SITTIG'S HANDBOOK OF TOXIC AND HAZARDOUS CHEMICALS AND CARCINOGENS

Sixth Edition

Volume 2: L–Z

SITTIG'S HANDBOOK OF TOXIC AND HAZARDOUS CHEMICALS AND CARCINOGENS

Sixth Edition

Volume 2: L–Z

Richard P. Pohanish

ELSEVIER

AMSTERDAM • BOSTON • HEIDELBERG • LONDON
NEW YORK • OXFORD • PARIS • SAN DIEGO
SAN FRANCISCO • SINGAPORE • SYDNEY • TOKYO
William Andrew is an imprint of Elsevier

William Andrew is an imprint of Elsevier
The Boulevard, Langford Lane, Kidlington, Oxford OX5 1GB, UK
225 Wyman Street, Waltham, MA 02451, USA

Fifth edition 2008
Sixth edition 2012

Notice
No responsibility is assumed by the publisher for any injury and/or damage to persons or property as a matter of products liability, negligence or otherwise, or from any use or operation of any methods, products, instructions or ideas contained in the material herein. Because of rapid advances in the medical sciences, in particular, independent verification of diagnoses and drug dosages should be made

British Library Cataloguing-in-Publication Data
A catalogue record for this book is available from the British Library

Library of Congress Cataloging-in-Publication Data
A catalog record for this book is available from the Library of Congress

ISBN-13: 978-1-4377-7869-4

Printed and bound in USA

11 12 13 14 15   10 9 8 7 6 5 4 3 2 1

NOTICE TO BE READ BY ALL USERS OF THIS PUBLICATION

Contents

Preface

For more than a quarter century, *Sittig's Handbook of Toxic and Hazardous Chemicals and Carcinogens* has continued to gather an ever-widening audience of users because it has been proven to be among the most reliable, easy to use, and essential reference works on hazardous materials. The 6th edition has been updated and expanded to keep pace with world events and to answer continuing and expanded need for information.

The 4th edition of *Sittig's Handbook of Toxic and Hazardous Chemicals and Carcinogens* was published in 2001, shortly before the tragic events of the morning of September 11, 2001. Following 9/11, the United States established the Department of Homeland Security and enacted laws such as the *Chemical Facilities Security Act of 2003* and released the DHS list of Chemicals of Concern, *Appendix to Chemical Facility Antiterrorism Standards; Final Rule, November 20, 2007*. These actions were prompted by concerns about infrastructure protection and the anticipation of another attack, possibly on the nation's chemical facilities or by using trucks or tank cars that transport highly dangerous and possibly lethal chemicals.

These facilities are found around the country in industrial parks, in seaports, and near the major population centers. Dangerous chemicals routinely travel along our highways, inland waterways, and on railcars that pass through the heart of major cities including Washington, D.C., just a short distance from Capitol Hill. Terrorist attacks on the US chemical industry have the potential to kill tens of thousands of Americans and seriously injure many more. In many instances, these attacks hold the potential for having a cascading effect across other infrastructures, particularly in the energy and transportation sectors. This is both because of the damage that can be caused by the attack, and the enormous expense and effort associated with the clean-up to an affected area in its aftermath.[83]

To put it more simply, using the same low tech/high concept approach that turned passenger planes into missiles, terrorists do not need to produce or amass chemical weapons or smuggle them into the United States in order to produce great damage.

"Commercial chemical incidents occur tens of thousands of times each year, often with devastating and exorbitantly expensive consequences. They are indiscriminate in their effects. Workers, companies, the public, emergency response organizations, and all levels of government pay the figurative and literal price. Yet, until now and with few exceptions, chemical incidents have been invisible. Perhaps it is due to their pervasiveness, or to the common tendency to overlook what is taken for granted."[84] This quote is from the highly publicized *600K Report* prepared by the Chemical Safety and Hazard Investigation Board (CSB), an independent, nonpartisan, quasi-legislative US government agency. The CSB described our nation's lack of definitive knowledge of the "big picture" surrounding chemical incidents as "... the industrial equivalent of two 737 airplanes 'crashing' year after year, killing all passengers (256 people) without anyone seeming to notice."[84]

More than 30 years ago, the United States Government Accounting Office (GAO) estimated that 62,000 chemicals were in commercial use. Today, that number has grown to beyond 82,000.

Each year, in the United States, over 2 billion tons of hazardous and toxic chemicals are manufactured. Including imports, more than 3 billion tons are transported employing 800,000 shipments each day. It is estimated that 1.3 billion tons are moved by truck and hundreds of billions of pounds of these hazardous materials are transported through populated areas. The average American household generates approximately 15 lb of hazardous waste per year. Nearly 5 million poisonings occur in the United States annually, resulting in thousands of deaths. Based on 2004, TRI data (publically released April 2006), over 4 billion pounds of toxic chemicals are released into the nation's environment each year, including 72 million pounds of recognized carcinogens from nearly 24,000 industrial facilities. The toxic chemicals problem in the United States; and, indeed, in all the world is frightening to many people. And, over the years, these fears are heightened by news stories about an oil field explosion in Mississippi (2006), a 48,000-lb chlorine release in Missouri (2002), Love Canal in New York, the Valley of the Drums in Kentucky, the Valley of Death in Brazil, major chemical spills, including Bhopal, India, terrorist attacks in Japan...and the like. All of these incidents generate emotional responses, often from people uninformed about science or technology. On the other hand, one encounters some industrialists who tell us that toxic chemicals are present in nature and that industrial contributions are just the price we have to pay for progress. There is little argument about the chemical industry's critical place in the nation's economy. The United States is the number one chemical producer in the world, generating more than $550 billion a year and employing more than 5 million people. So somewhere in between lies the truth—or at least an area in which we can function. Information is vital in a world where virtually every aspect of our lives is touched by chemical hazards.

Given the reality of problems related to chemical hazards, including accidents and spills, the advent of new threats to our way of life, and the challenges of communicating complex data, it is the goal of this book to provide data so that responsible decisions can be made by all who may have contact with chemicals in this reference work. With this in mind, the work can be used by those in the following professions:

- Chemicals manufacturers
- Emergency response personnel
- Protective safety equipment producers
- Environmental management
- Transportation managers
- Toxicologists
- Industrial hygienists
- Industrial safety engineers
- Lawyers
- Occupational doctors and nurses
- Chemists
- Industrial waste disposal operators
- Enforcement officials
- Special, technical, and university librarians
- Legislators
- Homeland security planners

The chemicals chosen for inclusion are officially recognized substances, defined as carcinogens, as belonging to some designated category of hazardous or toxic materials, with numerically defined safe limits in air in the workplace, ambient air, water, waste effluents. For the most part, these are materials of commerce that are heavily used and many are transported in bulk.

The 6th edition contains more regulated chemicals and expanded data on each material. Some material and appendices from the previous edition has been eliminated or moved to more appropriate sections. This was done to limit the work to a pair of handy volumes.

All of this has been done to make the work more relevant, more inclusive, and easier to use. The utility of the work has been enhanced by the addition of three appendices. Additionally there is a table that cross indexes the materials by chemical and trade names and CAS Registry Number.

Appendix 1: the list of oxidizing materials has been expanded. Appendix 2 contains many new confirmed and suspected carcinogens. Also, this edition allows the user to search the carcinogen list by name or CAS Number. Appendix 3 is a glossary of chemical, health, safety, medical, and environmental terms used in the handbook. The glossary was completely reviewed and many narrow interest medical terms were removed. More and new germane terms were added. The Introduction was replaced with the more accurate title, *How to Use this Book*. Following the use section is a revised *Key to the Abbreviations and Acronyms* used in the handbook.

In keeping with the broad changes originally initiated with the 4th edition, contents of the 6th edition are focused on the concept of "regulated chemicals." The carcinogen potential of each chemical was compared to listings and reports from eminent authorities as the International Agency for Research on Cancer (IARC) and the National Toxicology Progam (NTP).

The "Regulatory Authority and Advisory Bodies" section contains new items including, where available, EPA Gene-Tox Program findings, and many of the individual listings now contain useful advice sought after by the regulated community. As a result, the new volume should be even more practical for those users of specific chemicals and to those concerned with both adherence to and enforcement of regulations.

Data is furnished, to the extent currently available, in a uniform multisection uniform format to make it easy for users who must find information quickly and/or compare the data contained within records, in any or all of these important categories:

Chemical Description
Code Numbers (including CAS, DOT, RTECS, EC)
Synonyms
Regulatory Authority and Advisory Bodies (summary)
Description
Potential Exposure
Incompatibilities
Permissible Exposure Limits in Air
Determination in Air
Permissible Concentration in Water
Determination in Water
Routes of Entry
Short Term Exposure
Long Term Exposure
Points of Attack
Medical Surveillance
Decontamination (selected records)
First Aid
Decontamination (CWAs or WMDs)
Personal Protective Methods
Respirator Selection
Storage
Shipping
Spill Handling
Fire Extinguishing
Disposal Method Suggested
References
The 6th edition of Sittig has new and updated information in nearly every section, including the following: Synonyms, CAS Numbers, UN/NA & ERG (Emergency Response Guide) Number, EC Numbers (Annex I Index Numbers

added where assigned), Regulatory Authority and Advisory Bodies (added Rotterdam Convention Annex III [Chemicals Subject to the Prior Informed Consent Procedure (PIC)]; List of Stockholm Convention POPs: Annex A; European/ International Hazard Symbol, Risk phrases, Safety phrases; WGK (German Aquatic Hazard Class); Annex II Rotterdam Convention List information; hundreds of Department of the US Homeland Security Chemicals of Interest along with their Screening Threshold Quantities (STQs) [from the US Code of Federal regulations (6CFR Part 27 Appendix A]; California Proposition 65 Carcinogen and Reproduction Toxins; Description, Incompatibilities, Exposure Limits [now includes US Department of Energy (DOE) Protection Action Criteria (PACs)]. Short Term Exposure, Long Term Exposure, First Aid, Decontamination (especially, chemical warfare agents and weapons of mass destruction), Personal Protective Methods, Respirator Selection, Storage, Shipping, Spill Handling [more and updated Initial Isolation and Protective Distances (including both Imperial and Metric) from the *US DOT Emergency Response Guide*], Fire Extinguishing, Disposal. Specifically, additions include regulatory information, identifiers, chemical and physical properties, including explosive limits, NFPA (National Fire Protection Association)-type hazard ratings (based on NFPA-704 M Rating System), water solubility and hazard levels, exposure limits, odor thresholds, DOT isolation and protective distances, and full text of NIOSH respirator recommendations. Many records contain special warnings, including notes and reminders to Emergency Management Service (EMS) personnel, and other health-care professionals.

Although every effort has been made to produce an accurate and highly useful handbook, the author appreciates the need for constant improvement. Any comments, corrections, or advice from users of this book are welcomed by the author who asks that all correspondence be submitted in writing and mailed to the publisher who maintains a file for reprints and future editions.

A Brief history of this work
Sittig's Handbook of Toxic and Hazardous Chemicals and Carcinogens was first published 30 years ago. This work continues to provide first responders and occupational and environmental health and safety professionals with an accessible and portable reference source. The 6th edition of his handbook contains data on more than 2200 toxic and hazardous chemicals (up from nearly 600 in the first edition, nearly 800 in the second edition, nearly 1300 in the third edition, and 1500 in the 4th edition).

According to the Library of Congress, the history of the project is as follows: 1st edition published in 1981; 2nd edition published in 1985; 3rd edition published in 1991; 4th edition published in 2001; 5th edition published in 2008; 6th edition published in 2011.

Acknowledgments
The author would like to thank some individuals and institutions, without whose expertise and generous help, the 6th edition would not have been possible. In particular, the author wants to acknowledge the good work of the scientists and contract employees associated with NIOSH, US EPA, OSHA, ATSDR, ACGIH, IARC, DFG, CDC, TOXNET, NTP, AIHA, and many others who developed the various documents and databases that provided so much of the data that were compiled for this work. To each, the author is indebted. At the US Coast Guard Headquarters, the author wishes to thank the recently retired Alan Schneider, D.Sc., of the Marine Technical and Hazardous Materials Division.

How to Use This Book

Sittig's Handbook of Toxic and Hazardous Chemicals and Carcinogens focuses on critical data for more than 2200 commercially important and/or regulated and monitored substances, and many associated substances. Many of these chemicals or substances are found in the workplace; a few are found in the medical and research fields. Importance is defined by inclusion in official, regulatory, and advisory listings. Much of this information, found in US government sources, has been supplemented by a careful search of publications from various countries and other sources including United Nations and World Health Organization (WHO) publications.

This handbook is becoming more encyclopedic in nature. When one looks at most handbooks, one simply expects to find numerical data. Here, we have tried, wherever possible, to provide literature references to review documents which hopefully opens the door for users to a much broader field of published materials. It is recommended that this book be used as a guide. This book is not meant to be a substitute for workplace hazard communication programs required by regulatory bodies such as OSHA, and/or any other US, foreign, or international government agencies. If data are required for legal purposes, the original source documents and appropriate agencies, which are referenced, should be consulted.

In the pages which follow, the following categories of information will be discussed with reference to scope, sources, nomenclature employed, and the like. Omission of a category indicates a lack of available information.

Chemical name: Each record is arranged alphabetically by a chemical name used by regulatory and advisory bodies. In a very few cases the name may be a product name or trade name.

Formula: Generally, this has been limited to a commonly used one-line empirical or atomic formula. In the *Molecular Formula* field, the Hill system has been used showing number of carbons (if present), number of hydrogens (if present), and then alphabetically by element. Multiple carbon–carbon (double and triple) bonds have been displayed where appropriate.

Synonyms: This section contains scientific, product, trade, and other synonym names that are commonly used for each hazardous substance. Some of these names are registered trade names. Some are provided in other major languages other than English, including Spanish, French, and German. In some cases, "trivial" and nicknames (such as MEK for methyl ethyl ketone) have been included because they are commonly used in general communications and in the workplace. This section is important because the various "regulatory" lists published by federal, state, international, and advisory bodies and agencies do not always use the same name for a specific hazardous substance. Every attempt has been made to ensure the accuracy of the synonyms and trade names found in this volume, but errors are inevitable in compilations of this magnitude. Please note that this volume may not include the names of all products currently in commerce, particularly mixtures that may contain regulated chemicals.

The synonym index contains all synonym names listed in alphabetical order. It should be noted that organic chemical prefixes and interpolations, such as (α-) alpha-, (β-) beta-, (γ-) gamma-, delta- (δ-), (*o-*) *ortho-*, (*m-*) *meta-*; (*p-*) *para-*; *sec-* (*secondary-*), *trans-*, *cis-*, (*n-*) *normal-*, and numbers (1-; 1,2-), are not used when searching for a chemical name. In other words, these prefixes are not treated as part of the chemical name for the purposes of alphabetization. Users should use the substance name without the prefix. For example, to locate *n-*Butane, search for Butane; to locate 3,3$'$-Dichlorobenzidine, search for Dichlorobenzidine; and to locate α-Cyanotoluene or alpha-Cyanotoluene, search for Cyanotoluene.

CAS Number: The CAS number is a unique identifier assigned to each chemical registered with the Chemical Abstracts Service (CAS) of the American Chemical Society. This number is used to identify chemicals on the basis of their molecular structure. CAS numbers are given in the format nnn (...)-nn-n [two or more numeric characters (dash) two numeric characters (dash) followed by a single numeric check digit]. CAS numbers should always be used in conjunction with substance names to insure positive identification and avoid confusion with like-sounding names, i.e., benzene (71-43-2) and benzine (8032-32-4). This 6th edition contains some alternate CAS numbers that may now be considered related, retired, obsolete, and/or widely and incorrectly used in the literature. In this section, the first CAS number(s), before the word "alternate," is considered (based on several sources) to be the correct CAS number(s). Ultimately, it is the responsibility of the user to find and use the correct number.

RTECS®Number: The RTECS® numbers (Registry of Toxic Effects of Chemical Substances) are unique identifiers assigned and published by NIOSH. The RTECS® number in the format "AAnnnnnnn" (two alphabetic characters followed by seven numeric characters) may be useful for online searching for additional toxicologic information on specific substances. It can, for example, be used to provide access to the MEDLARS® computerized literature retrieval services of the National Library of Medicine (NLM) in Washington, DC. The RTECS number and the CAS number can serve to narrow down online searches.

DOT ID: The DOT hazard ID number is assigned to the substance by the US Department of Transportation (DOT). The DOT ID number format is "UNnnnn" or "NA nnnn." This ID number identifies substances regulated by DOT and must appear on shipping documents, the exterior of packages, and on specified containers. Identification numbers containing a UN prefix are also known as United Nations numbers and are authorized for use with all international shipments of hazardous materials. The "NA" prefix is used for shipments between Canada and the United States only, and may not be used for other international shipments.

EC Number: The European Commission number is a 7-digit identification code used by the European Union (EU) for commercially available chemical substances within the EU. An identification number from *European Inventory of Existing Commercial Chemical Substances*, published by the European Environment Agency, Copenhagen, Denmark. Use of these identification numbers for hazardous materials will (a) serve to verify descriptions of chemicals; (b) provide for rapid identification of materials when it might be inappropriate or confusing to require the display of lengthy chemical names on vehicles; (c) aid in speeding communication of information on materials from accident scenes and in the receipt of more accurate emergency response information; and (d) provide a means for quick access to immediate emergency response information in the *"North American Emergency Response Guidebook."*[31] In this latter volume, the various compounds have assigned "ID" numbers (or identification numbers) which correspond closely, but not always precisely, to the UN listing.[20] The EC number supercedes the outmoded EINECS, ELINCS, and NLP numbers. This section also includes Annex I, Index number for the Export and Import of Dangerous Chemicals found in Annex I of Regulation (EC) No. 689/2008.

Regulatory Authority and Advisory Bodies:
This section contains a listing of major regulatory and advisory lists containing the chemical of concern, including OSHA, US EPA, DFG, US DOT, ACGIH, IARC, NTP, WHMIS (Canada), and the EEC. Many law or regulatory references in this work have been abbreviated. For example, Title 40 of the Code of Federal Regulations, Part 261, subpart 32 has been abbreviated as 40CFR261.32. The symbol "§" may be used as well to designate a "section" or "part."

- European/International Hazard Symbols, Risk Phrases, and Safety Phrases. Explanation of these symbols and phrases can be found in the new Appendix 4. In the interim between the 6th and 7th edition, it is expected that the Globally Harmonized System of Classification and Labelling of Chemicals (GHS) will be phased in by many countries. Hazard statements are an essential element under the GHS, and will eventually replace the risk phrases (R-phrases) described earlier in the paragraph. In addition to hazard statements, containers and Material Safety Data Sheets (MSDS) will contain, where necessary, one or multiple pictograms, a signal word such as "Warning" or "Danger," and precautionary statements. The precautionary statements will indicate proper handling procedures aimed at protecting the user and other people who might come in contact with the substance during an accident or in the environment. The container and MSDS will also contain the name of the supplier, manufacturer, or importer. Each hazard statement contains a four-digit code, starting with the letter H (in the form Hxxx). Statements appear under various headings grouped together by code number. The purpose of the four-digit code is for reference only; however, following the code is exact *phrase* as it should appear on labels and MSDS.

- A carcinogen (the agency making such a determination, the nature of the carcinogenicity—whether human or animal and whether positive or suspected, are given in each case). These are frequently cited by IARC (International Agency for Research on Cancer),[12] and are classified as to their carcinogenic risk to humans by IARC as follows: Group 1: Human Carcinogen; Group 2A: Probable Human Carcinogen; Group 2B: Possible Human Carcinogen. Chemicals are classified as to their carcinogenic risk to humans by NTP as follows: Group K: Known Human Carcinogens; Group R: Reasonably Anticipated Human Carcinogens, or the NTP (US National Toxicology Programs).[10] It should be noted that the DFG have designated some substances as carcinogens not so classified by other agencies.

- A banned or severely restricted product as designated by the United Nations[13] or by the US EPA Office of Pesticide Programs under FIFRA (The Federal Insecticide, Fungicide, and Rodenticide Act).[14]

- A substance cited by the World Bank.[15]

- A substance with an air pollutant standard set or recommended by OSHA and/or NIOSH,[58] ACGIH,[1] DFG,[3] or HSE.[33] The OSHA limits are the enforceable pre-1989 PELs. The transitional limits that were vacated by court order have not been included. The NIOSH and ACGIH airborne limits are recommendations that do not carry the force of law.

- A substance whose allowable concentrations in workplace air are adopted or proposed by the American Conference of Government Industrial Hygienists (ACGIH),[1] DFG [Deutsche Forschungsgemeinschaft (German Research Society)].[3] Substances whose allowable concentrations in air and other safety considerations have been considered by OSHA and NIOSH.[2] Substances which have limits set in work-place air, in residential air, in water for domestic purposes, or in water for fishery purposes as set forth by the former USSRUNEP/IRPTC Project.[43]

- Substances that are specifically regulated by OSHA under 29CFR1910.1001 to 29CFR1910.1050.

- Highly hazardous chemicals, toxics, and reactives regulated by OSHA's *"Process Safety Management of Highly Hazardous Chemicals"* under 29CFR1910.119, Appendix A. Substances that are Hazardous Air Pollutants (Title I, Part A, § 112) as amended under 42USC7412. This list provided for regulating at least 189 specific substances using technology-based standards that employ Maximum Achievable Control Technology (MACT) standards; and, possibly health-based standards if required at a later time. § 112 of the Clean Air Act (CAA) requires emission control by the EPA on a source-by-source basis. Therefore, the emission of substances on this list does not necessarily mean that a firm is subject to regulation.

- Regulated Toxic Substances and Threshold Quantities for Accidental Release Prevention. These appear as Accidental Release Prevention/Flammable Substances, Clean Air Act (CAA) §112(r), Table 3, TQ (threshold quantity) in pounds and kilograms under 40 CFR68.130. The accidental release prevention regulations applies to stationary sources that have present more than a threshold quantity of a CAA § 112(r) regulated substance.
- Clean Air Act (CAA) Public Law 101–549, Title VI, *Protection of Stratospheric Ozone*, Subpart A, Appendix A, class I and Appendix B, Class II, Controlled Substances, (CFCs) Ozone-depleting substances under 40CFR82.
- Clean Water Act (CWA) Priority toxic water pollutants defined by the US Environmental Protection Agency for 65 pollutants and classes of pollutants which yielded 129 specific substances.[6]
- Chemicals designated by EPA as "Hazardous Substances"[4] under the Clean Water Act (CWA) 40CFR116.4, Table 116.4A.
- Clean Water Act (CWA) § 311 Hazardous Materials Discharge Reportable Quantities (RQs). This regulation establishes reportable quantities for substances designated as hazardous (see §116.4, above) and sets forth requirements for notification in the event of discharges into navigable waters. Source: 40CFR117.3, amended at 60FR30937.
- Clean Water Act (CWA) § 307 List of Toxic Pollutants. Source: 40CFR401.15.
- Clean Water Act (CWA) § 307 Priority Pollutant List. This list was developed from the List of Toxic Pollutants classes discussed above and includes substances with known toxic effects on human and aquatic life, and those known to be, or suspected of being, carcinogens, mutagens, or teratogens. Source: 40CFR423, Appendix A.
- Clean Water Act, § 313 Water Priority Chemicals. Source: 57FR41331.
- RCRA Maximum Concentration of Contaminants for the Toxicity Characteristic with Regulatory levels in mg/L. Source: 40CFR261.24.
- RCRA Hazardous Constituents. Source: 40CFR261, Appendix VIII. Substances listed have been shown, in scientific studies, to have carcinogenic, mutagenic, teratogenic, or toxic effects on humans and other life forms. This list also contains RCRA waste codes. The term "waste number not listed" appears when a RCRA number is NOT provided in Appendix VIII.

Characteristic Hazardous Wastes

Ignitability — A nonaqueous solution containing less than 24% alcohol by volume and having a closed cup flashpoint below 60°C/140°F using Pensky-Martens tester or equivalent. An ignitible compressed gas. A nonliquid capable of burning vigorously when ignited or causes fire by friction, moisture absorption, spontaneous chemical changes at standard pressure and temperature. An oxidizer. See §261.21.

Corrosivity — Liquids with a pH equal to or less than 2 or equal to or more than 12.5 or which corrode steel at a rate greater than 6.35 mm (0.25 in) per year at 55°C/130°F. See §261.22.

Reactivity — Unstable substances that undergo violent changes without detonating. Reacts violently with water or other substances to create toxic gases. Forms potentially explosive mixtures with air. See §261.23.

Toxicity — A waste that leaches specified amounts of metals, pesticides, or organic chemicals using Toxicity Characteristic Leaching Procedure (TCLP). See §261, Appendix II, and §268, Appendix I. **Listed Hazardous Wastes**.

"F" wastes — Hazardous wastes from nonspecific sources §261.31.

"K" Wastes — Hazardous wastes from specific sources §261.32.

"U" Wastes — Hazardous wastes from discarded commercial products, off-specification species, container residues §261.34. Covers some 455 compounds and their salts and some isomers of these compounds.

"P" Wastes — Acutely hazardous wastes from discarded commercial products, off-specification species, container residues §261.33. Covers some 203 compounds and their salts plus soluble cyanide salts.

Note: If a waste is not found on any of these lists, it may be found on state hazardous waste lists.

RCRA Maximum Concentration of Contaminants for the Toxicity Characteristic. Source: 40CFR261.24, Table I. These are listed with regulatory level in mg/L and "D" waste numbers representing the broad waste classes of ignitability, corrosivity, and reactivity.

EPA Hazardous Waste code(s), or RCRA number, appears in its own field. Acute hazardous wastes from commercial chemical products are identified with the prefix "P." Nonacutely hazardous wastes from commercial chemical products are identified with the prefix "U."

RCRA Universal Treatment Standards. Lists hazardous wastes that are banned from land disposal unless treated to meet standards established by the regulations. Treatment standard levels for wastewater (reported in mg/L) and nonwastewater [reported in mg/kg or mg/L TCLP (Toxicity Characteristic Leachability Procedure)] have been provided. Source: 40CFR268.48 and revision, 61FR15654.

RCRA Ground Water Monitoring List. Sets standards for owners and operators of hazardous waste treatment, storage, and disposal facilities, and contains test methods suggested by the EPA (see Report SW-846) followed by the Practical Quantitation Limit (PQL) shown in parentheses. The regulation applies only to the listed chemical; and, although both the test methods and PQL are provided, they are *advisory only*. Source: 40CFR264, Appendix IX.

Safe Drinking Water Act (SDWA) Maximum Contaminant Level Goals (MCLGs) for Organic Contaminants. Source: 40CFR141 and 40CFR141.50, amended 57FR31776.

- Maximum Contaminant Levels (MCLs) for Organic Contaminants. Source: 40CFR141.61.
- Maximum Contaminant Level Goals (MCLGs) for Inorganic Contaminants. Source: 40CFR141.51.
- Maximum Contaminant Levels (MCLs) for Inorganic Contaminants. Source: 40CFR141.62.
- Maximum Contaminant Levels for Inorganic Chemicals. The maximum contaminant level for arsenic applies only to community water systems. Compliance with the MCL for arsenic is calculated pursuant to §141.23. Source: 40CFR141.11.
- Secondary Maximum Contaminant Levels (SMCLs). Federal advisory standards for the states concerning substances that effect physical characteristics (i.e., smell, taste, color, etc.) of public drinking water systems. Source: 40CFR143.3.
- CERCLA Hazardous Substances ("RQ" Chemicals). From Consolidated List of Chemicals Subject to the Emergency Planning and Community Right-to-Know Act (EPCRA) and § 112(r) of the Clean Air Act, as Amended. Source: EPA 550-B-98-017 *Title III List of Lists*.
- Releases of CERCLA hazardous substances in quantities equal to or greater than their reportable quantity (RQ) are subject to reporting to the National Response Center under CERCLA. Such releases are also subject to state and local reporting under §304 of SARA Title III (EPCRA). CERCLA hazardous substances, and their reportable quantities, are listed in 40CFR302, Table 302.4. RQs are shown in pounds and kilograms for chemicals that are CERCLA hazardous substances. For metals listed under CERCLA (antimony, arsenic, beryllium, cadmium, chromium, copper, lead, nickel, selenium, silver, thallium, and zinc), no reporting of releases of the solid is required if the diameter of the pieces of solid metal released is 100 μm (0.004 in.) or greater. The RQs shown apply to smaller particles.
- EPCRA §302 Extremely Hazardous Substances (EHS). From Consolidated List of Chemicals Subject to the Emergency Planning and Community Right-to-Know Act (EPCRA) and § 112(r) of the Clean Air Act, as Amended. Source: EPA 550-B-98-017 *Title III List of*

Lists. The presence of Extremely Hazardous Substances in quantities in excess of the Threshold Planning Quantity (TPQ) requires certain emergency planning activities to be conducted. The Extremely Hazardous Substances and their TPQs are listed in 40CFR355, Appendices A & B. For chemicals that are solids, there may be two TPQs given (e.g., 500/10,000). In these cases, the lower quantity applies for solids in powder form with particle size less than 100 μm; or, if the substance is in solution or in molten form. Otherwise, the higher quantity (10,000 pounds in the example) TPQ applies.

- EPCRA §304 Reportable Quantities (RQs). In the event of a release or spill exceeding the reportable quantity, facilities are required to notify State Emergency Response Commissions (SERCs) and Local Emergency Planning Committees (LEPCs). From Consolidated List of Chemicals Subject to the Emergency Planning and Community Right-to-Know Act (EPCRA) and § 112(r) of the Clean Air Act, as Amended. Source: EPA 550-B-98-017 *Title III List of Lists*.
- EPCRA § 313 Toxic Chemicals. From Consolidated List of Chemicals Subject to the Emergency Planning and Community Right-to-Know Act (EPCRA) and § 112(r) of the Clean Air Act, as Amended. Source: EPA 550-B-98-017 *Title III List of Lists*. Chemicals on this list are reportable under §313 and §6607 of the Pollution Prevention Act. Some chemicals are reportable by category under §313. Category codes needed for reporting are provided for the EPCRA§313 categories. Information and Federal Register references have been provided where a chemical is subject to an administrative stay, and not reportable until further notice.
- From "*Toxic Chemical Release Inventory Reporting Form R and Instructions, Revised 2005 Version*," EPA document 260-B-06-001 was used for *de minimis* concentrations, toxic chemical categories.
- Chemicals which EPA has made the subject of Chemical Hazard Information Profiles or "CHIPS" review documents.
- Chemicals which NIOSH has made the subject of "Information Profile" review documents on "Current Intelligence Bulletins."
- Carcinogens identified by the National Toxicology Program of the US Department of Health and Human Services at Research Triangle Park, NC.[10]
- Substances regulated by EPA[7] under the major environmental laws: Clean Air Act, Clean Water Act, Safe Drinking Water Act, RCRA, CERCLA, EPCRA, etc. A more detailed list appears above. Substances with environmental standards set by some international bodies including those in Europe and Canada.[43]
- New to the 6th edition: United States Department of Homeland Security Chemicals of Interest from the

Federal Register, Appendix A, including all provisions of 6 CFR Part 27, including § 27.210(a)(1)(i). In developing the list, the DHS looked to existing expert sources of information including other federal regulations related to chemicals, including the following: Chemicals covered under the Environmental Protection Agency's Risk Management Program. Chemicals included in the Chemical Weapons Convention. Hazardous materials, such as gases, that are poisonous by inhalation. Explosives regulated by the Department of Transportation. The Department of Homeland Security has identified three security issues related to chemicals: *Release*—Toxic, flammable, or explosive chemicals or materials that, if released from a facility, have the potential for significant adverse consequences for human life or health. *Theft or Diversion*—Chemicals or materials that, if stolen or diverted, have the potential to be misused as weapons or easily converted into weapons using simple chemistry, equipment or techniques, in order to create significant adverse consequences for human life or health. *Sabotage or Contamination*—Chemicals or materials that, if mixed with readily available materials, have the potential to create significant adverse consequences for human life or health. Also considered were these security issues as well as to determine their potential future inclusion in Appendix A and/or coverage under Chemical Facility Anti-Terrorism Standards: *Critical to Government Mission*—Chemicals or facilities, the loss of which could create significant adverse consequences for national security or the ability of the government to deliver essential services, and *Critical to National Economy*—Chemicals or facilities, the loss of which could create significant adverse consequences for the national or regional economy.

- Chemicals on California's Proposition 65 List, revised as of January 7, 2011. The Safe Drinking Water and Toxic Enforcement Act of 1986 requires that the Governor revise and republish at least once per year the list of chemicals known to the State to cause cancer or reproductive toxicity.
- Also new in the 6th edition are the water hazard classifications from the German Federal Water Management Act on Water Hazard Classification, *Verwaltungsvorschrift Wassergefährdende Stoffe* (VwVwS). This law requires all chemical substances be evaluated for their detrimental impact on the physical, chemical, or biological characteristics of water. Substances can be classified as nonhazardous to water (*nwg, nicht wassergefährdende*) or assigned to one of three numeric water hazard classes, WGK: 1—low hazard to waters (low polluting to water), WGK: 2—hazard to waters (water pollutant), or WGK: 3—severe hazard to waters (severe pollutant). The English acronym for WGK is WHC (water hazard class). This work uses the German acronym "WGK" so there is no confusion as to its source. Material Safety

Data Sheets (MSDS) that use these water hazards also use the German acronym.

Description: This section contains a quick summary of physical properties of the substance including state (solid, liquid, or gas), color, odor description, molecular weight, density, boiling point, freezing/melting point, vapor pressure, flash point, autoignition temperature, explosion limits in air, Hazard Identification (based on NFPA-704 M Rating System) in the format: Health (ranked 1–4), Flammability (ranked 1–4), Reactivity (ranked 1–4) (see also below for a detailed explanation of the System and Fire Diamond), and solubility or miscibility in water. This section may also contain special and relevant comments about the substance. Terms in this section are also defined in the glossary.

Odor threshold: This is the lowest concentration in air that most humans can detect by smell. Some value ranges are reported. The value cannot be relied on to prevent overexposure, because human sensitivity to odors varies over wide limits, some chemicals cannot be smelled at toxic concentrations, odors can be masked by other odors, and some compounds rapidly deaden the sense of smell.

Molecular weight: The MW as calculated from the molecular formula using standard elemental molecular weights (e.g., carbon = 12.1).

Boiling point at 1 atm: The value is the temperature of a liquid when its vapor pressure is 1 atm. For example, when water is heated to 100°C/212°F its vapor pressure rises to 1 atm and the liquid boils. The boiling point at 1 atm indicates whether a liquid will boil and become a gas at any particular temperature and sea-level atmospheric pressure.

Melting/Freezing point: The melting/freezing point is the temperature at which a solid changes to liquid or a liquid changes to a solid. For example, liquid water changes to solid ice at 0°C/32°F. Some liquids solidify very slowly even when cooled below their melting/freezing point. When liquids are not pure (e.g., saltwater) their melting/freezing points are lowered slightly.

Flash point: This is defined as the lowest temperature at which vapors above a volatile combustible substance will ignite in air when exposed to a flame. Depending on the test method used, the values given are either Tag Closed Cup (cc) (ASTM D56) or Cleveland Open Cup (oc) (ASTM D93). The values, along with those in *Flammable Limits in Air* and *Autoignition temperature* below, give an indication of the relative flammability of the chemical. In general, the open cup value is slightly higher (perhaps 10–15°F higher) than the closed cup value. The flash points of flammable gases are often far below 0° (F or C) and these values are of little practical value, so the term "flammable gas" is often used instead of the flash point value.

Autoignition Temperature: This is the minimum temperature at which the material will ignite without a spark or flame being present. Values given are only approximate and may change substantially with changes in geometry, gas, or vapor concentrations, presence of catalysts, or other factors.

Flammable Limits in Air: The percent concentration in air (by volume) is given for the LEL (lower explosive flammable limit in air, % by volume) and UEL (upper explosive flammable limit in air, % by volume), at room temperature, unless other specified. The values along with those in "Flash point" and "Autoignition temperature" give an indication of the relative flammability of the chemical.

NFPA Hazard Classifications: The NFPA 704 Hazard Ratings (Classifications) are based on those found in *"Fire Protection Guide to Hazardous Materials,"* 2001 edition, National Fire Protection Association, Quincy, MA, ©1994. The classifications are defined in Table 1 below.

Table 1. Explanation of NFPA Hazard Classifications
Classification   Definition

HEALTH HAZARD (blue)

4   Materials which on very short exposure could cause death or major residual injury (even though prompt medical treatment was given), including those that are too dangerous to be approached without specialized protective equipment.

3   Materials which on short exposure could cause serious temporary or residual injury (even though prompt medical treatment was given), including those requiring protection from all bodily contact.

2   Materials that, on intense or continued (but not chronic) exposure, could cause temporary incapacitation or possible residual injury, including those requiring the use of protective clothing that has an independent air supply.

1   Materials which on exposure would cause irritation but only minor residual injury, including those requiring the use of an approved air-purifying respirator.

0   Materials that, on exposure under fire conditions, offer no hazard beyond that of ordinary combustible material.

FLAMMABILITY (red)

4   This degree includes flammable gases, pyrophoric liquids, and Class IA flammable liquids. Materials which will rapidly or completely vaporize at atmospheric pressure and normal ambient temperature, or which are readily dispersed in air and which will burn readily.

3   Includes Class IB and IC flammable liquids and materials that can be easily ignited under almost all normal temperature conditions.

2   Materials that must be moderately heated before ignition will occur and includes Class II and Class IIIA combustible liquids and solids and semisolids that readily give off ignitable vapors.

1   Materials that must be preheated before ignition will occur, such as Class IIIB combustible liquids, and solids and semisolids whose flash point exceeds 200°F/93.4°C, as well as most ordinary combustible materials.

0   Materials that will not burn.

REACTIVITY (yellow)

4   Materials that, in themselves, are readily capable of detonation, explosive decomposition, or explosive reaction at normal temperatures and pressures.

3   Materials that, in themselves, are capable of detonation, or explosive reaction, but require a strong initiating source or heating under confinement. This includes materials that are sensitive to thermal and mechanical shock at elevated temperatures and pressures and materials that react explosively with water.

2   Materials that are normally unstable and readily undergo violent chemical change, but are not capable of detonation. This includes materials that can undergo chemical change with rapid release of energy at normal temperatures and pressures. This also includes materials that may react violently with water or that may form potentially explosive mixtures in water.

1   Materials that are normally stable, but that may become unstable at elevated temperatures and pressures and materials that will react with water with some release of energy, but not violently.

0   Materials that are normally stable, even under fire exposure conditions, and that do not react with water.

OTHER (white)

₩   Materials which react so violently with water that a possible hazard results when they come in contact with water, as in a fire situation. Similar to Reactivity Classification 2.Oxy—Oxidizing material; any solid or liquid that readily yields oxygen or other oxidizing gas, or that readily reacts to oxidize combustible materials.

It should be noted that OSHA and DOT have differing definitions for the term "flammable liquid" and "combustible liquid." DOT defines a flammable liquid as one which, under specified procedures, has a flashpoint of 140°F/60°C or less. A combustible liquid is defined as "having a flashpoint above 140°F/60°C and below 200°F/93°C." OSHA defines a combustible liquid as having a flash point above 100°F/37.7°C.

Potential Exposure: A brief indication is given of the nature of exposure to each compound in the industrial environment. Where pertinent, some indications are given of background concentration and occurrence from other than industrial discharges such as water purification plants. Obviously in a volume of this size, this coverage must be very brief. It is of course recognized that nonoccupational exposures may be important as well. This 6th edition contains a brief summary called a Compound Description (Toxicity evaluation),[77] such as Agricultural Chemical, Mutagen, Tumorigen, Mutagen, Reproductive Effector, Primary Irritant, Human Data, etc. Compound descriptors define the types of toxicity data found in a record or uses or applications of the chemical if they are recognized by NIOSH. The Compound Descriptor does not represent an evaluation of the toxicity of a substance, nor are the descriptors all-inclusive (i.e., there may be some substances that should be, but are not, coded as belonging to certain application classes). The codes must be interpreted only in conjunction with the other information found in each record.[77]

Incompatibilities: Potentially hazardous incompatibilities of each substance are listed where available. Where a hazard with water exists, it is described. Reactivity with other materials are described including structural materials such as metal, wood, plastics, cement, and glass. The nature of the hazard, such as severe corrosion formation of a flammable gas, is described. This list is by no means complete or all inclusive. In some cases a very small quantity of material can act as a catalyst and produce violent reactions such as polymerization, disassociation, and condensation. Some chemicals can undergo rapid polymerization to form sticky, resinous materials, with the liberation of much heat. The containers may explode. For these chemicals the conditions under which the reaction can occur are given.

Permissible Exposure Limits in Air: The permissible exposure limit (PEL) has been cited as the federal standard where one exists. Inasmuch as OSHA has made the decision to enforce only pre-1989 PELs, we decided to use these values rather than the transitional limits that were vacated by court order. Except where otherwise noted, the PELs are 8-h work-shift time-weighted average (TWA) levels. Ceiling limits, Short-Term Exposure Limits (STELs), and TWAs that are averaged over other than full work shifts are noted.

The Short-Term Exposure Limit (STEL) values are derived from NIOSH,[58] ACGIH,[1] and HSE[33] publications. This value is the maximal concentration to which workers can be exposed for a period up to 15 min continuously without suffering from: irritation; chronic or irreversible tissue change; or narcosis of sufficient degree to increase accident proneness, impair self-rescue, or materially reduce work efficiency, provided that no more than four excursions per day are permitted, with at least 60 min between exposure periods, and provided that the daily TWA also is not exceeded. The "Immediately Dangerous to Life or Health" (IDLH) concentration represents a maximum level from which one could escape within 30 min without any impairing symptoms or any irreversible health effects. However, the 30-min period is meant to represent a MARGIN OF SAFETY and is NOT meant to imply that any person should stay in the work environments any longer than necessary. In fact, every effort should be made to exit immediately. The concentrations are reported in either parts per million (ppm) or milligrams per cubic meter (mg/m^3).

Most US specifications on permissible exposure limits in air have come from ACGIH[1] or NIOSH.[2] In the United Kingdom, the Health and Safety Executive has set forth Occupational Exposure Limits.[33] In Germany, the DFG has established Maximum Concentrations in the work-place[3] and the former USSR-UNEP/IRPTC project has set maximum allowable concentrations and tentative safe

exposure levels of harmful substance in work-place air and residential air for many substances.[43] This section also contains numerical values for allowable limits of various materials in ambient air[60] as assembled by the US EPA. Where available, this field contains legally enforceable airborne Permissible Exposure Limits (PELs) from OSHA. It also contains recommended airborne exposure limits from NIOSH, ACGIH, and international sources and special warnings when a chemical substance is a Special Health Hazard Substance. Each are described below. TLVs have not been developed as legal standards and the ACGIH does not advocate their use as such. The TLV is defined as the time-weighted average (TWA) concentration for a normal 8-h workday and a 40-h workweek, to which nearly all workers may be repeatedly exposed, day after day, without adverse effects. A ceiling value (TLV-C) is the concentration that should not be exceeded during any part of the working exposure. If instantaneous monitoring is not feasible, then the TLV-C can be assessed by sampling over a 15-min period except for those substances that may cause immediate irritation when exposures are short. As some people become ill after exposure to concentrations lower than the exposure limits, this value cannot be used to define exactly what is a "safe" or "dangerous" concentration. ACGIH threshold limit values (TLVs) are reprinted with permission of the American Conference of Governmental Industrial Hygienists, Inc., from the booklet entitled, *Threshold Limit Values for Chemical Substances and Physical Agents and Biological Exposure Indices.* This booklet is revised on an annual basis. No entry appears when the chemical is a mixture; it is possible to calculate the TLV for a mixture only when the TLV for each component of the mixture is known and the composition of the mixture by weight is also known. According to ACGIH, "Documentation of the Threshold Limit Values and Biological Exposure Indices, 7th Edition" is necessary to fully interpret and implement the TLVs.

OSHA Permissible Exposure Limits (PELs) are found in Tables Z-1, Z-2, and Z-3 of OSHA, "General Industry Air Contaminants Standard (29CFR1910.1000)" that were effective on July 1, 2001 and which are currently enforced by OSHA.

Unless otherwise noted, PELs are the Time-Weighted Average (TWA) concentrations that must not be exceeded during any 8-h shift of a 40-h workweek. An OSHA ceiling concentration must not be exceeded during any part of the workday; if instantaneous monitoring is not feasible, the ceiling must be assessed as a 15-min TWA exposure. In addition, there are a number of substances from Table Z-2 that have PEL ceiling values that must not be exceeded except for a maximum peak over a specified period (e.g., a 5-min maximum peak in any 2 h).

NIOSH Recommended Exposure Limits (RELs) are Time-Weighted Average (TWA) concentrations for up to a 10-h workday during a 40-h workweek. A ceiling REL should not be exceeded at any time. Exposure limits are usually expressed in units of parts per million (ppm), i.e., the parts of vapor (gas) per million parts of contaminated air by volume at 25°C/77°F and one atmosphere pressure. For a chemical that forms a fine mist or dust, the concentration is given in milligrams per cubic meter (mg/m^3).

Protective Action Criteria (PAC) are emergency exposure limits developed by the US Department of Energy (DOE) for 3388 chemicals in revision 26, published September 2010. These exposure limits can be used to estimate the consequences of the uncontrolled release of hazardous materials and to plan for emergency response. These PACs have been added to the 6th edition of Sittig because other well-established exposure limits in air are available for only a limited number of chemicals from other governmental and advisory sources. PAC values are given in parts per million (ppm) for volatile liquids and gases; in milligrams per cubic meter (mg/m^3) for solids, particulates, and nonvolatile liquids. Chemicals for which TEELs (Temporary Emergency Exposure Limits) are available have their values displayed using a regular (non-bold) font. Chemicals for which Acute Emergency Guideline Levels (AEGLs) and Emergency Response Planning Guidelines (ERPGs) have their values displayed in **bold** font. TEELs are intended for use until AEGLs or ERPGs are adopted for chemicals.

PAC Definitions:[SCAPA]

There are subtle difference in the definitions of AEGLs, ERPGs, and TEELs and major differences in how they are developed and issued. Differences in their definitions include: AEGLs pertain to the "general population, including susceptible individuals," but ERPGs and TEELs pertain to "nearly all individuals."

AEGLs are defined as the level "above which" certain health effects are expected, while ERPGs and TEELs are defined as the level "below which" certain health effects are not expected.

Acute Emergency Guideline Levels (AEGLs) are defined as follows:

- *AEGL-1:* the airborne concentration (expressed as ppm or mg/m^3) of a substance above which it is predicted that the general population, including susceptible individuals, could experience notable discomfort, irritation, or certain asymptomatic, nonsensory effects. However, these effects are not disabling and are transient and reversible upon cessation of exposure.
- *AEGL-2:* the airborne concentration (expressed as ppm or mg/m^3) of a substance above which it is predicted that the general population, including susceptible individuals, could experience irreversible or other serious, long-lasting adverse health effects or an impaired ability to escape.
- *AEGL-3:* the airborne concentration (expressed as ppm or mg/m^3) of a substance above which it is predicted that the general population, including susceptible individuals, could experience life-threatening adverse health effects or death.

Emergency Response Planning Guidelines (ERPGs) are defined as follows:

- *ERPG-1*: the maximum concentration in air below which it is believed nearly all individuals could be exposed for up to 1 h without experiencing other than mild transient adverse health effects or perceiving a clearly defined objectionable odor.
- *ERPG-2*: the maximum concentration in air below which it is believed nearly all individuals could be exposed for up to 1 h without experiencing or developing irreversible or other serious health effects or symptoms that could impair their abilities to take protective action.
- *ERPG-3*: the maximum concentration in air below which it is believed nearly all individuals could be exposed for up to 1 h without experiencing or developing life-threatening health effects.

Temporary Emergency Exposure Limits (TEELs) are defined as follows:

- *TEEL-0*: the threshold concentration below which most people will experience no adverse health effects.
- *TEEL-1:* the airborne concentration (expressed as ppm or mg/m^3) of a substance above which it is predicted that the general population, including susceptible individuals, could experience notable discomfort, irritation, or certain asymptomatic, nonsensory effects. However, these effects are not disabling and are transient and reversible upon cessation of exposure.
- *TEEL-2:* the airborne concentration (expressed as ppm or mg/m^3) of a substance above which it is predicted that the general population, including susceptible individuals, could experience irreversible or other serious, long-lasting adverse health effects or an impaired ability to escape.
- *TEEL-3:* the airborne concentration (expressed as ppm or mg/m^3) of a substance above which it is predicted that the general population, including susceptible individuals, could experience life-threatening adverse health effects or death.
- TEELs are intended for use until AEGLs or ERPGs are adopted for chemicals.

Additional information on PAC values, TEEL values, and links to other sources of information can be found on the webpage for the *Subcommittee for Consequence Assessment and Protective Action (SCAPA)*: http://orise.orau.gov/emi/scapa/teels.htm

The German MAK (DFG MAK) values are conceived and applied as 8-h time-weighted average (TWA) values.[3]

Short-Term Exposure Limits (15-min TWA): This field contains Short-Term Exposure Limits (STELs) from ACGIH, NIOSH, and OSHA. The parts of vapor (gas) per million parts of contaminated air by volume at 25°C/77°F and one atmosphere pressure is given. The limits are given in milligrams per cubic meter (mg/m^3) for chemicals that can form a fine mist or dust. Unless otherwise specified, the STEL is a 15-min TWA exposure that should not be exceeded at any time during the workday.

Determination in Air: The citations to analytical methods are drawn from various sources, such as the *NIOSH Manual of Analytical Methods*.[18] In addition, methods have been cited in the latest US Department of Health and Human Services publications including the *"NIOSH Pocket Guide to Chemical Hazards"* August, 2006.

Permissible Concentrations in Water: The permissible concentrations in water are drawn from various sources also, including: The National Academy of Sciences/National Research Council, Safe Drinking Water Committee Board on Toxicology and Environmental Health Hazards, *Drinking Water and Health,* 1980.[16]

The priority toxic pollutant criteria published by US EPA 1980.[6]

The multimedia environmental goals for environmental assessment study conducted by EPA.[32] Values are cited from this source when not available from other sources.

The US EPA has come forth with a variety of allowable concentration levels:

For allowable concentrations in "California List" wastes.[38] The California List consists of liquid hazardous wastes containing certain metals, free cyanides, polychlorinated biphenyls (PCBs), corrosives with a pH of less than or equal to 2.0, and liquid and nonliquid hazardous wastes containing halogenated organic compounds (HOCs).

For regulatory levels in leachates from landfills.[37]

For concentrations of various materials in effluents from the organic chemicals and plastics and synthetic fiber industries.[51]

For contaminants in drinking water.[36]

For National Primary and Secondary Drinking Water Regulations.[62]

In the form of health advisories for 16 pesticides,[47] 25 organics,[48] and 7 inorganics.[49]

For primary drinking water standards starting with a priority list of 8 Volatile Organic Chemicals.[40]

State drinking water standards and guidelines[61] as assembled by the US EPA.

Determination in Water: The sources of information in this field have been primarily US EPA publications including the test procedures for priority pollutant analysis[25] and later modifications.[42]

Routes of Entry: The toxicologically important routes of entry of each substance are listed. In other words, the way in which the people or experimental animals were exposed to the chemical is listed, e.g., eye contact, skin contact, inhalation, intraperitoneal, intravenous. Many of these are taken from the *NIOSH Pocket Guide,*[2] but are drawn from other sources as well.

Harmful Effects and Symptoms: These are primarily drawn from NIOSH, EPA publications, and New Jersey and New York State fact sheets on individual chemicals, and are supplemented from information from the draft criteria documents for priority toxic pollutants[26] and from other sources. The other sources include:

EPA Chemical Hazard Information Profiles (CHIPS) cited under individual entries.

NIOSH Information Profiles cited under individual entries. EPA Health and Environmental Effect Profiles cited under individual entries.

Particular attention has been paid to cancer as a "harmful effect" and special effort has been expended to include the latest data on carcinogenicity. See also "Regulatory Authority and Advisory Bodies" section.

Short Term Exposure: These are brief descriptions of the effects observed in humans when the vapor (gas) is inhaled, when the liquid or solid is ingested (swallowed), and when the liquid or solid comes in contact with the eyes or skin. The term LD_{50} signifies that about 50% of the animals given the specified dose by mouth will die. Thus, for a Grade 4 chemical (below 50 mg/kg), the toxic dose for 50% of animals weighing 70 kg (150 lb) is $70 \times 50 = 3500$ mg = 3.5 g, or less than 1 teaspoonful; it might be as little as a few drops. For a Grade 1 chemical (5–15 g/kg), the LD_{50} would be between a pint and a quart for a 150-lb man. All LD_{50} values have been obtained using small laboratory animals such as rodents, cats, and dogs. The substantial risks taken in using these values for estimating human toxicity are the same as those taken when new drugs are administered to humans for the first time.

Long Term Exposure: Where there is evidence that the chemical can cause cancer, mutagenic effects, teratogenic effects, or a delayed injury to vital organs such as the liver or kidney, a description of the effect is given.

Points of Attack: This category is based, in part, on the "Target Organs" in the *NIOSH Pocket Guide*[2] but the title has been changed as many of the points of attack are not organs (e.g., blood). This is human data unless otherwise noted.

Medical Surveillance: This information is often drawn from a NIOSH publication[27] but also from *New Jersey State Fact Sheets*[70] on individual chemicals. Where additional information is desired in areas of diagnosis, treatment, and medical control, the reader is referred to a private publication[28] which is adapted from the products of the NIOSH Standards Completion Program.

First Aid: Guides and guidance to first aid found in this work should not be construed as authorization to emergency personnel to perform the procedures or activities indicated or implied. Care of persons exposed to toxic chemicals must be directed by a physician or other recognized professional or authority. Simple first aid procedures are listed for response to eye contact, skin contact, inhalation, and ingestion of the toxic substance as drawn to a large extent from the *NIOSH Pocket Guide*[2] but supplemented by information from recent commercially available volumes in the United States,[29] in the United Kingdom, and in Japan[24] as well as from state fact sheets. They deal with exposure to the vapor (gas), liquid, or solid and include inhalation, ingestion (swallowing), and contact with eyes or skin. The instruction "Do NOT induce vomiting" is given if an unusual hazard is associated with the chemical being sucked into the lungs (aspiration) while the patient is vomiting.

"*Seek medical attention*" or "*Call a doctor*" is recommended in those cases where only competent medical personnel can treat the injury properly. In all cases of human exposure, seek medical assistance as soon as possible. In many cases, medical advice has been included for guidance only.

Personal Protective Methods: This information is drawn heavily from NIOSH publications[2, 77] and supplemented by information from the United States,[29] the United Kingdom, and Japan.[24] There are indeed other "personal protective methods" which space limitations prohibit describing here in full. One of these involves limiting the quantities of carcinogens to which a worker is exposed in the laboratory. The items listed are those recommended by (a) NIOSH and/or OSHA, (b) manufacturers, either in technical bulletins or in material safety data sheets (MSDS), (c) the Chemical Manufacturers Association (CMA), or (d) the National Safety Council (NSC), for use by personnel while responding to fire or accidental discharge of the chemical. They are intended to protect the lungs, eyes, and skin.

Respirator Selection: The 6th edition, like its predecessors, presents respirator selection with a full text description. For each line a maximum use concentration (in ppm, mg/m^3, µg/m^3, fibers/m^3, or mppcf) condition (e.g., escape) followed by the NIOSH code and full text related to respirator recommendations. All recommended respirators of a given class can be utilized at any concentration equal to or less than the class's listed maximum use concentration. Respirator selection should follow recommendations that provide the greatest degree of protection. Respirator codes found in the *NIOSH Pocket Guide* have been included to ease updating.

All respirators selected must be approved by NIOSH under the provisions of 42CFR84. The current listing of NIOSH/MSHA-certified respirators can be found in the *NIOSH Certified Equipment List*, which is available on www.cdc.gov/niosh/npptl/topics/respirators/cel (NIOSH Web site).

For firefighting, only self-contained breathing apparatuses with full facepieces operated in pressure-demand or other positive-pressure modes are recommended for all chemicals in the *NIOSH Pocket Guide*. In the case of chemical warfare agents, use only SCBA Respirator certified by NIOSH for CBRN environments. CBRN stands for "Chemical, Biological, Radiological, and Nuclear."

Pesticides are not identified as such in the respirator selection tables. For those substances that are pesticides, the recommended air-purifying respirator must be specifically approved by NIOSH/MSHA. Specific information on choosing the appropriate respirator will be provided on pesticide labels. Approved respirators will carry a "TC" number prefix, which signifies they have been tested and certified for a specific level of protection. New respirators may carry a "TC-84A" prefix in compliance with 42CFR84 for testing and certifying nonpowered, air-purifying, particulate-filter respirators. The new Part 84 respirators have passed a more demanding certification test than the old respirators (e.g., dust and mist [DM], dust, fume, and mist [DFM], spray paint, and pesticide) certified under 30CFR11. Additionally, a complete respirator protection program should be implemented including all requirements in 29CFR1910.134 and 42CFR84. At a minimum, a respirator protection program should include regular training, fit-testing, periodic environmental monitoring, maintenance inspection, and cleaning. The selection of the actual respirator to be used within the classes of recommended respirators depends on the particular use situation, and should only be made by a knowledgeable person. Remember, air-purifying respirators will not protect from oxygen-deficient atmospheres. For firefighting, only self-contained breathing apparatuses with full facepieces operated in pressure-demand or other positive-pressure modes are recommended for all chemicals in the *NIOSH Pocket Guide*.

Storage: The 6th edition now provides, as general guidance, a color-coded classification system similar to those often found in commerce and laboratories. It is the objective of any chemical storage classification system to prevent accidental combination of two or more incompatible materials that might be stored in the same space. To prevent an unwanted and possibly dangerous reaction, chemicals must be separated by space and/or physical barriers. Chemical storage areas should be appropriately labeled. Users must be careful to check the MSDS for both additional and specific information. Some chemical entries contain multiple storage codes because the chemical profile fits more than a single category.

Code	Hazard
Red	Flammables (flash point <100°F)
Blue	Health hazards/toxics/poisons
Yellow	Reactives/oxidizers
White	Contact hazards
Green*	General storage

*For general storage, the colors Gray and Orange are also used by some companies.

- Chemical containers that are not color coded should contain hazard information on the label.
- Check the MSDS to learn what personal protective equipment is required when using the substance
- *Red*: Flammability Hazard: Store in a flammable (liquid or materials) storage area or approved cabinet away from ignition sources and corrosive and reactive materials.
- *Blue*: Health Hazard/Toxics/Poisons: Store in a secure poison location.
- *Yellow*: Reactive Hazard; Store in a location separate from other materials, especially flammables and combustibles.
- *White*: Corrosive or Contact Hazard; Store separately in a corrosion-resistant location.
- *Green (or Gray or Orange)*: General storage may be used. Generally, for flammability, health, and corrosivity with an NFPA rating of no higher than "2."

- Chemicals with labels that are colored and diagonally striped may react with other chemicals in the same hazard class. See MSDS for more information.
- *Red Stripe*: Flammability Hazard: Store separately from all other flammable materials. Example: sodium metal.
- *Yellow Stripe*: Reactivity Hazard; Store separately in an area isolated from flammables, combustibles, or other yellow-coded materials. *Example*: reducing agents.
- *White stripe*: Contact Hazard; not compatible with materials in solid white category. Store separately. *Example*: Bases.

Other data in this field are drawn from, or based on, various resources, including the NFPA,[17] Japanese sources,[24] and publications such as the *Hazardous Substance Fact Sheets* published by the New Jersey Department of Health and Senior Services.[70]

Shipping: The shipping guidance offered herein does not replace the training requirements of the Department of Transportation and in no way guarantees that you will be in full compliance with the Department of Transportation Regulations. *Labeling*: This section refers to the type label or placard required by regulation on any container or packaging of the subject compound being shipped. In some cases a material may require more than one hazardous materials label. *Quantity limitation:* This section lists quantities of material that may be shipped on passenger aircraft, rail, and cargo aircraft. Materials in certain hazard classes may be shipped under the small quantities exception (see 49CFR173.4) with specific approval from the Associate Administrator for Hazardous Materials Safety, Department of Transportation. *Hazard class or division:* This number refers to the division number or hazard class that must appear on shipping papers. This information is drawn from DOT publications[19] as well as UN publications[20] and also NFPA publications.[17] The US Department of Transportation[19] has published listings of chemical substances which give a hazard classification and required labels. The US DOT listing now corresponds with the UN listing[20] and specifies first a hazard class of chemicals as defined in the following table, and then a packing group (I, II, or III) within each of the classes. These groups are variously defined depending on the hazard class but in general define materials presenting: I—a very severe risk (great danger); II—a serious risk (medium danger); and III—a relatively low risk (minor danger).

HAZARD CLASSIFICATION SYSTEM

The hazard class of dangerous goods is indicated either by its class (or division) number or name. For a placard corresponding to the primary hazard class of a material, the hazard class or division number must be displayed in the lower corner of the placard. However, no hazard class or division number may be displayed on a placard representing the subsidiary hazard of a material. For other than Class 7 or the OXYGEN placard, text indicating a hazard (e.g., "CORROSIVE") is not required. Text is shown only

in the United States. The hazard class or division number must appear on the shipping document after each shipping name.

Class 1—Explosives
Division 1.1: Explosives with a mass explosion hazard
Division 1.2: Explosives with a projection hazard
Division 1.3: Explosives with predominantly a fire hazard
Division 1.4: Explosives with no significant blast hazard
Division 1.5: Very insensitive explosives with a mass explosion hazard
Division 1.6: Extremely insensitive articles

Class 2—Gases
Division 2.1: Flammable gases
Division 2.2: Nonflammable, nontoxic* gases
Division 2.3: Toxic* gases

Class 3—Flammable liquids [and Combustible liquids (US)]

Class 4—Flammable solids; Spontaneously combustible materials; and Dangerous when wet materials/Water-reactive substances
Division 4.1: Flammable solids
Division 4.2: Spontaneously combustible materials
Division 4.3: Water-reactive substances/Dangerous when wet materials

Class 5—Oxidizing substances and Organic peroxides
Division 5.1: Oxidizing substances
Division 5.2: Organic peroxides

Class 6—Toxic* substances and Infectious substances
Division 6.1: Toxic* substances
Division 6.2: Infectious substances

Class 7—Radioactive materials

Class 8—Corrosive substances

Class 9—Miscellaneous hazardous materials/Products, Substances, or Organisms

*The words "poison" or "poisonous" are synonymous with the word "toxic."

Spill Handling: Spill or leak information provided is intended to be used only as a guide. The term *Issue warning* is used when the chemical is a poison, has a high flammability, is a water contaminant, is an air contaminant (so as to be hazardous to life), is an oxidizing material, or is corrosive. *Restrict access* is used for those chemicals that are unusually and immediately hazardous to personnel unless they are protected properly by appropriate protective clothing, eye protection, and respiratory protection equipment. *Evacuate area* is used primarily for unusually poisonous chemicals or those that ignite easily. *Mechanical containment* is used for water-insoluble chemicals that float and do not evaporate readily. *Should be removed* is used for chemicals that cannot be allowed to disperse because of potentially harmful effects on humans or on the ecological system in general. The term is not used unless there is a reasonable chance of preventing dispersal, after a discharge or leak, by chemical and physical treatment. *Chemical and physical treatment* is recommended for chemicals that can be removed by skimming, pumping, dredging, burning, neutralization, absorption, coagulation, or precipitation. The corrective response may also include the use of dispersing agents, sinking agents, and biological treatment. *Disperse and flush* is used for chemicals that can be made nonhazardous to humans by simple dilution with water. In a few cases the response is indicated even when the compound reacts with water because, when proper care is taken, dilution is still the most effective way of removing the primary hazard. This material safety data sheet information is drawn from a variety of sources including New Jersey Department of Health and Senior Services *Hazardous Substance Fact Sheets*[70] and EPA *Profiles on Extremely Hazardous Substances.*[82]

Fire Extinguishing: Fire information provided is intended to be used only as a guide. Certain extinguishing agents should not to be used because the listed agents react with the chemical and have the potential to create an additional hazard. In some cases they are listed because they are ineffective in putting out the fire. Many chemicals decompose or burn to give off toxic and irritating gases. Such gases may also be given off by chemicals that vaporize in the heat of a fire without either decomposing or burning. If no entry appears, the combustion products are thought to be similar to those formed by the burning of oil, gasoline, or alcohol; they include carbon monoxide (poisonous), carbon dioxide, and water vapor. The specific combustion products are usually not well known over the wide variety of conditions existing in fires; some may be hazardous. This information is drawn from NFPA publications,[17] FEMA,[78] and other sources. Any characteristic behavior that might increase significantly the hazard involved in a fire is described. The formation of flammable vapor clouds or dense smoke, the possibility of polymerization, and explosions is stated in this section and/or the incompatibility section. Unusual difficulty in extinguishing the fire is noted.

Disposal Method Suggested: The disposal methods for various chemical substances have been drawn from various sources, including government documents and a UN publication.[22, 79]

References: The general bibliography for this volume follows immediately. It includes both general reference sources and references dealing with analytical methods. The references at the end of individual chemical records are generally restricted to: references dealing only with that particular compound; and references which, in turn, contain bibliographies giving references to the original literature on toxicological and other behavior of the substance in question.

Key to Abbreviations and Acronyms

α — the Greek letter *alpha*; used as a prefix to denote the carbon atom in a straight chain compound to which the principal group is attached.

as- — prefix for asymmetric

ACGIH — American Conference of Governmental Industrial Hygienists

AEGL — Acute Emergency Guideline Level, developed by the EPA

AIHA — American Industrial Hygiene Association

approx. — approximately

asym- — prefix for asymmetric

atm. — atmosphere

ATSDR — Agency for Toxic Substances and Disease Registry

β — the Greek letter *beta*

BEI — Biological Exposure Indices (ACGIH in the USA; South Africa; New Zealand)

BLV — Biological Limit Values

BP — boiling point

$^{\circ}$C — degrees Centigrade

CAA — Clean Air Act

CAAA — Clean Air Act Amendments of 1990

CAMEO — Computer-Aided Management of Emergency Operations (NOAA)

carc. — carcinogen

CAS — Chemical Abstract Service

cc — cubic centimeter

cc — closed cup (Flash point)

CDC — US Center for Disease Control

CEPA — Canadian Environmental Protection Act

CERCLA — Comprehensive Environmental Response, Compensation, And Liability Act

CFCs — chlorofluorocarbons

CFR — *Code of Federal Regulations*

cis- — (Latin: on this side). Indicating one of two geometrical isomers in which certain atoms or groups are on the same side of a plane

CMA — Chemical Manufacturers Association

CPR — Cardiopulmonary Resuscitation

CWA — Clean Water Act

cyclo- — (Greek, circle). Cyclic, ring structure; as cyclohexane

Δ or δ — Greek letter *delta*

DFG — Deutsche Forschungs-gemeinschaft

DOE — US Department of Energy

DOT — US Department of Transportation

DOT ID — Department of Transportation Identification Numbers

EEC or EC — European Economic Community

EEGL — Emergency Exposure Guidance Level

EHS — Extremely Hazardous Substances

EINECS — European Inventory of Existing Commercial Chemical Substances

ε — Greek letter *epsilon*

EPA (US) — Environmental Protection Agency

EPCRA — Emergency Planning and Community Right-to-Know Act

ESLI — End of Service Life Indicator

EU — European Union

$^{\circ}$F — degrees Fahrenheit

FDA — Food and Drug Administration

FEMA — Federal Emergency Management Agency

FR — *Federal Register*

γ — Greek letter *gamma*

GHS — Globally Harmonized System of Classification and Labeling of Chemicals

h — hour(s)

HCFC — hydrochlorofluorocarbons

HCS — Hazard Communication Standard

HSDB — Hazardous Substances Data Bank

IARC — International Agency for Research on Cancer

IATA — International Air Transport Association

IDLH — Immediately Dangerous to Life or Health

ILO — International Labor Office

IRIS — Integrated Risk Information System (EPA)

iso- — (Greek, equal, alike). Usually denoting an isomer of a compound

kg — kilogram(s)

l — liter(s)

lb — pound(s)

LC_{50} — The concentration of a substance in air that kills 50% of the test population.

LC_{Lo} — The lowest concentration of a substance in air that has been shown to cause death in a test population.

LD_{50} — The dose of a substance administered by any route (other than inhalation) that causes death to 50% of the test population.

LD_{Lo} — The lowest dose of a substance administered by any route (other than inhalation) that has been shown to cause death in a test population.

LEL — Lower explosive (flammable) limit in air, % by volume at room temperature or other temperature as noted

LEPC — Local Emergency Planning Committees

LTEL — Long-Term Exposure Limit (UK)

m- — an abbreviation for "*meta-*," a prefix used to distinguish between isomers or nearly related compounds

m^3 — cubic meter

MACT	Maximum Achievable Control Technology (CAA)
MAK	airborne exposure limit "Maximale Arbeitsplatz-Konzentration" (maximum workplace concentration) from the Deutsche Forschungs-gemeinschaft (DFG), German Research Foundation.
MCL	Maximum Contaminant Level (SDWA)
MCLG	Maximum Contaminant Level Goal (SDWA)
mg	milligram(s)
μ	micro
μg	microgram(s)
min	minute(s)
mmHg	millimeters of mercury (non-SI symbol for pressure). Also known as "torr"
mppcf	million particles per cubic foot
MSDS	Material Safety Data Sheets
n-	abbreviation for "normal," referring to the arrangement of carbon atoms in a chemical molecule prefix for normal
N-	Symbol used in some chemical names, indicating that the next section of the name refers to a chemical group attached to a nitrogen atom. The bond to the nitrogen atom
NCI	National Cancer Institute (USA)
NFPA	National Fire Protection Association (USA)
NIOSH	National Institute for Occupational Safety and Health (USA)
NLM	National Library of Medicine (USA)
NOAA	National Oceanic and Atmospheric Administration (USA)
NOAEL	No Observed Adverse Effect Level
NOS	not otherwise specified
NPRI	National Pollutant Release Inventory (Canada)
NTIS	National Technical Information Service (USA)
NTP	National Toxicology Program (USA)
o-	*ortho-*, a prefix used to distinguish between isomers or nearly related compounds
OEL	Occupational Exposure Limit
ω	Greek letter *omega*
oc	open cup
OSHA	Occupational Safety and Health Administration (USA)
Oxy	Oxidizer or oxidizing agent
p-	an abbreviation for "*para-*," a prefix used to distinguish between isomers or nearly related compounds
PAC	Protective Action Criterion (US DOE)
PBB	polybrominated biphenyl
PCB	polychlorinated biphenyl
PE	polyethylene
PEL	Permissible Exposure Limit (USA, Japan, Mexico)

PNOS	Particulates Not Otherwise Specified
POTW	Publicly Owned Treatments Works
PP	polypropylene
ppb	parts per billion
PPE	Personal Protective Equipment
ppm	parts per million
PQL	Practical Quantitation Limit (RCRA)
prim-	prefix for primary
REL	Recommended Exposure Limits (NIOSH)
RQ	Reportable Quantity
RTECS	Registry of Toxic Effects of Chemical Substances
RTK	Right-to-Know
SARA	Superfund Amendments and Reauthorization Act
s. carc.	Suspected Carcinogen
SCAPA	Subcommittee on Consequence Assessment and Protective Actions (DOE)
SCBA	Self-Contained Breathing Apparatus
SDWA	Safe Drinking Water Act
sec-	prefix for secondary
SERC	State Emergency Response Commissions
SMCL	Secondary Maximum Contaminant Levels (SDWA)
STEL	Short-Term Exposure Limit
sym-	abbreviation for "symmetrical," referring to a particular arrangement of elements within a chemical molecule
t-	prefix for tertiary
TC_{Lo}	The lowest concentration of a substance in air that has been shown to produce toxic effects in a test population.
TD_{Lo}	The lowest dose of a substance that has been shown to produce toxic effects in a test population
TEEL	Temporary Emergency Exposure Limit
TRK	Technical Guiding Concentrations (DFG) for workplace control of carcinogens
tert-	abbreviation for "tertiary," referring to a particular arrangement of elements within a chemical molecule
TLV®	Threshold Limit Value (ACGIH)
TQ	Threshold Quantity
trans-	(Latin: across). Indicating that one of two geometrical isomers in which certain atoms or groups are on opposite sides of a plane
TRI	Toxic Release Inventory
TSCA	Toxic Substances Control Act
TWA	Time-Weighted Average. Often shown as TWA—8 h
UEL	Upper Explosive (flammable) Limit in air, % by volume at room temperature or other temperature as noted
UN	United Nations
unsym-	prefix for asymmetric

USDA	US Department of Agriculture	<	symbol for "less than"
USCG	US Coast Guard	≤	symbol for "less than or equal to"
VOCs	Volatile Organic Compounds	≥	symbol for "greater than or equal to"
WEEL	Workplace Environmental Exposure Level (AIHA)	°	degrees of temperature
		%	percent
WHMIS	Workplace Hazardous Materials Information System (Canada)		
>	symbol for "greater than"		

L

Lactonitrile   L:0050

Molecular Formula: C_3H_5NO
Common Formula: $CH_3CHOHCN$
Synonyms: Acetaldehyde cyanohydrin; 2-Hydroxypropionitrile; Lactonitrilo (Spanish); NSC-7764; Propionitrile, 2-hydroxy-
CAS Registry Number: 78-97-7
RTECS® Number: OD8225000
UN/NA & ERG Number: UN3276 (Nitriles, toxic, liquid, n.o.s.)/151
EC Number: 201-163-2
Regulatory Authority and Advisory Bodies
SUPERFUND/EPCRA 40CFR355, Extremely Hazardous Substances: TPQ = 1000 lb (454 kg).
Reportable Quantity (RQ): 1000 lb (454 kg).
European/International Regulations: not listed in Annex 1.
WGK (German Aquatic Hazard Class): No value assigned.
Description: Lactonitrile is a colorless to straw-colored liquid. Molecular weight = 71.09; Boiling point = 183°C (slight decomposition); Freezing/Melting point = −40°C; Flash point = 77°C. Hazard Identification (based on NFPA-704 M Rating System): Health 3, Flammability 2, Reactivity 0. Soluble in water.
Potential Exposure: This material is used as a solvent and as an intermediate in the production of ethyl lactate and lactic acid.
Incompatibilities: Alkalis form hydrogen cyanide. Violent reaction with strong oxidizers.
Permissible Exposure Limits in Air
Protective Action Criteria (PAC)
TEEL-0: 3.5 mg/m^3
PAC-1: 10 mg/m^3
PAC-2: 18 mg/m^3
PAC-3: 150 mg/m^3
Determination in Air:
NIOSH REL: (nitriles) 2 ppm, Ceiling Concentration, not to be exceeded in any 15-min work period.
Permissible Concentration in Water: See NIOSH Criteria Document 212 *Nitriles*.
Routes of Entry: Inhalation, ingestion, skin contact.
Harmful Effects and Symptoms
Short Term Exposure: Signs and symptoms of acute exposure to lactonitrile may include hypertension (high blood pressure) and tachycardia (rapid heart rate), followed by hypotension (low blood pressure) and bradycardia (slow heart rate). Cherry-red mucous membranes and blood, cardiac arrhythmias; other cardiac abnormalities are common. Cyanosis (blue tint to the skin and mucous membranes) may be present, following exposure to lactonitrile. Salivation, nausea, and vomiting may also occur. Tachypnea (rapid respiratory rate) may be followed by respiratory depression. Lung hemorrhage and pulmonary edema may occur. Headache, vertigo (dizziness), agitation, and giddiness may be followed by combative behavior, convulsions, paralysis, protruding eyeballs, dilated and unreactive pupils, and coma. Lactonitrile is irritating to the skin and mucous membranes. Lacrimation (tearing) and a burning sensation of the mouth and throat are common. *Warning:* Heart palpitations may occur within minutes after exposure. Caution is advised. Vital signs should be monitored closely. Symptoms may be delayed.
Long Term Exposure: May cause liver and kidney damage.
Points of Attack: Respiratory system, central nervous system, cardiovascular system, liver, kidneys.
Medical Surveillance: Liver function tests. Kidney function tests. See also NIOSH Criteria Document 212 *Nitriles*.
First Aid: If this chemical gets into the eyes, remove any contact lenses at once and irrigate immediately for at least 15 min, occasionally lifting upper and lower lids. Seek medical attention immediately. If this chemical contacts the skin, remove contaminated clothing and wash immediately with soap and water. Seek medical attention immediately. If this chemical has been inhaled, remove from exposure, begin rescue breathing (using universal precautions, including resuscitation mask) if breathing has stopped and CPR if heart action has stopped. Transfer promptly to a medical facility. When this chemical has been swallowed, get medical attention. Give large quantities of water and induce vomiting. Do not make an unconscious person vomit.
Personal Protective Methods: Wear protective gloves and clothing to prevent any reasonable probability of skin contact. Safety equipment suppliers/manufacturers can provide recommendations on the most protective glove/clothing material for your operation. All protective clothing (suits, gloves, footwear, headgear) should be clean, available each day, and put on before work. Contact lenses should not be worn when working with this chemical. Wear splash-proof chemical goggles and face shield unless full face-piece respiratory protection is worn. Employees should wash immediately with soap when skin is wet or contaminated. Provide emergency showers and eyewash. The NIOSH REL for nitriles is a Ceiling Concentration of 6 mg/m^3, not to be exceeded in any 15-min work period.
Respirator Selection: Follow regulations in OSHA 29CFR1910.134 or European Standard EN149. Use a NIOSH/MSHA- or European Standard EN149-approved respirator; or use an approved supplied-air respirator with a full face-piece operated in the positive-pressure mode, or with a full face-piece, hood, or helmet in the continuous-flow mode; or use a NIOSH/MSHA- or European Standard EN149-approved self-contained breathing apparatus with a full face-piece operated in pressure-demand or other positive-pressure mode.
Storage: Color Code—Blue: Health Hazard/Poison: Store in a secure poison location. Prior to working with this

chemical you should be trained on its proper handling and storage. Store in tightly closed containers in a cool, well-ventilated area away from oxidizers and alkalis. Store in a refrigerator under an inert atmosphere and protect from light for prolonged storage. Where possible, automatically pump liquid from drums or other storage containers to process containers.

Shipping: Nitriles, toxic, n.o.s. must be labeled "POISONOUS/TOXIC MATERIALS." This chemical falls in Hazard Class 6.1.

Spill Handling: Evacuate and restrict persons not wearing protective equipment from area of spill or leak until cleanup is complete. Remove all ignition sources. Ventilate area of spill or leak. Absorb liquids in vermiculite, dry sand, earth, peat, carbon, or a similar material and deposit in sealed containers. Keep this chemical out of a confined space, such as a sewer, because of the possibility of an explosion, unless the sewer is designed to prevent the buildup of explosive concentrations. It may be necessary to contain and dispose of this chemical as a hazardous waste. If material or contaminated runoff enters waterways, notify downstream users of potentially contaminated waters. Contact your local or federal environmental protection agency for specific recommendations. If employees are required to clean up spills, they must be properly trained and equipped. OSHA 1910.120(q) may be applicable.

Initial isolation and protective action distances: Distances shown are likely to be affected during the first 30 min after materials are spilled and could increase with time. If more than one tank car, cargo tank, portable tank, or large cylinder is involved in the incident is leaking, the protective action distance may need to be increased. You may need to seek emergency information from CHEMTREC at (800) 424-9300 or seek professional environmental engineering assistance from the US EPA Environmental Response Team at (908) 548-8730 (24-h response line).

Small spills (From a small package or a small leak from a large package)
First: Isolate in all directions (feet/meters)   100/30
Then: Protect persons downwind (miles/kilometers)
Day   0.1/0.1
Night   0.1/0.2

Large spills (From a large package or from many small packages)
First: Isolate in all directions (feet/meters)   200/60
Then: Protect persons downwind (miles/kilometers)
Day   0.3/0.5
Night   0.5/0.9

Fire Extinguishing: This chemical is a combustible liquid. Poisonous gases, including cyanide, are produced in fire. Use dry chemical, carbon dioxide, or alcohol foam extinguishers. Vapors are heavier than air and will collect in low areas. Vapors may travel long distances to ignition sources and flashback. Vapors in confined areas may explode when exposed to fire. Containers may explode in fire. Storage containers and parts of containers may rocket great distances, in many directions. If material or contaminated runoff enters waterways, notify downstream users of potentially contaminated waters. Notify local health and fire officials and pollution control agencies. From a secure, explosion-proof location, use water spray to cool exposed containers. If cooling streams are ineffective (venting sound increases in volume and pitch, tank discolors, or shows any signs of deforming), withdraw immediately to a secure position. If employees are expected to fight fires, they must be trained and equipped in OSHA 1910.156. The only respirators recommended for firefighting are self-contained breathing apparatuses that have full face-pieces and are operated in a pressure-demand or other positive-pressure mode.

Reference
US Environmental Protection Agency. (November 30, 1987). *Chemical Hazard Information Profile: Lactonitrile.* Washington, DC: Chemical Emergency Preparedness Program

Lead & inorganic compounds L:0100

Molecular Formula: Pb
Synonyms: C.I. 77575; C.I. Pigment metal 4; Glover; Haro mix CE-701; Haro mix CK-711; Haro mix MH-204; JMI Sloop; KS-4; Lead-S2; Lead element; Lead flake; Litharge; Omaha; PB-S 100; Plomo (Spanish); Plumbum
CAS Registry Number: 7439-92-1 (metallic lead)
RTECS® Number: OF7520000
UN/NA & ERG Number: UN3077 (Environmentally hazardous substances, solid, n.o.s.)/171; UN3089 (Metal powder, flammable, n.o.s.)/170
EC Number: 231-100-4
Regulatory Authority and Advisory Bodies
Carcinogenicity: IARC (inorganic): Human Inadequate Evidence, animal Sufficient Evidence, *probably carcinogenic to humans,* Group 2A; (organic) Animal Inadequate Evidence, Human Inadequate Evidence, *possibly carcinogenic to humans,* Group 2B; NTP: 11th Report on Carcinogens, 2004: Reasonably anticipated to be a human carcinogen; EPA: Sufficient evidence from animal studies; inadequate evidence or no useful data from epidemiologic studies.
US EPA Gene-Tox Program, Positive: Sperm morphology—human; Negative: *In vivo* cytogenetics—nonhuman bone marrow; Negative: *In vitro* cytogenetics—human lymphocyte; Inconclusive: Carcinogenicity—mouse/rat; Inconclusive: *In vivo* cytogenetics—human lymphocyte.
Air Pollutant Standard Set. See below, "Permissible Exposure Limits in Air" section.
OSHA, 29CFR1910 Specifically Regulated Chemicals (See 29 CFR 1910.1025).
Banned or Severely Restricted (Many countries, especially in food) (UN)[35], (Maine, Minnesota).[61]

Lead, metallic:
Clean Water Act: 40CFR423, Appendix A, Priority Pollutants; Section 313 Water Priority Chemicals (57FR41331, 9/9/92); 40CFR401.15 Section 307 Toxic Pollutants as lead and compounds.
US EPA Hazardous Waste Number (RCRA No.): D008.
RCRA Toxicity Characteristic (Section 261.24), Maximum. Concentration of Contaminants, regulatory level, 5.0 mg/L.
RCRA, 40CFR261, Appendix 8 Hazardous Constituents, as lead compounds, n.o.s., waste number not listed.
RCRA 40CFR268.48; 61FR15654, Universal Treatment Standards: Wastewater (mg/L), 0.69; Nonwastewater (mg/L), 0.37 TCLP.
RCRA 40CFR264, Appendix 9; TSD Facilities Ground Water Monitoring List. Suggested test method(s) (PQL µg/L): 6010 (40); 7420 (1000); 7421 (10).
Safe Drinking Water Act: MCL, zero; MCLG, zero; Regulated chemical (47 FR 9352).
Reportable Quantity (RQ): 10 lb (4.54 kg).
EPCRA Section 313 Form R *de minimis* concentration reporting level: 0.1%.
California Proposition 65 Chemical: Cancer; 10/1/92; Developmental/Reproductive toxin (male, female) 2/27/87.
Canada, WHMIS, Ingredients Disclosure List Concentration 0.1%.

Lead compounds:
Clean Water Act: 40CFR401.15 Section 307 Toxic Pollutants as lead and compounds.
RCRA, 40CFR261, Appendix 8 Hazardous Constituents, waste number not listed, as lead compounds, n.o.s.
EPCRA Section 313: Includes any unique chemical substance that contains lead as part of that chemical's infrastructure. Form R *de minimis* concentration reporting level: inorganic compounds 0.1%; organic compounds 1.0%.
US DOT Regulated Marine Pollutant (49CFR172.101, Appendix B) as lead compounds, soluble, n.o.s.
California Proposition 65 Chemical: Cancer; 10/1/92; Developmental/Reproductive toxin (male, female) 2/27/87
Canada, WHMIS, Ingredients Disclosure List Concentration 1.0% as inorganic compounds, n.o.s.
WGK (German Aquatic Hazard Class): Nonwater polluting agent.

Description: Inorganic lead includes lead oxides, metallic lead, lead salts, and organic salts, such as lead soaps, but excludes lead arsenate and organic lead compounds. Lead is a blue−gray metal which is very soft and malleable. Lustrous when freshly cut; tarnishes in moist air. Specific gravity $(H_2O:1) = 11.34$; Molecular weight $= 207.19$; Boiling point $= 1750°C$; Freezing/Melting point $= 327°C$. Hazard Identification (based on NFPA-704 M Rating System): Health 3, Flammability 0, Reactivity 0. Insoluble in water.

Potential Exposure: Compound Description: Tumorigen, Mutagen, Organometallic; Reproductive Effector; Human Data. Metallic lead is used for lining tanks, piping, and other equipment where pliability and corrosion resistance are required, such as in the chemical industry in handling corrosive gases and liquids used in the manufacture of sulfuric acid; in petroleum refining; and in halogenation, sulfonation, extraction, and condensation processes; and in building industry. It is also used as an ingredient in solder, a body filler in the automobile industry, and a shielding material for X-rays and atomic radiation, in manufacture of tetraethyl lead and organic and inorganic lead compounds, pigments for paints and varnishes, storage batteries, fling glass, vitreous enameling, ceramics as a glaze, litharge rubber, plastics, and electronic devices. Lead is utilized in metallurgy and may be added to bronze, brass, steel, and other alloys to improve their characteristics. It forms alloys with antimony, tin, copper, etc. It is also used in metallizing to provide protective coatings and as a heat treatment bath in wire drawing. Exposures to lead dust may occur during mining, smelting, refining, and to fume, during high temperature ($>500°C$) operations, such as welding or spray coating of metals with molten lead. There are numerous applications for lead compounds, some of the more common being in the plates of electric batteries and accumulators; as compounding agents in rubber manufacture; as ingredients in paints, glazes, enamels, glass, pigments; and in the chemical industry. In addition, to these usual levels of exposure from environmental media, there exist miscellaneous sources which are hazardous. The level of exposure resulting from contact is highly variable. Children with pica for paint chips or for soil may experience elevation in blood lead, ranging from marginal to sufficiently great to cause clinical illness. Certain adults may also be exposed to hazardous concentrations of lead in the workplace, notably in lead smelters and storage battery manufacturing plants. Again, the range of exposure is highly variable. Women in the workplace are more likely to experience adverse effects from lead exposure than men due to the fact that their hematopoietic system is more lead-sensitive than men. Because of health concerns, lead from gasoline, paints, and ceramic products, caulking, and pipe solder, has been dramatically reduced in recent years.

Incompatibilities: Lead dust is flammable in air and may explode when exposed to heat or flame. Reacts with sulfuric acid, hot concentrated nitric acid, boiling concentrated hydrochloric acid. Powdered lead can react (possibly violently) with strong oxidizers, ammonium nitrate, chlorine trifluoride, chemically active metals, concentrated hydrogen peroxide, sodium acetylide. Incompatible with sodium azide, disodium acetylide, hydrogen peroxide, active metals—sodium, potassium, zirconium. Lead is attacked by pure water and weak organic acids in the presence of oxygen. Attacked at room temperature by chlorine, fluorine.

Permissible Exposure Limits in Air
OSHA PEL: 0.050 mg[Pb]/m³ TWA. [*Note:* The PEL also applies to other inorganic lead compounds (as Pb).] Other OSHA requirements can be found in 29 CFR 1910.1025. The OSHA PEL (8-h TWA) for lead in "nonferrous foundries with less than 20 employees" is 0.075 mg/m³. OSHA

considers "lead" to mean metallic lead; all inorganic lead compounds (lead oxides and lead salts); and a class of organic compounds called "soaps;" all other lead compounds are excluded from this definition.

NIOSH REL: 0.050 mg[Pb]/m^3 TWA. Suspected carcinogen. Limit exposure to lowest feasible level. [*Note:* The REL also applies to other lead compounds (as Pb).] *Note:* NIOSH considers "lead" to mean metallic lead, lead oxides, and lead salts (including organic salts, such as lead "soaps" but excluding lead arsenate). Air concentrations should be maintained so that worker blood lead remains less than 0.060 mg[Pb]/100 g of whole blood. See *NIOSH Pocket Guide*, Appendix C.

ACGIH TLV®[1]: 0.05 mg[Pb]/m^3 TWA; BEI: 30 μg[Pb]/100 mL (blood). (Note: The TLV also applies to lead, inorganic compounds.) *Note:* women of child-bearing potential whose blood exceeds 10 μg[Pb]/dL are at risk of delivering a child with a blood [Pb] over the current CDC guideline of 10 μg[Pb]/dL and may cause birth defects. Confirmed animal carcinogen with unknown relevance to humans.

Protective Action Criteria (PAC)

NIOSH IDLH: 100 mg [Pb]/m^3

Metallic lead

TEEL-0: 0.05 mg/m^3
PAC-1: 0.15 mg/m^3
PAC-2: 0.25 mg/m^3
PAC-3: 100 mg/m^3

European OEL: 0.15 mg[Pb]/m^3 TWA (2002)

DFG MAK: BAT: 400 μg[Pb]/L (blood) not fixed; 100 μg [Pb]/L (blood) women age <45; Carcinogen Category 2; Germ Cell Mutagen Group 3A.

Austria: MAK 0.1 mg/m^3, 1999; Denmark: TWA 0.1 mg/m^3, 1999; Japan 0.1 mg/m^3, 1999; Norway: TWA 0.05 mg/m^3, 1999; Poland: MAC (TWA) 0.05 mg/m^3, 1999; Sweden: TWA 0.05 mg/m^3 (resp. dust), 1999; Sweden: TWA 0.1 mg/m^3 (total dust), 1999; the Netherlands: MAC-TGG 0.15 mg/m^3, 2003. The EPA requires lead in air not to exceed 1.5 μg/m^3 averaged over 3 months; the Czech Republic at 0.7 μg/m^3 on a daily average basis *and* at 2.0 μg/m^3 on a momentary basis. Several states have set guidelines or standards for lead in ambient air[60] ranging from 0.068 μg/m^3 (Massachusetts) to 0.357−1.5 μg/m^3 (North Dakota and Pennsylvania) to 2.5 μg/m^3 (New York) to 3.0 μg/m^3 (Connecticut) to 4.0 μg/m^3 (Nevada).

Determination in Air: Use NIOSH Analytical Method (IV) s #7082, 7105, 7300, 7301, 7303, 7700, 7701, 7702, 9100, 9102, 9105; OSHA Analytical Methods ID-121, ID-125G, ID-206.

Permissible Concentration in Water: The EPA limits lead in drinking water to 15 μg/L. Various organizations worldwide have set other standards for lead in drinking water as follows[35] (all in mg/L): Argentina 0.01; the Czech Republic 0.05; Germany 0.04; EEC 0.05; Japan 0.10; Mexico 0.05; former USSR-UNEP/IRPTC joint project 0.03; WHO 0.10. The states of Maine and Minnesota have set guidelines for lead in drinking water[61] at the level of 20 μg/L.

Determination in Water: Digestion followed by atomic absorption, or by colorimetric (dithizone) analysis, or by inductively coupled plasma (ICP) optical emission spectrometry. That gives total lead; dissolved lead may be determined by 0.45-μm filtration prior to such analysis.

Routes of Entry: Ingestion of dust, inhalation of dust or fume, skin and/or eye contact.

Harmful Effects and Symptoms

Short Term Exposure: The effects of exposure to fumes and dusts of inorganic lead may be slow to develop. Extremely high exposures can lead to seizures, coma, and death, but symptoms occur after weeks to months of exposure. The earliest symptoms may include tiredness, decreased physical fitness, fatigue, sleep disturbance, headache, aching bones and muscles, constipation, abdominal pains and decreased appetite, moodiness (mostly irritability and depression). Skin contact may cause irritation. Eyes contact may cause irritation. Ingestion of large amounts of lead may lead to seizures, coma, and death.

Long Term Exposure: Lead can accumulate in the body over a period of time. Therefore, long-term exposures to lower levels can result in a buildup of lead in the body and more severe symptoms. These may include high blood pressure, anemia, pale skin, a blue line at the gum margin, paralysis of forearm, wrist joint, and fingers, causing decreased hand-grip strength, abdominal pain, severe constipation, nausea, vomiting. Prolonged exposure to high enough levels may result in serious, permanent kidney and brain damage. If the nervous system is affected, usually due to very high exposures, the resulting effects include severe headache, mood and personality changes, retarded mental development, convulsions, coma, delirium, and death. In nonfatal cases, recovery is slow and not always complete. Alcohol ingestion and physical exertion may bring on symptoms. Lead is a probable teratogen in humans. Continuous exposure can result in decreased fertility in males and females. Elevated lead exposure of either parent before pregnancy can increase the chances of miscarriage or birth defects. Exposure of the mother during pregnancy can cause birth defects. May cause retarded development of the newborn.

Note: Lead is a cumulative poison. Increasing amounts can build up in the body, eventually reaching a point where symptoms and disability occur. Lead dust carried home on contaminated clothing can result in exposure and symptoms in other family members. Standards only protect for inhalation exposure. Extra precautions should be taken if skin exposure also occurs. The effects may be delayed. Medical observation is recommended. The biological half-life (PB) in the bones of humans = 10 years.

Points of Attack: Eyes, gastrointestinal tract, central nervous system, kidneys, blood, gingival tissue.

Medical Surveillance: OSHA mandated medical tests: whole blood (chemical/metabolite), blood lead level, hemoglobin, hematocrit, zinc protoporphyrin, blood urea nitrogen, serum creatinine, urinalysis (routine), microscopic examination. Prior to initial exposure, and annually for exposed person having blood lead readings exceeding 40 μg/100 g of whole blood, OSHA also requires a complete medical history, complete blood count and kidney function tests in addition to the tests listed above. OSHA defines "exposure" for these tests as air levels which average 30 μg of lead or more in a cubic meter of air. OSHA requires your employer to send the doctor a copy of the lead standard and provide one for you.

Note: Blood-lead level is a good indicator of total lead exposure. Current OSHA regulations require that if an individual has a blood-lead level greater than or equal to 0.050 mg lead per 100 mL blood, he or she must be removed from all exposures to lead and cannot return to the exposure environment until the blood level falls to 0.040 mg lead per 100 mL blood or less. The following tolerance levels for occupational exposures may also be useful: ACGIH BEI = 50 mg/L (blood); 150 mg/g creatinine (urine). DFG BAT = 70 mg/L (blood); 30 mg/L (blood) for women <45 years old.

NIOSH lists the following tests: whole blood (chemical/metabolite); blood lead level; biologic tissue/biopsy; complete blood count; nerve conduction studies; neurologic examination/electromyography; red blood cells/count; red blood cells/count: zinc protoporphyrin; urine (chemical/metabolite); urine (chemical/metabolite), end-of-shift; urine (chemical/metabolite), 24-h collection; Zinc Protoporphyrin; Zinc Protoporphyrin, after 1-month exposure.

First Aid: If this chemical gets into the eyes, remove any contact lenses at once and irrigate immediately for at least 15 min, occasionally lifting upper and lower lids. Seek medical attention immediately. If this chemical contacts the skin, remove contaminated clothing and wash immediately with soap and water. When this chemical has been swallowed, get medical attention.

Antidotes and special procedures: Persons with significant lead poisoning are sometimes treated with Ca EDTA while hospitalized. This "chelating" drug causes a rush of lead from the body organs into the blood and kidneys, and thus has its own hazards, and must be administered only by highly experienced medical personnel under controlled conditions and careful observation. Ca EDTA or similar drugs should never be used to prevent poisoning while exposure continues or without strict exposure control, as severe kidney damage can result.

Note to physician: For severe poisoning BAL [British Anti-Lewisite, dimercaprol, dithiopropanol ($C_3H_8OS_2$)] has been used to treat toxic symptoms of certain heavy metals poisoning. In the case of lead poisoning it may have SOME value. Although BAL is reported to have a large margin of safety, caution must be exercised, because toxic effects may be caused by excessive dosage. Most can be prevented by premedication with 1-ephedrine sulfate (CAS: 134-72-5).

Personal Protective Methods: Prevent skin contact: Any barrier that will prevent contamination from the dust. Safety equipment suppliers/manufacturers can provide recommendations on the most protective glove/clothing material for your operation. All protective clothing (suits, gloves, footwear, headgear) should be clean, available each day, and put on before work. Vacuum clothing with HEPA. Wear dust-proof chemical goggles and face shield unless full face-piece respiratory protection is worn. Employees should wash immediately with soap when skin is wet or contaminated. Provide emergency showers and eyewash. See also NIOSH 78-158 *Lead, inorganic dusts, and fumes.*

Respirator Selection: OSHA: *0.5 mg/m³:* 100XQ (APF = 10) [any air-purifying respirator with an N100, R100, or P100 filter (including N100, R100, and P100 filtering face-pieces) except quarter-mask respirators]. *1.25 mg/m³:* Sa:Cf (APF = 25) (any supplied-air respirator operated in a continuous-flow mode) or PaprHie (APF = 25) (any powered, air-purifying respirator with a high-efficiency particulate filter). *2.5 mg/m³:* 100F (APF = 50) (any air purifying, full-face-piece respirator with an N100, R100, or P100 filter) or SaT: Cf (APF = 50) (any supplied-air respirator that has a tight-fitting face-piece and is operated in a continuous-flow mode) or PaprTHie (APF = 50) (any powered, air-purifying respirator with a tight-fitting face-piece and a high-efficiency particulate filter) or SCBAF (APF = 50) (any self-contained breathing apparatus with a full face-piece) or SaF (APF = 50) (any supplied-air respirator with a full face-piece). *50 mg/m³:* Sa: Pd,Pp (APF = 1000) (any supplied-air respirator operated in a pressure-demand or other positive-pressure mode). *100 mg/m³:* SaF: Pd,Pp (APF = 2000) (any supplied-air respirator that has a full face-piece and is operated in a pressure-demand or other positive-pressure mode). *Emergency or planned entry into unknown concentrations or IDLH conditions:* SCBAF: Pd,Pp (APF = 10,000) (any self-contained breathing apparatus that has a full face-piece and is operated in a pressure-demand or other positive-pressure mode) or SaF: Pd,Pp: ASCBA (any supplied-air respirator that has a full face-piece and is operated in a pressure-demand or other positive-pressure mode in combination with an auxiliary self-contained breathing apparatus operated in a pressure-demand or other positive-pressure mode). *Escape:* 100F (APF = 50) (any air purifying, full-face-piece respirator with an N100, R100, or P100 filter) or SCBAE (any appropriate escape-type, self-contained breathing apparatus).

Storage: Color Code—Blue: Health Hazard/Poison: Store in a secure poison location. Prior to working with this chemical you should be trained on its proper handling and storage. Lead must be stored to avoid contact with oxidizers (such as perchlorates, peroxides, permanganates, chlorates, and nitrates) and chemically active metals (such as potassium, sodium, magnesium, and zinc), since violent reactions occur. Lead is regulated by OSHA Standard 1910. 1025. All requirements of the standard must be followed.

Shipping: For Metal powder, flammable, n.o.s. the required label is "SPONTANEOUSLY COMBUSTIBLE." They fall into Hazard Class 4.2 and Packing Group I.

Spill Handling: Evacuate persons not wearing protective equipment from area of spill or leak until cleanup is complete. Remove all ignition sources. Get all workers out of spill area. Put on necessary protective equipment including respirators. If spill is a solution, cover with absorbent and shovel into suitable container. If spill is in powder form, vacuum (HEPA) whenever possible to avoid raising dust by sweeping or blowing. Place in suitable container. Ventilate area after cleanup is complete. It may be necessary to contain and dispose of this chemical as a hazardous waste. If material or contaminated runoff enters waterways, notify downstream users of potentially contaminated waters. Contact your local or federal environmental protection agency for specific recommendations. If employees are required to clean up spills, they must be properly trained and equipped. OSHA 1910.120(q) may be applicable.

Fire Extinguishing: Lead powder is flammable when exposed to heat or flame. *Do not use water.* Use dry chemicals appropriate for extinguishing metal fires. Generally the problem is to fight a fire based on the surrounding combustible material since metallic lead is not combustible. Poisonous fumes, including lead, are produced in fire. If material or contaminated runoff enters waterways, notify downstream users of potentially contaminated waters. Notify local health and fire officials and pollution control agencies. From a secure, explosion-proof location, use water spray to cool exposed containers. If cooling streams are ineffective (venting sound increases in volume and pitch, tank discolors, or shows any signs of deforming), withdraw immediately to a secure position. If employees are expected to fight fires, they must be trained and equipped in OSHA 1910.156. The only respirators recommended for firefighting are self-contained breathing apparatuses that have full face-pieces and are operated in a pressure-demand or other positive-pressure mode.

Disposal Method Suggested: Recycle wherever possible. May be treated with HNO_3 to give lead nitrate; the lead may be precipitated as sulfide, which is then sent to a lead recovery plant.[22] Consult with environmental regulatory agencies for guidance on acceptable disposal practices. Generators of waste containing this contaminant ($\geq$100 kg/mo) must conform with EPA regulations governing storage, transportation, treatment, and waste disposal.

References

National Institute for Occupational Safety and Health. (1973). *Criteria for a Recommended Standard: Occupational Exposure to Inorganic Lead*, NIOSH Document No. 73-11010. Washington, DC

US Environmental Protection Agency. (1980). *Lead: Ambient Water Quality Criteria*. Washington, DC

US Environmental Protection Agency. (December 1979). *Status Assessment of Toxic Chemicals: Lead*, Report EPA-600/2-79-210h. Cincinnati, OH

US Environmental Protection Agency. (1978). *Reviews of the Environmental Effects of Pollutants: VII. Lead*. Report EPA-600/1-78-029. Cincinnati, OH

US Environmental Protection Agency. (1977). *Air Quality Criteria Document for Lead*, Report EPA-600/8-77-017. Research Triangle Park, NC

National Academy of Sciences. (1972). *Medical and Biologic Effects of Environmental Pollutants: Lead, Airborne Lead in Perspective*. Washington, DC

US Environmental Protection Agency. (April 30, 1980). *Lead: Health and Environmental Effects Profile No. 121*. Washington, DC: Office of Solid Waste

World Health Organization. (1977). *Lead: Environmental Health Criteria No. 3*. Geneva, Switzerland

Sax, N. I. (Ed.). (1981). *Dangerous Properties of Industrial Materials Report*, 1, 47−49.

US Department of Health and Human Services. (June 1999). *ATSDR ToxFAQs: Lead*. Atlanta, GA

New York State Department of Health. (June 1982). *Chemical Fact Sheet: Lead (Metallic and Inorganic Compounds)*. Version 3. Albany, NY: Bureau of Toxic Substance Assessment

New Jersey Department of Health and Senior Services. (September 2001). *Hazardous Substances Fact Sheet: Lead*. Trenton, NJ

Lead acetate   L:0110

Molecular Formula: $C_4H_6O_4Pb$
Common Formula: $Pb(OCOCH_3)_2$
Synonyms: Acetate de plomb (French); Acetato de plomo (Spanish); Acetic acid, lead(2 +) salt; Acetic acid, lead(II) salt; Arseniato de plomo (Spanish); Black marking ink, 105E; Bleiacetat (German); Dibasic lead acetate; Leadac; Lead(2+) acetate; Lead(II) acetate; Lead acetate trihydrate; Lead acetate(II), trihydrate; Lead diacetate; Multilayer dielectric ink HD; Neutral lead acetate; Normal lead acetate; Plumbous acetate; Salt of saturn; Sugar of lead
CAS Registry Number: 301-04-2; 15347-57-6; 6080-56-4 [Lead(II) acetate]
RTECS® Number: AI5250000
UN/NA & ERG Number: UN1616/151
EC Number: 206-104-4 [*Annex I Index No.:* 082-007-00-9]; 215-630-3 [*Annex I Index No.:* 082-007-00-9] (basic)
Regulatory Authority and Advisory Bodies
Carcinogenicity: IARC (*as organic lead compounds*): Animal Inadequate Evidence; Human Inadequate Evidence, *not classifiable as carcinogenic to humans*, Group 3.
FDA—over-the-counter drug.
OSHA, 29CFR1910 Specifically Regulated Chemicals (See 29 CFR 1910.1025).
Air Pollutant Standard Set. See below, "Permissible Exposure Limits in Air" section.
Clean Water Act: Section 311 Hazardous Substances/RQ 40CFR117.3 (same as CERCLA, see below); Section 313 Water Priority Chemicals (57FR41331, 9/9/92); 40CFR401.15 Section 307 Toxic Pollutants as lead and compounds.

US EPA Hazardous Waste Number (RCRA No.): U144.
RCRA, 40CFR261, Appendix 8 Hazardous Constituents.
Reportable Quantity (RQ): 10 lb (4.54 kg).
EPCRA Section 313 Form R *de minimis* concentration reporting level: 1.0%.
US DOT Regulated Marine Pollutant (49CFR172.101, Appendix B).
California Proposition 65 Chemical: Cancer 1/1/88.
Canada, WHMIS, Ingredients Disclosure List Concentration 0.1%.
European/International Regulations (*301-04-2; 15347-57-6*): Hazard Symbol: T, N; Risk phrases: R61; R33; R48/22; R50/53; Safety phrases: S53; S45; S60; S61 (see Appendix 4).
WGK (German Aquatic Hazard Class): 3—Severe hazard to waters.

Description: Lead acetate is a white, flaky crystalline substance with a slight odor of acetic acid. Commercial grades may be powdered granules, or brown or gray lumps. Molecular weight = 325.29; Boiling point = 280°C; Freezing/Melting point = 75°C (decomposes above 200°C). Hazard Identification (based on NFPA-704 M Rating System): Health 3, Flammability 0, Reactivity 0. Soluble in water, but absorbs carbon dioxide upon exposure to air, and may become insoluble in water.

Potential Exposure: Compound Description: Agricultural Chemical; Tumorigen, Organometallic, Mutagen; Reproductive Effector; Human Data. Lead acetate is used as a color additive in hair dyes, as a mordant in cotton dyes, in the lead coating of metals, as a drier in paints, varnishes and pigment inks, and in medicines such as astringents.

Incompatibilities: Reacts violently with strong oxidizers, bromates, strong acids, chemically active metals, phosphates, carbonates, phenols. Contact with strong acids forms acetic acid. Incompatible with strong bases, ammonia, amines, cresols, isocyanates, alkylene oxides, epichlorohydrin, sulfites, resorcinol, salicylic acid, chloral hydrate.

Permissible Exposure Limits in Air
Protective Action Criteria (PAC)
301-04-2
TEEL-0: 4 mg/m^3
PAC-1: 10 mg/m^3
PAC-2: 75 mg/m^3
PAC-3: 500 mg/m^3
Lead acetate(II) 6080-56-4
TEEL-0: 40 mg/m^3
PAC-1: 125 mg/m^3
PAC-2: 500 mg/m^3
PAC-3: 500 mg/m^3
As organic lead
ACGIH TLV®[1]: No listing for organic lead compounds.
Determination in Air: Use NIOSH Analytical Method (IV) s #7082, 7105, 7300, 7301, 7303, 7700, 7701, 7702, 9100, 9102, 9105; OSHA Analytical Methods ID-121, ID-125G, ID-206.
Permissible Concentration in Water: The EPA limits lead in drinking water to 15 µg/L. Various organizations

worldwide have set other standards for lead in drinking water as follows[35] (all in mg/L): Argentina 0.01; the Czech Republic 0.05; Germany 0.04; EEC 0.05; Japan 0.10; Mexico 0.05; former USSR-UNEP/IRPTC joint project 0.03; WHO 0.10. The states of Maine and Minnesota have set guidelines for lead in drinking water[61] at the level of 20 µg/L.
Determination in Water: Digestion followed by atomic absorption, or by colorimetric (dithizone) analysis, or by inductively coupled plasma (ICP) optical emission spectrometry. That gives total lead; dissolved lead may be determined by 0.45-µm filtration prior to such analysis.
Routes of Entry: Inhalation, ingestion. Passes through the skin.
Harmful Effects and Symptoms
Short Term Exposure: This material is poisonous. It emits toxic fumes of lead when heated to decomposition. Symptoms of exposure include irritation on the eyes, respiratory tract, and alimentary tract; headache, nausea, vomiting, colic, constipation, leg cramps; muscle weakness; paralysis, paresthesias, depression, dizziness, loss of consciousness; coma; and death. High levels of exposure to lead acetate may affect the central nervous system and blood, causing anemia, nervous disorders, kidney impairment.

Long Term Exposure: Repeated exposure may cause lead to accumulate in the body and may cause lead poisoning. See above for symptoms. May affect the blood, bone marrow, cardiovascular system, kidneys, nervous system including weakness and poor coordination in the arms and legs, hemolytic anemia, increase in blood pressure, paralysis, brain damage, behavioral effects including irritability, reduced memory, and disturbed sleep. Has been shown to cause kidney cancer in animals; possibly carcinogenic to humans. A probable teratogen. May cause reduced growth of offspring after birth while also decreasing fertility in males.

Points of Attack: Kidneys, brain, nervous system, blood.
Medical Surveillance: OSHA mandated medical tests: whole blood (chemical/metabolite), blood lead level, hemoglobin, hematocrit, zinc protoporphyrin, blood urea nitrogen, serum creatinine, urinalysis (routine), microscopic examination. Prior to initial exposure, and annually for exposed person having blood lead readings exceeding 40 µg/100 g of whole blood, OSHA also requires a complete medical history, complete blood count and kidney function tests in addition to the tests listed above. OSHA defines "exposure" for these tests as air levels which average 30 µg of lead or more in a cubic meter of air. OSHA requires your employer to send the doctor a copy of the lead standard and provide one for you.
Note: Blood-lead level is a good indicator of total lead exposure. Current OSHA regulations require that if an individual has a blood-lead level greater than or equal to 0.050 mg lead per 100 mL blood, he or she must be removed from all exposures to lead and cannot return to the

exposure environment until the blood level falls to 0.040 mg lead per 100 mL blood or less. The following tolerance levels for occupational exposures may also be useful: ACGIH BEI = 50 mg/L (blood); 150 mg/g creatinine (urine). DFG BAT = 70 mg/L (blood); 30 mg/L (blood) for women <45 years old.

NIOSH lists the following tests: whole blood (chemical/metabolite); Blood Lead Level; biologic tissue/biopsy; Complete blood count; Nerve Conduction Studies; Neurologic Examination/Electromyography; red blood cells/count; red blood cells/count: Zinc Protoporphyrin; urine (chemical/metabolite); urine (chemical/metabolite), end-of-shift; urine (chemical/metabolite), 24-h collection; Zinc Protoporphyrin; Zinc Protoporphyrin, after 1-month exposure.

First Aid: Skin Contact[52]: Flood all areas of body that have contacted the substance with water. Do not wait to remove contaminated clothing; do it under the water stream. Use soap to help assure removal. Isolate contaminated clothing when removed to prevent contact by others. *Eye Contact:* Remove any contact lenses at once. Immediately flush eyes well with copious quantities of water or normal saline for at least 20–30 min. Seek medical attention. *Inhalation:* Leave contaminated area immediately; breathe fresh air. Proper respiratory protection must be supplied to any rescuers. If coughing, difficult breathing, or any other symptoms develop, seek medical attention at once, even if symptoms develop many hours after exposure. *Ingestion:* Contact a physician, hospital, or poison center at once. If the victim is unconscious or convulsing, do not induce vomiting or give anything by mouth. Assure that the patient's airway is open and lay him on his side with his head lower than his body and transport immediately to a medical facility. If conscious and not convulsing, give a glass of water to dilute the substance. Vomiting should not be induced without a physician's advice.

Antidotes and special procedures for lead: Persons with significant lead poisoning are sometimes treated with Ca EDTA while hospitalized. This "chelating" drug causes a rush of lead from the body organs into the blood and kidneys, and thus has its own hazards, and must be administered only by highly experienced medical personnel under controlled conditions and careful observation. Ca EDTA or similar drugs should never be used to prevent poisoning while exposure continues or without strict exposure control, as severe kidney damage can result.

Note to physician: For severe poisoning BAL [British Anti-Lewisite, dimercaprol, dithiopropanol ($C_3H_8OS_2$)] has been used to treat toxic symptoms of certain heavy metals poisoning. In the case of lead poisoning it may have SOME value. Although BAL is reported to have a large margin of safety, caution must be exercised, because toxic effects may be caused by excessive dosage. Most can be prevented by premedication with 1-ephedrine sulfate (CAS: 134-72-5).

Personal Protective Methods: Avoid dust inhalation; wear NIOSH and US Bureau of Mines approved dust mask. Wear protective gloves and clothing to prevent any reasonable probability of skin contact. Use any barrier that will prevent contamination from the dust. Safety equipment suppliers/manufacturers can provide recommendations on the most protective glove/clothing material for your operation. All protective clothing (suits, gloves, footwear, headgear) should be clean, available each day, and put on before work. Contact lenses should not be worn when working with this chemical. Wear dust-proof chemical goggles and face shield unless full face-piece respiratory protection is worn. Employees should wash immediately with soap when skin is wet or contaminated. Provide emergency showers and eyewash. See also NIOSH 78-158 *Lead, inorganic dusts and fumes.*

Respirator Selection: OSHA: *0.5 mg/m³:* 100XQ (APF = 10) [any air-purifying respirator with an N100, R100, or P100 filter (including N100, R100, and P100 filtering face-pieces) except quarter-mask respirators]. *1.25 mg/m³:* Sa:Cf (APF = 25) (any supplied-air respirator operated in a continuous-flow mode) or PaprHie (APF = 25) (any powered, air-purifying respirator with a high-efficiency particulate filter). *2.5 mg/m³:* 100F (APF = 50) (any air purifying, full-face-piece respirator with an N100, R100, or P100 filter) or SaT:Cf (APF = 50) (any supplied-air respirator that has a tight-fitting face-piece and is operated in a continuous-flow mode) or PaprTHie (APF = 50) (any powered, air-purifying respirator with a tight-fitting face-piece and a high-efficiency particulate filter) or SCBAF (APF = 50) (any self-contained breathing apparatus with a full face-piece) or SaF (APF = 50) (any supplied-air respirator with a full face-piece). *50 mg/m³:* Sa: Pd,Pp (APF = 1000) (any supplied-air respirator operated in a pressure-demand or other positive-pressure mode). *100 mg/m³:* SaF: Pd,Pp (APF = 2000) (any supplied-air respirator that has a full face-piece and is operated in a pressure-demand or other positive-pressure mode). *Emergency or planned entry into unknown concentrations or IDLH conditions:* SCBAF: Pd,Pp (APF = 10,000) (any self-contained breathing apparatus that has a full face-piece and is operated in a pressure-demand or other positive-pressure mode) or SaF: Pd,Pp: ASCBA (any supplied-air respirator that has a full face-piece and is operated in a pressure-demand or other positive-pressure mode in combination with an auxiliary self-contained breathing apparatus operated in a pressure-demand or other positive-pressure mode). *Escape:* 100F (APF = 50) (any air purifying, full-face-piece respirator with an N100, R100, or P100 filter) or SCBAE (any appropriate escape-type, self-contained breathing apparatus).

Storage: Color Code—Blue: Health Hazard/Poison: Store in a secure poison location. Prior to working with this chemical you should be trained on its proper handling and storage. Store in a cool, dry place and keep tightly covered and avoid contact with oxidizers, strong acids, chemically active metals. A regulated, marked area should be established where this chemical is handled, used, or stored in compliance with OSHA Standard 1910.1045.

Shipping: This compound requires a shipping label of "POISONOUS/TOXIC MATERIALS." It falls in Hazard Class 6.1 and Packing Group III.

Spill Handling: Evacuate persons not wearing protective equipment from area of spill or leak until cleanup is complete. Remove all ignition sources. Dampen spilled material with water to avoid airborne dust then transfer material to vapor-tight plastic bags for eventual disposal. Collect powdered material in the most convenient and safe manner and deposit in sealed containers. Ventilate area after cleanup is complete. It may be necessary to contain and dispose of this chemical as a hazardous waste. If material or contaminated runoff enters waterways, notify downstream users of potentially contaminated waters. Contact your local or federal environmental protection agency for specific recommendations. If employees are required to clean up spills, they must be properly trained and equipped. OSHA 1910.120(q) may be applicable.

Fire Extinguishing: Lead acetate may burn but does not easily ignite. Use dry chemical, carbon dioxide, water spray, or foam extinguishers. Poisonous fumes of lead are produced in fire. If material or contaminated runoff enters waterways, notify downstream users of potentially contaminated waters. Notify local health and fire officials and pollution control agencies. From a secure, explosion-proof location, use water spray to cool exposed containers. If cooling streams are ineffective (venting sound increases in volume and pitch, tank discolors, or shows any signs of deforming), withdraw immediately to a secure position. If employees are expected to fight fires, they must be trained and equipped in OSHA 1910.156. The only respirators recommended for firefighting are self-contained breathing apparatuses that have full face-pieces and are operated in a pressure-demand or other positive-pressure mode.

Disposal Method Suggested: Convert to nitrate using nitric acid, evaporate, then saturate with H_2S, wash and dry the sulfide and ship to the supplier.[22] Consult with environmental regulatory agencies for guidance on acceptable disposal practices. Generators of waste containing this contaminant (≥ 100 kg/mo) must conform with EPA regulations governing storage, transportation, treatment, and waste disposal.

References

Sax, N. I. (Ed.). *Dangerous Properties of Industrial Materials Report*, 1, No. 4, 79–82 (1981) and 6, No. 2, 73–79 (1986)
New Jersey Department of Health and Senior Services. (April 2002). *Hazardous Substances Fact Sheet: Lead Acetate*. Trenton, NJ

Lead arsenate   L:0120

Molecular Formula: $AsHO_4Pb$; $As_2O_8Pb_3$
Common Formula: $PbHAsO_4$; $Pb_3(AsO_4)_2$
Synonyms: *3687-31-8*: Arseniate de plomb (French); Arseniato de plomo (Spanish); Arsenic acid, lead salt; Lead acetate acid; Plumbous arsenate
7784-40-9: Acid lead arsenate; Acid lead arsenite; Acid lead orthoarsenate; Arsenate of lead; Arseniato de plomo

(Spanish); Arsenic acid, lead(2+); Arsenic acid, lead(II); Arsenic acid, lead salt; Arsinette; Dibasic lead arsenate; Gypsine; Lead acid arsenate; Ortho L10 dust; Ortho L40 dust; Plumbous arsenate; Salt arsenate of lead; Schultenite; Security; Soprabel; Standard lead arsenate; Talbot
10102-48-4: Acid lead arsenate; Acid lead orthoarsenate; Arsenate of lead; Arseniato de plomo (Spanish); Arsenic acid, lead(2+) salt; Arsinette; Dibasic lead arsenate; Gypsine; Ortho L10 dust; Ortho L40 dust; Schultenite; Security; Soprabel; Standard lead arsenate; Talbot
CAS Registry Number: 3687-31-8; 7784-40-9 (dibasic lead arsenate); 7645-25-2 (PbH_3AsO_4)
RTECS® Number: CG0980000 ($PbHAsO_4$); CG0990000 [$Pb_3(AsO_4)_2$]; CG1000000 ($Pb_xH_3AsO_4$)
UN/NA & ERG Number: UN1617/151
EC Number: 232-064-2 [*Annex I Index No.:* 082-011-00-0] (lead hydrogen arsenate); 222-979-5 (trilead diarsenate)
Regulatory Authority and Advisory Bodies
Carcinogenicity: IARC: Animal Inadequate Evidence, Human Inadequate Evidence, *possibly carcinogenic to humans*, Group 2A, 1987; Carcinogenicity: NTP: 11th Report on Carcinogens, 2004: Reasonably anticipated to be a human carcinogen; EPA: Sufficient evidence from animal studies; inadequate evidence or no useful data from epidemiologic studies.
Air Pollutant Standard Set. See below, "Permissible Exposure Limits in Air" section.
Banned or Severely Restricted (In Agriculture in India, Japan) (UN).[13]
US EPA, FIFRA 1998 Status of Pesticides: Canceled.
OSHA, 29CFR1910 Specifically Regulated Chemicals (See 29 CFR 1910.1025 and 1910. 1018).
Clean Water Act: Section 311 Hazardous Substances/RQ 40CFR117.3 (same as CERCLA, see below); Section 313 Water Priority Chemicals (57FR41331, 9/9/92); 40CFR401.15 Section 307 Toxic Pollutants as lead and compounds.
RCRA, 40CFR261, Appendix 8 Hazardous Constituents, waste number not listed, as lead compounds, n.o.s.
Reportable Quantity (RQ): 1 lb (0.454 kg).
EPCRA Section 313 Form R *de minimis* concentration reporting level: 0.1%.
US DOT Regulated Marine Pollutant (49CFR172.101, Appendix B).
California Proposition 65 Chemical: Cancer 2/27/87; 10/1/92 (lead and compounds).
Canada, WHMIS, Ingredients Disclosure List Concentration 0.1%.
European/International Regulations: Hazard Symbol: T; Risk phrases: R45; R23/25; Safety phrases: S1/2; S20/21; S28; S45 as arsenic compounds.
WGK (German Aquatic Hazard Class): 3—Severe hazard to waters.
Description: Lead arsenate is an odorless, heavy, white powder, or crystals. Molecular weight = 347.12; Freezing/Melting point = (decomposes) approximately 280°C. Also

listed at 1042°C (decomposes). Hazard Identification (based on NFPA-704 M Rating System): Health 3, Flammability 0, Reactivity 0. Insoluble in cold water; soluble in hot water.

Potential Exposure: Compound Description: Agricultural Chemical; Tumorigen. Used as an insecticide, herbicide, and in manufacture of drugs and veterinary tapeworm medicine.

Incompatibilities: Violent reactions occur from contact with oxidizers, chemically active metals, strong acids. Acids and acid mists cause the release of arsine, a deadly gas. Decomposes above 270°C forming toxic fumes including arsenic and lead compounds.

Permissible Exposure Limits in Air
As arsenic
OSHA PEL: 0.01 mg[As]/m³ TWA; cancer hazard that can be inhaled, see 29CFR1910.1018.
NIOSH REL: 0.002 mg[As]/m³ [15 min] Ceiling Concentration. Limit exposure to lowest feasible level. See *NIOSH Pocket Guide*, Appendix A.
ACGIH TLV®[1]: TLV® [withdrawn].
DFG TRK: 0.10 mg[As]/m³; BAT: 1.30 μg[As]/L in urine/end-of-shift; Carcinogen Category; Germ Cell Mutagen Group 3A.
NIOSH IDLH: 5 mg[As]/m³.
Protective Action Criteria (PAC)
3687-31-8
TEEL-0: 0.06 mg/m³
PAC-1: 3 mg/m³
PAC-2: 20 mg/m³
PAC-3: 30 mg/m³
Dibasic, 7784-40-9
TEEL-0: 0.0463 mg/m³
PAC-1: 23.2 mg/m³
PAC-2: 23.2 mg/m³
PAC-3: 23.2 mg/m³
Arab Republic of Egypt: TWA 0.2 mg/m³, 1993; Australia: TWA 0.05 mg/m³, carcinogen, 1993; Belgium: TWA 0.2 mg/m³, 1993; Denmark: TWA 0.05 mg/m³, 1999; Finland: carcinogen, 1993; France: VME 0.2 mg/m³, 1993; Hungary: STEL 0.5 mg/m³, carcinogen, 1993; India: TWA 0.2 mg/m³, 1993; Norway: TWA 0.02 mg/m³, 1999; the Philippines: TWA 0.5 mg/m³, 1993; Poland: MAC (TWA) 0.01 mg/m³, 1999; Sweden: NGV 0.03 mg/m³, carcinogen, 1999; Switzerland: TWA 0.1 mg/m³, carcinogen, 1999; Thailand: TWA 0.5 mg/m³, 1993; Turkey: TWA 0.5 mg (As)/m³, 1993; Turkey: TWA 0.5 mg/m³, 1993; United Kingdom: TWA 0.1 mg/m³, carcinogen, 2000; Argentina, Bulgaria, Columbia, Jordan, South Korea, New Zealand, Singapore, Vietnam: ACGIH: TLV: Confirmed Human Carcinogen. Russia[43] set a MAC of 0.003 mg/m³ on an average daily basis for residential areas. Several states have set guidelines or standards for arsenic in ambient air[60]: 0.06 mg/m³ (California Prop. 65), 0.0002 μg/m³ (Rhode Island), 0.00023 μg/m³ (North Carolina), 0.024 μg/m³ (Pennsylvania), 0.05 μg/m³ (Connecticut), 0.07−0.39 μg/m³ (Montana), 0.67 μg/m³ (New York), 1.0 μg/m³ (South Carolina), 2.0 μg/m³ (North Dakota), 3.3 μg/m³ (Virginia), 5 μg/m³ (Nevada).
Lead:
OSHA PEL: 0.050 mg[Pb]/m³ TWA. [*Note*: The PEL also applies to other lead compounds (as Pb).] Other OSHA requirements can be found in 29 CFR 1910.1025. The OSHA PEL (8-h TWA) for lead in "nonferrous foundries with less than 20 employees" is 0.075 mg/m³. OSHA considers "lead" to mean metallic lead; all inorganic lead compounds (lead oxides and lead salts); and a class of organic compounds called "soaps;" all other lead compounds are excluded from this definition.
NIOSH REL: 0.050 mg[Pb]/m³ TWA. Suspected carcinogen. Limit exposure to lowest feasible level. [*Note*: The REL also applies to other lead compounds (as Pb).] *Note*: NIOSH considers "lead" to mean metallic lead, lead oxides, and lead salts (including organic salts, such as lead "soaps" but excluding lead arsenate). Air concentrations should be maintained so that worker blood lead remains less than 0.060 mg[Pb]/100 g of whole blood.
ACGIH TLV®[1]: 0.05 mg[Pb]/m³ TWA; BEI: 30 μg[Pb]/100 mL (blood). (Note: The TLV also applies to lead, inorganic compounds.) *Note:* women of child-bearing potential whose blood exceeds 10 μg[Pb]/dL are at risk of delivering a child with a blood [Pb] over the current CDC guideline of 10 μg[Pb]/dL and may cause birth defects. Confirmed animal carcinogen with unknown relevance to humans.
NIOSH IDLH: 100 mg [Pb]/m³.
DFG MAK: BAT: 400 μg[Pb]/L (blood) not fixed; 100 μg [Pb]/L (blood) women age <45; Carcinogen Category 2; Germ Cell Mutagen Group 3A.
Argentina[35] has set 0.15 mg/m³ as a TWA ambient air[60] ranging from 0.5 μg/m³ (New York) to 0.75 μg/m³ (South Carolina) to 0.5 μg/m³ (Florida and North Dakota) to 2.5 μg/m³ (Virginia) to 3.0 μg/m³ (Connecticut) to 4.0 μg/m³ (Nevada).

Permissible Concentration in Water: The EPA limits lead in drinking water to 15 μg/L. Various organizations worldwide have set other standards for lead in drinking water as follows[35] (all in mg/L): Argentina 0.01; the Czech Republic 0.05; Germany 0.04; EEC 0.05; Japan 0.10; Mexico 0.05; former USSR-UNEP/IRPTC joint project 0.03; WHO 0.10. The states of Maine and Minnesota have set guidelines for lead in drinking water[61] at the level of 20 μg/L.

Determination in Water: Digestion followed by atomic absorption, or by colorimetric (dithizone) analysis, or by inductively coupled plasma (ICP) optical emission spectrometry. That gives total lead; dissolved lead may be determined by 0.45-μm filtration prior to such analysis.

Routes of Entry: Inhalation, ingestion, skin and/or eye contact.

Harmful Effects and Symptoms
Short Term Exposure: Lead arsenate irritates the eyes, skin, and respiratory tract. Skin contact can cause burning sensation, itching, and rash. High exposure can cause poor

appetite; nausea, vomiting, and muscle cramps. May affect the heart, with abnormal EKG, gastrointestinal tract, and nervous system. Arsenic intoxication: nausea, diarrhea, inflammation of skin and mucous membranes; lead intoxication: abdominal pain, appetite loss, constipation, tiredness, weakness, nervousness, paresthesia. A rebuttable presumption against registration for pesticide uses was issued on October 18, 1978 by EPA on the basis of oncogenicity, teratogenicity, and mutagenicity.

Long Term Exposure: Lead arsenate is a carcinogen and has been shown to cause skin, lung, and liver cancer. Lead arsenate may also affect the gastrointestinal tract, nervous system, kidneys, and blood. Lead and certain lead compounds may be teratogens and cause reproductive damage in humans. See entry on "lead" for additional information on lead poisoning.

Points of Attack: Kidneys, blood, gingival tissue, lymphatic, skin, gastrointestinal system, central nervous system.

Medical Surveillance: Before first exposure and every 6 months thereafter, OSHA (1910.1025) requires your employer to provide: blood lead test, ZPP test (a special test for the effect of lead on blood cells). Examination of the nervous system. Prior to initial exposure, and annually for exposed person having blood lead readings exceeding 40 μg/100 g of whole blood, OSHA also requires a complete medical history, complete blood count and kidney function tests in addition to the tests listed above. OSHA defines "exposure" for these tests as air levels which average 30 μg of lead or more in a cubic meter of air. OSHA requires your employer to send the doctor a copy of the lead standard and provide one for you.

Note: Blood-lead level is a good indicator of total lead exposure. Current OSHA regulations require that if an individual has a blood-lead level greater than or equal to 0.050 mg lead per 100 mL blood, he or she must be removed from all exposures to lead and cannot return to the exposure environment until the blood level falls to 0.040 mg lead per 100 mL blood or less. The following tolerance levels for occupational exposures may also be useful: ACGIH BEI = 50 mg/L (blood); 150 mg/g creatinine (urine). DFG BAT = 70 mg/L (blood); 30 mg/L (blood) for women <45 years old.

Also seek prompt medical evaluation if health effects are noticed. With each visit, careful attention should be given to the inner nose, skin, nails, and nervous system. A test for serum arsenic is recommended. NIOSH recommends urine arsenic should not be greater than 50–100 μg/L of urine.

First Aid: If this chemical gets into the eyes, remove any contact lenses at once and irrigate immediately for at least 15 min, occasionally lifting upper and lower lids. Seek medical attention immediately. If this chemical contacts the skin, remove contaminated clothing and wash immediately with soap and water. Seek medical attention immediately. If this chemical has been inhaled, remove from exposure, begin rescue breathing (using universal precautions, including resuscitation mask) if breathing has stopped and CPR if

heart action has stopped. Transfer promptly to a medical facility. When this chemical has been swallowed, get medical attention. Give large quantities of water and induce vomiting. Do not make an unconscious person vomit.

Note to physician: For severe poisoning BAL [British Anti-Lewisite, dimercaprol, dithiopropanol ($C_3H_8OS_2$)] has been used to treat toxic symptoms of certain heavy metals poisoning including arsenic and may be of SOME value in the case of lead poisoning. Although BAL is reported to have a large margin of safety, caution must be exercised, because toxic effects may be caused by excessive dosage. Most can be prevented by premedication with 1-ephedrine sulfate (CAS: 134-72-5). For milder poisoning *penicillamine (not penicillin)* has been used, both with mixed success. Side effects occur with such treatment and it is never a substitute for controlling exposure. It can only be done under strict medical care.

Persons with significant lead poisoning can be treated with Ca EDTA while hospitalized. Since this drug causes a rush of lead from body organs into the blood and kidneys, and thus has its own hazards, it must be done by experienced medical persons under careful observation. It or other "chelating" drugs should never be used to prevent poisoning while exposure continues, as severe kidney damage can result.

Personal Protective Methods: Wear protective gloves and clothing to prevent any reasonable probability of skin contact. Use any barrier that will prevent contamination from the dust. Safety equipment suppliers/manufacturers can provide recommendations on the most protective glove/clothing material for your operation. All protective clothing (suits, gloves, footwear, headgear) should be clean, available each day, and put on before work. Contact lenses should not be worn when working with this chemical. Wear dust-proof chemical goggles and face shield unless full face-piece respiratory protection is worn. Employees should wash immediately with soap when skin is wet or contaminated. Provide emergency showers and eyewash. See also NIOSH Criteria Document #78-158, *Lead, inorganic dusts and fumes.*

Respirator Selection:

As inorganic arsenic

At concentrations above the NIOSH REL, or where there is no REL, at any detectable concentration: Sa (APF = 10) (any supplied-air respirator) or SCBAF (APF = 50) (any self-contained breathing apparatus with a full face-piece). *Emergency or planned entry into unknown concentrations or IDLH conditions:* SCBAF: Pd,Pp (APF = 10,000) (any self-contained breathing apparatus that has a full face-piece and is operated in a pressure-demand or other positive-pressure mode) or SaF: Pd,Pp: ASCBA (APF = 10,000) (any supplied-air respirator that has a full face-piece and is operated in a pressure-demand or other positive-pressure mode in combination with an auxiliary, self-contained breathing apparatus operated in a pressure-demand or other positive-pressure mode). *Escape:* GmFAg100 (APF = 50) [any air-purifying, full-face-piece respirator (gas mask) with a chin-style, front- or back-mounted acid gas canister having an N100, R100, or P100 filter] or SCBAE (any appropriate escape-type, self-contained breathing apparatus).

As lead

OSHA (lead): OSHA: *0.5 mg/m³:* 100XQ (APF = 10) [any air-purifying respirator with an N100, R100, or P100 filter (including N100, R100, and P100 filtering face-pieces) except quarter-mask respirators]. *1.25 mg/m³:* Sa:Cf (APF = 25) (any supplied-air respirator operated in a continuous-flow mode) or PaprHie (APF = 25) (any powered, air-purifying respirator with a high-efficiency particulate filter). *2.5 mg/m³:* 100F (APF = 50) (any air purifying, full-face-piece respirator with an N100, R100, or P100 filter) or SaT: Cf (APF = 50) (any supplied-air respirator that has a tight-fitting face-piece and is operated in a continuous-flow mode) or PaprTHie (APF = 50) (any powered, air-purifying respirator with a tight-fitting face-piece and a high-efficiency particulate filter) or SCBAF (APF = 50) (any self-contained breathing apparatus with a full face-piece) or SaF (APF = 50) (any supplied-air respirator with a full face-piece). *50 mg/m³:* Sa: Pd,Pp (APF = 1000) (any supplied-air respirator operated in a pressure-demand or other positive-pressure mode). *100 mg/m³:* SaF: Pd,Pp (APF = 2000) (any supplied-air respirator that has a full face-piece and is operated in a pressure-demand or other positive-pressure mode). *Emergency or planned entry into unknown concentrations or IDLH conditions:* SCBAF: Pd,Pp (APF = 10,000) (any self-contained breathing apparatus that has a full face-piece and is operated in a pressure-demand or other positive-pressure mode) or SaF: Pd,Pp: ASCBA (any supplied-air respirator that has a full face-piece and is operated in a pressure-demand or other positive-pressure mode in combination with an auxiliary self-contained breathing apparatus operated in a pressure-demand or other positive-pressure mode). *Escape:* 100F (APF = 50) (any air purifying, full-face-piece respirator with an N100, R100, or P100 filter) or SCBAE (any appropriate escape-type, self-contained breathing apparatus).

Storage: Color Code—Blue: Health Hazard/Poison: Store in a secure poison location. Prior to working with this chemical you should be trained on its proper handling and storage. Lead arsenate must be stored to avoid contact with oxidizers (such as perchlorates, peroxides, permanganates, chlorates and nitrates) and chemically active metals (such as potassium, sodium, magnesium, and zinc), since violent reactions occur. Avoid the presence of acids since arsine, a very deadly gas, is released in the presence of acid or acid mist. A regulated, marked area should be established where this chemical is handled, used, or stored in compliance with OSHA Standard 1910.1045. A regulated, marked area should be established where this chemical is handled, used, or stored in compliance with OSHA Standard 1910.1045.

Shipping: Lead arsenates require a shipping label of "POISONOUS/TOXIC MATERIALS." They fall in DOT Hazard Class 6.1 and Packing Group II.

Spill Handling: Evacuate persons not wearing protective equipment from area of spill or leak until cleanup is complete. Remove all ignition sources. Collect powdered material in the most convenient and safe manner and deposit in sealed containers. Ventilate area after cleanup is complete. It may be necessary to contain and dispose of this chemical as a hazardous waste. If material or contaminated runoff enters waterways, notify downstream users of potentially contaminated waters. Contact your local or federal environmental protection agency for specific recommendations. If employees are required to clean up spills, they must be properly trained and equipped. OSHA 1910.120(q) may be applicable.

Fire Extinguishing: Not flammable. Use dry chemical, carbon dioxide, water spray, or foam extinguishers. Protect against exposure to dust or fumes. Poisonous gases of lead and arsenic are produced in fire. If material or contaminated runoff enters waterways, notify downstream users of potentially contaminated waters. Notify local health and fire officials and pollution control agencies. From a secure, explosion-proof location, use water spray to cool exposed containers. If cooling streams are ineffective (venting sound increases in volume and pitch, tank discolors, or shows any signs of deforming), withdraw immediately to a secure position. If employees are expected to fight fires, they must be trained and equipped in OSHA 1910.156. The only respirators recommended for firefighting are self-contained breathing apparatuses that have full face-pieces and are operated in a pressure-demand or other positive-pressure mode.

Disposal Method Suggested: Long-term storage in large, weatherproof, and sift-proof storage bins or silos; may be disposed of by conversion to soluble salt, such as chloride, precipitation as sulfide, and return to supplier.[22]

References

New Jersey Department of Health and Senior Services. (September 2001). *Hazardous Substances Fact Sheet: Lead Arsenate.* Trenton, NJ

US Environmental Protection Agency, Special Review and Reregistration Division Office of Pesticide Programs. (1998). *Agency Status of Pesticides in Registration, Reregistration, and Special Review* (Rainbow Report). Washington, DC

Lead chloride   L:0130

Molecular Formula: Cl_2Pb

Synonyms: Cloruro de plomo (Spanish); Lead(2 +) chloride; Lead(II) chloride; Lead dichloride; Plumbous chloride

CAS Registry Number: 7758-95-4

RTECS® Number: OF9450000

UN/NA & ERG Number: UN2291/151

EC Number: 231-845-5

Regulatory Authority and Advisory Bodies

Carcinogenicity: IARC: Animal Inadequate Evidence, Human Inadequate Evidence, *possibly carcinogenic to humans*, Group 2A, 1987; Carcinogenicity: NTP: 11th Report on Carcinogens, 2004: Reasonably anticipated to be a human carcinogen; EPA: Sufficient evidence from animal studies; inadequate evidence or no useful data from epidemiologic studies.

OSHA, 29CFR1910 Specifically Regulated Chemicals (See CFR 1910.1025).

Clean Water Act: Section 311 Hazardous Substances/RQ 40CFR117.3 (same as CERCLA, see below); Section 313 Water Priority Chemicals (57FR41331, 9/9/92); 40CFR401.15 Section 307 Toxic Pollutants as lead and compounds.

RCRA, 40CFR261, Appendix 8 Hazardous Constituents, waste number not listed, as lead compounds, n.o.s.

Reportable Quantity (RQ): 10 lb (4.54 kg).

EPCRA Section 313 Form R *de minimis* concentration reporting level: 0.1%.

US DOT Regulated Marine Pollutant (49CFR172.101, Appendix B).

California Proposition 65 Chemical: Cancer 10/1/92 (lead and compounds).

Canada, WHMIS, Ingredients Disclosure List Concentration 0.1%.

European/International Regulations: not listed in Annex 1.

WGK (German Aquatic Hazard Class): No value assigned.

Description: Lead chloride is a white crystalline powder. Molecular weight = 278.00; Boiling point = 950°C; Freezing/Melting point = 501°C; Vapor pressure = 1 mmHg at 547°C. Hazard Identification (based on NFPA-704 M Rating System): Health 2, Flammability 0, Reactivity 0. Slightly soluble in cold water; more soluble in hot water.

Potential Exposure: Used to make lead salts, lead chromate pigments, as an analytical reagent for making other chemicals, making printed circuit boards, as a solder and flux.

Incompatibilities: Violent reaction with oxidizers, chemically active metals, and explosive with calcium + warming.

Permissible Exposure Limits in Air
Protective Action Criteria (PAC)
TEEL-0: 0.0671 mg/m^3
PAC-1: 0.201 mg/m^3
PAC-2: 7.5 mg/m^3
PAC-3: 134 mg/m^3
Lead
OSHA PEL: 0.050 mg[Pb]/m^3 TWA. [*Note*: The PEL also applies to other lead compounds (as Pb).] Other OSHA requirements can be found in 29 CFR 1910.1025. The OSHA PEL (8-h TWA) for lead in "nonferrous foundries with less than 20 employees" is 0.075 mg/m^3. OSHA considers "lead" to mean metallic lead; all inorganic lead compounds (lead oxides and lead salts); and a class of organic compounds called "soaps;" all other lead compounds are excluded from this definition.

NIOSH REL: 0.050 mg[Pb]/m^3 TWA. Suspected carcinogen. Limit exposure to lowest feasible level. [*Note*: The REL also applies to other lead compounds (as Pb).] *Note:* NIOSH considers "lead" to mean metallic lead, lead oxides, and lead salts (including organic salts, such as lead "soaps" but excluding lead arsenate). Air concentrations should be maintained so that worker blood lead remains less than 0.060 mg[Pb]/100 g of whole blood.

ACGIH TLV®[1]: 0.05 mg[Pb]/m^3 TWA; BEI: 30 μg[Pb]/100 mL (blood). (Note: The TLV also applies to lead, inorganic compounds.) *Note:* women of child-bearing potential whose blood exceeds 10 μg[Pb]/dL are at risk of delivering a child with a blood [Pb] over the current CDC guideline of 10 μg[Pb]/dL and may cause birth defects. Confirmed animal carcinogen with unknown relevance to humans.

The National primary and secondary ambient air quality standards for lead and its compounds (measured as elemental Pb) = 1 mg[Pb]/m^3, maximum arithmetic mean averaged over a calendar quarter. In addition, Russia[43] set a MAC of 0.0017 mg/m^3 in the ambient air of residential areas (on a daily average basis; about 6 times the same limits for lead compounds in general).

Determination in Air: Use NIOSH Analytical Method (IV) s #7082, 7105, 7300, 7301, 7303, 7700, 7701, 7702, 9100, 9102, 9105; OSHA Analytical Methods ID-121, ID-125G, ID-206.

Permissible Concentration in Water: 0.1 mg/L Pb in drinking water may cause chronic poisoning. The EPA limits lead in drinking water to 15 μg/L. Various organizations worldwide have set other standards for lead in drinking water as follows[35] (all in mg/L): Argentina 0.01; the Czech Republic 0.05; Germany 0.04; EEC 0.05; Japan 0.10; Mexico 0.05; former USSR-UNEP/IRPTC joint project 0.03; WHO 0.10. The states of Maine and Minnesota have set guidelines for lead in drinking water[61] at the level of 20 μg/L.

Determination in Water: Digestion followed by atomic absorption, or by colorimetric (dithizone) analysis, or by inductively coupled plasma (ICP) optical emission spectrometry. That gives total lead; dissolved lead may be determined by 0.45-μm filtration prior to such analysis.

Routes of Entry: Inhalation, skin. Can be absorbed through skin at chronically toxic levels.

Harmful Effects and Symptoms
Short Term Exposure: Lead chloride can irritate the eyes on contact. Inhalation can irritate the nose and throat. Skin contact can cause burning, itching, rash, and pigment changes. Ingestion of large amounts of lead may lead to seizures, coma, and death. The effects of exposure to fumes and dusts of inorganic lead may not develop quickly. Symptoms may include decreased physical fitness, fatigue, sleep disturbance, headache, aching bones and muscles, constipation, abdominal pains, and decreased appetite. These effects are reported to be reversible if exposure ceases. Inhalation of large amounts of lead may lead to seizures, coma, and death.

Note: Lead is a cumulative poison. Increasing amounts can build up in the body, eventually reaching a point where symptoms and disability occur. Lead dust carried home on contaminated clothing can result in exposure and symptoms in other family members. Standards only protect for inhalation exposure. Extra precautions should be taken if skin exposure also occurs.

Long Term Exposure: Lead chloride causes mutations. Such chemicals have a cancer risk. May damage the developing fetus. High levels can cause lead poisoning with symptoms of headache, irritability, disturbed sleep, tiredness, reduced memory, and personality changes. Higher levels can cause muscle or joint pains, weakness, and easy fatigue. Exposure can increase the risk of high blood pressure. May cause kidney and brain damage and anemia. Lead can accumulate in the body over a period of time. Therefore, long-term exposures to lower levels can result in a buildup of lead in the body and more severe symptoms. These may include anemia, pale skin, a blue line at the gum margin, decreased handgrip strength, abdominal pain, severe constipation, nausea, vomiting, and paralysis of the wrist joint. Prolonged exposure may also result in kidney and brain damage. If the nervous system is affected, usually due to very high exposures, the resulting effects include severe headache, convulsions, coma, delirium, and death. In nonfatal cases, recovery is slow and not always complete. Alcohol ingestion and physical exertion may bring on symptoms. Lead exposure increases the risk of high blood pressure. Continuous exposure can result in decreased fertility. Elevated lead exposure of either parent before pregnancy can increase the chances of miscarriage or birth defects. Exposure of the mother during pregnancy can cause birth defects.

Points of Attack: Blood, kidneys, brain, nervous system.

Medical Surveillance: Before first exposure and every 6 months thereafter, OSHA (1910.1025) requires your employer to provide: blood lead test, ZPP test (a special test for the effect of lead on blood cells). Examination of the nervous system. Prior to initial exposure, and annually for exposed person having blood lead readings exceeding 40 μg/100 g of whole blood, OSHA also requires a complete medical history, complete blood count and kidney function tests in addition to the tests listed above. OSHA defines "exposure" for these tests as air levels which average 30 μg of lead or more in a cubic meter of air. OSHA (under 1910.1020) requires your employer to send the doctor a copy of the lead standard and provide one for you.

Note: Blood-lead level is a good indicator of total lead exposure. Current OSHA regulations require that if an individual has a blood-lead level greater than or equal to 0.050 mg lead per 100 mL blood, he or she must be removed from all exposures to lead and cannot return to the exposure environment until the blood level falls to 0.040 mg lead per 100 mL blood or less. The following tolerance levels for occupational exposures may also be useful: ACGIH BEI = 50 mg/L (blood); 150 mg/g creatinine (urine). DFG BAT = 70 mg/L (blood); 30 mg/L (blood) for women <45 years old.

First Aid: If this chemical gets into the eyes, remove any contact lenses at once and irrigate immediately for at least 15 min, occasionally lifting upper and lower lids. Seek medical attention immediately. If this chemical contacts the skin, remove contaminated clothing and wash immediately with soap and water. Seek medical attention immediately. If this chemical has been inhaled, remove from exposure, begin rescue breathing (using universal precautions, including resuscitation mask) if breathing has stopped and CPR if heart action has stopped. Transfer promptly to a medical facility. When this chemical has been swallowed, get medical attention. Give large quantities of water and induce vomiting. Do not make an unconscious person vomit.

Note to physician: Administer saline cathartic and an enema. For relief of colic, administer antispasmodic (calcium gluconate, atropine, papaverine). Consider morphine sulfate for severe pain. Whole blood lead levels, circulating plasma/erythrocyte lead concentration ratio, urine ALA, and erythrocyte protoporphyrin fluorescent microscopy may all be useful in monitoring or assessing lead exposure. Chelating agents, such as edetate disodium calcium (Ca EDTA) and penicillamine (*not penicillin*), are generally useful in the therapy of acute lead intoxication.

Antidotes and special procedures for lead: Persons with significant lead poisoning are sometimes treated with Ca EDTA while hospitalized. This "chelating" drug causes a rush of lead from the body organs into the blood and kidneys, and thus has its own hazards, and must be administered only by highly experienced medical personnel under controlled conditions and careful observation. Ca EDTA or similar drugs should never be used to prevent poisoning while exposure continues or without strict exposure control, as severe kidney damage can result.

Note to physician: For severe poisoning BAL [British Anti-Lewisite, dimercaprol, dithiopropanol ($C_3H_8OS_2$)] has been used to treat toxic symptoms of certain heavy metals poisoning. In the case of lead poisoning it may have SOME value. Although BAL is reported to have a large margin of safety, caution must be exercised, because toxic effects may be caused by excessive dosage. Most can be prevented by premedication with 1-ephedrine sulfate (CAS: 134-72-5).

Personal Protective Methods: Wear protective gloves and clothing to prevent any reasonable probability of skin contact. Use any barrier that will prevent contamination from the dust. Safety equipment suppliers/manufacturers can provide recommendations on the most protective glove/clothing material for your operation. All protective clothing (suits, gloves, footwear, headgear) should be clean, available each day, and put on before work. Work clothing should be HEPA vacuumed before removal. Contact lenses should not be worn when working with this chemical. Wear dust-proof chemical goggles and face shield unless full face-piece respiratory protection is worn. Employees should wash immediately with soap when skin is wet or contaminated. Provide emergency showers and eyewash.

Respirator Selection: OSHA: *0.5 mg/m³*: 100XQ (APF = 10) [any air-purifying respirator with an N100, R100, or P100 filter (including N100, R100, and P100

filtering face-pieces) except quarter-mask respirators].
1.25 mg/m³: Sa:Cf (APF = 25) (any supplied-air respirator operated in a continuous-flow mode) or PaprHie (APF = 25) (any powered, air-purifying respirator with a high-efficiency particulate filter). *2.5 mg/m³:* 100F (APF = 50) (any air purifying, full-face-piece respirator with an N100, R100, or P100 filter) or SaT: Cf (APF = 50) (any supplied-air respirator that has a tight-fitting face-piece and is operated in a continuous-flow mode) or PaprTHie (APF = 50) (any powered, air-purifying respirator with a tight-fitting face-piece and a high-efficiency particulate filter) or SCBAF (APF = 50) (any self-contained breathing apparatus with a full face-piece) or SaF (APF = 50) (any supplied-air respirator with a full face-piece). *50 mg/m³:* Sa: Pd,Pp (APF = 1000) (any supplied-air respirator operated in a pressure-demand or other positive-pressure mode). *100 mg/m³:* SaF: Pd,Pp (APF = 2000) (any supplied-air respirator that has a full face-piece and is operated in a pressure-demand or other positive-pressure mode). *Emergency or planned entry into unknown concentrations or IDLH conditions:* SCBAF: Pd,Pp (APF = 10,000) (any self-contained breathing apparatus that has a full face-piece and is operated in a pressure-demand or other positive-pressure mode) or SaF: Pd,Pp: ASCBA (any supplied-air respirator that has a full face-piece and is operated in a pressure-demand or other positive-pressure mode in combination with an auxiliary self-contained breathing apparatus operated in a pressure-demand or other positive-pressure mode). *Escape:* 100F (APF = 50) (any air purifying, full-face-piece respirator with an N100, R100, or P100 filter) or SCBAE (any appropriate escape-type, self-contained breathing apparatus).

Storage: Color Code—Blue: Health Hazard/Poison: Store in a secure poison location. Prior to working with this chemical you should be trained on its proper handling and storage. A regulated, marked area should be established where this chemical is handled, used, or stored in compliance with OSHA Standard 1910.1045. Store in tightly closed containers in a cool, well-ventilated area away from oxidizers, chemically active metals, calcium, and heat. Lead is regulated by an OSHA Standard 1910.1025. All requirements of the standard must be followed.

Shipping: Lead compounds, soluble, n.o.s. require a label of "POISONOUS/TOXIC MATERIALS." Lead chloride falls in Hazard Class 6.1 and Packing Group III.

Spill Handling: Evacuate persons not wearing protective equipment from area of spill or leak until cleanup is complete. Remove all ignition sources. Collect powdered material in the most convenient and safe manner and deposit in sealed containers. Use vacuum or a wet method to reduce dust. Do not dry sweep. When vacuuming, a HEPA filter should be used, not a standard shop vac. Ventilate area after cleanup is complete. It may be necessary to contain and dispose of this chemical as a hazardous waste. If material or contaminated runoff enters waterways, notify downstream users of potentially contaminated waters. Contact your local

or federal environmental protection agency for specific recommendations. If employees are required to clean up spills, they must be properly trained and equipped. OSHA 1910.120(q) may be applicable.

Fire Extinguishing: This chemical may burn but does not easily ignite. Use dry chemical, carbon dioxide, water spray, or foam extinguishers. Poisonous fumes, including lead and chlorine, are produced in fire. If material or contaminated runoff enters waterways, notify downstream users of potentially contaminated waters. Notify local health and fire officials and pollution control agencies. From a secure, explosion-proof location, use water spray to cool exposed containers. If cooling streams are ineffective (venting sound increases in volume and pitch, tank discolors, or shows any signs of deforming), withdraw immediately to a secure position. If employees are expected to fight fires, they must be trained and equipped in OSHA 1910.156. The only respirators recommended for firefighting are self-contained breathing apparatuses that have full face-pieces and are operated in a pressure-demand or other positive-pressure mode.

Reference
New Jersey Department of Health and Senior Services. (April 2002). *Hazardous Substances Fact Sheet: Lead Chloride.* Trenton, NJ

Lead chromate   L:0140

Molecular Formula: CrO₄Pb
Common Formula: PbCrO₄
Synonyms: Canary chrome yellow 40-2250; Chrome green; Chrome yellow; Chromic acid, lead(2 +) salt (1:1); C.I. 77600; Cologne yellow; Crocoite; King's yellow; Lead chromate(VI); Leipzig yellow; Lemon yellow; Paris yellow; Plumbous chromate
CAS Registry Number: 7758-97-6; *(alt.)* 8049-64-7; *(alt.)* 11119-70-3; 18454-12-1 (lead chromate oxide)
RTECS® Number: GB2975000
UN/NA & ERG Number: UN3288 (Toxic solid, inorganic, n.o.s.)/151
EC Number: 231-846-0 [*Annex I Index No.:* 082-004-00-2]
Regulatory Authority and Advisory Bodies
Carcinogenicity: IARC (inorganic): Human Inadequate Evidence, animal Sufficient Evidence, *probably carcinogenic to humans,* Group 2A; NTP: 11th Report on Carcinogens, 2004: Known to be a human carcinogen; EPA [as Cr(VI)] *(inhalation):* Known human carcinogen; EPA *(oral):* Not Classifiable as to human carcinogenicity; NTP: Known to be a human carcinogen.
US EPA Gene-Tox Program, Positive: Carcinogenicity—mouse/rat; Cell transform.—SA7/SHE; Positive: *S. cerevisiae*—homozygosis.
OSHA, 29CFR1910 Specifically Regulated Chemicals (See 29 CFR 1910.1025)
Air Pollutant Standard Set. See below, "Permissible Exposure Limits in Air" section.

Clean Water Act: 40CFR401.15 Section 307 Toxic Pollutants as lead and compounds.

RCRA, 40CFR261, Appendix 8 Hazardous Constituents, waste number not listed, as lead compounds, n.o.s.

EPCRA Section 313: Includes any unique chemical substance that contains lead as part of that chemical's infrastructure. Form R *de minimis* concentration reporting level: inorganic compounds 0.1%; organic compounds 1.0%.

US DOT Regulated Marine Pollutant (49CFR172.101, Appendix B) as lead compounds, soluble, n.o.s.

California Proposition 65 Chemical: (*hexavalent chromium*) Cancer 2/27/87; Developmental/Reproductive toxin (male, female) 12/19/08; (lead and compounds) Cancer 10/1/92.

Canada, WHMIS, Ingredients Disclosure List Concentration 0.1%.

European/International Regulations: Hazard Symbol: T, N; Risk phrases: R45; R61; R62; R33; R50/53; Safety phrases: S53; S45; S60; S61 (see Appendix 4).

WGK (German Aquatic Hazard Class): No value assigned.

Description: Lead chromate is an orange or orange-yellow crystalline solid or powder. Molecular weight = 323.19; 546.38 (lead chromate oxide); Boiling point = (decomposes); Freezing/Melting point = 844°C. Hazard Identification (based on NFPA-704 M Rating System): Health 3, Flammability 0, Reactivity 0. Insoluble in water.

Potential Exposure: Compound Description: Agricultural Chemical; Tumorigen, Mutagen. Lead chromate is used to make paint pigments for wood and metal.

Incompatibilities: Oxidizers, active metals; hydrazine, sodium and potassium; organics at elevated temperature. Reacts with aluminum dinitronaphthalene, iron(III) hexacyanoferrate (IV).

Permissible Exposure Limits in Air

ACGIH TLV®[1]: 0.01 mg[Cr]/m^3 TWA, Confirmed Human Carcinogen.

ACGIH TLV®[1] (*as lead*): 0.05 mg[Pb]/m^3 TWA; BEI: 30 μg[Pb]/100 mL (blood). (Note: The TLV also applies to lead, inorganic compounds.) *Note:* women of child-bearing potential whose blood exceeds 10 μg[Pb]/dL are at risk of delivering a child with a blood [Pb] over the current CDC guideline of 10 μg[Pb]/dL and may cause birth defects. Confirmed animal carcinogen with unknown relevance to humans.

DFG MAK (lead chromate & lead chromate oxide): Carcinogen Category 3B.

Protective Action Criteria (PAC)

7758-97-6

TEEL-0: 0.0311 mg/m^3

PAC-1: 0.186 mg/m^3

PAC-2: 93.2 mg/m^3

PAC-3: 93.2 mg/m^3

As chromium(VI), inorganic insoluble compounds

OSHA PEL: 0.005 mg[Cr(VI)]/m^3 TWA Concentration. See 29CFR1910.1026.

NIOSH REL: 0.001 mg[Cr]/m^3 TWA, potential carcinogen, limit exposure to lowest feasible level. NIOSH considers all Cr(VI) compounds (including chromic acid, *tert*-butyl chromate, zinc chromate, and chromyl chloride) to be potential occupational carcinogens. See *NIOSH Pocket Guide*, Appendix A & C.

ACGIH TLV®[1]: 0.01 mg[Cr]/m^3 TWA, Confirmed Human Carcinogen.

DFG MAK: Danger of skin sensitization; Carcinogen Category 2; TRK: 0.05 mg[Cr]/m^3; 20 μg/L [Cr] in urine at end-of-shift.

NIOSH IDLH: 15 mg[Cr(VI)]/m^3.

As lead:

OSHA PEL: 0.050 mg[Pb]/m^3 TWA. [*Note:* The PEL also applies to other lead compounds (as Pb).] Other OSHA requirements can be found in 29 CFR 1910.1025. The OSHA PEL (8-h TWA) for lead in "nonferrous foundries with less than 20 employees" is 0.075 mg/m^3. OSHA considers "lead" to mean metallic lead; all inorganic lead compounds (lead oxides and lead salts); and a class of organic compounds called "soaps;" all other lead compounds are excluded from this definition.

NIOSH REL: 0.050 mg[Pb]/m^3 TWA. Suspected carcinogen. Limit exposure to lowest feasible level. [*Note:* The REL also applies to other lead compounds (as Pb).] *Note:* NIOSH considers "lead" to mean metallic lead, lead oxides, and lead salts (including organic salts, such as lead "soaps" but excluding lead arsenate). Air concentrations should be maintained so that worker blood lead remains less than 0.060 mg[Pb]/100 g of whole blood.

ACGIH TLV®[1]: 0.05 mg[Pb]/m^3 TWA; BEI: 30 μg[Pb]/100 mL (blood). (Note: The TLV also applies to lead, inorganic compounds.) *Note:* women of child-bearing potential whose blood exceeds 10 μg[Pb]/dL are at risk of delivering a child with a blood [Pb] over the current CDC guideline of 10 μg[Pb]/dL and may cause birth defects. Confirmed animal carcinogen with unknown relevance to humans.

NIOSH IDLH: 100 mg [Pb]/m^3.

DFG MAK: BAT: Carcinogen Category 3B.

Australia: TWA 0.05 mg/m^3, carcinogen, 1993; Austria: Suspected: carcinogen, 1999; Belgium: TWA 0.05 mg/m^3, carcinogen, 1993; Denmark: TWA 0.02 mg[Cr]/m^3, 1999; TWA 0.1 mg[Pb]/m^3, 1999; Finland: carcinogen, 1999; France: VME 0.05 mg[Cr]/m^3, VME 0.15 mg[Pb]/m^3, 1999; Japan 0.01 mg[Cr]/m^3, 1999; 0.1 mg[Pb]/m^3, 1999; Norway: TWA 0.1 mg[CrO$_3$]/m^3, 1999; Norway: TWA 0.05 mg[Pb]/m^3, 1999; Poland: TWA 0.1 mg/m^3; STEL 0.3 mg/m^3, 1999; Poland: MAC (TWA) 0.05 mg[Pb]/m^3, 1999; Sweden: TWA 0.02 mg[Cr]/m^3, carcinogen, 1999; Switzerland: MAK-W 0.05 mg[Cr]/m^3, carcinogen, 1999; United Kingdom: TWA 0.05 mg[Cr]/m^3, carcinogen, 2000; Argentina, Bulgaria, Columbia, Jordan, South Korea, New Zealand, Singapore, Vietnam: ACGIH TLV®: Suspected Human Carcinogen. Several states have set guidelines or standards for lead chromate in ambient air[60] ranging from 0.5 μg/m^3 (Connecticut and Virginia) to 1.0 μg/m^3 (Nevada).

Determination in Air: For lead use NIOSH Analytical Method (IV)s #7082, 7105, 7300, 7301, 7303, 7700, 7701,

7702, 9100, 9102, 9105; OSHA Analytical Methods ID-121, ID-125G, ID-206; #7024 for chromium, hexavalent.

Permissible Concentration in Water: The EPA limits lead in drinking water to 15 µg/L. Various organizations worldwide have set other standards for lead in drinking water as follows[35] (all in mg/L): Argentina 0.01; the Czech Republic 0.05; Germany 0.04; EEC 0.05; Japan 0.10; Mexico 0.05; former USSR-UNEP/IRPTC joint project 0.03; WHO 0.10. The states of Maine and Minnesota have set guidelines for lead in drinking water[61] at the level of 20 µg/L.

Determination in Water: Digestion followed by atomic absorption, or by colorimetric (dithizone) analysis, or by inductively coupled plasma (ICP) optical emission spectrometry. That gives total lead; dissolved lead may be determined by 0.45-µm filtration prior to such analysis.

Routes of Entry: Inhalation.

Harmful Effects and Symptoms

Short Term Exposure: Irritates the respiratory tract. Lead chromate is a carcinogen—handle with extreme caution. Lead poisoning symptoms may include poor appetite, colic, upset stomach, headache, irritability, muscle or joint pains, and weakness. Permanent kidney damage can result from long term or high exposure. Breathing lead chromate can cause a hole in the inner nose. Irritation of nose, throat, or bronchial tubes can also occur, with cough and/or wheezing. Skin contact with concentrated lead chromate can cause burns, deep ulcers, or an allergic skin rash.

Long Term Exposure: Lead chromate has been shown to cause kidney cancer. Repeated or prolonged contact may cause skin sensitization; dermatitis, irritation, chronic ulcers, eczema. Repeated or prolonged inhalation exposure may cause asthma. Lead chromate may affect the kidneys. May cause increased blood pressure. May cause reproductive toxicity and genetic damage in humans. May cause lead poisoning with symptoms of poor appetite, upset stomach, colic, headache, irritability, muscle or joint pains and weakness, constipation, disturbed sleep, and reduced memory. See also "Lead" entry.

Points of Attack: Kidneys, skin, lungs.

Medical Surveillance: NIOSH lists the following tests for chromates: Blood gas analysis, complete blood count, chest X-ray, electrocardiogram, liver function tests, pulmonary function tests, sputum cytology, urine (chemical/metabolite), urinalysis (routine), white blood cell count/differential. Before first exposure and every 6 months thereafter, OSHA (1910.1025) requires employers to provide: blood lead test, ZPP test (a special test for the effect of lead on blood cells). Examination of the nervous system. Prior to initial exposure, and annually for exposed person having blood lead readings exceeding 40 µg/100 g of whole blood, OSHA also requires a complete medical history, complete blood count and kidney function tests in addition to the tests listed above. OSHA defines "exposure" for these tests as air levels which average 30 µg of lead or more in a cubic meter of air. OSHA requires your employer to send the doctor a copy of the lead standard and provide one for you.

Note: Blood-lead level is a good indicator of total lead exposure. Current OSHA regulations require that if an individual has a blood-lead level greater than or equal to 0.050 mg lead per 100 mL blood, he or she must be removed from all exposures to lead and cannot return to the exposure environment until the blood level falls to 0.040 mg lead per 100 mL blood or less. The following tolerance levels for occupational exposures may also be useful: ACGIH BEI = 50 mg/L (blood); 150 mg/g creatinine (urine). DFG BAT = 70 mg/L (blood); 30 mg/L (blood) for women <45 years old.

First Aid: If this chemical gets into the eyes, remove any contact lenses at once and irrigate immediately for at least 15 min, occasionally lifting upper and lower lids. Seek medical attention immediately. If this chemical contacts the skin, remove contaminated clothing and wash immediately with soap and water. Seek medical attention immediately. If this chemical has been inhaled, remove from exposure, begin rescue breathing (using universal precautions, including resuscitation mask) if breathing has stopped and CPR if heart action has stopped. Transfer promptly to a medical facility. When this chemical has been swallowed, get medical attention. Give large quantities of water and induce vomiting. Do not make an unconscious person vomit.

Antidotes and special procedures for lead: Persons with significant lead poisoning are sometimes treated with Ca EDTA while hospitalized. This "chelating" drug causes a rush of lead from the body organs into the blood and kidneys, and thus has its own hazards, and must be administered only by highly experienced medical personnel under controlled conditions and careful observation. Ca EDTA or similar drugs should never be used to prevent poisoning while exposure continues or without strict exposure control, as severe kidney damage can result.

Note to physician: For severe poisoning BAL [British Anti-Lewisite, dimercaprol, dithiopropanol $(C_3H_8OS_2)$] has been used to treat toxic symptoms of certain heavy metals poisoning. In the case of lead poisoning it may have SOME value. Although BAL is reported to have a large margin of safety, caution must be exercised, because toxic effects may be caused by excessive dosage. Most can be prevented by premedication with l-ephedrine sulfate (CAS: 134-72-5).

Personal Protective Methods: Wear protective gloves and clothing to prevent any reasonable probability of skin contact: (*as chromic acid and chromates*) **8 h** (more than 8 h of resistance to breakthrough >0.1 µg/cm/min): polyethylene gloves, suits, boots; polyvinyl chloride gloves, suits, boots; Saranex™ coated suits; **4 h** (At least 4 but <8 h of resistance to breakthrough >0.1 µg/cm²/min): butyl rubber gloves, suits, boots; Viton™ gloves, suits. Safety equipment suppliers/manufacturers can provide recommendations on the most protective glove/clothing material for your operation. All protective clothing (suits, gloves, footwear, headgear) should be clean, available each day, and put on before work. Contact lenses should not be worn when working with this chemical. Wear dust-proof chemical goggles and face shield unless full face-piece respiratory protection is worn. Employees should wash immediately with soap when skin is wet or contaminated. Provide emergency showers and eyewash. Specific engineering controls are recommended in NIOSH Criteria Document #76-129 *Chromium(VI)* and NIOSH 78-158 *Lead, inorganic dusts and fumes*.

Respirator Selection:

As lead

OSHA: *0.5 mg/m³:* 100XQ (APF = 10) [any air-purifying respirator with an N100, R100, or P100 filter (including N100, R100, and P100 filtering face-pieces) except quarter-mask respirators]. *1.25 mg/m³:* Sa:Cf (APF = 25) (any supplied-air respirator operated in a continuous-flow mode) or PaprHie (APF = 25) (any powered, air-purifying respirator with a high-efficiency particulate filter). *2.5 mg/m³:* 100F (APF = 50) (any air purifying, full-face-piece respirator with an N100, R100, or P100 filter) or SaT: Cf (APF = 50) (any supplied-air respirator that has a tight-fitting face-piece and is operated in a continuous-flow mode) or PaprTHie (APF = 50) (any powered, air-purifying respirator with a tight-fitting face-piece and a high-efficiency particulate filter) or SCBAF (APF = 50) (any self-contained breathing apparatus with a full face-piece) or SaF (APF = 50) (any supplied-air respirator with a full face-piece). *50 mg/m³:* Sa: Pd,Pp (APF = 1000) (any supplied-air respirator operated in a pressure-demand or other positive-pressure mode). *100 mg/m³:* SaF: Pd,Pp (APF = 2000) (any supplied-air respirator that has a full face-piece and is operated in a pressure-demand or other positive-pressure mode). *Emergency or planned entry into unknown concentrations or IDLH conditions:* SCBAF: Pd,Pp (APF = 10,000) (any self-contained breathing apparatus that has a full face-piece and is operated in a pressure-demand or other positive-pressure mode) or SaF: Pd,Pp: ASCBA (any supplied-air respirator that has a full face-piece and is operated in a pressure-demand or other positive-pressure mode in combination with an auxiliary self-contained breathing apparatus operated in a pressure-demand or other positive-pressure mode). *Escape:* 100F (APF = 50) (any air purifying, full-face-piece respirator with an N100, R100, or P100 filter) or SCBAE (any appropriate escape-type, self-contained breathing apparatus).

As chromates:

NIOSH, as chromates: *at any concentrations above the NIOSH REL, or where there is no REL, at any detectable concentration:* SCBAF: Pd,Pp (APF = 10,000) (any self-contained breathing apparatus that has a full face-piece and is operated in a pressure-demand or other positive-pressure mode) or SaF: Pd,Pp: ASCBA (APF = 10,000) (any supplied-air respirator that has a full face-piece and is operated in a pressure-demand or other positive-pressure

mode in combination with an auxiliary, self-contained breathing apparatus operated in a pressure-demand or other positive-pressure mode). *Escape:* 100F (APF = 50) (any air purifying, full-face-piece respirator with an N100, R100, or P100 filter) or SCBAE (any appropriate escape-type, self-contained breathing apparatus).

Storage: Color Code—Blue: Health Hazard/Poison: Store in a secure poison location. Prior to working with this chemical you should be trained on its proper handling and storage. Lead chromate must be stored to avoid contact with oxidizers (such as perchlorates, peroxides, permanganates, chlorates, and nitrates) and chemically active metals (such as potassium, sodium, magnesium, zinc and ferric ferrocyanide), since violent reactions occur. A regulated, marked area should be established where this chemical is handled, used, or stored in compliance with OSHA Standard 1910.1045.

Shipping: Toxic solid, inorganic, n.o.s. requires a label of "POISONOUS/TOXIC MATERIALS." Lead chromate falls in Hazard Class 6.1 and Packing Group III.

Spill Handling: Evacuate persons not wearing protective equipment from area of spill or leak until cleanup is complete. Remove all ignition sources. Collect powdered material in the most convenient and safe manner and deposit in sealed containers. Ventilate area after cleanup is complete. It may be necessary to contain and dispose of this chemical as a hazardous waste. If material or contaminated runoff enters waterways, notify downstream users of potentially contaminated waters. Contact your local or federal environmental protection agency for specific recommendations. If employees are required to clean up spills, they must be properly trained and equipped. OSHA 1910.120(q) may be applicable.

Fire Extinguishing: Lead chromate may burn but does not easily ignite. NFPA recommends the use of water on fire. Poisonous gases, including lead and chromium, are produced in fire. If material or contaminated runoff enters waterways, notify downstream users of potentially contaminated waters. Notify local health and fire officials and pollution control agencies. From a secure, explosion-proof location, use water spray to cool exposed containers. If cooling streams are ineffective (venting sound increases in volume and pitch, tank discolors, or shows any signs of deforming), withdraw immediately to a secure position. If employees are expected to fight fires, they must be trained and equipped in OSHA 1910.156. The only respirators recommended for firefighting are self-contained breathing apparatuses that have full face-pieces and are operated in a pressure-demand or other positive-pressure mode.

References

Sax, N. I. (Ed.). (1981). *Dangerous Properties of Industrial Materials Report*, 1, No. 7, 65–66

New Jersey Department of Health and Senior Services. (September 2001). *Hazardous Substances Fact Sheet: Lead Chromate*. Trenton, NJ

Lead dioxide   L:0145

Molecular Formula: O_2Pb
Common Formula: PbO_2
Synonyms: Lead, brown; Lead(IV) oxide; Lead oxide, brown; Lead peroxide; Lead superoxide; Peroxyde de plomb (French)
CAS Registry Number: 1309-60-0
RTECS® Number: OG0700000
UN/NA & ERG Number: UN1872/141
EC Number: 215-174-5
Regulatory Authority and Advisory Bodies
Carcinogenicity: IARC: Animal Inadequate Evidence, Human Inadequate Evidence, *possibly carcinogenic to humans,* Group 2A, 1987; Carcinogenicity: NTP: 11th Report on Carcinogens, 2004: Reasonably anticipated to be a human carcinogen; EPA: Sufficient evidence from animal studies; inadequate evidence or no useful data from epidemiologic studies.
Air Pollutant Standard Set. See below, "Permissible Exposure Limits in Air" section.
Banned or Severely Restricted (various countries; in Pharmaceuticals) (UN).[13]
OSHA, 29CFR1910 Specifically Regulated Chemicals (See 29 CFR 1910.1025).
Clean Water Act: 40CFR401.15 Section 307 Toxic Pollutants as lead and compounds.
RCRA, 40CFR261, Appendix 8 Hazardous Constituents, waste number not listed, as lead compounds, n.o.s.
EPCRA Section 313: Includes any unique chemical substance that contains lead as part of that chemical's infrastructure. Form R *de minimis* concentration reporting level: inorganic compounds 0.1%; organic compounds 1.0%.
US DOT Regulated Marine Pollutant (49CFR172.101, Appendix B) as lead compounds, soluble, n.o.s.
California Proposition 65 Chemical: Cancer 10/1/92 (lead and compounds).
Canada, WHMIS, Ingredients Disclosure List Concentration 1.0% as lead inorganic compounds, n.o.s.
WGK (German Aquatic Hazard Class): No value assigned.
Description: Lead Dioxide is a dark brown crystalline solid or powder. Molecular weight = 239.19; Freezing/Melting point = (decomposes) 290°C. Hazard Identification (based on NFPA-704 M Rating System): Health 4, Flammability 1, Reactivity 3 (Oxidizer). Insoluble in water.
Potential Exposure: This material is used in electrodes for lead-acid batteries, in matches, explosives, and as a curing agent for polysulfide elastomers.
Incompatibilities: A powerful oxidizer. Violent reaction with many compounds, including reducing agents, chemically active metals, combustible materials.
Permissible Exposure Limits in Air
Protective Action Criteria (PAC)
1309-60-0
TEEL-0: 0.0577 mg/m³

PAC-1: 0.173 mg/m³
PAC-2: 0.3 mg/m³
PAC-3: 115 mg/m³
As inorganic lead
OSHA PEL: 0.050 mg[Pb]/m³ TWA. [*Note:* The PEL also applies to other lead compounds (as Pb).] Other OSHA requirements can be found in 29 CFR 1910.1025. The OSHA PEL (8-h TWA) for lead in "nonferrous foundries with less than 20 employees" is 0.075 mg/m³. OSHA considers "lead" to mean metallic lead; all inorganic lead compounds (lead oxides and lead salts); and a class of organic compounds called "soaps;" all other lead compounds are excluded from this definition. See 29CFR1910.1025.
NIOSH REL: 0.050 mg[Pb]/m³ TWA. Suspected carcinogen. Limit exposure to lowest feasible level [*Note:* The REL also applies to other lead compounds (as Pb).] *Note:* NIOSH considers "lead" to mean metallic lead, lead oxides, and lead salts (including organic salts, such as lead "soaps" but excluding lead arsenate). Air concentrations should be maintained so that worker blood lead remains less than 0.060 mg[Pb]/100 g of whole blood. See *NIOSH Pocket Guide,* Appendix C.
ACGIH TLV®[1]: 0.05 mg[Pb]/m³ TWA; BEI: 30 μg[Pb]/100 mL (blood). (Note: The TLV also applies to lead, inorganic compounds.) *Note:* women of child-bearing potential whose blood exceeds 10 μg[Pb]/dL are at risk of delivering a child with a blood [Pb] over the current CDC guideline of 10 μg[Pb]/dL and may cause birth defects. Confirmed animal carcinogen with unknown relevance to humans.
European OEL: 0.15 mg[Pb]/m³ TWA (2002).
NIOSH IDLH: 100 mg [Pb]/m³.
DFG MAK: BAT: 400 μg[Pb]/L (blood) not fixed; 100 μg [Pb]/L (blood) women age <45; Carcinogen Category 2; Germ Cell Mutagen Group 3A.
Determination in Air: Determination in Air: Use NIOSH Analytical Method (IV) s #7082, 7105, 7300, 7301, 7303, 7700, 7701, 7702, 9100, 9102, 9105; OSHA Analytical Methods ID-121, ID-125G, ID-206.
Permissible Concentration in Water: The EPA limits lead in drinking water to 15 μg/L. Various organizations worldwide have set other standards for lead in drinking water as follows[35] (all in mg/L): Argentina 0.01; the Czech Republic 0.05; Germany 0.04; EEC 0.05; Japan 0.10; Mexico 0.05; former USSR-UNEP/IRPTC joint project 0.03; WHO 0.10. The states of Maine and Minnesota have set guidelines for lead in drinking water[61] at the level of 20 μg/L.
Determination in Water: Digestion followed by atomic absorption, or by colorimetric (dithizone) analysis, or by inductively coupled plasma (ICP) optical emission spectrometry. That gives total lead; dissolved lead may be determined by 0.45-μm filtration prior to such analysis.
Routes of Entry: Inhalation.
Harmful Effects and Symptoms
Short Term Exposure: Lead dioxide can affect you when breathed in. Irritates eyes, skin, and respiratory tract. Lead

dioxide should be handled as a teratogen with extreme caution. Lead poisoning can cause poor appetite, colic, upset stomach, headaches, irritability, muscle or joint pains and weakness. Permanent kidney damage can result from high exposures.

Long Term Exposure: Repeated, prolonged, or high exposures may cause kidney damage. May cause lead poisoning with symptoms of poor appetite, upset stomach, colic, headache, irritability, muscle or joint pains and weakness, constipation, disturbed sleep, and reduced memory. See also "Lead" entry.

Points of Attack: Kidneys.

Medical Surveillance: Before first exposure and every 6 months thereafter, OSHA (1910.1025) requires your employer to provide: blood lead test, ZPP test (a special test for the effect of lead on blood cells). Examination of the nervous system. Prior to initial exposure, and annually for exposed person having blood lead readings exceeding 40 μg/100 g of whole blood, OSHA also requires a complete medical history, complete blood count and kidney function tests in addition to the tests listed above. OSHA defines "exposure" for these tests as air levels which average 30 μg of lead or more in a cubic meter of air. OSHA requires your employer to send the doctor a copy of the lead standard and provide one for you.

Note: Blood-lead level is a good indicator of total lead exposure. Current OSHA regulations require that if an individual has a blood-lead level greater than or equal to 0.050 mg lead per 100 mL blood, he or she must be removed from all exposures to lead and cannot return to the exposure environment until the blood level falls to 0.040 mg lead per 100 mL blood or less. The following tolerance levels for occupational exposures may also be useful: ACGIH BEI = 50 mg/L (blood); 150 mg/g creatinine (urine). DFG BAT = 70 mg/L (blood); 30 mg/L (blood) for women <45 years old.

First Aid: If this chemical gets into the eyes, remove any contact lenses at once and irrigate immediately for at least 15 min, occasionally lifting upper and lower lids. Seek medical attention immediately. If this chemical contacts the skin, remove contaminated clothing and wash immediately with soap and water. Seek medical attention immediately. If this chemical has been inhaled, remove from exposure, begin rescue breathing (using universal precautions, including resuscitation mask) if breathing has stopped and CPR if heart action has stopped. Transfer promptly to a medical facility. When this chemical has been swallowed, get medical attention. Give large quantities of water and induce vomiting. Do not make an unconscious person vomit.

Antidotes and special procedures for lead: Persons with significant lead poisoning are sometimes treated with Ca EDTA while hospitalized. This "chelating" drug causes a rush of lead from the body organs into the blood and kidneys, and thus has its own hazards, and must be administered only by highly experienced medical personnel under controlled conditions and careful observation. Ca EDTA or similar drugs should never be used to prevent poisoning while exposure continues or without strict exposure control, as severe kidney damage can result.

Note to physician: For severe poisoning BAL [British Anti-Lewisite, dimercaprol, dithiopropanol ($C_3H_8OS_2$)] has been used to treat toxic symptoms of certain heavy metals poisoning. In the case of lead poisoning it may have SOME value. Although BAL is reported to have a large margin of safety, caution must be exercised, because toxic effects may be caused by excessive dosage. Most can be prevented by premedication with 1-ephedrine sulfate (CAS: 134-72-5).

Personal Protective Methods: Wear protective gloves and clothing to prevent any reasonable probability of skin contact. Use any barrier that will prevent contamination from the dust. Safety equipment suppliers/manufacturers can provide recommendations on the most protective glove/clothing material for your operation. All protective clothing (suits, gloves, footwear, headgear) should be clean, available each day, and put on before work. Contact lenses should not be worn when working with this chemical. Wear dust-proof chemical goggles and face shield unless full face-piece respiratory protection is worn. Employees should wash immediately with soap when skin is wet or contaminated. Provide emergency showers and eyewash. See also NIOSH Criteria Document #78-158, *LEAD, inorganic dusts and fumes.*

Respirator Selection: OSHA: *0.5 mg/m³:* Any air-purifying respirator with an N100, R100, or P100 filter (including N100, R100, and P100 filtering face-pieces) except quarter-mask respirators. *1.25 mg/m³:* Sa:Cf (APF = 25) (any supplied-air respirator operated in a continuous-flow mode) or PaprHie (APF = 25) (any powered, air-purifying respirator with a high-efficiency particulate filter). *2.5 mg/m³:* 100F (APF = 50) (any air purifying, full-face-piece respirator with an N100, R100, or P100 filter) or SaT: Cf (APF = 50) (any supplied-air respirator that has a tight-fitting face-piece and is operated in a continuous-flow mode) or PaprTHie (APF = 50) (any powered, air-purifying respirator with a tight-fitting face-piece and a high-efficiency particulate filter) or SCBAF (APF = 50) (any self-contained breathing apparatus with a full face-piece) or SaF (APF = 50) (any supplied-air respirator with a full face-piece). *50 mg/m³:* Sa: Pd,Pp (APF = 1000) (any supplied-air respirator operated in a pressure-demand or other positive-pressure mode). *100 mg/m³:* SaF: Pd,Pp (APF = 2000) (any supplied-air respirator that has a full face-piece and is operated in a pressure-demand or other positive-pressure mode). *Emergency or planned entry into unknown concentrations or IDLH conditions:* SCBAF: Pd,Pp (APF = 10,000) (any self-contained breathing apparatus that has a full face-piece and is operated in a pressure-demand or other positive-pressure mode) or SaF: Pd,Pp: ASCBA (any supplied-air respirator that has a full face-piece and is operated in a pressure-demand or other positive-pressure mode in combination with an auxiliary self-contained breathing apparatus operated in a pressure-demand or other positive-pressure mode). *Escape:* 100F (APF = 50) (any air purifying, full-face-piece respirator with an N100, R100, or P100 filter) or SCBAE (any appropriate escape-type, self-contained breathing apparatus).

Storage: Color Code—Yellow: Reactive Hazard; Store in a location separate from other materials, especially flammables and combustibles. Prior to working with this chemical you should be trained on its proper handling and storage. Lead dioxide must be stored to avoid contact with oxidizers (such as perchlorates, peroxides, permanganates, chlorates and nitrates) chemically active metals (such as potassium, sodium, magnesium and zinc), since violent reactions occur. Store in tightly closed containers in a cool, well-ventilated area away from combustible materials, such as wood, paper, and oil. Lead dioxide is regulated by OSHA Standard 1910.1025. All requirements of the standard must be followed. See OSHA Standard 1910.104 and NFPA 43A *Code for the Storage of Liquid and Solid Oxidizers* for detailed handling and storage regulations.

Shipping: Lead dioxide requires a shipping label of "OXIDIZER." It falls in Hazard Class 5.1 and Packing Group III.

Spill Handling: Evacuate persons not wearing protective equipment from area of spill or leak until cleanup is complete. Remove all ignition sources. Collect powdered material in the most convenient and safe manner and deposit in sealed containers. Ventilate area after cleanup is complete. Keep combustibles (wood, paper, oil) away from spilled material. It may be necessary to contain and dispose of this chemical as a hazardous waste. If material or contaminated runoff enters waterways, notify downstream users of potentially contaminated waters. Contact your local or federal environmental protection agency for specific recommendations. If employees are required to clean up spills, they must be properly trained and equipped. OSHA 1910.120(q) may be applicable.

Fire Extinguishing: Lead dioxide does not burn but it will increase the intensity of a fire. Use extinguisher suitable for surrounding fire. Poisonous gases are produced in fire. If material or contaminated runoff enters waterways, notify downstream users of potentially contaminated waters. Notify local health and fire officials and pollution control agencies. From a secure, explosion-proof location, use water spray to cool exposed containers. If cooling streams are ineffective (venting sound increases in volume and pitch, tank discolors, or shows any signs of deforming), withdraw immediately to a secure position. If employees are expected to fight fires, they must be trained and equipped in OSHA 1910.156. The only respirators recommended for firefighting are self-contained breathing apparatuses that have full face-pieces and are operated in a pressure-demand or other positive-pressure mode.

Disposal Method Suggested: Conversion to soluble salt, precipitation as sulfide, and return to supplier.

Reference

New Jersey Department of Health and Senior Services. (September 2001). *Hazardous Substances Fact Sheet: Lead Dioxide.* Trenton, NJ

Lead fluoborate   L:0150

Molecular Formula: B_2F_8Pb
Common Formula: $Pb(BF_4)_2$
Synonyms: Borate(1-), tetrafluoro-, lead(2 +); Fluoborato de plomo (Spanish); Lead tetrafluoroborate; Tetrafluoro borate; Tetrafluoro borate(1-), lead(2 +)
CAS Registry Number: 13814-96-5
RTECS® Number: ED2700000
UN/NA & ERG Number: UN2291/151
EC Number: 237-486-0
Regulatory Authority and Advisory Bodies
Carcinogenicity: IARC: Animal Inadequate Evidence, Human Inadequate Evidence, *possibly carcinogenic to humans*, Group 2A, 1987; Carcinogenicity: NTP: 11th Report on Carcinogens, 2004: Reasonably anticipated to be a human carcinogen; EPA: Sufficient evidence from animal studies; inadequate evidence or no useful data from epidemiologic studies.

Air Pollutant Standard set. See below, "Permissible Exposure Limits in Air" section.

OSHA, 29CFR1910 Specifically Regulated Chemicals (See 29 CFR 1910.1025).

Hazardous Substance (EPA) (RQ = 5000/2270).[4]

Clean Water Act: Section 311 Hazardous Substances/RQ 40CFR117.3 (same as CERCLA, see below); Section 313 Water Priority Chemicals (57FR41331, 9/9/92); 40CFR401.15 Section 307 Toxic Pollutants as lead and compounds.

RCRA, 40CFR261, Appendix 8 Hazardous Constituents, waste number not listed, as lead compounds, n.o.s.

Reportable Quantity (RQ): 10 lb (4.54 kg).

EPCRA Section 313 Form R *de minimis* concentration reporting level: 0.1%.

US DOT Regulated Marine Pollutant (49CFR172.101, Appendix B).

California Proposition 65 Chemical: Cancer 10/1/92 (lead and compounds).

Canada, WHMIS, Ingredients Disclosure List Concentration 1.0% as lead inorganic compounds, n.o.s.

WGK (German Aquatic Hazard Class): No value assigned.

Description: Lead fluoborate is a nonflammable colorless liquid or crystalline powder with a faint odor. Specific gravity (H_2O:1) = 1.75 at 20°C; Molecular weight = 380.81. Hazard Identification (based on NFPA-704 M Rating System): Health 1, Flammability 0, Reactivity 0. Slightly soluble in water (decomposes).

Potential Exposure: This material is used in material finishing operations.

Incompatibilities: Aqueous solution is acidic. Incompatible with oxidizers, bases, active metals. Decomposes in water or alcohol. Attacks most metals, especially aluminum.

Permissible Exposure Limits in Air

Protective Action Criteria (PAC) *13814-96-5*

TEEL-0: 0.0919 mg/m^3

PAC-1: 0.276 mg/m^3

PAC-2: 0.459 mg/m^3

PAC-3: 184 mg/m^3

As inorganic lead

OSHA PEL: 0.050 mg[Pb]/m^3 TWA. [*Note*: The PEL also applies to other lead compounds (as Pb).] Other OSHA requirements can be found in 29 CFR 1910.1025. The OSHA PEL (8-h TWA) for lead in "nonferrous foundries with less than 20 employees" is 0.075 mg/m^3. OSHA considers "lead" to mean metallic lead; all inorganic lead compounds (lead oxides and lead salts); and a class of organic compounds called "soaps;" all other lead compounds are excluded from this definition. See 29CFR1910.1025.

NIOSH REL: 0.050 mg[Pb]/m^3 TWA. Suspected carcinogen. Limit exposure to lowest feasible level. [*Note*: The REL also applies to other lead compounds (as Pb).] *Note:* NIOSH considers "lead" to mean metallic lead, lead oxides, and lead salts (including organic salts, such as lead "soaps" but excluding lead arsenate). Air concentrations should be maintained so that worker blood lead remains less than 0.060 mg[Pb]/100 g of whole blood. See *NIOSH Pocket Guide*, Appendix C.

ACGIH TLV®[1]: 0.05 mg[Pb]/m^3 TWA; BEI: 30 μg[Pb]/100 mL (blood). (Note: The TLV also applies to lead, inorganic compounds.) *Note:* women of child-bearing potential whose blood exceeds 10 μg[Pb]/dL are at risk of delivering a child with a blood [Pb] over the current CDC guideline of 10 μg[Pb]/dL and may cause birth defects. Confirmed animal carcinogen with unknown relevance to humans.

European OEL: 0.15 mg[Pb]/m^3 TWA (2002).

NIOSH IDLH: 100 mg [Pb]/m^3.

DFG MAK: BAT: 400 μg[Pb]/L (blood) not fixed; 100 μg [Pb]/L (blood) women age <45; Carcinogen Category 2; Germ Cell Mutagen Group 3A.

Determination in Air: Use NIOSH Analytical Method (IV) s #7082, 7105, 7300, 7301, 7303, 7700, 7701, 7702, 9100, 9102, 9105; OSHA Analytical Methods ID-121, ID-125G, ID-206.

Permissible Concentration in Water: The EPA limits lead in drinking water to 15 μg/L. Various organizations worldwide have set other standards for lead in drinking water as follows[35] (all in mg/L): Argentina 0.01; the Czech Republic 0.05; Germany 0.04; EEC 0.05; Japan 0.10; Mexico 0.05; former USSR-UNEP/IRPTC joint project 0.03; WHO 0.10. The states of Maine and Minnesota have set guidelines for lead in drinking water[61] at the level of 20 μg/L.

Determination in Water: Digestion followed by atomic absorption, or by colorimetric (dithizone) analysis, or by inductively coupled plasma (ICP) optical emission spectrometry. That gives total lead; dissolved lead may be determined by 0.45-μm filtration prior to such analysis.

Routes of Entry: Inhalation, ingestion.

Harmful Effects and Symptoms

Short Term Exposure: Lead fluoborate can affect you when breathed in and if swallowed from food, drinks, or cigarettes. Contact can cause skin and eye irritation and burns. Irritates the respiratory tract. Can cause headache, irritability, mood changes, reduced memory, and disturbed sleep.

Long Term Exposure: Repeated exposure causes lead fluoborate build up in the body and lead to lead or fluoride poisoning. Low levels may cause tiredness, mood changes, headaches, stomach problems, and trouble sleeping. Higher levels may cause aching, weakness, and concentration or memory problems. May damage the nervous system causing numbness, "pins and needles" weakness in the hands and feet. Lead fluoborate can also cause serious permanent kidney, brain damage, and damage the blood cells, causing anemia. Lead fluoborate exposure increases risk of high blood pressure. Lead compounds have been determined to be teratogens and may also cause reproductive damage, such as reduced fertility and interference with menstrual cycles. Lead fluoborate should be handled as a teratogen with extreme caution.

Points of Attack: Kidneys, nervous system, brain, blood.

Medical Surveillance: Before first exposure and every 6 months thereafter, OSHA (1910.1025) requires your employer to provide: blood lead test, ZPP test (a special test for the effect of lead on blood cells). Examination of the nervous system. Prior to initial exposure, and annually for exposed person having blood lead readings exceeding 40 μg/100 g of whole blood, OSHA also requires a complete medical history, complete blood count and kidney function tests in addition to the tests listed above. OSHA defines "exposure" for these tests as air levels which average 30 μg of lead or more in a cubic meter of air. OSHA (under 1910.1020) requires your employer to send the doctor a copy of the lead standard and provide one for you.

Note: Blood-lead level is a good indicator of total lead exposure. Current OSHA regulations require that if an individual has a blood-lead level greater than or equal to 0.050 mg lead per 100 mL blood, he or she must be removed from all exposures to lead and cannot return to the exposure environment until the blood level falls to 0.040 mg lead per 100 mL blood or less. The following tolerance levels for occupational exposures may also be useful: ACGIH BEI = 50 mg/L (blood); 150 mg/g creatinine (urine). DFG BAT = 70 mg/L (blood); 30 mg/L (blood) for women <45 years old.

NIOSH lists the following tests for fluorides: chest X-ray, electrocardiogram, pulmonary function tests: forced vital capacity, forced expiratory volume (1 s); pelvic X-ray; sputum cytology; urine (chemical/metabolite); urine (chemical/metabolite) pre- and postshift; urinalysis (routine); complete blood count/differential.

First Aid: If this chemical gets into the eyes, remove any contact lenses at once and irrigate immediately for at least 15 min, occasionally lifting upper and lower lids. Seek medical attention immediately. If this chemical contacts the skin, remove contaminated clothing and wash immediately with soap and water. Seek medical attention immediately. If this chemical has been inhaled, remove from exposure, begin rescue breathing (using universal precautions, including resuscitation mask) if breathing has stopped and CPR if heart action has stopped. Transfer promptly to a medical facility. When this chemical has been swallowed, get medical attention. Give large quantities of water and induce vomiting. Do not make an unconscious person vomit.

Antidotes and special procedures for lead: Persons with significant lead poisoning are sometimes treated with Ca EDTA while hospitalized. This "chelating" drug causes a rush of lead from the body organs into the blood and kidneys, and thus has its own hazards, and must be administered only by highly experienced medical personnel under controlled conditions and careful observation. Ca EDTA or similar drugs should never be used to prevent poisoning while exposure continues or without strict exposure control, as severe kidney damage can result.

Note to physician: For severe poisoning BAL [British Anti-Lewisite, dimercaprol, dithiopropanol ($C_3H_8OS_2$)] has been used to treat toxic symptoms of certain heavy metals poisoning. In the case of lead poisoning it may have SOME value. Although BAL is reported to have a large margin of safety, caution must be exercised, because toxic effects may be caused by excessive dosage. Most can be prevented by premedication with 1-ephedrine sulfate (CAS: 134-72-5).

Personal Protective Methods: Wear protective gloves and clothing to prevent any reasonable probability of skin contact. Use any barrier that will prevent contamination from the dust. Safety equipment suppliers/manufacturers can provide recommendations on the most protective glove/clothing material for your operation. All protective clothing (suits, gloves, footwear, headgear) should be clean, available each day, and put on before work. Contact lenses should not be worn when working with this chemical. Wear splash-proof chemical goggles and face shield unless full face-piece respiratory protection is worn. Employees should wash immediately with soap when skin is wet or contaminated. Provide emergency showers and eyewash. See also NIOSH Criteria Document #78-158, *LEAD, inorganic dusts and fumes.*

Respirator Selection: OSHA: *0.5 mg/m³:* Any air-purifying respirator with an N100, R100, or P100 filter (including N100, R100, and P100 filtering face-pieces) except quarter-mask respirators. *1.25 mg/m³:* Sa:Cf (APF = 25) (any supplied-air respirator operated in a continuous-flow mode) or PaprHie (APF = 25) (any powered, air-purifying respirator with a high-efficiency particulate filter). *2.5 mg/m³:* 100F (APF = 50) (any air purifying, full-face-piece respirator with an N100, R100, or P100 filter) or SaT: Cf (APF = 50) (any supplied-air respirator that has a tight-fitting face-piece and is operated in a continuous-flow mode) or PaprTHie (APF = 50) (any powered, air-purifying respirator with a tight-fitting face-piece and a high-efficiency particulate filter) or SCBAF (APF = 50) (any self-contained breathing apparatus with a full face-piece) or SaF (APF = 50) (any supplied-air respirator with a full face-piece). *50 mg/m³:* Sa: Pd,Pp (APF = 1000) (any supplied-air respirator operated in a pressure-demand or other positive-pressure mode). *100 mg/m³:* SaF: Pd,Pp (APF = 2000) (any supplied-air respirator that has a full face-piece and is operated in a pressure-demand or other positive-pressure mode). *Emergency or planned entry into unknown concentrations or IDLH conditions:* SCBAF: Pd,Pp (APF = 10,000) (any self-contained breathing apparatus that has a full face-piece and is operated in a pressure-demand or other positive-pressure mode) or SaF: Pd,Pp: ASCBA (any supplied-air respirator that has a full face-piece and is operated in a pressure-demand or other positive-pressure mode in combination with an auxiliary self-contained breathing apparatus operated in a pressure-demand or other positive-pressure mode). *Escape:* 100F (APF = 50) (any air purifying, full-face-piece respirator with an N100, R100, or P100 filter) or SCBAE (any appropriate escape-type, self-contained breathing apparatus).

Storage: Color Code—Blue: Health Hazard/Poison: Store in a secure poison location. Prior to working with this chemical you should be trained on its proper handling and storage. Lead fluoborate must be stored to avoid contact with oxidizers (such as perchlorates, peroxides, permanganates, chlorates, and nitrates) and chemically active metals (such as potassium, sodium, magnesium, and zinc), since violent reactions occur. A regulated, marked area should be established where this chemical is handled, used, or stored in compliance with OSHA Standard 1910.1045. Lead is regulated by an OSHA Standard 1910.1025. All requirements of the standard must be followed.

Shipping: Lead compounds, soluble, n.o.s. require a shipping label of "POISONOUS/TOXIC MATERIALS." It falls in Hazard Class 6.1 and Packing Group III.

Spill Handling: Evacuate and restrict persons not wearing protective equipment from area of spill or leak until cleanup is complete. Remove all ignition sources. Absorb liquids in vermiculite, dry sand, earth, peat, carbon, or a similar material and deposit in sealed containers. Collect powdered material in the most convenient and safe manner and deposit in sealed containers. Ventilate and wash area after cleanup is complete. It may be necessary to contain and dispose of this chemical as a hazardous waste. If material or contaminated runoff enters waterways, notify downstream users of potentially contaminated waters. Contact your local or federal environmental protection agency for specific recommendations. If employees are required to clean up spills, they must be properly trained and equipped. OSHA 1910.120(q) may be applicable.

Fire Extinguishing: Lead fluoborate itself does not burn. Use any agent suitable for type of surrounding fire.

Poisonous fumes including lead and fluorine are in fire. Containers may explode in fire. Storage containers and parts of containers may rocket great distances, in many directions. If material or contaminated runoff enters waterways, notify downstream users of potentially contaminated waters. Notify local health and fire officials and pollution control agencies. From a secure, explosion-proof location, use water spray to cool exposed containers. If cooling streams are ineffective (venting sound increases in volume and pitch, tank discolors, or shows any signs of deforming), withdraw immediately to a secure position. If employees are expected to fight fires, they must be trained and equipped in OSHA 1910.156. The only respirators recommended for firefighting are self-contained breathing apparatuses that have full face-pieces and are operated in a pressure-demand or other positive-pressure mode.

References

Sax, N. I. (Ed.). (1981). *Dangerous Properties of Industrial Materials Report*, 1, No. 6, 79–80

New Jersey Department of Health and Senior Services. (November 1999). *Hazardous Substances Fact Sheet: Lead Fluoborate*. Trenton, NJ

Lead fluoride   L:0160

Molecular Formula: F_2Pb

Synonyms: Fluoruro de plomo (Spanish); Hydrofluoric acid, lead(2 +) salt; Hydrofluoric acid, lead(II) salt; Lead difluoride; Lead(2 +) fluoride; Lead(II) fluoride; Plomb fluorure (French); Plumbous fluoride

CAS Registry Number: 7783-46-2

RTECS® Number: OG1225000

UN/NA & ERG Number: UN3288 (Toxic solid, inorganic, n.o.s.)/151

EC Number: 231-998-8

Regulatory Authority and Advisory Bodies

Carcinogenicity: IARC: Animal Inadequate Evidence, Human Inadequate Evidence, *possibly carcinogenic to humans*, Group 2A, 1987; Carcinogenicity: NTP: 11th Report on Carcinogens, 2004: Reasonably anticipated to be a human carcinogen; EPA: Sufficient evidence from animal studies; inadequate evidence or no useful data from epidemiologic studies.

OSHA, 29CFR1910 Specifically Regulated Chemicals (See CFR 1910.1025).

Clean Water Act: Section 311 Hazardous Substances/RQ 40CFR117.3 (same as CERCLA, see below); Section 313 Water Priority Chemicals (57FR41331, 9/9/92); 40CFR401.15 Section 307 Toxic Pollutants as lead and compounds.

RCRA, 40CFR261, Appendix 8 Hazardous Constituents, waste number not listed, as lead compounds, n.o.s.

Reportable Quantity (RQ): 10 lb (4.54 kg).

EPCRA Section 313 Form R *de minimis* concentration reporting level: 0.1%.

US DOT Regulated Marine Pollutant (49CFR172.101, Appendix B).

California Proposition 65 Chemical: Cancer 10/1/92 (lead and compounds).

European/International Regulations: not listed in Annex 1.

WGK (German Aquatic Hazard Class): No value assigned.

Description: Lead fluoride is a white to colorless, odorless crystalline (rhombic, orthorhombic) solid. Molecular weight = 245.19; Boiling point = 1292°C; Freezing/Melting point = 825°C. Hazard Identification (based on NFPA-704 M Rating System): Health 1, Flammability 0, Reactivity 0. Slightly soluble in water.

Potential Exposure: Used to make other chemicals, underwater paints, electronic and optical parts (for growing single-crystal, solid-state lasers), in high-temperature dry-film lubricants, and for making special grades of glass.

Incompatibilities: Violent reaction with oxidizers, chemically active metals, calcium carbide. May ignite combustibles, such as wood, paper, oil, etc.

Permissible Exposure Limits in Air

Protective Action Criteria (PAC)

7783-46-2

TEEL-0: 0.0592 mg/m^3

PAC-1: 0.178 mg/m^3

PAC-2: 15 mg/m^3

PAC-3: 118 mg/m^3

As inorganic lead

OSHA PEL: 0.050 mg[Pb]/m^3 TWA. [*Note*: The PEL also applies to other lead compounds (as Pb).] Other OSHA requirements can be found in 29 CFR 1910.1025. The OSHA PEL (8-h TWA) for lead in "nonferrous foundries with less than 20 employees" is 0.075 mg/m^3. OSHA considers "lead" to mean metallic lead; all inorganic lead compounds (lead oxides and lead salts); and a class of organic compounds called "soaps;" all other lead compounds are excluded from this definition. See CFR1910.1025.

NIOSH REL: 0.050 mg[Pb]/m^3 TWA. Suspected carcinogen. Limit exposure to lowest feasible level. [*Note*: The REL also applies to other lead compounds (as Pb).] *Note:* NIOSH considers "lead" to mean metallic lead, lead oxides, and lead salts (including organic salts, such as lead "soaps" but excluding lead arsenate). Air concentrations should be maintained so that worker blood lead remains less than 0.060 mg[Pb]/100 g of whole blood. See *NIOSH Pocket Guide*, Appendix C.

ACGIH TLV®[1]: 0.05 mg[Pb]/m^3 TWA; BEI: 30 μg[Pb]/100 mL (blood). (Note: The TLV also applies to lead, inorganic compounds.) *Note:* women of child-bearing potential whose blood exceeds 10 μg[Pb]/dL are at risk of delivering a child with a blood [Pb] over the current CDC guideline of 10 μg[Pb]/dL and may cause birth defects. Confirmed animal carcinogen with unknown relevance to humans.

NIOSH IDLH: 100 mg [Pb]/m^3.

DFG MAK: BAT: 400 μg[Pb]/L (blood) not fixed; 100 μg [Pb]/L (blood) women age <45; Carcinogen Category 2; Germ Cell Mutagen Group 3A.

As inorganic fluorides
OSHA PEL: 3 ppm/2.5 mg[F]/m^3 TWA.
NIOSH REL: 3 ppm/2.5 mg[F]/m^3 TWA; 6 ppm/5 mg[F]/m^3, 15-min Ceiling Concentration.
ACGIH TLV$^{®[1]}$: 2.5 mg[F]/m^3 TWA; not classifiable as a human carcinogen; BEI: 3 mg[F]/g creatinine in urine *prior* to end-of-shift; 10 mg[F]/g creatinine in urine end-of-shift.
DFG MAK: 1 mg[F]/m^3, inhalable fraction [skin]; Peak Limitation Category II(4); Pregnancy Risk Group C; BAT: 7.0 mg[F]/g creatinine in urine at end-of-shift; 4.0 mg[F]/g creatinine in urine at the beginning of the next shift.
NIOSH IDLH: 250 mg[F]/m^3.

Determination in Air: Use NIOSH Analytical Method (IV) s #7082, 7105, 7300, 7301, 7303, 7700, 7701, 7702, 9100, 9102, 9105; OSHA Analytical Methods ID-121, ID-125G, ID-206.

Permissible Concentration in Water: 0.1 mg/L Pb in drinking water may cause chronic poisoning. The EPA limits lead in drinking water to 15 μg/L. Various organizations worldwide have set other standards for lead in drinking water as follows[35] (all in mg/L): Argentina 0.01; the Czech Republic 0.05; Germany 0.04; EEC 0.05; Japan 0.10; Mexico 0.05; former USSR-UNEP/IRPTC joint project 0.03; WHO 0.10. The states of Maine and Minnesota have set guidelines for lead in drinking water[61] at the level of 20 μg/L.

Determination in Water: Digestion followed by atomic absorption, or by colorimetric (dithizone) analysis, or by inductively coupled plasma (ICP) optical emission spectrometry. That gives total lead; dissolved lead may be determined by 0.45-μm filtration prior to such analysis.

Routes of Entry: Inhalation, skin. Lead can be absorbed through skin at chronically toxic levels.

Harmful Effects and Symptoms

Short Term Exposure: Contact can cause skin and eye irritation and burns. Inhalation can irritate the nose and throat. Lead fluoride can cause headache, irritability, mood changes, reduced memory, and disturbed sleep. The fluoride ion can cause protoplasmic poisoning at higher concentrations. See also entry on "Fluorides."

Note: Lead is a cumulative poison. Increasing amounts can build up in the body, eventually reaching a point where symptoms and disability occur. Lead dust carried home on contaminated clothing can result in exposure and symptoms in other family members. Standards only protect for inhalation exposure. Extra precautions should be taken if skin exposure also occurs.

Long Term Exposure: While lead fluoride has not been identified as a teratogen, or a reproductive hazard, lead and certain lead compounds have been determined to be teratogens and may also cause reproductive damage, such as reduced fertility and interfere with menstrual cycles. Handle with extreme caution. Lead can accumulate in the body over a period of time. Therefore, long-term exposures to lower levels can result in a buildup of lead in the body and more severe symptoms. These may include anemia, pale

skin, a blue line at the gum margin, decreased handgrip strength, abdominal pain, severe constipation, nausea, vomiting, and paralysis of the wrist joint. Prolonged exposure may also result in kidney and brain damage. If the nervous system is affected, usually due to very high exposures, the resulting effects include severe headache, convulsions, coma, delirium, and death. In nonfatal cases, recovery is slow and not always complete. Alcohol ingestion and physical exertion may bring on symptoms. Lead exposure increases the risk of high blood pressure.

Points of Attack: Blood, kidneys, brain, nervous system.

Medical Surveillance: Before first exposure and every 6 months thereafter, OSHA (1910.1025) requires your employer to provide: blood lead test, ZPP test (a special test for the effect of lead on blood cells). Examination of the nervous system. Prior to initial exposure, and annually for exposed person having blood lead readings exceeding 40 μg/100 g of whole blood, OSHA also requires a complete medical history, complete blood count and kidney function tests in addition to the tests listed above. OSHA defines "exposure" for these tests as air levels which average 30 μg of lead or more in a cubic meter of air. OSHA (1910.1020) requires your employer to send the doctor a copy of the lead standard and provide one for you.

Note: Blood-lead level is a good indicator of total lead exposure. Current OSHA regulations require that if an individual has a blood-lead level greater than or equal to 0.050 mg lead per 100 mL blood, he or she must be removed from all exposures to lead and cannot return to the exposure environment until the blood level falls to 0.040 mg lead per 100 mL blood or less. The following tolerance levels for occupational exposures may also be useful: ACGIH BEI = 50 mg/L (blood); 150 mg/g creatinine (urine). DFG BAT = 70 mg/L (blood); 30 mg/L (blood) for women <45 years old.

First Aid: If this chemical gets into the eyes, remove any contact lenses at once and irrigate immediately for at least 15 min, occasionally lifting upper and lower lids. Seek medical attention immediately. If this chemical contacts the skin, remove contaminated clothing and wash immediately with soap and water. Seek medical attention immediately. If this chemical has been inhaled, remove from exposure, begin rescue breathing (using universal precautions, including resuscitation mask) if breathing has stopped and CPR if heart action has stopped. Transfer promptly to a medical facility. When this chemical has been swallowed, get medical attention. Give large quantities of water and induce vomiting. Do not make an unconscious person vomit.

Note to physician: Administer saline cathartic and an enema. For relief of colic, administer antispasmodic (calcium gluconate, atropine, papaverine). Consider morphine sulfate for severe pain.

Whole blood lead levels, circulating plasma/erythrocyte lead concentration ratio, urine ALA, and erythrocyte protoporphyrin fluorescent microscopy may all be useful in monitoring or assessing lead exposure. Chelating agents, such as edetate disodium calcium (Ca EDTA) and penicillamine (*not penicillin*), are generally useful in the therapy of acute lead intoxication.

Antidotes and special procedures for lead: Persons with significant lead poisoning are sometimes treated with Ca EDTA while hospitalized. This "chelating" drug causes a rush of lead from the body organs into the blood and kidneys, and thus has its own hazards, and must be administered only by highly experienced medical personnel under controlled conditions and careful observation. Ca EDTA or similar drugs should never be used to prevent poisoning while exposure continues or without strict exposure control, as severe kidney damage can result.

Personal Protective Methods: Wear protective gloves and clothing to prevent any reasonable probability of skin contact. Use any barrier that will prevent contamination from the dust. Safety equipment suppliers/manufacturers can provide recommendations on the most protective glove/clothing material for your operation. All protective clothing (suits, gloves, footwear, headgear) should be clean, available each day, and put on before work. Work clothing should be HEPA vacuumed before removal. Contact lenses should not be worn when working with this chemical. Wear dust-proof chemical goggles and face shield unless full face-piece respiratory protection is worn. Employees should wash immediately with soap when skin is wet or contaminated. Provide emergency showers and eyewash.

Respirator Selection: OSHA: *0.5 mg/m³:* Any air-purifying respirator with an N100, R100, or P100 filter (including N100, R100, and P100 filtering face-pieces) except quarter-mask respirators. *1.25 mg/m³:* Sa:Cf (APF = 25) (any supplied-air respirator operated in a continuous-flow mode) or PaprHie (APF = 25) (any powered, air-purifying respirator with a high-efficiency particulate filter). *2.5 mg/m³:* 100F (APF = 50) (any air purifying, full-face-piece respirator with an N100, R100, or P100 filter) or SaT: Cf (APF = 50) (any supplied-air respirator that has a tight-fitting face-piece and is operated in a continuous-flow mode) or PaprTHie (APF = 50) (any powered, air-purifying respirator with a tight-fitting face-piece and a high-efficiency particulate filter) or SCBAF (APF = 50) (any self-contained breathing apparatus with a full face-piece) or SaF (APF = 50) (any supplied-air respirator with a full face-piece). *50 mg/m³:* Sa: Pd,Pp (APF = 1000) (any supplied-air respirator operated in a pressure-demand or other positive-pressure mode). *100 mg/m³:* SaF: Pd,Pp (APF = 2000) (any supplied-air respirator that has a full face-piece and is operated in a pressure-demand or other positive-pressure mode). *Emergency or planned entry into unknown concentrations or IDLH conditions:* SCBAF: Pd,Pp (APF = 10,000) (any self-contained breathing apparatus that has a full face-piece and is operated in a pressure-demand or other positive-pressure mode) or SaF: Pd,Pp: ASCBA (any supplied-air respirator that has a full face-piece and is operated in a pressure-demand or other positive-pressure mode in combination with an auxiliary self-contained breathing apparatus

operated in a pressure-demand or other positive-pressure mode). *Escape:* 100F (APF = 50) (any air purifying, full-face-piece respirator with an N100, R100, or P100 filter) or SCBAE (any appropriate escape-type, self-contained breathing apparatus).

As fluorides
NIOSH/OSHA *12.5 mg/m³:* Qm (APF = 25) (any quarter-mask respirator). *25 mg/m³:* 95XQ (APF = 10)* [any particulate respirator equipped with an N95, R95, or P95 filter (including N95, R95, and P95 filtering face-pieces) except quarter-mask respirators. The following filters may also be used: N99, R99, P99, N100, R100, P100] or SA* (any supplied-air respirator). *62.5 mg/m³:* Sa:Cf (APF = 25)*† (any supplied-air respirator operated in a continuous-flow mode) or PaprHie (APF = 25)* *if not present as a fume* (any powered, air-purifying respirator with a high-efficiency particulate filter). *125 mg/m³:* 100F (APF = 50)† [any particulate respirator equipped with an N95, R95, or P95 filter (including N95, R95, and P95 filtering face-pieces) except quarter-mask respirators. The following filters may also be used: N99, R99, P99, N100, R100, P100] or SCBAF (APF = 50) (any self-contained breathing apparatus with a full face-piece) or SaF (APF = 50) (any supplied-air respirator with a full face-piece). *250 mg/m³:* Sa: Pd,Pp (APF = 1000) (any supplied-air respirator operated in a pressure-demand or other positive-pressure mode). *Emergency or planned entry into unknown concentrations or IDLH conditions:* SCBAF: Pd,Pp (APF = 10,000) (any self-contained breathing apparatus that has a full faceplate and is operated in a pressure-demand or other positive-pressure mode) or SaF: Pd,Pp: ASCBA (APF = 10,000) (any supplied-air respirator that has a full face-piece and is operated in a pressure-demand or other positive-pressure mode in combination with an auxiliary, self-contained breathing apparatus operated in a pressure-demand or other positive-pressure mode). *Escape:* 100F (APF = 50)† [any particulate respirator equipped with an N95, R95, or P95 filter (including N95, R95, and P95 filtering face-pieces) except quarter-mask respirators. The following filters may also be used: N99, R99, P99, N100, R100, P100] or SCBAE (any appropriate escape-type, self-contained breathing apparatus).

*Substance reported to cause eye irritation or damage; may require eye protection.
†May need acid gas sorbent.

Storage: Color Code—Blue: Health Hazard/Poison: Store in a secure poison location. Prior to working with this chemical you should be trained on its proper handling and storage. Store in tightly closed containers in a cool, well-ventilated area. Lead is regulated by an OSHA Standard 1910.1025. All requirements of the standard must be followed.

Shipping: Toxic solid, inorganic, n.o.s. requires a label of "POISONOUS/TOXIC MATERIALS." Lead fluoride falls in Hazard Class 6.1.

Spill Handling: Evacuate persons not wearing protective equipment from area of spill or leak until cleanup is complete. Remove all ignition sources. Collect powdered material in the most convenient and safe manner and deposit in sealed containers. Use vacuum or a wet method to reduce dust. Do not dry sweep. When vacuuming, a HEPA filter should be used, not a standard shop vac. Ventilate area after cleanup is complete. It may be necessary to contain and dispose of this chemical as a hazardous waste. If material or contaminated runoff enters waterways, notify downstream users of potentially contaminated waters. Contact your local or federal environmental protection agency for specific recommendations. If employees are required to clean up spills, they must be properly trained and equipped. OSHA 1910.120(q) may be applicable.

Fire Extinguishing: This chemical does not burn. Use any extinguishing agent suitable for surrounding fire. Poisonous gases, including hydrogen fluoride, lead oxide, and lead fumes, are produced in fire. If material or contaminated run-off enters waterways, notify downstream users of potentially contaminated waters. Notify local health and fire officials and pollution control agencies. Containers may explode in fire. From a secure, explosion-proof location, use water spray to cool exposed containers. If cooling streams are ineffective (venting sound increases in volume and pitch, tank discolors, or shows any signs of deforming), withdraw immediately to a secure position. If employees are expected to fight fires, they must be trained and equipped in OSHA 1910.156. The only respirators recommended for firefighting are self-contained breathing apparatuses that have full face-pieces and are operated in a pressure-demand or other positive-pressure mode.

Reference

New Jersey Department of Health and Senior Services. (November, 1999). *Hazardous Substances Fact Sheet: Lead Fluoride.* Trenton, NJ

Lead iodide   L:0170

Molecular Formula: I_2Pb
Common Formula: PbI_2
Synonyms: Lead(2 +) iodide; Lead(II) iodide; Yoduro de plomo (Spanish)
CAS Registry Number: 10101-63-0
RTECS® Number: OG1515000
UN/NA & ERG Number: UN3288 (Toxic solid, inorganic, n.o.s.)/151
EC Number: 233-256-9
Regulatory Authority and Advisory Bodies
Carcinogenicity: IARC: Animal Inadequate Evidence, Human Inadequate Evidence, *possibly carcinogenic to humans*, Group 2A, 1987; Carcinogenicity: NTP: 11th Report on Carcinogens, 2004: Reasonably anticipated to be a human carcinogen; EPA: Sufficient evidence from animal studies; inadequate evidence or no useful data from epidemiologic studies.

Air Pollutant Standard Set. See below, "Permissible Exposure Limits in Air" section.
OSHA, 29CFR1910 Specifically Regulated Chemicals (See 29 CFR 1910.1025).
Hazardous Substance (EPA) (RQ = 5000/2270).[4]
Clean Water Act: Section 311 Hazardous Substances/RQ 40CFR117.3 (same as CERCLA, see below); Section 313 Water Priority Chemicals (57FR41331, 9/9/92); 40CFR401.15 Section 307 Toxic Pollutants as lead and compounds.
RCRA, 40CFR261, Appendix 8 Hazardous Constituents, waste number not listed, as lead compounds, n.o.s.
Reportable Quantity (RQ): 10 lb (4.54 kg).
EPCRA Section 313 Form R *de minimis* concentration reporting level: 0.1%.
US DOT Regulated Marine Pollutant (49CFR172.101, Appendix B).
California Proposition 65 Chemical: Cancer 10/1/92 (lead and compounds).
Canada, WHMIS, Ingredients Disclosure List Concentration 1.0% as lead, inorganic compounds, n.o.s.
WGK (German Aquatic Hazard Class): No value assigned.
Description: Lead Iodide is a heavy, bright-yellow, odorless powder. Hazard Identification (based on NFPA-704 M Rating System): Health 2, Flammability 0, Reactivity 0. Soluble in water.
Potential Exposure: Lead iodide is used in bronzing, gold pencils, mosaic gold, printing, and photography.
Incompatibilities: Contact with oxidizers or active metals may cause violent reaction.
Permissible Exposure Limits in Air
Protective Action Criteria (PAC) *10101-63-0*
TEEL-0: 0.111 mg/m^3
PAC-1: 0.334 mg/m^3
PAC-2: 0.556 mg/m^3
PAC-3: 222 mg/m^3
As iodides
ACGIH TLV®[1]: 0.01 ppm/0.1 mg/m^3, inhalable fraction and vapor, TWA.
As inorganic lead
OSHA PEL: 0.050 mg[Pb]/m^3 TWA. [*Note*: The PEL also applies to other lead compounds (as Pb).] Other OSHA requirements can be found in 29 CFR 1910.1025. The OSHA PEL (8-h TWA) for lead in "nonferrous foundries with less than 20 employees" is 0.075 mg/m^3. OSHA considers "lead" to mean metallic lead; all inorganic lead compounds (lead oxides and lead salts); and a class of organic compounds called "soaps;" all other lead compounds are excluded from this definition. See CFR1910.1025.
NIOSH REL: 0.050 mg[Pb]/m^3 TWA. Suspected carcinogen. Limit exposure to lowest feasible level. [*Note*: The REL also applies to other lead compounds (as Pb).] *Note:* NIOSH considers "lead" to mean metallic lead, lead oxides, and lead salts (including organic salts, such as lead "soaps" but excluding lead arsenate). Air concentrations should be maintained so that worker blood lead remains less than

0.060 mg[Pb]/100 g of whole blood. See *NIOSH Pocket Guide*, Appendix C.

ACGIH TLV®[1]: 0.05 mg[Pb]/m^3 TWA; BEI: 30 μg[Pb]/100 mL (blood). (Note: The TLV also applies to lead, inorganic compounds.) *Note:* women of child-bearing potential whose blood exceeds 10 μg[Pb]/dL are at risk of delivering a child with a blood [Pb] over the current CDC guideline of 10 μg[Pb]/dL and may cause birth defects. Confirmed animal carcinogen with unknown relevance to humans.

NIOSH IDLH: 100 mg [Pb]/m^3.

DFG MAK: BAT: 400 μg[Pb]/L (blood) not fixed; 100 μg [Pb]/L (blood) women age <45; Carcinogen Category 2; Germ Cell Mutagen Group 3A.

Determination in Air: Use NIOSH Analytical Method (IV) s #7082, 7105, 7300, 7301, 7303, 7700, 7701, 7702, 9100, 9102, 9105; OSHA Analytical Methods ID-121, ID-125G, ID-206.

Permissible Concentration in Water: The EPA limits lead in drinking water to 15 μg/L. Various organizations worldwide have set other standards for lead in drinking water as follows[35] (all in mg/L): Argentina 0.01; the Czech Republic 0.05; Germany 0.04; EEC 0.05; Japan 0.10; Mexico 0.05; former USSR-UNEP/IRPTC joint project 0.03; WHO 0.10. The states of Maine and Minnesota have set guidelines for lead in drinking water[61] at the level of 20 μg/L.

Determination in Water: Digestion followed by atomic absorption, or by colorimetric (dithizone) analysis, or by inductively coupled plasma (ICP) optical emission spectrometry. That gives total lead; dissolved lead may be determined by 0.45 μm filtration prior to such analysis.

Routes of Entry: Inhalation, ingestion.

Harmful Effects and Symptoms

Short Term Exposure: Can cause headache, irritability, reduced memory, and disturbed sleep. Lead poisoning can cause poor appetite, colic, upset stomach, headaches, irritability, muscle or joint pains, and weakness.

Long Term Exposure: Permanent kidney damage can result from long-term or high exposures. Repeated exposure may cause brain damage, and damage to the blood cells, leading to anemia. Higher levels can cause muscle and joint pains, weakness, and fatigue. May cause nerve damage. Lead exposure increases the risk of high blood pressure. Lead iodide should be handled as a teratogen, with extreme caution. Repeated exposure may cause lead poisoning and/or iodism. Symptoms of iodism can include running nose, headache, mucous membrane irritation, and skin rash.

Points of Attack: Kidneys, brain, nervous system, blood.

Medical Surveillance: Before first exposure and every 6 months thereafter, OSHA (1910.1025) requires your employer to provide blood lead test, ZPP test (a special test for the effect of lead on blood cells). Examination of the nervous system. Prior to initial exposure, and annually for exposed person having blood lead readings exceeding 40 μg/100 g of whole blood, OSHA also requires a complete medical history, complete blood count and kidney function tests in addition to the tests listed above. OSHA defines "exposure" for these tests as air levels which average 30 μg of lead or more in a cubic meter of air. OSHA (under 1910.1020) requires your employer to send the doctor a copy of the lead standard and provide one for you.

Note: Blood-lead level is a good indicator of total lead exposure. Current OSHA regulations require that if an individual has a blood-lead level greater than or equal to 0.050 mg lead per 100 mL blood, he or she must be removed from all exposures to lead and cannot return to the exposure environment until the blood level falls to 0.040 mg lead per 100 mL blood or less. The following tolerance levels for occupational exposures may also be useful: ACGIH BEI = 50 mg/L (blood); 150 mg/g creatinine (urine). DFG BAT = 70 mg/L (blood); 30 mg/L (blood) for women <45 years old.

First Aid: If this chemical gets into the eyes, remove any contact lenses at once and irrigate immediately for at least 15 min, occasionally lifting upper and lower lids. Seek medical attention immediately. If this chemical contacts the skin, remove contaminated clothing and wash immediately with soap and water. Seek medical attention immediately. If this chemical has been inhaled, remove from exposure, begin rescue breathing (using universal precautions, including resuscitation mask) if breathing has stopped and CPR if heart action has stopped. Transfer promptly to a medical facility. When this chemical has been swallowed, get medical attention. Give large quantities of water and induce vomiting. Do not make an unconscious person vomit.

Antidotes and special procedures for lead: Persons with significant lead poisoning are sometimes treated with Ca EDTA while hospitalized. This "chelating" drug causes a rush of lead from the body organs into the blood and kidneys, and thus has its own hazards, and must be administered only by highly experienced medical personnel under controlled conditions and careful observation. Ca EDTA or similar drugs should never be used to prevent poisoning while exposure continues or without strict exposure control, as severe kidney damage can result.

Note to physician: For severe poisoning BAL [British Anti-Lewisite, dimercaprol, dithiopropanol (C$_3$H$_8$OS$_2$)] has been used to treat toxic symptoms of certain heavy metals poisoning. In the case of lead poisoning it may have SOME value. Although BAL is reported to have a large margin of safety, caution must be exercised, because toxic effects may be caused by excessive dosage. Most can be prevented by premedication with 1-ephedrine sulfate (CAS: 134-72-5).

Personal Protective Methods: Wear protective gloves and clothing to prevent any reasonable probability of skin contact. Use any barrier that will prevent contamination from the dust. Safety equipment suppliers/manufacturers can provide recommendations on the most protective glove/clothing material for your operation. All protective clothing (suits, gloves, footwear, headgear) should be clean, available each day, and put on before work. Contact lenses should not be worn when working with this chemical. Wear dust-proof chemical goggles and face shield unless full face-piece respiratory protection is worn. Employees should wash immediately with soap when skin is wet or contaminated. Provide emergency showers and eyewash. See also NIOSH Criteria Document #78-158, *LEAD, inorganic dusts and fumes.*

Respirator Selection: OSHA: *0.5 mg/m³:* Any air-purifying respirator with an N100, R100, or P100 filter (including N100, R100, and P100 filtering face-pieces) except quarter-mask respirators. *1.25 mg/m³:* Sa:Cf (APF = 25) (any supplied-air respirator operated in a continuous-flow mode) or PaprHie (APF = 25) (any powered, air-purifying respirator with a high-efficiency particulate filter). *2.5 mg/m³:* 100F (APF = 50) (any air purifying, full-face-piece respirator with an N100, R100, or P100 filter) or SaT: Cf (APF = 50) (any supplied-air respirator that has a tight-fitting face-piece and is operated in a continuous-flow mode) or PaprTHie (APF = 50) (any powered, air-purifying respirator with a tight-fitting face-piece and a high-efficiency particulate filter) or SCBAF (APF = 50) (any self-contained breathing apparatus with a full face-piece) or SaF (APF = 50) (any supplied-air respirator with a full face-piece). *50 mg/m³:* Sa: Pd,Pp (APF = 1000) (any supplied-air respirator operated in a pressure-demand or other positive-pressure mode). *100 mg/m³:* SaF: Pd,Pp (APF = 2000) (any supplied-air respirator that has a full face-piece and is operated in a pressure-demand or other positive-pressure mode). *Emergency or planned entry into unknown concentrations or IDLH conditions:* SCBAF: Pd,Pp (APF = 10,000) (any self-contained breathing apparatus that has a full face-piece and is operated in a pressure-demand or other positive-pressure mode) or SaF: Pd,Pp: ASCBA (any supplied-air respirator that has a full face-piece and is operated in a pressure-demand or other positive-pressure mode in combination with an auxiliary self-contained breathing apparatus operated in a pressure-demand or other positive-pressure mode). *Escape:* 100F (APF = 50) (any air purifying, full-face-piece respirator with an N100, R100, or P100 filter) or SCBAE (any appropriate escape-type, self-contained breathing apparatus).

Storage: Color Code—Blue: Health Hazard/Poison: Store in a secure poison location. Prior to working with this chemical you should be trained on its proper handling and storage. Lead iodide must be stored to avoid contact with oxidizers (such as perchlorates, peroxides, permanganates, chlorates, and nitrates) and chemically active metals (such as potassium, sodium, magnesium, and zinc), since violent reactions occur. A regulated, marked area should be established where this chemical is handled, used, or stored in compliance with OSHA Standard 1910.1045. Lead iodide is regulated by an OSHA Standard, 1910.1025. All requirements of the standard must be followed.

Shipping: Toxic solid, inorganic, n.o.s. requires a label of "POISONOUS/TOXIC MATERIALS." Lead fluoride falls in Hazard Class 6.1.

Spill Handling: Evacuate persons not wearing protective equipment from area of spill or leak until cleanup is complete. Remove all ignition sources. Collect powdered material in the most convenient and safe manner and deposit in sealed containers. Ventilate area after cleanup is complete. It may be necessary to contain and dispose of this chemical as a hazardous waste. If material or contaminated runoff enters waterways, notify downstream users of potentially contaminated waters. Contact your local or federal environmental protection agency for specific recommendations. If employees are required to clean up spills, they must be properly trained and equipped. OSHA 1910.120(q) may be applicable.

Fire Extinguishing: Use dry chemical, carbon dioxide, water spray, or foam extinguishers. Poisonous fumes of lead and iodine are produced in fire. If material or contaminated runoff enters waterways, notify downstream users of potentially contaminated waters. Notify local health and fire officials and pollution control agencies. From a secure, explosion-proof location, use water spray to cool exposed containers. If cooling streams are ineffective (venting sound increases in volume and pitch, tank discolors, or shows any signs of deforming), withdraw immediately to a secure position. If employees are expected to fight fires, they must be trained and equipped in OSHA 1910.156. The only respirators recommended for firefighting are self-contained breathing apparatuses that have full face-pieces and are operated in a pressure-demand or other positive-pressure mode.

Reference
New Jersey Department of Health and Senior Services. (October 2004). *Hazardous Substances Fact Sheet: Lead Iodide.* Trenton, NJ

Lead phosphate   L:0180

Molecular Formula: $O_8P_2Pb_3$
Common Formula: $Pb_3(PO_4)_2$
Synonyms: Bleiphosphat (German); C.I. 77622; Fasfato de plomo (Spanish); Lead orthophosphate; Lead phosphate (3:2); Lead(2 +) phosphate; Lead(II) phosphate; Normal lead orthophosphate; Perlex paste; Phosphoric acid, lead salt; Phosphoric acid, lead(2 +) salt (2:3); Plumbous phosphate; Trilead phosphate; Trilead bis(orthophosphate)
CAS Registry Number: 7446-27-7
RTECS® Number: OG3675000
UN/NA & ERG Number: UN3288 (Toxic solid, inorganic, n.o.s.)/151
EC Number: 231-205-5 [*Annex I Index No.:* 082-006-00-3]
Regulatory Authority and Advisory Bodies
Carcinogenicity: IARC: Animal Inadequate Evidence, Human Inadequate Evidence, *possibly carcinogenic to humans*, Group 2A, 1987; Carcinogenicity: NTP: 11th Report on Carcinogens, 2004: Reasonably anticipated to be a human carcinogen; EPA: Sufficient evidence from animal studies; inadequate evidence or no useful data from epidemiologic studies.

Air Pollutant Standard Set. See below, "Permissible Exposure Limits in Air" section.

OSHA, 29CFR1910 Specifically Regulated Chemicals (See 29 CFR 1910.1025).

Clean Water Act: 40CFR401.15 Section 307 Toxic Pollutants as lead and compounds.

US EPA Hazardous Waste Number (RCRA No.): U145.

RCRA, 40CFR261, Appendix 8 Hazardous Constituents.

Reportable Quantity (RQ): 10 lb (4.54 kg).

EPCRA Section 313 Form R *de minimis* concentration reporting level: 0.1%.

Form R Toxic Chemical Category Code: N420.

US DOT Regulated Marine Pollutant (49CFR172.101, Appendix B).

California Proposition 65 Chemical: Cancer 4/1/88.

Canada, WHMIS, Ingredients Disclosure List Concentration 0.1%.

European/International Regulations: Hazard Symbol: T, N; Risk phrases: R61; R33; R48/22; R50/53; Safety phrases: S53; S45; S60; S61 (see Appendix 4).

WGK (German Aquatic Hazard Class): No value assigned.

Description: Lead phosphate is a white powder or colorless crystals. Molecular weight = 811.51; Freezing/Melting point = 1012°C. Hazard Identification (based on NFPA-704 M Rating System) (estimated): Health 3, Flammability 0, Reactivity 0. Insoluble in water.

Potential Exposure: Lead phosphate is used as a stabilizer in styrene and casein plastics.

Incompatibilities: Oxidizers, active metals.

Permissible Exposure Limits in Air

Protective Action Criteria (PAC)

7446-27-7

TEEL-0: 0.0653 mg/m^3

PAC-1: 0.196 mg/m^3

PAC-2: 30 mg/m^3

PAC-3: 131 mg/m^3

As inorganic lead

OSHA PEL: 0.05 mg[Pb]/m^3 TWA. [*Note:* The PEL also applies to other lead compounds (as Pb).] Other OSHA requirements can be found in 29 CFR 1910.1025. The OSHA PEL (8-h TWA) for lead in "nonferrous foundries with less than 20 employees" is 0.075 mg/m^3. OSHA considers "lead" to mean metallic lead; all inorganic lead compounds (lead oxides and lead salts); and a class of organic compounds called "soaps;" all other lead compounds are excluded from this definition. See 29CFR-1910.1025.

NIOSH REL: 0.05 mg[Pb]/m^3 TWA. Suspected carcinogen. Limit exposure to lowest feasible level. [*Note:* The REL also applies to other lead compounds (as Pb).] *Note:* NIOSH considers "lead" to mean metallic lead, lead oxides, and lead salts (including organic salts, such as lead "soaps" but excluding lead arsenate). Air concentrations should be maintained so that worker blood lead remains less than 0.060 mg[Pb]/100 g of whole blood. See *NIOSH Pocket Guide*, Appendix C.

ACGIH TLV®[1]: 0.05 mg[Pb]/m^3 TWA; BEI: 30 μg[Pb]/100 mL (blood). (Note: The TLV also applies to lead, inorganic compounds.) *Note:* women of child-bearing potential whose blood exceeds 10 μg[Pb]/dL are at risk of delivering a child with a blood [Pb] over the current CDC guideline of 10 μg[Pb]/dL and may cause birth defects. Confirmed animal carcinogen with unknown relevance to humans.

NIOSH IDLH: 100 mg [Pb]/m^3.

DFG MAK: BAT: 400 μg[Pb]/L (blood) not fixed; 100 μg [Pb]/L (blood) women age <45; Carcinogen Category 2; Germ Cell Mutagen Group 3A.

Determination in Air: Use NIOSH Analytical Method (IV) s #7082, 7105, 7300, 7301, 7303, 7700, 7701, 7702, 9100, 9102, 9105; OSHA Analytical Methods ID-121, ID-125G, ID-206.

Permissible Concentration in Water: The EPA limits lead in drinking water to 15 μg/L. Various organizations worldwide have set other standards for lead in drinking water as follows[35] (all in mg/L): Argentina 0.01; the Czech Republic 0.05; Germany 0.04; EEC 0.05; Japan 0.10; Mexico 0.05; former USSR-UNEP/IRPTC joint project 0.03; WHO 0.10. The states of Maine and Minnesota have set guidelines for lead in drinking water[61] at the level of 20 μg/L.

Determination in Water: Digestion followed by atomic absorption, or by colorimetric (dithizone) analysis, or by inductively coupled plasma (ICP) optical emission spectrometry. That gives total lead; dissolved lead may be determined by 0.45-μm filtration prior to such analysis.

Routes of Entry: Inhalation.

Harmful Effects and Symptoms

Short Term Exposure: Lead phosphate can cause headache, irritability, reduced memory, and disturbed sleeping patterns. Lead enters the body by breathing and from contaminated food, beverages, or cigarettes. Lead poisoning can cause poor appetite, colic, upset stomach, headaches, irritability, muscle or joint cramps, and weakness.

Long Term Exposure: Lead phosphate is a carcinogen, handle with extreme caution. Lead phosphate should be handled as a teratogen, with extreme caution. Permanent kidney damage can result from long term or high exposures. Lead exposure increases the risk of high blood pressure. High or repeated exposure may damage the nerves resulting in loss of coordination in the arms and legs. Can cause brain damage and anemia.

Points of Attack: Kidneys, blood, nervous system, brain.

Medical Surveillance: Before first exposure and every 6 months thereafter, OSHA (1910.1025) requires your employer to provide blood lead test, ZPP test (a special test for the effect of lead on blood cells). Examination of the nervous system. Prior to initial exposure, and annually for exposed person having blood lead readings exceeding 40 μg/100 g of whole blood, OSHA also requires a complete medical history, complete blood count and kidney function tests in addition to the tests listed above. OSHA defines "exposure" for these tests as air levels which average 30 μg of lead or more in a cubic meter of air. OSHA (under 1910.1020) requires your employer to send the doctor a copy of the lead standard and provide one for you.

Note: Blood-lead level is a good indicator of total lead exposure. Current OSHA regulations require that if an individual has a blood-lead level greater than or equal to 0.050 mg lead per 100 mL blood, he or she must be removed from all exposures to lead and cannot return to the exposure environment until the blood level falls to 0.040 mg lead per 100 mL blood or less. The following tolerance levels for occupational exposures may also be useful: ACGIH BEI = 50 mg/L (blood); 150 mg/g creatinine (urine). DFG BAT = 70 mg/L (blood); 30 mg/L (blood) for women <45 years old.

First Aid: If this chemical gets into the eyes, remove any contact lenses at once and irrigate immediately for at least 15 min, occasionally lifting upper and lower lids. Seek medical attention immediately. If this chemical contacts the skin, remove contaminated clothing and wash immediately with soap and water. Seek medical attention immediately. If this chemical has been inhaled, remove from exposure, begin rescue breathing (using universal precautions, including resuscitation mask) if breathing has stopped and CPR if heart action has stopped. Transfer promptly to a medical facility. When this chemical has been swallowed, get medical attention. Give large quantities of water and induce vomiting. Do not make an unconscious person vomit.

Antidotes and special procedures for lead: Persons with significant lead poisoning are sometimes treated with Ca EDTA while hospitalized. This "chelating" drug causes a rush of lead from the body organs into the blood and kidneys, and thus has its own hazards, and must be administered only by highly experienced medical personnel under controlled conditions and careful observation. Ca EDTA or similar drugs should never be used to prevent poisoning while exposure continues or without strict exposure control, as severe kidney damage can result.

Note to physician: For severe poisoning BAL [British Anti-Lewisite, dimercaprol, dithiopropanol ($C_3H_8OS_2$)] has been used to treat toxic symptoms of certain heavy metals poisoning. In the case of lead poisoning it may have SOME value. Although BAL is reported to have a large margin of safety, caution must be exercised, because toxic effects may be caused by excessive dosage. Most can be prevented by premedication with 1-ephedrine sulfate (CAS: 134-72-5).

Personal Protective Methods: Wear protective gloves and clothing to prevent any reasonable probability of skin contact. Use any barrier that will prevent contamination from the dust. Safety equipment suppliers/manufacturers can provide recommendations on the most protective glove/clothing material for your operation. All protective clothing (suits, gloves, footwear, headgear) should be clean, available each day, and put on before work. Work clothing should be HEPA vacuumed before removal. Contact lenses should not be worn when working with this chemical. Wear dust-proof chemical goggles and face shield unless full face-piece respiratory protection is worn. Employees should wash immediately with soap when skin is wet or contaminated. Provide emergency showers and eyewash. See also NIOSH Criteria Document #78-158, *LEAD, inorganic dusts and fumes.*

Respirator Selection: OSHA: *0.5 mg/m³:* Any air-purifying respirator with an N100, R100, or P100 filter (including N100, R100, and P100 filtering face-pieces) except quarter-mask respirators. *1.25 mg/m³:* Sa:Cf (APF = 25) (any supplied-air respirator operated in a continuous-flow mode) or PaprHie (APF = 25) (any powered, air-purifying respirator with a high-efficiency particulate filter). *2.5 mg/m³:* 100F (APF = 50) (any air purifying, full-face-piece respirator with an N100, R100, or P100 filter) or SaT: Cf (APF = 50) (any supplied-air respirator that has a tight-fitting face-piece and is operated in a continuous-flow mode) or PaprTHie (APF = 50) (any powered, air-purifying respirator with a tight-fitting face-piece and a high-efficiency particulate filter) or SCBAF (APF = 50) (any self-contained breathing apparatus with a full face-piece) or SaF (APF = 50) (any supplied-air respirator with a full face-piece). *50 mg/m³:* Sa: Pd,Pp (APF = 1000) (any supplied-air respirator operated in a pressure-demand or other positive-pressure mode). *100 mg/m³:* SaF: Pd,Pp (APF = 2000) (any supplied-air respirator that has a full face-piece and is operated in a pressure-demand or other positive-pressure mode). *Emergency or planned entry into unknown concentrations or IDLH conditions:* SCBAF: Pd,Pp (APF = 10,000) (any self-contained breathing apparatus that has a full face-piece and is operated in a pressure-demand or other positive-pressure mode) or SaF: Pd,Pp: ASCBA (any supplied-air respirator that has a full face-piece and is operated in a pressure-demand or other positive-pressure mode in combination with an auxiliary self-contained breathing apparatus operated in a pressure-demand or other positive-pressure mode). *Escape:* 100F (APF = 50) (any air purifying, full-face-piece respirator with an N100, R100, or P100 filter) or SCBAE (any appropriate escape-type, self-contained breathing apparatus).

Storage: Color Code—Blue: Health Hazard/Poison: Store in a secure poison location. Prior to working with this chemical you should be trained on its proper handling and storage. Store in tightly closed containers in a cool, well-ventilated area away from heat, oxidizers, strong acids. Lead is regulated by an OSHA Standard 1910.1025. All requirements of the standard must be followed. A regulated, marked area should be established where this chemical is handled, used, or stored in compliance with OSHA Standard 1910.1045.

Shipping: Toxic solid, inorganic, n.o.s. requires a label of "POISONOUS/TOXIC MATERIALS." Lead fluoride falls in Hazard Class 6.1.

Spill Handling: Evacuate persons not wearing protective equipment from area of spill or leak until cleanup is complete. Remove all ignition sources. Collect powdered material in the most convenient and safe manner and deposit in

sealed containers. Ventilate area after cleanup is complete. It may be necessary to contain and dispose of this chemical as a hazardous waste. If material or contaminated runoff enters waterways, notify downstream users of potentially contaminated waters. Contact your local or federal environmental protection agency for specific recommendations. If employees are required to clean up spills, they must be properly trained and equipped. OSHA 1910.120(q) may be applicable.

Fire Extinguishing: Use dry chemical, carbon dioxide, water spray, or extinguishers. Poisonous gases, including lead and phosphorus oxides, are produced in fire. If material or contaminated runoff enters waterways, notify downstream users of potentially contaminated waters. Notify local health and fire officials and pollution control agencies. From a secure, explosion-proof location, use water spray to cool exposed containers. If cooling streams are ineffective (venting sound increases in volume and pitch, tank discolors, or shows any signs of deforming), withdraw immediately to a secure position. If employees are expected to fight fires, they must be trained and equipped in OSHA 1910.156. The only respirators recommended for firefighting are self-contained breathing apparatuses that have full face-pieces and are operated in a pressure-demand or other positive-pressure mode.

Disposal Method Suggested: Consult with environmental regulatory agencies for guidance on acceptable disposal practices. Generators of waste containing this contaminant ($\geq$100 kg/mo) must conform with EPA regulations governing storage, transportation, treatment, and waste disposal.

Reference
New Jersey Department of Health and Senior Services. (October 2004). *Hazardous Substances Fact Sheet: Lead Phosphate*. Trenton, NJ

Lead stearate   L:0190

Molecular Formula: $C_{18}H_{36}O_2 \cdot xPb$
Synonyms: Bleistearat (German); Estearato de plomo (Spanish); Neutral lead stearate; Octadecanoic acid, Lead salt; Octadecanoic acid, lead(2 +) salt; Octadecanoic acid, lead(II) salt; Stearic acid, lead salt; Stearic acid, lead(2 +) Salt; Stearic acid, lead(II) salt; Steric acid, lead salt
CAS Registry Number: 7428-48-0 (stearic acid, lead salt); 1072-35-1 (lead distearate); 56189-09-4 [dioxobis(stearato) dilead]
RTECS®Number: WI4300000
UN/NA & ERG Number: UN2811 (toxic solid, organic, n.o.s.)/154
EC Number: 231-068-1 (stearic acid, lead salt); 214-005-2 (lead distearate); 260-043-8 [dioxobis(stearato)dilead]
Regulatory Authority and Advisory Bodies
Carcinogenicity (*as organic lead compounds*): IARC: Organic lead compounds are not classifiable as to their carcinogenicity to humans, (Group 3, 2004).

OSHA, 29CFR1910 Specifically Regulated Chemicals (See CFR 1910.1025).
Clean Water Act: Section 311 Hazardous Substances/RQ 40CFR117.3 (same as CERCLA, see below); Section 313 Water Priority Chemicals (57FR41331, 9/9/92); 40CFR401.15 Section 307 Toxic Pollutants as lead and compounds.
RCRA, 40CFR261, Appendix 8 Hazardous Constituents, waste number not listed, as lead compounds, n.o.s.
Reportable Quantity (RQ): 10 lb (4.54 kg).
EPCRA Section 313 Form R *de minimis* concentration reporting level: 1.0%.
US DOT Regulated Marine Pollutant (49CFR172.101, Appendix B).
California Proposition 65 Chemical: Cancer 10/1/92 (lead and compounds).
Canada, WHMIS, organic lead compounds are *not* included on the Ingredients Disclosure List.
WGK (German Aquatic Hazard Class): No value assigned.
Description: Lead stearate is an organic lead compound. It is a white powder with a slight fatty odor. Molecular weight = 1734.87; Freezing/Melting point = 116°C; Flash point = 232°C. Hazard Identification (based on NFPA-704 M Rating System): Health 2, Flammability 1, Reactivity 0. Insoluble in water.
Potential Exposure: It is used in extreme-pressure lubricants and as a drier in varnishes.
Incompatibilities: Oxidizers, strong acids. Dust may explode at high temperature or with source of ignition.
Permissible Exposure Limits in Air
As organic lead
ACGIH TLV®[1]: No listing for organic lead compounds.
Determination in Air: Use (*for inorganic lead*) NIOSH Analytical Method (IV) s #7082, 7105, 7300, 7301, 7303, 7700, 7701, 7702, 9100, 9102, 9105; OSHA Analytical Methods ID-121, ID-125G, ID-206.
Permissible Concentration in Water: 0.1 mg/L Pb in drinking water may cause chronic poisoning. The EPA limits lead in drinking water to 15 μg per liter. Various organizations worldwide have set other standards for lead in drinking water as follows[35] (all in mg/L): Argentina 0.01; the Czech Republic 0.05; Germany 0.04; EEC 0.05; Japan 0.10; Mexico 0.05; former USSR-UNEP/IRPTC joint project 0.03; WHO 0.10. The states of Maine and Minnesota have set guidelines for lead in drinking water[61] at the level of 20 μg/L.
Determination in Water: Digestion followed by atomic absorption, or by colorimetric (dithizone) analysis, or by inductively coupled plasma (ICP) optical emission spectrometry. That gives total lead; dissolved lead may be determined by 0.45-μm filtration prior to such analysis.
Routes of Entry: Ingestion, skin contact. Lead can be absorbed through skin at chronically toxic levels.
Harmful Effects and Symptoms
Short Term Exposure: Ingestion of large amounts of lead may lead to seizures, coma, and death. The effects of

exposure to fumes and dusts of inorganic lead may not develop quickly. Symptoms may include decreased physical fitness, fatigue, sleep disturbance, headache, aching bones and muscles, constipation, abdominal pains, and decreased appetite. These effects are reported to be reversible if exposure ceases. Inhalation of large amounts of lead may lead to seizures, coma, and death. Between 1 oz and 1 lb of lead stearate may be fatal.

Note: Lead is a cumulative poison. Increasing amounts can build up in the body, eventually reaching a point where symptoms and disability occur. Lead dust carried home on contaminated clothing can result in exposure and symptoms in other family members. Standards only protect for inhalation exposure. Extra precautions should be taken if skin exposure also occurs.

Long Term Exposure: While lead stearate has not been identified as a teratogen, or a reproductive hazard, lead and certain lead compounds have been determined to be teratogens and may also cause reproductive damage such as reduced fertility and interfere with menstrual cycles. Handle with extreme caution. Lead can accumulate in the body over a period of time. Therefore, long-term exposures to lower levels can result in a buildup of lead in the body and more severe symptoms. These may include anemia, pale skin, a blue line at the gum margin, decreased handgrip strength, abdominal pain, severe constipation, nausea, vomiting, and paralysis of the wrist joint. Prolonged exposure may also result in kidney and brain damage. If the nervous system is affected, usually due to very high exposures, the resulting effects include severe headache, convulsions, coma, delirium, and death. In nonfatal cases, recovery is slow and not always complete. Alcohol ingestion and physical exertion may bring on symptoms. Lead exposure increases the risk of high blood pressure.

Points of Attack: Blood, kidneys, brain, nervous system.

Medical Surveillance: for inorganic lead. Before first exposure and every 6 months thereafter, OSHA (1910.1025) requires your employer to provide blood lead test, ZPP test (a special test for the effect of lead on blood cells). Examination of the nervous system. Prior to initial exposure, and annually for exposed person having blood lead readings exceeding 40 μg/100 g of whole blood, OSHA also requires a complete medical history, complete blood count and kidney function tests in addition to the tests listed above. OSHA defines "exposure" for these tests as air levels which average 30 μg of lead or more in a cubic meter of air. OSHA (under 1910.1020) requires your employer to send the doctor a copy of the lead standard and provide one for you.

Note: Blood-lead level is a good indicator of total lead exposure. Current OSHA regulations require that if an individual has a blood-lead level greater than or equal to 0.050 mg lead per 100 mL blood, he or she must be removed from all exposures to lead and cannot return to the exposure environment until the blood level falls to 0.040 mg lead per 100 mL blood or less. The following tolerance levels for occupational exposures may also be

useful: ACGIH BEI = 50 mg/L (blood); 150 mg/g creatinine (urine). DFG BAT = 70 mg/L (blood); 30 mg/L (blood) for women <45 years old.

First Aid: If this chemical gets into the eyes, remove any contact lenses at once and irrigate immediately for at least 15 min, occasionally lifting upper and lower lids. Seek medical attention immediately. If this chemical contacts the skin, remove contaminated clothing and wash immediately with soap and water. Seek medical attention immediately. If this chemical has been inhaled, remove from exposure, begin rescue breathing (using universal precautions, including resuscitation mask) if breathing has stopped and CPR if heart action has stopped. Transfer promptly to a medical facility. When this chemical has been swallowed, get medical attention. Give large quantities of water and induce vomiting. Do not make an unconscious person vomit.

Note to physician: Administer saline cathartic and an enema. For relief of colic, administer antispasmodic (calcium gluconate, atropine, papaverine). Consider morphine sulfate for severe pain.

Whole blood lead levels, circulating plasma/erythrocyte lead concentration ratio, urine ALA, and erythrocyte protoporphyrin fluorescent microscopy may all be useful in monitoring or assessing lead exposure. Chelating agents, such as edetate disodium calcium (Ca EDTA) and penicillamine (*not penicillin*), are generally useful in the therapy of acute lead intoxication.

Antidotes and special procedures for lead: Persons with significant lead poisoning are sometimes treated with Ca EDTA while hospitalized. This "chelating" drug causes a rush of lead from the body organs into the blood and kidneys, and thus has its own hazards, and must be administered only by highly experienced medical personnel under controlled conditions and careful observation. Ca EDTA or similar drugs should never be used to prevent poisoning while exposure continues or without strict exposure control, as severe kidney damage can result.

Personal Protective Methods: Wear protective gloves and clothing to prevent any reasonable probability of skin contact. Use any barrier that will prevent contamination from the dust. Safety equipment suppliers/manufacturers can provide recommendations on the most protective glove/clothing material for your operation. All protective clothing (suits, gloves, footwear, headgear) should be clean, available each day, and put on before work. Work clothing should be HEPA vacuumed before removal. Contact lenses should not be worn when working with this chemical. Wear dust-proof chemical goggles and face shield unless full face-piece respiratory protection is worn. Employees should wash immediately with soap when skin is wet or contaminated. Provide emergency showers and eyewash.

Respirator Selection: for inorganic lead OSHA: *0.5 mg/m³:* Any air-purifying respirator with an N100, R100, or P100 filter (including N100, R100, and P100 filtering face-pieces) except quarter-mask respirators. *1.25 mg/m³:* Sa:Cf (APF = 25) (any supplied-air respirator operated in a continuous-flow mode) or PaprHie (APF = 25) (any powered, air-purifying respirator with a high-efficiency particulate filter). *2.5 mg/m³:* 100F (APF = 50) (any air purifying, full-face-piece respirator with an N100, R100, or P100 filter) or SaT: Cf (APF = 50) (any supplied-air respirator that has a tight-fitting face-piece and is operated in a continuous-flow mode) or PaprTHie (APF = 50) (any powered, air-purifying respirator with a tight-fitting face-piece and a high-efficiency particulate filter) or SCBAF (APF = 50) (any self-contained breathing apparatus with a full face-piece) or SaF (APF = 50) (any supplied-air respirator with a full face-piece). *50 mg/m³:* Sa: Pd,Pp (APF = 1000) (any supplied-air respirator operated in a pressure-demand or other positive-pressure mode). *100 mg/m³:* SaF: Pd,Pp (APF = 2000) (any supplied-air respirator that has a full face-piece and is operated in a pressure-demand or other positive-pressure mode). *Emergency or planned entry into unknown concentrations or IDLH conditions:* SCBAF: Pd, Pp (APF = 10,000) (any self-contained breathing apparatus that has a full face-piece and is operated in a pressure-demand or other positive-pressure mode) or SaF: Pd,Pp: ASCBA (any supplied-air respirator that has a full face-piece and is operated in a pressure-demand or other positive-pressure mode in combination with an auxiliary self-contained breathing apparatus operated in a pressure-demand or other positive-pressure mode). *Escape:* 100F (APF = 50) (any air purifying, full-face-piece respirator with an N100, R100, or P100 filter) or SCBAE (any appropriate escape-type, self-contained breathing apparatus).

Storage: Color Code—Blue: Health Hazard/Poison: Store in a secure poison location. Prior to working with this chemical you should be trained on its proper handling and storage. Store in tightly closed containers in a cool, well-ventilated area away from heat, oxidizers, strong acids. Dust may explode at high temperature. Lead is regulated by an OSHA Standard 1910.1025. All requirements of the standard must be followed.

Shipping: Toxic solids, organic, n.o.s. requires the label of "POISONOUS/TOXIC MATERIALS." They fall in Hazard Class 6.1 and Packing Group III.

Spill Handling: Evacuate persons not wearing protective equipment from area of spill or leak until cleanup is complete. Remove all ignition sources. Collect powdered material in the most convenient and safe manner and deposit in sealed containers. Use vacuum or a wet method to reduce dust. Do not dry sweep. When vacuuming, a HEPA filter should be used, not a standard shop vac. Ventilate area after cleanup is complete. It may be necessary to contain and dispose of this chemical as a hazardous waste. If material or contaminated runoff enters waterways, notify downstream users of potentially contaminated waters. Contact your local or federal environmental protection agency for specific recommendations. If employees are required to clean up spills, they must be properly trained and equipped. OSHA 1910.120(q) may be applicable.

Fire Extinguishing: This chemical is a noncombustible solid. Use dry chemical, carbon dioxide, water spray, or alcohol foam extinguishers. Poisonous fumes including lead are produced in fire. If material or contaminated runoff enters waterways, notify downstream users of potentially contaminated waters. Notify local health and fire officials and pollution control agencies. Containers may explode in fire. From a secure, explosion-proof location, use water spray to cool exposed containers. If cooling streams are ineffective (venting sound increases in volume and pitch, tank discolors, or shows any signs of deforming), withdraw immediately to a secure position. If employees are expected to fight fires, they must be trained and equipped in OSHA 1910.156. The only respirators recommended for firefighting are self-contained breathing apparatuses that have full face-pieces and are operated in a pressure-demand or other positive-pressure mode.

Reference
New Jersey Department of Health and Senior Services. (August 1999). *Hazardous Substances Fact Sheet: Lead Stearate.* Trenton, NJ

Lead subacetate   L:0200

Molecular Formula: $C_4H_{10}O_8Pb_3$

Synonyms: Basic lead acetate; Bis(acetato)tetrahydroxytrilead; Bis(aceto)dihydroxytrilead; BLA; Lead acetate, basic; Lead, bis(acetato-*O*)tetrahydroxytri-; Lead monosubacetate; Monobasic lead acetate; Subacetate lead; Subaceto de plomo (Spanish)

CAS Registry Number: 1335-32-6

RTECS® Number: OF8750000

UN/NA & ERG Number: UN1616/151

EC Number: 215-630-3 [*Annex I Index No.:* 082-007-00-9]

Regulatory Authority and Advisory Bodies

Carcinogenicity: Carcinogenicity (*as organic lead compounds*): IARC: Organic lead compounds are not classifiable as to their carcinogenicity to humans, (Group 3, 2004).

Clean Water Act: 40CFR401.15 Section 307 Toxic Pollutants as lead and compounds.

US EPA Hazardous Waste Number (RCRA No.): U146.

RCRA, 40CFR261, Appendix 8 Hazardous Constituents.

Reportable Quantity (RQ): 10 lb (4.54 kg).

EPCRA Section 313 Form R *de minimis* concentration reporting level: 1.0%.

US DOT Regulated Marine Pollutant (49CFR172.101, Appendix B).

California Proposition 65 Chemical: Cancer 10/1/89.

European/International Regulations: Hazard Symbol: T, N; Risk phrases: R61; R33; R40; R48/22; R50/53; Safety phrases: S53; S45; S60; S61 (see Appendix 4).

WGK (German Aquatic Hazard Class): No value assigned.

Description: Lead subacetate is a white, heavy powder or gelatinous solid. Molecular weight = 807.71. Hazard

Identification (based on NFPA-704 M Rating System): Health 3, Flammability 0, Reactivity 0. Soluble in water.

Permissible Exposure Limits in Air

Basic; subacetate

TEEL-0: 35 mg/m^3

PAC-1: 100 mg/m^3

PAC-2: 500 mg/m^3

PAC-3: 500 mg/m^3

As organic lead

ACGIH TLV®[1]: No listing for organic lead compounds.

Potential Exposure: Used as a decolorizing agent in sugar solutions and as an analytical chemical.

Incompatibilities: Oxidizers, strong acids.

Determination in Air: Use (*for inorganic lead*) NIOSH Analytical Method (IV) s #7082, 7105, 7300, 7301, 7303, 7700, 7701, 7702, 9100, 9102, 9105; OSHA Analytical Methods ID-121, ID-125G, ID-206.

Permissible Concentration in Water: 0.1 mg/L Pb in drinking water may cause chronic poisoning. The EPA limits lead in drinking water to 15 µg/L. Various organizations worldwide have set other standards for lead in drinking water as follows[35] (all in mg/L): Argentina 0.01; the Czech Republic 0.05; Germany 0.04; EEC 0.05; Japan 0.10; Mexico 0.05; former USSR-UNEP/IRPTC joint project 0.03; WHO 0.10. The states of Maine and Minnesota have set guidelines for lead in drinking water[61] at the level of 20 µg/L.

Determination in Water: Digestion followed by atomic absorption, or by colorimetric (dithizone) analysis, or by inductively coupled plasma (ICP) optical emission spectrometry. That gives total lead; dissolved lead may be determined by 0.45-µm filtration prior to such analysis.

Routes of Entry: Inhalation, skin. Lead can be absorbed through skin at chronically toxic levels.

Harmful Effects and Symptoms

Short Term Exposure: Lead subacetate can cause headache, irritability, mood changes, reduced memory, and disturbed sleep.

Note: Lead is a cumulative poison. Increasing amounts can build up in the body, eventually reaching a point where symptoms and disability occur. Lead dust carried home on contaminated clothing can result in exposure and symptoms in other family members. Standards only protect for inhalation exposure. Extra precautions should be taken if skin exposure also occurs.

Long Term Exposure: Lead subacetate may be a carcinogen in humans since it has been shown to cause kidney, lung, and brain cancer in animals. May be a teratogen in humans. May damage the testes. Repeated exposure may cause lead poisoning with symptoms of headache, irritability, disturbed sleep, tiredness, reduced memory, and personality changes. Higher levels can cause muscle or joint pains, weakness, disturbed sleep, and easy fatigue. Exposure can increase the risk of high blood pressure. May cause kidney and brain damage and anemia. Lead can accumulate in the body over a period of time. Therefore, long-term exposures to lower levels can result in a buildup of

lead in the body and more severe symptoms. These may include anemia, pale skin, a blue line at the gum margin, decreased handgrip strength, abdominal pain, severe constipation, nausea, vomiting, and paralysis of the wrist joint. Prolonged exposure may also result in kidney and brain damage. If the nervous system is affected, usually due to very high exposures, the resulting effects include severe headache, convulsions, coma, delirium, and death. In nonfatal cases, recovery is slow and not always complete. Alcohol ingestion and physical exertion may bring on symptoms. Lead exposure increases the risk of high blood pressure. Continuous exposure can result in decreased fertility. Elevated lead exposure of either parent before pregnancy can increase the chances of miscarriage or birth defects. Exposure of the mother during pregnancy can cause birth defects.

Points of Attack: Blood, kidneys, brain, nervous system.

Medical Surveillance: (*for inorganic lead*) Before first exposure and every 6 months thereafter, OSHA (1910.1025) requires your employer to provide blood lead test, ZPP test (a special test for the effect of lead on blood cells). Examination of the nervous system. Prior to initial exposure, and annually for exposed person having blood lead readings exceeding 40 µg/100 g of whole blood, OSHA also requires a complete medical history, complete blood count and kidney function tests in addition to the tests listed above. OSHA defines "exposure" for these tests as air levels which average 30 µg of lead or more in a cubic meter of air. OSHA (1910.1020) requires your employer to send the doctor a copy of the lead standard and provide one for you.

Note: Blood-lead level is a good indicator of total lead exposure. Current OSHA regulations require that if an individual has a blood-lead level greater than or equal to 0.050 mg lead per 100 mL blood, he or she must be removed from all exposures to lead and cannot return to the exposure environment until the blood level falls to 0.040 mg lead per 100 mL blood or less. The following tolerance levels for occupational exposures may also be useful: ACGIH BEI = 50 mg/L (blood); 150 mg/g creatinine (urine). DFG BAT = 70 mg/L (blood); 30 mg/L (blood) for women <45 years old.

First Aid: If this chemical gets into the eyes, remove any contact lenses at once and irrigate immediately for at least 15 min, occasionally lifting upper and lower lids. Seek medical attention immediately. If this chemical contacts the skin, remove contaminated clothing and wash immediately with soap and water. Seek medical attention immediately. If this chemical has been inhaled, remove from exposure, begin rescue breathing (using universal precautions, including resuscitation mask) if breathing has stopped and CPR if heart action has stopped. Transfer promptly to a medical facility. When this chemical has been swallowed, get medical attention. Give large quantities of water and induce vomiting. Do not make an unconscious person vomit.

Note to physician: Administer saline cathartic and an enema. For relief of colic, administer antispasmodic (calcium gluconate, atropine, papaverine). Consider morphine sulfate for severe pain.

Whole blood lead levels, circulating plasma/erythrocyte lead concentration ratio, urine ALA, and erythrocyte protoporphyrin fluorescent microscopy may all be useful in monitoring or assessing lead exposure. Chelating agents, such as edetate disodium calcium (Ca EDTA) and penicillamine (*not penicillin*), are generally useful in the therapy of acute lead intoxication.

Antidotes and special procedures for lead: Persons with significant lead poisoning are sometimes treated with Ca EDTA while hospitalized. This "chelating" drug causes a rush of lead from the body organs into the blood and kidneys, and thus has its own hazards, and must be administered only by highly experienced medical personnel under controlled conditions and careful observation. Ca EDTA or similar drugs should never be used to prevent poisoning while exposure continues or without strict exposure control, as severe kidney damage can result.

Personal Protective Methods: Wear protective gloves and clothing to prevent any reasonable probability of skin contact. Use any barrier that will prevent contamination from the dust. Safety equipment suppliers/manufacturers can provide recommendations on the most protective glove/clothing material for your operation. All protective clothing (suits, gloves, footwear, headgear) should be clean, available each day, and put on before work. Contact lenses should not be worn when working with this chemical. Wear dust-proof chemical goggles and face shield unless full face-piece respiratory protection is worn. Employees should wash immediately with soap when skin is wet or contaminated. Provide emergency showers and eyewash.

Respirator Selection: for inorganic lead OSHA: *0.5 mg/m³:* Any air-purifying respirator with an N100, R100, or P100 filter (including N100, R100, and P100 filtering face-pieces) except quarter-mask respirators. *1.25 mg/m³:* Sa:Cf (APF = 25) (any supplied-air respirator operated in a continuous-flow mode) or PaprHie (APF = 25) (any powered, air-purifying respirator with a high-efficiency particulate filter). *2.5 mg/m³:* 100F (APF = 50) (any air purifying, full-face-piece respirator with an N100, R100, or P100 filter) or SaT: Cf (APF = 50) (any supplied-air respirator that has a tight-fitting face-piece and is operated in a continuous-flow mode) or PaprTHie (APF = 50) (any powered, air-purifying respirator with a tight-fitting face-piece and a high-efficiency particulate filter) or SCBAF (APF = 50) (any self-contained breathing apparatus with a full face-piece) or SaF (APF = 50) (any supplied-air respirator with a full face-piece). *50 mg/m³:* Sa: Pd,Pp (APF = 1000) (any supplied-air respirator operated in a pressure-demand or other positive-pressure mode). *100 mg/m³:* SaF: Pd,Pp (APF = 2000) (any supplied-air respirator that has a full face-piece and is operated in a pressure-demand or other positive-pressure mode). *Emergency or planned entry into unknown concentrations or IDLH conditions:* SCBAF: Pd, Pp (APF = 10,000) (any self-contained breathing apparatus that has a full face-piece and is operated in a pressure-demand or other positive-pressure mode) or SaF: Pd,Pp: ASCBA (any supplied-air respirator that has a full face-piece and is operated in a pressure-demand or other positive-pressure mode in combination with an auxiliary self-contained breathing apparatus operated in a pressure-demand or other positive-pressure mode). *Escape:* 100F (APF = 50) (any air purifying, full-face-piece respirator with an N100, R100, or P100 filter) or SCBAE (any appropriate escape-type, self-contained breathing apparatus).

Storage: Color Code—Blue: Health Hazard/Poison: Store in a secure poison location. Prior to working with this chemical you should be trained on its proper handling and storage. Store in tightly closed containers in a cool, well-ventilated area away from heat, oxidizers, strong acids. Lead is regulated by an OSHA Standard 1910.1025. All requirements of the standard must be followed. A regulated, marked area should be established where this chemical is handled, used, or stored in compliance with OSHA Standard 1910.1045.

Shipping: Lead acetate requires a shipping label of "POISONOUS/TOXIC MATERIALS." Lead acetate falls in Hazard Class 6.1 and Packing Group III.

Spill Handling: Evacuate persons not wearing protective equipment from area of spill or leak until cleanup is complete. Remove all ignition sources. Collect powdered material in the most convenient and safe manner and deposit in sealed containers. Use vacuum or a wet method to reduce dust. Do not dry sweep. When vacuuming, a HEPA filter should be used, not a standard shop vac. Ventilate area after cleanup is complete. It may be necessary to contain and dispose of this chemical as a hazardous waste. If material or contaminated runoff enters waterways, notify downstream users of potentially contaminated waters. Contact your local or federal environmental protection agency for specific recommendations. If employees are required to clean up spills, they must be properly trained and equipped. OSHA 1910.120(q) may be applicable.

Fire Extinguishing: This chemical is a noncombustible solid. Use dry chemical, carbon dioxide, water spray, alcohol foam or polymer foam extinguishers. Poisonous gases, including lead oxide and carbon monoxide, are produced in fire. If material or contaminated runoff enters waterways, notify downstream users of potentially contaminated waters. Notify local health and fire officials and pollution control agencies. Containers may explode in fire. From a secure, explosion-proof location, use water spray to cool exposed containers. If cooling streams are ineffective (venting sound increases in volume and pitch, tank discolors, or shows any signs of deforming), withdraw immediately to a secure position. If employees are expected to fight fires, they must be trained and equipped in OSHA 1910.156. The only respirators recommended for firefighting are self-contained breathing apparatuses that have full face-pieces and are operated in a pressure-demand or other positive-pressure mode.

Disposal Method Suggested: Consult with environmental regulatory agencies for guidance on acceptable disposal practices. Generators of waste containing this contaminant (≥100 kg/mo) must conform with EPA regulations governing storage, transportation, treatment, and waste disposal.

Reference

New Jersey Department of Health and Senior Services. (July 1999). *Hazardous Substances Fact Sheet: Lead Subacetate*. Trenton, NJ

Lead sulfate   L:0210

Molecular Formula: O_4PbS

Common Formula: $PbSO_4$

Synonyms: Anglisite; Bleisulfat (German); C.I. 77630; C.I. Pigment white 3; Fast white; Freemans white lead; Lanarkite; Lead bottoms; Lead(2 +) sulfate(1:1); Lead(II) sulfate(1:1); Lead sulphate; Lead(2 +) sulphate(1:1); Lead (II) sulphate (1:1); Milk white; Mulhouse white; Sulfate de plomb (French); Sulfato de plomo (Spanish); Sulfuric acid, lead(2 +) salt(1:1); Sulfuric acid, lead(II) salt(1:1); White lead C.I. Pigment white

CAS Registry Number: 7446-14-2

RTECS® Number: OG4375000

UN/NA & ERG Number: UN1794 (with >3% free acid)/154

EC Number: 231-198-9

Regulatory Authority and Advisory Bodies

IARC: Animal Inadequate Evidence, Human Inadequate Evidence, *possibly carcinogenic to humans*, Group 2A, 1987; Carcinogenicity: NTP: 11th Report on Carcinogens, 2004: Reasonably anticipated to be a human carcinogen; EPA: Sufficient evidence from animal studies; inadequate evidence or no useful data from epidemiologic studies.

Air Pollutant Standard Set. See below, "Permissible Exposure Limits in Air" section.

OSHA, 29CFR1910 Specifically Regulated Chemicals (See 29 CFR 1910.1025).

Hazardous Substance (EPA) (RQ = 5000/2270).[4]

Clean Water Act: Section 311 Hazardous Substances/RQ 40CFR117.3 (same as CERCLA, see below); Section 313 Water Priority Chemicals (57FR41331, 9/9/92); 40CFR401.15 Section 307 Toxic Pollutants as lead and compounds.

RCRA, 40CFR261, Appendix 8 Hazardous Constituents, waste number not listed, as lead compounds, n.o.s.

Reportable Quantity (RQ): 10 lb (4.54 kg).

EPCRA Section 313 Form R *de minimis* concentration reporting level: 0.1%.

US DOT Regulated Marine Pollutant (49CFR172.101, Appendix B).

California Proposition 65 Chemical: Cancer 10/1/92 (lead and compounds).

Canada, WHMIS, Ingredients Disclosure List Concentration 1.0%.

European/International Regulations: not listed in Annex 1. WGK (German Aquatic Hazard Class): No value assigned.

Description: Lead sulfate is a heavy, white crystalline powder. Molecular weight = 303.25; Freezing/Melting point = 1170°C. Hazard Identification (based on NFPA-704 M Rating System): Health 3, Flammability 0, Reactivity 1. Soluble in water.

Potential Exposure: Lead sulfate is used in storage batteries and paint pigments. Used in the making of alloys, fast-drying oil varnishes, weighting fabrics, in lithography.

Incompatibilities: Contact with oxidizers and chemically active metals may cause violent reactions.

Permissible Exposure Limits in Air

Protective Action Criteria (PAC)

7446-14-2

TEEL-0: 0.0732 mg/m^3

PAC-1: 4 mg/m^3

PAC-2: 30 mg/m^3

PAC-3: 146 mg/m^3

As inorganic lead

OSHA PEL: 0.050 mg[Pb]/m^3 TWA. [*Note*: The PEL also applies to other lead compounds (as Pb).] Other OSHA requirements can be found in 29 CFR 1910.1025. The OSHA PEL (8-h TWA) for lead in "nonferrous foundries with less than 20 employees" is 0.075 mg/m^3. OSHA considers "lead" to mean metallic lead; all inorganic lead compounds (lead oxides and lead salts); and a class of organic compounds called "soaps;" all other lead compounds are excluded from this definition. See 29CFR1910.1025.

NIOSH REL: 0.050 mg[Pb]/m^3 TWA. Suspected carcinogen. Limit exposure to lowest feasible level. [*Note*: The REL also applies to other lead compounds (as Pb).] *Note*: NIOSH considers "lead" to mean metallic lead, lead oxides, and lead salts (including organic salts, such as lead "soaps" but excluding lead arsenate). Air concentrations should be maintained so that worker blood lead remains less than 0.060 mg[Pb]/100 g of whole blood. See *NIOSH Pocket Guide*, Appendix C.

ACGIH TLV®[1]: 0.05 mg[Pb]/m^3 TWA; BEI: 30 μg[Pb]/100 mL (blood). (Note: The TLV also applies to lead, inorganic compounds.) *Note:* women of child-bearing potential whose blood exceeds 10 μg[Pb]/dL are at risk of delivering a child with a blood [Pb] over the current CDC guideline of 10 μg[Pb]/dL and may cause birth defects. Confirmed animal carcinogen with unknown relevance to humans.

NIOSH IDLH: 100 mg [Pb]/m^3.

DFG MAK: BAT: 400 μg[Pb]/L (blood) not fixed; 100 μg [Pb]/L (blood) women age <45; Carcinogen Category 2; Germ Cell Mutagen Group 3A.

Determination in Air: Use NIOSH Analytical Method (IV) s #7082, 7105, 7300, 7301, 7303, 7700, 7701, 7702, 9100, 9102, 9105; OSHA Analytical Methods ID-121, ID-125G, ID-206.

Permissible Concentration in Water: The EPA limits lead in drinking water to 15 μg/L. Various organizations worldwide have set other standards for lead in drinking water as follows[35] (all in mg/L): Argentina 0.01; the Czech Republic 0.05; Germany 0.04; EEC 0.05; Japan 0.10; Mexico 0.05; former USSR-UNEP/IRPTC joint project 0.03; WHO 0.10. The states of Maine and Minnesota have set guidelines for lead in drinking water[61] at the level of 20 μg/L.

Determination in Water: Digestion followed by atomic absorption, or by colorimetric (dithizone) analysis, or by inductively coupled plasma (ICP) optical emission spectrometry. That gives total lead; dissolved lead may be determined by 0.45-μm filtration prior to such analysis.

Routes of Entry: Inhalation.

Harmful Effects and Symptoms

Short Term Exposure: This chemical is corrosive. Skin contact can cause severe irritation and burns, itching, rash, and pigment changes. Eye contact can cause severe irritation and burns. Inhalation can cause irritation of the respiratory tract. Ingestion of large amounts of lead may lead to seizures, coma, and death. The effects of exposure to fumes and dusts of inorganic lead may not develop quickly. Symptoms may include decreased physical fitness, fatigue, sleep disturbance, headache, aching bones and muscles, constipation, abdominal pain, and decreased appetite. These effects are reported to be reversible if exposure ceases. Inhalation of large amounts of lead may lead to seizures, coma, and death.

Note: Lead is a cumulative poison. Increasing amounts can build up in the body, eventually reaching a point where symptoms and disability occur. Lead dust carried home on contaminated clothing can result in exposure and symptoms in other family members. Standards only protect for inhalation exposure. Extra precautions should be taken if skin exposure also occurs.

Long Term Exposure: Highly irritating and corrosive substances can cause lung irritation that may lead to bronchitis. Lead can accumulate in the body over a period of time. Therefore, long-term exposures to lower levels can result in a buildup of lead in the body and more severe symptoms. These may include anemia, pale skin, a blue line at the gum margin, decreased handgrip strength, abdominal pain, severe constipation, nausea, vomiting, and paralysis of the wrist joint. Prolonged exposure may also result in kidney and brain damage. If the nervous system is affected, usually due to very high exposures, the resulting effects include severe headache, convulsions, coma, delirium, and death. In nonfatal cases, recovery is slow and not always complete. Alcohol ingestion and physical exertion may bring on symptoms. Lead exposure increases the risk of high blood pressure. Continuous exposure can result in decreased fertility. Elevated lead exposure of either parent before pregnancy can increase the chances of miscarriage or birth defects. Exposure of the mother during pregnancy can cause birth defects.

Points of Attack: Lungs, blood, kidneys, brain.

Medical Surveillance: Before first exposure and every 6 months thereafter, OSHA (1910. 1025) requires your employer to provide blood lead test, ZPP test (a special test for the effect of lead on blood cells). Examination of the nervous system. Prior to initial exposure, and annually for exposed person having blood lead readings exceeding 40 μg/100 g of whole blood, OSHA also requires a complete medical history, complete blood count and kidney function tests in addition to the tests listed above. OSHA defines "exposure" for these tests as air levels which average 30 μg of lead or more in a cubic meter of air. OSHA (under 1910.1020) requires your employer to send the doctor a copy of the lead standard and provide one for you.

Note: Blood-lead level is a good indicator of total lead exposure. Current OSHA regulations require that if an individual has a blood-lead level greater than or equal to 0.050 mg lead per 100 mL blood, he or she must be removed from all exposures to lead and cannot return to the exposure environment until the blood level falls to 0.040 mg lead per 100 mL blood or less. The following tolerance levels for occupational exposures may also be useful: ACGIH BEI = 50 mg/L (blood); 150 mg/g creatinine (urine). DFG BAT = 70 mg/L (blood); 30 mg/L (blood) for women <45 years old.

First Aid: If this chemical gets into the eyes, remove any contact lenses at once and irrigate immediately for at least 15 min, occasionally lifting upper and lower lids. Seek medical attention immediately. If this chemical contacts the skin, remove contaminated clothing and wash immediately with soap and water. Seek medical attention immediately. If this chemical has been inhaled, remove from exposure, begin rescue breathing (using universal precautions, including resuscitation mask) if breathing has stopped and CPR if heart action has stopped. Transfer promptly to a medical facility. When this chemical has been swallowed, get medical attention. Do not induce vomiting.

Note to physician: whole blood lead levels, circulating plasma/erythrocyte lead concentration ratio, urine ALA, and erythrocyte protoporphyrin fluorescent microscopy may all be useful in monitoring or assessing lead exposure. Chelating agents, such as edetate disodium calcium (Ca EDTA) and penicillamine (*not penicillin*), are generally useful in the therapy of acute lead intoxication.

Antidotes and special procedures for lead: Persons with significant lead poisoning are sometimes treated with Ca EDTA while hospitalized. This "chelating" drug causes a rush of lead from the body organs into the blood and kidneys, and thus has its own hazards, and must be administered only by highly experienced medical personnel under controlled conditions and careful observation. Ca EDTA or similar drugs should never be used to prevent poisoning while exposure continues or without strict exposure control, as severe kidney damage can result.

Note to physician: For severe poisoning BAL [British Anti-Lewisite, dimercaprol, dithiopropanol ($C_3H_8OS_2$)] has been used to treat toxic symptoms of certain heavy metals poisoning. In the case of lead poisoning it may have SOME value. Although BAL is reported to have a large margin of safety, caution must be exercised, because toxic effects may be caused by excessive dosage. Most can be prevented by premedication with 1-ephedrine sulfate (CAS: 134-72-5).

Personal Protective Methods: Wear protective gloves and clothing to prevent any reasonable probability of skin contact. Use any barrier that will prevent contamination from the dust. Safety equipment suppliers/manufacturers can provide recommendations on the most protective glove/clothing material for your operation. All protective clothing (suits, gloves, footwear, headgear) should be clean, available each day, and put on before work. Work clothing should be HEPA vacuumed before removal. Contact lenses should not be worn when working with this chemical. Wear dust-proof chemical goggles and face shield unless full face-piece respiratory protection is worn. Employees should wash immediately with soap when skin is wet or contaminated. Provide emergency showers and eyewash. See also NIOSH Criteria Document #78-158, *LEAD, inorganic dusts and fumes.*

Respirator Selection: OSHA: *0.5 mg/m³:* Any air-purifying respirator with an N100, R100, or P100 filter (including N100, R100, and P100 filtering face-pieces) except quarter-mask respirators. *1.25 mg/m³:* Sa:Cf (APF = 25) (any supplied-air respirator operated in a continuous-flow mode) or PaprHie (APF = 25) (any powered, air-purifying respirator with a high-efficiency particulate filter). *2.5 mg/m³:* 100F (APF = 50) (any air purifying, full-face-piece respirator with an N100, R100, or P100 filter) or SaT: Cf (APF = 50) (any supplied-air respirator that has a tight-fitting face-piece and is operated in a continuous-flow mode) or PaprTHie (APF = 50) (any powered, air-purifying respirator with a tight-fitting face-piece and a high-efficiency particulate filter) or SCBAF (APF = 50) (any self-contained breathing apparatus with a full face-piece) or SaF (APF = 50) (any supplied-air respirator with a full face-piece). *50 mg/m³:* Sa: Pd,Pp (APF = 1000) (any supplied-air respirator operated in a pressure-demand or other positive-pressure mode). *100 mg/m³:* SaF: Pd,Pp (APF = 2000) (any supplied-air respirator that has a full face-piece and is operated in a pressure-demand or other positive-pressure mode). *Emergency or planned entry into unknown concentrations or IDLH conditions:* SCBAF: Pd,Pp (APF = 10,000) (any self-contained breathing apparatus that has a full face-piece and is operated in a pressure-demand or other positive-pressure mode) or SaF: Pd,Pp: ASCBA (any supplied-air respirator that has a full face-piece and is operated in a pressure-demand or other positive-pressure mode in combination with an auxiliary self-contained breathing apparatus operated in a pressure-demand or other positive-pressure mode). *Escape:* 100F (APF = 50) (any air purifying, full-face-piece respirator with an N100, R100, or P100 filter) or SCBAE (any appropriate escape-type, self-contained breathing apparatus).

Storage: Color Code—White: Corrosive or Contact Hazard; Store separately in a corrosion-resistant location. Prior to working with this chemical you should be trained on its proper handling and storage. Lead sulfate must be stored to avoid contact with oxidizers (such as perchlorates, peroxides, permanganates, chlorates, and nitrates) and chemically active metals (such as potassium, sodium, magnesium, and zinc), since violent reactions occur. Lead is regulated by an OSHA Standard 1910.1025. All requirements of the standard must be followed. A regulated, marked area should be established where this chemical is handled, used, or stored in compliance with OSHA Standard 1910.1045.

Shipping: Lead sulfate with more than 3% free acid requires a shipping label of "CORROSIVE." It falls in Hazard Class 8 and Packing Group II.

Spill Handling: Evacuate persons not wearing protective equipment from area of spill or leak until cleanup is complete. Remove all ignition sources. Collect powdered material in the most convenient and safe manner and deposit in sealed containers. Ventilate area after cleanup is complete. It may be necessary to contain and dispose of this chemical as a hazardous waste. If material or contaminated runoff enters waterways, notify downstream users of potentially contaminated waters. Contact your local or federal environmental protection agency for specific recommendations. If employees are required to clean up spills, they must be properly trained and equipped. OSHA 1910.120(q) may be applicable.

Fire Extinguishing: Lead sulfate is not combustible. Use dry chemical, carbon dioxide, water spray, or foam extinguishers. Poisonous fumes including lead and sulfur oxides are produced in fire. If material or contaminated runoff enters waterways, notify downstream users of potentially contaminated waters. Notify local health and fire officials and pollution control agencies. From a secure, explosion-proof location, use water spray to cool exposed containers. If cooling streams are ineffective (venting sound increases in volume and pitch, tank discolors, or shows any signs of deforming), withdraw immediately to a secure position. If employees are expected to fight fires, they must be trained and equipped in OSHA 1910.156. The only respirators recommended for firefighting are self-contained breathing apparatuses that have full face-pieces and are operated in a pressure-demand or other positive-pressure mode.

References

New York State Department of Health. (March 1986). *Chemical Fact Sheet: Lead Sulfate.* Version 2. Albany, NY: Bureau of Toxic Substance Assessment
New Jersey Department of Health and Senior Services. (April, 2004). *Hazardous Substances Fact Sheet: Lead Sulphate.* Trenton, NJ

Lead sulfide

L:0220

Molecular Formula: PbS
Synonyms: C.I. 77640; Galena; Lead monosulfide; Natural lead sulfide; Plumbous sulfide; Sulfuro de plomo (Spanish)

CAS Registry Number: 1314-87-0
RTECS® Number: OG4550000
UN/NA & ERG Number: UN3288/165
EC Number: 215-246-6
Regulatory Authority and Advisory Bodies
Carcinogenicity: IARC: Animal Inadequate Evidence, Human Inadequate Evidence, *possibly carcinogenic to humans,* Group 2A, 1987; Carcinogenicity: NTP: 11th Report on Carcinogens, 2004: Reasonably anticipated to be a human carcinogen; EPA: Sufficient evidence from animal studies; inadequate evidence or no useful data from epidemiologic studies.
OSHA, 29CFR1910 Specifically Regulated Chemicals (See 29 CFR 1910.1025).
Air Pollutant Standard Set. See below, "Permissible Exposure Limits in Air" section.
Hazardous Substance (EPA) (RQ = 5000/2270).[4]
Section 261 Hazardous Constituents, waste number not listed, as lead compounds, n.o.s.
EPCRA Clean Water Act: 40CFR401.15 Section 307 Toxic Pollutants as lead and compounds.
RCRA Section 313: Includes any unique chemical substance that contains lead as part of that chemical's infrastructure. Form R *de minimis* concentration reporting level: inorganic compounds 0.1%; organic compounds 1.0%.
US DOT Regulated Marine Pollutant (49CFR172.101, Appendix B) as lead compounds, soluble, n.o.s.
California Proposition 65 Chemical: Cancer 10/1/92 (lead and compounds).
Canada, WHMIS, Ingredients Disclosure List Concentration 1.0%.
European/International Regulations: not listed in Annex 1.
WGK (German Aquatic Hazard Class): No value assigned.
Description: Lead sulfide is a silvery to black crystalline powder. Molecular weight = 239.25; Boiling point = 1281°C (sublimes); Freezing/Melting point = 1114°C. Hazard Identification (based on NFPA-704 M Rating System): Health 1, Flammability 0, Reactivity 0. Practically insoluble in water (0.000086 g/100-cc water at 13°C).
Potential Exposure: Lead sulfide is used in ceramics, infrared radiation detectors, and semiconductors.
Incompatibilities: Contact with oxidizers and chemically active metals may cause violent reactions. Sulfides react with acids to produce toxic and flammable vapors of hydrogen sulfide.
Permissible Exposure Limits in Air
Protective Action Criteria (PAC)
1314-87-0
TEEL-0: 0.0577 mg/m^3
PAC-1: 0.173 mg/m^3
PAC-2: 115 mg/m^3
PAC-3: 115 mg/m^3
As inorganic lead
OSHA PEL: 0.050 mg[Pb]/m^3 TWA. [*Note:* The PEL also applies to other lead compounds (as Pb).] Other OSHA

requirements can be found in 29 CFR 1910.1025. The OSHA PEL (8-h TWA) for lead in "nonferrous foundries with less than 20 employees" is 0.075 mg/m^3. OSHA considers "lead" to mean metallic lead; all inorganic lead compounds (lead oxides and lead salts); and a class of organic compounds called "soaps;" all other lead compounds are excluded from this definition. See 29CFR1910.1025.
NIOSH REL: 0.050 mg[Pb]/m^3 TWA. Suspected carcinogen. Limit exposure to lowest feasible level. [*Note:* The REL also applies to other lead compounds (as Pb).] *Note:* NIOSH considers "lead" to mean metallic lead, lead oxides, and lead salts (including organic salts, such as lead "soaps" but excluding lead arsenate). Air concentrations should be maintained so that worker blood lead remains less than 0.060 mg[Pb]/100 g of whole blood. See *NIOSH Pocket Guide*, Appendix C.
ACGIH TLV®[1]: 0.05 mg[Pb]/m^3 TWA; BEI: 30 μg[Pb]/100 mL (blood). (Note: The TLV also applies to lead, inorganic compounds.) *Note:* women of child-bearing potential whose blood exceeds 10 μg[Pb]/dL are at risk of delivering a child with a blood [Pb] over the current CDC guideline of 10 μg[Pb]/dL and may cause birth defects. Confirmed animal carcinogen with unknown relevance to humans.
NIOSH IDLH: 100 mg [Pb]/m^3.
DFG MAK: BAT: 400 μg[Pb]/L (blood) not fixed; 100 μg [Pb]/L (blood) women age <45; Carcinogen Category 2; Germ Cell Mutagen Group 3A.
In addition, Russia[43] set a MAC of 0.0017 mg/m^3 in the ambient air of residential areas (on a daily average basis; about 6 times the same limits for lead compounds in general).
Determination in Air: Use NIOSH Analytical Method (IV)s #7082, 7105, 7300, 7301, 7303, 7700, 7701, 7702, 9100, 9102, 9105; OSHA Analytical Methods ID-121, ID-125G, ID-206.
Permissible Concentration in Water: The EPA limits lead in drinking water to 15 μg/L. Various organizations worldwide have set other standards for lead in drinking water as follows[35] (all in mg/L): Argentina 0.01; the Czech Republic 0.05; Germany 0.04; EEC 0.05; Japan 0.10; Mexico 0.05; former USSR-UNEP/IRPTC joint project 0.03; WHO 0.10. The states of Maine and Minnesota have set guidelines for lead in drinking water[61] at the level of 20 μg/L.
Determination in Water: Digestion followed by atomic absorption, or by colorimetric (dithizone) analysis, or by inductively coupled plasma (ICP) optical emission spectrometry. That gives total lead; dissolved lead may be determined by 0.45-μm filtration prior to such analysis.
Routes of Entry: Inhalation.
Harmful Effects and Symptoms
Short Term Exposure: Lead sulfide is very slightly soluble in water, and sulfides react with water, forming hydrogen sulfide. Eye contact may cause eye irritation and damage. Ingestion of large amounts of lead may lead to seizures, coma, and death. The effects of exposure to fumes and dusts

of inorganic lead may not develop quickly. Symptoms may include decreased physical fitness, fatigue, sleep disturbance; headache, aching bones and muscles, constipation, abdominal pains, and decreased appetite. These effects are reported to be reversible if exposure ceases. Inhalation of large amounts of lead may lead to seizures, coma, and death. *Note:* Lead is a cumulative poison. Increasing amounts can build up in the body, eventually reaching a point where symptoms and disability occur. Lead dust carried home on contaminated clothing can result in exposure and symptoms in other family members. Standards only protect for inhalation exposure. Extra precautions should be taken if skin exposure also occurs.

Long Term Exposure: Lead can accumulate in the body over a period of time. Therefore, long-term exposures to lower levels can result in a buildup of lead in the body and more severe symptoms. These may include anemia, pale skin, a blue line at the gum margin, decreased handgrip strength, abdominal pain, severe constipation, nausea, vomiting, and paralysis of the wrist joint. Prolonged exposure may also result in kidney and brain damage. If the nervous system is affected, usually due to very high exposures, the resulting effects include severe headache, convulsions, coma, delirium, and death. In nonfatal cases, recovery is slow and not always complete. Alcohol ingestion and physical exertion may bring on symptoms. Lead exposure increases the risk of high blood pressure. Continuous exposure can result in decreased fertility. Elevated lead exposure of either parent before pregnancy can increase the chances of miscarriage or birth defects. Exposure of the mother during pregnancy can cause birth defects.

Points of Attack: Lungs, blood, kidneys, brain.

Medical Surveillance: Before first exposure and every 6 months thereafter, OSHA (1910.1025) requires your employer to provide blood lead test, ZPP test (a special test for the effect of lead on blood cells). Examination of the nervous system. Prior to initial exposure, and annually for exposed person having blood lead readings exceeding 40 μg/100 g of whole blood, OSHA also requires a complete medical history, complete blood count and kidney function tests in addition to the tests listed above. OSHA defines "exposure" for these tests as air levels which average 30 μg of lead or more in a cubic meter of air. OSHA (under 1910.1020) requires your employer to send the doctor a copy of the lead standard and provide one for you.

Note: Blood-lead level is a good indicator of total lead exposure. Current OSHA regulations require that if an individual has a blood-lead level greater than or equal to 0.050 mg lead per 100 mL blood, he or she must be removed from all exposures to lead and cannot return to the exposure environment until the blood level falls to 0.040 mg lead per 100 mL blood or less. The following tolerance levels for occupational exposures may also be useful: ACGIH BEI = 50 mg/L (blood); 150 mg/g creatinine (urine). DFG BAT = 70 mg/L (blood); 30 mg/L (blood) for women <45 years old.

First Aid: If this chemical gets into the eyes, remove any contact lenses at once and irrigate immediately for at least 15 min, occasionally lifting upper and lower lids. Seek medical attention immediately. If this chemical contacts the skin, remove contaminated clothing and wash immediately with soap and water. Seek medical attention immediately. If this chemical has been inhaled, remove from exposure, begin rescue breathing (using universal precautions, including resuscitation mask) if breathing has stopped and CPR if heart action has stopped. Transfer promptly to a medical facility. When this chemical has been swallowed, get medical attention.

Note to physician: whole blood lead levels, circulating plasma/erythrocyte lead concentration ratio, urine ALA, and erythrocyte protoporphyrin fluorescent microscopy may all be useful in monitoring or assessing lead exposure. Chelating agents, such as edetate disodium calcium (Ca EDTA) and penicillamine (*not penicillin*) are generally useful in the therapy of acute lead intoxication.

Antidotes and special procedures for lead: Persons with significant lead poisoning are sometimes treated with Ca EDTA while hospitalized. This "chelating" drug causes a rush of lead from the body organs into the blood and kidneys, and thus has its own hazards, and must be administered only by highly experienced medical personnel under controlled conditions and careful observation. Ca EDTA or similar drugs should never be used to prevent poisoning while exposure continues or without strict exposure control, as severe kidney damage can result.

Note to physician: For severe poisoning BAL [British Anti-Lewisite, dimercaprol, dithiopropanol ($C_3H_8OS_2$)] has been used to treat toxic symptoms of certain heavy metals poisoning. In the case of lead poisoning it may have SOME value. Although BAL is reported to have a large margin of safety, caution must be exercised, because toxic effects may be caused by excessive dosage. Most can be prevented by premedication with 1-ephedrine sulfate (CAS: 134-72-5).

Personal Protective Methods: Wear protective gloves and clothing to prevent any reasonable probability of skin contact. Use any barrier that will prevent contamination from the dust. Safety equipment suppliers/manufacturers can provide recommendations on the most protective glove/clothing material for your operation. All protective clothing (suits, gloves, footwear, headgear) should be clean, available each day, and put on before work. Work clothing should be HEPA vacuumed before removal. Contact lenses should not be worn when working with this chemical. Wear dust-proof chemical goggles and face shield unless full face-piece respiratory protection is worn. Employees should wash immediately with soap when skin is wet or contaminated. Provide emergency showers and eyewash. See also NIOSH Criteria Document #78-158, *LEAD, inorganic dusts and fumes.*

Respirator Selection: OSHA: *0.5 mg/m³:* Any air-purifying respirator with an N100, R100, or P100 filter (including N100, R100, and P100 filtering face-pieces) except quarter-mask respirators. *1.25 mg/m³:* Sa:Cf (APF = 25) (any supplied-air respirator operated in a continuous-flow mode) or PaprHie (APF = 25) (any powered, air-purifying respirator with a high-efficiency particulate filter). *2.5 mg/m³:* 100F (APF = 50) (any air purifying, full-face-piece respirator with an N100, R100, or P100 filter) or SaT: Cf (APF = 50) (any supplied-air respirator that has a tight-fitting face-piece and is operated in a continuous-flow mode) or PaprTHie (APF = 50) (any powered, air-purifying respirator with a tight-fitting face-piece and a high-efficiency particulate filter) or SCBAF (APF = 50) (any self-contained breathing apparatus with a full face-piece) or SaF (APF = 50) (any supplied-air respirator with a full face-piece). *50 mg/m³:* Sa: Pd,Pp (APF = 1000) (any supplied-air respirator operated in a pressure-demand or other positive-pressure mode). *100 mg/m³:* SaF: Pd,Pp (APF = 2000) (any supplied-air respirator that has a full face-piece and is operated in a pressure-demand or other positive-pressure mode). *Emergency or planned entry into unknown concentrations or IDLH conditions:* SCBAF: Pd,Pp (APF = 10,000) (any self-contained breathing apparatus that has a full face-piece and is operated in a pressure-demand or other positive-pressure mode) or SaF: Pd,Pp: ASCBA (any supplied-air respirator that has a full face-piece and is operated in a pressure-demand or other positive-pressure mode in combination with an auxiliary self-contained breathing apparatus operated in a pressure-demand or other positive-pressure mode). *Escape:* 100F (APF = 50) (any air purifying, full-face-piece respirator with an N100, R100, or P100 filter) or SCBAE (any appropriate escape-type, self-contained breathing apparatus).

Storage: Color Code—Blue: Health Hazard/Poison: Store in a secure poison location. Prior to working with this chemical you should be trained on its proper handling and storage. Lead sulfide must be stored to avoid contact with oxidizers (such as perchlorates, peroxides, permanganates, chlorates, and nitrates) and chemically active metals (such as potassium, sodium, magnesium, and zinc), since violent reactions occur. Store in tightly closed containers in a cool, well-ventilated area away from moisture and acids. Lead is regulated by an OSHA Standard 1910.1025. All requirements of the standard must be followed. A regulated, marked area should be established where this chemical is handled, used, or stored in compliance with OSHA Standard 1910.1045.

Shipping: Toxic solid, inorganic, n.o.s. requires a label of "POISONOUS/TOXIC MATERIALS." Lead fluoride falls in Hazard Class 6.1.

Spill Handling: Evacuate persons not wearing protective equipment from area of spill or leak until cleanup is complete. Remove all ignition sources. Collect powdered material in the most convenient and safe manner and deposit in sealed containers. Ventilate area after cleanup is complete. It may be necessary to contain and dispose of this chemical as a hazardous waste. If material or contaminated runoff enters waterways, notify downstream users of potentially contaminated waters. Contact your local or federal environmental protection agency for specific recommendations. If employees are required to clean up spills, they must be properly trained and equipped. OSHA 1910.120(q) may be applicable.

Fire Extinguishing: Use dry chemical, carbon dioxide, water spray, or foam extinguishers. Poisonous gases, including lead and sulfur oxides, are produced in fire. If material or contaminated runoff enters waterways, notify downstream users of potentially contaminated waters. Notify local health and fire officials and pollution control agencies. From a secure, explosion-proof location, use water spray to cool exposed containers. If cooling streams are ineffective (venting sound increases in volume and pitch, tank discolors, or shows any signs of deforming), withdraw immediately to a secure position. If employees are expected to fight fires, they must be trained and equipped in OSHA 1910.156. The only respirators recommended for firefighting are self-contained breathing apparatuses that have full face-pieces and are operated in a pressure-demand or other positive-pressure mode.

Reference
New Jersey Department of Health and Senior Services. (August 2005). *Hazardous Substances Fact Sheet: Lead Sulfide.* Trenton, NJ

Lead thiocyanate   L:0230

Molecular Formula: $C_2N_2PbS_2$
Common Formula: $Pb(SCN)_2$
Synonyms: Lead sulfocyanate; Lead(2 +) thiocyanate; Lead(II) thiocyanate; Thiocyanic acid, lead(2 +) salt; Thiocyanic acid, lead(II) salt
CAS Registry Number: 592-87-0
RTECS® Number: XL1538000
UN/NA & ERG Number: UN2291 (Lead compound, soluble, n.o.s.)/151
EC Number: 209-774-6
Regulatory Authority and Advisory Bodies
IARC (*as organic lead compounds*): Animal Inadequate Evidence; Human Inadequate Evidence, *not classifiable as carcinogenic to humans*, Group 3.
OSHA, 29CFR1910 Specifically Regulated Chemicals (See CFR 1910.1025).
Clean Air Act: Hazardous Air Pollutants (Title I, Part A, Section 112) as cyanide compounds.
Clean Water Act: Section 311 Hazardous Substances/RQ 40CFR117.3 (same as CERCLA, see below); Section 313 Water Priority Chemicals (57FR41331, 9/9/92); 40CFR401.15 Section 307 Toxic Pollutants as lead and compounds.
RCRA, 40CFR261, Appendix 8 Hazardous Constituents, waste number not listed, as lead compounds, n.o.s.
Reportable Quantity (RQ): 10 lb (4.54 kg).

EPCRA Section 313 Form R *de minimis* concentration reporting level: 0.1%.

US DOT Regulated Marine Pollutant (49CFR172.101, Appendix B).

California Proposition 65 Chemical: Cancer 10/1/92 (lead and compounds).

Canada, WHMIS, Ingredients Disclosure List Concentration 1.0% as lead compounds, n.o.s.

WGK (German Aquatic Hazard Class): No value assigned.

Description: Lead thiocyanate is a white or light yellow, odorless, crystalline powder. Odorless. Molecular weight = 323.36; Specific gravity (H_2O:1) = 3.82; Freezing/Melting point = 190°C (decomposes). Hazard Identification (based on NFPA-704 M Rating System): Health 3, Flammability 1, Reactivity 1. Slightly soluble in water; solubility = <0.05%.

Potential Exposure: An explosive, thermally unstable material. Used in making safety matches, primers for small-arms cartridges, pyrotechnic devices, and in dyes.

Incompatibilities: Oxidizers, strong acids. Contact with acids or acid fumes caused decomposition with fumes of cyanide. Will decompose in hot water.

Permissible Exposure Limits in Air

No TEEL available.

Determination in Air: Use NIOSH Analytical Method (IV) s #7082, 7105, 7300, 7301, 7303, 7700, 7701, 7702, 9100, 9102, 9105; OSHA Analytical Methods ID-121, ID-125G, ID-206.

Determination in Water: 0.1 mg/L Pb in drinking water may cause chronic poisoning. The EPA limits lead in drinking water to 15 μg/L. Various organizations worldwide have set other standards for lead in drinking water as follows[35] (all in mg/L): Argentina 0.01; the Czech Republic 0.05; Germany 0.04; EEC 0.05; Japan 0.10; Mexico 0.05; former USSR-UNEP/IRPTC joint project 0.03; WHO 0.10. The states of Maine and Minnesota have set guidelines for lead in drinking water[61] at the level of 20 μg/L.

Routes of Entry: Inhalation, skin. Lead can be absorbed through skin at chronically toxic levels.

Harmful Effects and Symptoms

Short Term Exposure: Skin and eye contact can cause irritation and burns. Lead thiocyanate can cause headache, irritability, mood changes, reduced memory, and disturbed sleep.

Note: Lead is a cumulative poison. Increasing amounts can build up in the body, eventually reaching a point where symptoms and disability occur. Lead dust carried home on contaminated clothing can result in exposure and symptoms in other family members. Standards only protect for inhalation exposure. Extra precautions should be taken if skin exposure also occurs.

Long Term Exposure: Repeated exposure can cause lead poisoning with symptoms of headache, irritability, disturbed sleep, tiredness, reduced memory, and personality changes. Higher levels can cause muscle or joint pains, weakness, disturbed sleep, and easy fatigue. Exposure can

increase the risk of high blood pressure. May cause kidney and brain damage and anemia. Lead can accumulate in the body over a period of time. Therefore, long-term exposures to lower levels can result in a build up of lead in the body and more severe symptoms. These may include anemia, pale skin, a blue line at the gum margin, decreased hand-grip strength, abdominal pain, severe constipation, nausea, vomiting, and paralysis of the wrist joint. Prolonged exposure may also result in kidney and brain damage. If the nervous system is affected, usually due to very high exposures, the resulting effects include severe headache, convulsions, coma, delirium, and death. In nonfatal cases, recovery is slow and not always complete. Alcohol ingestion and physical exertion may bring on symptoms. Lead exposure increases the risk of high blood pressure. Continuous exposure can result in decreased fertility. Elevated lead exposure of either parent before pregnancy can increase the chances of miscarriage or birth defects. Exposure of the mother during pregnancy can cause birth defects.

Points of Attack: Blood, kidneys, brain, nervous system.

Medical Surveillance: Before first exposure and every 6 months thereafter, OSHA (1910.1025) requires your employer to provide blood lead test, ZPP test (a special test for the effect of lead on blood cells). Examination of the nervous system. Prior to initial exposure, and annually for exposed person having blood lead readings exceeding 40 μg/100 g of whole blood, OSHA also requires a complete medical history, complete blood count and kidney function tests in addition to the tests listed above. OSHA defines "exposure" for these tests as air levels which average 30 μg of lead or more in a cubic meter of air. OSHA (1910.1020) requires your employer to send the doctor a copy of the lead standard and provide one for you.

Note: Blood-lead level is a good indicator of total lead exposure. Current OSHA regulations require that if an individual has a blood-lead level greater than or equal to 0.050 mg lead per 100 mL blood, he or she must be removed from all exposures to lead and cannot return to the exposure environment until the blood level falls to 0.040 mg lead per 100 mL blood or less. The following tolerance levels for occupational exposures may also be useful: ACGIH BEI = 50 mg/L (blood); 150 mg/g creatinine (urine). DFG BAT = 70 mg/L (blood); 30 mg/L (blood) for women <45 years old.

First Aid: If this chemical gets into the eyes, remove any contact lenses at once and irrigate immediately for at least 15 min, occasionally lifting upper and lower lids. Seek medical attention immediately. If this chemical contacts the skin, remove contaminated clothing and wash immediately with soap and water. Seek medical attention immediately. If this chemical has been inhaled, remove from exposure, begin rescue breathing (using universal precautions, including resuscitation mask) if breathing has stopped and CPR if heart action has stopped. Transfer promptly to a medical facility. When this chemical has been swallowed, get

medical attention. Give large quantities of water and induce vomiting. Do not make an unconscious person vomit.

Note to physician: Administer saline cathartic and an enema. For relief of colic, administer antispasmodic (calcium gluconate, atropine, papaverine). Consider morphine sulfate for severe pain.

Whole blood lead levels, circulating plasma/erythrocyte lead concentration ratio, urine ALA, and erythrocyte protoporphyrin fluorescent microscopy may all be useful in monitoring or assessing lead exposure. Chelating agents, such as edetate disodium calcium (Ca EDTA) and penicillamine (*not penicillin*), are generally useful in the therapy of acute lead intoxication.

Antidotes and special procedures for lead: Persons with significant lead poisoning are sometimes treated with Ca EDTA while hospitalized. This "chelating" drug causes a rush of lead from the body organs into the blood and kidneys, and thus has its own hazards, and must be administered only by highly experienced medical personnel under controlled conditions and careful observation. Ca EDTA or similar drugs should never be used to prevent poisoning while exposure continues or without strict exposure control, as severe kidney damage can result.

Personal Protective Methods: Wear protective gloves and clothing to prevent any reasonable probability of skin contact. Use any barrier that will prevent contamination from the dust. Safety equipment suppliers/manufacturers can provide recommendations on the most protective glove/clothing material for your operation. All protective clothing (suits, gloves, footwear, headgear) should be clean, available each day, and put on before work. Work clothing should be HEPA vacuumed before removal. Contact lenses should not be worn when working with this chemical. Wear dust-proof chemical goggles and face shield unless full face-piece respiratory protection is worn. Employees should wash immediately with soap when skin is wet or contaminated. Provide emergency showers and eyewash.

Respirator Selection: OSHA: *0.5 mg/m³:* Any air-purifying respirator with an N100, R100, or P100 filter (including N100, R100, and P100 filtering face-pieces) except quarter-mask respirators. *1.25 mg/m³:* Sa:Cf (APF = 25) (any supplied-air respirator operated in a continuous-flow mode) or PaprHie (APF = 25) (any powered, air-purifying respirator with a high-efficiency particulate filter). *2.5 mg/m³:* 100F (APF = 50) (any air purifying, full-face-piece respirator with an N100, R100, or P100 filter) or SaT: Cf (APF = 50) (any supplied-air respirator that has a tight-fitting face-piece and is operated in a continuous-flow mode) or PaprTHie (APF = 50) (any powered, air-purifying respirator with a tight-fitting face-piece and a high-efficiency particulate filter) or SCBAF (APF = 50) (any self-contained breathing apparatus with a full face-piece) or SaF (APF = 50) (any supplied-air respirator with a full face-piece). *50 mg/m³:* Sa: Pd,Pp (APF = 1000) (any supplied-air respirator operated in a pressure-demand or other positive-pressure mode). *100 mg/m³:* SaF: Pd,Pp (APF = 2000) (any supplied-air respirator that has a full face-piece and is operated in a pressure-demand or other positive-pressure mode).

Emergency or planned entry into unknown concentrations or IDLH conditions: SCBAF: Pd,Pp (APF = 10,000) (any self-contained breathing apparatus that has a full face-piece and is operated in a pressure-demand or other positive-pressure mode) or SaF: Pd,Pp: ASCBA (any supplied-air respirator that has a full face-piece and is operated in a pressure-demand or other positive-pressure mode in combination with an auxiliary self-contained breathing apparatus operated in a pressure-demand or other positive-pressure mode). *Escape:* 100F (APF = 50) (any air purifying, full-face-piece respirator with an N100, R100, or P100 filter) or SCBAE (any appropriate escape-type, self-contained breathing apparatus).

Storage: Color Code—Blue: Health Hazard/Poison: Store in a secure poison location. Prior to working with this chemical you should be trained on its proper handling and storage. Store in tightly closed containers in a cool, well-ventilated area away from hot water, heat, oxidizers, acids, acid fumes. Lead is regulated by an OSHA Standard 1910.1025. All requirements of the standard must be followed.

Shipping: Lead compounds, soluble, n.o.s. require a shipping label of "POISONOUS/TOXIC MATERIALS." It falls in Hazard Class 6.1 and Packing Group III.

Spill Handling: Evacuate persons not wearing protective equipment from area of spill or leak until cleanup is complete. Neutralize with lime or sodium bicarbonate. Remove all ignition sources. Collect powdered material in the most convenient and safe manner and deposit in sealed containers. Use vacuum or a wet method to reduce dust. Do not dry sweep. When vacuuming, a HEPA filter should be used, not a standard shop vac. Ventilate area after cleanup is complete. It may be necessary to contain and dispose of this chemical as a hazardous waste. If material or contaminated runoff enters waterways, notify downstream users of potentially contaminated waters. Contact your local or federal environmental protection agency for specific recommendations. If employees are required to clean up spills, they must be properly trained and equipped. OSHA 1910.120(q) may be applicable.

Fire Extinguishing: This chemical is slightly flammable when exposed to heat or flame. Use dry chemical, carbon dioxide, water spray, alcohol foam, or polymer foam extinguishers. Poisonous gases, including cyanides, nitrogen oxides, sulfur oxides, and lead oxide are produced in fire. If material or contaminated runoff enters waterways, notify downstream users of potentially contaminated waters. Notify local health and fire officials and pollution control agencies. Containers may explode in fire. From a secure, explosion-proof location, use water spray to cool exposed containers. If cooling streams are ineffective (venting sound increases in volume and pitch, tank discolors, or shows any signs of deforming), withdraw immediately to a secure position. If employees are expected to fight fires, they must be trained and equipped in OSHA 1910.156. The only respirators recommended for firefighting are self-contained breathing apparatuses that have full face-pieces and are operated in a pressure-demand or other positive-pressure mode.

Reference

New Jersey Department of Health and Senior Services. (June 1999). *Hazardous Substances Fact Sheet: Lead Thiocyanate.* Trenton, NJ

Leptophos   L:0240

Molecular Formula: $C_{13}H_{10}BrCl_2O_2PS$
Common Formula: $C_6H_5PS(OCH_3)OC_6H_2BrCl_2$
Synonyms: Abar; *O*-(4-Bromo-2,5-dichlorophenyl) *O*-methyl phenylphosphonothioate; *O*-(2,5-Dichloro-4-bromophenyl) *O*-methyl phenylthiophosphonate; Fosvel; K62-105; MBCP; *O*-Methyl *O*-(4-bromo-2,5-dichlorophenyl) phenyl thiophosphonate; NK711; Phenylphosphonothioic acid *O*-(4-bromo-2,5-bromo-2,5-dichlorophenyl) *O*-methyl ester; Phosphonothioic acid, phenyl-, *O*-(4-bromo-2,5-dichlorophenyl) *O*-methyl ester; Phosvel; PSL; V.C.S.; VCS-506; Velsicol 506; VelsicolVCS 506
CAS Registry Number: 21609-90-5
RTECS® Number: TB1720000
UN/NA & ERG Number: UN2783 (organophosphorus pesticides, solid, toxic)/152
EC Number: 244-472-8 [*Annex I Index No.:* 015-093-00-3]
Regulatory Authority and Advisory Bodies
Banned or Severely Restricted (many countries) (UN).[13]
SUPERFUND/EPCRA 40CFR355, Extremely Hazardous Substances: TPQ = 500/10,000 lb (227/4540 kg).
Reportable Quantity (RQ): 500 lb (227 kg).
EPCRA Section 313 Form R *de minimis* concentration reporting level: inorganic compounds 0.1%; organic compounds 1.0%.
US DOT Regulated Marine Pollutant (49CFR172.101, Appendix B).
US DOT 49CFR172.101, Inhalation Hazard Chemical as organophosphates.
European/International Regulations: Hazard Symbol: T, N; Risk phrases: R21; R25; R39/25; R50/53; Safety phrases: S1/2; S25; S36/37/39; S45; S60; S61 (see Appendix 4).
WGK (German Aquatic Hazard Class): No value assigned.
Description: Leptophos is a tan, waxy solid. Molecular weight = 412.07; Freezing/Melting point = about 70°C. Hazard Identification (based on NFPA-704 M Rating System): Health 3, Flammability 1, Reactivity 0. Practically insoluble in water; solubility = 0.028%.
Potential Exposure: Those involved in the manufacture, formulation, and application of this insecticide. Its use is not permitted in the United States.
Permissible Exposure Limits in Air
Protective Action Criteria (PAC)
TEEL-0: 6 mg/m^3

PAC-1: 15 mg/m^3
PAC-2: 30 mg/m^3
PAC-3: 30 mg/m^3
Determination in Air: OSHA versatile sampler-2; Toluene/ Acetone; Gas chromatography/Flame photometric detection for sulfur, nitrogen, or phosphorus; NIOSH Analytical Method (IV) Method #5600, Organophosphorus Pesticides.
Determination in Water: Octanol−water coefficient: Log $K_{ow} = 6.3$.
Routes of Entry: Inhalation, ingestion, skin contact.
Harmful Effects and Symptoms
Short Term Exposure: Organic phosphorus insecticides are absorbed by the skin as well as by the respiratory and gastrointestinal tracts. They are cholinesterase inhibitors. Symptoms of exposure include headache, giddiness, blurred vision, nervousness, weakness, nausea, cramps, diarrhea, and discomfort in the chest. Death may occur from failure of the respiratory center, paralysis of the respiratory muscles, intense bronchoconstriction, or all three. This material is highly toxic; it is capable of causing death or permanent injury by exposure during normal use. LD_{50} = (oral-rat) 30 mg/kg.
Long Term Exposure: Cholinesterase inhibitor; cumulative effect is possible. This chemical may damage the nervous system with repeated exposure, resulting in convulsions, respiratory failure. May cause liver damage.
Points of Attack: Respiratory system, lungs, central nervous system, cardiovascular system, skin, eyes, plasma and red blood cell cholinesterase.
Medical Surveillance: Before employment and at regular times after that, the following are recommended: plasma and red blood cell cholinesterase levels (tests for the enzyme poisoned by this chemical). If exposure stops, plasma levels return to normal in 1−2 weeks while red blood cell levels may be reduced for 1−3 months.
When cholinesterase enzyme levels are reduced by 25% or more below preemployment levels, risk of poisoning is increased, even if results are in lower ranges of "normal." Reassignment to work not involving organophosphate or carbamate pesticides is recommended until enzyme levels recover. If symptoms develop or overexposure occurs, repeat the above tests as soon as possible and get an examination of the nervous system. Also consider complete blood count. Consider chest X-ray following acute overexposure. Do not drink any alcoholic beverages before or during use. Alcohol promotes absorption of organic phosphates.
First Aid: If this chemical gets into the eyes, remove any contact lenses at once and irrigate immediately for at least 15 min, occasionally lifting upper and lower lids. Seek medical attention immediately. If this chemical contacts the skin, remove contaminated clothing and wash immediately with soap and water. Speed in removing material from skin is of extreme importance. Shampoo hair promptly if contaminated. Seek medical attention immediately. If this chemical has been inhaled, remove from exposure, begin rescue breathing (using universal precautions, including

resuscitation mask) if breathing has stopped and CPR if heart action has stopped. Transfer promptly to a medical facility. When this chemical has been swallowed, get medical attention. Give large quantities of water and induce vomiting. Do not make an unconscious person vomit. Keep victim quiet and maintain normal body temperature. Effects may be delayed; keep victim under observation.

Note to physician: 1,1′-trimethylenebis(4-formylpyridinium bromide)dioxime (a.k.a TMB-4 dibromide and TMV-4) has been used as an antidote for organophosphate poisoning.

Personal Protective Methods: Wear protective gloves and clothing to prevent any reasonable probability of skin contact. Safety equipment suppliers/manufacturers can provide recommendations on the most protective glove/clothing material for your operation. All protective clothing (suits, gloves, footwear, headgear) should be clean, available each day, and put on before work. Contact lenses should not be worn when working with this chemical. Wear dust-proof chemical goggles and face shield unless full face-piece respiratory protection is worn. Employees should wash immediately with soap when skin is wet or contaminated. Provide emergency showers and eyewash.

Respirator Selection: Follow regulations in OSHA 29CFR1910.134 or European Standard EN149. Use a NIOSH/MSHA- or European Standard EN149-approved respirator; or use an approved supplied-air respirator with a full face-piece operated in the positive-pressure mode, or with a full face-piece, hood, or helmet in the continuous-flow mode; or use a NIOSH/MSHA- or European Standard EN149-approved self-contained breathing apparatus with a full face-piece operated in pressure-demand or other positive-pressure mode.

Storage: Color Code—Blue: Health Hazard/Poison: Store in a secure poison location. Prior to working with this chemical you should be trained on its proper handling and storage. Store in tightly closed containers in a cool, well-ventilated area.

Shipping: Leptophos is classified as an organophosphorus pesticide, solid, toxic, n.o.s. This compound requires a shipping label of "POISONOUS/TOXIC MATERIALS." It falls in Hazard Class 6.1 and Packing Group II.

Spill Handling: Evacuate persons not wearing protective equipment from area of spill or leak until cleanup is complete. Remove all ignition sources. Collect powdered material in the most convenient and safe manner and deposit in sealed containers. Ventilate area after cleanup is complete. It may be necessary to contain and dispose of this chemical as a hazardous waste. If material or contaminated runoff enters waterways, notify downstream users of potentially contaminated waters. Contact your local or federal environmental protection agency for specific recommendations. If employees are required to clean up spills, they must be properly trained and equipped. OSHA 1910.120(q) may be applicable.

Fire Extinguishing: This material may burn but does not ignite readily. For small fires, use dry chemical, cargo dioxide, water spray, or foam. For large fires, use water spray, fog, or foam. Stay upwind; keep out of low areas. Move container from fire area if you can do it without risk. Fight fire from maximum distance. Dike fire control water for later disposal; do not scatter the material. Poisonous gases, including bromine, chlorine, phosphorus oxides, sulfur oxides, are produced in fire. If material or contaminated runoff enters waterways, notify downstream users of potentially contaminated waters. Notify local health and fire officials and pollution control agencies. From a secure, explosion-proof location, use water spray to cool exposed containers. If cooling streams are ineffective (venting sound increases in volume and pitch, tank discolors, or shows any signs of deforming), withdraw immediately to a secure position. If employees are expected to fight fires, they must be trained and equipped in OSHA 1910.156. The only respirators recommended for firefighting are self-contained breathing apparatuses that have full face-pieces and are operated in a pressure-demand or other positive-pressure mode.

Disposal Method Suggested: Small amounts may be treated with alkali and then burned in a landfill. Large quantities should be incinerated in a unit with effluent gas scrubbing.[22] In accordance with 40CFR165, follow recommendations for the disposal of pesticides and pesticide containers. Must be disposed properly by following package label directions or by contacting your local or federal environmental control agency or by contacting your regional EPA office.

Reference
US Environmental Protection Agency. (November 30, 1987). *Chemical Hazard Information Profile: Leptophos.* Washington, DC: Chemical Emergency Preparedness Program

Lewisite (Agents L-1, L-2, L-3, & HL—WMD)   L:0250

Molecular Formula: $C_2H_2AsCl_3$
Common Formula: $ClCH{=}CHAsCl_2$
Synonyms: (2-Chloroethenyl)arsenous dichloride; β-Chlorovinylbichloroarsine; (2-Chlorovinyl)dichloroarsine; 2-Chlorovinyldichloroarsine; Dichloro(2-chlorovinyl)arsine; L-1, L-2, L-3 (military designations); Levista (Spanish); Lewisite (arsenic compound)
Mustard-Lewisite: Agent HL; Sulfur mustard/Lewisite
CAS Registry Number: 541-25-3 (L-1); 40334-69-8 (L-2); 40334-70-1 (L-3); 1306-02-1
RTECS® Number: CH2975000
UN/NA & ERG Number: UN2810/153; UN3162 (Liquefied gas, poisonous, n.o.s.)
UN/NA & ERG Number: None assigned.
Regulatory Authority and Advisory Bodies
Department of Homeland Security Screening Threshold Quantity: *Theft hazard* CUM 100 g. (L1, L2, or L3).

Clean Air Act: Hazardous Air Pollutants (Title I, Part A, Section 112); List of high-risk pollutants (Section 63.74), as arsenic compounds.

Clean Water Act: Toxic Pollutant (Section 401.15) as arsenic and compounds.

RCRA, 40CFR261, Appendix 8 Hazardous Constituents, as arsenic compounds, n.o.s., waste number not listed.

SUPERFUND/EPCRA 40CFR355, Extremely Hazardous Substances: TPQ = 10 lb (4.54 kg).

Reportable Quantity (RQ): 1 lb (0.454 kg).

EPCRA Section 313 Form R *de minimis* concentration reporting level: 1.0%. (as organic arsenic compound) inhalation hazard.

US DOT Regulated Marine Pollutant (49CFR172.101, Appendix B) as arsenates.

US DOT 49CFR172.101, Inhalation Hazard Chemical.

WGK (German Aquatic Hazard Class): No value assigned (all above CAS numbers).

Description: Lewisite is a colorless, odorless liquid when pure. Industrially produced lewisite is an amber to dark brown oily liquid with an odor of geraniums. It turns violet to black or green with age. Odor is not a reliable indicator of the presence of toxic amount of vapor. Blister agent, lewisite (L) rapidly decomposes in relative humidity over 70%. Molecular weight = 207.32; Density (20°C) = 1.89 g/cm^3; Volatility = 2500 mg/m^3 at 20°C; Specific gravity = 1.89 at 20°C; Boiling point = 190°C (decomposes); Freezing/Melting point = 0.1°C. Flash point = None. Hazard Identification (based on NFPA-704 M Rating System): Health 4, Flammability 1, Reactivity 1. Sinks in water and is slightly soluble; solubility = 0.5 g/L.

Lewisite is a complex mixture of several *cis-* and *trans*-isomer compounds. In chemical-agent-grade lewisite, the L-1 isomer [2-Chlorovinylarsonous dichloride] generally predominates. The three homologues, L-1, L-2 [Bis(2-chlorovinyl)arsinous chloride], and L-3 [Tris(2-chlorovinyl)arsine] form from the catalyzed reaction of arsenic trichloride and acetylene. L-1 forms initially, but it continues to react with acetylene to form L-2 and L-3. L-1 is the vesicant agent. L-2 and L-3 are also toxic but considerably less than L-1.

Mustard—lewisite (military designation HL) is a liquid mixture of distilled mustard (HD) and lewisite (L) and has some properties of both. HL is both a blister agent (vesicant) and an alkylating agent (causes damage to the DNA of rapidly dividing cells). The mustard—lewisite mixture requires lower ambient temperatures before it will freeze; this property allows for improved ground dispersal and aerial spraying. Due to its low freezing point, the mixture remains a liquid in cold weather and at high altitudes. The mixture with the lowest freezing point consists of 63% Lewisite and 37% mustard. Mustard—Lewisite mixture may have a garlic odor. Odor however, should not be depended on to detect HL mixture.[NIOSH] Exposure to large amounts of HL may be fatal.

Potential Exposure: Those involved in the manufacture or use of this chemical warfare agent which is a vesicant. L-1,

L-2, and L-3 have been used as a blister-agent-type war gas.

Mustard—lewisite mixture was developed to achieve a lower freezing point for ground dispersal and aerial spraying.

Persistence of Chemical Agent: Agents L or HL: Summer: 1—3 days; Winter: May last for weeks.

Incompatibilities: Lewisite reacts with water and sweat; and, as it breaks down in water or sweat, it produces arsenic-containing materials which are less dangerous than lewisite but still dangerous. Heating causes lewisite to yield arsenic trichloride, tris-(2-chlorovinyl)arsine, and bis-(2-chlorovinyl)chloroarsine. Mustard—lewisite mixture is rapidly corrosive to brass at 65°C and will corrode steel at a rate of 0.0001 in. of steel per month at 65°C. It will hydrolyze into hydrochloric acid, thiodiglycol, and nonvesicant arsenic compounds.

Permissible Exposure Limits in Air The Surgeon General's Working Group (US Department of Health and Human Services) recommends (for the working place) 0.003 mg/m^3, Ceiling Concentration.

Protective Action Criteria (PAC) L-1; L-2; L-3
TEEL-0: 0.12 mg/m^3
PAC-1: 0.12 mg/m^3
PAC-2: 0.12 mg/m^3
PAC-3: 0.74 mg/m^3

Determination in Water: Lewisite dissolves rapidly in water and breaks down into toxic products that are much less dangerous than lewisite, but still poisonous, containing arsenic products. Warn pollution control authorities and advise shutting water intakes. Octanol—water coefficient: Log K_{ow} (estimated) = 2.56. Log $K_{benzene—water}$ = 0.15.

Routes of Entry: Inhalation, ingestion, skin contact. Absorbed through the skin.

Harmful Effects and Symptoms

Short Term Exposure: Lewisite is a cell irritant, blister agent, and systemic poison that can be absorbed through skin; a few drops can cause death. It produces an immediate searing sensation in the eye and permanent loss of sight *if not decontaminated within 1 min.* It produces an immediate and strong stinging sensation to the skin, followed by reddening within 30 min and blistering after about 13 h. This material causes pulmonary edema, diarrhea, restlessness, weakness, subnormal temperature, and low blood pressure. Inhalation of high concentrations may be fatal in as short a time as 10 min. Lethal dose in humans is 6 ppm (inhalation), 20 mg/kg (skin). Eye injury below 300 mg-min/m^3. Pulmonary edema, a medical emergency that can be delayed for several hours. This can cause death.

Long term Exposure: May cause sensitization and chronic lung impairment. It is a suspected carcinogen.

First Aid: If this chemical gets into the eyes, remove any contact lenses at once and irrigate immediately for at least 30 min, occasionally lifting upper and lower lids. Seek medical attention immediately. If this chemical contacts the skin, remove contaminated clothing and wash immediately

with soap and water. Speed in removing material from skin is of extreme importance. Seek medical attention immediately. If this chemical has been inhaled, remove from exposure, begin rescue breathing (using universal precautions, including resuscitation mask) if breathing has stopped and CPR if heart action has stopped. Transfer promptly to a medical facility. When this chemical has been swallowed, get medical attention. Give large quantities of water and induce vomiting. Do not make an unconscious person vomit. Medical observation is recommended for 24—48 h after breathing overexposure, as pulmonary edema may be delayed. As first aid for pulmonary edema, a doctor or authorized paramedic may consider administering a corticosteroid spray. Keep victim quiet and maintain normal body temperature. Effects may be delayed; keep victim under observation.

Note to physician: For severe poisoning BAL [British Anti-Lewisite, dimercaprol, dithiopropanol ($C_3H_8OS_2$)] has been used to treat toxic symptoms of certain heavy metals poisoning including arsenic. Although BAL is reported to have a large margin of safety, caution must be exercised, because toxic effects may be caused by excessive dosage. Most can be prevented by premedication with 1-ephedrine sulfate (CAS: 134-72-5). For milder poisoning *penicillamine (not penicillin)* has been used, both with mixed success. Side effects occur with such treatment and it is never a substitute for controlling exposure. It can only be done under strict medical care.

Decontamination: This is very important, and you have to decontaminate as soon as you can. Extra minutes before decontamination might make a big difference. If you do not have the equipment and training, do not enter the hot or the warm zone to rescue and decontaminate victims. If the victim cannot move, decontaminate without touching and without entering the hot or the warm zone. Use clean water from any source; if possible, use a hose (spray or fog to prevent injury to the victim) or other system so that you would not have to touch the victim; do not even wait for soap or for the victim to remove clothing, begin washing immediately. Immediately flush the eyes with water for at least 15 min. Use caution to avoid hypothermia in children and the elderly. Wash—strip—wash—evacuate upwind and uphill: the approach is to immediately wash with water, then have the victim (not the first responder) remove all the victim's clothing, then wash again (with soap if available), and then move away from the hot zone in an upwind and up-hill direction. Wash the victim with warm water and soap. There are differing guidelines for decontamination and more research is needed to identify the optimal decontamination method. The effect of lewisite and mustard—lewisite can be prevented by rapid topical application of 2,3-dimercaptopropanol, known as British anti-lewisite (BAL) which reacts with Lewisite to form a stable nontoxic cyclic product. Decontaminate with diluted household bleach (0.5%, or one part bleach to 200 parts water), but do not let any get in the victim's eyes, open wounds, or mouth.

Wash off the diluted bleach solution after 15 min. Be sure you have decontaminated the victims as much as you can before they leave the area so that they do not spread the Lewisite. Use the antidote "Anti-Lewisite." Use 5% solution of common bleach (sodium hypochlorite) or calcium hypochlorite solution (48 oz per 5 gallons of water) to decontaminate scissors used in clothing removal, clothes, and other items.

Personal Protective Methods: Wear protective gloves and clothing to prevent any reasonable probability of skin contact. Safety equipment suppliers/manufacturers can provide recommendations on the most protective glove/clothing material for your operation. All protective clothing (suits, gloves, footwear, headgear) should be clean, available each day, and put on before work. Contact lenses should not be worn when working with this chemical. Wear full face-piece respiratory protection. Employees should wash immediately with soap when skin is wet or contaminated. Provide emergency showers and eyewash.

Respirator Selection: *When used as a weapon, use SCBA Respirator Certified By NIOSH For CBRN Environments.* Follow regulations in OSHA 29CFR1910.134 or European Standard EN149. Use a NIOSH/MSHA- or European Standard EN149-approved respirator; or use an approved supplied-air respirator with a full face-piece operated in the positive-pressure mode, or with a full face-piece, hood, or helmet in the continuous-flow mode; or use a NIOSH/MSHA- or European Standard EN149-approved self-contained breathing apparatus with a full face-piece operated in pressure-demand or other positive-pressure mode.

Storage: Color Code—Red Stripe (UN3162): Flammability Hazard: Store separately from all other flammable materials. Color Code—Blue (UN2810): Health Hazard/Poison: Store in a secure poison location. Prior to working with this chemical you should be trained on its proper handling and storage. Store in tightly closed containers in a cool, well-ventilated area away from alkalis.

Shipping: Toxic, liquids, organic, n.o.s. [Inhalation hazard, Packing Group I, Zone A] requires a shipping label of "POISONOUS/TOXIC MATERIALS." Inhalation Hazard. Technical name required.

Spill Handling: BACK OFF! Isolate a wide area around the release and call for expert help. If in a building, evacuate and confine vapors by closing doors and shutting down HVAC systems. Evacuate persons not wearing protective equipment from area of spill or leak until cleanup is complete. Remove all ignition sources. Avoid contact with eyes and skin; avoid breathing vapors. Keep unnecessary people away; isolate hazard area and deny entry. Stay upwind; keep out of low areas. Ventilate closed spaces before entering them. Remove and isolate contaminated clothing at the site. Do not touch spilled material; stop leak if you can do so without risk. Use water spray to reduce vapors.

Initial isolation and protective action distances
Distances shown are likely to be affected during the first 30 min after materials are spilled and could increase with

timc. If more than one tank car, cargo tank, portable tank, or large cylinder involved in the incident is leaking, the protective action distance may need to be increased. You may need to seek emergency information from CHEMTREC at (800) 424-9300 or seek professional environmental engineering assistance from the US EPA Environmental Response Team at (908) 548-8730 (24-h response line).

Small spills (*From a small package or a small leak from a large package*)

Agents L-1, L-2, L-3, & HL when used as a weapon
First: Isolate in all directions (feet/meters)   100/30
Then: Protect persons downwind (miles/kilometers)
Day   0.1/0.2
Night   0.2/0.3

Large spills (*From a large package or from many small packages*)
First: Isolate in all directions (feet/meters)   300/100
Then: Protect persons downwind (miles/kilometers)
Day   0.3/0.5
Night   0.7/1.2

Small spills: Cover with vermiculite, diatomaceous earth, clay, or fine sand and neutralize as soon as possible using large amounts of alcoholic caustic, carbonate, or Decontaminating Solution No. 2 [DS2: (2% NaOH, 70% diethylenetriamine, 28% ethylene glycol monomethyl ether)]. Caution: acetylene given off. Place into containers for later disposal. *Large spills*: dike far ahead of spill for later disposal. Can be decontaminated by supertropical bleach or caustic soda. Ventilate area after cleanup is complete. It may be necessary to contain and dispose of this chemical as a hazardous waste. If material or contaminated runoff enters waterways, notify downstream users of potentially contaminated waters. Contact your local or federal environmental protection agency for specific recommendations. If employees are required to clean up spills, they must be properly trained and equipped. OSHA 1910.120(q) may be applicable.

Fire Extinguishing: Heating causes lewisite to yield arsenic compounds such as arsenic trichloride (A: 1570), tris-(2-chlorovinyl)arsine, and bis-(2-chlorovinyl)chloroarsine. Firefighting gear (including SCBA) does not provide adequate protection. Use unattended equipment whenever possible. If exposure occurs, remove and isolate gear immediately and thoroughly decontaminate personnel. Specially trained personnel operating from a safe distance can fight fires using foam or dry chemicals, or use fog streams to extinguish burning liquids. Poisonous gases, including chlorine and arsenic, are produced in fire. If material or contaminated runoff enters waterways, notify downstream users of potentially contaminated waters. Notify local health and fire officials and pollution control agencies. Containers may explode in fire. Storage containers and parts of containers may rocket great distances, in many directions. From a secure, explosion-proof location, use water spray from unattended equipment to cool exposed containers. If cooling streams are ineffective

(venting sound increases in volume and pitch, tank discolours, or shows any signs of deforming), withdraw immediately to a secure position. If employees are expected to fight fires, they must be trained and equipped in OSHA 1910.156. The only respirators recommended for firefighting are self-contained breathing apparatuses that have full face-pieces and are operated in a pressure-demand or other positive-pressure mode.

References
US Environmental Protection Agency. (November 30, 1987). *Chemical Hazard Information Profile: Lewisite.* Washington, DC: Chemical Emergency Preparedness Program
US Department of Health and Human Services. (November 2, 2006). *ATSDR Medical Management Guidelines for Blister Agents: Lewisite (L) (C₂H₂AsCl₃) and Mustard−Lewisite Mixture (HL).* Atlanta, GA
Schneider, A. L. (Ed.) (2007). *CHRIS + CD-ROM Version 2.0, United States Coast Guard Chemical Hazard Response Information System (COMDTINST 16465.12C).* Washington, DC: United States Coast Guard and the Department of Homeland Security
The Riegle Report: A Report of Chairman Donald W. Riegle, Jr. and Ranking Member Alfonse M. D'Amato of the Committee on Banking, Housing and Urban Affairs with Respect to Export Administration, United States Senate, 103rd Congress, 2d Session, (May 25, 1994).

Lindane   L:0260

Molecular Formula: $C_6H_6Cl_6$
Synonyms: Aalindan; Aficide; Agrisol G-20; Agrocide; Agronexit; Ameisenatod; Ameisenmittel (Merck); Aparasin; Aphtiria; Aplidal; Arbitex; BBH; Ben-Hex; Bentox 10; γ-Benzene hexachloride; Benzene hexachloride; Benzene hexachloride g isomer; Bexol; γ-BHC; BHC; Celanex; Chloresene; Codechine; 2,5-Cyclohexane, 1,2,3,4,5,6-hexachloro-, (1a,2a,3b,4a,5a,6b)-; DBH; Delsanex dairy fly spray; Detmol-extrakt; Detox 25; Devoran; DOL Granule; Drill Tox-Spezial Aglukon; Dual murganic RPB seed treatment; ENT7,796; Entomoxan; Exagama; Forlin; Fumite tecnalin smoke generators; Gallogama; Gamacid; Gamaphex; Gamene; Gammabenzene hexachlorocyclohexane (g isomer); gamma-BHC; gamma-Col; gamma HCH; gamma-HCH; Gammahexa; Gammahexane; Gammalex; Gammalin; Gammalin 20; Gammaphex; Gammasan 30; Gammaterr; Gammex; Gammexane; Gammexene; Gammopaz; Gexane; HCCH; γ-HCH; HCH; HCH BHC; Heclotox; Hexa; γ-Hexachloran; Hexachloran; γ-Hexachlorane; Hexachlorane; γ-Hexachlorobenzene; γ-1,2,3,4,5,6-Hexachlorocyclohexane; γ-Hexachlorocyclohexane; 1a,2a,3b,4a,5a,6b-Hexachlorocyclohexane; 1,2,3,4,5,6-

Hexachlorocyclohexane, *g* isomer; Hexachlorocyclohexane (*g* isomer); Hexachlorocyclohexane, 1,2,3,4,5,6-hexachlorcyclohexane; γ-Hexaclorobenzene; Hexaflow; Hexatox; Hexaverm; Hexicide; Hexyclan; HGI; Hortex; Inexit; Isotox; Jacutin; Kokotine; Kwell; Lendine; Lentox; Lidenal; Lindafor; Lindagam; Lindagrain; Lindagram; Lindagranox; γ-Lindane; Lindane; Lindapoudre; Lindatox; Lindosep; Lintox; Lorexane; Marstan fly spray; Mergamma 30; Milbol 49; Mist-O-Matic Lindex; Mszycol; NCI-C00204; Neo-Scabicidol; Nexen FB; Nexit; Nexit-Stark; Nexol-E; Nicochloran; Novigam; Omnitox; Ovadziak; Owadziak; Pedraczak; Pflanzol; Quellada; Rodesco insect powder; Sang gamma; Silvano; Silvano L; Spritz-Rapidin; Spruehpflanzol; Streunex; TAP85; TRI-6; Viton

CAS Registry Number: 58-89-9; *(alt)* 8007-42-9; *(alt)* 55963-79-6

RTECS® Number: GV4900000

UN/NA & ERG Number: UN2761/151

UN/NA & ERG Number: 200-401-2 [*Annex I Index No.:* 602-043-00-6]

Regulatory Authority and Advisory Bodies

Carcinogenicity: NTP: 11th Report on Carcinogens, 2004: Reasonably anticipated to be a human carcinogen; IARC: Animal Sufficient Evidence, Human No Adequate Data, *possibly carcinogenic to humans*, Group 2B.

NCI: Carcinogenesis Bioassay (feed); no evidence: mouse, rat; IARC: Animal Limited Evidence, 1987.

US EPA Gene-Tox Program, Positive: *S. cerevisiae* gene conversion; Positive/limited: Carcinogenicity—mouse/rat; Inconclusive: Host-mediated assay; *D. melanogaster* sex-linked lethal.

Air Pollutant Standard Set. See below, "Permissible Exposure Limits in Air" section.

US EPA, FIFRA, 1998 Status of Pesticides: Supported.

Clean Air Act: Hazardous Air Pollutants (Title I, Part A, Section 112).

Clean Water Act: Section 311 Hazardous Substances/RQ 40CFR117.3 (same as CERCLA, see below); 40CFR 423, Appendix A, Priority Pollutants; Section 313.

Water Priority Chemicals (57FR41331, 9/9/92); 40CFR 401.15 Section 307 Toxic Pollutants, as hexachlorocyclohexane.

US EPA Hazardous Waste Number (RCRA No.): U129.

RCRA Toxicity Characteristic (Section 261.24), Maximum. Concentration of Contaminants, regulatory level, 0.4 mg/L.

RCRA, 40CFR261, Appendix 8 Hazardous Constituents.

Safe Drinking Water Act: MCL, 0.0002 mg/L; MCLG, 0.0002 mg/L; Regulated chemical (47 FR 9352).

RCRA 40CFR264, Appendix 9; TSD Facilities Ground Water Monitoring List. Suggested test method(s) (PQL μg/L): 8080 (0.05).

RCRA 40CFR268.48; 61FR15654, Universal Treatment Standards: Wastewater (mg/L), 0.0017; Nonwastewater (mg/kg), 0.066.

SUPERFUND/EPCRA 40CFR355, Extremely Hazardous Substances: TPQ = 1000/10,000 lb (454/4540 kg).

Reportable Quantity (RQ): 1 lb (0.454 kg).

EPCRA Section 313 Form R *de minimis* concentration reporting level: 0.1%.

US DOT Regulated Marine Pollutant (49CFR172.101, Appendix B).

Rotterdam Convention Annex III [Chemicals Subject to the Prior Informed Consent Procedure (PIC)] (as lindane) See also HCH (mixed isomers) 608-73-1.

California Proposition 65 Chemical: *(Lindane and other hexachlorocyclohexane isomers)* Cancer 1/1/89.

List of Stockholm Convention POPs: Annex A (Elimination).

European/International Regulations: Hazard Symbol: T, N; Risk phrases: R20/21; R25; R48/22; R64; R50/53; Safety phrases: S1/2; S13; S36/37; S45; S60; S61 (see Appendix 4).

WGK (German Aquatic Hazard Class): 3—Severe hazard to waters.

Description: Lindane is a white to yellow, crystalline powder with a slight, musty odor (pure material is odorless). Molecular weight = 390.82; Specific gravity (H_2O:1) = 1.85; Boiling point = 323.3°C; Freezing/Melting point = 112.8°C; Vapor pressure = 0.00001 mmHg; Hazard Identification (based on NFPA-704 M Rating System): Health 3, Flammability 0, Reactivity 0. Insoluble in water. Noncombustible solid but may be dissolved in flammable liquids.

Potential Exposure: Compound Description: Agricultural Chemical; Tumorigen, Drug, Mutagen; Reproductive Effector; Human Data. Lindane has been used against insects in a wide range of applications including treatment of animals, buildings, humans for ectoparasites, clothes, water for mosquitoes, living plants, seeds, and soils. Some applications have been abandoned due to excessive residues, e.g., stored foodstuffs. Formulators, distributors, and users of lindane represent a special risk group. The major use of lindane in recent years has been to pretreat seeds. Thus, those engaged in treatment and planting can be exposed.

Incompatibilities: Lindane decomposes on contact with powdered iron, aluminum, zinc, and with alkalis producing trichlorobenzene. Corrosive to metals.

Permissible Exposure Limits in Air

OSHA PEL: 0.5 mg/m³ TWA [skin].

NIOSH REL: 0.5 mg/m³ TWA [skin].

ACGIH TLV®[1]: 0.5 mg/m³ TWA [skin]; confirmed animal carcinogen with unknown relevance to humans.

NIOSH IDLH: 50 mg/m³.

Protective Action Criteria (PAC)

TEEL-0: 0.5 mg/m³

PAC-1: 1.5 mg/m³

PAC-2: 50 mg/m³

PAC-3: 50 mg/m³

DFG MAK: 0.1 mg/m3, inhalable fraction; [skin]; Carcinogen Category 4; Pregnancy Risk Group C.

Austria: MAK 0.5 mg/m³, [skin], 1999; Denmark: TWA 0.5 mg/m³, [skin], 1999; Finland: TWA 0.5 mg/m³, [skin],

1999; France: VME 0.5 mg/m^3, [skin], 1999; Norway: TWA 0.5 mg/m^3, 1999; the Philippines: TWA 0.5 mg/m^3, [skin], 1993; Switzerland: MAK-W 0.5 mg/m^3, [skin], 1999; the Netherlands: MAC-TGG 4 µg/m^3, [skin], 2003; Turkey: TWA 0.5 mg/m^3, [skin], 1993; United Kingdom: TWA 0.1 mg/m^3, [skin], 2000; Argentina, Bulgaria, Columbia, Jordan, South Korea, New Zealand, Singapore, Vietnam: ACGIH TLV®: confirmed animal carcinogen with unknown relevance to humans. Several states have set guidelines or standards for lindane in ambient air[60] ranging from zero (North Dakota) to 0.068 µg/m^3 (Massachusetts) to 1.19 µg/m^3 (Kansas) to 1.2 µg/m^3 (Pennsylvania) to 1.67 µg/m^3 (New York) to 5.0 µg/m^3 (Connecticut, Florida, South Carolina) to 8.0 µg/m^3 (Virginia) to 12.0 µg/m^3 (Nevada).

Determination in Air: Collection on a filter, workup with isooctane, analysis by gas chromatography/flame ionization. Use NIOSH Analytical Method 5502.[18]

Permissible Concentration in Water: *To protect freshwater aquatic life:* 0.080 µg/L as a 24-h average, never to exceed 2.0 µg/L. *To protect saltwater aquatic life:* never to exceed 0.16 µg/L. *To protect human health:* preferably zero. An additional lifetime cancer risk of 1 in 100,000 is posed by a concentration of 0.186 µg/L.[6] Mexico[35] has set allowable limits on 0.2 µg/L in coastal waters and 2.0 µg/L in estuaries. WHO has set a limit of 3.0 µg/L in drinking water. The US EPA has set 120 µg/L as a long-term health advisory and 2 µg/L as a lifetime health advisory[47] for adults. The state of Maine has set a guideline of 4.0 µg/L for lindane in drinking water. EPA[62] has proposed a maximum level of 0.2 µg/L in drinking water.

Determination in Water: Methylene chloride extraction followed by gas chromatography with electron capture or halogen-specific detection (EPA Method 608) or gas chromatography plus mass spectrometry (EPA Method 625). Octanol−water coefficient: Log K_{ow} = >3.5.

Routes of Entry: Inhalation, skin absorption; ingestion, eye and/or skin contact. Fish Tox = 0.11310000 MATC (EXTRA HIGH).

Harmful Effects and Symptoms

Short Term Exposure: Lindane irritates the eyes and the respiratory tract and may affect the central nervous system. Symptoms of exposure include vomiting, faintness, tremor, restlessness, muscle spasms, unsteady gait, and convulsions may occur as a result of exposure. Elevated body temperature and pulmonary edema have been reported in children. Coma, respiratory failure, and death can result. Exposure to vapors of this compound or its thermal decomposition products may lead to headache, nausea, vomiting, and irritation of the eyes, nose, and throat. Lindane is a stimulant of the nervous system; causing violent convulsions that are rapid in onset and generally followed by death or recovery within 24 h. The probable human oral lethal dose is 50−500 mg/kg, or between 1 teaspoon and 1 oz for a 150-lb (70 kg) person. Human Tox = 0.20000 ppb MCL (EXTRA HIGH).

Long Term Exposure: Repeated or prolonged contact with skin may cause dermatitis. May damage the liver and kidneys. May damage the nerves in the arms and legs, possibly with weakness and poor coordination. May cause a serious drop in the blood cell count (aplastic anemia) or in the white blood cell count (agranulocytopenia). The Department of Health and Human Services has determined that HCH (hexachlorocyclohexanes) may reasonably be anticipated to be carcinogenic. Liver cancer has been seen in laboratory rodents that ate HCH for long periods of time. In animals, there is evidence that oral exposure to lindane during pregnancy results in an increased incidence of fetuses with extra ribs. However, ATSDR reports that animal studies have not shown birth defects in the babies of animals fed HCH during pregnancy. BCH has been detected in human breast milk.

Points of Attack: Eyes, central nervous system, blood, liver, kidneys, skin.

Medical Surveillance: NIOSH lists the following tests: whole blood (chemical/metabolite), blood serum, complete blood count, urine (chemical/metabolite). Blood test for lindane (may not be accurate longer than 1 week following last exposure). Liver and kidney function tests. See *Occupational Health Guidelines for Chemical Hazards.* NIOSH Pub Nos. 81-123; 88-118, Suppls. I−IV. 1981−1995.

First Aid: If this chemical gets into the eyes, remove any contact lenses at once and irrigate immediately for at least 15 min, occasionally lifting upper and lower lids. Seek medical attention immediately. If this chemical contacts the skin, remove contaminated clothing and wash immediately with soap and water. Seek medical attention immediately. If this chemical has been inhaled, remove from exposure, begin rescue breathing (using universal precautions, including resuscitation mask) if breathing has stopped and CPR if heart action has stopped. Transfer promptly to a medical facility. When this chemical has been swallowed, get medical attention. Give large quantities of water and induce vomiting. Do not make an unconscious person vomit.

Personal Protective Methods: Wear protective gloves and clothing to prevent any reasonable probability of skin contact. Safety equipment suppliers/manufacturers can provide recommendations on the most protective glove/clothing material for your operation. All protective clothing (suits, gloves, footwear, headgear) should be clean, available each day, and put on before work. Contact lenses should not be worn when working with this chemical. Wear dust-proof chemical goggles and face shield unless full face-piece respiratory protection is worn. Employees should wash immediately with soap when skin is wet or contaminated. Provide emergency showers and eyewash. Employees should wash immediately when skin is wet or contaminated. Work clothing should be changed daily if it is possible that clothing is contaminated. Remove nonimpervious clothing immediately if wet or contaminated. Provide emergency showers.

Respirator Selection: *up to 5 mg/m^3:* Any air-purifying half-mask respirator with organic vapor cartridge(s) in combination with an N95, R95, or P95 filter. The following filters may also be used: N99, R99, P99, N100, R100, P100; or Sa (APF = 10) (any supplied-air respirator). *Up to 12.5 mg/m³:* Sa:Cf* (APF = 25) (any supplied-air respirator operated in a continuous-flow mode) or PaprOvHie (APF = 25) (any powered air-purifying respirator with an organic vapor cartridge in combination with a high-efficiency particulate filter).* *Up to 25 mg/m³:* CcrFOv100 (APF = 50) [any air-purifying full-face-piece respirator equipped with organic vapor cartridge(s) in combination with an N100, R100, or P100 filter] or GmFOv100 (APF = 50) [any air-purifying, full-face-piece respirator (gas mask) with a chin-style, front- or back-mounted organic vapor canister having an N100, R100, or P100 filter] or PaprTOvHie* (APF = 50) [any powered, air-purifying respirator with a tight-fitting face-piece and organic vapor cartridge(s) in combination with a high-efficiency particulate filter] or SCBAF (APF = 50) (any self-contained breathing apparatus with a full face-piece) or SaF (APF = 50) (any supplied-air respirator with a full face-piece). *Up to 50 mg/m³:* SCBAF: Pd,Pp (APF = 10,000) (any self-contained breathing apparatus that has a full face-piece and is operated in a pressure-demand or other positive-pressure mode). *Emergency or planned entry into unknown concentrations or IDLH conditions:* SCBAF: Pd, Pp (APF = 10,000) (any self-contained breathing apparatus that has a full face-piece and is operated in a pressure-demand or other positive-pressure mode) or SaF: Pd,Pp: ASCBA (APF = 10,000) (any supplied-air respirator that has a full face-piece and is operated in a pressure-demand or other positive-pressure mode in combination with an auxiliary self-contained positive-pressure breathing apparatus). *Escape:* GmFOv100 (APF = 50) [any air-purifying, full-face-piece respirator (gas mask) with a chin-style, front- or back-mounted organic vapor canister having an N100, R100, or P100 filter] or SCBAE (any appropriate escape-type, self-contained breathing apparatus).

*Substance reported to cause eye irritation or damage; may require eye protection.

Storage: Color Code—Blue: Health Hazard/Poison: Store in a secure poison location. Prior to working with this chemical you should be trained on its proper handling and storage. Store in tightly sealed containers in a cool, dry place away from light and incompatible materials. Protect containers against physical damage. A regulated, marked area should be established where this chemical is handled, used, or stored in compliance with OSHA Standard 1910.1045.

Shipping: Lindane falls under the category of Organochlorine pesticides, solid, toxic, n.o.s. This compound requires a shipping label of "POISONOUS/TOXIC MATERIALS." It falls in Hazard Class 6.1 and Packing Group III.

Spill Handling: Evacuate persons not wearing protective equipment from area of spill or leak until cleanup is complete. Remove all ignition sources. Collect powdered material in the most convenient and safe manner and deposit in sealed containers. Ventilate area after cleanup is complete. It may be necessary to contain and dispose of this chemical as a hazardous waste. If material or contaminated runoff enters waterways, notify downstream users of potentially contaminated waters. Contact your local or federal environmental protection agency for specific recommendations. If employees are required to clean up spills, they must be properly trained and equipped. OSHA 1910.120(q) may be applicable.

Fire Extinguishing: Use dry chemical, carbon dioxide, water spray, or foam for small fires. Use water spray, fog, or foam for large fires. Move container from fire area if this can be done without risk. Use water to keep fire-exposed containers cool. Isolate hazard area and deny entry. Stay upwind and keep out of low areas. Ventilate closed spaces before entering. Wear positive breathing apparatus and special protective clothing. Fight fire from maximum distance, dike fire control water for later disposal. Poisonous gases, including phosgene and hydrogen chloride, are produced in fire. If material or contaminated runoff enters waterways, notify downstream users of potentially contaminated waters. Notify local health and fire officials and pollution control agencies. From a secure, explosion-proof location, use water spray to cool exposed containers. If cooling streams are ineffective (venting sound increases in volume and pitch, tank discolors, or shows any signs of deforming), withdraw immediately to a secure position. If employees are expected to fight fires, they must be trained and equipped in OSHA 1910.156. The only respirators recommended for firefighting are self-contained breathing apparatuses that have full face-pieces and are operated in a pressure-demand or other positive-pressure mode.

Disposal Method Suggested: For the disposal of lindane, a process has been developed involving destructive pyrolysis at 400–500°C with a catalyst mixture which contains 5–10% of either cupric chloride, ferric chloride, zinc chloride, or aluminum chloride on activated carbon. Consult with environmental regulatory agencies for guidance on acceptable disposal practices. Generators of waste containing this contaminant (≥100 kg/mo) must conform with EPA regulations governing storage, transportation, treatment, and waste disposal.

References

US Environmental Protection Agency. (1980). *Hexachlorocyclohexane: Ambient Water Quality Criteria.* Washington, DC

US Environmental Protection Agency. (April 30, 1980). *Gamma-Hexachlorocyclohexane: Health and Environmental Effects Profile No. 113.* Washington, DC: Office of Solid Waste

Sax, N. I. (Ed.). (1983). *Dangerous Properties of Industrial Materials Report*, 3, No. 1, 62–66

US Environmental Protection Agency. (November 30, 1987). *Chemical Hazard Information Profile: Lindane.* Washington, DC: Chemical Emergency Preparedness Program

New York State Department of Health. (May 1986). *Chemical Fact Sheet: Gamma-BHC.* Albany, NY: Bureau of Toxic Substance Assessment

US Department of Health and Human Services. (June 1999). *ATSDR ToxFAQs, Hexachlorocyclohexanes.* Atlanta, GA

US Environmental Protection Agency, Special Review and Reregistration Division Office of Pesticide Programs. (1998). *Agency Status of Pesticides in Registration, Reregistration, and Special Review* (Rainbow Report). Washington, DC

New Jersey Department of Health and Senior Services. (September 2001). *Hazardous Substances Fact Sheet: Lindane.* Trenton, NJ

Liquefied petroleum gas (LPG) L:0270

Molecular Formula: C_3H_8–C_3H_6–C_4H_{10}–C_4H_8 (mixture) A mixture of propane (C_3H_8) and butane (C_4H_{10}).

Synonyms: Bottled gas; Compressed petroleum gas; Gas de petroleo licuado (Spanish); Liquefied hydrocarbon gas; L.P. G.; Petroleum gas, liquefied; Propane-butane-(propylene); Pyrofax

CAS Registry Number: 68476-85-7

RTECS® Number: SE7545000

UN/NA & ERG Number: UN1075/115

EC Number: 270-704-2 [*Annex I Index No.:* 649-202-00-6]

Regulatory Authority and Advisory Bodies
Air Pollutant Standard Set. See below, "Permissible Exposure Limits in Air" section.

European/International Regulations: Hazard Symbol: F +, T; Risk phrases: R45; R46; R12; Safety phrases: S53; S45 (see Appendix 4).

WGK (German Aquatic Hazard Class): No value assigned.

Description: LPG is a colorless, noncorrosive, odorless gas when pure. A foul-smelling odorant is usually added. Shipped as a liquefied compressed gas (a mixture of propane, butanes, propylene, and butylenes). Boiling point ≥6°C; 42–44°C (propane); −1°C (butane); Flash point = (flammable gas): −104°C (propane); −60°C (butane). Explosive limits: LEL = 2.1%; UEL = 9.5% (propane); LEL = 1.9%; UEL = 8.5% (butane); Autoignition temperature = about 400°C: 466°C (propane); 405°C (butane). Hazard Identification (based on NFPA-704 M Rating System): Health 1, Flammability 4, Reactivity 0. Insoluble in water.

Potential Exposure: LPG is used as a fuel propellant, in metal cutting, and in the production of petrochemicals.

Incompatibilities: Forms explosive mixture with air. Contact with strong oxidizers and chlorine dioxide may cause fire and explosions. Attacks some plastics, rubber, and coatings.

Permissible Exposure Limits in Air
Conversion factor: 1 ppm = 1.72–2.37 mg/m^3 at 25°C & 1 atm.
OSHA PEL: 1000 ppm/1800 mg/m^3 TWA.
NIOSH REL: 1000 ppm/1800 mg/m^3 TWA.
ACGIH TLV®[1]: 1000 ppm TWA *as aliphatic hydrocarbon gas (C_1–C_4).*
NIOSH IDLH: 2000 ppm [LEL].
Protective Action Criteria (PAC)
TEEL-0: 1000 ppm
PAC-1: 2000 ppm
PAC-2: 2000 ppm
PAC-3: 2000 ppm
The Philippines: TWA 1000 ppm (1800 mg/m^3), 1993; Switzerland: MAK-W 1000 ppm (1800 mg/m^3), 1999; Turkey: TWA 1000 ppm (1800 mg/m^3), 1993; United Kingdom: TWA 1000 ppm (1750 mg/m^3); STEL 1250 ppm, 2000; the Netherlands: MAC-TGG 1800 mg/m^3, 2003; Argentina, Bulgaria, Columbia, Jordan, South Korea, New Zealand, Singapore, Vietnam: ACGIH TLV®: TWA 1000.

Routes of Entry: Inhalation, skin and/or eye contact (liquid).

Harmful Effects and Symptoms
Short Term Exposure: Liquefied petroleum gas can affect you when breathed in. Exposure to high levels can cause you to feel dizzy and lightheaded. Very high levels could cause suffocation and death from lack of oxygen. Contact with liquid liquefied petroleum gas can cause frostbite.

Long Term Exposure: Unknown at this time.

Points of Attack: Respiratory system, central nervous system.

Medical Surveillance: Consider the points of attack in pre-placement and periodic physical examinations.

First Aid: Skin Contact: Do not rub. Seek medical attention. *Breathing:* Remove the person from exposure. Begin rescue breathing (using universal precautions, including resuscitation mask) if breathing has stopped and CPR if heart action has stopped. Transfer promptly to a medical facility. If frostbite has occurred, seek medical attention immediately; do NOT rub the affected areas or flush them with water. In order to prevent further tissue damage, do NOT attempt to remove frozen clothing from frostbitten areas. If frostbite has NOT occurred, immediately and thoroughly wash contaminated skin with soap and water.

Personal Protective Methods: Wear appropriate personal protective clothing to prevent the skin from becoming frozen from contact with the evaporating liquid or from contact with vessels containing the liquid. Safety equipment suppliers/manufacturers can provide recommendations on the most protective glove/clothing material for your operation. All protective clothing (suits, gloves, footwear, headgear) should be clean, available each day, and put on before work. Contact lenses should not be worn when working with this chemical. Wear splash-proof chemical goggles and face shield when working with liquid a unless full facepiece respiratory protection is worn. Employees should wash immediately with soap when skin is wet or contaminated. Provide emergency showers and eyewash

Respirator Selection: 2000 ppm: Sa (APF = 10) (any supplied-air respirator) or SCBAF (APF = 50) (any self-contained breathing apparatus with a full face-piece). *Emergency or planned entry into unknown concentrations or IDLH conditions:* SCBAF: Pd,Pp (APF = 10,000) (any self-contained breathing apparatus that has a full face-piece and is operated in a pressure-demand or other positive-pressure mode) or SaF: Pd,Pp: ASCBA (APF = 10,000) (any supplied-air respirator that has a full face-piece and is operated in a pressure-demand or other positive-pressure mode in combination with an auxiliary, self-contained breathing apparatus operated in a pressure-demand or other positive-pressure mode). *Escape:* SCBAE (any appropriate escape-type, self-contained breathing apparatus).

Storage: Color Code—Red Stripe: Flammability Hazard: Store separately from all other flammable materials. Prior to working with this chemical you should be trained on its proper handling and storage. Before entering confined space where this chemical may be present, check to make sure that an explosive concentration does not exist. Store in tightly closed containers in a cool, well-ventilated area away from strong oxidizers (such as chlorine, bromine, and fluorine). Metal containers involving the transfer of 5 gallons or more of liquefied petroleum gas should be grounded and bonded. Drums must be equipped with self-closing valves, pressure vacuum bungs, and flame arresters. Sources of ignition, such as smoking and open flames, are prohibited where liquefied petroleum gas is handled, used, or stored. Wherever liquefied petroleum gas is used, handled, manufactured, or stored, use explosion-proof electrical equipment and fittings. Procedures for the handling, use, and storage of cylinders should be in compliance with OSHA 1910.101 and 1910.169, as with the recommendations of the Compressed Gas Association.

Shipping: This compound requires a shipping label of "FLAMMABLE GAS." It falls in Hazard Class 2.1.

Spill Handling: Evacuate and restrict persons not wearing protective equipment from area of spill or leak until cleanup is complete. Remove all ignition sources. Establish forced ventilation to keep levels below explosive limit. Ventilate area of leak to disperse the gas. Stop flow of gas. If source of leak is a cylinder and the leak cannot be stopped in place, remove the leaking cylinder to a safe place in the open air, and repair leak or allow cylinder to empty. Absorb liquids in vermiculite, dry sand, earth, peat, carbon, or a similar material and deposit in sealed containers. Keep this chemical out of a confined space, such as a sewer, because of the possibility of an explosion, unless the sewer is designed to prevent the buildup of explosive concentrations. It may be necessary to contain and dispose of this chemical as a hazardous waste. If material or contaminated runoff enters waterways, notify downstream users of potentially contaminated waters. Contact your local or federal environmental protection agency for specific recommendations. If employees are required to clean up spills, they must be properly trained and equipped. OSHA 1910.120(q) may be applicable.

Fire Extinguishing: Liquefied petroleum gas is a flammable gas. Use dry chemical, CO_2, water spray, or foam extinguishers. The flame may be invisible. Vapors may travel long distances to ignition sources and flashback. Vapors in confined areas may explode when exposed to fire. Containers may explode in fire. Storage containers and parts of containers may rocket great distances, in many directions. If material or contaminated runoff enters waterways, notify downstream users of potentially contaminated waters. Notify local health and fire officials and pollution control agencies. From a secure, explosion-proof location, use water spray to cool exposed containers. If cooling streams are ineffective (venting sound increases in volume and pitch, tank discolors, or shows any signs of deforming), withdraw immediately to a secure position. If employees are expected to fight fires, they must be trained and equipped in OSHA 1910.156. The only respirators recommended for firefighting are self-contained breathing apparatuses that have full face-pieces and are operated in a pressure-demand or other positive-pressure mode.

Disposal Method Suggested: Flaring using smokeless flare designs.

Reference

New Jersey Department of Health and Senior Services. (February 2001). *Hazardous Substances Fact Sheet: Liquefied Petroleum Gas.* Trenton, NJ

Lithium   L:0280

Molecular Formula: Li
Synonyms: Li; Lithium, elemental; Lithium metal; Lithium monohydride
CAS Registry Number: 7439-93-2
RTECS® Number: OJ5540000
UN/NA & ERG Number: UN1415/138; UN3089 (Metal powder, flammable, n.o.s.)/170
EC Number: 231-102-5 [*Annex I Index No.:* 003-001-00-4]
Regulatory Authority and Advisory Bodies
European/International Regulations: Hazard Symbol: F, C; Risk phrases: R14/15; R34; Safety phrases: S1/2; S8; S43; S45 (see Appendix 4).
WGK (German Aquatic Hazard Class): 2—Hazard to waters.

Description: Lithium is a silvery to grayish-white metal that turns yellow on exposure to air and/or moisture. Molecular weight = 6.94; Boiling point = 1336−1342°C; Freezing/Melting point = 181°C. Hazard Identification (based on NFPA-704 M Rating System): Health 3, Flammability 2, Reactivity 2W. Violent reaction with water.

Potential Exposure: Compound Description: Drug, Mutagen; Reproductive Effector. Lithium is used in inorganic syntheses, manufacture of storage batteries, heat transfer liquids, and metal alloys.

Incompatibilities: Violent reaction with water, forming flammable hydrogen gas and corrosive lithium hydroxide, a strong caustic solution. Heating may cause violent combustion or explosion. Finely divided particles or powdered form may ignite spontaneously in air. Contact with air forms corrosive fumes of lithium hydroxide. Violent reaction with oxidizers, acetonitrile, nitric acid, arsenic, bromobenzene, carbon tetrachloride, hydrocarbons, halogens, halons, sulfur, and many other substances. Forms impact- and friction-sensitive mixtures with bromobenzene, carbon tetrabromide, chloroform (weak explosion), iodoform, halogens, halocarbons, methyl dichloride, methyl diiodide, and other substances. Attacks plastics, rubber, ceramic materials, concrete, sand, and metal alloys: cobalt, iron, manganese, nickel.

Permissible Exposure Limits in Air
Protective Action Criteria (PAC)
TEEL-0: 4 mg/m^3
PAC-1: 12.5 mg/m^3
PAC-2: 75 mg/m^3
PAC-3: 400 mg/m^3

Determination in Air: Use OSHA Analytical Method ID-121; NIOSH Analytical Method #8005, Elements in blood or tissue.

Routes of Entry: Inhalation, ingestion, eye and/or skin contact.

Harmful Effects and Symptoms

Short Term Exposure: Corrosive. Contact can cause severe skin and eye burns. Inhalation can irritate the respiratory tract causing coughing, and/or shortness of breath. Higher exposures can cause pulmonary edema, a medical emergency that can be delayed for several hours. This can cause death. Exposure to lithium can cause loss of appetite, nausea, tremor, muscle twitches, apathy, convulsions, coma, and death.

Long Term Exposure: Exposure can cause loss of appetite, nausea, vomiting, diarrhea, and abdominal pain, headache, muscle weakness, loss of coordination, confusion, seizures, and coma. May affect the thyroid gland causing goiter. May affect the kidneys and heart function. Exposure may cause an allergy to develop, affecting the skin, blood vessels, and/or possibly the lungs.

Points of Attack: Lungs, thyroid, kidneys, heart, skin.

Medical Surveillance: Before beginning employment and at regular times after that, for those with frequent or potentially high exposures, the following are recommended: lung function tests. If symptoms develop or overexposure is suspected, the following may be useful: EKG, blood test for lithium. Thyroid function tests. Evaluation by a qualified allergist, including careful exposure history and special testing, may help diagnose skin allergy. Consider chest X-ray after acute overexposure.

First Aid: If this chemical gets into the eyes, remove any contact lenses at once and irrigate immediately for at least 15 min, occasionally lifting upper and lower lids. Seek medical attention immediately. If this chemical contacts the skin, remove contaminated clothing and wash immediately with soap and water. Seek medical attention immediately. If this chemical has been inhaled, remove from exposure, begin rescue breathing (using universal precautions, including resuscitation mask) if breathing has stopped and CPR if heart action has stopped. Transfer promptly to a medical facility. When this chemical has been swallowed, get medical attention. Give large quantities of water and induce vomiting. Do not make an unconscious person vomit. Medical observation is recommended for 24–48 h after breathing overexposure, as pulmonary edema may be delayed. As first aid for pulmonary edema, a doctor or authorized paramedic may consider administering a corticosteroid spray.

Personal Protective Methods: Wear protective gloves and clothing to prevent any reasonable probability of skin contact. Safety equipment suppliers/manufacturers can provide recommendations on the most protective glove/clothing material for your operation. All protective clothing (suits, gloves, footwear, headgear) should be clean, available each day, and put on before work. Contact lenses should not be worn when working with this chemical. Wear dust-proof chemical goggles and face shield unless full face-piece respiratory protection is worn. Employees should wash immediately with soap when skin is wet or contaminated. Provide emergency showers and eyewash.

Respirator Selection: Where there is potential for exposure to lithium, use a NIOSH/MSHA- or European Standard EN149-approved full face-piece respirator with a high-efficiency particulate filter. Greater protection is provided by a powdered-air purifying respirator. *Where there is potential for high exposures*, use a NIOSH/MSHA- or European Standard EN149-approved supplied-air respirator with a full face-piece operated in the positive-pressure mode, or with a full face-piece, hood, or helmet in the continuous-flow mode; or use a NIOSH/MSHA- or European Standard EN149-approved self-contained breathing apparatus with a full face-piece operated in pressure-demand or other positive-pressure mode.

Storage: (1) Color Code—Yellow Stripe: Reactivity Hazard; Store separately in an area isolated from flammables, combustibles, or other yellow coded materials. Reacts with body moisture, forming corrosive lithium hydroxide: (2) Color Code—White: Corrosive or Contact Hazard; Store separately in a corrosion-resistant location. (3) Color Code—Blue: Health Hazard/Poison: Store in a secure poison location. Prior to working with this chemical you should be trained on its proper handling and storage. Protect storage containers from physical damage. Lithium must be stored to avoid contact with water, halogenated compounds, carbon dioxide, oxidizers, strong acids, alcohols, metals, chlorinated arsenic compounds, hydrocarbons, sulfur, acetonitrile, nitrogen, and many other materials, since violent reactions occur. Store in tightly closed containers under kerosene or neutral oil in a cool, well-ventilated area. Sources of ignition, such as smoking and open flames, are prohibited where lithium is used, handled or stored in a manner that could create a potential fire or explosion hazard. Wherever lithium is used, handled, manufactured, or stored, use explosion-proof electrical equipment and fittings.

Shipping: This compound requires a shipping label of "DANGEROUS WHEN WET." It falls in Hazard Class 4.3 and Packing Group I. For Metal powder, flammable, n.o.s. the required label is "SPONTANEOUSLY COMBUSTIBLE." They fall into Hazard Class 4.2 and Packing Group II.

Spill Handling: Evacuate persons not wearing protective equipment from area of spill or leak until cleanup is complete. Remove all ignition sources. Collect powdered material in the most convenient and safe manner and deposit in sealed containers. Ventilate area after cleanup is complete. Keep lithium out of a confined space, such as a sewer, because of the possibility of an explosion, unless the sewer is designed to prevent the buildup of explosive concentrations. It may be necessary to contain and dispose of this chemical as a hazardous waste. If material or contaminated runoff enters waterways, notify downstream users of potentially contaminated waters. Contact your local or federal environmental protection agency for specific recommendations. If employees are required to clean up spills, they must be properly trained and equipped. OSHA 1910.120(q) may be applicable.

Fire Extinguishing: This chemical is a highly reactive, combustible solid that is difficult to extinguish. *Do not use water*, foam, dry chemical, halogenated hydrocarbons, CO_2 on fire or adjacent fire. Use Class D extinguishers, dry clay, dry graphite, limestone, or appropriate special metal fire-extinguishing powder. Poisonous gases, including lithium oxide, are produced in fire. If material or contaminated runoff enters waterways, notify downstream users of potentially contaminated waters. Notify local health and fire officials and pollution control agencies. From a secure, explosion-proof location, use water spray to cool exposed containers. If cooling streams are ineffective (venting sound increases in volume and pitch, tank discolors, or shows any signs of deforming), withdraw immediately to a secure position. If employees are expected to fight fires, they must be trained and equipped in OSHA 1910.156. The only respirators recommended for firefighting are self-contained breathing apparatuses that have full face-pieces and are operated in a pressure-demand or other positive-pressure mode.

Reference

New Jersey Department of Health and Senior Services. (November, 1999). *Hazardous Substances Fact Sheet: Lithium.* Trenton, NJ

Lithium aluminum hydride   L:0285

Molecular Formula: AlH_4Li
Common Formula: $LiAlH_4$

Synonyms: Aluminum lithium hydride; Lithium alanate; Lithium aluminohydride; Lithium aluminum tetrahydride; Lithium tetrahydroaluminate
CAS Registry Number: 16853-85-3
RTECS® Number: BD0100000
UN/NA & ERG Number: UN1410 (dry)/138; UN1411 (ether solution)/138
EC Number: 240-877-9 [*Annex I Index No.:* 001-002-00-4]
Regulatory Authority and Advisory Bodies
Canada, WHMIS, Ingredients Disclosure List Concentration 1.0%.
European/International Regulations: Hazard Symbol: F, C; Risk phrases: R15; R35; Safety phrases: S1/2; S7/8; S26; S36/37/39; S43; S45 (see Appendix 4).
WGK (German Aquatic Hazard Class): No value assigned.
Description: Lithium aluminum hydride is a white to gray powder. A combustible solid. Decomposes at 125°C. Hazard Identification (based on NFPA-704 M Rating System): Health 2, Flammability 3, Reactivity 2W. Reacts with water.
Potential Exposure: This material is used as a catalyst and as a specialty reducing agent in organic synthesis.
Incompatibilities: Combustible solid. Can ignite spontaneously in moist air or heat. Decomposes on heating at 125°C forming aluminum, lithium hydride, and flammable hydrogen gas. A strong reducing agent. Violent reaction with water, oxidizers, alcohols, acids, dimethylether, ethers, tetrahydrofuran, benzoyl peroxide, boron trifluoride etherate.
Permissible Exposure Limits in Air
Protective Action Criteria (PAC)
TEEL-0: 2.81 mg/m^3
PAC-1: 2.81 mg/m^3
PAC-2: 7.5 mg/m^3
PAC-3: 35 mg/m^3
Permissible Concentration in Water: No criteria set. Reacts violently with water.
Routes of Entry: Inhalation, ingestion, eyes and/or skin.
Harmful Effects and Symptoms
Short Term Exposure: Lithium aluminum hydride can affect you when breathed in. Contact can cause severe burns of the eyes and skin. Exposure can irritate the eyes, nose, throat, and lungs, causing coughing and sneezing. Higher exposures can cause pulmonary edema, a medical emergency that can be delayed for several hours. This can cause death. Exposure can cause loss of appetite, nausea, confusion, tremor, and muscle twitching. High exposure can cause coma and death.
Long Term Exposure: Can cause loss of appetite, nausea, vomiting, diarrhea and abdominal pain, headache, muscle weakness, loss of coordination, confusion, seizures, and coma. Can affect the thyroid gland causing goiter. May cause kidney damage.
Points of Attack: Thyroid gland, kidneys.
Medical Surveillance: For those with frequent or potentially high exposure, the following are recommended before beginning work and at regular times after that: lung

function tests. Thyroid function tests. If symptoms develop or overexposure is suspected, the following may be useful: serum lithium level. Consider chest X-ray after acute overexposure.

First Aid: If this chemical gets into the eyes, remove any contact lenses at once and irrigate immediately for at least 15 min, occasionally lifting upper and lower lids. Seek medical attention immediately. If this chemical contacts the skin, remove contaminated clothing and wash immediately with soap and water. Seek medical attention immediately. If this chemical has been inhaled, remove from exposure, begin rescue breathing (using universal precautions, including resuscitation mask) if breathing has stopped and CPR if heart action has stopped. Transfer promptly to a medical facility. When this chemical has been swallowed, get medical attention. Give large quantities of water and induce vomiting. Do not make an unconscious person vomit. Medical observation is recommended for 24–48 h after breathing overexposure, as pulmonary edema may be delayed. As first aid for pulmonary edema, a doctor or authorized paramedic may consider administering a corticosteroid spray.

Personal Protective Methods: Wear protective gloves and clothing to prevent any reasonable probability of skin contact. Safety equipment suppliers/manufacturers can provide recommendations on the most protective glove/clothing material for your operation. All protective clothing (suits, gloves, footwear, headgear) should be clean, available each day, and put on before work. Contact lenses should not be worn when working with this chemical. Wear dust-proof chemical goggles and face shield unless full face-piece respiratory protection is worn. Employees should wash immediately with soap when skin is wet or contaminated. Provide emergency showers and eyewash.

Respirator Selection: Where there is potential for exposures *over 5 mg/m³*, use a NIOSH/MSHA- or European Standard EN149-approved full face-piece respirator with a high-efficiency particulate filter. More protection is provided by a full face-piece respirator than by a half-mask respirator, and even greater protection is provided by a powered-air purifying respirator. Particulate filters must be checked every day before work for physical damage, such as rips or tears, and replaced as needed. *Where there is potential for high exposures*, use a NIOSH/MSHA- or European Standard EN149-approved supplied-air respirator with a full face-piece operated in the positive-pressure mode, or with a full face-piece, hood, or helmet in the continuous-flow mode; or use a NIOSH/MSHA- or European Standard EN149-approved self-contained breathing apparatus with a full face-piece operated in pressure-demand or other positive-pressure mode.

Storage: (1) Color Code—Red Stripe: Flammability Hazard: Do not store in the same area as other flammable materials. (2) Color Code—Yellow Stripe (*strong reducing agent*): Reactivity Hazard; Store separately in an area isolated from flammables, combustibles, or other yellow coded materials. Prior to working with this chemical you should be trained on its proper handling and storage. Lithium aluminum hydride must be stored to avoid contact with water, air, acids, alcohols, benzoyl peroxide, boron trifluoride etherate, ethers, tetrahydrofuran, and strong oxidizers (such as chlorine, bromine, and fluorine), since violent reactions occur. Sources of ignition, such as smoking and open flames are prohibited where lithium aluminum hydride is handled, used, or stored. Use only nonsparking tools and equipment, especially when opening and closing containers of lithium aluminum hydride. Wherever lithium aluminum hydride is used, handled, manufactured, or stored, use explosion-proof electrical equipment and fittings. Open only in inert atmospheres or very low humidity rooms.

Shipping: Lithium aluminum hydride (dry) requires a shipping label of "DANGEROUS WHEN WET." It falls in Hazard Class 4.3 and Packing Group I. Lithium aluminum hydride, ethereal, requires a shipping label of "DANGEROUS WHEN WET, FLAMMABLE LIQUID." It falls in Hazard Class 4.3 and Packing Group I.

Spill Handling: Evacuate persons not wearing protective equipment from area of spill or leak until cleanup is complete. Remove all ignition sources. Collect powdered material in the most convenient and safe manner and deposit in sealed containers. Ventilate area after cleanup is complete. It may be necessary to contain and dispose of this chemical as a hazardous waste. If material or contaminated runoff enters waterways, notify downstream users of potentially contaminated waters. Contact your local or federal environmental protection agency for specific recommendations. If employees are required to clean up spills, they must be properly trained and equipped. OSHA 1910.120(q) may be applicable.

Fire Extinguishing: This chemical is a combustible solid. *Do not use water*, CO₂, or foam. Use Class D extinguishers, dry sand, dry clay, dry limestone, dry graphite. Poisonous gases, including lithium hydride and aluminum oxide, are produced in fire. If material or contaminated runoff enters waterways, notify downstream users of potentially contaminated waters. Notify local health and fire officials and pollution control agencies. Containers may explode in fire. Storage containers and parts of containers may rocket great distances, in many directions. Fire may restart after it has been extinguished. From a secure, explosion-proof location, use water spray to cool exposed containers. If cooling streams are ineffective (venting sound increases in volume and pitch, tank discolors, or shows any signs of deforming), withdraw immediately to a secure position. If employees are expected to fight fires, they must be trained and equipped in OSHA 1910.156. The only respirators recommended for firefighting are self-contained breathing apparatuses that have full face-pieces and are operated in a pressure-demand or other positive-pressure mode.

Reference

New Jersey Department of Health and Senior Services. (November 1999). *Hazardous Substances Fact Sheet: Lithium Aluminum Hydride.* Trenton, NJ

Lithium carbonate   L:0290

Molecular Formula: Li$_2$CO$_3$
Synonyms: Camcolit; Candamide; Carbolith; Carbolithium; Carbonato de litio (Spanish); Carbonic acid, dilithium salt; Carbonic acid lithium salt; Ceglution; CP-15,467-61; Dilithium carbonate; Eskalith; Hypnorex; Limas; Liskonum; Litard; Lithane; Lithea; Lithicarb; Lithinate; Lithium phasal; Lithizine; Lithobid; Litho-carb; Lithonate; Lithotabs; Manialith; Neurolepsin; NSC 16895; PFI-lithium; PFL-lithium; Plenur; Priadel; Quilonorm; Quilonum retard
CAS Registry Number: 554-13-2
RTECS® Number: OJ5800000
UN/NA & ERG Number: UN2811 (toxic solid, organic, n.o.s.)/154
EC Number: 209-062-5
Regulatory Authority and Advisory Bodies
Listed in the TSCA inventory.
FDA—proprietary drug.
EPCRA Section 313 Form R *de minimis* concentration reporting level: 1.0%.
California Proposition 65 Developmental/Reproductive toxin 1/1/91.
WGK (German Aquatic Hazard Class): 1—Low hazard to waters.
Description: Lithium carbonate is a white hygroscopic powder. Molecular weight = 73.89; Boiling point = 1310°C (decomposes below BP); Freezing/Melting point = 618–735°C. Hazard Identification (based on NFPA-704 M Rating System): Health 1, Flammability 0, Reactivity 1. Slightly soluble in water.
Potential Exposure: Compound Description: Tumorigen, Drug, Mutagen; Reproductive Effector; Human Data. Lithium carbonate is used in treatment of manic-depressive psychoses, to make ceramics and porcelain glaze, varnishes, dyes, pharmaceuticals, coating of arc-welding electrodes, battery alloys, nucleonics, luminescent paints, glass ceramics, lubricating greases, in aluminum production.
Incompatibilities: The aqueous solution is a strong base. Reacts violently with acids and fluorine. Incompatible with oxidizers, moisture. Corrodes aluminum, copper, zinc.
Permissible Exposure Limits in Air
Protective Action Criteria (PAC)
TEEL-0: 0.2 mg/m^3
PAC-1: 0.6 mg/m^3
PAC-2: 4 mg/m^3
PAC-3: 200 mg/m^3
Routes of Entry: Inhalation.
Harmful Effects and Symptoms
Short Term Exposure: LD$_{50}$ = (oral-rat) 525 mg/kg. Contact causes skin, eye, and respiratory tract irritation. Inhalation can cause nausea, vomiting, diarrhea, and abdominal pain. Can cause lung irritation. Higher exposures can cause pulmonary edema, a medical emergency that can be delayed for several hours. This can cause death. Exposure can cause headache, muscle weakness, muscle twitching, blurred vision, loss of coordination, confusion, seizures, and coma. May affect the central nervous system; cardiovascular, and gastrointestinal systems if ingested.
Long Term Exposure: May cause damage to the developing fetus. May cause skin allergy and dermatitis. High exposure can cause enlarged thyroid (goiter). May damage the stomach, kidneys, and may affect the heart function. May cause reproductive toxicity in humans.
Points of Attack: Skin, thyroid, kidneys, heart.
Medical Surveillance: Blood tests for lithium level. Kidney function tests. Thyroid function tests. Consider chest X-ray following acute overexposure. Evaluation by a qualified allergist.
First Aid: If this chemical gets into the eyes, remove any contact lenses at once and irrigate immediately for at least 15 min, occasionally lifting upper and lower lids. Seek medical attention immediately. If this chemical contacts the skin, remove contaminated clothing and wash immediately with soap and water. Seek medical attention immediately. If this chemical has been inhaled, remove from exposure, begin rescue breathing (using universal precautions, including resuscitation mask) if breathing has stopped and CPR if heart action has stopped. Transfer promptly to a medical facility. When this chemical has been swallowed, get medical attention. Give large quantities of water and induce vomiting. Do not make an unconscious person vomit. Medical observation is recommended for 24–48 h after breathing overexposure, as pulmonary edema may be delayed. As first aid for pulmonary edema, a doctor or authorized paramedic may consider administering a corticosteroid spray.
Personal Protective Methods: Wear protective gloves and clothing to prevent any reasonable probability of skin contact. Safety equipment suppliers/manufacturers can provide recommendations on the most protective glove/clothing material for your operation. All protective clothing (suits, gloves, footwear, headgear) should be clean, available each day, and put on before work. Contact lenses should not be worn when working with this chemical. Wear dust-proof chemical goggles and face shield unless full face-piece respiratory protection is worn. Employees should wash immediately with soap when skin is wet or contaminated. Provide emergency showers and eyewash. Specific engineering controls are required for drug manufacture by the Food and Drug Administration. Refer to FDA regulation for Good Manufacturing Practices 21CFR210.
Respirator Selection: Follow regulations in OSHA 29CFR1910.134 or European Standard EN149. Use a NIOSH/MSHA- or European Standard EN149-approved respirator; or use an approved supplied-air respirator with a full face-piece operated in the positive-pressure mode, or with a full face-piece, hood, or helmet in the continuous-flow mode; or use a NIOSH/MSHA- or European Standard EN149-approved self-contained breathing apparatus with a

full face-piece operated in pressure-demand or other positive-pressure mode.

Storage: Color Code—Blue: Health Hazard/Poison: Store in a secure poison location. Prior to working with this chemical you should be trained on its proper handling and storage. Store in tightly closed containers in a cool, well-ventilated area away from moisture, fluorine, oxidizers, acids. Where possible, automatically transfer from drums or other storage containers to process containers.

Shipping: Label: "POISONOUS/TOXIC MATERIALS." It falls in Hazard Class 6.1 and Packing Group III.

Spill Handling: Evacuate persons not wearing protective equipment from area of spill or leak until cleanup is complete. Remove all ignition sources. Collect powdered material in the most convenient and safe manner and deposit in sealed containers. Ventilate area after cleanup is complete. It may be necessary to contain and dispose of this chemical as a hazardous waste. If material or contaminated runoff enters waterways, notify downstream users of potentially contaminated waters. Contact your local or federal environmental protection agency for specific recommendations. If employees are required to clean up spills, they must be properly trained and equipped. OSHA 1910.120(q) may be applicable.

Fire Extinguishing: This chemical is a combustible solid. Use dry chemical, carbon dioxide, water spray, or alcohol foam extinguishers. Poisonous gases are produced in fire. If material or contaminated runoff enters waterways, notify downstream users of potentially contaminated waters. Notify local health and fire officials and pollution control agencies. From a secure, explosion-proof location, use water spray to cool exposed containers. If cooling streams are ineffective (venting sound increases in volume and pitch, tank discolors, or shows any signs of deforming), withdraw immediately to a secure position. If employees are expected to fight fires, they must be trained and equipped in OSHA 1910.156. The only respirators recommended for firefighting are self-contained breathing apparatuses that have full face-pieces and are operated in a pressure-demand or other positive-pressure mode.

Reference
New Jersey Department of Health and Senior Services. (September 1998). *Hazardous Substances Fact Sheet: Lithium Carbonate.* Trenton, NJ

Lithium chromate   L:0300

Molecular Formula: $CrH_2O_4 \cdot 2Li$
Synonyms: Lithium chromate(VI); Chromic acid, dilithium salt; Chromium lithium oxide; Dilithium chromate
CAS Registry Number: 14307-35-8
RTECS® Number: GB2915000
UN/NA & ERG Number: UN3288 (Toxic solid, inorganic, n.o.s.)/151
EC Number: 238-244-7

Regulatory Authority and Advisory Bodies
Carcinogenicity: IARC: Human Sufficient Evidence; Animal Sufficient Evidence, *carcinogenic to humans,* Group 1, 1997; NTP: 11th Report on Carcinogens, 2004: Known to be a human carcinogen; EPA *(inhalation):* Known human carcinogen; EPA *(oral):* Not Classifiable as to human carcinogenicity; NTP: Known to be a human carcinogen.

California Proposition 65 Chemical: Cancer (developmental).

Air Pollutant Standard Set. See below, "Permissible Exposure Limits in Air" section.

Clean Air Act: Hazardous Air Pollutants (Title I, Part A, Section 112) as chromium compounds.

Clean Water Act: Section 311 Hazardous Substances/RQ 40CFR117.3 (same as CERCLA, see below); Toxic Pollutant (Section 401.15); Section 313 Water Priority Chemicals (57FR41331, 9/9/92).

RCRA, 40CFR261, Appendix 8 Hazardous Constituents, as chromium compounds, waste number not listed.

Reportable Quantity (RQ): 10 lb (4.54 kg).

EPCRA (Section 313): Includes any unique chemical substance that contains chromium as part of that chemical's infrastructure. Form R *de minimis* concentration reporting level: Chromium(VI) compounds: 0.1%.

California Proposition 65 Chemical: *(hexavalent chromium)* Cancer 2/27/87; Developmental/Reproductive toxin (male, female) 12/19/08.

Canada, WHMIS, Ingredients Disclosure List Concentration 0.1%.

European/International Regulations: Hazard Symbol: E, T +, N; Risk Phrases: R45; R46; R60; R61; R2; R8; R21; R25; R26; R34; R42/43; R48/23; R50/53; Safety phrases: S53; S45; S60; S61 (see Appendix 4).

WGK (German Aquatic Hazard Class): 3—Severe hazard to waters.

Description: Lithium chromate is a yellow crystalline powder. Molecular weight = 165.92; 129.87 (lithium chromate (VI)); 131.90 *(dilithium chromate);* Melting point = 75°C. Hazard Identification (based on NFPA-704 M Rating System): Health 2, Flammability 0, Reactivity 0. Soluble in water.

Potential Exposure: Used as a corrosion inhibitor, heat-transfer agent, and oxidizing agent in leather and metal finishing. Also used in photography, wood preservatives, batteries, safety matches, and cement.

Incompatibilities: Aqueous solution is caustic. An oxidizer; strong reaction with hydrazine, chromic acid, sulfur, reducing agents, combustibles, organic materials, acids. Attacks plastics and aluminum.

Permissible Exposure Limits in Air
Protective Action Criteria (PAC)
TEEL-0: 0.0125 mg/m^3
PAC-1: 1 mg/m^3
PAC-2: 7.5 mg/m^3
PAC-3: 37.5 mg/m^3

As chromium(VI) inorganic soluble compounds
OSHA PEL: 0.005 mg[Cr(VI)]/m^3 TWA Concentration. See 29CFR1910.1026.
NIOSH REL: 0.001 mg[Cr]/m^3 TWA, potential carcinogen, limit exposure to lowest feasible level. NIOSH considers all Cr(VI) compounds (including chromic acid, *tert*-butyl chromate, zinc chromate, and chromyl chloride) to be potential occupational carcinogens. See *NIOSH Pocket Guide*, Appendix A & C.
ACGIH TLV®[1]: 0.05 mg[Cr]/m^3 TWA, Confirmed Human Carcinogen; BEI issued.
NIOSH IDLH: 15 mg[Cr(VI)]/m^3.
DFG MAK: [skin] Danger of skin sensitization; Carcinogen Category 1; TRK: 0.05 mg[Cr]/m^3; 20 µg/L [Cr] in urine at end-of-shift.
Determination in Air: Use NIOSH Analytical Method (IV) #7024, Chromium.
Permissible Concentration in Water: For the protection of freshwater aquatic life: *Hexavalent chromium:* 0.29 µg/L as a 24-h average, never to exceed 21.0 µg/L. For the protection of saltwater aquatic life: Hexavalent chromium: 18 µg/L as a 24-h average, never to exceed 1260 µg/L. *To protect human health:* hexavalent chromium 50 µg/L according to EPA.[6]
US EPA[49] has set a long-term health advisory for adults of 0.84 mg/L and a lifetime health advisory of 0.12 mg/L (120 µg/L) for chromium. EPA's maximum drinking water level (MCL) is 0.1 mg/L.[62]
Germany, Canada, EEC, and WHO[35] have set a limit of 0.05 mg/L in drinking water.
The states of Maine and Minnesota have set guidelines for chromium in drinking water[61] as 50 µg/L for Maine and 120 µg/L for Minnesota.
Determination in Water: Total chromium may be determined by digestion followed by atomic absorption, or by colorimetry (diphenylcarbazide); or by inductively coupled plasma (CP) optical emission spectrometry. Chromium(VI) may be determined by extraction and atomic absorption or colorimetry (using diphenylhydrazide). Dissolved total Cr or Cr(VI) may be determined by 0.45-µm filtration followed by the above-cited methods.[49]
Routes of Entry: Inhalation, eye and/or skin.
Harmful Effects and Symptoms
Short Term Exposure: Contact causes severe skin and eye irritation and burns. Inhalation can irritate the nose and throat causing coughing and wheezing.
Long Term Exposure: Repeated exposure can cause loss of appetite, nausea, tremor, and convulsions. Lithium chromate is a hexavalent chromium compound and a possible human carcinogen. Handle with care. Related chromium compounds are teratogenic. Can irritate the lungs; may cause bronchitis to develop. May cause skin allergy. Repeated exposure may cause personality changes of depression, anxiety, or irritability. Prolonged exposure may cause deep slow-healing ulcers on the skin, and a sore or hole in nasal septum. May damage the liver and kidneys.

Points of Attack: Lungs, liver, kidneys, skin.
Medical Surveillance: NIOSH lists the following tests: Blood gas analysis, complete blood count, chest X-ray, electrocardiogram, liver function tests, pulmonary function tests, sputum cytology, urine (chemical/metabolite), urinalysis (routine), white blood cell count/differential. Lung function tests. Evaluation by a qualified allergist. Liver and kidney function tests. Check skin daily for blisters or little bumps, the first sign of "chrome ulcers." If not treated early, these can last for years following exposure.
First Aid: If this chemical gets into the eyes, remove any contact lenses at once and irrigate immediately for at least 15 min, occasionally lifting upper and lower lids. Seek medical attention immediately. If this chemical contacts the skin, remove contaminated clothing and wash immediately with soap and water. Seek medical attention immediately. If this chemical has been inhaled, remove from exposure, begin rescue breathing (using universal precautions, including resuscitation mask) if breathing has stopped and CPR if heart action has stopped. Transfer promptly to a medical facility. When this chemical has been swallowed, get medical attention. Give large quantities of water and induce vomiting. Do not make an unconscious person vomit.
Personal Protective Methods: Wear protective gloves and clothing to prevent any reasonable probability of skin contact. Safety equipment suppliers/manufacturers can provide recommendations on the most protective glove/clothing material for your operation. All protective clothing (suits, gloves, footwear, headgear) should be clean, available each day, and put on before work. Contact lenses should not be worn when working with this chemical. Wear dust-proof chemical goggles and face shield unless full face-piece respiratory protection is worn. Employees should wash immediately with soap when skin is wet or contaminated. Provide emergency showers and eyewash.
Respirator Selection: NIOSH, as chromates: *at any concentrations above the NIOSH REL, or where there is no REL, at any detectable concentration:* SCBAF: Pd,Pp (APF = 10,000) (any self-contained breathing apparatus that has a full face-piece and is operated in a pressure-demand or other positive-pressure mode) or SaF: Pd,Pp: ASCBA (APF = 10,000) (any supplied-air respirator that has a full face-piece and is operated in a pressure-demand or other positive-pressure mode in combination with an auxiliary, self-contained breathing apparatus operated in a pressure-demand or other positive-pressure mode). *Escape:* 100F (APF = 50) (any air purifying, full-face-piece respirator with an N100, R100, or P100 filter) or SCBAE (any appropriate escape-type, self-contained breathing apparatus).
Storage: Color Code—Blue: Health Hazard/Poison: Store in a secure poison location. Prior to working with this chemical you should be trained on its proper handling and storage. A regulated, marked area should be established where this chemical is handled, used, or stored in compliance with OSHA Standard 1910.1045. Store in tightly closed containers in a cool, well-ventilated area away from

acids, hydrazine, chromic acid, combustible materials, sulfur, aluminum, plastics, and reducing agents.

Shipping: Toxic solid, inorganic, n.o.s. requires a shipping label of "POISONOUS/TOXIC MATERIALS." It falls in Hazard Class 6.1 and Packing Group III.

Spill Handling: Evacuate persons not wearing protective equipment from area of spill or leak until cleanup is complete. Remove all ignition sources. Collect powdered material in the most convenient and safe manner and deposit in sealed containers. Ventilate area after cleanup is complete. It may be necessary to contain and dispose of this chemical as a hazardous waste. If material or contaminated runoff enters waterways, notify downstream users of potentially contaminated waters. Contact your local or federal environmental protection agency for specific recommendations. If employees are required to clean up spills, they must be properly trained and equipped. OSHA 1910.120(q) may be applicable.

Fire Extinguishing: This chemical is a noncombustible solid. Use extinguishing agents suitable for surrounding fire. Poisonous gases, including lithium oxide, are produced in fire. If material or contaminated runoff enters waterways, notify downstream users of potentially contaminated waters. Notify local health and fire officials and pollution control agencies. Containers may explode in fire. From a secure, explosion-proof location, use water spray to cool exposed containers. If cooling streams are ineffective (venting sound increases in volume and pitch, tank discolors, or shows any signs of deforming), withdraw immediately to a secure position. If employees are expected to fight fires, they must be trained and equipped in OSHA 1910.156. The only respirators recommended for firefighting are self-contained breathing apparatuses that have full face-pieces and are operated in a pressure-demand or other positive-pressure mode.

Reference
New Jersey Department of Health and Senior Services. (August 1998). *Hazardous Substances Fact Sheet: Lithium Chromate.* Trenton, NJ

Lithium hydride   L:0310

Molecular Formula: HLi
Common Formula: LiH
Synonyms: Hydrure de lithium (French); Hydruro de litio (Spanish); Lithium monohydride
CAS Registry Number: 7580-67-8
RTECS® Number: OJ6300000
UN/NA & ERG Number: UN1414/138; UN2805 (fused, solid)/138
EC Number: 231-484-3
Regulatory Authority and Advisory Bodies
Air Pollutant Standard Set. See below, "Permissible Exposure Limits in Air" section.
SUPERFUND/EPCRA 40CFR355, Extremely Hazardous Substances: TPQ = 100 lb (45.4 kg).

Reportable Quantity (RQ): 100 lb (45.4 kg).
Canada, WHMIS, Ingredients Disclosure List Concentration 1.0%.
European/International Regulations: not listed in Annex 1.
WGK (German Aquatic Hazard Class): 2—Hazard to waters.

Description: Lithium hydride is an off-white to grayish, translucent, odorless solid or white powder that darkens rapidly on exposure to light. Molecular weight = 7.95; Specific gravity (H_2O:1) = 0.78; Boiling point = 850°C (decomposes below BP); Freezing/Melting point = 689°C; Autoignition temperature = 200°C. Hazard Identification (based on NFPA-704 M Rating System): Health 3, Flammability 4, Reactivity 2. A combustible solid that can form airborne dust clouds which may explode on contact with flame, heat, or oxidizers.

Potential Exposure: Compound Description: Human Data. Lithium hydride is used in preparation of lithium aluminum hydride, as a desiccant, used in hydrogen generators and in organic synthesis as a reducing agent and condensing agent with ketones and acid esters, and reportedly used in thermonuclear weapons.

Incompatibilities: Incompatible with oxidizers, halogenated hydrocarbons; acids can cause fire and explosion. Reacts with water, forming caustic lithium hydroxide and flammable hydrogen gas; reaction may cause ignition. May ignite *spontaneously* in moist air and may reignite after fire is extinguished. Reacts with water to form hydrogen and lithium hydroxide. Powdered form and liquid oxygen form an explosive compound. Decomposes exothermically on contact with acids and upon heating to about 500°C, producing flammable hydrogen gas. Reacts with carboxylic acids, lower alcohols, chlorine, and ammonia (at 400°C), forming explosive hydrogen gas.

Permissible Exposure Limits in Air
OSHA PEL: 0.025 mg/m³ TWA.
NIOSH REL: 0.025 mg/m³ TWA.
ACGIH TLV®[1]: 0.025 mg/m³ TWA.
NIOSH IDLH: 0.5 mg/m³.
Protective Action Criteria (PAC)*
TEEL-0: 0.025 mg/m³
PAC-1: **0.025** mg/m³
PAC-2: **0.100** mg/m³
PAC-3: **0.500** mg/m³
*AEGLs (Acute Emergency Guideline Levels) & ERPGs (Emergency Response Planning Guideline) are in **bold face**.
DFG MAK: No numerical value established. Data may be available.
Australia: TWA 0.025 mg/m³, 1993; Austria: MAK 0.025 mg/m³, 1999; Belgium: TWA 0.025 mg/m³, 1993; Denmark: TWA 0.025 mg/m³, 1999; Finland: TWA 0.025 mg/m³; STEL 0.075 mg/m³, 1999; France: VME 0.025 mg/m³, 1999; the Netherlands: MAC-TGG 0.025 mg/m³, 2003; the Philippines: TWA 0.025 mg/m³, 1993; Poland: MAC (TWA) 0.025 mg/m³, 1999; Switzerland: MAK-W 0.025 mg/m³, 1999; Turkey: TWA 0.025 mg/m³,

1993; United Kingdom: TWA 0.025 mg/m^3, 2000; Argentina, Bulgaria, Columbia, Jordan, South Korea, New Zealand, Singapore, Vietnam: ACGIH TLV®: TWA 0.025 mg/m^3. Several states have set guidelines or standards for lithium hydride in ambient air[60] ranging from 0.4 μg/m^3 (Virginia) to 0.5 μg/m^3 (Connecticut) to 2.0 μg/m^3 (North Dakota) to 6.0 μg/m^3 (Nevada).

Determination in Air: No method available.

Permissible Concentration in Water: No criteria set, but EPA[32] has suggested a permissible ambient goal of 0.3 μg/L based on health effects.

Routes of Entry: Inhalation, ingestion, skin and/or eye contact.

Harmful Effects and Symptoms

Short Term Exposure: Lithium hydride is an alkaline-corrosive agent. Contact with eyes may result in severe damage to the cornea, conjunctiva, and blood vessels. Extreme caution is advised. Acute exposure to lithium hydride may result in irritation and burning of the skin, eyes, and mucous membranes. Increased salivation, dysphagia (difficulty in swallowing), abdominal pain, and spontaneous vomiting may occur. Stridor (high-pitched, noisy respirations), dyspnea (shortness of breath), and pulmonary edema are also common. Apathy and mental confusion may develop, with progression to coma and death.

Long Term Exposure: Lithium can cause loss of appetite, nausea, vomiting, diarrhea, and abdominal pain, headache, muscle weakness, loss of coordination, confusion, seizures, and coma. Can affect the thyroid gland causing goiter. May cause kidney damage.

Points of Attack: Eyes, skin, respiratory system, central nervous system, thyroid, kidneys.

Medical Surveillance: Consider the points of attack in preplacement and periodic physical examinations. Thyroid function tests. Kidney function tests. Lung function tests. Consider chest X-ray following acute overexposure.

First Aid: If this chemical gets into the eyes, remove any contact lenses at once and irrigate immediately for at least 15 min, occasionally lifting upper and lower lids. Seek medical attention immediately. If this chemical contacts the skin, remove contaminated clothing and wash immediately with soap and water. Seek medical attention immediately. If this chemical has been inhaled, remove from exposure, begin rescue breathing (using universal precautions, including resuscitation mask) if breathing has stopped and CPR if heart action has stopped. Transfer promptly to a medical facility. When this chemical has been swallowed, get medical attention. If victim is conscious, administer water or milk. Do not induce vomiting. Medical observation is recommended for 24—48 h after breathing overexposure, as pulmonary edema may be delayed. As first aid for pulmonary edema, a doctor or authorized paramedic may consider administering a corticosteroid spray.

Personal Protective Methods: Wear appropriate clothing to prevent any possibility of contact with air of >0.1 mg/m^3 content. Wear eye protection to prevent any possibility of eye contact. Provide emergency showers and eyewash if air containing >0.5 mg/m^3 is involved. Wear protective gloves and clothing to prevent any reasonable probability of skin contact. Safety equipment suppliers/manufacturers can provide recommendations on the most protective glove/clothing material for your operation. All protective clothing (suits, gloves, footwear, headgear) should be clean, available each day, and put on before work. Contact lenses should not be worn when working with this chemical. Employees should wash immediately with soap when skin is wet or contaminated.

Respirator Selection: up to 0.25 mg/m^3: 100XQ (any air-purifying respirator with a high-efficiency particulate filter) or Sa (APF = 10) (any supplied-air respirator). up to 0.5 mg/m^3: Sa:Cf* (APF = 25) (any supplied-air respirator operated in a continuous-flow mode) or 100F (APF = 50) (any air purifying, full-face-piece respirator with an N100, R100, or P100 filter) or PaprHie* (APF = 25) (any powered, air-purifying respirator with a high-efficiency particulate filter) or SCBAF (APF = 50) (any self-contained breathing apparatus with a full face-piece) or SaF (APF = 50) (any supplied-air respirator with a full face-piece). Emergency or planned entry into unknown concentrations or IDLH conditions: SCBAF: Pd,Pp (APF = 10,000) (any NIOSH/MSHA- or European Standard EN 149-approved self-contained breathing apparatus that has a full face-piece and is operated in a pressure-demand or other positive-pressure mode) or SaF: Pd,Pp: ASCBA (APF = 10,000) (any supplied-air respirator that has a full face-piece and is operated in a pressure-demand or other positive-pressure mode in combination with an auxiliary, self-contained breathing apparatus operated in a pressure-demand or other positive-pressure mode). Escape: 100 F (APF = 50) (any air-purifying, full-face-piece respirator with an N100, R100, or P100 filter) or SCBAE (any appropriate escape-type, self-contained breathing apparatus).

*Substance reported to cause eye irritation or damage; may require eye protection.

Storage: (1) Color Code—Red Stripe: Flammability Hazard: Store separately from all other flammable materials. (2) Color Code—Yellow Stripe (*strong reducing agent*): Reactivity Hazard; Store separately in an area isolated from flammables, combustibles, or other yellow coded materials. Prior to working with this chemical you should be trained on its proper handling and storage. Protect containers against physical damage. Store in isolated, well-ventilated, cool, dry area.

Shipping: Lithium hydride requires a shipping label of "DANGEROUS WHEN WET." It falls in Hazard Class 4.3 and Packing Group I.

Lithium hydride, fused solid, requires a shipping label of "DANGEROUS WHEN WET." It falls in Hazard Class 4.3 and Packing Group II.

Spill Handling: Evacuate persons not wearing protective equipment from area of spill or leak until cleanup is complete. Remove all ignition sources. Do not touch spilled materials.

Do not allow material to contact water. Shovel up *small spills* with noncombustible absorbent material. Confine *large spills* with dikes, sheets, or tarps to stop spreading. Ventilate area after cleanup is complete. It may be necessary to contain and dispose of this chemical as a hazardous waste. If material or contaminated runoff enters waterways, notify downstream users of potentially contaminated waters. Contact your local or federal environmental protection agency for specific recommendations. If employees are required to clean up spills, they must be properly trained and equipped. OSHA 1910.120(q) may be applicable.

Fire Extinguishing: In a fire, irritating alkali fumes may form. Lithium hydride can form airborne dust clouds which may explode on contact with flame, heat, or oxidizing materials. Additionally, spontaneous ignition occurs when nitrous oxide and lithium hydride are mixed. Lithium hydride also forms explosive mixtures with liquid oxygen. *Do not use water*, carbon dioxide, dry chemical, or halogenated extinguishing agents, such as carbon tetrachloride. Use Class D extinguishers, dry sand, dry graphite, dry limestone, or ground dolomite-based dry chemical extinguishers, such as "Lith-X." Wear protective goggles or face shield, rubberized gloves, flame-proof outer clothing, respirator, and high boots or shoes. *Large fires:* withdraw from area and let fire burn, as lithium hydride may continue to re-ignite. A fire, once started, cannot be extinguished by ordinary methods. If material or contaminated runoff enters waterways, notify downstream users of potentially contaminated waters. Notify local health and fire officials and pollution control agencies. From a secure, explosion-proof location, use water spray to cool exposed containers. If cooling streams are ineffective (venting sound increases in volume and pitch, tank discolors, or shows any signs of deforming), withdraw immediately to a secure position. If employees are expected to fight fires, they must be trained and equipped in OSHA 1910.156. The only respirators recommended for firefighting are self-contained breathing apparatuses that have full face-pieces and are operated in a pressure-demand or other positive-pressure mode.

Disposal Method Suggested: Lithium hydride may be mixed with sand, sprayed with butanol and then with water, neutralized and flushed to a sewer with water.

References
US Environmental Protection Agency. (September 1, 1976). *Chemical Hazard Information Profile: Lithium and Lithium Compounds.* Washington, DC
US Environmental Protection Agency. (November 30, 1987). *Chemical Hazard Information Profile: Lithium Hydride.* Washington, DC: Chemical Emergency Preparedness Program

Lithium nitrate   L:0320

Molecular Formula: $LiNO_3$
Synonyms: Nitric acid, lithium salt

CAS Registry Number: 7790-69-4
RTECS® Number: QU9200000
UN/NA & ERG Number: UN2722/140
EC Number: 232-218-9
Regulatory Authority and Advisory Bodies
WGK (German Aquatic Hazard Class): 1—Low hazard to waters.
Description: Lithium nitrate is a colorless deliquescent powder. Specific gravity (H_2O:1) = 2.38; Freezing/Melting point = 264°C. Hazard Identification (based on NFPA-704 M Rating System): Health 2, Flammability 1, Reactivity 2 (Oxidizer). Soluble in water.
Potential Exposure: Lithium nitrate is used in ceramics, pyrotechnics, salt baths, refrigeration systems, and rocket propellants.
Incompatibilities: May explode when exposed to sparks, shock, and heat. Violent reactions with combustible materials, oxidizers, organic materials, reducing agents, strong acids.
Permissible Exposure Limits in Air
Protective Action Criteria (PAC)
TEEL-0: 0.5 mg/m³
PAC-1: 1.5 mg/m³
PAC-2: 10 mg/m³
PAC-3: 50 mg/m³
Routes of Entry: Inhalation.
Harmful Effects and Symptoms
Short Term Exposure: Lithium nitrate can affect you when breathed in. Repeated heavy exposure may lead to lithium poisoning. This can cause loss of appetite, nausea, tremor, muscle twitches, apathy, convulsions, coma, and death. Contact with lithium nitrate water solution or powder can cause severe skin and eye burns. A very low sodium diet can increase your risk of health problems from exposure to lithium.
Long Term Exposure: Lithium can cause loss of appetite, nausea, vomiting, diarrhea and abdominal pain, headache, muscle weakness, loss of coordination, confusion, seizures, and coma. Can affect the thyroid gland causing goiter. May cause kidney damage.
Points of Attack: Thyroid, kidneys.
Medical Surveillance: If symptoms develop or overexposure is suspected, the following may be useful: serum lithium level. Thyroid function tests.
First Aid: If this chemical gets into the eyes, remove any contact lenses at once and irrigate immediately for at least 15 min, occasionally lifting upper and lower lids. Seek medical attention immediately. If this chemical contacts the skin, remove contaminated clothing and wash immediately with soap and water. Seek medical attention immediately. If this chemical has been inhaled, remove from exposure, begin rescue breathing (using universal precautions, including resuscitation mask) if breathing has stopped and CPR if heart action has stopped. Transfer promptly to a medical facility. When this chemical has been swallowed, get medical attention. Give large quantities of water and induce

vomiting. Do not make an unconscious person vomit. Medical observation is recommended for 24—48 h after breathing overexposure, as pulmonary edema may be delayed. As first aid for pulmonary edema, a doctor or authorized paramedic may consider administering a corticosteroid spray.

Personal Protective Methods: Wear protective gloves and clothing to prevent any reasonable probability of skin contact. Safety equipment suppliers/manufacturers can provide recommendations on the most protective glove/clothing material for your operation. All protective clothing (suits, gloves, footwear, headgear) should be clean, available each day, and put on before work. Contact lenses should not be worn when working with this chemical. Wear dust-proof goggles and face shield when working with powders or dust, unless full face-piece respiratory protection is worn. Wear splash-proof chemical goggles and face shield when working with liquid, unless full face-piece respiratory protection is worn. Employees should wash immediately with soap when skin is wet or contaminated. Provide emergency showers and eyewash.

Respirator Selection: Where there is potential for exposure to solid lithium nitrate use a NIOSH/MSHA- or European Standard EN149-approved respirator equipped with particulate (dust/fume/mist) filters. Particulate filters must be checked every day before work for physical damage, such as rips or tears, and replaced as needed. Where there is potential for high exposures to liquid lithium nitrate, use a NIOSH/MSHA- or European Standard EN149-approved supplied-air respirator with a full face-piece operated in the positive-pressure mode, or with a full face-piece, hood, or helmet in the continuous-flow mode; or use a NIOSH/MSHA- or European Standard EN149-approved self-contained breathing apparatus with a full face-piece operated in pressure-demand or other positive-pressure mode.

Storage: Color Code—Yellow: Reactive Hazard; Store in a location separate from other materials, especially flammables and combustibles. Prior to working with this chemical you should be trained on its proper handling and storage. Lithium nitrate must be stored to avoid contact with wood, paper, oil and heat, and oxidizers (such as perchlorates, peroxides, permanganates, chlorates, and nitrates), since violent reactions occur. Protect containers against physical damage, heat, shock, sparks. Store in tightly closed containers in a cool, well-ventilated area. Sources of ignition, such as smoking and open flames, are prohibited where this chemical is used, handled, or stored in a manner that could create a potential fire or explosion hazard. Wherever this chemical is used, handled, manufactured, or stored, use explosion-proof electrical equipment and fittings. See OSHA Standard 1910.104 and NFPA 43A *Code for the Storage of Liquid and Solid Oxidizers* for detailed handling and storage regulations.

Shipping: This compound requires a shipping label of "OXIDIZER." It falls in Hazard Class 5.1 and Packing Group III.

Spill Handling: Evacuate persons not wearing protective equipment from area of spill or leak until cleanup is complete. Remove all ignition sources. Collect powdered material in the most convenient and safe manner and deposit in sealed containers. Ventilate area after cleanup is complete. It may be necessary to contain and dispose of this chemical as a hazardous waste. If material or contaminated runoff enters waterways, notify downstream users of potentially contaminated waters. Contact your local or federal environmental protection agency for specific recommendations. If employees are required to clean up spills, they must be properly trained and equipped. OSHA 1910.120(q) may be applicable.

Fire Extinguishing: Use water only. Do not use dry chemical, carbon dioxide, or foam extinguishers. Poisonous gases, including nitrogen oxides, are produced in fire. If material or contaminated runoff enters waterways, notify downstream users of potentially contaminated waters. Notify local health and fire officials and pollution control agencies. From a secure, explosion-proof location, use water spray to cool exposed containers. If cooling streams are ineffective (venting sound increases in volume and pitch, tank discolors, or shows any signs of deforming), withdraw immediately to a secure position. If employees are expected to fight fires, they must be trained and equipped in OSHA 1910.156. The only respirators recommended for firefighting are self-contained breathing apparatuses that have full face-pieces and are operated in a pressure-demand or other positive-pressure mode.

Reference
New Jersey Department of Health and Senior Services. (September 1999). *Hazardous Substances Fact Sheet: Lithium Nitrate*. Trenton, NJ

Lomustine   L:0330

Molecular Formula: $C_9H_{16}ClN_3O_2$
Synonyms: Belustine; CCNU; Cecenu; CEENU; 1-(2-chloroethyl)-3-cyclohexyl-1-nitrosourea; (Chloro-2-ethyl)-1-cyclohexyl-3-nitrosourea; N-(2-Chloroethyl)-N'-cyclohexyl-N-nitrosourea; Chloroethylcyclohexylnitrosourea; CINU; ICIG1109; Lomustine; NCI-C04740; NSC-79037; RB 1509; SRI2200
CAS Registry Number: 13010-47-4
RTECS® Number: YS4900000
UN/NA & ERG Number: UN2811 (toxic solid, organic, n.o.s.)/154
EC Number: 235-859-2
Regulatory Authority and Advisory Bodies
Carcinogenicity: NCI: Carcinogenesis Bioassay (ipr); clear evidence: mouse, no evidence: rat, 1977; NTP: 7th Report on Carcinogens: Reasonably anticipated to be a human carcinogen; IARC: Animal Sufficient Evidence; Human Inadequate Evidence, Group 2A, 1999.

California Proposition 65 Chemical: Cancer as 1-(2-chloro-ethyl)-3-cyclohexyl-1-nitrosourea.

Regulatory Authority and Advisory Bodies
California Proposition 65 Chemical: Cancer 1/1/88; Developmental/Reproductive toxin 7/1/90.
WGK (German Aquatic Hazard Class): No value assigned.

Description: Lomustine is a pale yellow powder. Molecular weight = 233.73; Freezing/Melting point = 90°C. Insoluble in water.

Potential Exposure: Those involved in the manufacture, administration, or consumption of this antineoplastic (anti-cancer) agent.

Permissible Exposure Limits in Air
No standards or TEEL available.

Routes of Entry: Inhalation.

Harmful Effects and Symptoms
Long Term Exposure: There is sufficient evidence that 1-(2-chloroethyl)-3-cyclohexyl-1-nitrosourea (CCNU) is carcinogenic in experimental animals. CCNU caused lung cancers in rats following intraperitoneal or intravenous injection. In mice, intraperitoneal injections of CCNU resulted in a slight increase in the incidence of lymph system neoplasms. Applied to the skin of mice, CCNU did not induce skin tumors, but the duration of the experiment was inadequate. Evidence for the carcinogenicity of CCNU in humans is inadequate. In several reported cases, cancer patients who received CCNU developed leukemia. With one exception, all of these patients also had received other cytotoxic agents and/or irradiation.

Points of Attack: Blood.

Medical Surveillance: NIOSH lists the following tests: whole blood (chemical/metabolite), blood serum, biologic tissue/biopsy, urine (chemical/metabolite).

First Aid: If this chemical gets into the eyes, remove any contact lenses at once and irrigate immediately for at least 15 min, occasionally lifting upper and lower lids. Seek medical attention immediately. If this chemical contacts the skin, remove contaminated clothing and wash immediately with soap and water. Seek medical attention immediately. If this chemical has been inhaled, remove from exposure, begin rescue breathing (using universal precautions, including resuscitation mask) if breathing has stopped and CPR if heart action has stopped. Transfer promptly to a medical facility. When this chemical has been swallowed, get medical attention. Give large quantities of water and induce vomiting. Do not make an unconscious person vomit.

Personal Protective Methods: Wear protective gloves and clothing to prevent any reasonable probability of skin contact. Safety equipment suppliers/manufacturers can provide recommendations on the most protective glove/clothing material for your operation. All protective clothing (suits, gloves, footwear, headgear) should be clean, available each day, and put on before work. Contact lenses should not be worn when working with this chemical. Wear dust-proof chemical goggles and face shield unless full face-piece respiratory protection is worn. Employees should wash immediately with soap when skin is wet or contaminated. Provide emergency showers and eyewash.

Respirator Selection: Follow regulations in OSHA 29CFR1910.134 or European Standard EN149. Use a NIOSH/MSHA- or European Standard EN149-approved respirator; or use an approved supplied-air respirator with a full face-piece operated in the positive-pressure mode, or with a full face-piece, hood, or helmet in the continuous-flow mode; or use a NIOSH/MSHA- or European Standard EN149-approved self-contained breathing apparatus with a full face-piece operated in pressure-demand or other positive-pressure mode.

Storage: Color Code—Blue: Health Hazard/Poison: Store in a secure poison location. Prior to working with this chemical you should be trained on its proper handling and storage. Store in tightly closed containers in a cool, well-ventilated area. A regulated, marked area should be established where this chemical is handled, used, or stored in compliance with OSHA Standard 1910.1045.

Shipping: Toxic solid, organic, n.o.s. requires a shipping label of "POISONOUS/TOXIC MATERIALS." It falls in Hazard Class 6.1 and Packing Group III.

Spill Handling: Evacuate persons not wearing protective equipment from area of spill or leak until cleanup is complete. Remove all ignition sources. Collect powdered material in the most convenient and safe manner and deposit in sealed containers. Ventilate area after cleanup is complete. It may be necessary to contain and dispose of this chemical as a hazardous waste. If material or contaminated runoff enters waterways, notify downstream users of potentially contaminated waters. Contact your local or federal environmental protection agency for specific recommendations. If employees are required to clean up spills, they must be properly trained and equipped. OSHA 1910.120(q) may be applicable.

Fire Extinguishing: Use dry chemical, carbon dioxide, water spray, or alcohol foam extinguishers. Poisonous gases are produced in fire. If material or contaminated runoff enters waterways, notify downstream users of potentially contaminated waters. Notify local health and fire officials and pollution control agencies. From a secure, explosion-proof location, use water spray to cool exposed containers. If cooling streams are ineffective (venting sound increases in volume and pitch, tank discolors, or shows any signs of deforming), withdraw immediately to a secure position. If employees are expected to fight fires, they must be trained and equipped in OSHA 1910.156. The only respirators recommended for firefighting are self-contained breathing apparatuses that have full face-pieces and are operated in a pressure-demand or other positive-pressure mode.

M

Magnesium M:0100

Molecular Formula: Mg
Common Formula: Mg
Synonyms: Magnesium metal; Magnesium pellets; Magnesium powder; Magnesium ribbons; Magnesium scalpings; Magnesium shavings; Magnesium sheet; Magnesium turnings
CAS Registry Number: 7439-95-4
RTECS® Number: OM210000
UN/NA & ERG Number: UN1869/138; UN1418 (powder)/138
EC Number: 012-002-00-9 (pellets); 012-001-00-3 (powder, pyrophoric)
Regulatory Authority and Advisory Bodies
Department of Homeland Security Screening Threshold Quantity (pounds): *Theft hazard* 100 (ACG).
European/International Regulations: not listed in Annex 1.
WGK (German Aquatic Hazard Class): Nonwater polluting agent.
Description: Magnesium is a light, silvery-white metal in various forms and is a fire hazard. Molecular weight = 24.31; Boiling point = 1100°C; Freezing/Melting point = 649°C (pellets); 651°C (powder); Autoignition temperature = 473°C (powder). Explosive limits: LEL = 0.03 kg/m^3. Ignition temperature of dust cloud = 520°C; Minimum Explosive concentration = 0.020 oz/ft^3.[USBM] Relative explosion hazard of dust: Severe. Hazard Identification (based on NFPA-704 M Rating System): (*powder, turnings and ribbon*) Health 0, Flammability 2, Reactivity 2. Reacts with water; insoluble.
Potential Exposure: Magnesium alloyed with manganese, aluminum, thorium, zinc, cerium, and zirconium is used in aircraft, ships, automobiles, hand tools, etc., because of its lightness. Dow metal is the general name for a large group of alloys containing over 85% magnesium. Magnesium wire and ribbon are used for degassing valves in the radio industry and in various heating appliances; as a deoxidizer and desulfurizer in copper, brass, and nickel alloys; in chemical reagents; as a powder in the manufacture of flares, incendiary bombs, tracer bullets, and flashlight powders; in the nuclear energy process; and in a cement of magnesium oxide and magnesium chloride for floors. Magnesium is an essential element in human and animal nutrition and also in plants, where it is a component of all types of chlorophyll. It is the most abundant intracellular divalent cation in both plants and animals. It is an activator of many mammalian enzymes.
Incompatibilities: The substance is a strong reducing agent. Reacts violently with oxidizers, strong acids, acetylene, ammonium salts, arsenic, beryllium fluoride, carbon tetrachloride, carbonates, chloroform, cyanides, chlorinated hydrocarbons, ethylene oxide, hydrocarbons, metal oxides, methanol, phosphates, silver nitrate, sodium peroxide, sulfates, trichloroethylene, and many other substances, causing fire and explosion hazards. Finely powdered, chip or sheet form reacts with moisture or acids, evolving flammable hydrogen gas, causing fire and explosion hazard. Finely divided form is readily ignited by a spark or flame and splatters and burns at above 1260°C.
Permissible Exposure Limits in Air
Protective Action Criteria (PAC)
TEEL-0: 1.25 mg/m^3
PAC-1: 4 mg/m^3
PAC-2: 30 mg/m^3
PAC-3: 150 mg/m^3
No standards set for elemental magnesium.
Determination in Air: Filter collection and atomic absorption analysis.
Permissible Concentration in Water: The World Health Organization (WHO) has established European and International desirable limits ranging from 30 to 125 mg/L, depending on the sulfate concentration. If the sulfate exceeds 250 mg/L, the magnesium is limited to 30 mg/L. The WHO specifies an absolute maximum of 150 mg/L for magnesium in drinking water. In view of the fact that concentrations of magnesium in drinking water are less than those that impart astringent taste, they pose no health problem and are more likely to be beneficial; no limitation for reasons of health is needed. The USSR-UNEP/IRPTC joint project[43] set a MAC of 50 mg/L in water bodies used for fishery purposes.
Determination in Water: Magnesium in water can be determined by atomic-absorption spectrophotometry, with a sensitivity of 15 mg/L, and by photometry with a sensitivity of 100 µg/L.
Harmful Effects and Symptoms
Short Term Exposure: Irritates the eyes and skin. Inhaling dust can irritate the respiratory tract, causing coughing, wheezing, and/or shortness of breath. Magnesium and magnesium compounds are mild irritants to the conjunctiva and nasal mucosa, but are not specifically toxic. Inhalation may cause metal fume fever; the symptoms may be delayed for 4−12 h following exposure. On the skin, these hot particles are capable of producing second- and third-degree burns, but they respond to treatment as other thermal burns do. Metallic magnesium foreign bodies in the skin cause no unusual problems in humans. In animal experiments, however, they have caused "gas gangrene"—massive localized gaseous tumors with extensive necrosis. Magnesium salts at levels over 700 gm/L (especially magnesium sulfate) have a laxative effect, particularly on new users, although the human body can adapt to the effects of magnesium with time.
Long Term Exposure: Repeated exposure can cause an accumulation in the body, causing upset stomach.

Medical Surveillance: There is no special test.

First Aid: If this chemical gets into the eyes, remove any contact lenses at once and irrigate immediately for at least 15 min, occasionally lifting upper and lower lids. Seek medical attention immediately. If this chemical contacts the skin, remove contaminated clothing and wash immediately with soap and water. Seek medical attention immediately. If fragments have become imbedded in the skin and removal cannot be ensured by thorough scrubbing, medical attention for thorough removal is recommended. If this chemical has been inhaled, remove from exposure, begin rescue breathing (using universal precautions, including resuscitation mask) if breathing has stopped and CPR if heart action has stopped. Transfer promptly to a medical facility. When this chemical has been swallowed, get medical attention. Give large quantities of water and induce vomiting. Do not make an unconscious person vomit. The symptoms of metal fume fever may be delayed for 4−12 h following exposure: it may last less than 36 h.

Note to physician: In case of fume inhalation, treat pulmonary edema. Give prednisone or other corticosteroid orally to reduce tissue response to fume. Positive-pressure ventilation may be necessary. Treat metal fume fever with bed rest, analgesics, and antipyretics.

Personal Protective Methods: It burns with an intense flame; do not look directly at fire; wear adequate eye protection. Wear protective gloves and clothing to prevent any reasonable probability of skin contact. Safety equipment suppliers/manufacturers can provide recommendations on the most protective glove/clothing material for your operation. All protective clothing (suits, gloves, footwear, headgear) should be clean, available each day, and put on before work. Contact lenses should not be worn when working with this chemical. Wear dust-proof chemical goggles and face shield unless full face-piece respiratory protection is worn. Employees should wash immediately with soap when skin is wet or contaminated. Provide emergency showers and eyewash.

Respirator Selection: Where there is potential for exposures to magnesium, use a NIOSH/MSHA- or European Standard EN149-approved respirator equipped with particulate (dust/fume/mist) filters. Particulate filters must be checked every day before work for physical damage, such as rips or tears, and replaced as needed. *Where there is potential for high exposures*, use a NIOSH/MSHA- or European Standard EN149-approved supplied-air respirator with a full face-piece operated in the positive-pressure mode, or with a full face-piece, hood, or helmet in the continuous-flow mode; or use a NIOSH/MSHA- or European Standard EN149-approved self-contained breathing apparatus with a full face-piece operated in pressure-demand or other positive-pressure mode.

Storage: (1) Color Code—Red Stripe (*powder, turnings, and ribbon are flammable solids*): Flammability Hazard: Do not store in the same area as other flammable materials. (2) Color Code—Yellow Stripe (*strong reducing agent*): Reactivity Hazard; Store separately in an area isolated from flammables, combustibles, or other yellow coded materials. Magnesium must be stored to avoid contact with strong oxidizers (such as chlorine, bromine, and fluorine), strong acids (such as hydrochloric, sulfuric, and nitric), and chlorine trifluoride, since violent reactions occur. Store in tightly closed containers in a cool, well-ventilated area away from water. Use only nonsparking tools and equipment, especially when opening and closing containers of magnesium. Protect storage containers from physical damage.

Shipping: Magnesium pellets, turnings, or ribbons (UN1869) require a shipping label of "FLAMMABLE SOLID." It falls in Hazard Class 4.1 and Packing Group III. Magnesium powder requires a shipping label of "DANGEROUS WHEN WET, SPONTANEOUSLY COMBUSTIBLE." It falls in Hazard Class 4.3 and Packing Group III.

Magnesium granules, coated (particle size not <149 μm) require a shipping label of "DANGEROUS WHEN WET." It falls in Hazard Class 4.3 and Packing Group III.

Spill Handling: Evacuate persons not wearing protective equipment from area of spill or leak until cleanup is complete. Remove all ignition sources. Use a vacuum to reduce dust during cleanup. Do not dry sweep. Collect powdered material and deposit in sealed containers. Ventilate area after cleanup is complete. Keep magnesium out of a confined space, such as a sewer, because of the possibility of an explosion, unless the sewer is designed to prevent the buildup of explosive concentrations. It may be necessary to contain and dispose of this chemical as a hazardous waste. If material or contaminated runoff enters waterways, notify downstream users of potentially contaminated waters. Contact your local or federal environmental protection agency for specific recommendations. If employees are required to clean up spills, they must be properly trained and equipped. OSHA 1910.120(q) may be applicable.

Fire Extinguishing: Magnesium is a combustible solid or a flammable powder. It burns with an intense flame. Reacts violently with fire extinguishing agents, such as water, carbon dioxide, and powder. Dangerous when wet. It burns in a current of steam. Powders form explosive mixtures with air. Poisonous gas including magnesium oxide is produced in fire. Fire may restart after it has been extinguished. Use dry sand, MetL-X® powder, or G-1 graphite powder, soda ash, Class D extinguishers, or talc. *Do not use water* or hydrous agents. Fire may restart after it has been extinguished. If material or contaminated runoff enters waterways, notify downstream users of potentially contaminated waters. Notify local health and fire officials and pollution control agencies. From a secure, explosion-proof location, use water spray to cool exposed containers. If cooling streams are ineffective (venting sound increases in volume and pitch, tank discolors, or shows any signs of deforming), withdraw immediately to a secure position. If employees are expected to fight fires, they must be trained and equipped in OSHA 1910.156. The only respirators recommended for firefighting are self-contained breathing apparatuses that have full face-pieces and are operated in a pressure-demand or other positive-pressure mode.

References

Sax, N. I. (Ed.). (1984). *Dangerous Properties of Industrial Materials Report*, 4, No. 2, 79–81

New Jersey Department of Health and Senior Services. (September 1999). *Hazardous Substances Fact Sheet: Magnesium*. Trenton, NJ

Magnesium chlorate   M:0110

Molecular Formula: Cl_2MgO_6

Common Formula: $Mg(ClO_3)_2$

Synonyms: Chlorate salt of magnesium; Chloric acid, De-Fol-Ate®; Chloric acid, magnesium; E-Z-Off®; Magnesium dichlorate; Magnesium salt; Magron; MC defoliant; Ortho MC

CAS Registry Number: 10326-21-3

RTECS® Number: FO0175000

UN/NA & ERG Number: UN2723/140

EC Number: 233-711-1

Regulatory Authority and Advisory Bodies

Air Pollutant Standard Set. See below, "Permissible Exposure Limits in Air" section.

WGK (German Aquatic Hazard Class): 2—Hazard to waters.

Description: Magnesium chlorate is white crystalline solid. Molecular weight = 191.21; Boiling point = 120°C; Freezing/Melting point = 35°C. Soluble in water (reaction).

Potential Exposure: Used as a drying agent and defoliant.

Incompatibilities: A strong oxidizer. Violent reactions with arsenic, carbon, charcoal, copper, phosphorus, sulfur, magnesium oxide, metal sulfides (copper sulfide, arsenic sulfide, tin sulfide, fuels, and strong acids. Reacts with moisture.

Permissible Exposure Limits in Air: Russia[43] set a MAC of $0.1 \ mg/m^3$ in ambient air in residential areas both on a momentary and a daily average basis.

No standards or TEEL available.

Permissible Concentration in Water: Russia[43] set a MAC of 0.35 mg/L in water bodies used for fishery purposes.

Routes of Entry: Inhalation, ingestion, eye and/or skin contact.

Harmful Effects and Symptoms

Short Term Exposure: Magnesium chlorate can affect you when breathed in. Contact can irritate or even burn the skin and eyes. Inhaling the dust irritates the respiratory system. Exposure can interfere with the ability of the blood to carry oxygen, causing headaches, weakness, dizziness, trouble breathing, collapse, and possible death. Breathing the dust can irritate the air passages, cause sore throat and/or cough with phlegm.

Long Term Exposure: Repeated exposure can cause lung irritation; bronchitis may develop with coughing, phlegm, and/or shortness of breath. May affect the kidneys.

Points of Attack: Lungs, blood.

Medical Surveillance: Before beginning employment and at regular times after that, for those with frequent or potentially high exposures, the following are recommended: lung function tests. Kidney function tests. If symptoms develop or overexposure is suspected, the following may be useful: blood methemoglobin level.

First Aid: If this chemical gets into the eyes, remove any contact lenses at once and irrigate immediately for at least 15 min, occasionally lifting upper and lower lids. Seek medical attention immediately. If this chemical contacts the skin, remove contaminated clothing and wash immediately with soap and water. Seek medical attention immediately. If this chemical has been inhaled, remove from exposure, begin rescue breathing (using universal precautions, including resuscitation mask) if breathing has stopped and CPR if heart action has stopped. Transfer promptly to a medical facility. When this chemical has been swallowed, get medical attention. Give large quantities of water and induce vomiting. Do not make an unconscious person vomit.

Note to physician: Treat for methemoglobinemia. Spectrophotometry may be required for precise determination of levels of methemoglobin in urine.

Personal Protective Methods: Wear protective gloves and clothing to prevent any reasonable probability of skin contact. Safety equipment suppliers/manufacturers can provide recommendations on the most protective glove/clothing material for your operation. All protective clothing (suits, gloves, footwear, headgear) should be clean, available each day, and put on before work. Contact lenses should not be worn when working with this chemical. Wear dust-proof chemical goggles and face shield unless full face-piece respiratory protection is worn. Employees should wash immediately with soap when skin is wet or contaminated. Provide emergency showers and eyewash.

Respirator Selection: Where there is potential for exposure to magnesium chlorate, use a NIOSH/MSHA- or European Standard EN149-approved full face-piece respirator with a high-efficiency particulate filter. Greater protection is provided by a powered air-purifying respirator. Where there is potential for high exposure, use a NIOSH/MSHA- or European Standard EN149-approved supplied-air respirator with a full face-piece operated in the positive-pressure mode, or with a full face-piece, hood, or helmet in the continuous-flow mode; or use a NIOSH/MSHA- or European Standard EN149-approved self-contained breathing apparatus with a full face-piece operated in pressure-demand or other positive-pressure mode.

Storage: Color Code—Yellow: Reactive Hazard; Store in a location separate from other materials, especially flammables and combustibles. Prior to working with magnesium chlorate you should be trained on its proper handling and storage. Magnesium chlorate must be stored to avoid contact with aluminum, arsenic, carbon, copper, phosphorus, sulfur, magnesium oxide, metal sulfides, fuels, and strong acids, since violent reactions occur. Store in tightly closed containers in a cool, well-ventilated area away from flammable and combustible materials. See OSHA Standard 1910.104 and NFPA 43A *Code for the Storage of Liquid and Solid Oxidizers* for detailed handling and storage regulations.

Shipping: This compound requires a shipping label of "OXIDIZER." It falls in Hazard Class 5.1 and Packing Group II.

Spill Handling: Evacuate persons not wearing protective equipment from area of spill or leak until cleanup is complete. Remove all ignition sources. Collect powdered material in the most convenient and safe manner and deposit in sealed containers. Ventilate area after cleanup is complete. Spilled magnesium chlorate tends to become very sensitive to shock or friction and is an explosion hazard. Keep magnesium chlorate out of sewers since it can ignite flammable or combustible materials and thus cause a fire or explosion hazard. It may be necessary to contain and dispose of this chemical as a hazardous waste. If material or contaminated runoff enters waterways, notify downstream users of potentially contaminated waters. Contact your local or federal environmental protection agency for specific recommendations. If employees are required to clean up spills, they must be properly trained and equipped. OSHA 1910.120(q) may be applicable.

Fire Extinguishing: Extinguish fire using an agent suitable for type of surrounding fire. Magnesium chlorate itself does not burn but it will increase the intensity of a fire since it is an oxidizer. Containers may explode in fire; use water spray to keep fire-exposed containers cool. Poisonous gases, including chlorine and magnesium oxide, are produced in fire. If material or contaminated runoff enters waterways, notify downstream users of potentially contaminated waters. Notify local health and fire officials and pollution control agencies. From a secure, explosion-proof location, use water spray to cool exposed containers. If cooling streams are ineffective (venting sound increases in volume and pitch, tank discolors, or shows any signs of deforming), withdraw immediately to a secure position. If employees are expected to fight fires, they must be trained and equipped in OSHA 1910.156. The only respirators recommended for firefighting are self-contained breathing apparatuses that have full face-pieces and are operated in a pressure-demand or other positive-pressure mode.

Reference
New Jersey Department of Health and Senior Services. (September 1999). *Hazardous Substances Fact Sheet: Magnesium Chlorate.* Trenton, NJ

Magnesium hydride   M:0120

Molecular Formula: H_2Mg
Common Formula: MgH_2
Synonyms: Magnesium(II) hydride; Magnesium dihydride
CAS Registry Number: 7693-27-8; 60616-74-2

RTECS® Number: OM3560000
UN/NA & ERG Number: UN2010/138
EC Number: 231-705-3
Regulatory Authority and Advisory Bodies
WGK (German Aquatic Hazard Class): No value assigned.
Description: Magnesium hydride is a coarse, gray crystalline solid. Molecular weight = 26.33; Freezing/Melting point $\geq 200°C$; flash point = 110°C. Insoluble in water (dangerous reaction). Decomposes at 280°C in high vacuum.

Potential Exposure: Used in the production of hydrogen and magnesium alcoholates.

Incompatibilities: A strong reducing agent. Pyrophoric; the powder or dust may ignite spontaneously in air or in the presence of moisture. Contact with water or steam forms magnesium hydroxide, flammable hydrogen gas, and enough heat to ignite the hydrogen. Violent reaction with oxidizers, alcohols, halogens, chlorinated solvents. Incompatible with strong acids, acid chlorides. Store under nitrogen.

Permissible Exposure Limits in Air
No standards or TEEL available.

Routes of Entry: Inhalation, ingestion, skin and/or eye contact.

Harmful Effects and Symptoms
Short Term Exposure: Magnesium hydride can affect you when breathed in. Contact can irritate or even burn the skin and eyes. Breathing the dust can irritate the air passages, causing sore throat and/or cough with phlegm.

Long Term Exposure: Long-term effects are unknown at this time.

Points of Attack: Lungs.

Medical Surveillance: Before beginning employment and at regular times after that, for those with frequent or potentially high exposures, the following are recommended: lung function tests.

First Aid: If this chemical gets into the eyes, remove any contact lenses at once and irrigate immediately for at least 15 min, occasionally lifting upper and lower lids. Seek medical attention immediately. If this chemical contacts the skin, remove contaminated clothing and wash immediately with soap and water. Seek medical attention immediately. If this chemical has been inhaled, remove from exposure, begin rescue breathing (using universal precautions, including resuscitation mask) if breathing has stopped and CPR if heart action has stopped. Transfer promptly to a medical facility. When this chemical has been swallowed, get medical attention. Give large quantities of water and induce vomiting. Do not make an unconscious person vomit.

Personal Protective Methods: Wear protective gloves and clothing to prevent any reasonable probability of skin contact. Safety equipment suppliers/manufacturers can provide recommendations on the most protective glove/clothing material for your operation. All protective clothing (suits, gloves, footwear, headgear) should be clean, available each day, and put on before work. Contact lenses should not be

worn when working with this chemical. Wear dust-proof chemical goggles and face shield unless full face-piece respiratory protection is worn. Employees should wash immediately with soap when skin is wet or contaminated. Provide emergency showers and eyewash.

Respirator Selection: Where there is potential for exposures to magnesium hydride, use a NIOSH/MSHA- or European Standard EN149-approved full face-piece respirator with a particulate (dust/fume/mist) filter. Particulate filters must be checked every day before work for physical damage, such as rips or tears, and replaced as needed. *Where there is potential for high exposures*, use a NIOSH/MSHA- or European Standard EN149-approved supplied-air respirator with a full face-piece operated in the positive-pressure mode, or with a full face-piece, hood, or helmet in the continuous-flow mode; or use a NIOSH/MSHA- or European Standard EN149-approved self-contained breathing apparatus with a full face-piece operated in pressure-demand or other positive-pressure mode.

Storage: Color Code—Red Stripe: Flammability Hazard: Store separately from all other flammable materials. Store under nitrogen in tightly closed containers in cool (decomposes >250°C), well-ventilated area. Keep containers dry at all times. Magnesium hydride reacts violently with water, releasing caustic material, heat, and flammable gas. Sources of ignition, such as smoking and open flames, are prohibited where magnesium hydride is handled, used, or stored.

Shipping: Magnesium hydride requires a shipping label of "DANGEROUS WHEN WET." It falls in Hazard Class 4.3 and Packing Group I.

Spill Handling: Evacuate persons not wearing protective equipment from area of spill or leak until cleanup is complete. Remove all ignition sources. Use vacuum to reduce dust during cleanup. Do not use dry sweep. Collect powdered material in the safest manner and deposit in sealed containers. Ventilate area after cleanup is complete. Keep magnesium hydride out of sewers because of possibility of fire or explosion. It may be necessary to contain and dispose of this chemical as a hazardous waste. If material or contaminated runoff enters waterways, notify downstream users of potentially contaminated waters. Contact your local or federal environmental protection agency for specific recommendations. If employees are required to clean up spills, they must be properly trained and equipped. OSHA 1910.120(q) may be applicable.

Fire Extinguishing: Magnesium hydride is a flammable and reactive solid. Use dry chemical, sand, soda ash, or lime extinguishers. *Do not use water*, carbon dioxide, or foam. Magnesium hydride can catch fire spontaneously in air or in the presence of moisture. Fire may restart after it has been extinguished. Poisonous gases are produced in fire. If material or contaminated runoff enters waterways, notify downstream users of potentially contaminated waters. Notify local health and fire officials and pollution control agencies. From a secure, explosion-proof location, use water spray to cool exposed containers. If cooling streams are ineffective (venting sound increases in volume and pitch, tank discolors, or shows any signs of deforming), withdraw immediately to a secure position. If employees are expected to fight fires, they must be trained and equipped in OSHA 1910.156. The only respirators recommended for firefighting are self-contained breathing apparatuses that have full face-pieces and are operated in a pressure-demand or other positive-pressure mode.

Reference

New Jersey Department of Health and Senior Services. (August 1999). *Hazardous Substances Fact Sheet: Magnesium Hydride*. Trenton, NJ

Magnesium nitrate   M:0130

Molecular Formula: MgN_2O_6
Common Formula: $Mg(NO_3)_2$
Synonyms: Nitric acid, Magnesium salt; Nitromagnesite
CAS Registry Number: 10377-60-3;   10213-15-7 (hexahydrate)
RTECS® Number: OM3750000
UN/NA & ERG Number: UN1474/140
EC Number: 233-826-7 (see Appendix 4).
WGK (German Aquatic Hazard Class): 1—Low hazard to waters.

Description: Magnesium nitrate is white crystalline solid. Molecular weight = 148.33;   256.45 (hexahydrate); Freezing/Melting point = 129°C (dihydrate); 95−100°C (hexahydrate). It decomposes at 330°C. Hazard Identification (based on NFPA-704 M Rating System): Health 2, Flammability 0, Reactivity 2 OX. A strong oxidizer. Soluble in water.

Potential Exposure: Magnesium nitrate is used in fireworks and in the production of concentrated nitric acid.

Incompatibilities: A powerful oxidizer. Violent reaction with dimethylformamide, reducing agents, combustibles, fuels, organic and easily oxidizable matter.

Permissible Exposure Limits in Air
Protective Action Criteria (PAC)
TEEL-0: 10 mg/m^3
PAC-1: 30 mg/m^3
PAC-2: 50 mg/m^3
PAC-3: 250 mg/m^3

Routes of Entry: Inhalation, ingestion, skin and/or eye contact.

Harmful Effects and Symptoms
Short Term Exposure: Magnesium nitrate can affect you when breathed in. Contact can irritate or even burn the skin and eyes. The dust can irritate the eyes and air passages, causing sore throat and/or cough with phlegm. Exposure can interfere with the ability of the blood to carry oxygen, causing headaches, weakness, nausea, and a bluish color to the skin and lips (methemoglobinemia). Higher levels can cause trouble breathing, collapse, and even death.

Long Term Exposure: Repeated exposure can cause headache, weakness, and dizziness.

Medical Surveillance: If symptoms develop or overexposure is suspected, the following may be useful: blood methemoglobin level.

First Aid: If this chemical gets into the eyes, remove any contact lenses at once and irrigate immediately for at least 15 min, occasionally lifting upper and lower lids. Seek medical attention immediately. If this chemical contacts the skin, remove contaminated clothing and wash immediately with soap and water. Seek medical attention immediately. If this chemical has been inhaled, remove from exposure, begin rescue breathing (using universal precautions, including resuscitation mask) if breathing has stopped and CPR if heart action has stopped. Transfer promptly to a medical facility. When this chemical has been swallowed, get medical attention. Give large quantities of water and induce vomiting. Do not make an unconscious person vomit.

Note to physician: Treat for methemoglobinemia. Spectrophotometry may be required for precise determination of levels of methemoglobin in urine.

Personal Protective Methods: Wear protective gloves and clothing to prevent any reasonable probability of skin contact. Safety equipment suppliers/manufacturers can provide recommendations on the most protective glove/clothing material for your operation. All protective clothing (suits, gloves, footwear, headgear) should be clean, available each day, and put on before work. Contact lenses should not be worn when working with this chemical. Wear dust-proof chemical goggles and face shield unless full face-piece respiratory protection is worn. Employees should wash immediately with soap when skin is wet or contaminated. Provide emergency showers and eyewash.

Respirator Selection: Where there is potential for exposures to magnesium nitrate, use a NIOSH/MSHA- or European Standard EN149-approved full face-piece respirator with a with a high-efficiency particulate filter. Greater protection is provided by a powered air-purifying respirator. *Where there is potential for high exposures*, use a NIOSH/MSHA- or European Standard EN149-approved supplied-air respirator with a full face-piece operated in the positive-pressure mode, or with a full face-piece, hood, or helmet in the continuous-flow mode; or use a NIOSH/MSHA- or European Standard EN149-approved self-contained breathing apparatus with a full face-piece operated in pressure-demand or other positive-pressure mode.

Storage: Color Code—Yellow: Reactive Hazard (*strong oxidizer*); Store in a location separate from other materials, especially flammables and combustibles. Prior to working with this chemical you should be trained on its proper handling and storage. Magnesium nitrate must be stored to avoid contact with dimethyl formamide, fuels, and strong reducing agents, since violent reactions occur. Store in tightly closed containers in a cool, well-ventilated area away from flammable and combustible materials. Avoid storage on wood floors. See OSHA Standard 1910.104 and

NFPA 43A *Code for the Storage of Liquid and Solid Oxidizers* for detailed handling and storage regulations.

Shipping: This compound requires a shipping label of "OXIDIZER." It falls in Hazard Class 5.1 and Packing Group III.

Spill Handling: Evacuate persons not wearing protective equipment from area of spill or leak until cleanup is complete. Remove all ignition sources. Collect powdered material in the most convenient and safe manner and deposit in sealed containers. Ventilate area after cleanup is complete. Keep magnesium nitrate out of sewers since it can ignite flammable and combustible materials and thus cause a fire or explosion hazard. It may be necessary to contain and dispose of this chemical as a hazardous waste. If material or contaminated runoff enters waterways, notify downstream users of potentially contaminated waters. Contact your local or federal environmental protection agency for specific recommendations. If employees are required to clean up spills, they must be properly trained and equipped. OSHA 1910.120(q) may be applicable.

Fire Extinguishing: Extinguish fire using an agent suitable for type of surrounding fire. Magnesium nitrate itself does not burn but it will increase the intensity of a fire since it is an oxidizer. Use water spray to keep fire-exposed containers cool. Poisonous gases, including nitrogen oxides, are produced in fire. If material or contaminated runoff enters waterways, notify downstream users of potentially contaminated waters. Notify local health and fire officials and pollution control agencies. Containers may explode in fire. From a secure, explosion-proof location, use water spray to cool exposed containers. If cooling streams are ineffective (venting sound increases in volume and pitch, tank discolors, or shows any signs of deforming), withdraw immediately to a secure position. If employees are expected to fight fires, they must be trained and equipped in OSHA 1910.156. The only respirators recommended for firefighting are self-contained breathing apparatuses that have full face-pieces and are operated in a pressure-demand or other positive-pressure mode.

Reference
New Jersey Department of Health and Senior Services. (September 1999). *Hazardous Substances Fact Sheet: Magnesium Nitrate.* Trenton, NJ

Magnesium oxide M:0140

Molecular Formula: MgO
Synonyms: Akro-mag; Animag; Calcined brucite; Calcined magnesia; Calcined magnesite; Granmag; Magcal; Magchem 100; Maglite; Magnesia; Magnesia fume; Magnesia USTA; Magnesium oxide fume; Magox; Marmag; Periclase; Seawater magnesia
CAS Registry Number: 1309-48-4
RTECS® Number: OM3850000
UN/NA & ERG Number: UN1418 (powder)/138

EC Number: 215-171-9

Regulatory Authority and Advisory Bodies

Air Pollutant Standard Set. See below, "Permissible Exposure Limits in Air" section.

FDA—over-the-counter drug.

Water Pollution Standard Proposed (EPA).[32]

Canada, WHMIS, Ingredients Disclosure List Concentration 1.0%.

European/International Regulations: not listed in Annex 1.

WGK (German Aquatic Hazard Class): 1—Low hazard to waters.

Description: Magnesium oxide forms a finely divided white particulate dispersed in air. Molecular weight = 40.31; Boiling point = 3600°C; Freezing/Melting point = 2800°C. Hazard Identification (based on NFPA-704 M Rating System): Health 2, Flammability 0, Reactivity 1. Poor solubility in water.

Potential Exposure: Compound Description: Tumorigen; Human Data. Used in oil refining, pulp, and paper mills, in tire manufacturing, in the manufacture of refractory crucibles, fire bricks, magnesia cements, and boiler scale compounds. Exposure may occur when magnesium is burned, thermally cut, or welded upon.

Incompatibilities: Violent reaction with halogens, chlorine trifluoride, bromine pentalfluoride, phosphorous pentachloride, strong acids. May ignite and explode when heated with sublimed sulfur, magnesium powder, or aluminum powder.

Permissible Exposure Limits in Air

OSHA PEL: 15 mg/m^3 (*total particulate; fume*) TWA.

NIOSH: There is inadequate data to propose an exposure limit. See Appendix D of *The NIOSH Pocket Guide.*

ACGIH TLV®[1]: 10 mg/m^3 TWA (*inhalable fraction*); not classifiable as a human carcinogen.

NIOSH IDLH: 750 mg/m^3 (fume).

Protective Action Criteria (PAC)

TEEL-0: 15 mg/m^3

PAC-1: 30 mg/m^3

PAC-2: 50 mg/m^3

PAC-3: 500 mg/m^3

DFG MAK: 4 mg/m^3, inhalable fraction; 1.5 mg/m^3, respirable fraction; see sections V(f) and V(g). See V(h) for fume; Pregnancy Risk Group C.

Arab Republic of Egypt: TWA 10 mg/m^3, 1993; Australia: TWA 10 mg/m^3, 1993; Austria: MAK 6 mg/m^3, 1999; Belgium: TWA 10 mg/m^3, 1993; France: VME 10 mg/m^3, 1999; Hungary: TWA 5 mg/m^3; STEL 10 mg/m^3, 1993; Norway: TWA 10 mg/m^3, 1999; the Netherlands: MAC-TGG 10 mg/m^3, 2003; Poland: MAC (TWA) fume 5 mg/m^3, MAC (TWA) dust 10 mg/m^3, 1999; Russia: STEL 5 mg/m^3, 1993; Switzerland: MAK-W 6 mg/m^3, 1999; Turkey: TWA 15 mg/m^3, 1993; United Kingdom: TWA 10 mg[Mg]/m^3, total inhalable dust; TWA 4 mg[Mg]/m^3, fume and respirable dust, 2000; Argentina, Bulgaria, Columbia, Jordan, South Korea, New Zealand, Singapore, Vietnam: ACGIH TLV®: not classifiable as a human carcinogen. Several states have set guidelines or standards for magnesium oxide fume in ambient air[60] ranging from 100 μg/m^3 (North Dakota) to 160 μg/m^3 (Virginia) to 200 μg/m^3 (Connecticut) to 238 μg/m^3 (Nevada).

Determination in Air: Use NIOSH Analytical Method (IV) #7300 (Elements by ICP), #7301, #7303; OSHA Analytical Method #ID-125G.

Permissible Concentration in Water: No criteria set, but EPA[32] has suggested a permissible ambient goal of 138 μg/L for magnesium oxide based on health effects.

Routes of Entry: Inhalation of fume, eye and/or skin contact.

Harmful Effects and Symptoms

Short Term Exposure: Irritates the eyes and respiratory tract. Magnesium in the form of nascent magnesium oxide can cause metal fume fever with cough, chest pain, flu-like fever, if inhaled in sufficient quantity. The symptoms of metal fume fever may be delayed for 4—12 h following exposure.

Long Term Exposure: It has been noted that magnesium workers show a rise in serum magnesium, although no significant symptoms of ill health have been identified. Some investigators have reported higher incidence of digestive disorders and have related this to magnesium absorption, but the evidence is scant.

Points of Attack: Eyes, respiratory system.

Medical Surveillance: NIOSH lists the following test for fume exposure: whole blood (chemical/metabolite).

First Aid: If this chemical gets into the eyes, remove any contact lenses at once and irrigate immediately for at least 15 min, occasionally lifting upper and lower lids. Seek medical attention immediately. If this chemical contacts the skin, remove contaminated clothing and wash immediately with soap and water. Seek medical attention immediately. If this chemical has been inhaled, remove from exposure, begin rescue breathing (using universal precautions, including resuscitation mask) if breathing has stopped and CPR if heart action has stopped. Transfer promptly to a medical facility. When this chemical has been swallowed, get medical attention. Give large quantities of water and induce vomiting. Do not make an unconscious person vomit. The symptoms of metal fume fever may be delayed for 4—12 h following exposure: it may last less than 36 h.

Note to physician: In case of fume inhalation, treat for pulmonary edema. Give prednisone or other corticosteroid orally to reduce tissue response to fume. Positive-pressure ventilation may be necessary. Treat metal fume fever with bed rest, analgesics, and antipyretics.

Personal Protective Methods: Wear protective gloves and clothing to prevent any reasonable probability of skin contact. Safety equipment suppliers/manufacturers can provide recommendations on the most protective glove/clothing material for your operation. All protective clothing (suits, gloves, footwear, headgear) should be clean, available each day, and put on before work. Contact lenses should not be worn when working with this chemical. Wear dust-proof chemical goggles and face shield unless full face-piece

respiratory protection is worn. Employees should wash immediately with soap when skin is wet or contaminated. Provide emergency showers and eyewash.

Respirator Selection: as fume OSHA: *up to 150 mg/m³*: 95XQ [Any particulate respirator equipped with an N95, R95, or P95 filter (including N95, R95, and P95 filtering face-pieces) except quarter-mask respirators. The following filters may also be used: N99, R99, P99, N100, R100, P100]. Sa (APF = 10) (any supplied-air respirator). *Up to 375 mg/m³:* Sa:Cf (APF = 25) (any supplied-air respirator operated in a continuous-flow mode) or PAPRDMFu* (any powered, air-purifying respirator with a dust, mist, and fume filter). *Up to 750 mg/m³:* 100F (APF = 50) (any air-purifying, full-face-piece respirator with an N100, R100, or P100 filter) or PaprTHie (APF = 50) (any powered, air-purifying respirator with a tight-fitting face-piece and a high-efficiency particulate filter) SCBAF (APF = 50) (any self-contained breathing apparatus with a full face-piece) SaF (APF = 50) (any supplied-air respirator with a full face-piece). *Emergency or planned entry into unknown concentrations or IDLH conditions:* SCBAF: Pd,Pp (APF = 10,000) (any NIOSH/MSHA- or European Standard EN 149-approved self-contained breathing apparatus that has a full face-piece and is operated in a pressure-demand or other positive-pressure mode). SaF: Pd,Pp: ASCBA (APF = 10,000) (any supplied-air respirator that has a full face-piece and is operated in a pressure-demand or other positive-pressure mode in combination with an auxiliary, self-contained breathing apparatus operated in a pressure-demand or other positive-pressure mode). *Escape:* 100F (APF = 50) (any air-purifying, full-face-piece respirator with an N100, R100, or P100 filter) or SCBAE (any appropriate escape-type, self-contained breathing apparatus).

*Substance reported to cause eye irritation or damage; may require eye protection.

Storage: Color Code—Yellow Stripe (*strong reducing agent*): Reactivity Hazard; Store finely divided powder, chips, or shavings separately in detached fire-resistant building isolated from flammables, combustibles, or other yellow coded materials. Protect against physical damage. Prior to working with this chemical you should be trained on its proper handling and storage. Magnesium oxide must be stored to avoid contact with oxidizers (such as perchlorates, peroxides, permanganates, chlorates, and nitrates, and chlorine trifluoride), halogens, strong acids, since violent reactions occur. Store in tightly closed containers in a cool, well-ventilated, "No Smoking" area away from moisture. Use nonsparking-type tools and equipment, including explosion-proof ventilation. Containers of this material may be hazardous when empty since they retain product residues (dust, solids); observe all warnings and precautions listed for the product.

Shipping: Magnesium *powder* requires a shipping label of "DANGEROUS WHEN WET, SPONTANEOUSLY COMBUSTIBLE." It falls in Hazard Class 4.2 & 4.3 and Packing Group II.

Spill Handling: Evacuate persons not wearing protective equipment from area of spill or leak until cleanup is complete. Remove all ignition sources. Collect powdered material in the most convenient and safe manner and deposit in sealed containers. Ventilate area after cleanup is complete. It may be necessary to contain and dispose of this chemical as a hazardous waste. If material or contaminated runoff enters waterways, notify downstream users of potentially contaminated waters. Contact your local or federal environmental protection agency for specific recommendations. If employees are required to clean up spills, they must be properly trained and equipped. OSHA 1910.120(q) may be applicable.

Fire Extinguishing: This chemical is a noncombustible solid. May ignite and explode when heated with sublimed sulfur, magnesium powder, or aluminum powder. Use any extinguisher suitable for surrounding fire. Poisonous gases are produced in fire. If material or contaminated runoff enters waterways, notify downstream users of potentially contaminated waters. Notify local health and fire officials and pollution control agencies. From a secure, explosion-proof location, use water spray to cool exposed containers. If cooling streams are ineffective (venting sound increases in volume and pitch, tank discolors, or shows any signs of deforming), withdraw immediately to a secure position. If employees are expected to fight fires, they must be trained and equipped in OSHA 1910.156. The only respirators recommended for firefighting are self-contained breathing apparatuses that have full face-pieces and are operated in a pressure-demand or other positive-pressure mode.

Reference
New Jersey Department of Health and Senior Services. (January 2007). *Hazardous Substances Fact Sheet: Magnesium Oxide (Fume).* Trenton, NJ

Magnesium perchlorate   M:0150

Molecular Formula: Cl_2MgO_8
Common Formula: $Mg(ClO_4)_2$
Synonyms: Ammonium perchlorate, anhydrous; Ammonium perchlorate, hexahydride; Anhydrone®; Dehydrite®; Perchlorate de magnesium (French); Perchloric acid, Magnesium salt
CAS Registry Number: 10034-81-8
RTECS® Number: SC8925000
UN/NA & ERG Number: UN1475/140
EC Number: 233-108-3
Regulatory Authority and Advisory Bodies
WGK (German Aquatic Hazard Class): 1—Low hazard to waters.
Description: Magnesium perchlorate is a white crystalline solid. Molecular weight = 223.21; Freezing/Melting point ≥250°C (decomposes). Hazard Identification (based on NFPA-704 M Rating System): Health 1, Flammability 0, Reactivity 3 (Oxidizer). Soluble in water.

Potential Exposure: Magnesium perchlorate is used as a drying agent for gases and as an oxidizing agent.

Incompatibilities: A powerful oxidizer. Violent reaction with reducing agents, alkenes (above 220°C), ammonia gas, organic matter, ethylene oxide, powdered metals, phosphorus, dimethylsulfoxide, mineral acids, wet argon, hydrazines, alcohols, wet fluorobutane, butyl fluorides, organic materials. Forms explosive material with ethyl alcohol. Incompatible with many materials. Shock may cause magnesium perchlorate to explode.

Permissible Exposure Limits in Air
Protective Action Criteria (PAC)
TEEL-0: 6 mg/m^3
PAC-1: 15 mg/m^3
PAC-2: 125 mg/m^3
PAC-3: 500 mg/m^3

Routes of Entry: Inhalation.

Harmful Effects and Symptoms

Short Term Exposure: Magnesium perchlorate can affect you when breathed in. Contact can irritate or even burn the skin and eyes. Breathing the dust can irritate the air passages, causing sore throat and/or cough with phlegm. High levels can interfere with the ability of the blood to carry oxygen, causing headaches, dizziness, and a bluish color to the skin. Very high levels could cause death.

Long Term Exposure: Can irritate the lungs. Repeated exposure may cause bronchitis.

Points of Attack: Lungs, blood.

Medical Surveillance: For those with frequent or potentially high exposure the following are recommended before beginning work and at regular times after that: lung function tests. If symptoms develop or overexposure is suspected, the following maybe useful: blood methemoglobin level.

First Aid: If this chemical gets into the eyes, remove any contact lenses at once and irrigate immediately for at least 15 min, occasionally lifting upper and lower lids. Seek medical attention immediately. If this chemical contacts the skin, remove contaminated clothing and wash immediately with soap and water. Seek medical attention immediately. If this chemical has been inhaled, remove from exposure, begin rescue breathing (using universal precautions, including resuscitation mask) if breathing has stopped and CPR if heart action has stopped. Transfer promptly to a medical facility. When this chemical has been swallowed, get medical attention. Give large quantities of water and induce vomiting. Do not make an unconscious person vomit.

Note to physician: Treat for methemoglobinemia. Spectrophotometry may be required for precise determination of levels of methemoglobin in urine.

Personal Protective Methods: Wear protective gloves and clothing to prevent any reasonable probability of skin contact. Safety equipment suppliers/manufacturers can provide recommendations on the most protective glove/clothing material for your operation. All protective clothing (suits, gloves, footwear, headgear) should be clean, available each day, and put on before work. Contact lenses should not be worn when working with this chemical. Wear dust-proof chemical goggles and face shield unless full face-piece respiratory protection is worn. Employees should wash immediately with soap when skin is wet or contaminated. Provide emergency showers and eyewash.

Respirator Selection: Where there is potential for exposures to magnesium perchlorate, use a NIOSH/MSHA- or European Standard EN149-approved full face-piece respirator with a high-efficiency particulate filter. Greater protection is provided by a powered air-purifying respirator. *Where there is potential for high exposures*, use a NIOSH/MSHA- or European Standard EN149-approved supplied-air respirator with a full face-piece operated in the positive-pressure mode, or with a full face-piece, hood, or helmet in the continuous-flow mode; or use a NIOSH/MSHA- or European Standard EN149-approved self-contained breathing apparatus with a full face-piece operated in pressure-demand or other positive-pressure mode.

Storage: Color Code—Yellow: Reactive Hazard; Store in a location separate from other materials, especially flammables and combustibles. Prior to working with this chemical you should be trained on its proper handling and storage. Magnesium perchlorate must be stored to avoid contact with fuels, finely powdered metals, mineral acids, ammonia, ethylene oxide, phosphorus, dimethyl sulfoxide, and trimethyl phosphite, since violent reactions occur. Shock may cause magnesium perchlorate to explode. Store in tightly closed containers in a cool, well-ventilated area away from flammable and combustible materials. Do not store on wood floors. See OSHA Standard 1910.104 and NFPA 43A *Code for the Storage of Liquid and Solid Oxidizers* for detailed handling and storage regulations.

Shipping: This compound requires a shipping label of "OXIDIZER." It falls in Hazard Class 5.1 and Packing Group II.

Spill Handling: Evacuate persons not wearing protective equipment from area of spill or leak until cleanup is complete. Remove all ignition sources. Use vacuum to reduce dust during cleanup. Do not dry sweep. Collect powdered material in the safest manner and deposit in sealed containers. Ventilate area after cleanup is complete. It may be necessary to contain and dispose of this chemical as a hazardous waste. If material or contaminated runoff enters waterways, notify downstream users of potentially contaminated waters. Contact your local or federal environmental protection agency for specific recommendations. If employees are required to clean up spills, they must be properly trained and equipped. OSHA 1910.120(q) may be applicable.

Fire Extinguishing: Extinguish fire using an agent suitable for type of surrounding fire. Magnesium perchlorate itself does not burn but it will increase the intensity of a fire since it is an oxidizer. Poisonous gases, including chlorides and magnesium oxide, are produced in fire. If material or contaminated runoff enters waterways, notify downstream users of potentially contaminated waters.

Notify local health and fire officials and pollution control agencies. From a secure, explosion-proof location, use water spray to cool exposed containers. If cooling streams are ineffective (venting sound increases in volume and pitch, tank discolors, or shows any signs of deforming), withdraw immediately to a secure position. If employees are expected to fight fires, they must be trained and equipped in OSHA 1910.156. The only respirators recommended for firefighting are self-contained breathing apparatuses that have full face-pieces and are operated in a pressure-demand or other positive-pressure mode.

Reference
New Jersey Department of Health and Senior Services. (November 1999) *Hazardous Substances Fact Sheet: Magnesium Perchlorate*. Trenton, NJ

Magnesium peroxide   M:0160

Molecular Formula: MgO_2
Synonyms: IXPER 25M; Magnesium dioxide; Magnesium superoxol; Peromag; Peróxido de magnesio (Spanish); Peroxyde de magnésium (French)
CAS Registry Number: 14452-57-4 (MgO); 1335-26-8 (MgO_2)
RTECS® Number: OM4100000
UN/NA & ERG Number: UN1476/140
EC Number: 238-438-1
Regulatory Authority and Advisory Bodies
WGK (German Aquatic Hazard Class): 1—Low hazard to waters.
Description: Magnesium peroxide is a white, odorless crystalline solid. Molecular weight = 56.31 (MgO_2); 40.3 (MgO); Decomposes above 100°C. Insoluble in water.
Potential Exposure: Magnesium peroxide is used as a bleaching and oxidizing agent and in the manufacture of antacids and anti-infective drugs.
Incompatibilities: Powerful oxidizer. Dangerous fire risk with flammable and combustible materials. Violent reaction with acids. Keep away from moisture; causes the release of oxygen and heat.
Permissible Exposure Limits in Air
No standards or TEEL available.
Routes of Entry: Inhalation.
Harmful Effects and Symptoms
Short Term Exposure: Magnesium peroxide can affect you when breathed in. Contact can irritate or even burn the skin and eyes. Breathing the dust can irritate the air passages, causing sore throat and/or cough with phlegm.
Long Term Exposure: This chemical is a very irritating substance and it may cause lung effects or damage.
Points of Attack: Lungs.
Medical Surveillance: For those with frequent or potentially high exposure, the following are recommended before beginning work and at regular times after that: lung function tests.

First Aid: If this chemical gets into the eyes, remove any contact lenses at once and irrigate immediately for at least 15 min, occasionally lifting upper and lower lids. Seek medical attention immediately. If this chemical contacts the skin, remove contaminated clothing and wash immediately with soap and water. Seek medical attention immediately. If this chemical has been inhaled, remove from exposure, begin rescue breathing (using universal precautions, including resuscitation mask) if breathing has stopped and CPR if heart action has stopped. Transfer promptly to a medical facility. When this chemical has been swallowed, get medical attention. Give large quantities of water and induce vomiting. Do not make an unconscious person vomit.
Personal Protective Methods: Wear protective gloves and clothing to prevent any reasonable probability of skin contact. Safety equipment suppliers/manufacturers can provide recommendations on the most protective glove/clothing material for your operation. All protective clothing (suits, gloves, footwear, headgear) should be clean, available each day, and put on before work. Contact lenses should not be worn when working with this chemical. Wear dust-proof chemical goggles and face shield unless full face-piece respiratory protection is worn. Employees should wash immediately with soap when skin is wet or contaminated. Provide emergency showers and eyewash.
Respirator Selection: Color Code—Yellow: Reactive Hazard; Store in a location separate from other materials, especially flammables and combustibles. Where there is potential for exposures to magnesium peroxide, use a NIOSH/MSHA- or European Standard EN149-approved full face-piece respirator with a high-efficiency particulate filter. Greater protection is provided by a powered air-purifying respirator. *Where there is potential for high exposures*, use a NIOSH/MSHA- or European Standard EN149-approved supplied-air respirator with a full face-piece operated in the positive-pressure mode, or with a full face-piece, hood, or helmet in the continuous-flow mode; or use a NIOSH/MSHA- or European Standard EN149-approved self-contained breathing apparatus with a full face-piece operated in pressure-demand or other positive-pressure mode.
Storage: Color Code—Yellow: Reactive Hazard; Store in a location separate from other materials, especially flammables and combustibles. Prior to working with this chemical you should be trained on its proper handling and storage. Magnesium peroxide must be stored to avoid contact with acids, since violent reactions occur. Store in tightly closed containers in a cool, well-ventilated area away from flammable and combustible materials. Keep magnesium peroxide dry. In contact with moisture, it is a dangerous fire hazard because it releases oxygen and much heat. See OSHA Standard 1910.104 and NFPA 43A *Code for the Storage of Liquid and Solid Oxidizers* for detailed handling and storage regulations.
Shipping: This compound requires a shipping label of "OXIDIZER." It falls in Hazard Class 5.1 and Packing Group II.

Spill Handling: Evacuate persons not wearing protective equipment from area of spill or leak until cleanup is complete. Remove all ignition sources. Use a vacuum or a wet method to reduce dust during cleanup. Do not dry sweep. Collect powdered material in the most convenient and safe manner and deposit in sealed containers. Ventilate area after cleanup is complete. It may be necessary to contain and dispose of this chemical as a hazardous waste. If material or contaminated runoff enters waterways, notify downstream users of potentially contaminated waters. Contact your local or federal environmental protection agency for specific recommendations. If employees are required to clean up spills, they must be properly trained and equipped. OSHA 1910.120(q) may be applicable.

Fire Extinguishing: Extinguish fire using an agent suitable for type of surrounding fire. Magnesium peroxide itself does not burn but it will increase the intensity of a fire since it is an oxidizer. Poisonous gases, including magnesium oxides, are produced in fire. If material or contaminated runoff enters waterways, notify downstream users of potentially contaminated waters. Notify local health and fire officials and pollution control agencies. From a secure, explosion-proof location, use water spray to cool exposed containers. If cooling streams are ineffective (venting sound increases in volume and pitch, tank discolors, or shows any signs of deforming), withdraw immediately to a secure position. If employees are expected to fight fires, they must be trained and equipped in OSHA 1910.156. The only respirators recommended for firefighting are self-contained breathing apparatuses that have full face-pieces and are operated in a pressure-demand or other positive-pressure mode.

Reference
New Jersey Department of Health and Senior Services. (December 2000). *Hazardous Substances Fact Sheet: Magnesium Peroxide*. Trenton, NJ

Magnesium silicide   M:0170

Molecular Formula: Mg_2Si
Synonyms: Dimagnesium silicide
CAS Registry Number: 22831-39-6; 39404-03-0
RTECS® Number: OM4367000
UN/NA & ERG Number: UN2624/138
EC Number: 245-254-5
Regulatory Authority and Advisory Bodies
WGK (German Aquatic Hazard Class): No value assigned.
Description: Magnesium silicide is a slate-blue crystalline solid. Molecular weight = 76.71; Freezing/Melting point = 778°C. Hazard Identification (based on NFPA-704 M Rating System): Health 2, Flammability 0, Reactivity 1 ₩. Dangerous reaction with water.
Potential Exposure: Magnesium silicide is used in the semiconductor industry and to produce certain aluminum alloys.

Incompatibilities: Pyrophoric; mixtures with air are spontaneously explosive. Reacts with water; evolves explosive hydrogen and self-igniting toxic silane gas. Incompatible with mineral acids.
Permissible Exposure Limits in Air
No standards or TEEL available.
Determination in Air: No criteria set.
Routes of Entry: Inhalation, eye and/or skin contact.
Harmful Effects and Symptoms
Short Term Exposure: Contact can irritate the skin and/or eyes. Breathing the dust can irritate the air passages, causing cough with phlegm. In contact with moisture or acid or acid mist, a highly irritating and flammable gas (silane) is released.
Long Term Exposure: Unknown at this time.
First Aid: If this chemical gets into the eyes, remove any contact lenses at once and irrigate immediately for at least 15 min, occasionally lifting upper and lower lids. Seek medical attention immediately. If this chemical contacts the skin, remove contaminated clothing and wash immediately with soap and water. Seek medical attention immediately. If this chemical has been inhaled, remove from exposure, begin rescue breathing (using universal precautions, including resuscitation mask) if breathing has stopped and CPR if heart action has stopped. Transfer promptly to a medical facility. When this chemical has been swallowed, get medical attention. Give large quantities of water and induce vomiting. Do not make an unconscious person vomit.
Personal Protective Methods: Wear protective gloves and clothing to prevent any reasonable probability of skin contact. Safety equipment suppliers/manufacturers can provide recommendations on the most protective glove/clothing material for your operation. All protective clothing (suits, gloves, footwear, headgear) should be clean, available each day, and put on before work. Contact lenses should not be worn when working with this chemical. Wear dust-proof chemical goggles and face shield unless full face-piece respiratory protection is worn. Employees should wash immediately with soap when skin is wet or contaminated. Provide emergency showers and eyewash.
Respirator Selection: Where there is potential for exposures to magnesium silicide, use a NIOSH/MSHA- or European Standard EN149-approved full face-piece respirator equipped with particulate (dust/fume/mist) filters. More protection is provided by a full face-piece respirator than by a half-mask respirator, and even greater protection is provided by a powered air-purifying respirator. Particulate filters must be checked every day before work for physical damage, such as rips or tears, and replaced as needed. *Where there is potential for high exposures*, use a NIOSH/MSHA- or European Standard EN149-approved supplied-air respirator with a full face-piece operated in the positive-pressure mode, or with a full face-piece, hood, or helmet in the continuous-flow mode; or use a NIOSH/MSHA- or European Standard EN149-approved self-contained breathing apparatus with a full face-piece operated in pressure-demand or other positive-pressure mode.

Storage: Flammable solid. Color Code—Red: Flammability Hazard: Store in a flammable materials storage area. Prior to working with this chemical you should be trained on its proper handling and storage. Store in tightly closed containers in a cool, well-ventilated area away from acids. Keep dry at all times. Sources of ignition, such as smoking and open flames, are prohibited where magnesium silicide is handled, used, or stored.

Shipping: Magnesium silicide requires a shipping label of "DANGEROUS WHEN WET." It falls in Hazard Class 4.3 and Packing Group II.

Spill Handling: Evacuate persons not wearing protective equipment from area of spill or leak until cleanup is complete. Remove all ignition sources. Use HEPA vacuum; do not dry sweep. Collect powdered material in the most convenient and safe manner and deposit in sealed containers. Ventilate area after cleanup is complete. Keep magnesium silicide out of sewers because of possibility of fire or explosion. It may be necessary to contain and dispose of this chemical as a hazardous waste. If material or contaminated runoff enters waterways, notify downstream users of potentially contaminated waters. Contact your local or federal environmental protection agency for specific recommendations. If employees are required to clean up spills, they must be properly trained and equipped. OSHA 1910.120(q) may be applicable.

Fire Extinguishing: Magnesium silicide is a flammable solid. Forms flammable hydrogen or silane gas on contact with water. Use dry chemical, soda ash, or lime extinguishers. *Do not use water* or foam. Poisonous fumes of sodium monoxide are produced in a fire. Ignites spontaneously in air. Poisonous gases are produced in fire. If material or contaminated runoff enters waterways, notify downstream users of potentially contaminated waters. Notify local health and fire officials and pollution control agencies. From a secure, explosion-proof location, use water spray to cool exposed containers. If cooling streams are ineffective (venting sound increases in volume and pitch, tank discolors, or shows any signs of deforming), withdraw immediately to a secure position. If employees are expected to fight fires, they must be trained and equipped in OSHA 1910.156. The only respirators recommended for firefighting are self-contained breathing apparatuses that have full face-pieces and are operated in a pressure-demand or other positive-pressure mode.

Reference
New Jersey Department of Health and Senior Services. (December 2000). *Hazardous Substances Fact Sheet: Magnesium Silicide.* Trenton, NJ

Magnesium silicofluoride   M:0180

Molecular Formula: F_6MgSi
Common Formula: $MgSiF_6$
Synonyms: Eulava SM; Fluosilicate de magnesium (French); Magnesium fluorosillicate; Magnesium hexafluorosillicate; Magnesium hexahydrate; Magnesium silicofluoride; Silicate(2-), hexafluoro-
CAS Registry Number: 16949-65-8; 18972-56-0 (Hexafluorosilicate); *(alt.)* 1310-00-5
RTECS® Number: W8575000
UN/NA & ERG Number: UN2853/151
EC Number: 241-022-2 [*Annex I Index No.:* 009-018-00-3] (magnesium hexafluorosilicate)
Regulatory Authority and Advisory Bodies
Air Pollutant Standard Set. See below, "Permissible Exposure Limits in Air" section.
European/International Regulations (*16949-65-8; magnesium hexafluorosilicate*): Hazard Symbol: T; Risk phrases: R25; Safety phrases: S1/2; S24/25; S45 (see Appendix 4).
WGK (German Aquatic Hazard Class): 2—Hazard to waters.
Description: Magnesium silicofluoride is a white crystalline solid. Molecular weight = 274.52. Hazard Identification (based on NFPA-704 M Rating System): Health 3, Flammability 0, Reactivity 0. Soluble in water.
Potential Exposure: Magnesium silicofluoride is used as a concrete hardener and waterproofing agent. Hexafluoro is used as an agricultural chemical.
Incompatibilities: Strong acids.
Permissible Exposure Limits in Air
As fluorides
OSHA PEL: 3 ppm/2.5 mg[F]/m³ TWA.
NIOSH REL: 3 ppm/2.5 mg[F]/m³ TWA.
ACGIH TLV®[1]: 2.5 mg[F]/m³ TWA; not classifiable as a human carcinogen; BEI: 3 mg[F]/g creatinine in urine *prior* to end-of-shift; 10 mg[F]/g creatinine in urine end-of-shift
DFG MAK: 1 mg[F]/m³, inhalable fraction [skin]; Peak Limitation Category II(4); Pregnancy Risk Group C; BAT: 7.0 mg[F]/g creatinine in urine at end-of-shift; 4.0 mg[F]/g creatinine in urine at the beginning of the next shift.
NIOSH IDLH: 250 mg[F]/m³.
Arab Republic of Egypt: TWA 0.5 mg(Sb)/m³, 1993; Australia: TWA 0.5 mg(Sb)/m³, 1993; Australia: TWA 2.5 mg(F)/m³, 1993; Austria: MAK 0.5 mg(Sb)/m³, 1993; Austria: MAK 2.5 mg(F)/m³, 1999; Belgium: TWA 0.5 mg(Sb)/m³, 1993; Belgium: TWA 2.5 mg(F)/m³, 1993; Denmark: TWA 0.5 mg(Sb)/m³, 1999; the Netherlands: MAC-TGG 0.5 mg(Sb)/m³, 2003; Finland: TWA 0.5 mg(Sb)/m³, 1993; Finland: TWA 2.5 mg(F)/m³, 1993; France: VME 0.5 mg(Sb)/m³, 1993; France: VME 2.5 mg(F)/m³, 1999; Hungary STEL 0.5 mg(Sb)/m³, 1993; Hungary: TWA 1 mg(F)/m³, STEL 2 mg(F)/m³, 1993; Japan: 0.1 mg(Sb)/m³, 2B carcinogen, 1999; Norway: TWA 0.6 mg(F)/m³, 1999; the Philippines: TWA 0.5 mg(Sb)/m³, 1993; the Philippines: TWA 2.5 mg(F)/m³, 1993; Poland: MAC (time-weighted average) 0.5 mg(Sb)/m³, 1993; Poland: MAC (time-weighted average) 1 mg(HF)/m³; MAC (STEL) 3 mg(HF)/m³, 1999; Russia: STEL 0.3 mg/m³, 1993; Russia: TWA 0.2 mg(Sb)/m³, STEL 0.5 mg(Sb)/m³, 1993; Argentina, Bulgaria, Columbia, Jordan, South Korea, New Zealand, Singapore, Vietnam: ACGIH TLV®: TWA 2.5 mg

1648   Magnesium silicofluoride

(F)/m^3; United Kingdom: TWA 0.5 mg(Sb)/m^3; TWA 2.5 mg(F)/m^3, 2000.

Note: the OSHA PEL for fluorides (measured as fluorine) is 2.5 mg/m^3 TWA for an 8-h work-shift.

Routes of Entry: Inhalation, skin and/or eye contact.

Harmful Effects and Symptoms

Short Term Exposure: Magnesium silico-fluoride can affect you when breathed in. Skin contact may cause irritation, rash, and ulcers. Eye contact causes irritation with possible permanent damage.

Long Term Exposure: Repeated exposure can cause brittle bones, and muscle and ligament stiffness, with eventual crippling. Overexposure causes irritation of the throat and air passages. Repeated exposure may cause poor appetite, nausea, constipation, or diarrhea. May cause lung damage.

Points of Attack: Lungs.

Medical Surveillance: NIOSH lists the following tests: chest X-ray; electrocardiogram; pulmonary function tests: forced vital capacity, forced expiratory volume (1 s); pelvic X-ray; sputum cytology; urine (chemical/metabolite); urine (chemical/metabolite) pre- and postshift; urinalysis (routine); complete blood count/differential.

First Aid: If this chemical gets into the eyes, remove any contact lenses at once and irrigate immediately for at least 15 min, occasionally lifting upper and lower lids. Seek medical attention immediately. If this chemical contacts the skin, remove contaminated clothing and wash immediately with soap and water. Seek medical attention immediately. If this chemical has been inhaled, remove from exposure, begin rescue breathing (using universal precautions, including resuscitation mask) if breathing has stopped and CPR if heart action has stopped. Transfer promptly to a medical facility. When this chemical has been swallowed, get medical attention. Give large quantities of water and induce vomiting. Do not make an unconscious person vomit.

Personal Protective Methods: Wear protective gloves and clothing to prevent any reasonable probability of skin contact. Safety equipment suppliers/manufacturers can provide recommendations on the most protective glove/clothing material for your operation. All protective clothing (suits, gloves, footwear, headgear) should be clean, available each day, and put on before work. Contact lenses should not be worn when working with this chemical. Wear dust-proof chemical goggles and face shield unless full face-piece respiratory protection is worn. Employees should wash immediately with soap when skin is wet or contaminated. Provide emergency showers and eyewash. Specific engineering controls are recommended in NIOSH Criteria Document #76-103: *Inorganic Fluorides.*

Respirator Selection: NIOSH/OSHA as fluorine *1 ppm:* Sa (APF = 10) (any supplied-air respirator). *2.5 ppm:* Sa:Cf (APF = 25) (any supplied-air respirator operated in a continuous-flow mode). *5 ppm:* SCBAF (APF = 50) (any self-contained breathing apparatus with a full face-piece) or SaF (APF = 50) (any supplied-air respirator with a full face-piece). *25 ppm:* SaF: Pd,Pp (APF = 2000) (any supplied-air

respirator that has a full face-piece and is operated in a pressure-demand or other positive-pressure mode). *Emergency or planned entry into unknown concentrations or IDLH conditions:* SCBAF: Pd,Pp (APF = 10,000) (any self-contained breathing apparatus that has a full face-piece and is operated in a pressure-demand or other positive-pressure mode) or SaF: Pd,Pp: ASCBA (APF = 10,000) (any supplied-air respirator that has a full face-piece and is operated in a pressure-demand or other positive-pressure mode in combination with an auxiliary, self-contained breathing apparatus operated in a pressure-demand or other positive-pressure mode). *Escape:* GmFS end of service life indicator (ESLI) required [any air-purifying, full-face-piece respirator (gas mask) with a chin-style, front- or back-mounted canister protection against the compound of concern] or SCBAE (any appropriate escape-type, self-contained breathing apparatus).

Storage: Color Code—Blue: Health Hazard/Poison: Store in a secure poison location. Prior to working with this chemical you should be trained on its proper handling and storage. Store in tightly closed containers in a cool, well-ventilated area away from strong acids (such as hydrochloric, sulfuric, and nitric).

Shipping: This compound requires a shipping label of "POISONOUS/TOXIC MATERIALS." It falls in Hazard Class 6.1 and Packing Group III.

Spill Handling: Evacuate persons not wearing protective equipment from area of spill or leak until cleanup is complete. Remove all ignition sources. Collect powdered material in the most convenient and safe manner and deposit in sealed containers. Ventilate area after cleanup is complete. It may be necessary to contain and dispose of this chemical as a hazardous waste. If material or contaminated runoff enters waterways, notify downstream users of potentially contaminated waters. Contact your local or federal environmental protection agency for specific recommendations. If employees are required to clean up spills, they must be properly trained and equipped. OSHA 1910.120(q) may be applicable.

Fire Extinguishing: Use dry chemical, CO_2, water spray, or foam extinguishers. Poisonous gases and fumes are produced in a fire, including hydrogen fluoride, silicon tetrafluoride, and magnesium oxide fumes. If material or contaminated runoff enters waterways, notify downstream users of potentially contaminated waters. Notify local health and fire officials and pollution control agencies. From a secure, explosion-proof location, use water spray to cool exposed containers. If cooling streams are ineffective (venting sound increases in volume and pitch, tank discolors, or shows any signs of deforming), withdraw immediately to a secure position. If employees are expected to fight fires, they must be trained and equipped in OSHA 1910.156. The only respirators recommended for firefighting are self-contained breathing apparatuses that have full face-pieces and are operated in a pressure-demand or other positive-pressure mode.

Reference

New Jersey Department of Health and Senior Services. (January 2001). *Hazardous Substances Fact Sheet: Magnesium Silico-Fluoride.* Trenton, NJ

Malathion   M:0190

Molecular Formula: $C_{10}H_{19}O_6PS_2$

Synonyms: Agrichem greenfly spray; AI3-17034; All purpose garden insecticide; American Cyanamid 4,049; Banmite; *S*-[1,2-Bis(aethoxy-carbonyl)-aethyl]-*O,O*-dimethyl-dithiophosphat (German); *S*-[1,2-Bis(carbethoxy)ethyl] *O, O*-dimethyl dithiophosphate; *S*-[1,2-Bis(ethoxycarbonyl) ethyl] *O,O*-dimethyl phosphorodithioate; *S*-1,2-Bis(ethoxy-carbonyl)ethyl *O,O*-dimethyl thiophosphate; Butanedioic acid, [(dimethoxyphosphinothioyl)thio]-, diethyl ester; Calmathion; Carbethoxy malathion; Carbetovur; Carbetox; Chemathion; Cimexan; Compound 4049; Cromocide; Cython; Detmol 96%; Detmol MA; Detmol malathion; *S*-(1,2-Dicarbethoxyethyl) *O,O*-dimethyl phosphorodithioate; Dicarboethoxyethyl *O,O*-dimethyl phosphorodithioate; Diethyl [(dimethoxyphosphinothioyl)-thio]butanedioate; Diethyl (dimethoxyphosphinothioylthio)-succinate; Diethyl (dimethoxythiophosphorylthio)succinate; Diethyl mercapto-succinate, *O,O*-dimethyl dithiophosphate, *S*-ester; Diethyl mercaptosuccinate, *O,O*-dimethyl phosphorodithioate; Diethyl mercaptosuccinate, *O,O*-dimethyl thiophosphate; Diethyl mercaptosuccinate, *S*-ester with *O,O*-dimethyl phos-phorodithioate; [(Dimethoxyphosphinothioyl)thio]butane-dioic acid diethyl ester; *O,O*-Dimethyl *S*-(1,2-dicarbaethoxyaethyl)-dithiophosphat (German); *O,O*-Dimethyl *S*-(1,2-dicarbethoxyethyl) dithiophosphate; *O,O*-Dimethyl *S*-(1,2-dicarbethoxyethyl) phosphorodithioate; *O, O*-Dimethyl *S*-1,2-di(ethoxycarbamyl)ethyl phosphoro-dithioate; *O,O*-Dimethyl dithiophosphate diethyl mercapto-succinate; *O,O*-Dimethyl dithiophosphate of diethyl mercaptosuccinate; Dithiophosphate de *O,O*-dimethyle et de *S*-(1,2-dicarboethoxyethyle) (French); Duramitex; EL 4049; Emmatos; Emmatos extra; ENT 17,034; Ethiolacar; Etiol; Eveshield captan/malathion; Extermathion; Fisons greenfly and blackfly killer; FOG 3; Formal; Forthion; Fosfothion; Fosfotion; Greenfly aerosol spray; Hilthion (Indian); Karbofos; Kop-thion; Kypfos; Malacide; Malafor; Malagran; Malakill; Malamar; Malamar 50; Malasol; Malaspray; Malataf; Malathion 60; Malathion E50; Malathion LB concentrate; Malathion organophosphorous insecticide; Malathon; Malathyl; Malation (Spanish); Maldison (in Australia, New Zealand); Malmed; Malphos; Mercaptosuccinic acid diethyl ester; Mercaptothion; Moscarda; NCI-C00215; Oleophosphothion; Orthomalathion; PBI crop saver; Phosphothion; Prioderm; Sadofos; Sadophos; SF 60; Siptox I; Spray concentrate; STCC 4941156; Succinic acid, mercapto-, diethyl ester, *S*-ester with *O,O*-dimethyl phosphorodithioate; Sumitox; TAK; TM-4049; Vetiol; Zithiol

CAS Registry Number: 121-75-5

RTECS® Number: WM8400000

UN/NA & ERG Number: UN2783 (organophosphorus pesticides, solid, toxic)/152

EC Number: 204-497-7 [*Annex I Index No.:* 015-041-00-X]

Regulatory Authority and Advisory Bodies

Carcinogenicity: IARC: Animal Inadequate Evidence; Human No Adequate Data, *not classifiable as carcinogenic to humans,* Group 3, 1987; NCI: Carcinogenesis Bioassay (feed); no evidence: rat; IARC: Animal Inadequate Evidence.

US EPA Gene-Tox Program, Positive/dose response: *In vitro* SCE—human; Negative: Carcinogenicity—mouse/rat; Histidine reversion—Ames test; Negative: *D. melanogaster* sex-linked lethal; Negative: *In vitro* UDS—human fibro-blast; TRP reversion; Negative: *S. cerevisiae*—homozygosis; Inconclusive: *B. subtilis* rec assay; *E. coli* polA without S9.

Air Pollutant Standard Set. See below, "Permissible Exposure Limits in Air" section.

FDA—proprietary drug.

Clean Water Act: Section 311 Hazardous Substances/RQ 40CFR117.3 (same as CERCLA, see below).

Reportable Quantity (RQ): 100 lb (45.4 kg).

EPCRA Section 313 Form R *de minimis* concentration reporting level: 1.0%.

US DOT Regulated Marine Pollutant (49CFR172.101, Appendix B).

European/International Regulations: Hazard Symbol: Xn, N; Risk phrases: R22; R43; R50/53; Safety phrases: S2; 24; S37; S46; S60; S61 (see Appendix 4).

WGK (German Aquatic Hazard Class): 3—Severe hazard to waters.

Description: It is a deep-brown to yellow liquid with a garlic-like odor. Clear and colorless when pure; Freezing/Melting point = 3°C; Molecular weight = 330.38; Boiling point = 156−157°C; Vapor pressure = 8×10^{-6} mmHg at 20°C; Flash point ≥163°C. Hazard Identification (based on NFPA-704 M Rating System): Health 2, Flammability 1, Reactivity 0. Slightly soluble in water.

Potential Exposure: Compound Description: Agricultural Chemical; Tumorigen, Mutagen; Reproductive Effector; Human Data. Malathion is marketed as 99.6% technical grade liquid. Available formulations include wetable powders (25% and 50%), emulsifiable concentrates, dusts, and aerosols. Malathion is used as a broad spectrum insecticide and acaricide in the control of certain insect pests on fruits, vegetables, and ornamental plants. It has been used in the control of houseflies, mosquitoes, lice, and on farm and livestock animals.

Incompatibilities: Reacts violently with strong oxidizers, magnesium, alkaline pesticides. Attacks metals including iron, steel, tin plate, lead, copper, and some plastics, coatings, and rubbers.

Permissible Exposure Limits in Air

OSHA PEL: 15 mg/m³ (*total dust*) TWA [skin].

NIOSH REL: 10 mg/m³ TWA [skin].

ACGIH TLV®[1]: 1 mg/m³ TWA measured as inhalable fraction and vapor; [skin]; not classifiable as a human carcinogen; TLV-BEI$_A$ issued as Acetylcholinesterase-inhibiting pesticides.

NIOSH IDLH: 250 mg/m³.

Protective Action Criteria (PAC)*

TEEL-0: 1 mg/m³

PAC-1: **15** mg/m³

PAC-2: **120** mg/m³

PAC-3: **390** mg/m³

*AEGLs (Acute Emergency Guideline Levels) & ERPGs (Emergency Response Planning Guideline) are in **bold face**.

DFG MAK: 15 mg/m³ inhalable fraction TWA; Peak Limitation Category II(4); Pregnancy Risk Group D.

Arab Republic of Egypt: TWA 10 mg/m³, [skin], 1993; Australia: TWA 10 mg/m³, [skin], 1993; Austria: MAK: 10 mg/m³, 1999; Belgium: TWA 10 mg/m³, [skin], 1993; Denmark: TWA 5 mg/m³, [skin], 1999; Finland: TWA 10 mg/m³; STEL 20 mg/m³, 1999; France: VME 10 mg/m³, [skin], 1999; the Netherlands: MAC-TGG 10 mg/m³, [skin], 2003; Norway: TWA 5 mg/m³, 1999; the Philippines: TWA 15 mg/m³, [skin], 1993; Poland: MAC (TWA) 1 mg/m³, MAC (STEL) 10 mg/m³, 1999; Russia: STEL 0.5 mg/m³, [skin], 1993; Switzerland: MAK-W 10 mg/m³, [skin], 1999; Thailand: TWA 15 mg/m³, 1993; Turkey: TWA 15 mg/m³, [skin], 1993; United Kingdom: TWA 10 mg/m³, [skin], 2000; Argentina, Bulgaria, Columbia, Jordan, South Korea, New Zealand, Singapore, Vietnam: ACGIH TLV®: not classifiable as a human carcinogen. Argentina[35]: STEL 1.5 mg/m³. Russia[43] has set MAC values in the ambient air of residential areas at 0.015 mg/m³ on a momentary basis and 0.006 mg/m³ on a daily average basis. Several states have set guidelines or standards for malathion, in ambient air[60] ranging from 33.3 μg/m³ (New York) to 100 μg/m³ (Florida, North Dakota, South Carolina) to 160 μg/m³ (Virginia) to 200 μg/m³ (Connecticut) to 238 μg/m³ (Nevada).

Determination in Air: OSHA versatile sampler-2; Toluene/Acetone; Gas chromatography/Flame photometric detection for sulfur, nitrogen, or phosphorus; NIOSH Analytical Method (IV) #5600, Organophosphorus Pesticides; OSHA Analytical Method ID-62.

Permissible Concentration in Water: Russia[43] set a MAC of 0.05 mg/L in water bodies used for domestic purposes. The MAC in water bodies used for fishery purposes is zero. Several states have set guidelines for malathion in drinking water[61] ranging from 40 μg/L (Maine to 140 μg/L (Kansas) to 160 μg/L (California).

Determination in Water: Fish Tox = 0.28991000 ppb (EXTRA HIGH).

Routes of Entry: Inhalation of vapor, skin absorption, ingestion, and skin and/or eye contact.

Harmful Effects and Symptoms

Short Term Exposure: Malathion is representative of a general class called organophosphates. The effects caused by many short-term exposures during a week's time can be accumulated and felt as one intense response. Sometimes effects are not felt until hours or days after exposure. *Inhalation:* No effects were reported from exposures of up to 86 mg/m³ for 42 days. The only effect reported due to inhalation was the reduction in activity of an important nervous system enzyme. *Skin:* Important route of exposure during formulation and usage. Prolonged contact (hours) along with poor hygiene has resulted in irritation, as well as symptoms listed under ingestion. *Eyes:* Direct contact can lead to irritation and discomfort. *Ingestion:* Swallowing of malathion has caused severe poisoning and death. Swallowing of 1½–3 oz of a moisture (50% malathion) has caused severe poisoning with symptoms which include nausea, vomiting, headache, abdominal pain, diarrhea, difficulty in breathing, fall in blood pressure, muscle spasms, paralysis, loss of reflexes, convulsions, and coma. Between 3½ and 5 oz of a mixture (50% malathion) has caused death. Human Tox = 100.00000 ppb (VERY LOW).

Long Term Exposure: Prolonged, daily contact with exposed areas of skin has led to skin irritation and sensitization. May cause genetic changes (mutations). High or repeated exposure may damage the nerves, causing weakness, dizziness, and poor coordination in arms and legs. Repeated exposures may cause personality changes, depression, anxiety, or irritability.

Points of Attack: Eyes, skin, respiratory system, liver, blood cholinesterase, central nervous system, cardiovascular system, gastrointestinal tract.

Medical Surveillance: Before employment and at regular times after that, the following are recommended: plasma and red blood cell cholinesterase levels (tests for the enzyme poisoned by this chemical). If exposure stops, plasma levels return to normal in 1–2 weeks while red blood cell levels may be reduced for 1–3 months. When cholinesterase enzyme levels are reduced by 25% or more below preemployment levels, risk of poisoning is increased, even if results are in lower ranges of "normal." Reassignment to work not involving organophosphate or carbamate pesticides is recommended until enzyme levels recover. If symptoms develop or overexposure occurs, repeat the above tests as soon as possible and get an examination of the nervous system. Also consider complete blood count. Consider chest X-ray following acute overexposure. Preplacement and periodic medical examination shall include comprehensive initial or interim medical and work histories. A physical examination which shall be directed toward, but not limited to evidence of frequent headache, dizziness, nausea, tightness of the chest, dimness of vision, and difficulty in focusing the eyes. Determination, at the time of the preplacement examination, of a baseline or working baseline erythrocyte ChE activity. A judgment of the worker's physical ability to use negative or positive-pressure regulators as defined in 29 CFG 1910.134. Periodic examinations shall be made available on an annual basis or at some other interval determined by the responsible physician. Medical records shall be maintained for all workers engaged in the manufacture or formulation of malathion and such records shall be kept for at least 1 year after termination of employment. Pertinent medical information shall be available to medical representatives of the US Government, the employer, and the employees. Erythrocyte cholinesterase levels should be checked as noted above and as described in detail by NIOSH Criteria Document No. 76-205.

First Aid: If this chemical gets into the eyes, remove any contact lenses at once and irrigate immediately for at least 15 min, occasionally lifting upper and lower lids. Seek medical attention immediately. If this chemical contacts the skin, remove contaminated clothing and wash immediately with soap and water. Speed in removing material from skin is of extreme importance. Shampoo hair promptly if contaminated. Seek medical attention immediately. If this chemical has been inhaled, remove from exposure, begin rescue breathing (using universal precautions, including resuscitation mask) if breathing has stopped and CPR if heart action has stopped. Transfer promptly to a medical facility. When this chemical has been swallowed, get medical attention. Give large quantities of water and induce vomiting. Do not make an unconscious person vomit.

Note to physician or trained medical personnel: Administer atropine, 2 mg (0.030 g) intramuscularly or intravenously as soon as any local or systemic signs or symptoms of an intoxication are noted; repeat the administration of atropine every 3–8 min until signs of atropinization (mydriasis, dry mouth, rapid pulse, hot and dry skin) occur; initiate treatment in children with 1 mg of atropine. Watch respiration and remove bronchial secretions if they appear to be obstructing the airway; intubate if necessary. Give 2-PAM (Pralidoxime; Protopam), 2.5 g in 100 mL of sterile water or in 5% dextrose and water, intravenously, slowly, in 15–30 min; if sufficient fluid is not available, give 1 g of 2-PAM in 3 mL of distilled water by deep intramuscular injection; repeat this every half hour if respiration weakens or if muscle fasciculation or convulsions recur.

Personal Protective Methods: Wear protective gloves and clothing to prevent any reasonable probability of skin contact: **4 h:** Teflon™ gloves, suits, boots; 4H™ and Silver Shield™ gloves. Safety equipment suppliers/manufacturers can provide recommendations on the most protective glove/clothing material for your operation. All protective clothing (suits, gloves, footwear, headgear) should be clean, available each day, and put on before work. Contact lenses should not be worn when working with this chemical. Wear dust-proof chemical goggles and face shield unless full face-piece respiratory protection is worn. Employees should wash immediately with soap when skin is wet or contaminated. Provide emergency showers and eyewash. *Respirator Protection:* Engineering controls shall be used wherever feasible to maintain airborne malathion concentrations below the recommended work-place environment limit. Compliance with the work-place environmental limit by the use of respirators is allowed only when airborne malathion

concentrations exceed the work-place environmental limit because required engineering controls are being installed or tested, when nonroutine maintenance or repair is being accomplished, or during emergencies. When a respirator is thus permitted, it shall be selected and used in accordance with NIOSH requirements.

Respirator Selection: *100 mg/m³:* CcrOv95 (APF = 10) [any air-purifying half-mask respirator equipped with an organic vapor cartridge(s) in combination with an N95, R95, or P95 filter. The following filters may also be used: N99, R99, P99, N100, R100, P100]; or Sa (APF = 10) (any supplied-air respirator) SCBA (any self-contained breathing apparatus). *250 mg/m³:* Sa:Cf (APF = 25) (any supplied-air respirator operated in a continuous-flow mode) CcrFOv100 (APF = 50) [any air-purifying full-face-piece respirator equipped with organic vapor cartridge(s) in combination with an N100, R100, or P100 filter] or GmFOv100 (APF = 50) [any air-purifying, full-face-piece respirator (gas mask) with a chin-style, front- or back-mounted organic vapor canister having an N100, R100, or P100 filter] or PaprOvHie (APF = 25) (any powered air-purifying respirator with an organic vapor cartridge in combination with a high-efficiency particulate filter) SCBAF (APF = 50) (any self-contained breathing apparatus with full face-piece) SaF (APF = 50) (any supplied-air respirator with a full face-piece). *Emergency or planned entry into unknown concentrations or IDLH conditions:* SCBAF: Pd,Pp (APF = 10,000) (any self-contained breathing apparatus that has a full face-piece and is operated in a pressure-demand or other positive-pressure mode) SaF: Pd,Pp: ASCBA (APF = 10,000) (any supplied-air respirator that has a full face-piece and is operated in a pressure-demand or other positive-pressure mode in combination with an auxiliary self-contained breathing apparatus operated in a pressure-demand or other positive-pressure mode). *Escape:* GmFOv100 (APF = 50) [any air-purifying, full-face-piece respirator (gas mask) with a chin-style, front- or back-mounted organic vapor canister having an N100, R100, or P100 filter] or SCBAE (any appropriate escape-type, self-contained breathing apparatus).

Note: Substance reported to cause eye irritation or damage; may require eye protection.

Storage: Color Code—Blue: Health Hazard/Poison: Store in a secure poison location. Prior to working with this chemical you should be trained on its proper handling and storage. Store in tightly closed containers in a cool, well-ventilated, uninhabited area below 25°C. Store to avoid contact with oxidizers and alkaline pesticides. Store where possible leakage from containers cannot endanger the worker. Maintain regular inspection of containers for any leakage. Sources of ignition, such as smoking and open flames, are prohibited where this chemical is handled, used, or stored.

Shipping: This compound requires a shipping label of "POISONOUS/TOXIC MATERIALS." It falls in Hazard Class 6.1 and Packing Group III.

Spill Handling: Evacuate persons not wearing protective equipment from area of spill or leak until cleanup is complete. Remove all ignition sources. Ventilate the area of spill or leak. Absorb liquids in vermiculite, dry sand, earth, peat, carbon, or a similar material and deposit in sealed containers. Collect powdered material in the most convenient and safe manner and deposit in sealed containers. It may be necessary to contain and dispose of this chemical as a hazardous waste. If material or contaminated runoff enters waterways, notify downstream users of potentially contaminated waters. Contact your local or federal environmental protection agency for specific recommendations. If employees are required to clean up spills, they must be properly trained and equipped. OSHA 1910.120(q) may be applicable.

Fire Extinguishing: Malathion is combustible but ignites with difficulty. A fire should be extinguished with agents suitable to the surrounding combustibles. Poisonous gases, including sulfur dioxide and phosphoric acid, are produced in fire. If material or contaminated runoff enters waterways, notify downstream users of potentially contaminated waters. Notify local health and fire officials and pollution control agencies. Containers may explode in fire. From a secure, explosion-proof location, use water spray to cool exposed containers. If cooling streams are ineffective (venting sound increases in volume and pitch, tank discolors, or shows any signs of deforming), withdraw immediately to a secure position. If employees are expected to fight fires, they must be trained and equipped in OSHA 1910.156. The only respirators recommended for firefighting are self-contained breathing apparatuses that have full face-pieces and are operated in a pressure-demand or other positive-pressure mode.

Disposal Method Suggested: Malathion is reported to be "hydrolyzed almost instantly" at pH 12; 50%; hydrolysis at pH 0 requires 12 h. Alkaline hydrolysis under controlled conditions (0.5 N NaOH in ethanol) gives quantitative yields of $(CH_3O)_2P(S)SNa$, whereas hydrolysis in acidic media yields $(CH_3O)_2P(S)OH$. On prolonged contact with iron or iron-containing material, it is reported to break down and completely lose insecticidal activity. Incineration together with a flammable solvent in a furnace equipped with afterburner and scrubber is recommended. In accordance with 40CFR165, follow recommendations for the disposal of pesticides and pesticide containers. Must be disposed properly by following package label directions or by contacting your local or federal environmental control agency or by contacting your regional EPA office.

References

National Institute for Occupational Safety and Health. (June 1976). *Criteria for a Recommended Standard: Occupational Exposure to Malathion*, NIOSH Document No. 76-205. Washington, DC
New York State Department of Health. (March 1986). *Chemical Fact Sheet: Malathion*. Version 2. Albany, NY: Bureau of Toxic Substance Assessment
Sax, N. I. (Ed.). (1987). *Dangerous Properties of Industrial Materials Report*, 7, No. 5, 63–74
New Jersey Department of Health and Senior Services. (April 2004). *Hazardous Substances Fact Sheet: Malathion*. Trenton, NJ

Maleic acid   M:0200

Molecular Formula: $C_4H_4O_4$
Common Formula: HOOCCH = CHCOOH
Synonyms: *cis*-Butenedioic acid, (Z)-; Butenedioic acid, (Z)-; *cis*-Butenedioic anhydride; (Z)-1,2-Ethylenedicarboxylic acid; *cis*-1,2-Ethylenedicarboxylic acid; 1,2-Ethylenedicarboxylic acid, (Z); *cis*-1,2-Ethylenedicarboxylic acid, toxilic acid; Maleinic acid; Malenic acid; Toxilic acid
CAS Registry Number: 110-16-7
RTECS® Number: OM9625000
UN/NA & ERG Number: UN2215/156
EC Number: 203-742-5 [Annex I Index No.: 607-095-00-3]
Regulatory Authority and Advisory Bodies
Clean Water Act: Section 311 Hazardous Substances/RQ 40CFR117.3 (same as CERCLA, see below).
Reportable Quantity (RQ): 5000 lb (2270 kg).
Canada, WHMIS, Ingredients Disclosure List Concentration 1.0%.
European/International Regulations: Hazard Symbol: Xn; Risk phrases: R22; R36/37/38; R43; Safety phrases: S2; S24; S26; S28; S37; 46 (see Appendix 4).
WGK (German Aquatic Hazard Class): 1—Low hazard to waters.
Description: Maleic acid is a white crystalline solid with a faint, acidulous odor. Molecular weight = 116.08; Specific gravity (H_2O:1) = 1.59; Boiling point = (decomposes below BP at 135°C); Freezing/Melting point = 131°C; also listed at 138–139°C. Hazard Identification (based on NFPA-704 M Rating System): Health 1, Flammability 1, Reactivity 0. Highly soluble in water; solubility = >75%.
Potential Exposure: Compound Description: Mutagen, Primary Irritant. Maleic acid is used to make artificial resins, antihistamines, and to preserve (retard rancidity) fats and oils.
Incompatibilities: Oxidizers, strong bases, amines, reducing agents, alkali metals.
Permissible Exposure Limits in Air
Protective Action Criteria (PAC)
TEEL-0: 3 mg/m³
PAC-1: 7.5 mg/m³
PAC-2: 60 mg/m³
PAC-3: 300 mg/m³
Permissible Concentration in Water: Russia[43] set a MAC of 1.0 mg/L in water bodies used for domestic purposes.
Determination in Water: Octanol–water coefficient: Log K_{ow} = <– 0.6.
Routes of Entry: Inhalation, skin contact. Passes through the skin.

Harmful Effects and Symptoms

Short Term Exposure: Maleic acid can affect you when breathed in and by passing through your skin. Contact can cause severe eye burns leading to permanent damage. Contact can irritate the skin. Exposure may cause you to feel dizzy and lightheaded. Exposure can irritate the nose, throat, and lungs and cause coughing or shortness of breath. Very high exposures can cause pulmonary edema, a medical emergency that can be delayed for several hours. This can cause death.

Long Term Exposure: May cause lung irritation and bronchitis. May affect the kidneys.

Points of Attack: Lungs, kidneys.

Medical Surveillance: For those with frequent or potentially high exposure, the following are recommended before beginning work and at regular times after that: lung function tests. Kidney function tests. If symptoms develop or overexposure is suspected, the following maybe useful: consider chest X-ray after acute overexposure.

First Aid: If this chemical gets into the eyes, remove any contact lenses at once and irrigate immediately for at least 30 min, occasionally lifting upper and lower lids. Seek medical attention immediately. If this chemical contacts the skin, remove contaminated clothing and wash immediately with soap and water. Seek medical attention immediately. If this chemical has been inhaled, remove from exposure, begin rescue breathing (using universal precautions, including resuscitation mask) if breathing has stopped and CPR if heart action has stopped. Transfer promptly to a medical facility. When this chemical has been swallowed, get medical attention. If victim is conscious, administer water or milk. Do not induce vomiting. Medical observation is recommended for 24—48 h after breathing overexposure, as pulmonary edema may be delayed. As first aid for pulmonary edema, a doctor or authorized paramedic may consider administering a corticosteroid spray.

Personal Protective Methods: Wear protective gloves and clothing to prevent any reasonable probability of skin contact. Safety equipment suppliers/manufacturers can provide recommendations on the most protective glove/clothing material for your operation. For maleic acid (>70%) Natural rubber, Neoprene™, nitrile + PVC, nitrile, polyethylene, and Viton™ are among the recommended protective materials. All protective clothing (suits, gloves, footwear, headgear) should be clean, available each day, and put on before work. Contact lenses should not be worn when working with this chemical. Wear dust-proof chemical goggles and face shield unless full face-piece respiratory protection is worn. Employees should wash immediately with soap when skin is wet or contaminated.

Respirator Selection: Where there is potential for exposure to maleic acid as dust, mist, or fume, use a NIOSH/MSHA- or European Standard EN149-approved full face-piece respirator with a high-efficiency particulate filter. Greater protection is provided by a powered air-purifying respirator. Where there is potential for exposures to maleic acid as a liquid, or for high exposures, use a NIOSH/MSHA- or European Standard EN149-approved supplied air respirator with a full face-piece operated in the positive-pressure mode, or with a full face-piece, hood, or helmet in the continuous-flow mode; or use a NIOSH/MSHA- or European Standard EN149-approved self-contained breathing apparatus with a full face-piece operated in pressure-demand or other positive-pressure mode.

Storage: Color Code—White: Corrosive or Contact Hazard; Store separately in a corrosion-resistant location. Prior to working with this chemical you should be trained on its proper handling and storage. Maleic acid must be stored to avoid contact with oxidizers (such as perchlorates, peroxides, permanganates, chlorates, and nitrates), amines (such as aniline), and alkali metals (such as sodium and potassium), since violent reactions occur. Store in tightly closed containers in a cool, well-ventilated area away from moisture. Sources of ignition, such as smoking and open flames, are prohibited where this chemical is handled, used, or stored.

Shipping: This compound requires a shipping label of "CORROSIVE." It falls in Hazard Class 8 and Packing Group III.

Spill Handling: Evacuate persons not wearing protective equipment from area of spill or leak until cleanup is complete. Remove all ignition sources. For *small spills*, cover with soda ash or sodium bicarbonate, mix, add water, neutralize, and wash down drain with copious amounts of water. Use HEPA vacuum; do not dry sweep. Collect powdered material in the most convenient and safe manner and deposit in sealed containers. Ventilate area after cleanup is complete. It may be necessary to contain and dispose of this chemical as a hazardous waste. If material or contaminated runoff enters waterways, notify downstream users of potentially contaminated waters. Contact your local or federal environmental protection agency for specific recommendations. If employees are required to clean up spills, they must be properly trained and equipped. OSHA 1910.120(q) may be applicable.

Fire Extinguishing: This chemical may burn but does not easily ignite. Use dry chemical, carbon dioxide, water spray, polymer, or alcohol foam extinguishers. Poisonous gases are produced in fire. If material or contaminated runoff enters waterways, notify downstream users of potentially contaminated waters. Notify local health and fire officials and pollution control agencies. Containers may explode in fire. From a secure, explosion-proof location, use water spray to cool exposed containers. If cooling streams are ineffective (venting sound increases in volume and pitch, tank discolors, or shows any signs of deforming), withdraw immediately to a secure position. If employees are expected to fight fires, they must be trained and equipped in OSHA 1910.156. The only respirators recommended for firefighting are self-contained breathing apparatuses that have full face-pieces and are operated in a pressure-demand or other positive-pressure mode.

Disposal Method Suggested: Dissolve or mix the material with a combustible solvent and burn in a chemical incinerator equipped with an afterburner and scrubber. All federal, state, and local environmental regulations must be observed. *Liquid:* incinerate after mixing with a flammable solvent. Use afterburner for complete combustion. *Solid:* dissolve in a flammable solvent or package in paper and burn. See above.

References

Sax, N. I. (Ed.). (1987). *Dangerous Properties of Industrial Materials Report*, 7, No. 1, 61–65

New Jersey Department of Health and Senior Services. (November 1999). *Hazardous Substances Fact Sheet: Maleic Acid*. Trenton, NJ

Maleic anhydride   M:0210

Molecular Formula: $C_4H_2O_3$
Common Formula: $(CHCO)_2O$
Synonyms: Acido malico (Spanish); BM 10; *cis-*Butenedioic anhydride; Dihydro-2,5-dioxofuran; 2,5-Dihydrofuran-2,5-dione; 2,5-Furandione; 2,5-Furanedione; Maleic acid anhydride; Toxilic anhydride
CAS Registry Number: 108-31-6
RTECS® Number: ON3675000
UN/NA & ERG Number: UN2215/156
EC Number: 203-571-6 [*Annex I Index No.:* 607-096-00-9]
Regulatory Authority and Advisory Bodies
Air Pollutant Standard Set. See below, "Permissible Exposure Limits in Air" section.
Clean Air Act: Hazardous Air Pollutants (Title I, Part A, Section 112).
Clean Water Act: Section 311 Hazardous Substances/RQ 40CFR117.3 (same as CERCLA, see below); Section 313 Water Priority Chemicals (57FR41331, 9/9/92).
US EPA Hazardous Waste Number (RCRA No.): U147.
RCRA, 40CFR261, Appendix 8 Hazardous Constituents.
Reportable Quantity (RQ): 5000 lb (2270 kg).
EPCRA Section 313 Form R *de minimis* concentration reporting level: 1.0%.
Canada, WHMIS, Ingredients Disclosure List Concentration 0.1%.
European/International Regulations: Hazard Symbol: C; Risk phrases: R22; R34; R42/43; Safety phrases: S2; S22; S26; S36/37/39; S45 (see Appendix 4).
WGK (German Aquatic Hazard Class): 1—Low hazard to waters.
Description: Maleic anhydride is colorless needles, white lumps, or pellets with an irritating, choking odor. The odor threshold is 0.32 ppm. Molecular weight = 98.06; Specific gravity (H_2O:1) = 1.48; Boiling point = 202.2°C; Freezing/Melting point = 52.8°C; Vapor pressure = 0.2 mmHg at 20°C; Flash point = 102°C (cc); Autoignition temperature = 477°C. Explosive limits: LEL = 1.4%; UEL = 7.1%. Hazard Identification (based on NFPA-704 M Rating System):

Health 3, Flammability 1, Reactivity 1. Slightly reacts with and is soluble in water; solubility = 40%. Maleic anhydride may be transported as hot (70°C) liquid.
Potential Exposure: Compound Description: Tumorigen, Mutagen; Reproductive Effector; Primary Irritant. Maleic anhydride is used in unsaturated polyester resins, agricultural chemical and lubricating additives, in the manufacture of unsaturated polyester resins, in the manufacture of fumaric acid, in alkyd resin manufacture, and in the manufacture of pesticides (e.g., malathion, maleic hydrazide, and captan).
Incompatibilities: Reacts slowly with water (hydrolyzes) to form maleic acid, a medium strong acid. Reacts with strong oxidizers, oil, water, alkali metals, strong acids, strong bases. Violent reaction with alkali metals and amines above 66°C.
Permissible Exposure Limits in Air
Conversion factor: 1 ppm = 4.01 mg/m^3 at 25°C & 1 atm.
OSHA PEL: 0.25 ppm/1 mg/m^3/TWA.
NIOSH REL: 0.25 ppm/1 mg/m^3/TWA.
ACGIH TLV®[1]: 0.1 ppm/0.4 mg/m^3/TWA; danger of sensitizaton. Notice of intended change: 0.01 measured as inhalable fraction and vapor.
Protective Action Criteria (PAC)*
TEEL-0: 0.2 ppm
PAC-1: **0.2 ppm**
PAC-2: **2 ppm**
PAC-3: **20 ppm**
*AEGLs (Acute Emergency Guideline Levels) & ERPGs (Emergency Response Planning Guideline) are in **bold face**.
DFG MAK: 0.1 ppm/0.41 mg/m^3 TWA; Peak Limitation Category I(1), a momentary value of 0.2 mL/m^3/0.81 mg/m^3 should not be exceeded; danger of skin and airways; Pregnancy Risk Group C; danger of sensitization of airways and skin.
NIOSH IDLH: 10 mg/m^3; [danger of skin sensitization and airways]; Pregnancy Risk Group C.
Australia: TWA 0.25 ppm (1 mg/m^3), 1993; Austria: MAK 0.1 ppm (0.4 mg/m^3), 1999; Belgium: TWA 0.25 ppm (1 mg/m^3), 1993; Denmark: TWA 0.2 ppm (0.8 mg/m^3), 1999; Finland: TWA 0.1 ppm, ceiling 0.2 ppm, 1999; France: VLE 1 mg/m^3, 1999; Hungary: TWA 1 mg/m^3; STEL 2 mg/m^3, 1993; the Netherlands: MAC-TGG 0.4 mg/m^3, 2003; Norway: TWA 0.2 ppm (0.8 mg/m^3), 1999; the Philippines: TWA 0.25 ppm (1 mg/m^3), 1993; Poland: TWA 0.5 mg/m^3; STEL 1.0 mg/m^3, 1999; Russia: STEL 1 mg/m^3, 1993; Sweden: NGV 0.3 ppm (1.2 mg/m^3), KTV 0.6 ppm (2.5 mg/m^3), 1999; Switzerland: MAK-W 0.2 ppm (0.8 mg/m^3), KZV-(week) 0.4 ppm (1.6 mg/m^3), 1999; United Kingdom: TWA 1 mg/m^3; STEL 3 mg/m^3, 2000; Argentina, Bulgaria, Columbia, Jordan, South Korea, New Zealand, Singapore, Vietnam: ACGIH TLV®: not classifiable as a human carcinogen. The Czech Republic: TWA 1.0 mg/m^3 with the same value as an STEL.[35] Russia[35, 43] set a MAC of 0.2 mg/m^3 for ambient air in residential

areas on a momentary basis and 0.05 mg/m³ on a daily average basis. Several states have set guidelines or standards for maleic anhydride in ambient air[60] ranging from 0.14 µg/m³ (Massachusetts) to 3.3 µg/m³ (New York) to 10.0 µg/m³ (Florida and South Carolina) to 12.0 µg/m³ (North Carolina) to 17 µg/m³ (Virginia) to 20 µg/m³ (Connecticut) to 24.0 µg/m³ (Nevada) to 100.0 µg/m³ (North Carolina and North Dakota).

Determination in Air: Use NIOSH (IV), Method #3512; OSHA Analytical Method 25 or 86.

Routes of Entry: Inhalation, ingestion, eye and/or skin contact.

Harmful Effects and Symptoms

Short Term Exposure: Maleic anhydride severely irritates the eyes, skin, and respiratory tract. Higher exposures can cause pulmonary edema, a medical emergency that can be delayed for several hours. This can cause death. Inhalation may cause asthmatic reactions. The symptoms of asthma may be delayed for several hours and are aggravated by physical effort. Subacute inhalation of maleic anhydride can cause severe headaches, nosebleeds, nervousness, nausea, and temporary impairment of vision. It can also lead to conjunctivitis and corneal erosion. Maleic anhydride may be transported as hot liquid; skin contact causes burns.

Long Term Exposure: Repeated or prolonged skin contact may cause allergy and dermatitis. Repeated or prolonged inhalation exposure may cause bronchial asthma. Repeated exposure to concentrations above 1.25 ppm has caused asthmatic responses in workers. Allergies have developed so that lower concentrations of maleic anhydride can no longer be tolerated. An increased incidence of bronchitis and dermatitis has also been noted among workers with long-term exposure to maleic anhydride. Repeated exposure may cause photophobia (abnormal visual intolerance to light), double vision, bronchial asthma.

Points of Attack: Eyes, respiratory system, skin.

Medical Surveillance: Before beginning employment and at regular times after that, the following is recommended: lung function tests. These may be normal if the person is not having an attack at the time of the test. If symptoms develop or overexposure is suspected, the following may be useful: evaluation by a qualified allergist, including careful exposure history and special testing, may help diagnose skin allergy. Anyone who has developed symptoms of asthma due to contact with maleic anhydride should avoid all further contact with this chemical. Consider chest X-ray following acute overexposure. Lung function tests.

First Aid: If this chemical gets into the eyes, remove any contact lenses at once and irrigate immediately for at least 15 min, occasionally lifting upper and lower lids. Seek medical attention immediately. If this chemical contacts the skin, remove contaminated clothing and wash immediately with soap and water. Seek medical attention immediately. If this chemical has been inhaled, remove from exposure, begin rescue breathing (using universal precautions, including resuscitation mask) if breathing has stopped and CPR if heart action has stopped. Transfer promptly to a medical facility. When this chemical has been swallowed, get medical attention. If victim is conscious, administer water or milk. Do not induce vomiting. The symptoms of asthma may be delayed for several hours and are aggravated by physical effort. Rest and medical observation are highly recommended. Medical observation is recommended for 24–48 h after breathing overexposure, as pulmonary edema may be delayed. As first aid for pulmonary edema, a doctor or authorized paramedic may consider administering a corticosteroid spray.

Personal Protective Methods: Wear protective gloves and clothing to prevent any reasonable probability of skin contact. **8 h**: Responder™ suits. Safety equipment suppliers/manufacturers can provide recommendations on the most protective glove/clothing material for your operation. ACGIH recommends rubber, Neoprene™, nitrile, polyvinyl chloride and polystyrene as protective material. All protective clothing (suits, gloves, footwear, headgear) should be clean, available each day, and put on before work. Contact lenses should not be worn when working with this chemical. Wear dust-proof chemical goggles and face shield unless full face-piece respiratory protection is worn. Employees should wash immediately with soap when skin is wet or contaminated. Provide emergency showers and eyewash.

Respirator Selection: *10 mg/m³:* Sa:Cf (APF = 25) (any supplied-air respirator operated in a continuous-flow mode) or SCBAF (APF = 50) (any self-contained breathing apparatus with a full face-piece) or SaF (APF = 50) (any supplied-air respirator with a full face-piece). *Emergency or planned entry into unknown concentrations or IDLH conditions:* SCBAF: Pd,Pp (APF = 10,000) (any self-contained breathing apparatus that has a full face-piece and is operated in a pressure-demand or other positive-pressure mode) SaF: Pd,Pp: ASCBA (APF = 10,000) (any supplied-air respirator that has a full face-piece and is operated in a pressure-demand or other positive-pressure mode in combination with an auxiliary self-contained breathing apparatus operated in a pressure-demand or other positive-pressure mode). *Escape:* GmFOv100 (APF = 50) [any air-purifying, full-face-piece respirator (gas mask) with a chin-style, front- or back-mounted organic vapor canister having an N100, R100, or P100 filter] or SCBAE (any appropriate escape-type, self-contained breathing apparatus).

Note: May causes eye damage; eye protection needed.

Storage: Color Code—White: Corrosive or Contact Hazard; Store separately in a corrosion-resistant location. Maleic anhydride must be stored to avoid contact with water and strong oxidizers (such as chlorine and bromine), since violent reactions occur. Before entering confined space where this chemical may be present, check to make sure that an explosive concentration does not exist. Sources of ignition, such as smoking and open flames, are prohibited where this chemical is handled, used, or stored. Metal containers involving the transfer of 5 gallons or more of this chemical should be grounded and bonded. Drums must be equipped with self-closing valves, pressure vacuum bungs, and flame arresters. Use only nonsparking tools and equipment, especially when opening and closing containers of this chemical. Wherever this chemical is used, handled, manufactured, or stored, use explosion-proof electrical equipment and fittings.

Shipping: This compound requires a shipping label of "CORROSIVE." It falls in Hazard Class 8 and Packing Group III.

Spill Handling: Evacuate persons not wearing protective equipment from area of spill or leak until cleanup is complete. Remove all ignition sources. Use HEPA vacuum; do not dry sweep. Establish forced ventilation to keep levels below explosive limit. Collect powdered material in the most convenient and safe manner and deposit in sealed containers. Ventilate area after cleanup is complete. Keep maleic anhydride out of a confined space, such as a sewer, because of the possibility of an explosion, unless the sewer is designed to prevent the buildup of explosive concentrations. It may be necessary to contain and dispose of this chemical as a hazardous waste. If material or contaminated runoff enters waterways, notify downstream users of potentially contaminated waters. Contact your local or federal environmental protection agency for specific recommendations. If employees are required to clean up spills, they must be properly trained and equipped. OSHA 1910.120(q) may be applicable.

Fire Extinguishing: Maleic anhydride is a combustible solid. Dust clouds of the vapor of molten maleic anhydride are explosive on contact with spark or flame. Use CO_2 or alcohol foam extinguishers. Use of dry chemicals or water extinguishers may cause an explosion. Dust clouds or maleic anhydride or the vapors of molten maleic anhydride are explosive on contact with spark or flame. Poisonous gases are produced in fire. If material or contaminated runoff enters waterways, notify downstream users of potentially contaminated waters. Notify local health and fire officials and pollution control agencies. Containers may explode in fire. From a secure, explosion-proof location, use water spray to cool exposed containers. If cooling streams are ineffective (venting sound increases in volume and pitch, tank discolors, or shows any signs of deforming), withdraw immediately to a secure position. If employees are expected to fight fires, they must be trained and equipped in OSHA 1910.156. The only respirators recommended for firefighting are self-contained breathing apparatuses that have full face-pieces and are operated in a pressure-demand or other positive-pressure mode.

Disposal Method Suggested: Consult with environmental regulatory agencies for guidance on acceptable disposal practices. Generators of waste containing this contaminant (≥100 kg/mo) must conform with EPA regulations governing storage, transportation, treatment, and waste disposal. Controlled incineration: care must be taken that complete oxidation to nontoxic products occurs.

References
US Environmental Protection Agency. (August 1, 1978). *Chemical Hazard Information Profile: Maleic Anhydride.* Washington, DC
US Environmental Protection Agency. (April 30, 1980). *Maleic Anhydride, Health and Environmental Effects Profile No. 122.* Washington, DC: Office of Solid Waste
Sax, N. I. (Ed.). (1982). *Dangerous Properties of Industrial Materials Report*, 2, No. 3, 79–81
New Jersey Department of Health and Senior Services. (September 1999). *Hazardous Substances Fact Sheet: Maleic Anhydride.* Trenton, NJ

Maleic hydrazide   M:0220

Molecular Formula: $C_4H_4N_2O_2$
Common Formula: $C_4H_4O_2N_2$
Synonyms: BH dock killer; Bos MH; Burtolin; Chemform; De-cut; Desprout; 1,2-Dihydro-3,6-pyradazinedione; 1,2-Dihydro-3,6-pyridazinedione; 1,2-Dihydropyridazine-3,6-dione; Drexel-Super P; EC 300; ENT 18,870; Fair 30; Fair PS; Hydrazida maleica (Spanish); 6-Hydroxy-3(2H)-pyridazinone; KMH; MAH; Maintain 3; Malazide; Maleic acid hydrazide; Maleic hydrazide fungicide; Maleic hydrazine; Malein 30; Maleinsaurehydrazid (German); *N,N*-Maleoylhydrazine; Malzid; Mazide; MH; MH 30; MH 36 Bayer; MH 40; Regulox; Regulox 50W; Regulox W; Retard; Royal MH 30; Royal Slo-Gro; Slo-Gro; Sprout-stop; Stuntman; Sucker-stuff; Super de-sprout; Super sprout stop; 1,2,3-Tetrahydro-3,6-dioxopyridazine; Vondaldhyde; Vondrax
CAS Registry Number: 123-33-1
RTECS® Number: UR5950000
UN/NA & ERG Number: Not regulated
EC Number: 204-619-9
Regulatory Authority and Advisory Bodies
IARC: Animal Inadequate Evidence; Human No Adequate Data, *not classifiable as carcinogenic to humans*, Group 3, 1987.
Banned or Severely Restricted (Germany, Guatemala, United States) (UN).[13]
US EPA Hazardous Waste Number (RCRA No.): U148.
Reportable Quantity (RQ): 5000 lb (2270 kg).
European/International Regulations: not listed in Annex 1.
WGK (German Aquatic Hazard Class): No value assigned.
Description: Maleic Hydrazide is a crystalline solid. Molecular weight = 112.10; Freezing/Melting point = 292°C (decomposes at 260°C); also reported as >300°C. Hazard Identification (based on NFPA-704 M Rating System): Health 2, Flammability 1, Reactivity 0. Slightly soluble in cold water; more soluble in hot.
Potential Exposure: Those involved in the manufacture, formulation, and application of this plant growth retardant.
Incompatibilities: Strong oxidizers. Slightly corrosive to iron and zinc.

Permissible Exposure Limits in Air
Protective Action Criteria (PAC)
TEEL-0: 0.6 mg/m^3
PAC-1: 2 mg/m^3
PAC-2: 12.5 mg/m^3
PAC-3: 500 mg/m^3

Permissible Concentration in Water: A lowest-observed-adverse-effect-level (LOAEL) of 500 mg/kg/day has been calculated. On the basis of this, the US EPA has calculated a lifetime health advisory of 3.5 mg/L for an adult.

Routes of Entry: Inhalation, ingestion, skin and/eye contact.

Harmful Effects and Symptoms

Short Term Exposure: Irritation of eyes, skin and mucous membranes, tremors, muscle spasms, and skin sensitization are among the consequences of MH exposure. $LD_{50} =$ (oral-rat) 3800 mg/kg (slightly toxic).

Long Term Exposure: May cause liver damage and acute central nervous system effects. May cause mutations (genetic changes).

Points of Attack: Central nervous system, liver, skin.

Medical Surveillance: Liver function tests. Examination by a qualified allergist. Tests of the nervous system.

First Aid: Skin Contact[52]*:* Flood all areas of body that have contacted the substance with water. Do not wait to remove contaminated clothing; do it under the water stream. Use soap to help assure removal. Isolate contaminated clothing when removed to prevent contact by others. *Eye Contact:* Remove any contact lenses at once. Flush eyes well with copious quantities of water or normal saline for at least 20–23 min. Seek medical attention. *Inhalation:* Leave contaminated area immediately; breathe fresh air. Proper respiratory protection must be supplied to any rescuers. If coughing, difficult breathing, or any other symptoms develop, seek medical attention at once, even if symptoms develop many hours after exposure. *Ingestion:* If convulsions are not present, give a glass or two of water or milk to dilute the substance. Assure that the person's airway is unobstructed and contact a hospital or poison center immediately for advice on whether or not to induce vomiting.

Personal Protective Methods: Wear protective gloves and clothing to prevent any reasonable probability of skin contact. Safety equipment suppliers/manufacturers can provide recommendations on the most protective glove/clothing material for your operation. All protective clothing (suits, gloves, footwear, headgear) should be clean, available each day, and put on before work. Contact lenses should not be worn when working with this chemical. Wear dust-proof chemical goggles and face shield unless full face-piece respiratory protection is worn. Employees should wash immediately with soap when skin is wet or contaminated. Provide emergency showers and eyewash.

Respirator Selection: Follow regulations in OSHA 29CFR1910.134 or European Standard EN149. Use a NIOSH/MSHA- or European Standard EN149-approved respirator; or use an approved supplied-air respirator with a full face-piece operated in the positive-pressure mode, or with a full face-piece, hood, or helmet in the continuous-flow mode; or use a NIOSH/MSHA- or European Standard EN149-approved self-contained breathing apparatus with a full face-piece operated in pressure-demand or other positive-pressure mode.

Storage: Color Code—Green: General storage may be used. Prior to working with this chemical you should be trained on its proper handling and storage. Store in a refrigerator or a cool, dry place.

Spill Handling: Evacuate persons not wearing protective equipment from area of spill or leak until cleanup is complete. Remove all ignition sources. Dampen spilled material with acetone to avoid dust. Collect powdered material in the most convenient and safe manner and deposit in sealed containers. Ventilate area after cleanup is complete. It may be necessary to contain and dispose of this chemical as a hazardous waste. If material or contaminated runoff enters waterways, notify downstream users of potentially contaminated waters. Contact your local or federal environmental protection agency for specific recommendations. If employees are required to clean up spills, they must be properly trained and equipped. OSHA 1910.120(q) may be applicable.

Fire Extinguishing: Use dry chemical, carbon dioxide, water spray, or alcohol foam extinguishers. Poisonous gases are produced in fire. If material or contaminated run-off enters waterways, notify downstream users of potentially contaminated waters. Notify local health and fire officials and pollution control agencies. From a secure, explosion-proof location, use water spray to cool exposed containers. If cooling streams are ineffective (venting sound increases in volume and pitch, tank discolors, or shows any signs of deforming), withdraw immediately to a secure position. If employees are expected to fight fires, they must be trained and equipped in OSHA 1910.156. The only respirators recommended for firefighting are self-contained breathing apparatuses that have full face-pieces and are operated in a pressure-demand or other positive-pressure mode.

Disposal Method Suggested: Consult with environmental regulatory agencies for guidance on acceptable disposal practices. Generators of waste containing this contaminant ($\geq$100 kg/mo) must conform with EPA regulations governing storage, transportation, treatment, and waste disposal. In accordance with 40CFR165, follow recommendations for the disposal of pesticides and pesticide containers. Must be disposed properly by following package label directions or by contacting your local or federal environmental control agency or by contacting your regional EPA office.

Reference

U.S. Environmental Protection Agency. (August 1987). *Health Advisory: Maleic Hydrazide.* Washington, DC: Office of Drinking Water

Malononitrile   M:0230

Molecular Formula: $C_3H_2N_2$
Common Formula: $NCCH_2CN$
Synonyms: AI3-24285; Cyanoacetonitrile; α-Cyanoacetonitrile; Dicyanmethane; Dicyanomethane; Malonic acid dinitrile; Malonic dinitrile; Malonodinitrile; Malononitrilo (Spanish); Methane, dicyano-; Methylene cyanide; Methylenedinitrile; NSC 3769; Propanedinitrile; Propanedinitrite
CAS Registry Number: 109-77-3
RTECS® Number: OO3150000
UN/NA & ERG Number: UN2647/153
EC Number: 203-703-2 [*Annex I Index No.:* 608-009-00-7]
Listed on the TSCA inventory. Regulatory Authority and Advisory Bodies
Air Pollutant Standard Set. See below, "Permissible Exposure Limits in Air" section.
US EPA Hazardous Waste Number (RCRA No.): U149.
RCRA, 40CFR261, Appendix 8 Hazardous Constituents.
Superfund/EPCRA 40CFR355, Extremely Hazardous Substances: TPQ = 500/10,000 lb (227/4540 kg).
Reportable Quantity (RQ): 1000 lb (454 kg).
EPCRA Section 313 Form R *de minimis* concentration reporting level: 1.0%.
Canada, WHMIS, Ingredients Disclosure List Concentration 1.0%.
European/International Regulations: Hazard Symbol: T, N; Risk phrases: R23/24/25; R50/53; Safety phrases: S1/2; S23; S27; S45; S60; S61 (see Appendix 4).
WGK (German Aquatic Hazard Class): No value assigned.
Description: Malononitrile is a white powder or colorless, odorless crystalline substance. Molecular weight = 66.07; Specific gravity (H_2O:1) = 1.19; Boiling point = 218.9°C; Freezing/Melting point = 32.2°C; Flash point = 130°C (oc). Hazard Identification (based on NFPA-704 M Rating System): Health 3, Flammability 1, Reactivity 0. Soluble in water; solubility − 13%.
Potential Exposure: Compound Description: Primary Irritant. Malononitrile is used in organic synthesis, as a lubricating oil additive, for thiamine synthesis, for pteridine-type anti-cancer agent synthesis, and in the synthesis of photosensitizers, acrylic fibers, and dyestuffs. It has also been used in the treatment of various forms of mental illness. It has been used as a leaching agent for gold.
Incompatibilities: Strong bases. May polymerize violently on prolonged heating at 129°C or in contact with strong bases at lower temperatures.
Permissible Exposure Limits in Air
Conversion factor: 1 ppm = 2.70 mg/m³ at 25°C & 1 atm.
OSHA PEL: None.
NIOSH REL: 3 ppm/8 mg/m³ TWA.
Protective Action Criteria (PAC)*
TEEL-0: 3 ppm
PAC-1: 3 ppm

PAC-2: **4.9** ppm
PAC-3: **10** ppm
*AEGLs (Acute Emergency Guideline Levels) & ERPGs (Emergency Response Planning Guideline) are in **bold face**.
Connecticut has set a guideline or standard of 160 μg/m³ for malononitrile in ambient air.[60]
Determination in Air: Malononitrile may be determined in air by charcoal tube, Toluene, Gas chromatography/Flame ionization detection, NIOSH Nitriles Criteria Document.
Permissible Concentration in Water: Russia[43] set a MAC of 0.02 mg/L in water bodies used for domestic purposes.
Routes of Entry: Inhalation, skin absorption, ingestion, skin and/or eye contact.
Harmful Effects and Symptoms
Short Term Exposure: Metabolized by body to cyanide and thiocyanate; effects of inhalation of toxic fumes will be related to cyanide. Causes brain and heart damage related to lack of cellular oxygen. It is classified as extremely toxic. Probable oral lethal dose for humans is 5−50 mg/kg, or between 7 drops and 1 teaspoonful, for a 70-kg (150 lb) person. Symptoms of cyanide poisoning include rapid and irregular breathing, anxiety, confusion, odor of bitter almonds (on breath or vomitus), nausea, vomiting (if oral exposure), irregular heartbeat, a feeling of tightness in the chest, bright pink coloration of the skin, unconsciousness followed by convulsions, involuntary urination and defecation, paralysis, and respiratory arrest (heart will beat after breathing stops).
Medical Surveillance: See NIOSH Criteria Document *212 Nitriles*.
First Aid: If this chemical gets into the eyes, remove any contact lenses at once and irrigate immediately for at least 15 min, occasionally lifting upper and lower lids. Seek medical attention immediately. If this chemical contacts the skin, remove contaminated clothing and wash immediately with soap and water. Seek medical attention immediately. If this chemical has been inhaled, remove from exposure, begin rescue breathing (using universal precautions, including resuscitation mask) if breathing has stopped and CPR if heart action has stopped. Transfer promptly to a medical facility. When this chemical has been swallowed, get medical attention. Give large quantities of water and induce vomiting. Do not make an unconscious person vomit.
Personal Protective Methods: Wear protective gloves and clothing to prevent any reasonable probability of skin contact. Safety equipment suppliers/manufacturers can provide recommendations on the most protective glove/clothing material for your operation. All protective clothing (suits, gloves, footwear, headgear) should be clean, available each day, and put on before work. Contact lenses should not be worn when working with this chemical. Wear dust-proof chemical goggles and face shield unless full face-piece respiratory protection is worn. Employees should wash immediately with soap when skin is wet or contaminated. Provide emergency showers and eyewash.

Respirator Selection: NIOSH: *up to 80 mg/m³:* Sa (APF = 10) (any supplied-air respirator). *Up to 200 mg/m³:* Sa:Cf (APF = 25) (any supplied-air respirator operated in a continuous-flow mode). *Up to 400 mg/m³:* SCBAF (APF = 50) (any self-contained breathing apparatus with a full face-piece) SaF (APF = 50) (any supplied-air respirator with a full face-piece). *Up to 667 mg/m³:* SaF: Pd,Pp (APF = 2000) (any supplied-air respirator that has a full face-piece and is operated in a pressure-demand or other positive-pressure mode). Emergency or planned entry into unknown concentrations or IDLH conditions: SCBAF: Pd, Pp (APF = 10,000) (any self-contained breathing apparatus that has a full face-piece and is operated in a pressure-demand or other positive-pressure mode) SaF: Pd,Pp: ASCBA (APF = 10,000) (any supplied-air respirator that has a full face-piece and is operated in a pressure-demand or other positive-pressure mode in combination with an auxiliary self-contained breathing apparatus operated in a pressure-demand or other positive-pressure mode). Escape: GmFOv (APF = 50) [any air-purifying, full-face-piece respirator (gas mask) with a chin-style, front-or back-mounted organic vapor canister] or SCBAE (any appropriate escape-type, self-contained breathing apparatus).

Storage: Color Code—Blue: Health Hazard/Poison: Store in a secure poison location. Prior to working with this chemical you should be trained on its proper handling and storage. Store in a refrigerator under an inert atmosphere for prolonged storage. Keep away from strong bases. May polymerize violently on prolonged heating at 129°C or in contact with strong bases at lower temperatures. May spontaneously explode on storing above 70−80°C.

Shipping: This compound requires a shipping label of "POISONOUS/TOXIC MATERIALS." It falls in Hazard Class 6.1 and Packing Group II.

Spill Handling: Evacuate persons not wearing protective equipment from area of spill or leak until cleanup is complete. Remove all ignition sources. Avoid all skin contact, inhalation, and ingestion. Take up *small spills* with sand or other noncombustible material. Dike far ahead of *large spills* for later disposal. Do not touch spilled material; stop leak if you can do so without risk. Stay upwind and out of low areas. Ventilate area after cleanup is complete. It may be necessary to contain and dispose of this chemical as a hazardous waste. If material or contaminated runoff enters waterways, notify downstream users of potentially contaminated waters. Contact your local or federal environmental protection agency for specific recommendations. If employees are required to clean up spills, they must be properly trained and equipped. OSHA 1910.120(q) may be applicable.

Fire Extinguishing: Use chemical, carbon dioxide, water spray, fog, or foam. Move container from fire area if you can do it without risk. Poisonous gases, including nitrogen oxides and cyanide, are produced in fire. Wear positive pressure breathing apparatus and special protective clothing. When heated to decomposition, malononitrile emits highly toxic fumes. May explode spontaneously or polymerize violently on heating. If material or contaminated runoff enters waterways, notify downstream users of potentially contaminated waters. Notify local health and fire officials and pollution control agencies. Containers may explode in fire. From a secure, explosion-proof location, use water spray to cool exposed containers. If cooling streams are ineffective (venting sound increases in volume and pitch, tank discolors, or shows any signs of deforming), withdraw immediately to a secure position. If employees are expected to fight fires, they must be trained and equipped in OSHA 1910.156. The only respirators recommended for firefighting are self-contained breathing apparatuses that have full face-pieces and are operated in a pressure-demand or other positive-pressure mode.

Disposal Method Suggested: Generators of waste containing this contaminant (≥100 kg/mo) must conform with EPA regulations governing storage, transportation, treatment, and waste disposal.

References

U.S. Environmental Protection Agency. (April 30, 1980). *Malononitrile: Health and Environmental Effects Profile No. 123.* Washington, DC: Office of Solid Waste

National Institute for Occupational Safety and Health. (1978). *Criteria for a Recommended Standard. Occupational Exposure to Nitriles.* US DHEW (NIOSH) Report No. 78-212. Bethesda, MD

US Environmental Protection Agency. (November 30, 1987). *Chemical Hazard Information Profile: Malononitrile.* Washington, DC: Chemical Emergency Preparedness Program

Maneb   M:0240

Molecular Formula: $C_4H_6MnN_2S_4$

Synonyms: Aamangan; Akzo Chemie Maneb; BASF-Maneb Spritzpulver (German); Bavistin M, cosmic; Carbamic acid, ethylenebis(dithio-), manganese salt; Carbamodithioic acid, 1,2-ethanediylbis-, manganese salt; Chem neb; Chloroble M; Cleanacres; CR 3029; Delsene M flowable; Dithane M 22 special; EBDC; ENT 14,875; 1,2-Ethanediylbis(carbamodithioato)(2−)-manganese; 1,2-Ethanediylbiscarbamodithioic acid, manganese complex; 1,2-Ethanediylbiscarbamodithioic acid, manganese(2+) salt (1:1); 1,2-Ethanediylbismaneb, manganese(2+) salt (1:1); Ethylenebisdithiocarbamate manganese; *N,N′*-Ethylene bis (dithiocarbamate manganeux) (French); Ethylenebis(dithiocarbamato), manganese; Ethylenebis(dithiocarbamic acid), manganese salt; Ethylenebis(dithiocarbamic acid) manganous salt; 1,2-Ethylenediylbis(carbamodithioato)manganese; F 10; Griffin manex; Kypman 80; Lonocol M; Manam; Maneb 80; Maneba; Manebe (French); Manebe 80; Manebgan; Manesan; Manex; Mangan (II)-[*N,N′*-aethylen-bis (dithiocarbamate)] (German); Manganese ethylene-1,2-bis-dithiocarbamate; Manganese(II) ethylene di(dithiocarbamate);

Manganous ethylenebis(dithiocarbamate); Manoc; Manzate; Manzate D; Manzate Maneb fungicide; Manzeb; Manzin; M-Diphar; MEB; MNEBD; Multi-W, kascade; Nespor; Plantifog 160M; Polyram M; Remasan chloroble M; Rhodianehe; Sopranebe; Squadron and Quadrangle Manex; Superman Maneb F; Sup'r flo; Tersan-LSR; Trimangol; Trimangol 80; Trimanoc; Trithac; Tubothane; Unicrop Maneb; Vancide; Vancide Maneb 80; Vassgro Manex

CAS Registry Number: 12427-38-2; *(alt.)* 301-03-1; *(alt.)* 11004-49-2; *(alt.)* 12125-33-6; *(alt.)* 20316-06-7; *(alt.)* 28355-56-8; *(alt.)* 133317-06-3

RTECS® Number: OP0700000

UN/NA & ERG Number: UN2210 (with not <60% maneb)/135; UN2968 (preparation, stabilized against self-heating)/135

EC Number: 235-654-8 [*Annex I Index No.:* 006-077-00-7]

Regulatory Authority and Advisory Bodies
Carcinogenicity: IARC: Animal Inadequate Evidence; Human No Adequate Data, *not classifiable as carcinogenic to humans*, Group 3, 1987.
US EPA Gene-Tox Program, Positive: *S. cerevisiae*—homozygosis; Weakly Positive: *In vitro* UDS—human fibroblast; Negative: *S. cerevisiae* gene conversion.
US EPA, FIFRA, 1998 Status of Pesticides: Supported.
Air Pollutant Standard Set. See below, "Permissible Exposure Limits in Air" section
Banned or Severely Restricted (in agriculture) (former USSR-UNEP/IRPTC project).[13]
Carcinogenicity: (New Jersey) (See "References" Below) (California Prop. 65).
EPCRA Section 313 Form R *de minimis* concentration reporting level: 1.0%.
US DOT Regulated Marine Pollutant (49CFR172.101, appendix B).
California Proposition 65 Chemical: Cancer 1/1/90.
European/International Regulations: Hazard Symbol: Xn, N; Risk phrases: R20; R36; R43; R63; R50/53; Safety phrases: S2; S25; S36/37; S46; S60; S61 (see Appendix 4).
WGK (German Aquatic Hazard Class): No value assigned.

Description: Maneb is a yellow powder or crystalline solid with a faint odor. Molecular weight = 265.30; Freezing/Melting point = 130°C (decomposes below MP). Hazard Identification (based on NFPA-704 M Rating System): Health 3, Flammability 3, Reactivity 1. Moderately soluble in water.

Potential Exposure: Compound Description: Agricultural Chemical; Tumorigen, Mutagen; Reproductive Effector. Those involved in manufacture, formulation, and application of this broad spectrum fungicide. Some dithiocarbamates have been used as rubber components. DFG warns of danger of skin sensitization

Incompatibilities: Water, acid, oxidizing materials. Heat or contact with moisture or acids causes rapid decomposition and the generation of toxic and flammable hydrogen sulfide and carbon disulfide.

Permissible Exposure Limits in Air
OSHA PEL: 5 mg[Mn]/m^3 Ceiling Concentration.
NIOSH: 1 mg[Mn]/m^3 TWA; 3 mg[Mn]/m^3 STEL.
No TEEL available.
DFG MAK: Danger of skin sensitization.
Australia: TWA 5 mg[Mn]/m^3, 1993; Belgium: TWA 5 mg [Mn]/m^3, 1993; Denmark: TWA 2.5 mg[Mn]/m^3, 1999; Finland: TWA 2.5 mg[Mn]/m^3, 1999; Hungary: TWA 0.3 mg [Mn]/m^3, short-term exposure limit 0.6 mg[Mn]/m^3, 1993; Poland: MAC (TWA) 0.3 mg[Mn]/m^3, 1993; Russia: STEL 0.5 mg/m^3, 1993; United Kingdom: TWA 5 mg[Mn]/m^3, 2000.

Determination in Water: Use OSHA Analytical Method 107.

Permissible Concentration in Water: A no-adverse-effect-level in drinking water has been determined by MAS/NRC to be 0.035 mg/L. An acceptable daily intake (ADI) of 0.005 mg/kg/day has been calculated for maneb. The state of Maine has set a guideline of 10 μg/L for maneb in drinking water.[61]

Determination in Water: Fish Tox = 0.00193000 ppb (EXTRA HIGH).

Routes of Entry: Inhalation, ingestion, eye and/or skin contact.

Harmful Effects and Symptoms
Short Term Exposure: Maneb irritates the eyes, skin, and respiratory tract. Maneb is low in acute toxicity and does not present alarming properties during long-term administration to experimental animals, except at very high dosages. However, it is a material of concern because of evidence of mutagenic and teratogenic effects as well as the possibility of nitrosation to carcinogenic nitrosamines. A rebuttable presumption against registration of maneb for pesticide uses was issued by EPA on August 10, 1977 on the basis of oncogenicity, teratogenicity, and hazard to wildlife. Human Tox = 5.73770 ppb (HIGH).

Long Term Exposure: Repeated skin contact can cause skin sensitization and rash. High or repeated exposures may interfere with thyroid function (causing goiter), damage the central nervous system, affect liver function, or cause kidney damage.

Points of Attack: Skin, thyroid, liver, kidneys, central nervous system.

Medical Surveillance: If symptoms develop or overexposure is suspected, the following may be useful: examination of the nervous system. Thyroid function tests. Consider kidney and liver function tests with higher or repeated exposures. Examination by a qualified allergist.

First Aid: If this chemical gets into the eyes, remove any contact lenses at once and irrigate immediately for at least 15 min, occasionally lifting upper and lower lids. Seek medical attention immediately. If this chemical contacts the skin, remove contaminated clothing and wash immediately with soap and water. Seek medical attention immediately. If this chemical has been inhaled, remove from exposure, begin rescue breathing (using universal precautions,

including resuscitation mask) if breathing has stopped and CPR if heart action has stopped. Transfer promptly to a medical facility. When this chemical has been swallowed, get medical attention. Give large quantities of water and induce vomiting. Do not make an unconscious person vomit.

Personal Protective Methods: Wear protective gloves and clothing to prevent any reasonable probability of skin contact. Safety equipment suppliers/manufacturers can provide recommendations on the most protective glove/clothing material for your operation. All protective clothing (suits, gloves, footwear, headgear) should be clean, available each day, and put on before work. Contact lenses should not be worn when working with this chemical. Wear dust-proof chemical goggles and face shield unless full face-piece respiratory protection is worn. Employees should wash immediately with soap when skin is wet or contaminated. Provide emergency showers and eyewash.

Respirator Selection: Where there is potential for exposure to maneb, use a NIOSH/MSHA- or European Standard EN149-approved supplied-air respirator with a full face-piece operated in the positive-pressure mode, or with a full face-piece, hood, or helmet in the continuous-flow mode; or use a NIOSH/MSHA- or European Standard EN149-approved self-contained breathing apparatus with a full face-piece operated in pressure-demand or other positive-pressure mode.

Storage: Color Code—Red: Flammability Hazard: Store in a flammable liquid storage area or approved cabinet away from ignition sources and corrosive and reactive materials. Prior to working with this chemical you should be trained on its proper handling and storage. Maneb must be stored to avoid contact with water, since violent reactions occur. Store in tightly closed containers in a cool, well-ventilated area away from acids, moisture, heat, and oxidizing materials. Where possible, automatically transfer material from drums or other storage containers to process containers. A regulated, marked area should be established where this chemical is handled, used, or stored in compliance with OSHA Standard 1910.1045.

Shipping: Maneb, stabilized or maneb preparations, stabilized against self-heating requires a shipping label of "DANGEROUS WHEN WET." It falls in Hazard Class 4.3 and Packing Group III.

Maneb (with not <60% maneb) requires a shipping label of "SPONTANEOUSLY COMBUSTIBLE, DANGEROUS WHEN WET." It falls in Hazard Class 4.2 and Packing Group III.

Spill Handling: Evacuate persons not wearing protective equipment from area of spill or leak until cleanup is complete. Remove all ignition sources. Use HEPA vacuum or wet method to reduce dust during cleanup. Do not dry sweep. Collect powdered material in the most convenient and safe manner and deposit in sealed containers. Ventilate area after cleanup is complete. Keep maneb out of a confined space, such as a sewer, because of the possibility of an explosion, unless the sewer is designed to prevent the buildup of explosive concentrations. It may be necessary to contain and dispose of this chemical as a hazardous waste. If material or contaminated runoff enters waterways, notify downstream users of potentially contaminated waters. Contact your local or federal environmental protection agency for specific recommendations. If employees are required to clean up spills, they must be properly trained and equipped. OSHA 1910.120(q) may be applicable.

Fire Extinguishing: Use dry chemical, soda ash, sand, or lime extinguishers. *Do not use water*. Poisonous gases are produced in fire, including nitrogen oxides, hydrogen sulfide, sulfur oxides, and carbon disulfide. If material or contaminated runoff enters waterways, notify downstream users of potentially contaminated waters. Notify local health and fire officials and pollution control agencies. From a secure, explosion-proof location, use water spray to cool exposed containers. If cooling streams are ineffective (venting sound increases in volume and pitch, tank discolors, or shows any signs of deforming), withdraw immediately to a secure position. If employees are expected to fight fires, they must be trained and equipped in OSHA 1910.156. The only respirators recommended for firefighting are self-contained breathing apparatuses that have full face-pieces and are operated in a pressure-demand or other positive-pressure mode.

Disposal Method Suggested: Generators of waste containing this contaminant (≥100 kg/mo) must conform with EPA regulations governing storage, transportation, treatment, and waste disposal. Maneb is unstable in moisture and is hydrolyzed by acids and hot water. It decomposes at about 100°C but may spontaneously decompose vigorously when stored in bulk. Incineration is the preferred disposal means.[22] In accordance with 40CFR165, follow recommendations for the disposal of pesticides and pesticide containers. Must be disposed properly by following package label directions or by contacting your local or federal environmental control agency or by contacting your regional EPA office.

References

U.S. Environmental Protection Agency, Special Review and Reregistration Division Office of Pesticide Programs. (1998). *Agency Status of Pesticides in Registration, Reregistration, and Special Review* (Rainbow Report). Washington, DC

New Jersey Department of Health and Senior Services. (November 1999). *Hazardous Substances Fact Sheet: Maneb*. Trenton, NJ

Manganese (dust & fume)   M:0250

Molecular Formula: Mn
Synonyms: Colloidal manganese; Cutaval; Elemental manganese; JIS-G 1213; Manganese-55; Manganese element; Manganeso (Spanish); Tripart liquid; Tronamag

CAS Registry Number: 7439-96-5 (metal)
RTECS® Number: OO9275000
UN/NA & ERG Number: UN3077/171
EC Number: 231-105-1

Regulatory Authority and Advisory Bodies

EPA (*Mn and inorganic compounds*): Not classifiable as to human carcinogenicity.

Air Pollutant Standard Set. See below, "Permissible Exposure Limits in Air" section.

Clean Air Act: Hazardous Air Pollutants (Title I, Part A, Section 112), as manganese compounds.

Safe Drinking Water Act: SMCL, 0.05 mg/L.

EPCRA Section 313 Form R *de minimis* concentration reporting level: 1.0%.

Canada, WHMIS, Ingredients Disclosure List Concentration 1.0%.

European/International Regulations: not listed in Annex 1.

WGK (German Aquatic Hazard Class): Nonwater polluting agent.

Description: Manganese is a combustible, lustrous, brittle, silvery soft metal. The most important ore containing manganese is pyrolusite. Manganese may also be produced from ferrous scrap used in the production of electric and open-hearth steel. Molecular weight = 54.94; Specific gravity (H_2O:1) = 7.20 (metal); Boiling point = 1962°C; Freezing/Melting point = 123.8°C. Hazard Identification (based on NFPA-704 M Rating System): (*dust*): Health 0, Flammability 2, Reactivity 2. Insoluble in water.

Potential Exposure: Compound Description: Tumorigen, Mutagen, Human Data; Primary Irritant. Manganese is used in the production of steel, in the manufacture of welding rod coatings and fluxes, in iron and steel industry in steel alloys, e.g., ferromanganese, silicomanganese, manganin, spiegeleisen, and as an agent to reduce oxygen and sulfur content of molten steel. Other alloys may be formed with copper, zinc, and aluminum. Manganese and its compounds are utilized in the manufacture of dry cell batteries (MnO_2), paints, varnishes, inks, dyes, matches and firework, as a fertilizer, disinfectant, bleaching agent, laboratory reagent, drier for oils, an oxidizing agent in the chemical industry particularly in the synthesis of potassium permanganate, and as a decolorizer and coloring agent in the glass and ceramics industry. Exposure may occur during the mining, smelting, and refining of manganese, in the production of various materials, and in welding operations with manganese-coated rods. Manganese normally is ingested as a trace nutrient in food. The average human intake is approximately 10 mg/day.

Incompatibilities: Dust or powder may be pyrophoric or explosive in air. Reacts with water (slowly), steam, or acid producing flammable hydrogen gas. Reacts violently with concentrated hydrogen peroxide. Incompatible with nitrogen gas above 200°C. Oxidizers, nitric acid, nitrogen, finely divided aluminum and other metals, sulfur dioxide, carbon dioxide + heat may cause fire and explosions.

Permissible Exposure Limits in Air

OSHA PEL: 5 mg[Mn]/m³ Ceiling Concentration (inorganic compounds and fume).

NIOSH REL: 1 mg[Mn]/m³ TWA; 3 mg[Mn]/m³ STEL.

ACGIH TLV®[1]: 0.2 mg[Mn]/m³, inorganic compounds; *Notice of intended change*: 0.02 mg[Mn]/m³, respirable fraction; 0.2 mg[Mn]/m³, inhalable fraction; not classifiable as a human carcinogen.

NIOSH IDLH: 500 mg[Mn]/m³.

Protective Action Criteria (PAC)

TEEL-0: 0.2 mg/m³
PAC-1: 3 mg/m³
PAC-2: 5 mg/m³
PAC-3: 500 mg/m³

DFG MAK (*inorganic compounds and fume*): 0.5 mg[Mn]/m³ inhalable fraction (Mn and its inorganic compounds); Pregnancy Risk Group C.

Arab Republic of Egypt: TWA 5 mg/m³, 1993; Australia: TWA 1 mg/m³; STEL 3 mg/m³ (fume), 1993; Austria: MAK 5 mg/m³, 1999; Belgium: TWA 1 mg/m³; STEL 3 mg/m³ (fume), 1993; Denmark: TWA 2.5 mg[Mn]/m³, 1999; Finland: TWA 1 mg/m³, 1999; Finland: TWA 0.5 mg/m³, 1993; France: VME 1 mg/m³ (fume), 1999; the Netherlands: MAC-TGG 1 mg/m³, 2003; Japan: 0.3 mg/m³, respirable dust, 1999; Norway: TWA 1 mg/m³, 1999; the Philippines: TWA 5 mg/m³, 1993; Russia: STEL 0.2 mg/m³ (fume), 1993; Sweden: NGV 1 mg/m³, TGV 2.5 mg/m³ (respirable dust); NGV 2.5 mg/m³, TGV 5 mg/m³ (total dust), 1999; Switzerland: MAK-W 1 mg/m³ (fume); MAK-W 5 mg/m³, 1999; Thailand: TWA 5 mg/m³, 1993; Turkey: TWA 5 mg/m³ (fume), 1993; United Kingdom: TWA 1 mg[Mn]/m³; STEL 3 mg[Mn]/m³, fume; TWA 5 mg[Mn]/m³, 2000; Argentina, Bulgaria, Columbia, Jordan, South Korea, New Zealand, Singapore, Vietnam: ACGIH TLV®: TWA 0.2 mg. the Czech Republic 2.0 mg/m³ (6.0 mg/m³ as a ceiling value). Standards for ambient air in residential areas have been set as follows: the Czech Republic; momentary basis (mg/m³) 0.03; daily average basis (mg/m³) 0.01; Russia; momentary basis (mg/m³) 0.01; daily average basis (mg/m³) 0.001. Several states have set guidelines or standards for manganese in ambient air[60] ranging from 2.0 μg/m³ (Rhode Island) to 10 μg/m³ (North Dakota) to 17 μg/m³ (Virginia) to 20 μg/m³ (Connecticut and South Dakota) to 25 μg/m³ (Pennsylvania) to 119 μg/m³ (Nevada) to 300 μg/m³ (North Carolina).

Determination in Air: Use NIOSH Analytical Method, Elements by ICP, #7300; #7301; #7303; #9102; Elements in blood or tissue, #8005; Metals in urine, #8310; OSHA Analytical Method, ID-125G or ID-121.

Permissible Concentration in Water: There are a variety of maximum allowable concentrations set in various countries[35] (mg/L): the Czech Republic: 0.1 (drinking water); 0.2 (drinking water reserve); 0.5 (surface water); EEC 50.0 (drinking water); United States 0.05 (bottled water); Former USSR-UNEP/IRPTC joint project 0.1 (drinking and surface water); World Health Organization (WHO): 0.05

(maximum desirable); 0.10 (for esthetic quality); 0.50 (maximum permissible in drinking water). In addition, several states have set values for manganese in drinking water. These include a standard of 0.15 mg/L in Illinois and a guideline of 0.05 mg/L in Kansas.

Determination in Water: The manganese detection limit by direct flame atomization is 2 μg/L. However, solvent extraction is used for many determinations. Analytic conditions are more critical for the extraction of manganese than for most other metals, because many manganese-chelate complexes are unstable in solution. With pH control and immediate analysis after extraction, accurate determinations are possible. When the graphite furnace is used to increase sample atomization, the detection limit is lowered to 0.01 μg/L or ng/L according to NAS/NRC.

Routes of Entry: Inhalation of dust or fume, limited percutaneous absorption of liquids, ingestion.

Harmful Effects and Symptoms

Short Term Exposure: Manganese dust and fumes are irritants to the eyes and mucous membranes of the respiratory tract, and apparently are completely innocuous to the intact skin. Inhalation of dust may cause bronchitis and pneumonitis. The effects may be delayed.

Long Term Exposure: The substance may have effects on the lungs and nervous system resulting in bronchitis, pneumonitis, neurologic, and neuropsychiatric disorders (manganism). Animal tests show that this substance possibly causes toxic effects upon human reproduction. Chronic manganese poisoning has long been recognized as a clinical entity. The dust or fumes (manganous compounds) enter the respiratory tract and are absorbed into the blood stream. Manganese is then deposited in major body organs with a special predilection for the liver, spleen, and certain nerve cells of the brain and spinal cord. Among workers there is a very marked variation in individual susceptibility to manganese. Some workers have worked in heavy exposure for a lifetime and shown no signs of the disease; others have developed manganese intoxication within as little as 49 days of exposure. The early phase of chronic manganese poisoning is most difficult to recognize, but it is also important to recognize since early removal from the exposure may arrest the course of the disease. The onset is insidious, with apathy, anorexia, absthenia. Headache, hypersomnia, spasms, weakness of the legs, arthralgias, and irritability are frequently noted. Manganese psychosis follows with certain definitive features: unaccountable laughter, euphoria, impulsive acts, absent-mindedness, mental confusion, aggressiveness, and hallucinations. These symptoms usually disappear with the onset of true neurological disturbances, or may resolve completely with removal from manganese exposure. Progression of the disease presents a range of neurological manifestations that can vary widely among individuals affected. Speech disturbances are common: monotonous tone, inability to speak above a whisper, difficult articulation, incoherence, even complete muteness. The face may take on mask-like quality, and handwriting may be affected by micrographia. Disturbances in gait and balance occur, and frequently propulsion, retropropulsion, and lateropropulsion are affected, with no movement for protection when falling. Tremors are frequent, particularly of the tongue, arms, and legs. These will increase with intentional movements and are more frequent at night. Absolute detachment, broken by sporadic or spasmodic laughter, ensues, and as in extrapyramidal affections, there may be excessive salivation and excessive sweating. At this point the disease is indistinguishable from classical Parkinson's disease. Chronic manganese poisoning is not a fatal disease although it is extremely disabling. Manganese dust is not longer believed to be a causative factor in pneumonia. If there is any relationship at all, it appears to be as an aggravating factor to a pre-existing condition. Freshly formed fumes have been reported to cause fever and chills similar to metal fume fever.

Points of Attack: Respiratory system, central nervous system, lungs, blood, kidneys.

Medical Surveillance: NIOSH lists the following tests: whole blood (chemical/metabolite), biologic tissue/biopsy, complete blood count, chest X-ray, pulmonary function tests, urine (chemical/metabolite), urinalysis (routine). For those with frequent or potentially high exposure (half the TLV or greater), the following are recommended before beginning work and at regular times after that: A complete examination of the nervous system. Complete blood count. Lung function tests. These may be normal if the person is not having an attack at the time of the test. Kidney function tests. If symptoms develop or overexposure is suspected, the following may be useful: consider chest X-ray after acute overexposure. Liver function tests.

First Aid: If this chemical gets into the eyes, remove any contact lenses at once and irrigate immediately for at least 15 min, occasionally lifting upper and lower lids. Seek medical attention immediately. If this chemical contacts the skin, remove contaminated clothing and wash immediately with soap and water. Seek medical attention immediately. If this chemical has been inhaled, remove from exposure, begin rescue breathing (using universal precautions, including resuscitation mask) if breathing has stopped and CPR if heart action has stopped. Transfer promptly to a medical facility. When this chemical has been swallowed, get medical attention. Give large quantities of water and induce vomiting. Do not make an unconscious person vomit. The symptoms of metal fume fever may be delayed for 4—12 h following exposure: it may last less than 36 h. Medical observation is recommended.

Personal Protective Methods: Wear protective gloves and clothing to prevent any reasonable probability of skin contact. Safety equipment suppliers/manufacturers can provide recommendations on the most protective glove/clothing material for your operation. All protective clothing (suits, gloves, footwear, headgear) should be clean, available each day, and put on before work. Contact lenses should not be worn when working with this chemical. Wear dust-proof chemical goggles and face shield unless full face-piece

respiratory protection is worn. Employees should wash immediately with soap when skin is wet or contaminated. Provide emergency showers and eyewash.

Respirator Selection: Up to 10 mg/m³: 95XQ (APF = 10) [any particulate respirator equipped with an N95, R95, or P95 filter (including N95, R95, and P95 filtering face-pieces) except quarter-mask respirators. The following filters may also be used: N99, R99, P99, N100, R100, P100] or Sa (APF = 10) (any supplied-air respirator). *Up to 25 mg/m³:* Sa:Cf (APF = 25) (any supplied-air respirator operated in a continuous-flow mode) PaprHie (APF = 25) (any powered air-purifying respirator with a high-efficiency particulate filter). *Up to 50 mg/m³:* 100F (APF = 50) (any air-purifying, full-face-piece respirator with an N100, R100, or P100 filter) or SaT: Cf (APF = 50) (any supplied-air respirator that has a tight-fitting face-piece and is operated in a continuous-flow mode) or PaprTHie (APF = 50) (any powered, air-purifying respirator with a tight-fitting face-piece and a high-efficiency particulate filter) or SCBAF (APF = 50) (any self-contained breathing apparatus with a full face-piece) or SaF (APF = 50) (any supplied-air respirator with a full face-piece). *Up to 500 mg/m³:* Sa: Pd,Pp (APF = 1000) (any supplied-air respirator operated in a pressure-demand or other positive-pressure mode). *Emergency or planned entry into unknown concentrations or IDLH conditions:* SCBAF: Pd,Pp (APF = 10,000) (any self-contained breathing apparatus that has a full face-piece and is operated in a pressure-demand or other positive-pressure mode) Ascba (any supplied-air respirator that has a full face-piece and is operated in a pressure-demand or other positive-pressure mode in combination with an auxiliary self-contained positive-pressure breathing apparatus). *Escape:* 100F (APF = 50) (any air-purifying, full-face-piece respirator with an N100, R100, or P100 filter) or SCBAE (any appropriate escape-type, self-contained breathing apparatus).

Storage: Color Code—Green: General storage may be used. Prior to working with this chemical you should be trained on its proper handling and storage. Manganese must be stored to avoid contact with water and steam since flammable hydrogen gas is produced. Store in tightly closed containers in a cool, well-ventilated area away from oxidizers (such as perchlorates, peroxides, permanganates, chlorates and nitrates). Protect storage against physical damage.

Shipping: Manganese is not specifically cited in DOT's Performance-Oriented Packaging Standards.[19] However as an Environmentally hazardous solid, n.o.s. falls in Hazard Class 9 and Packing Group III.

Spill Handling: Evacuate persons not wearing protective equipment from area of spill or leak until cleanup is complete. Remove all ignition sources. Collect powdered material in the most convenient and safe manner and deposit in sealed containers. Ventilate area after cleanup is complete. It may be necessary to contain and dispose of this chemical as a hazardous waste. If material or contaminated runoff enters waterways, notify downstream users of potentially contaminated waters. Contact your local or federal environmental protection agency for specific recommendations. If employees are required to clean up spills, they must be properly trained and equipped. OSHA 1910.120(q) may be applicable.

Fire Extinguishing: Use dry chemicals appropriate for extinguishing metal fires. *Do not use water.* Poisonous gases are produced in fire. If material or contaminated runoff enters waterways, notify downstream users of potentially contaminated waters. Notify local health and fire officials and pollution control agencies. From a secure, explosion-proof location, use water spray to cool exposed containers. If cooling streams are ineffective (venting sound increases in volume and pitch, tank discolors, or shows any signs of deforming), withdraw immediately to a secure position. If employees are expected to fight fires, they must be trained and equipped in OSHA 1910.156. The only respirators recommended for firefighting are self-contained breathing apparatuses that have full face-pieces and are operated in a pressure-demand or other positive-pressure mode.

Disposal Method Suggested: Manganese metal-sanitary landfill. Manganese chloride or sulfate-chemical conversion to the oxide followed by land filling, or conversion to the sulfate for use in fertilizer. Consult with environmental regulatory agencies for guidance on acceptable disposal practices.

References

Illinois Institute for Environmental Quality. (September 1975). *Airborne Manganese Health Effects and Recommended Standard, Document No. 75-18.* Chicago, IL

National Academy of Sciences. (1973). *Manganese (in a Series on Medical and Biologic Effects of Environmental Pollutants).* Washington, DC

National Academy of Sciences. (1973). *Medical and Biologic Effects of Environmental Pollutants: Manganese.* Washington, DC

Sax, N. I. (Ed.). (1980). *Dangerous Properties of Industrial Materials Report*, 1, No. 2, 44–45

New Jersey Department of Health and Senior Services. (September 1999). *Hazardous Substances Fact Sheet: Manganese (Dust and Fume).* Trenton, NJ

Manganese dioxide   M:0260

Molecular Formula: MnO$_2$

Synonyms: Black manganese oxide; Bog manganese; Braunstein (German); Cement black; C.I. 77728; C.I. Pigment black 14; C.I. Pigment brown 8; Mangandioxid (German); Manganese binoxide; Manganese (bioxyd de) (French); Manganese black; Manganese (dioxyde de)

(French); Manganese peroxide; Manganese superoxide; Pyrolusite brown

CAS Registry Number: 1313-13-9

RTECS® Number: OP0350000

UN/NA & ERG Number: UN1479 (oxidizing solid, n.o.s.)/140

EC Number: 215-202-6 [*Annex I Index No.:* 025-001-00-3]

Regulatory Authority and Advisory Bodies

Air Pollutant Standard Set. See below, "Permissible Exposure Limits in Air" section.

Clean Air Act: Hazardous Air Pollutants (Title I, Part A, Section 112).

EPCRA Section 313: Includes any unique chemical substance that contains manganese as part of that chemical's infrastructure. Form R *de minimis* concentration reporting level: 1.0%.

Canada, WHMIS, Ingredients Disclosure List Concentration 1.0% as manganese compounds, n.o.s.

European/International Regulations: Hazard Symbol: Xn; Risk phrases R20/22; Safety phrases: S2; S25 (see Appendix 4).

WGK (German Aquatic Hazard Class): No value assigned.

Description: Manganese dioxide is a black crystalline solid. Molecular weight = 86.94; Freezing/Melting point = (decomposes) 553°C. Hazard Identification (based on NFPA-704 M Rating System): Health 3, Flammability 0, Reactivity 3 (Oxidizer). Insoluble in water.

Potential Exposure: Compound Description: Reproductive Effector; Human Data. Manganese dioxide is used as depolarizer for dry cell batteries, for production of manganese metal, as an oxidizing agent, laboratory reagent, and in making pyrotechnics and matches, in dry cell batteries.

Incompatibilities: A powerful oxidizer. Incompatible with strong acids, reducing agents, combustible materials (such as fuel, clothing and organic materials). Mixtures with calcium hydride is a heat- and friction-sensitive explosive. Vigorous reaction with hydrogen sulfide, diboron tetrafluoride, calcium hydride, chlorine trifluoride, hydrogen peroxide, hydroxyaluminum chloride, anilinium perchlorate. Decomposes when heated above 553°C producing manganese(III)oxide and oxygen, which increases fire hazard. Reacts violently with aluminum (thermite reaction), potassium azide, rubidium acetylide, in the presence of heat.

Permissible Exposure Limits in Air

OSHA PEL: 5 mg[Mn]/m^3 Ceiling Concentration (compounds and fume).

NIOSH: 1 mg[Mn]/m^3 TWA; 3 mg[Mn]/m^3 STEL.

ACGIH TLV®[1]: TWA 0.2 mg[Mn]/m^3, inorganic compounds.

Protective Action Criteria (PAC)

TEEL-0: 0.317 mg/m^3

PAC-1: 4.75 mg/m^3

PAC-2: 7.91 mg/m^3

PAC-3: 500 mg/m^3

DFG MAK: 0.5 mg[Mn]/m^3 inhalable fraction (Mn and its inorganic compounds); Pregnancy Risk Group C.

NIOSH IDLH: 500 mg[Mn]/m^3.

Australia: TWA 5 mg[Mn]/m^3, 1993; Belgium: TWA 5 mg[Mn]/m^3, 1993; Denmark: TWA 2.5 mg[Mn]/m^3, 1999; Finland: TWA 0.5 mg[Mn]/m^3, 1999; Hungary: TWA 0.3 mg[Mn]/m^3; STEL 0.6 mg[Mn]/m^3, 1993; Japan 0.3 mg[Mn]/m^3, respirable dust, 1999; Norway: TWA 2.5 mg[Mn]/m^3, 1999; Poland: MAC (TWA) 0.3 mg[Mn]/m^3, MAC 5 mg[Mn]/m^3, 1999; Sweden: NGV 1 mg[Mn]/m^3, TGV 2.5 mg[Mn]/m^3 (respirable dust), 1993; Sweden: NGV 2.5 mg[Mn]/m^3, TGV 5 mg[Mn]/m^3 (total dust), 1993; United Kingdom: TWA 5 mg[Mn]/m^3, 2000; Argentina, Bulgaria, Columbia, Jordan, South Korea, New Zealand, Singapore, Vietnam: ACGIH TLV®: TWA 0.2 mg[Mn]/m^3.

Routes of Entry: Inhalation, ingestion.

Harmful Effects and Symptoms

Short Term Exposure: Manganese dioxide can affect you when breathed in. Inhalation can cause irritation of the respiratory system. Manganese dioxide can cause a flu-like illness, with chills, fever, and aching. Chest congestion can occur with cough and shortness of breath. It can cause an asthma-like lung allergy. The effects may be delayed.

Long Term Exposure: Repeated exposure can cause permanent brain damage. Early symptoms include poor appetite, weakness, and sleepiness. Later effects include speech, balance, and personality changes. Later symptoms are identical to Parkinson's disease. High or repeated exposure may cause permanent lung damage, kidney and liver damage, and anemia. May affect the lungs and nervous system causing bronchitis, pneumonitis, neurologic and neuropsychiatric disorders (manganism). Animal tests show that this substance may cause toxic effects upon human reproduction.

Points of Attack: Lungs, blood, central nervous system, liver, kidneys.

Medical Surveillance: For those with frequent or potentially high exposure (half the TLV or greater), the following are recommended before beginning work and at regular times after that: complete examination of the nervous system. Complete blood count (CBC). Lung function tests. These may be normal if the person is not having an attack at the time of the test. Kidney function test. If symptoms develop or overexposure is suspected, the following may be useful: consider chest X-ray after acute overexposure. Liver function tests. Evaluation for brain effects. Positive and borderline individuals should be referred for neuropsychological testing.

First Aid: If this chemical gets into the eyes, remove any contact lenses at once and irrigate immediately for at least 15 min, occasionally lifting upper and lower lids. Seek medical attention immediately. If this chemical contacts the skin, remove contaminated clothing and wash immediately with soap and water. Seek medical attention immediately. If this chemical has been inhaled, remove from exposure, begin rescue breathing (using universal precautions, including resuscitation mask) if breathing has stopped and CPR if

heart action has stopped. Transfer promptly to a medical facility. When this chemical has been swallowed, get medical attention. Give large quantities of water and induce vomiting. Do not make an unconscious person vomit.

Personal Protective Methods: Wear protective gloves and clothing to prevent any reasonable probability of skin contact. Safety equipment suppliers/manufacturers can provide recommendations on the most protective glove/clothing material for your operation. All protective clothing (suits, gloves, footwear, headgear) should be clean, available each day, and put on before work. Contact lenses should not be worn when working with this chemical. Wear dust-proof chemical goggles and face shield unless full face-piece respiratory protection is worn. Employees should wash immediately with soap when skin is wet or contaminated. Provide emergency showers and eyewash.

Respirator Selection: Up to 10 mg/m^3: 95XQ (APF = 10) [any particulate respirator equipped with an N95, R95, or P95 filter (including N95, R95, and P95 filtering face-pieces) except quarter-mask respirators. The following filters may also be used: N99, R99, P99, N100, R100, P100] or Sa (APF = 10) (any supplied-air respirator). *Up to 25 mg/m^3:* Sa:Cf (APF = 25) (any supplied-air respirator operated in a continuous-flow mode) PaprHie (APF = 25) (any powered air-purifying respirator with a high-efficiency particulate filter). *Up to 50 mg/m^3:* 100F (APF = 50) (any air-purifying, full-face-piece respirator with an N100, R100, or P100 filter) or SaT: Cf (APF = 50) (any supplied-air respirator that has a tight-fitting face-piece and is operated in a continuous-flow mode) or PaprTHie (APF = 50) (any powered, air-purifying respirator with a tight-fitting face-piece and a high-efficiency particulate filter) or SCBAF (APF = 50) (any self-contained breathing apparatus with a full face-piece) or SaF (APF = 50) (any supplied-air respirator with a full face-piece). *Up to 500 mg/m^3:* Sa: Pd,Pp (APF = 1000) (any supplied-air respirator operated in a pressure-demand or other positive-pressure mode). *Emergency or planned entry into unknown concentrations or IDLH conditions:* SCBAF: Pd,Pp (APF = 10,000) (any self-contained breathing apparatus that has a full face-piece and is operated in a pressure-demand or other positive-pressure mode); Ascba (any supplied-air respirator that has a full face-piece and is operated in a pressure-demand or other positive-pressure mode in combination with an auxiliary self-contained positive-pressure breathing apparatus). *Escape:* 100F (APF = 50) (any air-purifying, full-face-piece respirator with an N100, R100, or P100 filter) or SCBAE (any appropriate escape-type, self-contained breathing apparatus).

Storage: Color Code—Yellow: Reactive Hazard (strong oxidizer); Store in a location separate from other materials, especially flammables and combustibles. Prior to working with this chemical you should be trained on its proper handling and storage. Manganese dioxide must be stored to avoid contact with heat and flammable materials and oxidizers (such as perchlorates, peroxides, permanganates, chlorates, and nitrates), since violent reactions occur. See also incompatibilities above. See OSHA Standard 1910.104 and NFPA 43A *Code for the Storage of Liquid and Solid Oxidizers* for detailed handling and storage regulations.

Shipping: Oxidizing solid, n.o.s. requires a shipping label of "OXIDIZER." It falls in Hazard Class 5.1 and Packing Group II.

Spill Handling: Evacuate persons not wearing protective equipment from area of spill or leak until cleanup is complete. Remove all ignition sources. Collect powdered material in the most convenient and safe manner and deposit in sealed containers. Ventilate area after cleanup is complete. It may be necessary to contain and dispose of this chemical as a hazardous waste. If material or contaminated runoff enters waterways, notify downstream users of potentially contaminated waters. Contact your local or federal environmental protection agency for specific recommendations. If employees are required to clean up spills, they must be properly trained and equipped. OSHA 1910.120(q) may be applicable.

Fire Extinguishing: Manganese dioxide itself does not burn, but it will intensify an existing fire. Use extinguishing agents suitable for surrounding fire. Poisonous gases are produced in fire. If material or contaminated runoff enters waterways, notify downstream users of potentially contaminated waters. Notify local health and fire officials and pollution control agencies. From a secure, explosion-proof location, use water spray to cool exposed containers. If cooling streams are ineffective (venting sound increases in volume and pitch, tank discolors, or shows any signs of deforming), withdraw immediately to a secure position. If employees are expected to fight fires, they must be trained and equipped in OSHA 1910.156. The only respirators recommended for firefighting are self-contained breathing apparatuses that have full face-pieces and are operated in a pressure-demand or other positive-pressure mode.

Reference
New Jersey Department of Health and Senior Services. (November 1999). *Hazardous Substances Fact Sheet: Manganese Dioxide.* Trenton, NJ

Manganese nitrate   M:0270

Molecular Formula: MnN$_2$O$_6$
Common Formula: Mn(NO$_3$)$_2$
Synonyms: Manganese dinitrate; Manganous nitrate; Manganous dinitrate; Nitric acid, manganese(2 +) salt
CAS Registry Number: 10377-66-9; 15710-66-4 (hydrate)
RTECS® Number: QU9780000
UN/NA & ERG Number: UN2724/140
EC Number: 233-828-8
Regulatory Authority and Advisory Bodies
Air Pollutant Standard Set. See below, "Permissible Exposure Limits in Air" section.
Clean Air Act: Hazardous Air Pollutants (Title I, Part A, Section 112).

EPCRA Section 313: Includes any unique chemical substance that contains manganese as part of that chemical's infrastructure. Form R *de minimis* concentration reporting level: 1.0%.

Canada, WHMIS, Ingredients Disclosure List Concentration 1.0% as manganese compounds, n.o.s.

WGK (German Aquatic Hazard Class): 1—Slightly water polluting (CAS: 10377-66-9).

Description: Manganese nitrate is a colorless to pink crystalline solid. Molecular weight = 178.96. Hazard Identification (based on NFPA-704 M Rating System): Health 3, Flammability 0, Reactivity 3 (Oxidizer). Highly soluble in water.

Potential Exposure: Manganese nitrate is used as a color agent in porcelain and ceramic manufacture, as a catalyst, and in the production of manganese dioxide.

Incompatibilities: A strong oxidizer. Violent reaction with reducing agents, organics, and combustible materials.

Permissible Exposure Limits in Air

OSHA PEL: $5 mg[Mn]/m^3$ Ceiling Concentration (compounds and fume).

NIOSH: $1 mg[Mn]/m^3$ TWA; $3 mg[Mn]/m^3$ STEL.

ACGIH TLV®[1]: TWA $0.2 mg[Mn]/m^3$, inorganic compounds.

Protective Action Criteria (PAC)

TEEL-0: $1.04 mg/m^3$

PAC-1: $15.7 mg/m^3$

PAC-2: $26.1 mg/m^3$

PAC-3: $500 mg/m^3$

Hydrate

TEEL-0: $0.717 mg/m^3$

PAC-1: $10.8 mg/m^3$

PAC-2: $17.9 mg/m^3$

PAC-3: $500 mg/m^3$

DFG MAK: $0.5 mg[Mn]/m^3$ inhalable fraction (Mn and its inorganic compounds); Pregnancy Risk Group C.

NIOSH IDLH: $500 mg[Mn]/m^3$.

Harmful Effects and Symptoms

Short Term Exposure: Manganese nitrate can affect you when breathed in. Eye or skin contact with the dust or concentrated liquid can cause irritation or burns. The dust or mist can irritate the nose, throat, and bronchial tubes, with cough and phlegm. Higher levels may irritate the lungs and could lead to a fluid buildup in the lungs (pulmonary edema). This can cause death. Overexposure can cause methemoglobinemia and anemia.

Long Term Exposure: Repeated exposure can cause permanent brain damage. Early symptoms include poor appetite, weakness, and sleepiness. Later effects include speech, balance, and personality changes. Later symptoms are identical to Parkinson's disease. High or repeated exposure may cause permanent lung damage, kidney and liver damage, and anemia. May affect the lungs and nervous system causing bronchitis, pneumonitis, neurologic and neuropsychiatric disorders (manganism).

Points of Attack: Blood, nervous system, lungs, liver, kidneys.

Medical Surveillance: For those with frequent or potentially high exposure (half the TLV or greater), the following are recommended before beginning work and at regular times after that: Complete examination of the nervous system. Complete blood count. Lung function and kidney function tests. If symptoms develop or overexposure is suspected, the following may be useful: consider chest X-ray after acute overexposure. Liver function tests. Evaluation for brain effects. Positive and borderline individuals should be referred for neuropsychological testing.

First Aid: If this chemical gets into the eyes, remove any contact lenses at once and irrigate immediately for at least 15 min, occasionally lifting upper and lower lids. Seek medical attention immediately. If this chemical contacts the skin, remove contaminated clothing and wash immediately with soap and water. Seek medical attention immediately. If this chemical has been inhaled, remove from exposure, begin rescue breathing (using universal precautions, including resuscitation mask) if breathing has stopped and CPR if heart action has stopped. Transfer promptly to a medical facility. When this chemical has been swallowed, get medical attention. Give large quantities of water and induce vomiting. Do not make an unconscious person vomit. Medical observation is recommended for 24—48 h after breathing overexposure, as pulmonary edema may be delayed. As first aid for pulmonary edema, a doctor or authorized paramedic may consider administering a corticosteroid spray.

Personal Protective Methods: Wear protective gloves and clothing to prevent any reasonable probability of skin contact. Safety equipment suppliers/manufacturers can provide recommendations on the most protective glove/clothing material for your operation. All protective clothing (suits, gloves, footwear, headgear) should be clean, available each day, and put on before work. Contact lenses should not be worn when working with this chemical. Wear dust-proof chemical goggles and face shield unless full face-piece respiratory protection is worn. Employees should wash immediately with soap when skin is wet or contaminated. Provide emergency showers and eyewash.

Respirator Selection: Up to $10 mg/m^3$: 95XQ (APF = 10) [any particulate respirator equipped with an N95, R95, or P95 filter (including N95, R95, and P95 filtering face-pieces) except quarter-mask respirators. The following filters may also be used: N99, R99, P99, N100, R100, P100] or Sa (APF = 10) (any supplied-air respirator). Up to $25 mg/m^3$: Sa:Cf (APF = 25) (any supplied-air respirator operated in a continuous-flow mode); PaprHie (APF = 25) (any powered air-purifying respirator with a high-efficiency particulate filter). Up to $50 mg/m^3$: 100F (APF = 50) (any air-purifying, full-face-piece respirator with an N100, R100, or P100 filter) or SaT: Cf (APF = 50) (any supplied-air respirator that has a tight-fitting face-piece and is operated in a continuous-flow mode) or PaprTHie (APF = 50) (any powered, air-purifying respirator with a tight-fitting face-piece and a high-efficiency particulate filter) or SCBAF (APF = 50) (any self-contained breathing apparatus with a full face-piece) or SaF (APF = 50) (any supplied-air respirator with a full face-piece). *Up to 500 mg/m³:* Sa: Pd,Pp (APF = 1000) (any supplied-air respirator operated in a pressure-demand or other positive-pressure mode). *Emergency or planned entry into unknown concentrations or IDLH conditions:* SCBAF: Pd,Pp (APF = 10,000) (any self-contained breathing apparatus that has a full face-piece and is operated in a pressure-demand or other positive-pressure mode); Ascba (any supplied-air respirator that has a full face-piece and is operated in a pressure-demand or other positive-pressure mode in combination with an auxiliary self-contained positive-pressure breathing apparatus). *Escape:* 100F (APF = 50) (any air-purifying, full-face-piece respirator with an N100, R100, or P100 filter) or SCBAE (any appropriate escape-type, self-contained breathing apparatus).

Storage: Color Code—Yellow: Reactive Hazard (*strong oxidizer*); Store in a location separate from other materials, especially flammables and combustibles. Prior to working with this chemical you should be trained on its proper handling and storage. Manganese nitrate must be stored to avoid contact with organic materials since violent reactions occur. Store in tightly closed containers in a cool, well-ventilated area. Where possible, automatically transfer material from drums or other storage containers to process containers. See OSHA Standard 1910.104 and NFPA 43A *Code for the Storage of Liquid and Solid Oxidizers* for detailed handling and storage regulations.

Shipping: This compound requires a shipping label of "OXIDIZER." It falls in Hazard Class 5.1 and Packing Group III.

Spill Handling: Evacuate persons not wearing protective equipment from area of spill or leak until cleanup is complete. Remove all ignition sources. Collect powdered material in the most convenient and safe manner and deposit in sealed containers. Ventilate area after cleanup is complete. Keep manganese nitrate out of a confined space, such as a sewer, because of the possibility of an explosion, unless the sewer is designed to prevent the buildup of explosive concentrations. It may be necessary to contain and dispose of this chemical as a hazardous waste. If material or contaminated runoff enters waterways, notify downstream users of potentially contaminated waters. Contact your local or federal environmental protection agency for specific recommendations. If employees are required to clean up spills, they must be properly trained and equipped. OSHA 1910.120(q) may be applicable.

Fire Extinguishing: Manganese nitrate may burn but does not readily ignite. Use dry chemical, CO_2, water spray, or alcohol foam extinguishers. Poisonous gases are produced in fire. If material or contaminated runoff enters waterways, notify downstream users of potentially contaminated waters. Notify local health and fire officials and pollution control agencies. From a secure, explosion-proof location, use water spray to cool exposed containers. If cooling streams are ineffective (venting sound increases in volume and pitch, tank discolors, or shows any signs of deforming), withdraw immediately to a secure position. If employees are expected to fight fires, they must be trained and equipped in OSHA 1910.156. The only respirators recommended for firefighting are self-contained breathing apparatuses that have full face-pieces and are operated in a pressure-demand or other positive-pressure mode.

Reference
New Jersey Department of Health and Senior Services. (February 2000). *Hazardous Substances Fact Sheet: Manganese Nitrate.* Trenton, NJ

Manganese, tricarbonyl methylcyclopentadienyl   M:0280

Molecular Formula: $C_9H_7MnO_3$
Common Formula: $C_5H_4(CH_3)$-$Mn(CO)_3$
Synonyms: Ak-33X; Antiknock-33; CI-2; Combustion improver-2; Manganese, (methylcyclopentadienyl)tricarbonyl-; Methylcyclopentadienyl manganese tricarbonyl; 2-Methylcyclopentadienyl manganese tricarbonyl; Methylcyklopentadientrikarbonylmanganium (German); MMT; TDS-1510; Tricarbonyl(methylcyclopentadienyl) manganese
CAS Registry Number: 12108-13-3
RTECS® Number: OP1450000
UN/NA & ERG Number: UN2810/153; UN3281 (Metal carbonyls, liquid, n.o.s.)/151
EC Number: 235-166-5
Regulatory Authority and Advisory Bodies
Air Pollutant Standard Set. See below, "Permissible Exposure Limits in Air" section.
Clean Air Act: Hazardous Air Pollutants (Title I, Part A, Section 112), as manganese compounds.
Superfund/EPCRA 40CFR355, Extremely Hazardous Substances: TPQ = 100 lb (45.4 kg).
Reportable Quantity (RQ): 100 lb (45.4 kg).
EPCRA Section 313 (as manganese compound) Form R *de minimis* concentration reporting level: 1.0%.
Canada, WHMIS, Ingredients Disclosure List Concentration 1.0%.
European/International Regulations: not listed in Annex 1.
WGK (German Aquatic Hazard Class): No value assigned.
Description: Manganese, tricarbonyl methylcyclopentadienyl is a dark orange liquid. Faintly pleasant, herb-like odor. Molecular weight = 218.10; Boiling point = 232°C; Freezing/Melting point = −18°C; Flash point = 74°C; 96°C. Hazard Identification (based on NFPA-704 M Rating System): Health 3, Flammability 0, Reactivity 0. Practically insoluble in water.
Potential Exposure: Compound Description: Organometallic, Primary Irritant. MMT is used as an octane improver in unleaded gasoline, other distillate fuels and fuel oils, and as a smoke abater in fuels.

Incompatibilities: May be air-reactive. Keep away from oxidizers, halogens. Light causes decomposition.

Permissible Exposure Limits in Air

OSHA PEL: None [5 mg[Mn]/m^3 Ceiling Concentration (vacated 6/30/1993)].

NIOSH REL: 0.2 mg[Mn]/m^3 TWA [skin].

ACGIH TLV$^{®[1]}$: 0.2 mg[Mn]/m^3 TWA [skin].

Protective Action Criteria (PAC)

TEEL-0: 0.2 mg/m^3

PAC-1: 0.6 mg/m^3

PAC-2: 0.6 mg/m^3

PAC-3: 7.5 mg/m^3

Australia: TWA 0.2 mg/m^3, [skin], 1993; Belgium: TWA 0.2 mg/m^3, [skin], 1993; Denmark: TWA 0.1 ppm (0.2 mg [Mn]/m^3), [skin], 1999; Finland: TWA 0.2 mg/m^3; STEL 0.6 mg/m^3, [skin], 1993; France: VME 0.2 mg/m^3, [skin], 1999; Switzerland: MAK-W 0.1 ppm (0.2 mg[Mn]/m^3), 1999; United Kingdom: TWA 0.2 mg[Mn]/m^3; STEL 0.6 mg[Mn]/m^3, [skin], 2000; the Netherlands: MAC-TGG 0.2 mg/m^3, [skin], 2003; Argentina, Bulgaria, Columbia, Jordan, South Korea, New Zealand, Singapore, Vietnam: ACGIH TLV$^®$: TWA 0.2 mg[Mn]/m^3 [skin]. States have set guidelines or standards for MCT in ambient air[60] ranging from 2.0 µg/m^3 (North Dakota) to 3.5 µg/m^3 (Virginia) to 4.0 µg/m^3 (Connecticut) to 5.0 µg/m^3 (Nevada). These compare to values for methyl cyclopentadienyl manganese tricarbonyl of 1.0 µg/m^3 (North Dakota) to 1.6 µg/m^3 (Virginia) to 2.0 µg/m^3 (Connecticut and Nevada).

Determination in Air: No method available.

Routes of Entry: Inhalation, ingestion, skin and/or eye contact. Absorbed through the skin.

Harmful Effects and Symptoms

Short Term Exposure: LD$_{50}$ = (oral-rat) 8 mg/kg. Irritates the eyes, skin, and respiratory tract. Medical observation is indicated. In concentrated form this compound is highly toxic by all routes of exposure. Approximately 5−15 mL, when spilled on the hand and wrist of a worker, produced toxic effects within 3−5 min. Giddiness, "thick tongue," nausea, and headache were reported to occur after skin exposure. Human exposure data are limited; the primary site of action is reported to be the central nervous system; may cause tissue lesions.

Long Term Exposure: Can cause kidney damage.

Points of Attack: Respiratory system, skin, central nervous system, kidneys.

Medical Surveillance: Examination of the nervous system. Kidney function tests.

First Aid: If this chemical gets into the eyes, remove any contact lenses at once and irrigate immediately for at least 15 min, occasionally lifting upper and lower lids. Seek medical attention immediately. If this chemical contacts the skin, remove contaminated clothing and wash immediately with soap and water. Seek medical attention immediately. If this chemical has been inhaled, remove from exposure, begin rescue breathing (using universal precautions, including resuscitation mask) if breathing has stopped and CPR if

heart action has stopped. Transfer promptly to a medical facility. When this chemical has been swallowed, get medical attention. Give large quantities of water and induce vomiting. Do not make an unconscious person vomit.

Personal Protective Methods: Wear protective gloves and clothing to prevent any reasonable probability of skin contact. Safety equipment suppliers/manufacturers can provide recommendations on the most protective glove/clothing material for your operation. All protective clothing (suits, gloves, footwear, headgear) should be clean, available each day, and put on before work. Contact lenses should not be worn when working with this chemical. Wear splash-proof chemical goggles and face shield unless full face-piece respiratory protection is worn. Employees should wash immediately with soap when skin is wet or contaminated. Provide emergency showers and eyewash.

Respirator Selection: *Up to 10 mg/m^3:* 95XQ (APF = 10) [any particulate respirator equipped with an N95, R95, or P95 filter (including N95, R95, and P95 filtering face-pieces) except quarter-mask respirators. The following filters may also be used: N99, R99, P99, N100, R100, P100] or Sa (APF = 10) (any supplied-air respirator). *Up to 25 mg/m^3:* Sa:Cf (APF = 25) (any supplied-air respirator operated in a continuous-flow mode); PaprHie (APF = 25) (any powered air-purifying respirator with a high-efficiency particulate filter). *Up to 50 mg/m^3:* 100F (APF = 50) (any air-purifying, full-face-piece respirator with an N100, R100, or P100 filter) or SaT: Cf (APF = 50) (any supplied-air respirator that has a tight-fitting face-piece and is operated in a continuous-flow mode) or PaprTHie (APF = 50) (any powered, air-purifying respirator with a tight-fitting face-piece and a high-efficiency particulate filter) or SCBAF (APF = 50) (any self-contained breathing apparatus with a full face-piece); or SaF (APF = 50) (any supplied-air respirator with a full face-piece). *Up to 500 mg/m^3:* Sa: Pd,Pp (APF = 1000) (any supplied-air respirator operated in a pressure-demand or other positive-pressure mode). *Emergency or planned entry into unknown concentrations or IDLH conditions:* SCBAF: Pd,Pp (APF = 10,000) (any self-contained breathing apparatus that has a full face-piece and is operated in a pressure-demand or other positive-pressure mode); Ascba (any supplied-air respirator that has a full face-piece and is operated in a pressure-demand or other positive-pressure mode in combination with an auxiliary self-contained positive-pressure breathing apparatus). *Escape:* 100F (APF = 50) (any air-purifying, full-face-piece respirator with an N100, R100, or P100 filter) or SCBAE (any appropriate escape-type, self-contained breathing apparatus).

Storage: Color Code—Blue: Health Hazard/Poison: Store in a secure poison location. Prior to working with this chemical you should be trained on its proper handling and storage. Store in tightly closed containers in a cool, well-ventilated area away from oxidizers. Where possible, automatically pump liquid from drums or other storage containers to process containers.

Shipping: Metal carbonyls, liquid, n.o.s. require a label of "POISONOUS/TOXIC MATERIALS." They fall in Hazard Class 6.1. and Packing Group II.

Spill Handling: Evacuate and restrict persons not wearing protective equipment from area of spill or leak until cleanup is complete. Remove all ignition sources. Ventilate area of spill or leak. Absorb liquids in vermiculite, dry sand, earth, peat, carbon, or a similar material and deposit in sealed containers. Large spills can be cleaned up using JP-4L or JP-5 (jet engine fuels). Isopropyl alcohol may be used to clean up moderate spills, while methyl ethyl ketone should be used for cleaning *small spills* and quick disconnects. Keep this chemical out of a confined space, such as a sewer, because of the possibility of an explosion, unless the sewer is designed to prevent the buildup of explosive concentrations. It may be necessary to contain and dispose of this chemical as a hazardous waste. If material or contaminated runoff enters waterways, notify downstream users of potentially contaminated waters. Contact your local or federal environmental protection agency for specific recommendations. If employees are required to clean up spills, they must be properly trained and equipped. OSHA 1910.120(q) may be applicable.

Initial isolation and protective action distances
Distances shown are likely to be affected during the first 30 min after materials are spilled and could increase with time. If more than one tank car, cargo tank, portable tank, or large cylinder involved in the incident is leaking, the protective action distance may need to be increased. You may need to seek emergency information from CHEMTREC at (800) 424-9300 or seek professional environmental engineering assistance from the US EPA Environmental Response Team at (908) 48-8730 (24-h response line).

Small spills (From a small package or a small leak from a large package)
First: Isolate in all directions (feet/meters)   500/150
Then: Protect persons downwind (miles/kilometers)
Day   0.9/1.5
Night   3.1/4.9

Large spills (From a large package or from many small packages)
First: Isolate in all directions (feet/meters)   3000/1000
Then: Protect persons downwind (miles/kilometers)
Day   7.0 + /11.0 +
Night   7.0 + /11.0 +

Fire Extinguishing: This chemical is a combustible liquid. Poisonous gases are produced in fire. Use dry chemical, carbon dioxide, or alcohol foam extinguishers. Vapors are heavier than air and will collect in low areas. Vapors may travel long distances to ignition sources and flashback. Vapors in confined areas may explode when exposed to fire. Containers may explode in fire. Storage containers and parts of containers may rocket great distances, in many directions. If material or contaminated runoff enters waterways, notify downstream users of potentially contaminated waters. Notify local health and fire officials and pollution

control agencies. From a secure, explosion-proof location, use water spray to cool exposed containers. If cooling streams are ineffective (venting sound increases in volume and pitch, tank discolors, or shows any signs of deforming), withdraw immediately to a secure position. If employees are expected to fight fires, they must be trained and equipped in OSHA 1910.156. The only respirators recommended for firefighting are self-contained breathing apparatuses that have full face-pieces and are operated in a pressure-demand or other positive-pressure mode.

References
US Environmental Protection Agency. (October 21, 1983). *Chemical Hazard Information Profile: Methylcyclopentadienyl Manganese Tricarbonyl.* Washington, DC
US Environmental Protection Agency. (November 30, 1987). *Chemical Hazard Information Profile: Manganese, Tricarbonyl Methylcyclopentadienyl.* Washington, DC: Chemical Emergency Preparedness Program

MCPA   M:0290

Molecular Formula: $C_9H_9ClO_3$
Common Formula: $H_3CC_6H_3ClOCH_2CO_2H$
Synonyms: Acetic acid (4-chloro-2-methylphenoxy)-; Acetic acid [(4-chloro-*o*-tolyl)-oxy]-; ACME MCPA amine 4; Agritox; Agroxone; Anicon kombi; Anicon M; BH MCPA; Bordermaster; Brominal M & plus; (4-Chloro-*o*-cresoxy)acetic acid; Chiptox; 4-Chloro-*o*-cresoxyacetic acid; (4-Chloro-2-methylphenoxy)acetic acid; 4-Chloro-2-methylphenoxyacetic acid; 4-Chloro-*o*-toloxyacetic acid; (4-Chloro-*o*-toloxy)acetic acid; [(4-Chloro-*o*-tolyl)oxy] acetic acid; Chwastox; Cornox M; Ded weed; Dicopur-M; Dicotex; Dow MCP amine weed killer; Emcepan; Empal; Hedapur M 52; Hedarex M; Hedonal M; Herbicide M; Hormotuho; Hornotuho; Kilsem; 4K-2M; Legumex DB; Leuna M; Leyspray; Linormone; M 40; 2M-4C; 2M-4CH; MCP; MCPA; Mephanac; Metaxon; Methoxone; Methylchlorophenoxyacetic acid; 2-Methyl-4-chlorophenoxyacetic acid; (2-Methyl-4-chlorophenoxy)acetic acid; 2-Methyl-4-chlorophenoxyessigsaeure (German); 2M-4KH; Okultin; Phenoxylene 50; Phenoxylene plus; Phenoxylene super; Razol dock killer; Rhonox; B-Selektonon M; Seppic MMD; U 46; U 46 M-Fluid; Vacate; Vesakontuho; Weedar; Weedar MCPA concentrate; Weedone MCPA ester; Weed RHAP; Zelan
CAS Registry Number: 94-74-6
RTECS® Number: AG1575000
UN/NA & ERG Number: UN2765 (Phenoxy pesticides, solid, toxic)/152
EC Number: 202-360-6 [*Annex I Index No.:* 607-051-00-3]
Regulatory Authority and Advisory Bodies
Carcinogenicity: IARC: Human Limited Evidence; Animal No Adequate Data, *possibly carcinogenic to humans*, Group 2B, 1987

US EPA Gene-Tox Program, Positive: *D. melanogaster* sex-linked lethal; *S. cerevisiae*—reversion; Negative: *D. melanogaster*—whole sex chrom. loss; Negative: *D. melanogaster*—nondisjunction; Host-mediated assay.

US EPA, FIFRA, 1998 Status of Pesticides: Supported.

EPCRA Section 313 Form R *de minimis* concentration reporting level: 0.1%.

European/International Regulations: Hazard Symbol: Xn; Risk phrases: R22; R38; R41; R50/53 Safety phrases: S2; S26; S37; S39; S60; S61 (see Appendix 4).

WGK (German Aquatic Hazard Class): 2—Hazard to waters.

Description: MCPA is a colorless crystalline solid. Molecular weight = 200.63; Freezing/Melting point = 118.8°C; Vapor pressure = 1.5×10^{-6} mmHg at 20°C. Insoluble in water.

Potential Exposure: Compound Description: Agricultural Chemical; Tumorigen, Mutagen; Reproductive Effector; Human Data; Primary Irritant. Those involved in the manufacture, formulation, and application of this postemergence herbicide, used for control of broadleaf weeds in agricultural applications.

Incompatibilities: A weak acid. Incompatible with alkalis.

Permissible Exposure Limits in Air

No TEEL available.

Poland: MAC (TWA) 1 mg/m³, MAC (STEL) 3 mg/m³, 1999

Determination in Air: No method available.

Permissible Concentration in Water: Russia set a MAC of 0.25 mg/L in surface water. The US EPA has determined a no-observed-adverse-effects-level (NOAEL) of 1.0 mg/kg/day from which they have calculated a long-term advisory of 0.35 mg/L (350 µg/L) for an adult. They have further calculated a lifetime health advisory of 0.0036 mg/L (3.6 µg/L) for an adult. In addition, Maine has set a guideline for MCPA in drinking water[61] of 2.5 µg/L.

Determination in Water: No method available. Fish Tox = 17986.45312000 ppb (VERY LOW). Octanol–water coefficient: Log K_{ow} = 3.29.

Routes of Entry: Inhalation, ingestion, skin and/or eye contact.

Harmful Effects and Symptoms

Short Term Exposure: Corrosive. Irritates the eyes, skin, and respiratory tract. This material is moderately toxic. LD_{50} = (oral-rat) 700 mg/kg. Human Tox = 4.00000 (HIGH). The approximate lethal dose to a 150-lb man is 3.3 tablespoonfuls (Sax).

Long Term Exposure: Animal tests show that this substance possibly causes toxic effects upon human reproduction. MCPA is classified as a chlorophenoxy-herbicide. These herbicides are a possible carcinogen to humans. May cause decreased blood pressure. May cause genetic changes.

Medical Surveillance: Monitor blood pressure.

First Aid: If this chemical gets into the eyes, remove any contact lenses at once and irrigate immediately for at least 15 min, occasionally lifting upper and lower lids. Seek medical attention immediately. If this chemical contacts the skin, remove contaminated clothing and wash immediately with soap and water. Seek medical attention immediately. If this chemical has been inhaled, remove from exposure, begin rescue breathing (using universal precautions, including resuscitation mask) if breathing has stopped and CPR if heart action has stopped. Transfer promptly to a medical facility. When this chemical has been swallowed, get medical attention. Give large quantities of water and induce vomiting. Do not make an unconscious person vomit.

Personal Protective Methods: Wear protective gloves and clothing to prevent any reasonable probability of skin contact. Safety equipment suppliers/manufacturers can provide recommendations on the most protective glove/clothing material for your operation. All protective clothing (suits, gloves, footwear, headgear) should be clean, available each day, and put on before work. Contact lenses should not be worn when working with this chemical. Wear dust-proof chemical goggles and face shield unless full face-piece respiratory protection is worn. Employees should wash immediately with soap when skin is wet or contaminated. Provide emergency showers and eyewash.

Respirator Selection: Follow regulations in OSHA 29CFR1910.134 or European Standard EN149. Use a NIOSH/MSHA- or European Standard EN149-approved respirator; or use an approved supplied-air respirator with a full face-piece operated in the positive-pressure mode, or with a full face-piece, hood, or helmet in the continuous-flow mode; or use a NIOSH/MSHA- or European Standard EN149-approved self-contained breathing apparatus with a full face-piece operated in pressure-demand or other positive-pressure mode.

Storage: Color Code—Blue: Health Hazard/Poison: Store in a secure poison location. Prior to working with this chemical you should be trained on its proper handling and storage. Store in tightly closed containers in a cool, well-ventilated area. A regulated, marked area should be established where this chemical is handled, used, or stored in compliance with OSHA Standard 1910.1045.

Shipping: This material falls into the category of Phenoxy pesticides, solid, toxic n.o.s. This compound requires a shipping label of "POISONOUS/TOXIC MATERIALS." It falls in Hazard Class 6.1 and Packing Group III.

Spill Handling: Evacuate persons not wearing protective equipment from area of spill or leak until cleanup is complete. Remove all ignition sources. Collect powdered material in the most convenient and safe manner and deposit in sealed containers. Ventilate area after cleanup is complete. It may be necessary to contain and dispose of this chemical as a hazardous waste. If material or contaminated runoff enters waterways, notify downstream users of potentially contaminated waters. Contact your local or federal environmental protection agency for specific recommendations. If employees are required to clean up spills, they must be properly trained and equipped. OSHA 1910.120(q) may be applicable. Also, can be precipitated with divalent cations. Carbon or peat can be used as sorbents.

Fire Extinguishing: Solid material not combustible. Liquid formulations containing organic solvents may be flammable. Use dry chemical, carbon dioxide, water spray, or alcohol foam extinguishers. Poisonous gases are produced in fire. If material or contaminated runoff enters waterways, notify downstream users of potentially contaminated waters. Notify local health and fire officials and pollution control agencies. From a secure, explosion-proof location, use water spray to cool exposed containers. If cooling streams are ineffective (venting sound increases in volume and pitch, tank discolors, or shows any signs of deforming), withdraw immediately to a secure position. If employees are expected to fight fires, they must be trained and equipped in OSHA 1910.156. The only respirators recommended for firefighting are self-contained breathing apparatuses that have full face-pieces and are operated in a pressure-demand or other positive-pressure mode.

Disposal Method Suggested: Incineration with added flammable solvent; incinerator equipped with fume scrubber.[22] In accordance with 40CFR165, follow recommendations for the disposal of pesticides and pesticide containers. Must be disposed properly by following package label directions or by contacting your local or federal environmental control agency or by contacting your regional EPA office.

References

US Environmental Protection Agency. (1975). *Initial Scientific and Minieconomic Review No. 21: MCPA.* Washington, DC: Office of Pesticide Programs
US Environmental Protection Agency. (August 1987). *Health Advisory: MCPA.* Washington, DC: Office of Drinking Water
Sax, N. I. (Ed.). (1988). *Dangerous Properties of Industrial Materials Report*, 8, No. 6, 35−41
US Environmental Protection Agency, Special Review and Reregistration Division Office of Pesticide Programs. (1998). *Agency Status of Pesticides in Registration, Reregistration, and Special Review* (Rainbow Report). Washington, DC

Mechlorethamine (Agent HN-2, WMD)
M:0300

Molecular Formula: $C_5H_{11}Cl_2N$
Common Formula: $CH_3N(CH_2CH_2Cl)_2$
Synonyms: N,N-Bis(2-chloroethyl)methylamine; Bis(β-chloroethyl)methylamine; Bis(2-chloroethyl)methylamine; Caryolysin; Chloramine; Chlorethazine; Chlormethine; 2-Chloro-N-(2-chloroethyl)-N-methylethanamine; Cloramin; Dichloramine; Dichloren (German); β,β-Dichlorodiethyl-N-methylamine; N,N-Di(chloroethyl)methylamine; Di(2-chloroethyl)methylamine; 2,2′-Dichloro-N-methyldiethylamine; Diethylamine, 2,2′-dichloro-N-methyl-(8CI); Embichin; ENT-25294; Ethanamine, 2-chloro-N-(2-chloroethyl)-N-methyl-; HN-2 (military designation); MBA; Mechlorethamine; N-Methyl-bis-chloraethylamin (German); N-Methyl-bis(β-chloroethyl)amine; N-Methyl-bis(2-chloroethyl)amine; Methylbis(β-chloroethyl)amine; Methylbis(2-chloroethyl)amine; N-Methyl-2,2′-dichlorodiethylamine; Methyldi(2-chloroethyl)amine; N-Methyllost (German); Mostaza de nitrogeno (Spanish); Mustargen; Mustine; Nitrogen Mustard-2; Nitrogen mustard agent HN-2; N-Lost (German); NSC 762; TL 146

CAS Registry Number: 51-75-2; 55-86-7 (hydrochloride); 126-85-2 (nitrogen mustard N-oxide); 302-70-5 (nitrogen mustard, N-oxide hydrochloride)
RTECS® Number: IA1750000
UN/NA & ERG Number: UN2810 (toxic liquids, organic, n.o.s.)/153
Regulatory Authority and Advisory Bodies
Department of Homeland Security Screening Threshold Quantity (pounds): *Theft hazard* 2.2 lb. (≥30.00% concentration) (hydrochloride); *Theft hazard* CUM 100 g. (*51-75-2*).
Carcinogenicity: IARC: Human Limited Evidence; Animal Sufficient Evidence, *probably carcinogenic to humans*, Group 2A, 1998; NTP (hydrochloride): Reasonably anticipated to be a human carcinogen.
Air Pollutant Standard Set. See below, "Permissible Exposure Limits in Air" section.
RCRA, 40CFR261, Appendix 8 Hazardous Constituents, waste number not listed.
Superfund/EPCRA 40CFR355, Extremely Hazardous Substances: TPQ = 10 lb (4.54 kg).
Reportable Quantity (RQ): 1 lb (0.454 kg).
EPCRA Section 313 Form R *de minimis* concentration reporting level: 0.1%.
US DOT 49CFR172.101, Inhalation Hazardous Chemical.
California Proposition 65 Chemical: (*hydrochloride*) Cancer 4/1/88; Developmental/Reproductive toxin 7/1/90.
WGK (German Aquatic Hazard Class): No value assigned.
Description: Nitrogen mustard is a pale yellow, oily, mobile liquid with a faint odor of herring. *Nitrogen mustards* are colorless when pure but are typically a yellow to brown oily substance. Odors are variably described as sweet, agreeable, slightly garlic- or mustard-like. NIOSH reports HN-2 as having "a fruity odor at high concentrations and a soapy or fishy odor at low concentrations." Molecular weight = 156.07; Specific gravity = 1.12 at 25°C; Boiling point = 87°C at 18 mmHg; 75°C at 10 mmHg [HN-2 decomposes before its boiling point is reached or condenses under all conditions; the reactions involved could generate enough heat to cause an explosion];[NIOSH] Freezing/Melting Point: −60°C; Vapor Pressure = 0.29 mmHg at 20°C. Hazard Identification (based on NFPA-704 M Rating System): Health 4, Flammability 1, Reactivity 0. This compound is very sparingly soluble in water; releases corrosive vapors on contact with water or steam.
Potential Exposure: Drug used in treatment of cancer. Sulfur mustards were formerly used as a gas warfare agent.

Nitrogen mustards have not previously been used in warfare.[NIOSH] Exposure to nitrogen mustard damages the eyes, skin, and respiratory tract and suppresses the immune system. Although the nitrogen mustards cause cellular changes within minutes of contact, the onset of pain and other symptoms is delayed. Exposure to large amounts can be fatal.[NIOSH]

Incompatibilities: HN-2 is not stable except as dry crystals. Polymerization of HN-2 results in components that present an explosion hazard in open air.[NIOSH] Avoid contact or contamination with oxidizers (e.g., nitrates, oxidizing acids; chlorine bleaches pool chlorine) which may result in ignition. Unstable in the presence of light and heat and forms dimers at temperatures above 50°C. Corrosive to ferrous alloys beginning at 65°C. Polymerizes slowly and so munitions would be effective for several years. When heated to decomposition emits hydrogen chloride and nitrogen oxide. Contact with metals may evolve flammable hydrogen gas.[NIOSH]

Note: Chlorinating agents destroy nitrogen mustards. Dry chlorinated lime and chloramines with a high content of active chlorine vigorously chlorinate nitrogen mustards to the carbon chain, giving low toxicity products. In the presence of water this interaction proceeds less actively. They are rapidly oxidized by peracids in aqueous solution at weakly alkaline pH. In acid solution the oxidation is much slower.

Permissible Exposure Limits in Air
The Surgeon General's Working Group (US Department of Health and Human Services) recommends (for the workplace) 0.003 mg/m³, Ceiling Concentration. Same for US military. Several states have set guidelines for nitrogen mustard in ambient air[60] and New York and South Carolina at zero.

Protective Action Criteria (PAC) HN-2*
TEEL-0: 0.001 mg/m³
PAC-1: 0.003 mg/m³
PAC-2: **0.022** mg/m³
PAC-3: **0.37** mg/m³
*AEGLs (Acute Emergency Guideline Levels) & ERPGs (Emergency Response Planning Guideline) are in **bold face**.
Hydrochloride
TEEL-0: 0.75 mg/m³
PAC-1: 2.5 mg/m³
PAC-2: 4 mg/m³
PAC-3: 4 mg/m³
DFG MAK: [skin] danger of skin sensitization; Carcinogen Category 2A; Germ Cell Mutation Category 2.

Determination in Water: A water contaminant. Octanol–water coefficient: Log K_{ow} = (estimated) 0.91 at pH 7.4; 1.53.[NIOSH]

Routes of Entry: Inhalation, ingestion, skin and/or eye contact (vapor or liquid).

Harmful Effects and Symptoms
Nitrogen mustard is a blister agent (vesicant) that causes delayed severe damage to the respiratory tract. It is an alkylating agent that damages the cells within the bone marrow that are necessary for making blood cells. Clinical effects do not appear until hours after exposure. Nitrogen mustards penetrate and bind quickly to cells of the body; their health effects develop slowly. The full extent of cellular injury may not be known for days. The sooner after exposure the health effects occur, the more likely it is that the patient/victim was exposed to a high concentration of HN-2. Similarly, the sooner the health effects occur, the more likely it is that they will progress and become severe.
Eye exposure: The eyes are the organs that are most sensitive to mustard vapor; eye injury may occur within 1–2 h after severe exposure, or 3–12 h after a mild to moderate exposure. *Inhalation exposure:* Airway injury may occur within 2–6 h after severe exposure and within 12–24 h after mild exposure. Skin exposure: The symptom-free (latent) period is 6–12 h in temperate conditions; hot, humid weather strikingly increases the action of nitrogen mustards. Some skin injury may appear as late as 48 h after exposure.[NIOSH]

Short Term Exposure: Extremely toxic and may damage the eyes, skin, and respiratory tract and suppress the immune system. Although this agent can cause cellular changes within minutes of contact, the onset of pain and other symptoms is delayed. Irritates and burns the eyes, with possible permanent damage. Skin contact can cause irritation, burns with itching and blisters. Inhalation can cause irritation of the respiratory tract with wheezing and coughing. Higher exposure can cause headache, nausea, vomiting, and dizziness. Symptoms of exposure include nausea and vomiting, bleeding, skin lesions, menstrual irregularities. Toxic doses as low as 400 mg/kg have been reported in humans. Blood clots may occur at site of intravenous injection, and tissue damage if outside vein. Powerful vesicant (causes blisters) when it contacts skin, mucous membranes, or eyes. *Delayed toxicity*—missed menstrual periods, alopecia (hair loss), hearing loss, tinnitus (ringing in ears), jaundice, impaired spermatogenesis and germinal aplasia, swelling, and hypersensitivity. High exposure can cause tinnitus (ringing in the ears) and possible hearing loss.

Long Term Exposure: Bone marrow suppression resulting in damage to the blood-forming (hematopoietic) system. Early signs of bone marrow suppression include a low white blood cell count, an increased risk for developing infections, a tendency for easy bruising and bleeding. May cause lymph node damage and a weakened immune system. It also causes liver and kidney damage, damage to the reproductive systems of both men and women leading to decreased fertility. It is mutagenic, toxic to the developing embryo, and carcinogenic. Following significant whole-body (systemic) absorption of nitrogen mustard, injury to the bone marrow, lymph nodes, and spleen may cause a drop in white blood cell counts (beginning on days 3–5), which can result in an increased risk for developing (life-threatening) infections. Counts of red blood cells and platelets may also fall due to bone marrow damage.

Points of Attack: Bone marrow, hearing.

Medical Surveillance: Complete blood count. Audiogram.

First Aid: There is no antidote for nitrogen mustard toxicity. Decontamination of all potentially exposed areas within minutes after exposure is the only effective means of decreasing tissue damage.[NIOSH] Because health effects due to nitrogen mustard may not occur until several hours after exposure, patients/victims should be observed in a hospital setting for at least 24 h. Gastric lavage is contraindicated following ingestion of this agent due to the risk of perforation of the esophagus or upper airway. If this chemical gets into the eyes, remove any contact lenses at once and irrigate immediately for at least 15 min, occasionally lifting upper and lower lids. Seek medical attention immediately. If this chemical contacts the skin, remove contaminated clothing and wash immediately with soap and water. Speed in removing material from skin is of extreme importance. Seek medical attention immediately. If this chemical has been inhaled, remove from exposure, begin rescue breathing (using universal precautions, including resuscitation mask) if breathing has stopped and CPR if heart action has stopped. Transfer promptly to a medical facility. When this chemical has been swallowed, get medical attention. Give large quantities of water and induce vomiting. Do not make an unconscious person vomit. Keep victim quiet and maintain normal body temperature. Effects may be delayed; keep victim under observation.

Personal Protective Methods: Wear Totally Encapsulating Chemical Protective (TECP) suit that provides protection against CBRN agents; chemical-resistant inner and outer gloves, chemical-resistant boots with a steel toe and shank, coveralls, long underwear, and a hard hat worn under the TECP suit are optional items. Take all necessary precautions to prevent any reasonable probability of skin contact. Safety equipment suppliers/manufacturers can provide recommendations on the most protective glove/clothing material for your operation. All protective clothing (suits, gloves, footwear, headgear) should be clean, available each day, and put on before work. Contact lenses should not be worn when working with this chemical. Wear splash-proof chemical goggles and face shield unless full face-piece respiratory protection is worn. Employees should wash immediately with soap when skin is wet or contaminated. Provide emergency showers and eyewash.

Decontamination: Decontamination of all potentially exposed areas within minutes after exposure is the only effective means of decreasing tissue damage.[NIOSH] Remove clothes and place contaminated clothes and personal belongings in a sealed double bag. Decontamination of mustard-exposed victims by either vapor or liquid should be performed within the first 2 min following the exposure to prevent tissue damage. If not accomplished within the first several minutes, decontamination should still be performed to ensure any residual liquid mustard is removed from the skin or clothes, or to ensure any trapped mustard vapor is removed with the clothing. Removing trapped mustard vapor will prevent vapor off-gassing or subsequent cross-contamination of other emergency responders/health-care providers or the health-care facility. Physical removal of the mustard agent, rather than detoxification or neutralization, is the most important principle in patient decontamination. Mustard is not detoxified by water alone and will remain in decontamination effluent (in dilute concentrations) if hydrolysis has not taken place.

(1) Patients exposed to vapor should be decontaminated by removing all clothing in a clean air environment and shampooing or rinsing the hair to prevent vapor off-gassing. (2) Patients exposed to liquid should be decontaminated by (a) washing in warm or hot water at least three times. Use liquid soap (dispose of container after use and replace), large volumes of water, and mild to moderate friction with a single-use sponge or washcloth in the first and second washes. Scrubbing of exposed skin with a brush is discouraged, because skin damage may occur which may enhance absorption. The third wash should be to rinse with large amounts of warm or hot water. Shampoo can be used to wash the hair. The rapid physical removal of a chemical agent is essential. If warm or hot water is not available, but cold water is, use cold water. Do not delay decontamination to obtain warm water. (b) Rinse the eyes, mucous membranes, or open wounds with sterile saline or water. (3) The health-care provider should (a) check the victim after the three washes to verify adequate decontamination before allowing entry to the medical treatment facility. If the washes were inadequate, repeat the entire process. (b) Be prepared to stabilize conventional injuries during the decontamination process. Careful decontamination can be a time-consuming process. The health-care provider may have to enter the contaminated area to treat the casualty during this process. Medical personnel should wear the proper PPE and evaluate the exposed workers.

Respirator Selection: When used as a weapon, use SCBA Respirator Certified by NIOSH for CBRN Environments. Where a potential exposure to the chemical exists, use a NIOSH-certified CBRN full-face-piece SCBA operated in a pressure-demand mode or a pressure-demand supplied air-hose respirator with an auxiliary escape bottle; or use a NIOSH/MSHA- or European Standard EN149-approved supplied-air respirator with a full face-piece operated in the positive-pressure mode, or with a full face-piece, hood, or helmet in the continuous-flow mode; or use a NIOSH/MSHA- or European Standard EN149-approved self-contained breathing apparatus (SCBA) with a full face-piece operated in pressure-demand or other positive-pressure mode.

Storage: Color Code—Blue: Health Hazard/Poison: Store in a secure poison location. Prior to working with this chemical you should be trained on its proper handling and storage. A regulated, marked area should be established where this chemical is handled, used, or stored. Store in tightly closed containers in a cool, well-ventilated area. Where possible, automatically pump liquid from drums or other storage containers to process containers.

Shipping: Toxic, liquids, organic, n.o.s. [Inhalation hazard, Packing Group I, Zone B] requires a shipping label of "POISONOUS/TOXIC MATERIALS." It falls in Hazard Class 6.1 and Packing Group III.

Spill Handling: Evacuate and restrict persons not wearing protective equipment from area of spill or leak until cleanup is complete. Remove all ignition sources. Ventilate area of spill or leak. Absorb liquids in vermiculite, dry sand, earth, peat, carbon, or a similar material and deposit in sealed containers. Keep this chemical out of a confined space, such as a sewer, because of the possibility of an explosion, unless the sewer is designed to prevent the buildup of explosive concentrations. It may be necessary to contain and dispose of this chemical as a hazardous waste. If material or contaminated runoff enters waterways, notify downstream users of potentially contaminated waters. Contact your local or federal environmental protection agency for specific recommendations. If employees are required to clean up spills, they must be properly trained and equipped. OSHA 1910.120(q) may be applicable.

HN-2, when used as a weapon
Initial isolation and protective action distances
Distances shown are likely to be affected during the first 30 min after materials are spilled and could increase with time. If more than one tank car, cargo tank, portable tank, or large cylinder involved in the incident is leaking, the protective action distance may need to be increased. You may need to seek emergency information from CHEMTREC at (800) 424-9300 or seek professional environmental engineering assistance from the US EPA Environmental Response Team at (908) 548-8730 (24-h response line).
Small spills (From a small package or a small leak from a large package)
HN-2, when used as a weapon
First: Isolate in all directions (feet/meters)   100/30
Then: Protect persons downwind (miles/kilometers)
Day   0.1/0.2
Night   0.1/0.2
Large spills (From a large package or from many small packages)
First: Isolate in all directions (feet/meters)   200/60
Then: Protect persons downwind (miles/kilometers)
Day   0.2/0.3
Night   0.3/0.5

Fire Extinguishing: Heated to decomposition emits hydrogen chloride and nitrogen oxide. *Small fires:* dry chemical, carbon dioxide, water spray, or foam. *Large fires:* water spray, fog, or foam. Stay upwind; keep out of low areas. Ventilate closed spaces before entering them. Wear positive pressure breathing apparatus and special protective clothing. Move container from fire area if you can do so without risk. Fight fire from maximum distance. Dike fire control water for later disposal; do not scatter the material. Containers

may explode in fire. Storage containers and parts of containers may rocket great distances, in many directions. If material or contaminated runoff enters waterways, notify downstream users of potentially contaminated waters. Notify local health and fire officials and pollution control agencies. From a secure, explosion-proof location, use water spray to cool exposed containers. If cooling streams are ineffective (venting sound increases in volume and pitch, tank discolors, or shows any signs of deforming), withdraw immediately to a secure position. If employees are expected to fight fires, they must be trained and equipped in OSHA 1910.156. The only respirators recommended for firefighting are self-contained breathing apparatuses that have full face-pieces and are operated in a pressure-demand or other positive-pressure mode.

Reference
US Environmental Protection Agency. (November 30, 1987). *Chemical Hazard Information Profile: Mechlorethamine.* Washington, DC: Chemical Emergency Preparedness Program

Melamine   M:0310

Molecular Formula: $C_3H_6N_6$
Common Formula: $NC(NH_2)NC(NH_2)NC(NH_2)$
Synonyms: Cyanuramide;   Cyanurotriamide; Cyanurotriamine;   1,3,5-Triazine-2,4,6-triamine;   2,4,6-Triaminotriazine
CAS Registry Number: 108-78-1; *(alt.)* 504-18-7; *(alt.)* 65544-34-5; *(alt.)* 67757-43-1; *(alt.)* 68379-55-5; *(alt.)* 70371-19-6; *(alt.)* 94977-27-2
RTECS® Number: OS0700000
EC Number: 203-615-4
Regulatory Authority and Advisory Bodies
Carcinogenicity: IARC: Animal Sufficient Evidence; Human No Adequate Data, *not classifiable as carcinogenic to humans*, Group 3, 1999; NCI: Carcinogenesis Studies (feed); clear evidence: rat; NCI: Carcinogenesis Studies (feed); no evidence: mouse.
US EPA Gene-Tox Program, Inconclusive: *D. melanogaster* sex-linked lethal.
WGK (German Aquatic Hazard Class): 1—Low hazard to waters.
Description: Melamine is a white crystalline solid. Molecular weight = 126.15; Specific gravity (H_2O:1) = 1.6; Freezing/Melting point = <250°C (decomposes); Flash point ≥300°C; Autoignition temperature ≥500°C. Hazard Identification (based on NFPA-704 M Rating System): Health 3, Flammability 0, Reactivity 0. Practically insoluble in water.
Potential Exposure: Compound Description: Tumorigen, Mutagen; Reproductive Effector; Primary Irritant. Manufactured from urea, melamine is used in the manufacture of plastics, melamine-formaldehyde resins, rubber, synthetic textiles, laminates, adhesives, and molding compounds.

Incompatibilities: Strong oxidizers, strong acids.

Permissible Exposure Limits in Air: Kansas[60] has set a guideline of 24.39 μg/m^3 for melamine in ambient air.

AIHA WEEL: 10 mg/m^3, inhalable TWA; 5 mg/m^3, respirable TWA.

Protective Action Criteria (PAC)

TEEL-0: 10 mg/m^3

PAC-1: 30 mg/m^3

PAC-2: 50 mg/m^3

PAC-3: 500 mg/m^3

Routes of Entry: Inhalation and ingestion.

Harmful Effects and Symptoms

Short Term Exposure: Irritates, eyes, skin, and mucous membranes. May cause dermatitis in humans.

Long Term Exposure: May cause kidney damage. There is limited evidence that melamine causes cancer in animals.

Points of Attack: Kidneys, skin.

Medical Surveillance: Kidney function tests.

First Aid: If this chemical gets into the eyes, remove any contact lenses at once and irrigate immediately for at least 15 min, occasionally lifting upper and lower lids. Seek medical attention immediately. If this chemical contacts the skin, remove contaminated clothing and wash immediately with soap and water. Seek medical attention immediately. If this chemical has been inhaled, remove from exposure, begin rescue breathing (using universal precautions, including resuscitation mask) if breathing has stopped and CPR if heart action has stopped. Transfer promptly to a medical facility. When this chemical has been swallowed, get medical attention. Give large quantities of water and induce vomiting. Do not make an unconscious person vomit.

Personal Protective Methods: Wear protective gloves and clothing to prevent any reasonable probability of skin contact. Safety equipment suppliers/manufacturers can provide recommendations on the most protective glove/clothing material for your operation. All protective clothing (suits, gloves, footwear, headgear) should be clean, available each day, and put on before work. Contact lenses should not be worn when working with this chemical. Wear dust-proof chemical goggles and face shield unless full face-piece respiratory protection is worn. Employees should wash immediately with soap when skin is wet or contaminated. Provide emergency showers and eyewash.

Respirator Selection: Follow regulations in OSHA 29CFR1910.134 or European Standard EN149. Use a NIOSH/MSHA- or European Standard EN149-approved respirator; or use an approved supplied-air respirator with a full face-piece operated in the positive-pressure mode, or with a full face-piece, hood, or helmet in the continuous-flow mode; or use a NIOSH/MSHA- or European Standard EN149-approved self-contained breathing apparatus with a full face-piece operated in pressure-demand or other positive-pressure mode.

Storage: Color Code—Blue: Health Hazard/Poison: Store in a secure poison location. Prior to working with this chemical you should be trained on its proper handling and storage. Store in tightly closed containers in a cool, well-ventilated area away from strong oxidizers and strong acids. Where possible, automatically pump liquid from drums or other storage containers to process containers. A regulated, marked area should be established where this chemical is handled, used, or stored in compliance with OSHA Standard 1910.1045.

Spill Handling: Evacuate persons not wearing protective equipment from area of spill or leak until cleanup is complete. Remove all ignition sources. Use HEPA vacuum or wet method to reduce dust. Collect powdered material in the most convenient and safe manner and deposit in sealed containers. Ventilate area after cleanup is complete. It may be necessary to contain and dispose of this chemical as a hazardous waste. If material or contaminated runoff enters waterways, notify downstream users of potentially contaminated waters. Contact your local or federal environmental protection agency for specific recommendations. If employees are required to clean up spills, they must be properly trained and equipped. OSHA 1910.120(q) may be applicable.

Fire Extinguishing: Melamine itself does not burn. Use any extinguishing agent suitable for surrounding fire. Poisonous gases, including carbon monoxide, hydrogen cyanide, nitrogen oxides, and ammonia, are produced in fire. If material or contaminated runoff enters waterways, notify downstream users of potentially contaminated waters. Notify local health and fire officials and pollution control agencies. From a secure, explosion-proof location, use water spray to cool exposed containers. If cooling streams are ineffective (venting sound increases in volume and pitch, tank discolors, or shows any signs of deforming), withdraw immediately to a secure position. If employees are expected to fight fires, they must be trained and equipped in OSHA 1910.156. The only respirators recommended for firefighting are self-contained breathing apparatuses that have full face-pieces and are operated in a pressure-demand or other positive-pressure mode.

References

US Environmental Protection Agency. (December 29, 1982). *Chemical Hazard Information Profile Draft Report: Melamine.* Washington, DC

Sax, N. I. (Ed.). (1988). *Dangerous Properties of Industrial Materials Report,* 8, No. 4, 40−41

New Jersey Department of Health and Senior Services. (June 1998). *Hazardous Substances Fact Sheet: Melamine.* Trenton, NJ

Melphalan   M:0320

Molecular Formula: $C_{13}H_{19}Cl_2N_2O_2$

Common   Formula: HOOCH(NH$_2$)CHCH$_2$C$_6$H$_4$N-(CH$_2$CH$_2$Cl)$_2$

Synonyms: Alanine nitrogen mustard; Alkeran; AT-290; 1-3-(*p*-[Bis(2-chloroethyl)amino]phenyl)alanine; *p,N*-Bis(2-chloroethyl)amino-l-phenylalanine; 3-[*p*-(*p*-[Bis(2-chloroethyl)

amino]phenyl)]-l-alanine; 4-[Bis(2-chloroethyl)amino]-l-phenylalanine; CB 3025; *p,N*-Di(chloroethyl)aminophenylalanine; *p*-Di(2-chloroethyl)amino-l-phenylalanine; 3-*p*-[Di(2-chloroethyl)amino]-phenyl-l-alanine; 2-(Diethoxyphosphinylimino)-4-methyl-1,3-dithiolane; I-PAM; L-Sarcolysin; Melfalano (Spanish); NCI-CO4853; NSC-8806; *i*-Phenylalanine nitrogen mustard; Phenylalanine nitrogen mustard; P-I-Sarcolysin; SK-15673

CAS Registry Number: 148-82-3
RTECS® Number: AY3675000
UN/NA & ERG Number: UN1851/151
EC Number: 205-726-3

Regulatory Authority and Advisory Bodies
Carcinogenicity: IARC: Human Sufficient Evidence; Animal Sufficient Evidence, Group 1, 1998; NTP: Known to be a human carcinogen.
US EPA Hazardous Waste Number (RCRA No.): U150.
RCRA, 40CFR261, Appendix 8 Hazardous Constituents.
Reportable Quantity (RQ): 1 lb (0.454 kg).
California Proposition 65 Chemical: Cancer 2/27/87; Developmental/Reproductive toxin 7/1/90.
WGK (German Aquatic Hazard Class): No value assigned.

Description: Melphalan forms solvated crystals from methanol. Molecular weight = 305.23; Freezing/Melting point = 182−183°C. (decomposes). Hazard Identification (based on NFPA-704 M Rating System): Health 3, Flammability 0, Reactivity 0.

Potential Exposure: An alkylating agent. Health-care workers may be exposed. As a drug it is an immunosuppressant, used in the treatment of multiple myeloma and cancer of the ovary. It is also used in investigation of other types of cancer and as an antineoplastic in animals. Human exposure to melphalan occurs principally during its use in cancer treatment. Melphalan is administered orally or intravenously. Adult dosage is 6 mg/day, 5 days per month. Has been used as a military poison gas (a nitrogen mustard, alkaline, crystals).

Permissible Exposure Limits in Air
No standards or TEEL available.

Harmful Effects and Symptoms
Short Term Exposure: LD_{50} = (oral-rat) 11.2 mg/kg. This material is highly toxic; local irritant of the skin, eyes, and mucous membranes. Skin rash, nausea, vomiting.

Long Term Exposure: A Confirmed Human Carcinogen capable of causing leukemia and Hodgkin's disease. High or repeated exposure causes marked bone marrow depression with possible anemia, neutropenia, and thrombocytopenia.

Points of Attack: Blood.
Medical Surveillance: Complete blood count (CBC).
First Aid: Skin Contact[52]: Flood all areas of body that have contacted the substance with water. Speed in removing material from skin is of extreme importance. Do not wait to remove contaminated clothing; do it under the water stream. Use soap to help assure removal. Isolate contaminated clothing when removed to prevent contact by others.

Eye Contact: Remove any contact lenses at once. Immediately flush eyes well with copious quantities of water or normal saline for at least 20−30 min. Seek medical attention.

Inhalation: Leave contaminated area immediately; move to the fresh air. Proper respiratory protection must be supplied to any rescuers. If coughing, difficult breathing or any other symptoms develop, seek medical attention at once, even if symptoms develop many hours after exposure.

Ingestion: Contact a physician, hospital or poison center at once. If the victim is unconscious or convulsing, do not induce vomiting or give anything by mouth. Assure that his airway is open and lay him on his side with his head lower than his body and transport immediately to a medical facility. If conscious and not convulsing, give a glass of water to dilute the substance. Vomiting should not be induced without a physician's advice.

Personal Protective Methods: Wear protective gloves and clothing to prevent any reasonable probability of skin contact. Safety equipment suppliers/manufacturers can provide recommendations on the most protective glove/clothing material for your operation. All protective clothing (suits, gloves, footwear, headgear) should be clean, available each day, and put on before work. Contact lenses should not be worn when working with this chemical. Wear dust-proof chemical goggles and face shield unless full face-piece respiratory protection is worn. Employees should wash immediately with soap when skin is wet or contaminated. Provide emergency showers and eyewash.

Respirator Selection: Follow regulations in OSHA 29CFR1910.134 or European Standard EN149. Use a NIOSH/MSHA- or European Standard EN149-approved respirator; or use an approved supplied-air respirator with a full face-piece operated in the positive-pressure mode, or with a full face-piece, hood, or helmet in the continuous-flow mode; or use a NIOSH/MSHA- or European Standard EN149-approved self-contained breathing apparatus with a full face-piece operated in pressure-demand or other positive-pressure mode.

Storage: Color Code—Blue: Health Hazard/Poison: Store in a secure poison location. Prior to working with this chemical you should be trained on its proper handling and storage. Store in tightly closed containers in a cool, well-ventilated area. A regulated, marked area should be established where this chemical is handled, used, or stored in compliance with OSHA Standard 1910.1045.

Shipping: This material falls into the category of Poisonous liquid, n.o.s. This compound requires a shipping label of "POISONOUS/TOXIC MATERIALS." It falls in Hazard Class 6.1.

Spill Handling: Evacuate persons not wearing protective equipment from area of spill or leak until cleanup is complete. Remove all ignition sources. Collect powdered material in the most convenient and safe manner and deposit in sealed containers. Ventilate area after cleanup is complete. It may be necessary to contain and dispose of this chemical

as a hazardous waste. If material or contaminated runoff enters waterways, notify downstream users of potentially contaminated waters. Contact your local or federal environmental protection agency for specific recommendations. If employees are required to clean up spills, they must be properly trained and equipped. OSHA 1910.120(q) may be applicable.

Fire Extinguishing: This chemical is a combustible solid. Use dry chemical, carbon dioxide, water spray, or alcohol foam extinguishers. Poisonous gases, including HCl and nitrogen oxides, are produced in fire. If material or contaminated runoff enters waterways, notify downstream users of potentially contaminated waters. Notify local health and fire officials and pollution control agencies. From a secure, explosion-proof location, use water spray to cool exposed containers. If cooling streams are ineffective (venting sound increases in volume and pitch, tank discolors, or shows any signs of deforming), withdraw immediately to a secure position. If employees are expected to fight fires, they must be trained and equipped in OSHA 1910.156. The only respirators recommended for firefighting are self-contained breathing apparatuses that have full face-pieces and are operated in a pressure-demand or other positive-pressure mode.

Disposal Method Suggested: Consult with environmental regulatory agencies for guidance on acceptable disposal practices. Generators of waste containing this contaminant ($\geq$100 kg/mo) must conform with EPA regulations governing storage, transportation, treatment, and waste disposal.

Mephosfolan   M:0330

Molecular Formula: $C_8H_{16}NO_3PS_2$
Synonyms: AC 47470; American cyanamid CL-47470; CL-47,470; Cyclic propylene(diethoxyphosphinyl) dithioimidocarbonate; Cytrolane; p,p-Diethyl cyclic propylene ester of phosphonodithioimidocarbonic acid; Diethyl (4-methyl-1,3-dithiolan-2-ylidene)phosphoroamidate; 2-(Diethoxyphosphinylimino)-4-methyl-1,3-dithiolane; EI-47470; ENT 25,991; (4-Methyl-1,3-dithiolan-2-ylidene) phosphoramidic acid, diethyl ester
CAS Registry Number: 950-10-7
RTECS® Number: JP1050000
UN/NA & ERG Number: UN2784 (organophosphorus pesticide, liquid, flammable, toxic)/131
EC Number: 213-447-3 [*Annex I Index No.*: 015-094-00-9]
Regulatory Authority and Advisory Bodies
Banned or Severely Restricted (in agriculture) (India) (UN).[13]
Superfund/EPCRA 40CFR355, Extremely Hazardous Substances: TPQ = 500 lb (227 kg).
Reportable Quantity (RQ): 500 lb (227 kg).
US DOT Regulated Marine Pollutant (49CFR172.101, Appendix B).
US DOT 49CFR172.101, Inhalation Hazard Chemical as organophosphates.

European/International Regulations: Hazard Symbol: T+, N; Risk phrases: R27/28; R51/53; Safety phrases: S1/2; S36/37/39; S45; S61 (see Appendix 4).
WGK (German Aquatic Hazard Class): No value assigned.
Description: Mephosfolan is a yellow to amber liquid. Molecular weight = 269.34; Boiling point = 120°C at 1.0 mmHg. Hazard Identification (based on NFPA-704 M Rating System): Health 4, Flammability 1, Reactivity 0. Moderately soluble in water.
Potential Exposure: Those involved in the production, formulation, and application of this insecticide and acaricide. An organophosphorus pesticide, liquid, poisonous, flammable, n.o.s.
Incompatibilities: Strong oxidizers, strong acids.
Permissible Exposure Limits in Air
Protective Action Criteria (PAC)
TEEL-0: 1.5 mg/m^3
PAC-1: 5 mg/m^3
PAC-2: 9 mg/m^3
PAC-3: 9 mg/m^3
Determination in Air: OSHA versatile sampler-2; Toluene/Acetone; Gas chromatography/Flame photometric detection for sulfur, nitrogen, or phosphorus; NIOSH Analytical Method (IV) Method #5600, Organophosphorus Pesticides.
Routes of Entry: Inhalation, ingestion, skin and/or eye contact.
Harmful Effects and Symptoms
Short Term Exposure: This is a highly to extremely toxic material. LD_{50} = (oral-rats) 9 mg/kg. Organic phosphorus insecticides are absorbed by the skin, as well as by the respiratory and gastrointestinal tracts. They are cholinesterase inhibitors. Symptoms of exposure include headache, giddiness, blurred vision, nervousness, weakness, nausea, cramps, diarrhea, and discomfort in the chest. Signs include sweating, tearing, salivation, vomiting, cyanosis, convulsions, coma, loss of reflexes, and loss of sphincter control.
Long Term Exposure: Cholinesterase inhibitor; cumulative effect is possible. This chemical may damage the nervous system with repeated exposure, resulting in convulsions, respiratory failure. May cause liver damage.
Points of Attack: Respiratory system, lungs, central nervous system; cardiovascular system, skin, eyes, plasma and red blood cell cholinesterase.
Medical Surveillance: Before employment and at regular times after that, the following are recommended: plasma and red blood cell cholinesterase levels (tests for the enzyme poisoned by this chemical). If exposure stops, plasma levels return to normal in 1–2 weeks while red blood cell levels may be reduced for 1–3 months. When cholinesterase enzyme levels are reduced by 25% or more below preemployment levels, risk of poisoning is increased, even if results are in lower ranges of "normal." Reassignment to work not involving organophosphate or carbamate pesticides is recommended until enzyme levels recover. If symptoms develop or overexposure occurs, repeat the above tests as soon as possible and get an

examination of the nervous system. Also consider complete blood count. Consider chest X-ray following acute overexposure. Do not drink any alcoholic beverages before or during use. Alcohol promotes absorption of organic phosphates.

First Aid: If this chemical gets into the eyes, remove any contact lenses at once and irrigate immediately for at least 15 min, occasionally lifting upper and lower lids. Seek medical attention immediately. If this chemical contacts the skin, remove contaminated clothing and wash immediately with soap and water. Speed in removing material from skin is of extreme importance. Shampoo hair promptly if contaminated. Seek medical attention immediately. If this chemical has been inhaled, remove from exposure, begin rescue breathing (using universal precautions, including resuscitation mask) if breathing has stopped and CPR if heart action has stopped. Transfer promptly to a medical facility. When this chemical has been swallowed, get medical attention. Give large quantities of water and induce vomiting. Do not make an unconscious person vomit.

Personal Protective Methods: Wear protective gloves and clothing to prevent any reasonable probability of skin contact. Safety equipment suppliers/manufacturers can provide recommendations on the most protective glove/clothing material for your operation. All protective clothing (suits, gloves, footwear, headgear) should be clean, available each day, and put on before work. Contact lenses should not be worn when working with this chemical. Wear splash-proof chemical goggles and face shield unless full face-piece respiratory protection is worn. Employees should wash immediately with soap when skin is wet or contaminated. Provide emergency showers and eyewash.

Respirator Selection: Follow regulations in OSHA 29CFR1910.134 or European Standard EN149. Use a NIOSH/MSHA- or European Standard EN149-approved respirator; or use an approved supplied-air respirator with a full face-piece operated in the positive-pressure mode, or with a full face-piece, hood, or helmet in the continuous-flow mode; or use a NIOSH/MSHA- or European Standard EN149-approved self-contained breathing apparatus with a full face-piece operated in pressure-demand or other positive-pressure mode.

Storage: Color Code—Blue: Health Hazard/Poison: Store in a secure poison location. Prior to working with this chemical you should be trained on its proper handling and storage. Store in tightly closed containers in a cool, well-ventilated area away from strong oxidizers, strong acids.

Shipping: This material falls into the category of Organophosphorus pesticides, liquid, toxic, flammable, n.o.s. This compound requires a shipping label of "POISONOUS/TOXIC MATERIALS." It falls in Hazard Class 6.1 and Packing Group I.

Spill Handling: Evacuate persons not wearing protective equipment from area of spill or leak until cleanup is complete. Remove all ignition sources. Use water spray to knock down vapors. Build dikes to contain material as

necessary. Avoid breathing vapors. Keep upwind. Avoid bodily contact with the material. Downwind evacuation should be considered. Collect powdered material in the most convenient and safe manner and deposit in sealed containers. Ventilate area after cleanup is complete. It may be necessary to contain and dispose of this chemical as a hazardous waste. If material or contaminated runoff enters waterways, notify downstream users of potentially contaminated waters. Contact your local or federal environmental protection agency for specific recommendations. If employees are required to clean up spills, they must be properly trained and equipped. OSHA 1910.120(q) may be applicable.

Fire Extinguishing: Combustible material. Vapors may travel to a source of ignition, and flash back. Use foam, carbon dioxide, or dry chemical. When heated to decomposition, this material emits very toxic fumes of nitrogen oxides, phosphorus oxides, and sulfur oxides. Use water spray to absorb vapors. Avoid breathing vapors. Keep upwind. Wear self-contained breathing apparatus. Avoid bodily contact with the material. Wear full protective clothing. If material or contaminated runoff enters waterways, notify downstream users of potentially contaminated waters. Notify local health and fire officials and pollution control agencies. Containers may explode in fire. From a secure, explosion-proof location, use water spray to cool exposed containers. If cooling streams are ineffective (venting sound increases in volume and pitch, tank discolors, or shows any signs of deforming), withdraw immediately to a secure position. If employees are expected to fight fires, they must be trained and equipped in OSHA 1910.156. The only respirators recommended for firefighting are self-contained breathing apparatuses that have full face-pieces and are operated in a pressure-demand or other positive-pressure mode.

Disposal Method Suggested: In accordance with 40CFR 165 recommendations for the disposal of pesticides and pesticide containers. Must be disposed properly by following package label directions or by contacting your local or federal environmental control agency or by contacting your regional EPA office.

References

Sax, N. I. (Ed.). (1983). *Dangerous Properties of Industrial Materials Report*, 3, No. 1, 72−74

US Environmental Protection Agency. (November 30, 1987). *Chemical Hazard Information Profile: Mephosfolan.* Washington, DC: Chemical Emergency Preparedness Program

Mercuric acetate   M:0340

Molecular Formula: $C_4H_6HgO_4$
Common Formula: $(CH_3COO)_2Hg$
Synonyms: Acetic acid, mercury(2 +) salt; Acetic acid, mercury(II) salt; bis(Acetyloxy)mercury; Diacetoxymercury; Mercuriacetate; Mercuric diacetate; Mercury acetate;

Mercury(2+) acetate; Mercury(II) acetate; Mercury diacetate; Mercuryl acetate
CAS Registry Number: 1600-27-7; *(alt.)* 6129-23-3; *(alt.)* 7619-62-7; *(alt.)* 19701-15-6
RTECS® Number: AI8575000
UN/NA & ERG Number: UN1629/151
EC Number: 216-491-1
Regulatory Authority and Advisory Bodies
Carcinogenicity: IARC: Human Inadequate Evidence, *possibly carcinogenic to humans*, Group 2B, 1993.
US EPA Gene-Tox Program, Negative: *In vivo* cytogenetics—mammalian oocyte.
Air Polutant Standard Set. See below, "Permissible Exposure Limits in Air" section.
Clean Water Act: 40CFR401.15 Section 307 Toxic Pollutants as mercury and compounds.
RCRA, 40CFR261, Appendix 8 Hazardous Constituents, waste number not listed, as mercury compounds, n.o.s.
Superfund/EPCRA 40CFR355, Extremely Hazardous Substances: TPQ = 500/10,000 lb (227/4540 kg).
Reportable Quantity (RQ): 500 lb (227 kg).
EPCRA Section 313 (as mercury compound). Form R *de minimis* concentration reporting level: 1.0%.
US DOT Regulated Marine Pollutant (49CFR172.101, Appendix B), severe pollutant.
California Proposition 65 Developmental/Reproductive toxin (mercury and mercury compounds) 7/1/90.
Canada, WHMIS, Ingredients Disclosure List Concentration 0.1%.
Rotterdam Convention Annex III [Chemicals Subject to the Prior Informed Consent Procedure (PIC)] (as mercury compounds, including inorganic mercury compounds, alkyl mercury compounds, and alkyloxylalkyl and aryl mercury compounds).
European/International Regulations: not listed in Annex 1.
WGK (German Aquatic Hazard Class): 3—Severe hazard to waters.
Description: Mercuric acetate is a white crystalline solid with a mild vinegar-like odor. Molecular weight = 318.69; Freezing/Melting point = 178−180°C (decomposes). Hazard Identification (based on NFPA-704 M Rating System): Health 3, Flammability 0, Reactivity 0. Soluble in water.
Potential Exposure: Compound Description: Tumorigen, Mutagen; Reproductive Effector. Mercuric acetate is used chiefly for mercuration of organic compounds, for the absorption of ethylene, as a chemical intermediate for phenylmercuric acetate, a mildewcide, and other organomercury compounds. It is used as a catalyst in organic synthesis and in the manufacture of pharmaceuticals.
Incompatibilities: Strong oxidizers, such as chlorine, acids. Light and heat cause decomposition.
Permissible Exposure Limits in Air
As mercury alkyl compounds
OSHA PEL: 0.1 mg[Hg]/m³ Ceiling Concentration.
NIOSH REL: Hg *(vapor)*: 0.05 mg[Hg]/m³ TWA [skin]; Other: 0.1 mg[Hg]/m³ Ceiling Concentration [skin].

ACGIH TLV®[1]: 0.01 mg[Hg]/m³ TWA; 0.03 mg/m³ (skin) STEL [skin].
Protective Action Criteria (PAC)
TEEL-0: 0.0159 mg/m³
PAC-1: 0.0477 mg/m³
PAC-2: 3.2 mg/m³
PAC-3: 3.2 mg/m³
DFG MAK: 0.1 mg[Hg]/m³; Peak Limitation Category II(8) danger of skin sensitization; Carcinogen Category 3B.
NIOSH IDLH: 10 mg[Hg]/m³.
Australia: TWA 0.05 mg[Hg]/m³, [skin], 1993; Belgium: TWA 0.05 mg[Hg]/m³, [skin], 1993; Denmark: TWA 0.05 mg[Hg]/m³, 1999; Finland: TWA 0.05 mg[Hg]/m³, 1993; France: VME 0.1 mg[Hg]/m³, [skin], 1999; Hungary: TWA 0.02 mg[Hg]/m³; STEL 0.04 mg[Hg]/m³, 1993; Japan: 0.05 mg[Hg]/m³, 1999; Norway: TWA 0.05 mg[Hg]/m³, 1999; the Philippines: TWA 0.05 mg[Hg]/m³, 1993; Poland: MAC (TWA) 0.05 mg[Hg]/m³, MAC (STEL) 0.15 mg[Hg]/m³, 1999; Russia: TWA 0.05 mg[Hg]/m³; STEL 0.01 mg[Hg]/m³, 1993; Sweden: NGV 0.05 mg[Hg]/m³, 1999; Thailand: STEL 0.05 mg[Hg]/m³, 1993; United Kingdom: TWA 0.025 mg[Hg]/m³, 2000; Argentina, Bulgaria, Columbia, Jordan, South Korea, ZEALAND, Singapore, Vietnam; ACGIH TLV®: TWA 0.01 mg[Hg]/m³; STEL 0.03 mg[Hg]/m³ [skin].
Determination in Air: Use NIOSH Analytical Method #6009; OSHA Analytical Method ID-140.
Permissible Concentration in Water: To protect freshwater aquatic life: 0.00057 µg/L as a 24-h average, never to exceed 0.0017 µg/L. *To protect saltwater aquatic life:* 0.025 µg/L as a 24-h average, never to exceed 3.7 µg/L. *To protect human health:* 0.144 µg/L (US EPA) set in 1979−1980.[6] These are the limits for inorganic mercury compounds in general.
Determination in Water: Total mercury is determined by flameless atomic absorption. Soluble mercury may be determined by 0.45-µm filtration followed by flameless atomic absorption.
Routes of Entry: Inhalation, ingestion, eye and/or skin contact. Absorbed through the skin.
Harmful Effects and Symptoms
Short Term Exposure: Mercuric acetate is corrosive to the eyes, skin, and respiratory tract. Corrosive on ingestion. Inhalation of the aerosol can cause pulmonary edema, a medical emergency that can be delayed for several hours. This can cause death. Signs and symptoms of acute exposure to mercuric acetate may be severe and include increased salivation, foul breath, inflammation and ulceration of the mucous membranes, abdominal pain, and bloody diarrhea. Oliguria (scanty urination), anuria (suppression of urine formation), and acute renal failure may be noted. Weak pulse, seizures, psychic disturbances, dyspnea (shortness of breath), chest pain, and circulatory collapse may be observed.
Long Term Exposure: Repeated or prolonged contact with skin may result in dermatitis (red inflamed skin) or may

cause skin to turn gray. Skin allergy may also occur. Repeated or prolonged exposure may cause death by hypovolemic shock, nephrotic syndrome, or kidney failure. Has been shown to be a teratogen in animals.

Points of Attack: Eyes, skin, central nervous system, peripheral nervous system, kidneys.

Medical Surveillance: NIOSH lists the following tests for inorganic mercury: whole blood (chemical/metabolite); whole blood (chemical/metabolite), end-of-shift; end-of-shift at end-of-work-week; biologic tissue/biopsy; nerve conduction studies; neurologic examination/electromyography; thyroid function test/thyroid profile; urine (chemical/metabolite); urine (chemical/metabolite) prior to shift, prior to next shift; urine (chemical/metabolite), sediment; urinalysis (routine). Before first exposure and every 6–12 months after, a complete medical history and examination is strongly recommended with eye examination. Consider lung function tests for persons with frequent exposure. Examination of the nervous system including handwriting. Routine urine test (UA). Urine test for mercury (should be less than 0.02 mg/L). Consider nerve conduction tests, urinary enzymes, and neurobehavioral test. After suspected illness or overexposure, repeat the tests above and get a blood test for mercury.

First Aid: If this chemical gets into the eyes, remove any contact lenses at once and irrigate immediately for at least 15 min, occasionally lifting upper and lower lids. Seek medical attention immediately. If this chemical contacts the skin, remove contaminated clothing and wash immediately with soap and water. Speed in removing material from skin is of extreme importance. Shampoo hair promptly if contaminated. Seek medical attention immediately. If this chemical has been inhaled, remove from exposure, begin rescue breathing (using universal precautions, including resuscitation mask) if breathing has stopped and CPR if heart action has stopped. Transfer promptly to a medical facility. When this chemical has been swallowed, get medical attention. If victim is conscious, administer water or milk. Do not induce vomiting. Medical observation is recommended for 24–48 h after breathing overexposure, as pulmonary edema may be delayed. As first aid for pulmonary edema, a doctor or authorized paramedic may consider administering a corticosteroid spray.

Antidotes and special procedures for medical personnel: The drug NAP (*n*-acetyl penicillamine) has been used to treat mercury poisoning, with mixed success.

Note to physician: For severe poisoning BAL [British Anti-Lewisite, dimercaprol, dithiopropanol ($C_3H_8OS_2$)] has been used to treat toxic symptoms of certain heavy metals poisoning including mercury. Although BAL is reported to have a large margin of safety, caution must be exercised, because toxic effects may be caused by excessive dosage. Most can be prevented by premedication with 1-ephedrine sulfate (CAS: 134-72-5).

Personal Protective Methods: Wear protective gloves and clothing to prevent any reasonable probability of skin contact. Safety equipment suppliers/manufacturers can provide recommendations on the most protective glove/clothing material for your operation. All protective clothing (suits, gloves, footwear, headgear) should be clean, available each day, and put on before work. Contact lenses should not be worn when working with this chemical. Wear dust-proof chemical goggles and face shield unless full face-piece respiratory protection is worn. Employees should wash immediately with soap when skin is wet or contaminated. Provide emergency showers and eyewash. Specific engineering controls are recommended in NIOSH Criteria Document #73-11024.

Respirator Selection: Mercury vapor: NIOSH: *Up to 0.5 mg/m³:* CcrS* (APF = 10) [any chemical cartridge respirator with cartridge(s) providing protection against the compound of concern] or Sa (APF = 10) (any supplied-air respirator). *Up to 1.25 mg/m³:* Sa:Cf (APF = 25) (any supplied-air respirator operated in a continuous-flow mode) or PaprS (APF = 25) [any powered, air-purifying respirator with cartridge(s) providing protection against the compound of concern]* (canister). *Up to 2.5 mg/m³:* CcrFS* (APF = 50) [any chemical cartridge respirator with a full face-piece and cartridge(s) providing protection against the compound of concern] or GmFS* (APF = 50) [any air-purifying, full-face-piece respirator (gas mask) with a chin-style, front- or back-mounted canister providing protection against the compound of concern] or SaT: Cf (APF = 50) (any supplied-air respirator that has a tight-fitting face-piece and is operated in a continuous-flow mode) or PaprTS (APF = 50) [any powered, air-purifying respirator with a tight-fitting face-piece and cartridge(s) providing protection against the compound of concern] or SCBAF (APF = 50) (any self-contained breathing apparatus with a full face-piece) or SaF (APF = 50) (any supplied-air respirator with a full face-piece). *Up to 10 mg/m³:* Sa: Pd,Pp (APF = 1000) (any supplied-air respirator operated in a pressure-demand or other positive-pressure mode). *Emergency or planned entry into unknown concentrations or IDLH conditions:* SCBAF: Pd,Pp (APF = 10,000) (any self-contained breathing apparatus that has a full face-piece and is operated in a pressure-demand or other positive-pressure mode) or SaF: Pd,Pp: ASCBA (any supplied-air respirator that has a full face-piece and is operated in a pressure-demand or other positive-pressure mode in combination with an auxiliary, self-contained breathing apparatus operated in a pressure-demand or other positive-pressure mode). *Escape:* GmFS* [any air-purifying, full-face-piece respirator (gas mask) with a chin-style, front- or back-mounted canister protection against the compound of concern] or SCBAE (any appropriate escape-type, self-contained breathing apparatus).
*End-of-service life indicator (ESLI) required.

Other mercury compounds: NIOSH/OSHA *Up to 1 mg/m³:* CcrS* (APF = 10) [any chemical cartridge respirator with cartridge(s) providing protection against the compound of concern] or Sa (APF = 10) (any supplied-air respirator). *Up to 2.5 mg/m³:* Sa:Cf (APF = 25) (any supplied-air respirator

operated in a continuous-flow mode) or PaprS (APF = 25) [any powered, air-purifying respirator with cartridge(s) providing protection against the compound of concern]* (canister). *Up to 5 mg/m³:* CcrFS* (APF = 50) [any chemical cartridge respirator with a full face-piece and cartridge(s) providing protection against the compound of concern] or GmFS* (APF = 50) [any air-purifying, full-face-piece respirator (gas mask) with a chin-style, front- or back-mounted canister providing protection against the compound of concern] or SaT: Cf (APF = 50) (any supplied-air respirator that has a tight-fitting face-piece and is operated in a continuous-flow mode) or PaprTS (APF = 50) [any powered, air-purifying respirator with a tight-fitting face-piece and cartridge(s) providing protection against the compound of concern] or SCBAF (APF = 50) (any self-contained breathing apparatus with a full face-piece) or SaF (APF = 50) (any supplied-air respirator with a full face-piece). *Up to 10 mg/m³:* Sa: Pd,Pp (APF = 1000) (any supplied-air respirator operated in a pressure-demand or other positive-pressure mode). *Emergency or planned entry into unknown concentrations or IDLH conditions:* SCBAF: Pd,Pp (APF = 10,000) (any self-contained breathing apparatus that has a full face-piece and is operated in a pressure-demand or other positive-pressure mode) or SaF: Pd,Pp: ASCBA (APF = 10,000) (any supplied-air respirator that has a full face-piece and is operated in a pressure-demand or other positive-pressure mode in combination with an auxiliary, self-contained breathing apparatus operated in a pressure-demand or other positive-pressure mode). *Escape:* GmFS* [any air-purifying, full-face-piece respirator (gas mask) with a chin-style, front- or back-mounted canister protection against the compound of concern] or SCBAE (any appropriate escape-type, self-contained breathing apparatus)*.
*End-of-service life indicator (ESLI) required.
Storage: Color Code—Blue: Health Hazard/Poison: Store in a secure poison location. Prior to working with this chemical you should be trained on its proper handling and storage. Store in tightly closed containers in a cool, well-ventilated area away from oxidizers (such as perchlorates, peroxides, permanganates, chlorates and nitrates), light, heat, and acids.
Shipping: This compound requires a shipping label of "POISONOUS/TOXIC MATERIALS." It falls in Hazard Class 6.1 and Packing Group II.
Spill Handling: Avoid breathing dusts. If material is not involved in fire, keep material out of water sources and sewers. Do not touch spilled material; stop leak if you can do so without risk. *Small wet spills:* absorb with sand or other noncombustible absorbent material and place into containers for later disposal. *Small dry spills:* with clean shovel place material into clean, dry container and cover; move containers from spill area. *Large spills:* Dike far ahead of spill for later disposal. Acute exposure to mercuric acetate may require decontamination and life support for the victim. Emergency personnel should wear protective clothing appropriate to the type and degree of contamination.

Air-purifying or supplied-air respiratory equipment should also be worn, as necessary. Rescue vehicles should carry supplies, such as plastic sheeting and disposable plastic bags to assist in preventing spread of contamination. Ventilate area after cleanup is complete. It may be necessary to contain and dispose of this chemical as a hazardous waste. If material or contaminated runoff enters waterways, notify downstream users of potentially contaminated waters. Contact your local or federal environmental protection agency for specific recommendations. If employees are required to clean up spills, they must be properly trained and equipped. OSHA 1910.120(q) may be applicable.
Fire Extinguishing: May burn but is difficult to ignite. Poisonous fumes including mercury are produced in fire. Extinguish fire using agent suitable for type of surrounding fire. Material itself does not burn or burns with difficulty. Use water in flooding quantities as fog. Use foam, carbon dioxide or dry chemical to extinguish fires. Avoid breathing dusts and fumes from burning material. Keep upwind. Avoid bodily contact with the material. Wear boots, protective gloves, and goggles. Do not handle broken packages without protective equipment. Wash away any material which may have contacted the body with copious amounts of water or soap and water. Wear self-contained breathing apparatus when fighting fires involving this material. If contact with the material is anticipated, wear full protective clothing. If material or contaminated runoff enters waterways, notify downstream users of potentially contaminated waters. Notify local health and fire officials and pollution control agencies. From a secure, explosion-proof location, use water spray to cool exposed containers. If cooling streams are ineffective (venting sound increases in volume and pitch, tank discolors, or shows any signs of deforming), withdraw immediately to a secure position. If employees are expected to fight fires, they must be trained and equipped in OSHA 1910.156. The only respirators recommended for firefighting are self-contained breathing apparatuses that have full face-pieces and are operated in a pressure-demand or other positive-pressure mode.
References
Sax, N. I. (Ed.). (1981). *Dangerous Properties of Industrial Materials Report*, 1, No. 3, 70
US Environmental Protection Agency. (November 30, 1987). *Chemical Hazard Information Profile: Mercuric Acetate.* Washington, DC: Chemical Preparedness Program
New Jersey Department of Health and Senior Services. (January 2000). *Hazardous Substances Fact Sheet: Mercuric Acetate*, Trenton, NJ

Mercuric bromide   M:0350

Molecular Formula: Br$_2$Hg
Common Formula: HgBr$_2$
Synonyms: Mercuric bromide, solid; Mercury bromide; Mercury(II) bromide (1:2)

CAS Registry Number: 7789-47-1
RTECS® Number: OV7415000
UN/NA & ERG Number: UN1634/154
EC Number: 232-169-3
Regulatory Authority and Advisory Bodies
Air Pollutant Standard Set. See below, "Permissible Exposure Limits in Air" section.
Clean Air Act: Hazardous Air Pollutants (Title I, Part A, Section 112).
Clean Water Act: 40CFR401.15 Section 307 Toxic Pollutants as mercury and compounds.
RCRA, 40CFR261, Appendix 8 Hazardous Constituents, waste number not listed, as mercury compounds, n.o.s.
EPCRA Section 313: Includes any unique chemical substance that contains mercury as part of that chemical's infrastructure. Form R *de minimis* concentration reporting level: 1.0%.
US DOT Regulated Marine Pollutant (49CFR172.101, Appendix B), severe pollutant.
California Proposition 65 Developmental/Reproductive toxin (mercury and mercury compounds) 7/1/90.
Canada, WHMIS, Ingredients Disclosure List Concentration 1.0%.
Rotterdam Convention Annex III [Chemicals Subject to the Prior Informed Consent Procedure (PIC)] (as mercury compounds, including inorganic mercury compounds, alkyl mercury compounds, and alkyloxylalkyl and aryl mercury compounds).
European/International Regulations: not listed in Annex 1.
WGK (German Aquatic Hazard Class): 3—Severe hazard to waters.
Description: Mercuric Bromide is a crystalline solid. Molecular weight = 360.41; Boiling point = 322°C; Freezing/Melting point = 236°C. Hazard Identification (based on NFPA-704 M Rating System): Health 3, Flammability 0, Reactivity 0. Slightly soluble in water.
Potential Exposure: This compound has applications in medicine.
Incompatibilities: Violent reaction with active metals, potassium, sodium. Store away from heat and light.
Permissible Exposure Limits in Air
OSHA PEL: 0.1 mg[Hg]/m^3 Ceiling Concentration.
NIOSH REL: Hg (*vapor*): 0.05 mg[Hg]/m^3 TWA [skin]; Other: 0.1 mg[Hg]/m^3 Ceiling Concentration [skin].
ACGIH TLV®[1]: 0.025 mg[Hg]/m^3 TWA [skin]; not classifiable as a carcinogen; BEI (preshift) 35 µg[Hg]/100 mL creatinine total inorganic Hg in urine; 15 µg[Hg]/L total inorganic Hg in blood; end-of-shift at end-of-work-week.
NIOSH IDLH: 10 mg[Hg]/m^3.
No TEEL available.
DFG MAK: 0.1 mg[Hg]/m^3; Peak Limitation Category II(8) danger of skin sensitization; Carcinogen Category 3B.
Determination in Air: Use NIOSH Analytical Method #6009; OSHA Analytical Method ID-140.
Permissible Concentration in Water: To protect freshwater aquatic life: 0.00057 µg/L as a 24-h average, never to

exceed 0.0017 µg/L. *To protect saltwater aquatic life:* 0.025 µg/L as a 24-h average, never to exceed 3.7 µg/L. *To protect human health:* 0.144 µg/L (US EPA) set in 1979–1980.[6] These are the limits for inorganic mercury compounds in general.
Determination in Water: Total mercury is determined by flameless atomic absorption. Soluble mercury may be determined by 0.45-µm filtration followed by flameless atomic absorption.
Routes of Entry: Inhalation, skin absorption, ingestion, skin and/or eye contact.
Harmful Effects and Symptoms
Short Term Exposure: Mercuric bromide can affect you when breathed and by passing through skin. Overexposures can cause kidney damage. Mercury poisoning can cause "shakes," irritability, sore gums, memory loss, increased saliva, personality changes, and permanent brain damage may result. Skin contact may cause burns or skin allergy, or gray skin color. Contact causes eye burns with permanent damage. Heating or use near acid releases toxic mercury vapors. Mercury can build up with permanent damage. Mercury can build up in the body. Health effects have been reported below NIOSH exposure levels.
Long Term Exposure: Repeated or prolonged contact with skin may result in dermatitis (red inflamed skin). Repeated or prolonged exposure may cause death by hypovolemic shock, nephrotic syndrome, or kidney failure.
Points of Attack: Eyes, skin, respiratory system, central nervous system, kidneys.
Medical Surveillance: NIOSH lists the following tests for inorganic mercury: whole blood (chemical/metabolite); whole blood (chemical/metabolite), end-of-shift; end-of-shift at end-of-work-week; biologic tissue/biopsy; nerve conduction studies; neurologic examination/electromyography; thyroid function test/thyroid profile; urine (chemical/metabolite); urine (chemical/metabolite) prior to shift, prior to next shift; urine (chemical/metabolite), sediment; urinalysis (routine). Before first exposure and every 6–12 months after, a complete medical history and examination is strongly recommended with examination of the nervous system including handwriting. Routine urine test (UA), urine test for mercury (should be less than 0.02 mg/L). Eye examination. After suspected illness or overexposure, repeat the test above and get a blood test for mercury. Consider chest x-ray after acute overexposure. Consider nerve conduction tests, urinary enzymes, and neurobehavioral testing.
First Aid: If this chemical gets into the eyes, remove any contact lenses at once and irrigate immediately for at least 15 min, occasionally lifting upper and lower lids. Seek medical attention immediately. If this chemical contacts the skin, remove contaminated clothing and wash immediately with soap and water. Seek medical attention immediately. If this chemical has been inhaled, remove from exposure, begin rescue breathing (using universal precautions, including resuscitation mask) if breathing has stopped and CPR if heart action has stopped. Transfer promptly to a medical

facility. When this chemical has been swallowed, get medical attention. Give large quantities of water and induce vomiting. Do not make an unconscious person vomit. Medical observation is recommended for 24—48 h after breathing overexposure, as pulmonary edema may be delayed. As first aid for pulmonary edema, a doctor or authorized paramedic may consider administering a corticosteroid spray.

Antidotes and special procedures for medical personnel: The drug NAP (*n*-acetyl penicillamine) has been used to treat mercury poisoning, with mixed success.

Note to physician: For severe poisoning BAL [British Anti-Lewisite, dimercaprol, dithiopropanol ($C_3H_8OS_2$)] has been used to treat toxic symptoms of certain heavy metals poisoning including mercury. Although BAL is reported to have a large margin of safety, caution must be exercised, because toxic effects may be caused by excessive dosage. Most can be prevented by premedication with 1-ephedrine sulfate (CAS: 134-72-5).

Personal Protective Methods: Wear protective gloves and clothing to prevent any reasonable probability of skin contact. Safety equipment suppliers/manufacturers can provide recommendations on the most protective glove/clothing material for your operation. All protective clothing (suits, gloves, footwear, headgear) should be clean, available each day, and put on before work. Contact lenses should not be worn when working with this chemical. Wear dust-proof chemical goggles and face shield unless full face-piece respiratory protection is worn. Employees should wash immediately with soap when skin is wet or contaminated. Provide emergency showers and eyewash.

Respirator Selection: Mercury vapor: NIOSH: *Up to 0.5 mg/m³:* CcrS* (APF = 10) [any chemical cartridge respirator with cartridge(s) providing protection against the compound of concern] or Sa (APF = 10) (any supplied-air respirator). *Up to 1.25 mg/m³:* Sa:Cf (APF = 25) (any supplied-air respirator operated in a continuous-flow mode) or PaprS (APF = 25) [any powered, air-purifying respirator with cartridge(s) providing protection against the compound of concern]* (canister). *Up to 2.5 mg/m³:* CcrFS* (APF = 50) [any chemical cartridge respirator with a full face-piece and cartridge(s) providing protection against the compound of concern] or GmFS* (APF = 50) [any air-purifying, full-face-piece respirator (gas mask) with a chin-style, front- or back-mounted canister providing protection against the compound of concern] or SaT: Cf (APF = 50) (any supplied-air respirator that has a tight-fitting face-piece and is operated in a continuous-flow mode) or PaprTS (APF = 50) [any powered, air-purifying respirator with a tight-fitting face-piece and cartridge(s) providing protection against the compound of concern] or SCBAF (APF = 50) (any self-contained breathing apparatus with a full face-piece) or SaF (APF = 50) (any supplied-air respirator with a full face-piece). *Up to 10 mg/m³:* Sa: Pd,Pp (APF = 1000) (any supplied-air respirator operated in a pressure-demand or other positive-pressure mode). *Emergency or planned entry into unknown concentrations or IDLH conditions:* SCBAF: Pd,Pp (APF = 10,000) (any self-contained breathing apparatus that has a full face-piece and is operated in a pressure-demand or other positive-pressure mode) or SaF: Pd,Pp: ASCBA (any supplied-air respirator that has a full face-piece and is operated in a pressure-demand or other positive-pressure mode in combination with an auxiliary, self-contained breathing apparatus operated in a pressure-demand or other positive-pressure mode). *Escape:* GmFS* [any air-purifying, full-face-piece respirator (gas mask) with a chin-style, front- or back-mounted canister protection against the compound of concern] or SCBAE (any appropriate escape-type, self-contained breathing apparatus).
*End-of-service life indicator (ESLI) required.

Other mercury compounds: NIOSH/OSHA *Up to 1 mg/m³:* CcrS* (APF = 10) [any chemical cartridge respirator with cartridge(s) providing protection against the compound of concern] or Sa (APF = 10) (any supplied-air respirator). *Up to 2.5 mg/m³:* Sa:Cf (APF = 25) (any supplied-air respirator operated in a continuous-flow mode) or PaprS (APF = 25) [any powered, air-purifying respirator with cartridge(s) providing protection against the compound of concern]* (canister). *Up to 5 mg/m³:* CcrFS* (APF = 50) [any chemical cartridge respirator with a full face-piece and cartridge(s) providing protection against the compound of concern] or GmFS* (APF = 50) [any air-purifying, full-face-piece respirator (gas mask) with a chin-style, front- or back-mounted canister providing protection against the compound of concern] or SaT: Cf (APF = 50) (any supplied-air respirator that has a tight-fitting face-piece and is operated in a continuous-flow mode) or PaprTS (APF = 50) [any powered, air-purifying respirator with a tight-fitting face-piece and cartridge(s) providing protection against the compound of concern] or SCBAF (APF = 50) (any self-contained breathing apparatus with a full face-piece) or SaF (APF = 50) (any supplied-air respirator with a full face-piece). *Up to 10 mg/m³:* Sa: Pd,Pp (APF = 1000) (any supplied-air respirator operated in a pressure-demand or other positive-pressure mode). *Emergency or planned entry into unknown concentrations or IDLH conditions:* SCBAF: Pd,Pp (APF = 10,000) (any self-contained breathing apparatus that has a full face-piece and is operated in a pressure-demand or other positive-pressure mode) or SaF: Pd,Pp: ASCBA (APF = 10,000) (any supplied-air respirator that has a full face-piece and is operated in a pressure-demand or other positive-pressure mode in combination with an auxiliary, self-contained breathing apparatus operated in a pressure-demand or other positive-pressure mode). *Escape:* GmFS* [any air-purifying, full-face-piece respirator (gas mask) with a chin-style, front- or back-mounted canister protection against the compound of concern] or SCBAE (any appropriate escape-type, self-contained breathing apparatus)*.
*End-of-service life indicator (ESLI) required.

Storage: Color Code—Blue: Health Hazard/Poison: Store in a secure poison location. Prior to working with this chemical you should be trained on its proper handling and

storage. Store in tightly closed containers in a cool, well-ventilated area away from light, heat, and acids. Mercuric bromide must be stored to avoid contact with sodium and potassium, since violent reactions occur.

Shipping: This compound requires a shipping label of "POISONOUS/TOXIC MATERIALS." It falls in Hazard Class 6.1 and Packing Group II.

Spill Handling: Evacuate persons not wearing protective equipment from area of spill or leak until cleanup is complete. Remove all ignition sources. Spills should be collected with special mercury vapor suppressants or special vacuums and deposited in sealed containers. Kits specific for cleanup of mercury spills should be available. Ventilate area after cleanup is complete. It may be necessary to contain and dispose of this chemical as a hazardous waste. If material or contaminated runoff enters waterways, notify downstream users of potentially contaminated waters. Contact your local or federal environmental protection agency for specific recommendations. If employees are required to clean up spills, they must be properly trained and equipped. OSHA 1910.120(q) may be applicable.

Fire Extinguishing: Mercuric bromide may burn but does not readily ignite. Use dry chemical, carbon dioxide, water spray, or foam extinguishers. Poisonous gases are produced in fire, including mercury and bromium. If material or contaminated runoff enters waterways, notify downstream users of potentially contaminated waters. Notify local health and fire officials and pollution control agencies. From a secure, explosion-proof location, use water spray to cool exposed containers. If cooling streams are ineffective (venting sound increases in volume and pitch, tank discolors, or shows any signs of deforming), withdraw immediately to a secure position. If employees are expected to fight fires, they must be trained and equipped in OSHA 1910.156. The only respirators recommended for firefighting are self-contained breathing apparatuses that have full face-pieces and are operated in a pressure-demand or other positive-pressure mode.

Reference
New Jersey Department of Health and Senior Services. (May 2001). *Hazardous Substances Fact Sheet: Mercuric Bromide*. Trenton, NJ

Mercuric chloride   M:0360

Molecular Formula: Cl_2Hg
Common Formula: $HgCl_2$
Synonyms: Bichloride of mercury; Bichlorure de mercure (French); Calochlor; Chlorure mercurique (French); Cloruro mercurico (Spanish); Corrosive mercury chloride; Fungchex; MC; Mercuric bichloride; Mercury bichloride; Mercury(2+) chloride; Mercury(II) chloride; Mercury perchloride; Mercury vichloride; NCI-C60173; Quecksilber chlorid (German); Perchloride of mercury; TL 898
CAS Registry Number: 7487-94-7

RTECS® Number: OV9100000
UN/NA & ERG Number: UN1624/154
EC Number: 231-299-8 [*Annex I Index No.:* 080-010-00-X]
Regulatory Authority and Advisory Bodies
Carcinogenicity: IARC: Animal Limited Evidence; Human Inadequate Evidence, *not classifiable as carcinogenic to humans*, Group 3, 1993; NCI: Carcinogenesis Studies (gavage); equivocal evidence: mouse, rat; EPA: Possible Human Carcinogen.
US EPA Gene-Tox Program, Positive: Cell transform.—SA7/SHE; *B. subtilis* rec assay; Negative: *N. crassa—aneuploidy.*
US EPA, FIFRA 1998 Status of Pesticides: Canceled.
Banned or Severely Restricted (in agriculture) (UK).[13]
FDA—over-the-counter drug.
Air Pollutant Standard Set. See below, "Permissible Exposure Limits in Air" section.
Clean Air Act: Hazardous Air Pollutants (Title I, Part A, Section 112).
Clean Water Act: 40CFR401.15 Section 307 Toxic Pollutants as mercury and compounds.
RCRA, 40CFR261, Appendix 8 Hazardous Constituents, waste number not listed, as mercury compounds, n.o.s.
EPCRA Section 313: Includes any unique chemical substance that contains mercury as part of that chemical's infrastructure. Form R *de minimis* concentration reporting level: 1.0%.
US DOT Regulated Marine Pollutant (49CFR172.101, Appendix B), severe pollutant.
California Proposition 65 Developmental/Reproductive toxin (mercury and mercury compounds) 7/1/90.
Canada, WHMIS, Ingredients Disclosure List Concentration 0.1%.
Rotterdam Convention Annex III [Chemicals Subject to the Prior Informed Consent Procedure (PIC)] (as mercury compounds, including inorganic mercury compounds, alkyl mercury compounds, and alkyloxylalkyl and aryl mercury compounds).
European/International Regulations: Hazard Symbol: T +, N; Risk phrases: R28; R34; R48/24/25; R50/53; Safety phrases: S1/2; S26; S36/37/39; S45; S60; S61 (see Appendix 4).
WGK (German Aquatic Hazard Class): 3—Severe hazard to waters.
Description: Mercuric chloride is an odorless white crystalline solid. Molecular weight = 271.49; Boiling point = 302°C; Freezing/Melting point = 276°C. Hazard Identification (based on NFPA-704 M Rating System): Health 3, Flammability 0, Reactivity 0. Soluble in water; solubility = 7.38% at 20°C.
Potential Exposure: Compound Description: Agricultural Chemical; Tumorigen, Drug, Organometallic, Mutagen, Reproductive Effector; Human Data; Primary Irritant. Mercuric chloride is used as dip for bulbs and tubers, for earthworm control; as repellent to ants, roaches, etc; in preserving wood and anatomical specimens; embalming,

browning, etching steel and iron; as a catalyst for organic synthesis; disinfectant, antiseptic, tanning; textile printing aid; manufacture of dyes; in agricultural chemicals; dry batteries; pharmaceuticals, and photographic chemicals.

Incompatibilities: Mercuric chloride may explode with friction or application on heat. Mixtures of mercuric chloride and sodium or potassium are shock sensitive and will explode on impact. Avoid contact with acids or acid fumes. Also avoid the presence of formats, sulfites, hypophosphites, phosphates, sulfide, albumin, gelatin, alkalies, alkaloid salts, ammonia, lime water, antimony, arsenic, bromides, borax, carbonates, reduced iron, copper, iron, lead, silver salts, infusions of cinchona, columbo, oak bark or senna, and tannic acid.

Permissible Exposure Limits in Air

OSHA PEL: 0.1 mg[Hg]/m^3 Ceiling Concentration.

NIOSH REL: Hg (*vapor*): 0.05 mg[Hg]/m^3 TWA [skin]; Other: 0.1 mg[Hg]/m^3 Ceiling Concentration [skin].

ACGIH TLV®[1]: 0.025 mg[Hg]/m^3 TWA [skin]; not classifiable as a carcinogen; BEI (preshift) 35 μg[Hg]/100 mL creatinine total inorganic Hg in urine; 15 μg[Hg]/L total inorganic Hg in blood; end-of-shift at end-of-work-week.

Protective Action Criteria (PAC)

TEEL-0: 0.0338 mg/m^3

PAC-1: 4 mg/m^3

PAC-2: 13.5 mg/m^3

PAC-3: 13.5 mg/m^3

DFG MAK: 0.1 mg[Hg]/m^3; Peak Limitation Category II(8) danger of skin sensitization; Carcinogen Category 3B.

NIOSH IDLH: 10 mg[Hg]/m^3.

Australia: TWA 0.05 mg[Hg]/m^3, [skin], 1993; Belgium: TWA 0.05 mg[Hg]/m^3, [skin], 1993; Denmark: TWA 0.05 mg[Hg]/m^3, [skin], 1999; Finland: TWA 1 mg[Hg]/m^3, 1999; France: VME 0.1 mg[Hg]/m^3, 1999; Hungary: TWA 0.02 mg[Hg]/m^3; STEL 0.04 mg[Hg]/m^3, 1993; Japan: 0.05 mg[Hg]/m^3, 1999; Norway: TWA 0.05 mg[Hg]/m^3, 1999; the Philippines: TWA 0.05 mg[Hg]/m^3, 1993; Poland: MAC (TWA) 0.05 mg[Hg]/m^3, MAC (STEL) 0.15 mg[Hg]/m^3, 1999; Russia: TWA 0.05 mg[Hg]/m^3; STEL 0.01 mg[Hg]/m^3, 1993; Sweden: NGV 0.05 mg[Hg]/m^3, [skin], 1999; Thailand STEL 0.05 mg[Hg]/m^3, 1993; United Kingdom: TWA 0.025 mg[Hg]/mg, 2000; Argentina, Bulgaria, Columbia, Jordan, South Korea, New Zealand, Singapore, Vietnam: ACGIH TLV®: not classifiable as a human carcinogen.

Determination in Air: Use NIOSH Analytical Method #6009; OSHA Analytical Method ID-140.

Permissible Concentration in Water: *To protect freshwater aquatic life:* 0.00057 μg/L as a 24-h average, never to exceed 0.0017 μg/L. *To protect saltwater aquatic life:* 0.025 μg/L as a 24-h average, never to exceed 3.7 μg/L. *To protect human health:* 0.144 μg/L (US EPA) set in 1979—1980.[6] These are the limits for inorganic mercury compounds in general. In addition, the former USSR-UNEP/IRTC project[43] set a MAC of 0.005 μg/L for water bodies used for domestic purposes.

Determination in Water: Total mercury is determined by flameless atomic absorption. Soluble mercury may be determined by 0.45-μm filtration followed by flameless atomic absorption. Octanol—water coefficient: Log K_{ow} = 0.12.

Routes of Entry: Inhalation, skin absorption, ingestion, skin and/or eye contact.

Harmful Effects and Symptoms

Short Term Exposure: Corrosive. The substance is corrosive to the eyes, the skin, and the respiratory tract. Corrosive on ingestion. Inhalation of its aerosol may cause lung edema. The substance may cause effects on the kidneys. Exposure far above OEL may result in death. The effects may be delayed. Medical observation is indicated. It is classified as extremely toxic. All forms of mercury are poisonous if absorbed. Probable oral lethal dose is 5—50 mg/kg and between 7 drops and 1 teaspoonful for a 150-lb person. Mercuric chloride is one of the most toxic salts of mercury. Material attacks the gastrointestinal tract and renal systems. Signs and symptoms of acute exposure or mercuric chloride may be severe and include increased salivation, foul breath, inflammation and ulceration of the mucous membranes, abdominal pain, and bloody diarrhea. Dermal exposure may result in dermatitis (red, inflamed skin) and burns. Oliguria (scanty urination), anuria (suppression of urine formation), and acute renal failure may be noted. Weak pulse, seizures, psychic disturbances, circulatory collapse, chest pain, and dyspnea (shortness of breath) may be observed.

Long Term Exposure: Repeated or prolonged contact with skin may result in dermatitis (red inflamed skin). Repeated or prolonged exposure may cause death by hypovolemic shock, nephrotic syndrome, or kidney failure.

Points of Attack: Eyes, skin, central nervous system, peripheral nervous system, kidneys.

Medical Surveillance: NIOSH lists the following tests for inorganic mercury: whole blood (chemical/metabolite); whole blood (chemical/metabolite), end-of-shift; end-of-shift at end-of-work-week; biologic tissue/biopsy; nerve conduction studies; neurologic examination/electromyography; thyroid function test/thyroid profile; urine (chemical/metabolite); urine (chemical/metabolite) prior to shift, prior to next shift; urine (chemical/metabolite), sediment; urinalysis (routine). Before first exposure and every 6—12 months after, a complete medical history and examination is strongly recommended with examination of the nervous system including handwriting. Routine urine test (UA). Urine test for mercury (should be less than 0.02 mg/L). Eye examination. Consider lung function tests for persons with frequent exposure. After suspected illness or overexposure, repeat the tests above and get a blood test for mercury. Consider chest X-ray after sudden overexposure. Consider nerve conduction tests, urinary enzymes, and neurobehavioral testing.

First Aid: If this chemical gets into the eyes, remove any contact lenses at once and irrigate immediately for at least 15 min, occasionally lifting upper and lower lids. Seek medical attention immediately. If this chemical contacts the skin, remove contaminated clothing and wash immediately with soap and water. Seek medical attention immediately. If this chemical has been inhaled, remove from exposure, begin rescue breathing (using universal precautions, including resuscitation mask) if breathing has stopped and CPR if heart action has stopped. Transfer promptly to a medical facility. When this chemical has been swallowed, get medical attention. Give large quantities of water and induce vomiting. Do not make an unconscious person vomit. Medical observation is recommended for 24–48 h after breathing overexposure, as pulmonary edema may be delayed. As first aid for pulmonary edema, a doctor or authorized paramedic may consider administering a corticosteroid spray.

Antidotes and special procedures for medical personnel: The drug NAP (*n*-acetyl penicillamine) has been used to treat mercury poisoning, with limited success.

Note to physician: For severe poisoning BAL [British Anti-Lewisite, dimercaprol, dithiopropanol ($C_3H_8OS_2$)] has been used to treat toxic symptoms of certain heavy metals poisoning including mercury. Although BAL is reported to have a large margin of safety, caution must be exercised, because toxic effects may be caused by excessive dosage. Most can be prevented by premedication with 1-ephedrine sulfate (CAS: 134-72-5).

Personal Protective Methods: Wear protective gloves and clothing to prevent any reasonable probability of skin contact. Safety equipment suppliers/manufacturers can provide recommendations on the most protective glove/clothing material for your operation. All protective clothing (suits, gloves, footwear, headgear) should be clean, available each day, and put on before work. Contact lenses should not be worn when working with this chemical. Wear dust-proof chemical goggles and face shield unless full face-piece respiratory protection is worn. Employees should wash immediately with soap when skin is wet or contaminated. Provide emergency showers and eyewash.

Respirator Selection: *Mercury vapor:* NIOSH: *Up to 0.5 mg/m³:* CcrS* (APF = 10) [any chemical cartridge respirator with cartridge(s) providing protection against the compound of concern] or Sa (APF = 10) (any supplied-air respirator). *Up to 1.25 mg/m³:* Sa:Cf (APF = 25) (any supplied-air respirator operated in a continuous-flow mode) or PaprS (APF = 25) [any powered, air-purifying respirator with cartridge(s) providing protection against the compound of concern]* (canister). *Up to 2.5 mg/m³:* CcrFS* (APF = 50) [any chemical cartridge respirator with a full face-piece and cartridge(s) providing protection against the compound of concern] or GmFS* (APF = 50) [any air-purifying, full-face-piece respirator (gas mask) with a chin-style, front- or back-mounted canister providing protection against the compound of concern] or SaT: Cf (APF = 50) (any supplied-air respirator that has a tight-fitting face-piece and is operated in a continuous-flow mode) or PaprTS (APF = 50) [any powered, air-purifying respirator with a

tight-fitting face-piece and cartridge(s) providing protection against the compound of concern] or SCBAF (APF = 50) (any self-contained breathing apparatus with a full face-piece) or SaF (APF = 50) (any supplied-air respirator with a full face-piece). *Up to 10 μg/m³:* Sa: Pd,Pp (APF = 1000) (any supplied-air respirator operated in a pressure-demand or other positive-pressure mode). Emergency or planned entry into unknown concentrations or IDLH conditions: SCBAF: Pd,Pp (APF = 10,000) (any self-contained breathing apparatus that has a full face-piece and is operated in a pressure-demand or other positive-pressure mode) or SaF: Pd,Pp: ASCBA (APF = 10,000) (any supplied-air respirator that has a full face-piece and is operated in a pressure-demand or other positive-pressure mode in combination with an auxiliary, self-contained breathing apparatus operated in a pressure-demand or other positive-pressure mode). Escape: GmFS* [any air-purifying, full-face-piece respirator (gas mask) with a chin-style, front- or back-mounted canister protection against the compound of concern] or SCBAE (any appropriate escape-type, self-contained breathing apparatus).*

*End-of-service life indicator (ESLI) required.

Other mercury compounds: NIOSH/OSHA *Up to 1 mg/m³:* CcrS* (APF = 10) [any chemical cartridge respirator with cartridge(s) providing protection against the compound of concern] or Sa (APF = 10) (any supplied-air respirator). *Up to 2.5 mg/m³:* Sa:Cf (APF = 25) (any supplied-air respirator operated in a continuous-flow mode) or PaprS (APF = 25) [any powered, air-purifying respirator with cartridge(s) providing protection against the compound of concern]* (canister). *Up to 5 mg/m³:* CcrFS (APF = 50) [any chemical cartridge respirator with a full face-piece and cartridge(s) providing protection against the compound of concern];* or GmFS* (APF = 50) [any air-purifying, full-face-piece respirator (gas mask) with a chin-style, front- or back-mounted canister providing protection against the compound of concern] or SaT: Cf (APF = 50) (any supplied-air respirator that has a tight-fitting face-piece and is operated in a continuous-flow mode) or PaprTS (APF = 50) [any powered, air-purifying respirator with a tight-fitting face-piece and cartridge(s) providing protection against the compound of concern] or SCBAF (APF = 50) (any self-contained breathing apparatus with a full face-piece) or SaF (APF = 50) (any supplied-air respirator with a full face-piece). *Up to 10 mg/m³:* Sa: Pd,Pp (APF = 1000) (any supplied-air respirator operated in a pressure-demand or other positive-pressure mode). *Emergency or planned entry into unknown concentrations or IDLH conditions:* SCBAF: Pd,Pp (APF = 10,000) (any self-contained breathing apparatus that has a full face-piece and is operated in a pressure-demand or other positive-pressure mode) or SaF: Pd,Pp: ASCBA (APF = 10,000) (any supplied-air respirator that has a full face-piece and is operated in a pressure-demand or other positive-pressure mode in combination with an auxiliary, self-contained breathing apparatus operated in a pressure-demand or other positive-pressure mode). *Escape:* GmFS*

[any air-purifying, full-face-piece respirator (gas mask) with a chin-style, front- or back-mounted canister protection against the compound of concern] or SCBAE (any appropriate escape-type, self-contained breathing apparatus).*
*End-of-service life indicator (ESLI) required.

Storage: Color Code—Blue: Health Hazard/Poison: Store in a secure poison location. Prior to working with this chemical you should be trained on its proper handling and storage. Mercuric chloride must be stored to avoid contact with potassium and sodium, since violent reactions occur. See also "Incompatibilities," above. Store in tightly closed containers in a cool, well-ventilated area.

Shipping: This compound requires a shipping label of "POISONOUS/TOXIC MATERIALS." It falls in Hazard Class 6.1 and Packing Group II.

Spill Handling: Evacuate persons not wearing protective equipment from area of spill or leak until cleanup is complete. Remove all ignition sources. Spills should be collected with special mercury vapor suppressants or special vacuums and deposited in sealed containers. Kits specific for cleanup of mercury spills should be available. Ventilate area after cleanup is complete. It may be necessary to contain and dispose of this chemical as a hazardous waste. If material or contaminated runoff enters waterways, notify downstream users of potentially contaminated waters. Contact your local or federal environmental protection agency for specific recommendations. If employees are required to clean up spills, they must be properly trained and equipped. OSHA 1910.120(q) may be applicable.

Fire Extinguishing: Not combustible. Extinguish with water spray, fog, foam, dry chemical, or carbon dioxide. Move container from fire area. Wear self-contained breathing apparatus and full body protective clothing. Poisonous gases are produced in fire. If material or contaminated runoff enters waterways, notify downstream users of potentially contaminated waters. Notify local health and fire officials and pollution control agencies. From a secure, explosion-proof location, use water spray to cool exposed containers. If cooling streams are ineffective (venting sound increases in volume and pitch, tank discolors, or shows any signs of deforming), withdraw immediately to a secure position. If employees are expected to fight fires, they must be trained and equipped in OSHA 1910.156. The only respirators recommended for firefighting are self-contained breathing apparatuses that have full face-pieces and are operated in a pressure-demand or other positive-pressure mode.

References
US Environmental Protection Agency. (November 30, 1987). *Chemical Hazard Information Profile: Mercuric Chloride.* Washington, DC: Chemical Emergency Preparedness Program
New York State Department of Health. (February 1986). *Chemical Fact Sheet: Mercuric Chloride.* Version 2. Albany, NY: Bureau of Toxic Substance Assessment
US Environmental Protection Agency, Special Review and Reregistration Division Office of Pesticide Programs.

(1998). *Agency Status of Pesticides in Registration, Reregistration, and Special Review* (Rainbow Report). Washington, DC
New Jersey Department of Health and Senior Services. (January 2000). *Hazardous Substances Fact Sheet: Mercuric Chloride.* Trenton, NJ

Mercuric cyanide   M:0370

Molecular Formula: C_2HgN_2
Common Formula: $Hg(CN)_2$
Synonyms: Cianuro mercurico (Spanish); Cyanure de mercure (French); Mercury(2+) cyanide; Mercury(II) cyanide
CAS Registry Number: 592-04-1
RTECS® Number: OW1515000
UN/NA & ERG Number: UN1636/154
EC Number: 209-741-6
Regulatory Authority and Advisory Bodies
Air Pollutant Standard Set. See below, "Permissible Exposure Limits in Air" section.
Clean Air Act: Hazardous Air Pollutants (Title I, Part A, Section 112).
Clean Water Act: 40CFR401.15 Section 307 Toxic Pollutants as mercury and compounds.
RCRA, 40CFR261, Appendix 8 Hazardous Constituents, waste number not listed, as mercury compounds, n.o.s.
EPCRA Section 313: Includes any unique chemical substance that contains mercury as part of that chemical's infrastructure. Form R *de minimis* concentration reporting level: 1.0%.
US DOT Regulated Marine Pollutant (49CFR172.101, Appendix B), severe pollutant.
California Proposition 65 Developmental/Reproductive toxin (mercury and mercury compounds) 7/1/90.
Canada, WHMIS, Ingredients Disclosure List Concentration 1.0%.
Rotterdam Convention Annex III [Chemicals Subject to the Prior Informed Consent Procedure (PIC)] (as mercury compounds, including inorganic mercury compounds, alkyl mercury compounds, and alkyloxylalkyl and aryl mercury compounds).
European/International Regulations: not listed in Annex 1.
WGK (German Aquatic Hazard Class): 3—Severe hazard to waters.

Description: Mercuric cyanide is an odorless, white crystalline solid; turns gray to dark brown when exposed to light. Molecular weight = 252.63; Decomposes at 319°C. Hazard Identification (based on NFPA-704 M Rating System): Health 3, Flammability 0, Reactivity 0. Soluble in water.

Potential Exposure: Mercuric cyanide is used in medicine, germicidal soaps, photography and in making cyanogen gas.

Incompatibilities: Violent reaction with fluorine, magnesium, sodium nitrite, acids. Heating or contact with acid releases toxic mercury and flammable HCN.

Permissible Exposure Limits in Air
Protective Action Criteria (PAC)
TEEL-0: 0.0315 mg/m^3
PAC-1: 1.5 mg/m^3
PAC-2: 12.6 mg/m^3
PAC-3: 12.6 mg/m^3

As inorganic mercury compound
OSHA PEL: 0.1 mg[Hg]/m^3 Ceiling Concentration.
NIOSH REL: Hg (*vapor*): 0.05 mg[Hg]/m^3 TWA [skin]; Other: 0.1 mg[Hg]/m^3 Ceiling Concentration [skin].
ACGIH TLV®[1]: 0.025 mg[Hg]/m^3 TWA [skin]; not classifiable as a carcinogen; BEI (preshift) 35 μg[Hg]/100 mL creatinine total inorganic Hg in urine; 15 μg[Hg]/L total inorganic Hg in blood; end-of-shift at end-of-work-week.
DFG MAK: 0.1 mg[Hg]/m^3; Peak Limitation Category II(8) danger of skin sensitization; Carcinogen Category 3B.
NIOSH IDLH: 10 mg[Hg]/m^3.

As cyanide compound (for reference)
OSHA PEL: 5 mg[CN]/m^3/4.7 ppm TWA.
NIOSH REL: 5 mg[CN]/m^3/4.7 ppm/10 min, Ceiling Concentration.
ACGIH TLV®[1]: 5 mg[CN]/m^3 [skin] Ceiling Concentration.
DFG MAK: 2 mg[CN]/m^3, inhalable fraction TWA; Peak Limitation Category II(1) [skin]; Pregnancy Risk Group: C.
NIOSH IDLH: 25 mg[CN]/m^3.

Determination in Air: Use NIOSH Analytical Method #6009; OSHA Analytical Method ID-140. See also NIOSH Analytical Method #7904, Cyanides.

Permissible Concentration in Water: *To protect freshwater aquatic life:* 0.00057 μg/L as a 24-h average, never to exceed 0.0017 μg/L. *To protect saltwater aquatic life:* 0.025 μg/L as a 24-h average, never to exceed 3.7 μg/L. *To protect human health:* 0.144 μg/L (US EPA) set in 1979–1980.[6] These are the limits for inorganic mercury compounds in general. See also the entry on "Cyanides."

Determination in Water
Total mercury is determined by flameless atomic absorption. Soluble mercury may be determined by 0.45-μm filtration followed by flameless atomic absorption.

Routes of Entry: Inhalation, skin absorption, ingestion, skin and/or eye contact.

Harmful Effects and Symptoms
Short Term Exposure: Mercuric cyanide can affect you when breathed and by passing through skin. Direct contact causes eye irritation and possible damage. Overexposures can cause kidney damage. Mercury poisoning can cause "shakes," irritability, sore gums, memory loss, increased saliva, metallic taste, personality changes, and brain damage. Skin contact can cause irritation, skin allergy, or a gray skin color. Mercury can build up in the body. Heating or contact with acid or acid mist causes the release of toxic mercury and cyanide vapors; may cause bronchitis and lung irritation. Overexposure to cyanide can cause sudden death.

Long Term Exposure: Repeated or prolonged contact with skin may cause dermatitis. Repeated or prolonged exposure may cause death by hypovolemic shock, nephrotic syndrome, or kidney failure. Organic mercury substances have been identified as human teratogens. Some related compounds damage the developing fetus and decrease fertility in males and females. It is unknown for certain if mercuric cyanide causes these effects and so care is indicated.

Points of Attack: Eyes, skin, respiratory system; central nervous system; kidneys.

Medical Surveillance: NIOSH lists the following tests for inorganic mercury: whole blood (chemical/metabolite); whole blood (chemical/metabolite), end-of-shift; end-of-shift at end-of-work-week; biologic tissue/biopsy; nerve conduction studies; neurologic examination/electromyography; thyroid function test/thyroid profile; urine (chemical/metabolite); urine (chemical/metabolite) prior to shift, prior to next shift; urine (chemical/metabolite), sediment; urinalysis (routine). Before first exposure and every 6–12 months after, a complete medical history and examination is strongly recommended with examination of the nervous system including handwriting. Routine urine test (UA). Urine test for mercury (should be less than 0.02 mg/L). Eye examination. After suspected illness or overexposure, repeat the tests above and get a blood test for mercury. Consider nerve conduction tests, urinary enzymes, and neurobehavioral testing.

First Aid: If this chemical gets into the eyes, remove any contact lenses at once and irrigate immediately for at least 15 min, occasionally lifting upper and lower lids. Seek medical attention immediately. If this chemical contacts the skin, remove contaminated clothing and wash immediately with soap and water. Seek medical attention immediately. If this chemical has been inhaled, remove from exposure, begin rescue breathing (using universal precautions, including resuscitation mask) if breathing has stopped and CPR if heart action has stopped. Transfer promptly to a medical facility. When this chemical has been swallowed, get medical attention. Give large quantities of water and induce vomiting. Do not make an unconscious person vomit.

Antidotes and special procedures for medical personnel: The drug NAP (*n*-acetyl penicillamine) has been used to treat mercury poisoning, with mixed success.

Note to physician: For severe poisoning BAL [British Anti-Lewisite, dimercaprol, dithiopropanol ($C_3H_8OS_2$)] has been used to treat toxic symptoms of certain heavy metals poisoning including mercury. Although BAL is reported to have a large margin of safety, caution must be exercised, because toxic effects may be caused by excessive dosage. Most can be prevented by premedication with 1-ephedrine sulfate (CAS: 134-72-5).

Personal Protective Methods: Wear protective gloves and clothing to prevent any reasonable probability of skin contact. Safety equipment suppliers/manufacturers can provide recommendations on the most protective glove/clothing material for your operation. All protective clothing (suits, gloves, footwear, headgear) should be clean, available each day, and put on before work. Contact lenses should not be

worn when working with this chemical. Wear dust-proof chemical goggles and face shield unless full face-piece respiratory protection is worn. Employees should wash immediately with soap when skin is wet or contaminated. Provide emergency showers and eyewash.

Respirator Selection: Mercury vapor: NIOSH: *Up to 0.5 mg/m³:* CcrS* (APF = 10) [any chemical cartridge respirator with cartridge(s) providing protection against the compound of concern] or Sa (APF = 10) (any supplied-air respirator). *Up to 1.25 mg/m³:* Sa:Cf (APF = 25) (any supplied-air respirator operated in a continuous-flow mode) or PaprS (APF = 25) [any powered, air-purifying respirator with cartridge(s) providing protection against the compound of concern]* (canister). *Up to 2.5 mg/m³:* CcrFS (APF = 50) [any chemical cartridge respirator with a full face-piece and cartridge(s) providing protection against the compound of concern];* or GmFS (APF = 50) [any air-purifying, full-face-piece respirator (gas mask) with a chin-style, front- or back-mounted canister providing protection against the compound of concern];* or SaT: Cf (APF = 50) (any supplied-air respirator that has a tight-fitting face-piece and is operated in a continuous-flow mode) or PaprTS (APF = 50) [any powered, air-purifying respirator with a tight-fitting face-piece and cartridge(s) providing protection against the compound of concern] or SCBAF (APF = 50) (any self-contained breathing apparatus with a full face-piece) or SaF (APF = 50) (any supplied-air respirator with a full face-piece). *Up to 10 mg/m³:* Sa: Pd,Pp (APF = 1000) (any supplied-air respirator operated in a pressure-demand or other positive-pressure mode). *Emergency or planned entry into unknown concentrations or IDLH conditions:* SCBAF: Pd,Pp (APF = 10,000) (any self-contained breathing apparatus that has a full face-piece and is operated in a pressure-demand or other positive-pressure mode) or SaF: Pd,Pp: ASCBA (any supplied-air respirator that has a full face-piece and is operated in a pressure-demand or other positive-pressure mode in combination with an auxiliary, self-contained breathing apparatus operated in a pressure-demand or other positive-pressure mode). *Escape:* GmFS* [any air-purifying, full-face-piece respirator (gas mask) with a chin-style, front- or back-mounted canister protection against the compound of concern] or SCBAE (any appropriate escape-type, self-contained breathing apparatus).*
*End-of-service life indicator (ESLI) required.

Other mercury compounds: NIOSH/OSHA *Up to 1 mg/m³:* CcrS (APF = 10) [any chemical cartridge respirator with cartridge(s) providing protection against the compound of concern]* or Sa (APF = 10) (any supplied-air respirator). *Up to 2.5 mg/m³:* Sa:Cf (APF = 25) (any supplied-air respirator operated in a continuous-flow mode) or PaprS (APF = 25) [any powered, air-purifying respirator with cartridge(s) providing protection against the compound of concern]* (canister). *Up to 5 mg/m³:* CcrFS (APF = 50) [any chemical cartridge respirator with a full face-piece and cartridge(s) providing protection against the compound of concern];* or GmFS (APF = 50) [any air-purifying,

full-face-piece respirator (gas mask) with a chin-style, front- or back-mounted canister providing protection against the compound of concern];* or SaT: Cf (APF = 50) (any supplied-air respirator that has a tight-fitting face-piece and is operated in a continuous-flow mode) or PaprTS (APF = 50) [any powered, air-purifying respirator with a tight-fitting face-piece and cartridge(s) providing protection against the compound of concern] or SCBAF (APF = 50) (any self-contained breathing apparatus with a full face-piece) or SaF (APF = 50) (any supplied-air respirator with a full face-piece). *Up to 10 mg/m³:* Sa: Pd,Pp (APF = 1000) (any supplied-air respirator operated in a pressure-demand or other positive-pressure mode). *Emergency or planned entry into unknown concentrations or IDLH conditions:* SCBAF: Pd,Pp (APF = 10,000) (any self-contained breathing apparatus that has a full face-piece and is operated in a pressure-demand or other positive-pressure mode) or SaF: Pd,Pp: ASCBA (APF = 10,000) (any supplied-air respirator that has a full face-piece and is operated in a pressure-demand or other positive-pressure mode in combination with an auxiliary, self-contained breathing apparatus operated in a pressure-demand or other positive-pressure mode). *Escape:* GmFS* [any air-purifying, full-face-piece respirator (gas mask) with a chin-style, front- or back-mounted canister protection against the compound of concern] or SCBAE (any appropriate escape-type, self-contained breathing apparatus).
*End-of-service life indicator (ESLI) required.

Storage: Color Code—Blue: Health Hazard/Poison: Store in a secure poison location. Prior to working with this chemical you should be trained on its proper handling and storage. Mercuric cyanide must be stored to avoid contact with fluorine, magnesium, and sodium nitrite, since violent reactions occur. Mercuric cyanide should not contact acid or heat because it will release flammable hydrogen cyanide gas. Store in tightly closed containers in a cool, well-ventilated area away from light. Protect containers from physical damage.

Shipping: This compound requires a shipping label of "POISONOUS/TOXIC MATERIALS." It falls in Hazard Class 6.1 and Packing Group II.

Spill Handling: Evacuate persons not wearing protective equipment from area of spill or leak until cleanup is complete. Remove all ignition sources. Spills should be collected with special mercury vapor suppressants or special vacuums and deposited in sealed containers. Kits specific for cleanup of mercury spills should be available. Ventilate area after cleanup is complete. It may be necessary to contain and dispose of this chemical as a hazardous waste. If material or contaminated runoff enters waterways, notify downstream users of potentially contaminated waters. Contact your local or federal environmental protection agency for specific recommendations. If employees are required to clean up spills, they must be properly trained and equipped. OSHA 1910.120(q) may be applicable.

Fire Extinguishing: Mercury cyanide may burn but does not readily ignite. Use dry chemical, CO_2, water spray, or foam extinguishers. Poisonous gases, including hydrogen cyanide, oxides of nitrogen, and mercury, are produced in fire. If material or contaminated runoff enters waterways, notify downstream users of potentially contaminated waters. Notify local health and fire officials and pollution control agencies. From a secure, explosion-proof location, use water spray to cool exposed containers. If cooling streams are ineffective (venting sound increases in volume and pitch, tank discolors, or shows any signs of deforming), withdraw immediately to a secure position. If employees are expected to fight fires, they must be trained and equipped in OSHA 1910.156. The only respirators recommended for firefighting are self-contained breathing apparatuses that have full face-pieces and are operated in a pressure-demand or other positive-pressure mode.

Disposal Method Suggested: Return to supplier for mercury recovery and deactivation.

Reference

New Jersey Department of Health and Senior Services. (February 2000). *Hazardous Substances Fact Sheet: Mercuric Cyanide.* Trenton, NJ

Mercuric iodide   M:0380

Molecular Formula: HgI_2

Synonyms: Hydrargyrum bijodatum (German); Mercuric iodide, red; Mercury biniodide; Mercury diiodide; Mercury (II) iodide; Red mercuric iodide

CAS Registry Number: 7774-29-0

RTECS® Number: OW5250000

UN/NA & ERG Number: UN1638 (solution and solid)/151

EC Number: 231-873-8

Regulatory Authority and Advisory Bodies

Air Pollutant Standard Set. See below, "Permissible Exposure Limits in Air" section.

Clean Air Act: Hazardous Air Pollutants (Title I, Part A, Section 112).

Clean Water Act: 40CFR401.15 Section 307 Toxic Pollutants as mercury and compounds.

RCRA, 40CFR261, Appendix 8 Hazardous Constituents, waste number not listed, as mercury compounds, n.o.s.

EPCRA Section 313: Includes any unique chemical substance that contains mercury as part of that chemical's infrastructure. Form R *de minimis* concentration reporting level: 1.0%.

US DOT Regulated Marine Pollutant (49CFR172.101, Appendix B), severe pollutant.

California Proposition 65 Developmental/Reproductive toxin (mercury and mercury compounds) 7/1/90.

Canada, WHMIS, Ingredients Disclosure List Concentration 0.1%.

Rotterdam Convention Annex III [Chemicals Subject to the Prior Informed Consent Procedure (PIC)] (as mercury compounds, including inorganic mercury compounds, alkyl mercury compounds, and alkyloxylalkyl and aryl mercury compounds).

European/International Regulations: not listed in Annex 1.

WGK (German Aquatic Hazard Class): 3—Severe hazard to waters.

Description: Mercuric iodide is a heavy, scarlet red, odorless, crystalline solid. It may be shipped as a red solution. It turns to a yellow powder at 127°C and red upon cooling. Molecular weight = 454.40; Boiling point = (sublimes) 354°C; Freezing/Melting point = 259°C. Hazard Identification (based on NFPA-704 M Rating System): Health 3, Flammability 0, Reactivity 0. Slightly soluble in water.

Potential Exposure: Mercuric iodide is used in medicine and in analytical chemistry.

Incompatibilities: Violent reaction with active metals, potassium, sodium, acids, chlorine trifluoride. Inorganic mercury compounds are incompatible with acetylene, ammonia, chlorine dioxide, azides, calcium (amalgam formation), sodium carbide, lithium, rubidium, copper

Permissible Exposure Limits in Air

Protective Action Criteria (PAC)

TEEL-0: 0.0566 mg/m^3

PAC-1: 0.0566 mg/m^3

PAC-2: 0.227 mg/m^3

PAC-3: 22.7 mg/m^3

As iodides

ACGIH TLV®[1]: 0.01 ppm/0.1 mg/m^3, inhalable fraction and vapor, TWA.

As inorganic mercury compound

OSHA PEL: 0.1 mg[Hg]/m^3 Ceiling Concentration.

NIOSH REL: Hg (*vapor*): 0.05 mg[Hg]/m^3 TWA [skin]; Other: 0.1 mg[Hg]/m^3 Ceiling Concentration [skin].

ACGIH TLV®[1]: 0.025 mg[Hg]/m^3 TWA [skin]; not classifiable as a carcinogen; BEI (preshift) 35 μg[Hg]/100 mL creatinine total inorganic Hg in urine; 15 μg[Hg]/L total inorganic Hg in blood; end-of-shift at end-of-work-week.

DFG MAK: 0.1 mg[Hg]/m^3; Peak Limitation Category II(8) danger of skin sensitization; Carcinogen Category 3B.

NIOSH IDLH: 10 mg[Hg]/m^3.

Determination in Air: Use NIOSH Analytical Method #6009; OSHA Analytical Method ID-140.

Permissible Concentration in Water: To protect freshwater aquatic life: 0.00057 μg/L as a 24-h average, never to exceed 0.0017 μg/L. To protect saltwater aquatic life: 0.025 μg/L as a 24-h average, never to exceed 3.7 μg/L. To protect human health: 0.144 μg/L (US EPA) set in 1979–1980.[6] These are the limits for inorganic mercury compounds in general.

Determination in Water: Total mercury is determined by flameless atomic absorption. Soluble mercury may be determined by 0.45-μm filtration followed by flameless atomic absorption.

Routes of Entry: Inhalation, skin absorption, ingestion, skin and/or eye contact.

Harmful Effects and Symptoms

Short Term Exposure: Mercuric iodide can affect you when breathed and by passing through skin. Overexposure can cause kidney damage. Mercury poisoning can cause "shakes," irritability, sore gums, memory loss, increased saliva, metallic taste, and personality changes. Permanent brain damage may result. Skin contact may cause skin burns, skin allergy, or a gray skin color. Eye contact can cause permanent damage. Heating or contact with acid or acid mist releases toxic mercury vapors. Mercury accumulates in the body. Health effects have been reported below exposure levels of 0.1 mg/m^3.

Long Term Exposure: Repeated or prolonged contact with skin may result in dermatitis (red inflamed skin). Repeated or prolonged exposure may cause death by hypovolemic shock, nephrotic syndrome, or kidney failure.

Points of Attack: Eyes, skin, respiratory system, central nervous system, kidneys.

Medical Surveillance: NIOSH lists the following tests for inorganic mercury: whole blood (chemical/metabolite); whole blood (chemical/metabolite), end-of-shift; end-of-shift at end-of-work-week; biologic tissue/biopsy; nerve conduction studies; neurologic examination/electromyography; thyroid function test/thyroid profile; urine (chemical/metabolite); urine (chemical/metabolite) prior to shift, prior to next shift; urine (chemical/metabolite), sediment; urinalysis (routine). Before first exposure and every 6–12 months after, a complete medical history and examination is strongly recommended with examination of the nervous system including handwriting. Routine urine test (UA). Urine test for mercury (should be less than 0.02 mg/L). Consider lung function tests for persons with frequent exposure. Eye examination. Consider nerve conduction tests, urinary enzymes and neurobehavioral testing. After suspected illness or overexposure, repeat the tests above and get a blood test for mercury. Consider chest X-ray after sudden exposure.

First Aid: If this chemical gets into the eyes, remove any contact lenses at once and irrigate immediately for at least 15 min, occasionally lifting upper and lower lids. Seek medical attention immediately. If this chemical contacts the skin, remove contaminated clothing and wash immediately with soap and water. Seek medical attention immediately. If this chemical has been inhaled, remove from exposure, begin rescue breathing (using universal precautions, including resuscitation mask) if breathing has stopped and CPR if heart action has stopped. Transfer promptly to a medical facility. When this chemical has been swallowed, get medical attention. Give large quantities of water and induce vomiting. Do not make an unconscious person vomit. Medical observation is recommended for 24–48 h after breathing overexposure, as pulmonary edema may be delayed. As first aid for pulmonary edema, a doctor or authorized paramedic may consider administering a corticosteroid spray.

Antidotes and special procedures for medical personnel: The drug NAP (*n*-acetyl penicillamine) has been used to treat mercury poisoning, with mixed success.

Note to physician: For severe poisoning BAL [British Anti-Lewisite, dimercaprol, dithiopropanol ($C_3H_8OS_2$)] has been used to treat toxic symptoms of certain heavy metals poisoning including mercury. Although BAL is reported to have a large margin of safety, caution must be exercised, because toxic effects may be caused by excessive dosage. Most can be prevented by premedication with 1-ephedrine sulfate (CAS: 134-72-5).

Personal Protective Methods: Wear protective gloves and clothing to prevent any reasonable probability of skin contact. Safety equipment suppliers/manufacturers can provide recommendations on the most protective glove/clothing material for your operation. All protective clothing (suits, gloves, footwear, headgear) should be clean, available each day, and put on before work. Contact lenses should not be worn when working with this chemical. Wear dust-proof chemical goggles and face shield unless full face-piece respiratory protection is worn. Employees should wash immediately with soap when skin is wet or contaminated. Provide emergency showers and eyewash.

Respirator Selection: Mercury vapor: NIOSH: *Up to 0.5 mg/m^3:* CcrS (APF = 10) [any chemical cartridge respirator with cartridge(s) providing protection against the compound of concern]* or Sa (APF = 10) (any supplied-air respirator). *Up to 1.25 mg/m^3:* Sa:Cf (APF = 25) (any supplied-air respirator operated in a continuous-flow mode) or PaprS (APF = 25) [any powered, air-purifying respirator with cartridge(s) providing protection against the compound of concern]* (canister). *Up to 2.5 mg/m^3:* CcrFS (APF = 50) [any chemical cartridge respirator with a full face-piece and cartridge(s) providing protection against the compound of concern]* or GmFS (APF = 50) [any air-purifying, full-face-piece respirator (gas mask) with a chin-style, front- or back-mounted canister providing protection against the compound of concern]* or SaT: Cf (APF = 50) (any supplied-air respirator that has a tight-fitting face-piece and is operated in a continuous-flow mode) or PaprTS (APF = 50) [any powered, air-purifying respirator with a tight-fitting face-piece and cartridge(s) providing protection against the compound of concern] or SCBAF (APF = 50) (any self-contained breathing apparatus with a full face-piece) or SaF (APF = 50) (any supplied-air respirator with a full face-piece). *Up to 10 mg/m^3:* Sa: Pd,Pp (APF = 1000) (any supplied-air respirator operated in a pressure-demand or other positive-pressure mode). *Emergency or planned entry into unknown concentrations or IDLH conditions:* SCBAF: Pd,Pp (APF = 10,000) (any self-contained breathing apparatus that has a full face-piece and is operated in a pressure-demand or other positive-pressure mode) or SaF: Pd,Pp: ASCBA (any supplied-air respirator that has a full face-piece and is operated in a pressure-demand or other positive-pressure mode in combination with an auxiliary, self-contained breathing apparatus operated in a pressure-demand or other positive-pressure mode). *Escape:* GmFS*

[any air-purifying, full-face-piece respirator (gas mask) with a chin-style, front- or back-mounted canister protection against the compound of concern] or SCBAE (any appropriate escape-type, self-contained breathing apparatus).*
*End-of-service life indicator (ESLI) required.
Other mercury compounds: NIOSH/OSHA *Up to 1 mg/m^3:* CcrS (APF = 10) [any chemical cartridge respirator with cartridge(s) providing protection against the compound of concern]* or Sa (APF = 10) (any supplied-air respirator). *Up to 2.5 mg/m^3:* Sa:Cf (APF = 25) (any supplied-air respirator operated in a continuous-flow mode) or PaprS (APF = 25) [any powered, air-purifying respirator with cartridge(s) providing protection against the compound of concern]* (canister). *Up to 5 mg/m^3:* CcrFS (APF = 50) [any chemical cartridge respirator with a full face-piece and cartridge(s) providing protection against the compound of concern]* or GmFS (APF = 50) [any air-purifying, full-face-piece respirator (gas mask) with a chin-style, front- or back-mounted canister providing protection against the compound of concern]* or SaT: Cf (APF = 50) (any supplied-air respirator that has a tight-fitting face-piece and is operated in a continuous-flow mode) or PaprTS (APF = 50) [any powered, air-purifying respirator with a tight-fitting face-piece and cartridge(s) providing protection against the compound of concern] or SCBAF (APF = 50) (any self-contained breathing apparatus with a full face-piece) or SaF (APF = 50) (any supplied-air respirator with a full face-piece). *Up to 10 mg/m^3:* Sa: Pd,Pp (APF = 1000) (any supplied-air respirator operated in a pressure-demand or other positive-pressure mode). *Emergency or planned entry into unknown concentrations or IDLH conditions:* SCBAF: Pd, Pp (APF = 10,000) (any self-contained breathing apparatus that has a full face-piece and is operated in a pressure-demand or other positive-pressure mode) or SaF: Pd,Pp: ASCBA (APF = 10,000) (any supplied-air respirator that has a full face-piece and is operated in a pressure-demand or other positive-pressure mode in combination with an auxiliary, self-contained breathing apparatus operated in a pressure-demand or other positive-pressure mode). *Escape:* GmFS* [any air-purifying, full-face-piece respirator (gas mask) with a chin-style, front- or back-mounted canister protection against the compound of concern] or SCBAE (any appropriate escape-type, self-contained breathing apparatus).*
*End-of-service life indicator (ESLI) required.
Storage: Color Code—Blue: Health Hazard/Poison: Store in a secure poison location. Prior to working with this chemical you should be trained on its proper handling and storage. Mercuric iodide must be stored to avoid contact with chlorine trifluoride, sodium, and potassium, since violent reactions occur. Store in tightly closed containers in a cool, well-ventilated area away from light, acids, and heat. Protect containers from physical damage.
Shipping: This compound requires a shipping label of "POISONOUS/TOXIC MATERIALS" (solution). It falls in Hazard Class 6.1 and Packing Group II.

Spill Handling: Evacuate persons not wearing protective equipment from area of spill or leak until cleanup is complete. Remove all ignition sources. Spills should be collected with special mercury vapor suppressants or special vacuums and deposited in sealed containers. Kits specific for cleanup of mercury spills should be available. Ventilate area after cleanup is complete. It may be necessary to contain and dispose of this chemical as a hazardous waste. If material or contaminated runoff enters waterways, notify downstream users of potentially contaminated waters. Contact your local or federal environmental protection agency for specific recommendations. If employees are required to clean up spills, they must be properly trained and equipped. OSHA 1910.120(q) may be applicable.
Fire Extinguishing: This chemical may burn but does not readily ignite. Use dry chemical, carbon dioxide, water spray, or foam extinguishers. Poisonous gases, including iodine, are produced in fire. If material or contaminated runoff enters waterways, notify downstream users of potentially contaminated waters. Notify local health and fire officials and pollution control agencies. From a secure, explosion-proof location, use water spray to cool exposed containers. If cooling streams are ineffective (venting sound increases in volume and pitch, tank discolors, or shows any signs of deforming), withdraw immediately to a secure position. If employees are expected to fight fires, they must be trained and equipped in OSHA 1910.156. The only respirators recommended for firefighting are self-contained breathing apparatuses that have full face-pieces and are operated in a pressure-demand or other positive-pressure mode.
Reference
New Jersey Department of Health and Senior Services. (February 2000). *Hazardous Substances Fact Sheet: Mercuric Iodide.* Trenton, NJ

Mercuric nitrate   M:0390

Molecular Formula: HgN_2O_6
Common Formula: $Hg(NO_3)_2$
Synonyms: Mercury(2+) nitrate (1:2); Mercury(II) nitrate (1:2); Mercury nitrate; Mercury pernitrate; Nitrate mercurique (French); Nitrato mercurico (Spanish); Nitric acid, mercury(2+) salt; Nitric acid, mercury(II) salt
CAS Registry Number: 10045-94-0
RTECS® Number: OW8225000
UN/NA & ERG Number: UN1625/154
EC Number: 233-152-3 [*Annex I Index No.:* 080-002-00-6] pesticide in the group of plant protection products; banned or severely restricted
Regulatory Authority and Advisory Bodies
Air Pollutant Standard Set. See below, "Permissible Exposure Limits in Air" section.
Clean Water Act: Section 311 Hazardous Substances/RQ 40CFR117.3 (same as CERCLA, see below); 40CFR 401.15 Section 307 Toxic Pollutants as mercury and

compounds; Section 313 Water Priority Chemicals (57FR41331, 9/9/92).

RCRA, 40CFR261, Appendix 8 Hazardous Constituents, waste number not listed, as mercury compounds, n.o.s. Reportable Quantity (RQ): 10 lb (4.54 kg).

EPCRA Section 313 (as mercury compound). Form R *de minimis* concentration reporting level: 1.0%.

US DOT Regulated Marine Pollutant (49CFR172.101, Appendix B), severe pollutant.

California Proposition 65 Developmental/Reproductive toxin (mercury and mercury compounds) 7/1/90.

Canada, WHMIS, Ingredients Disclosure List Concentration 1.0%.

Rotterdam Convention Annex III [Chemicals Subject to the Prior Informed Consent Procedure (PIC)] (as mercury compounds, including inorganic mercury compounds, alkyl mercury compounds, and alkyloxylalkyl and aryl mercury compounds).

European/International Regulations: Hazard Symbol: T+, N; Risk phrases: R6; R26; R48/23; R50/53; Safety phrases: S53; S45; S60; S61; as mercury.

WGK (German Aquatic Hazard Class): 3—Severe hazard to waters.

Description: Mercuric Nitrate is a white to yellowish crystalline solid with a nitric acid-like odor. Normally exists as hemihydrate or dihydrate. Molecular weight = 324.61; Boiling point = (decomposes); Freezing/Melting point = 70−79°C. Hazard Identification (based on NFPA-704 M Rating System): Health 3, Flammability 0, Reactivity 0. Soluble in water.

Potential Exposure: Mercuric nitrate is used in making other chemicals; in felt manufacture, and in making mercury fulminate.

Incompatibilities: A strong oxidizer. Reacts violently with combustibles, petroleum hydrocarbons, reducing agents. Reacts with acetylene, alcohol, phosphine and sulfur to form shock-sensitive compounds. Aqueous solution attacks most metals. Vigorous reaction with petroleum hydrocarbons. Incompatible with organic materials, acetylene, ethanol, phosphine, sulfur, hypophosphoric acid. Inorganic mercury compounds are incompatible with acetylene, ammonia, chlorine dioxide, azides, calcium (amalgam formation), sodium carbide, lithium, rubidium, copper. Decomposes in heat or on exposure to light, producing toxic fumes (mercury, nitrogen oxides).

Permissible Exposure Limits in Air
As inorganic mercury compounds
OSHA PEL: 0.1 mg[Hg]/m^3 Ceiling Concentration.
NIOSH REL: Hg (*vapor*): 0.05 mg[Hg]/m^3 TWA [skin]; Other: 0.1 mg[Hg]/m^3 Ceiling Concentration [skin].
ACGIH TLV®[1]: 0.025 mg[Hg]/m^3 TWA [skin]; not classifiable as a carcinogen; BEI (preshift) 35 μg[Hg]/100 mL creatinine total inorganic Hg in urine; 15 μg[Hg]/L total inorganic Hg in blood; end-of-shift at end-of-work-week.
Protective Action Criteria (PAC)
TEEL-0: 0.0405 mg/m^3

PAC-1: 0.2 mg/m^3
PAC-2: 1.62 mg/m^3
PAC-3: 16.2 mg/m^3
DFG MAK: 0.1 mg[Hg]/m^3; Peak Limitation Category II(8) danger of skin sensitization; Carcinogen Category 3B.
NIOSH IDLH: 10 mg[Hg]/m^3.
Australia: TWA 0.05 mg[Hg]/m^3, [skin], 1993; Belgium: TWA 0.05 mg[Hg]/m^3, [skin], 1993; Denmark: TWA 0.05 mg[Hg]/m^3, [skin], 1999; Finland: TWA 0.05 mg[Hg]/m^3, 1999; France: VME 0.1 mg[Hg]/m^3, [skin], 1999; Hungary: TWA 0.02 mg[Hg]/m^3; STEL 0.04 mg[Hg]/m^3, 1993; Japan: 0.05 mg[Hg]/m^3, 1999; Norway: TWA 0.05 mg[Hg]/m^3, 1999; the Philippines: TWA 0.05 mg[Hg]/m^3, 1993; Poland: MAC (TWA) 0.05 mg[Hg]/m^3, MAC (STEL) 0.15 mg[Hg]/m^3, 1999; Russia: TWA 0.05 mg[Hg]/m^3; STEL 0.01 mg[Hg]/m^3, 1993; Sweden: NGV 0.05 mg[Hg]/m^3, [skin], 1999; Thailand: STEL 0.05 mg[Hg]/m^3, 1993; United Kingdom: TWA 0.025 mg[Hg]/m^3, 2000; Argentina, Bulgaria, Columbia, Jordan, South Korea, New Zealand, Singapore, Vietnam: ACGIH TLV®: not classifiable as a human carcinogen.

Determination in Air: Use NIOSH Analytical Method #6009; OSHA Analytical Method ID-140.

Permissible Concentration in Water: To protect freshwater aquatic life: 0.00057 μg/L as a 24-h average, never to exceed 0.0017 μg/L. To protect saltwater aquatic life: 0.025 μg/L as a 24-h average, never to exceed 3.7 μg/L. To protect human health: 0.144 μg/L (US EPA) set in 1979−1980.[6] These are the limits for inorganic mercury compounds in general.

Determination in Water: Total mercury is determined by flameless atomic absorption. Soluble mercury may be determined by 0.45-μm filtration followed by flameless atomic absorption.

Routes of Entry: Inhalation, skin absorption, ingestion, skin and/or eye contact.

Harmful Effects and Symptoms
Short Term Exposure: Mercuric nitrate can affect you when breathed in and passed through skin (causing systemic poisoning). Eye contact causes ulceration of conjunctiva and cornea. Skin contact causes irritation and possible dermatitis. Acute systemic poisoning can be fatal within a few minutes. Death by uremic poisoning can be delayed 5−12 days. Overexposure can damage kidneys. Mercury poisoning can cause "shakes," irritability, sore gums, increased saliva, personality change and brain damage. Eye contact may cause burns. Heating or contact with acid or acid mist causes release of toxic mercury vapors and lung effects. Mercury accumulates in the body. Health effects have been reported below TLV levels of 0.1 mg/m^3. Acute poisoning has resulted from inhaling dust concentrations of 1.2−8.5 mg/m^3.

Long Term Exposure: Repeated or prolonged contact with skin may result in dermatitis (red inflamed skin), or cause skin to turn gray. Skin allergy may occur. Repeated or prolonged exposure may cause death by hypovolemic shock, nephrotic syndrome, or kidney failure. There is limited evidence that this chemical reduces fertility in males and females. Repeated exposures may cause brown staining of the eyes and may affect peripheral vision.

Points of Attack: Eyes, skin, respiratory system, central nervous system, kidneys.

Medical Surveillance: NIOSH lists the following tests for inorganic mercury: whole blood (chemical/metabolite); whole blood (chemical/metabolite), end-of-shift; end-of-shift at end-of-work-week; biologic tissue/biopsy; nerve conduction studies; neurologic examination/electromyography; thyroid function test/thyroid profile; urine (chemical/metabolite); urine (chemical/metabolite) prior to shift, prior to next shift; urine (chemical/metabolite), sediment; urinalysis (routine). Before first exposure and every 6–12 months after, a complete medical history and examination is strongly recommended with examination of the nervous system including handwriting. Routine urine test (UA). Urine test for mercury (should be less than 0.02 mg/L). Eye examination. After suspected illness or overexposure, repeat the tests above and get a blood test for mercury. Consider nerve conduction tests, urinary enzymes, and neurobehavioral testing. Eye examination.

First Aid: If this chemical gets into the eyes, remove any contact lenses at once and irrigate immediately for at least 15 min, occasionally lifting upper and lower lids. Seek medical attention immediately. If this chemical contacts the skin, remove contaminated clothing and wash immediately with soap and water. Seek medical attention immediately. If this chemical has been inhaled, remove from exposure, begin rescue breathing (using universal precautions, including resuscitation mask) if breathing has stopped and CPR if heart action has stopped. Transfer promptly to a medical facility. When this chemical has been swallowed, get medical attention. Give large quantities of water and induce vomiting. Do not make an unconscious person vomit.

Antidotes and special procedures for medical personnel: The drug NAP (*n*-acetyl penicillamine) has been used to treat mercury poisoning, with mixed success.

Note to physician: For severe poisoning BAL [British Anti-Lewisite, dimercaprol, dithiopropanol ($C_3H_8OS_2$)] has been used to treat toxic symptoms of certain heavy metals poisoning including mercury. Although BAL is reported to have a large margin of safety, caution must be exercised, because toxic effects may be caused by excessive dosage. Most can be prevented by premedication with 1-ephedrine sulfate (CAS: 134-72-5).

Personal Protective Methods: Wear protective gloves and clothing to prevent any reasonable probability of skin contact. Safety equipment suppliers/manufacturers can provide recommendations on the most protective glove/clothing material for your operation. All protective clothing (suits, gloves, footwear, headgear) should be clean, available each day, and put on before work. Contact lenses should not be worn when working with this chemical. Wear dust-proof chemical goggles and face shield unless full face-piece

respiratory protection is worn. Employees should wash immediately with soap when skin is wet or contaminated. Provide emergency showers and eyewash.

Respirator Selection: *Mercury vapor:* NIOSH: *Up to 0.5 mg/m³:* CcrS (APF = 10) [any chemical cartridge respirator with cartridge(s) providing protection against the compound of concern]* or Sa (APF = 10) (any supplied-air respirator). *Up to 1.25 mg/m³:* Sa:Cf (APF = 25) (any supplied-air respirator operated in a continuous-flow mode) or PaprS (APF = 25) [any powered, air-purifying respirator with cartridge(s) providing protection against the compound of concern]* (canister). *Up to 2.5 mg/m³:* CcrFS (APF = 50) [any chemical cartridge respirator with a full face-piece and cartridge(s) providing protection against the compound of concern]* or GmFS (APF = 50) [any air-purifying, full-face-piece respirator (gas mask) with a chin-style, front- or back-mounted canister providing protection against the compound of concern]* or SaT: Cf (APF = 50) (any supplied-air respirator that has a tight-fitting face-piece and is operated in a continuous-flow mode) or PaprTS (APF = 50) [any powered, air-purifying respirator with a tight-fitting face-piece and cartridge(s) providing protection against the compound of concern] or SCBAF (APF = 50) (any self-contained breathing apparatus with a full face-piece) or SaF (APF = 50) (any supplied-air respirator with a full face-piece). *Up to 10 mg/m³:* Sa: Pd,Pp (APF = 1000) (any supplied-air respirator operated in a pressure-demand or other positive-pressure mode). *Emergency or planned entry into unknown concentrations or IDLH conditions:* SCBAF: Pd,Pp (APF = 10,000) (any self-contained breathing apparatus that has a full face-piece and is operated in a pressure-demand or other positive-pressure mode) or SaF: Pd,Pp: ASCBA (APF = 10,000) (any supplied-air respirator that has a full face-piece and is operated in a pressure-demand or other positive-pressure mode in combination with an auxiliary, self-contained breathing apparatus operated in a pressure-demand or other positive-pressure mode). *Escape:* GmFS* [any air-purifying, full-face-piece respirator (gas mask) with a chin-style, front- or back-mounted canister protection against the compound of concern] or SCBAE (any appropriate escape-type, self-contained breathing apparatus).*

*End-of-service life indicator (ESLI) required.

Other mercury compounds: NIOSH/OSHA *Up to 1 mg/m³:* CcrS (APF = 10) [any chemical cartridge respirator with cartridge(s) providing protection against the compound of concern];* or Sa (APF = 10) (any supplied-air respirator). *Up to 2.5 mg/m³:* Sa:Cf (APF = 25) (any supplied-air respirator operated in a continuous-flow mode) or PaprS (APF = 25) [any powered, air-purifying respirator with cartridge(s) providing protection against the compound of concern]* (canister). *Up to 5 mg/m³:* CcrFS (APF = 50) [any chemical cartridge respirator with a full face-piece and cartridge(s) providing protection against the compound of concern]* or GmFS (APF = 50) [any air-purifying, full-face-piece respirator (gas mask) with a chin-style, front- or back-mounted canister providing protection against the compound of concern]* or SaT: Cf (APF = 50) (any supplied-air respirator that has a tight-fitting face-piece and is operated in a continuous-flow mode) or PaprTS (APF = 50) [any powered, air-purifying respirator with a tight-fitting face-piece and cartridge(s) providing protection against the compound of concern] or SCBAF (APF = 50) (any self-contained breathing apparatus with a full face-piece) or SaF (APF = 50) (any supplied-air respirator with a full face-piece). *Up to 10 mg/m³:* Sa: Pd,Pp (APF = 1000) (any supplied-air respirator operated in a pressure-demand or other positive-pressure mode). *Emergency or planned entry into unknown concentrations or IDLH conditions:* SCBAF: Pd, Pp (APF = 10,000) (any self-contained breathing apparatus that has a full face-piece and is operated in a pressure-demand or other positive-pressure mode) or SaF: Pd,Pp: ASCBA (APF = 10,000) (any supplied-air respirator that has a full face-piece and is operated in a pressure-demand or other positive-pressure mode in combination with an auxiliary, self-contained breathing apparatus operated in a pressure-demand or other positive-pressure mode). *Escape:* GmFS* [any air-purifying, full-face-piece respirator (gas mask) with a chin-style, front- or back-mounted canister protection against the compound of concern] or SCBAE (any appropriate escape-type, self-contained breathing apparatus).*

*End-of-service life indicator (ESLI) required.

Storage: Color Code—Blue: Health Hazard/Poison: Store in a secure poison location. Prior to working with this chemical you should be trained on its proper handling and storage. Mercuric nitrate must be stored to avoid contact with organic materials; acetylene, ethanol, phosphine, sulfur, and hypophosphoric acid, since violent reactions occur. See also "Incompatibilities." Do not store on wooden floors.

Shipping: This compound requires a shipping label of "POISONOUS/TOXIC MATERIALS." It falls in Hazard Class 6.1 and Packing Group II.

Spill Handling: Evacuate persons not wearing protective equipment from area of spill or leak until cleanup is complete. Remove all ignition sources. Spills should be collected with special mercury vapor suppressants or special vacuums and deposited in sealed containers. Kits specific for cleanup of mercury spills should be available. Ventilate area after cleanup is complete. It may be necessary to contain and dispose of this chemical as a hazardous waste. If material or contaminated runoff enters waterways, notify downstream users of potentially contaminated waters. Contact your local or federal environmental protection agency for specific recommendations. If employees are required to clean up spills, they must be properly trained and equipped. OSHA 1910.120(q) may be applicable.

Fire Extinguishing: Not combustible but enhances combustion of other substances. Mercuric nitrate is a dangerous fire risk on contact with combustible and organic materials (wood, paper, oils, etc). Use dry chemical, CO₂, water spray. Poisonous gases, including mercury and nitrogen oxides, are produced in fire. If material or contaminated runoff enters waterways, notify downstream users of potentially contaminated waters. Notify local health and fire officials and pollution control agencies. From a secure, explosion-proof location, use water spray to cool exposed containers. If cooling streams are ineffective (venting sound increases in volume and pitch, tank discolors, or shows any signs of deforming), withdraw immediately to a secure position. If employees are expected to fight fires, they must be trained and equipped in OSHA 1910.156. The only respirators recommended for firefighting are self-contained breathing apparatuses that have full face-pieces and are operated in a pressure-demand or other positive-pressure mode.

Reference

New Jersey Department of Health and Senior Services. (February 2000). *Hazardous Substances Fact Sheet: Mercuric Nitrate.* Trenton, NJ

Mercuric oxide   M:0400

Molecular Formula: HgO

Synonyms: C.I. 77760; Kankerex; Mercuric oxide, red; Mercuric oxide, yellow; Mercury monoxide; Mercury oxide; Oxido mercurico rojo (Spanish); Oxido mercurico amarillo (Spanish); Oxyde de mercure (French); Red mercuric oxide; Red oxide of mercury; Red precipitate; Santar; Yellow mercuric oxide; Yellow oxide of mercury; Yellow precipitate

CAS Registry Number: 21908-53-2; *(alt.)* 1344-45-2; *(alt.)* 8028-34-0

RTECS® Number: OW8750000

UN/NA & ERG Number: UN1641/151

EC Number: 244-654-7

Regulatory Authority and Advisory Bodies

Carcinogenicity: IARC: Human Inadequate Evidence, *not classifiable as carcinogenic to humans,* Group 3, 1993.

Banned or Severely Restricted (in agriculture) (EEC) (UK).[13]

Air Pollutant Standard Set. See below, "Permissible Exposure Limits in Air" section.

Clean Water Act: 40CFR401.15 Section 307 Toxic Pollutants as mercury and compounds.

RCRA, 40CFR261, Appendix 8 Hazardous Constituents, waste number not listed, as mercury compounds, n.o.s.

Superfund/EPCRA 40CFR355, Extremely Hazardous Substances: TPQ = 500/10,000 lb (227/4540 kg).

Reportable Quantity (RQ): 500 lb (227 kg).

EPCRA Section 313 (as mercury compound): Form R *de minimis* concentration reporting level: 1.0%.

US DOT Regulated Marine Pollutant (49CFR172.101, Appendix B), severe pollutant.

California Proposition 65 Developmental/Reproductive toxin (mercury and mercury compounds) 7/1/90.

Canada, WHMIS, Ingredients Disclosure List Concentration 0.1%.

Rotterdam Convention Annex III [Chemicals Subject to the Prior Informed Consent Procedure (PIC)] (as mercury compounds, including inorganic mercury compounds, alkyl mercury compounds and alkyloxylalkyl and aryl mercury compounds).

European/International Regulations: not listed in Annex 1.

WGK (German Aquatic Hazard Class): 3—Severe hazard to waters.

Description: Mercuric oxide is a red or orange-red heavy crystalline powder; yellow when finely powdered. Molecular weight = 216.59; Freezing/Melting point = 500°C (decomposes). Hazard Identification (based on NFPA-704 M Rating System): Health 3, Flammability 0, Reactivity 0. Insoluble in water.

Potential Exposure: Compound Description: Agricultural Chemical; Tumorigen; Reproductive Effector. Mercuric oxide is used for wound sealing and canker treatment of fruit and rubber trees; chemical intermediate for mercury salts; organic mercury compounds; chlorine monoxide; as an antiseptic in pharmaceuticals; component of dry cell batteries; pigment and glass modifier; fungicide; preservative in cosmetics; analytical reagent; formerly used in antifouling paints.

Incompatibilities: A powerful oxidizer. Decomposes on exposure to light, when heated above 500°C, producing highly toxic fumes including mercury and oxygen, which will add to the intensity of an existing fire. Violent reaction with combustible materials, other oxidizers, acetyl nitrate, aluminum, diboron tetrafluoride, reducing agents, phospham, hydrogen trisulfide (on ignition), hydrazine hydrate, hydrogen peroxide, hypophosphorous acid, acetyl nitrate, chlorine, magnesium (when heated), disulfur dichloride, alcohols, alkali metals (i.e., lithium, sodium, potassium, rubidium, cesium, francium). Forms heat- or impact-sensitive explosive mixtures with sulfur, phosphorus and other nonmetals, potassium, magnesium, sodium and other chemically active metals. Incompatible with strong bases and light.

Permissible Exposure Limits in Air

OSHA PEL: 0.1 mg[Hg]/m^3 Ceiling Concentration.

NIOSH REL: Hg (*vapor*): 0.05 mg[Hg]/m^3 TWA [skin]; Other: 0.1 mg[Hg]/m^3 Ceiling Concentration [skin].

ACGIH TLV®[1]: 0.025 mg[Hg]/m^3 TWA [skin]; not classifiable as a carcinogen; BEI (preshift) 35 µg[Hg]/100 mL creatinine total inorganic Hg in urine; 15 µg[Hg]/L total inorganic Hg in blood; end-of-shift at end-of-work-week.

NIOSH IDLH: 10 mg[Hg]/m^3.

Protective Action Criteria (PAC)

TEEL-0: 0.027 mg/m^3

PAC-1: 1.5 mg/m^3

PAC-2: 1.08 mg/m^3

PAC-3: 10.8 mg/m^3

DFG MAK: 0.1 mg[Hg]/m^3; Peak Limitation Category II(8) danger of skin sensitization; Carcinogen Category 3B.

Australia: TWA 0.05 mg[Hg]/m^3, [skin], 1993; Belgium: TWA 0.05 mg[Hg]/m^3, [skin], 1993; Denmark: TWA 0.05 mg[Hg]/m^3, [skin], 1999; Finland: TWA 0.05 mg[Hg]/m^3, 1999; France: VME 0.1 mg[Hg]/m^3, [skin], 1999; Hungary: TWA 0.02 mg[Hg]/m^3; STEL 0.04 mg[Hg]/m^3, 1993; Japan: 0.05 mg[Hg]/m^3, 1999; Norway: TWA 0.05 mg[Hg]/m^3, 1999; the Philippines: TWA 0.05 mg[Hg]/m^3, 1993; Poland: MAC (TWA) 0.05 mg[Hg]/m^3, MAC (STEL) 0.15 mg[Hg]/m^3, 1999; Russia: TWA 0.05 mg[Hg]/m^3; STEL 0.01 mg[Hg]/m^3, 1993; Sweden: NGV 0.05 mg [Hg]/m^3, [skin], 1999; Thailand: STEL 0.05 mg[Hg]/m^3, 1993; United Kingdom: LTEL 0.05 mg[Hg]/m^3; STEL 0.15 mg[Hg]/m^3, 1993; Argentina, Bulgaria, Columbia, Jordan, South Korea, New Zealand, Singapore, Vietnam: ACGIH TLV®: not classifiable as a human carcinogen. Russia set a MAC for ambient air in residential areas of 0.003 mg/m^3 on a daily average basis.

Determination in Air: Use NIOSH Analytical Method #6009; OSHA Analytical Method ID-140.

Permissible Concentration in Water: To protect freshwater aquatic life: 0.00057 µg/L as a 24-h average, never to exceed 0.0017 µg/L. *To protect saltwater aquatic life:* 0.025 µg/L as a 24-h average, never to exceed 3.7 µg/L. *To protect human health:* 0.144 µg/L (US EPA) set in 1979–1980.[6] These are the limits for inorganic mercury compounds in general.

Determination in Water: Total mercury is determined by flameless atomic absorption. Soluble mercury may be determined by 0.45-µm filtration followed by flameless atomic absorption.

Routes of Entry: Inhalation, skin absorption, ingestion, skin and/or eye contact.

Harmful Effects and Symptoms

Short Term Exposure: Mercuric oxide dust has a corrosive effect on eyes, skin, and respiratory tract. This material is highly toxic by ingestion, inhalation, or skin absorption. Very short exposure to small quantities may cause death or permanent injury. Following ingestion, mercuric oxide is readily converted to mercuric chloride, the most dangerous mercury compounds. Signs and symptoms of acute exposure to mercuric oxide may be severe and include increased salivation, foul breath, inflammation and ulceration of the mucous membranes, abdominal pain, and bloody diarrhea. Oliguria (scanty urination), anuria (suppression of urine formation), and acute renal failure may be noted. Weak pulse, seizures, psychic disturbances, circulatory collapse, chest pain, and dyspnea (shortness of breath) may be observed.

Long Term Exposure: Repeated or prolonged contact with skin may result in dermatitis and allergy. Repeated or prolonged exposure may cause brain damage and nervous system damage. Repeated or prolonged exposure may cause death by hypovolemic shock, nephrotic syndrome, and kidney failure. There is limited evidence that this chemical is a teratogen in animals. Can cause mercury to accumulate in the body and cause mercury poisoning. May cause permanent damage, such as gray colored skin, brown staining of the eyes, and decreased peripheral vision.

Points of Attack: Eyes, skin, respiratory system, central nervous system, kidneys.

Medical Surveillance: NIOSH lists the following tests for inorganic mercury: whole blood (chemical/metabolite); whole blood (chemical/metabolite), end-of-shift; end-of-shift at end-of-work-week; biologic tissue/biopsy; nerve conduction studies; neurologic examination/electromyography; thyroid function test/thyroid profile; urine (chemical/metabolite); urine (chemical/metabolite) prior to shift, prior to next shift; urine (chemical/metabolite), sediment; urinalysis (routine). Before first exposure and every 6–12 months after, a complete medical history and examination is strongly recommended with examination of the nervous system including handwriting. Routine urine test (UA), urine test for mercury (should be less than 0.02 mg/L). Eye examination. After suspected illness or overexposure, repeat the test above and get a blood test for mercury. Consider chest X-ray after acute overexposure. Consider nerve conduction tests, urinary enzymes and neurobehavioral testing. Evaluation by a qualified allergist. Eye examination. Consider chest X-ray following acute overexposure.

First Aid: Remove victims from exposure. Emergency personnel should avoid self-exposure to mercuric oxide. Evaluate vital signs, including pulse and respiratory rate, and note any trauma. If no pulse is detected, provide CPR. If not breathing, provide artificial respiration. If breathing is labored, administer oxygen or other respiratory support. Remove contaminated clothing as soon as possible. If eye exposure has occurred, remove any contact lenses at once; eyes must be flushed with lukewarm water for at least 15 min. Wash exposed skin areas for 15 min with soap and water. Obtain authorization and/or further instructions from the local hospital for administration of an antidote or performance of other invasive procedures in the event of inhalation or ingestion of HgO. Rush to a health-care facility.

Antidotes and special procedures for medical personnel: The drug NAP (*n*-acetyl penicillamine) has been used to treat mercury poisoning, with mixed success.

Note to physician: For severe poisoning BAL [British Anti-Lewisite, dimercaprol, dithiopropanol ($C_3H_8OS_2$)] has been used to treat toxic symptoms of certain heavy metals poisoning including mercury. Although BAL is reported to have a large margin of safety, caution must be exercised, because toxic effects may be caused by excessive dosage. Most can be prevented by premedication with 1-ephedrine sulfate (CAS: 134-72-5).

Personal Protective Methods: Wear protective gloves and clothing to prevent any reasonable probability of skin contact. Safety equipment suppliers/manufacturers can provide recommendations on the most protective glove/clothing material for your operation. All protective clothing (suits, gloves, footwear, headgear) should be clean, available each day, and put on before work. Contact lenses should not be worn when working with this chemical. Wear dust-proof chemical goggles and face shield unless full face-piece respiratory protection is worn. Employees should wash immediately with soap when skin is wet or contaminated. Provide emergency showers and eyewash.

Respirator Selection: *Mercury vapor:* NIOSH: *Up to 0.5 mg/m³:* CcrS (APF = 10) [any chemical cartridge respirator with cartridge(s) providing protection against the compound of concern]* or Sa (APF = 10) (any supplied-air respirator). *Up to 1.25 mg/m³:* Sa:Cf (APF = 25) (any supplied-air respirator operated in a continuous-flow mode) or PaprS (APF = 25) [any powered, air-purifying respirator with cartridge(s) providing protection against the compound of concern]* (canister). *Up to 2.5 mg/m³:* CcrFS (APF = 50) [any chemical cartridge respirator with a full face-piece and cartridge(s) providing protection against the compound of concern]* or GmFS (APF = 50) [any air-purifying, full-face-piece respirator (gas mask) with a chin-style, front- or back-mounted canister providing protection against the compound of concern]* or SaT: Cf (APF = 50) (any supplied-air respirator that has a tight-fitting face-piece and is operated in a continuous-flow mode) or PaprTS (APF = 50) [any powered, air-purifying respirator with a tight-fitting face-piece and cartridge(s) providing protection against the compound of concern] or SCBAF (APF = 50) (any self-contained breathing apparatus with a full face-piece) or SaF (APF = 50) (any supplied-air respirator with a full face-piece). *Up to 10 mg/m³:* Sa: Pd,Pp (APF = 1000) (any supplied-air respirator operated in a pressure-demand or other positive-pressure mode). *Emergency or planned entry into unknown concentrations or IDLH conditions:* SCBAF: Pd,Pp (APF = 10,000) (any self-contained breathing apparatus that has a full face-piece and is operated in a pressure-demand or other positive-pressure mode) or SaF: Pd,Pp: ASCBA (any supplied-air respirator that has a full face-piece and is operated in a pressure-demand or other positive-pressure mode in combination with an auxiliary, self-contained breathing apparatus operated in a pressure-demand or other positive-pressure mode). Escape: GmFS* [any air-purifying, full-face-piece respirator (gas mask) with a chin-style, front- or back-mounted canister protection against the compound of concern] or SCBAE (any appropriate escape-type, self-contained breathing apparatus).*
*End-of-service life indicator (ESLI) required.

Other mercury compounds: NIOSH/OSHA *Up to 1 mg/m³:* CcrS (APF = 10) [any chemical cartridge respirator with cartridge(s) providing protection against the compound of concern]* or Sa (APF = 10) (any supplied-air respirator). *Up to 2.5 mg/m³:* Sa:Cf (APF = 25) (any supplied-air respirator operated in a continuous-flow mode) or PaprS (APF = 25) [any powered, air-purifying respirator with cartridge(s) providing protection against the compound of concern]* (canister). *Up to 5 mg/m³:* CcrFS (APF = 50) [any chemical cartridge respirator with a full face-piece and cartridge(s) providing protection against the compound of concern]* or GmFS (APF = 50) [any air-purifying, full-face-piece respirator (gas mask) with a chin-style, front- or back-mounted

canister providing protection against the compound of concern]* or SaT: Cf (APF = 50) (any supplied-air respirator that has a tight-fitting face-piece and is operated in a continuous-flow mode) or PaprTS (APF = 50) [any powered, air-purifying respirator with a tight-fitting face-piece and cartridge(s) providing protection against the compound of concern] or SCBAF (APF = 50) (any self-contained breathing apparatus with a full face-piece) or SaF (APF = 50) (any supplied-air respirator with a full face-piece). *Up to 10 mg/m³*: Sa: Pd,Pp (APF = 1000) (any supplied-air respirator operated in a pressure-demand or other positive-pressure mode). Emergency or planned entry into unknown concentrations or IDLH conditions: SCBAF: Pd,Pp (APF = 10,000) (any self-contained breathing apparatus that has a full face-piece and is operated in a pressure-demand or other positive-pressure mode) or SaF: Pd,Pp: ASCBA (APF = 10,000) (any supplied-air respirator that has a full face-piece and is operated in a pressure-demand or other positive-pressure mode in combination with an auxiliary, self-contained breathing apparatus operated in a pressure-demand or other positive-pressure mode). *Escape:* GmFS* [any air-purifying, full-face-piece respirator (gas mask) with a chin-style, front- or back-mounted canister protection against the compound of concern] or SCBAE (any appropriate escape-type, self-contained breathing apparatus).*
*End-of-service life indicator (ESLI) required.

Storage: Color Code—Blue: Health Hazard/Poison: Store in a secure poison location. Prior to working with this chemical you should be trained on its proper handling and storage. Store in tightly closed containers in a cool, well-ventilated area away from chlorine, hydrogen peroxide, hypophosphorous acid, hydrazine hydrate, magnesium (when heated), disulfur dichloride; hydrogen trisulfide, reducing agents. See also "Incompatibilities."

Shipping: This compound requires a shipping label of "POISONOUS/TOXIC MATERIALS." It falls in Hazard Class 6.1 and Packing Group II.

Spill Handling: Evacuate persons not wearing protective equipment from area of spill or leak until cleanup is complete. Remove all ignition sources. Collect powdered material in the most convenient and safe manner and deposit in sealed containers. Spills should be collected with special mercury vapor suppressants or special vacuums. Kits specific for cleanup of mercury are available. Ventilate area after cleanup is complete. It may be necessary to contain and dispose of this chemical as a hazardous waste. If material or contaminated runoff enters waterways, notify downstream users of potentially contaminated waters. Contact your local or federal environmental protection agency for specific recommendations. If employees are required to clean up spills, they must be properly trained and equipped. OSHA 1910.120(q) may be applicable.

Fire Extinguishing: Not combustible but enhances combustion of other substances. For small fires, use dry chemical, carbon dioxide, water spray, or foam. For large fires, use water spray, fog, or foam. Wear full body protective clothing and self-contained breathing apparatus. Poisonous gases, including mercury fumes and mercury oxide, are produced in fire. If material or contaminated runoff enters waterways, notify downstream users of potentially contaminated waters. Notify local health and fire officials and pollution control agencies. From a secure, explosion-proof location, use water spray to cool exposed containers. If cooling streams are ineffective (venting sound increases in volume and pitch, tank discolors, or shows any signs of deforming), withdraw immediately to a secure position. If employees are expected to fight fires, they must be trained and equipped in OSHA 1910.156. The only respirators recommended for firefighting are self-contained breathing apparatuses that have full face-pieces and are operated in a pressure-demand or other positive-pressure mode.

Reference
US Environmental Protection Agency. (November 30, 1987). *Chemical Hazard Information Profile: Mercuric Oxide.* Washington, DC: Chemical Emergency Preparedness Program
New Jersey Department of Health and Senior Services. (February 2001). *Hazardous Substances Fact Sheet: Mercuric Oxide.* Trenton, NJ

Mercuric oxycyanide   M:0410

Molecular Formula: $C_2Hg_2N_2O$
Common Formula: $Hg(CN)_2 \cdot HgO$
Synonyms: Mercury cyanide oxide; Mercury oxycyanide
CAS Registry Number: 1335-31-5
RTECS® Number: OW1530000
UN/NA & ERG Number: UN1642/151
EC Number: 215-629-8 [*Annex I Index No.:* 080-006-00-8]
Regulatory Authority and Advisory Bodies
Air Pollutant Standard Set. See below, "Permissible Exposure Limits in Air" section.
Clean Air Act: Hazardous Air Pollutants (Title I, Part A, Section 112).
Clean Water Act: 40CFR401.15 Section 307 Toxic Pollutants as mercury and compounds.
RCRA, 40CFR261, Appendix 8 Hazardous Constituents, waste number not listed, as mercury compounds, n.o.s.
EPCRA Section 313: Includes any unique chemical substance that contains mercury as part of that chemical's infrastructure. Form R *de minimis* concentration reporting level: 1.0%.
US DOT Regulated Marine Pollutant (49CFR172.101, Appendix B), severe pollutant as mercury-based pesticides, liquid, flammable, toxic, n.o.s.; mercury-based pesticides, liquid, toxic, n.o.s.; mercury-based pesticides, solid, toxic, n.o.s.; mercury compounds, liquid, n.o.s.; mercury compounds, solid, n.o.s.; mercury(I) (mercurous) compounds (pesticides); mercury(II) (mercuric) compounds (pesticides).
California Proposition 65 Developmental/Reproductive toxin (mercury and mercury compounds) 7/1/90.

Canada, WHMIS, Ingredients Disclosure List Concentration 1.0% as mercury compounds, n.o.s.

Rotterdam Convention Annex III [Chemicals Subject to the Prior Informed Consent Procedure (PIC)] (as mercury compounds, including inorganic mercury compounds, alkyl mercury compounds, and alkyloxylalkyl and aryl mercury compounds).

European/International Regulations: Hazard Symbol: E, T, N; Risk phrases: R2; R23/24/25; R33; R50/53; Safety phrases: S1/2; S28; S36/37; S45; S60; S61 (see Appendix 4).

WGK (German Aquatic Hazard Class): 3—Severe hazard to waters.

Description: Mercuric oxycyanide is a white crystalline solid. It explodes instead of melting. Molecular weight = 721.9; Specific gravity (water = 1) = 4.43. Hazard Identification (based on NFPA-704 M Rating System): Health 3, Flammability 1, Reactivity 4. Soluble in water.

Potential Exposure: This material has been used in medicine as a topical antiseptic.

Incompatibilities: Mercuric oxycyanide is self reactive. Friction, heat, shock, and careless handling may cause an explosion. Contact with acid or acid mist causes the release of toxic mercury and hydrogen cyanide vapors.

Permissible Exposure Limits in Air
No TEEL available.

As organo mercury compounds
OSHA PEL: 0.01 mg/m^3 TWA; 0.04 mg/m^3 Ceiling Concentration.
NIOSH REL: 0.01 mg/m^3 TWA; 0.03 mg/m^3 STEL [skin].
ACGIH TLV®[1]: 0.01 mg/m^3 TWA; 0.03 mg/m^3 STEL [skin].
DFG MAK: 0.01 mg[Hg]/m^3 [skin] Danger of skin sensitization; Carcinogen Category 3 [skin] Danger of skin sensitization; Carcinogen Category 3.
NIOSH IDLH: 2 mg Hg/m^3.

As cyanides
OSHA PEL: 5 mg[CN]/m^3/4.7 ppm TWA.
NIOSH REL: 5 mg[CN]/m^3/4.7 ppm/10 min, Ceiling Concentration.
ACGIH TLV®[1]: 5 mg[CN]/m^3 [skin] Ceiling Concentration.
DFG MAK: 2 mg[CN]/m^3, inhalable fraction TWA; Peak Limitation Category II(1) [skin]; Pregnancy Risk Group: C.
NIOSH IDLH: 25 mg[CN]/m^3.

Determination in Air: Use NIOSH Analytical Method #7904, Cyanides.

Permissible Concentration in Water: *To protect freshwater aquatic life:* 0.00057 µg/L as a 24-h average, never to exceed 0.0017 µg/L. *To protect saltwater aquatic life:* 0.025 µg/L as a 24-h average, never to exceed 3.7 µg/L. *To protect human health:* 0.144 µg/L (US EPA) set in 1979–1980.[6] These are the limits for *inorganic* mercury compounds in general.

Determination in Water: Total mercury is determined by flameless atomic absorption. Soluble mercury may be determined by 0.45-µm filtration followed by flameless atomic absorption.

Routes of Entry: Inhalation, ingestion, eye and/or skin contact. Absorbed through the skin.

Harmful Effects and Symptoms

Short Term Exposure: LD$_{50}$ = (oral-rats)   26 mg/kg. Mercuric oxycyanide can affect you when breathed and by passing through skin. Overexposures can cause kidney damage. Mercury poisoning can cause "shakes," irritability, sore gums, memory loss, increased saliva, personality change, and even permanent brain damage. Skin contact can cause irritation, skin allergy, or a gray skin color. Eye contact causes irritation. Heating or use near acid can release toxic mercury and cyanide vapors. Health effects have been reported below permitted exposure levels.

Long Term Exposure: Repeated or prolonged contact with skin may result in dermatitis and allergy. Repeated or prolonged exposure may cause brain damage and nervous system damage. Repeated or prolonged exposure may cause death by hypovolemic shock, nephrotic syndrome, and kidney failure. Organic mercury substances have been identified as a teratogen in humans. Can cause mercury to accumulate in the body and cause mercury poisoning. May cause permanent damage, such as gray colored skin, brown staining of the eyes, and decreased peripheral vision.

Points of Attack: Eyes, skin, central nervous system, mucous membrane, peripheral nervous system, kidneys.

Medical Surveillance: NIOSH lists the following tests for inorganic mercury: whole blood (chemical/metabolite); whole blood (chemical/metabolite), end-of-shift; end-of-shift at end-of-work-week; biologic tissue/biopsy; nerve conduction studies; neurologic examination/electromyography; thyroid function test/thyroid profile; urine (chemical/metabolite); urine (chemical/metabolite) prior to shift, prior to next shift; urine (chemical/metabolite), sediment; urinalysis (routine). Before first exposure and every 6–12 months after, a complete medical history and exam is strongly recommended with examination of the nervous system including handwriting. Routine urine test (UA). Urine test for mercury (should be less than 0.02 mg/L). Eye examination. Kidney function tests. After suspected illness or overexposure, repeat the tests above and get a blood test for mercury.

First Aid: If this chemical gets into the eyes, remove any contact lenses at once and irrigate immediately for at least 15 min, occasionally lifting upper and lower lids. Seek medical attention immediately. If this chemical contacts the skin, remove contaminated clothing and wash immediately with soap and water. Seek medical attention immediately. If this chemical has been inhaled, remove from exposure, begin rescue breathing (using universal precautions, including resuscitation mask) if breathing has stopped and CPR if heart action has stopped. Transfer promptly to a medical facility. When this chemical has been swallowed, get medical attention. Give large quantities of water and induce vomiting. Do not make an unconscious person vomit.

Antidotes and special procedures for medical personnel: The drug NAP (*n*-acetyl penicillamine) has been used to treat mercury poisoning, with mixed success.

Note to physician: For severe poisoning BAL [British Anti-Lewisite, dimercaprol, dithiopropanol ($C_3H_8OS_2$)] has been used to treat toxic symptoms of certain heavy metals poisoning including mercury. Although BAL is reported to have a large margin of safety, caution must be exercised, because toxic effects may be caused by excessive dosage. Most can be prevented by premedication with 1-ephedrine sulfate (CAS: 134-72-5).

Personal Protective Methods: Wear protective gloves and clothing to prevent any reasonable probability of skin contact. Safety equipment suppliers/manufacturers can provide recommendations on the most protective glove/clothing material for your operation. All protective clothing (suits, gloves, footwear, headgear) should be clean, available each day, and put on before work. Contact lenses should not be worn when working with this chemical. Wear dust-proof chemical goggles and face shield unless full face-piece respiratory protection is worn. Employees should wash immediately with soap when skin is wet or contaminated. Provide emergency showers and eyewash. Specific engineering controls are recommended in NIOSH Criteria Document #73-11024.

Respirator Selection: Up to 0.1 mg/m³: Sa (APF = 10) (any supplied-air respirator). *Up to 0.25 mg/m³:* Sa:Cf (APF = 25) (any supplied-air respirator operated in a continuous-flow mode). *Up to 0.5 mg/m³:* SaT: Cf (APF = 50) (any supplied-air respirator that has a tight-fitting face-piece and is operated in a continuous-flow mode) or SCBAF (APF = 50) (any self-contained breathing apparatus with a full face-piece) or SaF (APF = 50) (any supplied-air respirator with a full face-piece). *Up to 2 mg/m³:* SA: PD, PP (any supplied-air respirator operated in a pressure-demand or other positive-pressure mode). *Emergency or planned entry into unknown concentrations or IDLH conditions:* SCBAF: Pd,Pp (APF = 10,000) (any NIOSH/MSHA- or European Standard EN 149-approved self-contained breathing apparatus that has a full face-piece and is operated in a pressure-demand or other positive-pressure mode) or SaF: Pd,Pp: ASCBA (APF = 10,000) (any supplied-air respirator that has a full face-piece and is operated in a pressure-demand or other positive-pressure mode in combination with an auxiliary self-contained breathing apparatus operated in a pressure-demand or other positive-pressure mode). *Escape:* SCBAE (any appropriate escape-type, self-contained breathing apparatus).

Storage: Color Code—Blue: Health Hazard/Poison: Store in a secure poison location. Prior to working with this chemical you should be trained on its proper handling and storage. Mercuric oxycyanide is self reactive. Friction, heat, and rough handling may cause an explosion. Store in tightly closed containers in a cool, well-ventilated area.

Shipping: Shipping of this material is FORBIDDEN *unless* it is desensitized. The desensitized material requires a shipping label of "POISONOUS/TOXIC MATERIALS." The desensitized material falls in DOT Hazard Class 6.1 and Packing Group II.

Spill Handling: Evacuate persons not wearing protective equipment from area of spill or leak until cleanup is complete. Remove all ignition sources. Spills should be collected with special mercury vapor suppressants or special vacuums and deposited in sealed containers. Kits specific for cleanup of mercury spills should be available. Ventilate area after cleanup is complete. It may be necessary to contain and dispose of this chemical as a hazardous waste. If material or contaminated runoff enters waterways, notify downstream users of potentially contaminated waters. Contact your local or federal environmental protection agency for specific recommendations. If employees are required to clean up spills, they must be properly trained and equipped. OSHA 1910.120(q) may be applicable.

Fire Extinguishing: Mercuric oxycyanide may burn but does not readily ignite. Use dry chemical, CO_2, water spray, or foam extinguishers. Poisonous gases, including cyanide gas and nitrogen oxides, are produced in fire. If material or contaminated runoff enters waterways, notify downstream users of potentially contaminated waters. Notify local health and fire officials and pollution control agencies. Containers may explode in fire. From a secure, explosion-proof location, use water spray to cool exposed containers. If cooling streams are ineffective (venting sound increases in volume and pitch, tank discolors, or shows any signs of deforming), withdraw immediately to a secure position. If employees are expected to fight fires, they must be trained and equipped in OSHA 1910.156. The only respirators recommended for firefighting are self-contained breathing apparatuses that have full face-pieces and are operated in a pressure-demand or other positive-pressure mode.

Reference

New Jersey Department of Health and Senior Services. (February 2000). *Hazardous Substances Fact Sheet: Mercuric Oxycyanide.* Trenton, NJ

Mercuric sulfate   M:0420

Molecular Formula: HgO_4S
Common Formula: $HgSO_4$
Synonyms: Mercury bisulfate; Mercury persulfate; Mercury (2+) sulfate (1:1); Mercury(II) sulfate (1:1); Sulfate mercurique (French); Sulfato mercurico (Spanish); Sulfuric acid, mercury(2+) salt (1:1); Sulfuric acid, mercury(II) salt (1:1)
CAS Registry Number: 7783-35-9
RTECS® Number: OX0500000
UN/NA & ERG Number: UN1645/151
EC Number: 231-992-5 [*Annex I Index No.:* 080-002-00-6]
Regulatory Authority and Advisory Bodies
Air Pollutant Standard Set. See below, "Permissible Exposure Limits in Air" section.

Clean Water Act: Section 311 Hazardous Substances/RQ 40CFR117.3 (same as CERCLA, see below); 40CFR 401.15 Section 307 Toxic Pollutants as mercury and compounds; Section 313 Water Priority Chemicals (57FR41331, 9/9/92). RCRA, 40CFR261, Appendix 8 Hazardous Constituents, waste number not listed, as mercury compounds, n.o.s. Reportable Quantity (RQ): 10 lb (4.54 kg).

EPCRA Section 313 (as mercury compound). Form R *de minimis* concentration reporting level: 1.0%.

US DOT Regulated Marine Pollutant (49CFR172.101, Appendix B), severe pollutant as mercuric sulphate.

California Proposition 65 Developmental/Reproductive toxin (mercury and mercury compounds) 7/1/90.

Canada, WHMIS, Ingredients Disclosure List Concentration 1.0%.

Rotterdam Convention Annex III [Chemicals Subject to the Prior Informed Consent Procedure (PIC)] (as mercury compounds, including inorganic mercury compounds, alkyl mercury compounds, and alkyloxylalkyl and aryl mercury compounds).

European/International Regulations: Hazard Symbol: T + , N; Risk phrases: R6; R26; R48/23; R50/53; Safety phrases: S53; S45; S60; S61 (see Appendix 4).

WGK (German Aquatic Hazard Class): 3—Severe hazard to waters.

Description: Mercuric sulfate is a white, odorless, crystalline solid. Molecular weight = 296.65; Freezing/Melting point = (decomposes). Hazard Identification (based on NFPA-704 M Rating System): Health 3, Flammability 0, Reactivity 0. Decomposes in water.

Potential Exposure: Mercuric sulfate is used in making other chemicals, as a battery electrolyte, and in extracting gold and silver from rock.

Incompatibilities: Contact with water produces insoluble basic mercuric subsulfate and sulfuric acid. Reacts with acids producing mercury vapors. Violent reaction with gaseous hydrogen chloride above 121°C. Decomposes in heat or on exposure to light, producing toxic fumes of mercury and sulfur oxides. Attacks magnesium, aluminum, zinc, iron, lead, copper.

Permissible Exposure Limits in Air
OSHA PEL: 0.1 mg[Hg]/m^3 Ceiling Concentration.
NIOSH REL: Hg (*vapor*): 0.05 mg[Hg]/m^3 TWA [skin]; Other: 0.1 mg[Hg]/m^3 Ceiling Concentration [skin].
ACGIH TLV®[1]: 0.025 mg[Hg]/m^3 TWA [skin]; not classifiable as a carcinogen; BEI (preshift) 35 μg[Hg]/100 mL creatinine total inorganic Hg in urine; 15 μg[Hg]/L total inorganic Hg in blood; end-of-shift at end-of-work-week.
ACGIH TLV®[1]: 0.1 mg/m^3 TWA *as persulfates.*
Protective Action Criteria (PAC)
TEEL-0: 0.037 mg/m^3
PAC-1: 0.037 mg/m^3
PAC-2: 0.148 mg/m^3
PAC-3: 14.8 mg/m^3
DFG MAK: 0.1 mg[Hg]/m^3; Peak Limitation Category II(8) danger of skin sensitization; Carcinogen Category 3B.

NIOSH IDLH: 10 mg[Hg]/m^3.
Australia: TWA 0.05 mg[Hg]/m^3, [skin], 1993; Belgium: TWA 0.05 mg[Hg]/m^3, [skin], 1993; Denmark: TWA 0.05 mg[Hg]/m^3, [skin], 1999; Finland: TWA 0.05 mg[Hg]/ m^3, 1999; France: VME 0.1 mg[Hg]/m^3, [skin], 1999; Hungary: TWA 0.02 mg[Hg]/m^3; STEL 0.04 mg[Hg]/m^3, 1993; Japan: 0.05 mg[Hg]/m^3, 1999; Norway: TWA 0.05 mg[Hg]/m^3, 1999; the Philippines: TWA 0.05 mg[Hg]/ m^3, 1993; Poland: MAC (TWA) 0.05 mg[Hg]/m^3, MAC (STEL) 0.15 mg[Hg]/m^3, 1999; Russia: TWA 0.05 mg[Hg]/ m^3; STEL 0.01 mg[Hg]/m^3, 1993; Sweden: NGV 0.05 mg [Hg]/m^3, [skin], 1999; Thailand: STEL 0.05 mg[Hg]/m^3, 1993; United Kingdom: LTEL 0.05 mg[Hg]/m^3; STEL 0.15 mg[Hg]/m^3, 1993; Argentina, Bulgaria, Columbia, Jordan, South Korea, New Zealand, Singapore, Vietnam: ACGIH TLV®: not classifiable as a human carcinogen.

Determination in Air: Use NIOSH Analytical Method #6009; OSHA Analytical Method ID-140.

Permissible Concentration in Water: To protect freshwater aquatic life: 0.00057 μg/L as a 24-h average, never to exceed 0.0017 μg/L. *To protect saltwater aquatic life:* 0.025 μg/L as a 24-h average, never to exceed 3.7 μg/L. *To protect human health:* 0.144 μg/L (US EPA) set in 1979−1980.[6] These are the limits for inorganic mercury compounds in general.

Determination in Water: Total mercury is determined by flameless atomic absorption. Soluble mercury may be determined by 0.45-μm filtration followed by flameless atomic absorption.

Routes of Entry: Inhalation, skin absorption, ingestion, skin and/or eye contact.

Harmful Effects and Symptoms
Short Term Exposure: Mercuric sulfate can affect you when breathed in and by passing through skin. Irritates the skin and is corrosive to the eyes and the respiratory tract. Corrosive if ingested. Inhalation of the aerosol can cause pulmonary edema, a medical emergency that can be delayed for several hours. This can cause death. Overexposure can cause kidney damage. Mercury poisoning can cause "shakes," irritability, sore gums, memory loss, increased saliva, personality change, and even permanent brain damage. Heating or contact with acid or acid "mist" causes release of toxic mercury vapors, and lung effects have been reported below permissible exposure levels. The substance may cause effects on the gastrointestinal tract. Very high exposure may result in death.

Long Term Exposure: Repeated or prolonged contact with skin may result in dermatitis and allergy. Repeated or prolonged exposure may cause brain damage and nervous system damage. Repeated or prolonged exposure may cause death by hypovolemic shock, nephrotic syndrome, and kidney failure. Organic mercury substances have been identified as a teratogen in humans. Can cause mercury to accumulate in the body and cause mercury poisoning. May cause permanent damage, such as gray colored skin, brown staining of the eyes, and decreased peripheral vision.

Points of Attack: Eyes, skin, respiratory system, central nervous system, kidneys.

Medical Surveillance: NIOSH lists the following tests for inorganic mercury: whole blood (chemical/metabolite); whole blood (chemical/metabolite), end-of-shift; end-of-shift at end-of-work-week; biologic tissue/biopsy; nerve conduction studies; neurologic examination/electromyography; thyroid function test/thyroid profile; urine (chemical/metabolite); urine (chemical/metabolite) prior to shift, prior to next shift; urine (chemical/metabolite), sediment; urinalysis (routine). Before first exposure and every 6−12 months after, a complete medical examination and history is strongly recommended, with examination of the nervous system including handwriting. Routine urine test (UA). Urine test for mercury (should be less than 0.02 mg/L). Eye examination. After suspected illness or overexposure, repeat the tests above and get a blood test for mercury. Consider nerve conduction tests, urinary enzymes, and neurobehavioral testing.

First Aid: If this chemical gets into the eyes, remove any contact lenses at once and irrigate immediately for at least 15 min, occasionally lifting upper and lower lids. Seek medical attention immediately. If this chemical contacts the skin, remove contaminated clothing and wash immediately with soap and water. Seek medical attention immediately. If this chemical has been inhaled, remove from exposure, begin rescue breathing (using universal precautions, including resuscitation mask) if breathing has stopped and CPR if heart action has stopped. Transfer promptly to a medical facility. When this chemical has been swallowed, get medical attention. Give large quantities of water and induce vomiting. Do not make an unconscious person vomit. Medical observation is recommended for 24−48 h after breathing overexposure, as pulmonary edema may be delayed. As first aid for pulmonary edema, a doctor or authorized paramedic may consider administering a corticosteroid spray.

Antidotes and special procedures for medical personnel: The drug NAP (N-acetyl penicillamine) has been used to treat mercury poisoning with limited success.

Note to physician: For severe poisoning BAL [British Anti-Lewisite, dimercaprol, dithiopropanol $(C_3H_8OS_2)$] has been used to treat toxic symptoms of certain heavy metals poisoning including mercury. Although BAL is reported to have a large margin of safety, caution must be exercised, because toxic effects may be caused by excessive dosage. Most can be prevented by premedication with 1-ephedrine sulfate (CAS: 134-72-5).

Personal Protective Methods: Wear protective gloves and clothing to prevent any reasonable probability of skin contact. Safety equipment suppliers/manufacturers can provide recommendations on the most protective glove/clothing material for your operation. All protective clothing (suits, gloves, footwear, headgear) should be clean, available each day, and put on before work. Contact lenses should not be worn when working with this chemical. Wear dust-proof chemical goggles and face shield unless full face-piece respiratory protection is worn. Employees should wash immediately with soap when skin is wet or contaminated. Provide emergency showers and eyewash.

Respirator Selection: Mercury vapor: NIOSH: *Up to 0.5 mg/m³:* CcrS (APF = 10) [any chemical cartridge respirator with cartridge(s) providing protection against the compound of concern]* or Sa (APF = 10) (any supplied-air respirator). *Up to 1.25 mg/m³:* Sa:Cf (APF = 25) (any supplied-air respirator operated in a continuous-flow mode) or PaprS (APF = 25) [any powered, air-purifying respirator with cartridge(s) providing protection against the compound of concern]* (canister). *Up to 2.5 mg/m³:* CcrFS (APF = 50) [any chemical cartridge respirator with a full face-piece and cartridge(s) providing protection against the compound of concern]* or GmFS (APF = 50) [any air-purifying, full-face-piece respirator (gas mask) with a chin-style, front- or back-mounted canister providing protection against the compound of concern]* or SaT: Cf (APF = 50) (any supplied-air respirator that has a tight-fitting face-piece and is operated in a continuous-flow mode) or PaprTS (APF = 50) [any powered, air-purifying respirator with a tight-fitting face-piece and cartridge(s) providing protection against the compound of concern] or SCBAF (APF = 50) (any self-contained breathing apparatus with a full face-piece) or SaF (APF = 50) (any supplied-air respirator with a full face-piece). *Up to 10 mg/m³:* Sa: Pd,Pp (APF = 1000) (any supplied-air respirator operated in a pressure-demand or other positive-pressure mode). *Emergency or planned entry into unknown concentrations or IDLH conditions:* SCBAF: Pd,Pp (APF = 10,000) (any self-contained breathing apparatus that has a full face-piece and is operated in a pressure-demand or other positive-pressure mode) or SaF: Pd,Pp: ASCBA (any supplied-air respirator that has a full face-piece and is operated in a pressure-demand or other positive-pressure mode in combination with an auxiliary, self-contained breathing apparatus operated in a pressure-demand or other positive-pressure mode). *Escape:* GmFS* [any air-purifying, full-face-piece respirator (gas mask) with a chin-style, front- or back-mounted canister protection against the compound of concern] or SCBAE (any appropriate escape-type, self-contained breathing apparatus).*
*End-of-service life indicator (ESLI) required.

Other mercury compounds: NIOSH/OSHA *Up to 1 mg/m³:* CcrS (APF = 10) [any chemical cartridge respirator with cartridge(s) providing protection against the compound of concern]* or Sa (APF = 10) (any supplied-air respirator). *Up to 2.5 mg/m³:* Sa:Cf (APF = 25) (any supplied-air respirator operated in a continuous-flow mode) or PaprS (APF = 25) [any powered, air-purifying respirator with cartridge(s) providing protection against the compound of concern]* (canister). *Up to 5 mg/m³:* CcrFS (APF = 50) [any chemical cartridge respirator with a full face-piece and cartridge(s) providing protection against the compound of concern]* or GmFS (APF = 50) [any air-purifying, full-face-piece respirator (gas mask) with a chin-style, front- or back-mounted canister providing protection against the compound of concern]* or SaT: Cf (APF = 50) (any supplied-air respirator that has a tight-fitting face-piece and is operated in a continuous-flow mode) or PaprTS (APF = 50) [any powered, air-purifying respirator with a tight-fitting face-piece and cartridge(s) providing protection against the compound of concern] or SCBAF (APF = 50) (any self-contained breathing apparatus with a full face-piece) or SaF (APF = 50) (any supplied-air respirator with a full face-piece). *Up to 10 mg/m³:* Sa: Pd,Pp (APF = 1000) (any supplied-air respirator operated in a pressure-demand or other positive-pressure mode). *Emergency or planned entry into unknown concentrations or IDLH conditions:* SCBAF: Pd,Pp (APF = 10,000) (any self-contained breathing apparatus that has a full face-piece and is operated in a pressure-demand or other positive-pressure mode) or SaF: Pd,Pp: ASCBA (APF = 10,000) (any supplied-air respirator that has a full face-piece and is operated in a pressure-demand or other positive-pressure mode in combination with an auxiliary, self-contained breathing apparatus operated in a pressure-demand or other positive-pressure mode). *Escape:* GmFS* [any air-purifying, full-face-piece respirator (gas mask) with a chin-style, front- or back-mounted canister protection against the compound of concern] or SCBAE (any appropriate escape-type, self-contained breathing apparatus).* *End-of-service life indicator (ESLI) required.

Storage: Color Code—Blue: Health Hazard/Poison: Store in a secure poison location. Prior to working with this chemical you should be trained on its proper handling and storage. Store in tightly closed containers in a cool, well-ventilated area away from light, water, and gaseous hydrogen chloride.

Shipping: This compound requires a shipping label of "POISONOUS/TOXIC MATERIALS." It falls in Hazard Class 6.1 and Packing Group II.

Spill Handling: Evacuate persons not wearing protective equipment from area of spill or leak until cleanup is complete. Remove all ignition sources. Spills should be collected with special mercury vapor suppressants or special vacuums and deposited in sealed containers. Kits specific for cleanup of mercury spills should be available. Ventilate area after cleanup is complete. It may be necessary to contain and dispose of this chemical as a hazardous waste. If material or contaminated runoff enters waterways, notify downstream users of potentially contaminated waters. Contact your local or federal environmental protection agency for specific recommendations. If employees are required to clean up spills, they must be properly trained and equipped. OSHA 1910.120(q) may be applicable.

Fire Extinguishing: This material is nonflammable. Use agent suitable to surrounding fire. Poisonous gases, including mercury and sulfur oxides, are produced in fire. If material or contaminated runoff enters waterways, notify downstream users of potentially contaminated waters. Notify local health and fire officials and pollution control agencies. From a secure, explosion-proof location, use water spray to cool exposed containers. If cooling streams are ineffective (venting sound increases in volume and pitch, tank discolors, or shows any signs of deforming), withdraw immediately to a secure position. If employees are expected to fight fires, they must be trained and equipped in OSHA 1910.156. The only respirators recommended for firefighting are self-contained breathing apparatuses that have full face-pieces and are operated in a pressure-demand or other positive-pressure mode.

Reference
New Jersey Department of Health and Senior Services. (February 2000). *Hazardous Substances Fact Sheet: Mercuric Sulfate.* Trenton, NJ

Mercury and inorganic compounds   M:0430

Molecular Formula: Hg
Synonyms: Colloidal mercury; Hydragyrum; Kwik; Liquid silver; Mercure (French); Mercury, metallic; Metallic mercury; NCI-C60399; Quecksilber (German); Quicksilver
CAS Registry Number: 7439-97-6; *(alt.)* 8030-64-6; *(alt.)* 51887-47-9; *(alt.)* 92355-34-5; *(alt.)* 92786-62-4; *(alt.)* 123720-03-6
RTECS® Number: OV4550000
UN/NA & ERG Number: UN2809/172
EC Number: 231-106-7 [*Annex I Index No.:* 080-001-00-0]
Regulatory Authority and Advisory Bodies
Carcinogenicity: IARC: Human Inadequate Evidence, animal Inadequate Evidence Group 3, 1993; EPA: Not Classifiable as to human carcinogenicity.
Banned or Severely Restricted (in agriculture) (many countries) (UN).[13]
Air Pollutant Standard Set. See below, "Permissible Exposure Limits in Air" section.
Clean Water Act: 40CFR401.15 Section 307 Toxic Pollutants as mercury and compounds; Section 313 Water Priority Chemicals (57FR41331, 9/9/92).
US EPA Hazardous Waste Number (RCRA No.): U151.
RCRA, 40CFR261, Appendix 8 Hazardous Constituents.
RCRA Toxicity Characteristic (Section 261.24), Maximum. Concentration of Contaminants, regulatory level, 0.2 mg/L.
RCRA 40CFR268.48; 61FR15654, Universal Treatment Standards: Wastewater (mg/L), 0.15; Nonwastewater (mg/L), 0.25 TCLP; Wastewater from retort, N/A; Nonwastewater from retort (mg/L), 0.20 TCLP.
RCRA 40CFR264, Appendix 9; TSD Facilities Ground Water Monitoring List. Suggested test method(s) (PQL µg/L): 7420 (2) as mercury, total dust.
Safe Drinking Water Act: MCL, 0.002 mg/L; MCLG, 0.002 mg/L.
Reportable Quantity (RQ): 1 lb (0.454 kg).
EPCRA Section 313 Form R *de minimis* concentration reporting level: 1.0%.
California Proposition 65 Developmental/Reproductive toxin (mercury and mercury compounds) 7/1/90.

Canada, WHMIS, Ingredients Disclosure List Concentration 0.1%.

Rotterdam Convention Annex III [Chemicals Subject to the Prior Informed Consent Procedure (PIC)] (as mercury compounds, including inorganic mercury compounds, alkyl mercury compounds, and alkyloxylalkyl and aryl mercury compounds).

European/International Regulations: not listed in Annex 1. WGK (German Aquatic Hazard Class): 3—Severe hazard to waters.

European/International Regulations: Hazard Symbol: T +, N; Risk phrases: R6; R26; R48/23; R50/53; Safety phrases: S53; S45; S60; S61.

Description: Mercury is a silvery, mobile, odorless liquid. Molecular weight = 200.59; Boiling point = 356−357°C; Freezing/Melting point = −39°C. Hazard Identification (based on NFPA-704 M Rating System): Health 1, Flammability 0, Reactivity 0. Insoluble in water.

Potential Exposure: Compound Description: Agricultural Chemical; Tumorigen, Drug, Mutagen; Reproductive Effector; Human Data. Mercury is used as a catalyst, in dental applications and pharmaceuticals, as a liquid cathode in cells for the electrolytic production of caustic and chlorine. It is used in electrical apparatus (lamps, rectifiers, and batteries) and in control instruments (switches, thermometers and barometers).

Incompatibilities: Heating mercury causes the formation of toxic mercury oxide fumes. Reacts violently with alkali metals, acetylene, azides, ammonia gas, chlorine, chlorine dioxide, many acids, most metals, ground mixtures of sodium carbide, and ethylene oxide. Contact with methyl azide forms shock- and spark-sensitive explosives. Attacks copper and many other metals, forming amalgams.

Permissible Exposure Limits in Air

OSHA PEL: 0.1 mg[Hg]/m^3 Ceiling Concentration.

NIOSH REL: (*Hg vapor*): 0.05 mg[Hg]/m^3 TWA [skin]; 0.1 mg[Hg]/m^3 Ceiling Concentration; [skin].

ACGIH TLV®[1]: 0.025 mg[Hg]/m^3 TWA [skin]; not classifiable as a carcinogen; BEI (preshift) 35 µg[Hg]/100 mL creatinine total inorganic Hg in urine; 15 µg[Hg]/L total inorganic Hg in blood; end-of-shift at end-of-work-week.

Protective Action Criteria (PAC) Mercury*

TEEL-0: 0.025 mg/m^3

PAC-1: 0.25 mg/m^3

PAC-2: **1.7** mg/m^3

PAC-3: **8.9** mg/m^3

*AEGLs (Acute Emergency Guideline Levels) & ERPGs (Emergency Response Planning Guideline) are in **bold face**.

Emergency Response Planning Guidelines (AIHA)

ERPG-1: Inappropriate

ERPG-2: 0.25 ppm

ERPG-3: 0.5 ppm

DFG MAK (*elemental and inorganic compounds*): 0.1 mg [Hg]/m^3; Peak Limitation Category II(8) danger of skin sensitization; Carcinogen Category 3B.

NIOSH IDLH: 10 mg[Hg]/m^3.

Arab Republic of Egypt: TWA 0.05 mg/m^3, 1993; Australia: TWA 0.1 mg/m^3, [skin], 1993; Austria: MAK 0.005 ppm (0.05 mg/m^3, 1999; Belgium: TWA 0.1 mg/m^3, [skin], 1993; Denmark: TWA 0.05 mg[Hg]/m^3, [skin], 1999; Finland: TWA 0.05 mg/m^3, 1993; France: VME 0.05 mg/m^3, [skin] (*vapor*), 1999; the Netherlands: MAC-TGG 0.05 mg/m^3, 2003; Japan: 0.05 mg/m^3, 1999; Norway: TWA 0.05 mg/m^3, 1999; Poland: MAC (TWA) vapors 0.025 mg/m^3, MAC (STEL) vapors 0.2 mg/m^3, 1999; Sweden: NGV 0.05 mg/m^3 (*vapor*), 1999; Switzerland: MAK-W 0.005 ppm (0.05 mg/m^3), [skin], 1999; Switzerland: MAK-W 0.01 mg/m^3, [skin], 1993; Turkey: TWA 0.1 mg/m^3, [skin], 1993; United Kingdom: TWA 0.025 mg[Hg]/m^3, 2000; Argentina, Bulgaria, Columbia, Jordan, South Korea, New Zealand, Singapore, Vietnam: ACGIH TLV®: not classifiable as a human carcinogen. Russia[35, 43] set a MAC of 0.0003 mg/m^3 for ambient air in residential areas. Several states have set guidelines or standards for mercury in ambient air[60] ranging from 0.024 µg/m^3 (Kansas) to 0.01−0.08 mg/m^3 (Montana) to 0.167 µg/m^3 (New York) to 0.2−2.0 µg/m^3 (Connecticut) to 0.24 µg/m^3 (Pennsylvania) to 0.25 µg/m^3 (South Carolina) to 0.5 µg/m^3 (North Dakota) to 0.5−1.0 µg/m^3 (Florida) to 0.8 µg/m^3 (Virginia) to 2.0 µg/m^3 (Nevada) to 3.0 µg/m^3 (North Carolina).

Determination in Air: Use NIOSH Analytical Method #6009; OSHA Analytical Method ID-140.

Permissible Concentration in Water: A variety of values have been set for mercury in drinking water in various parts of the world. These include[35]: the Czech Republic 0.001 mg/L; Canada 0.001 mg/L maximum acceptable concentration; Germany 0.004 mg/L EEC 0.001 mg/L; Japan 0.0005 mg/L; Mexico 0.005 mg/L; former USSR-UNEP/IRPTC joint project 0.0005 mg/L; USA 0.002 mg/L; WHO 0.001 mg/L. In addition, the US EPA[49] has set a lifetime health advisory at 0.0011 µg/L. And EPA has set a guideline of 0.002 µg/L for drinking water.[62] Beyond that several states have set guidelines for mercury in drinking water[61] including Maine at 2 µg/L and Minnesota at 3 µg/L.

Determination in Water: Total mercury is determined by flameless atomic absorption. Soluble mercury may be determined by 0.45-µm filtration followed by flameless atomic absorption.

Routes of Entry: Inhalation, skin absorption, eye and/or skin contact.

Harmful Effects and Symptoms

Short Term Exposure: Inhalation: Exposure to levels below 1 mg/m^3 has been shown to produce nonspecific symptoms, such as shyness, insomnia, anxiety, and loss of appetite. Higher levels (1−3 mg/m^3 for 2−5 h) may cause headache, salivation, metallic taste, chills, cough, fever, tremors, abdominal cramps, diarrhea, nausea, vomiting, tightness in the chest, difficult breathing, fatigue, lung irritation, and possible lung tissue damage. Symptoms may begin several hours after exposure and may last a week. Large doses may result in flu-like symptoms, which, in severe cases, may result in death due to pneumonia. *Lethal blood level in humans:* 0.4−22 mg/mL. *Skin:* Can be absorbed through the skin. Can cause irritation. Prolonged contact with skin can result in symptoms listed above. *Eyes:* Can cause eye irritation. *Ingestion:* Generally does not produce ill effects.

Long Term Exposure: May cause skin allergy. Mercury accumulates in the brain quickly during exposure but is released from the brain very slowly. This will result in a buildup in brain tissue over a long time. The liver and kidneys may also be damaged by mercury accumulation. It may cause headache, dizziness, restlessness, irritability, sleepiness, tremors, defective muscle control, increased salivation, loose teeth, irritation of the gums with a blue line between teeth and gums, loss of appetite, nausea, vomiting, diarrhea, liver damage, changes in urine, raised red areas and blisters of skin, impaired memory and possible permanent brain damage. Repeated exposure (usually more than 5 years) may cause clouding of the eyes, gray skin color. Frequency of complaints and severity of symptoms increase with levels of exposure, most notably above $0.1 mg/m^3$. However, many of these symptoms have been reported at levels below recommended limits due to the accumulation of mercury over long term. There is limited evidence that mercury may cause an increase in spontaneous abortions in exposed women.

Points of Attack: Eyes, skin, respiratory system, central nervous system, kidneys.

Medical Surveillance: NIOSH lists the following tests for inorganic mercury: whole blood (chemical/metabolite); whole blood (chemical/metabolite), end-of-shift; end-of-shift at end-of-work-week; biologic tissue/biopsy; nerve conduction studies; neurologic examination/electromyography; thyroid function test/thyroid profile; urine (chemical/metabolite); urine (chemical/metabolite) prior to shift, prior to next shift; urine (chemical/metabolite), sediment; urinalysis (routine). For those with frequent or potentially high exposure (half the TLV or greater, or significant skin contact), the following are recommended before beginning work and at regular times after that: examination of the nervous system (including handwriting test to detect early hand tremor). Urine mercury level (usually less than 0.02 mg/L). Kidney function tests. If symptoms develop or overexposure is suspected, the following may be useful: consider chest X-ray after acute overexposure. Evaluation by a qualified allergist, including careful exposure history and special testing, may help diagnose skin allergy.

First Aid: If this chemical gets into the eyes, remove any contact lenses at once and irrigate immediately for at least 15 min, occasionally lifting upper and lower lids. Seek medical attention immediately. If this chemical contacts the skin, remove contaminated clothing and wash immediately with soap and water. Seek medical attention immediately. If this chemical has been inhaled, remove from exposure, begin rescue breathing (using universal precautions, including resuscitation mask) if breathing has stopped and CPR if heart action has stopped. Transfer promptly to a medical facility. When this chemical has been swallowed, get medical attention. Give large quantities of water and do induce vomiting. Do not make an unconscious person vomit.

Antidotes and special procedures for medical personnel: The drug NAP (*n*-acetyl penicillamine) has been used to treat mercury poisoning, with mixed success.

Note to physician: For severe poisoning BAL [British Anti-Lewisite, dimercaprol, dithiopropanol ($C_3H_8OS_2$)] has been used to treat toxic symptoms of certain heavy metals poisoning including mercury. Although BAL is reported to have a large margin of safety, caution must be exercised, because toxic effects may be caused by excessive dosage. Most can be prevented by premedication with 1-ephedrine sulfate (CAS: 134-72-5).

Personal Protective Methods: Wear protective gloves and clothing to prevent any reasonable probability of skin contact. Safety equipment suppliers/manufacturers can provide recommendations on the most protective glove/clothing material for your operation. All protective clothing (suits, gloves, footwear, headgear) should be clean, available each day, and put on before work. Contact lenses should not be worn when working with this chemical. Wear chemical goggles and face shield unless full face-piece respiratory protection is worn. Employees should wash immediately with soap when skin is wet or contaminated. Provide emergency showers and eyewash.

Respirator Selection: Mercury vapor: NIOSH: *Up to $0.5 mg/m^3$:* CcrS (APF = 10) [any chemical cartridge respirator with cartridge(s) providing protection against the compound of concern]* or Sa (APF = 10) (any supplied-air respirator). *Up to $1.25 mg/m^3$:* Sa:Cf (APF = 25) (any supplied-air respirator operated in a continuous-flow mode) or PaprS (APF = 25) [any powered, air-purifying respirator with cartridge(s) providing protection against the compound of concern]* (canister). *Up to $2.5 mg/m^3$:* CcrFS (APF = 50) [any chemical cartridge respirator with a full face-piece and cartridge(s) providing protection against the compound of concern]* or GmFS (APF = 50) [any air-purifying, full-face-piece respirator (gas mask) with a chin-style, front- or back-mounted canister providing protection against the compound of concern]* or SaT: Cf (APF = 50) (any supplied-air respirator that has a tight-fitting face-piece and is operated in a continuous-flow mode) or PaprTS (APF = 50) [any powered, air-purifying respirator with a tight-fitting face-piece and cartridge(s) providing protection against the compound of concern] or SCBAF (APF = 50) (any self-contained breathing apparatus with a full face-piece) or SaF (APF = 50) (any supplied-air respirator with a full face-piece). *Up to $10 mg/m^3$:* Sa: Pd,Pp (APF = 1000) (any supplied-air respirator operated in a pressure-demand or other positive-pressure mode). *Emergency or planned entry into unknown concentrations or IDLH conditions:* SCBAF: Pd,Pp (APF = 10,000) (any self-contained breathing apparatus that has a full face-piece and is operated in a pressure-demand or other positive-pressure mode) or SaF: Pd,Pp: ASCBA (any supplied-air respirator that has a full face-piece and is operated in a pressure-demand or other positive-pressure mode in combination with an auxiliary, self-contained breathing apparatus operated in a pressure-demand or other positive-pressure mode). *Escape:* GmFS* [any air-purifying, full-face-piece respirator (gas mask) with a chin-style, front- or back-mounted canister protection against the compound of concern] or SCBAE (any appropriate escape-type, self-contained breathing apparatus).*

*End-of-service life indicator (ESLI) required.

Other mercury compounds: NIOSH/OSHA *Up to 1 mg/m³:* CcrS (APF = 10) [any chemical cartridge respirator with cartridge(s) providing protection against the compound of concern]* or Sa (APF = 10) (any supplied-air respirator). *Up to 2.5 mg/m³:* Sa:Cf (APF = 25) (any supplied-air respirator operated in a continuous-flow mode) or PaprS (APF = 25) [any powered, air-purifying respirator with cartridge(s) providing protection against the compound of concern]* (canister). *Up to 5 mg/m³:* CcrFS (APF = 50) [any chemical cartridge respirator with a full face-piece and cartridge(s) providing protection against the compound of concern]* or GmFS (APF = 50) [any air-purifying, full-face-piece respirator (gas mask) with a chin-style, front- or back-mounted canister providing protection against the compound of concern]* or SaT: Cf (APF = 50) (any supplied-air respirator that has a tight-fitting face-piece and is operated in a continuous-flow mode) or PaprTS (APF = 50) [any powered, air-purifying respirator with a tight-fitting face-piece and cartridge(s) providing protection against the compound of concern] or SCBAF (APF = 50) (any self-contained breathing apparatus with a full face-piece) or SaF (APF = 50) (any supplied-air respirator with a full face-piece). *Up to 10 mg/m³:* Sa: Pd,Pp (APF = 1000) (any supplied-air respirator operated in a pressure-demand or other positive-pressure mode). *Emergency or planned entry into unknown concentrations or IDLH conditions:* SCBAF: Pd, Pp (APF = 10,000) (any self-contained breathing apparatus that has a full face-piece and is operated in a pressure-demand or other positive-pressure mode) or SaF: Pd,Pp: ASCBA (APF = 10,000) (any supplied-air respirator that has a full face-piece and is operated in a pressure-demand or other positive-pressure mode in combination with an auxiliary, self-contained breathing apparatus operated in a pressure-demand or other positive-pressure mode). *Escape:* GmFS* [any air-purifying, full-face-piece respirator (gas mask) with a chin-style, front- or back-mounted canister protection against the compound of concern] or SCBAE (any appropriate escape-type, self-contained breathing apparatus).*

*End-of-service life indicator (ESLI) required.

Storage: Color Code—White: Corrosive or Contact Hazard; Store separately in a corrosion-resistant location. Prior to working with this chemical you should be trained on its proper handling and storage. Mercury must be stored to avoid contact with chlorine dioxide, nitric acid, nitrates, ethylene oxide, chlorine, and methylazide, since violent reactions occur. Store in tightly closed containers in a cool, well-ventilated area away from acetylene, ammonia, and nickel.

Shipping: This compound requires a shipping label of "CORROSIVE." It falls in Hazard Class 8 and Packing Group III.

Spill Handling: Evacuate persons not wearing protective equipment from area of spill or leak until cleanup is complete. Wearing protective equipment and clothing, clean up the spill with an industrial vacuum cleaner with a charcoal filter to absorb mercury vapor. For mercury spilled in cracks, cover with zinc dust to form an amalgam, or cover with calcium polysulfide with excess sulfur. Do not sweep or use compressed air to blow mercury droplets as it can increase air concentrations. Store contaminated or waste mercury in tightly covered or vapor-proof containers pending removal. It may be necessary to contain and dispose of this chemical as a hazardous waste. If material or contaminated runoff enters waterways, notify downstream users of potentially contaminated waters. Contact your local or federal environmental protection agency for specific recommendations. If employees are required to clean up spills, they must be properly trained and equipped. OSHA 1910.120(q) may be applicable.

Fire Extinguishing: Mercury is not combustible. Use agent suitable for surrounding fire. Poisonous gases, including Hg, are produced in fire. If material or contaminated runoff enters waterways, notify downstream users of potentially contaminated waters. Notify local health and fire officials and pollution control agencies. From a secure, explosion-proof location, use water spray to cool exposed containers. If cooling streams are ineffective (venting sound increases in volume and pitch, tank discolors, or shows any signs of deforming), withdraw immediately to a secure position. If employees are expected to fight fires, they must be trained and equipped in OSHA 1910.156. The only respirators recommended for firefighting are self-contained breathing apparatuses that have full face-pieces and are operated in a pressure-demand or other positive-pressure mode.

Disposal Method Suggested: Consult with environmental regulatory agencies for guidance on acceptable disposal practices. Generators of waste containing this contaminant (≥100 kg/mo) must conform with EPA regulations governing storage, transportation, treatment, and waste disposal. Accumulate for purification and re-use if possible. Mercury vapors may be adsorbed or treated with sulfide solutions and then sent to mercury recovery operations.[22]

References
National Institute for Occupational Safety and Health. (1973). *Criteria for a Recommended Standard: Occupational Exposure to Inorganic Mercury.* NIOSH Document No. 73-11024

US Environmental Protection Agency. (May 1977). *Toxicology of Metals. Vol. II: Mercury, Report EPA-600/1-77-022.* Research Triangle Park, NC, pp. 301—344

US Environmental Protection Agency. (1980). *Mercury: Ambient Water Quality Criteria.* Washington, DC

US Environmental Protection Agency. (December 1979). *Status Assessment of Toxic Chemicals: Mercury, Report EPA-600/2-79-210i.* Cincinnati, OH

US Environmental Protection Agency. (April 30, 1980). *Mercury, Health and Environmental Effects Profile No. 124.* Washington, DC: Office of Solid Waste

World Health Organization. (1976). *Mercury, Environmental Health Criteria No. 1.* Geneva, Switzerland

Sax, N. I. (Ed.). (1981). *Dangerous Properties of Industrial Materials Report* 1, No. 3, 70—72

US Public Health Service. (December 1988). *Toxicological Profile for Mercury.* Atlanta, GA: Agency for Toxic Substances and Disease Registry

New York State Department of Health. (February 1986). *Chemical Fact Sheet: Mercury (Metallic).* Version 2. Albany, NY: Bureau of Toxic Substance Assessment

New Jersey Department of Health and Senior Services. (February 2007). *Hazardous Substances Fact Sheet: Mercury.* Trenton, NJ

Mercury alkyl compounds   M:0440

Molecular Formula: CH_3ClHg
Common Formula: CH_3HgCl
Synonyms: *Methyl mercury chloride:* Caspan; Chloromethylmercury; Methylmercuric chloride; Methylmercury chloride; MMC; Monomethyl mercury chloride
Dimethyl mercury: Mercury dimethyl
CAS Registry Number: 115-09-3 (methyl mercury chloride); 593-74-8 (dimethyl mercury); 22967-92-6 (methyl mercury ion)
RTECS® Number: OW6320000 (methyl mercury ion); OW1225000 (methyl mercury chloride)
UN/NA & ERG Number: UN2025 (mercury compounds, solid, n.o.s.)/151; UN3024 (mercury compounds, liquid, n.o.s.)/131
EC Number: 204-064-2 (chloromethylmercury); 209-805-3 [*Annex I Index No.:* 080-007-00-3] (dimethylmercury)
Regulatory Authority and Advisory Bodies
Carcinogenicity: IARC: Human Inadequate Evidence, *possibly carcinogenic to humans*, Group 2B, 1993 (dimethyl mercury); EPA: Possible Human Carcinogen.
Maximum Contaminant Levels (Safe Drinking Water Act).
US EPA Gene-Tox Program, Negative: *In vivo* cytogenetics—mammalian oocyte; Inconclusive: Rodent dominant lethal (dimethyl mercury).
Air Pollutant Standard Set. See below, "Permissible Exposure Limits in Air" section.

Clean Air Act: Hazardous Air Pollutants (Title I, Part A, Section 112).
Clean Water Act: 40CFR401.15 Section 307 Toxic Pollutants as mercury and compounds.
RCRA, 40CFR261, Appendix 8 Hazardous Constituents, waste number not listed, as mercury compounds, n.o.s.
EPCRA Section 313: Includes any unique chemical substance that contains mercury as part of that chemical's infrastructure. Form R *de minimis* concentration reporting level: 1.0% (dimethyl mercury).
US DOT Regulated Marine Pollutant (49CFR172.101, Appendix B), severe pollutant.
California Proposition 65 Developmental/Reproductive toxin (mercury and mercury compounds) 7/1/90.
Canada, WHMIS, Ingredients Disclosure List Concentration 0.1%.
Rotterdam Convention Annex III [Chemicals Subject to the Prior Informed Consent Procedure (PIC)] (as mercury compounds, including inorganic mercury compounds, alkyl mercury compounds, and alkyloxylalkyl and aryl mercury compounds).
European/International Regulations (*593-74-8*): Hazard Symbol: T +, N; Risk phrases: R26/27/28; R33; R50/53; Safety phrases: S1/2; S13; S28; S36; S45; S60; S61 (see Appendix 4).
WGK (German Aquatic Hazard Class): No value assigned.
Note: All other mercury compounds are Hazard Class 3.
Description: Dimethyl mercury is a volatile colorless liquid with faint sweet odor. Molecular weight = 230.67; Boiling point = 96°C. Soluble in water. Methyl mercury chloride is a colorless, crystalline solid. Molecular weight = 251.08; Freezing/Melting point = 170°C. Practically insoluble in water; solubility = <0.1 mg/mL at 21°C.
Potential Exposure: Alkyl mercury compounds have been used as seed disinfectants and for fungicides. They have also been used in organic synthesis.
Incompatibilities: Strong oxidizers, such as chlorine. May be sensitive to light.
Permissible Exposure Limits in Air
OSHA PEL: 0.01 mg[Hg]/m³ TWA; 0.04 mg[Hg]/m³ Ceiling Concentration.
NIOSH REL: 0.01 mg[Hg]/m³ TWA; 0.03 mg[Hg]/m³ STEL [skin].
ACGIH TLV®[1]: 0.01 mg[Hg]/m³ TWA; 0.03 mg[Hg]/m³ STEL [skin].
NIOSH IDLH: 2 mg Hg/m³.
Dimethyl mercury
Protective Action Criteria (PAC)
TEEL-0: 0.0115 mg/m³
PAC-1: 0.0345 mg/m³
PAC-2: 0.046 mg/m³
PAC-3: 2.3 mg/m³
DFG MAK: 0.01 mg[Hg]/m³ [skin] Danger of skin sensitization; Carcinogen Category 3B.
Australia: TWA 0.01 mg[Hg]/m³; STEL 0.03 mg[Hg]/m³, [skin], 1993; Austria: MAK 0.01 mg[Hg]/m³, 1999;

Belgium: TWA 0.01 mg[Hg]/m^3; STEL 0.03 mg[Hg]/m^3, [skin], 1993; Denmark: TWA 0.01 mg[Hg]/m^3, [skin], 1999; Finland: TWA 0.01 mg[Hg]/m^3, [skin], 1999; France: VME 0.01 mg[Hg]/m^3, [skin], 1999; Norway: TWA 0.01 mg[Hg]/m^3, 1999; the Philippines: TWA 0.01 mg[Hg]/m^3, 1993; Poland: MAC (TWA) 0.01 mg[Hg]/m^3, MAC (STEL) 0.01 mg[Hg]/m^3, 1999; Sweden: NGV 0.01 mg [Hg]/m^3, [skin], 1999; Thailand: TWA 0.01 mg[Hg]/m^3; STEL 0.04 mg[Hg]/m^3, 1993; Turkey: TWA 0.01 mg[Hg]/ m^3, [skin], 1993; United Kingdom: TWA 0.01 mg[Hg]/m^3; STEL 0.03 mg[Hg]/m^3, [skin], 2000; Argentina, Bulgaria, Columbia, Jordan, South Korea, New Zealand, Singapore, Vietnam: ACGIH TLV$^®$: TWA 0.01; STEL 0.03 mg/m^3 [skin]. In addition, North Dakota has set guidelines for alkyl mercury compounds in ambient air[60] of 1−3 µg/m^3 (0.0001−0.0003 mg/m^3).

Determination in Air: No method available.

Permissible Concentration in Water: *Methyl mercury: To protect freshwater aquatic life:* 0.016 µg/L as a 24-h average, never to exceed 8.8 µg/L. *To protect saltwater aquatic life:* 0.025 µg/L as a 24-h average, never to exceed 2.8 µg/ L. *To protect human health:* 0.2 µg/L.[6]

Determination in Water: Total mercury is determined by flameless atomic absorption. Soluble mercury may be determined by 0.45-µm filtration followed by flameless atomic absorption.

Routes of Entry: Inhalation, ingestion, eye and/or skin contact. Absorbed through the skin.

Harmful Effects and Symptoms

Short Term Exposure: Alkyl mercury compounds can be absorbed through the skin. When deposited on the skin, they give no warning, and if contact is maintained, can cause second-degree burns. Sensitization may occur. Alkyl mercurials have very high toxicity. *Systemic:* The central nervous system, including the brain, is the principal target tissue for this group of toxic compounds. Severe poisoning may produce irreversible brain damage resulting in loss of higher functions. The effects of chronic poisoning with alkyl mercury compounds are progressive. In the early stages, there are fine tremors of the hands; and in some cases, of the face and arms.

Long Term Exposure: Repeated or prolonged contact with skin may result in dermatitis (red inflamed skin). Repeated or prolonged exposure may cause death by hypovolemic shock, nephrotic syndrome, or kidney failure. With repeated or continued exposure, tremors may become coarse and convulsive, scanning speech with moderate slurring and difficulty in pronunciation may also occur. The worker may then develop an unsteady gait of a spastic nature which can progress to severe ataxia of the arms and legs. Sensory disturbances including tunnel vision, blindness, and deafness are also common. A late symptom, constriction of the visual fields, is rarely reversible and may be associated with loss of understanding and reason which makes the victim completely out of touch with his environment. Severe cerebral effects have been seen in infants born to mothers who

had eaten large amounts of methylmercury-contaminated fish.

Points of Attack: Eyes, skin, central nervous system, peripheral nervous system, kidneys.

Medical Surveillance: Preplacement and periodic physical examinations should be concerned particularly with the skin, vision, central nervous system, and kidneys. Consideration should be given to the possible effects on the fetus of alkyl mercury exposure in the mother. Constriction of visual fields may be a useful diagnostic sign. Blood and urine levels of mercury have been studied, especially in the case of methylmercury. A precise correlation has not been found between exposure levels and concentrations. They may be of some value in indicating that exposure has occurred, however.

First Aid: If this chemical gets into the eyes, remove any contact lenses at once and irrigate immediately for at least 15 min, occasionally lifting upper and lower lids. Seek medical attention immediately. If this chemical contacts the skin, remove contaminated clothing and wash immediately with soap and water. Seek medical attention immediately. If this chemical has been inhaled, remove from exposure, begin rescue breathing (using universal precautions, including resuscitation mask) if breathing has stopped and CPR if heart action has stopped. Transfer promptly to a medical facility. When this chemical has been swallowed, get medical attention. Give large quantities of water and induce vomiting. Do not make an unconscious person vomit.

Antidotes and special procedures for medical personnel: The drug NAP (*n*-acetyl penicillamine) has been used to treat mercury poisoning, with mixed success.

Note to physician: For severe poisoning BAL [British Anti-Lewisite, dimercaprol, dithiopropanol (C$_3$H$_8$OS$_2$)] has been used to treat toxic symptoms of certain heavy metals poisoning including mercury. Although BAL is reported to have a large margin of safety, caution must be exercised, because toxic effects may be caused by excessive dosage. Most can be prevented by premedication with 1-ephedrine sulfate (CAS: 134-72-5).

Personal Protective Methods: Wear protective gloves and clothing to prevent any reasonable probability of skin contact. Safety equipment suppliers/manufacturers can provide recommendations on the most protective glove/clothing material for your operation. All protective clothing (suits, gloves, footwear, headgear) should be clean, available each day, and put on before work. Contact lenses should not be worn when working with this chemical. Wear chemical goggles and face shield unless full face-piece respiratory protection is worn. Employees should wash immediately with soap when skin is wet or contaminated. Provide emergency showers and eyewash.

Respirator Selection: *Up to 0.1 mg/m^3:* Sa (APF = 10) (any supplied-air respirator). *Up to 0.25 mg/m^3:* Sa:Cf (APF = 25) (any supplied-air respirator operated in a continuous-flow mode). *Up to 0.5 mg/m^3:* SaT: Cf (APF = 50) (any supplied-air respirator that has a tight-fitting face-piece and is operated in a continuous-flow mode) or SCBAF (APF = 50) (any self-contained breathing apparatus with a full face-piece) or SaF (APF = 50) (any supplied-air respirator with a full face-piece). *Up to 2 mg/m³:* SA: PD, PP (any supplied-air respirator operated in a pressure-demand or other positive-pressure mode). *Emergency or planned entry into unknown concentrations or IDLH conditions:* SCBAF: Pd,Pp (APF = 10,000) (any NIOSH/MSHA- or European Standard EN 149-approved self-contained breathing apparatus that has a full face-piece and is operated in a pressure-demand or other positive-pressure mode) or SaF: Pd,Pp: ASCBA (any supplied-air respirator that has a full face-piece and is operated in a pressure-demand or other positive-pressure mode in combination with an auxiliary self-contained breathing apparatus operated in a pressure-demand or other positive-pressure mode). *Escape:* SCBAE (any appropriate escape-type, self-contained breathing apparatus).

Storage: Color Code—Blue: Health Hazard/Poison: Store in a secure poison location. Prior to working with this chemical you should be trained on its proper handling and storage. These compounds should be stored in a refrigerator or a cool, dry place away from oxidizers. A regulated, marked area should be established where this chemical is handled, used, or stored in compliance with OSHA Standard 1910.1045.

Shipping: Mercury compounds n.o.s. require a shipping label of "POISONOUS/TOXIC MATERIALS." They fall in DOT Hazard Class 6.1.

Spill Handling: Evacuate persons not wearing protective equipment from area of spill or leak until cleanup is complete. Remove all ignition sources. Spills should be collected with special mercury vapor suppressants or special vacuums and deposited in sealed containers. Kits specific for cleanup of mercury spills should be available. Ventilate area after cleanup is complete. It may be necessary to contain and dispose of this chemical as a hazardous waste. If material or contaminated runoff enters waterways, notify downstream users of potentially contaminated waters. Contact your local or federal environmental protection agency for specific recommendations. If employees are required to clean up spills, they must be properly trained and equipped. OSHA 1910.120(q) may be applicable.

Fire Extinguishing: Use dry chemical, carbon dioxide, water spray, or alcohol foam extinguishers. Poisonous gases of chlorine and Hg are produced in fire. If material or contaminated runoff enters waterways, notify downstream users of potentially contaminated waters. Notify local health and fire officials and pollution control agencies. From a secure, explosion-proof location, use water spray to cool exposed containers. If cooling streams are ineffective (venting sound increases in volume and pitch, tank discolors, or shows any signs of deforming), withdraw immediately to a secure position. If employees are expected to fight fires, they must be trained and equipped in OSHA 1910.156. The only respirators recommended for firefighting are self-contained breathing apparatuses that have full face-pieces and are operated in a pressure-demand or other positive-pressure mode.

References
National Institute for Occupational Safety and Health. (October 1977). *Information Profiles on Potential Occupational Hazards: Organomercurials.* Rockville, MD, pp. 287–296, 678
US Environmental Protection Agency. (1979). *Mercury: Ambient Water Quality Criteria.* Washington, DC

Mercury thiocyanate   M:0450

Molecular Formula: $C_2HgN_2S_2$
Common Formula: $Hg(CNS)_2$
Synonyms: Bis(thyocyanato)-mercury; Mercuric sulfo cyanate, solid; Mercuric sulfocyanate; Mercuric sulfocyanide; Mercury dithiocyanate; Mercury thiocyanate; Tiocianato mercurico (Spanish)
CAS Registry Number: 592-85-8
RTECS® Number: XL1550000
UN/NA & ERG Number: UN1646/151
EC Number: 209-773-0
Regulatory Authority and Advisory Bodies
Air Pollutant Standard Set. See below, "Permissible Exposure Limits in Air" section.
Clean Air Act: Hazardous Air Pollutants (Title I, Part A, Section 112).
Clean Water Act: 40CFR401.15 Section 307 Toxic Pollutants as mercury and compounds.
RCRA, 40CFR261, Appendix 8 Hazardous Constituents, waste number not listed, as mercury compounds, n.o.s.
EPCRA Section 313: Includes any unique chemical substance that contains mercury as part of that chemical's infrastructure. Form R *de minimis* concentration reporting level: 1.0%; Category D1A.
California Proposition 65 Developmental/Reproductive toxin (mercury and mercury compounds) 7/1/90.
US DOT Regulated Marine Pollutant (49CFR172.101, Appendix B), severe pollutant as mercury-based pesticides, liquid, flammable, toxic, n.o.s.; mercury-based pesticides, liquid, toxic, n.o.s.; mercury-based pesticides, solid, toxic, n.o.s.; mercury compounds, liquid, n.o.s.; mercury compounds, solid, n.o.s.; mercury(I) (mercurous) compounds (pesticides); mercury(II) (mercuric) compounds (pesticides).
Canada, WHMIS, Ingredients Disclosure List Concentration 1.0%.
Rotterdam Convention Annex III [Chemicals Subject to the Prior Informed Consent Procedure (PIC)] (as mercury compounds, including inorganic mercury compounds, alkyl mercury compounds, and alkyloxylalkyl and aryl mercury compounds).
European/International Regulations: not listed in Annex 1.

WGK (German Aquatic Hazard Class): 3—Severe hazard to waters.

Description: Mercury thiocyanate is a white, odorless powder. Molecular weight = 316.79; Freezing/Melting point = about 165°C (decomposes). Slightly soluble in cold water.

Potential Exposure: Mercury thiocyanate is used in photography and fireworks.

Incompatibilities: Heat; expands to many times its original volume and then decomposes at freezing/melting point forming toxic fumes of sulfur oxides, mercury cyanide, and nitrogen oxides. Contact with acid or acid fumes causes release of toxic mercury and cyanide vapors.

Permissible Exposure Limits in Air

As organo mercury compounds

OSHA PEL: 0.01 mg/m³ TWA; 0.04 mg/m³ Ceiling Concentration.

NIOSH REL: 0.01 mg/m³ TWA; 0.03 mg/m³ STEL [skin].

ACGIH TLV®[1]: 0.01 mg/m³ TWA; 0.03 mg/m³ STEL [skin].

Protective Action Criteria (PAC)

TEEL-0: 0.0395 mg/m³

PAC-1: 0.0395 mg/m³

PAC-2: 0.158 mg/m³

PAC-3: 15.8 mg/m³

DFG MAK: 0.01 mg[Hg]/m³ [skin] Danger of skin sensitization; Carcinogen Category 3 [skin] Danger of skin sensitization; Carcinogen Category 3.

NIOSH IDLH: 2 mg Hg/m³.

Permissible Concentration in Water: *To protect freshwater aquatic life:* 0.00057 µg/L as a 24-h average, never to exceed 0.0017 µg/L. *To protect saltwater aquatic life:* 0.025 µg/L as a 24-h average, never to exceed 3.7 µg/L. *To protect human health:* 0.144 µg/L (US EPA) set in 1979–1980.[6] These are the limits for *inorganic* mercury compounds in general.

Determination in Water: Total mercury is determined by flameless atomic absorption. Soluble mercury may be determined by 0.45-µm filtration followed by flameless atomic absorption.

Routes of Entry: Inhalation, ingestion, eye and/or skin contact. Absorbed through the skin.

Harmful Effects and Symptoms

Short Term Exposure: Mercury thiocyanate can affect you when breathed in. High or repeated exposures can cause kidney damage. Mercury poisoning can cause "shakes," irritability, sore gums, memory loss, increased saliva, metallic taste, personality change, and/or brain damage. Skin and eye contact can cause irritation, allergy, and a gray skin color. Heating or contact with acid or acid "fumes" causes release of toxic mercury or cyanide vapors and lung effects. Health effects may occur below recommended exposure levels.

Long Term Exposure: Repeated or prolonged contact with skin may result in allergy, dermatitis, rash. Repeated or prolonged exposure may cause death by hypovolemic shock, nephrotic syndrome, or kidney failure. Related mercury compounds may damage the developing fetus and decrease fertility in males and females.

Points of Attack: Eyes, skin, central nervous system, peripheral nervous system, kidneys.

Medical Surveillance: NIOSH lists the following tests for inorganic mercury: whole blood (chemical/metabolite); whole blood (chemical/metabolite), end-of-shift; end-of-shift at end-of-work-week; biologic tissue/biopsy; nerve conduction studies; neurologic examination/electromyography; thyroid function test/thyroid profile; urine (chemical/metabolite); urine (chemical/metabolite) prior to shift, prior to next shift; urine (chemical/metabolite), sediment; urinalysis (routine). Before first exposure and every 6–12 months after, a complete medical history and examination is strongly recommended with examination of the nervous system including handwriting. Kidney function tests. Urine test for mercury (should be less than 0.02 mg/L). After suspected illness or overexposure, repeat the tests above and get a blood test for mercury. Examination by a qualified allergist.

First Aid: If this chemical gets into the eyes, remove any contact lenses at once and irrigate immediately for at least 15 min, occasionally lifting upper and lower lids. Seek medical attention immediately. If this chemical contacts the skin, remove contaminated clothing and wash immediately with soap and water. Seek medical attention immediately. If this chemical has been inhaled, remove from exposure, begin rescue breathing (using universal precautions, including resuscitation mask) if breathing has stopped and CPR if heart action has stopped. Transfer promptly to a medical facility. When this chemical has been swallowed, get medical attention. Give large quantities of water and induce vomiting. Do not make an unconscious person vomit.

Antidotes and special procedures for medical personnel: The drug NAP (n-acetyl penicillamine) has been used to treat mercury poisoning, with mixed success.

Note to physician: For severe poisoning BAL [British Anti-Lewisite, dimercaprol, dithiopropanol (C₃H₈OS₂)] has been used to treat toxic symptoms of certain heavy metals poisoning including mercury. Although BAL is reported to have a large margin of safety, caution must be exercised, because toxic effects may be caused by excessive dosage. Most can be prevented by premedication with 1-ephedrine sulfate (CAS: 134-72-5).

Personal Protective Methods: Wear protective gloves and clothing to prevent any reasonable probability of skin contact. Safety equipment suppliers/manufacturers can provide recommendations on the most protective glove/clothing material for your operation. All protective clothing (suits, gloves, footwear, headgear) should be clean, available each day, and put on before work. Contact lenses should not be worn when working with this chemical. Wear dust-proof chemical goggles and face shield unless full face-piece respiratory protection is worn. Employees should wash immediately with soap when skin is wet or contaminated. Provide emergency showers and eyewash.

Respirator Selection: *Up to 0.1 mg/m³:* Sa (APF = 10) (any supplied-air respirator). *Up to 0.25 mg/m³:* Sa:Cf (APF = 25) (any supplied-air respirator operated in a continuous-flow mode). *Up to 0.5 mg/m³:* SaT: Cf (APF = 50) (any supplied-air respirator that has a tight-fitting face-piece and is operated in a continuous-flow mode) or SCBAF (APF = 50) (any self-contained breathing apparatus with a full face-piece) or SaF (APF = 50) (any supplied-air respirator with a full face-piece). *Up to 2 mg/m³:* SA: PD, PP (any supplied-air respirator operated in a pressure-demand or other positive-pressure mode). *Emergency or planned entry into unknown concentrations or IDLH conditions:* SCBAF: Pd,Pp (APF = 10,000) (any NIOSH/MSHA- or European Standard EN 149-approved self-contained breathing apparatus that has a full face-piece and is operated in a pressure-demand or other positive-pressure mode) or SaF: Pd,Pp: ASCBA (APF = 10,000) (any supplied-air respirator that has a full face-piece and is operated in a pressure-demand or other positive-pressure mode in combination with an auxiliary self-contained breathing apparatus operated in a pressure-demand or other positive-pressure mode). *Escape:* SCBAE (any appropriate escape-type, self-contained breathing apparatus).

Storage: Color Code—Blue: Health Hazard/Poison: Store in a secure poison location. Prior to working with this chemical you should be trained on its proper handling and storage. Store in tightly closed containers in a cool, well-ventilated area away from light, heat, and acids, including fumes.

Shipping: This compound requires a shipping label of "POISONOUS/TOXIC MATERIALS." It falls in Hazard Class 6.1 and Packing Group II.

Spill Handling: Restrict persons not wearing protective equipment from area of spill until cleanup is complete. Spills should be collected with special mercury vapor suppressants or special vacuums. Kits specific for cleanup of mercury spills should be available. Ventilate area after cleanup is complete. It may be necessary to contain and dispose of this chemical as a hazardous waste. If material or contaminated runoff enters waterways, notify downstream users of potentially contaminated waters. Contact your local or federal environmental protection agency for specific recommendations. If employees are required to clean up spills, they must be properly trained and equipped. OSHA 1910.120(q) may be applicable.

Fire Extinguishing: Mercury thiocyanate may burn but does not readily ignite. Use dry chemical, CO₂, water spray, or foam extinguishers. Poisonous gases, including nitrogen oxides, sulfur oxides, mercury and cyanide, are produced in fire. If material or contaminated runoff enters waterways, notify downstream users of potentially contaminated waters. Notify local health and fire officials and pollution control agencies. From a secure, explosion-proof location, use water spray to cool exposed containers. If cooling streams are ineffective (venting sound increases in volume and pitch, tank discolors, or shows any signs of deforming),

withdraw immediately to a secure position. If employees are expected to fight fires, they must be trained and equipped in OSHA 1910.156. The only respirators recommended for firefighting are self-contained breathing apparatuses that have full face-pieces and are operated in a pressure-demand or other positive-pressure mode.

Reference
New Jersey Department of Health and Senior Services. (February 2000). *Hazardous Substances Fact Sheet: Mercury Thiocyanate.* Trenton, NJ

Mesitylene   M:0460

Molecular Formula: C₉H₁₂
Common Formula: C₆H₃(CH₃)₃
Synonyms: Benzene, 1,3,5-trimethyl-; *sym*-Trimethylbenzene; *symmetrical*-Trimethylbenzene; TMB; 1,3,5-Trimethylbenzene; Trimethyl benzol
CAS Registry Number: 108-67-8
RTECS® Number: OX6825000
UN/NA & ERG Number: UN2325/129
EC Number: 203-604-4 [*Annex I Index No.:* 601-025-00-5]
Regulatory Authority and Advisory Bodies
Air Pollutant Standard Set. See below, "Permissible Exposure Limits in Air" section.
Canada, WHMIS, Ingredients Disclosure List Concentration 0.1%.
Extremely Hazardous Substance (EPA-SARA, Dropped From Listing in 1988).
European/International Regulations: Hazard Symbol: Xi, N; Risk phrases: R10; R37; R51/53; Safety phrases: S2; S61 (see Appendix 4).
WGK (German Aquatic Hazard Class): No value assigned.
Description: Mesitylene is a clear, colorless liquid with a distinctive, aromatic odor. Molecular weight = 120.21; Specific gravity (H₂O:1) = 0.86; Boiling point = 165°C; Freezing/Melting point = −45°C; Vapor pressure = 2 mmHg at 20°C; Flash point = 50°C (cc); Autoignition temperature = 559°C. Hazard Identification (based on NFPA-704 M Rating System): Health 0, Flammability 2, Reactivity 0. Practically insoluble in water; solubility = 0.002%.
Potential Exposure: Compound Description: Mutagen, Human Data; Primary Irritant. Mesitylene is used as raw material in chemical synthesis and as ultraviolet stabilizer; as a paint thinner, solvent, and motor fuel component; as an intermediate in organic chemical manufacture.
Incompatibilities: Forms explosive mixture with air. Strong oxidizers cause a fire and explosion hazard. Violent reaction with nitric acid.
Permissible Exposure Limits in Air
Conversion factor: 1 ppm = 4.92 mg/m³ at 25°C & 1 atm.
OSHA PEL: None.
NIOSH REL: 25 ppm/125 mg/m³ TWA.

ACGIH TLV®[1]: 25 ppm TWA (lists a single CAS number for mixed isomers).

Protective Action Criteria (PAC)*

TEEL-0: 25 ppm

PAC-1: **140** ppm

PAC-2: **360** ppm

PAC-3: **500** ppm

*AEGLs (Acute Emergency Guideline Levels) & ERPGs (Emergency Response Planning Guideline) are in **bold face**.

DFG MAK: 20 ppm/100 mg/m^3; Pregnancy Risk Group C (all isomers of trimethylbenzene).

Denmark: TWA 25 ppm (120 mg/m^3), 1999; Japan: 25 ppm (120 mg/m^3), 1999; Norway: TWA 20 ppm (100 mg/m^3), 1999; Sweden: NGV 25 ppm (120 mg/m^3), KTV 35 ppm (170 mg/m^3), 1999; Switzerland: MAK-W 25 ppm (125 mg/m^3), 1999; the Netherlands: MAC-TGG 100 mg/m^3; Argentina, Bulgaria, Columbia, Jordan, South Korea, New Zealand, Singapore, Vietnam: ACGIH TLV®: TWA 25 ppm.

Determination in Air: Use OSHA Analytical Method PV-2091.

Permissible Concentration in Water: No method available.

Determination in Water: Octanol–water coefficient: Log K_{ow} = 3.41.

Harmful Effects and Symptoms

Short Term Exposure: Irritates the eyes, skin, and respiratory tract. Exposure can cause you to feel dizzy, light-headed, and to pass out. Symptoms of exposure can also include headache, drowsiness, fatigue, dizziness, nausea, a lack of coordination, vomiting, confusion. Acute-lowest toxic concentration for humans is 10 ppm, resulting in central nervous system effects. Liquid deposition in lungs causes chemical pneumonitis. Symptoms of exposure include nervousness, tension, anxiety, asthmatic bronchitis, and skin irritation.

Long Term Exposure: Repeated exposures can cause headaches, tiredness, and a feeling of nervous tension. Can affect the blood cells and the blood's clotting ability, hypochromic anemia. Delayed or chronic health hazard is possible asthmatic bronchitis with coughing and/or shortness of breath. The use of alcoholic beverages enhances the effect. May cause liver damage. The liquid destroys the skin's natural oils, causing drying and cracking.

Points of Attack: Eyes, skin, respiratory system, central nervous system, blood.

Medical Surveillance: Before beginning employment and at regular times after that, the following are recommended: lung function tests. Complete blood count and platelet count. If symptoms develop or overexposure is suspected, the following may be useful: liver function tests.

First Aid: If this chemical gets into the eyes, remove any contact lenses at once and irrigate immediately for at least 15 min, occasionally lifting upper and lower lids. Seek medical attention immediately. If this chemical contacts the skin, remove contaminated clothing and wash immediately with soap and water. Seek medical attention immediately. If this chemical has been inhaled, remove from exposure, begin rescue breathing (using universal precautions, including resuscitation mask) if breathing has stopped and CPR if heart action has stopped. Transfer promptly to a medical facility. When this chemical has been swallowed, get medical attention. Give large quantities of water and induce vomiting. Do not make an unconscious person vomit.

Personal Protective Methods: Wear protective gloves and clothing to prevent any reasonable probability of skin contact. Safety equipment suppliers/manufacturers can provide recommendations on the most protective glove/clothing material for your operation. All protective clothing (suits, gloves, footwear, headgear) should be clean, available each day, and put on before work. Contact lenses should not be worn when working with this chemical. Wear splash-proof chemical goggles and face shield unless full face-piece respiratory protection is worn. Employees should wash immediately with soap when skin is wet or contaminated. Provide emergency showers and eyewash.

Respirator Selection: Where there is potential for exposures *over 25 ppm*, use a NIOSH/MSHA- or European Standard EN149-approved respirator with an organic vapor cartridge/canister. More protection is provided by a full face-piece respirator than by a half-mask respirator, and even greater protection is provided by a powered air-purifying respirator. *Where there is potential for high exposures*, use a NIOSH/MSHA- or European Standard EN149-approved supplied-air respirator with a full face-piece operated in the positive-pressure mode, or with a full face-piece, hood, or helmet in the continuous-flow mode; or use a NIOSH/MSHA- or European Standard EN149-approved self-contained breathing apparatus with a full face-piece operated in pressure-demand or other positive-pressure mode.

Storage: Color Code—Red: Flammability Hazard: Store in a flammable liquid storage area or approved cabinet away from ignition sources and corrosive and reactive materials. Prior to working with this chemical you should be trained on its proper handling and storage. This chemical must be stored to avoid contact with oxidizers (such as perchlorates, peroxides, permanganates, chlorates, and nitrates) and strong oxidizers (such as chlorine, bromine, and fluorine), since violent reactions occur. Store in tightly closed containers in a cool, well-ventilated area away from heat. Sources of ignition, such as smoking and open flames are prohibited where this chemical is used, handled, or stored in a manner that could create a potential fire or explosion hazard. Metal containers involving the transfer of 5 gallons or more of this chemical should be grounded and bonded. Drums must be equipped with self-closing valves, pressure vacuum bungs, and flame arresters. Use only nonsparking tools and equipment, especially when opening and closing containers of this chemical.

Shipping: This compound requires a shipping label of "FLAMMABLE LIQUID." It falls in Hazard Class 3 and Packing Group II.

Spill Handling: Evacuate and restrict persons not wearing protective equipment from area of spill or leak until cleanup is complete. Remove all ignition sources. Ventilate area of spill or leak. Absorb liquids in vermiculite, dry sand, earth, peat, carbon, or a similar material and deposit in sealed containers. Keep this chemical out of a confined space, such as a sewer, because of the possibility of an explosion, unless the sewer is designed to prevent the buildup of explosive concentrations. It may be necessary to contain and dispose of this chemical as a hazardous waste. If material or contaminated runoff enters waterways, notify downstream users of potentially contaminated waters. Contact your local or federal environmental protection agency for specific recommendations. If employees are required to clean up spills, they must be properly trained and equipped. OSHA 1910.120(q) may be applicable.

Fire Extinguishing: This chemical is a flammable liquid. Poisonous gases are produced in fire. *Small fires:* dry chemical, carbon dioxide, water spray, or alcohol foam. *Large fires:* water spray, fog, or alcohol foam. Move container from fire area if you can do so without risk. Spray cooling water on containers that are exposed to flames until well after fire is out. For massive fire in cargo area, use unmanned hose holder or monitor nozzles; if this is impossible, withdraw from area and let fire burn. Withdraw immediately in case of rising sound from venting safety device or any discoloration of tank due to fire. Isolate for one-half mile in all directions if tank car or truck is involved in fire. Vapors are heavier than air and will collect in low areas. Vapors may travel long distances to ignition sources and flashback. Vapors in confined areas may explode when exposed to fire. Containers may explode in fire. Storage containers and parts of containers may rocket great distances, in many directions. If material or contaminated runoff enters waterways, notify downstream users of potentially contaminated waters. Notify local health and fire officials and pollution control agencies. From a secure, explosion-proof location, use water spray to cool exposed containers. If cooling streams are ineffective (venting sound increases in volume and pitch, tank discolors, or shows any signs of deforming), withdraw immediately to a secure position. If employees are expected to fight fires, they must be trained and equipped in OSHA 1910.156. The only respirators recommended for firefighting are self-contained breathing apparatuses that have full face-pieces and are operated in a pressure-demand or other positive-pressure mode.

Disposal Method Suggested: Dissolve or mix the material with a combustible solvent and burn in a chemical incinerator equipped with an afterburner and scrubber. All federal, state, and local environmental regulations must be observed.

References

US Environmental Protection Agency. (October 31, 1985). *Chemical Hazard Information Profile: Mesitylene.* Washington, DC: Chemical Emergency Preparedness Program

New Jersey Department of Health and Senior Services. (May 2003). *Hazardous Substances Fact Sheet: Trimethyl Benzene (mixed isomers).* Trenton, NJ

Mesityl oxide   M:0470

Molecular Formula: $C_6H_{10}O$
Common Formula: $CH_3COCH = C(CH_3)_2$
Synonyms: Isobutenyl methyl ketone; Isopropylideneacetone; Mesityloxid (German); Methyl isobutenyl ketone; 4-Methyl-3-pentene-2-one; 4-Methyl-3-penten-2-on (German); 2-Methyl-2-penten-4-one; 4-Methyl-3-penten-2-one; Oxyde de mesityle (French)
CAS Registry Number: 141-79-7
RTECS® Number: SB4200000
UN/NA & ERG Number: UN1229/129
EC Number: 205-502-5[*Annex I Index No.:* 606-009-00-1]
Regulatory Authority and Advisory Bodies
Air Pollutant Standard Set. See below, "Permissible Exposure Limits in Air" section.
European/International Regulations: Hazard Symbol: Xn; Risk phrases: R10; R20/21/22; Safety phrases: S2; S25 (see Appendix 4).
WGK (German Aquatic Hazard Class): 1—Low hazard to waters.
Description: Mesityl oxide is a clear, pale yellow, or colorless liquid with a strong peppermint odor. The odor threshold is 0.05 ppm. Molecular weight = 98.16; Specific gravity $(H_2O:1) = 0.87$; Boiling point = 130°C; Freezing/Melting point = −46.7°C; Vapor pressure = 9 mmHg at 20°C; Flash point = 30.6°C; Autoignition temperature = 343°C. Explosive limits: LEL = 1.4%; UEL = 7.2%. Hazard Identification (based on NFPA-704 M Rating System): Health 3, Flammability 3, Reactivity 1. Slightly soluble in water; solubility = 3%.[NIOSH]
Potential Exposure: Compound Description: Human Data; Primary Irritant. Mesityl oxide is used as a solvent for cellulose esters and ethers and other resins in lacquers and inks. It is used in paint and varnish removers and as an insect repellent.
Incompatibilities: Forms explosive mixture with air. May be able to form explosive peroxides. May react violently with nitric acid, aliphatic amines, alkanolamines, 2-aminoethanol, ethylene diamine, chlorosulfonic acid, oleum (fuming sulfuric acid). Not compatible with oxidizers, strong acids, strong bases, reducing agents, halogens. Dissolves some forms of plastics, resins, and rubber. Attacks copper.
Permissible Exposure Limits in Air
OSHA PEL: 25 ppm/100 mg/m³ TWA.
NIOSH REL: 10 ppm/40 mg/m³ TWA.
ACGIH TLV®[1]: 15 ppm/60 mg/m³ TWA; 25 ppm/100 mg/m³ STEL.
NIOSH IDLH: 1400 ppm [LEL].

Protective Action Criteria (PAC)
TEEL-0: 25 ppm
PAC-1: 25 ppm
PAC-2: 25 ppm
PAC-3: 1400 ppm
DFG MAK: 5 ppm/20 mg/m^3 TWA; Peak Limitation Category I(2) [skin]; Pregnancy Risk Group D.
Australia: TWA 15 ppm (60 mg/m^3); STEL 25 ppm, 1993; Austria: MAK 25 ppm (100 mg/m^3), 1999; Denmark: TWA 10 ppm (40 mg/m^3), 1999; Finland: TWA 25 ppm (100 mg/m^3); STEL 75 ppm (300 mg/m^3), [skin], 1999; France: VME 15 ppm (60 mg/m^3), 1999; Norway: TWA 10 ppm (40 mg/m^3), 1999; the Netherlands: MAC-TGG 60 mg/m^3, 2003; Poland: MAC (TWA) 20 mg/m^3, MAC (STEL) 100 mg/m^3, 1999; Russia: STEL 1 mg/m^3, [skin], 1993; Switzerland: MAK-W 15 ppm (60 mg/m^3), 1999; Turkey: TWA 25 ppm (100 mg/m^3), 1993; United Kingdom: TWA 15 ppm (61 mg/m^3); STEL 25 ppm, 2000; Argentina, Bulgaria, Columbia, Jordan, South Korea, New Zealand, Singapore, Vietnam: ACGIH TLV®: STEL 25 ppm. Several states have set guidelines or standards for mesityl oxide in ambient air[60] ranging from 0.6 to 1.0 mg/m^3 (North Dakota) to 0.8 mg/m^3 (Connecticut) to 1.0 mg/m^3 (Virginia) to 1.429 mg/m^3 (Nevada).

Determination in Air: Charcoal adsorption, workup with CS_2, analysis by gas chromatography/flame ionization. Use NIOSH Analytical Method #1301 for Ketones (II).[18]

Determination in Water: Octanol–water coefficient: Log $K_{ow} = 1.7$.

Routes of Entry: Inhalation, ingestion, skin and/or eye contact. Absorbed through the skin.

Harmful Effects and Symptoms

Short Term Exposure: Irritates and burns the eyes and skin. May cause permanent eye damage. Irritates the respiratory tract causing coughing, wheezing and shortness of breath. Exposure causes headache, sleepiness, dizziness, loss of coordination. Exposure far above OEL may result in narcosis, unconsciousness, coma.

Long Term Exposure: The liquid destroys the skin's natural oils. May affect the liver, kidneys, and lungs. May cause anemia.

Points of Attack: Eyes, skin, respiratory system, blood, liver, kidneys, lungs, central nervous system.

Medical Surveillance: Consider the points of attack in preplacement and periodic physical examinations. Liver and kidney function tests. Lung function tests. Complete blood count (CBC).

First Aid: If this chemical gets into the eyes, remove any contact lenses at once and irrigate immediately for at least 15 min, occasionally lifting upper and lower lids. Seek medical attention immediately. If this chemical contacts the skin, remove contaminated clothing and wash immediately with soap and water. Seek medical attention immediately. If this chemical has been inhaled, remove from exposure, begin rescue breathing (using universal precautions, including resuscitation mask) if breathing has stopped and CPR if heart action has stopped. Transfer promptly to a medical facility. When this chemical has been swallowed, get medical attention. Give large quantities of water and induce vomiting. Do not make an unconscious person vomit.

Personal Protective Methods: Wear protective gloves and clothing to prevent any reasonable probability of skin contact: **8 h:** Responder™ suits. Also, Viton™/chlorobutyl rubber is among the recommended protective materials. Safety equipment suppliers/manufacturers can provide recommendations on the most protective glove/clothing material for your operation. All protective clothing (suits, gloves, footwear, headgear) should be clean, available each day, and put on before work. Contact lenses should not be worn when working with this chemical. Wear splash-proof chemical goggles and face shield unless full face-piece respiratory protection is worn. Employees should wash immediately with soap when skin is wet or contaminated. Provide emergency showers and eyewash. For more information about engineering controls, see NIOSH Criteria Document 78-173, *Ketones.*

Respirator Selection: NIOSH: *250 ppm:* Sa:Cf (APF = 25) (any supplied-air respirator operated in a continuous-flow mode) or PaprOv (APF = 25) [any powered, air-purifying respirator with organic vapor cartridge(s)]. *500 ppm:* CcrFOv (APF = 50) [any chemical cartridge respirator with a full face-piece and organic vapor cartridge(s)] or GmFOv (APF = 50) [any air-purifying, full-face-piece respirator (gas mask) with a chin-style, front- or back-mounted acid gas canister] or PaprTOv (APF = 50) [any powered, air-purifying respirator with a tight-fitting face-piece and organic vapor cartridge(s)] or SCBAF (APF = 50) (any self-contained breathing apparatus with a full face-piece) or SaF (APF = 50) (any supplied-air respirator with a full face-piece). *1400 ppm:* SaF: Pd,Pp (APF = 2000) (any supplied-air respirator that has a full face-piece and is operated in a pressure-demand or other positive-pressure mode). *Emergency or planned entry into unknown concentrations or IDLH conditions:* SCBAF: Pd,Pp (APF = 10,000) (any self-contained breathing apparatus that has a full face-piece and is operated in a pressure-demand or other positive-pressure mode) or SaF: Pd,Pp: ASCBA (APF = 10,000) (any supplied-air respirator that has a full face-piece and is operated in a pressure-demand or other positive-pressure mode in combination with an auxiliary, self-contained breathing apparatus operated in a pressure-demand or other positive-pressure mode). *Escape:* GmFOv (APF = 50) [any air-purifying, full-face-piece respirator (gas mask) with a chin-style, front- or back-mounted organic vapor canister] or SCBAE (any appropriate escape-type, self-contained breathing apparatus).

Note: Substance causes eye irritation or damage; eye protection needed.

Storage: Color Code—Red: Flammability Hazard: Store in a flammable liquid storage area or approved cabinet away from ignition sources and corrosive and reactive materials.

Prior to working with this chemical you should be trained on its proper handling and storage. Before entering confined space where this chemical may be present, check to make sure that an explosive concentration does not exist. Store in tightly closed containers in a cool, well-ventilated area away from oxidizers, strong acids. See incompatibilities above. Where possible, automatically pump liquid from drums or other storage containers to process containers.

Shipping: This compound requires a shipping label of "FLAMMABLE LIQUID." It falls in Hazard Class 3 and Packing Group II.

Spill Handling: Evacuate and restrict persons not wearing protective equipment from area of spill or leak until cleanup is complete. Remove all ignition sources. Establish forced ventilation to keep levels below explosive limit. Absorb liquids in vermiculite, dry sand, earth, peat, carbon, or a similar material and deposit in sealed containers. Keep this chemical out of a confined space, such as a sewer, because of the possibility of an explosion, unless the sewer is designed to prevent the buildup of explosive concentrations. Oil-skimming equipment and sorbent foams can be applied to slick if done immediately. It may be necessary to contain and dispose of this chemical as a hazardous waste. If material or contaminated runoff enters waterways, notify downstream users of potentially contaminated waters. Contact your local or federal environmental protection agency for specific recommendations. If employees are required to clean up spills, they must be properly trained and equipped. OSHA 1910.120(q) may be applicable.

Fire Extinguishing: This chemical is a flammable liquid. Poisonous gases are produced in fire. Use dry chemical, carbon dioxide, alcohol foam, or polymer foam extinguishers. Water may be ineffective. Vapors are heavier than air and will collect in low areas. Vapors may travel long distances to ignition sources and flashback. Vapors in confined areas may explode when exposed to fire. Containers may explode in fire. Storage containers and parts of containers may rocket great distances, in many directions. If material or contaminated runoff enters waterways, notify downstream users of potentially contaminated waters. Notify local health and fire officials and pollution control agencies. From a secure, explosion-proof location, use water spray to cool exposed containers. If cooling streams are ineffective (venting sound increases in volume and pitch, tank discolors, or shows any signs of deforming), withdraw immediately to a secure position. If employees are expected to fight fires, they must be trained and equipped in OSHA 1910.156. The only respirators recommended for firefighting are self-contained breathing apparatuses that have full face-pieces and are operated in a pressure-demand or other positive-pressure mode.

Disposal Method Suggested: Dissolve or mix the material with a combustible solvent and burn in a chemical incinerator equipped with an afterburner and scrubber. All federal, state, and local environmental regulations must be observed.

References
National Institute for Occupational Safety and Health. (1978). *Criteria for a Recommended Standard: Occupational Exposure to Ketones.* NIOSH Document No. 78-173. Washington, DC
New Jersey Department of Health and Senior Services. (June 1999). *Hazardous Substances Fact Sheet: Mesityl Oxide.* Trenton, NJ

Metaldehyde   M:0480

Molecular Formula: $C_8H_{16}O_4$
Synonyms: Acetaldehyde, tetramer; Antimilace; Ariotox; Cekumeta; Halizan; META; Metacetaldehyde; Metaldehyd (German); Metason; Namekil; Slug-tox; 1,3,5,7-Tetroxocane, 2,4,6,8-tetramethyl-
CAS Registry Number: 108-62-3
RTECS® Number: XF9900000
UN/NA & ERG Number: UN1332/133
EC Number: 203-600-2 [*Annex I Index No.:* 605-005-00-7]
Regulatory Authority and Advisory Bodies
TSCA: 40CFR716.120(d)1 as aldehydes.
European/International Regulations: Hazard Symbol: Xn; Risk phrases: R11; R22; Safety phrases: S2; S13; S16; S25; S46 (see Appendix 4).
WGK (German Aquatic Hazard Class): No value assigned.
Description: Metaldehyde is a white crystalline powder with a mild menthol odor. Molecular weight = 176.24; Boiling point = 112−116°C; Freezing/Melting point = 47°C; Flash point = 36°C. Hazard Identification (based on NFPA-704 M Rating System): Health 3, Flammability 3, Reactivity 1. Insoluble in water.
Potential Exposure: It is used as a poison for slugs and snails, and as a fuel in small heaters.
Incompatibilities: Strong oxidizers.
Permissible Exposure Limits in Air
No standards or TEEL available.
Routes of Entry: Inhalation, ingestion, skin and/or eye contact.
Harmful Effects and Symptoms
Short Term Exposure: Contact can irritate the eyes, skin, and respiratory tract. Exposure can cause nausea, vomiting, diarrhea, abdominal pain, irritability, sleepiness, muscle twitching, convulsions, coma, and death.
Long Term Exposure: May cause kidney and liver damage. May damage the developing fetus.
Points of Attack: Kidneys, liver.
Medical Surveillance: Kidney function tests. Liver function tests.
First Aid: If this chemical gets into the eyes, remove any contact lenses at once and irrigate immediately for at least 15 min, occasionally lifting upper and lower lids. Seek medical attention immediately. If this chemical contacts the skin, remove contaminated clothing and wash immediately with soap and water. Seek medical attention immediately.

If this chemical has been inhaled, remove from exposure, begin rescue breathing (using universal precautions, including resuscitation mask) if breathing has stopped and CPR if heart action has stopped. Transfer promptly to a medical facility. When this chemical has been swallowed, get medical attention. Give large quantities of water and induce vomiting. Do not make an unconscious person vomit.

Personal Protective Methods: Wear protective gloves and clothing to prevent any reasonable probability of skin contact. Safety equipment suppliers/manufacturers can provide recommendations on the most protective glove/clothing material for your operation. All protective clothing (suits, gloves, footwear, headgear) should be clean, available each day, and put on before work. Contact lenses should not be worn when working with this chemical. Wear dust-proof chemical goggles and face shield unless full face-piece respiratory protection is worn. Employees should wash immediately with soap when skin is wet or contaminated. Provide emergency showers and eyewash.

Respirator Selection: Follow regulations in OSHA 29CFR1910.134 or European Standard EN149. Use a NIOSH/MSHA- or European Standard EN149-approved respirator; or use an approved supplied-air respirator with a full face-piece operated in the positive-pressure mode, or with a full face-piece, hood, or helmet in the continuous-flow mode; or use a NIOSH/MSHA- or European Standard EN149-approved self-contained breathing apparatus with a full face-piece operated in pressure-demand or other positive-pressure mode.

Storage: Color Code—Red: Flammability Hazard: Store in a flammable storage area. Prior to working with metaldehyde you should be trained on its proper handling and storage. Store in tightly closed containers in a cool, well-ventilated area away from strong oxidizers. Where possible, automatically transfer material from drums or other storage containers to process containers. Drums must be equipped with self-closing valves, pressure vacuum bungs, and flame arresters. Use only nonsparking tools and equipment, especially when opening and closing containers of this chemical. Wherever this chemical is used, handled, manufactured, or stored, use explosion-proof electrical equipment and fittings.

Shipping: Metaldehyde requires a shipping label of "FLAMMABLE SOLID." It falls in Hazard Class 4.1 and Packing Group III.

Spill Handling: Evacuate persons not wearing protective equipment from area of spill or leak until cleanup is complete. Remove all ignition sources. Use HEPA vacuum or wet method to reduce dust during cleanup. Do not dry sweep. Collect powdered material in the most convenient and safe manner and deposit in sealed containers. Ventilate area after cleanup is complete. It may be necessary to contain and dispose of this chemical as a hazardous waste. If material or contaminated runoff enters waterways, notify downstream users of potentially contaminated waters. Contact your local or federal environmental protection agency for specific recommendations. If employees are required to clean up spills, they must be properly trained and equipped. OSHA 1910.120(q) may be applicable.

Fire Extinguishing: This chemical is a flammable solid. Use dry chemical, carbon dioxide, water spray, or alcohol foam extinguishers. Fire may restart after it has been extinguished. Poisonous gases are produced in fire. If material or contaminated runoff enters waterways, notify downstream users of potentially contaminated waters. Notify local health and fire officials and pollution control agencies. From a secure, explosion-proof location, use water spray to cool exposed containers. If cooling streams are ineffective (venting sound increases in volume and pitch, tank discolors, or shows any signs of deforming), withdraw immediately to a secure position. If employees are expected to fight fires, they must be trained and equipped in OSHA 1910.156. The only respirators recommended for firefighting are self-contained breathing apparatuses that have full face-pieces and are operated in a pressure-demand or other positive-pressure mode.

Reference

New Jersey Department of Health and Senior Services. (June 1999). *Hazardous Substances Fact Sheet: Metaldehyde.* Trenton, NJ

Methacrylic acid   M:0490

Molecular Formula: $C_4H_6O_2$
Common Formula: $CH_2 = C(CH_3)COOH$
Synonyms: Acide methacrylique (French); Acido metacrilico (Spanish); Acido α-metacrilico (Spanish); Acrylic acid, 2-methyl-; Glacial methacrylic acid; Methacrylic acid, inhibited; α-Methyl-acrylic acid; Methacrylsaeure (German); 2-Methylpropenoic acid; 2-Methyl-2-propenoic acid; 2-Propenoic acid, 2-methyl-; Propionic acid, 2-methylene
CAS Registry Number: 79-41-4
RTECS® Number: OZ2975000
UN/NA & ERG Number: UN2531 (stabilized)/153
EC Number: 201-204-4 [*Annex I Index No.:* 607-088-00-5]
Regulatory Authority and Advisory Bodies
Air Pollutant Standard Set. See below, "Permissible Exposure Limits in Air" section.
Canada, WHMIS, Ingredients Disclosure List Concentration 1.0%.
European/International Regulations: Hazard Symbol: C; Risk phrases: R21/22; R35; Safety phrases: S1/2; S26; S36/37/39; S45 (see Appendix 4).
WGK (German Aquatic Hazard Class): 1—Low hazard to waters.

Description: Methacrylic acid is a colorless liquid. Molecular weight = 86.10; Specific gravity (H_2O:1) = 1.02 (liquid); Boiling point = 162.8°C; Freezing/Melting point = 16.1°C; Vapor pressure = 0.7 mmHg at 20°C; Flash point = 77°C (oc); 68°C (cc); Autoignition temperature = 68°C. Explosive limits: LEL = 1.6; UEL = 8.8. Hazard Identification (based on NFPA-704 M Rating System): Health 3, Flammability 2, Reactivity 2. Soluble in water; solubility = 9% at 25°C.

Potential Exposure: Compound Description: Drug, Mutagen; Human Data. Methacrylic acid is used in preparation of methacrylates and carboxylated polymers, in the production of the material or its alkyl esters, as monomers or comonomers for synthetic resins for the production of plastic sheets, moldings, and fibers.

Incompatibilities: Forms explosive mixture with air. A reducing agent; reacts with oxidizers. Incompatible with strong acids, caustics, ammonia, amines, isocyanates, alkylene oxides, epichlorohydrin. Will polymerize readily from heating above 59°F/15°C, or due to the presence of light, oxidizers (peroxides); or in the presence of traces of hydrochloric acid, with fire or explosion hazard. Attacks metals. *Note:* Typically contains 100 ppm of monomethyl ether hydroquinone (150-76-5) as an inhibitor to prevent polymerization.

Permissible Exposure Limits in Air
Conversion factor: 1 ppm = 3.52 mg/m^3 at 25°C & 1 atm.
OSHA PEL: None.
NIOSH REL: 20 ppm/70 mg/m^3 TWA [skin].
ACGIH TLV®[1]: 20 ppm/70 mg/m^3 TWA.
Protective Action Criteria (PAC)*
TEEL-0: 6.7 ppm
PAC-1: **6.7** ppm
PAC-2: **61** ppm
PAC-3: **220** ppm
*AEGLs (Acute Emergency Guideline Levels) & ERPGs (Emergency Response Planning Guideline) are in **bold face**.
DFG MAK: 5 ppm/18 mg/m^3 TWA; Peak Limitation Category I(2) [skin]; Pregnancy Risk Group C.
Australia: TWA 20 ppm (70 mg/m^3), 1993; Austria: MAK 20 ppm (70 mg/m^3), 1999; Belgium: TWA 20 ppm (70 mg/m^3), 1993; Denmark: TWA 20 ppm (70 mg/m^3), 1999; Finland: TWA 20 ppm, 1999; France: VME 20 ppm (70 mg/m^3), 1999; the Netherlands: MAC-TGG 70 mg/m^3, 2003; Russia: STEL 10 mg/m^3, [skin], 1993; Sweden: NGV 20 ppm (70 mg/m^3), KTV 30 ppm (100 mg/m^3), 1999; Switzerland: MAK-W 20 ppm (70 mg/m^3), 1999; United Kingdom: TWA 20 ppm (72 mg/m^3); STEL 40 ppm, 2000; Argentina, Bulgaria, Columbia, Jordan, South Korea, New Zealand, Singapore, Vietnam: ACGIH TLV®: TWA 20 ppm. Several states have set guidelines or standards for methacrylic acid in ambient air[60] ranging from 0.7 mg/m^3 (North Dakota) to 1.2 mg/m^3 (Virginia) to 1.4 mg/m^3 (Connecticut) to 1.67 mg/m^3 (Nevada).

Determination in Air: Use OSHA Analytical Method PV-2005.

Permissible Concentration in Water: Russia[43] set a MAC of 1.0 mg/L in water bodies used for domestic purposes.

Determination in Water: Octanol–water coefficient: Log K_{ow} = <1.0.

Routes of Entry: Inhalation, skin absorption, ingestion, skin and/or eye contact.

Harmful Effects and Symptoms
Short Term Exposure: Methacrylic acid can affect you when breathed in. A corrosive substance. Exposure can irritate the nose and throat. Methacrylic acid is a corrosive chemical and contact can burn the eyes, causing permanent damage. It can irritate and burn the skin. Higher exposures can cause pulmonary edema, a medical emergency that can be delayed for several hours. This can cause death.

Long Term Exposure: High or repeated exposure may damage the kidneys. Methacrylic acid can cause an allergic skin rash.

Points of Attack: Lungs, kidneys, skin.

Medical Surveillance: For those with frequent or potentially high exposure (half the TLV or greater), the following are recommended before beginning work and at regular times after that: lung function tests. If symptoms develop or overexposure is suspected, the following may be useful: kidney and liver function tests. Evaluation by a qualified allergist, including careful exposure history and special testing, may help diagnose skin allergy.

First Aid: If this chemical gets into the eyes, remove any contact lenses at once and irrigate immediately for at least 15 min, occasionally lifting upper and lower lids. Seek medical attention immediately. If this chemical contacts the skin, remove contaminated clothing and wash immediately with soap and water. Seek medical attention immediately. If this chemical has been inhaled, remove from exposure, begin rescue breathing (using universal precautions, including resuscitation mask) if breathing has stopped and CPR if heart action has stopped. Transfer promptly to a medical facility. When this chemical has been swallowed, get medical attention. If victim is conscious, administer water or milk. Do not induce vomiting. Medical observation is recommended for 24–48 h after breathing overexposure, as pulmonary edema may be delayed. As first aid for pulmonary edema, a doctor or authorized paramedic may consider administering a corticosteroid spray.

Personal Protective Methods: Wear protective gloves and clothing to prevent any reasonable probability of skin contact: **8 h**: butyl rubber gloves, suits, boots; Viton™ gloves, suits; 4H™ and Silver Shield™ gloves, Responder™ suits; Trychem 1000™ suits. Safety equipment suppliers/manufacturers can provide recommendations on the most protective glove/clothing material for your operation. All protective clothing (suits, gloves, footwear, headgear) should be clean, available each day, and put on before work. Contact lenses should not be worn when working with this chemical. Wear splash-proof chemical goggles and face shield unless full face-piece respiratory protection is worn. Employees should wash immediately with soap when skin is wet or contaminated. Provide emergency showers and eyewash.

Respirator Selection: Where there is potential for exposures *over 20 ppm*, use a NIOSH/MSHA- or European Standard EN149-approved supplied-air respirator with a full face-piece operated in the positive-pressure mode, or with a full face-piece, hood, or helmet in the continuous-flow mode; or use a NIOSH/MSHA- or European Standard EN149-approved self-contained breathing apparatus with a full face-piece operated in pressure-demand or other positive-pressure mode.

Storage: (1) Color Code—White: Corrosive or Contact Hazard; Store separately in a corrosion-resistant location. (2) Color Code—Blue: Health Hazard/Poison: Store in a secure poison location. (3) Color Code—Yellow Stripe (*strong reducing agent*): Reactivity Hazard; Store separately in a area isolated from flammables, combustibles, or other yellow coded materials. Prior to working with this chemical you should be trained on its proper handling and storage. Before entering confined space where this chemical may be present, check to make sure that an explosive concentration does not exist. Store in tightly closed containers in a cool, well-ventilated area away from oxidizers (such as perchlorates, peroxides, permanganates, chlorates, and nitrates). Methacrylic acid should be stored at temperatures below 15°C. Sources of ignition, such as smoking and open flames are prohibited where methacrylic acid is handled, used, or stored. Wherever methacrylic acid is used, handled, manufactured, or stored, use explosion-proof electrical equipment and fittings.

Shipping: Inhibited methacrylic acid requires a shipping label of "CORROSIVE." It falls in Hazard Class 8 and Packing Group III.

Spill Handling: Evacuate and restrict persons not wearing protective equipment from area of spill or leak until cleanup is complete. Remove all ignition sources. Establish forced ventilation to keep levels below explosive limit. Absorb liquids in vermiculite, dry sand, earth, peat, carbon, or a similar material and deposit in sealed containers. Using caution, neutralize remainder with aqueous sodium carbonate or lime. Then wash away with plenty of water. Keep this chemical out of a confined space, such as a sewer, because of the possibility of an explosion, unless the sewer is designed to prevent the buildup of explosive concentrations. It may be necessary to contain and dispose of this chemical as a hazardous waste. If material or contaminated runoff enters waterways, notify downstream users of potentially contaminated waters. Contact your local or federal environmental protection agency for specific recommendations. If employees are required to clean up spills, they must be properly trained and equipped. OSHA 1910.120(q) may be applicable.

Fire Extinguishing: This chemical is a combustible liquid. Poisonous gases are produced in fire. Use dry chemical, carbon dioxide, or alcohol foam extinguishers. Vapors are heavier than air and will collect in low areas. Vapors may travel long distances to ignition sources and flashback. Vapors in confined areas may explode when exposed to fire. Containers may explode in fire. Storage containers and parts of containers may rocket great distances, in many directions. If material or contaminated runoff enters waterways, notify downstream users of potentially contaminated waters. Notify local health and fire officials and pollution control agencies. From a secure, explosion-proof location, use water spray to cool exposed containers. If cooling streams are ineffective (venting sound increases in volume and pitch, tank discolors, or shows any signs of deforming), withdraw immediately to a secure position. If employees are expected to fight fires, they must be trained and equipped in OSHA 1910.156. The only respirators recommended for firefighting are self-contained breathing apparatuses that have full face-pieces and are operated in a pressure-demand or other positive-pressure mode.

Disposal Method Suggested: Dissolve or mix the material with a combustible solvent and burn in a chemical incinerator equipped with an afterburner and scrubber. All federal, state, and local environmental regulations must be observed.

Reference

New Jersey Department of Health and Senior Services. (August 2004). *Hazardous Substances Fact Sheet: Methacrylic Acid.* Trenton, NJ

Methacrylonitrile   M:0500

Molecular Formula: C_4H_6N
Common Formula: $CH_2 = CH(CH_3)CN$
Synonyms: AI3-52399; 2-Cyano-1-propene; 2-Cyanopropene-1; 2-Cyanopropene; Isopropene cyanide; Isopropenylnitrile; Metacrilonitrilo (Spanish); α-Methacrylonitrile; α-Methylacrylonitrile; Methyl acrylonitrile; 2-Methylacrylonitrile; 2-Methylpropenenitrile; 2-Methyl-2-propenenitrile; NSC 24145; 2-Propenenitrile, 2-methyl; Usafst-40
CAS Registry Number: 126-98-7
RTECS® Number: UD1400000
UN/NA & ERG Number: UN3079 (stabilized)/131
EC Number: 204-817-5 [*Annex I Index No.:* 608-010-00-2]
Regulatory Authority and Advisory Bodies
Department of Homeland Security Screening Threshold Quantity (pounds): *Release hazard* 10,000 (≥1.00% concentration).
Carcinogenicity: NCI: Carcinogenesis Studies (gavage); no evidence: mouse, rat; NTP: Toxicity studies, RPT#TOX-47, October 2000.
Air Pollutant Standard Set. See below, "Permissible Exposure Limits in Air" section.
OSHA 29CFR1910.119, Appendix A. Process Safety List of Highly Hazardous Chemicals, TQ = 250 lb.
Clean Air Act: Accidental Release Prevention/Flammable Substances, (Section 112[r], Table 3), TQ = 10,000 lb (4540 kg).
US EPA Hazardous Waste Number (RCRA No.): U152.
RCRA, 40CFR261, Appendix 8 Hazardous Constituents.
RCRA 40CFR268.48; 61FR15654, Universal Treatment Standards: Wastewater (mg/L), 0.24; Nonwastewater (mg/kg), 84.

RCRA Ground Water Monitoring List. Suggested test method(s) (PQL μg/L): 8015 (5); 8240 (5).
Superfund/EPCRA 40CFR355, Extremely Hazardous Substances: TPQ = 500 lb (227 kg).
Reportable Quantity (RQ): 1000 lb (454 kg).
EPCRA Section 313 Form R *de minimis* concentration reporting level: 1.0%.
Canada, WHMIS, Ingredients Disclosure List Concentration 1.0%.
European/International Regulations: Hazard Symbol: F, T; Risk phrases: R11; R23/24/25; R43; Safety phrases: S1/2; S9; S16; S18; S29; S45 (see Appendix 4).
WGK (German Aquatic Hazard Class): 1—Low hazard to waters.

Description: Methylacrylonitrile is a colorless liquid with an odor like bitter almonds. Odor threshold = 7.0 ppm. It is reported that methacrylonitrile cannot be detected by its smell even at concentrations which are already dangerous for humans. Hence, special attention must be given to ventilation and estimations of the amount of poison present and must be carried out frequently. Molecular weight = 67.10; Specific gravity (H_2O:1) = 0.80; Boiling point = 90.6°C; Freezing/Melting point = −35.8°C; Vapor pressure = 71 mmHg at 25°C; Flash point = 1.1°C (cc). Explosive limits: LEL = 2%; UEL = 6.8%. Hazard Identification (based on NFPA-704 M Rating System): Health 2, Flammability 3, Reactivity 2. Slightly soluble in water; solubility in water = 3%.

Potential Exposure: Compound Description: Tumorigen; Reproductive Effector; Primary Irritant. This material is used as a monomer in the preparation of polymeric coatings and elastomers.

Incompatibilities: Forms explosive mixture with air. Incompatible with aliphatic amines, alkanolamines, strong acids, strong oxidizers, alkali, light. Polymerization may occur due to elevated temperature, visible light, or contact with a concentrated alkali. Violent reaction with oxidizers.
Note: Typically contains 50 ppm of monoethyl ether hydroquinone (662-62-8) as an inhibitor to prevent polymerization.

Permissible Exposure Limits in Air
Conversion factor: 1 ppm = 2.74 mg/m³ at 25°C & 1 atm.
OSHA PEL: None.
NIOSH REL: 1 ppm/3 mg/m³ TWA [skin].
ACGIH TLV®[1]: 1 ppm/2.7 mg/m³ TWA [skin].
Protective Action Criteria (PAC)*
TEEL-0: 1 ppm
PAC-1: **1.0** ppm
PAC-2: **13** ppm
PAC-3: **25** ppm
*AEGLs (Acute Emergency Guideline Levels) & ERPGs (Emergency Response Planning Guideline) are in **bold face**.
Australia: TWA 1 ppm (3 mg/m³), [skin], 1993; Belgium: TWA 1 ppm (2.7 mg/m³), [skin], 1993; Denmark: TWA 1 ppm (3 mg/m³), [skin], 1999; Finland: TWA 1 ppm

(3 mg/m³); STEL 3 ppm (9 mg/m³), [skin], 1999; France: VME 1 ppm (3 mg/m³), [skin], 1999; Norway: TWA 1 ppm (3 mg/m³), 1999; Switzerland: MAK-W 1 ppm (3 mg/m³), [skin], 1993; the Netherlands: MAC-TGG 3 mg/m³, [skin], 2003; Argentina, Bulgaria, Columbia, Jordan, South Korea, New Zealand, Singapore, Vietnam: ACGIH TLV®: TWA 1 ppm [skin]. Several states have set guidelines or standards for methacrylonitrile in ambient air[60] ranging from 0.7 mg/m³ (North Dakota) to 1.2 mg/m³ (Virginia) to 1.4 mg/m³ (Connecticut) to 1.667 mg/m³ (Nevada).
Determination in Air: No measurement method listed.
Determination in Water: Octanol−water coefficient: Log K_{ow} = 0.7.
Routes of Entry: Inhalation, skin absorption, ingestion, skin and/or eye contact.

Harmful Effects and Symptoms
Short Term Exposure: Converted to cyanide in the body! A lacrimator (causes tearing); an insidious poison which causes delayed skin reactions. Very readily absorbed through skin. Highly toxic by all routes of exposure. Signs and symptoms of acute exposure to methylacrylonitrile may include hypertension (high blood pressure) and tachycardia (rapid heart rate), followed by hypotension (low blood pressure) and bradycardia (slow heart rate). Cherry-red mucous membranes and blood, cardiac arrhythmias, and other cardiac abnormalities are common. Cyanosis (blue tint to the skin and mucous membranes) is not a consistent finding. Tachypnea (rapid respiratory rate) may be followed by respiratory depression. Lung hemorrhage and pulmonary edema may also occur. Headache, vertigo (dizziness), agitation, and giddiness may be followed by combative behavior, convulsions, paralysis, protruding eyeballs, dilated and unreactive pupils, and coma. Methylacrylonitrile is irritating to the skin and mucous membranes. Lacrimation (tearing) and a burning sensation of the mouth and throat are common. Excessive salivation, nausea, and vomiting may also occur.
Long Term Exposure: May cause liver damage. May cause nervous system damage, causing weakness in the legs.
Points of Attack: Eyes, skin, liver, central nervous system.
Medical Surveillance: For those with frequent or potentially high exposure (half the TLV or greater, or significant skin contact), the following are recommended before beginning work and at regular times after that: examination of the nervous system. Liver function tests.
First Aid: If this chemical gets into the eyes, remove any contact lenses at once and irrigate immediately for at least 15 min, occasionally lifting upper and lower lids. Seek medical attention immediately. If this chemical contacts the skin, remove contaminated clothing and wash immediately with soap and water. Seek medical attention immediately. If this chemical has been inhaled, remove from exposure, begin rescue breathing (using universal precautions, including resuscitation mask) if breathing has stopped and CPR if heart action has stopped. Transfer promptly to a medical facility. When this chemical has been swallowed, get

medical attention. Give large quantities of water and induce vomiting. Do not make an unconscious person vomit.

Personal Protective Methods: Wear protective gloves and clothing to prevent any reasonable probability of skin contact. Safety equipment suppliers/manufacturers can provide recommendations on the most protective glove/clothing material for your operation. Butyl rubber is among the recommended protective materials. All protective clothing (suits, gloves, footwear, headgear) should be clean, available each day, and put on before work. Contact lenses should not be worn when working with this chemical. Wear splash-proof chemical goggles and face shield unless full face-piece respiratory protection is worn. Employees should wash immediately with soap when skin is wet or contaminated. Provide emergency showers and eyewash. See NIOSH Criteria Document 212 *Nitriles*.

Respirator Selection: Where there is potential for exposures *over 1 ppm*, use a NIOSH/MSHA- or European Standard EN149-approved supplied-air respirator with a full face-piece operated in the positive-pressure mode, or with a full face-piece, hood, or helmet in the continuous-flow mode; or use a NIOSH/MSHA- or European Standard EN149-approved self-contained breathing apparatus with a full face-piece operated in pressure-demand or other positive-pressure mode.

Storage: Color Code—Red: Flammability Hazard: Store in a flammable liquid storage area or approved cabinet away from ignition sources and corrosive and reactive materials. Prior to working with this chemical you should be trained on its proper handling and storage. Before entering confined space where this chemical may be present, check to make sure that an explosive concentration does not exist. Store in tightly closed containers in a cool, well-ventilated area away from heat. Sources of ignition, such as smoking and open flames, are prohibited where methylacrylonitrile is handled, used, or stored. Metal containers involving the transfer of 5 gallons or more of methylacrylonitrile should be grounded and bonded. Drums must be equipped with self-closing valves, pressure vacuum bungs, and flame arresters. Use only nonsparking tools and equipment, especially when opening and closing containers of methylacrylonitrile. Wherever methylacrylonitrile is used, handled, manufactured, or stored, use explosion-proof electrical equipment and fittings.

Shipping: Methacrylonitrile, inhibited, requires a shipping label of "FLAMMABLE LIQUID, POISONOUS/TOXIC MATERIALS." It falls in Hazard Class 3 and Packing Group I.

Spill Handling: Evacuate and restrict persons not wearing protective equipment from area of spill or leak until cleanup is complete. Remove all ignition sources. Establish forced ventilation to keep levels below explosive limit. Absorb liquids in vermiculite, dry sand, earth, peat, carbon, or a similar material and deposit in sealed containers. Oil-skimming equipment and sorbent foams can be applied to slick if done immediately. Keep this chemical out of a confined space, such as a sewer, because of the possibility of an explosion, unless the sewer is designed to prevent the buildup of explosive concentrations. It may be necessary to contain and dispose of this chemical as a hazardous waste. If material or contaminated runoff enters waterways, notify downstream users of potentially contaminated waters. Contact your local or federal environmental protection agency for specific recommendations. If employees are required to clean up spills, they must be properly trained and equipped. OSHA 1910.120(q) may be applicable.

Initial isolation and protective action distances
Distances shown are likely to be affected during the first 30 min after materials are spilled and could increase with time. If more than one tank car, cargo tank, portable tank, or large cylinder involved in the incident is leaking, the protective action distance may need to be increased. You may need to seek emergency information from CHEMTREC at (800) 424-9300 or seek professional environmental engineering assistance from the US EPA Environmental Response Team at (908) 48-8730 (24-h response line).

Small spills (From a small package or a small leak from a large package)
First: Isolate in all directions (feet/meters)   100/30
Then: Protect persons downwind (miles/kilometers)
Day   0.1/0.2
Night   0.1/0.2

Large spills (From a large package or from many small packages)
First: Isolate in all directions (feet/meters)   200/60
Then: Protect persons downwind (miles/kilometers)
Day   0.3/0.5
Night   0.5/0.8

Fire Extinguishing: This is a flammable liquid. Use dry chemical, CO_2, water spray, or alcohol foam extinguishers. Methacrylonitrile evolves flammable concentrations of vapor. Thus, at room temperatures, flammable concentrations are liable to be present. Also, the chemical will explode due to its tendency to polymerize violently. Vapors are heavier than air and will collect in low areas. Vapors may travel long distances to ignition sources and flashback. Vapors in confined areas may explode when exposed to fire. Containers may explode in fire. Storage containers and parts of containers may rocket great distances, in many directions. If material or contaminated runoff enters waterways, notify downstream users of potentially contaminated waters. Notify local health and fire officials and pollution control agencies. From a secure, explosion-proof location, use water spray to cool exposed containers. If cooling streams are ineffective (venting sound increases in volume and pitch, tank discolors, or shows any signs of deforming), withdraw immediately to a secure position. If employees are expected to fight fires, they must be trained and equipped in OSHA 1910.156. The only respirators recommended for firefighting are self-contained breathing apparatuses that have full face-pieces and are operated in a pressure-demand or other positive-pressure mode.

Disposal Method Suggested: Consult with environmental regulatory agencies for guidance on acceptable disposal practices. Generators of waste containing this contaminant (≥100 kg/mo) must conform with EPA regulations governing storage, transportation, treatment, and waste disposal. Add alcoholic NaOH, then oxidize with sodium hypochlorite. After reaction, flush to sewer with water.[24]

References
US Environmental Protection Agency. (November 30, 1987). *Chemical Hazard Information Profile: Methacrylonitrile.* Washington, DC: Chemical Emergency Preparedness Program
New Jersey Department of Health and Senior Services. (March 2000). *Hazardous Substances Fact Sheet: Methylacrylonitrile.* Trenton, NJ

Methallyl alcohol   M:0510

Molecular Formula: C_4H_8O
Common Formula: $CH_2 = C(CH_3)CH_2OH$
Synonyms: Isopropenyl carbinol; 2-Methyl-2-propen-1-ol; 2-Propen-1-ol, 2-methyl-
CAS Registry Number: 513-42-8
RTECS® Number: UD5250000
UN/NA & ERG Number: UN2614/129
EC Number: 208-161-0
Regulatory Authority and Advisory Bodies
WGK (German Aquatic Hazard Class): No value assigned.
Description: Methallyl alcohol is a colorless liquid with a pungent odor. Molecular weight = 72.12; Boiling point = 114°C; Flash point = 33°C. Hazard Identification (based on NFPA-704 M Rating System): Health 2, Flammability 3, Reactivity 0. Slightly soluble in water.
Potential Exposure: Used as an intermediate in organic synthesis.
Incompatibilities: Forms explosive mixture with air. Incompatible with strong acids, caustics, aliphatic amines, isocyanates, DMSO, oxidizers.
Permissible Exposure Limits in Air
No standards or TEEL available.
Determination in Air: Use NIOSH Analytical Method (IV) #1402, for Alcohols III.
Routes of Entry: Inhalation, ingestion, skin and/or eye contact. Passes through the skin.
Harmful Effects and Symptoms
Short Term Exposure: Methallyl alcohol can affect you when breathed in and by passing through your skin. Exposure to methallyl alcohol can cause irritation to eyes, nose, and throat. Contact can cause skin irritation.
Long Term Exposure: Similar allyl compounds cause liver damage. However, it is not known for certain that this chemical causes the same effects.
Points of Attack: Liver.
Medical Surveillance: Liver function tests.

First Aid: If this chemical gets into the eyes, remove any contact lenses at once and irrigate immediately for at least 15 min, occasionally lifting upper and lower lids. Seek medical attention immediately. If this chemical contacts the skin, remove contaminated clothing and wash immediately with soap and water. Seek medical attention immediately. If this chemical has been inhaled, remove from exposure, begin rescue breathing (using universal precautions, including resuscitation mask) if breathing has stopped and CPR if heart action has stopped. Transfer promptly to a medical facility. When this chemical has been swallowed, get medical attention. Give large quantities of water and induce vomiting. Do not make an unconscious person vomit.
Personal Protective Methods: Wear protective gloves and clothing to prevent any reasonable probability of skin contact. Safety equipment suppliers/manufacturers can provide recommendations on the most protective glove/clothing material for your operation. All protective clothing (suits, gloves, footwear, headgear) should be clean, available each day, and put on before work. Contact lenses should not be worn when working with this chemical. Wear splash-proof chemical goggles and face shield unless full face-piece respiratory protection is worn. Employees should wash immediately with soap when skin is wet or contaminated. Provide emergency showers and eyewash.
Respirator Selection: Where there is potential for exposures to methallyl alcohol, use a NIOSH/MSHA- or European Standard EN149-approved supplied-air respirator with a full face-piece operated in the positive-pressure mode, or with a full face-piece, hood, or helmet in the continuous-flow mode; or use a NIOSH/MSHA- or European Standard EN149-approved self-contained breathing apparatus with a full face-piece operated in pressure-demand or other positive-pressure mode.
Storage: Color Code—Red: Flammability Hazard: Store in a flammable liquid storage area or approved cabinet away from ignition sources and corrosive and reactive materials. Prior to working with this chemical you should be trained on its proper handling and storage. Store in tightly closed containers in a cool, well-ventilated area away from oxidizers and other incompatible materials. Metal containers involving the transfer of this chemical should be grounded and bonded. Drums must be equipped with self-closing valves. Sources of ignition, such as smoking and open flames, are prohibited where this chemical is used, handled, or stored in a manner that could create a potential fire or explosion hazard.
Shipping: This compound requires a shipping label of "FLAMMABLE LIQUID." It falls in Hazard Class 3 and Packing Group II.
Spill Handling: Evacuate and restrict persons not wearing protective equipment from area of spill or leak until cleanup is complete. Remove all ignition sources. Ventilate area of spill or leak. Absorb liquids in vermiculite, dry sand, earth, peat, carbon, or a similar material and deposit in sealed containers. Keep this chemical out of a confined space, such as

a sewer, because of the possibility of an explosion, unless the sewer is designed to prevent the buildup of explosive concentrations. It may be necessary to contain and dispose of this chemical as a hazardous waste. If material or contaminated runoff enters waterways, notify downstream users of potentially contaminated waters. Contact your local or federal environmental protection agency for specific recommendations. If employees are required to clean up spills, they must be properly trained and equipped. OSHA 1910.120(q) may be applicable.

Fire Extinguishing: Methallyl alcohol is a flammable liquid. Water may be ineffective. Alcohol foam is the recommended extinguishing agent. Acrid fumes and smoke are produced in fire. Vapors are heavier than air and will collect in low areas. Vapors may travel long distances to ignition sources and flashback. Vapors in confined areas may explode when exposed to fire. Containers may explode in fire. Storage containers and parts of containers may rocket great distances, in many directions. If material or contaminated runoff enters waterways, notify downstream users of potentially contaminated waters. Notify local health and fire officials and pollution control agencies. From a secure, explosion-proof location, use water spray to cool exposed containers. If cooling streams are ineffective (venting sound increases in volume and pitch, tank discolors, or shows any signs of deforming), withdraw immediately to a secure position. If employees are expected to fight fires, they must be trained and equipped in OSHA 1910.156. The only respirators recommended for firefighting are self-contained breathing apparatuses that have full face-pieces and are operated in a pressure-demand or other positive-pressure mode.

Disposal Method Suggested: Dissolve or mix the material with a combustible solvent and burn in a chemical incinerator equipped with an afterburner and scrubber. All federal, state, and local environmental regulations must be observed.

Reference

New Jersey Department of Health and Senior Services. (April 2000). *Hazardous Substances Fact Sheet: Methallyl Alcohol.* Trenton, NJ

Methamidophos   M:0520

Molecular Formula: $C_2H_8NO_2PS$
Common Formula: $CH_3OP(O)(NH_2)SCH_3$
Synonyms: Acephate-met; Bay 71625; Bayer 71628; Chevron 9006; Chevron ortho 9006; *O,S*-Dimethyl ester of amide of amidothioate; *O,S*-Dimethyl phosphoramidothioate; ENT 27,396; GS-13005; Hamidop; Metamidofos (Spanish); Metamidofos estrella; Monitor; MTD; MTD 600; NSC 190987; Ortho 9006; Pillaron; SRA 5172; Supracide; Tahmabon; Tamaron; Thiophosphorsaeure-*O,S*-dimethylesteramid (German)
CAS Registry Number: 10265-92-6

RTECS® Number: TB4970000
UN/NA & ERG Number: UN2783 (organophosphorus pesticides, solid, toxic)/152
EC Number: 233-606-0 [*Annex I Index No.:* 015-095-00-4]
Regulatory Authority and Advisory Bodies
Superfund/EPCRA 40CFR355, Extremely Hazardous Substances: TPQ = 100/10,000 lb (45.4/4540 kg).
Reportable Quantity (RQ): 100 lb (45.4 kg).
US DOT 49CFR172.101, Inhalation Hazard Chemical as organophosphates.
Rotterdam Convention Annex III [Chemicals Subject to the Prior Informed Consent Procedure (PIC)] [methamidophos (soluble liquid formulations of the substance that exceed 600 g active ingredient/l)].
European/International Regulations: Hazard Symbol: T +, N; Risk phrases: R27/28; R50/53; Safety phrases: S1/2; S28; S36/37; S45; S60; S61.
European/International Regulations: Hazard Symbol: T +, N; Risk phrases R24; R26/28; R50; Safety phrases: S1/2; S28; S36/37; S45; S61 (see Appendix 4).
WGK (German Aquatic Hazard Class): 3—Severe hazard to waters.
Description: Methamidophos is an off-white crystalline solid. Molecular weight = 141.12; Freezing/Melting point = 44.5°C; Vapor pressure = 8×10^{-4} mmHg at 20°C. Hazard Identification (based on NFPA-704 M Rating System): Health 3, Flammability 1, Reactivity 0. Slightly soluble in water.
Potential Exposure: Those involved in the manufacture, formulation, and application of this insecticide on vegetables and cotton.
Incompatibilities: Incompatible with strong acids or alkali. Attacks mild steel and copper-containing alloys (technical grade).
Permissible Exposure Limits in Air
Protective Action Criteria (PAC)
TEEL-0: 3.5 mg/m^3
PAC-1: 10 mg/m^3
PAC-2: 60 mg/m^3
PAC-3: 60 mg/m^3
Determination in Air: OSHA versatile sampler-2; Toluene/Acetone; Gas chromatography/Flame photometric detection for sulfur, nitrogen, or phosphorus; NIOSH Analytical Method (IV) Method #5600, Organophosphorus Pesticides.
Determination in Water: Fish Tox = 165.16992000 ppb (LOW).
Routes of Entry: Inhalation, ingestion, skin contact.
Harmful Effects and Symptoms
Short Term Exposure: Irritates the eyes. Organic phosphorus insecticides are absorbed by the skin, as well as by the respiratory and gastrointestinal tracts. They are cholinesterase inhibitors. Symptoms of exposure include headache, giddiness, blurred vision, nervousness, weakness, nausea, cramps, diarrhea, and discomfort in the chest. Signs include sweating, tearing, salivation, vomiting, cyanosis, convulsions, coma, loss of reflexes, and loss of sphincter control. Acute exposure

to methamidophos may produce the following signs and symptoms: pinpoint pupils, blurred vision, headache, dizziness, muscle spasms, and profound weakness. Vomiting, diarrhea, abdominal pain, seizures, and coma may also occur. The heart rate may decrease following oral exposure or increase following dermal exposure. Chest pain may be noted. Hypotension (low blood pressure) may be noted, although hypertension (high blood pressure) is not uncommon. Respiratory symptoms include dyspnea (shortness of breath), respiratory depression, and respiratory paralysis. Psychosis may occur. This material is highly toxic; LD_{50} = (oral-rat) 7.5 mg/kg. Human Tox = 7.00000 ppb (HIGH).

Long Term Exposure: The substance may have effects on the nervous system resulting in delayed neuropathy. Cholinesterase inhibitor; cumulative effect is possible. This chemical may damage the nervous system with repeated exposure, resulting in convulsions, respiratory failure. May cause liver damage.

Points of Attack: Respiratory system, lungs, central nervous system, cardiovascular system, skin, eyes, plasma and red blood cell cholinesterase.

Medical Surveillance: Before employment and at regular times after that, the following are recommended: plasma and red blood cell cholinesterase levels (tests for the enzyme poisoned by this chemical). If exposure stops, plasma levels return to normal in 1–2 weeks while red blood cell levels may be reduced for 13 months.

When cholinesterase enzyme levels are reduced by 25% or more below preemployment levels, risk of poisoning is increased, even if results are in lower ranges of "normal." Reassignment to work not involving organophosphate or carbamate pesticides is recommended until enzyme levels recover. If symptoms develop or overexposure occurs, repeat the above tests as soon as possible and get an examination of the nervous system. Also consider complete blood count. Consider chest X-ray following acute overexposure. Do not drink any alcoholic beverages before or during use. Alcohol promotes absorption of organic phosphates.

First Aid: If this chemical gets into the eyes, remove any contact lenses at once and irrigate immediately for at least 15 min, occasionally lifting upper and lower lids. Seek medical attention immediately. If this chemical contacts the skin, remove contaminated clothing and wash immediately with soap and water. Speed in removing material from skin is of extreme importance. Shampoo hair promptly if contaminated. Seek medical attention immediately. If this chemical has been inhaled, remove from exposure, begin rescue breathing (using universal precautions, including resuscitation mask) if breathing has stopped and CPR if heart action has stopped. Transfer promptly to a medical facility. When this chemical has been swallowed, get medical attention. Give large quantities of water and induce vomiting. Do not make an unconscious person vomit. Obtain authorization of an antidote or performance of other invasive procedures. The effects may be delayed. Medical observation recommended.

Note to physician: 1,1'-trimethylenebis(4-formylpyridinium bromide)dioxime (a.k.a TMB-4 dibromide and TMV-4) has been used as an antidote for organophosphate poisoning.

Personal Protective Methods: Wear protective gloves and clothing to prevent any reasonable probability of skin contact. Safety equipment suppliers/manufacturers can provide recommendations on the most protective glove/clothing material for your operation. All protective clothing (suits, gloves, footwear, headgear) should be clean, available each day, and put on before work. Contact lenses should not be worn when working with this chemical. Wear dust-proof chemical goggles and face shield unless full face-piece respiratory protection is worn. Employees should wash immediately with soap when skin is wet or contaminated. Provide emergency showers and eyewash. Do not drink any alcoholic beverages before or during use. Alcohol promotes absorption of organic phosphates.

Respirator Selection: Follow regulations in OSHA 29CFR1910.134 or European Standard EN149. Use a NIOSH/MSHA- or European Standard EN149-approved respirator; or use an approved supplied-air respirator with a full face-piece operated in the positive-pressure mode, or with a full face-piece, hood, or helmet in the continuous-flow mode; or use a NIOSH/MSHA- or European Standard EN149-approved self-contained breathing apparatus with a full face-piece operated in pressure-demand or other positive-pressure mode.

Storage: Color Code—Blue: Health Hazard/Poison: Store in a secure poison location. Prior to working with this chemical you should be trained on its proper handling and storage. Store in tightly closed containers in a cool, well-ventilated area

Shipping: This material falls into the class of "Organophosphorus Pesticides, solid, toxic, n.o.s." This compound requires a shipping label of "POISONOUS/TOXIC MATERIALS." It falls in Hazard Class 6.1 and Packing Group I.

Spill Handling: Evacuate persons not wearing protective equipment from area of spill or leak until cleanup is complete. Remove all ignition sources. Collect powdered material in the most convenient and safe manner and deposit in sealed containers. Ventilate area after cleanup is complete. It may be necessary to contain and dispose of this chemical as a hazardous waste. If material or contaminated runoff enters waterways, notify downstream users of potentially contaminated waters. Contact your local or federal environmental protection agency for specific recommendations. If employees are required to clean up spills, they must be properly trained and equipped. OSHA 1910.120(q) may be applicable.

Fire Extinguishing: This material may burn but does not ignite readily. For small fires, use dry chemical, carbon dioxide, water spray, or foam. For large fires, use water spray, fog, or foam. Stay upwind; keep out of low areas. Move containers from fire area if you can do it without risk. Fight fire from maximum distance. Dike fire control water for later disposal; do not scatter the material. Wear positive pressure breathing apparatus and special protective clothing. Poisonous gases, including nitrogen oxides, sulfur oxides, and phosphorus oxides, are produced in fire. If material or contaminated runoff enters waterways, notify downstream users of potentially contaminated waters. Notify local health and fire officials and pollution control agencies. Containers may explode in fire. From a secure, explosion-proof location, use water spray to cool exposed containers. If cooling streams are ineffective (venting sound increases in volume and pitch, tank discolors, or shows any signs of deforming), withdraw immediately to a secure position. If employees are expected to fight fires, they must be trained and equipped in OSHA 1910.156. The only respirators recommended for firefighting are self-contained breathing apparatuses that have full face-pieces and are operated in a pressure-demand or other positive-pressure mode.

Disposal Method Suggested: In accordance with 40CFR 165 recommendations for the disposal of pesticides and pesticide containers. Must be disposed properly by following package label directions or by contacting your local or federal environmental control agency or by contacting your regional EPA office.

Reference

US Environmental Protection Agency. (November 30, 1987). *Chemical Hazard Information Profile: Methamidophos.* Washington, DC: Chemical Emergency Preparedness Program

Methane   M:0530

Molecular Formula: CH_4
Synonyms: Biogas; Fire damp; Marsh gas; Metano (Spanish); Methyl hydride; Natural gas
CAS Registry Number: 74-82-8
RTECS® Number: PA1490000
UN/NA & ERG Number: UN1971 (compressed gas)/115; UN1972 (liquefied gas)/115
EC Number: 200-812-7 [*Annex I Index No.:* 601-001-00-4]
Regulatory Authority and Advisory Bodies
Department of Homeland Security Screening Threshold Quantity (pounds): *Release hazard* 10,000 (≥1.00% concentration).
Clean Air Act: Accidental Release Prevention/Flammable Substances, (Section 112[r], Table 3), TQ = 10,000 lb (4540 kg).
European/International Regulations: Hazard Symbol: F + ; Risk phrases: R12; Safety phrases: S2; S9; S16; S33 (see Appendix 4).
WGK (German Aquatic Hazard Class): Nonwater polluting agent.
Description: Methane is an odorless, colorless gas. Molecular weight = 16.05; Boiling point = −162°C; Freezing/Melting point = −183°C; Autoignition temperature = 537°C. Explosive limits: LEL = 5.0%;

UEL = 15.0%. Hazard Identification (based on NFPA-704 M Rating System): Health 1, Flammability 4, Reactivity 0. Insoluble in water. Natural gas consists primarily of methane (85%) with lesser amounts of ethane (9%), propane (3%), nitrogen (2%), and butane (1%).

Potential Exposure: Methane is used as a fuel and in the manufacture of organic chemicals, acetylene, hydrogen cyanide, and hydrogen. It may also be a cold liquid. Natural gas is used principally as a heating fuel. It is transported as a liquid under pressure. It is also used in the manufacture of various chemicals including acetaldehyde, acetylene, ammonia, carbon black, ethyl alcohol, formaldehyde, hydrocarbon fuels, hydrogenated oils, methyl alcohol, nitric acid, synthesis gas, and vinyl chloride. Helium can be extracted from certain types of natural gas.

Incompatibilities: Forms explosive mixture with air. Reacts violently with bromine pentafluoride, chlorine dioxide, nitrogen trifluoride, liquid oxygen and oxygen difluoride. In general, avoid contact with oxidizers.

Permissible Exposure Limits in Air Any loss of containment of Methane in a confined area can lower the oxygen content and cause suffocation. Oxygen content should be tested to ensure that it is at least 19% by volume.
OSHA PEL: Simple asphyxiant—inert gas and vapor.
ACGIH TLV®[1]: ACGIH TLV®[1]: 1000 ppm TWA *as aliphatic hydrocarbon gas (C_1-C_4).*
Protective Action Criteria (PAC)
TEEL-0: 1000 ppm
PAC-1: 3000 ppm
PAC-2: 5000 ppm
PAC-3: 200,000 ppm
Australia: asphyxiant, 1993; Belgium: asphyxiant, 1993; Hungary: asphyxiant, 1993; Switzerland: MAK-W 10,000 ppm (6700 mg/m³), 1999; United Kingdom: asphyxiant, 2000; Argentina, Bulgaria, Columbia, Jordan, South Korea, New Zealand, Singapore, Vietnam: ACGIH TLV®: Simple asphyxiant.
Routes of Entry: Inhalation.
Harmful Effects and Symptoms
Short Term Exposure: High levels can cause suffocation. Symptoms are due to a decrease in the concentration of oxygen available for breathing and include dizziness, difficult breathing, bluish color of the skin, and loss of consciousness. Any contact with liquid can cause freezing burns.
Long Term Exposure: No effects reported.
First Aid: If this chemical gets into the eyes, remove any contact lenses at once and irrigate immediately for at least 15 min, occasionally lifting upper and lower lids. Seek medical attention immediately. If this chemical contacts the skin, remove contaminated clothing and wash immediately with soap and water. Seek medical attention immediately. If this chemical has been inhaled, remove from exposure, begin rescue breathing (using universal precautions, including resuscitation mask) if breathing has stopped and CPR if heart action has stopped. Transfer promptly to a medical

facility. When this chemical has been swallowed, get medical attention. Give large quantities of water and induce vomiting. Do not make an unconscious person vomit. If frostbite has occurred, seek medical attention immediately; do *NOT* rub the affected areas or flush them with water. In order to prevent further tissue damage, do *NOT* attempt to remove frozen clothing from frostbitten areas. If frostbite has *NOT* occurred, immediately and thoroughly wash contaminated skin with soap and water.

Personal Protective Methods: Wear protective gloves and clothing to prevent any reasonable probability of skin contact. Safety equipment suppliers/manufacturers can provide recommendations on the most protective glove/clothing material for your operation. Polyethylene is among the recommended protective materials. All protective clothing (suits, gloves, footwear, headgear) should be clean, available each day, and put on before work. Contact lenses should not be worn when working with this chemical. Wear gas-proof chemical goggles and face shield unless full face-piece respiratory protection is worn. Employees should wash immediately with soap when skin is wet or contaminated. Provide emergency showers and eyewash. Where exposure to cold equipment, vapors, or liquid may occur, employees should be equipped with special clothing designed to prevent freezing of body tissues.

Respirator Selection: Exposure to methane is dangerous because it can replace oxygen and lead to suffocation. Only NIOSH/MSHA- or European Standard EN 149-approved self-contained breathing apparatus with a full face-piece operated in positive-pressure mode should be used in oxygen-deficient environments. Chemical cartridge respirators should not be used where methane exposure occurs. For high exposures use air supplied respirators.

Storage: Color Code—Red Stripe: Flammability Hazard: Store separately from all other flammable materials. Prior to working with this chemical you should be trained on its proper handling and storage. Before entering confined space where this chemical may be present, check to make sure that an explosive concentration does not exist. Methane must be stored to avoid contact with oxidizers (such as oxygen, chlorine, bromine, perchlorates, peroxides, nitrates, and permanganates), since violent reactions occur. Sources of ignition, such as smoking and open flames, are prohibited where methane is handled, used, or stored. Use only nonsparking tools and equipment, especially when opening and closing containers of methane. Wherever methane is used, handled, manufactured, or stored, use explosion-proof electrical equipment and fittings. Procedures for the handling, use, and storage of cylinders should be in compliance with OSHA 1910.101 and 1910.169, as with the recommendations of the Compressed Gas Association.

Shipping: This compound requires a shipping label of "FLAMMABLE GAS." It falls in Hazard Class 2.1.

Spill Handling: Evacuate and restrict persons not wearing protective equipment from area of spill or leak until cleanup is complete. Remove all ignition sources. Establish forced ventilation to keep levels below explosive limit. Stop the flow of gas if it can be done safely. If source of leak is a cylinder and the leak cannot be stopped in place, remove leaking cylinder to a safe place in the open air, and repair leak or allow cylinder to empty. Keep this chemical out of confined space, such as a sewer because of the possibility of explosion, unless the sewer is designed to prevent the buildup of explosive concentrations. It may be necessary to contain and dispose of this chemical as a hazardous waste. Contact your local or federal environmental protection agency for specific recommendations. If employees are required to clean up spills, they must be properly trained and equipped. OSHA 1910.120(q) may be applicable.

Fire Extinguishing: This chemical is a flammable gas. The flame may be invisible. Do not extinguish the fire unless the flow of gas can be stopped and any remaining gas is out of the line. Incomplete combustion of natural gas may produce carbon monoxide. Use water spray to disperse vapors. *Small fires:* use dry chemical or carbon dioxide extinguishers. *Large fires:* use water spray, fog, or foam. Specially trained personnel may use fog lines to cool exposures and let the fire burn itself out. Vapors are heavier than air and will collect in low areas. Vapors may travel long distances to ignition sources and flashback. Vapors in confined areas may explode when exposed to fire. Containers may explode in fire. Storage containers and parts of containers may rocket great distances, in many directions. If material or contaminated runoff enters waterways, notify downstream users of potentially contaminated waters. Notify local health and fire officials and pollution control agencies. From a secure, explosion-proof location, use water spray to cool exposed containers. If cooling streams are ineffective (venting sound increases in volume and pitch, tank discolors, or shows any signs of deforming), withdraw immediately to a secure position. If cylinders are exposed to excessive heat from fire or flame contact, withdraw immediately to a secure location. If employees are expected to fight fires, they must be trained and equipped in OSHA 1910.156. The only respirators recommended for firefighting are self-contained breathing apparatuses that have full face-pieces and are operated in a pressure-demand or other positive-pressure mode.

Disposal Method Suggested: Incineration (flaring).

References

New York State Department of Health. (March 1986). *Chemical Fact Sheet: Methane.* Albany, Bureau of Toxic Substance Assessment

New Jersey Department of Health and Senior Services. (June 2003). *Hazardous Substances Fact Sheet: Methane.* Trenton, NJ

Methidathion   M:0540

Molecular Formula: $C_6H_{11}N_2O_4PS_3$

Synonyms: Ciba-geigy GS 13005; *S*-(2,3-Dihydro-5-methoxy-2-oxo-1,4,4-thiadiazol-3-methyl); (*O,O*-Dimethyl)-S-[2-

methoxy-1,4,4-thiadiazole-5-(4H)-onyl-(4)-methyl]-dithio-phosphat (German); *O,O*-Dimethyl *S*-[2-methoxy-1,3,4-thiadiazole-5(4H)-on-4-ylmethyl] phosphorodithioate; *O, O*-Dimethyl phosphorodithioate *S*-ester with 4-(mercapto-methyl)-2-methoxy-δ-1,3,4-thiadiazolin-5-one; DMTP (Japan); ENT 27,193; Fisons NC 2964; Geigy 13005; GS-13005; *S*-([5-Methoxy-2-oxo-1,3,4-thiadiazol-3(2H)-yl]methyl) *O,O*-dimethyl phosphordithioate; Metidation (Spanish); Somonil; Surpracide; Ultracide

CAS Registry Number: 950-37-8
RTECS® Number: TE2100000
UN/NA & ERG Number: UN2783 (organophosphorus pesticides, solid, toxic)/152
EC Number: 213-449-4 [*Annex I Index No.:* 015-069-00-2]
Regulatory Authority and Advisory Bodies
Banned or Severely Restricted (Philippines) (UN).[13]
Superfund/EPCRA 40CFR355, Extremely Hazardous Substances: TPQ = 500/10,000 lb (227/4540 kg).
Reportable Quantity (RQ): 500 lb (227 kg).
US DOT Regulated Marine Pollutant (49CFR172.101, Appendix B).
US DOT 49CFR172.101, Inhalation Hazard Chemical as organophosphates.
European/International Regulations: Hazard Symbol: T + , N; Risk phrases: R21; R28; R50/53; Safety phrases: S1/2; S22; S28; S36/37; S45; S60; S61 (see Appendix 4).
WGK (German Aquatic Hazard Class): 3—Severe hazard to waters.
Description: Methidathion is a colorless crystalline solid. Molecular weight = 302.34; Freezing/Melting point = 39–40°C; Vapor pressure = 3.4×10^{-6} mmHg at 20°C. Very slightly soluble in water.
Potential Exposure: Those involved in the manufacture, formulation, and application of this nonsystemic insecticide.
Incompatibilities: None listed.
Permissible Exposure Limits in Air
Protective Action Criteria (PAC)
TEEL-0: 1 mg/m^3
PAC-1: 3 mg/m^3
PAC-2: 20 mg/m^3
PAC-3: 400 mg/m^3
Determination in Water: Fish Tox = 0.15019000 ppb (EXTRA HIGH).
Routes of Entry: Inhalation, ingestion, skin and/or eye contact. Absorbed through the skin.
Harmful Effects and Symptoms
Short Term Exposure: This material is poisonous to humans. Its toxic effects are by action on the nervous system. Organic phosphorus insecticides are absorbed by the skin, as well as by the respiratory and gastrointestinal tracts. They are cholinesterase inhibitors. Symptoms of exposure include headache, giddiness, blurred vision, nervousness, weakness, nausea, cramps, diarrhea, and discomfort in the chest. Signs include sweating, tearing, salivation, vomiting, cyanosis, convulsions, coma, loss of reflexes and loss of

sphincter control. Human volunteers ingesting 0.11 mg/kg/day for 6 weeks had no clinical effects. LD$_{50}$ = (oral-rat) 20 mg/kg (highly toxic). Symptoms are similar to parathion poisoning and may include nausea, vomiting, abdominal cramps, diarrhea, excessive salivation, headache, dizziness, giddiness, weakness, muscle twitching, difficult breathing, sensation of tightness of chest, blurring or dimness of vision, and loss of muscle coordination. Death may occur from failure of the respiratory center, paralysis of the respiratory muscles, intense bronchoconstriction, or all three. Human Tox = 1.05000 ppb (HIGH).
Long Term Exposure: Cholinesterase inhibitor; cumulative effect is possible. This chemical may damage the nervous system with repeated exposure, resulting in convulsions, respiratory failure. May cause liver damage.
Points of Attack: Respiratory system, lungs, central nervous system, cardiovascular system, skin, eyes, plasma and red blood cell cholinesterase.
Medical Surveillance: Before employment and at regular times after that, the following are recommended: plasma and red blood cell cholinesterase levels (tests for the enzyme poisoned by this chemical). If exposure stops, plasma levels return to normal in 1–2 weeks while red blood cell levels may be reduced for 1–3 months.
When cholinesterase enzyme levels are reduced by 25% or more below preemployment levels, risk of poisoning is increased, even if results are in lower ranges of "normal." Reassignment to work not involving organophosphate or carbamate pesticides is recommended until enzyme levels recover. If symptoms develop or overexposure occurs, repeat the above tests as soon as possible and get an examination of the nervous system. Also consider complete blood count. Consider chest X-ray following acute overexposure. Do not drink any alcoholic beverages before or during use. Alcohol promotes absorption of organic phosphates.
First Aid: If this chemical gets into the eyes, remove any contact lenses at once and irrigate immediately for at least 15 min, occasionally lifting upper and lower lids. Seek medical attention immediately. If this chemical contacts the skin, remove contaminated clothing and wash immediately with soap and water. Speed in removing material from skin is of extreme importance. Shampoo hair promptly if contaminated. Seek medical attention immediately. If this chemical has been inhaled, remove from exposure, begin rescue breathing (using universal precautions, including resuscitation mask) if breathing has stopped and CPR if heart action has stopped. Transfer promptly to a medical facility. When this chemical has been swallowed, get medical attention. Give large quantities of water and induce vomiting. Do not make an unconscious person vomit. Keep victim quiet and maintain normal body temperature. Effects may be delayed; keep victim under observation.
Personal Protective Methods: Wear protective gloves and clothing to prevent any reasonable probability of skin contact. Safety equipment suppliers/manufacturers can provide recommendations on the most protective glove/clothing

material for your operation. All protective clothing (suits, gloves, footwear, headgear) should be clean, available each day, and put on before work. Contact lenses should not be worn when working with this chemical. Wear dust-proof chemical goggles and face shield unless full face-piece respiratory protection is worn. Employees should wash immediately with soap when skin is wet or contaminated. Provide emergency showers and eyewash.

Respirator Selection: Follow regulations in OSHA 29CFR1910.134 or European Standard EN149. Use a NIOSH/MSHA- or European Standard EN149-approved respirator; or use an approved supplied-air respirator with a full face-piece operated in the positive-pressure mode, or with a full face-piece, hood, or helmet in the continuous-flow mode; or use a NIOSH/MSHA- or European Standard EN149-approved self-contained breathing apparatus with a full face-piece operated in pressure-demand or other positive-pressure mode.

Storage: Color Code—Blue: Health Hazard/Poison: Store in a secure poison location. Prior to working with this chemical you should be trained on its proper handling and storage. Store in tightly closed containers in a cool, well-ventilated area.

Shipping: This compound requires a shipping label of "POISONOUS/TOXIC MATERIALS." It falls in Hazard Class 6.1 and Packing Group II.

Spill Handling: Evacuate persons not wearing protective equipment from area of spill or leak until cleanup is complete. Remove all ignition sources. Collect powdered material in the most convenient and safe manner and deposit in sealed containers. Ventilate area after cleanup is complete. It may be necessary to contain and dispose of this chemical as a hazardous waste. If material or contaminated runoff enters waterways, notify downstream users of potentially contaminated waters. Contact your local or federal environmental protection agency for specific recommendations. If employees are required to clean up spills, they must be properly trained and equipped. OSHA 1910.120(q) may be applicable.

Fire Extinguishing: This material may burn but does not ignite readily. For small fires, use dry chemical, carbon dioxide, water spray, or foam. For large fires, use water spray, fog, or foam. Stay upwind; keep out of low areas. Move container from fire area if you can do so without risk. Fight fire from maximum distance. Dike fire control water for later disposal; do not scatter the material. Wear positive pressure breathing apparatus and special protective clothing. Use dry chemical, carbon dioxide, water spray, or alcohol foam extinguishers. Poisonous gases, including nitrogen oxides, sulfur oxides, and phosphorus oxides, are produced in fire. If material or contaminated runoff enters waterways, notify downstream users of potentially contaminated waters. Notify local health and fire officials and pollution control agencies. From a secure, explosion-proof location, use water spray to cool exposed containers. If cooling streams are ineffective (venting sound increases in volume and pitch, tank discolors, or shows any signs of deforming), withdraw immediately to a secure position. If employees are expected to fight fires, they must be trained and equipped in OSHA 1910.156. The only respirators recommended for firefighting are self-contained breathing apparatuses that have full face-pieces and are operated in a pressure-demand or other positive-pressure mode.

Disposal Method Suggested: Treat with strong alkali, mix with soil and bury in the case of small quantities.[22] For large quantities, use incineration with effluent gas scrubbing.

References

US Environmental Protection Agency. (November 30, 1987). *Chemical Hazard Information Profile: Methidathion.* Washington, DC: Chemical Emergency Preparedness Program

New Jersey Department of Health and Senior Services. (July 1999). *Hazardous Substances Fact Sheet: Methidathion.* Trenton, NJ

Methiocarb   M:0550

Molecular Formula: $C_{11}H_{15}NO_2S$
Common Formula: $C_6H_2(SCH_3)(CH_3)_2OCONHCH_3$
Synonyms: AI3-25726; B 37344; Bay 37344; Bay 5024; Bay 9026; Bayer 37344; Carbamic acid, methyl-, 3,5-dimethyl-4-(methylthio)phenyl ester; Carbamic acid, *N*-methyl-, 4-(methylthio)-3,5-xylyl ester; Carbamic acid, methyl-, 4-(methylthio)-3,5-xylyl ester; DCR 736; 3,5-Dimethyl-4-methylmercaptophenyl *N*-methylcarbamate; 3,5-Dimethyl-4-(methylthio)phenol methylcarbamate; 3,5-Dimethyl-4-(methylthio)phenyl methylcarbamate; 3,5-Dimethyl-4-methylthiophenyl *N*-methylcarbamate; Draza; Draza G micropellets; ENT 25,726; H 321; Mercaptodimethur; Mesurol; Methiocarbe; Methyl carbamic acid 4-(methylthio)-3,5-xylyl ester; 4-Methylmercapto-3,5-dimethylphenyl *N*-methylcarbamate; 4-Methylmercapto-3,5-xylyl methylcarbamate; 4-Methylthio-3,5-dimethylphenyl methylcarbamate; 4-(Methylthio)-3,5-xylyl *N*-methylcarbamate; 4-(Methylthio)-3,5-xylyl methylcarbamate; Metiocarb (Spanish); Metmercapturon; OMS-93; PBI Slug Gard; Phenol, 3,5-dimethyl-4-(methylthio)-, methylcarbamate; SD 9228
CAS Registry Number: 2032-65-7
RTECS® Number: FC5775000
UN/NA & ERG Number: UN2757/151
EC Number: 217-991-2 [*Annex I Index No.:* 006-023-00-2]
Regulatory Authority and Advisory Bodies
Clean Water Act: Section 311 Hazardous Substances/RQ 40CFR117.3 (same as CERCLA, see below).
US EPA Hazardous Waste Number (RCRA No.): P199.
RCRA, 40CFR261, Appendix 8 Hazardous Constituents.
RCRA 40CFR268.48; 61FR15654, Universal Treatment Standards: Wastewater (mg/L), 0.056; Nonwastewater (mg/kg), 1.4.

Superfund/EPCRA 40CFR355, Extremely Hazardous Substances: TPQ = 500/10,000 lb (227/4540 kg).

Reportable Quantity (RQ): 10 lb (4.54 kg).

EPCRA Section 313 Form R *de minimis* concentration reporting level: 1.0%.

US DOT Regulated Marine Pollutant (49CFR172.101, Appendix B) as mercaptodimethur.

European/International Regulations: Hazard Symbol: T, N; Risk phrases: R25; R50/53; Safety phrases: S1/2; S22; S37; S45; S60; S61 (see Appendix 4).

WGK (German Aquatic Hazard Class): 3—Severe hazard to waters.

Description: Methiocarb is a colorless crystalline powder. Molecular weight = 225.33; Freezing/Melting point = $117-118^{\circ}C$; Vapor pressure = 0.0001 mmHg. Hazard Identification (based on NFPA-704 M Rating System): Health 3, Flammability 1, Reactivity 0. Slightly soluble in water.

Potential Exposure: Those involved in the manufacture, formulation, and application of this nonsystemic acaricide and insecticide.

Permissible Exposure Limits in Air

Protective Action Criteria (PAC)

TEEL-0: 3 mg/m^3

PAC-1: 7.5 mg/m^3

PAC-2: 15 mg/m^3

PAC-3: 15 mg/m^3

Determination in Water: Fish Tox = 0.04597000 ppb (EXTRA HIGH).

Routes of Entry: Inhalation, ingestion, skin and/or eye contact.

Harmful Effects and Symptoms

Short Term Exposure: Contact irritates the skin and eyes. Inhalation will irritate the respiratory tract. As a carbamate insecticide, this compound is a reversible cholinesterase inhibitor and acts on the nervous system. It is classified as very toxic, and the probable oral lethal dose for humans is 50–500 mg/kg or between 1 teaspoon and 1 oz for a 150-lb adult. Symptoms include salivation, slowed heartbeat, spontaneous urination and defecation, labored breathing, headache, blurred vision, tremor, slight paralysis, and muscle twitching. Exposure to carbamate poisoning can also result in nausea, vomiting, diarrhea, abdominal pain, convulsions, coma, and death. Human Tox = 35.00000 ppb (INTERMEDIATE).

Long Term Exposure: The substance may have effects on the nervous system, liver.

Points of Attack: Central nervous system, liver, plasma and red blood cell cholinesterase.

Medical Surveillance: Before employment and at regular times after that, the following are recommended: plasma and red blood cell cholinesterase levels (tests for the enzyme poisoned by this chemical). If exposure stops, plasma levels return to normal in 1–2 weeks while red blood cell levels may be reduced for 1–3 months. When cholinesterase enzyme levels are reduced by 25% or more below preemployment levels, risk of poisoning is increased, even if results are in lower ranges of "normal." Reassignment to work not involving organophosphate or carbamate pesticides is recommended until enzyme levels recover. If symptoms develop or overexposure occurs, repeat the above tests as soon as possible and get an examination of the nervous system. Also consider complete blood count. Consider chest X-ray following acute overexposure. Do not drink any alcoholic beverages before or during use. Alcohol promotes absorption of organic phosphates.

First Aid: If this chemical gets into the eyes, remove any contact lenses at once and irrigate immediately for at least 15 min, occasionally lifting upper and lower lids. Seek medical attention immediately. If this chemical contacts the skin, remove contaminated clothing and wash immediately with soap and water. Speed in removing material from skin is of extreme importance. Shampoo hair promptly if contaminated. Seek medical attention immediately. If this chemical has been inhaled, remove from exposure, begin rescue breathing (using universal precautions, including resuscitation mask) if breathing has stopped and CPR if heart action has stopped. Transfer promptly to a medical facility. When this chemical has been swallowed, get medical attention. Give large quantities of water and induce vomiting. Do not make an unconscious person vomit. Keep victim quiet and maintain normal body temperature. Effects may be delayed; keep victim under observation.

Personal Protective Methods: Wear protective gloves and clothing to prevent any reasonable probability of skin contact. Safety equipment suppliers/manufacturers can provide recommendations on the most protective glove/clothing material for your operation. All protective clothing (suits, gloves, footwear, headgear) should be clean, available each day, and put on before work. Contact lenses should not be worn when working with this chemical. Wear dust-proof chemical goggles and face shield unless full face-piece respiratory protection is worn. Employees should wash immediately with soap when skin is wet or contaminated. Provide emergency showers and eyewash.

Respirator Selection: Follow regulations in OSHA 29CFR1910.134 or European Standard EN149. Use a NIOSH/MSHA- or European Standard EN149-approved respirator; or use an approved supplied-air respirator with a full face-piece operated in the positive-pressure mode, or with a full face-piece, hood, or helmet in the continuous-flow mode; or use a NIOSH/MSHA- or European Standard EN149-approved self-contained breathing apparatus with a full face-piece operated in pressure-demand or other positive-pressure mode.

Storage: Color Code—Blue: Health Hazard/Poison: Store in a secure poison location. Prior to working with this chemical you should be trained on its proper handling and storage. Store in tightly closed containers in a cool, well-ventilated area

Shipping: This material may be classified as a Carbamate pesticide, solid, toxic, n.o.s. This compound requires a shipping label of "POISONOUS/TOXIC MATERIALS." It falls in Hazard Class 6.1 and Packing Group I.

Spill Handling: Evacuate persons not wearing protective equipment from area of spill or leak until cleanup is complete. Remove all ignition sources. Use HEPA vacuum or wet method to reduce dust during cleanup. Do not dry sweep. Collect powdered material in the most convenient and safe manner and deposit in sealed containers. Ventilate area after cleanup is complete. It may be necessary to contain and dispose of this chemical as a hazardous waste. If material or contaminated runoff enters waterways, notify downstream users of potentially contaminated waters. Contact your local or federal environmental protection agency for specific recommendations. If employees are required to clean up spills, they must be properly trained and equipped. OSHA 1910.120(q) may be applicable.

Fire Extinguishing: This chemical does not burn. Poisonous gases, including nitrogen oxides and sulfur oxides, are produced in fire. Extinguish with dry chemical, carbon dioxide, water spray, fog, or foam. Keep unnecessary people away; isolate hazard area and deny entry. Stay upwind; keep out of low areas. Ventilate closed spaces before entering them. Wear positive pressure breathing apparatus and special protective clothing. Remove and isolate contaminated clothing at the site. Move container from fire area if you can do it without risk. Fight fire from maximum distance. Dike fire control water for later disposal; do not scatter the material. If material or contaminated runoff enters waterways, notify downstream users of potentially contaminated waters. Notify local health and fire officials and pollution control agencies. Containers may explode in fire. From a secure, explosion-proof location, use water spray to cool exposed containers. If cooling streams are ineffective (venting sound increases in volume and pitch, tank discolors, or shows any signs of deforming), withdraw immediately to a secure position. If employees are expected to fight fires, they must be trained and equipped in OSHA 1910.156. The only respirators recommended for firefighting are self-contained breathing apparatuses that have full face-pieces and are operated in a pressure-demand or other positive-pressure mode.

Disposal Method Suggested: In accordance with 40CFR 165 recommendations for the disposal of pesticides and pesticide containers. Must be disposed properly by following package label directions or by contacting your local or federal environmental control agency or by contacting your regional EPA office. Consult with environmental regulatory agencies for guidance on acceptable disposal practices. Generators of waste containing this contaminant (≥100 kg/mo) must conform with EPA regulations governing storage, transportation, treatment, and waste disposal. Remove material with contaminated soil and place in impervious containers. May be incinerated in a pesticide incinerator at the specified temperature/dwell-time combination. Any liquids, sludges, or solid residues generated should be disposed of in accordance with all applicable federal, state, and local pollution control requirements. If appropriate incineration facilities are not available, material may be buried in a chemical waste landfill. May be amenable to biological treatment at a municipal sewage treatment plant (Sax/DPIMR).

References

US Environmental Protection Agency. (November 30, 1987). *Chemical Hazard Information Profile: Methiocarb.* Washington, DC: Chemical Emergency Preparedness Program

New Jersey Department of Health and Senior Services. (November 1999). *Hazardous Substances Fact Sheet: Mercaptodimethur.* Trenton, NJ

Methomyl   M:0560

Molecular Formula: $C_5H_{10}N_2O_2S$

Synonyms: Acetimidic acid, thio-N-(methylcarbamoyl) oxy-, methyl ester; Acetimidothioic acid, methyl-N-(methylcarbamoyl) ester; Dupont insecticide 1179; ENT 27,341; Ethanimidothic acid, N-[(methylamino)carbonyl]; Fram fly kill; Improved blue Malrin sugar bait; Improved golden Malrin bait; Insecticide 1,179; Lannate; Lanox 216; Lanox 90; Mesomile; Methomex; Methyl N-([methylamino (carbonyl)oxy]ethanimido)thioate; S-Methyl N-(methylcarbamoyloxy)thioacetimidate; Methyl N-[methyl (carbamoyl) oxy]thioacetimidate; 2-Methylthio-propionaldehyd-o-(methylcarbamoyl)oxim (German); Metomilo (Spanish); Nu-bait II; Nudrin; Rentokil fram fly bait; Rentokill; Ridect; SD 14999; Sorex golden fly bait; 3-Thiabutan-2-one, O-(methylcarbamoyl) oxime; WL 18236

CAS Registry Number: 16752-77-5

RTECS® Number: AK2975000

UN/NA & ERG Number: UN2757/151

EC Number: 240-815-0 [*Annex I Index No.:* 006-045-00-2]

Regulatory Authority and Advisory Bodies

US EPA Gene-Tox Program, Negative: *D. melanogaster* sex-linked lethal; Negative: *In vitro* UDS—human fibroblast; TRP reversion; Negative: *S. cerevisiae*—homozygosis; Inconclusive: *B. subtilis* rec assay; *E. coli* polA without S9; Inconclusive: Histidine reversion—Ames test.

US EPA, FIFRA, 1998 Status of Pesticides: Supported.

Air Pollutant Standard Set. See below, "Permissible Exposure Limits in Air" section.

US EPA Hazardous Waste Number (RCRA No.): P066.

RCRA, 40CFR261, Appendix 8 Hazardous Constituents.

RCRA 40CFR268.48; 61FR15654, Universal Treatment Standards: Wastewater (mg/L), 0.028; Nonwastewater (mg/kg), 0.14.

Safe Drinking Water Act: Priority List (55 FR 1470).

Superfund/EPCRA 40CFR355, Extremely Hazardous Substances: TPQ = 500/10,000 lb (227/4540 kg).

Reportable Quantity (RQ): 100 lb (45.4 kg).

US DOT Regulated Marine Pollutant (49CFR172.101, Appendix B).

European/International Regulations: Hazard Symbol: T+, N; Risk phrases: R28; R50/53; Safety phrases: S1/2; S28; S36/37; S45; S60; S61 (see Appendix 4).

WGK (German Aquatic Hazard Class): No value assigned.

Description: Methomyl is a white crystalline solid with a slight sulfurous odor. A noncombustible solid that may be dissolved in flammable liquids that may alter physical properties listed here. Molecular weight = 162.23; Specific gravity (H$_2$O:1) = 1.29; Freezing/Melting point = 78−79°C; Vapor pressure: 0.00005 mmHg; also 0.0001 (EPA). Hazard Identification (based on NFPA-704 M Rating System): Health 3, Flammability 1, Reactivity 0. Moderately soluble in water; solubility 6% at 25°C.

Potential Exposure: Compound Description: Agricultural Chemical; Mutagen. Methomyl is a broad-spectrum insecticide used as insecticide in many vegetables, field crops, certain fruit crops, and ornamentals.

Incompatibilities: Keep away from strong bases, strong oxidizers. Heat causes decomposition forming toxic and irritating fumes including nitrogen oxides, sulfur oxides, hydrogen cyanide, methylisocyanate.

Permissible Exposure Limits in Air
OSHA PEL: None.
NIOSH REL: 2.5 mg/m^3 TWA.
ACGIH TLV®[1]: 2.5 mg/m^3 TWA not classifiable as a human carcinogen; TLV-BEI$_A$ issued for Acetylcholinesterase-inhibiting pesticides.
Protective Action Criteria (PAC)
TEEL-0: 2.5 mg/m^3
PAC-1: 7.5 mg/m^3
PAC-2: 10 mg/m^3
PAC-3: 200 mg/m^3
Australia: TWA 2.5 mg/m^3, 1993; Belgium: TWA 2.5 mg/m^3, 1993; Denmark: TWA 2.5 mg/m^3, [skin], 1999; France: VME 2.5 mg/m^3, [skin], 1999; Norway: TWA 2.5 mg/m^3, 1999; Switzerland: MAK-W 2.5 mg/m^3, [skin], 1999; United Kingdom: TWA 2.5 mg/m^3, 2000; the Netherlands: MAC-TGG 2.5 mg/m^3, [skin], 2003; Argentina, Bulgaria, Columbia, Jordan, South Korea, New Zealand, Singapore, Vietnam: ACGIH TLV®: not classifiable as a human carcinogen. Several states have set guidelines or standards for methomyl in ambient air[60] ranging from 25 μg/m^3 (North Dakota) to 40 μg/m^3 (Virginia) to 50 μg/m^3 (Connecticut) to 59.5 μg/m^3 (Nevada).

Determination in Air: OSHA versatile sampler-2; Reagent; High-pressure liquid chromatography/Ultraviolet detection; NIOSH Analytical Method (IV) #5601.

Permissible Concentration in Water: The US EPA has calculated a no-observed-adverse-effects-level (NOAEL) of 2.5 mg/kg/day from which a lifetime health advisory of 175 μg/L has been calculated. The state of Maine has set a guideline of 50 μg/L for methomyl in drinking water.[61]

Determination in Water: By high-performance liquid chromatography as described in EPA Health Advisory cited below. Fish Tox = 80.25646000 ppb (INTERMEDIATE). Octanol−water coefficient: Log K_{ow} = 0.63.

Routes of Entry: Inhalation, ingestion, skin and/or eye contact.

Harmful Effects and Symptoms
Short Term Exposure: Cholinesterase inhibitor. Irritates the eyes. May affect the nervous system resulting in respiratory failure and convulsions. Exposure may result in death. Methomyl has high oral toxicity, moderate inhalation toxicity, and low skin toxicity. The probable oral lethal dose for humans is between 7 drops and 1 teaspoon for a 150-lb adult. Death is due to respiratory arrest. Acute exposure to methomyl usually leads to a cholinergic crisis. Signs and symptoms may include increased salivation, lacrimation (tearing), spontaneous defecation, and spontaneous urination. Pinpoint pupils, blurred vision, tremor, muscle twitching, and loss of muscle coordination may occur. Mental confusion, convulsions, and coma may also be noted. Gastrointestinal effects include nausea, vomiting, diarrhea, and abdominal pain. Bradycardia (slow heart rate) occurs frequently. Dyspnea (shortness of breath), pulmonary edema, and respiratory arrest may also occur. Human Tox = 200.00000 ppb (VERY LOW).

Long Term Exposure: Cholinesterase inhibitor; cumulative effect is possible. Methomyl may damage the nervous system with repeated exposure, resulting in convulsions, respiratory failure. May cause liver damage. May cause anemia.

Points of Attack: Eyes, respiratory system, central nervous system, cardiovascular system, liver, kidneys, blood cholinesterase.

Medical Surveillance: Before employment and at regular times after that, the following are recommended: plasma and red blood cell cholinesterase levels (tests for the enzyme poisoned by this chemical). If exposure stops, plasma levels return to normal in 1−2 weeks while red blood cell levels may be reduced for 1−3 months. When cholinesterase enzyme levels are reduced by 25% or more below preemployment levels, risk of poisoning is increased, even if results are in lower ranges of "normal." Reassignment to work not involving carbamate pesticides is recommended until enzyme levels recover. If symptoms develop or overexposure occurs, repeat the above tests as soon as possible and get an examination of the nervous system. Also consider complete blood count. Consider chest X-ray following acute overexposure.

First Aid: If this chemical gets into the eyes, remove any contact lenses at once and irrigate immediately for at least 15 min, occasionally lifting upper and lower lids. Seek medical attention immediately. If this chemical contacts the skin, remove contaminated clothing and wash immediately with soap and water. Seek medical attention immediately. If this chemical has been inhaled, remove from exposure, begin rescue breathing (using universal precautions, including resuscitation mask) if breathing has stopped and CPR if heart action has stopped. Transfer promptly to a medical facility. When this chemical has been swallowed, get medical attention. Give large quantities of water and induce vomiting. Do not make an unconscious person vomit.

Medical observation is recommended for 24–48 h after breathing overexposure, as pulmonary edema may be delayed. As first aid for pulmonary edema, a doctor or authorized paramedic may consider administering a corticosteroid spray.

Personal Protective Methods: Wear protective gloves and clothing to prevent any reasonable probability of skin contact: **8 h:** Tychem 1000™ suits. Safety equipment suppliers/manufacturers can provide recommendations on the most protective glove/clothing material for your operation. All protective clothing (suits, gloves, footwear, headgear) should be clean, available each day, and put on before work. Contact lenses should not be worn when working with this chemical. Wear splash-proof chemical goggles and face shield when working with liquid, unless full face-piece respiratory protection is worn. Wear dust-proof goggles and face shield when working with powders or dust, unless full face-piece respiratory protection is worn. Employees should wash immediately with soap when skin is wet or contaminated. Provide emergency showers and eyewash.

Respirator Selection: Where there is potential for exposures over 2.5 mg/m^3, use a NIOSH/MSHA- or European Standard EN149-approved full face-piece respirator with a pesticide cartridge. Greater protection is provided by a powered air-purifying respirator. *Where there is potential for high exposures,* use a NIOSH/MSHA- or European Standard EN149-approved supplied-air respirator with a full face-piece operated in the positive-pressure mode, or with a full face-piece, hood, or helmet in the continuous-flow mode; or use a NIOSH/MSHA- or European Standard EN149-approved self-contained breathing apparatus with a full face-piece operated in pressure-demand or other positive-pressure mode.

Storage: Color Code—Blue: Health Hazard/Poison: Store in a secure poison location. Prior to working with this chemical you should be trained on its proper handling and storage. Store in tightly closed containers in a cool, well-ventilated area away from strong bases, strong oxidizers (such as chlorine, bromine and fluorine). Do not store in area where temperature is less than 32°F/0°C.

Shipping: This material may be classified under Carbamate pesticides, solid, toxic, n.o.s. This compound requires a shipping label of "POISONOUS/TOXIC MATERIALS." It falls in Hazard Class 6.1 and Packing Group I.

Spill Handling: Evacuate persons not wearing protective equipment from area of spill or leak until cleanup is complete. Remove all ignition sources. Keep unnecessary people away; isolate hazard areas and deny entry. Stay upwind and keep out of low areas. Do not touch spilled material, or handle broken packages without protective equipment. Do not breathe dust, vapors, or the fumes from burning material. Absorb liquids in vermiculite, dry sand, earth, or a similar material and deposit in sealed containers. Use HEPA vacuum or wet method to reduce dust during cleanup. Do not dry sweep. Collect powdered material in the most convenient manner and deposit in sealed containers. Ventilate area after cleanup is complete. It may be necessary to contain and dispose of this chemical as a hazardous waste. If material or contaminated runoff enters waterways, notify downstream users of potentially contaminated waters. Contact your local or federal environmental protection agency for specific recommendations. If employees are required to clean up spills, they must be properly trained and equipped. OSHA 1910.120(q) may be applicable.

Fire Extinguishing: Methomyl does not burn. Use any agent suitable for surrounding fire. Poisonous gases, including nitrogen oxides, sulfur oxides, hydrogen cyanide, methylisocyanate, are produced in fire. If material or contaminated runoff enters waterways, notify downstream users of potentially contaminated waters. Notify local health and fire officials and pollution control agencies. From a secure, explosion-proof location, use water spray to cool exposed containers. If cooling streams are ineffective (venting sound increases in volume and pitch, tank discolors, or shows any signs of deforming), withdraw immediately to a secure position. If employees are expected to fight fires, they must be trained and equipped in OSHA 1910.156. The only respirators recommended for firefighting are self-contained breathing apparatuses that have full face-pieces and are operated in a pressure-demand or other positive-pressure mode.

Disposal Method Suggested: Consult with environmental regulatory agencies for guidance on acceptable disposal practices. Generators of waste containing this contaminant (≥100 kg/mo) must conform with EPA regulations governing storage, transportation, treatment, and waste disposal. Alkaline hydrolysis leads to complete degradation to nontoxic products.[22] May be dissolved in water and sprayed into a furnace with effluent gas scrubbing also. In accordance with 40CFR165, follow recommendations for the disposal of pesticides and pesticide containers. Must be disposed properly by following package label directions or by contacting your local or federal environmental control agency, or by contacting your regional EPA office.

References

US Environmental Protection Agency. (April 30, 1980). *Methomyl, Health and Environmental Effects Profile No. 125.* Washington, DC: Office of Solid Waste

Sax, N. I. (Ed.). (1982). *Dangerous Properties of Industrial Materials Report* 2, No. 5, 79–81

US Environmental Protection Agency. (August 1987). *Health Advisory: Methomyl.* Washington, DC: Office of Drinking Water

US Environmental Protection Agency. (November 30, 1987). *Chemical Hazard Information Profile: Methomyl.* DC: Chemical Emergency Preparedness Program

US Environmental Protection Agency, Special Review and Reregistration Division Office of Pesticide Programs. (1998). *Agency Status of Pesticides in Registration, Reregistration, and Special Review* (Rainbow Report). Washington, DC

New Jersey Department of Health and Senior Services. (September 1999). *Hazardous Substances Fact Sheet: Methomyl*. Trenton, NJ

Methotrexate   M:0570

Molecular Formula: $C_{20}H_{22}N_8O_5$
Synonyms: Amethopterin; 4-Amino-4-deoxy-N^{10}-methyl-pteroyl glutamate; 4-Amino-4-deoxy-N^{10}-methylpteroyl-glutamic; 4-Amino-10-methylfolic acid; 4-Amino-N^{10}-methylpteroylglutamic acid; Antifolan; *N*-Bismethylpteroylglutamic acid; CL-14377; L-(+)-*N*-[*p*-([(2,4-Diamino-6-pteridinyl)methyl]methylamino)-benzoyl] glutamic acid; EMT 25,299; Emtexate; HDMTX; Methopterin; Methotextrate; Methylaminopterin; MTX; NCI-C04671; NSC-740; R 9985
CAS Registry Number: 59-05-2
RTECS® Number: MA1225000
UN/NA & ERG Number: UN2811 (toxic solid, organic, n.o.s.)/154
EC Number: 200-413-8
Regulatory Authority and Advisory Bodies
Carcinogenicity: IARC: Human Inadequate Evidence, animal Inadequate Evidence, *not classifiable as carcinogenic to humans*, Group 3, 1987; NCI: Carcinogenesis Studies (ipr); no evidence: mouse, rat.
California Proposition 65 Developmental/Reproductive toxin 1/1/89.
WGK (German Aquatic Hazard Class): No value assigned.
Description: Methotrexate is an orange-brown crystalline powder. Molecular weight = 454.50; Freezing/Melting point = 185−204°C (decomposes). Insoluble in water.
Potential Exposure: Methotrexate is an anti-cancer drug in tablet or injectable liquid form. It is also an insect chemosterilant.
Incompatibilities: Light and oxidizers.
Permissible Exposure Limits in Air
No standards or TEEL available.
Routes of Entry: Inhalation, skin and/or eyes.
Harmful Effects and Symptoms
Short Term Exposure: Irritates the eyes.
Long Term Exposure: Methotrexate causes mutations (genetic changes). Such chemicals may have a cancer or reproductive risk. Methotrexate is a probable teratogen in humans. There is limited evidence that methotrexate may affect sperm production in males. When taken as a medical drug, methotrexate can cause nausea, vomiting, loss of appetite, weight loss, bloody diarrhea, sores inside the mouth, hair loss, and skin rash. It can also damage the liver and kidneys. Methotrexate can damage bone marrow causing low blood cell count. It is not known if the effects can occur from work-place exposures.
Points of Attack: Blood, liver, kidneys.
Medical Surveillance: Before beginning employment and at regular times after that, for those with frequent or potentially high exposures, the following are recommended: complete blood count (CBC). Liver and kidney function tests.
First Aid: If this chemical gets into the eyes, remove any contact lenses at once and irrigate immediately for at least 15 min, occasionally lifting upper and lower lids. Seek medical attention immediately. If this chemical contacts the skin, remove contaminated clothing and wash immediately with soap and water. Seek medical attention immediately. If this chemical has been inhaled, remove from exposure, begin rescue breathing (using universal precautions, including resuscitation mask) if breathing has stopped and CPR if heart action has stopped. Transfer promptly to a medical facility. When this chemical has been swallowed, get medical attention. Give large quantities of water and induce vomiting. Do not make an unconscious person vomit.
Personal Protective Methods: Wear protective gloves and clothing to prevent any reasonable probability of skin contact. Safety equipment suppliers/manufacturers can provide recommendations on the most protective glove/clothing material for your operation. All protective clothing (suits, gloves, footwear, headgear) should be clean, available each day, and put on before work. Contact lenses should not be worn when working with this chemical. Wear dust-proof goggles when working with powders or dust, unless full face-piece respiratory protection is worn. Wear splash-proof chemical goggles when working with liquid, unless full face-piece respiratory protection is worn. Employees should wash immediately with soap when skin is wet or contaminated. Provide emergency showers and eyewash.
Respirator Selection: Where there is potential for exposures to methotrexate, use a NIOSH/MSHA- or European Standard EN149-approved supplied-air respirator with a full face-piece operated in the positive-pressure mode, or with a full face-piece, hood, or helmet in the continuous-flow mode; or use a NIOSH/MSHA- or European Standard EN149-approved self-contained breathing apparatus with a full face-piece operated in pressure-demand or other positive-pressure mode.
Storage: Color Code—Blue: Health Hazard/Poison: Store in a secure poison location. A regulated, marked area should be established where methotrexate is handled, used, or stored. Store in tightly closed containers in a cool, well-ventilated area away from light and oxidizers (such as perchlorates, peroxides, permanganates, chlorates, and nitrates). A regulated, marked area should be established where this chemical is handled, used, or stored in compliance with OSHA Standard 1910.1045.
Shipping: This material may be classed as Toxic solids, organic, n.o.s. which requires a shipping label of "POISONOUS/TOXIC MATERIALS." It falls in Hazard Class 6.1.
Spill Handling: Evacuate persons not wearing protective equipment from area of spill or leak until cleanup is complete. Remove all ignition sources. Collect powdered material in the most convenient and safe manner and deposit in

sealed containers. Ventilate area after cleanup is complete. It may be necessary to contain and dispose of this chemical as a hazardous waste. If material or contaminated runoff enters waterways, notify downstream users of potentially contaminated waters. Contact your local or federal environmental protection agency for specific recommendations. If employees are required to clean up spills, they must be properly trained and equipped. OSHA 1910.120(q) may be applicable.

Fire Extinguishing: Extinguish fire using an agent suitable for type of surrounding fire. Poisonous gases, including nitrogen oxides, are produced in fire. If material or contaminated runoff enters waterways, notify downstream users of potentially contaminated waters. Notify local health and fire officials and pollution control agencies. From a secure, explosion-proof location, use water spray to cool exposed containers. If cooling streams are ineffective (venting sound increases in volume and pitch, tank discolors, or shows any signs of deforming), withdraw immediately to a secure position. If employees are expected to fight fires, they must be trained and equipped in OSHA 1910.156. The only respirators recommended for firefighting are self-contained breathing apparatuses that have full face-pieces and are operated in a pressure-demand or other positive-pressure mode.

References

Sax, N. I. (Ed.). (1981). *Dangerous Properties of Industrial Materials Report* 1, No. 4, 82–83

New Jersey Department of Health and Senior Services. (April 1986). *Hazardous Substances Fact Sheet: Methotrexate.* Trenton, NJ

Methoxychlor   M:0580

Molecular Formula: C₁₆H₁₅Cl₃O₂
Common Formula: H₃COC₆H₄CH(CCl₃)C₆H₄OCH₃
Synonyms: Benzene, 1,1′-(2,2,2-trichloroethylidene)bis(4-methoxy-); 2,2-Bis(*p*-anisyl)-1,1,1-trichloroethane; 1,1-Bis(*p*-methoxyphenyl)-2,2,2-trichloroethane; 2,2-Bis(*p*-methoxyphenyl)-1,1,1-trichloroethane; Chemform; Dianisyltrichlorethane; 2,2-Di-*p*-anisyl-1,1,1-trichloroethane; Dimethoxy-DDT; *p,p′*-Dimethoxydiphenyltrichloroethane; Dimethoxy DT; Di(*p*-methoxyphenyl)-trichloro methyl methane; *p,p′*-DMDT; DMDT; ENT 1,716; Marlate; Marlate 50; Methoxide; Methoxo; *p,p′*-Methoxychlor; Methoxy DDT; 2,2-(*p*-Methoxyphenyl)-1,1,1-trichloroethane; Metox; Metoxicloro (Spanish); Moxie; NCI-C00497; 1,1,1-Trichloro-2,2-bis(*p*-anisyl)ethane; 1,1,1-Trichloro-2,2-bis(*p*-methoxyphenol)ethanol; 1,1,1-Trichloro-2,2-bis(4-methoxy-phenyl)aethane (German); 1,1,1-Trichloro-2,2-bis(*p*-methoxyphenyl)ethane; 1,1,1-Trichloro-2,2-di(*p*-methoxyphenyl)ethane; 1,1,1-Trichloro-2,2-di(4-methoxyphenyl)ethane; 1,1-(2,2,2-Trichloroethylidene)bis(4-methoxybenzene)

CAS Registry Number: 72-43-5
RTECS® Number: KJ3675000
UN/NA & ERG Number: UN2761/151
EC Number: 200-779-9
Regulatory Authority and Advisory Bodies
Carcinogenicity: NCI: Carcinogenesis Bioassay (feed); no evidence: mouse, rat; IARC: Human No Adequate Data, animal Inadequate Evidence, *not classifiable as carcinogenic to humans*, Group 3, 1987; EPA: Not Classifiable as to human carcinogenicity; NIOSH: Potential occupational carcinogen.
US EPA Gene-Tox Program, Positive/dose response: Cell transform.—BALB/c-3T3; Negative: Carcinogenicity—mouse/rat; SHE—clonal assay; Negative: Cell transform—RLV F344 rat embryo; Negative: Histidine reversion—Ames test; Negative: *D. melanogaster* sex-linked lethal; Negative: *In vitro* UDS—human fibroblast; TRP reversion; Negative: *S. cerevisiae*—homozygosis; Inconclusive: *B. subtilis* rec assay; *E. coli* polA without S9.
Banned or Severely Restricted (in agriculture) (several countries) (UN).[13]
US EPA, FIFRA, 1998 Status of Pesticides: Supported.
Air Pollutant Standard Set. See below, Permissible Exposure Limits in Air section.
Clean Air Act: Hazardous Air Pollutants (Title I, Part A, Section 112).
Clean Water Act: Section 311 Hazardous Substances/RQ 40CFR117.3 (same as CERCLA, see below); Section 313 Water Priority Chemicals (57FR41331, 9/9/92); Section 313 Water Priority Chemicals (57FR41331, 9/9/92).
US EPA Hazardous Waste Number (RCRA No.): U247.
RCRA Toxicity Characteristic (Section 261.24), Maximum. Concentration of Contaminants, regulatory level, 10.0 mg/L.
RCRA, 40CFR261, Appendix 8 Hazardous Constituents.
RCRA 40CFR268.48; 61FR15654, Universal Treatment Standards: Wastewater (mg/L), 0.25; Nonwastewater (mg/kg), 0.18.
RCRA 40CFR264, Appendix 9; TSD Facilities Ground Water Monitoring List. Suggested test method(s) (PQL µg/L): 8080 (2); 8270 (10).
Safe Drinking Water Act: MCL, 0.04 mg/L; MCLG, 0.04 mg/L; Regulated chemical (47 FR 9352); Priority List (55 FR 1470).
Reportable Quantity (RQ): 1 lb (0.45 kg).
EPCRA Section 313 Form R *de minimis* concentration reporting level: 1.0%.
WGK (German Aquatic Hazard Class): No value assigned.
Description: Methoxychlor is a colorless to tan solid with a slight fruity odor. Molecular weight = 345.66; Freezing/Melting point = 89°C. Hazard Identification (based on NFPA-704 M Rating System): Health 2, Flammability 1, Reactivity 0. Insoluble in water.
Potential Exposure: Compound Description: Agricultural Chemical; Tumorigen, Mutagen; Reproductive Effector; Human Data. Methoxychlor was introduced as an

insecticide in 1945. It is a close relative of DDT and has been used as an insecticide of very low mammalian toxicity for home and garden, on domestic animals for fly control, for elm bark-beetle vectors of Dutch elm disease, and for blackfly larvae in streams. Methoxychlor is registered for about 87 crops, alfalfa, nearly all fruits and vegetables, corn, wheat, rice, and other grains, beef and dairy cattle, and swine, goats and sheep, and for agricultural premises and outdoor fogging. Thus, those engaged in manufacture, formulation, and application of the material as well as people in application areas may be exposed.

Incompatibilities: Oxidizers.

Permissible Exposure Limits in Air
OSHA PEL: 15 mg/m^3 (*total dust*) TWA.
NIOSH REL: A potential occupational carcinogen. Limit exposure to lowest feasible concentration. See *NIOSH Pocket Guide*, Appendix A.
ACGIH TLV$^{®[1]}$: 10 mg/m^3 TWA not classifiable as a human carcinogen.
NIOSH IDLH: potential occupational carcinogen 5000 mg/m^3.
Protective Action Criteria (PAC)
TEEL-0: 15 mg/m^3
PAC-1: 30 mg/m^3
PAC-2: 250 mg/m^3
PAC-3: 500 mg/m^3
DFG MAK: 15 mg/m^3, inhalable fraction TWA; Peak Limitation Category II(8); Pregnancy Risk Group D.
Australia: TWA 10 mg/m^3, 1993; Austria: MAK 15 mg/m^3, 1999; Belgium: TWA 10 mg/m^3, 1993; Denmark: TWA 5 mg/m^3, 1999; Finland: TWA 10 mg/m^3; STEL 20 mg/m^3, 1999; France: VME 10 mg/m^3, 1999; Norway: TWA 5 mg/m^3, 1999; the Philippines: TWA 15 mg/m^3, 1993; the Netherlands: MAC-TGG 10 mg/m^3, 2003; Switzerland: MAK-W 10 mg/m^3, 1999; Thailand: TWA 15 mg/m^3, 1993; United Kingdom: TWA 10 mg/m^3, 2000; Argentina, Bulgaria, Columbia, Jordan, South Korea, New Zealand, Singapore, Vietnam: ACGIH TLV$^®$: not classifiable as a human carcinogen. Several states have set guidelines or standards for methoxychlor in ambient air[60] ranging from 23.8 μg/m^3 (Kansas) to 35.07 μg/m^3 (Pennsylvania) to 100 μg/m^3 (North Dakota) to 160 μg/m^3 (Virginia) to 200 μg/m^3 (Connecticut) to 238 μg/m^3 (Nevada).

Determination in Air: Use NIOSH II (4), Method #S-371; OSHA Analytical Method PV-2038.

Permissible Concentration in Water: The WHO[35] has recommended a limit of 30 μg/L of methoxychlor for drinking water. A US recommendation for bottled water for drinking purposes was 100 μg/L. Mexico[35] has set limits of 4.0 μg/L for methoxychlor in coastal waters, 40.0 μg/L in estuaries, and 35 μg/L in recovery waters used for drinking water supply. The EPA[47] has determined a lifetime health advisory of 340 μg/L for methoxychlor. More recently, EPA has proposed a guideline of 400 μg/L for drinking water.[62] Several states have set guidelines or standards for methoxychlor in drinking water.[61] These range from a

standard of 100 μg/L in Arizona and a guideline of 100 μg/L in Maine to a guideline of 340 μg/L for Minnesota.

Determination in Water: By liquid/liquid extraction followed by identification by gas chromatography.[47] Fish Tox: 0.11310000 ppb (EXTRA HIGH). Octanol−water coefficient: Log K_{ow} = >4.4.

Routes of Entry: Inhalation, ingestion, skin and/or eyes. Passes through the skin.

Harmful Effects and Symptoms
Short Term Exposure: Note: For application, methoxychlor is dissolved in organic or petroleum distillate solvents. These solvents may have poisonous effects in addition to those below. *Inhalation:* The results of accidental exposure and animal studies suggest that high levels may cause irritation to nose and throat and may cause headache, nausea, vomiting, staggering walk, drowsiness, convulsions, coma, and death. *Skin:* Absorbed in significant amounts especially when dissolved in organic solvents. Local irritation and numbing of affected area may be experienced. *Eyes:* May cause irritation. *Ingestion:* Symptoms are similar to those listed under inhalation. Ingestion of 5 oz a day for 6 weeks resulted in no observable symptoms. The least amount causing death has been reported as 1 lb. Exposure can cause anxiety, fatigue, nausea, vomiting, dizziness, confusion, weakness, "pins and needles" in extremities, muscle twitching, and tremor. Higher levels can cause convulsions, unconsciousness, and even death. Human Tox = 40.00000 ppb (INTERMEDIATE).

Long Term Exposure: Experiments with animals suggest that exposure to high levels for prolonged periods may cause excess salivation, tremors, seizures, and convulsions. These will generally go away when exposure stops. Methoxychlor has been shown to affect reproduction and to cause cancer at high exposure levels in some laboratory animals. Whether it does so in humans is not known. A potential occupational carcinogen. (NIOSH). May damage the liver and kidneys. Very high exposures may cause anemia.

Points of Attack: Central nervous system, liver, kidneys. Cancer site in animals: liver and ovaries.

Medical Surveillance: Liver and kidney function tests. Complete blood count (CBC).

First Aid: If this chemical gets into the eyes, remove any contact lenses at once and irrigate immediately for at least 15 min, occasionally lifting upper and lower lids. Seek medical attention immediately. If this chemical contacts the skin, remove contaminated clothing and wash immediately with soap and water. Seek medical attention immediately. If this chemical has been inhaled, remove from exposure, begin rescue breathing (using universal precautions, including resuscitation mask) if breathing has stopped and CPR if heart action has stopped. Transfer promptly to a medical facility. When this chemical has been swallowed, get medical attention. Give large quantities of water and induce vomiting. Do not make an unconscious person vomit.

Personal Protective Methods: Wear protective gloves and clothing to prevent any reasonable probability of skin

contact. Safety equipment suppliers/manufacturers can provide recommendations on the most protective glove/clothing material for your operation. All protective clothing (suits, gloves, footwear, headgear) should be clean, available each day, and put on before work. Contact lenses should not be worn when working with this chemical. Wear dust-proof chemical goggles and face shield unless full face-piece respiratory protection is worn. Employees should wash immediately with soap when skin is wet or contaminated. Provide emergency showers and eyewash.

Respirator Selection: NIOSH: *At any concentrations above the NIOSH REL, or where there is no REL, at any detectable concentration:* SCBAF: Pd,Pp (APF = 10,000) (any self-contained breathing apparatus that has a full face-piece and is operated in a pressure-demand or other positive-pressure mode) or SaF: Pd,Pp: ASCBA (any supplied-air respirator that has a full face-piece and is operated in a pressure-demand or other positive-pressure mode in combination with an auxiliary self-contained breathing apparatus operated in a pressure-demand or other positive-pressure mode). *Escape:* GmFOv100 (APF = 50) [any air-purifying, full-face-piece respirator (gas mask) with a chin-style, front- or back-mounted organic vapor canister having an N100, R100, or P100 filter] or SCBAE (any appropriate escape-type, self-contained breathing apparatus).

Storage: Color Code—Blue: Health Hazard/Poison: Store in a secure poison location. Protect containers from damage. Store in cool, dry area away from fire hazard and out of direct sunlight.

Shipping: This material may be classed under Organochlorine Pesticides, solid, toxic, n.o.s. This compound requires a shipping label of "POISONOUS/TOXIC MATERIALS." It falls in Hazard Class 6.1 and Packing Group III.

Spill Handling: Evacuate persons not wearing protective equipment from area of spill or leak until cleanup is complete. Remove all ignition sources. Use HEPA vacuum or wet method to reduce dust during cleanup. Do not dry sweep. Collect powdered material in the most convenient and safe manner and deposit in sealed containers. Ventilate area after cleanup is complete. It may be necessary to contain and dispose of this chemical as a hazardous waste. If material or contaminated runoff enters waterways, notify downstream users of potentially contaminated waters. Contact your local or federal environmental protection agency for specific recommendations. If employees are required to clean up spills, they must be properly trained and equipped. OSHA 1910.120(q) may be applicable.

Fire Extinguishing: This material is a combustible solid but difficult to ignite. Use agent suitable for surrounding fire. Use dry chemical, carbon dioxide, water spray, alcohol foam, or polymer foam extinguishers. Poisonous gases, including hydrogen chloride, are produced in fire. If material or contaminated runoff enters waterways, notify downstream users of potentially contaminated waters. Notify local health and fire officials and pollution control agencies.

From a secure, explosion-proof location, use water spray to cool exposed containers. If cooling streams are ineffective (venting sound increases in volume and pitch, tank discolors, or shows any signs of deforming), withdraw immediately to a secure position. If employees are expected to fight fires, they must be trained and equipped in OSHA 1910.156. The only respirators recommended for firefighting are self-contained breathing apparatuses that have full face-pieces and are operated in a pressure-demand or other positive-pressure mode.

Disposal Method Suggested: Consult with environmental regulatory agencies for guidance on acceptable disposal practices. Generators of waste containing this contaminant ($\geq$100 kg/mo) must conform with EPA regulations governing storage, transportation, treatment, and waste disposal.

References

Sax, N. I. (Ed.). (1987). *Dangerous Properties of Industrial Materials Report* 7, No. 5, 79−87

New York State Department of Health. (March 1986). *Chemical Fact Sheet: Methoxychlor.* Version 2. Albany, NY: Bureau of Toxic Substance Assessment

New Jersey Department of Health and Senior Services. (November 1999). *Hazardous Substances Fact Sheet: Methoxychlor.* Trenton, NJ

US Environmental Protection Agency, Special Review and Reregistration Division Office of Pesticide Programs. (1998). *Agency Status of Pesticides in Registration, Reregistration, and Special Review* (Rainbow Report). Washington, DC

Methoxyethyl acetate   M:0590

Molecular Formula: $C_5H_{10}O_3$
Common Formula: $CH_3COOCH_2CH_2OCH_3$
Synonyms: EGMEA; Ethylene glycol monomethyl ether acetate; Glycol monomethyl ether acetate; 2-Methoxyacetate ethanol; 2-Methoxyethyl acetate; 2-Methoxyethyl acrylate; Methyl Cellosolve® acetate; Methyl glycol acetate
CAS Registry Number: 110-49-6
RTECS® Number: KL5950000
UN/NA & ERG Number: UN1189/129
EC Number: 203-772-9 [*Annex I Index No.:* 607-036-00-1]
Regulatory Authority and Advisory Bodies
Air Pollutant Standard Set. See below, "Permissible Exposure Limits in Air" section.
California Proposition 65 Chemical: Developmental/Reproductive toxin 1/1/93.
Canada, WHMIS, Ingredients Disclosure List Concentration 0.1%.
European/International Regulations: Hazard Symbol: T; Risk phrases: R60; R61; R20/21/22; Safety phrases: S53; S45 (see Appendix 4).
WGK (German Aquatic Hazard Class): 1—Low hazard to waters.

Description: Methoxyethyl acetate is a colorless liquid with a mild, ether-like odor. Odor threshold = 0.33 ppm. Molecular weight = 118.15; Specific gravity (H_2O:1) = 1.01; Boiling point = 145°C; Freezing/Melting point = −65°C; Vapor pressure = 2 mmHg at 20°C; Flash point = 48.9°C (cc); Autoignition temperature = 380°C. Explosive limits: LEL = 1.7%; UEL = 8.2%. Hazard Identification (based on NFPA-704 M Rating System): Health 1, Flammability 2, Reactivity 0. Soluble in water.

Potential Exposure: Compound Description: Mutagen; Reproductive Effector; Hormone, Primary Irritant. Methoxyethyl acetate is used as a solvent for resins, oils, greases, and inks. It is also an ingredient of lacquers, paints, and adhesives.

Incompatibilities: Forms explosive mixture with air. Strong oxidants, strong bases, strong acids, and nitrates. May be able to form explosive peroxides.

Permissible Exposure Limits in Air
Conversion factor: 1 ppm = 4.83 mg/m³ at 25°C & 1 atm.
OSHA PEL: 25 ppm/120 mg/m³ TWA [skin].
NIOSH REL: 0.1 ppm/0.5 mg/m³ TWA [skin].
ACGIH TLV®[1]: 0.1 ppm/0.5 mg/m³ TWA [skin]; BEI issued.
NIOSH IDLH: 200 ppm.
Protective Action Criteria (PAC)
TEEL-0: 0.1 ppm
PAC-1: 0.3 ppm
PAC-2: 20 ppm
PAC-3: 200 ppm
DFG MAK: 1 ppm/4.9 mg/m³ (sum of the concentrations of EGMEA and its acetate in air); Peak Limitation Category II (8) [skin]; Pregnancy Risk Group B.
Australia: TWA 5 ppm (24 mg/m³), [skin], 1993; Austria: MAK 5 ppm (25 mg/m³), [skin], 1999; Belgium: TWA 5 ppm (24 mg/m³), [skin], 1993; Denmark: TWA 5 ppm (24 mg/m³), [skin], 1999; Finland: TWA 25 ppm (120 mg/m³); STEL 40 ppm (180 mg/m³), [skin], 1999; France: VME 5 ppm (24 mg/m³), [skin], 1999; Hungary: TWA 25 mg/m³; STEL 50 mg/m³, [skin], 1993; the Netherlands: MAC-TGG 1.5 mg/m³, [skin], 2003; Norway: TWA 5 ppm (22 mg/m³), 1999; the Philippines: TWA 25 ppm (120 mg/m³), [skin], 1993; Poland: MAC (TWA) 25 mg/m³, MAC (STEL) 100 mg/m³, 1999; Russia: TWA 5 ppm, 1993; Sweden: NGV 5 ppm (25 mg/m³), KTV 10 ppm (50 mg/m³), [skin], 1999; Switzerland: MAK-W 5 ppm (25 mg/m³), KZG-W 10 ppm (50 mg/m³), [skin], 1999; Turkey: TWA 25 ppm (120 mg/m³), [skin], 1993; United Kingdom: TWA 5 ppm (25 mg/m³), 2000; Argentina, Bulgaria, Columbia, Jordan, South Korea, New Zealand, Singapore, Vietnam: ACGIH TLV®: TWA 5 ppm [skin]. Several states have set guidelines or standards for methoxyethyl acetate in ambient air[60] ranging from 0.24 mg/m³ (North Dakota) to 0.4 mg/m³ (Virginia) to 0.48 mg/m³ (Connecticut) to 0.571 mg/m³ (Nevada).

Determination in Air: Use NIOSH Analytical Method (IV) #1451, Methyl cellosolve acetate; OSHA Analytical methods 53, 79.

Determination in Water: Octanol−water coefficient: Log K_{ow} = 0.1.

Routes of Entry: Inhalation, skin absorption, ingestion, skin and/or eye contact.

Harmful Effects and Symptoms
Short Term Exposure: 2-Methoxyethyl acetate can affect you when breathed in and by passing through your skin. Exposure to the vapor can irritate the eyes. High levels of exposure can cause headache, vomiting, dizziness, fatigue, confusion, lightheadedness, and unconsciousness. Higher exposure can cause irritation of the nose, throat, and lungs, causing coughing and/or shortness of breath. Higher exposures can cause pulmonary edema, a medical emergency that can be delayed for several hours. This can cause death. May affect the blood and central nervous system.

Long Term Exposure: Repeated high exposures can damage the brain and nervous system. May affect the blood, liver and kidneys. May cause low blood count (anemia). *In animals:* narcosis; reproductive, teratogenic effects.

Points of Attack: Eyes, respiratory system, kidneys, brain, central nervous system, peripheral nervous system, reproductive system, hematopoietic system.

Medical Surveillance: For those with frequent or potentially high exposure (half the TLV or greater), the following are recommended before beginning work and at regular times after that: lung function tests, liver function tests, kidney function tests. If symptoms develop or overexposure is suspected, the following may be useful: complete blood count (CBC). Examination of the nervous system. Consider chest X-ray following acute overexposure.

First Aid: If this chemical gets into the eyes, remove any contact lenses at once and irrigate immediately for at least 15 min, occasionally lifting upper and lower lids. Seek medical attention immediately. If this chemical contacts the skin, remove contaminated clothing and wash immediately with soap and water. Seek medical attention immediately. If this chemical has been inhaled, remove from exposure, begin rescue breathing (using universal precautions, including resuscitation mask) if breathing has stopped and CPR if heart action has stopped. Transfer promptly to a medical facility. When this chemical has been swallowed, get medical attention. Give large quantities of water and induce vomiting. Do not make an unconscious person vomit. Medical observation is recommended for 24−48 h after breathing overexposure, as pulmonary edema may be delayed. As first aid for pulmonary edema, a doctor or authorized paramedic may consider administering a corticosteroid spray.

Personal Protective Methods: Wear protective gloves and clothing to prevent any reasonable probability of skin contact: **8 h**: butyl rubber gloves, suits, boots; Tychem 1000™ suits. **4 h**: Saranex™ coated suits; 4H™ and Silver Shield™ gloves. Safety equipment suppliers/manufacturers can provide recommendations on the most protective glove/clothing material for your operation. All protective clothing (suits, gloves, footwear, headgear) should be clean, available each day, and put on before work. Contact lenses should not be worn when working with this chemical. Wear splash-proof chemical goggles and face shield unless full face-piece respiratory protection is worn. Employees should wash immediately with soap when skin is wet or contaminated. Provide emergency showers and eyewash.

Respirator Selection: NIOSH: *1 ppm:* Sa (APF = 10) (any supplied-air respirator). *2.5 ppm* Sa:Cf (APF = 25) (any supplied-air respirator operated in a continuous-flow mode). *5 ppm:* SCBAF (APF = 50) (any self-contained breathing apparatus with a full face-piece); saF (APF = 50) (any supplied-air respirator with a full face-piece). *100 ppm:* Sa: Pd, Pp (APF = 1000) (any supplied-air respirator operated in a pressure-demand or other positive-pressure mode). *200 ppm:* Sa: Pd,Pp (APF = 1000) (any supplied-air respirator that has a full face-piece and is operated in a pressure-demand or other positive-pressure mode). *Emergency or planned entry into unknown concentrations or IDLH conditions:* SCBAF: Pd,Pp (APF = 10,000) (any self-contained breathing apparatus that has a full face-piece and is operated in a pressure-demand or other positive-pressure mode) or SaF: Pd,Pp: ASCBA (APF = 10,000) (any supplied-air respirator that has a full face-piece and is operated in a pressure-demand or other positive-pressure mode in combination with an auxiliary self-contained breathing apparatus operated in a pressure-demand or other positive-pressure mode). *Escape:* GmFOv (APF = 50) [any air-purifying, full-face-piece respirator (gas mask) with a chin-style, front- or back-mounted organic vapor canister] or SCBAE (any appropriate escape-type, self-contained breathing apparatus).

Note: Substance reported to cause eye irritation or damage; may require eye protection.

Storage: Color Code—Red: Flammability Hazard: Store in a flammable liquid storage area or approved cabinet away from ignition sources and corrosive and reactive materials. Prior to working with this chemical you should be trained on its proper handling and storage. Before entering confined space where this chemical may be present, check to make sure that an explosive concentration does not exist. 2-Methoxyethyl acetate must be stored to avoid contact with oxidizers (such as perchlorates, peroxides, permanganates, chlorates, and nitrates) and strong caustics, since violent reactions occur. Sources of ignition, such as smoking and open flames are prohibited where 2-methoxyethyl acetate is used, handled, or stored in a manner that could create a potential fire or explosion hazard. Wherever 2-methoxyethyl acetate is used, handled, manufactured, or stored, use explosion-proof electrical equipment and fittings.

Shipping: This compound requires a shipping label of "FLAMMABLE LIQUID." It falls in Hazard Class 3 and Packing Group II.

Spill Handling: Evacuate and restrict persons not wearing protective equipment from area of spill or leak until cleanup is complete. Remove all ignition sources. Establish forced ventilation to keep levels below explosive limit. Absorb liquids in vermiculite, dry sand, earth, peat, carbon, or a similar material and deposit in sealed containers. Keep this chemical out of a confined space, such as a sewer, because of the possibility of an explosion, unless the sewer is designed to prevent the buildup of explosive concentrations. It may be necessary to contain and dispose of this chemical as a hazardous waste. If material or contaminated runoff enters waterways, notify downstream users of potentially contaminated waters. Contact your local or federal environmental protection agency for specific recommendations. If employees are required to clean up spills, they must be properly trained and equipped. OSHA 1910.120(q) may be applicable.

Fire Extinguishing: This chemical is a combustible liquid. Poisonous gases are produced in fire. Use dry chemical, carbon dioxide, alcohol, or polymer foam extinguishers. Vapors are heavier than air and will collect in low areas. Vapors may travel long distances to ignition sources and flashback. Vapors in confined areas may explode when exposed to fire. Containers may explode in fire. Storage containers and parts of containers may rocket great distances, in many directions. If material or contaminated runoff enters waterways, notify downstream users of potentially contaminated waters. Notify local health and fire officials and pollution control agencies. From a secure, explosion-proof location, use water spray to cool exposed containers. If cooling streams are ineffective (venting sound increases in volume and pitch, tank discolors, or shows any signs of deforming), withdraw immediately to a secure position. If employees are expected to fight fires, they must be trained and equipped in OSHA 1910.156. The only respirators recommended for firefighting are self-contained breathing apparatuses that have full face-pieces and are operated in a pressure-demand or other positive-pressure mode.

Disposal Method Suggested: Dissolve or mix the material with a combustible solvent and burn in a chemical incinerator equipped with an afterburner and scrubber. All federal, state, and local environmental regulations must be observed. Beware of possible presence of peroxides[22] in which case open burning may be used.

Reference

New Jersey Department of Health and Senior Services. (November 1999). *Hazardous Substances Fact Sheet: 2-Methoxyethyl Acetate.* Trenton, NJ

Methoxyethylmercuric acetate M:0600

Molecular Formula: $C_5H_{10}HgO_3$
Common Formula: $CH_3OCH_2CH_2HgOOCCH_3$
Synonyms: Acetato(2-methoxyethyl)mercury; Acetoxy(2-methoxyethyl)mercury; Cekusil Universal A®; Landisan®; MEMA; Mercuran; Mercury, acetoxy(2-methoxyethyl)-; Methoxyethylmercury acetate; 2-Methoxyethylmerkuriacetat (German); Panogen®; Panogen® M; Panogen® Metox; Radosan®
CAS Registry Number: 151-38-2

RTECS® Number: OV6300000
UN/NA & ERG Number: UN2777/151
EC Number: 205-790-2
Regulatory Authority and Advisory Bodies
Banned or Severely Restricted (in agriculture) (In UK).[13]
Air Pollutant Standard Set. See below, "Permissible Exposure Limits in Air" section.
Clean Water Act: 40CFR401.15 Section 307 Toxic Pollutants as mercury and compounds.
Superfund/EPCRA 40CFR355, Extremely Hazardous Substances: TPQ = 500/10,000 lb (227/4540 kg).
Reportable Quantity (RQ): 500 lb (227 kg).
US DOT Regulated Marine Pollutant (49CFR172.101, Appendix B).
California Proposition 65 Developmental/Reproductive toxin (mercury and mercury compounds) 7/1/90.
Canada, WHMIS, Ingredients Disclosure List Concentration 1.0% as mercury compounds.
Rotterdam Convention Annex III [Chemicals Subject to the Prior Informed Consent Procedure (PIC)] (as mercury compounds, including inorganic mercury compounds, alkyl mercury compounds and alkyloxylalkyl and aryl mercury compounds).
European/International Regulations: not listed in Annex 1.
WGK (German Aquatic Hazard Class): No value assigned.
Description: Methoxyethylmercuric acetate is a crystalline solid. Molecular weight = 318.74; Freezing/Melting point = 41°C. Hazard Identification (based on NFPA-704 M Rating System): Health 3, Flammability 1, Reactivity 0. Soluble in water.
Potential Exposure: Used as a pesticide in seed treatment for cotton and small grains. It is no longer approved for this use. It exhibits high fungicidal activity against leaf stripe of barley, stinking smut of wheat, snow mold of rye, against seedling diseases in beets and legumes, and for dressing "seed" potatoes, bulbs, and tubers. Not registered as a pesticide in the United States
Incompatibilities: Corrosive to iron and other metals.
Permissible Exposure Limits in Air
OSHA PEL: 0.01 mg[Hg]/m^3 TWA; 0.04 mg/m^3 Ceiling Concentration.
NIOSH REL: 0.01 mg[organomercury]/m^3 TWA; 0.03 mg/m^3 STEL [skin].
ACGIH TLV®[1]: 0.01 mg[Hg]/m^3 TWA; 0.03 mg/m^3 STEL [skin].
Protective Action Criteria (PAC)
TEEL-0: 0.0159 mg/m^3
PAC-1: 0.0477 mg/m^3
PAC-2: 0.0477 mg/m^3
PAC-3: 0.0477 mg/m^3
DFG MAK: 0.01 mg[Hg]/m^3 [skin] Danger of skin sensitization; Carcinogen Category 3.
NIOSH IDLH: 2 mg[Hg]/m^3.
In addition, North Dakota has set guidelines of 1–3 μg/m^3 (0.0001–0.0003 mg/m^3) for alkyl mercury compounds in ambient air.[60]

Determination in Air: No method available.
Permissible Concentration in Water: Presumably this material is covered in the Priority Toxic Pollutant Category. It is not specifically cited as an organomercurial as is methyl mercury.
Determination in Water: Total mercury is determined by flameless atomic absorption. Soluble mercury may be determined by 0.45- μm filtration followed by flameless atomic absorption.
Routes of Entry: Inhalation, ingestion, skin and/or eye contact.
Harmful Effects and Symptoms
Short Term Exposure: Highly toxic. Target organs are brain and central nervous system. Inhalation can cause lung damage; ingestion can cause kidney damage. Women of childbearing age should avoid exposure. Patients complain of headache; paresthesia of tongue, lips, fingers, and toes; a metallic taste in mouth; gastrointestinal disturbances; gas; and diarrhea. Nervous system symptoms may appear first after a relatively slight exposure or have a latency period: slight loss of coordination, loss of coordination of speech, writing, and gait. Uncoordination may progress to loss of ability to control voluntary movements. Irritability and bad temper may progress to mania. Stupor or coma may develop. Blisters or dermatitis may be present on skin. Symptoms persist for years even in cases of mild exposure.
Long Term Exposure: Repeated or prolonged contact with skin may result in dermatitis and allergy. Repeated or prolonged exposure may cause brain damage and nervous system damage. Repeated or prolonged exposure may cause death by hypovolemic shock, nephrotic syndrome, and kidney failure. Organic mercury substances have been identified as teratogen in humans. Can cause mercury to accumulate in the body and cause mercury poisoning. May cause permanent damage, such as gray colored skin, brown staining of the eyes, and decreased peripheral vision.
Points of Attack: Eyes, skin, central nervous system, peripheral nervous system, kidneys.
Medical Surveillance: Before first exposure and every 6–12 months after, a complete medical history and examination is strongly recommended with eye examination. Consider lung function tests for persons with frequent exposure. Examination of the nervous system. Routine urine test (UA). Urine test for mercury (should be less than 0.02 mg/L). Consider nerve conduction tests, urinary enzymes, and neurobehavioral test. After suspected illness or overexposure, repeat the tests above and get a blood test for mercury.
First Aid: If this chemical gets into the eyes, remove any contact lenses at once and irrigate immediately for at least 15 min, occasionally lifting upper and lower lids. Seek medical attention immediately. If this chemical contacts the skin, remove contaminated clothing and wash immediately with soap and water. Speed in removing material from skin is of extreme importance. Shampoo hair promptly if contaminated. Seek medical attention immediately. If this chemical has been inhaled, remove from exposure, begin

rescue breathing (using universal precautions, including resuscitation mask) if breathing has stopped and CPR if heart action has stopped. Transfer promptly to a medical facility. When this chemical has been swallowed, get medical attention. Give large quantities of water and induce vomiting. Do not make an unconscious person vomit.

Antidotes and special procedures for medical personnel: The drug NAP (*n*-acetyl penicillamine) has been used to treat mercury poisoning, with mixed success.

Note to physician: For severe poisoning BAL [British Anti-Lewisite, dimercaprol, dithiopropanol ($C_3H_8OS_2$)] has been used to treat toxic symptoms of certain heavy metals poisoning including mercury. Although BAL is reported to have a large margin of safety, caution must be exercised, because toxic effects may be caused by excessive dosage. Most can be prevented by premedication with 1-ephedrine sulfate (CAS: 134-72-5).

Personal Protective Methods: Wear protective gloves and clothing to prevent any reasonable probability of skin contact. Safety equipment suppliers/manufacturers can provide recommendations on the most protective glove/clothing material for your operation. All protective clothing (suits, gloves, footwear, headgear) should be clean, available each day, and put on before work. Contact lenses should not be worn when working with this chemical. Wear dust-proof chemical goggles and face shield unless full face-piece respiratory protection is worn. Employees should wash immediately with soap when skin is wet or contaminated. Provide emergency showers and eyewash.

Respirator Selection: Up to 0.1 mg/m³: Sa (APF = 10) (any supplied-air respirator). *Up to 0.25 mg/m³:* Sa:Cf (APF = 25) (any supplied-air respirator operated in a continuous-flow mode). *Up to 0.5 mg/m³:* SaT: Cf (APF = 50) (any supplied-air respirator that has a tight-fitting face-piece and is operated in a continuous-flow mode) or SCBAF (APF = 50) (any self-contained breathing apparatus with a full face-piece) or SaF (APF = 50) (any supplied-air respirator with a full face-piece). *Up to 2 mg/m³:* Sa: Pd,Pp (APF = 1000) (any supplied-air respirator operated in a pressure-demand or other positive-pressure mode). *Emergency or planned entry into unknown concentrations or IDLH conditions:* SCBAF: Pd,Pp (APF = 10,000) (any NIOSH/MSHA- or European Standard EN 149-approved self-contained breathing apparatus that has a full face-piece and is operated in a pressure-demand or other positive-pressure mode) or SaF: Pd,Pp: ASCBA (APF = 10,000) (any supplied-air respirator that has a full face-piece and is operated in a pressure-demand or other positive-pressure mode in combination with an auxiliary self-contained breathing apparatus operated in a pressure-demand or other positive-pressure mode). *Escape:* SCBAE (any appropriate escape-type, self-contained breathing apparatus).

Storage: Color Code—Blue: Health Hazard/Poison: Store in a secure poison location. Prior to working with this chemical you should be trained on its proper handling and storage. Store in tightly closed containers in a cool, well-ventilated area.

Shipping: Mercury-based pesticides, solid, toxic, requires a shipping label of "POISONOUS/TOXIC MATERIALS." It falls in Hazard Class 6.1 and Packing Group II.

Spill Handling: Evacuate persons not wearing protective equipment from area of spill or leak until cleanup is complete. Remove all ignition sources. Collect powdered material in the most convenient and safe manner and deposit in sealed containers. Spills should be collected with special mercury vapor suppressants or special vacuums. Kits specific for cleanup of mercury are available. Ventilate area after cleanup is complete. It may be necessary to contain and dispose of this chemical as a hazardous waste. If material or contaminated runoff enters waterways, notify downstream users of potentially contaminated waters. Contact your local or federal environmental protection agency for specific recommendations. If employees are required to clean up spills, they must be properly trained and equipped. OSHA 1910.120(q) may be applicable.

Fire Extinguishing: *Small fires*: dry chemical, carbon dioxide, water spray, or foam. *Large fires:* water spray, fog, or foam. Move container from fire area if you can do so without risk. Fight fire from maximum distance. Dike fire control water for later disposal; do not scatter the material. Runoff from fire control or dilution water may cause pollution. Poisonous gases, including Hg, are produced in fire. If material or contaminated runoff enters waterways, notify downstream users of potentially contaminated waters. Notify local health and fire officials and pollution control agencies. From a secure, explosion-proof location, use water spray to cool exposed containers. If cooling streams are ineffective (venting sound increases in volume and pitch, tank discolors, or shows any signs of deforming), withdraw immediately to a secure position. If employees are expected to fight fires, they must be trained and equipped in OSHA 1910.156. The only respirators recommended for firefighting are self-contained breathing apparatuses that have full face-pieces and are operated in a pressure-demand or other positive-pressure mode.

Disposal Method Suggested: In accordance with 40CFR 165 recommendations for the disposal of pesticides and pesticide containers. Must be disposed properly by following package label directions or by contacting your local or federal environmental control agency or by contacting your regional EPA office.

References

US Environmental Protection Agency. (November 30, 1987). *Chemical Hazard Information Profile: Methoxyethylmercuric Acetate.* Washington, DC: Chemical Emergency Preparedness Program

4-Methoxyphenol   M:0610

Molecular Formula: $C_7H_8O_2$
Common Formula: $HOC_6H_4OCH_3$

Synonyms: *p*-Guaicol; Hydroquinone monomethyl ether; Hydroquinone monomethyl ether and *p* hydroxyanisole; 1-Hydroxy-4-methoxybenzene; Mequinol; *p*-Methoxyphenol; MME; Mono methyl ether hydroquinone; Phenol, *p*-methoxy

CAS Registry Number: 150-76-5

RTECS® Number: SL7700000

EC Number: 205-769-8 [*Annex I Index No.:* 604-044-00-7]

Regulatory Authority and Advisory Bodies

Air Pollutant Standard Set. See below, "Permissible Exposure Limits in Air" section.

Canada, WHMIS, Ingredients Disclosure List Concentration 1.0%.

European/International Regulations: Hazard Symbol: Xn; Risk phrases: R22; R36; R43; Safety phrases: S2; S24/25; S26; S37/39; S46 (see Appendix 4).

WGK (German Aquatic Hazard Class): 1—Low hazard to waters.

Description: 4-Methoxyphenol is a colorless to white, waxy solid with an odor of caramel and phenol. A combustible solid. Molecular weight = 124.15; Boiling point = 243°C; Freezing/Melting point = 52−53°C; Flash point = 132°C. Hazard Identification (based on NFPA-704 M Rating System): Health 2, Flammability 1, Reactivity 0. Slightly soluble in water.

Potential Exposure: Compound Description: Mutagen, Primary Irritant. This compound is used in the manufacture of antioxidants, pharmaceuticals, plasticizers, and dyestuffs. It is used as a stabilizer and UV inhibitor in various polymers.

Incompatibilities: Strong oxidizers, strong bases, acid chlorides, acid anhydrides. Under certain conditions, a dust cloud can probably explode if ignited by a spark or flame.

Permissible Exposure Limits in Air

OSHA PEL: None.

NIOSH REL: 5 mg/m^3 TWA.

ACGIH TLV®[1]: 5 mg/m^3 TWA.

Protective Action Criteria (PAC).

TEEL-0: 5 mg/m^3

PAC-1: 15 mg/m^3

PAC-2: 100 mg/m^3

PAC-3: 500 mg/m^3

Australia: TWA 5 mg/m^3, 1993; Belgium: TWA 5 mg/m^3, 1993; Denmark: TWA mg/m^3, 1999; France: VME 5 mg/m^3, 1999; Norway: TWA 5 mg/m^3, 1999; United Kingdom: TWA 5 mg/m^3, 2000; the Netherlands: MAC-TGG 5 mg/m^3, 2003; Argentina, Bulgaria, Columbia, Jordan, South Korea, New Zealand, Singapore, Vietnam: ACGIH TLV®: TWA 5 mg/m^3.

Determination in Air: No NIOSH Analytical Method available. See OSHA Analytical Method 32.

Routes of Entry: Inhalation, skin absorption, ingestion, skin and/or eye contact. Absorbed through the skin.

Harmful Effects and Symptoms

Short Term Exposure: 4-Methoxyphenol can affect you when breathed in and by passing through your skin. Contact can cause severe eye burns and permanent damage. Irritates the skin, nose, throat, upper respiratory system. A central nervous system depressant.

Long Term Exposure: Prolonged contact can cause severe skin burns and scars. Repeated exposure may cause discoloration of the eye and skin. Damage to the vision may occur.

Points of Attack: Eyes, skin, respiratory system, central nervous system.

Medical Surveillance: If symptoms develop or overexposure is suspected, the following may be useful: examination of the eyes and vision, including a "slit-lamp" exam. Examination of the central nervous system.

First Aid: If this chemical gets into the eyes, remove any contact lenses at once and irrigate immediately for at least 15 min, occasionally lifting upper and lower lids. Seek medical attention immediately. If this chemical contacts the skin, remove contaminated clothing and wash immediately with soap and water. Seek medical attention immediately. If this chemical has been inhaled, remove from exposure, begin rescue breathing (using universal precautions, including resuscitation mask) if breathing has stopped and CPR if heart action has stopped. Transfer promptly to a medical facility. When this chemical has been swallowed, get medical attention. Give large quantities of water and induce vomiting. Do not make an unconscious person vomit.

Personal Protective Methods: Wear protective gloves and clothing to prevent any reasonable probability of skin contact. Safety equipment suppliers/manufacturers can provide recommendations on the most protective glove/clothing material for your operation. All protective clothing (suits, gloves, footwear, headgear) should be clean, available each day, and put on before work. Contact lenses should not be worn when working with this chemical. Wear dust-proof chemical goggles and face shield unless full face-piece respiratory protection is worn. Employees should wash immediately with soap when skin is wet or contaminated. Provide emergency showers and eyewash.

Respirator Selection: Where there is potential for exposures over 5 mg/m^3, use a NIOSH/MSHA- or European Standard EN149-approved full face-piece respirator with a high-efficiency particulate filter. Greater protection is provided by a powered air-purifying respirator. *Where there is potential for high exposures*, use a NIOSH/MSHA- or European Standard EN149-approved supplied-air respirator with a full face-piece operated in the positive-pressure mode; or with a full face-piece, hood, or helmet in the continuous-flow mode; or use a NIOSH/MSHA- or European Standard EN149-approved self-contained breathing apparatus with a full face-piece operated in pressure-demand or other positive-pressure mode.

Storage: Protect containers from physical damage. Store in tightly closed containers in a cool, well-ventilated area. Sources of ignition, such as smoking and open flames, are prohibited where 4-methoxyphenol is used, handled, or stored in a manner that could create a potential fire or explosion hazard.

Shipping: A combustible solid, but there are no label requirements or maximum shipping weights specified.

Spill Handling: Evacuate persons not wearing protective equipment from area of spill or leak until cleanup is complete. Remove all ignition sources. Ventilate area after cleanup is complete. It may be necessary to contain and dispose of this chemical as a hazardous waste. If material or contaminated runoff enters waterways, notify downstream users of potentially contaminated waters. Contact your local or federal environmental protection agency for specific recommendations. If employees are required to clean up spills, they must be properly trained and equipped. OSHA 1910.120(q) may be applicable.

Fire Extinguishing: 4-Methoxyphenol may burn but does not readily ignite. Use dry chemical or carbon dioxide extinguishers. Poisonous gases are produced in fire. If material or contaminated runoff enters waterways, notify downstream users of potentially contaminated waters. Notify local health and fire officials and pollution control agencies. From a secure, explosion-proof location, use water spray to cool exposed containers. If cooling streams are ineffective (venting sound increases in volume and pitch, tank discolors, or shows any signs of deforming), withdraw immediately to a secure position. If employees are expected to fight fires, they must be trained and equipped in OSHA 1910.156. The only respirators recommended for firefighting are self-contained breathing apparatuses that have full face-pieces and are operated in a pressure-demand or other positive-pressure mode.

Reference
New Jersey Department of Health and Senior Services. (March 2000). *Hazardous Substances Fact Sheet: 4-Methoxyphenol.* Trenton, NJ

Methyl acetate   M:0620

Molecular Formula: $C_3H_6O_2$
Common Formula: CH_3COOCH_3
Synonyms: Acetate de methyle (French); Acetic acid, methyl ester; Devoton; Methylacetat (German); Methyl acetic ester; Methyle (acetate de) (French); Methyl ethanoate; Tereton
CAS Registry Number: 79-20-9
RTECS® Number: AI9100000
UN/NA & ERG Number: UN1231/129
EC Number: 201-185-2 [*Annex I Index No.:* 607-021-00-X]
Regulatory Authority and Advisory Bodies
Air Pollutant Standard Set. See below, "Permissible Exposure Limits in Air" section.
Canada, WHMIS, Ingredients Disclosure List Concentration 1.0%.
European/International Regulations: not listed in Annex 1.
WGK (German Aquatic Hazard Class): 1—Low hazard to waters.

Description: Methyl acetate is a colorless liquid with a fruity odor. Odor threshold = 4.6 ppm. Molecular weight = 74.09; Specific gravity (H_2O:1) = 0.93; Boiling point = 57.2°C; Freezing/Melting point = −98°C; Vapor pressure = 173 mmHg at 20°C; Flash point = −10°C; Autoignition temperature = 455°C. The explosive limits are LEL = 3.1%; UEL = 16%. Hazard Identification (based on NFPA-704 M Rating System): Health 1, Flammability 3, Reactivity 0. Soluble in water; solubility = 25%.

Potential Exposure: Compound Description: Mutagen, Human Data; Primary Irritant. Methyl acetate is used as a solvent in lacquers and paint removers and as an intermediate in pharmaceutical manufacture.

Incompatibilities: Forms explosive mixture with air. A strong reducing agent. Incompatible water, acids, nitrates, strong oxidizers, alkalis. Attacks some plastics. Attacks many metals in presence of water. Reacts slowly with water, forming acetic acid and methanol. Decomposes in heat; on contact with air, bases, strong oxidizers, UV-light, causing fire and explosion hazard.

Permissible Exposure Limits in Air
Conversion factor: 1 ppm = 3.03 mg/m³ at 25°C & 1 atm.
OSHA PEL: 200 ppm/610 mg/m³ TWA.
NIOSH REL: 200 ppm/610 mg/m³ TWA; 250 ppm/760 mg/m³ STEL.
ACGIH TLV®[1]: 200 ppm/606 mg/m³ TWA; 250 ppm/757 mg/m³ STEL.
NIOSH IDLH: 3100 ppm [LEL].
Protective Action Criteria (PAC)
TEEL-0: 200 ppm
PAC-1: 250 ppm
PAC-2: 500 ppm
PAC-3: 3100 ppm
DFG MAK: 100 ppm/310 mg/m³ TWA; Peak Limitation Category I(4); Pregnancy Risk Group C.
Australia: TWA 200 ppm (610 mg/m³); STEL 250 ppm, 1993; Austria: MAK 200 ppm (610 mg/m³), 1999; Belgium: TWA 200 ppm (610 mg/m³); STEL 250 ppm (760 mg/m³), 1993; Denmark: TWA 150 ppm (455 mg/m³), 1999; Finland: TWA 200 ppm (610 mg/m³); STEL 250 ppm (765 mg/m³), 1993; France: VME 200 ppm (610 mg/m³), VLE 250 ppm (760 mg/m³), 1999; Hungary: TWA 200 mg/m³; STEL 500 mg/m³ [skin] 1993; the Netherlands: MAC-TGG 610 mg/m³, 2003; Norway: TWA 100 ppm (305 mg/m³), 1999; the Philippines: TWA 200 ppm (610 mg/m³), 1993; Poland: MAC (TWA) 100 mg/m³, MAC (STEL) 750 mg/m³, 1999; Russia: TWA 200 ppm; STEL 100 mg/m³, 1993; Sweden: TWA 150 ppm (450 mg/m³); STEL 300 ppm (900 mg/m³), 1999; Switzerland: MAK-W 200 ppm (610 mg/m³), KZG-W 400 ppm (1220 mg/m³), 1999; Turkey: TWA 200 ppm (610 mg/m³), 1993; United Kingdom: TWA 200 ppm (616 mg/m³); STEL 250 ppm (770 mg/m³), 2000; Argentina, Bulgaria, Columbia, Jordan, South Korea, New Zealand, Singapore, Vietnam: ACGIH TLV®: STEL 250 ppm. Several states have set guidelines or standards for methyl acetate in ambient air[60] ranging from 6.1−7.6

mg/m^3 (North Dakota) to 12.2 mg/m^3 (Connecticut) to 14.524 mg/m^3 (Nevada) to 15 mg/m^3 (Virginia).

Determination in Air: Use NIOSH Analytical Method (IV) #1458.

Permissible Concentration in Water: Russia[43] set a MAC of 0.1 mg/L in water bodies used for domestic purposes.

Determination in Water: Octanol−water coefficient: Log K_{ow} = 0.2.

Routes of Entry: Inhalation, ingestion, skin and/or eye contact.

Harmful Effects and Symptoms

Short Term Exposure: Methyl acetate can affect you when breathed in and by passing through your skin. Irritates the eyes and respiratory tract. May affect the central nervous system causing dizziness, lightheadedness, and unconsciousness. Higher exposures can cause pulmonary edema, a medical emergency that can be delayed for several hours. This can cause death.

Long Term Exposure: The liquid destroys the skin's natural oils. Repeated or high exposures may cause methanol poisoning, which can cause headaches, dizziness, coma, and affect the optic nerve, causing blindness. Death can occur.

Points of Attack: Eyes, skin, respiratory system, central nervous system.

Medical Surveillance: For those with frequent or potentially high exposures (half the TLV or greater), the following are recommended before beginning work and at regular times after that: lung function tests. Vision examination. If symptoms develop or overexposure is suspected, the following may be useful: consider chest X-ray following acute overexposure.

First Aid: If this chemical gets into the eyes, remove any contact lenses at once and irrigate immediately for at least 15 min, occasionally lifting upper and lower lids. Seek medical attention immediately. If this chemical contacts the skin, remove contaminated clothing and wash immediately with soap and water. Seek medical attention immediately. If this chemical has been inhaled, remove from exposure, begin rescue breathing (using universal precautions, including resuscitation mask) if breathing has stopped and CPR if heart action has stopped. Transfer promptly to a medical facility. When this chemical has been swallowed, get medical attention. Give large quantities of water and induce vomiting. Do not make an unconscious person vomit. Medical observation is recommended for 24−48 h after breathing overexposure, as pulmonary edema may be delayed. As first aid for pulmonary edema, a doctor or authorized paramedic may consider administering a corticosteroid spray.

Personal Protective Methods: Wear protective gloves and clothing to prevent any reasonable probability of skin contact: **8 h**: 4H™ and Silver Shield™ gloves. **4 h**: Teflon™ gloves, suits, boots. Also, butyl rubber is among the recommended protective materials. Safety equipment suppliers and manufacturers can provide recommendations on the most protective glove/clothing material for your operation. All protective clothing (suits, gloves, footwear, headgear) should be clean, available each day, and put on before work. Contact lenses should not be worn when working with this chemical. Wear splash-proof chemical goggles and face shield unless full face-piece respiratory protection is worn. Employees should wash immediately with soap when skin is wet or contaminated. Provide emergency showers and eyewash.

Respirator Selection: 2000 ppm: CcrOv (APF = 10) [any chemical cartridge respirator with organic vapor cartridge(s)] or Sa (APF = 10) (any supplied-air respirator); *3100 ppm:* Sa: Cf (APF = 25) (any supplied-air respirator operated in a continuous-flow mode) or CcrFOv (APF = 50) [any chemical cartridge respirator with a full face-piece and organic vapor cartridge(s)] or GmFOv (APF = 50) [any air-purifying, full-face-piece respirator (gas mask) with a chin-style, front- or back-mounted organic vapor canister] or PaprOv (APF = 25) [any powered, air-purifying respirator with organic vapor cartridge(s)] or SCBAF (APF = 50) (any self-contained breathing apparatus with a full face-piece) or SaF (APF = 50) (any supplied-air respirator with a full face-piece). *Emergency or planned entry into unknown concentrations or IDLH conditions:* SCBAF: Pd,Pp (APF = 10,000) (any self-contained breathing apparatus that has a full face-piece and is operated in a pressure-demand or other positive-pressure mode) or SaF: Pd,Pp: ASCBA (APF = 10,000) (any supplied-air respirator that has a full face-piece and is operated in a pressure-demand or other positive-pressure mode in combination with an auxiliary self-contained breathing apparatus operated in a pressure-demand or other positive-pressure mode). *Escape:* GmFOv (APF = 50) [any air-purifying, full-face-piece respirator (gas mask) with a chin-style, front- or back-mounted organic vapor canister] or SCBAE (any appropriate escape-type, self-contained breathing apparatus).

Note: Substance reported to cause eye irritation or damage; may require eye protection.

Storage: (1) Color Code—Red: Flammability Hazard: Store in a flammable liquid storage area or approved cabinet away from ignition sources and corrosive and reactive materials. (2) Color Code—Yellow Stripe (*strong reducing agent*): Reactivity Hazard; Store separately in an area isolated from flammables, combustibles, or other yellow coded materials. Prior to working with this chemical you should be trained on its proper handling and storage. Before entering confined space where this chemical may be present, check to make sure that an explosive concentration does not exist. Methyl acetate must be stored to avoid contact with strong oxidizers (such as chlorine, bromine, and fluorine) and strong acids (such as hydrochloric, sulfuric and nitric), since violent reactions occur. Store in tightly closed containers in a cool, well-ventilated area away from strong alkalis and nitrates. Sources of ignition, such as smoking and open flames, are prohibited where methyl acetate is used, handled, or stored in a manner that could create a potential fire or explosion hazard. Use only nonsparking tools and equipment, especially when opening and closing containers of methyl acetate.

Shipping: This compound requires a shipping label of "FLAMMABLE LIQUID." It falls in Hazard Class 3 and Packing Group II.

Spill Handling: Evacuate and restrict persons not wearing protective equipment from area of spill or leak until cleanup is complete. Remove all ignition sources. Establish forced ventilation to keep levels below explosive limit. Absorb liquids in vermiculite, dry sand, earth, peat, carbon, or a similar material and deposit in sealed containers. Keep this chemical out of a confined space, such as a sewer, because of the possibility of an explosion, unless the sewer is designed to prevent the buildup of explosive concentrations. It may be necessary to contain and dispose of this chemical as a hazardous waste. If material or contaminated runoff enters waterways, notify downstream users of potentially contaminated waters. Contact your local or federal environmental protection agency for specific recommendations. If employees are required to clean up spills, they must be properly trained and equipped. OSHA 1910.120(q) may be applicable.

Fire Extinguishing: This chemical is a combustible liquid. Poisonous gases are produced in fire. Use dry chemical, carbon dioxide, or alcohol foam extinguishers. Vapors are heavier than air and will collect in low areas. Vapors may travel long distances to ignition sources and flashback. Vapors in confined areas may explode when exposed to fire. Containers may explode in fire. Storage containers and parts of containers may rocket great distances, in many directions. If material or contaminated runoff enters waterways, notify downstream users of potentially contaminated waters. Notify local health and fire officials and pollution control agencies. From a secure, explosion-proof location, use water spray to cool exposed containers. If cooling streams are ineffective (venting sound increases in volume and pitch, tank discolors, or shows any signs of deforming), withdraw immediately to a secure position. If employees are expected to fight fires, they must be trained and equipped in OSHA 1910.156. The only respirators recommended for firefighting are self-contained breathing apparatuses that have full face-pieces and are operated in a pressure-demand or other positive-pressure mode.

Disposal Method Suggested: Dissolve or mix the material with a combustible solvent and burn in a chemical incinerator equipped with an afterburner and scrubber. All federal, state, and local environmental regulations must be observed.

Reference

New Jersey Department of Health and Senior Services. (June 2003). *Hazardous Substances Fact Sheet: Methyl Acetate.* Trenton, NJ

Methyl acetylene   M:0630

Molecular Formula: C_3H_4
Common Formula: $CH_3C = CH$

Synonyms: Acetylene, methyl-; Allylene; Propine; Propyne
CAS Registry Number: 74-99-7
RTECS® Number: UK4250000
UN/NA & ERG Number: UN1954/115
EC Number: 200-828-4
Regulatory Authority and Advisory Bodies
Department of Homeland Security Screening Threshold Quantity (pounds): *Release hazard* 10,000 ($\geq$1.00% concentration).
Air Pollutant Standard Set. See below, "Permissible Exposure Limits in Air" section.
Canada, WHMIS, Ingredients Disclosure List Concentration 1.0%.
European/International Regulations: not listed in Annex 1.
WGK (German Aquatic Hazard Class): Nonwater polluting agent.

Description: Methyl acetylene is a flammable, colorless gas with a sweet odor. Molecular weight = 40.07; Boiling point = $-23.3°C$; Freezing/Melting point = $-102.8°C$; Vapor pressure = 5.2 atm; Relative vapor density (air = 1) = 1.41. Explosive limits: LEL = 1.7%; UEL = 11.7%. Hazard Identification (based on NFPA-704 M Rating System): Health 2, Flammability 4, Reactivity 2. A fuel that is shipped as a liquefied compressed gas. Practically insoluble in water; solubility 0.4%.

Potential Exposure: Compound Description: Mutagen. This material may be used as a liquid rocket propellant, in admixture with propadiene as an industrial cutting fuel.

Incompatibilities: Forms explosive gas mixture with air. Can form explosive peroxide. Can decompose explosively on heating and at 4.5−5.6 atmospheres of pressure. Strong oxidizers may cause fire and explosions. Forms shock-sensitive compounds with copper, magnesium, silver, and their alloys. Copper or copper alloys containing more than 67% copper should not be used in handling equipment. Attacks some plastics, rubber, and coatings.

Permissible Exposure Limits in Air
Conversion factor = 1 ppm = 1.64 mg/m^3.
OSHA PEL: 1000 ppm/1650 mg/m^3 TWA.
NIOSH REL: 1000 ppm/1650 mg/m^3 TWA.
ACGIH TLV®[1]: 1000 ppm/1640 mg/m^3 TWA.
NIOSH IDLH: 1700 ppm [LEL].
Protective Action Criteria (PAC)
TEEL-0: 1000 ppm
PAC-1: 1700 ppm
PAC-2: 1700 ppm
PAC-3: 1700 ppm
Australia: TWA 1000 ppm (1650 mg/m^3); STEL 1250 ppm, 1993; Austria: MAK 1000 ppm (1650 mg/m^3), 1999; Belgium: TWA 1000 ppm (1640 mg/m^3); STEL 1250 ppm (2050 mg/m^3), 1993; Denmark: TWA 1000 ppm (1650 mg/m^3), 1999; Finland: TWA 1000 ppm (1650 mg/m^3); STEL 1250 ppm (2065 mg/m^3), 1999; France: VME 1000 ppm (1650 mg/m^3), 1999; the Netherlands: MAC-TGG 1650 mg/m^3, 2003; the Philippines: TWA 1000 ppm (1650 mg/m^3), 1993; Switzerland: MAK-W 1000 ppm (1650 mg/m^3), 1999;

Turkey: TWA 1000 ppm (1650 mg/m³), 1993; Argentina, Bulgaria, Columbia, Jordan, South Korea, New Zealand, Singapore, Vietnam: ACGIH TLV®: TWA 1000 ppm.

The Federal standard[58], the DFG MAK value[3], and the ACGIH TWA value is 1000 ppm (1650 mg/m³). Several states have set guidelines or standards for methyl acetylene in ambient air[60] ranging from 16.5–20.4 mg/m³ (North Dakota) to 30 mg/m³ (Virginia) to 33 mg/m³ (Connecticut) to 39.286 mg/m³ (Nevada).

Determination in Air: Gas collection bag; none; Gas chromatography/Flame ionization detection; NIOSH (II-5), Method #S8.

Routes of Entry: Inhalation, skin and/or eye contact (liquid).

Harmful Effects and Symptoms

Short Term Exposure: Irritates the respiratory tract. Higher exposures can cause pulmonary edema, a medical emergency that can be delayed for several hours. This can cause death. Rapid evaporation of the liquid may cause frostbite. Exposure can cause headache, dizziness, convulsions, and loss of consciousness.

Long Term Exposure: Can irritate the lungs. May cause bronchitis with coughing, phlegm, and/or shortness of breath.

Points of Attack: Central nervous system, lungs.

Medical Surveillance: Consider the points of attack in preplacement and periodic physical examinations. Consider chest X-ray following acute overexposure.

First Aid: If this chemical gets into the eyes, remove any contact lenses at once and irrigate immediately for at least 15 min, occasionally lifting upper and lower lids. Seek medical attention immediately. If this chemical contacts the skin, remove contaminated clothing and wash immediately with soap and water. Seek medical attention immediately. If this chemical has been inhaled, remove from exposure, begin rescue breathing (using universal precautions, including resuscitation mask) if breathing has stopped and CPR if heart action has stopped. Transfer promptly to a medical facility. When this chemical has been swallowed, get medical attention. Give large quantities of water and induce vomiting. Do not make an unconscious person vomit. Medical observation is recommended for 24–48 h after breathing overexposure, as pulmonary edema may be delayed. As first aid for pulmonary edema, a doctor or authorized paramedic may consider administering a corticosteroid spray. If frostbite has occurred, seek medical attention immediately; do NOT rub the affected areas or flush them with water. In order to prevent further tissue damage, do NOT attempt to remove frozen clothing from frostbitten areas. If frostbite has NOT occurred, immediately and thoroughly wash contaminated skin with soap and water.

Personal Protective Methods: Wear appropriate personal protective clothing to prevent the skin from becoming frozen from contact with the evaporating liquid or from contact with vessels containing the liquid. Safety equipment suppliers/manufacturers can provide recommendations on the most protective glove/clothing material for your operation. All protective clothing (suits, gloves, footwear, headgear) should be clean, available each day, and put on before work. Contact lenses should not be worn when working with this chemical. Wear gas-proof chemical goggles and face shield unless full face-piece respiratory protection is worn. Employees should wash immediately with soap when skin is wet or contaminated. Provide emergency showers and eyewash.

Respirator Selection: *Up to 1700 ppm:* Sa (APF = 10) (any supplied-air respirator) or SCBAF (APF = 50) (any self-contained breathing apparatus with a full face-piece). *Emergency or planned entry into unknown concentrations or IDLH conditions:* SCBAF: Pd,Pp (APF = 10,000) (any self-contained breathing apparatus that has a full face-piece and is operated in a pressure-demand or other positive-pressure mode) or SaF: Pd,Pp: ASCBA (APF = 10,000) (any supplied-air respirator that has a full face-piece and is operated in a pressure-demand or other positive-pressure mode in combination with an auxiliary self-contained breathing apparatus operated in a pressure-demand or other positive-pressure mode). *Escape:* GmFOv (APF = 50) [any air-purifying, full-face-piece respirator (gas mask) with a chin-style, front- or back-mounted organic vapor canister] or SCBAE (any appropriate escape-type, self-contained breathing apparatus).

Storage: Color Code—Red Stripe: Flammability Hazard: Store separately from all other flammable materials. May form peroxides in storage. Prior to working with this chemical you should be trained on its proper handling and storage. Before entering confined space where this chemical may be present, check to make sure that an explosive concentration does not exist. Store in tightly closed containers in a cool, well-ventilated area away from oxidizers. Where possible, automatically pump liquid from drums or other storage containers to process containers. Sources of ignition, such as smoking and open flames, are prohibited where this chemical is used, handled, or stored in a manner that could create a potential fire or explosion hazard. Wherever this chemical is used, handled, manufactured, or stored, use explosion-proof electrical equipment and fittings. Procedures for the handling, use and storage of cylinders should be in compliance with OSHA 1910.101 and 1910.169, as with the recommendations of the Compressed Gas Association.

Shipping: Compressed gases, flammable, n.o.s. require a label of "FLAMMABLE GAS." They fall in Hazard Class 2.1.

Spill Handling: If in a building, evacuate building and confine vapors by closing doors and shutting down HVAC systems. Restrict persons not wearing protective equipment from area of spill or leak until cleanup is complete. Remove all ignition sources. Establish forced ventilation to keep levels below explosive limit and to disperse the gas. Wear chemical protective suit with self-contained breathing apparatus to combat spills. Stay upwind and use water spray to "knock down" vapor; contain runoff. Stop the flow of gas, if it can be done safely from a distance. If source is a cylinder and the leak cannot be stopped in place, remove the leaking cylinder to a safe place, and repair leak or allow cylinder to empty. Keep this chemical out of confined spaces, such as a sewer, because of the possibility of explosion, unless the sewer is designed to prevent the buildup of explosive concentrations. If employees are required to clean up spills, they must be properly trained and equipped. OSHA 1910.120(q) may be applicable.

Fire Extinguishing: This chemical is a flammable gas. Poisonous gases are produced in fire. Do not extinguish the fire unless the flow of gas can be stopped and any remaining gas is out of the line. Specially trained personnel may use fog lines to cool exposures and let the fire burn itself out. Vapors are heavier than air and will collect in low areas. Vapors may travel long distances to ignition sources and flashback. Vapors in confined areas may explode when exposed to fire. Containers may explode in fire. Storage containers and parts of containers may rocket great distances, in many directions. If material or contaminated run-off enters waterways, notify downstream users of potentially contaminated waters. Notify local health and fire officials and pollution control agencies. From a secure, explosion-proof location, use water spray to cool exposed containers. If cooling streams are ineffective (venting sound increases in volume and pitch, tank discolors, or shows any signs of deforming), withdraw immediately to a secure position. If cylinders are exposed to excessive heat from fire or flame contact, withdraw immediately to a secure location. If employees are expected to fight fires, they must be trained and equipped in OSHA 1910.156. The only respirators recommended for firefighting are self-contained breathing apparatuses that have full face-pieces and are operated in a pressure-demand or other positive-pressure mode.

Note: Stopping the flow of gas is a better course of action than trying to extinguish the fire.[17] It may be dangerous to extinguish the flame and allow the gas to continue to flow, as an explosive mixture may be formed with air which may cause greater damage than the original fire if allowed to burn.

Disposal Method Suggested: Dissolve or mix the material with a combustible solvent and burn in a chemical incinerator equipped with an afterburner and scrubber. All federal, state, and local environmental regulations must be observed.

Reference
New Jersey Department of Health and Senior Services. (May 1999). *Hazardous Substances Fact Sheet: Methyl Acetylene*, Trenton, NJ

Methyl acetylene—propadiene mixture   M:0640

Molecular Formula: C_3H_4
Common Formula: $CH_3C \equiv CH$; $H_2C = C = CH_2$

Synonyms: Allene-methyl acetylene mixture; MAPP gas; Methyl acetylene-allene mixture; Propadiene and methylacetylene mixture; Propyne-allene mixture; Propyne-propadiene mixture
CAS Registry Number: 59355-75-8
RTECS® Number: UK4920000
UN/NA & ERG Number: UN1060 (stabilized)/116
UN/NA & ERG Number: None assigned.
Regulatory Authority and Advisory Bodies
Air Pollutant Standard Set. See below, "Permissible Exposure Limits in Air" section.
WGK (German Aquatic Hazard Class): No value assigned.
Description: This mixture of C_3H_4 isomers is a flammable, colorless gas with a strong, characteristic, foul odor. The odor threshold is 100 ppm. Boiling point = −38 to −20°C; Freezing/Melting point = −136°C; Relative vapor density (air = 1) = 1.48. Explosive limits: LEL = 3.4%; UEL = 10.8%. Hazard Identification (based on NFPA-704M Rating System): Health 0, Flammability 0, Reactivity 0. Insoluble in water. A fuel that is shipped as a liquefied compressed gas mixture containing 60−6.5% methylacetylene and propadiene; the balance is butane and propane.
Potential Exposure: This mixture is used as an industrial cutting fuel.
Incompatibilities: Strong oxidizers, copper alloys. Forms explosive compounds at high pressure in contact with alloys containing more than 67% copper.
Permissible Exposure Limits in Air
Conversion factor: 1 ppm = 1.64 mg/m³ at 25°C & 1 atm.
OSHA PEL: 1000 ppm/1800 mg/m³ TWA.
NIOSH REL: 1000 ppm/1800 mg/m³ TWA; 1250 ppm/2250 mg/m³ STEL.
ACGIH TLV®[1]: 1000 ppm/1640 mg/m³ TWA; 1250 ppm/2050 mg/m³ STEL.
NIOSH IDLH: 3,400 ppm [LEL].
Protective Action Criteria (PAC)
TEEL-0: 1000 ppm
PAC-1: 1250 ppm
PAC-2: 1250 ppm
PAC-3: 3400 ppm
Australia: TWA 1000 ppm (1800 mg/m³); STEL 1250 ppm, 1993; Belgium: TWA 1000 ppm (1640 mg/m³); STEL 1250 ppm (2050 mg/m³), 1993; Finland: TWA 1000 ppm (1800 mg/m³); STEL 1250 ppm (2250 mg/m³), 1999; the Philippines: TWA 1000 ppm (1800 mg/m³), 1993; Switzerland: MAK-W 1000 ppm (1800 mg/m³), 1999; the Netherlands: MAC-TGG 1800 mg/m³, 2003; Argentina, Bulgaria, Columbia, Jordan, South Korea, New Zealand, Singapore, Vietnam: ACGIH TLV®: STEL 1250 ppm. In addition, the state of North Dakota has set a guideline[60] of 18−22.5 mg/m³ for MAPP in ambient air.
Determination in Air: Gas collection bag; none; Gas chromatography/Flame ionization detection; NIOSH II (6), Method #S85.
Routes of Entry: Inhalation, skin and/or eye contact.

Harmful Effects and Symptoms
Short Term Exposure: Irritation of the respiratory system, excitement, confusion, anesthesia, liquid: frostbite.
Points of Attack: Respiratory system, central nervous system.
Medical Surveillance: Consider the points of attack in preplacement and periodic physical examinations.
First Aid: If this chemical gets into the eyes, remove any contact lenses at once and irrigate immediately for at least 15 min, occasionally lifting upper and lower lids. Seek medical attention immediately. If this chemical contacts the skin, remove contaminated clothing and wash immediately with soap and water. Seek medical attention immediately. If this chemical has been inhaled, remove from exposure, begin rescue breathing (using universal precautions, including resuscitation mask) if breathing has stopped and CPR if heart action has stopped. Transfer promptly to a medical facility. When this chemical has been swallowed, get medical attention. Give large quantities of water and induce vomiting. Do not make an unconscious person vomit. If frostbite has occurred, seek medical attention immediately; do NOT rub the affected areas or flush them with water. In order to prevent further tissue damage, do NOT attempt to remove frozen clothing from frostbitten areas. If frostbite has NOT occurred, immediately and thoroughly wash contaminated skin with soap and water.
Personal Protective Methods: Wear appropriate personal protective clothing to prevent the skin from becoming frozen from contact with the evaporating liquid or from contact with vessels containing the liquid. Safety equipment suppliers/manufacturers can provide recommendations on the most protective glove/clothing material for your operation. All protective clothing (suits, gloves, footwear, headgear) should be clean, available each day, and put on before work. Contact lenses should not be worn when working with this chemical. Wear gas-proof chemical goggles and face shield unless full face-piece respiratory protection is worn. Employees should wash immediately with soap when skin is wet or contaminated. Provide emergency showers and eyewash.
Respirator Selection: Up to 3400 ppm: Sa (APF = 10) (any supplied-air respirator) or SCBAF (APF = 50) (any self-contained breathing apparatus with a full face-piece). *Emergency or planned entry into unknown concentrations or IDLH conditions:* SCBAF: Pd,Pp (APF = 10,000) (any self-contained breathing apparatus that has a full face-piece and is operated in a pressure-demand or other positive-pressure mode) or SaF: Pd,Pp: ASCBA (APF = 10,000) (any supplied-air respirator that has a full face-piece and is operated in a pressure-demand or other positive-pressure mode in combination with an auxiliary, self-contained breathing apparatus operated in a pressure-demand or other positive-pressure mode). *Escape:* GmFOv (APF = 50) [any air-purifying, full-face-piece respirator (gas mask) with a chin-style, front- or back-mounted organic vapor canister] or SCBAE (any appropriate escape-type, self-contained breathing apparatus).

Storage: Color Code—Red Stripe: Flammability Hazard: Store separately from all other flammable materials. Prior to working with this chemical you should be trained on its proper handling and storage. Before entering confined space where this chemical may be present, check to make sure that an explosive concentration does not exist. Store in tightly closed containers in a cool, well-ventilated area away from oxidizers. Sources of ignition, such as smoking and open flames, are prohibited where this chemical is used, handled, or stored in a manner that could create a potential fire or explosion hazard. Wherever this chemical is used, handled, manufactured, or stored, use explosion-proof electrical equipment and fittings. Procedures for the handling, use and storage of cylinders should be in compliance with OSHA 1910.101 and 1910.169, as with the recommendations of the Compressed Gas Association.
Shipping: This compound requires a shipping label of "FLAMMABLE GAS." It falls in Hazard Class 2.1.
Spill Handling: If in a building, evacuate building and confine vapors by closing doors and shutting down HVAC systems. Restrict persons not wearing protective equipment from area of spill or leak until cleanup is complete. Remove all ignition sources. Establish forced ventilation to keep levels below explosive limit and to disperse the gas. Wear chemical protective suit with self-contained breathing apparatus to combat spills. Stay upwind and use water spray to "knock down" vapor; contain runoff. Stop the flow of gas, if it can be done safely from a distance. If source is a cylinder and the leak cannot be stopped in place, remove the leaking cylinder to a safe place, and repair leak or allow cylinder to empty. Keep this chemical out of confined spaces, such as a sewer, because of the possibility of explosion, unless the sewer is designed to prevent the buildup of explosive concentrations. If employees are required to clean up spills, they must be properly trained and equipped. OSHA 1910.120(q) may be applicable.
Fire Extinguishing: This chemical is a flammable gas. Poisonous gases are produced in fire. Do not extinguish the fire unless the flow of gas can be stopped and any remaining gas is out of the line. Specially trained personnel may use fog lines to cool exposures and let the fire burn itself out. Vapors are heavier than air and will collect in low areas. Vapors may travel long distances to ignition sources and flashback. Vapors in confined areas may explode when exposed to fire. Containers may explode in fire. Storage containers and parts of containers may rocket great distances, in many directions. If material or contaminated runoff enters waterways, notify downstream users of potentially contaminated waters. Notify local health and fire officials and pollution control agencies. From a secure, explosion-proof location, use water spray to cool exposed containers. If cooling streams are ineffective (venting sound increases in volume and pitch, tank discolors, or shows any signs of deforming), withdraw immediately to a secure position. If cylinders are exposed to excessive heat from fire or flame contact, withdraw immediately to a secure location.

If employees are expected to fight fires, they must be trained and equipped in OSHA 1910.156. The only respirators recommended for firefighting are self-contained breathing apparatuses that have full face-pieces and are operated in a pressure-demand or other positive-pressure mode.

Disposal Method Suggested: Dissolve or mix the material with a combustible solvent and burn in a chemical incinerator equipped with an afterburner and scrubber. All federal, state, and local environmental regulations must be observed.

Methyl acrylate   M:0650

Molecular Formula: $C_4H_6O_2$
Common Formula: $CH_2 = CHCOOCH_3$
Synonyms: Acrilato de metilo (Spanish); Acrylate de methyle (French); Acrylic acid methyl ester; Acrylsaeuremethylester (German); Curithane 103; Methoxycarbonylethylene; Methyl-acrylat (German); Methyl acrylate; Methyl propenate; Methyl 2-propenoate; Methyl propenoate; 2-Propenoic acid, methyl ester; Propenoic acid methyl ester
CAS Registry Number: 96-33-3
RTECS® Number: AT2800000
UN/NA & ERG Number: UN1919 (stabilized)/129
EC Number: 202-500-6 [*Annex I Index No.:* 607-034-00-0]
Regulatory Authority and Advisory Bodies
Carcinogenicity: IARC: Human No Adequate Data, animal Inadequate Evidence, *not classifiable as carcinogenic to humans*, Group 3, 1999; EPA: Not Classifiable as to human carcinogenicity.
Air Pollutant Standard Set. See below, "Permissible Exposure Limits in Air" section.
US EPA Hazardous Waste Number (RCRA No.): U328.
RCRA, 40CFR261, Appendix 8 Hazardous Constituents.
Reportable Quantity (RQ): 100 lb (45.4 kg).
EPCRA Section 313 Form R *de minimis* concentration reporting level: 1.0%.
Canada, WHMIS, Ingredients Disclosure List Concentration 1.0%.
European/International Regulations: Hazard Symbol: F, Xi; Risk phrases: R11; R36; R66; R67; Safety phrases: S2; S16; S26; S29; S33 (see Appendix 4).
WGK (German Aquatic Hazard Class): 2—Hazard to waters.
Description: Methyl acrylate is a clear, colorless liquid with a sharp, fruity odor. Molecular weight = 86.10; Specific gravity (H_2O:1) = 0.96; Boiling point = 80°C; Freezing/Melting point = −76.7°C; Flash point = −3°C (oc); Autoignition temperature = 468°C. Explosive limits: LEL = 2.8%; UEL = 25%. Hazard Identification (based on NFPA-704 M Rating System): Health 3, Flammability 3, Reactivity 2. Slightly soluble in water; solubility = 6%.
Potential Exposure: Compound Description: Tumorigen, Mutagen; Reproductive Effector; Human Data; Primary Irritant. Methyl acrylate is used in production of acrylates, copolymers, barrier resins; surfactants for shampoos; as a monomer in the manufacture of polymers for plastic films, textiles, paper, and leather coating resins. It is also used as a pesticide intermediate and in pharmaceutical manufacture.
Incompatibilities: Forms explosive mixture in air. Incompatible with nitrates, oxidizers, such as peroxides, strong alkalis. Polymerizes easily from heat, light, peroxides; usually contains an inhibitor, such as hydroquinone.
Permissible Exposure Limits in Air
Conversion factor: 1 ppm = 3.52 mg/m³ at 25°C & 1 atm.
OSHA PEL: 10 ppm/35 mg/m³ TWA [skin].
NIOSH REL: 10 ppm/35 mg/m³ TWA [skin].
ACGIH TLV®[1]: 2 ppm/7 mg/m³ TWA [skin]; danger of sensitization.
NIOSH IDLH: 250 ppm.
Protective Action Criteria (PAC)
TEEL-0: 2 ppm
PAC-1: 2 ppm
PAC-2: 7.5 ppm
PAC-3: 250 ppm
DFG MAK: 5 ppm/18 mg/m³, Peak Limitation Category I (1) danger of skin sensitization; Pregnancy Risk Group D.
Several states have set guidelines or standards for methyl acrylate in ambient air[60] ranging from 4.8 μg/m³ (Massachusetts) to 350 μg/m³ (North Dakota) to 500 μg/m³ (Virginia) to 700 μg/m³ (Connecticut) to 833 μg/m³ (Nevada).
Determination in Air: Charcoal adsorption, workup with CS_2, analysis by gas chromatography/flame ionization detection; NIOSH Analytical Method (IV) #1459; #2552; OSHA Analytical Method 92.
Permissible Concentration in Water: Russia[43] set a MAC of 0.02 mg/L in water bodies used for domestic purposes.
Determination in Water: Octanol−water coefficient: Log $K_{ow} = 0.79$.
Routes of Entry: Inhalation, skin absorption, ingestion, skin and/or eye contact.
Harmful Effects and Symptoms
Short Term Exposure: Methyl acrylate can affect you when breathed in and by passing through your skin. Irritates the eyes, skin, and respiratory tract. Eye and skin contact can cause severe irritation and burns. A lacrimator. Breathing high levels may cause severe lung irritation and may lead to fluid in the lungs (pulmonary edema). This can cause death. Prolonged contact can cause severe damage to the skin and eyes.
Long Term Exposure: Prolonged or repeated exposure to methyl acrylate may cause liver and kidney damage. Exposure may cause skin sensitization and allergy to develop.
Points of Attack: Eyes, skin, respiratory system.
Medical Surveillance: For those with frequent or potentially high exposure (half the TLV or greater), the following are recommended before beginning work and at regular times after that: lung function tests. If symptoms develop or overexposure is suspected, the following may be useful: evaluation by a qualified allergist, including careful

exposure history and special testing, may help diagnose skin allergy. Consider chest X-ray after acute overexposure. Liver and kidney function tests.

First Aid: If this chemical gets into the eyes, remove any contact lenses at once and irrigate immediately for at least 15 min, occasionally lifting upper and lower lids. Seek medical attention immediately. If this chemical contacts the skin, remove contaminated clothing and wash immediately with soap and water. Seek medical attention immediately. If this chemical has been inhaled, remove from exposure, begin rescue breathing (using universal precautions, including resuscitation mask) if breathing has stopped and CPR if heart action has stopped. Transfer promptly to a medical facility. When this chemical has been swallowed, get medical attention. Give large quantities of water and induce vomiting. Do not make an unconscious person vomit. Medical observation is recommended for 24–48 h after breathing overexposure, as pulmonary edema may be delayed. As first aid for pulmonary edema, a doctor or authorized paramedic may consider administering a corticosteroid spray.

Personal Protective Methods: Wear protective gloves and clothing to prevent any reasonable probability of skin contact: **8 h**: butyl rubber gloves, suits, boots; Tychem 1000™ suits. **4 h**: Teflon™ gloves, suits, boots. Also, polyvinyl alcohol is among the recommended protective materials. Safety equipment suppliers/manufacturers can provide recommendations on the most protective glove/clothing material for your operation. All protective clothing (suits, gloves, footwear, headgear) should be clean, available each day, and put on before work. Contact lenses should not be worn when working with this chemical. Wear splash-proof chemical goggles and face shield unless full face-piece respiratory protection is worn. Employees should wash immediately with soap when skin is wet or contaminated. Provide emergency showers and eyewash.

Respirator Selection: NIOSH: *Up to 100 ppm:* Sa (APF = 10) (any supplied-air respirator). *250 ppm:* Sa:Cf (APF = 25) (any supplied-air respirator operated in a continuous-flow mode) or SCBAF (APF = 50) (any self-contained breathing apparatus with a full face-piece) or SaF (APF = 50) (any supplied-air respirator with a full face-piece). *Emergency or planned entry into unknown concentrations or IDLH conditions:* SCBAF: Pd,Pp (APF = 10,000) (any self-contained breathing apparatus that has a full face-piece and is operated in a pressure-demand or other positive-pressure mode) or SaF: Pd,Pp: ASCBA (APF = 10,000) (any supplied-air respirator that has a full face-piece and is operated in a pressure-demand or other positive-pressure mode in combination with an auxiliary, self-contained breathing apparatus operated in a pressure-demand or other positive-pressure mode). *Escape:* GmFOv (APF = 50) [any air-purifying, full-face-piece respirator (gas mask) with a chin-style, front- or back-mounted organic vapor canister] or SCBAE (any appropriate escape-type, self-contained breathing apparatus).

Storage: Color Code—Red: Flammability Hazard: Store in a flammable liquid storage area or approved cabinet away from ignition sources and corrosive and reactive materials. Prior to working with this chemical you should be trained on its proper handling and storage. Before entering confined space where this chemical may be present, check to make sure that an explosive concentration does not exist. Methyl acrylate must be stored to avoid contact with oxidizers (such as perchlorates, peroxides, permanganates, chlorates, and nitrates), since violent reactions occur. Store in tightly closed containers in a cool, well-ventilated area away from heat and moisture. At temperatures above 20°C a violent reaction could take place. Methyl acrylate should always be used with an inhibitor. Check that the correct concentration of inhibitor is used; if it is not, a violent reaction could occur. Sources of ignition, such as smoking and open flames, are prohibited where methyl acrylate is handled, used, or stored. Metal containers involving the transfer of 5 gallons or more of methyl acrylate should be grounded and bonded. Drums must be equipped with self-closing valves, pressure vacuum bungs, and flame arresters. Use only non-sparking tools and equipment, especially when opening and closing containers of methyl acrylate. Wherever methyl acrylate is used, handled, manufactured, or stored, use explosion-proof electrical equipment and fittings.

Shipping: Methyl acrylate, inhibited, requires a shipping label of "FLAMMABLE LIQUID." It falls in Hazard Class 3 and Packing Group II.

Spill Handling: Evacuate and restrict persons not wearing protective equipment from area of spill or leak until cleanup is complete. Remove all ignition sources. Establish forced ventilation to keep levels below explosive limit. Absorb liquids in vermiculite, dry sand, earth, peat, carbon, or a similar material and deposit in sealed containers. Keep this chemical out of a confined space, such as a sewer, because of the possibility of an explosion, unless the sewer is designed to prevent the buildup of explosive concentrations. It may be necessary to contain and dispose of this chemical as a hazardous waste. If material or contaminated runoff enters waterways, notify downstream users of potentially contaminated waters. Contact your local or federal environmental protection agency for specific recommendations. If employees are required to clean up spills, they must be properly trained and equipped. OSHA 1910.120(q) may be applicable.

Fire Extinguishing: This chemical is a flammable liquid. Poisonous gases are produced in fire. Use dry chemical, carbon dioxide, or alcohol foam extinguishers. Vapors are heavier than air and will collect in low areas. Vapors may travel long distances to ignition sources and flashback. Vapors in confined areas may explode when exposed to fire. Containers may explode in fire. Storage containers and parts of containers may rocket great distances, in many directions. If material or contaminated runoff enters waterways, notify downstream users of potentially contaminated waters. Notify local health and fire officials and pollution control agencies. From a secure, explosion-proof location, use water spray to cool exposed containers. If cooling streams are ineffective (venting sound increases in volume and pitch, tank discolors, or shows any signs of deforming), withdraw immediately to a secure position. If employees are expected to fight fires, they must be trained and equipped in OSHA 1910.156. The only respirators recommended for firefighting are self-contained breathing apparatuses that have full face-pieces and are operated in a pressure-demand or other positive-pressure mode.

Disposal Method Suggested: Dissolve or mix the material with a combustible solvent and burn in a chemical incinerator equipped with an afterburner and scrubber. All federal, state, and local environmental regulations must be observed. Consult with environmental regulatory agencies for guidance on acceptable disposal practices. Generators of waste containing this contaminant (≥100 kg/mo) must conform with EPA regulations governing storage, transportation, treatment, and waste disposal.

Reference

New Jersey Department of Health and Senior Services. (March 2006). *Hazardous Substances Fact Sheet: Methyl Acrylate.* Trenton, NJ

Methylal   M:0660

Molecular Formula: $C_3H_8O_2$
Common Formula: $CH_3OCH_2OCH_3$
Synonyms: Anesthenyl;   Dimethoxymethane; Dimethylacetal formaldehyde; Formal; Formaldehyde dimethyl acetal; Methane, dimethoxy-; Methyl, dimethoxy-; Methylene dimethyl ether; Methyl formal
CAS Registry Number: 109-87-5
RTECS® Number: PA8750000
UN/NA & ERG Number: UN1234/127
EC Number: 203-714-2
Regulatory Authority and Advisory Bodies
Air Pollutant Standard Set. See below, "Permissible Exposure Limits in Air" section.
Canada, WHMIS, Ingredients Disclosure List Concentration 1.0%.
European/International Regulations: not listed in Annex 1.
WGK (German Aquatic Hazard Class): 1—Low hazard to waters.
Description: Methylal is a colorless liquid with a pungent odor. Molecular weight = 76.11; Specific gravity (H_2O:1) = 0.86; Boiling point = 43.9°C; Freezing/Melting point = −105°C; Vapor pressure 330 mmHg at 20°C; Flash point = −32°C; Autoignition temperature = 237°C. Explosive limits: LEL = 2.2%; UEL = 13.8%. Hazard Identification (based on NFPA-704 M Rating System): Health 2, Flammability 3, Reactivity 2. Soluble in water; solubility = 33%.
Potential Exposure: Compound Description: Primary Irritant. Methylal is used as a specialty fuel and as a solvent in adhesives and protective coatings.

Incompatibilities: Forms explosive mixture with air. Methylal may be able to form unstable and explosive peroxides. Heating may cause explosion. Strong reaction with strong oxidizers, acids, causing fire and explosion hazard. Hydrolyzes readily in presence of acids to generate aldehydes.
Permissible Exposure Limits in Air
Conversion factor: 1 ppm = 3.11 mg/m³ at 25°C & 1 atm.
OSHA PEL: 1000 ppm/3100 mg/m³ TWA.
NIOSH REL: 1000 ppm/3100 mg/m³ TWA.
ACGIH TLV®[1]: 1000 ppm; 3110 mg/m³ TWA.
NIOSH IDLH: 2200 ppm [LEL].
Protective Action Criteria (PAC)
TEEL-0: 1000 ppm
PAC-1: 2200 ppm
PAC-2: 2200 ppm
PAC-3: 2200 ppm
DFG MAK: 1000 ppm/3200 mg/m³ TWA; Peak Limitation II(2); Pregnancy Group C.
Australia: TWA 1000 ppm (3100 mg/m³), 1993; Austria: MAK 1000 ppm (3100 mg/m³), 1999; Belgium: TWA 1000 ppm (3110 mg/m³), 1993; Denmark: TWA 1000 ppm (3100 mg/m³), 1999; Finland: TWA 1000 ppm (3100 mg/m³); STEL 1250 ppm (3900 mg/m³), 1999; France: VME 1000 ppm (3100 mg/m³), 1999; the Netherlands: MAC-TGG 3100 mg/m³, 2003; the Philippines: TWA 1000 ppm (3100 mg/m³), 1993; Poland: MAC (TWA) 1000 mg/m³, MAC (STEL) 3500 mg/m³, 1999; Russia: STEL 10 mg/m³, 1993; Switzerland: MAK-W 1000 ppm (3100 mg/m³), 1999; Turkey: TWA 1000 ppm (3100 mg/m³), 1993; United Kingdom: TWA 1000 ppm (3160 mg/m³); STEL 1250 ppm, 2000; Argentina, Bulgaria, Columbia, Jordan, South Korea, New Zealand, Singapore, Vietnam: ACGIH TLV®: TWA 1000 ppm. Several states have set guidelines or standards for methylal in ambient air[60] ranging from 3.875−31.0 mg/m³ (North Dakota) to 62.0 mg/m³ (Connecticut) to 73.8 mg/m³ (Nevada).
Determination in Air: Use NIOSH Analytical Method #1611.[18]
Routes of Entry: Inhalation, ingestion, skin and/or eye contact.
Harmful Effects and Symptoms
Short Term Exposure: Methylal can affect you when breathed in. Irritates the eyes, skin, and respiratory tract. Exposure to high levels may cause you to feel dizzy, lightheaded, and to pass out. Methylal can irritate the eyes, nose, throat and skin. Very high levels may irritate the lungs. High exposure may damage the liver and kidneys.
Long Term Exposure: The liquid removes the skin's natural oils. Repeated exposure may damage the liver and kidneys.
Points of Attack: Skin, eyes, respiratory system, central nervous system.
Medical Surveillance: If symptoms develop or overexposure is suspected, the following may be useful: liver and kidney function tests. Lung function tests. Consider chest X-ray after acute overexposure.

First Aid: If this chemical gets into the eyes, remove any contact lenses at once and irrigate immediately for at least 15 min, occasionally lifting upper and lower lids. Seek medical attention immediately. If this chemical contacts the skin, remove contaminated clothing and wash immediately with soap and water. Seek medical attention immediately. If this chemical has been inhaled, remove from exposure, begin rescue breathing (using universal precautions, including resuscitation mask) if breathing has stopped and CPR if heart action has stopped. Transfer promptly to a medical facility. When this chemical has been swallowed, get medical attention. Give large quantities of water and induce vomiting. Do not make an unconscious person vomit.

Personal Protective Methods: Wear protective gloves and clothing to prevent any reasonable probability of skin contact. Safety equipment suppliers/manufacturers can provide recommendations on the most protective glove/clothing material for your operation. All protective clothing (suits, gloves, footwear, headgear) should be clean, available each day, and put on before work. Contact lenses should not be worn when working with this chemical. Wear splash-proof chemical goggles and face shield unless full face-piece respiratory protection is worn. Employees should wash immediately with soap when skin is wet or contaminated. Provide emergency showers and eyewash.

Respirator Selection: Up to 2,200 ppm: Sa (APF = 10) (any supplied-air respirator) or SCBAF (APF = 50) (any self-contained breathing apparatus with a full face-piece). Unknown concentrations or IDLH conditions: SCBAF: Pd, Pp (APF = 10,000) (any self-contained breathing apparatus that has a full face-piece and is operated in a pressure-demand or other positive-pressure mode) or SaF: Pd,Pp: ASCBA (APF = 10,000) (any supplied-air respirator that has a full face-piece and is operated in a pressure-demand or other positive-pressure mode in combination with an auxiliary, self-contained breathing apparatus operated in a pressure-demand or other positive-pressure mode). *Escape:* GmFOv (APF = 50) [any air-purifying, full-face-piece respirator (gas mask) with a chin-style, front- or back-mounted organic vapor canister] or SCBAE (any appropriate escape-type, self-contained breathing apparatus).

Storage: Color Code—Red: Flammability Hazard: Store in a flammable liquid storage area or approved cabinet away from ignition sources and corrosive and reactive materials. Prior to working with this chemical you should be trained on its proper handling and storage. Before entering confined space where this chemical may be present, check to make sure that an explosive concentration does not exist. Methylal must be stored to avoid contact with strong oxidizers, such as chlorine, chlorine dioxide, bromine, nitrates and permanganates, since violent reactions occur. Store in tightly closed containers in a cool, well-ventilated area away from heat. Sources of ignition, such as smoking and open flames, are prohibited where methylal is handled, used, or stored. Metal containers involving the transfer of 5 gallons or more of methylal should be grounded and bonded. Drums must be equipped with self-closing valves, pressure vacuum bungs, and flame arresters. Use only non-sparking tools and equipment, especially when opening and closing containers of methylal. Wherever methylal is used, handled, manufactured, or stored, use explosion-proof electrical equipment and fittings.

Shipping: This compound requires a shipping label of "FLAMMABLE LIQUID." It falls in Hazard Class 3 and Packing Group II.

Spill Handling: Evacuate and restrict persons not wearing protective equipment from area of spill or leak until cleanup is complete. Remove all ignition sources. Establish forced ventilation to keep levels below explosive limit. Absorb liquids in vermiculite, dry sand, earth, peat, carbon, or a similar material and deposit in sealed containers. Keep this chemical out of a confined space, such as a sewer, because of the possibility of an explosion, unless the sewer is designed to prevent the buildup of explosive concentrations. It may be necessary to contain and dispose of this chemical as a hazardous waste. If material or contaminated runoff enters waterways, notify downstream users of potentially contaminated waters. Contact your local or federal environmental protection agency for specific recommendations. If employees are required to clean up spills, they must be properly trained and equipped. OSHA 1910.120(q) may be applicable.

Fire Extinguishing: This chemical is a flammable liquid. Poisonous gases are produced in fire. Use dry chemical, carbon dioxide, or alcohol foam extinguishers. Vapors are heavier than air and will collect in low areas. Vapors may travel long distances to ignition sources and flashback. Vapors in confined areas may explode when exposed to fire. Containers may explode in fire. Storage containers and parts of containers may rocket great distances, in many directions. If material or contaminated runoff enters waterways, notify downstream users of potentially contaminated waters. Notify local health and fire officials and pollution control agencies. From a secure, explosion-proof location, use water spray to cool exposed containers. If cooling streams are ineffective (venting sound increases in volume and pitch, tank discolors, or shows any signs of deforming), withdraw immediately to a secure position. If employees are expected to fight fires, they must be trained and equipped in OSHA 1910.156. The only respirators recommended for firefighting are self-contained breathing apparatuses that have full face-pieces and are operated in a pressure-demand or other positive-pressure mode.

Disposal Method Suggested: Concentrated waste containing no peroxides: discharge liquid at a controlled rate near a pilot flame. *Concentrated waste containing peroxides:* perforation of containers of the waste from a safe distance followed by open burning.

Reference

New Jersey Department of Health and Senior Services. (April 2004). *Hazardous Substances Fact Sheet: Methylal.* Trenton, NJ

Methyl alcohol   M:0670

Molecular Formula: CH_4O
Common Formula: CH_3OH
Synonyms: Alcohol metilico (Spanish); Alcool methylique (French); Carbinol; Colonial spirit; Columbian spirit; Metanol (Spanish); Methyl alcohol; Methylalkohol (German); Methyl hydroxide; Methylol; Monohydroxymethane; Pyroxylic spirit; Wood alcohol; Wood naphtha; Wood spirit
CAS Registry Number: 67-56-1; *(alt.)* 54841-71-3
RTECS® Number: PC1400000
UN/NA & ERG Number: UN1230/131
EC Number: 200-659-6 [*Annex I Index No.:* 603-001-00-X]
Regulatory Authority and Advisory Bodies
US EPA Gene-Tox Program, Negative: SHE—clonal assay; Cell transform.—SA7/SHE; Negative: *N. crassa—aneuploidy*; *In vitro* SCE—nonhuman.
US EPA, FIFRA 1998 Status of Pesticides: Canceled.
Banned or Severely Restricted (Thailand and Sweden) (UN).[13]
Air Pollutant Standard Set. See below, "Permissible Exposure Limits in Air" section.
Clean Air Act: Hazardous Air Pollutants (Title I, Part A, Section 112); Accidental Release Prevention/Flammable Substances, (Section 112[r], Table 3), TQ = 5000 lb (2270 kg).
US EPA Hazardous Waste Number (RCRA No.): U154.
RCRA, 40CFR261, Appendix 8 Hazardous Constituents.
RCRA 40CFR268.48; 61FR15654, Universal Treatment Standards: Wastewater (mg/L), 5.6; Nonwastewater (mg/L/ TCLP), 0.75.
Reportable Quantity (RQ): 5000 lb (2270 kg).
EPCRA Section 313 Form R *de minimis* concentration reporting level: 1.0%.
Canada, WHMIS, Ingredients Disclosure List Concentration 1.0%; Class B2, D1B, D2B.
European/International Regulations: Hazard Symbol: F, T; Risk phrases: R11; R23/24/25; R39/23/24/25; Safety phrases: S1/2; S7; S16; S35/36; S45 (see Appendix 4).
WGK (German Aquatic Hazard Class): 1—Low hazard to waters.
Description: Methyl alcohol is a colorless, volatile liquid with a mild odor. Odor threshold = 100 ppm.[41] Molecular weight = 32.04; Specific gravity (H_2O:1) = 0.79; Boiling point = 63.9°C; Freezing/Melting point = −98°C; Vapor pressure = 96 mmHg at 20°C; 127 mmHg at 25°C; Flash point = 12°C; Autoignition temperature = 464°C. Explosive limit: LEL = 6.0%; UEL = 36.0%. Hazard Identification (based on NFPA-704 M Rating System): Health 1, Flammability 3, Reactivity 0. Soluble in water.
Potential Exposure: Compound Description: Drug, Mutagen; Reproductive Effector; Human Data; Primary Irritant. Methyl alcohol is used as a starting material in organic synthesis of chemicals, such as formaldehyde,

methacrylates, methyl amines, methyl halides, ethylene glycol, and pesticides, and as an industrial solvent for inks, resins, adhesives, and dyes. It is an ingredient in paint and varnish removers, cleaning and dewaxing preparations, spirit duplicating fluids, embalming fluids, antifreeze mixtures, and enamels, and is used in the manufacture of photographic film, plastics, celluloid, textile soaps, wood stains, coated fabrics, shatterproof glass, paper coating, waterproofing formulations, artificial leather, and synthetic indigo and other dyes. It has also been used as an extractant in many other processes, an antidetonant fuel-injection fluid for aircraft, a rubber accelerator, and a denaturant for ethyl alcohol.
Incompatibilities: Methanol reacts violently with strong oxidizers, causing a fire and explosion hazard.
Permissible Exposure Limits in Air
Conversion factor: 1 ppm = 1.31 mg/m^3 at 25°C & 1 atm.
OSHA PEL: 200 ppm/260 mg/m^3 TWA [skin].
NIOSH REL: 200 ppm/260 mg/m^3 TWA; 250 ppm/325 mg/ m^3 STEL [skin].
ACGIH TLV®[1]: 200 ppm/262 mg/m^3 TWA; 250 ppm/ 328 mg/m^3 STEL [skin]; BEI issued (2004).
Protective Action Criteria (PAC)*
TEEL-0: 200 ppm
PAC-1: **530** ppm
PAC-2: **2100** ppm
PAC-3: **7200** ppm
*AEGLs (Acute Emergency Guideline Levels) & ERPGs (Emergency Response Planning Guideline) are in **bold face**.
Emergency Response Planning Guidelines (AIHA)
ERPG-1: 200 ppm
ERPG-2: 1000 ppm
ERPG-3: 5000 ppm
DFG MAK: 200 ppm/270 mg/m^3 Peak Limitation Category II(4) [skin]; Pregnancy Risk Group C; BAT: 30 mg/L methanol in urine/end-of-shift; for long-term exposure, after several shifts.
NIOSH IDLH: 6000 ppm.
Arab Republic of Egypt: TWA 200 ppm (260 mg/m^3), [skin], 1993; Australia: TWA 200 ppm (260 mg/m^3); STEL 250 ppm, [skin], 1993; Austria: MAK 200 ppm (260 mg/ m^3), [skin], 1999; Belgium: TWA 200 ppm (262 mg/m^3); STEL 250 ppm, [skin], 1993; Denmark: TWA 200 ppm (260 mg/m^3), [skin], 1999; Finland: TWA 200 ppm (260 mg/m^3); STEL 250 ppm, [skin], 1999; France: VME 200 ppm, VLE 1000 ppm, 1999; Hungary: TWA 50 mg/m^3; STEL 100 mg/m^3, [skin], 1993; Japan 200 ppm (260 mg/ m^3), [skin], 1999; the Netherlands: MAC-TGG 260 mg/m^3, [skin], 2003; Norway: TWA 100 ppm (130 mg/m^3), 1999; the Philippines: TWA 200 ppm (260 mg/m^3), 1993; Poland: MAC (TWA) 100 mg/m^3, MAC (STEL) 300 mg/m^3, 1999; Russia: TWA 200 ppm; STEL 5 mg/m^3, [skin], 1993; Sweden: NGV 200 ppm (250 mg/m^3), KTV 250 ppm (350 mg/m^3), [skin], 1999; Thailand: TWA 200 ppm (260 mg/m^3), 1993; Turkey: TWA 200 ppm (260 mg/m^3),

1993; United Kingdom: TWA 200 ppm (255 mg/m^3); STEL 250 ppm, [skin], 2000; Argentina, Bulgaria, Columbia, Jordan, South Korea, New Zealand, Singapore, Vietnam: ACGIH TLV®: STEL 250 ppm [skin]. Russia[43] set a MAC of 1.0 mg/m^3 in ambient air in residential areas on a momentary basis and 0.5 mg/m^3 on a daily average basis. Several states have set guidelines or standards for methanol in ambient air[60] ranging from 0.036 mg/m^3 (Massachusetts) to 0.62 mg/m^3 (Kansas) to 2.6–3.1 mg/m^3 (North Dakota) to 4.3 mg/m^3 (Virginia) to 5.2 mg/m^3 (Connecticut and South Dakota) to 6.19 mg/m^3 (Nevada).

Determination in Air: Use NIOSH Analytical Method #2000, Methanol.[18]; #2549 Volatile organic compounds; OSHA Analytical Method 91.

Permissible Concentration in Water: EPA[32] has suggested a permissible ambient goal of 3600 μg/L based on health effects. Russia[43] set a MAC of 3 mg/L in water bodies used for domestic purposes and of 0.1 mg/L in water bodies used for fishery purposes.

Determination in Water: Octanol–water coefficient: Log K_{ow} = (estimated) <-0.77.

Routes of Entry: Inhalation of vapor, percutaneous absorption of liquid, ingestion, eye and/or skin contact.

Harmful Effects and Symptoms

Short Term Exposure: Irritates the eyes, skin, and respiratory tract. *Inhalation:* Below 500 ppm symptoms are rarely felt. Can cause headache, vomiting, irritation of the nose and throat, dilation of the pupils, feeling of intoxication, loss of muscle coordination, excessive sweating, bronchitis, and convulsions. Very high exposures may result in stupor, cramps, and visual difficulties, such as spotted vision, sensitivity to light, eye tenderness, and blindness. Recovery is not always complete and symptoms may recur without additional exposure. *Skin:* Can cause dry and cracked skin, irritation, and reddening. Skin absorption can be enough to contribute to symptoms described under inhalation. *Eyes:* Can cause irritation of eyes. *Ingestion:* Symptoms are similar to those under inhalation, plus damage to liver, kidneys, and heart. Nerve damage may occur causing loss of coordination and blindness. Recovery is not always complete. Death may occur. Usual fatal dose is about 100–250 mL but death from ingestion has occurred from as little as 30 mL (about 1 oz).

Long Term Exposure: Exposure to low levels may cause many of the symptoms listed above. Skin contact causes dryness and cracking. May cause liver damage. Because methyl alcohol is slowly eliminated from body, repeated low exposures may buildup to high levels, causing severe symptoms. Recovery is not always complete. Methanol has been found to be a teratogen (changes in the genetic material) in animals. Whether it does in humans is unknown.

Points of Attack: Eyes, skin, respiratory system, central nervous system, gastrointestinal tract.

Medical Surveillance: NIOSH lists the following tests: whole blood (chemical/metabolite); whole blood (chemical/metabolite) pH (Hydrogen ion concentration); whole blood (chemical/metabolite), pre- and postshift; blood plasma, bicarbonate, expired air, liver function tests; urine (chemical/metabolite); urine (chemical/metabolite), end-of-shift; urine (chemical/metabolite) end-of-work-week; urine (chemical/metabolite) prior to next shift.

First Aid: If this chemical gets into the eyes, remove any contact lenses at once and irrigate immediately for at least 15 min, occasionally lifting upper and lower lids. Seek medical attention immediately. If this chemical contacts the skin, remove contaminated clothing and wash immediately with soap and water. Seek medical attention immediately. If this chemical has been inhaled, remove from exposure, begin rescue breathing (using universal precautions, including resuscitation mask) if breathing has stopped and CPR if heart action has stopped. Transfer promptly to a medical facility. When this chemical has been swallowed, get medical attention. Give large quantities of water and induce vomiting. Do not make an unconscious person vomit.

Personal Protective Methods: Wear solvent-resistant gloves and clothing to prevent any reasonable probability of skin contact: **8 h:** butyl rubber gloves, suits, boots; Teflon™ gloves, suits, boots; Viton™ gloves, suits, Saranex™ coated suits; 4H™ and Silver Shield™ gloves, Responder™ suits, Trellchem™ HPS suits; Trychem 1000™ suits. Also, Viton™/chlorobutyl rubber, polyvinyl acetate, styrene/butadiene rubber, Viton™/Neoprene™, butyl/Neoprene™, and chlorinated polyethylene are among the recommended protective materials. Safety equipment suppliers/manufacturers can provide recommendations on the most protective glove/clothing material for your operation. All protective clothing (suits, gloves, footwear, headgear) should be clean, available each day, and put on before work. Contact lenses should not be worn when working with this chemical. Wear splash-proof chemical goggles and face shield unless full face-piece respiratory protection is worn. Employees should wash immediately with soap when skin is wet or contaminated. Provide emergency showers and eyewash.

Respirator Selection: 2000 ppm: Sa (APF = 10) (any supplied-air respirator). 5000 ppm: Sa:Cf (APF = 25) (any supplied-air respirator operated in a continuous-flow mode). 6000 ppm: SaT: Cf (APF = 50) (any supplied-air respirator that has a tight-fitting face-piece and is operated in a continuous-flow mode) or SCBAF (APF = 50) (any self-contained breathing apparatus with a full face-piece) or SaF (APF = 50) (any supplied-air respirator with a full face-piece). *Emergency or planned entry into unknown concentrations or IDLH conditions:* SCBAF: Pd,Pp (APF = 10,000) (any self-contained breathing apparatus that has a full face-piece and is operated in a pressure-demand or other positive-pressure mode) or SaF: Pd,Pp: ASCBA (any supplied-air respirator that has a full face-piece and is operated in a pressure-demand or other positive-pressure mode in combination with an auxiliary self-contained breathing apparatus operated in a pressure-demand or other positive-pressure mode). *Escape:* SCBAE (any appropriate escape-type, self-contained breathing apparatus).

Storage: Color Code—Red: Flammability Hazard: Store in a flammable liquid storage area or approved cabinet away from ignition sources and corrosive and reactive materials. Prior to working with this chemical you should be trained on its proper handling and storage. Before entering confined space where this chemical may be present, check to make sure that an explosive concentration does not exist. Methyl alcohol must be stored to avoid contact with strong oxidizers (such as chlorine, bromine, and fluorine). Store in tightly closed containers in a cool, well-ventilated area away from heat. Sources of ignition, such as smoking and open flames, are prohibited where methyl alcohol is handled, used, or stored. Metal containers involving the transfer of 5 gallons or more should be grounded and bonded. Drums must be equipped with self-closing valves, pressure vacuum bungs, and flame arresters. Use only nonsparking tools and equipment, especially when opening and closing containers of methyl alcohol.

Shipping: This compound requires a shipping label of "FLAMMABLE LIQUID, POISONOUS/TOXIC MATERIALS." It falls in Hazard Class 3 and Packing Group II.

Spill Handling: Evacuate and restrict persons not wearing protective equipment from area of spill or leak until cleanup is complete. Remove all ignition sources. Establish forced ventilation to keep levels below explosive limit. Absorb liquids in vermiculite, dry sand, earth, peat, carbon, or a similar material and deposit in sealed containers. Keep this chemical out of a confined space, such as a sewer, because of the possibility of an explosion, unless the sewer is designed to prevent the buildup of explosive concentrations. It may be necessary to contain and dispose of this chemical as a hazardous waste. If material or contaminated runoff enters waterways, notify downstream users of potentially contaminated waters. Contact your local or federal environmental protection agency for specific recommendations. If employees are required to clean up spills, they must be properly trained and equipped. OSHA 1910.120(q) may be applicable.

Fire Extinguishing: This chemical is a flammable liquid. Poisonous gases, including formaldehyde, are produced in fire. Use dry chemical, carbon dioxide, or alcohol foam extinguishers. Vapors are heavier than air and will collect in low areas. Vapors may travel long distances to ignition sources and flashback. Vapors in confined areas may explode when exposed to fire. Containers may explode in fire. Storage containers and parts of containers may rocket great distances, in many directions. If material or contaminated runoff enters waterways, notify downstream users of potentially contaminated waters. Notify local health and fire officials and pollution control agencies. From a secure, explosion-proof location, use water spray to cool exposed containers. If cooling streams are ineffective (venting sound increases in volume and pitch, tank discolors, or shows any signs of deforming), withdraw immediately to a secure position. If employees are expected to fight fires, they must be trained and equipped in OSHA 1910.156. The only respirators recommended for firefighting are self-contained breathing apparatuses that have full face-pieces and are operated in a pressure-demand or other positive-pressure mode.

Disposal Method Suggested: Consult with environmental regulatory agencies for guidance on acceptable disposal practices. Generators of waste containing this contaminant ($\geq$100 kg/mo) must conform with EPA regulations governing storage, transportation, treatment, and waste disposal. Incineration.

References

National Institute for Occupational Safety and Health. (1976). *Criteria for a Recommended Standard: Occupational Exposure to Methyl Alcohol*, NIOSH Document No. 76-148. Washington, DC

US Environmental Protection Agency. (July 11, 1977) (Revised 1979). *Chemical Hazard Information Profile: Methanol*. Washington, DC

US Environmental Protection Agency. (April 30, 1980). *Methyl Alcohol: Health and Environmental Effects Profile No. 126*. Washington, DC: Office of Solid Waste

Sax, N. I. (Ed.). (1985). *Dangerous Properties of Industrial Materials Report,* 5, No. 5, 58−64

New York State Department of Health. (February 1986). *Chemical Fact Sheet: Methyl alcohol.* Albany, NY: Bureau of Toxic Substance Assessment

US Environmental Protection Agency, Special Review and Reregistration Division Office of Pesticide Programs. (1998). *Agency Status of Pesticides in Registration, Reregistration, and Special Review* (Rainbow Report). Washington, DC

New Jersey Department of Health and Senior Services. (April 2002). *Hazardous Substances Fact Sheet: Methyl alcohol.* Trenton, NJ

Methylamine   M:0680

Molecular Formula: CH_5N
Common Formula: CH_3NH_2
Synonyms: Aminomethane; Carbinamine; Mercurialin; Methanamine; Methylamine; Metilamino (Spanish); Monomethylamine
CAS Registry Number: 74-89-5
RTECS® Number: PF6300000
UN/NA & ERG Number: UN1061 (anhydrous)/118; UN1235 (aqueous solution)/132
EC Number: 200-820-0 [*Annex I Index No.:* 612-001-00-9]
Regulatory Authority and Advisory Bodies
Department of Homeland Security Screening Threshold Quantity (pounds): *Release hazard* 10,000 ($\geq$1.00% concentration).
Air Pollutant Standard Set. See below, "Permissible Exposure Limits in Air" section.
OSHA 29CFR1910.119, Appendix A. Process Safety List of Highly Hazardous Chemicals, TQ = 1000 lb (450 kg).

Clean Air Act: Accidental Release Prevention/Flammable Substances, (Section 112[r], Table 3), TQ = 10,000 lb (4540 kg).

Clean Water Act: Section 311 Hazardous Substances/RQ 40CFR117.3 (same as CERCLA, see below).

Reportable Quantity (RQ): 100 lb (45.4 kg).

US DOT 49CFR172.101, Inhalation Hazardous Chemical.

Canada, WHMIS, Ingredients Disclosure List Concentration 0.1%.

European/International Regulations: Hazard Symbol: F +, Xn; Risk phrases: R12; R20; R37/38; R41; Safety phrases: S2; S16; S26; S39. (see Appendix 4).

WGK (German Aquatic Hazard Class): 2—Hazard to waters.

Description: Methylamine is a colorless gas with a fish- or ammonia-like odor; at low concentrations a fishy odor. Shipped as a liquefied compressed gas. The odor threshold is 3.2 ppm. Molecular weight = 31.07; Specific gravity (H_2O:1) = 0.7 (liquid); Boiling point = −6°C; Freezing/Melting point = −94°C; Relative vapor density (air = 1) = 1.08; Vapor pressure = 3 atm; Flash point = flammable gas; Autoignition temperature = 430°C. Explosive limits: LEL = 4.9%; UEL = 20.7%. Hazard Identification (based on NFPA-704 M Rating System): Health 3, Flammability 4, Reactivity 0. Soluble in water.

Potential Exposure: Compound Description: Agricultural Chemical; Mutagen, Natural Product; Primary Irritant. Methylamine is used in organic synthesis; a starting material for *N*-oleyltaurine, a surfactant; and *p-N*-methylaminophenol sulfate, a photographic developer. It has possible uses in solvent extraction systems in separation of aromatics from aliphatic hydrocarbons. It is also used in the synthesis of many different pharmaceuticals, pesticides, and rubber chemicals.

Incompatibilities: A medium strong base. Reacts violently with strong acids; mercury, strong oxidizers, nitromethane. Corrosive to copper, zinc alloys, aluminum, and galvanized surfaces.

Permissible Exposure Limits in Air

Conversion factor: 1 ppm = 1.27 mg/m^3 at 25°C & 1 atm.

OSHA PEL: 10 ppm/12 mg/m^3 TWA.

NIOSH REL: 10 ppm/12 mg/m^3 TWA.

ACGIH TLV®[1]: 5 ppm/6.4 mg/m^3; 15 ppm/19 mg/m^3 STEL.

Protective Action Criteria (PAC)*

TEEL-0: 10 ppm

PAC-1: **15** ppm

PAC-2: **64** ppm

PAC-3: **350** ppm

*AEGLs (Acute Emergency Guideline Levels) & ERPGs (Emergency Response Planning Guideline) are in **bold face**.

DFG MAK: 10 ppm/13 mg/m^3 TWA; Peak Limitation Category I(1); a momentary value of 10 mL/m^3/13 mg/m^3 should not be exceeded; Pregnancy Risk Group D.

NIOSH IDLH: 100 ppm.

Australia: TWA 10 ppm (12 mg/m^3), 1993; Austria: MAK 10 ppm (12 mg/m^3), 1999; Belgium: TWA 10 ppm (13 mg/m^3), 1993; Denmark: TWA 10 ppm (12 mg/m^3), [skin], 1999; Finland: STEL 10 ppm (12 mg/m^3), [skin], 1999; France: VLE 10 ppm (12 mg/m^3), 1999; Japan: 10 ppm (13 mg/m^3), 1999; the Netherlands: MAC-TGG 6.4 mg/m^3, 2003; Norway: TWA 10 ppm (12 mg/m^3), 1999; the Philippines: TWA 10 ppm (12 mg/m^3), 1993; Poland: MAC (TWA) 5 mg/m^3, MAC (STEL) 15 mg/m^3, 1999; Russia: TWA 10 ppm; STEL 1 mg/m^3, [skin], 1993; Sweden: NGV 10 ppm (13 mg/m^3), KTV 20 ppm (25 mg/m^3), [skin], 1999; United Kingdom: TWA 10 ppm (13 mg/m^3), 2000; Argentina, Bulgaria, Columbia, Jordan, South Korea, New Zealand, Singapore, Vietnam: ACGIH TLV®: STEL 15 ppm. Several states have set guidelines or standards for methylamine in ambient air[60] ranging from 28.6 μg/m^3 (Kansas) to 40.0 μg/m^3 (New York) to 120 μg/m^3 (Florida and North Dakota) to 200 μg/m^3 (Virginia) to 240 μg/m^3 (Connecticut) to 286 μg/m^3 (Nevada) to 300 μg/m^3 (South Carolina).

Determination in Air: Use OSHA Analytical Method 40.

Permissible Concentration in Water: Russia[43] set a MAC of 1.0 mg/L in water bodies used for drinking purposes.

Determination in Water: Octanol−water coefficient: Log K_{ow} = −0.69.

Routes of Entry: Inhalation, ingestion, skin absorption, eye and/or skin contact.

Harmful Effects and Symptoms

Short Term Exposure: Methylamine can affect you when breathed and by passing through your skin. Contact with the liquid can cause frostbite and severe burns of the eyes and skin. The vapor can irritate the eyes, nose, throat, and lungs. Higher levels can cause a buildup of fluid (pulmonary edema). This can cause death.

Long Term Exposure: Repeated or prolonged contact with skin may cause drying and cracking. May cause lung irritation and bronchitis.

Points of Attack: Eyes, skin, respiratory system.

Medical Surveillance: Before beginning employment and at regular times after that, the following are recommended: lung function test. If symptoms develop or overexposure is suspected, the following may be useful: consider chest X-ray after acute overexposure.

First Aid: If this chemical gets into the eyes, remove any contact lenses at once and irrigate immediately for at least 15 min, occasionally lifting upper and lower lids. Seek medical attention immediately. If this chemical contacts the skin, remove contaminated clothing and wash immediately with soap and water. Seek medical attention immediately. If this chemical has been inhaled, remove from exposure, begin rescue breathing (using universal precautions, including resuscitation mask) if breathing has stopped and CPR if heart action has stopped. Transfer promptly to a medical facility. When this chemical has been swallowed, get medical attention. Give large quantities of water and induce vomiting. Do not make an unconscious person vomit. Medical observation is recommended for 24—48 h after breathing overexposure, as pulmonary edema may be delayed. As first aid for pulmonary edema, a doctor or authorized paramedic may consider administering a corticosteroid spray.

Personal Protective Methods: Wear protective gloves and clothing to prevent any reasonable probability of skin contact: **8 h:** Responder™ suits; Trychem 1000™ suits. **4 h:** Teflon™ gloves, suits, boots. Also, nitrile rubber, Styrene-butadiene rubber, Neoprene™, Silvershield™, and PVC are among the recommended protective materials. Prevent possible skin freezing from direct liquid contact. Safety equipment suppliers/manufacturers can provide recommendations on the most protective glove/clothing material for your operation. All protective clothing (suits, gloves, footwear, headgear) should be clean, available each day, and put on before work. Contact lenses should not be worn when working with this chemical. Wear splash-proof chemical goggles and face shield when working with the liquid. Wear gas-proof chemical goggles and face shield when working with the gas, unless full face-piece respiratory protection is worn. Employees should wash immediately with soap when skin is wet or contaminated. Provide emergency showers and eyewash. Wear eye protection to prevent any possibility of eye contact. Employees should wash immediately when skin is wet or contaminated. Remove clothing immediately if wet or contaminated to avoid flammability hazard. Provide emergency showers and eyewash.

Respirator Selection: *100 ppm:* CcrFS (APF = 50) [any chemical cartridge respirator with a full face-piece and cartridge(s) providing protection against the compound of concern] or GmFS (APF = 50) [any air-purifying, full-face-piece respirator (gas mask) with a chin-style, front- or back-mounted canister providing protection against the compound of concern] or PaprS (APF = 25) [any powered, air-purifying respirator with cartridge(s) providing protection against the compound of concern] or SCBAF (APF = 50) (any self-contained breathing apparatus with a full face-piece) or SaF (APF = 50) (any supplied-air respirator with a full face-piece). *Emergency or planned entry into unknown concentrations or IDLH conditions:* SCBAF: Pd,Pp (APF = 10,000) (any self-contained breathing apparatus that has a full face-piece and is operated in a pressure-demand or other positive-pressure mode) or SaF: Pd,Pp: ASCBA (APF = 10,000) (any supplied-air respirator that has a full face-piece and is operated in a pressure-demand or other positive-pressure mode in combination with an auxiliary self-contained breathing apparatus operated in a pressure-demand or other positive-pressure mode). *Escape:* GmFS (APF = 50) [any air-purifying, full-face-piece respirator (gas mask) with a chin-style, front- or back-mounted canister providing protection against the compound of concern] or SCBAE (any appropriate escape-type, self-contained breathing apparatus).

Note: Substance causes eye irritation or damage; eye protection needed.

Storage: Color Code—Red Stripe: Flammability Hazard: Store separately from all other flammable materials. Prior to working with this chemical you should be trained on its proper handling and storage. Before entering confined space where this chemical may be present, check to make sure that an explosive concentration does not exist. Methylamine must be stored to avoid contact with mercury, flammable materials, and strong oxidizers, (such as chlorine dioxide, or bromine) since violent reactions occur. Store in tightly closed containers in a cool, well-ventilated area away from heat. Sources of ignition, such as smoking and open flames, are prohibited where methylamine is handled, used, or stored. Metal containers involving the transfer of 5 gallons or more of methylamine should be grounded and bonded. Drums must be equipped with self-closing valves, pressure vacuum bungs, and flame arresters. Use only nonsparking tools and equipment, especially when opening and closing containers of methylamine. Wherever methylamine is used, handled, manufactured, or stored, use explosion-proof electrical equipment and fittings. Procedures for the handling, use and storage of cylinders should be in compliance with OSHA 1910.101 and 1910.169, as with the recommendations of the Compressed Gas Association.

Shipping: Methylamine, *anhydrous,* requires a shipping label of "FLAMMABLE GAS." It falls in Hazard Class 2.1. Methylamine, *aqueous solution,* requires a shipping label of "FLAMMABLE LIQUID, CORROSIVE." It falls in Hazard Class 3 and Packing Group II.

Spill Handling: *Liquid:* Evacuate and restrict persons not wearing protective equipment from area of spill or leak until cleanup is complete. Remove all ignition sources. Establish forced ventilation to keep levels below explosive limit. Absorb liquids in vermiculite, dry sand, earth, peat, carbon, or a similar material and deposit in sealed containers. Alternatively, spread heavily with sodium bisulfate and sprinkle with water. Then drain into a sewer (see next sentence) with a large amount of water. Keep this chemical out of a confined space, such as a sewer, because of the possibility of an explosion, unless the sewer is designed to prevent the buildup of explosive concentrations. It may be necessary to contain and dispose of this chemical as a hazardous waste. If material or contaminated runoff enters waterways, notify downstream users of potentially contaminated waters. Contact your local or federal environmental protection agency for specific recommendations. If employees are required to clean up spills, they must be properly trained and equipped. OSHA 1910.120(q) may be applicable.

Gas: If in a building, evacuate building and confine vapors by closing doors and shutting down HVAC systems. Restrict persons not wearing protective equipment from area of spill or leak until cleanup is complete. Remove all ignition sources. Establish forced ventilation to keep levels below explosive limit and to disperse the gas. Wear chemical protective suit with self-contained breathing apparatus to combat spills. Stay upwind and use water spray to "knock down" vapor; contain runoff. Stop the flow of gas, if it can be done safely from a distance. If source is a cylinder and the leak cannot be stopped in place, remove the leaking cylinder to a safe place and repair leak or allow cylinder to empty. Keep this chemical out of confined spaces, such as a sewer, because of the possibility of explosion, unless the sewer is designed to prevent the buildup of explosive concentrations. If employees are required to clean up spills, they must be properly trained and equipped. OSHA 1910.120(q) may be applicable.

Fire Extinguishing: Methylamine is a flammable liquid or gas. If you are dealing with gas, stop the flow of gas if it can be done safely. Use water to keep fire-exposed containers cool and to protect people attempting shut-off. For water solutions, use water spray, CO_2, dry chemical, and alcohol foam extinguishers. Poisonous gases, including oxides of nitrogen, are produced in fire. Vapors are heavier than air and will collect in low areas. Vapors may travel long distances to ignition sources and flashback. Vapors in confined areas may explode when exposed to fire. Containers may explode in fire. Storage containers and parts of containers may rocket great distances, in many directions. If material or contaminated runoff enters waterways, notify downstream users of potentially contaminated waters. Notify local health and fire officials and pollution control agencies. From a secure, explosion-proof location, use water spray to cool exposed containers. If cooling streams are ineffective (venting sound increases in volume and pitch, tank discolors, or shows any signs of deforming), withdraw immediately to a secure position. If employees are expected to fight fires, they must be trained and equipped in OSHA 1910.156. The only respirators recommended for firefighting are self-contained breathing apparatuses that have full face-pieces and are operated in a pressure-demand or other positive-pressure mode.

Disposal Method Suggested: Controlled incineration (incinerator equipped with a scrubber or thermal unit to reduce nitrogen oxides emissions).

References

US Environmental Protection Agency. (May 1, 1978). *Chemical Hazard Information Profile: Methylamines.* Washington, DC

Sax, N. I. (Ed.). (1985). *Dangerous Properties of Industrial Materials Report*, 5, No. 4, 48−50

New Jersey Department of Health and Senior Services. (April 2004). *Hazardous Substances Fact Sheet: Methylamine*. Trenton, NJ

Methyl *n*-amyl ketone   M:0690

Molecular Formula: $C_7H_{14}O$
Common Formula: $CH_3COC_5H_{11}$
Synonyms: Amyl-methyl-cetone (French); Amyl methyl ketone; Butyl acetone; Methyl amyl ketone; Methyl-amyl-cetone (French); 2-Heptanone; 2-Ketoheptane; Methyl pentyl ketone; Pentyl methyl ketone

CAS Registry Number: 110-43-0
RTECS® Number: MJ5075000
UN/NA & ERG Number: UN1110/127
EC Number: 203-767-1 [*Annex I Index No.:* 606-024-00-3]
Regulatory Authority and Advisory Bodies
Air Pollutant Standard Set. See below, "Permissible Exposure Limits in Air" section.
Canada, WHMIS, Ingredients Disclosure List Concentration 1.0%.
European/International Regulations: Hazard Symbol: risk phrases: R10; R20/22; Safety phrases: S2; S22/25 (see Appendix 4).
WGK (German Aquatic Hazard Class): 1—Low hazard to waters.

Description: Methyl amyl ketone is a clear colorless liquid with a mild, banana-like odor. Molecular weight = 114.21; Specific gravity (H_2O:1) = 0.81; Boiling point = 151.7°C; Freezing/Melting point = −35.6°C; Vapor pressure = 3 mmHg at 20°C; Flash point = 39°C; Autoignition temperature = 393°C. Explosive limits: LEL = 1.1% at 66°C; UEL = 7.9% at 121°C. Hazard Identification (based on NFPA-704 M Rating System): Health 1, Flammability 2, Reactivity 0. Poor solubility in water; solubility = 0.4%.

Potential Exposure: Compound Description: Primary Irritant. Methyl amyl ketone is used as a solvent in metal roll coatings and in synthetic resin finishes; as a solvent for nitrocellulose in lacquers and as a relatively inert reaction medium.

Incompatibilities: Forms explosive mixture with air. Strong acids, alkalis, oxidizers. Attacks some forms of plastics.

Permissible Exposure Limits in Air
Conversion factor: 1 ppm = 4.67 mg/m³ at 25°C & 1 atm.
OSHA PEL: 100 ppm/465 mg/m³ TWA.
NIOSH REL: 100 ppm/465 mg/m³ TWA.
ACGIH TLV®[1]: 50 ppm/233 mg/m³ TWA.
NIOSH IDLH: 800 ppm.
Protective Action Criteria (PAC)
TEEL-0: 100 ppm
PAC-1: 100 ppm
PAC-2: 750 ppm
PAC-3: 800 ppm
Australia: TWA 50 ppm (235 mg/m³), 1993; Belgium: TWA 50 ppm (233 mg/m³), 1993; Denmark: TWA 50 ppm (230 mg/m³), 1999; Finland: TWA 50 ppm (230 mg/m³); STEL 75 ppm (345 mg/m³), 1999; France: VME 50 ppm (235 mg/m³), 1999; Norway: TWA 25 ppm (115 mg/m³), 1999; Sweden: TWA 25 ppm (120 mg/m³); STEL 50 ppm (250 mg/m³), 1999; the Netherlands: MAC-TGG 233 mg/m³, 2003; United Kingdom: LTEL 50 ppm (240 mg/m³), 1993; Argentina, Bulgaria, Columbia, Jordan, South Korea, New Zealand, Singapore, Vietnam: ACGIH TLV®: TWA 50 ppm. Several states have set guidelines or standards for methyl *n*-amyl ketone in ambient air[60] ranging from 2.35−4.65 mg/m³ (North Dakota) to 3.9 mg/m³ (Virginia) to 4.7 mg/m³ (Connecticut) to 5.595 mg/m³ (Nevada).

Determination in Air: Use NIOSH Analytical Methods #1301; #2553, Ketones II.[18]

Permissible Concentration in Water: No criteria set.

Routes of Entry: Inhalation, ingestion, skin and/or eye contact.

Harmful Effects and Symptoms

Short Term Exposure: Methyl *n*-amyl ketone can affect you when breathed in and by passing through your skin. Irritates the eyes and the respiratory tract. May affect the central nervous system. Breathing the vapor can cause dizziness and lightheadedness and can make you pass out.

Long Term Exposure: Causes skin irritation with cracking and drying; destroys the skin's natural oils. May cause liver and kidney damage. May affect the nervous system.

Points of Attack: Eyes, skin, respiratory system, central nervous system, peripheral nervous system.

Medical Surveillance: NIOSH lists the following tests: pulmonary function tests. If symptoms develop or overexposure is suspected, the following may be useful: liver and kidney function tests, examination of the nervous system.

First Aid: If this chemical gets into the eyes, remove any contact lenses at once and irrigate immediately for at least 15 min, occasionally lifting upper and lower lids. Seek medical attention immediately. If this chemical contacts the skin, remove contaminated clothing and wash immediately with soap and water. Seek medical attention immediately. If this chemical has been inhaled, remove from exposure, begin rescue breathing (using universal precautions, including resuscitation mask) if breathing has stopped and CPR if heart action has stopped. Transfer promptly to a medical facility. When this chemical has been swallowed, get medical attention. Give large quantities of water and induce vomiting. Do not make an unconscious person vomit.

Personal Protective Methods: Wear protective gloves and clothing to prevent any reasonable probability of skin contact: **4 h**: 4H™ and Silver Shield™ gloves. Safety equipment suppliers/manufacturers can provide recommendations on the most protective glove/clothing material for your operation. All protective clothing (suits, gloves, footwear, headgear) should be clean, available each day, and put on before work. Contact lenses should not be worn when working with this chemical. Wear splash-proof chemical goggles and face shield unless full face-piece respiratory protection is worn. Employees should wash immediately with soap when skin is wet or contaminated. Provide emergency showers and eyewash. See NIOSH Criteria Document 78-173, *Ketones.*

Respirator Selection: 800 ppm: CcrOv (APF = 10) [any chemical cartridge respirator with organic vapor cartridge(s)] or PaprOv (APF = 25) [any powered, air-purifying respirator with organic vapor cartridge(s)] or GmFOv (APF = 50) [any air-purifying, full-face-piece respirator (gas mask) with a chin-style, front- or back-mounted acid gas canister] or Sa (APF = 10) (any supplied-air respirator) or SCBAF (APF = 50) (any self-contained breathing apparatus with a full face-piece). *Emergency or planned entry into unknown concentrations or IDLH conditions:* SCBAF: Pd,Pp (APF = 10,000) (any self-contained breathing apparatus that has a full faceplate and is operated in a pressure-demand or other positive-pressure mode) or SaF: Pd,Pp: ASCBA (APF = 10,000) (any supplied-air respirator that has a full face-piece and is operated in a pressure-demand or other positive-pressure mode in combination with an auxiliary self-contained breathing apparatus operated in a pressure-demand or other positive-pressure mode). *Escape:* GmFOv (APF = 50) [any air-purifying, full-face-piece respirator (gas mask) with a chin-style, front- or back-mounted organic vapor canister] or SCBAE (any appropriate escape-type, self-contained breathing apparatus).

Note: Substance reported to cause eye irritation or damage; may require eye protection.

Storage: Color Code—Red: Flammability Hazard: Store in a flammable liquid storage area or approved cabinet away from ignition sources and corrosive and reactive materials. Prior to working with this chemical you should be trained on its proper handling and storage. Before entering confined space where this chemical may be present, check to make sure that an explosive concentration does not exist. Methyl (*n*-amyl) ketone must be stored to avoid contact with oxidizers, such as perchlorates, peroxides, chlorates, nitrites, and permanganates, since violent reaction occur. Store in tightly closed containers in a cool, well-ventilated area away from heat or flame. Sources of ignition, such as smoking and open flames, are prohibited where methyl (*n*-amyl) ketone is used, handled, or stored in a manner that could create a potential fire or explosion hazard.

Shipping: This compound requires a shipping label of "FLAMMABLE LIQUID." It falls in Hazard Class 3 and Packing Group II.

Spill Handling: Evacuate and restrict persons not wearing protective equipment from area of spill or leak until cleanup is complete. Remove all ignition sources. Establish forced ventilation to keep levels below explosive limit. Absorb liquids in vermiculite, dry sand, earth, peat, carbon, or a similar material and deposit in sealed containers. Keep this chemical out of a confined space, such as a sewer, because of the possibility of an explosion, unless the sewer is designed to prevent the buildup of explosive concentrations. It may be necessary to contain and dispose of this chemical as a hazardous waste. If material or contaminated runoff enters waterways, notify downstream users of potentially contaminated waters. Contact your local or federal environmental protection agency for specific recommendations. If employees are required to clean up spills, they must be properly trained and equipped. OSHA 1910.120(q) may be applicable.

Fire Extinguishing: This chemical is a combustible liquid. Poisonous gases are produced in fire. Use dry chemical, carbon dioxide, or alcohol foam extinguishers. Vapors are heavier than air and will collect in low areas. Vapors may travel long distances to ignition sources and flashback. Vapors in confined areas may explode when exposed to fire. Containers may explode in fire. Storage containers and parts of containers may rocket great distances, in many directions. If material or contaminated runoff enters waterways, notify downstream users of potentially contaminated waters. Notify local health and fire officials and pollution control agencies. From a secure, explosion-proof location, use water spray to cool exposed containers. If cooling streams are ineffective (venting sound increases in volume and pitch, tank discolors, or shows any signs of deforming), withdraw immediately to a secure position. If employees are expected to fight fires, they must be trained and equipped in OSHA 1910.156. The only respirators recommended for firefighting are self-contained breathing apparatuses that have full face-pieces and are operated in a pressure-demand or other positive-pressure mode.

Disposal Method Suggested: Dissolve or mix the material with a combustible solvent and burn in a chemical incinerator equipped with an afterburner and scrubber. All federal, state, and local environmental regulations must be observed.

References

National Institute for Occupational Safety and Health. (1978). *Criteria for a Recommended Standard: Occupational Exposure to Ketones*, NIOSH Document No. 78-173. Washington, DC

New Jersey Department of Health and Senior Services. (April 2004). *Hazardous Substances Fact Sheet: Methyl n-Amyl Ketone.* Trenton, NJ

N-Methylaniline   M:0700

Molecular Formula: C_7H_9N
Common Formula: $C_6H_5NH(CH_3)$
Synonyms: Anilinomethane; Benzenenamine, *N*-methyl-; MA; (Methylamino)benzene; *N*-Methylaminobenzene; Methyl aniline; Methylaniline (mono); *N*-Methylbenzenamine; 4,4′-Methylene-bis-(*N,N*-dimethylaniline)*;* *N*-Methylphenylamine; Methylphenyl amine; *N*-Monomethylaniline; Monomethyl aniline; *N*-Phenylmethylamine
CAS Registry Number: 100-61-8
RTECS® Number: BY4550000
UN/NA & ERG Number: UN2294/153
EC Number: 202-870-9 [*Annex I Index No.:* 612-015-00-5]
Regulatory Authority and Advisory Bodies
Air Pollutant Standard Set. See below, "Permissible Exposure Limits in Air" section.
Canada, WHMIS, Ingredients Disclosure List Concentration 1.0%.
European/International Regulations: Hazard Symbol: T, N; Risk phrases: R23/24/25; R33; R50/53; Safety phrases: S1/2; S28; S36/37; S45; S60; S61 (see Appendix 4).
WGK (German Aquatic Hazard Class): 3—Severe hazard to waters.

Description: *N*-Methylaniline is a yellow to light brown oily liquid with a weak, ammonia-like odor. Turns reddish-brown if left standing. The odor threshold is 1.7 ppm. Molecular weight = 107.17; Specific gravity (H_2O:1) = 0.99; Boiling point = 195.6°C; Freezing/Melting point = −57.2°C; Vapor pressure = 0.3 mmHg at 20°C; Flash point = 79.4°C. Hazard Identification (based on NFPA-704 M Rating System): Health 2, Flammability 2, Reactivity 0. Practically insoluble in water.
Potential Exposure: Compound Description: Mutagen. The material is used as an intermediate in organic synthesis, as a solvent and as an acid acceptor.
Incompatibilities: Reacts violently with strong acids, acid chlorides, acid anhydrides, strong oxidizers. Attacks some plastic.
Permissible Exposure Limits in Air
Conversion factor: 1 ppm = 4.38 mg/m^3 at 25°C & 1 atm.
OSHA PEL: 2 ppm/9 mg/m^3 [skin] TWA.
NIOSH REL: 0.5 ppm/2 mg/m^3 TWA [skin].
ACGIH TLV®[1]: 0.5 ppm/2.2 mg/m^3 TWA [skin].
NIOSH IDLH: 100 ppm.
Protective Action Criteria (PAC)
TEEL-0: 2 ppm
PAC-1: 2 ppm
PAC-2: 2 ppm
PAC-3: 100 ppm
DFG MAK: 0.5 ppm/2.2 mg/m^3 TWA; Peak Limitation Category II(2) [skin]; Pregnancy Risk Group D.
Arab Republic of Egypt: TWA 2 ppm (9 mg/m^3), [skin], 1993; Australia: TWA 0.5 ppm (2 mg/m^3), [skin], 1993; Austria: MAK 0.5 ppm (2 mg/m^3), [skin], 1999; Belgium: TWA 0.5 ppm (2.2 mg/m^3), [skin], 1993; Denmark: TWA 0.5 ppm (2.25 mg/m^3), [skin], 1999; France: VME 0.5 ppm (2 mg/m^3), [skin], 1999; the Netherlands: MAC-TGG 2 mg/m^3, 2003; Poland: MAC (TWA) 2 mg/m^3, 1999; Turkey: TWA 2 ppm (9 mg/m^3), [skin], 1993; United Kingdom: TWA 0.5 ppm (2.2 mg/m^3), [skin], 2000; Argentina, Bulgaria, Columbia, Jordan, South Korea, New Zealand, Singapore, Vietnam: ACGIH TLV®: TWA 0.5 ppm [skin]. Russia[43] set a MAC of 0.04 mg/m^3 (40 $\mu g/m^3$) in ambient air in residential areas both on a momentary and a daily average basis. Several states have set guidelines or standards for *n*-methylaniline in ambient air[60] ranging from 20 $\mu g/m^3$ (North Dakota) to 35 $\mu g/m^3$ (Virginia) to 40 $\mu g/m^3$ (Connecticut) to 47.6 $\mu g/m^3$ (Nevada).
Determination in Air: Bubbler; sodium hydroxide; Gas chromatography/Flame ionization detection; NIOSH Analytical Method (IV) #3511.
Determination in Water: Octanol−water coefficient: Log K_{ow} = 1.69.
Routes of Entry: Inhalation, skin absorption, ingestion, skin and/or eye contact.
Harmful Effects and Symptoms
Short Term Exposure: Methylaniline can affect you when breathed in and by passing through your skin. Irritates the eyes, skin, and respiratory tract. Exposure to methylaniline

can interfere with the ability of the blood to carry oxygen (methemoglobinemia). This can cause headaches, weakness, dizziness, dyspnea (breathing difficulty), cyanosis, a bluish color of the lips and nose. Higher exposure can cause shortness of breath, collapse, and even death. Higher exposures can cause pulmonary edema, a medical emergency that can be delayed for several hours. This can cause death. Exposure can damage the bladder, causing bloody urine.

Long Term Exposure: Repeated exposures may cause liver and kidney damage, and a low blood count (anemia). Repeated or prolonged contact may cause skin sensitization.

Points of Attack: Respiratory system, liver, kidneys, blood, central nervous system.

Medical Surveillance: NIOSH lists the following tests: whole blood (chemical/metabolite), methemoglobin, complete blood count, urinalysis (routine). Also, if symptoms develop or overexposure is suspected, the following may be considered: Liver and kidney function tests.

First Aid: If this chemical gets into the eyes, remove any contact lenses at once and irrigate immediately for at least 15 min, occasionally lifting upper and lower lids. Seek medical attention immediately. If this chemical contacts the skin, remove contaminated clothing and wash immediately with soap and water. Seek medical attention immediately. If this chemical has been inhaled, remove from exposure, begin rescue breathing (using universal precautions, including resuscitation mask) if breathing has stopped and CPR if heart action has stopped. Transfer promptly to a medical facility. When this chemical has been swallowed, get medical attention. Give large quantities of water and induce vomiting. Do not make an unconscious person vomit. Medical observation is recommended for 24−48 h after breathing overexposure, as pulmonary edema may be delayed. As first aid for pulmonary edema, a doctor or authorized paramedic may consider administering a corticosteroid spray.

Note to physician: Treat for methemoglobinemia. Spectrophotometry may be required for precise determination of levels of methemoglobin in urine.

Personal Protective Methods: Wear protective gloves and clothing to prevent any reasonable probability of skin contact. Safety equipment suppliers/manufacturers can provide recommendations on the most protective glove/clothing material for your operation. All protective clothing (suits, gloves, footwear, headgear) should be clean, available each day, and put on before work. Contact lenses should not be worn when working with this chemical. Wear splash-proof chemical goggles and face shield unless full face-piece respiratory protection is worn. Employees should wash immediately with soap when skin is wet or contaminated. Provide emergency showers and eyewash.

Respirator Selection: *5 ppm:* Sa (APF = 10) (any supplied-air respirator). *12.5 ppm:* Sa:Cf (APF = 25) (any supplied-air respirator operated in a continuous-flow mode). *25 ppm:* SaT: Cf (APF = 50) (any supplied-air respirator that has a tight-fitting face-piece and is operated in a continuous-flow mode) or SCBAF (APF = 50) (any self-contained breathing apparatus with a full face-piece) or SaF (APF = 50) (any supplied-air respirator with a full face-piece). *100 ppm:* SaF: Pd,Pp (APF = 2000) (any supplied-air respirator that has a full face-piece and is operated in a pressure-demand or other positive-pressure mode). *Emergency or planned entry into unknown concentrations or IDLH conditions:* SCBAF: Pd,Pp (APF = 10,000) (any self-contained breathing apparatus that has a full face-piece and is operated in a pressure-demand or other positive-pressure mode) or SaF: Pd,Pp: ASCBA (any supplied-air respirator that has a full face-piece and is operated in a pressure-demand or other positive-pressure mode in combination with an auxiliary self-contained breathing apparatus operated in a pressure-demand or other positive-pressure mode). *Escape:* GmFS (APF = 50) [any air-purifying, full-face-piece respirator (gas mask) with a chin-style, front- or back-mounted canister providing protection against the compound of concern] or SCBAE (any appropriate escape-type, self-contained breathing apparatus).

Storage: Color Code—Blue: Health Hazard/Poison: Store in a secure poison location. Prior to working with this chemical you should be trained on its proper handling and storage. Methylaniline must be stored to avoid contact with strong acids (such as hydrochloric, sulfuric, and nitric), since violent reactions occur. Store in tightly closed containers in a cool, well-ventilated area away from heat. Sources of ignition, such as smoking and open flames, are prohibited where methylaniline is used, handled, or stored in a manner that could create a potential fire or explosion hazard.

Shipping: This compound requires a shipping label of "POISONOUS/TOXIC MATERIALS." It falls in Hazard Class 6.1 and Packing Group III.

Spill Handling: Evacuate and restrict persons not wearing protective equipment from area of spill or leak until cleanup is complete. Remove all ignition sources. Ventilate area of spill or leak. Absorb liquids in vermiculite, dry sand, earth, peat, carbon, or a similar material and deposit in sealed containers. Keep this chemical out of a confined space, such as a sewer, because of the possibility of an explosion, unless the sewer is designed to prevent the build-up of explosive concentrations. It may be necessary to contain and dispose of this chemical as a hazardous waste. If material or contaminated runoff enters waterways, notify downstream users of potentially contaminated waters. Contact your local or federal environmental protection agency for specific recommendations. If employees are required to clean up spills, they must be properly trained and equipped. OSHA 1910.120(q) may be applicable.

Fire Extinguishing: This chemical is a combustible liquid. Poisonous gases, including nitrogen oxides, are produced in fire. Use dry chemical, carbon dioxide, or alcohol foam extinguishers. Vapors are heavier than air and will collect in low areas. Vapors may travel long distances to ignition sources and flashback. Vapors in confined areas may explode when exposed to fire. Containers may explode in fire. Storage containers and parts of containers may rocket great distances, in many directions. If material or contaminated runoff enters waterways, notify downstream users of potentially contaminated waters. Notify local health and fire officials and pollution control agencies. From a secure, explosion-proof location, use water spray to cool exposed containers. If cooling streams are ineffective (venting sound increases in volume and pitch, tank discolors, or shows any signs of deforming), withdraw immediately to a secure position. If employees are expected to fight fires, they must be trained and equipped in OSHA 1910.156. The only respirators recommended for firefighting are self-contained breathing apparatuses that have full face-pieces and are operated in a pressure-demand or other positive-pressure mode.

Disposal Method Suggested: Controlled incineration whereby oxides of nitrogen are removed from the effluent gas by scrubber, catalytic, or thermal device.

Reference

New Jersey Department of Health and Senior Services. (November 1999). *Hazardous Substances Fact Sheet: Methylaniline.* Trenton, NJ

Methyl benzoate   M:0710

Molecular Formula: $C_8H_8O_2$

Synonyms: Benzoato de metilo (Spanish); Benzoic acid, methyl ester; Essence of niobe; Methyl benzenecarboxylate; Niobe oil; Oil of niobe

CAS Registry Number: 93-58-3

RTECS® Number: DH3850000

UN/NA & ERG Number: UN2938/152

EC Number: 202-259-7

Regulatory Authority and Advisory Bodies

WGK (German Aquatic Hazard Class): 1—Low hazard to waters.

Description: Methyl Benzoate is a colorless, oily, transparent, liquid with a pleasant odor. Molecular weight = 136.16; Specific gravity (H_2O:1) = 1.09; Boiling point = 198.8°C; Freezing/Melting point = −13°C; Flash point = 83°C. Hazard Identification (based on NFPA-704 M Rating System): Health 0, Flammability 2, Reactivity 0. Insoluble in water.

Potential Exposure: Used as food additive and as a solvent for cellulose esters and ethers, resins, and rubber.

Incompatibilities: Incompatible with strong acids, strong bases, nitrates, oxidizers.

Permissible Exposure Limits in Air

Protective Action Criteria (PAC)

TEEL-0: 2 ppm

PAC-1: 6 ppm

PAC-2: 40 ppm

PAC-3: 75 ppm

Determination in Water: Octanol−water coefficient: Log K_{ow} = 2.1.

Routes of Entry: Inhalation, ingestion, skin and/or eye contact.

Harmful Effects and Symptoms

Short Term Exposure: Irritates the eyes, skin, and respiratory tract. Inhalation can cause coughing and wheezing. Swallowing the liquid may cause chemical pneumonia.

Long Term Exposure: May cause skin sensitization and allergy. May cause an asthma-like allergy. Can affect the nervous system causing tremors and muscle weakness.

Points of Attack: Skin, lungs, nervous system.

Medical Surveillance: Evaluation by a qualified allergist. Lung function tests. Examination of the nervous system.

First Aid: If this chemical gets into the eyes, remove any contact lenses at once and irrigate immediately for at least 15 min, occasionally lifting upper and lower lids. Seek medical attention immediately. If this chemical contacts the skin, remove contaminated clothing and wash immediately with soap and water. Seek medical attention immediately. If this chemical has been inhaled, remove from exposure, begin rescue breathing (using universal precautions, including resuscitation mask) if breathing has stopped and CPR if heart action has stopped. Transfer promptly to a medical facility. When this chemical has been swallowed, get medical attention. Give large quantities of water and induce vomiting. Do not make an unconscious person vomit.

Note to physician: Inhalation: Bronchodialators, decongestants, and oxygen may be used if necessary. Corticosteroids are useful for treating pneumonitis.

Personal Protective Methods: Wear protective gloves and clothing to prevent any reasonable probability of skin contact. Safety equipment suppliers/manufacturers can provide recommendations on the most protective glove/clothing material for your operation. All protective clothing (suits, gloves, footwear, headgear) should be clean, available each day, and put on before work. Contact lenses should not be worn when working with this chemical. Wear splash-proof chemical goggles and face shield unless full face-piece respiratory protection is worn. Employees should wash immediately with soap when skin is wet or contaminated. Provide emergency showers and eyewash.

Respirator Selection: Follow regulations in OSHA 29CFR1910.134 or European Standard EN149. Use a NIOSH/MSHA- or European Standard EN149-approved respirator; or use an approved supplied-air respirator with a full face-piece operated in the positive-pressure mode, or with a full face-piece, hood, or helmet in the continuous-flow mode; or use a NIOSH/MSHA- or European Standard EN149-approved self-contained breathing apparatus with a full face-piece operated in pressure-demand or other positive-pressure mode.

Storage: Color Code—Red: Flammability Hazard: Store in a flammable liquid storage area or approved cabinet away from ignition sources and corrosive and reactive materials. Prior to working with methyl benzoate you should be trained on its proper handling and storage. Store in tightly closed containers in a cool, well-ventilated area away from strong acids, strong bases, oxidizers, nitrates. Where possible, automatically pump liquid from drums or other storage containers to process containers. Drums must be equipped with self-closing valves, pressure vacuum bungs, and flame arresters. Use only nonsparking tools and equipment, especially when opening and closing containers of this chemical. Wherever this chemical is used, handled, manufactured, or stored, use explosion-proof electrical equipment and fittings.

Shipping: Methyl benzoate requires a shipping label of "POISONOUS/TOXIC MATERIALS." It falls in Hazard Class 6.1 and Packing Group III.

Spill Handling: Evacuate and restrict persons not wearing protective equipment from area of spill or leak until cleanup is complete. Remove all ignition sources. Ventilate area of spill or leak. Absorb liquids in vermiculite, dry sand, earth, peat, carbon, or a similar material and deposit in sealed containers. Keep this chemical out of a confined space, such as a sewer, because of the possibility of an explosion, unless the sewer is designed to prevent the buildup of explosive concentrations. It may be necessary to contain and dispose of this chemical as a hazardous waste. If material or contaminated runoff enters waterways, notify downstream users of potentially contaminated waters. Contact your local or federal environmental protection agency for specific recommendations. If employees are required to clean up spills, they must be properly trained and equipped. OSHA 1910.120(q) may be applicable.

Fire Extinguishing: This chemical is a combustible liquid. Poisonous gases are produced in fire. Use dry chemical, carbon dioxide, or alcohol foam extinguishers. Vapors are heavier than air and will collect in low areas. Vapors may travel long distances to ignition sources and flashback. Vapors in confined areas may explode when exposed to fire. Containers may explode in fire. Storage containers and parts of containers may rocket great distances, in many directions. If material or contaminated runoff enters waterways, notify downstream users of potentially contaminated waters. Notify local health and fire officials and pollution control agencies. From a secure, explosion-proof location, use water spray to cool exposed containers. If cooling streams are ineffective (venting sound increases in volume and pitch, tank discolors, or shows any signs of deforming), withdraw immediately to a secure position. If employees are expected to fight fires, they must be trained and equipped in OSHA 1910.156. The only respirators recommended for firefighting are self-contained breathing apparatuses that have full face-pieces and are operated in a pressure-demand or other positive-pressure mode.

Disposal Method Suggested: Dissolve or mix the material with a combustible solvent and burn in a chemical incinerator equipped with an afterburner and scrubber. All federal, state, and local environmental regulations must be observed.

Reference

New Jersey Department of Health and Senior Services. (May 1999). *Hazardous Substances Fact Sheet: Methyl Benzoate.* Trenton, NJ

Methyl bromide   M:0720

Molecular Formula: CH_3Br
Synonyms: Bromomethane; Brom-*o*-gas; Dawson 100; Dowfume; EDCO; Embafume; Halon 1001; Iscobrome; Kayafume; Methane, bromo-; Methogas; M-B-C fumigant; Monobromomethane; R 40BL; Rotox; Terabol; Terr-*o*-gas 100; Zytox
CAS Registry Number: 74-83-9
RTECS® Number: PA4900000
UN/NA & ERG Number: UN1062/123
EC Number: 200-813-2 [*Annex I Index No.:* 602-002-00-3]
Regulatory Authority and Advisory Bodies
Carcinogenicity: IARC: Human Inadequate Evidence, animal Limited Evidence, *not classifiable as carcinogenic to humans*, Group 3, 1999; NCI: Carcinogenesis Studies (inhalation); no evidence: mouse; EPA: Not Classifiable as to human carcinogenicity; NIOSH: Potential occupational carcinogen.
Carcinogenicity: (Animal Suspected) IARC[9]: DFG[3]: (ACGIH)[1] (suspected occupational carcinogen, NIOSH).
Toxic Chemical (World Bank).[15]
Air Pollutant Standard Set. See below, "Permissible Exposure Limits in Air" section.
OSHA 29CFR1910.119, Appendix A, Process Safety List of Highly Hazardous Chemicals, TQ = 2500 lb (1135 kg).
Clean Air Act: Hazardous Air Pollutants (Title I, Part A, Section 112); Stratospheric ozone protection (Title VI, Subpart A, Appendix A), Class I, Ozone Depletion Potential = 0.7.
Clean Water Act: 40CFR423, Appendix A, Priority Pollutants; Section 313 Water Priority Chemicals (57FR 41331, 9/9/92).
US EPA Hazardous Waste Number (RCRA No.): U029.
RCRA, 40CFR261, Appendix 8 Hazardous Constituents.
RCRA 40CFR268.48; 61FR15654, Universal Treatment Standards: Wastewater (mg/L), 0.11; Nonwastewater (mg/kg), 15.
RCRA 40CFR264, Appendix 9; TSD Facilities Ground Water Monitoring List. Suggested test method(s) (PQL µg/L): 8010 (20); 8240 (10).
Safe Drinking Water Act: Priority List (55 FR 1470).
Superfund/EPCRA 40CFR355, Extremely Hazardous Substances: TPQ = 1000 lb (454 kg).
Reportable Quantity (RQ): 1000 lb (454 kg).
EPCRA Section 313 Form R *de minimis* concentration reporting level: 1.0%.
US DOT Regulated Marine Pollutant (49CFR172.101, Appendix B). Only as methyl bromide and ethylene dibromide mixture, liquid.
US DOT 49CFR172.101, Inhalation Hazardous Chemical.
California Proposition 65 Developmental/Reproductive toxin (methyl bromide, as a structural fumigant) 1/1/93.
European/International Regulations: Hazard Symbol: T, N; Risk phrases: R23/25; R36/37/38; R48/20; R50; R59; R68;

Safety phrases: S1/2; S15; S27; S36/39; S38; S45; S59; S61 (see Appendix 4).

WGK (German Aquatic Hazard Class): 3—Severe hazard to waters.

Description: Methyl bromide is a colorless gas with a chloroform-like odor at high concentrations. A liquid below 3.3°C. Shipped as a liquefied compressed gas. Molecular weight = 94.95; Specific gravity (H₂O:1) = 1.73; Boiling point = 3.3°C; Freezing/Melting point = −93.7°C; Relative vapor density (air = 1) = 3.36; Vapor pressure = 1.9 atm; Flash point = practically nonflammable except in presence of a high-energy ignition sources; Autoignition temperature = 537°C. Explosive limits: LEL = 10.0%; UEL = 16.0%. Hazard Identification (based on NFPA-704 M Rating System): Health 3, Flammability 1, Reactivity 0. Slightly soluble in water; solubility = 2%.

Potential Exposure: Methyl bromide is used in fire extinguishers, as a fumigant in pest control, as a methylation agent in industry, and as an insect fumigant for soil, grain, warehouses, mills, ships, etc. It is also used as a chemical intermediate and a methylating agent, a refrigerant, a herbicide, a low-boiling solvent in aniline dye manufacture, for degreasing wool, for extracting oils from nuts, seeds, and flowers, and in ionization chambers. It is used as an intermediate in the manufacture of many drugs.

Incompatibilities: Attacks aluminum to form *spontaneously* flammable aluminum trimethyl. Incompatible with strong oxidizers, aluminum, dimethylsulfoxide, ethylene oxide, water. Attacks zinc, magnesium, alkali metals, and their alloys. Attacks some rubbers and coatings.

Permissible Exposure Limits in Air

Conversion factor: 1 ppm = 3.89 mg/m³ at 25°C & 1 atm.

OSHA PEL: 20 ppm/80 mg/m³ Ceiling Concentration [skin].

NIOSH REL: A potential occupational carcinogen. Limit exposure to lowest feasible concentration. See *NIOSH Pocket Guide*, Appendix A.

ACGIH TLV®[1]: 1 ppm/3.9 mg/m³ [skin]; not classifiable as a human carcinogen.

NIOSH IDLH: 250 ppm.

Protective Action Criteria (PAC)*

TEEL-0: 1 ppm

PAC-1: 30 ppm

PAC-2: **210** ppm

PAC-3: **740** ppm

*AEGLs (Acute Emergency Guideline Levels) & ERPGs (Emergency Response Planning Guideline) are in **bold face**.

DFG MAK: [skin], Carcinogen Category 3B.

Arab Republic of Egypt: TWA 5 ppm (20 mg/m³), [skin] 1993; Australia: TWA 5 ppm (20 mg/m³), [skin], 1993; Austria: Suspected: carcinogen, 1999; Denmark: TWA 5 ppm (20 mg/m³), [skin], 1999; Finland: TWA 5 ppm (20 mg/m³); STEL 10 ppm (40 mg/m³), [skin], 1999; France: VME 5 ppm (20 mg/m³), 1999; Hungary: TWA 10 mg/m³; STEL 20 mg/m³, [skin], 1993; the Netherlands:

MAC-TGG 1 mg/m³, [skin], 2003; Norway: TWA 5 ppm (20 mg/m³), 1999; the Philippines: TWA 20 ppm (80 mg/m³), [skin], 1993; Poland: MAC (TWA) 5 mg/m³, MAC (STEL) 40 mg/m³, 1999; Russia: STEL 1 mg/m³, 1993; Sweden: NGV 5 ppm (19 mg/m³), KTV 10 ppm (40 mg/m³), [skin], 1999; Switzerland: MAK-W 5 ppm (20 mg/m³), KZG-W 10 ppm (40 mg/m³), [skin], 1999; Thailand: TWA 20 ppm (80 mg/m³), 1993; Turkey: TWA 20 ppm (80 mg/m³), 1993; United Kingdom: TWA 5 ppm (20 mg/m³); STEL 15 ppm, [skin], 2000; Argentina, Bulgaria, Columbia, Jordan, South Korea, New Zealand, Singapore, Vietnam: ACGIH TLV®: not classifiable as a human carcinogen. The Czech Republic: MAC 1.0 mg/m³. Russia has set MAC value of 0.02 mg/m³ for the ambient air in residential areas on a momentary basis and 0.01 mg/m³ on a daily average basis. Several states have set guidelines or standards for methyl bromide in ambient air[60] ranging from 2.6 µg/m³ (Massachusetts) to 47.6 µg/m³ (Kansas) to 100 µg/m³ (South Carolina) to 200 µg/m³ (North Dakota) to 350 µg/m³ (Virginia) to 400 µg/m³ (Connecticut) to 476 µg/m³ (Nevada) to 480 µg/m³ (Pennsylvania).

Determination in Air: Charcoal adsorption, workup with CS₂, analysis by gas chromatography/flame ionization. Use NIOSH Analytical Method 2520;[18] OSHA Analytical Method PV-2040.

Permissible Concentration in Water: *To protect human health:* preferably zero. An additional lifetime cancer risk of 1 in 100,000 is posed by a concentration of 1.9 µg/L.[6] States which have set guidelines for methyl bromide in drinking water[61] include Arizona at 2.5 µg/L and Kansas at 0.19 µg/L.

Determination in Water: Inert gas purge followed by gas chromatography with halide specific detection (EPA Method 601) or gas chromatography plus mass spectrometry (EPA Method 624). Octanol—water coefficient: Log K_{ow} = 1.2.

Routes of Entry: Inhalation, percutaneous absorption, ingestion, skin and/or eye contact.

Harmful Effects and Symptoms

Short Term Exposure: Methyl bromide irritates the respiratory tract. Inhalation of the gas can cause pulmonary edema, a medical emergency that can be delayed for several hours. This can cause death. May affect the central nervous system causing psychological disturbances. Signs and symptoms of acute exposure to methyl bromide may be severe and include tremors, convulsions, brain hemorrhage, paralysis, coma, and permanent brain damage. Respiratory effects include cough, tachypnea (rapid respiratory rate), pulmonary edema, and respiratory collapse. Cyanosis (blue tint to the skin and mucous membranes), pallor, ventricular fibrillation, and circulatory collapse may also occur. Lethargy, profound weakness; headache, dizziness, mental confusion, slurring of speech, staggering gait, and blurred or double vision are often found. Gastrointestinal signs and symptoms include nausea, vomiting, abdominal pain, and anorexia. Oliguria (scanty urination), anuria (lack of urine formation), kidney hemorrhage, and kidney failure may occur. Contact with methyl bromide may cause dermatitis (red, inflamed skin) and conjunctivitis (red, inflamed eyes). *Inhalation:* A level of 35 ppm can cause nausea, vomiting, loss of appetite, headache, dizziness, drowsiness, and dimming of vision. These effects go away soon after exposure ceases. Headaches, dizziness, and weakness can be felt at 100 ppm and can last for months after exposure. Higher levels have caused coughing, nose and throat irritation, disturbed speech and walk, visual disturbances, twitching, numbness, paralysis, convulsions, and permanent nerve damage. Symptoms are often delayed 24–48 h. Exposures of 10,000 ppm for a few minutes can cause death. Can cause abdominal cramps and respiratory failure resulting in death. *Skin:* Contact with liquid can cause burning or tingling sensation, itching, redness, and swelling. Large amounts can cause blisters, numbness, or aching pain. Methyl bromide can be absorbed through the skin and cause symptoms described under inhalation. Death has occurred from skin absorption. *Eye:* Can cause irritation, tearing, reddening, or burning pain.

Ingestion: Can cause throat and stomach irritation as well as symptoms described under inhalation.

Note: Do not wear *ordinary rubber gloves* or *adhesive bandages* while using methyl bromide. It can dissolve rapidly through rubber or adhesive tape and cause severe symptoms.

Long Term Exposure: Levels between 20 and 35 ppm can cause symptoms as described under short-term inhalation. Symptoms can last months or years, or can be permanent. Repeated or prolonged contact with skin may cause dermatitis, lung damage, and broncho-spasms. Methyl bromide may affect the central nervous system causing paralysis, poor vision, psychological disorders, hallucinations, numbness in the arms and legs, and brain damage. May cause liver and kidney damage. Methyl bromide is a mutagen and may have a cancer risk. May damage the testes.

Points of Attack: Eyes, skin, respiratory system, central nervous system, brain. Cancer site in animals: lung, kidney, and fore-stomach.

Medical Surveillance: NIOSH lists the following tests: whole blood (chemical/metabolite); blood serum; chest X-ray, expired air, pulmonary function tests: forced vital capacity, forced expiratory volume (1 s); urine (chemical/metabolite). Evaluate the central nervous system; respiratory tract, and skin in preplacement and periodic examinations. Blood test for bromides (unexposed persons usually have serum levels of 5 mg/L or below). Kidney function tests. Evaluation for brain effects.

First Aid: If this chemical gets into the eyes, remove any contact lenses at once and irrigate immediately for at least 15 min, occasionally lifting upper and lower lids. Seek medical attention immediately. If this chemical contacts the skin, remove contaminated clothing and wash immediately with soap and water. Seek medical attention immediately. If this chemical has been inhaled, remove from exposure, begin rescue breathing (using universal precautions, including resuscitation mask) if breathing has stopped and CPR if heart action has stopped. Transfer promptly to a medical facility. When this chemical has been swallowed, get medical attention. Give large quantities of water and induce vomiting. Do not make an unconscious person vomit. Medical observation is recommended for 24–48 h after breathing overexposure, as pulmonary edema may be delayed. As first aid for pulmonary edema, a doctor or authorized paramedic may consider administering a corticosteroid spray. If frostbite has occurred, seek medical attention immediately; do *NOT* rub the affected areas or flush them with water. In order to prevent further tissue damage, do *NOT* attempt to remove frozen clothing from frostbitten areas. If frostbite has *NOT* occurred, immediately and thoroughly wash contaminated skin with soap and water.

Personal Protective Methods: Wear appropriate personal protective clothing to prevent the skin from becoming frozen from contact with the evaporating liquid or from contact with vessels containing the liquid: **8 h:** Responder™ suits; Trychem 1000™ suits. **4 h:** butyl rubber gloves, suits, boots; Neoprene™ rubber gloves, suits, boots; Teflon™ gloves, suits, boots. Also, Saranex™ and styrene-butadiene rubber are among the recommended protective materials. Safety equipment suppliers/manufacturers can provide recommendations on the most protective glove/clothing material for your operation. All protective clothing (suits, gloves, footwear, headgear) should be clean, available each day, and put on before work. Contact lenses should not be worn when working with this chemical. Wear eye protection to prevent any reasonable probability of eye contact. Employees should wash immediately with soap when skin is wet or contaminated. Provide emergency showers and eyewash.

Respirator Selection: NIOSH: *At concentrations above the NIOSH REL, or where there is no REL, at any detectable concentration:* SCBAF: Pd,Pp (APF = 10,000) (any NIOSH/MSHA- or European Standard EN 149-approved self-contained breathing apparatus that has a full face-piece and is operated in a pressure-demand or other positive-pressure mode) or SaF: Pd,Pp: ASCBA (APF = 10,000) (any supplied-air respirator that has a full face-piece and is operated in a pressure-demand or other positive-pressure mode in combination with an auxiliary, self-contained breathing apparatus operated in a pressure-demand or other positive-pressure mode). *Escape:* GmFOv (APF = 50) [any air-purifying, full-face-piece respirator (gas mask) with a chin-style, front- or back-mounted organic vapor canister] or SCBAE (any appropriate escape-type, self-contained breathing apparatus).

Storage: Poison gas. Color Code—Blue: Health Hazard/Poison: Store in a secure poison location. Prior to working with this chemical you should be trained on its proper handling and storage. Before entering confined space where this chemical may be present, check to make sure that an explosive concentration does not exist. Store in well-ventilated area away from direct sunlight. Maintain temperature below 40°C; avoid heat sources. Protect against physical damage. A regulated, marked area should be established where this chemical is handled, used, or stored in compliance with OSHA Standard 1910.1045.

Shipping: Methyl bromide requires a shipping label of "POISON GAS." It falls in Hazard Class 2.3. It is a violation of transportation regulations to refill compressed gas cylinders without the express written permission of the owner.

Spill Handling: If in a building, evacuate building and confine vapors by closing doors and shutting down HVAC systems. Restrict persons not wearing protective equipment from area of spill or leak until cleanup is complete. Remove all ignition sources. Ventilate area of spill or leak to disperse the gas. Wear chemical protective suit with self-contained breathing apparatus to combat spills. Stay upwind and use water spray to "knock down" vapor; contain runoff. Stop the flow of gas, if it can be done safely from a distance. If source is a cylinder and the leak cannot be stopped in place, remove the leaking cylinder to a safe place and repair leak or allow cylinder to empty. Keep this chemical out of confined spaces, such as a sewer, because of the possibility of explosion, unless the sewer is designed to prevent the buildup of explosive concentrations. If employees are required to clean up spills, they must be properly trained and equipped. OSHA 1910.120(q) may be applicable.

Initial isolation and protective action distances
Distances shown are likely to be affected during the first 30 min after materials are spilled and could increase with time. If more than one tank car, cargo tank, portable tank, or large cylinder involved in the incident is leaking, the protective action distance may need to be increased. You may need to seek emergency information from CHEMTREC at (800) 424-9300 or seek professional environmental engineering assistance from the US EPA Environmental Response Team at (908) 48-8730 (24-h response line).
Small spills (From a small package or a small leak from a large package)
First: Isolate in all directions (feet/meters)   100/30
Then: Protect persons downwind (miles/kilometers)
Day   0.1/0.2
Night   0.1/0.2
Large spills (From a large package or from many small packages)
First: Isolate in all directions (feet/meters)   500/150
Then: Protect persons downwind (miles/kilometers)
Day   0.4/0.6
Night   1.4/2.3

Fire Extinguishing: This chemical is a flammable gas but only in presence of a high energy ignition source. Nonflammable at room temperature. Poisonous gases, including hydrogen bromide and carbon monoxide, are produced in fire. Establish forced ventilation to keep levels below explosive limit. If liquid is spilled, evacuate area of spill; absorb liquid in vermiculite, dry sand, earth, or similar material and deposit in sealed containers for later disposal. If gas is leaked, do not extinguish the fire unless the flow of gas can be stopped and any remaining gas is out of the line. Specially trained personnel may use fog lines to cool exposures and let the fire burn itself out. Vapors are heavier than air and will collect in low areas. Vapors may travel long distances to ignition sources and flashback. Vapors in confined areas may explode when exposed to fire. Containers may explode in fire. Storage containers and parts of containers may rocket great distances, in many directions. If material or contaminated runoff enters waterways, notify downstream users of potentially contaminated waters. Notify local health and fire officials and pollution control agencies. From a secure, explosion-proof location, use water spray to cool exposed containers. If cooling streams are ineffective (venting sound increases in volume and pitch, tank discolors, or shows any signs of deforming), withdraw immediately to a secure position. If cylinders are exposed to excessive heat from fire or flame contact, withdraw immediately to a secure location. If employees are expected to fight fires, they must be trained and equipped in OSHA 1910.156. The only respirators recommended for firefighting are self-contained breathing apparatuses that have full face-pieces and are operated in a pressure-demand or other positive-pressure mode.

Disposal Method Suggested: Consult with environmental regulatory agencies for guidance on acceptable disposal practices. Generators of waste containing this contaminant (≥100 kg/mo) must conform with EPA regulations governing storage, transportation, treatment, and waste disposal. The recommended disposal procedure is to spray the gas into the fire box of an incinerator equipped with an afterburner and scrubber (alkali).[22]

References
US Environmental Protection Agency. (1980). *Halomethanes: Ambient Water Quality Criteria.* Washington, DC
US Environmental Protection Agency. (April 30, 1980). *Bromomethane: Health and Environmental Effects Profile No. 29.* Washington, DC: Office of Solid Waste
US Environmental Protection Agency. (February 20, 1985). *Chemical Hazard Information Profile: Methyl Bromide.* Washington, DC: Office of Toxic Substances
US Environmental Protection Agency. (November 30, 1987). *Chemical Hazard Information Profile: Methyl Bromide.* Washington, DC: Chemical Emergency Preparedness Program
New York State Department of Health. (March 1986). *Chemical Fact Sheet: Methyl Bromide.* Version 3. Albany, NY: Bureau of Toxic Substance Assessment
Sax, N. I. (Ed.). (1985). *Dangerous Properties of Industrial Materials Report,* 5, No. 6, 37—40
New Jersey Department of Health and Senior Services. (March 2006). *Hazardous Substances Fact Sheet: Methyl Bromide.* Trenton, NJ

Methyl-*tert*-butyl ether M:0730

Molecular Formula: $C_5H_{12}O$

Synonyms: Ether, *tert*-butyl methyl-; 2-Methoxy-2-methyl-propane; Methyl 1,1-dimethylethyl ether; Methyl *tert*-butyl ether; MTBE; Propane, 2-methoxy-2-methyl-

CAS Registry Number: 1634-04-4

RTECS® Number: KN5250000

UN/NA & ERG Number: UN2398/127

EC Number: 216-653-1 [*Annex I Index No.:* 603-181-00-X]

Regulatory Authority and Advisory Bodies

Carcinogenicity: IARC: Human Inadequate Evidence, animal Limited Evidence, *not classifiable as carcinogenic to humans*, Group 3.

Air Pollutant Standard Set. See below, "Permissible Exposure Limits in Air" section.

European/International Regulations: Hazard Symbol: T; Risk phrases: R10; R48/23; R62; R67; Safety phrases: S1/2; S36/37; R45 (see Appendix 4).

WGK (German Aquatic Hazard Class): 1—Low hazard to waters.

Description: Methyl-*tert*-butyl ether is a colorless liquid. Molecular weight = 88.17; Boiling point = 54°C; Freezing/Melting point = −109°C; Flash point = −10°C. Hazard Identification (based on NFPA-704 M Rating System): Health 2, Flammability 3, Reactivity 1. Soluble in water.

Potential Exposure: Compound Description: Tumorigen, Mutagen; Reproductive Effector. Used as an organic solvent, as an octane booster in unleaded gasolines, in making other chemicals, and in medicine to dissolve gall stones.

Incompatibilities: Forms explosive mixture with air. May be able to form unstable peroxides. Much less likely to form peroxides than other ethers. Incompatible with strong acids. Violent reaction with strong oxidizers. May accumulate static electrical charges and cause ignition of its vapors.

Permissible Exposure Limits in Air

OSHA PEL: None.

NIOSH REL: None.

ACGIH TLV®[1]: 50 ppm/180 mg/m³ TWA [skin]; confirmed animal carcinogen with unknown relevance to humans.

Protective Action Criteria (PAC)*

TEEL-0: 50 ppm

PAC-1: **50 ppm**

PAC-2: **570 ppm**

PAC-3: **5300 ppm**

*AEGLs (Acute Emergency Guideline Levels) & ERPGs (Emergency Response Planning Guideline) are in **bold face**.

DFG MAK: 50 ppm/180 mg/m³ TWA; Peak Limitation Category I(1.5); Carcinogen Category 3B; Pregnancy Risk Group C.

Russia: STEL 100 mg/m³, 1993; Sweden: TWA 50 ppm (180 mg/m³); STEL 75 ppm (250 mg/m³), 1999; United Kingdom: TWA 25 ppm (92 mg/m³); STEL 75 ppm (275 mg/m³), 2000; the Netherlands: MAC-TGG 180 mg/m³, 2003.

Determination in Air: Use NIOSH Analytical Method #1615, MTBE.

Routes of Entry: Inhalation, skin and/or eye contact. Absorbed through the skin.

Harmful Effects and Symptoms

Short Term Exposure: Irritates the eyes, skin, and respiratory tract. Exposure can cause difficulty concentrating and thinking. Higher levels can cause headache, nausea, dizziness, weakness and lightheadedness.

Long Term Exposure: May cause kidney damage.

Points of Attack: Kidneys.

Medical Surveillance: Kidney function tests. The use of alcoholic beverages should be avoided before or during use.

First Aid: If this chemical gets into the eyes, remove any contact lenses at once and irrigate immediately for at least 15 min, occasionally lifting upper and lower lids. Seek medical attention immediately. If this chemical contacts the skin, remove contaminated clothing and wash immediately with soap and water. Seek medical attention immediately. If this chemical has been inhaled, remove from exposure, begin rescue breathing (using universal precautions, including resuscitation mask) if breathing has stopped and CPR if heart action has stopped. Transfer promptly to a medical facility. When this chemical has been swallowed, get medical attention. Give large quantities of water and induce vomiting. Do not make an unconscious person vomit.

Personal Protective Methods: Wear protective gloves and clothing to prevent any reasonable probability of skin contact. Safety equipment suppliers/manufacturers can provide recommendations on the most protective glove/clothing material for your operation. All protective clothing (suits, gloves, footwear, headgear) should be clean, available each day, and put on before work. Contact lenses should not be worn when working with this chemical. Wear splash-proof chemical goggles and face shield unless full face-piece respiratory protection is worn. Employees should wash immediately with soap when skin is wet or contaminated. Provide emergency showers and eyewash. Do not drink any alcoholic beverages before or during use.

Respirator Selection: Follow regulations in OSHA 29CFR1910.134 or European Standard EN149. Use a NIOSH/MSHA- or European Standard EN149-approved respirator; or use an approved supplied-air respirator with a full face-piece operated in the positive-pressure mode, or with a full face-piece, hood, or helmet in the continuous-flow mode; or use a NIOSH/MSHA- or European Standard EN149-approved self-contained breathing apparatus with a full face-piece operated in pressure-demand or other positive-pressure mode.

Storage: Color Code—Red: Flammability Hazard: Store in a flammable liquid storage area or approved cabinet away from ignition sources and corrosive and reactive materials. Prior to working with MTBE you should be trained on its proper handling and storage. Store in tightly closed containers in a cool, well-ventilated area away from strong oxidizers, strong acids. Where possible, automatically pump liquid from drums or other storage containers to process containers. Drums must be equipped with self-closing valves, pressure vacuum bungs, and flame arresters. Use only nonsparking tools and equipment, especially when opening and closing containers of this chemical. Wherever this chemical is used, handled, manufactured, or stored, use explosion-proof electrical equipment and fittings.

Shipping: Methyl *tert*-butyl ether requires a shipping label of " FLAMMABLE LIQUID." It falls in Hazard Class 3 and Packing Group II.

Spill Handling: Evacuate and restrict persons not wearing protective equipment from area of spill or leak until cleanup is complete. Remove all ignition sources. Ventilate area of spill or leak. Absorb liquids in vermiculite, dry sand, earth, peat, carbon, or a similar material and deposit in sealed containers. Keep this chemical out of a confined space, such as a sewer, because of the possibility of an explosion, unless the sewer is designed to prevent the buildup of explosive concentrations. It may be necessary to contain and dispose of this chemical as a hazardous waste. If material or contaminated runoff enters waterways, notify downstream users of potentially contaminated waters. Contact your local or federal environmental protection agency for specific recommendations. If employees are required to clean up spills, they must be properly trained and equipped. OSHA 1910.120(q) may be applicable.

Fire Extinguishing: This chemical is a flammable liquid. Poisonous gases are produced in fire. Use dry chemical, carbon dioxide, or alcohol foam extinguishers. Solid streams of water may spread fire. Vapors are heavier than air and will collect in low areas. Vapors may travel long distances to ignition sources and flashback. Vapors in confined areas may explode when exposed to fire. Containers may explode in fire. Storage containers and parts of containers may rocket great distances, in many directions. If material or contaminated runoff enters waterways, notify downstream users of potentially contaminated waters. Notify local health and fire officials and pollution control agencies. From a secure, explosion-proof location, use water spray to cool exposed containers. If cooling streams are ineffective (venting sound increases in volume and pitch, tank discolors, or shows any signs of deforming), withdraw immediately to a secure position. If employees are expected to fight fires, they must be trained and equipped in OSHA 1910.156. The only respirators recommended for firefighting are self-contained breathing apparatuses that have full face-pieces and are operated in a pressure-demand or other positive-pressure mode.

Disposal Method Suggested: Dissolve or mix the material with a combustible solvent and burn in a chemical incinerator equipped with an afterburner and scrubber. All federal, state, and local environmental regulations must be observed.

Reference

New Jersey Department of Health and Senior Services. (March 2006). *Hazardous Substances Fact Sheet: Methyl-tert-Butyl Ether*. Trenton, NJ

Methyl *n*-butyl ketone   M:0740

Molecular Formula: $C_6H_{12}O$
Common Formula: $CH_3CO(CH_2)_3CH_3$
Synonyms: *n*-Butyl methyl ketone; Butyl methyl ketone; 2-Hexanone; Hexanone-2; Ketone, butyl methyl; MBK; Methyl butyl ketone; MNBK; Propylacetone
CAS Registry Number: 591-78-6
RTECS® Number: MP1400000
UN/NA & ERG Number: UN1224/127
EC Number: 209-731-1 [*Annex I Index No.:* 606-030-00-6]
Regulatory Authority and Advisory Bodies
Carcinogenicity: IARC: Human Inadequate Evidence, animal Limited Evidence, *not classifiable as carcinogenic to humans*, Group 3, 1999; EPA: Available data are inadequate for an assessment of human carcinogenic potential.
Air Pollutant Standard Set. See below, "Permissible Exposure Limits in Air" section.
California Proposition 65 Chemical: Developmental/Reproductive toxin (male) 8/7/09.
Canada, WHMIS, Ingredients Disclosure List Concentration 1.0%.
European/International Regulations: Hazard Symbol: T; Risk phrases: R10; R48/23; R62; R67; Safety phrases: S1/2; S36/37; S45 (see Appendix 4).
WGK (German Aquatic Hazard Class): No value assigned.
Description: MNBK is a colorless liquid with an acetone-like odor. The odor threshold is 0.08 ppm. Molecular weight = 100.18; Specific gravity (H_2O:1) = 0.81; Boiling point = 127.8°C; Freezing/Melting point = −57°C; Vapor pressure = 11 mmHg at 20°C; Flash point = 25°C (cc); Autoignition temperature = 423°C. Explosive limits: LEL = 1.2%; UEL = 8.0%. Hazard Identification (based on NFPA-704 M Rating System): Health 2, Flammability 3, Reactivity 0. Slightly soluble in water; solubility = 2%.
Potential Exposure: The material is used as a solvent.
Incompatibilities: Violent reaction with oxidizers. May form unstable peroxides. Attacks plastics.
Permissible Exposure Limits in Air
Conversion factor: 1 ppm = 4.10 mg/m³ at 25°C & 1 atm.
OSHA PEL: 100 ppm/410 mg/m³ TWA.
NIOSH REL: 1 ppm/4 mg/m³ TWA.
ACGIH TLV®[1]: 5 ppm/20 mg/m³ TWA; 10 ppm/40 mg/m³ STEL [skin].
DFG MAK: 5 ppm/21 mg/m³ TWA; Peak Limitation Category II(8) [skin]; BAT: 5 mg [hexane-2,5-dione +4,5-dihydroxy-2-hexanone]/L in urine at end-of-shift.
NIOSH IDLH: 1600 ppm.
Protective Action Criteria (PAC)

TEEL-0: 5 ppm
PAC-1: 10 ppm
PAC-2: 1500 ppm
PAC-3: 1600 ppm
Australia: TWA 5 ppm (20 mg/m^3), 1993; Austria: MAK 5 ppm (21 mg/m^3), 1999; Belgium: TWA 5 ppm (20 mg/m^3), [skin], 1993; Denmark: TWA 1 ppm (4 mg/m^3), [skin], 1999; Finland: TWA 5 ppm (20 mg/m^3); STEL 10 ppm (40 mg/m^3), [skin], 1999; France: VME 5 ppm (20 mg/m^3), VLE 8 ppm (35 mg/m^3), 1999; Hungary: TWA 20 mg/m^3; STEL 40 mg/m^3, [skin], 1993; the Netherlands: MAC-TGG 2 mg/m^3, 2003; Norway: TWA 1 ppm (4 mg/m^3), 1999; the Philippines: TWA 100 ppm (410 mg/m^3), 1993; Poland: MAC (TWA) 10 mg/m^3, MAC (STEL) 50 mg/m^3, 1999; Russia: TWA 5 ppm, 1993; Sweden: NGV 1 ppm (4 mg/m^3), KTV 2 ppm (8 mg/m^3), [skin], 1999; Turkey: TWA 100 ppm (410 mg/m^3), 1993; United Kingdom: TWA 5 ppm (21 mg/m^3), [skin], 2000; Argentina, Bulgaria, Columbia, Jordan, South Korea, New Zealand, Singapore, Vietnam: ACGIH TLV®: STEL 10 ppm [skin]. Several states have set guidelines or standards for MBK in ambient air[60] ranging from 54 μg/m^3 (Massachusetts) to 80 μg/m^3 (Connecticut) to 200 μg/m^3 (North Dakota) to 350 μg/m^3 (Virginia) to 476 μg/m^3 (Nevada).

Determination in Air: Use NIOSH Analytical Method #1300, Ketones, #2555; OSHA Analytical Method PV-2031.[18]

Determination in Water: Octanol−water coefficient: Log $K_{ow} = 1.4$.

Routes of Entry: Inhalation, skin absorption, ingestion, skin and/or eye contact.

Harmful Effects and Symptoms

Short Term Exposure: Methyl-*n*-butyl ketone can affect you when breathed in and by passing through your skin. Irritates the eyes, skin, and respiratory tract. May affect the nervous system. Exposure may result in unconsciousness. Overexposure can cause you to feel dizzy and lightheaded and to pass out.

Long Term Exposure: Repeated or prolonged contact with skin may cause drying and cracking. Exposure can damage the nerves, causing numbness and weakness, especially in the hands and feet. The vapor can irritate eyes, nose, and throat. There is limited evidence that this chemical may have reproductive effects. It may damage the male reproductive system. Handle with extreme caution. Exposure to other aliphatic monoketones (e.g., methyl ketone, methyl propyl ketone, methyl amyl ketone, hexyl ketone, etc.) may exacerbate the nerve damage caused by this chemical.

Points of Attack: Eyes, skin, respiratory system, central nervous system, peripheral nervous system.

Medical Surveillance: Before beginning employment and at regular times after that, the following are recommended: examination of the nervous system. Complete blood count. If symptoms develop or overexposure has occurred, the following may be useful: Nerve conduction tests should be considered. For men who suspect any problems conceiving a child, semen analysis and sperm count may be useful.

First Aid: If this chemical gets into the eyes, remove any contact lenses at once and irrigate immediately for at least 15 min, occasionally lifting upper and lower lids. Seek medical attention immediately. If this chemical contacts the skin, remove contaminated clothing and wash immediately with soap and water. Seek medical attention immediately. If this chemical has been inhaled, remove from exposure, begin rescue breathing (using universal precautions, including resuscitation mask) if breathing has stopped and CPR if heart action has stopped. Transfer promptly to a medical facility. When this chemical has been swallowed, get medical attention. Give large quantities of water and induce vomiting. Do not make an unconscious person vomit.

Personal Protective Methods: Wear protective gloves and clothing to prevent any reasonable probability of skin contact. Safety equipment suppliers/manufacturers can provide recommendations on the most protective glove/clothing material for your operation. Saranex™ is recommended in the literature. All protective clothing (suits, gloves, footwear, headgear) should be clean, available each day, and put on before work. Contact lenses should not be worn when working with this chemical. Wear splash-proof chemical goggles and face shield unless full face-piece respiratory protection is worn. Employees should wash immediately with soap when skin is wet or contaminated. Provide emergency showers and eyewash. See NIOSH Criteria Document 78-173, *Ketones.*

Respirator Selection: NIOSH: *10 ppm:* Sa (APF = 10) (any supplied-air respirator). *25 ppm:* Sa:Cf (APF = 25) (any supplied-air respirator operated in a continuous-flow mode). *50 ppm:* SaT: Cf (APF = 50) (any supplied-air respirator that has a tight-fitting face-piece and is operated in a continuous-flow mode) or SCBAF (APF = 50) (any self-contained breathing apparatus with a full face-piece) or SaF (APF = 50) (any supplied-air respirator with a full face-piece). *1600 ppm:* SaF: Pd,Pp (APF = 2000) (any supplied-air respirator that has a full face-piece and is operated in a pressure-demand or other positive-pressure mode). *Emergency or planned entry into unknown concentrations or IDLH conditions:* SCBAF: Pd,Pp (APF = 10,000) (any self-contained breathing apparatus that has a full face-piece and is operated in a pressure-demand or other positive-pressure mode) or SaF: Pd,Pp: ASCBA (APF = 10,000) (any supplied-air respirator that has a full face-piece and is operated in a pressure-demand or other positive-pressure mode in combination with an auxiliary, self-contained breathing apparatus operated in a pressure-demand or other positive-pressure mode). *Escape:* GmFOv (APF = 50) [any air-purifying, full-face-piece respirator (gas mask) with a chin-style, front- or back-mounted acid gas canister] or SCBAE (any appropriate escape-type, self-contained breathing apparatus).

Storage: Color Code—Red: Flammability Hazard: Store in a flammable liquid storage area or approved cabinet away from ignition sources and corrosive and reactive materials. Prior to working with this chemical you should be trained on its proper handling and storage. Methyl n-butyl ketone must be stored to avoid contact with strong oxidizers (such as peroxides, chlorates, perchlorates, permanganates, and nitrates) because violent reactions occur. Before entering confined space where this chemical may be present, check to make sure that an explosive concentration does not exist. Store in tightly closed containers in a cool, well-ventilated area away from heat, sparks, and flames. Sources of ignition, such as smoking and open flames, are prohibited where methyl n-butyl ketone is used, handled, or stored in a manner that could create a potential fire or explosion hazard. Metal containers used in the transfer of 5 gallons or more of methyl n-butyl ketone should be grounded and bonded. Drums must be equipped with self-closing valves, pressure vacuum bungs, and flame arresters. Use only non-sparking tools and equipment, especially when opening and closing containers of methyl n-butyl ketone.

Shipping: Ketones, liquid, n.o.s. must be labeled "FLAMMABLE LIQUID." This chemical falls in Hazard Class 3 and Packing Group II.

Spill Handling: Evacuate and restrict persons not wearing protective equipment from area of spill or leak until cleanup is complete. Remove all ignition sources. Establish forced ventilation to keep levels below explosive limit. Absorb liquids in vermiculite, dry sand, earth, peat, carbon, or a similar material and deposit in sealed containers. Keep this chemical out of a confined space, such as a sewer, because of the possibility of an explosion, unless the sewer is designed to prevent the buildup of explosive concentrations. It may be necessary to contain and dispose of this chemical as a hazardous waste. If material or contaminated runoff enters waterways, notify downstream users of potentially contaminated waters. Contact your local or federal environmental protection agency for specific recommendations. If employees are required to clean up spills, they must be properly trained and equipped. OSHA 1910.120(q) may be applicable.

Fire Extinguishing: This chemical is a flammable liquid. Poisonous gases are produced in fire. Use dry chemical, carbon dioxide, or alcohol foam extinguishers. Vapors are heavier than air and will collect in low areas. Vapors may travel long distances to ignition sources and flashback. Vapors in confined areas may explode when exposed to fire. Containers may explode in fire. Storage containers and parts of containers may rocket great distances, in many directions. If material or contaminated runoff enters waterways, notify downstream users of potentially contaminated waters. Notify local health and fire officials and pollution control agencies. From a secure, explosion-proof location, use water spray to cool exposed containers. If cooling streams are ineffective (venting sound increases in volume and pitch, tank discolors, or shows any signs of deforming), withdraw immediately to a secure position. If employees are expected to fight fires, they must be trained and equipped in OSHA 1910.156. The only respirators recommended for firefighting are self-contained breathing apparatuses that have full face-pieces and are operated in a pressure-demand or other positive-pressure mode.

Disposal Method Suggested: Dissolve or mix the material with a combustible solvent and burn in a chemical incinerator equipped with an afterburner and scrubber. All federal, state, and local environmental regulations must be observed.

References

National Institute for Occupational Safety and Health. (1978). *Criteria for a Recommended Standard: Occupational Exposure to Ketones*, NIOSH Document No. 78-173. Washington, DC

New Jersey Department of Health and Senior Services. (April 2004). *Hazardous Substances Fact Sheet: Methyl n-Butyl Ketone*. Trenton, NJ

Methyl chloride   M:0750

Molecular Formula: CH_3Cl
Synonyms: Artic; Chlor-methan (German); Chloromethane; Chlorure de methyle (French); Methylchlorid (German); Monochloromethane; Methane, chloro-
CAS Registry Number: 74-87-3
RTECS® Number: PA6300000
UN/NA & ERG Number: UN1063/115
EC Number: 200-817-4 [*Annex I Index No.:* 602-001-00-7]
Regulatory Authority and Advisory Bodies
Department of Homeland Security Screening Threshold Quantity (pounds): *Release hazard* 10,000 ($\geq$1.00% concentration).
Carcinogenicity: IARC: Human Inadequate Evidence, animal Inadequate Evidence, *not classifiable as carcinogenic to humans*, Group 3, 1999; EPA: Not Classifiable as to human carcinogenicity; NIOSH: Potential occupational carcinogen.
US EPA Gene-Tox Program, Positive: Histidine reversion—Ames test.
Air Pollutant Standard Set. See below, "Permissible Exposure Limits in Air" section.
OSHA 29CFR1910.119, Appendix A, Process Safety List of Highly Hazardous Chemicals, TQ = 15,000 lb.
Clean Air Act: Hazardous Air Pollutants (Title I, Part A, Section 112); Accidental Release Prevention/Flammable Substances, (Section 112[r], Table 3), TQ = 10,000 lb (4540 kg).
Clean Water Act: 40CFR423, Appendix A, Priority Pollutants; Section 313 Water Priority Chemicals (57FR 41331, 9/9/92).
US EPA Hazardous Waste Number (RCRA No.): U045.
RCRA, 40CFR261, Appendix 8 Hazardous Constituents.
Safe Drinking Water Act: Priority List (55 FR 1470).
RCRA 40CFR268.48; 61FR15654, Universal Treatment Standards: Wastewater (mg/L), 0.19; Nonwastewater (mg/kg), 30.

RCRA 40CFR264, Appendix 9; TSD Facilities Ground Water Monitoring List. Suggested test method(s) (PQL μg/L): 8010 (1); 8240 (10).

Reportable Quantity (RQ): 100 lb (45.4 kg).

EPCRA Section 313 Form R *de minimis* concentration reporting level: 1.0%.

California Proposition 65 Developmental/Reproductive toxin 3/10/00; male 8/7/09.

Canada, WHMIS, Ingredients Disclosure List Concentration 1.0%.

European/International Regulations: Hazard Symbol: F+, Xn; Risk phrases: R12; R40; R48/20; Safety phrases: S2; S9; S16; S33 (see Appendix 4).

WGK (German Aquatic Hazard Class): 2—Hazard to waters.

Description: Methyl chloride is a colorless gas with a faint, sweet odor which is not noticeable at dangerous concentrations. The odor threshold is 10 ppm. Shipped as a liquefied compressed gas. Molecular weight = 50.49; Specific gravity (H_2O:1) = 0.92; Boiling point = −24.4°C; Freezing/Melting point = −97.8°C; Relative vapor density (air = 1) = 1.78; Vapor pressure = 5.0 atm; Flash point = flammable gas; Autoignition temperature = 632°C. Explosive limits: LEL = 8.1%; UEL = 17.4%. Hazard Identification (based on NFPA-704 M Rating System): Health 1, Flammability 4, Reactivity 0. Slightly soluble in water (reaction); solubility 0.5%.

Potential Exposure: Compound Description: Tumorigen, Mutagen; Reproductive Effector; Human Data. Methyl chloride is used as a methylating and chlorinating agent in organic chemistry and in the production of silicones and tetramethyl lead. In petroleum refineries it is used as an extractant for greases, oils, and resins. Methyl chloride is also used as a solvent in the synthetic rubber industry, as a refrigerant, and as a propellant in polystyrene foam production. In the past it has been used as a local anesthetic (freezing). It is an intermediate in drug manufacture.

Incompatibilities: Violent reaction with chemically active metals, such as potassium, powdered aluminum, zinc, and magnesium. Reaction with aluminum trichloride, ethylene. Reacts with water (hydrolyzes) to form hydrochloric acid. Attacks many metals in the presence of moisture.

Permissible Exposure Limits in Air

OSHA PEL: 100 ppm TWA; 200 ppm Ceiling Concentration; 300 ppm [5 min max peak in any 3 h] [skin].

NIOSH REL: A potential occupational carcinogen. Limit exposure to lowest feasible concentration. See *NIOSH Pocket Guide*, Appendix A.

ACGIH TLV®[1]: 50 ppm/103 mg/m^3 TWA; 100 ppm/207 mg/m^3 STEL; [skin]; not classifiable as a human carcinogen.

Protective Action Criteria (PAC)*

TEEL-0: 100 ppm

PAC-1: 100 ppm

PAC-2: **910** ppm

PAC-3: **3000** ppm

*AEGLs (Acute Emergency Guideline Levels) & ERPGs (Emergency Response Planning Guideline) are in **bold face**.

DFG MAK: 50 ppm/100 mg/m^3 TWA; Peak Limitation Category II(2) [skin]; Pregnancy Risk Group B; Carcinogen Category 3.

NIOSH IDLH: 2000 ppm [Ca].

Arab Republic of Egypt: TWA 50 ppm (105 mg/m^3), 1993; Australia: TWA 50 ppm (105 mg/m^3); STEL 100 ppm, 1993; Austria: MAK 50 ppm (105 mg/m^3), Suspected: carcinogen, 1999; Belgium: TWA 50 ppm (103 mg/m^3); STEL 100 ppm (207 mg/m^3), 1993; Denmark: TWA 50 ppm (105 mg/m^3), 1999; Finland: TWA 50 ppm (105 mg/m^3); STEL 75 ppm (160 mg/m^3), 1999; France: VME 50 ppm (105 mg/m^3), VLE 100 ppm, carcinogen, 1999; the Netherlands: MAC-TGG 52 mg/m^3, 2003; Japan: 50 ppm (100 mg/m^3), 1999; Norway: TWA 25 ppm (50 mg/m^3), 1999; the Philippines: TWA 100 ppm (210 mg/m^3), 1993; Poland: MAC (TWA) 20 mg/m^3; STEL 160 mg/m^3, 1999; Russia: TWA 50 ppm; STEL 5 mg/m^3, 1993; Sweden: NGV 50 ppm (100 mg/m^3), KTV 100 ppm (200 mg/m^3), 1999; Switzerland: MAK-W 50 ppm (105 mg/m^3), KZG-W 100 ppm (210 mg/m^3); Thailand: TWA 100 mg/m^3; STEL 200 mg/m^3, 1993; Turkey: TWA 100 ppm (210 mg/m^3), 1993; United Kingdom: TWA 50 ppm (105 mg/m^3); STEL 100 ppm, 2000; Argentina, Bulgaria, Columbia, Jordan, South Korea, IN New Zealand, Singapore, Vietnam: ACGIH TLV®: STEL 100 ppm [skin]. Several states have set guidelines or standards for methyl chloride in ambient air[60] ranging from 1.6 μg/m^3 (Michigan) to 74.12 μg/m^3 (Kansas) to 1050−2050 μg/m^3 (North Dakota) to 1750 μg/m^3 (Virginia) to 2100 μg/m^3 (Connecticut and New York) to 2500 μg/m^3 (Nevada) to 2520 μg/m^3 (Pennsylvania).

Determination in Air: Use NIOSH Analytical Method (IV) #1001.

Permissible Concentration in Water: To protect human health: preferably zero. An additional lifetime cancer risk of 1 in 100,000 is posed by a concentration of 1.9 μg/L.[6] In addition, several states have set guidelines for methyl chloride in drinking water[61] ranging from 0.19 μg/L (Kansas) to 0.50 μg/L (Arizona).

Determination in Water: Gas chromatography (EPA Method 601) or gas chromatography and mass spectrometry (EPA Method #624). Octanol−water coefficient: Log K_{ow} = 0.9.

Routes of Entry: Inhalation, skin and/or eye contact (liquid).

Harmful Effects and Symptoms

Short Term Exposure: Inhalation: Illness has been reported at concentrations of 500 ppm. 10,000 ppm for 30 min has caused death. Can cause nausea, vomiting, painful neck, loss of appetite. More severe exposure may result in the above symptoms plus headache, diarrhea, dizziness, loss of coordination, tremors of hands and lips, drooping eyelids, and eye twitch. Very severe exposure may include the above plus burning sensation in mouth and throat, mustard-like taste, difficulty in swallowing, hallucinations, loss of memory, cold and clammy skin, rapid breathing, unconsciousness, coma and death. Onset of symptoms may be delayed several hours after exposure. Effects may last weeks or months. Higher exposures can cause pulmonary edema, a medical emergency that can be delayed for several hours. This can cause death. *Skin:* Contact with liquefied gas may cause freezing of skin, burns and permanent damage. Other symptoms are muscular pain, anemia, muscle weakness and fever. *Eyes:* Contact with the liquid can cause frostbite and severe burns, leading to permanent damage. May cause dimness of sight and abnormally dilated pupils. *Ingestion:* Ingestion of liquefied gas will cause freezing of mouth and throat.

Long Term Exposure: Long-term exposure may affect the testes, causing decreased production of male hormones and sperm. There is limited evidence that methyl chloride causes kidney cancer in animals. May damage the liver and kidneys. May cause brain damage. May cause blurred or double vision, and "drunken" behavior.

Points of Attack: Central nervous system, liver, kidneys, brain, reproductive system.

Medical Surveillance: NIOSH lists the following tests: whole blood (chemical/metabolite), expired air, urine (chemical/metabolite). For those with frequent or potentially high exposure (half the TLV or greater, or significant skin contact), the following are recommended before beginning work and at regular times after that: lung function tests. Examination of the nervous system. If symptoms develop or overexposure is suspected, the following may be useful: examination of the nervous system. Kidney function tests. Liver function tests. Examination for brain effects. Consider chest X-ray after acute overexposure.

First Aid: If this chemical gets into the eyes, remove any contact lenses at once and irrigate immediately for at least 15 min, occasionally lifting upper and lower lids. Seek medical attention immediately. If this chemical contacts the skin, remove contaminated clothing and wash immediately with soap and water. Seek medical attention immediately. If this chemical has been inhaled, remove from exposure, begin rescue breathing (using universal precautions, including resuscitation mask) if breathing has stopped and CPR if heart action has stopped. Transfer promptly to a medical facility. When this chemical has been swallowed, get medical attention. Give large quantities of water and induce vomiting. Do not make an unconscious person vomit. If frostbite has occurred, seek medical attention immediately; do NOT rub the affected areas or flush them with water. In order to prevent further tissue damage, do NOT attempt to remove frozen clothing from frostbitten areas. If frostbite has NOT occurred, immediately and thoroughly wash contaminated skin with soap and water. Medical observation is recommended for 24–48 h after breathing overexposure, as pulmonary edema may be delayed. As first aid for pulmonary edema, a doctor or authorized paramedic may consider administering a corticosteroid spray.

Personal Protective Methods: Wear appropriate personal protective clothing to prevent the skin from becoming frozen from contact with the evaporating liquid or from contact with vessels containing the liquid: **8 h:** Viton™ gloves, suits, Saranex™ coated suits, Barricade™ coated suits; Responder™ suits, Trellchem™ HPS suits; Trychem 1000™ suits. **4 h:** Teflon™ gloves, suits, boots. Prevent possible skin freezing from direct liquid contact. Safety equipment suppliers/manufacturers can provide recommendations on the most protective glove/clothing material for your operation. All protective clothing (suits, gloves, footwear, headgear) should be clean, available each day, and put on before work. Contact lenses should not be worn when working with this chemical. Wear eye protection to prevent any reasonable probability of eye contact. Employees should wash immediately with soap when skin is wet or contaminated. Provide emergency showers and eyewash.

Respirator Selection: NIOSH: *At any concentrations above the NIOSH REL, or where there is no REL, at any detectable concentration:* SCBAF: Pd,Pp (APF = 10,000) (any self-contained breathing apparatus that has a full face-piece and is operated in a pressure-demand or other positive-pressure mode) or SaF: Pd,Pp: ASCBA (any supplied-air respirator that has a full face-piece and is operated in a pressure-demand or other positive-pressure mode in combination with an auxiliary self-contained breathing apparatus operated in a pressure-demand or other positive-pressure mode). *Escape:* SCBAE (any appropriate escape-type, self-contained breathing apparatus).

Storage: Color Code—Red Stripe: Flammability Hazard: Store separately from all other flammable materials. Prior to working with this chemical you should be trained on its proper handling and storage. Before entering confined space where this chemical may be present, check to make sure that an explosive concentration does not exist. Methyl chloride must be stored to avoid contact with oxidizers (such as perchlorates, peroxides, chlorates, nitrates, and permanganates) or chemically active metals (such as sodium, potassium, powdered aluminum, zinc, and magnesium), since violent reactions occur. Store in tightly closed containers in a cool, well-ventilated area away from heat and direct sunlight. Sources of ignition, such as smoking and open flames, are prohibited where methyl chloride is used, handled, or stored in a manner that could create a potential fire or explosion hazard. Wherever methyl chloride is used, handled, manufactured, or stored, use explosion-proof electrical equipment and fittings. Procedures for the handling, use and storage of cylinders should be in compliance with OSHA 1910.101 and 1910.169, as with the recommendations of the Compressed Gas Association. A regulated, marked area should be established where this chemical is handled, used, or stored in compliance with OSHA Standard 1910.1045.

Shipping: This compound requires a shipping label of "FLAMMABLE GAS." It falls in Hazard Class 2.1.

Spill Handling: Evacuate and restrict persons not wearing protective equipment from area of spill or leak until cleanup is complete. Remove all ignition sources. Establish forced ventilation to keep levels below explosive limit. Stop the flow of gas if it can be done safely. If source of leak is a cylinder and the leak cannot be stopped in place, remove leaking cylinder to a safe place in the open air, and repair leak or allow cylinder to empty. Keep this chemical out of confined space, such as a sewer because of the possibility of explosion, unless the sewer is designed to prevent the buildup of explosive concentrations. It may be necessary to contain and dispose of this chemical as a hazardous waste. Contact your local or federal environmental protection agency for specific recommendations. If employees are required to clean up spills, they must be properly trained and equipped. OSHA 1910.120(q) may be applicable.

Fire Extinguishing: This chemical is a flammable gas. Poisonous gases, including hydrogen chloride, are produced in fire. Do not extinguish the fire unless the flow of gas can be stopped and any remaining gas is out of the line. Specially trained personnel may use fog lines to cool exposures and let the fire burn itself out. Vapors are heavier than air and will collect in low areas. Vapors may travel long distances to ignition sources and flashback. Vapors in confined areas may explode when exposed to fire. Containers may explode in fire. Storage containers and parts of containers may rocket great distances, in many directions. If material or contaminated runoff enters waterways, notify downstream users of potentially contaminated waters. Notify local health and fire officials and pollution control agencies. From a secure, explosion-proof location, use water spray to cool exposed containers. If cooling streams are ineffective (venting sound increases in volume and pitch, tank discolors, or shows any signs of deforming), withdraw immediately to a secure position. If cylinders are exposed to excessive heat from fire or flame contact, withdraw immediately to a secure location. If employees are expected to fight fires, they must be trained and equipped in OSHA 1910.156. The only respirators recommended for firefighting are self-contained breathing apparatuses that have full face-pieces and are operated in a pressure-demand or other positive-pressure mode.

Disposal Method Suggested: Consult with environmental regulatory agencies for guidance on acceptable disposal practices. Generators of waste containing this contaminant ($\geq$100 kg/mo) must conform with EPA regulations governing storage, transportation, treatment, and waste disposal. Controlled incineration with adequate scrubbing and ash disposal facilities.

References

US Environmental Protection Agency. (1980). *Halomethanes: Ambient Water Quality Criteria*, Report PB-296, 797. Washington, DC
National Institute for Occupational Safety and Health. (October 1977). *Information Profiles on Potential Occupational Hazards: Methyl Chloride*. Report PB-276-678. Rockville, MD, pp. 29–36
US Environmental Protection Agency. (April 30, 1980). *Chloromethane: Health and Environmental Effects Profile No. 48*. Washington, DC: Office of Solid Waste
New York State Department of Health. (March 1986). *Chemical Fact Sheet: Methyl Chloride*. Version 2. Albany, NY: Bureau of Toxic Substance Assessment
Sax, N. I. (Ed.). (1982). *Dangerous Properties of Industrial Materials Report*, 2, No. 4, 76–78
New Jersey Department of Health and Senior Services. (August 2004). *Hazardous Substances Fact Sheet: Methyl Chloride*. Trenton, NJ

Methyl 2-chloroacrylate   M:0760

Molecular Formula: $C_4H_5ClO_2$
Common Formula: $CH_2Cl = CHCOOCH_3$
Synonyms: 2-Chloroacrylate de méthyle (French); 2-Chloroacrylic acid, methyl ester; 2-Chloropropenoic acid, methyl ester; 2-Chloro-2-propenoic acid methyl ester; 2-Cloroacrilato de metilo (Spanish); Methyl-α-chloroacrylate
CAS Registry Number: 80-63-7
RTECS® Number: AS6380000
UN/NA & ERG Number: UN2924 (Flammable liquid, corrosive, n.o.s)/132
EC Number: 201-298-7
Regulatory Authority and Advisory Bodies
Superfund/EPCRA 40CFR355, Extremely Hazardous Substances: TPQ = 500 lb (227 kg).
Reportable Quantity (RQ): 500 lb (227 kg).
European/International Regulations: Hazard Symbol: C; Risk Phrases: R10; R34; Safety phrases: S9; S16; S25; S28; S33; S36/37/39 (see Appendix 4).
WGK (German Aquatic Hazard Class): No value assigned.
Description: Methyl 2-chloroacrylate is a colorless liquid. Molecular weight = 120.54; Boiling point = 52°C at 51 mmHg. Hazard Identification (based on NFPA-704 M Rating System): Health 3, Flammability 3, Reactivity 0. Slightly soluble in water; solubility = <1 mg/mL at 22°C.
Potential Exposure: Used to make acrylic high polymer with properties closely resembling those of polymethylmethacrylate. Monomer for specialty polymers (e.g., aircraft glazing).[EPA] Corrosive. Lacrimator.
Incompatibilities: May hydrolyze upon contact with moisture. Incompatible with nitrates. May be a polymerization hazard.
Permissible Exposure Limits in Air
Protective Action Criteria (PAC)
TEEL-0: 0.05 ppm
PAC-1: 0.15 ppm
PAC-2: 1.01 ppm
PAC-3: 7.5 ppm
Routes of Entry: Inhalation.
Harmful Effects and Symptoms
Short Term Exposure: Corrosive and a lacrimator. It is a skin, eye, and lung irritant. A trace on skin causes large blisters. Inhalation of high concentrations may cause rapid

breathing, headache, nausea, lethargy, convulsions, and death. Also, high exposures can cause pulmonary edema, a medical emergency that can be delayed for several hours. This can cause death.

Points of Attack: Lungs.

Medical Surveillance: Lung function tests. Consider chest X-ray following acute overexposure.

First Aid: If this chemical gets into the eyes, remove any contact lenses at once and irrigate immediately for at least 15 min, occasionally lifting upper and lower lids. Seek medical attention immediately. If this chemical contacts the skin, remove contaminated clothing and wash immediately with soap and water. Seek medical attention immediately. If this chemical has been inhaled, remove from exposure, begin rescue breathing (using universal precautions, including resuscitation mask) if breathing has stopped and CPR if heart action has stopped. Transfer promptly to a medical facility. When this chemical has been swallowed, get medical attention. Give large quantities of water and induce vomiting. Do not make an unconscious person vomit. Medical observation is recommended for 24–48 h after breathing overexposure, as pulmonary edema may be delayed. As first aid for pulmonary edema, a doctor or authorized paramedic may consider administering a corticosteroid spray.

Personal Protective Methods: Wear protective gloves and clothing to prevent any reasonable probability of skin contact. Safety equipment suppliers/manufacturers can provide recommendations on the most protective glove/clothing material for your operation. All protective clothing (suits, gloves, footwear, headgear) should be clean, available each day, and put on before work. Contact lenses should not be worn when working with this chemical. Wear splash-proof chemical goggles and face shield unless full face-piece respiratory protection is worn. Employees should wash immediately with soap when skin is wet or contaminated. Provide emergency showers and eyewash.

Respirator Selection: Follow regulations in OSHA 29CFR1910.134 or European Standard EN149. Use a NIOSH/MSHA- or European Standard EN149-approved respirator; or use an approved supplied-air respirator with a full face-piece operated in the positive-pressure mode, or with a full face-piece, hood, or helmet in the continuous-flow mode; or use a NIOSH/MSHA- or European Standard EN149-approved self-contained breathing apparatus with a full face-piece operated in pressure-demand or other positive-pressure mode.

Storage: Color Code—Red: Flammability Hazard: Store in a flammable liquid storage area or approved cabinet away from ignition sources and corrosive and reactive materials. Color Code—Blue: Health Hazard/Poison: Store in a secure poison location. Prior to working with this chemical you should be trained on its proper handling and storage. Store in tightly closed containers in a cool, well-ventilated area away from oxidizers and reducing agents. Where possible, automatically pump liquid from drums or other storage containers to process containers.

Shipping: This material requires a label of "FLAMMABLE LIQUID." It falls into Hazard Class 3(8), Packing Group III.

Spill Handling: Evacuate and restrict persons not wearing protective equipment from area of spill or leak until cleanup is complete. Remove all ignition sources. Ventilate area of spill or leak. Absorb liquids in vermiculite, dry sand; earth, peat, carbon, or a similar material and deposit in sealed containers. Keep this chemical out of a confined space, such as a sewer, because of the possibility of an explosion, unless the sewer is designed to prevent the buildup of explosive concentrations. It may be necessary to contain and dispose of this chemical as a hazardous waste. If material or contaminated runoff enters waterways, notify downstream users of potentially contaminated waters. Contact your local or federal environmental protection agency for specific recommendations. If employees are required to clean up spills, they must be properly trained and equipped. OSHA 1910.120(q) may be applicable.

Fire Extinguishing: Poisonous gases are produced in fire. *Small fires:* dry chemical, carbon dioxide, water spray, or foam. *Large fires:* water spray, fog, or foam. Move container from fire area if you can do so without risk. Fight fire from maximum distance. Stay upwind; keep out of low areas. Ventilate closed spaces before entering them. Wear positive pressure breathing apparatus and special protective clothing. Vapors are heavier than air and will collect in low areas. Vapors may travel long distances to ignition sources and flashback. Vapors in confined areas may explode when exposed to fire. Containers may explode in fire. Storage containers and parts of containers may rocket great distances, in many directions. If material or contaminated runoff enters waterways, notify downstream users of potentially contaminated waters. Notify local health and fire officials and pollution control agencies. From a secure, explosion-proof location, use water spray to cool exposed containers. If cooling streams are ineffective (venting sound increases in volume and pitch, tank discolors, or shows any signs of deforming), withdraw immediately to a secure position. If employees are expected to fight fires, they must be trained and equipped in OSHA 1910.156. The only respirators recommended for firefighting are self-contained breathing apparatuses that have full face-pieces and are operated in a pressure-demand or other positive-pressure mode.

Reference

US Environmental Protection Agency. (November 30, 1987). *Chemical Hazard Information Profile: Methyl 2-Chloroacrylate.* Washington, DC: Chemical Emergency Preparedness Program

Methyl chloroformate   M:0770

Molecular Formula: $C_2H_3ClO_2$

Synonyms: Carbonochloridic acid, methyl ester; Chlorameisensaeure methylester (German); Chlorocarbonate de methyle (French); Chlorocarbonic acid methyl ester; Chloroformic acid methyl ester; MCF; Methoxycarbonyl chloride

CAS Registry Number: 79-22-1

RTECS® Number: FG3675000

UN/NA & ERG Number: UN1238/155

EC Number: 201-187-3 [*Annex I Index No.:* 607-019-00-9]

Regulatory Authority and Advisory Bodies

Department of Homeland Security Screening Threshold Quantity (pounds): *Release hazard* 10,000 (≥1.00% concentration).

Air Pollutant Standard Set. See below, "Permissible Exposure Limits in Air" section.

OSHA 29CFR1910.119, Appendix A. Process Safety List of Highly Hazardous Chemicals, TQ = 500 lb (227 kg).

Clean Air Act: Accidental Release Prevention/Flammable Substances, (Section 112[r], Table 3), TQ = 5000 lb (2270 kg).

US EPA Hazardous Waste Number (RCRA No.): U156.

RCRA, 40CFR261, Appendix 8 Hazardous Constituents.

Superfund/EPCRA 40CFR355, Extremely Hazardous Substances: TPQ = 500 lb (227 kg).

Reportable Quantity (RQ): 1000 lb (454 kg).

EPCRA Section 313 Form R *de minimis* concentration reporting level: 1.0%.

US DOT 49CFR172.101, Inhalation Hazardous Chemical.

European/International Regulations: Hazard Symbol: F, T; Risk phrases: R11; R21/22; R26; R34; Safety phrases: S1/2; S14; S26; S28; S36/37/39; S45; S46; S63 (see Appendix 4). WGK (German Aquatic Hazard Class): 2—Hazard to waters.

Description: Methyl chloroformate is a colorless liquid with an unpleasant, acrid odor. This is a highly corrosive and flammable material. Molecular weight = 94.50; Specific gravity (H_2O:1) = 1.22; Boiling point = 71°C; Freezing/Melting point = −61°C; Flash point = 17°C; Autoignition temperature = 504°C. Flammability Limits: LEL = 6.7%; UEL = unknown. Hazard Identification (based on NFPA-704 M Rating System): Health 1, Flammability 3, Reactivity 1. Slightly soluble in water.

Potential Exposure: Compound Description: Human Data. Used in synthesis of pharmaceuticals; herbicides, plastics, and other organic chemicals; as a solvent in the photographic industry; as a chemical intermediate in the production of other chemicals. In WWI it was used as military tear-producing warfare agent.

Incompatibilities: Forms explosive mixture with air. Violent reaction with alkali metals, ethers. Incompatible with strong acids, strong bases, alcohols, oxidizers, dimethylsulfoxide, dimethyl formamide. Contact with water or moisture produces corrosive and poisonous hydrogen chloride gas. Corrodes metals in the presence of moisture. Attacks some plastics, rubber, and coatings.

Permissible Exposure Limits in Air

Protective Action Criteria (PAC)*

TEEL-0: 0.2 ppm

PAC-1: 0.3 ppm

PAC-2: **2.2** ppm

PAC-3: **6.7** ppm

*AEGLs (Acute Emergency Guideline Levels) & ERPGs (Emergency Response Planning Guideline) are in **bold face**.

DFG MAK: 0.2 ppm/0.78mg/m^3 TWA; Peak Limitation Category I(2); Pregnancy Risk Group C.

Determination in Air: No method available.

Determination in Water: No criteria have been established.

Routes of Entry: Inhalation, skin and/or eye contact.

Harmful Effects and Symptoms

Short Term Exposure: Highly corrosive, contact can irritate and burn the skin and eyes, with possible permanent damage. Inhalation irritates the respiratory tract causing coughing and/or shortness of breath. Higher exposures can cause pulmonary edema, a medical emergency that can be delayed for several hours. This can cause death. A concentration of 190 ppm has been lethal in 10 min.

Long Term Exposure: Can cause sensitization and skin allergy. Can cause lung irritation and bronchitis. After 2−3 inhalations, brief initial irritation may occur followed by massive symptoms (heavy cough) after 36 h. Relapse may occur in the following days with eventual recovery.

Points of Attack: Skin, lungs.

Medical Surveillance: Lung function tests. Examination by a qualified allergist. Consider chest X-ray following acute overexposure.

First Aid: If this chemical gets into the eyes, remove any contact lenses at once and irrigate immediately for at least 30 min, occasionally lifting upper and lower lids. Seek medical attention immediately. If this chemical contacts the skin, remove contaminated clothing and wash immediately with soap and water. Seek medical attention immediately. If this chemical has been inhaled, remove from exposure, begin rescue breathing (using universal precautions, including resuscitation mask) if breathing has stopped and CPR if heart action has stopped. Transfer promptly to a medical facility. When this chemical has been swallowed, get medical attention. If victim is conscious, administer water or milk. Do not induce vomiting. Medical observation is recommended for 24−48 h after breathing overexposure, as pulmonary edema may be delayed. As first aid for pulmonary edema, a doctor or authorized paramedic may consider administering a corticosteroid spray.

Personal Protective Methods: Wear protective gloves and clothing to prevent any reasonable probability of skin contact. Safety equipment suppliers/manufacturers can provide recommendations on the most protective glove/clothing material for your operation. All protective clothing (suits, gloves, footwear, headgear) should be clean, available each day, and put on before work. Contact lenses should not be worn when working with this chemical. Wear splash-proof chemical goggles and face shield unless full face-piece respiratory protection is worn. Employees should wash immediately with soap when skin is wet or contaminated. Provide emergency showers and eyewash.

Respirator Selection: Follow regulations in OSHA 29CFR1910.134 or European Standard EN149. Use a NIOSH/MSHA- or European Standard EN149-approved respirator; or use an approved supplied-air respirator with a full face-piece operated in the positive-pressure mode, or with a full face-piece, hood, or helmet in the continuous-flow mode; or use a NIOSH/MSHA- or European Standard EN149-approved self-contained breathing apparatus with a full face-piece operated in pressure-demand or other positive-pressure mode.

Storage: Color Code—Red: Flammability Hazard: Store in a flammable liquid storage area or approved cabinet away from ignition sources and corrosive and reactive materials. Prior to working with methyl chloroformate you should be trained on its proper handling and storage. Before entering confined space where this chemical may be present, check to make sure that an explosive concentration does not exist. Store in tightly closed containers in a cool, well-ventilated area away from all forms of moisture, oxidizers, strong acids, strong bases. See incompatibilities for other materials to avoid. Where possible, automatically pump liquid from drums or other storage containers to process containers. Drums must be equipped with self-closing valves, pressure vacuum bungs, and flame arresters. Use only nonsparking tools and equipment, especially when opening and closing containers of this chemical. Wherever this chemical is used, handled, manufactured, or stored, use explosion-proof electrical equipment and fittings.

Shipping: Methyl chloroformate requires a shipping label of "FLAMMABLE LIQUID, CORROSIVE." It falls in Hazard Class 6.1. Packing Group I.

Spill Handling: Evacuate and restrict persons not wearing protective equipment from area of spill or leak until cleanup is complete. Remove all ignition sources. Establish forced ventilation to keep levels below explosive limit. Absorb liquids in vermiculite, dry sand, earth, peat, carbon, or a similar material and deposit in sealed containers. Keep this chemical out of a confined space, such as a sewer, because of the possibility of an explosion, unless the sewer is designed to prevent the buildup of explosive concentrations. It may be necessary to contain and dispose of this chemical as a hazardous waste. If material or contaminated runoff enters waterways, notify downstream users of potentially contaminated waters. Contact your local or federal environmental protection agency for specific recommendations. If employees are required to clean up spills, they must be properly trained and equipped. OSHA 1910.120(q) may be applicable.

Initial isolation and protective action distances

Distances shown are likely to be affected during the first 30 min after materials are spilled and could increase with time. If more than one tank car, cargo tank, portable tank, or large cylinder involved in the incident is leaking, the protective action distance may need to be increased. You may need to seek emergency information from CHEMTREC at (800) 424-9300 or seek professional environmental engineering assistance from the US EPA Environmental Response Team at (908) 548-8730 (24-h response line).

Small spills (From a small package or a small leak from a large package)
First: Isolate in all directions (feet/meters) 100/30
Then: Protect persons downwind (miles/kilometers)
Day 0.2/0.3
Night 0.4/0.6

Large spills (From a large package or from many small packages)
First: Isolate in all directions (feet/meters) 500/150
Then: Protect persons downwind (miles/kilometers)
Day 0.8/1.2
Night 1.6/2.5

Fire Extinguishing: This chemical is a flammable liquid. Poisonous gases, including hydrogen chloride and phosgene, are produced in fire. *Do not use water.* Use dry chemical, carbon dioxide, alcohol foam, or polymer foam extinguishers. Vapors are heavier than air and will collect in low areas. Vapors may travel long distances to ignition sources and flashback. Vapors in confined areas may explode when exposed to fire. Containers may explode in fire. Storage containers and parts of containers may rocket great distances, in many directions. If material or contaminated runoff enters waterways, notify downstream users of potentially contaminated waters. Notify local health and fire officials and pollution control agencies. From a secure, explosion-proof location, use water spray to cool exposed containers. If cooling streams are ineffective (venting sound increases in volume and pitch, tank discolors, or shows any signs of deforming), withdraw immediately to a secure position. If employees are expected to fight fires, they must be trained and equipped in OSHA 1910.156. The only respirators recommended for firefighting are self-contained breathing apparatuses that have full face-pieces and are operated in a pressure-demand or other positive-pressure mode.

Disposal Method Suggested: Consult with environmental regulatory agencies for guidance on acceptable disposal practices. Generators of waste containing this contaminant ($\geq$100 kg/mo) must conform with EPA regulations governing storage, transportation, treatment, and waste disposal.

Reference
New Jersey Department of Health and Senior Services. (June 1999). *Hazardous Substances Fact Sheet: Methyl Chloroformate.* Trenton, NJ

Methyl chlorosilane   M:0780

Molecular Formula: CH_5ClSi
Common Formula: CH_3SiH_2Cl
Synonyms: Chloromethylsilane
CAS Registry Number: 993-00-0
RTECS® Number: VV2150000
UN/NA & ERG Number: UN2534/119
EC Number: 213-600-4

Regulatory Authority and Advisory Bodies

Department of Homeland Security Screening Threshold Quantity (pounds): *Theft hazard* 45 (≥45.00% concentration).

Canada, WHMIS, Ingredients Disclosure List Concentration 1.0%.

European/International Regulations: not listed in Annex 1. WGK (German Aquatic Hazard Class): 1—Low hazard to waters.

Description: Methyl chlorosilane is a colorless liquid. Flash point = −9°C. NJDHSS (New Jersey Department of Health and Senior Services) Hazard Rating: Health 3, Flammability 4, Reactivity 2W. Reacts with water.

Potential Exposure: Methyl chlorosilane is rarely used as a raw material, but it is released during the manufacture of silicones or siloxanes.

Incompatibilities: May form explosive gases with air. Contact with water, steam, or moisture, forms toxic and corrosive hydrogen chloride gas. Not compatible with strong bases, strong acids, oxidizers. Attacks metals, plastics, rubbers, and coatings.

Permissible Exposure Limits in Air

Protective Action Criteria (PAC)*

TEEL-0: 0.6 ppm

PAC-1: **1.8** ppm

PAC-2: **22** ppm

PAC-3: **100** ppm

*AEGLs (Acute Emergency Guideline Levels) & ERPGs (Emergency Response Planning Guideline) are in **bold face**.

Determination in Air: No method available.

Harmful Effects and Symptoms

Short Term Exposure: Methyl chlorosilane can affect you when breathed in. Methyl chlorosilane is a corrosive chemical and can cause severe eye and skin burns. Higher exposures can cause pulmonary edema, a medical emergency that can be delayed for several hours. This can cause death.

Long Term Exposure: Repeated exposure may cause bronchitis with cough, phlegm, and/or shortness of breath.

Points of Attack: Lungs.

Medical Surveillance: Before beginning employment and at regular times after that, for those with frequent or potentially high exposures, the following are recommended: lung function tests. If symptoms develop or overexposure is suspected, the following may be useful: consider chest X-ray after acute overexposure.

First Aid: If this chemical gets into the eyes, remove any contact lenses at once and irrigate immediately for at least 30 min, occasionally lifting upper and lower lids. Seek medical attention immediately. If this chemical contacts the skin, remove contaminated clothing and wash immediately with soap and water. Seek medical attention immediately. If this chemical has been inhaled, remove from exposure, begin rescue breathing (using universal precautions, including resuscitation mask) if breathing has stopped and CPR if heart action has stopped. Transfer promptly to a medical facility. When this chemical has been swallowed, get medical attention. If victim is conscious, administer water or milk. Do not induce vomiting. Medical observation is recommended for 24–48 h after breathing overexposure, as pulmonary edema may be delayed. As first aid for pulmonary edema, a doctor or authorized paramedic may consider administering a corticosteroid spray.

Personal Protective Methods: Wear protective gloves and clothing to prevent any reasonable probability of skin contact. Safety equipment suppliers/manufacturers can provide recommendations on the most protective glove/clothing material for your operation. All protective clothing (suits, gloves, footwear, headgear) should be clean, available each day, and put on before work. Contact lenses should not be worn when working with this chemical. Wear splash-proof chemical goggles and face shield unless full face-piece respiratory protection is worn. Employees should wash immediately with soap when skin is wet or contaminated. Provide emergency showers and eyewash.

Respirator Selection: Where there is potential for exposure to methyl chlorosilane, use a NIOSH/MSHA- or European Standard EN149-approved supplied-air respirator with a full face-piece operated in the positive-pressure mode, or with a full face-piece, hood, or helmet in the continuous-flow mode; or use a NIOSH/MSHA- or European Standard EN149-approved self-contained breathing apparatus with a full face-piece operated in pressure-demand or other positive-pressure mode.

Storage: Color Code—Red Stripe: Flammability Hazard: Store separately from all other flammable materials. Prior to working with this chemical you should be trained on its proper handling and storage. Methyl chlorosilane must be stored to avoid contact with water, steam, and moisture because toxic and corrosive chloride gases, including hydrogen chloride, can be produced. Sources of ignition, such as smoking and open flames, are prohibited where methyl chlorosilane is handled, used, or stored.

Shipping: Methylchlorosilane requires a shipping label of "POISON GAS, FLAMMABLE GAS, CORROSIVE." It falls in Hazard Class 2.3. It is a violation of transportation regulations to refill compressed gas cylinders without the express written permission of the owner.

Spill Handling: Evacuate and restrict persons not wearing protective equipment from area of spill or leak until cleanup is complete. Remove all ignition sources. *Do not use water* or wet method. Absorb liquids in vermiculite, dry sand, earth, peat, carbon, or a similar material and deposit in sealed containers. Ventilate and wash area (after neutralizing with sodium bicarbonate) after cleanup is complete. Keep this chemical out of a confined space, such as a sewer, because of the possibility of an explosion, unless the sewer is designed to prevent the buildup of explosive concentrations. It may be necessary to contain and dispose of this chemical as a hazardous waste. If material or contaminated runoff enters waterways, notify downstream users of potentially contaminated waters. Contact your local or federal environmental protection agency for specific recommendations. If employees are required to clean up spills, they must be properly trained and equipped. OSHA 1910.120(q) may be applicable.

Initial isolation and protective action distances

Distances shown are likely to be affected during the first 30 min after materials are spilled and could increase with time. If more than one tank car, cargo tank, portable tank, or large cylinder involved in the incident is leaking, the protective action distance may need to be increased. You may need to seek emergency information from CHEMTREC at (800) 424-9300 or seek professional environmental engineering assistance from the US EPA Environmental Response Team at (908) 548-8730 (24-h response line).

Small spills (From a small package or a small leak from a large package)

First: Isolate in all directions (feet/meters)   100/30
Then: Protect persons downwind (miles/kilometers)
Day   0.1/0.2
Night   0.4/0.6

Large spills (From a large package or from many small packages)

First: Isolate in all directions (feet/meters)   1000/300
Then: Protect persons downwind (miles/kilometers)
Day   1.0/1.5
Night   2.7/4.3/4.3

Fire Extinguishing: This chemical is a flammable and corrosive liquid. Poisonous gases, including chlorine and phosgene, are produced in fire. *Do not use water.* Use dry chemical or carbon dioxide extinguishers. Vapors are heavier than air and will collect in low areas. Vapors may travel long distances to ignition sources and flashback. Vapors in confined areas may explode when exposed to fire. Containers may explode in fire. Storage containers and parts of containers may rocket great distances, in many directions. If material or contaminated runoff enters waterways, notify downstream users of potentially contaminated waters. Notify local health and fire officials and pollution control agencies. From a secure, explosion-proof location, use water spray to cool exposed containers. If cooling streams are ineffective (venting sound increases in volume and pitch, tank discolors, or shows any signs of deforming), withdraw immediately to a secure position. If employees are expected to fight fires, they must be trained and equipped in OSHA 1910.156. The only respirators recommended for firefighting are self-contained breathing apparatuses that have full face-pieces and are operated in a pressure-demand or other positive-pressure mode.

Reference

New Jersey Department of Health and Senior Services. (December 1999). *Hazardous Substances Fact Sheet: Methyl Chlorosilane.* Trenton, NJ

Methyl cyanoacrylate   M:0790

Molecular Formula: $C_5H_5NO_2$
Common Formula: $CH_2=C(CN)COOCH_3$

Synonyms: Acrylic acid, 2-cyano-, methyl ester; Adhere®; 2-Cyanoacrylic acid methyl ester; Eastman 910 monomer®; Mecrylate; 2-Propenoic acid, 2-cyano-, methyl ester; Super Bonder®; Super Glue®

CAS Registry Number: 137-05-3

RTECS® Number: AS7000000

UN/NA & ERG Number: UN3334 (Aviation regulated liquid, n.o.s.)/171 (ICOA/IATA)

EC Number: 205-275-2 [*Annex I Index No.:* 607-235-00-3]

Regulatory Authority and Advisory Bodies

Air Pollutant Standard Set. See below, "Permissible Exposure Limits in Air" section.

Canada, WHMIS, Ingredients Disclosure List Concentration 1.0%.

European/International Regulations: Hazard Symbol: Xi; Risk phrases: R36/37/38; Safety phrases: S2; S23; 24/25; S26 (see Appendix 4).

WGK (German Aquatic Hazard Class): No value assigned.

Description: Methyl cyanoacrylate is a thick, clear liquid adhesive. The odor threshold is 2.2 ppm. Molecular weight = 111.11; Specific gravity (H_2O:1) = 1.10 at 27°C; Boiling point = 47−49°C at 1.8 mmHg; Vapor pressure = 0.2 mmHg at 20°C; Flash point = 78.9°C. Hazard Identification (based on NFPA-704 M Rating System): Health 2, Flammability 2, Reactivity 1. Insoluble in water.

Potential Exposure: Compound Description: Mutagen; Reproductive Effector; Primary Irritant. Methyl 2-cyanoacrylate is used in production of coatings and textiles; in the manufacture of quick-setting, high-strength, adhesive cements. Often found around the home; bonds eyes and skin in seconds. *Keep out of the reach of children.*

Incompatibilities: Reacts violently with water, bases, and peroxides. Contact with alcohols, water, amines, and alkalis can cause rapid polymerization.

Permissible Exposure Limits in Air

Conversion factor: 1 ppm = 4.54 mg/m^3 at 25°C & 1 atm.

OSHA PEL: None.

NIOSH REL: 2 ppm/8 mg/m^3 TWA; 4 ppm/16 mg/m^3 STEL.

ACGIH TLV®[1]: 0.2 ppm/1 mg/m^3 TWA.

Protective Action Criteria (PAC)

TEEL-0: 0.2 ppm

PAC-1: 4 ppm

PAC-2: 12.5 ppm

PAC-3: 12.5 ppm

DFG MAK: 2 ppm/9.2 mg/m^3 TWA; Peak Limitation Category I(1); Pregnancy Risk Group D.

Australia: TWA 2 ppm (8 mg/m^3); STEL 4 ppm, 1993; Austria: MAK 2 ppm (8 mg/m^3), 1999; Belgium: TWA 2 ppm (9.1 mg/m^3); STEL 4 ppm (18 mg/m^3), 1993; Denmark: TWA 2 ppm (8 mg/m^3), 1999; Finland: TWA 2 ppm (9 mg/m^3); STEL 4 ppm (18 mg/m^3), [skin], 1993; France: VME 2 ppm (8 mg/m^3), VLE 4 ppm (16 mg/m^3), 1999; Norway: TWA 2 ppm (8 mg/m^3), 1999; Sweden: NGV 2 ppm (9 mg/m^3), KTV 4 ppm (18 mg/m^3), 1999; Switzerland: MAK-W 2 ppm (9 mg/m^3), 1999; United

Kingdom: STEL 0.3 ppm (1.4 mg/m^3), 2000; Argentina, Bulgaria, Columbia, Jordan, South Korea, New Zealand, Singapore, Vietnam: ACGIH TLV$^®$: TWA 0.2 ppm. Several states have set guidelines or standards for methyl cyanoacrylate in ambient air[60] ranging from 48 μg/m^3 (Nevada) to 80–160 μg/m^3 (North Dakota) to 130 μg/m^3 (Virginia) to 160 μg/m^3 (Connecticut).

Determination in Air: Use OSHA Analytical Method 55.

Routes of Entry: Inhalation, ingestion, skin and/or eye contact.

Harmful Effects and Symptoms

Short Term Exposure: Methyl 2-cyanoacrylate can affect you when breathed in. Exposure can irritate the eyes, nose, and throat. Higher exposures can cause pulmonary edema, a medical emergency that can be delayed for several hours. This can cause death. Inhalation of vapor may cause an asthmatic reaction. Contact can irritate the eyes and skin. Capable of instantly gluing skin tissue.

Long Term Exposure: Methyl 2-cyanoacrylate may cause a skin allergy to develop. Once an allergy has developed, even very small future exposures will cause rash and itching. Repeated exposure may affect the liver and kidneys. May be able to cause lung damage.

Points of Attack: Eyes, skin, respiratory system.

Medical Surveillance: For those with frequent or potentially high exposure (half the TLV or greater), the following are recommended before beginning work and at regular times after that: lung function tests. If symptoms develop or overexposure is suspected, the following may be useful: liver and kidney function tests. Consider chest X-ray after acute overexposure. Evaluation by a qualified allergist, including careful exposure history and special testing, may help diagnose skin allergy.

First Aid: If this chemical gets into the eyes, remove any contact lenses at once and irrigate immediately for at least 15 min, occasionally lifting upper and lower lids. Seek medical attention immediately. If this chemical contacts the skin, remove contaminated clothing and wash immediately with plenty of soap and water. Remove cured adhesive with hot soapy water. Do not pull or scrape off adhesive as skin can also come off. Seek medical attention immediately. If this chemical has been inhaled, remove from exposure, begin rescue breathing (using universal precautions, including resuscitation mask) if breathing has stopped and CPR if heart action has stopped. Transfer promptly to a medical facility. When this chemical has been swallowed, get medical attention. Give large quantities of water and induce vomiting. Do not make an unconscious person vomit. Medical observation is recommended for 24–48 h after breathing overexposure, as pulmonary edema may be delayed. As first aid for pulmonary edema, a doctor or authorized paramedic may consider administering a corticosteroid spray.

Personal Protective Methods: Wear protective gloves and clothing to prevent any reasonable probability of skin contact. Safety equipment suppliers/manufacturers can provide recommendations on the most protective glove/clothing material for your operation. All protective clothing (suits, gloves, footwear, headgear) should be clean, available each day, and put on before work. Contact lenses should not be worn when working with this chemical. Wear splash-proof chemical goggles and face shield unless full face-piece respiratory protection is worn. Employees should wash immediately with soap when skin is wet or contaminated. Provide emergency showers and eyewash.

Respirator Selection: Where there is potential for exposures *over 2 ppm*, use a NIOSH/MSHA- or European Standard EN149-approved supplied-air respirator with a full face-piece operated in the positive-pressure mode, or with a full face-piece, hood, or helmet in the continuous-flow mode; or use a NIOSH/MSHA approve self-contained breathing apparatus with a full face-piece operated in pressure-demand or other positive-pressure mode.

Storage: Color Code—Red: Flammability Hazard: Store in a flammable liquid storage area or approved cabinet away from ignition sources and corrosive and reactive materials. Prior to working with this chemical you should be trained on its proper handling and storage. Methyl 2-cyanoacrylate must be stored to avoid contact with water, alkaline materials or peroxides, since violent reactions occur. Store in original containers under refrigerated conditions at 2–8°C. Where possible, automatically pump liquid from drums or other storage containers to process containers.

Shipping: AVIATION REGULATED LIQUID N.O.S. (Cyanoacrylate ester) Air: Packaging instructions (passenger or cargo): 906. Hazard Class 9.

Spill Handling: Evacuate and restrict persons not wearing protective equipment from area of spill or leak until cleanup is complete. Remove all ignition sources. Ventilate area of spill or leak. Absorb liquids in vermiculite, dry sand, earth, peat, carbon, or a similar material and deposit in sealed containers. Keep this chemical out of a confined space, such as a sewer, because of the possibility of an explosion, unless the sewer is designed to prevent the buildup of explosive concentrations. It may be necessary to contain and dispose of this chemical as a hazardous waste. If material or contaminated runoff enters waterways, notify downstream users of potentially contaminated waters. Contact your local or federal environmental protection agency for specific recommendations. If employees are required to clean up spills, they must be properly trained and equipped. OSHA 1910.120(q) may be applicable.

Fire Extinguishing: This chemical is a combustible liquid. Poisonous gases, including cyanide and nitrogen oxides, are produced in fire. Use dry chemical, carbon dioxide, or alcohol foam extinguishers. Avoid water unless you can flood the area. Water causes methyl 2-cyanoacrylate to polymerize and possibly to ignite spontaneously. Vapors are heavier than air and will collect in low areas. Vapors may travel long distances to ignition sources and flashback. Vapors in confined areas may explode when exposed to fire. Containers may explode in fire. Storage containers and parts of containers may rocket great distances, in many directions. If material or contaminated runoff enters waterways, notify downstream users of potentially contaminated waters. Notify local health and fire officials and pollution control agencies. From a secure, explosion-proof location, use water spray to cool exposed containers. If cooling streams are ineffective (venting sound increases in volume and pitch, tank discolors, or shows any signs of deforming), withdraw immediately to a secure position. If employees are expected to fight fires, they must be trained and equipped in OSHA 1910.156. The only respirators recommended for firefighting are self-contained breathing apparatuses that have full face-pieces and are operated in a pressure-demand or other positive-pressure mode.

Reference

New Jersey Department of Health and Senior Services. (February 2000). *Hazardous Substances Fact Sheet: Methyl 2-Cyanoacrylate*. Trenton, NJ

Methylcyclohexane   M:0800

Molecular Formula: C_7H_{14}
Common Formula: $C_6H_{11}CH_3$
Synonyms: Cyclohexane, methyl-; Cyclohexylmethane; Heptanaphthene; Hexahydrotoluene; MCH; Sextone B; Toluene hexahydride
CAS Registry Number: 108-87-2
RTECS® Number: GV6125000
UN/NA & ERG Number: UN2296/128
EC Number: 203-624-3 [*Annex I Index No.:* 601-018-00-7]
Regulatory Authority and Advisory Bodies
Air Pollutant Standard Set. See below, "Permissible Exposure Limits in Air" section.
Canada, WHMIS, Ingredients Disclosure List Concentration 1.0%.
European/International Regulations: Hazard Symbol: F; Xn, N; Risk phrases: R11; R38; R51/53; R38; R65; R67; Safety phrases: S2; S9; S16; S33; S61; S62 (see Appendix 4).
WGK (German Aquatic Hazard Class): 2—Hazard to waters.
Description: Methylcyclohexane, an alkene, is a colorless liquid with a faint benzene-like odor. The odor threshold is 630 ppm (this is above the OEL). Molecular weight = 98.21; Specific gravity (H_2O:1) = 0.77; Boiling point = 101°C; Freezing/Melting point = −127°C; Vapor pressure = 37 mmHg at 20°C; Flash point = −3.9°C (cc), −5.9°C (oc); Autoignition temperature = 258°C. Explosive limits: LEL = 1.2%; UEL = 6.7%. Hazard Identification (based on NFPA-704 M Rating System): Health 2, Flammability 3, Reactivity 0. Insoluble in water.
Potential Exposure: Compound Description: Primary Irritant. Methylcyclohexane is used as a solvent for

cellulose derivatives particularly with other solvents and as an organic intermediate in organic synthesis. A component of jet fuel.
Incompatibilities: Forms explosive mixture with air. Strong oxidizers may cause fire and explosions. Attacks some plastics, rubber, and coatings.
Permissible Exposure Limits in Air
OSHA PEL: 500 ppm/2000 mg/m³ TWA.
NIOSH REL: 400 ppm/1600 mg/m³ TWA.
ACGIH TLV®[1]: 400 ppm/1610 mg/m³ TWA.
NIOSH IDLH: 1200 ppm [LEL].
Temporary Emergency Exposure Limits (DOE
TEEL-0: 500 ppm
PAC-1: 1200 ppm
PAC-2: 1200 ppm
PAC-3: 1200 ppm
DFG MAK: 200 ppm/810 mg/m³ TWA; Peak Limitation Category II(2); Pregnancy Risk Group D
Australia: TWA 400 ppm (1600 mg/m³), 1993; Belgium: TWA 400 ppm (1610 mg/m³), 1993; Denmark: TWA 200 ppm (805 mg/m³), 1999; Finland: TWA 400 ppm (1600 mg/m³); STEL 500 ppm (2000 mg/m³), 1999; France: VME 400 ppm (1600 mg/m³), 1999; Japan: 400 ppm (1600 mg/m³), 1999; the Netherlands: MAC-TGG 1600 mg/m³, 2003; the Philippines: TWA 500 ppm (2000 mg/m³), 1993; Poland: MAC (TWA) 500 mg/m³, MAC (STEL) 2000 mg/m³, 1999; Russia: TWA 400 ppm; STEL 50 mg/m³, 1993; Switzerland: MAK-W 400 ppm (1600 mg/m³), KZG-W 800 ppm (3200 mg/m³), 1999; Turkey: TWA 500 ppm (2000 mg/m³), 1993; United Kingdom: LTEL 400 ppm (1600 mg/m³); STEL 500 ppm, 1993; Argentina, Bulgaria, Columbia, Jordan, South Korea, New Zealand, Singapore, Vietnam: ACGIH TLV®: TWA 400 ppm. Several states have set guidelines or standards for methyl cyclohexane in ambient air[60] ranging from 16−20 mg/m³ (North Dakota) to 27 mg/m³ (Virginia) to 32 mg/m³ (Connecticut) to 38.095 mg/m³ (Nevada).
Determination in Air: Charcoal adsorption, workup with CS_2, analysis by gas chromatography/flame ionization. Use NIOSH Analytical Method #1500 for Hydrocarbons, BP 36−126°C.[18]
Permissible Concentration in Water: Vermont has set a guideline of 28.6 mg/L for drinking water.[61]
Routes of Entry: Inhalation, ingestion, skin and/or eye contact.
Harmful Effects and Symptoms
Short Term Exposure: Methylcyclohexane can affect you when breathed in. Irritates the eyes, skin, and respiratory tract. May affect the central nervous system; exposure may cause dizziness, lightheadedness. High levels may act as an anesthetic. Unconsciousness and death may occur at higher levels. Ingesting the liquid may cause chemical pneumonitis.
Long Term Exposure: Prolonged or repeated skin contact can cause cracking and drying of exposed areas. May affect the liver and kidneys.

Points of Attack: Eyes, skin, respiratory system, central nervous system, kidneys, liver.

Medical Surveillance: Liver and kidney function tests.

First Aid: If this chemical gets into the eyes, remove any contact lenses at once and irrigate immediately for at least 15 min, occasionally lifting upper and lower lids. Seek medical attention immediately. If this chemical contacts the skin, remove contaminated clothing and wash immediately with soap and water. Seek medical attention immediately. If this chemical has been inhaled, remove from exposure, begin rescue breathing (using universal precautions, including resuscitation mask) if breathing has stopped and CPR if heart action has stopped. Transfer promptly to a medical facility. When this chemical has been swallowed, get medical attention. Give large quantities of water and induce vomiting. Do not make an unconscious person vomit.

Personal Protective Methods: Wear protective gloves and clothing to prevent any reasonable probability of skin contact. Safety equipment suppliers/manufacturers can provide recommendations on the most protective glove/clothing material for your operation. All protective clothing (suits, gloves, footwear, headgear) should be clean, available each day, and put on before work. Contact lenses should not be worn when working with this chemical. Wear splash-proof chemical goggles and face shield unless full face-piece respiratory protection is worn. Employees should wash immediately with soap when skin is wet or contaminated. Provide emergency showers and eyewash.

Respirator Selection: Up to 1200 ppm: Sa (APF = 10) (any supplied-air respirator) or SCBAF (APF = 50) (any self-contained breathing apparatus with a full face-piece). *Emergency or planned entry into unknown concentrations or IDLH conditions:* SCBAF: Pd,Pp (APF = 10,000) (any self-contained breathing apparatus that has a full face-piece and is operated in a pressure-demand or other positive-pressure mode) or SaF: Pd,Pp: ASCBA (any supplied-air respirator that has a full face-piece and is operated in a pressure-demand or other positive-pressure mode in combination with an auxiliary self-contained breathing apparatus operated in a pressure-demand or other positive-pressure mode). *Escape:* GmFOv (APF = 50) [any air-purifying, full-face-piece respirator (gas mask) with a chin-style, front- or back-mounted acid gas canister] or SCBAE (any appropriate escape-type, self-contained breathing apparatus).

Storage: Color Code—Red: Flammability Hazard: Store in a flammable liquid storage area or approved cabinet away from ignition sources and corrosive and reactive materials. Methylcyclohexane must be stored to avoid contact with strong oxidizers, (such as chlorine, bromine, chlorine oxide, nitrates and permanganates), since violent reactions occur. Before entering confined space where this chemical may be present, check to make sure that an explosive concentration does not exist. Store in tightly closed containers in a cool, well-ventilated area away from heat. Sources of ignition, such as smoking and open flames, are prohibited where methylcyclohexane is handled, used, or stored. Metal containers involving the transfer of 5 gallons or more of methylcyclohexane should be grounded and bonded. Drums must be equipped with self-closing valves, pressure vacuum bungs, and flame arresters. Use only nonsparking tools and equipment, especially when opening and closing containers of methylcyclohexane.

Shipping: This compound requires a shipping label of "FLAMMABLE LIQUID." It falls in Hazard Class 3 and Packing Group II.

Spill Handling: Evacuate and restrict persons not wearing protective equipment from area of spill or leak until cleanup is complete. Remove all ignition sources. Establish forced ventilation to keep levels below explosive limit. Absorb liquids in vermiculite, dry sand, earth, peat, carbon, or a similar material and deposit in sealed containers. Keep this chemical out of a confined space, such as a sewer, because of the possibility of an explosion, unless the sewer is designed to prevent the buildup of explosive concentrations. It may be necessary to contain and dispose of this chemical as a hazardous waste. If material or contaminated runoff enters waterways, notify downstream users of potentially contaminated waters. Contact your local or federal environmental protection agency for specific recommendations. If employees are required to clean up spills, they must be properly trained and equipped. OSHA 1910.120(q) may be applicable.

Fire Extinguishing: This chemical is a combustible liquid. Poisonous gases are produced in fire. Use dry chemical, carbon dioxide, or alcohol foam extinguishers. Vapors are heavier than air and will collect in low areas. Vapors may travel long distances to ignition sources and flashback. Vapors in confined areas may explode when exposed to fire. Containers may explode in fire. Storage containers and parts of containers may rocket great distances, in many directions. If material or contaminated runoff enters waterways, notify downstream users of potentially contaminated waters. Notify local health and fire officials and pollution control agencies. From a secure, explosion-proof location, use water spray to cool exposed containers. If cooling streams are ineffective (venting sound increases in volume and pitch, tank discolors, or shows any signs of deforming), withdraw immediately to a secure position. If employees are expected to fight fires, they must be trained and equipped in OSHA 1910.156. The only respirators recommended for firefighting are self-contained breathing apparatuses that have full face-pieces and are operated in a pressure-demand or other positive-pressure mode.

Disposal Method Suggested: Dissolve or mix the material with a combustible solvent and burn in a chemical incinerator equipped with an afterburner and scrubber. All federal, state, and local environmental regulations must be observed.

References

US Environmental Protection Agency. (1979). *Chemical Hazard Information Profile: Methylcyclohexane.* Washington, DC

New Jersey Department of Health and Senior Services. (April 2004). *Hazardous Substances Fact Sheet: Methyl Cyclohexane.* Trenton, NJ

Methylcyclohexanol   M:0810

Molecular Formula: $C_7H_{14}O$
Common Formula: $H_3CC_6H_{10}OH$
Synonyms: Hexahydrocresol; Hexahydromethyl phenol; Methylcyclohexanol; Methylcyclohexane
CAS Registry Number: 25639-42-3
RTECS® Number: GW0175000
UN/NA & ERG Number: UN2617/129
EC Number: 247-152-6
Regulatory Authority and Advisory Bodies
Air Pollutant Standard Set. See below, "Permissible Exposure Limits in Air" section.
Canada, WHMIS, Ingredients Disclosure List Concentration 1.0%.
European/International Regulations: not listed in Annex 1.
WGK (German Aquatic Hazard Class): 1—Low hazard to waters.
Description: 2-Methylcyclohexanol is a straw-colored liquid with a weak, menthol-like odor. The odor threshold is 490 ppm (higher than the OEL). Molecular weight = 114.21; Specific gravity (H_2O:1) = 0.92; Boiling point = 155−180°C (technical grade); Freezing/Melting point = −50°C; Vapor pressure = 2 mmHg at 30°C; Flash point = 65−68°C; Autoignition temperature = 296°C. Hazard Identification (based on NFPA-704 M Rating System): Health 0, Flammability 2, Reactivity 0. Slightly soluble in water.
Potential Exposure: Compound Description: Human Data. Methylcyclohexanol is used as a lacquer solvent, as a blending agent in textile soaps, and as an antioxidant in lubricants.
Incompatibilities: Forms explosive mixture with air. Contact with strong oxidizers may cause fire and explosions. Attacks some plastics, rubber, and coatings.
Permissible Exposure Limits in Air
Conversion factor: 1 ppm = 4.67 mg/m³ at 25°C & 1 atm.
OSHA PEL: 100 ppm/470 mg/m³ TWA.
NIOSH REL: 50 ppm/235 mg/m³ TWA.
ACGIH TLV®[1]: 50 ppm/234 mg/m³ TWA.
NIOSH IDLH: 500 ppm*
*The odor threshold is 490 ppm.
No TEEL available.
DFG MAK: MAK 50 ppm/235 mg/m³.
Australia: TWA 50 ppm (235 mg/m³), 1993; Belgium: TWA 50 ppm (234 mg/m³), 1993; Denmark: TWA 50 ppm (235 mg/m³), 1999; Finland: TWA 50 ppm (235 mg/m³); STEL 75 ppm (355 mg/m³), 1993; France: VME 50 ppm (235 mg/m³), 1999; Japan: 50 ppm (230 mg/m³), 1999; Norway: TWA 25 ppm (115 mg/m³), 1999; the Netherlands: MAC-TGG 50 mg/m³, [skin], 2003; the

Philippines: TWA 100 ppm (470 mg/m³), 1993; Poland: MAC (TWA) 50 mg/m³, MAC (STEL) 350 mg/m³, 1999; Russia: TWA 50 ppm, 1993; Switzerland: MAK-W 50 ppm (235 mg/m³), KZG-W 100 ppm (470 mg/m³), 1999; Turkey: TWA 100 ppm (470 mg/m³), 1993; United Kingdom: LTEL 50 ppm (235 mg/m³); STEL 75 ppm, 1993; United Kingdom: TWA 50 ppm (237 mg/m³); STEL 75 ppm, 2000; Argentina, Bulgaria, Columbia, Jordan, South Korea, New Zealand, Singapore, Vietnam: ACGIH TLV®: TWA 50 ppm. Several states have set guidelines or standards for methyl cyclohexanol in ambient air[60] ranging from 2.35−3.5 mg/m³ (North Dakota) to 3.9 mg/m³ (Virginia) to 4.7 mg/m³ (Connecticut) to 5.595 mg/m³ (Nevada).
Determination in Air: Charcoal tube; CH_2C_{12}; Gas chromatography/Flame ionization detection; NIOSH Analytical Method (IV) #1404. See also NIOSH: Methylcyclohexanol Method #S374.
Determination in Water: Octanol−water coefficient: Log $K_{ow} = 2.1$.
Routes of Entry: Inhalation, skin absorption, ingestion, skin and/or eye contact.
Harmful Effects and Symptoms
Short Term Exposure: Methylcyclohexanol can affect you when breathed in and by passing through your skin. Irritates the eyes and the skin. High levels of the vapor may cause irritation of eyes and upper respiratory tract. Repeated or prolonged exposure can cause headaches, irritation of the eyes, nose, and throat, and can also cause a skin rash. High exposures from skin contact or inhalation may cause damage to the heart, liver, kidneys, and lungs, and may result in death.
Long Term Exposure: Repeated or prolonged contact with skin may cause skin rash. Animal studies have shown this chemical to cause liver and kidney damage.
Points of Attack: Eyes, skin, respiratory system, central nervous system, kidneys, liver.
Medical Surveillance: If symptoms develop or overexposure has occurred, the following may be useful: lung function tests. Liver and kidney function tests.
First Aid: If this chemical gets into the eyes, remove any contact lenses at once and irrigate immediately for at least 15 min, occasionally lifting upper and lower lids. Seek medical attention immediately. If this chemical contacts the skin, remove contaminated clothing and wash immediately with soap and water. Seek medical attention immediately. If this chemical has been inhaled, remove from exposure, begin rescue breathing (using universal precautions, including resuscitation mask) if breathing has stopped and CPR if heart action has stopped. Transfer promptly to a medical facility. When this chemical has been swallowed, get medical attention. Give large quantities of water and induce vomiting. Do not make an unconscious person vomit.
Personal Protective Methods: Wear protective gloves and clothing to prevent any reasonable probability of skin contact. Safety equipment suppliers/manufacturers can provide

recommendations on the most protective glove/clothing material for your operation. All protective clothing (suits, gloves, footwear, headgear) should be clean, available each day, and put on before work. Contact lenses should not be worn when working with this chemical. Wear splash-proof chemical goggles and face shield unless full face-piece respiratory protection is worn. Employees should wash immediately with soap when skin is wet or contaminated. Provide emergency showers and eyewash.

Respirator Selection: 500 ppm: Sa (APF = 10) (any supplied-air respirator) or SCBAF (APF = 50) (any self-contained breathing apparatus with a full face-piece). *Emergency or planned entry into unknown concentrations or IDLH conditions:* SCBAF: Pd,Pp (APF = 10,000) (any self-contained breathing apparatus that has a full face-piece and is operated in a pressure-demand or other positive-pressure mode) or SaF: Pd,Pp: ASCBA (APF = 10,000) (any supplied-air respirator that has a full face-piece and is operated in a pressure-demand or other positive-pressure mode in combination with an auxiliary self-contained breathing apparatus operated in a pressure-demand or other positive-pressure mode). *Escape:* GmFOv (APF = 50) [any air-purifying, full-face-piece respirator (gas mask) with a chin-style, front- or back-mounted acid gas canister] or SCBAE (any appropriate escape-type, self-contained breathing apparatus).

Note: Substance reported to cause eye irritation or damage; may require eye protection.

Storage: Color Code—Red: Flammability Hazard: Store in a flammable liquid storage area or approved cabinet away from ignition sources and corrosive and reactive materials. Prior to working with this chemical you should be trained on its proper handling and storage. Methylcyclohexanol must be stored to avoid contact with strong oxidizers (such as peroxides, chlorates, perchlorates, nitrates, and permanganates), since violent reactions occur. Sources of ignition, such as smoking and open flames, are prohibited where methylcyclohexanol is used, handled, or stored in a manner that could create a potential fire or explosion hazard.

Shipping: This compound requires a shipping label of "FLAMMABLE LIQUID." It falls in Hazard Class 3 and Packing Group II.

Spill Handling: Evacuate and restrict persons not wearing protective equipment from area of spill or leak until cleanup is complete. Remove all ignition sources. Ventilate area of spill or leak. Absorb liquids in vermiculite, dry sand, earth, peat, carbon, or a similar material and deposit in sealed containers. Keep this chemical out of a confined space, such as a sewer, because of the possibility of an explosion, unless the sewer is designed to prevent the buildup of explosive concentrations. It may be necessary to contain and dispose of this chemical as a hazardous waste. If material or contaminated runoff enters waterways, notify downstream users of potentially contaminated waters. Contact your local or federal environmental protection agency for specific recommendations. If employees are required to clean up spills, they must be properly trained and equipped. OSHA 1910.120(q) may be applicable.

Fire Extinguishing: This chemical is a combustible liquid. Poisonous gases are produced in fire. Use dry chemical, carbon dioxide, or foam extinguishers. Vapors are heavier than air and will collect in low areas. Vapors may travel long distances to ignition sources and flashback. Vapors in confined areas may explode when exposed to fire. Containers may explode in fire. Storage containers and parts of containers may rocket great distances, in many directions. If material or contaminated runoff enters waterways, notify downstream users of potentially contaminated waters. Notify local health and fire officials and pollution control agencies. From a secure, explosion-proof location, use water spray to cool exposed containers. If cooling streams are ineffective (venting sound increases in volume and pitch, tank discolors, or shows any signs of deforming), withdraw immediately to a secure position. If employees are expected to fight fires, they must be trained and equipped in OSHA 1910.156. The only respirators recommended for firefighting are self-contained breathing apparatuses that have full face-pieces and are operated in a pressure-demand or other positive-pressure mode.

Disposal Method Suggested: Dissolve or mix the material with a combustible solvent and burn in a chemical incinerator equipped with an afterburner and scrubber. All federal, state, and local environmental regulations must be observed.

Reference

New Jersey Department of Health and Senior Services. (March 2000). *Hazardous Substances Fact Sheet: Methyl Cyclohexanol*. Trenton, NJ

2-Methylcyclohexanone   M:0820

Molecular Formula: C$_7$H$_{12}$O
Common Formula: H$_3$CC$_6$H$_9$O
Synonyms: Methyl anone; *o*-Methyl-cyclohexanon (German); Methylcyclohexanone; *o*-Methylcyclohexanone; 1-Methylcyclohexan-2-one
CAS Registry Number: 583-60-8
RTECS® Number: GW1750000
UN/NA & ERG Number: UN2297/128
EC Number: 209-513-6 [*Annex I Index No.:* 606-011-00-2]
Regulatory Authority and Advisory Bodies
Air Pollutant Standard Set. See below, "Permissible Exposure Limits in Air" section.
Canada, WHMIS, Ingredients Disclosure List Concentration 1.0%.
European/International Regulations: Hazard Symbol: Xn; Risk phrases: R10; R20; Safety phrases: S2; S25 (see Appendix 4).
WGK (German Aquatic Hazard Class): 1—Low hazard to waters.

Description: o-Methylcyclohexanone is a colorless liquid with a weak peppermint-like odor (also reported to be acetone-like). Molecular weight = 112.19; Specific gravity (H$_2$O:1) = 0.93; Boiling point = 162.7°C; Freezing/Melting point = −13.9°C; Flash point = 47.8°C. Hazard Identification (based on NFPA-704 M Rating System): Health 2, Flammability 2, Reactivity 0. Insoluble in water.

Potential Exposure: Methylcyclohexanone is used as a solvent in making varnish, plastics, and as a rust remover. Also used in the leather industry.

Incompatibilities: Reacts violently with aldehydes, strong oxidizers, strong acids. Contact with peroxides may form unstable heat- and shock-sensitive explosives.

Permissible Exposure Limits in Air
Conversion factor: 1 ppm = 4.59 mg/m^3 at 25°C & 1 atm.
OSHA PEL: 100 ppm/460 mg/m^3 TWA [skin].
NIOSH REL: 50 ppm/230 mg/m^3 TWA; 75 ppm/345 mg/m^3 STEL [skin].
ACGIH TLV®[1]: 50 ppm/229 mg/m^3 TWA; 75 ppm/344 mg/m^3 STEL [skin].
NIOSH IDLH: 600 ppm.
Protective Action Criteria (PAC)
TEEL-0: 50 ppm
PAC-1: 75 ppm
PAC-2: 125 ppm
PAC-3: 600 ppm
DFG MAK: 50 ppm/230 mg/m^3 [skin].
Australia: TWA 50 ppm (230 mg/m^3); STEL 75 ppm, [skin], 1993; Austria: MAK 50 ppm (230 mg/m^3), [skin], 1999; Belgium: TWA 50 ppm (229 mg/m^3); STEL 75 ppm (344 mg/m^3), [skin], 1993; Denmark: TWA 50 ppm (230 mg/m^3), [skin], 1999; Finland: TWA 50 ppm (230 mg/m^3); STEL 75 ppm (345 mg/m^3), [skin], 1999; France: VME 50 ppm (230 mg/m^3), [skin], 1999; Japan: 50 ppm (230 mg/m^3), [skin], 1999; the Netherlands: MAC-TGG 230 mg/m^3, [skin], 2003; Norway: TWA 25 ppm (115 mg/m^3), 1999; the Philippines: TWA 100 ppm (460 mg/m^3), [skin], 1993; Poland: MAC (TWA) 50 mg/m^3, MAC (STEL) 340 mg/m^3, 1999; Russia: TWA 50 ppm, 1993; Switzerland: MAK-W 50 ppm (230 mg/m^3), KZG-W 100 ppm (470 mg/m^3), [skin], 1999; Turkey: TWA 100 ppm (460 mg/m^3), [skin], 1993; United Kingdom: TWA 50 ppm (237 mg/m^3); STEL 75 ppm, 2000; Argentina, Bulgaria, Columbia, Jordan, South Korea, New Zealand, Singapore, Vietnam: ACGIH TLV®: STEL 75 ppm [skin]. Several states have set guidelines or standards[60] for methylcyclohexanone in ambient air ranging from 2.3−3.45 mg/m^3 (North Dakota) to 3.9 mg/m^3 (Virginia) to 4.6 mg/m^3 (Connecticut) to 5.476 mg/m^3 (Nevada).

Determination in Air: Collection in an adsorption tube, workup with acetone; analysis by gas chromatography/flame ionization. Use NIOSH Analytical Method #2521.[18]

Routes of Entry: Inhalation, ingestion, skin and/or eye contact. Passes through the skin.

Harmful Effects and Symptoms

Short Term Exposure: o-Methylcyclohexanone can affect you when breathed in and by passing through your skin. It can irritate the skin, eyes, nose, and throat. Breathing the vapor can cause headaches, dizziness, or lightheadedness. Contact can strongly irritate and even damage the eyes.

Long Term Exposure: Repeated or prolonged contact with skin may cause dermatitis to develop with thickening and cracking. May affect the liver, kidneys, and lungs.

Points of Attack: Skin, respiratory system, liver, kidneys, central nervous system.

Medical Surveillance: If symptoms develop or overexposure is suspected, the following may be useful: liver and kidney function tests. Lung function tests.

First Aid: If this chemical gets into the eyes, remove any contact lenses at once and irrigate immediately for at least 15 min, occasionally lifting upper and lower lids. Seek medical attention immediately. If this chemical contacts the skin, remove contaminated clothing and wash immediately with soap and water. Seek medical attention immediately. If this chemical has been inhaled, remove from exposure, begin rescue breathing (using universal precautions, including resuscitation mask) if breathing has stopped and CPR if heart action has stopped. Transfer promptly to a medical facility. When this chemical has been swallowed, get medical attention. Give large quantities of water and induce vomiting. Do not make an unconscious person vomit.

Personal Protective Methods: Wear protective gloves and clothing to prevent any reasonable probability of skin contact. Safety equipment suppliers/manufacturers can provide recommendations on the most protective glove/clothing material for your operation. All protective clothing (suits, gloves, footwear, headgear) should be clean, available each day, and put on before work. Contact lenses should not be worn when working with this chemical. Wear splash-proof chemical goggles and face shield unless full face-piece respiratory protection is worn. Employees should wash immediately with soap when skin is wet or contaminated. Provide emergency showers and eyewash.

Respirator Selection: *500 ppm:* Sa (APF = 10) (any supplied-air respirator). *600 ppm:* Sa:Cf (APF = 25) (any supplied-air respirator operated in a continuous-flow mode) or SCBAF (APF = 50) (any self-contained breathing apparatus with a full face-piece) or SaF (APF = 50) (any supplied-air respirator with a full face-piece). *Emergency or planned entry into unknown concentrations or IDLH conditions:* SCBAF: Pd,Pp (APF = 10,000) (any self-contained breathing apparatus that has a full face-piece and is operated in a pressure-demand or other positive-pressure mode) or SaF: Pd,Pp: ASCBA (APF = 10,000) (any supplied-air respirator that has a full face-piece and is operated in a pressure-demand or other positive-pressure mode in combination with an auxiliary self-contained breathing apparatus operated in a pressure-demand or other positive-pressure mode). *Escape:* GmFOv (APF = 50) [any air-purifying, full-face-piece respirator (gas mask) with a chin-style, front- or back-mounted acid gas canister] or SCBAE (any appropriate escape-type, self-contained breathing apparatus).

Note: Substance reported to cause eye irritation or damage; may require eye protection.

Storage: Color Code—Red: Flammability Hazard: Store in a flammable liquid storage area or approved cabinet away from ignition sources and corrosive and reactive materials. Prior to working with this chemical you should be trained on its proper handling and storage. *o*-methylcyclohexanone must be stored to avoid contact with strong oxidizers (such as chlorine, bromine, and fluorine) because violent reactions occur. Sources of ignition, such as smoking and open flames, are prohibited where *o*-Methylcyclohexanone is used, handled, or stored in a manner that could create a potential fire or explosion hazard. Metal containers involving the transfer of 5 gallons or more of this chemical should be grounded and bonded. Drums must be equipped with self-closing valves, pressure vacuum bungs, and flame arresters. Use only nonsparking tools and equipment, especially when opening and closing containers of this chemical. Wherever this chemical is used, handled, manufactured, or stored, use explosion-proof electrical equipment and fittings.

Shipping: This compound requires a shipping label of "FLAMMABLE LIQUID." It falls in Hazard Class 3 and Packing Group II.

Spill Handling: Evacuate and restrict persons not wearing protective equipment from area of spill or leak until cleanup is complete. Remove all ignition sources. Ventilate area of spill or leak. Absorb liquids in vermiculite, dry sand, earth, peat, carbon, or a similar material, and deposit in sealed containers. Keep this chemical out of a confined space, such as a sewer, because of the possibility of an explosion, unless the sewer is designed to prevent the buildup of explosive concentrations. It may be necessary to contain and dispose of this chemical as a hazardous waste. If material or contaminated runoff enters waterways, notify downstream users of potentially contaminated waters. Contact your local or federal environmental protection agency for specific recommendations. If employees are required to clean up spills, they must be properly trained and equipped. OSHA 1910.120(q) may be applicable.

Fire Extinguishing: This chemical is a combustible liquid. Poisonous gases are produced in fire. Use dry chemical, carbon dioxide, or foam extinguishers. Vapors are heavier than air and will collect in low areas. Vapors may travel long distances to ignition sources and flashback. Vapors in confined areas may explode when exposed to fire. Containers may explode in fire. Storage containers and parts of containers may rocket great distances, in many directions. If material or contaminated runoff enters waterways, notify downstream users of potentially contaminated waters. Notify local health and fire officials and pollution control agencies. From a secure, explosion-proof location, use water spray to cool exposed containers. If cooling streams are ineffective (venting sound increases in volume and pitch, tank discolors, or shows any signs of deforming), withdraw immediately to a secure position. If employees are expected to fight fires, they must be trained and equipped in OSHA 1910.156. The only respirators recommended for firefighting are self-contained breathing apparatuses that have full face-pieces and are operated in a pressure-demand or other positive-pressure mode.

Disposal Method Suggested: Dissolve or mix the material with a combustible solvent and burn in a chemical incinerator equipped with an afterburner and scrubber. All federal, state, and local environmental regulations must be observed.

Reference

New Jersey Department of Health and Senior Services. (December 2000). *Hazardous Substances Fact Sheet: o-Methylcyclohexanone.* Trenton, NJ

Methyl cyclopentane   M:0830

Molecular Formula: C_6H_{12}
Common Formula: $C_5H_9CH_3$
Synonyms: Cyclopentane, methyl-; MCP; Methylcyclopentane; Methylpentamethylene
CAS Registry Number: 96-37-7
RTECS® Number: GY4640000
UN/NA & ERG Number: UN2298/128
EC Number: 202-503-2
Regulatory Authority and Advisory Bodies
WGK (German Aquatic Hazard Class): No value assigned.
Description: Methyl cyclopentane is a colorless liquid with a sweet gasoline-like odor. Molecular weight = 84.18; Boiling point = 72°C; Freezing/Melting point = −142°C; Flash point = −7°C; Autoignition temperature = 258°C. Explosive limits: LEL = 1.0%; UEL = 8.35%. Hazard Identification (based on NFPA-704 M Rating System): Health 2, Flammability 3, Reactivity 0.
Potential Exposure: This material is used as a solvent, as a fuel, and in chemical synthesis.
Incompatibilities: Strong oxidizers; strong acids.
Permissible Exposure Limits in Air
Conversion factor: 1 ppm = 3.53 mg/m^3 at 25°C & 1 atm.
OSHA PEL: None.
NIOSH: 100 ppm/350 mg/m^3 TWA; STEL 510 ppm/1800 mg/m^3 [15 min]
ACGIH TLV®[1]: 500 ppm/1760 mg/m^3 TWA; 1000 ppm/3500 mg/m^3 STEL
Protective Action Criteria (PAC)
TEEL-0: 4 ppm
PAC-1: 12.5 ppm
PAC-2: 75 ppm
PAC-3: 4000 ppm
DFG MAK: 500 ppm/1800 mg/m^3 TWA; Peak Limitation Category II(2); Pregnancy Risk Group D.
Determination in Air: Charcoal tube; CS$_2$; Gas chromatography/Flame ionization detection; NIOSH Analytical Method (IV) #1500, for Hydrocarbons, BP 36−126°C.

Routes of Entry: Inhalation, ingestion, skin and/or eye contact.

Harmful Effects and Symptoms

Short Term Exposure: Methyl cyclopentane can affect you when breathed in. Exposure can cause you to feel dizzy, lightheaded, and to pass out. Higher levels can cause death. Exposure can irritate the eyes, nose, and throat. Contact can irritate the skin. Higher exposures can cause pulmonary edema, a medical emergency that can be delayed for several hours. This can cause death. Can cause central nervous system excitement followed by depression.

Long Term Exposure: May cause damage to the nervous system.

Points of Attack: Eyes, skin, respiratory system, central nervous system.

Medical Surveillance: Examination of the nervous system.

First Aid: If this chemical gets into the eyes, remove any contact lenses at once and irrigate immediately for at least 15 min, occasionally lifting upper and lower lids. Seek medical attention immediately. If this chemical contacts the skin, remove contaminated clothing and wash immediately with soap and water. Seek medical attention immediately. If this chemical has been inhaled, remove from exposure, begin rescue breathing (using universal precautions, including resuscitation mask) if breathing has stopped and CPR if heart action has stopped. Transfer promptly to a medical facility. When this chemical has been swallowed, get medical attention. Give large quantities of water and induce vomiting. Do not make an unconscious person vomit. Medical observation is recommended for 24—48 h after breathing overexposure, as pulmonary edema may be delayed. As first aid for pulmonary edema, a doctor or authorized paramedic may consider administering a corticosteroid spray.

Personal Protective Methods: Wear protective gloves and clothing to prevent any reasonable probability of skin contact. Safety equipment suppliers/manufacturers can provide recommendations on the most protective glove/clothing material for your operation. All protective clothing (suits, gloves, footwear, headgear) should be clean, available each day, and put on before work. Contact lenses should not be worn when working with this chemical. Wear splash-proof chemical goggles and face shield unless full face-piece respiratory protection is worn. Employees should wash immediately with soap when skin is wet or contaminated. Provide emergency showers and eyewash.

Respirator Selection: Where there is potential for exposure to liquid methyl cyclopentane, use a NIOSH/MSHA- or European Standard EN149-approved supplied-air respirator with a full face-piece operated in the positive-pressure mode, or with a full face-piece, hood, or helmet in the continuous-flow mode; or use a NIOSH/MSHA- or European Standard EN149-approved self-contained breathing apparatus with a full face-piece operated in pressure-demand or other positive-pressure mode.

Storage: Color Code—Red: Flammability Hazard: Store in a flammable liquid storage area or approved cabinet away from ignition sources and corrosive and reactive materials. Prior to working with this chemical you should be trained on its proper handling and storage. Before entering confined space where this chemical may be present, check to make sure that an explosive concentration does not exist. Methyl cyclopentane must be stored to avoid contact with oxidizers (such as perchlorates, peroxides, permanganates, chlorates and nitrates) and strong acids, since violent reactions occur. Sources of ignition, such as smoking and open flames, are prohibited where methyl cyclopentane is handled, used, or stored. Metal containers involving the transfer of 5 gallons or more of methyl cyclopentane should be grounded and bonded. Drums must be equipped with self-closing valves, pressure vacuum bungs, and flame arresters. Use only non-sparking tools and equipment, especially when opening and closing containers of methyl cyclopentane. Wherever methyl cyclopentane is used, handled, manufactured, or stored, use explosion-proof electrical equipment and fittings.

Shipping: This compound requires a shipping label of "FLAMMABLE LIQUID." It falls in Hazard Class 3 and Packing Group II.

Spill Handling: Evacuate and restrict persons not wearing protective equipment from area of spill or leak until cleanup is complete. Remove all ignition sources. Ventilate area of spill or leak. Absorb liquids in vermiculite, dry sand, earth, peat, carbon, or a similar material and deposit in sealed containers. Keep this chemical out of a confined space, such as a sewer, because of the possibility of an explosion, unless the sewer is designed to prevent the buildup of explosive concentrations. It may be necessary to contain and dispose of this chemical as a hazardous waste. If material or contaminated runoff enters waterways, notify downstream users of potentially contaminated waters. Contact your local or federal environmental protection agency for specific recommendations. If employees are required to clean up spills, they must be properly trained and equipped. OSHA 1910.120(q) may be applicable.

Fire Extinguishing: Methyl cyclopentane is a flammable liquid. Use dry chemical, carbon dioxide, alcohol foam, or polymer foam extinguishers. Water may be ineffective on fire. Poisonous gases are produced in fire. Establish forced ventilation to keep levels below explosive limit. Vapors are heavier than air and will collect in low areas. Vapors may travel long distances to ignition sources and flashback. Vapors in confined areas may explode when exposed to fire. Containers may explode in fire. Storage containers and parts of containers may rocket great distances, in many directions. If material or contaminated runoff enters waterways, notify downstream users of potentially contaminated waters. Notify local health and fire officials and pollution control agencies. From a secure, explosion-proof location, use water spray to cool exposed containers. If cooling streams are ineffective (venting sound increases in volume and pitch, tank discolors, or shows any signs of deforming), withdraw immediately to a secure position. If employees

are expected to fight fires, they must be trained and equipped in OSHA 1910.156. The only respirators recommended for firefighting are self-contained breathing apparatuses that have full face-pieces and are operated in a pressure-demand or other positive-pressure mode.

Disposal Method Suggested: Dissolve or mix the material with a combustible solvent and burn in a chemical incinerator equipped with an afterburner and scrubber. All federal, state, and local environmental regulations must be observed.

Reference

New Jersey Department of Health and Senior Services. (December 1999). *Hazardous Substances Fact Sheet: Methyl Cyclopentane.* Trenton, NJ

Methyl dichlorosilane   M:0840

Molecular Formula: CH_4Cl_2Si
Common Formula: CH_3SiHCl_2
Synonyms: Dichloromethylsilane; Monomethyl-dichlorosilane; Silane, dichloromethyl-
CAS Registry Number: 75-54-7
RTECS® Number: VV3500000
UN/NA & ERG Number: UN1242/139
EC Number: 200-877-1
Regulatory Authority and Advisory Bodies
Department of Homeland Security Screening Threshold Quantity (pounds): Sabotage/Contamination Hazard: A placarded amount (commercial grade).
US DOT 49CFR172.101, Inhalation Hazardous Chemical.
WGK (German Aquatic Hazard Class): 1—Low hazard to waters.
Description: Methyl dichlorosilane is a clear, straw-colored liquid. Molecular weight = 115.04; Boiling point = 41°C; Freezing/Melting point = −91°C; Flash point = −9°C; Autoignition temperature = 316°C. Explosive limits: LEL = 6.0%; UEL = 55.0%. Hazard Identification (based on NFPA-704 M Rating System): Health 3, Flammability 3, Reactivity 2W. Reacts with water.
Potential Exposure: This material is used to make siloxanes and other silicone polymer (polysiloxane) materials.
Incompatibilities: Forms explosive mixture with air. Reacts violently with water producing corrosive hydrochloric acid. Methyl dichlorosilane may spontaneously ignite on contact with air (even under inert gas) and on contact with potassium permanganate, lead(II) oxide, copper oxide, silver oxide. Violent reaction with oxidizers. Decomposes on contact with hot surfaces or flames producing toxic and corrosive fumes including silicon oxides, hydrogen chloride, and phosgene. Decomposes on contact with alkaline compounds producing highly flammable hydrogen gas. Corrodes many metals in presence of water. Attacks some plastics, rubber, and coatings.
Permissible Exposure Limits in Air
Protective Action Criteria (PAC)*

TEEL-0: 0.3 ppm
PAC-1: **0.90** ppm
PAC-2: **11** ppm
PAC-3: **50** ppm
*AEGLs (Acute Emergency Guideline Levels) & ERPGs (Emergency Response Planning Guideline) are in **bold face**.
Routes of Entry: Inhalation, ingestion, eye and/or skin contact.
Harmful Effects and Symptoms
Short Term Exposure: Methyl dichlorosilane is a corrosive chemical and contact can cause severe eye and skin burns leading to permanent damage. Exposure can irritate the lungs, causing coughing and/or shortness of breath. Higher exposures can cause a buildup of fluid in the lungs (pulmonary edema). This can cause death.
Long Term Exposure: May cause bronchitis with cough, phlegm, and/or shortness of breath.
Points of Attack: Lungs.
Medical Surveillance: For those with frequent or potentially high exposure the following are recommended before beginning work and at regular times after that: lung function tests. If symptoms develop or overexposure is suspected the following may be useful: consider chest X-ray after acute overexposure.
First Aid: If this chemical gets into the eyes, remove any contact lenses at once and irrigate immediately for at least 30 min, occasionally lifting upper and lower lids. Seek medical attention immediately. If this chemical contacts the skin, remove contaminated clothing and wash immediately with soap and water. Seek medical attention immediately. If this chemical has been inhaled, remove from exposure, begin rescue breathing (using universal precautions, including resuscitation mask) if breathing has stopped and CPR if heart action has stopped. Transfer promptly to a medical facility. When this chemical has been swallowed, get medical attention. If victim is conscious, administer water or milk. Do not induce vomiting. Medical observation is recommended for 24−48 h after breathing overexposure, as pulmonary edema may be delayed. As first aid for pulmonary edema, a doctor or authorized paramedic may consider administering a corticosteroid spray.
Personal Protective Methods: Wear protective gloves and clothing to prevent any reasonable probability of skin contact. Safety equipment suppliers/manufacturers can provide recommendations on the most protective glove/clothing material for your operation. All protective clothing (suits, gloves, footwear, headgear) should be clean, available each day, and put on before work. Contact lenses should not be worn when working with this chemical. Wear splash-proof chemical goggles and face shield unless full face-piece respiratory protection is worn. Employees should wash immediately with soap when skin is wet or contaminated. Provide emergency showers and eyewash.
Respirator Selection: Where there is potential for exposure to methyl dichlorosilane, use a NIOSH/MSHA- or

European Standard EN149-approved supplied-air respirator with a full face-piece operated in the positive-pressure mode, or with a full face-piece, hood, or helmet in the continuous-flow mode; or use a NIOSH/MSHA- or European Standard EN149-approved self-contained breathing apparatus with a full face-piece operated in pressure-demand or other positive-pressure mode.

Storage: (1) Color Code—Red: Flammability Hazard: Store in a flammable liquid storage area or approved cabinet away from ignition sources and corrosive and reactive materials. (2) Color Code—Blue: Health Hazard/Poison: Store in a secure poison location. (3) Color Code—White: Corrosive or Contact Hazard; Store separately in a corrosion-resistant location. Prior to working with this chemical you should be trained on its proper handling and storage. Before entering confined space where this chemical may be present, check to make sure that an explosive concentration does not exist. Methyl dichlorosilane must be stored to avoid contact with water, since violent reactions occur once hydrogen chloride is produced. Store in tightly closed containers in a cool, well-ventilated area away from heat. Sources of ignition, such as smoking and open flames, are prohibited where methyl dichlorosilane is handled, used, or stored. Metal containers involving the transfer of 5 gallons or more of methyl dichlorosilane should be grounded and bonded. Drums must be equipped with self-closing valves, pressure vacuum bungs, and flame arresters. Use only nonsparking tools and equipment, especially when opening and closing containers of methyl dichlorosilane. Wherever methyl dichlorosilane is used, handled, manufactured, or stored, use explosion-proof electrical equipment and fittings.

Shipping: Methyl dichlorosilane requires a shipping label of "DANGEROUS WHEN WET, CORROSIVE, FLAMMABLE LIQUID." It falls in Hazard Class 4.3 and Packing Group I.

Spill Handling: Evacuate and restrict persons not wearing protective equipment from area of spill or leak until cleanup is complete. Remove all ignition sources. Establish forced ventilation to keep levels below explosive limit. Absorb liquids in dry lime, dry sand, soda ash, or a similar material and deposit in sealed containers. Following cleanup, neutralize spill area by flushing with large quantities of water and then treat spill area with sodium bicarbonate. Keep this chemical out of a confined space, such as a sewer, because of the possibility of an explosion, unless the sewer is designed to prevent the buildup of explosive concentrations. It may be necessary to contain and dispose of this chemical as a hazardous waste. If material or contaminated runoff enters waterways, notify downstream users of potentially contaminated waters. Contact your local or federal environmental protection agency for specific recommendations. If employees are required to clean up spills, they must be properly trained and equipped. OSHA 1910.120(q) may be applicable.

Initial isolation and protective action distances
Distances shown are likely to be affected during the first 30 min after materials are spilled and could increase with time. If more than one tank car, cargo tank, portable tank, or large cylinder involved in the incident is leaking, the protective action distance may need to be increased. You may need to seek emergency information from CHEMTREC at (800) 424-9300 or seek professional environmental engineering assistance from the US EPA Environmental Response Team at (908) 548-8730 (24-h response line).

When spilled in water
Small spills (From a small package or a small leak from a large package)
First: Isolate in all directions (feet/meters)   100/30
Then: Protect persons downwind (miles/kilometers)
Day   0.1/0.2
Night   0.2/0.3
Large spills (From a large package or from many small packages)
First: Isolate in all directions (feet/meters)   200/60
Then: Protect persons downwind (miles/kilometers)
Day   0.5/0.8
Night   1.6/2.5

Fire Extinguishing: Methyl dichlorosilane is a flammable liquid and may ignite in air. Use dry chemical or CO_2 extinguishers. *Do not use water.* Fire may restart after it has been extinguished. Poisonous gases, including hydrogen chloride, silicon oxides, and phosgene, are produced in fire. Vapors are heavier than air and will collect in low areas. Vapors may travel long distances to ignition sources and flashback. Vapors in confined areas may explode when exposed to fire. Containers may explode in fire. Storage containers and parts of containers may rocket great distances, in many directions. If material or contaminated runoff enters waterways, notify downstream users of potentially contaminated waters. Notify local health and fire officials and pollution control agencies. From a secure, explosion-proof location, use water spray to cool exposed containers. If cooling streams are ineffective (venting sound increases in volume and pitch, tank discolors, or shows any signs of deforming), withdraw immediately to a secure position. If employees are expected to fight fires, they must be trained and equipped in OSHA 1910.156. The only respirators recommended for firefighting are self-contained breathing apparatuses that have full face-pieces and are operated in a pressure-demand or other positive-pressure mode.

Disposal Method Suggested: See "Spill Handling."
Reference
New Jersey Department of Health and Senior Services. (December 1999). *Hazardous Substances Fact Sheet: Methyl Dichlorosilane.* Trenton, NJ

4,4'- Methylenebis (2-chloroaniline)
M:0850

Molecular Formula: $C_{13}H_{12}Cl_2N_2$
Common Formula: $(C_6H_3ClNH_2)CH_2(C_6H_3ClNH_2)$

Synonyms: Aniline, 4,4'-methylenebis(2-chloro-); Benzenamine, 4,4'-methylenebis(2-chloro-); Bis amine; Bis (4-amino-3-chlorophenyl)methane; Bis(3-chloro-4-amino-phenyl) methane; BOCA; Cuamine MT; Curalin M; Curene 442; Cyanaset; DACPM; Diamet KH; Di-(4-amino-3-chlorophenyl)methane; 4,4'-Diamino-3,3'-dichlorodiphenyl-methane; 3,3'-Dichloro-4,4'-diaminodiphenylmethan (German); 3,3'-Dichloro-4,4'-diaminodiphenylmethane; MBOCA; Methylenebis(3-chloro-4-aminobenzene); *p,p'*-Methylenebis (α-chloroaniline); *p,p'*-Methylenebis(*o*-chloroaniline); 4,4'-Methylene(bis)-chloroaniline; 4,4'-Methylenebis(*o*-chloro-aniline); Methylene-4,4'-bis(*o*-chloroaniline); 4,4'-Methylenebis-2-chlorobenzenamine; 4,4'-Methylenebis (2-chloro-benzeneamine); Methylenebis(*o*-chloroaniline); *p,p'*-Metilenbis(*o*-cloroanilina) (Spanish); Millionate M; MOCA; Quodorole

CAS Registry Number: 101-14-4; *(alt.)* 29371-14-0; *(alt.)* 51065-07-7; *(alt.)* 78642-65-6

RTECS® Number: CY1050000

UN/NA & ERG Number: UN2811 (toxic solid, organic, n.o.s.)/154

EC Number: 202-918-9 [*Annex I Index No.:* 612-078-00-9]

Regulatory Authority and Advisory Bodies

Carcinogenicity: IARC: Animal Sufficient Evidence; Human Inadequate Evidence, Group 2A, 1993; NTP: 11th Report on Carcinogens, 2004: Reasonably anticipated to be a human carcinogen; NIOSH: Potential occupational carcinogen.

US EPA Gene-Tox Program, Positive: Carcinogenicity—mouse/rat; Positive: Cell transform.—RLV F344 rat embryo; Positive: Cell transform.—SA7/SHE; Mammalian micronucleus; Positive: Histidine reversion—Ames test; Positive/dose response: Cell transform.—BALB/c-3T3.

Banned or Severely Restricted (Sweden) (UN).[13]

Very Toxic Substance (World Bank).[15]

Air Pollutant Standard Set. See below, "Permissible Exposure Limits in Air" section.

Clean Air Act: Hazardous Air Pollutants (Title I, Part A, Section 112).

US EPA Hazardous Waste Number (RCRA No.): U158.

RCRA, 40CFR261, Appendix 8 Hazardous Constituents.

RCRA 40CFR268.48; 61FR15654, Universal Treatment Standards: Wastewater (mg/L), 0.50; Nonwastewater (mg/kg), 30.

Reportable Quantity (RQ): 10 lb (4.54 kg).

EPCRA Section 313 Form R *de minimis* concentration reporting level: 0.1%.

California Proposition 65 Chemical: Cancer 7/1/87.

Canada, WHMIS, Ingredients Disclosure List Concentration 0.1%.

European/International Regulations: Hazard Symbol: T, N; Risk phrases: R45; R22; R50/53; Safety phrases: S53; S45; S60; S61 (see Appendix 4).

WGK (German Aquatic Hazard Class): No value assigned.

Description: 4,4'-Methylenebis (2-chloroaniline) is a yellow to light gray-tan pellet and is also available in liquid form. Molecular weight = 267.17; Specific gravity (H_2O:1) = 1.44; Freezing/Melting point = 99−110°C; Vapor pressure = 0.00001 mmHg. Hazard Identification (based on NFPA-704 M Rating System): Health 1, Flammability 1, Reactivity 0. Slightly soluble in water.

Potential Exposure: Compound Description: Tumorigen, Mutagen. 4,4'-Methylenebis(2-chloroaniline) is used as a curing agent in the polyurethane industry for isocyanate containing polymers and in the production of solid elasto-meric parts. Other uses are in the manufacture of cross-linked urethane foams used in automobile seats and safety padded dashboards; it is also used in the manufacture of gun mounts, jet engine turbine blades, radar systems, and components in home appliances.

Incompatibilities: Reacts with chemically active metals (e.g., potassium, sodium, magnesium, and zinc). May ignite on contact with cellulose nitrate of high-surface area. Incompatible with acrolein, acrylonitrile, *tert*-butyl nitroacetylene, ethylene oxide, isopropyl chlorocarbonate, maleic anhydride, triisobutylaluminum.

Permissible Exposure Limits in Air

OSHA PEL: None.

NIOSH REL: 0.003 mg/m³ TWA [skin]; A potential occupational carcinogen. Limit exposure to lowest feasible concentration. See *NIOSH Pocket Guide*, Appendix A.

ACGIH TLV®[1]: 0.01 ppm/0.11 mg/m³ TWA; Suspected Human Carcinogen [skin] BEI issued.

Protective Action Criteria (PAC)

TEEL-0: 0.1 ppm

PAC-1: 1 ppm

PAC-2: 7.5 ppm

PAC-3: 40 ppm

DFG MAK: [skin] Carcinogen Category 2.

Australia: TWA 0.02 ppm (0.22 mg/m³), [skin], carcinogen, 1993; Austria: carcinogen, 1999; Belgium: TWA 0.02 ppm (0.22 mg/m³), [skin], carcinogen, 1993; Finland: TWA 0.02 ppm (0.2 mg/m³); STEL 0.06 ppm, [skin], 1993; Finland: TWA 0.02 ppm (0.2 mg/m³); STEL 0.06 ppm, carcinogen, 1999; France: VME 0.02 ppm (0.22 mg/m³), carcinogen, 1999; Japan: 0.005 mg/m³, [skin], 2A carcinogen, 1999; the Netherlands: MAC-TGG 0.02 mg/m³, 2003; Sweden: carcinogen, 1999; Switzerland: MAK-W 0.0 mg/m³, carcinogen, 1999; United Kingdom: TWA 0.005 mg/m³, [skin], carcinogen, 2000; Argentina, Bulgaria, Columbia, Jordan, South Korea, New Zealand, Singapore, Vietnam: ACGIH TLV®: Suspected Human Carcinogen. Several states have set guidelines or standards for MOCA in ambient air[60] ranging from zero (Nevada and North Dakota) to 0.015 μg/m³ (Connecticut) to 0.55 μg/m³ (Pennsylvania) to 1.0 μg/m³ (Rhode Island) to 2.2 μg/m³ (Virginia).

Determination in Air: Use NIOSH Analytical Method #8302, MBOCA in urine; OSHA Analytical Method 24; ID-71.

Determination in Water: Octanol−water coefficient: Log K_{ow} = 3.94.

Routes of Entry: Inhalation, skin absorption, ingestion, skin and/or eye contact. Passes through the skin.

Harmful Effects and Symptoms

Short Term Exposure: 4,4'-Methylenebis(2-chloroaniline) can affect you when breathed in and by passing through your skin. 4,4'-Methylenebis(2-chloroaniline) is a carcinogen; handle with extreme caution. Exposure can interfere with the ability of the blood to carry oxygen causing headaches, dizziness, nausea, and a bluish color to the skin and lips. High levels can cause trouble breathing, collapse, and death. Contact can irritate the eyes. High or repeated exposures may affect the kidneys, cause a low blood count, and cause bloody urine.

Long Term Exposure: May cause methemoglobinemia, anemia, kidney irritation. A potential occupational carcinogen. May affect the kidneys.

Points of Attack: Liver, blood, kidneys. Cancer site in animals: liver, lung, and bladder tumors.

Medical Surveillance: If symptoms develop or overexposure is suspected, the following may be useful: blood methemoglobin level. Complete blood count. Kidney function tests. Preplacement and periodic examinations should include a history of exposure to other carcinogens, alcohol and smoking habits, use of medications, and family history. Special attention should be given to liver size and function and to any changes in lung symptoms or X-rays.

First Aid: If this chemical gets into the eyes, remove any contact lenses at once and irrigate immediately for at least 15 min, occasionally lifting upper and lower lids. Seek medical attention immediately. If this chemical contacts the skin, remove contaminated clothing and wash immediately with soap and water. Seek medical attention immediately. If this chemical has been inhaled, remove from exposure, begin rescue breathing (using universal precautions, including resuscitation mask) if breathing has stopped and CPR if heart action has stopped. Transfer promptly to a medical facility. When this chemical has been swallowed, get medical attention. Give large quantities of water and induce vomiting. Do not make an unconscious person vomit.

Note to Physician: Treat for methemoglobinemia. Spectrophotometry may be required for precise determination of levels of methemoglobin in urine.

Personal Protective Methods: Wear protective gloves and clothing to prevent any reasonable probability of skin contact: **8 h**: Saranex™ coated suits, Barricade™ coated suits. **4 h**: 4H™ and Silver Shield™ gloves. Safety equipment suppliers/manufacturers can provide recommendations on the most protective glove/clothing material for your operation. All protective clothing (suits, gloves, footwear, headgear) should be clean, available each day, and put on before work. Contact lenses should not be worn when working with this chemical. Wear dust-proof chemical goggles and face shield unless full face-piece respiratory protection is worn. Employees should wash immediately with soap when skin is wet or contaminated. Provide emergency showers and eyewash.

Respirator Selection: *At any detectable concentration:* SCBAF: Pd,Pp (APF = 10,000) (any self-contained breathing apparatus that has a full faceplate and is operated in a pressure-demand or other positive-pressure mode) or SaF: Pd,Pp: ASCBA (APF = 10,000) (any supplied-air respirator that has a full face-piece and is operated in a pressure-demand or other positive-pressure mode in combination with an auxiliary self-contained breathing apparatus operated in a pressure-demand or other positive-pressure mode). *Escape:* GmFAg100 (APF = 50) [any air-purifying, full-face-piece respirator (gas mask) with a chin-style, front- or back-mounted acid gas canister having an N100, R100, or P100 filter] or SCBAE (any appropriate escape-type, self-contained breathing apparatus).

Storage: Color Code—Blue: Health Hazard/Poison: Store in a secure poison location. Prior to working with this chemical you should be trained on its proper handling and storage. 4,4'-Methylenebis(2-chloroaniline) must be stored to avoid contact with chemically active metals (such as potassium, sodium, magnesium and zinc), since violent reactions occur. Store in tightly closed containers in a cool, well-ventilated area. A regulated, marked area should be established where this chemical is handled, used, or stored in compliance with OSHA Standard 1910.1045.

Shipping: Toxic solids, organic, n.o.s. requires a shipping label of "POISONOUS/TOXIC MATERIALS." They fall in Hazard Class 6.1.

Spill Handling: Evacuate persons not wearing protective equipment from area of spill or leak until cleanup is complete. Remove all ignition sources. Use HEPA vacuum or wet method to reduce dust during cleanup. Do not dry sweep. Dampen spilled material with alcohol to avoid dust. Collect powdered material in the most convenient and safe manner and deposit in sealed containers. Ventilate area after cleanup is complete. It may be necessary to contain and dispose of this chemical as a hazardous waste. If material or contaminated runoff enters waterways, notify downstream users of potentially contaminated waters. Contact your local or federal environmental protection agency for specific recommendations. If employees are required to clean up spills, they must be properly trained and equipped. OSHA 1910.120(q) may be applicable.

Fire Extinguishing: This chemical is a combustible solid. Use dry chemical, carbon dioxide, water spray, or alcohol foam extinguishers. Poisonous gases, including chlorine and nitrogen oxides, are produced in fire. If material or contaminated runoff enters waterways, notify downstream users of potentially contaminated waters. Notify local health and fire officials and pollution control agencies. From a secure, explosion-proof location, use water spray to cool exposed containers. If cooling streams are ineffective (venting sound increases in volume and pitch, tank discolors, or shows any signs of deforming), withdraw immediately to a secure position. If employees are expected to fight fires, they must be trained and equipped in OSHA 1910.156. The only respirators recommended for firefighting are self-contained breathing apparatuses that have full face-pieces and are operated in a pressure-demand or other positive-pressure mode.

Disposal Method Suggested: Consult with environmental regulatory agencies for guidance on acceptable disposal practices. Generators of waste containing this contaminant ($\geq$100 kg/mo) must conform with EPA regulations governing storage, transportation, treatment, and waste disposal.

References

Lunch, A. L., O'Connor, G. B., Barnes, J. R., Killian, A. S., Jr., & Neeld, W. E., Jr. (1971). Methylene-bis-ortho-chloroaniline (MOCA): Evaluations of hazards and exposure control. *American Industrial Hygiene Association Journal*, 32, 802

New Jersey Department of Health and Senior Services. (April 2004). *Hazardous Substances Fact Sheet: 4,4'-Methylenebis(2-Chloroaniline)*. Trenton, NJ

Methylenebis(4-cyclohexyl isocyanate)   M:0860

Molecular Formula: $C_{15}H_{22}N_2O_2$
Common Formula: $OCNC_6H_{10}CH_2C_6H_{10}NCO$
Synonyms: Bis(4-isocyanatocyclohexyl)methane; Dicyclohexylmethane 4,4'-diisocyanate; DMDI; HMDI; Hydrogenated MDI; Nacconate H 12; Reduced MDI; Saturated MDI
CAS Registry Number: 5124-30-1
RTECS® Number: NQ9250000
UN/NA & ERG Number: Not regulated
EC Number: 225-863-2 [*Annex I Index No.*: 615-009-00-0]
Regulatory Authority and Advisory Bodies
Air Pollutant Standard Set. See below, "Permissible Exposure Limits in Air" section.
EPCRA Section 313 Form R *de minimis* concentration reporting level: 1.0%.
Canada, WHMIS, Ingredients Disclosure List Concentration 0.1%.
European/International Regulations: Hazard Symbol: T; Risk phrases: R23; R36/37/39; R42/43; Safety phrases: S1/2; S26; S28; S38; S45 (see Appendix 4).
WGK (German Aquatic Hazard Class): 1—Low hazard to waters.
Description: DMDI is a clear, colorless to light-yellow liquid. Molecular weight = 262.39; Specific gravity (H_2O:1) = 1.07 at 25°C; Freezing/Melting point = 19−23°C; Vapor pressure = 0.001 mmHg at 20°C; Flash point $\geq$202°C. Hazard Identification (based on NFPA-704 M Rating System): Health 2, Flammability 1, Reactivity 1. Insoluble in water; reactive.
Potential Exposure: Compound Description: Primary Irritant. Those involved in the manufacture of this compound or its use in the production of light-stable, non-yellowing polyurethane resins.

Incompatibilities: Violent reaction with alcohols, glycols. May slowly polymerize if heated above 122°F/50°C. Incompatible with moisture, heat, air, amines, amides, strong alkalis, and chemically active metals. Attacks some plastics, rubber, and coatings.
Permissible Exposure Limits in Air
Conversion factor: 1 ppm = 10.73 mg/m³ at 25°C & 1 atm.
OSHA PEL: None.
NIOSH REL: 0.01 ppm/0.11 mg/m³ Ceiling Concentration.
ACGIH TLV®[1]: 0.005 ppm/0.054 mg/m³ TWA.
Protective Action Criteria (PAC)
TEEL-0: 0.005 ppm
PAC-1: 0.005 ppm
PAC-2: 0.01 ppm
PAC-3: 0.15 ppm
Belgium: TWA 0.005 ppm (0.054 mg/m³), 1993; Denmark TWA 0.005 ppm (0.054 mg/m³), 1999; Norway: TWA 0.005 ppm (0.05 mg/m³), 1999; Sweden: TWA 0.005 ppm; STEL 0.01 ppm, 1999; United Kingdom: TWA 0.02 mg (NCO)/m³; STEL 0.07 mg[NCO]/m³, 2000; the Netherlands: MAC 0.1 1 mg/m³, [skin], 2003; Argentina, Bulgaria, Columbia, Jordan, South Korea, New Zealand, Singapore, Vietnam: ACGIH TLV®: TWA 0.005 ppm. Several states have set guidelines or standards for ambient air[60] ranging from 0.8 µg/m³ (Virginia) to 1.1 µg/m³ (North Dakota) to 0−3.0 µg/m³ (Nevada).
Determination in Air: Use NIOSH Analytical Method #5525; OSHA Analytical Method PV-2092.
Routes of Entry: Inhalation, ingestion, skin and/or eye contact.
Harmful Effects and Symptoms
Short Term Exposure: Methylene bis(4-cyclohexylisocyanate) can affect you when breathed in. Exposure can irritate the eyes, nose, and throat. Higher levels can irritate the lungs, causing a buildup of fluid (pulmonary edema). This can cause death. Contact can irritate and burn the eyes, causing permanent damage. Skin contact can cause blisters.
Long Term Exposure: Methylenebis(4-cyclohexylisocyanate) can cause an asthma-like allergy to develop. Future exposures can cause asthma attacks with shortness of breath, wheezing, cough, and/or chest tightness. Repeated exposures may permanently damage the lungs.
Points of Attack: Lungs.
Medical Surveillance: Before beginning employment and at regular times after that, the following are recommended: lung function tests. These may be normal at first if person is not having an attack at the time. If symptoms develop or overexposure is suspected, the following may be useful: consider chest X-ray after acute overexposure. Evaluation by a qualified allergist.
First Aid: If this chemical gets into the eyes, remove any contact lenses at once and irrigate immediately for at least 15 min, occasionally lifting upper and lower lids. Seek medical attention immediately. If this chemical contacts the skin, remove contaminated clothing and wash immediately with soap and water. Seek medical attention immediately.

If this chemical has been inhaled, remove from exposure, begin rescue breathing (using universal precautions, including resuscitation mask) if breathing has stopped and CPR if heart action has stopped. Transfer promptly to a medical facility. When this chemical has been swallowed, get medical attention. Give large quantities of water and induce vomiting. Do not make an unconscious person vomit. Medical observation is recommended for 24–48 h after breathing overexposure, as pulmonary edema may be delayed. As first aid for pulmonary edema, a doctor or authorized paramedic may consider administering a corticosteroid spray.

Personal Protective Methods: Wear protective gloves and clothing to prevent any reasonable probability of skin contact. Safety equipment suppliers/manufacturers can provide recommendations on the most protective glove/clothing material for your operation. All protective clothing (suits, gloves, footwear, headgear) should be clean, available each day, and put on before work. Contact lenses should not be worn when working with this chemical. Wear splash-proof chemical goggles and face shield unless full face-piece respiratory protection is worn. Employees should wash immediately with soap when skin is wet or contaminated. Provide emergency showers and eyewash.

Respirator Selection: NIOSH: *Up to 0.1 ppm:* Sa (APF = 10) (any supplied-air respirator).* *Up to 0.25 ppm:* Sa:Cf (APF = 25) (any supplied-air respirator operated in a continuous-flow mode.)* *Up to 0.5 ppm:* SCBAF (APF = 50) (any self-contained breathing apparatus with a full face-piece) or SaF (APF = 50) (any supplied-air respirator with a full face-piece). *Up to 1 ppm:* SaF: Pd,Pp (APF = 2000) (any supplied-air respirator that has a full face-piece and is operated in a pressure-demand or other positive-pressure mode). *Emergency or planned entry into unknown concentrations or IDLH conditions:* SCBAF: Pd, Pp (APF = 10,000) (any NIOSH/MSHA- or European Standard EN 149-approved self-contained breathing apparatus that has a full face-piece and is operated in a pressure-demand or other positive-pressure mode) or SaF: Pd,Pp: ASCBA (APF = 10,000) (any supplied-air respirator that has a full face-piece and is operated in a pressure-demand or other positive-pressure mode in combination with an auxiliary self-contained breathing apparatus operated in a pressure-demand or other positive-pressure mode). *Escape:* GmFOv (APF = 50) [any air-purifying, full-face-piece respirator (gas mask) with a chin-style, front-or back-mounted organic vapor canister] or SCBAE (any appropriate escape-type, self-contained breathing apparatus).

*Substance reported to cause eye irritation or damage; may require eye protection.

Storage: Color Code—Blue: Health Hazard/Poison: Store in a secure poison location. Prior to working with this chemical you should be trained on its proper handling and storage. Methylene bis(4-cyclohexylisocyanate) must be stored to avoid contact with alcohols, since violent reactions occur. Store in tightly closed containers in a cool, well-ventilated area away from moisture, heat, air, amines, strong bases, and chemically active metals.

Shipping: Not regulated.

Spill Handling: Evacuate and restrict persons not wearing protective equipment from area of spill or leak until cleanup is complete. Remove all ignition sources. Ventilate area of spill or leak. Use HEPA vacuum or wet method to reduce dust during cleanup. Do not dry sweep. Dampen spilled solid material with 60–70% ethanol to avoid airborne dust. Absorb liquids in vermiculite, dry sand, earth, peat, carbon, or a similar material, and deposit in sealed containers. Keep this chemical out of a confined space, such as a sewer, because of the possibility of an explosion, unless the sewer is designed to prevent the buildup of explosive concentrations. It may be necessary to contain and dispose of this chemical as a hazardous waste. If material or contaminated runoff enters waterways, notify downstream users of potentially contaminated waters. Contact your local or federal environmental protection agency for specific recommendations. If employees are required to clean up spills, they must be properly trained and equipped. OSHA 1910.120(q) may be applicable.

Fire Extinguishing: Methylene bis(4-cyclohexylisocyanate) may burn, but does not readily ignite. Use dry chemical, CO_2, or foam extinguishers. Poisonous gases, including hydrogen cyanide, ammonia, nitriles, and oxides of nitrogen, are produced in fire. Vapors are heavier than air and will collect in low areas. Vapors may travel long distances to ignition sources and flashback. Vapors in confined areas may explode when exposed to fire. Containers may explode in fire. Storage containers and parts of containers may rocket great distances, in many directions. If material or contaminated runoff enters waterways, notify downstream users of potentially contaminated waters. Notify local health and fire officials and pollution control agencies. From a secure, explosion-proof location, use water spray to cool exposed containers. If cooling streams are ineffective (venting sound increases in volume and pitch, tank discolors, or shows any signs of deforming), withdraw immediately to a secure position. If employees are expected to fight fires, they must be trained and equipped in OSHA 1910.156. The only respirators recommended for firefighting are self-contained breathing apparatuses that have full face-pieces and are operated in a pressure-demand or other positive-pressure mode.

Reference

New Jersey Department of Health and Senior Services. (January 2001). *Hazardous Substances Fact Sheet: Methylene bis(4-Cyclohexylisocyanate)*. Trenton, NJ

4,4'-Methylenebis(*N,N*-dimethyl) aniline   M:0870

Molecular Formula: $C_{17}H_{22}N_2$
Common Formula: $(CH_3)_2N\text{-}C_6H_4\text{-}CH_2\text{-}C_6H_4\text{-}N(CH_3)_2$

Synonyms: Aniline, 4,4′-methylenebis(*N,N*-dimethyl)-; Arnold's base; *tetra* Base; Benzenamine, 4,4′-methylenebis-(*N,N*-dimethyl-); *p,p′*-Bis(dimethylamino)diphenylmethane; 4,4′-Bis(dimethylamino)diphenylmethane; 4,4′-Bis(dimethyl-aminophenyl)methane; Bis[*p*-(*N,N*-dimethylamino)phenyl]methane; Bis[*p*-(dimethylamino)-phenyl]methane; Bis[4-(*N,N*-dimethylamino)-phenyl]methane; Bis[4-(dimethylamino)phenyl]methane; 4,4′-Methylene bis(*N,N*-dimethylaniline); Michler's base; Michler's hydride; Michler's methane; NCII-C01990; Reduced Michler's ketone; *N,N,N′,N′*-Tetramethyl-*p,p′*-diaminodiphenyl-methane; *N,N,N′,N′*-Tetramethyl-4,4′-diaminodiphenyl-methane; *p,p′*-Tetramethyldiamino-diphenylmethane; 4,4′-Tetramethyldiaminodiphenylmethane; Tetramethyldiamino-diphenylmethane

CAS Registry Number: 101-61-1
RTECS® Number: BY5250000
UN/NA & ERG Number: UN3143 (Dyes, solid, toxic, n.o.s. [or] Dye intermediates, solid, toxic, n.o.s.)/151
EC Number: 202-959-2 [*Annex I Index No.:* 612-201-00-6]

Regulatory Authority and Advisory Bodies
Carcinogenicity: EPA: Sufficient evidence from animal studies; inadequate evidence or no useful data from epidemiologic studies; NTP: Reasonably anticipated to be a human carcinogen.
Air Pollutant Standard Set. See below, "Permissible Exposure Limits in Air" section.
EPCRA Section 313 Form R *de minimis* concentration reporting level: 0.1%.
California Proposition 65 Chemical: Cancer 10/1/81.
Canada, WHMIS, Ingredients Disclosure List Concentration 0.1%.
European/International Regulations: Hazard Symbol: T, N; Risk phrases: R45; R50/53; Safety phrases: S53; S45; S60; S61 (see Appendix 4).
WGK (German Aquatic Hazard Class): No value assigned.

Description: 4,4′-Methylenebis(*N,N*-dimethyl)benzenamine is a yellow crystalline compound. Molecular weight = 254.41; Boiling point = 390°C; Freezing/Melting point = 90−91°C. Hazard Identification (based on NFPA-704 M Rating System): Health 2, Flammability 1, Reactivity 0. Insoluble in water.

Potential Exposure: 4,4′-methylenebis(N,N-dimethyl) benzenamine is used as an intermediate in dye manufacture and as an analytical reagent in the determination of lead.

Incompatibilities: Strong acids, oxidizers, acid chlorides, acid anhydrides.

Permissible Exposure Limits in Air
No TEEL available.
DFG MAK: Carcinogen Category 2.
The state of North Dakota has set a guideline of zero concentration for ambient air.[60]

Routes of Entry: Inhalation, ingestion, eye and/or skin contact. Passes through the skin.

Harmful Effects and Symptoms
Short Term Exposure: 4,4′-Methylenebis(*N,N′*-dimethyl) aniline can affect you when breathed in and by passing through your skin. 4,4′-Methylenebis(*N,N*-dimethyl)aniline is a carcinogen; handle with extreme caution. High exposure can interfere with the ability of the blood to carry oxygen (a condition called methemoglobinemia). This can cause headaches, dizziness, weakness, and a bluish color to the skin and lips. Higher levels can cause trouble breathing, collapse, and death.

Long Term Exposure: May be a carcinogen. It has been shown to cause liver and thyroid cancer in animals.

Medical Surveillance: If symptoms develop or overexposure is suspected, the following may be useful: blood test for methemoglobin.

First Aid: If this chemical gets into the eyes, remove any contact lenses at once and irrigate immediately for at least 15 min, occasionally lifting upper and lower lids. Seek medical attention immediately. If this chemical contacts the skin, remove contaminated clothing and wash immediately with soap and water. Seek medical attention immediately. If this chemical has been inhaled, remove from exposure, begin rescue breathing (using universal precautions, including resuscitation mask) if breathing has stopped and CPR if heart action has stopped. Transfer promptly to a medical facility. When this chemical has been swallowed, get medical attention. Give large quantities of water and induce vomiting. Do not make an unconscious person vomit.

Note to physician: Treat for methemoglobinemia. Spectrophotometry may be required for precise determination of levels of methemoglobin in urine.

Personal Protective Methods: Wear protective gloves and clothing to prevent any reasonable probability of skin contact. Safety equipment suppliers/manufacturers can provide recommendations on the most protective glove/clothing material for your operation. All protective clothing (suits, gloves, footwear, headgear) should be clean, available each day, and put on before work. Contact lenses should not be worn when working with this chemical. Wear dust-proof chemical goggles and face shield unless full face-piece respiratory protection is worn. Employees should wash immediately with soap when skin is wet or contaminated. Provide emergency showers and eyewash. Methylenebis(*N,N*-dimethyl)benzenamine. Wear protective gloves and clothing. Safety equipment suppliers/manufacturers can provide recommendations on the most protective glove/clothing material for your operation. All protective clothing (suits, gloves, footwear, headgear) should be clean, available each day, and put on before work.

Respirator Selection: At any exposure level, use a NIOSH/MSHA- or European Standard EN149-approved supplied-air-respirator with a full face-piece operated in the positive-pressure mode, or with a full face-piece, hood, or helmet in the continuous-flow mode; or use a NIOSH/MSHA- or European Standard EN149-approved self-contained breathing apparatus with a full face-piece operated in pressure-demand or other positive-pressure mode.

Storage: Color Code—Blue: Health Hazard/Poison: Store in a secure poison location. Prior to working with this chemical you should be trained on its proper handling and storage. 4,4′-Methylenebis(*N,N*-dimethyl)benzenamine must be stored to avoid contact with oxidizers (such as perchlorates, peroxides, permanganates, chlorates, and nitrates) and strong acids (such as hydrochloric, sulfuric, and nitric), since violent reactions occur. Store in tightly closed containers in a cool, well-ventilated area. A regulated, marked area should be established where this chemical is handled, used, or stored in compliance with OSHA Standard 1910.1045.

Shipping: Dyes, solid, toxic, n.o.s. [or] Dye intermediates, solid, toxic, n.o.s. require the label of "POISONOUS/TOXIC MATERIALS." It falls in Hazard Class 6.1 and Packing Group 1.

Spill Handling: Evacuate persons not wearing protective equipment from area of spill or leak until cleanup is complete. Remove all ignition sources. Dampen spilled material with 60–70% acetone to avoid airborne dust. Use HEPA vacuum or wet method to reduce dust during cleanup. Do not dry sweep. Collect powdered material in the most convenient and safe manner and deposit in sealed containers. Ventilate area after cleanup is complete. It may be necessary to contain and dispose of this chemical as a hazardous waste. If material or contaminated runoff enters waterways, notify downstream users of potentially contaminated waters. Contact your local or federal environmental protection agency for specific recommendations. If employees are required to clean up spills, they must be properly trained and equipped. OSHA 1910.120(q) may be applicable.

Fire Extinguishing: This chemical is a combustible solid. Use dry chemical, carbon dioxide, water spray, alcohol, or polymer foam extinguishers. Poisonous gases, including nitrogen oxides, are produced in fire. If material or contaminated runoff enters waterways, notify downstream users of potentially contaminated waters. Notify local health and fire officials and pollution control agencies. From a secure, explosion-proof location, use water spray to cool exposed containers. If cooling streams are ineffective (venting sound increases in volume and pitch, tank discolors, or shows any signs of deforming), withdraw immediately to a secure position. If employees are expected to fight fires, they must be trained and equipped in OSHA 1910.156. The only respirators recommended for firefighting are self-contained breathing apparatuses that have full face-pieces and are operated in a pressure-demand or other positive-pressure mode.

Reference
New Jersey Department of Health and Senior Services. (November 1999). *Hazardous Substances Fact Sheet: 4,4′-Methylenebis(N,N-Dimethyl)Benzenamine.* Trenton, NJ

Methylenebis(phenyl-isocyanate)
M:0880

Molecular Formula: C$_{15}$H$_{10}$N$_2$O$_2$
Common Formula: OCNC$_6$H$_4$CH$_2$C$_6$H$_4$NCO

Synonyms: AI3-15256; Benzene, 1,1′-methylenebis(4-isocyanato-); Bis(*p*-isocyanatophenyl)methane; Bis(1,4-isocyanatophenyl)methane; Bis(4-isocyanatophenyl) methane; Caradate 30; Desmodur 44; 4,4′-Diisocyanatodiphenylmethane; Di-(4-isocyanatophenyl)methane; *p,p*′-Diphenylmethane diisocyanate; 4,4′-Diphenylmethane diisocyanate; Diphenyl methane diisocyanate; Diphenylmethane *p,p*′-diisocyanate; Diphenylmethane 4,4′-diisocyanate; Diphenylmethane diisocyanate; Hylene M-50; Isocyanic acid, ester with diphenylmethane; Isocyanic acid, methylenedi-*p*-phenylene ester; Isonate 125M; Isonate 125 MF; MDI; MDR; Methyl bisphenylisocyanate; 1,1′-Methylenebis(4-isocyanatobenzene); 1,1-Methylenebis(4-isocyanatobenzene); Methylenebis(4-isocyanatobenzene); Methylenebis(*p*-phenylene isocyanate); Methylenebis(4-phenylene isocyanate); *p,p*′-Methylenebis(phenylisocyanate); 4,4′-Methylenebis(phenylisocyanate); Methylene bis(4-phenylisocyanate); Methylene bisphenylisocyanate; Methylenebis(*p*-phenylisocyanate); Methylenebis(4,4′-phenylisocyanate); Methylenebis(4-phenylisocyanate); 4,4′-Methylenedi(phenyldiisocyanate); 4,4′-Methylenedi-*p*-phenylene diisocyanate; Methylenedi(*p*-phenylene diisocyanate); Methylenedi-*p*-phenylene diisocyanate; 4,4′-Methylenedi(phenylene isocyanate); Methylene di(phenylene isocyanate); Methylenedi(*p*-phenylene isocyanate); Methylenedi-*p*-phenylene isocyanate; 4,4′-Methylene diphenylisocyanate; Metilenbis(fenilisocianato) (Spanish); Nacconate 300; NCI-C50668; Rubinate 44

CAS Registry Number: 101-68-8
RTECS® Number: NQ9350000
UN/NA & ERG Number: UN2811 (toxic solid, organic, n.o.s.)/154; 3082/171
EC Number: 202-966-0 [*Annex I Index No.:* 615-005-01-6]
Regulatory Authority and Advisory Bodies
Carcinogenicity: IARC: Human Inadequate Data, animal No Adequate Data*, not classifiable as carcinogenic to humans,* Group 3, 1999; EPA: Not Classifiable as to human carcinogenicity.

Air Pollutant Standard Set. See below, "Permissible Exposure Limits in Air" section.

Clean Air Act: Hazardous Air Pollutants (Title I, Part A, Section 112).

Reportable Quantity (RQ): 1 lb (0.454 kg).

EPCRA Section 313 Form R *de minimis* concentration reporting level: 1.0%.

US DOT 49CFR172.101, Inhalation Hazardous Chemical.

Canada, WHMIS, Ingredients Disclosure List Concentration 0.1%.

European/International Regulations: Hazard Symbol: Xn; Risk phrases: R20; R36/37/38; R40;R42/43; R48/20; Safety phrases: S1/2; S23; S23; S36/37; S45 (see Appendix 4).

WGK (German Aquatic Hazard Class): 1—Low hazard to waters.

Description: MDI is a white to light-yellow, odorless flakes. A liquid above 37°C. Molecular weight = 250.27; Specific gravity (H$_2$O:1) = 1.23 (solid at 25°C); 1.19 (liquid at 50°C); Boiling point = 313.9°C at 5 mmHg; Freezing/

Melting point = 37.2°C; Vapor pressure = 0.0000005 mmHg at 20°C; Flash point = 198.9°C (cc); Autoignition temperature = 240°C. Hazard Identification (based on NFPA-704 M Rating System): Health 2, Flammability 1, Reactivity 1. Slightly soluble in water (reaction).

Potential Exposure: Compound Description: Tumorigen, Mutagen; Reproductive Effector; Human Data; Primary Irritant. MDI is used in the production of polyurethane foams and plastics, polyurethane coatings, elastomers, and thermoplastic resins.

Incompatibilities: Violent reaction with oxidizers, strong alkalis, acids, alcohols, ammonia, amines, amides, glycols, caprolactum. Unstable above 100°F/38°C. Polymerizes at temperatures above 204°C. Reacts readily with water to form insoluble polyureas. Attacks some plastics, rubber, and coatings. Reacts with moisture.

Permissible Exposure Limits in Air

Conversion factor: 1 ppm = 10.24 mg/m^3 at 25°C & 1 atm.

OSHA PEL: 0.02 ppm/0.2 mg/m^3 Ceiling Concentration.

NIOSH REL: 0.005 ppm/0.05 mg/m^3 TWA; 0.2 mg/m^3/0.020 ppm/10 min, Ceiling Concentration.

ACGIH TLV®[1]: 0.005 ppm TWA, measured as inhalable fraction and vapor [skin]; BEI$_A$ issued as Acetylcholinesterase-inhibiting pesticides.

NIOSH IDLH: 75 mg/m^3.

Protective Action Criteria (PAC)*

TEEL-0: 0.051 mg/m^3

PAC-1: **0.2** mg/m^3

PAC-2: **2** mg/m^3

PAC-3: **2.5** mg/m^3

*AEGLs (Acute Emergency Guideline Levels) & ERPGs (Emergency Response Planning Guideline) are in **bold face**.

DFG MAK: 0.05 mg/m^3, inhalable fraction; danger of sensitization of the airways and skin; Peak Limitation Category I(1); a momentary value of 0.1 mg/m^3 should not be exceeded; Carcinogen Category 4; Pregnancy Risk Group C; BAT: 10 μg[4,4'-Diaminodiphenylmetane]/dL creatinine in urine/end-of-shift.

Austria: MAK 0.005 ppm (0.05 mg/m^3), Suspected: carcinogen, 1999; Belgium TWA 0.005 ppm (0.051 mg/m^3); STEL 0.02 ppm, 1993; Denmark TWA 0.005 ppm (0.05 mg/m^3), 1999; France: VME 0.01 ppm (0.1 mg/m^3), VLE 0.02 ppm (0.2 mg/m^3), 1999; Hungary TWA 0.05 mg/m^3; STEL 0.1 mg/m^3, 1993; Japan 0.05 mg/m^3, 1999; the Netherlands: MAC-TGG 0.05 mg/m^3, 2003; the Philippines: TWA 0.02 ppm (0.2 mg/m^3), 1993; Poland: MAC (TWA) 0.05 mg/m^3, MAC 0.2 mg/m^3, 1999; Russia: STEL 0.5 mg/m^3 [skin] 1993; Sweden: NGV 0.005 ppm, TGV 0.01 ppm, 1999; Switzerland: MAK- week 5 ppm (15 mg/m^3); STEL 25 ppm (75 mg/m^3), 1999; Thailand TWA 0.02 ppm (0.2 mg/m^3), 1993; United Kingdom TWA 0.02 mg[NCO]/m^3; STEL 0.07 mg[NCO]/m^3, 2000; Argentina, Bulgaria, Columbia, Jordan, South Korea, New Zealand, Singapore, Vietnam: ACGIH TLV®: TWA

0.005 ppm. Several states have set guidelines or standards for MDI in ambient air[60] ranging from 0.2 μg/m^3 (Rhode Island) to 0.67 μg/m^3 (New York) to 1.0 μg/m^3 (Connecticut) to 1.6 μg/m^3 (Virginia) to 2.0 μg/m^3 (North Dakota and South Carolina) to 5.0 μg/m^3 (Nevada).

Determination in Air: Use NIOSH Analytical Method (IV) #5521, #5522, #5525; OSHA Analytical Method 18, 33, 47.

Routes of Entry: Inhalation, ingestion, eye and/or skin contact.

Harmful Effects and Symptoms

Short Term Exposure: Inhalation: A lacrimator. Higher exposures can cause pulmonary edema, a medical emergency that can be delayed for several hours. This can cause death. Vapors are irritating at 0.001−0.026 ppm and may cause shortness of breath, asthma, sore throat, coughing, wheezing, chest tightness, depression, headache, nasal discharge, and insomnia. May cause allergic respiratory reactions. Symptoms may be delayed up to 8 h after exposure. *Skin:* Causes irritation, redness, and pain. May cause a rash. Irritation begins at levels of 0.05−0.1 ppm. May adhere firmly to skin. Attempts to remove it may increase or produce irritation. *Eyes:* Irritation at 0.05−0.1 ppm causing redness, pain, and blurred vision. *Ingestion:* Causes abdominal spasms and vomiting.

Long Term Exposure: Repeated or prolonged contact may cause skin sensitization and allergy. Repeated or prolonged inhalation exposure may cause asthma-like allergy. Prolonged exposure may lead to permanent breathing or respiratory problems.

Points of Attack: Eyes, skin, respiratory system.

Medical Surveillance: NIOSH lists the following tests: blood gas analysis; chest X-ray, electrocardiogram, pulmonary function tests: forced vital capacity, forced expiratory volume (1 sec); red blood cells/count; sputum cytology; urine (chemical/metabolite); white blood cell count/differential. Evaluation by a qualified allergist. Preplacement and periodic medical examinations should include chest X-ray, pulmonary function tests, and an evaluation of any respiratory disease or history of allergy. Periodic pulmonary function tests may be useful in detecting the onset of pulmonary sensitization.

First Aid: If this chemical gets into the eyes, remove any contact lenses at once and irrigate immediately for at least 15 min, occasionally lifting upper and lower lids. Seek medical attention immediately. If this chemical contacts the skin, remove contaminated clothing and wash immediately with soap and water. Seek medical attention immediately. If this chemical has been inhaled, remove from exposure, begin rescue breathing (using universal precautions, including resuscitation mask) if breathing has stopped and CPR if heart action has stopped. Transfer promptly to a medical facility. When this chemical has been swallowed, get medical attention. Give large quantities of water and induce vomiting. Do not make an unconscious person vomit. Medical observation is recommended for 24−48 h after breathing overexposure, as pulmonary edema may be delayed. As first aid for pulmonary edema, a doctor or authorized paramedic may consider administering a cortico-steroid spray.

Personal Protective Methods: Wear protective gloves and clothing to prevent any reasonable probability of skin contact: **8 h**: 4H™ and Silver Shield™ gloves; Barricade™ coated suits; Responder™ suits. Safety equipment suppliers/manufacturers can provide recommendations on the most protective glove/clothing material for your operation. All protective clothing (suits, gloves, footwear, headgear) should be clean, available each day, and put on before work. Contact lenses should not be worn when working with this chemical. Wear dust-proof chemical goggles and face shield unless full face-piece respiratory protection is worn. Employees should wash immediately with soap when skin is wet or contaminated. Provide emergency showers and eyewash.

Respirator Selection: NIOSH: *0.5 mg/m³:* Sa (APF = 10) (any supplied-air respirator) or SCBA (any self-contained breathing apparatus). *1.25 mg/m³:* Sa:Cf (APF = 25) (any supplied-air respirator operated in a continuous-flow mode). *2.5 mg/m³:* SCBAF (APF = 50) (any self-contained breathing apparatus with a full face-piece) or SaF (APF = 50) (any supplied-air respirator with a full face-piece). *75 mg/m³:* SaF: Pd,Pp (APF = 2000) (any supplied-air respirator that has a full face-piece and is operated in a pressure-demand or other positive-pressure mode). *Emergency or planned entry into unknown concentrations or IDLH conditions:* SCBAF: Pd,Pp (APF = 10,000) (any self-contained breathing apparatus that has a full face-piece and is operated in a pressure-demand or other positive-pressure mode) or SaF: Pd,Pp: ASCBA (APF = 10,000) (any supplied-air respirator that has a full face-piece and is operated in a pressure-demand or other positive-pressure mode in combination with an auxiliary, self-contained breathing apparatus operated in a pressure-demand or other positive-pressure mode). *Escape:* GmFOv100 (APF = 50) [Any air-purifying, full-face-piece respirator (gas mask) with a chin-style, front- or back-mounted organic vapor canister having an N100, R100, or P100 filter] or SCBAE (any appropriate escape-type, self-contained breathing apparatus).

Note: Substance reported to cause eye irritation or damage; may require eye protection.

Storage: Color Code—Blue: Health Hazard/Poison: Store in a secure poison location. Prior to working with this chemical you should be trained on its proper handling and storage. Store at temperatures indicated on labels, separately from acids, bases, amines, alcohols and ammonia, and with ventilation along the floor. Since MDI will react with moisture in the air, the storage area should be a dry place, away from all sources of fire or ignition.

Shipping: Toxic solids, organic, n.o.s. compound requires a shipping label of "POISONOUS/TOXIC MATERIALS." It falls in Hazard Class 6.1, Packing Group III.

Spill Handling: Evacuate persons not wearing protective equipment from area of spill or leak until cleanup is complete. Remove all ignition sources. Dampen spilled material with toluene to avoid dust. Collect powdered material in the most convenient and safe manner and deposit in sealed containers. Ventilate area after cleanup is complete. It may be necessary to contain and dispose of this chemical as a hazardous waste. If material or contaminated runoff enters waterways, notify downstream users of potentially contaminated waters. Contact your local or federal environmental protection agency for specific recommendations. If employees are required to clean up spills, they must be properly trained and equipped. OSHA 1910.120(q) may be applicable.

Fire Extinguishing: Use carbon dioxide, dry chemical, or halons. In case of large fires, water spray may be used to cool drums, taking care to prevent direct contact between water and MDI. Poisonous gases, including nitrogen oxides and hydrogen cyanide, are produced in fire. If material or contaminated runoff enters waterways, notify downstream users of potentially contaminated waters. Notify local health and fire officials and pollution control agencies. From a secure, explosion-proof location, use water spray to cool exposed containers. If cooling streams are ineffective (venting sound increases in volume and pitch, tank discolors, or shows any signs of deforming), withdraw immediately to a secure position. If employees are expected to fight fires, they must be trained and equipped in OSHA 1910.156. The only respirators recommended for firefighting are self-contained breathing apparatuses that have full face-pieces and are operated in a pressure-demand or other positive-pressure mode.

Disposal Method Suggested: Controlled incineration (oxides of nitrogen are removed from the effluent gas by scrubbers and/or thermal devices).

References

New York State Department of Health. (May 1986). *Chemical Fact Sheet 4,4'-Methylene Diphenyl Diisocyanate.* Version 2. Albany, NY: Bureau of Toxic Substance Assessment

US Environmental Protection Agency. (June 28, 1984). *Chemical Hazard Information Profile: Methylene Diphenyl Diisocyanate.* Washington, DC: Office of Toxic Substances

Methylene bromide   M:0890

Molecular Formula: CH_2Br_2
Synonyms: Bromuro de metileno (Spanish); Dibromomethane; Methane, dibromo-; Methylene dibromide
CAS Registry Number: 74-95-3
RTECS® Number: PA7350000
UN/NA & ERG Number: UN2664/160
EC Number: 200-824-2 [*Annex I Index No.:* 602-003-00-8]
Regulatory Authority and Advisory Bodies
Air Pollutant Standard Set. See below, "Permissible Exposure Limits in Air" section.
US EPA Hazardous Waste Number (RCRA No.): U068.

RCRA, 40CFR261, Appendix 8 Hazardous Constituents.
RCRA 40CFR268.48; 61FR15654, Universal Treatment Standards: Wastewater (mg/L), 0.11; Nonwastewater (mg/kg), 15.
RCRA 40CFR264, Appendix 9; TSD Facilities Ground Water Monitoring List. Suggested test method(s) (PQL μg/L): 8010 (15); 8240 (5).
Reportable Quantity (RQ): 1000 lb (454 kg).
EPCRA Section 313 Form R *de minimis* concentration reporting level: 1.0%.
Canada, WHMIS, Ingredients Disclosure List Concentration 0.1%.
European/International Regulations: Hazard Symbol: Xn, N; Risk phrases: R20; R52/53; Safety phrases: S2; S24; S61 (see Appendix 4).
WGK (German Aquatic Hazard Class): No value assigned.
Description: Methylene bromide is a colorless liquid with a sweet, pleasant odor. Molecular weight = 173.85; Boiling point = 96−97°C; Freezing/Melting point = −53°C. Hazard Identification (based on NFPA-704 M Rating System): Health 3, Flammability 0, Reactivity 1. Slightly soluble in water.
Potential Exposure: Compound Description: Mutagen; Human Data. Methylene bromide is used as a solvent and as a chemical intermediate.
Incompatibilities: Mixture with potassium forms a shock-sensitive explosive. Incompatible with oxidizers, aluminum, magnesium. The substance decomposes on contact with hot surfaces producing hydrogen bromide.
Permissible Exposure Limits in Air
Protective Action Criteria (PAC)
TEEL-0: 10 ppm
PAC-1: 30 ppm
PAC-2: 200 ppm
PAC-3: 1250 ppm
Russia: STEL: 10 mg/m^3, 1993.
Routes of Entry: Inhalation, ingestion, eye and/or skin contact.
Harmful Effects and Symptoms
Short Term Exposure: Methylene bromide can affect you when breathed in. Contact can irritate the eyes and skin. May affect the nervous system and blood, resulting in impaired functions and formation of carboxyhemoglobinemia. Exposure can cause you to feel dizzy, lightheaded and to pass out. High levels can cause death. Methylene bromide can cause the heart to beat irregularly or stop. This can cause death.
Long Term Exposure: Repeated skin contact can cause dryness and itching; removal of the skin's natural oils. May cause liver and kidney damage.
Points of Attack: Blood, kidneys, liver, skin.
Medical Surveillance: If symptoms develop or overexposure is suspected, the following may be useful: liver and kidney function tests. Serum bromine level. Carboxyhemoglobin level. Holter monitor (a special 24-h EKG to look for irregular heartbeat).

First Aid: If this chemical gets into the eyes, remove any contact lenses at once and irrigate immediately for at least 15 min, occasionally lifting upper and lower lids. Seek medical attention immediately. If this chemical contacts the skin, remove contaminated clothing and wash immediately with soap and water. Seek medical attention immediately. If this chemical has been inhaled, remove from exposure, begin rescue breathing (using universal precautions, including resuscitation mask) if breathing has stopped and CPR if heart action has stopped. Transfer promptly to a medical facility. When this chemical has been swallowed, get medical attention. Give large quantities of water and induce vomiting. Do not make an unconscious person vomit.
Personal Protective Methods: Wear protective gloves and clothing to prevent any reasonable probability of skin contact. Safety equipment suppliers/manufacturers can provide recommendations on the most protective glove/clothing material for your operation. Polyvinyl alcohol is among the recommended protective materials. All protective clothing (suits, gloves, footwear, headgear) should be clean, available each day, and put on before work. Contact lenses should not be worn when working with this chemical. Wear splash-proof chemical goggles and face shield unless full face-piece respiratory protection is worn. Employees should wash immediately with soap when skin is wet or contaminated. Provide emergency showers and eyewash.
Respirator Selection: Where there is potential for exposure to methylene bromide, use a NIOSH/MSHA- or European Standard EN149-approved supplied-air respirator with a full face-piece operated in the positive-pressure mode, or with a full face-piece, hood, or helmet in the continuous-flow mode; or use a NIOSH/MSHA- or European Standard EN149-approved self-contained breathing apparatus with a full face-piece operated in pressure-demand or other positive-pressure mode.
Storage: Color Code—Blue: Health Hazard/Poison: Store in a secure poison location. Prior to working with this chemical you should be trained on its proper handling and storage. Store in tightly closed container in a well-ventilated area away from potential high heat sources. Where possible, automatically pump liquid from drums or other storage containers to process containers.
Shipping: This compound requires a shipping label of "POISONOUS/TOXIC MATERIALS." It falls in Hazard Class 6.1 and Packing Group III.
Spill Handling: Evacuate and restrict persons not wearing protective equipment from area of spill or leak until cleanup is complete. Remove all ignition sources. Ventilate area of spill or leak. Use HEPA vacuum or wet method to reduce dust during cleanup. Do not dry sweep. Absorb liquids in vermiculite, dry sand, earth, peat, carbon, or a similar material and deposit in sealed containers. Keep this chemical out of a confined space, such as a sewer, because of the possibility of an explosion, unless the sewer is designed to prevent the buildup of explosive concentrations. It may be necessary to contain and dispose of this chemical as a

hazardous waste. If material or contaminated runoff enters waterways, notify downstream users of potentially contaminated waters. Contact your local or federal environmental protection agency for specific recommendations. If employees are required to clean up spills, they must be properly trained and equipped. OSHA 1910.120(q) may be applicable.

Fire Extinguishing: Methylene bromide may burn but does not readily ignite. Poisonous gases, including bromine and bromide, are produced in fire. Use dry chemical, carbon dioxide, alcohol, or polymer foam extinguishers. Containers may explode in fire. Storage containers and parts of containers may rocket great distances, in many directions. If material or contaminated runoff enters waterways, notify downstream users of potentially contaminated waters. Notify local health and fire officials and pollution control agencies. From a secure, explosion-proof location, use water spray to cool exposed containers. If cooling streams are ineffective (venting sound increases in volume and pitch, tank discolors, or shows any signs of deforming), withdraw immediately to a secure position. If employees are expected to fight fires, they must be trained and equipped in OSHA 1910.156. The only respirators recommended for firefighting are self-contained breathing apparatuses that have full face-pieces and are operated in a pressure-demand or other positive-pressure mode.

Disposal Method Suggested: Consult with environmental regulatory agencies for guidance on acceptable disposal practices. Generators of waste containing this contaminant ($\geq$100 kg/mo) must conform with EPA regulations governing storage, transportation, treatment, and waste disposal.

References

Sax, N. I. (Ed.). (1987). *Dangerous Properties of Industrial Materials Report,* 7, No. 2, 48–50

New Jersey Department of Health and Senior Services. (November 1999). *Hazardous Substances Fact Sheet: Methylene Bromide.* Trenton, NJ

Methylene chloride   M:0900

Molecular Formula: CH_2Cl_2
Synonyms: Aerothene MM; Chlorure de methylene (French); Cloruro de metileno (Spanish); DCM; Dichloromethane; Diclorometano (Spanish); Freon 30; Methane, dichloro-; Methane dichloride; Methylene bichloride; Methylene dichloride; Narkotil; NCI-C50102; R 30; Solaesthin; Solmethine
CAS Registry Number: 75-09-2
RTECS® Number: PA8050000
UN/NA & ERG Number: UN1593/160
EC Number: 200-838-9 [*Annex I Index No.:* 602-004-00-3]
Regulatory Authority and Advisory Bodies
Carcinogenicity: NCI: Carcinogenesis Studies (inhalation); clear evidence: mouse, rat; NTP: 11th Report on Carcinogens, 2004: Reasonably anticipated to be a human

carcinogen; IARC: Animal Sufficient Evidence; Human Inadequate Evidence, *possibly carcinogenic to humans,* Group 2B, 1999; EPA: Sufficient evidence from animal studies; inadequate evidence or no useful data from epidemiologic studies; US EPA Gene-Tox Program, Positive: Cell transform.—RLV F344 rat embryo; Positive: Histidine reversion—Ames test; Positive: *S. cerevisiae* gene conversion; *S. cerevisiae*—homozygosis; Positive: *S. cerevisiae*—reversion; Negative: *D. melanogaster* sex-linked lethal; OSHA Regulated carcinogen; NIOSH suspected human carcinogen; See *NIOSH Pocket Guide,* Appendix A.
US EPA, FIFRA 1998 Status of Pesticides: Canceled.
Air Pollutant Standard Set. See below, "Permissible Exposure Limits in Air" section.
Clean Air Act: Hazardous Air Pollutants (Title I, Part A, Section 112).
Clean Water Act: 40CFR423, Appendix A, Priority Pollutants.
US EPA Hazardous Waste Number (RCRA No.): U080.
RCRA, 40CFR261, Appendix 8 Hazardous Constituents.
Safe Drinking Water Act: Regulated chemical (47 FR 9352); MCL, 0.005 mg/L; MCLG, zero.
RCRA 40CFR268.48; 61FR15654, Universal Treatment Standards: Wastewater (mg/L), 0.089; Nonwastewater (mg/kg), 30.
RCRA 40CFR264, Appendix 9; TSD Facilities Ground Water Monitoring List. Suggested test method(s) (PQL μg/L): 8010 (5); 8240 (5).
Safe Drinking Water Act: MCL, 0.005 mg/L; MCLG, zero; Regulated chemical (47 FR 9352).
Reportable Quantity (RQ): 1000 lb (454 kg).
EPCRA Section 313 Form R *de minimis* concentration reporting level: 1.0%.
California Proposition 65 Chemical: Cancer 4/1/88.
Canada, WHMIS, Ingredients Disclosure List Concentration 0.1%.
European/International Regulations: Hazard Symbol: Xn; Risk phrases: R40; Safety phrases: S2; S23; S24/25; S36/37 (see Appendix 4).
WGK (German Aquatic Hazard Class): 2—Hazard to waters.
Description: Methylene chloride is a nonflammable, colorless liquid with a chloroform-like odor. A gas above 40°C/104°F. The odor is noticeable at 250 ppm. However, this level substantially exceeds the OSHA STEL and must not be relied upon as an adequate warning of unsafe concentrations. Molecular weight = 84.93; Specific gravity (H_2O:1) = 1.33; Boiling point = 40°C; Freezing/Melting point = −95°C; Relative vapor density (air = 1) = 2.91; Vapor pressure = 350 mmHg; Autoignition temperature = 556°C. Explosive limits: LEL = 13%; UEL = 23%. Hazard Identification (based on NFPA-704 M Rating System): Health 3, Flammability 1, Reactivity 0. Soluble in water; solubility = 2%.
Potential Exposure: Compound Description: Agricultural Chemical; Tumorigen, Drug, Mutagen; Reproductive

Effector; Human Data; Primary Irritant. Methylene chloride is used mainly as a low-temperature extractant of substances which are adversely affected by high temperature. It can be used as a solvent for oil, fats, waxes, bitumen, cellulose acetate, and esters. It is also used as a paint remover, as a degreaser, and in aerosol propellants.

Incompatibilities: Incompatible with strong oxidizers, caustics, chemically active metals, such as aluminum, magnesium powders, potassium, lithium, and sodium, and concentrated nitric acid causing fire and explosion hazard. Contact with hot surfaces or flames causes decomposition producing fumes of hydrogen chloride and phosgene gas. Attacks some forms of plastics, rubber, and coatings. Attacks metals in the presence of moisture.

Permissible Exposure Limits in Air

Conversion factor: 1 ppm = 3.47 mg/m^3 at 25°C & 1 atm.

OSHA PEL 25 ppm TWA; 125 ppm STEL, a potential occupational carcinogen, see 29CFR1910.1052; for Construction see 56FR57036.

NIOSH REL: A potential occupational carcinogen. Limit exposure to lowest feasible concentration. See *NIOSH Pocket Guide*, Appendix A.

ACGIH TLV®[1]: 50 ppm/174 mg/m^3 TWA; BEI: 0.2 mg [dichloromethane]/L in urine/end-of-shift; confirmed animal carcinogen with unknown relevance to humans.

NIOSH IDLH: potential occupational carcinogen, 2300 ppm.

Protective Action Criteria (PAC)*

TEEL-0: 25 ppm

PAC-1: **200** ppm

PAC-2: **560** ppm

PAC-3: **6900** ppm

*AEGLs (Acute Emergency Guideline Levels) & ERPGs (Emergency Response Planning Guideline) are in **bold face**.

DFG MAK: 100 ppm/350 mg/m^3; BAT: 5% [CO-Hb] in blood, at end-of-shift; 1 mg[dichloromethane]/L in blood/end-of-shift; Carcinogen Category 3A.

Australia: TWA 100 ppm (350 mg/m^3), carcinogen, 1993; Austria: MAK 100 ppm (360 mg/m^3), Suspected: carcinogen, 1999; Belgium: TWA 50 ppm (174 mg/m^3), carcinogen, 1993; Denmark: TWA 35 ppm (122 mg/m^3), [skin], 1999; Finland: TWA 100 ppm (350 mg/m^3); STEL 250 ppm (870 mg/m^3), 1999; France: VME 50 ppm (180 mg/m^3), VLE 100 ppm, continuous carcinogen, 1999; Hungary: STEL 10 mg/m^3, carcinogen, 1993; the Netherlands: MAC-TGG 350 mg/m^3, 2003; Norway: TWA 35 ppm (125 mg/m^3), 1999; the Philippines: TWA 500 ppm (1740 mg/m^3), 1993; Poland: MAC (TWA) 20 mg/m^3; STEL 50 mg/m^3, 1999; Russia: TWA 100 ppm; STEL 50 mg/m^3, 1993; Sweden: NGV 35 ppm (120 mg/m^3), KTV 70 ppm (250 mg/m^3), [skin] 1999; Switzerland: MAK-W 100 ppm (360 mg/m^3), KZG-W 500 ppm (1800 mg/m^3), 1999; Thailand: TWA 500 mg/m^3; STEL 1000 mg/m^3, 1993; Turkey: TWA 500 ppm (1740 mg/m^3), 1993; United Kingdom: TWA 100 ppm (350 mg/m^3); STEL 300 ppm, 2000; Argentina, Bulgaria, Columbia, Jordan, South Korea, New Zealand, Singapore, Vietnam: ACGIH TLV®: confirmed animal carcinogen with unknown relevance to humans; the Czech Republic: TWA 500 mg/m^3; Standards for methylene chloride in ambient air in residential areas have been set by Russia[43] at 8.8 mg/m^3 on a momentary basis and by the Czech Republic[35] at 3.0 mg/m^3 on a momentary basis and 1.0 mg/m^3 on a daily average basis. Several states have set guidelines or standards for methylene chloride in ambient air[60] ranging from 0.2 μg/m^3 (Rhode Island) to 2.4 μg/m^3 (Massachusetts and North Carolina) to 55.55 μg/m^3 (Kansas) to 1,167 μg/m^3 (New York) to 3500 μg/m^3 (South Dakota) to 3500−17,400 μg/m^3 (North Dakota) to 58,000 μg/m^3 (Virginia) to 7000 μg/m^3 (Connecticut) to 8333 μg/m^3 (Pennsylvania) to 8700 μg/m^3 (Indiana) to 8750 μg/m^3 (South Carolina).

Determination in Air: *Use* NIOSH Analytical Method #1005, Methylene chloride; #3800, Organic and Inorganic Gases, #2549, Volatile organic compounds; OSHA Analytical Methods 59 and 80.

Permissible Concentration in Water: *To protect human health:* preferably zero. An additional lifetime cancer risk of 1 in 100,000 results at a level of 1.9 μg/L.[6] A lifetime health advisory could not be calculated by EPA.[48] Several states have set standards and guidelines for methylene chloride in drinking water[61] ranging from standards of 2 μg/L (New Jersey) to 100 μg/L (New Mexico) and guidelines of 4.7 μg/L (Arizona) to 25 μg/L (Connecticut) to 40 μg/L (California) to 48 μg/L (Minnesota and Vermont) to 50 μg/L (Kansas) to 150 μg/L (Maine). Russia set a MAC[35] of 7.5 mg/L.

Determination in Water: Inert gas purge followed by gas chromatography with halide specific detection (EPA Method 601) or gas chromatography, plus mass spectrometry (EPA Method #624). Octanol−water coefficient: Log K_{ow} = 1.25

Routes of Entry: Inhalation of vapors, percutaneous absorption of liquid, ingestion, skin and/or eye contact.

Harmful Effects and Symptoms

Short Term Exposure: Irritates the eyes, skin, and respiratory tract. May affect the blood, causing the formation of methemoglobin and carboxyhemoglobin. Exposure can cause irregular heartbeat or cause heart to stop. This can cause death. *Inhalation:* Levels of 300−700 ppm for 3−5 hours has caused slight loss of muscle control and coordination. Effects of high concentrations include headaches, stupor, dizziness, fatigue, drunken behavior, chest pain, arm and leg pains, loss of feeling, loss of appetite, hot flashes and death. Higher exposures can cause pulmonary edema, a medical emergency that can be delayed for several hours. This can cause death. *Skin:* Contact is painful and highly irritating if confined on the skin by gloves or clothing. Absorbed slowly through the skin to cause symptoms listed under inhalation. *Eyes:* May cause pain, irritation, burns, and permanent damage. *Ingestion:* The liquid may cause chemical pneumonitis. Accidental ingestion of paint removers containing methylene chloride as the main ingredient have reportedly caused headache, nausea, vomiting, visual disturbance, presence of blood in the urine, and unconsciousness.

Note: Methylene chloride is changed to carbon monoxide in the human body. This is a particularly hazardous condition for those who have a history of heart trouble or those who are also exposed to carbon monoxide. See "Carbon monoxide" Fact Sheet. These persons should take extra precautions.

Long Term Exposure: Repeated or prolonged skin contact may cause drying and cracking. May be carcinogenic to humans. It has been shown to cause liver and lung cancer in animals. May affect the central nervous system and liver, causing degenerative brain disease and enlargement of the liver. See symptoms as above. Prolonged exposure can cause changes in blood, hallucinations, and decreased response to visual and auditory stimulation. Most of the effects will disappear after exposure stops. Methylene chloride caused genetic effects in certain bacteria and caused birth defects in chickens. In laboratory studies, methylene chloride has also been shown to cause tumors in mice and rats. Whether methylene chloride causes birth defects or tumors in humans is not known.

Points of Attack: Eyes, skin, cardiovascular system, central nervous system. Cancer site in animals: lung, liver, salivary and mammary gland.

Medical Surveillance: OSHA mandates the following tests: laboratory surveillance; carboxyhemoglobin; electrocardiogram (resting); hematocrit; liver function tests; cholesterol level. NIOSH lists the following tests: whole blood (chemical/metabolite); whole blood (chemical/metabolite), carboxyhemoglobin; whole blood (chemical/metabolite), carboxyhemoglobin, end-of-shift; whole blood (chemical/metabolite), carboxyhemoglobin, prior to next shift; whole blood (chemical/metabolite), end-of-shift; complete blood count; expired air, expired air, end-of-shift; liver function tests; urine (chemical/metabolite). If symptoms develop or overexposure is suspected, the following may be useful: Special 24-h EKG (Holter monitor) to look for irregular heartbeat. Blood carboxyhemoglobin (this must be done within a few hours after exposure). Consider chest X-ray after acute overexposure.

First Aid: If this chemical gets into the eyes, remove any contact lenses at once and irrigate immediately for at least 15 min, occasionally lifting upper and lower lids. Seek medical attention immediately. If this chemical contacts the skin, remove contaminated clothing and wash immediately with soap and water. Seek medical attention immediately. If this chemical has been inhaled, remove from exposure, begin rescue breathing (using universal precautions, including resuscitation mask) if breathing has stopped and CPR if heart action has stopped. Transfer promptly to a medical facility. When this chemical has been swallowed, get medical attention. Give large quantities of water and induce vomiting. Do not make an unconscious person vomit. Medical observation is recommended for 24—48 h after breathing overexposure, as

pulmonary edema may be delayed. As first aid for pulmonary edema, a doctor or authorized paramedic may consider administering a corticosteroid spray.

Personal Protective Methods: Wear protective gloves and clothing to prevent any reasonable probability of skin contact: **8 h:** polyvinyl alcohol gloves; 4H™ and Silver Shield™ gloves, Responder™ suits, Trellchem™ HPS suits; Trychem 1000™ suits. **4 h:** Teflon™ gloves, suits, boots; Barricade™ coated suits 4,4′-Methylenedianiline 101-77-9 Prevent skin contact. **8 h:** 4H™ and Silver Shield™ gloves. Safety equipment suppliers/manufacturers can provide recommendations on the most protective glove/clothing material for your operation. All protective clothing (suits, gloves, footwear, headgear) should be clean, available each day, and put on before work. Contact lenses should not be worn when working with this chemical. Wear splash-proof chemical goggles and face shield unless full face-piece respiratory protection is worn. Employees should wash immediately with soap when skin is wet or contaminated. Provide emergency showers and eyewash.

Respirator Selection: NIOSH: *At any concentrations above the NIOSH REL, or where there is no REL, at any detectable concentration:* SCBAF: Pd,Pp (APF = 10,000) (any self-contained breathing apparatus that has a full face-piece and is operated in a pressure-demand or other positive-pressure mode); or SaF: Pd,Pp: ASCBA (APF = 10,000) (any supplied-air respirator that has a full face-piece and is operated in a pressure-demand or other positive-pressure mode in combination with an auxiliary self-contained breathing apparatus operated in a pressure-demand or other positive-pressure mode). *Escape:* GmFOv (APF = 50) [any air-purifying, full-face-piece respirator (gas mask) with a chin-style, front- or back-mounted organic vapor canister] or SCBAE (any appropriate escape-type, self-contained breathing apparatus).

Storage: Color Code—Blue: Health Hazard/Poison: Store in a secure poison location. Prior to working with this chemical you should be trained on its proper handling and storage. Before entering confined space where this chemical may be present, check to make sure that an explosive concentration does not exist. Methylene chloride must be stored to avoid contact with strong oxidizers (such as perchlorates, peroxides, chlorates, nitrates, or permanganates), strong caustics, and chemically active metals (such as aluminum, magnesium powder, sodium, potassium, or lithium) because violent reactions occur. Store in tightly closed containers in a cool, well-ventilated area away from heat and moisture. A regulated, marked area should be established where this chemical is handled, used, or stored in compliance with OSHA Standard 1910.1045.

Shipping: This compound requires a shipping label of "POISONOUS/TOXIC MATERIALS." It falls in Hazard Class 6.1 and Packing Group III.

Spill Handling: Evacuate and restrict persons not wearing protective equipment from area of spill or leak until cleanup is complete. Remove all ignition sources. Establish forced ventilation to keep levels below explosive limit. Ventilate area of spill or leak. Absorb liquids in vermiculite, dry sand, earth, peat, carbon, or a similar material and deposit in sealed containers. Keep this chemical out of a confined space, such as a sewer, because of the possibility of an explosion, unless the sewer is designed to prevent the buildup of explosive concentrations. It may be necessary to contain and dispose of this chemical as a hazardous waste. If material or contaminated runoff enters waterways, notify downstream users of potentially contaminated waters. Contact your local or federal environmental protection agency for specific recommendations. If employees are required to clean up spills, they must be properly trained and equipped. OSHA 1910.120(q) may be applicable.

Fire Extinguishing: This chemical is not combustible, but may form a flammable mixture with air. Use an extinguishing agent suitable for surrounding fire. Poisonous gases, including carbon monoxide, hydrogen chloride, and phosgene, are produced in fire. Vapors are heavier than air and will collect in low areas. Vapors may travel long distances to ignition sources and flashback. Vapors in confined areas may explode when exposed to fire. Containers may explode in fire. Storage containers and parts of containers may rocket great distances, in many directions. If material or contaminated runoff enters waterways, notify downstream users of potentially contaminated waters. Notify local health and fire officials and pollution control agencies. From a secure, explosion-proof location, use water spray to cool exposed containers. If cooling streams are ineffective (venting sound increases in volume and pitch, tank discolors, or shows any signs of deforming), withdraw immediately to a secure position. If employees are expected to fight fires, they must be trained and equipped in OSHA 1910.156. The only respirators recommended for firefighting are self-contained breathing apparatuses that have full face-pieces and are operated in a pressure-demand or other positive-pressure mode.

Disposal Method Suggested: Consult with environmental regulatory agencies for guidance on acceptable disposal practices. Generators of waste containing this contaminant ($\geq$100 kg/mo) must conform with EPA regulations governing storage, transportation, treatment, and waste disposal. Incineration, preferably after mixing with another combustible fuel; care must be exercised to assure complete combustion to prevent the formation of phosgene; an acid scrubber is necessary to remove the halo acids produced.

References

National Institute for Occupational Safety and Health. (1976). *Criteria for a Recommended Standard: Occupational Exposure to Methylene Chloride.* NIOSH Document No. 76-138

US Environmental Protection Agency. (1980). *Halomethanes: Ambient Water Quality Criteria.* Washington, DC

US Environmental Protection Agency. (April 30, 1980). *Dichloromethane, Health and Environmental Effects Profile No. 74.* Washington, DC: Office of Solid Waste

Sax, N. I. (Ed.). *Dangerous Properties of Industrial Materials Report,* 1, No. 2, 45–47 (1980) and 6, No. 5, 51–52 (1986)

US Public Health Service. (December 1987). *Toxicological Profile for Methylene Chloride.* Atlanta, GA: Agency for Toxic Substances and Disease Registry

New York State Department of Health. (March 1986). *Chemical Fact Sheet: Methylene Chloride.* Version 2. Albany, NY: Bureau of Toxic Substance Assessment

US Environmental Protection Agency, Special Review and Reregistration Division Office of Pesticide Programs. (1998). *Agency Status of Pesticides in Registration, Reregistration, and Special Review* (Rainbow Report). Washington, DC

New Jersey Department of Health and Senior Services. (May 2001). *Hazardous Substances Fact Sheet: Methylene Chloride.* Trenton, NJ

Methyl ethyl ether   M:0910

Molecular Formula: C_3H_8O
Synonyms: Ether, ethyl methyl; Ethane, methoxy-; Ethyl methyl ether; Methane, ethoxy; Methoxyethane
CAS Registry Number: 540-67-0
RTECS® Number: KO0260000
UN/NA & ERG Number: UN1039/115
EC Number: None found.
Regulatory Authority and Advisory Bodies
WGK (German Aquatic Hazard Class): No value assigned.
Description: Methyl Ethyl Ether is a colorless liquid or gas at room temperature. Molecular weight = 60.11; Boiling point = 11°C; Flash point = −37°C; Autoignition temperature = 190°C. Explosive limits: LEL = 2.0%; UEL = 10.1%; Hazard Identification (based on NFPA-704 M Rating System): Health 1, Flammability 4, Reactivity 1. Soluble in water.
Potential Exposure: Used as a medicine and as an anesthetic.
Incompatibilities: Violent reaction with strong oxidizers, sulfuric and nitric acids. May be able to form explosive peroxides on standing.
Permissible Exposure Limits in Air
No standards or TEEL available.
Routes of Entry: Inhalation.
Harmful Effects and Symptoms
Short Term Exposure: Contact irritates the skin, eyes, and nose. Inhalation can cause lightheadedness and can reduce concentration. Higher levels of exposure may cause unconsciousness and even death.
Long Term Exposure: Repeated contact may cause skin dryness and cracking. Prolonged high exposure may affect the brain.
Points of Attack: Brain, skin.
Medical Surveillance: Evaluate for brain effect, such as memory, concentration, sleeping patterns and mood,

headache, and fatigue. Positive and borderline individuals should be referred for neuropsychological testing.

First Aid: If this chemical gets into the eyes, remove any contact lenses at once and irrigate immediately for at least 15 min, occasionally lifting upper and lower lids. Seek medical attention immediately. If this chemical contacts the skin, remove contaminated clothing and wash immediately with soap and water. Seek medical attention immediately. If this chemical has been inhaled, remove from exposure, begin rescue breathing (using universal precautions, including resuscitation mask) if breathing has stopped and CPR if heart action has stopped. Transfer promptly to a medical facility. When this chemical has been swallowed, get medical attention. Give large quantities of water and induce vomiting. Do not make an unconscious person vomit.

Personal Protective Methods: Wear protective gloves and clothing to prevent any reasonable probability of skin contact. Safety equipment suppliers/manufacturers can provide recommendations on the most protective glove/clothing material for your operation. All protective clothing (suits, gloves, footwear, headgear) should be clean, available each day, and put on before work. Contact lenses should not be worn when working with this chemical. Wear splash-proof chemical goggles and face shield unless full face-piece respiratory protection is worn. Employees should wash immediately with soap when skin is wet or contaminated. Provide emergency showers and eyewash.

Respirator Selection: Follow regulations in OSHA 29CFR1910.134 or European Standard EN149. Use a NIOSH/MSHA- or European Standard EN149-approved respirator; or use an approved supplied-air respirator with a full face-piece operated in the positive-pressure mode, or with a full face-piece, hood, or helmet in the continuous-flow mode; or use a NIOSH/MSHA- or European Standard EN149-approved self-contained breathing apparatus with a full face-piece operated in a pressure-demand or other positive-pressure mode.

Storage: Color Code—Red Stripe: Flammability Hazard: Store separately from all other flammable materials. Prior to working with methyl ethyl ether you should be trained on its proper handling and storage. Before entering confined space where this chemical may be present, check to make sure that an explosive concentration does not exist. Store in tightly closed containers in a cool, well-ventilated area away from oxidizers, strong acids. Where possible, automatically pump liquid from drums or other storage containers to process containers. Drums must be equipped with self-closing valves, pressure vacuum bungs, and flame arresters. Use only nonsparking tools and equipment, especially when opening and closing containers of this chemical. Wherever this chemical is used, handled, manufactured, or stored, use explosion-proof electrical equipment and fittings.

Shipping: Ethyl methyl ether requires a shipping label of "FLAMMABLE GAS." It falls in Hazard Class 2.1.

Spill Handling: *Liquid:* Evacuate and restrict persons not wearing protective equipment from area of spill or leak until cleanup is complete. Remove all ignition sources. Establish forced ventilation to keep levels below explosive limit. Absorb liquids in vermiculite, dry sand, earth, peat, carbon, or a similar material and deposit in sealed containers. Keep this chemical out of a confined space, such as a sewer, because of the possibility of an explosion, unless the sewer is designed to prevent the buildup of explosive concentrations. It may be necessary to contain and dispose of this chemical as a hazardous waste. If material or contaminated runoff enters waterways, notify downstream users of potentially contaminated waters. Contact your local or federal environmental protection agency for specific recommendations. If employees are required to clean up spills, they must be properly trained and equipped. OSHA 1910.120(q) may be applicable.

Gas: Evacuate and restrict persons not wearing protective equipment from area of spill or leak until cleanup is complete. Remove all ignition sources. Establish forced ventilation to keep levels below explosive limit and to disperse the gas. Stop the flow of gas if it can be done safely. If source of leak is a cylinder and the leak cannot be stopped in place, remove leaking cylinder to a safe place in the open air, and repair leak or allow cylinder to empty. Keep this chemical out of confined space, such as a sewer, because of the possibility of explosion, unless the sewer is designed to prevent the buildup of explosive concentrations. It may be necessary to contain and dispose of this chemical as a hazardous waste. Contact your local or federal environmental protection agency for specific recommendations. If employees are required to clean up spills, they must be properly trained and equipped. OSHA 1910.120(q) may be applicable.

Fire Extinguishing: *Liquid:* This chemical is a combustible liquid. Poisonous gases are produced in fire. Use alcohol foam extinguishers. Water may not be effective. Vapors are heavier than air and will collect in low areas. Vapors may travel long distances to ignition sources and flashback. Vapors in confined areas may explode when exposed to fire. Containers may explode in fire. Storage containers and parts of containers may rocket great distances, in many directions. If material or contaminated runoff enters waterways, notify downstream users of potentially contaminated waters. Notify local health and fire officials and pollution control agencies. From a secure, explosion-proof location, use water spray to cool exposed containers. If cooling streams are ineffective (venting sound increases in volume and pitch, tank discolors, or shows any signs of deforming), withdraw immediately to a secure position. If employees are expected to fight fires, they must be trained and equipped in OSHA 1910.156. The only respirators recommended for firefighting are self-contained breathing apparatuses that have full face-pieces and are operated in a pressure-demand or other positive-pressure mode.

Gas: Poisonous gases are produced in fire. Do not extinguish the fire unless the flow of gas can be stopped and any remaining gas is out of the line. Specially trained personnel may use fog lines to cool exposures and let the fire burn itself out. Vapors are heavier than air and will collect in low areas. Vapors may travel long distances to ignition sources and flashback. Vapors in confined areas may explode when exposed to fire. Containers may explode in fire. Storage containers and parts of containers may rocket great distances, in many directions. If material or contaminated runoff enters waterways, notify downstream users of potentially contaminated waters. Notify local health and fire officials and pollution control agencies. From a secure, explosion-proof location, use water spray to cool exposed containers. If cooling streams are ineffective (venting sound increases in volume and pitch, tank discolors, or shows any signs of deforming), withdraw immediately to a secure position. If cylinders are exposed to excessive heat from fire or flame contact, withdraw immediately to a secure location. If employees are expected to fight fires, they must be trained and equipped in OSHA 1910.156. The only respirators recommended for firefighting are self-contained breathing apparatuses that have full face-pieces and are operated in a pressure-demand or other positive-pressure mode.

Disposal Method Suggested: Dissolve or mix the material with a combustible solvent and burn in a chemical incinerator equipped with an afterburner and scrubber. All federal, state, and local environmental regulations must be observed.

Reference

New Jersey Department of Health and Senior Services. (November 1999). *Hazardous Substances Fact Sheet: Methyl Ethyl Ether.* Trenton, NJ

Methyl ethyl ketone   M:0920

Molecular Formula: C_4H_8O
Common Formula: $CH_3COCH_2CH_3$
Synonyms: Acetone, methyl-; Aethylmethylketon (German); Butanone; 2-Butanone; Butanone 2 (French); Ethyl methyl cetone (French); Ethyl methyl ketone; MEK; Methyl acetone; Methyl ketone; Ketone, ethyl methyl; Meetco; Metil etil cetona (Spanish)
CAS Registry Number: 78-93-3
RTECS® Number: EL6475000
UN/NA & ERG Number: UN1193/127
EC Number: 201-159-0 [*Annex I Index No.:* 606-002-00-3]
Regulatory Authority and Advisory Bodies
Carcinogenicity: EPA: Available data are inadequate for an assessment of human carcinogenic potential.; US EPA Gene-Tox Program, Inconclusive: *B. subtilis* rec assay.
US EPA, FIFRA 1998 Status of Pesticides: Canceled.
Air Pollutant Standard Set. See below, "Permissible Exposure Limits in Air" section.
Clean Air Act: Hazardous Air Pollutants (Title I, Part A, Section 112).
US EPA Hazardous Waste Number (RCRA No.): U159.
RCRA, 40CFR261, Appendix 8 Hazardous Constituents.

RCRA Toxicity Characteristic (Section 261.24), Maximum. Concentration of Contaminants, regulatory level, 200.0 mg/L.
RCRA 40CFR268.48; 61FR15654, Universal Treatment Standards: Wastewater (mg/L), 0.28; Nonwastewater (mg/kg), 36.
RCRA 40CFR264, Appendix 9; TSD Facilities Ground Water Monitoring List. Suggested test method(s) (PQL μg/L): 8015 (10); 8240 (100).
Safe Drinking Water Act: Priority List (55 FR 1470).
Reportable Quantity (RQ): 5000 lb (2270 kg).
EPCRA Section 313: Removed from the TRI list 2004 as a result of a court decision.
Canada, WHMIS, Ingredients Disclosure List Concentration 1.0%.
European/International Regulations: Hazard Symbol: F, Xi; Risk phrases: R11; R36; R66; R67; Safety phrases: S2; S9; S16 (see Appendix 4).
WGK (German Aquatic Hazard Class): 1—Low hazard to waters.
Description: MEK is a clear, colorless liquid with a fragrant, mint-like, moderately sharp odor. The odor threshold in air is 5.4 ppm. Molecular weight = 72.12; Specific gravity $(H_2O:1) = 0.81$; Boiling point = 79.4°C; Freezing/Melting point = −86.1°C; Vapor pressure = 78 mmHg at 20°C; Flash point = −9°C (cc); Autoignition temperature = 505°C. Explosive limits: LEL = 1.4% at 93°C; UEL = 11.4% at 93°C. Hazard Identification (based on NFPA-704 M Rating System): Health 1, Flammability 3, Reactivity 0. Soluble in water; solubility = 28%.
Potential Exposure: Compound Description: Mutagen; Reproductive Effector; Human Data; Hormone, Primary Irritant. MEK is used as a solvent in nitrocellulose coating and vinyl film manufacture; in smokeless powder manufacture; in cements and adhesives; and in the dewaxing of lubricating oils. It is also an intermediate in drug manufacture.
Incompatibilities: Forms explosive mixture with air. Violent reaction with strong oxidizers, amines, ammonia, inorganic acids, caustics, isocyanates, pyridines. Incompatible with potassium *tert*-butoxide, 2-propanol, chlorosulfonic acid, oleum. Attacks some plastics.
Permissible Exposure Limits in Air
Conversion factor: 1 ppm = 2.95 mg/m³ at 25°C & 1 atm.
OSHA PEL: 200 ppm/590 mg/m³ TWA.
NIOSH REL: 200 ppm/590 mg/m³ TWA; 300 ppm/885 mg/m³ STEL.
ACGIH TLV®[1]: 200 ppm/590 mg/m³ TWA; 300 ppm/885 mg/m³ STEL.
NIOSH IDLH: 3000 ppm.
Protective Action Criteria (PAC)
TEEL-0: 200 ppm.
PAC-1: 200 ppm.
PAC-2: 2700 ppm.
PAC-3: 4000 ppm.
DFG MAK: 200 ppm/590 mg/m³ [skin]; Peak Limitation Category I(1); Pregnancy Risk Group C; BAT: 5 mg/m³ in urine, at end-of-shift.

Australia: TWA 150 ppm (445 mg/m^3); STEL 300 ppm, 1993; Austria: MAK 200 ppm (590 mg/m^3), 1999; Belgium: TWA 200 ppm (590 mg/m^3); STEL 300 ppm (885 mg/m^3), 1993; Denmark: TWA 50 ppm (145 mg/m^3), [skin], 1999; Finland: STEL 100 ppm, 1993; France: VME 200 ppm (600 mg/m^3), 1999; Hungary: TWA 200 mg/m^3; STEL 600 mg/m^3, 1993; India: TWA 200 ppm (590 mg/m^3); STEL 300 ppm (885 mg/m^3), 1993; the Netherlands: MAC-TGG 590 mg/m^3, [skin], 2003; Norway: TWA 75 ppm (220 mg/m^3), 1999; the Philippines: TWA 200 ppm (590 mg/m^3), 1993; Poland: MAC (TWA) 200 mg/m^3, MAC (STEL) 850 mg/m^3, 1999; Russia: TWA 200 ppm; STEL 200 mg/m^3, 1993; Sweden: NGV 50 ppm (150 mg/m^3), KTV 100 ppm (300 mg/m^3), 1999; Switzerland: MAK-W 200 ppm (590 mg/m^3), KZG-V 400 ppm (1180 mg/m^3), 1999; Turkey: TWA 200 ppm (590 mg/m^3), 1993; United Kingdom: TWA 200 ppm (600 mg/m^3), STE 300 ppm, [skin], 2000; Argentina, Bulgaria, Columbia, Jordan, South Korea, New Zealand, Singapore, Vietnam: ACGIH TLV®: STEL 300 ppm. The Czech Republic: (ambient air in residential areas) have been set[35] at 0.3 mg/m^3 on a momentary and a daily average basis; Russia (in ambient air in residential areas): 0.1 mg/m^3 on a momentary basis. Several states have set guidelines or standards for methyl ethyl ketone in ambient air[60] ranging from 0.16 mg/m^3 (Massachusetts) to 1.967 mg/m^3 (New York) to 3.7–88.5 mg/m^3 (North Carolina) to 5.9 mg/m^3 (Florida) to 5.9–8.85 mg/m^3 (North Dakota) to 9.8 mg/m^3 (Virginia) to 11.8 mg/m^3 (Connecticut and South Dakota) to 14.048 mg/m^3 (Nevada) to 14.75 mg/m^3 (South Carolina).

Determination in Air: Use NIOSH Analytical Method #2500, Methyl ethyl ketone; #2555; #3800; #8002 in blood; #2549 Volatile organic compounds; OSHA Analytical Methods 16, 84, and 1004.

Permissible Concentration in Water: Russia[43] set a MAC of 1.0 mg/L in water bodies used for domestic purposes. The EPA[48] has set a lifetime health advisory at 170 μg/L. Several states have set guidelines for methyl ethyl ketone in drinking water[61] ranging from 60 μg/L (Massachusetts) to 170 μg/L (Arizona) to 172 μg/L (Minnesota) to 270 μg/L (New Jersey)[59] to 750 μg/L (Maine) to 1000 μg/L (Connecticut).

Determination in Water: Octanol–water coefficient: Log K_{ow} = 0.29.

Routes of Entry: Inhalation, ingestion, eye and/or skin contact.

Harmful Effects and Symptoms

Short Term Exposure: Irritates the eyes and the respiratory tract. May affect the central nervous system. *Inhalation:* Human exposures to levels of 350 ppm caused irritation of the nose and throat. Exposure can cause dizziness, lightheadedness, headache, nausea, blurred vision. Numbness in fingers, arms, and legs, accompanied by headache, nausea, vomiting, and fainting have occurred after exposure to levels of 300–600 ppm. *Skin:* Contact can irritate the skin causing rash and burning feeling. Liquid is absorbed readily and may cause numbing of fingers and arms. *Eyes:* Contact can irritate and cause burns and permanent damage. Exposure to levels of 200 ppm produced irritation. *Ingestion:* Can cause irritation of the mouth, throat, and stomach, the severity of which will be dependent upon amount swallowed. Symptoms of poisoning include nausea, vomiting, stomach pain, and diarrhea. Death can occur from ingestion of as little as 1 oz.

Long Term Exposure: Repeated exposure can cause drying and cracking of the skin. Has been implicated in certain nervous system and brain disorders characterized by weakness, fatigue, sleep disturbances, reduced coordination, heaviness in chest, and numbness of hands and feet. These symptoms may develop after 1 year of exposure to vapor concentrations of 50–200 ppm. Improvement is gradual and may take years after exposure is discontinued. Animal tests show that this chemical is a teratogen in animals and possibly causes toxic effects upon human reproduction.

Points of Attack: Eyes, skin, respiratory system, central nervous system.

Medical Surveillance: NIOSH lists the following tests: whole blood (chemical/metabolite), expired air, urine (chemical/metabolite); urine (chemical/metabolite), end-of-shift. If symptoms develop, or overexposure is suspected, examination of the nervous system is recommended. Special tests for nerve damage, called nerve conduction studies, may be useful.

First Aid: If this chemical gets into the eyes, remove any contact lenses at once and irrigate immediately for at least 15 min, occasionally lifting upper and lower lids. Seek medical attention immediately. If this chemical contacts the skin, remove contaminated clothing and wash immediately with soap and water. Seek medical attention immediately. If this chemical has been inhaled, remove from exposure, begin rescue breathing (using universal precautions, including resuscitation mask) if breathing has stopped and CPR if heart action has stopped. Transfer promptly to a medical facility. When this chemical has been swallowed, get medical attention. Give large quantities of water and induce vomiting. Do not make an unconscious person vomit.

Personal Protective Methods: Wear protective gloves and clothing to prevent any reasonable probability of skin contact. Safety equipment suppliers/manufacturers can provide recommendations on the most protective glove/clothing material for your operation. Butyl rubber and chlorobutyl rubber are among the recommended protective materials. All protective clothing (suits, gloves, footwear, headgear) should be clean, available each day, and put on before work. Contact lenses should not be worn when working with this chemical. Wear splash-proof chemical goggles and face shield unless full face-piece respiratory protection is worn. Employees should wash immediately with soap when skin is wet or contaminated. Provide emergency showers and eyewash. See NIOSH Criteria Document 78-173, *Ketones.*

Respirator Selection: *Up to 3000 ppm:* Sa:Cf (APF = 25) (any supplied-air respirator operated in a continuous-flow mode) or PaprOv (APF = 25) [any powered, air-purifying respirator with organic vapor cartridge(s)] or CcrOv (APF = 10) [any chemical cartridge respirator with a full face-piece and organic vapor cartridge(s)] or GmFOv (APF = 50) [any air-purifying, full-face-piece respirator (gas mask) with a chin-style, front- or back-mounted organic vapor canister] or SCBAF (APF = 50) (any self-contained breathing apparatus with a full face-piece) or SaF (APF = 50) (any supplied-air respirator with a full face-piece). *Emergency or planned entry into unknown concentrations or IDLH conditions: NIOSH:* SCBAF: Pd,Pp (APF = 10,000) (any NIOSH/MSHA- or European Standard EN 149-approved self-contained breathing apparatus that has a full face-piece and is operated in a pressure-demand or other positive-pressure mode) or SaF: Pd,Pp: ASCBA (APF = 10,000) (any supplied-air respirator that has a full face-piece and is operated in a pressure-demand or other positive-pressure mode in combination with an auxiliary self-contained breathing apparatus operated in a pressure-demand or other positive-pressure mode). *Escape:* GmFOv (APF = 50) [any air-purifying, full-face-piece respirator (gas mask) with a chin-style, front- or back-mounted organic vapor canister] or SCBAE (any appropriate escape-type, self-contained breathing apparatus).

Storage: Color Code—Red: Flammability Hazard: Store in a flammable liquid storage area or approved cabinet away from ignition sources and corrosive and reactive materials. Prior to working with this chemical you should be trained on its proper handling and storage. Before entering confined space where this chemical may be present, check to make sure that an explosive concentration does not exist. Methyl ethyl ketone must be stored to avoid contact with strong oxidizers (such as chlorine, bromine, and fluorine) since violent reactions occur. Store in tightly closed containers in a cool, well-ventilated area away from heat, sparks, or flame. Sources of ignition, such as smoking and open flames, are prohibited where methyl ethyl ketone is used, handled, or stored in a manner that could create a potential fire or explosion hazard. Metal containers involving the transfer of 5 gallons or more of methyl ethyl ketone should be grounded and bonded. Drums must be equipped with self-closing valves, pressure vacuum bungs, and flame arresters. Use only nonsparking tools and equipment, especially when opening and closing containers of methyl ethyl ketone.

Shipping: This compound requires a shipping label: "FLAMMABLE LIQUID." It falls in Hazard Class 3 and Packing Group II.

Spill Handling: Evacuate and restrict persons not wearing protective equipment from area of spill or leak until cleanup is complete. Remove all ignition sources. Establish forced ventilation to keep levels below explosive limit. Absorb liquids in vermiculite, dry sand, earth, peat, carbon, or a similar material and deposit in sealed containers. Keep this chemical out of a confined space, such as a sewer, because of the possibility of an explosion, unless the sewer is designed to prevent the buildup of explosive concentrations. It may be necessary to contain and dispose of this chemical as a hazardous waste. If material or contaminated runoff enters waterways, notify downstream users of potentially contaminated waters. Contact your local or federal environmental protection agency for specific recommendations. If employees are required to clean up spills, they must be properly trained and equipped. OSHA 1910.120(q) may be applicable.

Fire Extinguishing: This chemical is a flammable liquid. Poisonous gases are produced in fire. Use dry chemical, carbon dioxide, or alcohol foam extinguishers. Vapors are heavier than air and will collect in low areas. Vapors may travel long distances to ignition sources and flashback. Vapors in confined areas may explode when exposed to fire. Containers may explode in fire. Storage containers and parts of containers may rocket great distances, in many directions. If material or contaminated runoff enters waterways, notify downstream users of potentially contaminated waters. Notify local health and fire officials and pollution control agencies. From a secure, explosion-proof location, use water spray to cool exposed containers. If cooling streams are ineffective (venting sound increases in volume and pitch, tank discolors, or shows any signs of deforming), withdraw immediately to a secure position. If employees are expected to fight fires, they must be trained and equipped in OSHA 1910.156. The only respirators recommended for firefighting are self-contained breathing apparatuses that have full face-pieces and are operated in a pressure-demand or other positive-pressure mode.

Disposal Method Suggested: Consult with environmental regulatory agencies for guidance on acceptable disposal practices. Generators of waste containing this contaminant (≥100 kg/mo) must conform with EPA regulations governing storage, transportation, treatment, and waste disposal. Incineration.[22]

References

National Institute for Occupational Safety and Health. (1978). *Criteria for a Recommended Standard: Occupational Exposure to Ketones*, NIOSH Document No. 78-173. Washington, DC

US Environmental Protection Agency, Methyl Ethyl Ketone. (April 30, 1980). *Health and Environmental Effects Profile No. 128*. Washington, DC: Office of Solid Waste

Sax, N. I. (Ed.). (1981). *Dangerous Properties of Industrial Materials Report*, 1, No. 4, 85−87

New York State Department of Health. (March 1986). *Chemical Fact Sheet: 2-Butanone*. Albany, NY: Bureau of Toxic Substance Assessment

US Environmental Protection Agency, Special Review and Reregistration Division Office of Pesticide Programs. (1998). *Agency Status of Pesticides in Registration, Reregistration, and Special Review* (Rainbow Report). Washington, DC

New Jersey Department of Health and Senior Services. (August 2002). *Hazardous Substances Fact Sheet: Methyl Ethyl Ketone*. Trenton, NJ

Methyl ethyl ketoneperoxide   M:0930

Molecular Formula: $C_8H_{16}O_4$

Synonyms: 2-Butanone, peroxide; Butanox®; Hi-Point 90; Ketonox; Luperox®; Lupersol; MEKP; MEK peroxide; Methyl ethyl ketone hydroperoxide; NCI-C55447; Peroxido de metil etil cetona (Spanish); Quickset extra; Sprayset MEKP; the rmacure

CAS Registry Number: 1338-23-4

RTECS® Number: EL9450000

UN/NA & ERG Number: UN3101/146; UN3105/145; UN3107/145

EC Number: 215-661-2

Regulatory Authority and Advisory Bodies

Highly Reactive Substance and Explosive (World Bank).[15] Air Pollutant Standard Set. See below, "Permissible Exposure Limits in Air" section.

OSHA 29CFR1910.119, Appendix A, Process Safety List of Highly Hazardous Chemicals, TQ = 5000 lb (2270 kg).

US EPA Hazardous Waste Number (RCRA No.): U160.

RCRA, 40CFR261, Appendix 8 Hazardous Constituents.

Reportable Quantity (RQ): 10 lb (4.54 kg).

Canada, WHMIS, Ingredients Disclosure List Concentration 1.0%.

European/International Regulations: not listed in Annex 1.

WGK (German Aquatic Hazard Class): 1—Low hazard to waters.

Description: MEKP, an organic peroxide, is a colorless liquid. Molecular weight = 176.24; Specific gravity $(H_2O:1) = 1.12$ at 15°C; Boiling point = (decomposes) 117.8°C; Freezing/Melting point = about 60°C. Flash point = 52−93°C (oc) 60% MEKP.[52] Hazard Identification (based on NFPA-704 M Rating System): Health 3, Flammability 2, Reactivity 2. Insoluble in water.

Potential Exposure: Compound Description: Tumorigen, Human Data; Primary Irritant. MEKP is used as a curing agent for thermosetting polyester resins and as a crosslinking agent and catalyst in the production of other polymers.

Incompatibilities: Forms explosive mixture with air (flash point varies). MEKP may exist in several different structures; decomposition temperatures may vary. Pure substance is shock-sensitive. Explosive decomposition above 176°F/80°C (also reported at 230°F/110°C). Keep away from sources of ignition, heat, sunlight. A strong oxidizer. Violent reaction with strong acids, strong bases, reducing agents, combustible substances, organic materials, chemical accelerants, oxides of heavy metals, salts, trace contaminants, amines. May accumulate static electrical charges and cause ignition of its vapors. Commercial product is diluted with 40% dimethyl phthalate, cyclohexane peroxide, or diallyl phthalate to reduce sensitivity to shock.

Permissible Exposure Limits in Air

Conversion factor: 1 ppm = 7.21 mg/m^3 at 25°C & 1 atm.

OSHA PEL: None.

NIOSH REL: 0.2 ppm/1.5 mg/m^3 Ceiling Concentration.

ACGIH TLV®[1]: 0.2 ppm/1.5 mg/m^3 Ceiling Concentration.

Protective Action Criteria (PAC)

TEEL-0: 1 ppm

PAC-1: 3 ppm

PAC-2: 20 ppm

PAC-3: 20 ppm

DFG MAK: Produces very severe skin effects. See Section X(a).

Australia: TWA 0.2 ppm (1.5 mg/m^3), 1993; Belgium: STEL 0.2 ppm (1.5 mg/m^3), 1993; Denmark: TWA 1 mg/m^3, 1999; Finland: STEL 0.2 ppm (1.5 mg/m^3), [skin], 1999; France: VLE 0.2 ppm (1.5 mg/m^3), 1999; Norway: TWA 1 mg/m^3, 1999; the Netherlands: MAC 1.5 mg/m^3, 2003; Switzerland: MAK-W 0.2 ppm (1.5 mg/m^3), 1999; United Kingdom: STEL 0.2 ppm (1.5 mg/m^3), 2000; Argentina, Bulgaria, Columbia, Jordan, South Korea, New Zealand, Singapore, Vietnam: ACGIH TLV®: Ceiling Concentration 0.2 ppm. Several states have set guidelines or standards for MEK peroxide in ambient air[60] ranging from 11.0 μg/m^3 (Virginia) to 15.0 μg/m^3 (North Dakota) to 30.0 μg/m^3 (Connecticut and South Dakota) to 36.0 μg/m^3 (Nevada).

Determination in Air: Use NIOSH Analytical Method (IV) #3508; OSHA Analytical Method 77.

Routes of Entry: Inhalation, ingestion, skin and/or eye contact.

Harmful Effects and Symptoms

Short Term Exposure: Methyl ethyl ketone peroxide can affect you when breathed in and by passing through your skin. Contact can irritate the skin, and can cause burn and permanently damage the eyes. Exposure can irritate the nose and throat. Higher exposures can cause pulmonary edema, a medical emergency that can be delayed for several hours. This can cause death.

Long Term Exposure: Methyl ethyl ketone peroxide may affect the liver and kidneys. May cause lung irritation and bronchitis. May cause skin allergy.

Points of Attack: Eyes, skin, respiratory system, liver, kidneys.

Medical Surveillance: For those with frequent or potentially high exposure (half the TLV or greater) the following are recommended before beginning work and at regular times after that: lung function tests. If symptoms develop or overexposure is suspected, the following may be useful: liver and kidney function test. Evaluation by a qualified allergist.

First Aid: If this chemical gets into the eyes, remove any contact lenses at once and irrigate immediately for at least 15 min, occasionally lifting upper and lower lids. Seek medical attention immediately. If this chemical contacts the skin, remove contaminated clothing and wash immediately with soap and water. Seek medical attention immediately.

If this chemical has been inhaled, remove from exposure, begin rescue breathing (using universal precautions, including resuscitation mask) if breathing has stopped and CPR if heart action has stopped. Transfer promptly to a medical facility. When this chemical has been swallowed, get medical attention. Give large quantities of water and induce vomiting. Do not make an unconscious person vomit. Medical observation is recommended for 24−48 h after breathing overexposure, as pulmonary edema may be delayed. As first aid for pulmonary edema, a doctor or authorized paramedic may consider administering a corticosteroid spray.

Personal Protective Methods: Wear protective gloves and clothing to prevent any reasonable probability of skin contact. Safety equipment suppliers/manufacturers can provide recommendations on the most protective glove/clothing material for your operation. Butyl rubber, Neoprene, and Viton™ are among the recommended protective materials. All protective clothing (suits, gloves, footwear, headgear) should be clean, available each day, and put on before work. Contact lenses should not be worn when working with this chemical. Wear splash-proof chemical goggles and face shield unless full face-piece respiratory protection is worn. Employees should wash immediately with soap when skin is wet or contaminated. Provide emergency showers and eyewash.

Respirator Selection: Where there is potential for exposures *over 0.2 ppm*, use an NIOSH/MSHA- or European Standard EN 149-approved supplied-air respirator with a full face-piece operated in the positive-pressure mode, or with a full face-piece, hood, or helmet in the continuous-flow mode; or use an NIOSH/MSHA- or European Standard EN 149-approved self-contained breathing apparatus with a full face-piece operated in a pressure-demand or other positive mode.

Storage: Color Code—Yellow Stripe: Reactivity Hazard; Store separately in an area isolated from flammables, combustibles, or other yellow-coded materials. Prior to working with this chemical you should be trained on its proper handling and storage. Methyl ethyl ketone peroxide must be stored to avoid contact with heat, shock, and organics since violent reactions occur. Store in tightly closed containers in a cool, well-ventilated area away from organics. Use only nonsparking tools and equipment, especially when opening and closing containers of methyl ethyl ketone peroxide. Sources of ignition, such as smoking and open flames, are prohibited where methyl ethyl ketone peroxide is used, handled, or stored in a manner that could create a potential fire or explosion hazard. Wherever methyl ethyl ketone peroxide is used, handled, manufactured, or stored, use explosion-proof electrical equipment and fittings.

Shipping: Shipment of methyl ethyl ketone peroxide, in solution, with >9% by mass active oxygen is FORBIDDEN by any means. Solutions containing less than 9% active oxygen require a shipping label of "ORGANIC PEROXIDE." They fall in DOT Hazard Class 5.2 and Packing Group II.

Spill Handling: Evacuate and restrict persons not wearing protective equipment from area of spill or leak until cleanup is complete. Remove all ignition sources. Ventilate area of spill or leak. Absorb liquids in vermiculite, dry sand, earth, peat, carbon, or a similar material and deposit in sealed containers. Keep this chemical out of a confined space, such as a sewer, because of the possibility of an explosion, unless the sewer is designed to prevent the buildup of explosive concentrations. It may be necessary to contain and dispose of this chemical as a hazardous waste. If material or contaminated runoff enters waterways, notify downstream users of potentially contaminated waters. Contact your local or federal environmental protection agency for specific recommendations. If employees are required to clean up spills, they must be properly trained and equipped. OSHA 1910.120(q) may be applicable.

Fire Extinguishing: This chemical is a combustible liquid. Poisonous gases are produced in fire. Use dry chemical, carbon dioxide, alcohol or polymer foam extinguishers. Vapors are heavier than air and will collect in low areas. Vapors may travel long distances to ignition sources and flashback. Vapors in confined areas may explode when exposed to fire. Containers may explode in fire. Storage containers and parts of containers may rocket great distances, in many directions. If material or contaminated runoff enters waterways, notify downstream users of potentially contaminated waters. Notify local health and fire officials and pollution control agencies. From a secure, explosion-proof location, use water spray to cool exposed containers. If cooling streams are ineffective (venting sound increases in volume and pitch, tank discolors, or shows any signs of deforming), withdraw immediately to a secure position. If employees are expected to fight fires, they must be trained and equipped in OSHA 1910.156. The only respirators recommended for firefighting are self-contained breathing apparatuses that have full face-pieces and are operated in a pressure-demand or other positive-pressure mode.

Disposal Method Suggested: Consult with environmental regulatory agencies for guidance on acceptable disposal practices. Generators of waste containing this contaminant ($\geq$100 kg/mo) must conform with EPA regulations governing storage, transportation, treatment, and waste disposal. MEKP may be destroyed by adding 20% NaOH solution slowly in a quantity about 10 times the weight of MEKP. Incineration is recommended if NaOH treatment is not used. See the Sax (1985) reference cited below for details.

References

US Environmental Protection Agency. (1979). *Chemical Hazard Information Profile: Methyl Ethyl Ketone Peroxide.* Washington, DC

Na. Inst. for Occupational Safety and Health. (October 1977). *Information Profiles on Potential Occupational Hazards: Methyl Ethyl Ketone Peroxide*, Report PB-276,678. Rockville, MD, pp. 37−41

Sax, N. I. (Ed.). *Dangerous Properties of Industrial Materials Report*, 2, No. 6, 35—37 (1982) and 5, No. 4, 50—55 (1985)
New Jersey Department of Health and Senior Services. (October 1999). *Hazardous Substances Fact Sheet: Methyl Ethyl Ketone Peroxide*. Trenton, NJ

2-Methyl-5-ethyl pyridine   M:0940

Molecular Formula: $C_8H_{11}N$
Synonyms: Aldehydecollidine; Aldehydine; Collidine, aldehydecollidine; 3-Ethyl-6-methylpyridine; 5-Ethyl-2-methyl-pyridine; 5-Ethyl-2-picoline; MEP; 6-Methyl-3-ethylpyridine; Methyl ethyl pyridine; 2-Methyl-5-ethylpyridine
CAS Registry Number: 104-90-5
RTECS® Number: TJ6825000
UN/NA & ERG Number: UN2300/153
EC Number: 203-250-0
Regulatory Authority and Advisory Bodies
Air Pollutant Standard Set. See below, "Permissible Exposure Limits in Air" section.
Canada, WHMIS, Ingredients Disclosure List Concentration 1.0%.
European/International Regulations: not listed in Annex 1.
WGK (German Aquatic Hazard Class): 1—Low hazard to waters.
Description: 2-Methyl-5-ethyl pyridine is a colorless liquid with a sharp, aromatic odor. Molecular weight = 121.20; Boiling point = 128°C; Flash point = 39°C; also listed at 68°C; Autoignition temperature = 538°C. Explosive limits: LEL = 1.1%; UEL = 6.6%. Hazard Identification (based on NFPA-704 M Rating System): Health 2, Flammability 2, Reactivity 0. Slightly soluble in water.
Potential Exposure: In the manufacture of nicotinic acid, vinylpyridine monomer; as intermediates for insecticides, germicides, and textile chemicals.
Incompatibilities: Strong oxidizers, strong acids, aldehydes, strong bases, acid chlorides, chloroformates, isocyanates, phenols, cresols.
Permissible Exposure Limits in Air
No TEEL available.
Russia: MAC 2 mg/m³.[43]
Routes of Entry: Inhalation, ingestion, eye and/or skin contact. Passes through the skin.
Harmful Effects and Symptoms
Short Term Exposure: Irritates the eyes, skin, and respiratory tract. Higher exposures can cause pulmonary edema, a medical emergency that can be delayed for several hours. This can cause death. Exposure can cause headache, nausea, vomiting, and diarrhea. Moderately toxic by oral and dermal routes.
Long Term Exposure: May affect the central nervous system causing muscle weakness, loss of coordination, and loss of consciousness.

Points of Attack: Lungs, nervous system.
Medical Surveillance: Consider chest X-ray following acute overexposure. Examination of the nervous system.
First Aid: If this chemical gets into the eyes, remove any contact lenses at once and irrigate immediately for at least 15 min, occasionally lifting upper and lower lids. Seek medical attention immediately. If this chemical contacts the skin, remove contaminated clothing and wash immediately with soap and water. Seek medical attention immediately. If this chemical has been inhaled, remove from exposure, begin rescue breathing (using universal precautions, including resuscitation mask) if breathing has stopped and CPR if heart action has stopped. Transfer promptly to a medical facility. When this chemical has been swallowed, get medical attention. Give large quantities of water and induce vomiting. Do not make an unconscious person vomit. Medical observation is recommended for 24—48 h after breathing overexposure, as pulmonary edema may be delayed. As first aid for pulmonary edema, a doctor or authorized paramedic may consider administering a corticosteroid spray.
Personal Protective Methods: Wear protective gloves and clothing to prevent any reasonable probability of skin contact. Safety equipment suppliers/manufacturers can provide recommendations on the most protective glove/clothing material for your operation. All protective clothing (suits, gloves, footwear, headgear) should be clean, available each day, and put on before work. Contact lenses should not be worn when working with this chemical. Wear splash-proof chemical goggles and face shield unless full face-piece respiratory protection is worn. Employees should wash immediately with soap when skin is wet or contaminated. Provide emergency showers and eyewash.
Respirator Selection: Follow regulations in OSHA 29CFR1910.134 or European Standard EN149. Use a NIOSH/MSHA- or European Standard EN149-approved respirator; or use an approved supplied-air respirator with a full face-piece operated in the positive-pressure mode, or with a full face-piece, hood, or helmet in the continuous-flow mode; or use a NIOSH/MSHA- or European Standard EN149-approved self-contained breathing apparatus with a full face-piece operated in a pressure-demand or other positive-pressure mode.
Storage: Color Code—Blue: Health Hazard/Poison: Store in a secure poison location. Prior to working with this chemical you should be trained on its proper handling and storage. Before entering confined space where this chemical may be present, check to make sure that an explosive concentration does not exist. Store in tightly closed containers in a cool, well-ventilated area away from oxidizers. Where possible, automatically pump liquid from drums or other storage containers to process containers. Sources of ignition, such as smoking and open flames, are prohibited where this chemical is handled, used, or stored. Metal containers involving the transfer of 5 gallons or more of this chemical should be grounded and bonded. Drums must be

equipped with self-closing valves, pressure vacuum bungs, and flame arresters. Use only nonsparking tools and equipment, especially when opening and closing containers of this chemical. Wherever this chemical is used, handled, manufactured, or stored, use explosion-proof electrical equipment and fittings.

Shipping: This compound requires a shipping label of "POISONOUS/TOXIC MATERIALS." It falls in Hazard Class 6.1 and Packing Group III.

Spill Handling: Evacuate and restrict persons not wearing protective equipment from area of spill or leak until cleanup is complete. Remove all ignition sources. Establish forced ventilation to keep levels below explosive limit. Absorb liquids in vermiculite, dry sand, earth, peat, carbon, or a similar material and deposit in sealed containers. Oil-skimming equipment and sorbent foams can be applied to slick if done immediately. Keep this chemical out of a confined space, such as a sewer, because of the possibility of an explosion, unless the sewer is designed to prevent the buildup of explosive concentrations. It may be necessary to contain and dispose of this chemical as a hazardous waste. If material or contaminated runoff enters waterways, notify downstream users of potentially contaminated waters. Contact your local or federal environmental protection agency for specific recommendations. If employees are required to clean up spills, they must be properly trained and equipped. OSHA 1910.120(q) may be applicable.

Fire Extinguishing: This chemical is a flammable liquid. Poisonous gases, including nitrogen oxides, are produced in fire. Use dry chemical, carbon dioxide, alcohol or polymer foam extinguishers. Vapors are heavier than air and will collect in low areas. Vapors may travel long distances to ignition sources and flashback. Vapors in confined areas may explode when exposed to fire. Containers may explode in fire. Storage containers and parts of containers may rocket great distances, in many directions. If material or contaminated runoff enters waterways, notify downstream users of potentially contaminated waters. Notify local health and fire officials and pollution control agencies. From a secure, explosion-proof location, use water spray to cool exposed containers. If cooling streams are ineffective (venting sound increases in volume and pitch, tank discolors, or shows any signs of deforming), withdraw immediately to a secure position. If employees are expected to fight fires, they must be trained and equipped in OSHA 1910.156. The only respirators recommended for firefighting are self-contained breathing apparatuses that have full face-pieces and are operated in a pressure-demand or other positive-pressure mode.

Disposal Method Suggested: Dissolve or mix the material with a combustible solvent and burn in a chemical incinerator equipped with an afterburner and scrubber. All federal, state, and local environmental regulations must be observed.

References

Sax, N. I. (Ed.). (1982). *Dangerous Properties of Industrial Materials Report*, 2, No. 2, 54—55 and 3, No. 6, 48—49 (1983)

New Jersey Department of Health and Senior Services. (June 1999). *Hazardous Substances Fact Sheet: 2-Methyl-5-Ethylpyridine*. Trenton, NJ

Methyleugenol M:0945

Molecular Formula: $C_{11}H_{14}O_2$

Synonyms: 4-Allyl-1,2-dimethoxybenzene; 1-Allyl-3,4-dimethoxybenzene; 4-Allylveratrole; Benzene, 4-Allyl-1,2-dimethoxy-; Chavibetol methyl ether; 1,2-Dimethoxy-4-allylbenzene; 3,4-Dimethoxyallylbenzene; 1-(3,4-Dimethoxyphenyl)-2-propene; 3-(3,4-Dimethoxyphenyl)propene; 1,2-Dimethoxy-4-(2-propenyl)benzene; Eugenol methyl ether; 1,3,4-Eugenol methyl ether; Eugenyl methyl ether; Methyl eugenol ether; o-Methyleugenol; Veratrole, 4-allyl-; Veratrole methyl ether

CAS Registry Number: 93-15-2

RTECS®Number: CY2450000

UN/NA & ERG Number: UN2810

EC Number: 202-223-0

Regulatory Authority and Advisory Bodies

Carcinogenicity: NTP: 11th Report on Carcinogens, 2004: Reasonably anticipated to be a human carcinogen.

California Proposition 65 Chemical: Cancer 11/16/2001.

European/International Regulations: Hazard Symbol: Xn; Risk phrases: R22; R36/38; R40; R42/43; Safety phrases: S2; S26; S36/37/39 (see Appendix 4).

WGK (German Aquatic Hazard Class): 1—Low hazard to waters.

Description: Clear colorless to pale yellow liquid. Spicy, earthy odor. Bitter burning taste. This chemical is combustible. Molecular weight = 178.23; Specific gravity = 1.0396 at 20°C; Vapor density = >1.0; Boiling point = 249°C; 254.7°C at 760 mmHg; Melting point = −4°C; Vapor pressure = 0.02 mmHg at 20°C; 1 mmHg at 85°C; Flash point = 99°C; 109°C. Hazard Identification (based on NFPA-704 M Rating System): Health 2; Flammability 1; Reactivity 0. Practically insoluble in water; solubility = <1 mg/mL at 66°F; 500 mg/L.

Potential Exposure: Agricultural Chemical and Pesticide; Tumorigen; Mutagen; Primary Irritant. Methyleugenol is a naturally occurring substance found in the essential oils of several plant species. Methyleugenol is used as a flavoring agent in jellies, baked goods, nonalcoholic beverages, chewing gum, candy, pudding, relish, and ice cream. Methyleugenol has been used as an anesthetic in rodents. It is also used as an insect attractant in combination with insecticides.[NTP 2000]

Incompatibilities: When heated to decomposition, it emits acrid smoke, irritating fumes, and toxic fumes of carbon monoxide and carbon dioxide. Methyleugenol is incompatible with strong oxidizers; contact with reducing agents may cause the release of hydrogen gas.

Permissible Exposure Limits in Air: No OELs or TEEL available.

Determination in Air: No NIOSH or OSHA method available.

Permissible Concentration in Water: Not found.

Determination in Water: Bluegill (*Lepomis macrochirus*) LC_{50} = 8233 μg/L, Moderately Toxic; Rainbow trout, donaldson trout (*Oncorhynchus mykiss*) LC_{50} = 6450 μg/L, Moderately Toxic; octanol−water partition coefficient: Log K_{ow} = 3.0−3.5. Organic carbon−water partition coefficient: Log K_{oc} = 2.7.

Routes of Entry: Skin, eyes, inhalation, ingestion.

Harmful Effects and Symptoms

Short Term Exposure: This compound is a primary irritant and sensitizer and may cause contact dermatitis. Symptoms of exposure to this compound include nausea, vomiting, diarrhea, circulatory collapse, dizziness, rapid and shallow breathing, unconsciousness, convulsions, abdominal burning, dysuria, hematuria, tachycardia, bronchial irritation, anuria, pulmonary edema, bronchial pneumonia, and renal damage. LD_{50} = (oral-rat) 810 mg/kg; LC_{50} = (inhal.-rat) = >4800 mg/m³.

Long Term Exposure: This compound may be irritating to the skin and eyes.

Points of Attack: Liver, kidneys, skin, respiratory tract.

Medical Surveillance: Consideration should be given to the skin, eyes, and respiratory tract (lung function tests) in any placement or periodic examinations. Evaluation by a qualified allergist, including careful exposure history and special testing, may help diagnose skin or respiratory tract allergy.

First Aid: Eyes: Consideration should be given to the skin, eyes, and respiratory tract (lung function tests) in any placement or periodic examinations. Evaluation by a qualified allergist, including careful exposure history and special testing, may help diagnose skin or respiratory tract allergy.

Personal Protective Methods: Wear protective eyeglasses or chemical safety goggles as described in OSHA regulations 29CFR1910.133 or European Standard EN166. Recommended gloves: Ansell 5.109 (Latex); thickness: 0.18 mm; Breakthrough time: *15 min*; Edmont 29-870 (Neoprene); thickness: 0.51 mm; Breakthrough time: *135 min*; North Model F-091 (Viton); thickness: 0.41 mm; Breakthrough time: *480 min*.

Respirator Selection: Follow the regulations in OSHA 29CFR1910.134 or European Standard EN 149. Use a NIOSH/MSHA- or European Standard EN 149-approved respirator; or follow regulations in OSHA 29CFR1910.134 or European Standard EN 149. Use a NIOSH/MSHA- or European Standard EN 149-approved respirator; or use an approved half face respirator equipped with an organic vapor/acid gas cartridge (specific for organic vapors, HCl, acid gas, and SO₂) with a dust/mist filter.

Storage: Color Code—Green: General storage may be used. You should protect this material from exposure to light, and store it under ambient temperatures.

Shipping: Not regulated.

Spill Handling: First remove all sources of ignition. Then, use absorbent paper to pick up all liquid spill material. All contaminated clothing and absorbent paper should be sealed in a vapor-tight plastic bag for eventual disposal. Solvent wash all contaminated surfaces with 60−70% ethanol followed by washing with a soap and water solution. Do not reenter the contaminated area until the safety officer (or other responsible person) has verified that the area has been properly cleaned.

Fire Extinguishing: Wear a self-contained breathing apparatus in pressure-demand, NIOSH/MSHA- or European Standard EN 149-approved respirator, and full protective gear. During a fire or heated to decomposition, irritating and highly toxic gases may be generated and it emits acrid smoke, irritating fumes, and toxic fumes of carbon monoxide and carbon dioxide. Use water spray to keep fire-exposed containers cool. Containers may explode in the heat of a fire. Extinguishing media; cool containers with flooding quantities of water until well after fire is out. Fires involving this material can be controlled with a dry chemical, carbon dioxide, or Halon extinguisher. A water spray may also be used.

Disposal Method Suggested: Dissolve or mix the material with a combustible solvent and burn in a chemical incinerator equipped with an afterburner and scrubber. All federal, state, and local environmental regulations must be observed.

Reference

Hazardous Substances Data Bank [HSDB], US National Library of Medicine, TOXNET. *Eugenol.* <http://toxnet.nlm.nih.gov>

Methyl formate   M:0950

Molecular Formula: $C_2H_4O_2$

Common Formula: $HCOOCH_3$

Synonyms: Formiate de methyle (French); Formiato de metilo (Spanish); Formic acid, methyl ester; Methyle (formiate de) (French); Methylformiat (German); Methyl methanoate

CAS Registry Number: 107-31-3

RTECS® Number: LQ8925000

UN/NA & ERG Number: UN1243/129

EC Number: 203-481-7 [*Annex I Index No.:* 607-014-00-1]

Regulatory Authority and Advisory Bodies

Department of Homeland Security Screening Threshold Quantity (pounds): *Release hazard* 10,000 (≥1.00% concentration).

Air Pollutant Standard Set. See below, "Permissible Exposure Limits in Air" section.

Clean Air Act: Accidental Release Prevention/Flammable Substances, (Section 112[r], Table 3), TQ = 10,000 lb (4540 kg).

Canada, WHMIS, Ingredients Disclosure List Concentration 1.0%.

European/International Regulations: Hazard Symbol: F+, Xn; Risk phrases: R12; R20/22; R36/37; Safety phrases: S2; S9; S16; S24; S26; S33 (see Appendix 4).

WGK (German Aquatic Hazard Class): 1—Low hazard to waters.

Description: Methyl formate is a colorless liquid with a pleasant odor. Odor threshold = 2000 ppm. Molecular weight = 60.06; Specific gravity (H$_2$O:1) = 0.98; Boiling point = 3.7°C; Freezing/Melting point = −100°C; Vapor pressure = 476 mmHg at 20°C; Flash point = −18.9°C; Autoignition temperature = 449°C. Explosive limits: LEL = 4.5%; UEL = 23.0%. Hazard Identification (based on NFPA-704 M Rating System): Health 2, Flammability 4, Reactivity 0. Soluble in water; solubility = 30%.

Potential Exposure: Compound Description: Agricultural Chemical. Methyl formate is used as a solvent, as an intermediate in pharmaceutical manufacture, and as a fumigant.

Incompatibilities: Forms explosive mixture with air. Violent reaction with strong oxidizers. Reacts slowly with water to form methanol and formic acid.

Permissible Exposure Limits in Air

Conversion factor: 1 ppm = 2.46 mg/m^3 at 25°C & 1 atm.
OSHA PEL: 100 ppm/250 mg/m^3 TWA.
NIOSH REL: 100 ppm/250 mg/m^3 TWA; 150 ppm/375 mg/m^3 STEL.
ACGIH TLV®[1]: 100 ppm/246 mg/m^3 TWA; 150 ppm/368 mg/m^3 STEL.
NIOSH IDLH: 4500 ppm.
Protective Action Criteria (PAC)
TEEL-0: 100 ppm
PAC-1: 150 ppm
PAC-2: 750 ppm
PAC-3: 4500 ppm
DFG MAK: 50 ppm/120 mg/m^3 TWA; Peak Limitation Category II(4); [skin]; Pregnancy Risk Group C.
Australia: TWA 100 ppm (250 mg/m^3); STEL 150 ppm, 1993; Austria: MAK 100 ppm (250 mg/m^3), 1999; Belgium: TWA 100 ppm (246 mg/m^3); STEL 150 ppm (369 mg/m^3), 1993; Denmark: TWA 100 ppm (250 mg/m^3), 1999; Finland: TWA 100 ppm (250 mg/m^3); STEL 150 ppm (375 mg/m^3), 1999; France: VME 100 ppm (250 mg/m^3), 1999; the Netherlands: MAC-TGG 250 mg/m^3, 2003; Sweden: TWA 100 ppm (250 mg/m^3); STEL 150 ppm (350 mg/m^3), 1999; Switzerland: MAK-W 100 ppm (250 mg/m^3), KZG-W 200 ppm (500 mg/m^3), 1999; Turkey: TWA 100 ppm (250 mg/m^3), 1993; United Kingdom: TWA 100 ppm (250 mg/m^3); STEL 150 ppm, 2000; Argentina, Bulgaria, Columbia, Jordan, South Korea, New Zealand, Singapore, Vietnam: ACGIH TLV®: STEL 150 ppm. Several states have set guidelines or standards for methyl formate in ambient air[60] ranging from 2.50−3.75 mg/m^3 (North Dakota) to 4.2 mg/m^3 (Virginia) to 5.0 mg/m^3 (Connecticut) to 5.952 mg/m^3 (Nevada).

Determination in Air: Use NIOSH (II-5) Method #S-291. OSHA Analytical Method PV-2041.

Determination in Water: Octanol−water coefficient: Log K_{ow} = −0.2.

Routes of Entry: Inhalation, skin absorption, ingestion, skin and/or eye contact. Passes through the skin.

Harmful Effects and Symptoms

Short Term Exposure: Methyl formate can affect you when breathed in and by passing through your skin. Exposure can irritate the eyes, nose, and throat. Higher levels can irritate the lungs and cause a buildup of fluid (pulmonary edema). This can cause death. High levels attack the nervous system and cause you to become dizzy, lightheaded; and may cause unconsciousness and death.

Long Term Exposure: Prolonged or repeated contact can cause cracking and drying of the skin. Repeated exposure can irritate the lungs and may cause bronchitis to develop.

Points of Attack: Eyes, lungs, central nervous system.

Medical Surveillance: For those with frequent or potentially high exposure (half the TLV or greater), the following are recommended before beginning work and at regular times after that: lung function tests. If symptoms develop or overexposure is suspected, the following may be useful: consider chest X-ray after acute overexposure.

First Aid: If this chemical gets into the eyes, remove any contact lenses at once and irrigate immediately for at least 15 min, occasionally lifting upper and lower lids. Seek medical attention immediately. If this chemical contacts the skin, remove contaminated clothing and wash immediately with soap and water. Seek medical attention immediately. If this chemical has been inhaled, remove from exposure, begin rescue breathing (using universal precautions, including resuscitation mask) if breathing has stopped and CPR if heart action has stopped. Transfer promptly to a medical facility. When this chemical has been swallowed, get medical attention. Give large quantities of water and induce vomiting. Do not make an unconscious person vomit. Medical observation is recommended for 24−48 h after breathing overexposure, as pulmonary edema may be delayed. As first aid for pulmonary edema, a doctor or authorized paramedic may consider administering a corticosteroid spray.

Personal Protective Methods: Wear protective gloves and clothing to prevent any reasonable probability of skin contact. Safety equipment suppliers/manufacturers can provide recommendations on the most protective glove/clothing material for your operation. All protective clothing (suits, gloves, footwear, headgear) should be clean, available each day, and put on before work. Contact lenses should not be worn when working with this chemical. Wear splash-proof chemical goggles and face shield unless full face-piece respiratory protection is worn. Employees should wash immediately with soap when skin is wet or contaminated. Provide emergency showers and eyewash.

Respirator Selection: 1000 ppm: Sa (APF = 10) (any supplied-air respirator). 2500 ppm: Sa:Cf (APF = 25) (any supplied-air respirator operated in a continuous-flow mode). 4500 ppm: SCBAF (APF = 50) (any self-contained breathing apparatus with a full face-piece) or SaF (APF = 50) (any supplied-air respirator with a full face-piece). *Emergency or planned entry into unknown concentration or IDLH conditions:* SCBAF: Pd,Pp (APF = 10,000) (any self-contained breathing apparatus that has a full face-piece and is operated in a pressure-demand or other positive-pressure mode) or SaF: Pd,Pp: ASCBA (APF = 10,000) (any supplied-air respirator that has a full face-piece and is operated in a pressure-demand or other positive-pressure mode in combination with an auxiliary, self-contained breathing apparatus operated in a pressure-demand or other positive-pressure mode). *Escape:* GmFOv (APF = 50) [any air-purifying, full-face-piece respirator (gas mask) with a chin-style, front- or back-mounted acid gas canister] or SCBAE (any appropriate escape-type, self-contained breathing apparatus).

Note: Substance reported to cause eye irritation or damage; may require eye protection.

Storage: Color Code—Red: Flammability Hazard: Store in a flammable liquid storage area or approved cabinet away from ignition sources and corrosive and reactive materials. Prior to working with this chemical you should be trained on its proper handling and storage. Before entering confined space where this chemical may be present, check to make sure that an explosive concentration does not exist. Methyl formate must be stored to avoid contact with strong oxidizers, such as chlorine, bromine, chlorine dioxide, nitrates, and permanganates, since violent reactions occur. Store in tightly closed containers in a cool, well-ventilated area away from heat. Sources of ignition, such as smoking and open flames, are prohibited where methyl formate is handled, used, or stored. Metal containers involving the transfer of 5 gallons or more of methyl formate should be grounded and bonded. Drums must be equipped with self-closing valves, pressure vacuum bungs, and flame arresters. Use only nonsparking tools and equipment, especially when opening and closing containers of methyl formate. Wherever methyl formate is used, handled, manufactured, or stored, use explosion-proof electrical equipment and fittings.

Shipping: This compound requires a shipping label of "FLAMMABLE LIQUID." It falls in Hazard Class 3 and Packing Group I.

Spill Handling: Evacuate and restrict persons not wearing protective equipment from area of spill or leak until cleanup is complete. Remove all ignition sources. Establish forced ventilation to keep levels below explosive limit. Absorb liquids in vermiculite, dry sand, earth, peat, carbon, or a similar material and deposit in sealed containers. Keep this chemical out of a confined space, such as a sewer, because of the possibility of an explosion, unless the sewer is designed to prevent the buildup of explosive concentrations. It may be necessary to contain and dispose of this chemical as a hazardous waste. If material or contaminated runoff enters waterways, notify downstream users of potentially contaminated waters. Contact your local or federal environmental protection agency for specific recommendations. If employees are required to clean up spills, they must be properly trained and equipped. OSHA 1910.120(q) may be applicable.

Fire Extinguishing: This chemical is a flammable liquid. Poisonous gases are produced in fire. Use dry chemical, carbon dioxide, or alcohol foam extinguishers. Vapors are heavier than air and will collect in low areas. Vapors may travel long distances to ignition sources and flashback. Vapors in confined areas may explode when exposed to fire. Containers may explode in fire. Storage containers and parts of containers may rocket great distances, in many directions. If material or contaminated runoff enters waterways, notify downstream users of potentially contaminated waters. Notify local health and fire officials and pollution control agencies. From a secure, explosion-proof location, use water spray to cool exposed containers. If cooling streams are ineffective (venting sound increases in volume and pitch, tank discolors, or shows any signs of deforming), withdraw immediately to a secure position. If employees are expected to fight fires, they must be trained and equipped in OSHA 1910.156. The only respirators recommended for firefighting are self-contained breathing apparatuses that have full face-pieces and are operated in a pressure-demand or other positive-pressure mode.

Disposal Method Suggested: Incineration; atomizing in a suitable combustion chamber.

Reference

New Jersey Department of Health and Senior Services. (August 2004). *Hazardous Substances Fact Sheet: Methyl Formate.* Trenton, NJ

Methyl hydrazine   M:0960

Molecular Formula: CH_6N_2
Common Formula: CH_3NHNH_2
Synonyms: Hydrazine, methyl-; Hydrazomethane; *N*-Methyl hydrazine; 1-Methyl hydrazine; Metilhidrazina (Spanish); MMH; Monomethylhydrazine
CAS Registry Number: 60-34-4
RTECS® Number: MV5600000
UN/NA & ERG Number: UN1244/131
EC Number: 200-471-4
Regulatory Authority and Advisory Bodies
Department of Homeland Security Screening Threshold Quantity (pounds): *Release hazard* 15,000 (≥1.00% concentration).
Carcinogenicity: NIOSH: Suspected occupational carcinogen.
US EPA Gene-Tox Program, Positive: Histidine reversion—Ames test; TRP reversion; Positive/limited: Carcinogenicity—mouse/rat; Negative: Rodent dominant lethal; *In vitro* UDS—human fibroblast; Negative: *S. cerevisiae* gene conversion.
Air Pollutant Standard Set. See below, "Permissible Exposure Limits in Air" section.
OSHA 29CFR1910.119, Appendix A, Process Safety List of Highly Hazardous Chemicals, TQ = 100 lb (45 kg).

Clean Air Act: Hazardous Air Pollutants (Title I, Part A, Section 112); Accidental Release Prevention/Flammable Substances, (Section 112[r], Table 3), TQ = 15,000 lb (6810 kg).

US EPA Hazardous Waste Number (RCRA No.): P068.

RCRA, 40CFR261, Appendix 8 Hazardous Constituents.

Superfund/EPCRA 40CFR355, Extremely Hazardous Substances: TPQ = 500 lb (227 kg).

Reportable Quantity (RQ): 10 lb (4.54 kg).

EPCRA Section 313 Form R *de minimis* concentration reporting level: 1.0%.

US DOT 49CFR172.101, Inhalation Hazardous Chemical.

California Proposition 65 Chemical: Cancer (methyl hydrazine and its salts) 7/1/92.

Canada, WHMIS, Ingredients Disclosure List Concentration 0.1%.

European/International Regulations: not listed in Annex 1.

WGK (German Aquatic Hazard Class): No value assigned.

Description: Methyl hydrazine is a fuming, colorless liquid with an ammonia-like odor. The odor threshold is 1.3–1.7 ppm. Molecular weight = 46.09; Specific gravity (H_2O:1) = 0.87 at 25°C; Boiling point = 87.8°C; Freezing/Melting point = −52.2°C; Vapor pressure = 38 mmHg at 20°C; Flash point = −8.3°C; Autoignition temperature = 194°C. Explosive limits: LEL = 2.5%; UEL = 92.0%.[17] Hazard Identification (based on NFPA-704 M Rating System): Health 4, Flammability 3, Reactivity 2. Slightly soluble in water.

Potential Exposure: Compound Description: Tumorigen, Mutagen; Reproductive Effector. MMH has been used as the propellant in liquid propellant rockets; it is also used as a solvent and as an organic intermediate.

Incompatibilities: Forms explosive mixture with air. Highly reactive reducing agent and a medium strong base. May explode if heated or when in contact with metal oxides. Violent reaction with strong oxidizers, such as fluorine, chlorine, combustibles, nitric acid, hydrogen peroxide. Incompatible with acids, alcohols, glycols, isocyanates, phenols, cresols, porous materials, such as earth, asbestos, wood, and cloth. Oxides of iron or copper, manganese, lead, copper, or their alloys can lead to fire and explosions. Attacks cork, some plastics, coatings, and rubber.

Permissible Exposure Limits in Air

Conversion factor: 1 ppm = 1.89 mg/m³ at 25°C & 1 atm.

OSHA PEL: 0.2 ppm/0.35 mg/m³ Ceiling Concentration [skin].

NIOSH REL: 0.04 ppm/0.08 mg/m³ [120 min] Ceiling Concentration; A potential occupational carcinogen. Limit exposure to lowest feasible concentration. See *NIOSH Pocket Guide*, Appendix A.

ACGIH TLV®[1]: 0.01 ppm/0.019 mg/m³ [skin]; confirmed animal carcinogen with unknown relevance to humans.

NIOSH IDLH: potential occupational carcinogen 20 ppm.

Protective Action Criteria (PAC)*

TEEL-0: 0.01 ppm

PAC-1: 0.2 ppm

PAC-2: **0.9** ppm

PAC-3: **2.7** ppm

*AEGLs (Acute Emergency Guideline Levels) & ERPGs (Emergency Response Planning Guideline) are in **bold face**.

DFG MAK: [skin] danger of skin sensitization.

Australia: TWA 0.2 ppm (0.35 mg/m³), [skin], carcinogen, 1993; Belgium: STEL 0.2 ppm (0.38 mg/m³), [skin], Carcinogen 1993; Denmark: TWA 0.04 ppm (0.08 mg/m³), [skin], 1999; Finland: STEL 0.2 ppm (0.35 mg/m³), [skin], 1999; France: VME 0.2 ppm (0.35 mg/m³), 1999; Norway: TWA 0.08 mg/m³, 1999; the Netherlands: MAC 0.35 mg/m³, [skin], 2003; Poland: MAC (TWA) 0.02 mg/m³, MAC (STEL) 0.1 mg/m³, 1999; Switzerland: MAK-W 0.2 ppm (0.35 mg/m³), [skin], 1999; Thailand: TWA 0.2 ppm (0.35 mg/m³), 1993; Argentina, Bulgaria, Columbia, Jordan, South Korea, New Zealand, Singapore, Vietnam: ACGIH TLV®: confirmed animal carcinogen with unknown relevance to humans. Several states have set guidelines or standards for methyl hydrazine in ambient air[60] ranging from zero (Maryland, North Dakota, and Texas) to 0.88 µg/m³ (Pennsylvania) to 1.17 µg/m³ (New York) to 1.75 µg/m³ (South Carolina) to 3.5 µg/m³ (Virginia) to 8.0 µg/m³ (Nevada).

Determination in Air: Use NIOSH Analytical Method (IV) #3510.

Permissible Concentration in Water: No criteria set, but EPA[32] has suggested a permissible ambient goal of 5 µg/L based on health effects.

Determination in Water: Octanol–water coefficient: Log K_{ow} = −1.03.

Routes of Entry: Inhalation, skin absorption, ingestion, skin and/or eye contact.

Harmful Effects and Symptoms

Short Term Exposure: Corrosive to the eyes, skin, respiratory tract; and if ingested, may cause permanent damage. Inhalation can cause pulmonary edema, a medical emergency that can be delayed for several hours. This can cause death. May affect the central nervous system, liver, and blood, causing excitability, vomiting, tremors, convulsions, formation of methemoglobin, and death. Exposure at high concentrations may result in death. The effects may be delayed. Symptoms of acute exposure to methyl hydrazine may include facial numbness, facial swelling, and increased salivation. Headache, twitching, seizure, convulsions, and coma may also occur. Gastrointestinal signs and symptoms include anorexia, nausea, and vomiting. Methyl hydrazine is toxic to the liver, ruptures red blood cells; and may cause kidney damage. Methyl hydrazine vapors are extremely toxic and the liquid is corrosive to skin. Methyl hydrazine is the strongest convulsant and the most toxic of methyl-substituted hydrazine derivatives. It is more toxic than hydrazine. At high doses, it is a strong central nervous system poison.

Long Term Exposure: May damage the liver, kidneys, and blood, resulting in formation of methemoglobin. This substance causes liver cancer in animals and is possibly carcinogenic to humans.

Points of Attack: Eyes, skin, respiratory system, central nervous system, liver, blood, cardiovascular system. Cancer site in animals: lung, liver, blood vessels, and intestine.

Medical Surveillance: NIOSH lists the following tests: whole blood (chemical/metabolite), methemoglobin; pulmonary function tests. Before beginning employment and at regular times after that, the following is recommended: complete blood count. If symptoms develop or overexposure has occurred, the following may be useful: consider lung function tests, especially if lung symptoms are present. Test for kidney and liver function. Evaluation by a qualified allergist, including careful exposure history and special testing, may help diagnose skin allergy. Consider chest X-ray following acute overexposure.

First Aid: If this chemical gets into the eyes, remove any contact lenses at once and irrigate immediately for at least 15 min, occasionally lifting upper and lower lids. Seek medical attention immediately. If this chemical contacts the skin, remove contaminated clothing and wash immediately with soap and water. Seek medical attention immediately. If this chemical has been inhaled, remove from exposure, begin rescue breathing (using universal precautions, including resuscitation mask) if breathing has stopped and CPR if heart action has stopped. Transfer promptly to a medical facility. When this chemical has been swallowed, get medical attention. If victim is conscious, administer water or milk. Do not induce vomiting. Medical observation is recommended for 24–48 h after breathing overexposure, as pulmonary edema may be delayed. As first aid for pulmonary edema, a doctor or authorized paramedic may consider administering a corticosteroid spray.

Note to physician: Treat for methemoglobinemia. Spectrophotometry may be required for precise determination of levels of methemoglobin in urine.

Personal Protective Methods: Wear protective gloves and clothing to prevent any reasonable probability of skin contact: **8 h:** Responder™ suits; Trychem 1000™ suits. Also, Viton™, chlorobutyl rubber, and CR-39® may offer some protection. Safety equipment suppliers/manufacturers can provide recommendations on the most protective glove/clothing material for your operation. All protective clothing (suits, gloves, footwear, headgear) should be clean, available each day, and put on before work. Contact lenses should not be worn when working with this chemical. Wear splash-proof chemical goggles and face shield unless full face-piece respiratory protection is worn. Employees should wash immediately with soap when skin is wet or contaminated. Provide emergency showers and eyewash.

Respirator Selection: NIOSH: *At any concentrations above the NIOSH REL, or where there is no REL, at any detectable concentration:* SCBAF: Pd,Pp (APF = 10,000) (any NIOSH/MSHA- or European Standard EN 149-approved self-contained breathing apparatus that has a full face-piece and is operated in a pressure-demand or other positive-pressure mode) or SaF: Pd,Pp: ASCBA (APF = 10,000) (any supplied-air respirator that has a full face-piece and is operated in a pressure-demand or other positive-pressure mode in combination with an auxiliary self-contained breathing apparatus operated in a pressure-demand or other positive-pressure mode). *Escape:* SCBAE (any appropriate escape-type, self-contained breathing apparatus).

Storage: Color Code—Red: Flammability Hazard: Store in a flammable liquid storage area or approved cabinet away from ignition sources and corrosive and reactive materials. Prior to working with this chemical you should be trained on its proper handling and storage. Before entering confined space where this chemical may be present, check to make sure that an explosive concentration does not exist. Methyl hydrazine must be stored to avoid contact with oxides of iron and copper; manganese, lead, and copper alloys; porous materials (such as earth, asbestos, wood, and cloth); oxidizers (such as perchlorates, hydrogen peroxide, chlorates, nitrates, permanganates); and fuming nitric acid since violent reactions occur. Store in tightly closed containers in a cool, well-ventilated area away from heat and sparks. Sources of ignition, such as smoking and open flames, are prohibited where methyl hydrazine is handled, used, or stored. Metal containers involving the transfer of 5 gallons or more of methyl hydrazine should be grounded and bonded. Drums must be equipped with self-closing valves, pressure vacuum bungs, and flame arresters. Use only non-sparking tools and equipment, especially when opening and closing containers of methyl hydrazine. Wherever methyl hydrazine is used, handled, manufactured, or stored, use explosion-proof electrical equipment and fittings. A regulated, marked area should be established where this chemical is handled, used, or stored in compliance with OSHA Standard 1910.1045.

Shipping: This compound requires a shipping label of "POISONOUS/TOXIC MATERIALS, FLAMMABLE LIQUID, CORROSIVE." It falls in Hazard Class 6.1 and Packing Group I.

Spill Handling: Evacuate and restrict persons not wearing protective equipment from area of spill or leak until cleanup is complete. Remove all ignition sources. Establish forced ventilation to keep levels below explosive limit. Absorb liquids in vermiculite, dry sand, earth, peat, carbon, or a similar material and deposit in sealed containers. Keep this chemical out of a confined space, such as a sewer, because of the possibility of an explosion, unless the sewer is designed to prevent the buildup of explosive concentrations. It may be necessary to contain and dispose of this chemical as a hazardous waste. If material or contaminated runoff enters waterways, notify downstream users of potentially contaminated waters. Contact your local or federal environmental protection agency for specific recommendations. If employees are required to clean up spills, they must be properly trained and equipped. OSHA 1910.120(q) may be applicable.

Initial isolation and protective action distances
Distances shown are likely to be affected during the first 30 min after materials are spilled and could increase with time. If more than one tank car, cargo tank, portable tank, or large cylinder involved in the incident is leaking, the protective action distance may need to be increased. You may need to seek emergency information from CHEMTREC at (800) 424-9300 or seek professional environmental engineering assistance from the US EPA Environmental Response Team at (908) 48-8730 (24-h response line).
Small spills (From a small package or a small leak from a large package)
First: Isolate in all directions (feet/meters)   100/30
Then: Protect persons downwind (miles/kilometers)
Day   0.2/0.3
Night   0.4/0.6
Large spills (From a large package or from many small packages)
First: Isolate in all directions (feet/meters)   500/150
Then: Protect persons downwind (miles/kilometers)
Day   1.0/1.5
Night   1.5/2.4
Fire Extinguishing: This chemical is a flammable liquid. Poisonous gases, including nitrogen oxides, are produced in fire. For small fires, use dry chemical, carbon dioxide, water spray, and alcohol foam. For large fires, use water spray, fog, or foam. Keep unnecessary people away and isolate the hazardous area. Stay upwind and keep out of low-lying areas. Fire exposed containers should be kept cool with water. Use water spray to disperse vapors and protect responders attempting to stop a leak which has not ignited. Move container from fire area if it can be done without risk. Wear positive pressure breathing apparatus and special (full) protective clothing. No skin surface should be exposed. See "Isolation Distances" above. Vapors are heavier than air and will collect in low areas. Vapors may travel long distances to ignition sources and flashback. Vapors in confined areas may explode when exposed to fire. Containers may explode in fire. Storage containers and parts of containers may rocket great distances, in many directions. If material or contaminated runoff enters waterways, notify downstream users of potentially contaminated waters. Notify local health and fire officials and pollution control agencies. From a secure, explosion-proof location, use water spray to cool exposed containers. If cooling streams are ineffective (venting sound increases in volume and pitch, tank discolors, or shows any signs of deforming), withdraw immediately to a secure position. If employees are expected to fight fires, they must be trained and equipped in OSHA 1910.156. The only respirators recommended for firefighting are self-contained breathing apparatuses that have full face-pieces and are operated in a pressure-demand or other positive-pressure mode.
Disposal Method Suggested: Consult with environmental regulatory agencies for guidance on acceptable disposal practices. Generators of waste containing this contaminant

($\geq$100 kg/mo) must conform with EPA regulations governing storage, transportation, treatment, and waste disposal. There are two alternatives[24]: Dilute with water, neutralize with sulfuric acid, then flush to sewer with large volumes of water or incinerate with added flammable solvent in furnace equipped with afterburner and alkaline scrubber.
References
Nat. Inst. for Occupational Safety and Health. (1978). *Criteria for a Recommended Standard: Occupational Exposure to Hydrazines*, NIOSH Document No. 78-172. Washington, DC
Sax, N. I. (Ed.). *Dangerous Properties of Industrial Materials Report*, 2, No. 5, 86–90 (1982) and 5, No. 4, 55–59 (1985)
US Environmental Protection Agency. (November 30, 1987). *Chemical Hazard Information Profile: Methyl Hydrazine*. Washington, DC: Chemical Emergency Preparedness Program
New Jersey Department of Health and Senior Services. (July 2005). *Hazardous Substances Fact Sheet: Methyl Hydrazine*. Trenton, NJ

Methyl iodide   M:0970

Molecular Formula: CH_3I
Synonyms: Halon 10001; Iodomethane; Iodure de methyle (French); Jod-methan (German); Methyljodid (German); Methane, iodo-; Yoduro de metilo (Spanish)
CAS Registry Number: 74-88-4
RTECS® Number: PA9450000
UN/NA & ERG Number: UN2644/151
EC Number: 200-819-5 [*Annex I Index No.:* 602-005-00-9]
Regulatory Authority and Advisory Bodies
Carcinogenicity: IARC: Human No Adequate Data; Animal Limited Evidence, *not classifiable as carcinogenic to humans*, Group 3, 1999; NIOSH: Potential occupational carcinogen.
US EPA Gene-Tox Program, Positive: Carcinogenicity—mouse/rat; SHE—clonal assay; Positive: L5178Y cells *In vitro*—TK test; *E. coli* polA without S9; Positive: *S. cerevisiae*—homozygosis.
Air Pollutant Standard Set. See below, "Permissible Exposure Limits in Air" section.
OSHA 29CFR1910.119, Appendix A, Process Safety List of Highly Hazardous Chemicals, TQ = 7500 lb.
Clean Air Act: Hazardous Air Pollutants (Title I, Part A, Section 112).
US EPA Hazardous Waste Number (RCRA No.): U138.
RCRA, 40CFR261, Appendix 8 Hazardous Constituents.
RCRA 40CFR268.48; 61FR15654, Universal Treatment Standards: Wastewater (mg/L), 0.19; Nonwastewater (mg/kg), 65.
RCRA 40CFR264, Appendix 9; TSD Facilities Ground Water Monitoring List. Suggested test method(s) (PQL µg/L): 8010 (40); 8240 (5).

Reportable Quantity (RQ): 100 lb (45.4 kg).

EPCRA Section 313 Form R *de minimis* concentration reporting level: 1.0%.

US DOT 49CFR172.101, Inhalation Hazardous Chemical.

California Proposition 65 Chemical: Cancer 4/1/88.

Canada, WHMIS, Ingredients Disclosure List Concentration 0.1%.

European/International Regulations: Hazard Symbol: T; Risk phrases: R21; R23/25; R37/38;R40; Safety phrases: S1/2; S36/37; S38; S45 (see Appendix 4).

WGK (German Aquatic Hazard Class): No value assigned.

Description: Methyl iodide is a colorless liquid with a pungent, ether-like odor. Turns yellow, red, or brown on exposure to light and moisture. Molecular weight = 141.94; Specific gravity (H_2O:1) = 2.28; Boiling point = 42.8°C; Vapor pressure = 400 mmHg; Freezing/Melting point = −66.7°C. It is noncombustible. Hazard Identification (based on NFPA-704 M Rating System): Health 2, Flammability 1, Reactivity 0. Slightly soluble in water; solubility = 1%.

Potential Exposure: Compound Description: Tumorigen, Mutagen. Primary Irritant. Methyl iodide is used in fire extinguishers; as an intermediate in the manufacture of pharmaceuticals and some pesticides.

Incompatibilities: Violent reaction with strong oxidizers, strong bases, trialkylphosphines, silver chlorite, oxygen, sodium. Decomposes at 270°C.

Permissible Exposure Limits in Air

Conversion factor: 1 ppm = 5.80 mg/m^3 at 25°C & 1 atm.

OSHA PEL: 5 ppm/28 mg/m^3 TWA [skin].

NIOSH REL: 2 ppm/10 mg/m^3 TWA [skin]; A potential occupational carcinogen. Limit exposure to lowest feasible concentration. See *NIOSH Pocket Guide*, Appendix A.

ACGIH TLV®[1]: 2 ppm/12 mg/m^3 TWA [skin].

NIOSH IDLH: potential occupational carcinogen 100 ppm.

Protective Action Criteria (PAC)

TEEL-0: 5 ppm

PAC-1: 25 ppm

PAC-2: 50 ppm

PAC-3: 125 ppm

DFG MAK: [skin] Carcinogen Category 2.

Australia: TWA 2 ppm (10 mg/m^3), [skin], carcinogen, 1993; Austria: [skin], carcinogen, 1999; Belgium: TWA 2 ppm (12 mg/m^3), [skin], Carcinogen 1993; Denmark: TWA 1 ppm (5.6 mg/m^3), [skin], 1999; Finland: TWA 5 ppm (28 mg/m^3); STEL 10 ppm (56 mg/m^3), [skin], 1999; Norway: TWA 1 ppm (5 mg/m^3), 1999; the Netherlands: MAC-TGG 10 mg/m^3, [skin], 2003; Poland: MAC (TWA) 10 mg/m^3; MAC (STEL) 30 mg/m^3, 1999; Sweden: NGV 1 ppm (6 mg/m^3), KTV 5 ppm (30 mg/m^3), [skin], carcinogen, 1999; Switzerland: MAK-W 0.3 ppm (2 mg/m^3), [skin], carcinogen, 1999; United Kingdom: TWA 2 ppm (12 mg/m^3), [skin], 2000; Argentina, Bulgaria, Columbia, Jordan, South Korea, New Zealand, Singapore, Vietnam: ACGIH TLV®: TWA 2 ppm [skin].

Determination in Air: Use NIOSH Analytical Method #1014.[18]

Determination in Water: Octanol−water coefficient: Log K_{ow} = 1.63.

Routes of Entry: Inhalation, ingestion, eye and/or skin contact. Passes through the skin.

Harmful Effects and Symptoms

Short Term Exposure: Methyl iodide can affect you when breathed in and by passing through your skin. Methyl iodide is a carcinogen; handle with extreme caution. Contact can irritate the eyes and cause severe skin burns. It can cause brain damage leading to disorientation and psychotic behavior. Higher exposures can cause pulmonary edema, a medical emergency that can be delayed for several hours. This can cause death. Exposure may cause nausea, vomiting, diarrhea, dizziness, slurred speech, visual disturbances, irritability, loss of muscle control, drowsiness, delirium, serious mental disorders, coma, and death.

Long Term Exposure: Can cause lung irritation and bronchitis. May cause kidney damage. May affect the brain leading to disorientation and personality changes. Exposure can cause nausea, vomiting, vertigo (an illusion of movement), ataxia, slurred speech, drowsiness, dermatitis. A potential occupational carcinogen.

Points of Attack: Eyes, skin, respiratory system, kidneys, central nervous system.

Medical Surveillance: Before beginning employment and at regular times after that, the following are recommended: lung function tests; kidney function tests. If symptoms develop or overexposure is suspected, the following may be useful: consider urine test for iodine. Consider chest X-ray after acute overexposure. Evaluate for brain effects.

First Aid: If this chemical gets into the eyes, remove any contact lenses at once and irrigate immediately for at least 15 min, occasionally lifting upper and lower lids. Seek medical attention immediately. If this chemical contacts the skin, remove contaminated clothing and wash immediately with soap and water. Seek medical attention immediately. If this chemical has been inhaled, remove from exposure, begin rescue breathing (using universal precautions, including resuscitation mask) if breathing has stopped and CPR if heart action has stopped. Transfer promptly to a medical facility. When this chemical has been swallowed, get medical attention. Give large quantities of water and induce vomiting. Do not make an unconscious person vomit. Medical observation is recommended for 24−48 h after breathing overexposure, as pulmonary edema may be delayed. As first aid for pulmonary edema, a doctor or authorized paramedic may consider administering a corticosteroid spray.

Personal Protective Methods: Wear protective gloves and clothing to prevent any reasonable probability of skin contact: **8 h**: Viton™ gloves, suits; Responder™ suits; Trychem 1000™ suits. Safety equipment suppliers/manufacturers can provide recommendations on the most protective glove/clothing material for your operation. All protective clothing (suits, gloves, footwear, headgear) should be clean, available each day, and put on before work. Contact lenses should not be worn when working with this chemical. Wear splash-proof chemical goggles and face shield unless full face-piece respiratory protection is worn. Employees should wash immediately with soap when skin is wet or contaminated. Provide emergency showers and eyewash.

Respirator Selection: At any concentrations above the NIOSH REL, or where there is no REL, at any detectable concentration: SCBAF: Pd,Pp (APF = 10,000) (any NIOSH/MSHA- or European Standard EN 149-approved self-contained breathing apparatus that has a full face-piece and is operated in a pressure-demand or other positive-pressure mode) or SaF: Pd,Pp: ASCBA (APF = 10,000) (any supplied-air respirator that has a full face-piece and is operated in a pressure-demand or other positive-pressure mode in combination with an auxiliary self-contained breathing apparatus operated in a pressure-demand or other positive-pressure mode). *Escape:* GmFOv (APF = 50) [any air-purifying, full-face-piece respirator (gas mask) with a chin-style, front- or back-mounted organic vapor canister] or SCBAE (any appropriate escape-type, self-contained breathing apparatus).

Storage: Color Code—Blue: Health Hazard/Poison: Store in a secure poison location. Prior to working with this chemical you should be trained on its proper handling and storage. Store in tightly closed containers in a cool, well-ventilated area away from oxidizers. Where possible, automatically pump liquid from drums or other storage containers to process containers. A regulated, marked area should be established where this chemical is handled, used, or stored in compliance with OSHA Standard 1910.1045.

Shipping: This compound requires a shipping label of "POISONOUS/TOXIC MATERIALS." It falls in Hazard Class 6.1 and Packing Group I.

Spill Handling: Evacuate and restrict persons not wearing protective equipment from area of spill or leak until cleanup is complete. Remove all ignition sources. Ventilate area of spill or leak. Absorb liquids in vermiculite, dry sand, earth, peat, carbon, or a similar material and deposit in sealed containers. Keep this chemical out of a confined space, such as a sewer, because of the possibility of an explosion, unless the sewer is designed to prevent the buildup of explosive concentrations. It may be necessary to contain and dispose of this chemical as a hazardous waste. If material or contaminated runoff enters waterways, notify downstream users of potentially contaminated waters. Contact your local or federal environmental protection agency for specific recommendations. If employees are required to clean up spills, they must be properly trained and equipped. OSHA 1910.120(q) may be applicable.

Initial isolation and protective action distances
Distances shown are likely to be affected during the first 30 min after materials are spilled and could increase with time. If more than one tank car, cargo tank, portable tank, or large cylinder involved in the incident is leaking, the protective action distance may need to be increased. You may need to seek emergency information from CHEMTREC at (800) 424-9300 or seek professional environmental engineering assistance from the US EPA Environmental Response Team at (908) 48-8730 (24-h response line).

Small spills (From a small package or a small leak from a large package)
First: Isolate in all directions (feet/meters)   100/30
Then: Protect persons downwind (miles/kilometers)
Day   0.1/0.2
Night   0.1/0.2

Large spills (From a large package or from many small packages)
First: Isolate in all directions (feet/meters)   300/100
Then: Protect persons downwind (miles/kilometers)
Day   0.2/0.3
Night   0.5/0.8

Fire Extinguishing: Extinguish fire using an agent suitable for type of surrounding fire. Methyl iodide itself does not burn. Poisonous gases, including iodine, are produced in fire. Vapors are heavier than air and will collect in low areas. Vapors may travel long distances to ignition sources and flashback. Vapors in confined areas may explode when exposed to fire. Containers may explode in fire. Storage containers and parts of containers may rocket great distances, in many directions. If material or contaminated runoff enters waterways, notify downstream users of potentially contaminated waters. Notify local health and fire officials and pollution control agencies. From a secure, explosion-proof location, use water spray to cool exposed containers. If cooling streams are ineffective (venting sound increases in volume and pitch, tank discolors, or shows any signs of deforming), withdraw immediately to a secure position. If employees are expected to fight fires, they must be trained and equipped in OSHA 1910.156. The only respirators recommended for firefighting are self-contained breathing apparatuses that have full face-pieces and are operated in a pressure-demand or other positive-pressure mode.

Disposal Method Suggested: Consult with environmental regulatory agencies for guidance on acceptable disposal practices. Generators of waste containing this contaminant (≥100 kg/mo) must conform with EPA regulations governing storage, transportation, treatment, and waste disposal.

Reference
New Jersey Department of Health and Senior Services. (November 1999). *Hazardous Substances Fact Sheet: Methyl Iodide.* Trenton, NJ

Methyl isoamyl ketone   M:0980

Molecular Formula: $C_7H_{14}O$
Common Formula: $CH_3COCH_2CH_2CH(CH_3)_2$
Synonyms: 2-Hexanone, 5-methyl-; Isoamyl methyl ketone; Isopentyl methyl ketone; Ketone, methyl isoamyl; 5-Methylhexan-2-one; 2-Methyl-5-hexanone; 5-Methyl-2-hexanone; MIAK

CAS Registry Number: 110-12-3
RTECS® Number: MP3850000
UN/NA & ERG Number: UN2302/127
EC Number: 203-737-8 [*Annex I Index No.:* 606-026-00-4]
Regulatory Authority and Advisory Bodies
Air Pollutant Standard Set. See below, "Permissible Exposure Limits in Air" section.
Canada, WHMIS, Ingredients Disclosure List Concentration 1.0%.
European/International Regulations: Hazard Symbol: Xn; Risk phrases: R10; R20; Safety phrases: S2; S23; S24/25 (see Appendix 4).
WGK (German Aquatic Hazard Class): 1—Low hazard to waters.
Description: MIAK is a colorless liquid with a pleasant, fruity odor. Molecular weight = 114.21; Specific gravity (H_2O:1) = 0.81; Boiling point = 144°C; Freezing/Melting point = − 74°C; Vapor pressure = 5 mmHg at 20°C; Flash point = 36.1°C (cc); Autoignition temperature = 191°C. Explosive limits: LEL = 1.0% at 93.3°C; UEL = 8.2% at 93.3°C. Hazard Identification (based on NFPA-704 M Rating System): Health 1, Flammability 2, Reactivity 0. Slightly soluble in water; solubility = 0.5%.
Potential Exposure: Compound Description: Agricultural Chemical; Primary Irritant. MIAK is used as a solvent for cellulose esters, acrylics, and vinyl copolymers.
Incompatibilities: Reacts violently with strong oxidizers, strong bases, amines, and isocyanates. Attacks some plastic.
Permissible Exposure Limits in Air
OSHA PEL: 100 ppm/475 mg/m³ TWA.
NIOSH REL: 50 ppm/240 mg/m³ TWA.
ACGIH TLV®[1]: 50 ppm/234 mg/m³ TWA.
Protective Action Criteria (PAC)
TEEL-0: 100 ppm
PAC-1: 150 ppm
PAC-2: 1500 ppm
PAC-3: 1500 ppm
DFG MAK: 10 ppm/47 mg/m³ TWA; Peak Limitation Category I(2) [skin]; Pregnancy Risk Group D.
Australia: TWA 50 ppm (240 mg/m³), 1993; Belgium: TWA 50 ppm (234 mg/m³), 1993; Denmark: TWA 50 ppm (230 mg/m³), 1999; Finland: TWA 50 ppm (230 mg/m³); STEL 75 ppm (350 mg/m³), 1999; France: VME 50 ppm (240 mg/m³), 1999; Norway: TWA 25 ppm (115 mg/m³), 1999; Sweden: TWA 25 ppm (120 mg/m³); STEL 50 ppm (250 mg/m³), 1999; the Netherlands: MAC-TGG 233 mg/m³, 2003; United Kingdom: TWA 50 ppm (237 mg/m³); STEL 100 ppm, [skin], 2000; Argentina, Bulgaria, Columbia, Jordan, South Korea, New Zealand, Singapore, Vietnam: ACGIH TLV®: TWA 50 ppm. In addition, Several states have set guidelines or standards for MIAK in ambient air[60] ranging from 2.4 mg/m³ (North Dakota) to 4.0 mg/m³ (Virginia) to 4.6 mg/m³ (Connecticut) to 5.714 mg/m³ (Nevada).
Determination in Air: OSHA Analytical Method PV-2042.
Determination in Water: Octanol−water coefficient: Log K_{ow} = 1.72 (estimated).

Routes of Entry: Inhalation, ingestion, skin and/or eye contact.
Harmful Effects and Symptoms
Short Term Exposure: Methyl isoamyl ketone can affect you when breathed in and by passing through your skin. Exposure can cause you to become dizzy, lightheaded, and to pass out. Contact can irritate the eyes and skin. Repeated exposure can cause a skin rash. Methyl isoamyl ketone vapors can irritate the nose and throat. MIAK may affect the kidneys.
Long Term Exposure: Skin contact causes skin rash and drying and cracking. High exposures may cause liver damage and may affect the kidneys.
Points of Attack: Eyes, skin, respiratory system, central nervous system, liver, kidneys.
Medical Surveillance: If symptoms develop or overexposure is suspected, the following may be useful: liver and kidney function tests.
First Aid: If this chemical gets into the eyes, remove any contact lenses at once and irrigate immediately for at least 15 min, occasionally lifting upper and lower lids. Seek medical attention immediately. If this chemical contacts the skin, remove contaminated clothing and wash immediately with soap and water. Seek medical attention immediately. If this chemical has been inhaled, remove from exposure, begin rescue breathing (using universal precautions, including resuscitation mask) if breathing has stopped and CPR if heart action has stopped. Transfer promptly to a medical facility. When this chemical has been swallowed, get medical attention. Give large quantities of water and induce vomiting. Do not make an unconscious person vomit.
Personal Protective Methods: Wear protective gloves and clothing to prevent any reasonable probability of skin contact. Safety equipment suppliers/manufacturers can provide recommendations on the most protective glove/clothing material for your operation. All protective clothing (suits, gloves, footwear, headgear) should be clean, available each day, and put on before work. Contact lenses should not be worn when working with this chemical. Wear splash-proof chemical goggles and face shield unless full face-piece respiratory protection is worn. Employees should wash immediately with soap when skin is wet or contaminated. Provide emergency showers and eyewash. See NIOSH Criteria Document 78-173, *Ketones.*
Respirator Selection: NIOSH: *Up to 500 ppm:* CcrOv* (APF = 10) [any chemical cartridge respirator with organic vapor cartridge(s)] or Sa* (APF = 10) (any supplied-air respirator). *Up to 1250 ppm:* Sa:Cf* (APF = 25) (any supplied-air respirator operated in a continuous-flow mode) or PaprOv* (APF = 25) [any powered, air-purifying respirator with organic vapor cartridge(s)]. *Up to 2500 ppm:* CcrFOv (APF = 50) [any chemical cartridge respirator with a full face-piece and organic vapor cartridge(s)] or GmFOv (APF = 50) [any air-purifying, full-face-piece respirator (gas mask) with a chin-style, front- or back-mounted organic vapor canister] or PaprTOv* (APF = 50) [any powered,

air-purifying respirator with a tight-fitting face-piece and organic vapor cartridge(s)] or SaT: Cf* (APF = 50) (any supplied-air respirator that has a tight-fitting face-piece and is operated in a continuous-flow mode) or SCBAF (APF = 50) (any self-contained breathing apparatus with a full face-piece) or SaF (APF = 50) (any supplied-air respirator with a full face-piece). *Up to 5000 ppm:* SaF: Pd,Pp (APF = 2000) (any supplied-air respirator that has a full face-piece and is operated in a pressure-demand or other positive-pressure mode). *Emergency or planned entry into unknown concentrations or IDLH conditions:* SCBAF: Pd,Pp (APF = 10,000) (any NIOSH/MSHA- or European Standard EN 149-approved self-contained breathing apparatus that has a full face-piece and is operated in a pressure-demand or other positive-pressure mode) or SaF: Pd,Pp: ASCBA (APF = 10,000) (any supplied-air respirator that has a full face-piece and is operated in a pressure-demand or other positive-pressure mode in combination with an auxiliary self-contained breathing apparatus operated in a pressure-demand or other positive-pressure mode). *Escape:* GmFOv (APF = 50) [any air-purifying, full-face-piece respirator (gas mask) with a chin-style, front- or back-mounted organic vapor canister] or SCBAE (any appropriate escape-type, self-contained breathing apparatus).

*Substance reported to cause eye irritation or damage; may require eye protection.

Storage: Color Code—Red: Flammability Hazard: Store in a flammable liquid storage area or approved cabinet away from ignition sources and corrosive and reactive materials. Prior to working with this chemical you should be trained on its proper handling and storage. Before entering confined space where this chemical may be present, check to make sure that an explosive concentration does not exist. Store in tightly closed containers in a cool, well-ventilated area away from oxidizers (such as perchlorates, peroxides, permanganates, chlorates, and nitrates), strong oxidizers (such as chlorine, bromine, and fluorine), reducing agents, and aldehydes. Sources of ignition, such as smoking and open flames, are prohibited where methyl isoamyl ketone is handled, used, or stored. Metal containers involving the transfer of 5 gallons or more of methyl isoamyl ketone should be grounded and bonded. Drums must be equipped with self-closing valves, pressure vacuum bungs, and flame arresters.

Shipping: 5-Methylhexan-2-one compound requires a shipping label of "FLAMMABLE LIQUID." It falls in Hazard Class 3 and Packing Group II.

Spill Handling: Evacuate and restrict persons not wearing protective equipment from area of spill or leak until cleanup is complete. Remove all ignition sources. Establish forced ventilation to keep levels below explosive limit. Absorb liquids in vermiculite, dry sand, earth, peat, carbon, or a similar material and deposit in sealed containers. Keep this chemical out of a confined space, such as a sewer, because of the possibility of an explosion, unless the sewer is designed to prevent the buildup of explosive concentrations. It may be necessary to contain and dispose of this chemical as a hazardous waste. If material or contaminated runoff enters waterways, notify downstream users of potentially contaminated waters. Contact your local or federal environmental protection agency for specific recommendations. If employees are required to clean up spills, they must be properly trained and equipped. OSHA 1910.120(q) may be applicable.

Fire Extinguishing: This chemical is a combustible liquid. Poisonous gases are produced in fire. Use dry chemical, carbon dioxide, or alcohol foam extinguishers. Vapors are heavier than air and will collect in low areas. Vapors may travel long distances to ignition sources and flashback. Vapors in confined areas may explode when exposed to fire. Containers may explode in fire. Storage containers and parts of containers may rocket great distances, in many directions. If material or contaminated runoff enters waterways, notify downstream users of potentially contaminated waters. Notify local health and fire officials and pollution control agencies. From a secure, explosion-proof location, use water spray to cool exposed containers. If cooling streams are ineffective (venting sound increases in volume and pitch, tank discolors, or shows any signs of deforming), withdraw immediately to a secure position. If employees are expected to fight fires, they must be trained and equipped in OSHA 1910.156. The only respirators recommended for firefighting are self-contained breathing apparatuses that have full face-pieces and are operated in a pressure-demand or other positive-pressure mode.

Disposal Method Suggested: Dissolve or mix the material with a combustible solvent and burn in a chemical incinerator equipped with an afterburner and scrubber. All federal, state, and local environmental regulations must be observed.

Reference
New Jersey Department of Health and Senior Services. (October 1999). *Hazardous Substances Fact Sheet: Methyl Isoamyl Ketone.* Trenton, NJ

Methyl isobutyl carbinol   M:0990

Molecular Formula: $C_6H_{14}O$
Common Formula: $CH_3CHOHCH_2CH(CH_3)_2$
Synonyms: Alcool methyl amylique (French); Isobutylmethylcarbinol; Isobutylmethylmethanol; MAOH; Methyl amyl alcohol; Methylisobutyl carbinol; 2-Methyl-4-pentanol; 4-Methylpentanol-2; 4-Methyl-2-pentanol; MIBC; MIC; 3-MIC; 2-Pentanol, 4-methyl-
CAS Registry Number: 108-11-2
RTECS® Number: SA7350000
UN/NA & ERG Number: UN2053/129
EC Number: 203-551-7 [*Annex I Index No.:* 603-008-00-8]
Regulatory Authority and Advisory Bodies
Air Pollutant Standard Set. See below, "Permissible Exposure Limits in Air" section.
Canada, WHMIS, Ingredients Disclosure List Concentration 1.0%.

European/International Regulations: Hazard Symbol: Xi; Risk phrases: R10; R37; Safety phrases: S2; S24/25 (see Appendix 4).
WGK (German Aquatic Hazard Class): 1—Low hazard to waters.

Description: MIBC is a colorless liquid with a mild odor. The odor threshold is 0.52 ppm. Molecular weight = 102.20; Specific gravity (H_2O:1) = 0.81; Boiling point = 132.8°C; Freezing/Melting point = −90°C; Vapor pressure = 3 mmHg at 20°C; Flash point = 41°C. The explosive limits are LEL = 1.0%; UEL = 5.5%. Hazard Identification (based on NFPA-704 M Rating System): Health 2, Flammability 2, Reactivity 0. Slightly soluble in water; solubility = 2%.

Potential Exposure: MIBC is used as a solvent; in the formulation of brake fluids; as an intermediate in organic synthesis.

Incompatibilities: Forms explosive mixture with air. Contact with alkali metals produces hydrogen gas. Incompatible with strong oxidizers, strong acids, caustics, aliphatic amines, isocyanates.

Permissible Exposure Limits in Air
Conversion factor: 1 ppm = 4.18 mg/m³ at 25°C & 1 atm.
OSHA PEL: 25 ppm/100 mg/m³ TWA [skin].
NIOSH REL: 25 ppm/100 mg/m³ TWA; 40 ppm/165 mg/m³ STEL [skin].
ACGIH TLV®[1]: 25 ppm/104 mg/m³ TWA; 40 ppm/167 mg/m³ STEL [skin].
NIOSH IDLH: 400 ppm.
No TEEL available.
DFG MAK: 20 ppm/85 mg/m³ TWA; Peak Limitation Category I(1); Pregnancy Risk Group D.
Australia: TWA 25 ppm (100 mg/m³); STEL 40 ppm, [skin], 1993; Austria: MAK 25 ppm (50 mg/m³), [skin], 1999; Belgium: TWA 25 ppm (104 mg/m³); STEL 40 ppm, [skin] 1999; Denmark: TWA 25 ppm (100 mg/m³), [skin], 1999; Finland: TWA 25 ppm (100 mg/m³); STEL 40 ppm (170 mg/m³), [skin], 1999; France: VME 25 ppm (100 mg/m³), [skin], 1999; the Netherlands: MAC-TGG 100 mg/m³, [skin], 2003; Russia: STEL 10 mg/m³, [skin], 1993; Poland: MAC (TWA) 100 mg/m³; MAC (STEL) 160 mg/m³, 1999; Switzerland: TWA 25 ppm (100 mg/m³); STEL 125 ppm, [skin], 1993; Switzerland: MAK-W 25 ppm (100 mg/m³), KZG-W 125 ppm (500 mg/m³), [skin], 1999; United Kingdom: TWA 25 ppm (106 mg/m³); STEL 40 ppm, [skin], 2000; Argentina, Bulgaria, Columbia, Jordan, South Korea, New Zealand, Singapore, Vietnam: ACGIH TLV®: STEL 40 ppm [skin]. North Dakota has set a guideline of 1.0−1.65 mg/m³ for MIBC in ambient air.[60]

Determination in Air: Use NIOSH Analytical Method #1402, alcohols II; #1405; OSHA Analytical Method 7.[18]
Determination in Water: Octanol−water coefficient: Log K_{ow} = 1.4.
Routes of Entry: Inhalation, ingestion, eye and/or skin contact.

Harmful Effects and Symptoms
Short Term Exposure: MIBC can affect you when breathed in and by passing through your skin. Breathing the vapor can irritate the eyes, nose, and throat. Contact with the liquid can burn the eyes and can irritate the skin. Exposure to high concentrations can cause you to feel dizzy, lightheaded, and to pass out.

Long Term Exposure: Long-term contact can cause drying and cracking of the skin.

Points of Attack: Eyes, skin, central nervous system.
Medical Surveillance: NIOSH lists the following tests: whole blood (chemical/metabolite), expired air, pulmonary function tests; urine (chemical/metabolite), end-of-shift; urine (chemical/metabolite) end-of-work-week.

First Aid: If this chemical gets into the eyes, remove any contact lenses at once and irrigate immediately for at least 15 min, occasionally lifting upper and lower lids. Seek medical attention immediately. If this chemical contacts the skin, remove contaminated clothing and wash immediately with soap and water. Seek medical attention immediately. If this chemical has been inhaled, remove from exposure, begin rescue breathing (using universal precautions, including resuscitation mask) if breathing has stopped and CPR if heart action has stopped. Transfer promptly to a medical facility. When this chemical has been swallowed, get medical attention. Give large quantities of water and induce vomiting. Do not make an unconscious person vomit.

Personal Protective Methods: Wear protective gloves and clothing to prevent any reasonable probability of skin contact. Safety equipment suppliers/manufacturers can provide recommendations on the most protective glove/clothing material for your operation. All protective clothing (suits, gloves, footwear, headgear) should be clean, available each day, and put on before work. Contact lenses should not be worn when working with this chemical. Wear splash-proof chemical goggles and face shield unless full face-piece respiratory protection is worn. Employees should wash immediately with soap when skin is wet or contaminated. Provide emergency showers and eyewash.

Respiratory Selection: 250 ppm: Sa (APF = 10) (any supplied-air respirator). *400 ppm:* Sa:Cf (APF = 25) (any supplied-air respirator operated in a continuous-flow mode) or SCBAF (APF = 50) (any self-contained breathing apparatus with a full face-piece) or SaF (APF = 50) (any supplied-air respirator with a full face-piece). *At any concentrations above the NIOSH REL, or where there is no REL, at any detectable concentration:* SCBAF: Pd,Pp (APF = 10,000) (any NIOSH/MSHA- or European Standard EN 149-approved self-contained breathing apparatus that has a full face-piece and is operated in a pressure-demand or other positive-pressure mode) or SaF: Pd,Pp: ASCBA (APF = 10,000) (any supplied-air respirator that has a full face-piece and is operated in a pressure-demand or other positive-pressure mode in combination with an auxiliary self-contained breathing apparatus operated in a pressure-demand or other positive-pressure mode). *Escape:* GmFOv

(APF = 50) [any air-purifying, full-face-piece respirator (gas mask) with a chin-style, front- or back-mounted organic vapor canister] or SCBAE (any appropriate escape-type, self-contained breathing apparatus).

Storage: Color Code—Red: Flammability Hazard: Store in a flammable liquid storage area or approved cabinet away from ignition sources and corrosive and reactive materials. Prior to working with this chemical you should be trained on its proper handling and storage. Before entering confined space where this chemical may be present, check to make sure that an explosive concentration does not exist. Methyl amyl alcohol must be stored to avoid contact with peroxides, chlorates, perchlorates, permanganates, and nitrates since violent reactions occur. Store in tightly closed containers in a cool, well-ventilated area away from heat. Sources of ignition, such as smoking and open flames, are prohibited where methyl amyl alcohol is used, handled, or stored in a manner that could create a potential fire or explosion hazard.

Shipping: This compound requires a shipping label of "FLAMMABLE LIQUID." It falls in Hazard Class 3 and Packing Group II.

Spill Handling: Evacuate and restrict persons not wearing protective equipment from area of spill or leak until cleanup is complete. Remove all ignition sources. Establish forced ventilation to keep levels below explosive limit. Absorb liquids in vermiculite, dry sand, earth, peat, carbon, or a similar material and deposit in sealed containers. Keep this chemical out of a confined space, such as a sewer, because of the possibility of an explosion, unless the sewer is designed to prevent the buildup of explosive concentrations. It may be necessary to contain and dispose of this chemical as a hazardous waste. If material or contaminated runoff enters waterways, notify downstream users of potentially contaminated waters. Contact your local or federal environmental protection agency for specific recommendations. If employees are required to clean up spills, they must be properly trained and equipped. OSHA 1910.120(q) may be applicable.

Fire Extinguishing: This chemical is a combustible liquid. Poisonous gases are produced in fire. Use dry chemical, carbon dioxide, or alcohol foam extinguishers. Vapors are heavier than air and will collect in low areas. Vapors may travel long distances to ignition sources and flashback. Vapors in confined areas may explode when exposed to fire. Containers may explode in fire. Storage containers and parts of containers may rocket great distances, in many directions. If material or contaminated runoff enters waterways, notify downstream users of potentially contaminated waters. Notify local health and fire officials and pollution control agencies. From a secure, explosion-proof location, use water spray to cool exposed containers. If cooling streams are ineffective (venting sound increases in volume and pitch, tank discolors, or shows any signs of deforming), withdraw immediately to a secure position. If employees are expected to fight fires, they must be trained and equipped in OSHA 1910.156. The only respirators recommended for firefighting are self-contained breathing apparatuses that have full face-pieces and are operated in a pressure-demand or other positive-pressure mode.

Disposal Method Suggested: Incineration; other, more flammable solvent may be added.

Reference

New Jersey Department of Health and Senior Services. (April 2004). *Hazardous Substances Fact Sheet: Methyl n-Amyl Alcohol.* Trenton, NJ

Methyl isobutyl ketone   M:1000

Molecular Formula: $C_6H_{12}O$
Common Formula: $CH_3COCH_2CH(CH_3)_2$
Synonyms: Hexone; Isobutyl methyl ketone; Isopropylacetone; KTI COP Rinse I; KTI PMMA Rinse; Methyl-isobutyl-cetone (French); 2-Methyl-4-pentanone; 4-Methyl-2-pentanone; MIBK; MIK; 2-Pentanone, 4-methyl-; Metil isobutil cetona (Spanish); RN-10 E beam negative resist rinse; Shell MIBK
CAS Registry Number: 108-10-1
RTECS® Number: SA9275000
UN/NA & ERG Number: UN1245/127
EC Number: 203-550-1 [*Annex I Index No.:* 606-004-00-4]
Regulatory Authority and Advisory Bodies
Carcinogenicity: EPA: Available data are inadequate for an assessment of human carcinogenic potential.
Air Pollutant Standard Set. See below, "Permissible Exposure Limits in Air" section.
Clean Air Act: Hazardous Air Pollutants (Title I, Part A, Section 112).
US EPA Hazardous Waste Number (RCRA No.): U161.
RCRA, 40CFR261, Appendix 8 Hazardous Constituents.
RCRA 40CFR268.48; 61FR15654, Universal Treatment Standards: Wastewater (mg/L), 0.14; Nonwastewater (mg/kg), 33.
Safe Drinking Water Act: Priority List (55 FR 1470).
Reportable Quantity (RQ): 5000 lb (2270 kg).
EPCRA Section 313 Form R *de minimis* concentration reporting level: 1.0%.
Canada, WHMIS, Ingredients Disclosure List Concentration 1.0%.
European/International Regulations: Hazard Symbol: F, Xn; Risk phrases: R11; R20; R36/37; R66; Safety phrases: S2; S9; S16; S29 (see Appendix 4).
WGK (German Aquatic Hazard Class): 1—Low hazard to waters.
Description: MIBK is a colorless liquid with a pleasant, sweet, fruity odor. The odor threshold is 0.88 ppm. Molecular weight = 100.18; Specific gravity (H_2O:1) = 0.80; Boiling point = 116.7°C; Freezing/Melting point = −84.4°C; Vapor pressure = 16 mmHg at 20°C; Flash point = 18°C; Autoignition temperature: 450°C. Explosive limits: LEL = 1.2% at 93°C; UEL = 8.0% at 93°C. Hazard Identification (based on NFPA-704 M Rating

System): Health 2, Flammability 3, Reactivity 1. Slightly soluble in water; solubility = 2%.

Potential Exposure: Compound Description: Tumorigen; Reproductive Effector; Human Data; Primary Irritant. MIBK is used as a solvent; a denaturant; and as an extractant; in the manufacture of methyl amyl alcohol; as a solvent in paints, varnishes, and lacquers; as an alcohol denaturant; as a solvent in uranium extraction from fission products.

Incompatibilities: Able to form unstable and explosive peroxides on contact with air. Reacts violently with strong oxidizers, potassium *tert*-butoxide, strong acids, aliphatic amines, reducing agents.

Permissible Exposure Limits in Air

Conversion factor: 1 ppm = 4.10 mg/m^3 at 25°C & 1 atm.
OSHA PEL: 100 ppm/410 mg/m^3 TWA.
NIOSH REL: 50 ppm/205 mg/m^3 TWA; 75 ppm/300 mg/m^3 STEL.
ACGIH TLV®[1]: 20 ppm/82 mg/m^3 TWA; 75 ppm/307 mg/m^3 STEL; BEI issued.
NIOSH IDLH: 500 ppm.
Protective Action Criteria (PAC)
TEEL-0: 20 ppm
PAC-1: 75 ppm
PAC-2: 75 ppm
PAC-3: 500 ppm
DFG MAK: 20 ppm/83 mg/m^3 Peak Limitation Category I (2) [skin]; Pregnancy Risk Group C; BAT: 3.5 mg[hexone]/L in urine/end-of-shift.
Compound Description: Tumorigen; Reproductive Effector; Human Data; Primary Irritant.
Austria: MAK 100 ppm (400 mg/m^3), 1999; Denmark: TWA 25 ppm (100 mg/m^3), [skin], 1999; France: VME 50 ppm (205 mg/m^3), 1999; Japan: 50 ppm (105 mg/m^3), [skin], 1999; Norway: TWA 25 ppm (105 mg/m^3), 1999; Poland: MAC (TWA) 200 mg/m^3; MAC (STEL) 300 mg/m^3, 1999; Sweden: NGV 25 ppm (100 mg/m^3), KTV 50 ppm (200 mg/m^3), 1999; the Netherlands: MAC-TGG 104 mg/m^3, 2003. For ambient air in residential areas: The Czech Republic[35]: MAC of 0.2 mg/m^3 both on a momentary and a daily average basis; Russia[35]: MAC of 0.1 mg/m^3 on a once daily basis. Several states have set guidelines or standards for MIBK in ambient air[60] ranging from 0.28 mg/m^3 (Massachusetts) to 0.683 mg/m^3 (New York) to 2.05 mg/m^3 (Florida, North Dakota, South Carolina) to 2.50 mg/m^3 to 3.4 mg/m^3 (Virginia) to 4.1 mg/m^3 (Connecticut) to 4.76 mg/m^3 (Nevada) to 30.8 mg/m^3 (North Carolina).

Determination in Air: Charcoal adsorption, workup with CS_2; analysis by gas chromatography/flame ionization. See NIOSH (I) Method #1300, Ketones; #2555; OSHA Analytical Method 1004.[18]

Permissible Concentration in Water: The state of Massachusetts has set[61] a guideline of 40 µg/L (0.04 mg/L) for MIBK in drinking water. Russia has set a limit of 0.2 mg/L in surface water.

Routes of Entry: Inhalation, ingestion, skin and/or eye contact.

Harmful Effects and Symptoms

Short Term Exposure: Methyl isobutyl ketone can affect you when breathed in. Exposure to high concentrations can cause you to feel dizzy and lightheaded, and to pass out. Breathing the vapor may cause loss of appetite, nausea, vomiting, and diarrhea. Contact or the vapor can irritate the eyes, nose, mouth, throat. Contact can irritate the skin. Ingestion caused chemical pneumonitis.

Long Term Exposure: Long-term exposure may damage the liver and kidneys. Repeated or prolonged contact with skin may cause drying and cracking.

Points of Attack: Eyes, skin, respiratory system, central nervous system, liver, kidneys.

Medical Surveillance: If symptoms develop or overexposure is suspected, the following may be useful: liver and kidney function tests.

First Aid: If this chemical gets into the eyes, remove any contact lenses at once and irrigate immediately for at least 15 min, occasionally lifting upper and lower lids. Seek medical attention immediately. If this chemical contacts the skin, remove contaminated clothing and wash immediately with soap and water. Seek medical attention immediately. If this chemical has been inhaled, remove from exposure, begin rescue breathing (using universal precautions, including resuscitation mask) if breathing has stopped and CPR if heart action has stopped. Transfer promptly to a medical facility. When this chemical has been swallowed, get medical attention. Give large quantities of water and induce vomiting. Do not make an unconscious person vomit.

Personal Protective Methods: Wear protective gloves and clothing to prevent any reasonable probability of skin contact. Safety equipment suppliers/manufacturers can provide recommendations on the most protective glove/clothing material for your operation. Teflon™ and styrene—butadiene rubber are among the recommended protective materials. All protective clothing (suits, gloves, footwear, headgear) should be clean, available each day, and put on before work. Contact lenses should not be worn when working with this chemical. Wear splash-proof chemical goggles and face shield unless full face-piece respiratory protection is worn. Employees should wash immediately with soap when skin is wet or contaminated. Provide emergency showers and eyewash.

Respirator Selection: *500 ppm:* CcrOv (APF = 10) [any chemical cartridge respirator with organic vapor cartridge (s)] or GmFOv (APF = 50) [any air-purifying, full-face-piece respirator (gas mask) with a chin-style, front- or back-mounted organic vapor canister] or PaprOv (APF = 25) [any powered, air-purifying respirator with organic vapor cartridge(s)] or Sa (APF = 10) (any supplied-air respirator) or SCBAF (APF = 50) (any self-contained breathing apparatus with a full face-piece). *At any concentrations above the NIOSH REL, or where there is no REL, at any detectable concentration:* SCBAF: Pd,Pp (APF = 10,000) (any NIOSH/MSHA- or European Standard EN 149-approved self-contained breathing apparatus that has a full face-piece and is operated in a pressure-demand or other positive-pressure mode) or SaF: Pd,Pp: ASCBA (APF = 10,000) (any supplied-air respirator that has a full face-piece and is operated in a pressure-demand or other positive-pressure mode in combination with an auxiliary self-contained breathing apparatus operated in a pressure-demand or other positive-pressure mode). *Escape:* GmFOv (APF = 50) [any air-purifying, full-face-piece respirator (gas mask) with a chin-style, front- or back-mounted organic vapor canister] or SCBAE (any appropriate escape-type, self-contained breathing apparatus).

Storage: Color Code—Red: Flammability Hazard: Store in a flammable liquid storage area or approved cabinet away from ignition sources and corrosive and reactive materials. Prior to working with this chemical you should be trained on its proper handling and storage. Before entering confined space where this chemical may be present, check to make sure that an explosive concentration does not exist. Methyl isobutyl ketone must be stored to avoid contact with strong oxidizers because violent reactions occur. Store in tightly closed containers in a cool, well-ventilated area away from heat, sparks, and flames. Sources of ignition, such as smoking and open flames, are prohibited where methyl isobutyl ketone is used, handled, or stored in a manner that could crate a potential fire or explosion hazard. Metal containers involving the transfer of 5 gallons or more of methyl isobutyl ketone should be grounded and bonded. Drums must be equipped with self-closing valves, pressure vacuum bungs, and flame arresters. Use only nonsparking tools and equipment, especially when opening and closing containers of methyl isobutyl ketone.

Shipping: This compound requires a shipping label of "FLAMMABLE LIQUID." It falls in Hazard Class 3 and Packing Group II.

Spill Handling: Evacuate and restrict persons not wearing protective equipment from area of spill or leak until cleanup is complete. Remove all ignition sources. Establish forced ventilation to keep levels below explosive limit. Absorb liquids in vermiculite, dry sand, earth, peat, carbon, or a similar material and deposit in sealed containers. Oil-skimming equipment and sorbent foams can be applied to slick if done immediately. Keep this chemical out of a confined space, such as a sewer, because of the possibility of an explosion, unless the sewer is designed to prevent the buildup of explosive concentrations. It may be necessary to contain and dispose of this chemical as a hazardous waste. If material or contaminated runoff enters waterways, notify downstream users of potentially contaminated waters. Contact your local or federal environmental protection agency for specific recommendations. If employees are required to clean up spills, they must be properly trained and equipped. OSHA 1910.120(q) may be applicable.

Fire Extinguishing: This chemical is a flammable liquid. Poisonous gases are produced in fire. Use dry chemical, carbon dioxide, or alcohol foam extinguishers. Vapors are heavier than air and will collect in low areas. Vapors may travel long distances to ignition sources and flashback. Vapors in confined areas may explode when exposed to fire. Containers may explode in fire. Storage containers and parts of containers may rocket great distances, in many directions. If material or contaminated runoff enters waterways, notify downstream users of potentially contaminated waters. Notify local health and fire officials and pollution control agencies. From a secure, explosion-proof location, use water spray to cool exposed containers. If cooling streams are ineffective (venting sound increases in volume and pitch, tank discolors, or shows any signs of deforming), withdraw immediately to a secure position. If employees are expected to fight fires, they must be trained and equipped in OSHA 1910.156. The only respirators recommended for firefighting are self-contained breathing apparatuses that have full face-pieces and are operated in a pressure-demand or other positive-pressure mode.

Disposal Method Suggested: Consult with environmental regulatory agencies for guidance on acceptable disposal practices. Generators of waste containing this contaminant (≥100 kg/mo) must conform with EPA regulations governing storage, transportation, treatment, and waste disposal. Incineration.[22]

References

Nat. Inst. for Occupational Safety and Health. (1978). *Criteria for a Recommended Standard: Occupational Exposure to Ketones,* NIOSH Document No. 78-173. Washington, DC

US Environmental Protection Agency. (April 30, 1980). *Methyl Isobutyl Ketone: Health and Environmental Effects Profile No. 129.* Washington, DC: Office of Solid Waste

New Jersey Department of Health and Senior Services. (August 2005). *Hazardous Substances Fact Sheet: Methyl Isobutyl Ketone.* Trenton, NJ

Methyl isocyanate   M:1010

Molecular Formula: C_2H_3NO
Common Formula: CH_3NCO
Synonyms: Isocyanate de methyle (French); Isocianato de metilo (Spanish); Isocyanatomethane; Isocyanic acid, methyl ester; Isocyanate methane; Methane, isocyanato-; Methylcarbamyl amine; Methyl carbonimide; Methyl ester of isocyanic acid; Methyl isocyanat (German)
CAS Registry Number: 624-83-9
RTECS® Number: NQ9450000
UN/NA & ERG Number: UN2480/155
EC Number: 210-866-3 [*Annex I Index No.:* 615-001-00-7]
Regulatory Authority and Advisory Bodies
Department of Homeland Security Screening Threshold Quantity (pounds): *Release hazard* 10,000 (≥1.00% concentration).

Air Pollutant Standard Set. See below, "Permissible Exposure Limits in Air" section.

OSHA 29CFR1910.119, Appendix A. Process Safety List of Highly Hazardous Chemicals, TQ = 250 lb.

Clean Air Act: Hazardous Air Pollutants (Title I, Part A, Section 112); Accidental Release Prevention/Flammable Substances, (Section 112[r], Table 3), TQ = 10,000 lb (4540 kg).

US EPA Hazardous Waste Number (RCRA No.): P064.

RCRA, 40CFR261, Appendix 8 Hazardous Constituents.

Superfund/EPCRA 40CFR355, Extremely Hazardous Substances: TPQ = 500 lb (227 kg).

Reportable Quantity (RQ): 10 lb (4.54 kg).

EPCRA Section 313 Form R *de minimis* concentration reporting level: 1.0%.

US DOT 49CFR172.101, Inhalation Hazardous Chemical.

California Proposition 65 Chemical: Developmental/Reproductive toxin (female) 11/12/10.

Canada, WHMIS, Ingredients Disclosure List Concentration 0.1%.

European/International Regulations: Hazard Symbol: F, T + ; R11; R24/25; R26; R37/38; R42; R42/43; R63; Safety phrases: S1/2; S16; S26; S27/28; S36/37/39; S43; S45; S63 (see Appendix 4).

WGK (German Aquatic Hazard Class): No value assigned.

Description: Methyl isocyanate is a colorless liquid with a sharp odor which is a lachrymator (causes tears). Molecular weight = 57.06; Specific gravity (H_2O:1) = 0.96; Boiling point = 38.8−40°C; Freezing/Melting point = − 45°C; Vapor pressure = 348 mmHg at 20°C; Flash point = − 7°C; Autoignition temperature = 534°C. Explosive limits: LEL = 5.3%; UEL = 26°C. Hazard Identification (based on NFPA-704 M Rating System): Health 4, Flammability 3, Reactivity 2. Reacts with water; solubility = 10% at 15°C.

Potential Exposure: Methyl isocyanate is used in carbamates and as chemical intermediate; in the manufacture of a wide variety of pesticides; in the production of polyurethane foams and plastics. A release of this chemical was involved in the world's largest chemical accident, causing the death of thousands of industrial workers in 1984 in Bhopal, India.

Incompatibilities: Rapid reaction in the presence of acid, alkalis, amine, iron, tin, copper, their salts, or their catalysts (such as triphenylarsenic oxide, triethylphosphine, and tributyltin oxide). Exothermic reaction with water, producing carbon dioxide, highly flammable and air-reactive methylamine, dimethylurea, and/or trimethyl biuret. The reaction with water is slow at ≤ 20°C but violent at elevated temperatures and/or in the presence of acids and bases. Elevated temperatures may cause polymerization; usually contains inhibitors to prevent polymerization. Reacts with water, acids, alcohols, glycols, amines, amides, ammonia, caprolactum, caustics, strong oxidizers. Attacks some plastics, rubber, or coatings. Attacks some forms of plastic, rubber, and coatings.

Permissible Exposure Limits in Air

Conversion factor: 1 ppm = 2.34 mg/m³ at 25°C & 1 atm.

OSHA PEL: 0.02 ppm/0.05 mg/m³ TWA [skin].

NIOSH REL: 0.02 ppm/0.05 mg/m³ TWA [skin].

ACGIH TLV®[1]: 0.02 ppm/0.047 mg/m³ TWA [skin].

NIOSH IDLH: 3 ppm.

Protective Action Criteria (PAC)*

TEEL-0: 0.02 ppm

PAC-1: **0.025** ppm

PAC-2: **0.067** ppm

PAC-3: **0.2** ppm

*AEGLs (Acute Emergency Guideline Levels) & ERPGs (Emergency Response Planning Guideline) are in **bold face**.

DFG MAK: 0.01 ppm/0.024 mg/m³ TWA; Peak Limitation Category I(1); Pregnancy Risk Group D.

Austria: MAK 0.01 ppm (0.025 mg/m³), 1999; Belgium: TWA 0.02 ppm (0.047 mg/m³), [skin], 1993; Denmark: TWA 0.01 ppm (0.03 mg/m³), [skin], 1999; France: VME 0.02 ppm (0.05 mg/m³), [skin], 1999; Hungary: TWA 0.05 mg/m³; STEL 0.06 mg/m³, [skin], 1993; Norway: TWA 0.005 ppm (0.015 mg/m³), 1999; the Netherlands: MAC-TGG 0.05 mg/m³, [skin], 2003; Russia: STEL 0.05 mg/m³, [skin], 1993; Sweden: TWA 0.005 ppm, ceiling 0.01 ppm, 1999; Switzerland: MAK-W 0.01 ppm (0.025 mg/m³), KZG-W 0.02 ppm (0.05 mg/m³), 1999; United Kingdom: TWA 0.02 mg[NCO]/m³; STEL 0.07 mg [NCO]/m³, 2000; Argentina, Bulgaria, Columbia, Jordan, South Korea, New Zealand, Singapore, Vietnam: ACGIH TLV®: TWA 0.02 ppm [skin]. Several states have set guidelines or standards for methyl isocyanate in ambient air[60] ranging from 0.17 μg/m³ (New York) to 0.5 μg/m³ (North Dakota) to 0.8 μg/m³ (Virginia) to 1.0 μg/m³ (Connecticut and South Dakota).

Determination in Air: Use OSHA Analytical Method 54.

Routes of Entry: Inhalation, skin absorption, ingestion, skin and/or eye contact.

Harmful Effects and Symptoms

Short Term Exposure: Inhalation: Corrosive to the respiratory tract. Causes irritation of eyes, nose, throat, and lungs, bronchitis, cough, shortness of breath, increased secretions, chest pain, difficulty in breathing, and increased blood acidity. Allergic reactions may occur and trigger asthmatic response in sensitized individuals. Higher exposures can cause pulmonary edema, a medical emergency that can be delayed for several hours. This can cause death. Results of human volunteer experiments indicate the exposure for 1−5 min at 0.4 ppm causes no irritation of eyes, nose, or throat; at 2 ppm, irritation and tearing; at 4 ppm, stronger symptoms; and at 21 ppm exposure is unbearable; NIOSH reports that 3 ppm is immediately dangerous to life and health (IDLH). Improvement will occur in a few days if the dose is very low, and proper supportive therapy is given. At higher doses and longer durations of exposure, death may be immediate or delayed more than a month. *Skin:* Corrosive. Extremely irritating; can cause chemical burns. *Eyes:* Corrosive to the eyes. Levels of 2 ppm may cause irritation and tearing. Ulceration has occurred at high levels.

Very high levels may lead to permanent damage and blindness. *Ingestion:* Corrosive; causes vomiting, diarrhea, and abdominal pain.

Long Term Exposure: May cause allergic sensitization of skin and respiratory tract. Subsequent exposure even at extremely low levels may cause asthma attacks. May cause chronic lung disease and increased susceptibility to lung infection. May affect the lungs, causing tissue lesions. Exposure may cause miscarriages among pregnant women.

Points of Attack: Eyes, skin, respiratory system.

Medical Surveillance: NIOSH lists the following tests: Blood Gas Analysis; chest X-ray, electrocardiogram, pulmonary function tests: forced vital capacity, forced expiratory volume (1 s); sputum cytology; white blood cell count/differential. Before beginning employment and at regular times after that, the following are recommended: lung function tests. These may be normal at first if person is not having an attack at the time. Evaluation by a qualified allergist, including careful exposure history and special testing, may help diagnose skin allergy.

First Aid: Warning: Effects may be delayed for up to 15 h. If this chemical gets into the eyes, remove any contact lenses at once and irrigate immediately for at least 15 min, occasionally lifting upper and lower lids. Seek medical attention immediately. If this chemical contacts the skin, remove contaminated clothing and wash immediately with soap and water. Seek medical attention immediately. If this chemical has been inhaled, remove from exposure, begin rescue breathing (using universal precautions, including resuscitation mask) if breathing has stopped and CPR if heart action has stopped. Transfer promptly to a medical facility. When this chemical has been swallowed, get medical attention. Give large quantities of water and induce vomiting. Do not make an unconscious person vomit. Medical observation is recommended for 24–48 h after breathing overexposure, as pulmonary edema may be delayed. As first aid for pulmonary edema, a doctor or authorized paramedic may consider administering a corticosteroid spray.

Personal Protective Methods: Wear protective gloves and clothing to prevent any reasonable probability of skin contact: Prevent skin contact. **8 h:** polyvinyl alcohol gloves; Barricade™ coated suits; Responder™ suits, Trellchem™ HPS suits; Trychem 1000™ suits. Safety equipment suppliers/manufacturers can provide recommendations on the most protective glove/clothing material for your operation. All protective clothing (suits, gloves, footwear, headgear) should be clean, available each day, and put on before work. Contact lenses should not be worn when working with this chemical. Wear splash-proof chemical goggles and face shield unless full face-piece respiratory protection is worn. Employees should wash immediately with soap when skin is wet or contaminated. Provide emergency showers and eyewash.

Respirator Selection: *0.2 ppm:* Sa (APF = 10) (any supplied-air respirator). *0.5 ppm:* Sa:Cf (APF = 25) (any supplied-air respirator operated in a continuous-flow mode).

1 ppm: SCBAF (APF = 50) (any self-contained breathing apparatus with a full face-piece) or SaF (APF = 50) (any supplied-air respirator with a full face-piece). *3 ppm:* SaF: Pd,Pp (APF = 2000) (any supplied-air respirator that has a full face-piece and is operated in a pressure-demand or other positive-pressure mode). *Emergency or planned entry into unknown concentrations or IDLH conditions:* SCBAF: Pd,Pp (APF = 10,000) (any self-contained breathing apparatus that has a full face-piece and is operated in a pressure-demand or other positive-pressure mode) or SaF: Pd,Pp: ASCBA (APF = 10,000) (any supplied-air respirator that has a full face-piece and is operated in a pressure-demand or other positive-pressure mode in combination with an auxiliary self-contained breathing apparatus operated in a pressure-demand or other positive-pressure mode). *Escape:* GmFOv (APF = 50) [any air-purifying, full-face-piece respirator (gas mask) with a chin-style, front- or back-mounted acid gas canister] or SCBAE (any appropriate escape-type, self-contained breathing apparatus).

Note: Substance reported to cause eye irritation or damage; may require eye protection.

Storage: Color Code—Blue: Health Hazard/Poison: Store in a secure poison location. Prior to working with this chemical you should be trained on its proper handling and storage. Before entering confined space where this chemical may be present, check to make sure that an explosive concentration does not exist. Methyl isocyanate must be stored to avoid contact with water, acid, alkali, amines; or iron, tin, copper (or their salts); and certain other catalysts since violent reactions occur. Store in tightly closed containers in a cool, well-ventilated area away from water or heat. Sources of ignition, such as smoking and open flames, are prohibited where methyl isocyanate is handled, used, or stored. Metal containers involving the transfer of 5 gallons or more of methyl isocyanate should be grounded and bonded. Drums must be quipped with self-closing valves, pressure vacuum bungs, and flame arresters. Use only non-sparking tools and equipment, especially when opening and closing containers of methyl isocyanate. Wherever methyl isocyanate is used, handled, manufactured, or stored, use explosion-proof electrical equipment and fittings.

Shipping: This compound requires a shipping label of "POISONOUS/TOXIC MATERIALS, FLAMMABLE LIQUID." It falls in Hazard Class 6.1 and Packing Group I.

Spill Handling: Evacuate and restrict persons not wearing protective equipment from area of spill or leak until cleanup is complete. Remove all ignition sources. Establish forced ventilation to keep levels below explosive limit. For spills up to 5 gallons, cover with 6 in. of activated carbon for each 1 in. of liquid depth and let stand. Absorb liquids in vermiculite, dry sand, earth, peat, carbon, or a similar material and deposit in sealed containers. *Do not use water* or wet method. Keep this chemical out of a confined space, such as a sewer, because of the possibility of an explosion, unless the sewer is designed to prevent the buildup of explosive concentrations. It may be necessary to contain and dispose of this chemical as a hazardous waste. If material or contaminated runoff enters waterways, notify downstream users of potentially contaminated waters. Contact your local or federal environmental protection agency for specific recommendations. If employees are required to clean up spills, they must be properly trained and equipped. OSHA 1910.120(q) may be applicable.

Initial isolation and protective action distances
Distances shown are likely to be affected during the first 30 min after materials are spilled and could increase with time. If more than one tank car, cargo tank, portable tank, or large cylinder involved in the incident is leaking, the protective action distance may need to be increased. You may need to seek emergency information from CHEMTREC at (800) 424-9300 or seek professional environmental engineering assistance from the US EPA Environmental Response Team at (908) 48-8730 (24-h response line).
Small spills (From a small package or a small leak from a large package)
First: Isolate in all directions (feet/meters)   500/150
Then: Protect persons downwind (miles/kilometers)
Day   1.1/1.8
Night   3.3
Large spills (From a large package or from many small packages)
First: Isolate in all directions (feet/meters)   3000/1000
Then: Protect persons downwind (miles/kilometers)
Day   7.0+/11.0+
Night   7.0+/11.0+
Fire Extinguishing: This chemical is a flammable liquid. Use dry chemical, carbon dioxide, or alcohol foam. *Do not use water.* Material is extremely hazardous to health but areas may be entered with extreme care. Full protective clothing, including self-contained breathing apparatus (coat, pants, gloves, boots, and bands around legs, arms, and waist), should be provided. No skin surface should be exposed. Poisonous gases, including hydrogen cyanide, are produced in fire. Vapors are heavier than air and will collect in low areas. Vapors may travel long distances to ignition sources and flashback. Vapors in confined areas may explode when exposed to fire. Containers may explode in fire. Stay away from ends of tanks. Do not get water inside container. Storage containers and parts of containers may rocket great distances, in many directions. If material or contaminated runoff enters waterways, notify downstream users of potentially contaminated waters. Notify local health and fire officials and pollution control agencies. From a secure, explosion-proof location, use water spray to cool exposed containers. Spray cooling water on containers that are exposed to flames until well after fire is out. If cooling streams are ineffective (venting sound increases in volume and pitch, tank discolors, or shows any signs of deforming), withdraw immediately to a secure position. If employees are expected to fight fires, they must be trained and equipped in OSHA 1910.156. The only respirators recommended for firefighting are self-contained breathing

apparatuses that have full face-pieces and are operated in a pressure-demand or other positive-pressure mode.
Disposal Method Suggested: Consult with environmental regulatory agencies for guidance on acceptable disposal practices. Generators of waste containing this contaminant (≥100 kg/mo) must conform with EPA regulations governing storage, transportation, treatment, and waste disposal. Incineration in the presence of a flammable solvent.[22] A flue gas scrubber is recommended as well.

References
Sax, N. I. (Ed.). (1985). *Dangerous Properties of Industrial Materials Report*, 5, No. 2, 68–70
US Environmental Protection Agency. (November 30, 1987). *Chemical Hazard Information Profile: Methyl Isocyanate*. Washington, DC: Chemical Emergency Preparedness Program
New York State Department of Health. (May 1986). *Chemical Fact Sheet: Methyl Isocyanate*. Albany, NY: Bureau of Toxic Substance Assessment
New Jersey Department of Health and Senior Services. (April 2002). *Hazardous Substances Fact Sheet: Methyl Isocyanate*. Trenton, NJ

Methyl isopropyl ketone   M:1020

Molecular Formula: $C_5H_{10}O$
Common Formula: $CH_3COCH(CH_3)_2$
Synonyms: 2-Acetyl propane; Isopropyl methyl ketone; 3-Methyl-2-butanone; 3-Methyl butan-2-one; MIPK
CAS Registry Number: 563-80-4
RTECS® Number: EL9100000
UN/NA & ERG Number: UN2397/127
EC Number: 209-264-3 [*Annex I Index No.:* 606-007-00-0]
Regulatory Authority and Advisory Bodies
Air Pollutant Standard Set. See below, "Permissible Exposure Limits in Air" section.
Canada, WHMIS, Ingredients Disclosure List Concentration 1.0%.
European/International Regulations: Hazard Symbol: F; Risk phrases: R11; Safety phrases: S2; S9; S16; S33 (see Appendix 4).
WGK (German Aquatic Hazard Class): 1—Low hazard to waters.
Description: MIPK is a colorless liquid with an acetone-like odor. Molecular weight = 86.15; Specific gravity (H_2O:1) = 0.81; Boiling point = 93°C. Freezing point = −92°C; Vapor pressure = 42 mmHg at 20°C; Flash point = 6°C; Autoignition temperature = 475°C. Explosive limits: LEL = 1.2%; UEL = 8%. Hazard Identification (based on NFPA-704 M Rating System): Health 1, Flammability 4, Reactivity 0. Slightly soluble in water; solubility = 0.5% at 20°C.
Potential Exposure: Compound Description: Mutagen, Primary Irritant. This ketone is used as a solvent for nitrocellulose lacquers.

Incompatibilities: Violent reaction with oxidizers.

Permissible Exposure Limits in Air

Conversion factor: 1 ppm = 3.53 mg/m^3 at 25°C & 1 atm.

OSHA PEL: None.

NIOSH REL: 200 ppm/705 mg/m^3 TWA.

ACGIH TLV®[1]: 200 ppm/705 mg/m^3 TWA; Notice of intended change: 20 ppm.

Protective Action Criteria (PAC)

TEEL-0: 200 ppm

PAC-1: 200 ppm

PAC-2: 200 ppm

PAC-3: 600 ppm

Australia: TWA 200 ppm (705 mg/m^3), 1993; Belgium: TWA 200 ppm (705 mg/m^3), 1993; Denmark: TWA 200 ppm (705 mg/m^3), 1999; Finland: TWA 200 ppm (700 mg/m^3); STEL 250 ppm (875 mg/m^3), 1999; France: VME 200 ppm (705 mg/m^3), 1999; Norway: TWA 100 ppm (350 mg/m^3), 1999; Switzerland: MAK-W 200 ppm (720 mg/m^3), 1999; the Netherlands: MAC-TGG 705 mg/m^3, 2003; Argentina, Bulgaria, Columbia, Jordan, South Korea, New Zealand, Singapore, Vietnam: ACGIH TLV®: TWA 200 ppm. Several states have set guidelines or standards for MIPK in ambient air[60] ranging from 7.05 mg/m^3 (North Dakota) to 11.75 mg/m^3 (Virginia) to 14.1 mg/m^3 (Connecticut) to 16.786 mg/m^3 (Nevada).

Determination in Air: No NIOSH Analytical Method available.

Routes of Entry: Inhalation of vapor, skin absorption, ingestion, skin and/or eye contact.

Harmful Effects and Symptoms

Short Term Exposure: Eye, nose, throat, and skin irritation. In high concentrations, narcosis may be produced with symptoms of headache, nausea, vomiting, lightheadedness, dizziness, a lack of coordination, narcosis, and unconsciousness.

Long Term Exposure: Removes the skin's natural oils; causes drying and cracking.

Points of Attack: Eyes, skin, respiratory system.

Medical Surveillance: There is no special test for this chemical. However, if illness occurs or overexposure is suspected, medical attention is recommended.

First Aid: If this chemical gets into the eyes, remove any contact lenses at once and irrigate immediately for at least 15 min, occasionally lifting upper and lower lids. Seek medical attention immediately. If this chemical contacts the skin, remove contaminated clothing and wash immediately with soap and water. Seek medical attention immediately. If this chemical has been inhaled, remove from exposure, begin rescue breathing (using universal precautions, including resuscitation mask) if breathing has stopped and CPR if heart action has stopped. Transfer promptly to a medical facility. When this chemical has been swallowed, get medical attention. Give large quantities of water and induce vomiting. Do not make an unconscious person vomit.

Personal Protective Methods: Wear protective gloves and clothing to prevent any reasonable probability of skin contact. Safety equipment suppliers/manufacturers can provide recommendations on the most protective glove/clothing material for your operation. All protective clothing (suits, gloves, footwear, headgear) should be clean, available each day, and put on before work. Contact lenses should not be worn when working with this chemical. Wear splash-proof chemical goggles and face shield unless full face-piece respiratory protection is worn. Employees should wash immediately with soap when skin is wet or contaminated. Provide emergency showers and eyewash. See NIOSH Criteria Document 78-173, *Ketones.*

Respirator Selection: Where there is potential for exposures *over 200 ppm*, use a NIOSH/MSHA- or European Standard EN149-approved full-face-piece respirator with an organic vapor cartridge/canister. Greater protection is provided by a powered air-purifying respirator. *Where there is potential for high exposures*, use a NIOSH/MSHA- or European Standard EN149-approved supplied-air respirator with a full face-piece operated in the positive-pressure mode, or with a full face-piece, hood, or helmet in the continuous-flow mode; or use a NIOSH/MSHA- or European Standard EN149-approved self-contained breathing apparatus with a full face-piece operated in a pressure-demand or other positive-pressure mode.

Storage: Color Code—Red: Flammability Hazard: Store in a flammable liquid storage area or approved cabinet away from ignition sources and corrosive and reactive materials. Prior to working with this chemical you should be trained on its proper handling and storage. Before entering confined space where this chemical may be present, check to make sure that an explosive concentration does not exist. Store in tightly closed containers in a cool, well-ventilated area away from aldehydes. Sources of ignition, such as smoking and open flames, are prohibited where methyl isopropyl ketone is used, handled, or stored in a manner that could create a potential fire or explosion hazard. Use only non-sparking tools and equipment, especially when opening and closing containers of methyl isopropyl ketone. Wherever methyl isopropyl ketone is used, handled, manufactured, or stored, use explosion-proof electrical equipment and fittings.

Shipping: This compound requires a shipping label of "FLAMMABLE LIQUID." It falls in Hazard Class 3 and Packing Group II.

Spill Handling: Evacuate and restrict persons not wearing protective equipment from area of spill or leak until cleanup is complete. Remove all ignition sources. Establish forced ventilation to keep levels below explosive limit. Absorb liquids in vermiculite, dry sand, earth, peat, carbon, or a similar material and deposit in sealed containers. Keep this chemical out of a confined space, such as a sewer, because of the possibility of an explosion, unless the sewer is designed to prevent the buildup of explosive concentrations. It may be necessary to contain and dispose of this chemical as a hazardous waste. If material or contaminated runoff enters waterways, notify downstream users of potentially

contaminated waters. Contact your local or federal environmental protection agency for specific recommendations. If employees are required to clean up spills, they must be properly trained and equipped. OSHA 1910.120(q) may be applicable.

Fire Extinguishing: This chemical is a flammable liquid. Poisonous gases are produced in fire. Use dry chemical, carbon dioxide, or alcohol foam extinguishers. Vapors are heavier than air and will collect in low areas. Vapors may travel long distances to ignition sources and flashback. Vapors in confined areas may explode when exposed to fire. Containers may explode in fire. Storage containers and parts of containers may rocket great distances, in many directions. If material or contaminated runoff enters waterways, notify downstream users of potentially contaminated waters. Notify local health and fire officials and pollution control agencies. From a secure, explosion-proof location, use water spray to cool exposed containers. If cooling streams are ineffective (venting sound increases in volume and pitch, tank discolors, or shows any signs of deforming), withdraw immediately to a secure position. If employees are expected to fight fires, they must be trained and equipped in OSHA 1910.156. The only respirators recommended for firefighting are self-contained breathing apparatuses that have full face-pieces and are operated in a pressure-demand or other positive-pressure mode.

Disposal Method Suggested: Dissolve or mix the material with a combustible solvent and burn in a chemical incinerator equipped with an afterburner and scrubber. All federal, state, and local environmental regulations must be observed.

Reference

New Jersey Department of Health and Senior Services. (January 2001). *Hazardous Substances Fact Sheet: Methyl Isopropyl Ketone.* Trenton, NJ

Methyl isothiocyanate   M:1030

Molecular Formula: C_2H_3NS
Common Formula: CH_3NCS
Synonyms: AI3-28257; Di-Trapex; EP-161E; Isothiocyanate de methyle (French); Isothiocyanatomethane; Isothiocyanic acid, methyl ester; Methane, isothiocyanato-; Methyl-isothiocyanat (German); Methyl mustard; Methyl mustard oil; Methylsenfoel (German); MIC; MIT; MITC; Morton WP-161E; Trapex; Trapex-40; Trapexide; Vorlex; Vorlex 201; Vortex; WN 12
CAS Registry Number: 556-61-6
RTECS® Number: PA6925000
UN/NA & ERG Number: UN2477/131
EC Number: 209-132-5 [*Annex I Index No.:* 615-002-00-2]
Regulatory Authority and Advisory Bodies
Air Pollutant Standard Set. See below, "Permissible Exposure Limits in Air" section.

Superfund/EPCRA 40CFR355, Extremely Hazardous Substances: TPQ = 500 lb (227 kg).
Reportable Quantity (RQ): 500 lb (227 kg).
EPCRA Section 313 Form R *de minimis* concentration reporting level: 1.0%.
US DOT 49CFR172.101, Inhalation Hazardous Chemical.
European/International Regulations: Hazard Symbol: T, N; Risk phrases: R23/25; R34; R43; R50/53; Safety phrases: S1/2; S36/37; S38; S45; S60, S61 (see Appendix 4).
WGK (German Aquatic Hazard Class): 3—Severe hazard to waters.

Description: Methyl isothiocyanate is a crystalline solid with a horseradish odor. Molecular weight = 73.12; Boiling point = 119°C; Freezing/Melting point = 35−36°C; Flash point = 32°C. Hazard Identification (based on NFPA-704 M Rating System): Health 3, Flammability 3, Reactivity 0. Slightly soluble in water.

Potential Exposure: It is used as a soil fumigant. A mixture of methyl isothiocyanate and chlorinated C-3 hydrocarbons is used as a soil fumigant for control of weeds, fungi, insects, and nematodes.

Incompatibilities: Unstable and reactive; sensitive to oxygen and to light. Incompatible with oxidizers, strong acids, alcohols, strong bases, amines, water, heat, and cold. Attacks iron, zinc, and other metals.

Permissible Exposure Limits in Air
Protective Action Criteria (PAC)
TEEL-0: 1.5 mg/m^3
PAC-1: 4 mg/m^3
PAC-2: 33 mg/m^3
PAC-3: 500 mg/m^3
Russia[43] set a MAC of 0.1 mg/m^3 in work-place air.

Routes of Entry: Inhalation, ingestion, eye and/or skin contact.

Harmful Effects and Symptoms
Short Term Exposure: Highly irritating to the eyes, skin, and mucous membranes. Coughing, wheezing, and/or shortness of breath and other symptoms of extreme pulmonary irritation would be expected if vapors are inhaled. Very high exposures may cause pulmonary edema, a medical emergency that can be delayed for several hours. This can cause death. Exposure can cause headache, dizziness, depression, seizures, and even loss of consciousness. This material is very toxic; probable human oral lethal dose is 50−500 mg/kg or between 1 teaspoon and 1 oz for a 70-kg (150 lb) person. Human oral minimum lethal dose: approximately 1 g/kg.

Long Term Exposure: May cause skin allergy and bronchitis. May affect the thyroid gland.

Points of Attack: Lungs, skin, thyroid.

Medical Surveillance: Before beginning employment and at regular times after that, for those with frequent or potentially high exposures, the following are recommended: lung function tests. These may be normal if the person is not having an attack at the time. If symptoms develop or overexposure is suspected, the following may be useful: evaluation by a qualified allergist, including careful exposure history and special testing, may help diagnose skin allergy. Evaluation of thyroid function.

First Aid: If this chemical gets into the eyes, remove any contact lenses at once and irrigate immediately for at least 15 min, occasionally lifting upper and lower lids. Seek medical attention immediately. If this chemical contacts the skin, remove contaminated clothing and wash immediately with soap and water. Seek medical attention immediately. If this chemical has been inhaled, remove from exposure, begin rescue breathing (using universal precautions, including resuscitation mask) if breathing has stopped and CPR if heart action has stopped. Transfer promptly to a medical facility. When this chemical has been swallowed, get medical attention. Give large quantities of water and induce vomiting. Do not make an unconscious person vomit. Medical observation is recommended for 24–48 h after breathing overexposure, as pulmonary edema may be delayed. As first aid for pulmonary edema, a doctor or authorized paramedic may consider administering a corticosteroid spray.

Personal Protective Methods: Wear protective gloves and clothing to prevent any reasonable probability of skin contact. Safety equipment suppliers/manufacturers can provide recommendations on the most protective glove/clothing material for your operation. All protective clothing (suits, gloves, footwear, headgear) should be clean, available each day, and put on before work. Contact lenses should not be worn when working with this chemical. Wear dust-proof chemical goggles and face shield unless full face-piece respiratory protection is worn. Employees should wash immediately with soap when skin is wet or contaminated. Provide emergency showers and eyewash.

Respirator Selection: Where there is potential for exposure to methyl isothiocyanate, use a NIOSH/MSHA- or European Standard EN149-approved supplied-air respirator with a full face-piece operated in the positive-pressure mode, or with a full face-piece, hood, or helmet in the continuous-flow mode; or use a NIOSH/MSHA- or European Standard EN149-approved self-contained breathing apparatus with a full face-piece operated in a pressure-demand or other positive-pressure mode.

Storage: (1) Color Code—Red: Flammability Hazard: Store in a flammable liquid storage area or approved cabinet away from ignition sources and corrosive and reactive materials. (2) Color Code—Blue: Health Hazard/Poison: Store in a secure poison location. Prior to working with this chemical you should be trained on its proper handling and storage. Store in tightly closed containers in a cool, well-ventilated area. Sources of ignition, such as smoking and open flames, are prohibited where methyl isothiocyanate is used, handled, or stored in a manner that could create a potential fire or explosion hazard. Metal containers involving the transfer of 5 gallons or more of this chemical should be grounded and bonded. Drums must be equipped with self-closing valves, pressure vacuum bungs, and flame arresters. Use only nonsparking tools and equipment, especially when opening and closing containers of this chemical. Wherever this chemical is used, handled, manufactured, or stored, use explosion-proof electrical equipment and fittings.

Shipping: This compound requires a shipping label of "POISONOUS/TOXIC MATERIALS, FLAMMABLE LIQUID." It falls in Hazard Class 6.1 and Packing Group I. This material carries a plus sign (+), indicating that the designated proper shipping name and hazard class of the material must always be shown whether or not the material or its mixtures or solutions meet the definitions of the class.

Spill Handling: Evacuate and restrict persons not wearing protective equipment from area of spill or leak until cleanup is complete. Remove all ignition sources. Collect powdered material in the most convenient and safe manner and wash area following cleanup. Ventilate and wash spill areas when cleanup is complete. Keep this chemical out of a confined space, such as a sewer, because of the possibility of an explosion, unless the sewer is designed to prevent the buildup of explosive concentrations. It may be necessary to contain and dispose of this chemical as a hazardous waste. If material or contaminated runoff enters waterways, notify downstream users of potentially contaminated waters. Contact your local or federal environmental protection agency for specific recommendations. If employees are required to clean up spills, they must be properly trained and equipped. OSHA 1910.120(q) may be applicable.

Initial isolation and protective action distances
Distances shown are likely to be affected during the first 30 min after materials are spilled and could increase with time. If more than one tank car, cargo tank, portable tank, or large cylinder involved in the incident is leaking, the protective action distance may need to be increased. You may need to seek emergency information from CHEMTREC at (800) 424-9300 or seek professional environmental engineering assistance from the US EPA Environmental Response Team at (908) 548-8730 (24-h response line).

Small spills (From a small package or a small leak from a large package)
First: Isolate in all directions (feet/meters)   100/30
Then: Protect persons downwind (miles/kilometers)
Day   0.1/0.2
Night   0.1/0.2

Large spills (From a large package or from many small packages)
First: Isolate in all directions (feet/meters)   200/60
Then: Protect persons downwind (miles/kilometers)
Day   0.3/0.5
Night   0.5/0.8

Fire Extinguishing: Methyl isothiocyanate is a flammable solid. Poisonous gases are produced in fire, including cyanides, sulfur oxides, and nitrogen oxides. Use dry chemical, CO_2, water spray, or foam extinguishers. If material or contaminated runoff enters waterways, notify downstream users of potentially contaminated waters. Notify local health and fire officials and pollution control agencies. From a secure, explosion-proof location, use water spray to cool exposed containers. If cooling streams are ineffective (venting sound increases in volume and pitch, tank discolors, or shows any signs of deforming), withdraw immediately to a secure position. If employees are expected to fight fires, they must be trained and equipped in OSHA 1910.156. The only respirators recommended for firefighting are self-contained breathing apparatuses that have full face-pieces and are operated in a pressure-demand or other positive-pressure mode.

Disposal Method Suggested: In accordance with 40CFR 165 recommendations for the disposal of pesticides and pesticide containers. Must be disposed properly by following package label directions or by contacting your local or federal environmental control agency or by contacting your regional EPA office.

References

US Environmental Protection Agency. (November 30, 1987). *Chemical Hazard Information Profile: Methyl Isothiocyanate.* Washington, DC: Chemical Emergency Preparedness Program

New Jersey Department of Health and Senior Services. (November 1999). *Hazardous Substances Fact Sheet: Methyl Isothiocyanate.* Trenton, NJ

Methyl mercaptan   M:1040

Molecular Formula: CH_4S
Common Formula: CH_3SH
Synonyms: Mercaptan methylique (French); Mercaptomethane; Methanethiol; 1-Methanethiol; Methanthiol (German); Methyl sulfhydrate; Metilmercaptano (Spanish); Thiomethanol; Thiomethyl alcohol
CAS Registry Number: 74-93-1
RTECS® Number: PB4375000
UN/NA & ERG Number: UN1064/117
EC Number: 200-822-1 [*Annex I Index No.:* 016-021-00-3]
Regulatory Authority and Advisory Bodies
Department of Homeland Security Screening Threshold Quantity (pounds): *Release hazard* 10,000 (≥1.00% concentration).; *Theft hazard* 500 (>45.00% concentration).
Air Pollutant Standard Set. See below, "Permissible Exposure Limits in Air" section.
OSHA 29CFR1910.119, Appendix A. Process Safety List of Highly Hazardous Chemicals, TQ = 5000 lb (2270 kg).
Clean Air Act: Accidental Release Prevention/Flammable Substances, (Section 112[r], Table 3), TQ = 10,000 lb (4540 kg).
Clean Water Act: Section 311 Hazardous Substances/RQ 40CFR117.3 (same as CERCLA, see below).
US EPA Hazardous Waste Number (RCRA No.): U153.
RCRA, 40CFR261, Appendix 8 Hazardous Constituents.
Superfund/EPCRA 40CFR355, Extremely Hazardous Substances: TPQ = 500 lb (227 kg).
Reportable Quantity (RQ): 100 lb (45.4 kg).

EPCRA Section 313 Form R *de minimis* concentration reporting level: 1.0%. *Note:* Subject to an administrative stay under EPCRA Section 313. Not reportable until stay is lifted. See 8/22/94 (59 FR 43048).
US DOT Regulated Marine Pollutant (49CFR172.101, Appendix B).
US DOT 49CFR172.101, Inhalation Hazardous Chemical.
Canada, WHMIS, Ingredients Disclosure List Concentration 1.0%.
European/International Regulations: Hazard Symbol: F +, T, N; Risk phrases: R12; R23; R50/53; Safety phrases: S2; S16; S25; S60; S61 (see Appendix 4).
WGK (German Aquatic Hazard Class): 3—Severe hazard to waters.
Description: Methyl mercaptan is a colorless gas or white liquid with a disagreeable odor like garlic or rotten cabbage. The odor threshold = 0.0016 ppm. Shipped as a liquefied compressed gas. The odor threshold is 0.002 ppm. Molecular weight = 48.11; Specific gravity (H_2O:1) = 9.90 (liquid at 0°C); Relative density (H_2O:1): 0.9 (a liquid at 0°C); Boiling point = 6.1°C; Freezing/Melting point = −121.1°C; Vapor pressure = 1.7 atm; Relative vapor density (air = 1) = 1.66; Flash point = −17°C (liquid). Explosive limits: LEL = 3.9%; UEL = 21.8%. Hazard Identification (based on NFPA-704 M Rating System): Health 2, Flammability 4, Reactivity 0. Soluble in water; solubility = 2%.
Potential Exposure: Compound Description: Mutagen. Methyl mercaptan is used in methionine synthesis, and widely as an intermediate in pesticide manufacture.
Incompatibilities: Violent reaction with strong oxidizers, bleaches, copper, nickel and their alloys, aluminum. Reacts with acids producing flammable and toxic hydrogen sulfide.
Permissible Exposure Limits in Air
Conversion factor: 1 ppm = 1.97 mg/m^3 at 25°C & 1 atm.
OSHA PEL: 10 ppm/20 mg/m^3 Ceiling Concentration.
NIOSH REL: 0.5 ppm/1 mg/m^3 [15 min] Ceiling Concentration.
ACGIH TLV®[1]: 0.5 ppm/0.98 mg/m^3 TWA.
NIOSH IDLH: 150 ppm.
Protective Action Criteria (PAC)*
TEEL-0: 0.5 ppm
PAC-1: 6 ppm
PAC-2: **47** ppm
PAC-3: **68** ppm
*AEGLs (Acute Emergency Guideline Levels) & ERPGs (Emergency Response Planning Guideline) are in **bold face**.
DFG MAK: 0.5 ppm/1.0 mg/m^3 TWA; Peak Limitation Category II(2); Pregnancy Risk Group D.
Australia: TWA 0.5 ppm (1 mg/m^3), 1993; Austria: MAK 0.5 ppm (1 mg/m^3), 1999; Belgium: TWA 0.5 ppm (0.98 mg/m^3), 1993; Denmark: TWA 0.5 ppm (1 mg/m^3), 1999; Finland: TWA 0.5 ppm (1 mg/m^3); STEL 1.5 ppm (3 mg/m^3), 1999; France: VME 0.5 ppm (1 mg/m^3), 1999; Hungary: STEL 1 mg/m^3, 1993; the Netherlands:

MAC-TGG 1 mg/m^3, 2003; Norway: TWA 0.5 ppm (1 mg/m^3), 1999; the Philippines: TWA 10 ppm (20 mg/m^3), 1993; Poland: MAC (TWA) 1 mg/m^3; MAC (STEL) 2 mg/m^3, 1999; Russia: STEL 0.8 mg/m^3, 1993; Sweden: NGV 1 ppm, 1999; Switzerland: MAK-W 0.5 ppm (1 mg/m^3), KZG-W 1 ppm (2 mg/m^3), 1999; Thailand: TWA 10 ppm (20 mg/m^3), 1993; Turkey: TWA 10 ppm (20 mg/m^3), 1993; United Kingdom: TWA 0.5 ppm (1 mg/m^3), 2000; Argentina, Bulgaria, Columbia, Jordan, South Korea, New Zealand, Singapore, Vietnam: ACGIH TLV®: TWA 0.5 ppm. Russia[43] set a MAC of 9×10^{-6} mg/m^3 in ambient air in residential areas on a momentary basis. Several states have set guidelines or standards for methyl mercaptan in ambient air[60] ranging from 3.3 μg/m^3 (New York) to 10 μg/m^3 (Florida, North Dakota, South Carolina) to 16 μg/m^3 (Virginia) to 20 μg/m^3 (Connecticut) to 24 μg/m^3 (Nevada) to 50 μg/m^3 (North Carolina).

Determination in Air: Use NIOSH Analytical Method (IV) #2542, Mercaptans; OSHA Analytical Method 26.

Permissible Concentration in Water: EPA[32] has suggested a permissible ambient goal of 13.8 μg/L based on health effects. Russia[43] set a MAC of 0.2 μg/L in water bodies used for domestic purposes.

Routes of Entry: Inhalation, skin and/or eye contact (liquid).

Harmful Effects and Symptoms

Short Term Exposure: Irritates the eyes, skin, and respiratory tract. Skin contact can cause frostbite. Inhalation can cause pulmonary edema, a medical emergency that can be delayed for several hours. This can cause death. May affect the central nervous system. Signs and symptoms of acute exposure to methyl mercaptan may include fever, cough, shortness of breath, a feeling of tightness and burning in the chest, respiratory distress, respiratory paralysis, and respiratory failure/collapse. Headache, loss of the sense of smell, dizziness, staggering gait, and heightened emotions may occur. Gastrointestinal symptoms include difficulty in swallowing, redness of the tongue and pharynx, nausea, vomiting, abdominal pain, and diarrhea. Urinary disturbances may also be found.

Long Term Exposure: May cause liver and kidney damage. Repeated exposure may cause bronchitis. Memory loss, damage to the central and peripheral nervous systems, tremor, convulsions, and coma may also occur. May affect the blood cell causing anemia.

Points of Attack: Eyes, skin, respiratory system, liver, kidneys, central nervous system, blood.

Medical Surveillance: For those with frequent or potentially high exposure (half the TLV or greater) the following are recommended before beginning work and at regular times after that: lung function tests. If symptoms develop or overexposure is suspected, the following may be useful: consider chest X-ray after acute overexposure. Liver function tests. Complete blood count.

First Aid: If this chemical gets into the eyes, remove any contact lenses at once and irrigate immediately for at least 15 min, occasionally lifting upper and lower lids. Seek medical attention immediately. If this chemical contacts the skin, remove contaminated clothing and wash immediately with soap and water. Seek medical attention immediately. If this chemical has been inhaled, remove from exposure, begin rescue breathing (using universal precautions, including resuscitation mask) if breathing has stopped and CPR if heart action has stopped. Transfer promptly to a medical facility. When this chemical has been swallowed, get medical attention. Give large quantities of water and induce vomiting. Do not make an unconscious person vomit. Medical observation is recommended for 24—48 h after breathing overexposure, as pulmonary edema may be delayed. As first aid for pulmonary edema, a doctor or authorized paramedic may consider administering a corticosteroid spray. If frostbite has occurred, seek medical attention immediately; do *NOT* rub the affected areas or flush them with water. In order to prevent further tissue damage, do *NOT* attempt to remove frozen clothing from frostbitten areas. If frostbite has *NOT* occurred, immediately and thoroughly wash contaminated skin with soap and water.

Personal Protective Methods: Wear appropriate personal protective clothing to prevent the skin from becoming frozen from contact with the evaporating liquid or from contact with vessels containing the liquid: **8 h:** Barricade™ coated suits; Responder™ suits; Trychem 1000™ suits. Safety equipment suppliers/manufacturers can provide recommendations on the most protective glove/clothing material for your operation. All protective clothing (suits, gloves, footwear, headgear) should be clean, available each day, and put on before work. Contact lenses should not be worn when working with this chemical. Wear the proper chemical goggles (indirect vent, impact and slash resistant with liquids; nonvented, impact resistant with fumes, gases, or vapors) and face shield unless full face-piece respiratory protection is worn. Employees should wash immediately with soap when skin is wet or contaminated. Provide emergency showers and eyewash.

Respirator Selection: *5 ppm:* CcrOv (APF = 10) [any chemical cartridge respirator with organic vapor cartridge(s)] or Sa (APF = 10) (any supplied-air respirator). *12.5 ppm:* Sa:Cf (APF = 25) (any supplied-air respirator operated in a continuous-flow mode) or PaprOv (APF = 25) [any powered, air-purifying respirator with organic vapor cartridge(s)]. *25 ppm:* CcrFOv (APF = 50) [any air-purifying, full-face-piece respirator (gas mask) with a chin-style, front- or back-mounted acid gas canister] or GmFOv (APF = 50) [any air-purifying, full-face-piece respirator (gas mask) with a chin-style, front- or back-mounted organic vapor canister] or PaprTOv (APF = 50) [any powered, air-purifying respirator with a tight-fitting face-piece and organic vapor cartridge(s)] or SaT: Cf (APF = 50) (any supplied-air respirator that has a tight-fitting face-piece and is operated in a continuous-flow mode) or SCBAF (APF = 50) (any self-contained breathing apparatus with a full face-piece) or SaF (APF = 50) (any supplied-air respirator with a full face-piece). *Emergency or planned entry into unknown concentrations or IDLH conditions:* SCBAF: Pd,Pp (APF = 10,000) (any self-contained breathing apparatus that has a full face-piece and is operated in a pressure-demand or other positive-pressure mode) or SaF: Pd,Pp: ASCBA (APF = 10,000) (any supplied-air respirator that has a full face-piece and is operated in a pressure-demand or other positive-pressure mode in combination with an auxiliary self-contained breathing apparatus operated in a pressure-demand or other positive-pressure mode). *Escape:* GmFOv (APF = 50) [any air-purifying, full-face-piece respirator (gas mask) with a chin-style, front- or back-mounted organic vapor canister] or SCBAE (any appropriate escape-type, self-contained breathing apparatus).

Storage: Color Code—Red: Flammability Hazard: Store in a flammable liquid storage area or approved cabinet away from ignition sources and corrosive and reactive materials. Prior to working with this chemical you should be trained on its proper handling and storage. Before entering confined space where this chemical may be present, check to make sure that an explosive concentration does not exist. Methyl mercaptan must be stored to avoid contact with water, steam, or strong acids (such as hydrochloric, sulfuric, and nitric) because toxic flammable vapors will be released. It should not contact oxidizers (such as perchlorates, peroxides, permanganates, chlorates, and nitrates) since violent reactions occur. Store in tightly closed containers in a cool, well-ventilated area away from heat or sparks. Sources of ignition, such as smoking and open flames, are prohibited where methyl mercaptan is handled, used, or stored. Metal containers involving the transfer of 5 gallons or more of methyl mercaptan should be grounded and bonded. Drums must be equipped with self-closing valves, pressure vacuum bungs, and flame arresters. Use only nonsparking tools and equipment, especially when opening and closing containers of methyl mercaptan. Procedures for the handling, use, and storage of cylinders should be in compliance with OSHA 1910.101 and 1910.169, as with the recommendations of the Compressed Gas Association.

Shipping: Methyl mercaptan requires a shipping label of "POISON GAS, FLAMMABLE GAS." It falls in Hazard Class 2.3. It is a violation of transportation regulations to refill compressed gas cylinders without the express written permission of the owner.

Spill Handling: If in a building, evacuate building and confine vapors by closing doors and shutting down HVAC systems. Restrict persons not wearing protective equipment from area of spill or leak until cleanup is complete. Remove all ignition sources. Establish forced ventilation to keep levels below explosive limit and to disperse the gas. Wear chemical protective suit with self-contained breathing apparatus to combat spills. Stay upwind and use water spray to "knock down" vapor; contain runoff. Stop the flow of gas, if it can be done safely from a distance. If source is a cylinder and the leak cannot be stopped in place, remove the leaking cylinder to a safe place; and repair leak or allow

cylinder to empty. Keep this chemical out of confined spaces, such as a sewer, because of the possibility of explosion, unless the sewer is designed to prevent the buildup of explosive concentrations. If employees are required to clean up spills, they must be properly trained and equipped. OSHA 1910.120(q) may be applicable.

Initial isolation and protective action distances
Distances shown are likely to be affected during the first 30 min after materials are spilled and could increase with time. If more than one tank car, cargo tank, portable tank, or large cylinder involved in the incident is leaking, the protective action distance may need to be increased. You may need to seek emergency information from CHEMTREC at (800) 424-9300 or seek professional environmental engineering assistance from the US EPA Environmental Response Team at (908) 48-8730 (24-h response line).
Small spills (From a small package or a small leak from a large package)
First: Isolate in all directions (feet/meters)   100/30
Then: Protect persons downwind (miles/kilometers)
Day   0.1/0.2
Night   0.2/0.3
Large spills (From a large package or from many small packages)
First: Isolate in all directions (feet/meters)   600/200
Then: Protect persons downwind (miles/kilometers)
Day   0.8/1.2
Night   2.6/4.1

Fire Extinguishing: Keep unnecessary people away; isolate hazard area and deny entry. Stay upwind; keep out of low areas. Ventilate closed spaces before entering them. Wear positive pressure breathing apparatus and special protective clothing. Evacuate area endangered by gas. See isolation distances above. Combustion produces irritating sulfur dioxide. Very dangerous when exposed to heat, flame, or oxidizers. On decomposition it emits highly toxic fumes of sulfur oxides. Do not extinguish the fire unless the flow of gas can be stopped and any remaining gas is out of the line. Specially trained personnel may use fog lines to cool exposures and let the fire burn itself out. Vapors are heavier than air and will collect in low areas. Vapors may travel long distances to ignition sources and flashback. Vapors in confined areas may explode when exposed to fire. Containers may explode in fire. Storage containers and parts of containers may rocket great distances, in many directions. If material or contaminated runoff enters waterways, notify downstream users of potentially contaminated waters. Notify local health and fire officials and pollution control agencies. From a secure, explosion-proof location, use water spray to cool exposed containers. If cooling streams are ineffective (venting sound increases in volume and pitch, tank discolors, or shows any signs of deforming), withdraw immediately to a secure position. If cylinders are exposed to excessive heat from fire or flame contact, withdraw immediately to a secure location. If employees are expected to fight fires, they must be trained and equipped in

OSHA 1910.156. The only respirators recommended for firefighting are self-contained breathing apparatuses that have full face-pieces and are operated in a pressure-demand or other positive-pressure mode.

Disposal Method Suggested: Consult with environmental regulatory agencies for guidance on acceptable disposal practices. Generators of waste containing this contaminant ($\geq$100 kg/mo) must conform with EPA regulations governing storage, transportation, treatment, and waste disposal. Incineration followed by effective scrubbing of the effluent gas.

References

US Environmental Protection Agency. (November 30, 1987). *Chemical Hazard Information Profile: Methyl Mercaptans*. Washington, DC: Chemical Emergency Preparedness Program

New Jersey Department of Health and Senior Services. (January 2000). *Hazardous Substances Fact Sheet: Methyl Mercaptan*. Trenton, NJ

Methylmercuricdicyanamide   M:1050

Molecular Formula: $C_3H_6HgN_4$

Synonyms: Agrosol; Cyanoguanidine methyl mercury derivative; Cyano(methylmercury)guanidine; Guanidine, cyano-, methylmercury derivative; MEMA; Methylmercuric cyanoguanidine; Methylmercury dicyanandimide; Methylmercury dicyandiamide; Methylmerkuridikyandiamid (German); MMD; Morsodren; Morton EP-227; Morton soil drench; Pandrinox; Pano-Drench 4; Panodrin A-13; Panogen®; Panogen® 15; Panogen® 43; Panogen® PX; Panogen® turf fungicide; Panogen® turf spray; Panospray 30; R 8; R 8 fungicide

CAS Registry Number: 502-39-6

RTECS® Number: OW1750000

UN/NA & ERG Number: UN2777/151

EC Number: 207-935-5

Regulatory Authority and Advisory Bodies

Air Pollutant Standard Set. See below, "Permissible Exposure Limits in Air" section.

Clean Water Act: 40CFR401.15 Section 307 Toxic Pollutants as mercury and compounds.

Superfund/EPCRA 40CFR355, Extremely Hazardous Substances: TPQ = 500/10,000 lb (227/4540 kg).

Reportable Quantity (RQ): 500 lb (227 kg).

California Proposition 65 Chemical: Cancer 5/1/96; Developmental/Reproductive toxin 7/1/87.

California Proposition 65 Developmental/Reproductive toxin (mercury and mercury compounds) 7/1/90.

Canada, WHMIS, Ingredients Disclosure List Concentration 1.0% as mercury compounds.

Rotterdam Convention Annex III [Chemicals Subject to the Prior Informed Consent Procedure (PIC)] (as mercury compounds, including inorganic mercury compounds, alkyl mercury compounds and alkyloxylalkyl and aryl mercury compounds).

European/International Regulations: not listed in Annex 1.

WGK (German Aquatic Hazard Class): No value assigned.

Description: Methylmercuric dicyanamide is a crystalline solid. Molecular weight = 298.72; Freezing/Melting point = 156°C. Hazard Identification (based on NFPA-704 M Rating System): Health 3, Flammability 1, Reactivity 0. Soluble in water.

Potential Exposure: This material is used as a fungicide, a seed, soil, and turf treatment especially for cereals, sorghum, sugar beets, cotton, and flax. Not registered as a pesticide in the United States.

Permissible Exposure Limits in Air: As organo mercury compound.

OSHA PEL: 0.01 mg[Hg]/m³ TWA; 0.04 mg/m³ Ceiling Concentration.

NIOSH REL: 0.01 mg[organomercury]/m³ TWA; 0.03 mg/m³ STEL [skin].

ACGIH TLV®[1]: 0.01 mg[Hg]/m³ TWA; 0.03 mg/m³ STEL [skin].

NIOSH IDLH: 2 mg[Hg]/m³.

Protective Action Criteria (PAC)

TEEL-0: 0.0149 mg/m³

PAC-1: 0.0447 mg/m³

PAC-2: 2.98 mg/m³

PAC-3: 2.98 mg/m³

DFG MAK: 0.01 mg[Hg]/m³ [skin] Danger of skin sensitization; Carcinogen Category 3.

In addition, North Dakota has set a guideline of 1−3 μg/m³ for alkyl mercury compounds in ambient air.[60]

Determination in Air: No method available.

Permissible Concentration in Water: Methylmercury: To protect freshwater aquatic life: 0.016 μg/L as a 24-h average, never to exceed 8.8 μg/L. *To protect saltwater aquatic life:* 0.025 μg/L as a 24-h average, never to exceed 2.8 μg/L. *To protect human health:* 0.2 μg/L.[6]

Determination in Water: Total mercury is determined by flameless atomic absorption. Soluble mercury may be determined by 0.45 μm filtration followed by flameless atomic absorption.

Routes of Entry: Inhalation, ingestion, eye, and/or skin contact.

Harmful Effects and Symptoms

Short Term Exposure: Alkyl mercury compounds are primary skin irritants and may cause dermatitis. When deposited on the skin, they give no warning, and if contact is maintained, can cause second-degree burns and blisters. Sensitization may occur. In the case of ingestion there is nausea and abdominal pain. Vomiting and diarrhea may occur. Burning or prickling of the lips, tongue, and extremities. The patient may be confused, hallucinated, and irritated; have disturbed sleep; loss of muscular coordination; and memory loss. Visual fields may narrow concentrically; emotional instability may occur as well as inability to concentrate, with stupor and coma. Methylmercuric dicyanamide is extremely toxic to humans. The probable lethal dose for humans is 5–50 mg/kg of body weight (between 7 drops and 1 teaspoon for a 150-lb person). Humans may be poisoned by feeding on the flesh of animals that have ingested this fungicide. Eating treated seeds may also cause poisoning. The poisoning may show delayed manifestations on the nervous system. Patients frequently become gradually worse after their illness is recognized and exposure is stopped.

Long Term Exposure: Repeated or prolonged contact with skin may result in dermatitis (red inflamed skin). Repeated or prolonged exposure may cause death by hypovolemic shock, nephrotic syndrome, or kidney failure. The central nervous system, including the brain, is the principal target tissue for this group of toxic compounds. Severe poisoning may produce irreversible brain damage resulting in loss of higher functions. The effects of chronic poisoning with alkyl mercury compounds are progressive. In the early stages, there are fine tremors of the hands, and in some cases, of the face and arms. With continued exposure, tremors may become coarse and convulsive; scanning speech with moderate slurring and difficulty in pronunciation may also occur. The worker may then develop an unsteady gait of a spastic nature which can progress to severe ataxia of the arms and legs. Sensory disturbances, including tunnel vision, blindness, and deafness, are also common. A late symptom, constriction of the visual fields, is rarely reversible and may be associated with loss of understanding and reason which makes the victim completely out of touch with his environment. Severe cerebral effects have been seen in infants born to mothers who had eaten large amounts of methylmercury-contaminated fish.

Points of Attack: Eyes, skin, central nervous system, peripheral nervous system, kidneys.

Medical Surveillance: Before first exposure and every 6–12 months after, a complete medical history and examination is strongly recommended: eye examination. Consider lung function tests for persons with frequent exposure. Examination of the nervous system. Routine urine test (UA). Urine test for mercury (should be less than 0.02 mg/L). Consider nerve conduction tests, urinary enzymes and neurobehavioral test. After suspected illness or overexposure, repeat the tests above and get a blood test for mercury. Examination of the central nervous system and kidneys. Consideration should be given to the possible effects on the fetus of alkyl mercury exposure in the mother. Constriction of visual fields may be a useful diagnostic sign. Blood and urine levels of mercury have been studied, especially in the case of methylmercury.

First Aid: If this chemical gets into the eyes, remove any contact lenses at once and irrigate immediately for at least 15 min, occasionally lifting upper and lower lids. Seek medical attention immediately. If this chemical contacts the skin, remove contaminated clothing and wash immediately with soap and water. Seek medical attention immediately. If this chemical has been inhaled, remove from exposure, begin rescue breathing (using universal precautions, including resuscitation mask) if breathing has stopped and CPR if heart action has stopped. Transfer promptly to a medical facility. When this chemical has been swallowed, get medical attention. Give large quantities of water and induce vomiting. Do not make an unconscious person vomit. Keep victim quiet and maintain normal body temperature. Effects may be delayed; keep victim under observation.

Personal Protective Methods: Wear protective gloves and clothing to prevent any reasonable probability of skin contact. Safety equipment suppliers/manufacturers can provide recommendations on the most protective glove/clothing material for your operation. All protective clothing (suits, gloves, footwear, headgear) should be clean, available each day, and put on before work. Contact lenses should not be worn when working with this chemical. Wear dust-proof chemical goggles and face shield unless full face-piece respiratory protection is worn. Employees should wash immediately with soap when skin is wet or contaminated. Provide emergency showers and eyewash.

Respirator Selection: Up to 0.1 mg/m³: Sa (APF = 10) (any supplied-air respirator). *Up to 0.25 mg/m³:* Sa:Cf (APF = 25) (any supplied-air respirator operated in a continuous-flow mode). *Up to 0.5 mg/m³:* SaT: Cf (APF = 50) (any supplied-air respirator that has a tight-fitting face-piece and is operated in a continuous-flow mode) or SCBAF (APF = 50) (any self-contained breathing apparatus with a full face-piece) or SaF (APF = 50) (any supplied-air respirator with a full face-piece). *Up to 2 mg/m³:* SA: PD, PP (any supplied-air respirator operated in a pressure-demand or other positive-pressure mode). *Emergency or planned entry into unknown concentrations or IDLH conditions:* SCBAF: Pd,Pp (APF = 10,000) (any NIOSH/MSHA- or European Standard EN 149-approved self-contained breathing apparatus that has a full face-piece and is operated in a pressure-demand or other positive-pressure mode) or SaF: Pd,Pp: ASCBA (APF = 10,000) (any supplied-air respirator that has a full face-piece and is operated in a pressure-demand or other positive-pressure mode in combination with an auxiliary self-contained breathing apparatus operated in a pressure-demand or other positive-pressure mode). *Escape:* SCBAE (any appropriate escape-type, self-contained breathing apparatus).

Storage: Color Code—Blue: Health Hazard/Poison: Store in a secure poison location. Prior to working with this chemical you should be trained on its proper handling and storage. Store in tightly closed containers in a cool, well-ventilated area.

Shipping: Mercury-based pesticides, solid, toxic, require a shipping label of "POISONOUS/TOXIC MATERIALS." It falls in Hazard Class 6.1 and Packing Group II.

Spill Handling: Evacuate persons not wearing protective equipment from area of spill or leak until cleanup is complete. Remove all ignition sources. Collect powdered material in the most convenient and safe manner and deposit in sealed containers. Ventilate area after cleanup is complete.

It may be necessary to contain and dispose of this chemical as a hazardous waste. If material or contaminated runoff enters waterways, notify downstream users of potentially contaminated waters. Contact your local or federal environmental protection agency for specific recommendations. If employees are required to clean up spills, they must be properly trained and equipped. OSHA 1910.120(q) may be applicable.

Fire Extinguishing: This material may burn but does not ignite readily. Fire and runoff from fire control water may produce irritating or poisonous gases of mercury and nitrogen oxides. For small fires, use dry chemical, carbon dioxide, water spray, or foam. For large fires, use water spray, fog, or foam. Use dry chemical, carbon dioxide, water spray, or foam extinguishers. If material or contaminated runoff enters waterways, notify downstream users of potentially contaminated waters. Notify local health and fire officials and pollution control agencies. Containers may explode in fire. From a secure, explosion-proof location, use water spray to cool exposed containers. If cooling streams are ineffective (venting sound increases in volume and pitch, tank discolors, or shows any signs of deforming), withdraw immediately to a secure position. If employees are expected to fight fires, they must be trained and equipped in OSHA 1910.156. The only respirators recommended for firefighting are self-contained breathing apparatuses that have full face-pieces and are operated in a pressure-demand or other positive-pressure mode.

Disposal Method Suggested: In accordance with 40CFR 165 recommendations for the disposal of pesticides and pesticide containers. Must be disposed properly by following package label directions or by contacting your local or federal environmental control agency or by contacting your regional EPA office.

Reference

US Environmental Protection Agency. (November 30, 1987). *Chemical Hazard Information Profile: Methylmercuric Dicyanamide.* Washington, DC: Chemical Emergency Preparedness Program

Methyl methacrylate monomer   M:1060

Molecular Formula: $C_5H_8O_2$
Common Formula: $CH_2=C(CH_3)COOCH_3$
Synonyms: Acrylic acid, 2-methyl-, methyl ester; Diakon; MER; Metacrilato de metilo (Spanish); Methacrylate de methyle (French); Methacrylic acid Met; Methacrylic acid, methyl ester; Methacrylsaeuremethyl ester (German); Methyl ester of methacrylic acid; Methyl-methacrylat (German); Methyl methacrylate monomer; Methyl α-methylacrylate; Methyl 2-methyl-2-propenoate; Methyl 2-methylpropenoate; 2-Methyl propenoic acid, methyl ester; MMA; Monocite methacrylate monomer; NCI-C50680; Pegalan; 2-Propenoic acid, 2-methyl-, methyl ester

CAS Registry Number: 80-62-6
RTECS® Number: OZ5075000
UN/NA & ERG Number: UN1247/129
EC Number: 201-297-1 [*Annex I Index No.:* 607-035-00-6]
Regulatory Authority and Advisory Bodies
Carcinogenicity: IARC: Human Inadequate Evidence, Animal Lacks Carcinogenicity, *not classifiable as carcinogenic to humans*, Group 3, 1994; EPA: Evidence of noncarcinogenicity for humans; Not likely to be carcinogenic to humans; NCI: Carcinogenesis Studies (inhalation); no evidence: mouse, rat.
Air Pollutant Standard Set. See below, "Permissible Exposure Limits in Air" section.
Clean Air Act: Hazardous Air Pollutants (Title I, Part A, Section 112).
Clean Water Act: Section 311 Hazardous Substances/RQ 40CFR117.3 (same as CERCLA, see below); Section 313 Water Priority Chemicals (57FR41331, 9/9/92).
US EPA Hazardous Waste Number (RCRA No.): U162.
RCRA, 40CFR261, Appendix 8 Hazardous Constituents.
RCRA 40CFR268.48; 61FR15654, Universal Treatment Standards: Wastewater (mg/L), 0.14; Nonwastewater (mg/kg), 160.
RCRA 40CFR264, Appendix 9; TSD Facilities Ground Water Monitoring List. Suggested test method(s) (PQL μg/L): 8015 (2); 8240 (5).
Reportable Quantity (RQ): 1000 lb (454 kg).
EPCRA Section 313 Form R *de minimis* concentration reporting level: 1.0%.
US DOT Regulated Marine Pollutant (49CFR172.101, Appendix B).
Canada, WHMIS, Ingredients Disclosure List Concentration 1.0%.
European/International Regulations: Hazard Symbol: F,Xi; Risk phrases: R11; R37/38; R43; Safety phrases: S2; S24; S37; S46 (see Appendix 4).
WGK (German Aquatic Hazard Class): 1—Low hazard to waters.
Description: Methyl methacrylate is a colorless liquid with an acrid, fruity odor. The odor threshold is 0.05–0.083 ppm. Molecular weight = 100.13; Specific gravity (H_2O:1) = 0.94; Boiling point = 100°C; Freezing/Melting point = −48°C; Vapor pressure = 29 mmHg at 20°C; Flash point = 10°C; Autoignition temperature = 421°C. Explosive limits: LEL = 1.7%; UEL = 8.2%. Hazard Identification (based on NFPA-704 M Rating System): Health 2, Flammability 3, Reactivity 2. Slightly soluble in water; solubility = 1.5%.
Potential Exposure: Compound Description: Tumorigen; Mutagen; Reproductive Effector; Human Data; Primary Irritant. Virtually all of the methyl methacrylate monomer produced is used in the production of polymers, such as surface coating resins, plastics (Plexiglas® and Lucite®), ion exchange resins, and plastic dentures.
Incompatibilities: Reacts in air to form a heat-sensitive explosive product at 60°C. Incompatible with nitrates,

oxidizers, peroxides, strong acids, strong alkalis, oxidizers, reducing agents, amines, moisture. Contact with benzoyl peroxide may cause ignition, fire, and explosion. May polymerize if subjected to heat, polymerization catalysts (e.g. azoisobutyronitrile, dibenzoyl peroxide, di-*tert*-butyl peroxide, propionaldehyde), strong oxidizers, or ultraviolet light. Usually contains an inhibitor, such as hydroquinone.

Permissible Exposure Limits in Air
Conversion factor: 1 ppm = 4.09 mg/m^3 at 25°C & 1 atm.
OSHA PEL: 100 ppm/410 mg/m^3 TWA.
NIOSH REL: 100 ppm/410 mg/m^3 TWA.
ACGIH TLV®[1]: 50 ppm/205 mg/m^3 TWA; 100 ppm/ 410 mg/m^3 STEL, sensitizer, not classifiable as a human carcinogen.
NIOSH IDLH: 1000 ppm.
Protective Action Criteria (PAC)*
TEEL-0: 17 ppm
PAC-1: **17** ppm
PAC-2: **120** ppm
PAC-3: **570** ppm
*AEGLs (Acute Emergency Guideline Levels) & ERPGs (Emergency Response Planning Guideline) are in **bold face**.
DFG MAK: 50 ppm/210 mg/m^3 TWA; Peak Limitation Category I(2); danger of skin sensitization; Pregnancy Risk Group C.
Compound Description: Tumorigen, Mutagen; Reproductive Effector; Human Data; Primary Irritant.
Australia: TWA 100 ppm (410 mg/m^3), 1993; Austria: MAK 50 ppm (210 mg/m^3), 1999; Belgium: TWA 100 ppm (410 mg/m^3), 1993; Finland: TWA 100 ppm (410 mg/m^3); STEL 150 ppm (615 mg/m^3), 1999; France: VME 100 ppm (410 mg/m^3), VLE 200 ppm (820 mg/m^3), 1999; Hungary: TWA 50 mg/m^3; STEL 150 mg/m^3, 1993; the Netherlands: MAC-TGG 40 mg/m^3, 2003; the Philippines: TWA 100 ppm (410 mg/m^3), 1993; Poland: MAC (TWA) 50 mg/m^3; MAC (STEL) 400 mg/m^3, 1999; Russia: STEL 10 mg/m^3, 1993; Sweden: NGV 50 ppm (200 mg/m^3), KTV 150 ppm (600 mg/m^3), [skin], 1999; Switzerland: MAK-W 50 ppm (210 mg/m^3), KZG-W 100 ppm (420 mg/m^3), 1999; United Kingdom: TWA 50 ppm (208 mg/m^3); STEL 100 ppm, 2000; Argentina, Bulgaria, Columbia, Jordan, South Korea, New Zealand, Singapore, Vietnam: ACGIH TLV®: STEL 100 ppm. Russia has set a MAC value of 0.1 mg/m^3 for ambient air in residential areas on a momentary basis and also on a daily average basis. Several states have set guidelines or standards for methyl methacrylate in ambient air[60] ranging from 7.0 μg/m^3 (Massachusetts) to 1367 μg/m^3 (New York) to 4100 μg/m^3 (Florida and North Dakota) to 6800 μg/ m^3 (Virginia) to 8200 μg/m^3 (Connecticut) to 9742 μg/m^3 (Nevada) to 10,250 μg/m^3 (South Carolina).

Determination in Air: Use NIOSH Analytical Method (IV) #2537, Methyl methacrylate; OSHA Analytical Method 94.

Permissible Concentration in Water: A limit in drinking water has been set by Russia[43] at 0.01 mg/L. Maine[61] has set a guideline of 200 μg/L for methyl methacrylate in drinking water.

Routes of Entry: Inhalation, ingestion, skin and/or eye contact.

Harmful Effects and Symptoms

Short Term Exposure: Contact can irritate eyes and skin. Inhalation irritates the respiratory tract. High exposure can cause dizziness, lightheadedness narcosis, and unconsciousness.

Long Term Exposure: Repeated or prolonged contact may cause skin sensitization and allergy. Repeated or prolonged inhalation exposure may cause asthma. May affect the central nervous system and the peripheral nervous system. May affect the kidneys and liver. May damage the developing fetus.

Points of Attack: Central nervous system, kidneys, liver, skin.

Medical Surveillance: For those with frequent or potentially high exposure (half the TLV or greater) the following are recommended before beginning work and at regular times after that: lung function tests. If symptoms develop or overexposure is suspected, the following may be useful: examination of the nervous system. Evaluation by a qualified allergist, including careful exposure history and special testing, may help diagnose skin allergy. Liver and kidney function tests. Consider chest X-ray following acute overexposure.

First Aid: If this chemical gets into the eyes, remove any contact lenses at once and irrigate immediately for at least 15 min, occasionally lifting upper and lower lids. Seek medical attention immediately. If this chemical contacts the skin, remove contaminated clothing and wash immediately with soap and water. Seek medical attention immediately. If this chemical has been inhaled, remove from exposure, begin rescue breathing (using universal precautions, including resuscitation mask) if breathing has stopped and CPR if heart action has stopped. Transfer promptly to a medical facility. When this chemical has been swallowed, get medical attention. Give large quantities of water and induce vomiting. Do not make an unconscious person vomit.

Personal Protective Methods: Wear protective gloves and clothing to prevent any reasonable probability of skin contact: **8 h**: polyvinyl alcohol gloves; 4H™ and Silver Shield™ gloves; Barricade™ coated suits; Trellchem™ HPS suits. **4 h**: Teflon™ gloves, suits, boots. Safety equipment suppliers/manufacturers can provide recommendations on the most protective glove/clothing material for your operation. All protective clothing (suits, gloves, footwear, headgear) should be clean, available each day, and put on before work. Contact lenses should not be worn when working with this chemical. Wear splash-proof chemical goggles and face shield unless full face-piece respiratory protection is worn. Employees should wash immediately with soap when skin is wet or contaminated. Provide emergency showers and eyewash.

Respirator Selection: 1000 ppm: Sa:Cf (APF = 25) (any supplied-air respirator operated in a continuous-flow mode) or CcrFOv (APF = 50) [any chemical cartridge respirator with a full face-piece and organic vapor cartridge(s)] or GmFOv (APF = 50) [any air-purifying, full-face-piece respirator (gas mask) with a chin-style, front- or back-mounted acid gas canister] or PaprOv (APF = 25) [any powered, air-purifying respirator with organic vapor cartridge(s)] or SCBAF (APF = 50) (any self-contained breathing apparatus with a full face-piece) or SaF (APF = 50) (any supplied-air respirator with a full face-piece). *Emergency or planned entry into unknown concentrations or IDLH conditions:* SCBAF: PD, PP (any self-contained breathing apparatus that has a full face-piece and is operated in a pressure-demand or other positive-pressure mode) or SaF: Pd,Pp: ASCBA (APF = 10,000) (any supplied-air respirator that has a full face-piece and is operated in a pressure-demand or other positive-pressure mode in combination with an auxiliary self-contained breathing apparatus operated in a pressure-demand or other positive-pressure mode). *Escape:* GmFOv (APF = 50) [any air-purifying, full-face-piece respirator (gas mask) with a chin-style, front- or back-mounted organic vapor canister] or SCBAE (any appropriate escape-type, self-contained breathing apparatus).

Note: Substance causes eye irritation or damage; eye protection needed.

Storage: Color Code—Red: Flammability Hazard: Store in a flammable liquid storage area or approved cabinet away from ignition sources and corrosive and reactive materials. Prior to working with this chemical you should be trained on its proper handling and storage. Before entering confined space where this chemical may be present, check to make sure that an explosive concentration does not exist. Methyl methacrylate must be stored to avoid contact with oxidizers (such as nitrates, permanganates, perchlorates, chlorates, and peroxides), strong alkalis (such as sodium hydroxide and potassium hydroxide), and strong acids (such as nitric acid, hydrochloric acid, and sulfuric acid) since violent reactions occur. Store in tightly closed containers in a cool, well-ventilated area away from light, heat, and ionizing radiation because methyl methacrylate will react and release heat quickly causing an explosion. Store and use with an appropriate inhibitor. Lack of an appropriate inhibitor may cause an explosive reaction.

Shipping: This compound requires a shipping label of "FLAMMABLE LIQUID." It falls in Hazard Class 3 and Packing Group II.

Spill Handling: Evacuate and restrict persons not wearing protective equipment from area of spill or leak until cleanup is complete. Remove all ignition sources. Establish forced ventilation to keep levels below explosive limit. Absorb liquids in vermiculite, dry sand, earth, peat, carbon, or a similar material and deposit in sealed containers. Keep this chemical out of a confined space, such as a sewer, because of the possibility of an explosion, unless the sewer is designed to prevent the buildup of explosive concentrations. It may be necessary to contain and dispose of this chemical as a hazardous waste. If material or contaminated runoff enters waterways, notify downstream users of potentially

contaminated waters. Contact your local or federal environmental protection agency for specific recommendations. If employees are required to clean up spills, they must be properly trained and equipped. OSHA 1910.120(q) may be applicable.

Fire Extinguishing: This chemical is a flammable liquid. Poisonous gases, including carbon monoxide, are produced in fire. Use dry chemical, carbon dioxide, or alcohol foam extinguishers. Vapors are heavier than air and will collect in low areas. Vapors may travel long distances to ignition sources and flashback. Vapors in confined areas may explode when exposed to fire. Containers may explode in fire. Storage containers and parts of containers may rocket great distances, in many directions. If material or contaminated runoff enters waterways, notify downstream users of potentially contaminated waters. Notify local health and fire officials and pollution control agencies. From a secure, explosion-proof location, use water spray to cool exposed containers. If cooling streams are ineffective (venting sound increases in volume and pitch, tank discolors, or shows any signs of deforming), withdraw immediately to a secure position. If employees are expected to fight fires, they must be trained and equipped in OSHA 1910.156. The only respirators recommended for firefighting are self-contained breathing apparatuses that have full face-pieces and are operated in a pressure-demand or other positive-pressure mode.

Disposal Method Suggested: Consult with environmental regulatory agencies for guidance on acceptable disposal practices. Generators of waste containing this contaminant (≥100 kg/mo) must conform with EPA regulations governing storage, transportation, treatment, and waste disposal. Incineration may be allowed.

References

US Environmental Protection Agency. (April 30, 1980). *Methyl Methacrylate: Health and Environmental Effects Profile No. 130.* Washington, DC: Office of Solid Waste
Sax, N. I. (Ed.). (1986). *Dangerous Properties of Industrial Materials Report*, 6, No. 1, 86—90
New Jersey Department of Health and Senior Services. (July 2002). *Hazardous Substances Fact Sheet: Methyl Methacrylate.* Trenton, NJ

Methyl parathion   M:1070

Molecular Formula: $C_8H_{10}NO_5PS$
Synonyms: 8056HC; A-Gro; AI3-17292; Azofos; Azophos; Bay 11405; Bay E-601; Bladan M; Cekumethion; DALF; Devithion; *O,O*-Dimethyl *O*-(4-nitrophenyl)-monothiophosphat (German); Dimethyl-*p*-nitrophenyl monothiophosphate; Dimethyl *p*-nitrophenyl monothiophosphate; *O,O*-Dimethyl *O*-(*p*-nitrophenyl) phosphorothioate; *O,O*-Dimethyl *O*-(4-nitrophenyl) phosphorothioate; *O,O*-Dimethyl *O,p*-nitrophenyl phosphorothioate; *O,O*-Dimethyl *O*-4-nitrophenyl phosphorothioate; Dimethyl *p*-nitrophenyl

phosphorothionate; Dimethyl 4-nitrophenyl phosphorothionate; *O,O*-Dimethyl *O*-(*p*-nitrophenyl) thionophosphate; *O, O*-Dimethyl *O*-(*p*-nitrophenyl) thiophosphate; *O,O*-Dimethyl *O,p*-nitrophenyl thiophosphate; Dimethyl *p*-nitrophenyl thiophosphate; Dimethyl parathion; Drexel methyl parathion 4E; E 601; ENT 17,292; Folidoc; Folidol-80; Folidol M; Folidol M-40; Fosferno M 50; ME-Parathion; Meptox; Metacid 50; Metacide; Metafos (Pesticide); Metaphos; Methyl-E 605; Methyl fosferno; Methyl niran; Methylthiophos; Metilparationa (Spanish); Metron; M-Parathion; NCI-C02971; *p*-Nitrophenyldime thylthionophosphate; Nitrox; Nitrox 80; Oleovofotox; Parapest M-50; Parathion-methyl; Parathion metile; Partron M; Penncap M; Penncap MLS; Phenol, *p*-nitro-, *O*-ester with *O,O*-dimethyl phosphorothioate; Phosphorothioic acid, *O,O*-dimethyl *O*-(*p*-nitrophenyl) ester; Phosphorothioic acid, *O,O*-dimethyl *O*-(4-nitrophenyl) ester; Quinophos; Sinafid M-48; Sixty-three special E.C. insecticide; Thiophenit; Thylpar M-50; Toll; Vertac methyl parathion technisch 80%; Wofatox 50 EC

CAS Registry Number: 298-00-0

RTECS® Number: TG175000

UN/NA & ERG Number: UN2783 (organophosphorus pesticides, solid, toxic)/152; UN1967 (Parathion and compressed gas mixture)/123

EC Number: 206-050-1 [*Annex I Index No.:* 015-035-00-7]

Regulatory Authority and Advisory Bodies
Carcinogenicity: NCI: Carcinogenesis Bioassay (feed); no evidence: mouse, rat, 1979; IARC: Human No Adequate Data, Animal No Evidence, *not classifiable as carcinogenic to humans*, Group 3, 1987.

US EPA Gene-Tox Program, Positive: *S. cerevisiae*—homozygosis; Negative: Carcinogenicity—mouse/rat; *In vitro* UDS—human fibroblast; Negative: TRP reversion; Inconclusive: *B. subtilis* rec assay; *E. coli* polA without S9; Inconclusive: Histidine reversion—Ames test.

US EPA, FIFRA, 1998 Status of Pesticides: Supported.

Banned or Severely Restricted (several countries) (UN).[13]

Air Pollutant Standard Set. See below, "Permissible Exposure Limits in Air" section.

Clean Water Act: Section 311 Hazardous Substances/RQ 40CFR117.3 (same as CERCLA, see below).

US EPA Hazardous Waste Number (RCRA No.): P071.

RCRA, 40CFR261, Appendix 8 Hazardous Constituents.

RCRA 40CFR268.48; 61FR15654, Universal Treatment Standards: Wastewater (mg/L), 0.014; Nonwastewater (mg/kg), 4.6.

RCRA 40CFR264, Appendix 9; TSD Facilities Ground Water Monitoring List. Suggested test method(s) (PQL μg/L): 8140 (0.5); 8270 (10).

Superfund/EPCRA 40CFR355, Extremely Hazardous Substances: TPQ = 100/10,000 lb (45.4/4540 kg).

Reportable Quantity (RQ): 100 lb (45.4 kg).

EPCRA Section 313 Form R *de minimis* concentration reporting level: 1.0%.

US DOT 49CFR172.101, Inhalation Hazardous Chemical.

US DOT Regulated Marine Pollutant (49CFR172.101, Appendix B), severe pollutant as Parathion-methyl.

Rotterdam Convention Annex III [Chemicals Subject to the Prior Informed Consent Procedure (PIC)] [methyl parathion (emulsifiable concentrates (EC) at/or >19.5% active ingredient and dusts at/or >1.5% active ingredient)].

European/International Regulations: Hazard Symbol: T+, N; Risk phrases: R5; R10; R24; R26/28; R48/22; R50/53; Safety phrases: S1/2; S22-; S28; S36/37/39; S45; S60; 61 (see Appendix 4).

WGK (German Aquatic Hazard Class): 3—Severe hazard to waters.

Description: Methyl parathion is a white to yellow-brown, crystalline solid with a garlic odor. Molecular weight = 263.22; Specific gravity (H_2O:1) = 1.36; Boiling point = 147.8; Freezing/Melting point = 37.2°C. The commercial product in xylene is a tan liquid (80% methyl parathion/20% xylene); Vapor pressure = 0.00001 mmHg at 20°C; Flash point = 46°C. Hazard Identification (based on NFPA-704 M Rating System): Health 4, Flammability 1, Reactivity 2. Slightly soluble in water; solubility = 0.006% at 25°C.

Potential Exposure: A severely hazardous pesticide formulation. Compound Description: Agricultural Chemical; Tumorigen, Mutagen; Reproductive Effector. This material is used as an insecticide on over 50 crops, primarily cotton, and on several ornamentals.

Incompatibilities: Incompatible with oxidizers, strong bases, heat. Mixtures with magnesium, or endrin may be violent or explosive. Slightly decomposed by acid solutions. Rapidly decomposed by alkalis. Explosive risk when heated above 50°C. The liquid xylene solution decomposes violently at 120°C.

Permissible Exposure Limits in Air
OSHA PEL: None.

NIOSH REL: 0.2 mg/m³ TWA [skin].

ACGIH TLV®[1]: 0.2 mg/m³ TWA, measured as inhalable fraction and vapor [skin]; not classifiable as a human carcinogen; BEI_A issued as Acetylcholinesterase-inhibiting pesticides.

Protective Action Criteria (PAC)*

TEEL-0: 0.02 mg/m³

PAC-1: 0.06 mg/m³

PAC-2: **1.2** mg/m³

PAC-3: **3.5** mg/m³

*AEGLs (Acute Emergency Guideline Levels) & ERPGs (Emergency Response Planning Guideline) are in **bold face**.

Australia: TWA 0.2 mg/m³, [skin], 1993; Belgium: TWA 0.2 mg/m³, [skin], 1993; Denmark: TWA 0.2 mg/m³, [skin], 1999; Finland: TWA 0.2 mg/m³; STEL 0.6 mg/m³, [skin], 1999; France: VME 0.2 mg/m³, [skin], 1999; Hungary: TWA 0.1 mg/m³; STEL 0.2 mg/m³, [skin], 1993; Norway: TWA 0.2 mg/m³, 1999; the Netherlands: MAC-TGG 0.2 mg/m³, [skin], 2003; Russia: STEL 0.1 mg/m³, [skin], 1993; Switzerland: MAK-W 0.2 mg/m³, [skin], 1999;

United Kingdom: LTEL 0.2 mg/m^3; STEL 0.6 mg/m^3, [skin], 1993; Argentina, Bulgaria, Columbia, Jordan, South Korea, New Zealand, Singapore, Vietnam: ACGIH TLV®: not classifiable as a human carcinogen. Russia set a MAC of 0.008 mg/m^3 for ambient air in residential areas on a once daily basis. Several states have set guidelines or standards for methyl parathion in ambient air[60] ranging from 2.0 μg/m^3 (North Dakota) to 3.5 μg/m^3 (Virginia) to 4.0 μg/m^3 (Connecticut) to 5.0 μg/m^3 (Nevada).

Determination in Air: Use NIOSH Analytical Method (IV) #5600, Organophosphorus Pesticides; OSHA Analytical Method PV-2112, Methyl parathion.

Permissible Concentration in Water: Russia set a MAC of 0.02 mg/L in water bodies used for domestic purposes. Two states have set a guideline of 30 μg/L for methyl parathion in drinking water—California and Kansas.[61] A lifetime health advisory of 2 μg/L has been developed by EPA.

Determination in Water: Fish Tox = 5.59677000 ppb (HIGH). Octanol−water coefficient: Log K_{ow} = 2.7 (pure); 2.04 (V).

Routes of Entry: Inhalation, skin absorption, ingestion, skin and/or eye contact.

Harmful Effects and Symptoms

Short Term Exposure: Methyl parathion may affect the nervous system, causing convulsions, respiratory failure, and possible death. A cholinesterase inhibitor. Acute exposure to parathion-methyl may produce the following symptoms: pinpoint pupils, blurred vision, headache, dizziness, muscle spasms, and profound weakness. Vomiting, diarrhea, abdominal pain, seizures, and coma may also occur. High exposure may result in death. The heart rate may decrease following oral exposure or increase following dermal exposure. Hypotension (low blood pressure) may occur although hypertension (high blood pressure) is not uncommon. Chest pain may be noted. Respiratory symptoms include dyspnea (shortness of breath), respiratory depression, and respiratory paralysis. Psychosis may occur. Because this is a mutagen, handle it as a possible carcinogen—with extreme caution. Methyl parathion may damage the developing fetus. This material is extremely toxic; the probable oral lethal dose is 5−50 mg/kg or between 7 drops and 1 teaspoonful for a 150-lb person. Human Tox = 2.00000 ppb (HIGH).

Long Term Exposure: Cholinesterase inhibitor; cumulative effect is possible. This chemical may damage the nervous system with repeated exposure, resulting in convulsions, respiratory failure. May cause personality changes, depression, anxiety, irritability. May cause liver damage. May damage the developing fetus.

Points of Attack: Respiratory system, lungs, central nervous system, cardiovascular system, skin, eyes, plasma and red blood cell cholinesterase.

Medical Surveillance: Before employment and at regular times after that, the following are recommended: plasma and red blood cell cholinesterase levels (tests for the enzyme poisoned by this chemical). If exposure stops, plasma levels return to normal in 1−2 weeks while red blood cell levels may be reduced for 1−3 months.

When cholinesterase enzyme levels are reduced by 25% or more below preemployment levels, risk of poisoning is increased, even if results are in lower ranges of "normal." Reassignment to work not involving organophosphate or carbamate pesticides is recommended until enzyme levels recover. If symptoms develop or overexposure occurs, repeat the above tests as soon as possible and get an examination of the nervous system. Also, consider complete blood count. Consider chest X-ray following acute overexposure. Do not drink any alcoholic beverages before or during use. Alcohol promotes absorption of organic phosphates.

First Aid: If this chemical gets into the eyes, remove any contact lenses at once and irrigate immediately for at least 15 min, occasionally lifting upper and lower lids. Seek medical attention immediately. If this chemical contacts the skin, remove contaminated clothing and wash immediately with soap and water. Speed in removing material from skin is of extreme importance. Shampoo hair promptly if contaminated. Seek medical attention immediately. If this chemical has been inhaled, remove from exposure, begin rescue breathing (using universal precautions, including resuscitation mask) if breathing has stopped and CPR if heart action has stopped. Transfer promptly to a medical facility. When this chemical has been swallowed, get medical attention. Give large quantities of water and induce vomiting. Do not make an unconscious person vomit. Effects of exposure may be delayed. Medical monitoring is advised.

Personal Protective Methods: Wear protective gloves and clothing to prevent any reasonable probability of skin contact. Safety equipment suppliers/manufacturers can provide recommendations on the most protective glove/clothing material for your operation. All protective clothing (suits, gloves, footwear, headgear) should be clean, available each day, and put on before work. Contact lenses should not be worn when working with this chemical. Wear splash-proof chemical goggles and face shield unless full face-piece respiratory protection is worn. Employees should wash immediately with soap when skin is wet or contaminated. Provide emergency showers and eyewash.

Respirator Selection: NIOSH: *2 mg/m^3:* CcrOv95 (APF = 10) [any air-purifying half-mask respirator equipped with an organic vapor cartridge(s) in combination with an N95, R95, or P95 filter. The following filters may also be used: N99, R99, P99, N100, R100, P100] or Sa (APF = 10) (any supplied-air respirator). *5 mg/m^3:* Sa:Cf (APF = 25) (any supplied-air respirator operated in a continuous-flow mode) or PaprOvHie (APF = 25) (any powered air-purifying respirator with an organic vapor cartridge in combination with a high-efficiency particulate filter). *10 mg/m^3:* CcrFOv100 (APF = 50) [any air-purifying full-face-piece respirator equipped with organic vapor cartridge(s) in combination with an N100, R100, or P100 filter] or

GmFOv100 (APF = 50) [Any air-purifying, full-face-piece respirator (gas mask) with a chin-style, front- or back-mounted organic vapor canister having an N100, R100, or P100 filter] or PaprTOvHie (APF = 50) [any powered, air-purifying respirator with a tight-fitting face-piece and organic vapor cartridge(s) in combination with a high-efficiency particulate filter] or SaT: Cf (APF = 50) (any supplied-air respirator that has a tight-fitting face-piece and is operated in a continuous-flow mode) or SCBAF (APF = 50) (any self-contained breathing apparatus with full face-piece) or SaF (APF = 50) (any supplied-air respirator with a full face-piece). *200 mg/m³:* SaF: Pd,Pp (APF = 2000) (any supplied-air respirator that has a full face-piece and is operated in a pressure-demand or other positive-pressure mode). *Emergency or planned entry into unknown concentrations or IDLH conditions:* SCBAF: PD, PP (any self-contained breathing apparatus that has a full face-piece and is operated in a pressure-demand or other positive-pressure mode) or SaF: Pd,Pp: ASCBA (APF = 10,000) (any supplied-air respirator that has a full face-piece and is operated in a pressure-demand or other positive-pressure mode in combination with an auxiliary self-contained breathing apparatus operated in a pressure-demand or other positive-pressure mode). *Escape:* GmFOv100 (APF = 50) [any air-purifying, full-face-piece respirator (gas mask) with a chin-style, front- or back-mounted organic vapor canister having an N100, R100, or P100 filter] or SCBAE (any appropriate escape-type, self-contained breathing apparatus).

Storage: Color Code—Blue: Health Hazard/Poison: Store in a secure poison location. Store in tightly closed containers in a cool, well-ventilated area away from heat sources since violent reactions may occur. See incompatible material listed above. Sources of ignition, such as smoking and open flames, are prohibited where this chemical is handled, used, or stored. Metal containers involving the transfer of 5 gallons or more of this chemical should be grounded and bonded. Drums must be equipped with self-closing valves, pressure vacuum bungs, and flame arresters. Use only non-sparking tools and equipment, especially when opening and closing containers of this chemical. Wherever this chemical is used, handled, manufactured, or stored, use explosion-proof electrical equipment and fittings.

Shipping: Organophosphorus pesticides, solid, toxic, n.o.s. require a shipping label of "POISONOUS/TOXIC MATERIALS." It falls in Hazard Class 6.1. Methyl parathion, solid and liquid, carry the symbol "D." The letter "D" identifies proper shipping names that are appropriate for describing materials for *domestic* transportation but may be inappropriate for international transportation under the provisions of international regulations (e.g., IMO, ICAO). An alternate proper shipping name may be selected when either domestic or international transportation is involved. Methyl parathion, solid, requires a shipping label of "POISONOUS/TOXIC MATERIALS." It falls in Hazard Class 6.1 and Packing Group II. Methyl parathion, liquid, requires a shipping label of "POISONOUS/TOXIC

MATERIALS." It falls in Hazard Class 6.1 and Packing Group II.

Spill Handling: Do not touch spilled material; stop leak if you can do it without risk. Use water spray to reduce vapors. Evacuate and restrict persons not wearing protective equipment from area of spill or leak until cleanup is complete. Remove all ignition sources. Absorb liquids in vermiculite, dry sand, earth, peat, carbon, or a similar material and deposit in sealed containers. Collect powdered material in the most convenient and safe manner and deposit in sealed containers. Ventilate and wash area of spill or leak after cleanup is complete. Keep this chemical out of a confined space, such as a sewer, because of the possibility of an explosion, unless the sewer is designed to prevent the buildup of explosive concentrations. It may be necessary to contain and dispose of this chemical as a hazardous waste. If material or contaminated runoff enters waterways, notify downstream users of potentially contaminated waters. Contact your local or federal environmental protection agency for specific recommendations. If employees are required to clean up spills, they must be properly trained and equipped. OSHA 1910.120(q) may be applicable.

Fire Extinguishing: This chemical is a combustible solid. Often available and used in a flammable liquid solution. Poisonous gases, including nitrogen oxides, sulfur oxide, and phosphorus oxide, are produced in fire. Use dry chemical, carbon dioxide, or alcohol foam extinguishers. Vapors from the liquid are heavier than air and will collect in low areas. Vapors may travel long distances to ignition sources and flashback. Vapors in confined areas may explode when exposed to fire. Containers may explode in fire. Storage containers and parts of containers may rocket great distances, in many directions. If material or contaminated run-off enters waterways, notify downstream users of potentially contaminated waters. Notify local health and fire officials and pollution control agencies. From a secure, explosion-proof location, use water spray to cool exposed containers. If cooling streams are ineffective (venting sound increases in volume and pitch, tank discolors, or shows any signs of deforming), withdraw immediately to a secure position. If employees are expected to fight fires, they must be trained and equipped in OSHA 1910.156. The only respirators recommended for firefighting are self-contained breathing apparatuses that have full face-pieces and are operated in a pressure-demand or other positive-pressure mode.

Disposal Method Suggested: Consult with environmental regulatory agencies for guidance on acceptable disposal practices. Generators of waste containing this contaminant (≥100 kg/mo) must conform with EPA regulations governing storage, transportation, treatment, and waste disposal. Incineration (816°C, 0.5 s minimum for primary combustion; 1204°C, 1.0 s for secondary combustion) with adequate scrubbing and ash disposal facilities. In accordance with 40CFR165, follow recommendations for the disposal of pesticides and pesticide containers. Must be disposed properly by following package label directions or by contacting your local or federal environmental control agency or by contacting your regional EPA office.

References

National Institute for Occupational Safety and Health. (1977). *Criteria for a Recommended Standard: Occupational Exposure to Methyl Parathion*, NIOSH Document No. 77-106

Sax, N. I. (Ed.). (1986). *Dangerous Properties of Industrial Materials Report*, 6, No. 1, 90–97

US Environmental Protection Agency. (November 30, 1987) *Chemical Hazard Information Profile: Parathion-Methyl*. Washington, DC: Chemical Emergency Preparedness Program

US Environmental Protection Agency. (August 1987). *Health Advisory: Methyl Parathion*. Washington, DC: Office of Drinking Water

New Jersey Department of Health and Senior Services. (November 1999). *Hazardous Substances Fact Sheet: Methyl Parathion*. Trenton, NJ

US Environmental Protection Agency, Special Review and Reregistration Division Office of Pesticide Programs. (1998). *Agency Status of Pesticides in Registration, Reregistration, and Special Review* (Rainbow Report). Washington, DC

Methyl phenkapton   M:1080

Molecular Formula: $C_9H_{11}Cl_2O_2PS_3$

Synonyms: (2,5-Dichlorophenylthio)methanethiol *S*-ester with *O,O*-dimethyl phosphorodithioate; *S*-([(2,5-Dichlorophenyl)thio]methyl) *O,O*-dimethyl phosphorodithioate; *O,O*-Dimethyl *S*-(2,5-dichlorophenylthio)methyl phosphorodithioate; ENT 25,554; Geigy 30494; Methyl phencapton

CAS Registry Number: 3735-23-7

RTECS® Number: TD6125000

UN/NA & ERG Number: UN3018 (organophosphorus pesticide, liquid, toxic)/152

UN/NA & ERG Number: None assigned

Regulatory Authority and Advisory Bodies

Superfund/EPCRA 40CFR355, Extremely Hazardous Substances: TPQ = 500 lb (227 kg).

Reportable Quantity (RQ): 500 lb (227 kg).

US DOT 49CFR172.101, Inhalation Hazard Chemical as organophosphates.

WGK (German Aquatic Hazard Class): No value assigned.

Description: Methyl phenkapton is a liquid product. Molecular weight = 349.25. Hazard Identification (based on NFPA-704 M Rating System): Health 3, Flammability 1, Reactivity 0.

Potential Exposure: This material is an acaricide, insecticide. Not registered as a pesticide in the United States.

Incompatibilities: Strong oxidizers may cause release of toxic phosphorus oxides. Organophosphates, in the presence of strong reducing agents such as hydrides, may form highly toxic and flammable phosphine gas.

Permissible Exposure Limits in Air

Protective Action Criteria (PAC)

TEEL-0: 2 mg/m³

PAC-1: 6 mg/m³

PAC-2: 11 mg/m³

PAC-3: 100 mg/m³

Routes of Entry: Inhalation, ingestion, eye and/or skin contact.

Harmful Effects and Symptoms

Short Term Exposure: Organic phosphorus insecticides are absorbed by the skin, as well as by the respiratory and gastrointestinal tracts. They are cholinesterase inhibitors. Symptoms of exposure include headache, giddiness, blurred vision, nervousness, weakness, nausea, cramps, diarrhea, and discomfort in the chest. Signs include sweating, tearing, salivation, vomiting, cyanosis, convulsions, coma, loss of reflexes, and loss of sphincter control.

Long Term Exposure: Cholinesterase inhibitor; cumulative effect is possible. This chemical may damage the nervous system with repeated exposure, resulting in convulsions, respiratory failure. May cause liver damage.

Points of Attack: Respiratory system, lungs, central nervous system, cardiovascular system, skin, eyes, plasma and red blood cell cholinesterase.

Medical Surveillance: Before employment and at regular times after that, the following are recommended: plasma and red blood cell cholinesterase levels (tests for the enzyme poisoned by this chemical). If exposure stops, plasma levels return to normal in 1–2 weeks while red blood cell levels may be reduced for 1–3 months.

When cholinesterase enzyme levels are reduced by 25% or more below preemployment levels, risk of poisoning is increased, even if results are in lower ranges of "normal." Reassignment to work not involving organophosphate or carbamate pesticides is recommended until enzyme levels recover. If symptoms develop or overexposure occurs, repeat the above tests as soon as possible and get an examination of the nervous system. Also, consider complete blood count. Consider chest X-ray following acute overexposure. Do not drink any alcoholic beverages before or during use. Alcohol promotes absorption of organic phosphates.

First Aid: If this chemical gets into the eyes, remove any contact lenses at once and irrigate immediately for at least 15 min, occasionally lifting upper and lower lids. Seek medical attention immediately. If this chemical contacts the skin, remove contaminated clothing and wash immediately with soap and water. Speed in removing material from skin is of extreme importance. Shampoo hair promptly if contaminated. Seek medical attention immediately. If this chemical has been inhaled, remove from exposure, begin rescue breathing (using universal precautions, including resuscitation mask) if breathing has stopped and CPR if heart action has stopped. Transfer promptly to a medical facility. When this chemical has been swallowed, get medical attention. Give large quantities of water and induce vomiting. Do not make an unconscious person vomit. Keep victim quiet and maintain normal body temperature. Effects may be delayed; keep victim under observation.

Personal Protective Methods: Wear protective gloves and clothing to prevent any reasonable probability of skin contact. Safety equipment suppliers/manufacturers can provide recommendations on the most protective glove/clothing material for your operation. All protective clothing (suits, gloves, footwear, headgear) should be clean, available each day, and put on before work. Contact lenses should not be worn when working with this chemical. Wear splash-proof chemical goggles and face shield unless full face-piece respiratory protection is worn. Employees should wash immediately with soap when skin is wet or contaminated. Provide emergency showers and eyewash.

Respirator Selection: Follow regulations in OSHA 29CFR1910.134 or European Standard EN149. Use a NIOSH/MSHA- or European Standard EN149-approved respirator; or use an approved supplied-air respirator with a full face-piece operated in the positive-pressure mode, or with a full face-piece, hood, or helmet in the continuous-flow mode; or use a NIOSH/MSHA- or European Standard EN149-approved self-contained breathing apparatus with a full face-piece operated in a pressure-demand or other positive-pressure mode.

Storage: Color Code—Blue: Health Hazard/Poison: Store in a secure poison location. Prior to working with this chemical you should be trained on its proper handling and storage. Store in tightly closed containers in a cool, well-ventilated area away from oxidizers and reducing agents. Where possible, automatically pump liquid from drums or other storage containers to process containers.

Shipping: Organophosphorus pesticides, liquid, toxic, require a shipping label of "POISONOUS/TOXIC MATERIALS." It falls in Hazard Class 6.1 and Packing Group III.

Spill Handling: Stay upwind; keep out of low areas. Ventilate closed spaces before entering them. Do not touch spilled material; stop leak if you can do it without risk. Use water spray to reduce vapors. Evacuate and restrict persons not wearing protective equipment from area of spill or leak until cleanup is complete. Remove all ignition sources. Ventilate area of spill or leak. *Small spills:* take up with sand or other noncombustible absorbent material and place into containers for later disposal. *Small dry spills:* with clean shovel place material into clean, dry containers and cover; move containers from spill area. *Large spills:* dike far ahead of spill for later disposal. It may be necessary to contain and dispose of this chemical as a hazardous waste. If material or contaminated runoff enters waterways, notify downstream users of potentially contaminated waters. Contact your local or federal environmental protection agency for specific recommendations. If employees are required to clean up spills, they must be properly trained and equipped. OSHA 1910.120(q) may be applicable.

Initial isolation and protective action distances
Distances shown are likely to be affected during the first 30 min after materials are spilled and could increase with time. If more than one tank car, cargo tank, portable tank, or large cylinder involved in the incident is leaking, the protective action distance may need to be increased. You may need to seek emergency information from CHEMTREC at (800) 424-9300 or seek professional environmental engineering assistance from the US EPA Environmental Response Team at (908) 548-8730 (24-h response line).

Fire Extinguishing: This material may burn but does not ignite readily. Poisonous gases, including sulfur oxide, phosphorus oxide, and chlorine, are produced in fire. For small fires, use dry chemical, carbon dioxide, water spray, or foam. For large fires, use water spray, fog, or foam. Stay upwind; keep out of low areas. Move containers from fire area if you can do it without risk. Fight fire from maximum distance. Dike fire control water for later disposal; do not scatter the material. Wear positive pressure breathing apparatus and special protective clothing. Containers may explode in fire. Storage containers and parts of containers may rocket great distances, in many directions. If material or contaminated runoff enters waterways, notify downstream users of potentially contaminated waters. Notify local health and fire officials and pollution control agencies. From a secure, explosion-proof location, use water spray to cool exposed containers. If cooling streams are ineffective (venting sound increases in volume and pitch, tank discolors, or shows any signs of deforming), withdraw immediately to a secure position. If employees are expected to fight fires, they must be trained and equipped in OSHA 1910.156. The only respirators recommended for firefighting are self-contained breathing apparatuses that have full face-pieces and are operated in a pressure-demand or other positive-pressure mode.

Disposal Method Suggested: In accordance with 40CFR 165 recommendations for the disposal of pesticides and pesticide containers. Must be disposed properly by following package label directions or by contacting your local or federal environmental control agency or by contacting your regional EPA office.

Reference
US Environmental Protection Agency. (November 30, 1987). *Chemical Hazard Information Profile: Methyl Phenkapton.* Washington, DC: Chemical Emergency Preparedness Program

Methyl phosphonicdichloride M:1090

Molecular Formula: CH_3Cl_2OP
Common Formula: CH_3POCl_2
Synonyms: Dicloruro de metilfosfonico (Spanish); Methyl phosphonothioic dichloride, anhydrous; Phosphonic dichloride

CAS Registry Number: 676-97-1
RTECS® Number: TA1840000
UN/NA & ERG Number: UN9206/137
EC Number: 211-634-4

Regulatory Authority and Advisory Bodies
Department of Homeland Security Screening Threshold Quantity (pounds): Sabotage/Contamination Hazard: A placarded amount (commercial grade).
Superfund/EPCRA 40CFR355, Extremely Hazardous Substances: TPQ = 100 lb (45.4 kg).
Reportable Quantity (RQ): 100 lb (45.4 kg).
US DOT 49CFR172.101, Inhalation Hazardous Chemical.
WGK (German Aquatic Hazard Class): No value assigned.

Description: Methyl phosphonic dichloride is a low melting solid or colorless liquid. Molecular weight = 132.91; Boiling point = 162°C; Freezing/Melting point = 32°C; Flash point = ≥ 50°C (oc). Mixes violently with water.

Potential Exposure: Used as a chemical intermediate in pesticide manufacture.

Incompatibilities: Reacts violently with water, alcohols, forming hydrochloric acid/hydrogen chloride vapor. The reaction may be violent. Corrodes metals. Incompatible with strong oxidizers, alcohols, bases (including amines). May react violently, possibly explosively, when mixed with ethers and trace amounts of metal salts.

Permissible Exposure Limits in Air
Protective Action Criteria (PAC)
TEEL-0: 0.06 mg/m^3
PAC-1: 0.2 mg/m^3
PAC-2: 3.14 mg/m^3
PAC-3: 15 mg/m^3

Routes of Entry: Inhalation, ingestion, skin, and/or eye contact.

Harmful Effects and Symptoms
Short Term Exposure: Poisonous if inhaled or swallowed. Highly corrosive. Contact causes severe burns to skin and eyes. May cause permanent damage. Inhalation causes irritation. High levels of exposure can cause pulmonary edema, a medical emergency that can be delayed for several hours. This can cause death.

Long Term Exposure: May cause lung irritation or bronchitis to develop.

Points of Attack: Lungs, eyes, skin.

Medical Surveillance: Lung function tests. Consider chest X-ray following acute overexposure.

First Aid: If this chemical gets into the eyes, remove any contact lenses at once and irrigate immediately for at least 15 min, occasionally lifting upper and lower lids. Seek medical attention immediately. If this chemical contacts the skin, remove contaminated clothing and wash immediately with soap and water. Seek medical attention immediately. If this chemical has been inhaled, remove from exposure, begin rescue breathing (using universal precautions, including resuscitation mask) if breathing has stopped and CPR if heart action has stopped. Transfer promptly to a medical facility. When this chemical has been swallowed, get medical attention. If victim is conscious, administer water or milk. Do not induce vomiting. Keep victim quiet and maintain normal body temperature. Medical observation is recommended for 24–48 h after breathing overexposure, as pulmonary edema may be delayed. As first aid for pulmonary edema, a doctor or authorized paramedic may consider administering a corticosteroid spray.

Personal Protective Methods: Wear protective gloves and clothing to prevent any reasonable probability of skin contact. Safety equipment suppliers/manufacturers can provide recommendations on the most protective glove/clothing material for your operation. All protective clothing (suits, gloves, footwear, headgear) should be clean, available each day, and put on before work. Contact lenses should not be worn when working with this chemical. Wear dust-proof chemical goggles and face shield unless full face-piece respiratory protection is worn. Employees should wash immediately with soap when skin is wet or contaminated. Provide emergency showers and eyewash.

Respirator Selection: Follow regulations in OSHA 29CFR1910.134 or European Standard EN149. Use a NIOSH/MSHA- or European Standard EN149-approved respirator; or use an approved supplied-air respirator with a full face-piece operated in the positive-pressure mode, or with a full face-piece, hood, or helmet in the continuous-flow mode; or use a NIOSH/MSHA- or European Standard EN149-approved self-contained breathing apparatus with a full face-piece operated in a pressure-demand or other positive-pressure mode.

Storage: Color Code—White: Corrosive or Contact Hazard; Store separately in a corrosion-resistant location. Prior to working with this chemical you should be trained on its proper handling and storage. Store in tightly closed containers in a cool, well-ventilated area. Sources of ignition, such as smoking and open flames, are prohibited where this chemical is handled, used, or stored. Use only nonsparking tools and equipment, especially when opening and closing containers of this chemical. Wherever this chemical is used, handled, manufactured, or stored, use explosion-proof electrical equipment and fittings.

Shipping: This compound requires a shipping label of "POISONOUS/TOXIC MATERIALS, CORROSIVE." It falls in Hazard Class 8 and Packing Group I.

Spill Handling: Evacuate persons not wearing protective equipment from area of spill or leak until cleanup is complete. Remove all ignition sources. Do not breathe vapors. Do not get water inside container; stop leak if you can do so without risk. Do not touch spilled material. Use water spray to reduce vapors but do not put water on leak or spill area. Keep combustibles (wood, paper, oil, etc.) away from spilled material. Dike spill for later disposal; do not apply water unless directed to do so. Clean up only under supervision of an expert. Collect powdered material in the most convenient and safe manner and deposit in sealed containers. Ventilate area after cleanup is complete. It may be necessary to contain and dispose of this chemical as a hazardous waste. If material or contaminated runoff enters waterways, notify downstream users of potentially contaminated waters. Contact your local or federal environmental protection agency for specific recommendations. If employees are required to clean up spills, they must be properly trained and equipped. OSHA 1910.120(q) may be applicable.

Initial isolation and protective action distances

Distances shown are likely to be affected during the first 30 min after materials are spilled and could increase with time. If more than one tank car, cargo tank, portable tank, or large cylinder involved in the incident is leaking, the protective action distance may need to be increased. You may need to seek emergency information from CHEMTREC at (800) 424-9300 or seek professional environmental engineering assistance from the US EPA Environmental Response Team at (908) 548-8730 (24-h response line).

Small spills (From a small package or a small leak from a large package)

First: Isolate in all directions (feet/meters)   100/30
Then: Protect persons downwind (miles/kilometers)
Day   0.1/0.2
Night 0.1/0.2

Large spills (From a large package or from many small packages)

First: Isolate in all directions (feet/meters)   200/60
Then: Protect persons downwind (miles/kilometers)
Day   0.3/0.5
Night 0.4/0.6

Fire Extinguishing: This material may burn but does not ignite readily. *Small fires:* dry chemical or carbon dioxide. *Large fires:* flood fire area with water from a distance. Do not get solid stream of water on spilled material or inside open containers. Move container from fire area if you can do so without risk. Spray cooling water on containers that are exposed to flames until well after fire is out. Poisonous gases, including hydrogen chloride and sulfur dioxide, are produced in fire. If material or contaminated runoff enters waterways, notify downstream users of potentially contaminated waters. Notify local health and fire officials and pollution control agencies. Containers may explode in fire. From a secure, explosion-proof location, use water spray to cool exposed containers. If cooling streams are ineffective (venting sound increases in volume and pitch, tank discolors, or shows any signs of deforming), withdraw immediately to a secure position. If employees are expected to fight fires, they must be trained and equipped in OSHA 1910.156. The only respirators recommended for firefighting are self-contained breathing apparatuses that have full face-pieces and are operated in a pressure-demand or other positive-pressure mode.

Reference

US Environmental Protection Agency. (November 30, 1987). *Chemical Hazard Information Profile: Methyl Phosphonic Dichloride.* Washington, DC: Chemical Emergency Preparedness Program

Methyl propionate   M:1200

Molecular Formula: $C_4H_8O_2$

Synonyms: Methyl propanoate; Methyl propylate; Propanoic acid, methyl ester; Propionate de methyle (French); Propionato de metilo (Spanish)

CAS Registry Number: 554-12-1

RTECS® Number: UF5970000

UN/NA & ERG Number: UN1248/129

EC Number: 209-060-4 [*Annex I Index No.:* 607-027-00-2]

Regulatory Authority and Advisory Bodies

Canada, WHMIS, Ingredients Disclosure List Concentration 1.0%.

European/International Regulations: Hazard Symbol: F; Xn; Risk phrases: R11; R20; Safety phrases: S2; S16; S24; S29; S33 (see Appendix 4).

WGK (German Aquatic Hazard Class): 1—Low hazard to waters.

Description: Methyl propionate is a colorless liquid with a sweet, fruity, rum-like odor. Molecular weight = 88.12; Boiling point = 80°C; Freezing/Melting point = − 87°C; Flash point = − 2°C; Autoignition temperature = 469°C. Explosive limits: LEL = 2.5%; UEL = 13.0%; Hazard Identification (based on NFPA-704 M Rating System): Health 1, Flammability 3, Reactivity 0. Slightly soluble in water.

Potential Exposure: Compound Description: Natural Product; Primary Irritant. Used as a solvent; and in making paints, lacquers, and varnishes. Also, used in flavorings and fragrances.

Incompatibilities: Forms explosive mixture with air. Violent reactions with strong oxidizers, strong acids, strong bases. Keep away from heat and moisture.

Permissible Exposure Limits in Air

No TEEL available.

Russia: STEL 10 mg/m^3, 1993.

Routes of Entry: Inhalation, skin and/or eye contact.

Harmful Effects and Symptoms

Short Term Exposure: Contact can cause skin and eye irritation.

Long Term Exposure: Repeated high exposures may cause dizziness, lightheadedness, loss of coordination, and difficult breathing.

First Aid: If this chemical gets into the eyes, remove any contact lenses at once and irrigate immediately for at least 15 min, occasionally lifting upper and lower lids. Seek medical attention immediately. If this chemical contacts the skin, remove contaminated clothing and wash immediately with soap and water. Seek medical attention immediately. If this chemical has been inhaled, remove from exposure, begin rescue breathing (using universal precautions, including resuscitation mask) if breathing has stopped and CPR if heart action has stopped. Transfer promptly to a medical facility. When this chemical has been swallowed, get medical attention. Give large quantities of water and induce vomiting. Do not make an unconscious person vomit.

Personal Protective Methods: Wear protective gloves and clothing to prevent any reasonable probability of skin contact. Safety equipment suppliers/manufacturers can provide recommendations on the most protective glove/clothing material for your operation. All protective clothing (suits, gloves, footwear, headgear) should be clean, available each day, and put on before work. Contact lenses should not be worn when working with this chemical. Wear splash-proof chemical goggles and face shield unless full face-piece respiratory protection is worn. Employees should wash immediately with soap when skin is wet or contaminated. Provide emergency showers and eyewash.

Respirator Selection: Follow regulations in OSHA 29CFR1910.134 or European Standard EN149. Use a NIOSH/MSHA- or European Standard EN149-approved respirator; or use an approved supplied-air respirator with a full face-piece operated in the positive-pressure mode, or with a full face-piece, hood, or helmet in the continuous-flow mode; or use a NIOSH/MSHA- or European Standard EN149-approved self-contained breathing apparatus with a full face-piece operated in a pressure-demand or other positive-pressure mode.

Storage: Color Code—Red: Flammability Hazard: Store in a flammable liquid storage area or approved cabinet away from ignition sources and corrosive and reactive materials. Prior to working with methyl propionate you should be trained on its proper handling and storage. Before entering confined space where this chemical may be present, check to make sure that an explosive concentration does not exist. Store in tightly closed containers in a cool, well-ventilated area away from oxidizers, strong acids, strong bases. Where possible, automatically pump liquid from drums or other storage containers to process containers. Drums must be equipped with self-closing valves, pressure vacuum bungs, and flame arresters. Use only nonsparking tools and equipment, especially when opening and closing containers of this chemical. Wherever this chemical is used, handled, manufactured, or stored, use explosion-proof electrical equipment and fittings.

Shipping: Methyl propionate reqires a label of "FLAMMABLE LIQUID." It falls in Hazard Class 3 and Packing Group II.

Spill Handling: Evacuate and restrict persons not wearing protective equipment from area of spill or leak until cleanup is complete. Remove all ignition sources. Establish forced ventilation to keep levels below explosive limit. Ventilate area of spill or leak. Absorb liquids in vermiculite, dry sand, earth, peat, carbon, or a similar material and deposit in sealed containers. Keep this chemical out of a confined space, such as a sewer, because of the possibility of an explosion, unless the sewer is designed to prevent the buildup of explosive concentrations. It may be necessary to contain and dispose of this chemical as a hazardous waste. If material or contaminated runoff enters waterways, notify downstream users of potentially contaminated waters. Contact your local or federal environmental protection agency for specific recommendations. If employees are required to clean up spills, they must be properly trained and equipped. OSHA 1910.120(q) may be applicable.

Fire Extinguishing: This chemical is a combustible liquid. Poisonous gases are produced in fire. Use dry chemical, carbon dioxide, alcohol foam, or polymer foam extinguishers. Vapors are heavier than air and will collect in low areas. Vapors may travel long distances to ignition sources and flashback. Vapors in confined areas may explode when exposed to fire. Containers may explode in fire. Storage containers and parts of containers may rocket great distances, in many directions. If material or contaminated run-off enters waterways, notify downstream users of potentially contaminated waters. Notify local health and fire officials and pollution control agencies. From a secure, explosion-proof location, use water spray to cool exposed containers. If cooling streams are ineffective (venting sound increases in volume and pitch, tank discolors, or shows any signs of deforming), withdraw immediately to a secure position. If employees are expected to fight fires, they must be trained and equipped in OSHA 1910.156. The only respirators recommended for firefighting are self-contained breathing apparatuses that have full face-pieces and are operated in a pressure-demand or other positive-pressure mode.

Reference

New Jersey Department of Health and Senior Services. (May 1999). *Hazardous Substances Fact Sheet: Methyl Propionate.* Trenton, NJ

Methyl propyl ether   M:1210

Molecular Formula: $C_4H_{10}O$

Synonyms: Ether, methyl propyl; α-Methoxy propane; 1-Methoxypropane; Methyl *n*-propyl ether; Metopryl; Neothyl; Propane, 1-methoxy-

CAS Registry Number: 557-17-5

RTECS® Number: KO2280000

UN/NA & ERG Number: UN2612/127

EC Number: 209-158-7

Regulatory Authority and Advisory Bodies

WGK (German Aquatic Hazard Class): No value assigned.

Description: Methyl propyl ether is a clear, highly flammable, peroxidizable liquid. Molecular weight = 74.14; Boiling point = 39°C; Flash point ≤ −20°C. Hazard Identification (based on NFPA-704 M Rating System): Health 0, Flammability 3, Reactivity 0.

Potential Exposure: Used to make other chemicals.

Incompatibilities: Forms explosive mixture with air. May be able to form unstable and explosive peroxides. Violent reaction with strong oxidizers. Incompatible with strong acids.

Permissible Exposure Limits in Air

No standards or TEEL available.

Determination in Air: No methods listed.

Routes of Entry: Inhalation, eye and/or skin contact.

Harmful Effects and Symptoms

Short Term Exposure: Irritates the eyes, skin, and respiratory tract. Inhalation causes coughing and wheezing. High exposure can cause loss of appetite, headache, dizziness, followed by sleepiness and loss of consciousness.

Long Term Exposure: Causes drying and cracking of the skin.

First Aid: If this chemical gets into the eyes, remove any contact lenses at once and irrigate immediately for at least 15 min, occasionally lifting upper and lower lids. Seek medical attention immediately. If this chemical contacts the skin, remove contaminated clothing and wash immediately with soap and water. Seek medical attention immediately. If this chemical has been inhaled, remove from exposure, begin rescue breathing (using universal precautions, including resuscitation mask) if breathing has stopped and CPR if heart action has stopped. Transfer promptly to a medical facility. When this chemical has been swallowed, get medical attention. Give large quantities of water and induce vomiting. Do not make an unconscious person vomit.

Personal Protective Methods: Wear protective gloves and clothing to prevent any reasonable probability of skin contact. Safety equipment suppliers/manufacturers can provide recommendations on the most protective glove/clothing material for your operation. All protective clothing (suits, gloves, footwear, headgear) should be clean, available each day, and put on before work. Contact lenses should not be worn when working with this chemical. Wear splash-proof chemical goggles and face shield unless full face-piece respiratory protection is worn. Employees should wash immediately with soap when skin is wet or contaminated. Provide emergency showers and eyewash.

Respirator Selection: Follow regulations in OSHA 29CFR1910.134 or European Standard EN149. Use a NIOSH/MSHA- or European Standard EN149-approved respirator; or use an approved supplied-air respirator with a full face-piece operated in the positive-pressure mode, or with a full face-piece, hood, or helmet in the continuous-flow mode; or use a NIOSH/MSHA- or European Standard EN149-approved self-contained breathing apparatus with a full face-piece operated in a pressure-demand or other positive-pressure mode.

Storage: Color Code—Red: Flammability Hazard: Store in a flammable liquid storage area or approved cabinet away from ignition sources and corrosive and reactive materials. Prior to working with methyl propyl ether you should be trained on its proper handling and storage. Store in tightly closed containers in a cool, well-ventilated area away from oxidizers, heat, flames. Where possible, automatically pump liquid from drums or other storage containers to process containers. Drums must be equipped with self-closing valves, pressure vacuum bungs, and flame arresters. Use only nonsparking tools and equipment, especially when opening and closing containers of this chemical. Wherever this chemical is used, handled, manufactured, or stored, use explosion-proof electrical equipment and fittings.

Shipping: Methyl propyl ether requires a shipping label of "FLAMMABLE LIQUID." It falls in Hazard Class 3 and Packing Group II.

Spill Handling: Evacuate and restrict persons not wearing protective equipment from area of spill or leak until cleanup is complete. Remove all ignition sources. Ventilate area of spill or leak. Absorb liquids in vermiculite, dry sand, earth, peat, carbon, or a similar material and deposit in sealed containers. Keep this chemical out of a confined space, such as a sewer, because of the possibility of an explosion, unless the sewer is designed to prevent the buildup of explosive concentrations. It may be necessary to contain and dispose of this chemical as a hazardous waste. If material or contaminated runoff enters waterways, notify downstream users of potentially contaminated waters. Contact your local or federal environmental protection agency for specific recommendations. If employees are required to clean up spills, they must be properly trained and equipped. OSHA 1910.120(q) may be applicable.

Fire Extinguishing: This chemical is a flammable liquid. Poisonous gases are produced in fire. Use dry chemical, carbon dioxide, or alcohol foam extinguishers. Vapors are heavier than air and will collect in low areas. Vapors may travel long distances to ignition sources and flashback. Vapors in confined areas may explode when exposed to fire. Containers may explode in fire. Storage containers and parts of containers may rocket great distances, in many directions. If material or contaminated runoff enters waterways, notify downstream users of potentially contaminated waters. Notify local health and fire officials and pollution control agencies. From a secure, explosion-proof location, use water spray to cool exposed containers. If cooling streams are ineffective (venting sound increases in volume and pitch, tank discolors, or shows any signs of deforming), withdraw immediately to a secure position. If employees are expected to fight fires, they must be trained and equipped in OSHA 1910.156. The only respirators recommended for firefighting are self-contained breathing apparatuses that have full face-pieces and are operated in a pressure-demand or other positive-pressure mode.

Disposal Method Suggested: Dissolve or mix the material with a combustible solvent and burn in a chemical incinerator equipped with an afterburner and scrubber. All federal, state, and local environmental regulations must be observed.

Reference

New Jersey Department of Health and Senior Services. (June 1999). *Hazardous Substances Fact Sheet: Methyl Propyl Ether.* Trenton, NJ

Methyl propyl ketone   M:1220

Molecular Formula: $C_5H_{10}O$
Common Formula: $CH_3COCH_2CH_2CH_3$

Synonyms: Ethyl acetone; Methyl-*n*-propyl ketone; MPK; 2-Pentanone
CAS Registry Number: 107-87-9
RTECS® Number: SA7875000
UN/NA & ERG Number: UN1249/127
EC Number: 203-528-1
Regulatory Authority and Advisory Bodies
Air Pollutant Standard Set. See below, "Permissible Exposure Limits in Air" section.
Canada, WHMIS, Ingredients Disclosure List Concentration 1.0%.
European/International Regulations: not listed in Annex 1.
WGK (German Aquatic Hazard Class): 1—Low hazard to waters.
Description: MPK is a colorless to water-white liquid with a strong odor resembling acetone and ether. The odor threshold is 7.7 ppm. Molecular weight = 86.15; Specific gravity (H_2O:1) = 0.81; Boiling point = 102°C; Freezing/Melting point = −78°C; Vapor pressure = 27 mmHg at 20°C; Flash point = 7°C (cc); Autoignition temperature = 452°C. Explosive limits: LEL = 1.5%; UEL = 8.2%. Hazard Identification (based on NFPA-704 M Rating System): Health 2, Flammability 3, Reactivity 0. Slightly soluble in water; solubility = 6%.
Potential Exposure: Compound Description: Mutagen, Human Data; Primary Irritant. MPK is used as a solvent; as a synthetic food flavoring agent; and in organic synthesis; as a solvent replacement for diethyl ketone and acetone.
Incompatibilities: Reacts violently with strong oxidants, strong bases, amines, and isocyanates. Attacks some plastics.
Permissible Exposure Limits in Air
Conversion factor: 1 ppm = 3.52 mg/m³ at 25°C & 1 atm.
OSHA PEL: 200 ppm/700 mg/m³ TWA.
NIOSH REL: 150 ppm/530 mg/m³ TWA.
ACGIH TLV®[1]: 150 ppm/529 mg/m³ STEL.
NIOSH IDLH: 1500 ppm.
Protective Action Criteria (PAC)
TEEL-0: 150 ppm
PAC-1: 150 ppm
PAC-2: 150 ppm
PAC-3: 1500 ppm
DFG MAK: No numerical value established. Data may be available.
Australia: TWA 200 ppm (700 mg/m³), STEL 250 ppm, 1993; Austria: MAK 200 ppm (700 mg/m³), 1999; Belgium: TWA 200 ppm (705 mg/m³), STEL 250 ppm (881 mg/m³), 1993; Denmark: TWA 200 ppm (700 mg/m³), 1999; Finland: TWA 200 ppm (700 mg/m³), STEL 250 ppm (875 mg/m³), 1999; France: VME 200 ppm (700 mg/m³), 1999; the Netherlands: MAC-TGG 700 mg/m³, 2003; the Philippines: TWA 200 ppm (700 mg/m³), 1993; Poland: MAC (TWA) 100 mg/m³; MAC (STEL) 800 mg/m³, 1999; Russia: STEL 200 mg/m³, 1993; Switzerland: MAK-W 200 ppm (700 mg/m³), KZG-W 400 ppm (1400 mg/m³), 1999; Turkey: TWA 200 ppm

(700 mg/m³), 1993; United Kingdom: TWA 200 ppm (716 mg/m³), STEL 250 ppm, 2000; Argentina, Bulgaria, Columbia, Jordan, South Korea, New Zealand, Singapore, Vietnam: ACGIH TLV®: STEL 250 ppm.
Determination in Air: Use NIOSH Analytical Method #1300, Ketones, #2555 Ketones I.[18]
Determination in Water: Octanol−water coefficient: Log K_{ow} = <1.
Routes of Entry: Inhalation, ingestion, skin and/or eye contact.
Harmful Effects and Symptoms
Short Term Exposure: Methyl propyl ketone can affect you when breathed in and by passing through your skin. Exposure to the vapor can irritate the eyes, nose, throat, and lungs. Skin exposure to the liquid can cause a rash or burning feelings on contact. Contact irritates the eyes. Inhalation causes coughing, wheezing, and/or shortness of breath. Exposure to high concentrations can cause you to feel dizzy and lightheaded and to pass out.
Long Term Exposure: Long-term exposure can cause drying and cracking of the skin. Can irritate the lungs and may cause bronchitis to develop.
Points of Attack: Lungs, skin.
Medical Surveillance: NIOSH lists the following tests: chest X-ray; pulmonary function tests.
First Aid: If this chemical gets into the eyes, remove any contact lenses at once and irrigate immediately for at least 15 min, occasionally lifting upper and lower lids. Seek medical attention immediately. If this chemical contacts the skin, remove contaminated clothing and wash immediately with soap and water. Seek medical attention immediately. If this chemical has been inhaled, remove from exposure, begin rescue breathing (using universal precautions, including resuscitation mask) if breathing has stopped and CPR if heart action has stopped. Transfer promptly to a medical facility. When this chemical has been swallowed, get medical attention. Give large quantities of water and induce vomiting. Do not make an unconscious person vomit.
Personal Protective Methods: Wear protective gloves and clothing to prevent any reasonable probability of skin contact. Safety equipment suppliers/manufacturers can provide recommendations on the most protective glove/clothing material for your operation. All protective clothing (suits, gloves, footwear, headgear) should be clean, available each day, and put on before work. Contact lenses should not be worn when working with this chemical. Wear splash-proof chemical goggles and face shield unless full face-piece respiratory protection is worn. Employees should wash immediately with soap when skin is wet or contaminated. Provide emergency showers and eyewash.
Respirator Selection: NIOSH: *1500 ppm:* CcrFOv (APF = 50) [any air-purifying, full!-face-piece respirator (gas mask) with a chin-style, front- or back-mounted acid gas canister] or PaprOv (APF = 25) [any powered, air-purifying respirator with organic vapor cartridge(s)] or GmFOv (APF = 50) [any air-purifying, full-face-piece respirator

(gas mask) with a chin-style, front- or back-mounted organic vapor canister] or Sa (APF = 10) (any supplied-air respirator) or SCBAF (APF = 50) (any self-contained breathing apparatus with a full face-piece). *Emergency or planned entry into unknown concentrations or IDLH conditions:* SCBAF: Pd,Pp (APF = 10,000) (any self-contained breathing apparatus that has a full face-piece and is operated in a pressure-demand or other positive-pressure mode) or SaF: Pd,Pp: ASCBA (APF = 10,000) (any supplied-air respirator that has a full face-piece and is operated in a pressure-demand or other positive-pressure mode in combination with an auxiliary self-contained breathing apparatus operated in a pressure-demand or other positive-pressure mode). *Escape:* GmFOv (APF = 50) [any air-purifying, full-face-piece respirator (gas mask) with a chin-style, front- or back-mounted organic vapor canister] or SCBAE (any appropriate escape-type, self-contained breathing apparatus).

Note: Substance reported to cause eye irritation or damage; may require eye protection.

Storage: Color Code—Red: Flammability Hazard: Store in a flammable liquid storage area or approved cabinet away from ignition sources and corrosive and reactive materials. Prior to working with this chemical you should be trained on its proper handling and storage. Before entering confined space where this chemical may be present, check to make sure that an explosive concentration does not exist. Methyl propyl ketone must be stored to avoid contact with oxidizers since violent reactions occur. Store in tightly closed containers in a cool, well-ventilated area away from heat, sparks, or flame. Sources of ignition, such as smoking and open flames, are prohibited where methyl propyl ketone is handled, used, or stored. Metal containers used in the transfer of 5 gallons or more of methyl propyl ketone should be grounded and bonded. Drums must be equipped with self-closing valves, pressure vacuum bungs, and flame arresters. Use only nonsparking tools and equipment, especially when opening and closing containers of methyl propyl ketone. Wherever methyl propyl ketone is used, handled, manufactured, or stored, use explosion-proof electrical equipment and fittings.

Shipping: This compound requires a shipping label of "FLAMMABLE LIQUID." It falls in Hazard Class 3 and Packing Group II.

Spill Handling: Evacuate and restrict persons not wearing protective equipment from area of spill or leak until cleanup is complete. Remove all ignition sources. Establish forced ventilation to keep levels below explosive limit. Absorb liquids in vermiculite, dry sand, earth, peat, carbon, or a similar material and deposit in sealed containers. Keep this chemical out of a confined space, such as a sewer, because of the possibility of an explosion, unless the sewer is designed to prevent the buildup of explosive concentrations. It may be necessary to contain and dispose of this chemical as a hazardous waste. If material or contaminated runoff enters waterways, notify downstream users of potentially

contaminated waters. Contact your local or federal environmental protection agency for specific recommendations. If employees are required to clean up spills, they must be properly trained and equipped. OSHA 1910.120(q) may be applicable.

Fire Extinguishing: Methyl propyl ketone is a flammable liquid. Poisonous gases are produced in fire. Use dry chemical, carbon dioxide, alcohol foam, or polymer foam extinguishers. Vapors are heavier than air and will collect in low areas. Vapors may travel long distances to ignition sources and flashback. Vapors in confined areas may explode when exposed to fire. Containers may explode in fire. Storage containers and parts of containers may rocket great distances, in many directions. If material or contaminated runoff enters waterways, notify downstream users of potentially contaminated waters. Notify local health and fire officials and pollution control agencies. From a secure, explosion-proof location, use water spray to cool exposed containers. If cooling streams are ineffective (venting sound increases in volume and pitch, tank discolors, or shows any signs of deforming), withdraw immediately to a secure position. If employees are expected to fight fires, they must be trained and equipped in OSHA 1910.156. The only respirators recommended for firefighting are self-contained breathing apparatuses that have full face-pieces and are operated in a pressure-demand or other positive-pressure mode.

Disposal Method Suggested: Dissolve or mix the material with a combustible solvent and burn in a chemical incinerator equipped with an afterburner and scrubber. All federal, state, and local environmental regulations must be observed.

References

Nat. Inst. for Occupational Safety and Health. (1973). *Criteria for a Recommended Standard: Occupational Exposure to Ketones,* NIOSH Document No. 78-173. Washington, DC

US Environmental Protection Agency. (December 6, 1977). *Chemical Hazard Information Profile: 2-Pentanone.* Washington, DC

New Jersey Department of Health and Senior Services. (October 1999). *Hazardous Substances Fact Sheet: Methyl Propyl Ketone.* Trenton, NJ

Methyl silicate   M:1230

Molecular Formula: $C_4H_{12}O_4Si$
Common Formula: $(CH_3O)_4Si$
Synonyms: Methyl orthosilicate; Silicid acid, tetramethyl ester; Tetramethoxysilane; Tetramethyl ester of silicic acid; Tetramethyl silicate
CAS Registry Number: 681-84-5
RTECS® Number: VV9800000
UN/NA & ERG Number: UN2606/155
EC Number: 211-656-4

Regulatory Authority and Advisory Bodies

Air Pollutant Standard Set. See below, "Permissible Exposure Limits in Air" section.

US DOT 49CFR172.101, Inhalation Hazardous Chemical. Canada, WHMIS, Ingredients Disclosure List Concentration 1.0%.

European/International Regulations: not listed in Annex 1.

WGK (German Aquatic Hazard Class): No value assigned.

Description: Methyl silicate is a liquid. Molecular weight = 152.25; Specific gravity (H_2O:1) = 1.02; Boiling point = 121°C; Freezing/Melting point = −2°C; Vapor pressure = 12 mmHg at 25°C; Flash point = 20°C (cc). Insoluble in water.

Potential Exposure: Methyl silicate is used in coating screens of television picture tubes. It may be used in mold binders and in corrosion-resistant coatings, as well as in catalyst preparation and as a silicone intermediate.

Incompatibilities: Incompatible with oxidizers, water, moisture. Violent reaction with metal hexafluorides of rhenium, molybdenum, and tungsten.

Permissible Exposure Limits in Air

Conversion factor: 1 ppm = 6.23 mg/m³ at 25°C & 1 atm.

OSHA PEL: None.

NIOSH REL: 1 ppm/6 mg/m³ TWA.

ACGIH TLV®[1]: 1 ppm/6 mg/m³ TWA.

Protective Action Criteria (PAC)*

TEEL-0: 0.91 ppm

PAC-1: 0.91 ppm

PAC-2: **0.91** ppm

PAC-3: **1.4** ppm

*AEGLs (Acute Emergency Guideline Levels) & ERPGs (Emergency Response Planning Guideline) are in **bold face**.

DFG MAK: Danger of skin sensitization.

Australia: TWA 1 ppm (6 mg/m³), 1993; Belgium: TWA 1 ppm (6 mg/m³), 1993; Denmark: TWA 1 ppm (6 mg/m³), 1999; Finland: TWA 5 ppm (30 mg/m³), STEL 10 ppm (60 mg/m³), 1999; France: VME 1 ppm (6 mg/m³), 1999; Japan: 1 ppm (6 mg/m³), 1999; Norway: TWA 1 ppm (6 mg/m³), 1999; the Netherlands: MAC-TGG 6 mg/m³, 2003; United Kingdom: TWA 1 ppm (6.3 mg/m³), STEL 5 ppm (32 mg/m³), 2000; Argentina, Bulgaria, Columbia, Jordan, South Korea, New Zealand, Singapore, Vietnam: ACGIH TLV®: TWA 1 ppm.

Determination in Air: No method available.

Routes of Entry: Inhalation, ingestion, skin and/or eye contact.

Harmful Effects and Symptoms

Short Term Exposure: Methyl silicate can affect you when breathed in. Severely irritates and burns the eyes and skin. Exposure to the vapor can cause severe eye damage and cause permanent blindness. This can occur up to 12 h after exposure has ceased, even if no irritation is noticed at the time. Inhalation irritates the respiratory tract. Higher exposures can cause pulmonary edema, a medical emergency that can be delayed for several hours. This can cause death. Exposure to high levels can damage the lungs and kidneys.

Long Term Exposure: May cause kidney damage. Can irritate the lungs and may cause bronchitis to develop.

Points of Attack: Eyes, respiratory system, kidneys, liver.

Medical Surveillance: For those with frequent or potentially high exposure (half the TLV or greater), the following are recommended before beginning work and at regular times after that: lung function tests. If symptoms develop or overexposure is suspected, the following may be useful: examination of the eyes and vision; kidney function tests.

First Aid: If this chemical gets into the eyes, remove any contact lenses at once and irrigate immediately for at least 15 min, occasionally lifting upper and lower lids. Seek medical attention immediately. If this chemical contacts the skin, remove contaminated clothing and wash immediately with soap and water. Seek medical attention immediately. If this chemical has been inhaled, remove from exposure, begin rescue breathing (using universal precautions, including resuscitation mask) if breathing has stopped and CPR if heart action has stopped. Transfer promptly to a medical facility. When this chemical has been swallowed, get medical attention. Give large quantities of water and induce vomiting. Do not make an unconscious person vomit. Medical observation is recommended for 24−48 h after breathing overexposure, as pulmonary edema may be delayed. As first aid for pulmonary edema, a doctor or authorized paramedic may consider administering a corticosteroid spray.

Personal Protective Methods: Wear protective gloves and clothing to prevent any reasonable probability of skin contact. Safety equipment suppliers/manufacturers can provide recommendations on the most protective glove/clothing material for your operation. All protective clothing (suits, gloves, footwear, headgear) should be clean, available each day, and put on before work. Contact lenses should not be worn when working with this chemical. Wear splash-proof chemical goggles and face shield unless full face-piece respiratory protection is worn. Employees should wash immediately with soap when skin is wet or contaminated. Provide emergency showers and eyewash.

Respirator Selection: Where there is potential for exposures *over 1 ppm* to liquid methyl silicate, use a NIOSH/MSHA- or European Standard EN149-approved supplied-air respirator with a full face-piece operated in the positive-pressure mode, or with a full face-piece, hood, or helmet in the continuous-flow mode; or use a NIOSH/MSHA- or European Standard EN149-approved self-contained breathing apparatus with a full face-piece operated in a pressure-demand or other positive-pressure mode.

Where there is potential for high exposures, use a NIOSH/MSHA- or European Standard EN149-approved supplied-air respirator with a full face-piece operated in the positive-pressure mode, or with a full face-piece, hood, or helmet in the continuous-flow mode; or use a NIOSH/MSHA- or European Standard EN149-approved self-contained breathing apparatus with a full face-piece operated in a pressure-demand or other positive-pressure mode.

Storage: Color Code—Blue: Health Hazard/Poison: Store in a secure poison location. Prior to working with this chemical you should be trained on its proper handling and storage. Store in tightly closed containers in a cool, well-ventilated area away from water and moisture. Sources of ignition, such as smoking and open flames, are prohibited where methyl silicate is used, handled, or stored in a manner that could create a potential fire or explosion hazard.

Shipping: Methyl orthosilicate requires a shipping label of "POISONOUS/TOXIC MATERIALS, FLAMMABLE LIQUID." It falls in Hazard Class 6.1 and Packing Group I.

Spill Handling: Evacuate and restrict persons not wearing protective equipment from area of spill or leak until cleanup is complete. Remove all ignition sources. Ventilate area of spill or leak. Absorb liquids in vermiculite, dry sand, earth, peat, carbon, or a similar material and deposit in sealed containers. Keep this chemical out of a confined space, such as a sewer, because of the possibility of an explosion, unless the sewer is designed to prevent the buildup of explosive concentrations. It may be necessary to contain and dispose of this chemical as a hazardous waste. If material or contaminated runoff enters waterways, notify downstream users of potentially contaminated waters. Contact your local or federal environmental protection agency for specific recommendations. If employees are required to clean up spills, they must be properly trained and equipped. OSHA 1910.120(q) may be applicable.

Initial isolation and protective action distances
Distances shown are likely to be affected during the first 30 min after materials are spilled and could increase with time. If more than one tank car, cargo tank, portable tank, or large cylinder involved in the incident is leaking, the protective action distance may need to be increased. You may need to seek emergency information from CHEMTREC at (800) 424-9300 or seek professional environmental engineering assistance from the US EPA Environmental Response Team at (908) 48-8730 (24-h response line).

Small spills (From a small package or a small leak from a large package)
First: Isolate in all directions (feet/meters)   100/30
Then: Protect persons downwind (miles/kilometers)
Day   0.1/0.2
Night   0.1/0.2

Large spills (From a large package or from many small packages)
First: Isolate in all directions (feet/meters)   100/30
Then: Protect persons downwind (miles/kilometers)
Day   0.2/0.3
Night   0.3/0.5

Fire Extinguishing: This chemical is a combustible liquid. Poisonous gases, including silicon oxide, are produced in fire. Use dry chemical, carbon dioxide, alcohol or polymer foam extinguishers. *Do not use water.* Vapors are heavier than air and will collect in low areas. Vapors in confined areas may explode when exposed to fire. Containers may explode in fire. Storage containers and parts of containers may rocket great distances, in many directions. If material or contaminated runoff enters waterways, notify downstream users of potentially contaminated waters. Notify local health and fire officials and pollution control agencies. Containers may explode in fire. From a secure, explosion-proof location, use water spray to cool exposed containers. If cooling streams are ineffective (venting sound increases in volume and pitch, tank discolors, or shows any signs of deforming), withdraw immediately to a secure position. If employees are expected to fight fires, they must be trained and equipped in OSHA 1910.156. The only respirators recommended for firefighting are self-contained breathing apparatuses that have full face-pieces and are operated in a pressure-demand or other positive-pressure mode.

Reference
New Jersey Department of Health and Senior Services. (October 1999). *Hazardous Substances Fact Sheet: Methyl Silicate.* Trenton, NJ

α-Methylstyrene   M:1240

Molecular Formula: C_9H_{10}
Common Formula: $C_6H_5C(CH_3){=}CH_2$
Synonyms: AMS; Benzene, (1-methylethenyl)-; Isopropenyl benzene; 1-(Methylethyl) benzene; 1-Methyl-1-phenyl-ethene; 1-Methyl-1-phenyl-ethylene; Phenylpropylene; 2-Phenylpropylene; β-Phenylpropylene; Styrene, α-methyl-
CAS Registry Number: 98-83-9
RTECS® Number: WL5250000
UN/NA & ERG Number: UN2303/128
EC Number: 202-705-0 [*Annex I Index No.:* 601-027-00-6]
Regulatory Authority and Advisory Bodies
Carcinogenicity: NTP: Carcinogenesis studies; on test (pre-chronic studies), October 2000.
Air Pollutant Standard Set. See below, "Permissible Exposure Limits in Air" section.
Canada, WHMIS, Ingredients Disclosure List Concentration 1.0%.
European/International Regulations: Hazard Symbol: Xi, N; Risk phrases: R10; R36/37; R51/53; Safety phrases: S2; S61 (see Appendix 4).
WGK (German Aquatic Hazard Class): No value assigned.
Description: Methylstyrene is a colorless liquid with a characteristic odor. Molecular weight = 118.19; Specific gravity $(H_2O:1)$ = 0.91; Boiling point = 165.6°C; Freezing/Melting point = −23.3°C; Vapor pressure = 2 mmHg at 20°C; Flash point = 54°C; Autoignition temperature = 574°C. Explosive limits: LEL = 1.9%; UEL = 6.1%. Hazard Identification (based on NFPA-704 M Rating System): Health 1, Flammability 2, Reactivity 0. Insoluble in water.
Potential Exposure: Compound Description: Tumorigen, Mutagen, Human Data; Primary Irritant. Methylstyrene is used as an additive, a plasticizer, and as a copolymer; used in the production of modified polyester and alkyd resin formulations.

Incompatibilities: Incompatible with oxidizers, peroxides, halogens, catalysts for vinyl or ionic polymers; aluminum, iron chloride; copper. Methylstyrene may polymerize. Usually contains an inhibitor, such as *tert*-butyl catechol.

Permissible Exposure Limits in Air

Conversion factor: 1 ppm = 4.83 mg/m^3 at 25°C & 1 atm.

OSHA PEL: 100 ppm/480 mg/m^3 Ceiling Concentration.

NIOSH REL: 50 ppm/240 mg/m^3 TWA; 100 ppm/485 mg/m^3 STEL.

ACGIH TLV®[1]: 10 ppm/48 mg/m^3 TWA, confirmed animal carcinogen with unknown relevance to humans.

NIOSH IDLH: 700 ppm.

Protective Action Criteria (PAC)

TEEL-0: 10 ppm

PAC-1: 100 ppm

PAC-2: 100 ppm

PAC-3: 700 ppm

DFG MAK: 50 ppm/250 mg/m^3 TWA; Peak Limitation Category I(2); Pregnancy Risk Group D.

Compound Description: Tumorigen, Mutagen, Human Data; Primary Irritant.

Australia: TWA 50 ppm (240 mg/m^3); STEL 100 ppm, 1993; Austria: MAK 100 ppm (480 mg/m^3), 1999; Belgium: TWA 50 ppm (242 mg/m^3); STEL 100 ppm (484 mg/m^3), 1993; Denmark: TWA 50 ppm (240 mg/m^3), 1999; Finland: TWA 100 ppm (480 mg/m^3); STEL 150 ppm (720 mg/m^3), 1999; France: VME 50 ppm (240 mg/m^3), 1999; Norway: TWA 50 ppm (240 mg/m^3), 1999; the Netherlands: MAC-TGG 240 mg/m^3, 2003; Switzerland: MAK-W 50 ppm (240 mg/m^3), 1999; Thailand: TWA 100 ppm (480 mg/m^3), 1993; United Kingdom: STEL 100 ppm (491 mg/m^3), 2000; Argentina, Bulgaria, Columbia, Jordan, South Korea, New Zealand, Singapore, Vietnam: ACGIH TLV®: STEL 100 ppm. Russia[43] set a MAC of 0.04 mg/m^3 (40 μg/m^3) for ambient air in residential areas both on a momentary and a daily average basis. Several states have set guidelines or standards for methylstyrene in ambient air[60] ranging from 2.4–4.85 mg/m^3 (North Dakota) to 4.0 mg/m^3 (Virginia) to 4.8 mg/m^3 (Connecticut) to 5.714 mg/m^3 (Nevada).

Determination in Air: Use NIOSH Analytical Method 1501, aromatic hydrocarbons; OSHA Analytical Method 7.[18]

Permissible Concentration in Water: Russia[43] set a MAC of 0.1 mg/L in water bodies used for domestic purposes.

Determination in Water: Octanol−water coefficient: Log K_{ow} = 3.4.

Routes of Entry: Inhalation, ingestion, eye and/or skin contact. Passes through the skin.

Harmful Effects and Symptoms

Short Term Exposure: The substance irritates the eyes, skin, and the respiratory tract. Prolonged skin contact causes a burning sensation, redness, and blisters. Exposure can cause headache, dizziness, lightheadedness, and difficult breathing.

Long Term Exposure: Repeated or prolonged contact with skin may cause skin sensitization and allergy with itching and skin rash. May affect the central nervous system, kidneys, and liver. May cause brain effects or damage.

Points of Attack: Eyes, skin, respiratory system, central nervous system, liver, and kidneys.

Medical Surveillance: NIOSH lists the following tests: urine (chemical/metabolite). Consider the points of attack in preplacement and periodic physical examinations. Evaluation by a qualified allergist. Evaluate for brain effects. Evaluate nervous system. Liver and kidney function tests.

First Aid: If this chemical gets into the eyes, remove any contact lenses at once and irrigate immediately for at least 15 min, occasionally lifting upper and lower lids. Seek medical attention immediately. If this chemical contacts the skin, remove contaminated clothing and wash immediately with soap and water. Seek medical attention immediately. If this chemical has been inhaled, remove from exposure, begin rescue breathing (using universal precautions, including resuscitation mask) if breathing has stopped and CPR if heart action has stopped. Transfer promptly to a medical facility. When this chemical has been swallowed, get medical attention. Give large quantities of water and induce vomiting. Do not make an unconscious person vomit.

Personal Protective Methods: Wear protective gloves and clothing to prevent any reasonable probability of skin contact. Safety equipment suppliers/manufacturers can provide recommendations on the most protective glove/clothing material for your operation. All protective clothing (suits, gloves, footwear, headgear) should be clean, available each day, and put on before work. Contact lenses should not be worn when working with this chemical. Wear splash-proof chemical goggles and face shield unless full face-piece respiratory protection is worn. Employees should wash immediately with soap when skin is wet or contaminated. Provide emergency showers and eyewash.

Respirator Selection: *500 ppm:* CcrOv (APF = 10) [any chemical cartridge respirator with organic vapor cartridge(s)]; Sa (APF = 10) (any supplied-air respirator). *700 ppm:* Sa:Cf (APF = 25) (any supplied-air respirator operated in a continuous-flow mode) or CcrFOv (APF = 50) [any air-purifying, full-face-piece respirator (gas mask) with a chin-style, front- or back-mounted acid gas canister] or GmFOv (APF = 50) [any air-purifying, full-face-piece respirator (gas mask) with a chin-style, front- or back-mounted organic vapor canister] or PaprOv (APF = 25) [any powered, air-purifying respirator with organic vapor cartridge(s)] or SCBAF (APF = 50) (any self-contained breathing apparatus with a full face-piece) or SaF (APF = 50) (any supplied-air respirator with a full face-piece). *Emergency or planned entry into unknown concentrations or IDLH conditions:* SCBAF: Pd,Pp (APF = 10,000) (any self-contained breathing apparatus that has a full face-piece and is operated in a pressure-demand or other positive-pressure mode) or SaF: Pd,Pp: ASCBA (APF = 10,000) (any supplied-air respirator that has a full face-piece and is operated in a pressure-demand or other positive-pressure mode in combination with an auxiliary self-contained breathing apparatus operated in a pressure-demand or other positive-pressure mode). *Escape:* GmFOv (APF = 50) [any air-purifying, full-face-piece respirator (gas mask) with a chin-style, front- or back-mounted organic vapor canister] or SCBAE (any appropriate escape-type, self-contained breathing apparatus).

Note: Substance reported to cause eye irritation or damage; may require eye protection.

Storage: Color Code—Red: Flammability Hazard: Store in a flammable liquid storage area or approved cabinet away from ignition sources and corrosive and reactive materials. Prior to working with this chemical you should be trained on its proper handling and storage. Before entering confined space where this chemical may be present, check to make sure that an explosive concentration does not exist. Store in tightly closed containers in a cool, well-ventilated area away from oxidizers. Where possible, automatically pump liquid from drums or other storage containers to process containers.

Shipping: Isopropenylbenzene requires a shipping label of "FLAMMABLE LIQUID." It falls in Hazard Class 3 and Packing Group II.

Spill Handling: Evacuate and restrict persons not wearing protective equipment from area of spill or leak until cleanup is complete. Remove all ignition sources. Establish forced ventilation to keep levels below explosive limit. Absorb liquids in vermiculite, dry sand, earth, peat, carbon, or a similar material and deposit in sealed containers. Keep this chemical out of a confined space, such as a sewer, because of the possibility of an explosion, unless the sewer is designed to prevent the buildup of explosive concentrations. It may be necessary to contain and dispose of this chemical as a hazardous waste. If material or contaminated runoff enters waterways, notify downstream users of potentially contaminated waters. Contact your local or federal environmental protection agency for specific recommendations. If employees are required to clean up spills, they must be properly trained and equipped. OSHA 1910.120(q) may be applicable.

Fire Extinguishing: This chemical is a combustible and reactive liquid. Poisonous gases are produced in fire. Water may be ineffective. Use dry chemical, carbon dioxide, alcohol foam, or polymer foam extinguishers. Vapors are heavier than air and will collect in low areas. Vapors may travel long distances to ignition sources and flashback. Vapors in confined areas may explode when exposed to fire. Containers may explode in fire. Storage containers and parts of containers may rocket great distances, in many directions. If material or contaminated runoff enters waterways, notify downstream users of potentially contaminated waters. Notify local health and fire officials and pollution control agencies. From a secure, explosion-proof location, use water spray to cool exposed containers. If cooling streams are ineffective (venting sound increases in volume and pitch, tank discolors, or shows any signs of deforming),

withdraw immediately to a secure position. If employees are expected to fight fires, they must be trained and equipped in OSHA 1910.156. The only respirators recommended for firefighting are self-contained breathing apparatuses that have full face-pieces and are operated in a pressure-demand or other positive-pressure mode.

Disposal Method Suggested: Incineration, often by admixture with a more flammable solvent.[24]

Reference

New Jersey Department of Health and Senior Services. (June 1999). *Hazardous Substances Fact Sheet: Isopropenyl Benzene.* Trenton, NJ

Methyltetrahydrofuran   M:1250

Molecular Formula: $C_5H_{10}O$

Synonyms: Furan, tetrahydromethyl-; 2-Methyl-tetrahydrofuran; Methyltetrahydrofuran, 2-

CAS Registry Number: 96-47-9; 25265-68-3

RTECS® Number: LU6208000

UN/NA & ERG Number: UN2536/127

EC Number: 246-769-8

Regulatory Authority and Advisory Bodies

WGK (German Aquatic Hazard Class): 2—Water polluting (*CAS: 96-47-9*).

Description: Methyltetrahydrofuran is a colorless liquid with an ether-like odor. Boiling point = 80°C; Freezing/Melting point = − 136°C; Flash point = − 11°C. Hazard Identification (based on NFPA-704 M Rating System): Health 2, Flammability 3, Reactivity 1. Slightly soluble in water.

Potential Exposure: Used as a chemical intermediate and a solvent.

Incompatibilities: Strong oxidizers may cause fire and explosion.

Permissible Exposure Limits in Air

Protective Action Criteria (PAC)

96-47-9

TEEL-0: 6 ppm

PAC-1: 15 ppm

PAC-2: 125 ppm

PAC-3: 600 ppm

Routes of Entry: Inhalation, skin and/or eye contact.

Harmful Effects and Symptoms

Short Term Exposure: Contact irritates the skin and eyes. Irritates the respiratory tract.

Long Term Exposure: Closely related chemicals affect the nervous system although it is not known whether this chemical has this effect.

Medical Surveillance: There is no special test for this chemical. However, if illness occurs or overexposure is suspected, medical attention is recommended.

First Aid: If this chemical gets into the eyes, remove any contact lenses at once and irrigate immediately for at least

15 min, occasionally lifting upper and lower lids. Seek medical attention immediately. If this chemical contacts the skin, remove contaminated clothing and wash immediately with soap and water. Seek medical attention immediately. If this chemical has been inhaled, remove from exposure, begin rescue breathing (using universal precautions, including resuscitation mask) if breathing has stopped and CPR if heart action has stopped. Transfer promptly to a medical facility. When this chemical has been swallowed, get medical attention. Give large quantities of water and induce vomiting. Do not make an unconscious person vomit.

Personal Protective Methods: Wear protective gloves and clothing to prevent any reasonable probability of skin contact. Safety equipment suppliers/manufacturers can provide recommendations on the most protective glove/clothing material for your operation. All protective clothing (suits, gloves, footwear, headgear) should be clean, available each day, and put on before work. Contact lenses should not be worn when working with this chemical. Wear splash-proof chemical goggles and face shield unless full face-piece respiratory protection is worn. Employees should wash immediately with soap when skin is wet or contaminated. Provide emergency showers and eyewash.

Respirator Selection: Follow regulations in OSHA 29CFR1910.134 or European Standard EN149. Use a NIOSH/MSHA- or European Standard EN149-approved respirator; or use an approved supplied-air respirator with a full face-piece operated in the positive-pressure mode, or with a full face-piece, hood, or helmet in the continuous-flow mode; or use a NIOSH/MSHA- or European Standard EN149-approved self-contained breathing apparatus with a full face-piece operated in a pressure-demand or other positive-pressure mode.

Storage: Color Code—Red: Flammability Hazard: Store in a flammable liquid storage area or approved cabinet away from ignition sources and corrosive and reactive materials. Prior to working with methyltetrahydrofuran you should be trained on its proper handling and storage. Store in tightly closed containers in a cool, well-ventilated area. Where possible, automatically pump liquid from drums or other storage containers to process containers. Drums must be equipped with self-closing valves, pressure vacuum bungs, and flame arresters. Use only nonsparking tools and equipment, especially when opening and closing containers of this chemical. Wherever this chemical is used, handled, manufactured, or stored, use explosion-proof electrical equipment and fittings.

Shipping: Methyltetrahydrofuran requires a shipping label of "FLAMMABLE LIQUID." It falls in Hazard Class 3 and Packing Group II.

Spill Handling: Evacuate and restrict persons not wearing protective equipment from area of spill or leak until cleanup is complete. Remove all ignition sources. Ventilate area of spill or leak. Absorb liquids in vermiculite, dry sand, earth, peat, carbon, or a similar material and deposit in sealed containers. Keep this chemical out of a confined space, such as a sewer, because of the possibility of an explosion, unless the sewer is designed to prevent the buildup of explosive concentrations. It may be necessary to contain and dispose of this chemical as a hazardous waste. If material or contaminated runoff enters waterways, notify downstream users of potentially contaminated waters. Contact your local or federal environmental protection agency for specific recommendations. If employees are required to clean up spills, they must be properly trained and equipped. OSHA 1910.120(q) may be applicable.

Fire Extinguishing: This chemical is a combustible liquid. Poisonous gases are produced in fire. Use dry chemical, carbon dioxide, or alcohol foam extinguishers. Vapors are heavier than air and will collect in low areas. Vapors may travel long distances to ignition sources and flashback. Vapors in confined areas may explode when exposed to fire. Containers may explode in fire. Storage containers and parts of containers may rocket great distances, in many directions. If material or contaminated runoff enters waterways, notify downstream users of potentially contaminated waters. Notify local health and fire officials and pollution control agencies. From a secure, explosion-proof location, use water spray to cool exposed containers. If cooling streams are ineffective (venting sound increases in volume and pitch, tank discolors, or shows any signs of deforming), withdraw immediately to a secure position. If employees are expected to fight fires, they must be trained and equipped in OSHA 1910.156. The only respirators recommended for firefighting are self-contained breathing apparatuses that have full face-pieces and are operated in a pressure-demand or other positive-pressure mode.

Reference

New Jersey Department of Health and Senior Services. (June 1999). *Hazardous Substances Fact Sheet: Methyltetrahydrofuran.* Trenton, NJ

Methyl thiocyanate   M:1260

Molecular Formula: C_2H_3NS
Common Formula: CH_3CNS
Synonyms: Methyl rhodanate; Methylrhodanid (German); Methyl sulfocyanate; Methylthiokyanat; Thiocyanic acid, methyl ester
CAS Registry Number: 556-64-9
RTECS® Number: XL1575000
UN/NA & ERG Number: UN2810/153
EC Number: 209-134-6
Regulatory Authority and Advisory Bodies
Department of Homeland Security Screening Threshold Quantity (pounds): *Release hazard* 20,000.
Clean Air Act: Accidental Release Prevention/Flammable Substances, (Section 112[r], Table 3), TQ = 20,000 lb (9080 kg).
Superfund/EPCRA 40CFR355, Extremely Hazardous Substances: TPQ = 10,000 lb (4540 kg).

Reportable Quantity (RQ): 10,000 lb (4540 kg).
European/International Regulations: not listed in Annex 1.
WGK (German Aquatic Hazard Class): No value assigned.
Description: Methyl thiocyanate is a colorless liquid with an onion-like odor. Molecular weight = 73.12; Boiling point = 130−133°C; Freezing/Melting point = −51°C. Very slightly soluble in water.
Potential Exposure: It is used as an agricultural insecticide, a fumigant, and as a research chemical. No evidence of commercial production in the United States.
Incompatibilities: Incompatible with nitric acid. Violent reactions have occurred when mixed with chlorates, nitrates, nitric acid, peroxides, potassium chlorate, and sodium chlorate.
Permissible Exposure Limits in Air
Protective Action Criteria (PAC)
TEEL-0: 5 mg/m^3
PAC-1: 15 mg/m^3
PAC-2: 28.4 mg/m^3
PAC-3: 4 mg/m^3
Routes of Entry: Inhalation, ingestion, skin and/or eye contact.
Harmful Effects and Symptoms
Short Term Exposure: Prolonged skin absorption may produce various eruptions, runny nose, dizziness, cramps, nausea, vomiting, and mild or severe disturbances of the nervous system. This material is highly toxic if ingested. The ingestion of a concentrated solution may lead to vomiting. The principal systemic reaction is probably one of central nervous system depression, interrupted by periods of restlessness; abnormally fast and deep respiratory movements and convulsions. Death is usually due to respiratory arrest from paralysis of the medullary centers.
Long Term Exposure: May cause injury to the liver and kidneys.
Points of Attack: Liver, kidneys.
Medical Surveillance: Liver and kidney function tests.
First Aid: If this chemical gets into the eyes, remove any contact lenses at once and irrigate immediately for at least 15 min, occasionally lifting upper and lower lids. Seek medical attention immediately. If this chemical contacts the skin, remove contaminated clothing and wash immediately with soap and water. Seek medical attention immediately. If this chemical has been inhaled, remove from exposure, begin rescue breathing (using universal precautions, including resuscitation mask) if breathing has stopped and CPR if heart action has stopped. Transfer promptly to a medical facility. When this chemical has been swallowed, get medical attention. Give large quantities of water and induce vomiting. Do not make an unconscious person vomit.
Note: Because cyanide is probably largely responsible for poisonings, antidotal measures against cyanide should be instituted promptly. Use amyl nitrate capsules if symptoms develop. All area employees should be trained regularly in emergency measures for cyanide poisoning and in CPR. A cyanide antidote kit should be kept in the immediate work area and must be rapidly available. Kit ingredients should be replaced every 1−2 years to ensure freshness. Persons trained in the use of this kit, oxygen use, and CPR must be quickly available.
Personal Protective Methods: Wear protective gloves and clothing to prevent any reasonable probability of skin contact. Safety equipment suppliers/manufacturers can provide recommendations on the most protective glove/clothing material for your operation. All protective clothing (suits, gloves, footwear, headgear) should be clean, available each day, and put on before work. Contact lenses should not be worn when working with this chemical. Wear splash-proof chemical goggles and face shield unless full face-piece respiratory protection is worn. Employees should wash immediately with soap when skin is wet or contaminated. Provide emergency showers and eyewash.
Respirator Selection: (for cyanides) Up to 25 mg/m^3: Sa (APF = 10) (any supplied-air respirator) or SCBAF (APF = 50) (any self-contained breathing apparatus with full face-piece). *Emergency or planned entry into unknown concentrations or IDLH conditions:* SCBAF: Pd,Pp (APF = 10,000) (any self-contained breathing apparatus that has a full face-piece and is operated in a pressure-demand or other positive-pressure mode) or SaF: Pd,Pp: ASCBA (APF = 10,000) (any supplied-air respirator that has a full face-piece and is operated in a pressure-demand or other positive-pressure mode in combination with an auxiliary self-contained breathing apparatus operated in a pressure-demand or other positive-pressure mode). *Escape:* GmFS100 (APF = 50) [any air-purifying, full-face-piece respirator (gas mask) with a chin-style, front- or back-mounted canister providing protection against the compound of concern and having an N100, R100, or P100 filter] or SCBAE (any appropriate escape-type, self-contained breathing apparatus).
Storage: Color Code—Blue: Health Hazard/Poison: Store in a secure poison location. Prior to working with this chemical you should be trained on its proper handling and storage. Store in tightly closed containers in a cool, well-ventilated area away from oxidizers. Where possible, automatically pump liquid from drums or other storage containers to process containers.
Shipping: Toxic liquid, organic, n.o.s. require a label of "POISONOUS/TOXIC MATERIALS." It falls in Hazard Class 6.1.
Spill Handling: Evacuate and restrict persons not wearing protective equipment from area of spill or leak until cleanup is complete. Remove all ignition sources. Ventilate area of spill or leak. Absorb liquids in vermiculite, dry sand, earth, peat, carbon, or a similar material and deposit in sealed containers. Keep this chemical out of a confined space, such as a sewer, because of the possibility of an explosion, unless the sewer is designed to prevent the buildup of explosive concentrations. It may be necessary to contain and dispose of this chemical as a hazardous waste. If material or contaminated runoff enters waterways, notify downstream users

of potentially contaminated waters. Contact your local or federal environmental protection agency for specific recommendations. If employees are required to clean up spills, they must be properly trained and equipped. OSHA 1910.120(q) may be applicable.

Fire Extinguishing: Small fires: dry chemical, carbon dioxide, water spray, or foam. *Large fires:* water spray, fog, or foam. Move container from fire area if you can do so without risk. Fight fire from maximum distance. Dike fire control water for later disposal; do not scatter the material. Stay upwind; keep out of low areas. Ventilate closed spaces before entering them. Wear positive pressure breathing apparatus and special protective clothing. Poisonous gases, including nitrogen oxides, sulfur oxides, are produced in fire. Containers may explode in fire. Storage containers and parts of containers may rocket great distances, in many directions. If material or contaminated runoff enters waterways, notify downstream users of potentially contaminated waters. Notify local health and fire officials and pollution control agencies. From a secure, explosion-proof location, use water spray to cool exposed containers. If cooling streams are ineffective (venting sound increases in volume and pitch, tank discolors, or shows any signs of deforming), withdraw immediately to a secure position. If employees are expected to fight fires, they must be trained and equipped in OSHA 1910.156. The only respirators recommended for firefighting are self-contained breathing apparatuses that have full face-pieces and are operated in a pressure-demand or other positive-pressure mode.

Reference

US Environmental Protection Agency. (November 30, 1987). *Chemical Hazard Information Profile: Methyl Thiocyanate.* Washington, DC: Chemical Emergency Preparedness Program

Methyltrichlorosilane   M:1280

Molecular Formula: CH_3Cl_3Si
Synonyms: Methylsilicochloroform; Methylsilyl trichloride; KA 13; LS 40 (silane); Methylsilicon trichloride; Trichloromethylsilicon; Silane, trichloromethyl-; Trichloromethylsilane
CAS Registry Number: 75-79-6; *(alt.)* 175446-71-6
RTECS® Number: VV4550000
UN/NA & ERG Number: UN1250/155
EC Number: 200-902-6 [*Annex I Index No.:* 014-004-00-5]
Regulatory Authority and Advisory Bodies
Department of Homeland Security Screening Threshold Quantity (pounds): *Release hazard* 10,000 ($\geq 1.00\%$ concentration).
OSHA 29CFR1910.119, Appendix A. Process Safety List of Highly Hazardous Chemicals, TQ = 500 lb (227 kg).
US EPA Hazardous Waste Number (RCRA No.): U164.
RCRA, 40CFR261, Appendix 8 Hazardous Constituents.
Reportable Quantity (RQ): 10 lb (4.54 kg).

European/International Regulations: Hazard Symbol: F, Xi; Risk phrases: R11; 14; R36/37/38; Safety phrases: S2; S26; S39 (see Appendix 4).
WGK (German Aquatic Hazard Class): 1—Low hazard to waters.
Description: Methyltrichlorosilane is a colorless liquid with a sharp hydrochloric acid-like odor. Molecular weight = 149.48; Boiling point = 66°C; Freezing/Melting point = −90°C; Vapor pressure = 146.7 mmHg at 18°C; Flash point = −6.2°C; −8°C (cc); Autoignition temperature ≥404°C. Explosive limits: LEL = 7.6%; UEL ≥20%. Hazard Identification (based on NFPA-704 M Rating System): Health 3, Flammability 3, Reactivity 2W. Reacts with water.
Potential Exposure: Compound Description: Primary Irritant. Methyltrichlorosilane is used as an intermediate to make silicones; for making water repellants, electrical insulation, heat-resistant paints, and other products.
Incompatibilities: Water, steam, acids, alkalis, chemically active metals (potassium, sodium, magnesium, and zinc). Reacts violently with strong oxidizers. Reacts violently with water, moisture, and alkalis producing hydrogen chloride. Attacks metals, such as aluminum, magnesium.
Permissible Exposure Limits in Air
AIHA WEEL: 1 ppm, Ceiling Concentration.
Protective Action Criteria (PAC)*
TEEL-0: 0.2 ppm
PAC-1: **0.60** ppm
PAC-2: **7.3** ppm
PAC-3: **33** ppm
*AEGLs (Acute Emergency Guideline Levels) & ERPGs (Emergency Response Planning Guideline) are in **bold face**.
Routes of Entry: Inhalation, ingestion, skin and/or eye contact.
Harmful Effects and Symptoms
Short Term Exposure: If the eyes have come in contact with methyltrichlorosilane, then irritation, pain, swelling, corneal erosion, and blindness may result. Dermatitis (red, inflamed skin), severe burns, pain, and shock generally follow dermal exposure. As with other chlorosilanes, acute exposures may be highly toxic and may cause death or permanent injury after very short exposures to small quantities. Skin contact may produce severe burns with pain and risk of secondary infections. Ingestion may produce oral, esophageal, and stomach burns; intensity will vary from mild to very severe; gastrointestinal damage is rare but may occur. Signs and symptoms of acute ingestion of methyltrichlorosilane may include excessive salivation, intense thirst, difficulty in swallowing, chills, pain, and shock. Oral, esophageal, and stomach burns are common. Vomitus generally has a coffee-ground appearance. The potential for circulatory collapse is high following ingestion of methyltrichlorosilane. Acute inhalation exposure may result in hoarseness, laryngitis, a feeling of suffocation, dyspnea (shortness of breath), choking, respiratory tact irritation,

chest pain. Inhalation can cause pulmonary edema, a medical emergency that can be delayed for several hours. This can cause death. Sneezing, bleeding of the nose and gums, and ulceration of the nasal and oral mucosa may also occur.

Long Term Exposure: Highly irritating material; may affect the lungs, and bronchitis may develop. Renal toxicity has been observed in animals.

Points of Attack: Lungs, kidneys.

Medical Surveillance: For those with frequent or potentially high exposure the following are recommended before beginning work and at regular times after that: lung function tests. If symptoms develop or overexposure is suspected, the following may be useful: consider chest X-ray after acute overexposure, kidney function tests.

First Aid: If this chemical gets into the eyes, remove any contact lenses at once and irrigate immediately for at least 15 min, occasionally lifting upper and lower lids. Seek medical attention immediately. If this chemical contacts the skin, remove contaminated clothing and wash immediately with soap and water. Seek medical attention immediately. If this chemical has been inhaled, remove from exposure, begin rescue breathing (using universal precautions, including resuscitation mask) if breathing has stopped and CPR if heart action has stopped. Transfer promptly to a medical facility. When this chemical has been swallowed, get medical attention. If victim is conscious, administer water or milk. Do not induce vomiting. Medical observation is recommended for 24—48 h after breathing overexposure, as pulmonary edema may be delayed. As first aid for pulmonary edema, a doctor or authorized paramedic may consider administering a corticosteroid spray.

Personal Protective Methods: Wear protective gloves and clothing to prevent any reasonable probability of skin contact. Safety equipment suppliers/manufacturers can provide recommendations on the most protective glove/clothing material for your operation. All protective clothing (suits, gloves, footwear, headgear) should be clean, available each day, and put on before work. Contact lenses should not be worn when working with this chemical. Wear splash-proof chemical goggles and face shield unless full face-piece respiratory protection is worn. Employees should wash immediately with soap when skin is wet or contaminated. Provide emergency showers and eyewash.

Respirator Selection: Where there is potential for exposure to methyltrichlorosilane, use a NIOSH/MSHA- or European Standard EN149-approved supplied-air respirator with a full face-piece operated in the positive-pressure mode, or with a full face-piece, hood, or helmet in the continuous-flow mode; or use a NIOSH/MSHA- or European Standard EN149-approved self-contained breathing apparatus with a full face-piece operated in a pressure-demand or other positive-pressure mode.

Storage: Color Code—Red: Flammability Hazard: Store in a flammable liquid storage area or approved cabinet away from ignition sources and corrosive and reactive materials. Prior to working with this chemical you should be trained on its proper handling and storage. Before entering confined space where this chemical may be present, check to make sure that an explosive concentration does not exist. Methyltrichlorosilane must be stored to avoid contact with water, acids, chemically active metals (such as potassium, sodium, magnesium, and zinc), and alkalis since violent reactions occur and hydrogen chloride is produced. Store in tightly closed containers in a cool, well-ventilated area away from heat. Sources of ignition, such as smoking and open flames, are prohibited where methyltrichlorosilane is handled, used, or stored. Metal containers involving the transfer of 5 gallons or more of methyltrichlorosilane should be grounded and bonded. Drums must be equipped with self-closing valves, pressure vacuum bungs, and flame arresters. Use only nonsparking tools and equipment, especially when opening and closing containers of methyltrichlorosilane. Wherever methyltrichlorosilane is used, handled, manufactured, or stored, use explosion-proof electrical equipment and fittings.

Shipping: This compound requires a shipping label of "FLAMMABLE LIQUID, CORROSIVE." It falls in Hazard Class 3 and Packing Group I.

Spill Handling: Evacuate and restrict persons not wearing protective equipment from area of spill or leak until cleanup is complete. Remove all ignition sources. Spills can be neutralized by flushing with large quantities of water followed by treatment with sodium bicarbonate. Provide adequate protection against generated hydrogen chloride. Do not allow water to get into container since resulting pressure could cause container to rupture. Protect against potentially violent reaction with water. Avoid breathing vapors and contact with skin. Establish forced ventilation to keep levels below explosive limit. Absorb liquids in vermiculite, dry sand, earth, peat, carbon, or a similar material and deposit in sealed containers. Keep this chemical out of a confined space, such as a sewer, because of the possibility of an explosion, unless the sewer is designed to prevent the buildup of explosive concentrations. It may be necessary to contain and dispose of this chemical as a hazardous waste. If material or contaminated runoff enters waterways, notify downstream users of potentially contaminated waters. Contact your local or federal environmental protection agency for specific recommendations. If employees are required to clean up spills, they must be properly trained and equipped. OSHA 1910.120(q) may be applicable.

Initial isolation and protective action distances

Distances shown are likely to be affected during the first 30 min after materials are spilled and could increase with time. If more than one tank car, cargo tank, portable tank, or large cylinder involved in the incident is leaking, the protective action distance may need to be increased. You may need to seek emergency information from CHEMTREC at (800) 424-9300 or seek professional environmental engineering assistance from the US EPA Environmental Response Team at (908) 548-8730 (24-h response line).

Small spills (From a small package or a small leak from a large package)
when spilled in water
First: Isolate in all directions (feet/meters)   100/30
Then: Protect persons downwind (miles/kilometers)
Day   0.1/0.2
Night   0.2/0.3
Large spills (From a large package or from many small packages)
First: Isolate in all directions (feet/meters)   200/60
Then: Protect persons downwind (miles/kilometers)
Day   0.4/0.6
Night   1.3/2.0
Fire Extinguishing: This chemical is a flammable liquid. Poisonous gases are produced in fire. Dry chemical or carbon dioxide may be used for small fires. *Do not use water* or hydrous agents. However, water may be used for large fires if firefighters are protected from violent reactions of methyltrichlorosilane with water. Water may be used to keep containers cool. Self-contained breathing apparatus is required as combustion/decomposition yields acid gases/pulmonary irritants. Corrosion-resistant protective clothing, as well as appropriate foot, hand, arm, head, eye, and face protection are required where contact is possible. Vapors are heavier than air and will collect in low areas. Vapors may travel long distances to ignition sources and flashback. Vapors in confined areas may explode when exposed to fire. Containers may explode in fire. Storage containers and parts of containers may rocket great distances, in many directions. If material or contaminated runoff enters waterways, notify downstream users of potentially contaminated waters. Notify local health and fire officials and pollution control agencies. From a secure, explosion-proof location, use water spray to cool exposed containers. If cooling streams are ineffective (venting sound increases in volume and pitch, tank discolors, or shows any signs of deforming), withdraw immediately to a secure position. If employees are expected to fight fires, they must be trained and equipped in OSHA 1910.156. The only respirators recommended for firefighting are self-contained breathing apparatuses that have full face-pieces and are operated in a pressure-demand or other positive-pressure mode.
Disposal Method Suggested: Consult with environmental regulatory agencies for guidance on acceptable disposal practices. Generators of waste containing this contaminant ($\geq$100 kg/mo) must conform with EPA regulations governing storage, transportation, treatment, and waste disposal.
References
US Environmental Protection Agency. (November 30, 1987). *Chemical Hazard Information Profile: Methyltrichlorosilane.* Washington, DC: Chemical Emergency Preparedness Program
New Jersey Department of Health and Senior Services. (January 2000). *Hazardous Substances Fact Sheet: Methyl Trichlorosilane.* Trenton, NJ

Methyl vinyl ketone　　　　M:1290

Molecular Formula: C_4H_6O
Common Formula: $CH_2{=}CHCOCH_3$
Synonyms: Acetyl ethylene; 3-Butene-2-one; Methylene acetone; Methyl-vinyl-cetone (French); Methylvinylketon (German); Metil vinil cetona (Spanish); MVK; γ-Oxo-α-butylene; Vinyl methyl ketone
CAS Registry Number: 78-94-4
RTECS® Number: EM9800000
UN/NA & ERG Number: UN1251 (stabilized)/131
EC Number: 201-160-6
Regulatory Authority and Advisory Bodies
Air Pollutant Standard Set. See below, "Permissible Exposure Limits in Air" section.
Superfund/EPCRA 40CFR355, Extremely Hazardous Substances: TPQ = 10 lb (4.54 kg).
Reportable Quantity (RQ): 10 lb (4.54 kg).
US DOT 49CFR172.101, Inhalation Hazardous Chemical.
Canada, WHMIS, Ingredients Disclosure List Concentration 1.0%.
European/International Regulations: not listed in Annex 1.
WGK (German Aquatic Hazard Class): 3—Severe hazard to waters.
Description: Methyl vinyl ketone is a colorless liquid with a pungent odor. The odor threshold is 0.5 mg/m^3. Molecular weight = 70.01; Boiling point = 81°C; Flash point = -7°C; Autoignition temperature = 491°C. Explosive limits: LEL = 2.1%; UEL = 15.6%. Hazard Identification (based on NFPA-704 M Rating System): Health 3, Flammability 3, Reactivity 2.
Potential Exposure: Compound Description: Mutagen, Primary Irritant. Methyl vinyl ketone is used as an alkylating agent, a starting material for plastics; and an intermediate in the synthesis of steroids and vitamin A.
Incompatibilities: Forms explosive mixture with air. Heat or shock may cause explosive polymerization. Violent reaction with strong oxidizers.
Permissible Exposure Limits in Air
ACGIH TLV®[1]: 0.2 ppm/0.6 mg/m^3 [skin] danger of skin sensitization, Ceiling Concentration.
Protective Action Criteria (PAC)*
TEEL-0: 0.05 ppm
PAC-1: **0.17** ppm
PAC-2: **1.2** ppm
PAC-3: **2.4** ppm
*AEGLs (Acute Emergency Guideline Levels) & ERPGs (Emergency Response Planning Guideline) are in **bold face**.
DFG MAK: [skin] danger of skin sensitization.
Russia: STEL 0.1 mg/m^3, [skin], 1993.
Routes of Entry: Inhalation, ingestion, skin, and/or eye contact.
Harmful Effects and Symptoms
Short Term Exposure: Warning: Methyl vinyl ketone is easily absorbed through the skin, causing general poisoning;

inhalation has central nervous system depressant effects. It is irritating to mucous membranes and respiratory tract and to the skin; it is a lachrymator and can cause eye injury. Liquid or high concentration of vapors causes blistering of the skin. Similar to other ketones, it can cause sore throat, sneezing, coughing, and salivation. Inhalation may cause nausea and vomiting; inhalation of high concentrations can cause headache, dizziness, fainting, tremor, uncoordination, lowered body temperature, depressed respiratory and heart rate, gasping, coma, and death. Direct aspiration of liquid into lungs can cause chemical pneumonia.

First Aid: If this chemical gets into the eyes, remove any contact lenses at once and irrigate immediately for at least 15 min, occasionally lifting upper and lower lids. Seek medical attention immediately. If this chemical contacts the skin, remove contaminated clothing and wash immediately with soap and water. Seek medical attention immediately. If this chemical has been inhaled, remove from exposure, begin rescue breathing (using universal precautions, including resuscitation mask) if breathing has stopped and CPR if heart action has stopped. Transfer promptly to a medical facility. When this chemical has been swallowed, get medical attention. Give large quantities of water and induce vomiting. Do not make an unconscious person vomit. Medical observation is recommended for 24−48 h after breathing overexposure, as pulmonary edema may be delayed. As first aid for pulmonary edema, a doctor or authorized paramedic may consider administering a corticosteroid spray.

Personal Protective Methods: For emergency situations, wear protective gloves and clothing to prevent any reasonable probability of skin contact. Safety equipment suppliers/ manufacturers can provide recommendations on the most protective glove/clothing material for your operation. All protective clothing (suits, gloves, footwear, headgear) should be clean, available each day, and put on before work. Contact lenses should not be worn when working with this chemical. Wear splash-proof chemical goggles and face shield unless full face-piece respiratory protection is worn. Employees should wash immediately with soap when skin is wet or contaminated. Provide emergency showers and eyewash.

Respirator Selection: Follow regulations in OSHA 29CFR1910.134 or European Standard EN149. Use a NIOSH/MSHA- or European Standard EN149-approved respirator; or use an approved supplied-air respirator with a full face-piece operated in the positive-pressure mode, or with a full face-piece, hood, or helmet in the continuous-flow mode; or use a NIOSH/MSHA- or European Standard EN149-approved self-contained breathing apparatus with a full face-piece operated in a pressure-demand or other positive-pressure mode.

Storage: Color Code—Blue: Health Hazard/Poison: Store in a secure poison location. Prior to working with this chemical you should be trained on its proper handling and storage. Before entering confined space where this chemical may be present, check to make sure that an explosive concentration does not exist. Protect against physical damage. Outside or detached storage is preferred. Inside storage should be in a standard flammable liquids storage room. Separate from oxidizing materials. MVK vapors are uninhibited and may form polymers in the flame arresters of storage tanks, resulting in stoppage of vent.

Shipping: Methyl vinyl ketone, stabilized, requires a shipping label of "POISONOUS/TOXIC MATERIALS, FLAMMABLE LIQUID, CORROSIVE." It falls in Hazard Class 6.1 and Packing Group I.

Spill Handling: Evacuate and restrict persons not wearing protective equipment from area of spill or leak until cleanup is complete. Remove all ignition sources. Establish forced ventilation to keep levels below explosive limit. Absorb liquids in vermiculite, dry sand, earth, peat, carbon, or a similar material and deposit in sealed containers. Keep this chemical out of a confined space, such as a sewer, because of the possibility of an explosion, unless the sewer is designed to prevent the buildup of explosive concentrations. It may be necessary to contain and dispose of this chemical as a hazardous waste. If material or contaminated runoff enters waterways, notify downstream users of potentially contaminated waters. Contact your local or federal environmental protection agency for specific recommendations. If employees are required to clean up spills, they must be properly trained and equipped. OSHA 1910.120(q) may be applicable.

Initial isolation and protective action distances

Distances shown are likely to be affected during the first 30 min after materials are spilled and could increase with time. If more than one tank car, cargo tank, portable tank, or large cylinder involved in the incident is leaking, the protective action distance may need to be increased. You may need to seek emergency information from CHEMTREC at (800) 424-9300 or seek professional environmental engineering assistance from the US EPA Environmental Response Team at (908) 48-8730 (24-h response line).

Small spills (From a small package or a small leak from a large package)
First: Isolate in all directions (feet/meters)   500/150
Then: Protect persons downwind (miles/kilometers)
Day   1.0/1.5
Night   2.3./3.6

Large spills (From a large package or from many small packages)
First: Isolate in all directions (feet/meters)   3000/1000
Then: Protect persons downwind (miles/kilometers)
Day   7.0 + /11.0 +
Night   7.0 + /11.0 +

Fire Extinguishing: This chemical is a flammable liquid. Poisonous gases are produced in fire. Use dry chemical, alcohol foam, or carbon dioxide. Water spray may be ineffective as an extinguishing agent. *Small fires:* dry chemical, carbon dioxide, and foam. *Large fires:* fog or foam. Move container form fire area if you can do so without risk. Dike fire control water for later disposal; do not scatter the material. Spray cooling water on containers that are exposed to flames until well after fire is out. Wear positive pressure breathing apparatus and special protective clothing. *See above for isolation distances.* Vapors are heavier than air and will collect in low areas. Vapors may travel long distances to ignition sources and flashback. Vapors in confined areas may explode when exposed to fire. Containers may explode in fire. Storage containers and parts of containers may rocket great distances, in many directions. If material or contaminated runoff enters waterways, notify downstream users of potentially contaminated waters. Notify local health and fire officials and pollution control agencies. From a secure, explosion-proof location, use water spray to cool exposed containers. If cooling streams are ineffective (venting sound increases in volume and pitch, tank discolors, or shows any signs of deforming), withdraw immediately to a secure position. If employees are expected to fight fires, they must be trained and equipped in OSHA 1910.156. The only respirators recommended for firefighting are self-contained breathing apparatuses that have full face-pieces and are operated in a pressure-demand or other positive-pressure mode.

Reference
US Environmental Protection Agency. (November 30, 1987). *Chemical Hazard Information Profile: Methyl Vinyl Ketone.* Washington, DC: Chemical Emergency Preparedness Program

2-Methyl-5-vinylpyridine   M:1300

Molecular Formula: C_8H_9N
Synonyms: 5-Ethenyl-2-methylpyridine; MVP; 5-Vinyl-2-picoline
CAS Registry Number: 140-76-1
RTECS® Number: UT2975000
UN/NA & ERG Number: UN3073 (stabilized)/131
EC Number: 205-432-5
Regulatory Authority and Advisory Bodies
Air Pollutant Standard Set. See below, "Permissible Exposure Limits in Air" section.
Superfund/EPCRA 40CFR355, Extremely Hazardous Substances: TPQ = 500 lb (227 kg).
Reportable Quantity (RQ): 500 lb (227 kg).
Canada, WHMIS, Ingredients Disclosure List Concentration 1.0%.
European/International Regulations: not listed in Annex 1.
WGK (German Aquatic Hazard Class): No value assigned.
Description: 2-Methyl-5-vinylpyridine is a clear to faintly opalescent liquid. Molecular weight = 119.18; Boiling point = 181°C. Hazard Identification (based on NFPA-704 M Rating System): Health 2, Flammability 2, Reactivity 0; Flash point = 74°C.
Potential Exposure: 2-Methyl-5-vinylpyridine is used as a monomer for resins, oil additive, ore flotation agent, and dye acceptor.

Incompatibilities: Strong oxidizers.
Permissible Exposure Limits in Air
Protective Action Criteria (PAC)
TEEL-0: 0.35 mg/m^3
PAC-1: 1 mg/m^3
PAC-2: 1.9 mg/m^3
PAC-3: 40 mg/m^3
Russia[43] set a MAC of 2.0 mg/m^3 in work-place air.
Routes of Entry: Inhalation, ingestion, skin and/or eye contact.
Harmful Effects and Symptoms
Short Term Exposure: This material is poisonous by ingestion, inhalation, and absorption through the skin. Vapors may cause dizziness or suffocation.
Long Term Exposure: May cause liver and kidney damage.
First Aid: If this chemical gets into the eyes, remove any contact lenses at once and irrigate immediately for at least 15 min, occasionally lifting upper and lower lids. Seek medical attention immediately. If this chemical contacts the skin, remove contaminated clothing and wash immediately with soap and water. Seek medical attention immediately. If this chemical has been inhaled, remove from exposure, begin rescue breathing (using universal precautions, including resuscitation mask) if breathing has stopped and CPR if heart action has stopped. Transfer promptly to a medical facility. When this chemical has been swallowed, get medical attention. Give large quantities of water and induce vomiting. Do not make an unconscious person vomit.
Personal Protective Methods: Wear protective gloves and clothing to prevent any reasonable probability of skin contact. Safety equipment suppliers/manufacturers can provide recommendations on the most protective glove/clothing material for your operation. All protective clothing (suits, gloves, footwear, headgear) should be clean, available each day, and put on before work. Contact lenses should not be worn when working with this chemical. Wear splash-proof chemical goggles and face shield unless full face-piece respiratory protection is worn. Employees should wash immediately with soap when skin is wet or contaminated. Provide emergency showers and eyewash.
Respirator Selection: Follow regulations in OSHA 29CFR1910.134 or European Standard EN149. Use a NIOSH/MSHA- or European Standard EN149-approved respirator; or use an approved supplied-air respirator with a full face-piece operated in the positive-pressure mode, or with a full face-piece, hood, or helmet in the continuous-flow mode; or use a NIOSH/MSHA- or European Standard EN149-approved self-contained breathing apparatus with a full face-piece operated in a pressure-demand or other positive-pressure mode.
Storage: Color Code—Blue: Health Hazard/Poison: Store in a secure poison location. Prior to working with this chemical you should be trained on its proper handling and storage. Store in tightly closed containers in a cool, well-ventilated area away from oxidizers. Where possible,

automatically pump liquid from drums or other storage containers to process containers.

Shipping: Vinylpyridines require a label of "POISONOUS/ TOXIC MATERIALS, FLAMMABLE LIQUID, CORROSIVE." They fall in Hazard Class 6.1 and Packing Group II.

Spill Handling: Evacuate and restrict persons not wearing protective equipment from area of spill or leak until cleanup is complete. Remove all ignition sources. Ventilate area of spill or leak. Absorb liquids in vermiculite, dry sand, earth, peat, carbon, or a similar material and deposit in sealed containers. Keep this chemical out of a confined space, such as a sewer, because of the possibility of an explosion, unless the sewer is designed to prevent the buildup of explosive concentrations. It may be necessary to contain and dispose of this chemical as a hazardous waste. If material or contaminated runoff enters waterways, notify downstream users of potentially contaminated waters. Contact your local or federal environmental protection agency for specific recommendations. If employees are required to clean up spills, they must be properly trained and equipped. OSHA 1910.120(q) may be applicable.

Fire Extinguishing: This chemical is a flammable liquid. Poisonous gases, including nitrogen oxides, are produced in fire. Use dry chemical, carbon dioxide, or alcohol foam extinguishers. Vapors are heavier than air and will collect in low areas. Vapors may travel long distances to ignition sources and flashback. Vapors in confined areas may explode when exposed to fire. Containers may explode in fire. Storage containers and parts of containers may rocket great distances, in many directions. If material or contaminated runoff enters waterways, notify downstream users of potentially contaminated waters. Notify local health and fire officials and pollution control agencies. From a secure, explosion-proof location, use water spray to cool exposed containers. If cooling streams are ineffective (venting sound increases in volume and pitch, tank discolors, or shows any signs of deforming), withdraw immediately to a secure position. If employees are expected to fight fires, they must be trained and equipped in OSHA 1910.156. The only respirators recommended for firefighting are self-contained breathing apparatuses that have full face-pieces and are operated in a pressure-demand or other positive-pressure mode.

Reference

US Environmental Protection Agency. (November 30, 1987). *Chemical Hazard Information Profile: Pyridine, 2-Methyl-5-Vinyl.* Washington, DC: Chemical Emergency Preparedness Program

Metolachlor   M:1310

Molecular Formula: $C_{15}H_{22}ClNO_2$
Synonyms: 2-Aethyl-6-methyl-*N*-(1-methyl-2-methoxy-aethyl)-chloracetanilid (German); Bicep; CGA-24705;

α-Chlor-6′-aethyl-*N*-(2-methoxy-1-methylaethyl)-acet-*o*-toluidin (German); 2-Chloro-6′-cthyl-*N*-(2 methoxy-1-methy-lethyl)acet-*o*-toluidide; α-Chloro-2′-ethyl-6′-methyl-*N*-(1-methyl-2-methoxyethyl)-acetanilide; 2-Chloro-*N*-(2-ethyl-6-methylphenyl)-*N*-(2-methoxy-1-methylethyl) acetamide; 2-Chloro-*N*-(6-ethyl-*o*-tolyl)-*N*-(2-methoxy-1-methylethyl)-acetamide; Codal®; Cotoran® Multi®; Dual®; 2-Ethyl-6-methyl-1-*N*-(2-methoxy-1-methylethyl)chloroacetanilide; Metelilachlor; Milocep; Ontrack 8E®; Primagram®; Primextra®

CAS Registry Number: 51218-45-2; *(alt.)* 63150-68-5; *(alt.)* 94449-58-8
RTECS® Number: AN3430000
UN/NA & ERG Number: UN2902 (Pesticides, liquid, toxic, n.o.s.)/151
EC Number: 257-060-8
Regulatory Authority and Advisory Bodies
Carcinogenicity: EPA: Possible Human Carcinogen.
WGK (German Aquatic Hazard Class): No value assigned.
Description: Metolachlor is a colorless or tan to brown, oily liquid with a slightly sweet odor. Molecular weight = 283.83; Boiling point = 100°C at 0.001 mmHg. It is stable to about 300°C. Hazard Identification (based on NFPA-704 M Rating System): Health 1, Flammability 0, Reactivity 0. Slightly soluble in water.
Potential Exposure: Compound Description: Agricultural Chemical; Mutagen; Reproductive Effector; Primary Irritant. It is a selective herbicide used for weed control in corn and for controlling grasses in a variety of crops including cotton and peanuts.
Incompatibilities: Oxidizers, strong acids, nitrates.
Permissible Exposure Limits in Air
No standards or TEEL available.
Permissible Concentration in Water: The US EPA has set a lifetime health advisory of 10 μg/L. Several states have set guidelines for metolachlor in drinking water ranging from 1.0 μg/L (Illinois) to 17.5 μg/L (Kansas) to 25 μ/mL (Wisconsin).
Determination in Water: Extraction with methylene chloride followed by separation by gas chromatography and measurement using a nitrogen-phosphorus detector. Fish Tox = 1117.14617000 ppb (VERY LOW).
Harmful Effects and Symptoms
Short Term Exposure: Irritates the eyes and skin. The acute oral LD_{50} for rats is 2780 mg/kg (slightly toxic). Signs of human intoxication from metolachlor and/or its formulations (presumably following acute deliberate or accidental exposures) include abdominal cramps, anemia, ataxia, dark urine, methemoglobinemia, cyanosis, hypothermia, collapse, convulsions, diarrhea, gastrointestinal irritation, jaundice, weakness, nausea, shock, sweating, vomiting, CNS depression, dizziness, dyspnea, liver damage, nephritis, cardiovascular failure, skin irritation, dermatitis, sensitization dermatitis, eye and mucous membrane irriation, corneal opacity, and adverse reproductive effects. Human Tox = 100.00000 ppb (VERY LOW).

Long Term Exposure: May cause tumors.

Points of Attack: Blood.

Medical Surveillance: Test for methemoglobinemia. Complete blood count (CBC).

First Aid: If this chemical gets into the eyes, remove any contact lenses at once and irrigate immediately for at least 15 min, occasionally lifting upper and lower lids. Seek medical attention immediately. If this chemical contacts the skin, remove contaminated clothing and wash immediately with soap and water. Seek medical attention immediately. If this chemical has been inhaled, remove from exposure, begin rescue breathing (using universal precautions, including resuscitation mask) if breathing has stopped and CPR if heart action has stopped. Transfer promptly to a medical facility. When this chemical has been swallowed, get medical attention. Give large quantities of water and induce vomiting. Do not make an unconscious person vomit.

Note to physician: Treat for methemoglobinemia. Spectrophotometry may be required for precise determination of levels of methemoglobin in urine.

Personal Protective Methods: Wear protective gloves and clothing to prevent any reasonable probability of skin contact. Safety equipment suppliers/manufacturers can provide recommendations on the most protective glove/clothing material for your operation. All protective clothing (suits, gloves, footwear, headgear) should be clean, available each day, and put on before work. Contact lenses should not be worn when working with this chemical. Wear splash-proof chemical goggles and face shield unless full face-piece respiratory protection is worn. Employees should wash immediately with soap when skin is wet or contaminated. Provide emergency showers and eyewash.

Respirator Selection: Follow regulations in OSHA 29CFR1910.134 or European Standard EN149. Use a NIOSH/MSHA- or European Standard EN149-approved respirator; or use an approved supplied-air respirator with a full face-piece operated in the positive-pressure mode, or with a full face-piece, hood, or helmet in the continuous-flow mode; or use a NIOSH/MSHA- or European Standard EN149-approved self-contained breathing apparatus with a full face-piece operated in a pressure-demand or other positive-pressure mode.

Storage: Color Code—Blue: Health Hazard/Poison: Store in a secure poison location. Prior to working with this chemical you should be trained on its proper handling and storage. Store in tightly closed containers in a cool, well-ventilated area away from oxidizers. Where possible, automatically pump liquid from drums or other storage containers to process containers.

Shipping: Pesticides, liquid, toxic, n.o.s. require the label of "POISONOUS/TOXIC MATERIALS." It falls in Hazard Class 6.1.

Spill Handling: Evacuate and restrict persons not wearing protective equipment from area of spill or leak until cleanup is complete. Remove all ignition sources. Ventilate area of spill or leak. Absorb liquids in vermiculite, dry sand, earth, peat, carbon, or a similar material and deposit in sealed containers. Keep this chemical out of a confined space, such as a sewer, because of the possibility of an explosion, unless the sewer is designed to prevent the buildup of explosive concentrations. It may be necessary to contain and dispose of this chemical as a hazardous waste. If material or contaminated runoff enters waterways, notify downstream users of potentially contaminated waters. Contact your local or federal environmental protection agency for specific recommendations. If employees are required to clean up spills, they must be properly trained and equipped. OSHA 1910.120(q) may be applicable.

Fire Extinguishing: This chemical is a combustible liquid. Poisonous gases, including nitrogen oxides and chlorine, are produced in fire. Use dry chemical, carbon dioxide, or alcohol foam extinguishers. Vapors are heavier than air and will collect in low areas. Vapors in confined areas may explode when exposed to fire. Containers may explode in fire. Storage containers and parts of containers may rocket great distances, in many directions. If material or contaminated runoff enters waterways, notify downstream users of potentially contaminated waters. Notify local health and fire officials and pollution control agencies. From a secure, explosion-proof location, use water spray to cool exposed containers. If cooling streams are ineffective (venting sound increases in volume and pitch, tank discolors, or shows any signs of deforming), withdraw immediately to a secure position. If employees are expected to fight fires, they must be trained and equipped in OSHA 1910.156. The only respirators recommended for firefighting are self-contained breathing apparatuses that have full face-pieces and are operated in a pressure-demand or other positive-pressure mode.

Disposal Method Suggested: In accordance with 40CFR 165 recommendations for the disposal of pesticides and pesticide containers. Must be disposed properly by following package label directions or by contacting your local or federal environmental control agency or by contacting your regional EPA office.

Reference

US Environmental Protection Agency. (August 1987). *Health Advisory.* Washington, DC: Office of Drinking Water

Metolcarb   M:1320

Molecular Formula: $C_9H_{11}NO_2$

Common Formula: $C_6H_4(CH_3)OCONHCH_3$

Synonyms: Carbamic acid, methyl-, 3-methylphenyl ester; Carbamic acid, methyl-, 3-tolyl ester; *m*-Cresyl ester of *N*-methylcarbamic acid; *m*-Cresyl methyl carbamate; *m*-Cresyl methylcarbamate; Dicresyl; Dicresyl *N*-methylcarbamate; DRC 3341; Kumiai; Metacrate; Metolcarb; Methylcarbamic acid *m*-toyl ester; *m*-Methylphenyl methylcarbamate; 3-Methylphenyl *N*-methylcarbamate; Metolcarb; MTMC; S 1065; *m*-Tolyester kyseliny methyl karbaminove;

m-Tolyl *N* methylcarbamate; 3-Tolyl *N*-methylcarbamate; Tsumacide; Tsumaunka

CAS Registry Number: 1129-41-5

RTECS® Number: FC8050000

UN/NA & ERG Number: UN2757(carbamate pesticides, solid, toxic)/151

EC Number: 214-446-0 [*Annex I Index No.:* 006-056-00-2]

Regulatory Authority and Advisory Bodies

US EPA Hazardous Waste Number (RCRA No.): P190.

Superfund/EPCRA [40CFR 302 and 355, F R: 8/16/06,Vol 71, No. 158] Reportable Quantity (RQ): 1000 lb (454 kg).

RCRA, 40CFR261, Appendix 8 Hazardous Constituents.

RCRA 40CFR268.48; 61FR15654, Universal Treatment Standards: Wastewater (mg/L), 0.056; Nonwastewater (mg/kg), 1.4.

Superfund/EPCRA 40CFR355, Extremely Hazardous Substances: TPQ = 100/10,000 lb (45.4/4540 kg).

Reportable Quantity (RQ): 1000 lb (454 kg).

European/International Regulations: Hazard Symbol: Xn, N; Risk phrases: R22; R51/53 Safety phrases: S2; S61 (see Appendix 4).

WGK (German Aquatic Hazard Class): No value assigned.

Description: Metolcarb is a colorless crystalline solid. Molecular weight = 165.21; Freezing/Melting point = 74−75°C. Hazard Identification (based on NFPA-704 M Rating System): Health 3, Flammability 0, Reactivity 0.

Potential Exposure: Metolcarb is an insecticide used for the control of rice leafhoppers, plant-hoppers, codling moth, citrus mealy bug, onion thrips, fruit flies, bollworms, and aphids. Not registered as a pesticide in the United States.

Permissible Exposure Limits in Air

Protective Action Criteria (PAC)

TEEL-0: 1 mg/m^3

PAC-1: 3 mg/m^3

PAC-2: 4.8 mg/m^3

PAC-3: 200 mg/m^3

Routes of Entry: Inhalation, ingestion, skin and/or eye contact.

Harmful Effects and Symptoms

Short Term Exposure: Metolcarb is a carbamate insecticide. Signs and symptoms of poisoning by carbamates are similar to those for organic phosphorus compounds. Symptoms of poisoning by organic phosphorus compounds include headache, giddiness, nervousness, blurred vision, weakness, nausea, cramps, diarrhea, and discomfort in the chest. Signs include sweating, myosis, tearing, salivation and other excessive respiratory tract secretion, vomiting, cyanosis, uncontrollable muscle twitches followed by muscular weakness, convulsions, coma, loss of reflexes, and loss of muscular control. Metolcarb exhibits high oral and skin toxicity, and moderate inhalation toxicity. Some carbamates appear to be carcinogenic, teratogenic, and/or mutagenic. Carbamates are cholinesterase inhibitors.

Long Term Exposure: Cholinesterase inhibitor; cumulative effect is possible. This chemical may damage the nervous system with repeated exposure, resulting in convulsions, respiratory failure. May cause liver damage.

Points of Attack: Respiratory system, lungs, central nervous system, cardiovascular system, skin, eyes, plasma and red blood cell cholinesterase.

Medical Surveillance: Before employment and at regular times after that, the following are recommended: plasma and red blood cell cholinesterase levels (tests for the enzyme poisoned by this chemical). If exposure stops, plasma levels return to normal in 1−2 weeks while red blood cell levels may be reduced for 1−3 months.

When cholinesterase enzyme levels are reduced by 25% or more below preemployment levels, risk of poisoning is increased, even if results are in lower ranges of "normal." Reassignment to work not involving organophosphate or carbamate pesticides is recommended until enzyme levels recover. If symptoms develop or overexposure occurs, repeat the above tests as soon as possible and get an examination of the nervous system. Also, consider complete blood count. Consider chest X-ray following acute overexposure. Do not drink any alcoholic beverages before or during use. Alcohol promotes absorption of organic phosphates.

First Aid: If this chemical gets into the eyes, remove any contact lenses at once and irrigate immediately for at least 15 min, occasionally lifting upper and lower lids. Seek medical attention immediately. If this chemical contacts the skin, remove contaminated clothing and wash immediately with soap and water. Speed in removing material from skin is of extreme importance. Shampoo hair promptly if contaminated. Seek medical attention immediately. If this chemical has been inhaled, remove from exposure, begin rescue breathing (using universal precautions, including resuscitation mask) if breathing has stopped and CPR if heart action has stopped. Transfer promptly to a medical facility. When this chemical has been swallowed, get medical attention. Give large quantities of water and induce vomiting. Do not make an unconscious person vomit. Keep victim quiet and maintain normal body temperature. Carefully observe victim since effects may be delayed.

Personal Protective Methods: Wear protective gloves and clothing to prevent any reasonable probability of skin contact. Safety equipment suppliers/manufacturers can provide recommendations on the most protective glove/clothing material for your operation. All protective clothing (suits, gloves, footwear, headgear) should be clean, available each day, and put on before work. Contact lenses should not be worn when working with this chemical. Wear splash-proof chemical goggles and face shield unless full face-piece respiratory protection is worn. Employees should wash immediately with soap when skin is wet or contaminated. Provide emergency showers and eyewash.

Respirator Selection: Follow regulations in OSHA 29CFR1910.134 or European Standard EN149. Use a NIOSH/MSHA- or European Standard EN149-approved respirator; or use an approved supplied-air respirator with a

full face-piece operated in the positive-pressure mode, or with a full face-piece, hood, or helmet in the continuous-flow mode; or use a NIOSH/MSHA- or European Standard EN149-approved self-contained breathing apparatus with a full face-piece operated in a pressure-demand or other positive-pressure mode.

Storage: Color Code—Blue: Health Hazard/Poison: Store in a secure poison location. Prior to working with this chemical you should be trained on its proper handling and storage. Store in tightly closed containers in a cool, well-ventilated area away from oxidizers. Where possible, automatically pump liquid from drums or other storage containers to process containers.

Shipping: Carbamate pesticides, solid, toxic, require a label of "POISONOUS/TOXIC MATERIALS." It falls in Hazard Class 6.1 and Packing Group III.

Spill Handling: Evacuate persons not wearing protective equipment from area of spill or leak until cleanup is complete. Remove all ignition sources. Collect powdered material in the most convenient and safe manner and deposit in sealed containers. Ventilate area after cleanup is complete. It may be necessary to contain and dispose of this chemical as a hazardous waste. If material or contaminated runoff enters waterways, notify downstream users of potentially contaminated waters. Contact your local or federal environmental protection agency for specific recommendations. If employees are required to clean up spills, they must be properly trained and equipped. OSHA 1910.120(q) may be applicable.

Fire Extinguishing: Solid carbamate pesticides may burn but do not ignite readily. For small fires, use dry chemical, carbon dioxide, water spray, and foam. For large fires, use water spray, fog, or foam. Dike fire control water for later disposal. Stay upwind and keep out of low areas. Wear positive pressure breathing apparatus and special protective clothing. Move container from fire area if you can do it without risk. Fight fire from maximum distance. Dike fire control water for later disposal; do not scatter the material. Poisonous gases, including nitrogen oxides, are produced in fire. If material or contaminated runoff enters waterways, notify downstream users of potentially contaminated waters. Notify local health and fire officials and pollution control agencies. From a secure, explosion-proof location, use water spray to cool exposed containers. If cooling streams are ineffective (venting sound increases in volume and pitch, tank discolors, or shows any signs of deforming), withdraw immediately to a secure position. If employees are expected to fight fires, they must be trained and equipped in OSHA 1910.156. The only respirators recommended for firefighting are self-contained breathing apparatuses that have full face-pieces and are operated in a pressure-demand or other positive-pressure mode.

Disposal Method Suggested: Consult with environmental regulatory agencies for guidance on acceptable disposal practices. Generators of waste containing this contaminant (≥100 kg/mo) must conform with EPA regulations governing storage, transportation, treatment, and waste disposal. In accordance with 40CFR165, follow recommendations for the disposal of pesticides and pesticide containers. Must be disposed properly by following package label directions or by contacting your local or federal environmental control agency or by contacting your regional EPA office.

Reference
US Environmental Protection Agency. (November 30, 1987). *Chemical Hazard Information Profile: Metolcarb.* Washington, DC: Chemical Emergency Preparedness Program

Metribuzin   M:1330

Molecular Formula: $C_8H_{14}N_4OS$

Synonyms: 4-Amino-6-*tert*-butyl-3-(methylthio)-1,2,4-triazin-5-one; 4-Amino-6-*tert*-butyl-3-methylthio-As-triazin-5-one; 4-Amino-6-(1,1-dimethylethyl)-3-(methylthio)-1,2,4-triazin-5-(4H)-one; As-triazin-5(4H)-one,4-amino-6-*tert*-butyl-3-(methylthio)-; Bay 61597; Bay DIC 1468; Bayer 6159H; Bayer 6443H; Bayer 94337; DIC 1468; Lexone; Lexoneex; Metribuzina (Spanish); Sencor; Sencoral; Sencorer; Sencorex; 1,2,4-Triazin-5-(4H)-one, 4-Amino-6-(1,1-dimethylethyl)-3-(methylthio)-

CAS Registry Number: 21087-64-9

RTECS® Number: XZ2990000

UN/NA & ERG Number: UN2763 (triazine pesticide, solid, poisonous)/151

EC Number: 244-209-7 [*Annex I Index No.:* 606-034-00-8]

Regulatory Authority and Advisory Bodies
EPA: Not Classifiable as to human carcinogenicity.

Air Pollutant Standard Set. See below, "Permissible Exposure Limits in Air" section.

EPCRA Section 313 Form R *de minimis* concentration reporting level: 1.0%.

Safe Drinking Water Act: Priority List (55 FR 1470).

European/International Regulations: Hazard Symbol: Xn, N; Risk phrases: 22; R50/53; Safety phrases: S2; S60; S61 (see Appendix 4).

WGK (German Aquatic Hazard Class): No value assigned.

Description: Metribuzin is a colorless crystalline solid with a mild sulfurous odor; Freezing/Melting point = 125−127°C; Vapor pressure = 4×10^{-7} mmHg. Slightly soluble in water. Available in different concentrations (4%, 50%, 75%).

Potential Exposure: Those involved in manufacture, formulation, and application of this herbicide.

Incompatibilities: None reported.

Permissible Exposure Limits in Air
OSHA PEL: None.

NIOSH REL: 5 mg/m³ TWA.

ACGIH TLV®[1]: 5 mg/m³ TWA, not classifiable as a human carcinogen.

No TEEL available.

Guidelines or standards for metribuzin in ambient air[60] have been set by several states ranging from 50 $\mu g/m^3$ (North Dakota) to 100 $\mu g/m^3$ (Connecticut) to 119 $\mu g/m^3$ (Nevada).

Determination in Air: No method available.

Permissible Concentration in Water: The US EPA has set a lifetime health advisory of 175 $\mu g/L$. Several states have set guidelines for metribuzin in drinking water[61] ranging from 1.0 $\mu g/L$ (Illinois) to 25 $\mu g/L$ (Wisconsin) to 175 $\mu g/L$ (Kansas).

Determination in Water: Solvent extraction with methylene chloride followed by exchange to acetone; separation by gas chromatography and measurement with a thermionic bead detector. Fish Tox = 7683.76758000 ppb (VERY LOW).

Routes of Entry: Inhalation, ingestion, skin and/or eye contact.

Harmful Effects and Symptoms

Short Term Exposure: Metribuzin can affect you when breathed in and by passing through your skin. Acute poisoning can cause difficult breathing and drowsiness. High exposures may cause upset stomach, fatigue, and depression of the central nervous system, causing poor coordination, tremors, and weakness. Human Tox = 200.00000 ppb (VERY LOW).

Long Term Exposure: Repeated or high exposure may cause liver enzyme changes, goiter, and may affect thyroid function.

Points of Attack: Central nervous system, thyroid, liver.

Medical Surveillance: If symptoms develop or overexposure is suspected, the following may be useful: thyroid function tests.

First Aid: If this chemical gets into the eyes, remove any contact lenses at once and irrigate immediately for at least 15 min, occasionally lifting upper and lower lids. Seek medical attention immediately. If this chemical contacts the skin, remove contaminated clothing and wash immediately with soap and water. Seek medical attention immediately. If this chemical has been inhaled, remove from exposure, begin rescue breathing (using universal precautions, including resuscitation mask) if breathing has stopped and CPR if heart action has stopped. Transfer promptly to a medical facility. When this chemical has been swallowed, get medical attention. Give large quantities of water and induce vomiting. Do not make an unconscious person vomit.

Personal Protective Methods: Wear protective gloves and clothing to prevent any reasonable probability of skin contact. Safety equipment suppliers/manufacturers can provide recommendations on the most protective glove/clothing material for your operation. All protective clothing (suits, gloves, footwear, headgear) should be clean, available each day, and put on before work. Contact lenses should not be worn when working with this chemical. Wear dust-proof chemical goggles and face shield unless full face-piece respiratory protection is worn. Employees should wash immediately with soap when skin is wet or contaminated. Provide emergency showers and eyewash.

Respirator Selection: Where there is potential for exposures *over 5 mg/m³*, use a NIOSH/MSHA- or European Standard EN149-approved full-face-piece respirator with a pesticide cartridge. Greater protection is provided by a powered air-purifying respirator. *Where there is potential for high exposures*, use a NIOSH/MSHA- or European Standard EN149-approved supplied-air respirator with a full face-piece operated in the positive-pressure mode, or with a full face-piece hood, or helmet in the continuous-flow mode; or use a NIOSH/MSHA- or European Standard EN149-approved self-contained breathing apparatus with a full face-piece operated in a pressure-demand or other positive-pressure mode.

Storage: Color Code—Blue: Health Hazard/Poison: Store in a secure poison location. Store in tightly closed containers in a cool, dry area.

Shipping: Triazine pesticides, solid, toxic, require a shipping label of "POISONOUS/TOXIC MATERIALS." It falls in Hazard Class 6.1 and Packing Group III.

Spill Handling: Evacuate persons not wearing protective equipment from area of spill or leak until cleanup is complete. Remove all ignition sources. Collect powdered material in the most convenient and safe manner and deposit in sealed containers. Ventilate area after cleanup is complete. It may be necessary to contain and dispose of this chemical as a hazardous waste. If material or contaminated runoff enters waterways, notify downstream users of potentially contaminated waters. Contact your local or federal environmental protection agency for specific recommendations. If employees are required to clean up spills, they must be properly trained and equipped. OSHA 1910.120(q) may be applicable.

Fire Extinguishing: Extinguish fire using an agent suitable for type of surrounding fire. Metribuzin itself does not burn. Poisonous gases, including nitrogen oxides and sulfur oxides, are produced in fire. If material or contaminated runoff enters waterways, notify downstream users of potentially contaminated waters. Notify local health and fire officials and pollution control agencies. From a secure, explosion-proof location, use water spray to cool exposed containers. If cooling streams are ineffective (venting sound increases in volume and pitch, tank discolors, or shows any signs of deforming), withdraw immediately to a secure position. If employees are expected to fight fires, they must be trained and equipped in OSHA 1910.156. The only respirators recommended for firefighting are self-contained breathing apparatuses that have full face-pieces and are operated in a pressure-demand or other positive-pressure mode.

Disposal Method Suggested: In accordance with 40CFR165, follow recommendations for the disposal of pesticides and pesticide containers. Must be disposed properly by following package label directions or by contacting your local or federal environmental control agency or by contacting your regional EPA office.

References
US Environmental Protection Agency. (August 1987). *Health Advisory: Metribuzin.* Washington, DC: Office of Drinking Water
New Jersey Department of Health and Senior Services. (January 2001). *Hazardous Substances Fact Sheet: Metribuzin.* Trenton, NJ

Metronidazole   M:1340

Molecular Formula: $C_6H_9N_3O_3$
Synonyms: Acromona; Anagiardil; Atrivyl; Bayer 5360; Bexon; Clont; Cont; Danizol; Deflamon-wirkstoff; Efloran; Elyzol; Entizol; 1-(β-Ethylol)-2-methyl-5-nitro-3-azapyrrole; Eumin; Flagemona; Flagesol; Flagil; Flagyl; Giatricol; Gineflavir; 1-(β-Hydroxyethyl)-2-methyl-5-nitroimidazole; 1-(2-Hydroxy-1-ethyl)-2-methyl-5-nitroimidazole; 1-(2-Hydroxyethyl)-2-methyl-5-nitroimidazole; 1-Hydroxyethyl-2-methyl-5-nitroimidazole; Klion; Meronidal; 2-Methyl-1-(2-hydroxyethyl)-5-nitroimidazole; 2-Methyl-3-(2-hydroxyethyl)-4-nitroimidazole; Metronidaz; Metronidazol; Metronidazolo; Monagyl; Nalox; Neo-Tric; NIDA; Novonidazol; NSC-50364; Orvagil; 1-(β-Oxyethyl)-2-methyl-5-nitroimidazole; RP 8823; Sanatrichom; SC 10295; Trichazol; Trichocide; Trichomol; Trichomonacid "Pharmachim"; Trichopol; Tricom; Tricowas B; Trikojol; Trimeks; Trivazol; Vagilen; Vagimid; Vertisal
CAS Registry Number: 443-48-1
RTECS® Number: NI5600000
UN/NA & ERG Number: UN3249 (Medicine, solid, toxic, n.o.s.)/151
EC Number: 207-136-1
Regulatory Authority and Advisory Bodies
Carcinogenicity: IARC: Animal Sufficient Evidence; Human Inadequate Evidence, *possibly carcinogenic to humans*, Group 2B; NTP: Reasonably anticipated to be a human carcinogen.
California Proposition 65 Chemical: Cancer 1/1/88.
WGK (German Aquatic Hazard Class): No value assigned.
Description: Metronidazole is an odorless, white, yellow, or cream-colored crystalline solid. Darkens on exposure to light. Molecular weight = 171.18; Freezing/Melting point = 158−160°C. Soluble in water.
Potential Exposure: Metronidazole is an orally administered drug for the treatment of infections due to *entamoeba histolytica*, *trichomonas vaginalis*, *giardia lamblia* and has also been used for treating Vincent's infection. It can be used as a trichomonacide in veterinary medicine. One firm has petitioned EPA to use metronidazole as a disinfectant for cooling tower water.
Permissible Exposure Limits in Air
No standards or TEEL available.
Determination in Water: Octanol−water coefficient: Log K_{ow} = <−0.1.

Harmful Effects and Symptoms
Short Term Exposure: Symptoms of exposure include headache, anorexia, nausea, occasional vomiting, diarrhea, and rash.
Long Term Exposure: There is evidence that this substance is carcinogenic in animals. Possibly carcinogenic to humans.
First Aid: Skin Contact[52]*:* Flood all areas of body that have contacted the substance with water. Do not wait to remove contaminated clothing; do it under the water stream. Use soap to help assure removal. Isolate contaminated clothing when removed to prevent contact by others. *Eye Contact:* Remove any contact lenses at once. Flush eyes well with copious quantities of water or normal saline for at least 20−30 min. Seek medical attention. *Inhalation:* Leave area immediately; breathe fresh air. Proper respiratory protection must be supplied to any rescuers. If coughing, difficult breathing, or any other symptoms develop, seek medical attention at once, even if symptoms develop many hours after exposure. *Ingestion:* If convulsions are not present, give a glass or two of water or milk to dilute the substance. Assure that the person's airway is unobstructed and contact a hospital or poison center immediately for advice on whether or not to induce vomiting.
Personal Protective Methods: Wear protective gloves and clothing to prevent any reasonable probability of skin contact. Safety equipment suppliers/manufacturers can provide recommendations on the most protective glove/clothing material for your operation. All protective clothing (suits, gloves, footwear, headgear) should be clean, available each day, and put on before work. Contact lenses should not be worn when working with this chemical. Wear dust-proof chemical goggles and face shield unless full face-piece respiratory protection is worn. Employees should wash immediately with soap when skin is wet or contaminated. Provide emergency showers and eyewash.
Respirator Selection: Follow regulations in OSHA 29CFR1910.134 or European Standard EN149. Use a NIOSH/MSHA- or European Standard EN149-approved respirator; or use an approved supplied-air respirator with a full face-piece operated in the positive-pressure mode, or with a full face-piece, hood, or helmet in the continuous-flow mode; or use a NIOSH/MSHA- or European Standard EN149-approved self-contained breathing apparatus with a full face-piece operated in a pressure-demand or other positive-pressure mode.
Storage: Color Code—Blue: Health Hazard/Poison: Store in a secure poison location. Prior to working with this chemical you should be trained on its proper handling and storage. Store in a refrigerator. A regulated, marked area should be established where this chemical is handled, used, or stored in compliance with OSHA Standard 1910.1045.
Shipping: Medicine, solid, toxic, n.o.s. require a shipping label of "POISONOUS/TOXIC MATERIALS." It falls in Hazard Class 6.1.
Spill Handling: Evacuate persons not wearing protective equipment from area of spill or leak until cleanup is

complete. Remove all ignition sources. Dampen spilled material with water to avoid dust. Use HEPA vacuum or wet method to reduce dust during cleanup. Do not dry sweep. Collect powdered material in the most convenient and safe manner and deposit in sealed containers. Ventilate area after cleanup is complete. It may be necessary to contain and dispose of this chemical as a hazardous waste. If material or contaminated runoff enters waterways, notify downstream users of potentially contaminated waters. Contact your local or federal environmental protection agency for specific recommendations. If employees are required to clean up spills, they must be properly trained and equipped. OSHA 1910.120(q) may be applicable.

Fire Extinguishing: This chemical is a combustible solid. Use dry chemical, carbon dioxide, water spray, or alcohol foam extinguishers. Poisonous gases are produced in fire. If material or contaminated runoff enters waterways, notify downstream users of potentially contaminated waters. Notify local health and fire officials and pollution control agencies. From a secure, explosion-proof location, use water spray to cool exposed containers. If cooling streams are ineffective (venting sound increases in volume and pitch, tank discolors, or shows any signs of deforming), withdraw immediately to a secure position. If employees are expected to fight fires, they must be trained and equipped in OSHA 1910.156. The only respirators recommended for firefighting are self-contained breathing apparatuses that have full face-pieces and are operated in a pressure-demand or other positive-pressure mode.

Reference

New Jersey Department of Health and Senior Services. (September 2001). *Hazardous Substances Fact Sheet: Metronidazole.* Trenton, NJ

Mevinphos   M:1350

Molecular Formula: $C_7H_{13}O_6P$

Synonyms: AI3-22374; Apavinphos; 2-Butenoic acid, 3-[(dimethoxyphosphinyl)oxy]-, methyl ester; 2-Carbomethoxy-1-methylvinyl dimethyl phosophate, α-isomer; α-2-Carbomethoxy-1-methylvinyl dimethyl phosphate; (α-2-Carbomethoxy-1-methylvinyl) dimethyl phosphate; 2-Carbomethoxy-1-methylvinyl dimethyl phosphate; 2-Carbomethoxy-1-propen-2-yl dimethyl phosphate; Caswell No. 160B; CMDP; Compound 2046; Crotonic acid, 3-hydroxy-, methyl ester, dimethyl phosphate; Crotonic acid, 3-hydroxy-, methyl ester, dimethyl phosphate, (E)-; 3-[(Dimethoxyphosphinyl)oxy]-2-butenoic acid methyl ester; O,O-Dimethyl O-(2-carbomethoxy-1-methylvinyl) phosphate; O,O-Dimethyl 1-carbomethoxy-1-propen-2-yl phosphate; Dimethyl-1-carbomethoxy-1-propen-2-yl phosphate; Dimethyl (2-methoxycarbonyl-1-methylvinyl) phosphate; Dimethyl methoxycarbonylpropenyl phosphate; Dimethyl (1-methoxycarboxypropen-2-yl) phosphate; O,O-Dimethyl O-(1-methyl-2-carboxyvinyl) phosphate; Dimethyl

phosphate of methyl 3-hydroxy-*cis*-crotonate; Duraphos; ENT 22,374; EPA pesticide chemical code 015801; Gesfid; Gestid; 3-Hydroxycrotonic acid methyl ester dimethyl phosphate; Menite; (*cis*-2-Methoxycarbonyl-1-methylvinyl) dimethyl phosphate; *cis*-2-Methoxycarbonyl-1-methylvinyl dimethylphosphate; 2-Methoxycarbonyl-1-methylvinyl dimethyl phosphate; 1-Methoxycarbonyl-1-propen-2-yl dimethyl phosphate; Methyl 3-[(dimethoxyphosphinyl)oxy]-2-butenoate; Methyl-3-[(dimethoxyphosphinyl)oxy]-2-butenoate, α-isomer; Methyl 3-(dimethoxyphosphinyloxy)crotonate; Methyl 3-hydroxy-α-crotonate dimethyl phosphate; Methyl 3-hydroxycrotonate dimethyl phosphate ester; Methyl-3-hydroxy-α-crotonate, dimethyl phosphate ester; Mevinfos (Spanish); NSC 46470; PD 5; *cis*-Phosdrin; Phosdrin; Phosfene; Phosphene; Phosphoric acid, dimethyl ester, with methyl 3-hydroxycrotonate; Phosphoric acid, (1-methoxy-carboxypropen-2-yl) dimethyl ester

CAS Registry Number: 7786-34-7

RTECS® Number: GQ5250000

UN/NA & ERG Number: UN3018 (organophosphorus pesticide, liquid, toxic)/152

EC Number: 232-095-1 [*Annex I Index No.:* 015-020-00-5]

Regulatory Authority and Advisory Bodies

Banned or Severely Restricted (India, Norway) (UN).[13]

US EPA, FIFRA 1998 Status of Pesticides: RED completed.

Very Toxic Substance (World Bank).[15]

Clean Water Act: Section 311 Hazardous Substances/RQ 40CFR117.3 (same as CERCLA, see below).

Superfund/EPCRA 40CFR355, Extremely Hazardous Substances: TPQ = 500 lb (227 kg).

Reportable Quantity (RQ): 10 lb (4.54 kg).

EPCRA Section 313 Form R *de minimis* concentration reporting level: 1.0%.

US DOT 49CFR172.101, Inhalation Hazard Chemical as organophosphates.

European/International Regulations: Hazard Symbol: T +, N; Risk phrases: R27/28; R50/53; Safety phrases: S1/2; S23; S28; S36/37; S45; S60; S61 (see Appendix 4).

WGK (German Aquatic Hazard Class): 3—Severe hazard to waters.

Description: Mevinphos is a pale yellow to orange high-boiling liquid with a weak odor. The carrier solvent may change the physical properties listed here. Molecular weight = 224.17; Specific gravity (H_2O:1) = 1.25; Boiling point = decomposes; Freezing/Melting point = 7°C (*trans*-); 21°C (*cis*-); Vapor pressure: 0.0001 mmHg at 20°C; Flash point = 30°C (oc); 79.5°C (oc).[ICSC] Hazard Identification (based on NFPA-704 M Rating System): Health 4, Flammability 1, Reactivity 0. Soluble in water. Commercial product is a mixture of the *cis*- and *trans*-isomers. Insecticide that may be absorbed on a dry carrier.

Potential Exposure: Compound Description: Agricultural Chemical; Mutagen; Human Data. Those engaged in the manufacture, formulation, and application of this contact and systemic insecticide and acaricide.

Incompatibilities: Decomposes in heat (below boiling point at 300°C) producing phosphoric acid and phosphorus oxides fumes. Strong oxidizers may cause release of toxic phosphorus oxides. Organophosphates, in the presence of strong reducing agents such as hydrides, may form highly toxic and flammable phosphine gas. Keep away from alkaline materials. Corrosive to cast iron, some stainless steels, and brass. Attacks some forms of plastics, rubber, and coatings.

Permissible Exposure Limits in Air

Conversion factor: 1 ppm = 9.17 mg/m^3 at 25°C & 1 atm.

OSHA PEL: 0.1 mg/m^3 TWA [skin].

NIOSH REL: 0.01 ppm/0.1 mg/m^3 TWA; 0.03 ppm/0.3 mg/m^3 STEL [skin].

ACGIH TLV®[1]: 0.01 mg/m^3 measured as inhalable fraction and vapor TWA [skin]; not classifiable as a human carcinogen; BEI$_A$ issued as Acetylcholinesterase-inhibiting pesticides.

NIOSH IDLH: 4 ppm.

Protective Action Criteria (PAC)

TEEL-0: 0.01 mg/m^3

PAC-1: 0.3 mg/m^3

PAC-2: 4 mg/m^3

PAC-3: 36.6 mg/m^3

DFG MAK: 0.01 ppm/0.093 mg/m^3 TWA; Peak Limitation Category II(2) [skin].

Arab Republic of Egypt: TWA 0.01 ppm (0.1 mg/m^3) [skin] 1993; Australia: TWA 0.01 ppm (0.1 mg/m^3); STEL 0.03 ppm [skin] 1993; Austria: MAK 0.01 ppm (0.1 mg/m^3) [skin] 1999; Belgium: TWA 0.01 ppm (0.09 mg/m^3); STEL 0.03 ppm [skin] 1993; Denmark: TWA 0.01 ppm (0.1 mg/m^3) [skin] 1999; Finland: TWA 0.001 ppm, 1999; France: VME 0.01 ppm (0.1 mg/m^3) [skin] 1999; the Netherlands: MAC-TGG 0.1 mg/m^3 [skin] 2003; Norway: TWA 25 ppm (125 mg/m^3), 1999; the Philippines: TWA 0.1 mg/m^3 [skin] 1993; Switzerland: MAK-W 0.01 ppm (0.1 mg/m^3) [skin] 1999; Thailand: TWA 0.1 mg/m^3, 1993; United Kingdom: TWA 0.01 ppm (0.09 mg/m^3); STEL 0.03 ppm, 2000; Argentina, Bulgaria, Columbia, Jordan, South Korea, New Zealand, Singapore, Vietnam: ACGIH TLV®: STEL 0.27 mg/m^3 (skin). Several states have set guidelines or standards for mevinphos in ambient air[60] ranging from 1−3 μg/m^3 (North Dakota) to 1.6 μg/m^3 (Virginia) to 2.0 μg/m^3 (Connecticut and Nevada).

Determination in Air: Use NIOSH Analytical Method (IV) #5600, Organophosphorus Pesticides.

Permissible Concentration in Water: No criteria set. This chemical is highly toxic to aquatic life.

Determination in Water: Fish Tox = 0.96183000 ppb (EXTRA HIGH).

Routes of Entry: Inhalation, skin absorption, ingestion, skin and/or eye contact.

Short Term Exposure: Cholinesterase inhibitor. Mevinphos may affect the nervous system, causing convulsions, respiratory failure. This material is super toxic; the probable oral lethal dose for humans is less than 5 mg/kg or a taste (less than 7 drops) for a 150-lb person. It has direct and immediate effects whether it is swallowed, inhaled, or absorbed through the skin. Symptoms include nausea, vomiting, abdominal cramps, diarrhea, excessive salivation, headache, giddiness, dizziness, runny nose, tightness in the chest, blurring and dimming of vision, slurring of speech, twitching of muscles, mental confusion, disorientation, troubled breathing, blueing of skin, convulsions, coma, and death. Human Tox = 1.75000 ppb (HIGH).

Long Term Exposure: Cholinesterase inhibitor; cumulative effect is possible. This chemical may damage the nervous system with repeated exposure, resulting in convulsions, respiratory failure. May cause liver damage.

Points of Attack: Respiratory system, lungs, central nervous system, cardiovascular system, skin, eyes, plasma and red blood cell cholinesterase.

Medical Surveillance: NIOSH lists the following tests: Blood serum; Cholinesterase, Blood Serum, Red blood cells/count. Before employment and at regular times after that, the following are recommended: plasma and red blood cell cholinesterase levels (tests for the enzyme poisoned by this chemical). If exposure stops, plasma levels return to normal in 1−2 weeks while red blood cell levels may be reduced for 1−3 months. When cholinesterase enzyme levels are reduced by 25% or more below preemployment levels, risk of poisoning is increased, even if results are in lower ranges of "normal." Reassignment to work not involving organophosphate or carbamate pesticides is recommended until enzyme levels recover. If symptoms develop or overexposure occurs, repeat the above tests as soon as possible and get an examination of the nervous system. Also, consider complete blood count. Consider chest X-ray following acute overexposure. Do not drink any alcoholic beverages before or during use. Alcohol promotes absorption of organic phosphates.

First Aid: If this chemical gets into the eyes, remove any contact lenses at once and irrigate immediately for at least 15 min, occasionally lifting upper and lower lids. Seek medical attention immediately. If this chemical contacts the skin, remove contaminated clothing and wash immediately with soap and water. Speed in removing material from skin is of extreme importance. Shampoo hair promptly if contaminated. Seek medical attention immediately. If this chemical has been inhaled, remove from exposure, begin rescue breathing (using universal precautions, including resuscitation mask) if breathing has stopped and CPR if heart action has stopped. Transfer promptly to a medical facility. When this chemical has been swallowed, get medical attention. Give large quantities of water and induce vomiting. Do not make an unconscious person vomit.

Personal Protective Methods: Wear protective gloves and clothing to prevent any reasonable probability of skin contact. Safety equipment suppliers/manufacturers can provide recommendations on the most protective glove/clothing material for your operation. All protective clothing (suits, gloves, footwear, headgear) should be clean, available each day, and put on before work. Contact lenses should not be

worn when working with this chemical. Wear dust-proof chemical goggles and face shield unless full face-piece respiratory protection is worn. Employees should wash immediately with soap when skin is wet or contaminated. Provide emergency showers and eyewash.

Respirator Selection: 0.1 ppm: Sa (APF = 10) (any supplied-air respirator). *0.25 ppm:* Sa:Cf (APF = 25) (any supplied-air respirator operated in a continuous-flow mode). *0.5 ppm:* SaT: Cf (APF = 50) (any supplied-air respirator that has a tight-fitting face-piece and is operated in a continuous-flow mode) or SCBAF (APF = 50) (any self-contained breathing apparatus with a full face-piece) or SaF (APF = 50) (any supplied-air respirator with a full face-piece). *4 ppm:* Sa: Pd,Pp (APF = 1000) (any supplied-air respirator operated in a pressure-demand or other positive-pressure mode). *Emergency or planned entry into unknown concentrations or IDLH conditions:* SCBAF: Pd,Pp (APF = 10,000) (any self-contained breathing apparatus that has a full face-piece and is operated in a pressure-demand or other positive-pressure mode) or SaF: Pd,Pp: ASCBA (any supplied-air respirator that has a full face-piece and is operated in a pressure-demand or other positive-pressure mode in combination with an auxiliary self-contained breathing apparatus operated in a pressure-demand or other positive-pressure mode). *Escape:* GmFOv100 (APF = 50) [any air-purifying, full-face-piece respirator (gas mask) with a chin-style, front- or back-mounted organic vapor canister having an N100, R100, or P100 filter] or SCBAE (any appropriate escape-type, self-contained breathing apparatus).

Storage: Color Code—Blue: Health Hazard/Poison: Store in a secure poison location. Prior to working with this chemical you should be trained on its proper handling and storage. Store in tightly closed containers in a cool, well-ventilated area away from oxidizers.

Shipping: Organophosphorus pesticides, liquid, toxic, require a shipping label of "POISONOUS/TOXIC MATERIALS." It falls in Hazard Class 6.1 and Packing Group I.

Spill Handling: Evacuate and restrict persons not wearing protective equipment from area of spill or leak until cleanup is complete. Remove all ignition sources. Ventilate area of spill or leak. Absorb liquids in vermiculite, dry sand, earth, peat, carbon, or a similar material and deposit in sealed containers. Keep this chemical out of a confined space, such as a sewer, because of the possibility of an explosion, unless the sewer is designed to prevent the buildup of explosive concentrations. It may be necessary to contain and dispose of this chemical as a hazardous waste. If material or contaminated runoff enters waterways, notify downstream users of potentially contaminated waters. Contact your local or federal environmental protection agency for specific recommendations. If employees are required to clean up spills, they must be properly trained and equipped. OSHA 1910.120(q) may be applicable.

Fire Extinguishing: This chemical is a combustible liquid. Poisonous gases, including phosphorus oxides, are produced in fire. Use dry chemical, carbon dioxide, or alcohol foam extinguishers. Vapors are heavier than air and will collect in low areas. Vapors may travel long distances to ignition sources and flashback. Vapors in confined areas may explode when exposed to fire. Containers may explode in fire. Storage containers and parts of containers may rocket great distances, in many directions. If material or contaminated runoff enters waterways, notify downstream users of potentially contaminated waters. Notify local health and fire officials and pollution control agencies. From a secure, explosion-proof location, use water spray to cool exposed containers. If cooling streams are ineffective (venting sound increases in volume and pitch, tank discolors, or shows any signs of deforming), withdraw immediately to a secure position. If employees are expected to fight fires, they must be trained and equipped in OSHA 1910.156. The only respirators recommended for firefighting are self-contained breathing apparatuses that have full face-pieces and are operated in a pressure-demand or other positive-pressure mode.

Disposal Method Suggested: Mevinphos is 50% hydrolyzed in aqueous solutions at an unspecified temperature in 1.4 h at pH 11, 35 days at pH 7, and 120 days at pH 6. Decomposition is rapidly accomplished by lime sulfur. Mevinphos may also be incinerated. In accordance with 40CFR165, follow recommendations for the disposal of pesticides and pesticide containers. Must be disposed properly by following package label directions or by contacting your local or federal environmental control agency or by contacting your regional EPA office.

References

Sax, N. I. (Ed.). (1986). *Dangerous Properties of Industrial Materials Report*, 6, No. 1, 97−101

US Environmental Protection Agency. (November 30, 1987). *Chemical Hazard Information Profile: Mevinphos.* Washington, DC: Chemical Emergency Preparedness Program

US Environmental Protection Agency, Special Review and Reregistration Division Office of Pesticide Programs. (1998). *Agency Status of Pesticides in Registration, Reregistration, and Special Review* (Rainbow Report). Washington, DC

New Jersey Department of Health and Senior Services. (March 2007). *Hazardous Substances Fact Sheet: Mevinphos.* Trenton, NJ

Mexacarbate M:1360

Molecular Formula: $C_{12}H_{18}N_2O_2$
Synonyms: Carbamate, 4-dimethylamino-3,5-xylyln-methyl-; Carbamic acid, methyl-, 4-(dimethylamino)-3,5-xylyl ester; Carbamic acid, methyl-, methylcarbamate (ester); 4-(Dimethylamine)-3,5-xylyl *N*-methylcarbamate;

4-(Dimethylamino)-3,5-dimethylphenol methylcarbamate (ester); 4-(Dimethylamino)-3,5-dimethylphenyl N-methyl-carbamate; 4-(Dimethylamino)-3,5-xylenol, methylcarbamate (ester); 4-(N,N-Dimethylamino)-3,5-xylyl N-methylcarbamate; 4-Dimethylamino-3,5-xylyl N-methylcarbamate; 4-Dimethylamino-3,5-xylyl methylcarbamate; 5-Dimethylphenol methylcarbamate ester; DowCo® 139; ENT 25766; Methylcarbamic acid, 4-(dimethylamino)-3,5-xylyl ester; Methyl 4-dimethylamino-3,5-xylyl carbamate; Methyl-4-dimethylamino-3,5-xylyl ester of carbamic acid; Mexacarbato (Spanish); NCI-C00544; OMS-47; Phenol, 4-(dimethylamino)-3,5- dimethyl-methylcarbamate (ester); 3,5-Xylenol, 4-(dimethylamino)-, methylcarbamate; Zactran; Zectane; Zectran; Zextran

CAS Registry Number: 315-18-4
RTECS® Number: FC0700000
UN/NA & ERG Number: UN2757(carbamate pesticides, solid, toxic)/151
EC Number: 206-249-3 [*Annex I Index No.:* 006-054-00-1]

Regulatory Authority and Advisory Bodies
Carcinogenicity: NCI: Carcinogenesis Bioassay (feed); no evidence: mouse, rat, 1979; IARC: Human No Adequate Data, Animal No Evidence, *not classifiable as carcinogenic to humans*, Group 3.
Clean Water Act: Section 311 Hazardous Substances/RQ 40CFR117.3 (same as CERCLA, see below).
US EPA Hazardous Waste Number (RCRA No.): P128.
RCRA, 40CFR261, Appendix 8 Hazardous Constituents.
RCRA 40CFR268.48; 61FR15654, Universal Treatment Standards: Wastewater (mg/L), 0.056; Nonwastewater (mg/kg), 1.4.
Superfund/EPCRA 40CFR355, Extremely Hazardous Substances: TPQ = 500/10,000 lb (227/4540 kg).
Reportable Quantity (RQ): 1000 lb (454 kg).
US DOT Regulated Marine Pollutant (49CFR172.101, Appendix B).
European/International Regulations: Hazard Symbol: T +, N; Risk phrases: R1; R28; R50/53; Safety phrases: S1/2; S36/37; S45; S60; S61 (see Appendix 4).
WGK (German Aquatic Hazard Class): No value assigned.

Description: Mexacarbate is an odorless, white to tan crystalline solid. Molecular weight = 222.32; Freezing/Melting point = 85°C; Vapor pressure = 0.1 mmHg at 20°C. Hazard Identification (based on NFPA-704 M Rating System): Health 3, Flammability 1, Reactivity 0. Insoluble in water.

Potential Exposure: It is an insecticide for nonagricultural uses, e.g., lawn and turf, flowers, gardens, vines, forest lands, woody shrubs and trees; and also a molluscicide. It is not produced or used commercially in the United States.

Incompatibilities: Alkalis, strong oxidizers.

Permissible Exposure Limits in Air
Protective Action Criteria (PAC)
TEEL-0: 2.5 mg/m³
PAC-1: 7.5 mg/m³
PAC-2: 14 mg/m³
PAC-3: 14 mg/m³

Determination in Water: Fish Tox = 31.03658000 ppb (INTERMEDIATE).

Routes of Entry: Inhalation, ingestion, skin, and/or contact. Absorbed through the skin.

Harmful Effects and Symptoms
Short Term Exposure: Extremely toxic: probable oral lethal dose for humans is 5–50 mg/kg, between 7 drops and 1 teaspoonful for a 70-kg person (150 lb). Poisonous; may be fatal if inhaled, swallowed, or absorbed through skin. Contact may cause burns to skin and eyes. Symptoms of carbamate poisoning resemble those of parathion. This material is similar to carbaryl; symptoms of carbaryl exposure include nausea, vomiting, abdominal cramps, diarrhea, excessive salivation, sweating, lassitude, and weakness. Runny nose and sensation of tightness in chest may occur with inhalation exposures. Blurring or dimness of vision, tearing, eye muscle spasm, loss of muscle coordination, slurring of speech, and twitching of muscles may also occur.

Long Term Exposure: Cholinesterase inhibitor; cumulative effect is possible. Mexacarbate may damage the nervous system with repeated exposure, resulting in convulsions, respiratory failure. May cause liver damage.

Points of Attack: Respiratory system, lungs, central nervous system, cardiovascular system, skin, eyes, plasma and red blood cell cholinesterase.

Medical Surveillance: Before employment and at regular times after that, the following are recommended: plasma and red blood cell cholinesterase levels (tests for the enzyme poisoned by this chemical). If exposure stops, plasma levels return to normal in 1–2 weeks while red blood cell levels may be reduced for 1–3 months. When cholinesterase enzyme levels are reduced by 25% or more below preemployment levels, risk of poisoning is increased, even if results are in lower ranges of "normal." Reassignment to work not involving organophosphate or carbamate pesticides is recommended until enzyme levels recover. If symptoms develop or overexposure occurs, repeat the above tests as soon as possible and get an examination of the nervous system. Also, consider complete blood count. Consider chest X-ray following acute overexposure. Do not drink any alcoholic beverages before or during use. Alcohol promotes absorption of organic phosphates.

First Aid: If this chemical gets into the eyes, remove any contact lenses at once and irrigate immediately for at least 15 min, occasionally lifting upper and lower lids. Seek medical attention immediately. If this chemical contacts the skin, remove contaminated clothing and wash immediately with soap and water. Speed in removing material from skin is of extreme importance. Shampoo hair promptly if contaminated. Seek medical attention immediately. If this chemical has been inhaled, remove from exposure, begin rescue breathing (using universal precautions, including resuscitation mask) if breathing has stopped and CPR if heart action has stopped. Transfer promptly to a medical

facility. When this chemical has been swallowed, get medical attention. Give large quantities of water and induce vomiting. Do not make an unconscious person vomit.

Personal Protective Methods: Wear protective gloves and clothing to prevent any reasonable probability of skin contact. Safety equipment suppliers/manufacturers can provide recommendations on the most protective glove/clothing material for your operation. All protective clothing (suits, gloves, footwear, headgear) should be clean, available each day, and put on before work. Contact lenses should not be worn when working with this chemical. Wear dust-proof chemical goggles and face shield unless full face-piece respiratory protection is worn. Employees should wash immediately with soap when skin is wet or contaminated. Provide emergency showers and eyewash.

Respirator Selection: Follow regulations in OSHA 29CFR1910.134 or European Standard EN149. Use a NIOSH/MSHA- or European Standard EN149-approved respirator; or use an approved supplied-air respirator with a full face-piece operated in the positive-pressure mode, or with a full face-piece, hood, or helmet in the continuous-flow mode; or use a NIOSH/MSHA- or European Standard EN149-approved self-contained breathing apparatus with a full face-piece operated in a pressure-demand or other positive-pressure mode.

Storage: Color Code—Blue: Health Hazard/Poison: Store in a secure poison location. Store in a cool, dry place or a refrigerator.

Shipping: Carbamate pesticide, solid, toxic, requires a shipping label of "POISONOUS/TOXIC MATERIALS." It falls in Hazard Class 6.1 and Packing Group I.

Spill Handling: Evacuate persons not wearing protective equipment from area of spill or leak until cleanup is complete. Remove all ignition sources. Use HEPA vacuum or wet method to reduce dust during cleanup. Do not dry sweep. Collect powdered material in the most convenient and safe manner and deposit in sealed containers. Ventilate area after cleanup is complete. It may be necessary to contain and dispose of this chemical as a hazardous waste. If material or contaminated runoff enters waterways, notify downstream users of potentially contaminated waters. Contact your local or federal environmental protection agency for specific recommendations. If employees are required to clean up spills, they must be properly trained and equipped. OSHA 1910.120(q) may be applicable.

Fire Extinguishing: This chemical is a combustible solid. *Small fires:* dry chemical, carbon dioxide, water spray, or foam. *Large fires:* water spray, fog, or foam. Dike fire control water for later disposal; do not scatter the material. Poisonous gases, including nitrogen oxides, are produced in fire. If material or contaminated runoff enters waterways, notify downstream users of potentially contaminated waters. Notify local health and fire officials and pollution control agencies. From a secure, explosion-proof location, use water spray to cool exposed containers. If cooling streams are ineffective (venting sound increases in volume and pitch, tank discolors, or shows any signs of deforming), withdraw immediately to a secure position. If employees are expected to fight fires, they must be trained and equipped in OSHA 1910.156. The only respirators recommended for firefighting are self-contained breathing apparatuses that have full face-pieces and are operated in a pressure-demand or other positive-pressure mode.

Disposal Method Suggested: Consult with environmental regulatory agencies for guidance on acceptable disposal practices. Generators of waste containing this contaminant ($\geq$100 kg/mo) must conform with EPA regulations governing storage, transportation, treatment, and waste disposal. In accordance with 40CFR165, follow recommendations for the disposal of pesticides and pesticide containers. Must be disposed properly by following package label directions or by contacting your local or federal environmental control agency or by contacting your regional EPA office.

References

US Environmental Protection Agency. (November 30, 1987). *Chemical Hazard Information Profile: Mexacarbate.* Washington, DC: Chemical Emergency Preparedness Program

New Jersey Department of Health and Senior Services. (December 1999). *Hazardous Substances Fact Sheet: Mexacarbate.* Trenton, NJ

Mica   M:1370

Molecular Formula: $Al_6H_4K_2O_{24}Si_6$
Common Formula: $K_2Al_4(Al_2Si_6O_{20})(OH)_4$
Synonyms: Amber mica; Biotite; Fluorophlogopite; Lepidolite; Margarite; Mica silicate; Muscovite; Phlogopite; Roscoelite, suzorite mica; Zimmwaldite
CAS Registry Number: 12001-26-2
RTECS® Number: VV8760000
UN/NA & ERG Number: Not regulated.
Regulatory Authority and Advisory Bodies
Air Pollutant Standard Set. See below, "Permissible Exposure Limits in Air" section.
Canada, WHMIS, Ingredients Disclosure List Concentration 1.0%.

Description: Mica (Muscovite) takes the form of a colorless, odorless solid that separates into flakes or thin sheets of hydrous silicates. Molecular weight = 797 (approx.); Specific gravity (H_2O:1) = 2.6−3.2. Insoluble in water.

Potential Exposure: Mica is used as reinforcing filler for plastics, substitute for asbestos; for insulation in electrical equipment; used in the manufacture of roofing shingles, wallpaper, and paint.

Incompatibilities: Silicates react with lithium.
Permissible Exposure Limits in Air
OSHA PEL: 20 mppcf, <1% crystalline silica TWA.

NIOSH REL: 3 mg/m^3 respirable dust; containing <1% quartz TWA.

ACGIH TLV$^{®[1]}$: 3 mg/m^3 respirable fraction TWA.

NIOSH IDLH: 1500 mg/m^3.

Protective Action Criteria (PAC)

TEEL-0: 3 mg/m^3

PAC-1: 9 mg/m^3

PAC-2: 15 mg/m^3

PAC-3: 500 mg/m^3

Australia: TWA 2.5 mg/m^3, 1993; Belgium: TWA 3 mg/m^3, 1993; Switzerland: MAK-W 3 mg/m^3, 1999; the Netherlands: MAC-TGG 5 mg/m^3 (total dust); MAC-TGG 2.5 mg/m^3 (respirable dust), 2003; United Kingdom: TWA 10 mg/m^3, total inhalable dust; TWA 0.8 mg/m^3, respirable dust, 2000; Argentina, Bulgaria, Columbia, Jordan, South Korea, New Zealand, Singapore, Vietnam: ACGIH TLV$^®$: TWA 3 mg/m^3, respirable fraction.

Determination in Air: Use NIOSH Analytical Method (IV) #0600, Particulates NOR (respiratory).

Routes of Entry: Inhalation, skin and/or eye contact.

Harmful Effects and Symptoms

Short Term Exposure: Unknown at this time.

Long Term Exposure: Pneumoconiosis, cough, dyspnea, weakness, weight loss. Repeated heavy exposure can irritate the lungs. After years of high exposure, lung scarring (fibrosis) may result. This causes an abnormal chest X-ray and may cause cough and a shortness of breath.

Points of Attack: Lungs.

Medical Surveillance: If symptoms develop or overexposure is suspected, the following may be useful: lung function tests; chest X-ray.

First Aid: If this chemical gets into the eyes, remove any contact lenses at once and irrigate immediately. If a person breathes in large amounts of this chemical, move the exposed person to fresh air at once.

Personal Protective Methods: Wear protective gloves and clothing to prevent any reasonable probability of skin contact. Safety equipment suppliers/manufacturers can provide recommendations on the most protective glove/clothing material for your operation. All protective clothing (suits, gloves, footwear, headgear) should be clean, available each day, and put on before work. Contact lenses should not be worn when working with this chemical. Wear dust-proof chemical goggles and face shield unless full face-piece respiratory protection is worn. Employees should wash immediately with soap when skin is wet or contaminated. Provide emergency showers and eyewash.

Respirator Selection: NIOSH: *Up to 15 mg/m^3*: Qm (APF = 25) (any quarter-mask respirator). *Up to 30 mg/m^3:* Any particulate respirator equipped with an N95, R95, or P95 filter (including N95, R95, and P95 filtering face-pieces) except quarter-mask respirators. The following filters may also be used: N99, R99, P99, N100, R100, P100; or Sa (APF = 10) (any supplied-air respirator). *Up to 75 mg/m^3*: Sa:Cf (APF = 25) (any supplied-air respirator operated in a continuous-flow mode) or PaprHie

(APF = 25) (any powered, air-purifying respirator with a high-efficiency particulate filter). *Up to 150 mg/m^3*: 100F (APF = 50) (any air-purifying, full-face-piece respirator with an N100, R100, or P100 filter) or SaT: Cf (APF = 50) (any supplied-air respirator that has a tight-fitting face-piece and is operated in a continuous-flow mode) or PaprTHie (APF = 50) (any powered, air-purifying respirator with a tight-fitting face-piece and a high-efficiency particulate filter) or SCBAF (APF = 50) (any self-contained breathing apparatus with a full face-piece) or SaF (APF = 50) (any supplied-air respirator with a full face-piece). *Up to 1500 mg/m^3:* Sa: Pd,Pp (APF = 1000) (any supplied-air respirator operated in a pressure-demand or other positive-pressure mode). *Emergency or planned entry into unknown concentrations or IDLH conditions:* SCBAF: Pd,Pp (APF = 10,000) (any self-contained breathing apparatus that has a full faceplate and is operated in a pressure-demand or other positive-pressure mode) or SaF: Pd,Pp: ASCBA (APF = 10,000) (any supplied-air respirator that has a full face-piece and is operated in a pressure-demand or other positive-pressure mode in combination with an auxiliary, self-contained breathing apparatus operated in a pressure-demand or other positive-pressure mode). *Escape:* 100F APF = 50 (any air-purifying, full-face-piece respirator with an N100, R100, or P100 filter) or SCBAE (any appropriate escape-type, self-contained breathing apparatus).

Storage: Color Code—Green: General storage may be used. Prior to working with this chemical you should be trained on its proper handling and storage. Store in tightly closed containers in a cool, well-ventilated area.

Spill Handling: Evacuate persons not wearing protective equipment from area of spill or leak until cleanup is complete. Remove all ignition sources. Collect powdered material in the most convenient and safe manner and deposit in sealed containers. Ventilate area after cleanup is complete. It may be necessary to contain and dispose of this chemical as a hazardous waste. If material or contaminated runoff enters waterways, notify downstream users of potentially contaminated waters. Contact your local or federal environmental protection agency for specific recommendations. If employees are required to clean up spills, they must be properly trained and equipped. OSHA 1910.120(q) may be applicable.

Fire Extinguishing: Extinguish fire using an agent suitable for type of surrounding fire. Mica silica itself does not burn. Poisonous gases are produced in fire. If employees are expected to fight fires, they must be trained and equipped in OSHA 1910.156. The only respirators recommended for firefighting are self-contained breathing apparatuses that have full face-pieces and are operated in a pressure-demand or other positive-pressure mode.

Disposal Method Suggested: Landfill.

Reference

New Jersey Department of Health and Senior Services. (January 1996). *Hazardous Substances Fact Sheet: Silica, Mica.* Trenton, NJ

Michlers ketone   M:1380

Molecular Formula: $C_{17}H_{20}N_2O$

Common Formula: $(CH_3)_2N-C_6H_4-CO-C_6H_4-N(CH_3)_2$

Synonyms: Benzophenone, 4,4'-bis(dimethylamino)-; *p,p'*-Bis(dimethylamino)benzophenone; 4,4'-Bis(dimethylamino)benzophenone; Bis(4-dimethylaminophenyl) ketone; Bis[*p*-(*N,N*-dimethylamino)phenyl] ketone; Cetona de michler (Spanish); Methanone, bis[4-(dimethylamino)phenyl]-; *p,p'*-Michler's ketone; NCI-C02006; *N,N,N',N'*-Tetramethyl-4,4'-diaminobenzophenone; Tetramethyldiaminobenzophenone

CAS Registry Number: 90-94-8

RTECS® Number: DJ0250000

UN/NA & ERG Number: UN1602 (Dye intermediate, liquid, toxic, n.o.s.)/151

EC Number: 202-027-5 [*Annex I Index No.:* 606-073-00-0]

Regulatory Authority and Advisory Bodies

Carcinogenicity: IARC: Animal Sufficient Evidence; Human Inadequate Evidence, *possibly carcinogenic to humans*, Group 2B; NTP: Reasonably anticipated to be a human carcinogen; NCI: Carcinogenesis Studies (feed); clear evidence: rat, mouse, 1979.

EPCRA Section 313 Form R *de minimis* concentration reporting level: 0.1%.

California Proposition 65 Chemical: Cancer 1/1/88.

Canada, WHMIS, Ingredients Disclosure List Concentration 1.0%.

European/International Regulations: Hazard Symbol: T; Risk phrases: R45; R41; R68; Safety phrases: S53; S45 (see Appendix 4).

WGK (German Aquatic Hazard Class): No value assigned.

Description: Michler's ketone is a blue powder or white to green-colored leaflet material. Molecular weight = 268.37; Boiling point ≥360°C (decomposition); Freezing/Melting point = 172−176°C; Flash point = 220°C; Autoignition temperature = 480°C. Hazard Identification (based on NFPA-704 M Rating System): Health 3, Flammability 1, Reactivity 0. Insoluble in water.

Potential Exposure: Mutagen. Animal Carcinogen. Michler's ketone is a dye intermediate and derivative of dimethylaniline. It is also used in antifreeze formulations, cosmetics, cleaning compounds, heat transfer fluids; as a chemical intermediate in the synthesis of at least 13 dyes and pigments, especially auramine derivatives.

Incompatibilities: Ketones can react violently with oxidizers and strong reducing agents, aldehydes, nitric acid. Contact with hydrogen peroxide may form heat- and shock-sensitive explosives.

Permissible Exposure Limits in Air

Protective Action Criteria (PAC)

TEEL-0: 1 mg/m^3

PAC-1: 3.5 mg/m^3

PAC-2: 25 mg/m^3

PAC-3: 40 mg/m^3

DFG MAK: Carcinogen Category 2.

Routes of Entry: Inhalation and skin adsorption.

Harmful Effects and Symptoms

Irritates the eyes, skin, and mucous membranes. Absorbed through the skin.

Short Term Exposure: May have a narcotic or an anesthetic effect.

Points of Attack: Gastrointestinal or liver cancer.

Long Term Exposure: A potential occupational carcinogen.

First Aid: *Skin Contact*[52]*:* Flood all areas of body that have contacted the substance with water. Do not wait to remove contaminated clothing; do it under the water stream. Use soap to help assure removal. Isolate contaminated clothing when removed to prevent contact by others. *Eye Contact:* Remove any contact lenses at once. Immediately flush eyes well with copious quantities of water or normal saline for at least 20−30 min. Seek medical attention. *Inhalation:* Leave contaminated area immediately; breathe fresh air. Proper respiratory protection must be supplied to any rescuers. If coughing, difficult breathing, or any other symptoms develop, seek medical attention at once, even if symptoms develop many hours after exposure. *Ingestion:* If unconscious or convulsing, do not induce vomiting or give anything by mouth. Assure that victim's airway is open and lay him on his side with his head lower than his body and transport at once to a medical facility. If conscious and not convulsing, give a glass of water to dilute the substance. If medical advice is not readily available, do not induce vomiting, and rush the victim to the nearest medical facility.

Personal Protective Methods: Wear protective gloves and clothing to prevent any reasonable probability of skin contact. Safety equipment suppliers/manufacturers can provide recommendations on the most protective glove/clothing material for your operation. All protective clothing (suits, gloves, footwear, headgear) should be clean, available each day, and put on before work. Contact lenses should not be worn when working with this chemical. Wear splash-proof chemical goggles and face shield unless full face-piece respiratory protection is worn. Employees should wash immediately with soap when skin is wet or contaminated. Provide emergency showers and eyewash. See NIOSH Criteria Document 78-173, *Ketones*.

Respirator Selection: Follow regulations in OSHA 29CFR1910.134 or European Standard EN149. Use a NIOSH/MSHA- or European Standard EN149-approved respirator; or use an approved supplied-air respirator with a full face-piece operated in the positive-pressure mode, or with a full face-piece, hood, or helmet in the continuous-flow mode; or use a NIOSH/MSHA- or European Standard EN149-approved self-contained breathing apparatus with a full face-piece operated in a pressure-demand or other positive-pressure mode.

Storage: Color Code—Blue: Health Hazard/Poison: Store in a secure poison location. Prior to working with this chemical you should be trained on its proper handling and storage. Store in a refrigerator or a cool, dry place away from peroxides, aldehydes, strong acids. Where possible, automatically pump liquid from drums or other storage containers to process containers. A regulated, marked area should be established where this chemical is handled, used, or stored in compliance with OSHA Standard 1910.1045.

Shipping: Dye intermediates, liquid, toxic, n.o.s. require a label of "POISONOUS/TOXIC MATERIALS." It falls in Hazard Class 6.1, Packing Group III.

Spill Handling: Evacuate and restrict persons not wearing protective equipment from area of spill or leak until cleanup is complete. Remove all ignition sources. Ventilate area of spill or leak. Absorb liquids in vermiculite, dry sand, earth, peat, carbon, or a similar material and deposit in sealed containers. Keep this chemical out of a confined space, such as a sewer, because of the possibility of an explosion, unless the sewer is designed to prevent the buildup of explosive concentrations. It may be necessary to contain and dispose of this chemical as a hazardous waste. If material or contaminated runoff enters waterways, notify downstream users of potentially contaminated waters. Contact your local or federal environmental protection agency for specific recommendations. If employees are required to clean up spills, they must be properly trained and equipped. OSHA 1910.120(q) may be applicable.

Fire Extinguishing: This chemical is a combustible liquid. Poisonous gases, including nitrogen oxides, are produced in fire. Use dry chemical, carbon dioxide, or alcohol foam extinguishers. Vapors are heavier than air and will collect in low areas. Vapors may travel long distances to ignition sources and flashback. Vapors in confined areas may explode when exposed to fire. Containers may explode in fire. Storage containers and parts of containers may rocket great distances, in many directions. If material or contaminated runoff enters waterways, notify downstream users of potentially contaminated waters. Notify local health and fire officials and pollution control agencies. From a secure, explosion-proof location, use water spray to cool exposed containers. If cooling streams are ineffective (venting sound increases in volume and pitch, tank discolors, or shows any signs of deforming), withdraw immediately to a secure position. If employees are expected to fight fires, they must be trained and equipped in OSHA 1910.156. The only respirators recommended for firefighting are self-contained breathing apparatuses that have full face-pieces and are operated in a pressure-demand or other positive-pressure mode.

Mineral oil   M:1385

Synonyms: Adepsine oil; Heavy mineral oil mist; Mist of white mineral oil; Cutting oil; Heat-treating oil; Hydraulic oil; Cable oil; Lubricating oil; Paraffin oil mist; White mineral oil mist

CAS Registry Number: 8012-95-1; *(alt.)* 39355-35-6; *(alt.)* 79956-36-8; *(alt.)* 83046-05-3

RTECS® Number: PY8030000

EC Number: 232-384-2 (paraffin oils)

Regulatory Authority and Advisory Bodies

Carcinogenicity: IARC (*highly refined*): Human No Adequate Data, Animal No Evidence, *not classifiable as carcinogenic to humans*, Group 3; IARC (*untreated and poorly refined*): Human Sufficient Evidence; Animal Limited Evidence, *carcinogenic to humans*, Group 1; NTP (*untreated and poorly refined*): Known to be a human carcinogen.

US EPA, FIFRA, 1998 Status of Pesticides: Supported.

FDA—over-the-counter drug.

Air Pollutant Standard Set. See below, "Permissible Exposure Limits in Air" section.

Canada, WHMIS, Ingredients Disclosure List Concentration 1.0%.

European/International Regulations: Not listed in Annex 1.

WGK (German Aquatic Hazard Class): No value assigned.

Description: Mineral oil mist is a colorless, oily liquid aerosol dispersed in air with an odor like burned lubricating oil. The odor threshold is 1.0 ppm. Specific gravity (H_2O:1) = 0.865 at 60°C; Boiling point = 250−360°C; Vapor pressure = <0.5 mmHg at 20°C; Flash point = 193°C; Autoignition temperature = 260−371°C. Hazard Identification (based on NFPA-704 M Rating System): Health 0, Flammability 1, Reactivity 0. Insoluble in water.

Potential Exposure: Compound Description: Tumorigen, Human Data; Primary Irritant. Mineral oil is used in cosmetics, pharmaceutical bases, food, and fiber production; as carriers and bases; as a lubricating oil; and as a solvent for inks in the printing industry. Oil mist would be encountered in quenching of hot metal parts and in metal machining operations.

Incompatibilities: Strong oxidizers, nitric acid.

Permissible Exposure Limits in Air

As oil mist (mineral)

OSHA PEL: 5 mg/m^3 TWA.

NIOSH REL: 5 mg/m^3 TWA; 10 mg/m^3 STEL.

ACGIH TLV®[1]: 5 mg/m^3 inhalable fraction from highly refined and pure mineral oil, TWA.

NIOSH IDLH: 2500 mg/m^3.

Protective Action Criteria (PAC)

TEEL-0: 5 mg/m^3

PAC-1: 10 mg/m^3

PAC-2: 10 mg/m^3

PAC-3: 500 mg/m^3

Australia: TWA 5 mg/m^3; STEL 10 mg/m^3, 1993; Belgium: TWA 5 mg/m^3; STEL 10 mg/m^3, 1993; Finland: TWA 5 mg/m^3, 1999; Hungary: STEL 5 mg/m^3, carcinogen, 1993; Japan: 3 mg/m^3, Group 1 carcinogen, 1999; the Philippines: TWA 5 mg/m^3, 1993; Poland: MAC (TWA) 5 mg/m^3; MAC (STEL) 10 mg/m^3, 1999; Russia: STEL 5 mg/m^3, 1993; the Netherlands: MAC-TGG 5 mg/m^3, 2003; Sweden: NGV 3 mg/m^3; STEL 5 mg/m^3, 1993; Switzerland: MAK-W 5 mg/m^3, 1999; United Kingdom: LTEL 5 mg/m^3; STEL 10 mg/m^3, 1993; Argentina,

Bulgaria, Columbia, Jordan, South Korea, New Zealand, Singapore, Vietnam: ACGIH TLV®: STEL 10 mg/m³. Several states have set guidelines or standards for mineral oil mist in ambient air[60] ranging from 16.7 µg/m³ (New York) to 25 µg/m³ (South Carolina) to 50 µg/m³ (Florida) to 80 µg/m³ (Virginia) to 100 µg/m³ (Nevada).

Determination in Air: Use NIOSH Analytical Method (IV) #5026, Oil mist, mineral; #5524, Metalworking Fluids

Permissible Concentration in Water: The EEC[35] set a MAC of 10 µg/L in drinking water. Russia set a MAC of 6.5 mg/L in surface water for fishery purposes.

Routes of Entry: Inhalation, skin and/or eye contact.

Harmful Effects and Symptoms

Short Term Exposure: Inhalation: May irritate nose and throat. If taken into lungs may cause coughing and swelling of lung tissue; may cause pneumonia. Similarity to kerosene indicates that headache, nausea, ringing in the ears, weakness, confusion, drowsiness, coma, and death may occur. *Skin:* May cause redness and swelling if not promptly removed. *Eyes:* May cause severe irritation if not promptly removed. *Ingestion:* Will cause burning sensation in mouth, throat, and stomach if swallowed. Vomiting, diarrhea, and belching may follow. If liquid gets into lungs, it may cause rapid breathing, blue skin coloration, rapid heartbeat, and fever with rapid onset of chemical pneumonia and possible secondary infection. Death may result from as little as half a liquid ounce in the absence of lung involvement.

Note: Food grades are highly purified and are of low toxicity.

Long Term Exposure: Prolonged contact may cause skin irritation; acne-like rash may develop. May cause skin allergy with itching and rash. There is some evidence that some substances referred to as "mineral oils" may be carcinogens. However, this information is unclear at this time and mineral oils (except for food grades) should be treated with caution.

Points of Attack: Eyes, skin, respiratory system.

Medical Surveillance: If symptoms develop or overexposure is suspected, the following may be useful: consider chest X-ray after acute overexposure. Special tests of the sputum to look for oil droplets. Evaluation by a qualified allergist, including careful exposure history and special testing, may help diagnose skin allergy.

First Aid: If this chemical gets into the eyes, remove any contact lenses at once and irrigate immediately for at least 15 min, occasionally lifting upper and lower lids. Seek medical attention immediately. If this chemical contacts the skin, remove contaminated clothing and wash immediately with soap and water. Seek medical attention immediately. If this chemical has been inhaled, remove from exposure, begin rescue breathing (using universal precautions, including resuscitation mask) if breathing has stopped and CPR if heart action has stopped. Transfer promptly to a medical facility.

Personal Protective Methods: Wear protective gloves and clothing to prevent any reasonable probability of skin contact. Safety equipment suppliers/manufacturers can provide recommendations on the most protective glove/clothing material for your operation. All protective clothing (suits, gloves, footwear, headgear) should be clean, available each day, and put on before work. Contact lenses should not be worn when working with this chemical. Wear splash-proof chemical goggles and face shield unless full face-piece respiratory protection is worn. Employees should wash immediately with soap when skin is wet or contaminated. Provide emergency showers and eyewash.

Respirator Selection: 50 mg/m³: 100XQ (APF = 10) [Any air-purifying respirator with an N100, R100, or P100 filter (including N100, R100, and P100 filtering face-pieces) except quarter-mask respirators] or Sa (APF = 10) (any supplied-air respirator). *125 mg/m³:* Sa:Cf (APF = 25) (any supplied-air respirator operated in a continuous-flow mode) or PaprHie (APF = 25) (any powered, air-purifying respirator with a high-efficiency particulate filter). *250 mg/m³:* 100F (APF = 50) (any air-purifying, full-face-piece respirator with an N100, R100, or P100 filter) or SaT: Cf (APF = 50) (any supplied-air respirator that has a tight-fitting face-piece and is operated in a continuous-flow mode) or PaprTHie (APF = 50) (any powered, air-purifying respirator with a tight-fitting face-piece and a high-efficiency particulate filter) or SCBAF (APF = 50) (any self-contained breathing apparatus with a full face-piece) or SaF (APF = 50) (any supplied-air respirator with a full face-piece). *2500 mg/m³:* Sa: Pd,Pp (APF = 1000) (any supplied-air respirator operated in a pressure-demand or other positive-pressure mode). *Emergency or Planned Entry into Unknown Concentrations or IDLH Conditions* SCBAF: Pd, Pp (APF = 10,000) (any self-contained breathing apparatus that has a full face-piece and is operated in a pressure-demand or other positive-pressure mode) or SaF: Pd,Pp: ASCBA (APF = 10,000) (any supplied-air respirator that has a full face-piece and is operated in a pressure-demand or other positive-pressure mode in combination with an auxiliary self-contained breathing apparatus operated in a pressure-demand or other positive-pressure mode). *Escape:* 100F (APF = 50) (any air-purifying, full-face-piece respirator with an N100, R100, or P100 filter) or SCBAE (any appropriate escape-type, self-contained breathing apparatus).

Storage: Color Code—Green: General storage may be used. Prior to working with this chemical you should be trained on its proper handling and storage. Store in tightly closed containers in a cool, well-ventilated area away from oxidizers.

Spill Handling: Evacuate and restrict persons not wearing protective equipment from area of spill or leak until cleanup is complete. Remove all ignition sources. Ventilate area of spill or leak. Absorb liquids in vermiculite, dry sand, earth, peat, carbon, or a similar material and deposit in sealed containers. Keep this chemical out of a confined space, such as a sewer, because of the possibility of an explosion, unless the sewer is designed to prevent the buildup of explosive

concentrations. It may be necessary to contain and dispose of this chemical as a hazardous waste. If material or contaminated runoff enters waterways, notify downstream users of potentially contaminated waters. Contact your local or federal environmental protection agency for specific recommendations. If employees are required to clean up spills, they must be properly trained and equipped. OSHA 1910.120(q) may be applicable.

Fire Extinguishing: This chemical is a combustible liquid. Poisonous gases are produced in fire. Use dry chemical, carbon dioxide, or alcohol foam extinguishers. Vapors are heavier than air and will collect in low areas. Vapors may travel long distances to ignition sources and flashback. Vapors in confined areas may explode when exposed to fire. Containers may explode in fire. Storage containers and parts of containers may rocket great distances, in many directions. If material or contaminated runoff enters waterways, notify downstream users of potentially contaminated waters. Notify local health and fire officials and pollution control agencies. From a secure, explosion-proof location, use water spray to cool exposed containers. If cooling streams are ineffective (venting sound increases in volume and pitch, tank discolors, or shows any signs of deforming), withdraw immediately to a secure position. If employees are expected to fight fires, they must be trained and equipped in OSHA 1910.156. The only respirators recommended for firefighting are self-contained breathing apparatuses that have full face-pieces and are operated in a pressure-demand or other positive-pressure mode.

Disposal Method Suggested: Dissolve or mix the material with a combustible solvent and burn in a chemical incinerator equipped with an afterburner and scrubber. All federal, state, and local environmental regulations must be observed.

References
New York State Department of Health. (March 1986). *Chemical Fact Sheet: Mineral Seal Oil.* Albany, NY: Bureau of Toxic Substance Assessment

Sax, N. I. (Ed.). (1980). *Dangerous Properties of Industrial Materials Report,* 1, No. 2, 47–48

New Jersey Department of Health and Senior Services. (June 2001). *Hazardous Substances Fact Sheet: Oil Mist, Mineral.* Trenton, NJ

Mirex   M:1390

Molecular Formula: $C_{10}Cl_{12}$
Synonyms: Bichlorendo; CG-1283; Dechlorane 4070; 1,1a,2,2,3,3a,4,5,5,5a,5b,6-Dodecachlorooctahydro-1,3,4-metheno-1H-cyclobuta(c,d)pentalene; Dodecachlorooctahydro-1,3,4-metheno-2H-cyclobuta(c,d)pentalene; Dodecachloropentacyclodecane; ENT 25,719; Ferriamicide; 1,2,3,4,5,5-Hexachloro-1,3-cyclopentadiene dimer; Hexachlorocyclopentadienedimer; HRS 1276; NCI-C06428; Perchlorodihomocubane; Perchloropentacyclodecane

CAS Registry Number: 2385-85-5
RTECS® Number: PC8225000
UN/NA & ERG Number: UN2761/151
EC Number: 219-196-6 [*Annex I Index No.:* 602-077-00-1]
Regulatory Authority and Advisory Bodies
Carcinogenicity: IARC: Animal Sufficient Evidence, Human Inadequate Evidence, *possibly carcinogenic to humans,* Group 2B; NTP: Reasonably anticipated to be a human carcinogen.
Banned or Severely Restricted (several countries) (UN).[13]
Persistent Organic Pollutants (UN).
Air Pollutant Standard Set. See below, "Permissible Exposure Limits in Air" section.
California Proposition 65 Chemical: Cancer 1/188.
List of Stockholm Convention POPs: Annex A (Elimination).
European/International Regulations: Hazard Symbol: Xn, N; Risk phrases: R21/22; R40; R50/53; R62; R63; R64; Safety phrases: S2; S13; S36/37; S46; S60; S61 (see Appendix 4).
WGK (German Aquatic Hazard Class): No value assigned.
Description: Mirex is a snow-white crystalline solid. Molecular weight = 545.50; Boiling point = (decomposes) 485°C; Vapor pressure = 8×10^{-7} mm at 20°C. Hazard Identification (based on NFPA-704 M Rating System): Health 2, Flammability 1, Reactivity 0. Insoluble in water.
Potential Exposure: Those involved in the manufacture, formulation, and application of the insecticide (particularly effective against fire ants). Also used as a fire retardant in plastics. Not produced in the United States but may be found in imported products.
Incompatibilities: Strong oxidizers, dichromates.
Permissible Exposure Limits in Air
Protective Action Criteria (PAC)
TEEL-0: 0.1 mg/m^3
PAC-1: 0.3 mg/m^3
PAC-2: 2 mg/m^3
PAC-3: 100 mg/m^3
Several states have set guidelines or standards for mirex in ambient air[60] ranging from zero (Massachusetts) to 0.03 μg/m^3 (New York) to 0.88 μg/m^3 (Pennsylvania) to 4500 μg/m^3 (South Carolina).
Permissible Concentration in Water: 0.001 mg/L for protection of aquatic life (Sax—see reference below).
Determination in Water: Fish Tox = 1.98564000 ppb (HIGH).
Routes of Entry: Inhalation, ingestion, skin and/or eye contact. Passes through the skin.
Harmful Effects and Symptoms
Short Term Exposure: Inhalation: Can irritate the respiratory tract. This compound is moderately toxic (the LD_{50} value for rats is 300 mg/kg). *Skin:* Can cause irritation, burning sensation, and rash. *Eyes:* Can cause irritation. *Ingestion:* No cases of human toxicity reported. Possible symptoms include nausea, vomiting, restlessness, tremor, weight loss, nervous system and liver abnormalities, skin

rash, and reproductive failure. Exposure can cause nausea and vomiting, headache, dizziness, muscular weakness, fatigue, convulsions, and unconsciousness.

Long Term Exposure: May damage the developing fetus. May cause damage to the testes. May damage the liver and cause anemia. High exposure can cause arrhythmia (irregular heartbeat) and may cause death. Mirex has caused cataracts, liver and thyroid cancer, and birth defects in both rats and mice. Whether it does so in humans is not known.

Points of Attack: Blood, liver, nervous system.

Medical Surveillance: Complete blood count (CBC), liver function tests; EKG, examination of the nervous system.

First Aid: If this chemical gets into the eyes, remove any contact lenses at once and irrigate immediately for at least 15 min, occasionally lifting upper and lower lids. Seek medical attention immediately. If this chemical contacts the skin, remove contaminated clothing and wash immediately with soap and water. Speed in removing material from skin is of extreme importance. Seek medical attention immediately. If this chemical has been inhaled, remove from exposure, begin rescue breathing (using universal precautions, including resuscitation mask) if breathing has stopped and CPR if heart action has stopped. Transfer promptly to a medical facility. When this chemical has been swallowed, get medical attention. Give large quantities of water and induce vomiting. Do not make an unconscious person vomit.

Note to physician: Gastric lavage or catharsis may be useful. High urine organic chlorine is indicative of exposure but not of severity.

Personal Protective Methods: Wear protective gloves and clothing to prevent any reasonable probability of skin contact. Safety equipment suppliers/manufacturers can provide recommendations on the most protective glove/clothing material for your operation. All protective clothing (suits, gloves, footwear, headgear) should be clean, available each day, and put on before work. Contact lenses should not be worn when working with this chemical. Wear dust-proof chemical goggles and face shield unless full face-piece respiratory protection is worn. Employees should wash immediately with soap when skin is wet or contaminated. Provide emergency showers and eyewash.

Respirator Selection: Pesticide respirators may be used to limit exposure. Follow regulations in OSHA 29CFR1910.134 or European Standard EN149. Use a NIOSH/MSHA- or European Standard EN149-approved respirator; or use an approved supplied-air respirator with a full face-piece operated in the positive-pressure mode, or with a full face-piece, hood, or helmet in the continuous-flow mode; or use a NIOSH/MSHA- or European Standard EN149-approved self-contained breathing apparatus with a full face-piece operated in a pressure-demand or other positive-pressure mode.

Storage: Color Code—Blue: Health Hazard/Poison: Store in a secure poison location. Prior to working with this chemical you should be trained on its proper handling and storage. Store in a cool area in closed containers away from oxidizers and dichromates. A regulated, marked area should be established where this chemical is handled, used, or stored in compliance with OSHA Standard 1910.1045.

Shipping: Organochlorine pesticides, solid toxic, require a shipping label of "POISONOUS/TOXIC MATERIALS." It falls in Hazard Class 6.1 and Packing Group III.

Spill Handling: Evacuate persons not wearing protective equipment from area of spill or leak until cleanup is complete. Remove all ignition sources. Use HEPA vacuum; do not use wet method. Collect powdered material in the most convenient and safe manner and deposit in sealed containers. Ventilate area after cleanup is complete. It may be necessary to contain and dispose of this chemical as a hazardous waste. If material or contaminated runoff enters waterways, notify downstream users of potentially contaminated waters. Contact your local or federal environmental protection agency for specific recommendations. If employees are required to clean up spills, they must be properly trained and equipped. OSHA 1910.120(q) may be applicable.

Fire Extinguishing: Mirex is a noncombustible solid. Use dry chemical, carbon dioxide, water spray, or foam extinguishers. Poisonous gases are produced in fire. If material or contaminated runoff enters waterways, notify downstream users of potentially contaminated waters. Notify local health and fire officials and pollution control agencies. From a secure, explosion-proof location, use water spray to cool exposed containers. If cooling streams are ineffective (venting sound increases in volume and pitch, tank discolors, or shows any signs of deforming), withdraw immediately to a secure position. If employees are expected to fight fires, they must be trained and equipped in OSHA 1910.156. The only respirators recommended for firefighting are self-contained breathing apparatuses that have full face-pieces and are operated in a pressure-demand or other positive-pressure mode.

Disposal Method Suggested: High-temperature incineration is recommended. In accordance with 40CFR165, follow recommendations for the disposal of pesticides and pesticide containers. Must be disposed properly by following package label directions or by contacting your local or federal environmental control agency or by contacting your regional EPA office.

References

Sax, N. I. (Ed.). *Dangerous Properties of Industrial Materials Report*, 1, No. 2, 48 (1980) and 7, No. 5, 88–91 (1987)

New York State Department of Health. (March 1986). *Chemical Fact Sheet: Mirex.* Albany, NY: Bureau of Toxic Substance Assessment

New Jersey Department of Health and Senior Services. (July 1999). *Hazardous Substances Fact Sheet: Mirex.* Trenton, NJ

Mitomycin C   M:1400

Molecular Formula: $C_{15}H_{18}N_4O_5$

Synonyms: Ametycin; 7-Amino-9-α-methoxymitosane; 7-Amino-9-α-methoxymitosane; Azirino (2′,3′: 3,4) pyrrolo (1,2-a) indole-4,7-dione, 6-amino-8-([(aminocarbonyl) oxy] methyl)-1,1a, 2,8,8a,8b-hexahydro-8-α-methoxy-5-methyl-, [1aS-(1a-a, 8b, 8a-a, 8ba)]; MIT-C; MITO-C; Mitocin-C; Mitomycin; Mitomycin-C; Mitomycinum; MMC; Mutamycin; Mytomycin; NCI-C04706; NSC 26980

CAS Registry Number: 50-07-7

RTECS® Number: CN0700000

DOT ID and ERG Number: UN3249 (medicine, solid, toxic, n.o.s)/151

EC Number: 200-008-6

Regulatory Authority and Advisory Bodies

Carcinogenicity: IARC: Animal Sufficient Evidence; Human Inadequate Evidence, *possibly carcinogenic to humans*, Group 2B, 1987; NCI: Carcinogenesis Studies (ipr); clear evidence: rat; no evidence: mouse, 1975.

US EPA Hazardous Waste Number (RCRA No.): U010.

RCRA, 40CFR261, Appendix 8 Hazardous Constituents.

Superfund/EPCRA 40CFR355, Extremely Hazardous Substances: TPQ = 500/10,000 lb (227/4540 kg).

Reportable Quantity (RQ): 10 lb (4.54 kg).

California Proposition 65 Chemical: Cancer 4/1/88.

WGK (German Aquatic Hazard Class): 3—Severe hazard to waters.

Description: Mitomycin is a blue-violet crystalline solid. Molecular weight = 334.37; Freezing/Melting point = ≥ 360°C. Soluble in water.

Potential Exposure: This compound is an antitumor antibiotic complex. This drug is usually injected intravenously.

Incompatibilities: Keep away from heat.

Permissible Exposure Limits in Air

Protective Action Criteria (PAC)

TEEL-0: 4 mg/m^3

PAC-1: 12.5 mg/m^3

PAC-2: 23 mg/m^3

PAC-3: 23 mg/m^3

Routes of Entry: Inhalation. It is not known if this chemical penetrates the unbroken skin.

Harmful Effects and Symptoms

Short Term Exposure: Contact may irritate and damage the eyes. Toxic doses as low as 750 mg/kg have been reported in humans. The major toxic effect is myelosuppression, characterized by marked leukopenia and thrombocytopenia; this may be delayed and cumulative. Interstitial pneumonia and glomerular damage resulting in kidney failure are unusual but well-documented complications. Administration of mitomycin has been recognized as causing pneumonitis, alveolitis, and pulmonary fibrosis. Administration of mitomycin can cause kidney damage. Kidney toxicity was observed in 1−5% of patients. Depressed immune conditions were also noted. Headaches,

blurred vision, confusion, drowsiness, fatigue, diarrhea, and pain have been occasionally noted as symptoms of mitomycin exposure. These do not appear to be dose related by intravenous administration. Fever and anorexia occur in 15% of patients. Hair loss, sloughing of skin, and loss of feeling occur in approximately 4% of patients. Labored breathing, cough, and pneumonia occur in some cases. Renal toxicity is sometimes observed.

Long Term Exposure: Causes mutations. Causes cancer in animals. There is limited evidence that Mitomycin C is a teratogen in animals. Can damage the bone marrow and cause kidney damage.

Points of Attack: Blood, kidneys.

Medical Surveillance: Before beginning employment and at regular times after that, for those with frequent or potentially high exposures, the following are recommended: complete blood count and platelet count. Kidney function tests.

First Aid: If this chemical gets into the eyes, remove any contact lenses at once and irrigate immediately for at least 15 min, occasionally lifting upper and lower lids. Seek medical attention immediately. If this chemical contacts the skin, remove contaminated clothing and wash immediately with soap and water. Seek medical attention immediately. If this chemical has been inhaled, remove from exposure, begin rescue breathing (using universal precautions, including resuscitation mask) if breathing has stopped and CPR if heart action has stopped. Transfer promptly to a medical facility. When this chemical has been swallowed, get medical attention. Give large quantities of water and induce vomiting. Do not make an unconscious person vomit. Keep victim quiet and maintain normal body temperature.

Personal Protective Methods: Wear protective gloves and clothing to prevent any reasonable probability of skin contact. Safety equipment suppliers/manufacturers can provide recommendations on the most protective glove/clothing material for your operation. All protective clothing (suits, gloves, footwear, headgear) should be clean, available each day, and put on before work. Contact lenses should not be worn when working with this chemical. Wear dust-proof chemical goggles and face shield unless full face-piece respiratory protection is worn. Employees should wash immediately with soap when skin is wet or contaminated. Provide emergency showers and eyewash.

Respirator Selection: Where there is potential for exposure to mitomycin C, use a NIOSH/MSHA- or European Standard EN149-approved supplied-air respirator with a full face-piece operated in the positive-pressure mode, or with a full face-piece, hood, or helmet in the continuous-flow mode; or use a NIOSH/MSHA- or European Standard EN149-approved self-contained breathing apparatus with a full face-piece operated in pressure demand or other positive-pressure mode.

Storage: Color Code—Blue: Health Hazard/Poison: Store in a secure poison location. Prior to working with this chemical you should be trained on its proper handling and storage. Store in tightly closed containers in a cool, well-ventilated area away from heat (temperatures over 40°C). If you are required to work in a "sterile" environment, you require special training. A regulated, marked area should be established where this chemical is handled, used, or stored in compliance with OSHA Standard 1910.1045.

Shipping: Medicine, solid, toxic, n.o.s. requires a shipping label of "POISONOUS/TOXIC MATERIALS." It falls in Hazard Class 6.1.

Spill Handling: Evacuate persons not wearing protective equipment from area of spill or leak until cleanup is complete. Remove all ignition sources. Collect powdered material in the most convenient and safe manner and deposit in sealed containers. Ventilate area after cleanup is complete. It may be necessary to contain and dispose of this chemical as a hazardous waste. If material or contaminated runoff enters waterways, notify downstream users of potentially contaminated waters. Contact your local or federal environmental protection agency for specific recommendations. If employees are required to clean up spills, they must be properly trained and equipped. OSHA 1910.120(q) may be applicable.

Fire Extinguishing: This chemical is a combustible solid. Use dry chemical, carbon dioxide, water spray, or alcohol foam extinguishers. Poisonous gases, including oxides of nitrogen, are produced in fire. If material or contaminated runoff enters waterways, notify downstream users of potentially contaminated waters. Notify local health and fire officials and pollution control agencies. From a secure, explosion-proof location, use water spray to cool exposed containers. If cooling streams are ineffective (venting sound increases in volume and pitch, tank discolors, or shows any signs of deforming), withdraw immediately to a secure position. If employees are expected to fight fires, they must be trained and equipped in OSHA 1910.156. The only respirators recommended for firefighting are self-contained breathing apparatuses that have full face-pieces and are operated in a pressure-demand or other positive-pressure mode.

Disposal Method Suggested: Consult with environmental regulatory agencies for guidance on acceptable disposal practices. Generators of waste containing this contaminant ($\geq$100 kg/mo) must conform with EPA regulations governing storage, transportation, treatment, and waste disposal.

References

US Environmental Protection Agency. (November 30, 1987). *Chemical Hazard Information Profile: Mitomycin C.* Washington, DC: Chemical Emergency Preparedness Program

New Jersey Department of Health and Senior Services. (June 2000). *Hazardous Substances Fact Sheet: Mitomycin C.* Trenton, NJ

Molybdenum   M:1410

Molecular Formula: Mo
Synonyms: Elemental molybdenum; Molybdate; Molybdenum metal

CAS Registry Number: 7439-98-7
RTECS® Number: QA4680000 (elemental)
UN/NA & ERG Number: No citation for metal.
EC Number: 231-107-2
Regulatory Authority and Advisory Bodies
Air Pollutant Standard Set. See below, "Permissible Exposure Limits in Air" section.
Canada, WHMIS, Ingredients Disclosure List Concentration 1.0%.
European/International Regulations: not listed in Annex 1.
WGK (German Aquatic Hazard Class): Nonwater polluting agent.

Description: Molybdenum is a silvery-white metal or dark gray or black powder with a metallic luster. Molecular weight = 95.94; Boiling point = 4825°C; Freezing/Melting point = 2652°C. Hazard Identification (based on NFPA-704 M Rating System): Health 1, Flammability 0, Reactivity 0. Molybdenite is the only important commercial source. This ore is often associated with copper ore. Molybdenum is insoluble in water and soluble in hot concentrated nitric and sulfuric acid.

Potential Exposure: Compound Description: Mutagen; Reproductive Effector. Most of the molybdenum produced is used in alloys: steel, stainless steel; tool steel; case iron; steel mill rolls; manganese, nickel, chromium, and tungsten. The metal is used in electronic parts (contacts, spark plugs, X-ray tubes, filaments, screens, and grids for radios), induction heating elements, electrodes for glass melting, and metal spraying applications. Molybdenum compounds are utilized as lubricants; as pigments for printing inks; lacquers, paints, for coloring rubber animal fibers, leather, and as a mordant; as catalysts for hydrogenation cracking; alkylation, and reforming in the petroleum industry; in Fischer–Tropsch synthesis; in ammonia production; and in various oxidation–reduction and organic cracking reactions; as a coating for quartz glass; in vitreous enamels to increase adherence to steel; in fertilizers, particularly for legumes; in electroplating to form protective coatings; and in the production of tungsten. Hazardous exposures may occur during high-temperature treatment in the fabrication and production of molybdenum products, spraying applications, or through loss of catalyst. MoO_3 sublimes above 800°C.

Incompatibilities: *Soluble compounds:* alkali metals, sodium, potassium, molten magnesium. *Insoluble compounds:* Violent reaction with oxidizers, nitric acid, sulfuric acid. Forms explosive mixture with potassium nitrate. Metallic Mo is a combustible solid in the form of dust or powder and is potentially explosive.

Permissible Exposure Limits in Air
Mo, Metal and insoluble compounds
OSHA PEL: 15 mg[Mo]/m^3 total dust TWA. [*Note:* The PEL also applies to other insoluble molybdenum compounds (as Mo).]
NIOSH: See Appendix D of the *NIOSH Pocket Guide.*
ACGIH TLV®[1]: 10 mg[Mo]/m^3 inhalable fraction; 3 mg [Mo]/m^3 respirable fraction.

NIOSH IDLH: 5000 mg[Mo]/m^3.
Protective Action Criteria (PAC)
Elemental
TEEL-0: 15 mg/m^3
PAC-1: 15 mg/m^3
PAC-2: 15 mg/m^3
PAC-3: 500 mg/m^3
Mo, soluble compounds
OSHA PEL: 5 mg[Mo]/m^3 TWA.
NIOSH: See Appendix D of the *NIOSH Pocket Guide*.
ACGIH TLV®[1]: 0.5 mg[Mo]/m^3 respirable fraction TWA confirmed animal carcinogen with unknown relevance to humans.
DFG MAK: No numerical value established. Data may be available; testing for carcinogenic effects.
NIOSH IDLH: 1000 mg[Mo]/m^3.
Russia[43] established a MAC of 2 mg/m^3 for *soluble compounds, aerosol condensates*; 4 mg/m^3 for *soluble compounds as dusts*; MAC of 6 mg/m^3 for *insoluble compounds*. Several states have set guidelines or standards for molybdenum in ambient air[60] ranging from 100–200 µg/m^3 (Connecticut) to 119 µg/m^3 (Nevada) to 160 µg/m^3 (Virginia).
Determination in Air: Use NIOSH Analytical Method #7300, Elements by ICP (Nitric/perchloric acid ashing); #7301, Elements by ICP (Aqua regia ashing); #7303, Elements by ICP [Hot Block (HCl/HNO$_3$ Digestion)]; #9102, Elements on wipes; #8310, Metals in urine, #8005, Elements in blood or tissue; OSHA Analytical Methods ID-121 and ID-125G.
Permissible Concentration in Water: Russia[43] has established a molybdenum limit of 0.5 mg/L in bodies used for domestic purposes. EPA[32] has suggested a permissible ambient goal of 70 µg/L based on health effects.
Determination in Water: With atomic-absorption spectrophotometry, a detection limit of 20 µg/L is attainable by direct aspiration into the flame, necessitating concentration for ordinary determinations. When the graphite furnace is used to increase sample atomization, the detection limit is lowered to 0.5 µg/L or. Neutron activation may be used at even lower detection limits, according to US EPA.
Routes of Entry: Inhalation, ingestion, skin and/or eye contact.
Harmful Effects and Symptoms
Short Term Exposure: Contact irritates the skin and eyes. Inhalation can irritate the respiratory tract causing coughing and wheezing.
Long Term Exposure: Can cause headache, fatigue, loss of appetite, muscle and joint pain. Uric acid levels may be elevated which can lead to gout. May damage the liver and kidneys. May cause low blood count (anemia). In animals: irritation of eyes, nose, throat; anorexia, diarrhea, weight loss; listlessness; liver, kidney damage. Soluble compounds (e.g., sodium molybdate) and freshly generated molybdenum fumes are considerably more toxic. Inhalation of high concentrations of molybdenum trioxide dust is highly

irritating to animals and has caused weight loss, diarrhea, loss of muscular coordination, and a high mortality rate. Molybdenum trioxide dust is more toxic than the fumes. Large oral doses of ammonium molybdate in rabbits caused some fetal deformities.
Points of Attack: *Soluble compounds:* Eyes, respiratory system, kidneys, blood. *Insoluble compounds:* Eyes, respiratory system, liver, kidneys.
Medical Surveillance: NIOSH lists the following tests for molybdenum and compounds: whole blood (chemical/metabolite); biologic tissue/biopsy; urine (chemical/metabolite). Preemployment and periodic physical examinations should evaluate any irritant effects to the eyes or respiratory tract and the general health of the worker. Liver and kidney function tests. Complete blood count (CBC). Uric acid level. Molybdenum is considered to be an essential trace element in many species, including humans. However, excessive intake of molybdenum may produce signs of copper deficiency. The normal intake of copper in the diet may be sufficient to combat or prevent systemic toxic effects due to molybdenum poisoning.
First Aid: If this chemical gets into the eyes, remove any contact lenses at once and irrigate immediately for at least 15 min, occasionally lifting upper and lower lids. Seek medical attention immediately. If this chemical contacts the skin, remove contaminated clothing and wash immediately with soap and water. Seek medical attention immediately. If this chemical has been inhaled, remove from exposure, begin rescue breathing (using universal precautions, including resuscitation mask) if breathing has stopped and CPR if heart action has stopped. Transfer promptly to a medical facility. When this chemical has been swallowed, get medical attention. Give large quantities of water and induce vomiting. Do not make an unconscious person vomit.
Personal Protective Methods: Wear protective gloves and clothing to prevent any reasonable probability of skin contact. Safety equipment suppliers/manufacturers can provide recommendations on the most protective glove/clothing material for your operation. All protective clothing (suits, gloves, footwear, headgear) should be clean, available each day, and put on before work. Contact lenses should not be worn when working with this chemical. Wear eye protection (chemical goggles and face shield unless full face-piece respiratory protection is worn). Employees should wash immediately with soap when skin is wet or contaminated.
Respirator Selection: *For Insoluble Compounds:* OSHA *Up to 75 mg/m^3:* Qm* (APF = 25) (any quarter-mask respirator). *Up to 150 mg/m^3:* Any particulate respirator equipped with an N95, R95, or P95 filter (including N95, R95, and P95 filtering face-pieces) except quarter-mask respirators. The following filters may also be used: N99, R99, P99, N100, R100, P100; or Sa (APF = 10) (any supplied-air respirator). *Up to 375 mg/m^3:* Sa:Cf (APF = 25) (any supplied-air respirator operated in a continuous-flow mode) or PaprHie* (APF = 25) (any powered, air-purifying respirator with a high-efficiency particulate filter). *Up to 750 mg/m^3:*

100F (APF = 50) (any air-purifying, full-face-piece respirator with an N100, R100, or P100 filter) or SaT: Cf (APF = 50) (any supplied-air respirator that has a tight-fitting face-piece and is operated in a continuous-flow mode) or PaprTHie (APF = 50) (any powered, air-purifying respirator with a tight-fitting face-piece and a high-efficiency particulate filter) or SCBAF (APF = 50) (any self-contained breathing apparatus with a full face-piece) or SaF (APF = 50) (any supplied-air respirator with a full face-piece). *Up to 5000 mg/m³:* Sa: Pd,Pp (APF = 1000) (any supplied-air respirator operated in a pressure-demand or other positive-pressure mode). *Emergency or planned entry into unknown concentrations or IDLH conditions:* SCBAF: Pd,Pp (APF = 10,000) (any self-contained breathing apparatus that has a full faceplate and is operated in a pressure-demand or other positive-pressure mode) or SaF: Pd,Pp: ASCBA (any supplied-air respirator that has a full face-piece and is operated in a pressure-demand or other positive-pressure mode in combination with an auxiliary, self-contained breathing apparatus operated in a pressure-demand or other positive-pressure mode). *Escape:* 100F (APF = 50) (any air-purifying, full-face-piece respirator with an N100, R100, or P100 filter) or SCBAE (any appropriate escape-type, self-contained breathing apparatus).
*If not present as a fume.
For Soluble Compounds: *25 mg/m³:* Qm* (APF = 25) (any quarter-mask respirator). *50 mg/m³:* Any particulate respirator equipped with an N95, R95, or P95 filter (including N95, R95, and P95 filtering face-pieces) except quarter-mask respirators. The following filters may also be used: N99, R99, P99, N100, R100, P100*; or Sa* (APF = 10) (any supplied-air respirator). *125 mg/m³:* Sa:Cf* (APF = 25) (any supplied-air respirator operated in a continuous-flow mode) or PAPRDM, if not present as a fume (any powered, air-purifying respirator with a dust and mist filter).* *250 mg/m³:* 100F (APF = 50) (any air-purifying, full-face-piece respirator with an N100, R100, or P100 filter) or SaT: Cf (APF = 50) (any supplied-air respirator that has a tight-fitting face-piece and is operated in a continuous-flow mode) or PaprTHie* (APF = 50) (any powered, air-purifying respirator with a tight-fitting face-piece and a high-efficiency particulate filter); or SCBAF (APF = 50) (any self-contained breathing apparatus with a full face-piece) or SaF (APF = 50) (any supplied-air respirator with a full face-piece). *1000 mg/m³:* SaF: Pd,Pp (APF = 2000) (any supplied-air respirator that has a full face-piece and is operated in a pressure-demand or other positive-pressure mode). *Emergency or planned entry into unknown concentrations or IDLH conditions:* SCBAF: Pd,Pp (APF = 10,000) (any self-contained breathing apparatus that has a full faceplate and is operated in a pressure-demand or other positive-pressure mode) or SaF: Pd,Pp: ASCBA (APF = 10,000) (any supplied-air respirator that has a full face-piece and is operated in a pressure-demand or other positive-pressure mode in combination with an auxiliary, self-contained breathing apparatus operated in a pressure-

demand or other positive-pressure mode). *Escape:* 100F (APF = 50) (any air-purifying, full-face-piece respirator with an N100, R100, or P100 filter) or SCBAE (any appropriate escape-type, self-contained breathing apparatus).
*Substance reported to cause eye irritation or damage; may require eye protection.

Storage: Color Code—Green: General storage may be used. Prior to working with this chemical you should be trained on its proper handling and storage. Molybdenum must be stored to avoid contact with strong oxidizers (such as chlorine, bromine, and fluorine) since violent reactions occur. Store in tightly closed containers in a cool, well-ventilated area away from bromine, trifluoride, fluorine, chlorine trifluoride, and lead dioxide.

Spill Handling: Evacuate persons not wearing protective equipment from area of spill or leak until cleanup is complete. Remove all ignition sources. Use HEPA vacuum or wet method to reduce dust during cleanup. Do not dry sweep. Collect powdered material in the most convenient and safe manner and deposit in sealed containers. Ventilate area after cleanup is complete. It may be necessary to contain and dispose of this chemical as a hazardous waste. If material or contaminated runoff enters waterways, notify downstream users of potentially contaminated waters. Contact your local or federal environmental protection agency for specific recommendations. If employees are required to clean up spills, they must be properly trained and equipped. OSHA 1910.120(q) may be applicable.

Fire Extinguishing: Use dry chemicals appropriate for extinguishing metal fires, such as dry sand, dolomite, or graphite. *Do not use water.* Poisonous gases are produced in fire. Dust or powdered molybdenum may cause a dust explosion. If material or contaminated runoff enters waterways, notify downstream users of potentially contaminated waters. Notify local health and fire officials and pollution control agencies. From a secure, explosion-proof location, use water spray to cool exposed containers. If cooling streams are ineffective (venting sound increases in volume and pitch, tank discolors, or shows any signs of deforming), withdraw immediately to a secure position. If employees are expected to fight fires, they must be trained and equipped in OSHA 1910.156. The only respirators recommended for firefighting are self-contained breathing apparatuses that have full face-pieces and are operated in a pressure-demand or other positive-pressure mode.

Disposal Method Suggested: Recovery is indicated whenever possible. Processes for recovery of Molybdenum from scrap, flue dusts, spent catalysts, and other industrial wastes have been developed.

References

US Environmental Protection Agency. (November 1975). *Molybdenum: A Toxicological Appraisal*, Report EPA-600/1-75-004. Research Triangle Park, NC: Health Effects Research Laboratory

US Environmental Protection Agency. (May 1977). *Toxicology of Metals, Vol. II: Molybdenum*, Report EPA-600/1-77-022. Research Triangle Park, NC, pp. 345—357

New Jersey Department of Health and Senior Services. (November 1999). *Hazardous Substances Fact Sheet: Molybdenum*. Trenton, NJ

Molybdenum trioxide   M:1420

Molecular Formula: MoO_3

Synonyms: MO 1202T; Molybdena; Molybdenum oxide; Molybdenum(VI) oxide; Molybdenum oxide (MoO_3); Molybdenum(VI) oxide; Molybdenum(VI) trioxide; Molybdic acid anhydride; Molybdic anhydride; Trioxido de molibdeno (Spanish)

CAS Registry Number: 1313-27-5

RTECS® Number: QA4725000

DOT ID and ERG Number: UN2811 (toxic solid, organic, n.o.s.)/154

EC Number: 215-204-7 [*Annex I Index No.:* 042-001-00-9]

Regulatory Authority and Advisory Bodies

Air Pollutant Standard Set. See below, "Permissible Exposure Limits in Air" section.

EPCRA Section 313 Form R *de minimis* concentration reporting level: 1.0%.

Safe Drinking Water Act: Priority List (55 FR 1470) as molybdenum.

Canada, WHMIS, Ingredients Disclosure List Concentration 1.0%.

European/International Regulations: Hazard Symbol: Xn; Risk phrases: R36/37; R40; Safety phrases: S2; S22; S36/37 (see Appendix 4).

WGK (German Aquatic Hazard Class): 1—Low hazard to waters.

Description: Molybdenum trioxide is an odorless, white crystalline powder that turns yellow when heated; Freezing/Melting point = 795°C. Hazard Identification (based on NFPA-704 M Rating System): Health 2, Flammability 0, Reactivity 0. Slightly soluble in water.

Potential Exposure: Molybdenum trioxide is used in agriculture; in the manufacture of metallic molybdenum, ceramic glazes, enamels, pigments, and in analytical chemistry.

Incompatibilities: Explodes on contact with molten magnesium. Violent reaction with strong oxidizers, such as chlorine trifluoride, bromine pentafluoride. Not compatible with strong acids, active metals (sodium, potassium, lithium).

Permissible Exposure Limits in Air

OSHA PEL: 5 mg[Mo]/m³ TWA *soluble compounds*.

NIOSH: See Appendix D of the *NIOSH Pocket Guide*.

ACGIH TLV®[1]: 0.5 mg[Mo]/m³ TWA *soluble compounds*, confirmed animal carcinogen with unknown relevance to humans.

NIOSH IDLH: 1000 mg[Mo]/m³.

Protective Action Criteria (PAC)

TEEL-0: 0.75 mg/m³

PAC-1: 0.75 mg/m³

PAC-2: 0.75 mg/m³

PAC-3: 500 mg/m³

DFG MAK: Carcinogen Category 3B.

Determination in Air: Use NIOSH Analytical Method #7300, Elements by ICP (Nitric/perchloric acid ashing); #7301, Elements by ICP (Aqua regia ashing); #7303, Elements by ICP [Hot block (HCl/HNO₃ Digestion)]; #9102, Elements on wipes; #8310, Metals in urine, #8005, Elements in blood or tissue; OSHA Analytical Methods ID-121 and ID-125G.

Permissible Concentration in Water: No specific values set for MoO_3; see this section in the entry on "Molybdenum and Compounds."

Routes of Entry: Inhalation, ingestion, skin and/or eye contact.

Harmful Effects and Symptoms

Short Term Exposure: Irritation of the skin and eyes. Dust or mist can irritate the respiratory tract causing cough and/or tightness in the chest. Can cause anorexia, weight loss, headache, muscle and joint aches, listlessness, hair loss, lack of muscular coordination, diarrhea, liver and kidney damage.

Long Term Exposure: Can irritate the lungs; bronchitis may develop. May cause anemia. May affect the liver and kidneys.

Points of Attack: Lungs, blood, liver and kidneys.

Medical Surveillance: NIOSH lists the following tests for molybdenum and compounds: whole blood (chemical/metabolite); biologic tissue/biopsy; urine (chemical/metabolite). If symptoms develop or overexposure is suspected, the following may be useful: tests for liver and kidney function, complete blood count; uric acid level; lung function tests.

First Aid: If this chemical gets into the eyes, remove any contact lenses at once and irrigate immediately for at least 15 min, occasionally lifting upper and lower lids. Seek medical attention immediately. If this chemical contacts the skin, remove contaminated clothing and wash immediately with soap and water. Seek medical attention immediately. If this chemical has been inhaled, remove from exposure, begin rescue breathing (using universal precautions, including resuscitation mask) if breathing has stopped and CPR if heart action has stopped. Transfer promptly to a medical facility. When this chemical has been swallowed, get medical attention. Give large quantities of water and induce vomiting. Do not make an unconscious person vomit.

Personal Protective Methods: Wear protective gloves and clothing to prevent any reasonable probability of skin contact. Safety equipment suppliers/manufacturers can provide recommendations on the most protective glove/clothing material for your operation. All protective clothing (suits, gloves, footwear, headgear) should be clean, available each day, and put on before work. Contact lenses should not be worn when working with this chemical. Wear dust-proof chemical goggles and face shield unless full face-piece

respiratory protection is worn. Employees should wash immediately with soap when skin is wet or contaminated. Provide emergency showers and eyewash.

Respirator Selection: *For Soluble Compounds: 25 mg/m³:* Qm (APF = 25) (any quarter-mask respirator).* *50 mg/m³:* Any particulate respirator equipped with an N95, R95, or P95 filter (including N95, R95, and P95 filtering face-pieces) except quarter-mask respirators. The following filters may also be used: N99, R99, P99, N100, R100, P100*; or Sa (APF = 10) (any supplied-air respirator).* *125 mg/m³:* Sa:Cf (APF = 25) (any supplied-air respirator operated in a continuous-flow mode)*; or PAPRDM, if not present as a fume (any powered, air-purifying respirator with a dust and mist filter).* *250 mg/m³:* 100F (APF = 50) (any air-purifying, full-face-piece respirator with an N100, R100, or P100 filter) or SaT: Cf (APF = 50) (any supplied-air respirator that has a tight-fitting face-piece and is operated in a continuous-flow mode) or PaprTHie (APF = 50) (any powered, air-purifying respirator with a tight-fitting face-piece and a high-efficiency particulate filter)*; or SCBAF (APF = 50) (any self-contained breathing apparatus with a full face-piece) or SaF (APF = 50) (any supplied-air respirator with a full face-piece). *1000 mg/m³:* SaF: Pd,Pp (APF = 2000) (any supplied-air respirator that has a full face-piece and is operated in a pressure-demand or other positive-pressure mode). *Emergency or planned entry into unknown concentrations or IDLH conditions:* SCBAF: Pd,Pp (APF = 10,000) (any self-contained breathing apparatus that has a full faceplate and is operated in a pressure-demand or other positive-pressure mode) or SaF: Pd,Pp: ASCBA (APF = 10,000) (any supplied-air respirator that has a full face-piece and is operated in a pressure-demand or other positive-pressure mode in combination with an auxiliary, self-contained breathing apparatus operated in a pressure-demand or other positive-pressure mode). *Escape:* 100F (APF = 50) (any air-purifying, full-face-piece respirator with an N100, R100, or P100 filter) or SCBAE (any appropriate escape-type, self-contained breathing apparatus).
*Substance reported to cause eye irritation or damage; may require eye protection.

Storage: Color Code—Blue: Health Hazard/Poison: Store in a secure poison location. Prior to working with this chemical you should be trained on its proper handling and storage. Molybdenum trioxide must be stored to avoid contact with strong acids (such as hydrochloric, sulfuric, and nitric); alkalis, sodium, potassium, and molten magnesium since violent reactions occur. Store in tightly closed containers in a cool, well-ventilated area.

Shipping: Toxic solids, organic, n.o.s. compound require a shipping label of "POISONOUS/TOXIC MATERIALS." It falls in Hazard Class 6.1, Packing Group III.

Spill Handling: Evacuate persons not wearing protective equipment from area of spill or leak until cleanup is complete. Remove all ignition sources. Use HEPA vacuum or wet method to reduce dust during cleanup. Do not dry sweep. Collect powdered material in the most convenient and safe manner and deposit in sealed containers. Ventilate area after cleanup is complete. It may be necessary to contain and dispose of this chemical as a hazardous waste. If material or contaminated runoff enters waterways, notify downstream users of potentially contaminated waters. Contact your local or federal environmental protection agency for specific recommendations. If employees are required to clean up spills, they must be properly trained and equipped. OSHA 1910.120(q) may be applicable.

Fire Extinguishing: Extinguish fire using an agent suitable for type of surrounding fire. Molybdenum trioxide itself does not burn. Poisonous gases are produced in fire. If material or contaminated runoff enters waterways, notify downstream users of potentially contaminated waters. Notify local health and fire officials and pollution control agencies. From a secure, explosion-proof location, use water spray to cool exposed containers. If cooling streams are ineffective (venting sound increases in volume and pitch, tank discolors, or shows any signs of deforming), withdraw immediately to a secure position. If employees are expected to fight fires, they must be trained and equipped in OSHA 1910.156. The only respirators recommended for firefighting are self-contained breathing apparatuses that have full face-pieces and are operated in a pressure-demand or other positive-pressure mode.

References

Sax, N. I. (Ed.). (1988). *Dangerous Properties of Industrial Materials Report*, 8, No. 3, 73−78
New Jersey Department of Health and Senior Services. (July 1999). *Hazardous Substances Fact Sheet: Molybdenum Trioxide*. Trenton, NJ

Monocrotophos   M:1430

Molecular Formula: $C_7H_{14}NO_5P$
Common Formula: $(CH_3O)_2P(O)OC(CH_3)=CHCONHCH_3$
Synonyms: Apadrin; Azodrin; Biloborn; Bilobran; C 1414; Ciba 1414; Crisodin; Crisodrin; Crotonamide, 3-hydroxy-*N*-methyl-, dimethylphosphate, (*E*)-; Crotonamide, 3-Hydroxy-*N*-methyl-, dimethylphosphate, *cis*-; 3-(Dimethoxyphosphinyloxy)-*N*-methyl-*cis*-crotonamide; *O,O*-Dimethyl-*O*-(2-*N*-methyl-carbamoyl-1-methyl)-vinyl-phosphat (German); *O,O*-Dimethyl *O*-(2-*N*-methylcarbamoyl-1-methylvinyl) phosphate; (*E*)-Dimethyl 1-methyl-3-(methylamino)-3-oxo-1-propenyl phosphate; Dimethyl 1-methyl-2-(methylcarbamoyl)vinyl phosphate, *cis*-; Dimethyl phosphate ester of 3-hydroxy-*N*-methyl-*cis*-crotonamide; Dimethyl phosphate of 3-hydroxy-*N*-methyl-*cis*-crotonamine; ENT 27,129; Glore Phos 36; 3-Hydroxy-*N*-methyl-*cis*-crotonamide dimethyl phosphate; 3-Hydroxy-*N*-methylcrotonamide dimethyl phosphate; *cis*-1-Methyl-2-methyl carbamoyl vinyl phosphate; Monocron; Monocrotofos (Spanish); Monodrin; Nuvacron; Phosphate de dimethyle et de 2-methylcarbamoyl 1-methyl vinyle (French); Phosphoric acid, dimethyl

ester, with *cis*-3-hydroxy-*N*-methylcrotonamide; Pillardin; Plantdrin; SD 9129; Shell SD 9129; Susvin; Ulvair

CAS Registry Number: 6923-22-4

RTECS® Number: TC4375000

UN/NA & ERG Number: UN2783 (organophosphorus pesticides, solid, toxic)/152

EC Number: 230-042-7 [*Annex I Index No.:* 015-072-00-9]

Regulatory Authority and Advisory Bodies

US EPA Gene-Tox Program, Positive: *In vitro* UDS—human fibroblast; *S. cerevisiae*—homozygosis; Negative: *D. melanogaster* sex-linked lethal; TRP reversion; Inconclusive: *B. subtilis* rec assay; *E. coli* polA without S9; Inconclusive: Histidine reversion—Ames test.

US EPA, FIFRA 1998 Status of Pesticides: Canceled.

Air Pollutant Standard Set. See below, "Permissible Exposure Limits in Air" section.

Superfund/EPCRA 40CFR355, Extremely Hazardous Substances: TPQ = 10/10,000 lb (4.54/4540 kg).

Reportable Quantity (RQ): 10 lb (4.54 kg).

US DOT Regulated Marine Pollutant (49CFR172.101, Appendix B).

US DOT 49CFR172.101, Inhalation Hazard Chemical as organophosphates.

Rotterdam Convention Annex III [Chemicals Subject to the Prior Informed Consent Procedure (PIC)] [monocrotophos and dustable powder formulations containing a combination of monocrotophos (solid pesticide and soluble liquid formulations of the substance that exceed 600 g active ingredient/L)].

European/International Regulations: Hazard Symbol: T+, N; Risk phrases: R24; R28; R50/53; R68; Safety phrases: S1/2; S36/37; SS45-S60; S61 (see Appendix 4).

WGK (German Aquatic Hazard Class): No value assigned.

Description: Monocrotophos is a colorless to reddish-brown solid with a mild ester odor. Molecular weight = 223.19; Specific gravity (H$_2$O:1) = 1.3; Boiling point = 125°C; Freezing/Melting point = 53.9°C (pure); 25–30°C (the reddish brown technical product); Vapor pressure = 7×10^{-6} mm at 20°C; Flash point = >93°C. Hazard Identification (based on NFPA-704 M Rating System): Health 3, Flammability 1, Reactivity 0. Soluble in water. Commercially available as a water-miscible solution.

Potential Exposure: The liquid form is a severely hazardous pesticide formulation. Those involved in the manufacture, formulation, and application of this insecticide.

Incompatibilities: Alkaline pesticides. Attacks black iron, drum steel, stainless steel, brass.

Permissible Exposure Limits in Air

OSHA PEL: None.

NIOSH REL: 0.25 mg/m^3 TWA.

ACGIH TLV®[1]: 0.05 mg/m^3 TWA, inhalable fraction and vapor; [skin], not classifiable as a human carcinogen; BEI$_A$ releases as Acetylcholinesterase-inhibiting pesticides.

Protective Action Criteria (PAC)

TEEL-0: 0.05 mg/m^3

PAC-1: 0.15 mg/m^3

PAC-2: 0.63 mg/m^3

PAC-3: 25 mg/m^3

Several states have set guidelines or standards for monocrotophos in ambient air[60] ranging from 2.5 μg/m^3 (North Dakota) to 40 μg/m^3 (Virginia) to 5.0 μg/m^3 (Connecticut) to 6.0 μg/m^3 (Nevada). Australia: TWA 0.25 mg/m^3, 1993; Belgium: TWA 0.25 mg/m^3, 1993; Denmark: TWA 0.25 mg/m^3, 1999; France: VME 0.25 mg/m^3 1999; Switzerland: MAK-W 0.25 mg/m^3, 1999; the Netherlands: MAC-TGG 0.25 mg/m^3, 2003; Argentina, Bulgaria, Columbia, Jordan, South Korea, New Zealand, Singapore, Vietnam: ACGIH TLV®: not classifiable as a human carcinogen.

Determination in Air: OSHA versatile sampler-2; Toluene/Acetone; Gas chromatography/Flame photometric detection for sulfur, nitrogen, or phosphorus; NIOSH Analytical Method (IV) #5600, Organophosphorus Pesticides; OSHA Analytical Method PV-2045, Monocrotophos.

Determination in Water: Fish Tox = 728.00039000 ppb (VERY LOW).

Routes of Entry: Inhalation, ingestion, skin contact.

Harmful Effects and Symptoms

Short Term Exposure: Monocrotophos is a highly toxic, direct acting, water-soluble cholinesterase inhibitor which appears to be capable of penetrating through the skin but is excreted rapidly and does not accumulate in the body. Acute exposure to monocrotophos may result in the following signs and symptoms: pinpoint pupils; blurred vision; headache, dizziness, muscle spasms; and profound weakness. Vomiting, diarrhea, abdominal pain, seizures, and coma may also occur. The heart rate may decrease following oral exposure or increase following dermal exposure. Hypotension (low blood pressure) may occur although hypertension (high blood pressure) is not uncommon. Chest pain may be noted. Dyspnea (shortness of breath) may lead to respiratory collapse. Giddiness is common. Monocrotophos acts on the nervous system. Extremely toxic; probable oral lethal dose to humans 5–50 mg/kg or between 7 drops and 1 teaspoon for a 70-kg (150 lb) person. Repeated inhalation or skin contact with this material may, without symptoms, progressively increase susceptibility to poisoning. Monocrotophos may cause mutations. Handle with extreme caution. In animals: possible teratogenic effects. Human Tox = 0.35000 ppb (EXTRA HIGH).

Long Term Exposure: Cholinesterase inhibitor; cumulative effect is possible. This chemical may damage the nervous system with repeated exposure, resulting in convulsions, respiratory failure. May cause liver damage. May cause personality changes with depression, anxiety, irritability.

Points of Attack: Eyes, respiratory system, central nervous system, cardiovascular system, blood cholinesterase, reproductive system.

Medical Surveillance: Before employment and at regular times after that, the following are recommended: plasma and red blood cell cholinesterase levels (tests for the enzyme poisoned by this chemical). If exposure stops, plasma levels return to normal in 1–2 weeks while red blood cell levels may be reduced for 1–3 months. When cholinesterase enzyme levels are reduced by 25% or more below preemployment levels, risk of poisoning is increased, even if results are in lower ranges of "normal." Reassignment to work not involving organophosphate or carbamate pesticides is recommended until enzyme levels recover. If symptoms develop or overexposure occurs, repeat the above tests as soon as possible and get an examination of the nervous system. Also, consider complete blood count. Consider chest X-ray following acute overexposure. Do not drink any alcoholic beverages before or during use. Alcohol promotes absorption of organic phosphates.

First Aid: If this chemical gets into the eyes, remove any contact lenses at once and irrigate immediately for at least 15 min, occasionally lifting upper and lower lids. Seek medical attention immediately. If this chemical contacts the skin, remove contaminated clothing and wash immediately with soap and water. Speed in removing material from skin is of extreme importance. Shampoo hair promptly if contaminated. Seek medical attention immediately. If this chemical has been inhaled, remove from exposure, begin rescue breathing (using universal precautions, including resuscitation mask) if breathing has stopped and CPR if heart action has stopped. Transfer promptly to a medical facility. When this chemical has been swallowed, get medical attention. Give large quantities of water and induce vomiting. Do not make an unconscious person vomit. Obtain authorization and/or further instructions from the local hospital for administration of an antidote or performance of other invasive procedures.

Personal Protective Methods: Wear protective gloves and clothing to prevent any reasonable probability of skin contact. Safety equipment suppliers/manufacturers can provide recommendations on the most protective glove/clothing material for your operation. All protective clothing (suits, gloves, footwear, headgear) should be clean, available each day, and put on before work. Contact lenses should not be worn when working with this chemical. Wear splash-proof chemical goggles and face shield when working with liquid, or wear dust-proof goggles and face shield when working with powders or dust unless full face-piece respiratory protection is worn. Employees should wash immediately with soap when skin is wet or contaminated. Provide emergency showers and eyewash.

Respirator Selection: Follow regulations in OSHA 29CFR1910.134 or European Standard EN149. Use a NIOSH/MSHA- or European Standard EN149-approved respirator; or use an approved supplied-air respirator with a full face-piece operated in the positive-pressure mode, or with a full face-piece, hood, or helmet in the continuous-flow mode; or use a NIOSH/MSHA- or European Standard EN149-approved self-contained breathing apparatus with a full face-piece operated in a pressure-demand or other positive-pressure mode.

Storage: Color Code—Blue: Health Hazard/Poison: Store in a secure poison location. Prior to working with this chemical you should be trained on its proper handling and storage. Store in tightly closed containers in a cool, well-ventilated area. Sources of ignition, such as smoking and open flames, are prohibited where this chemical is handled, used, or stored. Metal containers involving the transfer of 5 gallons or more of this chemical should be grounded and bonded. Drums must be equipped with self-closing valves, pressure vacuum bungs, and flame arresters. Use only non-sparking tools and equipment, especially when opening and closing containers of this chemical. Wherever this chemical is used, handled, manufactured, or stored, use explosion-proof electrical equipment and fittings.

Shipping: Organophosphorus pesticides, solid, toxic, n.o.s. require a shipping label of "POISONOUS/TOXIC MATERIALS." It falls in Hazard Class 6.1 and Packing Group II.

Spill Handling: Evacuate persons not wearing protective equipment from area of spill or leak until cleanup is complete. Remove all ignition sources. As with other organophosphorus pesticides, do not touch spilled material; stop leak if you can do it without risk. Use water spray to reduce vapors. *Small spills*: take up with vermiculite, dry sand, earth, or other noncombustible absorbent material and place into containers for later disposal. *Small dry spills*: with clean shovel place material into clean, dry container and cover; move containers from spill area. Use HEPA vacuum or wet method to reduce dust during cleanup. Do not dry sweep. *Large spills*: dike far ahead of spill for later disposal. Keep unnecessary people away; isolate hazard area and deny entry. Stay upwind; keep out of low areas. Ventilate area after cleanup is complete. It may be necessary to contain and dispose of this chemical as a hazardous waste. If material or contaminated runoff enters waterways, notify downstream users of potentially contaminated waters. Contact your local or federal environmental protection agency for specific recommendations. If employees are required to clean up spills, they must be properly trained and equipped. OSHA 1910.120(q) may be applicable.

Fire Extinguishing: This chemical is a combustible liquid or solid. Like other organophosphorus pesticides, extinguish with dry chemical, carbon dioxide, water spray, fog, or foam. Move container from fire area if you can do so without risk. Fight fire from maximum distance. Dike fire control water for later disposal; do not scatter the material. Wear positive pressure breathing apparatus and full protective clothing. Poisonous gases, including nitrogen oxides and phosphorus oxides, are produced in fire. If material or contaminated runoff enters waterways, notify downstream users of potentially contaminated waters. Notify local health and fire officials and pollution control agencies. Container may explode in heat of fire. From a secure, explosion-proof location, use water spray to cool exposed containers. If cooling streams are ineffective (venting sound increases in volume and pitch, tank discolors, or shows any signs of deforming), withdraw immediately to a secure position. If employees are expected to fight fires, they must be trained and equipped in OSHA 1910.156. The only respirators recommended for firefighting are self-contained breathing apparatuses that have full face-pieces and are operated in a pressure-demand or other positive-pressure mode.

Disposal Method Suggested: Small amounts may be hydrolyzed with water.[22] Incineration in a unit with effluent gas scrubbing is recommended for larger amounts. In accordance with 40CFR165, follow recommendations for the disposal of pesticides and pesticide containers. Must be disposed properly by following package label directions or by contacting your local or federal environmental control agency or by contacting your regional EPA office.

References

US Environmental Protection Agency. (November 30, 1987). *Chemical Hazard Information Profile: Monocrotophos*. Washington, DC: Chemical Emergency Preparedness Program

US Environmental Protection Agency, Special Review and Reregistration Division Office of Pesticide Programs. (1998). *Agency Status of Pesticides in Registration, Reregistration, and Special Review* (Rainbow Report). Washington, DC

New Jersey Department of Health and Senior Services. (November 1999). *Hazardous Substances Fact Sheet: Monocrotophos*. Trenton, NJ

Morpholine   M:1440

Molecular Formula: C_4H_9NO

Synonyms: Diethyleneimide oxide; Diethylene imidoxide; Diethylene oximide; N,N-Dimethylacetamide; Diethylenimide oxide; *p*-Isoxazine, tetrahydro-; 1-Oxa-4-azacyclohexane; 2H-1,4-Oxazine, tetrahydro-; Tetrahydro-1,4-isoxazine; Tetrahydro-1,4-oxazine; Tetrahydro-2H-1,4-oxazine; Tetrahydro-*p*-oxazine

CAS Registry Number: 110-91-8

RTECS® Number: QD6475000

UN/NA & ERG Number: UN2054/132

EC Number: 203-815-1 [*Annex I Index No.:* 613-028-00-9]

Regulatory Authority and Advisory Bodies

Carcinogenicity: IARC: Human, No Adequate Data; Animal, Inadequate Evidence, *not classifiable as carcinogenic to humans*, Group 3, 1999.

US EPA Gene-Tox Program, Positive: Host-mediated assay. Air Pollutant Standard Set. See below, "Permissible Exposure Limits in Air" section.

Canada, WHMIS, Ingredients Disclosure List Concentration 1.0%.

European/International Regulations: Hazard Symbol: C; Risk phrases: R10; R20/21/22; R34; Safety phrases: S1/2; S23; S36; S45 (see Appendix 4).

WGK (German Aquatic Hazard Class): 2—Hazard to waters.

Description: Morpholine is a colorless liquid with a weak ammonia or fish-like odor. The odor threshold is 0.01 ppm. Molecular weight = 87.14; Specific gravity (H_2O:1) = 1.007; Boiling point = 128.9°C; Freezing/Melting point = −5°C; Vapor pressure = 6 mmHg at 20°C; Flash point = 37°C. Autoignition temperature = 310°C. Explosive limits: LEL = 1.4%; UEL = 11.2%. Hazard Identification (based on NFPA-704 M Rating System): Health 3, Flammability 3, Reactivity 0. Soluble in water.

Potential Exposure: Compound Description: Tumorigen, Mutagen. Primary Irritant. Morpholine is used as a separating agent for volatile amines; an intermediate for textile lubricants; in the synthesis of rubber accelerators and pharmaceuticals. It is also used as a solvent; as a boiler water additive; and in the formulation of waxes, polishers, and cleaners.

Incompatibilities: Strong acids, strong oxidizers, metals, nitro compounds. Corrosive to metals; attacks copper and its compounds.

Permissible Exposure Limits in Air

Conversion factor: 1 ppm = 3.56 mg/m³ at 25°C & 1 atm.

OSHA PEL: 20 ppm/70 mg/m³ TWA [skin].

NIOSH REL: 20 ppm/70 mg/m³ TWA; 30 ppm/105 mg/m³ STEL [skin].

ACGIH TLV®[1] 20 ppm/71 mg/m³ TWA [skin].

DFG MAK: 10 ppm/36 mg/m³ TWA; Peak Limitation Category I(2); [skin]; Pregnancy Risk Group D.

NIOSH IDLH: 1400 ppm [LEL].

Protective Action Criteria (PAC)

TEEL-0: 20 ppm

PAC-1: 30 ppm

PAC-2: 30 ppm

PAC-3: 1400 ppm

Australia: TWA 20 ppm (70 mg/m³); STEL 30 ppm, [skin], 1993; Austria: MAK 20 ppm (70 mg/m³), [skin], 1999; Belgium: TWA 20 ppm (71 mg/m³); STEL 30 ppm, [skin] 1999; Denmark: TWA 20 ppm (70 mg/m³), [skin], 1999; Finland: TWA 20 ppm (70 mg/m³); STEL 30 ppm (105 mg/m³), [skin], 1999; France: VME 20 ppm (70 mg/m³), VLE 30 ppm (105 mg/m³), 1999; Hungary: STEL 10 mg/m³, [skin], 1993; Norway: TWA 20 ppm (70 mg/m³), 1999; the Netherlands: MAC-TGG 36 mg/m³, [skin], 2003; the Philippines: TWA 20 ppm (70 mg/m³), [skin], 1993; Poland: MAC (TWA) 70 mg/m³; MAC (STEL) 100 mg/m³, 1999; Russia: TWA 0.5 mg/m³; STEL 1.5 mg/m³, [skin], 1993; Sweden: NGV 20 ppm (70 mg/m³), KTV 30 ppm (110 mg/m³), [skin], 1999; Switzerland: MAK-W 20 ppm (70 mg/m³), KZG-W 40 ppm (140 mg/m³), [skin], 1999; United Kingdom: TWA 20 ppm (72 mg/m³); STEL 30 ppm, [skin], 2000; Argentina, Bulgaria, Columbia, Jordan, South Korea, New Zealand, Singapore, Vietnam: ACGIH TLV®: not classifiable as a human carcinogen. Russia set a MAC of 0.01 mg/m³ for ambient air in residential areas (10 μg/m³) on a once-daily basis. Several states have set guidelines or standards for morpholine in ambient air[60] ranging from 0.7–1.05 mg/m^3 (North Dakota) to 1.15 mg/m^3 (Virginia) to 1.4 mg/m^3 (Connecticut) to 1.6667 mg/m^3 (Nevada).

Determination in Air: Use NIOSH (II-3) Method #S-150.

Permissible Concentration in Water: Russia[35][43] set a MAC of 0.04 mg/L for water bodies used for domestic purposes.

Determination in Water: Octanol–water coefficient: Log $K_{ow} = -0.9$.

Routes of Entry: Inhalation, skin absorption, ingestion, and skin and/or eye contact.

Harmful Effects and Symptoms

Short Term Exposure: The substance is corrosive to the eyes, skin, and the respiratory tract. Contact can cause redness, swelling, and burns. Exposure can cause corneal (eye) swelling resulting in blurring vision. Inhalation can cause pulmonary edema, a medical emergency that can be delayed for several hours. This can cause death.

Long Term Exposure: Morpholine may damage the liver and kidneys. Can cause a skin rash.

Points of Attack: Eyes, skin, respiratory system, liver, kidneys.

Medical Surveillance: For those with frequent or potentially high exposure (half the TLV or greater) the following are recommended before beginning work and at regular times after that: lung function tests. If symptoms develop or overexposure is suspected, the following may be useful: liver and kidney function tests. Consider chest X-ray following acute overexposure.

First Aid: If this chemical gets into the eyes, remove any contact lenses at once and irrigate immediately for at least 15 min, occasionally lifting upper and lower lids. Seek medical attention immediately. If this chemical contacts the skin, remove contaminated clothing and wash immediately with soap and water. Seek medical attention immediately. If this chemical has been inhaled, remove from exposure, begin rescue breathing (using universal precautions, including resuscitation mask) if breathing has stopped and CPR if heart action has stopped. Transfer promptly to a medical facility. When this chemical has been swallowed, get medical attention. Give large quantities of water and induce vomiting. Do not make an unconscious person vomit. Medical observation is recommended for 24–48 h after breathing overexposure, as pulmonary edema may be delayed. As first aid for pulmonary edema, a doctor or authorized paramedic may consider administering a corticosteroid spray.

Personal Protective Methods: Wear protective gloves and clothing to prevent any reasonable probability of skin contact: **8 h**: butyl rubber gloves, suits, boots; 4H™ and Silver Shield™ gloves. Also, polyvinyl alcohol and Viton are among the recommended protective materials. Safety equipment suppliers/manufacturers can provide recommendations on the most protective glove/clothing material for your operation. All protective clothing (suits, gloves, footwear, headgear) should be clean, available each day, and put on

before work. Contact lenses should not be worn when working with this chemical. Wear splash-proof chemical goggles and face shield unless full face-piece respiratory protection is worn. Employees should wash immediately with soap when skin is wet or contaminated. Provide emergency showers and eyewash.

Respirator Selection: *550 ppm:* Sa:Cf (APF = 25) (any supplied-air respirator operated in a continuous-flow mode) or PaprOv (APF = 25) [any powered, air-purifying respirator with organic vapor cartridge(s)]. *1000 ppm:* CcrFOv (APF = 50) [any chemical cartridge respirator with a full face-piece and organic vapor cartridge(s)] or GmFOv (APF = 50) [any air-purifying, full-face-piece respirator (gas mask) with a chin-style, front- or back-mounted acid gas canister] or PaprTOv (APF = 50) [any powered, air-purifying respirator with a tight-fitting face-piece and organic vapor cartridge(s)] or SCBAF (APF = 50) (any self-contained breathing apparatus with a full face-piece) or SaF (APF = 50) (any supplied-air respirator with a full face-piece). *1400 ppm:* SaF: Pd,Pp (APF = 2000) (any supplied-air respirator that has a full face-piece and is operated in a pressure-demand or other positive-pressure mode). *Emergency or planned entry into unknown concentrations or IDLH conditions:* SCBAF: Pd,Pp (APF = 10,000) (any self-contained breathing apparatus that has a full face-piece and is operated in a pressure-demand or other positive-pressure mode) or SaF: Pd,Pp: ASCBA (any supplied-air respirator that has a full face-piece and is operated in a pressure-demand or other positive-pressure mode in combination with an auxiliary self-contained breathing apparatus operated in a pressure-demand or other positive-pressure mode). *Escape:* GmFOv (APF = 50) [any air-purifying, full-face-piece respirator (gas mask) with a chin-style, front- or back-mounted organic vapor canister] or SCBAE (any appropriate escape-type, self-contained breathing apparatus).

Note: Substance causes eye irritation or damage; eye protection needed.

Storage: Color Code—Red: Flammability Hazard: Store in a flammable liquid storage area or approved cabinet away from ignition sources and corrosive and reactive materials. Prior to working with this chemical you should be trained on its proper handling and storage. Before entering confined space where this chemical may be present, check to make sure that an explosive concentration does not exist. Morpholine must be stored to avoid contact with strong acids (such as nitric acid) and strong oxidizers (such as chlorine, chlorine dioxide, bromine, nitrates, and permanganates) since violent reactions occur.

Shipping: This compound requires a shipping label of "FLAMMABLE LIQUID." It falls in Hazard Class 3 and Packing Group II.

Spill Handling: Evacuate and restrict persons not wearing protective equipment from area of spill or leak until cleanup is complete. Remove all ignition sources. Establish forced ventilation to keep levels below explosive limit. Absorb liquids in vermiculite, dry sand, earth, peat, carbon, or a similar material and deposit in sealed containers. Keep this chemical out of a confined space, such as a sewer, because of the possibility of an explosion, unless the sewer is designed to prevent the buildup of explosive concentrations. It may be necessary to contain and dispose of this chemical as a hazardous waste. If material or contaminated runoff enters waterways, notify downstream users of potentially contaminated waters. Contact your local or federal environmental protection agency for specific recommendations. If employees are required to clean up spills, they must be properly trained and equipped. OSHA 1910.120(q) may be applicable.

Fire Extinguishing: This chemical is a combustible liquid. Poisonous gases, including nitrogen oxides, are produced in fire. Use dry chemical, carbon dioxide, or alcohol foam extinguishers. Vapors are heavier than air and will collect in low areas. Vapors may travel long distances to ignition sources and flashback. Vapors in confined areas may explode when exposed to fire. Containers may explode in fire. Storage containers and parts of containers may rocket great distances, in many directions. If material or contaminated runoff enters waterways, notify downstream users of potentially contaminated waters. Notify local health and fire officials and pollution control agencies. From a secure, explosion-proof location, use water spray to cool exposed containers. If cooling streams are ineffective (venting sound increases in volume and pitch, tank discolors, or shows any signs of deforming), withdraw immediately to a secure position. If employees are expected to fight fires, they must be trained and equipped in OSHA 1910.156. The only respirators recommended for firefighting are self-contained breathing apparatuses that have full face-pieces and are operated in a pressure-demand or other positive-pressure mode.

Disposal Method Suggested: Controlled incineration (incinerator equipped with a scrubber or thermal unit to reduce nitrogen oxides emissions).[22]

References

US Environmental Protection Agency. (November 16, 1977). *Chemical Hazard Information Profile: Morpholine.* Washington, DC

Sax, N. I. (Ed.). (1981). *Dangerous Properties of Industrial Materials Report,* 1, No. 8, 82−84

New Jersey Department of Health and Senior Services. (January 2000). *Hazardous Substances Fact Sheet: Morpholine.* Trenton, NJ

Muscimol   M:1450

Molecular Formula: $C_4H_6N_2O_2$
Synonyms: Agarin; 5-Aminomethyl-3-hydroxyisoxazole; 5-(Aminomethyl)-3-isoxazolol; 5-(Aminomethyl)-3-(2H)isoxazolone; 5-Hydroxy-5-aminomethylisoxazole; Muscimol
CAS Registry Number: 2763-96-4
RTECS® Number: NY3325000
UN/NA & ERG Number: Not regulated.

EC Number: 220-430-4
Regulatory Authority and Advisory Bodies
US EPA Hazardous Waste Number (RCRA No.): P007.
RCRA, 40CFR261, Appendix 8 Hazardous Constituents.
Superfund/EPCRA 40CFR355, Extremely Hazardous Substances, TPQ = 500/10,000 lb (227/4550 kg).
Reportable Quantity (RQ): 1000 lb(455 kg).
European/International Regulations: not listed in Annex 1.
WGK (German Aquatic Hazard Class): 3—Severe hazard to waters.
Description: Muscimol is a crystalline solid. Molecular weight = 114.12; Freezing/Melting point = 175°C (decomposes).
Potential Exposure: Formerly used as a sedative and as an antiemetic; and for experimental laboratory purposes.
Permissible Exposure Limits in Air
Protective Action Criteria (PAC)
TEEL-0: 3.5 mg/m^3
PAC-1: 10 mg/m^3
PAC-2: 17 mg/m^3
PAC-3: 20 mg/m^3
Routes of Entry: Inhalation, ingestion, skin and/or eye contact.
Harmful Effects and Symptoms
Short Term Exposure: Muscimol is a potent central nervous system depressant. Initially, drowsiness, stupor, or sleep is followed by mild nausea and vomiting. Muscle spasms in extremities, various emotional changes and distorted perceptions of space and time, but only rarely hallucinations. This compound is a natural constituent of amanita mushrooms and is extremely toxic. It is a potent central nervous system depressant and is believed to be responsible for most of the nervous system effects that result from eating this mushroom. The lowest toxic dose in humans has been reported at 109 mg/kg.
Long Term Exposure: May be neurotoric.
Points of Attack: Central nervous system.
Medical Surveillance: Examination of the nervous system. Examination for brain effects.
First Aid: If this chemical gets into the eyes, remove any contact lenses at once and irrigate immediately for at least 15 min, occasionally lifting upper and lower lids. Seek medical attention immediately. If this chemical contacts the skin, remove contaminated clothing and wash immediately with soap and water. Speed in removing material from skin is of extreme importance. Seek medical attention immediately. If this chemical has been inhaled, remove from exposure, begin rescue breathing (using universal precautions, including resuscitation mask) if breathing has stopped and CPR if heart action has stopped. Transfer promptly to a medical facility. When this chemical has been swallowed, get medical attention. Give large quantities of water and induce vomiting. Do not make an unconscious person vomit.
Personal Protective Methods: Wear protective gloves and clothing to prevent any reasonable probability of skin

contact. Safety equipment suppliers/manufacturers can provide recommendations on the most protective glove/clothing material for your operation. All protective clothing (suits, gloves, footwear, headgear) should be clean, available each day, and put on before work. Contact lenses should not be worn when working with this chemical. Wear dust-proof chemical goggles and face shield unless full face-piece respiratory protection is worn. Employees should wash immediately with soap when skin is wet or contaminated. Provide emergency showers and eyewash.

Respirator Selection: Follow regulations in OSHA 29CFR1910.134 or European Standard EN149. Use a NIOSH/MSHA- or European Standard EN149-approved respirator; or use an approved supplied-air respirator with a full face-piece operated in the positive-pressure mode, or with a full face-piece, hood, or helmet in the continuous-flow mode; or use a NIOSH/MSHA- or European Standard EN149-approved self-contained breathing apparatus with a full face-piece operated in a pressure-demand or other positive-pressure mode.

Storage: Color Code—Blue: Health Hazard/Poison: Store in a secure poison location. Prior to working with this chemical you should be trained on its proper handling and storage. Store in tightly closed containers in a cool, well-ventilated area away from oxidizers and reducing agents. Where possible, automatically pump liquid from drums or other storage containers to process containers.

Shipping: Not regulated.

Spill Handling: Evacuate persons not wearing protective equipment from area of spill or leak until cleanup is complete. Remove all ignition sources. Collect powdered material in the most convenient and safe manner and deposit in sealed containers. Ventilate area after cleanup is complete. It may be necessary to contain and dispose of this chemical as a hazardous waste. If material or contaminated runoff enters waterways, notify downstream users of potentially contaminated waters. Contact your local or federal environmental protection agency for specific recommendations. If employees are required to clean up spills, they must be properly trained and equipped. OSHA 1910.120(q) may be applicable.

Fire Extinguishing: Use dry chemical, carbon dioxide, water spray, or alcohol foam extinguishers. Poisonous gases, including nitrogen oxides, are produced in fire. If material or contaminated runoff enters waterways, notify downstream users of potentially contaminated waters. Notify local health and fire officials and pollution control agencies. From a secure, explosion-proof location, use water spray to cool exposed containers. If cooling streams are ineffective (venting sound increases in volume and pitch, tank discolors, or shows any signs of deforming), withdraw immediately to a secure position. If employees are expected to fight fires, they must be trained and equipped in OSHA 1910.156. The only respirators recommended for firefighting are self-contained breathing apparatuses that have full face-pieces and are operated in a pressure-demand or other positive-pressure mode.

Disposal Method Suggested: Consult with environmental regulatory agencies for guidance on acceptable disposal practices. Generators of waste containing this contaminant ($\geq$100 kg/mo) must conform with EPA regulations governing storage, transportation, treatment, and waste disposal.

References
Sax, N. I. (Ed.). (1982). *Dangerous Properties of Industrial Materials Report*, 2, No. 3, 81
US Environmental Protection Agency. (November 30, 1987). *Chemical Hazard Information Profile: Muscimol.* Washington, DC: Chemical Emergency Preparedness Program

Mustard gas (Agents H, HD & HT, WMD)   M:1460

Molecular Formula: $C_4H_8Cl_2S$; $C_4(H_8)Cl_2(S)$
Common Formula: $(ClCH_2CH_2)_2S$
Synonyms: Bis(β-chloroethyl) sulfide; Bis(2-chloroethyl) sulfide; 1-Chloro-2-(β-chloroethylthio)ethane; β,β'-Dichlorodiethyl sulfide; 2,2'-Dichlorodiethyl sulfide; Dichloro diethyl sulfide; β,β'-Dichloroethyl sulfide; 2,2'-Dichloroethyl sulfide; Di-2-chloroethyl sulfide; Distilled mustard (HD); Ethane, 1,1'-thiobis-2-chloro-; Gas mostaza (Spanish); H and HD (military designations); Iprit; Kampstoff lost; Lost (German); Pyro (Agent HD); Senfgas; Sesquimustard; S-Lost (German); Sulfide, bis(2-chloroethyl); Sulfur mustard; Sulfur mustard gas; 1,1'-Thiobis(2-chloroethane); Yellow cross gas; Yellow cross liquid; S-Yperite; Yperite
HT: Mixture of BIS(2-chloroethyl)sulfide and BIS[2-(2-chloroethylthio)ethyl]ether; Mustard-T mixture
CAS Registry Number: 505-60-2; 39472-40-7, 68157-62-0; *(alt.)* 69020-37-7; 6392-89-8 (HT); 3563-36-8 (HD)
RTECS® Number: WQ090000
UN/NA & ERG Number: UN2810/153
Regulatory Authority and Advisory Bodies
Department of Homeland Security Screening Threshold Quantity: *Theft hazard* CUM 100 g. [*505-60-2; 3563-36-8 (HD)*].
Carcinogenicity: NTP: 11th Report on Carcinogens, 2004: Known to be a human carcinogen; IARC: Human Sufficient Evidence; Animal Limited Evidence, *carcinogenic to humans*, Group 1, 1998; HT (Agent T) should be treated as a suspect carcinogen due to its similarity to Mustard Agent HD.
US EPA Gene-Tox Program, Positive: *D. melanogaster*—whole sex chrom. loss; Positive: *D. melanogaster*—reciprocal translocation; Positive: L5178Y cells *In vitro*—TK test; *N. crassa*—reversion; Positive: *D. melanogaster* sex-linked lethal; Positive/dose response: TRP reversion; Positive/limited: Carcinogenicity—mouse/rat.
RCRA, 40CFR261, Appendix 8 Hazardous Constituents, waste number not listed.

Superfund/EPCRA 40CFR355, Extremely Hazardous Substances: TPQ = 500 lb (227 kg).

Reportable Quantity (RQ): 1 lb (0.454 kg).

EPCRA Section 313 Form R *de minimis* concentration reporting level: 0.1%.

US DOT 49CFR172.101, Inhalation Hazard Chemical.

California Proposition 65 Chemical: Cancer 2/27/87.

WGK (German Aquatic Hazard Class): No value assigned (all above CAS numbers).

Description: Mustard gas, a chlorinated sulfur compound (s), is an oily, yellow to black liquid (clear when pure) with a sweet, burnt garlic or horseradish-like odor. The odor threshold for HD is 0.0006 mg/m^3. Molecular weight = 159.08; Specific gravity (H$_2$O = 1) = 1.27 at 20°C; Volatility = 610 mg/m^3 at 20°C; 920 mg/m^3 at 25°C; Boiling point = (decomposes) 215−217°C; Freezing/Melting point = 13−14°C; Vapor pressure = 0.072 mmHg at 20°C; 0.11 mmHg at 25°C; Vapor density (air = 1) = 5.5; Flash point (ignited by large explosive charges) = 104°C. Hazard Identification (based on NFPA-704 M Rating System): Health 4, Flammability 1, Reactivity 1W. Slightly soluble in water (reaction). Soluble in acetone, CH3(C1), tetrachloroethane, ethylbenzoate, and ether. Thickened is essentially the same as HD except for viscosity, which uses HD K125 (acryloid copolymer, 5%) as a thickener. K125 is not known to be hazardous except in a finely-divided, powder form. The viscosity of HD is between 1000 and 1200 centistokes at 25°C.

HT is a mixture of 60% HD and 40% Agent HT [(bis-(2-chloroethylthio ethyl)ether], a closely related vesicant with a lower freezing point). Agent HT is essentially the same as Agent HD, but HT is more stable. HT is a clear, yellowish, highly viscous liquid. It has a longer duration of effectiveness and has a lower garlic-like odor similar to HD. Although volatility is low, vapors can reach hazardous levels in warm weather. Specific gravity (water = 1) = 1.2361 at 25°C; Boiling point: 120°C at 0.02 torr; 174°C at 2.0 torr; Freezing/Melting point = 9.6−9.9°C; 13−14°C[NIOSH]; Volatility (mg/L) = 4.1 × 10^{-4} at 25°C; Viscosity (centistoke) = 14.7 at 25°C; Vapor pressure = 0.025 mmHg at 0°C; 0.090 mmHg at 30°C. Vapor density (air = 1) = 9.08; Flash point = 100°C. Practically insoluble in water; solubility = 0.8 g/L at 20°C.

Potential Exposure: Compound Description: Tumorigen, Mutagen; Reproductive Effector; Human Data; Primary Irritant. Mustard gas is used as an alkylating agent. It has also been used as a chemical warfare agent, causing delayed casualties. It is a vesicant and blister agent in chemical warfare (especially during World War I, military designation H or HD). Mustard gas is used as a model compound in biological studies. Mustard gas has been tested as an antineoplastic agent, but its clinical use as a tumor inhibitor has been minimal.

Incompatibilities: Sulfur mustard is stable at ambient temperatures. Reacts with oxidizers (vigorous), strong acids, acid fumes, strong alkalis, oxygen, water, steam, and other forms of moisture. On contact with acid or acid fumes, it emits highly toxic fumes of oxides of sulfur and chlorine. Rapidly corrosive to brass at 65°C. Will corrode steel at a rate of 0.0001 in/month at 65°C. Contact with metals may evolve flammable hydrogen gas. HD reacts with water; will hydrolyze, forming HCI and thiodiglycol. When heated to decomposition (between 149°C and 177°C), highly toxic fumes of hydrochloric acid, oxides of sulfur and chlorine are emitted.

Persistence of Chemical Agent: HD: Summer: 3 days to 1 week; Winter: May last for weeks.

Permissible Exposure Limits in Air

IDLH: 0.7 mg/m^3.

STEL: 0.003 mg/m^3.

WPL (Worker population limit): 0.0004 mg/m^3.

GPL (General population limit): 0.00002 mg/m^3.

AEGL (Acute Exposure Guidelines): 0.06 ppm/0.4 mg/m^3 (10 min); 0.001 ppm/0.008 mg/m^3 (8 h).

Protective Action Criteria (PAC) HD, 505-60-2*

TEEL-0: 0.0035 ppm

PAC-1: **0.01** ppm

PAC-2: **0.02** ppm

PAC-3: **0.32** ppm

*AEGLs (Acute Emergency Guideline Levels) & ERPGs (Emergency Response Planning Guideline) are in **bold face**.

DFG MAK: [skin] Carcinogen Category 1 as bis(β-chloroethyl)sulfide.

Finland: carcinogen, 1999; France: carcinogen, 1993; Sweden: carcinogen, 1999; Switzerland: carcinogen, 1999; United Kingdom: carcinogen, 2000.

Determination in Air: Available monitoring equipment for agent HD is the M8/M9 Detector paper, blue bank tube; M256/M256A1 kits; bubbler. Depot Area Air Monitoring System (DAMMS); automated Continuous Air Monitoring System (ACMS); CAM-M1, Hydrogen Flame Photometric Emission Detector (HYFED); and the Minature Chemical Agent Monitor (MINICAM).

Determination in Water: Ecotoxicology: Log K_{ow} (estimated) = 2.41: Log $K_{benzene-water}$ = 0.15. Bulk mustard can persist for decades in soil or water. When exposed to sea water, mustard forms a thick outer "crust" over a core of mustard which allows the mustard to be brought to the surface where it can injure unsuspecting fishermen, who may snare plastic lumps of mustard gas in their nets. Mustard and a number of its hydrolysis products are oxidized (air, oxygen, hypochlorite, hydrogen peroxide, nitric acid, potassium permanganate, and chromic acid) to give the less toxic sulfoxide and sulfone analogs. Mustard and its hydrolysis products do not significantly degrade in sunlight and are stable at less than 49°C.

Routes of Entry: Ingestion, skin and/or eye contact.

Harmful Effects and Symptoms

Sulfur mustard causes severe, delayed burns to the eyes, skin, and respiratory tract. Sulfur mustard damages cells within minutes of contact; however, the onset of pain and other health effects is delayed until hours after exposure. Large exposures to sulfur mustard may be fatal.[CDC]

Short Term Exposure: Contact with the liquid or exposure to high vapor concentrations can cause severe burns and permanent eye damage. There is no pain on contact but hours later redness, swelling, and pain occur. Blindness may result. Inhalation can cause pulmonary edema, a medical emergency that can be delayed for several hours. This can cause death. In $1-12$ h there may be coughing, swollen eyelids, reddened skin, and severe itching. There may be swelling and destruction of tissue in the respiratory tract and exposed skin. Ingestion may cause nausea and destruction of tissue in the respiratory tract and exposed skin. Ingestion may cause nausea and vomiting. The median lethal dosage is 1500 mg-min/m^3 for inhalation and $10,000$ mg-min/m^3 for skin absorption (masked personnel). The median incapacitating dosage is 200 mg-min/m^3 for eye injury and 2000 mg-min/m^3 for skin absorption (masked personnel). Wet skin absorbs more material than dry skin. May cause death or permanent injury after very short exposure to small quantities. It is a blistering gas and is highly irritating to eyes, skin, and lungs. Pulmonary lesions are often fatal. HD is a vesicant (causing blisters) and alkylating agent producing cytotoxic action on the hematopoietic (blood-forming) tissues which are especially sensitive. The rate of detoxification of HD in the body is very slow and repeated exposures produce a cumulative effect. Median doses of HD in humans are: LD_{50} (skin) $=$ 100 mg/kg; ICt_{50} (skin) $= 2000$ mg-min/m^3 at $21-27°C$ (humid environment); $= 1000$ mg-min/m^3 at $32°C$ (dry environment); ICt_{50} (eyes) $= 200$ mg-min/m^3; ICt_{50} (inhalation) $= 1500$ mg-min/m^3 (Ct unchanged with time) LD_{50} (oral-rat) $= 0.7$ mg/kg. Maximum safe Ct for skin and eyes are 5 and 2 mg-min/m^3, respectively.

Acute physiological action, local: HD affects both the eyes and the skin. *Skin* damage occurs after percutaneous resorption. Being lipid soluble, HD can be reabsorbed into all organs. Skin penetration is rapid without skin irritation. Swelling (blisters) and reddening (erythema) of the skin occurs after a latency period of $4-24$ h following the exposure, depending on degree of exposure and individual sensitivity. The skin healing process is very slow. Tender skin, mucous membrane, and perspiration-covered skin are more sensitive to the effects of HD. HD's effect on the skin, however, is less than on the eyes. Local action on the eyes produces severe necrotic damage and loss of eyesight. Exposure of eyes to HD vapor or aerosol produces lacrimation, photophobia, and inflammation of the conjunctiva and cornea.

Acute physiological action, systemic: Occur primarily through inhalation and ingestion. The HD vapor or aerosol is less toxic to the skin or eyes than the liquid form. When inhaled, the upper respiratory tract (nose, throat, trachea) is inflamed after a few hours latency period, accompanied by sneezing, coughing, and bronchitis, loss of appetite, diarrhea, fever, and apathy. Exposure to nearly lethal dose of

HD can produce injury to bone marrow, lymph nodes, and spleen, as indicated by a drop in WBC count and, therefore, results in increased susceptibility to local and systemic infections. Ingestion of HD will produce severe stomach pains, vomiting, and bloody stools after a 15- to 20-min latency period.

Long Term Exposure: HD can cause sensitization, chronic bronchitis, and lung impairment (cough, shortness of breath, chest pain); and can cause cancer of the mouth, throat, respiratory tract, skin, and leukemia. It may also cause birth defects. Mustard gas is a carcinogen.

Points of Attack: Lungs, eyes.

Medical Surveillance: Lung function tests. Examination of the eyes, including slit lamp. Consider chest X-ray following acute overexposure.

First Aid: There is no antidote for sulfur mustard toxicity. If this chemical gets into the eyes instantly flush with water. A delay of seconds can cause permanent damage. *Inhalation*: Remove from the source *immediately*. If breathing has stopped, give artificial respiration. If breathing is difficult, administer oxygen. *Seek medical attention immediately. Eye contact:* Speed in decontaminating the eyes is absolutely essential. Remove person from the liquid source; flush the eyes immediately with water by tilting the head to the side, pulling the eyelids apart with the fingers, and pouring water slowly into the eyes. Do not cover eyes with bandages but, if necessary, protect eyes by means of dark or opaque goggles. *Transfer the patient to a medical facility immediately. Ingestion:* Do not induce vomiting. Give victim milk to drink. Seek medical attention immediately. *Skin contact:* Don respiratory protective mask and gloves; remove victim from agent source immediately. Flush skin and clothes with solution of sodium hypochlorite or liquid household bleach (see decontamination procedure below) within 1 min. Cut and remove contaminated clothing, flush contaminated skin area again with sodium hypochlorite solution, then wash contaminated skin area with soap and water. If shower facilities are available, wash thoroughly and transfer to medical facility. If the skin becomes contaminated with a thickened agent, blot/wipe off the material immediately with an absorbent pad/paper towel prior to using decontaminating solution.

Decontamination: This is very important, and you have to decontaminate as soon as you can. Extra minutes before decontamination might make a big difference. If you do not have the equipment and training, do not enter the hot zone to rescue and decontaminate victims. If the victim cannot move, decontaminate without touching and without entering the hot zone. Use clean water from any source; if possible, use a hose (spray or fog to prevent injury to the victim) or other system so that you would not have to touch the victim; do not even wait for soap or for the victim to remove clothing, begin washing immediately. Immediately flush the eyes with water for at least 15 min. Wash—strip—wash—evacuate upwind and uphill: The approach is to immediately wash with water, then have the victim (not the first responder) remove all the victim's clothing, then wash again (with soap if available); and then move away from the hot zone in an upwind and uphill direction. Wash the victim with warm water and soap. Decontaminate with diluted household bleach (10%, or one part bleach to nine parts water), but do not let anything to get into the victim's eyes, open wounds, or mouth. Wash off the diluted bleach solution after 15 min. Be sure that you have decontaminated the victims as much as you can before they leave the area so that they do not spread the agent. Use 5% solution of common bleach (sodium hypochlorite) or calcium hypochlorite solution (48 oz per 5 gallons of water) to decontaminate scissors used in clothing removal, clothes, and other items.

Personal Protective Methods: Protective gloves, mandatory: Butyl, toxicological agent, protective gloves (M3, M4, glove set). As a minimum, chemical goggles will be worn. For splash hazards use goggles and face shield. Full protective clothing will consist of the M3 butyl rubber suit with hood, M2A1 butyl boots; M3 gloves; impregnated underwear; M9 series mask and coveralls (if desired); or the Demilitarization Protective Ensemble (DPE). For general lab work, gloves and lab coat shall be worn with M9 or M17 mask readily available. In addition, when handling contaminated lab animals, a daily clean smock, foot covers, and head covers are required. *Ventilation*: Local Exhaust: Mandatory. Must be filtered or scrubbed. *Special:* Chemical laboratory hoods shall have an average inward face velocity of 100 linear feet per minute (lfpm) plus or minus 10% with the velocity at any point not deviating from the average face velocity by more than 20%. Laboratory hoods shall be located so that cross-drafts do not exceed 20% of the inward face velocity. A visual performance test utilizing smoke-producing devices shall be performed in assessing the ability of the hood to contain agent HD. Other: recirculation of exhaust air from agent areas is prohibited. No connection between agent areas and other areas through ventilation system is permitted. Emergency backup power is necessary. Hoods should be tested semi-annually or after modification or maintenance operations. Operations should be performed with 20 cm inside hoods.

Swatch Test Results for Level A Suits and Chemical Protective Gloves for HD (Mustard gas)

Item	Breakthrough
25-mil chemical protective gloves	360 min
Kappler Suit Model 42483	150 min
TYCHEM 10,000 Pkg Style No. 12645	330 min
Trellchem HPS suit	>480 min
Ready 1 Limited Use Suit: Model 91	125 min
First Team XE HazMat suit	385 min
Commander Ultrapro Suit, Style 79102	280 min
Kappler Suit Model 50660	435 min
TYCHEM Style No. 11645	>480 min
Trellchem TLU suit	>480 min
Chemturion Suit: Model 13	110 min
Chempruf II BETEX Suit	125 min
Commander Brigade: F91	>480 min

Respirator Selection: When used as a weapon, use SCBA Respirator Certified By NIOSH For CBRN Environments. Do not use chemical cartridge or canister respirators. Less than or equal to 0.003: Protective mask not required provided that: as an 8-h TWA (a) Continuous real-time monitoring (with alarm capability is conducted in the work area at the 0.003 mg/m^3 level of detection. (b) M9, M17 or M40 mask is available and donned if ceiling concentrations exceed 0.003 mg/m^3. (c) Exposure has been limited to the extent practicable by engineering controls (remote operations, ventilation, and process isolation) or work practices. *If these conditions are not met then the following applies:* full face-piece chemical canister, air-purifying respirators. (The M9, M17, or M40 series or other certified equivalent masks are acceptable for this purpose in conjunction with the M3 toxicological agent protective (TAP) suit for dermal protection.) *Greater than 0.003 as an 8-h TWA*: The Demilitarization Protective Ensemble (DPE), 30 mil, may be used with prior approval from the AMC Field Safety Activity. Use time for the 30 mil DPE must be restricted to 2 h or less. *Note*: When 30 mil DPE is not available, the M9 or M40 series mask with Level A protective ensemble including impregnated innerwear can be used. However, use time shall be restricted to the extent operationally feasible and may not exceed 1 h. As an additional precaution, the cuffs of the sleeves and the legs of the M3 suit shall be taped to the gloves and boots, respectively, to reduce aspiration.

Storage: Color Code—Blue: Health Hazard/Poison: Store in a secure poison location. Stable at ambient temperatures. Decomposition temperature is 149–177°C. Mustard is a persistent agent depending on pH and moisture and has been known to remain active for up to 3 years in soil. A regulated, marked area should be established where this chemical is handled, used, or stored in compliance with OSHA Standard 1910.1045. During handling, the "buddy" (two-man) system will be used. Containers should be periodically inspected for leaks either visually or using a detector kit, prior to transferring the containers from storage to work areas. Stringent control over all personnel handling HD must be exercised. Chemical showers, eyewash stations, and personal cleanliness facilities must be provided. Each worker will wash their hands before meals and shower thoroughly with special attention given to hair, face, neck, and hands using plenty of soap before leaving at the end of the workday. No smoking, eating, or drinking is permitted at the work site. Decontaminating equipment shall be conveniently located. Exits must be designed to permit rapid evacuation. HD should be stored in containers made of glass for Research, Development, Test, and Evaluation (RDTE) for small quantities or one-ton steel containers for large quantities. Agent shall be double-contained in liquid-tight containers when in storage.

Shipping: Toxic liquids, organic, n.o.s. [Inhalation hazard, Packing Group I, Zone B] require a shipping label of "POISONOUS/TOXIC MATERIALS." Bis-(2-chloroethyl) sulfide UN 2810, Inhalation Hazard. It falls in Hazard Class 6.1 and Packing Group III.[NIOSH]

Spill Handling: Evacuate and restrict persons not wearing protective equipment from area of spill or leak until cleanup is complete. Remove all ignition sources. Ventilate area of spill or leak. To clean up a spill, cover with up to 15% calcium hypochloride. Place in container; neutralize after 12 h if necessary. Keep sparks, flames, and other sources of ignition away. Keep material out of water sources and sewers. Attempt to stop leak if it is without hazard. Use water spray to knock down vapors. Avoid breathing vapors and bodily contact with the material. Keep upwind. Wash away any material which may have contacted the body with copious amounts of soap and water. Downwind evacuation must be considered. Absorb liquids in vermiculite, dry sand, earth, peat, carbon, or a similar material and deposit in sealed containers. Keep this chemical out of a confined space, such as a sewer, because of the possibility of an explosion, unless the sewer is designed to prevent the buildup of explosive concentrations. It may be necessary to contain and dispose of this chemical as a hazardous waste. If material or contaminated runoff enters waterways, notify downstream users of potentially contaminated waters. Contact your local or federal environmental protection agency for specific recommendations. If employees are required to clean up spills, they must be properly trained and equipped. OSHA 1910.120(q) may be applicable.

If spills or leaks of THD (Thickened HD) occur, follow the same procedures as those for HD, but dissolve the THD in acetone prior to introducing any decontaminating solution. Containment of THD is generally not necessary. Spilled THD can be carefully scraped off the contaminated surface and placed in a fully removable head drum with a high density, polyethylene lining. The THD can then be decontaminated, after it has been dissolved in acetone, using the same procedures used for HD. Contaminated surfaces should be treated with acetone, then decontaminated using the same procedures as those used for HD. Handling the THD requires careful observation of the "stringers" (elastic, thread-like attachments) formed when the agents are transferred or dispensed. These stringers must be broken cleanly before moving the contaminating device or dispensing device to another location, or unwanted contamination of a working surface will result.

Initial isolation and protective action distances

Distances shown are likely to be affected during the first 30 min after materials are spilled and could increase with time. If more than one tank car, cargo tank, portable tank, or large cylinder involved in the incident is leaking, the protective action distance may need to be increased. You may need to seek emergency information from CHEMTREC at (800) 424-9300 or seek professional environmental engineering assistance from the US EPA Environmental Response Team at (908) 548-8730 (24-h response line).

Small spills (From a small package or a small leak from a large package)

H, HD, when used as a weapon
First: Isolate in all directions (feet/meters)   100/30
Then: Protect persons downwind (miles/kilometers)
Day   0.1/0.2
Night   0.1/0.2

Large spills (From a large package or from many small packages)
First: Isolate in all directions (feet/meters)   200/60
Then: Protect persons downwind (miles/kilometers)
Day   0.2/0.3
Night   0.3/0.5

Recommended field procedures (H, HD): The mustard should be contained using vermiculite, diatomaceous earth, clay, or fine sand, and neutralized as soon as possible using copious amounts of 5.25% sodium hypochlorite solution. Scoop all material and place in an approved DOT container. Cover the contents of the drum with decontaminating solution as above. The exterior of the drum shall be decontaminated and then labeled per IAW, EPA, and DOT regulations. All leaking containers shall be over-packed with vermiculite placed between the interior and exterior containers. Decontaminate and label in accordance with IAW, EPA, and DOT regulations. Dispose of the material used to decontaminate exterior of drum in accordance with IAW, federal, state, and local regulations. Conduct general area monitoring with an approved monitor to confirm that the atmospheric concentrations do not exceed the airborne exposure limit. If 5.25% Sodium Hypochlorite solution is not available, then the following decontaminates may be used instead and are listed in the order of preference: Calcium Hypochlorite Decontaminating Solution No. 2 [DS2: (2% NaOH, 70% diethylenetriamine, 28% ethylene glycol monomethyl ether)]; and Super Tropical Bleach Slurry (STB).

Warning: Pure, undiluted calcium hypochlorite (HTH) will burn on contact with liquid blister agent.

Recommended laboratory procedures: A minimum of 65 g of decontamination fluid per gram of HD is allowed to agitate for a minimum of 1 h. Agitation is not necessary following the first hour if a single phase is obtained. At the end of 24 h, the resulting solution shall be adjusted to a pH between 10 and 11. Test for presence of active chlorine by use of acidic potassium iodide solution to give free iodine color. Place 3 mL of the decontaminate in a test tube. Add several crystals of potassium Iodine and swirl to dissolve. Add 3 mL of 50 wt.% sulfuric acid: water and swirl. IMMEDIATE iodine color indicates the presence of active chlorine. If negative, add additional 5.25% sodium hypochlorite solution to the decontamination solution, wait for 2 h, then test again for active chlorine. Continue procedure until positive chlorine is given by solution. A 10-wt.% calcium hypochlorite (HTH) mixture may be substituted for sodium hypochlorite. Use 65 g of decon/gram of HD and continue the test as described for sodium hypochlorite. Scoop up all material and place in approved DOT containers. Cover the contents of the drum with decontaminating solution as above. The exterior of the drum shall be decontaminated and then labeled in accordance with IAW, EPA, and DOT regulations. All leaking containers shall be overpacked with vermiculite placed between the interior and exterior containers. Decontaminate and label in accordance with IAW, EPA, and DOT regulations. Dispose of the material used to decontaminate exterior of drum in accordance with IAW federal, state, and local regulations. Conduct general area monitoring with an approved monitor to confirm that the atmospheric concentrations do not exceed the airborne exposure limits. *Note*: Surfaces contaminated with HD, and then rinse decontaminated, may evolve sufficient mustard vapor to produce a physiological response.

Recommended field procedures (HT): HT should be contained using vermiculite, diatomaceous earth, clay, or fine sand, and neutralized as soon as possible using copious amounts of alcoholic caustic, carbonate, or Decontaminating Solution No. 2 [DS2: (2% NaOH, 70% diethylenetriamine, 28% ethylene glycol monomethyl ether)]. Caution must be exercised when using these decontaminates since acetylene will be given off. Household bleach can also be used if accompanied by stirring to allow contact. Scoop up all contaminated material and place in approved DOT containers. Cover the contents with additional decontaminant. All leaking containers will be overpacked with vermiculite placed between the interior and exterior containers. Decontaminate the outside of the container and label according to DOT and EPA requirements. Dispose of the material according to waste disposal methods provided below. Dispose of decontaminate according to federal, state, and local laws. Conduct general area monitoring with an approved monitor to confirm that the atmospheric concentrations do not exceed the airborne exposure limit. *Warning:* Never use dry High Test Hypochlorite (HTH) or Super Tropical Bleach (STB) since they will react violently with HT and may burst into flames.

Recommended laboratory procedures: A minimum of 65 g of decontamination solution per gram of HT is allowed to agitate for a minimum of 1 h. Agitation is not necessary following the first hour if a single phase is obtained. At the end of 24 h, the resulting solution will be adjusted to a pH between 10 and 11. Test for presence of active chlorine by use of acidic potassium iodide solution to give free iodine color. Place 3 mL of the decontaminate in a test tube. Add several crystals of potassium iodine and swirl to dissolve. Add 3 mL of 50 wt.% sulfuric acid: water and swirl. Immediate iodine color shows the presence of active chlorine. If negative, add additional 5.25% sodium hypochlorite solution to the decontamination solution, wait for 2 h, then test again for active chlorine. Continue procedure until positive chlorine is given by solution. Scoop up all material and place in approved DOT containers. Cover the contents with additional decontaminate as above. The exterior of the container will be decontaminated and labeled according to EPA and DOT regulations. All leaking containers will be overpacked with vermiculite placed between the interior and

exterior containers. Decontaminate and label according to EPA and DOT regulations. Dispose of the material according to waste disposal methods provided below. Dispose of decontaminate according to federal, state, and local regulations. Conduct general area monitoring with an approved monitor to confirm that the atmospheric concentrations do not exceed the airborne exposure limits.

A 10-wt.% calcium hypochlorite mixture may be substituted for sodium hypochlorite. Use 65 g of decon/gram of HT and continue the test as described for sodium hypochlorite. *Note:* Surfaces contaminated with HT, then rinse decontaminated, may evolve sufficient HT vapor to produce a physiological response. HT on laboratory glassware may be oxidized by vigorous reaction with concentrated nitric acid.

Waste disposal method: All neutralized material should be collected, contained, and thermally decomposed in EPA-approved incinerators that will filter or scrub toxic by-products from effluent air before discharge to the atmosphere. Any contaminated materials or protective clothing should be decontaminated using HTH or bleach and analyzed to assure it is free of detectable contamination (3 ×) level. Contaminated clothes and personal belongings should be placed in a sealed double bag and placed inside properly labeled drums and held for shipment back to the DA issue point.

Note: Several states define decontaminated surety material as a RCRA hazardous waste.

Fire Extinguishing: This chemical may burn but does not easily ignite. HD will hydrolyze, forming HCl and thiodiglycol. Poisonous gases, including oxides of sulfur and chlorides, are produced in fire. Extinguish with water, foam, dry chemical, or carbon dioxide. Protective clothing and self-contained breathing apparatus are required in the presence of mustard gas. Vapors are heavier than air and will collect in low areas. Containers may explode in fire. Storage containers and parts of containers may rocket great distances, in many directions. If material or contaminated runoff enters waterways, notify downstream users of potentially contaminated waters. Notify local health and fire officials and pollution control agencies. From a secure, explosion-proof location, use water spray to cool exposed containers. If cooling streams are ineffective (venting sound increases in volume and pitch, tank discolors, or shows any signs of deforming), withdraw immediately to a secure position. If employees are expected to fight fires, they must be trained and equipped in OSHA 1910.156. The only respirators recommended for firefighting are self-contained breathing apparatuses that have full face-pieces and are operated in a pressure-demand or other positive-pressure mode.

Disposal Method Suggested: All decontaminated material should be collected, contained and chemically decontaminated or thermally decomposed in an EPA-approved incinerator, which will filter or scrub toxic by-products from effluent air before discharge to the atmosphere. Any contaminated protective clothing should be decontaminated using calcium hypochlorite (HTH) or bleach and analyzed to assure it is free of detectable contamination (3X) level.

Contaminated clothes and personal belongings should be placed in a sealed double bag and subsequently placed inside properly labeled drums and held for shipment back to the DA issue point. Decontamination of waste or excess material shall be accomplished in accordance with the procedures outlined above with the following exceptions: (a) HD on laboratory glassware may be oxidized by its vigorous reaction with concentrated nitric acid. (b) Open pit burning or burying of HD or items containing or contaminated with HD in any quantity is prohibited.

Note: Several states define decontaminated surety material as a RCRA Hazardous Waste.

References

US Environmental Protection Agency. (November 30, 1987). *Chemical Hazard Information Profile: Mustard Gas.* Washington, DC: Chemical Emergency Preparedness Program USAEHA Technical Guide No. 173. *Occupational Health Guidelines for the Evaluation and Control of Occupational Exposure to Mustard Agents H, HD, and HT*

New Jersey Department of Health and Senior Services. (March 2006). *Hazardous Substances Fact Sheet: Mustard Gas.* Trenton, NJ

Schneider, A. L., et al. (2007). *CHRIS + CD-ROM Version 2.0 (United Coast Guard Chemical Hazard Response Information System (COMDTINST 16465.12C).* Washington, DC: United States Coast Guard and the Department of Homeland Security

The Riegle Report: A Report of Chairman Donald W. Riegle, Jr. and Ranking Member Alfonse M. D'Amato of the Committee on Banking, Housing and Urban Affairs with Respect to Export Administration, United States Senate, 103rd Congress, 2d Session, (May 25, 1994)

Belmont, R. B. (June 1998). *TR Tests of Level A Suits— Protection against Chemical and Biological Warfare Agents and Simulants: Executive Summary.* Aberdeen Proving Ground, MD: CBRD-EN (US Army Chemical and Biological Defense Command)

N

Naled   N:0100

Molecular Formula: $C_4H_7Br_2Cl_2O_4P$

Synonyms: AI3-24988; Arthodibrom; Bromchlophos; Bromex; BRP; Dibrom; *O*-(1,2-Dibrom-2,2-dichloraethyl)-*O,O*-dimethyl-phosphat (German); 1,2-Dibromo-2,2-dichloroethyl dimethyl phosphate; *O,O*-Dimethyl *O*-(1,2-dibromo-2,2-dichloroethyl) phosphate; Dimethyl 1,2-dibromo-2,2-dichloroethyl phosphate; *O,O*-Dimethyl *O*-2,2-dichloro-1,2-dibromoethyl phosphate; ENT 24,988; Ethanol, 1,2-dibromo-2,2-dichloro-, dimethyl phosphate; Hibrom; OMS 75; Ortho 4355; Orthodibrom; Orthodibromo; Phosphate de *O,O*-dimethyle et de *O*-(1,2-dibromo-2-dichlorethyle) (French); Phosphoric acid, 1,2-dibromo-2,2-dichloroethyl dimethyl ester

CAS Registry Number: 300-76-5; *(alt.)* 53095-31-1

RTECS® Number: TB9450000

UN/NA & ERG Number: UN3018 (organophosphorus pesticide, liquid, toxic)/152; UN2783 (organophosphorus pesticides, solid, toxic)/152

EC Number: 206-098-3 [*Annex I Index No.:* 015-055-00-6]

Regulatory Authority and Advisory Bodies
US EPA, FIFRA, 1998 Status of Pesticides: Supported.
Air Pollutant Standard Set. See below, "Permissible Exposure Limits in Air" section.
Clean Water Act: Section 311 Hazardous Substances/RQ 40CFR117.3 (same as CERCLA, see below).
Reportable Quantity (RQ): 10 lb (4.54 kg).
EPCRA Section 313 Form R *de minimis* concentration reporting level: 1.0%.
US DOT Regulated Marine Pollutant (49CFR172.101, Appendix B).
US DOT 49CFR172.101, Inhalation Hazard Chemical as organophosphates.
European/International Regulations: Hazard Symbol: Xn, N; Risk phrases: R21/22; R36/38; R50; Safety phrases: S2; S36/37; S61 (see Appendix 4).
WGK (German Aquatic Hazard Class): No value assigned.

Description: Naled is a white crystalline solid (when pure) or light straw-colored liquid (above 26.7°C) with a slightly pungent insecticide odor. Molecular weight = 380.80; Specific gravity (H_2O:1) = 1.96 at 25°C; Boiling point = (decomposes) 110°C; Freezing/Melting point = 26.7°C; Vapor pressure = 2×10^{-4} mmHg at 20°C. Hazard Identification (based on NFPA-704 M Rating System): Health 1, Flammability 0, Reactivity 1. Insoluble in water.

Potential Exposure: Compound Description: Agricultural Chemical; Mutagen, Primary Irritant. Those involved in the manufacture, formulation, and application of this insecticide, fungicide, bactericide, acaricide. Also used in cooling towers, veterinary medicine, pulp and paper mill systems, hospitals, swimming pools, and bathrooms.

Incompatibilities: Incompatible with oxidizers. Hydrolyzed in presence of water. Degraded by sunlight. Decomposes when heated and on contact with acids, acid fumes, bases, producing fumes of hydrogen chloride, hydrogen bromide, phosphorus oxides. Reacts with acids, strong oxidizers in sunlight. Reacts with water. Corrosive to metals. Attacks some plastics, rubber, and coatings.

Permissible Exposure Limits in Air
OSHA PEL: 3 mg/m³ TWA.
NIOSH REL: 3 mg/m³ TWA [skin].
ACGIH TLV®[1]: 0.1 mg/m³ inhalable fraction and vapor [skin, sensitizer]; not classifiable as a human carcinogen; BEI_A issued as Acetylcholinesterase-inhibiting pesticides.
NIOSH IDLH: 200 mg/m³.
No TEEL available.
DFG MAK: 1 mg/m³, measured as the inhalable fraction TWA; Peak Limitation Category II(2); [skin] danger of skin sensitization; Pregnancy Risk Group C.
Australia: TWA 3 mg/m³, [skin], 1993; Austria: MAK 3 mg/m³, 1999; Belgium: TWA 3 mg/m³, [skin], 1993; Finland: TWA 3 mg/m³; STEL 6 mg/m³, 1999; France: VME 3 mg/m³, [skin], 1999; Norway: TWA 3 mg/m³, 1999; the Netherlands: MAC-TGG 3 mg/m³, 2003; Thailand: TWA 3 mg/m³, 1993; United Kingdom: TWA 3 mg/m³; STEL 6 mg/m³, 2000; Argentina, Bulgaria, Columbia, Jordan, South Korea, New Zealand, Singapore, Vietnam: ACGIH TLV®: not classifiable as a human carcinogen. Several states have set guidelines or standards for dibrom in ambient air[60] ranging from 30 µg/m³ (North Dakota) to 50 µg/m³ (Virginia) to 60 µg/m³ (Connecticut) to 71 µg/m³ (Nevada).

Determination in Air: No method available.

Permissible Concentration in Water: Mexico[35] has set maximum permissible concentration values of 3.0 µg/L for coastal waters and 0.03 mg/L (30 µg/L) for estuaries.

Determination in Water: Fish Tox = 10.17349000 ppb (INTERMEDIATE).

Routes of Entry: Inhalation, skin absorption, ingestion, eye and/or skin contact. Absorbed by the skin.

Harmful Effects and Symptoms
Short Term Exposure: Irritates the eyes, skin, and respiratory tract. May affect the nervous system causing convulsions, respiratory failure. Organic phosphorus insecticides are absorbed by the skin, as well as by the respiratory and gastrointestinal tracts. They are cholinesterase inhibitors. Symptoms of exposure include headache, giddiness, blurred vision, nervousness, weakness, nausea, cramps, diarrhea, and discomfort in the chest. Signs include sweating, tearing, salivation, vomiting, cyanosis, convulsions, coma, loss of reflexes, and loss of sphincter control. High exposure can result in death. Highly toxic; a probable human lethal dose may be between 1 teaspoon and 1 oz. Human Tox = 14.00000 ppb (INTERMEDIATE).

Long Term Exposure: May cause skin allergy. Cholinesterase inhibitor; cumulative effect is possible. This chemical may damage the nervous system with repeated exposure, resulting in convulsions, respiratory failure. May cause liver damage.

Points of Attack: Respiratory system, central nervous system, cardiovascular system, skin, eyes, blood cholinesterase.

Medical Surveillance: NIOSH lists the following tests: cholinesterase: blood plasma, red blood cells/count; cholinesterase: blood serum, red blood cells/count; cholinesterase: red blood cells/count. Before employment and at regular times after that, the above tests are recommended. Cholinesterase levels tests for the enzyme poisoned by this chemical. If exposure stops, plasma levels return to normal in 1−2 weeks while red blood cell levels may be reduced for 1−3 months. When cholinesterase enzyme levels are reduced by 25% or more below preemployment levels, risk of poisoning is increased, even if results are in lower ranges of "normal." Reassignment to work not involving organophosphate or carbamate pesticides is recommended until enzyme levels recover. If symptoms develop or overexposure occurs, repeat the above tests as soon as possible and get an examination of the nervous system. Also consider complete blood count. Consider chest X-ray following acute overexposure. Evaluation by a qualified allergist. Do not drink any alcoholic beverages before or during use. Alcohol promotes absorption of organic phosphates.

First Aid: If this chemical gets into the eyes, remove any contact lenses at once and irrigate immediately for at least 15 min, occasionally lifting upper and lower lids. Seek medical attention immediately. If this chemical contacts the skin, remove contaminated clothing and wash immediately with soap and water. Speed in removing material from skin is of extreme importance. Shampoo hair promptly if contaminated. Seek medical attention immediately. If this chemical has been inhaled, remove from exposure, begin rescue breathing (using universal precautions, including resuscitation mask) if breathing has stopped and CPR if heart action has stopped. Transfer promptly to a medical facility. When this chemical has been swallowed, get medical attention. Give large quantities of water and induce vomiting. Do not make an unconscious person vomit. Effects may be delayed. Medical observation is recommended.

Note to physician: 1−4 mg atropine; maintenance: 2-mg doses at intervals of 15−60 min.

Personal Protective Methods: Wear protective gloves and clothing to prevent any reasonable probability of skin contact. Safety equipment suppliers/manufacturers can provide recommendations on the most protective glove/clothing material for your operation. All protective clothing (suits, gloves, footwear, headgear) should be clean, available each day, and put on before work. Contact lenses should not be worn when working with this chemical. Wear splash-proof chemical goggles and face shield when working with liquid, unless full face-piece respiratory protection is worn. Wear dust-proof goggles and face shield when working with

powders or dust, unless full face-piece respiratory protection is worn. Employees should wash immediately with soap when skin is wet or contaminated. Provide emergency showers and eyewash.

Respirator Selection: $30 \ mg/m^3$: 95XQ (APF = 10) [any particulate respirator equipped with an N95, R95, or P95 filter (including N95, R95, and P95 filtering face-pieces) except quarter-mask respirators. The following filters may also be used: N99, R99, P99, N100, R100, P100] or Sa (APF = 10) (any supplied-air respirator). $75 \ mg/m^3$: Sa:Cf (APF = 25) (any supplied-air respirator operated in a continuous-flow mode) or PaprHie (APF = 25) (any powered, air-purifying respirator with a high-efficiency particulate filter). $150 \ mg/m^3$: 100F (APF = 50) (any air-purifying, full-face-piece respirator with an N100, R100, or P100 filter) or SaT: Cf (APF = 50) (any supplied-air respirator that has a tight-fitting face-piece and is operated in a continuous-flow mode) or PaprTHie (APF = 50) (any powered, air-purifying respirator with a tight-fitting face-piece and a high-efficiency particulate filter) or SCBAF (APF = 50) (any self-contained breathing apparatus with a full face-piece). $200 \ mg/m^3$: Sa: Pd,Pp (APF = 1000) (any supplied-air respirator operated in a pressure-demand or other positive-pressure mode). *Emergency or planned entry into unknown concentrations or IDLH conditions:* SCBAF: Pd,Pp (APF = 10,000) (any self-contained breathing apparatus that has a full face-piece and is operated in a pressure-demand or other positive-pressure mode) or SaF: Pd,Pp: ASCBA (APF = 10,000) (any supplied-air respirator that has a full face-piece and is operated in a pressure-demand or other positive-pressure mode in combination with an auxiliary self-contained breathing apparatus operated in a pressure-demand or other positive-pressure mode). *Escape:* 100F (APF = 50) (any air-purifying, full-face-piece respirator with an N100, R100, or P100 filter) or SCBAE (any appropriate escape-type, self-contained breathing apparatus).

Storage: Color Code—Blue: Health Hazard/Poison: Store in a secure poison location. Color Code—Red: Flammability Hazard: Store in a flammable liquid storage area or approved cabinet away from ignition sources and corrosive and reactive materials. Prior to working with this chemical you should be trained on its proper handling and storage. It must be stored to avoid contact with strong oxidizers, such as chlorine and chlorine dioxide, since violent reactions occur. Keep away from acids, acid fumes, sunlight, heat, and water.

Shipping: Organophosphorus pesticides, liquid, toxic, require a shipping label of "POISONOUS/TOXIC MATERIALS." They fall in DOT Hazard Class 6.1 and Naled in Packing Group III.

Spill Handling: *Dry:* Evacuate persons not wearing protective equipment from area of spill or leak until cleanup is complete. Remove all ignition sources. Use HEPA vacuum or wet method to reduce dust during cleanup. Do not dry sweep. Collect powdered material in the most convenient and safe manner and deposit in sealed containers. Ventilate area after cleanup is complete. It may be necessary to contain and dispose of this chemical as a hazardous waste. If material or contaminated runoff enters waterways, notify downstream users of potentially contaminated waters. Contact your local or federal environmental protection agency for specific recommendations. If employees are required to cleanup spills, they must be properly trained and equipped. OSHA 1910.120(q) may be applicable.

Liquid: Evacuate and restrict persons not wearing protective equipment from area of spill or leak until cleanup is complete. Remove all ignition sources. Ventilate area of spill or leak. Absorb liquids in vermiculite, dry sand, earth, peat, carbon, or a similar material and deposit in sealed containers. Keep this chemical out of a confined space, such as a sewer, because of the possibility of an explosion, unless the sewer is designed to prevent the buildup of explosive concentrations. It may be necessary to contain and dispose of this chemical as a hazardous waste. If material or contaminated runoff enters waterways, notify downstream users of potentially contaminated waters. Contact your local or federal environmental protection agency for specific recommendations. If employees are required to clean up spills, they must be properly trained and equipped. OSHA 1910.120(q) may be applicable.

Fire Extinguishing: This chemical is noncombustible. Use agents suitable for surrounding fire. Poisonous gases, including hydrogen chloride, hydrogen bromide, phosphorus oxides, are produced in fire. If material or contaminated runoff enters waterways, notify downstream users of potentially contaminated waters. Notify local health and fire officials and pollution control agencies. From a secure, explosion-proof location, use water spray to cool exposed containers. If cooling streams are ineffective (venting sound increases in volume and pitch, tank discolors, or shows any signs of deforming), withdraw immediately to a secure position. If employees are expected to fight fires, they must be trained and equipped in OSHA 1910.156. The only respirators recommended for firefighting are self-contained breathing apparatuses that have full face-pieces and are operated in a pressure-demand or other positive-pressure mode.

Disposal Method Suggested: This pesticide is more stable to hydrolysis than dichlorvos (50% hydrolysis at pH 9 at 37.5°C in 301 min). It is unstable in alkaline conditions, in presence of iron, and is degraded by sunlight. About 10% hydrolysis per day is obtained in ambient water. Incineration is recommended for large amounts.[22] In accordance with 40CFR165, follow recommendations for the disposal of pesticides and pesticide containers. Must be disposed properly by following package label directions or by contacting your local or federal environmental control agency or by contacting your regional EPA office.

References

US Environmental Protection Agency. (August 1976). *Investigation of Selected Potential Environmental Contaminants: Haloalkyl Phosphates*, Report EPA 560/2/76-007. Washington, DC

Sax, N. I. (Ed.). (1985). *Dangerous Properties of Industrial Materials Report*, 5, No. 3, 44−47

New Jersey Department of Health and Senior Services. (December 1998). *Hazardous Substances Fact Sheet: Dimethyl-1,2-Dibromo-2,2-Dichloroethyl Phosphate.* Trenton, NJ

Naphthas   N:0110

Synonyms: coal tar naphtha: Coal tar naphtha; Crude solvent coal tar naphtha; High solvent naphtha; Naphtha
Petroleum naphtha: Aliphatic petroleum naphtha; Petroleum naphtha; Petroleum spirit; rubber solvent
VM&P naphtha: Ligroin; Painters naphtha; Petroleum ether; Petroleum spirit; Refined solvent naphtha; Varnish makers' & painters' naphtha

CAS Registry Number: 8002-05-9 (petroleum distillate; petroleum asphalt); 8030-30-6; *(alt.)* 50813-73-5; *(alt.)* 54847-97-1 (naphtha, low boiling point); 121448-83-7 (coal tar naphtha); 8032-32-4 (ligroin or VM&P naphtha); *(alt.)* 8031-06-9; 64475-85-0 (naphtha, petroleum spirits); 64742-89-8 [Naphtha (petroleum) light aliphatic; rubber solvent]

RTECS® Number: SE7449000 [petroleum distillates (naphtha)]; DE3030000 (coal tar naphtha); OI6180000 (VM&P naphtha)

UN/NA & ERG Number: UN1136 (coal tar distillate)/128; UN1268 (petroleum distillates, n.o.s. or petroleum products, n.o.s.)/128; UN1300 (turpentine substitute)/128; UN1993/128

EC Number: 232-298-5 [*Annex I Index No.:* 649-049-00-5] (crude solvent coal tar naphtha); 232-443-2 [*Annex I Index No.:* 649-262-00-3] (naphtha, low boiling point); 232-453-7 [*Annex I Index No.:* 649-263-00-9] (ligroin or VM&P naphtha)

Regulatory Authority and Advisory Bodies

Carcinogenicity: IARC (*8002-05-9*): Human, No Adequate Data; Animal, Inadequate Evidence, *not classifiable as carcinogenic to humans,* Group 3.

Air Pollutant Standard Set. See below, "Permissible Exposure Limits in Air" section.

Canada, WHMIS, Ingredients Disclosure List Concentration 1.0% VM&P naphtha.

European/International Regulations (*8002-05-9*): Hazard Symbol: T; Risk phrases: R45; Safety phrases: S53; S45; (*8030-30-6; Naptha, low boiling point* and *8032-32-4 ligroin or VM&P naphtha*): Hazard Symbol: T; Risk phrases: R45; R46; R65; Safety phrases: S53; S45 (see Appendix 4).

WGK (German Aquatic Hazard Class): 1—Slightly water polluting (*CAS: 8002-05-9*); 1—Slightly water polluting (*CAS: 8030-30-6; 8032-32-4*).

Description: Naphthas derived from both petroleum and coal tar are included in this group. *Petroleum naphthas* are colorless liquids with a gasoline- or kerosene-like odor. A mixture of paraffins (C5−C13) may contain a small amount of aromatic hydrocarbons and are termed "close-cut" fractions. "Medium-range" and "wide-range" fractions are made up of 40−80% aliphatic hydrocarbons, 25−50% naphthenic hydrocarbons, 0−10% benzene, and 0−20% other aromatic hydrocarbons. Molecular weight = (approx.) 99 [petroleum distillates (naphtha)]; Specific gravity (H_2O:1) = 0.63−0.66 [petroleum distillates (naphtha)]; Boiling point = 35−60°C; Freezing/Melting point = −73°C; Vapor pressure = (approx.) 40 mmHg [petroleum distillates (naphtha)]; Flash point = −40 to −66°C [petroleum distillates (naphtha)]; Autoignition temperature = 288°C. Explosive Limits: LEL = 1.1%; UEL = 5.9%. Hazard Identification (based on NFPA-704 M Rating System): Health 1, Flammability 4, Reactivity 0. Insoluble in water.

Coal tar naphtha is a mixture of aromatic hydrocarbons, principally toluene, xylene, and cumene. Coal tar naphtha is a reddish-brown, mobile liquid with an aromatic odor. Shipped as a molten solid. Benzene is present in appreciable amounts in those coal tar naphthas with low boiling points. Molecular weight = (approx.) 110; Specific gravity (H_2O:1) = 0.89−0.97; Boiling point = 160−220°C; Vapor pressure = <5 mmHg at 20°C; Flash point = 38−43°C. Explosive Limits: LEL = 1.0%; UEL—unknown. Hazard Identification (based on NFPA-704 M Rating System): (coal tar) Health 2, Flammability 2, Reactivity 0. Insoluble in water.

VM&P naphtha is a clear to yellowish liquid with a pleasant, aromatic odor: Molecular weight = (approx) 87−114; Specific gravity (H_2O:1) = 0.73−0.76; Boiling point = 100−177°C; Vapor pressure = 2−20 mmHg at 20°C; Flash points = −2 to 29°C; Autoignition temperature = 232°C. Explosive limits: vary somewhat but typical values are LEL = 0.9%; UEL = 6.7%. Hazard Identification (based on NFPA-704 M Rating System): (VM&P) Health 1, Flammability 3, Reactivity 0. See also Stoddard Solvent.

Potential Exposure: Compound Description [*Petroleum distillates (naphtha)*]: Human Data; Primary Irritant; (*coal tar naphtha*) Agricultural Chemical; Tumorigen, Human Data; Primary Irritant; *(VM&P naphtha)* Reproductive Effector; Primary Irritant. Naphthas are used as organic solvents for dissolving or softening rubber, oils, greases, bituminous paints, varnishes, and plastics. The less flammable fractions are used in dry cleaning. The heavy naphthas are used as a vehicle for various pesticides. Coal tar naphthas are used as quick-drying paint solvent, in the manufacture of floor coverings, resin solution, varnish; VM&P naphtha is used as a solvent for lacquers and varnishes and as a rapid-dry paint thinner.

Incompatibilities: Strong oxidizers.

Permissible Exposure Limits in Air
Naphtha, petroleum distillates
Conversion factor: 1 ppm = 4.05 mg/m³ at 25°C & 1 atm.
OSHA PEL: 500 ppm/2000 mg/m³ TWA.
NIOSH REL: 350 mg/m³ TWA; 1800 mg/m³ [15 min] Ceiling Concentration.
NIOSH IDLH: 1100 ppm [LEL].

Protective Action Criteria (PAC)
Petroleum asphalt 8002-05-9
TEEL-0: 350 mg/m³
PAC-1: 350 mg/m³
PAC-2: 500 mg/m³
PAC-3: 500 mg/m³
Naphtha, coal tar
Conversion factor: 1 ppm = 4.50 mg/m³ (approx.) at 25°C & 1 atm.
OSHA PEL: 100 ppm/400 mg/m³ TWA.
NIOSH REL: 100 ppm/400 mg/m³ TWA.
NIOSH IDLH: 1000 ppm [LEL].
Australia: TWA 480 mg/m³, 1993; Denmark: TWA 25 ppm (145 mg/m³), 1999; Hungary: TWA 300 mg/m³; STEL 800 mg/m³, 1993; the Philippines: TWA 100 ppm (400 mg/m³), 1993; Poland: MAC (TWA) 300 mg/m³; MAC (STEL) 900 mg/m³, 1999; Russia: STEL 100 mg/m³, 1993; Switzerland: MAK-W 300 mg/m³ (1100 mg/m³), 1999; Turkey: TWA 100 ppm (400 mg/m³), 1993
VM&P naphtha, legroin, paint solvent 8032-32-4:
OSHA PEL: None.
NIOSH REL: 350 mg/m³ TWA; 1800 mg/m³ [15 min] Ceiling Concentration.
ACGIH withdrawn; confirmed animal carcinogen with unknown relevance to humans.
Protective Action Criteria (PAC)
TEEL-0: 75 ppm
PAC-1: 75 ppm
PAC-2: 400 ppm
PAC-3: 1100 ppm
Poland: TWA 500 mg/m³; STEL 1500 mg/m³, 1999. In addition, Several states have set guidelines or standards for naphtha in ambient air[60] ranging from zero (Nevada) to 60−27,000 μg/m³ (Connecticut) to 225 μg/m³ (Virginia).
64742-89-8 [naphtha (petroleum) light aliphatic; rubber solvent]
Protective Action Criteria (PAC)
TEEL-0: 87.9 ppm
PAC-1: 87.9 ppm
PAC-2: 453 ppm
PAC-3: 1100 ppm

Determination in Air: Use NIOSH Analytical Method (IV) #1550, Naphthas, OSHA Analytical Method 48.

Routes of Entry: Inhalation of vapor, ingestion, skin and/or eye contact. Percutaneous absorption of liquid is important in development of systemic effects if benzene is present.

Harmful Effects and Symptoms

Short Term Exposure: The naphthas are irritating to the skin conjunctiva and the mucous membranes of the upper respiratory tract. Skin "chapping" and photosensitivity may develop after repeated contact with the liquid. If confined against skin by clothing, the naphthas may cause skin burn. Exposure can cause dizziness, lightheadedness, and unconsciousness. Petroleum naphtha has a lower order of toxicity than that derived from coal tar, where the major hazard is brought about by the aromatic hydrocarbon content.

Sufficient quantities of both naphthas cause central nervous system depression. Symptoms include eye irritation followed by headache and nausea. In severe cases, dizziness, convulsions, and unconsciousness occasionally result. Symptoms of anorexia and nervousness have been reported to persist for several months following an acute overexposure, but this appears to be rare. One fraction, hexane, has been reported to have been associated with peripheral neuropathy. If benzene is present, coal tar naphthas may produce blood changes, such as leukopenia, aplastic anemia, or leukemia. The kidneys and spleen have also been affected in animal experiments. At vapor concentrations up to 450 ppm, petroleum naphtha inhalations may produce slight throat irritation. At 880 pm, definite throat irritation is observed. Vapors may also irritate the nose. High concentrations may produce difficulty in breathing, blue coloration of skin, excitement, and dizziness. Inhalation of vapors in the absence of oxygen is immediately life-threatening. A vapor concentration up to 450 ppm causes mild, temporary irritation; at 880 ppm more severe irritation may be experienced. Human Tox = 14.00000 ppb (INTERMEDIATE).

Long Term Exposure: Irritates the eyes and upper respiratory system. Coal tar naphtha may contain benzene, a cancer-causing agent in humans. Exposure may cause nervous system and kidney damage. Some coal tar naphthas contain other substances that can cause blood cell damage. Longer exposure may cause drying and cracking of the skin, and make the skin sunburn more easily. Swallowing the liquid may cause chemical pneumonia.

Points of Attack: Eyes, skin, respiratory system, central nervous system, liver, kidneys.

Medical Surveillance: Preplacement and periodic medical examinations should include the central nervous system. If benzene exposure is present, workers should have a periodic complete blood count (CBC) including hematocrits, hemoglobin, white blood cell count and differential count, mean corpuscular volume and platelet count, reticulocyte count, serum bilirubin determination, and urinary phenol in the preplacement examination and at 3-month intervals. There are no specific diagnostic tests for naphtha exposure but urinary phenols may indicate exposure to benzene and aromatic hydrocarbons. It should be noted that benzene content of vapor may be higher than predicted by content in the liquid.

First Aid: If this chemical gets into the eyes, remove any contact lenses at once and irrigate immediately for at least 15 min, occasionally lifting upper and lower lids. Seek medical attention immediately. If this chemical contacts the skin, remove contaminated clothing and wash immediately with soap and water. Seek medical attention immediately. If this chemical has been inhaled, remove from exposure, begin rescue breathing (using universal precautions, including resuscitation mask) if breathing has stopped and CPR if heart action has stopped. Transfer promptly to a medical facility. When this chemical has been swallowed, get immediate medical attention.

Note to physician: Inhalation: Bronchodilators, decongestants, and oxygen may be used if necessary. Corticosteroids are useful for treating pneumonitis.

Personal Protective Methods: Wear protective gloves and clothing to prevent any reasonable probability of skin contact. Safety equipment suppliers/manufacturers can provide recommendations on the most protective glove/clothing material for your operation. Nitrile, Neoprene™ rubber, Silvershield™, chlorinated polyethylene, styrene-butadiene rubber, and polyvinyl alcohol are among the recommended protective materials. All protective clothing (suits, gloves, footwear, headgear) should be clean, available each day, and put on before work. Contact lenses should not be worn when working with this chemical. Wear splash or dust-proof chemical goggles and face shield unless full face-piece respiratory protection is worn. Employees should wash immediately with soap when skin is wet or contaminated. Provide emergency showers and eyewash.

Respirator Selection: *Coal tar naphthas: Up to 1000 ppm:* Sa:Cf (APF = 25) (any supplied-air respirator operated in a continuous-flow mode); CcrFOv (APF = 50) [any chemical cartridge respirator with a full face-piece and organic vapor cartridge(s)] or GmFOv (APF = 50) [any air-purifying, full-face-piece respirator (gas mask) with a chin-style, front- or back-mounted organic vapor canister] or PaprOv (APF = 25) [any powered, air-purifying respirator with organic vapor cartridge(s)] or SCBAF (APF = 50) (any self-contained breathing apparatus with a full face-piece) or SaF (APF = 50) (any supplied-air respirator with a full face-piece). *Emergency or planned entry into unknown concentrations or IDLH conditions:* SCBAF: Pd,Pp (APF = 10,000) (any self-contained breathing apparatus that has a full face-piece and is operated in a pressure-demand or other positive-pressure mode) or SaF: Pd,Pp: ASCBA (APF = 10,000) (any supplied-air respirator that has a full face-piece and is operated in a pressure-demand or other positive-pressure mode in combination with an auxiliary, self-contained breathing apparatus operated in a pressure-demand or other positive-pressure mode). *Escape:* GmFOv (APF = 50) [any air-purifying, full-face-piece respirator (gas mask) with a chin-style, front- or back-mounted organic vapor canister] or SCBAE (any appropriate escape-type, self-contained breathing apparatus).

Note: Causes eye irritation or damage; may require eye protection.

Petroleum naphtha: NIOSH: *Up to 850 ppm:* Sa (APF = 10) (any supplied-air respirator). *Up to 1100 ppm:* Sa:Cf (APF = 25) (any supplied-air respirator operated in a continuous-flow mode);* SCBAF (APF = 50) (any self-contained breathing apparatus with a full face-piece); SaF (APF = 50) (any supplied-air respirator with a full face-piece). *Emergency or planned entry into unknown concentrations or IDLH conditions:* SCBAF: Pd,Pp (APF = 10,000) (any self-contained breathing apparatus that has a full face-piece and is operated in a pressure-demand or other positive-pressure mode) or SaF: Pd,Pp: ASCBA

(APF = 10,000) (any supplied-air respirator that has a full face-piece and is operated in a pressure-demand or other positive-pressure mode in combination with an auxiliary, self-contained breathing apparatus operated in a pressure-demand or other positive-pressure mode). *Escape:* GmFOv (APF = 50) [any air-purifying, full-face-piece respirator (gas mask) with a chin-style, front- or back-mounted organic vapor canister] or SCBAE (any appropriate escape-type, self-contained breathing apparatus).

*Substance reported to cause eye irritation or damage; may require eye protection.

VM & P naphtha: 3500 mg/m³: CcrOv (APF = 10) [any chemical cartridge respirator with organic vapor cartridge(s)] or Sa (APF = 10) (any supplied-air respirator). *8750 mg/m³:* Sa:Cf (APF = 25) (any supplied-air respirator operated in a continuous-flow mode) or PaprOv (APF = 25) [any powered, air-purifying respirator with organic vapor cartridge (s)]. *17,500 mg/m³:* CcrFOv (APF = 50) [any air-purifying, full-face-piece respirator (gas mask) with a chin-style, front- or back-mounted acid gas canister] or GmFOv (APF = 50) [any air-purifying, full-face-piece respirator (gas mask) with a chin-style, front- or back-mounted organic vapor canister] or PaprTOv (APF = 50) [any powered, air-purifying respirator with a tight-fitting face-piece and organic vapor cartridge(s)] or SCBAF (APF = 50) (any self-contained breathing apparatus with a full face-piece) or SaF (APF = 50) (any supplied-air respirator with a full face-piece). *Emergency or planned entry into unknown concentrations or IDLH conditions:* SCBAF: Pd,Pp (APF = 10,000) (any self-contained breathing apparatus that has a full face-piece and is operated in a pressure-demand or other positive-pressure mode) or SaF: Pd,Pp: ASCBA (APF = 10,000) (any supplied-air respirator that has a full face-piece and is operated in a pressure-demand or other positive-pressure mode in combination with an auxiliary self-contained breathing apparatus operated in a pressure-demand or other positive-pressure mode). *Escape:* GmFOv (APF = 50) [any air-purifying, full-face-piece respirator (gas mask) with a chin-style, front- or back-mounted organic vapor canister] or SCBAE (any appropriate escape-type, self-contained breathing apparatus).

Storage: Color Code—Red: Flammability Hazard: Store in a flammable liquid storage area or approved cabinet away from ignition sources and corrosive and reactive materials. Naphthas must be stored to avoid contact with strong oxidizers (such as chlorine, bromine, and fluorine), since violent reactions occur. Before entering confined space where this chemical may be present, check to make sure that an explosive concentration does not exist. Store in tightly closed containers in a cool, well-ventilated area away from heat. Sources of ignition, such as smoking and open flames are prohibited where naphthas are handled, used, or stored. Metal containers involving the transfer of 5 gallons or more of naphthas should be grounded and bonded. Drums must be equipped with self-closing valves, pressure vacuum bungs, and flame arresters. Use only nonsparking tools and equipment, especially when opening and closing containers of naphtha. Wherever naphtha is used, handled, manufactured, or stored, use explosion-proof electrical equipment and fittings.

Shipping: This compound requires a shipping label of "FLAMMABLE LIQUID." It falls in DOT Hazard Class 3 and Packing Group II.

Spill Handling: Evacuate and restrict persons not wearing protective equipment from area of spill or leak until cleanup is complete. Remove all ignition sources. Establish forced ventilation to keep levels below explosive limit. Absorb liquids in vermiculite, dry sand, earth, peat, carbon, or a similar material and deposit in sealed containers. Keep this chemical out of a confined space, such as a sewer, because of the possibility of an explosion, unless the sewer is designed to prevent the buildup of explosive concentrations. It may be necessary to contain and dispose of this chemical as a hazardous waste. If material or contaminated runoff enters waterways, notify downstream users of potentially contaminated waters. Contact your local or federal environmental protection agency for specific recommendations. If employees are required to clean up spills, they must be properly trained and equipped. OSHA 1910.120(q) may be applicable.

Fire Extinguishing: These chemicals are flammable liquids. Poisonous gases are produced in fire. Use dry chemical, carbon dioxide, or alcohol foam extinguishers. Use water for cooling only. Vapors are heavier than air and will collect in low areas. Vapors may travel long distances to ignition sources and flashback. Vapors in confined areas may explode when exposed to fire. Containers may explode in fire. Storage containers and parts of containers may rocket great distances, in many directions. If material or contaminated runoff enters waterways, notify downstream users of potentially contaminated waters. Notify local health and fire officials and pollution control agencies. From a secure, explosion-proof location, use water spray to cool exposed containers. If cooling streams are ineffective (venting sound increases in volume and pitch, tank discolors, or shows any signs of deforming), withdraw immediately to a secure position. If employees are expected to fight fires, they must be trained and equipped in OSHA 1910.156. The only respirators recommended for firefighting are self-contained breathing apparatuses that have full face-pieces and are operated in a pressure-demand or other positive-pressure mode.

Disposal Method Suggested: Dissolve or mix the material with a combustible solvent and burn in a chemical incinerator equipped with an afterburner and scrubber. All federal, state, and local environmental regulations must be observed.

References

National Institute for Occupational Safety and Health. (1977). *Criteria for a Recommended Standard: Occupational Exposure to Refined Petroleum,* NIOSH Document No. 77-192. Washington, DC

New Jersey Department of Health and Senior Services. (December 1985). *Hazardous Substances Fact Sheet: Benzine.* Trenton, NJ

New Jersey Department of Health and Senior Services. (March 1986). *Hazardous Substances Fact Sheet: Coal Tar Naphtha.* Trenton, NJ

New Jersey Department of Health and Senior Services. (August 1998). *Hazardous Substances Fact Sheet: VM&P Naphtha.* Trenton, NJ

New York State Department of Health. (May 1986). *Chemical Fact Sheet: VM&P Naphtha.* Version 2. Albany, NY: Bureau of Toxic Substance Assessment

US Environmental Protection Agency. (September 28, 1984). *Chemical Hazard Information Profile: Naphtha (Petroleum) Solvents.* Washington, DC: Office of Toxic Substances

US Environmental Protection Agency, Special Review and Reregistration Division Office of Pesticide Programs. (1998). *Agency Status of Pesticides in Registration, Reregistration, and Special Review* (Rainbow Report). Washington, DC

New Jersey Department of Health and Senior Services. (January 2001). *Hazardous Substances Fact Sheet: Naphtha (Coal Tar).* Trenton, NJ

Naphthalene   N:0120

Molecular Formula: $C_{10}H_8$

Synonyms: Agitene 141/super; Albocarbon; Camphor tar; Dezodorator; Mighty 150; Moth balls; Moth flakes; Naftaleno (Spanish); Naphthaline; Napthalene, moletn; Napthalin; Napthaline; Napthene; NCI-C52904; Tar camphor

CAS Registry Number: 91-20-3

RTECS® Number: QJ0525000

UN/NA & ERG Number: UN1334 (crude and refined)/133; UN2304 (molten)/133

EC Number: 202-049-5 [*Annex I Index No.:* 601-052-00-2] (naphthalene, crude or refined)

Regulatory Authority and Advisory Bodies

Carcinogenicity: NCI: Carcinogenesis Studies (inhalation); clear evidence: rat; equivocal evidence: mouse; NTP: Carcinogenesis Studies (inhalation); some evidence: mouse; NTP: 11th Report on Carcinogens, 2004: Reasonably anticipated to be a human carcinogen; IARC: Animal Sufficient Evidence; Human Inadequate Evidence, *possibly carcinogenic to humans*, Group 2B, 2002; EPA: Possible Human Carcinogen; Cannot be Determined.

US EPA Gene-Tox Program, Negative: Cell transform.—mouse embryo; Negative: Cell transform.—RLV F344 rat embryo; Negative: Histidine reversion—Ames test.

US EPA, FIFRA, 1998 Status of Pesticides: Supported.

Air Pollutant Standard Set. See below, "Permissible Exposure Limits in Air" section.

Clean Air Act: Hazardous Air Pollutants (Title I, Part A, Section 112).

Clean Water Act: Section 311 Hazardous Substances/RQ 40CFR117.3 (same as CERCLA, see below); 40CFR401.15 Section 307 Toxic Pollutants.

US EPA Hazardous Waste Number (RCRA No.): U165.

RCRA, 40CFR261, Appendix 8 Hazardous Constituents.

RCRA 40CFR268.48; 61FR15654, Universal Treatment Standards: Wastewater (mg/L), 0.059; Nonwastewater (mg/kg), 5.6.

RCRA 40CFR264, Appendix 9; TSD Facilities Ground Water Monitoring List. Suggested test method(s) (PQL µg/L): 8100 (200); 8270 (10).

Safe Drinking Water Act: Regulated chemical (47 FR 9352).

Reportable Quantity (RQ): 100 lb (45.4 kg).

EPCRA Section 313 Form R *de minimis* concentration reporting level: 1.0%.

US DOT Regulated Marine Pollutant (49CFR172.101, Appendix B).

California Proposition 65 Chemical: Cancer 4/19/02.

Canada, WHMIS, Ingredients Disclosure List Concentration 1.0%.

European/International Regulations: Hazard Symbol: Xn, N; Risk phrases: R22; R40; R50/53; Safety phrases: S2; S36/37; S46; S60; S61 (see Appendix 4).

WGK (German Aquatic Hazard Class): 3—Severe hazard to waters.

Description: Naphthalene is a colorless to brown crystalline solid with a characteristic "moth ball" odor. Shipped as a molten solid. Odor threshold = 0.038 ppm. Molecular weight = 128.18; Specific gravity (H_2O:1) = 1.15; Boiling point = 217.8°C; Freezing/Melting point = 74−80°C; Vapor pressure = 0.08 mmHg at 20°C; Flash point = 78.9°C; Autoignition temperature = 540°C. Explosive limits: LEL = 0.9%; UEL = 5.9%. Hazard Identification (based on NFPA-704 M Rating System): Health 2, Flammability 2, Reactivity 0. Practically insoluble in water; solubility = 0.003%.

Potential Exposure: Compound Description: Agricultural Chemical; Tumorigen, Mutagen; Reproductive Effector; Human Data; Primary Irritant. Naphthalene is used as a chemical intermediate or feedstock for synthesis of phthalic, anthranilic, hydroxyl (naphthols), amino (naphthylamines), and sulfonic compounds, which are used in the manufacture of various dyes and in the preparation of phthalic anhydride, 1-naphthyl-*N*-methyl carbonate, and β-naphthol. Naphthalene is also used in the manufacture of hydronaphthalenes, synthetic resins, lampblack, smokeless powder, and celluloid. Naphthalene has been used as a moth repellent.

Approximately 100 million people worldwide have G6PD deficiency which would make them more susceptible to hemolytic anemia on exposure to naphthalene. At present, more than 80 variants of this enzyme deficiency have been identified. The incidence of this deficiency is 0.1% in

American and European Caucasians but can range as high as 20% in American blacks and greater than 50% in certain Jewish groups. Newborn infants have a similar sensitivity to the hemolytic effects of naphthalene, even without G6PD deficiency.

Incompatibilities: Violent reactions with chromium(III) oxide, dinitrogen pentoxide, chromic anhydride, and strong oxidizers.

Permissible Exposure Limits in Air
OSHA PEL: 10 ppm/50 mg/m^3 TWA.
NIOSH REL: 10 ppm/50 mg/m^3 TWA; 15 ppm/75 mg/m^3 STEL.
ACGIH TLV®[1]: 10 ppm/52 mg/m^3 TWA; 15 ppm/79 mg/m^3 STEL [skin].
DFG MAK: [skin] Carcinogen Category 3B.
NIOSH IDLH: 250 ppm.
Protective Action Criteria (PAC)
TEEL-0: 10 ppm
PAC-1: 15 ppm
PAC-2: 15 ppm
PAC-3: 250 ppm
Arab Republic of Egypt: TWA 10 ppm (50 mg/m^3), 1993; Austria: MAK 10 ppm (50 mg/m^3), 1999; Belgium: TWA 10 ppm (52 mg/m^3); STEL 15 ppm (79 mg/m^3), 1993; Denmark: TWA 10 ppm (50 mg/m^3), 1999; Finland: TWA 10 ppm (50 mg/m^3); STEL 20 ppm (100 mg/m^3), 1999; France: VME 10 ppm (50 mg/m^3), 1999; Hungary: TWA 40 mg/m^3; STEL 80 mg/m^3, [skin], 1993; the Netherlands: MAC-TGG 50 mg/m^3, 2003; Norway: TWA 10 ppm (50 mg/m^3), 1999; the Philippines: TWA 10 ppm (50 mg/m^3), 1993; Poland: MAC (TWA) 20 mg/m^3; MAC (STEL) 75 mg/m^3, 1999; Russia: STEL 20 mg/m^3, 1993; Switzerland: MAK-W 10 ppm (50 mg/m^3), 1999; United Kingdom: TWA 10 ppm (53 mg/m^3); STEL 15 ppm (80 mg/m^3), 2000; Argentina, Bulgaria, Columbia, Jordan, South Korea, New Zealand, Singapore, Vietnam: ACGIH TLV®: STEL 15 ppm [skin]. Russia[43] set a MAC of 0.003 mg/m^3 (3 μg/m^3) in ambient air in residential areas. Several states have set guidelines or standards for naphthalene in ambient air[60] ranging from 14 μg/m^3 (Massachusetts) to 166.7 μg/m^3 (New York) to 500 μg/m^3 (Florida) to 500−750 μg/m^3 (North Dakota) to 800 μg/m^3 (Virginia) to 1000 μg/m^3 (Connecticut) to 1190 μg/m^3 (Nevada) to 1250 μg/m^3 (South Carolina).

Determination in Air: Use NIOSH Analytical Methods #1501; OSHA Analytical Method 35.

Permissible Concentration in Water: *To protect freshwater aquatic life:* 2300 μg/L on an acute toxicity basis and 620 μg/L on a chronic basis. *To protect saltwater aquatic life:* 2350 μg/L on an acute toxicity basis. For the protection of human health from the toxic properties of naphthalene ingested through water and through contaminated aquatic organisms, no ambient water criterion has been set due to insufficient data.[6] Russia[43] set a MAC of 0.004 mg/L in water bodies used for fishery purposes. Kansas has set a guideline of 143 μg/L for naphthalene in drinking water.[61]

Determination in Water: Methylene chloride extraction followed by high-pressure liquid chromatography with fluorescence or UV detection; or gas chromatography (EPA Method 610) or gas chromatography plus mass spectrometry (EPA Method 625). Octanol−water coefficient: Log $K_{ow} = 3.3$.

Routes of Entry: Inhalation of vapor or dust, skin absorption, ingestion, skin and/or eye contact.

Harmful Effects and Symptoms
Short Term Exposure: Irritates the eyes, skin, and respiratory tract. High levels cause headache, fatigue, confusion, excitement, malaise, nausea, and vomiting. *Inhalation:* Levels above 10 ppm may cause headache, nausea, excessive sweating, and vomiting. *Skin:* May cause irritation. If hypersensitive to naphthalene, severe irritation may occur. *Eyes:* Levels above 15 ppm may cause irritation. Direct contact may cause severe irritation, injury to the cornea, and a blurring of vision. *Ingestion:* Ingestion of ½ g (1/60 oz) may cause nausea, vomiting, abdominal pain, irritation of the bladder, and brown or black coloration of the urine. The symptoms usually disappear after a few days. Animal studies indicate that the probable lethal dose for an adult is 5−15 g (1/16−½ oz).

Long Term Exposure: Repeated exposure or ingestion may cause clouding of the eye (cataract). Inhalation of levels above 10 ppm may cause headaches, nausea, vomiting, and a feeling of general discomfort. Chronic skin problems are rare, except in cases of hypersensitivity. May cause skin allergy, kidney, and liver damage. May damage the red blood cells causing anemia.

Points of Attack: Eyes, skin, blood, liver, kidneys, central nervous system.

Medical Surveillance: NIOSH lists the following tests: blood plasma, hemoglobin; complete blood count; liver function tests; red blood cells/count, RBC hemolysis; urine (chemical/metabolite); urine (chemical/metabolite), hemoglobin; urinalysis (routine); white blood cell count/differential. For those with frequent or potentially high exposure (half the TLV or greater, or significant skin contact), the following is recommended before beginning work and at regular times after that: examination of the eyes and vision. If symptoms develop or overexposure is suspected, the following may be useful: liver and kidney function tests. Complete blood count (CBC). Evaluation by a qualified allergist, including careful exposure history and special testing, may help diagnose skin allergy.

First Aid: If this chemical gets into the eyes, remove any contact lenses at once and irrigate immediately for at least 15 min, occasionally lifting upper and lower lids. Seek medical attention immediately. If this chemical contacts the skin, remove contaminated clothing and wash immediately with soap and water. Seek medical attention immediately. If this chemical has been inhaled, remove from exposure, begin rescue breathing (using universal precautions, including resuscitation mask) if breathing has stopped and CPR if heart action has stopped. Transfer promptly to a medical

facility. When this chemical has been swallowed, get medical attention. Give large quantities of water and induce vomiting. Do not make an unconscious person vomit.

Personal Protective Methods: Wear protective gloves and clothing to prevent any reasonable probability of skin contact. Safety equipment suppliers/manufacturers can provide recommendations on the most protective glove/clothing material for your operation. All protective clothing (suits, gloves, footwear, headgear) should be clean, available each day, and put on before work. Contact lenses should not be worn when working with this chemical. Wear dust-proof chemical goggles and face shield unless full face-piece respiratory protection is worn. Employees should wash immediately with soap when skin is wet or contaminated. Provide emergency showers and eyewash.

Respirator Selection: Up to 2.5 mg/m^3: Qm (APF = 25) (any quarter-mask respirator). Up to 5 mg/m^3: 95XQ (APF = 10) [any particulate respirator equipped with an N95, R95, or P95 filter (including N95, R95, and P95 filtering face-pieces) except quarter-mask respirators. The following filters may also be used: N99, R99, P99, N100, R100, P100] or any supplied-air respirator.* Up to 12.5 mg/m^3: Sa:Cf (any supplied-air respirator operated in a continuous-flow mode)*; or PaprHie* (APF = 25) (any powered, air-purifying respirator with a high-efficiency particulate filter). Up to 25 mg/m^3: 100F (APF = 50) (any air-purifying, full-face-piece respirator with an N100, R Sa:Cf* (any supplied-air respirator operated in a continuous-flow mode) 100, or P100 filter) or any powered, air-purifying respirator with a tight-fitting face-piece and a high-efficiency particulate filter*; or any self-contained breathing apparatus with a full face-piece; or any supplied-air respirator with a full face-piece. *Emergency or planned entry into unknown concentrations or IDLH conditions:* SCBAF: Pd,Pp (APF = 10,000) (any self-contained breathing apparatus that has a full face-piece and is operated in a pressure-demand or other positive-pressure mode) or SaF: Pd,Pp: ASCBA (APF = 10,000) (any supplied-air respirator that has a full face-piece and is operated in a pressure-demand or other positive-pressure mode in combination with an auxiliary, self-contained breathing apparatus operated in a pressure-demand or other positive-pressure mode). *Escape:* GmFOv100 [any air-purifying, full-face-piece respirator (gas mask) with a chin-style, front- or back-mounted organic vapor canister having an N100, R100, or P100 filter] or SCBAE (any appropriate escape-type, self-contained breathing apparatus).

Note: Substance reported to cause eye irritation or damage; may require eye protection.

Storage: Color Code—Red: Flammability Hazard: Store in a flammable storage area. Naphthalene must be stored to avoid contact with chromium(III) oxide, dinitrogen pentoxide, and strong oxidizers (such as chlorine, bromine, and fluorine), since violent reactions occur. Before entering confined space where this chemical may be present, check to make sure that an explosive concentration does not exist. Store in tightly closed containers in a cool, well-ventilated area. Sources of ignition, such as smoking and open flames

are prohibited where naphthalene is used, handled, or stored in a manner that could create a potential fire or explosion hazard. Metal containers involving the transfer of 5 gallons or more of naphthalene should be grounded and bonded. Drums must be equipped with self-closing valves, pressure vacuum bungs, and flame arresters. Liquid naphthalene must avoid contact with water.

Shipping: Naphthalene, crude or Naphthalene, refined, requires a shipping label of "FLAMMABLE SOLID." It falls in Hazard Class 4.1 and Packing Group III.

Spill Handling: Evacuate persons not wearing protective equipment from area of spill or leak until cleanup is complete. Remove all ignition sources. Establish forced ventilation to keep levels below explosive limit. Use HEPA vacuum or wet method to reduce dust during cleanup. Do not dry sweep. Collect powdered material in the most convenient and safe manner and deposit in sealed containers. Ventilate area after cleanup is complete. Keep naphthalene out of a confined space, such as a sewer, because of the possibility of an explosion, unless the sewer is designed to prevent the buildup of explosive concentrations. It may be necessary to contain and dispose of this chemical as a hazardous waste. If material or contaminated runoff enters waterways, notify downstream users of potentially contaminated waters. Contact your local or federal environmental protection agency for specific recommendations. If employees are required to clean up spills, they must be properly trained and equipped. OSHA 1910.120(q) may be applicable.

Fire Extinguishing: This chemical is a combustible solid but is not easily ignited. Use dry chemical, carbon dioxide, water spray, or foam extinguishers. Use caution when using water spray or foam directly on molten naphthalene as extensive foaming may occur. Poisonous gases are produced in fire. If material or contaminated runoff enters waterways, notify downstream users of potentially contaminated waters. Notify local health and fire officials and pollution control agencies. From a secure, explosion-proof location, use water spray to cool exposed containers. If cooling streams are ineffective (venting sound increases in volume and pitch, tank discolors, or shows any signs of deforming), withdraw immediately to a secure position. If employees are expected to fight fires, they must be trained and equipped in OSHA 1910.156. The only respirators recommended for firefighting are self-contained breathing apparatuses that have full face-pieces and are operated in a pressure-demand or other positive-pressure mode.

Disposal Method Suggested: Dissolve or mix the material with a combustible solvent and burn in a chemical incinerator equipped with an afterburner and scrubber. All federal, state, and local environmental regulations must be observed. Consult with environmental regulatory agencies for guidance on acceptable disposal practices. Generators of waste containing this contaminant ($\geq$100 kg/mo) must conform with EPA regulations governing storage, transportation, treatment, and waste disposal.

References
US Environmental Protection Agency. (1980). *Naphthalene: Ambient Water Quality Criteria.* Washington, DC
National Institute for Occupational Safety and Health. (1977). *Profiles on Occupational Hazards for Criteria Document Priorities: Naphthalene*, Report PB-274,073. Cincinnati, OH. pp. 269–273
US Environmental Protection Agency. (April 30, 1980). *Naphthalene, Health and Environmental Effects Profile No. 131.* Washington, DC: Office of Solid Waste
Sax, N. I. (Ed.). (1985). *Dangerous Properties of Industrial Materials Report,* 5, No. 4, 71–74
New York State Department of Health. (March 1986). *Chemical Fact Sheet: Naphthalene.* Version 2. Albany, NY: Bureau of Toxic Substance Assessment
US Environmental Protection Agency, Special Review and Reregistration Division Office of Pesticide Programs. (1998). *Agency Status of Pesticides in Registration, Reregistration, and Special Review* (Rainbow Report). Washington, DC
New Jersey Department of Health and Senior Services. (October 2004). *Hazardous Substances Fact Sheet: Naphthalene.* Trenton, NJ

Naphthenic acid   N:0130

Molecular Formula: $R_2C\text{-}CR_2\text{-}CR_2\text{-}CR_2\text{-}CR\text{-}(CH_2)_n\text{-}COOH$ (where $n = 2–6$); $CnH_{2n}1COOH$
Synonyms: Acido naftalico (Spanish); Agenap; Agenap HMW-H; Cyclopentane carboxylic acid; Naphid; Sunaptic acid B; Sunaptic acid C
CAS Registry Number: 1338-24-5
RTECS® Number: QK8750000
UN/NA & ERG Number: UN3077/171
EC Number: 215-662-8
Regulatory Authority and Advisory Bodies
Clean Water Act: Section 311 Hazardous Substances/RQ 40CFR117.3 (same as CERCLA, see below).
Reportable Quantity (RQ): 100 lb (45.4 kg).
US DOT Regulated Marine Pollutant (49CFR172.101, Appendix B).
European/International Regulations: not listed in Annex 1.
WGK (German Aquatic Hazard Class): No value assigned.
Description: Naphthenic acid is a gold to black, odorless liquid. Molecular weight = (approx.) 200–250 (mixture); Specific gravity (H_2O:1) = 0.982 at 20°C; Boiling point = 132–243°C; Freezing/Melting point = 31°C; Flash point = 149°C (oc). Explosive limits: LEL = 0.9–1%; UEL = unknown. Hazard Identification (based on NFPA-704 M Rating System): Health 2, Flammability 2, Reactivity 0; Slightly soluble in water.
Potential Exposure: Used to make metallic naphthenates for paint dryers and cellulose preservatives. It is also used as a solvent, detergent, rubber reclaiming agent. Used in catalysts, cutting oils, drilling compounds, rust inhibitors, surfactants, emulsions, grease, and wood preservatives.
Incompatibilities: Incompatible with sulfuric acid, caustics, ammonia, aliphatic amines, alkanolamines, isocyanates, alkylene oxides, epichlorohydrin, strong oxidizers. Corrosive to metals.
Permissible Exposure Limits in Air
Protective Action Criteria (PAC)
Lead salt
TEEL-0: 1.25 mg/m³
PAC-1: 4 mg/m³
PAC-2: 30 mg/m³
PAC-3: 500 mg/m³
Determination in Air: No method established.
Harmful Effects and Symptoms
Short Term Exposure: Vapors irritate the eyes, skin, and respiratory tract. Contact with the skin causes reddening.
Long Term Exposure: Repeated or prolonged contact causes dry and cracked skin and may cause contact dermatitis. May cause liver and central nervous system damage.
Medical Surveillance: There are no special tests.
First Aid: If this chemical gets into the eyes, remove any contact lenses at once and irrigate immediately for at least 15 min, occasionally lifting upper and lower lids. Seek medical attention immediately. If this chemical contacts the skin, remove contaminated clothing and wash immediately with soap and water. Seek medical attention immediately. If this chemical has been inhaled, remove from exposure, begin rescue breathing (using universal precautions, including resuscitation mask) if breathing has stopped and CPR if heart action has stopped. Transfer promptly to a medical facility. When this chemical has been swallowed, get medical attention. Give large quantities of water and induce vomiting. Do not make an unconscious person vomit.
Personal Protective Methods: Wear protective gloves and clothing to prevent any reasonable probability of skin contact. Safety equipment suppliers/manufacturers can provide recommendations on the most protective glove/clothing material for your operation. All protective clothing (suits, gloves, footwear, headgear) should be clean, available each day, and put on before work. Contact lenses should not be worn when working with this chemical. Wear splash-proof chemical goggles and face shield unless full face-piece respiratory protection is worn. Employees should wash immediately with soap when skin is wet or contaminated. Provide emergency showers and eyewash.
Respirator Selection: Follow regulations in OSHA 29CFR1910.134 or European Standard EN149. Use a NIOSH/MSHA- or European Standard EN149-approved respirator; or use an approved supplied-air respirator with a full face-piece operated in the positive-pressure mode, or with a full face-piece, hood, or helmet in the continuous-flow mode; or use a NIOSH/MSHA- or European Standard EN149-approved self-contained breathing apparatus with a full face-piece operated in pressure-demand or other positive-pressure mode.

Storage: Color Code—Green: General storage may be used. Prior to working with this chemical you should be trained on its proper handling and storage. Store in tightly closed containers in a cool, well-ventilated area away from metals and other incompatible materials listed above. Where possible, automatically pump liquid from drums or other storage containers to process containers. Sources of ignition, such as smoking and open flames are prohibited where this chemical is handled, used, or stored. Metal containers involving the transfer of 5 gallons or more of this chemical should be grounded and bonded. Drums must be equipped with self-closing valves, pressure vacuum bungs, and flame arresters. Use only nonsparking tools and equipment, especially when opening and closing containers of this chemical. Wherever this chemical is used, handled, manufactured, or stored, use explosion-proof electrical equipment and fittings.

Shipping: The name of this material is not on the DOT list of materials[19] for label and packaging standards. However, based on regulations, it may be classified[52] as an Environmentally hazardous substances, liquid, n.o.s. It falls in Hazard Class 9 and Packing Group III.[20, 21]

Spill Handling: Evacuate and restrict persons not wearing protective equipment from area of spill or leak until cleanup is complete. Remove all ignition sources. Ventilate area of spill or leak. Absorb liquids in vermiculite, dry sand, earth, peat, carbon, or a similar material and deposit in sealed containers. Keep this chemical out of a confined space, such as a sewer, because of the possibility of an explosion, unless the sewer is designed to prevent the buildup of explosive concentrations. It may be necessary to contain and dispose of this chemical as a hazardous waste. If material or contaminated runoff enters waterways, notify downstream users of potentially contaminated waters. Contact your local or federal environmental protection agency for specific recommendations. If employees are required to clean up spills, they must be properly trained and equipped. OSHA 1910.120(q) may be applicable.

Fire Extinguishing: This chemical may burn but does not easily ignite. Poisonous gases are produced in fire. Use dry chemical, carbon dioxide, or foam extinguishers. Vapors are heavier than air and will collect in low areas. Vapors may travel long distances to ignition sources and flashback. Vapors in confined areas may explode when exposed to fire. Containers may explode in fire. Storage containers and parts of containers may rocket great distances, in many directions. If material or contaminated runoff enters waterways, notify downstream users of potentially contaminated waters. Notify local health and fire officials and pollution control agencies. From a secure, explosion-proof location, use water spray to cool exposed containers. If cooling streams are ineffective (venting sound increases in volume and pitch, tank discolors, or shows any signs of deforming), withdraw immediately to a secure position. If employees are expected to fight fires, they must be trained and equipped in OSHA 1910.156. The only respirators recommended for firefighting are self-contained breathing apparatuses that have full face-pieces and are operated in a pressure-demand or other positive-pressure mode.

Reference
New Jersey Department of Health and Senior Services. (June1999). *Hazardous Substances Fact Sheet: Naphthenic Acid.* Trenton, NJ

2-Naphthol   N:0140

Molecular Formula: $C_{10}H_8O$
Synonyms: β-Hydroxynaphthalene; 2-Naphthalenol; 2-Naphthol; β-Naphthol
CAS Registry Number: 135-19-3
RTECS® Number: QL2975000
UN/NA & ERG Number: UN2811 (toxic solid, organic, n.o.s.)/154
EC Number: 205-182-7 [*Annex I Index No.:* 604-007-00-5]
Regulatory Authority and Advisory Bodies
US EPA, FIFRA 1998 Status of Pesticides: Canceled.
Air Pollutant Standard Set. See below, "Permissible Exposure Limits in Air" section.
Canada, WHMIS, Ingredients Disclosure List Concentration 1%.
European/International Regulations: Hazard Symbol: Xn, N; Risk phrases: R20/22; R50; Safety phrases: S2; S24/25; S61 (see Appendix 4).
WGK (German Aquatic Hazard Class): 2—Hazard to waters.

Description: 2-Naphthol is a white, crystalline solid with a slight phenolic odor. Darkens in air and on exposure to light. Molecular weight = 144.18; Boiling point = 285−286°C; Freezing/Melting point = 121−123°C; Flash point = 153°C. Hazard Identification (based on NFPA-704 M Rating System): Health 2, Flammability 1, Reactivity 0; (1-naphthol) Health 3, Flammability 1, Reactivity 1. Slightly soluble in water.

Potential Exposure: Compound Description: Drug, Mutagen, Primary Irritant. Those involved in rubber antioxidant production, synthesis of dyes, leather processing, fungicides, pharmaceuticals, and perfumes. Used as an antioxidant for fats, oils, as an antiseptic, in insecticides.

Incompatibilities: Oxidizers, iron salts, 2,3-dimethyl-1-phenyl-3-pyrazolin-5-one (antipyrine), camphor, phenol, menthol, urethane.

Permissible Exposure Limits in Air
Protective Action Criteria (PAC)
TEEL-0: 0.0025 mg/m^3
PAC-1: 0.0075 mg/m^3
PAC-2: 0.05 mg/m^3
PAC-3: 500 mg/m^3
Russia: STEL 0.1 mg/m^3, 1993.
Russia[43] set a MAC of 0.006 mg/m^3 in ambient air in residential areas on a momentary basis and 0.003 mg/m^3 on a daily average basis.

Permissible Concentration in Water: Russia[43] set a MAC of 0.4 mg/L in water bodies used for domestic purposes.

Routes of Entry: Inhalation, skin absorption.

Harmful Effects and Symptoms

Short Term Exposure: Irritates the skin and eyes. Ingestion can cause nephritis, lens opacity, vomiting, diarrhea, abdominal pain, circulatory collapse, convulsion, hemolytic anemia, and death.

Long Term Exposure: May cause kidney damage, anemia. May cause mutations.

Points of Attack: Eyes, skin, blood, liver, kidneys, central nervous system.

Medical Surveillance: Complete blood count (CBC), kidney damage.

First Aid: If this chemical gets into the eyes, remove any contact lenses at once and irrigate immediately for at least 15 min, occasionally lifting upper and lower lids. Seek medical attention immediately. If this chemical contacts the skin, remove contaminated clothing and wash immediately with soap and water. Seek medical attention immediately. If this chemical has been inhaled, remove from exposure, begin rescue breathing (using universal precautions, including resuscitation mask) if breathing has stopped and CPR if heart action has stopped. Transfer promptly to a medical facility. When this chemical has been swallowed, get medical attention. Give large quantities of water and induce vomiting. Do not make an unconscious person vomit.

Personal Protective Methods: Wear protective gloves and clothing to prevent any reasonable probability of skin contact. Safety equipment suppliers/manufacturers can provide recommendations on the most protective glove/clothing material for your operation. All protective clothing (suits, gloves, footwear, headgear) should be clean, available each day, and put on before work. Contact lenses should not be worn when working with this chemical. Wear dust-proof chemical goggles and face shield unless full face-piece respiratory protection is worn. Employees should wash immediately with soap when skin is wet or contaminated. Provide emergency showers and eyewash.

Respirator Selection: Self-contained breathing apparatus.

Storage: Color Code—Blue (*1-naphthol*): Health Hazard/Poison: Store in a secure poison location. Color Code—Green (2-naphthol): General storage may be used. Prior to working with this chemical you should be trained on its proper handling and storage. Store in a cool, dark place.

Shipping: Toxic solids, organic, n.o.s. compound requires a shipping label of "POISONOUS/TOXIC MATERIALS." It falls in Hazard Class 6.1, Packing Group III.

Spill Handling: Evacuate persons not wearing protective equipment from area of spill or leak until cleanup is complete. Remove all ignition sources. Use HEPA vacuum or wet method to reduce dust during cleanup. Do not dry sweep. Collect powdered material in the most convenient and safe manner and deposit in sealed containers. Ventilate area after cleanup is complete. It may be necessary to contain and dispose of this chemical as a hazardous waste. If material or contaminated runoff enters waterways, notify downstream users of potentially contaminated waters. Contact your local or federal environmental protection agency for specific recommendations. If employees are required to clean up spills, they must be properly trained and equipped. OSHA 1910.120(q) may be applicable.

Fire Extinguishing: This chemical is a combustible solid. Use dry chemical, carbon dioxide, water spray, or alcohol foam extinguishers. Poisonous gases are produced in fire. If material or contaminated runoff enters waterways, notify downstream users of potentially contaminated waters. Notify local health and fire officials and pollution control agencies. From a secure, explosion-proof location, use water spray to cool exposed containers. If cooling streams are ineffective (venting sound increases in volume and pitch, tank discolors, or shows any signs of deforming), withdraw immediately to a secure position. If employees are expected to fight fires, they must be trained and equipped in OSHA 1910.156. The only respirators recommended for firefighting are self-contained breathing apparatuses that have full face-pieces and are operated in a pressure-demand or other positive-pressure mode.

Disposal Method Suggested: Mix with flammable solvent and atomize into an incinerator.

References

Sax, N. I. (Ed.). *Dangerous Properties of Industrial Materials Report,* 2, No. 3, 81−83 (1982) & 3, No. 6, 49−52 (1983) & 8, No. 3, 79−86 (1988)

US Environmental Protection Agency, Special Review and Reregistration Division Office of Pesticide Programs. (1998). *Agency Status of Pesticides in Registration, Reregistration, and Special Review* (Rainbow Report). Washington, DC

1,4-Naphthoquinone   N:0150

Molecular Formula: $C_{10}H_6O_2$

Synonyms: 1,4-Dihydro-1,4-diketonaphthalene; Naftoquinona (Spanish); 1,4-Naphthalenedione; α-Naphthoquinone

CAS Registry Number: 130-15-4

RTECS® Number: QL7175000

UN/NA & ERG Number: UN2811 (toxic solid, organic, n.o.s.)/154

EC Number: 204-977-6

Regulatory Authority and Advisory Bodies

Air Pollutant Standard Set. See below, "Permissible Exposure Limits in Air" section.

US EPA Hazardous Waste Number (RCRA No.): U166.

RCRA, 40CFR261, Appendix 8 Hazardous Constituents.

RCRA 40CFR264, Appendix 9; TSD Facilities Ground Water Monitoring List. Suggested test method(s) (PQL µg/L): 8270 (10).

Reportable Quantity (RQ): 5000 lb (2270 kg).

European/International Regulations: not listed in Annex 1.

WGK (German Aquatic Hazard Class): 3—Severe hazard to waters.

Description: 1,4-Naphthoquinone is a yellow to greenish-yellow crystalline solid with a pungent odor. Odor threshold = 0.02 ppm. Molecular weight = 158.16; Sublimation point = 100°C; Freezing/Melting point = 123−126°C. Hazard Identification (based on NFPA-704 M Rating System): Health 2, Flammability 1, Reactivity 0. Slightly soluble in water.

Potential Exposure: Compound Description: Agricultural Chemical; Drug, Tumorigen, Mutagen; Reproductive Effector. 1,4-Naphthoquinone is used as a polymerization regulator for rubber and polyester resins, in the synthesis of dyes and pharmaceuticals, and as a fungicide and algicide.

Incompatibilities: Oxidizers.

Permissible Exposure Limits in Air
Protective Action Criteria (PAC)
TEEL-0: 7.5 mg/m^3
PAC-1: 25 mg/m^3
PAC-2: 75 mg/m^3
PAC-3: 75 mg/m^3
Russia: STEL 0.1 mg/m^3, [skin] 1993.
Russia[43] set a MAC of 0.005 mg/m^3 in ambient air in residential areas both on a momentary and a daily average basis.

Routes of Entry: Inhalation, ingestion, skin and/or eye contact.

Harmful Effects and Symptoms
Long Term Exposure: The most consistent findings reported in the literature for health effects of 1,4-naphthoquinone involve hematological changes, irritant and allergenic activity, and inhibition of biochemical oxidation processes. One study found 1,4-naphthoquinone to be oncogenic. Some evidence of inhibition of *in vitro* endocrine function and of nerve activity was reported.

First Aid: If this chemical gets into the eyes, remove any contact lenses at once and irrigate immediately for at least 15 min, occasionally lifting upper and lower lids. Seek medical attention immediately. If this chemical contacts the skin, remove contaminated clothing and wash immediately with soap and water. Seek medical attention immediately. If this chemical has been inhaled, remove from exposure, begin rescue breathing (using universal precautions, including resuscitation mask) if breathing has stopped and CPR if heart action has stopped. Transfer promptly to a medical facility. When this chemical has been swallowed, get medical attention. Give large quantities of water and induce vomiting. Do not make an unconscious person vomit.

Personal Protective Methods: Wear protective gloves and clothing to prevent any reasonable probability of skin contact. Safety equipment suppliers/manufacturers can provide recommendations on the most protective glove/clothing material for your operation. All protective clothing (suits, gloves, footwear, headgear) should be clean, available each day, and put on before work. Contact lenses should not be worn when working with this chemical. Wear dust-proof chemical goggles and face shield unless full face-piece respiratory protection is worn. Employees should wash immediately with soap when skin is wet or contaminated. Provide emergency showers and eyewash.

Respirator Selection: Follow regulations in OSHA 29CFR1910.134 or European Standard EN149. Use a NIOSH/MSHA- or European Standard EN149-approved respirator; or use an approved supplied-air respirator with a full face-piece operated in the positive-pressure mode, or with a full face-piece, hood, or helmet in the continuous-flow mode; or use a NIOSH/MSHA- or European Standard EN149-approved self-contained breathing apparatus with a full face-piece operated in pressure-demand or other positive-pressure mode.

Storage: Color Code—Blue: Health Hazard/Poison: Store in a secure poison location. Prior to working with this chemical you should be trained on its proper handling and storage. Store in tightly closed containers in a cool, well-ventilated area away from oxidizers.

Shipping: Toxic solids, organic, n.o.s. compound require a shipping label of "POISONOUS/TOXIC MATERIALS." It falls in Hazard Class 6.1, Packing Group III.

Spill Handling: Evacuate persons not wearing protective equipment from area of spill or leak until cleanup is complete. Remove all ignition sources. Use HEPA vacuum or wet method to reduce dust during cleanup. Do not dry sweep. Collect powdered material in the most convenient and safe manner and deposit in sealed containers. Ventilate area after cleanup is complete. It may be necessary to contain and dispose of this chemical as a hazardous waste. If material or contaminated runoff enters waterways, notify downstream users of potentially contaminated waters. Contact your local or federal environmental protection agency for specific recommendations. If employees are required to clean up spills, they must be properly trained and equipped. OSHA 1910.120(q) may be applicable.

Fire Extinguishing: This chemical is a combustible solid. Use dry chemical, carbon dioxide, water spray, or alcohol foam extinguishers. Poisonous gases are produced in fire. If material or contaminated runoff enters waterways, notify downstream users of potentially contaminated waters. Notify local health and fire officials and pollution control agencies. From a secure, explosion-proof location, use water spray to cool exposed containers. If cooling streams are ineffective (venting sound increases in volume and pitch, tank discolors, or shows any signs of deforming), withdraw immediately to a secure position. If employees are expected to fight fires, they must be trained and equipped in OSHA 1910.156. The only respirators recommended for firefighting are self-contained breathing apparatuses that have full face-pieces and are operated in a pressure-demand or other positive-pressure mode.

Disposal Method Suggested: Consult with environmental regulatory agencies for guidance on acceptable disposal practices. Generators of waste containing this contaminant

(≥100 kg/mo) must conform with EPA regulations governing storage, transportation, treatment, and waste disposal.

References

US Environmental Protection Agency. (April 30, 1980). *1,4-Naphthoquinone, Health and Environmental Effects Profile No. 132*. Washington, DC: Office of Solid Waste

Sax, N. I. (Ed.). (1984). *Dangerous Properties of Industrial Materials Report*, 4, No. 2, 81–83

1-Naphthylamine   N:0160

Molecular Formula: $C_{10}H_9N$
Common Formula: $C_{10}H_7NH_2$
Synonyms: 1-Aminonaphthalene; C.I. azoic diazo component 114; Fast garnet B base; α-Naftilamina (Spanish); 1-Naftilamina (Spanish); 1-Naphthalenamine; Naphthalidine; 1-Naphthylamin (German); α-Naphthylamine; Naphthylamine; Naphthylamine-a; Napthalidine
CAS Registry Number: 134-32-7
RTECS® Number: QM1400000
UN/NA & ERG Number: UN2077/53
EC Number: 205-138-7 [*Annex I Index No.:* 612-020-00-2]
Regulatory Authority and Advisory Bodies
Carcinogenicity: IARC: Human, Inadequate Evidence; Animal, Inadequate Evidence, *not classifiable as carcinogenic to humans*, Group 3, 1997; OSHA: Potential human carcinogen; NIOSH: Potential occupational carcinogen.
US EPA Gene-Tox Program, Positive: Cell transform.—SA7/SHE; Host-mediated assay; Positive: *E. coli* polA with S9; Histidine reversion—Ames test; Negative: SHE—clonal assay; Cell transform.—RLV F344 rat embryo; Negative: *E. coli* polA without S9; *In vitro* UDS—human fibroblast; Negative: *S. cerevisiae*—homozygosis; Negative/limited: Carcinogenicity—mouse/rat; Inconclusive: Mammalian micronucleus; *N. crassa*—forward mutation; Inconclusive: Sperm morphology—mouse.
OSHA, 29CFR1910 Specifically Regulated Chemicals (See CFR 1910.1004).
US EPA Hazardous Waste Number (RCRA No.): U167.
RCRA, 40CFR261, Appendix 8 Hazardous Constituents.
RCRA 40CFR264, Appendix 9; TSD Facilities Ground Water Monitoring List. Suggested test method(s) (PQL μg/L): 8270 (10).
Reportable Quantity (RQ): 100 lb (45.4 kg).
EPCRA Section 313 Form R *de minimis* concentration reporting level: 0.1%.
California Proposition 65 Chemical: Cancer 10/1/89.
Canada, WHMIS, Ingredients Disclosure List Concentration 1.0%.
European/International Regulations: Hazard Symbol: Xn, N; Risk phrases: R22; R51/53; Safety phrases: S2; S24; S6 (see Appendix 4).
WGK (German Aquatic Hazard Class): 2—Hazard to waters.
Description: α-Naphthylamine exists as white needle-like crystals which turn red on exposure to air. Has a weak ammonia-like odor. Molecular weight = 143.20; Specific gravity (H_2O:1) = 1.12; Boiling point = 300.6°C; Freezing/Melting point = 44.4°C; Vapor pressure = 1 mmHg at 104°C; Flash point = 157.2°C; Autoignition temperature = 460°C. Hazard Identification (based on NFPA-704 M Rating System): Health 2, Flammability 1, Reactivity 0. Practically insoluble in water; solubility = 0.002%; 1 mg/mL at 20°C.[NTP]

Potential Exposure: Compound Description: Tumorigen, Mutagen. α-Naphthylamine is used as an intermediate in dye production, for manufacturing herbicides and antioxidants, in the manufacture of condensation colors, rubber, and in the synthesis of many chemicals, such as α-naphthol, sodium naphthionate, *o*-naphthionic acid, Neville and Winther's acid, sulfonated naphthylamines, α-naphthylthiourea (a rodenticide), and *N*-phenyl-α-naphthylamine.

Incompatibilities: Oxidizes in air. Incompatible with nitrous acid, oxidizers, nitrates, organic anhydrides, isocyanates, aldehydes.

Permissible Exposure Limits in Air
OSHA: Cancer Suspect Agent, see 29CFR1910.1003.
NIOSH REL: A potential occupational carcinogen. Limit exposure to lowest feasible concentration. See *NIOSH Pocket Guide*, Appendix A.
ACGIH TLV®[1]: Confirmed Human Carcinogen.
Protective Action Criteria (PAC)
TEEL-0: 0.6 mg/m³
PAC-1: 1.5 mg/m³
PAC-2: 12.5 mg/m³
PAC-3: 350 mg/m³
Austria: carcinogen, 1999; Poland: MAC (TWA) 0 mg/m³, MAC (STEL) 0 mg/m³, 1999; Sweden: carcinogen, 1993. α-Naphthylamine is included in the Federal standard for carcinogens; all contact with it should be avoided.[63] Several states have set guidelines or standards for 1-naphthylamine in ambient air[60] ranging from zero in South Carolina to 0.03 μg/m³ (New York).

Determination in Air: Use NIOSH Analytical Method #5518, Naphthylamines; OSHA Analytical Method 93.

Determination in Water: Octanol–water coefficient: Log K_{ow} = 2.3.

Routes of Entry: Inhalation, ingestion, skin and/or eye contact. Percutaneous absorption.

Harmful Effects and Symptoms

Short Term Exposure: 1-Naphthylamine can affect you when breathed in and by passing through your skin. 1-Naphthylamine should be handled as a carcinogen, with extreme caution. High exposure can cause the skin and lips to turn blue. This reduces the ability of the blood cells to carry oxygen to body organs (methemoglobinemia). Higher levels can cause breathing difficulties, collapse, and possible death.

Long Term Exposure: Some related chemicals can damage the liver and/or cause skin allergies. It is not known whether 1-naphthylamine has these effects. A report of excess bladder cancer among individuals who worked with this chemical has been published. This may be due to contamination with 2-naphthylamine, a known human carcinogen.

Points of Attack: Bladder, skin. *Cancer site:* bladder.

Medical Surveillance: OSHA mandates the following tests or information: *Increased Risk:* reduced immunologic competence, steroid treatment, pregnancy, cigarette smoking. *NIOSH lists the following tests: *increased risk:* reduced immunologic competence, steroid treatment, pregnancy, cigarette smoking, cystoscopy, urinalysis (routine). Before beginning employment and at regular times after that, the following are recommended: urine cytology (a test for abnormal cells in the urine). A urine test for 1-naphthylamine can also be done to determine whether this cancer agent is entering the body. For accuracy, it should be done shortly after exposure. If symptoms develop or overexposure is suspected, the following may be useful: test for methemoglobin (most accurate a few hours after exposure; the blood sample must be promptly tested).

Code of Federal Regulations. 29 CFR Part 1910.1000. Subpart Z. Air Contaminants. US Government Printing Office. (July 1, 2004).

First Aid: If this chemical gets into the eyes, remove any contact lenses at once and irrigate immediately for at least 15 min, occasionally lifting upper and lower lids. Seek medical attention immediately. If this chemical contacts the skin, remove contaminated clothing and wash immediately with soap and water. Seek medical attention immediately. If this chemical has been inhaled, remove from exposure, begin rescue breathing (using universal precautions, including resuscitation mask) if breathing has stopped and CPR if heart action has stopped. Transfer promptly to a medical facility. When this chemical has been swallowed, get medical attention. Give large quantities of water and induce vomiting. Do not make an unconscious person vomit.

Note to physician: Treat for methemoglobinemia. Spectrophotometry may be required for precise determination of levels of methemoglobin in urine.

Personal Protective Methods: Wear protective gloves and clothing to prevent any reasonable probability of skin contact. Safety equipment suppliers/manufacturers can provide recommendations on the most protective glove/clothing material for your operation. All protective clothing (suits, gloves, footwear, headgear) should be clean, available each day, and put on before work. Contact lenses should not be worn when working with this chemical. Wear dust-proof chemical goggles and face shield unless full face-piece respiratory protection is worn. Employees should wash immediately with soap when skin is wet or contaminated. Provide emergency showers and eyewash.

Respirator Selection: At any detectable concentration: SCBAF: Pd,Pp (APF = 10,000) (any NIOSH/MSHA- or European Standard EN 149-approved self-contained breathing apparatus that has a full face-piece and is operated in a pressure-demand or other positive-pressure mode) or SaF: Pd, Pp: ASCBA (APF = 10,000) (any supplied-air respirator that has a full face-piece and is operated in a pressure-demand or other positive-pressure mode in combination with an auxiliary, self-contained breathing apparatus operated in a pressure-demand or other positive-pressure mode). *Escape:* 100F (APF = 50) (any air-purifying, full-face-piece respirator with an N100, R100, or P100 filter) or SCBAE (any appropriate escape-type, self-contained breathing apparatus).

Storage: Color Code—Blue: Health Hazard/Poison: Store in a secure poison location. Prior to working with this chemical you should be trained on its proper handling and storage. Store in tightly closed containers in a cool, well-ventilated area away from light. A regulated, marked area should be established where this chemical is handled, used, or stored in compliance with OSHA Standard 1910.1045.

Shipping: alpha-Naphthylamine requires a shipping label of "POISONOUS/TOXIC MATERIALS." It falls in Hazard Class 6.1 and Packing Group III.

Spill Handling: Evacuate persons not wearing protective equipment from area of spill or leak until cleanup is complete. Remove all ignition sources. Use HEPA vacuum or wet method to reduce dust during cleanup. Do not dry sweep. Collect powdered material in the most convenient and safe manner and deposit in sealed containers. Ventilate area after cleanup is complete. It may be necessary to contain and dispose of this chemical as a hazardous waste. If material or contaminated runoff enters waterways, notify downstream users of potentially contaminated waters. Contact your local or federal environmental protection agency for specific recommendations. If employees are required to clean up spills, they must be properly trained and equipped. OSHA 1910.120(q) may be applicable.

Fire Extinguishing: 1-Naphthylamine may burn but does not readily ignite. Use dry chemical, CO_2, water spray, or foam extinguishers, however, water or foam may cause frothing. Poisonous gases, including nitrogen oxides, are produced in fire. If material or contaminated runoff enters waterways, notify downstream users of potentially contaminated waters. Notify local health and fire officials and pollution control agencies. From a secure, explosion-proof location, use water spray to cool exposed containers. If cooling streams are ineffective (venting sound increases in volume and pitch, tank discolors, or shows any signs of deforming), withdraw immediately to a secure position. If employees are expected to fight fires, they must be trained and equipped in OSHA 1910.156. The only respirators recommended for firefighting are self-contained breathing apparatuses that have full face-pieces and are operated in a pressure-demand or other positive-pressure mode.

Disposal Method Suggested: Controlled incineration whereby oxides of nitrogen are removed from the effluent gas by scrubber, catalyst, or thermal device. Consult with environmental regulatory agencies for guidance on acceptable disposal practices. Generators of waste containing this contaminant (≥100 kg/mo) must conform with EPA regulations governing storage, transportation, treatment, and waste disposal.

References
Sax, N. I. (Ed.). (1984). *Dangerous Properties of Industrial Materials Report*, 4, No. 3, 79–82

New Jersey Department of Health and Senior Services. (April 2004). *Hazardous Substances Fact Sheet: 1-Naphthylamine*. Trenton, NJ

2-Naphthylamine   N:0170

Molecular Formula: $C_{10}H_9N$

Common Formula: $C_{10}H_7NH_2$

Synonyms: 2-Aminonaphthalene; C.I. 37270; Fast scarlet base B; NA; β-Naftilamina (Spanish); 2-Naftilamina (Spanish); 2-Naphthalenamine; β-Naphthamin (German); 2-Naphthylamin (German); 2-Naphthylamine; 6-Naphthylamine; Naphthylamine-b; 2-Naphthylamine mustard

CAS Registry Number: 91-59-8

RTECS® Number: QM2100000

UN/NA & ERG Number: UN1650/153

EC Number: 202-080-4 [*Annex I Index No.:* 612-022-00-3]

Regulatory Authority and Advisory Bodies

Carcinogenicity: IARC: Human Sufficient Evidence; Animal Sufficient Evidence, *carcinogenic to humans*, Group 1, 1998; Carcinogenicity: NTP: 11th Report on Carcinogens, 2004: Known to be a human carcinogen; NIOSH: Potential occupational carcinogen; OSHA: Potential human carcinogen.

US EPA Gene-Tox Program, Positive: Carcinogenicity—mouse/rat; SHE—clonal assay; Positive: Cell transform.—mouse embryo; Positive: Cell transform.—RLV F344 rat embryo; Positive: Cell transform.—SA7/SHE; Host-mediated assay; Positive: Mammalian micronucleus; *N. crassa*—forward mutation; Positive: *E. coli* polA with S9; Histidine reversion—Ames test; Positive: *S. cerevisiae* gene conversion; Negative: Cell transform.—BALB/c-3T3; *E. coli* polA without S9; Negative: Sperm morphology—mouse; *S. cerevisiae*—homozygosis; Inconclusive: *D. melanogaster* sex-linked lethal; Positive: CHO gene mutation.

Banned or Severely Restricted (several countries) (UN).[35]

Very Toxic Substance (World Bank).[15]

OSHA, 29CFR1910 Specifically Regulated Chemicals (See CFR 1910.1009).

US EPA Hazardous Waste Number (RCRA No.): U168.

RCRA, 40CFR261, Appendix 8 Hazardous Constituents.

RCRA 40CFR268.48; 61FR15654, Universal Treatment Standards: Wastewater (mg/L), 0.52; Nonwastewater (mg/kg), N/A.

RCRA 40CFR264, Appendix 9; TSD Facilities Ground Water Monitoring List. Suggested test method(s) (PQL μg/ L): 8270 (10).

Reportable Quantity (RQ): 10 lb (4.54 kg).

EPCRA Section 313 Form R *de minimis* concentration reporting level: 0.1%.

California Proposition 65 Chemical: Cancer 2/27/87.

European/International Regulations: Hazard Symbol: T, N; Risk phrases: R45; R22; R51/53; Safety phrases: S53; S45; S61 (see Appendix 4).

WGK (German Aquatic Hazard Class): No value assigned.

Description: 2-Naphthylamine is a white to red crystal with a faint, aromatic odor. Darkens in air to a reddish-purple color. Molecular weight = 143.20; Specific gravity (H_2O:1) = 1.06 at 97.8°C; Boiling point = 294°C; Freezing/Melting point = 109–110°C; Vapor pressure = 1 mmHg at 104°C; Flash point = 157°C. Soluble in hot water.

Potential Exposure: Compound Description: Tumorigen, Mutagen. 2-Naphthylamine is presently used only for research purposes. It is present as an impurity in α-naphthylamine. It is used as an intermediate in the preparation of other compounds. 2-Naphthylamine was widely used in the manufacture of dyestuffs, as an antioxidant for rubber, and in rubber-coated cables.

Incompatibilities: Strong oxidizers. A weak base. Incompatible with nitrous acid.

Permissible Exposure Limits in Air

OSHA: Cancer Suspect Agent, see 29CFR1910.1003.

NIOSH REL: A potential occupational carcinogen. Limit exposure to lowest feasible concentration. See *NIOSH Pocket Guide*, Appendix A.

ACGIH TLV®[1]: Exposure by all routes should be carefully controlled to levels as low as possible; Confirmed Human Carcinogen.

No TEEL available.

DFG MAK: [skin]; Carcinogen Category 1.

Australia: carcinogen, 1993; Austria [skin] carcinogen, 1999; Belgium: carcinogen, 1993; Finland: carcinogen, 1999; France: VME 0.001 ppm (0.005 mg/m³), continuous; carcinogen, 1993; Poland: MAC (TWA) 0 mg/m³, MAC (STEL) 0 mg/m³, 1999; Sweden: carcinogen, 1999; Switzerland: [skin] carcinogen, 1999; United Kingdom: carcinogen, 2000; Argentina, Bulgaria, Columbia, Jordan, South Korea, New Zealand, Singapore, Vietnam: ACGIH TLV®: Confirmed Human Carcinogen. 2-Naphthylamine is included in the federal standard for carcinogens; all contact with it should be avoided. ACGIH states that β-naphthylamine is a human carcinogen without an assigned TLV. Several states have set guidelines of standards for 2-naphthylamine in ambient air[60] ranging from zero (North Dakota, New York and South Carolina) to 3.0 μg/m³ (Virginia) to 19.07 μg/m³ (Pennsylvania).

Determination in Air: Use NIOSH Analytical Method #5518, Naphthylamines; OSHA Analytical Method 93.

Permissible Concentration in Water: No criteria set, but EPA[32] has suggested an ambient level goal based on health effects of 291 μg/L.

Routes of Entry: Inhalation, ingestion, skin and/or eye contact. Percutaneous absorption.

Harmful Effects and Symptoms

Short Term Exposure: 2-Naphthylamine is irritating to the eyes and skin; has produced contact dermatitis. High levels can interfere with the blood's ability to carry oxygen (methemoglobinemia). Higher levels can cause breathing difficulties, collapse, and even death.

Long Term Exposure: 2-Naphthylamine is a known human bladder carcinogen with a latent period of about 16 years. The symptoms are frequent urination, dysuria, and hematuria. Acute poisoning leads to methemoglobinemia or acute hemorrhagic cystitis. 2-Naphthylamine is carcinogenic, producing urinary bladder carcinomas in hamsters, dogs, and nonhuman primates, and hepatomas in mice, after oral administration. Epidemiological studies have shown that occupational exposure to 2-naphthylamine, either alone or when present as an impurity in other compounds, is causally associated with bladder cancer.

Points of Attack: Bladder, skin. *Cancer site:* bladder.

Medical Surveillance: OSHA mandates the following tests or information: *Increased Risk:* reduced immunologic competence, steroid treatment, pregnancy, cigarette smoking. *NIOSH lists the following tests: *increased risk:* reduced immunologic competence, steroid treatment, pregnancy, cigarette smoking, cystoscopy, urinalysis (routine). Preplacement and periodic examinations should include an evaluation of exposure to other carcinogens, use of alcohol, smoking, and medications, and family history. Special attention should be given on a regular basis to urine sediment and cytology. If red cells or positive smears are seen, cystoscopy should be done at once. The general health of exposed persons should also be evaluated in periodic examinations.

First Aid: If this chemical gets into the eyes, remove any contact lenses at once and irrigate immediately for at least 15 min, occasionally lifting upper and lower lids. Seek medical attention immediately. If this chemical contacts the skin, remove contaminated clothing and wash immediately with soap and water. Speed in removing material from skin is of extreme importance. Seek medical attention immediately. If this chemical has been inhaled, remove from exposure, begin rescue breathing (using universal precautions, including resuscitation mask) if breathing has stopped and CPR if heart action has stopped. Transfer promptly to a medical facility. When this chemical has been swallowed, get medical attention. Give large quantities of water and induce vomiting. Do not make an unconscious person vomit. *Note to physician:* Treat for methemoglobinemia. Spectrophotometry may be required for precise determination of levels of methemoglobin in urine.

Personal Protective Methods: Wear protective gloves and clothing to prevent any reasonable probability of skin contact. Safety equipment suppliers/manufacturers can provide recommendations on the most protective glove/clothing material for your operation. All protective clothing (suits, gloves, footwear, headgear) should be clean, available each day, and put on before work. Contact lenses should not be worn when working with this chemical. Wear dust-proof chemical goggles and face shield unless full face-piece respiratory protection is worn. Employees should wash immediately with soap when skin is wet or contaminated. Provide emergency showers and eyewash.

Respirator Selection: NIOSH: *At any detectable concentration:* SCBAF: Pd,Pp (APF = 10,000) (any NIOSH/MSHA- or European Standard EN 149-approved self-contained breathing apparatus that has a full face-piece and is operated in a pressure-demand or other positive-pressure mode) or SaF: Pd,Pp: ASCBA (APF = 10,000) (any supplied-air respirator that has a full face-piece and is operated in a pressure-demand or other positive-pressure mode in combination with an auxiliary, self-contained breathing apparatus operated in a pressure-demand or other positive-pressure mode). *Escape:* 100F (APF = 50) (any air-purifying, full-face-piece respirator with an N100, R100, or P100 filter) or SCBAE (any appropriate escape-type, self-contained breathing apparatus).

Storage: Color Code—Blue: Health Hazard/Poison: Store in a secure poison location. Prior to working with this chemical you should be trained on its proper handling and storage. Store away from heat and light in a refrigerator or a cool, dry place away from incompatible materials. A regulated, marked area should be established where this chemical is handled, used, or stored in compliance with OSHA Standard 1910.1045.

Shipping: This compound requires a shipping label of "POISONOUS/TOXIC MATERIALS." It falls in Hazard Class 6.1 and Packing Group II.

Spill Handling: Evacuate persons not wearing protective equipment from area of spill or leak until cleanup is complete. Remove all ignition sources. Dampen spilled material with 60–70% acetone to avoid airborne dust. Transfer to vapor-tight plastic bags for eventual disposal. Collect powdered material in the most convenient and safe manner and deposit in sealed containers. Ventilate area after cleanup is complete. It may be necessary to contain and dispose of this chemical as a hazardous waste. If material or contaminated runoff enters waterways, notify downstream users of potentially contaminated waters. Contact your local or federal environmental protection agency for specific recommendations. If employees are required to clean up spills, they must be properly trained and equipped. OSHA 1910.120(q) may be applicable.

Fire Extinguishing: This chemical is a combustible solid. Use dry chemical, carbon dioxide, water spray, or alcohol foam extinguishers. Poisonous gases, including nitrogen oxides, are produced in fire. If material or contaminated runoff enters waterways, notify downstream users of potentially contaminated waters. Notify local health and fire officials and pollution control agencies. From a secure, explosion-proof location, use water spray to cool exposed containers. If cooling streams are ineffective (venting sound increases in volume and pitch, tank discolors, or shows any signs of deforming), withdraw immediately to a secure position. If employees are expected to fight fires, they must be trained and equipped in OSHA 1910.156. The only respirators recommended for firefighting are self-contained breathing apparatuses that have full face-pieces and are operated in a pressure-demand or other positive-pressure mode.

Disposal Method Suggested: Controlled incineration whereby oxides of nitrogen are removed from the effluent gas by scrubber, catalyst, or thermal device.[22] Consult with environmental regulatory agencies for guidance on acceptable disposal practices. Generators of waste containing this contaminant ($\geq$100 kg/mo) must conform with EPA regulations governing storage, transportation, treatment, and waste disposal.

Reference

Sax, N. I. (Ed.). *Dangerous Properties of Industrial Materials Report*, 2, No. 2, 56−58 (1982) & 3, No. 6, 52−56 (1983)

Neochromium trivalene   N:0180

Molecular Formula: $CrHO_5S$
Common Formula: $Cr(OH)SO_4$
Synonyms: Basic chromic sulfate; Basic chromic sulphate; Basic chromium sulfate; Basic chromium sulphate; Chromium hydroxide sulfate; Chromium sulfate; Chromium sulfate, basic; Chromium sulphate; Koreon; Monobasic chromium sulfate; Monobasic chromium sulphate; Sulfuric acid, chromium salt, basic
CAS Registry Number: 64093-79-4
RTECS® Number: QO6800000
UN/NA & ERG Number: UN2240/154
Regulatory Authority and Advisory Bodies
Carcinogenicity: IARC: Human Inadequate Evidence; Animal Inadequate Evidence, *not classifiable as carcinogenic to humans*, Group 3, 1997; EPA (insoluble salts): Not Classifiable as to human carcinogenicity; Cannot be Determined.
Air Pollutant Standard Set. See below, "Permissible Exposure Limits in Air" section.
Clean Air Act: Hazardous Air Pollutants (Title I, Part A, Section 112).
Clean Water Act: Toxic Pollutant (Section 401.15); 40CFR401.15 Section 307 Toxic Pollutants as chromium and compounds.
RCRA, 40CFR261, Appendix 8 Hazardous Constituents, waste number not listed.
EPCRA (Section 313): Includes any unique chemical substance that contains chromium as part of that chemical's infrastructure.
Form R *de minimis* concentration reporting level: Chromium(III) compounds: 1.0%.
Canada, WHMIS, Ingredients Disclosure List Concentration 1.0% as Chromium(III) compounds, n.o.s.; National Pollutant Release Inventory (NPRI); CEPA Priority Substance List as chromium compounds.
WGK (German Aquatic Hazard Class): No value assigned.
Description: Neochromium is a violet or green powder. Molecular weight = 165.07; Boiling point = 98°C;

Freezing/Melting point = 67°C. Hazard Identification (based on NFPA-704 M Rating System): Health 3, Flammability 0, Reactivity 0.
Potential Exposure: Neochromium is used in papermaking, photography, dyeing, printing, and tanning.
Permissible Exposure Limits in Air
OSHA PEL: 0.5 mg[Cr]/m^3 TWA.
NIOSH REL: 0.5 mg[Cr]/m^3 TWA limit exposures to lowest feasible concentration.
ACGIH TLV®[1]: 0.5 mg[Cr]/m^3 TWA; not classifiable as a human carcinogen.
NIOSH IDLH: 25 mg Cr(III)/m^3.
No TEEL available.
DFG MAK: [skin] danger of skin sensitization.
Determination in Air: Use NIOSH Analytical Method #7300, Elements by ICP (Nitric/perchloric Acid Ashing); #7301, Elements by ICP (Aqua regia ashing); #7303, Elements by ICP [Hot block (HCl/HNO$_3$ Digestion)]; #9102, Elements on wipes; #8310, Metals in urine, #8005, Elements in blood or tissue; OSHA Analytical Methods ID-121 and ID-125G.
Permissible Concentration in Water: For the protection of freshwater aquatic life: *Trivalent chromium:* not to exceed e[1.08 In (hardness) + 3.48] μg/L. For the protection of saltwater aquatic life: *Trivalent chromium:* 10,300 μg/L on an acute toxicity basis. *To protect human health: Trivalent chromium:* 170 μg/L; Hexavalent chromium 50 μg/L according to EPA.[6] For chromium, EPA[49] has set a long-term health advisory of 0.84 mg/L for adults and a lifetime health advisory of 0.12 mg/L (120 μg/L). EPA's maximum drinking water level (MCL) is 0.1 mg/L.[62] Germany, Canada, EEC, and WHO[35] have set a limit of 0.05 mg/L in drinking water. The states of Maine and Minnesota have set guidelines for chromium in drinking water[61] as 50 μg/L for Maine and 120 μg/L for Minnesota.
Routes of Entry: Inhalation, ingestion, skin and/or eye contact.
Harmful Effects and Symptoms
Short Term Exposure: Neochromium can affect you when breathed in. Skin and eye contact can cause severe burns. Overexposure can irritate the nose, throat, and bronchial tubes. Higher exposures can cause pulmonary edema, a medical emergency that can be delayed for several hours. This can cause death. LD$_{50}$ (oral-rat) = 1500 mg/kg.
Long Term Exposure: Skin allergy sometimes occurs with itching, redness, and/or an eczema-like rash. If this happens, future contact can trigger symptoms. Breathing neochromium may cause a sore or hole in the nasal septum. Can irritate the lungs; bronchitis may develop. In addition, RTECS[9] states neochromium is tumorigenic, that is, it facilitates the action of known carcinogens.
Points of Attack: Eyes, skin.
Medical Surveillance: For those with frequent or potentially high exposure (half the TLV or greater), the following are recommended before beginning work and at regular times after that: lung function tests. If symptoms develop or

overexposure is suspected, the following may be useful: consider chest X-ray after acute overexposure. Evaluation by a qualified allergist, including careful exposure history and special testing, may help diagnose skin allergy. NIOSH lists the following tests [chromium(III) compounds]: whole blood (chemical/metabolite), biologic tissue/biopsy, chest X-ray, pulmonary function tests, red blood cells/count, urine (chemical/metabolite) [end-of-shift] [end-of-shift at end-of-work-week] [end-of-work-week] [pre- and postshift].

First Aid: If this chemical gets into the eyes, remove any contact lenses at once and irrigate immediately for at least 15 min, occasionally lifting upper and lower lids. Seek medical attention immediately. If this chemical contacts the skin, remove contaminated clothing and wash immediately with soap and water. Seek medical attention immediately. If this chemical has been inhaled, remove from exposure, begin rescue breathing (using universal precautions, including resuscitation mask) if breathing has stopped and CPR if heart action has stopped. Transfer promptly to a medical facility. When this chemical has been swallowed, get medical attention. Give large quantities of water and do not induce vomiting. Medical observation is recommended for 24−48 h after breathing overexposure, as pulmonary edema may be delayed. As first aid for pulmonary edema, a doctor or authorized paramedic may consider administering a corticosteroid spray.

Personal Protective Methods: Wear protective gloves and clothing to prevent any reasonable probability of skin contact. Safety equipment suppliers/manufacturers can provide recommendations on the most protective glove/clothing material for your operation. All protective clothing (suits, gloves, footwear, headgear) should be clean, available each day, and put on before work. Contact lenses should not be worn when working with this chemical. Wear dust-proof chemical goggles and face shield unless full face-piece respiratory protection is worn. Employees should wash immediately with soap when skin is wet or contaminated. Provide emergency showers and eyewash.

Respirator Selection: *Up to 2.5 mg/m³:* Qm* (APF = 25) (any quarter-mask respirator). *Up to 5 mg/m³:* 95XQ* (APF = 10) [any particulate respirator equipped with an N95, R95, or P95 filter (including N95, R95, and P95 filtering face-pieces) except quarter-mask respirators. The following filters may also be used: N99, R99, P99, N100, R100, P100]; or Sa* (APF = 10) (any supplied-air respirator). *Up to 12.5 mg/m³:* Sa:Cf* (any supplied-air respirator operated in a continuous-flow mode) or PaprHie* (any powered air-purifying respirator with a high-efficiency particulate filter). *Up to 25 mg/m³:* 100F (APF = 50) (any air-purifying, full-face-piece respirator with an N100, R100, or P100 filter) or PaprTHie* (any powered, air-purifying respirator with a tight-fitting face-piece and a high-efficiency particulate filter) or SCBAF (any self-contained breathing apparatus with a full face-piece) or SaF (any supplied-air respirator with a full face-piece). *Emergency or planned entry into unknown concentrations or IDLH conditions:* SCBAF: Pd,Pp

(APF = 10,000) (any self-contained breathing apparatus that has a full face-piece and is operated in a pressure-demand or other positive-pressure mode) or SaF: Pd,Pp: ASCBA (APF = 10,000) (any supplied-air respirator that has a full face-piece and is operated in a pressure-demand or other positive-pressure mode in combination with an auxiliary, self-contained breathing apparatus operated in a pressure-demand or other positive-pressure mode). *Escape:* 100F (APF = 50) (any air-purifying, full-face-piece respirator with an N100, R100, or P100 filter) or SCBAE (any appropriate escape-type, self-contained breathing apparatus).

Note: Substance reported to cause eye irritation or damage; may require eye protection.

Storage: Color Code—White: Corrosive or Contact Hazard; Store separately in a corrosion-resistant location. Prior to working with this chemical you should be trained on its proper handling and storage. Store in tightly closed containers in a cool, dry, well-ventilated area. A regulated, marked area should be established where this chemical is handled, used, or stored in compliance with OSHA Standard 1910.1045.

Shipping: This chemical requires a "CORROSIVE" label. It falls in Hazard Class 8 and Packing Group I.[19]

Spill Handling: Evacuate persons not wearing protective equipment from area of spill or leak until cleanup is complete. Remove all ignition sources. Use HEPA vacuum or wet method to reduce dust during cleanup. Do not dry sweep. Collect powdered material in the most convenient and safe manner and deposit in sealed containers. Ventilate area after cleanup is complete. It may be necessary to contain and dispose of this chemical as a hazardous waste. If material or contaminated runoff enters waterways, notify downstream users of potentially contaminated waters. Contact your local or federal environmental protection agency for specific recommendations. If employees are required to clean up spills, they must be properly trained and equipped. OSHA 1910.120(q) may be applicable.

Fire Extinguishing: Neochromium may burn but does not readily ignite. Use dry chemical, carbon dioxide, water spray, or alcohol foam extinguishers. Poisonous gases, including sulfur oxides are produced in fire. If material or contaminated runoff enters waterways, notify downstream users of potentially contaminated waters. Notify local health and fire officials and pollution control agencies. From a secure, explosion-proof location, use water spray to cool exposed containers. If cooling streams are ineffective (venting sound increases in volume and pitch, tank discolors, or shows any signs of deforming), withdraw immediately to a secure position. If employees are expected to fight fires, they must be trained and equipped in OSHA 1910.156. The only respirators recommended for firefighting are self-contained breathing apparatuses that have full face-pieces and are operated in a pressure-demand or other positive-pressure mode.

Reference

New Jersey Department of Health and Senior Services. (August 2004). *Hazardous Substances Fact Sheet: Neochromium.* Trenton, NJ

Neon   N:0190

Molecular Formula: Ne
Synonyms: Neon, elemental
CAS Registry Number: 7440-01-9
RTECS® Number: QP4450000
UN/NA & ERG Number: UN1065 (compressed); UN1913 (liquid)
EC Number: 231-110-9
Description: Neon is a colorless, odorless, tasteless gas or liquid. Molecular weight = 20.18; Boiling point = −246°C; Freezing/Melting point = −249°C. Slightly soluble in water.
Potential Exposure: Neon is used in photoelectric bulbs and certain light tubes, in the electronic industry, in lasers, in plasma studies, and in other research.
Permissible Exposure Limits in Air
ACGIH TLV®[1]: Simple asphyxiant.
Protective Action Criteria (PAC)
TEEL-0: 65,000 ppm
PAC-1: 65,000 ppm
PAC-2: 230,000 ppm
PAC-3: 400,000 ppm
Australia: asphyxiant, 1993; Belgium: asphyxiant, 1993; Hungary: asphyxiant, 1993; Switzerland: asphyxiant, 1999; United Kingdom: asphyxiant, 2000; Argentina, Bulgaria, Columbia, Jordan, South Korea, New Zealand, Singapore, Vietnam: ACGIH TLV®: Simple asphyxiant. Large amounts of neon will decrease the amount of available oxygen. Oxygen content should be tested to ensure that it is at least 19% by volume.
Routes of Entry: Inhalation.
Harmful Effects and Symptoms
Short Term Exposure: Neon can affect you when breathed in. Exposure can cause you to feel dizzy and lightheaded. Very high levels can cause you to pass out and could cause suffocation from lack of oxygen. Contact with liquefied neon could cause frostbite. Before entering a confined space, be certain that sufficient oxygen exists.
First Aid: If this chemical gets into the eyes, remove any contact lenses at once and irrigate immediately for at least 15 min, occasionally lifting upper and lower lids. Seek medical attention immediately. If this chemical contacts the skin, remove contaminated clothing and wash immediately with soap and water. Seek medical attention immediately. If this chemical has been inhaled, remove from exposure, begin rescue breathing (using universal precautions, including resuscitation mask) if breathing has stopped and CPR if heart action has stopped. Transfer promptly to a medical facility. When this chemical has been swallowed, get medical attention. Give large quantities of water and induce vomiting. Do not make an unconscious person vomit. If frostbite has occurred, seek medical attention immediately; do *NOT* rub the affected areas or flush them with water. In order to prevent further tissue damage, do *NOT* attempt to remove frozen clothing from frostbitten areas. If frostbite has *NOT* occurred, immediately and thoroughly wash contaminated skin with soap and water.
Personal Protective Methods: Wear protective gloves and clothing to prevent any reasonable probability of skin contact. Safety equipment suppliers/manufacturers can provide recommendations on the most protective glove/clothing material for your operation. All protective clothing (suits, gloves, footwear, headgear) should be clean, available each day, and put on before work. Contact lenses should not be worn when working with this chemical. Wear splash-proof chemical goggles and face shield when working with liquid, unless full face-piece respiratory protection is worn. Eye protection is required in laser operations to avoid eye burning. Employees should wash immediately with soap when skin is wet or contaminated. Provide emergency showers and eyewash (gloves, footwear, headgear) which should be clean, available each day, and put on before work. Where exposure to cold equipment, vapors, or liquid may occur, employees should be provided with special clothing designed to prevent the freezing of body tissues.
Respirator Selection: Exposure to neon is dangerous because it can replace oxygen and lead to suffocation. Only NIOSH/MSHA- or European Standard EN 149-approved self-contained breathing apparatus with a full face-piece operated in positive-pressure mode should be used in oxygen-deficient environments.
Storage: Color Code—Green: General storage may be used. Prior to working with this chemical you should be trained on its proper handling and storage. Storage areas should be dry and well ventilated. Protect cylinders of neon from physical damage. Procedures for the handling, use, and storage of cylinders should be in compliance with OSHA 1910.101 and 1910.169, as with the recommendations of the Compressed Gas Association.
Shipping: Neon, *compressed,* and Neon, *refrigerated liquid (cryogenic liquid),* require a shipping label of "NONFLAMMABLE GAS." Both fall in DOT Hazard Class 2.2.
Spill Handling: If in a building, evacuate building and confine vapors by closing doors and shutting down HVAC systems. Restrict persons not wearing protective equipment from area of spill or leak until cleanup is complete. Remove all ignition sources. Ventilate area of spill or leak to disperse the gas. Wear chemical protective suit with self-contained breathing apparatus to combat spills. Stay upwind and use water spray to "knock down" vapor; contain runoff. Stop the flow of gas, if it can be done safely from a distance. If source is a cylinder and the leak cannot be stopped in place, remove the leaking cylinder to a safe place and repair leak or allow cylinder to empty. Keep this chemical out of confined spaces, such as a sewer, because of the possibility of explosion, unless the sewer is designed to prevent the buildup of explosive concentrations. If employees are required to clean up spills, they must be properly trained and equipped. OSHA 1910.120(q) may be applicable.

Fire Extinguishing: This chemical is a nonflammable gas, but containers may explode in fire. Do not extinguish the fire unless the flow of gas can be stopped and any remaining gas is out of the line. Vapors are heavier than air and will collect in low areas. Containers may explode in fire. Storage containers and parts of containers may rocket great distances, in many directions. From a secure, explosion-proof location, use water spray to cool exposed containers. If cooling streams are ineffective (venting sound increases in volume and pitch, tank discolors, or shows any signs of deforming), withdraw immediately to a secure position. If employees are expected to fight fires, they must be trained and equipped in OSHA 1910.156. The only respirators recommended for firefighting are self-contained breathing apparatuses that have full face-pieces and are operated in a pressure-demand or other positive-pressure mode.

Disposal Method Suggested: Venting to atmosphere.

Reference

New Jersey Department of Health and Senior Services. (September 2001). *Hazardous Substances Fact Sheet: Neon.* Trenton, NJ

Neopentane   N:0200

Molecular Formula: C_5H_{12}
Common Formula: $CH_3C(CH_3)_2CH_3$
Synonyms: 2,2-Dimethylpropane; Dimethylpropane; Neopentane; *tert*-Pentane; Propane, 2,2-dimethyl-; Tetramethylmethane; 1,1,1-Trimethylethane
CAS Registry Number: 463-82-1
RTECS® Number: TY1190000
UN/NA & ERG Number: UN2044/115
EC Number: 207-343-7 [*Annex I Index No.:* 601-005-00-6]

Regulatory Authority and Advisory Bodies

Department of Homeland Security Screening Threshold Quantity (pounds): *Release hazard* 10,000 ($\geq$1.00% concentration).

Air Pollutant Standard Set. See below, "Permissible Exposure Limits in Air" section.

Clean Air Act: Accidental Release Prevention/Flammable Substances, (Section 112[r], Table 3), TQ = 10,000 lb (4540 kg).

Canada, WHMIS, Ingredients Disclosure List (neopentane not listed) *n*-pentane Concentration 1%.

European/International Regulations: Hazard Symbol: F +, N; Risk phrases: R12; R51/53; Safety phrases: S2; S9; S16; S33; S61 (see Appendix 4).

WGK (German Aquatic Hazard Class): 1—Low hazard to waters.

Description: Neopentane is a volatile liquid. Molecular weight = 72.17; Boiling point = 9°C; Flash point = (Gas) − 7°C. Explosive limits: LEL = 1.4%; UEL = 7.5%; Autoignition temperature = 450°C. Hazard Identification (based on NFPA-704 M Rating System): Health 0, Flammability 4, Reactivity 0. Insoluble in water.

Potential Exposure: Neopentane is used as a gasoline blending component and for making butyl rubber. A research chemical. Reacts with strong oxidizers, causing fire and explosion hazard. Attacks some plastics, rubbers, and coatings.

Permissible Exposure Limits in Air

OSHA gives limits for *n*-pentane as do several states, but they do not single out "neopentane." *This is shown for reference only.*

OSHA PEL: 1000 ppm/2950 mg/m^3 TWA.
NIOSH REL: 120 ppm/350 mg/m^3 TWA.
ACGIH TLV®[1]: 600 ppm/1770 TWA.
NIOSH IDLH: 1500 ppm [LEL].
Protective Action Criteria (PAC)
TEEL-0: 610 ppm
PAC-1: 610 ppm
PAC-2: 610 ppm
PAC-3: 1500 ppm
DFG MAK: 1000 ppm/3000 mg/m^3 TWA; Peak Limitation Category II(2); Pregnancy Risk Group C.
Austria: MAK 600 ppm (1899 mg/m^3), 1999; Denmark: TWA 500 ppm (1500 mg/m^3), 1999; Argentina, Bulgaria, Columbia, Jordan, South Korea, New Zealand, Singapore, Vietnam: ACGIH TLV®: TWA 600 ppm.
United Kingdom[33] TWA 600 ppm (1800 mg/m^3); STEL of 750 ppm (2250 mg/m^3) (all isomers of pentane).

Determination in Air: Charcoal tube; CS2; Gas chromatography/Flame ionization detection; NIOSH Analytical Method (IV) #1500, Hydrocarbons.

Routes of Entry: Inhalation, ingestion, skin and/or eye contact.

Harmful Effects and Symptoms

Short Term Exposure: Exposure can cause skin irritation and severe burns with redness, itching. May affect the central nervous systems. Skin contact with the undiluted material for 5 h causes blisters; for 1 h it causes irritation, itching, erythema, pigmentation, swelling, burning, and pain. Skin contact can cause frostbite. Inhalation can cause headache, dizziness, and suffocation. The toxicity is via the inhalation route. It is narcotic in high concentrations. Ingesting the liquid may cause chemical pneumonia (aspiration).

Long Term Exposure: Repeated or prolonged contact with skin may cause dermatitis. See above.

Points of Attack: Eyes, skin, respiratory system, central nervous system.

Medical Surveillance: Lung function tests. Consider chest X-ray following acute overexposure.

First Aid: If this chemical gets into the eyes, remove any contact lenses at once and irrigate immediately for at least 15 min, occasionally lifting upper and lower lids. Seek medical attention immediately. If this chemical contacts the skin, remove contaminated clothing and wash immediately with soap and water. Seek medical attention immediately. If this chemical has been inhaled, remove from exposure, begin rescue breathing (using universal precautions,

including resuscitation mask) if breathing has stopped and CPR if heart action has stopped. Transfer promptly to a medical facility. When this chemical has been swallowed, get medical attention. Give large quantities of water and induce vomiting. Do not make an unconscious person vomit. If frostbite has occurred, seek medical attention immediately; do *NOT* rub the affected areas or flush them with water. In order to prevent further tissue damage, do *NOT* attempt to remove frozen clothing from frostbitten areas. If frostbite has *NOT* occurred, immediately and thoroughly wash contaminated skin with soap and water.

Note to physician: Inhalation: bronchodilators, decongestants, and oxygen may be used if necessary. Corticosteroids are useful for treating pneumonitis.

Personal Protective Methods: Wear appropriate personal protective clothing to prevent the skin from becoming frozen from contact with the evaporating liquid or from contact with vessels containing the liquid. Safety equipment suppliers/manufacturers can provide recommendations on the most protective glove/clothing material for your operation. All protective clothing (suits, gloves, footwear, headgear) should be clean, available each day, and put on before work. Contact lenses should not be worn when working with this chemical. Wear gas-proof chemical goggles and face shield unless full face-piece respiratory protection is worn. Employees should wash immediately with soap when skin is wet or contaminated. Provide emergency showers and eyewash.

Respirator Selection: NIOSH (as pentane): *1200 ppm:* Sa (APF = 10) (any supplied-air respirator). *1500 ppm:* Sa:Cf (APF = 25) (any supplied-air respirator operated in a continuous-flow mode) or SCBAF (APF = 50) (any self-contained breathing apparatus with a full face-piece) or SaF (APF = 50) (any supplied-air respirator with a full face-piece). *Emergency or planned entry into unknown concentrations or IDLH conditions:* SCBAF: Pd,Pp (APF = 10,000) (any self-contained breathing apparatus that has a full face-piece and is operated in a pressure-demand or other positive-pressure mode); or SaF: Pd,Pp: ASCBA (APF = 10,000) (any supplied-air respirator that has a full face-piece and is operated in a pressure-demand or other positive-pressure mode in combination with an auxiliary self-contained breathing apparatus operated in a pressure-demand or other positive-pressure mode). *Escape:* GmFOv (APF = 50) [any air-purifying, full-face-piece respirator (gas mask) with a chin-style, front- or back-mounted organic vapor canister] or SCBAE (any appropriate escape-type, self-contained breathing apparatus).

Storage: Color Code—Red: Flammability Hazard: Store in a flammable liquid storage area or approved cabinet away from ignition sources and corrosive and reactive materials. Prior to working with this chemical you should be trained on its proper handling and storage. Before entering confined space where this chemical may be present, check to make sure that an explosive concentration does not exist. Store in tightly closed containers in a cool, well-ventilated area

away from oxidizers and heat. Sources of ignition, such as smoking and open flames, are prohibited where this chemical is handled, used, or stored. Drums must be equipped with self-closing valves, pressure vacuum bungs, and flame arresters. Use only nonsparking tools and equipment, especially when opening and closing containers of this chemical. Wherever this chemical is used, handled, manufactured, or stored, use explosion-proof electrical equipment and fittings. Procedures for the handling, use, and storage of cylinders should be in compliance with OSHA 1910.101 and 1910.169, as with the recommendations of the Compressed Gas Association.

Shipping: Pentanes require a label of "FLAMMABLE LIQUID." They fall in Hazard Class 3 and Packing Group I.

Spill Handling: Evacuate and restrict persons not wearing protective equipment from area of spill or leak until cleanup is complete. Remove all ignition sources. Establish forced ventilation to keep levels below explosive limit. Stop the flow of gas if it can be done safely. If source of leak is a cylinder and the leak cannot be stopped in place, remove leaking cylinder to a safe place in the open air, and repair leak or allow cylinder to empty. Keep this chemical out of confined space, such as a sewer, because of the possibility of explosion, unless the sewer is designed to prevent the buildup of explosive concentrations. It may be necessary to contain and dispose of this chemical as a hazardous waste. Contact your local or federal environmental protection agency for specific recommendations. If employees are required to clean up spills, they must be properly trained and equipped. OSHA 1910.120(q) may be applicable.

Fire Extinguishing: This chemical is a flammable gas. Poisonous gases are produced in fire. Do not extinguish the fire unless the flow of gas can be stopped and any remaining gas is out of the line. Specially trained personnel may use fog lines to cool exposures and let the fire burn itself out. Vapors are heavier than air and will collect in low areas. Vapors may travel long distances to ignition sources and flashback. Vapors in confined areas may explode when exposed to fire. Containers may explode in fire. Storage containers and parts of containers may rocket great distances, in many directions. If material or contaminated runoff enters waterways, notify downstream users of potentially contaminated waters. Notify local health and fire officials and pollution control agencies. From a secure, explosion-proof location, use water spray to cool exposed containers. If cooling streams are ineffective (venting sound increases in volume and pitch, tank discolors, or shows any signs of deforming), withdraw immediately to a secure position. If cylinders are exposed to excessive heat from fire or flame contact, withdraw immediately to a secure location. If employees are expected to fight fires, they must be trained and equipped in OSHA 1910.156. The only respirators recommended for firefighting are self-contained breathing apparatuses that have full face-pieces and are operated in a pressure-demand or other positive-pressure mode.

Disposal Method Suggested: Dissolve or mix the material with a combustible solvent and burn in a chemical incinerator equipped with an afterburner and scrubber. All federal, state, and local environmental regulations must be observed.
Reference
New Jersey Department of Health and Senior Services. (February 1999). *Hazardous Substances Fact Sheet: Dimethylpropane.* Trenton, NJ

Nerve agents

Nerve agents consist of a group of very toxic organophosphate chemicals specifically designed for military warfare. Other organophosphate chemicals include commercial insecticides, such as Malathion®. These chemicals all cause similar effects on the human body by disrupting how nerves communicate and control muscles, glands, and organs. Though they cause similar effects, nerve agents are more toxic than commercial insecticides—so smaller amounts can cause effects of concern. Most of the nerve agents exist as liquids but some (such as GB) volatilize into the air on their own. VX is the least likely to become airborne, but in conditions involving explosions, it could vaporize and spread in the air. The major chemical warfare agents in this category and their code names are listed below along with their record number for quick access.
Cyclosarin, agent GF see C:1795.
Sarin, agent GB see S:0130.
Soman, agent GD see S:0565.
Tabun, agent GA see T:0110.
VX, agent VX see V:0250.

Niacinamide   N:0210

Molecular Formula: $C_6H_6N_2O$
Synonyms: Acidamide; Nicotimanide; Nicotine acid amide; 3-Pyridine carboxamide; 3-Pyridine carboxylic acid amide; VI-Nicotyl; VI-Nictyl; Vitamin B_3; Vitamin Pp
CAS Registry Number: 98-92-0
RTECS® Number: QS3675000
EC Number: 202-713-4
Regulatory Authority and Advisory Bodies
WGK (German Aquatic Hazard Class): 1—Low hazard to waters.
Description: Niacinamide is a white crystalline powder or forms colorless needle-like crystals. Molecular weight = 122.14; Boiling point = 155°C; Freezing/Melting point = 129–130°C. Hazard Identification (based on NFPA-704 M Rating System): Health 1, Flammability 0, Reactivity 0. Highly soluble in water.
Potential Exposure: Compound Description: Mutagen. Used as a dietary supplement and food additive.
Permissible Exposure Limits in Air
Protective Action Criteria (PAC)

TEEL-0: 6 mg/m³
PAC-1: 20 mg/m³
PAC-2: 150 mg/m³
PAC-3: 500 mg/m³
Routes of Entry: Ingestion.
Harmful Effects and Symptoms
Short Term Exposure: These symptoms reported from cases of medical treatment or self-prescribed massive vitamin dosage. The sudden onset of nausea, vomiting, and fatigue have been reported following increased dose from 4 to 9 g daily. From animal studies, the lethal human dose has been estimated to be about ½ lb.
Long Term Exposure: The recommended dietary supplement is 10–20 mg per day. This level causes no adverse effects. Levels of 4–9 g/day may cause nausea and vomiting as well as serious changes in liver tissue and enzymes. May cause mutations.
First Aid: If this chemical gets into the eyes, remove any contact lenses at once and irrigate immediately for at least 15 min, occasionally lifting upper and lower lids. Seek medical attention immediately. If this chemical contacts the skin, remove contaminated clothing and wash immediately with soap and water. Seek medical attention immediately. If this chemical has been inhaled, remove from exposure, begin rescue breathing (using universal precautions, including resuscitation mask) if breathing has stopped and CPR if heart action has stopped. Transfer promptly to a medical facility. When this chemical has been swallowed, get medical attention. Give large quantities of water and induce vomiting. Do not make an unconscious person vomit.
Personal Protective Methods: Wear protective gloves and clothing to prevent any reasonable probability of skin contact. Safety equipment suppliers/manufacturers can provide recommendations on the most protective glove/clothing material for your operation. All protective clothing (suits, gloves, footwear, headgear) should be clean, available each day, and put on before work. Contact lenses should not be worn when working with this chemical. Wear dust-proof chemical goggles and face shield unless full face-piece respiratory protection is worn. Employees should wash immediately with soap when skin is wet or contaminated. Provide emergency showers and eyewash.
Respirator Selection: Follow regulations in OSHA 29CFR1910.134 or European Standard EN149. Use a NIOSH/MSHA- or European Standard EN149-approved respirator; or use an approved supplied-air respirator with a full face-piece operated in the positive-pressure mode, or with a full face-piece, hood, or helmet in the continuous-flow mode; or use a NIOSH/MSHA- or European Standard EN149-approved self-contained breathing apparatus with a full face-piece operated in pressure-demand or other positive-pressure mode.
Storage: Color Code—Green: General storage may be used. Prior to working with this chemical you should be trained on its proper handling and storage.

Spill Handling: Evacuate persons not wearing protective equipment from area of spill or leak until cleanup is complete. Remove all ignition sources. Use HEPA vacuum or wet method to reduce dust during cleanup. Do not dry sweep. Collect powdered material in the most convenient and safe manner and deposit in sealed containers. Ventilate area after cleanup is complete. It may be necessary to contain and dispose of this chemical as a hazardous waste. If material or contaminated runoff enters waterways, notify downstream users of potentially contaminated waters. Contact your local or federal environmental protection agency for specific recommendations. If employees are required to clean up spills, they must be properly trained and equipped. OSHA 1910.120(q) may be applicable.

Fire Extinguishing: This chemical is a noncombustible solid. Use dry chemical, carbon dioxide, water spray, or alcohol foam extinguishers. Poisonous gases are produced in fire. If material or contaminated runoff enters waterways, notify downstream users of potentially contaminated waters. Notify local health and fire officials and pollution control agencies. From a secure, explosion-proof location, use water spray to cool exposed containers. If cooling streams are ineffective (venting sound increases in volume and pitch, tank discolors, or shows any signs of deforming), withdraw immediately to a secure position. If employees are expected to fight fires, they must be trained and equipped in OSHA 1910.156. The only respirators recommended for firefighting are self-contained breathing apparatuses that have full face-pieces and are operated in a pressure-demand or other positive-pressure mode.

References
New York State Department of Health. (March 1986). *Chemical Fact Sheet: Niacinamide.* Albany, NY: Bureau of Toxic Substance Assessment

Nickel & compounds   N:0220

Molecular Formula: Ni
Synonyms: metal: Alloy 725; Alloy 732; Alloy 735; Alloy 762; Alloy 770; C.I. 77775; FM 1208; HCA 1; Metallic Nickel; Ni; Ni 0901S (Harshaw); Ni 233; Ni 270; Ni 4303T; Nickel, Elemental; Nickel 0901 S; Nickel 200; Nickel 201; Nickel 203; Nickel 204; Nickel 205; Nickel 211; Nickel 212; Nickel 213; Nickel 222; Nickel 223; Nickel 225; Nickel 229; Nickel 233; Nickel 270; Nickel 4303 T; Nickel Sponge; Niklad 794-A; NP 2; Raney Alloy; Raney Nickel; RCH 55/5; Synonyms of other nickel compounds vary depending upon the specific compound.
CAS Registry Number: 7440-02-0
RTECS® Number: QR5950000 [*Annex I Index No.:* 028-002-00-7]
UN/NA & ERG Number: UN3077/171; metal powder, in bulk, may be pyrophoric: UN3089 (Metal powder, flammable, n.o.s.)/170; UN2881(Nickel catalyst, dry)/135 Nickel catalyst, dry

EC Number: 231-111-4 [*Annex I Index No.:* 028-002-00-7]
Regulatory Authority and Advisory Bodies
Carcinogenicity: IARC (*Ni compounds*) Human Sufficient Evidence; Animal Sufficient Evidence, *carcinogenic to humans,* Group 1; (*elemental*): Animal Sufficient Evidence; Human Inadequate Evidence, *possibly carcinogenic to humans,* Group 2B, (compounds) *carcinogenic to humans,* Group 1, 1997; NTP (*elemental and Ni compounds, soluble and insoluble*): 11th Report on Carcinogens, 2004: Reasonably anticipated to be a human carcinogen; NIOSH (*elemental and Ni compounds, soluble and insoluble*): Potential occupational carcinogen; US EPA Gene-Tox Program, Positive: Carcinogenicity—mouse/rat.
Very Toxic Substance (World Bank).[15]
Air Pollutant Standard Set. See below, "Permissible Exposure Limits in Air" section.
Clean Air Act: Hazardous Air Pollutants (Title I, Part A, Section 112) as nickel compounds.
Clean Water Act: 40CFR401.15 Section 307 Toxic Pollutants as nickel and compounds; Section 313 Water Priority Chemicals (57FR41331, 9/9/92).
RCRA, 40CFR261, Appendix 8 Hazardous Constituents, waste number not listed.
RCRA 40CFR268.48; 61FR15654, Universal Treatment Standards: Wastewater (mg/L), 3.98; Nonwastewater (mg/L), 5.0 TCLP.
RCRA 40CFR264, Appendix 9; TSD Facilities Ground Water Monitoring List. Suggested test method(s) (PQL μg/L): total dust 6010 (50); 7520 (400).
Safe Drinking Water Act: Regulated chemical (47 FR 9352).
Reportable Quantity (RQ): 100 lb (45.4 kg).
EPCRA Section 313 Form R *de minimis* concentration reporting level: 0.1%.
Dropped from listing of Extremely Hazardous Substance (EPCRA) in 1988.
Canada, WHMIS, Ingredients Disclosure List Concentration 0.1% as elemental nickel.
Nickel compounds:
Clean Air Act: Hazardous Air Pollutants (Title I, Part A, Section 112).
Clean Water Act: 40CFR401.15 Section 307 Toxic Pollutants as nickel and compounds.
RCRA, 40CFR261, Appendix 8 Hazardous Constituents, waste number not listed, as nickel compounds, n.o.s.
EPCRA Section 313: Includes any unique chemical substance that contains nickel as part of that chemical's infrastructure. Form R *de minimis* concentration reporting level: 0.1.
California Proposition 65 Chemical: Cancer 1/1/89; nickel compounds 5/7/04; Nickel refinery dust from the pyrometallurgical process 10/1/87.
Canada, WHMIS, Ingredients Disclosure List Concentration (most listed compounds are 0.1%). Nickel, water-insoluble compounds, n.o.s and Nickel, water-soluble inorganic compounds, n.o.s. are 1%. See list.

European/International Regulations: Hazard Symbol (*powder*): T,N; Risk phrases: R40; R43; R48/23; R52/53; Safety phrases: S2; S36/37/39; S45; S61 (see Appendix 4).
WGK (German Aquatic Hazard Class): 2—Water polluting (*nickel metal; nickel powder, particle size <0.1 mm*); 1—Slightly water polluting (*nickel, particle size >0.1 mm*).

Description: Nickel metal is a hard, ductile, magnetic metal with a silver-white color. Molecular weight = 58.71; Boiling point = 2837°C; Freezing/Melting point = 1555°C. Hazard Identification (based on NFPA-704 M Rating System) (*powder*): Health 2, Flammability 4, Reactivity 1; (*metal*) Health 1, Flammability 0, Reactivity 0. Insoluble in water. It occurs free in meteorites and in ores combined with sulfur, antimony, or arsenic. Processing and refining of nickel is accomplished by either the Oxford (sodium sulfide and electrolysis) or the Mond (nickel carbonyl) processes. In the latter, impure nickel powder is reacted with carbon monoxide to form gaseous nickel carbonyl which is then treated to deposit high-purity metallic nickel.

Potential Exposure: Compound Description: Tumorigen, Mutagen; Reproductive Effector. Nickel is used as an alloy additive in steel manufacture and in the production of coins and other utensils. Nickel forms alloys with copper, manganese, zinc, chromium, iron, molybdenum, etc. Stainless steel is the most widely used nickel alloy. An important nickel—copper alloy is Monel metal, which contains 66% nickel and 32% copper and has excellent corrosion resistance properties. Permanent magnets are alloys chiefly of nickel, cobalt, aluminum, and iron. Elemental nickel is used in electroplating, anodizing aluminum casting operations for machine parts and in coinage, in the manufacture of acid-resisting and magnetic alloys, magnetic tapes, surgical and dental instruments, nickel-cadmium batteries, nickel soaps in crankcase oil, in ground-coat enamels, colored ceramics, and glass. It is used as a catalyst in the hydrogenation synthesis of acrylic esters for plastics. Exposure to nickel may also occur during mining, smelting, and refining operations. The route by which most people in the general population receive the largest portion of daily nickel intake is through food. Based on the available data from composite diet analysis, between 300 and 600 μg nickel/day is ingested. Fecal nickel analysis, a more accurate measure of dietary nickel intake, suggests about 300 μg/day. The highest level of nickel observed in water was 75 μg/L. Average drinking water levels are about 5 μg/L. A typical consumption of 2 L daily would yield an additional 10 μg of nickel, of which up to 1 μg would be absorbed.

Incompatibilities: Nickel dust is a spontaneously flammable solid and a dangerous fire hazard.

Permissible Exposure Limits in Air

OSHA PEL (*elemental, soluble & insoluble compounds*): 1 mg[Ni]/m³ TWA.
NIOSH REL (*elemental, soluble & insoluble compounds*): 0.015 mg[Ni]/m³ TWA; A potential occupational carcinogen. Limit exposure to lowest feasible concentration. See *NIOSH Pocket Guide*, Appendix A.

ACGIH TLV®[1] (*elemental*): 1.5 mg[Ni]/m³ inhalable fraction TWA; not suspected as a human carcinogen; (*inorganic, insoluble compounds*) 0.2 mg[Ni]/m³ inhalable fraction TWA, confirmed human carcinogen; (*inorganic, soluble compounds*) 0.1 mg[Ni]/m³ inhalable fraction TWA, confirmed human carcinogen.
NIOSH IDLH: 10 mg[Ni]/m³.
Protective Action Criteria (PAC)
TEEL-0: 1 mg/m³
PAC-1: 4.5 mg/m³
PAC-2: 10 mg/m³
PAC-3: 10 mg/m³
DFG MAK (*elemental & nickel compounds*): Inhalable fraction, sensitization of the respiratory tract and skin; Carcinogen Category 1; DFG TRK: *As inhalable dusts/aerosols from nickel metal:* 0.50 μg[Ni]/L in urine, after several shifts; Carcinogen Category 1 DFG TRK: 0.50 μg [Ni]/L in urine, after several shifts; Category 1, human carcinogen, as inhalable dusts/aerosols from nickel metal, nickel carbonate arising in production and processing.
Arab Republic of Egypt: TWA 0.1 mg/m³, 1993; Australia: TWA 1 mg/m³, 1993; Austria: carcinogen, 1999; Denmark: TWA 0.5 mg/m³, 1999; Finland: TWA 0.1 mg/m³ [skin] carcinogen, 1999; France: VME 1 mg/m³, continuous carcinogen, 1999; Hungary: STEL 0.005 mg[Ni]/m³, Carcinogen (insoluble compounds), 1993; the Netherlands: MAC-TGG 1 mg/m³, 2003; the Philippines: TWA 1 mg/m³, 1993; Poland: MAC (TWA) 0.25 mg/m³, 1999; Russia: STEL 0.05 mg/m³, 1993; Sweden: NGV 0.5 mg/m³ (*dust*), 1999; Switzerland: MAK-W 0.5 mg/m³, carcinogen, 1999; Thailand: TWA 1 mg/m³, 1993; United Kingdom: TWA 0.1 mg/m³, 2000; Argentina, Bulgaria, Columbia, Jordan, South Korea, New Zealand, Singapore, Vietnam: ACGIH TLV®: Confirmed Human Carcinogen. Russia[35, 43] set a MAC of 0.0002 mg/m³ in ambient air in residential areas for soluble nickel salts. Various states have set guidelines or standards for nickel in ambient air[60] ranging from 0.002 μg/m³ (Rhode Island) to 0.002–0.018 μg/m³ (Massachusetts) to 0.0303 μg/m³ (Kansas) to 0.13–0.70 μg/m³ (Montana) to 0.24 μg/m³ (Pennsylvania) to 0.5 μg/m³ (North Carolina and South Carolina) to 1.0 μg/m³ (North Dakota) to 2.0 μg/m³ (Nevada) to 3.3 μg/m³ (New York) to 5.0 μg/m³ (Connecticut) to 10.0 μg/m³ (Virginia).

Determination in Air: Use NIOSH Analytical Method #7300, Elements by ICP (Nitric/perchloric acid ashing); #7301, Elements by ICP (Aqua regia ashing); #7303, Elements by ICP [Hot block (HCl/HNO₃ Digestion)]; #9102, Elements on wipes; #8310, Metals in urine, #8005, Elements in blood or tissue; OSHA Analytical Methods ID-121 and ID-125G.

Permissible Concentration in Water: To protect freshwater aquatic life: e[0.76 ln (hardness) + 1.06] as a 24-h average, never to exceed e[0.76 ln (hardness) + 4.02] at any time. *To protect saltwater aquatic life:* 7.1 μg/L as a 24-h average, never to exceed 140 μg/L. *To protect human health:* 13.4 μg/L.[6] A lifetime health advisory of 150 μg/L has been developed by EPA.[49] Mexico[35] has set a limit of 0.1 mg/L in estuaries and 0.008 mg/L in coastal waters. The Czech Republic[35] has set a limit of 0.1 mg/L in surface waters and 0.05 mg/L in drinking water reserves. States which have set guidelines for nickel in drinking water[61] include Minnesota at 150 µg/L and Kansas at 1000 µg/L.

Determination in Water: Digestion followed by atomic absorption, or by colorimetric (heptoxime) determination, or by inductively coupled plasma (ICP) optical emission spectrometry. This gives total nickel; dissolved nickel may be determined by the same method preceded by 0.45-µm filtration.

Routes of Entry: Inhalation, ingestion, skin and/or eye contact.

Harmful Effects and Symptoms

Short Term Exposure: Nickel dusts and fumes can affect you when breathed in. Can cause irritation of the eyes and skin. Skin contact may cause skin allergy, with itching, redness and later, rash. Lung allergy occasionally occurs with asthma-type effects. Fumes from heated nickel can cause pneumonia-like illness with cough and shortness of breath. Higher exposures can cause pulmonary edema, a medical emergency that can be delayed for several hours or days. This can cause death.

Long Term Exposure: May cause skin sensitization and allergy. May cause skin allergy. May cause allergic asthma, pneumonitis. Breathing nickel dust and fume can cause a sore or hole in the nasal septum. May damage the kidneys and affect liver function. Nickel is a carcinogen and may damage the developing fetus. Occupational exposure to nickel refinery dust contains nickel subsulfide and is associated with lung cancer. Handle with extreme caution.

Points of Attack: Nasal cavities, lungs, skin, liver, kidneys. *Cancer site:* lung, throat, and nasal cavity.

Medical Surveillance: NIOSH lists the following tests: Blood gas analysis; whole blood (chemical/metabolite); blood plasma; blood plasma, end-of-shift; blood serum; biologic tissue/biopsy; chest X-ray, electrocardiogram, pulmonary function tests; pre- and postshift; sputum cytology; urine (chemical/metabolite); urine (chemical/metabolite), end-of-shift; white blood cell count/differential. Before beginning employment and at regular times after that, the following are recommended: lung function tests. These may be normal if the person is not having an attack at the time of the test. Urine or plasma test for nickel (unexposed persons have urine levels less than 10 µg/L). If symptoms develop or overexposure is suspected, the following may be useful: daily urine nickel for several days (persons with urine nickel over 100 µg need medical attention). Consider chest X-ray for acute overexposure. Evaluation by a qualified allergist, including careful exposure history and special testing, may help diagnose skin allergy. Liver and kidney function tests.

First Aid: If this chemical gets into the eyes, remove any contact lenses at once and irrigate immediately for at least 15 min, occasionally lifting upper and lower lids. Seek medical attention immediately. If this chemical contacts the skin, remove contaminated clothing and wash immediately with soap and water. Seek medical attention immediately. If this chemical has been inhaled, remove from exposure, begin rescue breathing (using universal precautions, including resuscitation mask) if breathing has stopped and CPR if heart action has stopped. Transfer promptly to a medical facility. When this chemical has been swallowed, get medical attention. Give large quantities of water and induce vomiting. Do not make an unconscious person vomit.

Note to physician: Inhalation: Bronchodilators, decongestants, and oxygen may be used if necessary. Corticosteroids are useful for treating pneumonitis.

Note to physician: For severe poisoning BAL [British Anti-Lewisite, dimercaprol, dithiopropanol ($C_3H_8OS_2$)] has been used to treat toxic symptoms of certain heavy metals poisoning including nickel. Although BAL is reported to have a large margin of safety, caution must be exercised, because toxic effects may be caused by excessive dosage. Most can be prevented by premedication with 1-ephedrine sulfate (CAS: 134-72-5).

Personal Protective Methods: Wear protective gloves and clothing to prevent any reasonable probability of skin contact. Safety equipment suppliers/manufacturers can provide recommendations on the most protective glove/clothing material for your operation. All protective clothing (suits, gloves, footwear, headgear) should be clean, available each day, and put on before work. Contact lenses should not be worn when working with this chemical. Wear dust-proof chemical goggles and face shield unless full face-piece respiratory protection is worn. Employees should wash immediately with soap when skin is wet or contaminated. Provide emergency showers and eyewash. Specific engineering controls are recommended by NIOSH Criteria Document #77-164, *Inorganic Nickel.*

Respirator Selection: At concentrations above the NIOSH REL, or where there is no REL, at any detectable concentration: SCBAF: Pd,Pp (APF = 10,000) (any NIOSH/MSHA- or European Standard EN 149-approved self-contained breathing apparatus that has a full face-piece and is operated in a pressure-demand or other positive-pressure mode) or SaF: Pd,Pp: ASCBA (APF = 10,000) (any supplied-air respirator that has a full face-piece and is operated in a pressure-demand or other positive-pressure mode in combination with an auxiliary, self-contained breathing apparatus operated in a pressure-demand or other positive-pressure mode). *Escape:* 100F (APF = 50) (any air-purifying, full-face-piece respirator with an N100, R100, or P100 filter.

Storage: Pyrophoric (dry nickel powder, Raney nickel, and nickel catalyst) are fire hazards. Prior to working with these chemicals, you should be trained on its proper handling and storage. Finely divided nickel must be stored to avoid contact with strong acids (such as hydrochloric, sulfuric and nitric), since violent reactions occur. Store in tightly closed containers in a cool, well-ventilated area away from acids, fluorine, ammonia, phosphorus, sulfur, selenium, hydrazine, and performic acid. Sources of ignition, such as smoking and open flames, are prohibited where pulverized nickel is handled, used, or stored. Wherever pulverized nickel is used, handled, manufactured, or stored, use explosion-proof electrical equipment and fittings. A regulated, marked area should be established where this chemical is handled, used, or stored in compliance with OSHA Standard 1910.1045.

Shipping: Dry powder, in bulk, requires a shipping label of "SPONTANEOUSLY COMBUSTIBLE." It falls in Hazard Class 4.2 and Packing Group II.

The name of this material is not on the DOT list of materials[19] for label and packaging standards. However, based on regulations, it may be classified[52] as an Environmentally hazardous substances, liquid, n.o.s. It falls in Hazard Class 9 and Packing Group III.[20,21]

Spill Handling: Evacuate persons not wearing protective equipment from area of spill or leak until cleanup is complete. Remove all ignition sources. Use dry chemicals, sand, water spray, or foam. Do not dry sweep. Collect powdered material in the most convenient and safe manner and deposit in sealed containers. Ventilate area after cleanup is complete. Keep nickel catalyst spills out of confined spaces, such as a sewers, because of the possibility of an explosion or fire. It may be necessary to contain and dispose of this chemical as a hazardous waste. If material or contaminated runoff enters waterways, notify downstream users of potentially contaminated waters. Contact your local or federal environmental protection agency for specific recommendations. If employees are required to clean up spills, they must be properly trained and equipped. OSHA 1910.120(q) may be applicable.

Fire Extinguishing: Nickel dust is flammable. Use dry chemical, soda ash, or lime extinguishers. Poisonous gases are produced in fire, including nickel carbonyl. Dry nickel catalyst may spontaneously ignite and the fire may restart after it has been extinguished. If material or contaminated runoff enters waterways, notify downstream users of potentially contaminated waters. Notify local health and fire officials and pollution control agencies. From a secure, explosion-proof location, use water spray to cool exposed containers. If cooling streams are ineffective (venting sound increases in volume and pitch, tank discolors, or shows any signs of deforming), withdraw immediately to a secure position. If employees are expected to fight fires, they must be trained and equipped in OSHA 1910.156. The only respirators recommended for firefighting are self-contained breathing apparatuses that have full face-pieces and are operated in a pressure-demand or other positive-pressure mode.

Disposal Method Suggested: Nickel compounds—encapsulation followed by disposal in a chemical waste landfill. However, nickel from various industrial wastes may also be recovered and recycled as described in the literature.

References
National Institute for Occupational Safety and Health. (1977). *Criteria for a Recommended Standard: Occupational Exposure to Inorganic Nickel*, NIOSH Document No. 77-64
National Academy of Sciences. (1975). *Report on Medical and Biological Effects of Environmental Pollutants: Nickel*. Washington, DC
US Environmental Protection Agency. (1980). *Nickel: Ambient Water Quality Criteria*. Washington, DC
US Environmental Protection Agency. (April 30, 1980). *Nickel, Health and Environmental Effects Profile No. 133*. Washington, DC: Office of Solid Waste
Sax, N. I. (Ed.). *Dangerous Properties of Industrial Materials Report,* 1, No. 1, 50−51 (1980) and 3, No. 3, 76−80 (1983)
US Public Health Service. (October 31, 1987). *Toxicological Profile for Nickel*. Atlanta, GA: Agency for Toxic Substances and Disease Registry
US Environmental Protection Agency. (October 31, 1985). *Chemical Hazard Information Profile: Nickel*. Washington, DC: Chemical Emergency Preparedness Program
New York State Department of Health. (March 1986). *Chemical Fact Sheet: Nickel Metal and Soluble Nickel Compounds*. Albany, NY: Bureau of Toxic Substance Assessment
New Jersey Department of Health and Senior Services. (March 2007). *Hazardous Substances Fact Sheet: Nickel*. Trenton, NJ

Nickel ammonium sulfate   N:0230

Molecular Formula: $H_{20}N_2NiO_{14}S_2$
Common Formula: $NiSO_4 \cdot (NH_4)_2SO_4 \cdot 6H_2O$
Synonyms: Ammonium disulfatonickelate (II); Ammonium nickel sulfate; Nickel ammonium sulphate; Sulfato de niquel y amonio (Spanish); Sulfuric acid, ammonium nickel (2+) salt (2:2:1); Sulfuric acid, ammonium nickel(II) salt (2:2:1)
CAS Registry Number: 15699-18-0
RTECS® Number: WS6050000
UN/NA & ERG Number: UN3288 (Toxic solid, inorganic, n.o.s.)/151
EC Number: 239-793-5
Regulatory Authority and Advisory Bodies
Carcinogenicity: IARC: (compounds) *carcinogenic to humans, carcinogenic to humans*, Group 1, 1997; NTP: 11th Report on Carcinogens, 2004: Reasonably anticipated to be a human carcinogen; NIOSH: Potential occupational carcinogen.
Air Pollutant Standard Set. See below, "Permissible Exposure Limits in Air" section.
Water Pollution Standard Proposed (EPA).[6, 49]
Clean Air Act: Hazardous Air Pollutants (Title I, Part A, Section 112) as nickel compounds.
Clean Water Act: Section 311 Hazardous Substances/RQ 40CFR117.3 (same as CERCLA, see below); 40CFR401.15

Section 307 Toxic Pollutants as nickel and compounds; Section 313 Water Priority Chemicals (57FR41331, 9/9/92). RCRA, 40CFR261, Appendix 8 Hazardous Constituents, waste number not listed, as nickel compounds, n.o.s. Reportable Quantity (RQ): 100 lb (45.4 kg). EPCRA Section 313 Form R *de minimis* concentration reporting level: 0.1%. California Proposition 65 Chemical: Cancer (as nickel compounds) 5/7/07. Canada, WHMIS, Ingredients Disclosure List Concentration 0.1%. European/International Regulations: not listed in Annex 1. WGK (German Aquatic Hazard Class): No value assigned.

Description: Nickel ammonium sulfate is a green, odorless powder. Molecular weight = 394.4. Hazard Identification (based on NFPA-704 M Rating System): Health 3, Flammability 0, Reactivity 1. Soluble in water.

Potential Exposure: This material is used in electroplating.

Incompatibilities: Violent reaction with strong acids. Incompatible with nickel nitrate, sulfur, selenium, wood, organics, and other combustibles.

Permissible Exposure Limits in Air

OSHA PEL (*elemental, soluble & insoluble compounds*): 1 mg[Ni]/m^3 TWA.

NIOSH REL (*elemental, soluble & insoluble compounds*): 0.015 mg[Ni]/m^3 TWA; A potential occupational carcinogen. Limit exposure to lowest feasible concentration. See *NIOSH Pocket Guide*, Appendix A.

ACGIH TLV®[1] (*inorganic, soluble compounds*) 0.1 mg [Ni]/m^3 inhalable fraction TWA, confirmed human carcinogen.

NIOSH IDLH: 10 mg[Ni]/m^3.

Protective Action Criteria (PAC)

TEEL-0: 0.489 mg/m^3

PAC-1: 1.47 mg/m^3

PAC-2: 2.5 mg/m^3

PAC-3: 48.9 mg/m^3

DFG MAK (*elemental & nickel compounds*): Inhalable fraction, sensitization of the respiratory tract and skin; Carcinogen Category 1; DFG TRK: *As inhalable dusts/ aerosols from nickel metal*: 0.50 μg[Ni]/L in urine, after several shifts; Carcinogen Category 1.

Determination in Air: Use NIOSH Analytical Method #7300, Elements by ICP (Nitric/Perchloric Acid Ashing); #7301, Elements by ICP (Aqua Regia Ashing); #7303, Elements by ICP [Hot Block (HCl/HNO$_3$ Digestion)]; #9102, Elements on Wipes; #8310, Metals in urine, #8005, Elements in blood or tissue; OSHA Analytical Methods ID-121 and ID-125G.

Permissible Concentration in Water: The EPA[6] has set a limit of 13.4 μg/L to protect human health. A lifetime health advisory of 150 μg/L has more recently been promulgated by EPA.[49] See also this section in the entry on "Nickel and Soluble Compounds."

Routes of Entry: Inhalation, ingestion, skin and/or eye contact.

Harmful Effects and Symptoms

Short Term Exposure: Nickel ammonium sulfate can affect you when breathed in. Eye and skin contact may cause irritation and burns. Lung damage may result from a single high exposure or lower repeated exposure. Higher exposures can cause pulmonary edema, a medical emergency that can be delayed for several hours. This can cause death. This is sometimes delayed for 1−2 days after exposure.

Long Term Exposure: Skin contact may cause skin allergy with itching, redness, and later, rash. Lung allergy occasionally occurs, with asthma-type effects. High or repeated lower exposures may cause scarring of the lungs, and may damage the heart, liver, or kidneys.

Points of Attack: Lungs, kidneys, heart, liver, skin.

Medical Surveillance: NIOSH lists the following tests: blood gas analysis; whole blood (chemical/metabolite); blood plasma; blood plasma, end-of-shift; blood serum; biologic tissue/biopsy; chest X-ray, electrocardiogram, pulmonary function tests; pre- and postshift; sputum cytology; urine (chemical/metabolite); urine (chemical/metabolite), end-of-shift; white blood cell count/differential. Before beginning employment and at regular times after that, the following are recommended: lung function tests. These may be normal if the person is not having an attack at the time of the test. Urine or plasma test for nickel (unexposed persons have urine levels less than 10 μg/L). If symptoms develop or overexposure is suspected, the following may be useful: daily urine nickel for several days (person's urine nickel over 100 μg needs medical attention). Lung function tests. Consider chest X-ray following acute overexposure. Liver and kidney function tests. Evaluation by a qualified allergist.

First Aid: If this chemical gets into the eyes, remove any contact lenses at once and irrigate immediately for at least 15 min, occasionally lifting upper and lower lids. Seek medical attention immediately. If this chemical contacts the skin, remove contaminated clothing and wash immediately with soap and water. Seek medical attention immediately. If this chemical has been inhaled, remove from exposure, begin rescue breathing (using universal precautions, including resuscitation mask) if breathing has stopped and CPR if heart action has stopped. Transfer promptly to a medical facility for medical observation up to 2 days and tests for urine nickel. When this chemical has been swallowed, get medical attention. Give large quantities of water and induce vomiting. Do not make an unconscious person vomit. Medical observation is recommended for 24−48 h after breathing overexposure, as pulmonary edema may be delayed. As first aid for pulmonary edema, a doctor or authorized paramedic may consider administering a corticosteroid spray.

Note to physician: For severe poisoning BAL [British Anti-Lewisite, dimercaprol, dithiopropanol (C$_3$H$_8$OS$_2$)] has been used to treat toxic symptoms of certain heavy metals poisoning including nickel. Although BAL is reported to have a large margin of safety, caution must be exercised, because toxic effects may be caused by excessive dosage. Most can be prevented by premedication with 1-ephedrine sulfate (CAS: 134-72-5).

Personal Protective Methods: Wear protective gloves and clothing to prevent any reasonable probability of skin contact. Safety equipment suppliers/manufacturers can provide recommendations on the most protective glove/clothing material for your operation. All protective clothing (suits, gloves, footwear, headgear) should be clean, available each day, and put on before work. Contact lenses should not be worn when working with this chemical. Wear dust-proof chemical goggles and face shield unless full face-piece respiratory protection is worn. Employees should wash immediately with soap when skin is wet or contaminated. Provide emergency showers and eyewash. See NIOSH Criteria Document #77-164, *Inorganic Nickel.*

Respirator Selection: *At concentrations above the NIOSH REL, or where there is no REL, at any detectable concentration:* SCBAF: Pd,Pp (APF = 10,000) (any NIOSH/MSHA- or European Standard EN 149-approved self-contained breathing apparatus that has a full face-piece and is operated in a pressure-demand or other positive-pressure mode) or SaF: Pd,Pp: ASCBA (APF = 10,000) (any supplied-air respirator that has a full face-piece and is operated in a pressure-demand or other positive-pressure mode in combination with an auxiliary, self-contained breathing apparatus operated in a pressure-demand or other positive-pressure mode). *Escape:* 100F (APF = 50) (any air-purifying, full-face-piece respirator with an N100, R100, or P100 filter).

Storage: Color Code—Blue: Health Hazard/Poison: Store in a secure poison location. Prior to working with this chemical you should be trained on its proper handling and storage. Nickel ammonium sulfate must be stored to avoid contact with strong acids (such as hydrochloric, sulfuric and nitric), since violent reactions occur. Store in tightly closed containers in a cool, well-ventilated area away from sulfur.

Shipping: Toxic solid, inorganic, n.o.s. materials require a shipping label of "POISONOUS/TOXIC MATERIALS." They fall in Hazard Class 6.1.

Spill Handling: Evacuate persons not wearing protective equipment from area of spill or leak until cleanup is complete. Remove all ignition sources. Use HEPA vacuum or wet method to reduce dust during cleanup. Do not dry sweep. Collect powdered material in the most convenient and safe manner and deposit in sealed containers. Ventilate area after cleanup is complete. It may be necessary to contain and dispose of this chemical as a hazardous waste. If material or contaminated runoff enters waterways, notify downstream users of potentially contaminated waters. Contact your local or federal environmental protection agency for specific recommendations. If employees are required to clean up spills, they must be properly trained and equipped. OSHA 1910.120(q) may be applicable.

Fire Extinguishing: This chemical may burn but does not easily ignite. Use dry chemical, carbon dioxide, water spray, or foam extinguishers. Poisonous gases, including nitrogen oxides and sulfur oxides, are produced in fire. If material or contaminated runoff enters waterways, notify downstream users of potentially contaminated waters. Notify local health and fire officials and pollution control agencies. From a secure, explosion-proof location, use water spray to cool exposed containers. If cooling streams are ineffective (venting sound increases in volume and pitch, tank discolors, or shows any signs of deforming), withdraw immediately to a secure position. If employees are expected to fight fires, they must be trained and equipped in OSHA 1910.156. The only respirators recommended for firefighting are self-contained breathing apparatuses that have full face-pieces and are operated in a pressure-demand or other positive-pressure mode.

References
Sax, N. I. (Ed.). (1985). *Dangerous Properties of Industrial Materials Report,* 5, No. 4, 74–76
New Jersey Department of Health and Senior Services. (January 1996). *Hazardous Substances Fact Sheet: Nickel Ammonium Sulfate.* Trenton, NJ

Nickel carbonyl   N:0240

Molecular Formula: C_4NiO_4
Common Formula: $Ni(CO)_4$
Synonyms: Nickel carbonyle (French); Nickel tetracarbonyl; Nickel tetracarbonyle (French); Niquel carbonilo (Spanish); Tetracarbonyl nickel
CAS Registry Number: 13463-39-3
RTECS® Number: QR6300000
UN/NA & ERG Number: UN1259/131
EC Number: 236-669-2 [*Annex I Index No.:* 028-001-00-1]
Regulatory Authority and Advisory Bodies
Department of Homeland Security Screening Threshold Quantity (pounds): *Release hazard* 10,000 ($\geq$1.00% concentration).
Carcinogenicity: IARC[9]: *possibly carcinogenic to humans,* Group B2; NTP: 11th Report on Carcinogens, 2004: Known to be a human carcinogen; EPA: Sufficient evidence from animal studies; inadequate evidence or no useful data from epidemiologic studies; NIOSH: Potential occupational carcinogen.
Very Toxic Substance (World Bank).[15]
Air Pollutant Standard Set. See below, "Permissible Exposure Limits in Air" section.
OSHA 29CFR1910.119, Appendix A, Process Safety List of Highly Hazardous Chemicals, TQ = 150 lb.
Clean Air Act: Hazardous Air Pollutants (Title I, Part A, Section 112) as nickel compounds; Accidental Release Prevention/Flammable Substances, (Section 112[r], Table 3), TQ = 1000 lb (454.0 kg).
Clean Water Act: 40CFR401.15 Section 307 Toxic Pollutants as nickel and compounds.
US EPA Hazardous Waste Number (RCRA No.): P073.

RCRA, 40CFR261, Appendix 8 Hazardous Constituents.
Reportable Quantity (RQ): 10 lb (4.54 kg).
EPCRA Section 313 Form R *de minimis* concentration reporting level: 0.1%.
SUPERFUND/EPCRA 40CFR355, Extremely Hazardous Substances: TPQ = 1 lb (0.454 kg).
US DOT Marine Pollutant (49CFR, Subchapter 172.101, Appendix B.
US DOT 49CFR172.101, Inhalation Hazardous Chemical.
California Proposition 65 Chemical: Cancer 10/1/87; Developmental/Reproductive toxin 9/1/96.
Canada, WHMIS, Ingredients Disclosure List Concentration 0.1%.
European/International Regulations: Hazard Symbol: F, T+, N; Risk phrases: R61; R11; R26; R40; R50/53; Safety phrases: S53; S45; S60; S61 (see Appendix 4).
WGK (German Aquatic Hazard Class): No value assigned.

Description: Nickel carbonyl is a colorless, highly volatile, flammable liquid with a musty odor. The odor threshold is 1.3 ppm. It decomposes above room temperature producing carbon monoxide and finely divided nickel. Molecular weight = 170.75; Specific gravity (H_2O:1) = 1.32 at 17°C; Boiling point = 43.3°C; Freezing/Melting point = −19°C; Vapor pressure = 315 mmHg at 20°C; Flash point < −24°C (cc); Autoignition temperature: 60°C. Explosive limits: LEL = 2%; UEL = 34%. Hazard Identification (based on NFPA-704 M Rating System): Health 4, Flammability 3, Reactivity 3. Insoluble in water.

Potential Exposure: Compound Description: Tumorigen; Reproductive Effector; Human Data. Nickel carbonyl is used as an intermediate product in the refining of nickel. The primary use for nickel carbonyl is in the production of nickel by the Mond process. Impure nickel powder is reacted with carbon monoxide to form gaseous nickel carbonyl which is then treated to deposit high-purity metallic nickel and release carbon monoxide. Other uses include gas plating, the production of nickel products, in chemical synthesis as a catalyst, particularly for oxo reactions (addition reaction of hydrogen and carbon monoxide with unsaturated hydrocarbons to form oxygen-function compounds), e.g., synthesis of acrylic esters, and as a reactant.

Incompatibilities: May spontaneously ignite on contact with air. In the presence of air, oxidizes and forms a deposit which becomes peroxidized; this tends to decompose and ignite. May explode when heated above 60°C. Decomposes on contact with acids, producing carbon monoxide. Violent reaction with oxidizers; may cause fire and explosions. Vapor may promote the ignition of mixtures of combustible vapors (such as gasoline) and air. Attacks some plastics, rubber, and coatings. Store under inert gas blanket.

Permissible Exposure Limits in Air
Conversion factor: 1 ppm = 6.98 mg/m³ at 25°C & 1 atm.
OSHA PEL: 0.0001 ppm/0.007 mg[Ni]/m³ TWA.
NIOSH REL: 0.0001 ppm/0.007 mg[Ni]/m³ TWA, potential carcinogen, limit occupational exposure to lowest feasible level; See *NIOSH Pocket Guide*, Appendix A.

ACGIH TLV®[1]: 0.05 ppm/0.12 mg[Ni]/m³ TWA.
NIOSH IDLH: 2 ppm.
Protective Action Criteria (PAC)*
TEEL-0: 0.001 ppm
PAC-1: 0.005 ppm
PAC-2: **0.036** ppm
PAC-3: **0.16** ppm
*AEGLs (Acute Emergency Guideline Levels) & ERPGs (Emergency Response Planning Guideline) are in **bold face**.
DFG MAK: [skin] Carcinogen Category 2.
Compound Description: Tumorigen; Reproductive Effector; Human Data.
Arab Republic of Egypt: TWA 0.007 ppm (0.001 mg/m³), 1993; Australia: TWA 0.05 ppm (0.1 mg/m³), 1993; Austria: carcinogen, 1999; Belgium: carcinogen, 1993; Denmark: TWA 0.001 ppm (0.007 mg[Ni]/m³), [skin], 1999; Finland: TWA 0.001 ppm (0.007 mg/m³); STEL 0.003 ppm, carcinogen, 1999; France: VME 0.05 ppm (0.23 mg[Ni]/m³), 1999; the Netherlands: MAC-TGG 0.35 mg/m³, 2003; Japan: 0.001 ppm (0.007 mg/m³), 1999; Norway: TWA 0.001 ppm (0.007 mg/m³), 1999; the Philippines: TWA 0.001 ppm (0.007 mg/m³), 1993; Russia: TWA 0.001 ppm; STEL 0.0005 mg/m³, carcinogen, 1993; Sweden: NGV 0.001 ppm (0.007 mg/m³), carcinogen, 1999; Switzerland: MAK-W 0.05 ppm (0.35 mg/m³), [skin], carcinogen, 1999; Thailand: TWA 0.001 ppm (0.007 mg/m³), 1993; Turkey: TWA 0.001 ppm (0.007 mg/m³), 1993; United Kingdom: STEL 0.1 ppm (0.24 mg/m³), 2000; Argentina, Bulgaria, Columbia, Jordan, South Korea, New Zealand, Singapore, Vietnam: ACGIH TLV®: TWA 0.05 (Ni)ppm. The Czech Republic[35]: TWA 0.01 mg/m³; ceiling value 0.02 mg/m³. Several states have set guidelines or standards for nickel carbonyl in ambient air[60] ranging from 1.17 μg/m³ (New York) to 1.75 μg/m³ (Connecticut and South Carolina) to 5.0 μg/m³ (Virginia) to 8.0 μg/m³ (Nevada).

Determination in Air: Use NIOSH Analytical Method #6007, Nickel carbonyl. Charcoal tube (low Ni); HNO_3; Graphite furnace atomic absorption spectrometry; NIOSH Analytical Method (IV) #6007.

Permissible Concentration in Water: No criteria set, but EPA[32] has suggested a permissible ambient goal of 1.4 μg/L based on health effects.

Routes of Entry: It may be possible for appreciable amounts of the liquid to be absorbed through the skin; also ingestion and eye and skin contact.

Harmful Effects and Symptoms
Short Term Exposure: Irritates and burns the eyes and skin. Irritates the respiratory tract. May affect the central nervous system and the kidneys, causing tissue lesions. Medical observation is advised for 3 days or more; delayed lung effects may occur. Probable oral lethal dose for a human is between 50 and 500 mg/kg, between one teaspoon and 1 oz per 150-lb person. Nickel carbonyl has also been estimated to be lethal in humans at atmospheric exposures of 30 ppm for 20 min. Autopsies show congestion, collapse, and tissue

destruction as well as hemorrhage in the brain. Dermatitis, recurrent asthmatic attacks, and increased number of white blood cells (eosinophils) in respiratory tract are acute health hazards. Acute exposure to nickel carbonyl may result in dizziness, giddiness, weakness, convulsions, hallucinations, delirium, nausea, vomiting, and diarrhea. Following inhalation, respiration will initially be rapid, accompanied by a nonproductive cough and followed by pain and tightness in the chest. Pulmonary edema, cerebral edema, and hepatic (liver) degeneration may also occur. Vapor is irritating to the eyes, nose, and throat. Nickel contact dermatitis is the most common reaction to nickel carbonyl. Nickel itch may begin with a burning sensation and itching, often followed by erythema (redness) and nodular eruptions.

Long Term Exposure: Repeated or prolonged inhalation exposure may cause skin and lung sensitization and asthma. A potential occupational carcinogen. In animals: reproductive, teratogenic effects. Permanent lung damage may occur following a single high exposure or lower repeated exposure. High or repeated exposures may cause damage to the heart muscle, liver and/or kidney damage.

Points of Attack: Lungs, paranasal sinus, central nervous system, reproductive system, liver, kidney, heart.

Medical Surveillance: NIOSH lists the following tests: Blood gas analysis; blood plasma; chest X-ray, electrocardiogram, expired air, 2H; pulmonary function tests; forced vital capacity, forced expiratory volume (1 s); sputum cytology; urine (chemical/metabolite), urinalysis (routine); white blood cell count/differential. Before beginning employment and at regular times after that, the following are recommended: lung function tests. Urine or plasma test for nickel (unexposed persons have urine levels under 10 µg/L). If symptoms develop or overexposure is suspected, the following may be useful: daily urine nickel for several days. Lung function tests. Chest X-ray (persons with urine levels over 100 µg/L need medical observation). Evaluation by a qualified allergist, including careful exposure history and special testing, may help diagnose skin allergy. Liver and kidney function tests.

First Aid: If this chemical gets into the eyes, remove any contact lenses at once and irrigate immediately for at least 15 min, occasionally lifting upper and lower lids. Seek medical attention immediately. If this chemical contacts the skin, remove contaminated clothing and wash immediately with soap and water. Seek medical attention immediately. If this chemical has been inhaled, remove from exposure, begin rescue breathing (using universal precautions, including resuscitation mask) if breathing has stopped and CPR if heart action has stopped. Transfer promptly to a medical facility. When this chemical has been swallowed, get medical attention. Give large quantities of water and induce vomiting. Do not make an unconscious person vomit. Medical observation is advised for 3 days or more; delayed lung effects including pulmonary edema may occur.

Note to physician: For severe poisoning BAL [British Anti-Lewisite, dimercaprol, dithiopropanol ($C_3H_8OS_2$)] has been used to treat toxic symptoms of certain heavy metals poisoning including nickel. Although BAL is reported to have a large margin of safety, caution must be exercised, because toxic effects may be caused by excessive dosage. Most can be prevented by premedication with 1-ephedrine sulfate (CAS: 134-72-5).

Personal Protective Methods: Wear protective gloves and clothing to prevent any reasonable probability of skin contact. Safety equipment suppliers/manufacturers can provide recommendations on the most protective glove/clothing material for your operation. All protective clothing (suits, gloves, footwear, headgear) should be clean, available each day, and put on before work. Contact lenses should not be worn when working with this chemical. Wear splash-proof chemical goggles and face shield unless full face-piece respiratory protection is worn. Employees should wash immediately with soap when skin is wet or contaminated. Provide emergency showers and eyewash. See NIOSH Criteria Document #77-164, *Inorganic Nickel.*

Respirator Selection: At any detectable concentration: SCBAF: Pd,Pp (APF = 10,000) (any NIOSH/MSHA- or European Standard EN 149-approved self-contained breathing apparatus that has a full face-piece and is operated in a pressure-demand or other positive-pressure mode) or SaF: Pd,Pp: ASCBA (APF = 10,000) (any supplied-air respirator that has a full face-piece and is operated in a pressure-demand or other positive-pressure mode in combination with an auxiliary, self-contained breathing apparatus operated in a pressure-demand or other positive-pressure mode). *Escape:* 100 F (APF = 50) (any air-purifying, full-face-piece respirator with an N100, R100, or P100 filter) or SCBAE (any appropriate escape-type, self-contained breathing apparatus).

Storage: (1) Color Code—Red: Flammability Hazard: Store in a flammable liquid storage area or approved cabinet away from ignition sources and corrosive and reactive materials. (2) Color Code—Blue: Health Hazard/Poison: Store in a secure poison location. Prior to working with this chemical you should be trained on its proper handling and storage. A regulated, marked area should be established where this chemical is handled, used, or stored in compliance with OSHA Standard 1910.1045. Before entering confined space where this chemical may be present, check to make sure that an explosive concentration does not exist. Nickel carbonyl must be stored to avoid contact with strong oxidizers (such as chlorine, bromine, and fluorine), since violent reactions occur. Store in tightly closed containers in a cool, well-ventilated area away from strong acids (such as hydrochloric, sulfuric, and nitric). Sources of ignition, such as smoking and open flames, are prohibited where nickel carbonyl is handled, used, or stored. Metal containers involving the transfer of 5 gallons or more of nickel carbonyl should be grounded and bonded. Drums must be equipped with self-closing valves, pressure vacuum bungs, and flame arresters. Use only nonsparking tools and equipment, especially when opening and closing containers of nickel carbonyl. Wherever nickel carbonyl is used, handled, manufactured, or stored, use explosion-proof electrical equipment and fittings.

Shipping: Nickel carbonyl requires a shipping label of "POISONOUS/TOXIC MATERIALS; FLAMMABLE LIQUID." It falls in Hazard Class 6.1 and Packing Group I. A US DOT Severe Marine Pollutant.

Spill Handling: Evacuate and restrict persons not wearing protective equipment from area of spill or leak until cleanup is complete. Remove all ignition sources. Establish forced ventilation to keep levels below explosive limit. Absorb liquids in vermiculite, dry sand, earth, peat, carbon, or a similar material and deposit in sealed containers. Keep this chemical out of a confined space, such as a sewer, because of the possibility of an explosion, unless the sewer is designed to prevent the buildup of explosive concentrations. It may be necessary to contain and dispose of this chemical as a hazardous waste. If material or contaminated runoff enters waterways, notify downstream users of potentially contaminated waters. Contact your local or federal environmental protection agency for specific recommendations. If employees are required to clean up spills, they must be properly trained and equipped. OSHA 1910.120(q) may be applicable.

Initial isolation and protective action distances

Distances shown are likely to be affected during the first 30 min after materials are spilled and could increase with time. If more than one tank car, cargo tank, portable tank, or large cylinder involved in the incident is leaking, the protective action distance may need to be increased. You may need to seek emergency information from CHEMTREC at (800) 424-9300 or seek professional environmental engineering assistance from the US EPA Environmental Response Team at (908) 548-8730 (24-h response line).

Small spills (From a small package or a small leak from a large package)

First: Isolate in all directions (feet/meters)   500/150
Then: Protect persons downwind (miles/kilometers)
Day   0.9/1.5
Night   3.1/4.9

Large spills (From a large package or from many small packages)

First: Isolate in all directions (feet/meters)   3000/1000
Then: Protect persons downwind (miles/kilometers)
Day   7.0 + /11.0+
Night   7.0+ /11.0+

Fire Extinguishing: This chemical is a flammable liquid. Poisonous gases, including carbon monoxide, are produced in fire. Use dry chemical, carbon dioxide, or foam extinguishers. Solid streams of water may be ineffective. Material is too dangerous to health to expose fire fighters. A few whiffs of the vapor could cause death. If liquid or vapor penetrates fire fighter's protective gear it will cause fatality. Normal full protective gear available to the average fire department will not provide adequate inhalation or skin protection. Vapors are heavier than air and will collect in low areas. Vapors may travel long distances to ignition sources and flashback. Vapors in confined areas may explode when exposed to fire. Containers may explode in fire. Storage containers and parts of containers may rocket great distances, in many directions. If material or contaminated runoff enters waterways, notify downstream users of potentially contaminated waters. Notify local health and fire officials and pollution control agencies. From a secure, explosion-proof location, use water spray to cool exposed containers. If cooling streams are ineffective (venting sound increases in volume and pitch, tank discolors, or shows any signs of deforming), withdraw immediately to a secure position. If employees are expected to fight fires, they must be trained and equipped in OSHA 1910.156. The only respirators recommended for firefighting are self-contained breathing apparatuses that have full face-pieces and are operated in a pressure-demand or other positive-pressure mode.

Disposal Method Suggested: Incineration in admixture with a flammable solvent. Also, nickel carbonyl used in metalizing operations may be recovered and recycled. Consult with environmental regulatory agencies for guidance on acceptable disposal practices. Generators of waste containing this contaminant ($\geq$100 kg/mo) must conform with EPA regulations governing storage, transportation, treatment, and waste disposal.

References

Sax, N. I. (Ed.). *Dangerous Properties of Industrial Materials Report,* 5, No. 4, 76−82 (1985) and 8, No. 6, 8−16 (1988)

US Environmental Protection Agency. (November 30, 1987). *Chemical Hazard Information Profile: Nickel Carbonyl.* Washington, DC: Chemical Emergency Preparedness Program

New York State Department of Health. (March 1986). *Chemical Fact Sheet: Nickel Carbonyl.* Version 2. Albany, NY: Bureau of Toxic Substance Assessment

New Jersey Department of Health and Senior Services. (February 2001). *Hazardous Substances Fact Sheet: Nickel Carbonyl.* Trenton, NJ

Nickel chloride   N:0250

Molecular Formula: Cl_2Ni
Common Formula: $NiCl_2$
Synonyms: Cloruro de niquel (Spanish); Nickel(2+) chloride; Nickel(2+) chloride (1:2); Nickel(II) chloride; Nickel (II) chloride (1:2); Nickel chloride (ous); Nickelous chloride
CAS Registry Number: 7718-54-9; 37211-05-5
7791-20-0 (hexahydrate)
RTECS® Number: QR6475000
UN/NA & ERG Number: UN3288 (Toxic solid, inorganic, n.o.s.)/151
EC Number: 231-743-0 [Annex I Index No.: 028-011-00-6] (37211-05-5)

Regulatory Authority and Advisory Bodies

Carcinogenicity: IARC: (compounds) *carcinogenic to humans, carcinogenic to humans*, Group 1, 1997; NTP: 11th Report on Carcinogens, 2004: Reasonably anticipated to be a human carcinogen; NIOSH: Potential occupational carcinogen.

Air Pollutant Standard Set. See below, "Permissible Exposure Limits in Air" section.

Clean Air Act: Hazardous Air Pollutants (Title I, Part A, Section 112) as nickel compounds.

Clean Water Act: Section 311 Hazardous Substances/RQ 40CFR117.3 (same as CERCLA, see below); 40CFR401.15 Section 307 Toxic Pollutants as nickel and compounds; Section 313 Water Priority Chemicals (57FR41331, 9/9/92). RCRA, 40CFR261, Appendix 8 Hazardous Constituents, waste number not listed, as nickel compounds, n.o.s.

Reportable Quantity (RQ): 100 lb (45.4 kg).

EPCRA Section 313 Form R *de minimis* concentration reporting level: 0.1%.

Canada, WHMIS, Ingredients Disclosure List Concentration 0.1%.

European/International Regulations: Hazard Symbol: T, N; Risk phrases (*7718-54-9*): R49; R61; R23/25; R38; R42/43; R48/23; R68; R50/53; Safety phrases: S53; S45; S60; S61 (see Appendix 4).

WGK (German Aquatic Hazard Class): 3—Highly water polluting (CAS: 7718-54-9).

Description: Nickel chloride appears as green or brown scales, or sparkling golden-yellow powder. Molecular weight = 129.60; Sublimation temperature = $973°C$; Freezing/Melting point = $1000°C$. Hazard Identification (based on NFPA-704 M Rating System): Health 3, Flammability 0, Reactivity 0; (*hexahydrate*) Health 3, Flammability 0, Reactivity 0. Soluble in water.

Potential Exposure: Nickel chloride is used in electroplating and ink manufacturing.

Incompatibilities: Strong acids, potassium, sulfur. Forms an impact-sensitive mixture with potassium.

Permissible Exposure Limits in Air

OSHA PEL: 1 mg[Ni]/m^3 TWA.

NIOSH REL: 0.015 mg[Ni]/m^3 TWA; A potential occupational carcinogen. Limit exposure to lowest feasible concentration. See *NIOSH Pocket Guide*, Appendix A.

ACGIH TLV[1]: 0.1 mg[Ni]/m^3 TWA, inhalable fraction, confirmed human carcinogen.

NIOSH IDLH: 10 mg [Ni]/m^3.

Protective Action Criteria (PAC)

TEEL-0: 0.221 mg/m^3
PAC-1: 0.221 mg/m^3
PAC-2: 0.221 mg/m^3
PAC-3: 22.1 mg/m^3

Hexahydrate
TEEL-0: 0.405 mg/m^3
PAC-1: 15 mg/m^3
PAC-2: 40.5 mg/m^3
PAC-3: 40.5 mg/m^3

DFG MAK: Inhalable fraction, sensitization of the respiratory tract and skin, Carcinogen Category 1.

NIOSH IDLH: 10 mg[Ni]/m^3.

Russia set a MAC level of 0.005 mg/m^3 for nickel salts and aerosols.

Determination in Air: Use NIOSH Analytical Method #7300, Elements by ICP (Nitric/Perchloric Acid Ashing); #7301, Elements by ICP (Aqua Regia Ashing); #7303, Elements by ICP [Hot Block (HCl/HNO_3 Digestion)]; #9102, Elements on Wipes; #8310, Metals in urine, #8005, Elements in blood or tissue; OSHA Analytical Methods ID-121 and ID-125G.

Permissible Concentration in Water: The EPA[6] has set a limit of 13.4 µg/L on nickel to protect human health. A lifetime health advisory of 150 µg/L has more recently been promulgated by EPA.[49]

Routes of Entry: Inhalation, ingestion, skin and/or eye contact.

Harmful Effects and Symptoms

Short Term Exposure: Nickel chloride can affect you when breathed in. Exposure can irritate and inflame the air passages and sinuses, causing cough, phlegm, shortness of breath, and ulceration. Contact can irritate and burn the eyes or skin.

Long Term Exposure: Repeated or prolonged contact with skin may cause dermatitis. Repeated or prolonged contact may cause skin sensitization. Repeated or prolonged inhalation exposure may cause asthma-like allergy. Once allergy develops, even low future exposures can trigger symptoms. Repeated exposure can cause lung scarring and may affect the kidneys. Nickel chloride may cause mutations. Handle with extreme caution.

Points of Attack: Skin, lungs, kidneys.

Medical Surveillance: NIOSH lists the following tests: blood gas analysis; whole blood (chemical/metabolite); blood plasma; blood plasma, end-of-shift; blood serum; biologic tissue/biopsy; chest X-ray, electrocardiogram, pulmonary function tests; pre- and postshift; sputum cytology; urine (chemical/metabolite); urine (chemical/metabolite), end-of-shift; white blood cell count/differential. If symptoms develop or overexposure is suspected, the following may be useful: evaluation by a qualified allergist, including careful exposure history and special testing, may help diagnose skin allergy. Consider lung function tests if lung symptoms are present. Kidney function tests. Chest X-ray. Daily testing for urine nickel for several days. Persons with urine nickel over 100 µg/L require immediate medical attention.

First Aid: If this chemical gets into the eyes, remove any contact lenses at once and irrigate immediately for at least 15 min, occasionally lifting upper and lower lids. Seek medical attention immediately. If this chemical contacts the skin, remove contaminated clothing and wash immediately with soap and water. Seek medical attention immediately. If this chemical has been inhaled, remove from exposure, begin rescue breathing (using universal precautions, including resuscitation mask) if breathing has stopped and CPR if

heart action has stopped. Transfer promptly to a medical facility. When this chemical has been swallowed, get medical attention. Give large quantities of water and induce vomiting. Do not make an unconscious person vomit.

Note to physician: For severe poisoning BAL [British Anti-Lewisite, dimercaprol, dithiopropanol ($C_3H_8OS_2$)] has been used to treat toxic symptoms of certain heavy metals poisoning including nickel. Although BAL is reported to have a large margin of safety, caution must be exercised, because toxic effects may be caused by excessive dosage. Most can be prevented by premedication with 1-ephedrine sulfate (CAS: 134-72-5).

Personal Protective Methods: Wear protective gloves and clothing to prevent any reasonable probability of skin contact. Safety equipment suppliers/manufacturers can provide recommendations on the most protective glove/clothing material for your operation. All protective clothing (suits, gloves, footwear, headgear) should be clean, available each day, and put on before work. Contact lenses should not be worn when working with this chemical. Wear splash-proof chemical goggles and face shield when working with liquid, unless full face-piece respiratory protection is worn. Wear dust-proof goggles and face shield when working with powders or dust, unless full face-piece respiratory protection is worn. Employees should wash immediately with soap when skin is wet or contaminated. Provide emergency showers and eyewash. See NIOSH Criteria Document #77-164, *Inorganic Nickel.*

Respirator Selection: At concentrations above the NIOSH REL, or where there is no REL, at any detectable concentration: SCBAF: Pd,Pp (APF = 10,000) (any NIOSH/MSHA- or European Standard EN 149-approved self-contained breathing apparatus that has a full face-piece and is operated in a pressure-demand or other positive-pressure mode) or SaF: Pd,Pp: ASCBA (APF = 10,000) (any supplied-air respirator that has a full face-piece and is operated in a pressure-demand or other positive-pressure mode in combination with an auxiliary, self-contained breathing apparatus operated in a pressure-demand or other positive-pressure mode). *Escape:* 100F (APF = 50) (any air-purifying, full-face-piece respirator with an N100, R100, or P100 filter).

Storage: Color Code—Blue: Health Hazard/Poison: Store in a secure poison location. Prior to working with this chemical you should be trained on its proper handling and storage. Nickel chloride must be stored to avoid contact with strong acids (such as hydrochloric, sulfuric, and nitric), since violent reactions occur. Store in tightly closed containers in a cool, well-ventilated area away from potassium and sulfur. Mixtures of nickel chloride and potassium will produce a strong explosion on impact.

Shipping: Toxic solid, inorganic, n.o.s. materials require a shipping label of "POISONOUS/TOXIC MATERIALS." They fall in Hazard Class 6.1.

Spill Handling: Evacuate and restrict persons not wearing protective equipment from area of spill or leak until cleanup is complete. Remove all ignition sources. Ventilate area of spill or leak. Absorb liquids in vermiculite, dry sand, earth, peat, carbon, or a similar material and deposit in sealed containers. Keep this chemical out of a confined space, such as a sewer, because of the possibility of an explosion, unless the sewer is designed to prevent the buildup of explosive concentrations. It may be necessary to contain and dispose of this chemical as a hazardous waste. If material or contaminated runoff enters waterways, notify downstream users of potentially contaminated waters. Contact your local or federal environmental protection agency for specific recommendations. If employees are required to clean up spills, they must be properly trained and equipped. OSHA 1910.120(q) may be applicable.

Fire Extinguishing: Extinguish fire using an agent suitable for type of surrounding fire. Nickel chloride itself may burn but does not easily ignite. Poisonous gases, including chlorine and nickel carbonyl, are produced in fire. If material or contaminated runoff enters waterways, notify downstream users of potentially contaminated waters. Notify local health and fire officials and pollution control agencies. From a secure, explosion-proof location, use water spray to cool exposed containers. If cooling streams are ineffective (venting sound increases in volume and pitch, tank discolors, or shows any signs of deforming), withdraw immediately to a secure position. If employees are expected to fight fires, they must be trained and equipped in OSHA 1910.156. The only respirators recommended for firefighting are self-contained breathing apparatuses that have full face-pieces and are operated in a pressure-demand or other positive-pressure mode.

Disposal Method Suggested: Recycle or disposal in a chemical waste landfill is recommended.[22]

Reference

New Jersey Department of Health and Senior Services. (June 2002). *Hazardous Substances Fact Sheet: Nickel Chloride.* Trenton, NJ

Nickel cyanide   N:0260

Molecular Formula: C_2N_2Ni
Common Formula: $Ni(CN)_2$
Synonyms: Cianuro de niquel (Spanish); Nickel(2+) cyanide; Nickel(II) cyanide; Nickel cyanide, solid
CAS Registry Number: 557-19-7
RTECS® Number: QR6495000
UN/NA & ERG Number: UN1653/151
EC Number: 209-160-8 [*Annex I Index No.:* 028-034-00-1]
Regulatory Authority and Advisory Bodies
Carcinogenicity: IARC: (compounds) *carcinogenic to humans, carcinogenic to humans,* Group 1, 1997; NTP: 11th Report on Carcinogens, 2004: Reasonably anticipated to be a human carcinogen; NIOSH: Potential occupational carcinogen.

Air Pollutant Standard Set. See below, "Permissible Exposure Limits in Air" section.

Clean Air Act: Hazardous Air Pollutants (Title I, Part A, Section 112) as nickel compounds.

Clean Water Act: 40CFR401.15 Section 307 Toxic Pollutants as nickel and compounds.

US EPA Hazardous Waste Number (RCRA No.): P074.

RCRA, 40CFR261, Appendix 8 Hazardous Constituents.

Reportable Quantity (RQ): 10 lb (4.54 kg).

EPCRA Section 313 Form R *de minimis* concentration reporting level: 0.1%.

US DOT Regulated Marine Pollutant (49CFR172.101, Appendix B).

Canada, WHMIS, Ingredients Disclosure List Concentration 1.0%.

European/International Regulations: Hazard Symbol: T, N; Risk phrases: R49; R32; R42/43; R48/23;R50/53; Safety phrases:S53; S45; S60; S61 (see Appendix 4).

WGK (German Aquatic Hazard Class): No value assigned.

Description: Nickel cyanide is a yellowish-brown plate or powder that may change to a green color by absorbing moisture. It has a weak almond odor like cyanide. Molecular weight = 110.75; Hazard Identification (based on NFPA-704 M Rating System): Health 2, Flammability 0, Reactivity 0. Insoluble in water.

Potential Exposure: Nickel cyanide is used in metallurgy and electroplating, and in making other chemicals.

Incompatibilities: Acids, active metals. Violent reaction with magnesium. Heat or acid contact can cause release of toxic cyanide.

Permissible Exposure Limits in Air

OSHA PEL (*elemental, soluble & insoluble compounds*): 1 mg[Ni]/m^3 TWA.

NIOSH REL (*elemental, soluble & insoluble compounds*): 0.015 mg[Ni]/m^3 TWA; A potential occupational carcinogen. Limit exposure to lowest feasible concentration. See *NIOSH Pocket Guide*, Appendix A.

ACGIH TLV®[1] (*inorganic, soluble compounds*) 0.1 mg[Ni]/m^3 inhalable fraction TWA, confirmed human carcinogen.

NIOSH IDLH: 10 mg[Ni]/m^3.

Protective Action Criteria (PAC)

TEEL-0: 0.377 mg/m^3

PAC-1: 0.5 mg/m^3

PAC-2: 3.5 mg/m^3

PAC-3: 18.9 mg/m^3

DFG MAK (*elemental & nickel compounds*): Inhalable fraction, sensitization of the respiratory tract and skin; Carcinogen Category 1; DFG TRK: *As inhalable dusts/aerosols from nickel metal:* 0.50 μg[Ni]/L in urine, after several shifts; Carcinogen Category 1.

Russia set a MAC level of 0.005 mg/m^3 for nickel salts and aerosols.

As cyanides

NIOSH recommends a level of 5 mg/m^3 not to be exceeded during any 10-min work period. ACGIH recommends a TWA of 5 mg/m^3.

Determination in Air: Use NIOSH Analytical Method #7300, Elements by ICP (Nitric/Perchloric Acid Ashing); #7301, Elements by ICP (Aqua Regia Ashing); #7303, Elements by ICP [Hot Block (HCl/HNO$_3$ Digestion)]; #9102, Elements on Wipes; #8310, Metals in urine, #8005, Elements in blood or tissue; OSHA Analytical Methods ID-121 and ID-125G.

Permissible Concentration in Water: Again one should consider the restrictions imposed both by nickel and by cyanide. However, the lifetime health advisories developed by EPA[49] are 154 μg/L for cyanide and 150 μg/L for nickel, which are nearly the same.

Routes of Entry: Inhalation, ingestion, skin and/or eye contact.

Harmful Effects and Symptoms

Short Term Exposure: Nickel cyanide can affect you when breathed in. Eye contact may cause irritation. Skin contact may cause skin allergy, with itching and redness, and later, rash. Fumes can cause pneumonia-like illness, with coughing and/or shortness of breath. Lung damage may result from a single high exposure or lower repeated exposures. Higher exposures can cause pulmonary edema, a medical emergency that can be delayed for several hours. This can cause death. Lung allergy occasionally occurs with asthma-type effect. High or repeated lower exposures may damage the heart, liver, or kidneys.

Long Term Exposure: Lung allergy occasionally occurs with asthma-type effect. Skin contact can cause allergy. High or repeated lower exposures may damage the lungs, with scarring of the lung tissue, and may cause damage to the heart, liver, or kidneys.

Points of Attack: Lungs, liver, kidneys.

Medical Surveillance: NIOSH lists the following tests: blood gas analysis; whole blood (chemical/metabolite); blood plasma; blood plasma, end-of-shift; blood serum; biologic tissue/biopsy; chest X-ray, electrocardiogram, pulmonary function tests; pre- and postshift; sputum cytology; urine (chemical/metabolite); urine (chemical/metabolite), end-of-shift; white blood cell count/differential. Before beginning employment and at regular times after that, the following are recommended: lung function tests, urine or plasma test for nickel (unexposed persons have urine levels less than 10 μg/L). If symptoms develop or overexposure is suspected, the following may be useful: daily urine nickel for several days (persons with urine nickel over 100 μg need medical attention). Consider chest X-ray after acute overexposure. Evaluation by a qualified allergist, including careful exposure history and special testing, may help diagnose skin allergy. Liver and kidney function tests. Consider chest X-ray following acute overexposure.

First Aid: If this chemical gets into the eyes, remove any contact lenses at once and irrigate immediately for at least 15 min, occasionally lifting upper and lower lids. Seek medical attention immediately. If this chemical contacts the skin, remove contaminated clothing and wash immediately with soap and water. Seek medical attention immediately. If this chemical has been inhaled, remove from exposure, begin rescue breathing (using universal precautions, including resuscitation mask) if breathing has stopped and CPR if heart action has stopped. Transfer promptly to a medical facility for medical observation up to 2 days and test for urine nickel. When this chemical has been swallowed, get medical attention. Give large quantities of water and induce vomiting. Do not make an unconscious person vomit. Medical observation is recommended for 24—48 h after breathing overexposure, as pulmonary edema may be delayed. As first aid for pulmonary edema, a doctor or authorized paramedic may consider administering a corticosteroid spray.

Use amyl nitrate capsules if symptoms develop. All area employees should be trained regularly in emergency measures for cyanide poisoning and in CPR. A cyanide antidote kit should be kept in the immediate work area and must be rapidly available. Kit ingredients should be replaced every 1—2 years to ensure freshness. Persons trained in the use of this kit, oxygen use, and CPR must be quickly available.

Note to physician: For severe poisoning BAL [British Anti-Lewisite, dimercaprol, dithiopropanol ($C_3H_8OS_2$)] has been used to treat toxic symptoms of certain heavy metals poisoning including nickel. Although BAL is reported to have a large margin of safety, caution must be exercised, because toxic effects may be caused by excessive dosage. Most can be prevented by premedication with 1-ephedrine sulfate (CAS: 134-72-5).

Personal Protective Methods: Wear protective gloves and clothing to prevent any reasonable probability of skin contact. Safety equipment suppliers/manufacturers can provide recommendations on the most protective glove/clothing material for your operation. All protective clothing (suits, gloves, footwear, headgear) should be clean, available each day, and put on before work. Contact lenses should not be worn when working with this chemical. Wear dust-proof chemical goggles and face shield unless full face-piece respiratory protection is worn. Employees should wash immediately with soap when skin is wet or contaminated. Provide emergency showers and eyewash. See NIOSH Criteria Document #77-164, *Inorganic Nickel.*

Respirator Selection:
Nickel: At concentrations above the NIOSH REL, or where there is no REL, at any detectable concentration: SCBAF: Pd,Pp (APF = 10,000) (any NIOSH/MSHA- or European Standard EN 149-approved self-contained breathing apparatus that has a full face-piece and is operated in a pressure-demand or other positive-pressure mode) or SaF: Pd,Pp: ASCBA (APF = 10,000) (any supplied-air respirator that has a full face-piece and is operated in a pressure-demand or other positive-pressure mode in combination with an auxiliary, self-contained breathing apparatus operated in a pressure-demand or other positive-pressure mode). *Escape:* 100F (APF = 50) (any air-purifying, full-face-piece respirator with an N100, R100, or P100 filter).

Cyanides: Up to 25 mg/m^3: Sa (APF = 10) (any supplied-air respirator) or SCBAF (APF = 50) (any self-contained breathing apparatus with full face-piece). *Emergency or planned entry into unknown concentrations or IDLH conditions:* SCBAF: Pd,Pp (APF = 10,000) (any self-contained breathing apparatus that has a full face-piece and is operated in a pressure-demand or other positive-pressure mode) or SaF: Pd,Pp: ASCBA (APF = 10,000) (any supplied-air respirator that has a full face-piece and is operated in a pressure-demand or other positive-pressure mode in combination with an auxiliary self-contained breathing apparatus operated in a pressure-demand or other positive-pressure mode). *Escape:* GmFS100 (APF = 50) [any air-purifying, full-face-piece respirator (gas mask) with a chin-style, front- or back-mounted canister providing protection against the compound of concern and having an N100, R100, or P100 filter] or SCBAE (any appropriate escape-type, self-contained breathing apparatus).

Storage: Color Code—Blue: Health Hazard/Poison: Store in a secure poison location. Prior to working with this chemical you should be trained on its proper handling and storage. Nickel cyanide must be stored to avoid contact with magnesium, since violent reactions occur. Store in tightly closed containers in a cool, well-ventilated area away from acids since heat and acids can release toxic cyanide.

Shipping: Nickel cyanide requires a shipping label of "POISONOUS/TOXIC MATERIALS." It falls in Hazard Class 6.1 and Packing Group II.

Spill Handling: Evacuate persons not wearing protective equipment from area of spill or leak until cleanup is complete. Remove all ignition sources. Use HEPA vacuum or wet method to reduce dust during cleanup. Do not dry sweep. Collect powdered material in the most convenient and safe manner and deposit in sealed containers. Ventilate area after cleanup is complete. It may be necessary to contain and dispose of this chemical as a hazardous waste. If material or contaminated runoff enters waterways, notify downstream users of potentially contaminated waters. Contact your local or federal environmental protection agency for specific recommendations. If employees are required to clean up spills, they must be properly trained and equipped. OSHA 1910.120(q) may be applicable.

Fire Extinguishing: Extinguish fire using an agent suitable for type of surrounding fire. Nickel cyanide itself does not burn. Poisonous gases, including cyanide, are produced in fire. If material or contaminated runoff enters waterways, notify downstream users of potentially contaminated waters. Notify local health and fire officials and pollution control agencies. From a secure, explosion-proof location, use water spray to cool exposed containers. If cooling streams are ineffective (venting sound increases in volume and pitch, tank discolors, or shows any signs of deforming), withdraw immediately to a secure position. If employees are expected to fight fires, they must be trained and equipped in OSHA 1910.156. The only respirators recommended for firefighting are self-contained breathing apparatuses that have full face-pieces and are operated in a pressure-demand or other positive-pressure mode.

Disposal Method Suggested: Consult with environmental regulatory agencies for guidance on acceptable disposal practices. Generators of waste containing this contaminant ($\geq$100 kg/mo) must conform with EPA regulations governing storage, transportation, treatment, and waste disposal.

Reference

New Jersey Department of Health and Senior Services. (July 2001). *Hazardous Substances Fact Sheet: Nickel Cyanide.* Trenton, NJ

Nickel hydroxide   N:0270

Molecular Formula: H_2NiO_2
Synonyms: Hidroxido niquel (Spanish); Nickel black; Nickel dihydroxide; Nickel(2+) hydroxide; Nickel(II) hydroxide; Nickelic hydroxide; Nickelous hydroxide
CAS Registry Number: 12054-48-7; *(alt)*12125-56-3
RTECS® Number: QR7040000
UN/NA & ERG Number: UN3077/171
EC Number: 235-008-5 [*Annex I Index No.:* 028-008-00-X]
Regulatory Authority and Advisory Bodies
Carcinogenicity: IARC: (compounds) *carcinogenic to humans*, Group 1, 1997; NTP: 11th Report on Carcinogens, 2004: Reasonably anticipated to be a human carcinogen; NIOSH: Potential occupational carcinogen.
Air Pollutant Standard Set. See below, "Permissible Exposure Limits in Air" section.
Clean Air Act: Hazardous Air Pollutants (Title I, Part A, Section 112) as nickel compounds.
Clean Water Act: Section 311 Hazardous Substances/RQ 40CFR117.3 (same as CERCLA, see below); 40CFR401.15 Section 307 Toxic Pollutants as nickel and compounds; Section 313 Water Priority Chemicals (57FR41331, 9/9/92). RCRA, 40CFR261, Appendix 8 Hazardous Constituents, waste number not listed, as nickel compounds, n.o.s.
Reportable Quantity (RQ): 10 lb (4.54 kg).
EPCRA Section 313 Form R *de minimis* concentration reporting level: 0.1%.
California Proposition 65 Chemical: Cancer 10/1/89.
Canada, WHMIS, Ingredients Disclosure List Concentration 0.1%.
European/International Regulations: Hazard Symbol (12054-48-7): T, N; Risk phrases: R49; R61; R20/22; R42/43; R48/23; R68; R50/53; Safety phrases:S53; S45; S60; S61 (see Appendix 4).
WGK (German Aquatic Hazard Class): 3—Severe hazard to waters.
Description: Nickel hydroxide is a light, apple-green powder. Molecular weight = 92.73; Freezing/Melting point = 230°C (decomposes); Autoignition temperature = 400°C.[CHRIS] Hazard Identification (based on NFPA-704 M

Rating System): Health 2, Flammability 0, Reactivity 0. Insoluble in water.
Potential Exposure: It may be found in the workplace as a dust, liquid, or acid solution. This compound may be used in nickel plating operations.
Incompatibilities: Incompatible with strong acids. Aqueous solution may be acidic.
Permissible Exposure Limits in Air
OSHA PEL (*elemental, soluble & insoluble compounds*): 1 mg[Ni]/m³ TWA.
NIOSH REL (*elemental, soluble & insoluble compounds*): 0.015 mg[Ni]/m³ TWA; A potential occupational carcinogen. Limit exposure to lowest feasible concentration. See *NIOSH Pocket Guide*, Appendix A.
ACGIH TLV®[1] (*inorganic, insoluble compounds*) 0.2 mg [Ni]/m³ inhalable fraction TWA, confirmed human carcinogen.
NIOSH IDLH: 10 mg[Ni]/m³.
Protective Action Criteria (PAC)
12054-48-7
TEEL-0: 0.0237 mg/m³
PAC-1: 0.06 mg/m³
PAC-2: 0.4 mg/m³
PAC-3: 15.8 mg/m³
DFG MAK (*elemental & nickel compounds*): Inhalable fraction, sensitization of the respiratory tract and skin; Carcinogen Category 1; DFG TRK: *As inhalable dusts/aerosols from nickel metal*: 0.50 µg[Ni]/L in urine, after several shifts; Carcinogen Category 1.
Determination in Air: Use NIOSH Analytical Method #7300, Elements by ICP (Nitric/Perchloric Acid Ashing); #7301, Elements by ICP (Aqua Regia Ashing); #7303, Elements by ICP [Hot Block (HCl/HNO₃ Digestion)]; #9102, Elements on Wipes; #8310, Metals in urine, #8005, Elements in blood or tissue; OSHA Analytical Methods ID-121 and ID-125G.
Permissible Concentration in Water: The EPA[6] has set a limit on nickel to protect human health of 13.4 µg/L. A lifetime health advisory of 150 µg/L has more recently been promulgated by EPA.[49]
Routes of Entry: Inhalation, ingestion, skin and/or eye contact.
Harmful Effects and Symptoms
Short Term Exposure: Nickel hydroxide can affect you when breathed in. Irritates and burns the eyes and skin on contact. Irritates the respiratory tract causing phlegm and/or shortness of breath. Lung damage may result from a single high or lower repeated exposures. Higher exposures can cause pulmonary edema, a medical emergency that can be delayed for several hours. This can cause death.
Long Term Exposure: Lung allergy occasionally occurs with asthma-type effects. High or repeated lower exposure may cause lung scarring and may damage the heart, liver, or kidneys. Skin contact may cause skin allergy, with itching, redness, and later rash. Nickel hydroxide is a carcinogen; handle with extreme caution.

Points of Attack: Skin, lungs, kidneys.

Medical Surveillance: NIOSH lists the following tests: blood gas analysis; whole blood (chemical/metabolite); blood plasma; blood plasma, end-of-shift; blood serum; biologic tissue/biopsy; chest X-ray, electrocardiogram, pulmonary function tests; pre- and postshift; sputum cytology; urine (chemical/metabolite); urine (chemical/metabolite), end-of-shift; white blood cell count/differential. Before beginning employment and at regular times after that, the following are recommended: lung function tests. Urine or plasma test for nickel (unexposed persons have urine levels less than 10 µg/L). If symptoms develop or overexposure is suspected, the following may be useful: daily urine nickel levels for several days (persons with urine nickel over 100 µg need medical attention). Consider chest X-ray after acute overexposure. Evaluation by a qualified allergist, including careful exposure history and special testing, may help diagnose skin allergy. Kidney function tests.

First Aid: If this chemical gets into the eyes, remove any contact lenses at once and irrigate immediately for at least 15 min, occasionally lifting upper and lower lids. Seek medical attention immediately. If this chemical contacts the skin, remove contaminated clothing and wash immediately with soap and water. Seek medical attention immediately. If this chemical has been inhaled, remove from exposure, begin rescue breathing (using universal precautions, including resuscitation mask) if breathing has stopped and CPR if heart action has stopped. Transfer promptly to a medical facility. When this chemical has been swallowed, get medical attention. Give large quantities of water and do not induce vomiting. Medical observation is recommended for 24–48 h after breathing overexposure, as pulmonary edema may be delayed. As first aid for pulmonary edema, a doctor or authorized paramedic may consider administering a corticosteroid spray.

Note to physician: For severe poisoning BAL [British Anti-Lewisite, dimercaprol, dithiopropanol ($C_3H_8OS_2$)] has been used to treat toxic symptoms of certain heavy metals poisoning including nickel. Although BAL is reported to have a large margin of safety, caution must be exercised, because toxic effects may be caused by excessive dosage. Most can be prevented by premedication with 1-ephedrine sulfate (CAS: 134-72-5).

Personal Protective Methods: Wear protective gloves and clothing to prevent any reasonable probability of skin contact. Safety equipment suppliers/manufacturers can provide recommendations on the most protective glove/clothing material for your operation. All protective clothing (suits, gloves, footwear, headgear) should be clean, available each day, and put on before work. Contact lenses should not be worn when working with this chemical. Wear dust-proof chemical goggles and face shield unless full face-piece respiratory protection is worn. Employees should wash immediately with soap when skin is wet or contaminated. Provide emergency showers and eyewash.

Respirator Selection: At concentrations above the NIOSH REL, or where there is no REL, at any detectable concentration: SCBAF: Pd,Pp (APF = 10,000) (any NIOSH/MSHA- or European Standard EN 149-approved self-contained breathing apparatus that has a full face-piece and is operated in a pressure-demand or other positive-pressure mode) or SaF: Pd,Pp: ASCBA (APF = 10,000) (any supplied-air respirator that has a full face-piece and is operated in a pressure-demand or other positive-pressure mode in combination with an auxiliary, self-contained breathing apparatus operated in a pressure-demand or other positive-pressure mode). *Escape:* 100F (APF = 50) (any air-purifying, full-face-piece respirator with an N100, R100, or P100 filter).

Storage: Color Code—Blue: Health Hazard/Poison: Store in a secure poison location. Prior to working with this chemical you should be trained on its proper handling and storage. Store in tightly closed containers away from strong acids. A regulated, marked area should be established where this chemical is handled, used, or stored in compliance with OSHA Standard 1910.1045.

Shipping: The name of this material is not on the DOT list of materials[19] for label and packaging standards. However, based on regulations, it may be classified[52] as an Environmentally hazardous substance, liquid, n.o.s. It falls in Hazard Class 9 and Packing Group III.[20, 21]

Spill Handling: Evacuate persons not wearing protective equipment from area of spill or leak until cleanup is complete. Remove all ignition sources. Use HEPA vacuum or wet method to reduce dust during cleanup. Do not dry sweep. Collect powdered material in the most convenient and safe manner and deposit in sealed containers. Ventilate area after cleanup is complete. It may be necessary to contain and dispose of this chemical as a hazardous waste. If material or contaminated runoff enters waterways, notify downstream users of potentially contaminated waters. Contact your local or federal environmental protection agency for specific recommendations. If employees are required to clean up spills, they must be properly trained and equipped. OSHA 1910.120(q) may be applicable.

Fire Extinguishing: Nickel hydroxide is combustible but not easy to ignite. Use dry chemicals, carbon dioxide, water spray, or foam extinguishers. Poisonous gases, including nickel, are produced in fire. If material or contaminated runoff enters waterways, notify downstream users of potentially contaminated waters. Notify local health and fire officials and pollution control agencies. From a secure, explosion-proof location, use water spray to cool exposed containers. If cooling streams are ineffective (venting sound increases in volume and pitch, tank discolors, or shows any signs of deforming), withdraw immediately to a secure position. If employees are expected to fight fires, they must be trained and equipped in OSHA 1910.156. The only respirators recommended for firefighting are self-contained breathing apparatuses that have full face-pieces and are operated in a pressure-demand or other positive-pressure mode.

Disposal Method Suggested: Recover and recycle where possible or dispose of in a chemical waste landfill.[22]
Reference
New Jersey Department of Health and Senior Services. (March 2002). *Hazardous Substances Fact Sheet: Nickel Hydroxide.* Trenton, NJ

Nickel nitrate   N:0280

Molecular Formula: N_2NiO_6
Common Formula: $Ni(NO_3)_2$
Synonyms: Nickel(2+) nitrate (1:2); Nickel(II) nitrate (1:2); Nickel nitrate hexahydride; Nickel nitrate (ous); Nickelous nitrate; Nitric acid, nickel(2+) salt; Nitric acid, nickel(II) salt
CAS Registry Number: 13138-45-9; 13478-00-7 (hexahydrate); 14216-75-2 (nickel dinitrate)
RTECS® Number: OR7200000
UN/NA & ERG Number: UN2725/140
EC Number: 236-068-5 [*Annex I Index No.:* 028-012-00-1]; 238-076-4 [*Annex I Index No.:* 028-012-00-1] (nickel dinitrate)
Regulatory Authority and Advisory Bodies
Carcinogenicity: IARC: (compounds) *carcinogenic to humans, carcinogenic to humans,* Group 1, 1997; NTP: 11th Report on Carcinogens, 2004: Reasonably anticipated to be a human carcinogen; NIOSH: Potential occupational carcinogen.
Air Pollutant Standard Set. See below, "Permissible Exposure Limits in Air" section.
Clean Air Act: Hazardous Air Pollutants (Title I, Part A, Section 112) as nickel compounds.
Clean Water Act: 40CFR401.15 Section 307 Toxic Pollutants as nickel and compounds.
RCRA, 40CFR261, Appendix 8 Hazardous Constituents, waste number not listed, as nickel compounds, n.o.s.
EPCRA Section 313 Form R *de minimis* concentration reporting level: 0.1%.
European/International Regulations: Hazard Symbol: O, T, N; Risk phrases: R49; R61; R8; R20/22; R38; R41; R42/43; R48/23; R68; R50/53; Safety phrases: S53; S45; S60; S61 (see Appendix 4).
WGK (German Aquatic Hazard Class): 3—Severe hazard to waters.
14216-75-2
Clean Air Act: Hazardous Air Pollutants (Title I, Part A, Section 112) as nickel compounds.
Clean Water Act: Section 311 Hazardous Substances/RQ 40CFR117.3 (same as CERCLA, see below); 40CFR401.15 Section 307 Toxic Pollutants as nickel and compounds; Section 313 Water Priority Chemicals (57FR41331, 9/9/92).
RCRA, 40CFR261, Appendix 8 Hazardous Constituents, waste number not listed, as nickel compounds, n.o.s.
Reportable Quantity (RQ): 100 lb (45.4 kg).

EPCRA Section 313 Form R *de minimis* concentration reporting level: 0.1%.
Canada, WHMIS, Ingredients Disclosure List Concentration 0.1%.
European/International Regulations: Hazard Symbol: O, T, N; Risk phrases: R49; R61; R8; R20/22; R38; R41; R42/43; R48/23; R68; R50/53; Safety phrases: S53; S45; S60; S61 (see Appendix 4).
WGK (German Aquatic Hazard Class): 3—Severe hazard to waters.
Description: Nickel nitrate is a green powder. Molecular weight = 182.73; Boiling point = 137°C; Freezing/Melting point = 55−57°C. Hazard Identification (based on NFPA-704 M Rating System): Health 2, Flammability 0, Reactivity 3 (Oxidizer). Highly soluble in water.
Potential Exposure: Nickel nitrate is used in electroplating, nickel catalyst production, and in the manufacture of brown ceramic colors.
Incompatibilities: A powerful oxidizer. Incompatible with strong acids, sulfur, combustibles, organics, and other easily oxidizable materials.
Permissible Exposure Limits in Air
OSHA PEL (*elemental, soluble & insoluble compounds*): 1 mg[Ni]/m^3 TWA.
NIOSH REL (*elemental, soluble & insoluble compounds*): 0.015 mg[Ni]/m^3 TWA; A potential occupational carcinogen. Limit exposure to lowest feasible concentration. See *NIOSH Pocket Guide,* Appendix A.
ACGIH TLV®[1] (*inorganic, soluble compounds*) 0.1 mg [Ni]/m^3 inhalable fraction TWA, confirmed human carcinogen.
NIOSH IDLH: 10 mg[Ni]/m^3.
Protective Action Criteria (PAC)
TEEL-0: 0.311 mg/m^3
PAC-1: 1.5 mg/m^3
PAC-2: 12.5 mg/m^3
PAC-3: 31.1 mg/m^3
Hexahydrate
TEEL-0: 0.314 mg/m^3
PAC-1: 0.93 mg/m^3
PAC-2: 12.5 mg/m^3
PAC-3: 31 mg/m^3
DFG MAK (*elemental & nickel compounds*): Inhalable fraction, sensitization of the respiratory tract and skin; Carcinogen Category 1; DFG TRK: *As inhalable dusts/ aerosols from nickel metal*: 0.50 μg[Ni]/L in urine, after several shifts; Carcinogen Category 1.
Russia set a MAC level of 0.005 mg/m^3 for nickel salts and aerosols.
Determination in Air: Use NIOSH Analytical Method #7300, Elements by ICP (Nitric/Perchloric Acid Ashing); #7301, Elements by ICP (Aqua Regia Ashing); #7303, Elements by ICP [Hot Block (HCl/HNO$_3$ Digestion)]; #9102, Elements on Wipes; #8310, Metals in urine, #8005, Elements in blood or tissue; OSHA Analytical Methods ID-121 and ID-125G.

Permissible Concentration in Water: The EPA[6] has set a limit to protect human health of 13.4 µg/L. A lifetime health advisory of 150 µg/L has more recently been promulgated by EPA.[49] See also this section in the entry on "Nickel and Soluble Compounds."

Routes of Entry: Inhalation, ingestion, skin and/or eye contact.

Harmful Effects and Symptoms

Short Term Exposure: Nickel nitrate can affect you when breathed in. Irritates and burns the eyes and skin. Irritates the respiratory tract causing coughing and phlegm.

Long Term Exposure: Skin contact may cause skin allergy with itching, redness, and rash. Nickel nitrate may cause mutations. There is limited evidence that nickel nitrate may decrease fertility in males. Handle with extreme caution. Exposure may also cause lung allergy (asthma) to nickel. Once allergy asthma develops, even low future exposures can cause symptoms.

Points of Attack: Skin, lungs.

Medical Surveillance: NIOSH lists the following tests: blood gas analysis; whole blood (chemical/metabolite); blood plasma; blood plasma, end-of-shift; blood serum; biologic tissue/biopsy; chest X-ray, electrocardiogram, pulmonary function tests; pre- and postshift; sputum cytology; urine (chemical/metabolite); urine (chemical/metabolite), end-of-shift; white blood cell count/differential. If symptoms develop or overexposure is suspected, the following may be useful: evaluation by a qualified allergist, including careful exposure history and special testing, may help diagnose skin allergy. Consider lung function tests if lung symptoms are present.

First Aid: If this chemical gets into the eyes, remove any contact lenses at once and irrigate immediately for at least 15 min, occasionally lifting upper and lower lids. Seek medical attention immediately. If this chemical contacts the skin, remove contaminated clothing and wash immediately with soap and water. Seek medical attention immediately. If this chemical has been inhaled, remove from exposure, begin rescue breathing (using universal precautions, including resuscitation mask) if breathing has stopped and CPR if heart action has stopped. Transfer promptly to a medical facility. When this chemical has been swallowed, get medical attention. Give large quantities of water and induce vomiting. Do not make an unconscious person vomit.

Note to physician: For severe poisoning BAL [British Anti-Lewisite, dimercaprol, dithiopropanol ($C_3H_8OS_2$)] has been used to treat toxic symptoms of certain heavy metals poisoning including nickel. Although BAL is reported to have a large margin of safety, caution must be exercised, because toxic effects may be caused by excessive dosage. Most can be prevented by premedication with 1-ephedrine sulfate (CAS: 134-72-5).

Personal Protective Methods: Wear protective gloves and clothing to prevent any reasonable probability of skin contact. Safety equipment suppliers/manufacturers can provide recommendations on the most protective glove/clothing material for your operation. All protective clothing (suits, gloves, footwear, headgear) should be clean, available each day, and put on before work. Contact lenses should not be worn when working with this chemical. Wear dust-proof chemical goggles and face shield unless full face-piece respiratory protection is worn. Employees should wash immediately with soap when skin is wet or contaminated. Provide emergency showers and eyewash. See NIOSH Criteria Document #77-164, *Inorganic Nickel.*

Respirator Selection: At concentrations above the NIOSH REL, or where there is no REL, at any detectable concentration: SCBAF: Pd,Pp (APF = 10,000) (any NIOSH/MSHA- or European Standard EN 149-approved self-contained breathing apparatus that has a full face-piece and is operated in a pressure-demand or other positive-pressure mode) or SaF: Pd,Pp: ASCBA (APF = 10,000) (any supplied-air respirator that has a full face-piece and is operated in a pressure-demand or other positive-pressure mode in combination with an auxiliary, self-contained breathing apparatus operated in a pressure-demand or other positive-pressure mode). *Escape:* 100F (APF = 50) (any air-purifying, full-face-piece respirator with an N100, R100, or P100 filter).

Storage: Color Code—Yellow: Reactive Hazard; Store in a location separate from other materials, especially flammables and combustibles. Color Code—Blue: Health Hazard/Poison: Store in a secure poison location. Prior to working with this chemical you should be trained on its proper handling and storage. Store in tightly closed containers in a cool, well-ventilated area away from strong acids, sulfur, combustibles, organics, or other readily oxidizable materials. Avoid storage on wood floors. See OSHA Standard 1910.104 and NFPA 43A *Code for the Storage of Liquid and Solid Oxidizers* for detailed handling and storage regulations.

Shipping: This compound requires a shipping label of "OXIDIZER." It falls in Hazard Class 5.1 and Packing Group III.

Spill Handling: Evacuate persons not wearing protective equipment from area of spill or leak until cleanup is complete. Remove all ignition sources. Use HEPA vacuum or wet method to reduce dust during cleanup. Do not dry sweep. Collect powdered material in the most convenient and safe manner and deposit in sealed containers. Ventilate area after cleanup is complete. It may be necessary to contain and dispose of this chemical as a hazardous waste. If material or contaminated runoff enters waterways, notify downstream users of potentially contaminated waters. Contact your local or federal environmental protection agency for specific recommendations. If employees are required to clean up spills, they must be properly trained and equipped. OSHA 1910.120(q) may be applicable.

Fire Extinguishing: Nickel nitrate is an oxidizer. It increases the flammability of any combustible substance. Extinguish fire using an agent suitable for type of surrounding fire. Nickel nitrate itself does not burn. Poisonous gases, including nitrogen oxides, are produced in fire. If material or contaminated runoff enters waterways, notify downstream users of potentially contaminated waters. Notify local health and fire officials and pollution control agencies. From a secure, explosion-proof location, use water spray to cool exposed containers. If cooling streams are ineffective (venting sound increases in volume and pitch, tank discolors, or shows any signs of deforming), withdraw immediately to a secure position. If employees are expected to fight fires, they must be trained and equipped in OSHA 1910.156. The only respirators recommended for firefighting are self-contained breathing apparatuses that have full face-pieces and are operated in a pressure-demand or other positive-pressure mode.

Disposal Method Suggested: Consult with environmental regulatory agencies for guidance on acceptable disposal practices. Generators of waste containing this contaminant (≥100 kg/mo) must conform with EPA regulations governing storage, transportation, treatment, and waste disposal.

References

Sax, N. I. (Ed.). (1985). *Dangerous Properties of Industrial Materials Report*, 5, No. 6, 64–67

New Jersey Department of Health and Senior Services. (July 2001). *Hazardous Substances Fact Sheet: Nickel Nitrate*. Trenton, NJ

Nickel sulfate   N:0290

Molecular Formula: $NiSO_4$
Synonyms: NCI-C60344; Nickelous sulfate; Nickel(2+) sulfate(1:1); Nickel(II) sulfate; Nickel sulphate; Sulfato de niquel (Spanish); Sulfuric acid, nickel(2+) salt; Sulfuric acid, nickel(II) salt
CAS Registry Number: 7786-81-4; 10101-97-0 (hexahydrate)
RTECS® Number: QR9350000
UN/NA & ERG Number: UN3288 (Toxic solid, inorganic, n.o.s.)/151
EC Number: 232-104-9 [*Annex I Index No.:* 028-009-00-5]
Regulatory Authority and Advisory Bodies
Carcinogenicity: Carcinogenicity: IARC: (compounds) *carcinogenic to humans, carcinogenic to humans*, Group 1, 1997; NTP: 11th Report on Carcinogens, 2004: Reasonably anticipated to be a human carcinogen; NIOSH: Potential occupational carcinogen.
US EPA Gene-Tox Program, Positive: Cell transform.—SA7/SHE; Inconclusive: Carcinogenicity—mouse/rat.
Air Pollutant Standard Set. See below, "Permissible Exposure Limits in Air" section.
Clean Air Act: Hazardous Air Pollutants (Title I, Part A, Section 112) as nickel compounds.
Clean Water Act: Section 311 Hazardous Substances/RQ 40CFR117.3 (same as CERCLA, see below); 40CFR401.15 Section 307 Toxic Pollutants as nickel and compounds; Section 313 Water Priority Chemicals (57FR41331, 9/9/92).

RCRA, 40CFR261, Appendix 8 Hazardous Constituents, waste number not listed, as nickel compounds, n.o.s.
Reportable Quantity (RQ): 100 lb (45.4 kg)
EPCRA Section 313 Form R *de minimis* concentration reporting level: 0.1%.
Canada, WHMIS, Ingredients Disclosure List Concentration 0.1%.
European/International Regulations: not listed in Annex 1.
WGK (German Aquatic Hazard Class): No value assigned.
Description: Nickel sulfate is a blue to blue-green crystalline solid with a sweet taste. Molecular weight = 154.77; Freezing/Melting point = 848°C (decomposes). Hazard Identification (based on NFPA-704 M Rating System): Health 2, Flammability 0, Reactivity 0; (hexahysrate) Health 3, Flammability 0, Reactivity 0. Soluble in water.
Potential Exposure: Compound Description: Tumorigen, Mutagen, Organometallic; Reproductive Effector; Human Data; Primary Irritant. Nickel sulfate is used in plating baths, and as an intermediate in the production of nickel ammonium sulphate, as a mordant in dyeing, and in printing textiles, coatings, and ceramics.
Incompatibilities: The aqueous solution is a weak acid. Sulfates may react violently with aluminum, magnesium.
Permissible Exposure Limits in Air
OSHA PEL (*elemental, soluble & insoluble compounds*): 1 mg[Ni]/m³ TWA.
NIOSH REL (*elemental, soluble & insoluble compounds*): 0.015 mg[Ni]/m³ TWA; A potential occupational carcinogen. Limit exposure to lowest feasible concentration. See *NIOSH Pocket Guide*, Appendix A.
ACGIH TLV®[1] (*inorganic, soluble compounds*) 0.1 mg[Ni]/m³ inhalable fraction TWA, confirmed human carcinogen.
NIOSH IDLH: 10 mg[Ni]/m³.
Protective Action Criteria (PAC)
TEEL-0: 0.264 mg/m³
PAC-1: 0.791 mg/m³
PAC-2: 1.32 mg/m³
PAC-3: 26.4 mg/m³
Hexahydrate
TEEL-0: 0.448 mg/m³
PAC-1: 1.34 mg/m³
PAC-2: 10 mg/m³
PAC-3: 44.8 mg/m³
DFG MAK (*elemental & nickel compounds*): Inhalable fraction, sensitization of the respiratory tract and skin; Carcinogen Category 1; DFG TRK: *As inhalable dusts/aerosols from nickel metal*: 0.50 μg[Ni]/L in urine, after several shifts; Carcinogen Category 1.
Arab Republic of Egypt: TWA 0.1 mg[Ni]/m³, 1993; Australia: TWA 1 mg[Ni]/m³, 1993; Belgium: TWA 1 mg[Ni]/m³ (insoluble compounds), 1993; Denmark: TWA 0.1 mg[Ni]/m³, 1999; Finland: TWA 0.1 mg[Ni]/m³, [skin], carcinogen, 1999; France: VME 1 mg[Ni]/m³, 1999; Hungary: STEL 0.005 mg[Ni]/m³, Carcinogen (insoluble compounds), 1993; the Philippines: TWA 1 mg[Ni]/m³,

1993; Poland: MAC (TWA) 0.25 mg[Ni]/m^3, 1999; Russia: STEL 0.05 mg[Ni]/m^3, 1993; Sweden: NGV 0.1 mg[Ni]/m^3, carcinogen, 1999; Switzerland: MAK-W 0.5 mg[Ni]/m^3 (insoluble compounds), 1999; Switzerland: MAK-W 0.5 mg[Ni]/m^3, carcinogen, 1999; Thailand: TWA 1 mg[Ni]/m^3, 1993; United Kingdom: TWA 0.1 mg[Ni]/m^3, 2000; Argentina, Bulgaria, Columbia, Jordan, South Korea, New Zealand, Singapore, Vietnam: ACGIH TLV®: TWA 0.1 mg[Ni]/m^3.

Determination in Air: Use NIOSH Analytical Method #7300, Elements by ICP (Nitric/Perchloric Acid Ashing); #7301, Elements by ICP (Aqua Regia Ashing); #7303, Elements by ICP [Hot Block (HCl/HNO$_3$ Digestion)]; #9102, Elements on Wipes; #8310, Metals in urine, #8005, Elements in blood or tissue; OSHA Analytical Methods ID-121 and ID-125G.

Permissible Concentration in Water: The EPA[6] has set a limit to protect human health of 13.4 μg/L. A lifetime health advisory of 150 μg/L has more recently been promulgated by EPA.[49] See also this section in the entry on "Nickel and Soluble Compounds."

Routes of Entry: Inhalation, ingestion, skin and/or eye contact.

Harmful Effects and Symptoms

Short Term Exposure: Nickel sulfate can affect you when breathed in. It may also cause infertility in males. High or repeated exposures can scar the lungs. Higher exposures can cause pulmonary edema, a medical emergency that can be delayed for several hours. This can cause death.

Long Term Exposure: Repeated or prolonged contact may cause skin sensitization and allergy, with itching, redness, and later rash. Repeated or prolonged inhalation exposure may cause asthma-like lung allergy. Lungs may be scarred by repeated or prolonged exposure to the aerosol. Nickel sulfate may cause mutations. Handle with extreme caution. Carcinogenic to humans. Animal tests show that this substance possibly causes toxic effects upon human reproduction. May cause kidney damage.

Points of Attack: Skin, lungs, kidneys.

Medical Surveillance: NIOSH lists the following tests: blood gas analysis; whole blood (chemical/metabolite); blood plasma; blood plasma, end-of-shift; blood serum; biologic tissue/biopsy; chest X-ray, electrocardiogram, pulmonary function tests; pre- and postshift; sputum cytology; urine (chemical/metabolite); urine (chemical/metabolite), end-of-shift; white blood cell count/differential. Before beginning employment and at regular times after that, the following are recommended: lung function test. Urine or plasma test for nickel (unexposed persons have urine levels less than 10 μg/L). If symptoms develop or overexposure is suspected, the following may be useful: daily urine nickel for several days (persons with urine nickel over 100 μg/L need medical attention). Lung function tests. Consider chest X-ray after acute overexposure. Evaluation by a qualified allergist, including careful exposure history and special testing, may help diagnose skin allergy. Kidney function tests.

First Aid: If this chemical gets into the eyes, remove any contact lenses at once and irrigate immediately for at least 15 min, occasionally lifting upper and lower lids. Seek medical attention immediately. If this chemical contacts the skin, remove contaminated clothing and wash immediately with soap and water. Seek medical attention immediately. If this chemical has been inhaled, remove from exposure, begin rescue breathing (using universal precautions, including resuscitation mask) if breathing has stopped and CPR if heart action has stopped. Transfer promptly to a medical facility. When this chemical has been swallowed, get medical attention. Give large quantities of water and induce vomiting. Do not make an unconscious person vomit. Medical observation is recommended for 24−48 h after breathing overexposure, as pulmonary edema may be delayed. As first aid for pulmonary edema, a doctor or authorized paramedic may consider administering a corticosteroid spray.

Note to physician: For severe poisoning BAL [British Anti-Lewisite, dimercaprol, dithiopropanol (C$_3$H$_8$OS$_2$)] has been used to treat toxic symptoms of certain heavy metals poisoning including nickel. Although BAL is reported to have a large margin of safety, caution must be exercised, because toxic effects may be caused by excessive dosage. Most can be prevented by premedication with 1-ephedrine sulfate (CAS: 134-72-5).

Personal Protective Methods: Wear protective gloves and clothing to prevent any reasonable probability of skin contact. Safety equipment suppliers/manufacturers can provide recommendations on the most protective glove/clothing material for your operation. All protective clothing (suits, gloves, footwear, headgear) should be clean, available each day, and put on before work. Contact lenses should not be worn when working with this chemical. Wear dust-proof chemical goggles and face shield unless full face-piece respiratory protection is worn. Employees should wash immediately with soap when skin is wet or contaminated. Provide emergency showers and eyewash. See NIOSH Criteria Document #77-164, *Inorganic Nickel.*

Respirator Selection: *At concentrations above the NIOSH REL, or where there is no REL, at any detectable concentration:* SCBAF: Pd,Pp (APF = 10,000) (any NIOSH/MSHA- or European Standard EN 149-approved self-contained breathing apparatus that has a full face-piece and is operated in a pressure-demand or other positive-pressure mode) or SaF: Pd,Pp: ASCBA (APF = 10,000) (any supplied-air respirator that has a full face-piece and is operated in a pressure-demand or other positive-pressure mode in combination with an auxiliary, self-contained breathing apparatus operated in a pressure-demand or other positive-pressure mode). *Escape:* 100F (APF = 50) (any air-purifying, full-face-piece respirator with an N100, R100, or P100 filter).

Storage: Color Code—Blue: Health Hazard/Poison: Store in a secure poison location. Prior to working with this chemical you should be trained on its proper handling and storage. Nickel sulfate must be stored to avoid contact with strong acids (such as hydrochloric, sulfuric, and nitric); wood, and other combustibles, since violent reactions occur. A regulated, marked area should be established where nickel sulfate is handled, used, or stored.

Shipping: Toxic solid, inorganic, n.o.s. materials require a shipping label of "POISONOUS/TOXIC MATERIALS." They fall in Hazard Class 6.1.

Spill Handling: Evacuate persons not wearing protective equipment from area of spill or leak until cleanup is complete. Remove all ignition sources. Use HEPA vacuum or wet method to reduce dust during cleanup. Do not dry sweep. Collect powdered material in the most convenient and safe manner and deposit in sealed containers. Ventilate area after cleanup is complete. It may be necessary to contain and dispose of this chemical as a hazardous waste. If material or contaminated runoff enters waterways, notify downstream users of potentially contaminated waters. Contact your local or federal environmental protection agency for specific recommendations. If employees are required to clean up spills, they must be properly trained and equipped. OSHA 1910.120(q) may be applicable.

Fire Extinguishing: Nickel sulfate is not flammable. Use agent suitable for surrounding fire. Poisonous gases, including oxides of sulfur, are produced in fire. If material or contaminated runoff enters waterways, notify downstream users of potentially contaminated waters. Notify local health and fire officials and pollution control agencies. From a secure, explosion-proof location, use water spray to cool exposed containers. If cooling streams are ineffective (venting sound increases in volume and pitch, tank discolors, or shows any signs of deforming), withdraw immediately to a secure position. If employees are expected to fight fires, they must be trained and equipped in OSHA 1910.156. The only respirators recommended for firefighting are self-contained breathing apparatuses that have full face-pieces and are operated in a pressure-demand or other positive-pressure mode.

Disposal Method Suggested: Consult with environmental regulatory agencies for guidance on acceptable disposal practices. Generators of waste containing this contaminant ($\geq$100 kg/mo) must conform with EPA regulations governing storage, transportation, treatment, and waste disposal.

References

Sax, N. I. (Ed.). (1985). *Dangerous Properties of Industrial Materials Report,* 5, No. 6, 68–71

New Jersey Department of Health and Senior Services. (April 2003). *Hazardous Substances Fact Sheet: Nickel Sulfate.* Trenton, NJ

Nicotine   N:0300

Molecular Formula: $C_{10}H_{14}N_2$

Synonyms: Black leaf; Campbell's nico-soap; Destruxol orchard spray; Di-tetrahydronicotyrine; Emo-Nib; ENT 3,424; Flux MAAG; Fumeto-Bac; Mach-Nic; 1-Methyl-2-(3-pyridyl) pyrrolidine; 3-(N-Methylpyrrolidino)pyridine; (s)-3-(1-Methyl-2-pyrrolidinyl)pyridine; 3-(1-Methyl-2-pyr-rolidinyl)pyridine; (−)-3-(1-Methyl-2-pyrrolidyl)pyridine; 1-3-(1-Methyl-2-pyrrolidyl) pyridine; 3-(1-Methyl-2-pyrro-lidyl) pyridine; Niagra P.A. dust; Nicocide; Nicodust; Nicofume; Nicotina (Spanish); 1-Nicotine; Nicotine alka-loid; Nikotin (German); Orthon-4 dust; Orthon-5 dust; Pyridine, 3-(1-methyl-2-pyrrolidinyl)-; Pyridine, (s)-3-(1-methyl-2-pyrrolidinyl)- and salts; Pyridine, 3-(tetrahydro-1-methylpyrrol-2-yl); β-Pyridyl-α-N-methylpyrrolidine; Tendust; Xl all insecticide

CAS Registry Number: 54-11-5; *(alt.)* 6912-85-2; *(alt.)* 16760-37-5

RTECS® Number: QS5250000

UN/NA & ERG Number: UN1654/151

EC Number: 200-193-3 [*Annex I Index No.:* 614-001-00-4]

Regulatory Authority and Advisory Bodies

US EPA Gene-Tox Program, Negative: *N. crassa—aneu-ploidy*; Histidine reversion—Ames test.

US EPA, FIFRA, 1998 Status of Pesticides: Supported.

Banned or Severely Restricted (in agriculture) (Germany, Hungary) (UN).[13]

Air Pollutant Standard Set. See below, "Permissible Exposure Limits in Air" section.

US EPA Hazardous Waste Number (RCRA No.): No. P075.

RCRA, 40CFR261, Appendix 8 Hazardous Constituents.

SUPERFUND/EPCRA 40CFR355, Extremely Hazardous Substances: TPQ = 100 lb (45.4 kg).

Reportable Quantity (RQ): 100 lb (45.4 kg).

EPCRA Section 313 Form R *de minimis* concentration reporting level: 1.0%.

California Proposition 65 Developmental/Reproductive toxin 4/1/90.

Canada, WHMIS, Ingredients Disclosure List Concentration 1.0%.

European/International Regulations: Hazard Symbol: T+, N; Risk phrases: R25; R27; R51/53; Safety phrases: S1/2; S36/37; S45; S61 (see Appendix 4).

WGK (German Aquatic Hazard Class): No value assigned.

Description: Nicotine is a pale yellow to dark brown, oily liquid with a slight, fishy odor when warm. It is also available as a powder. Molecular weight = 162.23; Specific gravity (H_2O:1) = 1.01; Boiling point = 250°C; Freezing/Melting point = − 78.9°C; Vapor pressure = 0.08 mmHg at 20°C; Flash point = 95°C (cc); Autoignition temperature = 240°C. Explosive limits: LEL = 0.7%; UEL = 4.0%. Hazard Identification (based on NFPA-704 M Rating System): Health 4, Flammability 1, Reactivity 0. Slightly soluble in water.

Potential Exposure: Compound Description: Agricultural Chemical; Drug, Mutagen; Reproductive Effector; Human Data; Natural Product. Nicotine is used in some drugs and in tanning. At one time, nicotine was used in the United States as an insecticide and fumigant; however, it is no longer produced or used in the United States for this purpose.

Incompatibilities: Contact with strong oxidizers may be violent. Incompatible with strong acids. Attacks some forms of plastics, rubber, and coatings. Nicotine decomposes on heating, producing nitrogen oxides, carbon monoxide, and other highly toxic fumes.

Permissible Exposure Limits in Air

OSHA PEL: 0.5 mg/m^3 TWA [skin].

NIOSH REL: 0.5 mg/m^3 TWA [skin].

ACGIH TLV®[1]: 0.5 mg/m^3 TWA [skin].

NIOSH IDLH: 5 mg/m^3.

Protective Action Criteria (PAC)

TEEL-0: 0.5 mg/m^3

PAC-1: 1.5 mg/m^3

PAC-2: 3.5 mg/m^3

PAC-3: 5 mg/m^3

DFG MAK: [skin]

Arab Republic of Egypt: TWA 0.5 mg/m^3, [skin], 1993; Australia: TWA 0.5 mg/m^3, [skin], 1993; Austria: MAK 0.07 ppm, 1999; Belgium: TWA 0.5 mg/m^3, [skin], 1993; Denmark: TWA 0.5 mg/m^3, [skin], 1999; Finland: TWA 0.5 mg/m^3; STEL 1.5 mg/m^3, [skin], 1999; France: VME 0.5 mg/m^3, [skin], 1999; the Netherlands: MAC-TGG 0.5 mg/m^3, [skin], 2003; Norway: TWA 0.5 mg/m^3, 1999; the Philippines: TWA 0.5 mg/m^3, [skin], 1993; Poland: MAC (TWA) 0.5 mg/m^3; MAC (STEL) 1.5 mg/m^3, 1999; Switzerland: MAK-W 0.07 ppm (0.5 mg/m^3), KZG-W 0.14 ppm, [skin], 1999; Thailand: TWA 0.5 mg/m^3, 1993; Turkey: TWA 0.5 mg/m^3, [skin], 1993; United Kingdom: TWA 0.5 mg/m^3; STEL 1.5 mg/m^3, [skin], 2000; Argentina, Bulgaria, Columbia, Jordan, South Korea, New Zealand, Singapore, Vietnam: ACGIH TLV®: TWA 0.5 mg/m^3 [skin]. Several states have set guidelines or standards for nicotine in ambient air[60] ranging from 8.0 μg/m^3 (Virginia) to 10.0 μg/m^3 (Connecticut) to 12.0 μg/m^3 (Nevada).

Determination in Air: Use NIOSH Analytical Method (IV) #2544, #2551.

Determination in Water: Octanol–water coefficient: Log $K_{ow} = 1.17$.

Routes of Entry: Inhalation, ingestion, skin and/or eye contact.

Harmful Effects and Symptoms

Nicotine affects the nervous system and the heart. Exposure to relatively small amounts can rapidly be fatal.[NIOSH]

Short Term Exposure: Irritates the eyes and skin. Even small exposures can cause increased heart rate, increased blood fat levels, and change vital hormone levels. May affect the cardiovascular system and central nervous system, resulting in convulsions and respiratory failure. Nicotine is classified as super toxic. Probable oral lethal dose in humans is less than 5 mg/kg or a taste (less than 7 drops) for a 70-kg (150 lb) person. It may be assumed that ingestion of 40–60 mg of nicotine is lethal to humans. There is a fundamental difference between acute toxicity from use of nicotine as an insecticide instead of from ingestion and chronic toxicity that may be caused by prolonged exposure to small doses as occurs in smoking. Maternal smoking during pregnancy is associated with increased risk of spontaneous abortion, low birth weight, and stillbirth. Acute exposure to nicotine may result in headache, dizziness, confusion, agitation, restlessness, lethargy, seizures, and coma. Victims may experience hypertension (high blood pressure), tachycardia (rapid heart rate), and tachypnea (rapid respirations), followed by hypotension (low blood pressure), bradycardia (slow heart rate), and respiratory depression. Cardiac arrhythmias may also occur. Gastrointestinal effects include nausea, vomiting, abdominal pain, or burning sensation, and diarrhea. Increased salivation, lacrimation (tearing), and sweating may be noted. High levels, far above the OEL, may result in death.

Long Term Exposure: Animal tests show that this substance possibly causes toxic effects upon human reproduction. Nicotine was found as a co-carcinogen in animals.

Points of Attack: Central nervous system, cardiovascular system, lungs, GI tract, reproduction system. Has been shown to be a teratogen in animals; may be a teratogen in humans. Causes fat deposits in the arteries (reducing blood supply to many body organs). This increases the risk of heart attack, stroke, and many other poor circulation problems. Chronic high blood pressure can also result.

Medical Surveillance: NIOSH lists the following tests: blood plasma, during exposure, pre- and postshift, urine (chemical/metabolite). Before beginning employment and at regular times after that, the following is recommended: blood test for nicotine (only accurate shortly after exposure); consider test to evaluate typical exposures as well as for suspected overexposure or if symptoms are present. Even those who have smoked for a long time can reduce the risk of developing health problems by stopping.

First Aid: If this chemical gets into the eyes, remove any contact lenses at once and irrigate immediately for at least 15 min, occasionally lifting upper and lower lids. Seek medical attention immediately. If this chemical contacts the skin, remove contaminated clothing and wash immediately with soap and water. Seek medical attention immediately. If this chemical has been inhaled, remove from exposure, begin rescue breathing (using universal precautions, including resuscitation mask) if breathing has stopped and CPR if heart action has stopped. Transfer promptly to a medical facility. When this chemical has been swallowed, get medical attention. Give large quantities of water and induce vomiting. Do not make an unconscious person vomit.

Personal Protective Methods: Wear protective gloves and clothing to prevent any reasonable probability of skin contact. Safety equipment suppliers/manufacturers can provide recommendations on the most protective glove/clothing material for your operation. All protective clothing (suits, gloves, footwear, headgear) should be clean, available each day, and put on before work. Contact lenses should not be worn when working with this chemical. Wear splash-proof chemical goggles and face shield unless full face-piece respiratory protection is worn. Employees should wash immediately with soap when skin is wet or contaminated. Provide emergency showers and eyewash. Possibility of skin contact. Wear eye protection to prevent any possibility of eye contact. Employees should wash immediately when skin is wet or contaminated. Remove nonimpervious clothing immediately if wet or contaminated. Provide emergency showers and eyewash.

Respirator Selection: 5 mg/m^3: Sa (APF = 10) (any supplied-air respirator) or SCBAF (APF = 50) (any self-contained breathing apparatus with a full face-piece). *Emergency or planned entry into unknown concentrations or IDLH conditions:* SCBAF: Pd,Pp (APF = 10,000) (any self-contained breathing apparatus that has a full face-piece and is operated in a pressure-demand or other positive-pressure mode) or SaF: Pd,Pp: ASCBA (APF = 10,000) (any supplied-air respirator that has a full face-piece and is operated in a pressure-demand or other positive-pressure mode in combination with an auxiliary self-contained breathing apparatus operated in a pressure-demand or other positive-pressure mode). *Escape:* GmFOv (APF = 50) [any air-purifying, full-face-piece respirator (gas mask) with a chin-style, front- or back-mounted acid gas canister] or SCBAE (any appropriate escape-type, self-contained breathing apparatus).

Note: Substance reported to cause eye irritation or damage; may require eye protection.

Storage: Color Code—Blue: Health Hazard/Poison: Store in a secure poison location. Prior to working with this chemical you should be trained on its proper handling and storage. Before entering confined space where this chemical may be present, check to make sure that an explosive concentration does not exist. Nicotine must be stored to avoid contact with strong oxidizers (such as chlorine, bromine and fluorine), strong acids (such as hydrochloric, sulfuric and nitric), since violent reactions occur. Where possible, automatically pump liquid from drums or other storage containers to process containers. Sources of ignition, such as smoking and open flames, are prohibited where this chemical is handled, used, or stored. Metal containers involving the transfer of 5 gallons or more of this chemical should be grounded and bonded. Drums must be equipped with self-closing valves, pressure vacuum bungs, and flame arresters. Use only nonsparking tools and equipment, especially when opening and closing containers of this chemical. Wherever this chemical is used, handled, manufactured, or stored, use explosion-proof electrical equipment and fittings.

Shipping: This compound requires a shipping label of "POISONOUS/TOXIC MATERIALS." It falls in Hazard Class 6.1 and Packing Group II.

Spill Handling: Evacuate and restrict persons not wearing protective equipment from area of spill or leak until cleanup is complete. Remove all ignition sources. Establish forced ventilation to keep levels below explosive limit. Absorb liquids in vermiculite, dry sand, earth, peat, carbon, or a similar material and deposit in sealed containers. Keep this chemical out of a confined space, such as a sewer, because of the possibility of an explosion, unless the sewer is designed to prevent the buildup of explosive concentrations. It may be necessary to contain and dispose of this chemical as a hazardous waste. If material or contaminated runoff enters waterways, notify downstream users of potentially contaminated waters. Contact your local or federal environmental protection agency for specific recommendations. If employees are required to clean up spills, they must be properly trained and equipped. OSHA 1910.120(q) may be applicable.

Fire Extinguishing: Nicotine is a combustible liquid. Extinguish with foam, dry chemical, or carbon dioxide extinguishers. Poisonous gases, including oxides of nitrogen and carbon monoxide, are produced in fire. Water may cause frothing if it gets below surface of liquid and turns to steam. However, water fog gently applied to surface will cause frothing which will extinguish the fire. Vapors are heavier than air and will collect in low areas. Vapors may travel long distances to ignition sources and flashback. Vapors in confined areas may explode when exposed to fire. Containers may explode in fire. Storage containers and parts of containers may rocket great distances, in many directions. If material or contaminated runoff enters waterways, notify downstream users of potentially contaminated waters. Notify local health and fire officials and pollution control agencies. From a secure, explosion-proof location, use water spray to cool exposed containers. If cooling streams are ineffective (venting sound increases in volume and pitch, tank discolors, or shows any signs of deforming), withdraw immediately to a secure position. If employees are expected to fight fires, they must be trained and equipped in OSHA 1910.156. The only respirators recommended for firefighting are self-contained breathing apparatuses that have full face-pieces and are operated in a pressure-demand or other positive-pressure mode.

Disposal Method Suggested: Dissolve or mix the material with a combustible solvent and burn in a chemical incinerator equipped with an afterburner and scrubber. All federal, state, and local environmental regulations must be observed.[22] In accordance with 40CFR165, follow recommendations for the disposal of pesticides and pesticide containers. Must be disposed properly by following package label directions or by contacting your local or federal environmental control agency or by contacting your regional EPA office. Consult with environmental regulatory agencies for guidance on acceptable disposal practices. Generators of waste containing this contaminant ($\geq$ 100 kg/mo) must conform with EPA regulations governing storage, transportation, treatment, and waste disposal.

References

Sax, N. I. (Ed.). *Dangerous Properties of Industrial Materials Report*, 1, No. 8, 84–85 (1981) and 5, No. 4, 82–85 (1985)

US Environmental Protection Agency. (November 30, 1987). *Chemical Hazard Information Profile: Nicotine.* Washington, DC: Chemical Emergency Preparedness Program

New Jersey Department of Health and Senior Services. (March 2000). *Hazardous Substances Fact Sheet: Nicotine.* Trenton, NJ

Nicotine sulfate   N:0310

Molecular Formula: $C_{10}H_{18}N_2O_8S_2$
Common Formula: $C_{10}H_{14}N_2 \cdot 2H_2SO_4$
Synonyms: ENT 2,435; 1-1-Methyl-2-(3-pyridyl)-pyrrolidine sulfate; (S)-3-(1-Methyl-2-pyrrolidinyl)pyridine sulfate (2:1); 1-3-(1-Methyl-2-pyrrolidinyl)pyridine sulfate; Nicotine sulfate (2:1); Nicotine sulphate; Nicotine sulphate (2:1); Nikotinsulfat (German); Pyridine, 3-(1-methyl-2-pyrrolidinyl)-, (S)-, sulfate (2:1); Pyrrolidine, 1-methyl-2-(3-pyridyl)-, sulfate; Sulfate de nicotine (French); Sulfato de nicotina (Spanish)
CAS Registry Number: 65-30-5
RTECS® Number: QS9625000
UN/NA & ERG Number: UN1658/151
EC Number: 200-606-7
Regulatory Authority and Advisory Bodies
US EPA, FIFRA 1998 Status of Pesticides: Canceled.
Banned or Severely Restricted (in agriculture) (East Germany, New Zealand, former USSR) (UN).[13]
Air Pollutant Standard Set. See below, "Permissible Exposure Limits in Air" section.
SUPERFUND/EPCRA 40CFR355, Extremely Hazardous Substances: TPQ = 100/10,000 lb (45.4/4540 kg).
Reportable Quantity (RQ): 100 lb (45.4 kg).
EPCRA Section 313 Form R *de minimis* concentration reporting level: 1.0%.
European/International Regulations (*nicotine salts*): Hazard Symbol: T+, N; Risk phrases: R26/27/28; Safety phrases: S1/2; S13; S28; S45 (see Appendix 4).
WGK (German Aquatic Hazard Class): No value assigned.
Description: Nicotine sulfate is a white crystalline solid. Molecular weight = 420.58; Autoignition temperature = 244°C. Hazard Identification (based on NFPA-704 M Rating System): Health 4, Flammability 1, Reactivity 0. Soluble in water.
Potential Exposure: Compound Description: Agricultural Chemical; Reproductive Effector. It is used as an insecticide and in veterinary medicine as an anthelmintic and external parasiticide.
Incompatibilities: Oxidizing materials.
Permissible Exposure Limits in Air
Protective Action Criteria (PAC)
TEEL-0: 4 mg/m^3
PAC-1: 9 mg/m^3
PAC-2: 9 mg/m^3
PAC-3: 9 mg/m^3
Russia: STEL 0.1 mg/m^3, 1993.
Russia set a MAC in ambient air in residential areas of 0.005 mg/m^3 on a once-daily basis and 0.001 mg/m^3 on a daily average basis.

Routes of Entry: Inhalation, ingestion, skin and/or eye contact.
Harmful Effects and Symptoms
Short Term Exposure: The liquid irritates the eyes and skin. Inhalation irritates nose and throat. May affect the central nervous system, causing convulsions and respiratory failure. Exposure at high concentrations may result in death. Onset of acute poisoning is rapid. Symptoms include nausea, salivation, abdominal pain, vomiting, diarrhea, cold sweat, headache, dizziness, disturbed hearing and vision, mental confusion, marked weakness, faintness and prostration, lowered blood pressure, difficult breathing, and weak, rapid and irregular pulse. It is classified as super toxic. Probable oral lethal dose in humans is less than 5 mg/kg (less than 7 drops) for a 70-kg (150 lb) person. Death is possible from respiratory failure caused by paralysis of the respiratory muscles.
Long Term Exposure: Animal tests show that this substance possibly causes toxic effects upon human reproduction.
First Aid: If this chemical gets into the eyes, remove any contact lenses at once and irrigate immediately for at least 15 min, occasionally lifting upper and lower lids. Seek medical attention immediately. If this chemical contacts the skin, remove contaminated clothing and wash immediately with soap and water. Seek medical attention immediately. If this chemical has been inhaled, remove from exposure, begin rescue breathing (using universal precautions, including resuscitation mask) if breathing has stopped and CPR if heart action has stopped. Transfer promptly to a medical facility. When this chemical has been swallowed, get medical attention. Give large quantities of water and induce vomiting. Do not make an unconscious person vomit.
Personal Protective Methods: Wear protective gloves and clothing to prevent any reasonable probability of skin contact. Safety equipment suppliers/manufacturers can provide recommendations on the most protective glove/clothing material for your operation. All protective clothing (suits, gloves, footwear, headgear) should be clean, available each day, and put on before work. Contact lenses should not be worn when working with this chemical. Wear dust-proof chemical goggles and face shield unless full face-piece respiratory protection is worn. Employees should wash immediately with soap when skin is wet or contaminated. Provide emergency showers and eyewash.
Respirator Selection: Follow regulations in OSHA 29CFR1910.134 or European Standard EN149. Use a NIOSH/MSHA- or European Standard EN149-approved respirator; or use an approved supplied-air respirator with a full face-piece operated in the positive-pressure mode, or with a full face-piece, hood, or helmet in the continuous-flow mode; or use a NIOSH/MSHA- or European Standard EN149-approved self-contained breathing apparatus with a full face-piece operated in pressure-demand or other positive-pressure mode.
Storage: Color Code—Blue: Health Hazard/Poison: Store in a secure poison location. Prior to working with this

chemical you should be trained on its proper handling and storage. Store in tightly closed containers in a cool, well-ventilated area away from oxidizers. Metal containers involving the transfer of 5 gallons or more of this chemical should be grounded and bonded. Drums must be equipped with self-closing valves, pressure vacuum bungs, and flame arresters. Use only nonsparking tools and equipment, especially when opening and closing containers of this chemical. Wherever this chemical is used, handled, manufactured, or stored, use explosion-proof electrical equipment and fittings.

Shipping: This compound requires a shipping label of "POISONOUS/TOXIC MATERIALS." It falls in Hazard Class 6.1 and Packing Group II.

Spill Handling: Solid: Evacuate persons not wearing protective equipment from area of spill or leak until cleanup is complete. Remove all ignition sources. Use HEPA vacuum or wet method to reduce dust during cleanup. Do not dry sweep. Collect powdered material in the most convenient and safe manner and deposit in sealed containers. Ventilate area after cleanup is complete. It may be necessary to contain and dispose of this chemical as a hazardous waste. If material or contaminated runoff enters waterways, notify downstream users of potentially contaminated waters. Contact your local or federal environmental protection agency for specific recommendations. If employees are required to clean up spills, they must be properly trained and equipped. OSHA 1910.120(q) may be applicable.

Solution: Evacuate and restrict persons not wearing protective equipment from area of spill or leak until cleanup is complete. Remove all ignition sources. Ventilate area of spill or leak. To clean up, do not touch spilled material; stop leak if possible. Use water spray to reduce vapors. *Small spills:* take up with sand or other noncombustible absorbent material and place into containers for disposal. *Small dry spill:* with clean shovel place material into clean, dry container and cover; move containers from spill area. *Large spills:* dike far ahead of spill for later disposal. Keep this chemical out of a confined space, such as a sewer, because of the possibility of an explosion, unless the sewer is designed to prevent the buildup of explosive concentrations. It may be necessary to contain and dispose of this chemical as a hazardous waste. If material or contaminated runoff enters waterways, notify downstream users of potentially contaminated waters. Contact your local or federal environmental protection agency for specific recommendations. If employees are required to clean up spills, they must be properly trained and equipped. OSHA 1910.120(q) may be applicable.

Fire Extinguishing: As for nicotine, extinguish with alcohol foam, dry chemical, or carbon dioxide. Large fires can be extinguished with water spray, fog, or foam. Wear positive pressure breathing apparatus and special protective clothing. Dike fire control water; do not scatter the material. Poisonous gases are produced in fire. If material or contaminated runoff enters waterways, notify downstream users of

potentially contaminated waters. Notify local health and fire officials and pollution control agencies. From a secure, explosion-proof location, use water spray to cool exposed containers. If cooling streams are ineffective (venting sound increases in volume and pitch, tank discolors, or shows any signs of deforming), withdraw immediately to a secure position. If employees are expected to fight fires, they must be trained and equipped in OSHA 1910.156. The only respirators recommended for firefighting are self-contained breathing apparatuses that have full face-pieces and are operated in a pressure-demand or other positive-pressure mode.

Disposal Method Suggested: In accordance with 40CFR165, follow recommendations for the disposal of pesticides and pesticide containers. Must be disposed properly by following package label directions or by contacting your local or federal environmental control agency or by contacting your regional EPA office.

References

Sax, N. I. (Ed.). (1985). *Dangerous Properties of Industrial Materials Report,* 5, No. 4, 88−90

US Environmental Protection Agency. (November 30, 1987). *Chemical Hazard Information Profile: Nicotine Sulfate.* Washington, DC: Chemical Preparedness Program New Jersey Department of Health and Senior Services. (April 2002). *Hazardous Substances Fact Sheet: Nicotine Sulfate.* Trenton, NJ

Nitrapyrin   N:0320

Molecular Formula: $C_6H_3Cl_4N$

Synonyms: 2-Chloro-6-(trichloromethyl)pyridine; 2-Chloro-6-trichloromethylpyridine; 4-Chloro-6-(trichloromethyl)-pyridine; Dowco-163®; Nitrapyrine; *N*-Serve®; *N*-Serve nitrogen stabilizer®; Pyridine, 2-chloro-6-(trichloromethyl)-

CAS Registry Number: 1929-82-4

RTECS® Number: US7525000

UN/NA & ERG Number: UN3077/171

EC Number: 217-682-2 [*Annex I Index No.:* 006-057-00-8]

Regulatory Authority and Advisory Bodies

Carcinogenicity: (Animal Positive) NCI.[9]

Air Pollutant Standard Set. See below, "Permissible Exposure Limits in Air" section.

EPCRA Section 313 Form R *de minimis* concentration reporting level: 1.0%.

California Proposition 65 Chemical: Cancer 10/5/05; Developmental/Reproductive toxin 4/30/99.

European/International Regulations: Hazard Symbol: Xn, N; Risk phrases: R22; R51/53; Safety phrases: S2; S24; S61 (see Appendix 4).

WGK (German Aquatic Hazard Class): 2—Hazard to waters.

Description: Nitrapyrin is a colorless crystalline solid with a mild, sweet odor. Molecular weight = 230.90; Freezing/Melting point = 62.8°C; Vapor pressure = 0.0028 mmHg at 22.8°C. Insoluble in water.

Potential Exposure: Compound Description: Agricultural Chemical; Mutagen; Reproductive Effector. Nitrapyrin is used as a fertilizer additive; to stabilize nitrogen.

Incompatibilities: Aluminum, magnesium.

Permissible Exposure Limits in Air

OSHA PEL: 15 mg/m^3 TWA (total dust); 5 mg/m^3 TWA, respirable fraction.

NIOSH REL: 10 mg/m^3 TWA (total dust); 5 mg/m^3 TWA, respirable fraction TWA; 20 mg/m^3 respirable fraction STEL.

ACGIH TLV®[1]: 10 mg/m^3 TWA; 20 mg/m^3 STEL; not classifiable as a human carcinogen.

Protective Action Criteria (PAC)

TEEL-0: 15 mg/m^3

PAC-1: 20 mg/m^3

PAC-2: 20 mg/m^3

PAC-3: 400 mg/m^3

Austria: MAK 10 mg/m^3; STEL 20 mg/m^3, 1993; Belgium: TWA 10 mg/m^3; STEL 20 mg/m^3, 1993; France: VME 10 mg/m^3, 1999; Switzerland: MAK-W 10 mg/m^3, 1999; United Kingdom: TWA 10 mg/m^3; STEL 20 mg/m^3 2000; the Netherlands: MAC-TGG 10 mg/m^3, 2003; Argentina, Bulgaria, Columbia, Jordan, South Korea, New Zealand, Singapore, Vietnam: ACGIH TLV®: STEL 20 mg/m^3. Several states have set guidelines for nitrapyrin in ambient air[60] ranging from 100 to 200 µg/m^3 (North Dakota) to 160 µg/m^3 (Virginia) to 200 µg/m^3 (Connecticut) to 238 µg/m^3 (Nevada).

Determination in Air: No NIOSH Analytical Method available.

Determination in water: Fish Tox = 314.70281000 ppb (LOW).

Routes of Entry: Inhalation, ingestion, skin and/or eye contact.

Harmful Effects and Symptoms

Short Term Exposure: Nitrapyrin can affect you when breathed in and by passing through your skin. Exposure can irritate the eyes, nose, and throat. High levels may cause you to feel dizzy, lightheaded, and to pass out. Contact can irritate and may damage the eyes and skin. No adverse effects noted in ingestion studies with animals. Human Tox = 210.00000 ppb (VERY LOW).

Long Term Exposure: There may be damage to the liver and kidneys. Repeated exposure to nitrapyrin may cause symptoms of headaches, dizziness, loss of appetite, and trouble sleeping.

Medical Surveillance: If symptoms develop or overexposure is suspected, the following may be useful: liver and kidney function tests.

First Aid: If this chemical gets into the eyes, remove any contact lenses at once and irrigate immediately for at least 15 min, occasionally lifting upper and lower lids. Seek medical attention immediately. If this chemical contacts the skin, remove contaminated clothing and wash immediately with soap and water. Seek medical attention immediately. If this chemical has been inhaled, remove from exposure, begin rescue breathing (using universal precautions, including resuscitation mask) if breathing has stopped and CPR if heart action has stopped. Transfer promptly to a medical facility. When this chemical has been swallowed, get medical attention. Give large quantities of water and induce vomiting. Do not make an unconscious person vomit.

Personal Protective Methods: Wear protective gloves and clothing to prevent any reasonable probability of skin contact. Safety equipment suppliers/manufacturers can provide recommendations on the most protective glove/clothing material for your operation. All protective clothing (suits, gloves, footwear, headgear) should be clean, available each day, and put on before work. Contact lenses should not be worn when working with this chemical. Wear dust-proof chemical goggles and face shield unless full face-piece respiratory protection is worn. Employees should wash immediately with soap when skin is wet or contaminated. Provide emergency showers and eyewash.

Respirator Selection: Where there is potential for exposures over *10 mg/m^3*, use a NIOSH/MSHA- or European Standard EN149-approved full face-piece respirator with a high-efficiency particulate filter. Greater protection is provided by a powered air-purifying respirator. *Where there is potential for high exposures,* use a NIOSH/MSHA- or European Standard EN149-approved supplied-air respirator with a full face-piece operated in the positive-pressure mode, or with a full face-piece, hood, or helmet in the continuous-flow mode; or use a NIOSH/MSHA- or European Standard EN149-approved self-contained breathing apparatus with a full face-piece operated in pressure-demand or other positive-pressure mode.

Storage: Color Code—Green: General storage may be used. Prior to working with this chemical you should be trained on its proper handling and storage. Store in tightly closed containers in a cool, well-ventilated area away from aluminum, magnesium, or their alloys. Do not store in unlined containers. A regulated, marked area should be established where this chemical is handled, used, or stored in compliance with OSHA Standard 1910.1045.

Shipping: The name of this material is not on the DOT list of materials[19] for label and packaging standards. However, based on regulations, it may be classified[52] as an Environmentally hazardous substances, solid, n.o.s. This chemical requires a shipping label of "CLASS 9." It falls in Hazard Class 9 and Packing Group III.[20, 21]

Spill Handling: Evacuate persons not wearing protective equipment from area of spill or leak until cleanup is complete. Remove all ignition sources. Use HEPA vacuum or wet method to reduce dust during cleanup. Do not dry sweep. Collect powdered material in the most convenient and safe manner and deposit in sealed containers. Ventilate area after cleanup is complete. It may be necessary to contain and dispose of this chemical as a hazardous waste. If material or contaminated runoff enters waterways, notify downstream users of potentially contaminated waters. Contact your local or federal environmental protection agency for specific recommendations. If employees are required to clean up spills, they must be properly trained and equipped. OSHA 1910.120(q) may be applicable.

Fire Extinguishing: Extinguish fire using an agent suitable for type of surrounding fire. Nitrapyrin itself does not burn. Poisonous gases, including oxides of nitrogen and chloride ion, are produced in fire. If material or contaminated runoff enters waterways, notify downstream users of potentially contaminated waters. Notify local health and fire officials and pollution control agencies. From a secure, explosion-proof location, use water spray to cool exposed containers. If cooling streams are ineffective (venting sound increases in volume and pitch, tank discolors, or shows any signs of deforming), withdraw immediately to a secure position. If employees are expected to fight fires, they must be trained and equipped in OSHA 1910.156. The only respirators recommended for firefighting are self-contained breathing apparatuses that have full face-pieces and are operated in a pressure-demand or other positive-pressure mode.

Disposal Method Suggested: The manufacturer of this nitrification inhibitor suggests that unwanted quantities can be disposed of by burial in a sanitary landfill. In accordance with 40CFR165, follow recommendations for the disposal of pesticides and pesticide containers. Must be disposed properly by following package label directions or by contacting your local or federal environmental control agency or by contacting your regional EPA office.

Reference
New Jersey Department of Health and Senior Services. (April 2000). *Hazardous Substances Fact Sheet: Nitrapyrin.* Trenton, NJ

Nitrates   N:0330

Molecular Formula: $N_vM_vO_{vx3}$
Common Formula: $M_v(NO_3)_v$
Synonyms: Vary with specific compounds
CAS Registry Number: 14797-55-8
RTECS® Number: WC5600000 (sodium nitrate)
UN/NA & ERG Number: UN3218 (nitrates, inorganic, aqueous solution, n.o.s.)/140; UN1477 (nitrates, inorganic, n.o.s.)/140
Regulatory Authority and Advisory Bodies
Air Pollutant Standard Set. See below, "Permissible Exposure Limits in Air" section.
Nitrate and nitrite:
Safe Drinking Water Act: MCL, 10 mg/L as nitrogen; MCLG, 10 mg/L as nitrogen.
Nitrate compounds (water dissociable).
EPCRA Section 313: Reportable only when in aqueous solution. Form R *de minimis* concentration reporting level: 1.0%.
US DOT Regulated Marine Pollutant (49CFR172.101, Appendix B), as nitrates, inorganic, n.o.s.

WGK (German Aquatic Hazard Class): No value assigned (CAS: 14797-55-8). May vary by compound.
Description: Nitrates are salts with varying properties depending on the specific compounds. Organic nitrates are called "nitro" compounds. Hazard Identification (based on NFPA-704 M Rating System): Hazard Identification (based on NFPA-704 M Rating System) (*estimated*): Health 2, Flammability 0, Reactivity 0. Oxidizers.
Potential Exposure: Among the major point sources of nitrogen entry into water bodies are municipal and industrial wastewaters, septic tanks, and feedlot discharges. Diffuse sources of nitrogen include farm-site fertilizer and animal wastes, lawn fertilizer, leachate from waste disposal in dumps or sanitary landfill, atmospheric fallout, nitric oxide and nitrite discharges from automobile exhausts and other combustion processes, and losses from natural sources, such as mineralization of soil organic matter.
Incompatibilities: Nearly all nitrates are powerful oxidizers. Possible violent reactions with reducing agents, combustible materials, organic materials, finely divided (powdered) metals, may form explosive mixtures or cause fire and explosions. Many nitrates are flammable by spontaneous chemical reaction. Some may explode by spontaneous chemical reaction, when exposed to elevating temperatures and/or shock. Some nitrates are high explosives, especially when confined and/or exposed to heat and/or fire.
Permissible Exposure Limits in Air
Protective Action Criteria (PAC) *nitrate(s)*
TEEL-0: 10 mg/m^3
PAC-1: 30 mg/m^3
PAC-2: 50 mg/m^3
PAC-3: 250 mg/m^3
Permissible Concentration in Water: Nitrite in water at concentrations less than 1000 mg/L is not of serious concern as a direct toxicant. It is a health hazard because of its conversion to nitrite. Nitrite is directly toxic when it reacts with hemoglobin to form methemoglobin which causes methemoglobinemia. It also reacts readily under appropriate conditions with secondary amines and similar nitrogenous compounds to form *N*-nitroso compounds, many of which are potent carcinogens. Epidemiological evidence on the occurrence of methemoglobinemia in infants tends to confirm a value near 10 mg/L nitrate as nitrogen as a maximum concentration level for water with no observed adverse health effects, but there is little margin of safety in this value. The EPA[49] has published a health advisory for nitrate/nitrite which confirms the 10 mg/L level. The state of Maine[61] has set a guideline at 10 mg/L as well. EPA[62] has recently set a guideline for drinking water of 10 mg/L also.
Determination in Water: A variety of methods for determination of nitrate exists, but none is particularly precise, accurate, or sensitive in the milligram per liter concentration range. Further development and standardization of analytical methodology will be required if standard routine determinations are to be considered reliable within the

range required for proper control and assessment of health effects.

Most standard procedures for nitrate determination in the milligram per liter range are spectrophotometric. Traditionally, three types of fractions of nitrate have been used as bases: nitration of a phenolic substance to a colored derivative, oxidation of an organic substance to a colored product, and reduction of the nitrate to nitrite or ammonia, followed by reaction of the reduced nitrogenous materials to give colored substances. In addition, direct spectrophotometric determination based on ultraviolet absorption of nitrate at 273 nm is possible and becoming established. Electrochemical determination with the use of a nitrate electrode may also be feasible but is subject to numerous interferences.

Routes of Entry: Inhalation, ingestion, skin and/or eye contact.

Harmful Effects and Symptoms

Short Term Exposure: High dosages, if swallowed, may lead to serious and possibly fatal effects. Symptoms of exposure include headache, dizziness, abdominal cramps, vomiting, bloody diarrhea, weakness, convulsions, collapse, and death.

Long Term Exposure: Two health hazards are related to the consumption of water containing large concentrations of nitrate (or nitrite): induction of methemoglobinemia, particularly in infants, and possible formation for carcinogenic nitrosamines. Acute toxicity of nitrate occurs as a result of reduction to nitrite, a process that can occur under specific conditions in the stomach, as well as in the saliva. Nitrite acts in the blood to oxidize the hemoglobin to methemoglobin, which does not perform as an oxygen carrier to the tissues. Consequently, anoxia and death may ensue. According to Sax/Lewis, "... small, repeated doses may lead to weakness, general depression, headache, and mental impairment. Also, there is some implication of increased cancer incidence among those exposed." Acute nitrate toxicity is almost always seen in infants rather than adults. This increased susceptibility of infants has been attributed to high intake per unit weight, to the presence of nitrate-reducing bacteria in the upper gastrointestinal tract, to the condition of the mucosa, and to greater ease of oxidation of fetal hemoglobin."

Points of Attack: Blood.

First Aid: See specific entries.

Note to physician: Treat for methemoglobinemia. Spectrophotometry may be required for precise determination of levels of methemoglobin in urine.

Personal Protective Methods: See specific entries.

Respirator Selection: See specific entries.

Storage: Color Code—Yellow: Reactive Hazard; Store in a location separate from other materials, especially flammables and combustibles. Prior to working with nitrates chemicals you should be trained on its proper handling and storage. See statement about incompatibilities above and specific entries. See OSHA Standard 1910.104 and NFPA 43A *Code for the Storage of Liquid and Solid Oxidizers* for detailed handling and storage regulations.

Shipping: Nitrates, inorganic, n.o.s. and Nitrates, inorganic, aqueous solution, n.o.s. require a shipping label of "OXIDIZER." They fall in DOT Hazard Class 5.1 and Packing Group II or III.

Spill Handling: *Dry material:* Evacuate persons not wearing protective equipment from area of spill or leak until cleanup is complete. Remove all ignition sources. Use HEPA vacuum or wet method to reduce dust during cleanup. Do not dry sweep. Collect powdered material in the most convenient and safe manner and deposit in sealed containers. Ventilate area after cleanup is complete. It may be necessary to contain and dispose of this chemical as a hazardous waste. If material or contaminated runoff enters waterways, notify downstream users of potentially contaminated waters. Contact your local or federal environmental protection agency for specific recommendations. If employees are required to clean up spills, they must be properly trained and equipped. OSHA 1910.120(q) may be applicable.

Liquid: Evacuate and restrict persons not wearing protective equipment from area of spill or leak until cleanup is complete. Remove all ignition sources. Ventilate area of spill or leak. Absorb liquids in vermiculite, dry sand, earth, peat, carbon, or a similar material and deposit in sealed containers. Keep this chemical out of a confined space, such as a sewer, because of the possibility of an explosion, unless the sewer is designed to prevent the buildup of explosive concentrations. It may be necessary to contain and dispose of this chemical as a hazardous waste. If material or contaminated runoff enters waterways, notify downstream users of potentially contaminated waters. Contact your local or federal environmental protection agency for specific recommendations. If employees are required to clean up spills, they must be properly trained and equipped. OSHA 1910.120(q) may be applicable.

Fire Extinguishing: Nitrates will increase the activity of an existing fire. Many will explode spontaneously. Poisonous gases, including nitrogen oxides, are produced in fire. If material or contaminated runoff enters waterways, notify downstream users of potentially contaminated waters. Notify local health and fire officials and pollution control agencies. From a secure, explosion-proof location, use water spray to cool exposed containers. If cooling streams are ineffective (venting sound increases in volume and pitch, tank discolors, or shows any signs of deforming), withdraw immediately to a secure position. If employees are expected to fight fires, they must be trained and equipped in OSHA 1910.156. The only respirators recommended for firefighting are self-contained breathing apparatuses that have full face-pieces and are operated in a pressure-demand or other positive-pressure mode.

Disposal Method Suggested: *Sodium nitrate:* the material is diluted to the recommended provisional limit in water.

(The pH is adjusted to between 6.5 and 9.1 and then the material can be discharged into sewers or natural streams.)

References

Sax, N. I. (Ed.). (1984). *Dangerous Properties of Industrial Materials Report*, 4, No. 2, 29–32

National Academy of Sciences. (1978). *Medical and Biologic Effects of Environmental Pollutants: Nitrates*. Washington, DC

World Health Organization. (1978). *Nitrates, Nitrites and N-Nitroso Compounds*, Environmental Health Criteria No. 5. Geneva

Nitric acid   N:0340

Molecular Formula: HNO$_3$

Synonyms: Acide nitrique (French); Acido nitrico (Spanish); Aluminum etch 16-1-1-2; Aluminum etch 82-3-15-0; Aluminum etch II; Aluminum etch III; Aqua fortis; Aqua regia; Azotic acid; Chrome etch KTI; Copper, brass brite dip 1127; Copper, brass brite dip 127; Copperlite RD-25; C-P 8 solution; Doped poly etch; Freckle etch; Hydrogen nitrate; Kovar bright dip (412X); Kovar bright dip (RDX-555); KTI aluminum etch I; KTI chrome etch; MAE etchants; Mixed acid etch (5-2-2); Mixed acid etch (6-1-1); NF solder stripper 3114-B; Nital; Nitraline; Nitric acid, red fuming; Nitric acid, white fuming; Nitrous fumes; Nitryl hydroxide; Passivation solution; Patclin 958; Poly etch 95%; Red fuming nitric acid; RFNA; RT-2 stripping solution; Salpetersaure (German); Silicon etch solution; Solder strip NP-A; Stress relief etch; Wet K-etch; WFNA; White fuming nitric acid

CAS Registry Number: 7697-37-2

RTECS® Number: QU5775000

UN/NA & ERG Number: UN2031 (other than red fuming, with >70% nitric acid)/157; UN2032 (fuming)/157

EC Number: 231-714-2 [*Annex I Index No.:* 007-004-00-1]

Regulatory Authority and Advisory Bodies

Department of Homeland Security Screening Threshold Quantity (pounds): *Release hazard* 15,000 (≥80.00% concentration); *Theft hazard* 400 (≥68.00% concentration).

US EPA Gene-Tox Program, Negative: Cell transform.—SA7/SHE

Air Pollutant Standard Set. See below, "Permissible Exposure Limits in Air" section.

OSHA 29CFR1910.119, Appendix A. Process Safety List of Highly Hazardous Chemicals, TQ = 500 lb (227 kg) (≥94.5%).

Clean Water Act: Section 311 Hazardous Substances/RQ 40CFR117.3 (same as CERCLA, see below); Section 313 Water Priority Chemicals (57FR41331, 9/9/92).

SUPERFUND/EPCRA 40CFR355, Extremely Hazardous Substances: TPQ = 1000 lb (454 kg).

Reportable Quantity (RQ): 1000 lb (454 kg).

EPCRA Section 313 Form R *de minimis* concentration reporting level: 1.0%.

US DOT 49CFR172.101, Inhalation Hazardous Chemical. Canada, WHMIS, Ingredients Disclosure List Concentration 1.0%.

European/International Regulations: Hazard Symbol: O, C; Risk phrases: R8; R35; Safety phrases: S1/2; S26; S36; S45 (see Appendix 4).

WGK (German Aquatic Hazard Class): 1—Low hazard to waters.

Description: Nitric acid is a colorless to light brown fuming liquid with an acrid, suffocating odor. Fuming nitric acid is a reddish fuming liquid. Fumes in moist air. Often used in an aqueous solution. Fuming nitric acid is concentrated nitric acid that contains dissolved nitrogen dioxide. Nitric acid is a solution of nitrogen dioxide, NO$_2$, in water and so-called fuming nitric acid contains an excess of NO$_2$ and is yellow to brownish-red in color. Molecular weight = 63.02; Specific gravity (H$_2$O:1) = 1.50 at 25°C; Boiling point = 82.8°C; Freezing/Melting point = −42°C (monohydrate); −19°C (trihydrate); Vapor pressure = 48 mmHg. Soluble in water.

Potential Exposure: Compound Description: Mutagen; Reproductive Effector; Human Data. Nitric acid is the second most important industrial acid and its production represents the sixth largest chemical industry in the United States. Nitric acid is used in chemicals, explosives, fertilizers, steel pickling, metal cleaning. The largest use of nitric acid is in the production of fertilizers. Almost 15% of the production goes into the manufacture of explosives, with the remaining 10% distributed among a variety of uses, such as etching, bright-dipping, electroplating, photoengraving, production of rocket fuel, and pesticide manufacture.

Incompatibilities: A strong oxidizer and strong acid. Reacts violently with combustible and reducing agents, carbides, hydrogen sulfide, turpentine, charcoal, alcohol, powdered metals, strong bases. Heat causes decomposition producing nitrogen oxides. Attacks some plastics. Corrosive to metals.

Permissible Exposure Limits in Air

Conversion factor: 1 ppm = 2.58 mg/m^3 at 25°C & 1 atm.

OSHA PEL: 2 ppm/5 mg/m^3 TWA.

NIOSH REL: 2 ppm/5 mg/m^3 TWA; 4 ppm/10 mg/m^3 STEL.

ACGIH TLV®[1]: 2 ppm/5.2 mg/m^3 TWA; 4 ppm/10 mg/m^3 STEL.

NIOSH IDLH: 25 ppm.

Protective Action Criteria (PAC)*

TEEL-0: 0.53 ppm

PAC-1: **0.53** ppm

PAC-2: **24** ppm

PAC-3: **92** ppm

*AEGLs (Acute Emergency Guideline Levels) & ERPGs (Emergency Response Planning Guideline) are in **bold face**.

DFG MAK: 2 ppm/5.2 mg/m^3 TWA; Peak Limitation Category I(1); Pregnancy Risk Group D.

Arab Republic of Egypt: TWA 2 ppm (5 mg/m^3), 1993; Australia: TWA 2 ppm (5 mg/m^3); STEL 4 ppm, 1993;

Austria: MAK 2 ppm (5 mg/m^3), 1999; Belgium: TWA 2 ppm (5.2 mg/m^3); STEL 4 ppm (10 mg/m^3), 1993; Denmark: TWA 2 ppm (5 mg/m^3), 1999; Finland: TWA 2 ppm (5 mg/m^3); STEL 5 ppm (13 mg/m^3), [skin], 1999; France: VME 2 ppm (5 mg/m^3), VLE 5 ppm (10 mg/m^3), 1999; Hungary: STEL 5 mg/m^3, 1993; Japan: 2 ppm (5.2 mg/m^3), 1999; Norway: TWA 2 ppm (5 mg/m^3), 1999; the Philippines: TWA 2 ppm (5 mg/m^3), 1993; Poland: MAC (TWA) 5 mg/m^3; MAC (STEL) 10 mg/m^3, 1999; Russia: TWA 2 ppm; STEL 2 mg/m^3, [skin], 1993; Sweden: NGV 2 ppm (5 mg/m^3), KTV 5 ppm (13 mg/m^3), 1999; Thailand: TWA 2 ppm (5 mg/m^3), 1993; Turkey: TWA 2 ppm (5 mg/m^3), 1993; United Kingdom: LTEL 2 ppm (5 mg/m^3); STEL 4 ppm (10 mg/m^3), 1993; Argentina, Bulgaria, Columbia, Jordan, South Korea, New Zealand, Singapore, Vietnam: ACGIH TLV®: STEL 4 ppm. Russia set a MAC of 0.4 mg/m^3 for nitric acid in ambient air in residential areas on a once-daily basis and 0.15 mg/m^3 on a daily average basis. Several states have set guidelines or standards for nitric acid in ambient air[60] ranging from 50−100 µg/m^3 (North Dakota) to 80 µg/m^3 (Virginia) to 100 µg/m^3 (Connecticut, Florida, New York and South Dakota) to 119 µg/m^3 (Nevada) to 125 µg/m^3 (South Carolina) to 1000 µg/m^3 (North Carolina).

Determination in Air: Use NIOSH Analytical Method #7903, acids, inorganic; OSHA Analytical Methods ID-127 and ID-165SG.

Permissible Concentration in Water: The EEC set a MAC of 50 mg NO$_3$/L in drinking water and a guideline level of 25 mg NO$_3$/L.[35] Russia[43] set a MAC of 0.4 mg/L in water bodies used for domestic purposes.

Routes of Entry: Inhalation, ingestion, skin and/or eye contact.

Harmful Effects and Symptoms

Short Term Exposure: Corrosive to the eyes, skin, and respiratory tract. Corrosive if ingested. Inhalation can cause pulmonary edema, a medical emergency that can be delayed for several hours. This can cause death. This compound is a primary irritant, and causes burns and ulceration of all tissues and membranes that it contacts. This includes burns to the eyes and skin by contact; burns to the mouth, throat, esophagus, and stomach by ingestion; and the entire respiratory tract by inhalation. Circulatory collapse and shock are often the immediate causes of death. The approximate minimum lethal dose is 5 mL for a 150-lb person. Persons with skin, eye, or cardiopulmonary disorders are at a greater risk. Signs and symptoms of acute ingestion of nitric acid may be severe and include increased salivation, intense thirst, difficulty in swallowing, chills, pain, and shock. Oral, esophageal, and stomach burns are common. Vomitus generally has a coffee-ground appearance. The potential for circulatory collapse is high following ingestion of nitric acid. Acute inhalation exposure may result in sneezing, hoarseness, choking, laryngitis, dyspnea (shortness of breath), respiratory tract irritation, and chest pain. Bleeding of nose and gums, ulceration of the nasal and oral mucosa, chronic bronchitis, and pneumonia may also

occur. If the eyes have come in contact with nitric acid, irritation, pain, swelling, corneal erosion, and blindness may occur. Dermal exposure may result in severe burns, pain, and dermatitis (red, inflamed skin).

Long Term Exposure: The mists or vapors may cause erosion of the teeth. May affect the lungs.

Points of Attack: Eyes, skin, respiratory system, teeth.

Medical Surveillance: NIOSH recommends that workers subject to nitric acid exposure have comprehensive preplacement and annual medical examinations including a 14″ × 17″ posterior−anterior chest X-ray; pulmonary function tests: forced vital capacity, forced expiratory volume (1 s); and a visual examination of the teeth for evidence of dental erosion.

First Aid: If this chemical gets into the eyes, remove any contact lenses at once and irrigate immediately for at least 15 min, occasionally lifting upper and lower lids. Seek medical attention immediately. If this chemical contacts the skin, remove contaminated clothing and wash immediately with soap and water. Seek medical attention immediately. If this chemical has been inhaled, remove from exposure, begin rescue breathing (using universal precautions, including resuscitation mask) if breathing has stopped and CPR if heart action has stopped. Transfer promptly to a medical facility. When this chemical has been swallowed, get medical attention. If victim is conscious, administer water or milk. Do not induce vomiting. Medical observation is recommended for 24−48 h after breathing overexposure, as pulmonary edema may be delayed. As first aid for pneumonitis or pulmonary edema, a doctor or authorized paramedic may consider administering a corticosteroid spray.

Personal Protective Methods: Wear protective gloves and clothing to prevent any reasonable probability of skin contact. Safety equipment suppliers/manufacturers can provide recommendations on the most protective glove/clothing material for your operation. Nitric acid: Neoprene™ (Up to 30%); Natural rubber or polyvinyl chloride (Up to 70%); Neoprene™/natural rubber or Saranex™ (more than 70%). For fuming nitric acid the following are recommended: Viton™, nitrile, Neoprene™, natural rubber, chlorobutyl, or Neoprene™/natural rubber. All protective clothing (suits, gloves, footwear, headgear) should be clean, available each day, and put on before work. Contact lenses should not be worn when working with this chemical. Wear splash-proof chemical goggles and face shield unless full face-piece respiratory protection is worn. Employees should wash immediately with soap when skin is wet or contaminated. Provide emergency showers and eyewash. Possibility of skin contact with liquids of pH < 2.5 or repeated or prolonged contact with liquids of pH > 2.5. Wear eye protection to prevent any possibility of eye contact. Employees should wash immediately when skin is wet or contaminated. Remove nonimpervious clothing immediately if wet or contaminated. Provide emergency showers and eyewash if liquids of pH < 2.5 are involved.

Respirator Selection: *25 ppm:* Sa:Cf (APF = 25) (any supplied-air respirator operated in a continuous-flow mode)

or CcrFS (APF = 50) [any chemical cartridge respirator with a full face-piece and cartridge(s) providing protection against the compound of concern] or GmFS (APF = 50) [any air-purifying, full-face-piece respirator (gas mask) with a chin-style, front- or back-mounted canister providing protection against the compound of concern] or SCBAF (APF = 50) (any self-contained breathing apparatus with a full face-piece) or SaF (APF = 50) (any supplied-air respirator with a full face-piece). *Emergency or planned entry into unknown concentrations or IDLH conditions:* SCBAF: Pd,Pp (APF = 10,000) (any self-contained breathing apparatus that has a full face-piece and is operated in a pressure-demand or other positive-pressure mode) or SaF: Pd,Pp: SCBA (any supplied-air respirator that has a full face-piece and is operated in a pressure-demand or other positive-pressure mode in combination with an auxiliary self-contained breathing apparatus operated in a pressure-demand or other positive-pressure mode). *Escape:* GmFS (APF = 50) [any air-purifying, full-face-piece respirator (gas mask) with a chin-style, front- or back-mounted canister providing protection against the compound of concern] or SCBAE (any appropriate escape-type, self-contained breathing apparatus).
Note: Substance reported to cause eye irritation or damage; may require eye protection. Only nonoxidizable sorbents are allowed (not charcoal).

Storage: Color Code—White: Corrosive or Contact Hazard; Store separately in a corrosion-resistant location. Prior to working with this chemical you should be trained on its proper handling and storage. Nitric acid must be stored to avoid contact with metallic powders, carbides, hydrogen sulfide, turpentine, or strong bases because violent reactions occur. Store in tightly closed containers in a cool, well-ventilated area away from heat. Heat may cause containers to burst and result in an escape of poisonous gases.

Shipping: Nitric acid other than red fuming, with >70% nitric acid or Nitric acid other than red fuming, with not >70% nitric acid require a shipping label of "CORROSIVE." They fall in DOT Hazard Class 8. Packing Groups are as follows: (with >70% nitric acid) is Group I; (with not >70% nitric acid) is Group II.
Nitric acid, red fuming, requires a shipping label of "CORROSIVE, OXIDIZER, POISONOUS/TOXIC MATERIALS." It falls in Hazard Class 8 and Packing Group I. Red fuming nitric acid carries a plus sign (+) symbol, indicating that the designated proper shipping name and hazard class of the material must always be shown whether or not the material or its mixtures or solutions meet the definitions of the class.

Spill Handling: Keep unnecessary people away. Do not touch spilled material; stop leak if you can do so without risk. Isolate the hazard area and deny entry. Stay upwind and keep out of low areas. Ventilate closed spaces before entering them. Ventilate area of spill or leak. Remove all ignition sources. Keep combustibles (wood, paper, oil, etc.) away from spilled material. Use water spray to reduce vapors; do not get water inside container. *Small spills:* flush area with flooding amounts of water. *Large spills:* dike far ahead of spill for later disposal. Flush with copious quantities of water and neutralize with alkaline material (such as soda ash, lime, etc.). Absorb liquids in vermiculite, dry sand, earth, peat, carbon, or a similar material and deposit in sealed containers. It may be necessary to contain and dispose of this chemical as a hazardous waste. If material or contaminated runoff enters waterways, notify downstream users of potentially contaminated waters. Contact your local or federal environmental protection agency for specific recommendations. If employees are required to clean up spills, they must be properly trained and equipped. OSHA 1910.120(q) may be applicable.

Initial isolation and protective action distances (UN 2032 fuming nitric acid)
Distances shown are likely to be affected during the first 30 min after materials are spilled and could increase with time. If more than one tank car, cargo tank, portable tank, or large cylinder involved in the incident is leaking, the protective action distance may need to be increased. You may need to seek emergency information from CHEMTREC at (800) 424-9300 or seek professional environmental engineering assistance from the US EPA Environmental Response Team at (908) 548-8730 (24-h response line).
Small spills (From a small package or a small leak from a large package)
Nitric acid, fuming and red fuming
First: Isolate in all directions (feet/meters)   100/30
Then: Protect persons downwind (miles/kilometers)
Day   0.1/0.2
Night   0.2/0.3
Large spills (From a large package or from many small packages)
First: Isolate in all directions (feet/meters)   500/150
Then: Protect persons downwind (miles/kilometers)
Day   0.4/0.6
Night   0.7/1.2

Fire Extinguishing: Firefighting gear (including SCBA) may not provide adequate protection. If exposure occurs, remove and isolate gear immediately and thoroughly decontaminate personnel. Nitric acid is noncombustible. However, it can increase the flammability of combustible, organic, and readily oxidizable materials, or even cause ignition of some of these with water. Poisonous gases, including nitrogen oxides, are produced in fire. Use water spray. *Small fires:* water, dry chemical, or soda ash. *Large fires:* flood fire area with water. Move container from fire area if you can do so without risk. For massive fire in cargo area, use unmanned hose holder or monitor nozzles; if this is impossible, withdraw from area and let fire burn. Vapors are heavier than air and will collect in low areas. Containers may explode in fire. Storage containers and parts of containers may rocket great distances, in many directions. If material or contaminated runoff enters waterways, notify downstream users of potentially contaminated waters. Notify local health and fire officials and pollution control agencies. From a secure, explosion-proof location, use water spray to cool exposed containers. If cooling streams are ineffective (venting sound increases in volume and pitch, tank discolors, or shows any signs of deforming), withdraw immediately to a secure position. If employees are expected to fight fires, they must be trained and equipped in OSHA 1910.156. The only respirators recommended for firefighting are self-contained breathing apparatuses that have full face-pieces and are operated in a pressure-demand or other positive-pressure mode.

Disposal Method Suggested: Soda ash-slaked lime is added to form the neutral solution of nitrate of sodium and calcium. This solution can be discharged after dilution with water.[22] Also, nitric acid can be recovered and reused in some cases as with acrylic fiber spin solutions. Consult with environmental regulatory agencies for guidance on acceptable disposal practices. Generators of waste containing this contaminant ($\geq$100 kg/mo) must conform with EPA regulations governing storage, transportation, treatment, and waste disposal.

References

National Institute for Occupational Safety and Health. (1976). *Criteria for a Recommended Standard: Occupational Exposure to Nitric Acid*, NIOSH Document No. 76–141

Sax, N. I. (Ed.). *Dangerous Properties of Industrial Materials Report*, 1, No. 5, 71–72 (1981) and 5, No. 3, 64–67 (1985)

US Environmental Protection Agency. (November 30, 1987). *Chemical Hazard Information Profile: Nitric Acid.* Washington, DC: Chemical Emergency Preparedness Program

New York State Department of Health. (March 1986). *Chemical Fact Sheet: Nitric Acid.* Version 2. NY: Bureau of Toxic Substance Assessment

New Jersey Department of Health and Senior Services. (May 2001). *Hazardous Substances Fact Sheet: Nitric Acid.* Trenton, NJ

Nitric oxide   N:0350

Molecular Formula: NO

Synonyms: Bioxyde d'azote (French); Monoxido de nitrogeno (Spanish); Nitric oxide; Nitrogen monoxide; NO (military designation); Oxido nitrico (Spanish); Oxyde nitrique (French); Stickmonoxyd (German)

CAS Registry Number: 10102-43-9; *(alt.)* 51005-21-1; *(alt.)* 90452-29-2; *(alt.)* 90880-94-7

RTECS® Number: QX525000

UN/NA & ERG Number: UN1660/124

EC Number: 233-271-0

Regulatory Authority and Advisory Bodies

Department of Homeland Security Screening Threshold Quantity (pounds): *Release hazard* 10,000 ($\geq$1.00% concentration); *Theft hazard* 15 ($\geq$3.83% concentration).

Air Pollutant Standard Set. See below, "Permissible Exposure Limits in Air" section.

OSHA 29CFR1910.119, Appendix A. Process Safety List of Highly Hazardous Chemicals, TQ = 250 lb.

Clean Air Act: Accidental Release Prevention/Flammable Substances, (Section 112[r], Table 3), TQ = 10,000 lb (4540 kg).

US EPA Hazardous Waste Number (RCRA No.): P076.

RCRA, 40CFR261, Appendix 8 Hazardous Constituents.

SUPERFUND/EPCRA 40CFR355, Extremely Hazardous Substances: TPQ = 100 lb (45.4 kg).

Reportable Quantity (RQ): 10 lb (4.54 kg).

US DOT 49CFR172.101, Inhalation Hazardous Chemical.

Canada, WHMIS, Ingredients Disclosure List Concentration 1.0%.

European/International Regulations: not listed in Annex 1.

WGK (German Aquatic Hazard Class): 1—Low hazard to waters.

Description: Nitric oxide is a colorless gas with a sharp, sweet odor; brown at high concentration in air. Odor threshold = 0.3–1.0 ppm. Shipped as a nonliquefied compressed gas. Molecular weight = 30.01; Boiling point = −151.6°C; Freezing/Melting point = −163.9°C; Relative vapor density (air = 1) = 1.04; Vapor Pressure = 34.2 atm. Slightly soluble in water; solubility = 5%.

Potential Exposure: Nitric oxide is used in the manufacture of nitric acid; it is also used in the bleaching of rayon; it is a raw material for nitrosyl halide preparation.

Incompatibilities: Explosive reaction with nitrogen trichloride, ozone, carbon disulfide, pentacarbonyl iron, chlorine monoxide. Incompatible with halogens, combustibles, metals, oil, alcohols, chlorinated hydrocarbons (e.g., trichloroethylene), reducing agents (such as NH_3), oxygen, fluorine, metals. Reacts with water to form nitric acid. Rapidly converted in air to nitrogen dioxide.

Permissible Exposure Limits in Air

Conversion factor: 1 ppm = 1.88 mg/m^3 at 25°C & 1 atm.

OSHA PEL: 25 ppm/30 mg/m^3 TWA.

NIOSH REL: 25 ppm/30 mg/m^3 TWA.

ACGIH TLV®[1]: 25 ppm/31 mg/m^3 TWA; BEI$_M$ issued for Methemoglobin inducers.

NIOSH IDLH: 100 ppm

Protective Action Criteria (PAC)*

TEEL-0: 0.5 ppm

PAC-1: **0.50** ppm

PAC-2: **12** ppm

PAC-3: **20** ppm

*AEGLs (Acute Emergency Guideline Levels) & ERPGs (Emergency Response Planning Guideline) are in **bold face**.

DFG MAK: 0.5 ppm/0.63 mg/m^3 TWA; Peak Limitation Category II(2); Pregnancy Risk Group D.

Austria: MAK 25 ppm (30 mg/m^3), 1999; Denmark: TWA 25 ppm (30 mg/m^3), 1999; France: VME 25 ppm (30 mg/m^3), 1999; Norway: TWA 25 ppm (30 mg/m^3), 1999; Poland: MAC (TWA) 5 mg/m^3; MAC (STEL) 10 mg/m^3, 1999;

Sweden: NGV 25 ppm (30 mg/m^3), KTV 50 ppm (60 mg/m^3), 1999; Thailand: TWA 25 ppm (30 mg/m^3), 1993; the Netherlands: MAC-TGG 30 mg/m^3, 2003; Argentina, Bulgaria, Columbia, Jordan, South Korea, New Zealand, Singapore, Vietnam: ACGIH TLV®: TWA 25 ppm. Several states have set guidelines or standards for nitric oxide in ambient air[60] ranging from 0.3 mg/m^3 (North Dakota) to 0.5 mg/m^3 (Virginia) to 0.6 mg/m^3 (Connecticut) to 0.714 mg/m^3 (Nevada).

Determination in Air: Use NIOSH Analytical Method (IV) #6014, Nitric oxide and nitrogen dioxide, 6014; OSHA Analytical Method #ID-109.

Routes of Entry: Inhalation, ingestion, skin and/or eye contact.

Harmful Effects and Symptoms

Short Term Exposure: Irritates the eyes, skin. Strong respiratory tract irritant. High levels can interfere with the blood's ability to carry oxygen (methemoglobinemia). Nitric oxide forms acids in the respiratory system which are irritating and can cause congestion in the lungs or pulmonary edema, a medical emergency that can be delayed for several hours. This can cause death. Nitric oxide can cause death or permanent injury after a very short exposure to small quantities. Can cause unconsciousness. Concentrations of 60−150 ppm cause immediate irritation of the nose and throat with coughing and burning in the throat and chest 6−24 h after exposure; labored breathing and unconsciousness may result. Concentrations of 100−150 ppm are dangerous for a short exposure of 30−60 min. Concentrations of 200−700 ppm may be fatal after very short exposure. Nitric oxide can cause death due to blockage of gas exchange in lungs. Initially, symptoms include slight coughing, fatigue, and nausea at high concentrations; coughing, choking, headache, nausea, abdominal pain, and shortness of breath are seen. Latent symptoms are uneasiness, restlessness, rapid and shallow breathing, bluing of skin, lips and fingernail beds, anxiety, mental confusion; and finally loss of consciousness.

Long Term Exposure: Can irritate the lungs; bronchitis may develop. Can cause headache, nausea, vomiting, fatigue, mental confusion, unconsciousness, and death. Increased blood methemoglobin levels.

Points of Attack: Eyes, skin, respiratory system, blood, central nervous system.

Medical Surveillance: NIOSH lists the following tests: chest X-ray, electrocardiogram, on workers over 40 years, expired air, pulmonary function tests: forced vital capacity, forced expiratory volume (1 s). Consider the points of attack in preplacement and periodic physical examinations. Blood methemoglobin levels.

First Aid: If this chemical gets into the eyes, remove any contact lenses at once and irrigate immediately for at least 15 min, occasionally lifting upper and lower lids. Seek medical attention immediately. If this chemical contacts the skin, remove contaminated clothing and wash immediately with soap and water. Seek medical attention immediately. If this chemical has been inhaled, remove from exposure, begin rescue breathing (using universal precautions, including resuscitation mask) if breathing has stopped and CPR if heart action has stopped. Transfer promptly to a medical facility. When this chemical has been swallowed, get medical attention. Give large quantities of water and induce vomiting. Do not make an unconscious person vomit. Effects may be delayed; keep victim under observation. Medical observation is recommended for 24−48 h after breathing overexposure, as pulmonary edema may be delayed. As first aid for pulmonary edema, a doctor or authorized paramedic may consider administering a corticosteroid spray.

Note to physician: Treat for methemoglobinemia. Spectrophotometry may be required for precise determination of levels of methemoglobin in urine.

Personal Protective Methods: Wear protective gloves and clothing to prevent any reasonable probability of skin contact. Safety equipment suppliers/manufacturers can provide recommendations on the most protective glove/clothing material for your operation. All protective clothing (suits, gloves, footwear, headgear) should be clean, available each day, and put on before work. Contact lenses should not be worn when working with this chemical. Wear nonvented, impact resistant, gas-proof chemical goggles and face shield unless full face-piece respiratory protection is worn. Employees should wash immediately with soap when skin is wet or contaminated. Provide emergency showers and eyewash; pressure, pressure-demand, full face-piece self-contained breathing apparatus (SCBA) or pressure-demand supplied air respirator with escape SCBA and a fully-encapsulating, chemical-resistant suit.

Respirator Selection: NIOSH/OSH: *1000 ppm:* Sa:Cf (APF = 25) (any supplied-air respirator operated in a continuous-flow mode) or CcrFS (APF = 50) [any chemical cartridge respirator with a full face-piece and cartridge(s) providing protection against the compound of concern] or PaprS (APF = 25) [any powered, air-purifying respirator with cartridge(s) providing protection against the compound of concern] or GmFS (APF = 50) [any air-purifying, full-face-piece respirator (gas mask) with a chin-style, front- or back-mounted canister providing protection against the compound of concern] or Sa (APF = 10) (any supplied-air respirator) or SCBAF (APF = 50) (any self-contained breathing apparatus with a full face-piece). *Emergency or planned entry into unknown concentrations or IDLH conditions:* SCBAF: Pd,Pp (APF = 10,000) (any self-contained breathing apparatus that has a full face-piece and is operated in a pressure-demand or other positive-pressure mode) or SaF: Pd,Pp: ASCBA (APF = 10,000) (any supplied-air respirator that has a full face-piece and is operated in a pressure-demand or other positive-pressure mode in combination with an auxiliary self-contained breathing apparatus operated in a pressure-demand or other positive-pressure mode). *Escape:* GmFS (APF = 50) [any air-purifying, full-face-piece respirator (gas mask) with a chin-style, front- or back-mounted canister providing protection against the compound of concern] or SCBAE (any appropriate escape-type, self-contained breathing apparatus).

Note: Only nonoxidizable sorbents are allowed (not charcoal).

Storage: Color Code—Yellow Stripe: Reactivity Hazard; Store separately in an area isolated from flammables, combustibles, or other yellow coded materials. Prior to working with this chemical you should be trained on its proper handling and storage. High concentrations cause a deficiency of oxygen with the risk of unconsciousness or death. Check that oxygen content is at least 19% before entering storage or spill area. Procedures for the handling, use, and storage of cylinders should be in compliance with OSHA 1910.101 and 1910.169, as with the recommendations of the Compressed Gas Association.

Shipping: Nitric oxide requires a shipping label of "POISON GAS, OXIDIZER, CORROSIVE." It falls in Hazard Class 2.3. It is a violation of transportation regulations to refill compressed gas cylinders without the express written permission of the owner.

Spill Handling: If in a building, evacuate building and confine vapors by closing doors and shutting down HVAC systems. Restrict persons not wearing protective equipment from area of spill or leak until cleanup is complete. Remove all ignition sources. Ventilate area of spill or leak to disperse the gas. Wear chemical protective suit with self-contained breathing apparatus to combat spills. Stay upwind and use water spray to "knock down" vapor; contain runoff. Stop the flow of gas, if it can be done safely from a distance. If source is a cylinder and the leak cannot be stopped in place, remove the leaking cylinder to a safe place, and repair leak or allow cylinder to empty. Keep this chemical out of confined spaces, such as a sewer, because of the possibility of explosion, unless the sewer is designed to prevent the buildup of explosive concentrations. If employees are required to clean up spills, they must be properly trained and equipped. OSHA 1910.120(q) may be applicable.

Initial isolation and protective action distances

Distances shown are likely to be affected during the first 30 min after materials are spilled and could increase with time. If more than one tank car, cargo tank, portable tank, or large cylinder involved in the incident is leaking, the protective action distance may need to be increased. You may need to seek emergency information from CHEMTREC at (800) 424-9300 or seek professional environmental engineering assistance from the US EPA Environmental Response Team at (908) 548-8730 (24-h response line).

Nitric oxide or nitric oxide, compressed

Small spills (From a small package or a small leak from a large package)

First: Isolate in all directions (feet/meters)   100/30
Then: Protect persons downwind (miles/kilometers)
Day   0.1/0.2
Night   0.4/0.6

Large spills (From a large package or from many small packages)

First: Isolate in all directions (feet/meters)   300/100
Then: Protect persons downwind (miles/kilometers)
Day   0.4/0.6
Night   1.4/2.3

Fire Extinguishing: Nitric oxide itself does not burn. Will react with water or steam to produce heat and corrosive fumes. When heated to decomposition, highly toxic fumes of nitrogen oxides are emitted. May ignite other combustible materials (wood, paper, oil, etc.). Mixture with fuels may explode. Do not extinguish the fire unless the flow of gas can be stopped and any remaining gas is out of the line. Specially trained personnel may use fog lines to cool exposures and let the fire burn itself out. Vapors are heavier than air and will collect in low areas. Vapor explosion and poison hazard indoors, outdoors, or in sewers. Containers may explode in fire. Storage containers and parts of containers may rocket great distances, in many directions. If material or contaminated runoff enters waterways, notify downstream users of potentially contaminated waters. Notify local health and fire officials and pollution control agencies. From a secure, explosion-proof location, use water spray to cool exposed containers. If cooling streams are ineffective (venting sound increases in volume and pitch, tank discolors, or shows any signs of deforming), withdraw immediately to a secure position. If cylinders are exposed to excessive heat from fire or flame contact, withdraw immediately to a secure location. If employees are expected to fight fires, they must be trained and equipped in OSHA 1910.156. The only respirators recommended for firefighting are self-contained breathing apparatuses that have full face-pieces and are operated in a pressure-demand or other positive-pressure mode.

Disposal Method Suggested: Incineration with added hydrocarbon fuel, controlled so as to produce elemental nitrogen, CO_2, and water. Consult with environmental regulatory agencies for guidance on acceptable disposal practices. Generators of waste containing this contaminant ($\geq$100 kg/mo) must conform with EPA regulations governing storage, transportation, treatment, and waste disposal.

References

National Institute for Occupational Safety and Health. (1976). *Criteria for a Recommended Standard: Occupational Exposure to Oxides of Nitrogen*, NIOSH Document No. 76-149. Washington, DC

World Health Organization. (1977). *Oxides of Nitrogen, Environmental Health Criteria No. 4*. Geneva, Switzerland

Sax, N. I. (Ed.). (1981). *Dangerous Properties of Industrial Materials Report*, 1, No. 5, 73—74

US Environmental Protection Agency. (November 30, 1987). *Chemical Hazard Information Profile: Nitric Oxide*. Washington, DC: Chemical Emergency Preparedness Program

New Jersey Department of Health and Senior Services. (August, 1999). *Hazardous Substances Fact Sheet: Nitric Oxide*. Trenton, NJ

Nitrilotriacetic acid   N:0360

Molecular Formula: $C_6H_9NO_6$

Common Formula: $N(CH_2COOH)_3$

Synonyms: Acetic acid, nitrilotri-; Acido nitrilotriacetico (Spanish); Aminotriacetic acid; *N,N*-Bis(carboxymethyl) glycine; Chel 300; Complexon I; Glycine, *N,N*-Bis(carboxymethyl)-; Hampshire NTA acid; Komplexon I; NCI-C02766; Nitrilo-2,2',2''-triacetic acid; NTA; Titriplex I; Tri (carboxymethyl)amine; Triglycine; Triglycollamic acid; Trilon A; α,α',α''-Trimethylaminetricarboxylic acid; Versene NTA acid

CAS Registry Number: 139-13-9; *(alt.)* 26627-44-1; *(alt.)* 26627-45-2; *(alt.)* 80751-51-5

RTECS® Number: AJ0175000

UN/NA & ERG Number: UN2811 (toxic solid, organic, n.o.s.)/154

EC Number: 205-355-7

Regulatory Authority and Advisory Bodies

Carcinogenicity: NCI: Carcinogenesis Bioassay (feed); clear evidence: mouse, rat; NTP: 11th Report on Carcinogens, 2002: Reasonably anticipated to be a human carcinogen; IARC: Animal Sufficient Evidence; Human Inadequate Evidence, *possibly carcinogenic to humans*, Group 2B, 1999.

US EPA Gene-Tox Program, Positive: Carcinogenicity—mouse/rat; Negative: *D. melanogaster* sex-linked lethal.

EPCRA Section 313 Form R *de minimis* concentration reporting level: 0.1%.

California Proposition 65 Chemical: Cancer 1/1/88.

WGK (German Aquatic Hazard Class): 2—Hazard to waters.

Description: Nitrilotriacetic acid is a crystalline compound. Molecular weight = 191.16; Boiling point = 167°C; Freezing/Melting point = 242°C (decomposition). Hazard Identification (based on NFPA-704 M Rating System): Health 1, Flammability 1, Reactivity 0. Slightly soluble in water.

Potential Exposure: Compound Description: Tumorigen, Mutagen. Nitrilotriacetic acid (NTA) was used as a phosphate replacement in laundry detergents in the late 1960s. In 1971, the use of NTA was discontinued. The possibility of resumed use arose in 1980. NTA is now used in laundry detergents in states where phosphates are banned. NTA is also used as a boiler feed-water additive at a maximum use level of 5 ppm of trisodium salt. Currently, the remaining nondetergent uses of NTA are for water treatment, textile treatment; metal plating and cleaning; and pulp and paper processing.

Incompatibilities: Aqueous solution is an acid. Violent reaction with strong bases.

Permissible Exposure Limits in Air

Protective Action Criteria (PAC)

TEEL-0: 35 mg/m^3

PAC-1: 100 mg/m^3

PAC-2: 500 mg/m^3

PAC-3: 500 mg/m^3

DFG MAK: avoid simultaneous exposure to Iron (Fe) compounds.

Permissible Concentration in Water: No criteria set for drinking water; NTA addition to boiler feed-water is limited to 5 ppm by FDA. NTA levels in the US drinking water prior to NTA's discontinued use in detergents was estimated by EPA to have ranged from 0.20 to 24.5 μg/L. NTA is rapidly degraded under aerobic conditions at temperatures above 5°C, and biodegradation does not lead to the formation of persistent intermediates.

Determination in Water: Octanol−water coefficient: Log $K_{ow} = -3.75$.

Routes of Entry: Inhalation, ingestion, skin and/or eye contact.

Harmful Effects and Symptoms

Short Term Exposure: Irritates the eyes and skin. This material is moderately toxic. The LD50 (oral-rat) is 1470 mg/kg.

Long Term Exposure: A Confirmed Human Carcinogen.

First Aid: Skin Contact: Flood all areas of body that have contacted the substance with water. Do not wait to remove contaminated clothing; do it under the water stream. Use soap to help assure removal. Isolate contaminated clothing when removed to prevent contact by others.[52] *Eye Contact:* Remove any contact lenses at once. Flush eyes well with copious quantities of water or normal saline for at least 20−30 min. Seek medical attention. *Inhalation:* Leave contaminated area immediately; breathe fresh air. Proper respiratory protection must be supplied to any rescuers. If coughing, difficult breathing or any other symptoms develop, seek medical attention at once, even if symptoms develop many hours after exposure. *Ingestion:* If convulsions are not present, give a glass or two of water or milk to dilute the substance. Assure that the person's airway is unobstructed and contact a hospital or poison center immediately for advice on whether or not to induce vomiting.

Personal Protective Methods: Wear protective gloves and clothing to prevent any reasonable probability of skin contact. Safety equipment suppliers/manufacturers can provide recommendations on the most protective glove/clothing material for your operation. All protective clothing (suits, gloves, footwear, headgear) should be clean, available each day, and put on before work. Contact lenses should not be worn when working with this chemical. Wear dust-proof chemical goggles and face shield unless full face-piece respiratory protection is worn. Employees should wash immediately with soap when skin is wet or contaminated. Provide emergency showers and eyewash.

Respirator Selection: Specific respirator(s) have not been recommended by NIOSH. However, based on potential carcinogenicity the following might be considered:

At any detectable concentration: SCBAF: Pd,Pp (APF = 10,000) (any NIOSH/MSHA- or European Standard EN 149-approved self-contained breathing apparatus that has a full face-piece and is operated in a pressure-demand or other positive-pressure mode) or SaF: Pd,Pp: ASCBA (APF = 10,000) (any supplied-air respirator that has a full face-piece and is operated in a pressure-demand or other positive-pressure mode in combination with an auxiliary, self-contained breathing apparatus operated in a pressure-demand or other positive-pressure mode). *Escape:* 100 F (APF = 50) (any air-purifying, full-face-piece respirator with an N100, R100, or P100 filter) or SCBAE (any appropriate escape-type, self-contained breathing apparatus).

Storage: Color Code—Blue: Health Hazard/Poison: Store in a secure poison location. Prior to working with this chemical you should be trained on its proper handling and storage. Store in a refrigerator or in a cool, dry place. A regulated, marked area should be established where this chemical is handled, used, or stored in compliance with OSHA Standard 1910.1045.

Shipping: Toxic solids, organic, n.o.s. requires a shipping label of "POISONOUS/TOXIC MATERIALS." It falls in Hazard Class 6.1 and Packing Group III.

Spill Handling: Evacuate persons not wearing protective equipment from area of spill or leak until cleanup is complete. Remove all ignition sources. Dampen spilled material with toluene to avoid dust. Or, use HEPA vacuum or wet method to reduce dust during cleanup. Do not dry sweep. Collect powdered material in the most convenient and safe manner and deposit in sealed containers. Ventilate area after cleanup is complete. It may be necessary to contain and dispose of this chemical as a hazardous waste. If material or contaminated runoff enters waterways, notify downstream users of potentially contaminated waters. Contact your local or federal environmental protection agency for specific recommendations. If employees are required to clean up spills, they must be properly trained and equipped. OSHA 1910.120(q) may be applicable.

Fire Extinguishing: This chemical is a combustible solid. Use dry chemical, carbon dioxide, water spray, or alcohol foam extinguishers. Poisonous gases are produced in fire. If material or contaminated runoff enters waterways, notify downstream users of potentially contaminated waters. Notify local health and fire officials and pollution control agencies. From a secure, explosion-proof location, use water spray to cool exposed containers. If cooling streams are ineffective (venting sound increases in volume and pitch, tank discolors, or shows any signs of deforming), withdraw immediately to a secure position. If employees are expected to fight fires, they must be trained and equipped in OSHA 1910.156. The only respirators recommended for firefighting are self-contained breathing apparatuses that have full face-pieces and are operated in a pressure-demand or other positive-pressure mode.

Disposal Method Suggested: Dissolve or mix the material with a combustible solvent and burn in a chemical incinerator equipped with an afterburner and scrubber. All federal, state, and local environmental regulations must be observed.

5-Nitroacenaphthene   N:0370

Molecular Formula: $C_{12}H_9O_2$
Synonyms: 1-Amino-2-methoxy-5-nitrobenzene; 2-Amino-1-methoxy-4-nitrobenzene; 3-Amino-4-methoxynitrobenzene; 2-Amino-4-nitroanisole; *o*-Anisidine nitrate; *o*-Anisidine, 5-nitro-; Azoamine scarlet; Azoamine scarlet K; Azogene Ecarlate R; Azoic diazo component 13 base; Benzenamine, C.I. 37130; Benzenamine, 2-methoxy-5-nitro-; C.I. 37130; C.I. Azoic diazo component 13; Fast scarlet R; 1-Methoxy-2-amino-4-nitrobenzene; 2-Methoxy-5-nitro-; 2-Methoxy-5-nitroaniline; 2-Methoxy-5-nitrobenzenamine; NCI-C01934; 3-Nitro-6-methoxyaniline; 5-Nitro-2-methoxyaniline
CAS Registry Number: 602-87-9
RTECS® Number: AB1060000
EC Number: 210-025-0 [*Annex I Index No.:* 609-037-00-2]
Regulatory Authority and Advisory Bodies
Carcinogenicity: IARC: Animal Sufficient Evidence; Human Inadequate Evidence, *possibly carcinogenic to humans*, Group 2B, 1987; NCI: Carcinogenesis Bioassay (feed); clear evidence: mouse, rat, 1987
EPCRA Section 313 Form R *de minimis* concentration reporting level: 1.0%.
California Proposition 65 Chemical: Cancer 4/1/88.
European/International Regulations: Hazard Symbol: T; Risk phrases: R45; Safety phrases: S53; S45 (see Appendix 4).
WGK (German Aquatic Hazard Class): No value assigned.
Description: 5-Nitroacenaphthene is a yellow crystalline solid. Molecular weight = 199.22; Freezing/Melting point = 103−104°C. Hazard Identification (based on NFPA-704 M Rating System): Health 1, Flammability 1, Reactivity 0.
Potential Exposure: Used in organic synthesis.
Incompatibilities: 5-Nitroacenaphthene is an aromatic hydrocarbon (nitro compound). It may be flammable or explosive. Incompatible with strong oxidizers, alkalis which may increase the thermal sensitivity of the substance.
Permissible Exposure Limits in Air
No TEEL available.
DFG MAK: Carcinogen Category 2.
Routes of Entry: Inhalation, ingestion, skin and/or eye contact. Absorbed through the skin.
Harmful Effects and Symptoms
Short Term Exposure: May reduce the blood's ability to carry oxygen (methemoglobinemia) with cyanosis, fatigue, dizziness, headache. May affect the central nervous system.
Long Term Exposure: Little is known aside from the fact that it is a carcinogen. May cause liver and kidney damage. Poisoning may cause anemia, cyanosis, fatigue, insomnia, weight loss.
First Aid: *Skin Contact:* Flood all areas of body that have contacted the substance with water. Do not wait to remove contaminated clothing; do it under the water stream. Use

soap to help assure removal. Isolate contaminated clothing when removed to prevent contact by others.[52] *Eye Contact:* Remove any contact lenses at once. Flush eyes well with copious quantities of water or normal saline for at least 20–30 min. Seek medical attention. *Inhalation:* Leave contaminated area immediately; breathe fresh air. Proper respiratory protection must be supplied to any rescuers. If coughing, difficult breathing or any other symptoms develop, seek medical attention at once, even if symptoms develop many hours after exposure. *Ingestion:* If convulsions are not present, give a glass or two of water or milk to dilute the substance. Assure that the person's airway is unobstructed and contact a hospital or poison center immediately for advice on whether or not to induce vomiting.

Personal Protective Methods: Wear protective gloves and clothing to prevent any reasonable probability of skin contact. Safety equipment suppliers/manufacturers can provide recommendations on the most protective glove/clothing material for your operation. Teflon™, Silvershield™, Viton™, Viton™/chlorobutyl rubber, chlorinated polyethylene, and polyvinyl alcohol are among the recommended protective materials. All protective clothing (suits, gloves, footwear, headgear) should be clean, available each day, and put on before work. Contact lenses should not be worn when working with this chemical. Wear dust-proof chemical goggles and face shield unless full face-piece respiratory protection is worn. Employees should wash immediately with soap when skin is wet or contaminated. Provide emergency showers and eyewash.

Respirator Selection: Follow regulations in OSHA 29CFR1910.134 or European Standard EN149. Use a NIOSH/MSHA- or European Standard EN149-approved respirator; or use an approved supplied-air respirator with a full face-piece operated in the positive-pressure mode, or with a full face-piece, hood, or helmet in the continuous-flow mode; or use a NIOSH/MSHA- or European Standard EN149-approved self-contained breathing apparatus with a full face-piece operated in pressure-demand or other positive-pressure mode.

Storage: Color Code—Green: General storage may be used. Prior to working with this chemical you should be trained on its proper handling and storage. Store in a refrigerator or a cool, dry place. A regulated, marked area should be established where this chemical is handled, used, or stored in compliance with OSHA Standard 1910.1045.

Spill Handling: Evacuate persons not wearing protective equipment from area of spill or leak until cleanup is complete. Remove all ignition sources. Dampen spilled material with acetone. Collect powdered material in the most convenient and safe manner and deposit in sealed containers. Ventilate area after cleanup is complete. It may be necessary to contain and dispose of this chemical as a hazardous waste. If material or contaminated runoff enters waterways, notify downstream users of potentially contaminated waters. Contact your local or federal environmental protection agency for specific recommendations. If employees are required to clean up spills, they must be properly trained and equipped. OSHA 1910.120(q) may be applicable.

Fire Extinguishing: This chemical is a combustible solid. Use dry chemical, carbon dioxide, water spray, or alcohol foam extinguishers. Poisonous gases, including nitrogen oxides, are produced in fire. If material or contaminated runoff enters waterways, notify downstream users of potentially contaminated waters. Notify local health and fire officials and pollution control agencies. From a secure, explosion-proof location, use water spray to cool exposed containers. If cooling streams are ineffective (venting sound increases in volume and pitch, tank discolors, or shows any signs of deforming), withdraw immediately to a secure position. If employees are expected to fight fires, they must be trained and equipped in OSHA 1910.156. The only respirators recommended for firefighting are self-contained breathing apparatuses that have full face-pieces and are operated in a pressure-demand or other positive-pressure mode.

Disposal Method Suggested: Careful incineration in an incinerator equipped with afterburner and scrubbers.[22]

p-Nitroaniline   N:0380

Molecular Formula: $C_6H_6N_2O_2$
Common Formula: $H_2NC_6H_4NO_2$
Synonyms: *p*-Aminonitrobenzene; 1-Amino-4-nitrobenzene; Aniline, *p*-nitro-; Aniline, 4-nitro-; Azoamine red ZH; Azofix red GG salt; Azoic diazo component 37; Benzenamine, 4-nitro-; C.I. 37035; C.I. azoic diazo component 37; C.I. Developer 17; Developer P; Devol red GG; Diazo fast red GG; Fast red 2G base; Fast red 2G salt; Fast red base; Fast red base 2J; Fast red base GG; Fast red GG base; Fast red GG salt; Fast red MP base; Fast red P base; Fast red P salt; Fast red salt 2J; Fast red salt GG; Naphtoelan red GG base; NCI-C60786; Nitoraniline-*p*; 4-Nitranbine; *p*-Nitraniline; Nitrazol CF extra; *p*-Nitroanilina (Spanish); *p*-Nitroaniline; 4-Nitroaniline; 4-Nitrobenzenamine; *p*-Nitrophenylamine; PNA; Red 2G base; Shinnippon fast red GG base
CAS Registry Number: 100-01-6
RTECS® Number: BY7000000
UN/NA & ERG Number: UN1661/153
EC Number: 202-810-1 [*Annex I Index No.:* 612-012-00-9]
Regulatory Authority and Advisory Bodies
Carcinogenicity: NCI: Carcinogenesis Studies (gavage); equivocal evidence: rat.
US EPA Gene-Tox Program, Negative: Sperm morphology—mouse.
Air Pollutant Standard Set. See below, "Permissible Exposure Limits in Air" section.
OSHA 29CFR1910.119, Appendix A. Process Safety List of Highly Hazardous Chemicals, TQ = 5000 lb (2270 kg).
US EPA Hazardous Waste Number (RCRA No.): P077.
RCRA, 40CFR261, Appendix 8 Hazardous Constituents.

RCRA 40CFR268.48; 61FR15654, Universal Treatment Standards: Wastewater (mg/L), 0.028; Nonwastewater (mg/kg), 28.

Reportable Quantity (RQ): 5000 lb (2270 kg).

EPCRA Section 313 Form R *de minimis* concentration reporting level: 1.0%.

Canada, WHMIS, Ingredients Disclosure List Concentration 1.0%.

European/International Regulations (*all isomers*): Hazard Symbol: T, N; Risk phrases: R23/24/25; R33; R52/53; Safety phrases: S1/2; S28; S36/37; S45; S61 (see Appendix 4).

WGK (German Aquatic Hazard Class): 2—Hazard to waters.

Description: *p*-Nitroaniline consists of yellow crystals with a pungent, faint ammonia-like odor. Molecular weight = 138.14; Specific gravity (H_2O:1) = 1.42; Boiling point = 332.2°C; Freezing/Melting point = 146.1°C; Vapor pressure = 0.000002 mmHg at 20°C; Flash point = 198.9°C; Autoignition temperature = 510°C. Hazard Identification (based on NFPA-704 M Rating System): Health 3, Flammability 1, Reactivity 0. Slightly soluble in water.

Potential Exposure: Compound Description: Tumorigen, Mutagen; Reproductive Effector. *p*-Nitroaniline is used as an intermediate in the manufacture of dyes, antioxidants, pharmaceuticals, and pesticides.

Incompatibilities: A combustible liquid. A strong oxidizer. Violent reaction with reducing agents, strong acids, combustibles, strong oxidizers. May explode on heating. May result in spontaneous heating of organic materials in the presence of moisture or strong oxidizers.

Permissible Exposure Limits in Air
OSHA PEL: 1 ppm/6 mg/m^3 TWA [skin].
NIOSH REL: 3 mg/m^3 TWA [skin].
ACGIH TLV®[1]: 3 mg/m^3 TWA [skin]; not classifiable as a human carcinogen; BEI$_M$ issued for Methemoglobin inducers.
DFG MAK: [skin]; Carcinogen Category 3A.
NIOSH IDLH: 300 mg/m^3.
Protective Action Criteria (PAC)
TEEL-0: 6 mg/m^3
PAC-1: 9 mg/m^3
PAC-2: 300 mg/m^3
PAC-3: 300 mg/m^3
Arab Republic of Egypt: TWA 1 ppm (6 mg/m^3), [skin], 1993; Australia: TWA 3 mg/m^3, [skin], 1993; Austria: MAK 1 ppm (6 mg/m^3), [skin], 1999; Belgium: TWA 3 mg/m^3, [skin], 1993; Denmark: TWA 0.5 ppm (3 mg/m^3), [skin], 1999; Finland: TWA 1 ppm (6 mg/m^3); STEL 3 ppm (18 mg/m^3), [skin], 1993; France: VME 3 mg/m^3, [skin], 1999; the Netherlands: MAC-TGG 6 mg/m^3, [skin], 2003; Japan: 3 mg/m^3, [skin], 1999; Norway: TWA 1 mg/m^3, 1999; the Philippines: TWA 1 ppm (6 mg/m^3), [skin], 1993; Poland: MAC (TWA) 3 mg/m^3; MAC (STEL) 10 mg/m^3, 1999; Russia: STEL 0.1 mg/m^3, [skin], 1993; Switzerland: MAK-W 0.5 ppm (3 mg/m^3), [skin], 1999; Turkey: TWA

1 ppm (6 mg/m^3), [skin], 1993; United Kingdom: TWA 6 mg/m^3, [skin], 2000; Argentina, Bulgaria, Columbia, Jordan, South Korea, New Zealand, Singapore, Vietnam: ACGIH TLV®: not classifiable as a human carcinogen. Several states have set guidelines or standards for *p*-nitroaniline in ambient air[60] ranging from zero (Connecticut) to 6.0 µg/m^3 (New York) to 7.143 µg/m^3 (Kansas) to 15 µg/m^3 (South Carolina) to 30 µg/m^3 (Florida and North Dakota) to 50 µg/m^3 (Virginia) to 71 µg/m^3 (Nevada).

Determination in Air: Use NIOSH Analytical Method (IV) #5033.

Determination in Water: Octanol−water coefficient: Log $K_{ow} = 2.7$.

Routes of Entry: Inhalation, ingestion, skin and/or eye contact.

Harmful Effects and Symptoms
Short Term Exposure: *p*-Nitroaniline can affect you when breathed in and by passing through your skin. Exposure by skin contact or breathing can interfere with the ability of the blood to carry oxygen (methegloblnemia) and kidney impairment. This can cause headaches, dizziness, cyanosis (a blue color to the skin and lips), trouble breathing, and even collapse and death.

Long Term Exposure: Because this is a mutagen, handle it as a possible cancer-causing substance with extreme caution. May affect the blood, causing the formation of methemoglobin, low blood count (anemia), kidney and liver damage. People with "G-6-P-D deficiency" may be at higher risk for developing health problems following exposure.

Points of Attack: Respiratory system, blood, heart, liver, kidneys.

Medical Surveillance: NIOSH lists the following tests: whole blood (chemical/metabolite), Methemoglobin; Complete blood count; liver function tests. Before beginning employment and at regular times after that, for those with frequent or potentially high exposures, the following is recommended: kidney function tests. Also, tests for the condition called "G-6-P-D deficiency."

First Aid: If this chemical gets into the eyes, remove any contact lenses at once and irrigate immediately for at least 15 min, occasionally lifting upper and lower lids. Seek medical attention immediately. If this chemical contacts the skin, remove contaminated clothing and wash immediately with soap and water. Seek medical attention immediately. If this chemical has been inhaled, remove from exposure, begin rescue breathing (using universal precautions, including resuscitation mask) if breathing has stopped and CPR if heart action has stopped. Transfer promptly to a medical facility. When this chemical has been swallowed, get medical attention. Give large quantities of water and induce vomiting. Do not make an unconscious person vomit.

Note to physician: Treat for methemoglobinemia. Spectrophotometry may be required for precise determination of levels of methemoglobine in urine.

Personal Protective Methods: Wear protective gloves and clothing to prevent any reasonable probability of skin

contact. Safety equipment suppliers/manufacturers can provide recommendations on the most protective glove/clothing material for your operation. All protective clothing (suits, gloves, footwear, headgear) should be clean, available each day, and put on before work. Contact lenses should not be worn when working with this chemical. Wear dust-proof chemical goggles and face shield unless full face-piece respiratory protection is worn. Employees should wash immediately with soap when skin is wet or contaminated. Provide emergency showers and eyewash.

Respirator Selection: NIOSH: *Up to 30 mg/m³:* Sa (APF = 10) (any supplied-air respirator).* *Up to 75 mg/m³:* Sa:Cf (APF = 25) (any supplied-air respirator operated in a continuous-flow mode).* *Up to 150 mg/m³:* SCBAF (APF = 50) (any self-contained breathing apparatus with a full face-piece) or SaF (APF = 50) (any supplied-air respirator with a full face-piece). *Up to 300 mg/m³:* SaF: Pd,Pp (APF = 2000) (any supplied-air respirator that has a full face-piece and is operated in a pressure-demand or other positive-pressure mode). *Emergency or planned entry into unknown concentrations or IDLH conditions:* SCBAF: Pd, Pp (APF = 10,000) (any self-contained breathing apparatus that has a full face-piece and is operated in a pressure-demand or other positive-pressure mode) or SaF: Pd,Pp; ASCBA (any supplied-air respirator that has a full face-piece and is operated in a pressure-demand or other positive-pressure mode in combination with an auxiliary self-contained positive-pressure breathing apparatus). *Escape:* GmFOv100 [any air-purifying, full-face-piece respirator (gas mask) with a chin-style, front- or back-mounted organic vapor canister having an N100, R100, or P100 filter] or SCBAE (any appropriate escape-type, self-contained breathing apparatus).

*Substance reported to cause eye irritation or damage; may require eye protection.

Storage: Color Code—Blue: Health Hazard/Poison: Store in a secure poison location. Prior to working with this chemical you should be trained on its proper handling and storage. *p*-Nitroaniline must be stored to avoid contact with oxidizers (such as perchlorates, peroxides, permanganates, chlorates and nitrates), reducers, and strong acids (such as hydrochloric, sulfuric, and nitric), since violent reactions occur. Store in tightly closed containers in a cool, well-ventilated area. Protect storage containers from physical damage. Wherever *p*-nitroaniline is used, handled, manufactured, or stored, use explosion-proof electrical equipment and fittings.

Shipping: Nitroanilines require a shipping label of "POISONOUS/TOXIC MATERIALS." They fall in DOT Hazard Class 6.1 and Packing Group II.

Spill Handling: Evacuate persons not wearing protective equipment from area of spill or leak until cleanup is complete. Remove all ignition sources. Use HEPA vacuum or wet method to reduce dust during cleanup. Do not dry sweep. Collect powdered material in the most convenient and safe manner and deposit in sealed containers. Ventilate area after cleanup is complete. It may be necessary to contain and dispose of this chemical as a hazardous waste. If material or contaminated runoff enters waterways, notify downstream users of potentially contaminated waters. Contact your local or federal environmental protection agency for specific recommendations. If employees are required to clean up spills, they must be properly trained and equipped. OSHA 1910.120(q) may be applicable.

Fire Extinguishing: This chemical is a combustible solid. Use dry chemical, carbon dioxide, water spray, or alcohol foam extinguishers. Poisonous gases, including sulfur dioxide and nitrogen oxides, are produced in fire. If material or contaminated runoff enters waterways, notify downstream users of potentially contaminated waters. Notify local health and fire officials and pollution control agencies. From a secure, explosion-proof location, use water spray to cool exposed containers. If cooling streams are ineffective (venting sound increases in volume and pitch, tank discolors, or shows any signs of deforming), withdraw immediately to a secure position. If employees are expected to fight fires, they must be trained and equipped in OSHA 1910.156. The only respirators recommended for firefighting are self-contained breathing apparatuses that have full face-pieces and are operated in a pressure-demand or other positive-pressure mode.

Disposal Method Suggested: Incineration (982°C, 2.0 s minimum) with scrubbing for nitrogen oxides abatement. Consult with environmental regulatory agencies for guidance on acceptable disposal practices. Generators of waste containing this contaminant (≥100 kg/mo) must conform with EPA regulations governing storage, transportation, treatment, and waste disposal.

Reference

New Jersey Department of Health and Senior Services. (April 2004). *Hazardous Substances Fact Sheet: p-Nitroaniline.* Trenton, NJ

5-Nitro-*o*-anisidine   N:0390

Molecular Formula: $C_7H_8N_2O_3$
Common Formula: $C_6H_3(OCH_3)(NH_2)(NO_2)$
Synonyms: 1-Amino-2-methoxy-5-nitrobenzene; 2-Amino-1-methoxy-4-nitrobenzene; 3-Amino-4-methoxynitrobenzene; 2-Amino-4-nitroanisole; *o*-Anisidine nitrate; *o*-Anisidine, 5-nitro-; Azoamine scarlet ; Azoamine scarlet K; Azogene Ecarlate R; Azoic diazo component 13 base; Benzenamine, C.I. 37130; Benzenamine, 2-methoxy-5-nitro-; C.I. 37130; C.I. azoic diazo component 13; Fast scarlet R; 1-Methoxy-2-amino-4-nitrobenzene; 2-Methoxy-5-nitro-; 2-Methoxy-5-nitroaniline; 2-Methoxy-5-nitrobenzenamine; NCI-C01934; 3-Nitro-6-methoxyaniline; 5-Nitro-2-methoxyaniline
CAS Registry Number: 99-59-2
RTECS® Number: BZ7175000

UN/NA & ERG Number: UN3143 (Dyes, solid, toxic, n.o.s.)/151

EC Number: 202-770-5

Regulatory Authority and Advisory Bodies

Carcinogenicity: IARC: Animal Limited Evidence; Human No Evidence, *not classifiable as carcinogenic to humans,* Group 3, 1987; NCI: Carcinogenesis Bioassay (feed); clear evidence: mouse, rat, 1978. 5-Nitro-*o*-anisidine was removed from the NTP 6th Report on Carcinogens as a substance "reasonably anticipated to be a human carcinogen" in 1991, when NTP concluded there was insufficient evidence of carcinogenicity. Delisted from the California Proposition 65 list as of December 8, 2006.

Air Pollutant Standard Set. See below, "Permissible Exposure Limits in Air" section.

EPCRA Section 313 Form R *de minimis* concentration reporting level: 1.0%.

California Proposition 65 Chemical: Cancer 10/1/89; Delisted 12/8/06.

Canada, WHMIS, Ingredients Disclosure List Concentration 0.1%.

European/International Regulations: not listed in Annex 1.

WGK (German Aquatic Hazard Class): 2—Hazard to waters.

Description: 5-Nitro-*o*-anisidine is an orange-red crystalline compound. Molecular weight = 168.17; Freezing/Melting point = 118°C. Hazard Identification (based on NFPA-704 M Rating System): Health 2, Flammability 1, Reactivity 0.

Potential Exposure: 5-Nitro-*o*-anisidine is a chemical intermediate in the production of C.I. Pigment red 23, which is used as a colorant for commodities, such as printing inks, interior latex paints; lacquers, rubber, plastics, floor coverings; paper coating; and textiles. It is also used with other C.I. coupling components to produce various hues of red, brown, yellow, and violet on cotton, silk, acetate, and nylon.

Permissible Exposure Limits in Air

No standards or TEEL available.

Incompatibilities: Strong acids, acid chlorides, acid anhydrides, chloroformates.

Permissible Exposure Limits in Air: No OELs have been established, but this chemical can be absorbed through the skin. Several states have set guidelines or standards for 5-nitro-*o*-ansidine in ambient air[60] ranging from zero (North Dakota) to 0.08 μg/m^3 (Rhode Island).

Routes of Entry: Inhalation, ingestion, skin and/or eye contact.

Harmful Effects and Symptoms

Short Term Exposure: High levels can interfere with the blood's ability to carry oxygen (methemoglobinemia) causing dizziness, cyanosis. Higher levels can cause trouble breathing, collapse, and even death.

Long Term Exposure: This compound is a proved carcinogen in experimental animals. Related chemicals can cause allergic skin rash, and irritate the nose, throat, and lungs. It is not known whether this substance can cause these effects.

Points of Attack: Blood. *Cancer site:* bladder.

Medical Surveillance: Consider the points of attack in pre-placement and periodic physical examinations. Blood hemoglobin level. Complete blood count.

First Aid: If this chemical gets into the eyes, remove any contact lenses at once and irrigate immediately for at least 15 min, occasionally lifting upper and lower lids. Seek medical attention immediately. If this chemical contacts the skin, remove contaminated clothing and wash immediately with soap and water. Seek medical attention immediately. If this chemical has been inhaled, remove from exposure, begin rescue breathing (using universal precautions, including resuscitation mask) if breathing has stopped and CPR if heart action has stopped. Transfer promptly to a medical facility. When this chemical has been swallowed, get medical attention. Give large quantities of water and induce vomiting. Do not make an unconscious person vomit.

Personal Protective Methods: Wear protective gloves and clothing to prevent any reasonable probability of skin contact. Safety equipment suppliers/manufacturers can provide recommendations on the most protective glove/clothing material for your operation. All protective clothing (suits, gloves, footwear, headgear) should be clean, available each day, and put on before work. Contact lenses should not be worn when working with this chemical. Wear dust-proof chemical goggles and face shield unless full face-piece respiratory protection is worn. Employees should wash immediately with soap when skin is wet or contaminated. Provide emergency showers and eyewash.

Respirator Selection: Follow regulations in OSHA 29CFR1910.134 or European Standard EN149. Use a NIOSH/MSHA- or European Standard EN149-approved respirator; or use an approved supplied-air respirator with a full face-piece operated in the positive-pressure mode, or with a full face-piece, hood, or helmet in the continuous-flow mode; or use a NIOSH/MSHA- or European Standard EN149-approved self-contained breathing apparatus with a full face-piece operated in pressure-demand or other positive-pressure mode.

Storage: Color Code—Blue: Health Hazard/Poison: Store in a secure poison location. Prior to working with this chemical you should be trained on its proper handling and storage. Store in tightly closed containers in a cool, well-ventilated area away from strong acids, acid chlorides, acid anhydrides, and chloroformates. Where possible, automatically pump liquid from drums or other storage containers to process containers. Sources of ignition, such as smoking and open flames, are prohibited where this chemical is handled, used, or stored. Metal containers involving the transfer of 5 gallons or more of this chemical should be grounded and bonded. Drums must be equipped with self-closing valves, pressure vacuum bungs, and flame arresters. Use only nonsparking tools and equipment, especially when opening and closing containers of this chemical. Wherever this chemical is used, handled, manufactured, or stored, use explosion-proof electrical equipment and fittings.

A regulated, marked area should be established where this chemical is handled, used, or stored in compliance with OSHA Standard 1910.1045.

Shipping: Dye intermediates, solid, toxic, n.o.s. require a shipping label of "POISONOUS/TOXIC MATERIALS." They fall in Hazard Class 6.1.

Spill Handling: Evacuate persons not wearing protective equipment from area of spill or leak until cleanup is complete. Remove all ignition sources. Use HEPA vacuum or wet method to reduce dust during cleanup. Do not dry sweep. Collect powdered material in the most convenient and safe manner and deposit in sealed containers. Ventilate area after cleanup is complete. It may be necessary to contain and dispose of this chemical as a hazardous waste. If material or contaminated runoff enters waterways, notify downstream users of potentially contaminated waters. Contact your local or federal environmental protection agency for specific recommendations. If employees are required to clean up spills, they must be properly trained and equipped. OSHA 1910.120(q) may be applicable.

Fire Extinguishing: This chemical is a combustible solid. Use dry chemical, carbon dioxide, water spray, or alcohol foam extinguishers. Poisonous gases, including nitrogen oxides, are produced in fire. If material or contaminated runoff enters waterways, notify downstream users of potentially contaminated waters. Notify local health and fire officials and pollution control agencies. From a secure, explosion-proof location, use water spray to cool exposed containers. If cooling streams are ineffective (venting sound increases in volume and pitch, tank discolors, or shows any signs of deforming), withdraw immediately to a secure position. If employees are expected to fight fires, they must be trained and equipped in OSHA 1910.156. The only respirators recommended for firefighting are self-contained breathing apparatuses that have full face-pieces and are operated in a pressure-demand or other positive-pressure mode.

Disposal Method Suggested: Incineration (982°C, 2.0 s minimum) with scrubbing for nitrogen oxides abatement.

References

US Environmental Protection Agency. (1979). *Chemical Hazard Information Profile: 5-Nitro-o-Anisidine.* Washington, DC

New Jersey Department of Health and Senior Services. (September 2004). *Hazardous Substances Fact Sheet: 5-Nitro-o-anisidine.* Trenton, NJ

Nitrobenzene   N:0400

Molecular Formula: $C_6H_5NO_2$

Synonyms: Benzene, nitro-; Essence of mirbane; Essence of myrbane; Mirbane oil; NCI-C60082; Nitrobenceno (Spanish); Nitrobenzol; Nitrobenzol, L; Nitro, liquid; Oil of mirbane; Oil of myrbane

CAS Registry Number: 98-95-3

RTECS® Number: DA6475000

UN/NA & ERG Number: UN1662/152

EC Number: 202-716-0 [*Annex I Index No.:* 609-003-00-7]

Regulatory Authority and Advisory Bodies

Department of Homeland Security Screening Threshold Quantity (pounds): *Explosive hazard*; *Theft hazard* 100 (ACG concentration).

Carcinogenicity: NTP: 11th Report on Carcinogens, 2004: Reasonably anticipated to be a human carcinogen; IARC: Animal Sufficient Evidence; Human Inadequate Evidence, *possibly carcinogenic to humans*, Group 2B, 1996; EPA: Likely to produce cancer in humans; NTP: Reasonably anticipated to be a human carcinogen.

US EPA TSCA Section 8(e) Risk Notification, 8EHQ-0293-8703; 8EHQ-0293-8723; 8EHQ-0293-8724; 8EHQ-0892-9102; 8EHQ-0892-9103.

Air Pollutant Standard Set. See below, "Permissible Exposure Limits in Air" section.

Water Pollution Standard Proposed (EPA)[6] (Russia)[35,43] (Maine, Kansas).[61]

Clean Air Act: Hazardous Air Pollutants (Title I, Part A, Section 112).

Clean Water Act: Section 311 Hazardous Substances/RQ 40CFR117.3 (same as CERCLA, see below); 40CFR401.15 Section 307 Toxic Pollutants; Section 313 Water Priority Chemicals (57FR41331, 9/9/92).

US EPA Hazardous Waste Number (RCRA No.): U169.

RCRA, 40CFR261, Appendix 8 Hazardous Constituents.

RCRA 40CFR268.48; 61FR15654, Universal Treatment Standards: Wastewater (mg/L), 0.068; Nonwastewater (mg/kg), 14.

RCRA 40CFR264, Appendix 9; TSD Facilities Ground Water Monitoring List. Suggested test method(s) (PQL µg/L): 8090 (40); 8270 (10).

Safe Drinking Water Act: Regulated chemical (47 FR 9352).

SUPERFUND/EPCRA 40CFR355, Extremely Hazardous Substances: TPQ = 10,000 lb (4540 kg).

Reportable Quantity (RQ): 1000 lb (454 kg).

EPCRA Section 313 Form R *de minimis* concentration reporting level: 1.0%.

California Proposition 65 Chemical: Cancer 8/26/97; Developmental/Reproductive toxin (male) 3/30/10.

Canada, WHMIS, Ingredients Disclosure List Concentration 1.0%.

European/International Regulations: Hazard Symbol: T, N; Risk phrases: R23/24/25; R40; R48/23/24; R51/53; R62; Safety phrases: S1/2; S28; S36/37; S45; S61 (see Appendix 4).

WGK (German Aquatic Hazard Class): 2—Hazard to waters.

Description: Nitrobenzene is a pale yellow to dark brown oily liquid whose odor resembles bitter almonds (or black paste shoe polish). The odor threshold is 0.044; 0.02 ppm (NJ). Molecular weight = 123.12; Specific gravity ($H_2O:1$) = 1.20; Boiling point = 210°C; Freezing/Melting point = 5.6°C; Vapor pressure = 0.3 mmHg at 25°C; Flash point = 88°C (cc); Autoignition temperature = 480°C. Explosive limits: LEL = 1.8% at 93°C; UEL: 40%. Hazard Identification (based on NFPA-704 M Rating System): Health 3, Flammability 2, Reactivity 1. Practically insoluble in water; solubility = 0.2%.

Potential Exposure: Compound Description: Agricultural Chemical; Tumorigen, Mutagen; Reproductive Effector; Human Data; Natural Product; Primary Irritant. Nitrobenzene is used in the manufacture of explosives and aniline dyes and as solvent and intermediate. It is also used in floor polishes, leather dressings and polishing compounds, paint solvents, and to mask other unpleasant odors. Substitution reactions with nitrobenzene are used to form *m*-derivatives. Pregnant women may be especially at risk with respect to nitrobenzene as with many other chemical compounds, due to transplacental passage of the agent. Individuals with glucose-6-phosphate dehydrogenase deficiency may also be special risk groups. Additionally, because alcohol ingestion or chronic alcoholism can lower the lethal or toxic dose of nitrobenzene, individuals consuming alcoholic beverages may be at risk.

Incompatibilities: Concentrated nitric acid, nitrogen tetroxide, caustics, phosphorus pentachloride, chemically active metals, such as tin or zinc. Violent reaction with strong oxidizers and reducing agents. Attacks many plastics. Forms thermally unstable compounds with many organic and inorganic compounds.

Permissible Exposure Limits in Air
Conversion factor: 1 ppm = 5.04 mg/m^3 at 25°C & 1 atm.
OSHA PEL: 1 ppm/5 mg/m^3 TWA [skin].
NIOSH REL: 1 ppm/5 mg/m^3 TWA [skin].
ACGIH TLV®[1]: 1 ppm TWA [skin]; BEI: 5 mg[total *p*-nitrophenol]/g creatinine in urine/end-of-shift at end-of-work-week; 1.5% methemoglobin in blood, end-of-shift. Confirmed animal carcinogen with unknown relevance to humans.
NIOSH IDLH: 200 ppm.
Protective Action Criteria (PAC)
TEEL-0: 1 ppm
PAC-1: 2.5 ppm
PAC-2: 19.9 ppm
PAC-3: 200 ppm
DFG MAK: [skin] Carcinogen Category 3B; BAT: 100 μg [Aniline, released from aniline−hemoglobin conjugate]/L in blood, for long-term exposure, after several shifts (sampling time) [skin].
Arab Republic of Egypt: TWA 1 ppm (5 mg/m^3), [skin], 1993; Australia: TWA 1 ppm (5 mg/m^3), [skin], 1993; Austria: MAK 1 ppm (5 mg/m^3), [skin], 1999; Belgium: TWA 1 ppm (5 mg/m^3), [skin], 1993; Denmark: TWA 1 ppm (5 mg/m^3), [skin], 1999; Finland: TWA 1 ppm (5 mg/m^3); STEL 3 ppm (15 mg/m^3), [skin], 1999; France: VME 1 ppm (5 mg/m^3), 1999; the Netherlands: MAC-TGG 5 mg/m^3, [skin], 2003; Japan: 1 ppm (5 mg/m^3), [skin], 1999; Norway: TWA 1 ppm (5 mg/m^3), 1999; Poland: MAC (TWA) 3 mg/m^3; MAC (STEL) 10 mg/m^3, 1999;

Russia: TWA 1 ppm; STEL 3 mg/m^3, [skin], 1993; Sweden: NGV 1 ppm (5 mg/m^3), KTV 2 ppm (10 mg/m^3), [skin], 1999; Switzerland: MAK-W 1 ppm (5 mg/m^3), KZG-W 2 ppm (10 mg/m^3), [skin], 1999; Turkey: TWA 1 ppm (5 mg/m^3), [skin], 1993; United Kingdom: TWA 1 ppm (5.1 mg/m^3); STEL 2 ppm (10 mg/m^3), [skin], 2000; New Zealand, Singapore, Vietnam: ACGIH TLV®: confirmed animal carcinogen with unknown relevance to humans. The Czech Republic: TWA 1 ppm (5 mg/m^3).[35] Russia[43] has set 0.008 mg/m^3 (8 μg/m^3) as an MAC for ambient air in residential areas. Several states have set guidelines or standards for nitrobenzene in ambient air[60] ranging from 6.8 μg/m^3 (Massachusetts) to 16.7 μg/m^3 (New York) to 25 μg/m^3 (North Carolina and South Carolina) to 50 μg/m^3 (North Dakota) to 80 μg/m^3 (Virginia) to 100 μg/m^3 (Connecticut) to 119 μg/m^3 (Nevada).

Determination in Air: Use NIOSH Analytical Method (IV) #2017. See also Method #2005.

Permissible Concentration in Water: *To protect freshwater aquatic life:* 27,000 μg/L on an acute toxicity basis. *To protect saltwater aquatic life:* 6680 μg/L on an acute toxicity basis. To protect humans—30 μg/L based on organoleptic considerations and 19,800 μg/L based on toxicity considerations.[6] Guidelines for nitrobenzene in drinking water have been set by Maine at 1.4 μg/L and by Kansas at 5.0 μg/L.[61]

Determination in Water: Methylene chloride extraction followed by exchange to toluene, gas chromatography with flame ionization detection (EPA Method 609) or gas chromatography plus mass spectrometry (EPA Method 625). Octanol−water coefficient: Log K_{ow} = 1.86.

Routes of Entry: Inhalation, ingestion, skin and/or eye contact. Penetrates the skin.

Harmful Effects and Symptoms
Short Term Exposure: *Inhalation:* Has caused headache and nausea at 3−6 ppm. 40 ppm may cause intoxication. Symptoms due to decreased ability of blood to carry oxygen may include blue coloration of lips, fingernails, and earlobes; headache, dizziness, loss of coordination; labored breathing; rapid heartbeat; vomiting, coma, and death. Symptoms may be delayed up to 4 h. *Skin:* Easily absorbed through the skin and contributes significantly to symptoms listed under inhalation. May also cause irritation and allergic sensitization. Death has been reported from skin absorption. *Eyes:* May cause irritation and damage to the cornea. *Ingestion:* May cause symptoms listed under inhalation and include burning of throat, abdominal pain, bloody diarrhea, and enlarged spleen and liver. Death has resulted from as little as 0.4 mL (0.05 liq. oz), about 8 drops. High levels of exposure can interfere with the blood's ability to carry oxygen (methemoglobinemia). Signs and symptoms of acute exposure to nitrobenzene may be severe and include cyanosis (blue tint to the skin and mucous membranes), tachycardia (rapid heart rate), Hypotension (low blood pressure), and cardiac arrhythmias. Respiratory depression and

respiratory failure may also occur. Headache, lethargy, weakness, vertigo (dizziness), severe depression, and coma may be noted. Gastrointestinal symptoms include nausea and vomiting. Urine and vomitus may have the odor of bitter almonds.

Long Term Exposure: Occupational exposure to 40 ppm for 6 months has caused intoxication and anemia. Can cause skin allergy. Can cause jaundice, liver and spleen damage, fatigue, bladder distress, nerve damage. May affect the blood forming organs, causing anemia. Exposure may affect vision (acuity and contraction of fields).

Points of Attack: Blood, liver, kidneys, cardiovascular system, skin.

Medical Surveillance: NIOSH lists the following tests: whole blood (chemical/metabolite); whole blood (chemical/metabolite), carboxyhemoglobin; whole blood (chemical/metabolite), methemoglobin; whole blood (chemical/metabolite), methemoglobin, end-of-shift; complete blood count; urine (chemical/metabolite); urine (chemical/metabolite), end-of-shift at end-of-work-week; urine (chemical/metabolite), end-of-work-week. Preemployment and periodic examinations should be concerned particularly with a history of dyscrasias, reactions to medications, alcohol intake, eye disease, skin, and cardiovascular status. Liver and renal functions should be evaluated periodically, as well as blood and general health. Follow methemoglobin levels until normal in all cases of suspected cyanosis. The metabolites in urine, *p*-nitro- and *p*-aminophenol, can be used as evidence of exposure. Liver function tests.

Note: Alcohol ingestion increases the toxic effects of nitrobenzene. Persons with blood, heart, liver or lung diseases should not work with this substance.

First Aid: If this chemical gets into the eyes, remove any contact lenses at once and irrigate immediately for at least 15 min, occasionally lifting upper and lower lids. Seek medical attention immediately. If this chemical contacts the skin, remove contaminated clothing and wash immediately with soap and water. Seek medical attention immediately. If this chemical has been inhaled, remove from exposure, begin rescue breathing (using universal precautions, including resuscitation mask) if breathing has stopped and CPR if heart action has stopped. Transfer promptly to a medical facility. When this chemical has been swallowed, get medical attention. Give large quantities of water and induce vomiting. Do not make an unconscious person vomit.

Note to physician: Treat for methemoglobinemia. Spectrophotometry may be required for precise determination of levels of methemoglobin in urine.

Personal Protective Methods: Wear protective gloves and clothing to prevent any reasonable probability of skin contact. Safety equipment suppliers/manufacturers can provide recommendations on the most protective glove/clothing material for your operation. Chlorinated polyethylene, polyvinyl alcohol, teflon, Viton™/chlorobutyl, and Silvershield™ are recommended. All protective clothing (suits, gloves, footwear, headgear) should be clean, available each day, and put on before work. Contact lenses should not be worn when working with this chemical. Wear splash-proof chemical goggles and face shield unless full face-piece respiratory protection is worn. Employees should wash immediately with soap when skin is wet or contaminated. Provide emergency showers and eyewash.

Respirator Selection: 10 ppm: CcrOv (APF = 10) [any chemical cartridge respirator with organic vapor cartridge(s)] Sa (APF = 10) (any supplied-air respirator). *25 ppm:* Sa:Cf (APF = 25) (any supplied-air respirator operated in a continuous-flow mode) or PaprOv (APF = 25) [any powered, air-purifying respirator with organic vapor cartridge(s)]. *50 ppm:* CcrFOv (APF = 50) [any air-purifying, full-face-piece respirator (gas mask) with a chin-style, front- or back-mounted acid gas canister] or GmFOv (APF = 50) [any air-purifying, full-face-piece respirator (gas mask) with a chin-style, front- or back-mounted organic vapor canister] or PaprTOv (APF = 50) [any powered, air-purifying respirator with a tight-fitting face-piece and organic vapor cartridge(s)] or SCBAF (APF = 50) (any self-contained breathing apparatus with a full face-piece) or SaF (APF = 50) (any supplied-air respirator with a full face-piece). *200 ppm:* SaF: Pd,Pp (APF = 2000) (any supplied-air respirator that has a full face-piece and is operated in a pressure-demand or other positive-pressure mode). *Emergency or planned entry into unknown concentrations or IDLH conditions:* SCBAF: Pd,Pp (APF = 10,000) (any self-contained breathing apparatus that has a full face-piece and is operated in a pressure-demand or other positive-pressure mode) or SaF: Pd,Pp: ASCBA (APF = 10,000) (any supplied-air respirator that has a full face-piece and is operated in a pressure-demand or other positive-pressure mode in combination with an auxiliary self-contained breathing apparatus operated in a pressure-demand or other positive-pressure mode). *Escape:* GmFOv (APF = 50) [any air-purifying, full-face-piece respirator (gas mask) with a chin-style, front- or back-mounted organic vapor canister] or SCBAE (any appropriate escape-type, self-contained breathing apparatus).

Note: Substance reported to cause eye irritation or damage; may require eye protection.

Storage: Color Code—Blue: Health Hazard/Poison: Store in a secure poison location. Prior to working with this chemical you should be trained on its proper handling and storage. Before entering confined space where this chemical may be present, check to make sure that an explosive concentration does not exist. Nitrobenzene must be stored to avoid contact with strong acids (such as hydrochloric, sulfuric, and nitric) and chemically active metals (such as potassium, sodium, magnesium, and zinc) caustic nitrogen tetroxide or silver perchlorate, since violent reactions occur. Sources of ignition, such as smoking and open flames are prohibited where this chemical is handled, used, or stored. Metal containers involving the transfer of 5 gallons or more of this chemical should be grounded and bonded. Drums must be equipped with self-closing valves, pressure vacuum bungs, and flame arresters.

Use only nonsparking tools and equipment, especially when opening and closing containers of this chemical. Wherever this chemical is used, handled, manufactured, or stored, use explosion-proof electrical equipment and fittings.

Shipping: This compound requires a shipping label of "POISONOUS/TOXIC MATERIALS." It falls in Hazard Class 6.1 and Packing Group II.

Spill Handling: Evacuate and restrict persons not wearing protective equipment from area of spill or leak until cleanup is complete. Remove all ignition sources. Establish forced ventilation to keep levels below explosive limit. Absorb liquids in vermiculite, dry sand, earth, peat, carbon, or a similar material and deposit in sealed containers. Keep this chemical out of a confined space, such as a sewer, because of the possibility of an explosion, unless the sewer is designed to prevent the buildup of explosive concentrations. It may be necessary to contain and dispose of this chemical as a hazardous waste. If material or contaminated runoff enters waterways, notify downstream users of potentially contaminated waters. Contact your local or federal environmental protection agency for specific recommendations. If employees are required to clean up spills, they must be properly trained and equipped. OSHA 1910.120(q) may be applicable.

Fire Extinguishing: This chemical is a combustible liquid. Poisonous gases, including nitrogen oxides, are produced in fire. Use dry chemical, carbon dioxide, or alcohol foam extinguishers. Vapors are heavier than air and will collect in low areas. Vapors may travel long distances to ignition sources and flashback. Vapors in confined areas may explode when exposed to fire. Containers may explode in fire. Storage containers and parts of containers may rocket great distances, in many directions. If material or contaminated runoff enters waterways, notify downstream users of potentially contaminated waters. Notify local health and fire officials and pollution control agencies. From a secure, explosion-proof location, use water spray to cool exposed containers. If cooling streams are ineffective (venting sound increases in volume and pitch, tank discolors, or shows any signs of deforming), withdraw immediately to a secure position. If employees are expected to fight fires, they must be trained and equipped in OSHA 1910.156. The only respirators recommended for firefighting are self-contained breathing apparatuses that have full face-pieces and are operated in a pressure-demand or other positive-pressure mode.

Disposal Method Suggested: Incineration (982°C, 2.0 s minimum) with scrubbing for nitrogen oxides abatement.[22] Consult with environmental regulatory agencies for guidance on acceptable disposal practices. Generators of waste containing this contaminant (≥100 kg/mo) must conform with EPA regulations governing storage, transportation, treatment, and waste disposal.

References

National Institute for Occupational Safety and Health. (October 1977). *Information Profiles on Potential Occupational Hazards: Nitrobenzenes*, Report PB-276, 678. Rockville, MD. 198–211

US Environmental Protection Agency. (1979). *Chemical Hazard Information Profile: Nitrobenzene*. Washington, DC
US Environmental Protection Agency. (1980). *Nitrobenzene: Ambient Water Quality Criteria.* Washington, DC
US Environmental Protection Agency. (April 30, 1980). *Nitrobenzene: Health and Environmental Effects Profile No. 134.* Washington, DC: Office of Solid Waste
US Environmental Protection Agency. (November 30, 1987). *Chemical Hazard Information Profile: Nitrobenzene.* Washington, DC: Chemical Emergency Preparedness Program
Sax, N. I. (Ed.). (1985). *Dangerous Properties of Industrial Materials Report*, 5, No. 6, 77–81
New York State Department of Health. (March 1986). *Chemical Fact Sheet: Nitrobenzene.* Albany, NY: Bureau of Toxic Substance Assessment
New Jersey Department of Health and Senior Services. (April 2004). *Hazardous Substances Fact Sheet: Nitrobenzene.* Trenton, NJ

4-Nitrobiphenyl N:0410

Molecular Formula: $C_{12}H_9NO_2$
Common Formula: $C_6H_5C_6H_4NO_2$
Synonyms: BA 2794; 1,1′-Biphenyl, 4-nitro-; Biphenyl, 4-nitro-; *p*-Nitrobiphenyl; *p*-Nitrodiphenyl; 4-Nitrodiphenyl; *p*-Nitrofenol (Spanish); 4-Nitrofenol (Spanish); 1-Nitro-4-phenylbenzene; *p*-Phenylnitrobenzene; 4-Phenylnitrobenzene; PNB
CAS Registry Number: 92-93-3
RTECS® Number: DV5600000
UN/NA & ERG Number: UN2811 (toxic solid, organic, n.o.s.)/154
EC Number: 202-204-7 [*Annex I Index No.:* 609-039-00-3]
Regulatory Authority and Advisory Bodies
Carcinogenicity: IARC: Animal Inadequate Evidence; Human No Evidence, *not classifiable as carcinogenic to humans*, Group 3, 1987; NIOSH: Potential occupational carcinogen.
US EPA Gene-Tox Program, Positive: SHE—clonal assay; Cell transform.—SA7/SHE; Positive: Host-mediated assay; *E. coli* polA without S9; Positive: Histidine reversion—Ames test; Positive/limited: Carcinogenicity—mouse/rat; Negative: *N. crassa*—aneuploidy; *S. cerevisiae*—homozygosis.
OSHA, 29CFR1910 Specifically Regulated Chemicals (See CFR 1910.1003).
Air Pollutant Standard Set. See below, "Permissible Exposure Limits in Air" section.
Clean Air Act: Hazardous Air Pollutants (Title I, Part A, Section 112).
Reportable Quantity (RQ): 1 lb (0.454 kg).
EPCRA Section 313 Form R *de minimis* concentration reporting level: 0.1%.

California Proposition 65 Chemical: Cancer 4/1/88.

Canada, WHMIS, Ingredients Disclosure List Concentration 0.1%.

European/International Regulations: Hazard Symbol: T, N; Risk phrases: R 45; R51/53; Safety phrases: S53; S45; S61 (see Appendix 4).

WGK (German Aquatic Hazard Class): No value assigned.

Description: 4-Nitrobiphenyl exists as yellow plates or needles. Molecular weight = 199.22; Boiling point = 340°C; Freezing/Melting point = 113.9°C; Flash point = 143.3°C. Hazard Identification (based on NFPA-704 M Rating System): Health 2, Flammability 1, Reactivity 0. Insoluble in water.

Potential Exposure: Compound Description: Tumorigen, Mutagen. 4-Nitrobiphenyl was formerly used in the synthesis of 4-aminodiphenyl. It is presently used only for research purposes; there are no commercial uses.

Incompatibilities: Strong reducers, strong oxidizers.

Permissible Exposure Limits in Air

OSHA: Cancer Suspect Agent [skin], see Code of Federal Regulations 29CFR1910.1003.

NIOSH REL: A potential occupational carcinogen [skin]; Limit exposure to lowest feasible concentration. See *NIOSH Pocket Guide*, Appendix A.

ACGIH TLV[R][1]: [skin] Suspected Human Carcinogen *as 4-nitrodiphenyl.*

NIOSH IDLH: Not determined. Potential occupational carcinogen.

Protective Action Criteria (PAC)

TEEL-0: 0.25 mg/m^3

PAC-1: 0.75 mg/m^3

PAC-2: 5 mg/m^3

PAC-3: 500 mg/m^3

DFG MAK: [skin] Carcinogen Category 2.

Australia: carcinogen, 1993; Austria: [skin], carcinogen, 1999; Belgium: carcinogen, 1993; Finland: carcinogen, 1999; Norway: TWA 0.01 mg/m^3, 1999; Sweden: carcinogen, 1999; Switzerland: [skin], carcinogen, 1999; United Kingdom: carcinogen, 2000; Argentina, Bulgaria, Columbia, Jordan, South Korea, New Zealand, Singapore, Vietnam: ACGIH TLV[R]: Suspected Human Carcinogen. Several states have set guidelines or standards for nitrobiphenyl in ambient air[60] ranging from zero (New York, North Dakota, South Carolina, Virginia) to 2.77 μg/m^3 (Pennsylvania).

Determination in Air: Collection on a glass fiber filter in series with silica gel, elution with 2-propanol, analysis by gas chromatography/flame ionization detection; NIOSH (II-4) P&CAM Method #273; OSHA Analytical Method PV-2082.

Permissible Concentration in Water: No criteria set, but EPA[32] has suggested a permissible ambient goal of 890 μg/L based on health effects.

Determination in Water: Octanol—water coefficient: Log K_{ow} = 3.8.

Routes of Entry: Inhalation, ingestion, skin and/or eye contact. Percutaneous absorption.

Harmful Effects and Symptoms

Short Term Exposure: 4-Nitrobiphenyl can affect you when breathed in and by passing through your skin. Other health effects are not well known at this time, but contact with biphenyls can cause irritation of the skin and eyes; and may cause liver nerve damage, disturbed sleep, headache, lethargy (drowsiness or indifference), dizziness, dyspnea (breathing difficulty), ataxia, weakness, methemoglobinemia, urinary burning, acute hemorrhagic cystitis.

Long Term Exposure: 4-Nitrobiphenyl is a potential occupational carcinogen. Handle with extreme caution. May cause liver damage. Related compounds have caused damage to the nerves of the arms and legs.

Points of Attack: Bladder, blood, liver. Cancer site in animals: bladder.

Medical Surveillance: OSHA mandates tests and information on the following: *Increased Risk:* reduced immunologic competence, steroid treatment, pregnancy, cigarette smoking. NIOSH lists the following tests: *increased risk:* reduced immunologic competence, steroid treatment, pregnancy, cigarette smoking, urine (chemical/metabolite) placement, and periodic examinations should include an evaluation of exposure to other carcinogens, as well as an evaluation of smoking, or use of alcohol and medications and of family history. Special attention should be given on a regular basis to urine sediment and cytology. If red cells or positive smears are seen, cystoscopy should be done at once. The general health of exposed persons should also be evaluated in periodic examinations. Liver function tests. Complete blood count (CBC).

First Aid: If this chemical gets into the eyes, remove any contact lenses at once and irrigate immediately for at least 15 min, occasionally lifting upper and lower lids. Seek medical attention immediately. If this chemical contacts the skin, remove contaminated clothing and wash immediately with soap and water. Seek medical attention immediately. If this chemical has been inhaled, remove from exposure, begin rescue breathing (using universal precautions, including resuscitation mask) if breathing has stopped and CPR if heart action has stopped. Transfer promptly to a medical facility. When this chemical has been swallowed, get medical attention. Give large quantities of water and induce vomiting. Do not make an unconscious person vomit.

Note to physician: Treat for methemoglobinemia. Spectrophotometry may be required for precise determination of levels of methemoglobin in urine.

Personal Protective Methods: Wear protective gloves and clothing to prevent any reasonable probability of skin contact. Safety equipment suppliers/manufacturers can provide recommendations on the most protective glove/clothing material for your operation. All protective clothing (suits, gloves, footwear, headgear) should be clean, available each day, and put on before work. Contact lenses should not be worn when working with this chemical. Wear dust-proof chemical goggles and face shield unless full face-piece respiratory protection is worn. Employees should wash

immediately with soap when skin is wet or contaminated. Provide emergency showers and eyewash.

Respirator Selection: At any detectable concentration: SCBAF: Pd,Pp (APF = 10,000) (any NIOSH/MSHA- or European Standard EN 149-approved self-contained breathing apparatus that has a full face-piece and is operated in a pressure-demand or other positive-pressure mode) or SaF: Pd,Pp: ASCBA (APF = 10,000) (any supplied-air respirator that has a full face-piece and is operated in a pressure-demand or other positive-pressure mode in combination with an auxiliary, self-contained breathing apparatus operated in a pressure-demand or other positive-pressure mode). *Escape:* 100F (APF = 50) (any air-purifying, full-face-piece respirator with an N100, R100, or P100 filter) or SCBAE (any appropriate escape-type, self-contained breathing apparatus).

Storage: Color Code—Blue: Health Hazard/Poison: Store in a secure poison location. Prior to working with this chemical you should be trained on its proper handling and storage. Store in tightly closed containers in a cool, well-ventilated area away from heat and flame. A regulated, marked area should be established where this chemical is handled, used, or stored in compliance with OSHA Standard 1910.1045.

Shipping: Toxic solids, organic, n.o.s. requires a shipping label of "POISONOUS/TOXIC MATERIALS." It falls in Hazard Class 6.1.

Spill Handling: Evacuate persons not wearing protective equipment from area of spill or leak until cleanup is complete. Remove all ignition sources. Use HEPA vacuum or wet method to reduce dust during cleanup. Do not dry sweep. Collect powdered material in the most convenient and safe manner and deposit in sealed containers. Ventilate area after cleanup is complete. It may be necessary to contain and dispose of this chemical as a hazardous waste. If material or contaminated runoff enters waterways, notify downstream users of potentially contaminated waters. Contact your local or federal environmental protection agency for specific recommendations. If employees are required to clean up spills, they must be properly trained and equipped. OSHA 1910.120(q) may be applicable.

Fire Extinguishing: This chemical is a combustible solid. Use dry chemical, carbon dioxide, water spray, or alcohol foam extinguishers. Poisonous gases are produced in fire. If material or contaminated runoff enters waterways, notify downstream users of potentially contaminated waters. Notify local health and fire officials and pollution control agencies. From a secure, explosion-proof location, use water spray to cool exposed containers. If cooling streams are ineffective (venting sound increases in volume and pitch, tank discolors, or shows any signs of deforming), withdraw immediately to a secure position. If employees are expected to fight fires, they must be trained and equipped in OSHA 1910.156. The only respirators recommended for firefighting are self-contained breathing apparatuses that have full face-pieces and are operated in a pressure-demand or other positive-pressure mode.

Disposal Method Suggested: Incineration (982°C, 2.0 s minimum) with scrubbing for nitrogen oxides abatement.

Reference

New Jersey Department of Health and Senior Services. (February 2000). *Hazardous Substances Fact Sheet: 4-Nitrobiphenyl.* Trenton, NJ

Nitrocellulose   N:0420

Molecular Formula: $C_{12}H_{16}(ONO_2)_4O_6$

Synonyms: Box toe gum; Celloidin; Cellulose nitrate solution; Collodion cotton; Gun cotton; Nitrocellulose; Nitrocellulose gum; Nitrocellulose solution; Nitrocellulose, with plasticizer; Nitrocotton; Nitron; Nixon N/C; NT; Pyroxylin solution; Synpor; Tsapolak 964; Xyloidin

CAS Registry Number: 9004-70-0

RTECS® Number: QW0970000

UN/NA & ERG Number: UN2059 [nitrocellulose, solution, flammable with not >12.6% nitrogen, by mass, and not >55% nitrocellulose]/127; UN2555 [nitrocellulose with water with not <25% water, by mass]/113; UN2556 [nitrocellulose with alcohol with not <25% alcohol by mass, and with not >12.6% nitrogen, by dry mass]/113; UN2557 [nitrocellulose, with not >12.6% nitrogen, by dry mass, or nitrocellulose mixture with pigment or nitrocellulose mixture with plasticizer or nitrocellulose mixture with pigment and plasticizer]/133; UN0340 [Nitrocellulose, dry or wetted with <25% water (or alcohol), by mass]/112; UN0341 [nitrocellulose, unmodified or plasticized with <18% plasticizing substance, by mass]/112; UN0343 [nitrocellulose, plasticized with not <18% plasticizing substance, by mass]/112; UN0342 [nitrocellulose, wetted with not <25% alcohol, by mass]/112; 3270 (Nitrocellulose membrane filters)/133.

Regulatory Authority and Advisory Bodies

Department of Homeland Security Screening Threshold Quantity (pounds): *Release hazard: explosives* 5000 commercial grade); *Theft hazard* 400 (commercial grade).

European/International Regulations: not listed in Annex 1.

WGK (German Aquatic Hazard Class): No value assigned.

Description: Nitrocellulose is a pulpy, cotton-like solid, or a colorless liquid solution. Molecular weight = 504.31; Boiling point = about 34°C; Flash point = 13−27°C (wet with alcohol); Autoignition temperature = 170°C. Hazard Identification (based on NFPA-704 M Rating System): Health 2, Flammability 3, Reactivity 3 (Oxidizer). Insoluble in water.

Potential Exposure: It is used in making explosives, rocket propellants, and celluloid.

Incompatibilities: Strong acids, alkaline materials, oxidizers. Do not allow to become dry. Dry material is a shock-sensitive explosive.

Permissible Exposure Limits in Air
Protective Action Criteria (PAC)
TEEL-0: 20 mg/m^3
PAC-1: 60 mg/m^3
PAC-2: 400 mg/m^3
PAC-3: 500 mg/m^3

Harmful Effects and Symptoms

Short Term Exposure: Only those associated with the flammable and explosive nature of this flammable and reactive material. However, it may be wetted with alcohol, ether, or other dangerous liquid material that can be irritating to the eyes, nose, and throat. If inhaled will cause dizziness, difficult breathing, or loss of consciousness.

First Aid: If this chemical gets into the eyes, remove any contact lenses at once and irrigate immediately for at least 15 min, occasionally lifting upper and lower lids. Seek medical attention immediately. If this chemical contacts the skin, remove contaminated clothing and wash immediately with soap and water. Seek medical attention immediately. If this chemical has been inhaled, remove from exposure, begin rescue breathing (using universal precautions, including resuscitation mask) if breathing has stopped and CPR if heart action has stopped. Transfer promptly to a medical facility. When this chemical has been swallowed, get medical attention. Give large quantities of water and induce vomiting. Do not make an unconscious person vomit.

Personal Protective Methods: Wear protective gloves and clothing to prevent any reasonable probability of skin contact. Safety equipment suppliers/manufacturers can provide recommendations on the most protective glove/clothing material for your operation. All protective clothing (suits, gloves, footwear, headgear) should be clean, available each day, and put on before work. Contact lenses should not be worn when working with this chemical. Wear dust-proof chemical goggles and face shield unless full face-piece respiratory protection is worn. Wear splash-proof chemical goggles when working with liquid, unless full face-piece respiratory protection is worn. Employees should wash immediately with soap when skin is wet or contaminated. Provide emergency showers and eyewash.

Respirator Selection: Engineering controls must be effective to ensure that exposure to nitrocellulose does not occur. See respirator for solvent used as a wetting agent.

Storage: Color Code—Red: Flammability Hazard: Store in a flammable materials storage area. Color Code—Yellow: Reactive Hazard; Store in a location separate from other materials, especially flammables and combustibles. Prior to working with this chemical you should be trained on its proper handling and storage. Nitrocellulose must be stored to avoid contact with oxidizers, strong acids (such as hydrochloric, sulfuric, and nitric) and alkaline materials (such as sodium hydroxide and potassium hydroxide), since violent reactions occur. Store nitrocellulose away from high temperatures and direct sunlight. Do not allow material to become dry. Sources of ignition, such as smoking and open flames, are prohibited where nitrocellulose is handled, used, or stored. Wherever nitrocellulose is used, handled, manufactured, or stored, use explosion-proof electrical equipment and fittings.

Shipping: UN2555, UN2556, and UN2557 require a shipping label of "FLAMMABLE SOLID." They fall in DOT Hazard Class 4.1 and Packing Group II.
UN0342 and UN0343 require a shipping label of "EXPLOSIVE." They fall in DOT Hazard Class 1.3C and Packing Group II.
UN2059 requires a shipping label of "FLAMMABLE LIQUID." It falls in Hazard Class 3 and Packing Group III.

Spill Handling: Restrict persons not wearing protective equipment from area of spill or leak until cleanup is complete. Remove all ignition sources. Ventilate area of spill or leak. Immediately flood spill area with water, collect material, and deposit in sealed containers. Keep nitrocellulose out of a confined space, such as a sewer, because of the possibility of an explosion, unless the sewer is designed to prevent the buildup of explosive concentrations. It may be necessary to contain and dispose of this chemical as a hazardous waste. If material or contaminated runoff enters waterways, notify downstream users of potentially contaminated waters. Contact your local or federal environmental protection agency for specific recommendations. If employees are required to clean up spills, they must be properly trained and equipped. OSHA 1910.120(q) may be applicable.

Fire Extinguishing[17]: Presents an unusually severe fire hazard; when dry, ignites readily and burns explosively. Should never be kept for any appreciable time in any dry fibrous state. Unstabilized product decomposes gradually at relatively low temperature, with evolution of copious volumes of toxic and flammable gases and rapid heat generation. In prolonged storage and aging of nitrocellulose plastics, camphor is lost with deterioration and the decomposition temperature may be lowered to 40°C. The resulting flameless decomposition is self-sustaining and accelerative, presenting the added hazard of dangerous pressures in building structures.

Use extreme caution in approaching fires involving this material as it may explode. No attempt should be made to fight advanced fires, except for remote activation of installed fire extinguishing equipment and/or with unmanned fixed turrets and hose nozzles. The surrounding areas should be evacuated. Fires should be approached from upwind and self-contained breathing apparatus used. Since cellulose nitrate supplies its own oxygen, prompt cooling with a large quantity of water is essential; water applied through spray nozzles is effective if fused quickly and in sufficient volume, in a manner to wet the entire exposed surface. Poisonous gases, including oxides of nitrogen (and possibly hydrogen cyanide and carbon monoxide), are produced in fire. If material or contaminated runoff enters waterways, notify downstream users of potentially contaminated waters. Notify local health and fire officials and pollution control agencies. From a secure, explosion-proof location, use water spray to cool exposed containers. If cooling streams are ineffective (venting sound increases in volume and pitch, tank discolors, or shows any signs of deforming), withdraw immediately to a secure position. If employees are expected to fight fires, they must be trained and equipped in OSHA 1910.156. The only respirators recommended for firefighting are self-contained breathing apparatuses that have full face-pieces and are operated in a pressure-demand or other positive-pressure mode.

Reference

New Jersey Department of Health and Senior Services. (April 2001). *Hazardous Substances Fact Sheet: Nitrocellulose*. Trenton, NJ

p-Nitrochlorobenzene   N:0430

Molecular Formula: $C_6H_4ClNO_2$
Common Formula: p-$ClC_6H_4NO_2$
Synonyms: Benzene, 1-chloro-4-nitro-; 1-Chlor-4-nitrobenzol (German); *p*-Chloronitrobenzene; 1-Chloro-4-nitrobenzene; 4-Chloro-1-nitrobenzene; 4-Chloronitrobenzene; *p*-Nitrochlorobenzol (German); *p*-Nitroclorobenzene; PNCB
CAS Registry Number: 100-00-5
RTECS® Number: CZ1050000
UN/NA & ERG Number: UN1578/152
EC Number: 202-809-6 [*Annex I Index No.:* 610-005-00-5]
Regulatory Authority and Advisory Bodies
Carcinogenicity: IARC: Animal Inadequate Evidence; Human Inadequate Evidence, *not classifiable as carcinogenic to humans*, Group 3, 1996; NIOSH: Potential occupational carcinogen.
Air Pollutant Standard Set. See below, "Permissible Exposure Limits in Air" section.
California Proposition 65 Chemical: Cancer 10/29/99.
Canada, WHMIS, Ingredients Disclosure List Concentration 1.0%.
European/International Regulations: Hazard Symbol: T, N; Risk phrases: R23/24/25; R40; R48/20/21/22; R68; R51/53; Safety phrases: S1/2; S28; S36/37; S45; S61 (see Appendix 4). WGK (German Aquatic Hazard Class): 2—Hazard to waters.
Description: *p*-Nitrochlorobenzene is a yellow crystalline solid with a sweet odor. Molecular weight = 157.56; Specific gravity (H_2O:1) = 1.52; Boiling point = 242°C; Freezing/Melting point = 83.3°C; Vapor pressure = 0.2 mmHg at 20°C; Flash point = 127.2°C (cc). Hazard Identification (based on NFPA-704 M Rating System): Health 2, Flammability 1, Reactivity 3. Shock and heat sensitive. Insoluble in water.
Potential Exposure: Compound Description: Tumorigen, Mutagen; Reproductive Effector. *p*-Nitrochlorobenzene (PNCB) is used as an intermediate in pesticide (parathion) manufacture, drug (phenacetin and acetaminophen) manufacture, in dye making, and in rubber and antioxidant manufacture.

Incompatibilities: A strong oxidizer. Reacts violently with oxidizers, combustibles, alkalis, sodium methoxide, and reducing materials.
Permissible Exposure Limits in Air
OSHA PEL: 1 mg/m³ TWA [skin].
NIOSH REL: A potential occupational carcinogen [skin]; Limit exposure to lowest feasible concentration. See *NIOSH Pocket Guide*, Appendix A. BEI_M issued for Methemoglobin inducers.
ACGIH TLV®[1]: 0.1 ppm/0.64 mg/m³ TWA [skin]; confirmed animal carcinogen with unknown relevance to humans.
NIOSH IDLH: potential occupational carcinogen 100 mg/m³.
Protective Action Criteria (PAC)
TEEL-0: 1 mg/m³
PAC-1: 6 mg/m³
PAC-2: 40 mg/m³
PAC-3: 100 mg/m³
DFG MAK: [skin] Carcinogen Category 3B.
Australia: TWA 0.1 ppm (0.6 mg/m³), [skin], 1993; Austria Suspected: carcinogen, 1999; Belgium: TWA 0.1 ppm (0.64 mg/m³), [skin], 1993; Denmark: TWA 0.1 ppm (0.64 mg/m³), [skin], 1999; Finland: TWA 1 mg/m³; STEL 3 mg/m³, [skin], 1999; Hungary: TWA 1 mg/m³; STEL 2 mg/m³, [skin], 1993; Japan: 0.1 ppm (0.64 mg/m³), [skin], 1999; the Netherlands: MAC-TGG 1 mg/m³, [skin], 2003; Norway: TWA 1 mg/m³, 1999; the Philippines: TWA 1 mg/m³, [skin], 1993; Poland: MAC (TWA) 1 mg/m³; MAC (STEL) 3 mg/m³, 1999; Russia: TWA 0.1 ppm; STEL 1 mg/m³, [skin], 1993; Switzerland: MAK-W 1 mg/m³, KZG-W 2 mg/m³, [skin], 1999; United Kingdom: TWA 1 mg/m³; STEL 2 mg/m³, [skin], 2000; Argentina, Bulgaria, Columbia, Jordan, South Korea, New Zealand, Singapore, Vietnam: ACGIH TLV®: confirmed animal carcinogen with unknown relevance to humans. Russia[43] set a MAC of 0.004 mg/m³ in ambient air of residential areas both on a momentary and a daily average basis. Several states have set guidelines or standards for 4-chloronitrobenzene in ambient air[60] ranging from 3.3 μg/m³ (New York) to 5 μg/m³ (South Carolina) to 10 μg/m³ (Florida) to 20 μg/m³ (Connecticut) to 24 μg/m³ (Nevada) to 30 μg/m³ (North Dakota) to 50 μg/m³ (Virginia) to 83.33 μg/m³ (Kansas).
Determination in Air: Use NIOSH Analytical Method (IV) #2005[18], Nitrobenzene.
Determination in Water: Octanol—water coefficient: Log K_{ow} = 2.4.
Permissible Concentration in Water: A limit of 0.05 mg/L has been set in the former USSR.[43]
Routes of Entry: Inhalation, ingestion, skin and/or eye contact.
Harmful Effects and Symptoms
Short Term Exposure: Irritates the eyes and skin on contact. Inhalation can cause irritation of the respiratory tract with coughing and wheezing. High levels can interfere with the body's ability to carry oxygen (methemoglobinemia)

causing cyanosis, headache, dizziness, fatigue. Can also cause nausea, vomiting, dyspnea. Higher levels can cause trouble breathing, collapse, and death.

Long Term Exposure: Repeated or prolonged contact may cause skin sensitization and allergy. Can affect the nervous system. May damage the liver and kidneys. May cause methemoglobinemia, hemoglobinuria, anemia, spleen, bone marrow changes, reproductive effects. Potential occupational carcinogen. In animals: hematuria (blood in the urine).

Points of Attack: Blood, liver, kidneys, cardiovascular system, spleen, bone marrow, reproductive system. Cancer site in animals: vascular and liver.

Medical Surveillance: NIOSH lists the following tests: whole blood (chemical/metabolite), Methemoglobin; complete blood count. Consider the points of attack in preplacement and periodic physical examinations. Evaluation by a qualified allergist. Liver and kidney function tests. Alcohol consumption may increase liver damage caused by PNCB.

First Aid: If this chemical gets into the eyes, remove any contact lenses at once and irrigate immediately for at least 15 min, occasionally lifting upper and lower lids. Seek medical attention immediately. If this chemical contacts the skin, remove contaminated clothing and wash immediately with soap and water. Seek medical attention immediately. If this chemical has been inhaled, remove from exposure, begin rescue breathing (using universal precautions, including resuscitation mask) if breathing has stopped and CPR if heart action has stopped. Transfer promptly to a medical facility. When this chemical has been swallowed, get medical attention. Give large quantities of water and induce vomiting. Do not make an unconscious person vomit.

Note to physician: Treat for methemoglobinemia. Spectrophotometry may be required for precise determination of levels of methemoglobin in urine.

Personal Protective Methods: Wear protective gloves and clothing to prevent any reasonable probability of skin contact. Safety equipment suppliers/manufacturers can provide recommendations on the most protective glove/clothing material for your operation. All protective clothing (suits, gloves, footwear, headgear) should be clean, available each day, and put on before work. Contact lenses should not be worn when working with this chemical. Wear dust-proof chemical goggles and face shield unless full face-piece respiratory protection is worn. Employees should wash immediately with soap when skin is wet or contaminated. Provide emergency showers and eyewash.

Respirator Selection: NIOSH: *At any detectable concentration:* SCBAF: Pd,Pp (APF = 10,000) (any NIOSH/MSHA- or European Standard EN 149-approved self-contained breathing apparatus that has a full face-piece and is operated in a pressure-demand or other positive-pressure mode); or SaF: Pd,Pp: ASCBA (APF = 10,000) (any supplied-air respirator that has a full face-piece and is operated in a pressure-demand or other positive-pressure mode in combination with an auxiliary, self-contained breathing apparatus operated in a pressure-demand or other positive-pressure mode). *Escape:* 100F (APF = 50) (any air-purifying, full-face-piece respirator with an N100, R100, or P100 filter) or SCBAE (any appropriate escape-type, self-contained breathing apparatus).

Storage: Color Code—Blue: Health Hazard/Poison: Store in a secure poison location. Prior to working with this chemical you should be trained on its proper handling and storage. Store in a refrigerator and protect from shock, heat sources, light, oxidizers, reducing agents, alkalis, and sodium methoxide. Sources of ignition, such as smoking and open flames, are prohibited where this chemical is handled, used, or stored. Metal containers involving the transfer of 5 gallons or more of this chemical should be grounded and bonded. Drums must be equipped with self-closing valves, pressure vacuum bungs, and flame arresters. Use only nonsparking tools and equipment, especially when opening and closing containers of this chemical. Wherever this chemical is used, handled, manufactured, or stored, use explosion-proof electrical equipment and fittings. A regulated, marked area should be established where this chemical is handled, used, or stored in compliance with OSHA Standard 1910.1045.

Shipping: Chloronitrobenzenes [meta or para, solid] require a shipping label of "POISONOUS/TOXIC MATERIALS." They fall in DOT Hazard Class 6.1 and Packing Group II. A plus sign (+) symbol indicates that the designated proper shipping name and hazard class of the material must always be shown whether or not the material or its mixtures or solutions meet the definitions of the class.

Spill Handling: Evacuate persons not wearing protective equipment from area of spill or leak until cleanup is complete. Remove all ignition sources. Use HEPA vacuum or wet method to reduce dust during cleanup. Do not dry sweep. Collect powdered material in the most convenient and safe manner and deposit in sealed containers. Ventilate area after cleanup is complete. It may be necessary to contain and dispose of this chemical as a hazardous waste. If material or contaminated runoff enters waterways, notify downstream users of potentially contaminated waters. Contact your local or federal environmental protection agency for specific recommendations. If employees are required to clean up spills, they must be properly trained and equipped. OSHA 1910.120(q) may be applicable.

Fire Extinguishing: This chemical is a combustible solid. It does not readily ignite. Use dry chemical, carbon dioxide, water spray; alcohol foam or polymer foam extinguishers. Poisonous gases, including nitrogen oxides and hydrogen chloride, are produced in fire. If material or contaminated runoff enters waterways, notify downstream users of potentially contaminated waters. Notify local health and fire officials and pollution control agencies. Containers may explode in fire. From a secure, explosion-proof location, use water spray to cool exposed containers. If cooling streams are ineffective (venting sound increases in volume and pitch, tank discolors, or shows any signs of deforming),

withdraw immediately to a secure position. If employees are expected to fight fires, they must be trained and equipped in OSHA 1910.156. The only respirators recommended for firefighting are self-contained breathing apparatuses that have full face-pieces and are operated in a pressure-demand or other positive-pressure mode.

Disposal Method Suggested: Incineration (816°C, 0.5 s for primary combustion; 1204°C, 1.0 s for secondary combustion). The formation of elemental chlorine can be prevented through injection of steam or methane into the combustion process. Nitrogen oxides may be abated through the use of thermal or catalytic devices.

References

US Environmental Protection Agency. (June 13, 1983). *Chemical Hazard Information Profile Draft Report: 4-Chloronitrobenzene*. Washington, DC

New Jersey Department of Health and Senior Services. (July 1999). *Hazardous Substances Fact Sheet: p-Nitrochlorobenzene*. Trenton, NJ

New Jersey Department of Health and Senior Services. (January 2007). *Hazardous Substances Fact Sheet: Chloronitrobenzenes (mixed isomers)*. Trenton, NJ

Nitrocyclohexane N:0440

Molecular Formula: $C_6H_{11}NO_2$
Synonyms: Cyclohexane, nitro-
CAS Registry Number: 1122-60-7
RTECS® Number: GV6600000
UN/NA & ERG Number: UN2810/153
EC Number: 214-354-0
Regulatory Authority and Advisory Bodies
Air Pollutant Standard Set. See below, "Permissible Exposure Limits in Air" section.
SUPERFUND/EPCRA 40CFR355, Extremely Hazardous Substances: TPQ = 500 lb (227 kg).
Reportable Quantity (RQ): 500 lb (227 kg).
European/International Regulations: not listed in Annex I (see Appendix 4).
WGK (German Aquatic Hazard Class): No value assigned.
Description: Nitrocyclohexane is a highly flammable, colorless liquid. Molecular weight = 129.18; Boiling point = 206°C (decomposition); Freezing/Melting point = −34°C; Flash point = 88°C. Hazard Identification (based on NFPA-704 M Rating System): Health 3, Flammability 2, Reactivity 2 (Oxidizer).
Potential Exposure: Used in organic synthesis.
Incompatibilities: A nitro compound; forms explosive mixture with air. Incompatible with strong oxidizers, alkalis, and metal oxides. This chemical is highly reactive and may be heat- and shock-sensitive and a fire and explosive hazard.
Permissible Exposure Limits in Air
Protective Action Criteria (PAC)
TEEL-0: 0.3 mg/m^3

PAC-1: 0.75 mg/m^3
PAC-2: 1.5 mg/m^3
PAC-3: 60 mg/m^3
Russia[43] set a MAC of 1.0 mg/m^3 in work-place air.
Permissible Concentration in Water: Russia[43] set a MAC of 0.1 mg/L in water bodies used for domestic purposes.
Routes of Entry: Inhalation, ingestion, skin and/or eye contact.
Harmful Effects and Symptoms
The LD_{50} = (oral-mouse) 250 mg/kg (moderately toxic).
Short Term Exposure: Insufficient data are available on the effect of this substance on human health; therefore, utmost care must be taken. May be absorbed through the skin. May be an irritant to the eyes, skin, and respiratory system. Similar chemical can cause cyanosis due to formation of methemoglobin.
Long Term Exposure: Similar chemicals can cause kidney and liver damage.
Points of Attack: Most of the above information is based on similar nitro compounds of aromatic hydrocarbons. May cause blood, liver, and kidney effects.
Medical Surveillance: There is no special test for this chemical. However, if illness occurs or overexposure is suspected, medical attention is recommended. The following might be considered: blood methemoglobin level, complete blood count (CBC), liver and kidney function tests.
First Aid: If this chemical gets into the eyes, remove any contact lenses at once and irrigate immediately for at least 15 min, occasionally lifting upper and lower lids. Seek medical attention immediately. If this chemical contacts the skin, remove contaminated clothing and wash immediately with soap and water. Speed in removing material from skin is of extreme importance. Seek medical attention immediately. If this chemical has been inhaled, remove from exposure, begin rescue breathing (using universal precautions, including resuscitation mask) if breathing has stopped and CPR if heart action has stopped. Transfer promptly to a medical facility. When this chemical has been swallowed, get medical attention. Give large quantities of water and induce vomiting. Do not make an unconscious person vomit. Keep victim quiet and maintain normal body temperature. Effects may be delayed; keep victim under observation.
Note to physician: Treat for methemoglobinemia. Spectrophotometry may be required for precise determination of levels of methemoglobin in urine.
Personal Protective Methods: Wear protective gloves and clothing to prevent any reasonable probability of skin contact. Safety equipment suppliers/manufacturers can provide recommendations on the most protective glove/clothing material for your operation. All protective clothing (suits, gloves, footwear, headgear) should be clean, available each day, and put on before work. Contact lenses should not be worn when working with this chemical. Wear splash-proof chemical goggles and face shield unless full face-piece respiratory protection is worn. Employees should wash

immediately with soap when skin is wet or contaminated. Provide emergency showers and eyewash; pressure, pressure-demand, full face-piece self-contained breathing apparatus (SCBA) or pressure-demand supplied air respirator with escape SCBA and a fully-encapsulating, chemical-resistant suit.

Respirator Selection: Follow regulations in OSHA 29CFR1910.134 or European Standard EN149. Use a NIOSH/MSHA- or European Standard EN149-approved respirator; or use an approved supplied-air respirator with a full face-piece operated in the positive-pressure mode, or with a full face-piece, hood, or helmet in the continuous-flow mode; or use a NIOSH/MSHA- or European Standard EN149-approved self-contained breathing apparatus with a full face-piece operated in pressure-demand or other positive-pressure mode.

Storage: Color Code—Blue: Health Hazard/Poison: Store in a secure poison location. Prior to working with this chemical you should be trained on its proper handling and storage. Store in tightly closed containers in a cool, well-ventilated area away from oxidizers, alkalis, metal oxides. Where possible, automatically pump liquid from drums or other storage containers to process containers. Sources of ignition, such as smoking and open flames, are prohibited where this chemical is handled, used, or stored. Metal containers involving the transfer of 5 gallons or more of this chemical should be grounded and bonded. Drums must be equipped with self-closing valves, pressure vacuum bungs, and flame arresters. Use only nonsparking tools and equipment, especially when opening and closing containers of this chemical. Wherever this chemical is used, handled, manufactured, or stored, use explosion-proof electrical equipment and fittings.

Shipping: Toxic, liquids, organic, n.o.s. [Inhalation hazard, Packing Group I, Zone B] requires a shipping label of "POISONOUS/TOXIC MATERIALS." It falls in Hazard Class 6.1 and Packing Group III.

Spill Handling: Evacuate and restrict persons not wearing protective equipment from area of spill or leak until cleanup is complete. Remove all ignition sources. Ventilate area of spill or leak. Do not touch spilled material; stop leak if you can do so without risk. *Large fires:* water spray, fog, or foam. Move container from fire area if you can do it without risk. Stay upwind; keep out of low areas. Wear positive pressure breathing apparatus and special protective clothing. Absorb liquids in vermiculite, dry sand, earth, peat, carbon, or a similar material and deposit in sealed containers. Keep this chemical out of a confined space, such as a sewer, because of the possibility of an explosion, unless the sewer is designed to prevent the build-up of explosive concentrations. It may be necessary to contain and dispose of this chemical as a hazardous waste. If material or contaminated runoff enters waterways, notify downstream users of potentially contaminated waters. Contact your local or federal environmental protection agency for specific recommendations. If employees are required to clean up spills, they must be properly trained and equipped. OSHA 1910.120(q) may be applicable.

Fire Extinguishing: This chemical is a combustible liquid. Poisonous gases, including nitrogen oxides, are produced in fire. *Small fires:* dry chemical, carbon dioxide, water spray, or foam. *Large fires:* water spray, fog, or foam. Move container from fire area if you can do it without risk. Stay upwind; keep out of low areas. Vapors are heavier than air and will collect in low areas. Vapors may travel long distances to ignition sources and flashback. Vapors in confined areas may explode when exposed to fire. Containers may explode in fire. Storage containers and parts of containers may rocket great distances, in many directions. If material or contaminated runoff enters waterways, notify downstream users of potentially contaminated waters. Notify local health and fire officials and pollution control agencies. From a secure, explosion-proof location, use water spray to cool exposed containers. If cooling streams are ineffective (venting sound increases in volume and pitch, tank discolors, or shows any signs of deforming), withdraw immediately to a secure position. If employees are expected to fight fires, they must be trained and equipped in OSHA 1910.156. The only respirators recommended for firefighting are self-contained breathing apparatuses that have full face-pieces and are operated in a pressure-demand or other positive-pressure mode.

Disposal Method Suggested: Dissolve or mix the material with a combustible solvent and burn in a chemical incinerator equipped with an afterburner and scrubber. All federal, state, and local environmental regulations must be observed.

References

US Environmental Protection Agency. (November 30, 1987). *Chemical Hazard Information Profile: Nitrocyclohexane.* Washington, DC: Chemical Emergency Preparedness Program

New Jersey Department of Health and Senior Services. (April 2002). *Hazardous Substances Fact Sheet: Nitrocyclohexane.* Trenton, NJ

Nitroethane   N:0450

Molecular Formula: $C_2H_5NO_2$
Common Formula: $CH_3CH_2NO_2$
Synonyms: Ethane, nitro-; Nitroetano (Spanish)
CAS Registry Number: 79-24-3
RTECS® Number: KI5600000
UN/NA & ERG Number: UN2842/129
EC Number: 201-188-9 [*Annex I Index No.:* 609-035-00-1]
Regulatory Authority and Advisory Bodies
US EPA Gene-Tox Program, Inconclusive: Mammalian micronucleus.
Air Pollutant Standard Set. See below, "Permissible Exposure Limits in Air" section.

Canada, WHMIS, Ingredients Disclosure List Concentration 1.0%.

European/International Regulations: Hazard Symbol: Xn; Risk phrases: R10; R20/22; Safety phrases: S2; S9; S25; S41 (see Appendix 4).

WGK (German Aquatic Hazard Class): 2—Hazard to waters.

Description: Nitroethane is a colorless, oily liquid with a mild, fruity odor. The odor threshold is 163 ppm.[41] Molecular weight = 75.08; Specific gravity (H_2O:1) = 1.05; Boiling point = 113.9°C; Freezing/Melting point = −90°C; Vapor pressure = 21 mmHg at 25°C; Flash point = 28°C. Begins to decompose at 300°C; Autoignition temperature = 414°C. Explosive limits: LEL = 3.4%; UEL = Unknown. Hazard Identification (based on NFPA-704 M Rating System): Health 1, Flammability 3, Reactivity 3. Slightly soluble in water; solubility = 5%.

Potential Exposure: Compound Description: Mutagen; Human Data. Nitroethane is used as solvent for polymers, cellulose esters, vinyl, waxes, fats, dyestuffs, and alkyd resins, as a stabilizer. It has been used as a rocket propellant. It is used as an intermediate in pharmaceutical manufacture and in pesticide manufacture.

Incompatibilities: A nitroparaffin, nitroethane forms explosive mixture with air. Explodes when heated or when shocked; in confined area, with elevated temperatures. A strong reducing agent. Violent reaction with oxidizers, hydrocarbons, other combustibles, amines, metal oxides. Forms shock-sensitive compounds with strong acids, strong alkalis. Attacks some plastics and coatings.

Permissible Exposure Limits in Air

OSHA PEL: 100 ppm/310 mg/m³ TWA.

NIOSH REL: 100 ppm/310 mg/m³ TWA.

ACGIH TLV®[1]: 100 ppm/307 mg/m³ TWA.

NIOSH IDLH: 1000 ppm.

Protective Action Criteria (PAC)

TEEL-0: 100 ppm

PAC-1: 100 ppm

PAC-2: 200 ppm

PAC-3: 1000 ppm

DFG MAK: 100 ppm/310 mg/m³ TWA; Peak Limitation Category II(4); Pregnancy Risk Group D.

Australia: TWA 100 ppm (310 mg/m³), 1993; Austria: MAK 100 ppm (310 mg/m³), 1999; Belgium: TWA 100 ppm (307 mg/m³), 1993; Denmark: TWA 100 ppm (310 mg/m³), 1999; Finland: TWA 100 ppm (310 mg/m³); STEL 150 ppm (465 mg/m³), 1999; France: VME 100 ppm (310 mg/m³), 1999; the Netherlands: MAC-TGG 60 mg/m³, 2003; the Philippines: TWA 100 ppm (310 mg/m³), 1993; Poland: MAC (TWA) 30 mg/m³, MAC (STEL) 240 mg/m³, 1999; Russia: STEL 30 mg/m³, 1993; Sweden: TWA 20 ppm (60 mg/m³); STEL 50 ppm (150 mg/m³), 1999; Switzerland: MAK-W 100 ppm (310 mg/m³), 1999; Turkey: TWA 100 ppm (310 mg/m³), 1993; United Kingdom: TWA 100 ppm (312 mg/m³), 2000; Argentina, Bulgaria, Columbia, Jordan, South Korea, New Zealand,

Singapore, Vietnam: ACGIH TLV®: TWA 100 ppm. Several states have set limits in ambient air ranging from 3.1 mg/m³ (North Dakota) to 5.2 mg/m³ (Virginia) to 6.2 mg/m³ (Connecticut) to 7.38 mg/m³ (Nevada).

Determination in Air: Use NIOSH Analytical Method (IV) #2526.

Permissible Concentration in Water: Russia[43] set a MAC of 1.0 mg/L in water bodies used for domestic purposes.

Determination in Water: Octanol−water coefficient: Log K_{ow} = 0.18.

Routes of Entry: Inhalation, ingestion, skin and/or eye contact.

Harmful Effects and Symptoms

Short Term Exposure: Irritates the eyes, skin, and respiratory tract. Inhalation can cause coughing and wheezing. High exposure could cause headache, dizziness, and unconsciousness. Higher exposures can cause pulmonary edema, a medical emergency that can be delayed for several hours. This can cause death. LD_{50} = (oral-rat) 250 mg/kg (moderately toxic).

Long Term Exposure: May cause liver and kidney damage. Can cause dermatitis, drying, and cracking skin. Can cause lung irritation; bronchitis may develop.

Points of Attack: Skin, respiratory system, central nervous system, kidneys, liver.

Medical Surveillance: Consider the points of attack in pre-placement and periodic physical examinations. Liver and kidney function tests. Consider chest X-ray following acute overexposure.

First Aid: If this chemical gets into the eyes, remove any contact lenses at once and irrigate immediately for at least 15 min, occasionally lifting upper and lower lids. Seek medical attention immediately. If this chemical contacts the skin, remove contaminated clothing and wash immediately with soap and water. Seek medical attention immediately. If this chemical has been inhaled, remove from exposure, begin rescue breathing (using universal precautions, including resuscitation mask) if breathing has stopped and CPR if heart action has stopped. Transfer promptly to a medical facility. When this chemical has been swallowed, get medical attention. Give large quantities of water and induce vomiting. Do not make an unconscious person vomit. Medical observation is recommended for 24−48 h after breathing overexposure, as pulmonary edema may be delayed. As first aid for pulmonary edema, a doctor or authorized paramedic may consider administering a corticosteroid spray.

Personal Protective Methods: Wear protective gloves and clothing to prevent any reasonable probability of skin contact. Safety equipment suppliers/manufacturers can provide recommendations on the most protective glove/clothing material for your operation. Butyl rubber and polyvinyl alcohol are among the recommended protective materials. All protective clothing (suits, gloves, footwear, headgear) should be clean, available each day, and put on before work. Contact lenses should not be worn when working with this chemical. Wear splash-proof chemical goggles and face shield unless full face-piece respiratory protection is worn. Employees should wash immediately with soap when skin is wet or contaminated. Provide emergency showers and eyewash.

Respirator Selection: 1000 ppm: SCBAF (APF = 50) (any self-contained breathing apparatus with a full face-piece) or SaF (APF = 50) (any supplied-air respirator with a full face-piece). *Emergency or planned entry into unknown concentrations or IDLH conditions:* SCBAF: Pd,Pp (APF = 10,000) (any self-contained breathing apparatus that has a full face-piece and is operated in a pressure-demand or other positive-pressure mode) or Sa: Pd,Pp (APF = 1000): ASCBA (any supplied-air respirator that has a full face-piece and is operated in a pressure-demand or other positive-pressure mode in combination with an auxiliary self-contained breathing apparatus operated in a pressure-demand or other positive-pressure mode). *Escape:* SCBAE (any appropriate escape-type, self-contained breathing apparatus).

Storage: (1) Color Code—Red: Flammability Hazard: Store in a flammable liquid storage area or approved cabinet away from ignition sources and corrosive and reactive materials. (2) Color Code—Yellow Stripe (*strong reducing agent*): Reactivity Hazard; Store separately in an area isolated from flammables, combustibles, or other yellow coded materials. Prior to working with this chemical you should be trained on its proper handling and storage. Before entering confined space where this chemical may be present, check to make sure that an explosive concentration does not exist. Store in an explosion-proof refrigerator away from oxidizers, strong acids, amines, alkalis, hydrocarbons, combustibles, metal oxides, strong bases, and reducing agents. Where possible, automatically pump liquid from drums or other storage containers to process containers. Sources of ignition, such as smoking and open flames are prohibited where this chemical is handled, used, or stored. Metal containers involving the transfer of 5 gallons or more of this chemical should be grounded and bonded. Drums must be equipped with self-closing valves, pressure vacuum bungs, and flame arresters. Use only nonsparking tools and equipment, especially when opening and closing containers of this chemical. Wherever this chemical is used, handled, manufactured, or stored, use explosion-proof electrical equipment and fittings.

Shipping: This compound requires a shipping label of "FLAMMABLE LIQUID." It falls in Hazard Class 3 and Packing Group II.

Spill Handling: Evacuate and restrict persons not wearing protective equipment from area of spill or leak until cleanup is complete. Remove all ignition sources. Establish forced ventilation to keep levels below explosive limit. Absorb liquids in vermiculite, dry sand, earth, peat, carbon, or a similar material, and deposit in sealed containers. Keep this chemical out of a confined space, such as a sewer, because of the possibility of an explosion, unless the sewer is designed to prevent the buildup of explosive concentrations. It may be necessary to contain and dispose of this chemical as a hazardous waste. If material or contaminated runoff enters waterways, notify downstream users of potentially contaminated waters. Contact your local or federal environmental protection agency for specific recommendations. If employees are required to clean up spills, they must be properly trained and equipped. OSHA 1910.120(q) may be applicable.

Fire Extinguishing: This chemical is a flammable liquid. Poisonous gases, including nitrogen oxides and carbon monoxide, are produced in fire. Do not use dry chemical powder. Use carbon dioxide, or alcohol, or polymer foam extinguishers. Water may be ineffective. Vapors are heavier than air and will collect in low areas. Vapors may travel long distances to ignition sources and flashback. Vapors in confined areas may explode when exposed to fire. Containers may explode in fire. Storage containers and parts of containers may rocket great distances, in many directions. If material or contaminated runoff enters waterways, notify downstream users of potentially contaminated waters. Notify local health and fire officials and pollution control agencies. From a secure, explosion-proof location, use water spray to cool exposed containers. If cooling streams are ineffective (venting sound increases in volume and pitch, tank discolors, or shows any signs of deforming), withdraw immediately to a secure position. If employees are expected to fight fires, they must be trained and equipped in OSHA 1910.156. The only respirators recommended for firefighting are self-contained breathing apparatuses that have full face-pieces and are operated in a pressure-demand or other positive-pressure mode.

Disposal Method Suggested: Incineration: large quantities of material may require nitrogen oxide removal by catalytic or scrubbing processes.[22]

Reference
New Jersey Department of Health and Senior Services. (June 1999). *Hazardous Substances Fact Sheet: Nitroethane.* Trenton, NJ

Nitrofen   N:0460

Molecular Formula: $C_{12}H_7Cl_2NO_3$
Common Formula: $O_2NC_6H_4OC_6H_3Cl_3$
Synonyms: Benzenamine, 4-Ethoxy-*N*-(5-nitro-2-furanyl) methylene-; Benzene, 2,4-dichloro-1-(4-nitrophenoxy)-; 2′,4′-Dichloro-4′-nitrodiphenyl ether; 2,4-Dichloro-1-(4-nitrophenoxy)benzene; 4-(2,4-Dichlorophenoxy) nitrobenzene; 2,4-Dichlorophenyl-4-nirtophenylaether (German); 2,4-Dichlorophenyl *p*-nitrophenyl ether; 2,4-Dichlorophenyl 4-nitrophenyl ether; Ether, 2,4-dichlorophenyl *p*-nitrophenyl; FW 925; Mezotox; NCI-C00420; Niclofen; NIP; Nitrochlor; 4′-Nitro-2,4-dichlorodiphenyl ether; 4-Nitro-2′,4′-dichlorodiphenyl ether; Nitrofene (French); Nitrophen;

Nitrophene; Preparation 125; TOK; TOK-2; TOK E; TOK E 25; TOK E 40; Tokkom; Tokkorn; TOK WP-50; Trizilin
CAS Registry Number: 1836-75-5; *(alt.)* 51274-07-8
RTECS® Number: KN8400000
UN/NA & ERG Number: UN2765 (Phenoxy pesticides, solid, toxic)/152
EC Number: 217-406-0 [*Annex I Index No.:* 609-040-00-9]
Regulatory Authority and Advisory Bodies
Carcinogenicity: IARC (*technical grade*): Animal Sufficient Evidence; Human No Adequate Data, *possibly carcinogenic to humans*, Group 2B, 1987; NTP (*technical grade*): 11th Report on Carcinogens, 2004: Reasonably anticipated to be a human carcinogen; NCI: Carcinogenesis Bioassay (feed); clear evidence: mouse, rat, 1978; (feed); clear evidence: mouse, 1979; (feed); no evidence: rat, 1979.
US EPA Gene-Tox Program, Positive: Carcinogenicity—mouse/rat; Inconclusive: Mammalian micronucleus.
Banned or Severely Restricted (many countries) (UN).[13]
Air Pollutant Standard Set. See below, "Permissible Exposure Limits in Air" section.
EPCRA Section 313 Form R *de minimis* concentration reporting level: 0.1%.
US DOT Regulated Marine Pollutant (49CFR172.101, Appendix B).
California Proposition 65 Chemical: Cancer 1/1/88.
WGK (German Aquatic Hazard Class): No value assigned.
Description: Nitrofen is a crystalline solid. Molecular weight = 284.10; Freezing/Melting point = 70−71°C; Vapor pressure = 1.2×10^{-7} mmHg at 20°C. Hazard Identification (based on NFPA-704 M Rating System): Health 2, Flammability 3, Reactivity 0. Very slightly soluble in water.
Potential Exposure: Compound Description: Agricultural Chemical; Drug, Tumorigen, Mutagen; Reproductive Effector; Human Data. Nitrofen is a contact herbicide used for pre- and postemergency control of annual grasses and broadleaf weeds on a variety of food and ornamental crops. Occupational exposure to nitrofen, primarily through inhalation and dermal contact, may occur among workers at production facilities. Field handlers of the herbicide are subject to inhalation exposure during application procedures.
Permissible Exposure Limits in Air
No TEEL available.
Finland: carcinogen, 1999; Russia: STEL 1 mg/m³, 1993; United Kingdom: carcinogen, TWA 1 ppm (7.1 mg/m³), 2000. Russia set a MAC of 0.02 mg/m³ in ambient air in residential areas on a once-daily basis and 0.01 mg/m³ on an average daily basis. Pennsylvania has set a guideline for nitrofen in ambient air[60] of 0.75 μg/m³.
Permissible Concentration in Water: Nitrofen presumably falls under the EPA Priority Toxic Pollutant category of haloethers[6] but specific limits have not been set. Russia has set a limit of 4.0 mg/L in surface water.
Determination in water: Fish Tox = 92.54948000 ppb (INTERMEDIATE).
Routes of Entry: Inhalation, ingestion, skin and/or eye contact.

Harmful Effects and Symptoms
Short Term Exposure: Toxic by ingestion. Severe eye irritant. Causes skin irritation on contact. Inhalation can cause irritation to the respiratory tract. May cause difficult breathing; fatigue, and loss of appetite.
Long Term Exposure: Long term exposure may cause damage to the blood cells, causing low white cell (leukocyte) count, reduced hemoglobin, and reduced serum cholinesterase and erythrocyte catalase activities. May cause liver damage. May affect the nervous system.
Points of Attack: Blood, liver, kidneys, nervous system.
Medical Surveillance: Complete blood count (CBC). Liver function tests. Examination of the nervous system and interview for brain effects.
First Aid: Skin Contact: Flood all areas of body that have contacted the substance with water. Do not wait to remove contaminated clothing; do it under the water stream. Use soap to help assure removal. Isolate contaminated clothing when removed to prevent contact by others.[52]
Eye Contact: Remove any contact lenses at once. Immediately flush eyes well with copious quantities of water or normal saline for at least 20−30 min. Seek medical attention.
Inhalation: Leave contaminated area immediately; breathe fresh air. Proper respiratory protection must be supplied to any rescuers. If coughing, difficult breathing, or any other symptoms develop, seek medical attention at once, even if symptoms develop many hours after exposure.
Ingestion: Contact a physician, hospital, or poison center at once. If the victim is unconscious or convulsing, do not induce vomiting or give anything by mouth. Assure that the patient's airway is open and lay him on his side with his head lower that his body and transport immediately to a medical facility. If conscious and not convulsing, give a glass of water to dilute the substance. Vomiting should not be induced without a physician's advice.
Personal Protective Methods: Wear protective gloves and clothing to prevent any reasonable probability of skin contact. Safety equipment suppliers/manufacturers can provide recommendations on the most protective glove/clothing material for your operation. All protective clothing (suits, gloves, footwear, headgear) should be clean, available each day, and put on before work. Contact lenses should not be worn when working with this chemical. Wear dust-proof chemical goggles and face shield unless full face-piece respiratory protection is worn. Employees should wash immediately with soap when skin is wet or contaminated. Provide emergency showers and eyewash.
Respirator Selection: Follow regulations in OSHA 29CFR1910.134 or European Standard EN149. Use a NIOSH/MSHA- or European Standard EN149-approved respirator; or use an approved supplied-air respirator with a full face-piece operated in the positive-pressure mode, or with a full face-piece, hood, or helmet in the continuous-flow mode; or use a NIOSH/MSHA- or European Standard EN149-approved self-contained breathing apparatus with a

full face-piece operated in pressure-demand or other positive-pressure mode.

Storage: Color Code—Blue: Health Hazard/Poison: Store in a secure poison location. Prior to working with this chemical you should be trained on its proper handling and storage. Store in a refrigerator or a cool, dry place. A regulated, marked area should be established where this chemical is handled, used, or stored in compliance with OSHA Standard 1910.1045.

Shipping: Phenoxy pesticides, solid, toxic, require a shipping label of "POISONOUS/TOXIC MATERIALS." It falls in Hazard Class 6.1 and Packing Group II.

Spill Handling: Evacuate persons not wearing protective equipment from area of spill or leak until cleanup is complete. Remove all sources of ignition and dampen spilled material with 60−70% acetone to avoid airborne dust. Use HEPA vacuum or wet method to reduce dust during cleanup. Do not dry sweep. Collect powdered material in the most convenient and safe manner and deposit in sealed containers. Ventilate area after cleanup is complete. It may be necessary to contain and dispose of this chemical as a hazardous waste. If material or contaminated runoff enters waterways, notify downstream users of potentially contaminated waters. Contact your local or federal environmental protection agency for specific recommendations. If employees are required to clean up spills, they must be properly trained and equipped. OSHA 1910.120(q) may be applicable.

Fire Extinguishing: This chemical is a combustible solid, but does not readily ignite. Use dry chemical, carbon dioxide, water spray, or foam extinguishers. Poisonous gases, including nitrogen oxides and chlorine, are produced in fire. If material or contaminated runoff enters waterways, notify downstream users of potentially contaminated waters. Notify local health and fire officials and pollution control agencies. From a secure, explosion-proof location, use water spray to cool exposed containers. If cooling streams are ineffective (venting sound increases in volume and pitch, tank discolors, or shows any signs of deforming), withdraw immediately to a secure position. If employees are expected to fight fires, they must be trained and equipped in OSHA 1910.156. The only respirators recommended for firefighting are self-contained breathing apparatuses that have full face-pieces and are operated in a pressure-demand or other positive-pressure mode.

Disposal Method Suggested: Small quantities may be landfilled but large quantities should be incinerated.[22] In accordance with 40CFR165, follow recommendations for the disposal of pesticides and pesticide containers. Must be disposed properly by following package label directions or by contacting your local or federal environmental control agency or by contacting your regional EPA office.

Reference

New Jersey Department of Health and Senior Services. (September 2001). *Hazardous Substances Fact Sheet: Nitrofen.* Trenton, NJ

Nitrogen N:0470

Molecular Formula: N_2
Synonyms: Liquid nitrogen; Nitrogen, compressed; Nitrogen, cryogenic liquid; Nitrogen gas; Nitrogen, refrigerated liquid
CAS Registry Number: 7727-37-9
RTECS® Number: QW9700000
UN/NA & ERG Number: UN1066 (compressed)/121; UN1977 (refrigerated liquid)/120
EC Number: 231-783-9
Regulatory Authority and Advisory Bodies
US EPA, FIFRA 1998 Status of Pesticides: Active registration.
WGK (German Aquatic Hazard Class): Nonwater polluting agent.
Description: Nitrogen is a nonflammable, stable, odorless, cryogenic liquid or a compressed gas. Molecular weight = 28.02; Boiling point = −196°C; Freezing/Melting point = −210°C. Hazard Identification (based on NFPA-704 M Rating System): Health 2, Flammability 0, Reactivity 0. Slightly soluble in water.
Potential Exposure: Nitrogen is present in the air we breathe. Health effects may occur at concentrations above 80%. It has many medical and industrial uses including the quick freezing of food. The gas is used for purging, heat treating, food freezing, annealing, cooling, oil recovery, in the inert blanketing of sensitive materials and as a reactant in chemical synthesis of ammonia.
Incompatibilities: Containers may explode when heated.
Permissible Exposure Limits in Air: Before entering an enclosed space where nitrogen may be present, oxygen content should be tested to ensure that it is at least 19% by volume.
Determination in Air:
OSHA PEL: Simple asphyxiant—inert gases and vapors.
NIOSH REL: Simple asphyxiant—inert gases and vapors.
ACGIH TLV®[1]: Simple asphyxiant.
Protective Action Criteria (PAC)
TEEL-0: 796,000 ppm
PAC-1: 796,000 ppm
PAC-2: 832,000 ppm
PAC-3: 869,000 ppm
United Kingdom: asphyxiant, 2000; Argentina, Bulgaria, Columbia, Jordan, South Korea, New Zealand, Singapore, Vietnam: ACGIH TLV®: Simple asphyxiant: nitrogen is a simple asphyxiant inert gas or vapor; *oxygen* content should be at least 19%.
Routes of Entry: Inhalation.
Harmful Effects and Symptoms
Short Term Exposure: *Inhalation:* No significant toxic effects except as an asphyxiant; that is, it may threaten life if levels are so high as to reduce oxygen levels below 19%. Since nitrogen is odorless, colorless, and tasteless, there may not be adequate warning of high levels. Symptoms of lack of

oxygen may include nausea, drowsiness, blue coloration of skin and lips, unconsciousness, and death. *Skin, Eyes, Ingestion:* Liquid may cause frostbite and freezing burns

Long Term Exposure: No information is known at this time.

First Aid: *Inhalation:* Move person to fresh air. Give oxygen or artificial respiration as necessary. *Skin:* Remove liquid-soaked clothing after allowing to thaw. If frostbite has occurred, seek medical attention immediately; do *NOT* rub the affected areas or flush them with water. In order to prevent further tissue damage, do *NOT* attempt to remove frozen clothing from frostbitten areas. If frostbite has *NOT* occurred, immediately and thoroughly wash contaminated skin with soap and water. Seek medical attention. *Eyes:* Seek immediate medical attention if contact with liquid occurs. *Ingestion:* Seek medical attention as necessary.

Personal Protective Methods: Wear appropriate personal protective clothing to prevent the skin from becoming frozen from contact with the evaporating liquid or from contact with vessels containing the liquid. Safety equipment suppliers/manufacturers can provide recommendations on the most protective glove/clothing material for your operation. All protective clothing (suits, gloves, footwear, headgear) should be clean, available each day, and put on before work. Contact lenses should not be worn when working with this chemical. Wear splash-proof chemical goggles and face shield unless full face-piece respiratory protection is worn. Employees should wash immediately with soap when skin is wet or contaminated. Provide emergency showers and eyewash. If vapors, or liquid may occur, employees should be provided with special clothing designed to prevent the freezing of body tissues. All protective clothing (suits, gloves, footwear, headgear) should be clean, available each day, and put on before work.

Respirator Selection: Exposure to nitrogen is dangerous because it can replace oxygen and lead to suffocation. Only NIOSH/MSHA- or European Standard EN 149-approved self-contained breathing apparatus with a full face-piece operated in positive-pressure mode should be used in oxygen-deficient environments.

Storage: Color Code—Green: General storage may be used. Prior to working with this chemical you should be trained on its proper handling and storage. A regulated, marked area should be established where nitrogen is handled, used, or stored. Store liquid containers and cylinders in cool, well-ventilated areas. Use only in well-ventilated areas. Cylinders must be secured and protected against damage. Procedures for the handling, use and storage of cylinders should be in compliance with OSHA 1910.101 and 1910.169, as with the recommendations of the Compressed Gas Association.

Shipping: Nitrogen, *compressed*, or nitrogen, *refrigerated liquid [cryogenic liquid]*, requires a shipping label of "NONFLAMMABLE GAS." They fall in DOT Hazard Class 2.2.

Spill Handling: If liquid nitrogen is spilled or leaked, take the following steps: Restrict persons not wearing protective equipment from area of spill or leak until cleanup is complete. Ventilate the area of spill or leak. Stop the leak or move the container to a safe area and allow the liquid to evaporate. If nitrogen gas is leaked, take the following steps: Restrict persons not wearing protective equipment from area of leak until cleanup is complete. Stop flow of gas. If source of leak is a cylinder and the leak cannot be stopped in place, remove the leaking cylinder to a safe place in the open air, and repair leak, or allow cylinder to empty. Absorb liquids in vermiculite, dry sand, earth, peat, carbon, or a similar material, and deposit in sealed containers. It may be necessary to contain and dispose of this chemical as a hazardous waste. If material or contaminated runoff enters waterways, notify downstream users of potentially contaminated waters. Contact your local or federal environmental protection agency for specific recommendations. If employees are required to clean up spills, they must be properly trained and equipped. OSHA 1910.120(q) may be applicable.

Fire Extinguishing: Extinguish fire using an agent suitable for type of surrounding fire. Nitrogen itself does not burn. Vapors are heavier than air and will collect in low areas. Containers may explode in fire. Storage containers and parts of containers may rocket great distances, in many directions. If material or contaminated runoff enters waterways, notify downstream users of potentially contaminated waters. Notify local health and fire officials and pollution control agencies. From a secure, explosion-proof location, use water spray to cool exposed containers. If cooling streams are ineffective (venting sound increases in volume and pitch, tank discolors, or shows any signs of deforming), withdraw immediately to a secure position. If employees are expected to fight fires, they must be trained and equipped in OSHA 1910.156. The only respirators recommended for firefighting are self-contained breathing apparatuses that have full face-pieces and are operated in a pressure-demand or other positive-pressure mode.

Disposal Method Suggested: Vent to atmosphere.

References

New York State Department of Health. (March 1986). *Chemical Fact Sheet: Nitrogen.* Albany, NY: Bureau of Toxic Substance Assessment

New Jersey Department of Health and Senior Services. (September 2004). *Hazardous Substances Fact Sheet: Nitrogen.* Trenton, NJ

Nitrogen dioxide   N:0480

Molecular Formula: NO_2,
Common Formula: N_2O_4 (nitrogen tetroxide)
Synonyms: Dinitrogen dioxide; Dinitrogen dioxide, di-; Dinitrogen tetroxide (N_2O_4); Dioxido de nitrogeno (Spanish); Nitrogen peroxide; Nitrogen tetroxide

CAS Registry Number: 10102-44-0; 10544-72-6 (dinitrogen tetroxide)

RTECS® Number: QW9800000

UN/NA & ERG Number: UN1067 (dinitrogen tetroxide)/124

EC Number: 233-272-6 [*Annex I Index No.:* 007-002-00-0]; 234-126-4 [*Annex I Index No.:* 007-002-00-0] (dinitrogen tetroxide)

Regulatory Authority and Advisory Bodies

Department of Homeland Security Screening Threshold Quantity (pounds): *Theft hazard* 15 ($\geq 3.80\%$ concentration) (*dinitrogen tetroxide*).

Toxic Substance (World Bank).[15]

Air Pollutant Standard Set. See below, "Permissible Exposure Limits in Air" section.

OSHA 29CFR1910.119, Appendix A. Process Safety List of Highly Hazardous Chemicals, TQ = 250 lb.

Clean Water Act: Section 311 Hazardous Substances/RQ 40CFR117.3 (same as CERCLA, see below).

US EPA Hazardous Waste Number (RCRA No.): P078.

RCRA, 40CFR261, Appendix 8 Hazardous Constituents.

SUPERFUND/EPCRA 40CFR355, Extremely Hazardous Substances: TPQ = 100 lb (45.4 kg).

Reportable Quantity (RQ): 10 lb (4.54 kg).

US DOT 49CFR172.101, Inhalation Hazardous Chemical.

Canada, WHMIS, Ingredients Disclosure List Concentration 1.0%.

European/International Regulations (*nitrogen dioxide*; *nitrogen tetroxide*): Hazard Symbol: T + , N; Risk phrases: R8; R26; R34; Safety phrases: S1/2; S9; S26; S28; S36/37/39; S45 (see Appendix 4).

WGK (German Aquatic Hazard Class): 1—Slightly water polluting (*nitrogen dioxide, nitrogen tetroxide*).

Description: Nitrogen dioxide (and nitrogen tetroxide, the solid dimer) is a dark brown gas (above 21°C) or a yellow, fuming liquid or colorless solid with a pungent, acrid odor. The solid form is colorless below about −11°C; it is found structurally as N_2O_4. The odor threshold is 5 pm. Molecular weight = 46.01; Specific gravity (H_2O:1) = 1.44 (liquid at 20°C); Boiling point = 21°C; Freezing/Melting point = −9.4°C; Vapor pressure = 720 mmHg. Decomposes at 160°C. Hazard Identification (based on NFPA-704 M Rating System): Health 3, Flammability 0, Reactivity ₩ (oxidizer). Decomposes (reacts) with water.

Potential Exposure: Compound Description: Tumorigen, Mutagen; Reproductive Effector; Human Data. Nitrogen dioxide is found in automotive and diesel emissions. Nitrogen dioxide is an industrial chemical used as an intermediate in nitric and sulfuric acid manufacture, in the nitration of organic compounds, and as an oxidizer in liquid propellant rocket fuel combinations. It is also used in firefighting, welding, and brazing.

Incompatibilities: A strong oxidizer. Reacts violently with combustible matter, chlorinated hydrocarbons, ammonia, carbon disulfide, reducing materials. Reacts with water, forming nitric acid and nitric oxide. Attacks steel in the presence of moisture.

Permissible Exposure Limits in Air

Conversion factor: 1 ppm = 1.88 mg/m³ at 25°C & 1 atm.

OSHA PEL: 5 ppm/9 mg/m³ Ceiling Concentration.

NIOSH REL: 1 ppm/1.8 mg/m³ STEL.

ACGIH TLV®[1]: 3 ppm/5.6 mg/m³ TWA; 5 ppm/9.4 mg/m³ STEL, not classifiable as a human carcinogen.

Protective Action Criteria (PAC)*

TEEL-0: 0.5 ppm

PAC-1: **0.50** ppm

PAC-2: **12** ppm

PAC-3: **20** ppm

*AEGLs (Acute Emergency Guideline Levels) & ERPGs (Emergency Response Planning Guideline) are in **bold face**.

10544-72-6 (dinitrogen tetroxide)

TEEL-0: 0.3 ppm

PAC-1: **0.94** ppm

PAC-2: **23** ppm

PAC-3: **38** ppm

*AEGLs (Acute Emergency Guideline Levels) & ERPGs (Emergency Response Planning Guideline) are in **bold face**.

DFG MAK: 0.5 ppm/0.95 mg/m³ TWA; Peak Limitation Category I(1); Pregnancy Risk Group D; Carcinogen Category 3B.

NIOSH IDLH: 20 ppm.

Arab Republic of Egypt: TWA 3 ppm (6 mg/m³), 1993; Austria: MAK 3 ppm (6 mg/m³), 1999; Denmark: TWA 3 ppm (5.6 mg/m³), 1999; Finland: TWA 3 ppm (6 mg/m³); STEL 6 ppm (12 mg/m³), 1999; France: VLE 3 ppm (6 mg/m³), 1999; Japan: pending, 1999; the Netherlands: MAC-TGG 4 mg/m³, 2003; the Philippines: TWA 5 ppm (9 mg/m³), 1993; Poland: MAC (TWA) 5 mg/m³; MAC (STEL) 10 mg/m³, 1999; Sweden: NGV 2 ppm (4 mg/m³), TGV 5 ppm (13 mg/m³), 1999; Switzerland: MAK-W 3 ppm (6 mg/m³), KZG-W 6 ppm (12 mg/m³), 1999; Thailand: TWA 5 ppm (9 mg/m³), 1993; Turkey: TWA 5 ppm (9 mg/m³), 1993; United Kingdom: TWA 3 ppm (5.7 mg/m³); STEL 5 ppm (9.6 mg/m³), 2000; Argentina, Bulgaria, Columbia, Jordan, South Korea, New Zealand, Singapore, Vietnam: ACGIH TLV®: STEL 5 ppm. Russia[43] set a MAC of 0.085 mg/m³ (85 µg/m³) for ambient air in residential areas. Several states have set guidelines or standards for nitrogen dioxide in ambient air[60] ranging from 100 µg/m³ (Arizona and Connecticut) to 143 µg/m³ (Nevada).

Protective Action Criteria (PAC)*

10544-72-6 (dinitrogen tetroxide)

TEEL-0: 0.3 ppm

PAC-1: **0.94** ppm

PAC-2: **23** ppm

PAC-3: **38** ppm

*AEGLs (Acute Emergency Guideline Levels) & ERPGs (Emergency Response Planning Guideline) are in **bold face**.

Determination in Air: Use NIOSH Analytical Method (IV) #6014, Nitric oxide and nitrogen dioxide; OSHA Analytical Methods ID-109 and ID-182.

Routes of Entry: Inhalation, ingestion, skin and/or eye contact.

Harmful Effects and Symptoms

Short Term Exposure: Nitrogen dioxide and its vapors irritate the eyes, skin, and respiratory tract. Inhalation exposure can cause pulmonary edema, a medical emergency that can be delayed for several hours. This can cause death. Acute exposure to nitrogen dioxide may be severe and result in a weak, rapid pulse; cyanosis (blue tint to the skin and mucous membranes); and circulatory collapse. Cough, dyspnea (shortness of breath), bronchitis, pneumonitis, and pulmonary edema may occur following inhalation exposure. Gastrointestinal symptoms include nausea and abdominal pain. Fatigue, lethargy, restlessness, fever, anxiety, headache, mental confusion, and loss of consciousness may also occur. Contact with the skin and mucous membranes may result in severe irritation and burns. When liquid nitrogen dioxide contacts the skin, frostbite will result. *Inhalation:* 10−20 ppm can cause mild irritation of the nose and throat. 25−50 ppm can cause an inflammation of the lungs, such as bronchitis or pneumonia. Levels above 100 ppm can cause death. Only highly concentrated fumes cause immediate symptoms, such as coughing, choking, headache, nausea, and stomach or chest pain. However, exposures to less concentrated fumes may produce these symptoms after 5−72 h. Rapid and shallow breathing, bluish coloration in skin, and unconsciousness may develop along with lung irritation or congestion. *Skin:* Can cause severe irritation and burns. *Eyes:* Levels of 10−20 ppm can cause irritation. Higher vapor concentration can cause eye injury. Contact with liquid can cause severe chemical burns. *Ingestion:* Can cause burns in mouth, throat, and stomach.

Long Term Exposure: Can cause headache, weakness, loss of sleep and appetite, sores in nose and mouth, nausea, and erosion of teeth. Exposure to 0.4−2.7 ppm for 4−6 years has been associated with emphysema and bronchitis. Genetic changes have been shown in experimental animals; possibly causes toxic effects on human reproduction. Nitrogen dioxide may affect the immune system, resulting in a decreased resistance to infection.

Points of Attack: Respiratory system, lungs, cardiovascular system.

Medical Surveillance: NIOSH lists the following tests: Electrocardiogram, expired air, pulmonary function tests: forced vital capacity, forced expiratory volume (1 s); sputum cytology; white blood cell count/differential. Preplacement and periodic examinations should be concerned particularly with the skin, eyes, and with significant pulmonary and heart diseases. Smoking history should be known. Methemoglobin studies may be of interest if exposure to nitric oxide is present. In the case of nitric acid vapor mist exposure, dental effects may be present.

First Aid: If this chemical gets into the eyes, remove any contact lenses at once and irrigate immediately for at least 15 min, occasionally lifting upper and lower lids. Seek medical attention immediately. If this chemical contacts the skin, remove contaminated clothing and wash immediately with soap and water. Seek medical attention immediately. If this chemical has been inhaled, remove from exposure, begin rescue breathing (using universal precautions, including resuscitation mask) if breathing has stopped and CPR if heart action has stopped. Transfer promptly to a medical facility. When this chemical has been swallowed, get medical attention. Give large quantities of water and do not induce vomiting. Do not make an unconscious person vomit. Medical observation is recommended for 24−48 h after breathing overexposure, as pulmonary edema may be delayed. As first aid for pulmonary edema, a doctor or authorized paramedic may consider administering a corticosteroid spray. If frostbite has occurred, seek medical attention immediately; do *NOT* rub the affected areas or flush them with water. In order to prevent further tissue damage, do *NOT* attempt to remove frozen clothing from frostbitten areas. If frostbite has *NOT* occurred, immediately and thoroughly wash contaminated skin with soap and water.

Note to physician: Inhalation: bronchodilators, decongestants and oxygen may be used if necessary. Corticosteroids are useful for treating pneumonitis. Treat for methemoglobinemia. Spectrophotometry may be required for precise determination of levels of methemoglobin in urine.

Personal Protective Methods: Wear protective gloves and clothing to prevent any reasonable probability of skin or eye contact. Safety equipment suppliers/manufacturers can provide recommendations on the most protective glove/clothing material for your operation. All protective clothing (suits, gloves, footwear, headgear) should be clean, available each day, and put on before work. Contact lenses should not be worn when working with this chemical. Individuals should be equipped with supplied air respirators with full-face-piece or chemical goggles, and enclosed areas should be properly ventilated before entering. An observer equipped with appropriate respiratory protection should be outside the area and standing by to supply any aid needed. Employees should wash immediately with soap when skin is wet or contaminated. Provide emergency showers and eyewash, where nitrogen oxides may accumulate (for example, silos).

Respirator Selection: NIOSH: *20 ppm:* Sa:Cf (APF = 25)* (any supplied-air respirator operated in a continuous-flow mode) SCBAF (APF = 50) (any self-contained breathing apparatus with a full face-piece) or SaF (APF = 50) (any supplied-air respirator with a full face-piece). *Emergency or planned entry into unknown concentrations or IDLH conditions:* SCBAF: Pd,Pp (APF = 10,000) (any self-contained breathing apparatus that has a full face-piece and is operated in a pressure-demand or other positive-pressure mode) or SaF: Pd,Pp: ASCBA (APF = 10,000) (any supplied-air respirator that has a full face-piece and is operated in a pressure-demand or other positive-pressure mode in combination with an auxiliary self-contained breathing apparatus operated in a pressure-demand or other positive-pressure mode). *Escape:* GmFS* [any air-purifying, full-face-piece respirator (gas mask) with a chin-style, front- or back-mounted canister protection against the compound of concern] or SCBAE (any appropriate escape-type, self-contained breathing apparatus).

*Only nonoxidizable sorbents are allowed (NOT charcoal).

Storage: Color Code—Yellow Stripe: Reactivity Hazard; Store separately in an area isolated from flammables, combustibles, or other yellow coded materials. Prior to working with this chemical you should be trained on its proper handling and storage. Protect containers from physical damage. Store separately from combustible, organic, and readily oxidizable materials. Transfer facilities should be outdoors.

Shipping: Nitrogen dioxide, liquefied, requires a shipping label of "POISON GAS, OXIDIZER." It falls in Hazard Class 2.3 and Packing Group I. It is a violation of transportation regulations to refill compressed gas cylinders without the express written permission of the owner.

Spill Handling: Keep unnecessary people away; isolate hazard area and deny entry. Stay upwind; keep out of low areas. Ventilate closed spaces before entering them. Evacuate area endangered by gas. For water spills, neutralize with agricultural lime (slaked lime), crushed limestone, or sodium bicarbonate. For an air spill, apply water spray or mist to knock down vapors. Vapor knockdown water is corrosive or toxic and should be diked for containment. Keep combustibles (wood, paper, oil, etc.) away from spilled material. Stop leak if you can do so without risk. Use water spray to reduce vapor but do not put water on leak or spill area. Isolate area until gas has dispersed. It may be necessary to contain and dispose of this chemical as a hazardous waste. If material or contaminated runoff enters waterways, notify downstream users of potentially contaminated waters. Contact your local or federal environmental protection agency for specific recommendations. If employees are required to clean up spills, they must be properly trained and equipped. OSHA 1910.120(q) may be applicable.

Initial isolation and protective action distances

Distances shown are likely to be affected during the first 30 min after materials are spilled and could increase with time. If more than one tank car, cargo tank, portable tank, or large cylinder involved in the incident is leaking, the protective action distance may need to be increased. You may need to seek emergency information from CHEMTREC at (800) 424-9300 or seek professional environmental engineering assistance from the US EPA Environmental Response Team at (908) 548-8730 (24-h response line).

Small spills (From a small package or a small leak from a large package)

Dinitrogen tetroxide nitrogen dioxide

First: Isolate in all directions (feet/meters)   100/30
Then: Protect persons downwind (miles/kilometers)
Day   0.1/0.2
Night   0.2/0.3

Large spills (From a large package or from many small packages)

First: Isolate in all directions (feet/meters)   1250/400

Then: Protect persons downwind (miles/kilometers)
Day   0.7/1.1
Night   1.9/3.1

Dinitrogen tetroxide and nitric oxide mixture; nitric oxide and dinitrogen tetroxide mixture; nitric oxide and nitrogen dioxide mixture; nitric oxide and nitrogen tetroxide mixture; nitrogen dioxide and nitric oxide mixture; nitrogen tetroxide and nitric oxide mixture

First: Isolate in all directions (feet/meters)   100/30
Then: Protect persons downwind (miles/kilometers)
Day   0.1/0.2
Night   0.4/0.6

Large spills (From a large package or from many small packages)

First: Isolate in all directions (feet/meters)   300/100
Then: Protect persons downwind (miles/kilometers)
Day   0.4/0.6
Night   1.4/2.3

Fire Extinguishing: NO_2 is nonflammable but supports combustion. Wearing proper equipment, shut off flow of gas. Use water spray to keep containers cool and to also direct escaping gas away from personnel attempting to shut off leak. Do not extinguish the fire unless the flow of gas can be stopped and any remaining gas is out of the line. Specially trained personnel may use fog lines to cool exposures and let the fire burn itself out. Vapors are heavier than air and will collect in low areas. Vapors may travel long distances to ignition sources and flashback. Vapors in confined areas may explode when exposed to fire. Containers may explode in fire. Storage containers and parts of containers may rocket great distances, in many directions. If material or contaminated runoff enters waterways, notify downstream users of potentially contaminated waters. Notify local health and fire officials and pollution control agencies. From a secure, explosion-proof location, use water spray to cool exposed containers. If cooling streams are ineffective (venting sound increases in volume and pitch, tank discolors, or shows any signs of deforming), withdraw immediately to a secure position. If cylinders are exposed to excessive heat from fire or flame contact, withdraw immediately to a secure location. If employees are expected to fight fires, they must be trained and equipped in OSHA 1910.156. The only respirators recommended for firefighting are self-contained breathing apparatuses that have full face-pieces and are operated in a pressure-demand or other positive-pressure mode.

Disposal Method Suggested: Destroy this chemical by incineration with the addition of hydrocarbon fuel, controlled in such a way that combustion products are elemental nitrogen, CO_2, and water. Consult with environmental regulatory agencies for guidance on acceptable disposal practices. Generators of waste containing this contaminant (≥ 100 kg/mo) must conform with EPA regulations governing storage, transportation, treatment, and waste disposal.

References

National Institute for Occupational Safety and Health. (1976). *Criteria for a Recommended Standard:*

Occupational Exposure to Oxides of Nitrogen. NIOSH Document No. 76-149

World Health Organization. (1977). *Oxides of Nitrogen, Environmental Health Criteria No. 4.* Geneva, Switzerland

Sax, N. I. (Ed.). *Dangerous Properties of Industrial Materials Report,* 1, No. 5, 74−76 (1981) and 5, No. 6, 81−83 (1985)

US Environmental Protection Agency. (November 30, 1987). *Chemical Hazard Information Profile: Nitrogen Dioxide.* Washington, DC: Chemical Emergency Preparedness Program

New York State Department of Health. (March 1986). *Chemical Fact Sheet: Nitrogen Dioxide.* Version 30. Albany, NY: Bureau of Toxic Substance Assessment

Nitrogen oxides   N:0490

Molecular Formula: NO (nitric oxide); NO_2 (nitrogen dioxide); N_2O_3 (nitrogen trioxide); N_2O_4 (dinitrogen tetroxide); N_2O_5 (nitrogen pentoxide)

Synonyms: nitrogen oxides; Oxides of nitrogen

CAS Registry Number: 10024-97-2 (nitrous oxide); 10102-43-9 (nitric monoxide); 10544-73-7 (nitrogen trioxide); 10102-03-1 (dinitrogen pentoxide)

UN/NA & ERG Number: UN1660 (nitric oxide)/124; UN2201 (Nitrous oxide, refrigerated liquid)/122; UN2421 (Nitrogen trioxide)/124; UN1067 (dinitrogen tetroxide)/124

EC Number: 233-032-0 (nitrous oxide); 233-271-0 9 (nitrogen monoxide); 234-128-5 (dinitrogen trioxide); 233-264-2 (dinitrogen pentoxide)

Regulatory Authority and Advisory Bodies

Department of Homeland Security Screening Threshold Quantity (pounds): Sabotage/Contamination Hazard: A placarded amount (commercial grade). (*Nitrogen trioxide*).

See entries for specific compound.

US DOT 49CFR172.101, Inhalation Hazardous Chemical (nitrogen dioxide, nitric oxide, nitrogen tetroxide, nitrogen trioxide).

European/International Regulations: not listed in Annex 1.

WGK (German Aquatic Hazard Class): 1—Slightly water polluting (for above CAS number except dinitrogen pentoxide; no hazard value assigned).

Description: Nitrogen oxides are colorless (NO, N_2O) to brick red (NO_2) gases with little or no odor or an irritating odor (NO_2). When frozen they appear to be white to bluish-white snow. Molecular weight- = varies by entry; Boiling point = $-52°C$ (*nitric oxide*); $21°C$ (*nitrogen dioxide*); $47°C$ (*nitrogen pentoxide*); Freezing/Melting point = $-164°C$ (*nitric oxide*); $-9°C$ (*nitrogen dioxide*); $30°C$ (*nitrogen pentoxide*). Hazard Identification (based on NFPA-704 M Rating System) (*nitric oxide*): Health 4, Flammability 0, Reactivity 1; (*nitrogen dioxide*): Health 2, Flammability 0, Reactivity 0. They decompose in water.

Potential Exposure: See entries for specific compound.

Incompatibilities: Stability and reactivity are variable depending on specific compound. All are strong oxidizers that enhance the combustion of easily oxidizer materials, reducing agents, combustibles, organics.

Permissible Exposure Limits in Air

Protective Action Criteria (PAC) *see also nitrous oxide* entry

TEEL-0: 50 ppm

PAC-1: 150 ppm

PAC-2: 10,000 ppm

PAC-3: 20,000 ppm

Nitric oxide; nitrogen dioxide; nitrogen tetroxide

TEEL-0: 0.5 ppm

PAC-1: 0.5 ppm

PAC-2: 12 ppm

PAC-3: 20 ppm

Nitrogen trioxide

TEEL-0: 5 ppm

PAC-1: 15 ppm

PAC-2: 100 ppm

PAC-3: 500 ppm

See entries for specific compound. The US EPA has set national ambient air quality standards of 0.05 ppm (100 $\mu g/m^3$) for nitrogen oxides as an annual arithmetic mean value.

Harmful Effects and Symptoms

Short Term Exposure: Serious health hazards. Corrosive to the eyes and skin. Harmful if inhaled. See "Nitric oxide," "Nitrogen dioxide," etc.

First Aid: If any of these chemicals gets into the eyes, remove any contact lenses at once and irrigate immediately for at least 15 min, occasionally lifting upper and lower lids. Seek medical attention immediately. If this chemical contacts the skin, remove contaminated clothing and wash immediately with soap and water. Seek medical attention immediately. If this chemical has been inhaled, remove from exposure, begin rescue breathing (using universal precautions, including resuscitation mask) if breathing has stopped and CPR if heart action has stopped. Transfer promptly to a medical facility. When this chemical has been swallowed, get medical attention. Give large quantities of water and induce vomiting. Do not make an unconscious person vomit. Medical observation is recommended for 24−48 h after breathing overexposure, as pulmonary edema may be delayed. As first aid for pulmonary edema, a doctor or authorized paramedic may consider administering a corticosteroid spray.

Personal Protective Methods: Wear protective gloves and clothing to prevent any reasonable probability of skin contact. Safety equipment suppliers/manufacturers can provide recommendations on the most protective glove/clothing material for your operation. All protective clothing (suits, gloves, footwear, headgear) should be clean, available each day, and put on before work. Contact lenses should not be worn when working with this chemical. Wear gas-proof chemical goggles and face shield unless full face-piece respiratory protection is worn. Employees should wash

immediately with soap when skin is wet or contaminated. Provide emergency showers and eyewash.

Respirator Selection: Follow regulations in OSHA 29CFR1910.134 or European Standard EN149. Use a NIOSH/MSHA- or European Standard EN149-approved respirator; or use an approved supplied-air respirator with a full face-piece operated in the positive-pressure mode, or with a full face-piece, hood, or helmet in the continuous-flow mode; or use a NIOSH/MSHA- or European Standard EN149-approved self-contained breathing apparatus with a full face-piece operated in pressure-demand or other positive-pressure mode.

Storage: Color Code—Yellow: Reactive Hazard; Store in a location separate from other materials, especially flammables and combustibles. Prior to working with nitrogen oxides you should be trained on its proper handling and storage. Store in tightly closed containers in a cool, well-ventilated area away from oxidizable materials. Outside or detached storage is preferred. Do not put on wooden floors. See NFPA 43A *Code for the Storage of Liquid and Solid Oxidizers.*

Shipping: See separate entries for specific compound. See each chemical.

Spill Handling: See "Nitric oxide", "Nitrogen dioxide", etc. Runoff of less volatile nitrogen oxides may contain nitric acid.

Fire Extinguishing: See "Nitric oxide," "Nitrogen dioxide," etc. Nitrogen oxides enhance the activity of an existing fire.

References

National Academy of Sciences. (1977). *Medical and Biologic Effects of Environmental Pollutants: Nitrogen Oxides.* Washington, DC

US Environmental Protection Agency. (1993). *Air Quality Criteria for Oxides of Nitrogen.* Research Triangle Park, NC: Criteria and Assessment Office

Nitrogen trifluoride   N:0500

Molecular Formula: F_3N
Common Formula: NF_3
Synonyms: Nitrogen fluoride; Trifluoramine; Trifluorammonia
CAS Registry Number: 7783-54-2
RTECS® Number: QX1925000
UN/NA & ERG Number: UN2451/122
EC Number: 232-007-1
Regulatory Authority and Advisory Bodies
Air Pollutant Standard Set. See below, "Permissible Exposure Limits in Air" section.
OSHA 29CFR1910.119, Appendix A. Process Safety List of Highly Hazardous Chemicals, TQ = 5000 lb (2270 kg).
US DOT 49CFR172.101, Inhalation Hazardous Chemical.

Canada, WHMIS, Ingredients Disclosure List Concentration 1.0%.
European/International Regulations: not listed in Annex 1.
WGK (German Aquatic Hazard Class): No value assigned.

Description: Nitrogen trifluoride is a colorless gas with a moldy odor. Shipped as a nonliquefied compressed gas. Molecular weight = 71.01; Boiling point = −129°C; Freezing/Melting point = −206.7°C; Relative vapor density (air = 1) 2.46 at 20°C; Vapor pressure = >1 atm; Hazard Identification (based on NFPA-704 M Rating System): Health 3, Flammability 0, Reactivity ₩. Slightly soluble in water (reactive).

Potential Exposure: This material has been used in chemical synthesis and as an oxidizer for high-energy fuels (as an oxidizer in rocket propellant combinations).

Incompatibilities: A powerful oxidizer. Reacts with oil, grease, reducing agents and other oxidizable materials, combustibles, organics, ammonia, carbon monoxide, methane, hydrogen, hydrogen sulfide, activated charcoal, diborane, water.

Permissible Exposure Limits in Air
Conversion factor: 1 ppm = 2.90 mg/m^3 at 25°C & 1 atm.
OSHA PEL: 10 ppm/29 mg/m^3 TWA.
NIOSH REL: 10 ppm/29 mg/m^3 TWA.
ACGIH TLV®[1]: 10 ppm/29 mg/m^3 TWA; BEI_M issued for Methemoglobin inducers.
NIOSH IDLH: 1000 ppm.
Protective Action Criteria (PAC)*
TEEL-0: 10 ppm
PAC-1: **200** ppm
PAC-2: **530** ppm
PAC-3: **860** ppm
*AEGLs (Acute Emergency Guideline Levels) & ERPGs (Emergency Response Planning Guideline) are in **bold face**.
United Kingdom: TWA 10 ppm (30 mg/m^3); STEL 15 ppm (44 mg/m^3), 2000; United Kingdom: TWA 2.5 $mg[F]/m^3$, 2000; the Netherlands: MAC-TGG 29 mg/m^3, 2003. Several states have set guidelines or standards for NF_3 in ambient air[60] ranging from 0.3 mg/m^3 (North Dakota) to 0.5 mg/m^3 (Virginia) to 0.58 mg/m^3 (Connecticut) to 0.714 mg/m^3 (Nevada).

Determination in Air: No NIOSH Analytical Method available.

Permissible Concentration in Water: Fluoride guidelines are 1.8 mg/L in Arizona, 2.4 mg/L in Maine, and 4.0 mg/L according to EPA.

Routes of Entry: Inhalation.

Harmful Effects and Symptoms

Short Term Exposure: Insufficient data are available on the effect of this substance on human health; therefore, utmost care must be taken. May be corrosive to eyes, skin, and respiratory tract. In animals: anoxia, cyanosis, methemoglobinemia, weakness, dizziness, headache, liver, kidney injury. (NIOSH).

Long Term Exposure: However, a Japanese source[24] states that NF_3 is corrosive to tissue and that teeth and bones are affected on long inhalation. See also, "Fluoride," above.

Points of Attack: Blood, liver, kidneys.

Medical Surveillance: NIOSH lists the following tests: whole blood (chemical/metabolite), methemoglobin; complete blood count. Liver and kidney function tests. Urine fluoride test (levels above 3−4 mg/L at the end of exposure represent increased exposure).

First Aid: If this chemical gets into the eyes, remove any contact lenses at once and irrigate immediately for at least 15 min, occasionally lifting upper and lower lids. Seek medical attention immediately. If this chemical contacts the skin, remove contaminated clothing and wash immediately with soap and water. Seek medical attention immediately. If this chemical has been inhaled, remove from exposure, begin rescue breathing (using universal precautions, including resuscitation mask) if breathing has stopped and CPR if heart action has stopped. Transfer promptly to a medical facility. When this chemical has been swallowed, get medical attention. Give large quantities of water and induce vomiting. Do not make an unconscious person vomit. If frostbite has occurred, seek medical attention immediately; do NOT rub the affected areas or flush them with water. In order to prevent further tissue damage, do NOT attempt to remove frozen clothing from frostbitten areas. If frostbite has NOT occurred, immediately and thoroughly wash contaminated skin with soap and water.

Note to physician: Treat for methemoglobinemia. Spectrophotometry may be required for precise determination of levels of methemoglobin in urine.

Personal Protective Methods: Wear appropriate personal protective clothing to prevent the skin from becoming frozen from contact with the evaporating liquid or from contact with vessels containing the liquid. No protective devices other than respirators are indicated by NIOSH. Rubber gloves, face shield, and overalls are suggested by others, however.[24] Specific engineering controls are recommended in NIOSH Criteria Document #76-103: *Inorganic fluorides.*

Respirator Selection: *Up to 100 ppm:* CcrS (APF = 10) [any chemical cartridge respirator with cartridge(s) providing protection against the compound of concern] or Sa (APF = 10) (any supplied-air respirator). *Up to 250 ppm:* Sa:Cf (APF = 25) (any supplied-air respirator operated in a continuous-flow mode) or PaprS (APF = 25) [any powered, air-purifying respirator with cartridge(s) providing protection against the compound of concern]. *Up to 500 ppm:* CcrFS (APF = 50) [any chemical cartridge respirator with a full face-piece and cartridge(s) providing protection against the compound of concern] or GmFS (APF = 50) [any air-purifying, full-face-piece respirator (gas mask) with a chin-style, front- or back-mounted canister providing protection against the compound of concern] or PaprTS (APF = 50) [any powered, air-purifying respirator with a tight-fitting face-piece and cartridge(s) providing protection against the compound of concern]; or *SaT: Cf (APF = 50) (any

supplied-air respirator that has a tight-fitting face-piece and is operated in a continuous-flow mode); or *SCBAF (APF = 50) (any self-contained breathing apparatus with a full face-piece) or SAF (any supplied-air respirator with a full face-piece). *Up to 1000 ppm:* SaF: Pd,Pp (APF = 2000) (any supplied-air respirator that has a full face-piece and is operated in a pressure-demand or other positive-pressure mode). *Emergency or planned entry into unknown concentrations or IDLH conditions:* SCBAF: Pd,Pp (APF = 10,000) (any self-contained breathing apparatus that has a full face-piece and is operated in a pressure-demand or other positive-pressure mode) or SaF: Pd,Pp: ASCBA (any supplied-air respirator that has a full face-piece and is operated in a pressure-demand or other positive-pressure mode in combination with an auxiliary self-contained breathing apparatus operated in a pressure-demand or other positive-pressure mode). *Escape:* GmFS (APF = 50) [any air-purifying, full-face-piece respirator (gas mask) with a chin-style, front- or back-mounted canister providing protection against the compound of concern] or SCBAE (any appropriate escape-type, self-contained breathing apparatus).

*Substance reported to cause eye irritation or damage; may require eye protection.

Storage: Color Code—Yellow Stripe: Reactivity Hazard; Store separately in an area isolated from flammables, combustibles, or other yellow coded materials. Prior to working with this chemical you should be trained on its proper handling and storage. High concentrations cause a deficiency of oxygen with the risk of unconsciousness or death. Check that oxygen content is at least 19% before entering storage or spill area. Procedures for the handling, use, and storage of cylinders should be in compliance with OSHA 1910.101 and 1910.169, as with the recommendations of the Compressed Gas Association. See OSHA Standard 1910.104 and NFPA 43A *Code for the Storage of Liquid and Solid Oxidizers* for detailed handling and storage regulations.

Shipping: Nitrogen trifluoride, compressed, requires a shipping label of "POISON GAS, OXIDIZER." It falls in Hazard Class 2.3. It is a violation of transportation regulations to refill compressed gas cylinders without the express written permission of the owner.

Spill Handling: If in a building, evacuate building and confine vapors by closing doors and shutting down HVAC systems. Restrict persons not wearing protective equipment from area of spill or leak until cleanup is complete. Remove all ignition sources. Ventilate area of spill or leak to disperse the gas. Wear chemical protective suit with self-contained breathing apparatus to combat spills. Stay upwind and use water spray to "knock down" vapor; contain runoff. Stop the flow of gas, if it can be done safely from a distance. If source is a cylinder and the leak cannot be stopped in place, remove the leaking cylinder to a safe place, and repair leak or allow cylinder to empty. Keep this chemical out of confined spaces, such as a sewer, because of the possibility of explosion, unless the sewer is designed to prevent the buildup of explosive concentrations. If employees are required to clean up spills, they must be properly trained and equipped. OSHA 1910.120(q) may be applicable.

Fire Extinguishing: This material is a nonflammable gas. Poisonous gases, including fluorine, are produced in fire. Do not extinguish the fire unless the flow of gas can be stopped and any remaining gas is out of the line. Specially trained personnel may use fog lines to cool exposures and let the fire burn itself out. Vapors are heavier than air and will collect in low areas. Containers may explode in fire. Storage containers and parts of containers may rocket great distances, in many directions. If material or contaminated runoff enters waterways, notify downstream users of potentially contaminated waters. Notify local health and fire officials and pollution control agencies. From a secure, explosion-proof location, use water spray to cool exposed containers. If cooling streams are ineffective (venting sound increases in volume and pitch, tank discolors, or shows any signs of deforming), withdraw immediately to a secure position. If cylinders are exposed to excessive heat from fire or flame contact, withdraw immediately to a secure location. If employees are expected to fight fires, they must be trained and equipped in OSHA 1910.156. The only respirators recommended for firefighting are self-contained breathing apparatuses that have full face-pieces and are operated in a pressure-demand or other positive-pressure mode.

Disposal Method Suggested: Vent into large volume of concentrated reducing agent (bisulfites, ferrous salts, or hypo) solution, then neutralize and flush to sewer with large volumes of water.

References

National Institute for Occupational Safety and Health. (1975). *Criteria for a Recommended Standard: Occupational Exposure to Inorganic Fluorides.* NIOSH Document No. 75-103. Washington, DC

New Jersey Department of Health and Senior Services. (April 2000). *Hazardous Substances Fact Sheet: Nitrogen Dioxide.* Trenton, NJ

Nitroglycerin   N:0510

Molecular Formula: $C_3H_5N_3O_9$
Common Formula: $C_3H_5(NO_3)_3$
Synonyms: Angibid; Anginine; Angiolingual; Angorin; Blasting gelatin; Blasting oil; Cardmist; Glonoin; Glucor nitro; Glycerol nitric acid triester; Glycerol (trinitrate de) (French); Glycerol trinitrate; Glyceryl nitrate; Glyceryl trinitrate; GTN; Klavi kordal; Lenitral; Myocon; Myoglycerin; NG; Niglycon; Niong; Nitora; Nitric acid triester of glycerol; Nitrin; Nitrine; Nitrine-TDC; Nitro-dur; Nitroglicerina (Spanish); Nitroglycerine; Nitroglycerol; Nitroglyn; Nitrol; Nitrolan; Nitrolent; Nitroletten; Nitrolingual; Nitrolowe; Nitronet; Nitrong; Nitrorectal; Nitro-Span; Nitrostabilin; Nitrostat; Nitrozell retard; NK-843; NTG; Perglottal; 1,2,3-Propanetriyl nitrate; 1,2,3-Propanetrol, trinitrate; Pyro-glycerine; SK-106N; SNG; Soup; Spirit of glonoin; Spirit of glyceryl trinitrate; Spirit of trinitroglycerin; Temponitrin; TNG; Trinitrin; Trinitroglycerin; Trinitroglycerol

CAS Registry Number: 55-63-0
RTECS® Number: QX2100000
UN/NA & ERG Number: UN1204 (solution in alcohol with not >1% nitroglycerin)/127; UN3064 (solution in alcohol, with >1% but not >5% nitroglycerin)/127; UN0144 (solution in alcohol, with >1% but not >10% nitrogylcerin)/112; UN0143 (desensitized with not <40% nonvolatile, water-insoluble phlegmatizer, by mass)/112; UN3319 (nitroglycerin mixture with more than 2% but not >10% Nitroglycerin, desensitized)/113; UN3343 (Nitroglycerin mixture, desensitized, liquid, flammable, n.o.s., with not >30% Nitroglycerin)/113; UN3357 (Nitroglycerin mixture, desensitized, liquid, n.o.s., with not >30% Nitroglycerin)/113

EC Number: 200-240-8 [*Annex I Index No.:* 603-034-00-X]
Regulatory Authority and Advisory Bodies
Department of Homeland Security Screening Threshold Quantity (pounds): *Release hazard* 5000 commercial grade); *Theft hazard* 400 (commercial grade).
FDA—proprietary drug.
Air Pollutant Standard Set. See below, "Permissible Exposure Limits in Air" section.
US EPA Hazardous Waste Number (RCRA No.): P081.
RCRA, 40CFR261, Appendix 8 Hazardous Constituents.
Reportable Quantity (RQ): 10 lb (4.54 kg).
EPCRA Section 313 Form R *de minimis* concentration reporting level: 1.0%.
European/International Regulations: Hazard Symbol: E, T +, N; Risk phrases: R3; R26/27/28; R33; R51/53; Safety phrases: S1/2; S28; S33; S35; S36/37; S45; S61 (see Appendix 4).
WGK (German Aquatic Hazard Class): No value assigned.
Description: Nitroglycerin is a pale yellow liquid or crystalline solid (below 13°C). Molecular weight = 227.11; Specific gravity (H_2O:1) = 1.6; Boiling point = begins to decompose at 50–60°C; explodes at 261°C; Freezing/Melting point = 13.3°C; Vapor pressure = 0.0003 mmHg at 20°C; Flash point = explodes; Autoignition temperature = 261°C. Hazard Identification (based on NFPA-704 M Rating System): Health 2, Flammability 2, Reactivity 4. Practically insoluble in water.
Potential Exposure: Compound Description: Tumorigen, Drug, Mutagen; Reproductive Effector; Human Data; Primary Irritant. An explosive ingredient in dynamite (20–40%) with ethylene glycol dinitrate (80–60%). It is also used in making other explosives, rocket propellants, and medicine (vasodilator).
Incompatibilities: Heat, ozone, shock, acids. An OSHA Class A Explosive (1910.109). Heating may cause violent combustion or explosion. May explosively decompose on shock, friction, or concussion. Reacts with ozone causing explosion hazard.

Permissible Exposure Limits in Air

Conversion factor: 1 ppm = 9.29 mg/m^3 at 25°C & 1 atm.
OSHA PEL: 0.2 ppm/2 mg/m^3 Ceiling Concentration [skin].
NIOSH REL: 0.1 mg/m^3 STEL [skin].
ACGIH TLV®[1]: 0.05 ppm/0.46 mg/m^3 [skin].
NIOSH IDLH: 75 mg/m^3.
Protective Action Criteria (PAC)
TEEL-0: 0.1 ppm
PAC-1: 0.1 ppm
PAC-2: 2 ppm
PAC-3: 75 ppm
DFG MAK: [skin] Carcinogen Category 3B.
Arab Republic of Egypt: TWA 0.02 ppm (0.2 mg/m^3), [skin], 1993; Australia: TWA 0.05 ppm (0.5 mg/m^3), [skin], 1993; Austria: MAK 0.05 ppm (0.5 mg/m^3), [skin], 1999; Belgium: TWA 0.05 ppm (0.46 mg/m^3), [skin], 1993; Denmark: TWA 0.02 ppm (0.2 mg/m^3), [skin], 1999; Finland: TWA 0.1 ppm (0.9 mg/m^3); STEL 0.3 ppm (3 mg/m^3), [skin], 1999; France: VME 0.1 ppm (1 mg/m^3), [skin], 1999; the Netherlands: MAC-TGG 0.5 mg/m^3, [skin], 2003; Japan: STEL 0.05 ppm (0.46 mg/m^3), [skin], 1999; Norway: TWA 0.03 ppm (0.27 mg/m^3), 1999; the Philippines: TWA 0.2 ppm (2 mg/m^3), [skin], 1993; Poland: MAC (TWA) 0.5 mg/m^3; MAC (STEL) 1 mg/m^3, 1999; Russia: STEL 0.5 ppm, 1993; Sweden: NGV 0.03 ppm (0.3 mg/m^3), KTV 0.1 ppm (0.9 mg/m^3), [skin], 1999; Switzerland: MAK-W 0.05 ppm (0.5 mg/m^3), KZG-W 0.1 ppm (1 mg/m^3), [skin], 1999; Thailand: TWA 0.2 ppm (2 mg/m^3), 1993; Turkey: TWA 0.2 ppm (2 mg/m^3), [skin], 1993; United Kingdom: TWA 0.2 ppm (1.9 mg/m^3); STEL 0.2 ppm, [skin], 2000; Argentina, Bulgaria, Columbia, Jordan, South Korea, New Zealand, Singapore, Vietnam: ACGIH TLV®: TWA 0.05 ppm [skin]. Several states have set guidelines or standards for nitroglycerin in ambient air[60] ranging from 1.67 µg/m^3 (Nevada) to 5.0 µg/m^3 (Florida, North Dakota and South Carolina) to 8.0 µg/m^3 (Virginia) to 10.0 µg/m^3 (Connecticut) to 12.0 µg/m^3 (Nevada).

Determination in Air: Use NIOSH Analytical Method (IV) #2507. See also OSHA Analytical Method #43.

Routes of Entry: Inhalation, ingestion, skin and/or eye contact.

Harmful Effects and Symptoms

Short Term Exposure: Nitroglycerin can affect you when breathed and by passing through skin. Irritates the eyes. May affect the cardiovascular system and blood, causing lowered blood pressure, circulatory collapse, and interfere with the blood's ability to carry oxygen resulting in cyanosis, the formation of methemoglobin, trouble breathing, and even death. Exposure can cause headaches, nausea, and lightheadedness.

Long Term Exposure: After repeated exposure to nitroglycerin, a marked tolerance develops. Returning to work after a short absence from exposure can cause headaches and other symptoms, and may lead to sudden death. Angina (chest pain) and heart attacks can occur when exposure stops suddenly. Repeated or prolonged contact may cause skin sensitization and allergy.

Points of Attack: Cardiovascular system, blood, skin, central nervous system.

Medical Surveillance: NIOSH lists the following tests: whole blood (chemical/metabolite); blood plasma; complete blood count, electrocardiogram, EKG (immediately, if any chest discomfort is felt). Blood methemoglobin level.

First Aid: If this chemical gets into the eyes, remove any contact lenses at once and irrigate immediately for at least 15 min, occasionally lifting upper and lower lids. Seek medical attention immediately. If this chemical contacts the skin, remove contaminated clothing and wash immediately with soap and water. Seek medical attention immediately. If this chemical has been inhaled, remove from exposure, begin rescue breathing (using universal precautions, including resuscitation mask) if breathing has stopped and CPR if heart action has stopped. Transfer promptly to a medical facility. When this chemical has been swallowed, get medical attention. Give large quantities of water and induce vomiting. Do not make an unconscious person vomit.

Note to physician: Treat for methemoglobinemia. Spectrophotometry may be required for precise determination of levels of methemoglobin in urine.

Personal Protective Methods: Wear protective gloves and clothing to prevent any reasonable probability of skin contact. Safety equipment suppliers/manufacturers can provide recommendations on the most protective glove/clothing material for your operation. All protective clothing (suits, gloves, footwear, headgear) should be clean, available each day, and put on before work. Contact lenses should not be worn when working with this chemical. Wear splash-proof chemical goggles and face shield when working with liquid, unless full face-piece respiratory protection is worn. Wear dust-proof goggles and face shield when working with powders or dust, unless full face-piece respiratory protection is worn. Employees should wash immediately with soap when skin is wet or contaminated. Provide emergency showers and eyewash. See NIOSH Criteria Document #78-187 *Occupational Exposure to Nitroglycerin and Ethylene Glycol Dinitrate.*

Respirator Selection: NIOSH: *Up to 1 mg/m^3:* Sa (APF = 10) (any supplied-air respirator).* *Up to 2.5 mg/m^3:* Sa:Cf (APF = 25) (any supplied-air respirator operated in a continuous-flow mode).* *Up to 5 mg/m^3:* SaT: Cf (APF = 50) (any supplied-air respirator that has a tight-fitting face-piece and is operated in a continuous-flow mode) or *SCBAF (APF = 50) (any self-contained breathing apparatus with a full face-piece) or SaF (APF = 50) (any supplied-air respirator with a full face-piece). *Up to 75 mg/m^3:* SaF: Pd,Pp (APF = 2000) (any supplied-air respirator that has a full face-piece and is operated in a pressure-demand or other positive-pressure mode). *Emergency or planned entry into unknown concentrations or IDLH conditions:* SCBAF: Pd,Pp (APF = 10,000) (any self-contained

breathing apparatus that has a full face-piece and is operated in a pressure-demand or other positive-pressure mode) or SaF: Pd,Pp: ASCBA (APF = 10,000) (any supplied-air respirator that has a full face-piece and is operated in a pressure-demand or other positive-pressure mode in combination with an auxiliary, self-contained breathing apparatus operated in a pressure-demand or other positive-pressure mode). *Escape:* GmFOv100 (APF = 50) [any air-purifying, full-face-piece respirator (gas mask) with a chin-style, front- or back-mounted organic vapor canister having an N100, R100, or P100 filter] or SCBAE (any appropriate escape-type, self-contained breathing apparatus).
*Substance reported to cause eye irritation or damage; may require eye protection.

Storage: Explosive. Color Code—Red: Flammability Hazard: Store in a flammable liquid storage area or approved cabinet away from ignition sources and corrosive and reactive materials. Prior to working with this chemical you should be trained on its proper handling and storage. Nitroglycerin must be stored to avoid contact with heat, flames, mechanical shock or ozone, since violent reactions occur. Store in tightly closed containers in a cool, well-ventilated area away from strong acids (such as hydrochloric, sulfuric, and nitric). Sources of ignition, such as smoking and open flames, are prohibited where nitroglycerin is handled, used, or stored. Nitroglycerin has a special shipping regulation by DOT and therefore requires specific handling procedures. Metal containers involving the transfer of 5 gallons or more of nitroglycerin should be grounded and bonded. Drums must be equipped with self-closing valves, pressure vacuum bungs, and flame arresters. Use only nonsparking tools and equipment, especially when opening and closing containers of nitroglycerin. Wherever nitroglycerin is used, handled, manufactured, or stored, use explosion-proof electrical equipment and fittings. Nitroglycerin is often found mixed with ethylene glycol dinitrate. Also see entry on "Ethylene glycol dinitrate."

Shipping: Nitroglycerin, solution in alcohol, with >1% but not >5% nitroglycerin requires a shipping label of "FLAMMABLE LIQUID." It falls in Hazard Class 3 and Packing Group II.
Nitroglycerin solution in alcohol with not >1% nitroglycerin requires a shipping label of "FLAMMABLE LIQUID." It falls in Hazard Class 3 and Packing Group II.
Nitroglycerin, desensitized with not <40% nonvolatile, water-insoluble phlegmatizer, by mass. It falls in Hazard Class 1.1D (subsidiary hazard: 6.1) and Packing Group II.
Nitroglycerin, solution in alcohol, with >1% but not >10% nitroglycerin falls in Hazard Class 1.1D.

Spill Handling: Evacuate and restrict persons not wearing protective equipment from area of spill or leak until cleanup is complete. Do not touch or disturb spilled material. Consult an expert trained for this kind of emergency. Do NOT wash away into sewer (extra personal protection: complete protective clothing including self-contained breathing apparatus). It may be necessary to contain and

dispose of this chemical as a hazardous waste. If material or contaminated runoff enters waterways, notify downstream users of potentially contaminated waters. Contact your local or federal environmental protection agency for specific recommendations. If employees are required to clean up spills, they must be properly trained and equipped. OSHA 1910.120(q) may be applicable.

Fire Extinguishing: This chemical is an explosive and flammable liquid. In case of fire, evacuate area. Combustion in an enclosed space can result in explosion. Isolate area around fire and call for expert help. Consider letting fire burn. Poisonous gases, including nitrogen oxides, are produced in fire. Vapors are heavier than air and will collect in low areas. Vapors in confined areas may explode when exposed to fire. Containers may explode in fire. Storage containers and parts of containers may rocket great distances, in many directions. If material or contaminated runoff enters waterways, notify downstream users of potentially contaminated waters. Notify local health and fire officials and pollution control agencies. From a secure, explosion-proof location, use water spray to cool exposed containers. If cooling streams are ineffective (venting sound increases in volume and pitch, tank discolors, or shows any signs of deforming), withdraw immediately to a secure position. If employees are expected to fight fires, they must be trained and equipped in OSHA 1910.156. The only respirators recommended for firefighting are self-contained breathing apparatuses that have full face-pieces and are operated in a pressure-demand or other positive-pressure mode.

Disposal Method Suggested: Do not wash into sewer. Consult with environmental regulatory agencies for guidance on acceptable disposal practices. Generators of waste containing this contaminant (≥ 100 kg/mo) must conform with EPA regulations governing storage, transportation, treatment, and waste disposal.

References
Sax, N. I. (Ed.). (1981). *Dangerous Properties of Industrial Materials Report*, 1, No. 4, 89−90
New Jersey Department of Health and Senior Services. (July 2001). *Hazardous Substances Fact Sheet: Nitroglycerin.* Trenton, NJ

Nitromethane   N:0520

Molecular Formula: CH_3NO_2
Synonyms: Methan, nitro-; Nitrocarbol
CAS Registry Number: 75-52-5
RTECS® Number: PA9800000
UN/NA & ERG Number: UN1261/129
EC Number: 200-876-6 [*Annex I Index No.:* 609-036-00-7]
Regulatory Authority and Advisory Bodies
Department of Homeland Security Screening Threshold Quantity (pounds): *Theft hazard* 400 (commercial grade).
Carcinogenicity: IARC: Animal Sufficient Evidence; Human No Adequate Data, *possibly carcinogenic to*

humans, Group 2B, 2000; NCI: Carcinogenesis Studies (inhalation); clear evidence: mouse, rat; NTP: 11th Report on Carcinogens, 2004: Reasonably anticipated to be a human carcinogen;

Air Pollutant Standard Set. See below, "Permissible Exposure Limits in Air" section.

OSHA 29CFR1910.119, Appendix A, Process Safety List of Highly Hazardous Chemicals, TQ = 2500 lb (1135 kg).

Carcinogenicity: (New Jersey).

California Proposition 65 Chemical: Cancer 5/1/97.

Canada, WHMIS, Ingredients Disclosure List Concentration 1.0%.

European/International Regulations: Hazard Symbol: Xn; Risk phrases: R5; R10; R22; Safety phrases: S2; S41 (see Appendix 4).

WGK (German Aquatic Hazard Class): 2—Hazard to waters.

Description: Nitromethane is a highly flammable and explosive colorless liquid with a strong, disagreeable odor. The odor threshold is below 200 ppm. Molecular weight = 671.05; Specific gravity (H_2O:1) = 1.14; Boiling point = 101.1°C; Freezing/Melting point = -28.9°C; Vapor pressure = 28 mmHg at 20°C; Flash point = 35°C (cc); Autoignition temperature = 417°C. Explosive limits: LEL = 7.3%; UEL = 63%. Hazard Identification (based on NFPA-704 M Rating System): Health 2, Flammability 3, Reactivity 4 (Possible detonation). Slightly soluble in water.

Potential Exposure: Compound Description: Tumorigen. Nitromethane is used in the production of the fumigant, chloropicrin. It is best known as racing car fuel. It is also used as a solvent and as an intermediate in the pharmaceutical industry.

Incompatibilities: May explode from heat, shock, friction, or concussion. Reacts with alkalis, strong acids, metallic oxides. Detonates or reacts violently with strong oxidizers, strong reducing agents, formaldehyde, copper, copper alloys, lead, lead alloys, hydrocarbons, and other combustibles, causing fire and explosion hazard. Forms shock sensitive mixture when contaminated with acids, amines, bases, metal oxides, hydrocarbons, and other combustible materials.

Permissible Exposure Limits in Air

Conversion factor: 1 ppm = 2.50 mg/m³ at 25°C & 1 atm.

OSHA PEL: 100 ppm/250 mg/m³ TWA.

NIOSH REL: see *NIOSH Pocket Guide*, Appendix D.

ACGIH TLV®[1]: 20 ppm/50 mg/m³ TWA, confirmed animal carcinogen with unknown relevance to humans.

NIOSH IDLH: 750 ppm.

Protective Action Criteria (PAC)

TEEL-0: 20 ppm

PAC-1: 60 ppm

PAC-2: 750 ppm

PAC-3: 750 ppm

DFG MAK: [skin] Carcinogen Category 3B.

Austria: MAK 100 ppm (250 mg/m³), 1999; Denmark: TWA 100 ppm (250 mg/m³), 1999; Finland: TWA 100 ppm (250 mg/m³); STEL 150 ppm (375 mg/m³), 1999; France: VME 100 ppm (250 mg/m³), 1999; Norway: TWA 50 ppm (125 mg/m³), 1999; the Philippines: TWA 100 ppm (250 mg/m³), 1993; the Netherlands: MAC-TGG 50 mg/m³, 2003; Sweden: NGV 20 ppm (50 mg/m³), KTV 50 ppm (125 mg/m³), 1999; Switzerland: MAK-W 100 ppm (250 mg/m³), 1999; Turkey: TWA 100 ppm (250 mg/m³), 1993; United Kingdom: TWA 100 ppm (254 mg/m³); STEL 150 ppm, 2000; New Zealand, Singapore, Vietnam: ACGIH TLV®: confirmed animal carcinogen with unknown relevance to humans. Several states have set guidelines or standards for nitromethane in ambient air[60] ranging from 2.5 mg/m³ (North Dakota) to 4.0 mg/m³ (Virginia) to 5.0 mg/m³ (Connecticut) to 5.952 mg/m³ (Nevada).

Determination in Air: Use NIOSH Analytical Method (IV) #2527, Nitromethane.

Permissible Concentration in Water: Russia[43] set a MAC of 0.005 mg/L in water bodies used for domestic purposes.

Routes of Entry: Inhalation, ingestion, skin and/or eye contact.

Harmful Effects and Symptoms

Short Term Exposure: Irritates the skin and eyes on contact. Inhaling nitromethane irritates the nose and throat causing mild pulmonary irritation with coughing and wheezing. May affect the central nervous system, causing CNS depression, weakness, muscular incoordination, convulsions.

Long Term Exposure: Repeated or prolonged contact may cause dry and cracked skin. May affect or damage the peripheral nervous system. May cause kidney and liver damage. May cause anorexia, nausea, vomiting, and diarrhea.

Points of Attack: Eyes, skin, central nervous system, kidneys, liver.

Medical Surveillance: Consider the points of attack in preplacement and periodic physical examinations.

First Aid: If this chemical gets into the eyes, remove any contact lenses at once and irrigate immediately for at least 15 min, occasionally lifting upper and lower lids. Seek medical attention immediately. If this chemical contacts the skin, remove contaminated clothing and wash immediately with soap and water. Seek medical attention immediately. If this chemical has been inhaled, remove from exposure, begin rescue breathing (using universal precautions, including resuscitation mask) if breathing has stopped and CPR if heart action has stopped. Transfer promptly to a medical facility. When this chemical has been swallowed, get medical attention. Give large quantities of water and induce vomiting. Do not make an unconscious person vomit.

Personal Protective Methods: Wear protective gloves and clothing to prevent any reasonable probability of skin contact. Safety equipment suppliers/manufacturers can provide recommendations on the most protective glove/clothing material for your operation. Butyl rubber; Neoprene™, polyethylene are among the recommended protective materials. All protective clothing (suits, gloves, footwear, headgear) should be clean, available each day, and put on before work. Contact lenses should not be worn when working with this chemical. Wear splash-proof chemical goggles and face shield unless full face-piece respiratory protection is worn. Employees should wash immediately with soap when skin is wet or contaminated. Provide emergency showers and eyewash.

Respirator Selection: OSHA: *750 ppm:* Sa:Cf (APF = 25) (any supplied-air respirator operated in a continuous-flow mode) or SCBAF (APF = 50) (any self-contained breathing apparatus with a full face-piece) or SaF (APF = 50) (any supplied-air respirator with a full face-piece). *Emergency or planned entry into unknown concentrations or IDLH conditions:* SCBAF: Pd,Pp (APF = 10,000) (any self-contained breathing apparatus that has a full face-piece and is operated in a pressure-demand or other positive-pressure mode) or SaF: Pd,Pp: ASCBA (APF = 10,000) (any supplied-air respirator that has a full face-piece and is operated in a pressure-demand or other positive-pressure mode in combination with an auxiliary self-contained breathing apparatus operated in a pressure-demand or other positive-pressure mode). *Escape:* SCBAE (any appropriate escape-type, self-contained breathing apparatus).

Note: Substance causes eye irritation or damage; eye protection needed.

Storage: (1) Color Code—Red Stripe: Flammability Hazard: Do not store in the same area as other flammable materials. (2) Color Code—Red: Flammability Hazard: Store in a flammable liquid storage area or approved cabinet away from ignition sources and corrosive and reactive materials. Prior to working with this chemical you should be trained on its proper handling and storage. Before entering confined space where this chemical may be present, check to make sure that an explosive concentration does not exist. Store in an explosion-proof refrigerator and protect from heat, oxidizers, strong acids, strong bases, formaldehyde, amines, hydrocarbons, combustibles, metallic oxides, reducing agents, copper and its alloys, lead and its alloys, light. Sources of ignition, such as smoking and open flames, are prohibited where this chemical is handled, used, or stored. Metal containers involving the transfer of this chemical should be grounded and bonded. Use only nonsparking tools and equipment, especially when opening and closing containers of this chemical. Wherever this chemical is used, handled, manufactured, or stored, use explosion-proof electrical equipment and fittings. A regulated, marked area should be established where this chemical is handled, used, or stored in compliance with OSHA Standard 1910.1045.

Shipping: This compound requires a shipping label of "FLAMMABLE LIQUID." It falls in Hazard Class 3 and Packing Group II.

Spill Handling: Evacuate and restrict persons not wearing protective equipment from area of spill or leak until cleanup is complete. Remove all ignition sources. Establish forced ventilation to keep levels below explosive limit. Absorb liquids in vermiculite, dry sand, earth, peat, carbon, or a similar material and deposit in sealed containers. Follow by washing surfaces well first with alcohol, then with soap and water. Keep this chemical out of a confined space, such as a sewer, because of the possibility of an explosion, unless the sewer is designed to prevent the buildup of explosive concentrations. It may be necessary to contain and dispose of this chemical as a hazardous waste. If material or contaminated runoff enters waterways, notify downstream users of potentially contaminated waters. Contact your local or federal environmental protection agency for specific recommendations. If employees are required to clean up spills, they must be properly trained and equipped. OSHA 1910.120(q) may be applicable.

Fire Extinguishing: This chemical is a flammable liquid. Poisonous gases, including nitrogen oxides, are produced in fire. Use dry chemical, carbon dioxide, or alcohol foam extinguishers. Vapors are heavier than air and will collect in low areas. Vapors may travel long distances to ignition sources and flashback. Vapors in confined areas may explode when exposed to fire. Containers may explode in fire. Storage containers and parts of containers may rocket great distances, in many directions. If material or contaminated runoff enters waterways, notify downstream users of potentially contaminated waters. Notify local health and fire officials and pollution control agencies. From a secure, explosion-proof location, use water spray to cool exposed containers. If cooling streams are ineffective (venting sound increases in volume and pitch, tank discolors, or shows any signs of deforming), withdraw immediately to a secure position. If employees are expected to fight fires, they must be trained and equipped in OSHA 1910.156. The only respirators recommended for firefighting are self-contained breathing apparatuses that have full face-pieces and are operated in a pressure-demand or other positive-pressure mode. Water may be ineffective. Alcohol foam is recommended.[17]

Disposal Method Suggested: Incineration: large quantities of material may require nitrogen oxide removal by catalytic or scrubbing processes.[22]

References

National Institute for Occupational Safety and Health. (April 1978). *Information Profiles on Potential Occupational Hazards: Classes of Chemicals: Nitroparaffins*, NIOSH Publication No. TR-78-518. Rockville, MD. pp. 199—210

New Jersey Department of Health and Senior Services. (August 1999). *Hazardous Substances Fact Sheet: Nitromethane*. Trenton, NJ

Nitrophenols   N:0530

Molecular Formula: $C_6H_5NO_3$
Common Formula: $NO_2C_6H_4OH$
Synonyms: (2-nitrophenol; o-isomer): 2-Hydroxynitrobenzene; *o*-Nitrofenol (Spanish); *o*-Nitrophenol; Orthonitrophenol;

Phenol, *o*-nitro-; Phenol, 2-nitro- *(3-nitrophenol; m-isomer):* *m*-Hydroxynitrobenzene; 3-Hydroxynitrobenzene; *m*-Nitrofenol (Spanish); 3-Nitrophenol; Phenol, 3-nitro- *(4-nitrophenol; p-isomer):* Degradation product of parathion; 4-Hydroxynitrobenzene; NCI-C55992; Niphen; *p*-Nitrofenol (Spanish); 4-Nitrofenol (Spanish); 4-Nitrophenol; Paranitrophenol (French, German); Phenol, *p*-nitro; Phenol, 4-nitro-; PNP

CAS Registry Number: 88-75-5 (2-nitrophenol; *o*-isomer); 554-84-7 (3-nitrophenol; *m*-isomer); 100-02-7 (4-nitrophenol; *p*-isomer); 25154-55-6 (mixed isomers)

RTECS® Number: SM2100000 (2-nitrophenol; *o*-isomer); SM1925000 (3-nitrophenol; *m*-isomer); SM2275000 (4-nitrophenol; *p*-isomer)

UN/NA & ERG Number: UN1663 (nitrophenols)/153

EC Number: 201-857-5 (2-nitrophenol); 209-073-5 (3-nitrophenol); 202-811-7 *[Annex I Index No.:* 609-015-00-2] (4-nitrophenol)

Regulatory Authority and Advisory Bodies
Air Pollutant Standard Set. See below, "Permissible Exposure Limits in Air" section.

Mixed isomers:
Clean Water Act: Section 311 Hazardous Substances/RQ 40CFR117.3 (same as CERCLA, see below); 40CFR401.15 Section 307 Toxic Pollutants as nitrophenols.
Reportable Quantity (RQ): 100 lb (45.4 kg).

m-isomer:
Clean Water Act: 40CFR401.15 Section 307 Toxic Pollutants as nitrophenols.
Reportable Quantity (RQ): 100 lb (45.4 kg).
Canada, WHMIS, Ingredients Disclosure List Concentration 1.0%, nitrophenols (*o*-; *m*-; *p*-).

p-isomer:
Carcinogenicity: NCI: Carcinogenesis Studies (derm); no evidence: mouse.
US EPA Gene-Tox Program, Positive: *S. cerevisiae* gene conversion; Negative: Histidine reversion—Ames test; Inconclusive: Host-mediated assay.
US EPA, FIFRA 1998 Status of Pesticides: RED completed.
Clean Air Act: Hazardous Air Pollutants (Title I, Part A, Section 112).
Clean Water Act: 40CFR401.15 Section 307 Toxic Pollutants as nitrophenols; Section 313 Water Priority Chemicals (57FR41331, 9/9/92).
US EPA Hazardous Waste Number (RCRA No.): U170.
RCRA, 40CFR261, Appendix 8 Hazardous Constituents.
RCRA 40CFR268.48; 61FR15654, Universal Treatment Standards: Wastewater (mg/L), 0.12; Nonwastewater (mg/kg), 29.
RCRA 40CFR264, Appendix 9; TSD Facilities Ground Water Monitoring List. Suggested test method(s) (PQL µg/L): 8040 (10); 8270 (50).
Reportable Quantity (RQ): 100 lb (45.4 kg).
EPCRA Section 313 Form R *de minimis* concentration reporting level: 1.0%.

o-isomer:
Clean Water Act: 40CFR401.15 Section 307 Toxic Pollutants as nitrophenols; Section 313 Water Priority Chemicals (57FR41331, 9/9/92).
RCRA 40CFR268.48; 61FR15654, Universal Treatment Standards: Wastewater (mg/L), 0.028; Nonwastewater (mg/kg), 13.
RCRA 40CFR264, Appendix 9; TSD Facilities Ground Water Monitoring List. Suggested test method(s) (PQL µg/L): 8040 (5); 8270 (10).
Reportable Quantity (RQ): 100 lb (45.4 kg).
EPCRA Section 313 Form R *de minimis* concentration reporting level: 1.0%.
European/International Regulations *(p-isomer)*: Hazard Symbol: Xn; Risk phrases: R20/21/22; R33; Safety phrases: S2; S28 (see Appendix 4).
WGK (German Aquatic Hazard Class): 2—Water polluting *(p-isomer)*

Description: There are three isomers of nitrophenol. The isomer of greatest concern, and the subject of *ATSDR Toxicology Profile*, is the *p*-isomer (4-nitrophenol). The meta-form is produced from *m*-nitroaniline, and the *o*- and *p*-isomers are produced by nitration of phenol. They are colorless to slightly yellowish crystals with an aromatic to sweetish odor. Molecular weight = 139.12 (*p*-isomer); Boiling point = (*o*-isomer) 215°C; (*m*-isomer) 194°C; (*p*-isomer) 279°C (decomposes); Freezing/Melting point = (*p*-isomer) 113−115°C (sublimes); (*o*-isomer) 45°C; 97°C (*m*-isomer); Flash point = (*o*-, *m*-) 102°C; (*p*-isomer) 169°C; Autoignition temperature = (*p*-isomer) 283°C. NFPA 704 M Hazard Identification (*m*-, *o*-, *p*-): Health 3, Flammability 1, Reactivity 2. Not soluble in water.

Potential Exposure: Compound Description (*p*-isomer): Agricultural Chemical; Tumorigen, Mutagen. Nitrophenols are used as intermediates in production of dyes, photochemicals, pesticides, pharmaceuticals, and in leather tanning.

Incompatibilities: Nitrophenols are strong oxidizers. Reacts violently with combustible and reducing agents. Contact with potassium hydroxide forms an explosive mixture. May explode on heating.

Permissible Exposure Limits in Air
Protective Action Criteria (PAC)
o-isomer and m-isomer
TEEL-0: 1.25 mg/m^3
PAC-1: 4 mg/m^3
PAC-2: 30 mg/m^3
PAC-3: 150 mg/m^3
Protective Action Criteria (PAC)
p-isomer 100-02-7 and mixed isomers
TEEL-0: 6 mg/m^3
PAC-1: 20 mg/m^3
PAC-2: 75 mg/m^3
PAC-3: 75 mg/m^3
Russia set a MAC for 4-nitrophenols in ambient air in residential areas of 0.003 mg/m^3 on a once-daily basis. Guidelines for *p*-nitrophenol in ambient air[60] have been

set by South Carolina at zero and by New York at 0.03 $\mu g/m^3$.

Permissible Concentration in Water: *To protect freshwater aquatic life:* 230 $\mu g/L$ on an acute toxicity basis. *To protect saltwater aquatic life:* 4580 $\mu g/L$ on an acute toxicity basis. *To protect humans:* 0.06 $\mu g/L$ for 2- and 3-nitrophenol and 0.02 mg/L for 4-nitrophenol. In addition, guidelines for nitrophenols in drinking water have been set[61] by Maine at 83 $\mu g/L$ and by Kansas at 290 $\mu g/L$.

Determination in Water: Methylene chloride extraction followed by gas chromatography with flame ionization or electron capture detection (EPA Method 604) or gas chromatography plus mass spectrometry (EPA Method 625).

Routes of Entry: Inhalation, ingestion, skin and/or eye contact. Percutaneous absorption of liquid.

Harmful Effects and Symptoms

Short Term Exposure: Nitrophenols can affect you when breathed in and by passing through your skin. Nitrophenols can irritate and burn the skin and eyes with possible eye damage. Vapors can cause irritation of the respiratory tract with coughing and/or shortness of breath. High levels of vapors may cause metabolism increase and rapid heartbeat. High levels can lower the ability of the blood to carry oxygen (methemoglobinemia), leading to cyanosis (a bluish color to the skin and lips), headaches, dizziness, and collapse. Higher levels can cause death. Exposure can cause headache, upset stomach, dizziness, weakness, confusion, fever, breathing trouble, a slow pulse, fall in blood pressure, convulsions (fits), and death.

Long Term Exposure: Nitrophenols may damage the kidney and liver. Nitrophenols can irritate the lungs; bronchitis may develop. High or repeated exposure may affect the nervous system.

Points of Attack: Liver, kidneys, blood.

Medical Surveillance: If symptoms develop or overexposure is suspected, the following may be useful: A blood test for methemoglobin level, liver function tests, kidney function tests, nervous system tests.

First Aid: If this chemical gets into the eyes, remove any contact lenses at once and irrigate immediately for at least 15 min, occasionally lifting upper and lower lids. Seek medical attention immediately. If this chemical contacts the skin, remove contaminated clothing and wash immediately with soap and water. Seek medical attention immediately. If this chemical has been inhaled, remove from exposure, begin rescue breathing (using universal precautions, including resuscitation mask) if breathing has stopped and CPR if heart action has stopped. Transfer promptly to a medical facility. When this chemical has been swallowed, get medical attention. Give large quantities of water and induce vomiting. Do not make an unconscious person vomit.

Note to physician: Treat for methemoglobinemia. Spectrophotometry may be required for precise determination of levels of methemoglobin in urine.

Personal Protective Methods: Wear protective gloves and clothing to prevent any reasonable probability of skin contact. Safety equipment suppliers/manufacturers can provide recommendations on the most protective glove/clothing material for your operation. Butyl rubber is recommended by a manufacturer and other authorities as a protective material. All protective clothing (suits, gloves, footwear, headgear) should be clean, available each day, and put on before work. Contact lenses should not be worn when working with this chemical. Wear dust-proof chemical goggles and face shield unless full face-piece respiratory protection is worn. Employees should wash immediately with soap when skin is wet or contaminated. Provide emergency showers and eyewash.

Respirator Selection: Where there is potential for exposure to nitrophenols, use a NIOSH/MSHA- or European Standard EN149-approved full face-piece respirator with a high-efficiency particulate filter. Greater protection is provided by a powered air-purifying respirator. *Where there is potential for high exposures,* use a NIOSH/MSHA- or European Standard EN149-approved supplied-air respirator with a full face-piece operated in the positive-pressure mode, or with a full face-piece, hood, or helmet in the continuous-flow mode; or use a NIOSH/MSHA- or European Standard EN149-approved self-contained breathing apparatus with a full face-piece operated in pressure-demand or other positive-pressure mode.

Storage: Color Code—Blue: Health Hazard/Poison: Store in a secure poison location. Prior to working with this chemical you should be trained on its proper handling and storage. Store in tightly closed containers in a cool, well-ventilated area. Nitrophenols must be stored to avoid contact with reducing agents, oxidizers, combustibles, organic materials, and strong bases, since violent reactions occur. Where possible, automatically transfer material from drums or other storage containers to process containers. Sources of ignition, such as smoking and open flames, are prohibited where this chemical is handled, used, or stored. Metal containers involving the transfer of this chemical should be grounded and bonded. Use only nonsparking tools and equipment, especially when opening and closing containers of this chemical. Wherever this chemical is used, handled, manufactured, or stored, use explosion-proof electrical equipment and fittings.

Shipping: Nitrophenols (*o*-; *m*-; *p*-) require a shipping label of "POISONOUS/TOXIC MATERIALS." They fall in DOT Hazard Class 6.1 and Packing Group III.

Spill Handling: Evacuate persons not wearing protective equipment from area of spill or leak until cleanup is complete. Remove all ignition sources. Use HEPA vacuum or wet method to reduce dust during cleanup. Do not dry sweep. Collect powdered material in the most convenient and safe manner and deposit in sealed containers. Ventilate area after cleanup is complete. It may be necessary to contain and dispose of this chemical as a hazardous waste. If material or contaminated runoff enters waterways, notify downstream users of potentially contaminated waters. Contact your local or federal environmental protection agency for specific recommendations. If employees are required to clean up spills, they must be properly trained and equipped. OSHA 1910.120(q) may be applicable.

Fire Extinguishing: This chemical is a combustible solid. Use dry chemical, carbon dioxide, water spray, or alcohol foam extinguishers. Poisonous gases, including nitrogen oxides, are produced in fire. If material or contaminated runoff enters waterways, notify downstream users of potentially contaminated waters. Notify local health and fire officials and pollution control agencies. From a secure, explosion-proof location, use water spray to cool exposed containers. If cooling streams are ineffective (venting sound increases in volume and pitch, tank discolors, or shows any signs of deforming), withdraw immediately to a secure position. If employees are expected to fight fires, they must be trained and equipped in OSHA 1910.156. The only respirators recommended for firefighting are self-contained breathing apparatuses that have full face-pieces and are operated in a pressure-demand or other positive-pressure mode.

Disposal Method Suggested: Controlled incineration—care must be taken to maintain complete combustion at all times. Incineration of large quantities may require scrubbers to control the emission of nitrogen oxides. In accordance with 40CFR165, follow recommendations for the disposal of pesticides and pesticide containers. Must be disposed properly by following package label directions or by contacting your local or federal environmental control agency or by contacting your regional EPA office. Consult with environmental regulatory agencies for guidance on acceptable disposal practices. Generators of waste containing this contaminant (≥ 100 kg/mo) must conform with EPA regulations governing storage, transportation, treatment, and waste disposal.

References

National Institute for Occupational Safety and Health. (October 1977). *Information Profiles on Potential Occupational Hazards: Nitrophenols*, Report No. PB-276, 678. Rockville, MD. pp. 212–226

US Environmental Protection Agency. (1980). *Nitrophenols: Ambient Water Quality Criteria*. Washington, DC

US Environmental Protection Agency. (April 30, 1980). *4-Nitrophenol: Health and Environmental Effects Profile No. 135*. Washington, DC: Office of Solid Waste

US Environmental Protection Agency. (April 30, 1980). *Nitrophenols: Health and Environmental Effects Profile No. 136*. Washington, DC: Office of Solid Waste

Sax, N. I. (Ed.). *Dangerous Properties of Industrial Materials Report*, 5, No. 3, 67–70 (1985) (2-Nitrophenol) and 1, No. 6, 89–90 (1981) and 6, No. 3, 63–66 (3-Nitrophenol) and 3, No. 3, 82–85 (1983) (4-Nitrophenol)

US Environmental Protection Agency, Special Review and Reregistration Division Office of Pesticide Programs. (1998). *Agency Status of Pesticides in Registration, Reregistration, and Special Review* (Rainbow Report). Washington, DC

New Jersey Department of Health and Senior Services. (February 2000). *Hazardous Substances Fact Sheet: 2-Nitrophenol*. Trenton, NJ

New Jersey Department of Health and Senior Services. (February 2000). *Hazardous Substances Fact Sheet: 3-Nitrophenol*. Trenton, NJ

New Jersey Department of Health and Senior Services. (September 2004). *Hazardous Substances Fact Sheet: 4-Nitrophenol*. Trenton, NJ

1-Nitropropane   N:0540

Molecular Formula: $C_3H_7NO_2$
Common Formula: $CH_3CH_2CH_2NO_2$
Synonyms: α-Nitropropane; Nitropropane; 1-Nitropropano (Spanish); 1-NP; Propane, 1-nitro-; Propane, nitro-
CAS Registry Number: 108-03-2
RTECS® Number: TZ5075000
UN/NA & ERG Number: UN2608/129
EC Number: 203-544-9 [*Annex I Index No.:* 609-001-00-6]
Regulatory Authority and Advisory Bodies
Air Pollutant Standard Set. See below, "Permissible Exposure Limits in Air" section.
Canada, WHMIS, Ingredients Disclosure List Concentration 1.0%.
European/International Regulations: Hazard Symbol: Xn; Risk phrases: R10; R20/21/22; Safety phrases: S2; S9 (see Appendix 4).
WGK (German Aquatic Hazard Class): 1—Low hazard to waters.

Description: 1-Nitropropane is a colorless liquid with a mild, fruity odor. Odor threshold = 140 ppm. Molecular weight = 89.11; Specific gravity (H_2O:1) = 1.00; Boiling point = 131.7°C; Freezing/Melting point = −107.8°C; Vapor pressure = 8 mmHg at 20°C; Flash point = 35.6°C (cc); Autoignition temperature = 421°C. Explosive limits: LEL = 2.2%; UEL—unknown. Hazard Identification (based on NFPA-704 M Rating System): Health 1, Flammability 3, Reactivity 2. Slightly soluble in water; solubility = 1%.
Note: Technical products measurably contaminated with 2-Nitropropane.

Potential Exposure: Compound Description: Mutagen, Human Data; Primary Irritant. 1-Nitropropane is used as a solvent for polymers, as a stabilizer, and in organic synthesis. *Note:* Technical products measurably contaminated with 2-Nitropropane, see also "2-Nitropropane."

Incompatibilities: 1-Nitropropane, a nitroparaffin is incompatible with reducing agents, nitrates, strong bases, amines, strong acids, oxidizers, hydrocarbons and other combustible materials, metal oxides. May explode on heating.

Permissible Exposure Limits in Air
OSHA PEL: 25 ppm/90 mg/m^3 TWA.
NIOSH REL: 25 ppm/90 mg/m^3 TWA.

ACGIH TLV®[1]: 25 ppm/91 mg/m³ TWA, not classifiable as a human carcinogen.

NIOSH IDLH: 1000 ppm.

Temporary Emergency Exposure Limits (DOE)

TEEL-0: 25 ppm

PAC-1: 75 ppm

PAC-2: 125 ppm

PAC-3: 1000 ppm

DFG MAK: 25 ppm/92 mg/m³ TWA; Peak Limitation Category I(4); Pregnancy Risk Group D.

Austria: MAK 25 ppm (90 mg/m³), 1999; Denmark: TWA 5 ppm (18 mg/m³), 1999; Finland: TWA 25 ppm (90 mg/m³); STEL 40 ppm (150 mg/m³), 1999; France: VME 25 ppm (90 mg/m³), 1999; Norway: TWA 20 ppm (70 mg/m³), 1999; the Philippines: TWA 25 ppm (90 mg/m³), 1993; the Netherlands: MAC-TGG 90 mg/m³, 2003; Switzerland: MAK-W 5 ppm (18 mg/m³), carcinogen, 1999; Turkey: TWA 25 ppm (90 mg/m³), 1993; United Kingdom: TWA 25 ppm (93 mg/m³), 2000; Argentina, Bulgaria, Columbia, Jordan, South Korea, New Zealand, Singapore, Vietnam: ACGIH TLV®: not classifiable as a human carcinogen. Several states have set guidelines or standards for 1-nitropropane in ambient air[60] ranging from 0.3 mg/m³ (New York) to 0.9 mg/m³ (Florida and North Dakota) to 1.5 mg/m³ (Virginia) to 1.8 mg/m³ (Connecticut) to 2.143 mg/m³ (Nevada) to 2.25 mg/m³ (South Carolina).

Determination in Air: Use OSHA Analytical Method 46.

Permissible Concentration in Water: Russia[43] set a MAC of 1.0 mg/L in water bodies used for domestic purposes.

Routes of Entry: Inhalation, ingestion, skin and/or eye contact.

Harmful Effects and Symptoms

Short Term Exposure: Irritates the eyes, skin, and respiratory tract. Inhalation can cause coughing, wheezing, and/or shortness of breath. High levels can interfere with the ability of the blood to carry oxygen (methemoglobinemia), causing cyanosis, headaches, fatigue, dizziness nausea, vomiting, diarrhea. Higher levels can cause collapse and death.

Long Term Exposure: In animals: liver, kidney damage.

Points of Attack: Eyes, central nervous system, liver, kidneys.

Medical Surveillance: Liver and kidney function tests. Blood methemoglobin level.

First Aid: If this chemical gets into the eyes, remove any contact lenses at once and irrigate immediately for at least 15 min, occasionally lifting upper and lower lids. Seek medical attention immediately. If this chemical contacts the skin, remove contaminated clothing and wash immediately with soap and water. Seek medical attention immediately. If this chemical has been inhaled, remove from exposure, begin rescue breathing (using universal precautions, including resuscitation mask) if breathing has stopped and CPR if heart action has stopped. Transfer promptly to a medical facility. When this chemical has been swallowed, get medical attention. Give large quantities of water and induce vomiting. Do not make an unconscious person vomit.

Note to physician: Treat for methemoglobinemia. Spectrophotometry may be required for precise determination of levels of methemoglobin in urine.

Personal Protective Methods: Wear protective gloves and clothing to prevent any reasonable probability of skin contact. Safety equipment suppliers/manufacturers can provide recommendations on the most protective glove/clothing material for your operation. Butyl rubber, Teflon™, and Silvershield™ are among the recommended protective materials. All protective clothing (suits, gloves, footwear, headgear) should be clean, available each day, and put on before work. Contact lenses should not be worn when working with this chemical. Wear splash-proof chemical goggles and face shield unless full face-piece respiratory protection is worn. Employees should wash immediately with soap when skin is wet or contaminated. Provide emergency showers and eyewash.

Respirator Selection: *Up to 250 ppm:* Sa (APF = 10) (any supplied-air respirator).* *Up to 625 ppm:* Sa:Cf (APF = 25) (any supplied-air respirator operated in a continuous-flow mode).* *Up to 1000 ppm:* SCBAF (APF = 50) (any self-contained breathing apparatus with a full face-piece) or SaF (APF = 50) (any supplied air respirator with a full face-piece). *Emergency or planned entry into unknown concentrations or IDLH conditions:* SCBAF: Pd,Pp (APF = 10,000) (any self-contained breathing apparatus that has a full face-piece and is operated in a pressure-demand or other positive-pressure mode) or SaF: Pd,Pp: ASCBA (APF = 10,000) (any supplied-air respirator that has a full face-piece and is operated in a pressure-demand or other positive-pressure mode in combination with an auxiliary, self-contained breathing apparatus operated in a pressure-demand or other positive-pressure mode). *Escape:* SCBAE (any appropriate escape-type, self-contained breathing apparatus).

*Substance reported to cause eye irritation or damage; may require eye protection.

Storage: Color Code—Red: Flammability Hazard: Store in a flammable liquid storage area or approved cabinet away from ignition sources and corrosive and reactive materials. Prior to working with this chemical you should be trained on its proper handling and storage. Before entering confined space where this chemical may be present, check to make sure that an explosive concentration does not exist.

Shipping: This compound requires a shipping label of "FLAMMABLE LIQUID." It falls in Hazard Class 3 and Packing Group II.

Spill Handling: Evacuate and restrict persons not wearing protective equipment from area of spill or leak until cleanup is complete. Remove all ignition sources. Establish forced ventilation to keep levels below explosive limit. Absorb liquids in vermiculite, dry sand, earth, peat, carbon, or a similar material and deposit in sealed containers. Keep this chemical out of a confined space, such as a sewer, because of the possibility of an explosion, unless the sewer is designed to prevent the buildup of explosive concentrations. It may be necessary to contain and dispose of this chemical as a hazardous waste. If material or contaminated runoff enters waterways, notify downstream users of potentially contaminated waters. Contact your local or federal environmental protection agency for specific recommendations. If employees are required to clean up spills, they must be properly trained and equipped. OSHA 1910.120(q) may be applicable.

Fire Extinguishing: This chemical is a flammable liquid. Poisonous gases, including nitrogen oxides, are produced in fire. Use dry chemical, carbon dioxide, alcohol foam, or polymer foam extinguishers. Vapors are heavier than air and will collect in low areas. Vapors may travel long distances to ignition sources and flashback. Vapors in confined areas may explode when exposed to fire. Containers may explode in fire. Storage containers and parts of containers may rocket great distances, in many directions. If material or contaminated runoff enters waterways, notify downstream users of potentially contaminated waters. Notify local health and fire officials and pollution control agencies. From a secure, explosion-proof location, use water spray to cool exposed containers. If cooling streams are ineffective (venting sound increases in volume and pitch, tank discolors, or shows any signs of deforming), withdraw immediately to a secure position. If employees are expected to fight fires, they must be trained and equipped in OSHA 1910.156. The only respirators recommended for firefighting are self-contained breathing apparatuses that have full face-pieces and are operated in a pressure-demand or other positive-pressure mode.

Reference

New Jersey Department of Health and Senior Services. (September 1999). *Hazardous Substances Fact Sheet: 1-Nitropropane.* Trenton, NJ

2-Nitropropane   N:0550

Molecular Formula: $C_3H_7NO_2$
Common Formula: $CH_3CH(NO_2)CH_3$
Synonyms: Dimethylnitromethane; Isonitropropane; Nipar S-20; Nipars-20 solvent; Nipar S-30 solvent; Nitroisopropane; β-Nitropropane; *sec*-Nitropropane; 2-Nitropropano (Spanish); 2-NP; Propane, 2-nitro-
CAS Registry Number: 79-46-9
RTECS® Number: TZ5250000
UN/NA & ERG Number: UN2608/129
EC Number: 201-209-1 [*Annex I Index No.:* 609-002-00-1]
Regulatory Authority and Advisory Bodies
Carcinogenicity: IARC: Animal Sufficient Evidence; Human Inadequate Evidence, *possibly carcinogenic to humans,* Group 2B, 1999; NTP: 11th Report on Carcinogens, 2004: Reasonably anticipated to be a human carcinogen; NIOSH: Potential occupational carcinogen.

US EPA Gene-Tox Program, Positive: Carcinogenicity—mouse/rat; Inconclusive: Mammalian micronucleus.
Air Pollutant Standard Set. See below, "Permissible Exposure Limits in Air" section.
Clean Air Act: Hazardous Air Pollutants (Title I, Part A, Section 112).
US EPA Hazardous Waste Number (RCRA No.): U171.
RCRA, 40CFR261, Appendix 8 Hazardous Constituents.
Reportable Quantity (RQ): 10 lb (4.54 kg).
EPCRA Section 313 Form R *de minimis* concentration reporting level: 0.1%.
California Proposition 65 Chemical: Cancer 1/1/88.
Canada, WHMIS, Ingredients Disclosure List Concentration 0.1%.
European/International Regulations: Hazard Symbol: T; Risk phrases: R45; R10; R20/22; Safety phrases: S53; S45 (see Appendix 4).
WGK (German Aquatic Hazard Class): No value assigned.
Description: 2-Nitropropane is a colorless liquid. The odor threshold is 300 ppm. Molecular weight = 89.11; Specific gravity (H_2O:1) = 0.99; Boiling point = 120.5°C; Freezing/Melting point = −92.8°C; Vapor pressure: 13 mmHg at 20°C; Flash point = 23.9°C; Autoignition temperature = 428°C. Explosive limits: LEL = 2.6%; UEL = 11.0%. Hazard Identification (based on NFPA-704 M Rating System): Health 1, Flammability 3, Reactivity 2. Slightly soluble in water; solubility = 2%.
Potential Exposure: Compound Description: Agricultural Chemical; Tumorigen, Mutagen; Reproductive Effector; Human Data. 2-Nitropropane is used as a solvent for polymers, organic compounds; cellulose, esters; gums, vinyl resins; waxes, epoxy resins, fats, dyes, and chlorinated rubber; as a stabilizer. Its combustion properties have made it useful as a rocket propellant and as a gasoline and diesel fuel additive. 2-Nitropropane also has limited use as a paint and varnish remover. It serves as an intermediate in organic synthesis of some pharmaceuticals, dyes, insecticides, and textile chemicals.
Incompatibilities: 1-Nitropropane, a nitroparaffinin, forms explosive mixture with air. Contact with heavy metal oxides may cause decomposition. Mixtures with hydrocarbons are extremely flammable. Attacks some plastics, rubber, and coatings. May explode on heating. Violent reaction with strong bases, strong acids, and metal oxides. Shock-sensitive compounds are formed with acids, amines, inorganic bases, and heavy metal oxides. Incompatible with strong oxidizers, combustible materials. 2-Nitropropane reacts with activated carbon causing decomposition. *This reaction may occur in activated carbon respirator filters.*
Permissible Exposure Limits in Air
OSHA PEL: 25 ppm/90 mg/m³ TWA
NIOSH REL: A potential occupational carcinogen [skin]; Limit exposure to lowest feasible concentration. See *NIOSH Pocket Guide,* Appendix A.
ACGIH TLV®[1]: 10 ppm/36 ppm TWA, confirmed animal carcinogen with unknown relevance to humans.

NIOSH IDLH: 100 ppm.
Protective Action Criteria (PAC)
TEEL-0: 25 ppm
PAC-1: 25 ppm
PAC-2: 25 ppm
PAC-3: 100 ppm
DFG MAK: [skin] Carcinogen Category 2.
Austria carcinogen, 1999; Denmark: TWA 5 ppm (18 mg/m³), 1999; France: carcinogen, 1993; Norway: TWA 10 ppm (35 mg/m³), 1999; the Philippines: TWA 25 ppm (90 mg/m³), 1993; Sweden: NGV 5 ppm (18 mg/m³), TGV 10 ppm (35 mg/m³), 1999; the Netherlands: MAC-TGG 0.036 mg/m³, 2003; Turkey: TWA 25 ppm (90 mg/m³), 1993; United Kingdom: TWA 5 ppm (19 mg/m³), carcinogen, 2000; Argentina, Bulgaria, Columbia, Jordan, South Korea, New Zealand, Singapore, Vietnam: ACGIH TLV®: confirmed animal carcinogen with unknown relevance to humans. Brazil[35] has set a TWA of 20 ppm (70 mg/m³). Several states have set guidelines or standards for 2-nitropropane in ambient air[60] ranging from zero in North Dakota to 0.2 μg/m³ (Rhode Island) to 21.67 μg/m³ (Pennsylvania) to 350 μg/m³ (Virginia) to 360 μg/m³ (Connecticut) to 2143 μg/m³ (Nevada).

Determination in Air: Use NIOSH Analytical Method (IV) #2528; OSHA Analytical Method 46 (which supercedes Method 15).

Permissible Concentration in Water: Russia[43] set a MAC in water bodies used for domestic purposes of 1.0 mg/L.

Determination in Water: Octanol−water coefficient: Log $K_{ow} = 0.9$.

Routes of Entry: Inhalation, ingestion, skin and/or eye contact.

Harmful Effects and Symptoms

Short Term Exposure: The vapor irritates the eyes and respiratory tract. 2-Nitropropane can affect you when breathed in. 2-Nitropropane may cause mutations. Handle with extreme caution. Exposure can cause headaches, dizziness, nausea, vomiting, and diarrhea. At levels, causing these symptoms, severe liver damage can occur that can cause death. Higher exposures can cause pulmonary edema, a medical emergency that can be delayed for several hours. This can cause death. Exposure may also damage the kidneys, heart, and may interfere with the ability of the blood to carry oxygen (methemoglobinemia); this can cause weakness, trouble breathing and cyanosis, a bluish color to the skin and lips. Exposure to high levels may cause liver damage. Exposure to very high levels may result in death.

Long Term Exposure: May affect the liver, kidneys, heart, and nervous system. Based on animal tests, this chemical is a potential occupational carcinogen. Animal tests show that this substance may cause mutations and affect human reproduction.

Points of Attack: Eyes, skin, respiratory system, central nervous system, kidneys, liver, heart. Cancer site in animals: liver.

Medical Surveillance: For those with frequent or potentially high exposure (half the TLV or greater), the following are recommended before beginning work and at regular times after that: Liver function tests. If symptoms develop or overexposure is suspected, the following may be useful: consider chest X-ray after acute overexposure. Kidney function tests. Blood methemoglobin level.

First Aid: If this chemical gets into the eyes, remove any contact lenses at once and irrigate immediately for at least 15 min, occasionally lifting upper and lower lids. Seek medical attention immediately. If this chemical contacts the skin, remove contaminated clothing and wash immediately with soap and water. Seek medical attention immediately. If this chemical has been inhaled, remove from exposure, begin rescue breathing (using universal precautions, including resuscitation mask) if breathing has stopped and CPR if heart action has stopped. Transfer promptly to a medical facility. When this chemical has been swallowed, get medical attention. Give large quantities of water and induce vomiting. Do not make an unconscious person vomit.

Note to physician: Treat for methemoglobinemia. Spectrophotometry may be required for precise determination of levels of methemoglobin in urine.

Personal Protective Methods: Wear protective gloves and clothing to prevent any reasonable probability of skin contact. Safety equipment suppliers/manufacturers can provide recommendations on the most protective glove/clothing material for your operation. Butyl rubber and polyvinyl chloride are recommended. All protective clothing (suits, gloves, footwear, headgear) should be clean, available each day, and put on before work. Contact lenses should not be worn when working with this chemical. Wear splash-proof chemical goggles and face shield unless full face-piece respiratory protection is worn. Employees should wash immediately with soap when skin is wet or contaminated. Provide emergency showers and eyewash.

Respirator Selection: NIOSH: *At any detectable concentration:* SCBAF: Pd,Pp (APF = 10,000) (any NIOSH/MSHA- or European Standard EN 149-approved self-contained breathing apparatus that has a full face-piece and is operated in a pressure-demand or other positive-pressure mode) or SaF: Pd,Pp: ASCBA (APF = 10,000) (any supplied-air respirator that has a full face-piece and is operated in a pressure-demand or other positive-pressure mode in combination with an auxiliary, self-contained breathing apparatus operated in a pressure-demand or other positive-pressure mode). *Escape:* SCBAE (any appropriate escape-type, self-contained breathing apparatus).

Storage: Color Code—Red: Flammability Hazard: Store in a flammable liquid storage area or approved cabinet away from ignition sources and corrosive and reactive materials. Prior to working with this chemical you should be trained on its proper handling and storage. Before entering confined space where this chemical may be present, check to make sure that an explosive concentration does not exist. 2-Nitropropane must be stored to avoid contact with strong bases and strong acids (such as hydrochloric, sulfuric, and nitric) and metal oxides, since violent reactions occur. Store in tightly closed containers in a cool, well-ventilated area. Protect storage containers from physical damage. Sources of ignition, such as smoking and open flames, are prohibited where this chemical is handled, used, or stored. Metal containers involving the transfer of this chemical should be grounded and bonded. Drums must be equipped with self-closing valves, pressure vacuum bungs, and flame arresters. Use only nonsparking tools and equipment, especially when opening and closing containers of 2-nitropropane. A regulated, marked area should be established where this chemical is handled, used, or stored in compliance with OSHA Standard 1910.1045.

Shipping: This compound requires a shipping label of "FLAMMABLE LIQUID." It falls in Hazard Class 3 and Packing Group II.

Spill Handling: Evacuate and restrict persons not wearing protective equipment from area of spill or leak until cleanup is complete. Remove all ignition sources. Establish forced ventilation to keep levels below explosive limit. Absorb liquids in vermiculite, dry sand, earth, peat, carbon, or a similar material and deposit in sealed containers. Keep this chemical out of a confined space, such as a sewer, because of the possibility of an explosion, unless the sewer is designed to prevent the buildup of explosive concentrations. It may be necessary to contain and dispose of this chemical as a hazardous waste. If material or contaminated runoff enters waterways, notify downstream users of potentially contaminated waters. Contact your local or federal environmental protection agency for specific recommendations. If employees are required to clean up spills, they must be properly trained and equipped. OSHA 1910.120(q) may be applicable.

Fire Extinguishing: This chemical is a flammable liquid. Poisonous gases are produced in fire. Use dry chemical, carbon dioxide, or alcohol foam extinguishers. Vapors are heavier than air and will collect in low areas. Vapors may travel long distances to ignition sources and flashback. Vapors in confined areas may explode when exposed to fire. Containers may explode in fire. Storage containers and parts of containers may rocket great distances, in many directions. If material or contaminated runoff enters waterways, notify downstream users of potentially contaminated waters. Notify local health and fire officials and pollution control agencies. From a secure, explosion-proof location, use water spray to cool exposed containers. If cooling streams are ineffective (venting sound increases in volume and pitch, tank discolors, or shows any signs of deforming), withdraw immediately to a secure position. If employees are expected to fight fires, they must be trained and equipped in OSHA 1910.156. The only respirators recommended for firefighting are self-contained breathing apparatuses that have full face-pieces and are operated in a pressure-demand or other positive-pressure mode.

Disposal Method Suggested: Incineration: large quantities of material may require nitrogen oxide removal by catalytic or scrubbing processes.[22] Dilute with pure kerosene and burn with care as it is potentially explosive. Consult with environmental regulatory agencies for guidance on acceptable disposal practices. Generators of waste containing this contaminant (≥ 100 kg/mo) must conform with EPA regulations governing storage, transportation, treatment, and waste disposal.

References

US Environmental Protection Agency. (September 27, 1977). *Chemical Hazard Information Profile: 2-Nitropropane*. Washington, DC

National Institute for Occupational Safety and Health. (April 1978). *Information Profiles on Potential Occupational Hazards: Classes of Chemicals: Nitroparaffins*. NIOSH Publication No. RT 78-518. Rockville, MD. pp. 199–210

National Institute for Occupational Safety and Health. (April 25, 1977). *2-Nitropropane, Current Intelligence Bulletin No. 17*. Rockville, MD

Sax, N. I. (Ed.). *Dangerous Properties of Industrial Materials Report*, 2, No. 2, 58–59 (1982) and 4, No. 1, 92–94 (1984)

New Jersey Department of Health and Senior Services. (July 2001). *Hazardous Substances Fact Sheet: 2-Nitropropane*. Trenton, NJ

N-Nitrosodi-*n*-butylamine N:0560

Molecular Formula: $C_8H_{18}N_2O$

Synonyms: 1-Butanamine, *n*-butyl-*N*-nitroso-; Butylamine, *N*-nitrosodi-; *n*-Butyl-*N*-nitroso-1-butamine; DBN; DBNA; Dibutylamine, *N*-nitroso-; Di-*n*-butylnitrosamin (German); *N,N*-Di-*n*-butylnitrosamine; Di-*n*-butylnitrosamine; Dibutylnitrosamine; *N,N*-Dibutylnitrosoamine; NDBA; *N*-Nitroso-di-*n*-butylamine; Nitrosodibutylamine

CAS Registry Number: 924-16-3

RTECS® Number: EJ4025000

UN/NA & ERG Number: UN3082/171

EC Number: 213-101-1

Regulatory Authority and Advisory Bodies

Carcinogenicity: IARC: Animal Sufficient Evidence; Human Insufficient Evidence, *possibly carcinogenic to humans*, Group 2B, 1978; EPA: Sufficient evidence from animal studies; inadequate evidence or no useful data from epidemiologic studies; NTP: Reasonably anticipated to be a human carcinogen.

US EPA Hazardous Waste Number (RCRA No.): U172.

RCRA, 40CFR261, Appendix 8 Hazardous Constituents.

RCRA 40CFR268.48; 61FR15654, Universal Treatment Standards: Wastewater (mg/L), 0.40; Nonwastewater (mg/kg), 17.

RCRA 40CFR264, Appendix 9; TSD Facilities Ground Water Monitoring List. Suggested test method(s) (PQL μg/L): 8270 (10).

Reportable Quantity (RQ): 10 lb (4.54 kg).

EPCRA Section 313 Form R de minimis concentration reporting level: 0.1%.

California Proposition 65 Chemical: Cancer 10/1/87.

Canada, WHMIS, Ingredients Disclosure List Concentration 0.1%.

European/International Regulations: not listed in Annex 1.

WGK (German Aquatic Hazard Class): No value assigned.

Description: *N*-Nitrosodi-*n*-butylamine is a yellow, oily liquid. Molecular weight = 158.28; Boiling point = 235°C; 116°C at 14 mmHg. Hazard Identification (based on NFPA-704 M Rating System): Health 1, Flammability 0, Reactivity 0. Slightly soluble in water.

Potential Exposure: This chemical is a carcinogenic nitrosamine. It is primarily used in research.

Incompatibilities: Light sensitive.

Permissible Exposure Limits in Air

DFG MAK: [skin] Carcinogen Category 2.

There are no established numerical OELs. However, this chemical is a carcinogen and exposure should be reduced to the lowest possible level.

Determination in Air: Use NIOSH Analytical Method (IV) #2522, Nitrosamines; OSHA Analytical Method 27.

Routes of Entry: Inhalation.

Harmful Effects and Symptoms

Short Term Exposure: Irritates the skin and respiratory tract. Some related nitrosamines can cause headaches, abdominal cramps, weakness, and dizziness. It is unknown if this chemical causes all of these same symptoms.

Long Term Exposure: May be a human carcinogen; causes urinary bladder, esophagus, and liver cancer in animals. This chemical is toxic to the fetus and causes fetal death in animals. Repeated exposure may cause liver damage.

Points of Attack: Liver.

Medical Surveillance: Liver function tests.

First Aid: If this chemical gets into the eyes, remove any contact lenses at once and irrigate immediately for at least 15 min, occasionally lifting upper and lower lids. Seek medical attention immediately. If this chemical contacts the skin, remove contaminated clothing and wash immediately with soap and water. Seek medical attention immediately. If this chemical has been inhaled, remove from exposure, begin rescue breathing (using universal precautions, including resuscitation mask) if breathing has stopped and CPR if heart action has stopped. Transfer promptly to a medical facility. When this chemical has been swallowed, get medical attention. Give large quantities of water and induce vomiting. Do not make an unconscious person vomit.

Personal Protective Methods: Wear protective gloves and clothing to prevent any reasonable probability of skin contact. Safety equipment suppliers/manufacturers can provide recommendations on the most protective glove/clothing material for your operation. All protective clothing (suits, gloves, footwear, headgear) should be clean, available each day, and put on before work. Contact lenses should not be worn when working with this chemical. Wear splash-proof chemical goggles and face shield unless full face-piece respiratory protection is worn. Employees should wash immediately with soap when skin is wet or contaminated. Provide emergency showers and eyewash.

Respirator Selection: Specific respirator(s) have not been recommended by NIOSH. However, based on potential carcinogenicity the following might be considered: *At any detectable concentration:* SCBAF: Pd,Pp (APF = 10,000) (any NIOSH/MSHA- or European Standard EN 149-approved self-contained breathing apparatus that has a full face-piece and is operated in a pressure-demand or other positive-pressure mode) or SaF: Pd,Pp: ASCBA (APF = 10,000) (any supplied-air respirator that has a full face-piece and is operated in a pressure-demand or other positive-pressure mode in combination with an auxiliary, self-contained breathing apparatus operated in a pressure-demand or other positive-pressure mode). *Escape:* 100 F (APF = 50) (any air-purifying, full-face-piece respirator with an N100, R100, or P100 filter) or SCBAE (any appropriate escape-type, self-contained breathing apparatus).

Storage: Color Code—Green: General storage may be used. Prior to working with this chemical you should be trained on its proper handling and storage. Store in tightly closed containers in a cool, well-ventilated area away from oxidizers. Where possible, automatically pump liquid from drums or other storage containers to process containers. Sources of ignition, such as smoking and open flames are prohibited where this chemical is handled, used, or stored. Metal containers involving the transfer of this chemical should be grounded and bonded. Drums must be equipped with self-closing valves, pressure vacuum bungs, and flame arresters. Use only nonsparking tools and equipment, especially when opening and closing containers of this chemical. Wherever this chemical is used, handled, manufactured, or stored, use explosion-proof electrical equipment and fittings. A regulated, marked area should be established where this chemical is handled, used, or stored in compliance with OSHA Standard 1910.1045.

Shipping: The name of this material is not on the DOT list of materials[19] for label and packaging standards. However, based on regulations, it may be classified[52] as an Environmentally hazardous substances, liquid, n.o.s. It falls in Hazard Class 9 and Packing Group III.[20, 21]

Spill Handling: Evacuate and restrict persons not wearing protective equipment from area of spill or leak until cleanup is complete. Remove all ignition sources. Ventilate area of spill or leak. Absorb liquids in vermiculite, dry sand, earth, peat, carbon, or a similar material and deposit in sealed containers. Keep this chemical out of a confined space, such as a sewer, because of the possibility of an explosion, unless the sewer is designed to prevent the buildup of explosive concentrations. It may be necessary to contain and dispose of this chemical as a hazardous waste. If material or contaminated runoff enters waterways, notify downstream users of potentially contaminated waters. Contact your local or federal environmental protection agency for specific recommendations. If employees are required to clean up spills, they must be properly trained and equipped. OSHA 1910.120(q) may be applicable.

Fire Extinguishing: This chemical is a combustible liquid. Poisonous gases, including nitrogen oxides, are produced in fire. Use dry chemical, carbon dioxide, or alcohol foam extinguishers. Vapors are heavier than air and will collect in low areas. Vapors may travel long distances to ignition sources and flashback. Vapors in confined areas may explode when exposed to fire. Containers may explode in fire. Storage containers and parts of containers may rocket great distances, in many directions. If material or contaminated runoff enters waterways, notify downstream users of potentially contaminated waters. Notify local health and fire officials and pollution control agencies. From a secure, explosion-proof location, use water spray to cool exposed containers. If cooling streams are ineffective (venting sound increases in volume and pitch, tank discolors, or shows any signs of deforming), withdraw immediately to a secure position. If employees are expected to fight fires, they must be trained and equipped in OSHA 1910.156. The only respirators recommended for firefighting are self-contained breathing apparatuses that have full face-pieces and are operated in a pressure-demand or other positive-pressure mode.

Disposal Method Suggested: Consult with environmental regulatory agencies for guidance on acceptable disposal practices. Generators of waste containing this contaminant (≥ 100 kg/mo) must conform with EPA regulations governing storage, transportation, treatment, and waste disposal. Under 40 CFR 261.5 small quantity generators of this waste may qualify for partial exclusion from hazardous waste regulations.

Reference

New Jersey Department of Health and Senior Services. (January 2007). *Hazardous Substances Fact Sheet: N-nitrosodi-N-butylamine.* Trenton, NJ

N-Nitrosodiethylamine   N:0570

Molecular Formula: $C_4H_{10}N_2O$
Synonyms: DANA; DEN; DENA; Diaethylnitrosamin (German); Diethylamine, *N*-nitroso-; Diethylnitrosamide; Diethylnitrosamine; *N,N*-Diethylnitrosoamine; Diethylnitrosoamine; Ethanamine, *n*-ethyl-*N*-nitroso-; *n*-Ethyl-*N*-nitrosoethanamine; NDEA; *N*-Nitrosodiaethylamin (German); *N*-Nitroso-*N,N*-diethylamine; Nitrosodiethylamine; *N*-Nitrosodietilamina (Spanish)
CAS Registry Number: 55-18-5
RTECS® Number: IA3500000
UN/NA & ERG Number: UN3082/171
EC Number: 200-226-1
Regulatory Authority and Advisory Bodies
Carcinogenicity: IARC: Animal Sufficient Evidence; Human Insufficient Evidence, Group 2A, 1998; EPA: Sufficient evidence from animal studies; inadequate evidence or no useful data from epidemiologic studies;

NTP: Reasonably anticipated to be a human carcinogen; NCI: Carcinogenesis Studies (ipr); clear evidence: mouse, rat 1975.

US EPA Hazardous Waste Number (RCRA No.): U174.
RCRA, 40CFR261, Appendix 8 Hazardous Constituents.
RCRA 40CFR268.48; 61FR15654, Universal Treatment Standards: Wastewater (mg/L), 0.40; Nonwastewater (mg/kg), 28.
RCRA 40CFR264, Appendix 9; TSD Facilities Ground Water Monitoring List. Suggested test method(s) (PQL μg/L): 8270 (10).
Reportable Quantity (RQ): 1 lb (0.454 kg).
EPCRA Section 313 Form R *de minimis* concentration reporting level: 0.1%.
California Proposition 65 Chemical: Cancer 10/1/87.
Canada, WHMIS, Ingredients Disclosure List Concentration 0.1%.
European/International Regulations: not listed in Annex 1.
WGK (German Aquatic Hazard Class): No value assigned.
Description: *N*-Nitrosodiethylamine is a yellow liquid. Molecular weight = 102.16; Boiling point = 177°C. Hazard Identification (based on NFPA-704 M Rating System): Health 2, Flammability 1, Reactivity 0. Soluble in water.
Potential Exposure: An additive in gasoline and lubricants; an antioxidant and stabilizer in plastics. Used in research.
Incompatibilities: Oxidizers, reducing agents (may form hydrazine), hydrogen bromide. Light sensitive; rapidly decomposes.
Permissible Exposure Limits in Air
No TEEL available.
DFG MAK: [skin] Carcinogen Category 2.
There are no established numerical OELs. However, this chemical is a carcinogen and exposure should be reduced to the lowest possible level.
Determination in Air: Use NIOSH Analytical Method (IV) #2522, Nitrosamines; OSHA Analytical Method 27.
Routes of Entry: Inhalation, ingestion, skin and/or eye contact.
Harmful Effects and Symptoms
Short Term Exposure: Contact can irritate the skin and eyes. The vapors cause respiratory tract irritation.
Long Term Exposure: Potential human carcinogen. Prolonged or repeated exposure may cause liver damage. May damage the developing fetus.
Points of Attack: Liver.
Medical Surveillance: Liver function tests.
First Aid: If this chemical gets into the eyes, remove any contact lenses at once and irrigate immediately for at least 15 min, occasionally lifting upper and lower lids. Seek medical attention immediately. If this chemical contacts the skin, remove contaminated clothing and wash immediately with soap and water. Seek medical attention immediately. If this chemical has been inhaled, remove from exposure, begin rescue breathing (using universal precautions, including resuscitation mask) if breathing has stopped and CPR if heart action has stopped. Transfer promptly to a medical

facility. When this chemical has been swallowed, get medical attention. Give large quantities of water and induce vomiting. Do not make an unconscious person vomit.

Personal Protective Methods: Wear protective gloves and clothing to prevent any reasonable probability of skin contact. Safety equipment suppliers/manufacturers can provide recommendations on the most protective glove/clothing material for your operation. Plastic, latex, or Neoprene™ gloves may be effective protection. All protective clothing (suits, gloves, footwear, headgear) should be clean, available each day, and put on before work. Contact lenses should not be worn when working with this chemical. Wear splash-proof chemical goggles and face shield unless full face-piece respiratory protection is worn. Employees should wash immediately with soap when skin is wet or contaminated. Provide emergency showers and eyewash.

Respirator Selection: Specific respirator(s) have not been recommended by NIOSH. However, based on potential carcinogenicity the following might be considered: *At any detectable concentration:* SCBAF: Pd,Pp (APF = 10,000) (any NIOSH/MSHA- or European Standard EN 149-approved self-contained breathing apparatus that has a full face-piece and is operated in a pressure-demand or other positive-pressure mode) or SaF: Pd,Pp: ASCBA (APF = 10,000) (any supplied-air respirator that has a full face-piece and is operated in a pressure-demand or other positive-pressure mode in combination with an auxiliary, self-contained breathing apparatus operated in a pressure-demand or other positive-pressure mode). *Escape:* 100 F (APF = 50) (any air-purifying, full-face-piece respirator with an N100, R100, or P100 filter) or SCBAE (any appropriate escape-type, self-contained breathing apparatus).

Storage: Color Code—Green: General storage may be used. Prior to working with this chemical you should be trained on its proper handling and storage. Store in tightly closed containers in a cool, dark, well-ventilated area away from light, oxidizers, reducing agents, hydrogen bromide. Where possible, automatically pump liquid from drums or other storage containers to process containers. Sources of ignition, such as smoking and open flames, are prohibited where this chemical is handled, used, or stored. Metal containers involving the transfer of 5 gallons or more of this chemical should be grounded and bonded. Drums must be equipped with self-closing valves, pressure vacuum bungs, and flame arresters. Use only nonsparking tools and equipment, especially when opening and closing containers of this chemical. Wherever this chemical is used, handled, manufactured, or stored, use explosion-proof electrical equipment and fittings. A regulated, marked area should be established where this chemical is handled, used, or stored in compliance with OSHA Standard 1910.1045.

Shipping: The name of this material is not on the DOT list of materials[19] for label and packaging standards. However, based on regulations, it may be classified[52] as an Environmentally hazardous substances, liquid, n.o.s. It falls in Hazard Class 9 and Packing Group III.[20, 21]

Spill Handling: Evacuate and restrict persons not wearing protective equipment from area of spill or leak until cleanup is complete. Remove all ignition sources. Ventilate area of spill or leak. Absorb liquids in vermiculite, dry sand, earth, peat, carbon, or a similar material and deposit in sealed containers. Keep this chemical out of a confined space, such as a sewer, because of the possibility of an explosion, unless the sewer is designed to prevent the buildup of explosive concentrations. It may be necessary to contain and dispose of this chemical as a hazardous waste. If material or contaminated runoff enters waterways, notify downstream users of potentially contaminated waters. Contact your local or federal environmental protection agency for specific recommendations. If employees are required to clean up spills, they must be properly trained and equipped. OSHA 1910.120(q) may be applicable.

Fire Extinguishing: This chemical is a combustible liquid. Poisonous gases, including nitrogen oxides, are produced in fire. Use dry chemical, carbon dioxide, alcohol foam, or polymer foam extinguishers. Vapors are heavier than air and will collect in low areas. Vapors may travel long distances to ignition sources and flashback. Vapors in confined areas may explode when exposed to fire. Containers may explode in fire. Storage containers and parts of containers may rocket great distances, in many directions. If material or contaminated runoff enters waterways, notify downstream users of potentially contaminated waters. Notify local health and fire officials and pollution control agencies. From a secure, explosion-proof location, use water spray to cool exposed containers. If cooling streams are ineffective (venting sound increases in volume and pitch, tank discolors, or shows any signs of deforming), withdraw immediately to a secure position. If employees are expected to fight fires, they must be trained and equipped in OSHA 1910.156. The only respirators recommended for firefighting are self-contained breathing apparatuses that have full face-pieces and are operated in a pressure-demand or other positive-pressure mode.

Disposal Method Suggested: Consult with environmental regulatory agencies for guidance on acceptable disposal practices. Generators of waste containing this contaminant (≥100 kg/mo) must conform with EPA regulations governing storage, transportation, treatment, and waste disposal.

Reference
New Jersey Department of Health and Senior Services. (August 1999). *Hazardous Substances Fact Sheet: N-Nitrosodiethylamine.* Trenton, NJ

N-Nitrosodimethylamine   N:0580

Molecular Formula: $C_2H_6N_2O$
Common Formula: $(CH_3)_2NN=O$
Synonyms: Dimethylamine, *N*-nitroso-; Dimethylnitrosamin (German); Dimethylnitrosamine; *N,N*-Dimethylnitrosoamine;

DMN; DMNA; Methanamine, *n*-methyl-*N*-nitroso-; *n*-Methyl-*N*-nitrosomethan amine; NDMA; *N*-Nitroso-*N,N*-dimethylamine; Nitrosodimethylamine; *N*-Nitrosodimetilamina (Spanish)
CAS Registry Number: 62-75-9
RTECS® Number: IQ0525000
UN/NA & ERG Number: UN2810/153
EC Number: 200-549-8 [*Annex I Index No.:* 612-077-00-3]
Regulatory Authority and Advisory Bodies
Carcinogenicity: NTP: 11th Report on Carcinogens, 2004: Reasonably anticipated to be a human carcinogen; IARC: Animal Sufficient Evidence; Human No Adequate Data, Group 2A, 1987; EPA: Sufficient evidence from animal studies; inadequate evidence or no useful data from epidemiologic studies; OSHA: Potential human carcinogen; NIOSH: Potential occupational carcinogen.
US EPA Gene-Tox Program, Positive: Carcinogenicity—mouse/rat; Positive: *D. melanogaster*—reciprocal translocation; Positive: Host-mediated assay; L5178Y cells *in vitro*—TK test; Positive: Mammalian micronucleus; *N. crassa*—forward mutation; Positive: *N. crassa*—reversion; *E. coli* polA with S9; Positive: Histidine reversion—Ames test; Positive: *D. melanogaster* sex-linked lethal; Positive: *In vitro* UDS in rat liver; V79 cell culture-gene mutation; Positive: TRP reversion; *S. cerevisiae* gene conversion; Positive: *S. cerevisiae*—forward mutation; *S. cerevisiae*—homozygosis; Positive: *S. pombe*—forward mutation; Positive/dose response: *In vitro* cytogenetics—nonhuman; Positive/dose response: *In vitro* SCE—nonhuman; *In vivo* SCE—nonhuman; Positive/dose response: *In vitro* UDS—human fibroblast; Negative: Cytogenetics—male germ cell; Sperm morphology—mouse; Negative: UDS in mouse germ cells; Inconclusive: SHE—clonal assay; Cell transform.—RLV F344 rat embryo; Inconclusive: Rodent dominant lethal; Mouse spot test; Inconclusive: *E. coli* polA without S9; Positive: CHO gene mutation.
OSHA, 29CFR1910 Specifically Regulated Chemicals (See CFR 1910.1016).
Clean Air Act: Hazardous Air Pollutants (Title I, Part A, Section 112).
Clean Water Act: Section 313 Water Priority Chemicals (57FR41331, 9/9/92).
US EPA Hazardous Waste Number (RCRA No.): P082.
RCRA, 40CFR261, Appendix 8 Hazardous Constituents.
RCRA 40CFR268.48; 61FR15654, Universal Treatment Standards: Wastewater (mg/L), 0.40; Nonwastewater (mg/kg), 2.3.
RCRA 40CFR264, Appendix 9; TSD Facilities Ground Water Monitoring List. Suggested test method(s) (PQL μg/L): 8270 (10).
SUPERFUND/EPCRA 40CFR355, Extremely Hazardous Substances: TPQ = 1000 lb (454 kg).
Reportable Quantity (RQ): 10 lb (4.54 kg).
EPCRA Section 313 Form R *de minimis* concentration reporting level: 0.1%.
California Proposition 65 Chemical: Cancer 10/1/87.

California Proposition 65 Chemical: Cancer (methylhydrazine and its salts) 7/1/92.
European/International Regulations: Hazard Symbol: T+, N; Risk phrases: R45; R25; R26; R48/25; R51/53; Safety phrases: S48/25; S51/53; S53; S45; S61 (see Appendix 4).
WGK (German Aquatic Hazard Class): No value assigned.
Description: N-Nitrosodimethylamine is a yellow oily liquid with a faint, characteristic odor. Molecular weight = 74.10; Boiling point = 152°C; Flash point = 61°C. Hazard Identification (based on NFPA-704 M Rating System): Health 2, Flammability 1, Reactivity 1. Soluble in water.
Potential Exposure: Compound Description: Tumorigen, Mutagen; Reproductive Effector; Human Data. Nitrosodimethylamine was formerly used in the production of rocket fuels. Presently used as an antioxidant, as an additive for lubricants, and as a softener of copolymers. It is used as an intermediate for 1,1-dimethylhydrazine.
Incompatibilities: Oxidants, especially peracids. Sensitive to UV light. Should be stored in dark bottles.
Permissible Exposure Limits in Air
OSHA PEL: Cancer suspect agent. Exposure of workers to this chemical is to be controlled through the required use of engineering controls, work practice, and personal protective equipment, including respirators. See 29CFR1910.1003.
NIOSH REL: A potential occupational carcinogen. Limit exposure to lowest feasible concentration.
ACGIH TLV®[1]: Exposures by all routes should be carefully controlled to levels as low as possible [skin]. Confirmed animal carcinogen with unknown relevance to humans.
Protective Action Criteria (PAC)
TEEL-0: 3.5 mg/m^3
PAC-1: 10 mg/m^3
PAC-2: 19 mg/m^3
PAC-3: 100 mg/m^3
DFG MAK: [skin] Carcinogen Category 2
Austria: carcinogen, 1999; France: carcinogen, 1993; Switzerland: MAK-W 0.001 mg/m^3, carcinogen, 1999; United Kingdom: carcinogen, 2000; the Netherlands: MAC-TGG 0.001 mg/m^3, 2003; Argentina, Bulgaria, Columbia, Jordan, South Korea, New Zealand, Singapore, Vietnam: ACGIH TLV®: confirmed animal carcinogen with unknown relevance to humans.
Determination in Air: Use NIOSH IV; Method #2522; OSHA Analytical Methods 6 and 27.
Permissible Concentration in Water: *To protect freshwater aquatic life:* 5850 μg/L on an acute toxicity basis for nitrosamines as a class. *To protect saltwater aquatic life:* 3,300,000 μg/L on an acute toxicity basis for nitrosamines as a class. For protection of human health: preferably zero. An additional lifetime cancer risk of 1 in 100,000 is posed by a concentration of 0.014 μg/L.[6] The states of Kansas and Minnesota have set guidelines for DMNA in drinking water[61], Kansas at 0.0014 μg/L and Minnesota at 0.014 μg/L.
Determination in Water: Methylene chloride extraction followed by gas chromatography with nitrogen–phosphorus or

reductive Hall detectors (EPA Method 607); or gas chromatography plus mass spectrometry (EPA Method 625).

Routes of Entry: Inhalation, skin absorption, ingestion, skin and/or eye contact.

Harmful Effects and Symptoms

Short Term Exposure: Irritates the eyes, skin, and respiratory tract. Symptoms of exposure include nausea, vomiting, and malaise. Extremely high toxicity. The lowest lethal oral dose in humans has been reported at 10 mg/kg/80 week intermittent exposure.

Long Term Exposure: Chronic exposure may cause liver disease with jaundice and swelling with low platelet count and cirrhosis. The effects may be delayed. Based on animal tests, this substance may be a potential carcinogen in humans.

Points of Attack: Liver.

Medical Surveillance: OSHA mandates the following tests: *increased risk:* reduced immunologic competence, steroid treatment, pregnancy, cigarette smoking. NIOSH lists the tests: *increased risk:* reduced immunologic competence, steroid treatment, pregnancy, cigarette smoking, liver function tests, pulmonary function tests. Based on human experience and on animal studies, preplacement and periodic examinations should include a history of exposure to other carcinogens, alcohol and smoking habits, medications, and family history. Special attention should be given to liver size and function, and to any changes in lung symptoms or X-rays. Renal function should be followed. Sputum and urine cytology may be useful.

First Aid: If this chemical gets into the eyes, remove any contact lenses at once and irrigate immediately for at least 15 min, occasionally lifting upper and lower lids. Seek medical attention immediately. If this chemical contacts the skin, remove contaminated clothing and wash immediately with soap and water. Seek medical attention immediately. If this chemical has been inhaled, remove from exposure, begin rescue breathing (using universal precautions, including resuscitation mask) if breathing has stopped and CPR if heart action has stopped. Transfer promptly to a medical facility. When this chemical has been swallowed, get medical attention. Give large quantities of water and induce vomiting. Do not make an unconscious person vomit. Medical observation is recommended for because the symptoms of jaundice may be delayed. Keep victim quiet and maintain normal body temperature.

Personal Protective Methods: Wear protective gloves and clothing to prevent any reasonable probability of skin contact. Safety equipment suppliers/manufacturers can provide recommendations on the most protective glove/clothing material for your operation. All protective clothing (suits, gloves, footwear, headgear) should be clean, available each day, and put on before work. Contact lenses should not be worn when working with this chemical. Wear splash-proof chemical goggles and face shield unless full face-piece respiratory protection is worn. Employees should wash immediately with soap when skin is wet or contaminated. Provide emergency showers and eyewash.

Respirator Selection: At any detectable concentration: SCBAF: Pd,Pp (APF = 10,000) (any NIOSH/MSHA- or European Standard EN 149-approved self-contained breathing apparatus that has a full face-piece and is operated in a pressure-demand or other positive-pressure mode) or SaF: Pd,Pp: ASCBA (APF = 10,000) (any supplied-air respirator that has a full face-piece and is operated in a pressure-demand or other positive-pressure mode in combination with an auxiliary, self-contained breathing apparatus operated in a pressure-demand or other positive-pressure mode). *Escape:* 100 F (APF = 50) (any air-purifying, full-face-piece respirator with an N100, R100, or P100 filter) or SCBAE (any appropriate escape-type, self-contained breathing apparatus). See also Appendix E *NIOSH Pocket Guide.*

Storage: Color Code—Blue: Health Hazard/Poison: Store in a secure poison location. Prior to working with this chemical you should be trained on its proper handling and storage. Store in a refrigerator in brown bottles, and protect from oxidizers and prolonged exposure to light. Where possible, automatically pump liquid from drums or other storage containers to process containers. Sources of ignition, such as smoking and open flames, are prohibited where this chemical is handled, used, or stored. Metal containers involving the transfer of 5 gallons or more of this chemical should be grounded and bonded. Drums must be equipped with self-closing valves, pressure vacuum bungs, and flame arresters. Use only nonsparking tools and equipment, especially when opening and closing containers of this chemical. Wherever this chemical is used, handled, manufactured, or stored, use explosion-proof electrical equipment and fittings. A regulated, marked area should be established where this chemical is handled, used, or stored in compliance with OSHA Standard 1910.1045.

Shipping: Toxic, liquids, organic, n.o.s. require a shipping label of "POISONOUS/TOXIC MATERIALS." *N*-Nitrosodimethylamine falls in DOT Hazard Class 6.1 and Packing Group I.

Spill Handling: Evacuate and restrict persons not wearing protective equipment from area of spill or leak until cleanup is complete. Remove all ignition sources. Ventilate area of spill or leak. Absorb liquids in celite, vermiculite, dry sand, earth, peat, carbon, or a similar material and deposit in sealed containers. Keep this chemical out of a confined space, such as a sewer, because of the possibility of an explosion, unless the sewer is designed to prevent the buildup of explosive concentrations. It may be necessary to contain and dispose of this chemical as a hazardous waste. If material or contaminated runoff enters waterways, notify downstream users of potentially contaminated waters. Contact your local or federal environmental protection agency for specific recommendations. If employees are required to clean up spills, they must be properly trained and equipped. OSHA 1910.120(q) may be applicable.

Fire Extinguishing: This chemical is a combustible liquid. Poisonous gases, including nitrogen oxides, are produced in fire. Use dry chemical, carbon dioxide, or alcohol foam extinguishers. Vapors are heavier than air and will collect in low areas. Vapors may travel long distances to ignition sources and flashback. Vapors in confined areas may explode when exposed to fire. Containers may explode in fire. Storage containers and parts of containers may rocket great distances, in many directions. If material or contaminated runoff enters waterways, notify downstream users of potentially contaminated waters. Notify local health and fire officials and pollution control agencies. From a secure, explosion-proof location, use water spray to cool exposed containers. If cooling streams are ineffective (venting sound increases in volume and pitch, tank discolors, or shows any signs of deforming), withdraw immediately to a secure position. If employees are expected to fight fires, they must be trained and equipped in OSHA 1910.156. The only respirators recommended for firefighting are self-contained breathing apparatuses that have full face-pieces and are operated in a pressure-demand or other positive-pressure mode.

Disposal Method Suggested: Pour over soda ash, neutralize with HCl, then flush to drain with large volumes of water. Consult with environmental regulatory agencies for guidance on acceptable disposal practices. Generators of waste containing this contaminant (≥ 100 kg/mo) must conform with EPA regulations governing storage, transportation, treatment, and waste disposal.

References

US Environmental Protection Agency. (1979). *Chemical Hazard Information Profile: N-Nitroso Compounds.* Washington, DC
US Environmental Protection Agency. (1980). *Nitrosamines: Ambient Water Quality Criteria.* Washington, DC
US Environmental Protection Agency. (April 30, 1980). *Dimethylnitrosamine: Health and Environmental Effects Profile No. 86.* Washington, DC: Office of Solid Waste
Sax, N. I. (Ed.). *Dangerous Properties of Industrial Materials Report*, 1, No. 2, 50–51 (1980) and 2, No. 6, 65–69 (1982)
US Public Health Service. (December 1988). *Toxicological Profile for N-Nitrosodimethylamine.* Atlanta, GA: Agency for Toxic Substances and Disease Registry
US Environmental Protection Agency. (November 30, 1987). *Chemical Hazard Information Profile: N-Nitrosodimethylamine.* Washington, DC: Chemical Emergency Preparedness Program

N-Nitrosodiphenylamine   N:0590

Molecular Formula: $C_{12}H_{10}N_2O$
Common Formula: $C_6H_5N(NO)C_6H_5$
Synonyms: Benzenamine, *N*-nitroso-*N*-phenyl-; Curetard A; Delac J; Diphenylamine, *N*-nitrosoamine; Diphenylnitrosamin (German); *N,N*-Diphenylnitrosamine; Diphenylnitrosamine; *N,N*-Diphenyl-*N*-nitrosoamine;   NCI-C02880;   NDPA;

NDPHA; *N*-Nitrosodifenilamina (Spanish); *N*-Nitroso-*N*-diphenylamine; Nitrosodiphenylamine; *N*-Nitroso-*N*-phenyla-niline; Nitrous diphenylamide; Redax; Retarder J; TJB; Valcatard; Vulcalent A; Vulcatard A; Vultrol
CAS Registry Number: 86-30-6
RTECS® Number: JJ9800000
UN/NA & ERG Number: UN2811 (toxic solid, organic, n.o.s.)/154
EC Number: 201-663-0
Regulatory Authority and Advisory Bodies
Carcinogenicity: NCI: Carcinogenesis Bioassay (feed); clear evidence: rat; no evidence: mouse; IARC: Animal Limited Evidence; Human No Adequate Data, *not classifiable as carcinogenic to humans*, Group 3, 1987; EPA: Sufficient evidence from animal studies; inadequate evidence or no useful data from epidemiologic studies.
US EPA Gene-Tox Program, Positive: SHE—clonal assay; Cell transform.—RLV F344 rat embryo; Positive/limited: Carcinogenicity—mouse/rat; Negative: Cell transform.—BALB/c-3T3; Host-mediated assay; Negative: *E. coli* polA with S9; Histidine reversion—Ames test; Negative: Sperm morphology—mouse; *In vitro* UDS—human fibroblast; Negative: *In vitro* UDS in rat liver; V79 cell culture—gene mutation; Negative: *S. cerevisiae*—homozygosis; Inconclusive: L5178Y cells *In vitro*—TK test; Mammalian micronucleus; Inconclusive: *E. coli* polA without S9; *In vitro* SCE—nonhuman; Inconclusive: *D. melanogaster* sex-linked lethal.
Clean Water Act: Section 313 Water Priority Chemicals (57FR41331, 9/9/92).
RCRA Land Ban Waste.
RCRA 40CFR268.48; 61FR15654, Universal Treatment Standards: Wastewater (mg/L), 0.92; Nonwastewater (mg/kg), 13.
RCRA 40CFR264, Appendix 9; TSD Facilities Ground Water Monitoring List. Suggested test method(s) (PQL µg/L): 8270 (10).
Reportable Quantity (RQ): 100 lb (45.4 kg).
EPCRA Section 313 Form R *de minimis* concentration reporting level: 1.0%.
California Proposition 65 Chemical: Cancer 4/1/88.
As nitrosamines:
Clean Water Act: 40CFR401.15 Section 307 Toxic Pollutants as nitrosamines.
RCRA, 40CFR261, Appendix 8 Hazardous Constituents, waste number not listed.
WGK (German Aquatic Hazard Class): 2—Hazard to waters.
Description: *N*-Nitrosodiphenylamine is a yellow to orange-brown crystalline solid. Molecular weight = 198.24; Boiling point = 268°C (estimated); Freezing/Melting point = 67°C. Hazard Identification (based on NFPA-704 M Rating System): Health 1, Flammability 2, Reactivity 0. Slightly soluble in water.
Potential Exposure: Compound Description: Tumorigen, Mutagen. Primary Irritant. *N*-Nitrosodiphenylamine is not a

naturally occurring substance; it is a man-made chemical that is no longer produced in the United States. It was used in the manufacture of plastics, resins, rubber, and synthetic textiles to help control processes involved in making rubber products, such as tires and mechanical goods; however, in the early 1980s, the US manufacturers stopped producing *N*-nitrosodiphenylamine because new and more efficient chemicals were found to replace its uses. In addition, the use of *N*-nitrosodiphenylamine had several undesirable side effects which do not occur with the replacement chemicals.

Incompatibilities: Oxidizing materials.

Permissible Exposure Limits in Air
Protective Action Criteria (PAC)
TEEL-0: 7.5 mg/m^3
PAC-1: 25 mg/m^3
PAC-2: 150 mg/m^3
PAC-3: 500 mg/m^3
DFG MAK: Carcinogen Category 3 B.
There are no established numerical OELs. However, this chemical is a carcinogen and exposure should be reduced to the lowest possible level.

Determination in Air: Use OSHA Analytical Method 23.

Permissible Concentration in Water: Russia set a MAC of 0.1 mg/L in surface water. Two states have set guidelines for nitrosodiphenylamine in drinking water.[60] They are Kansas at 71.0 µg/L, and Minnesota at 71.1 µg/L.

Routes of Entry: Inhalation, ingestion, skin and/or eye contact.

Harmful Effects and Symptoms

Short Term Exposure: Information is not available regarding effects of brief exposures to *N*-nitrosodiphenylamine on human health. Very little is known about the health effects of brief exposures to *N*-nitrosodiphenylamine in experimental animals, other than that relatively high doses by ingestion are required to produce death. It is not known if exposure to *N*-nitrosodiphenylamine by breathing or skin contact can affect the health of humans or animals, but ingestion of *N*-nitrosodiphenylamine has been shown to have adverse health effects in animals. Exposure of humans to *N*-nitrosodiphenylamine should be minimized.

Long Term Exposure: Long-term exposure of experimental animals to *N*-nitrosodiphenylamine by ingestion produced inflammation and cancer of the bladder. It is not known whether these effects or birth defects would occur in humans if they were exposed to *N*-nitrosodiphenylamine.

Points of Attack: Cancer site in animals: bladder.

Medical Surveillance: Although the presence of the chemical in blood and urine can be detected by chemical analysis, this analysis has not been used as a test for human exposure or to predict potential health effects.

First Aid: Skin Contact: Flood all areas of body that have contacted the substance with water. Do not wait to remove contaminated clothing; do it under the water stream. Use soap to help assure removal. Isolate contaminated clothing when removed to prevent contact by others.[52] *Eye Contact:* Remove any contact lenses at once. Immediately flush eyes well with copious quantities of water or normal saline for at least 20–30 min. Seek medical attention. *Inhalation:* Leave contaminated area immediately; breathe fresh air. Proper respiratory protection must be supplied to any rescuers. If coughing, difficult breathing, or any other symptoms develop, seek medical attention at once, even if symptoms develop many hours after exposure. *Ingestion:* Contact a physician, hospital, or poison center at once. If the victim is unconscious or convulsing, do not induce vomiting or give anything by mouth. Assure that the patient's airway is open and lay him on his side with his head lower than his body and transport immediately to a medical facility. If conscious and not convulsing, give a glass of water to dilute the substance. Vomiting should not be induced without a physician's advice.

Personal Protective Methods: Wear protective gloves and clothing to prevent any reasonable probability of skin contact. Safety equipment suppliers/manufacturers can provide recommendations on the most protective glove/clothing material for your operation. All protective clothing (suits, gloves, footwear, headgear) should be clean, available each day, and put on before work. Contact lenses should not be worn when working with this chemical. Wear dust-proof chemical goggles and face shield unless full face-piece respiratory protection is worn. Employees should wash immediately with soap when skin is wet or contaminated. Provide emergency showers and eyewash.

Respirator Selection: Follow regulations in OSHA 29CFR1910.134 or European Standard EN149. Use a NIOSH/MSHA- or European Standard EN149-approved respirator; or use an approved supplied-air respirator with a full face-piece operated in the positive-pressure mode, or with a full face-piece, hood, or helmet in the continuous-flow mode; or use a NIOSH/MSHA- or European Standard EN149-approved self-contained breathing apparatus with a full face-piece operated in pressure-demand or other positive-pressure mode.

Storage: Color Code—Blue: Health Hazard/Poison: Store in a secure poison location. Prior to working with this chemical you should be trained on its proper handling and storage. Store in a refrigerator under an inert atmosphere for prolonged storage. Store in tightly closed containers in a cool, well-ventilated area away from oxidizers. Where possible, automatically transfer from drums or other storage containers to process containers. Sources of ignition, such as smoking and open flames, are prohibited where this chemical is handled, used, or stored. Metal containers involving the transfer of 5 gallons or more of this chemical should be grounded and bonded. Use only nonsparking tools and equipment, especially when opening and closing containers of this chemical. Wherever this chemical is used, handled, manufactured, or stored, use explosion-proof electrical equipment and fittings. A regulated, marked area should be established where this chemical is handled, used, or stored in compliance with OSHA Standard 1910.1045.

Shipping: Toxic solids, organic, n.o.s. require a shipping label of "POISONOUS/TOXIC MATERIALS." It falls in Hazard Class 6.1.

Spill Handling: Evacuate persons not wearing protective equipment from area of spill or leak until cleanup is complete. Use HEPA vacuum or wet method to reduce dust during cleanup. Do not dry sweep[52]: remove all sources of ignition and dampen spilled material with 60−70% acetone to avoid airborne dust, then transfer material to a suitable container. Ventilate the spill area and use absorbent paper dampened with 60−70% acetone to pick up remaining material. Wash surfaces well with soap and water. Seal all wastes in vapor-tight plastic bags for eventual disposal. It may be necessary to contain and dispose of this chemical as a hazardous waste. If material or contaminated runoff enters waterways, notify downstream users of potentially contaminated waters. Contact your local or federal environmental protection agency for specific recommendations. If employees are required to clean up spills, they must be properly trained and equipped. OSHA 1910.120(q) may be applicable.

Fire Extinguishing: This chemical is a combustible solid. Use dry chemical, carbon dioxide, water spray, or alcohol foam extinguishers. Poisonous gases, including nitrogen oxides, are produced in fire. If material or contaminated runoff enters waterways, notify downstream users of potentially contaminated waters. Notify local health and fire officials and pollution control agencies. From a secure, explosion-proof location, use water spray to cool exposed containers. If cooling streams are ineffective (venting sound increases in volume and pitch, tank discolors, or shows any signs of deforming), withdraw immediately to a secure position. If employees are expected to fight fires, they must be trained and equipped in OSHA 1910.156. The only respirators recommended for firefighting are self-contained breathing apparatuses that have full face-pieces and are operated in a pressure-demand or other positive-pressure mode.

Disposal Method Suggested: Burn in admixture with flammable solvent in furnace equipped with afterburner and scrubber.[22]

References

US Public Health Service. (October 1987). *Toxicological Profile for Nitrosodiphenylamine.* Atlanta, GA: Agency for Toxic Substances & Disease Registry

New Jersey Department of Health and Senior Services. (February 2000). *Hazardous Substances Fact Sheet: N-Nitrosodiphenylamine.* Trenton, NJ

p-Nitrosodiphenylamine   N:0600

Molecular Formula: $C_{12}H_{10}N_2O$

Synonyms: Benzenamine, 4-nitroso-*N*-phenyl-; Diphenylamine, 4-nitroso-; Naugard TKB; NCI-C02244; *p*-Nitrosodifenilamina (Spanish); 4-Nitrosodiphenylamine; *p*-Nitroso-*N*-phenylaniline; 4-Nitroso-*N*-phenylaniline; *p*-Phenylaminonitrosobenzene; *N*-Phenyl-*p*-nitrosoaniline; TKB

CAS Registry Number: 156-10-5

RTECS® Number: JK0175000

UN/NA & ERG Number: UN3077/171

EC Number: 205-848-7

Regulatory Authority and Advisory Bodies

Carcinogenicity: IARC: Animal Insufficient Evidence; Human No Adequate Data, *not classifiable as carcinogenic to humans*, Group 3, 1982, 1998.

EPCRA Section 313 Form R *de minimis* concentration reporting level: 1.0%.

California Proposition 65 Chemical: Cancer 1/1/88.

European labeling: Hazard symbol: Xn (Harmful); Possible carcinogen; Risk Phrases: R40; Safety Phrases: S36/37; S45; S53 (see Appendix 4).

WGK (German Aquatic Hazard Class): No value assigned.

Description: TKB is a black powder or a green plate-like material with a bluish luster. Molecular weight = 198.2; Freezing/Melting point = 145°C. Hazard Identification (based on NFPA-704 M Rating System): Health 3, Flammability 0, Reactivity 0. Slightly soluble in water.

Potential Exposure: Used as a chemical intermediate for dyes and pharmaceuticals, in making monomers, and vulcanizing rubber.

Incompatibilities: Oxidizers.

Permissible Exposure Limits in Air

Protective Action Criteria (PAC)

TEEL-0: 0.1 mg/m^3

PAC-1: 0.3 mg/m^3

PAC-2: 2 mg/m^3

PAC-3: 150 mg/m^3

This chemical is a potential occupational carcinogen and all exposure should be reduced to the lowest possible level.

Determination in Air: Use NIOSH IV; Method #2522; OSHA Analytical Method 27.

Permissible Concentration in Water: Russia set a MAC of 0.1 mg/L in surface water. Two states have set guidelines for nitrosodiphenylamine in drinking water.[60] They are Kansas at 71.0 µg/L and Minnesota at 71.1 µg/L.

Routes of Entry: Inhalation, ingestion, skin and/or eyes. Absorbed through the skin.

Harmful Effects and Symptoms

Short Term Exposure: Contact can cause eye irritation.

Long Term Exposure: There is limited evidence that this chemical causes bladder cancer in animals.

First Aid: If this chemical gets into the eyes, remove any contact lenses at once and irrigate immediately for at least 15 min, occasionally lifting upper and lower lids. Seek medical attention immediately. If this chemical contacts the skin, remove contaminated clothing and wash immediately with soap and water. Seek medical attention immediately. If this chemical has been inhaled, remove from exposure, begin rescue breathing (using universal precautions, including resuscitation mask) if breathing has stopped and CPR if heart action has stopped. Transfer promptly to a medical

facility. When this chemical has been swallowed, get medical attention. Give large quantities of water and induce vomiting. Do not make an unconscious person vomit.

Personal Protective Methods: Wear protective gloves and clothing to prevent any reasonable probability of skin contact. Safety equipment suppliers/manufacturers can provide recommendations on the most protective glove/clothing material for your operation. All protective clothing (suits, gloves, footwear, headgear) should be clean, available each day, and put on before work. Contact lenses should not be worn when working with this chemical. Wear dust-proof chemical goggles and face shield unless full face-piece respiratory protection is worn. Employees should wash immediately with soap when skin is wet or contaminated. Provide emergency showers and eyewash.

Respirator Selection: Follow regulations in OSHA 29CFR1910.134 or European Standard EN149. Use a NIOSH/MSHA- or European Standard EN149-approved respirator; or use an approved supplied-air respirator with a full face-piece operated in the positive-pressure mode, or with a full face-piece, hood, or helmet in the continuous-flow mode; or use a NIOSH/MSHA- or European Standard EN149-approved self-contained breathing apparatus with a full face-piece operated in pressure-demand or other positive-pressure mode.

Storage: Color Code—Green: General storage may be used. Prior to working with this chemical you should be trained on its proper handling and storage. Store in tightly closed containers in a cool, well-ventilated area away from oxidizers. Where possible, automatically transfer material from drums or other storage containers to process containers. Sources of ignition, such as smoking and open flames, are prohibited where this chemical is handled, used, or stored. Metal containers involving the transfer of this chemical should be grounded and bonded. Use only nonsparking tools and equipment, especially when opening and closing containers of this chemical. Wherever this chemical is used, handled, manufactured, or stored, use explosion-proof electrical equipment and fittings. A regulated, marked area should be established where this chemical is handled, used, or stored in compliance with OSHA Standard 1910.1045.

Shipping: This material may be classed as an Environmentally hazardous solid, n.o.s. It falls in Hazard Class 9 and "CLASS 9." Packing Group III.

Spill Handling: Evacuate persons not wearing protective equipment from area of spill or leak until cleanup is complete. Remove all ignition sources. Use HEPA vacuum or wet method to reduce dust during cleanup. Do not dry sweep. Collect powdered material in the most convenient and safe manner and deposit in sealed containers. Ventilate area after cleanup is complete. It may be necessary to contain and dispose of this chemical as a hazardous waste. If material or contaminated runoff enters waterways, notify downstream users of potentially contaminated waters. Contact your local or federal environmental protection agency for specific recommendations. If employees are required to clean up spills, they must be properly trained and equipped. OSHA 1910.120(q) may be applicable.

Fire Extinguishing: This chemical is a combustible solid. Use dry chemical, carbon dioxide, water spray, or foam extinguishers. Poisonous gases, including nitrogen oxides, are produced in fire. If material or contaminated runoff enters waterways, notify downstream users of potentially contaminated waters. Notify local health and fire officials and pollution control agencies. From a secure, explosion-proof location, use water spray to cool exposed containers. If cooling streams are ineffective (venting sound increases in volume and pitch, tank discolors, or shows any signs of deforming), withdraw immediately to a secure position. If employees are expected to fight fires, they must be trained and equipped in OSHA 1910.156. The only respirators recommended for firefighting are self-contained breathing apparatuses that have full face-pieces and are operated in a pressure-demand or other positive-pressure mode.

Reference

New Jersey Department of Health and Senior Services. (January 2000). *Hazardous Substances Fact Sheet: p-Nitrosodiphenylamine.* Trenton, NJ

N-Nitrosodipropylamine   N:0610

Molecular Formula: $C_6H_{14}N_2O$
Common Formula: $C_3H_7N(NO)C_3H_7$
Synonyms: Dipropylamine, *N*-nitroso-; Di-*N*-propylnitrosamine; Dipropylnitrosamine; DPN; DPNA; NDPA; *N*-Nitroso-*N*-dipropylamine; *N*-Nitrosodipropylamine; Nitrosodipropylamine; *N*-Nitroso-*N*-propylpropanamine; 1-Propanamine, *N*-nitroso-*N*-propyl-
CAS Registry Number: 621-64-7
RTECS® Number: JL9700000
UN/NA & ERG Number: UN3082/171
EC Number: 210-698-0 [*Annex I Index No.:* 612-098-00-8]
Regulatory Authority and Advisory Bodies

Carcinogenicity: IARC: Animal Sufficient Evidence; Human Insufficient Evidence, *possibly carcinogenic to humans*, Group 2B; EPA: Sufficient evidence from animal studies; inadequate evidence or no useful data from epidemiologic studies; NTP: Reasonably anticipated to be a human carcinogen; NCI: Carcinogenesis Bioassay (feed); clear evidence: rat; no evidence: mouse.

US EPA Gene-Tox Program, Positive: SHE—clonal assay; Cell transform.—RLV F344 rat embryo; Positive/limited: Carcinogenicity—mouse/rat; Negative: Cell transform.—BALB/c-3T3; Host-mediated assay; Negative: *E. coli* polA with S9; Histidine reversion—Ames test; Negative: Sperm morphology—mouse; *In vitro* UDS—human fibroblast; Negative: *In vitro* UDS in rat liver; V79 cell culture-gene mutation; Negative: *S. cerevisiae*—homozygosis; Inconclusive: L5178Y cells *In vitro*-TK test; Mammalian micronucleus; Inconclusive: *E. coli* polA without S9; *In vitro* SCE—nonhuman; Inconclusive: *D. melanogaster* sex-linked lethal.

Clean Water Act: Section 313 Water Priority Chemicals (57FR41331, 9/9/92).

US EPA Hazardous Waste Number (RCRA No.): U111.

RCRA, 40CFR261, Appendix 8 Hazardous Constituents.

RCRA 40CFR268.48; 61FR15654, Universal Treatment Standards: Wastewater (mg/L), 0.40; Nonwastewater (mg/kg), 14.

RCRA 40CFR264, Appendix 9; TSD Facilities Ground Water Monitoring List. Suggested test method(s) (PQL μg/L): 8270 (10).

Reportable Quantity (RQ): 10 lb (4.54 kg).

EPCRA Section 313 Form R *de minimis* concentration reporting level: 0.1%.

California Proposition 65 Chemical: Cancer 1/1/88.

As nitrosamines:

Clean Water Act: 40CFR401.15 Section 307 Toxic Pollutants as nitrosamines.

RCRA, 40CFR261, Appendix 8 Hazardous Constituents, waste number not listed.

European/International Regulations: Hazard Symbol: T, N; Risk phrases: R45; R22; R51/53; Safety phrases: S53; S45; S61.

WGK (German Aquatic Hazard Class): No value assigned.

Description: *N*-nitrosodi-*N*-propylamine is a yellow liquid. Molecular weight = 198.24; Boiling point = 206°C; Freezing/Melting point = 144°C. Slightly soluble in water.

Potential Exposure: Compound Description: Tumorigen, Mutagen. Primary Irritant. *N*-nitrosodi-*N*-propylamine is used in the manufacture of plastics, resins, rubber, and synthetic textiles. There is no evidence that *N*-nitrosodi-*N*-propylamine exists naturally in soil, air, food, or water. Small quantities of *N*-nitrosodi-*N*-propylamine are inadvertently produced during some manufacturing processes as an impurity in some commercially available dinitroaniline based weed killers, and during the manufacture of some rubber products. However, according to Sax, some similar *N*-nitroso compounds are formed in the environment and absorbed from precursors in food, water, or air, from tobacco, and from naturally occurring compounds.

Incompatibilities: Oxidizers. Sensitive to UV light.

Permissible Exposure Limits in Air

Protective Action Criteria (PAC)

TEEL-0: 0.06 mg/m^3

PAC-1: 0.2 mg/m^3

PAC-2: 1.25 mg/m^3

PAC-3: 200 mg/m^3

DFG MAK: [skin] Carcinogen Category 2.

There are no established numerical OELs. However, this chemical is a carcinogen and exposure should be reduced to the lowest possible level.

Determination in Air: Use OSHA Analytical Method 23.

Routes of Entry: *N*-nitrosodi-*N*-propylamine can enter the body by breathing air that contains *N*-nitrosodi-*N*-propylamine or by eating food or drinking water contaminated with *N*-nitrosodi-*N*-propylamine. *N*-Nitrosodi-*N*-propylamine is not likely to get into your body unless you eat certain foods or drink alcoholic beverages, or are exposed to it at a waste disposal site by breathing *N*-nitrosodi-*N*-propylamine vapors. It is not known whether *N*-nitrosodi-*N*-propylamine can enter the body by direct skin contact with wastes, pesticides, or soil containing *N*-nitrosodi-*N*-propylamine. Experiments with animals suggest that if *N*-nitrosodi-*N*-propylamine enters the body, it will be broken down into other compounds and will leave the body in the urine.

Harmful Effects and Symptoms

Short Term Exposure: Irritates the eyes and skin. The effects of brief or long-term exposures to *N*-nitrosodi-*N*-propylamine on human health have not been studied in depth. Little is known about the health effects of brief exposures to *N*-nitrosodi-*N*-propylamine in experimental animals except that eating or drinking certain amounts of this chemical can cause liver disease and death.

Long Term Exposure: Long-term exposure of experimental animals to *N*-nitrosodi-*N*-propylamine in food or drinking water produces cancer of the liver, esophagus, and nasal cavities. Although human studies are not available, the animal evidence indicates that it is reasonable to expect that exposure to *N*-nitrosodi-*N*-propylamine by eating or drinking could cause liver disease and cancer in humans. It is not known whether other effects, such as birth defects, occur in animals or could occur in humans if they were exposed to *N*-nitrosodi-*N*-propylamine by eating or drinking.

Points of Attack: Liver.

Medical Surveillance: The presence of *N*-nitrosodi-*N*-propylamine in blood and urine can be detected by chemical analysis, but this analysis is not routinely available and has not been used as a test for human exposure or to predict potential health effects. Liver function tests.

First Aid: If this chemical gets into the eyes, remove any contact lenses at once and irrigate immediately for at least 15 min, occasionally lifting upper and lower lids. Seek medical attention immediately. If this chemical contacts the skin, remove contaminated clothing and wash immediately with soap and water. Seek medical attention immediately. If this chemical has been inhaled, remove from exposure, begin rescue breathing (using universal precautions, including resuscitation mask) if breathing has stopped and CPR if heart action has stopped. Transfer promptly to a medical facility. When this chemical has been swallowed, get medical attention. Give large quantities of water and induce vomiting. Do not make an unconscious person vomit.

Personal Protective Methods: Wear protective gloves and clothing to prevent any reasonable probability of skin contact. Safety equipment suppliers/manufacturers can provide recommendations on the most protective glove/clothing material for your operation. All protective clothing (suits, gloves, footwear, headgear) should be clean, available each day, and put on before work. Contact lenses should not be worn when working with this chemical. Wear splash-proof chemical goggles and face shield unless full face-piece

respiratory protection is worn. Employees should wash immediately with soap when skin is wet or contaminated. Provide emergency showers and eyewash.

Respirator Selection: Follow regulations in OSHA 29CFR1910.134 or European Standard EN149. Use a NIOSH/MSHA- or European Standard EN149-approved respirator; or use an approved supplied-air respirator with a full face-piece operated in the positive-pressure mode, or with a full face-piece, hood, or helmet in the continuous-flow mode; or use a NIOSH/MSHA- or European Standard EN149-approved self-contained breathing apparatus with a full face-piece operated in pressure-demand or other positive-pressure mode.

Storage: Color Code—Green: General storage may be used. Prior to working with this chemical you should be trained on its proper handling and storage. Store in tightly closed containers in a cool, well-ventilated area away from oxidizers. Sources of ignition, such as smoking and open flames, are prohibited where this chemical is handled, used, or stored. Metal containers involving the transfer of this chemical should be grounded and bonded. Use only nonsparking tools and equipment, especially when opening and closing containers of this chemical. Wherever this chemical is used, handled, manufactured, or stored, use explosion-proof electrical equipment and fittings. A regulated, marked area should be established where this chemical is handled, used, or stored in compliance with OSHA Standard 1910.1045.

Shipping: The name of this material is not on the DOT list of materials[19] for label and packaging standards. However, based on regulations, it may be classified[52] as an Environmentally hazardous substances, liquid, n.o.s. It falls in Hazard Class 9 and Packing Group III.[20, 21]

Spill Handling: Evacuate and restrict persons not wearing protective equipment from area of spill or leak until cleanup is complete. Remove all ignition sources. Ventilate area of spill or leak. Absorb liquids in vermiculite, dry sand, earth, peat, carbon, or a similar material and deposit in sealed containers. Keep this chemical out of a confined space, such as a sewer, because of the possibility of an explosion, unless the sewer is designed to prevent the buildup of explosive concentrations. It may be necessary to contain and dispose of this chemical as a hazardous waste. If material or contaminated runoff enters waterways, notify downstream users of potentially contaminated waters. Contact your local or federal environmental protection agency for specific recommendations. If employees are required to clean up spills, they must be properly trained and equipped. OSHA 1910.120(q) may be applicable.

Fire Extinguishing: This chemical is a combustible liquid. Poisonous gases, including nitrogen oxides, are produced in fire. Use dry chemical, carbon dioxide, or alcohol foam extinguishers. Vapors are heavier than air and will collect in low areas. Vapors may travel long distances to ignition sources and flashback. Vapors in confined areas may explode when exposed to fire. Containers may explode in fire. Storage containers and parts of containers may rocket great distances, in many directions. If material or contaminated runoff enters waterways, notify downstream users of potentially contaminated waters. Notify local health and fire officials and pollution control agencies. From a secure, explosion-proof location, use water spray to cool exposed containers. If cooling streams are ineffective (venting sound increases in volume and pitch, tank discolors, or shows any signs of deforming), withdraw immediately to a secure position. If employees are expected to fight fires, they must be trained and equipped in OSHA 1910.156. The only respirators recommended for firefighting are self-contained breathing apparatuses that have full face-pieces and are operated in a pressure-demand or other positive-pressure mode.

Disposal Method Suggested: *N*-Nitrosodi-*N*-propylamine may be destroyed by high temperature incineration in an incinerator equipped with a nitrogen oxide scrubber. Chemical treatment methods may also be used to destroy *N*-nitrosodi-*N*-propylamine. These methods involve (a) denitrosation by reaction with 3% hydrobromic acid in glacial acetic acid; (b) oxidation by reaction with potassium permanganate-sulfuric acid; or (c) extraction of the nitrosamine from the waste using dichloromethane and subsequent reaction with triethyloxonium tetrafluoroborate (TOEF). Consult with environmental regulatory agencies for guidance on acceptable disposal practices. Generators of waste containing this contaminant ($\geq$ 100 kg/mo) must conform with EPA regulations governing storage, transportation, treatment, and waste disposal.

References
US Public Health Service. (December 1988). *Toxicological Profile for N-nitroso-di-N-Propylamine*. Atlanta, GA: Agency for Toxic Substances & Disease Registry
New Jersey Department of Health and Senior Services. (December 2006). *Hazardous Substances Fact Sheet: N-Nitrosodi-N-Phenylamine*. Trenton, NJ

N-Nitroso-*N*-ethyl urea   N:0620

Molecular Formula: $C_3H_7N_3O_2$
Synonyms: AENH (German); Aethylnitroso-harnstoff (German); ENU; *N*-Ethyl-*N*-nitrosocarbamide; *N*-Ethyl-*N*-nitrosourea; 1-Ethyl-1-nitrosourea; Ethyl-1-nitrosourea; Ethylnitrosourea; NEU; Nitrosoethylurea; NSC 45403; Urea, *N*-ethyl-*N*-nitroso-; Urea, 1-ethyl-1-nitroso-
CAS Registry Number: 759-73-9
RTECS® Number: YT3150000
UN/NA & ERG Number: UN3077/171
EC Number: 212-072-2
Regulatory Authority and Advisory Bodies
Carcinogenicity: IARC: Animal Sufficient Evidence; Human Limited Evidence, Group 2A, 1998; NTP: Reasonably anticipated to be a human carcinogen.
US EPA Hazardous Waste Number (RCRA No.): U176.
RCRA, 40CFR261, Appendix 8 Hazardous Constituents.
Reportable Quantity (RQ): 1 lb (0.454 kg).

EPCRA Section 313 Form R *de minimis* concentration reporting level: 0.1%.

California Proposition 65 Chemical: Cancer 1/1/87.

WGK (German Aquatic Hazard Class): No value assigned.

Description: *N*-Nitroso-*N*-ethyl urea is a pale yellow, crystalline powder; Freezing/Melting point = 103–104°C (decomposes). Hazard Identification (based on NFPA-704 M Rating System): Health 2, Flammability 1, Reactivity 0. Soluble in water.

Potential Exposure: Used as an anti-cancer drug.

Possible Exposure Limits in Air:

DFG MAK: [skin]

No OELs have been established. However this chemical is a carcinogen and exposure should be reduced to the lowest possible level.

Routes of Entry: Inhalation, ingestion, skin and/or eye contact.

Harmful Effects and Symptoms

Short Term Exposure: Contact can cause irritation of the eyes and skin. Inhalation can cause respiratory tract irritation. Higher exposure can cause headache, drowsiness, fatigue, weakness, and loss of appetite.

Long Term Exposure: A potential occupational carcinogen. This chemical has been shown to be a teratogen in animals. High or repeated exposure may cause liver damage.

Points of Attack: Liver.

Medical Surveillance: Liver function tests.

First Aid: If this chemical gets into the eyes, remove any contact lenses at once and irrigate immediately for at least 15 min, occasionally lifting upper and lower lids. Seek medical attention immediately. If this chemical contacts the skin, remove contaminated clothing and wash immediately with soap and water. Seek medical attention immediately. If this chemical has been inhaled, remove from exposure, begin rescue breathing (using universal precautions, including resuscitation mask) if breathing has stopped and CPR if heart action has stopped. Transfer promptly to a medical facility. When this chemical has been swallowed, get medical attention. Give large quantities of water and induce vomiting. Do not make an unconscious person vomit.

Personal Protective Methods: Wear protective gloves and clothing to prevent any reasonable probability of skin contact. Safety equipment suppliers/manufacturers can provide recommendations on the most protective glove/clothing material for your operation. All protective clothing (suits, gloves, footwear, headgear) should be clean, available each day, and put on before work. Contact lenses should not be worn when working with this chemical. Wear dust-proof chemical goggles and face shield unless full face-piece respiratory protection is worn. Employees should wash immediately with soap when skin is wet or contaminated. Provide emergency showers and eyewash. A Class I, Type B, biological safety hood should be used for handling and mixing in a laboratory environment. Specific engineering controls are required for drug manufacture by the Food and Drug Administration. See 21CFR210.

Respirator Selection: Specific respirator(s) have not been recommended by NIOSH. However, based on potential carcinogenicity the following might be considered: *At any detectable concentration:* SCBAF: Pd,Pp (APF = 10,000) (any NIOSH/MSHA- or European Standard EN 149-approved self-contained breathing apparatus that has a full face-piece and is operated in a pressure-demand or other positive-pressure mode) or SaF: Pd,Pp: ASCBA (APF = 10,000) (any supplied-air respirator that has a full face-piece and is operated in a pressure-demand or other positive-pressure mode in combination with an auxiliary, self-contained breathing apparatus operated in a pressure-demand or other positive-pressure mode). *Escape:* 100 F (APF = 50) (any air-purifying, full-face-piece respirator with an N100, R100, or P100 filter) or SCBAE (any appropriate escape-type, self-contained breathing apparatus).

Storage: Color Code—Green: General storage may be used. Prior to working with this chemical you should be trained on its proper handling and storage. Store in tightly closed containers in a cool, well-ventilated area. Where possible, automatically transfer this material from storage containers to process containers. Sources of ignition, such as smoking and open flames, are prohibited where this chemical is handled, used, or stored. Metal containers involving the transfer of this chemical should be grounded and bonded. Use only nonsparking tools and equipment, especially when opening and closing containers of this chemical. Wherever this chemical is used, handled, manufactured, or stored, use explosion-proof electrical equipment and fittings. A regulated, marked area should be established where this chemical is handled, used, or stored in compliance with OSHA Standard 1910.1045.

Shipping: Environmentally hazardous substances, solid, n.o.s. It falls in Hazard Class 9 and "CLASS 9." Packing Group III.

Spill Handling: Evacuate persons not wearing protective equipment from area of spill or leak until cleanup is complete. Remove all ignition sources. Use HEPA vacuum or wet method to reduce dust during cleanup. Do not dry sweep. Collect powdered material in the most convenient and safe manner and deposit in sealed containers. Ventilate area after cleanup is complete. It may be necessary to contain and dispose of this chemical as a hazardous waste. If material or contaminated runoff enters waterways, notify downstream users of potentially contaminated waters. Contact your local or federal environmental protection agency for specific recommendations. If employees are required to clean up spills, they must be properly trained and equipped. OSHA 1910.120(q) may be applicable.

Fire Extinguishing: Use dry chemical, carbon dioxide, water spray, or foam extinguishers. Poisonous gases, including nitrogen oxides and carbon monoxide, are produced in fire. If material or contaminated runoff enters waterways, notify downstream users of potentially contaminated waters. Notify local health and fire officials and pollution control agencies. From a secure, explosion-proof location, use water spray to cool exposed containers. If cooling streams are ineffective (venting sound increases in volume and pitch, tank discolors, or shows any signs of deforming), withdraw immediately to a secure position. If employees are expected to fight fires, they must be trained and equipped in OSHA 1910.156. The only respirators recommended for firefighting are self-contained breathing apparatuses that have full face-pieces and are operated in a pressure-demand or other positive-pressure mode.

Disposal Method Suggested: Consult with environmental regulatory agencies for guidance on acceptable disposal practices. Generators of waste containing this contaminant (≥ 100 kg/mo) must conform with EPA regulations governing storage, transportation, treatment, and waste disposal.

Reference

New Jersey Department of Health and Senior Services. (April 2002). *Hazardous Substances Fact Sheet: N-Nitroso-N-Ethylurea.* Trenton, NJ

N-Nitrosomethylvinylamine   N:0630

Molecular Formula: $C_3H_6N_2O$
Synonyms: Ethenamine, *N*-methyl-*N*-nitroso-; Ethylene, *N*-methyl-*N*-nitroso-;   *N*-Methyl-*N*-nitrosovinylamine; Methylvinylnitrosamine; Methylvinylnitrosamine (German); MVNA;   *N*-nitroso-*N*-methylvinyl   amine;   NMVA; Vinylamine, *N*-methyl-*N*-nitroso-
CAS Registry Number: 4549-40-0
RTECS® Number: YZ0875000
UN/NA & ERG Number: UN3082/171
Regulatory Authority and Advisory Bodies
Carcinogenicity: IARC: Animal   Sufficient   Evidence; Human Limited Evidence, *possibly carcinogenic to humans,* Group 2B, 1987; NTP: Reasonably anticipated to be a human carcinogen.
US EPA Hazardous Waste Number (RCRA No.): P084.
RCRA, 40CFR261, Appendix 8 Hazardous Constituents.
Reportable Quantity (RQ): 10 lb (4.54 kg).
EPCRA Section 313 Form R *de minimis* concentration reporting level: 0.1%.
California Proposition 65 Chemical: Cancer 1/1/88.
Canada, WHMIS, Ingredients Disclosure List Concentration 0.1%.
European/International Regulations: not listed in Annex 1.
WGK (German Aquatic Hazard Class): No value assigned.
Description: *N*-Nitrosomethylvinylamine is a yellow liquid. Molecular weight = 86.11. Hazard Identification (based on NFPA-704 M Rating System): Health 3, Flammability 1, Reactivity 0. Decomposes in water.
Potential Exposure: This chemical is not manufactured but occurs as a chemical reaction byproduct found in the dye, automotive, rubber, and leather industries.
Incompatibilities: Strong oxidizers, light.
Permissible Exposure Limits in Air
No TEEL available.

DFG MAK: [skin].
There are no established numerical OELs. However this chemical is a carcinogen and exposure should be reduced to the lowest possible level.
Determination in Air: T-Sorb; Methanol/CH_2Cl_2; Gas chromatography/Flame   ionization   detection;   NIOSH   IV; Method #2522. See also OSHA Analytical Method #27.
Routes of Entry: Inhalation, eye and/or skin contact.
Harmful Effects and Symptoms
Short Term Exposure: Irritates the eyes and skins. Some related nitrosamines can cause headaches, abdominal cramps, weakness, and dizziness. It is unknown if this chemical causes all of these same symptoms.
Long Term Exposure: A potential occupational carcinogen. Has been shown to be a teratogen in animals. May cause liver and kidney damage.
Points of Attack: Liver and kidneys.
Medical Surveillance: Liver and kidney function tests.
First Aid: If this chemical gets into the eyes, remove any contact lenses at once and irrigate immediately for at least 15 min, occasionally lifting upper and lower lids. Seek medical attention immediately. If this chemical contacts the skin, remove contaminated clothing and wash immediately with soap and water. Seek medical attention immediately. If this chemical has been inhaled, remove from exposure, begin rescue breathing (using universal precautions, including resuscitation mask) if breathing has stopped and CPR if heart action has stopped. Transfer promptly to a medical facility. When this chemical has been swallowed, get medical attention. Give large quantities of water and induce vomiting. Do not make an unconscious person vomit.
Personal Protective Methods: Wear protective gloves and clothing to prevent any reasonable probability of skin contact. Safety equipment suppliers/manufacturers can provide recommendations on the most protective glove/clothing material for your operation. All protective clothing (suits, gloves, footwear, headgear) should be clean, available each day, and put on before work. Contact lenses should not be worn when working with this chemical. Wear splash-proof chemical goggles and face shield unless full face-piece respiratory protection is worn. Employees should wash immediately with soap when skin is wet or contaminated. Provide emergency showers and eyewash.
Respirator Selection: Specific respirator(s) have not been recommended by NIOSH. However, based on potential carcinogenicity the following might be considered:
At any detectable concentration: SCBAF: Pd,Pp (APF = 10,000) (any NIOSH/MSHA- or European Standard EN 149-approved self-contained breathing apparatus that has a full face-piece and is operated in a pressure-demand or other positive-pressure mode) or SaF: Pd,Pp: ASCBA (APF = 10,000) (any supplied-air respirator that has a full face-piece and is operated in a pressure-demand or other positive-pressure mode in combination with an auxiliary, self-contained breathing apparatus operated in a pressure-demand or other positive-pressure mode). *Escape:* 100 F

(APF = 50) (any air-purifying, full-face-piece respirator with an N100, R100, or P100 filter) or SCBAE (any appropriate escape-type, self-contained breathing apparatus).

Storage: Color Code—Blue: Health Hazard/Poison: Store in a secure poison location. However, storage is not likely. Prior to working with this chemical you should be trained on its proper handling and storage. Store in tightly closed containers in a cool, well-ventilated area away from oxidizers, light. Sources of ignition, such as smoking and open flames are prohibited where this chemical is handled, used, or stored. Use only nonsparking tools and equipment, especially when opening and closing containers of this chemical. Wherever this chemical is used, handled, manufactured, or stored, use explosion-proof electrical equipment and fittings. A regulated, marked area should be established where this chemical is handled, used, or stored in compliance with OSHA Standard 1910.1045.

Shipping: The name of this material is not on the DOT list of materials[19] for label and packaging standards. However, based on regulations, it may be classified[52] as an Environmentally hazardous substances, liquid, n.o.s. It falls in Hazard Class 9 and Packing Group III.[20, 21]

Spill Handling: Evacuate and restrict persons not wearing protective equipment from area of spill or leak until cleanup is complete. Remove all ignition sources. Ventilate area of spill or leak. Absorb liquids in vermiculite, dry sand, earth, peat, carbon, or a similar material and deposit in sealed containers. Keep this chemical out of a confined space, such as a sewer, because of the possibility of an explosion, unless the sewer is designed to prevent the buildup of explosive concentrations. It may be necessary to contain and dispose of this chemical as a hazardous waste. If material or contaminated runoff enters waterways, notify downstream users of potentially contaminated waters. Contact your local or federal environmental protection agency for specific recommendations. If employees are required to clean up spills, they must be properly trained and equipped. OSHA 1910.120(q) may be applicable.

Fire Extinguishing: This chemical is a combustible liquid. Poisonous gases, including nitrogen oxides, are produced in fire. Use dry chemical, carbon dioxide, or alcohol foam extinguishers. Vapors are heavier than air and will collect in low areas. Vapors may travel long distances to ignition sources and flashback. Vapors in confined areas may explode when exposed to fire. Containers may explode in fire. Storage containers and parts of containers may rocket great distances, in many directions. If material or contaminated runoff enters waterways, notify downstream users of potentially contaminated waters. Notify local health and fire officials and pollution control agencies. From a secure, explosion-proof location, use water spray to cool exposed containers. If cooling streams are ineffective (venting sound increases in volume and pitch, tank discolors, or shows any signs of deforming), withdraw immediately to a secure position. If employees are expected to fight fires, they must be trained and equipped in OSHA 1910.156. The only

respirators recommended for firefighting are self-contained breathing apparatuses that have full face-pieces and are operated in a pressure-demand or other positive-pressure mode.

Disposal Method Suggested: Consult with environmental regulatory agencies for guidance on acceptable disposal practices. Generators of waste containing this contaminant (≥100 kg/mo) must conform with EPA regulations governing storage, transportation, treatment, and waste disposal.

Reference

New Jersey Department of Health and Senior Services. (March 2006). *Hazardous Substances Fact Sheet: N-Nitrosomethylvinylamine*. Trenton, NJ

N-Nitrosopiperidine   N:0640

Molecular Formula: $C_5H_{10}N_2O$
Synonyms: Hexahydro-*N*-nitrosopyridine; Nitrosopiperidin (German); 1-Nitrosopiperidine; N.N-PIP; No-PIP; NPIP; Piperidine, 1-nitroso
CAS Registry Number: 100-75-4
RTECS® Number: TN2100000
UN/NA & ERG Number: UN3082/171
EC Number: 202-886-6
Regulatory Authority and Advisory Bodies
Carcinogenicity: IARC: Animal Sufficient Evidence; Human Limited Evidence, *possibly carcinogenic to humans*, Group 2B, 1987; NTP: Reasonably anticipated to be a human carcinogen.
US EPA Hazardous Waste Number (RCRA No.): U179.
RCRA, 40CFR261, Appendix 8 Hazardous Constituents.
RCRA 40CFR268.48; 61FR15654, Universal Treatment Standards: Wastewater (mg/L), 0.013; Nonwastewater (mg/kg), 35.
RCRA 40CFR264, Appendix 9; TSD Facilities Ground Water Monitoring List. Suggested test method(s) (PQL µg/L): 8270 (10).
Reportable Quantity (RQ): 10 lb (4.54 kg).
EPCRA Section 313 Form R *de minimis* concentration reporting level: 0.1%.
California Proposition 65 Chemical: Cancer 1/1/88.
WGK (German Aquatic Hazard Class): No value assigned.
Description: *N*-Nitrosopiperidine is a clear, yellow, oily liquid. Molecular weight = 114.17; Boiling point = 217°C; Flash point = 93°C. Hazard Identification (based on NFPA-704 M Rating System): Health 3, Flammability 1, Reactivity 0. Soluble in water.
Potential Exposure: It is found in some foods and tobacco smoke. Used as a research chemical.
Incompatibilities: Oxidizers. Light may cause decomposition.
Permissible Exposure Limits in Air
No TEEL available.
DFG MAK: [skin] Carcinogen Category 2.

There are no established numerical OELs. However, this chemical is a carcinogen and exposure should be reduced to the lowest possible level.

Determination in Air: Use NIOSH IV; Method #2522; OSHA Analytical Method 27.

Routes of Entry: Inhalation, ingestion, eye and/or skin contact.

Harmful Effects and Symptoms

Short Term Exposure: Can irritate and may cause permanent damage to the eyes; inflammation in the pigmented area and damage to the cornea with clouded patches of vision.

Long Term Exposure: A potential carcinogen. Prolonged exposure may cause liver damage.

Points of Attack: Eyes, liver.

Medical Surveillance: Examination of the eyes. Liver function tests.

First Aid: If this chemical gets into the eyes, remove any contact lenses at once and irrigate immediately for at least 15 min, occasionally lifting upper and lower lids. Seek medical attention immediately. If this chemical contacts the skin, remove contaminated clothing and wash immediately with soap and water. Seek medical attention immediately. If this chemical has been inhaled, remove from exposure, begin rescue breathing (using universal precautions, including resuscitation mask) if breathing has stopped and CPR if heart action has stopped. Transfer promptly to a medical facility. When this chemical has been swallowed, get medical attention. Give large quantities of water and induce vomiting. Do not make an unconscious person vomit.

Personal Protective Methods: Wear protective gloves and clothing to prevent any reasonable probability of skin contact. Safety equipment suppliers/manufacturers can provide recommendations on the most protective glove/clothing material for your operation. All protective clothing (suits, gloves, footwear, headgear) should be clean, available each day, and put on before work. Contact lenses should not be worn when working with this chemical. Wear splash-proof chemical goggles and face shield unless full face-piece respiratory protection is worn. Employees should wash immediately with soap when skin is wet or contaminated. Provide emergency showers and eyewash.

Respirator Selection: Specific respirator(s) have not been recommended by NIOSH. However, based on potential carcinogenicity, and where the potential exists for exposure, the following might be considered:

At any detectable concentration: SCBAF: Pd,Pp (APF = 10,000) (any NIOSH/MSHA- or European Standard EN 149-approved self-contained breathing apparatus that has a full face-piece and is operated in a pressure-demand or other positive-pressure mode) or SaF: Pd,Pp: ASCBA (APF = 10,000) (any supplied-air respirator that has a full face-piece and is operated in a pressure-demand or other positive-pressure mode in combination with an auxiliary, self-contained breathing apparatus operated in a pressure-demand or other positive-pressure mode). *Escape:* 100 F

(APF = 50) (any air-purifying, full-face-piece respirator with an N100, R100, or P100 filter) or SCBAE (any appropriate escape-type, self-contained breathing apparatus).

Storage: Color Code—Blue: Health Hazard/Poison: Store in a secure poison location. Prior to working with this chemical you should be trained on its proper handling and storage. Store in tightly closed containers in a cool, well-ventilated area away from oxidizers. Where possible, automatically pump liquid from drums or other storage containers to process containers. Sources of ignition, such as smoking and open flames, are prohibited where this chemical is handled, used, or stored. Metal containers involving the transfer of 5 gallons or more of this chemical should be grounded and bonded. Drums must be equipped with self-closing valves, pressure vacuum bungs, and flame arresters. Use only nonsparking tools and equipment, especially when opening and closing containers of this chemical. Wherever this chemical is used, handled, manufactured, or stored, use explosion-proof electrical equipment and fittings. A regulated, marked area should be established where this chemical is handled, used, or stored in compliance with OSHA Standard 1910.1045.

Shipping: The name of this material is not on the DOT list of materials[19] for label and packaging standards. However, based on regulations, it may be classified[52] as an Environmentally hazardous substances, liquid, n.o.s. It falls in Hazard Class 9 and Packing Group III.[20, 21]

Spill Handling: Evacuate and restrict persons not wearing protective equipment from area of spill or leak until cleanup is complete. Remove all ignition sources. Ventilate area of spill or leak. Absorb liquids in vermiculite, dry sand, earth, peat, carbon, or a similar material and deposit in sealed containers. Keep this chemical out of a confined space, such as a sewer, because of the possibility of an explosion, unless the sewer is designed to prevent the buildup of explosive concentrations. It may be necessary to contain and dispose of this chemical as a hazardous waste. If material or contaminated runoff enters waterways, notify downstream users of potentially contaminated waters. Contact your local or federal environmental protection agency for specific recommendations. If employees are required to clean up spills, they must be properly trained and equipped. OSHA 1910.120(q) may be applicable.

Fire Extinguishing: This chemical is a combustible liquid. Poisonous gases, including nitrogen oxides, are produced in fire. Use dry chemical, carbon dioxide, or alcohol foam extinguishers. Vapors are heavier than air and will collect in low areas. Vapors may travel long distances to ignition sources and flashback. Vapors in confined areas may explode when exposed to fire. Containers may explode in fire. Storage containers and parts of containers may rocket great distances, in many directions. If material or contaminated runoff enters waterways, notify downstream users of potentially contaminated waters. Notify local health and fire officials and pollution control agencies. From a secure, explosion-proof location, use water spray to cool exposed containers. If cooling streams are ineffective (venting sound increases in volume and pitch, tank discolors, or shows any signs of deforming), withdraw immediately to a secure position. If employees are expected to fight fires, they must be trained and equipped in OSHA 1910.156. The only respirators recommended for firefighting are self-contained breathing apparatuses that have full face-pieces and are operated in a pressure-demand or other positive-pressure mode.

Disposal Method Suggested: Consult with environmental regulatory agencies for guidance on acceptable disposal practices. Generators of waste containing this contaminant (≥100 kg/mo) must conform with EPA regulations governing storage, transportation, treatment, and waste disposal. Under 40 CFR 261.5 small quantity generators of this waste may qualify for partial exclusion from hazardous waste regulations.

Reference

New Jersey Department of Health and Senior Services. (March 2006). *Hazardous Substances Fact Sheet: N-Nitrosopiperidine.* Trenton, NJ

N-Nitrosopyrrolidine   N:0650

Molecular Formula: $C_4H_8N_2O$
Synonyms: *N*-Nitrosopyrrolidin   (German);   1-Nitrosopyrrolidine;   N-N-PYR;   No-PYR;   NPYR; Tetrahydro-*N*-nitrosopyrrole
CAS Registry Number: 930-55-2
RTECS® Number: UY1575000
UN/NA & ERG Number: UN3082/171
EC Number: 213-218-8
Regulatory Authority and Advisory Bodies
Carcinogenicity: IARC: Animal Sufficient Evidence; Human Limited Evidence, *possibly carcinogenic to humans*, Group 2B, 1987; NTP: Reasonably anticipated to be a human carcinogen; EPA: Sufficient evidence from animal studies; inadequate evidence or no useful data from epidemiologic studies.
US EPA Hazardous Waste Number (RCRA No.): U180.
RCRA, 40CFR261, Appendix 8 Hazardous Constituents.
RCRA 40CFR268.48; 61FR15654, Universal Treatment Standards: Wastewater (mg/L), 0.013; Nonwastewater (mg/kg), 35.
RCRA 40CFR264, Appendix 9; TSD Facilities Ground Water Monitoring List. Suggested test method(s) (PQL µg/L): 8270 (10).
Reportable Quantity (RQ): 1 lb (0.454 kg).
California Proposition 65 Chemical: Cancer 10/1/87.
WGK (German Aquatic Hazard Class): No value assigned.
Description: *N*-Nitrosopyrrolidine is a yellow liquid. Molecular weight = 100.14; Boiling point = 105°C at 20 mmHg; Flash point = 83°C. Hazard Identification (based on NFPA-704 M Rating System): Health 3, Flammability 2, Reactivity 0. Soluble in water.

Potential Exposure: *N*-Nitrosopyrrolidine is a research chemical.
Incompatibilities: Oxidizers.
Permissible Exposure Limits in Air
No TEEL available.
DFG MAK: [skin] Carcinogen Category 2.
There are no established numerical OELs. However, this chemical is a carcinogen and exposure should be reduced to the lowest possible level.
Determination in Air: Use NIOSH IV; Method #2522; OSHA Analytical Method 27.
Routes of Entry: Inhalation, ingestion, skin and/or eye contact.
Harmful Effects and Symptoms
Short Term Exposure: Contact can irritate the skin and eyes.
Long Term Exposure: A potential occupational carcinogen. May cause damage to the liver and kidneys.
Points of Attack: Liver, kidneys.
Medical Surveillance: Liver and kidney function tests.
First Aid: If this chemical gets into the eyes, remove any contact lenses at once and irrigate immediately for at least 15 min, occasionally lifting upper and lower lids. Seek medical attention immediately. If this chemical contacts the skin, remove contaminated clothing and wash immediately with soap and water. Seek medical attention immediately. If this chemical has been inhaled, remove from exposure, begin rescue breathing (using universal precautions, including resuscitation mask) if breathing has stopped and CPR if heart action has stopped. Transfer promptly to a medical facility. When this chemical has been swallowed, get medical attention. Give large quantities of water and induce vomiting. Do not make an unconscious person vomit.
Personal Protective Methods: Wear protective gloves and clothing to prevent any reasonable probability of skin contact. Safety equipment suppliers/manufacturers can provide recommendations on the most protective glove/clothing material for your operation. All protective clothing (suits, gloves, footwear, headgear) should be clean, available each day, and put on before work. Contact lenses should not be worn when working with this chemical. Wear splash-proof chemical goggles and face shield unless full face-piece respiratory protection is worn. Employees should wash immediately with soap when skin is wet or contaminated. Provide emergency showers and eyewash.
Respirator Selection: Specific respirator(s) have not been recommended by NIOSH. However, based on potential carcinogenicity, and where the potential exists for exposure, the following might be considered:
At any detectable concentration: SCBAF: Pd,Pp (APF = 10,000) (any NIOSH/MSHA- or European Standard EN 149-approved self-contained breathing apparatus that has a full face-piece and is operated in a pressure-demand or other positive-pressure mode) or SaF: Pd,Pp: ASCBA (APF = 10,000) (any supplied-air respirator that has a full face-piece and is operated in a pressure-demand or other

positive-pressure mode in combination with an auxiliary, self-contained breathing apparatus operated in a pressure-demand or other positive-pressure mode). *Escape:* 100 F (APF = 50) (any air-purifying, full-face-piece respirator with an N100, R100, or P100 filter) or SCBAE (any appropriate escape-type, self-contained breathing apparatus).

Storage: Color Code—Blue: Health Hazard/Poison: Store in a secure poison location. Prior to working with this chemical you should be trained on its proper handling and storage. Store in tightly closed containers in a cool, well-ventilated area away from oxidizers. Where possible, automatically pump liquid from drums or other storage containers to process containers. Sources of ignition, such as smoking and open flames, are prohibited where this chemical is handled, used, or stored. Metal containers involving the transfer of 5 gallons or more of this chemical should be grounded and bonded. Drums must be equipped with self-closing valves, pressure vacuum bungs, and flame arresters. Use only nonsparking tools and equipment, especially when opening and closing containers of this chemical. Wherever this chemical is used, handled, manufactured, or stored, use explosion-proof electrical equipment and fittings. A regulated, marked area should be established where this chemical is handled, used, or stored in compliance with OSHA Standard 1910.1045.

Shipping: The name of this material is not on the DOT list of materials[19] for label and packaging standards. However, based on regulations, it may be classified[52] as an Environmentally hazardous substances, liquid, n.o.s. It falls in Hazard Class 9 and Packing Group III.[20, 21]

Spill Handling: Evacuate and restrict persons not wearing protective equipment from area of spill or leak until cleanup is complete. Remove all ignition sources. Ventilate area of spill or leak. Absorb liquids in vermiculite, dry sand, earth, peat, carbon, or a similar material and deposit in sealed containers. Keep this chemical out of a confined space, such as a sewer, because of the possibility of an explosion, unless the sewer is designed to prevent the buildup of explosive concentrations. It may be necessary to contain and dispose of this chemical as a hazardous waste. If material or contaminated runoff enters waterways, notify downstream users of potentially contaminated waters. Contact your local or federal environmental protection agency for specific recommendations. If employees are required to clean up spills, they must be properly trained and equipped. OSHA 1910.120(q) may be applicable.

Fire Extinguishing: This chemical is a combustible liquid. Poisonous gases, including nitrogen oxides, are produced in fire. Use dry chemical, carbon dioxide, alcohol foam, or polymer foam extinguishers. Vapors are heavier than air and will collect in low areas. Vapors may travel long distances to ignition sources and flashback. Vapors in confined areas may explode when exposed to fire. Containers may explode in fire. Storage containers and parts of containers may rocket great distances, in many directions. If material or contaminated runoff enters waterways, notify downstream users of potentially contaminated waters. Notify local health and fire officials and pollution control agencies. From a secure, explosion-proof location, use water spray to cool exposed containers. If cooling streams are ineffective (venting sound increases in volume and pitch, tank discolors, or shows any signs of deforming), withdraw immediately to a secure position. If employees are expected to fight fires, they must be trained and equipped in OSHA 1910.156. The only respirators recommended for firefighting are self-contained breathing apparatuses that have full face-pieces and are operated in a pressure-demand or other positive-pressure mode.

Disposal Method Suggested: Under 40 CFR 261.5 small quantity generators of this waste may qualify for partial exclusion from hazardous waste regulations. Consult with environmental regulatory agencies for guidance on acceptable disposal practices. Generators of waste containing this contaminant ($\geq$100 kg/mo) must conform with EPA regulations governing storage, transportation, treatment, and waste disposal.

Reference
New Jersey Department of Health and Senior Services. (July 1999). *Hazardous Substances Fact Sheet: N-Nitrosopyrrolidine.* Trenton, NJ

Nitrotoluenes   N:0660

Molecular Formula: $C_7H_7NO_2$
Common Formula: $CH_3C_6H_4NO_2$
Synonyms: *m-isomer:* Benzene, 1-methyl-3-nitro-; *m*-Nitrotoluene; *m*-Methylnitrobenzene; 3-Methylnitrobenzene; MNT; 3-Nitrotoluene; Nitrotoluene, 3-; *m*-Nitrotolueno (Spanish); 3-Nitrotoluol
Mixed isomers: Mixo-nitrotoluene
ortho-isomer: Benzene, 1-methyl-2-nitro-; *o*-Methylnitrobenzene; 2-Methylnitrobenzene; 2-Nitrotoluene; Nitrotoluene, 2-; *o*-Nitrotolueno (Spanish); *o*-Nitrotoluol; ONT; Orthonitrotoluene
California Proposition 65 Chemical: Cancer 5/15/98
para-isomer: Benzene, 1-methyl-4-nitro-; *p*-Methylnitrobenzene; 4-Methylnitrobenzene; NCI-C60537; 4-Nitrotoluene; Nitrotoluene, 4-; *p*-Nitrotolueno (Spanish); *p*-Nitrotoluol; 4-Nitrotoluol; *p*-Nitrotoluene; PNT
CAS Registry Number: 88-72-2 (*o*-isomer); *(alt.)* 57158-05-1; 99-08-1 (*m*-isomer); 99-99-0 (*p*-isomer); 1321-12-6 (mixed isomers)
RTECS® Number: XT3150000 (*o*-isomer); XT2975000 (*m*-isomer); XT3325000 (*p*-isomer)
UN/NA & ERG Number: UN1664/152
EC Number: 201-853-3 [*Annex I Index No.:* 609-065-00-5]
Regulatory Authority and Advisory Bodies
Carcinogenicity: (*m*-isomer) IARC: Animal Inadequate Evidence; Human Inadequate Evidence, *not classifiable as carcinogenic to humans*, Group 3, 1996; NTP; NCI: Carcinogenesis Studies (feed); clear evidence: mouse, rat;

NTP: Carcinogenesis studies; on test (2-year studies), October 2000; (*p*-isomer) IARC: Animal Inadequate Evidence; Human Inadequate Evidence, *not classifiable as carcinogenic to humans*, Group 3, 1996; NCI: Carcinogenesis Studies (feed); equivocal evidence: rat; test completed (peer review), October 2000.

NTP: Toxicity studies, (all isomers) Report No: TOX-23, October 2000; (*o*-isomer) Report No: TOX-44, October 2000.

All isomers:

Air Pollutant Standard Set. See below, "Permissible Exposure Limits in Air" section.

Clean Water Act: Section 311 Hazardous Substances/RQ 40CFR117.3 (same as CERCLA, see below).

Reportable Quantity (RQ): 1000 lb (454 kg).

Canada, WHMIS, Ingredients Disclosure List Concentration 1.0% all isomers.

European/International Regulations: Hazard Symbol: T, N; Risk phrases: R45; R46; R22; R62; R51/53; Safety phrases: S51/53 S53; S45; S61 (see Appendix 4).

WGK (German Aquatic Hazard Class): 3—Highly water polluting (*o*-isomer); 2—Water polluting (*m*- and *p*-isomers).

Description: Nitrotoluene is formed in 3 isomeric forms. The *o*- and *m*- forms are yellow liquids or solids. The *p*-form is a pale yellow crystalline solid. All have weak aromatic odors. The odor thresholds are 0.05 mg/L (*o*-isomer); 1.74 ppm (*m*-isomer). Molecular weight = 137.15 (all isomers); Specific gravity (H_2O:1) = 1.16 (all isomer); Boiling point = 222°C (*o*-isomer); 232°C (*m*-isomer); 238° (*p*-isomer); Freezing/Melting point = − 4°C (*o*-isomer); 16°C (*m*-isomer); 52°C (*p*-isomer); Vapor pressure = 0.1 mmHg (all isomers); Flash point = 101−106°C (all isomers). Explosive limits: LEL = 2.2; UEL—unknown. Hazard Identification (based on NFPA-704 M Rating System): (*o*-, *p*-) Health 3, Flammability 1, Reactivity 1. Insoluble in water.

Potential Exposure: Compound Description (all isomers): Tumorigen, Mutagen; Reproductive Effector. The nitrotoluenes are used in the production of toluidines and other dye intermediates. All isomers are used in manufacture of agriculture and rubber chemicals and in various dyes.

Incompatibilities: Decomposes on contact with strong oxidizers, strong acids, reducing agents, strong bases, ammonia, amines producing toxic fumes, causing fire and explosion hazard. Heat above 190°C may cause explosive decomposition. Attacks some plastics, rubbers, and coatings.

Permissible Exposure Limits in Air

Conversion factor: 1 ppm = 5.61 mg/m³ (all isomers) at 25°C & 1 atm.

m-isomer

OSHA PEL: 5 ppm/30 mg/m³ TWA [skin].

NIOSH REL: 2 ppm/11 mg/m³ TWA [skin].

ACGIH TLV®[1]: 2 ppm/11 mg/m³ TWA [skin]; BEI$_M$ issued for Methemoglobin inducers.

Protective Action Criteria (PAC)

TEEL-0: 5 ppm

PAC-1: 5 ppm

PAC-2: 40 ppm

PAC-3: 200 ppm

NIOSH IDLH: 200 ppm

DFG MAK: [skin] Carcinogen Category 3B.

Australia: TWA 2 ppm (11 mg/m³), [skin], 1993; Austria: MAK 2 ppm (11 mg/m³), [skin], 1999; Belgium: TWA 2 ppm (11 mg/m³), [skin], 1993; Denmark: TWA 2 ppm (12 mg/m³), 1999; France: VME 2 ppm (11 mg/m³), [skin], 1999; Norway: TWA 1 ppm (5.5 mg/m³), 1999; Poland: MAC (TWA) 3 mg/m³; MAC (STEL) 9 mg/m³, 1999; the Netherlands: MAC-TGG 6 mg/m³, [skin], 2003; Switzerland: MAK-W 2 ppm (11 mg/m³), KZG-W 4 ppm (22 mg/m³), [skin], 1999; United Kingdom: TWA 5 ppm (29 mg/m³); STEL 10 ppm, [skin], 2000; Argentina, Bulgaria, Columbia, Jordan, South Korea, New Zealand, Singapore, Vietnam: ACGIH TLV®: TWA 2 ppm [skin].

o-isomer

OSHA PEL: 5 ppm/30 mg/m³ TWA [skin].

NIOSH REL: 2 ppm/11 mg/m³ TWA [skin].

ACGIH TLV®[1]: 2 ppm/11 mg/m³ TWA [skin]; BEI$_M$ issued for Methemoglobin inducers.

NIOSH IDLH: 200 ppm.

DFG MAK: [skin] Carcinogen Category 3B.

Australia: TWA 2 ppm (11 mg/m³), [skin], 1993; Austria: MAK 2 ppm (11 mg/m³), [skin], 1999; Belgium: TWA 2 ppm (11 mg/m³), [skin], 1993; Denmark: TWA 2 ppm (12 mg/m³), 1999; Norway: TWA 1 ppm (5.5 mg/m³), 1999; Sweden: NGV 1 ppm (6 mg/m³), KTV 2 ppm (11 mg/m³), [skin], 1999; Switzerland: [skin], carcinogen, 1999; United Kingdom: TWA 5 ppm (29 mg/m³); STEL 10 ppm, [skin], 2000; Argentina, Bulgaria, Columbia, Jordan, South Korea, New Zealand, Singapore, Vietnam: ACGIH TLV®: TWA 2 ppm [skin].

p-isomer

OSHA PEL: 5 ppm/30 mg/m³ TWA [skin].

NIOSH REL: 2 ppm/11 mg/m³ TWA [skin].

ACGIH TLV®[1]: 2 ppm/11 mg/m³ TWA [skin]; BEI$_M$ issued for Methemoglobin inducers.

DFG MAK: [skin] Carcinogen Category 3B.

NIOSH IDLH: 200 ppm.

Australia: TWA 2 ppm (11 mg/m³), [skin], 1993; Austria: MAK 2 ppm (11 mg/m³), [skin], 1999; Belgium: TWA 2 ppm (11 mg/m³), [skin], 1993; Denmark: TWA 2 ppm (12 mg/m³), 1999; Norway: TWA 1 ppm (5.5 mg/m³), 1999; Poland: MAC (TWA) 3 mg/m³; MAC (STEL) 9 mg/m³, 1999; Sweden: NGV 1 ppm (6 mg/m³), KTV 2 ppm (11 mg/m³), [skin], 1999; Switzerland: MAK-W 2 ppm (11 mg/m³), KZG-W 4 ppm (22 mg/m³), [skin], 1999; United Kingdom: TWA 5 ppm (29 mg/m³); STEL 10 ppm, [skin], 2000; Argentina, Bulgaria, Columbia, Jordan, South Korea, New Zealand, Singapore, Vietnam: ACGIH TLV®: TWA 2 ppm [skin]. In addition, Several states have set guidelines or standards for *p*-nitrotoluene in ambient air[60] ranging from 5.5 $\mu g/m^3$ (South Carolina) to 36.7 $\mu g/m^3$ (New York) to 110.0 $\mu g/m^3$ (Florida). Further, Several states have set guidelines or standards for *m*-nitrotoluene in ambient air[60] ranging from 37.0 $\mu g/m^3$ (New York) to 110.0 $\mu g/m^3$ (North Dakota) to 220.0 $\mu g/m^3$ (Connecticut) to 262.0 $\mu g/m^3$ (Nevada).

Determination in Air: Use NIOSH Analytical Method (IV) #2005, Nitrobenzenes.

Permissible Concentration in Water: No criteria set, but EPA[32] has suggested a permissible ambient goal of 414 $\mu g/L$ based on health effects.

Determination in Water: Octanol–water coefficient: Log $K_{ow} = 2.3 - 2.5$ (all isomers).

Routes of Entry: Inhalation, skin absorption, ingestion, skin and/or eye contact.

Harmful Effects and Symptoms

Short Term Exposure: Irritates the eyes, skin, and respiratory tract. Inhalation can cause coughing and wheezing. Exposure can cause headache, flushing of the skin, rapid heartbeat, nausea, vomiting, weakness, irritability, convulsions, coma, and death. High levels may affect the blood, causing formation of methemoglobinemia. Symptoms include anoxia, cyanosis, headache, weakness, dizziness, ataxia, dyspnea (breathing difficulty), tachycardia, vomiting.

Long Term Exposure: May damage the liver, kidneys, and blood; anemia may develop. May cause damage to the testes.

Points of Attack: Blood, skin, gastrointestinal tract, cardiovascular system, central nervous system, liver, kidneys.

Medical Surveillance: NIOSH lists the following tests (all isomers): whole blood (chemical/metabolite), Methemoglobin, complete blood count, urinalysis (chemical/metabolite). Consider the points of attack in preplacement and periodic physical examinations. Liver and kidney function tests.

First Aid: If this chemical gets into the eyes, remove any contact lenses at once and irrigate immediately for at least 15 min, occasionally lifting upper and lower lids. Seek medical attention immediately. If this chemical contacts the skin, remove contaminated clothing and wash immediately with soap and water. Seek medical attention immediately. If this chemical has been inhaled, remove from exposure, begin rescue breathing (using universal precautions, including resuscitation mask) if breathing has stopped and CPR if heart action has stopped. Transfer promptly to a medical facility. When this chemical has been swallowed, get medical attention. Give large quantities of water and induce vomiting. Do not make an unconscious person vomit.

Note to physician: Treat for methemoglobinemia. Spectrophotometry may be required for precise determination of levels of methemoglobin in urine.

Personal Protective Methods: Wear protective gloves and clothing to prevent any reasonable probability of skin contact. Safety equipment suppliers/manufacturers can provide recommendations on the most protective glove/clothing material for your operation. For *p*-nitrotoluene, Butyl rubber and polycarbonate are among the recommended protective materials. All protective clothing (suits, gloves, footwear, headgear) should be clean, available each day, and put on before work. Contact lenses should not be worn when working with this chemical. Wear splash-proof chemical goggles and face shield unless full face-piece respiratory protection is worn. Employees should wash immediately with soap when skin is wet or contaminated. Provide emergency showers and eyewash.

Respirator Selection: *20 ppm:* Sa (APF = 10) (any supplied-air respirator) or SCBA (any self-contained breathing apparatus). *50 ppm:* Sa:Cf (APF = 25) (any supplied-air respirator operated in a continuous-flow mode). *100 ppm:* SaT: Cf (APF = 50) (any supplied-air respirator that has a tight-fitting face-piece and is operated in a continuous-flow mode) or SCBAF (APF = 50) (any self-contained breathing apparatus with a full face-piece) or SaF (APF = 50) (any supplied-air respirator with a full face-piece). *200 ppm:* SaF: Pd,Pp (APF = 2000) (any supplied-air respirator that has a full face-piece and is operated in a pressure-demand or other positive-pressure mode). *Emergency or planned entry into unknown concentrations or IDLH conditions:* SCBAF: Pd,Pp (APF = 10,000) (any self-contained breathing apparatus that has a full face-piece and is operated in a pressure-demand or other positive-pressure mode) or SaF: Pd,Pp: ASCBA (APF = 10,000) (any supplied-air respirator that has a full face-piece and is operated in a pressure-demand or other positive-pressure mode in combination with an auxiliary self-contained breathing apparatus operated in a pressure-demand or other positive-pressure mode). *Escape:* GmFOv100 (APF = 50) [any air-purifying, full-face-piece respirator (gas mask) with a chin-style, front- or back-mounted organic vapor canister having an N100, R100, or P100 filter] or SCBAE (any appropriate escape-type, self-contained breathing apparatus).

Note: Substance reported to cause eye irritation or damage; may require eye protection.

Storage: Color Code—Blue: Health Hazard/Poison: Store in a secure poison location. Prior to working with this chemical you should be trained on its proper handling and storage. Before entering confined space where this chemical may be present, check to make sure that an explosive concentration does not exist. Store in tightly closed containers in a cool, well-ventilated area away from combustibles, reducing agents, strong oxidants, strong bases, strong acids. Where possible, automatically transfer materials from drums or other storage containers to process containers. Sources of ignition, such as smoking and open flames, are prohibited where this chemical is handled, used, or stored. Metal containers involving the transfer of 5 gallons or more of this chemical should be grounded and bonded. Drums must be equipped with self-closing valves, pressure vacuum bungs, and flame arresters. Use only nonsparking tools and equipment, especially when opening and closing containers of this chemical. Wherever this chemical is used, handled, manufactured, or stored, use explosion-proof electrical equipment and fittings.

Shipping: Nitrotoluenes, *liquid* (*o*-; *m*-; *p*-) require a shipping label of "POISONOUS/TOXIC MATERIALS." They fall in DOT Hazard Class 6.1 and Packing Group II. Nitrotoluenes, *solid* (*m*- or *p*-) require a shipping label of "POISONOUS/TOXIC MATERIALS." They fall in DOT Hazard Class 6.1 and Packing Group II.

Spill Handling: Solid: Evacuate persons not wearing protective equipment from area of spill or leak until cleanup is complete. Remove all ignition sources. Establish forced ventilation to keep levels below explosive limit. With the solid (*p*-isomer) isomer, dampen spilled material with alcohol to avoid dust. Collect powdered material in the most convenient and safe manner and deposit in sealed containers. Ventilate area after cleanup is complete. It may be necessary to contain and dispose of this chemical as a hazardous waste. If material or contaminated runoff enters waterways, notify downstream users of potentially contaminated waters. Contact your local or federal environmental protection agency for specific recommendations. If employees are required to clean up spills, they must be properly trained and equipped. OSHA 1910.120(q) may be applicable.

Liquid: Evacuate and restrict persons not wearing protective equipment from area of spill or leak until cleanup is complete. Remove all ignition sources. Ventilate area of spill or leak. Absorb liquids in vermiculite, dry sand, earth, peat, carbon, or a similar material and deposit in sealed containers. Keep this chemical out of a confined space, such as a sewer, because of the possibility of an explosion, unless the sewer is designed to prevent the buildup of explosive concentrations. It may be necessary to contain and dispose of this chemical as a hazardous waste. If material or contaminated runoff enters waterways, notify downstream users of potentially contaminated waters. Contact your local or federal environmental protection agency for specific recommendations. If employees are required to clean up spills, they must be properly trained and equipped. OSHA 1910.120(q) may be applicable.

Fire Extinguishing: Solid: This chemical is a combustible solid. Use dry chemical, carbon dioxide, water spray, or alcohol foam extinguishers. Poisonous gases, including nitrogen oxides, are produced in fire. If material or contaminated runoff enters waterways, notify downstream users of potentially contaminated waters. Notify local health and fire officials and pollution control agencies. From a secure, explosion-proof location, use water spray to cool exposed containers. If cooling streams are ineffective (venting sound increases in volume and pitch, tank discolors, or shows any signs of deforming), withdraw immediately to a secure position. If employees are expected to fight fires, they must be trained and equipped in OSHA 1910.156. The only respirators recommended for firefighting are self-contained breathing apparatuses that have full face-pieces and are operated in a pressure-demand or other positive-pressure mode. *Liquid:* This chemical is a combustible liquid. Poisonous gases, including nitrogen oxides, are produced in fire. Use

dry chemical, carbon dioxide, or alcohol foam extinguishers. Vapors are heavier than air and will collect in low areas. Vapors may travel long distances to ignition sources and flashback. Vapors in confined areas may explode when exposed to fire. Containers may explode in fire. Storage containers and parts of containers may rocket great distances, in many directions. If material or contaminated runoff enters waterways, notify downstream users of potentially contaminated waters. Notify local health and fire officials and pollution control agencies. From a secure, explosion-proof location, use water spray to cool exposed containers. If cooling streams are ineffective (venting sound increases in volume and pitch, tank discolors, or shows any signs of deforming), withdraw immediately to a secure position. If employees are expected to fight fires, they must be trained and equipped in OSHA 1910.156. The only respirators recommended for firefighting are self-contained breathing apparatuses that have full face-pieces and are operated in a pressure-demand or other positive-pressure mode.

Disposal Method Suggested: Controlled incineration—care must be taken to maintain complete combustion at all times. Incineration of large quantities may require scrubbers to control the emission of nitrogen oxides.

References

National Institute for Occupational Safety and Health. (October 1977). *Information Profiles on Potential Occupational Hazards: Nitrotoluenes*, Report PB-276-678. Rockville, MD. pp. 227–240
Sax, N. I. (Ed.). (1983). *Dangerous Properties of Industrial Materials Report*, 3, No. 3, 85–88 (*p*-Nitrotoluene)
New Jersey Department of Health and Senior Services. (August 1999). *Hazardous Substances Fact Sheet: Nitrotoluenes.* Trenton, NJ

5-Nitro-*o*-toluidine   N:0670

Molecular Formula: $C_7H_8N_2O_2$
Synonyms: AI3-01557; Amarthol fast scarlet G base; Amarthol fast scarlet G salt; 1-Amino-2-methyl-5-nitrobenzene; 2-Amino-4-nitrotoluene; Azoene fast scarlet GC base; Azoene fast scarlet GC salt; Azofix scarlet G salt; Azogene fast scarlet G; Azoic diazo component 12; Benzenamine, 2-methyl-5-nitro-; C.I. 37105; C.I. azoic diazo component 12; Dainichi fast scarlet G base; Daito scarlet base G; Devol scarlet B; Devol scarlet G salt; Diabase scarlet G; Diazo fast scarlet G; Fast red SG base; Fast scarlet base G; Fast scarlet base J; Fast scarlet G; Fast scarlet G base; Fast scarlet GC base; Fast scarlet G salt; Fast scarlet J salt; Fast scarlet M 4NT base; Fast scarlet T base; Hiltonil fast scarlet G base; Hiltonil fast scarlet GC base; Hiltonil fast scarlet G salt; Kayaku scarlet G base; Lake scarlet G base; Lithosol orange R base; 2-Methyl-5-nitroaniline; 6-Methyl-3-nitroaniline; 2-Methyl-5-nitrobenzenamine; 2-Methyl-5-nitrobenzeneamine; Mitsui scarlet G base; Naphthanil scarlet G base; Naphtoelan fast scarlet G base; Naphtoelan fast scarlet

G salt; NCI-C01843; 4-Nitro-2-aminotoluene; 3-Nitro-6-methylaniline; 5-Nitro-2-methylaniline; *p*-Nitro-*o*-toluidina (Spanish); *p*-Nitro-*o*-toluidine; 5-Nitro-2-toluidine; NSC 8947; PNOT; scarlet base Ciba II; scarlet base IRGA II; scarlet base NSP; scarlet G base; Sugai fast scarlet G base; Symulon scarlet G base; *o*-Toluidine, 5-nitro-
CAS Registry Number: 99-55-8
RTECS® Number: XU8225000
UN/NA & ERG Number: UN2660/153
EC Number: 202-765-8 [*Annex I Index No.:* 612-210-00-5]
Regulatory Authority and Advisory Bodies
IARC: Animal Inadequate Evidence; Human Inadequate Evidence, *not classifiable as carcinogenic to humans*, Group 3.
US EPA Hazardous Waste Number (RCRA No.): U181.
RCRA, 40CFR261, Appendix 8 Hazardous Constituents.
RCRA 40CFR268.48; 61FR15654, Universal Treatment Standards: Wastewater (mg/L), 0.32; Nonwastewater (mg/kg), 28.
RCRA 40CFR264, Appendix 9; TSD Facilities Ground Water Monitoring List. Suggested test method(s) (PQL µg/L): 8270 (10).
Reportable Quantity (RQ): 100 lb (45.4 kg).
EPCRA Section 313 Form R *de minimis* concentration reporting level: 1.0%.
European/International Regulations: Hazard Symbol: T, N; Risk phrases: R23/24/25; R40; 52/53; Safety phrases: S1/2; S28; S36/37; S45; S61 (see Appendix 4).
WGK (German Aquatic Hazard Class): 3—Severe hazard to waters.
Description: 5-Nitro-*o*-toluidine is a yellow, crystalline solid. Molecular weight = 152.17; Freezing/Melting point = 107°C.
Incompatibilities: Oxidizers, strong acids, acid chlorides, acid anhydrides, chloroformates.
Permissible Exposure Limits in Air
ACGIH TLV®[1]: 1 mg/m^3 inhalable fraction TWA; confirmed animal carcinogen with unknown relevance to humans.
Protective Action Criteria (PAC)
TEEL-0: 1 mg/m^3
PAC-1: 3 mg/m^3
PAC-2: 150 mg/m^3
PAC-3: 250 mg/m^3
DFG MAK: Carcinogen Category 2
Determination in Air: Use NIOSH Analytical Method (IV) #2005, Nitrobenzenes.
Routes of Entry: Inhalation, ingestion, eye and/or skin contact.
Harmful Effects and Symptoms
Short Term Exposure: Contact may irritate the skin and eyes. High levels of exposure may interfere with the blood's ability to carry oxygen (methemoglobinemia) causing headache, fatigue, dizziness, and cyanosis (a blue color to the skin and lips). Higher levels can cause trouble breathing, collapse, and death.

Long Term Exposure: There is limited evidence that this chemical causes liver cancer in animals.
Points of Attack: Liver, blood.
Medical Surveillance: Blood methemoglobin levels. Complete blood count (CBC). Liver function disease.
First Aid: If this chemical gets into the eyes, remove any contact lenses at once and irrigate immediately for at least 15 min, occasionally lifting upper and lower lids. Seek medical attention immediately. If this chemical contacts the skin, remove contaminated clothing and wash immediately with soap and water. Seek medical attention immediately. If this chemical has been inhaled, remove from exposure, begin rescue breathing (using universal precautions, including resuscitation mask) if breathing has stopped and CPR if heart action has stopped. Transfer promptly to a medical facility. When this chemical has been swallowed, get medical attention. Give large quantities of water and induce vomiting. Do not make an unconscious person vomit.
Note to physician: Treat for methemoglobinemia. Spectrophotometry may be required for precise determination of levels of methemoglobin in urine.
Personal Protective Methods: Wear protective gloves and clothing to prevent any reasonable probability of skin contact. Safety equipment suppliers/manufacturers can provide recommendations on the most protective glove/clothing material for your operation. All protective clothing (suits, gloves, footwear, headgear) should be clean, available each day, and put on before work. Contact lenses should not be worn when working with this chemical. Wear dust-proof chemical goggles and face shield unless full face-piece respiratory protection is worn. Employees should wash immediately with soap when skin is wet or contaminated. Provide emergency showers and eyewash.
Respirator Selection: Follow regulations in OSHA 29CFR1910.134 or European Standard EN149. Use a NIOSH/MSHA- or European Standard EN149-approved respirator; or use an approved supplied-air respirator with a full face-piece operated in the positive-pressure mode, or with a full face-piece, hood, or helmet in the continuous-flow mode; or use a NIOSH/MSHA- or European Standard EN149-approved self-contained breathing apparatus with a full face-piece operated in pressure-demand or other positive-pressure mode.
Storage: Color Code—Blue: Health Hazard/Poison: Store in a secure poison location. Prior to working with this chemical you should be trained on its proper handling and storage. Store in tightly closed containers in a cool, well-ventilated area away from oxidizers, strong acids, acid chlorides, acid anhydrides, chloroformates. Where possible, automatically pump liquid from drums or other storage containers to process containers. Sources of ignition, such as smoking and open flames, are prohibited where this chemical is handled, used, or stored. Metal containers involving the transfer of 5 gallons or more of this chemical should be grounded and bonded. Drums must be equipped with self-closing valves, pressure vacuum bungs, and flame arresters.

Use only nonsparking tools and equipment, especially when opening and closing containers of this chemical. Wherever this chemical is used, handled, manufactured, or stored, use explosion-proof electrical equipment and fittings.

Shipping: Nitrotoluidines (mono) require a label of "POISONOUS/TOXIC MATERIALS." They fall in Hazard Class 6.1 and Packing Group III.

Spill Handling: Evacuate persons not wearing protective equipment from area of spill or leak until cleanup is complete. Remove all ignition sources. Use HEPA vacuum or wet method to reduce dust during cleanup. Do not dry sweep. Collect powdered material in the most convenient and safe manner and deposit in sealed containers. Ventilate area after cleanup is complete. It may be necessary to contain and dispose of this chemical as a hazardous waste. If material or contaminated runoff enters waterways, notify downstream users of potentially contaminated waters. Contact your local or federal environmental protection agency for specific recommendations. If employees are required to cleanup spills, they must be properly trained and equipped. OSHA 1910.120(q) may be applicable.

Fire Extinguishing: This chemical may burn but does not easily ignite. Use dry chemical, carbon dioxide, water spray, alcohol foam, or polymer foam extinguishers. Poisonous gases, including nitrogen oxides, are produced in fire. If material or contaminated runoff enters waterways, notify downstream users of potentially contaminated waters. Notify local health and fire officials and pollution control agencies. From a secure, explosion-proof location, use water spray to cool exposed containers. If cooling streams are ineffective (venting sound increases in volume and pitch, tank discolors, or shows any signs of deforming), withdraw immediately to a secure position. If employees are expected to fight fires, they must be trained and equipped in OSHA 1910.156. The only respirators recommended for firefighting are self-contained breathing apparatuses that have full face-pieces and are operated in a pressure-demand or other positive-pressure mode.

Disposal Method Suggested: Consult with environmental regulatory agencies for guidance on acceptable disposal practices. Generators of waste containing this contaminant ($\geq$100 kg/mo) must conform with EPA regulations governing storage, transportation, treatment, and waste disposal.

Reference

New Jersey Department of Health and Senior Services. (August 1999). *Hazardous Substances Fact Sheet: 5-Nitro-o-toluidine*. Trenton, NJ

Nitrous oxide   N:0680

Molecular Formula: N_2O
Synonyms: Dinitrogen monoxide; Factitious air; Hyponitrous acid anhydride; Laughing gas; Nitrogen Oxide
CAS Registry Number: 10024-97-2

RTECS® Number: QX1350000
UN/NA & ERG Number: UN1070 (compressed)/122; UN2201 (refrigerated liquid)/122
EC Number: 233-032-0
Regulatory Authority and Advisory Bodies
Carcinogenicity: IARC: Animal Inadequate Evidence, Human Inadequate Evidence, *not classifiable as carcinogenic to humans*, Group 3, 1987.
US EPA Gene-Tox Program, Positive: V79 cell culture—gene mutation; Negative: Sperm morphology—mouse; Inconclusive: Carcinogenicity—mouse/rat.
Air Pollutant Standard Set. See below, "Permissible Exposure Limits in Air" section.
California Proposition 65 Developmental/Reproductive toxin 8/1/08.
Canada, WHMIS, Ingredients Disclosure List Concentration 0.1%.
European/International Regulations: not listed in Annex 1.
WGK (German Aquatic Hazard Class): 1—Low hazard to waters.
Description: Nitrous oxide is a colorless gas with a slightly sweet odor. Shipped as a liquefied compressed gas. Molecular weight = 44.02; Boiling point = −88°C; Freezing/Melting point = −91.1°C; Relative density of the vapor/air mixture (air = 1) = 1.53 at 20°C; Vapor pressure = 51.3 atm. Slightly soluble in water; solubility = 0.1% at 25°C. Hazard Identification (based on NFPA-704 M Rating System): Health 2, Flammability 0, Reactivity 0.
Potential Exposure: Compound Description: Tumorigen, Drug, Mutagen; Reproductive Effector; Human Data. Used as an anesthetic in dentistry and surgery, as a gas in food aerosols, such as whipped cream, in the manufacture of nitrites, in rocket fuels, firefighting, and diesel emissions. Large amounts of nitrous oxide will decrease the amount of available oxygen. Oxygen should be routinely tested to ensure that it is at least 19% by volume.
Incompatibilities: Violent reactions with organic peroxides, hydrazine, hydrogen, hydrogen sulfide, lithium, boron, lithium hydride, sodium, aluminum, phosphine. This chemical is a strong oxidizer above 300°C and self-explodes at high temperature. May form explosive mixtures with ammonia, carbon monoxide, hydrogen sulphide, oil, grease and fuels.
Permissible Exposure Limits in Air
Conversion factor: 1 ppm = 1.80 mg/m³ at 25°C & 1 atm.
OSHA PEL: none.
NIOSH REL: 25 ppm/46 mg/m³ TWA over the time exposed; [*Note:* REL for exposure to waste anesthetic gas].
ACGIH TLV®[1]: 50 ppm TWA; not classifiable as a human carcinogen.
Protective Action Criteria (PAC)
TEEL-0: 50 ppm
PAC-1: 150 ppm
PAC-2: 10,000 ppm
PAC-3: 20,000 ppm
DFG MAK: 100 ppm/180 mg/m³ TWA; Peak Limitation Category II(2); Pregnancy Risk Group C.

Denmark: TWA 50 ppm (90 mg/m³), 1999; Finland: TWA 100 ppm, 1999; Norway: TWA 100 ppm (180 mg/m³), 1999; Sweden: NGV 100 ppm (180 mg/m³), KTV 500 ppm (900 mg/m³), 1999; Switzerland: MAK-W 100 ppm (200 mg/m³), KZG-W 200 ppm (400 mg/m³), 1999; United Kingdom: TWA 100 ppm (183 mg/m³), 2000; the Netherlands: MAC-TGG 152 mg/m³, 2003; Argentina, Bulgaria, Columbia, Jordan, South Korea, New Zealand, Singapore, Vietnam: ACGIH TLV®: not classifiable as a human carcinogen. The recommended ACGIH TLV is 50 ppm. Connecticut[60] has set a guideline for nitrous oxide in ambient air of 1.34 mg/m³.

Determination in Air: Use NIOSH Analytical Method (IV) #6600, Nitrous oxide; OSHA Analytical Methods ID-166.

Routes of Entry: Inhalation, skin and/or eye contact (liquid).

Harmful Effects and Symptoms

Short Term Exposure: *Inhalation:* May cause dizziness and difficult breathing. Excessive exposure may cause headaches, nausea, fatigue, and irritability. Loss of consciousness may result from exposure to concentrations of 400,000−800,000 ppm. Anesthetic grades are composed of 80% nitrous oxide with 20% oxygen. High concentrations may cause a deficiency of oxygen in the air. *Skin, Eyes:* Liquid may cause frostbite and freezing burns. Vapors are nonirritating. *Ingestion:* Liquid may cause frostbite and freezing burns of the mouth and throat.

Long Term Exposure: Increased incidence of liver and kidney disease, neurological disease, and spontaneous abortion have been reported. Nitrous oxide has been shown to cause birth defects in rats. Repeated exposure can damage the nervous system, causing numbness and weakness in the arms and legs. May damage the bone marrow and affect blood cell production. May be a teratogen in humans.

Points of Attack: Respiratory system, central nervous system, blood, reproductive system.

Medical Surveillance: Before beginning employment and regular times after that, for those with frequent or potentially high exposures, the following is recommended: examination of the nervous system. If symptoms develop or overexposure is suspected, the following may be useful: consider nerve conduction studies, complete blood count (CBC), examination of the nervous system. Check the work-place air to make certain that the oxygen level is at least 19%.

First Aid: If this chemical gets into the eyes, remove any contact lenses at once and irrigate immediately for at least 15 min, occasionally lifting upper and lower lids. Seek medical attention immediately. If this chemical contacts the skin, remove contaminated clothing and wash immediately with soap and water. Seek medical attention immediately. If this chemical has been inhaled, remove from exposure, begin rescue breathing (using universal precautions, including resuscitation mask) if breathing has stopped and CPR if heart action has stopped. Transfer promptly to a medical facility. When this chemical has been swallowed, get

medical attention. Give large quantities of water and induce vomiting. Do not make an unconscious person vomit. If frostbite has occurred, seek medical attention immediately; do NOT rub the affected areas or flush them with water. In order to prevent further tissue damage, do NOT attempt to remove frozen clothing from frostbitten areas. If frostbite has NOT occurred, immediately and thoroughly wash contaminated skin with soap and water.

Personal Protective Methods: Wear appropriate personal protective clothing to prevent the skin from becoming frozen from contact with the evaporating liquid or from contact with vessels containing the liquid. Safety equipment suppliers/manufacturers can provide recommendations on the most protective glove/clothing material for your operation. All protective clothing (suits, gloves, footwear, headgear) should be clean, available each day, and put on before work. Contact lenses should not be worn when working with this chemical. Wear gas-proof goggles, unless full face-piece respiratory protection is worn. Employees should wash immediately with soap when skin is wet or contaminated. Provide emergency showers and eyewash.

Respirator Selection: Exposure to nitrous oxide is dangerous because it can replace oxygen and lead to suffocation. Where there is potential for exposure *over 25 ppm*, use a NIOSH/MSHA- or European Standard EN149-approved supplied-air respirator with a full face-piece operated in the positive-pressure mode, or with a full face-piece, hood, or helmet in the continuous-flow mode; or use a NIOSH/MSHA- or European Standard EN149-approved self-contained breathing apparatus with a full face-piece operated in pressure-demand or other positive-pressure mode.

Storage: Color Code—Yellow: Reactive Hazard; Store in a location separate from other materials, especially flammables and combustibles. Prior to working with this chemical you should be trained on its proper handling and storage. Nitrous oxide must be stored to avoid contact with organic peroxides, ammonia, carbon monoxide, hydrogen, hydrogen sulfide, and phosphine, since violent reactions occur. Cylinders of nitrous oxide should be stored in a cool, preferably fire-resistant area, away from heat sources.

Shipping: Nitrous oxide requires a shipping label of "NONFLAMMABLE GAS, OXIDIZER." Nitrous oxide, *refrigerated liquid*, requires a shipping label of "NONFLAMMABLE GAS." Both fall in DOT Hazard Class 2.2.

Spill Handling: If in a building, evacuate building and confine vapors by closing doors and shutting down HVAC systems. Restrict persons not wearing protective equipment from area of spill or leak until cleanup is complete. Remove all ignition sources. Ventilate area of spill or leak to disperse the gas. Wear chemical protective suit with self-contained breathing apparatus to combat spills. Stay upwind and use water spray to "knock down" vapor; contain runoff. Stop the flow of gas, if it can be done safely from a distance. If source is a cylinder and the leak cannot be stopped in place, remove the leaking cylinder to a safe place and repair leak or allow cylinder to empty. If flow cannot be stopped, allow it to flow into a mixture of caustic soda and slaked lime and dispose of the resulting material in a hood. Keep this chemical out of confined spaces, such as a sewer, because of the possibility of explosion, unless the sewer is designed to prevent the buildup of explosive concentrations. If employees are required to clean up spills, they must be properly trained and equipped. OSHA 1910.120(q) may be applicable.

Fire Extinguishing: Nitrous oxide is an oxidizer and will increase the intensity of any fire. Nitrous oxide self-explodes at high temperature. Extinguish fire using an agent suitable for type of surrounding fire. Nitrous oxide itself does not burn. Do not extinguish the fire unless the flow of gas can be stopped and any remaining gas is out of the line. Specially trained personnel may use fog lines to cool exposures and let the fire burn itself out. Vapors are heavier than air and will collect in low areas. Vapors in confined areas may explode when exposed to fire. Containers may explode in fire. Storage containers and parts of containers may rocket great distances, in many directions. If material or contaminated runoff enters waterways, notify downstream users of potentially contaminated waters. Notify local health and fire officials and pollution control agencies. From a secure, explosion-proof location, use water spray to cool exposed containers. If cooling streams are ineffective (venting sound increases in volume and pitch, tank discolors, or shows any signs of deforming), withdraw immediately to a secure position. If cylinders are exposed to excessive heat from fire or flame contact, withdraw immediately to a secure location. If employees are expected to fight fires, they must be trained and equipped in OSHA 1910.156. The only respirators recommended for firefighting are self-contained breathing apparatuses that have full face-pieces and are operated in a pressure-demand or other positive-pressure mode.

Disposal Method Suggested: Disperse in atmosphere or spray on dry soda ash/lime with great care; then flush to sewer.

References

Sax, N. I. (Ed.). (1981). *Dangerous Properties of Industrial Materials Report*, 1, No. 7, 66–67

New Jersey Department of Health and Senior Services. (September 2004). *Hazardous Substances Fact Sheet: Nitrous Oxide.* Trenton, NJ

New York State Department of Health. (May 1986). *Chemical Fact Sheet: Nitrous Oxide.* Albany, NY: Bureau of Toxic Substance Assessment

Nonane   N:0685

Molecular Formula: C_9H_{20}
Common Formula: $CH_3(CH_2)_7CH_3$
Synonyms: *N*.Nonane; Nonyl hydride; Shellsol 140
CAS Registry Number: 111-84-2

RTECS® Number: RA6115000
UN/NA & ERG Number: UN1920/128
EC Number: 203-913-4
Regulatory Authority and Advisory Bodies
Air Pollutant Standard Set. See below, "Permissible Exposure Limits in Air" section.
Canada, WHMIS, Ingredients Disclosure List Concentration 1.0%.
European/International Regulations: not listed in Annex 1.
WGK (German Aquatic Hazard Class): No value assigned.
Description: Nonane is a colorless liquid. Odor threshold = 47 ppm. Molecular weight = 128.29; Specific gravity (H_2O:1) = 0.72; Boiling point = 150.6°C; Freezing/Melting point = −51°C; Vapor pressure 3 mmHg at 20°C; Flash point = 31°C; Autoignition temperature = 205°C. Explosive limits: LEL = 0.8%; UEL = 2.9%. Hazard Identification (based on NFPA-704 M Rating System): Health 0, Flammability 3, Reactivity 0. Insoluble in water.
Potential Exposure: Compound Description: Drug, Primary Irritant. Nonane is used in the synthesis of biodegradable detergents as a distillation chaser and as an ingredient in Stoddard solvent and gasoline.
Incompatibilities: Strong oxidizers.
Permissible Exposure Limits in Air
Conversion factor: 1 ppm = 5.25 mg/m^3 at 25°C & 1 atm.
OSHA PEL: None.
NIOSH REL: 200 ppm/1050 mg/m^3 TWA.
ACGIH TLV®[1]: 200 ppm TWA.
Protective Action Criteria (PAC)
TEEL-0: 200 ppm
PAC-1: 200 ppm
PAC-2: 350 ppm
PAC-3: 350 ppm
Denmark: TWA 200 ppm (1050 mg/m^3), 1999; Finland: TWA 200 ppm (1050 mg/m^3); STEL 250 ppm (1315 mg/m^3), 1999; France: VME 200 ppm (1050 mg/m^3), 1999; Japan: 200 ppm (1050 mg/m^3), 1999; Norway: TWA 100 ppm (525 mg/m^3), 1999; Sweden: NGV 150 ppm (800 mg/m^3), KTV 200 ppm (1100 mg/m^3), 1999; Switzerland: MAK-W 200 ppm (1050 mg/m^3), 1999; the Netherlands: MAC-TGG 1050 mg/m^3, 2003; Argentina, Bulgaria, Columbia, Jordan, South Korea, New Zealand, Singapore, Vietnam: ACGIH TLV®: TWA 200 ppm. Several states have set guidelines or standards for nonane in ambient air[60] ranging from 10.5–13.0 mg/m^3 (North Dakota) to 17.5 mg/m^3 (Virginia) to 21.0 mg/m^3 (Connecticut) to 25.0 mg/m^3 (Nevada).
Determination in Air: No NIOSH Analytical Method available.
Routes of Entry: Inhalation, ingestion, skin and/or eye contact.
Harmful Effects and Symptoms
Short Term Exposure: Nonane can affect you when breathed in. Narcotic at high concentrations; may affect the central nervous system. Exposure to high levels can cause headache, drowsiness, dizziness, confusion, nausea, tremor,

lack of coordination, and unconsciousness. Irritates the eyes, skin, respiratory tract. Swallowing the liquid may cause chemical pneumonitis.

Long Term Exposure: Prolonged contact can cause drying and cracking of the skin. Repeated exposure can affect the liver.

Points of Attack: Eyes, skin, respiratory system, central nervous system, liver.

Medical Surveillance: If symptoms develop or overexposure is suspected, the following may be useful: liver function tests.

First Aid: If this chemical gets into the eyes, remove any contact lenses at once and irrigate immediately for at least 15 min, occasionally lifting upper and lower lids. Seek medical attention immediately. If this chemical contacts the skin, remove contaminated clothing and wash immediately with soap and water. Seek medical attention immediately. If this chemical has been inhaled, remove from exposure, begin rescue breathing (using universal precautions, including resuscitation mask) if breathing has stopped and CPR if heart action has stopped. Transfer promptly to a medical facility. When this chemical has been swallowed, get medical attention. Give large quantities of water and induce vomiting. Do not make an unconscious person vomit.

Note to physician: Inhalation: bronchodilators, decongestants, and oxygen may be used if necessary. Corticosteroids are useful for treating pneumonitis.

Personal Protective Methods: Wear protective gloves and clothing to prevent any reasonable probability of skin contact. Safety equipment suppliers/manufacturers can provide recommendations on the most protective glove/clothing material for your operation. All protective clothing (suits, gloves, footwear, headgear) should be clean, available each day, and put on before work. Contact lenses should not be worn when working with this chemical. Wear splash-proof chemical goggles and face shield unless full face-piece respiratory protection is worn. Employees should wash immediately with soap when skin is wet or contaminated. Provide emergency showers and eyewash. Wear solvent-resistant gloves and clothing. Safety equipment suppliers/manufacturers can provide recommendations on the most protective glove/clothing material for your operation. All protective clothing (suits, gloves, footwear, headgear) should be clean, available each day, and put on before work.

Respirator Selection: Where there is potential for exposures *over 200 ppm*, use a NIOSH/MSHA- or European Standard EN149-approved full face-piece respirator with an organic vapor cartridge/canister. Greater protection is provided by a powered air-purifying respirator. Where there is potential for high exposure, use a NIOSH/MSHA- or European Standard EN149-approved supplied-air respirator with a full face-piece operated in the positive-pressure mode, or with a full face-piece hood, or helmet in the continuous-flow mode; or use a NIOSH/MSHA- or European Standard EN149-approved self-contained breathing apparatus with a full face-piece operated in pressure demand or other positive-pressure mode.

Storage: Color Code—Red: Flammability Hazard: Store in a flammable liquid storage area or approved cabinet away from ignition sources and corrosive and reactive materials. Prior to working with this chemical you should be trained on its proper handling and storage. Before entering confined space where this chemical may be present, check to make sure that an explosive concentration does not exist. Store in tightly closed containers in a cool, well-ventilated area. Sources of ignition, such as smoking and open flames, are prohibited where nonane is handled, used, or stored. Metal containers involving the transfer of 5 gallons or more of nonane should be grounded and bonded. Drums must be equipped with self-closing valves, pressure vacuum bungs, and flame arresters. Wherever nonane is used, handled, manufactured, or stored, use explosion-proof electrical equipment and fittings. Use only nonsparking tools and equipment, especially when opening and closing containers of nonane.

Shipping: Nonanes require a shipping label of "FLAMMABLE LIQUID." It falls in Hazard Class 3 and Packing Group II.

Spill Handling: Evacuate and restrict persons not wearing protective equipment from area of spill or leak until cleanup is complete. Remove all ignition sources. Establish forced ventilation to keep levels below explosive limit. Absorb liquids in vermiculite, dry sand, earth, peat, carbon, or a similar material and deposit in sealed containers. Keep this chemical out of a confined space, such as a sewer, because of the possibility of an explosion, unless the sewer is designed to prevent the buildup of explosive concentrations. It may be necessary to contain and dispose of this chemical as a hazardous waste. If material or contaminated runoff enters waterways, notify downstream users of potentially contaminated waters. Contact your local or federal environmental protection agency for specific recommendations. If employees are required to clean up spills, they must be properly trained and equipped. OSHA 1910.120(q) may be applicable.

Fire Extinguishing: This chemical is a flammable liquid. Poisonous gases are produced in fire. Use dry chemical, carbon dioxide, or foam extinguishers. Vapors are heavier than air and will collect in low areas. Vapors may travel long distances to ignition sources and flashback. Vapors in confined areas may explode when exposed to fire. Containers may explode in fire. Storage containers and parts of containers may rocket great distances, in many directions. If material or contaminated runoff enters waterways, notify downstream users of potentially contaminated waters. Notify local health and fire officials and pollution control agencies. From a secure, explosion-proof location, use water spray to cool exposed containers. If cooling streams are ineffective (venting sound increases in volume and pitch, tank discolors, or shows any signs of deforming), withdraw immediately to a secure position. If employees

are expected to fight fires, they must be trained and equipped in OSHA 1910.156. The only respirators recommended for firefighting are self-contained breathing apparatuses that have full face-pieces and are operated in a pressure-demand or other positive-pressure mode.

Disposal Method Suggested: Dissolve or mix the material with a combustible solvent and burn in a chemical incinerator equipped with an afterburner and scrubber. All federal, state, and local environmental regulations must be observed.

References

National Institute for Occupational Safety and Health. (1977). *Criteria for a Recommended Standard: Occupational Exposure to Alkanes.* NIOSH Document No. 77-151. Washington, DC

New Jersey Department of Health and Senior Services. (February 2000). *Hazardous Substances Fact Sheet: Nonane.* Trenton, NJ

Nonyl trichlorosilane   N:0690

Molecular Formula: $C_9H_{19}Cl_3Si$
Common Formula: $C_9H_{19}SiCl_3$
Synonyms: Silane, nonyltrichloro-; Silane, trichlorononyl-; Trichlorononylsilane
CAS Registry Number: 5283-67-0
RTECS® Number: VV4660000
UN/NA & ERG Number: UN1799/156
EC Number: 226-113-7
Regulatory Authority and Advisory Bodies
Department of Homeland Security Screening Threshold Quantity (pounds): Sabotage/Contamination Hazard: A placarded amount (commercial grade).
Carcinogenicity: IARC: Animal Inadequate Evidence, Human Inadequate Evidence, *not classifiable as carcinogenic to humans,* Group 3 (see Appendix 4).
WGK (German Aquatic Hazard Class): 1—Low hazard to waters.
Description: Nonyl trichlorosilane is a clear fuming liquid with an irritating odor. Molecular weight = 261.73. Hazard Identification (based on NFPA-704 M Rating System): Health 3, Flammability 1, Reactivity 2W. Reacts with water.
Potential Exposure: Used in silicone (polysiloxane) manufacture.
Incompatibilities: Chlorosilanes can self-ignite in air. They react with water or steam, producing heat and fumes of HCl. Forms a self-igniting compound with ammonia.
Protective Action Criteria (PAC)*
TEEL-0: 0.2 ppm
PAC-1: **0.60** ppm
PAC-2: **7.3** ppm
PAC-3: **33** ppm
*AEGLs (Acute Emergency Guideline Levels) & ERPGs (Emergency Response Planning Guideline) are in **bold face**.

Routes of Entry: Inhalation, ingestion, skin and/or eye contact.
Harmful Effects and Symptoms
Short Term Exposure: Nonyl trichlorosilane can affect you when breathed in. Nonyl trichlorosilane is a corrosive chemical and can cause severe eye burns leading to permanent damage. Contact can cause severe skin burns. Exposure can irritate the eyes, nose, and throat. Exposure can irritate the lungs, causing coughing and/or shortness of breath. Higher exposures can cause pulmonary edema, a medical emergency that can be delayed for several hours. This can cause death.
Long Term Exposure: May cause lung irritation; bronchitis may develop.
Points of Attack: Lungs.
Medical Surveillance: For those with frequent or potentially high exposure the following are recommended before beginning work and at regular times after that: lung function tests. If symptoms develop or overexposure is suspected, the following may be useful: consider chest X-ray after acute overexposure.
First Aid: If this chemical gets into the eyes, remove any contact lenses at once and irrigate immediately for at least 30 min, occasionally lifting upper and lower lids. Seek medical attention immediately. If this chemical contacts the skin, remove contaminated clothing and wash immediately with soap and water. Seek medical attention immediately. If this chemical has been inhaled, remove from exposure, begin rescue breathing (using universal precautions, including resuscitation mask) if breathing has stopped and CPR if heart action has stopped. Transfer promptly to a medical facility. When this chemical has been swallowed, get medical attention. If victim is *conscious,* administer water or milk. Do not induce vomiting. Medical observation is recommended for 24—48 h after breathing overexposure, as pulmonary edema may be delayed. As first aid for pulmonary edema, a doctor or authorized paramedic may consider administering a corticosteroid spray.
Personal Protective Methods: Wear protective gloves and clothing to prevent any reasonable probability of skin contact. Safety equipment suppliers/manufacturers can provide recommendations on the most protective glove/clothing material for your operation. All protective clothing (suits, gloves, footwear, headgear) should be clean, available each day, and put on before work. Contact lenses should not be worn when working with this chemical. Wear splash-proof chemical goggles and face shield unless full face-piece respiratory protection is worn. Employees should wash immediately with soap when skin is wet or contaminated. Provide emergency showers and eyewash.
Respirator Selection: Where there is potential for exposure to nonyl trichlorosilane, use a NIOSH/MSHA- or European Standard EN149-approved supplied-air respirator with a full face-piece operated in the positive-pressure mode, or with a full face-piece, hood, or helmet in the continuous-flow mode; or use a NIOSH/MSHA- or European Standard

EN149-approved self-contained breathing apparatus with a full face-piece operated in pressure-demand or other positive-pressure mode.

Storage: Color Code—White: Corrosive or Contact Hazard; Store separately in a corrosion-resistant location. Prior to working with this chemical you should be trained on its proper handling and storage. Store in tightly closed containers in a cool, well-ventilated area away from combustible materials, such as wood, paper, or oil.

Shipping: Nonyl trichlorosilane requires a shipping label of "CORROSIVE." It falls in Hazard Class 8 and Packing Group II.

Spill Handling: Evacuate and restrict persons not wearing protective equipment from area of spill or leak until cleanup is complete. Remove all ignition sources. Ventilate area of spill or leak. Absorb liquids in vermiculite, dry sand, earth, peat, carbon, or a similar material and deposit in sealed containers. Keep this chemical out of a confined space, such as a sewer, because of the possibility of an explosion, unless the sewer is designed to prevent the buildup of explosive concentrations. It may be necessary to contain and dispose of this chemical as a hazardous waste. If material or contaminated runoff enters waterways, notify downstream users of potentially contaminated waters. Contact your local or federal environmental protection agency for specific recommendations. If employees are required to clean up spills, they must be properly trained and equipped. OSHA 1910.120(q) may be applicable.

Initial isolation and protective action distances

Distances shown are likely to be affected during the first 30 min after materials are spilled and could increase with time. If more than one tank car, cargo tank, portable tank, or large cylinder involved in the incident is leaking, the protective action distance may need to be increased. You may need to seek emergency information from CHEMTREC at (800) 424-9300 or seek professional environmental engineering assistance from the US EPA Environmental Response Team at (908) 548-8730 (24-h response line).

Small spills (From a small package or a small leak from a large package)

when spilled in water

First: Isolate in all directions (feet/meters)   100/30
Then: Protect persons downwind (miles/kilometers)
Day   0.1/0.2
Night   0.1/0.2

Large spills (From a large package or from many small packages)

First: Isolate in all directions (feet/meters)   200/60
Then: Protect persons downwind (miles/kilometers)
Day   0.3/0.5
Night   1.0/1.5

Fire Extinguishing: Nonyl trichlorosilane may burn but does not readily ignite. Use dry chemical, CO_2, or foam extinguishers. Poisonous gases, including hydrogen chloride and chlorine, are produced in fire. Vapors are heavier than air and will collect in low areas. Containers may explode in fire. Storage containers and parts of containers may rocket great distances in many directions. If material or contaminated runoff enters waterways, notify downstream users of potentially contaminated waters. Notify local health and fire officials and pollution control agencies. From a secure, explosion-proof location, use water spray to cool exposed containers. If cooling streams are ineffective (venting sound increases in volume and pitch, tank discolors, or shows any signs of deforming), withdraw immediately to a secure position. If employees are expected to fight fires, they must be trained and equipped in OSHA 1910.156. The only respirators recommended for firefighting are self-contained breathing apparatuses that have full face-pieces and are operated in a pressure-demand or other positive-pressure mode.

Reference

New Jersey Department of Health and Senior Services. (February 2000). *Hazardous Substances Fact Sheet: Nonyl Trichlorosilane.* Trenton, NJ

Norbormide   N:0700

Molecular Formula: $C_{33}H_{25}N_3O_3$

Synonyms: Compound S-6,999; ENT 51,762; 5 (α-Hydroxy-α-2-pyridylbenzyl)-7-(α-2-pyridylbenzylidene)-5-norborene-2,3-dicarboxide; MCN 1025; Norbormida (Spanish); Raticate®; Raticide®; S-6,999; Shoxin

CAS Registry Number: 991-42-4

RTECS® Number: RB8750000

UN/NA & ERG Number: UN2588/151

EC Number: 213-589-6 [*Annex I Index No.:* 650-004-00-7]

Regulatory Authority and Advisory Bodies

SUPERFUND/EPCRA 40CFR355, Extremely Hazardous Substances: TPQ = 100/10,000 lb (45.4/4540 kg).

Reportable Quantity (RQ): 100 lb (45.4 kg).

European/International Regulations: Hazard Symbol: Xn; Risk phrases: R22; Safety phrases: S2 (see Appendix 4).

WGK (German Aquatic Hazard Class): No value assigned.

Description: Norbormide is a white crystalline powder. Molecular weight = 511.61; Freezing/Melting point = 190−198°C. Hazard Identification (based on NFPA-704 M Rating System): Health 3, Flammability 0, Reactivity 0. Insoluble in water.

Potential Exposure: This material is used as a selective rat poison.

Incompatibilities: Alkalis.

Permissible Exposure Limits in Air

Protective Action Criteria (PAC)

TEEL-0: 0.75 mg/m^3

PAC-1: 2 mg/m^3

PAC-2: 3.8 mg/m^3

PAC-3: 3.8 mg/m^3

Routes of Entry: Inhalation, ingestion, skin and/or eye contact.

Harmful Effects and Symptoms

Short Term Exposure: Moderately to highly toxic to humans. Probable human lethal dose is 50–500 mg/kg, or 1 teaspoon to 1 pint for a 150-lb person. Exposure may cause a transient decrease in temperature and blood pressure.

First Aid: If this chemical gets into the eyes, remove any contact lenses at once and irrigate immediately for at least 15 min, occasionally lifting upper and lower lids. Seek medical attention immediately. If this chemical contacts the skin, remove contaminated clothing and wash immediately with soap and water. Seek medical attention immediately. If this chemical has been inhaled, remove from exposure, begin rescue breathing (using universal precautions, including resuscitation mask) if breathing has stopped and CPR if heart action has stopped. Transfer promptly to a medical facility. When this chemical has been swallowed, get medical attention. Give large quantities of water and induce vomiting. Do not make an unconscious person vomit.

Personal Protective Methods: Wear protective gloves and clothing to prevent any reasonable probability of skin contact. Safety equipment suppliers/manufacturers can provide recommendations on the most protective glove/clothing material for your operation. All protective clothing (suits, gloves, footwear, headgear) should be clean, available each day, and put on before work. Contact lenses should not be worn when working with this chemical. Wear dust-proof chemical goggles and face shield unless full face-piece respiratory protection is worn. Employees should wash immediately with soap when skin is wet or contaminated. Provide emergency showers and eyewash.

Respirator Selection: Follow regulations in OSHA 29CFR1910.134 or European Standard EN149. Use a NIOSH/MSHA- or European Standard EN149-approved respirator; or use an approved supplied-air respirator with a full face-piece operated in the positive-pressure mode, or with a full face-piece, hood, or helmet in the continuous-flow mode; or use a NIOSH/MSHA- or European Standard EN149-approved self-contained breathing apparatus with a full face-piece operated in pressure-demand or other positive-pressure mode.

Storage: Color Code—Blue: Health Hazard/Poison: Store in a secure poison location. Prior to working with this chemical you should be trained on its proper handling and storage.

Shipping: Pesticides, solid, toxic, n.o.s. require a shipping label of "POISONOUS/TOXIC MATERIALS." It falls in Hazard Class 6.1 and Packing Group III.

Spill Handling: Evacuate and restrict persons not wearing protective equipment from area of spill or leak until cleanup is complete. Remove all ignition sources. Ventilate area of spill or leak. *Small wet spills:* Absorb liquids in vermiculite, dry sand, earth, peat, carbon, or a similar material and deposit in sealed containers. *Small dry spills:* with clean shovel place material into clean, dry container and cover; move containers from spill area. *Large spills*: dike far ahead of spill for later disposal. Keep this chemical out of a confined space, such as a sewer, because of the possibility of an explosion, unless the sewer is designed to prevent the buildup of explosive concentrations. It may be necessary to contain and dispose of this chemical as a hazardous waste. If material or contaminated runoff enters waterways, notify downstream users of potentially contaminated waters. Contact your local or federal environmental protection agency for specific recommendations. If employees are required to clean up spills, they must be properly trained and equipped. OSHA 1910.120(q) may be applicable.

Fire Extinguishing: Use dry chemical, carbon dioxide, water spray, or alcohol foam extinguishers. Poisonous gases are produced in fire. If material or contaminated runoff enters waterways, notify downstream users of potentially contaminated waters. Notify local health and fire officials and pollution control agencies. From a secure, explosion-proof location, use water spray to cool exposed containers. If cooling streams are ineffective (venting sound increases in volume and pitch, tank discolors, or shows any signs of deforming), withdraw immediately to a secure position. If employees are expected to fight fires, they must be trained and equipped in OSHA 1910.156. The only respirators recommended for firefighting are self-contained breathing apparatuses that have full face-pieces and are operated in a pressure-demand or other positive-pressure mode.

Disposal Method Suggested: Small amounts may be treated with alkali and then landfilled. Large amounts should be incinerated.[22] In accordance with 40CFR165, follow recommendations for the disposal of pesticides and pesticide containers. Must be disposed properly by following package label directions or by contacting your local or federal environmental control agency or by contacting your regional EPA office.

Reference

US Environmental Protection Agency. (November 30, 1987). *Chemical Hazard Information Profile: Norbormide.* Washington, DC: Chemical Emergency Preparedness Program

O

Octafluorocyclobutane O:0100

Molecular Formula: C_4F_8
Synonyms: Cyclobutane, cyclooctafluorobutane; Freon C-318®; Octafluoro-; Perfluorocyclobutane
CAS Registry Number: 115-25-3
RTECS® Number: GU1779500
UN/NA & ERG Number: UN1976/126
EC Number: 204-075-2
Regulatory Authority and Advisory Bodies
Air Pollutant Standard Set. See below, "Permissible Exposure Limits in Air" section.
WGK (German Aquatic Hazard Class): No value assigned.
Description: Octafluorocyclobutane is a colorless gas. Molecular weight = 200.03; Boiling point = −6°C; Freezing/Melting point = −41°C. Hazard Identification (based on NFPA-704 M Rating System): Health 2, Flammability 0, Reactivity 0. Slightly soluble (slight hydrolysis).
Potential Exposure: This material is used as a refrigerant.
Permissible Exposure Limits in Air
Protective Action Criteria (PAC)
TEEL-0: 25,000 ppm
PAC-1: 65,000 ppm
PAC-2: 230,000 ppm
PAC-3: 300,000 ppm
Routes of Entry: Inhalation, skin and/or eye contact.
Harmful Effects and Symptoms
Short Term Exposure: Octafluorocyclobutane can affect you when inhaled. Inhalation can irritate the lungs, causing coughing and shortness of breath. Higher exposures can cause pulmonary edema, a medical emergency that can be delayed for several hours. This can cause death. High levels can cause you to feel dizzy, lightheaded, and to pass out. Very high levels could cause death.
Long Term Exposure: Similar chemicals can cause irregular heartbeat, which could lead to death. Highly irritating chemicals may affect the lungs; may cause bronchitis and lung damage.
Points of Attack: Lungs, heart.
Medical Surveillance: Before beginning employment and at regular times after that, for those with frequent or potentially high exposures, the following are recommended: lung function tests. If symptoms develop or overexposure is suspected, the following may be useful: special 24-h EKG (Holter Monitor) to look for irregular heartbeat. Consider chest X-ray after acute overexposure.
First Aid: If this chemical gets into the eyes, remove any contact lenses at once and irrigate immediately for at least 15 min, occasionally lifting upper and lower lids. Seek medical attention immediately. If this chemical contacts the skin, remove contaminated clothing and wash immediately with soap and water. Seek medical attention immediately. If this chemical has been inhaled, remove from exposure, begin rescue breathing (using universal precautions, including resuscitation mask) if breathing has stopped and CPR if heart action has stopped. Transfer promptly to a medical facility. When this chemical has been swallowed, get medical attention. Give large quantities of water and induce vomiting. Do not make an unconscious person vomit. Medical observation is recommended for 24−48 h after breathing overexposure, as pulmonary edema may be delayed. As first aid for pulmonary edema, a doctor or authorized paramedic may consider administering a corticosteroid spray.
Personal Protective Methods: Wear protective gloves and clothing to prevent any reasonable probability of skin contact. Safety equipment suppliers/manufacturers can provide recommendations on the most protective glove/clothing material for your operation. All protective clothing (suits, gloves, footwear, headgear) should be clean, available each day, and put on before work. Contact lenses should not be worn when working with this chemical. Wear gas-proof goggles unless full face-piece respiratory protection is worn. Employees should wash immediately with soap when skin is wet or contaminated. Provide emergency showers and eyewash.
Respirator Selection: Where there is potential for exposures to octafluorocyclobutane, use NIOSH/MSHA- or European Standard EN 149-approved supplied-air respirator with a full face-piece operated in the positive-pressure mode, or with the full face-piece, hood, or helmet in the continuous-flow mode; or use a NIOSH/MSHA- or European Standard EN149-approved self-contained breathing apparatus with a full face-piece operated in a pressure-demand or other positive-pressure mode.
Storage: Color Code—Green: General storage may be used. Prior to working with this chemical you should be trained on its proper handling and storage. Store in tightly closed containers in a cool, well-ventilated area away from potential high heat sources. Protect cylinders from physical damage. Procedures for the handling, use, and storage of cylinders should be in compliance with OSHA 1910.101 and 1910.169, as with the recommendations of the Compressed Gas Association.
Shipping: This compound requires a shipping label of "NONFLAMMABLE GAS." Octafluorocyclobutane falls in DOT Hazard Class 2.2.
Spill Handling: If in a building, evacuate building and confine vapors by closing doors and shutting down HVAC systems. Restrict persons not wearing protective equipment from area of spill or leak until cleanup is complete. Remove all ignition sources. Ventilate area of spill or leak to disperse the gas. Wear chemical protective suit with self-contained breathing apparatus to combat spills. Stay upwind and use water spray to "knock down" vapor; contain runoff.

Sittig's Handbook of Toxic and Hazardous Chemicals and Carcinogens.
DOI: 10.1016/B978-1-4377-7869-4.00014-X

Stop the flow of gas, if it can be done safely from a distance. If source is a cylinder and the leak cannot be stopped in place, remove the leaking cylinder to a safe place; and repair leak or allow cylinder to empty. Keep this chemical out of confined spaces, such as a sewer, because of the possibility of explosion, unless the sewer is designed to prevent the buildup of explosive concentrations. If employees are required to clean up spills, they must be properly trained and equipped. OSHA 1910.120(q) may be applicable.

Fire Extinguishing: Extinguish fire using an agent suitable for type of surrounding fire. Octafluorocyclobutane itself does not burn. Poisonous gases, including fluorine, are produced in fire. Do not extinguish the fire unless the flow of gas can be stopped and any remaining gas is out of the line. Specially trained personnel may use fog lines to cool exposures and let the fire burn itself out. Vapors are heavier than air and will collect in low areas. Containers may explode in fire. Storage containers and parts of containers may rocket great distances, in many directions. If material or contaminated runoff enters waterways, notify downstream users of potentially contaminated waters. Notify local health and fire officials and pollution control agencies. From a secure, explosion-proof location, use water spray to cool exposed containers. If cooling streams are ineffective (venting sound increases in volume and pitch, tank discolors, or shows any signs of deforming), withdraw immediately to a secure position. If cylinders are exposed to excessive heat from fire or flame contact, withdraw immediately to a secure location. If employees are expected to fight fires, they must be trained and equipped in OSHA 1910.156. The only respirators recommended for firefighting are self-contained breathing apparatuses that have full face-pieces and are operated in a pressure-demand or other positive-pressure mode.

Reference

New Jersey Department of Health and Senior Services. (February 2000). *Hazardous Substances Fact Sheet: Octafluorocyclobutane.* Trenton, NJ

Octamethyl diphosphoramide O:0110

Molecular Formula: $C_8H_{24}N_4O_3P_2$
Common Formula: $C_8H_{24}N_4P_2O_3$
Synonyms: Bis(bisdimethylamino)phosphonousanhydride; Bis(bisdimethylaminophosphonous)anhydride; Bis-bisdimethylaminophosphonous anhydride; Bis(bisdimethylamino) phosphoric anhydride; Bis-*N,N,N′,N′*-tetramethylphosphorodiamidic anhydride; Diphosphoramide, octamethyl-; ENT 17,291; Letha laire G-59; Octamethyl-diphosphorsaeuretetramid (German); Octamethylpyrophosphoramide; Octamethyl pyrophosphortetramide; Octamethyl tetramido pyrophosphate; Octametilpirofosforamida (Spanish); OMPA; Ompacide; Ompatox; Ompax; Pestox; Pestox 3; Pestox III; Pyrophosphoric acid octamethyltetraamide; Pyrophosphorytetrakisdimethylamide; Schradan; Schradane

(French); Systam; Systophos; Sytam; Tetrakisdimethylaminophosphonous anhydride; Tetrakisdimethylaminophosphoric anhydride
CAS Registry Number: 152-16-9
RTECS® Number: UX5950000
UN/NA & ERG Number: UN3018 (organophosphorus pesticide, liquid, toxic)/152
EC Number: 205-801-0 [*Annex I Index No.:* 015-026-00-8]
Regulatory Authority and Advisory Bodies
Banned or Severely Restricted (in agriculture) (Russia, USA) (UN).[13]
Air Pollutant Standard Set. See below, "Permissible Exposure Limits in Air" section.
US EPA Hazardous Waste Number (RCRA No.): P085.
RCRA, 40CFR261, Appendix 8 Hazardous Constituents.
SUPERFUND/EPCRA 40CFR355, Extremely Hazardous Substances: TPQ = 100 lb (45.4 kg).
Reportable Quantity (RQ): 100 lb (45.4 kg).
European/International Regulations: Hazard Symbol: T+; Risk phrases: R27/28; Safety phrases: S1/2; S36/37; S38; S45 (see Appendix 4).
WGK (German Aquatic Hazard Class): No value assigned.
Description: OMPA is a dark brown viscous liquid. Molecular weight = 286.30; Boiling point = 120−125°C at 0.5 mmHg; Freezing/Melting point = 14−20°C. Hazard Identification (based on NFPA-704 M Rating System): Health 3, Flammability 1, Reactivity 0. Soluble in water.
Potential Exposure: Material is used as a systemic insecticide for plants and as an acaricide. Not registered as a pesticide in the United States.
Incompatibilities: Acids.
Permissible Exposure Limits in Air
Protective Action Criteria (PAC)
TEEL-0: 0.15 mg/m^3
PAC-1: 0.5 mg/m^3
PAC-2: 0.8 mg/m^3
PAC-3: 3.5 mg/m^3
Russia[35, 43] set a MAC of 0.02 mg/m^3 in work-place air and a MAC of 0.002 mg/m^3 for ambient air in residential areas on a momentary basis and 0.0004 mg/m^3 on an average daily basis.
Routes of Entry: Inhalation, ingestion, skin and/or eye contact.
Harmful Effects and Symptoms
Short Term Exposure: Acute exposure to OMPA may produce the following signs and symptoms: pinpoint pupils, blurred vision, headache, dizziness, muscle spasms, and profound weakness. Vomiting, diarrhea, abdominal pain, seizures, and coma may also occur. The heart rate may decrease following oral exposure or increase following dermal exposure. Hypotension (low blood pressure) and chest pain may be noted. Hypertension (high blood pressure) is not uncommon. Respiratory symptoms include dyspnea, respiratory depression, and respiratory paralysis. Psychosis may occur. This material is extremely toxic; probable oral lethal dose in humans is 5−50 mg/kg, between 7 drops and

1 teaspoonful for a 150-lb person. It is highly toxic when inhaled. Material is a cholinesterase inhibitor. It is similar in action to other organophosphorus pesticides in its toxicity. It is slightly less toxic than parathion. Gastrointestinal, neurologic, and respiratory symptoms may accompany poisoning with this material. High doses may cause a toxic psychosis similar to acute alcoholism.

Note: Persons taking the following drugs may be at greater risk: Phenobarbital and phenaglycodol together; glutethimide, chlorpromazine hydrochloride; or mepromabate. These drugs appear to enhance the toxicity of the material markedly.

Long Term Exposure: A cholinesterase inhibitor; cumulative effect is possible. This chemical may damage the nervous system with repeated exposure, resulting in convulsions, respiratory failure. May cause liver damage.

Points of Attack: Respiratory system, lungs, central nervous system, cardiovascular system, skin, eyes, plasma and red blood cell cholinesterase.

Medical Surveillance: Before employment and at regular times after that, the following are recommended: plasma and red blood cell cholinesterase levels (tests for the enzyme poisoned by this chemical). If exposure stops, plasma levels return to normal in 1−2 weeks while red blood cell levels may be reduced for 1−3 months. When cholinesterase enzyme levels are reduced by 25% or more below preemployment levels, risk of poisoning is increased, even if results are in lower ranges of "normal." Reassignment to work not involving organophosphate or carbamate pesticides is recommended until enzyme levels recover. If symptoms develop or overexposure occurs, repeat the above tests as soon as possible and get an examination of the nervous system. Also, consider complete blood count. Consider chest X-ray following acute overexposure. Do not drink any alcoholic beverages before or during use. Alcohol promotes absorption of organic phosphates.

First Aid: If this chemical gets into the eyes, remove any contact lenses at once and irrigate immediately for at least 15 min, occasionally lifting upper and lower lids. Seek medical attention immediately. If this chemical contacts the skin, remove contaminated clothing and wash immediately with soap and water. Speed in removing material from skin is of extreme importance. Shampoo hair promptly if contaminated. Seek medical attention immediately. If this chemical has been inhaled, remove from exposure, begin rescue breathing (using universal precautions, including resuscitation mask) if breathing has stopped and CPR if heart action has stopped. Transfer promptly to a medical facility. When this chemical has been swallowed, get medical attention. Give large quantities of water and induce vomiting. Do not make an unconscious person vomit. Obtain authorization and/or further instructions from the local hospital for administration of an antidote or performance of other invasive procedures.

Note to physician: 1,1′-trimethylenebis(4-formylpyridinium bromide)dioxime (a.k.a TMB-4 dibromide and TMV-4) has been used as an antidote for organophosphate poisoning.

Personal Protective Methods: Wear protective gloves and clothing to prevent any reasonable probability of skin contact. Safety equipment suppliers/manufacturers can provide recommendations on the most protective glove/clothing material for your operation. All protective clothing (suits, gloves, footwear, headgear) should be clean, available each day, and put on before work. Contact lenses should not be worn when working with this chemical. Wear splash-proof chemical goggles and face shield unless full face-piece respiratory protection is worn. Employees should wash immediately with soap when skin is wet or contaminated. Provide emergency showers and eyewash; pressure, pressure-demand, full face-piece self-contained breathing apparatus (SCBA) or pressure-demand supplied-air respirator with escape SCBA and a fully-encapsulating, chemical-resistant suit.

Respirator Selection: Follow regulations in OSHA 29CFR1910.134 or European Standard EN149. Use a NIOSH/MSHA- or European Standard EN149-approved respirator; or use an approved supplied-air respirator with a full face-piece operated in the positive-pressure mode, or with a full face-piece, hood, or helmet in the continuous-flow mode; or use a NIOSH/MSHA- or European Standard EN149-approved self-contained breathing apparatus with a full face-piece operated in a pressure-demand or other positive-pressure mode.

Storage: Color Code—Blue: Health Hazard/Poison: Store in a secure poison location. Prior to working with this chemical you should be trained on its proper handling and storage. Store in tightly closed containers in a cool, well-ventilated area away from acids. Where possible, automatically transfer from drums or other storage containers to process containers.

Shipping: This chemical is technically Organophosphorus pesticides, liquid, toxic; its effects are very similar and requires a shipping label of "POISONOUS/TOXIC MATERIALS." This falls in DOT Hazard Class 6.1 and Packing Group I.

Spill Handling: Use water spray to knock down vapors. Attempt to stop leak if it can be done without hazard. Avoid breathing vapors. Keep upwind. Avoid bodily contact with material. Do not handle broken packages without protective equipment. Wash away any material which may have contacted the body with copious amounts of water or soap and water. Ventilate area of spill or leak. Absorb liquids in vermiculite, dry sand, earth, peat, carbon, or a similar material and deposit in sealed containers. Keep this chemical out of a confined space, such as a sewer, because of the possibility of an explosion, unless the sewer is designed to prevent the buildup of explosive concentrations. It may be necessary to contain and dispose of this chemical as a hazardous waste. If material or contaminated runoff enters waterways, notify downstream users of potentially contaminated waters. Contact your local or federal environmental protection agency for specific recommendations. If employees are required to clean up spills, they must be properly trained and equipped. OSHA 1910.120(q) may be applicable.

Fire Extinguishing: This chemical is a noncombustible liquid. Poisonous gases, including nitrogen oxides and potassium oxides, are produced in fire. Do not extinguish fire unless flow can be stopped. Use water in flooding quantities as fog. Solid streams of water may be ineffective. Cool all affected containers with flooding quantities of water. Apply water from as far a distance as possible. Use alcohol foam, carbon dioxide, or dry chemical. Vapors are heavier than air and will collect in low areas. Containers may explode in fire. Storage containers and parts of containers may rocket great distances, in many directions. If material or contaminated runoff enters waterways, notify downstream users of potentially contaminated waters. Notify local health and fire officials and pollution control agencies. From a secure, explosion-proof location, use water spray to cool exposed containers. If cooling streams are ineffective (venting sound increases in volume and pitch, tank discolors, or shows any signs of deforming), withdraw immediately to a secure position. If employees are expected to fight fires, they must be trained and equipped in OSHA 1910.156. The only respirators recommended for firefighting are self-contained breathing apparatuses that have full face-pieces and are operated in a pressure-demand or other positive-pressure mode.

Disposal Method Suggested: In accordance with 40CFR165, follow recommendations for the disposal of pesticides and pesticide containers. Must be disposed properly by following package label directions or by contacting your local or federal environmental control agency or by contacting your regional EPA office. Consult with environmental regulatory agencies for guidance on acceptable disposal practices. Generators of waste containing this contaminant ($\geq$100 kg/mo) must conform with EPA regulations governing storage, transportation, treatment, and waste disposal.

Reference

US Environmental Protection Agency. (November 30, 1987). *Chemical Hazard Information Profile: Octamethyl Phosphoramide.* Washington, DC: Chemical Emergency Preparedness Program

Octane   O:0120

Molecular Formula: C_8H_{18}
Synonyms: n-Octane; normal-Octane
CAS Registry Number: 111-65-9 (n-); 540-84-1 (iso-)
RTECS® Number: RG8400000
UN/NA & ERG Number: UN1262/128
EC Number: (n-) 203-892-1 [*Annex I Index No.:* 601-009-00-8]; (iso-) 208-759-1[*Annex I Index No.:* 601-009-00-8]
Regulatory Authority and Advisory Bodies
Carcinogenicity: EPA (*isooctane; oral*): Inadequate Information to assess carcinogenic potential.

Air Pollutant Standard Set. See below, "Permissible Exposure Limits in Air" section.
Canada, WHMIS, Ingredients Disclosure List Concentration 1.0%.
European/International Regulations (n- and *iso*-): Hazard Symbol: F, Xn, N; Risk phrases: R11; R38; R65; R67; R50/53; Safety phrases: S2; S9; S16; S29; S33; S60; S61; S62 (see Appendix 4).
WGK (German Aquatic Hazard Class): 2—Water polluting (*octane and isomers*).

Description: Octane is a colorless liquid with a gasoline-like odor. The odor threshold is 4 ppm[41] and 48 ppm (New Jersey Fact Sheet). Molecular weight = 114.26; Specific gravity (H_2O:1) = 0.70; Boiling point = 125.6°C; Freezing/Melting point = −56.7°C; Vapor pressure = 10 mmHg at 20°C; Flash point = 13°C (cc); Autoignition temperature = 206°C. Explosive limits: LEL = 1.0%; UEL = 6.5%. Hazard Identification (based on NFPA-704 M Rating System): Health 0, Flammability 3, Reactivity 0. Practically insoluble in water; solubility = 7×10^{-5}.

Potential Exposure: Octane is used as a solvent; as a fuel; as an intermediate in organic synthesis; and in azeotropic distillations.

Incompatibilities: Reacts with strong oxidizers, causing fire and explosion hazard. Attacks some forms of plastics, rubber, and coatings.

Permissible Exposure Limits in Air
Conversion factor: 1 ppm = 4.67 mg/m^3 at 25°C & 1 atm.
OSHA PEL (*n-octane only*): 500 ppm/2350 mg/m^3 TWA; *Construction Industry:* 400 ppm/1900 mg/m^3 TWA.
NIOSH REL (*n-octane only*): 75 ppm/350 mg/m^3 TWA; 385 ppm/1800 mg/m^3 [15-min] Ceiling Concentration.
ACGIH TLV®[1] (*all isomers*): 300 ppm/1401 mg/m^3 TWA.
NIOSH IDLH: 1000 ppm.
Protective Action Criteria (PAC)
n-isomer
TEEL-0: 300 ppm
PAC-1: 300 ppm
PAC-2: 385 ppm
PAC-3: 1000 ppm
iso-isomer
TEEL-0: 300 ppm
PAC-1: 300 ppm
PAC-2: 300 ppm
PAC-3: 1000 ppm
DFG MAK (*n-isomer*): 500 ppm/2400 mg/m^3 TWA; Peak Limitation Category II(2); Pregnancy Risk Group D.
Australia: TWA 300 ppm (1450 mg/m^3); STEL 375 ppm, 1993; Austria: MAK 300 ppm (1400 mg/m^3), 1999; Belgium: TWA 300 ppm (1400 mg/m^3); STEL 375 ppm (1750 mg/m^3), 1993; Denmark: TWA 200 ppm (935 mg/m^3), 1999; Finland: TWA 300 ppm (1400 mg/m^3); STEL 375 ppm (1750 mg/m^3), 1999; France: VME 300 ppm (1450 mg/m^3), 1999; Japan: 300 ppm (1400 mg/m^3), 1999; the Netherlands: MAC-TGG 1450 mg/m^3, 2003; Norway: TWA 150 ppm (725 mg/m^3), 1999; the Philippines: TWA 500 ppm

(2350 mg/m^3), 1993; Poland: MAC (TWA) 1000 mg/m^3; MAC (STEL) 1800 mg/m^3, 1999; Russia: TWA 300 ppm, 1993; Sweden: NGV 200 ppm (900 mg/m^3), KTV 300 ppm (1400 mg/m^3), 1999; Switzerland: MAK-W 300 ppm (1400 mg/m^3), KZG-W 600 ppm (2800 mg/m^3), 1999; Turkey: TWA 400 ppm (1900 mg/m^3), 1993; United Kingdom: LTEL 300 ppm (1450 mg/m^3); STEL 375 ppm, 1993; Argentina, Bulgaria, Columbia, Jordan, South Korea, New Zealand, Singapore, Vietnam: ACGIH TLV®: TWA 300 ppm. Several states have set guidelines or standards for octane in ambient air[60] ranging from 7.0 mg/m^3 (Connecticut) to 14.5–18.0 mg/m^3 (North Dakota) to 24.0 mg/m^3 (Virginia) to 34.524 mg/m^3 (Nevada).

Determination in Air: Use NIOSH Analytical Method (IV) #1500, for Hydrocarbons, BP 36-126°C; #2549, Volatile organic compounds; OSHA Analytical Method 7.

Determination in Water: Octanol-water coefficient: Log K_{ow} = 4.5.

Routes of Entry: Inhalation, ingestion, skin and/or eye contact.

Harmful Effects and Symptoms

Short Term Exposure: Octane can affect you when breathed in. Irritates the eyes, skin, and respiratory tract. Skin contact can cause rash and a burning sensation. Swallowing the liquid may cause aspiration into the lungs and chemical pneumonitis. Exposure to high concentrations of vapor can cause lightheadedness, dizziness, confusion, and may cause you to pass out.

Long Term Exposure: Repeated or prolonged contact can result in dry, cracked skin.

Points of Attack: Eyes, skin, respiratory system, central nervous system.

Medical Surveillance: Consider the points of attack in pre-placement and periodic physical examinations.

First Aid: If this chemical gets into the eyes, remove any contact lenses at once and irrigate immediately for at least 15 min, occasionally lifting upper and lower lids. Seek medical attention immediately. If this chemical contacts the skin, remove contaminated clothing and wash immediately with soap and water. Seek medical attention immediately. If this chemical has been inhaled, remove from exposure, begin rescue breathing (using universal precautions, including resuscitation mask) if breathing has stopped and CPR if heart action has stopped. Transfer promptly to a medical facility. When this chemical has been swallowed, get medical attention. Do NOT induce vomiting. Give victim nothing to drink.

Note to physician: Inhalation: bronchodilators, decongestants, and oxygen may be used if necessary. Corticosteroids are useful for treating pneumonitis.

Personal Protective Methods: Wear protective gloves and clothing to prevent any reasonable probability of skin contact. Safety equipment suppliers/manufacturers can provide recommendations on the most protective glove/clothing material for your operation. Neoprene™, Nitrile/PVC, and Nitrile are recommended. All protective clothing (suits, gloves, footwear, headgear) should be clean, available each day, and put on before work. Contact lenses should not be worn when working with this chemical. Wear splash-proof chemical goggles and face shield unless full face-piece respiratory protection is worn. Employees should wash immediately with soap when skin is wet or contaminated. Provide emergency showers and eyewash.

Respirator Selection: NIOSH: *Up to 750 ppm:* Sa* (APF = 10) (any supplied-air respirator). *Up to 1000 ppm:* Sa:Cf* (APF = 25) (any supplied-air respirator operated in a continuous-flow mode); or SCBAF (APF = 50) (any self-contained breathing apparatus with a full face-piece) or SaF (APF = 50) (any supplied-air respirator with a full face-piece). *Emergency or planned entry into unknown concentrations or IDLH conditions:* SCBAF: Pd,Pp (APF = 10,000) (any self-contained breathing apparatus that has a full face-piece and is operated in a pressure-demand or other positive-pressure mode) or SaF: Pd,Pp: ASCBA (APF = 10,000) (any supplied-air respirator that has a full face-piece and is operated in a pressure-demand or other positive-pressure mode in combination with an auxiliary self-contained breathing apparatus operated in a pressure-demand or other positive-pressure mode). *Escape:* GmFOv (APF = 50) [any air-purifying, full-face-piece respirator (gas mask) with a chin-style, front- or back-mounted organic vapor canister] or SCBAE (any appropriate escape-type, self-contained breathing apparatus).

*Substance reported to cause eye irritation or damage; may require eye protection.

Storage: Color Code—Red: Flammability Hazard: Store in a flammable liquid storage area or approved cabinet away from ignition sources and corrosive and reactive materials. Prior to working with this chemical you should be trained on its proper handling and storage. Before entering confined space where this chemical may be present, check to make sure that an explosive concentration does not exist. Octane must be stored to avoid contact with strong oxidizers (such as chlorine and bromine) because violent reactions occur. Store in tightly closed containers in a cool, well-ventilated area away from heat. Sources of ignition, such as smoking and open flames, are prohibited where octane is used, handled, or stored. Metal containers used in the transfer of 5 gallons or more of octane should be grounded and bonded. Drums must be equipped with self-closing valves, pressure vacuum bungs, and flame arresters. Use only nonsparking tools and equipment, especially when opening and closing containers of octane.

Shipping: Octanes require a shipping label of "FLAMMABLE LIQUID." It falls in DOT Hazard Class 3 and Packing Group II.

Spill Handling: Evacuate and restrict persons not wearing protective equipment from area of spill or leak until cleanup is complete. Remove all ignition sources. Establish forced ventilation to keep levels below explosive limit. Absorb liquids in vermiculite, dry sand, earth, peat, carbon, or a similar material and deposit in sealed containers. Keep this chemical out of a confined space, such as a sewer, because of the possibility of an explosion, unless the sewer is designed to prevent the buildup of explosive concentrations. It may be necessary to contain and dispose of this chemical as a hazardous waste. If material or contaminated runoff enters waterways, notify downstream users of potentially contaminated waters. Contact your local or federal environmental protection agency for specific recommendations. If employees are required to clean up spills, they must be properly trained and equipped. OSHA 1910.120(q) may be applicable.

Fire Extinguishing: This chemical is a flammable liquid. Poisonous gases, including carbon monoxide, are produced in fire. Use dry chemical, carbon dioxide, or foam extinguishers. Vapors are heavier than air and will collect in low areas. Vapors may travel long distances to ignition sources and flashback. Vapors in confined areas may explode when exposed to fire. Containers may explode in fire. Storage containers and parts of containers may rocket great distances, in many directions. If material or contaminated runoff enters waterways, notify downstream users of potentially contaminated waters. Notify local health and fire officials and pollution control agencies. From a secure, explosion-proof location, use water spray to cool exposed containers. If cooling streams are ineffective (venting sound increases in volume and pitch, tank discolors, or shows any signs of deforming), withdraw immediately to a secure position. If employees are expected to fight fires, they must be trained and equipped in OSHA 1910.156. The only respirators recommended for firefighting are self-contained breathing apparatuses that have full face-pieces and are operated in a pressure-demand or other positive-pressure mode.

Disposal Method Suggested: Dissolve or mix the material with a combustible solvent and burn in a chemical incinerator equipped with an afterburner and scrubber. All federal, state, and local environmental regulations must be observed.

References

National Institute for Occupational Safety and Health. (1977). *Criteria for a Recommended Standard: Occupational Exposure to Alkanes*, NIOSH Document No. 77-151. Washington, DC

New Jersey Department of Health and Senior Services. (January 2000). *Hazardous Substances Fact Sheet: Octane.* Trenton, NJ

Octyl phenol

O:0130

Molecular Formula: $C_{14}H_{22}O$
Common Formula: $C_8H_{17}C_6H_4OH$
Synonyms: Diisobutyl phenol
CAS Registry Number: 1322-69-6; 27193-28-8
RTECS® Number: SM5775000
EC Number: 248-310-7

Regulatory Authority and Advisory Bodies
FDA: 21CFR§175.105
WGK (German Aquatic Hazard Class): No value assigned.
Description: Octyl phenol is a white to pink crystalline solid. Molecular weight = 206.36; Boiling point = 280−302°C; Freezing/Melting point = 72−74°C; Flash point = about 149°C. Insoluble in water.
Potential Exposure: As a fuel oil stabilizer; as an intermediate for resins; in fungicides, bactericides, dyestuffs, adhesives, antioxidants, nonionic surfactants; in plasticizers and rubber; in adhesives; and in food packaging.
Incompatibilities: Oxidizers, such as chlorine and dichromate.
Permissible Exposure Limits in Air
No standards or TEEL available.
Routes of Entry: Inhalation, ingestion, skin and/or eye contact.
Harmful Effects and Symptoms
Short Term Exposure: Inhalation: May cause irritation to lungs and throat. *Skin:* May cause irritation. *Eyes:* May cause severe irritation. *Ingestion:* Moderately toxic; may cause digestive upset. LD_{50} = (ip-mouse) 25 mg/kg.
Long Term Exposure: May cause liver and/or kidney damage. May be a mutagenic.
Points of Attack: Liver, kidneys.
Medical Surveillance: Liver and kidney function tests.
First Aid: If this chemical gets into the eyes, remove any contact lenses at once and irrigate immediately for at least 15 min, occasionally lifting upper and lower lids. Seek medical attention immediately. If this chemical contacts the skin, remove contaminated clothing and wash immediately with soap and water. Seek medical attention immediately. If this chemical has been inhaled, remove from exposure, begin rescue breathing (using universal precautions, including resuscitation mask) if breathing has stopped and CPR if heart action has stopped. Transfer promptly to a medical facility. When this chemical has been swallowed, get medical attention. Give large quantities of water and induce vomiting. Do not make an unconscious person vomit.
Personal Protective Methods: Wear protective gloves and clothing to prevent any reasonable probability of skin contact. Safety equipment suppliers/manufacturers can provide recommendations on the most protective glove/clothing material for your operation. All protective clothing (suits, gloves, footwear, headgear) should be clean, available each day, and put on before work. Contact lenses should not be worn when working with this chemical. Wear splash-proof chemical goggles and face shield unless full face-piece respiratory protection is worn. Employees should wash immediately with soap when skin is wet or contaminated. Provide emergency showers and eyewash.
Respirator Selection: Follow regulations in OSHA 29CFR1910.134 or European Standard EN149. Use a NIOSH/MSHA- or European Standard EN149-approved respirator; or use an approved supplied-air respirator with a

full face-piece operated in the positive-pressure mode, or with a full face-piece, hood, or helmet in the continuous-flow mode; or use a NIOSH/MSHA- or European Standard EN149-approved self-contained breathing apparatus with a full face-piece operated in a pressure-demand or other positive-pressure mode.

Storage: Color Code—Blue: Health Hazard/Poison: Store in a secure poison location. Prior to working with this chemical you should be trained on its proper handling and storage.

Spill Handling: Evacuate and restrict persons not wearing protective equipment from area of spill or leak until cleanup is complete. Remove all ignition sources. Ventilate area of spill or leak. Absorb liquids in vermiculite, dry sand, earth, peat, carbon, or a similar material and deposit in sealed containers. Keep this chemical out of a confined space, such as a sewer, because of the possibility of an explosion, unless the sewer is designed to prevent the buildup of explosive concentrations. It may be necessary to contain and dispose of this chemical as a hazardous waste. If material or contaminated runoff enters waterways, notify downstream users of potentially contaminated waters. Contact your local or federal environmental protection agency for specific recommendations. If employees are required to clean up spills, they must be properly trained and equipped. OSHA 1910.120(q) may be applicable.

Fire Extinguishing: This chemical is a combustible liquid. Will ignite at about 149°C. Poisonous gases are produced in fire. Use dry chemical, carbon dioxide, or alcohol foam extinguishers. Vapors are heavier than air and will collect in low areas. Vapors may travel long distances to ignition sources and flashback. Vapors in confined areas may explode when exposed to fire. Containers may explode in fire. Storage containers and parts of containers may rocket great distances, in many directions. If material or contaminated runoff enters waterways, notify downstream users of potentially contaminated waters. Notify local health and fire officials and pollution control agencies. From a secure, explosion-proof location, use water spray to cool exposed containers. If cooling streams are ineffective (venting sound increases in volume and pitch, tank discolors, or shows any signs of deforming), withdraw immediately to a secure position. If employees are expected to fight fires, they must be trained and equipped in OSHA 1910.156. The only respirators recommended for firefighting are self-contained breathing apparatuses that have full face-pieces and are operated in a pressure-demand or other positive-pressure mode.

Disposal Method Suggested: Dissolve or mix the material with a combustible solvent and burn in a chemical incinerator equipped with an afterburner and scrubber. All federal, state, and local environmental regulations must be observed.

Reference
New York State Department of Health. (April 1986). *Chemical Fact Sheet: Octyl Phenol.* Albany, NY: Bureau of Toxic Substance Assessment

Osmium & Osmium tetroxide   O:0140

Molecular Formula: Os; O_4Os
Synonyms: Milas' reagent; Osmic acid anhydride; Osmium oxide (OsO_4); Osmium(IV) oxide; Tetroxido de osmio (Spanish)
CAS Registry Number: 7440-04-2 (elemental); 20816-12-0 (tetroxide)
RTECS® Number: RN1100000 (elemental); RN1140000 (tetroxide)
UN/NA & ERG Number: UN2471 (tetroxide)/154
EC Number: 231-114-0; 244-058-7 [*Annex I Index No.:* 076-001-00-5] (tetroxide)
Regulatory Authority and Advisory Bodies
US EPA Gene-Tox Program (*tetroxide*), Positive: *B. subtilis* rec assay; Inconclusive: *D. melanogaster* sex-linked lethal.
Air Pollutant Standard Set. See below, "Permissible Exposure Limits in Air" section.
OSHA 29CFR1910.119, Appendix A, Process Safety List of Highly Hazardous Chemicals, TQ = 100 lb (45 kg).
Superfund/EPCRA 40CFR355, Extremely Hazardous Substances: Dropped From Listing In 1988.
US EPA Hazardous Waste Number (RCRA No.): P087.
RCRA, 40CFR261, Appendix 8 Hazardous Constituents.
Reportable Quantity (RQ): 1000 lb (454 kg).
EPCRA Section 313 Form R *de minimis* concentration reporting level: 1.0%.
Canada, WHMIS, Ingredients Disclosure List Concentration 0.1% (tetroxide).
European/International Regulations (*tetroxide*): Hazard Symbol: T+, N; Risk phrases: R26/27/28; R34; Safety phrases: S1/2; S7/9; S26; S45 (see Appendix 4).
WGK (German Aquatic Hazard Class): Nonwater polluting agent. (metal); no value assigned to the tetroxide.

Description: Osmium is a blue-white metal. It is found in platinum ores and in the naturally occurring alloy osmiridium. Osmium when heated in air or when the finely divided form is exposed to air at room temperature, oxidizes to form the tetroxide (OsO_4), osmic acid.
Osmium tetraoxide is a colorless, crystalline solid or pale-yellow mass with an unpleasant, acrid, chlorine-like odor. A liquid above 41°C. Odor threshold = 0.0019 ppm. Molecular weight = 190.20; 254.20 (OsO_4); Specific gravity (H_2O:1) = 5.10 (OsO_4); Boiling point = 130°C (sublimes well below the BP) (OsO_4); Freezing/Melting point = 41°C (OsO_4); Vapor pressure = 7 mmHg at 20°C (OsO_4). Slightly soluble in water; solubility = 6% at 25°C (OsO_4). Hazard Identification (based on NFPA-704 M Rating System): Health 4, Flammability 0, Reactivity 3 (Oxidizer).
Potential Exposure: Compound Description (tetroxide): Drug, Mutagen; Reproductive Effector; Human Data. Osmium may be alloyed with platinum metals, iron, cobalt, and nickel, and it forms compounds with tin and zinc. The alloy with iridium is used in the manufacture of fountain pen points, engraving tool, record player needles, electrical

contacts, compass needles, fine machine bearings, and parts for watch and lock mechanisms. The metal is a catalyst in the synthesis of ammonia and in the dehydrogenation of organic compounds. It is also used as a stain for histological examination of tissues. Osmium tetroxide is used as an oxidizing agent, catalyst, and as a fixative for tissues in electron microscopy. Other osmium compounds find use in photography. Osmium no longer is used in incandescent lights or in fingerprinting.

Incompatibilities: Osmium tetroxide is a strong oxidizer. Reacts with combustibles and reducing materials. Reacts with hydrochloric acid to form toxic chlorine gas. Forms unstable compounds with alkalis.

Permissible Exposure Limits in Air

Conversion factor: 1 ppm = 10.40 mg/m^3 (tetroxide) at 25°C & 1 atm.

OSHA PEL: 0.002 mg[Os]/m^3 TWA.

NIOSH REL: 0.0002 ppm/0.002 mg[Os]/m^3 TWA; 0.006 mg[Os]/m^3/0.0006 ppm STEL.

ACGIH TLV®[1]: 0.0002 ppm/0.0016 mg[Os]/m^3 TWA; 0.0047 mg[Os]/m^3/0.0006 ppm STEL.

NIOSH IDLH: 1 mg[Os]/m^3.

Protective Action Criteria (PAC)

TEEL-0: 0.25 mg/m^3

PAC-1: 0.75 mg/m^3

PAC-2: 6 mg/m^3

PAC-3: 30 mg/m^3

Tetroxide

TEEL-0: 0.0002 ppm

PAC-1: 0.0006 ppm

PAC-2: 0.0084 ppm

PAC-3: 4 ppm

Arab Republic of Egypt: TWA 0.0002 ppm (0.002 mg[Os]/m^3), 1993; Australia: TWA 0.0002 ppm (0.002 mg[Os]/m^3); STEL 0.0006 ppm, 1993; Austria: MAK 0.0002 ppm (0.002 mg[Os]/m^3), 1999; Belgium: TWA 0.0002 ppm (0.0016 mg[Os]/m^3); STEL 0.0006 ppm, 1993; Denmark: TWA 0.0002 ppm (0.002 mg[Os]/m^3), 1999; Finland: TWA 0.0002 mg[Os]/m^3; STEL 0.002 mg[Os]/m^3, 1999; France: VME 0.0002 ppm (0.002 mg[Os]/m^3), 1999; the Netherlands: MAC-TGG 0.002 mg[Os]/m^3, 2003; Norway: TWA 0.0002 ppm (0.002 mg[Os]/m^3), 1999; the Philippines: TWA 0.002 mg[Os]/m^3, 1993; Switzerland: MAK-W 0.0002 ppm (0.002 mg[Os]/m^3), KZG-W 0.0004 ppm, 1999; United Kingdom: TWA 0.0002 ppm (0.002 mg[Os]/m^3); STEL 0.0006 ppm, 2000; Argentina, Bulgaria, Columbia, Jordan, South Korea, New Zealand, Singapore, Vietnam: ACGIH TLV®: STEL 0.0006 (Os) ppm. There is presently no PEL for elemental, metallic osmium. Several states have set guidelines or standards for osmium tetroxide in ambient air[60] ranging from zero in North Dakota to 0.04 μg/m^3 (Connecticut) to 1.0 μg/m^3 (Nevada) to 3000 μg/m^3 (Virginia).

Determination in Air: No NIOSH Analytical Method available.

Determination in Water: Osmium tetraoxide may be hazardous to the environment; crustacea, in particular, may be at risk.

Routes of Entry: Inhalation, ingestion, skin and/or eye contact.

Harmful Effects and Symptoms

Short Term Exposure: Osmium metal is innocuous, but persons engaged in the production of the metal may be exposed to acids and chlorine vapors. By contrast, osmium tetroxide vapors are poisonous and extremely irritating to the eyes; even in low concentrations they may cause weeping and persistent conjunctivitis. Longer exposure can cause redness, swelling of the eye tissue, blurred vision, and may result in damage to the cornea and permanent loss of vision. Contact with skin may cause discoloration (green or black), dermatitis and ulceration. *Inhalation*: fumes are extremely irritating to the respiratory system, causing tracheitis, bronchitis, bronchial spasm, and difficulty in breathing which may last several hours. Longer exposures can cause serious inflammatory lesions of the lungs (bronchopneumonia with suppuration and gangrene). Higher exposures can cause pulmonary edema, a medical emergency that can be delayed for several hours. This can cause death. Exposure to high concentrations may result in death. Slight kidney damage was seen in rabbits inhaling lethal concentrations of vapor for 30 min. Some fatty degeneration of renal tubules was seen in one fatal human case along with bronchopneumonia following an accidental overexposure.

Long Term Exposure: Repeated or prolonged contact with skin may cause dermatitis. The substance may have effects on the kidneys. Repeated exposure may cause lung irritation; bronchitis may develop. There is limited evidence that osmium tetroxide may cause mutations.

Points of Attack: Eyes, respiratory system, lungs, skin, kidneys.

Medical Surveillance: Consider the skin, eyes, respiratory tract, and renal function in preplacement or periodic examinations. Lung function tests. Complete eye examination. Urinalysis. Chest X-ray.

First Aid: If this chemical gets into the eyes, remove any contact lenses at once and irrigate immediately for at least 15 min, occasionally lifting upper and lower lids. Seek medical attention immediately. If this chemical contacts the skin, remove contaminated clothing and wash immediately with soap and water. Seek medical attention immediately. If this chemical has been inhaled, remove from exposure, begin rescue breathing (using universal precautions, including resuscitation mask) if breathing has stopped and CPR if heart action has stopped. Transfer promptly to a medical facility. When this chemical has been swallowed, get medical attention. Give nothing to drink. Do not make an unconscious person vomit. Medical observation is recommended for 24–48 h after breathing overexposure, as pulmonary edema may be delayed. As first aid for pulmonary edema, a doctor or authorized paramedic may consider administering a corticosteroid spray.

Personal Protective Methods: Avoid all contact. Wear protective gloves and clothing to prevent any reasonable probability of skin contact. Safety equipment suppliers/manufacturers can provide recommendations on the most protective glove/clothing material for your operation. All protective clothing (suits, gloves, footwear, headgear) should be clean, available each day, and put on before work. Contact lenses should not be worn when working with this chemical. Wear dust-proof chemical goggles and face shield unless full face-piece respiratory protection is worn. Employees should wash immediately with soap when skin is wet or contaminated. Provide emergency showers and eyewash.

Respirator Selection: Up to 0.1 mg/m³: CcrFS100 (APF = 50) [Any air-purifying full-face-piece respirator equipped with cartridge(s) providing protection against the compound of concern in combination with an N100, R100, or P100 filter] or GmFS100 (APF = 50) [any air-purifying, full-face-piece respirator (gas mask) with a chin-style, front- or back-mounted canister providing protection against the compound of concern and having an N100, R100, or P100 filter] or SCBAF (APF = 50) (any self-contained breathing apparatus with a full face-piece); SaF (APF = 50) (any supplied-air respirator with a full face-piece). *Up to 1 mg/m³:* SaF: Pd,Pp (APF = 2000) (any supplied-air respirator that has a full face-piece and is operated in a pressure-demand or other positive-pressure mode). *Emergency or planned entry into unknown concentrations or IDLH conditions:* SCBAF: Pd,Pp (APF = 10,000) (any self-contained breathing apparatus that has a full face-piece and is operated in a pressure-demand or other positive-pressure mode) or SaF: Pd,Pp: ASCBA (APF = 10,000) (any supplied-air respirator that has a full face-piece and is operated in a pressure-demand or other positive-pressure mode in combination with an auxiliary self-contained breathing apparatus operated in a pressure-demand or other positive-pressure mode). *Escape:* GmFS100 (APF = 50) [any air-purifying, full-face-piece respirator (gas mask) with a chin-style, front- or back-mounted canister providing protection against the compound of concern and having an N100, R100, or P100 filter] or SCBAE (any appropriate escape-type, self-contained breathing apparatus).

Storage: Color Code—Blue: Health Hazard/Poison (*osmium tetroxide*): Store in a secure poison location. Prior to working with this chemical you should be trained on its proper handling and storage. Store refrigerated in tightly closed containers away from hydrochloric acid, reducing agents and easily oxidized materials. Where possible, automatically pump liquid from drums or other storage containers to process containers. Sources of ignition, such as smoking and open flames, are prohibited where this chemical is handled, used, or stored. Metal containers involving the transfer of 5 gallons or more of this chemical should be grounded and bonded. Drums must be equipped with self-closing valves, pressure vacuum bungs, and flame arresters.

Use only nonsparking tools and equipment, especially when opening and closing containers of this chemical. Wherever this chemical is used, handled, manufactured, or stored, use explosion-proof electrical equipment and fittings. See OSHA Standard 1910.104 and NFPA 43A *Code for the Storage of Liquid and Solid Oxidizers* for detailed handling and storage regulations.

Shipping: Osmium tetroxide requires a shipping label of "POISONOUS/TOXIC MATERIALS." It falls in Hazard Class 6.1 and Packing Group I.

Spill Handling: Evacuate persons not wearing protective equipment from area of spill or leak until cleanup is complete. Remove all ignition sources. Collect powdered material in the most convenient and safe manner and deposit in sealed containers. Ventilate area after cleanup is complete. It may be necessary to contain and dispose of this chemical as a hazardous waste. If material or contaminated runoff enters waterways, notify downstream users of potentially contaminated waters. Contact your local or federal environmental protection agency for specific recommendations. If employees are required to clean up spills, they must be properly trained and equipped. OSHA 1910.120(q) may be applicable.

Fire Extinguishing: Osmium tetraoxide may burn but does not easily ignite. *Small fires:* dry chemical, carbon dioxide, water spray, or foam extinguishers. *Large fires:* water spray, fog, or foam. Move container from fire area if you can do it without risk. Fight fire from maximum distance. Dike fire control water for later disposal; do not scatter the material. Contact with easily oxidized organic materials may cause fires and explosions. Poisonous gases, including osmium, are produced in fire. If material or contaminated runoff enters waterways, notify downstream users of potentially contaminated waters. Notify local health and fire officials and pollution control agencies. Containers may explode in fire. From a secure, explosion-proof location, use water spray to cool exposed containers. If cooling streams are ineffective (venting sound increases in volume and pitch, tank discolors, or shows any signs of deforming), withdraw immediately to a secure position. If employees are expected to fight fires, they must be trained and equipped in OSHA 1910.156. The only respirators recommended for firefighting are self-contained breathing apparatuses that have full face-pieces and are operated in a pressure-demand or other positive-pressure mode.

Disposal Method Suggested: Consult with environmental regulatory agencies for guidance on acceptable disposal practices. Generators of waste containing this contaminant (≥100 kg/mo) must conform with EPA regulations governing storage, transportation, treatment, and waste disposal.

References

US Environmental Protection Agency. (Oct. 31, 1985). *Chemical Hazard Information Profile: Osmium Tetroxide.* Washington, DC: Chemical Emergency Preparedness Program

New Jersey Department of Health and Senior Services. (August 2002). *Hazardous Substances Fact Sheet: Osmium Tetroxide*. Trenton, NJ

Ouabain   O:0150

Molecular Formula: $C_{29}H_{44}O_{12}$
Synonyms: Acocantherin; Astrobain; Gratibain; Gratus strophanthin; G-Strophanthin; Ouabagenin-l-rhamnosid (German); Ouabaine; Oubain; Purostrophan; Quabagenin-l-rhamnoside; Strophanthin G; Strophoperm
CAS Registry Number: 630-60-4; 11018-89-6 (octahydrate)
RTECS® Number: RN3675000
UN/NA & ERG Number: UN1544 (Alkaloids, solid, n.o.s.)/151
EC Number: 211-139-3 [*Annex I Index No.:* 614-025-00-5]
Regulatory Authority and Advisory Bodies
SUPERFUND/EPCRA 40CFR355, Extremely Hazardous Substances: TPQ = 100/10,000 lb (45.4/4540 kg).
Reportable Quantity (RQ): 100 lb (45.4 kg).
Canada: WHMIS, Class D1A; Not on DSL or NDSL lists.
European/International Regulations (*630-60-4*): Hazard Symbol: T; Risk phrases: R23/25; R33; Safety phrases: S1/2; S45 (see Appendix 4).
WGK (German Aquatic Hazard Class): 3—Severe hazard to waters.
Description: Ouabain is a white crystalline solid. Molecular weight = 584.652; Freezing/Melting point = 190−200°C. Slightly soluble in waer.
Potential Exposure: Compound Description: Drug, Mutagen, Natural Product. Ouabain, similar to digitoxin, is used to produce rapid digitalization in acute congestive heart failure. Also recommended in treatment of atrial or nodal paroxysmal tachycardia and atrial flutter; enzyme inhibitor.
Permissible Exposure Limits in Air
Protective Action Criteria (PAC)
TEEL-0: 1.5 mg/m^3
PAC-1: 5 mg/m^3
PAC-2: 8.3 mg/m^3
PAC-3: 12.5 mg/m^3
Routes of Entry: Inhalation, ingestion, skin and/or eye contact.
Harmful Effects and Symptoms
Short Term Exposure: Upon exposure to ouabain, symptoms of heart failure occur, with marked increase in serum potassium. Signs may include confusion, vomiting, coma, convulsions, and respiratory failure. It is classified as extremely toxic. Probable oral lethal dose in humans is less than 5 mg/kg or a taste (less than 7 drops) for a 70-kg (150 lb) person. Exposure may result in respiratory and cardiac failure; and/or hyperalkemia. LD$_{50}$ = (oral-mouse) ≥ 500 mg/kg.
Note: Patients with frequent premature ventricular heartbeat or who have received any preparation of digitalis during preceding 3 weeks are prone to toxicity.

Points of Attack: Heart.
Medical Surveillance: EKG.
First Aid: If this chemical gets into the eyes, remove any contact lenses at once and irrigate immediately for at least 15 min, occasionally lifting upper and lower lids. Seek medical attention immediately. If this chemical contacts the skin, remove contaminated clothing and wash immediately with soap and water. Seek medical attention immediately. If this chemical has been inhaled, remove from exposure, begin rescue breathing (using universal precautions, including resuscitation mask) if breathing has stopped and CPR if heart action has stopped. Transfer promptly to a medical facility. When this chemical has been swallowed, get medical attention. Give large quantities of water and induce vomiting. Do not make an unconscious person vomit.
Personal Protective Methods: For emergency situations, wear protective gloves and clothing to prevent any reasonable probability of skin contact. Safety equipment suppliers/manufacturers can provide recommendations on the most protective glove/clothing material for your operation. All protective clothing (suits, gloves, footwear, headgear) should be clean, available each day, and put on before work. Contact lenses should not be worn when working with this chemical. Wear dust-proof chemical goggles and face shield unless full face-piece respiratory protection is worn. Employees should wash immediately with soap when skin is wet or contaminated. Provide emergency showers and eyewash; pressure, pressure-demand, full face-piece self-contained breathing apparatus (SCBA) or pressure-demand supplied-air respirator with escape SCBA and a fully-encapsulating, chemical-resistant suit.
Respirator Selection: Follow regulations in OSHA 29CFR1910.134 or European Standard EN149. Use a NIOSH/MSHA- or European Standard EN149-approved respirator; or use an approved supplied-air respirator with a full face-piece operated in the positive-pressure mode, or with a full face-piece, hood, or helmet in the continuous-flow mode; or use a NIOSH/MSHA- or European Standard EN149-approved self-contained breathing apparatus with a full face-piece operated in a pressure-demand or other positive-pressure mode.
Storage: Color Code—Blue: Health Hazard/Poison: Store in a secure poison location. Prior to working with this chemical you should be trained on its proper handling and storage.
Shipping: Alkaloids, solid, n.o.s. require a label of "POISONOUS/TOXIC MATERIALS." They fall in Hazard Class 6.1 and ergotamine tartrate in Packing Group I.
Spill Handling: Evacuate persons not wearing protective equipment from area of spill or leak until cleanup is complete. Remove all ignition sources. Keep combustibles (wood, paper, oil, etc.) away from spilled material. Do not touch spilled material. *Small spills:* absorb with sand or other noncombustible absorbent material and place into containers for later disposal. *Small dry spills:* with clean shovel place material into clean, dry container and cover;

move container from spill area. *Large spills*: dike far ahead of spill for later disposal. Ventilate area after cleanup is complete. It may be necessary to contain and dispose of this chemical as a hazardous waste. If material or contaminated runoff enters waterways, notify downstream users of potentially contaminated waters. Contact your local or federal environmental protection agency for specific recommendations. If employees are required to clean up spills, they must be properly trained and equipped. OSHA 1910.120(q) may be applicable.

Fire Extinguishing: For small fires, use dry chemical, carbon dioxide, water spray, or foam. For large fires, use water spray, fog, or foam. Cool containers that are exposed to flames with water from the side until well after fire is out. For massive fires use unmanned hose holder or monitor nozzles; if this is impossible, withdraw and let fire burn. Wear self-contained (positive pressure if available) breathing apparatus and full protective clothing. Poisonous gases are produced in fire. If material or contaminated runoff enters waterways, notify downstream users of potentially contaminated waters. Notify local health and fire officials and pollution control agencies. From a secure, explosion-proof location, use water spray to cool exposed containers. If cooling streams are ineffective (venting sound increases in volume and pitch, tank discolors, or shows any signs of deforming), withdraw immediately to a secure position. If employees are expected to fight fires, they must be trained and equipped in OSHA 1910.156. The only respirators recommended for firefighting are self-contained breathing apparatuses that have full face-pieces and are operated in a pressure-demand or other positive-pressure mode.

Reference

US Environmental Protection Agency. (November 30, 1987). *Chemical Hazard Information Profile: Ouabain*. Washington, DC: Chemical Emergency Preparedness Program

Oxalic acid   O:0160

Molecular Formula: $C_2H_2O_4$; $C_2H_6O_6$
Common Formula: HOOCCOOH · $_2H_2O$
Synonyms: Acide oxalique (French); Ethanedioic acid; NCI-C55209; Oxalic acid dihydrate; Oxalsaeure (German)
CAS Registry Number: 144-62-7
RTECS® Number: RO2450000
UN/NA & ERG Number: UN1759/154
EC Number: 205-634-3 [*Annex I Index No.:* 607-006-00-8]
Regulatory Authority and Advisory Bodies
US EPA, FIFRA 1998 Status of Pesticides: RED completed.
Air Pollutant Standard Set. See below, "Permissible Exposure Limits in Air" section.
Canada, WHMIS, Ingredients Disclosure List Concentration 0.1%.

European/International Regulations: Hazard Symbol: Xn; Risk phrases: R21/22; Safety phrases: S2; S24/25 (see Appendix 4).
WGK (German Aquatic Hazard Class): 1—Low hazard to waters.

Description: Oxalic acid is a colorless, odorless powder or granular solid. The anhydrous form $(COOH)_2$ is an odorless, white solid; the solution is a colorless liquid. Molecular weight = 90.04; Specific gravity (H_2O:1) = 1.90; Sublimation point = 150−157°C; Freezing/Melting point (decomposes): 101.7°C; 190°C (anhydrous). Hazard Identification (based on NFPA-704 M Rating System): Health 3, Flammability 1, Reactivity 0. Moderately soluble in water.

Potential Exposure: Compound Description: Agricultural Chemical; Reproductive Effector; Human Data; Primary Irritant. Oxalic acid is used in textile finishing, paint stripping; metal and equipment cleaning; as an intermediate; as an analytic reagent and in the manufacture of dyes, inks, bleaches, paint removers; varnishes, wood, and metal cleansers; dextrin, cream of tartar, celluloid, oxalates, tartaric acid, purified methyl alcohol, glycerol, and stable hydrogen cyanide. It is also used in the photographic, ceramic, metallurgic, rubber, leather, engraving, pharmaceutical, paper, and lithographic industries.

Incompatibilities: The aqueous solution is a medium strong acid. Incompatible with strong oxidizers, silver compounds, strong alkalis, chlorites. Contact with some silver compounds forms explosive materials.

Permissible Exposure Limits in Air
OSHA PEL: 1 mg/m³ TWA.
NIOSH REL: 1 mg/m³ TWA; 2 mg/m³ STEL.
ACGIH TLV®[1]: 1 mg/m³ TWA; 2 mg/m³ STEL.
NIOSH IDLH: 500 mg/m³.
Protective Action Criteria (PAC)
TEEL-0: 1 mg/m³
PAC-1: 2 mg/m³
PAC-2: 40 mg/m³
PAC-3: 500 mg/m³
Arab Republic of Egypt: TWA 1 mg/m³, 1993; Australia: TWA 1 mg/m³; STEL 2 mg/m³, 1993; Austria: MAK 1 mg/m³, 1999; Denmark: TWA 1 mg/m³, 1999; Finland: TWA 1 mg/m³; STEL 3 mg/m³, 1999; France: VME 1 mg/m³, 1999; the Netherlands: MAC-TGG 1 mg/m³, 2003; Poland: MAC (TWA) 1 mg/m³; MAC (STEL) 2 mg/m³, 1999; Sweden: NGV 1 mg/m³, KTV 2 mg/m³, 1999; Switzerland: MAK-W 1 mg/m³, 1999; United Kingdom: TWA 1 mg/m³; STEL 0.2 mg/m³, 2000; Argentina, Bulgaria, Columbia, Jordan, South Korea, New Zealand, Singapore, Vietnam: ACGIH TLV®: STEL 2 mg/m³. Several states have set guidelines or standards for oxalic acid in ambient air[60] ranging from 3.3 μg/m³ (New York) to 10.0 μg/m³ (Florida and South Carolina) to 10−20 μg/m³ (North Dakota) to 16.0 μg/m³ (Virginia) to 20.0 μg/m³ (Connecticut) to 24.0 μ/m³ (Nevada).

Determination in Air: No NIOSH Analytical Method available.

Determination in Water: Octanol-water coefficient: Log $K_{ow} = <-1$ (estimated).

Routes of Entry: Inhalation, ingestion, skin and/or eye contact.

Harmful Effects and Symptoms

Short Term Exposure: Corrosive to the eyes, skin, and respiratory tract. Corrosive on ingestion. Higher exposures can cause pulmonary edema, a medical emergency that can be delayed for several hours. This can cause death. Also, very high exposure (far above the OEL) can cause death. *Inhalation:* Contact with dust or mist can cause irritation, burns, and sores of the nose and throat. *Skin:* Contact with solid or solution may cause severe burns. *Eyes:* Contact with solid or solution may cause severe burns and eye damage. *Ingestion:* Onset of symptoms is usually rapid and includes burning and erosion of mouth, throat, and stomach tissue; nausea; vomiting, including vomiting blood; abdominal pain; diarrhea and bloody stools; numbness of fingers and toes; shock; collapse and convulsions; and kidney damage. Ingestion of 5 g (1/6 oz) can be fatal.

Long Term Exposure: Prolonged skin contact can cause irritation and slowly healing sores, pain, and discoloration (blue color) in fingers and localized tissue damage; gangrene may develop. Contact with dust or mist may cause inflammation and irritation of the nose and throat. May affect the kidneys and urinary stone formation. Repeated exposure may cause irritability, headache, and weakness.

Points of Attack: Eyes, skin, respiratory system, kidneys.

Medical Surveillance: NIOSH lists the following tests: urine (chemical/metabolite). If symptoms develop or overexposure is suspected, the following may be useful: kidney function tests. Examination of the blood vessels in exposed areas. For an acute overexposure, consider testing serum calcium level.

First Aid: If this chemical gets into the eyes, remove any contact lenses at once and irrigate immediately for at least 15 min, occasionally lifting upper and lower lids. Seek medical attention immediately. If this chemical contacts the skin, remove contaminated clothing and wash immediately with soap and water. Seek medical attention immediately. If this chemical has been inhaled, remove from exposure, begin rescue breathing (using universal precautions, including resuscitation mask) if breathing has stopped and CPR if heart action has stopped. Transfer promptly to a medical facility. When this chemical has been swallowed, get medical attention. If victim is *conscious*, administer water or milk. Do not induce vomiting. Medical observation is recommended for 24–48 h after breathing overexposure, as pulmonary edema may be delayed. As first aid for pulmonary edema, a doctor or authorized paramedic may consider administering a corticosteroid spray.

Personal Protective Methods: Wear protective gloves and clothing to prevent any reasonable probability of skin contact. Safety equipment suppliers/manufacturers can provide recommendations on the most protective glove/clothing material for your operation. Natural rubber, Neoprene™,

Nitrile/PVC, nitrile, and polyethylene are recommended. All protective clothing (suits, gloves, footwear, headgear) should be clean, available each day, and put on before work. Contact lenses should not be worn when working with this chemical. Wear splash-proof chemical goggles and face shield when working with liquid unless full face-piece respiratory protection is worn. Wear dust-proof goggles and face shield when working with powders or dust unless full face-piece respiratory protection is worn. Employees should wash immediately with soap when skin is wet or contaminated. Provide emergency showers and eyewash.

Respirator Selection: NIOSH: *25 mg/m³:* Sa:Cf* (APF = 25) (any supplied-air respirator operated in a continuous-flow mode) or PAPRDM* (any powered, air-purifying respirator with a dust and mist filter). *50 mg/m³:* 100F (APF = 50) (any air-purifying, full-face-piece respirator with an N100, R100, or P100 filter) or SCBAF (APF = 50) (any self-contained breathing apparatus with a full face-piece) or SaF (APF = 50) (any supplied-air respirator with a full face-piece). *500 mg/m³:* SaF: Pd,Pp (APF = 2000) (any supplied-air respirator that has a full face-piece and is operated in a pressure-demand or other positive-pressure mode). *Emergency or planned entry into unknown concentrations or IDLH conditions:* SCBAF: Pd,Pp (APF = 10,000) (any self-contained breathing apparatus that has a full face-piece and is operated in a pressure-demand or other positive-pressure mode) or SaF: Pd,Pp: ASCBA (APF = 10,000) (any supplied-air respirator that has a full face-piece and is operated in a pressure-demand or other positive-pressure mode in combination with an auxiliary self-contained breathing apparatus operated in a pressure-demand or other positive-pressure mode). *Escape:* 100F (APF = 50) (any air-purifying, full-face-piece respirator with an N100, R100, or P100 filter) or SCBAE (any appropriate escape-type, self-contained breathing apparatus).

*Substance causes eye irritation or damage; eye protection needed.

Storage: Color Code—White: Corrosive or Contact Hazard; Store separately in a corrosion-resistant location. Prior to working with this chemical you should be trained on its proper handling and storage. Oxalic acid must be stored to avoid contact with silver or strong oxidizers (such as chlorine and bromine) because violent reactions occur. Store in tightly closed containers in a cool, well-ventilated area away from heat. Sources of ignition, such as smoking and open flames, are prohibited where oxalic acid is used, handled, or stored in a manner that could create a potential fire or explosion hazard.

Shipping: Corrosive solids, n.o.s. require a shipping label of "CORROSIVE." Oxalic acid falls in DOT Hazard Class 8 and Packing Group II.

Spill Handling: Evacuate persons not wearing protective equipment from area of spill or leak until cleanup is complete. Remove all ignition sources. Wearing proper protective clothing and equipment, cover with soda ash or sodium bicarbonate. Collect powdered material in the most convenient and safe manner and deposit in sealed containers. Ventilate area after cleanup is complete. It may be necessary to contain and dispose of this chemical as a hazardous waste. If material or contaminated runoff enters waterways, notify downstream users of potentially contaminated waters. Contact your local or federal environmental protection agency for specific recommendations. If employees are required to clean up spills, they must be properly trained and equipped. OSHA 1910.120(q) may be applicable.

Fire Extinguishing: This chemical is a combustible solid. Use dry chemical, carbon dioxide, or water spray, or alcohol foam extinguishers. Poisonous gases, including carbon monoxide and formic acid, are produced in fire. If material or contaminated runoff enters waterways, notify downstream users of potentially contaminated waters. Notify local health and fire officials and pollution control agencies. From a secure, explosion-proof location, use water spray to cool exposed containers. If cooling streams are ineffective (venting sound increases in volume and pitch, tank discolors, or shows any signs of deforming), withdraw immediately to a secure position. If employees are expected to fight fires, they must be trained and equipped in OSHA 1910.156. The only respirators recommended for firefighting are self-contained breathing apparatuses that have full face-pieces and are operated in a pressure-demand or other positive-pressure mode.

Disposal Method Suggested: Pretreatment involves chemical reaction with limestone or calcium oxide forming calcium oxalate. This may then be incinerated utilizing particulate collection equipment to collect calcium oxide for recycling.

References

National Institute for Occupational Safety and Health. (October 1977). *Information Profiles on Potential Hazards: Oxalic Acid*, Report PB-276,678. Rockville, MD, pp. 42−46

New York State Department of Health. (March 1986). *Chemical Fact Sheet: Oxalic Acid*. Version 2. Albany, NY: Bureau of Toxic Substance Assessment

New Jersey Department of Health and Senior Services. (February 2000). *Hazardous Substances Fact Sheet: Oxalic Acid*. Trenton, NJ

Oxamyl   O:0170

Molecular Formula: $C_7H_{13}N_3O_3S$
Common Formula: $(CH_3)_2NCOC(SCH_3) = NOCONHCH_3$
Synonyms: D-1410; 2-(Dimethylamino)-*N*-([(methylamino)carbonyl]oxy)2-oxoethanimidothioic acid methyl ester; 2-Dimethylamino-1-(methylamino)glyoxal-*O*-methylcarbamoyl monoxime; *N,N*-Dimethyl-α-methylcarbamoyloxyimino-α-(methylthio)acetamide; *N,N*-Dimethyl-*N*-[(methylcarbamoyl)oxy]-1-thiooxamimidic acid methyl ester; DPX 1410; Insecticide-nemacide 1410; Methyl 2-(dimethylamino)-*N*-[((methylamino)carbonyl)oxy]-2-oxoethanimidothioate; *S*-Methyl 1-(dimethylcarbamoyl)-*N*-[(methylcarbamoyl)oxy]thioformimidate; Methyl 1-(dimethylcarbamoyl)-*N*-(methylcarbamoyloxy)thioformimidate; Methyl *N,N*′-dimethyl-*N*-[(methylcarbamoyl)oxy]-1-thiooxamimidate; Oxamyl carbamate insecticide; Thioxamyl; Vydate; Vydate 10G; Vydate insecticide/nematicide; Vydate L; Vydate Oxamyl insecticide/nematocide

CAS Registry Number: 23135-22-0
RTECS® Number: RP2300000
UN/NA & ERG Number: UN2757/151; 2991/151
EC Number: 245-445-3 [*Annex I Index No.:* 006-059-00-9]

Regulatory Authority and Advisory Bodies
US EPA Hazardous Waste Number (RCRA No.): P194.
Superfund/EPCRA [40CFR 302 and 355, F R: 8/16/06, Vol 71, No. 158] Reportable Quantity (RQ): 100 lb (45.4 kg).
RCRA, 40CFR261, Appendix 8 Hazardous Constituents.
RCRA 40CFR268.48; 61FR15654, Universal Treatment Standards: Wastewater (mg/L), 0.056; Nonwastewater (mg/kg), 0.28.
SUPERFUND/EPCRA 40CFR355, Extremely Hazardous Substances: TPQ = 100/10,000 lb (45.4/4540 kg).
Reportable Quantity (RQ): 100 lb (45.4 kg).
European/International Regulations: Hazard Symbol: T+, N; Risk phrases: R21; R26/28; R51; 53; Safety phrases: S1/2; S36/37; S45; S61 (see Appendix 4).
WGK (German Aquatic Hazard Class): 3—Severe hazard to waters.

Description: Oxamyl is a white crystalline solid with a sulfur- or garlic-like odor. Molecular weight = 219.29; Freezing/Melting point = 101°C. Hazard Identification (based on NFPA-704 M Rating System): Health 4, Flammability 1, Reactivity 0. Slightly soluble in water.

Potential Exposure: Used as an insecticide, nematicide, and acaricide on many field crops, vegetables, fruits, and ornamentals.

Permissible Exposure Limits in Air
Protective Action Criteria (PAC)
TEEL-0: 0.35 mg/m^3
PAC-1: 1 mg/m^3
PAC-2: 1.7 mg/m^3
PAC-3: 15 mg/m^3

Permissible Concentration in Water: A lifetime health advisory of 175 μg/L has been developed by EPA.[47] In addition, Massachusetts has set a guideline of 50 μg/L for oxamyl in drinking water.[61]

Determination in Water: Fish Tox = 707.10678000 ppb MATC (VERY LOW).

Routes of Entry: Inhalation, ingestion, skin and/or eye contact.

Harmful Effects and Symptoms
Short Term Exposure: Contact can cause skin and eye irritation. Acute exposure to oxamyl usually leads to a cholinergic crisis. Signs and symptoms may include increased salivation, lacrimation (tearing), perspiration, spontaneous defecation, and spontaneous urination. Pinpoint pupils,

blurred vision, tremor, muscle twitching, mental confusion, convulsions, and coma may occur. Gastrointestinal symptoms include abdominal pain, diarrhea, nausea, and vomiting. Bradycardia (slow heart rate) is common. Dyspnea (shortness of breath) and pulmonary edema may also occur. Classified by the World Health Organization as highly hazardous. Has also been rated as extremely to super-toxic. Acute oral exposure (ingestion) to oxamyl has caused death. Oxamyl is a potent cholinesterase inhibitor.

Long Term Exposure: Cholinesterase inhibitor; cumulative effect is possible. This chemical may damage the nervous system causing numbness and/or weakness in the hands and feet. Repeated exposure may cause personality changes with depression, anxiety, and irritability. May cause liver damage. Human Tox = 200.00000 ppb MCL (VERY LOW).

Points of Attack: Respiratory system, lungs, central nervous system, cardiovascular system, skin, eyes, liver, plasma and red blood cell cholinesterase.

Medical Surveillance: Before employment and at regular times after that, the following are recommended: plasma and red blood cell cholinesterase levels (tests for the enzyme poisoned by this chemical). If exposure stops, plasma levels return to normal in 1–2 weeks while red blood cell levels may be reduced for 1–3 months. When cholinesterase enzyme levels are reduced by 25% or more below preemployment levels, risk of poisoning is increased, even if results are in lower ranges of "normal." Reassignment to work not involving carbamate pesticides is recommended until enzyme levels recover. If symptoms develop or overexposure occurs, repeat the above tests as soon as possible and get an examination of the nervous system. Also, consider complete blood count. Consider chest X-ray following acute overexposure. Consider liver function tests.

First Aid: If this chemical gets into the eyes, remove any contact lenses at once and irrigate immediately for at least 15 min, occasionally lifting upper and lower lids. Seek medical attention immediately. If this chemical contacts the skin, remove contaminated clothing and wash immediately with soap and water. Seek medical attention immediately. If this chemical has been inhaled, remove from exposure, begin rescue breathing (using universal precautions, including resuscitation mask) if breathing has stopped and CPR if heart action has stopped. Transfer promptly to a medical facility. When this chemical has been swallowed, get medical attention. Give large quantities of water and induce vomiting. Do not make an unconscious person vomit. Obtain authorization and/or further instructions from the local hospital for administration of an antidote or performance of other invasive procedures. Transport to a the health-care facility.

Personal Protective Methods: Wear protective gloves and clothing to prevent any reasonable probability of skin contact. Safety equipment suppliers/manufacturers can provide recommendations on the most protective glove/clothing material for your operation. All protective clothing (suits, gloves, footwear, headgear) should be clean, available each day, and put on before work. Contact lenses should not be worn when working with this chemical. Wear dust-proof chemical goggles and face shield unless full face-piece respiratory protection is worn. Employees should wash immediately with soap when skin is wet or contaminated. Provide emergency showers and eyewash.

Respirator Selection: Follow regulations in OSHA 29CFR1910.134 or European Standard EN149. Use a NIOSH/MSHA- or European Standard EN149-approved respirator; or use an approved supplied-air respirator with a full face-piece operated in the positive-pressure mode, or with a full face-piece, hood, or helmet in the continuous-flow mode; or use a NIOSH/MSHA- or European Standard EN149-approved self-contained breathing apparatus with a full face-piece operated in a pressure-demand or other positive-pressure mode.

Storage: Color Code—Blue: Health Hazard/Poison: Store in a secure poison location. Prior to working with this chemical you should be trained on its proper handling and storage. Store in tightly closed containers in a cool, well-ventilated area away from heat and light. Where possible, automatically transfer material from drums or other storage containers to process containers.

Shipping: Carbamate pesticide, solid, toxic, n.o.s. requires a shipping label of "POISONOUS/TOXIC MATERIALS." It falls in Hazard Class 6.1.

Spill Handling: Evacuate persons not wearing protective equipment from area of spill or leak until cleanup is complete. Remove all ignition sources. Collect powdered material in the most convenient and safe manner and deposit in sealed containers. Ventilate area after cleanup is complete. It may be necessary to contain and dispose of this chemical as a hazardous waste. If material or contaminated runoff enters waterways, notify downstream users of potentially contaminated waters. Contact your local or federal environmental protection agency for specific recommendations. If employees are required to clean up spills, they must be properly trained and equipped. OSHA 1910.120(q) may be applicable.

Fire Extinguishing: Extinguish fire using agent suitable for type of surrounding fire, as the material itself does not burn or burns with difficulty. Use water in flooding quantities as a fog. Use alcohol foam, carbon dioxide, or dry chemical. Move container from fire area. Fight fire from maximum distance. Dike fire control water for later disposal do not scatter the material. Wear positive pressure breathing apparatus and special protective clothing. Poisonous gases, including nitrogen oxides and sulfur oxides, are produced in fire. If material or contaminated runoff enters waterways, notify downstream users of potentially contaminated waters. Notify local health and fire officials and pollution control agencies. From a secure, explosion-proof location, use water spray to cool exposed containers. If cooling streams are ineffective (venting sound increases in volume and pitch, tank discolors, or shows any signs of deforming),

withdraw immediately to a secure position. If employees are expected to fight fires, they must be trained and equipped in OSHA 1910.156. The only respirators recommended for firefighting are self-contained breathing apparatuses that have full face-pieces and are operated in a pressure-demand or other positive-pressure mode.

Disposal Method Suggested: Consult with environmental regulatory agencies for guidance on acceptable disposal practices. Generators of waste containing this contaminant ($\geq$100 kg/mo) must conform with EPA regulations governing storage, transportation, treatment, and waste disposal. Small quantities may be treated with alkali and buried in a landfill.[22] In accordance with 40CFR165, follow recommendations for the disposal of pesticides and pesticide containers. Must be disposed properly by following package label directions or by contacting your local or federal environmental control agency or by contacting your regional EPA office.

References

US Environmental Protection Agency. (November 30, 1987). *Chemical Hazard Information Profile: Oxamyl.* Washington, DC: Chemical Emergency Preparedness Program

New Jersey Department of Health and Senior Services. (July 1999). *Hazardous Substances Fact Sheet: Oxamyl.* Trenton, NJ

4,4'-Oxydianiline   O:0180

Molecular Formula: $C_{12}H_{12}N_2O$
Common Formula: $H_2NC_6H_4OC_6H_4NH_2$
Synonyms: *p*-Aminophenyl ether; 4-Aminophenyl ether; Aniline, 4,4'-oxydi-; Benzenamine, 4,4'-oxybis-; Bis(*p*-aminophenyl) ether; Bis(4-aminophenyl) ether; 4,4-Diaminodiphenyl ether; Diaminodiphenyl ether; 4,4'-Diaminofenol eter (Spanish); 4,4'-Diaminophenyl ether; NCI-C50146; Oxybis(4-aminobenzene); *p,p'*-Oxybis(aniline); 4,4'-Oxybis(aniline); *p,p'*-Oxydianiline; 4,4'-Oxydianiline; 4,4'-Oxydiphenylamine; Oxydi-*p*-phenylenediamine
CAS Registry Number: 101-80-4
RTECS® Number: BY7900000
UN/NA & ERG Number: UN2811 (toxic solid, organic, n.o.s.)/154
EC Number: 202-977-0 [*Annex I Index No.:* 612-199-00-7]
Regulatory Authority and Advisory Bodies
Carcinogenicity: IARC: Animal Sufficient Evidence; Human Limited Evidence, *possibly carcinogenic to humans*, Group 2B, 1987; NCI: Carcinogenesis Studies (feed); clear evidence: mouse, rat; NTP: Reasonably anticipated to be a human carcinogen.
Air Pollutant Standard Set. See below, "Permissible Exposure Limits in Air" section.
EPCRA Section 313 Form R *de minimis* concentration reporting level: 0.1%.

European/International Regulations: Hazard Symbol: T, N; Risk phrases: R45; R46; R 23/24/25; R62; R51/53; Safety phrases: S53; S45; S61 (see Appendix 4).
WGK (German Aquatic Hazard Class): No value assigned.
Description: 4,4'-Oxydianiline is a white crystalline solid, or a beige powder. Molecular weight = 200.26; Boiling point $\geq$300°C (sublimes); Freezing/Melting point = 186−187°C; Flash point = 219°C. Very slightly soluble in water.
Potential Exposure: Intermediate in the manufacture of high-temperature-resistant, straight polyimide and poly (esterimide) resins capable of withstanding temperatures of up to 480°C for short periods or 260°C for prolonged periods of time. Some *p*-phenylenediamine compounds have been used as rubber components, and DFG warns of danger of skin sensitization.
Incompatibilities: Strong oxidizers.
Permissible Exposure Limits in Air
Protective Action Criteria (PAC)
TEEL-0: 0.5 mg/m^3
PAC-1: 1.5 mg/m^3
PAC-2: 10 mg/m^3
PAC-3: 300 mg/m^3
DFG MAK: Dangerous skin sensitization; Carcinogen Category 2.
Russia set a MAC of 5.0 mg/m^3 in work-place air.
Routes of Entry: Inhalation, ingestion, skin and/or eye contact.
Harmful Effects and Symptoms
Short Term Exposure: Irritation of the skin and/or eyes. Poisonous. LD$_{50}$ = (oral-rat) 725 mg/kg.
Long Term Exposure: It has caused liver disease and retinopathy in rats. A potential occupational carcinogen.
Points of Attack: Liver.
Medical Surveillance: Liver tests.
First Aid: *Skin Contact:* Flood all areas of body that have contacted the substance with water. Do not wait to remove contaminated clothing; do it under the water stream. Use soap to help assure removal. Isolate contaminated clothing when removed to prevent contact by others.[52] *Eye Contact:* Remove any contact lenses at once. Immediately flush eyes well with copious quantities of water or normal saline for at least 20−30 min. Seek medical attention.
Inhalation: Leave contaminated area immediately; breathe fresh air. Proper respiratory protection must be supplied to any rescuers. If coughing, difficult breathing, or any other symptoms develop, seek medical attention at once, even if symptoms develop many hours after exposure. *Ingestion:* Contact a physician, hospital, or poison center at once. If the victim is unconscious or convulsing, do not induce vomiting or give anything by mouth. Assure that the patient's airway is open and lay him on his side with his head lower than his body and transport immediately to a medical facility. If conscious and not convulsing, give a glass of water to dilute the substance. Vomiting should not be induced without a physician's advice.

Personal Protective Methods: Wear protective gloves and clothing to prevent any reasonable probability of skin contact. Safety equipment suppliers/manufacturers can provide recommendations on the most protective glove/clothing material for your operation. All protective clothing (suits, gloves, footwear, headgear) should be clean, available each day, and put on before work. Contact lenses should not be worn when working with this chemical. Wear dust-proof chemical goggles and face shield unless full face-piece respiratory protection is worn. Employees should wash immediately with soap when skin is wet or contaminated. Provide emergency showers and eyewash.

Respirator Selection: Follow regulations in OSHA 29CFR1910.134 or European Standard EN149. Use a NIOSH/MSHA- or European Standard EN149-approved respirator; or use an approved supplied-air respirator with a full face-piece operated in the positive-pressure mode, or with a full face-piece, hood, or helmet in the continuous-flow mode; or use a NIOSH/MSHA- or European Standard EN149-approved self-contained breathing apparatus with a full face-piece operated in a pressure-demand or other positive-pressure mode.

Storage: Color Code—Blue: Health Hazard/Poison: Store in a secure poison location. Prior to working with this chemical you should be trained on its proper handling and storage. Store in a refrigerator under an inert atmosphere. Protect from exposure to light and oxidizing agents. A regulated, marked area should be established where this chemical is handled, used, or stored in compliance with OSHA Standard 1910.1045.

Shipping: Toxic solids, organic, n.o.s. require a shipping label of "POISONOUS/TOXIC MATERIALS." It falls in Hazard Class 6.1 and Packing Group III.

Spill Handling: Evacuate persons not wearing protective equipment from area of spill or leak until cleanup is complete. Remove all ignition sources. Remove all sources of ignition and dampen spilled material with 60–70% acetone to avoid airborne dust and collect powdered material in the most convenient and safe manner and deposit in sealed containers. Ventilate area after cleanup is complete. It may be necessary to contain and dispose of this chemical as a hazardous waste. If material or contaminated runoff enters waterways, notify downstream users of potentially contaminated waters. Contact your local or federal environmental protection agency for specific recommendations. If employees are required to clean up spills, they must be properly trained and equipped. OSHA 1910.120(q) may be applicable.

Fire Extinguishing: This chemical is a combustible solid. Use dry chemical, carbon dioxide, water spray, or alcohol foam extinguishers. Poisonous gases, including nitrogen oxides, are produced in fire. If material or contaminated runoff enters waterways, notify downstream users of potentially contaminated waters. Notify local health and fire officials and pollution control agencies. From a secure, explosion-proof location, use water spray to cool exposed containers. If cooling streams are ineffective (venting sound increases in volume and pitch, tank discolors, or shows any signs of deforming), withdraw immediately to a secure position. If employees are expected to fight fires, they must be trained and equipped in OSHA 1910.156. The only respirators recommended for firefighting are self-contained breathing apparatuses that have full face-pieces and are operated in a pressure-demand or other positive-pressure mode.

Disposal Method Suggested: Incineration with provision for nitrogen oxides removal from flue gases.

9,10-Oxydiphenoxarsine   O:0190

Molecular Formula: $C_{24}H_{16}As_2O_3$

Synonyms: Bis(10-phenoxarsinl) oxide; Bis(phenoxarsin-10-yl) ether; Bis(10-phenoxarsyl) oxide; 10,10′-Bis(phenoxyarsinyl) oxide; Bis(10-phenoxyarsinyl) oxide; DID 47; Diphenoxarsin-10-yl oxide; OBPA; 10-10′-Oxidiphenoxarsine; 10-10′-Oxybisphenoxyarsine; Phenoxaksine oxide; PXO; SA 546; Vinadine; Vinyzene; Vinyzene BP 5; Vinyzene BP 5-2; Vinyzene (Pesticide); Vinyzene SB 1

CAS Registry Number: 58-36-6

RTECS® Number: SP6800000

UN/NA & ERG Number: Not regulated

EC Number: 200-377-3

Regulatory Authority and Advisory Bodies
Carcinogenicity: NTP: 11th Report on Carcinogens, 2004: Known to be a human carcinogen; IARC: Human Sufficient Evidence, 1980; Animal Limited Evidence, *carcinogenic to humans*, Group 1, 1987.

Air Pollutant Standard Set. See below, "Permissible Exposure Limits in Air" section.

SUPERFUND/EPCRA 40CFR355, Extremely Hazardous Substances: TPQ = 500/10,000 lb (227/4540 kg).

Reportable Quantity (RQ): 500 lb (227 kg).

As arsenic compound:
Carcinogenicity: NTP: 11th Report on Carcinogens, 2004: Known to be a human carcinogen; IARC: Human Sufficient Evidence, 1980; Animal Limited Evidence, *carcinogenic to humans*, Group 1, 1987.

Clean Air Act: Hazardous Air Pollutants (Title I, Part A, Section 112) as arsenic compounds.

Clean Water Act: Toxic Pollutant (Section 401.15) as arsenic and compounds.

RCRA, 40CFR261, Appendix 8 Hazardous Constituents, waste number not listed.

Reportable Quantity (RQ): 1 lb (0.454 kg).

EPCRA (Section 313): Includes any unique chemical substance that contains arsenic as part of that chemical's infrastructure. Form R *de minimis* concentration reporting level: organics 1.0%.

US DOT Regulated Marine Pollutant (49CFR172.101, Appendix B) as arsenates, liquid, n.o.s.; arsenates, solid, n.o.s.; arsenical pesticides, liquid, toxic, flammable, n.o.s.

Canada, WHMIS, Ingredients Disclosure List Concentration 1.0% as arsenic, water-soluble compounds, n.o.s.
Canada: Priority Substance List & Restricted Substances/ Ocean Dumping FORBIDDEN (CEPA), National Pollutant Release Inventory (NPRI) (arsenic compounds).
European/International Regulations: Hazard Symbol: T, N; Risk phrases: R45; R23/25; R50/53; Safety phrases: S53; S45; S60; S61 (see Appendix 4).
WGK (German Aquatic Hazard Class): No value assigned.
Description: 10-10'-Oxydiphenoxarsine is a colorless crystalline, organometallic solid or a dense yellow liquid. Molecular weight = 502.23; Boiling point = 233°C; Freezing/Melting point = 185°C. Decomposition temperature = 380°C. Flash point = <32°C. Hazard Identification (based on NFPA-704 M Rating System): Health 1, Flammability 0, Reactivity 1. Practically insoluble in water.
Incompatibilities: A reducing agent; reacts violently with strong oxidizers. Keep away from strong acids and strong bases.
Potential Exposure: This material is used primarily for fungicidal and bactericidal protection of plastics. It is an organoarsenic and a heavy metal compound.

Permissible Exposure Limits in Air
Arsenic, organic compounds
OSHA PEL: 0.5 mg[As]/m^3 TWA.
NIOSH REL: Not established. See NIOSH Pocket Guide, Appendix A.
ACGIH TLV$^{®[1]}$: 0.01 mg[As]/m^3 TWA; Confirmed Human Carcinogen; BEI established.
Protective Action Criteria (PAC)
TEEL-0: 1.68 mg/m^3
PAC-1: 2 mg/m^3
PAC-2: 14 mg/m^3
PAC-3: 14 mg/m^3
Permissible Concentration in Water: While not specifically citing the compounds, EPA[6] gives a desirable level of zero for arsenic in water.
Routes of Entry: Inhalation, ingestion, skin and/or eye contact.
Harmful Effects and Symptoms
Short Term Exposure: Irritates the eyes and skin. Contact with eyes may be severe. Ingestion causes nausea, vomiting, and diarrhea. Arsenic compounds are acutely poisonous by ingestion. In severe cases, there may be bloody vomitus and stools and the victim may suffer collapse and shock with weak, rapid pulse; cold sweats; coma; and death. LD$_{50}$ = (oral-rat) 40 mg/kg.
Long Term Exposure: Arsenic compounds are recognized carcinogens of the skin, lungs, and liver. Ingestion or inhalation may result in chronic poisoning. Symptoms may include disturbances of the digestive system, loss of appetite, cramps, nausea, constipation, diarrhea. May cause liver damage, resulting in jaundice. May cause disturbances of the blood, kidneys, and nervous system. May cause skin abnormalities including itching, pigmentation, and even cancerous changes.

Points of Attack: Blood, kidneys, skin, nervous system.
Medical Surveillance: Kidney function tests. Examination of the nervous system.
First Aid: If this chemical gets into the eyes, remove any contact lenses at once and irrigate immediately for at least 15 min, occasionally lifting upper and lower lids. Seek medical attention immediately. If this chemical contacts the skin, remove contaminated clothing and wash immediately with soap and water. Seek medical attention immediately. If this chemical has been inhaled, remove from exposure, begin rescue breathing (using universal precautions, including resuscitation mask) if breathing has stopped and CPR if heart action has stopped. Transfer promptly to a medical facility. When this chemical has been swallowed, get medical attention. Give large quantities of water and induce vomiting. Do not make an unconscious person vomit.
Note to physician: For severe poisoning BAL [British Anti-Lewisite, Dimercaprol, dithiopropanol (C$_3$H$_8$OS$_2$)] has been used to treat toxic symptoms of certain heavy metals poisoning—including arsenic. Although BAL is reported to have a large margin of safety, caution must be exercised because toxic effects may be caused by excessive dosage. Most can be prevented by premedication with 1-ephedrine sulfate (CAS: 134-72-5). For milder poisoning *penicillamine* (*not penicillin*) has been used, both with mixed success. Side effects occur with such treatment and it is never a substitute for controlling exposure. It can only be done under strict medical care.
Personal Protective Methods: Wear protective gloves and clothing to prevent any reasonable probability of skin contact. Safety equipment suppliers/manufacturers can provide recommendations on the most protective glove/clothing material for your operation. All protective clothing (suits, gloves, footwear, headgear) should be clean, available each day, and put on before work. Contact lenses should not be worn when working with this chemical. Wear dust-proof chemical goggles and face shield unless full face-piece respiratory protection is worn. Employees should wash immediately with soap when skin is wet or contaminated. Provide emergency showers and eyewash.
Respirator Selection: At any concentrations above the NIOSH REL, or where there is no REL, at any detectable concentration: SCBAF: Pd,Pp (APF = 10,000) (any self-contained breathing apparatus that has a full faceplate and is operated in a pressure-demand or other positive-pressure mode) or SaF: Pd,Pp: ASCBA (APF = 10,000) (any supplied-air respirator that has a full face-piece and is operated in a pressure-demand or other positive-pressure mode in combination with an auxiliary self-contained breathing apparatus operated in a pressure-demand or other positive-pressure mode). *Escape:* GmFAg100 (APF = 50) [any air-purifying, full-face-piece respirator (gas mask) with a chin-style, front- or back-mounted acid gas canister having an N100, R100, or P100 filter] or SCBAE (any appropriate escape-type, self-contained breathing apparatus).

Storage: (1) Color Code—Red: Flammability Hazard (*dissolved in a flammable carrier solvent*): Store in a flammable liquid storage area or approved cabinet away from ignition sources and corrosive and reactive materials. (2) Color Code—Blue (*carcinogen*): Health Hazard/Poison: Store in a secure poison location. Prior to working with this chemical you should be trained on its proper handling and storage. Store in tightly closed containers in a cool, well-ventilated area away from oxidizers. Where possible, automatically transfer material from drums or other storage containers to process containers. Sources of ignition, such as smoking and open flames, are prohibited where this chemical is handled, used, or stored. Metal containers involving the transfer of this chemical should be grounded and bonded. Use only nonsparking tools and equipment, especially when opening and closing containers of this chemical. Wherever this chemical is used, handled, manufactured, or stored, use explosion-proof electrical equipment and fittings. A regulated, marked area should be established where this chemical is handled, used, or stored in compliance with OSHA Standard 1910.1045.

Shipping: Not regulated.

Spill Handling: Evacuate persons not wearing protective equipment from area of spill or leak until cleanup is complete. Remove all ignition sources. Keep unnecessary people away; isolate hazard area and deny entry. Stay upwind; keep out of low areas. Do not touch spilled material; stop leak if you can do so without risk. *Small spills:* absorb with sand or other noncombustible absorbent material and place into containers for later disposal. *Small dry spills:* with clean shovel place material into clean, dry container and cover; move containers from spill area. *Large spills:* dike far ahead of spill for later disposal. Ventilate area after cleanup is complete. It may be necessary to contain and dispose of this chemical as a hazardous waste. If material or contaminated runoff enters waterways, notify downstream users of potentially contaminated waters. Contact your local or federal environmental protection agency for specific recommendations. If employees are required to clean up spills, they must be properly trained and equipped. OSHA 1910.120(q) may be applicable.

Fire Extinguishing: Small fires: dry chemical, carbon dioxide, water spray, or foam. *Large fires:* water spray, fog, or foam. Move container from fire area if you can do so without risk. Stay upwind; keep out of low areas. Wear self-contained (positive pressure if available) breathing apparatus and full protective clothing. Poisonous gases, including arsenic, are produced in fire. If material or contaminated runoff enters waterways, notify downstream users of potentially contaminated waters. Notify local health and fire officials and pollution control agencies. From a secure, explosion-proof location, use water spray to cool exposed containers. If cooling streams are ineffective (venting sound increases in volume and pitch, tank discolors, or shows any signs of deforming), withdraw immediately to a secure position. If employees are expected to fight fires, they must be trained and equipped in OSHA 1910.156. The only respirators recommended for firefighting are self-contained breathing apparatuses that have full face-pieces and are operated in a pressure-demand or other positive-pressure mode.

Reference
US Environmental Protection Agency. (November 30, 1987). *Chemical Hazard Information Profile: 10,10-Oxydiphenoxarsine.* Washington, DC: Chemical Emergency Preparedness Program

Oxydisulfoton   O:0200

Molecular Formula: $C_8H_{19}OsPS_3$
Synonyms: BAY 23323; *O,O*-Diethyl *S*-[2-(ethylsulfinyl) ethyl] phosphorodithioate; *O,O*-Diethyl *S*-[(ethylsulfinyl) ethyl] phosphorodithioate; Disulfoton disulfide; Disulfoton sulfoxide; Disyston sulfoxide; Ethylthiomelton sulfoxide
CAS Registry Number: 2497-07-6
RTECS® Number: TD8600000
UN/NA & ERG Number: UN3018 (organophosphorus pesticide, liquid, toxic)/152
EC Number: 219-679-1 [*Annex I Index No.:* 015-096-00-X]
Regulatory Authority and Advisory Bodies
Very Toxic Substance (World Bank).[15]
SUPERFUND/EPCRA 40CFR355, Extremely Hazardous Substances: TPQ = 500 lb (227 kg).
Reportable Quantity (RQ): 500 lb (227 kg).
European/International Regulations: Hazard Symbol: T+, N; Risk phrases: R24; R28; R50/53; Safety phrases: S1/2; S28; S36/37; S45; 60; S61 (see Appendix 4).
WGK (German Aquatic Hazard Class): No value assigned.
Description: Oxydisulfoton is a liquid. Molecular weight = 290.42. Hazard Identification (based on NFPA-704 M Rating System): Health 4, Flammability 1, Reactivity 0.
Potential Exposure: This material is an agricultural insecticide.
Incompatibilities: Strong oxidizers may cause release of toxic phosphorus oxides. Organophosphates, in the presence of strong reducing agents such as hydrides, may form highly toxic and flammable phosphine gas. Keep away from alkaline materials.
Permissible Exposure Limits in Air
Protective Action Criteria (PAC)
TEEL-0: 0.6 mg/m^3
PAC-1: 2 mg/m^3
PAC-2: 3.5 mg/m^3
PAC-3: 3.5 mg/m^3
Routes of Entry: Inhalation, ingestion, skin and/or eye contact.
Harmful Effects and Symptoms
Short Term Exposure: Organic phosphorus insecticides are absorbed by the skin, as well as by the respiratory and gastrointestinal tracts. They are cholinesterase inhibitors. Symptoms include the following: *mild exposure:* headache,

loss of appetite, nausea, dizziness; *moderate exposure:* abdominal cramps, diarrhea, salivation, excessive tearing, muscular cramps; *severe exposure:* fever, blue lips, lack of sphincter control, coma, heart shock, difficult breathing.

Long Term Exposure: Cholinesterase inhibitor; cumulative effect is possible. This chemical may damage the nervous system with repeated exposure, resulting in convulsions, respiratory failure. May cause liver damage.

Points of Attack: Respiratory system, lungs, central nervous system, cardiovascular system, skin, eyes, plasma and red blood cell cholinesterase.

Medical Surveillance: Before employment and at regular times after that, the following are recommended: plasma and red blood cell cholinesterase levels (tests for the enzyme poisoned by this chemical). If exposure stops, plasma levels return to normal in 1−2 weeks while red blood cell levels may be reduced for 1−3 months. When cholinesterase enzyme levels are reduced by 25% or more below preemployment levels, risk of poisoning is increased, even if results are in lower ranges of "normal." Reassignment to work not involving organophosphate or carbamate pesticides is recommended until enzyme levels recover. If symptoms develop or overexposure occurs, repeat the above tests as soon as possible and get an examination of the nervous system. Also, consider complete blood count. Consider chest X-ray following acute overexposure. Do not drink any alcoholic beverages before or during use. Alcohol promotes absorption of organic phosphates.

First Aid: If this chemical gets into the eyes, remove any contact lenses at once and irrigate immediately for at least 15 min, occasionally lifting upper and lower lids. Seek medical attention immediately. If this chemical contacts the skin, remove contaminated clothing and wash immediately with soap and water. Speed in removing material from skin is of extreme importance. Shampoo hair promptly if contaminated. Seek medical attention immediately. If this chemical has been inhaled, remove from exposure, begin rescue breathing (using universal precautions, including resuscitation mask) if breathing has stopped and CPR if heart action has stopped. Transfer promptly to a medical facility. When this chemical has been swallowed, get medical attention. Give large quantities of water and induce vomiting. Do not make an unconscious person vomit. Keep victim quiet and maintain normal body temperature. Effects may be delayed; keep victim under observation.

Personal Protective Methods: Wear protective gloves and clothing to prevent any reasonable probability of skin contact. Safety equipment suppliers/manufacturers can provide recommendations on the most protective glove/clothing material for your operation. All protective clothing (suits, gloves, footwear, headgear) should be clean, available each day, and put on before work. Contact lenses should not be worn when working with this chemical. Wear splash-proof chemical goggles and face shield unless full face-piece respiratory protection is worn. Employees should wash

immediately with soap when skin is wet or contaminated. Provide emergency showers and eyewash.

Respirator Selection: Follow regulations in OSHA 29CFR1910.134 or European Standard EN149. Use a NIOSH/MSHA- or European Standard EN149-approved respirator; or use an approved supplied-air respirator with a full face-piece operated in the positive-pressure mode, or with a full face-piece, hood, or helmet in the continuous-flow mode; or use a NIOSH/MSHA- or European Standard EN149-approved self-contained breathing apparatus with a full face-piece operated in a pressure-demand or other positive-pressure mode.

Storage: Color Code—Blue: Health Hazard/Poison: Store in a secure poison location. Prior to working with this chemical you should be trained on its proper handling and storage.

Shipping: Organophosphorus pesticides, liquid, toxic, require a shipping label of "POISONOUS/TOXIC MATERIALS." It falls in Hazard Class 6.1 and Packing Group III.

Spill Handling: Evacuate and restrict persons not wearing protective equipment from area of spill or leak until cleanup is complete. Remove all ignition sources. Ventilate area of spill or leak. Do not touch spilled material. *Small spills:* take up with sand or other noncombustible absorbent material and place into containers for later disposal. *Small dry spills:* with clean shovel place material into clean, dry container and cover; move containers from spill area. *Large spills:* dike far ahead of spill for later disposal. Keep this chemical out of a confined space, such as a sewer, because of the possibility of an explosion, unless the sewer is designed to prevent the buildup of explosive concentrations. It may be necessary to contain and dispose of this chemical as a hazardous waste. If material or contaminated runoff enters waterways, notify downstream users of potentially contaminated waters. Contact your local or federal environmental protection agency for specific recommendations. If employees are required to clean up spills, they must be properly trained and equipped. OSHA 1910.120(q) may be applicable.

Fire Extinguishing: Extinguish with dry chemical, carbon dioxide, water spray, fog, or foam. Fight fire from maximum distance. Dike fire control water for later disposal; do not scatter the material. Poisonous gases may be generated from the fire or runoff water. Vapors are heavier than air and will collect in low areas. Vapors may travel long distances to ignition sources and flashback. Vapors in confined areas may explode when exposed to fire. Containers may explode in fire. Storage containers and parts of containers may rocket great distances, in many directions. If material or contaminated runoff enters waterways, notify downstream users of potentially contaminated waters. Notify local health and fire officials and pollution control agencies. From a secure, explosion-proof location, use water spray to cool exposed containers. If cooling streams are ineffective (venting sound increases in volume and pitch, tank discolors, or shows any signs of deforming), withdraw immediately to a secure position. If employees are expected to fight fires, they must be trained and equipped in OSHA 1910.156. The only respirators recommended for firefighting are self-contained breathing apparatuses that have full face-pieces and are operated in a pressure-demand or other positive-pressure mode.

Disposal Method Suggested: In accordance with 40CFR 165 recommendations for the disposal of pesticides and pesticide containers. Must be disposed properly by following package label directions or by contacting your local or federal environmental control agency or by contacting your regional EPA office.

Reference

US Environmental Protection Agency. (November 30, 1987). *Chemical Hazard Information Profile: Oxydisulfoton.* Washington, DC: Chemical Emergency Preparedness Program

Oxygen   O:0210

Molecular Formula: O_2

Synonyms: Liquid oxygen; LOX; Oxygen, liquid

CAS Registry Number: 7782-44-7

RTECS® Number: RS2060000

UN/NA & ERG Number: UN1072 (compressed)/122; UN1073 (refrigerated liquid)/122

EC Number: 231-956-9 [*Annex I Index No.:* 008-001-00-8]

Regulatory Authority and Advisory Bodies

US EPA Gene-Tox Program, Positive: V79 cell culture-gene mutation.

European/International Regulations: Hazard Symbol: O; Risk phrases: R8; Safety phrases: S2; S17 (see Appendix 4).

WGK (German Aquatic Hazard Class): Nonwater polluting agent.

Description: Oxygen is a colorless odorless gas or a bluish cryogenic liquid. Molecular weight = 32.00; Boiling point = −183°C; Freezing/Melting point = −219°C. Hazard Identification (based on NFPA-704 M Rating System): Health 3, Flammability 0, Reactivity 0. Oxidizer. Slightly soluble in water. Liquid sinks and boils in water.

Potential Exposure: Compound Description: Mutagen; Reproductive Effector; Human Data. Compressed oxygen is used in various oxidation processes, for feedstock; and enrichment purposes; as a medicinal gas; a chemical intermediate; in oxyacetylene welding; in metallurgy. Liquid oxygen is used as a rocket fuel. Oxygen is naturally present at a concentration of 21% in breathing air.

Incompatibilities: A strong oxidizer. Reacts violently with nearly every element, combustibles, organics, and reducing materials.

Permissible Exposure Limits in Air

No standards or TEEL available.

Minimum acceptable breathing air contains 19% oxygen.

Determination in Air: Use NIOSH Analytical Method #6601, Oxygen.

Routes of Entry: Inhalation, skin and/or eye contact.

Harmful Effects and Symptoms

Short Term Exposure: *Note:* Increased pressure speeds up the toxic effects of oxygen. Drugs and chemical can also effects toxicity either positively or negatively. *Inhalation:* The air we breathe contains 21% oxygen. Normal activity requires a minimum of 19% oxygen. Breathing up to 50% oxygen at normal pressure produced no symptoms. Breathing 100% oxygen at normal pressure produced no symptoms after 12 h, but after 24 h has caused weakness, dizziness, burning in the nose and throat; fatigue, pain in joints and muscles; numbness and tingling in the arms and legs; palpitations, headache, cough, nasal congestion; ear disturbances; nausea, vomiting, loss of appetite; sore throat, fever, and swelling of the mucous membranes. Pressure levels greater than 3 times normal may cause convulsions and coma. Higher exposures can cause pulmonary edema, a medical emergency that can be delayed for several hours. This can cause death. *Skin:* Contact with liquid oxygen may cause freezing burns and tissue damage. *Eyes:* Liquid oxygen may cause freezing burns and tissue damage. *Ingestion:* Liquid oxygen may cause freezing burns in the mouth and throat.

Long Term Exposure: Breathing of 50−100% oxygen at normal pressure even intermittently, over a prolonged period can cause lung damage.

Points of Attack: Lungs.

Medical Surveillance: Before beginning employment and at regular times after that, for those with frequent or potentially high exposures to pure oxygen, the following are recommended: lung function tests. If symptoms develop or overexposure is suspected, the following may be useful: consider chest X-ray after acute overexposure to pure oxygen.

First Aid: *Eye Contact:* With liquid oxygen—immediately remove any contact lenses and flush with large amounts of water for at least 15 min, occasionally lifting upper and lower lids. Seek medical attention immediately. *Skin Contact*: with liquid oxygen—if frostbite has occurred, seek medical attention immediately; do *NOT* rub the affected areas or flush them with water. In order to prevent further tissue damage, do *NOT* attempt to remove frozen clothing from frostbitten areas. If frostbite has *NOT* occurred, immediately and thoroughly wash contaminated skin with soap and water. Seek medical attention. *Breathing Pure Oxygen or Gases* >40% O_2; Remove the person from exposure. Begin rescue breathing (using universal precautions, including resuscitation mask) if breathing has stopped and CPR if heart action has stopped. Transfer promptly to a medical facility. Medical observation is recommended for 24−48 h after breathing overexposure, as pulmonary edema may be delayed.

Personal Protective Methods: Wear appropriate personal protective clothing to prevent the skin from becoming frozen from contact with the evaporating liquid or from contact with vessels containing the liquid. Safety equipment suppliers/manufacturers can provide recommendations on the most protective glove/clothing material for your operation. All protective clothing (suits, gloves, footwear, headgear) should be clean, available each day, and put on before work. Contact lenses should not be worn when working with this chemical. Wear splash-proof chemical goggles and face shield when working with liquefied oxygen. Wear gas-proof chemical goggles and face shield unless full facepiece respiratory protection is worn. Employees should wash immediately with soap when skin is wet or contaminated. Provide emergency showers and eyewash. All protective clothing (suits, gloves, footwear, headgear) should be clean, available each day, and put on before work. Specific engineering controls are required by OSHA. See OSHA Standard 1910.104.

Respirator Selection: Entering into oxygen enriched atmospheres (greater than 21% oxygen) is highly dangerous. Consult *NFPA Code 53M.*

Storage: Color Code—Yellow: Reactive Hazard; Store in a location separate from other materials, especially flammables and combustibles. Prior to working with this chemical you should be trained on its proper handling and storage. Liquid oxygen must be stored to avoid contact with organic and combustible materials (such as oil, grease, and coal dust) since violent reactions occur. Open storage is preferred. Oxygen gas evaporating from liquid or from oxygen enriched environments is easily absorbed into clothing and any source of ignition (such as a static spark) can cause flash burning. Compressed oxygen cylinders must be securely stored separately from fuel cylinders. Liquid oxygen tanks should be stored outdoors. Sources of ignition, such as smoking and open flames, are prohibited where oxygen is used, handled, or stored in a manner that could create a potential fire or explosion hazard. See OSHA Standard 1910.104 and NFPA 43A *Code for the Storage of Liquid and Solid Oxidizers* for detailed handling and storage regulations. Procedures for the handling, use, and storage of cylinders should be in compliance with OSHA 1910.101 and 1910.169, as with the recommendations of the Compressed Gas Association.

Shipping: Oxygen, *compressed,* or oxygen, *refrigerated liquid (cryogenic liquid)*, requires a shipping label of "NONFLAMMABLE GAS, OXIDIZER." They fall in DOT Hazard Class 2.2.

Spill Handling: Evacuate and restrict persons not wearing protective equipment from area of spill or leak until cleanup is complete. Remove all ignition sources. *Liquid:* Do NOT absorb liquid in saw-dust or similar combustible absorbents. This is an oxidizer and a dangerous fire and explosion risk. Use only water. Do not use chemical or carbon dioxide extinguishers. NEVER direct water jet on liquid. Allow liquid oxygen spills to evaporate. *Gas:* Ventilate area of leak to disperse the gas. Stop flow of gas. If source of leak is a cylinder and the leak cannot be stopped in place, remove the leaking cylinder to a safe place in the open air, and repair leak or allow cylinder to empty. If liquid oxygen is spilled or leaked, take the following steps: Keep combustibles (wood, paper, oil, etc.) away from spill. It may be necessary to contain and dispose of this chemical as a hazardous waste. If material or contaminated runoff enters waterways, notify downstream users of potentially contaminated waters. Contact your local or federal environmental protection agency for specific recommendations. If employees are required to clean up spills, they must be properly trained and equipped. OSHA 1910.120(q) may be applicable.

Fire Extinguishing: Oxygen is not flammable but supports combustion and greatly increases the intensity of any fire. Mixtures of liquid oxygen and any fuel are highly explosive. Do not extinguish the fire unless the flow of gas can be stopped and any remaining gas is out of the line. Specially trained personnel may use fog lines to cool exposures and let the fire burn itself out. For a large fire, evacuate danger area and consult an expert. For small fire, extinguish using an agent suitable for type of surrounding fire. Oxygen itself does not burn. Do not use chemical or carbon dioxide on *liquid* oxygen. Containers may explode in fire. Storage containers and parts of containers may rocket great distances, in many directions. If material or contaminated runoff enters waterways, notify downstream users of potentially contaminated waters. Notify local health and fire officials and pollution control agencies. From a secure, explosion-proof location, use water spray to cool exposed containers. If cooling streams are ineffective (venting sound increases in volume and pitch, tank discolors, or shows any signs of deforming), withdraw immediately to a secure position. If cylinders are exposed to excessive heat from fire or flame contact, withdraw immediately to a secure location. If employees are expected to fight fires, they must be trained and equipped in OSHA 1910.156. The only respirators recommended for firefighting are self-contained breathing apparatuses that have full face-pieces and are operated in a pressure-demand or other positive-pressure mode.

Disposal Method Suggested: Vent to atmosphere.

References
New Jersey Department of Health and Senior Services. (April 2004). *Hazardous Substances Fact Sheet: Oxygen.* Trenton, NJ
New York State Department of Health. (February 1986). *Chemical Fact Sheet: Oxygen (compressed or liquefied).* Albany, NY: Bureau of Toxic Substance Assessment

Oxygen difluoride   O:0220

Molecular Formula: F_2O
Common Formula: OF_2
Synonyms: Difluorine monoxide; Fluorine monoxide; Fluorine oxide; Oxygen fluoride

CAS Registry Number: 7783-41-7
RTECS® Number: RS2100000
UN/NA & ERG Number: UN2190/124
EC Number: 231-996-7

Regulatory Authority and Advisory Bodies

Department of Homeland Security Screening Threshold Quantity (pounds): *Theft hazard* 15 (≥0.09% concentration).

Very Toxic Substance (World Bank).[15]

Air Pollutant Standard Set. See below, "Permissible Exposure Limits in Air" section.

OSHA 29CFR1910.119, Appendix A, Process Safety List of Highly Hazardous Chemicals, TQ = 100 lb (45 kg).

US DOT 49CFR172.101, Inhalation Hazardous Chemical.

Canada, WHMIS, Ingredients Disclosure List Concentration 1.0%.

European/International Regulations: not listed in Annex 1.

WGK (German Aquatic Hazard Class): No value assigned.

Description: Oxygen difluoride is a colorless gas with a foul, peculiar odor. Shipped as a nonliquefied compressed gas. Molecular weight = 54.00; Boiling point = −145.6°C; Freezing/Melting point = −223.9°C; Vapor pressure = >1 atm at 20°C; Relative vapor density (air = 1) = 1.88. Slightly soluble in water (slowly reactive); solubility = 0.02%.

Potential Exposure: Compound Description: Reproductive Effector; Human Data. Oxygen difluoride may be used as an oxidant in missile propellant systems.

Incompatibilities: A strong oxidizer. Explodes on contact with steam. Violent reaction with reducing agents, combustible materials, chlorine, bromine, iodine, platinum, metal oxides, moist air, hydrogen sulfide (explosive in ambient air), hydrocarbons, water. Attacks mercury.

Permissible Exposure Limits in Air

Conversion factor: 1 ppm = 2.21 mg/m^3 at 25°C & 1 atm.

OSHA PEL: 0.05 ppm/0.1 mg/m^3 TWA.

NIOSH REL: 0.05 ppm/0.1 mg/m^3 Ceiling Concentration.

ACGIH TLV®[1]: 0.05 ppm/0.11 mg/m^3 Ceiling Concentration.

NIOSH IDLH: 0.5 ppm.

Protective Action Criteria (PAC)*

TEEL-0: 0.05 ppm

PAC-1: 0.1 ppm

PAC-2: **0.83** ppm

PAC-3: **2.5** ppm

*AEGLs (Acute Emergency Guideline Levels) & ERPGs (Emergency Response Planning Guideline) are in **bold face**.

Arab Republic of Egypt: TWA 0.05 ppm (0.1 mg/m^3), 1993; Austria: MAK 2.5 mg[F]/m^3, 1999; Denmark: TWA 0.05 ppm (0.1 mg/m^3), 1999; Finland: STEL 0.05 ppm (0.1 mg/m^3), 1999; Norway: TWA 0.05 ppm (0.1 mg/m^3), 1999; the Philippines: TWA 0.05 ppm (0.1 mg/m^3), 1993; Poland: MAC (TWA) 1 mg[HF]/m^3; MAC (STEL) 3 mg [HF]/m^3, 1999; the Netherlands: MAC 0.1 mg/m^3, 2003;

United Kingdom: TWA 2.5 mg[F]/m^3, 2000; New Zealand, Singapore, Vietnam: ACGIH TLV®: Ceiling Concentration 0.05 ppm. Several states have set guidelines or standards for oxygen difluoride in ambient air[60] ranging from 1.0 μg/m^3 (North Dakota) to 2.0 μg/m^3 (Connecticut and Nevada) to 80,000 μg/m^3 (Virginia).

Determination in Air: No NIOSH Analytical Method available.

Routes of Entry: Inhalation, ingestion, skin and/or eye contact.

Harmful Effects and Symptoms

Short Term Exposure: Gas under pressure is corrosive to eyes, skin, and respiratory tract; causes burns. Higher exposures can cause pulmonary edema, a medical emergency that can be delayed for several hours. This can cause death. Exposure at low levels may result in severe, intractable headaches.

Points of Attack: Lungs, eyes.

Medical Surveillance: NIOSH lists the following tests: chest X-ray; electrocardiogram; pulmonary function tests: forced vital capacity, forced expiratory volume (1 sec); sputum cytology; white blood cell count/differential.

First Aid: If this chemical gets into the eyes, remove any contact lenses at once and irrigate immediately for at least 15 min, occasionally lifting upper and lower lids. Seek medical attention immediately. If this chemical contacts the skin, remove contaminated clothing and wash immediately with soap and water. Seek medical attention immediately. If this chemical has been inhaled, remove from exposure, begin rescue breathing (using universal precautions, including resuscitation mask) if breathing has stopped and CPR if heart action has stopped. Transfer promptly to a medical facility. When this chemical has been swallowed, get medical attention. Give large quantities of water and induce vomiting. Do not make an unconscious person vomit. Medical observation is recommended for 24−48 h after breathing overexposure, as pulmonary edema may be delayed. As first aid for pulmonary edema, a doctor or authorized paramedic may consider administering a corticosteroid spray. If frostbite has occurred, seek medical attention immediately; do NOT rub the affected areas or flush them with water. In order to prevent further tissue damage, do NOT attempt to remove frozen clothing from frostbitten areas. If frostbite has NOT occurred, immediately and thoroughly wash contaminated skin with soap and water.

Personal Protective Methods: Wear protective gloves and clothing to prevent any reasonable probability of skin contact. Safety equipment suppliers/manufacturers can provide recommendations on the most protective glove/clothing material for your operation. All protective clothing (suits, gloves, footwear, headgear) should be clean, available each day, and put on before work. Contact lenses should not be worn when working with this chemical. Wear gas-proof chemical goggles and face shield unless full face-piece respiratory protection is worn. Employees should wash

immediately with soap when skin is wet or contaminated. Provide emergency showers and eyewash.

Respirator Selection: *Up to 0.5 ppm:* Sa (APF = 10) (any supplied-air respirator) or SCBAF (APF = 50) (any self-contained breathing apparatus with a full face-piece). *Emergency or planned entry into unknown concentrations or IDLH conditions:* SCBAF: Pu,Pp (APF = 10,000) (any self-contained breathing apparatus that has a full face-piece and is operated in a pressure-demand or other positive-pressure mode) or SaF: Pd,Pp: ASCBA (APF = 10,000) (any supplied-air respirator that has a full face-piece and is operated in a pressure-demand or other positive-pressure mode in combination with an auxiliary self-contained breathing apparatus operated in a pressure-demand or other positive-pressure mode). *Escape:* GmFS (APF = 50)* [any air-purifying, full-face-piece respirator (gas mask) with a chin-style, front- or back-, mounted canister providing protection against the compound of concern] or SCBAE (any appropriate escape-type, self-contained breathing apparatus).

*Only nonoxidizable sorbents are allowed (not charcoal).

Storage: Color Code—Yellow Stripe: Reactivity Hazard; Store separately in an area isolated from flammables, combustibles, or other yellow-coded materials. Prior to working with this chemical you should be trained on its proper handling and storage. Protect cylinder containers against physical damage. Do not use wood pallets. Preferably handle behind body shield, in outdoor, or open protective fences. Wear long rubber gloves, goggles, protective clothing, and self-contained breathing apparatus. See 29 CFR 1910.101 for specific regulations on storage of compressed gas. See OSHA Standard 1910.104 and NFPA 43A *Code for the Storage of Liquid and Solid Oxidizers* for detailed handling and storage regulations.

Shipping: Oxygen difluoride requires a shipping label of "POISON GAS, OXIDIZER, CORROSIVE." It falls in Hazard Class 2.3. It is a violation of transportation regulations to refill compressed gas cylinders without the express written permission of the owner.

Spill Handling: If in a building, evacuate building and confine vapors by closing doors and shutting down HVAC systems. Restrict persons not wearing protective equipment from area of spill or leak until cleanup is complete. Remove all ignition sources. Ventilate area of spill or leak to disperse the gas. Wear chemical protective suit with self-contained breathing apparatus to combat spills. Stay upwind and use water spray to "knock down" vapor; contain runoff. Stop the flow of gas, if it can be done safely from a distance. If source is a cylinder and the leak cannot be stopped in place, remove the leaking cylinder to a safe place; and repair leak or allow cylinder to empty. Keep this chemical out of confined spaces, such as a sewer, because of the possibility of explosion, unless the sewer is designed to prevent the buildup of explosive concentrations. If employees are required to clean up spills, they must be properly trained and equipped. OSHA 1910.120(q) may be applicable.

Initial isolation and protective action distances

Distances shown are likely to be affected during the first 30 min after materials are spilled and could increase with time. If more than one tank car, cargo tank, portable tank, or large cylinder involved in the incident is leaking, the protective action distance may need to be increased. You may need to seek emergency information from CHEMTREC at (800) 424-9300 or seek professional environmental engineering assistance from the US EPA Environmental Response Team at (908) 548-8730 (24-h response line).

Small spills (From a small package or a small leak from a large package)

Oxygen difluoride, oxygen difluoride, compressed

First: Isolate in all directions (feet/meters) 2500/800
Then: Protect persons downwind (miles/kilometers)
Day   3.3/5.3
Night 7.0+/11.0+

Large spills (From a large package or from many small packages)

First: Isolate in all directions (feet/meters) 3000/1000
Then: Protect persons downwind (miles/kilometers)
Day   7.0+/11.0+
Night 7.0+/11.0+

Fire Extinguishing: This chemical is a nonflammable gas but a strong oxidizer that can increase the intensity of a fire. Poisonous gases are produced in fire. Do not extinguish the fire unless the flow of gas can be stopped and any remaining gas is out of the line. Specially trained personnel may use fog lines to cool exposures and let the fire burn itself out. Vapors are heavier than air and will collect in low areas. Containers may explode in fire. Storage containers and parts of containers may rocket great distances, in many directions. If material or contaminated runoff enters waterways, notify downstream users of potentially contaminated waters. Notify local health and fire officials and pollution control agencies. From a secure, explosion-proof location, use water spray to cool exposed containers. If cooling streams are ineffective (venting sound increases in volume and pitch, tank discolors, or shows any signs of deforming), withdraw immediately to a secure position. If cylinders are exposed to excessive heat from fire or flame contact, withdraw immediately to a secure location. If employees are expected to fight fires, they must be trained and equipped in OSHA 1910.156. The only respirators recommended for firefighting are self-contained breathing apparatuses that have full face-pieces and are operated in a pressure-demand or other positive-pressure mode.

Disposal Method Suggested: Spray or sift on a thick layer of a (1:1) mixture of dry soda ash and slaked lime behind a shield. After mixing, spray water from an atomizer with great precaution. Transfer slowly into a large amount of water. Neutralize and drain into the sewer with sufficient water.

Ozone   O:0230

Molecular Formula: O_3
Synonyms: Oxygen mol (O_3); Ozono (Spanish); Triatomic oxygen
CAS Registry Number: 10028-15-6
RTECS® Number: RS8225000
UN/NA & ERG Number: UN1955/123
EC Number: 233-069-2
Regulatory Authority and Advisory Bodies
Carcinogenicity: NCI: Carcinogenesis Studies (inhalation); equivocal evidence: mouse; no evidence: rat; NTP: Carcinogenesis Studies (inhalation); some evidence: mouse.
US EPA Gene-Tox Program, Negative: *In vivo* cytogenetics—nonhuman bone marrow; Negative: *In vivo* SCE—nonhuman; Inconclusive: *In vivo* cytogenetics—nonhuman lymphocyte; Inconclusive: *In vitro* cytogenetics—human lymphocyte; Inconclusive: *In vivo* cytogenetics—human lymphocyte; Inconclusive: Cytogenetics—male germ cell.
Air Pollutant Standard Set. See below, "Permissible Exposure Limits in Air" section.
OSHA 29CFR1910.119, Appendix A, Process Safety List of Highly Hazardous Chemicals, TQ = 100 lb (45 kg).
SUPERFUND/EPCRA 40CFR355, Extremely Hazardous Substances: TPQ = 100 lb (45.4 kg).
Reportable Quantity (RQ): 100 lb (45.4 kg).
EPCRA Section 313 Form R *de minimis* concentration reporting level: 1.0%.
SARA Hazard Classes: Sudden Release of pressure hazard; Fire Hazard; Acute Health Hazard; Chronic Health Hazard.
Canada, WHMIS, Ingredients Disclosure List Concentration 1.0%; Category DIA, C. D2B.
WGK (German Aquatic Hazard Class): No value assigned.
Description: Ozone is a colorless to blue gas with a very pungent, characteristic, sulfur-like odor, associated with electrical sparks; condenses to a blue-black liquid or crystalline solid. The odor threshold is 0.045 ppm. Molecular weight = 48.00; Boiling point = $-112°C$; Freezing/Melting point = $-193°C$; Vapor pressure = >1 atm at 20°C; Hazard Identification (based on NFPA-704 M Rating System): Health 2, Flammability 0, Reactivity 3. Slightly soluble in water; solubility = 0.001% at 0°C.
Potential Exposure: Compound Description: Tumorigen, Mutagen; Reproductive Effector; Human Data; Natural Product; Primary Irritant. Ozone is found naturally in the atmosphere as a result of the action of solar radiation and electrical storms. It is also formed around electrical sources, such as X-ray or ultraviolet generators, electric arcs; mercury vapor lamps; linear accelerators; and electrical discharges. Ozone is used as an oxidizing agent in the organic chemical industry (e.g., production of azelaic acid); as a disinfectant for air, mold and bacteria inhibitor for food in cold storage rooms, and for water (e.g., public water supplies; swimming pools; sewage treatment); for bleaching textiles; waxes, flour, mineral oils and their derivatives;

paper pulp; starch and sugar; for aging liquor and wood; for processing certain perfumes; vanillin and camphor; in treating industrial wastes; in the rapid drying of varnishes and printing inks; and in the deodorizing of feathers.
Incompatibilities: A powerful oxidizer. A severe explosion hazard when exposed to shock or heat. Spontaneously decomposes to oxygen under ordinary conditions; heating increases oxygen production. Reacts with all reducing agents, combustibles, organic, and inorganic oxidizable materials, and can form products that are highly explosive. Incompatible with alkenes, aniline, benzene, bromine, ether, ethylene, hydrogen bromide, nitric oxide, stibine. Attacks metals except gold and platinum.
Permissible Exposure Limits in Air
OSHA PEL: 0.1 ppm/0.2 mg/m³ TWA.
NIOSH REL: 0.1 ppm/0.2 mg/m³ Ceiling Concentration.
ACGIH TLV®[1]: (*heavy work*) 0.05 ppm/0.1 mg/m³ TWA; (*moderate work*) 0.08 ppm/0.16 mg/m³ TWA; (*light work*) 0.01 ppm/0.2 mg/m³ TWA; (*light, moderate, or heavy workload* ≤2 h) 0.2 ppm/0.4 mg/m³ TWA; not classifiable as a human carcinogen.
NIOSH IDLH: 5 ppm.
Protective Action Criteria (PAC)
TEEL-0: 0.1 ppm
PAC-1: 0.15 ppm
PAC-2: 1 ppm
PAC-3: 5 ppm
DFG MAK: Carcinogen Category 3B.
Arab Republic of Egypt: TWA 0.1 ppm (0.02 mg/m³), 1993; Australia: TWA 0.1 ppm (0.2 mg/m³); STEL 0.3 ppm, 1993; Austria: MAK 0.1 ppm (0.2 mg/m³), 1999; Belgium: STEL 0.1 ppm (0.2 mg/m³), 1993; Denmark: TWA 0.1 ppm (0.2 mg/m³), 1999; Finland: TWA 0.1 ppm (0.2 mg/m³); STEL 0.3 ppm (0.6 mg/m³), 1999; France: VME 0.1 ppm (0.2 mg/m³), VLE 0.2 ppm (0.4 mg/m³), 1999; Hungary: TWA 0.2 mg/m³; STEL 0.4 mg/m³, 1993; Japan: 0.1 ppm (0.2 mg/m³), 1999; Norway: TWA 0.1 ppm (0.2 mg/m³), 1999; the Philippines: TWA 0.1 ppm (0.2 mg/m³), 1993; Poland: MAC (TWA) 0.1 mg/m³; MAC (STEL) 0.6 mg/m³, 1999; Russia: TWA 0.1 ppm; STEL 0.1 mg/m³, 1993; Sweden: NGV 0.1 ppm (0.2 mg/m³), TGV 0.3 ppm (0.6 mg/m³), 1999; Switzerland: MAK-W 0.1 ppm (0.2 mg/m³), KZG-W 0.2 ppm (0.4 mg/m³), 1999; Turkey: TWA 0.1 ppm (0.2 mg/m³), 1993; United Kingdom: STEL 0.2 ppm (0.4 mg/m³), 2000; Argentina, Bulgaria, Columbia, Jordan, South Korea, New Zealand, Singapore, Vietnam: ACGIH TLV®: not classifiable as a human carcinogen.
The US EPA has set a national ambient air quality standard of 0.12 ppm (240 μg/m³) for ozone on a 1-h average basis and a standard of 0.08 ppm (160 μg/m³) for total photochemical oxidants (expressed as ozone) on a 1-h average basis (to be exceeded not more than once a year). The Czech Republic[35] has set a TWA of 0.1 mg/m³ in workplace air with a ceiling of 0.2 mg/m³. Russia set a MAC of 0.16 mg/m³ for ambient air in residential areas on a once-daily basis and 0.03 mg/m³ on an average daily basis. State

limits for ozone in ambient air[60] range from 0.005 mg/m^3 in Nevada to 0.235 mg/m^3 in Connecticut.

Determination in Air: Use OSHA Analytical Method ID-214.

Routes of Entry: Inhalation, ingestion, skin and/or eye contact.

Harmful Effects and Symptoms

Short Term Exposure: Irritates the eyes and respiratory tract. Eye exposure may result in conjunctivitis (red, inflamed eyes). Inhalation of the gas can cause pulmonary edema, a medical emergency that can be delayed for several hours. This can cause death. Inhalation of the gas may cause asthma-like reactions. The liquid may cause frostbite. May affect the central nervous system, causing headache and impaired concentration and performance. Signs and symptoms of acute exposure to ozone may be severe and include irritation and burns of the skin, eyes, and mucous membranes. An increased respiratory rate, shallow breathing, cough, dyspnea (shortness of breath), bronchitis, pulmonary edema, and pulmonary hemorrhage may occur. Tachycardia (rapid heart rate) and hypotension (low blood pressure) may be observed. Neurologic effects include fatigue, dizziness, drowsiness, headache, exhilaration, and depression. Nausea, vomiting, and anorexia may occur. A level of 0.2 ppm for 3 h did not produce symptoms. Levels of 0.3 ppm may cause tightness in chest and throat, dry throat; and irritation of throat and lungs within 30 min. Levels of 0.5 ppm and above produce a sulfur-like odor and may cause headache, drowsiness, loss of coordination, and accumulation of fluid in the lungs. Levels near 10 ppm may result in immediate, severe irritation of throat and lungs, excessive sweating, continual coughing, decreased blood pressure, weak and rapid pulse, and severe chemical pneumonia. Death may occur from prolonged exposures at 2 ppm or short exposures at 10 ppm.

Note: There may be a delay in onset of breathing difficulties for up to 6 h.

Ozone is highly toxic via inhalation or by contact of liquid to skin, eyes, or mucous membranes. It is capable of causing acute to chronic lung damage, burns, and death or permanent injury. Ozone can be toxic at a concentration of 100 ppm for 1 min. Ozone is capable of causing death from pulmonary edema. It increases sensitivity of the lungs to bronchoconstrictors and allergens; increases susceptibility to and severity of lung bacterial and viral infections.

Long Term Exposure: Repeated or prolonged exposure may cause lung damage and/or chronic respiratory disease. Ozone may damage the developing fetus.

Points of Attack: Eyes, respiratory system, lungs.

Medical Surveillance: NIOSH lists the following tests: chest X-ray; electrocardiogram; pulmonary function tests: forced vital capacity, forced expiratory volume (1 sec); sputum cytology; white blood cell count/differential. Preemployment and periodic physical examinations should be concerned especially with significant respiratory diseases. Eye irritation may also be important.

First Aid: If this chemical gets into the eyes, remove any contact lenses at once and irrigate immediately for at least 15 min, occasionally lifting upper and lower lids. Seek medical attention immediately. If this chemical contacts the skin, remove contaminated clothing and wash immediately with soap and water. Seek medical attention immediately. If this chemical has been inhaled, remove from exposure, begin rescue breathing (using universal precautions, including resuscitation mask) if breathing has stopped and CPR if heart action has stopped. Transfer promptly to a medical facility. Administer 100% O$_2$. Medical observation is recommended for 24–48 h after breathing overexposure, as pulmonary edema may be delayed. As first aid for pulmonary edema, a doctor or authorized paramedic may consider administering a corticosteroid spray.

Personal Protective Methods: In areas of excessive concentration, gas masks with proper canister and full face-piece or goggles or the use of supplied-air respirators is recommended. Wear protective gloves and clothing to prevent any reasonable probability of skin contact. Safety equipment suppliers/manufacturers can provide recommendations on the most protective glove/clothing material for your operation. All protective clothing (suits, gloves, footwear, headgear) should be clean, available each day, and put on before work. Contact lenses should not be worn when working with this chemical. Wear gas-proof chemical goggles and face shield unless full face-piece respiratory protection is worn. Employees should wash immediately with soap when skin is wet or contaminated. Provide emergency showers and eyewash.

Respirator Selection: *Up to 1 ppm:* CcrS (APF = 10) [any chemical cartridge respirator with cartridge(s) providing protection against the compound of concern] or Sa (APF = 10) (any supplied-air respirator). *Up to 2.5 ppm:* Sa:Cf (APF = 25) (any supplied-air respirator operated in a continuous-flow mode) or PaprS (APF = 25) [any powered, air-purifying respirator with cartridge(s) providing protection against the compound of concern]. *Up to 5 ppm:* CcrFS (APF = 50) [any chemical cartridge respirator with a full face-piece and cartridge(s) providing protection against the compound of concern] or GmFS (APF = 50) [any air-purifying, full-face-piece respirator (gas mask) with a chin-style, front- or back-mounted canister providing protection against the compound of concern] or SaT: Cf (APF = 50) (any supplied-air respirator that has a tight-fitting face-piece and is operated in a continuous-flow mode) or SCBAF (APF = 50) (any self-contained breathing apparatus with a full face-piece) or SaF (APF = 50) (any supplied-air respirator with a full face-piece). *Emergency or planned entry into unknown concentrations or IDLH conditions:* SCBAF: Pd,Pp (APF = 10,000) (any self-contained breathing apparatus that has a full face-piece and is operated in a pressure-demand or other positive-pressure mode) or SaF: Pd,Pp: ASCBA (APF = 10,000) (any supplied-air respirator that has a full face-piece and is operated in a pressure-demand or other positive-pressure mode in combination with an auxiliary self-contained breathing apparatus operated in a pressure-demand or other positive-pressure mode). *Escape:* GmFS (APF = 50) [any air-purifying, full-face-piece respirator (gas mask) with a chin-style, front- or back-, mounted canister providing protection against the compound of concern] or SCBAE (any appropriate escape-type, self-contained breathing apparatus).

Storage: Color Code—Red Stripe: Flammability Hazard: Store separately from all other flammable materials. Prior to working with this chemical you should be trained on its proper handling and storage. Store in a cool, dry, well-ventilated place. Detached storage preferred, away from combustible material and reducing agents. See OSHA Standard 1910.104 and NFPA 43A *Code for the Storage of Liquid and Solid Oxidizers* for detailed handling and storage regulations.

Shipping: Compressed gas, toxic, n.o.s. (Ozone) must be labeled "POISON GAS, OXIDIZER." It falls in Hazard Class 2.3. It is a violation of transportation regulations to refill compressed gas cylinders without the express written permission of the owner.

Spill Handling: If in a building, evacuate building and confine vapors by closing doors and shutting down HVAC systems. Restrict persons not wearing protective equipment from area of spill or leak until cleanup is complete. Remove all ignition sources. Ventilate area of spill or leak to disperse the gas. Wear chemical protective suit with self-contained breathing apparatus to combat spills. Stay upwind and use water spray to "knock down" vapor, contain runoff. Stop the flow of gas, if it can be done safely from a distance. If source is a cylinder and the leak cannot be stopped in place, remove the leaking cylinder to a safe place; and repair leak or allow cylinder to empty. Absorb liquids in vermiculite, dry sand, earth, or a similar material and deposit in sealed containers. Keep this chemical out of confined spaces, such as a sewer, because of the possibility of explosion, unless the sewer is designed to prevent the buildup of explosive concentrations. If employees are required to clean up spills, they must be properly trained and equipped. OSHA 1910.120(q) may be applicable.

Fire Extinguishing: Not flammable; however, ozone may react with any combustible substance to cause fire or explosion. Use dry chemical, carbon dioxide, water spray, or alcohol foam extinguishers. Vapors are heavier than air and will collect in low areas. Containers may explode in fire. Storage containers and parts of containers may rocket great distances, in many directions. If material or contaminated runoff enters waterways, notify downstream users of potentially contaminated waters. Notify local health and fire officials and pollution control agencies. From a secure, explosion-proof location, use water spray to cool exposed containers. If cooling streams are ineffective (venting sound increases in volume and pitch, tank discolors, or shows any signs of deforming), withdraw immediately to a secure position. If cylinders are exposed to excessive heat from fire or flame contact, withdraw immediately to a secure location. If employees are expected to fight fires, they must be trained and equipped in OSHA 1910.156. The only respirators recommended for firefighting are self-contained breathing apparatuses that have full face-pieces and are operated in a pressure-demand or other positive-pressure mode.

Disposal Method Suggested: Vent to atmosphere.

References

National Institute for Occupational Safety and Health. (October 1977). *Information Profiles on Potential Occupational Hazards: Ozone*, Report PB-276.678. Rockville, MD, pp. 47—50

National Academy of Sciences. (1977). *Medical and Biologic Effects of Environmental Pollutants: Ozone and Other Photochemical Oxidants*. Washington, DC

US Environmental Protection Agency. (1978). *Air Quality Criteria for Ozone and Other Photochemical Oxidants*, Report EPA-600/8-78-004. Research Triangle Park, NC

US Environmental Protection Agency. (November 30, 1987). *Chemical Hazard Information Profile: Ozone.* Washington, DC: Chemical Emergency Preparedness Program

New York State Department of Health. (March 1986). *Chemical Fact Sheet: Ozone.* Albany, NY: Bureau of Toxic Substance Assessment

Sax, N. I. (Ed.). (1980). *Dangerous Properties of Industrial Materials Report*, 1, No. 2, 52—53

New Jersey Department of Health and Senior Services. (June 2003). *Hazardous Substances Fact Sheet: Ozone.* Trenton, NJ

P

Pancreatin P:0050

Synonyms: Beef viokase; Diastase vera; Donnazyme; Entozyme; Ilozyme; Intrazyme; Pancreatic extract; Pancrex-V; Pankreon; Pankrotanon; Panteric; Stamyl; Viobin; Viokase; Zypanar

CAS Registry Number: 8057-43-0; 9002-16-8; 8049-47-6; 9046-39-3

RTECS® Number: RT9033000

Regulatory Authority and Advisory Bodies

WGK (German Aquatic Hazard Class): No value assigned.

Description: Pancreatin is a yellowish to cream-colored amorphous powder with a strong odor. Slightly soluble in water. Hazard Identification (based on NFPA-704 M Rating System): Health 1, Flammability 1, Reactivity 0.

Potential Exposure: It is an enzyme found in the pancreas and is used in medicines and in treating leather and textiles.

Incompatibilities: Alcohols, acids.

Permissible Exposure Limits in Air

No standards or TEEL available.

Routes of Entry: Inhalation, ingestion, skin and/or eye contact.

Harmful Effects and Symptoms

Short Term Exposure: Pancreatin can affect you when breathed in. Exposure can cause an asthma-like lung reaction, with rapidly occurring symptoms of wheezing and shortness of breath. A second type of lung reaction, with fatigue, shortness of breath, and possibly fever, can occur hours after exposure. This can lead to scars in the lungs. Once allergy develops, even low exposures can trigger symptoms. Other proteolytic enzymes similar to Pancreatin can cause severe eye irritation and irritate the tongue, mouth, and cause nosebleeds and skin sores.

Long Term Exposure: Repeated breathing exposure may cause changes in lung function, even without symptoms. May cause skin allergy, with rash and itching. May cause asthma-like allergy. Can cause severe allergic lung reaction with chills, fever, chest tightness, cough, and/or shortness of breath. Repeated attacks may lead to permanent lung scarring. Once allergy develops, even low exposures can trigger symptoms.

Points of Attack: Lungs.

Medical Surveillance: Before beginning employment and at regular times after that, for those with frequent or potentially high exposures, the following are recommended: lung function tests. These may be normal if the person is not having an attack at the time of the test. If symptoms develop or overexposure is suspected, the following may be useful: consider chest X-ray after acute overexposure.

First Aid: If this chemical gets into the eyes, remove any contact lenses at once and irrigate immediately for at least 15 min, occasionally lifting upper and lower lids. Seek medical attention immediately. If this chemical contacts the skin, remove contaminated clothing and wash immediately with soap and water. Seek medical attention immediately. If this chemical has been inhaled, remove from exposure, begin rescue breathing (using universal precautions, including resuscitation mask) if breathing has stopped and CPR if heart action has stopped. Transfer promptly to a medical facility. When this chemical has been swallowed, get medical attention. Give large quantities of water and induce vomiting. Do not make an unconscious person vomit. Medical observation for up to 8 h after breathing exposure is recommended, as symptoms may be delayed.

Personal Protective Methods: Wear protective gloves and clothing to prevent any reasonable probability of skin contact. Safety equipment suppliers/manufacturers can provide recommendations on the most protective glove/clothing material for your operation. All protective clothing (suits, gloves, footwear, headgear) should be clean, available each day, and put on before work. Contact lenses should not be worn when working with this chemical. Wear dust-proof chemical goggles and face shield unless full face-piece respiratory protection is worn. Employees should wash immediately with soap when skin is wet or contaminated. Provide emergency showers and eyewash.

Respirator Selection: Where there is potential for exposure to pancreatin, use a NIOSH/MSHA- or European Standard EN149-approved full face-piece respirator with a high-efficiency particulate filter. Greater protection is provided by a powered air-purifying respirator. Particulate filters must be checked every day before work for physical damage, such as rips or tears, and replaced as needed. *Where there is potential for high exposures*, use a NIOSH/MSHA- or European Standard EN149-approved supplied-air respirator with a full face-piece operated in the positive-pressure mode, or with a full face-piece, hood, or helmet in the continuous-flow mode; or use a NIOSH/MSHA- or European Standard EN149-approved self-contained breathing apparatus with a full face-piece operated in pressure-demand or other positive-pressure mode.

Storage: Color Code—Green: General storage may be used. Prior to working with this chemical you should be trained on its proper handling and storage. Store in tightly closed containers in a cool, well-ventilated area away from alcohol and acids.

Spill Handling: Evacuate persons not wearing protective equipment from area of spill or leak until cleanup is complete. Remove all ignition sources. Collect powdered material in the most convenient and safe manner and deposit in sealed containers. Ventilate area after cleanup is complete. It may be necessary to contain and dispose of this chemical as a hazardous waste. If material or contaminated runoff enters waterways, notify downstream users of potentially contaminated waters. Contact your local or federal environmental protection agency for specific recommendations.

Sittig's Handbook of Toxic and Hazardous Chemicals and Carcinogens.
DOI: 10.1016/B978-1-4377-7869-4.00015-1

If employees are required to clean up spills, they must be properly trained and equipped. OSHA 1910.120(q) may be applicable.

Fire Extinguishing: This chemical is a combustible solid but does not easily ignite. Use dry chemical, carbon dioxide, water spray, or alcohol foam extinguishers. Poisonous gases are produced in fire. If material or contaminated run-off enters waterways, notify downstream users of potentially contaminated waters. Notify local health and fire officials and pollution control agencies. From a secure, explosion-proof location, use water spray to cool exposed containers. If cooling streams are ineffective (venting sound increases in volume and pitch, tank discolors, or shows any signs of deforming), withdraw immediately to a secure position. If employees are expected to fight fires, they must be trained and equipped in OSHA 1910.156. The only respirators recommended for firefighting are self-contained breathing apparatuses that have full face-pieces and are operated in a pressure-demand or other positive-pressure mode.

Reference

New Jersey Department of Health and Senior Services. (February 1987). *Hazardous Substances Fact Sheet: Pancreatin.* Trenton, NJ

Paraffin wax P:0100

Molecular Formula: C_nH_{2n+2}

Synonyms: Hard paraffin; Paraffin; Paraffin, *n*-; Paraffin fume

CAS Registry Number: 8002-74-2; 71808-29-2 (waxes, petroleum, clay-treated, reaction)

RTECS® Number: RV0350000

UN/NA & ERG Number: Not regulated.

EC Number: 232-315-6

Regulatory Authority and Advisory Bodies

FDA—over-the-counter drug.

US EPA TSCA Section 8(e) Risk Notification, 8EHQ-0892-9311.

Air Pollutant Standard Set. See below, "Permissible Exposure Limits in Air" section.

WGK (German Aquatic Hazard Class): Nonwater polluting agent.

Description: Paraffin wax is a white, somewhat translucent solid and consists of a mixture of solid aliphatic hydrocarbons. It may be obtained from petroleum and consists of a mixture of high-molecular-weight hydrocarbons (e.g., $C_{36}H_{74}$). Specific gravity (H_2O:1) = 0.88−0.92; Freezing/Melting point = 46−68°C; Flash point = 198.9°C; Autoignition temperature = 245°C. Hazard Identification (based on NFPA-704 M Rating System): Health 1, Flammability 0, Reactivity 1. Insoluble in water.

Potential Exposure: Compound Description: Tumorigen, Primary Irritant. Paraffin is used in the manufacture of paraffin paper, candles, food package material, varnishes, floor polishes, and cosmetics. It is also used in waterproofing and extracting of essential oils from flowers for perfume.

Incompatibilities: Strong oxidizers.

Permissible Exposure Limits in Air

Paraffin wax fume

OSHA PEL: None.

NIOSH REL: 2 mg/m³ TWA.

ACGIH TLV®[1]: 2 mg/m³ TWA.

Protective Action Criteria (PAC)

8002-74-2

TEEL-0: 2 mg/m³

PAC-1: 6 mg/m³

PAC-2: 100 mg/m³

PAC-3: 500 mg/m³

71808-29-2

TEEL-0: 10 mg/m³

PAC-1: 30 mg/m³

PAC-2: 50 mg/m³

PAC-3: 250 mg/m³

Australia: TWA 2 mg/m³ (fume), 1993; Belgium: TWA 2 mg/m³ (fume), 1993; Denmark: TWA 2 mg/m³ (fume), 1999; Finland: TWA 1 mg/m³ (fume), 1999; France: VME 2 mg/m³ (fume), 1999; Norway: TWA 2 mg/m³, 1999; Switzerland: MAK-W 2 mg/m³ (fume), 1999; the Netherlands: MAC-TGG 2 mg/m³, 2003; Argentina, Bulgaria, Columbia, Jordan, South Korea, New Zealand, Singapore, Vietnam: ACGIH TLV®: TWA 2 mg/m³. Several states have set guidelines or standards for paraffin wax fume in ambient air[60] ranging from 20−60 μg/m³ (North Dakota) to 35 μg/m³ (Virginia) to 40 μg/m³ (Connecticut) to 48 μg/m³ (Nevada).

Determination in Air: Use OSHA Analytical Method PV-2047.

Routes of Entry: Inhalation, eye contact.

Harmful Effects and Symptoms

Short Term Exposure: Irritates eyes and respiratory system. Inhalation can cause nausea. Occasionally sensitivity reactions have been reported. Chronic exposure can produce chronic dermatitis, wax boils, folliculitis, comedones, melanoderma, papules, and hyperkeratoses.

Long Term Exposure: Fume can cause lung damage, and paraffins contain carcinogens. Carcinoma of the scrotum in pressmen exposed to crude petroleum wax has been documented. Other malignant lesions of an exposed area in employees working with finished paraffin are less well documented. Carcinoma of the scrotum, occurring in workmen exposed 10 years or more, began as a hyperkeratotic nevus like a lesion and developed into a squamous cell carcinoma. The lesions can metastasize to regional inguinal and pelvic lymph nodes. Paraffinoma has been reported from use of paraffin for cosmetic purposes. Summarize by calling it an equivocal tumorigenic agent.

Points of Attack: Eyes, skin, respiratory system.

Medical Surveillance: Medical examinations should be concerned especially with the skin. Surveillance should be continued indefinitely.

First Aid: If this chemical gets into the eyes, remove any contact lenses at once and irrigate immediately for at least 15 min, occasionally lifting upper and lower lids. Seek medical attention immediately. If this chemical contacts the skin, remove contaminated clothing and wash immediately with soap and water. Seek medical attention immediately. If this chemical has been inhaled, remove from exposure, begin rescue breathing (using universal precautions, including resuscitation mask) if breathing has stopped and CPR if heart action has stopped. Transfer promptly to a medical facility. When this chemical has been swallowed, get medical attention. Give large quantities of water and induce vomiting. Do not make an unconscious person vomit.

Personal Protective Methods: Wear protective gloves and clothing to prevent any reasonable probability of skin contact. Safety equipment suppliers/manufacturers can provide recommendations on the most protective glove/clothing material for your operation. All protective clothing (suits, gloves, footwear, headgear) should be clean, available each day, and put on before work. Contact lenses should not be worn when working with this chemical. Wear chemical goggles and face shield unless full face-piece respiratory protection is worn. Employees should wash immediately with soap when skin is wet or contaminated. Provide emergency showers and eyewash.

Respirator Selection: Follow regulations in OSHA 29CFR1910.134 or European Standard EN149. Use a NIOSH/MSHA- or European Standard EN149-approved respirator; or use an approved supplied-air respirator with a full face-piece operated in the positive-pressure mode, or with a full face-piece, hood, or helmet in the continuous-flow mode; or use a NIOSH/MSHA- or European Standard EN149-approved self-contained breathing apparatus with a full face-piece operated in pressure-demand or other positive-pressure mode.

Storage: Color Code—Green: General storage may be used. Prior to working with this chemical you should be trained on its proper handling and storage. Store in tightly closed containers in a cool, well-ventilated area away from oxidizers.

Spill Handling: Evacuate and restrict persons not wearing protective equipment from area of spill or leak until cleanup is complete. Remove all ignition sources. Ventilate area of spill or leak. Absorb liquids in vermiculite, dry sand, earth, peat, carbon, or a similar material and deposit in sealed containers. Keep this chemical out of a confined space, such as a sewer, because of the possibility of an explosion, unless the sewer is designed to prevent the buildup of explosive concentrations. It may be necessary to contain and dispose of this chemical as a hazardous waste. If material or contaminated runoff enters waterways, notify downstream users of potentially contaminated waters. Contact your local or federal environmental protection agency for specific recommendations. If employees are required to clean up spills, they must be properly trained and equipped. OSHA 1910.120(q) may be applicable.

Fire Extinguishing: Paraffin wax is a combustible solid. Use dry chemical, carbon dioxide, water spray, or alcohol foam extinguishers. Poisonous gases are produced in fire. If material or contaminated runoff enters waterways, notify downstream users of potentially contaminated waters. Notify local health and fire officials and pollution control agencies. From a secure, explosion-proof location, use water spray to cool exposed containers. If cooling streams are ineffective (venting sound increases in volume and pitch, tank discolors, or shows any signs of deforming), withdraw immediately to a secure position. If employees are expected to fight fires, they must be trained and equipped in OSHA 1910.156. The only respirators recommended for firefighting are self-contained breathing apparatuses that have full face-pieces and are operated in a pressure-demand or other positive-pressure mode.

Disposal Method Suggested: Dissolve or mix the material with a combustible solvent and burn in a chemical incinerator equipped with an afterburner and scrubber. All federal, state, and local environmental regulations must be observed.

Reference

Sax, N. I. (Ed.). (1981). *Dangerous Properties of Industrial Materials Report*, 1, No. 7, 69–70

Paraformaldehyde   P:0120

Molecular Formula: $(CH_2O)_x$; $C_3H_6O_4$

Synonyms: Aldacide; Flo-more; Formagene; Formaldehyde polymer; Granuform; Paraform; Paraform 3; Paraformaldehido (Spanish); Polyformaldehyde; Polymerized formaldehyde; Polyoxymethylene; Polyoxymethylene glycol; Triformol; Trioxymethylene

CAS Registry Number: 30525-89-4; 110-88-3 (1,3,5-trioxane)

RTECS® Number: RV0540000 (paraformaldehyde); YM1400000 (trioxane)

UN/NA & ERG Number: UN2213/133

EC Number: 203-812-5 [*Annex I Index No.:* 605-002-00-0] (1,3,5-trioxane)

Regulatory Authority and Advisory Bodies

Air Pollutant Standard Set. See below, "Permissible Exposure Limits in Air" section.

US EPA, FIFRA, 1998 Status of Pesticides: Supported.

Clean Water Act: Section 311 Hazardous Substances/RQ 40CFR117.3 (same as CERCLA, see below).

Reportable Quantity (RQ): 1000 lb (454 kg).

Canada, WHMIS, Ingredients Disclosure List Concentration 1.0%.

European/International Regulations (*110-88-3*): Hazard Symbol: F, Xn; Risk phrases: R11, R37; R63; Safety phrases: S2; S36/37; S46 (see Appendix 4).

WGK (German Aquatic Hazard Class): 2—Hazard to waters.

Description: Paraformaldehyde is a white crystalline solid with an irritating odor. The term "trioxane" applies specifically to this trimer $(CH_2O)_3$, but paraformaldehyde is applied both to trioxane and other low polymers or oligomers of formaldehyde. Boiling point = 115°C (trioxane); Freezing/Melting point = 64°C (trioxane); 120–180°C (decomposition/paraformaldehyde); Flash points = 45°C (trioxane); 70°C (paraformaldehyde); Autoignition temperature = 300°C. Explosive limits: LEL = 3.6%; UEL = 28.7% (trioxane); LEL = 7.0%; UEL = 73.0% (paraformaldehyde). Hazard Identification (based on NFPA-704 M Rating System): Health 3, Flammability 3, Reactivity 1. Slightly soluble in water.

Potential Exposure: Compound Description: Agricultural Chemical; Mutagen, Primary Irritant. Paraformaldehyde is used in polyacetal resin manufacture, as a food additive, and as an odorless fuel.

Incompatibilities: Dust forms an explosive mixture with air. Decomposes on contact with oxidizers, strong acids, acid fumes, and bases; with elevated temperatures, forms formaldehyde. May explode when heated. May explode on impact if peroxide contamination develops. Mixtures with hydrogen peroxide or liquid oxygen are explosives sensitive to heat, shock, or contact with lead.

Permissible Exposure Limits in Air
Protective Action Criteria (PAC)
TEEL-0: 4 mg/m^3
PAC-1: 12.5 mg/m^3
PAC-2: 75 mg/m^3
PAC-3: 100 mg/m^3
No TEEL available for *trioxane*.
The Netherlands: 2 ppm/3 mg/m^3 ceiling value[57]; Japan: 5 ppm TWA.

Routes of Entry: Inhalation, ingestion, skin and/or eye contact.

Harmful Effects and Symptoms

Short Term Exposure: Exposure can irritate the eyes, nose, throat, and skin. Exposure can irritate the lungs, causing coughing and/or shortness of breath. Higher exposures can cause pulmonary edema, a medical emergency that can be delayed for several hours. This can cause death. See also "Formaldehyde"; this chemical forms formaldehyde when heated.

Long Term Exposure: Repeated or prolonged contact may cause skin and lung sensitization, resulting in allergies. Paraformaldehyde may cause mutations. Handle with extreme caution. May cause kidney damage. Testing has not been completed to determine the carcinogenicity of paraformaldehyde. However, the limited studies to date indicate that these substances have chemical reactivity and mutagenicity similar to acetaldehyde and malonaldehyde. Therefore, NIOSH recommends that careful consideration should be given to reducing exposures to this aldehyde. Further information can be found in the *NIOSH Current Intelligence Bulletin 55: Carcinogenicity of Acetaldehyde and Malonaldehyde, and Mutagenicity of Related Low-Molecular-Weight Aldehydes* [DHHS (NIOSH), Publication No. 91-112].

Points of Attack: Lungs, kidneys, skin.

Medical Surveillance: Before beginning employment and at regular times after that, for those with frequent or potentially high exposures, the following are recommended: lung function tests. These may be normal if person is not having an attack at the time. If symptoms develop or overexposure is suspected, the following may be useful: consider chest X-ray after acute overexposure. Evaluation by a qualified allergist, including careful exposure history and special testing, may help diagnose skin allergy. Kidney function tests.

First Aid: If this chemical gets into the eyes, remove any contact lenses at once and irrigate immediately for at least 30 min, occasionally lifting upper and lower lids. Seek medical attention immediately. If this chemical contacts the skin, remove contaminated clothing and wash immediately with soap and water. Seek medical attention immediately. If this chemical has been inhaled, remove from exposure, begin rescue breathing (using universal precautions, including resuscitation mask) if breathing has stopped and CPR if heart action has stopped. Transfer promptly to a medical facility. When this chemical has been swallowed, get medical attention. Give large quantities of water and induce vomiting. Do not make an unconscious person vomit. Medical observation is recommended for 24–48 h after breathing overexposure, as pulmonary edema may be delayed. As first aid for pulmonary edema, a doctor or authorized paramedic may consider administering a corticosteroid spray.

Personal Protective Methods: Wear protective gloves and clothing to prevent any reasonable probability of skin contact. Safety equipment suppliers/manufacturers can provide recommendations on the most protective glove/clothing material for your operation. All protective clothing (suits, gloves, footwear, headgear) should be clean, available each day, and put on before work. Contact lenses should not be worn when working with this chemical. Wear dust-proof chemical goggles and face shield unless full face-piece respiratory protection is worn. Employees should wash immediately with soap when skin is wet or contaminated. Provide emergency showers and eyewash.

Respirator Selection: Where there is potential for exposure to Paraformaldehyde, use a NIOSH/MSHA- or European Standard EN149-approved full face-piece respirator with a high-efficiency particulate filter. Greater protection is provided by a powered air-purifying respirator. *Where there is potential for high exposures,* use a NIOSH/MSHA- or European Standard EN149-approved supplied-air respirator with a full face-piece operated in the positive-pressure mode, or with a full face-piece, hood, or helmet in the continuous-flow mode; or use a NIOSH/MSHA- or European Standard EN149-approved self-contained breathing apparatus with a full face-piece operated in pressure-demand or other positive-pressure mode.

Storage: Color Code—Red: Flammability Hazard: Store in a flammable materials storage area. Prior to working with this chemical you should be trained on its proper handling and storage. Before entering confined space where this chemical may be present, check to make sure that an explosive concentration is not a danger. Paraformaldehyde must be stored to avoid contact with oxidizers (such as perchlorates, peroxides, permanganates, chlorates, and nitrates), strong acids (such as hydrochloric, sulfuric, and nitric), and alkaline materials (such as potassium or sodium hydroxide), since violent reactions occur. Sources of ignition, such as smoking and open flames, are prohibited where paraformaldehyde is used, handled, or stored in a manner that could create a potential fire or explosion hazard. Protect containers from physical damage. Store in tightly closed containers in a cool, well-ventilated area away from areas of high humidity.

Shipping: Paraformaldehyde requires a shipping label of "FLAMMABLE SOLID." It falls in Hazard Class 4.1 and Packing Group III.

Spill Handling: Evacuate persons not wearing protective equipment from area of spill or leak until cleanup is complete. Remove all ignition sources. Establish forced ventilation to keep levels below explosive limit. Collect powdered material in the most convenient and safe manner and deposit in sealed containers. Ventilate area after cleanup is complete. Keep paraformaldehyde out of a confined space, such as a sewer, because of the possibility of an explosion, unless the sewer is designed to prevent the buildup of explosive concentrations. It may be necessary to contain and dispose of this chemical as a hazardous waste. If material or contaminated runoff enters waterways, notify downstream users of potentially contaminated waters. Contact your local or federal environmental protection agency for specific recommendations. If employees are required to clean up spills, they must be properly trained and equipped. OSHA 1910.120(q) may be applicable.

Fire Extinguishing: This chemical is a flammable solid. Dust can form an explosive mixture with air. Use dry chemical, carbon dioxide, water spray, or alcohol foam extinguishers. Poisonous gases are produced in fire, including formaldehyde. If material or contaminated runoff enters waterways, notify downstream users of potentially contaminated waters. Notify local health and fire officials and pollution control agencies. From a secure, explosion-proof location, use water spray to cool exposed containers. If cooling streams are ineffective (venting sound increases in volume and pitch, tank discolors, or shows any signs of deforming), withdraw immediately to a secure position. If employees are expected to fight fires, they must be trained and equipped in OSHA 1910.156. The only respirators recommended for firefighting are self-contained breathing apparatuses that have full face-pieces and are operated in a pressure-demand or other positive-pressure mode.

Disposal Method Suggested: Dissolve or mix the material with a combustible solvent and burn in a chemical incinerator equipped with an afterburner and scrubber. All federal, state, and local environmental regulations must be observed.

References

Sax, N. I. (Ed.). (1983). *Dangerous Properties of Industrial Materials Report*, 3, No. 3, 90–92

New Jersey Department of Health and Senior Services. (February 2000). *Hazardous Substances Fact Sheet: Paraformaldehyde*. Trenton, NJ

Paraldehyde   P:0130

Molecular Formula: $C_6H_{12}O_3$
Common Formula: $(CH_3CHO)_3$
Synonyms: A13-03115; *p*-Acetaldehyde; Acetaldehyde, trimer; DEANo. 2585; Elaldehyde; NSC9799; Paraacetaldehyde; Paracetaldehyde; Paral; Paraldehido (Spanish); Paraldehyd (German); Paraldehyde draught; Paraldehyde enema; PCHO; Poral; Triacetaldehyde (French); 2,4,6-Trimethyl-1,3,5-trioxacyclohexane; 1,3,5-Trimethyl-2,4,6-trioxane; 2,4,6-Trimethyl-*s*-trioxane; 2,4,6-Trimethyl-1,3,5-trioxane; *s*-Trimethyltrioxymethylene; *S*-Trioxane, 2,4,6-trimethyl
CAS Registry Number: 123-63-7
RTECS® Number: YK0525000
UN/NA & ERG Number: UN1264/129
EC Number: 204-639-8 [*Annex I Index No.:* 605-004-00-1]
Regulatory Authority and Advisory Bodies
Air Pollutant Standard Set. See below, "Permissible Exposure Limits in Air" section.
US EPA Hazardous Waste Number (RCRA No.): U182.
RCRA, 40CFR261, Appendix 8 Hazardous Constituents.
Reportable Quantity (RQ): 1000 lb (454 kg).
EPCRA Section 313 Form R *de minimis* concentration reporting level: 1.0%.
Canada, WHMIS, Ingredients Disclosure List Concentration 0.1%.
European/International Regulations: Hazard Symbol: F; Risk phrases: R10; Safety phrases: S2; S29. (See Appendix 4).
WGK (German Aquatic Hazard Class): 1—Low hazard to waters.
Description: Paraldehyde is a colorless liquid with a pleasant odor. Molecular weight = 132.18; Boiling point = 125°C; Freezing/Melting point = 12°C; Flash point = 36°C; Autoignition temperature = 237°C. Explosive Limits: LEL = 1.3%; UEL — unknown. Hazard Identification (based on NFPA-704 M Rating System): Health 2, Flammability 3, Reactivity 1. Slightly soluble in water.
Potential Exposure: Compound Description: Drug, Human Data; Primary Irritant. Paraldehyde is used primarily in medicine. It is used as a hypnotic agent, in delirium tremens, and in treatment of psychiatric states characterized by excitement when drugs given over a long period of time. It is also administered for intractable pain which

does not respond to opiates and for basal and obstetrical anesthesia. It is effective against experimentally induced convulsions and has been used in emergency therapy of tetanus, eclampsia, status epilepticus, and poisoning by convulsant drugs. Since it is used primarily in medicine, the chance of accidental human exposure or environmental contamination is low. However, paraldehyde decomposes to acetaldehyde and acetic acid; these compounds have been found to be toxic. In this case, occupational exposure or environmental contamination is possible. Since paraldehyde is prepared from acetaldehyde by polymerization in the presence of an acid catalyst, there exists a potential for adverse effects, although none have been reported in the available literature. It is also used in the manufacture of organic compounds.

Incompatibilities: Forms explosive mixture with air. Incompatible with strong oxidants, strong acids, alkalis, ammonia, amines, iodides, hydrocyanic acid. Violent reaction with liquid oxygen. Contact with acids form acetaldehyde. Attacks rubber and plastics.

Permissible Exposure Limits in Air
Protective Action Criteria (PAC)
TEEL-0: 15 mg/m^3
PAC-1: 40 mg/m^3
PAC-2: 300 mg/m^3
PAC-3: 300 mg/m^3
Russia[43] set a MAC of 5.0 mg/m^3 in work-place air.

Routes of Entry: Inhalation, ingestion, skin and/or eye contact.

Harmful Effects and Symptoms
Short Term Exposure: Paraldehyde can affect you when breathed in and by passing through your skin. Contact can cause severe eye irritation or burns with possible permanent damage, and irritates the skin. Overexposure can cause poor coordination and make you sleepy. Higher exposures can cause pulmonary edema, a medical emergency that can be delayed for several hours. This can cause death.

Long Term Exposure: High or repeated exposure can damage the liver and kidneys. Can irritate the lungs; bronchitis may develop. Repeated exposure may cause fatigue, tremors, changes in speech, personality changes, and/or poor memory.

Points of Attack: Liver, kidneys, lungs, brain.

Medical Surveillance: Liver and kidney function tests. Examine for brain effects. Consider chest X-ray following acute overexposure.

First Aid: If this chemical gets into the eyes, remove any contact lenses at once and irrigate immediately for at least 15 min, occasionally lifting upper and lower lids. Seek medical attention immediately. If this chemical contacts the skin, remove contaminated clothing and wash immediately with soap and water. Seek medical attention immediately. If this chemical has been inhaled, remove from exposure, begin rescue breathing (using universal precautions, including resuscitation mask) if breathing has stopped and CPR if heart action has stopped. Transfer promptly to a medical facility. When this chemical has been swallowed, get medical attention. Give large quantities of water and induce vomiting. Do not make an unconscious person vomit. Medical observation is recommended for 24–48 h after breathing overexposure, as pulmonary edema may be delayed. As first aid for pulmonary edema, a doctor or authorized paramedic may consider administering a corticosteroid spray.

Personal Protective Methods: Wear protective gloves and clothing to prevent any reasonable probability of skin contact. Safety equipment suppliers/manufacturers can provide recommendations on the most protective glove/clothing material for your operation. All protective clothing (suits, gloves, footwear, headgear) should be clean, available each day, and put on before work. Contact lenses should not be worn when working with this chemical. Wear splash-proof chemical goggles and face shield unless full face-piece respiratory protection is worn. Employees should wash immediately with soap when skin is wet or contaminated. Provide emergency showers and eyewash.

Respirator Selection: Where there is potential for exposures to paraldehyde, use a NIOSH/MSHA- or European Standard EN149-approved supplied-air respirator with a full face-piece operated in the positive-pressure mode, or with a full face-piece, hood, or helmet in the continuous-flow mode; or use a NIOSH/MSHA- or European Standard EN149-approved self-contained breathing apparatus with a full face-piece operated in pressure-demand or other positive-pressure mode.

Storage: Color Code—Red: Flammability Hazard: Store in a flammable liquid storage area or approved cabinet away from ignition sources and corrosive and reactive materials. Prior to working with this chemical you should be trained on its proper handling and storage. Before entering confined space where this chemical may be present, check to make sure that an explosive concentration is not a danger. Paraldehyde must be stored to avoid contact with oxidizers (such as perchlorates, peroxides, permanganates, chlorates, and nitrates), liquid oxygen, alkalis, and nitric acid, since violent reactions occur. Store in tightly closed containers in a cool, well-ventilated area. Sources of ignition, such as smoking and open flames, are prohibited where paraldehyde is handled, used, or stored. Metal containers involving the transfer of 5 gallons or more of paraldehyde should be grounded and bonded. Drums must be equipped with self-closing valves, pressure vacuum bungs, and flame arresters. Use only nonsparking tools and equipment, especially when opening and closing containers of paraldehyde.

Shipping: This compound requires a shipping label of "FLAMMABLE LIQUID." It falls in Hazard Class 3 and Packing Group II.

Spill Handling: Evacuate and restrict persons not wearing protective equipment from area of spill or leak until cleanup is complete. Remove all ignition sources. Establish forced ventilation to keep levels below explosive limit. Absorb liquids in vermiculite, dry sand, earth, peat, carbon, or a similar material and deposit in sealed containers. Keep this chemical out of a confined space, such as a sewer, because of the possibility of an explosion, unless the sewer is designed to prevent the buildup of explosive concentrations. It may be necessary to contain and dispose of this chemical as a hazardous waste. If material or contaminated runoff enters waterways, notify downstream users of potentially contaminated waters. Contact your local or federal environmental protection agency for specific recommendations. If employees are required to clean up spills, they must be properly trained and equipped. OSHA 1910.120(q) may be applicable.

Fire Extinguishing: This chemical is a flammable liquid. Poisonous gases, including formaldehyde, are produced in fire. Use dry chemical, carbon dioxide, alcohol foam, or polymer foam extinguishers. Vapors are heavier than air and will collect in low areas. Vapors may travel long distances to ignition sources and flashback. Vapors in confined areas may explode when exposed to fire. Containers may explode in fire. Storage containers and parts of containers may rocket great distances, in many directions. If material or contaminated runoff enters waterways, notify downstream users of potentially contaminated waters. Notify local health and fire officials and pollution control agencies. From a secure, explosion-proof location, use water spray to cool exposed containers. If cooling streams are ineffective (venting sound increases in volume and pitch, tank discolors, or shows any signs of deforming), withdraw immediately to a secure position. If employees are expected to fight fires, they must be trained and equipped in OSHA 1910.156. The only respirators recommended for firefighting are self-contained breathing apparatuses that have full face-pieces and are operated in a pressure-demand or other positive-pressure mode.

Disposal Method Suggested: Incineration in added solvent. Consult with environmental regulatory agencies for guidance on acceptable disposal practices. Generators of waste containing this contaminant ($\geq$100 kg/mo) must conform with EPA regulations governing storage, transportation, treatment, and waste disposal.

References

US Environmental Protection Agency. (April 30, 1980). *Paraldehyde, Health and Environmental Effects Profile No. 140.* Washington, DC: Office of Solid Waste

Sax, N. I. (Ed.). *Dangerous Properties of Industrial Materials Report*, 5, No. 6, 87–90 (1985) and 8, No. 6, 74–79 (1988)

New Jersey Department of Health and Senior Services. (January 2000). *Hazardous Substances Fact Sheet: Paraldehyde.* Trenton, NJ

Paraoxon   P:0140

Molecular Formula: $C_{10}H_{14}NO_6P$

Synonyms: Chinorta; Diaethyl-*p*-nitrophenylphosphorsaeureester (German); *O,O'*-Diethyl-*p*-nitrophenylphosphat

(German); *O,O*-Diethyl *O,p*-nitrophenyl phosphate; *O,O*-Diethyl *p*-nitrophenyl phosphate; Diethyl-*p*-nitrophenyl phosphate; Diethyl paraoxon; *O,O*-Diethylphosphoric acid *O,p*-nitrophenyl ester; E 600; ENT16,087; Ester 25; Ethyl-*p*-nitrophenyl ethylphosphate; Ethyl paraoxon; Eticol; Fosfakol; HC2072; Mintaco; Mintacol; Miotisal; Miotisal A; *O,p*-Nitrofenilfosfato de *O,O*-dietilo (Spanish); *p*-Nitrophenyl diethylphosphate; Oxyparathion; Paraoxone; Paroxan; Pestox 101; Phosphacol; Phosphoric acid, diethyl *p*-nitrophenyl ester; Phosphoric acid, diethyl 4-nitrophenyl ester; Soluglacit; TS219

CAS Registry Number: 311-45-5

RTECS® Number: TC2275000

UN/NA & ERG Number: UN3278 (organophosphorus compound, toxic n.o.s.)/151

EC Number: 206-221-0

Regulatory Authority and Advisory Bodies

US EPA Hazardous Waste Number (RCRA No.): P041.

RCRA, 40CFR261, Appendix 8 Hazardous Constituents.

Reportable Quantity (RQ): 100 lb (45.4 kg).

US DOT Regulated Marine Pollutant (49CFR172.101, Appendix B).

US DOT 49CFR172.101, Inhalation Hazard Chemical as organophosphates.

European/International Regulations: Hazard Symbol: T +, N; Risk phrases: R27/28; R50/53; Safety phrases: S1/2; S28; S36/37; S45; S60; S61 (see Appendix 4).

WGK (German Aquatic Hazard Class): No value assigned.

Description: Paraoxon is an odorless, reddish-yellow oil. Molecular weight = 275.22; Boiling point = 170°C at 1 mmHg; Freezing/Melting point = 189°C; Vapor pressure = 9×10^{-5} mmHg at 25°C. Slightly soluble in water; solubility = 3640 mg/L at 20°C.

Potential Exposure: An organophosphate insecticide. It has been used as a medication.

Incompatibilities: Decomposes in alkaline materials.

Permissible Exposure Limits in Air

No standards or TEEL available.

This chemical can be absorbed through the skin, thereby increasing exposure.

Determination in Air: Use NIOSH Analytical Method (IV) Method #5600, Organophosphorus Pesticides.

Determination in Water: Octanol—water coefficient: Log $K_{ow} = 1.59$.

Routes of Entry: Inhalation, ingestion, skin and/or eye contact.

Harmful Effects and Symptoms

Short Term Exposure: Can cause rapid organophosphate poisoning. Organic phosphorus insecticides are absorbed by the skin as well as by the respiratory and gastrointestinal tracts. They are cholinesterase inhibitors. Symptoms of exposure include headache, giddiness, blurred vision, nervousness, weakness, nausea, cramps, diarrhea, and discomfort in the chest. Signs include sweating, tearing, salivation, vomiting, cyanosis, convulsions, coma, loss of reflexes, and loss of sphincter control. LD_{50} = (oral-rat) 1.8 mg/kg.

Long Term Exposure: Cholinesterase inhibitor; cumulative effect is possible. This chemical may damage the nervous system with repeated exposure, resulting in convulsions, respiratory failure. Repeated exposure may cause personality changes, including depression, anxiety, irritability. May cause liver damage.

Points of Attack: Respiratory system, lungs, central nervous system, cardiovascular system, skin, eyes, plasma and red blood cell cholinesterase.

Medical Surveillance: Before employment and at regular times after that, the following are recommended: plasma and red blood cell cholinesterase levels (tests for the enzyme poisoned by this chemical). If exposure stops, plasma levels return to normal in 1–2 weeks while red blood cell levels may be reduced for 1–3 months. When cholinesterase enzyme levels are reduced by 25% or more below preemployment levels, risk of poisoning is increased, even if results are in lower ranges of "normal." Reassignment to work not involving organophosphate or carbamate pesticides is recommended until enzyme levels recover. If symptoms develop or overexposure occurs, repeat the above tests as soon as possible and get an examination of the nervous system. Also, consider complete blood count. Consider chest X-ray following acute overexposure. Do not drink any alcoholic beverages before or during use. Alcohol promotes absorption of organic phosphates. Refer to the NIOSH Criteria Documents #78-174 and #76-147 on manufacturing, formulating, and working safely with pesticides.

First Aid: If this chemical gets into the eyes, remove any contact lenses at once and irrigate immediately for at least 15 min, occasionally lifting upper and lower lids. Seek medical attention immediately. If this chemical contacts the skin, remove contaminated clothing and wash immediately with soap and water. Speed in removing material from skin is of extreme importance. Shampoo hair promptly if contaminated. Seek medical attention immediately. If this chemical has been inhaled, remove from exposure, begin rescue breathing (using universal precautions, including resuscitation mask) if breathing has stopped and CPR if heart action has stopped. Transfer promptly to a medical facility. When this chemical has been swallowed, get medical attention. Give large quantities of water and induce vomiting. Do not make an unconscious person vomit.

Personal Protective Methods: Wear protective gloves and clothing to prevent any reasonable probability of skin contact. Safety equipment suppliers/manufacturers can provide recommendations on the most protective glove/clothing material for your operation. All protective clothing (suits, gloves, footwear, headgear) should be clean, available each day, and put on before work. Contact lenses should not be worn when working with this chemical. Wear splash-proof chemical goggles and face shield unless full face-piece respiratory protection is worn. Employees should wash immediately with soap when skin is wet or contaminated. Provide emergency showers and eyewash.

Respirator Selection: Follow regulations in OSHA 29CFR1910.134 or European Standard EN149. Use a NIOSH/MSHA- or European Standard EN149-approved respirator; or use an approved supplied-air respirator with a full face-piece operated in the positive-pressure mode, or with a full face-piece, hood, or helmet in the continuous-flow mode; or use a NIOSH/MSHA- or European Standard EN149-approved self-contained breathing apparatus with a full face-piece operated in pressure-demand or other positive-pressure mode.

Storage: Color Code—Blue: Health Hazard/Poison: Store in a secure poison location. Prior to working with this chemical you should be trained on its proper handling and storage. Store in tightly closed containers in a cool, well-ventilated area away from strong oxidizers. Where possible, automatically transfer material from drums or other storage containers to process containers. Sources of ignition, such as smoking and open flames, are prohibited where this chemical is handled, used, or stored. Metal containers involving the transfer of this chemical should be grounded and bonded. Wherever this chemical is used, handled, manufactured, or stored, use explosion-proof electrical equipment and fittings.

Shipping: Organophosphorus compound, toxic n.o.s. require a shipping label of "POISONOUS/TOXIC MATERIALS." It falls in Hazard Class 6.1.

Spill Handling: Evacuate and restrict persons not wearing protective equipment from area of spill or leak until cleanup is complete. Remove all ignition sources. Ventilate area of spill or leak. Absorb liquids in vermiculite, dry sand, earth, peat, carbon, or a similar material and deposit in sealed containers. Keep this chemical out of a confined space, such as a sewer, because of the possibility of an explosion, unless the sewer is designed to prevent the buildup of explosive concentrations. It may be necessary to contain and dispose of this chemical as a hazardous waste. If material or contaminated runoff enters waterways, notify downstream users of potentially contaminated waters. Contact your local or federal environmental protection agency for specific recommendations. If employees are required to clean up spills, they must be properly trained and equipped. OSHA 1910.120(q) may be applicable.

Initial isolation and protective action distances: Distances shown are likely to be affected during the first 30 min after materials are spilled and could increase with time. If more than one tank car, cargo tank, portable tank, or large cylinder involved in the incident is leaking, the protective action distance may need to be increased. You may need to seek emergency information from CHEMTREC at (800) 424-9300 or seek professional environmental engineering assistance from the US EPA Environmental Response Team at (908) 548-8730 (24-h response line).

Small spills (From a small package or a small leak from a large package)

First: Isolate in all directions (feet/meters)   100/30

Then: Protect persons downwind (miles/kilometers)

Day   0.3/0.4
Night   0.8/1.2
Large spills (From a large package or from many small packages)
First: Isolate in all directions (feet/meters)   600/200
Then: Protect persons downwind (miles/kilometers)
Day   1.6/2.6
Night   2.8/4.5

Fire Extinguishing: This chemical is a combustible liquid, but does not readily ignite. Poisonous gases, including oxides of phosphorus and nitrogen, are produced in fire. Use dry chemical, carbon dioxide, or alcohol foam extinguishers. Vapors are heavier than air and will collect in low areas. Vapors in confined areas may explode when exposed to fire. Containers may explode in fire. Storage containers and parts of containers may rocket great distances, in many directions. If material or contaminated runoff enters waterways, notify downstream users of potentially contaminated waters. Notify local health and fire officials and pollution control agencies. From a secure, explosion-proof location, use water spray to cool exposed containers. If cooling streams are ineffective (venting sound increases in volume and pitch, tank discolors, or shows any signs of deforming), withdraw immediately to a secure position. If employees are expected to fight fires, they must be trained and equipped in OSHA 1910.156. The only respirators recommended for firefighting are self-contained breathing apparatuses that have full face-pieces and are operated in a pressure-demand or other positive-pressure mode.

Disposal Method Suggested: In accordance with 40CFR165, follow recommendations for the disposal of pesticides and pesticide containers. Must be disposed properly by following package label directions or by contacting your local or federal environmental control agency or by contacting your regional EPA office.

Reference

New Jersey Department of Health and Senior Services. (May 2000). *Hazardous Substances Fact Sheet: Paraoxon.* Trenton, NJ

Paraquat (paraquat dichloride) P:0150

Molecular Formula: $C_{12}H_{14}Cl_2N_2$
Common Formula: $C_{12}H_{14}N_2Cl_2$
Synonyms: AH 501; AI3-61943; 4,4′-Bipyridinium, 1,1′-dimethyl-, dichloride; Bipyridinium, 1,1′-dimethyl-4,4′-, dichloride; Cekuquat; *para*-COL; Crisquat; Dextrone; Dextrone-X; *N,N′*-Dimethyl-4,4′-bipyridinium dichloride; 1,1′-Dimethyl-4, 4′-bipyridinium dichloride; *N,N′*-Dimethyl-4,4′-bipyridylium dichloride; 1,1′-Dimethyl-4,4′-bipyridynium dichloride; 1,1-Dimethyl-4,4-dipyridilium dichloride; 4,4′-Dimethyldipyridyl dichloride; 1,1′-Dimethyl-4,4′-dipyridylium chloride; *N,N′*-Dimethyl-4,4′-dipyridylium dichloride; 1,1′-Dimethyl-4,4′-dipyridylium

dichloride; Dimethyl violgen chloride; Dimethyl violgen chloride; Esgram; Gamixel; Goldquat 276; Gramoxone; Gramoxone D; Gramoxone dichloride; Gramoxone S; Gramoxone W; Herboxone; Methyl violgen; Methyl violgen (reduced); Methyl violgen chloride; Methyl violgen dichloride; NSC263500; NSC 88126; OK622; Paraquat chloride; *ortho*-Paraquat Cl; Paraquat Cl; Paraquat dichloride; Paraquat dichloride bipyridylnium herbicide; Pathclear; Pillarquat; Pillarxone; PP 148; Sweep; Terraklene; Toxer total; Violgen, methyl-; Weedol

Note: Paraquat is a cation ($C_{12}H_{14}N_2$ ++; 1,1-Dimethyl-4,4-bipyridinium ion); the commercial product is the dichloride salt of paraquat.[NIOSH]

CAS Registry Number: 4685-14-7 (cation); 1910-42-5 (dichloride); *(alt.)* 3765-78-4; *(alt.)* 57593-74-5; *(alt.)* 65982-50-5; *(alt.)* 136338-65-3; *(alt.)* 205105-68-6; *(alt.)* 247050-57-3; 4032-26-2 (diiodide); 2074-50-2 (dimethylsulfate; methosulfat)

RTECS® Number: DW1960000 (cation); DW2275000 (dichloride); DW2280000 (diiodide); DW2010000 (dimethylsulfate)

UN/NA & ERG Number: UN2781

EC Number: 225-141-7; 217-615-7 [*Annex I Index No.:* 613-090-00-7]; 223-714-6 [*Annex I Index No.:* 613-089-00-1] (diquat dichloride); 218-196-3 [*Annex I Index No.:* 613-090-00-7] (paraquat-dimethylsulfate)

Regulatory Authority and Advisory Bodies

Carcinogenicity: EPA (*dichloride*): Possible Human Carcinogen.

US EPA Gene-Tox Program, Positive: *S. cerevisiae* gene conversion.

US EPA, FIFRA 1998 Status of Pesticides: RED completed.

Banned or Severely Restricted (several countries) (UN).[13]

Air Pollutant Standard Set. See below, "Permissible Exposure Limits in Air" section.

SUPERFUND/EPCRA 40CFR355, Extremely Hazardous Substances: TPQ = 100/10,000 lb   (45.4/4540 kg) (dichloride).

Reportable Quantity (RQ): 10 lb (4.54 kg) (dichloride).

EPCRA Section 313 Form R *de minimis* concentration reporting level: 1.0%.

European/International Regulations (*dichloride*): Hazard Symbol: T, N; Risk phrases: R24/25; R26; R36/37/38; R48/25; R50/53; Safety phrases: S1/2; S22; S28; S36/37/39; S45; S60; 61 (see Appendix 4).

WGK (German Aquatic Hazard Class): No value assigned.

Description: Paraquat is a yellow solid with a faint, ammonia-like odor. Molecular weight = 186.28; Specific gravity (H_2O:1) = 1.25; Boiling point = decomposes; Freezing/Melting point = 298°C (decomposes). Hazard Identification (based on NFPA-704 M Rating System): (paraquat) Health 4, Flammability 0, Reactivity 0. Soluble in water. Paraquat dichloride is a quaternary ion which is usually used as the dichloride salt. Molecular weight = 257.18. Highly soluble in water.

Potential Exposure: Compound Description: Agricultural Chemical; Drug, Mutagen; Reproductive Effector; Human Data; Primary Irritant. Those engaged in the manufacture, formulation, and application of this herbicide. Classified for restricted use: limited to use by a certified applicator, or those under applicator's direct supervision.

Incompatibilities: Strong oxidizers, alkylaryl-sulfonate wetting agents; strong bases (hydrolysis). Corrosive to metals. Decomposes in the presence of ultraviolet light. Decomposes in heat (see physical properties, above) and in the presence of UV light, producing nitrogen oxides, hydrogen chloride.

Permissible Exposure Limits in Air

OSHA PEL (*cation, dichloride, methosulfate*): 0.5 mg/m^3 respirable dust TWA [skin].

NIOSH REL (*dichloride*): 0.1 mg/m^3 respirable fraction TWA [skin].

ACGIH TLV$^{®[1]}$ (*cation*): 0.5 mg/m^3/; 0.1 mg/m^3 respirable fraction TWA.

NIOSH IDLH: 1 mg/m^3.

Protective Action Criteria (PAC) (*dichloride*)

TEEL-0: 0.1 mg/m^3

PAC-1: 0.1 mg/m^3

PAC-2: 0.15 mg/m^3

PAC-3: 1 mg/m^3

4685-14-7

TEEL-0: 0.1 mg/m^3

PAC-1: 0.25 mg/m^3

PAC-2: 1 mg/m^3

PAC-3: 1 mg/m^3

DFG MAK (*dichloride*): 0.1 mg/m^3 measured as the inhalable fraction TWA; Peak Limitation Category I(1); [skin].

Austria: MAK 0.1 mg/m^3, [skin], 1999; Denmark: TWA 0.1 mg/m^3, [skin], 1999; Switzerland: MAK-W 0.1 mg/m^3, KZG-W 0.2 mg/m^3, [skin], 1999; United Kingdom: TWA 0.08 mg/m^3, respirable dust, 2000; the Netherlands: MAC-TGG 0.1 mg/m^3, 2003; Argentina, Bulgaria, Columbia, Jordan, South Korea, New Zealand, Singapore, Vietnam: ACGIH TLV$^®$: TWA 0.5 mg/m^3.

Several states have set guidelines or standards for paraquat in ambient air[60] ranging from 0.33 μg/m^3 (New York) to 0.50 μg/m^3 (South Carolina) to 1.0 μg/m^3 (Florida) to 1.6 μg/m^3 (Virginia) to 2.0 μg/m^3 (Connecticut and Nevada).

Determination in Air: Use NIOSH Analytical Method (IV) #5003, Paraquat.

Permissible Concentration in Water: A lifetime health advisory of 3.0 μg/L has been derived by EPA (See "References" Below). In addition, the state of Maine[61] has set a guideline of 17.0 μg/L for paraquat in drinking water.

Determination in Water: Fish Tox = 2115.13304000 ppb (VERY LOW); Octanol—water coefficient: Log K_{ow} = −4.2.

Routes of Entry: Inhalation, ingestion, skin and/or eye contact. Absorbed through the skin.

Harmful Effects and Symptoms

Short Term Exposure: Irritates the eyes, skin, and respiratory tract. Inhalation can cause pulmonary edema, a medical emergency that can be delayed for several hours. This can cause death. Effects occur in two stages, immediate and delayed. Caution is advised. Exposure to paraquat may be fatal; there is no effective antidote. Signs and symptoms of acute exposure to paraquat may be severe and include nausea, vomiting, diarrhea, and abdominal pain. A burning sensation of the mouth and esophagus with possible ulceration may occur following ingestion. Eye exposure may result in corneal opacification (cloudiness). Dermatitis and nail atrophy may occur following dermal contact. Delayed effects include transient reversible liver injury, acute renal failure, and progressive pulmonary fibrosis with associated dyspnea (shortness of breath) and pulmonary edema. Absorbed through the skin and can lead to symptoms as listed in the following paragraph. In addition, can cause fingernail discoloration and damage (which returns to normal when exposure stops), irritation, redness, swelling, and burning. Exposure through ingestion may cause burning of the mouth and throat, nausea, vomiting, abdominal pain, diarrhea, and damage to the kidneys, heart, and liver. Lung damage, leading to death, may occur. One-half ounce of a 20% solution has caused death. Ingestion can also cause lung hemorrhage and fibrosis. The substance may cause effects on the lungs, kidneys, liver, cardiovascular system, and gastrointestinal tract, resulting in impaired functions, tissue lesions.

Long Term Exposure: Repeated or prolonged contact with skin may cause damage and possible loss of the fingernails, and can lead to dry and cracking skin with blistering. Repeated or prolonged exposure to the aerosol can cause lung irritation, lung damage; bronchitis may develop. Can cause scarring of the lungs leading to breathlessness. Can damage the liver, kidneys, and affect the heart. Human Tox = 3.15000 ppb Health Advisory (HIGH).

Points of Attack: Eyes, skin, respiratory system, heart, liver, kidneys, gastrointestinal tract.

Medical Surveillance: NIOSH lists the following tests: chest X-ray; liver function tests; pulmonary function tests: forced vital capacity, forced expiratory volume (1s); urine (chemical/metabolite); urinalysis (routine). Consider the points of attack in preplacement and periodic physical examinations. Kidney function tests. EKG. Chemical users should be cautioned about the use of alcohol which can increase liver damage.

First Aid: If this chemical gets into the eyes, remove any contact lenses at once and irrigate immediately for at least 15 min, occasionally lifting upper and lower lids. Seek medical attention immediately. If this chemical contacts the skin, remove contaminated clothing and wash immediately with soap and water. Seek medical attention immediately. If this chemical has been inhaled, remove from exposure, begin rescue breathing (using universal precautions, including resuscitation mask) if breathing has stopped and CPR if heart action has stopped. Transfer promptly to a medical facility. When this chemical has been swallowed, get medical attention. Give large quantities of water, or bentonite clay in water, or activated charcoal in water; and induce vomiting. Do not make an unconscious person vomit. Medical observation is recommended for 24–48 h after breathing overexposure, as pulmonary edema may be delayed. As first aid for pulmonary edema, a doctor or authorized paramedic may consider administering a corticosteroid spray. Obtain authorization and/or further instructions from the local hospital for performance of other invasive procedures. Rush to a health-care facility.

Personal Protective Methods: Wear protective gloves and clothing to prevent any reasonable probability of skin contact. Safety equipment suppliers/manufacturers can provide recommendations on the most protective glove/clothing material for your operation. All protective clothing (suits, gloves, footwear, headgear) should be clean, available each day, and put on before work. Contact lenses should not be worn when working with this chemical. Wear dust-proof chemical goggles and face shield unless full face-piece respiratory protection is worn. Employees should wash immediately with soap when skin is wet or contaminated. Provide emergency showers and eyewash.

Respirator Selection: NIOSH: *Up to 1 mg/m³:* CcrOv95* (APF = 10) [any air-purifying half-mask respirator with organic vapor cartridge(s) in combination with an N95, R95, or P95 filter. The following filters may also be used: N99, R99, P99, N100, R100, P100]; or PaprOvHie* (APF = 25) (any powered air-purifying respirator with an organic vapor cartridge in combination with a high-efficiency particulate filter); or Sa* (APF = 10) (any supplied-air respirator); or SCBAF (APF = 50) (any self-contained breathing apparatus with a full face-piece). *Emergency or planned entry into unknown concentrations or IDLH conditions:* SCBAF: Pd,Pp (APF = 10,000) (any NIOSH/MSHA- or European Standard EN 149-approved self-contained breathing apparatus that has a full face-piece and is operated in a pressure-demand or other positive-pressure mode) or SaF: Pd,Pp: ASCBA (APF = 10,000) (any supplied-air respirator that has a full face-piece and is operated in a pressure-demand or other positive-pressure mode in combination with an auxiliary, self-contained breathing apparatus operated in a pressure-demand or other positive-pressure mode). *Escape:* GmFOv100 (APF = 50) [any air-purifying, full-face-piece respirator (gas mask) with a chin-style, front- or back-mounted organic vapor canister having an N100, R100, or P100 filter] or SCBAE (any appropriate escape-type, self-contained breathing apparatus). *Substance reported to cause eye irritation or damage; may require eye protection.

Storage: Color Code—Blue: Health Hazard/Poison: Store in a secure poison location. Prior to working with this chemical you should be trained on its proper handling and storage. This chemical is inactivated by inert clays and anionic surfactants. Store in tightly closed containers in a cool, well-ventilated area away from oxidizers, alkylaryl-sulfonate wetting agents, light. Where possible, automatically pump material from drums or other storage containers to process containers. Sources of ignition, such as smoking and open flames, are prohibited where this chemical is handled, used, or stored. Metal containers involving the transfer of this chemical should be grounded and bonded. Use only nonsparking tools and equipment, especially when opening and closing containers of this chemical. Wherever this chemical is used, handled, manufactured, or stored, use explosion-proof electrical equipment and fittings.

Shipping: Bipyridilium pesticides, solid, toxic, require a shipping label of "POISONOUS/TOXIC MATERIALS." They fall in DOT Hazard Class 6.1 and Paraquat is in Packing Group II.

Spill Handling: Evacuate persons not wearing protective equipment from area of spill or leak until cleanup is complete. Poisonous gases, including nitrogen oxides, are produced in fire. Remove all ignition sources. Stay upwind; keep out of low areas. Ventilate closed spaces before entering them. Remove and isolate contaminated clothing at the site. If water pollution occurs, notify appropriate authorities. Do not touch spilled material; stop leak if you can do so without risk. Use water spray to reduce vapors. *Small spills:* absorb with sand or other noncombustible absorbent material and place into containers for later disposal. *Small dry spills:* with clean shovel place material into clean, dry container and cover; move containers from spill area. *Large spills:* dike far ahead of spill for later disposal. Ventilate area after cleanup is complete. It may be necessary to contain and dispose of this chemical as a hazardous waste. If material or contaminated runoff enters waterways, notify downstream users of potentially contaminated waters. Contact your local or federal environmental protection agency for specific recommendations. If employees are required to clean up spills, they must be properly trained and equipped. OSHA 1910.120(q) may be applicable.

Fire Extinguishing: This chemical is a combustible solid. Procedures for bipyridilium pesticides are as follows. *Small fires:* dry chemical, carbon dioxide, water spray, or foam. *Large fires:* water spray, fog, or foam. Move container from fire area if you can do so without risk. Fight fire from maximum distance. Dike fire control water for later disposal; do not scatter the material. Poisonous gases are produced in fire. If material or contaminated runoff enters waterways, notify downstream users of potentially contaminated waters. Notify local health and fire officials and pollution control agencies. Containers may explode in fire. From a secure, explosion-proof location, use water spray to cool exposed containers. If cooling streams are ineffective (venting sound increases in volume and pitch, tank discolors, or shows any signs of deforming), withdraw immediately to a secure position. If employees are expected to fight fires, they must be trained and equipped in OSHA 1910.156. The only respirators recommended for firefighting are self-contained breathing apparatuses that have full face-pieces and are operated in a pressure-demand or other positive-pressure mode.

Disposal Method Suggested: Paraquat is rapidly inactivated in soil. It is also inactivated by anionic surfactants.

Therefore, an effective and environmentally safe disposal method would be to mix the product with ordinary household detergent and bury the mixture in clay soil. In accordance with 40CFR165, follow recommendations for the disposal of pesticides and pesticide containers. Must be disposed properly by following package label directions or by contacting your local or federal environmental control agency or by contacting your regional EPA office. Consult with environmental regulatory agencies for guidance on acceptable disposal practices. Generators of waste containing this contaminant ($\geq$100 kg/mo) must conform with EPA regulations governing storage, transportation, treatment, and waste disposal.

References

Pasi, A. (1978). *The Toxicology of Paraquat, Diquat and Morfamquat.* Bern, Switzerland: H. Huber

US Environmental Protection Agency. (November 30, 1987). *Chemical Hazard Information Profile: Paraquat.* Washington, DC: Chemical Emergency Preparedness Program

Sax, N. I. (Ed.). (1988). *Dangerous Properties of Industrial Materials Report,* 8, No. 2, 67−72

US Environmental Protection Agency. (August 1987). *Health Advisory: Paraquat.* Washington, DC: Office of Drinking Water

New York State Department of Health. (February 1986). *Chemical Fact Sheet: Paraquat* (Version 2 and Version 3). Albany, NY: Bureau of Toxic Substance Assessment

New Jersey Department of Health and Senior Services. (September 1999). *Hazardous Substances Fact Sheet: Paraquat.* Trenton, NJ

US Environmental Protection Agency, Special Review and Reregistration Division Office of Pesticide Programs. (1998). *Agency Status of Pesticides in Registration, Reregistration, and Special Review* (Rainbow Report). Washington, DC

Paraquat methosulfate   P:0160

Molecular Formula: $C_{14}H_{20}N_2O_8S_2$
Common Formula: $C_{12}H_{14}N_2(CH_3SO_4)_2$
Synonyms: 4,4-Bipyridinium, 1,1′-dimethyl-, bis(methyl sulfate); 1,1′-Dimethyl-4,4′-bipyridyniumdimethylsulfate; 1,1′-Dimethyl-4,4′-dipyridynium di(methyl sulfate); Gramoxone methyl sulfate; Paraqiat I; Paraquat bis(methyl sulfate); Paraquat dimethosulfate; Paraquat dimethyl sulphate; Paraquat dimethyl sulfate; Paraquat methsulfate bipyridylnium herbicide; PP 910
CAS Registry Number: 2074-50-2
RTECS® Number: DW2010000
UN/NA & ERG Number: UN2781/151
EC Number: 218-196-3 [*Annex I Index No.:* 613-090-00-7]
Regulatory Authority and Advisory Bodies
Banned or Severely Restricted (Hungary) (UN).[13]

SUPERFUND/EPCRA 40CFR355, Extremely Hazardous Substances: TPQ = 10/10,000 lb (4.54/4540 kg).
Reportable Quantity (RQ): 10 lb (4.54 kg).
European/International Regulations (*includes dichloride*): Hazard Symbol: T, N; Risk phrases: R24/25; R26; R36/37/38; R48/25; R50/53; Safety phrases: S1/2; S22; S28; S36/37/39; S45; S60; 61 (see Appendix 4).
WGK (German Aquatic Hazard Class): No value assigned.
Description: Paraquat methosulfate is a white to yellow crystalline solid. Molecular weight = 408.48; Freezing/Melting point = 175−180°C (decomposition). Hazard Identification (based on NFPA-704 M Rating System): Health 4, Flammability 0, Reactivity 0. Soluble in water.
Potential Exposure: Those who might be involved in the manufacture or use of this contact herbicide and desiccant.
Incompatibilities: Strong oxidizers.
Permissible Exposure Limits in Air
Protective Action Criteria (PAC)
TEEL-0: 0.75 mg/m^3
PAC-1: 2 mg/m^3
PAC-2: 15 mg/m^3
PAC-3: 40 mg/m^3
A MAC of 0.01 mg/m^3 has been set in Bulgaria according to the EPA Profile (see "References," below).
Determination in Air: Use NIOSH Analytical Method (IV) #5003, Paraquat.
Permissible Concentration in Water: Paraquat: A lifetime health advisory of 3.0 μg/L has been derived by EPA (see "References," below). In addition, the state of Maine[61] has set a guideline of 17.0 μg/L for paraquat in drinking water.
Routes of Entry: Inhalation, ingestion, skin and/or eye contact.
Harmful Effects and Symptoms
Short Term Exposure: Contact causes irritation. Inhalation causes nosebleeds, headaches, coughing, and a sore throat. Higher exposures can cause pulmonary edema, a medical emergency that can be delayed for several hours. This can cause death. Swallowing causes burning in mouth, throat, and abdomen; vomiting, bloody vomitus; diarrhea with bloody stools; and headaches. It can cause death by shock and/or pulmonary damage. The fatal dose is estimated to be 6 g of paraquat ion. Exposure may cause renal tubular damage and liver dysfunction. Death may occur in 24 h or less.
Long Term Exposure: Liver and kidney damage.
Points of Attack: Lungs, liver, kidneys.
Medical Surveillance: NIOSH lists the following tests (paraquat dichloride): chest X-ray; liver function tests; pulmonary function tests: forced vital capacity, forced expiratory volume (1 s); urine (chemical/metabolite); urinalysis (routine).
First Aid: If this chemical gets into the eyes, remove any contact lenses at once and irrigate immediately for at least 15 min, occasionally lifting upper and lower lids. Seek medical attention immediately. If this chemical contacts the skin, remove contaminated clothing and wash immediately with soap and water. Seek medical attention immediately.

If this chemical has been inhaled, remove from exposure, begin rescue breathing (using universal precautions, including resuscitation mask) if breathing has stopped and CPR if heart action has stopped. Transfer promptly to a medical facility. When this chemical has been swallowed, get medical attention. Give large quantities of water and induce vomiting. Do not make an unconscious person vomit. Medical observation is recommended for 24–48 h after breathing overexposure, as pulmonary edema may be delayed. As first aid for pulmonary edema, a doctor or authorized paramedic may consider administering a corticosteroid spray.

Personal Protective Methods: Wear protective gloves and clothing to prevent any reasonable probability of skin contact. Safety equipment suppliers/manufacturers can provide recommendations on the most protective glove/clothing material for your operation. All protective clothing (suits, gloves, footwear, headgear) should be clean, available each day, and put on before work. Contact lenses should not be worn when working with this chemical. Wear dust-proof chemical goggles and face shield unless full face-piece respiratory protection is worn. Employees should wash immediately with soap when skin is wet or contaminated. Provide emergency showers and eyewash.

Respirator Selection: Follow regulations in OSHA 29CFR1910.134 or European Standard EN149. Use a NIOSH/MSHA- or European Standard EN149-approved respirator; or use an approved supplied-air respirator with a full face-piece operated in the positive-pressure mode, or with a full face-piece, hood, or helmet in the continuous-flow mode; or use a NIOSH/MSHA- or European Standard EN149-approved self-contained breathing apparatus with a full face-piece operated in pressure-demand or other positive-pressure mode.

The following is for reference: NIOSH (for *paraquat dichloride*): *Up to 1 mg/m³:* CcrOv95* (APF = 10) [any air-purifying half-mask respirator equipped with an organic vapor cartridge(s) in combination with an N95, R95, or P95 filter. The following filters may also be used: N99, R99, P99, N100, R100, P100]; or PaprOvHie* (APF = 25) (any powered air-purifying respirator with an organic vapor cartridge in combination with a high-efficiency particulate filter); or Sa* (APF = 10) (any supplied-air respirator); or SCBAF (APF = 50) (any self-contained breathing apparatus with a full face-piece). *Emergency or planned entry into unknown concentrations or IDLH conditions:* SCBAF: Pd, Pp (APF = 10,000) (any NIOSH/MSHA- or European Standard EN 149-approved self-contained breathing apparatus that has a full face-piece and is operated in a pressure-demand or other positive-pressure mode) or SaF: Pd,Pp: ASCBA (APF = 10,000) (any supplied-air respirator that has a full face-piece and is operated in a pressure-demand or other positive-pressure mode in combination with an auxiliary, self-contained breathing apparatus operated in a pressure-demand or other positive pressure mode). *Escape:* GmFOv100 (APF = 50) [any air-purifying, full-face-piece respirator (gas mask) with a chin-style, front- or back-mounted organic vapor canister having an N100, R100, or P100 filter] or SCBAE (any appropriate escape-type, self-contained breathing apparatus).

*Substance reported to cause eye irritation or damage; may require eye protection.

Storage: Color Code—Blue: Health Hazard/Poison: Store in a secure poison location. Prior to working with this chemical you should be trained on its proper handling and storage. Store in tightly closed containers in a cool, well-ventilated area away from oxidizers. Where possible, automatically transfer material from drums or other storage containers to process containers. Sources of ignition, such as smoking and open flames, are prohibited where this chemical is handled, used, or stored. Metal containers involving the transfer of this chemical should be grounded and bonded. Use only nonsparking tools and equipment, especially when opening and closing containers of this chemical. Wherever this chemical is used, handled, manufactured, or stored, use explosion-proof electrical equipment and fittings.

Shipping: Bipyridilium pesticides, solid, toxic, require a shipping label of "POISONOUS/TOXIC MATERIALS." They fall in DOT Hazard Class 6.1 and Paraquat is in Packing Group III.

Spill Handling: Remove all ignition sources. Keep unnecessary people away; isolate hazard area and deny entry. Stay upwind; keep out of low areas. Do not touch spilled material; stop leak if you can do so without risk. *Small spills:* absorb with sand or other noncombustible absorbent material and place into containers for later disposal. *Small dry spills:* with clean shovel place material into clean, dry container and cover; move containers from spill area. *Large spills:* dike far ahead of spill for later disposal. Ventilate area after cleanup is complete. It may be necessary to contain and dispose of this chemical as a hazardous waste. If material or contaminated runoff enters waterways, notify downstream users of potentially contaminated waters. Contact your local or federal environmental protection agency for specific recommendations. If employees are required to clean up spills, they must be properly trained and equipped. OSHA 1910.120(q) may be applicable.

Fire Extinguishing: This chemical is a noncombustible solid. Use dry chemical, carbon dioxide, water spray, or alcohol foam extinguishers. Poisonous gases are produced in fire, including nitrogen oxides and sulfur oxides. If material or contaminated runoff enters waterways, notify downstream users of potentially contaminated waters. Notify local health and fire officials and pollution control agencies. From a secure, explosion-proof location, use water spray to cool exposed containers. If cooling streams are ineffective (venting sound increases in volume and pitch, tank discolors, or shows any signs of deforming), withdraw immediately to a secure position. If employees are expected to fight fires, they must be trained and equipped in OSHA 1910.156. The only respirators recommended for firefighting are self-contained breathing apparatuses that have full face-pieces and are operated in a pressure-demand or other positive-pressure mode.

Disposal Method Suggested: In accordance with 40CFR165, follow recommendations for the disposal of pesticides and pesticide containers. Must be disposed properly by following package label directions or by contacting your local or federal environmental control agency or by contacting your regional EPA office.

Reference

US Environmental Protection Agency. (November 30, 1987). *Chemical Hazard Information Profile: Paraquat Methosulfate.* Washington, DC: Chemical Emergency Preparedness Program

Parathion   P:0170

Molecular Formula: $C_{10}H_{14}NO_5PS$

Synonyms: AAT; AATP; ACC 3422; Alkron; Alleron; American cyanamid 3422; Aphamite; Aralo; B 404; BAY E-605; Bayer E-605; Bladan F; Compound 3422; Corothion; Corthion; Corthione; Danthion; *O,O*-Diethyl *O*-(*p*-nitrophenyl) phosphorothioate; *O,O*-Diethyl *O*-(4-nitrophenyl) phosphorothioate; *O,O*-Diethyl *O,p*-nitrophenyl phosphorothioate; Diethyl *p*-nitrophenyl phosphorothionate; Diethyl 4-nitrophenyl phosphorothionate; Diethyl *p*-nitrophenyl thionophosphate; *O,O*-Diethyl *O,p*-nitrophenyl thiophosphate; Diethyl parathion; DNTP:; DPP; Drexel parathion 8E; E 605; E 605 F; Ecatox; Ekatin WF & WF ULV; Ekatox; ENT15,108; Ethlon; Ethyl parathion; Etilon; Folidol; Folidol E; Folidol E-605; Folidol E&E 605; Folidol oil; Fosfermo; Fosferno; Fosfex; Fosfive; Fosova; Fostern; Fostox; Gearphos; Genithion; Kalphos; Kypthion; Lethalaire G-54; Lirothion; Murfos; Murphos; NCI-C00226; Niran; Niran E-4; Nitrostigmin (German); Nitrostigmine; NIUIF 100; Nourithion; Oleofos 20; Oleoparathene; Oleoparathion; OMS 19; Orthophos; PAC; Pacol; Panthion; Paradust; Paramar; Paramar 50; Paraphos; Parathene; Parathion-ethyl; Parathion thiophos; Parationa (Spanish); Parawet; Penncap E; Pestox plus; Pethion; Phoskil; Phosphorothioic acid, *O,O*-diethyl *O*-(*p*-nitrophenyl) ester; Phosphorothioic acid, *O,O*-diethyl *O*-(4-nitrophenyl) ester; Phosphostigmine; Pleoparaphene; RB; Rhodiasol; Rhodiatox; Rhodiatrox; Selephos; SNP; Soprathion; Stathion; STCC4921469; Sulphos; Super rodiatox; T-47; Thiomex; Thiophos; Thiophos 3422; Tiofos; TOX 47; Toxol (3); Vapophos; Vitrex

CAS Registry Number: 56-38-2; *(alt.)* 8057-70-3; *(alt.)* 11111-91-4; *(alt.)* 110616-89-2

RTECS® Number: TF4550000

UN/NA & ERG Number: UN2783 (Methyl parathion, solid)/152; UN1967 (Parathion and compressed gas mixture)/123; UN3018 (organophosphorus pesticide, liquid, toxic)/152

EC Number: 200-271-7 [*Annex I Index No.:* 015-034-00-1]

Regulatory Authority and Advisory Bodies

Carcinogenicity: IARC: Animal Inadequate Evidence; Human No Adequate Data, *not classifiable as carcinogenic to humans*, Group 3, 1987; EPA: Possible Human Carcinogen; NCI: Carcinogenesis Bioassay (feed); clear evidence: rat; no evidence: mouse.

US EPA Gene-Tox Program, Negative: *In vitro* UDS—human fibroblast; TRP reversion; Negative: *S. cerevisiae*—homozygosis; Inconclusive: *B. subtilis* rec assay; *E. coli* polA without S9; Inconclusive: Histidine reversion—Ames test; Inconclusive: *D. melanogaster* sex-linked lethal Banned or Severely Restricted (many countries) (UN).[13]

US EPA, FIFRA, 1998 Status of Pesticides: Supported.

Very Toxic Substance (World Bank).[15]

Air Pollutant Standard Set. See below, "Permissible Exposure Limits in Air" section.

Clean Water Act: Section 311 Hazardous Substances/RQ 40CFR117.3 (same as CERCLA, see below); Section 313 Water Priority Chemicals (57FR41331, 9/9/92).

US EPA Hazardous Waste Number (RCRA No.): P089.

RCRA, 40CFR261, Appendix 8 Hazardous Constituents.

RCRA 40CFR268.48; 61FR15654, Universal Treatment Standards: Wastewater (mg/L), 0.014; Nonwastewater (mg/kg), 4.6.

RCRA 40CFR264, Appendix 9; TSD Facilities Ground Water Monitoring List. Suggested test method(s) (PQL μg/L): 8270 (10).

Safe Drinking Water Act: Priority List (55 FR 1470) as parathion degradation.

Superfund/EPCRA 40CFR355, Extremely Hazardous Substances: TPQ = 100 lb (45.4 kg).

Reportable Quantity (RQ): 10 lb (4.54 kg).

EPCRA Section 313 Form R *de minimis* concentration reporting level: 1.0%.

US DOT Regulated Marine Pollutant (49CFR172.101, Appendix B), severe pollutant.

US DOT 49CFR172.101, Inhalation Hazard Chemical as organophosphates.

Rotterdam Convention Annex III [Chemicals Subject to the Prior Informed Consent Procedure (PIC)] [parathion (all formulations—aerosols, dustable powder (DP), emulsifiable concentrate (EC), granules (GR), and wettable powders (WP)—of this substance are included, *except* capsule suspensions (CS))].

European/International Regulations: Hazard Symbol: T +, N; Risk phrases: R24; R26/28; R48/25; R50/53; Safety phrases: S1/2; S28; S36/37; S45; S60; S61 (see Appendix 4).

WGK (German Aquatic Hazard Class): 3—Severe hazard to waters.

Description: Parathion is a clear liquid when fresh; pale yellow to dark-brown liquid with a garlic-like odor. Commercial formulations use carrier solvents that may change the physical properties shown. Molecular weight = 291.28; Specific gravity $(H_2O:1) = 1.27$; Boiling point = 375°C; 157−162°C at 6 mmHg (for CW agent); Freezing/Melting point = 6.1°C; also listed at 2.9°C for chemical warfare (CW) agent; Flash point = 195°C (oc). Hazard Identification (based on NFPA-704 M Rating System): Health 4, Flammability 0, Reactivity 0. Slightly soluble in water; solubility = 24 mg/L.

Potential Exposure: A severely hazardous pesticide formulation. Those engaged in the manufacture, formulation, and application of this broad-spectrum insecticide. This material has also been used as a chemical warfare agent.

Incompatibilities: Strong oxidizers may cause release of toxic phosphorus oxides. Organophosphates, in the presence of strong reducing agents such as hydrides, may form highly toxic and flammable phosphine gas. Keep away from alkaline materials. Attacks some plastics, rubbers, and coatings. Rapidly hydrolyzed by alkalis.

Permissible Exposure Limits in Air
OSHA PEL: 0.1 mg/m^3 TWA [skin].
NIOSH REL: 0.05 mg/m^3 TWA [skin].
ACGIH TLV®[1]: 0.05 mg/m^3 TWA, inhalable fraction and vapor [skin]; not classifiable as a human carcinogen; BEI; 0.05 mg[creatinine]/g in urine, end-of-shift.
NIOSH IDLH: 10 mg/m^3.
Protective Action Criteria (PAC)*
TEEL-0: 0.1 mg/m^3
PAC-1: 0.15 mg/m^3
PAC-2: **1.5** mg/m^3
PAC-3: **2** mg/m^3
AEGLs (Acute Emergency Guideline Levels) & ERPGs (Emergency Response Planning Guideline) are in **bold face.**
DFG MAK: 0.1 mg/m^3 measured as the inhalable fraction TWA; Peak Limitation Category II(8) [skin]; Pregnancy Risk Group D; BAT: 100 μg[*p*-nitrophenol]/L in urine after several shifts (sampling time).
Arab Republic of Egypt: TWA 0.1 mg/m^3, [skin], 1993; Australia: TWA 0.1 mg/m^3, [skin], 1993; Austria: MAK 0.1 mg/m^3, [skin], 1999; Belgium: TWA 0.1 mg/m^3, [skin], 1993; Denmark: TWA 0.1 mg/m^3, [skin], 1999; Finland: TWA 0.1 mg/m^3, short-term exposure limit 0.3 mg/m^3, [skin], 1999; France: VME 0.1 mg/m^3, [skin], 1999; the Netherlands: MAC-TGG 0.1 mg/m^3, [skin], 2003; Japan: 0.1 mg/m^3, [skin], 1999; Norway: TWA 0.05 mg/m^3, 1999; the Philippines: TWA 0.1 mg/m^3, [skin], 1993; Russia: STEL 0.05 mg/m^3, 1993; Thailand: TWA 0.11 mg/m^3, 1993; Turkey: TWA 0.1 mg/m^3, [skin], 1993; United Kingdom: TWA 0.1 mg/m^3, short-term exposure limit 0.3 mg/m^3, [skin], 2000; Argentina, Bulgaria, Columbia, Jordan, South Korea, New Zealand, Singapore, Vietnam: ACGIH TLV®: not classifiable as a human carcinogen. Several states have set guidelines or standards for parathion in ambient air[60] ranging from 0.238 μg/m^3 (Kansas) to 0.33 μg/m^3 (New York) to 0.5 μg/m^3 (South Carolina) to 1.0 μg/m^3 (North Dakota) to 1.6 μg/m^3 (Virginia) to 1.87 μg/m^3 (Pennsylvania) to 2.0 μg/m^3 (Connecticut and Nevada).

Determination in Air: Use NIOSH Analytical Method (IV) #5600, Organophosphorus pesticides; OSHA Analytical Method ID-62.

Permissible Concentration in Water: Russia set a MAC[35] of 3.0 μg/L in surface water and Mexico has set maximum permissible concentrations of 1.0 μg/L in coastal waters and 10.0 μg/L in estuaries. Several states have set guidelines for parathion in drinking water[61] ranging from 8.6 μg/L in Maine to 30.0 μg/L in California and Kansas.

Determination in Water: Fish Tox = 0.26514000 ppb MATC (EXTRA HIGH); Octanol−water coefficient: Log K_{ow} = 3.15−3.8.

Routes of Entry: Inhalation, ingestion, skin and/or eye contact. Absorbed through the skin.

Harmful Effects and Symptoms
Short Term Exposure: Parathion irritates the eyes, skin, and respiratory tract. A cholinesterase inhibitor. Acute exposure to parathion may produce the following signs and symptoms: pinpoint pupils, blurred vision, headache, dizziness, muscle spasms, and profound weakness. Vomiting, diarrhea, abdominal pain, seizures, and coma may also occur. The heart rate may decrease following oral exposure or increase following dermal exposure. Hypotension (low blood pressure) is not uncommon. Respiratory symptoms include dyspnea (shortness of breath), respiratory depression, and respiratory paralysis. Psychosis may occur. This material is extremely toxic; the probable oral lethal dose is 5−50 mg/kg or between 7 drops and 1 teaspoonful for a 150-lb person. As little as 1 drop can endanger life if splashed in the eye. Toxicity is highest by inhalation. People at special risk are those with a history of glaucoma, cardiovascular disease, hepatic disease, renal disease, or central nervous system abnormalities. Some additional details on short-term exposure to parathion are as follows: *Inhalation:* Occasional human exposures at concentrations of 0.1−0.8 mg/m^3 did not give rise to any symptoms. Occasional human exposure at 1.5−2.0 mg/m^3 resulted in nausea and vomiting. Higher exposures can give rise to dizziness, blurred vision, wheezing, excessive salivation, and muscle and abdominal cramps. An estimated 10−20 mg (1/1500 oz) may cause death. *Skin:* However, many human poisonings have occurred through extensive skin contact at unspecified levels. This is the greatest hazard for some workers. Symptoms of poisoning include nausea, vomiting, weakness, blurring of vision, and muscle cramps. NIOSH lists the following symptoms of exposure: irritation of the eyes, skin, respiratory system; miosis; rhinorrhea (discharge of thin nasal mucus); headache; chest tightness; wheezing, laryngeal spasm; salivation, cyanosis, anorexia, nausea, vomiting, abdominal cramps; diarrhea; sweating; muscle fasciculation; weakness, paralysis; giddiness, confusion, ataxia; convulsions, coma; low blood pressure; cardiac irregular/irregularities.

Long Term Exposure: Cholinesterase inhibitor; cumulative effect is possible. This chemical may damage the nervous system with repeated exposure, resulting in convulsions, respiratory failure. May cause liver damage. Human Tox = 0.23100 ppb Health Advisory (EXTRA HIGH).

Points of Attack: Respiratory system, central nervous system, cardiovascular system, eyes, skin, blood cholinesterase.

Medical Surveillance: NIOSH lists the following tests: blood serum; cholinesterase: whole blood (chemical/metabolite); cholinesterase: blood plasma; cholinesterase: blood plasma, red blood cells/count; urine (chemical/metabolite); urine (chemical/metabolite), end-of-shift; urine (chemical/metabolite), end-of-workweek. NIOSH recommends that medical surveillance, including preemployment and periodic examinations, shall be made available to workers who may be occupationally exposed to parathion. Biologic monitoring is also recommended as an additional safety measure. Before employment and at regular times after that, the following are recommended: plasma and red blood cell cholinesterase levels (tests for the enzyme poisoned by this chemical). If exposure stops, plasma levels return to normal in 1−2 weeks while red blood cell levels may be reduced for 1−3 months. Do not drink any alcoholic beverages before or during use. Alcohol promotes absorption of organic phosphates. When cholinesterase enzyme levels are reduced by 25% or more below preemployment levels, risk of poisoning is increased, even if results are in lower ranges of "normal." Reassignment to work not involving organophosphate or carbamate pesticides is recommended until enzyme levels recover. If symptoms develop or overexposure occurs, repeat the above tests as soon as possible and get an examination of the nervous system. Also, consider complete blood count. Consider chest X-ray following acute overexposure. Do not drink any alcoholic beverages before or during use. Alcohol promotes absorption of organic phosphates.

First Aid: If this chemical gets into the eyes, remove any contact lenses at once and irrigate immediately for at least 15 min, occasionally lifting upper and lower lids. Seek medical attention immediately. If this chemical contacts the skin, remove contaminated clothing and wash immediately with soap and water. Seek medical attention immediately. If this chemical has been inhaled, remove from exposure, begin rescue breathing (using universal precautions, including resuscitation mask) if breathing has stopped and CPR if heart action has stopped. Transfer promptly to a medical facility. When this chemical has been swallowed, get medical attention. Give large quantities of water and induce vomiting. Do not make an unconscious person vomit. Medical observation is recommended for 24−48 h after breathing overexposure, as pulmonary edema may be delayed. As first aid for pulmonary edema, a doctor or authorized paramedic may consider administering a corticosteroid spray.

Personal Protective Methods: Wear protective gloves and clothing to prevent any reasonable probability of skin contact. Safety equipment suppliers/manufacturers can provide recommendations on the most protective glove/clothing material for your operation. All protective clothing (suits, gloves, footwear, headgear) should be clean, available each day, and put on before work. Contact lenses should not be worn when working with this chemical. Wear splash-proof chemical goggles and face shield unless full face-piece respiratory protection is worn. Employees should wash immediately with soap when skin is wet or contaminated. Provide emergency showers and eyewash.

Respirator Selection: NIOSH: *0.5 mg/m³*: CcrOv95 (APF = 10) [any air-purifying half-mask respirator with organic vapor cartridge(s) in combination with an N95, R95, or P95 filter. The following filters may also be used: N99, R99, P99, N100, R100, P100] or Sa (APF = 10) (any supplied-air respirator). *1.25 mg/m³*: Sa:Cf (APF = 25) (any supplied-air respirator operated in a continuous-flow mode) or PaprOvHie (APF = 25) (any powered air-purifying respirator with an organic vapor cartridge in combination with a high-efficiency particulate filter). *2.5 mg/m³*: CcrFOv100 (APF = 50) [any air-purifying full-face-piece respirator equipped with organic vapor cartridge(s) in combination with an N100, R100, or P100 filter] or SaT: Cf (APF = 50) (any supplied-air respirator that has a tight-fitting face-piece and is operated in a continuous-flow mode) or PaprTOvHie (APF = 50) [any powered, air-purifying respirator with a tight-fitting face-piece and organic vapor cartridge(s) in combination with a high-efficiency particulate filter] or SCBAF (APF = 50) (any self-contained breathing apparatus with full face-piece) or SaF (APF = 50) (any supplied-air respirator with a full face-piece). *10 mg/m³*: Sa: Pd,Pp (APF = 1000) (any supplied-air respirator operated in a pressure-demand or other positive-pressure mode). *Emergency or planned entry into unknown concentrations or IDLH conditions:* SCBAF: Pd,Pp (APF = 10,000) (any self-contained breathing apparatus that has a full face-piece and is operated in a pressure-demand or other positive-pressure mode) or SaF: Pd,Pp: ASCBA (APF = 10,000) (any supplied-air respirator that has a full face-piece and is operated in a pressure-demand or other positive-pressure mode in combination with an auxiliary self-contained breathing apparatus operated in a pressure-demand or other positive-pressure mode). *Escape:* GmFOv100 (APF = 50) [any air-purifying, full-face-piece respirator (gas mask) with a chin-style, front- or back-mounted organic vapor canister having an N100, R100, or P100 filter] or SCBAE (any appropriate escape-type, self-contained breathing apparatus).

Storage: Color Code—Blue: Health Hazard/Poison: Store in a secure poison location. Prior to working with this chemical you should be trained on its proper handling and storage. Store where possible leakage from containers will not endanger the worker. Maintain regular inspection of containers for leakage. Store in tightly closed containers in a cool, well-ventilated area away from oxidizers and alkaline material. Where possible, automatically pump liquid from drums or other storage containers to process containers. Sources of ignition, such as smoking and open flames, are prohibited where this chemical is handled, used, or stored. Metal containers involving the transfer of 5 gallons or more of this chemical should be grounded and bonded. Drums must be equipped with self-closing valves, pressure vacuum bungs, and flame arresters. Use only nonsparking tools and equipment, especially when opening and closing containers of this chemical. Wherever this chemical is used, handled, manufactured, or stored, use explosion-proof electrical equipment and fittings.

Shipping: Organophosphorus pesticides, liquid, toxic, require a shipping label of "POISONOUS/TOXIC MATERIALS." Parathion falls in DOT Hazard Class 6.1 and Packing Group I.

Spill Handling: Evacuate and restrict persons not wearing protective equipment from area of spill or leak until cleanup is complete. Remove all ignition sources. Ventilate area of spill or leak. Absorb liquids in vermiculite, dry sand, earth, peat, carbon, or a similar material and deposit in sealed containers. Keep this chemical out of a confined space, such as a sewer, because of the possibility of an explosion, unless the sewer is designed to prevent the buildup of explosive concentrations. It may be necessary to contain and dispose of this chemical as a hazardous waste. If material or contaminated runoff enters waterways, notify downstream users of potentially contaminated waters. Contact your local or federal environmental protection agency for specific recommendations. If employees are required to clean up spills, they must be properly trained and equipped. OSHA 1910.120(q) may be applicable.

Fire Extinguishing: This chemical is not very combustible. Poisonous gases, including carbon monoxide, sulfur oxides, phosphorous oxides, nitrogen oxides, are produced in fire. Use dry chemical, carbon dioxide, or alcohol foam extinguishers. Vapors are heavier than air and will collect in low areas. Vapors in confined areas may explode when exposed to fire. Containers may explode in fire. Storage containers and parts of containers may rocket great distances, in many directions. If material or contaminated runoff enters waterways, notify downstream users of potentially contaminated waters. Notify local health and fire officials and pollution control agencies. From a secure, explosion-proof location, use water spray to cool exposed containers. If cooling streams are ineffective (venting sound increases in volume and pitch, tank discolors, or shows any signs of deforming), withdraw immediately to a secure position. If employees are expected to fight fires, they must be trained and equipped in OSHA 1910.156. The only respirators recommended for firefighting are self-contained breathing apparatuses that have full face-pieces and are operated in a pressure-demand or other positive-pressure mode.

Disposal Method Suggested: Consult with environmental regulatory agencies for guidance on acceptable disposal practices. Generators of waste containing this contaminant ($\geq$100 kg/mo) must conform with EPA regulations governing storage, transportation, treatment, and waste disposal. In accordance with 40CFR165, follow recommendations for the disposal of pesticides and pesticide containers. Must be disposed properly by following package label directions or by contacting your local or federal environmental control agency or by contacting your regional EPA office. One manufacturer recommends the use of a detergent in a 5% trisodium phosphate solution for parathion disposal and cleanup problems. For parathion disposal in general, however, the recommended method is incineration (816°C, 0.5 s minimum for primary combustion; 1204°C, 1.0 s for secondary combustion) with adequate scrubbing and ash disposal facilities.[22]

References

National Institute for Occupational Safety and Health. (1976). *Criteria for a Recommended Standard: Occupational Exposure to Parathion,* NIOSH Document No. 76-190

Sax, N. I. (Ed.). (1983). *Dangerous Properties of Industrial Materials Report,* 3, No. 3, 92−97

US Environmental Protection Agency. (November 30, 1987). *Chemical Hazard Information Profile: Parathion.* Washington, DC: Chemical Emergency Preparedness Program

New York State Department of Health. (March 1986). *Chemical Fact Sheet: Parathion* (Version 2 and Version 3). Albany, NY: Bureau of Toxic Substance Assessment

US Environmental Protection Agency, Special Review and Reregistration Division Office of Pesticide Programs. (1998). *Agency Status of Pesticides in Registration, Reregistration, and Special Review* (Rainbow Report). Washington, DC

Paris green   P:0180

Molecular Formula: $C_4H_6As_6Cu_4O_{16}$

Synonyms: Acetoarsenito de cuivre (French); Acetoarsenito de cobre (Spanish); Basle green; C.I. 77410; C.I. Pigment green 21; Copper acetoarsenite; Cupric acetoarsenite; Emerald green; ENT884; French green; Imperial green; King's green; Meadow green; Mineral green; Mitis green; Moss green; Mountain green; Neuwied green; New green; Paris green; Parrot green; Patent green; O-P-G bait; Powder green; Schweinfurtergruen (German); Schweinfurt green; Sowbug & cutworm bait; Swedish green; Vienna green

CAS Registry Number: 12002-03-8

RTECS® Number: GL6475000

UN/NA & ERG Number: UN1585/151

Regulatory Authority and Advisory Bodies

Clean Air Act: Hazardous Air Pollutants (Title I, Part A, Section 112); List of high-risk pollutants (Section 63.74) as arsenic compounds.

Clean Water Act: Section 311 Hazardous Substances/RQ 1 lb (0.454 kg); Toxic Pollutant (Section 401.15) as copper and compounds; Section 313 Water Priority Chemicals (57FR41331, 9/9/92).

Safe Drinking Water Act 47FR9352 Regulated chemical: MCL, 0.05 mg/L (Section 141.11) applies only to community water systems (arsenic).

RCRA 40CFR264, Appendix 9; TSD Facilities Ground Water Monitoring List. Suggested test method(s) (PQL μg/L): 6010 (60); 7210 (200) *Note:* All species in the ground water that contain copper are included.

Superfund/EPCRA 40CFR355, Extremely Hazardous Substances: TPQ = 500/10,000 lb (227/4540 kg).

Reportable Quantity (RQ): 500 lb (227 kg).

EPCRA Section 313 Form R *de minimis* concentration reporting level: 1.0%.

US DOT Regulated Marine Pollutant (49CFR172.101, Appendix B) as arsenates, liquid, n.o.s.; arsenates, solid, n.o.s.; arsenical pesticides liquid, toxic, flammable, n.o.s.

California Proposition 65 Chemical: Cancer 2/27/87.

Canada: Priority Substance List & Restricted Substances/ Ocean Dumping FORBIDDEN (CEPA), National Pollutant Release Inventory (NPRI) (arsenic compounds).

Canada, WHMIS, Ingredients Disclosure List Concentration 1.0% arsenic, water-soluble compounds.

European/International Regulations: Hazard Symbol: T, N; Risk phrases: R45; R23/25; R50/53; Safety phrases: S53; S45; S60; S61 (see Appendix 4).

WGK (German Aquatic Hazard Class): 3—Severe hazard to waters.

Description: Paris green (copper acetoarsenite) is an odorless emerald green crystalline powder which decomposes upon heating. Molecular weight = 1013.78. Hazard Identification (based on NFPA-704 M Rating System): Health 3, Flammability 1, Reactivity 0. Insoluble in water.

Potential Exposure: This material is used primarily as an insecticide; it may be used as a wood preservative and a pigment, particularly for ships and submarines; and also finds use as an anthelmintic.

Incompatibilities: Can react vigorously with oxidizers. Emits highly toxic arsenic fumes on contact with acid or acid fumes; and in elevated temperatures.

Permissible Exposure Limits in Air

Arsenic, organic compounds

OSHA PEL: 0.5 mg[As]/m^3 TWA.

NIOSH REL: Not established. See NIOSH Pocket Guide, Appendix A.

ACGIH TLV®[1]: 0.01 mg[As]/m^3 TWA; Confirmed Human Carcinogen; BEI established.

Protective Action Criteria (PAC)

Paris green; cupric acetoarsenite

TEEL-0: 1.13 mg/m^3

PAC-1: 3.38 mg/m^3

PAC-2: 22 mg/m^3

PAC-3: 22 mg/m^3

As arsenic, organic compounds

TEEL-0: 0.5 mg/m^3

PAC-1: 1.5 mg/m^3

PAC-2: 2.5 mg/m^3

PAC-3: 350 mg/m^3

Arab Republic of Egypt: TWA 0.2 mg/m^3, 1993; Australia: TWA 0.05 mg/m^3, carcinogen, 1993; Belgium: TWA 0.2 mg/m^3, 1993; Denmark: TWA 0.05 mg/m^3, 1999; Finland: carcinogen, 1993; France: VME 0.2 mg/m^3, 1993; Hungary: STEL 0.5 mg/m^3, carcinogen, 1993; India: TWA 0.2 mg/m^3, 1993; Norway: TWA 0.02 mg/m^3, 1999; the Philippines: TWA 0.5 mg/m^3, 1993; Poland: MAC (TWA)

0.01 mg/m^3, 1999; Sweden: NGV 0.03 mg/m^3, carcinogen, 1999; Switzerland: TWA 0.1 mg/m^3, carcinogen, 1999; Thailand: TWA 0.5 mg/m^3, 1993; Turkey: TWA 0.5 mg (As)/m^3; TWA 0.5 mg/m^3, 1993; United Kingdom: TWA 0.1 mg/m^3, carcinogen, 2000; Argentina, Bulgaria, Columbia, Jordan, South Korea, New Zealand, Singapore, Vietnam: ACGIH: TLV: Confirmed Human Carcinogen. Russia[43] set a MAC of 0.003 mg/m^3 on a daily average basis for residential areas. Several states have set guidelines or standards for arsenic in ambient air[60]: 0.06 mg/m^3 (California Prop. 65), 0.0002 µg/m^3 (Rhode Island), 0.00023 µg/m^3 (North Carolina), 0.024 µg/m^3 (Pennsylvania), 0.05 µg/m^3 (Connecticut), 0.07−0.39 µg/m^3 (Montana), 0.67 µg/m^3 (New York), 1.0 µg/m^3 (South Carolina), 2.0 µg/m^3 (North Dakota), 3.3 µg/m^3 (Virginia), 5 µg/m^3 (Nevada).

Determination in Air: NIOSH Analytical Methods (inorganic arsenic): #7300, #7301, #7303, #7900, #9102; OSHA Analytical Methods ID-105. The American Conference of Government Industrial Hygienists (ACGIH) Method 803 measures total particulate arsenic in air.

Permissible Concentration in Water: EPA[6] recommends a zero concentration of arsenic for human health reasons but has set a guideline of 50 µg/L[61] for drinking water.

Determination in Water: *For arsenic:* The atomic absorption graphite furnace technique is often used for measurement of total arsenic in water. It has also been standardized by EPA. Total arsenic may be determined by digestion followed by silver diethyldithiocarbamate; an alternative is atomic absorption; another is inductively coupled plasma (ICP) optical emission spectrometry. See OSHA Analytical Method #ID-105 for arsenic.

Routes of Entry: Inhalation, ingestion, skin and/or eye contact.

Harmful Effects and Symptoms

Short Term Exposure: It may cause eye and respiratory tract irritation. Industrial exposure may cause dermatitis. This material is extremely toxic; the probable oral lethal dose for humans is 5−50 mg/kg or between 7 drops and 1 teaspoonful for a 150-lb person. Some absorption may occur through the skin and by inhalation, but most poisonings result from ingestion. Symptoms usually appear ½ to 1 h after ingestion, but may be delayed. Causes gastric disturbance, tremors, muscular cramps, and nervous collapse, which may lead to death. Symptoms of exposure also include a sweetish, metallic taste and garlicky odor; difficulty in swallowing; abdominal pain; vomiting and diarrhea; dehydration; rapid heartbeat; dizziness and headache; and eventually coma; sometimes convulsions; and death.

Long Term Exposure: May cause liver damage. Arsenic compounds may cause blood, kidneys, and nervous system damage, and skin abnormalities may develop. See also entries for "Arsenic" and "Copper."

Points of Attack: Liver, kidneys, blood, skin.

Medical Surveillance: Liver and kidney function tests. Blood tests including CBC.

First Aid: If this chemical gets into the eyes, remove any contact lenses at once and irrigate immediately for at least 15 min, occasionally lifting upper and lower lids. Seek medical attention immediately. If this chemical contacts the skin, remove contaminated clothing and wash immediately with soap and water. Seek medical attention immediately. If this chemical has been inhaled, remove from exposure, begin rescue breathing (using universal precautions, including resuscitation mask) if breathing has stopped and CPR if heart action has stopped. Transfer promptly to a medical facility. When this chemical has been swallowed, get medical attention. Give large quantities of water and induce vomiting. Do not make an unconscious person vomit.

Note to physician: For severe poisoning BAL [British Anti-Lewisite, dimercaprol, dithiopropanol $(C_3H_8OS_2)$] has been used to treat toxic symptoms of certain heavy metal poisoning—including arsenic. Although BAL is reported to have a large margin of safety, caution must be exercised, because toxic effects may be caused by excessive dosage. Most can be prevented by premedication with 1-ephedrine sulfate (CAS: 134-72-5). For milder poisoning *penicillamine (not penicillin)* has been used, both with mixed success. Side effects occur with such treatment and it is never a substitute for controlling exposure. It can only be done under strict medical care.

Personal Protective Methods: Wear protective gloves and clothing to prevent any reasonable probability of skin contact. Safety equipment suppliers/manufacturers can provide recommendations on the most protective glove/clothing material for your operation. All protective clothing (suits, gloves, footwear, headgear) should be clean, available each day, and put on before work. Contact lenses should not be worn when working with this chemical. Wear dust-proof chemical goggles and face shield unless full face-piece respiratory protection is worn. Employees should wash immediately with soap when skin is wet or contaminated. Provide emergency showers and eyewash. Specific engineering controls are required under OSHA 1910.1018, *Inorganic Arsenic.* See also NIOSH Criteria Document #75-149, *"Inorganic Arsenic."*

Respirator Selection: *Copper dusts and mists: 5 mg/m³:* Qm (APF = 25) (any quarter-mask respirator). *10 mg/m³:* Any particulate respirator equipped with an N95, R95, or P95 filter (including N95, R95, and P95 filtering facepieces) except quarter-mask respirators. The following filters may also be used: N99, R99, P99, N100, R100, P100; or Sa (APF = 10) (any supplied-air respirator). *25 mg/m³:* Sa:Cf (APF = 25) (any supplied-air respirator operated in a continuous-flow mode); PaprHie (APF = 25) (any powered, air-purifying respirator with a high-efficiency particulate filter). *50 mg/m³:* 100F (APF = 50) (any air-purifying, full-face-piece respirator with an N100, R100, or P100 filter) or PaprTHie (APF = 50) (any powered, air-purifying respirator with a tight-fitting face-piece and a high-efficiency particulate filter) or SCBAF (APF = 50) (any self-contained breathing apparatus with a full face-piece) or SaF

(APF = 50) (any supplied-air respirator with a full face-piece). *100 mg/m³:* SaF: Pd,Pp (APF = 2000) (any supplied-air respirator that has a full face-piece and is operated in a pressure-demand or other positive-pressure mode). *Emergency or planned entry into unknown concentrations or IDLH conditions:* SCBAF: Pd,Pp (APF = 10,000) (any self-contained breathing apparatus that has a full face-piece and is operated in a pressure-demand or other positive-pressure mode) or SaF: Pd,Pp: ASCBA (APF = 10,000) (any supplied-air respirator that has a full face-piece and is operated in a pressure-demand or other positive-pressure mode in combination with an auxiliary self-contained breathing apparatus operated in a pressure-demand or other positive-pressure mode). *Escape:* 100F (APF = 50) (any air-purifying, full-face-piece respirator with an N100, R100, or P100 filter) or SCBAE (any appropriate escape-type, self-contained breathing apparatus).

Note: Substance reported to cause eye irritation or damage; may require eye protection.

Arsenic: At concentrations above the NIOSH REL, or where there is no REL, at any detectable concentration: Sa (APF = 10) (any supplied-air respirator) or SCBAF (APF = 50) (any self-contained breathing apparatus with a full face-piece). *Emergency or planned entry into unknown concentrations or IDLH conditions:* SCBAF: Pd,Pp (APF = 10,000) (any self-contained breathing apparatus that has a full face-piece and is operated in a pressure-demand or other positive-pressure mode) or SaF: Pd,Pp: ASCBA (APF = 10,000) (any supplied-air respirator that has a full face-piece and is operated in a pressure-demand or other positive-pressure mode in combination with an auxiliary, self-contained breathing apparatus operated in a pressure-demand or other positive-pressure mode). *Escape:* GmFAg100 (APF = 50) [any air-purifying, full-face-piece respirator (gas mask) with a chin-style, front- or back-mounted acid gas canister having an N100, R100, or P100 filter] or SCBAE (any appropriate escape-type, self-contained breathing apparatus).

Storage: Color Code—Blue: Health Hazard/Poison: Store in a secure poison location. Prior to working with copper acetoarsenite you should be trained on its proper handling and storage. Store in tightly closed containers in a cool, well-ventilated area, away from strong bases, strong acids, and moisture. Where possible, automatically pump liquid from drums or other storage containers to process containers.

Shipping: Copper acetoarsenite requires a shipping label of "POISONOUS/TOXIC MATERIALS." Copper acetoarsenite falls in Hazard Class 6.1 and Packing Group II.

Spill Handling: Evacuate persons not wearing protective equipment from area of spill or leak until cleanup is complete. Remove all ignition sources. Stay upwind; keep out of low areas. Do not touch spilled material. Take up *small spills* with sand or other noncombustible absorbent material and place in sealed containers for later disposal. For *small dry spills,* use a clean shovel to place material in clean, dry container. For *large spills*, dike far ahead of spill for later disposal. Use water spray to knock down dust. Ventilate area after cleanup is complete. It may be necessary to contain and dispose of this chemical as a hazardous waste. If material or contaminated runoff enters waterways, notify downstream users of potentially contaminated waters. Contact your local or federal environmental protection agency for specific recommendations. If employees are required to clean up spills, they must be properly trained and equipped. OSHA 1910.120(q) may be applicable.

Fire Extinguishing: This chemical may burn but does not easily ignite. Use dry chemical, carbon dioxide, water spray, or alcohol foam extinguishers. Poisonous fumes are produced in fire, including arsenic oxide and copper. If material or contaminated runoff enters waterways, notify downstream users of potentially contaminated waters. Notify local health and fire officials and pollution control agencies. From a secure, explosion-proof location, use water spray to cool exposed containers. If cooling streams are ineffective (venting sound increases in volume and pitch, tank discolors, or shows any signs of deforming), withdraw immediately to a secure position. If employees are expected to fight fires, they must be trained and equipped in OSHA 1910.156. The only respirators recommended for firefighting are self-contained breathing apparatuses that have full face-pieces and are operated in a pressure-demand or other positive-pressure mode.

Disposal Method Suggested: Consult with environmental regulatory agencies for guidance on acceptable disposal practices. Generators of waste containing this contaminant (≥100 kg/mo) must conform with EPA regulations governing storage, transportation, treatment, and waste disposal. In accordance with 40CFR165, follow recommendations for the disposal of pesticides and pesticide containers. Must be disposed properly by following package label directions or by contacting your local or federal environmental control agency or by contacting your regional EPA office.

References

New Jersey Department of Health and Senior Services. (January 1999). *Hazardous Substances Fact Sheet: Copper Acetoarsenite.* Trenton, NJ

US Environmental Protection Agency. (November 30, 1987). *Chemical Hazard Information Profile: Paris Green.* Washington, DC: Chemical Emergency Preparedness Program

Pentaborane   P:0190

Molecular Formula: B_5H_9

Synonyms: Dihydropentaborane (9); Pentaborane (9); Pentaborane undecahydride; Pentaborano (Spanish); (9)-Pentaboron nonahydride; Pentaboron nonahydride; Pentaboron undecahydride; Stable pentaborane

CAS Registry Number: 19624-22-7

RTECS® Number: RY8925000

UN/NA & ERG Number: UN1380/135

EC Number: 243-194-4

Regulatory Authority and Advisory Bodies

Very Toxic Substance (World Bank).[15]

Air Pollutant Standard Set. See below, "Permissible Exposure Limits in Air" section.

Reportable Quantity (RQ): 500 lb (227 kg).

Superfund/EPCRA 40CFR355, Extremely Hazardous Substances: TPQ = 500 lb (227 kg).

US DOT 49CFR172.101, Inhalation Hazardous Chemical.

Canada, WHMIS, Ingredients Disclosure List Concentration 1.0%.

European/International Regulations: not listed in Annex 1.

WGK (German Aquatic Hazard Class): No value assigned.

Description: Pentaborane is a colorless, volatile liquid with an unpleasant, sweetish odor, like sour milk. The odor threshold is 0.8 ppm. Molecular weight = 63.14; Specific gravity (H_2O:1) = 0.62; Boiling point = 60°C; Freezing/Melting point = −47°C; Vapor pressure = 171 mmHg at 25°C; Flash point = 30°C; Autoignition temperature: about 35°C. Explosive limits: LEL = 0.42%; UEL = 98%. Hazard Identification (based on NFPA-704 M Rating System): Health 4, Flammability 4, Reactivity 2. Reacts with water.

Potential Exposure: Pentaborane is used in rocket propellants and in gasoline additives.

Incompatibilities: Reacts on contact with oxidizers, halogens, water, halogenated hydrocarbons. May ignite *spontaneously* in moist air, decomposes at 150°C. Corrosive to natural rubber. Hydrolyzes slowly with heat in water to form boric acid. Contact with solvents, such as ketones, ethers, esters, forms shock-sensitive compounds. Corrosive to natural rubber and some synthetic rubber; some lubricants.

Permissible Exposure Limits in Air

Conversion factor: 1 ppm = 2.58 mg/m³ at 25°C & 1 atm.

OSHA PEL: 0.005 ppm/0.01 mg/m³ TWA.

NIOSH REL: 0.005 ppm/0.01 mg/m³ TWA; 0.015 ppm/0.03 mg/m³ STEL.

ACGIH TLV®[1]: 0.005 ppm/0.013 mg/m³ TWA; 0.015 ppm/0.039 mg/m³ STEL.

NIOSH IDLH: 1 ppm.

Protective Action Criteria (PAC)*

TEEL-0: 0.005 mg/m³

PAC-1: 0.015 mg/m³

PAC-2: **0.14** mg/m³

PAC-3: **0.70** mg/m³

*AEGLs (Acute Emergency Guideline Levels) & ERPGs (Emergency Response Planning Guideline) are in **bold face**.

DFG MAK: 0.005 ppm/0.013 mg/m³ TWA; Peak Limitation Category II(2).

Arab Republic of Egypt: TWA 0.1 mg/m³, [skin], 1993; Australia: TWA 0.1 mg/m³, [skin], 1993; Austria: MAK 0.1 mg/m³, [skin], 1999; Belgium: TWA 0.1 mg/m³, [skin], 1993; Denmark: TWA 0.1 mg/m³, [skin], 1999; Finland: TWA 0.1 mg/m³, short-term exposure limit 0.3 mg/m³,

[skin], 1999; France: VME 0.1 mg/m^3, [skin], 1999; the Netherlands: MAC-TGG 0.1 mg/m^3, [skin], 2003; Japan: 0.1 mg/m^3, [skin], 1999; Norway: TWA 0.05 mg/m^3, 1999; the Philippines: TWA 0.1 mg/m^3, [skin], 1993; Russia: STEL 0.05 mg/m^3, 1993; Thailand: TWA 0.11 mg/m^3, 1993; Turkey: TWA 0.1 mg/m^3, [skin], 1993; United Kingdom: TWA 0.1 mg/m^3, short-term exposure limit 0.3 mg/m^3, [skin], 2000; Argentina, Bulgaria, Columbia, Jordan, South Korea, New Zealand, Singapore, Vietnam: ACGIH TLV®: not classifiable as a human carcinogen. Several states have set guidelines or standards for pentaborane in ambient air[60] ranging from 0.16 μg/m^3 (Virginia) to 0.2 μg/m^3 (Connecticut) to 1.0−3.0 μg/m^3 (North Dakota) to 2.0 μg/m^3 (Nevada).

Determination in Air: No Analytical Method available.

Routes of Entry: Inhalation, ingestion, skin and/or eye contact.

Harmful Effects and Symptoms

Short Term Exposure: Causes severe irritation to the respiratory tract. Pentaborane may affect the central nervous system, resulting in visual disturbances, poor judgment, behavioral changes, loss of recent memory, nausea, vomiting, drowsiness, and difficulty in focusing. Inhalation of higher concentrations may cause headache, dizziness, nervous excitation, muscular pain, muscle incoordination, cramps, tremors, convulsions, and coma. Death can occur by central nervous system poisoning.

Points of Attack: Central nervous system, eyes, skin.

Medical Surveillance: Preemployment and periodic physical examinations to determine the status of the workers' general health should be performed. These examinations should be concerned especially with any history of central nervous system disease, personality or behavioral changes, as well as liver, kidney, or pulmonary disease of any significant nature. Chest X-ray, blood, liver, and renal function studies may be helpful.

First Aid: If this chemical gets into the eyes, remove any contact lenses at once and irrigate immediately for at least 15 min, occasionally lifting upper and lower lids. Seek medical attention immediately. If this chemical contacts the skin, remove contaminated clothing and wash immediately with soap and water. Seek medical attention immediately. If this chemical has been inhaled, remove from exposure, begin rescue breathing (using universal precautions, including resuscitation mask) if breathing has stopped and CPR if heart action has stopped. Transfer promptly to a medical facility. When this chemical has been swallowed, get medical attention. Give large quantities of water and induce vomiting. Do not make an unconscious person vomit.

Personal Protective Methods: Wear protective gloves and clothing to prevent any reasonable probability of skin contact. Safety equipment suppliers/manufacturers can provide recommendations on the most protective glove/clothing material for your operation. All protective clothing (suits, gloves, footwear, headgear) should be clean, available each day, and put on before work. Contact lenses should not be worn when working with this chemical. Wear splash-proof chemical goggles and face shield unless full face-piece respiratory protection is worn. Employees should wash immediately with soap when skin is wet or contaminated. Provide emergency showers and eyewash.

Respirator Selection: *0.05 ppm:* Sa (APF = 10) (any supplied-air respirator). *0.125 ppm:* Sa:Cf (APF = 25) (any supplied-air respirator operated in a continuous-flow mode). *0.25 ppm:* SaT: Cf (APF = 50) (any supplied-air respirator that has a tight-fitting face-piece and is operated in a continuous-flow mode) or SCBAF (APF = 50) (any self-contained breathing apparatus with a full face-piece) or SaF (APF = 50) (any supplied-air respirator with a full face-piece). *1 ppm:* Sa: Pd,Pp (APF = 1000) (any supplied-air respirator operated in a pressure-demand or other positive-pressure mode). *Emergency or planned entry into unknown concentrations or IDLH conditions:* SCBAF: Pd,Pp (APF = 10,000) (any self-contained breathing apparatus that has a full face-piece and is operated in a pressure-demand or other positive-pressure mode) or SaF: Pd,Pp: ASCBA (any supplied-air respirator that has a full face-piece and is operated in a pressure-demand or other positive-pressure mode in combination with an auxiliary self-contained breathing apparatus operated in a pressure-demand or other positive-pressure mode). *Escape:* GmFS (APF = 50) [any air-purifying, full-face-piece respirator (gas mask) with a chin-style, front- or back-mounted canister providing protection against the compound of concern] or SCBAE (any appropriate escape-type, self-contained breathing apparatus).

Note: Make certain respirator has no exposed rubber gaskets.

Storage: Color Code—Red: Flammability Hazard: Store in a flammable materials storage area. Prior to working with this chemical you should be trained on its proper handling and storage. Before entering confined space where this chemical may be present, check to make sure that an explosive concentration is not a danger. Store in tightly closed containers in a cool, well-ventilated area away from oxidizers, halogens, water, halogenated hydrocarbons. Where possible, automatically pump liquid from drums or other storage containers to process containers. Sources of ignition, such as smoking and open flames, are prohibited where this chemical is handled, used, or stored. Metal containers involving the transfer of 5 gallons or more of this chemical should be grounded and bonded. Drums must be equipped with self-closing valves, pressure vacuum bungs, and flame arresters. Use only nonsparking tools and equipment, especially when opening and closing containers of this chemical. Wherever this chemical is used, handled, manufactured, or stored, use explosion-proof electrical equipment and fittings.

Shipping: This compound requires a shipping label of "SPONTANEOUSLY COMBUSTIBLE, POISONOUS/TOXIC MATERIALS." It falls in Hazard Class 4.2 and Packing Group I. A plus sign (+) indicates that the designated proper shipping name and hazard class of the material must always be shown whether or not the material or its mixtures or solutions meet the definitions of the class. **Spill Handling:** Evacuate and restrict persons not wearing protective equipment from area of spill or leak until cleanup is complete. Remove all ignition sources. Establish forced ventilation to keep levels below explosive limit. Do not touch spilled material, stop leak if you can do it without risk. For spills, dike for later disposal and do not apply water unless directed to do so. Clean up only under supervision of an expert. Keep unnecessary people away; isolate hazard area and deny entry. Stay upwind; keep out of low areas. Ventilate closed spaces before entering them. Avoid breathing vapors, and keep upwind. Avoid bodily contact with the material. Do not handle broken packages without protective equipment. Wash away any material which may have contacted the body with copious amounts of water or soap and water. Keep this chemical out of a confined space, such as a sewer, because of the possibility of an explosion, unless the sewer is designed to prevent the buildup of explosive concentrations. It may be necessary to contain and dispose of this chemical as a hazardous waste. If material or contaminated runoff enters waterways, notify downstream users of potentially contaminated waters. Contact your local or federal environmental protection agency for specific recommendations. If employees are required to clean up spills, they must be properly trained and equipped. OSHA 1910.120(q) may be applicable.

Initial isolation and protective action distances
Distances shown are likely to be affected during the first 30 min after materials are spilled and could increase with time. If more than one tank car, cargo tank, portable tank, or large cylinder involved in the incident is leaking, the protective action distance may need to be increased. You may need to seek emergency information from CHEMTREC at (800) 424-9300 or seek professional environmental engineering assistance from the US EPA Environmental Response Team at (908) 548-8730 (24-h response line).
Small spills (From a small package or a small leak from a large package)
First: Isolate in all directions (feet/meters)   200/60
Then: Protect persons downwind (miles/kilometers)
Day   0.4/0.6
Night 1.4/2.3
Large spills (From a large package or from many small packages)
First: Isolate in all directions (feet/meters)   1250/400
Then: Protect persons downwind (miles/kilometers)
Day   2.9/4.7
Night 5.5/8.9
Fire Extinguishing: This chemical is a combustible liquid. Poisonous gases, including boron, are produced in fire. Ignites spontaneously in air. Reacts violently with halogenated extinguishing agents. Fires tend to reignite. If material is on fire or involved in fire, do not extinguish unless flow can be stopped. *Do not use water.* Extinguish *small fires* with dry chemical or carbon dioxide. For *large fires* withdraw and let burn. Move container from fire area if you can do it without risk. Cool containers that are exposed to flames with water from the side until well after fire is out. *For massive fire* in cargo area, use unmanned hose holder or monitor nozzles; if this is impossible, withdraw from area and let fire burn. Wear positive pressure breathing apparatus and full protective clothing. If fire becomes uncontrollable or container is exposed to direct flame—evacuate for a radius of 1500 feet. If material is leaking (not on fire), downwind evacuation must be considered. Vapors are heavier than air and will collect in low areas. Vapors may travel long distances to ignition sources and flashback. Vapors in confined areas may explode when exposed to fire. Containers may explode in fire. Storage containers and parts of containers may rocket great distances, in many directions. If material or contaminated runoff enters waterways, notify downstream users of potentially contaminated waters. Notify local health and fire officials and pollution control agencies. From a secure, explosion-proof location, use water spray to cool exposed containers. If cooling streams are ineffective (venting sound increases in volume and pitch, tank discolors, or shows any signs of deforming), withdraw immediately to a secure position. If employees are expected to fight fires, they must be trained and equipped in OSHA 1910.156. The only respirators recommended for firefighting are self-contained breathing apparatuses that have full face-pieces and are operated in a pressure-demand or other positive-pressure mode.
Disposal Method Suggested: Incineration with aqueous scrubbing of exhaust gases to remove B_2O_3 particulates.
Reference
US Environmental Protection Agency. (November 30, 1987). *Chemical Hazard Information Profile: Pentaborane.* Washington, DC: Chemical Emergency Preparedness Program

Pentachlorobenzene   P:0200

Molecular Formula: C_6HCl_5
Synonyms: Benzene, pentachloro-; 1,2,3,4,5-Pentachlorobenzene; QCB
CAS Registry Number: 608-93-5
RTECS® Number: DA6640000
UN/NA & ERG Number: UN3077/171
EC Number: 210-172-0 [*Annex I Index No.:* 602-074-00-5]
Regulatory Authority and Advisory Bodies
Carcinogenicity: EPA: Not Classifiable as to human carcinogenicity.
TSCA: Subject to a proposed or final SNUR; Subject to a Section 4 test rule.
US EPA Hazardous Waste Number (RCRA No.): U183.
RCRA, 40CFR261, Appendix 8 Hazardous Constituents.
RCRA 40CFR268.48; 61FR15654, Universal Treatment Standards: Wastewater (mg/L), 0.055; Nonwastewater (mg/kg), 10.

RCRA 40CFR264, Appendix 9; TSD Facilities Ground Water Monitoring List. Suggested test method(s) (PQL µg/L): 8270 (10).

Reportable Quantity (RQ): 10 lb (4.54 kg).

List of Stockholm Convention POPs: Annex A (Elimination); Annex C (Unintentional production and release).

European/International Regulations: Hazard Symbol: Risk phrases: F, Xn, N; R11; R22; R50/53; Safety phrases: S2; S41; S46; S50; S60; S61 (see Appendix 4).

WGK (German Aquatic Hazard Class): 3—Severe hazard to waters.

Description: Pentachlorobenzene is a colorless crystalline solid with a pleasant aroma. Molecular weight = 250.32; Boiling point = 277°C; Freezing/Melting point = 86°C. Insoluble in water; solubility 0.65 mg/L.

Potential Exposure: Pentachlorobenzene is used primarily as a precursor in the synthesis of the fungicide pentachloronitrobenzene, and as a flame retardant. Drug/Therapeutic Agent; Fungicide; bactericide; wood preservative; industrial insecticides; organochlorine; Reproductive Effect; Tumor data.

Incompatibilities: Polychlorinated hydrocarbons can react with oxidizers and may react violently with aluminum, liquid oxygen, potassium, sodium.

Permissible Exposure Limits in Air

Protective Action Criteria (PAC)

TEEL-0: 0.04 mg/m^3

PAC-1: 0.125 mg/m^3

PAC-2: 0.75 mg/m^3

PAC-3: 400 mg/m^3

Determination in Air: Filter/XAD-2; workup with hexane; GC/ECD; NIOSH Analytical Method (IV) #5517, Polychlorobenzenes.

Determination in Water: A persistent organic pollutant. Log K_{ow} = 4.75–5.75.

Permissible Concentration in Water: The US EPA has set a criterion of 0.5 µg/L based on toxicity studies.

Routes of Entry: Inhalation, ingestion, skin and/or eye contact.

Harmful Effects and Symptoms

Short Term Exposure: May affect the nervous system. LD$_{50}$= (oral-rat) 1080 mg/kg.

Long Term Exposure: Pentachlorobenzene may affect the liver and kidneys, causing tissue lesions. Limited animal studies have produced developmental effects and decreased body weights in fetuses.

Points of Attack: Liver, kidneys.

Medical Surveillance: Liver and kidney function tests.

First Aid: Skin Contact[52]*:* Flood all areas of body that have contacted the substance with water. Do not wait to remove contaminated clothing; do it under the water stream. Use soap to help assure removal. Isolate contaminated clothing when removed to prevent contact by others. *Eye Contact:* Remove any contact lenses at once. Flush eyes well with copious quantities of water or normal saline for at least 20–30 min. Seek medical attention. *Inhalation:* Leave contaminated area immediately; breathe fresh air. Proper respiratory protection must be supplied to any rescuers. If coughing, difficult breathing, or any other symptoms develop, seek medical attention at once, even if symptoms develop many hours after exposure. *Ingestion:* If convulsions are not present, give a glass or two of water or milk to dilute the substance. Assure that the person's airway is unobstructed and contact a hospital or poison center immediately for advice on whether or not to induce vomiting.

Personal Protective Methods: Wear protective gloves and clothing to prevent any reasonable probability of skin contact. Safety equipment suppliers/manufacturers can provide recommendations on the most protective glove/clothing material for your operation. All protective clothing (suits, gloves, footwear, headgear) should be clean, available each day, and put on before work. Contact lenses should not be worn when working with this chemical. Wear dust-proof chemical goggles and face shield unless full face-piece respiratory protection is worn. Employees should wash immediately with soap when skin is wet or contaminated. Provide emergency showers and eyewash.

Respirator Selection: Follow regulations in OSHA 29CFR1910.134 or European Standard EN149. Use a NIOSH/MSHA- or European Standard EN149-approved respirator; or use an approved supplied-air respirator with a full face-piece operated in the positive-pressure mode, or with a full face-piece, hood, or helmet in the continuous-flow mode; or use a NIOSH/MSHA- or European Standard EN149-approved self-contained breathing apparatus with a full face-piece operated in pressure-demand or other positive-pressure mode.

Storage: Color Code—Blue: Health Hazard/Poison: Store in a secure poison location. Prior to working with this chemical you should be trained on its proper handling and storage. Store in a refrigerator or a cool, dry place. Use only nonsparking tools and equipment, especially when opening and closing containers of this chemical. Wherever this chemical is used, handled, manufactured, or stored, use explosion-proof electrical equipment and fittings.

Shipping: The name of this material is not in the DOT list of materials[19] for label and packaging standards. However, based on regulations, it may be classified[52] as an Environmentally hazardous substances, solid, n.o.s. This chemical requires a shipping label of "CLASS 9." It falls in Hazard Class 9 and Packing Group III.[20, 21]

Spill Handling: Evacuate persons not wearing protective equipment from area of spill or leak until cleanup is complete. Remove all ignition sources.[52] Dampen spilled material with alcohol to avoid dust, then transfer material to a suitable container. Use absorbent dampened with alcohol to pick up remaining material. Wash surfaces well with soap and water. Seal all wastes in vapor-tight plastic bags for eventual disposal. Ventilate area after cleanup is complete. It may be necessary to contain and dispose of this chemical as a hazardous waste. If material or contaminated runoff enters waterways, notify downstream users of potentially contaminated waters. Contact your local or federal environmental protection agency for specific recommendations. If employees are required to clean up spills, they must be properly trained and equipped. OSHA 1910.120(q) may be applicable.

Fire Extinguishing: This chemical is a combustible solid. Use dry chemical, carbon dioxide, water spray, or alcohol foam extinguishers. Poisonous gases are produced in fire, including chlorine. If material or contaminated runoff enters waterways, notify downstream users of potentially contaminated waters. Notify local health and fire officials and pollution control agencies. From a secure, explosion-proof location, use water spray to cool exposed containers. If cooling streams are ineffective (venting sound increases in volume and pitch, tank discolors, or shows any signs of deforming), withdraw immediately to a secure position. If employees are expected to fight fires, they must be trained and equipped in OSHA 1910.156. The only respirators recommended for firefighting are self-contained breathing apparatuses that have full face-pieces and are operated in a pressure-demand or other positive-pressure mode.

Disposal Method Suggested: Consult with environmental regulatory agencies for guidance on acceptable disposal practices. Generators of waste containing this contaminant (≥100 kg/mo) must conform with EPA regulations governing storage, transportation, treatment, and waste disposal. Incineration after mixing with another combustible fuel. Care must be exercised to assure complete combustion to prevent the formation of phosgene. An acid scrubber is necessary to remove the halo acids produced.

References

US Environmental Protection Agency. (April 30, 1980). *Pentachlorobenzene, Health and Environmental Effects Profile No. 141.* Washington, DC: Office of Solid Waste
US EPA. (1980). Chlorinated Benzenes: Ambient Water Quality Criteria. Washington, DC
Sax, N. I. (Ed.). (1986). *Dangerous Properties of Industrial Materials Report*, 6, No. 1, 105–107

Pentachloroethane   P:0210

Molecular Formula: C_2HCl_5
Common Formula: CCl_3CHCl_2
Synonyms: Ethane pentachloride; Ethane, pentachloro-; NCI-C53894; Pentachloraethan (German); Pentachlorethane (French); Pentacloroetano (Spanish); Pentalin
CAS Registry Number: 76-01-7
RTECS® Number: KI6300000
UN/NA & ERG Number: UN1669/151
EC Number: 200-925-1 [*Annex I Index No.:* 602-017-00-4]
Regulatory Authority and Advisory Bodies
Carcinogenicity: NCI: Carcinogenesis Studies (gavage); clear evidence: mouse; equivocal evidence: rat; IARC: Animal Limited Evidence; Human No Adequate Data, *not classifiable as carcinogenic to humans*, Group 3, 1999.

Air Pollutant Standard Set. See below, "Permissible Exposure Limits in Air" section.
US EPA Hazardous Waste Number (RCRA No.): U184.
RCRA, 40CFR261, Appendix 8 Hazardous Constituents.
RCRA 40CFR268.48; 61FR15654, Universal Treatment Standards: Wastewater (mg/L), 0.055; Nonwastewater (mg/kg), 6.0.
RCRA 40CFR264, Appendix 9; TSD Facilities Ground Water Monitoring List. Suggested test method(s) (PQL μg/L): 8240 (5); 8270 (10).
Reportable Quantity (RQ): 10 lb (4.54 kg).
EPCRA Section 313 Form R *de minimis* concentration reporting level: 1.0%.
US DOT Regulated Marine Pollutant (49CFR172.101, Appendix B).
Canada, WHMIS, Ingredients Disclosure List Concentration 1.0%.
European/International Regulations: Hazard Symbol: T, N; Risk phrases: R40; R48/23; R51/53; Safety phrases: S1/2; S23; S36/37; S45; S61 (see Appendix 4).
WGK (German Aquatic Hazard Class): No value assigned.
Description: Pentachloroethane is a colorless, heavy, nonflammable liquid with a sweetish chloroform- or camphor-like odor. Molecular weight = 202.28; Boiling point = 162°C; Freezing/Melting point = −29°C; Flash point = 75°C (cc). Hazard Identification (based on NFPA-704 M Rating System): Health 3, Flammability 2, Reactivity 0. Practically insoluble in water.
Potential Exposure: Compound Description: Tumorigen, Drug, Mutagen. Pentachloroethane is used in the manufacture of tetrachloroethylene and as a solvent for cellulose acetate, certain cellulose ethers, resins, and gums. It is also used as a drying agent for timber by immersion at temperatures greater than 100°C.
Incompatibilities: May self-ignite. Violent reaction with alkali metals (i.e., lithium, sodium, potassium, rubidium, cesium, francium) will produce spontaneous explosive chloroacetylenes. Shock- and friction-sensitive material formed by mixture with potassium.
Permissible Exposure Limits in Air
NIOSH REL: Handle with caution; See *NIOSH Pocket Guide*, Appendix C.
Protective Action Criteria (PAC)
TEEL-0: 46 mg/m³
PAC-1: 126 mg/m³
PAC-2: 500 mg/m³
PAC-3: 500 mg/m³
DFG MAK: 5 ppm/42 mg/m³ TWA; Peak Limitation Category II(2).
Rumania: TWA 30 mg/m³; Yugoslavia: MAC 5 ppm/40 mg/m³.
Determination in Air: Use NIOSH Analytical Method (IV) #2517.
Permissible Concentration in Water: To protect freshwater aquatic life: 7240 μg/L on an acute toxicity basis and 1100 μg/L on a chronic basis. To protect saltwater aquatic

life: 390 µg/L on an acute toxicity basis and 231 µg/L on a chronic basis. To protect human health—no criteria derived due to insufficient data.[6]

Routes of Entry: Inhalation, ingestion, skin and/or eye contact.

Harmful Effects and Symptoms

Short Term Exposure: Pentachloroethane is an irritant and a strong central nervous system depressant. Pentachloroethane has a strong narcotic effect. Symptoms include prompt nausea, vomiting, abdominal pain with diarrhea, headaches, dizziness, confusion, drowsiness, and occasionally, convulsions. Visual disturbances may arise followed by coma and possible death from respiratory arrest or circulatory collapse. Death may occur by respiratory arrest or circulatory collapse. Occasionally, sudden death may occur due to ventricular fibrillation. Other effects may include weight gain, edema, loss of appetite, jaundice, and pain (due to enlarged liver). The chemical is very toxic with a probable oral lethal dose of 50–500 mg/kg or between 1 teaspoon and 1 oz for a 150-lb person. In animals: irritation of eyes, skin; weakness, restlessness, irregular/irregularities of respiration, muscle incoordination; liver, kidney, lung changes (NIOSH).

Long Term Exposure: Exposure to this material may result in injury to the liver, lungs, and kidneys.

Points of Attack: Eyes, skin, respiratory system, central nervous system, liver, kidneys.

First Aid: If this chemical gets into the eyes, remove any contact lenses at once and irrigate immediately for at least 15 min, occasionally lifting upper and lower lids. Seek medical attention immediately. If this chemical contacts the skin, remove contaminated clothing and wash immediately with soap and water. Seek medical attention immediately. If this chemical has been inhaled, remove from exposure, begin rescue breathing (using universal precautions, including resuscitation mask) if breathing has stopped and CPR if heart action has stopped. Transfer promptly to a medical facility. When this chemical has been swallowed, get medical attention. Give large quantities of water and induce vomiting. Do not make an unconscious person vomit. Keep victim quiet and maintain normal body temperature. Effects may be delayed; keep victim under observation.

Personal Protective Methods: Wear protective gloves and clothing to prevent any reasonable probability of skin contact. Safety equipment suppliers/manufacturers can provide recommendations on the most protective glove/clothing material for your operation. All protective clothing (suits, gloves, footwear, headgear) should be clean, available each day, and put on before work. Contact lenses should not be worn when working with this chemical. Wear splash-proof chemical goggles and face shield unless full face-piece respiratory protection is worn. Employees should wash immediately with soap when skin is wet or contaminated. Provide emergency showers and eyewash.

Respirator Selection: Follow regulations in OSHA 29CFR1910.134 or European Standard EN149. Use a NIOSH/MSHA- or European Standard EN149-approved respirator; or use an approved supplied-air respirator with a full face-piece operated in the positive-pressure mode, or with a full face-piece, hood, or helmet in the continuous-flow mode; or use a NIOSH/MSHA- or European Standard EN149-approved self-contained breathing apparatus with a full face-piece operated in pressure-demand or other positive-pressure mode.

Storage: Color Code—Blue: Health Hazard/Poison: Store in a secure poison location. Prior to working with this chemical you should be trained on its proper handling and storage. Store in a refrigerator away from alkalis, reactive metals, water. Where possible, automatically pump liquid from drums or other storage containers to process containers. Sources of ignition, such as smoking and open flames, are prohibited where this chemical is handled, used, or stored. Metal containers involving the transfer of 5 gallons or more of this chemical should be grounded and bonded. Drums must be equipped with self-closing valves, pressure vacuum bungs, and flame arresters. Use only nonsparking tools and equipment, especially when opening and closing containers of this chemical. Wherever this chemical is used, handled, manufactured, or stored, use explosion-proof electrical equipment and fittings. A regulated, marked area should be established where this chemical is handled, used, or stored in compliance with OSHA Standard 1910.1045.

Shipping: This compound requires a shipping label of "POISONOUS/TOXIC MATERIALS." It falls in Hazard Class 6.1 and Packing Group II.

Spill Handling: Evacuate and restrict persons not wearing protective equipment from area of spill or leak until cleanup is complete. Remove all ignition sources. Ventilate area of spill or leak. Absorb liquids in vermiculite, dry sand, earth, peat, carbon, or a similar material and deposit in sealed containers. Keep this chemical out of a confined space, such as a sewer, because of the possibility of an explosion, unless the sewer is designed to prevent the buildup of explosive concentrations. It may be necessary to contain and dispose of this chemical as a hazardous waste. If material or contaminated runoff enters waterways, notify downstream users of potentially contaminated waters. Contact your local or federal environmental protection agency for specific recommendations. If employees are required to clean up spills, they must be properly trained and equipped. OSHA 1910.120(q) may be applicable.

Fire Extinguishing: This chemical is a combustible liquid. Poisonous gases, including chlorine, are produced in fire. Fires should be extinguished using water, carbon dioxide, or dry chemical. Move container from fire area if you can do so without risk. Fight fire from maximum distance. Dike fire control water for later disposal; do not scatter the material. Keep unnecessary people away; isolate hazard area and deny entry. Stay upwind; keep out of low areas. Ventilate closed spaces before entering them. Wear positive pressure breathing apparatus and special protective clothing. Remove and isolate contaminated clothing at the site.

Vapors are heavier than air and will collect in low areas. Vapors may travel long distances to ignition sources and flashback. Vapors in confined areas may explode when exposed to fire. Containers may explode in fire. Storage containers and parts of containers may rocket great distances, in many directions. If material or contaminated runoff enters waterways, notify downstream users of potentially contaminated waters. Notify local health and fire officials and pollution control agencies. From a secure, explosion-proof location, use water spray to cool exposed containers. If cooling streams are ineffective (venting sound increases in volume and pitch, tank discolors, or shows any signs of deforming), withdraw immediately to a secure position. If employees are expected to fight fires, they must be trained and equipped in OSHA 1910.156. The only respirators recommended for firefighting are self-contained breathing apparatuses that have full face-pieces and are operated in a pressure-demand or other positive-pressure mode.

Disposal Method Suggested: Consult with environmental regulatory agencies for guidance on acceptable disposal practices. Generators of waste containing this contaminant ($\geq$100 kg/mo) must conform with EPA regulations governing storage, transportation, treatment, and waste disposal. Incineration after mixing with another combustible fuel. Care must be exercised to assure complete combustion to prevent the formation of phosgene. An acid scrubber is necessary to remove the halo acids produced.[22]

References

National Institute for Occupational Safety and Health. (1977). *Profiles on Occupational Hazards for Criteria Document Priorities: Pentachloroethane*, Report PB-274,073. Cincinnati, OH, pp. 303–305

US Environmental Protection Agency. (1980). *Chlorinated Ethanes: Ambient Water Quality Criteria*. Washington, DC

US Environmental Protection Agency. (January 4, 1983). *Chemical Hazard Information Profile Draft Reports: Pentachloroethane*. Washington, DC

US Environmental Protection Agency. (October 31, 1985). *Chemical Hazard Information Profile: Pentachloroethane*. Washington, DC: Chemical Emergency Preparedness Program

Pentachloronaphthalene   P:0220

Molecular Formula: $C_{10}H_3C1_5$
Synonyms: Halowax 1013; Naphthalene, pentachloro-; 1,2,3,4,5-Pentachloronaphthalene
CAS Registry Number: 1321-64-8
RTECS® Number: QK0300000
EC Number: 215-320-8 [*Annex I Index No.:* 602-041-00-5]
Regulatory Authority and Advisory Bodies
US EPA GENETOX PROGRAM: Positive: Histidine reversion—Ames test, 1988.
Air Pollutant Standard Set. See below, "Permissible Exposure Limits in Air" section.

Canada, WHMIS, Ingredients Disclosure List Concentration 1.0%.
European/International Regulations: Hazard Symbol: Xn, N; Risk phrases: R21/22; R36/38; R50/53; Safety phrases: S2 (see Appendix 4).
WGK (German Aquatic Hazard Class): 3—Severe hazard to waters.
Description: Pentachloronaphthalene is a pale yellow or white solid powder with an aromatic odor. Molecular weight = 300.38; Specific gravity (H_2O:1) = 1.67; Boiling point = 335.6°C. Freezing/Freezing/Melting point = 120°C; Vapor pressure = <1 mmHg at 25°C. Hazard Identification (based on NFPA-704 M Rating System): Health 3, Flammability 0, Reactivity 0. Insoluble in water.
Potential Exposure: Compound Description: Tumorigen, Mutagen. Used in electric wire insulation, in additives to specialized lubricants, and as a fire- and water-proofing agent.
Incompatibilities: Violent reaction with strong oxidizers, aluminum, liquid oxygen, potassium, sodium. Heat may contribute to instability.
Permissible Exposure Limits in Air
OSHA PEL: 0.5 mg/m^3 TWA [skin].
NIOSH REL: 0.5 mg/m^3 TWA [skin].
ACGIH TLV®[1]: 0.5 mg/m^3 TWA [skin].
No TEEL available.
DFG MAK: 0.5 mg/m^3 TWA [skin].
Australia: TWA 0.5 mg/m^3, 1993; Austria: MAK 0.5 mg/m^3, [skin], 1999; Belgium: TWA 0.5 mg/m^3, 1993; Denmark: TWA 0.5 mg/m^3, [skin], 1999; France: VME 0.5 mg/m^3, 1999; Norway: TWA 0.5 mg/m^3, 1999; the Netherlands: MAC-TGG 0.5 mg/m^3, 2003; Poland: MAC (TWA) 0.5 mg/m^3, MAC (STEL) 1.5 mg/m^3, 1999; Sweden: NGV 0.2 mg/m^3, KTV 0.3 mg/m^3, [skin], 1999; Switzerland: MAK-W 0.5 mg/m^3, KZG-W 2.5 mg/m^3, [skin], 1999; Turkey: TWA 0.5 mg/m^3, [skin], 1993; Argentina, Bulgaria, Columbia, Jordan, South Korea, New Zealand, Singapore, Vietnam: ACGIH TLV®: TWA 0.5 mg/m^3 [skin].
When skin contact also occurs, you may be overexposed, even though air levels are less than the limit listed above.
Determination in Air: Use NIOSH II(2), Method #S9.
Determination in Water: Octanol–water coefficient: Log K_{ow} = 8.73–9.13 (ICSC).
Routes of Entry: Inhalation, skin absorption, ingestion, skin and/or eye contact.
Harmful Effects and Symptoms
Short Term Exposure: Irritates the eyes, skin, and respiratory tract. Skin rash may occur if contaminated skin is exposed to sunlight. Can affect the nervous system, causing headache, fatigue, dizziness, vertigo (an illusion of movement), anorexia (loss of appetite).
Long Term Exposure: Repeated or prolonged contact with skin may cause acne-like rash (chloracne), pruritus. May affect the liver, causing jaundice; liver necrosis.
Points of Attack: Skin, liver, central nervous system.

Medical Surveillance: NIOSH lists the following tests: Liver function tests.

First Aid: If this chemical gets into the eyes, remove any contact lenses at once and irrigate immediately for at least 15 min, occasionally lifting upper and lower lids. Seek medical attention immediately. If this chemical contacts the skin, remove contaminated clothing and wash immediately with soap and water. Seek medical attention immediately. If this chemical has been inhaled, remove from exposure, begin rescue breathing (using universal precautions, including resuscitation mask) if breathing has stopped and CPR if heart action has stopped. Transfer promptly to a medical facility. When this chemical has been swallowed, get medical attention. Give large quantities of water and induce vomiting. Do not make an unconscious person vomit.

Personal Protective Methods: Wear protective gloves and clothing to prevent any reasonable probability of skin contact. Safety equipment suppliers/manufacturers can provide recommendations on the most protective glove/clothing material for your operation. All protective clothing (suits, gloves, footwear, headgear) should be clean, available each day, and put on before work. Contact lenses should not be worn when working with this chemical. Wear dust-proof chemical goggles and face shield unless full face-piece respiratory protection is worn. Employees should wash immediately with soap when skin is wet or contaminated. Provide emergency showers and eyewash.

Respirator Selection: Up to 5 mg/m³: Sa* (APF = 10) (any supplied-air respirator); or SCBAF (APF = 50) (any self-contained breathing apparatus with a full face-piece). *Emergency or planned entry into unknown concentrations or IDLH conditions:* SCBAF: Pd,Pp (APF = 10,000) (any NIOSH/MSHA- or European Standard EN 149-approved self-contained breathing apparatus that has a full face-piece and is operated in a pressure-demand or other positive-pressure mode) or SaF: Pd,Pp: ASCBA (APF = 10,000) (any supplied-air respirator that has a full face-piece and is operated in a pressure-demand or other positive-pressure mode in combination with an auxiliary, self-contained breathing apparatus operated in a pressure-demand or other positive-pressure mode). *Escape:* GmFOv100 (APF = 50) (any air-purifying, full-face-piece respirator (gas mask) with a chin-style, front- or back-mounted organic vapor canister having an N100, R100, or P100 filter) or SCBAE (any appropriate escape-type, self-contained breathing apparatus).

*Substance reported to cause eye irritation or damage; may require eye protection.

Storage: Color Code—Blue: Health Hazard/Poison: Store in a secure poison location. Prior to working with this chemical you should be trained on its proper handling and storage. Store in tightly closed containers in a cool, well-ventilated area away from strong oxidizers. Where possible, automatically transfer material from drums or other storage containers to process containers. Sources of ignition, such as smoking and open flames, are prohibited where this chemical is handled, used, or stored. Metal containers involving the transfer of this chemical should be grounded and bonded. Wherever this chemical is used, handled, manufactured, or stored, use explosion-proof electrical equipment and fittings.

Spill Handling: Evacuate persons not wearing protective equipment from area of spill or leak until cleanup is complete. Remove all ignition sources. Use HEPA vacuum or wet method to reduce dust during cleanup. Do not dry sweep. Collect powdered material in the most convenient and safe manner and deposit in sealed containers. Ventilate area after cleanup is complete. It may be necessary to contain and dispose of this chemical as a hazardous waste. If material or contaminated runoff enters waterways, notify downstream users of potentially contaminated waters. Contact your local or federal environmental protection agency for specific recommendations. If employees are required to clean up spills, they must be properly trained and equipped. OSHA 1910.120(q) may be applicable.

Fire Extinguishing: This chemical is a combustible solid. Use dry chemical, carbon dioxide, water spray, or alcohol foam extinguishers. Poisonous gases are produced in fire. If material or contaminated runoff enters waterways, notify downstream users of potentially contaminated waters. Notify local health and fire officials and pollution control agencies. From a secure, explosion-proof location, use water spray to cool exposed containers. If cooling streams are ineffective (venting sound increases in volume and pitch, tank discolors, or shows any signs of deforming), withdraw immediately to a secure position. If employees are expected to fight fires, they must be trained and equipped in OSHA 1910.156. The only respirators recommended for firefighting are self-contained breathing apparatuses that have full face-pieces and are operated in a pressure-demand or other positive-pressure mode.

Reference

New Jersey Department of Health and Senior Services. (December 1999). *Hazardous Substances Fact Sheet: Pentachloronaphthalene.* Trenton, NJ

Pentachloronitrobenzene   P:0230

Molecular Formula: $C_6Cl_5NO_2$

Synonyms: Avicol (Pesticide); Bartilex; Batrilex; Benzene, pentachloronitro-; Botrilex; Brassicol; Brassicol 75; Brassicol earthcide; Brassicol super; Chinozan; Fartox; Folosan; Fomac 2; Fungichlor; GC 3944-3-4; Kobu; Kobutol; KP 2; Marisan forte; NCI-C00419; Nitropentachlorobenzene; Olipsan; Olpisan; PCNB; Pentachlornitrobenzol (German); Pentachloronitrobenzene; Pentagen; Phomasan; PKHNB; Quinosan; Quintocene; Quintoceno (Spanish); Quintozene; RTU1010; Saniclor 30; Terrachlor; Terraclor; Terraclor30 G; Terrafun; Tilcarex; Tripcnb; Tritisan

CAS Registry Number: 82-68-8; *(alt.)* 39378-26-2

RTECS® Number: DA6650000

UN/NA & ERG Number: UN2811 (toxic solid, organic, n.o.s.)/154

EC Number: 201-435-0 [*Annex I Index No.:* 609-043-00-5] (quintozene)

Regulatory Authority and Advisory Bodies

Carcinogenicity: IARC: Animal Limited Evidence; Human No Adequate Data, *not classifiable as carcinogenic to humans*, Group 3, 1987; NCI: Carcinogenesis Studies (feed); no evidence: mouse; NTP: Carcinogenesis Studies (feed); no evidence: mouse.

US EPA Gene-Tox Program, Negative: Host-mediated assay; *In vitro* UDS—human fibroblast; Negative: TRP reversion; *S. cerevisiae*—homozygosis; Negative/limited: Carcinogenicity—mouse/rat; Inconclusive: *B. subtilis* rec assay; *E. coli* polA without S9; Inconclusive: Histidine reversion—Ames test; Inconclusive: *D. melanogaster* sex-linked lethal.

Banned or Severely Restricted (Germany, US) (UN).[13]

US EPA, FIFRA, 1998 Status of Pesticides: Supported.

Air Pollutant Standard Set. See below, "Permissible Exposure Limits in Air" section.

US EPA Hazardous Waste Number (RCRA No.): U185.

RCRA, 40CFR261, Appendix 8 Hazardous Constituents.

RCRA 40CFR268.48; 61FR15654, Universal Treatment Standards: Wastewater (mg/L), 0.055; Nonwastewater (mg/kg), 4.8.

RCRA 40CFR264, Appendix 9; TSD Facilities Ground Water Monitoring List. Suggested test method(s) (PQL μg/L): 8270 (10).

Reportable Quantity (RQ): 100 lb (45.4 kg).

EPCRA Section 313 Form R *de minimis* concentration reporting level: 1.0%.

European/International Regulations: Hazard Symbol: Xi, N; Risk phrases: R43; R50/53; Safety phrases: S2; S13; S24; S37; S60; S61 (see Appendix 4).

WGK (German Aquatic Hazard Class): No value assigned.

Description: Pentachloronitrobenzene forms colorless needles. Technical-grade PCNB contains an average of 97.8% PCNB, 1.8% hexachlorobenzene (HCB), 0.4% 2,3,4,5-tetra-chloronitrobenzene (TCNB), and less than 0.1% pentachlorobenzene. Molecular weight = 295.32; Boiling point = 328°C; Freezing/Melting point = 146°C; Vapor pressure = 1×10^{-4} mbar at 25°C. Hazard Identification (based on NFPA-704 M Rating System): Health 3, Flammability 0, Reactivity 0. Practically insoluble in water.

Potential Exposure: Compound Description: Agricultural Chemical; Tumorigen, Mutagen; Reproductive Effector. Those engaged in the manufacture, formulation, and application of this soil fungicide and seed treatment chemical.

Incompatibilities: Alkalis.

Permissible Exposure Limits in Air

ACGIH TLV®[1]: 0.5 mg/m^3 TWA: not classifiable as a human carcinogen.

Protective Action Criteria (PAC)

TEEL-0: 0.5 mg/m^3

PAC-1: 1.5 mg/m^3

PAC-2: 100 mg/m^3

PAC-3: 500 mg/m^3

Denmark: TWA 0.5 mg/m^3, 1999; the Netherlands: MAC-TGG 0.5 mg/m^3, 2003; Russia[35,43] set a MAC of 0.5 mg/m^3 in work-place air and MAC values for ambient air in residential areas of 0.01 mg/m^3 on a momentary basis and 0.006 mg/m^3 on a daily average basis. A guideline in ambient air has been set[60] in Pennsylvania at 2.47 μg/m^3.

Determination in Water: Fish Tox = 64.49856000 MATC (INTERMEDIATE).

Routes of Entry: Inhalation, ingestion, skin and/or eye contact.

Harmful Effects and Symptoms

Short Term Exposure: May cause skin and eye irritation; sensitization with erythema, itching, and edema. A rebuttable presumption against registration of PCNB for pesticidal uses was issued on October 13, 1977 by EPA on the basis of oncogenicity.

Long Term Exposure: There is limited evidence that this compound is an animal carcinogen. Human Tox = 2.10000 ppb (HIGH).

First Aid: Skin Contact[52]: Flood all areas of body that have contacted the substance with water. Do not wait to remove contaminated clothing; do it under the water stream. Use soap to help assure removal. Isolate contaminated clothing when removed to prevent contact by others. *Eye Contact:* Remove any contact lenses at once. Flush eyes well with copious quantities of water or normal saline for at least 20−30 min. Seek medical attention. *Inhalation:* Leave contaminated area immediately; breathe fresh air. Proper respiratory protection must be supplied to any rescuers. If coughing, difficult breathing, or any other symptoms develop, seek medical attention at once, even if symptoms develop many hours after exposure. *Ingestion:* If convulsions are not present, give a glass or two of water or milk to dilute the substance. Assure that the person's airway is unobstructed and contact a hospital or poison center immediately for advice on whether or not to induce vomiting.

Personal Protective Methods: Wear protective gloves and clothing to prevent any reasonable probability of skin contact. Safety equipment suppliers/manufacturers can provide recommendations on the most protective glove/clothing material for your operation. All protective clothing (suits, gloves, footwear, headgear) should be clean, available each day, and put on before work. Contact lenses should not be worn when working with this chemical. Wear dust-proof chemical goggles and face shield unless full face-piece respiratory protection is worn. Employees should wash immediately with soap when skin is wet or contaminated. Provide emergency showers and eyewash.

Respirator Selection: Follow regulations in OSHA 29CFR1910.134 or European Standard EN149. Use a NIOSH/MSHA- or European Standard EN149-approved respirator; or use an approved supplied-air respirator with a full face-piece operated in the positive-pressure mode, or with a full face-piece, hood, or helmet in the continuous-flow mode; or use a NIOSH/MSHA- or European Standard EN149-approved self-contained breathing apparatus with a full face-piece operated in pressure-demand or other positive-pressure mode.

Storage: Color Code—Blue: Health Hazard/Poison: Store in a secure poison location. Prior to working with this chemical you should be trained on its proper handling and storage. Store in a refrigerator or a cool, dry place away from strong bases. A regulated, marked area should be established where this chemical is handled, used, or stored in compliance with OSHA Standard 1910.1045.

Shipping: Toxic solids, organic, n.o.s. requires a shipping label of "POISONOUS/TOXIC MATERIALS." It falls in Hazard Class 6.1 and Packing Group III.

Spill Handling: Evacuate persons not wearing protective equipment from area of spill or leak until cleanup is complete. Remove all ignition sources. Dampen spilled material with alcohol to avoid dust, then transfer material to a suitable container. Use absorbent dampened with alcohol to pick up remaining material. Wash surfaces well with soap and water. Seal all wastes in vapor-tight plastic bags for eventual disposal. Ventilate area after cleanup is complete. It may be necessary to contain and dispose of this chemical as a hazardous waste. If material or contaminated runoff enters waterways, notify downstream users of potentially contaminated waters. Contact your local or federal environmental protection agency for specific recommendations. If employees are required to clean up spills, they must be properly trained and equipped. OSHA 1910.120(q) may be applicable.

Fire Extinguishing: This chemical is a combustible solid. Use dry chemical, carbon dioxide, water spray, or alcohol foam extinguishers. Poisonous gases are produced in fire. If material or contaminated runoff enters waterways, notify downstream users of potentially contaminated waters. Notify local health and fire officials and pollution control agencies. From a secure, explosion-proof location, use water spray to cool exposed containers. If cooling streams are ineffective (venting sound increases in volume and pitch, tank discolors, or shows any signs of deforming), withdraw immediately to a secure position. If employees are expected to fight fires, they must be trained and equipped in OSHA 1910.156. The only respirators recommended for firefighting are self-contained breathing apparatuses that have full face-pieces and are operated in a pressure-demand or other positive-pressure mode.

Disposal Method Suggested: Consult with environmental regulatory agencies for guidance on acceptable disposal practices. Generators of waste containing this contaminant (≥100 kg/mo) must conform with EPA regulations governing storage, transportation, treatment, and waste disposal. In accordance with 40CFR165, follow recommendations for the disposal of pesticides and pesticide containers. Must be disposed properly by following package label directions or by contacting your local or federal environmental control agency or by contacting your regional EPA office. It has been observed that the product decomposes readily when burned with polyethylene. The compound is highly stable in soil in general, as would be expected on the basis of the polychlorinated aromatic structure.[22]

References

US Environmental Protection Agency. (April 30, 1980). *Pentachloronitrobenzene, Health and Environmental Effects Profile No. 142*. Washington, DC: Office of Solid Waste

Sax, N. I. (Ed.). (1985). *Dangerous Properties of Industrial Materials Report*, 5, No. 3, 11−16

US Environmental Protection Agency, Special Review and Reregistration Division Office of Pesticide Programs. (1998). *Agency Status of Pesticides in Registration, Reregistration, and Special Review* (Rainbow Report). Washington, DC

New Jersey Department of Health and Senior Services. (January 2007). *Hazardous Substances Fact Sheet: Chloronitrobenzenes (mixed isomers)*. Trenton, NJ

Pentachlorophenol   P:0240

Molecular Formula: C_6HCl_5O
Common Formula: C_6Cl_5OH
Synonyms: Chem-tol; Chlon; Chlorophen; Cryptogil ol; Dowcide 7; Dowcide 7; Dowcide EC-7; Dowcide G; Dow pentachlorophenol DP-2 antimicrobial; Dura treet II; Durotox; EP30; Fungifen; Glaze penta; Grundier arbezol; 1-Hydroxypentachlorobenzene; Lauxtol; Lauxtol A; Liroprem; NCI-C54933; NCI-C55378; NCI-C56655; PCP; Penchlorol; Penta; Pentachlorofenol; Pentachlorophenate; 2,3,4,5,6-Pentachlorophenol; Pentachlorophenol, Dowcide EC-7; Pentachlorophenol, DP-2; Pentachlorophenol, technical; Pentachlorphenol (German); Pentaclorofenol (Spanish); Pentacon; Penta-Kil; Pentasol; Penwar; Peratox; Permacide; Permagard; Permasan; Permatox DP-2; Permatox penta; Permite; Phenol, pentachloro-; Pol nu; Preventol P; Priltox; Santobrite; Santophen; Santophen 20; Sinituho; Term-i-trol; Thompson's wood fix; Weedone; Woodtreat A
CAS Registry Number: 87-86-5
RTECS® Number: SM6300000
UN/NA & ERG Number: UN3155/154
EC Number: 201-778-6 [*Annex I Index No.:* 604-002-00-8]
Regulatory Authority and Advisory Bodies
Carcinogenicity: IARC: Animal Sufficient Evidence; Human Inadequate Evidence, *possibly carcinogenic to humans*, Group 2B, 1991; EPA: Sufficient evidence from animal studies; inadequate evidence or no useful data from epidemiologic studies; NCI: Carcinogenesis Studies (feed); clear evidence: mouse; (feed); equivocal evidence: rat.
US EPA Gene-Tox Program, Positive: Cell transform.—SA7/SHE; *S. cerevisiae* gene conversion; Positive: *S. cerevisiae*—forward mutation; Negative: Host-mediated assay; Mouse spot test; Negative: Histidine reversion—Ames test; *S. cerevisiae*—homozygosis.

US EPA, FIFRA, 1998 Status of Pesticides: Supported. Banned or Severely Restricted (several countries) (UN).[13] Air Pollutant Standard Set. See below, "Permissible Exposure Limits in Air" section.

Clean Air Act: Hazardous Air Pollutants (Title I, Part A, Section 112).

Clean Water Act: Section 311 Hazardous Substances/RQ 40CFR117.3 (same as CERCLA, see below); 40CFR401.15 Section 307 Toxic Pollutants; 40CFR423, Appendix A, Priority Pollutants; Section 313 Water Priority Chemicals (57FR41331, 9/9/92).

US EPA Hazardous Waste Number (RCRA No.): D037.

RCRA, 40CFR261, Appendix 8 Hazardous Constituents.

RCRA Toxicity Characteristic (Section 261.24), Maximum. Concentration of Contaminants, regulatory level, 100 mg/L.

RCRA 40CFR268.48; 61FR15654, Universal Treatment Standards: Wastewater (mg/L), 0.089; Nonwastewater (mg/kg), 7.4.

RCRA 40CFR264, Appendix 9; TSD Facilities Ground Water Monitoring List. Suggested test method(s) (PQL μg/L): 8040 (5); 8270 (50).

Safe Drinking Water Act: MCL, 0.001 mg/L; MCLG, zero; Regulated chemical (47 FR 9352).

Reportable Quantity (RQ): 10 lb (4.54 kg)

EPCRA Section 313 Form R *de minimis* concentration reporting level: 1.0%.

US DOT Regulated Marine Pollutant (49CFR172.101, Appendix B), severe pollutant.

Rotterdam Convention Annex III [Chemicals Subject to the Prior Informed Consent Procedure (PIC)] (as pentachlorophenol and its salts and esters).

California Proposition 65 Chemical: Cancer 1/1/90.

European/International Regulations: Hazard Symbol: T+, N; Risk phrases: R24/25; R26; R36/37/38; R40; R50/53; Safety phrases: S1/2; S22; S36/37; S45; S52; S60; S61 (see Appendix 4).

WGK (German Aquatic Hazard Class): 3—Severe hazard to waters.

Description: Pentachlorophenol is a colorless to white crystalline solid. It has a benzene-like odor; pungent when hot. The odor threshold in water is 1600 μg/L and the taste threshold in water is 30 μg/L. Molecular weight = 266.32; Specific gravity (H_2O:1) = 1.98; Boiling point = 308.9°C (decomposes); Freezing/Melting point = 190°C (anhydrous); Vapor pressure = 0.0001 mmHg at 25°C. Hazard Identification (based on NFPA-704 M Rating System): Health 3, Flammability 0, Reactivity 0. Practically insoluble in water; solubility in water = 0.001 at 20°C.

Potential Exposure: Compound Description: Agricultural Chemical; Tumorigen, Mutagen; Reproductive Effector; Human Data; Primary Irritant. Pentachlorophenol (PCP) is a commercially produced bactericide, fungicide, and slimicide used primarily for the preservation of wood, wood products, and other materials. As a chlorinated hydrocarbon, its biological properties have also resulted in its use as an herbicide and molluscicide. Two groups can be expected to encounter the largest exposures. One involves the small number of employees involved in the manufacture of PCP. All of these are presently under industrial health surveillance programs. The second and larger group are the formulators and wood theaters. Exposure, hygiene and industrial health practices can be expected to vary from the small theaters to the larger companies. The principal use as a wood preservative results in both point source water contamination at manufacturing and wood preservation sites and, conceivably, nonpoint source water contamination through runoff wherever there are PCP-treated lumber products exposing PCP to soil.

Incompatibilities: Reacts violently with strong oxidizers, acids, alkalis, and water.

Permissible Exposure Limits in Air

OSHA PEL: 0.5 mg/m³ TWA [skin].

NIOSH REL: 0.5 mg/m³ TWA [skin].

ACGIH TLV®[1]: 0.5 mg/m³ TWA [skin]; BEI: 2 mg[total PCP]/g creatinine in urine/prior to last shift of workweek; 5 mg [free PCP]/L in plasma/end-of-shift; confirmed animal carcinogen with unknown relevance to humans.

NIOSH IDLH: 2.5 mg/m³.

Protective Action Criteria (PAC)

TEEL-0: 0.5 mg/m³

PAC-1: 2 mg/m³

PAC-2: 2.5 mg/m³

PAC-3: 2.5 mg/m³

DFG MAK: [skin] Carcinogen Category 2.

Australia: TWA 0.5 mg/m³, [skin], 1993; Austria: [skin], carcinogen, 1999; Belgium: TWA 0.5 mg/m³, [skin], 1993; Denmark: TWA 0.005 ppm (0.05 mg/m³), [skin], 1999; Finland: TWA 0.5 mg/m³; STEL 1.5 mg/m³, [skin], 1999; France: VME 0.5 mg/m³, [skin], continuous carcinogen, 1999; Hungary: TWA 0.2 mg/m³; STEL 0.4 mg/m³, [skin], 1993; the Netherlands: MAC-TGG 0.06 mg/m³, [skin], 2003; Norway: TWA 0.05 ppm (0.5 mg/m³), 1999; Poland: MAC (TWA) 0.5 mg/m³; MAC (STEL) 1.5 mg/m³, 1999; Russia: STEL 0.1 mg/m³, [skin], 1993; Sweden: NGV 0.5 mg/m³, KTV 1.5 mg/m³, [skin], 1999; Switzerland: MAK-W 0.05 ppm (0.5 mg/m³); STEL 0.1 ppm, [skin], 1999; Turkey: TWA 0.5 mg/m³, [skin], 1993; United Kingdom: TWA 0.5 mg/m³; STEL 1.5 mg/m³, [skin], 2000; Argentina, Bulgaria, Columbia, Jordan, South Korea, New Zealand, Singapore, Vietnam: ACGIH TLV®: confirmed animal carcinogen with unknown relevance to humans. The notation "skin" is added to indicate the possibility of cutaneous absorption. Russia[43] set a MAC of 0.1 mg/m³ in work-place air and a MAC in ambient basis. Several states have set guidelines or standards for Pentachlorophenol in ambient air[60] ranging from zero (North Carolina) to 0.034 μg/m³ (Massachusetts) to 1.67 μg/m³ (New York) to 5.0 μg/m³ (North Dakota and South Carolina) to 8.0 μg/m³ (Virginia) to 10.0 μg/m³ (Connecticut and South Dakota) to 12.0 μg/m³ (Nevada and Pennsylvania) to 25.64 μg/m³ (Kansas).

Determination in Air: Use NIOSH (IV) Analytical Method #5512, Pentachlorophenol; in blood, #8001; in urine, #8303; OSHA Analytical Method 39.

Determination in Water: Octanol−water coefficient: Log $K_{ow} = 5.0$.

Permissible Concentration in Water: *To protect freshwater aquatic life:* 55 μg/L on an acute toxicity basis and 3.2 μg/L on a chronic basis. *To protect saltwater aquatic life:* 53 μg/L on an acute basis and 34 μg/L on a chronic basis. To protect human health—1.010 μg/L is the criteria set by EPA based on toxicity data. A value of 30 μg/L is set on an organoleptic basis.[6] More recently, EPA[47] has developed a lifetime health advisory of 220 μg/L. WHO[35] has set a limit of 10 μg/L on pentachlorophenol in drinking water. Russia[43] set a MAC of 300 μg/L in water bodies used for domestic purposes. More recently, EPA has set a guideline of 200 μg/L for drinking water.[62] Several states have set guidelines for Pentachlorophenol in drinking water[61] ranging from 6 μg/L (Maine) to 30 μg/L (California) to 200 μg/L (Arizona) to 220 μg/L (Kansas and Minnesota).

Determination in Water: Methylene chloride extraction followed by gas chromatography with electron capture or halogen specific detection (EPA Method 608) or gas chromatography plus mass spectrometry (EPA Method 625). Fish Tox = 23.89351000 MATC (INTERMEDIATE) ppb.

Routes of Entry: Inhalation, ingestion, skin and/or eye contact. Absorbed by the skin.

Harmful Effects and Symptoms

Short Term Exposure: Pentachlorophenol irritates the eyes, skin, and respiratory tract. May affect the cardiovascular system. *Inhalation:* Levels of 1 mg/m^3 can cause severe irritation of the nose, throat, and lungs. Higher exposures can cause pulmonary edema, a medical emergency that can be delayed for several hours. This can cause death. Breathing dust or particulates tainted with pentachlorophenol can give rise to sneezing. *Skin:* A 0.04% solution can cause pain and inflammation at point of contact. Chloracne, a skin disorder, has been observed in workers in pentachlorophenol manufacturing plants and wood preserving operations. Profuse sweating and elevated temperature are symptoms of poisoning due to prolonged contact. Excessive skin exposure has caused human death. *Eyes:* Levels of 1 mg/m^3 may be irritating and excessive contact can lead to loss of sight due to corneal damage. *Ingestion:* The lethal human dose is approximately equal to 1 teaspoon for a 150-lb person. Ingestion of 4−8 oz followed by prompt emergency treatment still produced symptoms of poisoning which included rapid breathing followed by a decrease in breathing rate, abdominal pain, reduced blood pressure, excessive and slurred speech, and weakness.

Long Term Exposure: Irritation of eyes, throat, nose, and upper lungs has been reported by individuals using pentachlorophenol as an insecticide for periods of a few years. Chemical acne has been associated with prolonged exposure to this compound. May affect the central nervous system, kidneys, liver, lungs. May be a carcinogen in humans. May damage the developing fetus. There is limited evidence that pentachlorophenol is a teratogen in animals. Tumors have been detected in experimental animals. Human Tox = 1.00000 ppb MCL (HIGH).

Points of Attack: Eyes, skin, respiratory system, cardiovascular system, liver, kidneys, central nervous system. Cancer site in animals: liver.

Medical Surveillance: NIOSH lists the following tests: whole blood (chemical/metabolite); blood plasma; blood plasma, end-of-shift; blood serum; urine (chemical/metabolite); urine (chemical/metabolite), end-of-workweek; urine (chemical/metabolite), prior-to-shift; urine (chemical/metabolite), prior-to-last-shift-of-workweek. If symptoms develop or overexposure is suspected, the following may be useful: Urine test for pentachlorophenol. Liver and kidney function tests. Refer to the NIOSH Criteria Documents #78-174 and #76-147, *Manufacturing, formulating, and working safely with pesticides.*

First Aid: If this chemical gets into the eyes, remove any contact lenses at once and irrigate immediately for at least 15 min, occasionally lifting upper and lower lids. Seek medical attention immediately. If this chemical contacts the skin, remove contaminated clothing and wash immediately with soap and water. Seek medical attention immediately. If this chemical has been inhaled, remove from exposure, begin rescue breathing (using universal precautions, including resuscitation mask) if breathing has stopped and CPR if heart action has stopped. Transfer promptly to a medical facility. When this chemical has been swallowed, get medical attention. Give large quantities of water and induce vomiting. Do not make an unconscious person vomit. Medical observation is recommended for 24−48 h after breathing overexposure, as pulmonary edema may be delayed. As first aid for pulmonary edema, a doctor or authorized paramedic may consider administering a corticosteroid spray.

Personal Protective Methods: Wear protective gloves and clothing to prevent any reasonable probability of skin contact. Safety equipment suppliers/manufacturers can provide recommendations on the most protective glove/clothing material for your operation. Nitrile, polyvinyl chloride, and Tychem® (from E.I. du Pont de Nemours & Company) are among the recommended protective materials. All protective clothing (suits, gloves, footwear, headgear) should be clean, available each day, and put on before work. Contact lenses should not be worn when working with this chemical. Wear dust-proof chemical goggles and face shield unless full face-piece respiratory protection is worn. Employees should wash immediately with soap when skin is wet or contaminated. Provide emergency showers and eyewash.

Respirator Selection: *2.5 mg/m^3:* CcrOv95 (APF = 10) [any air-purifying half-mask respirator with organic vapor cartridge(s) in combination with an N95, R95, or P95 filter. The following filters may also be used: N99, R99, P99, N100, R100, P100] or PaprOvHie (APF = 25) (any powered air-purifying respirator with an organic vapor cartridge in combination with a high-efficiency particulate filter) or Sa

(APF = 10) (any supplied-air respirator) or SCBAF (APF = 50) (any self-contained breathing apparatus with a full face-piece). *Emergency or planned entry into unknown concentrations or IDLH conditions:* SCBAF: Pd,Pp (APF = 10,000) (any self-contained breathing apparatus that has a full face-piece and is operated in a pressure-demand or other positive-pressure mode) or SaF: Pd,Pp: ASCBA (APF = 10,000) (any supplied-air respirator that has a full face-piece and is operated in a pressure-demand or other positive-pressure mode in combination with an auxiliary self-contained breathing apparatus operated in a pressure-demand or other positive-pressure mode). *Escape:* GmFOv100 (APF = 50) (any air-purifying, full-face-piece respirator (gas mask) with a chin-style, front- or back-mounted organic vapor canister having an N100, R100, or P100 filter) or SCBAE (any appropriate escape-type, self-contained breathing apparatus).

Note: Substance reported to cause eye irritation or damage; may require eye protection.

Storage: Color Code—Blue: Health Hazard/Poison: Store in a secure poison location. Prior to working with this chemical you should be trained on its proper handling and storage. Pentachlorophenol must be stored to avoid contact with strong oxidizers (such as chlorine, bromine, and fluorine) because violent reactions occur. Where possible, automatically transfer material from drums or other storage containers to process containers. Sources of ignition, such as smoking and open flames, are prohibited where this chemical is handled, used, or stored. Metal containers involving the transfer of this chemical should be grounded and bonded. A regulated, marked area should be established where this chemical is handled, used, or stored in compliance with OSHA Standard 1910.1045.

Shipping: Pentachlorophenol requires a shipping label of "POISONOUS/TOXIC MATERIALS." It falls in Hazard Class 6.1 and Packing Group II.

Spill Handling: Evacuate persons not wearing protective equipment from area of spill or leak until cleanup is complete. Remove all ignition sources. Collect powdered material in the most convenient and safe manner and deposit in sealed containers. Ventilate area after cleanup is complete. It may be necessary to contain and dispose of this chemical as a hazardous waste. If material or contaminated runoff enters waterways, notify downstream users of potentially contaminated waters. Contact your local or federal environmental protection agency for specific recommendations. If employees are required to clean up spills, they must be properly trained and equipped. OSHA 1910.120(q) may be applicable.

Fire Extinguishing: Extinguish fire using an agent suitable for type of surrounding fire (the material itself does not burn). Poisonous gases, including hydrogen chloride, dioxines, and chlorinated phenols, are produced in fire. If material or contaminated runoff enters waterways, notify downstream users of potentially contaminated waters. Notify local health and fire officials and pollution control agencies. From a secure, explosion-proof location, use water spray to cool exposed containers. If cooling streams are ineffective (venting sound increases in volume and pitch, tank discolors, or shows any signs of deforming), withdraw immediately to a secure position. If employees are expected to fight fires, they must be trained and equipped in OSHA 1910.156. The only respirators recommended for firefighting are self-contained breathing apparatuses that have full face-pieces and are operated in a pressure-demand or other positive-pressure mode.

Disposal Method Suggested: Consult with environmental regulatory agencies for guidance on acceptable disposal practices. Generators of waste containing this contaminant (≥100 kg/mo) must conform with EPA regulations governing storage, transportation, treatment, and waste disposal. In accordance with 40CFR165, follow recommendations for the disposal of pesticides and pesticide containers. Must be disposed properly by following package label directions or by contacting your local or federal environmental control agency or by contacting your regional EPA office. Incineration (600−900°C) coupled with adequate scrubbing and ash disposal facilities.[22] Alternatively pentachlorophenol in wastewaters, for example, may be recovered and recycled.

References

US Environmental Protection Agency. (1980). *Pentachlorophenol: Ambient Water Quality Criteria.* Washington, DC

Rao, K. R. (Ed.). (1978). *Pentachlorophenol: Chemistry, Pharmacology and Environmental Toxicology.* Proceedings of a Symposium, Pensacola, FL, June 1977, New York, Plenum Press

US Environmental Protection Agency. (April 30, 1980). *Pentachlorophenol, Health and Environmental Effects Profile No. 143.* Washington, DC: Office of Solid Waste

Sax, N. I. (Ed.). *Dangerous Properties of Industrial Materials Report*, 3, No. 4, 73−77 (1983) and 4, No. 3, 24−26 (1984)

US Public Health Service. (December 1988). *Toxicological Profile for Pentachlorophenol.* Atlanta, GA: Agency for Toxic Substances and Disease Registry

US Environmental Protection Agency. (October 31, 1985). *Chemical Hazard Information Profile: Pentachlorophenol.* Washington, DC: Chemical Emergency Preparedness Program

New York State Department of Health. (1986). *Chemical Fact Sheet: Pentachlorophenol* (Version 2 and Version 3). Albany, NY: Bureau of Toxic Substance Assessment

US Environmental Protection Agency, Special Review and Reregistration Division Office of Pesticide Programs. (1998). *Agency Status of Pesticides in Registration, Reregistration, and Special Review* (Rainbow Report). Washington, DC

New Jersey Department of Health and Senior Services. (August 2002). *Hazardous Substances Fact Sheet: Pentachlorophenol.* Trenton, NJ

Pentaerythritol   P:0250

Molecular Formula: $C_5H_{12}O_4$
Common Formula: $C(CH_2OH)_4$
Synonyms: 2,2-bis(Hydroxymethyl)-1,3-propanediol;
Methane tetramethylol; Monopentaerythritol; PE;
Pentaerythrite; Tetrahydroxymethylmethane; Tetramethyl-
olmethane
CAS Registry Number: 115-77-5
RTECS® Number: RZ2490000
EC Number: 204-104-9
Regulatory Authority and Advisory Bodies
Air Pollutant Standard Set. See below, "Permissible
Exposure Limits in Air" section.
WGK (German Aquatic Hazard Class): 1—Low hazard to
waters.
Description: Pentaerythritol is a white crystalline solid.
Molecular weight = 136.17; Specific gravity (H_2O:1) = 1.38;
Boiling point = (sublimes) 276°C at 30 mmHg; Freezing/
Melting point = (sublimes) 261°C; Vapor pressure =
8×10^{-8} mmHg at 25°C. Hazard Identification (based on
NFPA-704 M Rating System): Health 1, Flammability 2,
Reactivity 0. Slightly soluble in water; 6% at 15°C.
Potential Exposure: Pentaerythritol is used in coatings and
stabilizers; in the formation of alkyd resins and varnishes. It
is used as an intermediate in the manufacture of plasticizers,
explosives (PETN), and pharmaceuticals.
Incompatibilities: Organic acids, oxidizers. Explosive com-
pound is formed when a mixture of PE and thiophosphoryl
chloride is heated.
Permissible Exposure Limits in Air
OSHA PEL: 15 mg/m³ (total inhalable dust) TWA; 5 mg/
m³ TWA, respirable fraction.
NIOSH REL: 10 mg/m³ (total inhalable dust) TWA; 5 mg/
m³ TWA, respirable fraction.
ACGIH TLV®[1]: 10 mg/m³ TWA.
Protective Action Criteria (PAC)
TEEL-0: 15 mg/m³
PAC-1: 30 mg/m³
PAC-2: 50 mg/m³
PAC-3: 500 mg/m³
Australia: TWA 10 mg/m³, 1993; Belgium: TWA 10 mg/
m³, 1993; Finland: TWA 10 mg/m³; STEL 20 mg/m³, 1999;
France: VME 10 mg/m³, 1999; United Kingdom: TWA
10 mg/m³; STEL 20 mg/m³, total inhalable dust; TWA
4 mg/m³, respirable dust, 2000; the Netherlands: MAC-
TGG 10 mg/m³ (total dust), 2003; the Netherlands: MAC-
TGG 5 mg/m³ (respirable dust), 2003; Argentina, Bulgaria,
Columbia, Jordan, South Korea, New Zealand, Singapore,
Vietnam: ACGIH TLV®: TWA 10 mg/m³. States which
have set guidelines or standards for pentaerythritol in ambient air[60] include Virginia at 80 μg/m³ and Connecticut at
300 μg/m³.
Determination in Air: Use NIOSH IV Method #0500, total
dust; Method #0600 (respirable dust), Particulates NOR.

Permissible Concentration in Water: Russia[43] set a MAC
of 0.1 mg/L in water bodies used for domestic purposes.
Routes of Entry: Inhalation, ingestion, skin and/or eye
contact.
Harmful Effects and Symptoms
Short Term Exposure: Irritates the eyes and respiratory
system. Feeding studies using human volunteers showed
that 85% of Pentaerythritol fed was eliminated unchanged
in the urine within 30 h. There are, in general, no significant
effects on health by common routes of exposure even at
abnormal use concentrations.
Points of Attack: Eyes, respiratory system.
Medical Surveillance: There is no special test for this
chemical. However, if illness occurs or overexposure is suspected, medical attention is recommended.
First Aid: If this chemical gets into the eyes, remove any
contact lenses at once and irrigate immediately for at least
15 min, occasionally lifting upper and lower lids. Seek
medical attention immediately. If this chemical contacts
the skin, remove contaminated clothing and wash immediately with soap and water. Seek medical attention immediately. If this chemical has been inhaled, remove from
exposure, begin rescue breathing (using universal precautions, including resuscitation mask) if breathing has
stopped and CPR if heart action has stopped. Transfer
promptly to a medical facility. When this chemical has
been swallowed, get medical attention. Give large quantities of water and induce vomiting. Do not make an unconscious person vomit.
Personal Protective Methods: Wear protective gloves and
clothing to prevent any reasonable probability of skin contact. Safety equipment suppliers/manufacturers can provide
recommendations on the most protective glove/clothing
material for your operation. All protective clothing (suits,
gloves, footwear, headgear) should be clean, available each
day, and put on before work. Contact lenses should not be
worn when working with this chemical. Wear dust-proof
chemical goggles and face shield unless full face-piece
respiratory protection is worn. Employees should wash
immediately with soap when skin is wet or contaminated.
Provide emergency showers and eyewash.
Respirator Selection: Use dust respirator.
Storage: Color Code—Green: General storage may be used.
Prior to working with this chemical you should be trained
on its proper handling and storage. Store in tightly closed
containers in a cool, well-ventilated area away from organic
acids, oxidizers.
Shipping: Not regulated.
Spill Handling: Evacuate persons not wearing protective
equipment from area of spill or leak until cleanup is complete. Remove all ignition sources. Collect powdered material in the most convenient and safe manner and deposit in
sealed containers. Ventilate area after cleanup is complete.
It may be necessary to contain and dispose of this chemical
as a hazardous waste. If material or contaminated runoff
enters waterways, notify downstream users of potentially

contaminated waters. Contact your local or federal environmental protection agency for specific recommendations. If employees are required to clean up spills, they must be properly trained and equipped. OSHA 1910.120(q) may be applicable.

Fire Extinguishing: This chemical is a combustible solid. Use dry chemical, carbon dioxide, water spray, or alcohol foam extinguishers. Poisonous gases are produced in fire. If material or contaminated runoff enters waterways, notify downstream users of potentially contaminated waters. Notify local health and fire officials and pollution control agencies. From a secure, explosion-proof location, use water spray to cool exposed containers. If cooling streams are ineffective (venting sound increases in volume and pitch, tank discolors, or shows any signs of deforming), withdraw immediately to a secure position. If employees are expected to fight fires, they must be trained and equipped in OSHA 1910.156. The only respirators recommended for firefighting are self-contained breathing apparatuses that have full face-pieces and are operated in a pressure-demand or other positive-pressure mode.

Pentaerythritol tetranitrate   P:0255

Molecular Formula: $C_5H_8N_4O_{12}$
Common Formula: $C(CH_2ONO_2)_4$
Synonyms: Angicap; Angitet; Cardiacap; 1,3-Dinitrato-2,2-bis(nitratomethyl)propane; Nitropenta; Pentaerythrite tetranitrate; Pentaerithrityl tetranitrate; Pentaerithrityltetranitrat (German); Pentrita (Spanish); PETN; 1,3-Propanediol,2,2-bis[(nitrooxy) methyl]-, dinitrate (ester); Tetranitrato de pentaeritritilo (Spanish); Tétranitrate de pentaerithrityle (French); Vasitol; Vasodiatol
CAS Registry Number: 78-11-5; *(alt.)* 103842-90-6; *(alt.)* 108736-71-6
RTECS®Number: RZ2620000
UN/NA & ERG Number: UN3344 (Pentaerythrite tetranitrate mixture, desensitized, solid, n.o.s. with >10% but not >20% PETN, by mass)/113; UN0150 (Pentaerythrite tetranitrate, wetted or Pentaerythritol tetranitrate, wetted, or PETN, wetted or Pentaerythrite tetranitrate, or Pentaerythritol tetranitrate or PETN, desensitized)/112
EC Number: 201-084-3
Regulatory Authority and Advisory Bodies
Department of Homeland Security Screening Threshold Quantity (pounds): *Release hazard* 5000 (commercial grade); *Theft hazard* 400 (commercial grade).
Not listed under California Proposition 65.
Chemicals Subject to TSCA 12(b) Export Notification Requirements, Section 4 (1%).
European/International Regulations: Hazard Symbol: E; Risk phrases: R3; Safety phrases: S2; S35 (see Appendix 4). WGK (German Aquatic Hazard Class): No value assigned.

Description: Pentaerythrite tetranitrate (PETN) is a high explosive, especially when dry. PETN is a sand-like, white crystalline solid. Practically odorless. Molecular weight = 316.17; Boiling point = 205−215°C (explodes); Vapor pressure = negligible; 1.04×10^{-10} mmHg at 25°C; 1.035×10^{-10} mmHg at 25°C. Density 1.75−1.77 g/cm³. Freezing/Melting point = 138−140°C. Autoignition temperature = (explodes) 210°C. Hazard Identification (based on NFPA-704 M Rating System): Health 2; Flammability, 2; Reactivity, 4. Solubility in water = 4.3 mg/100 g at 25°C. The principal hazard from PETN is blast from sudden and abrupt explosion; not from ruptured or bursting container fragments or rocketing projectiles.
Potential Exposure: Compound Description: Tumorigen, Drug, Mutagen, Human Data. First introduced following WWII, PETN shares the same chemical family as nitroglycerine. It is 70% more powerful than TNT.[NYT] Used in the manufacture of fuses for detonation and explosive specialties, including the plastic explosive, Semtex, and in blasting caps. PETN is also used as a medical vasodilator to lower blood pressure by widening blood vessels to improve blood flow. PRTN has been used in terrorism attempts in 2001 by the so-called "shoe bomber," in 2009 by the "underwear bomber," and most recently in October 2010, hidden in printer cartridges being shipped internationally by passenger jet.
Incompatibilities: Normally stable, PETN does not easily explode when dropped or set on fire (rapid heating can cause detonation when heated to 210°C). Nevertheless, PETN is a dangerous high explosive and a strong oxidizer. PETN normally requires a blasting cap or other kind of detonator but may decompose explosively from concussion, shock, friction, static charges. Contact with reducing agents (e.g., zinc, alkaline metals) may cause explosion. Keep away from combustible materials, other oxidizers (e.g., nitrates, permanganates). Contact with sulfur trioxide may cause detonation. May explode in the presence of strong bases (i.e., sodium or potassium hydroxide). May react with heavy metals.
Permissible Exposure Limits in Air
Protective Action Criteria (PAC)
TEEL-0: 0.015 mg/m³
PAC-1: 0.05 mg/m³
PAC-2: 0.35 mg/m³
PAC-3: 500 mg/m³
If material is mixed with TNT, see entry T:0920.
Determination in Air: NIOSH method not established. If material is mixed with TNT, see entry T:0920.
Determination in Water: Octanol−water coefficient: Log $K_{ow} = 1.60$.
Routes of Entry: Inhalation of dust, fume, or vapor; ingestion of dust; percutaneous absorption from dust, skin and/or eye contact.
Harmful Effects and Symptoms
Short Term Exposure: Corrosive and highly irritating. Target organs are the skin and heart. Contact irritates the

eyes, nose, and throat. Can penetrate the skin. Exposure can cause red eyes, dizziness, irritability, headache, convulsions, nausea, vomiting.

PETN is a vasodilator and can affect the cardiovascular system, resulting in widening of blood vessels and the lowering of blood pressure. Medical care is advised.

Long Term Exposure: Lowest published toxic dose (oral, man): 1669 mg/kg/8 year—continuous. [British Journal of Dermatology (87, 498, 1972)].

Medical Surveillance: If material is mixed with TNT, see entry T:0920.

First Aid: If this chemical gets into the eyes, remove any contact lenses at once and irrigate immediately for at least 15 min, occasionally lifting upper and lower lids. Seek medical attention immediately. If this chemical contacts the skin, remove contaminated clothing and wash immediately with soap and water. Seek medical attention immediately. If this chemical has been inhaled, remove from exposure, begin rescue breathing (using universal precautions, including resuscitation mask) if breathing has stopped and CPR if heart action has stopped. Transfer promptly to a medical facility. When this chemical has been swallowed, get medical attention. Give large quantities of water and induce vomiting. Do not make an unconscious person vomit.

Personal Protective Methods: Wear protective gloves and clothing to prevent any reasonable probability of skin contact. Safety equipment suppliers/manufacturers can provide recommendations on the most protective glove/clothing material for your operation. Polyvinyl chloride is among the recommended protective materials. All protective clothing (suits, gloves, footwear, headgear) should be clean, available each day, and put on before work. Contact lenses should not be worn when working with this chemical. Wear splash-proof chemical goggles and face shield unless full-face-piece respiratory protection is worn. Employees should wash immediately with soap when skin is wet or contaminated. Provide emergency showers and eyewash. The Webster skin test (colorimetric tests with alcoholic sodium hydroxide) or indicator soap should be used to make sure workers have washed all PETN off their skins.

Respirator Selection: 95XQ (APF = 10) [any particulate respirator equipped with an N95, R95, or P95 filter (including N95, R95, and P95 filtering face-pieces) except quarter-mask respirators. The following filters may also be used: N99, R99, P99, N100, R100, P100]; or any air-purifying full-face-piece respirator equipped with an N95, R95, or P95 filter. The following filters may also be used: N99, R99, P99, N100, R100, P100; or PaprHie (APF = 25) (any powered, air-purifying respirator with a high-efficiency particulate filter) or PaprTHie (APF = 50) (any powered, air-purifying respirator with a tight-fitting face-piece and a high-efficiency particulate filter) or any supplied-air respirator with a full face-piece that is operated in a pressure-demand or other positive-pressure mode. *Emergency or planned entry into unknown concentrations or IDLH conditions:* SaF: Pd,Pp: ASCBA (any supplied-air respirator that has a full face-piece and is operated in a pressure-demand or other positive-pressure mode in combination with an auxiliary self-contained breathing apparatus operated in a pressure-demand or other positive-pressure mode). *Escape:* 100F (APF = 50) (any air-purifying, full-face-piece respirator with an N100, R100, or P100 filter) or SCBAE (any appropriate escape-type, self-contained breathing apparatus).

Storage: An explosive and a strong oxidizer. Color Code—Red Stripe: Flammability Hazard: Store in a explosion-proof refrigerator and keep away from reducing agents.[52] Store in a cool, dark place in a airtight container, separately from all other flammable materials. Unless specified by manufacturer, store between 15°C and 30°C. Protect from light. Prior to working with this chemical you should be trained on its proper handling and storage. Keep material wet with water and treat as an explosive. Keep away from heat, sources of ignition, metal, nitric acid, and reducing materials. Protect containers from shock. Use only nonsparking tools and equipment, especially when opening and closing containers of this chemical. A regulated, marked area should be established where this chemical is handled, used, or stored in compliance with OSHA Standard 1910.1045.

Shipping: Pentaerythrite tetranitrate mixture, desensitized, solid, n.o.s. with >10% but not >20% PETN, by mass. Must be labeled "FLAMMABLE SOLID." This falls into Hazard Class 4.1 and Packing Group II. Pentaerythrite tetranitrate, wetted or Pentaerythritol tetranitrate, wetted, or PETN, wetted or Pentaerythrite tetranitrate, or Pentaerythritol tetranitrate or PETN, desensitized must be labeled "EXPLOSIVE 1.1D." It falls in Hazard Class 1.1D and Packing Group II. Air transport in passenger and cargo planes is FORBIDDEN.

Spill Handling: Seek expert help or contact manufacturer. Evacuate persons not wearing protective equipment from area of spill or leak until cleanup is complete. Remove all ignition sources. Dampen spilled material with water to avoid dust. Do not wash material to sewer. If material or contaminated runoff enters waterways, notify downstream users of potentially contaminated waters. Collect waste material in the most convenient and safe manner and deposit in sealed containers. Ventilate area after cleanup is complete. It may be necessary to contain and dispose of this chemical as a hazardous waste. Contact your local or federal environmental protection agency for specific recommendations. If employees are required to clean up spills, they must be properly trained and equipped. OSHA 1910.120(q) may be applicable.

Fire Extinguishing: Seek expert help or contact manufacturer. This chemical is a dangerously explosive solid and a strong oxidizer. It will increase the activity of an existing fire. The major hazard from PETN is the blast from sudden and abrupt explosion; not necessarily from ruptured or bursting container fragments or rocketing projectiles. When heated to decomposition, this material emits highly toxic nitrogen oxide fumes. If material is on fire and conditions permit, do not extinguish. Evacuate area and let burn. Cool exposures using unattended monitors. If fire must be extinguished, use any agent appropriate for the burning material. If material or contaminated runoff enters waterways, notify downstream users of potentially contaminated waters. Notify local health and fire officials and pollution control agencies. If cooling streams are ineffective (venting sound increases in volume and pitch, tank discolors, or shows any signs of deforming), withdraw immediately to a secure position. If employees are expected to fight fires, they must be trained and equipped in OSHA 1910.156. The only respirators recommended for firefighting are self-contained breathing apparatuses that have full face-pieces and are operated in a pressure-demand or other positive-pressure mode.

References

New Jersey Department of Health and Senior Services. (October 2001). *Hazardous Substances Fact Sheet: Pentaerythrite Tetranitrate*. Trenton, NJ

Chang, K. (November 31, 2010). The New York Times news article, *Explosive on Planes Was Used in Past Plots*. New York, NY

Pentane   P:0260

Molecular Formula: C_5H_{12}
Synonyms: Amyl hydride; Normalpentane; *n*-Pentane; *normal*-Pentane; *n*-Pentano (Spanish); Skellysolve-A
CAS Registry Number: 109-66-0
RTECS® Number: RZ9450000
UN/NA & ERG Number: UN1265/128
EC Number: 203-692-4 [*Annex I Index No.:* 601-006-00-1]
Regulatory Authority and Advisory Bodies
Department of Homeland Security Screening Threshold Quantity (pounds): *Release hazard* 10,000 (≥1.00% concentration).
Air Pollutant Standard Set. See below, "Permissible Exposure Limits in Air" section.
Clean Air Act: Accidental Release Prevention/Flammable Substances, (Section 112[r], Table 3), TQ = 10,000 lb (4540 kg).
Canada, WHMIS, Ingredients Disclosure List Concentration 1.0%.
European/International Regulations: Hazard Symbol: F, Xn, N; Risk phrases: R12; R51/53; R65; R66; R67; Safety phrases: S2; S9; S16; S29; S33 (see Appendix 4).
WGK (German Aquatic Hazard Class): 2—Hazard to waters.
Description: Pentane is a colorless liquid. Gas above 36°C. Gasoline-like odor. Molecular weight = 72.17; Specific gravity (H_2O:1) = 0.63; Boiling point = 36°C; Freezing/Melting point = −129°C; Vapor pressure = 420 mmHg at 25°C; Flash point = −49.4°C (cc); Autoignition temperature = 260°C; also listed at 284 and 309°C.

Explosive limits: LEL = 1.4%; UEL = 7.8%. Hazard Identification (based on NFPA-704 M Rating System): Health 1, Flammability 4, Reactivity 0. Slightly soluble in water.
Potential Exposure: Compound Description: Drug, Human Data. Pentane is used in the manufacture of ice, low-temperature thermometers, in solvent extraction processes, as a blowing agent in plastics, as a fuel, as a chemical intermediate (e.g., amylchlorides).
Incompatibilities: Reacts with strong oxidizers. Attacks some plastics, rubbers, and coatings.
Permissible Exposure Limits in Air
Conversion factor: 1 ppm = 2.95 mg/m³ at 25°C & 1 atm.
OSHA PEL: 1000 ppm/2950 mg/m³ TWA.
NIOSH REL: 120 ppm/350 mg/m³ TWA; 610 ppm/1800 mg/m³ [15 min] Ceiling Concentration.
ACGIH TLV®[1]: 600 ppm/1770 TWA.
NIOSH IDLH: 1500 ppm [LEL].
Protective Action Criteria (PAC)
TEEL-0: 120 ppm
PAC-1: 120 ppm
PAC-2: 610 ppm
PAC-3: 1500 ppm
DFG MAK: 1000 ppm/3000 mg/m³ TWA; Peak Limitation Category II(2); Pregnancy Risk Group C.
Australia: TWA 600 ppm (1800 mg/m³); STEL 750 ppm, 1993; Austria: MAK 600 ppm (1899 mg/m³), 1999; Belgium: TWA 600 ppm (1770 mg/m³); STEL 750 ppm (2210 mg/m³), 1993; Denmark: TWA 500 ppm (1500 mg/m³), 1999; Finland: TWA 500 ppm (1500 mg/m³); STEL 625 ppm (1800 mg/m³), 1999; France: VME 600 ppm (1800 mg/m³), 1999; Hungary: TWA 500 mg/m³; STEL 1500 mg/m³, 1993; the Netherlands: MAC-TGG 1800 mg/m³, 2003; the Philippines: TWA 1000 ppm (2950 mg/m³), 1993; Poland: MAC (TWA) 1800 mg/m³; MAC (STEL) 2300 mg/m³, 1999; Russia: TWA 300 ppm; STEL 300 mg/m³, 1993; Sweden: NGV 600 ppm (1800 mg/m³), KTV 750 ppm (2000 mg/m³), 1999; Switzerland: MAK-W 600 ppm (1800 mg/m³), 1999; Turkey: TWA 1000 ppm (2950 mg/m³), 1993; United Kingdom: LTEL 600 ppm (1800 mg/m³); STEL 750 ppm, 1993; Argentina, Bulgaria, Columbia, Jordan, South Korea, New Zealand, Singapore, Vietnam: ACGIH TLV®: TWA 600 ppm. Russia[43] set a MAC of 100 mg/m³ in ambient air in residential areas on a momentary basis and 25 mg/m³ on a daily average basis. Several states have set guidelines or standards for pentane in ambient air[60] ranging from 7.0 mg/m³ (Connecticut) to 18.0−22.5 mg/m³ (North Dakota) to 30.0 mg/m³ (Virginia) to 42.857 mg/m³ (Nevada).
Determination in Air: Use NIOSH Analytical Method #1500, Hydrocarbons, BP 36-126°C; #2549, Volatile organic compounds.[18]
Determination in Water: Octanol−water coefficient: Log K_{ow} = 3.4.
Routes of Entry: Inhalation, ingestion, skin and/or eye contact.

Harmful Effects and Symptoms

Short Term Exposure: Pentane can affect you when breathed in. Exposure can cause lightheadedness and dizziness and may cause you to pass out. It may damage the nervous system, causing numbness, "pins and needles," and weakness in the arms and legs. Skin contact may cause rash and a burning sensation.

Long Term Exposure: Repeated or prolonged contact can cause dry, cracked skin.

Points of Attack: Eyes, skin, respiratory system, and central nervous system.

Medical Surveillance: If symptoms develop or overexposure is suspected, the following may be useful: examination of the nervous system. Nerve conduction studies should be considered.

First Aid: If this chemical gets into the eyes, remove any contact lenses at once and irrigate immediately for at least 15 min, occasionally lifting upper and lower lids. Seek medical attention immediately. If this chemical contacts the skin, remove contaminated clothing and wash immediately with soap and water. Seek medical attention immediately. If this chemical has been inhaled, remove from exposure, begin rescue breathing (using universal precautions, including resuscitation mask) if breathing has stopped and CPR if heart action has stopped. Transfer promptly to a medical facility. When this chemical has been swallowed, get medical attention. Give large quantities of water and induce vomiting. Do not make an unconscious person vomit.

Personal Protective Methods: Wear protective gloves and clothing to prevent any reasonable probability of skin contact. Safety equipment suppliers/manufacturers can provide recommendations on the most protective glove/clothing material for your operation. Viton is among the recommended protective materials. All protective clothing (suits, gloves, footwear, headgear) should be clean, available each day, and put on before work. Contact lenses should not be worn when working with this chemical. Wear splash-proof chemical goggles and face shield unless full face-piece respiratory protection is worn. Employees should wash immediately with soap when skin is wet or contaminated. Provide emergency showers and eyewash.

Respirator Selection: NIOSH: *1200 ppm:* Sa (APF = 10) (any supplied-air respirator). *1500 ppm:* Sa:Cf (APF = 25) (any supplied-air respirator operated in a continuous-flow mode) or SCBAF (APF = 50) (any self-contained breathing apparatus with a full face-piece) or SaF (APF = 50) (any supplied-air respirator with a full face-piece). *Emergency or planned entry into unknown concentrations or IDLH conditions:* SCBAF: Pd,Pp (APF = 10,000) (any self-contained breathing apparatus that has a full face-piece and is operated in a pressure-demand or other positive-pressure mode) or SaF: Pd,Pp: ASCBA (APF = 10,000) (any supplied-air respirator that has a full face-piece and is operated in a pressure-demand or other positive-pressure mode in combination with an auxiliary self-contained breathing apparatus operated in a pressure-demand or other positive-pressure

mode). *Escape:* GmFOv (APF = 50) [any air-purifying, full-face-piece respirator (gas mask) with a chin-style, front- or back-mounted organic vapor canister] or SCBAE (any appropriate escape-type, self-contained breathing apparatus).

Storage: Color Code—Red: Flammability Hazard: Store in a flammable liquid storage area or approved cabinet away from ignition sources and corrosive and reactive materials. Prior to working with this chemical you should be trained on its proper handling and storage. Before entering confined space where this chemical may be present, check to make sure that an explosive concentration is not a danger. Pentane must be stored to avoid contact with strong oxidizers (such as chlorine and bromine) because violent reactions occur. Store in tightly closed containers in a cool, well-ventilated area away from heat. Sources of ignition, such as smoking and open flames, are prohibited where pentane is handled, used, or stored. Metal containers involving the transfer of 5 gallons or more of pentane should be grounded and bonded. Drums must be equipped with self-closing valves, pressure vacuum bungs, and flame arresters. Use only nonsparking tools and equipment, especially when opening and closing containers of pentane.

Shipping: Pentanes require a label of "FLAMMABLE LIQUID." They fall in Hazard Class 3 and Packing Group I.

Spill Handling: Evacuate and restrict persons not wearing protective equipment from area of spill or leak until cleanup is complete. Remove all ignition sources. Establish forced ventilation to keep levels below explosive limit. Absorb liquids in vermiculite, dry sand, earth, peat, carbon, or a similar material and deposit in sealed containers. Oil-skimming equipment and sorbent foams can be applied to slick if done immediately. Keep this chemical out of a confined space, such as a sewer, because of the possibility of an explosion, unless the sewer is designed to prevent the buildup of explosive concentrations. It may be necessary to contain and dispose of this chemical as a hazardous waste. If material or contaminated runoff enters waterways, notify downstream users of potentially contaminated waters. Contact your local or federal environmental protection agency for specific recommendations. If employees are required to clean up spills, they must be properly trained and equipped. OSHA 1910.120(q) may be applicable.

Fire Extinguishing: This chemical is a flammable liquid. Poisonous gases are produced in fire. Use dry chemical, carbon dioxide, or alcohol foam extinguishers. Vapors are heavier than air and will collect in low areas. Vapors may travel long distances to ignition sources and flashback. Vapors in confined areas may explode when exposed to fire. Containers may explode in fire. Storage containers and parts of containers may rocket great distances, in many directions. If material or contaminated runoff enters waterways, notify downstream users of potentially contaminated waters. Notify local health and fire officials and pollution control agencies. From a secure, explosion-proof location, use water spray to cool exposed containers. If cooling streams are ineffective (venting sound increases in volume and pitch, tank discolors, or shows any signs of deforming), withdraw immediately to a secure position. If employees are expected to fight fires, they must be trained and equipped in OSHA 1910.156. The only respirators recommended for firefighting are self-contained breathing apparatuses that have full face-pieces and are operated in a pressure-demand or other positive-pressure mode.

Disposal Method Suggested: Dissolve or mix the material with a combustible solvent and burn in a chemical incinerator equipped with an afterburner and scrubber. All federal, state, and local environmental regulations must be observed.

References

National Institute for Occupational Safety and Health. (1977). *Criteria for a Recommended Standard: Occupational Exposure to Alkanes*, NIOSH Document No. 77-151. Washington, DC

New Jersey Department of Health and Senior Services. (February 2000). *Hazardous Substances Fact Sheet: Pentane*. Trenton, NJ

2,4-Pentanedione   P:0270

Molecular Formula: $C_5H_8O_2$
Common Formula: $CH_3COCH_2COCH_3$
Synonyms: Acetoacetone; Acetyl acetone; Diacetylmethane; Pentane-2,4-dione; 2-Propanone, acetyl
CAS Registry Number: 123-54-6; 81235-32-7
RTECS® Number: SA1925000
UN/NA & ERG Number: UN2310/131
EC Number: 204-634-0 [*Annex I Index No.:* 606-029-00-0]
Regulatory Authority and Advisory Bodies
Canada, WHMIS, Ingredients Disclosure List Concentration 1.0%.
European/International Regulations (*23-54-6*): Hazard Symbol: Xn; Risk phrases: R10; R22; Safety phrases: S2; S21; S23; S24/25 (see Appendix 4).
WGK (German Aquatic Hazard Class): 1—Low hazard to waters.
Description: 2,4-Pentanedione is a colorless to yellowish liquid with a sour, rancid odor. The odor threshold is 0.01 ppm. Molecular weight = 100.13; Boiling point = 139°C; Freezing/Melting point = −23°C; Flash point = 34°C; Autoignition temperature = 340°C. Explosive limits: LEL = 2.4%; UEL = 11.6%. Hazard Identification (based on NFPA-704 M Rating System): Health 2, Flammability 2, Reactivity 0. Soluble in water.
Potential Exposure: Acetoacetic acid derivative. Compound Description: Agricultural Chemical; Mutagen; Reproductive Effector; Primary Irritant. 2,4-Pentanedione is used in gasoline and lubricant additives, fungicides, insecticides, and colors manufacture; as a chemical intermediate and in the manufacture of metal chelates.

Incompatibilities: Oxidizing material, bases, reducing agents, halogens, aliphatic amines, alkanolamines, organic acids, isocyanates. Light may cause polymerization.
Permissible Exposure Limits in Air
ACGIH TLV®[1]: (2010 Notice of intended change) 25 ppm TWA [skin].
Protective Action Criteria (PAC)
TEEL-0: 20 ppm
PAC-1: 50 ppm
PAC-2: 100 ppm
PAC-3: 100 ppm
DFG MAK: 20 ppm/83 mg/m³ TWA; Peak Limitation Category II(2) [skin]; Pregnancy Risk Group C.
Routes of Entry: Inhalation, ingestion, skin and/or eye contact.
Harmful Effects and Symptoms
Short Term Exposure: Irritates the eyes, skin, and respiratory tract. Eye irritation may be severe. May affect the nervous system. If inhaled, will cause dizziness, coughing, headaches, convulsions, loss of consciousness, and possible death. In addition, to neuropathy, 2,4-pentanedione causes thymic atrophy; it complexes with and inhibits the activities of oxidizing enzymes; it causes minor to severe eye injury and minor to moderate skin irritation in animals; and it has caused contact urticaria and allergic contact dermatitis in humans.
Long Term Exposure: Repeated or prolonged contact may cause skin sensitization and allergy. High exposure may affect the brain. May affect the lungs, thymus, central nervous system. There is limited evidence of reproductive damage and mutations.
Points of Attack: Skin, brain, lungs, central nervous system, thymus.
Medical Surveillance: Evaluation by a qualified allergist. Evaluation of brain effects. Thymus function tests. Consider chest X-ray following acute overexposure.
First Aid: If this chemical gets into the eyes, remove any contact lenses at once and irrigate immediately for at least 15 min, occasionally lifting upper and lower lids. Seek medical attention immediately. If this chemical contacts the skin, remove contaminated clothing and wash immediately with soap and water. Seek medical attention immediately. If this chemical has been inhaled, remove from exposure, begin rescue breathing (using universal precautions, including resuscitation mask) if breathing has stopped and CPR if heart action has stopped. Transfer promptly to a medical facility. When this chemical has been swallowed, get medical attention. Give large quantities of water and induce vomiting. Do not make an unconscious person vomit.
Personal Protective Methods: Wear protective gloves and clothing to prevent any reasonable probability of skin contact. Safety equipment suppliers/manufacturers can provide recommendations on the most protective glove/clothing material for your operation. All protective clothing (suits, gloves, footwear, headgear) should be clean, available each day, and put on before work. Contact lenses should not be

worn when working with this chemical. Wear splash-proof chemical goggles and face shield unless full face-piece respiratory protection is worn. Employees should wash immediately with soap when skin is wet or contaminated. Provide emergency showers and eyewash.

Respirator Selection: Follow regulations in OSHA 29CFR1910.134 or European Standard EN149. Use a NIOSH/MSHA- or European Standard EN149-approved respirator; or use an approved supplied-air respirator with a full face-piece operated in the positive-pressure mode, or with a full face-piece, hood, or helmet in the continuous-flow mode; or use a NIOSH/MSHA- or European Standard EN149-approved self-contained breathing apparatus with a full face-piece operated in pressure-demand or other positive-pressure mode.

Storage: Color Code—Red: Flammability Hazard: Store in a flammable liquid storage area or approved cabinet away from ignition sources and corrosive and reactive materials. Prior to working with this chemical you should be trained on its proper handling and storage. Before entering confined space where this chemical may be present, check to make sure that an explosive concentration is not a danger. Store in stainless steel containers away from oxidizers, reducing agents, bases. Where possible, automatically pump liquid from drums or other storage containers to process containers. Sources of ignition, such as smoking and open flames, are prohibited where this chemical is handled, used, or stored. Metal containers involving the transfer of 5 gallons or more of this chemical should be grounded and bonded. Drums must be equipped with self-closing valves, pressure vacuum bungs, and flame arresters. Use only nonsparking tools and equipment, especially when opening and closing containers of this chemical. Wherever this chemical is used, handled, manufactured, or stored, use explosion-proof electrical equipment and fittings.

Shipping: Pentane-2,4-dione requires a shipping label of "FLAMMABLE LIQUID." It falls in Hazard Class 3 and Packing Group III.

Spill Handling: Evacuate and restrict persons not wearing protective equipment from area of spill or leak until cleanup is complete. Remove all ignition sources. Establish forced ventilation to keep levels below explosive limit. Absorb liquids in vermiculite, dry sand, earth, peat, carbon, or a similar material and deposit in sealed containers. Keep this chemical out of a confined space, such as a sewer, because of the possibility of an explosion, unless the sewer is designed to prevent the buildup of explosive concentrations. It may be necessary to contain and dispose of this chemical as a hazardous waste. If material or contaminated runoff enters waterways, notify downstream users of potentially contaminated waters. Contact your local or federal environmental protection agency for specific recommendations. If employees are required to clean up spills, they must be properly trained and equipped. OSHA 1910.120(q) may be applicable.

Fire Extinguishing: This chemical is a flammable liquid. Poisonous gases are produced in fire. Use dry chemical, carbon dioxide, alcohol foam, or polymer foam extinguishers. Vapors are heavier than air and will collect in low areas. Vapors may travel long distances to ignition sources and flashback. Vapors in confined areas may explode when exposed to fire. Containers may explode in fire. Storage containers and parts of containers may rocket great distances, in many directions. If material or contaminated runoff enters waterways, notify downstream users of potentially contaminated waters. Notify local health and fire officials and pollution control agencies. From a secure, explosion-proof location, use water spray to cool exposed containers. If cooling streams are ineffective (venting sound increases in volume and pitch, tank discolors, or shows any signs of deforming), withdraw immediately to a secure position. If employees are expected to fight fires, they must be trained and equipped in OSHA 1910.156. The only respirators recommended for firefighting are self-contained breathing apparatuses that have full face-pieces and are operated in a pressure-demand or other positive-pressure mode.

Disposal Method Suggested: Dissolve or mix the material with a combustible solvent and burn in a chemical incinerator equipped with an afterburner and scrubber. All federal, state, and local environmental regulations must be observed.

References

Sax, N. I. (Ed.). (1981). *Dangerous Properties of Industrial Materials Report*, 1, No. 7, 25−26 (as Acetylacetone)
US Environmental Protection Agency. (August 25, 1988). *Chemical Hazard Information Profile Draft Report: Pentanedione*. Washington, DC: Office of Toxic Substances
New Jersey Department of Health and Senior Services. (August 1999). *Hazardous Substances Fact Sheet: Pentane-2,4-dione*. Trenton, NJ

1-Pentene   P:0280

Molecular Formula: C_5H_{10}
Common Formula: $CH_3(CH_2)_2CH{=}CH_2$
Synonyms: Amylene; α-*n*-Amylene; Pentene; Pentylene; Propylethylene
CAS Registry Number: 109-67-1; *(alt.)* 25377-72-4
RTECS®Number: SB2179000
UN/NA & ERG Number: UN1108/128
EC Number: 246-916-6
Regulatory Authority and Advisory Bodies
Department of Homeland Security Screening Threshold Quantity (pounds): *Release hazard* 10,000 (≥1.00% concentration).
Air Pollutant Standard Set. See below, "Permissible Exposure Limits in Air" section.
WGK (German Aquatic Hazard Class): No value assigned.
Description: Pentene is a colorless liquid. Molecular weight = 70.15; Boiling point = 30°C; Vapor pressure = 60.8 mmHg at 20°C; Flash point = −18°C; Autoignition

temperature = 276°C. Explosive limits: LEL = 1.5%; UEL = 8.7%. Hazard Identification (based on NFPA-704 M Rating System): Health 1, Flammability 4, Reactivity 1.

Potential Exposure: Workers in petroleum refineries and petrochemical plants.

Incompatibilities: Strong oxidants.

Permissible Exposure Limits in Air
Protective Action Criteria (PAC)
109-67-1
TEEL-0: 300 ppm
PAC-1: 750 ppm
PAC-2: 6000 ppm
PAC-3: 75,000 ppm
Russia[43] set a MAC for amylene in ambient air of residential areas at 1.5 mg/m³ both on a momentary and a daily average basis.

Routes of Entry: Inhalation, ingestion, skin and/or eye contact.

Harmful Effects and Symptoms

Short Term Exposure: Simple asphyxiant. Narcotic in high concentrations. May affect central nervous system. Moderately toxic by oral and inhalation routes.

Long Term Exposure: There is no special test for this chemical. However, if illness occurs or overexposure is suspected, medical attention is recommended.

Points of Attack: Skin, eyes, respiratory tract, and nervous system.

First Aid: If this chemical gets into the eyes, remove any contact lenses at once and irrigate immediately for at least 15 min, occasionally lifting upper and lower lids. Seek medical attention immediately. If this chemical contacts the skin, remove contaminated clothing and wash immediately with soap and water. Seek medical attention immediately. If this chemical has been inhaled, remove from exposure, begin rescue breathing (using universal precautions, including resuscitation mask) if breathing has stopped and CPR if heart action has stopped. Transfer promptly to a medical facility. When this chemical has been swallowed, get medical attention. Give large quantities of water and induce vomiting. Do not make an unconscious person vomit.

Personal Protective Methods: Use protective gloves and safety goggles. Wear protective gloves and clothing to prevent any reasonable probability of skin contact. Safety equipment suppliers/manufacturers can provide recommendations on the most protective glove/clothing material for your operation. All protective clothing (suits, gloves, footwear, headgear) should be clean, available each day, and put on before work. Contact lenses should not be worn when working with this chemical. Wear splash-proof chemical goggles and face shield unless full face-piece respiratory protection is worn. Employees should wash immediately with soap when skin is wet or contaminated. Provide emergency showers and eyewash.

Respirator Selection: Follow regulations in OSHA 29CFR1910.134 or European Standard EN149. Use a NIOSH/MSHA- or European Standard EN149-approved respirator; or use an approved supplied-air respirator with a full face-piece operated in the positive-pressure mode, or with a full face-piece, hood, or helmet in the continuous-flow mode; or use a NIOSH/MSHA- or European Standard EN149-approved self-contained breathing apparatus with a full face-piece operated in pressure-demand or other positive-pressure mode.

Storage: Color Code—Red: Flammability Hazard: Store in a flammable liquid storage area or approved cabinet away from ignition sources and corrosive and reactive materials. Prior to working with this chemical you should be trained on its proper handling and storage. Store in tightly closed containers in a cool, well-ventilated area away from oxidizers. Before entering confined space where this chemical may be present, check to make sure that an explosive concentration is not a danger. Where possible, automatically pump liquid from drums or other storage containers to process containers. Sources of ignition, such as smoking and open flames, are prohibited where this chemical is handled, used, or stored. Metal containers involving the transfer of 5 gallons or more of this chemical should be grounded and bonded. Drums must be equipped with self-closing valves, pressure vacuum bungs, and flame arresters. Use only non-sparking tools and equipment, especially when opening and closing containers of this chemical. Wherever this chemical is used, handled, manufactured, or stored, use explosion-proof electrical equipment and fittings.

Shipping: 1-Pentene requires a shipping label of "FLAMMABLE LIQUID." It falls in Hazard Class 3 and Packing Group I.

Spill Handling: Evacuate and restrict persons not wearing protective equipment from area of spill or leak until cleanup is complete. Remove all ignition sources. Establish forced ventilation to keep levels below explosive limit. Absorb liquids in vermiculite, dry sand, earth, peat, carbon, or a similar material and deposit in sealed containers. Keep this chemical out of a confined space, such as a sewer, because of the possibility of an explosion, unless the sewer is designed to prevent the buildup of explosive concentrations. It may be necessary to contain and dispose of this chemical as a hazardous waste. If material or contaminated runoff enters waterways, notify downstream users of potentially contaminated waters. Contact your local or federal environmental protection agency for specific recommendations. If employees are required to clean up spills, they must be properly trained and equipped. OSHA 1910.120(q) may be applicable.

Fire Extinguishing: This chemical is a flammable liquid. Poisonous gases are produced in fire. Use dry chemical, carbon dioxide, or alcohol foam extinguishers. Vapors are heavier than air and will collect in low areas. Vapors may travel long distances to ignition sources and flashback. Vapors in confined areas may explode when exposed to fire. Containers may explode in fire. Storage containers and parts of containers may rocket great distances, in many directions. If material or contaminated runoff enters water-ways, notify downstream users of potentially contaminated waters. Notify local health and fire officials and pollution control agencies. From a secure, explosion-proof location, use water spray to cool exposed containers. If cooling streams are ineffective (venting sound increases in volume and pitch, tank discolors, or shows any signs of deforming), withdraw immediately to a secure position. If employees are expected to fight fires, they must be trained and equipped in OSHA 1910.156. The only respirators recommended for firefighting are self-contained breathing apparatuses that have full face-pieces and are operated in a pressure-demand or other positive-pressure mode.

Disposal Method Suggested: Dissolve or mix the material with a combustible solvent and burn in a chemical incinerator equipped with an afterburner and scrubber. All federal, state, and local environmental regulations must be observed.

References

Sax, N. I. (Ed.). *Dangerous Properties of Industrial Materials Report*, 2, No. 6, 69–71 (1982) and 3, No. 2, 56–57 (1983)

New Jersey Department of Health and Senior Services. (March 2007). *Hazardous Substances Fact Sheet: n-Pentene.* Trenton, NJ

Peracetic acid   P:0290

Molecular Formula: $C_2H_4O_3$
Common Formula: CH_3COOOH
Synonyms: Acetic peroxide; Acetyl hydroperoxide; Acide peracetique (French); Acido peracetico (Spanish); Desoxon 1; Estosteril; Ethaneperoxoic acid; Hydrogen peroxide and peroxyacetic acid mixture; Hydroperoxide, acetyl; Monoperacetic acid; Osbon AC; Oxymaster; PAA; Peroxyacetic acid; Proxitane; Proxitane 4002
CAS Registry Number: 79-21-0
RTECS® Number: SD8750000
UN/NA & ERG Number: UN3105 [Organic peroxide type D, liquid (stabilized)]/145; UN3109 [Organic Peroxide Type F, Liquid (with ≤17% Peracetic Acid with ≤26% Hydrogen Peroxide)]/145
EC Number: 201-186-8 [*Annex I Index No.:* 607-094-00-8]
Regulatory Authority and Advisory Bodies
Department of Homeland Security Screening Threshold Quantity (pounds): *Release hazard* 10,000 (≥1.00% concentration).
US EPA, FIFRA 1998 Status of Pesticides: RED completed.
Highly Reactive Substance and Explosive (World Bank).[15]
Clean Air Act: Accidental Release Prevention/Flammable Substances, (Section 112[r], Table 3), TQ = 10,000 lb (4540 kg).
Superfund/EPCRA 40CFR355, Extremely Hazardous Substances: TPQ = 500 lb (227 kg).

Reportable Quantity (RQ): 500 lb (227 kg).
EPCRA Section 313 Form R *de minimis* concentration reporting level: 1.0%.
Canada, WHMIS, Ingredients Disclosure List Concentration 1.0%.
European/International Regulations: Hazard Symbol: O, C, N; Risk phrases: R7; R10; R20/21/22; R35; R50; Safety phrases: S1/2; S3/7; S14; S36/37/39; S45; S61 (see Appendix 4).
WGK (German Aquatic Hazard Class): 2—Hazard to waters.
Description: Peracetic acid is a colorless liquid. Transported and stored in diluted solution with acetic acid and hydrogen peroxide to prevent explosion. Molecular weight = 76.06; Specific gravity (H_2O:1) = 1.2; Boiling point = 105°C (violent decomposition at 110°C); Freezing/Melting point = 0.1°C; Flash point = 41.3°C (oc); 56°C (32% in dilute acetic acid and <6% hydrogen peroxide); Autoignition temperature = 198°C; Hazard Identification (based on NFPA-704 M Rating System): Health 3, Flammability 2, Reactivity 2 (Oxidizer). Soluble in water.
Potential Exposure: Compound Description: Agricultural Chemical; Tumorigen, Primary Irritant. This compound is used as polymerization initiator, curing agent, and cross-linking agent; as bactericide and fungicide, especially in food processing; a reagent in making caprolactam and glycerol; an oxidant for preparing epoxy compounds; a bleaching agent; a sterilizing agent; and a polymerization catalyst for polyester resins.
Incompatibilities: This material is a powerful oxidizer. Thermally unstable, it decomposes violently at 110°C. Concentrated material is shock- and friction-sensitive. May explode if concentration exceeds 56% of carrier, due to evaporation. Isolate from other stored material, particularly accelerators, oxidizers, organic or combustible materials, olefins, hydrogen peroxide, acetic anhydride, reducing substances. Keep away from acids, alkalis, heavy metals, organic materials.
Permissible Exposure Limits in Air
Protective Action Criteria (PAC)*
TEEL-0: 0.15 mg/m^3
PAC-1: **0.52** mg/m^3
PAC-2: **1.6** mg/m^3
PAC-3: **15** mg/m^3
*AEGLs (Acute Emergency Guideline Levels) & ERPGs (Emergency Response Planning Guideline) are in **bold face**.
DFG MAK: Carcinogen Category 3B; See section X(a).
Routes of Entry: Inhalation, ingestion, skin and/or eye contact.
Harmful Effects and Symptoms
Short Term Exposure: Eye contact can cause severe irritation and burns; may cause permanent damage. Irritates the respiratory tract. Contact may burn the skin. Higher exposures can cause pulmonary edema, a medical emergency that can be delayed for several hours. This can cause death.

Signs and symptoms of acute ingestion of peracetic acid may include corrosion of mucous membranes of mouth, throat, and esophagus with immediate pain and dysphagia (difficulty in swallowing); ingestion may cause gastrointestinal tract irritation. This is a very toxic compound. The probable oral lethal dose for humans is 50–500 mg/kg or between 1 teaspoon and 1 oz for a 150-lb person.

Long Term Exposure: There is limited evidence that peracetic acid causes cancer in animals. It may cause cancer of the lungs. High or repeated exposure may affect the liver and kidneys.

Points of Attack: Liver, kidneys, lungs.

Medical Surveillance: Liver and kidney function tests. Consider chest X-ray following acute overexposure.

First Aid: If this chemical gets into the eyes, remove any contact lenses at once and irrigate immediately for at least 15 min, occasionally lifting upper and lower lids. Seek medical attention immediately. If this chemical contacts the skin, remove contaminated clothing and wash immediately with soap and water. Seek medical attention immediately. If this chemical has been inhaled, remove from exposure, begin rescue breathing (using universal precautions, including resuscitation mask) if breathing has stopped and CPR if heart action has stopped. Transfer promptly to a medical facility. When this chemical has been swallowed, get medical attention. If victim is *conscious*, administer water or milk. Do not induce vomiting. Medical observation is recommended for 24–48 h after breathing overexposure, as pulmonary edema may be delayed. As first aid for pulmonary edema, a doctor or authorized paramedic may consider administering a corticosteroid spray.

Personal Protective Methods: Wear protective gloves and clothing to prevent any reasonable probability of skin contact. Safety equipment suppliers/manufacturers can provide recommendations on the most protective glove/clothing material for your operation. Butyl rubber and Viton are recommended. All protective clothing (suits, gloves, footwear, headgear) should be clean, available each day, and put on before work. Contact lenses should not be worn when working with this chemical. Wear splash-proof chemical goggles and face shield unless full face-piece respiratory protection is worn. Employees should wash immediately with soap when skin is wet or contaminated. Provide emergency showers and eyewash.

Respirator Selection: *Where there is no REL, at any detectable concentration:* Sa:Cf (APF = 25) (any supplied-air respirator operated in a continuous-flow mode) or PaprOv (APF = 25) [any powered, air-purifying respirator with organic vapor cartridge(s)]; CcrFOv (APF = 50) [any air-purifying, full-face-piece respirator (gas mask) with a chin-style, front- or back-mounted acid gas canister] or GmFOv (APF = 50) [any air-purifying, full-face-piece respirator (gas mask) with a chin-style, front- or back-mounted organic vapor canister] or SCBAF (APF = 50) (any self-contained breathing apparatus with a full face-piece) or SaF (APF = 50) (any supplied-air respirator with a full face-

piece). *Emergency or planned entry into unknown concentrations or IDLH conditions:* SCBAF: Pd,Pp (APF = 10,000) (any self-contained breathing apparatus that has a full face-piece and is operated in a pressure-demand or other positive-pressure mode) or SaF: Pd,Pp: ASCBA (APF = 10,000) (any supplied-air respirator that has a full face-piece and is operated in a pressure-demand or other positive-pressure mode in combination with an auxiliary self-contained breathing apparatus operated in a pressure-demand or other positive-pressure mode). *Escape:* GmFOv (APF = 50) [any air-purifying, full-face-piece respirator (gas mask) with a chin-style, front- or back-mounted organic vapor canister] or SCBAE (any appropriate escape-type, self-contained breathing apparatus).

Storage: Color Code—Red Stripe: Flammability Hazard: Do not store in the same area as other flammable materials. Prior to working with this chemical you should be trained on its proper handling and storage. Keep in a cool, well-ventilated area, separated from organic and combustible materials. Where possible, automatically pump liquid from drums or other storage containers to process containers. Sources of ignition, such as smoking and open flames, are prohibited where this chemical is handled, used, or stored. Metal containers involving the transfer of 5 gallons or more of this chemical should be grounded and bonded. Drums must be equipped with self-closing valves, pressure vacuum bungs, and flame arresters. Use only nonsparking tools and equipment, especially when opening and closing containers of this chemical. Wherever this chemical is used, handled, manufactured, or stored, use explosion-proof electrical equipment and fittings. See OSHA Standard 1910.104 and NFPA 43A *Code for the Storage of Liquid and Solid Oxidizers* for detailed handling and storage regulations.

Shipping: Organic peroxide type D, liquid, requires a shipping label of "ORGANIC PEROXIDE." They fall in DOT Hazard Class 5.2 and Packing Group I. Organic peroxide type F, liquid, requires a shipping label of "Organic Peroxide Type F, Liquid (with <= 17% Peracetic Acid with <= 26% Hydrogen Peroxide)." They fall in DOT Hazard Class 5.2 and Packing Group II. Shipment of solutions with >43% peracetic acid is FORBIDDEN under any conditions.

Spill Handling: Evacuate and restrict persons not wearing protective equipment from area of spill or leak until cleanup is complete. Remove all ignition sources. Ventilate area of spill or leak. Avoid breathing vapors. Do not touch the spilled material; shut off all ignition sources and stop the leak if this can be done without risk. Absorb liquids in vermiculite, dry sand, earth, peat, carbon, or a similar material and deposit in sealed containers. Do not use spark-generating metals or organic materials for sweeping up or handling spilled material. Dispose of the absorbed peroxyacetic acid solution, in small quantities at a time, by placing it on the ground in a remote outdoor area and igniting with a long torch. Empty containers should be washed with a 10% sodium hydroxide solution. Keep this chemical out of a confined space, such as a sewer, because of the possibility of an explosion, unless the sewer is designed to prevent the buildup of explosive concentrations. It may be necessary to contain and dispose of this chemical as a hazardous waste. If material or contaminated runoff enters waterways, notify downstream users of potentially contaminated waters. Contact your local or federal environmental protection agency for specific recommendations. If employees are required to clean up spills, they must be properly trained and equipped. OSHA 1910.120(q) may be applicable.

Fire Extinguishing: This chemical is a combustible liquid and a powerful oxidizer that can increase the activity of an existing fire. It explodes at 230°F/110°C and is shock sensitive, particularly if organic solvents are used in place of acetic acid as a carrier. Poisonous gases are produced in fire. Fight fires from an explosion-resistant location. In advanced or massive fires, area should be evacuated. For small fires: use dry chemical, carbon dioxide, water spray, or foam. For large fires: flood area with water. If fire occurs in the vicinity of this compound, water should be used to keep containers cool. Cleanup and salvage operations should not be attempted until all of the peroxyacetic acid solution has cooled completely. Keep unnecessary people away; wear self-contained breathing apparatus and full protective clothing. Vapors are heavier than air and will collect in low areas. Vapors may travel long distances to ignition sources and flashback. Vapors in confined areas may explode when exposed to fire. Containers may explode in fire. Storage containers and parts of containers may rocket great distances, in many directions. If material or contaminated runoff enters waterways, notify downstream users of potentially contaminated waters. Notify local health and fire officials and pollution control agencies. From a secure, explosion-proof location, use water spray to cool exposed containers. If cooling streams are ineffective (venting sound increases in volume and pitch, tank discolors, or shows any signs of deforming), withdraw immediately to a secure position. If employees are expected to fight fires, they must be trained and equipped in OSHA 1910.156. The only respirators recommended for firefighting are self-contained breathing apparatuses that have full face-pieces and are operated in a pressure-demand or other positive-pressure mode.

References
US Environmental Protection Agency. (November 30, 1987). *Chemical Hazard Information Profile: Peracetic Acid.* Washington, DC: Chemical Emergency Preparedness Program
New York State Department of Health. (April 1986). *Chemical Fact Sheet: Peracetic Acid.* Albany, NY: Bureau of Toxic Substance Assessment
New Jersey Department of Health and Senior Services. (October 2004). *Hazardous Substances Fact Sheet: Peroxyacetic Acid.* Trenton, NJ

Perchloromethyl mercaptan   P:0300

Molecular Formula: CCl_4S
Common Formula: CCl_3SCl
Synonyms: Clairsit; Mercaptan methylique perchlore (French); Perchloromethanethiol; PMM; Trichloromethane sulfenyl chloride; Trichloromethylsulfenyl chloride; Trichloromethyl sulfur chloride; Trichloromethylsulphenyl chloride PCV
CAS Registry Number: 594-42-3
RTECS® Number: PB0370000
UN/NA & ERG Number: UN1670 (inhalation zone B)/157
EC Number: 209-840-4
Regulatory Authority and Advisory Bodies
Department of Homeland Security Screening Threshold Quantity (pounds): *Release hazard* 10,000 (≥1.00% concentration).
Air Pollutant Standard Set. See below, "Permissible Exposure Limits in Air" section.
OSHA 29CFR1910.119, Appendix A, Process Safety List of Highly Hazardous Chemicals, TQ = 150 lb (67.5 kg).
US DOT 49CFR172.101, Inhalation Hazardous Chemical.
Canada, WHMIS, Ingredients Disclosure List Concentration 1.0%.
European/International Regulations: not listed in Annex 1.
WGK (German Aquatic Hazard Class): No value assigned.
Description: Perchloromethyl mercaptan is a pale yellow oily liquid with a foul-smelling, unbearable, acrid odor. Molecular weight = 185.87; Specific gravity (H_2O:1) = 1.69; Boiling point = (decomposes) 147.2°C; Vapor pressure = 3 mmHg at 25°C. Hazard Identification (based on NFPA-704 M Rating System): Health 3, Flammability 0, Reactivity 0. Insoluble in water.
Potential Exposure: Compound Description: Primary Irritant. Perchloromethyl mercaptan is used as an intermediate for the synthesis of dyes and fungicides, such as Captan and Folpet. It has been considered as a warfare tear gas because of its highly irritant properties.
Incompatibilities: Water contact forms HCl, sulfur, and carbon dioxide. Reacts with alkalies, amines, hot water, alcohols, oxidizers, reducing agents, iron, and steel. Attacks most metals.
Permissible Exposure Limits in Air
Conversion factor: 1 ppm = 7.60 mg/m³ at 25°C & 1 atm.
OSHA PEL: 0.1 ppm/0.8 mg/m³ TWA.
NIOSH REL: 0.1 ppm/0.8 mg/m³ TWA.
ACGIH TLV®[1]: 0.1 ppm/0.76 mg/m³ TWA.
NIOSH IDLH: 10 ppm.
Protective Action Criteria (PAC)*
TEEL-0: 0.013 ppm
PAC-1: **0.013** ppm
PAC-2: **0.3** ppm
PAC-3: **0.9** ppm
*AEGLs (Acute Emergency Guideline Levels) & ERPGs (Emergency Response Planning Guideline) are in **bold face**.

Australia: TWA 0.1 ppm (0.8 mg/m^3), 1993; Austria: MAK 0.1 ppm (0.8 mg/m^3), 1999; Belgium: TWA 0.1 ppm (0.76 mg/m^3), 1993; Denmark: TWA 0.1 ppm (0.8 mg/m^3), 1999; Finland: STEL 0.1 ppm (0.8 mg/m^3), 1999; France: VME 0.1 ppm (0.8 mg/m^3), 1999; Norway: TWA 0.1 ppm (0.8 mg/m^3), 1999; the Netherlands: MAC-TGG 0.8 mg/m^3, 2003; Switzerland: MAK-W 0.1 ppm (0.8 mg/m^3), KZG-W 0.2 ppm, 1999; Turkey: TWA 0.1 ppm (0.8 mg/m^3), 1993; Argentina, Bulgaria, Columbia, Jordan, South Korea, New Zealand, Singapore, Vietnam: ACGIH TLV®: TWA 0.1 ppm. Several states have set guidelines or standards for PMM in ambient air[60] ranging from 8.0 μg/m^3 (North Dakota) to 13.0 μg/m^3 (Virginia) to 16.0 μg/m^3 (Connecticut) to 19.0 μg/m^3 (Nevada).

Determination in Air: Sample collection by charcoal tube, analysis by gas liquid chromatography.

Routes of Entry: Inhalation, ingestion, skin and/or eye contact.

Harmful Effects and Symptoms

Short Term Exposure: Irritates the eyes, skin, and respiratory tract. Higher exposures can cause pulmonary edema, a medical emergency that can be delayed for several hours. This can cause death. Signs and symptoms of acute exposure to perchloromethyl mercaptan may lead to liver, heart, and kidney damage. Respiratory effects include coughing, dyspnea (shortness of breath), painful breathing, and lung congestion. Tachycardia (rapid heart rate) is often observed. Nausea, vomiting, abdominal cramping, and diarrhea may also occur. Contact with Perchloromethyl mercaptan may result in severe dermatitis (red, inflamed skin), conjunctivitis (red, inflamed eyes), and burns with ulceration and severe pain. May cause death or permanent injury after short exposure to small quantities. Brief exposure to lower concentrations may produce central nervous system depression and lung, liver, and heart congestion. Severe exposures may be fatal. May be absorbed through the skin in quantities sufficient to cause general toxic effects. Ingestion may cause damage to mucous membranes and result in pain and burning of the mouth and throat, nausea, vomiting, cramps, and diarrhea. In severe cases, tissue ulceration and CNS depression may occur.

Medical Surveillance: NIOSH lists the following tests: Blood Gas Analysis; chest X-ray, electrocardiogram, liver function tests; pulmonary function tests; pulmonary function tests: forced vital capacity, forced expiratory volume (1 s); sputum cytology; urinalysis (routine); white blood cell count/differential.

First Aid: If this chemical gets into the eyes, remove any contact lenses at once and irrigate immediately for at least 15 min, occasionally lifting upper and lower lids. Seek medical attention immediately. If this chemical contacts the skin, remove contaminated clothing and wash immediately with soap and water. Seek medical attention immediately. If this chemical has been inhaled, remove from exposure, begin rescue breathing (using universal precautions, including resuscitation mask) if breathing has stopped and CPR if heart action has stopped. Transfer promptly to a medical facility. When this chemical has been swallowed, get medical attention. Give large quantities of water and induce vomiting. Do not make an unconscious person vomit.

Personal Protective Methods: Wear protective gloves and clothing to prevent any reasonable probability of skin contact. Safety equipment suppliers/manufacturers can provide recommendations on the most protective glove/clothing material for your operation. All protective clothing (suits, gloves, footwear, headgear) should be clean, available each day, and put on before work. Contact lenses should not be worn when working with this chemical. Wear splash-proof chemical goggles and face shield unless full face-piece respiratory protection is worn. Employees should wash immediately with soap when skin is wet or contaminated. Provide emergency showers and eyewash.

Respirator Selection: *Up to 1 ppm:* CcrOv* (APF = 10) [any chemical cartridge respirator with organic vapor cartridge(s)]; Sa* (APF = 10) (any supplied-air respirator). *Up to 2.5 ppm:* Sa:Cf* (APF = 25) (any supplied-air respirator operated in a continuous-flow mode); PaprOv* (APF = 25) [any powered, air-purifying respirator with organic vapor cartridge(s)]. *Up to 5 ppm:* CcrFOv (APF = 50) [any chemical cartridge respirator with a full face-piece and organic vapor cartridge(s)] or GmFOv (APF = 50) [any air-purifying, full-face-piece respirator (gas mask) with a chin-style, front- or back-mounted organic vapor canister] or PaprTOv* (APF = 50) [any powered, air-purifying respirator with a tight-fitting face-piece and organic vapor cartridge(s)]; or SaT: Cf* (APF = 50) (any supplied-air respirator that has a tight-fitting face-piece and is operated in a continuous-flow mode); or SCBAF (APF = 50) (any self-contained breathing apparatus with a full face-piece) or SaF (APF = 50) (any supplied-air respirator with a full face-piece). *Up to 10 ppm:* SaF: Pd,Pp (APF = 2000) (any supplied-air respirator that has a full face-piece and is operated in a pressure-demand or other positive-pressure mode). *Emergency or planned entry into unknown concentrations or IDLH conditions:* SCBAF: Pd,Pp (APF = 10,000) (any NIOSH/MSHA- or European Standard EN 149-approved self-contained breathing apparatus that has a full face-piece and is operated in a pressure-demand or other positive-pressure mode) or SaF: Pd,Pp: ASCBA (APF = 10,000) (any supplied-air respirator that has a full face-piece and is operated in a pressure-demand or other positive-pressure mode in combination with an auxiliary, self-contained breathing apparatus operated in a pressure-demand or other positive-pressure mode). *Escape:* GmFOv100 (APF = 50) [any air-purifying, full-face-piece respirator (gas mask) with a chin-style, front- or back-mounted organic vapor canister having an N100, R100, or P100 filter] or SCBAE (any appropriate escape-type, self-contained breathing apparatus). *Substance reported to cause eye irritation or damage; may require eye protection.

Storage: Color Code—Blue: Health Hazard/Poison: Store in a secure poison location. Prior to working with this

chemical you should be trained on its proper handling and storage. Store in a cool, dry place. Protect from moisture, metals, oxidizing and reducing agents. Where possible, automatically pump liquid from drums or other storage containers to process containers. Sources of ignition, such as smoking and open flames, are prohibited where this chemical is handled, used, or stored. Metal containers involving the transfer of 5 gallons or more of this chemical should be grounded and bonded. Drums must be equipped with self-closing valves, pressure vacuum bungs, and flame arresters. Use only nonsparking tools and equipment, especially when opening and closing containers of this chemical. Wherever this chemical is used, handled, manufactured, or stored, use explosion-proof electrical equipment and fittings.

Shipping: Perchloromethyl mercaptan requires a shipping label of "POISONOUS/TOXIC MATERIALS." It falls in Hazard Class 6.1 and Packing Group I.

Spill Handling: Evacuate and restrict persons not wearing protective equipment from area of spill or leak until cleanup is complete. Remove all ignition sources. Ventilate area of spill or leak. Absorb liquids in vermiculite, dry sand, earth, peat, carbon, or a similar material and deposit in sealed containers. Keep this chemical out of a confined space, such as a sewer, because of the possibility of an explosion, unless the sewer is designed to prevent the buildup of explosive concentrations. It may be necessary to contain and dispose of this chemical as a hazardous waste. If material or contaminated runoff enters waterways, notify downstream users of potentially contaminated waters. Contact your local or federal environmental protection agency for specific recommendations. If employees are required to clean up spills, they must be properly trained and equipped. OSHA 1910.120(q) may be applicable. It may be necessary to seek emergency assistance.

Initial isolation and protective action distances
Distances shown are likely to be affected during the first 30 min after materials are spilled and could increase with time. If more than one tank car, cargo tank, portable tank, or large cylinder involved in the incident is leaking, the protective action distance may need to be increased. You may need to seek emergency information from CHEMTREC at (800) 424-9300 or seek professional environmental engineering assistance from the US EPA Environmental Response Team at (908) 548-8730 (24-h response line).

Small spills (From a small package or a small leak from a large package)
First: Isolate in all directions (feet/meters)   100/30
Then: Protect persons downwind (miles/kilometers)
Day   0.2/0.3
Night   0.2/0.3

Large spills (From a large package or from many small packages)
First: Isolate in all directions (feet/meters)   300/100
Then: Protect persons downwind (miles/kilometers)
Day   0.5/0.8
Night   0.9/1.5

Fire Extinguishing: This compound is neither flammable nor a serious fire hazard, although it will support combustion. Fight small fires with dry chemical, carbon dioxide, water spray, or foam, and large fires with water spray, fog, or foam. Move containers containing this compound away from fire area if possible. Fight fire from maximum distance. Dike fire control water for later disposal; do not scatter the material. Positive pressure breathing apparatus and special protective clothing should be worn. Poisonous gases, including chlorine and sulfur oxides, are produced in fire. Vapors are heavier than air and will collect in low areas. Containers may explode in fire. Storage containers and parts of containers may rocket great distances, in many directions. If material or contaminated runoff enters waterways, notify downstream users of potentially contaminated waters. Notify local health and fire officials and pollution control agencies. From a secure, explosion-proof location, use water spray to cool exposed containers. If cooling streams are ineffective (venting sound increases in volume and pitch, tank discolors, or shows any signs of deforming), withdraw immediately to a secure position. If employees are expected to fight fires, they must be trained and equipped in OSHA 1910.156. The only respirators recommended for firefighting are self-contained breathing apparatuses that have full face-pieces and are operated in a pressure-demand or other positive-pressure mode.

Disposal Method Suggested: Incineration together with a flammable solvent in a furnace equipped with afterburner and scrubber.

References
US Environmental Protection Agency. (November 30, 1987). *Chemical Hazard Information Profile: Perchloromethyl Mercaptan.* Washington, DC: Chemical Emergency Preparedness Program
New Jersey Department of Health and Senior Services. (February 2000). *Hazardous Substances Fact Sheet: Perchloromethyl Mercaptan.* Trenton, NJ

Perchloryl fluoride   P:0310

Molecular Formula: $ClFO_3$
Common Formula: ClO_3F
Synonyms: Chlorine oxyfluoride; Chlorine fluoride oxide; Trioxychlorofluoride
CAS Registry Number: 7616-94-6
RTECS® Number: SD1925000
UN/NA & ERG Number: UN3083/124
EC Number: 231-526-0
Regulatory Authority and Advisory Bodies
Department of Homeland Security Screening Threshold Quantity (pounds): *Theft hazard* 45 ($\geq$25.67% concentration).
Air Pollutant Standard Set. See below, "Permissible Exposure Limits in Air" section.

OSHA 29CFR1910.119, Appendix A. Process Safety List of Highly Hazardous Chemicals, TQ = 5000 lb (2270 kg). US DOT 49CFR172.101, Inhalation Hazardous Chemical. Canada, WHMIS, Ingredients Disclosure List Concentration 1.0%.

European/International Regulations: not listed in Annex 1. WGK (German Aquatic Hazard Class): No value assigned.

Description: Perchloryl fluoride is a colorless gas with a characteristic sweet odor. Shipped as a liquefied compressed gas. Molecular weight = 102.45; Boiling point = −46.7°C; Freezing/Melting point = −147.8°C; Relative vapor density (air = 1) = 3.64; Vapor pressure = 10.5 atm at 25°C; Relative vapor density (air = 1) = 3.64. Hazard Identification (based on NFPA-704 M Rating System): Health 3, Flammability 2, Reactivity 3 (Oxidizer). Slightly soluble in water; solubility = 0.06% at 20°C.

Potential Exposure: Perchloryl fluoride has been used as a liquid oxidant in rocket propellant combinations, as an insulating gas in high-voltage electrical systems, as a fluorinating agent in organic synthesis.

Incompatibilities: A strong oxidizer. Violent reaction with benzene, calcium hydride, combustibles, olefins, strong bases, sulfur, sulfuric acid, amines, reducing agents, alcohols. Contact with carbonaceous materials (such as charcoal) or finely divided metals (such as powdered magnesium, aluminum, zinc) are a fire and explosion hazard. Attacks some plastics, rubber, and coatings.

Permissible Exposure Limits in Air
Conversion factor: 1 ppm = 4.19 mg/m³ at 25°C & 1 atm.
OSHA PEL: 3 ppm/13.5 mg/m³ TWA.
NIOSH REL: 3 ppm/14 mg/m³ TWA; 6 ppm/28 mg/m³ STEL.
ACGIH TLV®[1]: 3 ppm/13 mg/m³ TWA; 6 ppm/25 mg/m³ STEL.
NIOSH IDLH: 100 ppm.
Protective Action Criteria (PAC)*
TEEL-0: 1.5 ppm
PAC-1: **1.5 ppm**
PAC-2: **4.0 ppm**
PAC-3: **12 ppm**
*AEGLs (Acute Emergency Guideline Levels) & ERPGs (Emergency Response Planning Guideline) are in **bold face**.
Australia: TWA 3 ppm (14 mg/m³); STEL 6 ppm, 1993; Austria: MAK 2.5 mg[F]/m³, 1999; Belgium: TWA 3 ppm (13 mg/m³); STEL 6 ppm (25 mg/m³), 1993; Denmark: TWA 3 ppm (14 mg/m³), 1999; Finland: TWA 3 ppm (14 mg/m³); STEL 6 ppm (25 mg/m³), 1999; France: VME 3 ppm (14 mg/m³), 1999; Norway: TWA 3 ppm (14 mg/m³), 1999; the Netherlands: MAC-TGG 14 mg/m³, 2003; Sweden: NGV 2 mg[F]/m³, 1999; Switzerland: MAK-W 3 ppm (13 mg/m³), 1999; United Kingdom: TWA 3 ppm (13 mg/m³); STEL 6 ppm (26 mg/m³), 2000; Argentina, Bulgaria, Columbia, Jordan, South Korea, New Zealand, Singapore, Vietnam: ACGIH TLV®: STEL 6 ppm.
Several states have set guidelines or standards for perchloryl fluoride in ambient air[60] ranging from 140−280 μg/m³ (North Dakota) to 230 μg/m³ (Virginia) to 270 μg/m³ (Connecticut) to 333 μg/m³ (Nevada).

Determination in Air: Sample collection by impinger or fritted bubbler; analysis by ion-specific electrode.

Routes of Entry: Inhalation, ingestion, skin and/or eye contact. Absorbed through the skin.

Harmful Effects and Symptoms
Short Term Exposure: Irritates the respiratory tract. May affect the blood, causing the destruction of red blood cells; formation of methemoglobin. Cyanosis and anemia may result. The liquid may cause frostbite.

Long Term Exposure: May cause anemia. Repeated high exposures can cause deposits of fluorides in the bones (fluorosis), which may cause pain, disability, and mottling of the teeth. Repeated exposure may cause nausea, vomiting, loss of appetite, diarrhea, or constipation.

Points of Attack: Respiratory system, skin, blood.

Medical Surveillance: NIOSH lists the following tests: Blood Gas Analysis; whole blood (chemical/metabolite), Methemoglobin; Complete blood count; chest X-ray; pulmonary function tests: forced vital capacity, forced expiratory volume (1 s); sputum cytology; urine (chemical/metabolite); white blood cell count/differential. Consider the points of attack in preplacement and periodic physical examinations. Fluoride level in urine (use NIOSH #8308). Levels higher than 4 mg/L may indicate overexposure.

First Aid: If this chemical gets into the eyes, remove any contact lenses at once and irrigate immediately for at least 15 min, occasionally lifting upper and lower lids. Seek medical attention immediately. If this chemical contacts the skin, remove contaminated clothing and wash immediately with soap and water. Seek medical attention immediately. If this chemical has been inhaled, remove from exposure, begin rescue breathing (using universal precautions, including resuscitation mask) if breathing has stopped and CPR if heart action has stopped. Transfer promptly to a medical facility. When this chemical has been swallowed, get medical attention. Give large quantities of water and induce vomiting. Do not make an unconscious person vomit. If frostbite has occurred, seek medical attention immediately; do NOT rub the affected areas or flush them with water. In order to prevent further tissue damage, do NOT attempt to remove frozen clothing from frostbitten areas. If frostbite has NOT occurred, immediately and thoroughly wash contaminated skin with soap and water.

Personal Protective Methods: Wear appropriate personal protective clothing to prevent the skin from becoming frozen from contact with the evaporating liquid or from contact with vessels containing the liquid. Safety equipment suppliers/manufacturers can provide recommendations on the most protective glove/clothing material for your operation. All protective clothing (suits, gloves, footwear, headgear) should be clean, available each day, and put on before work. Contact lenses should not be worn when working with this chemical. Wear eye protection to prevent any reasonable probability of eye contact. Employees should wash

immediately with soap when skin is wet or contaminated. Provide emergency showers and eyewash.

Respirator Selection: *Up to 30 ppm:* Sa (APF = 10) (any supplied-air respirator). *Up to 75 ppm:* Sa:Cf* (APF = 25) (any supplied-air respirator operated in a continuous-flow mode). *Up to 100 ppm:* SCBAF (APF = 50) (any self-contained breathing apparatus with a full face-piece) or SaF (APF = 50) (any supplied-air respirator with a full face-piece). *Emergency or planned entry into unknown concentrations or IDLH conditions:* SCBAF: Pd,Pp (APF = 10,000) (any self-contained breathing apparatus that has a full face-piece and is operated in a pressure-demand or other positive-pressure mode) or SaF: Pd,Pp: ASCBA (APF = 10,000) (any supplied-air respirator that has a full face-piece and is operated in a pressure-demand or other positive-pressure mode in combination with an auxiliary self-contained breathing apparatus operated in a pressure-demand or other positive-pressure mode). *Escape:* GmFS100 (APF = 50) [any air-purifying, full-face-piece respirator (gas mask) with a chin-style, front- or back-mounted canister providing protection against the compound of concern and having an N100, R100, or P100 filter] or SCBAE (any appropriate escape-type, self-contained breathing apparatus).

*Substance reported to cause eye irritation or damage; may require eye protection.

Storage: (1) Color Code—Yellow: Reactive Hazard; Store in a location separate from other materials, especially flammables and combustibles. (2) Color Code—Blue: Health Hazard/Poison: Store in a secure poison location. Prior to working with this chemical you should be trained on its proper handling and storage. Store in a cool, well-ventilated area away from incompatible materials listed above. Procedures for the handling, use and storage of cylinders should be in compliance with OSHA 1910.101 and 1910.169, as with the recommendations of the Compressed Gas Association. See OSHA Standard 1910.104 and NFPA 43A *Code for the Storage of Liquid and Solid Oxidizers* for detailed handling and storage regulations.

Shipping: Perchloryl fluoride requires a shipping label of "POISON GAS, OXIDIZER." It falls in Hazard Class 2.3. It is a violation of transportation regulations to refill compressed gas cylinders without the express written permission of the owner.

Spill Handling: If in a building, evacuate building and confine vapors by closing doors and shutting down HVAC systems. Restrict persons not wearing protective equipment from area of spill or leak until cleanup is complete. Remove all ignition sources. Ventilate area of spill or leak to disperse the gas. Wear chemical protective suit with self-contained breathing apparatus to combat spills. Stay upwind and use water spray to "knock down" vapor; contain runoff. Stop the flow of gas, if it can be done safely from a distance. If source is a cylinder and the leak cannot be stopped in place, remove the leaking cylinder to a safe place; and repair leak or allow cylinder to empty. Keep this chemical out of confined spaces, such as a sewer, because of the possibility of explosion, unless the sewer is designed to prevent the buildup of explosive concentrations. If employees are required to clean up spills, they must be properly trained and equipped. OSHA 1910.120(q) may be applicable.

Initial isolation and protective action distances
Distances shown are likely to be affected during the first 30 min after materials are spilled and could increase with time. If more than one tank car, cargo tank, portable tank, or large cylinder involved in the incident is leaking, the protective action distance may need to be increased. You may need to seek emergency information from CHEMTREC at (800) 424-9300 or seek professional environmental engineering assistance from the US EPA Environmental Response Team at (908) 548-8730 (24-h response line).

Small spills (From a small package or a small leak from a large package)
First: Isolate in all directions (feet/meters)   100/30
Then: Protect persons downwind (miles/kilometers)
Day   0.1/0.2
Night   0.4/0.6

Large spills (From a large package or from many small packages)
First: Isolate in all directions (feet/meters)   1500/500
Then: Protect persons downwind (miles/kilometers)
Day   2.0/3.2
Night   5.2/8.4

Fire Extinguishing: Nonflammable gas, but will support combustion and add to the intensity of an existing fire. Poisonous gases, including fluorine, fluorine oxides, chlorine, and chlorine oxides, are produced in fire. Vapors are heavier than air and will collect in low areas. Containers may explode in fire. Storage containers and parts of containers may rocket great distances, in many directions. If material or contaminated runoff enters waterways, notify downstream users of potentially contaminated waters. Notify local health and fire officials and pollution control agencies. From a secure, explosion-proof location, use water spray to cool exposed containers. If cooling streams are ineffective (venting sound increases in volume and pitch, tank discolors, or shows any signs of deforming), withdraw immediately to a secure position. If cylinders are exposed to excessive heat from fire or flame contact, withdraw immediately to a secure location. If employees are expected to fight fires, they must be trained and equipped in OSHA 1910.156. The only respirators recommended for firefighting are self-contained breathing apparatuses that have full face-pieces and are operated in a pressure-demand or other positive-pressure mode.

Disposal Method Suggested: Incineration together with flammable solvent in furnace equipped with afterburner and scrubber.

Reference
New Jersey Department of Health and Senior Services. (November 2001). *Hazardous Substances Fact Sheet: Perchloryl Fluoride*. Trenton, NJ

Persulfates

See "Potassium Persulfate" as good example of this class of compounds.

Phenanthrene   P:0320

Molecular Formula: C_4H_{10}
Synonyms: Coal tar pitch volatiles: Phenanthrene; Phenanthren (German); Phenantrin
CAS Registry Number: 85-01-8
RTECS® Number: VB2600000
UN/NA & ERG Number: Not regulated.
EC Number: 201-581-5
Regulatory Authority and Advisory Bodies
Carcinogenicity: IARC: Animal Inadequate Evidence; Human No Adequate Data, *not classifiable as carcinogenic to humans*, Group 3, 1987; EPA: Not Classifiable as to human carcinogenicity.
OSHA, 29CFR1910 Specifically Regulated Chemicals (CFR1910.1002) as coal tar pitch volatiles.
Clean Water Act: Section 307 Toxic Pollutants, 40CFR401.15 (effluent limitations); 40CFR413.02, Total Toxic Organics, 40CFR423, Priority Pollutants, as polynuclear aromatic hydrocarbons (PAH).
RCRA 40CFR258, Appendix 2.
RCRA 40CFR268.48; 61FR15654, Universal Treatment Standards: Wastewater (mg/L), 0.059; Nonwastewater (mg/kg), 3.4.
RCRA, 40CFR264, Appendix 9, Ground Water Monitoring List, Suggested Testing Methods (PQL µg/L): 8100 (200); 8270 (10).
Superfund/EPCRA 40CFR302.4, Appendix A, Reportable Quantity (RQ): 100 lb (45.4 kg).
Canada, WHMIS, Ingredients Disclosure List Concentration: 1% as phenanthrene; 0.1% as coal tar pitch volatiles; DSL list.
Mexico, Drinking Water, Criteria (Ecological): 0.02 mg/L; wastewater: organic toxic pollutant.
European/International Regulations: Hazard Symbol: Xn, N; Risk phrases: R22; R40; R50/53; Safety phrases: S29; S36/37; S61 (see Appendix 4).
WGK (German Aquatic Hazard Class): No value assigned.
Description: Phenanthrene is a white[2] crystalline substance with a weak aromatic odor. Polynuclear aromatic hydrocarbons (PAHs) are compounds containing multiple benzene rings and are also called polycyclic aromatic hydrocarbons. Molecular weight = 178.22.[2] Molecular weight = 178.24; Boiling point = 340°C at 760 mmHg[2]; Freezing/Melting point = 100°C[2]; Flash point = 171°C. Hazard Identification (based on NFPA-704 M Rating System): Health 2, Flammability 0, Reactivity 0. Insoluble in water.
Potential Exposure: Mutagen. Used for making dyes, other chemicals; explosives, pharmaceuticals; in biological research.

Incompatibilities: Oxidizers.
Permissible Exposure Limits in Air: No specific standards have been established for phenanthrene.
OSHA PEL: 0.2 mg/m³ TWA [1910.1002] (benzene-soluble fraction). OSHA defines "coal tar pitch volatiles" in 29 CFR 1910.1002 as the fused polycyclic hydrocarbons that volatilize from the distillation residues of coal, petroleum (excluding asphalt), wood, and other organic matter.
NIOSH REL: 0.1 mg/m³ (cyclohexane-extractable fraction). NIOSH considers coal tar products (i.e., coal tar, coal tar pitch, or creosote) to be potential occupational carcinogens.
ACGIH TLV®[1]: 0.2 mg/m³ TWA (as benzene-soluble aerosol); Confirmed Human Carcinogen.
NIOSH IDLH: 80 mg/m³.
Protective Action Criteria (PAC)
TEEL-0: 2 mg/m³
PAC-1: 6 mg/m³
PAC-2: 40 mg/m³
PAC-3: 500 mg/m³
DFG MAK: [skin].
Several states have set guidelines or standards for coal tar pitch volatiles in ambient air[60] ranging from zero (North Carolina) to 0.0161 µg/m³ (Kansas) to 0.48 µg/m³ (Pennsylvania) to 2.0 µg/m³ (Connecticut and Virginia) to 5.0 µg/m³ (Nevada).
Determination in Air: Use NIOSH Analytical Method #5506 polynuclear aromatic hydrocarbons by HPLC; NIOSH Analytical Method #5515, Polynuclear aromatic hydrocarbons by GC; OSHA Analytical Method ID-58.
Routes of Entry: Inhalation, ingestion, skin and/or eye contact.
Harmful Effects and Symptoms
Short Term Exposure: Skin contact can cause irritation. A skin photosensitizer, contaminated skin exposed to sunlight can develop rash, skin burns, and blisters. Irritates the eyes and respiratory tract.
Long Term Exposure: May cause skin allergy. If allergy develops, very low future exposure can cause itching and skin rash.
Points of Attack: Skin, respiratory system, bladder, liver, kidneys.
Medical Surveillance: NIOSH lists: complete blood count; chest X-ray; pulmonary function tests: Forced Vital Capacity; Forced Expiratory Volume (1 s); photopatch testing; sputum cytology; urinalysis (routine); cytology, hematuria.[2]
First Aid: If this chemical gets into the eyes, remove any contact lenses at once and irrigate immediately for at least 15 min, occasionally lifting upper and lower lids. Seek medical attention immediately. If this chemical contacts the skin, remove contaminated clothing and wash immediately with soap and water. Seek medical attention immediately. If this chemical has been inhaled, remove from exposure, begin rescue breathing (using universal precautions, including resuscitation mask) if breathing has stopped and CPR if heart action has stopped. Transfer promptly to a medical

facility. When this chemical has been swallowed, get medical attention. Give large quantities of water and induce vomiting. Do not make an unconscious person vomit.

Personal Protective Methods: Wear protective gloves and clothing to prevent any reasonable probability of skin contact. Safety equipment suppliers/manufacturers can provide recommendations on the most protective glove/clothing material for your operation. All protective clothing (suits, gloves, footwear, headgear) should be clean, available each day, and put on before work. Contact lenses should not be worn when working with this chemical. Wear dust-proof chemical goggles and face shield unless full face-piece respiratory protection is worn. Employees should wash immediately with soap when skin is wet or contaminated. Provide emergency showers and eyewash.

Respirator Selection: NIOSH: *At any detectable concentration over 0.1 mg/m³:* SCBAF: Pd,Pp (APF = 10,000) (any NIOSH/MSHA- or European Standard EN 149-approved self-contained breathing apparatus that has a full face-piece and is operated in a pressure-demand or other positive-pressure mode) or SaF: Pd,Pp: ASCBA (APF = 10,000) (any supplied-air respirator that has a full face-piece and is operated in a pressure-demand or other positive-pressure mode in combination with an auxiliary, self-contained breathing apparatus operated in a pressure-demand or other positive-pressure mode). *Escape:* GmFOv100 (APF = 50) [any air-purifying, full-face-piece respirator (gas mask) with a chin-style, front- or back-mounted organic vapor canister having an N100, R100, or P100 filter] or SCBAE (any appropriate escape-type, self-contained breathing apparatus).

Storage: Color Code—Blue: Health Hazard/Poison: Store in a secure poison location. Prior to working with this chemical you should be trained on its proper handling and storage. Store in tightly closed containers in a cool, well-ventilated area away from strong oxidizers. Where possible, automatically transfer material from drums or other storage containers to process containers. Sources of ignition, such as smoking and open flames, are prohibited where this chemical is handled, used, or stored. Metal containers involving the transfer of this chemical should be grounded and bonded. Wherever this chemical is used, handled, manufactured, or stored, use explosion-proof electrical equipment and fittings.

Shipping: Not regulated. However, the "Acridine" standard may be used for this chemical. The required label is "POISONOUS/TOXIC MATERIALS." It would fall in Hazard Class 6.1 and Packing Group III.

Spill Handling: Evacuate persons not wearing protective equipment from area of spill or leak until cleanup is complete. Remove all ignition sources. Collect powdered material in the most convenient and safe manner and deposit in sealed containers. Ventilate area after cleanup is complete. It may be necessary to contain and dispose of this chemical as a hazardous waste. If material or contaminated runoff enters waterways, notify downstream users of potentially contaminated waters. Contact your local or federal environmental protection agency for specific recommendations. If employees are required to clean up spills, they must be properly trained and equipped. OSHA 1910.120(q) may be applicable.

Fire Extinguishing: This chemical is a combustible solid. Use dry chemical, carbon dioxide, water spray, or alcohol foam extinguishers. Poisonous gases are produced in fire. If material or contaminated runoff enters waterways, notify downstream users of potentially contaminated waters. Notify local health and fire officials and pollution control agencies. From a secure, explosion-proof location, use water spray to cool exposed containers. If cooling streams are ineffective (venting sound increases in volume and pitch, tank discolors, or shows any signs of deforming), withdraw immediately to a secure position. If employees are expected to fight fires, they must be trained and equipped in OSHA 1910.156. The only respirators recommended for firefighting are self-contained breathing apparatuses that have full face-pieces and are operated in a pressure-demand or other positive-pressure mode.

Disposal Method Suggested: Consult with environmental regulatory agencies for guidance on acceptable disposal practices. Generators of waste containing this contaminant (≥100 kg/mo) must conform with EPA regulations governing storage, transportation, treatment, and waste disposal.

References

US EPA. (April 1975). *Identification of Organic Compounds in Effluents from Industrial Sources*, EPA-560/3-75-002. Washington, DC

New Jersey Department of Health and Senior Services. (August 1999). *Hazardous Substances Fact Sheet: Phenanthrene*. Trenton, NJ

Phenazopyridine & phenazo-pyridine hydrochloride　　　P:0330

Molecular Formula: $C_{11}H_{12}ClN_5$

Synonyms: AP; 2,6-Diamino-3-phenylazopyridine; Diridone; DPP; Gastracid; Gastrotest; Mallophene; NC150; Phenazodine; Phenylazo; 3-(Phenylazo)-2,6-pyridinediamine; Pirid; Pyrazofen; Pyridacil; Pyridium; Pyripyridium; Sedural; Uridinal; Urodine; W 1655

Hydrochloride: Azodine; Azodium; Azodyne; Azo gantrisin; Azo gastanol; Azo-mandelamine; Azomine; Azo-standard; Azo-stat; Azotrex; Baridium; Bisteril; Cystopyrin; Cystural; 2,6-Diamino-3-phenylazopyridine hydrochloride; 2,6-Diamino-3-(phenylazo)pyridine monohydrochloride; Di-azo; Diridone; Dolonil; Eucistin; Giracid; Mallofeen; Mallophene; NC150; NCI-C01672; Nefrecil; PAP; PDP; Phenazo; Phenazodine; Phenazopyridine hydrochloride; Phenazopyridinium chloride; β-Phenylazo-α,α′-diamino-pyridine hydrochloride; 3-Phenylazo-2,6-diaminopyridine hydrochloride; Phenylazodiaminopyridine hydrochloride;

Phenylazo-α,α'-diaminopyridine monohydrochloride; 3-(Phenylazo)-2,6-pyridinediamine, hydrochloride; Phenylazopyridine hydrochloride; Phenyl-idium; Phenylidium 200; Pirid; Piridacil; Pyrazodine; Pyrazofen; Pyredal; Pyridacil; Pyridenal; Pyridene; Pyridiate; Pyridium; Pyridivite; Pyripyridium; Pyrizin; Sedural; Suladyne; Sulodyne; Thiosulfil-A forte; Urazium; Uridinal; Uriplex; Urobiotic-250; Urodine; Urofeen; Uromide; Urophenyl; Uropyridin; Uropyrine; Utostan; Vestin; W 1655

CAS Registry Number: 94-78-0; 136-40-3 (hydrochloride)
RTECS® Number: US7875000 (hydrochloride); US7700000
UN/NA & ERG Number: UN2811 (toxic solid, organic, n.o.s.)/154
EC Number: 202-363-2; 205-243-8 (hydrochloride)
Regulatory Authority and Advisory Bodies
Carcinogenicity: (hydrochloride) IARC: Animal Sufficient Evidence; Human Limited Evidence, *possibly carcinogenic to humans*, Group 2B, 1987; NCI: Carcinogenesis Studies (feed); clear evidence: mouse; (feed); clear evidence: rat; NTP: Reasonably anticipated to be a human carcinogen.
California Proposition 65 Chemical: Cancer 1/1/88; 1/1/88 (hydrochloride).
European/International Regulations: not listed in Annex 1.
WGK (German Aquatic Hazard Class): No value assigned.
Description: Phenazopyridine is a red crystalline compound. Molecular weight = 213.27; 249.73 (hydrochloride); Freezing/Melting point = (fine base) 139°C; 233–238°C (hydrochloride). Slightly soluble in water.
Potential Exposure: Phenazopyridine hydrochloride has been used for 50 years as an analgesic drug either alone or in combination with other drugs to reduce pain associated with urinary tract infection. Also used as a local anesthetic. Exposure to phenazopyridine hydrochloride occurs during manufacture and formulation.
Incompatibilities: None listed.
Permissible Exposure Limits in Air
No standards or TEEL available.
Routes of Entry: Inhalation, ingestion, skin and/or eye contact.
Harmful Effects and Symptoms
Short Term Exposure: Can affect you if swallowed. Symptoms of exposure include diarrhea, nausea, or vomiting; dehydration, decreased urine volume. May affect the development of red blood cells, causing cyanosis and methemoglobinemia and changes in blood sodium levels.
Long Term Exposure: May affect the kidneys. Phenazopyridine hydrochloride was tested in mice and rats by oral administration. In female mice, it significantly increased the incidence of hepatocellular adenomas and carcinomas. In male and female rats, it induced tumors of the colon and rectum. Symptoms of exposure include deeply stained vomitus and urine, methemoglobinemia, Heinz body anemia, hepatic enlargement, abnormal renal function. May cause mutations.
Points of Attack: Blood, kidneys.

Medical Surveillance: Complete blood count (CBC). Kidney function tests.
First Aid: *Skin Contact*[52]: Flood all areas of body that have contacted the substance with water. Do not wait to remove contaminated clothing; do it under the water stream. Use soap to help assure removal. Isolate contaminated clothing when removed to prevent contact by others. *Eye Contact:* Remove any contact lenses at once. Flush eyes well with copious quantities of water or normal saline for at least 20–30 min. Seek medical attention. *Inhalation:* Leave contaminated area immediately; breathe fresh air. Proper respiratory protection must be supplied to any rescuers. If coughing, difficult breathing, or any other symptoms develop, seek medical attention at once, even if symptoms develop many hours after exposure. *Ingestion:* If convulsions are not present, give a glass or two of water or milk to dilute the substance. Assure that the person's airway is unobstructed and contact a hospital or poison center immediately for advice on whether or not to induce vomiting.
Note to physician: Treat for methemoglobinemia. Spectrophotometry may be required for precise determination of levels of methemoglobin in urine.
Personal Protective Methods: Wear protective gloves and clothing to prevent any reasonable probability of skin contact. Safety equipment suppliers/manufacturers can provide recommendations on the most protective glove/clothing material for your operation. All protective clothing (suits, gloves, footwear, headgear) should be clean, available each day, and put on before work. Contact lenses should not be worn when working with this chemical. Wear dust-proof chemical goggles and face shield unless full face-piece respiratory protection is worn. Employees should wash immediately with soap when skin is wet or contaminated. Provide emergency showers and eyewash.
Respirator Selection: Follow regulations in OSHA 29CFR1910.134 or European Standard EN149. Use a NIOSH/MSHA- or European Standard EN149-approved respirator; or use an approved supplied-air respirator with a full face-piece operated in the positive-pressure mode, or with a full face-piece, hood, or helmet in the continuous-flow mode; or use a NIOSH/MSHA- or European Standard EN149-approved self-contained breathing apparatus with a full face-piece operated in pressure-demand or other positive-pressure mode.
Storage: Color Code—Blue: Health Hazard/Poison: Store in a secure poison location. Prior to working with this chemical you should be trained on its proper handling and storage. Store in a refrigerator or a cool, dry place. Protect from air and light. A regulated, marked area should be established where this chemical is handled, used, or stored in compliance with OSHA Standard 1910.1045.
Shipping: Toxic solids, organic, n.o.s. requires a shipping label of "POISONOUS/TOXIC MATERIALS." It falls in Hazard Class 6.1 and Packing Group III.

Spill Handling: Evacuate persons not wearing protective equipment from area of spill or leak until cleanup is complete. Remove all ignition sources. Collect powdered material in the most convenient and safe manner and deposit in sealed containers. Ventilate area after cleanup is complete. It may be necessary to contain and dispose of this chemical as a hazardous waste. If material or contaminated runoff enters waterways, notify downstream users of potentially contaminated waters. Contact your local or federal environmental protection agency for specific recommendations. If employees are required to clean up spills, they must be properly trained and equipped. OSHA 1910.120(q) may be applicable.

Fire Extinguishing: This chemical is a combustible solid. Use dry chemical, carbon dioxide, water spray, or alcohol foam extinguishers. Poisonous gases are produced in fire, including hydrogen chloride and nitrogen oxides. If material or contaminated runoff enters waterways, notify downstream users of potentially contaminated waters. Notify local health and fire officials and pollution control agencies. From a secure, explosion-proof location, use water spray to cool exposed containers. If cooling streams are ineffective (venting sound increases in volume and pitch, tank discolors, or shows any signs of deforming), withdraw immediately to a secure position. If employees are expected to fight fires, they must be trained and equipped in OSHA 1910.156. The only respirators recommended for firefighting are self-contained breathing apparatuses that have full face-pieces and are operated in a pressure-demand or other positive-pressure mode.

Phenol   P:0340

Molecular Formula: C_6H_6O
Common Formula: C_6H_5OH
Synonyms: Acide carbolique (French); Benzene, hydroxy-; Benzenol; Carbolic acid; Carbolsaure (German); ENT 1814; Fenol (Spanish); Hydroxybenzene; Monohydroxybenzene; Monophenol; NCI-C50124; Oxybenzene; Phenic acid; Phenole (German); Phenyl alcohol; Phenyl hydrate; Phenyl hydroxide; Phenylic acid; Phenylic alcohol
CAS Registry Number: 108-95-2
RTECS® Number: SJ3325000
UN/NA & ERG Number: UN1671 (solid)/153; UN2312/153 (molten); UN2821 (solution)/153
EC Number: 203-632-7 [*Annex I Index No.:* 604-001-00-2]
Regulatory Authority and Advisory Bodies
Carcinogenicity: NCI: Carcinogenesis Bioassay (oral); no evidence: mouse, rat; IARC: Animal Inadequate Evidence; Human Inadequate Evidence, *not classifiable as carcinogenic to humans*, Group 3, 1999; EPA: Available data are inadequate for an assessment of human carcinogenic potential; Not Classifiable as to human carcinogenicity.
US EPA Gene-Tox Program, Negative: *N. crassa*—reversion.

US EPA, FIFRA, 1998 Status of Pesticides: Supported.
Air Pollutant Standard Set. See below, "Permissible Exposure Limits in Air" section.
Clean Air Act: Hazardous Air Pollutants (Title I, Part A, Section 112).
Clean Water Act: Section 311 Hazardous Substances/RQ 40CFR117.3 (same as CERCLA, see below); 40CFR401.15 Section 307 Toxic Pollutants; 40CFR423, Appendix A, Priority Pollutants; Section 313 Water Priority Chemicals (57FR41331, 9/9/92).
US EPA Hazardous Waste Number (RCRA No.): U188.
RCRA, 40CFR261, Appendix 8 Hazardous Constituents.
RCRA 40CFR268.48; 61FR15654, Universal Treatment Standards: Wastewater (mg/L), 0.039; Nonwastewater (mg/kg), 6.2.
RCRA 40CFR264, Appendix 9; TSD Facilities Ground Water Monitoring List. Suggested test method(s) (PQL µg/L): 8040 (1); 8270 (10).
Superfund/EPCRA 40CFR355, Extremely Hazardous Substances: TPQ = 500/10,000 lb (227/4540 kg).
Reportable Quantity (RQ): 1000 lb (454 kg).
EPCRA Section 313 Form R *de minimis* concentration reporting level: 1.0%.
Canada, WHMIS, Ingredients Disclosure List Concentration 1.0%.
European/International Regulations: Hazard Symbol: T; Risk phrases: R23/24/25; R34; R48/21/22/23; R68; Safety phrases: S1/2; S24/25; S26; S28; S36/37/39; S45 (see Appendix 4).
WGK (German Aquatic Hazard Class): 2—Hazard to waters.

Description: Phenol is a colorless to light pink crystalline solid with a sweet, acrid odor. Phenol liquefies by mixing with about 8% water. The odor threshold in air is 0.04 ppm and in water is 7.9 ppm. Molecular weight = 94.12; Boiling point = 182°C; Freezing/Melting point = 42.8°C; Relative vapor density (air = 1) = 1.00 at 20°C; Vapor pressure = 0.4 mmHg at 25°C; Relative vapor density (air = 1): 3.2; Flash point = 79.4°C (cc); Autoignition temperature = 715°C. Explosive limits: LEL = 1.3%; UEL = 8.6%. Hazard Identification (based on NFPA-704 M Rating System) (*liquid, crystals*): Health 4, Flammability 2, Reactivity 1 (Corrosive). Soluble in water; solubility = 9% at 25°C.

Potential Exposure: Compound Description: Agricultural Chemical; Tumorigen, Mutagen; Reproductive Effector; Human Data; Primary Irritant. Phenol is used as a pharmaceutical, in the production of fertilizer; coke, illuminating gas; lampblack, paints, paint removers; rubber, asbestos goods; wood preservatives; synthetic resins; textiles, drugs, pharmaceutical preparations; perfumes, bakelite, and other plastics (phenol formaldehyde resins); polymer intermediates (caprolactam, bisphenol-A, and adipic acid). Phenol also finds wide use as a disinfectant and veterinary drug.

Incompatibilities: The aqueous solution is a weak acid. Violent reaction with strong oxidizers, calcium hypochlorite, aluminum chloride, acids. Reacts with metals.

Permissible Exposure Limits in Air
Conversion factor: 1 ppm = 3.85 mg/m^3 at 25°C & 1 atm.
OSHA PEL: 5 ppm/19 mg/m^3 TWA [skin].
NIOSH REL: 5 ppm/19 mg/m^3 TWA [skin]; 15.6 ppm/60 mg/m^3/15 min Ceiling Concentration.
ACGIH TLV®[1]: 5 ppm/19 mg/m^3 TWA [skin], not classifiable as a human carcinogen; BEI: 250 mg[total phenol]/g creatinine in urine/end-of-shift.
NIOSH IDLH: 250 ppm.
Protective Action Criteria (PAC)*
TEEL-0: 5 ppm
PAC-1: **15 ppm**
PAC-2: **23 ppm**
PAC-3: **200 ppm**
*AEGLs (Acute Emergency Guideline Levels) & ERPGs (Emergency Response Planning Guideline) are in **bold face**.
DFG MAK; [skin], Carcinogen Category: 3B.
Arab Republic of Egypt: TWA 5 ppm (19 mg/m^3), [skin], 1993; Australia: TWA 5 ppm (19 mg/m^3), [skin], 1993; Austria: MAK 5 ppm (19 mg/m^3), [skin], 1999; Belgium: TWA 5 ppm (19 mg/m^3), [skin], 1993; Denmark: TWA 1 ppm (4 mg/m^3), [skin], 1999; Finland: TWA 5 ppm (19 mg/m^3); STEL 10 ppm (38 mg/m^3), [skin], 1999; France: VME 5 ppm (19 mg/m^3), [skin], 1999; the Netherlands: MAC-TGG 8 mg/m^3, [skin], 2003; Japan: 5 ppm (19 mg/m^3), [skin], 1999; Norway: TWA 1 ppm (4 mg/m^3), 1999; the Philippines: TWA 5 ppm (10 mg/m^3), [skin], 1993; Poland: MAC (TWA) 10 mg/m^3; MAC (STEL) 20 mg/m^3, 1999; Russia: TWA 5 ppm; STEL 0.3 mg/m^3, [skin], 1993; Sweden: NGV 1 ppm (4 mg/m^3), KTV 2 ppm (8 mg/m^3), [skin], 1999; Switzerland: MAK-W 5 ppm (19 mg/m^3), KZG-W 10 ppm (38 mg/m^3), [skin], 1999; Thailand: TWA 5 ppm (19 mg/m^3), 1993; Turkey: TWA 5 ppm (19 mg/m^3), [skin], 1993; United Kingdom: TWA 5 ppm (20 mg/m^3); STEL 10 ppm, [skin], 2000; Argentina, Bulgaria, Columbia, Jordan, South Korea, New Zealand, Singapore, Vietnam: ACGIH TLV®: not classifiable as a human carcinogen. Russia[35, 43] has also set a MAC of 0.01 mg/m^3 (10 μg/m^3) for ambient air in residential areas both on a momentary and a daily average basis. Many states have set guidelines or standards for phenol in ambient air[60] ranging, for example, from 10.0 μg/m^3 (New York) to 45.23 μg/m^3 (Kansas) to 52.0 μg/m^3 (Massachusetts) to 95.0 μg/m^3 (Indiana) to 190 μg/m^3 (Florida, North Dakota, South Carolina) to 315.0 μg/m^3 (Virginia) to 380.0 μg/m^3 (Connecticut, North Dakota, South Dakota) to 452.0 μg/m^3 (Nevada) to 456.0 μg/m^3 (Pennsylvania) to 95.0 μg/m^3 (North Carolina).
Determination in Air: Use NIOSH Analytical Method (IV) #2546, Cresols and Phenol, OSHA Analytical Method 32.
Permissible Concentration in Water: To protect freshwater aquatic life: 10,200 μg/L, based on acute toxicity data and 2560 μg/L, based on chronic toxicity data. To protect saltwater aquatic life: 5800 μg/L, based on acute toxicity data. For the protection of human health from phenol ingested through water and through contaminated aquatic organisms, the concentration in water should not exceed 3500 μg/L. For the prevention of adverse effects due to the organoleptic properties of chlorinated phenols inadvertently formed during water purification processes, the phenol concentration in water should not exceed 300 μg/L.[6] The Czech Republic[35] set a MAC of 0.2 mg/L for surface water and a MAC of 0.05 mg/L for drinking water. The EEC set a MAC of 0.5 μg/L in drinking water. Mexico has set maximum permissible concentrations of 1.0 μg/L in receiving waters used for drinking water supply; 1.0 mg/L in receiving water for recreational use; 0.1 mg/L in estuaries; and 0.01 mg/L in coastal waters. Russia[35, 43] set a MAC of 0.001 mg/L in drinking water. States which have set guidelines for phenol in drinking water[61] include California at 1.0 μg/L and Kansas at 300 μg/L.
Determination in Water: Methylene chloride extraction followed by gas chromatography with flame ionization or electron capture detection (EPA Method 604) or gas chromatography plus mass spectrometry (EPA Method 625). Octanol–water coefficient: Log K_{ow} = 1.46.
Routes of Entry: Inhalation, ingestion, skin and/or eye contact. Absorbed through the skin.
Harmful Effects and Symptoms
Short Term Exposure: Phenol and its vapor are corrosive to the eyes, skin, and respiratory tract. Eye contact can cause severe and painful burns and permanent damage. Skin contact may cause severe and painful burns, which promptly become anesthetized (numb) to touch, but deep damage and local gangrene can result. Significant skin contact or inhalation can cause death within minutes. Ulceration may follow. Inhalation can cause pulmonary edema, a medical emergency that can be delayed for several hours. This can cause death. May affect the central nervous system, heart, liver, and kidneys, causing convulsions, coma, cardiac disorders, respiratory failure, collapse. Signs and symptoms of acute exposure to phenol may be severe, and range from tachycardia (rapid heart rate) and tachypnea (rapid respiratory rate) to hypotension (low blood pressure), weak pulse, cardiac failure, pulmonary edema, and respiratory arrest. Cardiac arrhythmias may be noted. Weakness, headache, dizziness, tinnitus (ringing in the ears), delirium, and shock are common. Seizures may often be followed by coma. Pallor, profuse sweating, dilated pupils, and a profound drop in body temperature may occur. Gastrointestinal effects may include nausea, abdominal pain, bloody vomitus, and bloody diarrhea. Renal insufficiency may lead to hematuria (bloody urine). Toxic hazard rating is very toxic: probable oral lethal dose (human) is 50–500 mg/kg. Ingestion of 1 g has been lethal to humans. Lethal amounts may be absorbed through skin or inhaled. Industrial contact can cause chronic poisoning with kidney and liver damage.
Long Term Exposure: Repeated or prolonged contact with skin may cause dermatitis. The substance may damage the liver and kidneys and have an effect on the pancreas and heart muscle. May affect the central nervous system and

cause nerve and/or brain damage. Phenol causes mutations and may cause reproductive damage in humans; and may be a cancer risk.

Points of Attack: Eyes, skin, respiratory system, liver, kidneys.

Medical Surveillance: NIOSH lists the following tests: liver function tests; urine (chemical/metabolite); urine (chemical/metabolite), last 2 h of 8-h exposure; urine (chemical/metabolite), end-of-shift; urine (chemical/metabolite), pre- and postshift; urinalysis. Urinary phenol (See also NIOSH #8305 *Phenol and p-cresol in urine*). These tests should be repeated if overexposure is suspected. Interview for brain effects.

First Aid: If this chemical gets into the eyes, remove any contact lenses at once and irrigate immediately for at least 15 min, occasionally lifting upper and lower lids. Seek medical attention immediately. If this chemical contacts the skin, remove contaminated clothing and wash immediately with soap and water. If concentrated phenol gets on a large area of the skin, immediately rush victim to shower and use at full blast; remove all contaminated clothing; scrub the contaminated area with *soap* for at least 10 min—*water alone may be harmful.* If polyethyleneglycol-300 is available, swab exposed area with cotton soaked in it. Seek medical attention immediately. If this chemical has been inhaled, remove from exposure, begin rescue breathing (using universal precautions, including resuscitation mask) if breathing has stopped and CPR if heart action has stopped. Transfer promptly to a medical facility. When this chemical has been swallowed, get medical attention. Rinse mouth. Give plenty of water and/or vegetable oil to drink. Do not allow the consumption of alcohol. Induce vomiting. Do not make an unconscious person vomit. Medical observation is recommended for 24–48 h after breathing overexposure, as pulmonary edema may be delayed. As first aid for pulmonary edema, a doctor or authorized paramedic may consider administering a corticosteroid spray.

Personal Protective Methods: Wear protective gloves and clothing to prevent any reasonable probability of skin contact. Safety equipment suppliers/manufacturers can provide recommendations on the most protective glove/clothing material for your operation. For phenol <30%, sealed chemical materials with good to excellent resistance: polyethylene. For phenol >70%, sealed chemical materials with good to excellent resistance: butyl rubber; Neoprene™, Teflon™, Viton™, Silvershield™. Also, polyethylene offers limited protection. All protective clothing (suits, gloves, footwear, headgear) should be clean, available each day, and put on before work. Contact lenses should not be worn when working with this chemical. Wear dust-proof chemical goggles and face shield unless full face-piece respiratory protection is worn. Employees should wash immediately with soap when skin is wet or contaminated. Provide emergency showers and eyewash.

Respirator Selection: *50 ppm:* CcrOv95 (APF = 10) [any air-purifying half-mask respirator with organic vapor cartridge(s) in combination with an N95, R95, or P95 filter.

The following filters may also be used: N99, R99, P99, N100, R100, P100]; Sa (APF = 10) (any supplied-air respirator). *125 ppm:* Sa:Cf (APF = 25) (any supplied-air respirator operated in a continuous-flow mode) or PaprOvHie (APF = 25) (any air-purifying full-face-piece respirator equipped with an organic vapor cartridge in combination with a high-efficiency particulate filter). *250 ppm:* CcrFOv100 (APF = 50) [air-purifying full-face-piece respirator equipped with organic vapor cartridge(s) in combination with an N100, R100, or P100 filter] or GmFOv100 (APF = 50) [any air-purifying, full-face-piece respirator (gas mask) with a chin-style, front- or back-mounted organic vapor canister having an N100, R100, or P100 filter] or PaprTOvHie (APF = 50) [any powered, air-purifying respirator with a tight-fitting face-piece and organic vapor cartridge(s) in combination with a high-efficiency particulate filter] or SCBAF (APF = 50) (any self-contained breathing apparatus with a full face-piece) or SaF (APF = 50) (any supplied-air respirator with a full face-piece). *Emergency or planned entry into unknown concentrations or IDLH conditions:* SCBAF: Pd,Pp (APF = 10,000) (any self-contained breathing apparatus that has a full face-piece and is operated in a pressure-demand or other positive-pressure mode) or SaF: Pd,Pp: ASCBA (APF = 10,000) (any supplied-air respirator that has a full face-piece and is operated in a pressure-demand or other positive-pressure mode in combination with an auxiliary self-contained breathing apparatus operated in a pressure-demand or other positive-pressure mode). *Escape:* GmFOv100 (APF = 50) [any air-purifying, full-face-piece respirator (gas mask) with a chin-style, front- or back-mounted organic vapor canister having an N100, R100, or P100 filter] or SCBAE (any appropriate escape-type, self-contained breathing apparatus).

Storage: Color Code—White stripe: Contact Hazard; Store separately; not compatible with materials in solid white category. Prior to working with this chemical you should be trained on its proper handling and storage. Before entering confined space where this chemical may be present, check to make sure that an explosive concentration is not a danger. Phenol must be stored to avoid contact with calcium hypochlorite and other strong oxidizers (such as chlorine and bromine), since violent reactions occur. Store in tightly closed containers in a cool, well-ventilated area away from heat. Where possible, automatically pump liquid from drums or other storage containers to process containers. Sources of ignition, such as smoking and open flames, are prohibited where this chemical is handled, used, or stored. Metal containers involving the transfer of 5 gallons or more of this chemical should be grounded and bonded. Drums must be equipped with self-closing valves, pressure vacuum bungs, and flame arresters. Use only nonsparking tools and equipment, especially when opening and closing containers of this chemical. Wherever this chemical is used, handled, manufactured, or stored, use explosion-proof electrical equipment and fittings.

Shipping: *Molten phenol* requires a shipping label of "POISONOUS/TOXIC MATERIALS." It falls in Hazard Class 6.1 and Packing Group II. *Solid phenol* requires a shipping label of "POISONOUS/TOXIC MATERIALS." It falls in Hazard Class 6.1 and Packing Group II. *Phenol solutions* require a shipping label of "POISONOUS/TOXIC MATERIALS." They fall in DOT Hazard Class 6.1 and Packing Group II or III.

Spill Handling: Remove all ignition sources. Spills must be disposed of immediately by properly protected personnel; no others should remain in area. Flush with flooding quantities of water, then use caustic soda solution for neutralization. Remove and isolate contaminated clothing at the site. Establish forced ventilation to keep levels below explosive limit. Collect powdered material in the most convenient and safe manner and deposit in sealed containers. Absorb liquids in vermiculite, dry sand, earth, or similar material and deposit in sealed containers for later disposal. It may be necessary to contain and dispose of this chemical as a hazardous waste. If material or contaminated runoff enters waterways, notify downstream users of potentially contaminated waters. Contact your local or federal environmental protection agency for specific recommendations. If employees are required to clean up spills, they must be properly trained and equipped. OSHA 1910.120(q) may be applicable.

Fire Extinguishing: This chemical is a combustible solid or liquid. Flammable vapors are produced when phenol is heated. *Small fires:* dry chemical, carbon dioxide, water spray, or alcohol foam. *Large fires:* water spray, fog, or foam; use water spray to cool containers in fire area. Move container from fire area if it can be done without risk; fight fire from maximum distance. Dike fire control water for later disposal; do not scatter the material. Poisonous gases are produced in fire. If material or contaminated runoff enters waterways, notify downstream users of potentially contaminated waters. Notify local health and fire officials and pollution control agencies. Vapors are heavier than air and will collect in low areas. Vapors in confined areas may explode when exposed to fire. Containers may explode in fire. From a secure, explosion-proof location, use water spray to cool exposed containers. If cooling streams are ineffective (venting sound increases in volume and pitch, tank discolors, or shows any signs of deforming), withdraw immediately to a secure position. If employees are expected to fight fires, they must be trained and equipped in OSHA 1910.156. The only respirators recommended for firefighting are self-contained breathing apparatuses that have full face-pieces and are operated in a pressure-demand or other positive-pressure mode.

Disposal Method Suggested: Consult with environmental regulatory agencies for guidance on acceptable disposal practices. Generators of waste containing this contaminant ($\geq$100 kg/mo) must conform with EPA regulations governing storage, transportation, treatment, and waste disposal. Incineration.

References

National Institute for Occupational Safety and Health. (1976). *Criteria for a Recommended Standard: Occupational Exposure to Phenol*, NIOSH Document No. 76-196

US Environmental Protection Agency. (1980). *Phenol: Ambient Water Quality Criteria*. Washington, DC

US Environmental Protection Agency. (April 30, 1980). *Phenol: Health and Environmental Effects Profile No. 144.* Washington, DC: Office of Solid Waste

Sax, N. I. (Ed.). (1983). *Dangerous Properties of Industrial Materials Report*, 3, No. 4, 77–84

US Public Health Service. (December 1988). *Toxicological Profile for Phenol*. Atlanta, GA: Agency for Toxic Substances and Disease Registry

US Environmental Protection Agency. (November 30, 1987). *Chemical Hazard Information Profile: Phenol*. Washington, DC: Chemical Emergency Preparedness Program

New York State Department of Health. (April 1986). *Chemical Fact Sheet: Phenol*. Albany, NY: Bureau of Toxic Substance Assessment

New Jersey Department of Health and Senior Services. (June 2001). *Hazardous Substances Fact Sheet: Phenol*. Trenton, NJ

Phenol, 3-(1-methylethyl)-, methylcarbamate   P:0350

Molecular Formula: $C_{11}H_{15}NO_2$

Synonyms: Carbamic acid, methyl-, *m*-cumenyl ester; Compound 10854; *m*-Cumenol methylcarbamate; *m*-Cumenyl methylcarbamate; ENT 25,500; ENT25,543; H 5727; H 8757; Hercules AC5727; Hercules 5727; HIP; *m*-Isopropylphenol *N*-methylcarbamate; *m*-Isopropylphenol methylcarbamate; 3-Isopropylphenol *N*-methylcarbamate; 3-Isopropylphenol methylcarbamate; *m*-Isopropylphenyl *N*-methylcarbamate; 3-Isopropylphenyl methylcarbamate; Methylcarbamic acid *m*-cumenyl ester; 3-(1-Methylethyl) phenol methylcarbamate; *N*-Methyl-*m*-isopropylphenyl carbamate; *N*-Methyl-3-isopropylphenyl carbamate; OMS-15; *m*-Psopropylphenyl methylcarbamate; UC 10854; Union Carbide UC10,854

CAS Registry Number: 64-00-6

RTECS® Number: FB7875000

UN/NA & ERG Number: UN2757/151

EC Number: 200-572-3

Regulatory Authority and Advisory Bodies

Reportable Quantity (RQ): 10 lb (4.54 kg).

RCRA 40CFR268.48; 61FR15654, Universal Treatment Standards: Wastewater (mg/L), 0.056; Nonwastewater (mg/kg), 1.4 as *m*-cumenyl methylcarbamate. Superfund/EPCRA 40CFR355, Extremely Hazardous Substances: TPQ = 500/10,000 lb (227/4540 kg).

Reportable Quantity (RQ): 10 lb (4.54 kg).

European/International Regulations: not listed in Annex 1.

WGK (German Aquatic Hazard Class): No value assigned.

Description: Phenol, 3-(1-methylethyl)-, methylcarbamate is a white, crystalline, odorless solid. Molecular weight = 193.27; Freezing/Melting point = 73°C. Soluble in water; 270 mg/L at 25°C.

Potential Exposure: Used as a carbamate insecticide.

Incompatibilities: Strong alkalies.

Permissible Exposure Limits in Air

Protective Action Criteria (PAC)*

TEEL-0: 3 mg/m^3

PAC-1: 10 mg/m^3

PAC-2: 16 mg/m^3

PAC-3: 16 mg/m^3

Routes of Entry: Inhalation, ingestion, skin and/or eye contact. Absorbed through the skin.

Harmful Effects and Symptoms

Short Term Exposure: Irritates the eyes and respiratory tract causing coughing, wheezing, shortness of breath. Inhalation or skin contact can cause rapid, severe carbamate poisoning with headache, dizziness, blurred vision, nervousness, weakness, nausea, cramps, diarrhea, and discomfort in the chest. Signs also include sweating, tearing, salivation, vomiting, cyanosis, convulsions, coma, loss of reflexes and loss of sphincter control, death. LD_{50} = (oral-guinea pig) 10 mg/kg.

Long Term Exposure: May affect the nervous system. Cholinesterase inhibitor; cumulative effect is possible. This chemical may damage the nervous system with repeated exposure, resulting in convulsions, respiratory failure. May cause liver damage.

Points of Attack: Respiratory system, lungs, central nervous system, cardiovascular system, skin, eyes, plasma and red blood cell cholinesterase.

Medical Surveillance: Before employment and at regular times after that, the following are recommended: plasma and red blood cell cholinesterase levels (tests for the enzyme poisoned by this chemical). If exposure stops, plasma levels return to normal in 1−2 weeks while red blood cell levels may be reduced for 1−3 months. When cholinesterase enzyme levels are reduced by 25% or more below preemployment levels, risk of poisoning is increased, even if results are in lower ranges of "normal." Reassignment to work not involving organophosphate or carbamate pesticides is recommended until enzyme levels recover. If symptoms develop or overexposure occurs, repeat the above tests as soon as possible and get an examination of the nervous system. Also, consider complete blood count. Consider chest X-ray following acute overexposure. Do not drink any alcoholic beverages before or during use. Alcohol promotes absorption of organic phosphates. Refer to the NIOSH Criteria Documents #78-174 and #76-147 on manufacturing, formulating, and working safely with pesticides.

First Aid: If this chemical gets into the eyes, remove any contact lenses at once and irrigate immediately for at least 15 min, occasionally lifting upper and lower lids. Seek medical attention immediately. If this chemical contacts the skin, remove contaminated clothing and wash immediately with soap and water. Seek medical attention immediately. If this chemical has been inhaled, remove from exposure, begin rescue breathing (using universal precautions, including resuscitation mask) if breathing has stopped and CPR if heart action has stopped. Transfer promptly to a medical facility. When this chemical has been swallowed, get medical attention. Give large quantities of water and induce vomiting. Do not make an unconscious person vomit.

Personal Protective Methods: Wear protective gloves and clothing to prevent any reasonable probability of skin contact. Safety equipment suppliers/manufacturers can provide recommendations on the most protective glove/clothing material for your operation. All protective clothing (suits, gloves, footwear, headgear) should be clean, available each day, and put on before work. Contact lenses should not be worn when working with this chemical. Wear dust-proof chemical goggles and face shield unless full face-piece respiratory protection is worn. Employees should wash immediately with soap when skin is wet or contaminated. Provide emergency showers and eyewash.

Respirator Selection: Follow regulations in OSHA 29CFR1910.134 or European Standard EN149. Use a NIOSH/MSHA- or European Standard EN149-approved respirator; or use an approved supplied-air respirator with a full face-piece operated in the positive-pressure mode, or with a full face-piece, hood, or helmet in the continuous-flow mode; or use a NIOSH/MSHA- or European Standard EN149-approved self-contained breathing apparatus with a full face-piece operated in pressure-demand or other positive-pressure mode.

Storage: Color Code—Blue: Health Hazard/Poison: Store in a secure poison location. Prior to working with this chemical you should be trained on its proper handling and storage. Store in tightly closed containers in a cool, well-ventilated area away from strong alkaline materials. Where possible, automatically transfer material from drums or other storage containers to process containers. Sources of ignition, such as smoking and open flames, are prohibited where this chemical is handled, used, or stored. Metal containers involving the transfer of this chemical should be grounded and bonded. Wherever this chemical is used, handled, manufactured, or stored, use explosion-proof electrical equipment and fittings.

Shipping: Carbamate pesticides, solid, toxic, require a shipping label of "POISONOUS/TOXIC MATERIALS." Phenol, 3-(methylethyl)-, Methylcarbamate falls in Hazard Class 6.1 and Packing Group III.

Spill Handling: Evacuate persons not wearing protective equipment from area of spill or leak until cleanup is complete. Remove all ignition sources. Collect powdered material in the most convenient and safe manner and deposit in sealed containers. Ventilate area after cleanup is complete. It may be necessary to contain and dispose of this chemical

as a hazardous waste. If material or contaminated runoff enters waterways, notify downstream users of potentially contaminated waters. Contact your local or federal environmental protection agency for specific recommendations. If employees are required to clean up spills, they must be properly trained and equipped. OSHA 1910.120(q) may be applicable.

Fire Extinguishing: This chemical is a noncombustible solid. Use extinguishing agents suitable for surrounding fire. Poisonous gases are produced in fire, including nitrogen oxides. If material or contaminated runoff enters waterways, notify downstream users of potentially contaminated waters. Notify local health and fire officials and pollution control agencies. From a secure, explosion-proof location, use water spray to cool exposed containers. If cooling streams are ineffective (venting sound increases in volume and pitch, tank discolors, or shows any signs of deforming), withdraw immediately to a secure position. If employees are expected to fight fires, they must be trained and equipped in OSHA 1910.156. The only respirators recommended for firefighting are self-contained breathing apparatuses that have full face-pieces and are operated in a pressure-demand or other positive-pressure mode.

Disposal Method Suggested: In accordance with 40CFR165, follow recommendations for the disposal of pesticides and pesticide containers. Must be disposed properly by following package label directions or by contacting your local or federal environmental control agency or by contacting your regional EPA office.

Reference

New Jersey Department of Health and Senior Services. (September 1999). *Hazardous Substances Fact Sheet: Phenol, 3-(methylethyl)-, Methylcarbamate.* Trenton, NJ

Phenothiazine   P:0360

Molecular Formula: $C_{12}H_9NS$
Common Formula: $S(C_6H_4)_2NH$
Synonyms: AFI-tiazin; Agrazine; Antiverm; Biverm; Contaverm; Dibenzoparathiazine; Dibenzothiazine; Dibenzo-1,4-thiazine; ENT 38; Feeno; Fenoverm; Fentiazin; Helmetina; Lethelmin; Nemazene; Nemazine; Orimon; Padophene; Penthazine; Phenegic; Phenosan; Phenoverm; Phenovis; Phenoxur; Phenthiazine; Reconox; Souframine; Thiodiphenylamin (German); Vermitin; Wurm-thional; XL-50
CAS Registry Number: 92-84-2
RTECS®Number: SN5075000
EC Number: 202-196-5
Regulatory Authority and Advisory Bodies
US EPA Gene-Tox Program, Positive: Cell transform.—SA7/SHE.
US EPA, FIFRA 1998 Status of Pesticides: Canceled.
Air Pollutant Standard Set. See below, "Permissible Exposure Limits in Air" section.

WGK (German Aquatic Hazard Class): 1—Low hazard to waters.
Description: Phenothiazine is a greenish-yellow to greenish-gray crystalline substance with a slight odor and taste. Molecular weight = 199.28; Boiling point = (decomposes) 371°C; Freezing/Melting point = 185°C (sublimes). Hazard Identification (based on NFPA-704 M Rating System): Health 2, Flammability 1, Reactivity 0. Insoluble in water.
Potential Exposure: Compound Description: Agricultural Chemical; Drug, Mutagen; Reproductive Effector; Human Data; Primary Irritant. Phenothiazine is used as an insecticide; as a base for the manufacture of tranquilizers; as anthelmintic in medicine and veterinary medicine; it is used widely as an intermediate in pharmaceutical manufacture; polymerization inhibitor, antioxidant.
Incompatibilities: Organosulfides are incompatible with strong acids and acid fumes; elevated temperatures; sulfur oxides and nitrogen oxides can be produced. Contact with strong reducing agents, azo and diazo compounds, halocarbons, isocyanates can generate heat and may form explosive hydrogen gas.
Permissible Exposure Limits in Air
OSHA PEL: None.
NIOSH REL: 5 mg/m^3 TWA [skin].
ACGIH: 5 mg/m^3 TWA [skin].
No TEEL available.
Australia: TWA 5 mg/m^3, [skin], 1993; Belgium: TWA 5 mg/m^3, [skin], 1993; Denmark: TWA 5 mg/m^3, [skin], 1999; Finland: TWA 5 mg/m^3; STEL 10 mg/m^3, [skin], 1999; France: VME 5 mg/m^3, [skin], 1999; Norway: TWA 5 mg/m^3, 1999; the Philippines: TWA 5 mg/m^3, [skin], 1993; the Netherlands: MAC-TGG 5 mg/m^3, [skin], 2003; Argentina, Bulgaria, Columbia, Jordan, South Korea, New Zealand, Singapore, Vietnam: ACGIH TLV®: 5 mg/m^3 [skin]. Several states have set guidelines or standards for Phenothiazine is ambient air[60] ranging from 50 μg/m^3 (North Dakota) to 80 μg/m^3 (Virginia) to 100 μg/m^3 (Connecticut) to 119 μg/m^3 (Nevada).
Determination in Air: No NIOSH Analytical Method available.
Determination in Water: Octanol—water coefficient: Log $K_{ow} = 4.2$.
Routes of Entry: Inhalation, ingestion, skin and/or eye contact. Skin absorption.
Harmful Effects and Symptoms
Short Term Exposure: Phenothiazine can affect you when breathed in and by passing through your skin. Exposure can irritate the skin and eyes. Exposure can cause an inflammation in the eye (keratitis). This can also be made worse by sunlight (photosensitization) and cause a severe skin reaction with rash and color changes. Can cause a severe allergic liver reaction. High levels of exposure may affect the blood cells, causing hemolytic anemia and toxic liver degeneration. Exposure may affect the nervous system, causing muscle twitching and shaking. May affect heart rhythm, causing irregular heartbeat.

Long Term Exposure: Repeated or prolonged contact with skin may cause dermatitis and allergy. Can cause kidney and liver damage. Repeated or prolonged contact may cause skin sensitization as well as skin photophobia (abnormal visual intolerance to light). There is limited evidence that this chemical may damage the developing fetus. Several related phenothiazine compounds have been associated with human teratogenic effects.

Points of Attack: Skin, cardiovascular system, liver, kidneys, heart.

Medical Surveillance: For those with frequent or potentially high exposure (half the TLV or greater, or significant skin contact), the following are recommended before beginning work and at regular times after that: examination of the nervous system and eyes. Liver function tests, especially bile salts. Complete blood count. Evaluation by a qualified allergist, including careful exposure history and special testing, may help diagnose skin allergy. EKG.

First Aid: If this chemical gets into the eyes, remove any contact lenses at once and irrigate immediately for at least 15 min, occasionally lifting upper and lower lids. Seek medical attention immediately. If this chemical contacts the skin, remove contaminated clothing and wash immediately with soap and water. Seek medical attention immediately. If this chemical has been inhaled, remove from exposure, begin rescue breathing (using universal precautions, including resuscitation mask) if breathing has stopped and CPR if heart action has stopped. Transfer promptly to a medical facility. When this chemical has been swallowed, get medical attention. Give large quantities of water and induce vomiting. Do not make an unconscious person vomit.

Personal Protective Methods: Wear protective gloves and clothing to prevent any reasonable probability of skin contact. Safety equipment suppliers/manufacturers can provide recommendations on the most protective glove/ clothing material for your operation. All protective clothing (suits, gloves, footwear, headgear) should be clean, available each day, and put on before work. Contact lenses should not be worn when working with this chemical. Wear dust-proof chemical goggles and face shield unless full face-piece respiratory protection is worn. Employees should wash immediately with soap when skin is wet or contaminated. Provide emergency showers and eyewash.

Respirator Selection: Where there is potential for exposures *over 5 mg/m³*, use a NIOSH/MSHA- or European Standard EN149-approved full face-piece respirator with a high-efficiency particulate filter. Greater protection is provided by a powered air-purifying respirator. *Where there is potential for high exposures*, use a NIOSH/MSHA- or European Standard EN149-approved supplied-air respirator with a full face-piece operated in the positive-pressure mode, or with a full face-piece, hood, or helmet in the continuous-flow mode; or use a NIOSH/MSHA- or European Standard EN149-approved self-contained breathing apparatus with a full face-piece operated in pressure-demand or other positive-pressure mode.

Storage: Color Code—Green: General storage may be used. Prior to working with this chemical you should be trained on its proper handling and storage. Store in tightly closed containers in a cool, well-ventilated area away from strong acids (such as hydrochloric, sulfuric, and nitric) since toxic fumes can result.

Shipping: Not regulated.

Spill Handling: Evacuate persons not wearing protective equipment from area of spill or leak until cleanup is complete. Remove all ignition sources. Collect powdered material in the most convenient and safe manner and deposit in sealed containers. Ventilate area after cleanup is complete. It may be necessary to contain and dispose of this chemical as a hazardous waste. If material or contaminated runoff enters waterways, notify downstream users of potentially contaminated waters. Contact your local or federal environmental protection agency for specific recommendations. If employees are required to clean up spills, they must be properly trained and equipped. OSHA 1910.120(q) may be applicable.

Fire Extinguishing: Phenothiazine may burn, but does not readily ignite. Use dry chemical, carbon dioxide, water spray, or alcohol foam extinguishers. Poisonous gases are produced in fire, including sulfur oxides and nitrogen oxides. If material or contaminated runoff enters waterways, notify downstream users of potentially contaminated waters. Notify local health and fire officials and pollution control agencies. From a secure, explosion-proof location, use water spray to cool exposed containers. If cooling streams are ineffective (venting sound increases in volume and pitch, tank discolors, or shows any signs of deforming), withdraw immediately to a secure position. If employees are expected to fight fires, they must be trained and equipped in OSHA 1910.156. The only respirators recommended for firefighting are self-contained breathing apparatuses that have full face-pieces and are operated in a pressure-demand or other positive-pressure mode.

Disposal Method Suggested: Dissolve in combustible solvent and spray into incinerator equipped with afterburner and scrubber. In accordance with 40CFR165, follow recommendations for the disposal of pesticides and pesticide containers. Must be disposed properly by following package label directions or by contacting your local or federal environmental control agency or by contacting your regional EPA office.

References

New Jersey Department of Health and Senior Services. (May 2000). *Hazardous Substances Fact Sheet: Phenothiazine.* Trenton, NJ

US Environmental Protection Agency, Special Review and Reregistration Division Office of Pesticide Programs. (1998). *Agency Status of Pesticides in Registration, Reregistration, and Special Review* (Rainbow Report). Washington, DC

Phenyl dichloroarsine (Agent PD, WMD)   P:0370

Molecular Formula: $C_6H_5AsCl_2$

Synonyms: Arsine, dichlorophenyl-; Arsonous dichloride, phenyl-; Diclorofenilarsina (Spanish); Dichlorophenylarsine; PD (military designation); Phenylarsinedichloride; Phenylarsonous dichloride; Phenyl arsonous dichloride; Phenyldichloroarsine

CAS Registry Number: 696-28-6

RTECS®Number: CH5425000

UN/NA & ERG Number: UN1556/152

EC Number: 211-791-9

Regulatory Authority and Advisory Bodies

Air Pollutant Standard Set. See below, "Permissible Exposure Limits in Air" section.

US EPA Hazardous Waste Number (RCRA No.): P036.

RCRA, 40CFR261, Appendix 8 Hazardous Constituents.

Superfund/EPCRA 40CFR355, Extremely Hazardous Substances: TPQ = 500 lb (227 kg).

Reportable Quantity (RQ): 1 lb (0.454 kg).

US DOT 49CFR172.101, Inhalation Hazardous Chemical.

As arsenic compounds

Clean Air Act: Hazardous Air Pollutants (Title I, Part A, Section 112) as arsenic compounds.

Clean Water Act: Toxic Pollutant (Section 401.15) as arsenic and compounds.

RCRA, 40CFR261, Appendix 8 Hazardous Constituents, waste number not listed.

Reportable Quantity (RQ): 1 lb (0.454 kg).

EPCRA (Section 313): Includes any unique chemical substance that contain arsenic as part of that chemical's infrastructure; Form R *de minimis* concentration reporting level: organics 1.0%.

US DOT Regulated Marine Pollutant (49CFR172.101, Appendix B).

Canada: Priority Substance List & Restricted Substances/Ocean Dumping FORBIDDEN (CEPA), National Pollutant Release Inventory (NPRI) (arsenic compounds).

European/International Regulations: not listed in Annex 1.

WGK (German Aquatic Hazard Class): No value assigned.

Description: Phenyldichloroarsine is a colorless or light yellow liquid or gas. Odorless. Molecular weight = 222.93; Boiling point = 257°C; Freezing/Melting point = −19°C; Flash point = 16°C. Hazard Identification (based on NFPA-704 M Rating System): Health 4, Flammability 2, Reactivity 2W. Reaction with water; insoluble.

Potential Exposure: It is used in organic synthesis and as a solvent. PD has been used as a military tear gas, vesicant, and blister agent.

Incompatibilities: Contact with water forms HCl. Heat produces fumes of arsenic and chlorine. Attacks some metals in the presence of moisture.

Permissible Exposure Limits in Air

Arsenic, organic compounds

OSHA PEL: 0.5 mg[As]/m³ TWA.

NIOSH REL: Not established. See NIOSH Pocket Guide, Appendix A.

ACGIH TLV®[1]: 0.01 mg[As]/m³ TWA; Confirmed Human Carcinogen; BEI established.

Phenyl dichloroarsine; dichlorophenylarsine

Protective Action Criteria (PAC)*

TEEL-0: 0.061 mg/m³

PAC-1: 0.061 mg/m³

PAC-2: **0.061** mg/m³

PAC-3: **0.18** mg/m³

AEGLs (Acute Emergency Guideline Levels) & ERPGs (Emergency Response Planning Guideline) are in **bold face**.

As arsenic, organic compounds

TEEL-0: 0.5 mg/m³

PAC-1: 1.5 mg/m³

PAC-2: 2.5 mg/m³

PAC-3: 350 mg/m³

NIOSH IDLH: 5 mg[As]/m³.

Determination in Air: Filter; Acid; Hydride generation atomic absorption spectrometry; NIOSH Analytical Method (IV) #7900. See also #7300, Elements (arsenic).

Permissible Concentration in Water: EPA[6] recommends a zero concentration of arsenic for human health reasons but has set a guideline of 50 μg/L[61] for drinking water.

Determination in Water: When phenyldichloroarsine mixes with water it breaks down into hydrochloric acid and arsenicals. *For arsenic:* The atomic absorption graphite furnace technique is often used for the measurement of total arsenic in water. It has also been standardized by EPA. Total arsenic may be determined by digestion followed by silver diethyldithiocarbamate; an alternative is atomic absorption; another is inductively coupled plasma (ICP) optical emission spectrometry. See OSHA Analytical Method #ID-105 for arsenic.

Routes of Entry: Inhalation, ingestion, skin and/or eye contact.

Harmful Effects and Symptoms

Short Term Exposure: Phenyldichloroarsine reacts with many enzymes which damages the body. Phenyldichloroarsine will blind you and blister your skin severely, and with enough it will kill. You will know that you have been exposed when you feel immediate pain and you begin to vomit violently. Contact may cause burns to skin and eyes. Strong irritant to eyes, skin, and tissue. Corrosive if swallowed. Poisonous; may be fatal if inhaled, swallowed, or absorbed through skin. Vomiting and blistering are among symptoms of exposure. The median lethal dosage is 2600 mg-min/m³. The mean incapacitating dosage is 16 mg-min/m³ as a vomiting agent and 1800 mg-min/m³ as a blistering agent. 633 mg-min/m³ produces eye injury.

Long Term Exposure: In animals: kidney damage; muscle tremor, seizure; possible gastrointestinal tract; reproductive effects; possible liver damage.

Points of Attack: Skin, respiratory system, kidneys, central nervous system, liver, gastrointestinal tract, reproductive system.

Medical Surveillance: Kidney function tests. Lung function tests. Consider chest X-ray following acute overexposure.

First Aid: If this chemical gets into the eyes, remove any contact lenses at once and irrigate immediately for at least 15 min, occasionally lifting upper and lower lids. Seek medical attention immediately. If this chemical contacts the skin, remove contaminated clothing and wash immediately with soap and water. Speed in removing material from skin is of extreme importance. Seek medical attention immediately. If this chemical has been inhaled, remove from exposure, begin rescue breathing (using universal precautions, including resuscitation mask) if breathing has stopped and CPR if heart action has stopped. Transfer promptly to a medical facility. When this chemical has been swallowed, get medical attention. Give large quantities of water and induce vomiting. Do not make an unconscious person vomit. Keep victim quiet and maintain normal body temperature. Effects may be delayed; keep victim under observation.

Note to physician: For severe poisoning BAL [British anti-lewisite, dimercaprol, dithiopropanol ($C_3H_8OS_2$)] has been used to treat toxic symptoms of certain heavy metal poisoning—including arsenic. Although BAL is reported to have a large margin of safety, caution must be exercised, because toxic effects may be caused by excessive dosage. Most can be prevented by premedication with 1-ephedrine sulfate (CAS: 134-72-5). For milder poisoning *penicillamine (not penicillin)* has been used, both with mixed success. Side effects occur with such treatment and it is never a substitute for controlling exposure. It can only be done under strict medical care.

Decontamination: This is very important, and you have to decontaminate as soon as you can. Extra minutes before decontamination might make a big difference. If you do not have the equipment and training do not enter the hot or the warm zone to rescue and decontaminate victims. If the victim cannot move, decontaminate without touching and without entering the hot or the warm zone. Use clean water from any source; if possible, use a hose (spray or fog to prevent injury to the victim) or other system so that you would not have to touch the victim; do not even wait for soap or for the victim to remove clothing, begin washing immediately. Immediately flush the eyes with water for at least 15 min. Use caution to avoid hypothermia in children and the elderly. Wash—strip—wash—evacuate upwind and uphill: The approach is to immediately wash with water, then have the victim (not the first responder) remove all the victim's clothing, then wash again (with soap if available) and then move away from the hot zone in an upwind and uphill direction. Wash the victim with warm water and soap. Decontaminate with diluted household bleach (0.5%, or one part bleach to 200 parts water), but do not let any get in the victim's eyes, open wounds, or mouth. Wash off the diluted bleach solution after 15 min. Be sure you have decontaminated the victims as much as you can before they leave the area so that they do not spread the phenyldichloroarsine. Use the antidote "Anti-Lewisite." See "First Aid" above. Use 5% solution of common bleach (sodium hypochlorite) or calcium hypochlorite solution (48 oz per 5 gallons of water) to decontaminate scissors used in clothing removal, clothes and other items.

Personal Protective Methods: Wear protective gloves and clothing to prevent any reasonable probability of skin contact. Safety equipment suppliers/manufacturers can provide recommendations on the most protective glove/clothing material for your operation. All protective clothing (suits, gloves, footwear, headgear) should be clean, available each day, and put on before work. Contact lenses should not be worn when working with this chemical. Wear splash-proof chemical goggles and face shield unless full face-piece respiratory protection is worn. Employees should wash immediately with soap when skin is wet or contaminated. Provide emergency showers and eyewash.

Respirator Selection: Follow regulations in OSHA 29CFR1910.134 or European Standard EN149. Use a NIOSH/MSHA- or European Standard EN149-approved respirator; or use an approved supplied-air respirator with a full face-piece operated in the positive-pressure mode, or with a full face-piece, hood, or helmet in the continuous-flow mode; or use a NIOSH/MSHA- or European Standard EN149-approved self-contained breathing apparatus with a full face-piece operated in pressure-demand or other positive-pressure mode.

Storage: Color Code—Blue: Health Hazard/Poison: Store in a secure poison location. Prior to working with this chemical you should be trained on its proper handling and storage. Store in tightly closed containers in a cool, well-ventilated area away from oxidizers. Where possible, automatically pump liquid from drums or other storage containers to process containers. Sources of ignition, such as smoking and open flames, are prohibited where this chemical is handled, used, or stored. Metal containers involving the transfer of 5 gallons or more of this chemical should be grounded and bonded. Drums must be equipped with self-closing valves, pressure vacuum bungs, and flame arresters. Use only nonsparking tools and equipment, especially when opening and closing containers of this chemical. Wherever this chemical is used, handled, manufactured, or stored, use explosion-proof electrical equipment and fittings.

Shipping: Arsenic compounds, liquid, n.o.s. requires a shipping label of "POISONOUS/TOXIC MATERIALS." Phenyl dichloroarsine falls in DOT Hazard Class 6.1 and Packing Group I.

Spill Handling: Evacuate and restrict persons not wearing protective equipment from area of spill or leak until cleanup is complete. Remove all ignition sources. Ventilate area of spill or leak. Absorb liquids in vermiculite, dry sand, earth, peat, carbon, or a similar material and deposit in sealed containers. Keep this chemical out of a confined space, such as a sewer, because of the possibility of an explosion, unless the sewer is designed to prevent the buildup of explosive concentrations. It may be necessary to contain and dispose of this chemical as a hazardous waste. If material or contaminated runoff enters waterways, notify downstream users of potentially contaminated waters. Contact your local or federal environmental protection agency for specific recommendations. If employees are required to clean up spills, they must be properly trained and equipped. OSHA 1910.120(q) may be applicable.

Initial isolation and protective action distances
Distances shown are likely to be affected during the first 30 min after materials are spilled and could increase with time. If more than one tank car, cargo tank, portable tank, or large cylinder involved in the incident is leaking, the protective action distance may need to be increased. You may need to seek emergency information from CHEMTREC at (800) 424-9300 or seek professional environmental engineering assistance from the US EPA Environmental Response Team at (908) 548-8730 (24-h response line).

PD, when used as a weapon
Small spills (From a small package or a small leak from a large package)
First: Isolate in all directions (feet/meters)   100/30
Then: Protect persons downwind (miles/kilometers)
Day   0.1/0.2
Night   0.1/0.2
Large spills (From a large package or from many small packages)
First: Isolate in all directions (feet/meters)   100/30
Then: Protect persons downwind (miles/kilometers)
Day   0.1/0.2
Night   0.1/0.2
Fire Extinguishing: Phenyl dichloroarsine may burn but does not ignite readily. Burned phenyl dichloroarsine is safer and better than the unburned product. In case of fire, evacuate the area. If there is some reason that you have to put out the fire—for example, there are things you cannot let burn nearby—use unattended equipment. You can fight phenyl dichloroarsine fires with water streams, water fog; ordinary foam; universal foam; and, for confined fires, carbon dioxide. Remember that phenyl dichloroarsine breaks down in water, forming toxic hydrochloric acid and arsenic chemicals. Wear positive pressure breathing apparatus. Move container from fire area if you can do it without risk. Fight fire from maximum distance. Dike fire control water for later disposal; do not scatter the material. Poisonous gases, including arsenic and chlorine, are produced in fire. Water produces corrosive chlorine fumes. Vapors are heavier than air and will collect in low areas. Vapors in confined areas may explode when exposed to fire. Containers may explode in fire. Storage containers and parts of containers may rocket great distances, in many directions. If material or contaminated runoff enters waterways, notify downstream users of potentially contaminated waters. Notify local health and fire officials and pollution control

agencies. From a secure, explosion-proof location, use water spray to cool exposed containers. If cooling streams are ineffective (venting sound increases in volume and pitch, tank discolors, or shows any signs of deforming), withdraw immediately to a secure position. If employees are expected to fight fires, they must be trained and equipped in OSHA 1910.156. The only respirators recommended for firefighting are self-contained breathing apparatuses that have full face-pieces and are operated in a pressure-demand or other positive-pressure mode.
Disposal Method Suggested: Consult with environmental regulatory agencies for guidance on acceptable disposal practices. Generators of waste containing this contaminant ($\geq$100 kg/mo) must conform with EPA regulations governing storage, transportation, treatment, and waste disposal. In accordance with 40CFR165, follow recommendations for the disposal of pesticides and pesticide containers. Must be disposed properly by following package label directions or by contacting your local or federal environmental control agency or by contacting your regional EPA office.

References
US Environmental Protection Agency. (November 30, 1987). *Chemical Hazard Information Profile: Phenyl Dichloroarsine*. Washington, DC: Chemical Emergency Preparedness Program
Schneider, A. L., (Ed.) (2007). *CHRIS + CD-ROM Version 2.0, United States Coast Guard Chemical Hazard Response Information System (COMDTINST 16465.12C)*. Washington, DC: United States Coast Guard and the Department of Homeland Security

m-Phenylenediamine   P:0380

Molecular Formula: $C_6H_8N_2$
Synonyms: AI3-52607; 3-Aminoaniline; *m*-Aminoaniline; *meta*-Aminoaniline; Aminoaniline, *meta*-; Apco 2330; *meta*-Benzenediamine; *m*-Benzenediamine; 1,3-Benzenediamine; Benzene, 1,3-diamino-; C.I. 76025; C.I. Developer 11; Developer 11; Developer C; Developer H; Developer M; *meta*-Diaminobenzene; *m*-Diaminobenzene; 1,3-Diaminobenzene; Direct brown BR; Direct brown GG; 1,3-Fenilendiamina (Spanish); *m*-Fenilendiamina (Spanish); Metaphenylenediamine; 3-Phenylenediamine; *m*-Phenylenediamine; Phenylenediamine, *meta*-
CAS Registry Number: 108-45-2
RTECS®Number: SS7700000
UN/NA & ERG Number: UN1673/153
EC Number: 203-584-7 [*Annex I Index No.:* 612-147-00-3] (*m*-)
Regulatory Authority and Advisory Bodies
Carcinogenicity: IARC: Animal Inadequate Evidence; Human No Adequate Data, *not classifiable as carcinogenic to humans*, Group 3, 1987.

US EPA Gene-Tox Program, Positive: SHE—clonal assay; Histidine reversion—Ames test; Inconclusive: Rodent dominant lethal.

EPCRA Section 313 Form R *de minimis* concentration reporting level: 1.0%.

Canada, WHMIS, Ingredients Disclosure List Concentration 0.1%.

European/International Regulations: Hazard Symbol: T, N; Risk phrases: R20/21; R25; R36; R40; R43; R50/53; R68; Safety phrases: S1/2; S28; S36/37; S45; S60; S61 (see Appendix 4).

WGK (German Aquatic Hazard Class): 2—Hazard to waters.

Description: *m*-Phenylenediamine is a colorless to white crystalline substance that turns red upon exposure to air. Molecular weight = 108.16; Boiling point = 287°C; Freezing/Melting point = 64.4°C; Flash point = 187°C; Autoignition temperature = 555°C. Hazard Identification (based on NFPA-704 M Rating System): Health 0, Flammability 1, Reactivity 0. Soluble in water.

Potential Exposure: Compound Description: Tumorigen, Mutagen; Reproductive Effector; Primary Irritant. Used in making various dyes; as a curing agent for epoxy resin; rubber, textile fibers; urethanes, corrosion inhibitors; adhesives; in photographic and analytical procedures and processes.

Incompatibilities: Reacts violently with strong oxidizers, strong acids, acid chlorides, acid anhydrides, chloroformates. Heat and light contribute to instability. Keep away from metals.

Permissible Exposure Limits in Air
ACGIH TLV®[1]: 0.1 mg/m^3 TWA [skin]; not classifiable as a human carcinogen.
Protective Action Criteria (PAC)
TEEL-0: 0.1 mg/m^3
PAC-1: 0.3 mg/m^3
PAC-2: 10 mg/m^3
PAC-3: 125 mg/m^3
DFG MAK: [skin]; Carcinogen Category 3B.
Austria [skin], Suspected: carcinogen, 1999; Denmark: TWA 0.1 mg/m^3, 1999; Norway: TWA 0.1 mg/m^3, 1999; Switzerland: MAK-W 0.1 mg/m^3, [skin], 1999; the Netherlands: MAC-TGG 0.1 mg/m^3, 2003; Argentina, Bulgaria, Columbia, Jordan, South Korea, New Zealand, Singapore, Vietnam: ACGIH TLV®: not classifiable as a human carcinogen.

Determination in Air: Use OSHA Analytical Method 87.

Routes of Entry: Inhalation, ingestion, skin and/or eye contact.

Harmful Effects and Symptoms
Short Term Exposure: Contact can irritate the eyes and skin. Irritates the respiratory tract, causing coughing, wheezing, and/or shortness of breath. May affect the blood, causing the formation of methemoglobin and cyanosis with blue coloration of the skin and lips, headache, fatigue, dizziness. High levels can cause troubled breathing, collapse, and death. The *p*-isomer is more toxic and a more severe irritant than the *m*-isomers. LD$_{50}$ = (oral-rat) 720−1600 mg/kg.

Long Term Exposure: Repeated or prolonged contact may cause skin sensitization and allergy. If allergy develops, very low future exposure can cause itching and a skin rash. Repeated or prolonged inhalation exposure may cause asthma-like allergy. May cause kidney and liver impairment. Exposure may cause anemia.

Points of Attack: Respiratory system, skin, lungs, liver, kidneys, blood.

Medical Surveillance: Consider the points of attack in preplacement and periodic physical examinations. Complete blood count (CBC). Examination of the eyes and vision. Evaluation by a qualified allergist. Lung function tests. Blood methemoglobin levels. Liver and kidney function tests.

First Aid: If this chemical gets into the eyes, remove any contact lenses at once and irrigate immediately for at least 15 min, occasionally lifting upper and lower lids. Seek medical attention immediately. If this chemical contacts the skin, remove contaminated clothing and wash immediately with soap and water. Seek medical attention immediately. If this chemical has been inhaled, remove from exposure, begin rescue breathing (using universal precautions, including resuscitation mask) if breathing has stopped and CPR if heart action has stopped. Transfer promptly to a medical facility. When this chemical has been swallowed, get medical attention. Give large quantities of water and induce vomiting. Do not make an unconscious person vomit.

Note to physician: Treat for methemoglobinemia. Spectrophotometry may be required for precise determination of levels of methemoglobin in urine.

Personal Protective Methods: Wear protective gloves and clothing to prevent any reasonable probability of skin contact. Safety equipment suppliers/manufacturers can provide recommendations on the most protective glove/clothing material for your operation. All protective clothing (suits, gloves, footwear, headgear) should be clean, available each day, and put on before work. Contact lenses should not be worn when working with this chemical. Wear dust-proof chemical goggles and face shield unless full face-piece respiratory protection is worn. Employees should wash immediately with soap when skin is wet or contaminated. Provide emergency showers and eyewash.

Respirator Selection: Where there is potential for exposure to this chemical of more than 0.1 mg/m^3, use a NIOSH/MSHA- or European Standard EN149-approved supplied-air respirator with a full face-piece operated in the positive-pressure mode, or with a full face-piece, hood, or helmet in the continuous-flow mode; or use a NIOSH/MSHA- or European Standard EN149-approved self-contained breathing apparatus with a full face-piece operated in pressure-demand or other positive-pressure mode.

Storage: Color Code—Blue: Health Hazard/Poison: Store in a secure poison location. Prior to working with this chemical you should be trained on its proper handling and

storage. Store in tightly closed containers in a cool, well-ventilated area away from strong oxidizers. Where possible, automatically transfer material from drums or other storage containers to process containers. Sources of ignition, such as smoking and open flames, are prohibited where this chemical is handled, used, or stored. Metal containers involving the transfer of this chemical should be grounded and bonded. Wherever this chemical is used, handled, manufactured, or stored, use explosion-proof electrical equipment and fittings.

Shipping: Phenylenediamines require a shipping label of "POISONOUS/TOXIC MATERIALS." They fall in DOT Hazard Class 6.1 and Packing Group III.

Spill Handling: Evacuate persons not wearing protective equipment from area of spill or leak until cleanup is complete. Remove all ignition sources. Collect powdered material in the most convenient and safe manner and deposit in sealed containers. Ventilate area after cleanup is complete. It may be necessary to contain and dispose of this chemical as a hazardous waste. If material or contaminated runoff enters waterways, notify downstream users of potentially contaminated waters. Contact your local or federal environmental protection agency for specific recommendations. If employees are required to clean up spills, they must be properly trained and equipped. OSHA 1910.120(q) may be applicable.

Fire Extinguishing: This chemical is a combustible solid. Use dry chemical, carbon dioxide, water spray, alcohol foam or polymer foam extinguishers. Poisonous gases are produced in fire, including nitrogen oxides. If material or contaminated runoff enters waterways, notify downstream users of potentially contaminated waters. Notify local health and fire officials and pollution control agencies. From a secure, explosion-proof location, use water spray to cool exposed containers. If cooling streams are ineffective (venting sound increases in volume and pitch, tank discolors, or shows any signs of deforming), withdraw immediately to a secure position. If employees are expected to fight fires, they must be trained and equipped in OSHA 1910.156. The only respirators recommended for firefighting are self-contained breathing apparatuses that have full face-pieces and are operated in a pressure-demand or other positive-pressure mode.

Disposal Method Suggested: Controlled incineration whereby oxides of nitrogen are removed from the effluent gas by scrubber, catalytic or thermal device.[22]

Reference
New Jersey Department of Health and Senior Services. (August 1999). *Hazardous Substances Fact Sheet: m-Phenylenediamine.* Trenton, NJ

o-Phenylenediamine   P:0390

Molecular Formula: $C_6H_8N_2$
Synonyms: AI3-24343; 2-Aminoaniline; *o*-Benzenediamine; 1,2-Benzenediamine; C.I. 76010; C.I. Oxidation base 16;

o-Diaminobenzene; 1,2-Diaminobenzene; 1,2-Fenilendiamina (Spanish); *o*-Fenilendiamina (Spanish); OPDA; Orthamine; *o*-Phenylenediamine; Phenylenediamine, ortho-; PODA
CAS Registry Number: 95-54-5
RTECS®Number: SS7875000
UN/NA & ERG Number: UN1673/153
EC Number: 202-430-6 [*Annex I Index No.:* 612-145-00-2] *(o-)*

Regulatory Authority and Advisory Bodies
US EPA Gene-Tox Program, Negative: Rodent dominant lethal; *N. crassa—aneuploidy.*
RCRA 40CFR268.48; 61FR15654, Universal Treatment Standards: Wastewater (mg/L), 0.056; Nonwastewater (mg/kg), 5.6.
EPCRA Section 313 Form R *de minimis* concentration reporting level: 1.0%.
California Proposition 65 Chemical: Cancer 5/15/98.
European/International Regulations: Hazard Symbol: T, N; Risk phrases: R20/21; R25; R36; R40; R43; R50/53; R68; Safety phrases: S1/2; S28; S36/37; S45; S60; S61 (see Appendix 4).
WGK (German Aquatic Hazard Class): 3—Severe hazard to waters.

Description: *o*-Phenylenediamine is a white to brownish crystalline substance that turns red upon exposure to air. Molecular weight = 108.16; Boiling point = 257°C; Freezing/Melting point = 104°C; Flash point = 156°C. Hazard Identification (based on NFPA-704 M Rating System): Health 2, Flammability 1, Reactivity 0. Slightly soluble in water.

Potential Exposure: Used as an intermediate in the making of dyes; pesticides, pharmaceuticals, and rubber chemicals; in making fungicides and other chemicals; in photographic and analytical procedures and processes.

Incompatibilities: Reacts violently with strong oxidizers, strong acids, acid chlorides, acid anhydrides, chloroformates. Heat and light contribute to instability. Keep away from metals.

Permissible Exposure Limits in Air
ACGIH TLV®[1]: 0.1 mg/m³ TWA [skin]; confirmed animal carcinogen with unknown relevance to humans.
Protective Action Criteria (PAC)
TEEL-0: 0.1 mg/m³
PAC-1: 0.3 mg/m³
PAC-2: 200 mg/m³
PAC-3: 500 mg/m³
DFG MAK: [skin] Danger of skin sensitization; Carcinogen Category 3B.
Austria: carcinogen, 1999; Denmark: TWA 0.1 mg/m³, 1999; Norway: TWA 0.1 mg/m³, 1999; Switzerland: MAK-W 0.1 mg/m³, [skin], carcinogen, 1999.

Determination in Air: Use OSHA Analytical Method 87.
Routes of Entry: Inhalation, ingestion, skin and/or eye contact.
Harmful Effects and Symptoms
Short Term Exposure: Contact can irritate the eyes and skin. Eye contact may cause permanent damage. Irritates

the respiratory tract, causing coughing, wheezing, and/or shortness of breath. May affect the blood, causing the formation of methemoglobin and cyanosis with blue coloration of the skin and lips. Can cause stomach ache, headache, fatigue, dizziness, shaking, and convulsions. The *p*-isomer is more toxic and a more severe irritant than the *o*-isomers.

Long Term Exposure: Repeated or prolonged contact may cause skin sensitization and allergy. If allergy develops, very low future exposure can cause itching and a skin rash. Repeated or prolonged inhalation exposure may cause asthma-like allergy. May cause kidney and liver impairment. Exposure may cause anemia.

Points of Attack: Respiratory system, skin, lungs, liver, kidneys, blood.

Medical Surveillance: Consider the points of attack in pre-placement and periodic physical examinations. Complete blood count (CBC). Examination of the eyes and vision. Evaluation by a qualified allergist. Lung function tests. Blood methemoglobin levels. Liver and kidney function tests.

First Aid: If this chemical gets into the eyes, remove any contact lenses at once and irrigate immediately for at least 15 min, occasionally lifting upper and lower lids. Seek medical attention immediately. If this chemical contacts the skin, remove contaminated clothing and wash immediately with soap and water. Seek medical attention immediately. If this chemical has been inhaled, remove from exposure, begin rescue breathing (using universal precautions, including resuscitation mask) if breathing has stopped and CPR if heart action has stopped. Transfer promptly to a medical facility. When this chemical has been swallowed, get medical attention. Give large quantities of water and induce vomiting. Do not make an unconscious person vomit.

Note to physician: Treat for methemoglobinemia. Spectrophotometry may be required for precise determination of levels of methemoglobin in urine.

Personal Protective Methods: Wear protective gloves and clothing to prevent any reasonable probability of skin contact. Safety equipment suppliers/manufacturers can provide recommendations on the most protective glove/clothing material for your operation. All protective clothing (suits, gloves, footwear, headgear) should be clean, available each day, and put on before work. Contact lenses should not be worn when working with this chemical. Wear dust-proof chemical goggles and face shield unless full face-piece respiratory protection is worn. Employees should wash immediately with soap when skin is wet or contaminated. Provide emergency showers and eyewash.

Respirator Selection: Where there is potential for exposure to this chemical of more than 0.1 mg/m^3, use a NIOSH/MSHA- or European Standard EN149-approved supplied-air respirator with a full face-piece operated in the positive-pressure mode, or with a full face-piece, hood, or helmet in the continuous-flow mode; or use a NIOSH/MSHA- or European Standard EN149-approved self-contained breathing apparatus with a full face-piece operated in pressure-demand or other positive-pressure mode.

Storage: Color Code—Blue: Health Hazard/Poison: Store in a secure poison location. Prior to working with this chemical you should be trained on its proper handling and storage. Store in tightly closed containers in a cool, well-ventilated area away from strong oxidizers. Where possible, automatically transfer material from drums or other storage containers to process containers. Sources of ignition, such as smoking and open flames, are prohibited where this chemical is handled, used, or stored. Metal containers involving the transfer of this chemical should be grounded and bonded. Wherever this chemical is used, handled, manufactured, or stored, use explosion-proof electrical equipment and fittings.

Shipping: Phenylenediamines require a shipping label of "POISONOUS/TOXIC MATERIALS." They fall in DOT Hazard Class 6.1 and Packing Group III.

Spill Handling: Evacuate persons not wearing protective equipment from area of spill or leak until cleanup is complete. Remove all ignition sources. Collect powdered material in the most convenient and safe manner and deposit in sealed containers. Ventilate area after cleanup is complete. It may be necessary to contain and dispose of this chemical as a hazardous waste. If material or contaminated runoff enters waterways, notify downstream users of potentially contaminated waters. Contact your local or federal environmental protection agency for specific recommendations. If employees are required to clean up spills, they must be properly trained and equipped. OSHA 1910.120(q) may be applicable.

Fire Extinguishing: This chemical is a combustible solid. Use dry chemical, carbon dioxide, water spray, alcohol foam or polymer foam extinguishers. Poisonous gases are produced in fire, including nitrogen oxides. If material or contaminated runoff enters waterways, notify downstream users of potentially contaminated waters. Notify local health and fire officials and pollution control agencies. From a secure, explosion-proof location, use water spray to cool exposed containers. If cooling streams are ineffective (venting sound increases in volume and pitch, tank discolors, or shows any signs of deforming), withdraw immediately to a secure position. If employees are expected to fight fires, they must be trained and equipped in OSHA 1910.156. The only respirators recommended for firefighting are self-contained breathing apparatuses that have full face-pieces and are operated in a pressure-demand or other positive-pressure mode.

Disposal Method Suggested: Controlled incineration whereby oxides of nitrogen are removed from the effluent gas by scrubber, catalytic or thermal device.[22]

Reference

New Jersey Department of Health and Senior Services. (August 1999). *Hazardous Substances Fact Sheet: o-Phenylenediamine.* Trenton, NJ

p-Phenylenediamine   P:0400

Molecular Formula: $C_6H_8N_2$
Common Formula: $H_2NC_6H_4NH_2$
Synonyms: 4-Aminoaniline; *p*-Aminoaniline; BASF Ursol D; *p*-Benzenediamine; 1,4-Benzenediamine; Benzofur D; C.I. 76060; C.I. Developer 13; C.I. Oxidation base 10; Developer PF; *p*-Diaminobenzene; 1,4-Diaminobenzene; 1,4-Diaminobenzol; Durafur black R; 1,4-Fenilendiamina (Spanish); *p*-Fenilendiamina (Spanish); Fouramine D; Fourrine 1; Fourrine D; Fur black 41867; Fur brown 41866; Furro D; Fur yellow; Futramine D; Nako H; Orsin; Pelagol D; Pelagol grey D; Peltol D; 1,4-Phenylenediamine; Phenylene diamine, para-; PPD; Renal PF; Santoflex IC; Tertral D; Ursol D; Vulkanox 4020; Zoba black D
Hydrochloride:
1,4-Aminoaniline dihydrochloride; 1,4-Benzenediamine dihydrochloride; 1,4-Phenylenediame dihydrochloride
CAS Registry Number: 106-50-3; 624-18-0 (dihydrochloride)
RTECS® Number: SS8050000
UN/NA & ERG Number: UN1673/153
EC Number: 203-404-7 [*Annex I Index No.:* 612-028-00-6]; 210-834-9 [*Annex I Index No.:* 612-029-00-1] (*p*-phenylenediamine dihydrochloride)
Regulatory Authority and Advisory Bodies
Carcinogenicity: IARC: Animal Inadequate Evidence; Human No Adequate Data, *not classifiable as carcinogenic to humans*, Group 3, 1978.
US EPA Gene-Tox Program, Positive: Cell transform.—RLV F344 rat embryo; Positive: Cell transform.—SA7/SHE; *D. melanogaster* sex-linked lethal; Negative: Carcinogenicity—mouse/rat; Rodent dominant lethal; Negative: *N. crassa—aneuploidy*; Sperm morphology—mouse; Inconclusive: Mammalian micronucleus.
Banned or Severely Restricted (several countries) (UN).[13]
Air Pollutant Standard Set. See below, "Permissible Exposure Limits in Air" section.
Clean Air Act: Hazardous Air Pollutants (Title I, Part A, Section 112).
RCRA 40CFR264, Appendix 9; TSD Facilities Ground Water Monitoring List. Suggested test method(s) (PQL µg/L): 8270 (10).
Reportable Quantity (RQ): 1 lb (0.454 kg).
EPCRA Section 313 Form R *de minimis* concentration reporting level: 1.0%.
Canada, WHMIS, Ingredients Disclosure List Concentration 0.1%.
European/International Regulations (*includes dihydrochloride*): Hazard Symbol: T, N; Risk phrases: R23/24/25; R36; R43; R50/53; Safety phrases: S1/2; S28; S36/37; S45; S60; S61 (see Appendix 4).
WGK (German Aquatic Hazard Class): 3—Severe hazard to waters.

Description: *p*-Phenylenediamines are white to slightly red crystalline solids. They have been described as gray "light brown" which may result from exposure to air. Molecular weight = 108.16; Specific gravity (H_2O:1) = 1.1; Boiling point = 267°C (sublimes); Freezing/Melting point = 146°C; Vapor pressure = <1 mmHg at 25°C; Flash point = 156°C; Autoignition temperature = 400°C. Explosive limits: LEL = 1.5%; UEL—unknown. Hazard Identification (based on NFPA-704 M Rating System): Health 3, Flammability 1, Reactivity 0. Slightly soluble in water; solubility = 4% at 25°C.
Potential Exposure: Compound Description: Agricultural Chemical; Tumorigen, Mutagen, Human Data; Primary Irritant. Compound Description: Agricultural Chemical; Tumorigen, Mutagen, Human Data; Primary Irritant. *p*-Phenylenediamine has been used in dyestuff manufacture, in hair dyes, in photographic developers, in synthetic fibers, in polyurethanes, and as a monomer and in the manufacture of improved tire cords. Also used as a gasoline additive and in making antioxidants.
Incompatibilities: A strong reducing agent. Reacts violently with strong oxidizers, strong acids, acid chlorides, acid anhydrides, chloroformates, and strong bases. Incompatible with organic anhydrides, isocyanates, aldehydes. Heat and light contribute to instability. Keep away from metals.
Permissible Exposure Limits in Air
OSHA PEL: 0.1 mg/m³ TWA [skin].
NIOSH REL: 0.1 mg/m³ TWA [skin].
ACGIH TLV®[1]: 0.1 mg/m³ TWA; not classifiable as a human carcinogen.
NIOSH IDLH: 25 mg/m³.
Protective Action Criteria (PAC)
TEEL-0: 0.1 mg/m³
PAC-1: 6 mg/m³
PAC-2: 25 mg/m³
PAC-3: 25 mg/m³
Dihydrochloride
TEEL-0: 0.6 mg/m³
PAC-1: 1.5 mg/m³
PAC-2: 12.5 mg/m³
PAC-3: 60 mg/m³
DFG MAK: 0.1 mg/m³, measured as the inhalable fraction TWA; Peak Limitation Category II(2); [skin], danger of skin sensitization; Carcinogen Category 3; Pregnancy Risk Group C.
Australia: TWA 0.1 mg/m³, [skin], 1993; Austria: MAK 0.1 mg/m³, [skin], Suspected: carcinogen, 1999; Belgium: TWA 0.1 mg/m³, [skin], 1993; Denmark: TWA 0.1 mg/m³, [skin], 1999; Finland: TWA 0.1 mg/m³, STEL 0.3 mg/m³, [skin], 1999; France: VME 0.1 mg/m³, [skin], 1999; the Netherlands: MAC-TGG 0.1 mg/m³, [skin], 2003; the Philippines: TWA 0.1 mg/m³, [skin], 1993; Poland: MAC (TWA) 0.1 mg/m³; MAC (STEL) 0.3 mg/m³, 1999; Russia: STEL 0.05 mg/m³, 1993; Sweden: NGV 0.1 mg/m³, KTV 0.3 mg/m³, [skin], 1999; United Kingdom: TWA 0.1 mg/m³, 2000; Argentina, Bulgaria, Columbia, Jordan,

South Korea, New Zealand, Singapore, Vietnam: ACGIH TLV®: not classifiable as a human carcinogen. Several states have set guidelines or standards for *p*-phenylenediamine in ambient air[60] ranging from 0.33 μg/m^3 (New York) to 1.0 μg/m^3 (Florida, North Dakota, South Carolina) to 1.6 μg/m^3 (Virginia) to 2.0 μg/m^3 (Connecticut and Nevada).

Determination in Air: Use OSHA Analytical Method 87.

Permissible Concentration in Water: Russia[35, 43] set a MAC of 0.1 mg/L in surface water.

Routes of Entry: Inhalation, ingestion, skin and/or eye contact. Absorbed through the skin.

Harmful Effects and Symptoms

Short Term Exposure: Contact can severely irritate and burn the eyes and skin. May cause permanent eye damage. Irritates the respiratory tract, causing coughing, wheezing, and/or shortness of breath. Inhalation of dust may irritate the pharynx and larynx; bronchial asthmatic reactions. Swelling of mouth and throat may be observed following ingestion. Exposure can cause abdominal pain, nausea, high blood pressure, dizziness, seizures, and even coma. May affect the blood, causing the formation of methemoglobin and cyanosis with blue coloration of the skin and lips, headache, fatigue, dizziness. High levels can cause troubled breathing, collapse, and death. The *p*-isomer is more toxic and a more severe irritant than the *o*- and *m*-isomers.

Long Term Exposure: Repeated or prolonged contact may cause skin sensitization and allergy. If allergy develops, very low future exposure can cause itching and a skin rash. Repeated or prolonged inhalation exposure may cause asthma-like allergy. May cause kidney and liver impairment. Repeated high exposure can cause cataracts. Exposure may cause anemia.

Points of Attack: Respiratory system, skin, lungs, liver, kidneys, blood.

Medical Surveillance: Consider the points of attack in preplacement and periodic physical examinations. Complete blood count (CBC). Examination of the eyes and vision. Evaluation by a qualified allergist. Lung function tests. Blood methemoglobin levels. Liver and kidney function tests.

First Aid: If this chemical gets into the eyes, remove any contact lenses at once and irrigate immediately for at least 15 min, occasionally lifting upper and lower lids. Seek medical attention immediately. If this chemical contacts the skin, remove contaminated clothing and wash immediately with soap and water. Seek medical attention immediately. If this chemical has been inhaled, remove from exposure, begin rescue breathing (using universal precautions, including resuscitation mask) if breathing has stopped and CPR if heart action has stopped. Transfer promptly to a medical facility. When this chemical has been swallowed, get medical attention. Give large quantities of water or a slurry of activated charcoal in water; and induce vomiting. Do not make an unconscious person vomit.

Note to physician: Treat for methemoglobinemia. Spectrophotometry may be required for precise determination of levels of methemoglobin in urine.

Personal Protective Methods: Wear protective gloves and clothing to prevent any reasonable probability of skin contact. Safety equipment suppliers/manufacturers can provide recommendations on the most protective glove/clothing material for your operation. All protective clothing (suits, gloves, footwear, headgear) should be clean, available each day, and put on before work. Contact lenses should not be worn when working with this chemical. Wear dust-proof chemical goggles and face shield unless full face-piece respiratory protection is worn. Employees should wash immediately with soap when skin is wet or contaminated. Provide emergency showers and eyewash.

Respirator Selection: *Up to 2.5 mg/m^3:* Sa:Cf (APF = 25) (any supplied-air respirator operated in a continuous-flow mode). *Up to 5 mg/m^3:* SCBAF (APF = 50) (any self-contained breathing apparatus with a full face-piece) or SaF (APF = 50) (any supplied-air respirator with a full face-piece). *Up to 25 mg/m^3:* SaF: Pd,Pp (APF = 2000) (any supplied-air respirator that has a full face-piece and is operated in a pressure-demand or other positive-pressure mode). *Emergency or planned entry into unknown concentrations or IDLH conditions:* SCBAF: Pd,Pp (APF = 10,000) (any self-contained breathing apparatus that has a full face-piece and is operated in a pressure-demand or other positive-pressure mode) or SaF: Pd,Pp: ASCBA (APF = 10,000) (any supplied-air respirator that has a full face-piece and is operated in a pressure-demand or other positive-pressure mode in combination with an auxiliary self-contained breathing apparatus operated in a pressure-demand or other positive-pressure mode). *Escape:* GmFS100 (APF = 50) [any air-purifying, full-face-piece respirator (gas mask) with a chin-style, front- or back-mounted canister providing protection against the compound of concern and having an N100, R100, or P100 filter] or SCBAE (any appropriate escape-type, self-contained breathing apparatus).

Note: Causes eye irritation and damage; eye protection needed.

Storage: (1) Color Code—Yellow Stripe (*strong reducing agent*): Reactivity Hazard; Store separately in an area isolated from flammables, combustibles, or other yellow-coded materials. Prior to working with this chemical you should be trained on its proper handling and storage. Before entering confined space where this chemical may be present, check to make sure that an explosive concentration is not a danger. Store in tightly closed containers in a cool, well-ventilated area away from oxidizers, strong acids, acid chlorides, acid anhydrides, chloroformates, and metals. Where possible, automatically transfer material from drums or other storage containers to process containers.

Shipping: Phenylenediamines require a shipping label of "POISONOUS/TOXIC MATERIALS." They fall in DOT Hazard Class 6.1 and Packing Group III.

Spill Handling: Evacuate persons not wearing protective equipment from area of spill or leak until cleanup is complete. Remove all ignition sources. Collect powdered material in the most convenient and safe manner and deposit in

sealed containers. Establish forced ventilation to keep levels below explosive limit. It may be necessary to contain and dispose of this chemical as a hazardous waste. If material or contaminated runoff enters waterways, notify downstream users of potentially contaminated waters. Contact your local or federal environmental protection agency for specific recommendations. If employees are required to clean up spills, they must be properly trained and equipped. OSHA 1910.120(q) may be applicable.

Fire Extinguishing: This chemical is a combustible solid. Use dry chemical, carbon dioxide, water spray, alcohol foam or polymer foam extinguishers. Poisonous gases, including nitrogen oxides, are produced in fire. If material or contaminated runoff enters waterways, notify downstream users of potentially contaminated waters. Notify local health and fire officials and pollution control agencies. From a secure, explosion-proof location, use water spray to cool exposed containers. If cooling streams are ineffective (venting sound increases in volume and pitch, tank discolors, or shows any signs of deforming), withdraw immediately to a secure position. If employees are expected to fight fires, they must be trained and equipped in OSHA 1910.156. The only respirators recommended for firefighting are self-contained breathing apparatuses that have full face-pieces and are operated in a pressure-demand or other positive-pressure mode.

Disposal Method Suggested: Controlled incineration whereby oxides of nitrogen are removed from the effluent gas by scrubber, catalytic or thermal device.[22]

References

US Environmental Protection Agency. (June 1, 1978). *Chemical Hazard Information Profile: Phenylenediamines.* Washington, DC

New Jersey Department of Health and Senior Services. (September, 1999). *Hazardous Substances Fact Sheet: p-Phenylenediamine.* Trenton, NJ

Phenyl glycidyl ether   P:0410

Molecular Formula: $C_9H_{10}O_2$
Synonyms: 1,2-Epoxy-3-phenoxypropane; 2,3-Epoxypropylphenyl ether; Glycidyl phenyl ether; PGE; Phenol-glycidaether (German); Phenol glycidyl ether; 3-Phenoxy-1,2-epoxypropane; Phenoxypropene oxide; Phenoxypropylene oxide; Phenyl-2,3-epoxypropyl ether
CAS Registry Number: 122-60-1
RTECS®Number: TZ3675000
UN/NA & ERG Number: UN2810/153
EC Number: 204-557-2 [*Annex I Index No.:* 603-067-00-X]
Regulatory Authority and Advisory Bodies
Carcinogenicity: IARC: Animal Sufficient Evidence; Human No Adequate Data, *possibly carcinogenic to humans,* Group 2B, 1999; NIOSH: Potential occupational carcinogen.

US EPA Gene-Tox Program, Positive: SHE—focus assay.
Air Pollutant Standard Set. See below, "Permissible Exposure Limits in Air" section.
California Proposition 65 Chemical: Cancer 10/1/90; male 8/7/09.
Canada, WHMIS, Ingredients Disclosure List Concentration 0.1%.
European/International Regulations: Hazard Symbol: T, N; Risk phrases: R45; R20; R37/38; R43; R68; R52/53; Safety phrases: S53; S45; S61 (see Appendix 4).
WGK (German Aquatic Hazard Class): 3—Severe hazard to waters.

Description: Phenyl glycidyl ether is a colorless liquid with an unpleasant sweet odor. Molecular weight = 150.19; Specific gravity (H_2O:1) = 1.11; Boiling point = 245°C; Freezing/Melting point = 3.3°C; Vapor pressure = 0.01 mmHg at 20°C; Flash point = 120°C. Explosive limits: LEL = 1.1%; UEL—unknown. Hazard Identification (based on NFPA-704 M Rating System): Health 2, Flammability 1, Reactivity 0. Slightly soluble in water; solubility = 0.24% at 20°C.

Potential Exposure: Compound Description: Tumorigen, Mutagen; Reproductive Effector; Primary Irritant. PGE is used to increase storage time and stability of halogenated compounds; as a reactive diluent in uncured epoxy resins to reduce the viscosity of the uncured system for ease in casting, adhesive, and laminating applications. NIOSH once estimated that 8000 workers are potentially exposed to PGE.

Incompatibilities: Strong oxidizers, amines, strong acids, strong bases, and curing agents. PGE can presumably form explosive peroxides.

Permissible Exposure Limits in Air
Conversion factor: 1 ppm = 6.14 mg/m^3 at 25°C & 1 atm.
OSHA PEL: 10 ppm/60 mg/m^3 TWA.
NIOSH REL: 1 ppm/6 mg/m^3 [15 min] Ceiling Concentration; A potential occupational carcinogen [skin]; Limit exposure to lowest feasible concentration. See *NIOSH Pocket Guide*, Appendix A.
ACGIH TLV®[1]: 0.1 ppm TWA [skin] danger of skin sensitization; confirmed animal carcinogen with unknown relevance to humans.
No TEEL available.
DFG MAK: [skin], danger of skin sensitization; Carcinogen Category 2.
NIOSH IDLH: potential occupational carcinogen 100 ppm.
Australia: TWA 1 ppm (6 mg/m^3), 1993; Austria: [skin], carcinogen, 1999; Belgium: TWA 1 ppm (6.1 mg/m^3), 1993; Denmark: TWA 1 ppm (5 mg/m^3), 1999; Finland; STEL 10 ppm (60 mg/m^3), [skin], 1999; France: VME 1 ppm (6 mg/m^3), 1999; Norway: TWA 1 ppm (5 mg/m^3), 1999; the Philippines: TWA 10 ppm (62 mg/m^3), 1993; Poland: MAC (TWA) 0.6 mg/m^3; MAC (STEL) 3 mg/m^3, 1999; Sweden: NGV 10 ppm (60 mg/m^3), KTV 15 ppm (90 mg/m^3), 1999; Switzerland: MAK-W 1 ppm (6 mg/m^3), [skin], carcinogen, 1999; Turkey: TWA 10 ppm (60 mg/m^3), 1993; United

Kingdom: TWA 1 ppm (6.2 mg/m^3), 2000; Argentina, Bulgaria, Columbia, Jordan, South Korea, New Zealand, Singapore, Vietnam: ACGIH TLV®: confirmed animal carcinogen with unknown relevance to humans. Several states have set guidelines or standards for PGE in ambient air[60] ranging from 60 μg/m^3 (Connecticut and North Dakota) to 100 μg/m^3 (Virginia) to 143 μg/m^3 (Nevada).

Determination in Air: Use NIOSH Analytical Method (IV), Phenyl glycidyl ether.

Determination in Water: Octanol–water coefficient: Log $K_{ow} = 1.1$.

Routes of Entry: Inhalation, ingestion, skin and/or eye contact.

Harmful Effects and Symptoms

Short Term Exposure: The substance irritates the eyes, skin, and upper respiratory tract. Eye and skin irritation may be severe. Exposure could cause lowering of consciousness, with headache, loss of concentration, dizziness, and unconsciousness.

Long Term Exposure: Repeated or prolonged contact may cause skin sensitization and allergy. Possible hematopoietic, reproductive effects. A potential occupational carcinogen.

Points of Attack: Eyes, skin, central nervous system, hematopoietic system, reproductive system. The liquid destroys the skin's natural oils, causing dermatitis. Cancer site in animals: nasal cavity.

Medical Surveillance: NIOSH lists the following tests: pulmonary function tests.

First Aid: If this chemical gets into the eyes, remove any contact lenses at once and irrigate immediately for at least 15 min, occasionally lifting upper and lower lids. Seek medical attention immediately. If this chemical contacts the skin, remove contaminated clothing and wash immediately with soap and water. Seek medical attention immediately. If this chemical has been inhaled, remove from exposure, begin rescue breathing (using universal precautions, including resuscitation mask) if breathing has stopped and CPR if heart action has stopped. Transfer promptly to a medical facility. When this chemical has been swallowed, get medical attention. Give large quantities of water and induce vomiting. Do not make an unconscious person vomit.

Personal Protective Methods: Wear protective gloves and clothing to prevent any reasonable probability of skin contact. Safety equipment suppliers/manufacturers can provide recommendations on the most protective glove/clothing material for your operation. Butyl rubber, natural rubber, and polyvinyl alcohol are among the recommended protective materials. All protective clothing (suits, gloves, footwear, headgear) should be clean, available each day, and put on before work. Contact lenses should not be worn when working with this chemical. Wear splash-proof chemical goggles and face shield unless full face-piece respiratory protection is worn. Employees should wash immediately with soap when skin is wet or contaminated. Provide emergency showers and eyewash.

Respirator Selection: At any detectable concentration: SCBAF: Pd,Pp (APF = 10,000) (any NIOSH/MSHA- or European Standard EN 149-approved self-contained breathing apparatus that has a full face-piece and is operated in a pressure-demand or other positive-pressure mode) or SaF: Pd,Pp: ASCBA (APF = 10,000) (any supplied-air respirator that has a full face-piece and is operated in a pressure-demand or other positive-pressure mode in combination with an auxiliary, self-contained breathing apparatus operated in a pressure-demand or other positive-pressure mode). *Escape:* GmFOv (APF = 50) [any air-purifying, full-face-piece respirator (gas mask) with a chin-style, front- or back-mounted organic vapor canister] or SCBAE (any appropriate escape-type, self-contained breathing apparatus).

Storage: Color Code—Blue: Health Hazard/Poison: Store in a secure poison location. Prior to working with this chemical you should be trained on its proper handling and storage. Before entering confined space where this chemical may be present, check to make sure that an explosive concentration is not a danger. Store in a refrigerator or in a cool, dry place. Protect from exposure to acids, bases, oxidizers, and curing agents. Where possible, automatically pump liquid from drums or other storage containers to process containers. Sources of ignition, such as smoking and open flames, are prohibited where this chemical is handled, used, or stored. Metal containers involving the transfer of 5 gallons or more of this chemical should be grounded and bonded. Drums must be equipped with self-closing valves, pressure vacuum bungs, and flame arresters. Use only nonsparking tools and equipment, especially when opening and closing containers of this chemical. Wherever this chemical is used, handled, manufactured, or stored, use explosion-proof electrical equipment and fittings. A regulated, marked area should be established where this chemical is handled, used, or stored in compliance with OSHA Standard 1910.1045.

Shipping: Toxic, liquids, organic, n.o.s. require a shipping label of "POISONOUS/TOXIC MATERIALS." *N*-Nitrosodimethylamine falls in DOT Hazard Class 6.1 and Packing Group III.

Spill Handling: Evacuate and restrict persons not wearing protective equipment from area of spill or leak until cleanup is complete. Remove all ignition sources. Establish forced ventilation to keep levels below explosive limit. Absorb liquids in vermiculite, dry sand, earth, peat, carbon, or a similar material and deposit in sealed containers. Follow by washing surfaces well, first with 60–70% ethanol; then with soap and with 60–70% ethanol; then with soap and water. Keep this chemical out of a confined space, such as a sewer, because of the possibility of an explosion, unless the sewer is designed to prevent the buildup of explosive concentrations. It may be necessary to contain and dispose of this chemical as a hazardous waste. If material or contaminated runoff enters waterways, notify downstream users of potentially contaminated waters. Contact your local or federal environmental protection agency for specific recommendations. If employees are required to clean up spills, they must be properly trained and equipped. OSHA 1910.120(q) may be applicable.

Fire Extinguishing: This chemical is a combustible liquid. Poisonous gases are produced in fire. Use dry chemical, carbon dioxide, or alcohol foam extinguishers. Vapors are heavier than air and will collect in low areas. Vapors may travel long distances to ignition sources and flashback. Vapors in confined areas may explode when exposed to fire. Containers may explode in fire. Storage containers and parts of containers may rocket great distances, in many directions. If material or contaminated runoff enters waterways, notify downstream users of potentially contaminated waters. Notify local health and fire officials and pollution control agencies. From a secure, explosion-proof location, use water spray to cool exposed containers. If cooling streams are ineffective (venting sound increases in volume and pitch, tank discolors, or shows any signs of deforming), withdraw immediately to a secure position. If employees are expected to fight fires, they must be trained and equipped in OSHA 1910.156. The only respirators recommended for firefighting are self-contained breathing apparatuses that have full face-pieces and are operated in a pressure-demand or other positive-pressure mode.

Disposal Method Suggested: Concentrated waste containing no peroxides—discharge liquid at a controlled rate near a pilot flame. Concentrated waste containing peroxides—perforation of a container of the waste from a safe distance followed by open burning.

References

US Environmental Protection Agency. (1979). *Chemical Hazard Information Profile: Phenyl Glycidyl Ether.* Washington, DC

National Institute for Occupational Safety and Health. (1978). *Criteria for a Recommended Standard: Occupational Exposure to Glycidyl Ethers,* NIOSH Document No. 78-166. Washington, DC

National Institute for Occupational Safety and Health. (October 1977). *Information Profiles on Potential Occupational Hazards: Glycidyl Ethers,* Report PB-276,678. Rockville, MD, pp. 116—123

Phenylhydrazine   P:0420

Molecular Formula: $C_6H_8N_2$
Common Formula: $C_6H_5NHNH_2$
Synonyms: Fenilhidrazina (Spanish); Hydrazine-benzene; Hydrazinobenzene; Monophenylhydrazine
Hydrochloride: Cloruro de fenilhidrazinio (Spanish); Phenylhydrazine monohydrochloride; Phenylhydrazin hydrochlorid (German); Phenylhydrazinium chloride
CAS Registry Number: 100-63-0; 59-88-1 (hydrochloride)
RTECS® Number: MV8925000; MV9000000 (hydrochloride)
UN/NA & ERG Number: UN2572/53

EC Number: 202-873-5 [*Annex I Index No.:* 612-023-00-9]; 200-444-7 (phenylhydrazinium chloride) [*Annex I Index No.:* 612-023-00-9]

Regulatory Authority and Advisory Bodies
Carcinogenicity: NIOSH: Potential occupational carcinogen.
Air Pollutant Standard Set. See below, "Permissible Exposure Limits in Air" section.
California Proposition 65 Chemical: (*Phenylhydrazine and its salts*) Cancer 7/1/92.
Canada, WHMIS, Ingredients Disclosure List Concentration 0.1%.
Hydrochloride:
Superfund/EPCRA 40CFR355, Extremely Hazardous Substances: TPQ = 1000/10,000 lb (454/4540 kg) (hydrochloride).
Reportable Quantity (RQ): 1000 lb (454 kg) (hydrochloride).
Canada, WHMIS, Ingredients Disclosure List Concentration 0.1%.
European/International Regulations: Hazard Symbol (*includes hydrochloride*): T, N; Risk phrases: R45; R23/24/25; R36/38; R43; R48/23/24/25; R68; R50; Safety phrases: S53; S45; S61 (see Appendix 4).
WGK (German Aquatic Hazard Class): 3—Severe hazard to waters.

Description: Phenylhydrazine is a colorless to pale yellow liquid or solid with a weak aromatic odor. The hydrochloride is a white to tan solid with a weak odor. Molecular weight = 108.16; Specific gravity (H_2O:1) = 1.10 (base); Boiling point = (decomposes) 243.3°C (base); Freezing/Melting point = 19.4°C (base); 24°C (hemihydrate); 243—246°C (hydrochloride); Vapor pressure = 0.04 mmHg at 25°C; Flash point = 88°C (cc) (base); Autoignition temperature = 174°C (base). Explosive limits: LEL = 1.1%; UEL—unknown. Hazard Identification (based on NFPA-704 M Rating System): Health 3, Flammability 2, Reactivity 2; (*hydrochloride*) Hazard Identification (based on NFPA-704 M Rating System): Health 2, Flammability 0, Reactivity 0. Slightly soluble in water.

Potential Exposure: Compound Description: Tumorigen, Mutagen; Reproductive Effector. Phenylhydrazine is a widely used reagent in conjunction with sugars, aldehydes, and ketones. In addition, it is used in the synthesis of dyes; pharmaceuticals, such as antipyrin, cryogenin, and pyramidone; and other organic chemicals. The hydrochloride salt is used in the treatment of polycythemia vera.

Incompatibilities: Phenylhydrazine is very reactive with carbonyl compounds, strong oxidizers, strong bases, alkali metals, ammonia, lead dioxide (violent). Attacks copper salts, nickel, and chromates.

Permissible Exposure Limits in Air
Conversion factor: 1 ppm = 4.42 mg/m^3 at 25°C & 1 atm.
OSHA PEL: 5 ppm/22 mg/m^3 TWA [skin].
NIOSH REL: 0.14 ppm/0.6 mg/m^3 [120 min] Ceiling Concentration [skin]; A potential occupational carcinogen

[skin]; Limit exposure to lowest feasible concentration. See *NIOSH Pocket Guide*, Appendix A.

ACGIH TLV®[1]: 0.1 ppm/0.44 mg/m³ TWA [skin]; confirmed animal carcinogen with unknown relevance to humans.

NIOSH IDLH: 15 ppm (potential occupational carcinogen).

Protective Action Criteria (PAC)

Phenylhydrazine

TEEL-0: 0.1 ppm

PAC-1: 0.3 ppm

PAC-2: 2 ppm

PAC-3: 15 ppm

Phenylhydrazine hydrochloride

TEEL-0: 50 mg/m³

PAC-1: 150 mg/m³

PAC-2: 250 mg/m³

PAC-3: 250 mg/m³

DFG MAK: [skin], danger of skin sensitization; Carcinogen Category 3B.

Arab Republic of Egypt: TWA 5 ppm (20 mg/m³), [skin], 1993; Australia: TWA 5 ppm (20 mg/m³); STEL 10 ppm, [skin], carcinogen, 1993; Austria: MAK 5 ppm (22 mg/m³), [skin], Suspected: carcinogen, 1999; Belgium: TWA 5 ppm (22 mg/m³); STEL 10 ppm, [skin], carcinogen, 1993; Denmark: TWA 0.1 ppm (0.6 mg/m³), [skin], 1999; Finland: STEL 5 ppm (22 mg/m³), [skin], 1999; Norway: TWA 0.6 mg/m³, 1999; the Philippines: TWA 5 ppm (22 mg/m³), 1993; Poland: MAC (TWA) 20 mg/m³, 1999; Switzerland: MAK-W 5 ppm (22 mg/m³), [skin], 1999; Turkey: TWA 5 ppm (22 mg/m³), [skin], 1993; United Kingdom: CHAN, 2000; Argentina, Bulgaria, Columbia, Jordan, South Korea, New Zealand, Singapore, Vietnam: ACGIH TLV®: confirmed animal carcinogen with unknown relevance to humans. The Czech Republic[35]: MAC 1.0 mg/m³. Several states have set guidelines or standards for phenylhydrazine in ambient air[60] ranging from zero (North Dakota) to 66.7 μg/m³ (New York) to 200.0 μg/m³ (Connecticut, Florida, South Carolina, and Virginia) to 476.0 μg/m³ (Nevada).

Determination in Air: Use NIOSH Analytical Method (IV) #3518.

Permissible Concentration in Water: Russia[35, 43] set a MAC of 0.01 mg/L of phenylhydrazine in water bodies used for domestic purposes.

Determination in Water: Octanol−water coefficient: Log K_{ow} = 1.25.

Routes of Entry: Inhalation, ingestion, skin and/or eye contact. Absorbed through the skin.

Harmful Effects and Symptoms

Short Term Exposure: The dust and fumes can irritate and burn the eyes and skin. Inhalation can cause irritation, coughing, and difficult breathing. This material is poisonous if swallowed or if fumes are inhaled. Phenylhydrazine is a chronic poison. High levels can cause cyanosis and methemoglobinemia. Higher levels can cause troubled breathing, collapse, and even death. Exposure can cause headache, nausea, vomiting, lightheadedness, nervousness, shaking, seizures, and coma.

Long Term Exposure: Repeated or prolonged contact may cause skin irritation, dermatitis, sensitization, and allergy. May affect the blood, causing red cell damage, cyanosis, hemolytic anemia, kidney and liver damage, vascular thrombosis. A potential occupational carcinogen; it may cause leukemia.

Points of Attack: Blood, respiratory system, liver, kidneys, skin. Cancer site in animals: lungs, liver, blood vessels, and intestine. May affect the bone marrow and cause leukemia.

Medical Surveillance: NIOSH lists the following tests: blood plasma, hemoglobin; complete blood count; liver function tests; pulmonary function tests; urine (chemical/metabolite), hemoglobin; urinalysis (routine); white blood cell count/differential. Evaluation by a qualified allergist.

First Aid: If this chemical gets into the eyes, remove any contact lenses at once and irrigate immediately for at least 15 min, occasionally lifting upper and lower lids. Seek medical attention immediately. If this chemical contacts the skin, remove contaminated clothing and wash immediately with soap and water. Seek medical attention immediately. If this chemical has been inhaled, remove from exposure, begin rescue breathing (using universal precautions, including resuscitation mask) if breathing has stopped and CPR if heart action has stopped. Transfer promptly to a medical facility. When this chemical has been swallowed, get medical attention. Give large quantities of water and induce vomiting. Do not make an unconscious person vomit.

Note to physician: Treat for methemoglobinemia. Spectrophotometry may be required for precise determination of levels of methemoglobin in urine.

Personal Protective Methods: Wear protective gloves and clothing to prevent any reasonable probability of skin contact. Safety equipment suppliers/manufacturers can provide recommendations on the most protective glove/clothing material for your operation. All protective clothing (suits, gloves, footwear, headgear) should be clean, available each day, and put on before work. Contact lenses should not be worn when working with this chemical. Wear splash- or dust-proof chemical goggles and face shield unless full face-piece respiratory protection is worn. Employees should wash immediately with soap when skin is wet or contaminated. Provide emergency showers and eyewash.

Respirator Selection: *At any detectable concentration:* SCBAF: Pd,Pp (APF = 10,000) (any NIOSH/MSHA- or European Standard EN 149-approved self-contained breathing apparatus that has a full face-piece and is operated in a pressure-demand or other positive-pressure mode) or SaF: Pd,Pp: ASCBA (APF = 10,000) (any supplied-air respirator that has a full face-piece and is operated in a pressure-demand or other positive-pressure mode in combination with an auxiliary, self-contained breathing apparatus operated in a pressure-demand or other positive-pressure mode).

Escape: SCBAE (any appropriate escape-type, self-contained breathing apparatus).

Storage: (1) Color Code—Red Stripe (*100-63-0*): Flammability Hazard: Do not store in the same area as other flammable materials. (2) Color Code—Blue (*hydrochloride, 59-88-1*): Health Hazard: Store in a secure poison location. Prior to working with this chemical you should be trained on its proper handling and storage. Store in a refrigerator under an inert atmosphere and protect from exposure to light, strong bases, ammonia, oxidizers, metal salts. Where possible, automatically pump liquid from drums or other storage containers to process containers. Sources of ignition, such as smoking and open flames, are prohibited where this chemical is handled, used, or stored. Metal containers involving the transfer of 5 gallons or more of this chemical should be grounded and bonded. Drums must be equipped with self-closing valves, pressure vacuum bungs, and flame arresters. Use only nonsparking tools and equipment, especially when opening and closing containers of this chemical. Wherever this chemical is used, handled, manufactured, or stored, use explosion-proof electrical equipment and fittings. A regulated, marked area should be established where this chemical is handled, used, or stored in compliance with OSHA Standard 1910.1045.

Shipping: Phenylhydrazine requires a shipping label of "POISONOUS/TOXIC MATERIALS." It falls in Hazard Class 6.1 and Packing Group II.

Spill Handling: Evacuate and restrict persons not wearing protective equipment from area of spill or leak until cleanup is complete. Remove all ignition sources. Ventilate area of spill or leak. Avoid contact with solid and dust. Restrict access. Disperse and flush. Keep unnecessary people away; isolate hazard area and deny entry. Stay upwind; keep out of low areas. Do not touch spilled material; stop leak if you can do it without risk. *Small liquid spills:* take up with sand or other noncombustible absorbent material and place into containers for later disposal. *Small dry spills:* collect powdered material in the most convenient and safest manner and deposit in sealed containers; move container from spill area. *Large spills:* dike far ahead of spill for later disposal. Keep this chemical out of a confined space, such as a sewer, because of the possibility of an explosion, unless the sewer is designed to prevent the buildup of explosive concentrations. It may be necessary to contain and dispose of this chemical as a hazardous waste. If material or contaminated runoff enters waterways, notify downstream users of potentially contaminated waters. Contact your local or federal environmental protection agency for specific recommendations. If employees are required to clean up spills, they must be properly trained and equipped. OSHA 1910.120(q) may be applicable.

Fire Extinguishing: This chemical is a combustible liquid or solid, but does not easily ignite. Poisonous gases, including nitrogen oxides and hydrogen chloride, are produced in fire. Use dry chemical, carbon dioxide, alcohol foam, or polymer foam extinguishers. Vapors are heavier than air and will collect in low areas. Vapors in confined areas may explode when exposed to fire. Containers may explode in fire. Storage containers and parts of containers may rocket great distances, in many directions. If material or contaminated runoff enters waterways, notify downstream users of potentially contaminated waters. Notify local health and fire officials and pollution control agencies. From a secure, explosion-proof location, use water spray to cool exposed containers. If cooling streams are ineffective (venting sound increases in volume and pitch, tank discolors, or shows any signs of deforming), withdraw immediately to a secure position. If employees are expected to fight fires, they must be trained and equipped in OSHA 1910.156. The only respirators recommended for firefighting are self-contained breathing apparatuses that have full face-pieces and are operated in a pressure-demand or other positive-pressure mode.

Disposal Method Suggested: Controlled incineration whereby oxides of nitrogen are removed from the effluent gas by scrubber, catalytic or thermal device.

References

National Institute for Occupational Safety and Health. (1978). *Criteria for a Recommended Standard: Occupational Exposure to Hydrazines*, NIOSH Document No. 78-172. Washington, DC

US Environmental Protection Agency. (November 30, 1987). *Chemical Hazard Information Profile: Phenylhydrazine Hydrochloride*. Washington, DC: Chemical Emergency Preparedness Program

New Jersey Department of Health and Senior Services. (September 1999). *Hazardous Substances Fact Sheet: Phenylhydrazine Hydrochloride*. Trenton, NJ

Phenyl isocyanate   P:0430

Molecular Formula: C_7H_5NO
Common Formula: C_6H_5NCO
Synonyms: Carbanil; Fenylisokyanat; Isocyanic acid, Phenyl ester; Karbanil; Mondur P; Phenylcarbimide; Phenyl carbonimide
CAS Registry Number: 103-71-9
RTECS®Number: DA3675000
UN/NA & ERG Number: UN2487/155
EC Number: 203-137-6
Regulatory Authority and Advisory Bodies
Air Pollutant Standard Set. See below, "Permissible Exposure Limits in Air" section.
US DOT 49CFR172.101, Inhalation Hazardous Chemical.
European/International Regulations: Hazard Symbol: T; Risk phrases: R10; R22; R26; R34; R37; R42/43; Safety phrases: S/23; S/26; S28; S36/37/39; S38; S45; S61 (see Appendix 4).
WGK (German Aquatic Hazard Class): 2—Hazard to waters.
Description: Phenyl isocyanate is a colorless liquid with an irritating odor. Molecular weight = 119.30; Boiling point = 160°C; Freezing/Melting point = −30°C; Flash point = 56°C. Hazard Identification (based on NFPA-704 M

Rating System): Health 1, Flammability 2, Reactivity 0. Decomposes in water.

Potential Exposure: Compound Description: Mutagen. Phenyl isocyanate is used as a laboratory reagent and in organic synthesis.

Incompatibilities: Forms explosive mixture with air. Violent reaction with strong oxidizers. Isocyanates are incompatible with acids, caustics, ammonia, amines, amides, alcohols, glycols, caprolactum solution, water.

Permissible Exposure Limits in Air
Protective Action Criteria (PAC)*
TEEL-0: 0.006 mg/m^3
PAC-1: **0.020** mg/m^3
PAC-2: **0.15** mg/m^3
PAC-3: **0.24** mg/m^3
*AEGLs (Acute Emergency Guideline Levels) & ERPGs (Emergency Response Planning Guideline) are in **bold face**.
DFG MAK: Danger of skin and airway sensitization.
Russia[43] set a MAC of 0.5 mg/m^3 in work-place air.

Routes of Entry: Inhalation, ingestion, skin and/or eye contact. Absorbed through the skin.

Harmful Effects and Symptoms

Short Term Exposure: Phenyl isocyanate can affect you when breathed in and by passing through your skin. Exposure can strongly irritate the skin, nose, throat, and lungs. Higher levels may cause a buildup of fluid in the lungs (pulmonary edema). This can cause death. LD_{50} = (oral-rat) 800 mg/kg.

Long Term Exposure: Phenyl isocyanate can cause an asthma-like lung allergy to develop, with cough, shortness of breath, and wheezing. It can also cause an allergic skin rash.

Points of Attack: Lungs, skin.

Medical Surveillance: For those with frequent or potentially high exposure, the following are recommended before beginning work and at regular times after that: lung function tests. These may be normal if the person is not having an attack at the time of the test. If symptoms develop or overexposure is suspected, the following may be useful: evaluation by a qualified allergist, including careful exposure history and special testing, may help diagnose skin allergy. Consider chest X-ray after acute overexposure.

First Aid: If this chemical gets into the eyes, remove any contact lenses at once and irrigate immediately for at least 15 min, occasionally lifting upper and lower lids. Seek medical attention immediately. If this chemical contacts the skin, remove contaminated clothing and wash immediately with soap and water. Seek medical attention immediately. If this chemical has been inhaled, remove from exposure, begin rescue breathing (using universal precautions, including resuscitation mask) if breathing has stopped and CPR if heart action has stopped. Transfer promptly to a medical facility. When this chemical has been swallowed, get medical attention. Give large quantities of water and induce vomiting. Do not make an unconscious person vomit. Medical observation is recommended for 24—48 h after breathing overexposure, as pulmonary edema may be delayed. As first aid for pulmonary edema, a doctor or authorized paramedic may consider administering a corticosteroid spray.

Personal Protective Methods: Wear protective gloves and clothing to prevent any reasonable probability of skin contact. Safety equipment suppliers/manufacturers can provide recommendations on the most protective glove/clothing material for your operation. All protective clothing (suits, gloves, footwear, headgear) should be clean, available each day, and put on before work. Contact lenses should not be worn when working with this chemical. Wear splash-proof chemical goggles and face shield unless full face-piece respiratory protection is worn. Employees should wash immediately with soap when skin is wet or contaminated. Provide emergency showers and eyewash.

Respirator Selection: Where there is potential for exposures to phenyl isocyanate, as a NIOSH/MSHA- or European Standard EN149-approved supplied-air respirator with a full face-piece operated in the positive-pressure mode, or with a full face-piece, hood, or helmet in the continuous-flow mode; or use a NIOSH/MSHA- or European Standard EN149-approved self-contained breathing apparatus with a full face-piece operated in pressure-demand or other positive-pressure mode.

Storage: Color Code—Blue: Health Hazard/Poison: Store in a secure poison location. Prior to working with this chemical you should be trained on its proper handling and storage. Store in tightly closed containers in a cool, well-ventilated area. Where possible, automatically pump liquid from drums or other storage containers to process containers. Sources of ignition, such as smoking and open flames, are prohibited where this chemical is handled, used, or stored. Metal containers involving the transfer of 5 gallons or more of this chemical should be grounded and bonded. Drums must be equipped with self-closing valves, pressure vacuum bungs, and flame arresters. Use only nonsparking tools and equipment, especially when opening and closing containers of this chemical. Wherever this chemical is used, handled, manufactured, or stored, use explosion-proof electrical equipment and fittings.

Shipping: Phenyl isocyanate requires a shipping label of "POISONOUS/TOXIC MATERIALS, FLAMMABLE LIQUID." It falls in Hazard Class 6.1 and Packing Group I. A plus sign (+) indicates that the designated proper shipping name and hazard class of the material must always be shown whether or not the material or its mixtures or solutions meet the definitions of the class.

Spill Handling: Evacuate and restrict persons not wearing protective equipment from area of spill or leak until cleanup is complete. Remove all ignition sources. Ventilate area of spill or leak. Absorb liquids in vermiculite, dry sand, earth, peat, carbon, or a similar material and deposit in sealed containers. Keep this chemical out of a confined space, such as a sewer, because of the possibility of an explosion, unless the sewer is designed to prevent the buildup of explosive

concentrations. It may be necessary to contain and dispose of this chemical as a hazardous waste. If material or contaminated runoff enters waterways, notify downstream users of potentially contaminated waters. Contact your local or federal environmental protection agency for specific recommendations. If employees are required to clean up spills, they must be properly trained and equipped. OSHA 1910.120(q) may be applicable.

Initial isolation and protective action distances
Distances shown are likely to be affected during the first 30 min after materials are spilled and could increase with time. If more than one tank car, cargo tank, portable tank, or large cylinder involved in the incident is leaking, the protective action distance may need to be increased. You may need to seek emergency information from CHEMTREC at (800) 424-9300 or seek professional environmental engineering assistance from the US EPA Environmental Response Team at (908) 548-8730 (24-h response line).

Small spills (From a small package or a small leak from a large package)
First: Isolate in all directions (feet/meters)   100/30
Then: Protect persons downwind (miles/kilometers)
Day   0.3/0.5
Night   0.4/0.6

Large spills (From a large package or from many small packages)
First: Isolate in all directions (feet/meters)   500/150
Then: Protect persons downwind (miles/kilometers)
Day   1.0/1.5
Night   1.6/2.5

Fire Extinguishing: Phenyl isocyanate may burn, but does not readily ignite. Poisonous gases are produced in fire, including oxides of nitrogen. Containers may explode in fire. Use dry chemical, CO_2, water spray; or foam extinguishers. Vapors are heavier than air and will collect in low areas. Vapors may travel long distances to ignition sources and flashback. Vapors in confined areas may explode when exposed to fire. Containers may explode in fire. Storage containers and parts of containers may rocket great distances, in many directions. If material or contaminated runoff enters waterways, notify downstream users of potentially contaminated waters. Notify local health and fire officials and pollution control agencies. From a secure, explosion-proof location, use water spray to cool exposed containers. If cooling streams are ineffective (venting sound increases in volume and pitch, tank discolors, or shows any signs of deforming), withdraw immediately to a secure position. If employees are expected to fight fires, they must be trained and equipped in OSHA 1910.156. The only respirators recommended for firefighting are self-contained breathing apparatuses that have full face-pieces and are operated in a pressure-demand or other positive-pressure mode.

Reference
New Jersey Department of Health and Senior Services. (November 2000). *Hazardous Substances Fact Sheet: Phenyl Isocyanate.* Trenton, NJ

Phenyl mercaptan   P:0440

Molecular Formula: C_6H_6S
Common Formula: C_6H_5SH
Synonyms: Benzenethiol; Mercaptobenzene; Phenol, thio-; Phenylmercaptan; Phenylthiol; Thiophenol
CAS Registry Number: 108-98-5
RTECS® Number: DC0525000
UN/NA & ERG Number: UN2337/131
EC Number: 203-635-3
Regulatory Authority and Advisory Bodies
Air Pollutant Standard Set. See below, "Permissible Exposure Limits in Air" section.
US EPA Hazardous Waste Number (RCRA No.): P014.
RCRA, 40CFR261, Appendix 8 Hazardous Constituents.
Superfund/EPCRA 40CFR355, Extremely Hazardous Substances: TPQ = 500 lb (227 kg).
Reportable Quantity (RQ): 100 lb (45.4 kg).
US DOT 49CFR172.101, Inhalation Hazardous Chemical.
Canada, WHMIS, Ingredients Disclosure List Concentration 0.1%.
European/International Regulations: not listed in Annex 1.
WGK (German Aquatic Hazard Class): 3—Severe hazard to waters.

Description: Phenyl mercaptan is a water-white liquid with a repulsive, penetrating, garlic-like odor. The odor threshold is 0.0003 ppm. Molecular weight = 110.18; Specific gravity (H_2O:1) = 1.08; Boiling point = 126°C; Freezing/Melting point = −9.4°C; Vapor pressure = 1 mmHg at 18°C; Flash point = 55.6°C. Explosive limits: LEL = 1.2%; UEL—unknown. Hazard Identification (based on NFPA-704 M Rating System): Health 3, Flammability 3, Reactivity 0. Insoluble in water.

Potential Exposure: Compound Description: Reproductive Effector; Primary Irritant. Phenyl mercaptan is used as a chemical intermediate in pesticide manufacture; as a mosquito larvicide. It is used in solvent formulations for the removal of polysulfide sealants.

Incompatibilities: Strong acids, strong bases, calcium hypochlorite, alkali metals. Oxidizes on exposure to air; supplied under nitrogen. At normal room temperature may vaporize forming explosive mixtures with air.

Permissible Exposure Limits in Air
OSHA PEL: None.
NIOSH REL: 0.1 ppm/0.5 mg/m³ [15 min] Ceiling Concentration.
ACGIH TLV®[1]: 0.1 ppm/0.45 mg/m³ TWA [skin].
Protective Action Criteria (PAC)*
TEEL-0: 0.1 ppm
PAC-1: 0.1 ppm
PAC-2: **0.53** ppm
PAC-3: **1.6** ppm
*AEGLs (Acute Emergency Guideline Levels) & ERPGs (Emergency Response Planning Guideline) are in **bold face**.

Australia: TWA 0.5 ppm (2 mg/m^3), 1993; Belgium: TWA 0.5 ppm (2.3 mg/m^3), 1993; Denmark: TWA 0.5 ppm (2.3 mg/m^3), 1999; Finland: STEL 0.5 ppm (2.6 mg/m^3), 1993; France: VME 0.5 ppm (2 mg/m^3), 1999; Norway: TWA 0.5 ppm (2 mg/m^3), 1999; Switzerland: MAK-W 0.5 ppm (2.3 mg/m^3), 1999; the Netherlands: MAC-TGG 2 mg/m^3, 2003; Argentina, Bulgaria, Columbia, Jordan, South Korea, New Zealand, Singapore, Vietnam: ACGIH TLV®: TWA 0.5 ppm.

Several states have set guidelines or standards for benzenethiol in ambient air[60] ranging from 20 μg/m^3 (North Dakota) to 35 μg/m^3 (Virginia) to 40 μg/m^3 (Connecticut) to 48 μg/m^3 (Nevada).

Determination in Air: Use OSHA Analytical Method PV-2075.

Determination in Water: Octanol–water coefficient: Log K_{ow} = 2.52.

Routes of Entry: Inhalation, ingestion, skin and/or eye contact. Absorbed through the skin.

Harmful Effects and Symptoms

Short Term Exposure: Benzenethiol can affect you when breathed in and by passing through your skin. Irritates the eyes, skin, and respiratory tract. Benzenethiol can severely burn the eyes, causing permanent damage. Higher exposures can cause pulmonary edema, a medical emergency that can be delayed for several hours. This can cause death. Exposure can cause weakness, dizziness, cough, wheezing, dyspnea (breathing difficulty). Higher levels can cause restlessness and irritability followed by paralysis and death. High or repeated exposure can cause liver, kidney, or lung damage. Acute exposure to thiophenol may result in cough, troubled breathing, irritation of the lungs, and pneumonitis. Nausea, vomiting, and diarrhea are often seen. May affect the nervous system.

Long Term Exposure: Repeated or prolonged contact with skin may cause dermatitis. May cause lung, kidney, liver, spleen damage.

Points of Attack: Eyes, skin, respiratory system, central nervous system, kidneys, liver, spleen.

Medical Surveillance: If symptoms develop or overexposure is suspected, tests of the following may be helpful: Kidney and liver function. Lung function tests.

First Aid: If this chemical gets into the eyes, remove any contact lenses at once and irrigate immediately for at least 30 min, occasionally lifting upper and lower lids. If available, flush eyes with large amounts of 0.5% silver nitrate, followed immediately by very large amounts of water. Continue water for 15 min. Seek medical attention immediately. If this chemical contacts the skin, remove contaminated clothing and wash immediately with soap and water. Speed in removing material from skin is of extreme importance. Seek medical attention immediately. If this chemical has been inhaled, remove from exposure, begin rescue breathing (using universal precautions, including resuscitation mask) if breathing has stopped and CPR if heart action has stopped. Transfer promptly to a medical facility. When this chemical has been swallowed, get medical attention. Give large quantities of water and induce vomiting. Do not make an unconscious person vomit. Medical observation is recommended for 24–48 h after breathing overexposure, as pulmonary edema may be delayed. As first aid for pneumonitis or pulmonary edema, a doctor or authorized paramedic may consider administering a corticosteroid spray.

Personal Protective Methods: Wear protective gloves and clothing to prevent any reasonable probability of skin contact. Safety equipment suppliers/manufacturers can provide recommendations on the most protective glove/clothing material for your operation. All protective clothing (suits, gloves, footwear, headgear) should be clean, available each day, and put on before work. Contact lenses should not be worn when working with this chemical. Wear dust-proof chemical goggles and face shield unless full face-piece respiratory protection is worn. Employees should wash immediately with soap when skin is wet or contaminated. Provide emergency showers and eyewash.

Respirator Selection: NIOSH: *Up to 1 ppm:* CcrOv (APF = 10) [any chemical cartridge respirator with organic vapor cartridge(s)] or Sa (APF = 10) (any supplied-air respirator). *Up to 2.5 ppm:* Sa:Cf (APF = 25) (any supplied-air respirator operated in a continuous-flow mode) or PaprOv (APF = 25) [any powered, air-purifying respirator with organic vapor cartridge(s)]. *Up to 5 ppm:* CcrFOv (APF = 50) [any chemical cartridge respirator with a full face-piece and organic vapor cartridge(s)]; GmFOv (APF = 50) [any air-purifying, full-face-piece respirator (gas mask) with a chin-style, front- or back-mounted organic vapor canister]; PaprTOv (APF = 50) [any powered, air-purifying respirator with a tight-fitting face-piece and organic vapor cartridge(s)] or SCBAF (APF = 50) (any self-contained breathing apparatus with a full face-piece) or SaF (APF = 50) (any supplied-air respirator with a full face-piece). *Emergency or planned entry into unknown concentrations or IDLH conditions:* SCBAF: Pd,Pp (APF = 10,000) (any NIOSH/MSHA- or European Standard EN 149-approved self-contained breathing apparatus that has a full face-piece and is operated in a pressure-demand or other positive-pressure mode) or SaF: Pd,Pp: ASCBA (APF = 10,000) (any supplied-air respirator that has a full face-piece and is operated in a pressure-demand or other positive-pressure mode in combination with an auxiliary self-contained breathing apparatus operated in a pressure-demand or other positive-pressure mode). *Escape:* GmFOv (APF = 50) [any air-purifying, full-face-piece respirator (gas mask) with a chin-style, front- or back-mounted organic vapor canister] or SCBAE (any appropriate escape-type, self-contained breathing apparatus).

Storage: (1) Color Code—Red: Flammability Hazard: Store in a flammable liquid storage area or approved cabinet away from ignition sources and corrosive and reactive materials. (2) Color Code—Blue: Health Hazard/Poison: Store in a secure poison location. Prior to working with this chemical you should be trained on its proper handling and storage. Store in airtight containers in a cool, well-ventilated area. Sources of ignition, such as smoking and open flames, should be prohibited where benzenethiol is handled, used, or stored. Use only nonsparking tools and equipment, especially when opening and closing containers of benzenethiol.

Shipping: This compound requires a shipping label of "POISONOUS/TOXIC MATERIALS, FLAMMABLE LIQUID." It falls in Hazard Class 6.1 and Packing Group I.

Spill Handling: Evacuate and restrict persons not wearing protective equipment from area of spill or leak until cleanup is complete. Remove all ignition sources. Ventilate area of spill or leak. Absorb liquids in vermiculite, dry sand, earth, peat, carbon, or a similar material and deposit in sealed containers. Keep this chemical out of a confined space, such as a sewer, because of the possibility of an explosion, unless the sewer is designed to prevent the buildup of explosive concentrations. It may be necessary to contain and dispose of this chemical as a hazardous waste. If material or contaminated runoff enters waterways, notify downstream users of potentially contaminated waters. Contact your local or federal environmental protection agency for specific recommendations. If employees are required to clean up spills, they must be properly trained and equipped. OSHA 1910.120(q) may be applicable.

Initial isolation and protective action distances

Distances shown are likely to be affected during the first 30 min after materials are spilled and could increase with time. If more than one tank car, cargo tank, portable tank, or large cylinder involved in the incident is leaking, the protective action distance may need to be increased. You may need to seek emergency information from CHEMTREC at (800) 424-9300 or seek professional environmental engineering assistance from the US EPA Environmental Response Team at (908) 548-8730 (24-h response line).

Small spills (From a small package or a small leak from a large package)

First: Isolate in all directions (feet/meters)   100/30
Then: Protect persons downwind (miles/kilometers)
Day   0.1/0.2
Night   0.1/0.2

Large spills (From a large package or from many small packages)

First: Isolate in all directions (feet/meters)   100/30
Then: Protect persons downwind (miles/kilometers)
Day   0.2/0.3
Night   0.3/0.5

Fire Extinguishing: This chemical is a flammable liquid. Poisonous gases, including sulfur dioxide, are produced in fire. Use dry chemical, carbon dioxide, or alcohol foam extinguishers. Vapors are heavier than air and will collect in low areas. Vapors may travel long distances to ignition sources and flashback. Vapors in confined areas may explode when exposed to fire. Containers may explode in fire. Storage containers and parts of containers may rocket great distances, in many directions. If material or

contaminated runoff enters waterways, notify downstream users of potentially contaminated waters. Notify local health and fire officials and pollution control agencies. From a secure, explosion-proof location, use water spray to cool exposed containers. If cooling streams are ineffective (venting sound increases in volume and pitch, tank discolors, or shows any signs of deforming), withdraw immediately to a secure position. If employees are expected to fight fires, they must be trained and equipped in OSHA 1910.156. The only respirators recommended for firefighting are self-contained breathing apparatuses that have full face-pieces and are operated in a pressure-demand or other positive-pressure mode.

Disposal Method Suggested: Consult with environmental regulatory agencies for guidance on acceptable disposal practices. Generators of waste containing this contaminant ($\geq$100 kg/mo) must conform with EPA regulations governing storage, transportation, treatment, and waste disposal. Dissolve in flammable solvent and burn in furnace equipped with afterburner and alkaline scrubber.[22]

References

US Environmental Protection Agency. (November 30, 1987). *Chemical Hazard Information Profile: Thiophenol.* Washington, DC: Chemical Emergency Preparedness Program

New Jersey Department of Health and Senior Services. (July 2004). *Hazardous Substances Fact Sheet: Benzenethiol.* Trenton, NJ

Phenylmercury acetate   P:0450

Molecular Formula: $C_8H_8HgO_2$
Common Formula: $C_6H_5HgOOCCH_3$
Synonyms: Acetate phenylmercurique (French); (Aceato) phenylmercury; Acetato fenilmercurio (Spanish); Acetic acid, phenylmercury derivitive; Agrosan; Agrosand; Agrosan GN 5; Algimycin; Antimucin WDR; Benzene, (acetoxymercuri)-; Benzene, (acetoxymercurio); Bufen; Cekusil; Celmer; Ceresan; Ceresan universal; Ceresol; Contra creme; Dynacide; Femma; FMA; Fungitox OR; Gallotox; HL-331; Hong kien; Hostaquick; Kwiksan; Leytosan; Liquiphene; Mercuriphenyl acetate; Mercury(II) acetate, phenyl; Mercury (acetoxy)phenyl-; Mergamma; Mersolite; Mersolite 8; Metasol 30; Norforms; Nymerate; Pamisan; Phenmad; Phenomercury acetate; Phenyl-murcuriacetate; Phenylmercuric acetate; Phenylquecksilberacetat (German); Phix; PMA; PMAC; PM acetate; PMAL; PMAS; Purasan-SC-10; Puraturf 10; Quicksan; Sanitized SPG; SC-110; Scutl; Seedtox; Shimmerex; Sporkil; Tag; Tag 331; Tag HL 331; Tag fungicide; Trigosan; Ziarnik
CAS Registry Number: 62-38-4; *(alt.)* 1337-06-0; *(alt.)* 61840-45-7; *(alt.)* 64684-45-3
RTECS® Number: OV6475000

UN/NA & ERG Number: UN1674/151
EC Number: 200-532-5 [*Annex I Index No.:* 080-011-00-5]
Regulatory Authority and Advisory Bodies
Carcinogenicity: IARC: Human Inadequate Evidence, *possibly carcinogenic to humans*, Group 2B, 1993.
US EPA Gene-Tox Program, Positive: *D. melanogaster*—whole sex chrom. loss; Positive: *D. melanogaster*—nondisjunction; *B. subtilis* rec assay.
US EPA, FIFRA 1998 Status of Pesticides: Canceled.
Banned or Severely Restricted (several countries) (UN).[13]
Air Pollutant Standard Set. See below, "Permissible Exposure Limits in Air" section.
US EPA Hazardous Waste Number (RCRA No.): P092.
RCRA, 40CFR261, Appendix 8 Hazardous Constituents.
Superfund/EPCRA 40CFR355, Extremely Hazardous Substances: TPQ = 500/10,000 lb (227/4540 kg).
Reportable Quantity (RQ): 100 lb (45.4 kg).
US DOT Regulated Marine Pollutant (49CFR172.101, Appendix B), severe pollutant.
California Proposition 65 Chemical: Reproductive toxin.
Canada, WHMIS, Ingredients Disclosure List Concentration 0.1%.
European/International Regulations: Hazard Symbol: T, N; Risk phrases: R25; R34; R48/24/25; R50/53; Safety phrases: S1/2; S3; S24/25; S37; S45; S60; S61 (see Appendix 4).
WGK (German Aquatic Hazard Class): 3—Severe hazard to waters.
Description: Phenylmercury acetate is a white or yellow crystalline solid. Molecular weight = 336.75; Freezing/Melting point = 152°C; Flash point = >38°C. Hazard Identification (based on NFPA-704 M Rating System): Health 3, Flammability 1, Reactivity 0. Slightly soluble in water.
Potential Exposure: Compound Description: Agricultural Chemical; Tumorigen, Organometallic, Mutagen; Reproductive Effector; Primary Irritant. Phenylmercury acetate is used as an antiseptic, fungicide; for fungal and bacterial control; herbicide and control of crabgrass; mildewcide for paints; slimicide in paper mills. It was also used in contraceptive gels and foams.
Incompatibilities: Strong oxidizers, halogens.
Permissible Exposure Limits in Air
As organo mercury compound
OSHA PEL: 0.01 mg/m^3 TWA; 0.04 mg/m^3 Ceiling Concentration.
NIOSH REL: 0.01 mg/m^3 TWA; 0.03 mg/m^3 STEL [skin].
ACGIH TLV®[1]: 0.01 mg/m^3 TWA; 0.03 mg/m^3 STEL [skin].
NIOSH IDLH: 2 mg Hg/m^3.
Protective Action Criteria (PAC)
TEEL-0: 0.168 mg/m^3
PAC-1: 2.5 mg/m^3
PAC-2: 16.8 mg/m^3
PAC-3: 16.8 mg/m^3
DFG MAK: 0.01 mg[Hg]/m^3 [skin] Danger of skin sensitization; Carcinogen Category 3.

Australia: TWA 0.05 mg[Hg]/m^3, [skin], 1993; Belgium: TWA 0.05 mg[Hg]/m^3, [skin], 1993; Denmark: TWA 0.05 mg[Hg]/m^3, [skin], 1999; Finland: TWA 1 mg[Hg]/m^3, 1999; France: VME 0.1 mg[Hg]/m^3, [skin], 1999; Hungary: TWA 0.02 mg[Hg]/m^3; STEL 0.04 mg[Hg]/m^3, 1993; Japan: 0.05 mg[Hg]/m^3, 1999; Norway: TWA 0.05 mg[Hg]/m^3, 1999; the Philippines: TWA 0.05 mg[Hg]/m^3, 1993; Poland: MAC (TWA) 0.05 mg[Hg]/m^3; MAC (STEL) 0.15 mg[Hg]/m^3, 1999; Russia: TWA 0.05 mg[Hg]/m^3; STEL 0.01 mg[Hg]/m^3, 1993; Sweden: NGV 0.05 mg[Hg]/m^3, [skin], 1999; Thailand: STEL 0.05 mg[Hg]/m^3, 1993; United Kingdom: LTEL 0.05 mg[Hg]/m^3; STEL 0.15 mg[Hg]/m^3, 1993; Argentina, Bulgaria, Columbia, Jordan, South Korea, New Zealand, Singapore, Vietnam: ACGIH TLV®: TWA 0.1 mg[Hg]/m^3 [skin].
Permissible Concentration in Water: *To protect freshwater aquatic life:* 0.00057 µg/L as a 24-h average, never to exceed 0.0017 µg/L. *To protect saltwater aquatic life:* 0.025 µg/L as a 24-h average, never to exceed 3.7 µg/L. *To protect human health:* 0.144 µg/L (US EPA) set in 1979–1980.[6] These are the limits for inorganic mercury compounds in general.
Determination in Water: Total mercury is determined by flameless atomic absorption. Soluble mercury may be determined by 0.45 µm filtration followed by flameless atomic absorption.
Routes of Entry: Inhalation, ingestion, skin and/or eye contact. Absorbed through the skin.
Harmful Effects and Symptoms
Short Term Exposure: Irritates the eyes, skin, and respiratory tract. Overexposure affects the kidneys, causing renal function failure. Extremely toxic. The probable oral lethal dose for humans is 5–50 mg/kg, between 7 drops and 1 teaspoonful for a 70-kg (150 lb) person. Symptoms arising from acute exposure may occur at varying intervals up to several weeks following exposure. Ingestion of mercurial fungicide-treated grain resulted in gastrointestinal irritation with nausea, vomiting, abdominal pain, and diarrhea. Alkylmercurials produce severe neurologic toxicity, such as loss of feeling in lips, tongue, and extremities; confusion, hallucinations, irritability, sleep disturbances; staggering walk; memory loss; slurred speech; auditory defects; emotional instability; and inability to concentrate. It is also a strong skin irritant; erythema and blistering may result 6–12 h after exposure. Phenylmercury acetate, at sufficient concentration, is expected to be injurious to the eye externally. Mercury poisoning can cause "shakes," irritability, sore gums; increased saliva; personality change and brain damage. Skin contact can cause burns, skin allergy, and a gray skin color. Heating or contact with acid or acid "fumes" releases toxic mercury vapors.
Long Term Exposure: Mercury accumulates in the body. Repeated or prolonged contact with skin may cause dermatitis. May affect the nervous system, causing nervous disorders. Based on animal tests, phenylmercuric acetate should

be handled as a teratogen—with extreme caution. It also may cause mutations.

Points of Attack: Eyes, skin, central nervous system, peripheral nervous system, kidneys.

Medical Surveillance: Before first exposure and every 6–12 months after, a complete medical history and examination is strongly recommended with: examination of the nervous system, including handwriting. Routine urine test (UA). Urine test for mercury (should be less than 0.02 mg/L). Consider lung function tests for persons with frequent exposures. After suspected illness or overexposure, repeat the above tests and get a blood test for mercury. Consider chest X-ray after acute overexposure.

First Aid: If this chemical gets into the eyes, remove any contact lenses at once and irrigate immediately for at least 15 min, occasionally lifting upper and lower lids. Seek medical attention immediately. If this chemical contacts the skin, remove contaminated clothing and wash immediately with soap and water. Seek medical attention immediately. If this chemical has been inhaled, remove from exposure, begin rescue breathing (using universal precautions, including resuscitation mask) if breathing has stopped and CPR if heart action has stopped. Transfer promptly to a medical facility. When this chemical has been swallowed, get medical attention. Give large quantities of water and induce vomiting. Do not make an unconscious person vomit. Keep victim quiet and maintain normal body temperature. Effects may be delayed; keep victim under observation.

Antidotes and Special Procedures for medical personnel: The drug NAP (*n*-acetyl penicillamine) has been used to treat mercury poisoning, with mixed success.

Note to physician: For severe poisoning BAL [British Anti-Lewisite, dimercaprol, dithiopropanol ($C_3H_8OS_2$)] has been used to treat toxic symptoms of certain heavy metal poisoning—including mercury. Although BAL is reported to have a large margin of safety, caution must be exercised, because toxic effects may be caused by excessive dosage. Most can be prevented by premedication with 1-ephedrine sulfate (CAS: 134-72-5).

Personal Protective Methods: Wear protective gloves and clothing to prevent any reasonable probability of skin contact. Safety equipment suppliers/manufacturers can provide recommendations on the most protective glove/clothing material for your operation. All protective clothing (suits, gloves, footwear, headgear) should be clean, available each day, and put on before work. Contact lenses should not be worn when working with this chemical. Wear dust-proof chemical goggles and face shield unless full face-piece respiratory protection is worn. Employees should wash immediately with soap when skin is wet or contaminated. Provide emergency showers and eyewash. Specific engineering controls are recommended in NIOSH Criteria Document #73-11024.

Respirator Selection: *Up to 0.1 mg/m³:* Sa (APF = 10) (any supplied-air respirator). *Up to 0.25 mg/m³:* Sa:Cf (APF = 25) (any supplied-air respirator operated in a continuous-flow mode). *Up to 0.5 mg/m³:* SaT: Cf (APF = 50) (any supplied-air respirator that has a tight-fitting face-piece and is operated in a continuous-flow mode) or SCBAF (APF = 50) (any self-contained breathing apparatus with a full face-piece) or SaF (APF = 50) (any supplied-air respirator with a full face-piece). *Up to 2 mg/m³:* SA: PD, PP (any supplied-air respirator operated in a pressure-demand or other positive-pressure mode). *Emergency or planned entry into unknown concentrations or IDLH conditions:* SCBAF: Pd,Pp (APF = 10,000) (any NIOSH/MSHA- or European Standard EN 149-approved self-contained breathing apparatus that has a full face-piece and is operated in a pressure-demand or other positive-pressure mode) or SaF: Pd,Pp: ASCBA (APF = 10,000) (any supplied-air respirator that has a full face-piece and is operated in a pressure-demand or other positive-pressure mode in combination with an auxiliary self-contained breathing apparatus operated in a pressure-demand or other positive-pressure mode). *Escape:* SCBAE (any appropriate escape-type, self-contained breathing apparatus).

Storage: Color Code—Blue: Health Hazard/Poison: Store in a secure poison location. Prior to working with this chemical you should be trained on its proper handling and storage. Store in tightly closed containers in a cool, well-ventilated area away from strong oxidizers (such as chlorine, bromine, and fluorine). Sources of ignition, such as smoking and open flames, are prohibited where phenylmercuric acetate is used, handled, or stored in a manner that could create a potential fire or explosion hazard.

Shipping: Phenylmercuric acetate requires a shipping label of "POISONOUS/TOXIC MATERIALS." It falls in Hazard Class 6.1 and Packing Group II.

Spill Handling: Evacuate persons not wearing protective equipment from area of spill or leak until cleanup is complete. Remove all ignition sources. Spills should be collected with special mercury vapor suppressants or special vacuums. Kits specific for cleanup of mercury spills should be available. Ventilate area after cleanup is complete. It may be necessary to contain and dispose of this chemical as a hazardous waste. If material or contaminated runoff enters waterways, notify downstream users of potentially contaminated waters. Contact your local or federal environmental protection agency for specific recommendations. If employees are required to clean up spills, they must be properly trained and equipped. OSHA 1910.120(q) may be applicable.

Fire Extinguishing: Use dry chemical, foam, or carbon dioxide on solution. Use water as necessary, but run-off should be limited and controlled to prevent it from entering streams of water supplies. Materials are extremely hazardous to health, but areas may be entered with extreme care. Full protective clothing, including self-contained breathing apparatus; rubber gloves; boots and bands around legs, arms, and waist, should be provided. No skin should be exposed. Poisonous gases are produced in fire, including mercury. If material or contaminated runoff enters waterways, notify downstream users of potentially contaminated waters. Notify local health and fire officials and pollution control agencies. From a secure, explosion-proof location, use water spray to cool exposed containers. If cooling streams are ineffective (venting sound increases in volume and pitch, tank discolors, or shows any signs of deforming), withdraw immediately to a secure position. If employees are expected to fight fires, they must be trained and equipped in OSHA 1910.156. The only respirators recommended for firefighting are self-contained breathing apparatuses that have full face-pieces and are operated in a pressure-demand or other positive-pressure mode.

Disposal Method Suggested: Consult with environmental regulatory agencies for guidance on acceptable disposal practices. Generators of waste containing this contaminant (≥ 100 kg/mo) must conform with EPA regulations governing storage, transportation, treatment, and waste disposal. React to produce soluble nitrate form, precipitate as mercuric sulfide. Return to supplier.

References

US Environmental Protection Agency. (November 30, 1987). *Chemical Hazard Information Profile: Phenylmercury Acetate*. Washington, DC: Chemical Emergency Preparedness Program

New Jersey Department of Health and Senior Services. (February 2000). *Hazardous Substances Fact Sheet: Phenylmercuric Acetate*. Trenton, NJ

N-Phenyl-β-naphthylamine   P:0460

Molecular Formula: $C_{16}H_{13}N$
Common Formula: $C_{10}H_7NHC_6H_5$
Synonyms: 2-Anilinonaphthalene; β-Naphthylphenylamine; PBNA;   2-Phenylaminonaphthalene;   Phenyl-β-naphthylamine
CAS Registry Number: 135-88-6
RTECS®Number: QV4550000
UN/NA & ERG Number: UN2811 (toxic solid, organic, n.o.s.)/154
EC Number: 205-223-9 [*Annex I Index No.:* 612-135-00-8]
Regulatory Authority and Advisory Bodies
Carcinogenicity: IARC: Animal Limited Evidence; Human Inadequate Evidence, *not classifiable as carcinogenic to humans*, Group 3, 1978; NCI: Carcinogenesis Studies (feed); equivocal evidence: mouse; no evidence: rat; NIOSH) (*since metabolized to* β-*naphthylamine*): Potential occupational carcinogen.
Banned or Severely Restricted (Sweden) (UN).[13]
Air Pollutant Standard Set. See below, "Permissible Exposure Limits in Air" section.
Canada, WHMIS, Ingredients Disclosure List Concentration 0.1%.
European/International Regulations: Hazard Symbol: Xn, N; Risk phrases: R36/38; R40; R43; R51/53; Safety phrases: S2; S26; S36/37; S61 (see Appendix 4).

WGK (German Aquatic Hazard Class): No value assigned.
Description: Phenyl-β-naphthylamine is a light gray powder. A combustible solid. Molecular weight = 219.30; Specific gravity (H_2O:1) = 1.24; Boiling point = 396°C; Freezing/Melting point = 107.8°C. Insoluble in water.
Potential Exposure: Compound Description: Tumorigen, Mutagen. Phenyl-β-naphthylamine is used as a rubber antioxidant, as an inhibitor for butadiene, a stabilizer in lubricants, and an intermediate in chemical synthesis.
Incompatibilities: Incompatible with oxidizers, strong acids, organic anhydrides, isocyanates, aldehydes.
Permissible Exposure Limits in Air
OSHA PEL: None.
NIOSH REL: A potential occupational carcinogen* [skin]; Limit exposure to lowest feasible concentration. See *NIOSH Pocket Guide*, Appendix A. [*Note:* Since metabolized to β-naphthylamine].
ACGIH TLV®[1]: not classifiable as a human carcinogen.
No TEEL available.
DFG MAK: Carcinogen Category 3B.
Australia: carcinogen, 1993; Austria: Suspected: carcinogen, 1999; Belgium: carcinogen, 1993; Finland: carcinogen, 1999; Poland: MAC (TWA) 0.03 mg/m³, 1999; Sweden: carcinogen, 1999; Argentina, Bulgaria, Columbia, Jordan, South Korea, New Zealand, Singapore, Vietnam: ACGIH TLV®: not classifiable as a human carcinogen.
Several states have set guidelines or standards for this compound in ambient air[60] ranging from zero (North Dakota) to 3.0 μg/m³ (Virginia) to 45.0 μg/m³ (Pennsylvania).
Determination in Air: Use OSHA Analytical Method 96.
Determination in Water: Octanol−water coefficient: Log $K_{ow} = 4.4$.
Routes of Entry: Inhalation, ingestion, skin and/or eye contact.
Harmful Effects and Symptoms
Short Term Exposure: The main problem with this compound is that phenyl-β-naphthylamine, a known carcinogen, is both a contaminant in, and a metabolic product of PBNA. Phenyl-β-naphthylamine can affect you when breathed in and by passing through your skin. Contact can cause skin irritation and rash. Phenyl-β-naphthylamine should be handled as a carcinogen—with extreme caution. Exposure can affect the ability of the blood to carry oxygen (methemoglobinemia), causing cyanosis, a bluish skin color. Higher levels can cause headache and dizziness. Very high levels can cause death. LD_{50} = (oral-mouse) 1450 kg/mg.
Long Term Exposure: Repeated or prolonged contact may cause skin irritation, sensitization, allergy; hypersensitivity to sunlight. A potential occupational carcinogen. Can cause methemoglobinemia (see above); anemia may result.
Points of Attack: Eyes, skin, bladder. *Cancer site:* bladder; in animals: lung, pancreas.
Medical Surveillance: If symptoms develop or overexposure is suspected, the following may be useful: methemoglobin level. Complete blood count (CBC).

First Aid: If this chemical gets into the eyes, remove any contact lenses at once and irrigate immediately for at least 15 min, occasionally lifting upper and lower lids. Seek medical attention immediately. If this chemical contacts the skin, remove contaminated clothing and wash immediately with soap and water. Seek medical attention immediately. If this chemical has been inhaled, remove from exposure, begin rescue breathing (using universal precautions, including resuscitation mask) if breathing has stopped and CPR if heart action has stopped. Transfer promptly to a medical facility. When this chemical has been swallowed, get medical attention. Give large quantities of water and induce vomiting. Do not make an unconscious person vomit.

Note to physician: Treat for methemoglobinemia. Spectrophotometry may be required for precise determination of levels of methemoglobin in urine.

Personal Protective Methods: Wear protective gloves and clothing to prevent any reasonable probability of skin contact. Safety equipment suppliers/manufacturers can provide recommendations on the most protective glove/clothing material for your operation. All protective clothing (suits, gloves, footwear, headgear) should be clean, available each day, and put on before work. Contact lenses should not be worn when working with this chemical. Wear dust-proof chemical goggles and face shield unless full face-piece respiratory protection is worn. Employees should wash immediately with soap when skin is wet or contaminated. Provide emergency showers and eyewash.

Respirator Selection: *At any detectable concentration:* SCBAF: Pd,Pp (APF = 10,000) (any NIOSH/MSHA- or European Standard EN 149-approved self-contained breathing apparatus that has a full face-piece and is operated in a pressure-demand or other positive-pressure mode) or SaF: Pd,Pp: ASCBA (APF = 10,000) (any supplied-air respirator that has a full face-piece and is operated in a pressure-demand or other positive-pressure mode in combination with an auxiliary, self-contained breathing apparatus operated in a pressure-demand or other positive-pressure mode). *Escape:* GmFOv100 (APF = 50) [any air-purifying, full-face-piece respirator (gas mask) with a chin-style, front- or back-mounted organic vapor canister having an N100, R100, or P100 filter] or SCBAE (any appropriate escape-type, self-contained breathing apparatus).

Storage: Color Code—Blue: Health Hazard/Poison: Store in a secure poison location. Prior to working with this chemical you should be trained on its proper handling and storage. Store in tightly closed containers in a cool, well-ventilated area away from heat and oil. A regulated, marked area should be established where *N*-phenyl-β-naphthylamine is handled, used, or stored.

Shipping: Toxic solids, organic, n.o.s. requires a shipping label of "POISONOUS/TOXIC MATERIALS." It falls in Hazard Class 6.1 and Packing Group III. As a hazardous substance, solid, n.o.s., this imposes no label requirements or maximum on shipping weights.

Spill Handling: Evacuate persons not wearing protective equipment from area of spill or leak until cleanup is complete. Remove all ignition sources. Collect powdered material in the most convenient and safe manner and deposit in sealed containers. Ventilate area after cleanup is complete. It may be necessary to contain and dispose of this chemical as a hazardous waste. If material or contaminated runoff enters waterways, notify downstream users of potentially contaminated waters. Contact your local or federal environmental protection agency for specific recommendations. If employees are required to clean up spills, they must be properly trained and equipped. OSHA 1910.120(q) may be applicable.

Fire Extinguishing: This chemical is a combustible solid. Use dry chemical, carbon dioxide, water spray, or alcohol foam extinguishers. Poisonous gases are produced in fire, including nitrogen oxides. If material or contaminated runoff enters waterways, notify downstream users of potentially contaminated waters. Notify local health and fire officials and pollution control agencies. From a secure, explosion-proof location, use water spray to cool exposed containers. If cooling streams are ineffective (venting sound increases in volume and pitch, tank discolors, or shows any signs of deforming), withdraw immediately to a secure position. If employees are expected to fight fires, they must be trained and equipped in OSHA 1910.156. The only respirators recommended for firefighting are self-contained breathing apparatuses that have full face-pieces and are operated in a pressure-demand or other positive-pressure mode.

References

National Institute for Occupational Safety and Health. (December 17, 1976). *Metabolic Precursors of a Known Human Carcinogen, β-Naphthylamine*, Current Intelligence Bulletin No. 16. Rockville, MD

New Jersey Department of Health and Senior Services. (January 2007). *Hazardous Substances Fact Sheet: N-Phenyl-beta-Naphthylamine*. Trenton, NJ

o-Phenylphenol   P:0470

Molecular Formula: $C_{12}H_{10}O$
Common Formula: $C_6H_5-C_6H_4OH$
Synonyms: Anthrapole 73; 2-Biphenylol; *o*-Biphenylol; (1,1′-Biphenyl)-2-ol; *o*-Biphenylol; *o*-Diphenylol; Dowicide 1; Dowcide 1 antimicrobial; *o*-Fenilfenol (Spanish); 2-Hydroxybiphenyl; *o*-Hydroxybiphenyl; 2-Hydroxy-1,1′-biphenyl; *o*-Hydroxydiphenyl; 2-Hydroxydiphenyl; Invalon OP; Kiwiydiphenyl; Nectryl; Orthophenylphenol; Orthoxenol; *o*-Phenylphenol; 2-Phenylphenol; Preventol O extra; Remol TRF; Tetrosin OE; Tetrosin OE-N; Torsite; Tumescal OPE; *o*-Xenol
CAS Registry Number: 90-43-7
RTECS® Number: DV5775000
UN/NA & ERG Number: UN3143 Dyes, solid, toxic, n.o.s. [or] Dye intermediates, solid, toxic, n.o.s./151

EC Number: 201-993-5 [*Annex I Index No.:* 604-020-00-6]

Regulatory Authority and Advisory Bodies

Carcinogenicity: NCI: Carcinogenesis Studies (derm); no evidence: rat; IARC: Animal Limited Evidence; Human Inadequate Evidence, *not classifiable as carcinogenic to humans*, Group 3, 1999.

US EPA, FIFRA, 1998 Status of Pesticides: Supported.

EPCRA Section 313 Form R *de minimis* concentration reporting level: 1.0%.

California Proposition 65 Chemical: Cancer 8/4/00.

Canada, WHMIS, Ingredients Disclosure List Concentration 1.0%.

European/International Regulations: Hazard Symbol: Xi, N; Risk phrases: R36/37/38; R50; Safety phrases: S2; S22; S61 (see Appendix 4).

WGK (German Aquatic Hazard Class): 2—Hazard to waters.

Description: *o*-Phenylphenol is a white to buff-colored crystalline solid. Molecular weight = 170.22; Boiling point = 286°C; Freezing/Melting point = 57°C; Flash point = 124°C; Autoignition temperature = 530°C. Hazard Identification (based on NFPA-704 M Rating System): Health 3, Flammability 1, Reactivity 0. Slightly soluble in water.

Potential Exposure: Compound Description: Agricultural Chemical; Tumorigen, Mutagen; Reproductive Effector; Primary Irritant. *o*-Phenylphenol is used in the manufacture of plastics, resins, rubber, as agricultural chemical; in making fungicides; as an intermediate in making dye stuffs and rubber chemicals; a germicide; used in food packaging.

Incompatibilities: Strong bases, strong oxidizers.

Permissible Exposure Limits in Air

Protective Action Criteria (PAC)

TEEL-0: 60 mg/m^3

PAC-1: 150 mg/m^3

PAC-2: 500 mg/m^3

PAC-3: 500 mg/m^3

DFG MAK: No numerical value established. Data may be available.

Routes of Entry: Inhalation, ingestion, skin and/or eye contact.

Harmful Effects and Symptoms

Short Term Exposure: Irritates the eyes, skin, and respiratory tract. High exposures may affect the kidney, liver, and lungs; gastrointestinal tract; cardiovascular system, causing respiratory failure. *Inhalation:* Dusts can cause irritation of the nose, throat, and lungs. *Skin:* Can cause severe irritation and burns. Concentrations of 0.5% or higher of the sodium form can cause irritation. *Eyes:* Can cause severe irritation, burns, and damage to cornea, especially the sodium form. *Ingestion:* Based on studies of phenol, can cause burning sensation and pain in mouth and throat, sores, abdominal pain, nausea, vomiting, diarrhea, and skin rash. Larger doses may also cause muscle weakness, irregular rapid breathing, blue coloration of the skin, shock, unconsciousness, collapse, and death. Based on animal studies, 5 oz would be lethal to a 150-lb healthy adult.

Long Term Exposure: May cause kidney damage. Prolonged skin contact may cause severe irritation, sores, and skin allergy. Very irritating substances may affect the lungs; bronchitis may develop.

Points of Attack: Lungs, kidneys, skin.

Medical Surveillance: Before beginning employment and at regular times after that, for those with frequent or potentially high exposures, the following are recommended: lung function tests. Consider chest X-ray following acute overexposure.

First Aid: If this chemical gets into the eyes, remove any contact lenses at once and irrigate immediately for at least 15 min, occasionally lifting upper and lower lids. Seek medical attention immediately. If this chemical contacts the skin, remove contaminated clothing and wash immediately with soap and water. Seek medical attention immediately. If this chemical has been inhaled, remove from exposure, begin rescue breathing (using universal precautions, including resuscitation mask) if breathing has stopped and CPR if heart action has stopped. Transfer promptly to a medical facility. When this chemical has been swallowed, get medical attention. Give large quantities of water and induce vomiting. Do not make an unconscious person vomit.

Personal Protective Methods: Wear protective gloves and clothing to prevent any reasonable probability of skin contact. Safety equipment suppliers/manufacturers can provide recommendations on the most protective glove/clothing material for your operation. All protective clothing (suits, gloves, footwear, headgear) should be clean, available each day, and put on before work. Contact lenses should not be worn when working with this chemical. Wear dust-proof chemical goggles and face shield unless full face-piece respiratory protection is worn. Employees should wash immediately with soap when skin is wet or contaminated. Provide emergency showers and eyewash.

Respirator Selection: Where there is potential for exposures to *o*-phenylphenol, use a NIOSH/MSHA- or European Standard EN149-approved full face-piece respirator with a high-efficiency particulate filter. Greater protection is provided by a powered air-purifying respirator. *Where there is potential for high exposures*, use a NIOSH/MSHA- or European Standard EN149-approved supplied-air respirator with a full face-piece operated in the positive-pressure mode, or with a full face-piece, hood, or helmet in the continuous-flow mode; or use a NIOSH/MSHA- or European Standard EN149-approved self-contained breathing apparatus with a full face-piece operated in pressure-demand or other positive-pressure mode.

Storage: Color Code—Blue: Health Hazard/Poison: Store in a secure poison location. Prior to working with this chemical you should be trained on its proper handling and storage. Store in tightly closed containers in a cool, well-ventilated area away from water. Sources of ignition, such as smoking and open flames, are prohibited where *o*-phenylphenol is used, handled, or stored in a manner that could create a potential fire or explosion hazard. A regulated, marked area should be established where this chemical is handled, used, or stored in compliance with OSHA Standard 1910.1045.

Shipping: Dye intermediates, solid, toxic, n.o.s. requires a shipping label of "POISONOUS/TOXIC MATERIALS." They fall in Hazard Class 6.1.

Spill Handling: Evacuate persons not wearing protective equipment from area of spill or leak until cleanup is complete. Remove all ignition sources. Collect powdered material in the most convenient and safe manner and deposit in sealed containers. Ventilate area after cleanup is complete. It may be necessary to contain and dispose of this chemical as a hazardous waste. If material or contaminated runoff enters waterways, notify downstream users of potentially contaminated waters. Contact your local or federal environmental protection agency for specific recommendations. If employees are required to clean up spills, they must be properly trained and equipped. OSHA 1910.120(q) may be applicable.

Fire Extinguishing: This chemical is a combustible solid. Use dry chemical, carbon dioxide, water spray, or alcohol foam extinguishers. Poisonous gases are produced in fire. If material or contaminated runoff enters waterways, notify downstream users of potentially contaminated waters. Notify local health and fire officials and pollution control agencies. From a secure, explosion-proof location, use water spray to cool exposed containers. If cooling streams are ineffective (venting sound increases in volume and pitch, tank discolors, or shows any signs of deforming), withdraw immediately to a secure position. If employees are expected to fight fires, they must be trained and equipped in OSHA 1910.156. The only respirators recommended for firefighting are self-contained breathing apparatuses that have full face-pieces and are operated in a pressure-demand or other positive-pressure mode.

Disposal Method Suggested: In accordance with 40CFR165, follow recommendations for the disposal of pesticides and pesticide containers. Must be disposed properly by following package label directions or by contacting your local or federal environmental control agency or by contacting your regional EPA office.

References

New York State Department of Health. (April 1986). *Chemical Fact Sheet ortho-Phenylphenol.* Albany, NY: Bureau of Toxic Substance Assessment
New Jersey Department of Health and Senior Services. (December 2000). *Hazardous Substances Fact Sheet: o-Phenylphenol.* Trenton, NJ

Phenylphosphine   P:0480

Molecular Formula: C_6H_7P
Common Formula: $C_6H_5PH_2$
Synonyms: Fenylfosfin; PF; Phosphaniline
CAS Registry Number: 638-21-1

RTECS® Number: SZ2100000
EC Number: 211-325-4
Regulatory Authority and Advisory Bodies
Air Pollutant Standard Set. See below, "Permissible Exposure Limits in Air" section.
California Proposition 65 Developmental/Reproductive toxin 8/7/09.
Canada, WHMIS, Ingredients Disclosure List Concentration 1.0%.
European/International Regulations: not listed in Annex 1.
WGK (German Aquatic Hazard Class): No value assigned.
Description: Polyphosphinate is a clear, colorless liquid with a foul odor. Molecular weight = 110.10; Specific gravity (H_2O:1) = 1.001 at 15°C; Boiling point = 160°C. Insoluble in water; reacts.
Potential Exposure: Polyphosphinate is used as an intermediate or a chemical reagent. Polyphosphinate compounds are used as catalysts and antioxidants disproportionate, when heated to give phosphonic acid derivatives plus PF.
Incompatibilities: A strong reducing agent. Reacts violently with strong oxidizers. Water reactive; spontaneously combustible in high concentrations in moist air. Potential exposure to gaseous phenylphosphine and phosphorus oxides when heated above 200°C.
Permissible Exposure Limits in Air
Conversion factor: 1 ppm = 4.50 mg/m³ at 25°C & 1 atm.
OSHA PEL: None.
NIOSH REL: 0.05 ppm/0.25 mg/m³ Ceiling Concentration.
ACGIH TLV®[1]: 0.05 ppm/0.23 mg/m³ Ceiling Concentration.
Protective Action Criteria (PAC)
TEEL-0: 0.0025 ppm
PAC-1: 0.0075 ppm
PAC-2: 0.05 ppm
PAC-3: 4 ppm
Australia: TWA 0.05 ppm (0.25 mg/m³), 1993; Belgium: STEL 0.05 ppm (0.23 mg/m³), 1993; Denmark: TWA 0.05 ppm (0.25 mg/m³), 1999; Finland: STEL 0.05 ppm (0.25 mg/m³), 1999; France: VLE 0.05 ppm (0.25 mg/m³), 1999; Switzerland: MAK-W 0.05 ppm (0.25 mg/m³), 1999; the Netherlands: MAC 0.25 mg/m³, 2003; Argentina, Bulgaria, Columbia, Jordan, South Korea, New Zealand, Singapore, Vietnam: ACGIH TLV®: Ceiling Concentration 0.05 ppm. Several states have set guidelines or standards for PF in ambient air[60] ranging from 2.0 μg/m³ (Virginia) to 2.5 μg/m³ (North Dakota) to 6.0 μg/m³ (Nevada).
Determination in Air: No method available.
Routes of Entry: Inhalation, ingestion, skin and/or eye contact.
Harmful Effects and Symptoms
Short Term Exposure: A level of 0.6 ppm is a threshold effect level for laboratory animals; hypersensitivity to sound and touch and mild hyperemia developed above this level. Above 2.2 ppm, chronic effects developed including decreases in red blood cells, dermatitis, and severe testicular degeneration (which was, however, reversible).[53] This

material is highly toxic by inhalation and ingestion; mild respiratory irritant; emits toxic fumes of phosphorus oxides when heated to decomposition.

Symptoms of exposure include mild respiratory irritation, dyspnea, nausea, vomiting, diarrhea, thirst, sensation of pressure in the chest, back pains, chills, stupor, and fainting with marked pulmonary edema. Phenylphosphine can affect you when breathed in. Exposure can cause nausea, loss of appetite, shaking (tremor), irritation of the eyes, and flushed skin.

Long Term Exposure: Repeated exposure can cause skin rash. Phenylphosphine can damage the blood cells. In animals: blood changes; anemia, testicular degeneration; loss of appetite; diarrhea, lacrimation (discharge of tears), hind leg tremor; dermatitis.

Points of Attack: Blood, central nervous system, skin, reproductive system.

Medical Surveillance: If symptoms develop or overexposure is suspected, the following may be useful: complete blood count. Examination of the nervous system.

First Aid: If this chemical gets into the eyes, remove any contact lenses at once and irrigate immediately for at least 15 min, occasionally lifting upper and lower lids. Seek medical attention immediately. If this chemical contacts the skin, remove contaminated clothing and wash immediately with soap and water. Seek medical attention immediately. If this chemical has been inhaled, remove from exposure, begin rescue breathing (using universal precautions, including resuscitation mask) if breathing has stopped and CPR if heart action has stopped. Transfer promptly to a medical facility. When this chemical has been swallowed, get medical attention. Give large quantities of water and induce vomiting. Do not make an unconscious person vomit.

Personal Protective Methods: Wear protective gloves and clothing to prevent any reasonable probability of skin contact. Safety equipment suppliers/manufacturers can provide recommendations on the most protective glove/clothing material for your operation. All protective clothing (suits, gloves, footwear, headgear) should be clean, available each day, and put on before work. Contact lenses should not be worn when working with this chemical. Wear splash-proof chemical goggles and face shield unless full face-piece respiratory protection is worn. Employees should wash immediately with soap when skin is wet or contaminated. Provide emergency showers and eyewash.

Respirator Selection: Where there is potential for exposures *over 0.05* ppm, use a NIOSH/MSHA- or European Standard EN149-approved supplied-air respirator with a full face-piece operated in the positive-pressure mode, or with a full face-piece, hood, or helmet in the continuous-flow mode; or use a NIOSH/MSHA- or European Standard EN149-approved self-contained breathing apparatus with a full face-piece operated in pressure-demand or other positive-pressure mode.

Storage: Pyrophoric. Color Code—Yellow Stripe (*strong reducing agent*): Reactivity Hazard; Store separately in an area isolated from flammables, combustibles, or other yellow-coded materials. Prior to working with this chemical you should be trained on its proper handling and storage. Store in tightly closed containers in a cool, well-ventilated area.

Spill Handling: Evacuate and restrict persons not wearing protective equipment from area of spill or leak until cleanup is complete. Remove all ignition sources. Ventilate area of spill or leak. Absorb liquids in vermiculite, dry sand, earth, peat, carbon, or a similar material and deposit in sealed containers. Keep this chemical out of a confined space, such as a sewer, because of the possibility of an explosion, unless the sewer is designed to prevent the buildup of explosive concentrations. It may be necessary to contain and dispose of this chemical as a hazardous waste. If material or contaminated runoff enters waterways, notify downstream users of potentially contaminated waters. Contact your local or federal environmental protection agency for specific recommendations. If employees are required to clean up spills, they must be properly trained and equipped. OSHA 1910.120(q) may be applicable.

Fire Extinguishing: This chemical is a flammable liquid. Poisonous gases, including phosphorus oxides, are produced in fire. Use dry chemical, carbon dioxide, or alcohol foam extinguishers. Vapors are heavier than air and will collect in low areas. Vapors may travel long distances to ignition sources and flashback. Vapors in confined areas may explode when exposed to fire. Containers may explode in fire. Storage containers and parts of containers may rocket great distances, in many directions. If material or contaminated runoff enters waterways, notify downstream users of potentially contaminated waters. Notify local health and fire officials and pollution control agencies. From a secure, explosion-proof location, use water spray to cool exposed containers. If cooling streams are ineffective (venting sound increases in volume and pitch, tank discolors, or shows any signs of deforming), withdraw immediately to a secure position. If employees are expected to fight fires, they must be trained and equipped in OSHA 1910.156. The only respirators recommended for firefighting are self-contained breathing apparatuses that have full face-pieces and are operated in a pressure-demand or other positive-pressure mode.

Reference

New Jersey Department of Health and Senior Services. (July 2001). *Hazardous Substances Fact Sheet: Phenylphosphine.* Trenton, NJ

Phenylthiourea   P:0490

Molecular Formula: $C_7H_8N_2S$
Common Formula: $C_6H_5NCHCSNH_2$
Synonyms: NCI-C02017; Phenylthiocarbamide; *N*-phenylthiourea; α-Phenylthiourea; Phenyl-2-thiourea; 1-Phenylthiourea; PTC; PTU

CAS Registry Number: 103-85-5
RTECS® Number: YU1400000
UN/NA & ERG Number: UN2767/151
EC Number: 203-151-2
Regulatory Authority and Advisory Bodies
US EPA Hazardous Waste Number (RCRA No.): P093.
RCRA, 40CFR261, Appendix 8 Hazardous Constituents.
Superfund/EPCRA 40CFR355, Extremely Hazardous Substances: TPQ = 100/10,000 lb (45.4/4540 kg).
Reportable Quantity (RQ): 100 lb (45.4 kg).
European/International Regulations: Hazard Symbol: Xn, N; Risk phrases: R36/38; R43; R51/53; Safety phrases: S28; S36/37-45 (see Appendix 4).
WGK (German Aquatic Hazard Class): No value assigned.
Description: *N*-phenylthiourea is a colorless crystalline solid. Molecular weight = 152.23; Freezing/Melting point = 148−154°C. Soluble in water.
Potential Exposure: Used as a repellent for rats, rabbits, and weasels; in the manufacture of rodenticides and in medical genetics.
Incompatibilities: Incompatible with oxidizers, strong bases, and acids. Contact with acids or acid fumes produces toxic fumes of sulfur oxide.
Permissible Exposure Limits in Air
Protective Action Criteria (PAC)
TEEL-0: 0.6 mg/m^3
PAC-1: 1.5 mg/m^3
PAC-2: 3 mg/m^3
PAC-3: 3 mg/m^3
Routes of Entry: Inhalation, ingestion, skin and/or eye contact.
Harmful Effects and Symptoms
Short Term Exposure: Irritates the eyes, skin, and respiratory tract. High exposures can cause lung irritation, coughing, and/or shortness of breath. Higher exposures can cause pulmonary edema, a medical emergency that can be delayed for several hours. This can cause death. Exposure may result in vomiting, difficult breathing, noisy breathing, cyanosis, and low body temperature. It is classified as extremely toxic. The probable oral lethal dose is 5−50 mg/kg or between 7 drops and 1 teaspoon for a 70-kg (150 lb) person.
Long Term Exposure: Not tested for long-term health effects. May cause methemoglobinemia, cyanosis, and anemia. Phenylthiourea is reported to be similar to ANTU.
Points of Attack: Lungs.
Medical Surveillance: Lung function tests. Blood methemogloblin level. Completed blood count (CBC). Consider chest X-ray following acute overexposure.
First Aid: If this chemical gets into the eyes, remove any contact lenses at once and irrigate immediately for at least 15 min, occasionally lifting upper and lower lids. Seek medical attention immediately. If this chemical contacts the skin, remove contaminated clothing and wash immediately with soap and water. Seek medical attention immediately. If this chemical has been inhaled, remove from exposure,

begin rescue breathing (using universal precautions, including resuscitation mask) if breathing has stopped and CPR if heart action has stopped. Transfer promptly to a medical facility. When this chemical has been swallowed, get medical attention. Give large quantities of water and induce vomiting. Do not make an unconscious person vomit. Medical observation is recommended for 24−48 h after breathing overexposure, as pulmonary edema may be delayed. As first aid for pulmonary edema, a doctor or authorized paramedic may consider administering a corticosteroid spray.
Note to physician: Treat for methemoglobinemia. Spectrophotometry may be required for precise determination of levels of methemoglobin in urine.
Personal Protective Methods: Wear protective gloves and clothing to prevent any reasonable probability of skin contact. Safety equipment suppliers/manufacturers can provide recommendations on the most protective glove/clothing material for your operation. All protective clothing (suits, gloves, footwear, headgear) should be clean, available each day, and put on before work. Contact lenses should not be worn when working with this chemical. Wear dust-proof chemical goggles and face shield unless full face-piece respiratory protection is worn. Employees should wash immediately with soap when skin is wet or contaminated. Provide emergency showers and eyewash.
Respirator Selection: Follow regulations in OSHA 29CFR1910.134 or European Standard EN149. Use a NIOSH/MSHA- or European Standard EN149-approved respirator; or use an approved supplied-air respirator with a full face-piece operated in the positive-pressure mode, or with a full face-piece, hood, or helmet in the continuous-flow mode; or use a NIOSH/MSHA- or European Standard EN149-approved self-contained breathing apparatus with a full face-piece operated in pressure-demand or other positive-pressure mode.
Storage: Color Code—Blue: Health Hazard/Poison: Store in a secure poison location. Prior to working with this chemical you should be trained on its proper handling and storage. Store in a refrigerator or a cool, dry place.
Shipping: Phenylurea pesticides, solid, toxic, n.o.s. requires a shipping label of "POISONOUS/TOXIC MATERIALS." It falls in Hazard Class 6.1 and Packing Group I.
Spill Handling: Evacuate persons not wearing protective equipment from area of spill or leak until cleanup is complete. Remove all ignition sources. Dampen spilled material with alcohol to avoid dust or use HEPA vacuum or wet method to reduce dust during cleanup. Do not dry sweep. Collect powdered material in the most convenient and safe manner and deposit in sealed containers. Ventilate area after cleanup is complete. It may be necessary to contain and dispose of this chemical as a hazardous waste. If material or contaminated runoff enters waterways, notify downstream users of potentially contaminated waters. Contact your local or federal environmental protection agency for specific recommendations. If employees are required to clean up

spills, they must be properly trained and equipped. OSHA 1910.120(q) may be applicable.

Fire Extinguishing: This chemical is a combustible solid. Use dry chemical, carbon dioxide, water spray, alcohol foam or polymer foam extinguishers. Poisonous gases are produced in fire, including nitrogen oxides and sulfur oxides. If material or contaminated runoff enters waterways, notify downstream users of potentially contaminated waters. Notify local health and fire officials and pollution control agencies. From a secure, explosion-proof location, use water spray to cool exposed containers. If cooling streams are ineffective (venting sound increases in volume and pitch, tank discolors, or shows any signs of deforming), withdraw immediately to a secure position. If employees are expected to fight fires, they must be trained and equipped in OSHA 1910.156. The only respirators recommended for firefighting are self-contained breathing apparatuses that have full face-pieces and are operated in a pressure-demand or other positive-pressure mode.

Disposal Method Suggested: In accordance with 40CFR165, follow recommendations for the disposal of pesticides and pesticide containers. Must be disposed properly by following package label directions or by contacting your local or federal environmental control agency or by contacting your regional EPA office.

References

US Environmental Protection Agency. (November 30, 1987). *Chemical Hazard Information Profile: Phenylthiourea.* Washington, DC: Chemical Emergency Preparedness Program

New Jersey Department of Health and Senior Services. (August 1999). *Hazardous Substances Fact Sheet: Phenylthiourea.* Trenton, NJ

Phenyl trichlorosilane   P:0500

Molecular Formula: $C_6H_5Cl_3Si$

Synonyms: Phenylsilicon trichloride; Phenyl trichlorosilane; Silicon phenyl trichloride; Silane, trichlorophenyl-; Trichlorophenylsilane; Tricloro(fenil)silano (Spanish)

CAS Registry Number: 98-13-5

RTECS® Number: VV6650000

UN/NA & ERG Number: UN1804/156

EC Number: 202-640-8

Regulatory Authority and Advisory Bodies

Department of Homeland Security Screening Threshold Quantity (pounds): Sabotage/Contamination Hazard: A placarded amount (commercial grade).

Superfund/EPCRA 40CFR355, Extremely Hazardous Substances: TPQ = 500 lb (227 kg).

Reportable Quantity (RQ): 500 lb (227 kg).

US DOT 49CFR172.101, Inhalation Hazardous Chemical.

WGK (German Aquatic Hazard Class): 1—Low hazard to waters.

Description: Phenyl trichlorosilane is a colorless to light yellow liquid. Molecular weight = 211.55; Specific gravity = 1.32 at 25°C; Boiling point = 202°C; Flash point = 80.6°C; 91°C (oc). Hazard Identification (based on NFPA-704 M Rating System): Health 3, Flammability 2, Reactivity 2W. Water reactive.

Potential Exposure: Phenyl trichlorosilane is used to make silicones for water repellants, insulating resins, heat-resistant paints, and as a laboratory reagent.

Incompatibilities: May spontaneously ignite in air above flash point. Contact with water, steam, or moisture forms hydrogen chloride. Trichlorosilanes may react violently with strong oxidants, strong acids, bases, amines, alcohols, acetone, ammonia. Attacks many metals in the presence of water, releasing explosive hydrogen gas.

Permissible Exposure Limits in Air

Protective Action Criteria (PAC)

TEEL-0: 0.2 ppm

PAC-1: 0.6 ppm

PAC-2: 7.3 ppm

PAC-3: 33 ppm

Routes of Entry: Inhalation, ingestion, skin and/or eye contact.

Harmful Effects and Symptoms

Short Term Exposure: Corrosive to the eyes, skin, and respiratory tract. Eye contact may damage the corneas and cause blindness. Inhalation may cause throat to swell, causing suffocation; and may cause pulmonary edema, a medical emergency that can be delayed for several hours. This can cause death. Highly toxic; may cause death or permanent injury after short inhalation exposure to small quantity. Chemical burns to all exposed membranes and tissues with severe tissue destruction. *Delayed:* after oral exposure, stomach and intestines may perforate to be obstructed by scar tissue. Ingestion may cause mild to moderately severe oral and esophageal burns, with severe burns occurring in stomach. Perforations and peritonitis may occur. Severe irritation may produce spontaneous vomiting. Viscid white or blood-stained foamy mucus and threads of tissue may appear in mouth.

Long Term Exposure: Many highly irritating substances can cause lung damage; bronchitis may develop.

Points of Attack: Lungs.

Medical Surveillance: For those with frequent or potentially high exposure, the following are recommended before beginning work and at regular times after that: lung function tests. If symptoms develop or overexposure is suspected, the following may be useful: lung function tests. Consider chest X-ray after acute overexposure.

First Aid: If this chemical gets into the eyes, remove any contact lenses at once and irrigate immediately for at least 30 min, occasionally lifting upper and lower lids. Seek medical attention immediately. If this chemical contacts the skin, remove contaminated clothing and wash immediately with soap and water. Seek medical attention immediately. If this chemical has been inhaled, remove from exposure,

begin rescue breathing (using universal precautions, including resuscitation mask) if breathing has stopped and CPR if heart action has stopped. Transfer promptly to a medical facility. When this chemical has been swallowed, get medical attention. If victim is *conscious*, administer water or milk; then give demulcents, such as milk, cornstarch, and water. Do not induce vomiting.

Medical observation is recommended for 24—48 h after breathing overexposure, as pulmonary edema may be delayed. As first aid for pulmonary edema, a doctor or authorized paramedic may consider administering a corticosteroid spray.

Personal Protective Methods: Wear protective gloves and clothing to prevent any reasonable probability of skin contact. Safety equipment suppliers/manufacturers can provide recommendations on the most protective glove/clothing material for your operation. All protective clothing (suits, gloves, footwear, headgear) should be clean, available each day, and put on before work. Contact lenses should not be worn when working with this chemical. Wear splash-proof chemical goggles and face shield unless full face-piece respiratory protection is worn. Employees should wash immediately with soap when skin is wet or contaminated. Provide emergency showers and eyewash.

Respirator Selection: Where there is potential for exposure to phenyl trichlorosilane, use a NIOSH/MSHA- or European Standard EN149-approved supplied-air respirator with a full face-piece operated in the positive-pressure mode, or with a full face-piece, hood, or helmet in the continuous-flow mode; or use a NIOSH/MSHA- or European Standard EN149-approved self-contained breathing apparatus with a full face-piece operated in pressure-demand or other positive-pressure mode.

Storage: Color Code—White: Corrosive or Contact Hazard; Store separately in a corrosion resistant location. Prior to working with this chemical you should be trained on its proper handling and storage. Store in tightly closed containers in a cool, well-ventilated area away from water at temperatures below 50°C. Phenyl trichlorosilane can give off corrosive hydrogen chloride gas on contact with water, steam, or moisture. Sources of ignition, such as smoking and open flames, are prohibited where Phenyl trichlorosilane is used, handled, or stored in a manner that could create a potential fire or explosion hazard.

Shipping: Phenyl trichlorosilane requires a shipping label of "CORROSIVE." It falls in Hazard Class 8 and Packing Group II.

Spill Handling: Evacuate and restrict persons not wearing protective equipment from area of spill or leak until cleanup is complete. Remove all ignition sources. Ventilate area of spill or leak. Absorb liquids in vermiculite, dry sand, earth, peat, carbon, or a similar material and deposit in sealed containers. Keep this chemical out of a confined space, such as a sewer, because of the possibility of an explosion, unless the sewer is designed to prevent the buildup of explosive concentrations. It may be necessary to contain and dispose of this chemical as a hazardous waste. If material or contaminated runoff enters waterways, notify downstream users of potentially contaminated waters. Contact your local or federal environmental protection agency for specific recommendations. If employees are required to clean up spills, they must be properly trained and equipped. OSHA 1910.120(q) may be applicable.

Initial isolation and protective action distances
Distances shown are likely to be affected during the first 30 min after materials are spilled and could increase with time. If more than one tank car, cargo tank, portable tank, or large cylinder involved in the incident is leaking, the protective action distance may need to be increased. You may need to seek emergency information from CHEMTREC at (800) 424-9300 or seek professional environmental engineering assistance from the US EPA Environmental Response Team at (908) 548-8730 (24-h response line).
Small spills (From a small package or a small leak from a large package)
when spilled in water
First: Isolate in all directions (feet/meters)   100/30
Then: Protect persons downwind (miles/kilometers)
Day   0.1/0.2
Night   0.1/0.2
Large spills (From a large package or from many small packages)
First: Isolate in all directions (feet/meters)   200/60
Then: Protect persons downwind (miles/kilometers)
Day   0.3/0.5
Night   1.0/1.5

Fire Extinguishing: This chemical is a combustible and corrosive liquid. Poisonous gases, including chlorine and hydrogen chloride, are produced in fire. Use dry chemical, carbon dioxide. *Do not use water* or hydrous agents. Full protective clothing, including self-contained breathing apparatus; coat, pants, gloves, boots; and bands around legs, arms, and waist, should be provided. No skin surface should be exposed. Move container from fire area if you can do so without risk. Vapors are heavier than air and will collect in low areas. Vapors may travel long distances to ignition sources and flashback. Vapors in confined areas may explode when exposed to fire. Containers may explode in fire. Storage containers and parts of containers may rocket great distances, in many directions. If material or contaminated runoff enters waterways, notify downstream users of potentially contaminated waters. Notify local health and fire officials and pollution control agencies. From a secure, explosion-proof location, use water spray to cool exposed containers. Do not get water inside containers. If cooling streams are ineffective (venting sound increases in volume and pitch, tank discolors, or shows any signs of deforming), withdraw immediately to a secure position. If employees are expected to fight fires, they must be trained and equipped in OSHA 1910.156. The only respirators recommended for firefighting are self-contained breathing apparatuses that have full face-pieces and are operated in a pressure-demand or other positive-pressure mode.

References
US Environmental Protection Agency. (November 30, 1987). *Chemical Hazard Information Profile: Trichlorophenylsilane*. Washington, DC: Chemical Emergency Preparedness Program
New Jersey Department of Health and Senior Services. (May 2000). *Hazardous Substances Fact Sheet: Phenyl Trichlorosilane*. Trenton, NJ

Phenytoin   P:0510

Molecular Formula: $C_{15}H_{12}N_2O_2$
Synonyms: AI3-52498; Aleviatin; Antisacer; Auranile; Causoin; Citrullamon; Citrulliamon; Comital; Comitoina; Convul; Danten; Dantinal; Dantoinal; Dantoinal klinos; Dantoine; Denyl; Didan TDC 250; Difenilhidantoina (Spanish); Difenin; Difhydan; Dihycon; di-Hydan; Dihydantoin; di-Lan; Dilantin acid; Dilantine; Dillantin; Dintion; Diphantoin; Diphedal; Diphenine; Diphentoin; Diphentyn; Diphenylan; 5,5-Diphenylhydantoin; Diphenylhydantoin; 5,5-Diphenylimidazolidin-2,4-dione; 5,5-Diphenyl-2,4-imida zolidinedione; Diphenylhydantoine; di-Phetine; Ditoinate; DPH; EKKO; EKKO Capsules; Enkelfel; Elepsindon; Epamin; Epanutin; Epasmir 5; Epdantoine simple; Epelin; Epilan; Epilantin; Epinat; Epised; Eptal; Eptoin; Fenantoin; Fenidantoin S; Fenitoina; Fenylepsin; Fenytoine; Gerot-epilan-D; Hidan; Hidantilo; Hidantina; Hidantina senosian; Hidantina vitoria; Hidantomin; Hydantoin; Hydantoin, 5,5-diphenyl-; Hydantoinal; Ictalis simple; Idantoin 2,4-imidazolidine-dione, 5,5-diphenyl-; Kessodanten; Labopal; Lehydan; Lepitoin; Lepsin; Minetoin; NCI-C55765; Neos-hidantoina; Neosidantoina; Novantoina; OM-hidantoine simple; OM-hydantoine; Oxylan; Phanantin; Phanatine; Phenatine; Phenatoine; Phenitoin; Ritmenal; Saceril; Sanepil; Silantin; Sodanthon; Sodantoin; Solantoin; Sylantoic; Tacosal; Thilophenyl; Toin; Toin unicelles; Zentronal; Zentropil
CAS Registry Number: 57-41-0; 630-93-3 (sodium salt)
RTECS® Number: MU1050000
UN/NA & ERG Number: UN3249 (Medicines, toxic, solid, n.o.s.)/151
EC Number: 200-328-6; 211-148-2 (phenytoin sodium)
Regulatory Authority and Advisory Bodies
Carcinogenicity: IARC: Human Limited Evidence, animal Sufficient Evidence, *possibly carcinogenic to humans*, Group 2B, 1987; NTP: Reasonably anticipated to be a human carcinogen.
Clean Air Act: Accidental Release Prevention/Flammable Substances, (Section 112[r], Table 3), TQ = 15,000 lb (6810 kg).
US EPA Hazardous Waste Number (RCRA No.): U098.
RCRA, 40CFR261, Appendix 8 Hazardous Constituents.
EPCRA Section 313 Form R *de minimis* concentration reporting level: 0.1%.

Superfund/EPCRA 40CFR355, Extremely Hazardous Substances: TPQ = 1000 lb (454 kg).
Reportable Quantity (RQ): 10 lb (4.54 kg).
California Proposition 65 Chemical: Cancer 1/1/88; Reproductive toxin 7/1/87; Cancer 1/1/88 (sodium salt).
European/International Regulations: not listed in Annex 1.
WGK (German Aquatic Hazard Class): No value assigned.
Description: Phenytoin is a crystalline compound. Molecular weight = 252.29; Freezing/Melting point = 295−298°C; Ignition temperature = 585°C. May react with water.
Potential Exposure: Phenytoin is a pharmaceutical used in the treatment of grand mal epilepsy, Parkinson's syndrome; and in veterinary medicine. Human exposure to phenytoin occurs principally during its use as a drug. Figures on the number of patients using phenytoin are not available, but phenytoin is given to a major segment of those individuals with epilepsy. The oral dose rate is initially 100 mg given 3 times per day and can gradually increase by 100 mg every 2−4 weeks until the desired therapeutic response is obtained. The intravenous dose is 200−350 mg/day.
Incompatibilities: Strong acids, strong oxidizers, water.
Permissible Exposure Limits in Air
No standards or TEEL available.
Routes of Entry: Inhalation, ingestion, skin and/or eye contact.
Harmful Effects and Symptoms
Short Term Exposure: Symptoms of exposure include blurred vision, hyperactivity, confusion, drowsiness, nausea, vomiting, epigastric pain, swelling of gums, fever, liver and kidney damage.
Long Term Exposure: Phenytoin is carcinogenic in mice after oral administration or by intraperitoneal injection, producing lymphomas and leukemias.
Points of Attack: Liver, kidneys.
Medical Surveillance: Liver and kidney function tests.
First Aid: Skin Contact[52]: Flood all areas of body that have contacted the substance with water. Do not wait to remove contaminated clothing; do it under the water stream. Use soap to help assure removal. Isolate contaminated clothing when removed to prevent contact by others. *Eye Contact:* Remove any contact lenses at once. Flush eyes well with copious quantities of water or normal saline for at least 20−30 min. Seek medical attention. *Inhalation:* Leave contaminated area immediately; breathe fresh air. Proper respiratory protection must be supplied to any rescuers. If coughing, difficult breathing, or any other symptoms develop, seek medical attention at once, even if symptoms develop many hours after exposure. *Ingestion:* If convulsions are not present, give a glass or two of water or milk to dilute the substance. Assure that the person's airway is unobstructed and contact a hospital or poison center immediately for advice on whether or not to induce vomiting.
Personal Protective Methods: Wear protective gloves and clothing to prevent any reasonable probability of skin contact. Safety equipment suppliers/manufacturers can provide

recommendations on the most protective glove/clothing material for your operation. All protective clothing (suits, gloves, footwear, headgear) should be clean, available each day, and put on before work. Contact lenses should not be worn when working with this chemical. Wear dust-proof chemical goggles and face shield unless full face-piece respiratory protection is worn. Employees should wash immediately with soap when skin is wet or contaminated. Provide emergency showers and eyewash.

Respirator Selection: Specific respirator(s) have not been recommended by NIOSH. However, based on potential carcinogenicity, and where the potential exists for exposure, the following might be considered:

At any detectable concentration: SCBAF: Pd,Pp (APF = 10,000) (any NIOSH/MSHA- or European Standard EN 149-approved self-contained breathing apparatus that has a full face-piece and is operated in a pressure-demand or other positive-pressure mode) or SaF: Pd,Pp: ASCBA (APF = 10,000) (any supplied-air respirator that has a full face-piece and is operated in a pressure-demand or other positive-pressure mode in combination with an auxiliary, self-contained breathing apparatus operated in a pressure-demand or other positive-pressure mode). *Escape:* 100 F (APF = 50) (any air-purifying, full-face-piece respirator with an N100, R100, or P100 filter) or SCBAE (any appropriate escape-type, self-contained breathing apparatus).

Storage: Color Code—Blue: Health Hazard/Poison: Store in a secure poison location. Prior to working with this chemical you should be trained on its proper handling and storage. Store in a cool, dry place or in a refrigerator. Protection from air, light, and moisture is recommended for long term storage.[52] A regulated, marked area should be established where this chemical is handled, used, or stored in compliance with OSHA Standard 1910.1045.

Shipping: Medicine, solid, toxic, n.o.s. requires a shipping label of "POISONOUS/TOXIC MATERIALS." This compound falls in Hazard Class 6.1 and Packing Group III.

Spill Handling: Evacuate persons not wearing protective equipment from area of spill or leak until cleanup is complete. Remove all ignition sources[52]: Dampen spilled material with alcohol to avoid dust, then transfer material to a suitable container. Use absorbent dampened with alcohol to pick up remaining material. Wash surfaces well with soap and water. Collect powdered material in the most convenient and safe manner and deposit in sealed containers. Ventilate area after cleanup is complete. It may be necessary to contain and dispose of this chemical as a hazardous waste. If material or contaminated runoff enters waterways, notify downstream users of potentially contaminated waters. Contact your local or federal environmental protection agency for specific recommendations. If employees are required to clean up spills, they must be properly trained and equipped. OSHA 1910.120(q) may be applicable.

Fire Extinguishing: This chemical is a combustible solid. Use dry chemical, carbon dioxide, water spray, or alcohol foam extinguishers. Poisonous gases, including nitrogen oxides, are produced in fire. If material or contaminated runoff enters waterways, notify downstream users of potentially contaminated waters. Notify local health and fire officials and pollution control agencies. From a secure, explosion-proof location, use water spray to cool exposed containers. If cooling streams are ineffective (venting sound increases in volume and pitch, tank discolors, or shows any signs of deforming), withdraw immediately to a secure position. If employees are expected to fight fires, they must be trained and equipped in OSHA 1910.156. The only respirators recommended for firefighting are self-contained breathing apparatuses that have full face-pieces and are operated in a pressure-demand or other positive-pressure mode.

Reference

New Jersey Department of Health and Senior Services. (April 2001). *Hazardous Substances Fact Sheet: Phenytoin.* Trenton, NJ

Phorate   P:0520

Molecular Formula: $C_7H_{17}O_2PS_3$

Synonyms: Aastar; AC3911; American cyanamid 3,911; *O,O*-Diaethyl-*S*-(aethylthio-methyl)-dithiophosphat (German); *O,O*-Diethyl *S*-ethylmercaptomethyl dithiophosphonate; *O,O*-Diethyl *S*-ethylthiomethyl dithiophosphonate; *O,O*-Diethyl *S*-(ethylthio)methyl phosphorodithioate; *O,O*-Diethyl *S*-[(ethylthio)methyl] phosphorodithioate; *O,O*-Diethylethylthiomethyl phosphorodithioate; *O,O*-Diethyl *S*-ethylthiomethyl thiothionophosphate; Dithiophosphatede *O,O*-diethyle et d'ethylthiomethyle (French); EL3911; ENT 24,042; Experimental insecticide 3911; Forato (Spanish); Geomet; Gramtox; Granutox; L11/6; Methanethiol, ethylthio-*S*-ester with *O,O*-diethyl phosphorodithioate; Phorat (German); Phorate-10G; Rampart; Terrathion granules; the met®; Thimet®; Vegfru; Vergfru Foratox

CAS Registry Number: 298-02-2

RTECS® Number: TD9450000

UN/NA & ERG Number: UN3018 (organophosphorus pesticide, liquid, toxic)/152

EC Number: 206-052-2 [*Annex I Index No.:* 015-033-00-6]

Regulatory Authority and Advisory Bodies

US EPA Gene-Tox Program, Negative: *D. melanogaster* sex-linked lethal; Negative: *In vitro* UDS—human fibroblast; TRP reversion; Negative: *S. cerevisiae*—homozygosis; Inconclusive: *B. subtilis* rec assay; *E. coli* polA without S9; Inconclusive: Histidine reversion—Ames test.

US EPA, FIFRA, 1998 Status of Pesticides: Supported.

Banned or Severely Restricted (Malaysia) (UN).[13]

Very Toxic Substance (World Bank).[15]

Air Pollutant Standard Set. See below, "Permissible Exposure Limits in Air" section.

US EPA Hazardous Waste Number (RCRA No.): P094.

RCRA, 40CFR261, Appendix 8 Hazardous Constituents.

RCRA 40CFR268.48; 61FR15654, Universal Treatment Standards: Wastewater (mg/L), 0.021; Nonwastewater (mg/kg), 4.6.

RCRA 40CFR264, Appendix 9; TSD Facilities Ground Water Monitoring List. Suggested test method(s) (PQL μg/L): 8140 (2); 8270 (10).

Superfund/EPCRA 40CFR355, Extremely Hazardous Substances: TPQ = 10 lb (4.54 kg).

Reportable Quantity (RQ): 10 lb (4.54 kg).

US DOT Regulated Marine Pollutant (49CFR172.101, Appendix B), severe pollutant.

US DOT 49CFR172.101, Inhalation Hazard Chemical as organophosphates.

European/International Regulations: Hazard Symbol: T+, N; Risk phrases: R27/28; R50/53 Safety phrases: S1/2; S28; S36/37; S45; S60; S61 (see Appendix 4).

WGK (German Aquatic Hazard Class): No value assigned.

Description: Phorate is a clear mobile liquid with a skunk-like odor. Molecular weight = 260.39; Specific gravity (H_2O:1) = 1.16 at 25°C; Boiling point = 118–120°C at 0.8 mm; Freezing/Melting point = −42.8°C; Vapor pressure = 0.0008 at 20°C; Flash point = 160°C. Hazard Identification (based on NFPA-704 M Rating System): Health 4, Flammability 1, Reactivity 0. Practically insoluble in water; solubility = 0.005%.

Potential Exposure: Compound Description: Agricultural Chemical; Mutagen; Reproductive Effector; Human Data; Primary Irritant. Those engaged in the manufacture, formulation, and application of this systemic and contact insecticide and acaricide. It is also used as a soil insecticide.

Incompatibilities: Water, alkalis. Hydrolyzed in the presence of moisture and by alkalis; may produce toxic oxides of phosphorus and sulfur. Strong oxidizers may cause release of toxic phosphorus oxides. Organophosphates, in the presence of strong reducing agents such as hydrides, may form highly toxic and flammable phosphine gas. Keep away from alkaline materials.

Permissible Exposure Limits in Air

OSHA PEL: None.

NIOSH REL: 0.05 mg/m³ TWA; 0.2 mg/m³ STEL [skin].

ACGIH TLV®[1]: 0.05 mg/m³ TWA, inhalable fraction and vapor; [skin] not classifiable as a human carcinogen; BEI$_A$ issued for Acetylcholinesterase inhibiting pesticides.

Protective Action Criteria (PAC)*

TEEL-0: 0.04 mg/m³

PAC-1: 0.04 mg/m³

PAC-2: **0.040** mg/m³

PAC-3: **0.12** mg/m³

AEGLs (Acute Emergency Guideline Levels) & ERPGs (Emergency Response Planning Guideline) are in **bold face**.

Australia: TWA 0.05 mg/m³; STEL 0.2 mg/m³, [skin], 1993; Belgium: TWA 0.05 mg/m³, [skin], 1993; Denmark: TWA 0.05 mg/m³, [skin], 1999; France: VME 0.05 mg/m³, [skin], 1999; Norway: TWA 0.05 mg/m³, 1999; Switzerland: MAK-W 0.05 mg/m³, [skin], 1999; United Kingdom: TWA 0.05 mg/m³; STEL 0.2 mg/m³, [skin], 2000; the Netherlands: MAC-TGG 0.05 mg/m³, [skin], 2003; Argentina, Bulgaria, Columbia, Jordan, South Korea, New Zealand, Singapore, Vietnam: ACGIH TLV®: STEL 0.2 mg/m³ [skin]. Several states have set guidelines or standards for Phorate in ambient air[60] ranging from 0.5–2.0 μg/m³ (North Dakota) to 0.8 μg/m³ (Virginia) to 1.0 μg/m³ (Connecticut and Nevada).

Determination in Air: Use NIOSH Analytical Method (IV) #5600, Organophosphorus pesticides.

Permissible Concentration in Water: Maine[61] has set a guideline for phorate in drinking water of 0.2 μg/L.

Determination in Water: Fish Tox = 0.13505000 ppb (EXTRA HIGH). Octanol–water coefficient: Log K_{ow} = 3.88.

Routes of Entry: Inhalation, ingestion, skin and/or eye contact. Absorbed through the skin.

Harmful Effects and Symptoms

Short Term Exposure: Acute exposure to phorate may produce the following signs and symptoms: pinpoint pupils, blurred vision, headache, dizziness, muscle spasms, and profound weakness. Vomiting, diarrhea, abdominal pain, seizures, and coma may also occur. The heart rate may decrease following oral exposure or increase following dermal exposure. Chest pain may be noted. Hypotension (low blood pressure) may occur, although hypertension (high blood pressure) is not uncommon. Dyspnea (shortness of breath) may be followed by respiratory collapse. Giddiness is common. This material is one of the most toxic organophosphorus insecticides. It is a cholinesterase inhibitor that acts on the nervous system; and produces toxicity similar to parathion. The probable oral lethal dose for humans is less than 5 mg/kg, i.e., a taste (less than 7 drops) for a 70-kg (150 lb) person. LD$_{50}$ = (oral-rat) 37 mg/kg.

Long Term Exposure: Cholinesterase inhibitor; cumulative effect is possible. This chemical may damage the nervous system with repeated exposure; resulting in convulsions, respiratory failure. May cause liver damage. Human Tox = 3.50000 ppm (HIGH).

Points of Attack: Respiratory system, lungs, central nervous system, cardiovascular system, skin, eyes, plasma and red blood cell cholinesterase.

Medical Surveillance: Before employment and at regular times after that, the following are recommended: plasma and red blood cell cholinesterase levels (tests for the enzyme poisoned by this chemical). If exposure stops, plasma levels return to normal in 1–2 weeks while red blood cell levels may be reduced for 1–3 months. Do not drink any alcoholic beverages before or during use. Alcohol promotes absorption of organic phosphates.

When cholinesterase enzyme levels are reduced by 25% or more below preemployment levels, risk of poisoning is increased, even if results are in lower ranges of "normal." Reassignment to work not involving organophosphate or carbamate pesticides is recommended until enzyme levels recover. If symptoms develop or overexposure occurs, repeat the above tests as soon as possible and get an examination of the nervous system. Also consider complete blood count. Consider chest X-ray following acute overexposure. Do not drink any alcoholic beverages before or during use. Alcohol promotes absorption of organic phosphates.

First Aid: If this chemical gets into the eyes, remove any contact lenses at once and irrigate immediately for at least 15 min, occasionally lifting upper and lower lids. Seek medical attention immediately. If this chemical contacts the skin, remove contaminated clothing and wash immediately with soap and water. Speed in removing material from skin is of extreme importance. Seek medical attention immediately. If this chemical has been inhaled, remove from exposure, begin rescue breathing (using universal precautions, including resuscitation mask) if breathing has stopped and CPR if heart action has stopped. Transfer promptly to a medical facility. When this chemical has been swallowed, get medical attention. Give large quantities of water and induce vomiting. Do not make an unconscious person vomit.

Personal Protective Methods: Wear protective gloves and clothing to prevent any reasonable probability of skin contact. Safety equipment suppliers/manufacturers can provide recommendations on the most protective glove/clothing material for your operation. All protective clothing (suits, gloves, footwear, headgear) should be clean, available each day, and put on before work. Contact lenses should not be worn when working with this chemical. Wear splash-proof chemical goggles and face shield unless full face-piece respiratory protection is worn. Employees should wash immediately with soap when skin is wet or contaminated. Provide emergency showers and eyewash.

Respirator Selection: Not available according to NIOSH.

The following is included for reference: NIOSH: *(parathion) 0.5 mg/m^3:* CcrOv95 (APF = 10) [any air-purifying half-mask respirator with organic vapor cartridge(s) in combination with an N95, R95, or P95 filter. The following filters may also be used: N99, R99, P99, N100, R100, P100] or Sa (APF = 10) (any supplied-air respirator). *1.25 mg/m^3:* Sa:Cf (APF = 25) (any supplied-air respirator operated in a continuous-flow mode) or PaprOvHie (APF = 25) (any powered air-purifying respirator with an organic vapor cartridge in combination with a high-efficiency particulate filter). *2.5 mg/m^3:* CcrFOv100 (APF = 50) [any air-purifying full-face-piece respirator equipped with organic vapor cartridge(s) in combination with an N100, R100, or P100 filter] or SaT: Cf (APF = 50) (any supplied-air respirator that has a tight-fitting face-piece and is operated in a continuous-flow mode) or PaprTOvHie (APF = 50) [any powered, air-purifying respirator with a tight-fitting face-piece and organic vapor cartridge(s) in combination with a high-efficiency particulate filter] or SCBAF (APF = 50) (any self-contained breathing apparatus with full face-piece) or SaF (APF = 50) (any supplied-air respirator with a full face-piece). *10 mg/m^3:* Sa: Pd,Pp (APF = 1000) (any supplied-air respirator operated in a pressure-demand or other

positive-pressure mode). *Emergency or planned entry into unknown concentrations or IDLH conditions:* SCBAF: Pd, Pp (APF = 10,000) (any self-contained breathing apparatus that has a full face-piece and is operated in a pressure-demand or other positive-pressure mode) or SaF: Pd,Pp: ASCBA (APF = 10,000) (any supplied-air respirator that has a full face-piece and is operated in a pressure-demand or other positive-pressure mode in combination with an auxiliary self-contained breathing apparatus operated in a pressure-demand or other positive-pressure mode). *Escape:* GmFOv100 (APF = 50) [any air-purifying, full-face-piece respirator (gas mask) with a chin-style, front- or back-mounted organic vapor canister having an N100, R100, or P100 filter] or SCBAE (any appropriate escape-type, self-contained breathing apparatus).

Storage: Color Code—Blue: Health Hazard/Poison: Store in a secure poison location. Prior to working with this chemical you should be trained on its proper handling and storage. Store in tightly closed containers in a cool, well-ventilated area away from water and alkalis. Where possible, automatically pump liquid from drums or other storage containers to process containers. Sources of ignition, such as smoking and open flames, are prohibited where this chemical is handled, used, or stored. Metal containers involving the transfer of 5 gallons or more of this chemical should be grounded and bonded. Drums must be equipped with self-closing valves, pressure vacuum bungs, and flame arresters. Use only non-sparking tools and equipment, especially when opening and closing containers of this chemical. Wherever this chemical is used, handled, manufactured, or stored, use explosion-proof electrical equipment and fittings.

Shipping: Organophosphorus pesticides, liquid, toxic, n.o.s. requires a shipping label of "POISONOUS/TOXIC MATERIALS." It falls in Hazard Class 6.1 and Packing Group I.

Spill Handling: Evacuate and restrict persons not wearing protective equipment from area of spill or leak until cleanup is complete. Remove all ignition sources. Ventilate area of spill or leak. Stay upwind; keep out of low areas. Ventilate closed spaces before entering them. Remove and isolate contaminated clothing at the site. Do not touch spilled material; stop leak if you can do it without risk. Use water spray to reduce vapors. *Small spills:* take up with sand or other noncombustible absorbent material and place into containers for later disposal. *Large spills:* dike far ahead of spill for later disposal. Keep this chemical out of a confined space, such as a sewer, because of the possibility of an explosion, unless the sewer is designed to prevent the buildup of explosive concentrations. It may be necessary to contain and dispose of this chemical as a hazardous waste. If material or contaminated runoff enters waterways, notify downstream users of potentially contaminated waters. Contact your local or federal environmental protection agency for specific recommendations. If employees are required to clean up spills, they must be properly trained and equipped. OSHA 1910.120(q) may be applicable.

Fire Extinguishing: This chemical is a combustible liquid. Poisonous gases, including nitrogen oxides, phosphorous oxides, sulfur oxides, are produced in fire. Use dry chemical, carbon dioxide, or alcohol foam extinguishers. Wear positive pressure self-contained breathing apparatus. Move container from fire area if you can do it without risk. Fight fire from maximum distance. Dike fire control water for later disposal; do not scatter the material. Vapors are heavier than air and will collect in low areas. Vapors may travel long distances to ignition sources and flashback. Vapors in confined areas may explode when exposed to fire. Containers may explode in fire. Storage containers and parts of containers may rocket great distances, in many directions. If material or contaminated runoff enters waterways, notify downstream users of potentially contaminated waters. Notify local health and fire officials and pollution control agencies. From a secure, explosion-proof location, use water spray to cool exposed containers. If cooling streams are ineffective (venting sound increases in volume and pitch, tank discolors, or shows any signs of deforming), withdraw immediately to a secure position. If employees are expected to fight fires, they must be trained and equipped in OSHA 1910.156. The only respirators recommended for firefighting are self-contained breathing apparatuses that have full face-pieces and are operated in a pressure-demand or other positive-pressure mode.

Disposal Method Suggested: In accordance with 40CFR165, follow recommendations for the disposal of pesticides and pesticide containers. Must be disposed properly by following package label directions or by contacting your local or federal environmental control agency or by contacting your regional EPA office. Consult with environmental regulatory agencies for guidance on acceptable disposal practices. Generators of waste containing this contaminant ($\geq$100 kg/mo) must conform with EPA regulations governing storage, transportation, treatment, and waste disposal.

References

US Environmental Protection Agency. (April 30, 1980). *Phorate: Health and Environmental Effects Profile No. 145.* Washington, DC: Office of Solid Waste

US Environmental Protection Agency. (November 30, 1987). *Chemical Hazard Information Profile: Phorate.* Washington, DC: Chemical Emergency Preparedness Program

New Jersey Department of Health and Senior Services. (September 2001). *Hazardous Substances Fact Sheet: Phorate.* Trenton, NJ

Phosacetim   P:0530

Molecular Formula: $C_{14}H_{13}Cl_2N_2O_2PS$
Synonyms: Acetimidoylphosphoramidothioic acid *O,O*-bis (*p*-chlorophenyl) ester; BAY 33819; Bayer 33819; *O,O*-Bis (*p*-chlorophenyl) acetimidoyl phosphoramidothioate; *O,O*-Bis (4-chlorophenyl) *N*-acetimidoyl phosphoramidothioate; *O,O*-Bis(4-chlorophenyl) 1-iminoethyl phosphoramidothioate; *O,O*-Bis(4-chlorophenyl)-1-iminoethylphosphoramidothioic acid; DRC-714; Gophacide; (1-Iminoethyl)phosphoramidothioic acid, *O,O*-bis(4-chlorophenyl) ester; Phosazetim; Phosphonodithio-imidocarbonic acid, acetimidoyl-, *O,O*-bis (*p*-chlorophenyl) ester; Phosphonodithioimidocarbonic acid, (1-iminoethyl)-, *O,O*-bis(*p*-chlorophenyl) ester
CAS Registry Number: 4104-14-7
RTECS® Number: TB4725000
UN/NA & ERG Number: UN2783 (organophosphorus pesticides, solid, toxic)/152
EC Number: 223-874-7 [*Annex I Index No.:* 015-092-00-8]
Regulatory Authority and Advisory Bodies
Banned or Severely Restricted (East Germany, Philippines) (UN).[13]
Very Toxic Substance (World Bank).[15]
Superfund/EPCRA 40CFR355, Extremely Hazardous Substances: TPQ = 100/10,000 lb (45.4/4540 kg).
Reportable Quantity (RQ): 100 lb (45.4 kg).
US DOT 49CFR172.101, Inhalation Hazard Chemical as organophosphates.
European/International Regulations: Hazard Symbol: T+, N; Risk phrases: R27/28; R50/53; Safety phrases: S1/2; S28; S36/37; S45; S60; S61 (see Appendix 4).
WGK (German Aquatic Hazard Class): No value assigned.
Description: Phosacetim is a crystalline solid. Molecular weight = 375.22. Hazard Identification (based on NFPA-704 M Rating System): Health 4 Flammability 1, Reactivity 0.
Potential Exposure: Used as a rodenticide.
Incompatibilities: Strong oxidizers, nitrates. May hydrolyze on contact with moisture.
Permissible Exposure Limits in Air
Protective Action Criteria (PAC)
TEEL-0: 0.75 mg/m^3
PAC-1: 2 mg/m^3
PAC-2: 3.7 mg/m^3
PAC-3: 3.7 mg/m^3
Determination in Air: Use NIOSH Analytical Method (IV) #5600, Organophosphorus Pesticides.
Routes of Entry: Inhalation, ingestion, skin and/or eye contact. Absorbed by the skin.
Harmful Effects and Symptoms
Short Term Exposure: Organic phosphorus insecticides are absorbed by the skin as well as by the respiratory and gastrointestinal tracts. They are cholinesterase inhibitors. Symptoms of exposure include headache, giddiness, blurred vision, nervousness, weakness, nausea, cramps, diarrhea, and discomfort in the chest. Signs include sweating, tearing, salivation, vomiting, cyanosis, convulsions, coma, loss of reflexes, and loss of sphincter control. Highly toxic. LD$_{50}$ oral rat is 3.7 mg/kg.
Long Term Exposure: Cholinesterase inhibitor; cumulative effect is possible. This chemical may damage the nervous

system with repeated exposure, resulting in convulsions and respiratory failure. May cause liver damage.

Points of Attack: Respiratory system, lungs, central nervous system, cardiovascular system, skin, eyes, plasma and red blood cell cholinesterase.

Medical Surveillance: Before employment and at regular times after that, the following are recommended: plasma and red blood cell cholinesterase levels (tests for the enzyme poisoned by this chemical). If exposure stops, plasma levels return to normal in 1–2 weeks while red blood cell levels may be reduced for 1–3 months.

When cholinesterase enzyme levels are reduced by 25% or more below preemployment levels, risk of poisoning is increased, even if results are in lower ranges of "normal." Reassignment to work not involving organophosphate or carbamate pesticides is recommended until enzyme levels recover. If symptoms develop or overexposure occurs, repeat the above tests as soon as possible and get an examination of the nervous system. Also consider complete blood count. Consider chest X-ray following acute overexposure. Do not drink any alcoholic beverages before or during use. Alcohol promotes absorption of organic phosphates.

First Aid: If this chemical gets into the eyes, remove any contact lenses at once and irrigate immediately for at least 15 min, occasionally lifting upper and lower lids. Seek medical attention immediately. If this chemical contacts the skin, remove contaminated clothing and wash immediately with soap and water. Speed in removing material from skin is of extreme importance. Seek medical attention immediately. If this chemical has been inhaled, remove from exposure, begin rescue breathing (using universal precautions, including resuscitation mask) if breathing has stopped and CPR if heart action has stopped. Transfer promptly to a medical facility. When this chemical has been swallowed, get medical attention. Give large quantities of water and induce vomiting. Do not make an unconscious person vomit.

Personal Protective Methods: Wear protective gloves and clothing to prevent any reasonable probability of skin contact. Safety equipment suppliers/manufacturers can provide recommendations on the most protective glove/clothing material for your operation. All protective clothing (suits, gloves, footwear, headgear) should be clean, available each day, and put on before work. Contact lenses should not be worn when working with this chemical. Wear dust-proof chemical goggles and face shield unless full face-piece respiratory protection is worn. Employees should wash immediately with soap when skin is wet or contaminated. Provide emergency showers and eyewash.

Respirator Selection: Follow regulations in OSHA 29CFR1910.134 or European Standard EN149. Use a NIOSH/MSHA- or European Standard EN149-approved respirator; or use an approved supplied-air respirator with a full face-piece operated in the positive-pressure mode, or with a full face-piece, hood, or helmet in the continuous-flow mode; or use a NIOSH/MSHA- or European Standard

EN149-approved self-contained breathing apparatus with a full face-piece operated in pressure-demand or other positive-pressure mode.

Storage: Color Code—Blue: Health Hazard/Poison: Store in a secure poison location. Prior to working with this chemical you should be trained on its proper handling and storage.

Shipping: Organophosphorus pesticides, solid, toxic, requires a shipping label of "POISONOUS/TOXIC MATERIALS." It falls in Hazard Class 6.1 and Packing Group I.

Spill Handling: Evacuate persons not wearing protective equipment from area of spill or leak until cleanup is complete. Remove all ignition sources. Collect powdered material in the most convenient and safe manner and deposit in sealed containers. Ventilate area after cleanup is complete. It may be necessary to contain and dispose of this chemical as a hazardous waste. If material or contaminated runoff enters waterways, notify downstream users of potentially contaminated waters. Contact your local or federal environmental protection agency for specific recommendations. If employees are required to clean up spills, they must be properly trained and equipped. OSHA 1910.120(q) may be applicable.

Fire Extinguishing: This chemical is a flammable solid. Use dry chemical, carbon dioxide, water spray, or alcohol foam extinguishers. Poisonous gases are produced in fire, including oxides of phosphorus, sulfur, nitrogen, and chlorine. If material or contaminated runoff enters waterways, notify downstream users of potentially contaminated waters. Notify local health and fire officials and pollution control agencies. From a secure, explosion-proof location, use water spray to cool exposed containers. If cooling streams are ineffective (venting sound increases in volume and pitch, tank discolors, or shows any signs of deforming), withdraw immediately to a secure position. If employees are expected to fight fires, they must be trained and equipped in OSHA 1910.156. The only respirators recommended for firefighting are self-contained breathing apparatuses that have full face-pieces and are operated in a pressure-demand or other positive-pressure mode.

Disposal Method Suggested: Consult with environmental regulatory agencies for guidance on acceptable disposal practices. Generators of waste containing this contaminant (≥100 kg/mo) must conform with EPA regulations governing storage, transportation, treatment, and waste disposal. In accordance with 40CFR165, follow recommendations for the disposal of pesticides and pesticide containers. Must be disposed properly by following package label directions or by contacting your local or federal environmental control agency or by contacting your regional EPA office.

Reference

US Environmental Protection Agency. (November 30, 1987). *Chemical Hazard Information Profile: Phosacetim.* Washington, DC: Chemical Emergency Preparedness Program

Phosfolan

P:0540

Molecular Formula: $C_7H_{14}NO_3PS_2$

Synonyms: AC 47031; American cyanamid 47031; C.I. 47031; Cyclic ethylene(diethoxyphosphinothioyl)-dithioimidocarbonate; Cyclic ethylene *p,p*-diethylphosphono dithioimidocarbonate; Cylan; Cyolane; Cyolane insecticide; (Diethoxyphosphinyl)dithioimidocarbonic acid cyclic ethylene ester; 2-(Diethoxyphosphinylimino)-1,3-dithiolan; 2-(Diethoxyphosphinylimino)-1,3-dithiolane; *p,p*-Diethyl cyclic ethylene ester of phosphonodithioimidocarbonate; *p,p*-Diethyl cyclic ethylene ester of phosphonodithioimidocarbonic acid; Diethyl 1,3-dithiolan-2-ylidenephosphoramidate; EI 47031; ENT 25,830; 1,2-Ethanedithiol, cyclic ester with *p,p*-diethyl phosphonodithioimidocarbonate; 1,2-Ethanedithiol, cyclic ester with phosphonodithioimidocarbonic acid *p,p*-diethyl ester; Imidocarbonic acid, phosphonodithio-, cyclic ethylene *p,p*-diethyl ester; Phosphoroamidic acid, 1,3-dithiolan-2-ylidene-, diethyl ester

CAS Registry Number: 947-02-4

RTECS® Number: NJ6475000

UN/NA & ERG Number: UN2783 (organophosphorus pesticides, solid, toxic)/152

EC Number: 213-423-2 [*Annex I Index No.:* 015-111-00-X]

Regulatory Authority and Advisory Bodies

Superfund/EPCRA 40CFR355, Extremely Hazardous Substances: TPQ = 100/10,000 lb (45.4/4540 kg).

Reportable Quantity (RQ): 100 lb (45.4 kg).

US DOT 49CFR172.101, Inhalation Hazard Chemical as organophosphates.

European/International Regulations: Hazard Symbol: T+, N; Risk phrases: R27/28; Safety phrases: S1/2; S28; S36/37; S45 (see Appendix 4).

WGK (German Aquatic Hazard Class): No value assigned.

Description: Phosfolan is a colorless to yellow solid. Molecular weight = 255.31; Boiling point = 115−118°C at 0.001 mm; Freezing/Melting point = 37−45°C. Hazard Identification (based on NFPA-704 M Rating System): Health 4, Flammability 1, Reactivity 0. Soluble in water.

Potential Exposure: Those involved in the manufacture, formulation, and application of this insecticide.

Incompatibilities: Incompatible with nitrates and water. May hydrolyze upon contact with water, steam, and moisture, and produce toxic oxides of phosphorus, nitrogen, sulfur, and chlorine.

Permissible Exposure Limits in Air

Protective Action Criteria (PAC)

TEEL-0: 1.5 mg/m^3

PAC-1: 5 mg/m^3

PAC-2: 9 mg/m^3

PAC-3: 9 mg/m^3

Routes of Entry: Inhalation, ingestion, skin and/or eye contact. Absorbed through the skin.

Harmful Effects and Symptoms

Short Term Exposure: Similar to parathion in health hazards. Death may result due to respiratory arrest as a result of paralysis of respiratory muscles and intense bronchoconstriction. Also considered a cholinesterase inhibitor. Symptoms similar to parathion include nausea, vomiting, abdominal cramps, diarrhea, excessive salivation, headache, giddiness, dizziness, tightness in the chest, blurring or dimness of vision, tearing, loss of muscle coordination, slurring of speech, twitching of muscles, drowsiness, difficulty in breathing, respiratory rales, and random jerky movements.

Long Term Exposure: Cholinesterase inhibitor; cumulative effect is possible. This chemical may damage the nervous system with repeated exposure, resulting in convulsions, respiratory failure. May cause liver damage.

Points of Attack: Respiratory system, lungs, central nervous system, cardiovascular system, skin, eyes, plasma and red blood cell cholinesterase.

Medical Surveillance: Before employment and at regular times after that, the following are recommended: plasma and red blood cell cholinesterase levels (tests for the enzyme poisoned by this chemical). If exposure stops, plasma levels return to normal in 1−2 weeks while red blood cell levels may be reduced for 1−3 months. When cholinesterase enzyme levels are reduced by 25% or more below preemployment levels, risk of poisoning is increased, even if results are in lower ranges of "normal." Reassignment to work not involving organophosphate or carbamate pesticides is recommended until enzyme levels recover. If symptoms develop or overexposure occurs, repeat the above tests as soon as possible and get an examination of the nervous system. Also consider complete blood count. Consider chest X-ray following acute overexposure. Do not drink any alcoholic beverages before or during use. Alcohol promotes absorption of organic phosphates.

First Aid: If this chemical gets into the eyes, remove any contact lenses at once and irrigate immediately for at least 15 min, occasionally lifting upper and lower lids. Seek medical attention immediately. If this chemical contacts the skin, remove contaminated clothing and wash immediately with soap and water. Speed in removing material from skin is of extreme importance. Seek medical attention immediately. If this chemical has been inhaled, remove from exposure, begin rescue breathing (using universal precautions, including resuscitation mask) if breathing has stopped and CPR if heart action has stopped. Transfer promptly to a medical facility. When this chemical has been swallowed, get medical attention. Give large quantities of water and induce vomiting. Do not make an unconscious person vomit. Keep victim quiet and maintain normal body temperature. Effects may be delayed; keep victim under observation.

Personal Protective Methods: Wear protective gloves and clothing to prevent any reasonable probability of skin contact. Safety equipment suppliers/manufacturers can provide recommendations on the most protective glove/clothing material for your operation. All protective clothing (suits, gloves, footwear, headgear) should be clean, available each day, and put on before work. Contact lenses should not be worn when working with this chemical. Wear dust-proof chemical goggles and face shield unless full face-piece respiratory protection is worn. Employees should wash immediately with soap when skin is wet or contaminated. Provide emergency showers and eyewash.

Respirator Selection: Follow regulations in OSHA 29CFR1910.134 or European Standard EN149. Use a NIOSH/MSHA- or European Standard EN149-approved respirator; or use an approved supplied-air respirator with a full face-piece operated in the positive-pressure mode, or with a full face-piece, hood, or helmet in the continuous-flow mode; or use a NIOSH/MSHA- or European Standard EN149-approved self-contained breathing apparatus with a full face-piece operated in pressure-demand or other positive-pressure mode.

Storage: Color Code—Blue: Health Hazard/Poison: Store in a secure poison location. Prior to working with this chemical you should be trained on its proper handling and storage. Store in tightly closed containers in a cool, well-ventilated area away from oxidizers, nitrates, and other incompatible materials listed above. Where possible, automatically transfer material from other storage containers to process containers.

Shipping: Organophosphorus pesticides, solid, toxic, require a shipping label of "POISONOUS/TOXIC MATERIALS." It falls in Hazard Class 6.1 and Packing Group I.

Spill Handling: Evacuate persons not wearing protective equipment from area of spill or leak until cleanup is complete. Remove all ignition sources. Collect powdered material in the most convenient and safe manner and deposit in sealed containers. Ventilate area after cleanup is complete. It may be necessary to contain and dispose of this chemical as a hazardous waste. If material or contaminated runoff enters waterways, notify downstream users of potentially contaminated waters. Contact your local or federal environmental protection agency for specific recommendations. If employees are required to clean up spills, they must be properly trained and equipped. OSHA 1910.120(q) may be applicable.

Fire Extinguishing: This chemical is a combustible solid. Extinguish with dry chemical, carbon dioxide, water spray, foam, or fog. Fight fire from maximum distance. Dike fire control water for later disposal; do not scatter the material. Wear positive pressure breathing apparatus and special protective clothing. Poisonous gases are produced in fire. If material or contaminated runoff enters waterways, notify downstream users of potentially contaminated waters. Notify local health and fire officials and pollution control agencies. From a secure, explosion-proof location, use water spray to cool exposed containers. If cooling streams are ineffective (venting sound increases in volume and pitch, tank discolors, or shows any signs of deforming), withdraw immediately to a secure position. If employees are expected to fight fires, they must be trained and equipped in OSHA 1910.156. The only respirators recommended for firefighting are self-contained breathing apparatuses that have full face-pieces and are operated in a pressure-demand or other positive-pressure mode.

Disposal Method Suggested: Consult with environmental regulatory agencies for guidance on acceptable disposal practices. Generators of waste containing this contaminant ($\geq$100 kg/mo) must conform with EPA regulations governing storage, transportation, treatment, and waste disposal. In accordance with 40CFR165, follow recommendations for the disposal of pesticides and pesticide containers. Must be disposed properly by following package label directions or by contacting your local or federal environmental control agency or by contacting your regional EPA office.

Reference

US Environmental Protection Agency. (November 30, 1987). *Chemical Hazard Information Profile: Phosfolan.* Washington, DC: Chemical Emergency Preparedness Program

Phosgene (Agents CG & DP, WMD) P:0550

Molecular Formula: CCl_2O
Common Formula: $COCl_2$
Synonyms: Carbone (oxychlorure de) (French); Carbon dichloride oxide; Carbon oxychloride; Carbonic dichloride; Carbon oxychloride; Carbonylchlorid (German); Carbonyl chloride; Carbonyl dichloride; CG (military designation); Chloroformyl chloride; Combat gas; Diphosgene; DP (military designation for diphosgene); Fosgeno (Spanish); NCI-C60219; Phosgen (German); Trichloroacetyl chloride (diphosgene)

CAS Registry Number: 75-44-5; 503-38-8 (diphosgene)
RTECS® Number: SY5600000
UN/NA & ERG Number: UN1076/125
EC Number: 200-870-3 [*Annex I Index No.:* 006-002-00-8]; 207-965-9 (diphosgene or trichloromethyl chloroformate)

Regulatory Authority and Advisory Bodies

Department of Homeland Security Screening Threshold Quantity (pounds): *Release hazard* 500 (1.00% concentration); *Theft hazard* 15 ($\geq$0.17% concentration).

Carcinogenicity: EPA: Inadequate Information to assess carcinogenic potential.

Air Pollutant Standard Set. See below, "Permissible Exposure Limits in Air" section.

Clean Air Act: Hazardous Air Pollutants (Title I, Part A, Section 112); List of high-risk pollutants (Section 63.74); Accidental Release Prevention/Flammable Substances, (Section 112[r], Table 3), TQ = 500 lb (227 kg).

Clean Water Act: Section 311 Hazardous Substances/RQ 40CFR117.3 (same as CERCLA, see below); Section 313 Water Priority Chemicals (57FR41331, 9/9/92).

US EPA Hazardous Waste Number (RCRA No.): P095.
RCRA, 40CFR261, Appendix 8 Hazardous Constituents.
Superfund/EPCRA 40CFR355, Extremely Hazardous Substances: TPQ = 10 lb (4.54 kg).
Reportable Quantity (RQ): 10 lb (4.54 kg).
EPCRA Section 313 Form R *de minimis* concentration reporting level: 1.0%.
US DOT 49CFR172.101, Inhalation Hazardous Chemical.
Canada, WHMIS, Ingredients Disclosure List Concentration 1.0%.
European/International Regulations: Hazard Symbol *(75-44-5)*: T + ; Risk phrases: R26; R34; Safety phrases: S1/2; S9; S26; S36/37/39; S45 (see Appendix 4).
WGK (German Aquatic Hazard Class): 2—Hazard to waters.

Description: Phosgene (CG) is a colorless, noncombustible gas. It is shipped as a liquefied compressed gas in steel cylinders. At low concentrations CG has a sweet (not pleasant) odor like newly mown hay, green corn, or moldy hay. In higher concentrations, it is poisonous with an odor that is suffocating, irritating, and pungent. The odor is only detectable for a short amount of time when CG is initially released and odor should not be regarded as a reliable indicator of overexposure. A fuming liquid below 8.3°C/47°F. Shipped as a liquefied compressed gas. The odor threshold is between 1.5 and 6 mg/m^3. A choking agent, phosgene (CG), rapidly decomposes in relative humidity over 70%. Molecular weight = 98.92; Boiling point = 8.2°C; Freezing/Melting point = −118°C; Relative vapor density (air = 1) = 3.48; Vapor pressure = 1.61 atm at 25°C. Hazard Identification (based on NFPA-704 M Rating System): Health 4, Flammability 0, Reactivity 1. Reacts with water (slightly soluble).

Potential Exposure: Compound Description: Human Data. Phosgene can be deadly at a concentration as low as 2 ppm. Phosgene is used as an intermediate in the manufacture of many industrial chemicals, including dyes and plastics; in the making of dyestuffs based on triphenylmethane, coal tar, and urea. It is also used in the organic synthesis of isocyanates and their derivatives, carbonic acid esters (polycarbonates), and acid chlorides. Other applications include its utilization in metallurgy; and in the manufacture of some insecticides and pharmaceuticals. Exposure to phosgene may occur during arc welding and in fires involving vinyl chloride; released from household paint removers and degreasers when they are used in the presence of heat. Phosgene (CG) has been used as a military choking, pulmonary agent since WW I, and has become a staple of chemical arsenals in many countries. *Persistence of Chemical Agent:* Phosgene (CG & DO): Summer: 1−10 min; Winter: 10 min to 1 h.

Incompatibilities: Moisture, alkalis, ammonia, alcohols, copper. Reacts slowly in water to form hydrochloric acid and carbon dioxide. Violent reaction with strong oxidizers, amines, aluminum. Attacks many metals in the presence of water. Attacks plastic, rubber, and coatings.

Permissible Exposure Limits in Air
Conversion factor: 1 ppm = 4.05 mg/m^3 at 25°C & 1 atm.
OSHA PEL: 0.1 ppm/0.4 mg/m^3 TWA.
NIOSH REL: 0.1 ppm/0.4 mg/m^3 TWA; 0.2 ppm/0.8 mg/m^3 [15 min] Ceiling Concentration.
ACGIH TLV®[1]: 0.1 ppm/0.4 mg/m^3 TWA.
Protective Action Criteria (PAC) **CG***
TEEL-0: 0.1 ppm
PAC-1: 0.1 ppm
PAC-2: **0.3** ppm
PAC-3: **0.75** ppm
*AEGLs (Acute Emergency Guideline Levels) & ERPGs (Emergency Response Planning Guideline) are in **bold face**.
NIOSH IDLH: 2 ppm.
NIOSH IDLH: 2 ppm.
Emergency Response Planning Guidelines (AIHA)
ERPG-1: Inappropriate
ERPG-2: 0.2 ppm
ERPG-3: 1 ppm
DFG MAK: 0.02 ppm; 0.082 mg/m^3 TWA; Peak Limitation Category I(2); Pregnancy Risk Group C.
Austria: MAK 0.1 ppm (0.4 mg/m^3), 1999; Denmark: TWA 0.05 ppm (0.2 mg/m^3), 1999; Finland: STEL 0.05 ppm (0.2 mg/m^3), [skin], 1999; France: VLE 0.1 ppm (0.4 mg/m^3), 1999; Japan: 0.1 ppm (0.4 mg/m^3), 1999; the Netherlands: MAC-TGG 0.08 mg/m^3, 2003; the Philippines: TWA 0.1 ppm (0.1 mg/m^3), 1993; Poland: MAC (TWA) 0.5 mg/m^3; MAC (STEL) 1.5 mg/m^3, 1999; Sweden: TGV 0.05 ppm (0.2 mg/m^3), 1999; Switzerland: MAK-W 0.1 ppm (0.4 mg/m^3), KZG-W 0.2 ppm (0.8 mg/m^3), 1999; Thailand: TWA 0.1 ppm (0.4 mg/m^3), 1993; Turkey: TWA 0.1 ppm (0.4 mg/m^3), 1993; United Kingdom: TWA 0.02 ppm (0.08 mg/m^3); STEL 0.06 ppm, 2000; Argentina, Bulgaria, Columbia, Jordan, South Korea, New Zealand, Singapore, Vietnam: ACGIH TLV®: TWA 0.1 ppm. The Czech Republic has set a TWA of 0.5 mg/m^3 and a ceiling value of 1.0 mg/m^3 in work-place air, and MAC in ambient air of 0.01 mg/m^3 and 0.003 mg/m^3 on a daily average basis. Several states have set guidelines or standards for phosgene in ambient air[60] ranging from zero (North Carolina) to 1.33 μg/m^3 (New York) to 4.0 μg/m^3 (Florida, North Dakota, South Carolina) to 7.0 μg/m^3 (Virginia) to 8.0 μg/m^3 (Connecticut) to 10.0 μg/m^3 (Nevada).

Determination in Air: Use OSHA Analytical Method 61.
Determination in Water: Octanol−water coefficient: Log K_{ow} = −0.71.
Routes of Entry: Inhalation, ingestion, skin and/or eye contact.

Harmful Effects and Symptoms
Short Term Exposure: Acute exposure to phosgene may result in severe irritation and burns of the skin, eyes, mucous membranes, and respiratory passages. Cough, dyspnea (shortness of breath), pain in the chest, and severe pulmonary edema may also occur. Cyanosis and anxiety may be observed.

Note: The detection of the odor of phosgene at any time indicates the need for immediate, corrective action or withdrawal. *Inhalation:* Both immediate and delayed symptoms may be felt. Immediate symptoms of irritation to mouth, throat, and eyes, tearing, coughing, and difficult breathing are felt at levels of 5 ppm and above. Delayed effects are the accumulation of fluid in the lungs and death; if proper, rapid treatment is not obtained. The length of delay depends on the dose but may be between 2 and 15 h. Death may result from short exposures to high levels (30 ppm, 17 min) or long exposures to low levels (3 ppm, 3 h). Phosgene is particularly dangerous at low levels because lethal doses may be inhaled without warning symptoms. *Skin:* Contact with skin may lead to severe chemical burns. Liquid may cause frostbite. *Eyes:* Eye irritation begins at 3–5 ppm. Severe and permanent damage may result. Liquid phosgene is more hazardous than vapor. Liquid may cause frostbite. *Ingestion:* Expected symptoms may include severe irritation and chemical burns of the mouth, throat, lungs, and digestive tract.

Long Term Exposure: Even low levels can cause permanent lung damage, emphysema, bronchitis, pulmonary fibrosis.

Points of Attack: Respiratory system, lungs, skin, eyes.

Medical Surveillance: Preemployment medical examinations should include chest X-rays and baseline pulmonary function tests. Consider chest X-ray following acute overexposure. The eyes and skin should be examined. Smoking history should be known. Periodic pulmonary function studies should be done. Workers who are known to have inhaled phosgene should remain under medical observation for at least 24 h to insure that delayed symptoms do not occur.

First Aid: If this chemical gets into the eyes, remove any contact lenses at once and irrigate immediately. If this chemical contacts the skin, flush with water immediately. If a person breathes in large amounts of this chemical, move the exposed person to fresh air at once and perform artificial respiration. When this chemical has been swallowed, get medical attention. Do not induce vomiting. Medical observation is recommended for 24–48 h after breathing overexposure, as pulmonary edema may be delayed. As first aid for pulmonary edema, a doctor or authorized paramedic may consider administering a corticosteroid spray. If frostbite has occurred, seek medical attention immediately; do NOT rub the affected areas or flush them with water. In order to prevent further tissue damage, do NOT attempt to remove frozen clothing from frostbitten areas. If frostbite has NOT occurred, immediately and thoroughly wash contaminated skin with soap and water.

Decontamination: Decontaminate as soon as possible. This is extremely important. If you do not have the equipment and training, do not enter the hot zone to rescue and/or decontaminate victims. If the victim cannot move, begin the decontamination process without touching and without entering the hot zone. Use clean water from any source; if possible, use a hose (spray or fog to prevent injury to the victim) or other system so that you would not have to touch the victim; do not even wait for soap or for the victim to remove clothing, begin washing immediately. Immediately flush the eyes with water for at least 15 min. Wash–strip–wash–evacuate upwind and uphill: The approach is to immediately wash with water, then have the victim (not the first responder) remove all the victim's clothing, then wash again (with soap if available); and subsequently move away from the hot zone in an upwind and uphill direction. Wash the victim with warm water and soap. Decontaminate with diluted household bleach (10%, or one part bleach to nine parts water), but do not let any of the bleach solution get in the victim's eyes, open wounds, or mouth. Rinse off the diluted bleach solution after 15 min. In order to prevent spreading the agent, be certain the victims have been decontaminated as much as possible before they leave the decontamination area. If you get any amount of the agent on yourself, decontaminate immediately. Even if you think you are not contaminated, be sure to thoroughly shower and change clothes as soon as you can after the incident.

Personal Protective Methods: Where liquid phosgene is encountered, protective clothing should be supplied which is impervious to phosgene. Where gas is encountered above safe limits, full-face gas masks with phosgene canisters or supplied-air respirators should be used. Because of the potentially serious consequences of acute overexposure and the poor warning properties of the gas to the human senses, automatic continuous monitors with alarm systems are strongly recommended. Wear appropriate clothing to prevent any reasonable probability of skin contact. Wear eye protection to prevent any possibility of eye contact. Employees should wash immediately when skin is wet or contaminated. Remove nonimpervious clothing immediately if wet or contaminated. Provide emergency showers.

Respirator Selection: *1 ppm:* Sa (APF = 10) (any supplied-air respirator). *2 ppm:* SCBAF (APF = 50) (any self-contained breathing apparatus with a full face-piece) or SaF (APF = 50) (any supplied-air respirator with a full face-piece). *Emergency or planned entry into unknown concentrations or IDLH conditions:* SCBAF: Pd,Pp (APF = 10,000) (any self-contained breathing apparatus that has a full face-piece and is operated in a pressure-demand or other positive-pressure mode) or SaF: Pd,Pp: ASCBA (any supplied-air respirator that has a full face-piece and is operated in a pressure-demand or other positive-pressure mode in combination with an auxiliary self-contained breathing apparatus operated in a pressure-demand or other positive-pressure mode). *Escape:* GmFS (APF = 50) [any air-purifying, full-face-piece respirator (gas mask) with a chin-style, front- or back-mounted canister providing protection against the compound of concern] or SCBAE (any appropriate escape-type, self-contained breathing apparatus).

Note: Substance reported to cause eye irritation or damage; may require eye protection.

Storage: Color Code—Yellow Stripe: Reactivity Hazard; Store separately in an area isolated from flammables, combustibles, or other yellow-coded materials. Color Code—Blue: Health Hazard/Poison: Store in a secure poison location. Prior to working with this chemical you should be trained on its proper handling and storage. Phosgene must be stored to avoid contact with water, moisture, or steam, since violent reactions occur. Store in tightly closed, steel containers in an isolated area away from the work area and separated from all other materials, as well as sunlight. Although phosgene in anhydrous equipment is not corrosive to ordinary metals; in the presence of moisture, use monel, tantalum, or glass-lined storage containers. Phosgene should be stored away from heating and cooling ducts. Containers should be frequently inspected for leaks. Procedures for the handling, use, and storage of cylinders should be in compliance with OSHA 1910.101 and 1910.169 with the recommendations of the Compressed Gas Association.

Shipping: Phosgene requires a shipping label of "POISON GAS, CORROSIVE." It falls in Hazard Class 2.3. It is a violation of transportation regulations to refill compressed gas cylinders without the express written permission of the owner.

Special precautions: Cylinders must be transported in a secure upright position, in a well-ventilated truck.

Spill Handling: Evacuate persons not wearing protective equipment from area of spill or leak until cleanup is complete. Ventilate area of leak to disperse the gas. Stop flow of gas. If source of leak is a cylinder and the leak cannot be stopped in place, remove the leaking cylinder to a safe place in the open air, and repair leak or allow cylinder to empty. Absorb liquids in vermiculite, dry sand, earth, or a similar material and deposit in sealed containers. Phosgene may be neutralized by covering it with sodium bicarbonate or an equal mixture of soda ash and slaked lime. After mixing, spray very carefully with water. Transfer slowly to a larger container of water. *Do not use water* directly on spill. It may be necessary to contain and dispose of this chemical as a hazardous waste. If material or contaminated runoff enters waterways, notify downstream users of potentially contaminated waters. Contact your local or federal environmental protection agency for specific recommendations. If employees are required to clean up spills, they must be properly trained and equipped. OSHA 1910.120(q) may be applicable.

Initial isolation and protective action distances
Distances shown are likely to be affected during the first 30 min after materials are spilled and could increase with time. If more than one tank car, cargo tank, portable tank, or large cylinder involved in the incident is leaking, the protective action distance may need to be increased. You may need to seek emergency information from CHEMTREC at (800) 424-9300 or seek professional environmental engineering assistance from the US EPA Environmental Response Team at (908) 548-8730 (24-h response line).

CG, when used as a weapon
Small spills (From a small package or a small leak from a large package)
First: Isolate in all directions (feet/meters)   600/200
Then: Protect persons downwind (miles/kilometers)
Day   0.7/1.1
Night   2.5/4.1
Large spills (From a large package or from many small packages)
First: Isolate in all directions (feet/meters)   3000/1000
Then: Protect persons downwind (miles/kilometers)
Day   4.7/7.5
Night   7.0+ /11.0+

DP, when used as a weapon
Small spills (From a small package or a small leak from a large package)
First: Isolate in all directions (feet/meters)   100/30
Then: Protect persons downwind (miles/kilometers)
Day   0.2/0.3
Night   0.5/0.8
Large spills (From a large package or from many small packages)
First: Isolate in all directions (feet/meters)   600/200
Then: Protect persons downwind (miles/kilometers)
Day   0.7/1.1
Night   1.6/2.5

Phosgene
Small spills (From a small package or a small leak from a large package)
First: Isolate in all directions (feet/meters)   300/100
Then: Protect persons downwind (miles/kilometers)
Day   0.4/0.6
Night   1.6/2.5
Large spills (From a large package or from many small packages)
First: Isolate in all directions (feet/meters)   500/150
Then: Protect persons downwind (miles/kilometers)
Day   2.0/3.2
Night   6.1/9.7

Diphosgene
Small spills (From a small package or a small leak from a large package)
First: Isolate in all directions (feet/meters)   100/30
Then: Protect persons downwind (miles/kilometers)
Day   0.1/0.2
Night   0.1/0.2
Large spills (From a large package or from many small packages)
First: Isolate in all directions (feet/meters)   100/30
Then: Protect persons downwind (miles/kilometers)
Day   0.2/0.3
Night   0.3/0.5

Fire Extinguishing: Phosgene may burn, but does not easily ignite. For small fires, use dry chemical or carbon dioxide. Use water spray, fog, or foam for larger fires. Move container from fire area if you can do so without risk. Stay away from the ends of tanks and cool exposed containers with water until well after the fire is out. Isolate the area until gas has dispersed. Poisonous gases are produced in fire, including hydrogen chloride, carbon monoxide, and chlorine fumes. If material or contaminated runoff enters waterways, notify downstream users of potentially contaminated waters. Notify local health and fire officials and pollution control agencies. From a secure, explosion-proof location, use water spray to cool exposed containers. Do not get water inside containers. If cooling streams are ineffective (venting sound increases in volume and pitch, tank discolors, or shows any signs of deforming), withdraw immediately to a secure position. If employees are expected to fight fires, they must be trained and equipped in OSHA 1910.156. The only respirators recommended for firefighting are self-contained breathing apparatuses that have full face-pieces and are operated in a pressure-demand or other positive-pressure mode.

Diphosgene (DP) will not burn. However, it is possible that a DP tank may be adjacent to a fire. In a fire, a storage tank will heat and the tank may overpressurize and explode, so evacuate the area. When heated, DP breaks down to toxic phosgene, which breaks down into chlorine and hydrogen chloride gases. The danger from a heated DP tank is too great to risk a manned firefighting effort; if possible, an unattended fire monitor aimed at the upper part of the diphosgene tank will cool the tank and may prevent tank failure. In general, it is best to use a spray or fog pattern rather than a solid stream, to avoid spreading the burning fuel around.

Disposal Method Suggested: Phosgene may be neutralized by covering it with sodium bicarbonate or an equal mixture of soda ash and slaked lime. After mixing, spray carefully with water. Transfer slowly to a larger container of water. *Do not use water* directly on spill. Pass controlled discharges of phosgene through 10% NaOH solution in a scrubbing tower.[22] Consult with environmental regulatory agencies for guidance on acceptable disposal practices. Generators of waste containing this contaminant ($\geq$100 kg/mo) must conform with EPA regulations governing storage, transportation, treatment, and waste disposal.

References
National Institute for Occupational Safety and Health. (1976). *Criteria for a Recommended Standard: Occupational Exposure to Phosgene*, NIOSH Document No. 76-137
US Environmental Protection Agency. (June 13, 1977). *Chemical Hazard Information Profile: Phosgene.* Washington, DC
Sax, N. I. (Ed.). (1983). Dangerous Properties of Industrial Materials Report, 3, No. 97−99
US Environmental Protection Agency. (November 30, 1987). *Chemical Hazard Information Profile: Phosgene.* Washington, DC: Chemical Emergency Preparedness Program
New York State Department of Health. (March 1986). *Chemical Fact Sheet: Phosgene.* Albany, NY: Bureau of Toxic Substance Assessment
New Jersey Department of Health and Senior Services. (April 2004). *Hazardous Substances Fact Sheet: Phosgene.* Trenton, NJ
Schneider, A. L., et al. (2007). *CHRIS + CD-ROM Version 2.0, United Coast Guard Chemical Hazard Response Information System (COMDTINST 16465.12C).* Washington, DC: United States Coast Guard and the Department of Homeland Security
The Riegle Report: A Report of Chairman Donald W. Riegle, Jr. and Ranking Member Alfonse M. D'Amato of the Committee on Banking, Housing and Urban Affairs with Respect to Export Administration, United States Senate, 103rd Congress, 2d Session, (May 25, 1994)
Besch, T., Moss, C., the Battlebook Project Team, et al. (1991). *MEDICAL NBC Battlebook* (304 p), Tech Guide 244. Ft. Leonard Wood, MO: The US Army Center for Health Promotion and Preventive Medicine (USACHPPM)

Phosgene oxime (Agent CX, WMD) P:0555

Molecular Formula: $CHCl_2NO$
Synonyms: Carbonyl chloride oxime; CX; Dichlorformaldehyde-oxime; Dichloroformaldoxime; 1,2-Dichloroformoxime; Dichloroformoxime; Dichlormethylenhydroxylamine; Dichloroximinomethane; Kohlensauredichloridoxime (German)
CAS Registry Number: 1794-86-1
RTECS® Number: Not established
UN/NA & ERG Number: 2811/154
Regulatory Authority and Advisory Bodies
Report any release of WMD to National Response Center 1-800-424-8802.
While not a mandated "Federally listed" waste, CX is more toxic than most RCRA listed chemicals. However, GF is a "listed" hazardous waste in some states where it may have been stockpiled by the military.
WGK (German Aquatic Class): No value assigned.
Description: Phosgene oxime (military designation CX) is a noncombustible urticant (nettle agent, blister agent) with a short (seconds to minutes) latency period. CX is a colorless, low-melting point (crystalline, white powder) solid or as a liquid (liquid above 39°C; solid below 35°C). On hot days (or at body temperature) it can appear as a yellowish-brown liquid. It has a high vapor pressure (the vapor pressure of the solid is high enough to produce symptoms), slowly decomposes at normal temperatures. It has an intense, disagreeable, penetrating, and violently irritating, peppery odor. Odor detectable at less than 0.3 ppm. Molecular weight = 113.93 Da; Freezing/Melting point = 35−40°C; boiling point = 129°C (with decomposition); Vapor density = 3.9; Vapor pressure = 11.2 mmHg at 20°C (solid); 13 mmHg at 40°C (liquid); Volatility = 1800 mg/m^3 at

$20^{\circ}C$; 76,000 mg/m^3 at $40^{\circ}C$; Latent heat of vaporization = 101 cal/g at $40^{\circ}C$; Decomposition temperature = $<128^{\circ}C$. Solubility in water = dissolves slowly and completely. High solubility in organic solvents.

History of the chemical: CX was invented in Germany in 1929; it is among the least well-studied chemical warfare agents; therefore, detailed information is limited. Although it is believed that CX was never used on the battlefield, it was after WW II that the military tested concentrated phosgene oxime. These tests revealed that CX was a highly effective and painful chemical warfare agent. Phosgene oxime is of military interest because it easily penetrates garments and rubber much more quickly than do other chemical agents. It is possible that Iraq used CX in the Iran–Iraq war, and North Korea may have produced and stocked quantities of this chemical agent. Phosgene oxime (CX) gives off dangerous gas. It does take very little of this gas to damage a victim's lungs. Terrorists might put CX in an exploding bomb in order to break up the solid into an aerosol that can penetrate the skin and lungs. Once exposed, the victim feels immediate pain and the need to escape. Phosgene oxime (CX) easily penetrates fabrics and rubbers and cause great pain and skin damage (without blisters) in less than a minute. Soon after contact the skin dies. Recovery may take up to 6 months. Eye contact can cause blindness. Inhalation attacks the lungs, resulting in damage that can be permanent.

Potential Exposure: There is no industrial use for Phosgene oxime (CX) and because of its extreme instability, the pure material is not likely to be used in military operations.[FM 3-9] CX is especially dangerous when mixed with other chemicals such as nerve agents. It burns away the skin making it more permeable to any other "added" agents. No other chemical agent is capable of producing immediate extreme pain followed by rapid local tissue death (necrosis). Post World War II studies indicate that concentrations below 8% cause no or inconsistent effects.[ATSDR]

Persistence of Chemical Agent: Soil: about 2 h. *Material surfaces and water:* relatively nonpersistent.

Incompatibilities: Phosgene oxime (CX) is among the most important halogenated oximes. CX reacts with water, sweat, and heat, forming hydrochloric acid. CX may be an oxidizer, and it may ignite combustibles (e.g., wood, paper, oil, or clothing). CX is incompatible with strong acids and bases; hydrides and other strong reducing agents; strong oxidizing acids, peroxides, and hydroperoxides. Not hydrolyzed by dilute acids; reacts violently in basic solutions forming carbon dioxide, hydrogen chloride, and hydroxylamine. Hydrolysis products include HCl and methylarsenic oxide. CX quickly penetrates rubber and clothing. CX decomposes when in contact with many metals; it is corrosive to most metals, and contact with metals may cause the release flammable hydrogen gas. Traces of many metals cause it to decompose; however, it corrodes most metals.

Permissible Exposure Limits in Air: Conversion factor = 1 ppm = 4.66 mg/m^3 at $77^{\circ}C$.

The immediately dangerous to life and health (IDLH) concentration of CX has not been defined.
Protective Action Criteria (PAC) **CG***
TEEL-0: 0.0075 mg/m^3
PAC-1: **0.028** mg/m^3
PAC-2: **0.083** mg/m^3
PAC-3: **13** mg/m^3
*AEGLs (Acute Emergency Guideline Levels) & ERPGs (Emergency Response Planning Guideline) are in **bold face**.

Determination in Air:
According to the USAMRICD, do NOT depend on the following for the detection of Agent CX: M272 water testing kit, MINICAMS, ICAD, M21 remote sensing alarm, CAM, ACAMS, DAAMS, and M8A1 automatic chemical-agent detector alarm are incapable of detecting CX. Likewise, M8 and M9 paper should not be depended upon to detect Agent CX.

The M256A1 detector ticket reacts to the presence of CX, but the detection threshold is not known with certainty. Liquid detection: The portable M256A1 has a response time of up to 15 min. The following detectors are listed for Agent CX detection in the *Guide for the Selection of Chemical Detection Equipment for Emergency First Responders, 3rd Edition*, published by the US Department of Homeland Security: Chemical Agent Detector C2 Kit (021330) for vapor, liquids, aerosols: Start-up time (based on experience) 1–5 min; response time (regardless of experience) 20–25 min (Anachemia Canada, Inc.); and M256A1 (T503) (063230COM) Chemical Agent Detector Simulator Training Kits for vapor only: Start-up time (based on experience) 1–5 min; response time (regardless of experience) 20–25 min (Anachemia Canada, Inc.)

The following detectors have the capacity to detect CX at the threshold limits given[USAMRICD].
Liquid: M18A2   0.5 mg/min^3
Air: M90 (M90-D1-C)   0.15 mg/min^3
M93A1 Fox   10–100 µg/L

Permissible Concentration in Water: Do NOT use the M272 water testing kit.[USAMRICD]

Determination in Water: Contact pollution control authorities and advise shutting water intakes. CX dissolves in water and breaks down into toxic products that are much less dangerous than phosgene oxime, but still poisonous. Do not allow people to drink water containing even the breakdown products.

Routes of Entry: Skin, eye contact, inhalation.

Harmful Effects and Symptoms
Phosgene oxime (CX) is a rapid-acting casualty agent. The effects of phosgene oxime vapor and liquid on the skin, eyes, and lungs are almost instantaneous, causing immediate pain upon contact with the liquid and when the vapors are inhaled. Little is known about how CX works but it eats its way through protective clothing, the skin, and eventually reaches the blood. Phosgene oxime (CX) will cause blindness, kill skin horribly, and with enough of it on the skin a

victim could die. *Eyes:* The eyes will immediately burn, the eyelids will swell, and the victim's cornea will scar, causing permanent damage to the eyes with possible blindness. *Lungs:* The effects are immediate: The victim will sneeze and cough, with runny nose. The lungs will fill with fluid causing pulmonary edema, a medical emergency that can be delayed for several hours. This can cause death. *Skin:* Immediate stinging pain will be felt. The skin reddens and will eventually blister (up to 12 h after exposure). *Ingestion:* It is difficult to understand how one might swallow Phosgene oxime (CX), but it possibly may happen. Animal studies show that the victim's stomach and intestines would swell and bleed.

Short Term Exposure: Pain and local tissue destruction occur immediately on contact with skin, eyes, and mucous membranes. Phosgene oxime is rapidly absorbed from the skin and eyes and may result in systemic toxicity. Phosgene oxime causes redness, wheals (hives), and urticaria on the skin, but does not produce a fluid-filled blister (vesication). Despite the lack of initial blister formation, phosgene oxime produces more tissue damage than the blister agents. Known as a "nettle gas," CX produces immediate pain varying from a mild prickling to almost intolerable pain similar to a severe bee sting. Phosgene oxime has no antidote. Treatment is similar to that of the mustard agents. It causes violent irritation to the mucous membranes of the nose and eyes. Even at low temperature it has sufficient vapor pressure to produce tearing. When CX comes in contact to the skin, the area turns pale in 30 s and develops a red ring around the area. A wheal forms in about 30 min; the blanched area turns brown in 24 h and a scab forms in about a week. The scab usually falls off in about 3 weeks. Itching may be present throughout healing, which in some cases may be delayed beyond 2 months.[Army FM 3-9 and CDC] The LD$_{50}$ for skin exposure is estimated as 25 mg/kg. LCt$_{50}$ = no accurate data available; the estimated LCt$_{50}$ by inhalation is 1500−2000 mg-min/m^3; ICt$_{50}$ (respiratory) 25 mg-min/m^3. Inhaled phosgene oxime is extremely irritating to the upper airways and causes pulmonary edema. Irritation occurs with exposures to 0.2 mg-min/m^3 and becomes unbearable at 3 mg-min/m^3. The estimated LCt$_{50}$ (the product of concentration times time that is lethal to 50% of the exposed population by inhalation) is 1500 to 2000 mg-min/m^3.[ATSDR]

Long Term Exposure: May cause permanent injury, including lung damage and blindness. Information is unavailable about the carcinogenicity, developmental toxicity, or reproductive toxicity from chronic or repeated exposure to phosgene oxime.

Points of Attack: Skin, mucous membranes of the nose and eyes.

Medical Surveillance: There are no specific tests to confirm exposure. Extreme pain may persist for days. Patients/victims should be observed for signs of whole-body (systemic) toxicity, including accumulation of fluid in the lungs (pulmonary edema). Gastric lavage is contraindicated

following ingestion of this agent due to the risk of perforation of the esophagus or upper airway.[NIOSH]

First Aid: A note to first responders: You cannot help the victim if you kill yourself. *Do not enter an area contaminated with either CX vapor or liquid unprotected.* The only option to your own survival is by wearing "Level A" protection. You must act quickly.

Remove the victim to fresh air without exposing yourself—do not touch the victim! Only when the victim is outside the hot and warm zones and decontaminated you can help. If breathing is difficult, give oxygen. Do not make the victim vomit if the victim has swallowed Phosgene oxime. If the victim is conscious, give the victim milk. Get the victim to a doctor or medical facility as soon as possible, even if symptoms do not appear serious; symptoms are often delayed.

Decontamination: *Note to first responders:* Depending on the dose and the equipment available to you: Evacuate the area and shut down heating, ventilation, and air conditioning systems to prevent further spread of Phosgene oxime (CX). Call for medical and hazmat assistance immediately. If you get CX on yourself, decontaminate immediately. Even if you think you are not contaminated, be sure to thoroughly shower and change clothes as soon as you can after the incident. Because of the rapid reaction of CX with the skin, decontamination may not be entirely effective once pain occurs. Nevertheless, decontaminate as rapidly as possible by flushing the area with large amounts of water to remove any agent that has not reacted with the skin. Wash the victim with warm water and soap. *Bleach does not work with CX.* Be certain you have decontaminated the victims as much as you can before they leave the area so that they do not spread the Phosgene oxime (CX).

Personal Protective Methods: Use Level A protection. CX quickly penetrates rubber and clothing. Corrosive to most metals.

Respirator Selection: CX quickly penetrates rubber and clothing. Corrosive to most metals.

Storage: Stable in steel containers.

Shipping: Shipping Name: Toxic solids, organic, n.o.s. Hazardous Class or Division: 6.1. Subsidiary Hazardous Class or Division: Label Poison (Toxic). Packing Group III.

Spill Handling: Immediately evacuate everyone, including yourself. Immediately call for medical and hazmat assistance. Notify police, federal authorities, medical, hazmat, and emergency authorities. Immediately decontaminate victim. Do not touch the victim or allow Phosgene oxime (CX) to touch your skin or eyes. If possible, ventilate the area. If response personnel must walk through the spilled agent, wear the appropriate Level A protection. Keep combustibles (e.g., wood, paper, and oil) away from the spilled agent. Use water spray to reduce aerosols or divert aerosol cloud drift. Avoid allowing water runoff to contact the spilled agent. Do not direct water at the spill or the source of the leak. Stop the leak if it is possible to do so without risk to personnel. Prevent entry into waterways, sewers, basements, or confined areas. Isolate the area until aerosol has dispersed.[NIOSH]

Initial isolation and protective action distances

Distances shown are likely to be affected during the first 30 min after materials are spilled and could increase with time. If more than one tank car, cargo tank, portable tank, or large cylinder involved in the incident is leaking, the protective action distance may need to be increased. You may need to seek emergency information from CHEMTREC at (800) 424-9300 or seek professional environmental engineering assistance from the US EPA Environmental Response Team at (908) 548-8730 (24-h response line).

If a tank, rail car, or tank truck is involved in a fire, isolate it for 0.5 mile (800 m) in all directions; also, consider initial evacuation for 0.5 mile (800 m) in all directions.

CX, when used as a weapon

Small spills [involving the release of approximately 52.83 gallons (200 L) or less]

First: Isolate in all directions (feet/meters)   100/30
Then: Protect persons downwind (miles/kilometers)
Day   0.1/0.2
Night   0.3/0.5

Large spills [involving quantities greater than 52.83 gallons (200 L)]

First: Isolate in all directions (feet/meters)   300/90
Then: Protect persons downwind (miles/kilometers)
Day   0.6/1.0
Night   1.9/3.1

Fire Extinguishing: Phosgene oxime (CX) is noncombustible; it burns weakly, if at all. CX may decompose upon heating producing corrosive and/or toxic gases. Containers may explode when heated. In case of fire, evacuate the area, including yourself. CX may be an oxidizer, and it may ignite combustible materials (e.g., wood, paper, oil, or clothing). If there is some reason that you have to put out the fire—for example, there are things nearby that cannot be allowed to burn—use unattended equipment, then evacuate everyone immediately, including yourself. If you *must* extinguish a Phosgene oxime (CX) fire, use water streams, water fog, alcohol foam, universal foam, and, for confined fires, carbon dioxide. (For small fires, use dry chemical, carbon dioxide, or water spray. For large fires, use dry chemical, carbon dioxide, alcohol-resistant foam, or water spray.) Vapors are heavier than air and will collect in low areas. Keep out of these areas; stay upwind. Hazardous concentrations may spread along the ground and collect and stay in poorly ventilated, low-lying, or confined areas (e.g., sewers, basements, and tanks). If material or contaminated runoff enters waterways, notify downstream users of potentially contaminated waters. Notify local health and fire officials and pollution control agencies. Move containers from the fire area if it is possible to do so without risk to personnel. Dike fire control water for later disposal; do not scatter the material. For fire involving tanks or car/trailer loads, fight the fire from maximum distance or use unmanned hose holders or monitor nozzles. Do not get water inside containers. Cool containers with flooding quantities of water until well after fire is out. Withdraw immediately in case of rising sound from venting safety devices or discoloration of tanks. Always stay away from tanks engulfed in fire. Run-off from fire control or dilution water may be corrosive and/or toxic, and it may cause pollution. If the situation allows, control and properly dispose of run-off (effluent).

Disposal Method Suggested: Seek expert advice from armed services (see Reference section), Center for Disease Control headquarters in Atlanta, GA.

References

US Army Field Manual (DA FM) 3-9 (PCN 320 008457 00); US Navy Publication No P-467; US Air Force Manual No 355-7; *Potential Military Chemical/Biological Agents and Compounds.* Washington, DC: Headquarters, Department of the Army, Headquarters, Department of the Navy, Headquarters, Department of the Air Force, December 1990

Sidell, R. (1997). *Medical Aspects of Chemical and Biological Warfare*, Borden Institute, Walter Reed Army Medical Center, Washington, DC; Office of The Surgeon General, United States Army, Falls Church, Virginia; United States Army Medical Department Center and School, Fort Sam Houston, Texas; United States Army Medical Research and Materiel Command, Fort Detrick, Frederick, Maryland, Uniformed Services University of the Health Sciences, Bethesda, MD[MACBW]

CDC/NIOSH, *The Emergency Response Safety and Health Database,* <http://www.cdc.gov/NIOSH/ershdb/EmergencyResponseCard_29750015.html>

Fatha, A. A., Arcilesi, R. D., Peterson, J. C., Lattin, C. H., Well, C. Y., & McClintock, J. A. (January 2007). *Guide for the Selection of Chemical Detection Equipment for Emergency First Responders, 3rd Edition, Guide 100-6.* Washington, DC: US Department of Homeland Security

Phosmet   P:0560

Molecular Formula: $C_{11}H_{12}NO_4PS_2$

Synonyms: APPA; Decemthion; Decemthion P-6; *O,O*-Dimethyl phthalimidiomethyl dithiophosphate; *O,O*-Dimethyl *S*-(*N*-phthalimidomethyl) dithiophosphate; *O,O*-Dimethyl *S*-phthalimidomethyl phosphorodithioate; ENT25,705; Fosmet (Spanish); Ftalophos; Imidan; Kemolate; *N*-(Mercaptomethyl)phthalimide *S*-(*O,O*-dimethyl phosphorodithioate); Percolate; Phosphorodithioic acid, *S*-[(1,3-dihydro-1,3-dioxo-isoindol-2-yl)methyl] *O,O*-dimethyl ester; Phosphorodithioic acid, *O,O*-dimethyl ester, *S*-ester with *N*-(mercaptomethyl) phthalimide; Phthalimide, *N*-(mercaptomethyl)-, *S*-ester with *O,O*-dimethyl phosphorodithioate; Phthalimido-*O,O*-dimethyl phosphorodithioate; Phthalimidomethyl *O,O*-dimethyl phosphorodithioate; Phthalophos; PMP; Prolate; R 1504; Smidan; Stauffer R 1504

CAS Registry Number: 732-11-6

RTECS® Number: TE2275000

UN/NA & ERG Number: UN2783 (organophosphorus pesticides, solid, toxic)/152

EC Number: 211-987-4 [*Annex I Index No.:* 015-101-00-5]

Regulatory Authority and Advisory Bodies

US EPA, FIFRA, 1998 Status of Pesticides: Supported.

Air Pollutant Standard Set. Scc below, "Permissible Exposure Limits in Air" section.

Superfund/EPCRA 40CFR355, Extremely Hazardous Substances: TPQ = 10/10,000 lb (4.54/4540 kg).

Reportable Quantity (RQ): 1 lb (0.454 kg).

US DOT MARINE POLLUTANT (49CFR, Subchapter 172.101, Appendix B).

US DOT 49CFR172.101, Inhalation Hazard Chemical as organophosphates.

European/International Regulations (not listed in Annex I, but the following may apply): Hazard Symbol: T+, N; Risk phrases: R27/28; R50/53; Safety phrases: S1/2; S28; S36/37; S45; S60; S61 (see Appendix 4).

WGK (German Aquatic Hazard Class): No value assigned.

Description: Phosmet is a white crystalline solid. Molecular weight = 317.33; Boiling point = (decomposes below BP) >100°C; Freezing/Melting point = 72°C. Hazard; Vapor pressure = 4.9×10^{-7} mbar at 20°C. Identification (based on NFPA-704 M Rating System): Health 3, Flammability 1, Reactivity 2. Slightly soluble in water.

Potential Exposure: Compound Description: Agricultural Chemical; Mutagen; Reproductive Effector; Human Data. Used as an organophosphorus insecticide and acaricide.

Incompatibilities: Not compatible with other pesticides under alkaline conditions. Contact with water, steam, or moisture forms phthalic acids. Slightly corrosive to metals in the presence of moisture.

Permissible Exposure Limits in Air

Protective Action Criteria (PAC)

TEEL-0: 0.025 mg/m^3

PAC-1: 0.075 mg/m^3

PAC-2: 0.54 mg/m^3

PAC-3: 40 mg/m^3

Russia has set a ceiling value of 0.3 mg/m^3 in work-place air. Russia has also set a MAC of 0.009 mg/m^3 in ambient air in residential area on a once-daily basis and 0.004 mg/m^3 on a daily average basis.

Determination in Air: NIOSH Analytical Method (IV) Method #5600, Organophosphorus pesticides.

Permissible Concentration in Water: Russia[36] set a MAC of 0.2 mg/L in surface water.

Determination in Water: Fish Tox: 4.41815000 ppb MATC (HIGH). Octanol–water coefficient: Log K_{ow} = 2.8.

Routes of Entry: Inhalation, ingestion, skin and/or eye contact. Absorbed by the skin.

Harmful Effects and Symptoms

Short Term Exposure: Irritates the eyes and skin on contact. This material is a highly toxic organophosphate; the probable oral lethal dose for humans is 50–500 mg/kg or between 1 teaspoon and 1 oz for a 150-lb person. It is a cholinesterase inhibitor and has central nervous system effects. Oral lethal doses in humans have been reported at 50 mg/kg. Acute exposure to phosmet may produce the following signs and symptoms: pinpoint pupils, blurred vision, headache, dizziness, muscle spasms, and profound weakness. Vomiting, diarrhea, abdominal pain, seizures, and coma may also occur. The heart rate may decrease following oral exposure or increase following dermal exposure. Chest pain may be noted. Hypotension (low blood pressure) may occur, although hypertension (high blood pressure) is not uncommon. Dyspnea (shortness of breath) may be followed by respiratory collapse. Giddiness is common.

Long Term Exposure: Cholinesterase inhibitor; cumulative effect is possible. This chemical may damage the nervous system with repeated exposure, resulting in convulsions, respiratory failure. May cause liver damage. Human Tox = 7.00000 ppm Health Advisory (HIGH).

Points of Attack: Respiratory system, lungs, central nervous system, cardiovascular system, skin, eyes, plasma and red blood cell cholinesterase.

Medical Surveillance: Before employment and at regular times after that, the following are recommended: plasma and red blood cell cholinesterase levels (tests for the enzyme poisoned by this chemical). If exposure stops, plasma levels return to normal in 1–2 weeks while red blood cell levels may be reduced for 1–3 months.

When cholinesterase enzyme levels are reduced by 25% or more below preemployment levels, risk of poisoning is increased, even if results are in lower ranges of "normal." Reassignment to work not involving organophosphate or carbamate pesticides is recommended until enzyme levels recover. If symptoms develop or overexposure occurs, repeat the above tests as soon as possible and get an examination of the nervous system. Also consider complete blood count. Consider chest X-ray following acute overexposure. Liver function tests. Do not drink any alcoholic beverages before or during use. Alcohol promotes absorption of organic phosphates.

First Aid: If this chemical gets into the eyes, remove any contact lenses at once and irrigate immediately for at least 15 min, occasionally lifting upper and lower lids. Seek medical attention immediately. If this chemical contacts the skin, remove contaminated clothing and wash immediately with soap and water. Speed in removing material from skin is of extreme importance. Shampoo hair promptly if contaminated. Seek medical attention immediately. If this chemical has been inhaled, remove from exposure, begin rescue breathing (using universal precautions, including resuscitation mask) if breathing has stopped and CPR if heart action has stopped. Transfer promptly to a medical facility. When this chemical has been swallowed, get medical attention. Give large quantities of water and/or slurry of activated charcoal in water; and induce vomiting. Do not make an unconscious person vomit. Obtain authorization and/or further instructions from the local hospital for administration of an antidote or performance of other invasive procedures. Transport to a health-care facility.

Personal Protective Methods: Wear protective gloves and clothing to prevent any reasonable probability of skin contact. Safety equipment suppliers/manufacturers can provide recommendations on the most protective glove/clothing material for your operation. All protective clothing (suits, gloves, footwear, headgear) should be clean, available each day, and put on before work. Contact lenses should not be worn when working with this chemical. Wear dust-proof chemical goggles and face shield unless full face-piece respiratory protection is worn. Employees should wash immediately with soap when skin is wet or contaminated. Provide emergency showers and eyewash.

Respirator Selection: Follow regulations in OSHA 29CFR1910.134 or European Standard EN149. Use a NIOSH/MSHA- or European Standard EN149-approved respirator; or use an approved supplied-air respirator with a full face-piece operated in the positive-pressure mode, or with a full face-piece, hood, or helmet in the continuous-flow mode; or use a NIOSH/MSHA- or European Standard EN149-approved self-contained breathing apparatus with a full face-piece operated in pressure-demand or other positive-pressure mode.

Storage: Color Code—Blue: Health Hazard/Poison: Store in a secure poison location. Prior to working with this chemical you should be trained on its proper handling and storage. Store in tightly closed containers in a cool, well-ventilated area away from other pesticide, alkaline conditions, water, and other forms of moisture. Where possible, automatically transfer material from drums or other storage containers to process containers. Sources of ignition, such as smoking and open flames, are prohibited where this chemical is handled, used, or stored. Metal containers involving the transfer of this chemical should be grounded and bonded. Use only nonsparking tools and equipment, especially when opening and closing containers of this chemical. Wherever this chemical is used, handled, manufactured, or stored, use explosion-proof electrical equipment and fittings.

Shipping: Organophosphorus pesticides, solid, toxic, require a shipping label of "POISONOUS/TOXIC MATERIALS." It falls in Hazard Class 6.1 and Packing Group III.

Spill Handling: Evacuate persons not wearing protective equipment from area of spill or leak until cleanup is complete. Remove all ignition sources. As for other organophosphorus pesticides stay upwind; keep out of low areas. Ventilate closed spaces before entering them. Remove and isolate contaminated clothing at the site. Do not touch spilled material. Use water spray to reduce vapors. Take up *small spills* with sand or other noncombustible absorbent material and place in containers for later disposal. Take up small, dry spills with clean shovel and place in clean, dry container. Dike far ahead of *large spills* for later disposal. Ventilate area after cleanup is complete. It may be necessary to contain and dispose of this chemical as a hazardous waste. If material or contaminated runoff enters waterways, notify downstream users of potentially contaminated waters. Contact your local or federal environmental protection agency for specific recommendations. If employees are required to clean up spills, they must be properly trained and equipped. OSHA 1910.120(q) may be applicable.

Fire Extinguishing: This material may burn but does not ignite readily. For small fires, use dry chemical, carbon dioxide, water spray, or foam. For large fires, use water spray, fog, or foam. Stay upwind; keep out of low areas. Move container from fire area if you can do it without risk. Fight fire from maximum distance. Dike fire control water for later disposal; do not scatter material. Wear positive pressure breathing apparatus and special protective clothing. Poisonous gases are produced in fire, including nitrogen oxides, phosphorous oxides, and sulfur oxides. If material or contaminated runoff enters waterways, notify downstream users of potentially contaminated waters. Notify local health and fire officials and pollution control agencies. Containers may explode in fire. From a secure, explosion-proof location, use water spray to cool exposed containers. If cooling streams are ineffective (venting sound increases in volume and pitch, tank discolors, or shows any signs of deforming), withdraw immediately to a secure position. If employees are expected to fight fires, they must be trained and equipped in OSHA 1910.156. The only respirators recommended for firefighting are self-contained breathing apparatuses that have full face-pieces and are operated in a pressure-demand or other positive-pressure mode.

Disposal Method Suggested: Consult with environmental regulatory agencies for guidance on acceptable disposal practices. Generators of waste containing this contaminant ($\geq$100 kg/mo) must conform with EPA regulations governing storage, transportation, treatment, and waste disposal. In accordance with 40CFR165, follow recommendations for the disposal of pesticides and pesticide containers. Must be disposed properly by following package label directions or by contacting your local or federal environmental control agency or by contacting your regional EPA office. Small amounts may be decomposed with hypochlorite. For large amounts, incineration with effective gas scrubbing is recommended.[22]

References

US Environmental Protection Agency. (November 30, 1987). *Chemical Hazard Information Profile: Phosmet.* Washington, DC: Chemical Emergency Preparedness Program

New Jersey Department of Health and Senior Services. (March 1999). *Hazardous Substances Fact Sheet: Decemthion.* Trenton, NJ

Phosphamidon   P:0570

Molecular Formula: $C_{10}H_{19}ClNO_5P$
Common Formula: $(CH_3O)_2POOC(CH_3){=}C(Cl)CON(CH_2CH_3)_2$

Synonyms: Apamidon; C 570; C-570; (2-Chlor-3-diaethyla-mino-methyl-3-oxo-prop-1-en-yl)-dimethylphosphat (German); 2-Chloro-3-(diethylamino)-1-methyl-3-oxo-1-propenyl-dimethyl phosphate; 2-Chloro-2-diethylcarbamoyl-1-methylvinyl dimethyl phosphate; 1-Chloro-diethylcarba-moyl-1-propen-2-yl dimethyl phosphate; Ciba 570; Crophosphate; Dimecron; Dimecron 100; *O,O*-Dimethyl *O*-[2-chloro-2-(*N,N*-diethylcarbamoyl)-1-methylvinyl] phosphate; Dimethyl 2-chloro-2-diethylcarbamoyl-1-methylvi-nylphosphate; Dimethyl diethylamido-1-chlorocrotonyl(2) phosphate; *O,O*-Dimethyl-*O*-(1-methyl-2-chlor-2-*N*, *N*-diethyl-carbamoyl)-vinyl-phosphat (German); [*O,O*-Dimethyl *O*-[1-methyl-(2-chloro-2-diethylcarbamoyl)vinyl] phosphate; Dimethyl phosphate of 2-chloro-*N,N*-diethyl-3-hydroxycrotonamide; Dimethyl phosphate ester with 2-chloro-*N,N*-diethyl-3-hydroxycrotonamide; Dimonex; Dixon; ENT25,515; Fosfamidon (Spanish); Fosfamidone; Foszfamidon; ML 97; Merkon phosphamidone; NCI-C00588; OMS 1325; OR1191; Phosphamidon; Phosphate de dimethyle et de(2-chloro-2-diethylcarbamoyl-1-methyl-vinyle) (French); Phosphoric acid, 2-chloro-3-(diethylami-no)-1-methyl-3-oxo-1-propenyl dimethyl ester; Phosphoric acid, dimethyl ester, with 2-chloro-*N,N*-diethyl-3-hydroxycrotonamide

CAS Registry Number: 13171-21-6; 23783-98-4 [(Z) isomer)]; 297-99-4 [(E) isomer]

RTECS® Number: TC2800000

UN/NA & ERG Number: UN3018 (organophosphorus pesticide, liquid, toxic)/152

EC Number: 236-116-5 [*Annex I Index No.:* 015-022-00-6] (phosphamidon)

Regulatory Authority and Advisory Bodies

Carcinogenicity: NCI: Carcinogenesis Bioassay (feed); equivocal evidence: rat; no evidence: mouse.

US EPA, FIFRA 1998 Status of Pesticides: Canceled.

Very Toxic Substance (World Bank).[15]

Superfund/EPCRA 40CFR355, Extremely Hazardous Substances: TPQ = 100 lb (45.4 kg).

Reportable Quantity (RQ): 100 lb (45.4 kg).

US DOT Regulated Marine Pollutant (49CFR172.101, Appendix B), severe pollutant.

US DOT 49CFR172.101, Inhalation Hazard Chemical as organophosphates.

Rotterdam Convention Annex III [Chemicals Subject to the Prior Informed Consent Procedure (PIC)] (soluble liquid formulations of the substance that exceeds 1000 g active ingredient).

European/International Regulations: Hazard Symbol: T + , N; Risk phrases: R24; R28; R50/53; R68; Safety phrases: S1/2; S23; S36/37; S45-S60; S61 (see Appendix 4).

WGK (German Aquatic Hazard Class): 3—Highly water polluting (CAS: 13171-21-6).

Description: Phosphamidon is a pale yellow oily liquid. Molecular weight = 299.72; Specific gravity (H_2O:1) = 1.22; Boiling point = 162°C at 1.5 mmHg; Freezing/Melting point = −45°C. Hazard Identification (based on

NFPA-704 M Rating System): Health 3, Flammability 1, Reactivity 0. Soluble in water.

Potential Exposure: Compound Description: Agricultural Chemical; Tumorigen, Mutagen; Reproductive Effector. This material is used as an insecticide on citrus, cotton, and deciduous fruit and nuts. It is also an acaricide.

Incompatibilities: Strong oxidizers may cause release of toxic phosphorus oxides. Organophosphates, in the presence of strong reducing agents such as hydrides, may form highly toxic and flammable phosphine gas. Keep away from alkaline materials. Attacks metals, such as aluminum, iron, tin.

Permissible Exposure Limits in Air

Protective Action Criteria (PAC)

TEEL-0: 0.06 mg/m^3

PAC-1: 0.15 mg/m^3

PAC-2: 0.3 mg/m^3

PAC-3: 60 mg/m^3

Determination in Water: Fish Tox = 1445.67277000 ppb (EXTRA LOW); FISH STV (Sediment Toxicity Value): LOW; Octanol−water coefficient: Log K_{ow} = <0.9.

Routes of Entry: Inhalation, ingestion, skin and/or eye contact.

Harmful Effects and Symptoms

Short Term Exposure: Irritates the eyes. May affect the nervous system, causing convulsions, respiratory failure, and death. Higher exposures can cause pulmonary edema, a medical emergency that can be delayed for several hours. This can cause death. This material is extremely toxic; the probable oral lethal dose for humans is 5−50 mg/kg or between 7 drops and 1 teaspoonful for a 150-lb person. It is a cholinesterase inhibitor. Acute exposure to phosphamidon may produce pinpoint pupils, blurred vision, headache, dizziness, muscle spasms, and profound weakness. Vomiting, diarrhea, abdominal pain, seizures, and coma may also occur. The heart rate may decrease following oral exposure or increase following dermal exposure. Hypotension (low blood pressure) may occur, although hypertension (high blood pressure) is not uncommon. Chest pain may be noted. Respiratory effects include dyspnea (shortness of breath), respiratory depression, and respiratory paralysis. Psychosis may occur.

Long Term Exposure: Cholinesterase inhibitor; cumulative effect is possible. This chemical may damage the nervous system with repeated exposure, resulting in convulsions, respiratory failure. May cause liver damage. Human Tox = 0.14000 ppm (EXTRA HIGH).

Points of Attack: Respiratory system, lungs, central nervous system, cardiovascular system, skin, eyes, plasma and red blood cell cholinesterase.

Medical Surveillance: Before employment and at regular times after that, the following are recommended: plasma and red blood cell cholinesterase levels (tests for the enzyme poisoned by this chemical). If exposure stops, plasma levels return to normal in 1−2 weeks while red blood cell levels may be reduced for 1−3 months.

When cholinesterase enzyme levels are reduced by 25% or more below preemployment levels, risk of poisoning is increased, even if results are in lower ranges of "normal." Reassignment to work not involving organophosphate or carbamate pesticides is recommended until enzyme levels recover. If symptoms develop or overexposure occurs, repeat the above tests as soon as possible and get an examination of the nervous system. Also consider complete blood count. Consider chest X-ray following acute overexposure. Do not drink any alcoholic beverages before or during use. Alcohol promotes absorption of organic phosphates.

First Aid: If this chemical gets into the eyes, remove any contact lenses at once and irrigate immediately for at least 15 min, occasionally lifting upper and lower lids. Seek medical attention immediately. If this chemical contacts the skin, remove contaminated clothing and wash immediately with soap and water. Speed in removing material from skin is of extreme importance. Shampoo hair promptly if contaminated. Seek medical attention immediately. If this chemical has been inhaled, remove from exposure, begin rescue breathing (using universal precautions, including resuscitation mask) if breathing has stopped and CPR if heart action has stopped. Transfer promptly to a medical facility. When this chemical has been swallowed, get medical attention. Give large quantities of water and induce vomiting. Do not make an unconscious person vomit. Obtain authorization and/or further instructions from the local hospital for administration of an antidote or performance of other invasive procedures. Transport to a health-care facility. Medical observation is recommended for 24—48 h after breathing overexposure, as pulmonary edema may be delayed. As first aid for pulmonary edema, a doctor or authorized paramedic may consider administering a corticosteroid spray.

Personal Protective Methods: Wear protective gloves and clothing to prevent any reasonable probability of skin contact. Safety equipment suppliers/manufacturers can provide recommendations on the most protective glove/clothing material for your operation. All protective clothing (suits, gloves, footwear, headgear) should be clean, available each day, and put on before work. Contact lenses should not be worn when working with this chemical. Wear splash-proof chemical goggles and face shield unless full face-piece respiratory protection is worn. Employees should wash immediately with soap when skin is wet or contaminated. Provide emergency showers and eyewash.

Respirator Selection: Follow regulations in OSHA 29CFR1910.134 or European Standard EN149. Use a NIOSH/MSHA- or European Standard EN149-approved respirator; or use an approved supplied-air respirator with a full face-piece operated in the positive-pressure mode, or with a full face-piece, hood, or helmet in the continuous-flow mode; or use a NIOSH/MSHA- or European Standard EN149-approved self-contained breathing apparatus with a full face-piece operated in pressure-demand or other positive-pressure mode.

Storage: Color Code—Blue: Health Hazard/Poison: Store in a secure poison location. Prior to working with this chemical you should be trained on its proper handling and storage. Store in tightly closed containers in a cool, well-ventilated area away from alkalis.

Shipping: This compound requires a shipping label of "POISONOUS/TOXIC MATERIALS." It falls in Hazard Class 6.1 and Packing Group II.

Spill Handling: Evacuate persons not wearing protective equipment from area of spill or leak until cleanup is complete. Remove all ignition sources. Stay upwind; keep out of low areas. Ventilate closed spaces before entering them. Do not touch spilled material; stop leak if you can do so without risk. Use water spray to reduce vapors. *Small spills:* absorb with sand or other noncombustible absorbent material and place into containers for later disposal. *Large spills:* dike far ahead of spill for later disposal. It may be necessary to contain and dispose of this chemical as a hazardous waste. If material or contaminated runoff enters waterways, notify downstream users of potentially contaminated waters. Contact your local or federal environmental protection agency for specific recommendations. If employees are required to clean up spills, they must be properly trained and equipped. OSHA 1910.120(q) may be applicable.

Fire Extinguishing: This material may burn, but does not ignite readily. For small fires, use dry chemical, carbon dioxide, water spray, or foam. For large fires: use water spray, fog, or foam. Stay upwind; keep out of low areas. Move containers from fire area if you can do it without risk. Fight fire from maximum distance. Dike fire control water for later disposal; do not scatter the material. Poisonous gases, including phosphorous oxides, hydrogen chloride, and nitrogen oxides, are produced in fire. Vapors are heavier than air and will collect in low areas. Vapors may travel long distances to ignition sources and flashback. Vapors in confined areas may explode when exposed to fire. Containers may explode in fire. Storage containers and parts of containers may rocket great distances, in many directions. If material or contaminated runoff enters waterways, notify downstream users of potentially contaminated waters. Notify local health and fire officials and pollution control agencies. From a secure, explosion-proof location, use water spray to cool exposed containers. If cooling streams are ineffective (venting sound increases in volume and pitch, tank discolors, or shows any signs of deforming), withdraw immediately to a secure position. If employees are expected to fight fires, they must be trained and equipped in OSHA 1910.156. The only respirators recommended for firefighting are self-contained breathing apparatuses that have full face-pieces and are operated in a pressure-demand or other positive-pressure mode.

Disposal Method Suggested: Small quantities may be treated with alkali followed by landfill disposal. Large quantities should be incinerated with effluent gas scrubbing.[22] In accordance with 40CFR165, follow recommendations for the disposal of pesticides and pesticide containers. Must be disposed properly by following package label directions or by contacting your local or federal environmental control agency or by contacting your regional EPA office.

References
US Environmental Protection Agency. (November 30, 1987). *Chemical Hazard Information Profile: Phosphamidon*. Washington, DC: Chemical Emergency Preparedness Program

New Jersey Department of Health and Senior Services. (September 1999). *Hazardous Substances Fact Sheet: Phosphamidon*. Trenton, NJ

Phosphine   P:0580

Molecular Formula: H$_3$P
Common Formula: PH$_3$
Synonyms: Celphos; Delicia; Detia gas-EX-B; Fosfamia (Spanish); Hydrogen phosphide; Phosphorous trihydride; Phosphorous hydride; Phosphorated hydrogen; Phosphorwasserstoff (German); Phostoxin
CAS Registry Number: 7803-51-2
RTECS® Number: SY7525000
UN/NA & ERG Number: UN2199/119
EC Number: 232-260-8 [Annex I Index No.: 015-181-00-1]
Regulatory Authority and Advisory Bodies
Department of Homeland Security Screening Threshold Quantity (pounds): *Release hazard* 10,000 (≥1.00% concentration). (1% concentration; *Theft hazard* 15 (≥0.67% concentration).
Department of Homeland Security Screening Threshold Quantity (pounds): Sabotage/Contamination Hazard: A placarded amount (commercial grade).
Carcinogenicity: EPA: Not Classifiable as to human carcinogenicity.
Banned or Severely Restricted (several countries) (UN).[13]
Very Toxic Substance (World Bank).[15]
Air Pollutant Standard Set. See below, "Permissible Exposure Limits in Air" section.
Clean Air Act: Hazardous Air Pollutants (Title I, Part A, Section 112); Accidental Release Prevention/Flammable Substances, (Section 112[r], Table 3), TQ = 5000 lb (2270 kg).
US EPA Hazardous Waste Number (RCRA No.): P096.
RCRA, 40CFR261, Appendix 8 Hazardous Constituents.
Superfund/EPCRA 40CFR355, Extremely Hazardous Substances: TPQ = 500 lb (227 kg).
Reportable Quantity (RQ): 100 lb (45.4 kg).
EPCRA Section 313 Form R *de minimis* concentration reporting level: 1.0%.
US DOT 49CFR172.101, Inhalation Hazardous Chemical.
European/International Regulations: Hazard Symbol: F, T, N; Risk phrases: R12; R17; R26; R34; R50; Safety phrases: S1/2; S28; S36/37; S45; S61; S63 (see Appendix 4).
WGK (German Aquatic Hazard Class): 2—Hazard to waters.

Description: Phosphine is a colorless gas that is shipped as liquefied compressed gas. Odorless when pure, it has the odor of garlic or the foul odor of decaying fish. The level at which humans detect the odor of phosphine (odor threshold) does not provide sufficient warning of dangerous concentrations. Phosphine presents an additional hazard in that it ignites at very low temperatures. Shipped as a liquefied compressed gas. The pure compound is odorless. The odor threshold is 0.14 ppm. Molecular weight = 34.00; Specific gravity (H$_2$O:1) = 0.8; Boiling point = −88°C; Freezing/Melting point = −133°C; Relative vapor density (air = 1): 1.18; Vapor pressure = 41.3 atm; >760 mmHg at 20°C; Flash point = (flammable gas) 104°C; Autoignition temperature (depends on concentration and diluent) = 38−100°C. Explosive limits: LEL = 1.8%; UEL = unknown. Hazard Identification (based on NFPA-704 M Rating System): Health 4, Flammability 4, Reactivity 2. Slightly soluble in water; solubility = 25 mL/100 mL at 17°C.
Potential Exposure: Compound Description: Agricultural Chemical; Mutagen; Human Data. Phosphine is used as a fumigant; in the semiconductor industry, as a doping agent for electronic components (to introduce phosphorus into silicon crystals); in chemical synthesis; used as a polymerization initiator; as an intermediate for some flame retardants. Also, exposures may occur when acid or water comes in contact with metallic phosphides (aluminum phosphide, calcium phosphide). These two phosphides are used as insecticides or rodenticides for grain, and phosphine is generated during grain fumigation. When phosphine toxicity is suspected, but phosphine exposure is not obvious, one should suspect transdermal contamination and/or ingestion of phosphides. Phosphine may also evolve during the generation of acetylene from impure calcium carbide, as well as during metal shaving; sulfuric acid tank cleaning; rustproofing, ferrosilicon, phosphoric acid; and yellow phosphorus explosive handling.
Incompatibilities: Phosphine reacts with acids, air, copper, moisture, oxidizers, oxygen, chlorine, nitrogen oxides, metal nitrates, halogens, halogenated hydrocarbons, copper, and many other substances, causing fire and explosion hazard. Extremely explosive; may ignite spontaneously on contact with air at or about 100°C. Attacks many metals.
Permissible Exposure Limits in Air
Conversion factor: 1 ppm = 1.39 mg/m^3 at 25°C & 1 atm.
OSHA PEL: 0.3 ppm/0.4 mg/m^3 TWA.
NIOSH REL: 0.3 ppm/0.4 mg/m^3 TWA; 1 ppm/1 mg/m^3 STEL.
ACGIH TLV®[1]: 0.3 ppm/0.42 mg/m^3 TWA; 1 ppm/1.4 mg/m^3 STEL.
NIOSH IDLH: 50 ppm.
Protective Action Criteria (PAC) Phosphine*
TEEL-0: 0.3 ppm
PAC-1: 1 ppm
PAC-2: **2.0** ppm
PAC-3: **3.6** ppm

*AEGLs (Acute Emergency Guideline Levels) & ERPGs (Emergency Response Planning Guideline) are in **bold face**.
Emergency Response Planning Guidelines (AIHA)
ERPG-1: Inappropriate
ERPG-2: 0.5 ppm
ERPG-3: 3.5 ppm
DFG MAK: 0.1 ppm/0.14 mg/m^3 TWA; Peak Limitation Category II(2); Pregnancy Risk Group C.
Arab Republic of Egypt: TWA 0.3 ppm (0.4 mg/m^3), 1993; Australia: TWA 0.3 ppm (0.4 mg/m^3); STEL 1 ppm (1 mg/m^3), 1993; Austria: MAK 0.1 ppm (0.15 mg/m^3), 1999; Belgium: TWA 0.3 ppm (0.42 mg/m^3); STEL 1 ppm (1.4 mg/m^3), 1993; Denmark: TWA 0.1 ppm (0.15 mg/m^3), 1999; Finland: TWA 0.1 ppm (0.15 mg/m^3); STEL 0.3 ppm (0.4 mg/m^3), 1999; France: VME 0.1 ppm (0.13 mg/m^3), VLE 0.3 ppm (0.4 mg/m^3), 1999; the Netherlands: MAC-TGG 0.4 mg/m^3, 2003; Norway: TWA 0.1 ppm (0.15 mg/m^3), 1999; the Philippines: TWA 0.3 ppm (0.4 mg/m^3), 1993; Poland: MAC (TWA) 0.1 mg/m^3; MAC (STEL) 0.8 mg/m^3, 1999; Russia: STEL 0.1 mg/m^3, 1993; Sweden: NGV 0.3 ppm (0.4 mg/m^3), KTV 1 ppm (1.4 mg/m^3), 1999; Switzerland: MAK-W 0.1 ppm (0.15 mg/m^3), KZG-W 0.2 ppm (0.3 mg/m^3), 1999; Thailand: TWA 0.3 ppm (0.4 mg/m^3), 1993; Turkey: TWA 0.3 ppm (0.4 mg/m^3), 1993; United Kingdom: STEL 0.3 ppm (intermittent 0.42 mg/m^3), 2000; Argentina, Bulgaria, Columbia, Jordan, South Korea, New Zealand, Singapore, Vietnam: ACGIH TLV®: STEL 1 ppm. Several states have set guidelines or standards for phosphine in ambient air[60] ranging from 1.33 μg/m^3 (New York) to 4.0 μg/m^3 (Florida and North Dakota) to 6.7 μg/m^3 (Virginia) to 8.0 μg/m^3 (Connecticut) to 10.0 μg/m^3 (Nevada and North Dakota) to 130 μg/m^3 (North Carolina).

Determination in Air: Use NIOSH Analytical Method #1003, Phosphine, OSHA Analytical Method ID-180. See also NIOSH Analytical Method #6002.

Permissible Concentration in Water: No criteria set, but EPA[32] has suggested a permissible ambient goal of 5.5 μg/L based on health effects.

Determination in Water: Phosphine cannot be used to contaminate water supplies; it breaks down in water. Octanol–water coefficient: Log K_{ow} = (estimated) −0.27.

Routes of Entry: Phosphine can be absorbed into the body by inhalation. Direct contact with phosphine liquid may cause frostbite.

Harmful Effects and Symptoms

Short Term Exposure: Severe irritation to the respiratory tract. Higher exposures can cause pulmonary edema, a medical emergency that can be delayed for several hours. This can cause death. Contact with the liquid may cause frostbite. May affect the central nervous system, cardiovascular system, heart, gastrointestinal tract, liver, and kidneys. Phosphine is a super-toxic gas with a probable oral lethal dose of 5 mg/kg or 7 drops for a 150-lb person. An air concentration of 3 ppm is safe for long-term exposure,

500 ppm is lethal in 30 min, and concentration of 1000 ppm is lethal after a few breaths. Acute exposure to phosphine usually results in headache, cough, tightness and pain in the chest, shortness of breath, dizziness, lethargy, and stupor. Fatigue, muscle pain, chills, tremors, loss of coordination, seizures, and coma may be seen. Pulmonary edema and cardiac arrhythmias are common. Gastrointestinal symptoms include nausea, vomiting, abdominal pain, and diarrhea. Renal (kidney) damage, hepatic (liver) damage; and jaundice may also occur.

Long Term Exposure: Chronic poisoning may cause toothache, swelling of the jaw, spontaneous fractures of bones. May cause anemia. May damage the liver and kidneys. The effects are cumulative. Can irritate the lungs; bronchitis may develop.

Points of Attack: Respiratory system, liver.

Medical Surveillance: NIOSH lists the following tests: Blood Gas Analysis; chest X-ray, electrocardiogram, pulmonary function tests: forced vital capacity, forced expiratory volume (1 s); sputum cytology; white blood cell count/differential. For those with frequent or potentially high exposure (half the TLV or greater), the following are recommended before beginning work and at regular times after that: lung function tests. If symptoms develop or overexposure is suspected, the following may be useful: consider chest X-ray after acute overexposure. Liver function tests.

First Aid: If this chemical gets into the eyes, remove any contact lenses at once and irrigate immediately for at least 15 min, occasionally lifting upper and lower lids. Seek medical attention immediately. If this chemical contacts the skin, remove contaminated clothing and wash immediately with soap and water. Seek medical attention immediately. If this chemical has been inhaled, remove from exposure, begin rescue breathing (using universal precautions, including resuscitation mask) if breathing has stopped and CPR if heart action has stopped. Transfer promptly to a medical facility. When this chemical has been swallowed, if phosphides have been ingested, **do not induce emesis**. Phosphides will release phosphine in the stomach; therefore, watch for signs similar to those produced by phosphine inhalation. Administer a slurry of activated charcoal at 1 g/kg (usual adult dose: 60–90 g; child dose: 25–50 g). A soda can and a straw may be of assistance when offering charcoal to a child. Do not make an unconscious person vomit. Medical observation is recommended for 24–48 h after breathing overexposure, as pulmonary edema may be delayed. As first aid for pulmonary edema, a doctor or authorized paramedic may consider administering a corticosteroid spray.

Decontamination: This is very important. The rapid physical removal of a chemical agent is essential. If you do not have the equipment and training, do not enter the hot or the warm zone to rescue and/or decontaminate victims. Medical personnel should wear the proper PPE. If the victim cannot move, decontaminate without touching and without entering the hot or the warm zone. Metallic phosphides on clothes, skin, or hair can off-gas phosphine after contact with water or moisture, so a risk of secondary contamination may be present. Have the victim remove clothing; and seal contaminated clothes and personal belongings in a sealed double bag. For skin exposure to the metallic phosphides, scrape or brush all visible particles from the skin and hair. Use clean water from any source; if possible, use a hose (spray or fog to prevent injury to the victim) or other system to avoid touching the victim. Do not wait for soap or for the victim to remove clothing, begin washing immediately. Do not delay decontamination to obtain warm water; time is of the essence; use cold water instead. Immediately flush the eyes with water for at least 15 min. Use caution to avoid hypothermia in children and the elderly. Persons exposed only to phosphine gas do not pose substantial risks of secondary contamination. Vomitus containing phosphides can also off-gas phosphine. Rinse the eyes, mucous membranes, or open wounds with sterile saline or water and then move away from the hot zone in an upwind and uphill direction.

Personal Protective Methods: Wear protective gloves and clothing to prevent any reasonable probability of skin contact. Safety equipment suppliers/manufacturers can provide recommendations on the most protective glove/clothing material for your operation. All protective clothing (suits, gloves, footwear, headgear) should be clean, available each day, and put on before work. Contact lenses should not be worn when working with this chemical. Wear gas-proof chemical goggles and face shield unless full face-piece respiratory protection is worn. Employees should wash immediately with soap when skin is wet or contaminated. Provide emergency showers and eyewash.

Respirator Selection: Up to 3 ppm: Sa (APF = 10) (any supplied-air respirator). *Up to 7.5 ppm:* Sa:Cf (APF = 25) (any supplied-air respirator operated in a continuous-flow mode). *Up to 15 ppm:* GmFS (APF = 50) [any air-purifying, full-face-piece respirator (gas mask) with a chin-style, front- or back-mounted canister providing protection against the compound of concern]; SCBAF (APF = 50) (any self-contained breathing apparatus with a full face-piece) or SaF (APF = 50) (any supplied-air respirator with a full face-piece). *Up to 50 ppm:* Sa: Pd,Pp (APF = 1000) (any supplied-air respirator operated in a pressure-demand or other positive-pressure mode). *Emergency or planned entry into unknown concentrations or IDLH conditions:* SCBAF: Pd, Pp (APF = 10,000) (any self-contained breathing apparatus that has a full face-piece and is operated in a pressure-demand or other positive-pressure mode) or SaF: Pd,Pp: ASCBA (APF = 10,000) (any supplied-air respirator that has a full face-piece and is operated in a pressure-demand or other positive-pressure mode in combination with an auxiliary self-contained breathing apparatus operated in a pressure-demand or other positive-pressure mode). *Escape:* GmFS (APF = 50) [any air-purifying, full-face-piece respirator (gas mask) with a chin-style, front- or back-mounted canister providing protection against the compound of concern] or SCBAE (any appropriate escape-type, self-contained breathing apparatus).

Storage: Color Code—Red Stripe: Flammability Hazard: Store separately from all other flammable materials. Prior to working with this chemical you should be trained on its proper handling and storage. Before entering confined space where this chemical may be present, check to make sure that an explosive concentration is not a danger. Phosphine must be stored to avoid contact with oxidizers (such as perchlorates, peroxides; permanganates, chlorates, and nitrates), strong acids (such as hydrochloric, sulfuric, and nitric), oxygen, and halogenated hydrocarbons, since violent reactions occur. Store in tightly closed containers from physical damage. Use only nonsparking tools and equipment, especially when opening and closing containers of phosphine. Procedures for the handling, use, and storage of cylinders should be in compliance with OSHA 1910.101 and 1910.169, as with the recommendations of the Compressed Gas Association.

Shipping: Phosphine requires a shipping label of "POISON GAS, FLAMMABLE GAS." It falls in Hazard Class 2.3 and Packing Group I. It is a violation of transportation regulations to refill compressed gas cylinders without the express written permission of the owner.

Spill Handling: Evacuate and restrict persons not wearing protective equipment from area of spill or leak until cleanup is complete. Remove all ignition sources. Establish forced ventilation to keep levels below explosive limit. Stop the flow of gas if it can be done safely. If source of leak is a cylinder and the leak cannot be stopped in place, remove leaking cylinder to a safe place in the open air, and repair leak or allow cylinder to empty. Keep this chemical out of confined space, such as a sewer, because of the possibility of explosion, unless the sewer is designed to prevent the buildup of explosive concentrations. It may be necessary to contain and dispose of this chemical as a hazardous waste. Contact your local or federal environmental protection agency for specific recommendations. If employees are required to clean up spills, they must be properly trained and equipped. OSHA 1910.120(q) may be applicable.

Initial isolation and protective action distances

Distances shown are likely to be affected during the first 30 min after materials are spilled and could increase with time. If more than one tank car, cargo tank, portable tank, or large cylinder involved in the incident is leaking, the protective action distance may need to be increased. You may need to seek emergency information from CHEMTREC at (800) 424-9300 or seek professional environmental engineering assistance from the US EPA Environmental Response Team at (908) 548-8730 (24-h response line).

Small spills (From a small package or a small leak from a large package)

First: Isolate in all directions (feet/meters)   3000/1000
Then: Protect persons downwind (miles/kilometers)
Day   0.4/0.6
Night   1.5/2.4

Large spills (From a large package or from many small packages)
First: Isolate in all directions (feet/meters)   2500/800
Then: Protect persons downwind (miles/kilometers)
Day   2.7/4.4
Night   5.6/8.9

Fire Extinguishing: This chemical is a flammable gas. Poisonous gases and mists including oxides of phosphorus and phosphonic acid are produced in fire. Specially trained personnel may use fog lines to cool exposures and let the fire burn itself out. If material is on fire or involved in a fire, do not extinguish unless flow can be stopped; use water in flooding quantities as fog; cool all affected containers with flooding quantities of water; apply water from as far a distance as possible; solid streams of water may be ineffective; use "alcohol" foam, carbon dioxide or dry chemical. Wear full protective clothing including self-contained breathing apparatus; rubber gloves; boots and bands around legs, arms, and waist. No skin surface should be exposed. For massive fires in cargo areas, use unmanned hose holders or monitor nozzles. Move containers from fire area. The gas is heavier than air and may travel along the ground to an ignition source. Container may explode in heat of fire. Vapors are heavier than air and will collect in low areas. Vapors may travel long distances to ignition sources and flashback. Vapors in confined areas may explode when exposed to fire. Containers may explode in fire. Storage containers and parts of containers may rocket great distances, in many directions. If material or contaminated run-off enters waterways, notify downstream users of potentially contaminated waters. Notify local health and fire officials and pollution control agencies. From a secure, explosion-proof location, use water spray to cool exposed containers. If cooling streams are ineffective (venting sound increases in volume and pitch, tank discolors, or shows any signs of deforming), withdraw immediately to a secure position. If cylinders are exposed to excessive heat from fire or flame contact, withdraw immediately to a secure location. If employees are expected to fight fires, they must be trained and equipped in OSHA 1910.156. The only respirators recommended for firefighting are self-contained breathing apparatuses that have full face-pieces and are operated in a pressure-demand or other positive-pressure mode.

Disposal Method Suggested: Consult with environmental regulatory agencies for guidance on acceptable disposal practices. Generators of waste containing this contaminant ($\geq$100 kg/mo) must conform with EPA regulations governing storage, transportation, treatment, and waste disposal. In accordance with 40CFR165, follow recommendations for the disposal of pesticides and pesticide containers. Must be disposed properly by following package label directions or by contacting your local or federal environmental control agency or by contacting your regional EPA office. Controlled discharges of Phosphine may be passed through 10% NAOH solution in a scrubbing tower. The product may be discharged to a sewer.[22]

References
US Environmental Protection Agency. (November 30, 1987). *Chemical Hazard Information Profile: Phosphine*. Washington, DC: Chemical Emergency Preparedness Program
Sax, N. I. (Ed.). (1986). Dangerous Properties of Industrial Materials Report, 6, No. 2, 103—107
New Jersey Department of Health and Senior Services. (April 2004). *Hazardous Substance Fact Sheet: Phosphine*. Trenton, NJ

Phosphoric acid, ortho-   P:0590

Molecular Formula: H_3O_4P
Common Formula: H_3PO_4
Synonyms: Acide phosphorique (French); Acido fosforico Spanish); Decon 4512; Evits; Orthophosphoric acid; *o*-Phosphoric acid; Phosphorsaeureloesungen (German); Sonac; WC-Reiniger; White phosphoric acid
CAS Registry Number: 7664-38-2
RTECS® Number: TB6300000
UN/NA & ERG Number: UN1805/154
EC Number: 231-633-2 [*Annex I Index No.:* 015-011-00-6]
Regulatory Authority and Advisory Bodies
US EPA Gene-Tox Program, Negative: Cell transform.—SA7/SHE.
US EPA, FIFRA 1998 Status of Pesticides: RED completed.
Air Pollutant Standard Set. See below, "Permissible Exposure Limits in Air" section.
FDA—over-the-counter and proprietary drug.
Clean Water Act: Section 311 Hazardous Substances/RQ 40CFR117.3 (same as CERCLA, see below); Section 313 Water Priority Chemicals (57FR41331, 9/9/92).
Reportable Quantity (RQ): 5000 lb (2270 kg).
EPCRA Section 313 Form R *de minimis* concentration reporting level: 1.0%.
Canada, WHMIS, Ingredients Disclosure List Concentration 1.0%.
European/International Regulations (*orthophosphoric acid*): Hazard Symbol: C; Risk phrases: R34; Safety phrases: S1/2; S26; S45 (see Appendix 4).
WGK (German Aquatic Hazard Class): 1—Low hazard to waters.

Description: Phosphoric acid is a colorless, odorless, crystalline solid, or thick syrupy liquid. Physical state is strength and temperature dependent. Molecular weight = 98.00; Specific gravity (H_2O:1) = 1.87 (pure); 1.33 (50% solution) at 25°C; Boiling point = Decomposes below BP at 212.8°C; Freezing/Melting point = 42.2°C; Vapor pressure = 0.03 mmHg at 20°C. Hazard Identification (based on NFPA-704 M Rating System): Health 3, Flammability 0, Reactivity 2 (Corrosive). Highly soluble in water.

Potential Exposure: Compound Description: Mutagen, Human Data; Primary Irritant. Phosphoric acid is used in the manufacture of fertilizers, phosphate salts; polyphosphates, detergents, activated carbon; animal feed; ceramics, dental cement; pharmaceuticals, soft drinks; gelatin, rust inhibitors; wax, and rubber latex. Exposure may also occur during electropolishing, engraving, photoengraving, lithographing, metal cleaning; sugar refining; and water-treating.

Incompatibilities: The substance is a medium strong acid. Incompatible with strong caustics; most metals. Readily attacks and reacts with metals forming flammable hydrogen gas. *Do not mix with solutions containing bleach or ammonia.* Violently polymerizes on contact with azo compounds; epoxides, and other polymerizable compounds. Decomposes on contact with metals, alcohols, aldehydes, cyanides, ketones, phenols, esters, sulfides, halogenated organics; producing toxic fumes.

Permissible Exposure Limits in Air
OSHA PEL: 1 mg/m^3 TWA.
NIOSH REL: 1 mg/m^3 TWA; 3 mg/m^3 STEL.
ACGIH TLV®[1]: 1 mg/m^3 TWA; 3 mg/m^3 STEL.
NIOSH IDLH: 1000 mg/m^3.
Protective Action Criteria (PAC)
TEEL-0: 1 mg/m^3
PAC-1: 3 mg/m^3
PAC-2: 500 mg/m^3
PAC-3: 500 mg/m^3
DFG MAK: 2 mg/m^3, inhalable fraction TWA; Peak Limitation Category I(2); Pregnancy risk Group C.
Arab Republic of Egypt: TWA 1 mg/m^3, 1993; Australia: TWA 1 mg/m^3; STEL 3 mg/m^3, 1993; Austria: MAK 1 mg/m^3, 1999; Belgium: TWA 1 mg/m^3; STEL 3 mg/m^3, 1993; Denmark: TWA 1 mg/m^3, 1999; Finland: TWA 1 mg/m^3; STEL 3 mg/m^3 [skin]1999; France: VME 1 mg/m^3, VLE 3 mg/m^3, 1999; Japan 1 mg/m^3, 1999; the Netherlands: MAC-TGG 1 mg/m^3, 2003; Poland: MAC (TWA) 1 mg/m^3; MAC (STEL) 3 mg/m^3, 1999; Sweden: NGV 1 mg/m^3, KTV 3 mg/m^3, 1999; Switzerland: MAK-W 1 mg/m^3, 1999; Thailand: TWA 1 mg/m^3, 1993; United Kingdom: STEL 2 mg/m^3, 2000; Argentina, Bulgaria, Columbia, Jordan, South Korea, New Zealand, Singapore, Vietnam: ACGIH TLV®: STEL 3 mg/m^3. Several states have set guidelines or standards for phosphoric acid in ambient air[60] ranging from 1.4 $\mu g/m^3$ (Massachusetts) to 10−30 $\mu g/m^3$ (North Dakota) to 10−33 $\mu g/m^3$ (Virginia) to 20.0 $\mu g/m^3$ (Connecticut) to 24.0 $\mu g/m^3$ (Nevada) to 25.0 $\mu g/m^3$ (South Carolina).

Determination in Air: Use NIOSH Analytical Method (IV) #7903, Inorganic Acids; OSHA Analytical Method ID-165-SG.

Determination in Water: Inhalation, ingestion, skin and/or eye contact.

Harmful Effects and Symptoms
Short Term Exposure: Corrosive to the eyes, skin, and respiratory tract. Eye contact may cause permanent damage. Inhalation can cause pulmonary edema, a medical

emergency that can be delayed for several hours. This can cause death. Solid is especially irritating to skin in the presence of moisture. Corrosive if swallowed. May cause pain in the throat and stomach, nausea, vomiting, and intense thirst. Severe exposures may result in shock with clammy skin, weak and rapid pulse; shallow breathing; reduced urine output; and death. 1−5 mg/m^3 may cause irritation of nose and throat. 4−11 mg/m^3 may cause coughing. Inhalation of acid mist can cause lung irritation. 1−5 mg/m^3 may cause irritation of nose and throat. 4−11 mg/m^3 may cause coughing.

Long Term Exposure: Repeated or prolonged skin exposure may cause irritation, drying, cracking, and dermatitis. Can cause bronchitis to develop.

Points of Attack: Eyes, skin, respiratory system.

Medical Surveillance: Before beginning employment and at regular times after that, the following is recommended: lung function tests. If symptoms develop or overexposure is suspected, the following may be useful: consider chest X-ray after acute overexposure.

First Aid: If this chemical gets into the eyes, remove any contact lenses at once and irrigate immediately for at least 15 min, occasionally lifting upper and lower lids. Seek medical attention immediately. If this chemical contacts the skin, remove contaminated clothing and wash immediately with soap and water. Seek medical attention immediately. If this chemical has been inhaled, remove from exposure, begin rescue breathing (using universal precautions, including resuscitation mask) if breathing has stopped and CPR if heart action has stopped. Transfer promptly to a medical facility. When this chemical has been swallowed, get medical attention. If victim is *conscious*, administer water or milk. Do not induce vomiting. Medical observation is recommended for 24−48 h after breathing overexposure, as pulmonary edema may be delayed. As first aid for pulmonary edema, a doctor or authorized paramedic may consider administering a corticosteroid spray.

Personal Protective Methods: Wear protective gloves and clothing to prevent any reasonable probability of skin contact. Safety equipment suppliers/manufacturers can provide recommendations on the most protective glove/clothing material for your operation. Natural rubber, Neoprene™, nitrile + pvc, nitrile, Saranex™, and polyvinyl chloride are among the recommended protective materials. All protective clothing (suits, gloves, footwear, headgear) should be clean, available each day, and put on before work. Contact lenses should not be worn when working with this chemical. Wear splash or dust-proof chemical goggles and face shield unless full face-piece respiratory protection is worn. Employees should wash immediately with soap when skin is wet or contaminated. Provide emergency showers and eyewash.

Respirator Selection: 25 mg/m^3: Sa:Cf (APF = 25) (any supplied-air respirator operated in a continuous-flow mode). 50 mg/m^3: 100F (APF = 50) (any air-purifying, full-face-piece respirator with an N100, R100, or P100 filter) or

SCBAF (APF = 50) (any self-contained breathing apparatus with a full face-piece) or SaF (APF = 50) (any supplied-air respirator with a full face-piece). *1000 mg/m³:* SaF: Pd,Pp (APF = 2000) (any supplied-air respirator that has a full face-piece and is operated in a pressure-demand or other positive-pressure mode). *Emergency or planned entry into unknown concentrations or IDLH conditions:* SCBAF: Pd, Pp (APF = 10,000) (any self-contained breathing apparatus that has a full face-piece and is operated in a pressure-demand or other positive-pressure mode) or SaF: Pd,Pp: ASCBA (APF = 10,000) (any supplied-air respirator that has a full face-piece and is operated in a pressure-demand or other positive-pressure mode in combination with an auxiliary self-contained breathing apparatus operated in a pressure-demand or other positive-pressure mode). *Escape:* 100F (APF = 50) (any air-purifying, full-face-piece respirator with an N100, R100, or P100 filter) or SCBAE (any appropriate escape-type, self-contained breathing apparatus).

Note: Substance reported to cause eye irritation or damage; may require eye protection.

Storage: Color Code—White: Corrosive or Contact Hazard; Store separately in a corrosion-resistant location. Prior to working with this chemical you should be trained on its proper handling and storage. Phosphoric acid must be stored to avoid contact with metals, aldehydes, cyanides, mercaptans, and sulfides, because violent reactions occur.

Shipping: This compound requires a shipping label of "CORROSIVE." It falls in Hazard Class 8 and Packing Group III.

Spill Handling: Evacuate persons not wearing protective equipment from area of spill or leak until cleanup is complete. Remove all ignition sources. Collect powdered material in the most convenient and safe manner and deposit in sealed containers. Ventilate area after cleanup is complete. It may be necessary to contain and dispose of this chemical as a hazardous waste. If material or contaminated runoff enters waterways, notify downstream users of potentially contaminated waters. Contact your local or federal environmental protection agency for specific recommendations. If employees are required to clean up spills, they must be properly trained and equipped. OSHA 1910.120(q) may be applicable.

Fire Extinguishing: Phosphoric acid is a noncombustible solid or liquid. Contact with common metals may form flammable hydrogen gas. Use extinguishing agent suitable for surrounding fire. Use water only to keep fire-exposed containers cool and to flush away spills. Poisonous gases are produced in fire, including oxides of phosphorus. If material or contaminated runoff enters waterways, notify downstream users of potentially contaminated waters. Notify local health and fire officials and pollution control agencies. From a secure, explosion-proof location, use water spray to cool exposed containers. If cooling streams are ineffective (venting sound increases in volume and pitch, tank discolors, or shows any signs of deforming), withdraw immediately to a secure position.

If employees are expected to fight fires, they must be trained and equipped in OSHA 1910.156. The only respirators recommended for firefighting are self-contained breathing apparatuses that have full face-pieces and are operated in a pressure-demand or other positive-pressure mode.

Disposal Method Suggested: Add slowly to solution of soda ash and slaked lime with stirring, then flush to sewer with large volumes of water.

References
Sax, N. I. (Ed.). (1983). Dangerous Properties of Industrial Materials Report, 3, No. 4, 84−87

New York State Department of Health. (April 1986). *Chemical Fact Sheet: Phosphoric Acid* (Version 2). Albany, NY: Bureau of Toxic Substance Assessment

US Environmental Protection Agency, Special Review and Reregistration Division Office of Pesticide Programs. (1998). *Agency Status of Pesticides in Registration, Reregistration, and Special Review* (Rainbow Report). Washington, DC

New Jersey Department of Health and Senior Services. (April 2004). *Hazardous Substances Fact Sheet: Phosphoric Acid.* Trenton, NJ

Phosphorous acid   P:0600

Molecular Formula: H₃O₃P
Common Formula: H₃PO₃
Synonyms: Orthophosphorus acid; Phosphonic acid; Phosphorous acid; Phosphorus trihydroxide; Trihydroxy
CAS Registry Number: 13598-36-2; 10294-56-1
RTECS® Number: SZ6475500
UN/NA & ERG Number: UN2834/154
EC Number: 237-066-7 [*Annex I Index No.:* 015-157-00-0]
Regulatory Authority and Advisory Bodies
Canada, WHMIS, Ingredients Disclosure List Concentration 1.0%.

European/International Regulations: Hazard Symbol (*13598-36-2; 10294-56-1*): C; Risk phrases: R22; R35; Safety phrases: S26; S36/3739; S45 (see Appendix 4).
WGK (German Aquatic Hazard Class): 1—Low hazard to waters.

Description: Phosphonic acid is a white to yellow crystalline solid. Molecular weight = 82.00; Boiling point = about 200°C (decomposes); Freezing/Melting point = 74°C. Hazard Identification (based on NFPA-704 M Rating System): Health 3, Flammability 0, Reactivity 1. Soluble in water.

Potential Exposure: Used in chemicals manufacture; a laboratory chemical.

Incompatibilities: Incompatible with aliphatic amines, alkanolamines, alkylene oxides; aromatic amines; amides, ammonia, ammonium hydroxide; bases, calcium oxide; epichlorohydrin, isocyanates. Unless it is stored in airtight containers it readily absorbs oxygen forming orthophosphoric acid. Attacks some metals.

Permissible Exposure Limits in Air
Protective Action Criteria (PAC)
TEEL-0: 0.006 mg/m^3
PAC-1: 0.015 mg/m^3
PAC-2: 0.125 mg/m^3
PAC-3: 500 mg/m^3.

Routes of Entry: Inhalation, ingestion, skin and/or eye contact.

Harmful Effects and Symptoms

Short Term Exposure: Orthophosphorous acid can affect you when breathed in. Contact can cause severe eye burns leading to permanent eye damage. Inhalation can irritate the eyes, skin, and respiratory tract.

Long Term Exposure: Highly irritating substances can cause lung effects; bronchitis may develop.

Points of Attack: Lungs.

Medical Surveillance: Before beginning employment and at regular times after that, for those with frequent or potentially high exposures, the following are recommended: lung function tests. Consider chest X-ray following acute overexposure.

First Aid: If this chemical gets into the eyes, remove any contact lenses at once and irrigate immediately for at least 15 min, occasionally lifting upper and lower lids. Seek medical attention immediately. If this chemical contacts the skin, remove contaminated clothing and wash immediately with soap and water. Seek medical attention immediately. If this chemical has been inhaled, remove from exposure, begin rescue breathing (using universal precautions, including resuscitation mask) if breathing has stopped and CPR if heart action has stopped. Transfer promptly to a medical facility. When this chemical has been swallowed, get medical attention. If victim is *conscious*, administer water or milk. Do not induce vomiting.

Personal Protective Methods: Wear protective gloves and clothing to prevent any reasonable probability of skin contact. Safety equipment suppliers/manufacturers can provide recommendations on the most protective glove/clothing material for your operation. All protective clothing (suits, gloves, footwear, headgear) should be clean, available each day, and put on before work. Contact lenses should not be worn when working with this chemical. Wear splash-proof chemical goggles and face shield when working with liquid, unless full face-piece respiratory protection is worn. Wear dust-proof goggles and face shield when working with powders or dust, unless full face-piece respiratory protection is worn. Employees should wash immediately with soap when skin is wet or contaminated. Provide emergency showers and eyewash.

Respirator Selection: Where there is potential for exposures to solid ortho phosphorous acid, use a NIOSH/MSHA- or European Standard EN149-approved full face-piece respirator with a high-efficiency particulate filter. Greater protection is provided by a powered air-purifying respirator. Where there is potential for high exposures of liquid ortho phosphorous acid, use a NIOSH/MSHA- or European Standard EN149-approved supplied-air respirator with a full face-piece operated in the positive-pressure mode, or with a full face-piece, hood, or helmet in the continuous-flow mode; or use a NIOSH/MSHA- or European Standard EN149-approved self-contained breathing apparatus with a full face-piece operated in the pressure-demand or other positive-pressure mode.

Storage: Color Code—White: Corrosive or Contact Hazard; Store separately in a corrosion-resistant location. Prior to working with this chemical you should be trained on its proper handling and storage. o-phosphorous acid must be stored in airtight containers away from incompatible materials listed above. Avoid contact with metals.

Shipping: This compound requires a shipping label of "CORROSIVE." It falls in Hazard Class 8 and Packing Group III.

Spill Handling: Evacuate persons not wearing protective equipment from area of spill or leak until cleanup is complete. Remove all ignition sources. Restrict persons not wearing protective equipment from area of spill or leak until cleanup is complete. Absorb liquids in vermiculite, dry sand, earth, or a similar material and deposit in sealed containers. Collect powdered material in the most convenient and safe manner and deposit in sealed containers. It may be necessary to contain and dispose of this chemical as a hazardous waste. If material or contaminated runoff enters waterways, notify downstream users of potentially contaminated waters. Contact your local or federal environmental protection agency for specific recommendations. If employees are required to clean up spills, they must be properly trained and equipped. OSHA 1910.120(q) may be applicable.

Fire Extinguishing: This chemical is a combustible solid or liquid. Use dry chemical, carbon dioxide, water spray, or alcohol foam extinguishers. Poisonous gases are produced in fire, including phosphorus oxides. If material or contaminated runoff enters waterways, notify downstream users of potentially contaminated waters. Notify local health and fire officials and pollution control agencies. From a secure, explosion-proof location, use water spray to cool exposed containers. If cooling streams are ineffective (venting sound increases in volume and pitch, tank discolors, or shows any signs of deforming), withdraw immediately to a secure position. If employees are expected to fight fires, they must be trained and equipped in OSHA 1910.156. The only respirators recommended for firefighting are self-contained breathing apparatuses that have full face-pieces and are operated in a pressure-demand or other positive-pressure mode.

Reference
New Jersey Department of Health and Senior Services. (February 2001). *Hazardous Substances Fact Sheet: Phosphorous Acid, ortho.* Trenton, NJ

Phosphorus   P:0610

Molecular Formula: P; P$_4$
Synonyms: Bonide blue death rat killer; Common sense cockroach and rat preparations; Exolite 405; Exolit

LPKN275; Exolit VPK-N 361; Fosforo blanco (Spanish); Gelber phosphor (German); Phosphore blanc (French); Phosphorous yellow; Phosphorus-31; Phosphorus elemental, white; Rat-NIP; Red phosphorus; RP (military designation); ST CC4916140; Tetraphosphor (German); Weiss phosphor (German); White phosphorus; Yellow phosphorus

CAS Registry Number: 7723-14-0 (white, red); 12185-10-3 (yellow)

RTECS® Number: TH3500000

UN/NA & ERG Number: UN1381 (Phosphorus, white, dry; under water, in solution; Phosphorus, yellow, dry; yellow, under water; in solution)/136; UN2447 (Phosphorus, white, molten)/136

EC Number: 231-768-7 [*Annex I Index No.:* 015-001-00-1]

Regulatory Authority and Advisory Bodies

Department of Homeland Security Screening Threshold Quantity (pounds): *Theft hazard* 400 (Commercial grade).

Carcinogenicity: EPA: Not Classifiable as to human carcinogenicity.

Air Pollutant Standard Set. See below, "Permissible Exposure Limits in Air" section.

Clean Air Act: Hazardous Air Pollutants (Title I, Part A, Section 112).

Clean Water Act: Section 311 Hazardous Substances/RQ 40CFR117.3 (same as CERCLA, see below); Section 313 Water Priority Chemicals (57FR41331, 9/9/92).

Superfund/EPCRA 40CFR355, Extremely Hazardous Substances: TPQ = 100 lb (45.4 kg).

Reportable Quantity (RQ): 1 lb (0.454 kg).

EPCRA Section 313 (yellow or white) Form R *de minimis* concentration reporting level: 1.0%.

US DOT Regulated Marine Pollutant (49CFR172.101, Appendix B), severe pollutant, white, yellow dry, molten or in solution.

Canada, WHMIS, Ingredients Disclosure List Concentration 1.0%.

European/International Regulations (*7723-14-0*): Hazard Symbol: F,T,C,N; Risk phrases: R17; R26/28; R35; R50; Safety phrases: S1/2; S5; S26; S38; S45; S61 (see Appendix 4).

WGK (German Aquatic Hazard Class): No value assigned.

Description: Phosphorus is a white to yellow, soft, waxy solid with acrid fumes in air. *White/yellow* phosphorus is either a yellow or colorless, volatile, crystalline solid which darkens when exposed to light and ignites in air to form white fumes and greenish light. It has a garlic-like odor. Usually shipped or stored in water. Molecular weight = 123.88; Boiling point = 280°C; Freezing/Melting point = (decomposes) 44°C; Vapor pressure = 0.026 mmHg at 20°C; 0.181 mmHg at 44.1°C; Autoignition temperature = 30°C. Hazard Identification (based on NFPA-704 M Rating System): (*white, red powder*) Health 4, Flammability 4, Reactivity 2. Insoluble in water.

Red phosphorus is a brick red, reddish-brown, or violet amorphous powder, frequently contaminated with a small amount of the yellow. Molecular weight = 30.97; Boiling

point = 280°C (with ignition at 200°C); Freezing/Melting point = 416°C (sublimes); Flash point = 260°C; Autoignition temperature = 260°C. Hazard Identification (based on NFPA-704 M Rating System): Health 1, Flammability 1, Reactivity 1. Insoluble in water.

Potential Exposure: Compound Description (white): Agricultural Chemical; Reproductive Effector; Human Data. White or yellow phosphorus is handled away from air so that exposure is usually limited. Phosphorus was at one time used for the production of matches or "lucifers" but has long since been replaced due to its chronic toxicity. It is used in the manufacture of munitions including tracer bullets, pyrotechnics, explosives, smoke bombs; and other incendiary agents; (because it spontaneously catches fire in air) and as a smoke agent (because it produces clouds of irritating white smoke). Phosphorus is used artificial fertilizers; rodenticides, phosphor bronze alloys; semiconductors, Electro-luminescent coating; and chemicals, such as phosphoric and metallic phosphides. RP is used as a choking/pulmonary agent.

Incompatibilities: Phosphorus spontaneously ignites on contact with air, producing toxic phosphorus oxide fumes. Reacts with strong bases, releasing toxic phosphine gas. Phosphorus reacts violently with oxidizers, halogens, some metals, nitrites, sulfur, and many other compounds, causing a fire and explosion hazard. *White/yellow* reacts with air, halogens, halides, sulfur, oxidizers, alkali hydroxides (forming gas), and metals (forming reactive phosphides). *Red* is a combustible solid. Friction or contact with oxidizers can cause ignition. Incompatible with many other substances. Forms gas and phosphoric acid on contact with moisture. Opened packages of red phosphorus should be stored under inert gas blanket.

Permissible Exposure Limits in Air

OSHA PEL (*yellow*): 0.1 mg/m^3 TWA.

NIOSH REL (*yellow*): 0.1 mg/m^3 TWA.

ACGIH TLV®[1] (*yellow*): 0.02 ppm/0.1 mg/m^3 TWA.

NIOSH IDLH: 5 mg/m^3.

Protective Action Criteria (PAC) RP

TEEL-0: 0.05 mg/m^3

PAC-1: 0.15 mg/m^3

PAC-2: 3 mg/m^3

PAC-3: 5 mg/m^3

DFG MAK (*White/yellow/red*): 0.05 mg/m^3, inhalable fraction TWA; Peak Limitation Category II(2); Pregnancy Risk Group C; *red:* No numerical value established.

Arab Republic of Egypt: TWA 0.1 mg/m^3, 1993; Australia: TWA 0.1 mg/m^3, 1993; Belgium: TWA 0.1 mg/m^3, 1993; Denmark: TWA 0.1 mg/m^3, 1999; Finland: STEL 0.1 mg/m^3, [skin], 1999; France: VME 0.1 mg/m^3, VLE 0.3 mg/m^3, 1999; Hungary: TWA 0.3 mg/m^3; STEL 0.06 mg/m^3, 1993; the Netherlands: MAC-TGG 0.1 mg/m^3, 2003; the Philippines: TWA 0.1 mg/m^3, 1993; Poland: MAC (TWA) 0.3 mg/m^3, 1993; Russia: STEL 0.03 mg/m^3, 1993; Switzerland: MAK-W 0.1 mg/m^3, KZG-W 0.2 mg/m^3, 1999; Thailand: TWA 0.1 mg/m^3, 1993; Turkey: TWA 0.1 mg/m^3,

1993; United Kingdom: LTEL 0.1 mg/m³; STEL 0.3 mg/m³, 1993; Argentina, Bulgaria, Columbia, Jordan, South Korea, New Zealand, Singapore, Vietnam: ACGIH TLV®: TWA 0.1 mg/m³. The Czech Republic[35]: TWA 0.03 mg/m³; 0.06 mg/m³ STEL. Several states have set guidelines or standards for yellow phosphorus in ambient air[60] ranging from 0.33 µg/m³ (New York) to 1.0 µg/m³ (Florida) to 1.6 µg/m³ (Virginia) to 2.0 µg/m³ (Connecticut and Nevada) to 10.0–30.0 µg/m³ (North Dakota).

Determination in Air: Use NIOSH Analytical Method (IV) #7905.

Permissible Concentration in Water: EPA[32] has suggested a permissible ambient goal of 1.4 µg/L based on health effects. Russia[35, 43] set a MAC of 0.1 µg/L in water bodies used for domestic purposes and zero in surface water used for fishery purposes.

Determination in Water: Octanol–water coefficient: Log K_{ow} = (estimated) −0.27.

Routes of Entry: Inhalation, ingestion, skin and/or eye contact. Yellow phosphorus can be absorbed through the skin.

Harmful Effects and Symptoms

Short Term Exposure: Phosphorus is corrosive to the eyes, skin, and respiratory tract. Eye contact may lead to a total destruction of the eyes. Victims may experience spontaneous hemorrhaging of phosphorus-contaminated skin and mucous membranes. Sudden death, possibly due to irregular heartbeat, may occur after relatively minor (10–15%) burns. *Yellow:* Fumes are irritating to the respiratory tract and cause severe ocular irritation. On contact with the skin it may ignite and produce severe skin burns with blistering. Very high exposure may cause severe or fatal poisoning. *Red:* Irritates eyes. Corrosive if ingested. Inhalation can cause pulmonary edema, a medical emergency that can be delayed for several hours. This can cause death. May affect the kidneys, liver. Exposure may result in death. Phosphorus is classified as super toxic. The probable lethal dose is less than 5 mg/kg (a taste or less than 7 drops) for a 70-kg (150 lb) person. Signs and symptoms of acute exposure to phosphorus may be severe and occur in three stages. The first stage will involve burns, pain, shock, intense thirst; nausea, vomiting, diarrhea, severe abdominal pain; and "smoking stools." The breath and feces may have garlicky odor. The second stage will be a symptom-free period of several days in which the patient appears to be recovering. The third stage may be severe and include nausea, bloody vomitus; diarrhea (may be bloody), jaundice, liver enlargement with tenderness; renal damage; hematuria (bloody urine), and either oliguria (little urine formation) or anuria (no urine formation). Headache, convulsions, delirium, coma, cardiac arrhythmias; and cardiovascular collapse may also occur. If phosphorus contacts the eyes, then severe irritation and burns, blepharospasm (spasmodic winking), lacrimation (tearing), and photophobia (heightened sensitivity to light) may occur.

Long Term Exposure: Phosphorus may affect the bones, causing bone degeneration (especially the jaw bone, known as "phossy" jaw), dental pain; salivation, jaw pain and swelling. This process can extend into one or both eye sockets. Repeated low exposure can cause low blood count (anemia), weight loss; and bronchitis. May cause jaundice; liver and kidney damage; cachexia. May cause nervous system damage.

Points of Attack: Respiratory system, liver, kidneys, jaw, teeth, blood, eyes, skin.

Medical Surveillance: NIOSH lists the following tests: Complete Blood Count, anemia; Dental X-ray/Examination; liver function tests. Also consider EKG. Special consideration should be given to the skin, eyes, jaws, teeth, respiratory tract; and liver. Preplacement medical and dental examination with X-rays of teeth is highly recommended in the case of yellow phosphorus exposure. Poor dental hygiene may increase the risk in yellow phosphorus exposures, and any required dental work should be completed before workers are assigned to areas of possible exposure. Workers experiencing any jaw injury, tooth extraction; or any abnormal dental conditions should be removed from areas of exposure and observed. X-ray examinations may show necrosis; however, in order to prevent full development of sequestra, the disease should be diagnosed in earlier stages. Liver function should be evaluated periodically.

First Aid: There is no antidote for white phosphorus toxicity. If this chemical gets into the eyes, remove any contact lenses at once and irrigate immediately for at least 30 min, occasionally lifting upper and lower lids. Seek medical attention immediately. If this chemical contacts the skin, remove contaminated clothing and brush all traces of dry chemical from skin. Submerge burning phosphorus (yellow) in water or 1% copper sulfate solution if embedded in skin, or wash exposed area with large amounts of water. Seek medical attention immediately. Skin burns from yellow phosphorus should be observed for 1–3 days for possible delayed effects. If this chemical has been inhaled, remove from exposure, begin rescue breathing (using universal precautions, including resuscitation mask) if breathing has stopped and CPR if heart action has stopped. Transfer promptly to a medical facility. When this chemical has been swallowed, get medical attention. Give large quantities of water and induce vomiting. Do not make an unconscious person vomit. Medical observation is recommended for 24–48 h after breathing overexposure, as pulmonary edema may be delayed. As first aid for pulmonary edema, a doctor or authorized paramedic may consider administering a corticosteroid spray.

Personal Protective Methods: Wear protective gloves and clothing to prevent any reasonable probability of skin contact. Safety equipment suppliers/manufacturers can provide recommendations on the most protective glove/clothing material for your operation. All protective clothing (suits, gloves, footwear, headgear) should be clean, available each day, and put on before work. Contact lenses should not be worn when working with this chemical. Wear dust-proof chemical goggles and face shield unless full face-piece

respiratory protection is worn. Employees should wash immediately with soap when skin is wet or contaminated. Provide emergency showers and eyewash.

Respirator Selection: *1 mg/m³:* Sa (APF = 10) (any supplied-air respirator). *2.5 mg/m³:* Sa:Cf (APF = 25) (any supplied-air respirator operated in a continuous-flow mode). *5 mg/m³:* SCBAF (APF = 50) (any self-contained breathing apparatus with a full face-piece) or SaF (APF = 50) (any supplied-air respirator with a full face-piece). *Emergency or planned entry into unknown concentrations or IDLH conditions:* SCBAF: Pd,Pp (APF = 10,000) (any self-contained breathing apparatus that has a full face-piece and is operated in a pressure-demand or other positive-pressure mode) or SaF: Pd,Pp: ASCBA (APF = 10,000) (any supplied-air respirator that has a full face-piece and is operated in a pressure-demand or other positive-pressure mode in combination with an auxiliary self-contained breathing apparatus operated in a pressure-demand or other positive-pressure mode). *Escape:* SCBAE (any appropriate escape-type, self-contained breathing apparatus).

Note: Substance causes eye irritation or damage; eye protection needed.

Storage: Color Code—Red Stripe (*7723-14-0*): Flammability Hazard: Do not store in the same area as other flammable materials. Color Code—Yellow Stripe (*strong reducing agent*): Reactivity Hazard; Store separately in an area isolated from flammables, combustibles, or other yellow-coded materials. Prior to working with this chemical you should be trained on its proper handling and storage. Phosphorus must be stored in a cool, well-ventilated area away from heat, direct sunlight; air, organic materials; oxidizers (such as perchlorates, peroxides, permanganates, chlorates, and nitrates), since violent reactions occur. Always store away from alkaline materials because of the extreme fire hazard and because poisonous gas is produced. Always store yellow phosphorus under water and protect it from physical damage. Opened packages of red phosphorus should be stored under inert gas blanket. Sources of ignition, such as smoking and open flames, are prohibited where phosphorus is used, handled, or stored in a manner that could create a potential fire or explosion hazard. Use only nonsparking tools and equipment, especially when opening and closing containers of this chemical. Sources of ignition, such as smoking and open flames, are prohibited where this chemical is used, handled, or stored in a manner that could create a potential fire or explosion hazard. Wherever this chemical is used, handled, manufactured, or stored, use explosion-proof electrical equipment and fittings.

Shipping: Phosphorus white, molten, requires a shipping label of "SPONTANEOUSLY COMBUSTIBLE, POISONOUS/TOXIC MATERIALS." It falls in Hazard Class 4.2 and Packing Group III. Subsidiary Hazardous Class or Division 6.1.

Spill Handling: Evacuate persons not wearing protective equipment from area of spill or leak until cleanup is complete. Remove all ignition sources. Keep spilled material wet and cover with wet sand or dirt. Collect solidified material in the most convenient and safe manner and cover with water in sealed containers. Ventilate area after cleanup is complete. Keep phosphorus out of a confined space, such as a sewer, because of the possibility of an explosion, unless the sewer is designed to prevent the buildup of explosive concentrations. It may be necessary to contain and dispose of this chemical as a hazardous waste. If material or contaminated runoff enters waterways, notify downstream users of potentially contaminated waters. Contact your local or federal environmental protection agency for specific recommendations. If employees are required to clean up spills, they must be properly trained and equipped. OSHA 1910.120(q) may be applicable.

Fire Extinguishing: Phosphorus (white/yellow) is a flammable solid which ignites spontaneously in moist air. Combustion in a confined space will deplete oxygen causing asphyxiation. Poisonous gases are produced in fire, including oxides of phosphorus; and phosphoric acid if water is present. Fire may restart after it has been extinguished. *Small fires:* dry chemical, sand, water spray; or foam. *Large fires:* water spray, fog, or foam. Cool containers that are exposed to flames with water from the side until well after fire is out. *White/Yellow:* Deluge with water, taking care not to scatter, until fire is extinguished and phosphorus has solidified, then cover with wet sand or dirt. *Red:* Flood with water and when fire is extinguished, cover with wet sand or dirt. Extreme caution should be used during cleanup of the more hazardous white phosphorus. *White/Yellow:* Ignites at approximately 30°C in air; ignition temperature is higher when air is dry. *Black:* Does not catch fire spontaneously. *Red:* Catches fire when heated in air to approximately 260°C and burns with formation of the pentoxide. If material or contaminated runoff enters waterways, notify downstream users of potentially contaminated waters. Notify local health and fire officials and pollution control agencies. From a secure, explosion-proof location, use water spray to cool exposed containers. If cooling streams are ineffective (venting sound increases in volume and pitch, tank discolors, or shows any signs of deforming), withdraw immediately to a secure position. If employees are expected to fight fires, they must be trained and equipped in OSHA 1910.156. The only respirators recommended for firefighting are self-contained breathing apparatuses that have full face-pieces and are operated in a pressure-demand or other positive-pressure mode.

Disposal Method Suggested: Controlled incineration followed by alkaline scrubbing and particulate removal equipment.

References
Sax, N. I. (Ed.). (1983). Dangerous Properties of Industrial Materials Report, 3, No. 4, 90–93
US Environmental Protection Agency. (November 30, 1987). *Chemical Hazard Information Profile: Phosphorus*. Washington, DC: Chemical Emergency Preparedness Program

New Jersey Department of Health and Senior Services. (October 2002). *Hazardous Substances Fact Sheet: Phosphorus*. Trenton, NJ

Phosphorus oxychloride   P:0620

Molecular Formula: Cl_3OP
Common Formula: $POCl_3$
Synonyms: Fosforoxychlorid; Oxicloruro de fosforo (Spanish); Oxychlorid fosforecny; Phosphoric chloride; Phosphorus chloride oxide; Phosphorus oxytrichloride; Phosphoryl chloride; Phosphoryl trichloride
CAS Registry Number: 10025-87-3
RTECS® Number: TH4897000
UN/NA & ERG Number: UN1810/137
EC Number: 233-046-7 [*Annex I Index No.:* 015-009-00-5]
Regulatory Authority and Advisory Bodies
Department of Homeland Security Screening Threshold Quantity (pounds): *Release hazard* 5000 ($\geq$1.00% concentration). (1% concentration); *Theft hazard* 220 ($\geq$80.00% concentration).
Sabotage/Contamination Hazard: A placarded amount (commercial grade).
Air Pollutant Standard Set. See below, "Permissible Exposure Limits in Air" section.
Clean Air Act: Accidental Release Prevention/Flammable Substances, (Section 112[r], Table 3), TQ = 5000 lb (2270 kg).
Clean Water Act: Section 311 Hazardous Substances/RQ 40CFR117.3 (same as CERCLA, see below).
Superfund/EPCRA 40CFR355, Extremely Hazardous Substances: TPQ = 500 lb (227 kg).
Reportable Quantity (RQ): 1000 lb (454 kg).
US DOT 49CFR172.101, Inhalation Hazardous Chemical.
Canada, WHMIS, Ingredients Disclosure List Concentration 1.0%.
European/International Regulations: Hazard Symbol: T + , C; Risk phrases: R14; R22; R26; R35; R48/23; Safety phrases: S1/2; S7/8; S26; S36/37/39; S45 (see Appendix 4).
WGK (German Aquatic Hazard Class): 1—Low hazard to waters.
Description: Phosphorus oxychloride is a clear, colorless to yellow, fuming, oily liquid with a pungent and musty odor. Molecular weight = 153.32; Specific gravity (H_2O:1) = 1.65 at 25°C; Boiling point = 105.6°C; Freezing/Melting point = 1.25°C; Vapor pressure = 40 mmHg. Hazard Identification (based on NFPA-704 M Rating System): Health 4, Flammability 0, Reactivity 3W (Corrosive). Reacts with water; decomposes with heat.
Potential Exposure: Phosphorus oxychloride is used in the manufacture of pesticides, pharmaceuticals, plasticizers, gasoline additives; and hydraulic fluids.
Incompatibilities: A powerful oxidizer. Violently decomposes in water, forming heat and hydrochloric and phosphoric acids. Violent reaction with alcohols, phenols, amines, reducing agents; combustible materials; carbon disulfide; dimethylformamide, and many other materials. Rapid corrosion of metals, except nickel and lead.
Permissible Exposure Limits in Air
Conversion factor: 1 ppm = 6.27 mg/m^3 at 25°C & 1 atm.
OSHA PEL: None.
NIOSH REL: 0.1 ppm/0.6 mg/m^3 TWA; 0.5 ppm/3 mg/m^3 STEL.
ACGIH TLV®[1]: 0.1 ppm/0.63 mg/m^3 TWA.
Protective Action Criteria (PAC)*
TEEL-0: 0.1 ppm
PAC-1: 0.479 ppm
PAC-2: 0.479 ppm
PAC-3: **0.85** ppm
*AEGLs (Acute Emergency Guideline Levels) & ERPGs (Emergency Response Planning Guideline) are in **bold face**.
DFG MAK: 0.2 ppm/1.3 mg/m^3 TWA; Peak Limitation Category I(1); Pregnancy Risk Group C.
Australia: TWA 0.1 ppm (0.6 mg/m^3); STEL 0.5 ppm, 1993; Austria: MAK 0.2 ppm (1 mg/m^3), 1999; Belgium: TWA 0.1 ppm (0.63 mg/m^3); STEL 0.5 ppm, 1993; Denmark: TWA 0.1 ppm (0.6 mg/m^3), 1999; Finland: STEL 0.5 ppm (3 mg/m^3), [skin], 1999; France: VME 0.1 ppm (0.6 mg/m^3), 1999; the Netherlands: MAC-TGG 0.6 mg/m^3, 2003; Russia: STEL 0.05 mg/m^3, [skin], 1993; Switzerland: MAK-W 0.1 ppm (0.6 mg/m^3), KZG-W 0.2 ppm (1.2 mg/m^3), 1999; Turkey: TWA 0.5 ppm (3 mg/m^3), 1993; United Kingdom: TWA 0.2 ppm (1.3 mg/m^3); STEL 0.6 ppm, 2000; Argentina, Bulgaria, Columbia, Jordan, South Korea, New Zealand, Singapore, Vietnam: ACGIH TLV®: TWA 0.1 ppm. Several states have set guidelines or standards for $POCl_2$ in ambient air[60] ranging from 6.0–30.0 µg/m^3 (North Dakota) to 12.0 µg/m^3 (Connecticut) to 14.0 µg/m^3 (Nevada).
Determination in Air: No method available.
Permissible Concentration in Water: No criteria set. ($POCl_2$ decomposes in water).
Routes of Entry: Inhalation, ingestion, skin and/or eye contact.
Harmful Effects and Symptoms
Short Term Exposure: Corrosive to the eyes, skin, and respiratory tract. Eye contact can cause permanent damage. Inhalation of the vapors can cause pulmonary edema, a medical emergency that can be delayed for several hours. This can cause death. This material is toxic by inhalation and ingestion and is strongly irritating to skin and tissues. It causes burns of the mucous membranes of the mouth and digestive tract; and may be fatal. Symptoms include burns and extensive reddening of eyes, pains in throat; coughing, labored breathing with a shortness of breath; dizziness, headache, weakness, nausea, vomiting, chest pain; bronchitis, bronchopneumonia, kidney, and liver damage.
Long Term Exposure: May cause nephritis; kidney damage. May cause liver damage.

Points of Attack: Eyes, skin, respiratory system, central nervous system, kidneys, liver.

Medical Surveillance: Lung function tests. Test for liver and kidney functions. Consider chest X-ray following acute overexposure.

First Aid: If this chemical gets into the eyes, remove any contact lenses at once and irrigate immediately for at least 15 min, occasionally lifting upper and lower lids. Seek medical attention immediately. If this chemical contacts the skin, remove contaminated clothing and wash immediately with soap and water. Seek medical attention immediately. If this chemical has been inhaled, remove from exposure, begin rescue breathing (using universal precautions, including resuscitation mask) if breathing has stopped and CPR if heart action has stopped. Transfer promptly to a medical facility. When this chemical has been swallowed, get medical attention. If victim is *conscious*, administer water or milk. Do not induce vomiting. Medical observation is recommended for 24—48 h after breathing overexposure, as pulmonary edema may be delayed. As first aid for pneumonitis or pulmonary edema, a doctor or authorized paramedic may consider administering a corticosteroid spray.

Personal Protective Methods: Wear protective gloves and clothing to prevent any reasonable probability of skin contact. Safety equipment suppliers/manufacturers can provide recommendations on the most protective glove/clothing material for your operation. All protective clothing (suits, gloves, footwear, headgear) should be clean, available each day, and put on before work. Contact lenses should not be worn when working with this chemical. Wear splash-proof chemical goggles and face shield unless full face-piece respiratory protection is worn. Employees should wash immediately with soap when skin is wet or contaminated. Provide emergency showers and eyewash.

Respirator Selection: Follow regulations in OSHA 29CFR1910.134 or European Standard EN149. Use a NIOSH/MSHA- or European Standard EN149-approved respirator; or use an approved supplied-air respirator with a full face-piece operated in the positive-pressure mode, or with a full face-piece, hood, or helmet in the continuous-flow mode; or use a NIOSH/MSHA- or European Standard EN149-approved self-contained breathing apparatus with a full face-piece operated in pressure-demand or other positive-pressure mode.

Storage: (1) Color Code—White stripe: Contact Hazard; Store separately; not compatible with materials in solid white category. (2) Color Code—White: Corrosive or Contact Hazard; Store separately in a corrosion resistant location. Prior to working with this chemical you should be trained on its proper handling and storage. Store in tightly closed containers in a cool, well-ventilated area away from incompatible materials listed above. Where possible, automatically pump liquid from drums or other storage containers to process containers. Sources of ignition, such as smoking and open flames, are prohibited where this chemical is handled, used, or stored. Metal containers involving the transfer of 5 gallons or more of this chemical should be grounded and bonded. Drums must be equipped with self-closing valves, pressure vacuum bungs, and flame arresters. Use only nonsparking tools and equipment, especially when opening and closing containers of this chemical. Wherever this chemical is used, handled, manufactured, or stored, use explosion-proof electrical equipment and fittings.

Shipping: This compound requires a shipping label of "CORROSIVE, POISONOUS/TOXIC MATERIALS." It falls in Hazard Class 8 and Packing Group II.

Spill Handling: Keep material out of water sources and sewers; build dikes to contain flow as necessary; use water spray to knock down vapors; *Do not use water* on material itself; and neutralize spilled material with crushed limestone, soda ash; or lime. *For a land spill*, dig a pit, pond, lagoon, or holding area to contain liquid or solid material; dike surface flow using soil, sand bags; foamed polyurethane; or foamed concrete; absorb bulk liquid with fly ash or cement powder; neutralize with agricultural lime (slaked lime), crushed limestone; or sodium bicarbonate. *For a water spill*, neutralize with agricultural lime (slaked lime), crushed limestone; or sodium bicarbonate; use mechanical dredges or lifts to remove immobilized masses of pollutants and precipitates; adjust pH to neutral (pH 7). *For air spills* apply water spray or mist to knock down vapors; vapor knock down water is corrosive or toxic and should be diked for containment. Stop leak if you can do so without risk. Do not touch spilled material. Keep combustibles (wood, paper, oil, etc.) away from spilled material. Clean up only under supervision of an expert. Keep this chemical out of a confined space, such as a sewer, because of the possibility of an explosion, unless the sewer is designed to prevent the buildup of explosive concentrations. It may be necessary to contain and dispose of this chemical as a hazardous waste. If material or contaminated runoff enters waterways, notify downstream users of potentially contaminated waters. Contact your local or federal environmental protection agency for specific recommendations. If employees are required to clean up spills, they must be properly trained and equipped. OSHA 1910.120(q) may be applicable.

Initial isolation and protective action distances

Distances shown are likely to be affected during the first 30 min after materials are spilled and could increase with time. If more than one tank car, cargo tank, portable tank, or large cylinder involved in the incident is leaking, the protective action distance may need to be increased. You may need to seek emergency information from CHEMTREC at (800) 424-9300 or seek professional environmental engineering assistance from the US EPA Environmental Response Team at (908) 548-8730 (24-h response line).

Small spills (From a small package or a small leak from a large package)

when spilled on land

First: Isolate in all directions (feet/meters)   100/30
Then: Protect persons downwind (miles/kilometers)

Day   0.2/0.3
Night   0.4/0.6
Large spills (From a large package or from many small packages)
First: Isolate in all directions (feet/meters)   300/100
Then: Protect persons downwind (miles/kilometers)
Day   0.7/1.1
Night   1.3/2.0
when spilled in water
First: Isolate in all directions (feet/meters)   100/30
Then: Protect persons downwind (miles/kilometers)
Day   0.1/0.2
Night   0.2/0.3
Large spills (From a large package or from many small packages)
First: Isolate in all directions (feet/meters)   200/60
Then: Protect persons downwind (miles/kilometers)
Day   0.5/0.8
Night   1.4/2.3

Fire Extinguishing: This chemical reacts violently with moisture producing hydrochloric and phosphoric acids. Poisonous gases, including chlorides and phosphorus oxides, are produced in fire. *Do not use water* unless used in flooding quantities to control a large fire by wetting down combustibles burning in vicinity of this material. Use dry chemical, carbon dioxide, or dry sand; *Do not use water* on material itself. Use water spray to absorb vapors and cool all affected containers with flooding quantities of water. Apply water from as far a distance as possible. Avoid breathing vapors; keep upwind. Wear self-contained breathing apparatus. Avoid bodily contact with the material. Wear boots, protective gloves, and goggles. Do not handle broken packages without protective equipment. Wash away any material which may have contacted the body with copious amounts of water or soap and water. If contact with the material is anticipated, wear full protective clothing. Keep unnecessary people away; isolate hazard area and deny entry. Vapors are heavier than air and will collect in low areas. Containers may explode in fire. Storage containers and parts of containers may rocket great distances, in many directions. If material or contaminated runoff enters waterways, notify downstream users of potentially contaminated waters. Notify local health and fire officials and pollution control agencies. From a secure, explosion-proof location, use water spray to cool exposed containers. If cooling streams are ineffective (venting sound increases in volume and pitch, tank discolors, or shows any signs of deforming), withdraw immediately to a secure position. If employees are expected to fight fires, they must be trained and equipped in OSHA 1910.156. The only respirators recommended for firefighting are self-contained breathing apparatuses that have full face-pieces and are operated in a pressure-demand or other positive-pressure mode.

Disposal Method Suggested: Pour onto sodium bicarbonate. Spray with aqueous ammonia and add crushed ice. Neutralize and pour into drain with running water. In accordance with 40CFR165, follow recommendations for the disposal of pesticides and pesticide containers. Must be disposed properly by following package label directions or by contacting your local or federal environmental control agency or by contacting your regional EPA office.

References
Sax, N. I. (Ed.). (1983). Dangerous Properties of Industrial Materials Report, 3, No. 4, 87–88
US Environmental Protection Agency. (November 30, 1987). *Chemical Hazard Information Profile: Phosphorus Oxychloride*. Washington, DC: Chemical Emergency Preparedness Program
New Jersey Department of Health and Senior Services. (September 2001). *Hazardous Substances Fact Sheet: Phosphorus Oxychloride*. Trenton, NJ

Phosphorus pentachloride   P:0630

Molecular Formula: Cl_5P
Common Formula: PCl_5
Synonyms: Pentacloruro de fosforo (Spanish); Phosphore (pentachlorure de) (French); Phosphoric chloride; Phosphorpentachlorid (German); Phosphorus perchloride
CAS Registry Number: 10026-13-8
RTECS® Number: TB6125000
UN/NA & ERG Number: UN1806/137
EC Number: 233-060-3 [*Annex I Index No.:* 015-008-00-X]
Regulatory Authority and Advisory Bodies
Department of Homeland Security Screening Threshold Quantity (pounds): Sabotage/Contamination Hazard: A placarded amount (commercial grade).
Air Pollutant Standard Set. See below, "Permissible Exposure Limits in Air" section.
Superfund/EPCRA 40CFR355, Extremely Hazardous Substances: TPQ = 500 lb (227 kg).
Reportable Quantity (RQ): 500 lb (227 kg).
US DOT 49CFR172.101, Inhalation Hazardous Chemical.
Canada, WHMIS, Ingredients Disclosure List Concentration 1.0%.
European/International Regulations: Hazard Symbol: C; Risk phrases: R14; R22; R26; R34; R48/20; Safety phrases: S1/2; S7/8; S26; S36/37/39; S45 (see Appendix 4).
WGK (German Aquatic Hazard Class): No value assigned.

Description: Phosphorus pentachloride is a pale yellow, fuming solid with an odor like hydrochloric acid. Molecular weight = 208.22; Specific gravity (H_2O:1) = 3.60 at 25°C; Boiling point = (sublimes) 160°C; Freezing/Melting point = (sublimes) 162°C; Vapor pressure = 1 mmHg at 55.6°C. Hazard Identification (based on NFPA-704 M Rating System): Health 3, Flammability 0, Reactivity 2W. Reacts violently with water.

Potential Exposure: Phosphorus pentachloride is used as a chlorinating and dehydrating agent and as a catalyst. It is used in the manufacture of agricultural chemicals;

chlorinated compounds; gasoline additives, plasticizers, and surfactants; and in pharmaceutical manufacture.

Incompatibilities: A powerful oxidizer. Reacts with water (violent), magnesium oxide, chemically active metals, such as sodium and potassium, alkalis, amines. Hydrolyzes in water (even in humid air) to form hydrochloric acid and phosphoric acid. Corrosive to many metals, forming flammable and explosive hydrogen gas. Attacks plastic and rubber.

Permissible Exposure Limits in Air

OSHA PEL: 1 mg/m³ TWA.

NIOSH REL: 1 mg/m³ TWA.

ACGIH TLV®[1]: 0.1 ppm/1 mg/m³ TWA.

NIOSH IDLH: 70 mg/m³.

Protective Action Criteria (PAC)

TEEL-0: 1 mg/m³

PAC-1: 3 mg/m³

PAC-2: 20 mg/m³

PAC-3: 70 mg/m³

DFG MAK: 1 mg/m³, measured as the inhalable fraction TWA; Peak Limitation Category I(1); Pregnancy Risk Group C.

Australia: TWA 0.1 ppm (1 mg/m³), 1993; Austria: MAK 1 mg/m³, 1999; Belgium: TWA 0.1 ppm (0.85 mg/m³), 1993; Denmark: TWA 1 mg/m³, 1999; Finland: STEL 1 mg/m³, [skin], 1999; France: VME 0.1 ppm (1 mg/m³), 1999; Japan 0.1 ppm (0.85 mg/m³), 1999; the Netherlands: MAC-TGG 1 mg/m³, 2003; Norway: TWA 0.1 mg/m³, 1999; the Philippines: TWA 1 mg/m³, 1993; Poland: MAC (TWA) 0.3 mg/m³; MAC (STEL) 0.9 mg/m³, 1999; Russia: TWA 0.1 ppm; STEL 0.2 mg/m³, [skin], 1993; Switzerland: MAK-W 1 mg/m³, KZG-W 2 mg/m³, 1999; Thailand: TWA 1 mg/m³, 1993; Turkey: TWA 1 mg/m³, 1993; United Kingdom: TWA 0.1 ppm (0.87 mg/m³), 2000; Argentina, Bulgaria, Columbia, Jordan, South Korea, New Zealand, Singapore, Vietnam: ACGIH TLV®: TWA 0.1 ppm Several states have set guidelines or standards for PCl₅ in ambient air[60] ranging from 10.0 µg/m³ (North Dakota) to 16.0 µg/m³ (Virginia) to 20.0 µg/m³ (Connecticut) to 24.0 µg/m³ (Nevada).

Determination in Air: Use NIOSH II(5), Method #S-257.

Routes of Entry: Inhalation, ingestion, skin and/or eye contact.

Harmful Effects and Symptoms

Short Term Exposure: The substance is corrosive to the eyes, skin, and respiratory tract. Corrosive if swallowed. Higher exposures can cause pulmonary edema, a medical emergency that can be delayed for several hours. Can cause death by pulmonary edema or circulation shock. Fumes cause irritation of eyes and respiratory passages. Upon ingestion, immediate pain in the mouth and throat, abdominal pain; nausea, vomiting of mucoid and "coffee-ground" material, intense thirst; clammy skin, weak and rapid pulse; shallow respiration, and circulatory shock occur.

Long Term Exposure: Repeated or prolonged contact with skin may cause dermatitis. May cause lung irritation; bronchitis may develop. May cause liver and kidney damage.

Points of Attack: Respiratory system, lungs, eyes, skin, liver, kidneys.

Medical Surveillance: NIOSH lists the following tests: chest X-ray; pulmonary function tests: forced vital capacity, forced expiratory volume (1 s). Consider the points of attack in preplacement and periodic physical examinations. Lung function tests. Liver and kidney function tests. Consider chest X-ray following acute overexposure.

First Aid: If this chemical gets into the eyes, remove any contact lenses at once and irrigate immediately for at least 15 min, occasionally lifting upper and lower lids. Seek medical attention immediately. If this chemical contacts the skin, remove contaminated clothing and wash immediately with soap and water. Seek medical attention immediately. If this chemical has been inhaled, remove from exposure, begin rescue breathing (using universal precautions, including resuscitation mask) if breathing has stopped and CPR if heart action has stopped. Transfer promptly to a medical facility. When this chemical has been swallowed, get medical attention. If victim is *conscious*, administer water or milk. Do not induce vomiting. Medical observation is recommended for 24–48 h after breathing overexposure, as pulmonary edema may be delayed. As first aid for pulmonary edema, a doctor or authorized paramedic may consider administering a corticosteroid spray.

Personal Protective Methods: Wear protective gloves and clothing to prevent any reasonable probability of skin contact. Safety equipment suppliers/manufacturers can provide recommendations on the most protective glove/clothing material for your operation. All protective clothing (suits, gloves, footwear, headgear) should be clean, available each day, and put on before work. Contact lenses should not be worn when working with this chemical. Wear dust-proof chemical goggles and face shield unless full face-piece respiratory protection is worn. Employees should wash immediately with soap when skin is wet or contaminated. Provide emergency showers and eyewash.

Respirator Selection: Up to 10 mg/m³: Sa* (APF = 10) (any supplied-air respirator). *Up to 25 mg/m³:* Sa:Cf* (APF = 25) (any supplied-air respirator operated in a continuous-flow mode). *Up to 50 mg/m³:* SCBAF (APF = 50) (any self-contained breathing apparatus with a full face-piece); or SaF (APF = 50) (any supplied-air respirator with a full face-piece). *Up to 70 mg/m³:* SaF: Pd,Pp (APF = 2000) (any supplied-air respirator that has a full face-piece and is operated in a pressure-demand or other positive-pressure mode). *Emergency or planned entry into unknown concentrations or IDLH conditions:* SCBAF: Pd, Pp (APF = 10,000) (any NIOSH/MSHA- or European Standard EN 149-approved self-contained breathing apparatus that has a full face-piece and is operated in a pressure-demand or other positive-pressure mode) or SaF: Pd,Pp: ASCBA (APF = 10,000) (any supplied-air respirator that has a full face-piece and is operated in a pressure-demand or other positive-pressure mode in combination with an auxiliary, self-contained breathing apparatus operated in a pressure-demand or other positive-pressure mode). *Escape:* GmFOv100 (APF = 50) [any air-purifying, full-face-piece respirator (gas mask) with a chin-style, front- or back-mounted organic vapor canister having an N100, R100, or P100 filter] or SCBAE (any appropriate escape-type, self-contained breathing apparatus).

*Substance reported to cause eye irritation or damage; may require eye protection.

Storage: (1) Color Code—White: Corrosive or Contact Hazard; Store separately in a corrosion resistant location. (2) Color Code—Yellow Stripe (strong oxidizer): Reactivity Hazard; Store separately in an area isolated from flammables, combustibles, or other yellow-coded materials. Prior to working with this chemical you should be trained on its proper handling and storage. Store in tightly closed containers in a cool, well-ventilated area away from all other combustible and oxidizable materials, and moisture. Where possible, automatically transfer material storage containers to process containers. Sources of ignition, such as smoking and open flames, are prohibited where this chemical is handled, used, or stored. Metal containers involving the transfer of this chemical should be grounded and bonded. Use only nonsparking tools and equipment, especially when opening and closing containers of this chemical. Wherever this chemical is used, handled, manufactured, or stored, use explosion-proof electrical equipment and fittings.

Shipping: This compound requires a shipping label of "CORROSIVE." It falls in Hazard Class 8 and Packing Group II.

Spill Handling: Evacuate persons not wearing protective equipment from area of spill or leak until cleanup is complete. Remove all ignition sources. Keep material out of water sources and sewers. Use water spray to knock down vapors. *Do not use water* on material itself; neutralize spilled material with crushed limestone, soda ash, or lime. Avoid breathing vapors; keep upwind. Avoid bodily contact with the materials. Do not handle broken packages without protective equipment. Wash away any materials which may have contacted the body with copious amounts of water or soap and water. Ventilate area after cleanup is complete. It may be necessary to contain and dispose of this chemical as a hazardous waste. If material or contaminated runoff enters waterways, notify downstream users of potentially contaminated waters. Contact your local or federal environmental protection agency for specific recommendations. If employees are required to clean up spills, they must be properly trained and equipped. OSHA 1910.120(q) may be applicable.

Initial isolation and protective action distances
Distances shown are likely to be affected during the first 30 min after materials are spilled and could increase with time. If more than one tank car, cargo tank, portable tank, or large cylinder involved in the incident is leaking, the protective action distance may need to be increased. You may need to seek emergency information from CHEMTREC at (800) 424-9300 or seek professional environmental engineering assistance from the US EPA Environmental Response Team at (908) 548-8730 (24-h response line).

Small spills (From a small package or a small leak from a large package)
when spilled in water
First: Isolate in all directions (feet/meters)   100/30
Then: Protect persons downwind (miles/kilometers)
Day   0.1/0.2
Night 0.2/0.3
Large spills (From a large package or from many small packages)
First: Isolate in all directions (feet/meters)   100/30
Then: Protect persons downwind (miles/kilometers)
Day   0.3/0.5
Night 1.0/1.5

Fire Extinguishing: A noncombustible solid. If material is involved in fire then use dry chemical, carbon dioxide, or dry sand. *Do not use water* on material itself. If large quantities of combustibles are involved, use water in flooding quantities (i.e., spray or fog), and use water spray to absorb vapors. Avoid breathing vapors; keep upwind. Wear self-contained breathing apparatus. Avoid bodily contact with the material; wear boots, protective gloves, and goggles. Poisonous gases are produced in fire, including hydrogen chloride and phosphorus oxide. If material or contaminated runoff enters waterways, notify downstream users of potentially contaminated waters. Notify local health and fire officials and pollution control agencies. Containers may explode in fire. From a secure, explosion-proof location, use water spray to cool exposed containers. If cooling streams are ineffective (venting sound increases in volume and pitch, tank discolors, or shows any signs of deforming), withdraw immediately to a secure position. If employees are expected to fight fires, they must be trained and equipped in OSHA 1910.156. The only respirators recommended for firefighting are self-contained breathing apparatuses that have full face-pieces and are operated in a pressure-demand or other positive-pressure mode.

Disposal Method Suggested: Decompose with water, forming phosphoric and hydrochloric acids. Neutralize acids and dilute if necessary for discharge into the sewer system.

References
US Environmental Protection Agency. (November 30, 1987). *Chemical Hazard Information Profile: Phosphorus Pentachloride.* Washington, DC: Chemical Emergency Preparedness Program
New Jersey Department of Health and Senior Services. (August 1999). *Hazardous Substances Fact Sheet: Phosphorus Pentachloride.* Trenton, NJ

Phosphorus pentasulfide   P:0640

Molecular Formula: P_2S_5; P_4S_{10}
Common Formula: P_4S_{10}
Synonyms: Pentasulfure de phosphore (French); Phosphoric sulfide; Phosphorus pentasulfide; Phosphorus persulfide;

Phosphorus sulfide; Sulphur phosphide; Thiophosphoric anhydride

CAS Registry Number: 1314-80-3

RTECS® Number: TH4375000

UN/NA & ERG Number: UN1340 (free from yellow or white phosphorus)/139

EC Number: 215-242-4 [*Annex I Index No.:* 015-104-00-1]

Regulatory Authority and Advisory Bodies

Department of Homeland Security Screening Threshold Quantity (pounds): Sabotage/Contamination Hazard: A placarded amount (commercial grade).

Air Pollutant Standard Set. See below, "Permissible Exposure Limits in Air" section.

US EPA Hazardous Waste Number (RCRA No.): U189.

RCRA, 40CFR261, Appendix 8 Hazardous Constituents.

Reportable Quantity (RQ): 100 lb (45.4 kg).

Canada, WHMIS, Ingredients Disclosure List Concentration 1.0%.

European/International Regulations: Hazard Symbol: F, Xn, N; Risk phrases: R11; R20/22; R29; Safety phrases: S2; S61 (see Appendix 4).

WGK (German Aquatic Hazard Class): 3—Severe hazard to waters.

Description: Phosphorus pentasulfide is a greenish-gray to yellow, crystalline solid with an odor of rotten eggs. The odor threshold is 0.005 ppm. Molecular weight = 222.24 (P_2S_5); 444.6 (P_4S_{10}); Specific gravity (H_2O:1) = 2.09; Boiling point = 513.8°C; Freezing/Melting point = 286°C; Vapor pressure = 1 mmHg at 300°C; Autoignition temperature (dry air) = 142°C. Hazard Identification (based on NFPA-704 M Rating System): Health 1, Flammability 2, Reactivity W. Reacts with water.

Potential Exposure: Compound Description: Primary Irritant. Phosphorus pentasulfide is used as an intermediate in the manufacture of lubricant additives, insecticides, flotation agents, lubricating oil, ignition compounds, and matches. It is also used to introduce sulfur into rubber, and organic chemicals, such as pharmaceuticals.

Incompatibilities: Flammable solid; water reactive. Violent reaction with water, alcohols, strong oxidizers, acids, alkalis. Reaction with water to produce heat, hydrogen sulfide, sulfur dioxide, and phosphoric acid. May self-ignite in moist air.

Permissible Exposure Limits in Air

OSHA PEL: 1 mg/m³ TWA.

NIOSH REL: 1 mg/m³ TWA; 3 mg/m³ STEL.

ACGIH TLV®[1]: 1 mg/m³ TWA; 3 mg/m³ STEL.

NIOSH IDLH: 250 mg/m³.

Protective Action Criteria (PAC)

TEEL-0: 1 mg/m³

PAC-1: 3 mg/m³

PAC-2: 50 mg/m³

PAC-3: 250 mg/m³

Australia: TWA 1 mg/m³; STEL 3 mg/m³, 1993; Austria: MAK 1 mg/m³, 1999; Belgium: TWA 1 mg/m³; STEL 3 mg/m³, 1993; Denmark: TWA 1 mg/m³, 1999; Finland: STEL 1 mg/m³, 1999; France: VME 1 mg/m³, 1999; the Netherlands: MAC-TGG 1 mg/m³, 2003; the Philippines: TWA 1 mg/m³, 1993; Poland: TWA 1 mg/m³; STEL 3 mg/m³, 1999; Switzerland: MAK-W 1 mg/m³, KZG-W 2 mg/m³, 1999; Thailand: TWA 1 mg/m³, 1993; Turkey: TWA 1 mg/m³, 1993; United Kingdom: TWA 1 mg/m³; STEL 3 mg/m³, 2000; Argentina, Bulgaria, Columbia, Jordan, South Korea, New Zealand, Singapore, Vietnam: ACGIH TLV®: STEL 3 mg/m³. Several states have set guidelines or standards for P_2S_5 in ambient air[60] ranging from 10 μg/m³ (North Dakota) to 16 μg/m³ (Virginia) to 20 μg/m³ (Connecticut) to 24 μg/m³ (Nevada).

Determination in Air: No method available.

Routes of Entry: Inhalation, ingestion, skin and/or eye contact.

Harmful Effects and Symptoms

Short Term Exposure: Severely irritates the eyes, skin, and respiratory tract. Inhalation of fumes produced by phosphorus compounds may cause irritation of pulmonary tissues with resultant acute pulmonary edema. The hazards of phosphorus pentasulfide are the same as for hydrogen sulfide to which it rapidly hydrolyzes in the presence of moisture. Symptoms include apnea, coma, convulsions, lacrimation (discharge of tears), photophobia (abnormal visual intolerance to light), kerato-conjunctivitis, corneal vesiculation, respiratory system irritation, dizziness, headaches, fatigue, irritability, insomnia, gastrointestinal disturbances.

Long Term Exposure: Chronic exposure may lead to lung irritation, cough, bronchitis, and pneumonia.

Points of Attack: Respiratory system, lungs, central nervous system, eyes, skin.

Medical Surveillance: NIOSH lists the following tests: chest X-ray; pulmonary function tests: forced vital capacity, forced expiratory volume (1 s). Consider the points of attack in preplacement and periodic physical examinations. Lung function tests. Consider chest X-ray following acute overexposure.

First Aid: If this chemical gets into the eyes, remove any contact lenses at once and irrigate immediately for at least 15 min, occasionally lifting upper and lower lids. Seek medical attention immediately. If this chemical contacts the skin, remove contaminated clothing and wash immediately with soap and water. Seek medical attention immediately. If this chemical has been inhaled, remove from exposure, begin rescue breathing (using universal precautions, including resuscitation mask) if breathing has stopped and CPR if heart action has stopped. Transfer promptly to a medical facility. When this chemical has been swallowed, get medical attention. Give large quantities of water and induce vomiting. Do not make an unconscious person vomit. Medical observation is recommended for 24−48 h after breathing overexposure, as pulmonary edema may be delayed. As first aid for pulmonary edema, a doctor or authorized paramedic may consider administering a corticosteroid spray.

Personal Protective Methods: Wear protective gloves and clothing to prevent any reasonable probability of skin

contact. Safety equipment suppliers/manufacturers can provide recommendations on the most protective glove/clothing material for your operation. All protective clothing (suits, gloves, footwear, headgear) should be clean, available each day, and put on before work. Contact lenses should not be worn when working with this chemical. Wear dust-proof chemical goggles and face shield unless full face-piece respiratory protection is worn. Employees should wash immediately with soap when skin is wet or contaminated. Provide emergency showers and eyewash.

Respirator Selection: *Up to 10 mg/m³* Sa (APF = 10) (any supplied-air respirator).* *Up to 25 mg/m³*: Sa:Cf (APF = 25) (any supplied-air respirator operated in a continuous-flow mode).* *Up to 50 mg/m³*: SCBAF (APF = 50) (any self-contained breathing apparatus with a full face-piece) or SaF (APF = 50) (any supplied-air respirator with a full face-piece). *Up to 250 mg/m³*: SaF: Pd,Pp (APF = 2000) (any supplied-air respirator that has a full face-piece and is operated in a pressure-demand or other positive-pressure mode). *Emergency or planned entry into unknown concentrations or IDLH conditions:* GmFS100 (APF = 50) [Any air-purifying, full-face-piece respirator (gas mask) with a chin-style, front- or back-mounted canister providing protection against the compound of concern and having an N100, R100, or P100 filter] or SCBAE (any appropriate escape-type, self-contained breathing apparatus).

*Substance reported to cause eye irritation or damage; may require eye protection.

Storage: Color Code—Red: Flammability Hazard: Store in a flammable materials storage area. Prior to working with this chemical you should be trained on its proper handling and storage. Store in tightly closed containers in a cool, well-ventilated area away from moisture, water, alcohols, strong oxidizers, acids, alkalis. Where possible, automatically transfer material from drums or other storage containers to process containers. Sources of ignition, such as smoking and open flames, are prohibited where this chemical is handled, used, or stored. Metal containers involving the transfer of this chemical should be grounded and bonded. Wherever this chemical is used, handled, manufactured, or stored, use explosion-proof electrical equipment and fittings.

Shipping: This compound requires a shipping label of "DANGEROUS WHEN WET, FLAMMABLE SOLID." It falls in Hazard Class 4.3 and Packing Group II.

Spill Handling: Evacuate persons not wearing protective equipment from area of spill or leak until cleanup is complete. Remove all ignition sources. Collect powdered material in the most convenient and safe manner and deposit in sealed containers. Ventilate area after cleanup is complete. It may be necessary to contain and dispose of this chemical as a hazardous waste. If material or contaminated runoff enters waterways, notify downstream users of potentially contaminated waters. Contact your local or federal environmental protection agency for specific recommendations. If employees are required to clean up spills, they must be properly trained and equipped. OSHA 1910.120(q) may be applicable.

Fire Extinguishing: This chemical is a combustible solid. Use dry chemical, carbon dioxide, sand. Consider the use of sodium chloride-base extinguisher suitable for metal fires. *Do not use water.* Poisonous gases are produced in fire, including phosphorus oxides and fluorine. If material or contaminated runoff enters waterways, notify downstream users of potentially contaminated waters. Notify local health and fire officials and pollution control agencies. Containers may explode in fire. From a secure, explosion-proof location, use water spray to cool exposed containers. If cooling streams are ineffective (venting sound increases in volume and pitch, tank discolors, or shows any signs of deforming), withdraw immediately to a secure position. If employees are expected to fight fires, they must be trained and equipped in OSHA 1910.156. The only respirators recommended for firefighting are self-contained breathing apparatuses that have full face-pieces and are operated in a pressure-demand or other positive-pressure mode.

Disposal Method Suggested: Consult with environmental regulatory agencies for guidance on acceptable disposal practices. Generators of waste containing this contaminant (≥100 kg/mo) must conform with EPA regulations governing storage, transportation, treatment, and waste disposal. Decompose with water, forming phosphoric acid, sulfuric acid, and hydrogen sulfide. Provisions must be made for scrubbing hydrogen sulfide emissions. The acids may then be neutralized and diluted slowly to solution of soda ash and slaked lime with stirring, then flush to sewer with large volumes of water.

Reference
Sax, N. I. (Ed.). (1983). *Dangerous Properties of Industrial Materials Report*, 3, No. 4, 89–90

Phosphorus pentoxide   P:0650

Molecular Formula: O_5P_2
Common Formula: P_2O_5
Synonyms: Diphosphorus pentoxide; Pentoxido de fosforo (Spanish); Phosphoric anhydride; Phosphorus(V) oxide; Phosphorus(5+) oxide; Phosphorus pentaoxide; Phosphorus oxide; POX
CAS Registry Number: 1314-56-3
RTECS®Number: TH3945000
UN/NA & ERG Number: UN1807/137
EC Number: 215-236-1 [*Annex I Index No.:* 015-010-00-0]
Regulatory Authority and Advisory Bodies
Air Pollutant Standard Set. See below, "Permissible Exposure Limits in Air" section.
US EPA, FIFRA 1998 Status of Pesticides: Active registration.
Superfund/EPCRA 40CFR355, Extremely Hazardous Substances: TPQ = 10 lb (4.54 kg).

Reportable Quantity (RQ): 1 lb (0.454 kg).

Canada, WHMIS, Ingredients Disclosure List Concentration 1.0%.

European/International Regulations: Hazard Symbol: C; Risk phrases: R35; Safety phrases: S1/2; S22; S26; S45 (see Appendix 4).

WGK (German Aquatic Hazard Class): 1—Low hazard to waters.

Description: Phosphorus pentoxide is a white crystalline solid. Molecular weight = 141.94; Freezing/Melting point = 340°C; it begins to sublime at 360°C. Hazard Identification (based on NFPA-704 M Rating System): Health 3, Flammability 0, Reactivity 2₩ (Corrosive). Reacts with water.

Potential Exposure: This material is used as an intermediate in organic synthesis; as a catalyst, condensing agent, dehydrating agent; in the preparation of acrylate esters, surfactants, sugar refining; in medicine, fire extinguishing, and special glasses.

Incompatibilities: Reacts violently with water, forming highly corrosive phosphoric acid. Phosphorus pentoxide reacts violently with the following: perchloric acid, ammonia, hydrofluoric acid, oxidizers, hydrogen fluoride, formic acid, oxygen difluoride, potassium, sodium, propargyl alcohol, calcium oxide, inorganic bases, sodium hydroxide, and chlorine trifluoride. Attacks many metals in the presence of water.

Permissible Exposure Limits in Air

Protective Action Criteria (PAC)*

TEEL-0: 1 mg/m^3

PAC-1: **1** mg/m^3

PAC-2: **10** mg/m^3

PAC-3: **50** mg/m^3

*AEGLs (Acute Emergency Guideline Levels) & ERPGs (Emergency Response Planning Guideline) are in **bold face**.

DFG MAK: 2 mg/m^3, measured as inhalable fraction TWA; Peak Limitation Category I(2); Pregnancy Risk Group C.

Austria: MAK 1 mg/m^3, 1999; Denmark: ceiling 1 mg/m^3, 1999; France: VME 1 mg/m^3, 1999; Hungary: TWA 1 mg/m^3; STEL 2 mg/m^3, 1993; Norway: TWA 1 mg/m^3, 1999; Poland: MAC (TWA) 1 mg/m^3; STEL 3 mg/m^3, 1999; Russia: STEL 1 mg/m^3, [skin], 1993; the Netherlands: MAC-TGG 1 mg/m^3, 2003; Turkey: TWA 1 mg/m^3, 1993; United Kingdom: STEL 2 mg/m^3, 2000.

Permissible Concentration in Water: No criteria set (reacts with water).

Routes of Entry: Inhalation, ingestion, skin and/or eye contact.

Harmful Effects and Symptoms

Short Term Exposure: Highly corrosive to the eyes, skin, and respiratory tract. Eye contact may lead to a total destruction of the eyes. Particles in contact with eyes react vigorously and even a small amount may cause permanent burns. Contact with the skin will cause severe burns. Corrosive if ingested; will damage the gastrointestinal tract.

Higher exposures can cause pulmonary edema, a medical emergency that can be delayed for several hours. This can cause death. Signs and symptoms of acute exposure to phosphorus pentoxide may include severe burns, pain, shock, intense thirst, nausea, vomiting, diarrhea, severe abdominal pain; and "smoking stools." The breath and feces may have a garlicky odor. A symptom-free period of several days may follow. Exposure to phosphorus pentoxide may also result in bloody vomitus and diarrhea, jaundice, liver enlargement with tenderness, renal damage, hematuria (bloody urine), and either oliguria (scanty urination) or anuria (suppression of urine formation). Headache, convulsions, delirium, coma, cardiac arrhythmias, and cardiovascular collapse may occur. If phosphorus pentoxide contacts the eyes, severe irritation and burns, blepharospasm (spasmodic winking), lacrimation (tearing), and photophobia (heightened sensitivity to light) may occur. Victims may experience spontaneous hemorrhaging of phosphorus pentoxide-contaminated skin and mucous membranes.

Long Term Exposure: Highly corrosive materials can cause lung damage; bronchitis may develop.

Points of Attack: Lungs, skin, eyes,

Medical Surveillance: Lung function tests. Consider chest X-ray following acute overexposure.

First Aid: If this chemical gets into the eyes, remove any contact lenses at once and irrigate immediately for at least 15 min, occasionally lifting upper and lower lids. Seek medical attention immediately. If this chemical contacts the skin, remove contaminated clothing and wash immediately with soap and water. Seek medical attention immediately. If this chemical has been inhaled, remove from exposure, begin rescue breathing (using universal precautions, including resuscitation mask) if breathing has stopped and CPR if heart action has stopped. Transfer promptly to a medical facility. When this chemical has been swallowed, get medical attention. If victim is *conscious*, administer water or milk. Do not induce vomiting. Obtain authorization and/or further instructions from the local hospital for administration of an antidote or performance of other invasive procedures. Rush to a health-care facility.

Personal Protective Methods: Wear protective gloves and clothing to prevent any reasonable probability of skin contact. Safety equipment suppliers/manufacturers can provide recommendations on the most protective glove/clothing material for your operation. All protective clothing (suits, gloves, footwear, headgear) should be clean, available each day, and put on before work. Contact lenses should not be worn when working with this chemical. Wear dust-proof chemical goggles and face shield unless full face-piece respiratory protection is worn. Employees should wash immediately with soap when skin is wet or contaminated. Provide emergency showers and eyewash. *Never pour water into this substance*; always add POX slowly to water when diluting or dissolving.

Respirator Selection: Follow regulations in OSHA 29CFR1910.134 or European Standard EN149. Use a

NIOSH/MSHA- or European Standard EN149-approved respirator; or use an approved supplied-air respirator with a full face-piece operated in the positive-pressure mode, or with a full face-piece, hood, or helmet in the continuous-flow mode; or use a NIOSH/MSHA- or European Standard EN149-approved self-contained breathing apparatus with a full face-piece operated in a pressure-demand or other positive-pressure mode.

Storage: Color Code—White: Corrosive or Contact Hazard; Store separately in a corrosion-resistant location. Prior to working with this chemical you should be trained on its proper handling and storage. Store in tightly closed containers in a cool, well-ventilated area away from incompatible materials listed above.

Shipping: This compound requires a shipping label of "CORROSIVE." It falls in Hazard Class 8 and Packing Group II.

Spill Handling: Keep unnecessary people away. Stay upwind. Keep out of low areas. Ventilate closed spaces before entering them. Stop leak if possible without risk. Do not touch spilled material. Use water spray to reduce vapors, but do not put water on leak or spill. Keep combustibles away from spilled material. Dike spilled area and keep water away from spill. Clean up requires supervision by an expert. Ventilate area after cleanup is complete. It may be necessary to contain and dispose of this chemical as a hazardous waste. If material or contaminated runoff enters waterways, notify downstream users of potentially contaminated waters. Contact your local or federal environmental protection agency for specific recommendations. If employees are required to clean up spills, they must be properly trained and equipped. OSHA 1910.120(q) may be applicable.

Fire Extinguishing: Does not support combustion. For small fires, use dry chemical, carbon dioxide, or sand. *Do not use water* or hydrous extinguishers. Wear positive pressure breathing apparatus and special protective clothing. Keep combustibles away from spilled material. Poisonous gases, including phosphorus oxides, are produced in fire. If material or contaminated runoff enters waterways, notify downstream users of potentially contaminated waters. Notify local health and fire officials and pollution control agencies. *For large fires,* flood fire area with water from a distance. Do not get solid stream of water on spilled material or in open containers. From a secure, explosion-proof location, use water spray to cool exposed containers long after flames have been extinguished. If cooling streams are ineffective (venting sound increases in volume and pitch, tank discolors, or shows any signs of deforming), withdraw immediately to a secure position. If employees are expected to fight fires, they must be trained and equipped in OSHA 1910.156. The only respirators recommended for firefighting are self-contained breathing apparatuses that have full face-pieces and are operated in a pressure-demand or other positive-pressure mode.

Disposal Method Suggested: Decompose with water, forming phosphoric and hydrochloric acids. The acids may then be neutralized and diluted slowly to solution of soda ash and slaked lime with stirring then flush to sewer with large volumes of water.

References
US Environmental Protection Agency. (November 30, 1987). *Chemical Hazard Information Profile: Phosphorus Pentoxide.* Washington, DC: Chemical Emergency Preparedness Program
US Environmental Protection Agency, Special Review and Reregistration Division Office of Pesticide Programs. (1998). *Agency Status of Pesticides in Registration, Reregistration, and Special Review* (Rainbow Report). Washington, DC

Phosphorus trichloride   P:0660

Molecular Formula: Cl₃P
Common Formula: PCl₃
Synonyms: Chloride of phosphorus; Phosphore (trichlorure de) (French); Phosphorous chloride; Phosphortrichlorid (German); Phosphorus chloride; Trichloro; Tricloruro de fosforo (Spanish)
CAS Registry Number: 7719-12-2
RTECS® Number: TH3675000
UN/NA & ERG Number: UN1809/137
EC Number: 231-749-3 [Annex I Index No.: 015-007-00-4]
Regulatory Authority and Advisory Bodies
Department of Homeland Security Screening Threshold Quantity (pounds): *Release hazard* 15,000 (1% concentration); *Theft hazard* 45 (≥3.48% concentration); Sabotage/Contamination Hazard: A placarded amount (commercial grade).
Air Pollutant Standard Set. See below, "Permissible Exposure Limits in Air" section.
Clean Air Act: Accidental Release Prevention/Flammable Substances, (Section 112[r], Table 3), TQ = 15,000 lb (6810 kg).
Clean Water Act: Section 311 Hazardous Substances/RQ 40CFR117.3 (same as CERCLA, see below).
Superfund/EPCRA 40CFR355, Extremely Hazardous Substances: TPQ = 1000 lb (454 kg).
Reportable Quantity (RQ): 1000 lb (454 kg).
US DOT 49CFR172.101, Inhalation Hazardous Chemical.
Canada, WHMIS, Ingredients Disclosure List Concentration 1.0%.
European/International Regulations: Hazard Symbol: T +, C; Risk phrases: R14; R26/28; R48/20; Safety phrases: S1/2; S7/8; S26; S36/37/39; S45 (see Appendix 4).
WGK (German Aquatic Hazard Class): 1—Low hazard to waters.
Description: Phosphorus trichloride is a colorless to yellow, fuming liquid with an odor like hydrochloric acid. Molecular weight = 137.32; Specific gravity (H₂O:1) = 1.58; Boiling point = 76.1°C; Freezing/Melting

point = $-112.2°C$; Vapor pressure = 100 mmHg at 20°C. Hazard Identification (based on NFPA-704 M Rating System): Health 3, Flammability 0, Reactivity 2W. Reacts with water.

Potential Exposure: Phosphorus trichloride is used as an intermediate and as a chlorinating agent and catalyst; in the manufacture of agricultural chemicals, pharmaceuticals, chlorinated compounds, dyes, gasoline additives, acetyl cellulose, phosphorus oxychloride, plasticizers, saccharin, and surfactants.

Incompatibilities: Violent reaction with alcohols, phenols and bases, water, when in contact with combustible organics, chemically active metals: sodium, potassium, aluminum, strong nitric acid. Violent reaction with water, producing heat and hydrochloric and phosphorous acids. Attacks most metals. Attacks plastics, rubber, and coatings.

Permissible Exposure Limits in Air
Conversion factor: 1 ppm = $5.62 mg/m^3$ at 25°C & 1 atm.
OSHA PEL: 0.5 ppm/3 mg/m^3 TWA.
NIOSH REL: 0.2 ppm/1.5 mg/m^3 TWA; 0.5 ppm/3 mg/m^3 STEL.
ACGIH TLV®[1]: 0.2 ppm/1.1 mg/m^3 TWA; 0.5 ppm/2.8 mg/m^3 STEL.
NIOSH IDLH: 25 ppm.
Protective Action Criteria (PAC)*
TEEL-0: 0.34 ppm
PAC-1: **0.34 ppm**
PAC-2: **2 ppm**
PAC-3: **5.6 ppm**
*AEGLs (Acute Emergency Guideline Levels) & ERPGs (Emergency Response Planning Guideline) are in **bold face**.
DFG MAK: 0.5 ppm/2.8 mg/m^3 TWA; Peak Limitation Category I(1); Pregnancy Risk Group C.
Arab Republic of Egypt: TWA 0.5 ppm (3 mg/m^3), 1993; Australia: TWA 0.2 ppm (1.5 mg/m^3); STEL 0.5 ppm (3 mg/m^3), 1993; Austria: MAK 0.25 ppm (1.5 mg/m^3), 1999; Belgium: TWA 0.2 ppm (1.1 mg/m^3); STEL 0.5 ppm (2.8 mg/m^3), 1993; Denmark: TWA 0.2 ppm (1.2 mg/m^3), 1999; Finland: STEL 0.5 ppm (3 mg/m^3), [skin], 1999; France: VME 0.2 ppm (1.5 mg/m^3), 1999; the Netherlands: MAC-TGG 1.5 mg/m^3, 2003; Japan 0.2 ppm (1.1 mg/m^3), 1999; Norway: TWA 0.2 ppm (1.5 mg/m^3), 1999; the Philippines: TWA 0.5 ppm (3 mg/m^3), 1993; Poland: MAC (TWA) 3 mg/m^3, 1999; Russia: TWA 0.2 ppm; STEL 0.2 mg/m^3, [skin], 1993; Turkey: TWA 0.5 ppm (3 mg/m^3), 1993; United Kingdom: TWA 0.2 ppm (1.1 mg/m^3); STEL 0.5 ppm, 2000; Argentina, Bulgaria, Columbia, Jordan, South Korea, New Zealand, Singapore, Vietnam: ACGIH TLV®: STEL 0.5 ppm. Several states have set guidelines or standards for PCl$_3$ in ambient air[60] ranging from 15–30 μg/m^3 (North Dakota) to 30 μg/m^3 (Connecticut) to 36 μg/m^3 (Nevada).

Determination in Air: Use NIOSH Analytical Method 6402.

Routes of Entry: Inhalation, ingestion, skin and/or eye contact.

Harmful Effects and Symptoms
Short Term Exposure: Corrosive to the eyes, skin, and respiratory tract. Corrosive if swallowed. Inhalation can cause pulmonary edema, a medical emergency that can be delayed for several hours. Very high levels of exposure to vapors can cause death. This material is highly toxic; it may cause death or permanent injury. Exposure may cause dizziness, headache, anorexia, respiratory difficulties, nausea, and vomiting. It can also cause liver and lung disturbances. Occupational exposure has caused coughs, bronchitis, pneumonia, and conjunctivitis.

Long Term Exposure: Highly corrosive materials can cause lung damage; bronchitis may develop.
Points of Attack: Eyes, skin, respiratory system.
Medical Surveillance: NIOSH lists the following tests: chest X-ray, electrocardiogram, pulmonary function tests: forced vital capacity, forced expiratory volume (1 s); sputum cytology; white blood cell count/differential.
First Aid: If this chemical gets into the eyes, remove any contact lenses at once and irrigate immediately for at least 30 min, occasionally lifting upper and lower lids. Seek medical attention immediately. If this chemical contacts the skin, remove contaminated clothing and wash immediately with soap and water. Seek medical attention immediately. If this chemical has been inhaled, remove from exposure, begin rescue breathing (using universal precautions, including resuscitation mask) if breathing has stopped and CPR if heart action has stopped. Transfer promptly to a medical facility. When this chemical has been swallowed, get medical attention. If victim is *conscious*, administer water or milk. Do not induce vomiting. Medical observation is recommended for 24–48 h after breathing overexposure, as pulmonary edema may be delayed. As first aid for pulmonary edema, a doctor or authorized paramedic may consider administering a corticosteroid spray.

Personal Protective Methods: Wear protective gloves and clothing to prevent any reasonable probability of skin contact. Safety equipment suppliers/manufacturers can provide recommendations on the most protective glove/clothing material for your operation. All protective clothing (suits, gloves, footwear, headgear) should be clean, available each day, and put on before work. Contact lenses should not be worn when working with this chemical. Wear splash-proof chemical goggles and face shield unless full face-piece respiratory protection is worn. Employees should wash immediately with soap when skin is wet or contaminated. Provide emergency showers and eyewash.

Respirator Selection: *10 ppm:* SCBAF (APF = 50) (any self-contained breathing apparatus with a full face-piece) or SaF (APF = 50) (any supplied-air respirator with a full face-piece). *25 ppm:* SaF: Pd,Pp (APF = 2000) (any supplied-air respirator that has a full face-piece and is operated in a pressure-demand or other positive-pressure mode). *Emergency or planned entry into unknown concentrations or IDLH conditions:* SCBAF: Pd,Pp (APF = 10,000) (any self-contained breathing apparatus that has a full face-piece and is operated in a pressure-demand or other positive-pressure mode) or SaF: Pd,Pp: ASCBA (APF = 10,000) (any supplied-air respirator that has a full face-piece and is operated in a pressure-demand or other positive-pressure mode in combination with an auxiliary self-contained breathing apparatus operated in a pressure-demand or other positive-pressure mode). *Escape:* GmFS (APF = 50) [any air-purifying, full-face-piece respirator (gas mask) with a chin-style, front- or back-mounted canister providing protection against the compound of concern] or SCBAE (any appropriate escape-type, self-contained breathing apparatus). *Note:* Substance causes eye irritation or damage; eye protection needed.

Storage: Color Code—Blue: Health Hazard/Poison: Store in a secure poison location. Prior to working with this chemical you should be trained on its proper handling and storage. Phosphorus trichloride must be stored to avoid contact with acetic acid, aluminum, chromyl chloride, fluorine, alcohol, nitric acid, sodium, potassium, water, hydroxylamine, and lead dioxide since violent reactions occur. Store in tightly closed containers in a cool, well-ventilated area away from water and moisture. Phosphorus trichloride corrodes most metals and will attack some forms of plastics, rubber, and coatings.

Shipping: This compound requires a shipping label of "POISONOUS/TOXIC MATERIALS, CORROSIVE." It falls in Hazard Class 6.1 and Packing Group I. A plus sign (+) indicates that the designated proper shipping name and hazard class of the material must always be shown whether or not the material or its mixtures or solutions meet the definitions of the class.

Spill Handling: Evacuate and restrict persons not wearing protective equipment from area of spill or leak until cleanup is complete. Remove all ignition sources. Ventilate area of spill or leak. Absorb liquids in vermiculite, dry sand, earth, peat, carbon, or a similar material and deposit in sealed containers. Keep this chemical out of a confined space, such as a sewer, because of the possibility of an explosion, unless the sewer is designed to prevent the buildup of explosive concentrations. It may be necessary to contain and dispose of this chemical as a hazardous waste. If material or contaminated runoff enters waterways, notify downstream users of potentially contaminated waters. Contact your local or federal environmental protection agency for specific recommendations. If employees are required to clean up spills, they must be properly trained and equipped. OSHA 1910.120(q) may be applicable.

Initial isolation and protective action distances
Distances shown are likely to be affected during the first 30 min after materials are spilled and could increase with time. If more than one tank car, cargo tank, portable tank, or large cylinder involved in the incident is leaking, the protective action distance may need to be increased. You may need to seek emergency information from CHEMTREC at (800) 424-9300 or seek professional environmental

engineering assistance from the US EPA Environmental Response Team at (908) 548-8730 (24-h response line).
Small spills (From a small package or a small leak from a large package)
when spilled on land
First: Isolate in all directions (feet/meters)   100/30
Then: Protect persons downwind (miles/kilometers)
Day   0.2/0.3
Night   0.4/0.6
Large spills (From a large package or from many small packages)
First: Isolate in all directions (feet/meters)   500/150
Then: Protect persons downwind (miles/kilometers)
Day   0.9/1.5
Night   1.9/3.1
when spilled in water
First: Isolate in all directions (feet/meters)   100/30
Then: Protect persons downwind (miles/kilometers)
Day   0.1/0.2
Night   0.2/0.3
Large spills (From a large package or from many small packages)
First: Isolate in all directions (feet/meters)   200/60
Then: Protect persons downwind (miles/kilometers)
Day   0.5/0.8
Night   1.7/2.7

Fire Extinguishing: This material may burn but does not easily ignite. Use carbon dioxide or dry chemical on fires involving phosphorus trichloride. *Do not use water.* Poisonous gases are produced in fire, including hydrogen chloride and phosphoric acid. If material or contaminated runoff enters waterways, notify downstream users of potentially contaminated waters. Notify local health and fire officials and pollution control agencies. From a secure, explosion-proof location, use water spray to cool exposed containers. Spray cooling water on containers that are exposed to flames until well after fire is out. Do not get water inside containers. If cooling streams are ineffective (venting sound increases in volume and pitch, tank discolors, or shows any signs of deforming), withdraw immediately to a secure position. If employees are expected to fight fires, they must be trained and equipped in OSHA 1910.156. The only respirators recommended for firefighting are self-contained breathing apparatuses that have full face-pieces and are operated in a pressure-demand or other positive-pressure mode.

Disposal Method Suggested: Decompose with water, forming phosphoric and hydrochloric acids. The acids may then be neutralized and diluted slowly to solution of soda ash and slaked lime with stirring, then flush to sewer with large volumes of water.

References
Sax, N. I. (Ed.). (1983). *Dangerous Properties of Industrial Materials Report*, 3, No. 4, 93–94
US Environmental Protection Agency. (November 30, 1987). *Chemical Hazard Information Profile: Phosphorus*

Trichloride. Washington, DC: Chemical Emergency Preparedness Program

New Jersey Department of Health and Senior Services. (March 2001). *Hazardous Substances Fact Sheet: Phosphorous Trichloride.* Trenton, NJ

Phthalic anhydride   P:0670

Molecular Formula: $C_8H_4O_3$

Synonyms: Anhidrido ftalico (Spanish); Anhydride phthalique (French); Araldite HT 901; 1,2-Benzenedicarboxylic anhydride; 1,2-Benzenedicarboxylic acid anhydride; 1,2-Dioxophthalan phthalandione; 1,3-Dioxophthalan; Esen; HT 901; 1,3-Isobenzofurandione; NCI-C03601; PAN; Phthalandione; 1,3-Phthalandione; Phthalanhydride; Phthalic acid anhydride; Phthalsaeureanhydrid (German); Retarder AK; Retarder esen; Retarder PD; TGL 6525; Vulkalent B/C

CAS Registry Number: 85-44-9

RTECS® Number: TI3150000

UN/NA & ERG Number: UN2214/156

EC Number: 201-607-5 [*Annex I Index No.:* 607-009-00-4]

Regulatory Authority and Advisory Bodies

Carcinogenicity: NCI: Carcinogenesis Bioassay (feed); no evidence: mouse, rat.

US EPA Gene-Tox Program, Negative: Carcinogenicity—mouse/rat.

Air Pollutant Standard Set. See below, "Permissible Exposure Limits in Air" section.

Clean Air Act: Hazardous Air Pollutants (Title I, Part A, Section 112).

US EPA Hazardous Waste Number (RCRA No.): U190.

RCRA, 40CFR261, Appendix 8 Hazardous Constituents.

RCRA 40CFR268.48; 61FR15654, Universal Treatment Standards: Wastewater (mg/L), 0.055; Nonwastewater (mg/kg), 28.

Reportable Quantity (RQ): 5000 lb (2270 kg).

EPCRA Section 313 Form R *de minimis* concentration reporting level: 1.0%.

Canada, WHMIS, Ingredients Disclosure List Concentration 0.1%.

European/International Regulations: Hazard Symbol: Xi; Risk phrases: R22; R37/38; R41; R42/43; Safety phrases: S2; S23; S24/25; S26; S37/39; S46 (see Appendix 4).

WGK (German Aquatic Hazard Class): 1—Low hazard to waters.

Description: Phthalic Anhydride is moderately flammable, white solid (flake) or a clear, colorless, mobile liquid (molten) with a characteristic, acrid, choking odor. The odor threshold is 0.05 ppm. Molecular weight = 148.12 (flake); 1.20 (molten); Boiling point = (sublimes) 295°C; Freezing/Melting point = 130.6°C; Vapor pressure = 0.0015 mmHg at 20°C; Flash point = 151.7°C (cc); Autoignition temperature = 570°C. Explosive limits: LEL = 1.7%;

UEL = 10.5%. Hazard Identification (based on NFPA-704 M Rating System): Health 3, Flammability 1, Reactivity 0. Slightly soluble in water; slow reaction.

Potential Exposure: Compound Description: Tumorigen, Mutagen; Reproductive Effector; Primary Irritant. Phthalic anhydride is used in plasticizers, in the manufacture of phthaleins, benzoic acid, alkyd and polyester resins, synthetic indigo, and phthalic acid; which is used as a plasticizer for vinyl resins. To a lesser extent, it is used in the production of alizarin, dye, anthranilic acid, anthraquinone, diethyl phthalate, dimethyl phthalate, erythrosine, isophthalic acid, methylaniline, phenolphthalein, phthalamide, sulfathalidine, and terephthalic acid. It has also found uses as a pesticide intermediate.

Incompatibilities: Dust forms an explosive mixture with air. Strong acids, caustics, ammonia, amines, strong oxidizers, water. Converted to phthalic acid in hot water. Reacts violently with copper oxide or sodium nitrite + heat.

Permissible Exposure Limits in Air

Conversion factor: 1 ppm = 6.06 mg/m³ at 25°C & 1 atm.

OSHA PEL: 2 ppm/12 mg/m³ TWA.

NIOSH REL: 1 ppm/6 mg/m³ TWA.

ACGIH TLV®[1]: 1 ppm/6.1 mg/m³ TWA; danger of sensitization; not classifiable as a human carcinogen.

NIOSH IDLH: 60 mg/m³.

Protective Action Criteria (PAC)

TEEL-0: 12 mg/m³

PAC-1: 12 mg/m³

PAC-2: 12 mg/m³

PAC-3: 60 mg/m³

DFG MAK: Danger of skin sensitization; No numerical value established. Data may be available.

Australia: TWA 1 ppm (6 mg/m³), 1993; Austria: MAK 1 mg/m³, 1999; Belgium: TWA 1 ppm (6.1 mg/m³), 1993; Denmark: TWA 2 mg/m³, 1999; Finland: TWA 0.2 mg/m³, 1999; France: VLE 6 mg/m³, 1999; Hungary: TWA 1 mg/m³; STEL 2 mg/m³, 1993; the Netherlands: MAC-TGG 1 mg/m³, 2003; Norway: TWA 2 mg/m³, 1999; the Philippines: TWA 2 ppm (12 mg/m³), 1993; Poland: MAC (TWA) 1 mg/m³; STEL 2 mg/m³ (vapors and aerosols), 1999; Russia: STEL 1 mg/m³, [skin], 1993; Sweden: NGV 2 mg/m³, TGV 3 mg/m³, 1999; Switzerland: MAK-W 1 mg/m³, KZG-W 2 mg/m³, 1999; United Kingdom: TWA 4 mg/m³; STEL 12 mg/m³, 2000; Argentina, Bulgaria, Columbia, Jordan, South Korea, New Zealand, Singapore, Vietnam: ACGIH TLV®: not classifiable as a human carcinogen. Russia[43] set a MAC of 0.1 mg/m³ for ambient air in residential areas (100 µg/m³). Several states have set guidelines or standards for phthalic anhydride in ambient air[60] ranging from 0.82 µg/m³ (Massachusetts) to 60–240 µg/m³ (North Dakota) to 100 µg/m³ (Virginia) to 120 µg/m³ (Connecticut) to 143 µg/m³ (Nevada).

Determination in Air: Use NIOSH II(3), Method #S179; OSHA Analytical Method 90.

Determination in Water: Octanol–water coefficient: Log K_{ow} = 1.58.

Routes of Entry: Inhalation, ingestion, skin and/or eye contact.

Harmful Effects and Symptoms

Short Term Exposure: Irritates the eyes, skin, and respiratory tract. *Inhalation:* May cause irritation of nose, throat, and mouth with coughing, sneezing, shortness of breath, and excessive discharge and bleeding from nose. Studies suggest that this will occur at about 4 ppm. *Skin:* Rapid chemical burns may occur on contact with wet skin. Molten material may cause severe burns unless removed immediately. *Eyes:* May cause severe irritation and chemical burns on contact or at dust levels above 5 ppm. *Ingestion:* May cause severe irritation to mouth and throat. Animal studies suggest that death may occur from ingestion of 4–8 oz.

Long Term Exposure: May cause irritation of nose, mouth, throat, and lungs. Repeated or prolonged contact may cause conjunctivitis, nasal ulcer bleeding. Allergy may develop in sensitive individuals which can lead to bronchial asthma. Repeated or prolonged skin contact may cause dermatitis, skin sensitization, and allergy. In animals: liver, kidney damage.

Points of Attack: Eyes, skin, respiratory system, liver, kidneys.

Medical Surveillance: NIOSH lists the following tests: chest X-ray; pulmonary function tests: forced vital capacity, forced expiratory volume (1 s); urine (chemical/metabolite). Lung function tests may be normal if person is not having an attack at the time. If symptoms develop or overexposure is suspected, the following may be useful: evaluation by a qualified allergist, including careful exposure history and special testing, may help diagnose skin allergy.

First Aid: If this chemical gets into the eyes, remove any contact lenses at once and irrigate immediately for at least 15 min, occasionally lifting upper and lower lids. Seek medical attention immediately. If this chemical contacts the skin, remove contaminated clothing and wash immediately with soap and water. Seek medical attention immediately. If this chemical has been inhaled, remove from exposure, begin rescue breathing (using universal precautions, including resuscitation mask) if breathing has stopped and CPR if heart action has stopped. Transfer promptly to a medical facility. When this chemical has been swallowed, get medical attention. If victim is *conscious,* administer water or milk. Do not induce vomiting.

Personal Protective Methods: Wear protective gloves and clothing to prevent any reasonable probability of skin contact. Safety equipment suppliers/manufacturers can provide recommendations on the most protective glove/clothing material for your operation. All protective clothing (suits, gloves, footwear, headgear) should be clean, available each day, and put on before work. Contact lenses should not be worn when working with this chemical. Wear splash-proof chemical goggles and face shield when working with liquid, or wear dust-proof goggles when working with powders or dust unless full face-piece respiratory protection is worn.

Employees should wash immediately with soap when skin is wet or contaminated. Provide emergency showers and eyewash.

Respirator Selection: NIOSH: *30 mg/m³:* Qm (APF = 25) (any quarter-mask respirator). *60 mg/m³:* 95 XQ [any particulate respirator equipped with an N95, R95, or P95 filter (including N95, R95, and P95 filtering face-pieces) except quarter-mask respirators. The following filters may also be used: N99, R99, P99, N100, R100, P100] or 95F (APF = 10) (any air-purifying full-face-piece respirator equipped with an N95, R95, or P95 filter. The following filters may also be used: N99, R99, P99, N100, R100, P100) or PaprHie (APF = 25)* (any powered air-purifying respirator with a high-efficiency particulate filter) or SA* (any supplied-air respirator) or SCBAF (APF = 50) (any self-contained breathing apparatus with a full face-piece). *Emergency or planned entry into unknown concentrations or IDLH conditions:* SCBAF: Pd,Pp (APF = 10,000) (any self-contained breathing apparatus that has a full face-piece and is operated in a pressure-demand or other positive-pressure mode) or SaF: Pd,Pp: ASCBA (APF = 10,000) (any supplied-air respirator that has a full face-piece and is operated in a pressure-demand or other positive-pressure mode in combination with an auxiliary self-contained breathing apparatus operated in a pressure-demand or other positive-pressure mode). *Escape:* 100F (APF = 50) (any air-purifying, full-face-piece respirator with an N100, R100, or P100 filter) or SCBAE (any appropriate escape-type, self-contained breathing apparatus).

*Substance reported to cause eye irritation or damage; may require eye protection.

Storage: Color Code—White Stripe (*flake*): Contact Hazard; not compatible with materials in solid white category. Color Code—White: Corrosive or Contact Hazard; Store separately in a corrosion-resistant location. Prior to working with this chemical you should be trained on its proper handling and storage. Before entering confined space where this chemical may be present, check to make sure that an explosive concentration is not a danger. Phthalic anhydride must be stored to avoid contact with strong oxidizers (such as chlorine and bromine) since violent reactions occur. Sources of ignition (such as smoking and open flames) are prohibited where phthalic anhydride is used, handled, or stored in a manner that could create a potential fire or explosion hazard.

Shipping: This compound requires a shipping label of "CORROSIVE." It falls in Hazard Class 8 and Packing Group III.

Spill Handling: Evacuate persons not wearing protective equipment from area of spill or leak until cleanup is complete. Remove all ignition sources. Establish forced ventilation to keep levels below explosive limit. Collect powdered material in the most convenient and safe manner and deposit in sealed containers. Ventilate area after cleanup is complete. It may be necessary to contain and dispose of this chemical as a hazardous waste. If material or contaminated runoff enters waterways, notify downstream users of potentially contaminated waters. Contact your local or federal environmental protection agency for specific recommendations. If employees are required to clean up spills, they must be properly trained and equipped. OSHA 1910.120(q) may be applicable.

Fire Extinguishing: This chemical is a combustible solid. Use dry chemical, carbon dioxide, water spray, or alcohol foam extinguishers. Poisonous gases are produced in fire. If material or contaminated runoff enters waterways, notify downstream users of potentially contaminated waters. Notify local health and fire officials and pollution control agencies. From a secure, explosion-proof location, use water spray to cool exposed containers. If cooling streams are ineffective (venting sound increases in volume and pitch, tank discolors, or shows any signs of deforming), withdraw immediately to a secure position. If employees are expected to fight fires, they must be trained and equipped in OSHA 1910.156. The only respirators recommended for firefighting are self-contained breathing apparatuses that have full face-pieces and are operated in a pressure-demand or other positive-pressure mode.

Disposal Method Suggested: Dissolve or mix the material with a combustible solvent and burn in a chemical incinerator equipped with an afterburner and scrubber. All federal, state, and local environmental regulations must be observed. Consult with environmental regulatory agencies for guidance on acceptable disposal practices. Generators of waste containing this contaminant ($\geq$100 kg/mo) must conform with EPA regulations governing storage, transportation, treatment, and waste disposal.

References
US Environmental Protection Agency. (April 30, 1980). *Phthalic Anhydride, Health and Environmental Effects Profile No. 147.* Washington, DC: Office of Solid Waste New York State Department of Health. (March 1986). *Chemical Fact Sheet: Phthalic Anhydride.* Version 2. Albany, NY: Bureau of Toxic Substance Assessment New Jersey Department of Health and Senior Services. (August 2001). *Hazardous Substances Fact Sheet: Phthalic Anhydride.* Trenton, NJ

m-Phthalodinitrile   P:0680

Molecular Formula: $C_8H_4N_2$
Common Formula: $C_6H_4(CN)_2$
Synonyms: 1,3-Benzenedicarbonitrile;   *m*-Benzenedicarbonitrile;   *m*-Dicyanobenzene;   1,3-Dicyanobenzene;   Isophthalodinitrile;   *m*-PDN; Phthalonitrile, *m*-Dicyanobenzene
CAS Registry Number: 626-17-5
RTECS® Number: CZ1900000
UN/NA & ERG Number: UN3276 (Nitriles, toxic, liquid, n.o.s.)/151

EC Number: 210-933-7
Regulatory Authority and Advisory Bodies
Air Pollutant Standard Set. See below, "Permissible Exposure Limits in Air" section.
Canada, WHMIS, Ingredients Disclosure List Concentration 1.0%.
European/International Regulations: not listed in Annex 1.
WGK (German Aquatic Hazard Class): 1—Low hazard to waters.

Description: *m*-Phthalodinitrile is a needle-like, colorless to white, flaky solid with an almond-like odor. Molecular weight = 128.14; Specific gravity (H_2O:1) = 4.42; Boiling point = sublimes; Freezing/Melting point = 162°C (sublimes); Vapor pressure = 0.01 mmHg at 20°C. Hazard Identification (based on NFPA-704 M Rating System): Health 2, Flammability 0, Reactivity 0. Slightly soluble in water.

Potential Exposure: Compound Description: Drug, Primary Irritant. This material is used as an intermediate in the manufacture of polyurethane paints and varnishes; in pharmaceuticals, for synthetic fibers, agricultural chemicals, rubber chemicals; an intermediate for phthalocyanine pigments and dyes, and for high-temperature lubricants and coatings; it may be used to produce phthalate esters.

Incompatibilities: Combustible solid. The dust is a severe explosion hazard. Strong oxidizers (e.g., chlorine, bromine, fluorine).

Permissible Exposure Limits in Air
OSHA PEL: None.
NIOSH REL: 5 mg/m^3 TWA.
ACGIH TLV®[1]: 5 mg/m^3, measured as inhalable fraction and vapor, TWA.
No TEEL available.
Australia: TWA 5 mg/m^3, 1993; Belgium: TWA 5 mg/m^3, 1993; Denmark: TWA 5 mg/m^3, 1999; Finland: TWA 5 mg/m^3; STEL 20 mg/m^3, [skin], 1999; France: VME 5 mg/m^3, 1999; Norway: TWA 5 mg/m^3, 1999; Switzerland: MAK-W 5 mg/m^3, 1999; the Netherlands: MAC-TGG 5 mg/m^3, 2003; Argentina, Bulgaria, Columbia, Jordan, South Korea, New Zealand, Singapore, Vietnam: ACGIH TLV®: TWA 5 mg/m^3. Several states have set guidelines or standards for *m*-phthalodinitrile in ambient air[60] ranging from 50 µg/m^3 (North Dakota) to 100 µg/m^3 (Connecticut) to 119 µg/m^3 (Nevada).

Determination in Air: No method available.
Routes of Entry: Inhalation, ingestion, skin and/or eye contact. Absorbed by the skin.

Harmful Effects and Symptoms
Short Term Exposure: *m*-Phthalodinitrile can affect you when breathed in and by passing through your skin. High exposure may cause headache, nausea, weakness, confusion and may cause you to pass out. In animals: irritation of the eyes, skin.
Long Term Exposure: May cause headaches and nausea. No permanent effects are known at this time.
Points of Attack: Eyes, skin, central nervous system.

First Aid: If this chemical gets into the eyes, remove any contact lenses at once and irrigate immediately for at least 15 min, occasionally lifting upper and lower lids. Seek medical attention immediately. If this chemical contacts the skin, remove contaminated clothing and wash immediately with soap and water. Seek medical attention immediately. If this chemical has been inhaled, remove from exposure, begin rescue breathing (using universal precautions, including resuscitation mask) if breathing has stopped and CPR if heart action has stopped. Transfer promptly to a medical facility. When this chemical has been swallowed, get medical attention. Give large quantities of water and induce vomiting. Do not make an unconscious person vomit.

Personal Protective Methods: Wear protective gloves and clothing to prevent any reasonable probability of skin contact. Safety equipment suppliers/manufacturers can provide recommendations on the most protective glove/clothing material for your operation. All protective clothing (suits, gloves, footwear, headgear) should be clean, available each day, and put on before work. Contact lenses should not be worn when working with this chemical. Wear dust-proof chemical goggles and face shield unless full face-piece respiratory protection is worn. Employees should wash immediately with soap when skin is wet or contaminated. Provide emergency showers and eyewash. See NIOSH Criteria Document 212 *Nitriles*.

Respirator Selection: Where there is potential for exposures *over 5 mg/m³*, use a NIOSH/MSHA- or European Standard EN149-approved full-face-piece respirator with a high-efficiency particulate filter. Greater protection is provided by a powered air-purifying respirator. *Where there is potential for high exposures*, use a NIOSH/MSHA- or European Standard EN149-approved supplied-air respirator with a full face-piece operated in the positive-pressure mode, or with a full face-piece, hood, or helmet in the continuous-flow mode; or use a MSHA/NIIOSH approved self-contained breathing apparatus with a full face-piece operated in a pressure-demand or other positive-pressure mode.

Storage: Color Code—Blue: Health Hazard/Poison: Store in a secure poison location. Prior to working with this chemical you should be trained on its proper handling and storage. *m*-Phthalodinitrile is incompatible with strong oxidizers, such as chlorine, bromine, and fluorine. Store in tightly closed containers in a cool, well-ventilated area away from open flames and high temperatures. Sources of ignition, such as smoking and open flames, are prohibited where *m*-phthalodinitrile is used, handled, or stored in a manner that could create potential fire of explosion hazard. *m*-Phthalodinitrile dust is a severe explosion hazard. Use explosion proof equipment when handling *m*-phthalodinitrile.

Spill Handling: Evacuate persons not wearing protective equipment from area of spill or leak until cleanup is complete. Remove all ignition sources. Collect powdered material in the most convenient and safe manner and deposit in sealed containers. Ventilate area after cleanup is complete. It may be necessary to contain and dispose of this chemical as a hazardous waste. If material or contaminated runoff enters waterways, notify downstream users of potentially contaminated waters. Contact your local or federal environmental protection agency for specific recommendations. If employees are required to clean up spills, they must be properly trained and equipped. OSHA 1910.120(q) may be applicable.

Initial isolation and protective action distances: Distances shown are likely to be affected during the first 30 min after materials are spilled and could increase with time. If more than one tank car, cargo tank, portable tank, or large cylinder involved in the incident is leaking, the protective action distance may need to be increased. You may need to seek emergency information from CHEMTREC at (800) 424-9300 or seek professional environmental engineering assistance from the US EPA Environmental Response Team at (908) 548-8730 (24-h response line).

Small spills (From a small package or a small leak from a large package)
First: Isolate in all directions (feet/meters)   100/30
Then: Protect persons downwind (miles/kilometers)
Day   0.1/0.1
Night   0.1/0.2

Large spills (From a large package or from many small packages)
First: Isolate in all directions (feet/meters)   200/60
Then: Protect persons downwind (miles/kilometers)
Day   0.3/0.5
Night   0.5/0.9

Fire Extinguishing: This chemical is a combustible solid. Use dry chemical, carbon dioxide, water spray, or alcohol foam extinguishers. Poisonous gases are produced in fire, including ammonia, hydrogen cyanide, and nitrogen oxides. If material or contaminated runoff enters waterways, notify downstream users of potentially contaminated waters. Notify local health and fire officials and pollution control agencies. From a secure, explosion-proof location, use water spray to cool exposed containers. If cooling streams are ineffective (venting sound increases in volume and pitch, tank discolors, or shows any signs of deforming), withdraw immediately to a secure position. If employees are expected to fight fires, they must be trained and equipped in OSHA 1910.156. The only respirators recommended for firefighting are self-contained breathing apparatuses that have full face-pieces and are operated in a pressure-demand or other positive-pressure mode.

Disposal Method Suggested: React with alcoholic NaOH; after 1 h, evaporate alcohol and add calcium hypochlorite; after 24 h flush into sewer with large volumes of water.

Reference
New Jersey Department of Health and Senior Services. (November 2000). *Hazardous Substances Fact Sheet: m-Phthalodinitrile*. Trenton, NJ

Phylloquinone   P:0690

Molecular Formula: $C_{31}H_{46}O_2$

Synonyms: Antihemorrhagic vitamin; Aqua mephyton; Combinal K_1; Kativ N; Kephton; Kinadion; Konakion; Mephyton; 2-Methyl-3-phythyl-1,4-naphthochinon (German); 2-Methyl-3-(3,7,11,15-tetramethyl-2-hexadecenyl)-1,4-naphthalenedione; Monodion; Phyllochinon (German); α-Phylloquinone; *trans*-Phylloquinone; Phytomenadione; Phytonadione; Vitamin K_1

CAS Registry Number: 84-80-0

RTECS® Number: QJ5800000

EC Number: 201-564-2

Regulatory Authority and Advisory Bodies

Superfund/EPCRA 40CFR355, Extremely Hazardous Substances: Dropped From Listing in 1988).

European/International Regulations: not listed in Annex 1.

WGK (German Aquatic Hazard Class): No value assigned.

Description: Phylloquinone is an odorless yellow viscous oil or crystals. Molecular weight = 450.77; Boiling point = 140−145°C at 0.001 mm; Freezing/Melting point = −4°C; −20° C. Hazard Identification (based on NFPA-704 M Rating System): Health 1, Flammability 0, Reactivity 0. Insoluble in water.

Potential Exposure: Phylloquinone is a dietary component essential for normal biosynthesis of several factors required for clotting of blood, as a therapeutic drug used to correct bleeding tendency, and as a food supplement.

Incompatibilities: Phylloquinone decomposes in sunlight and is destroyed by alkali hydroxides and reducing agents.

Permissible Exposure Limits in Air

No standards or TEEL available.

Routes of Entry: Inhalation, ingestion, skin and/or eye contact.

Harmful Effects and Symptoms

Short Term Exposure: High oral toxicity. Intravenous injection can cause toxic responses and occasionally death. Rapid intravenous administration of phylloquinone has produced flushing, irregular breathing, and chest pains. In newborns, it can cause hemolytic anemia and hemoglobinuria. In patients who have severe liver disease, administration of large doses of menadione or phylloquinone may further depress function of liver. Individuals resistant to coumarin may have unusual sensitivity to the antidotal effects of vitamin K.

First Aid: If this chemical gets into the eyes, remove any contact lenses at once and irrigate immediately for at least 15 min, occasionally lifting upper and lower lids. Seek medical attention immediately. If this chemical contacts the skin, remove contaminated clothing and wash immediately with soap and water. Seek medical attention immediately. If this chemical has been inhaled, remove from exposure, begin rescue breathing (using universal precautions, including resuscitation mask) if breathing has stopped and CPR if heart action has stopped. Transfer promptly to a medical facility. When this chemical has been swallowed, get medical attention. Give large quantities of water and induce vomiting. Do not make an unconscious person vomit. Keep victim quiet and maintain normal body temperature.

Personal Protective Methods: Wear protective gloves and clothing to prevent any reasonable probability of skin contact. Safety equipment suppliers/manufacturers can provide recommendations on the most protective glove/clothing material for your operation. All protective clothing (suits, gloves, footwear, headgear) should be clean, available each day, and put on before work. Contact lenses should not be worn when working with this chemical. Wear splash-proof chemical goggles and face shield unless full face-piece respiratory protection is worn. Employees should wash immediately with soap when skin is wet or contaminated. Provide emergency showers and eyewash.

Respirator Selection: Follow regulations in OSHA 29CFR1910.134 or European Standard EN149. Use a NIOSH/MSHA- or European Standard EN149-approved respirator; or use an approved supplied-air respirator with a full face-piece operated in the positive-pressure mode, or with a full face-piece, hood, or helmet in the continuous-flow mode; or use a NIOSH/MSHA- or European Standard EN149-approved self-contained breathing apparatus with a full face-piece operated in a pressure-demand or other positive-pressure mode.

Storage: Color Code—Green: General storage may be used. Prior to working with this chemical you should be trained on its proper handling and storage. Store in tightly closed containers in a cool, well-ventilated area away from alkali hydroxides and reducing agents. Where possible, automatically pump liquid from drums or other storage containers to process containers. Sources of ignition, such as smoking and open flames, are prohibited where this chemical is handled, used, or stored. Metal containers involving the transfer of 5 gallons or more of this chemical should be grounded and bonded. Drums must be equipped with self-closing valves, pressure vacuum bungs, and flame arresters. Use only nonsparking tools and equipment, especially when opening and closing containers of this chemical. Wherever this chemical is used, handled, manufactured, or stored, use explosion-proof electrical equipment and fittings.

Spill Handling: Keep unnecessary people away; isolate hazard area and deny entry. Stay upwind; keep out of low areas. Wear self-contained (positive pressure if available) breathing apparatus and full protective clothing. Shut off ignition sources; no flares, smoking, or flames in hazard area. Keep combustibles (wood, paper, oil, etc.) away from spilled material. Do not touch spilled material. *Small spills:* absorb with sand or other noncombustible absorbent material and place into containers for later disposal. *Large spills:* dike far ahead of spill for later disposal. Ventilate area of spill or leak. Absorb liquids in vermiculite, dry sand, earth, peat, carbon, or a similar material and deposit in sealed containers. Keep this chemical out of a confined space, such as a sewer, because of the possibility of an explosion, unless the sewer is designed to prevent the buildup of explosive concentrations. It may be necessary to contain and dispose of this chemical as a hazardous waste. If material or contaminated runoff enters waterways, notify downstream users of potentially contaminated waters. Contact your local or federal environmental protection agency for specific recommendations. If employees are required to clean up spills, they must be properly trained and equipped. OSHA 1910.120(q) may be applicable.

Fire Extinguishing: Irritating fumes are produced in fire. Use dry chemical, carbon dioxide, or alcohol foam extinguishers. Containers may explode in fire. Storage containers and parts of containers may rocket great distances, in many directions. If material or contaminated runoff enters waterways, notify downstream users of potentially contaminated waters. Notify local health and fire officials and pollution control agencies. From a secure, explosion-proof location, use water spray to cool exposed containers. If cooling streams are ineffective (venting sound increases in volume and pitch, tank discolors, or shows any signs of deforming), withdraw immediately to a secure position. If employees are expected to fight fires, they must be trained and equipped in OSHA 1910.156. The only respirators recommended for firefighting are self-contained breathing apparatuses that have full face-pieces and are operated in a pressure-demand or other positive-pressure mode.

Reference
US Environmental Protection Agency. (October 31, 1985). *Chemical Hazard Information Profile: Phylloquinone.* Washington, DC: Chemical Emergency Preparedness Program

Physostigmine   P:0700

Molecular Formula: $C_{15}H_{21}N_3O_2$
Synonyms: Calabarine; Erserine; Eserine; Eserolein; Fisostigmina (Spanish); Methylcarbamate (ester); Methylcarbamic acid, ester with eseroline; Physostol
CAS Registry Number: 57-47-6
RTECS® Number: TJ2100000
UN/NA & ERG Number: UN2757(carbamate pesticides, solid, toxic)/151 see "Potential Exposure"
EC Number: 200-332-8 [*Annex I Index No.:* 614-020-00-8]
Regulatory Authority and Advisory Bodies
US EPA Hazardous Waste Number (RCRA No.): P204.
RCRA, 40CFR261, Appendix 8 Hazardous Constituents.
RCRA 40CFR268.48; 61FR15654, Universal Treatment Standards: Wastewater (mg/L), 0.056; Nonwastewater (mg/kg), 1.4.
Superfund/EPCRA 40CFR355, Extremely Hazardous Substances: TPQ = 100/10,000 lb (45.4/4540 kg).
Reportable Quantity (RQ): 100 lb (45.4 kg).
European/International Regulations: Hazard Symbol: T+, N; Risk phrases: R26/28; R1/2; Safety phrases: S25; S45 (see Appendix 4).

WGK (German Aquatic Hazard Class): 3—Severe hazard to waters.
Description: Physostigmine is an odorless white crystalline solid. Molecular weight = 275.39; Freezing/Melting point = 86−87°C (unstable form); 105−106°C (stable form). Hazard Identification (based on NFPA-704 M Rating System): Health 3, Flammability 1, Reactivity 1. Slightly soluble in water.
Potential Exposure: Material is used as a cholinergic (anticholinesterase) agent and as a veterinary medication. Listed as a carbamate pesticide; however, physostigmine is not registered in the United States as such.
Incompatibilities: Light and heat.
Permissible Exposure Limits in Air
Protective Action Criteria (PAC)
TEEL-0: 0.75 mg/m^3
PAC-1: 2.5 mg/m^3
PAC-2: 4.5 mg/m^3
PAC-3: 4.5 mg/m^3
Routes of Entry: Inhalation, ingestion, skin and/or eye contact.
Harmful Effects and Symptoms
Short Term Exposure: Super toxic. Probable oral lethal dose is less than 5 mg/kg for a 70-kg (150 lb) person. Material is a cholinesterase inhibitor. Effects of exposure may involve the respiratory, gastrointestinal, cardiovascular, and central nervous systems. Death occurs due to respiratory paralysis or impaired cardiac function. Time to death may vary from 5 min to 24 h, in severely poisoned patients, depending on factors, such as the dose and route. General symptoms include: increased secretions, fatigability and generalized weakness; involuntary twitching; severe weakness of skeletal muscles. Symptoms of exposure to material by major organ system: gastrointestinal: lack of appetite, nausea and vomiting, abdominal cramps, and diarrhea. Central nervous system: confusion, uncoordination, slurred speech, loss of reflexes, rapid irregular breathing, generalized convulsions, and coma. Cardiovascular: slowed heartbeat resulting in hypotension and fall in cardiac output.
Long Term Exposure: Cholinesterase inhibitor; cumulative effect is possible. This chemical may damage the nervous system with repeated exposure, resulting in convulsions, respiratory failure. May cause liver damage.
Points of Attack: Respiratory system, lungs, central nervous system, cardiovascular system, skin, eyes, plasma and red blood cell cholinesterase.
Medical Surveillance: Before employment and at regular times after that, the following are recommended: plasma and red blood cell cholinesterase levels (tests for the enzyme poisoned by this chemical). If exposure stops, plasma levels return to normal in 1−2 weeks while red blood cell levels may be reduced for 1−3 months. When cholinesterase enzyme levels are reduced by 25% or more below preemployment levels, risk of poisoning is increased, even if results are in lower ranges of "normal." Reassignment to work not involving carbamate pesticides is

recommended until enzyme levels recover. If symptoms develop or overexposure occurs, repeat the above tests as soon as possible and get an examination of the nervous system. Also, consider complete blood count. Consider chest X-ray following acute overexposure.

First Aid: If this chemical gets into the eyes, remove any contact lenses at once and irrigate immediately for at least 15 min, occasionally lifting upper and lower lids. Seek medical attention immediately. If this chemical contacts the skin, remove contaminated clothing and wash immediately with soap and water. Seek medical attention immediately. If this chemical has been inhaled, remove from exposure, begin rescue breathing (using universal precautions, including resuscitation mask) if breathing has stopped and CPR if heart action has stopped. Transfer promptly to a medical facility. When this chemical has been swallowed, get medical attention. Give large quantities of water and induce vomiting. Do not make an unconscious person vomit.

Personal Protective Methods: Wear protective gloves and clothing to prevent any reasonable probability of skin contact. Safety equipment suppliers/manufacturers can provide recommendations on the most protective glove/clothing material for your operation. All protective clothing (suits, gloves, footwear, headgear) should be clean, available each day, and put on before work. Contact lenses should not be worn when working with this chemical. Wear splash-proof chemical goggles and face shield unless full face-piece respiratory protection is worn. Employees should wash immediately with soap when skin is wet or contaminated. Provide emergency showers and eyewash.

Respirator Selection: Follow regulations in OSHA 29CFR1910.134 or European Standard EN149. Use a NIOSH/MSHA- or European Standard EN149-approved respirator; or use an approved supplied-air respirator with a full face-piece operated in the positive-pressure mode, or with a full face-piece, hood, or helmet in the continuous-flow mode; or use a NIOSH/MSHA- or European Standard EN149-approved self-contained breathing apparatus with a full face-piece operated in a pressure-demand or other positive-pressure mode.

Storage: Color Code—Blue: Health Hazard/Poison: Store in a secure poison location. Prior to working with this chemical you should be trained on its proper handling and storage. Store in tightly closed containers in a cool, well-ventilated area.

Shipping: Carbamate pesticides, solid, toxic, require a shipping label of "POISONOUS/TOXIC MATERIALS." It falls in Hazard Class 6.1 and Packing Group I.

Spill Handling: Evacuate persons not wearing protective equipment from area of spill or leak until cleanup is complete. As for other carbamate pesticides, avoid breathing dusts and fumes from burning materials. Keep upwind. Avoid bodily contact with the material. Wash away any material which may have contacted the body with copious amounts of water or soap and water. Remove all ignition sources. Collect powdered material in the most convenient and safe manner and deposit in sealed containers. Ventilate area after cleanup is complete. It may be necessary to contain and dispose of this chemical as a hazardous waste. If material or contaminated runoff enters waterways, notify downstream users of potentially contaminated waters. Contact your local or federal environmental protection agency for specific recommendations. If employees are required to clean up spills, they must be properly trained and equipped. OSHA 1910.120(q) may be applicable.

Fire Extinguishing: As for other carbamate pesticides, extinguish fire using agent suitable for type of surrounding fire (material itself burns with difficulty). Use water in flooding quantities as fog. Use alcohol foam, carbon dioxide, or dry chemical. Wear self-contained breathing apparatus when fighting fires. Poisonous gases are produced in fire, including nitrogen oxides. If material or contaminated runoff enters waterways, notify downstream users of potentially contaminated waters. Notify local health and fire officials and pollution control agencies. From a secure, explosion-proof location, use water spray to cool exposed containers. If cooling streams are ineffective (venting sound increases in volume and pitch, tank discolors, or shows any signs of deforming), withdraw immediately to a secure position. If employees are expected to fight fires, they must be trained and equipped in OSHA 1910.156. The only respirators recommended for firefighting are self-contained breathing apparatuses that have full face-pieces and are operated in a pressure-demand or other positive-pressure mode.

Disposal Method Suggested: In accordance with 40CFR165, follow recommendations for the disposal of pesticides and pesticide containers. Must be disposed properly by following package label directions or by contacting your local or federal environmental control agency or by contacting your regional EPA office.

Reference
US Environmental Protection Agency. (November 30, 1987). *Chemical Hazard Information Profile: Physostigmine.* Washington, DC: Chemical Emergency Preparedness Program

Picloram   P:0710

Molecular Formula: $C_6H_3Cl_3N_2O_2$
Synonyms: Amdon; Amdon grazon; 4-Aminotrichloropicolinic acid; 4-Amino-3,5,6-trichloro-2-picolinic acid; 4-Amino-3,5,6-trichloropicolinic acid; 4-Amino-3,5,6-trichloro-2-pyridinecarboxylic acid; 4-Amino-3,5,6-trichloropyridine-2-carboxylic acid; 4-Amino-3,5,6-trichlorpicolinsaeure (German); ATCP; Borolin; K-Pin; NCI-C00237; NSC 233899; Picolinic acid, 4-Amino-3,5,6-trichloro-; 2-Pyridine carboxylic acid, 4-amino-3,5,6-trichloro-; Tordon; Tordon 10K; Tordon 22K; Tordon 101 mixture; 3,5,6-Trichloro-4-aminopicolinic acid
CAS Registry Number: 1918-02-1

RTECS® Number: TJ7525000
UN/NA & ERG Number: UN2588/151
EC Number: 217-636-1

Regulatory Authority and Advisory Bodies

Carcinogenicity: NCI: Carcinogenesis Bioassay (feed); equivocal evidence: rat; no evidence: mouse; IARC: Animal Limited Evidence; Human No Available Data, *not classifiable as carcinogenic to humans,* Group 3, 1991.

US EPA, FIFRA 1998 Status of Pesticides: RED completed. Banned or Severely Restricted (Sweden) (UN).[13]

Air Pollutant Standard Set. See below, "Permissible Exposure Limits in Air" section.

Safe Drinking Water Act: MCL, 0.5 mg/L; MCLG, 0.5 mg/L; Regulated chemical (47 FR 9352) as pichloram.

EPCRA Section 313 Form R *de minimis* concentration reporting level: 1.0%.

WGK (German Aquatic Hazard Class): No value assigned.

Description: Picloram is a colorless powder with a chlorine-like odor. Molecular weight = 241.46; Freezing/Melting point = (decomposes) 219.8°C; Vapor pressure = 6×10^{-7} mmHg at 36.7°C. Poor solubility in water.

Potential Exposure: Compound Description: Agricultural Chemical; Tumorigen, Mutagen; Reproductive Effector. Those involved in the manufacture, formulation, or application of the herbicide.

Incompatibilities: This material is acidic. Reacts with hot concentrated alkali (hydrolyzes), strong bases. Attacks some metals.

Permissible Exposure Limits in Air

OSHA PEL: 15 mg/m^3 TWA, total dust; 5 mg/m^3 TWA, respirable fraction.

NIOSH REL: See Appendix D of the *NIOSH Pocket Guide.*
ACGIH TLV®[1]: 10 mg/m^3 TWA; not classifiable as a human carcinogen.

No TEEL available.

Australia: TWA 10 mg/m^3; STEL 20 mg/m^3, 1993; Belgium: TWA 10 mg/m^3; STEL 20 mg/m^3, 1993; Denmark: TWA 10 mg/m^3, 1999; Finland: TWA 10 mg/m^3; STEL 20 mg/m^3, 1999; France: VME 10 mg/m^3, 1999; Russia: STEL 2 mg/m^3, 1993; Switzerland: MAK-W 10 mg/m^3, 1999; United Kingdom: TWA 10 mg/m^3; STEL 20 mg/m^3, 2000; the Netherlands: MAC-TGG 10 mg/m^3, 2003; Argentina, Bulgaria, Columbia, Jordan, South Korea, New Zealand, Singapore, Vietnam: ACGIH TLV®: not classifiable as a human carcinogen.

Russia set a MAC of 0.03 mg/m^3 for ambient air on a once-daily basis and 0.02 mg/m^3 on a daily average basis. Several states have set guidelines or standards for picloram in ambient air[60] ranging from 0.1–0.2 mg/m^3 (North Dakota) to 0.16 mg/m^3 (Virginia) to 0.2 mg/m^3 (Connecticut) to 0.238 mg/m^3 (Nevada).

Determination in Air: Filter; none; Gravimetric; NIOSH IV, Particulates NOR: Method #0500, total dust; Method #0600 (respirable dust).

Permissible Concentration in Water: The US EPA has set a lifetime health advisory of 0.49 mg/L. Russia set a MAC of 10.0 mg/L in surface water. States which have set guidelines for Picloram in drinking water[61] include Kansas at 0.175 mg/L and Maine at 0.3 mg/L.

Determination in Water: Fish Tox = 703.55815000 ppb (VERY LOW).

Routes of Entry: Inhalation, ingestion, skin and/or eye contact.

Harmful Effects and Symptoms

Short Term Exposure: Irritates the eyes, skin, and respiratory tract. Exposure can cause nausea.

Long Term Exposure: Picloram should be handled as a carcinogen—with extreme caution. It may damage the testes. May affect the kidneys and liver. In animals: liver, kidney changes. Human Tox = 500.00000 ppb (VERY LOW).

Points of Attack: Eyes, skin, respiratory system, liver, kidneys. Cancer site in animals: liver, uterus, pituitary gland.

Medical Surveillance: Liver and kidney function tests.

First Aid: If this chemical gets into the eyes, remove any contact lenses at once and irrigate immediately for at least 15 min, occasionally lifting upper and lower lids. Seek medical attention immediately. If this chemical contacts the skin, remove contaminated clothing and wash immediately with soap and water. Seek medical attention immediately. If this chemical has been inhaled, remove from exposure, begin rescue breathing (using universal precautions, including resuscitation mask) if breathing has stopped and CPR if heart action has stopped. Transfer promptly to a medical facility. When this chemical has been swallowed, get medical attention. Give large quantities of water and induce vomiting. Do not make an unconscious person vomit.

Personal Protective Methods: Wear protective gloves and clothing to prevent any reasonable probability of skin contact. Safety equipment suppliers/manufacturers can provide recommendations on the most protective glove/clothing material for your operation. All protective clothing (suits, gloves, footwear, headgear) should be clean, available each day, and put on before work. Contact lenses should not be worn when working with this chemical. Wear dust-proof chemical goggles and face shield unless full face-piece respiratory protection is worn. Employees should wash immediately with soap when skin is wet or contaminated. Provide emergency showers and eyewash.

Respirator Selection: Where there is potential for exposures over *10 mg/m^3*, use a NIOSH/MSHA- or European Standard EN149-approved supplied-air respirator with a full face-piece operated in the positive-pressure mode, or with a full face-piece, hood, or helmet in the continuous-flow mode; or use a NIOSH/MSHA- or European Standard EN149-approved self-contained breathing apparatus with a full face-piece operated in a pressure-demand or other positive-pressure mode.

Storage: Color Code—Blue: Health Hazard/Poison: Store in a secure poison location. Prior to working with this chemical you should be trained on its proper handling and

storage. Store in tightly closed containers in a cool, well-ventilated area away from acids, bases, and metals. A regulated, marked area should be established where this chemical is handled, used, or stored. The use of picloram has been restricted. Be sure that your operation follows regulations. A regulated, marked area should be established where this chemical is handled, used, or stored in compliance with OSHA Standard 1910.1045.

Shipping: Pesticides, solid, toxic, n.o.s. require a shipping label of "POISONOUS/TOXIC MATERIALS." It falls in Hazard Class 6.1 and Packing Group III.

Spill Handling: Evacuate persons not wearing protective equipment from area of spill or leak until cleanup is complete. Remove all ignition sources. Collect powdered material in the most convenient and safe manner and deposit in sealed containers. Ventilate area after cleanup is complete. It may be necessary to contain and dispose of this chemical as a hazardous waste. If material or contaminated runoff enters waterways, notify downstream users of potentially contaminated waters. Contact your local or federal environmental protection agency for specific recommendations. If employees are required to clean up spills, they must be properly trained and equipped. OSHA 1910.120(q) may be applicable.

Fire Extinguishing: This chemical is a combustible solid. Use dry chemical, carbon dioxide, water spray, or alcohol foam extinguishers. Poisonous gases are produced in fire, including hydrogen chloride and nitrogen oxides. If material or contaminated runoff enters waterways, notify downstream users of potentially contaminated waters. Notify local health and fire officials and pollution control agencies. From a secure, explosion-proof location, use water spray to cool exposed containers. If cooling streams are ineffective (venting sound increases in volume and pitch, tank discolors, or shows any signs of deforming), withdraw immediately to a secure position. If employees are expected to fight fires, they must be trained and equipped in OSHA 1910.156. The only respirators recommended for firefighting are self-contained breathing apparatuses that have full face-pieces and are operated in a pressure-demand or other positive-pressure mode.

Disposal Method Suggested: This chlorinated brush killer is usually formulated with 2,4-D and the disposal problems are similar. Incineration at 1000°C for 2 s is required for thermal decomposition. Alternatively, the free acid can be precipitated from its solutions by addition of a mineral acid. The concentrated acid can then be incinerated and the dilute residual solution disposed in an area where several years' persistence in the soil can be tolerated.

References

US Environmental Protection Agency. (August 1987). *Health Advisory: Picloram.* Washington, DC: Office of Drinking Water

US Environmental Protection Agency, Special Review and Reregistration Division Office of Pesticide Programs. (1998). *Agency Status of Pesticides in Registration, Reregistration, and Special Review* (Rainbow Report). Washington, DC

New Jersey Department of Health and Senior Services. (May 2001). *Hazardous Substances Fact Sheet: Picloram.* Trenton, NJ

Picolines   P:0720

Molecular Formula: C_6H_7N

Synonyms: 2-Picoline (o-isomer): AI3-2409; AI3-24109; α-Methylpyridine; 2-Methylpyridine; Metilpiridina (Spanish); NSC 3409; α-Picoline; *o*-Picoline; 2-Picoline; Picoline; Pyridine, 2-methyl-; Pyridine, methyl-

3-Picoline   (m-isomer): β-Picoline;   *m*-Picoline; β-Methylpyridine; *m*-Methylpyridine; 3-Methylpyridine; Pyridine, 3-methyl

4-Picoline   (p-isomer): γ-Picoline;   *p*-Picoline; γ-Methylpyridine; *p*-Methylpyridine; 4-Methylpyridine; Pyridine, 4-methyl

CAS Registry Number: 109-06-8 (2-Picoline); 108-99-6 (3-Picoline); 108-89-4 (4-Picoline); 1333-41-1 (mixed isomers)

RTECS® Number: TJ4900000 (2-Picoline); TJ5000000 (3-Picoline); UT5425000 (4-Picoline)

UN/NA & ERG Number: UN2313/129

EC Number: 203-643-7 [*Annex I Index No.:* 613-036-00-2] (2-Picoline); 203-636-9 (3-Picoline); 203-626-4 [*Annex I Index No.:* 613-037-00-8] (4-Picoline); 215-588-6 (methylpyridine or mixed isomers)

Regulatory Authority and Advisory Bodies
(2-Picoline)

Air Pollutant Standard Set. See below, "Permissible Exposure Limits in Air" section.

o-isomer:
US EPA Hazardous Waste Number (RCRA No.): U191.
RCRA, 40CFR261, Appendix 8 Hazardous Constituents.
Reportable Quantity (RQ): 5000 lb (2270 kg).
EPCRA Section 313 Form R *de minimis* concentration reporting level: 1.0%.
Canada, WHMIS, Ingredients Disclosure List Concentration 1.0%.
European/International Regulations (*2-picoline*): Hazard Symbol: Xn; Risk phrases: R10; R20/21/22; R36/37; Safety phrases: S2; S26; S36; (*4-picoline*) Hazard Symbol: T; Risk phrases: R10; R20/22; R24; R36/37/38; Safety phrases: S1/2; S26; S36; S45 (see Appendix 4).
WGK (German Aquatic Hazard Class): 1—Low hazard to waters (*all isomers*).

Description: Picolines are colorless liquids with a strong, unpleasant, pyridine-like odor. Odor threshold is 0.023 ppm. "Picoline" is often used as mixed isomers and physical data for the other isomers are listed. The *o*-isomer is the most heavily regulated (see above). Molecular weight = 93.14; Specific gravity ($H_2O:1$) = 0.96 at 25°C;

Boiling point = 129°C (*o*-isomer); 143–144°C (*m*-isomer); 145°C (*p*-isomer); Freezing/Melting point = −70°C (*o*-isomer); −18°C (*m*-isomer); 3.7°C (*p*-isomer); Flash point = 39°C (oc) (*o*-isomer); 37.8°C (cc) (*m*-isomer); 57°C (*p*-isomer); Autoignition temperature = 535°C (*o*-isomer). Explosive limits (*o*-isomer): LEL = 1.4%; UEL = 8.6%. Hazard Identification (based on NFPA-704 M Rating System): (*2-isomer*) Health 3, Flammability 2, Reactivity 0; (*3- and 4-isomer*s) Health 2, Flammability 2, Reactivity 0. Soluble in water.

Potential Exposure: Compound Description (*o*-isomer): Mutagen, Primary Irritant; (*m*-isomer): Tumorigen, Primary Irritant. Picolines are used as intermediates in pharmaceutical manufacture, pesticide manufacture, and in the manufacture of dyes and rubber chemicals. It is also used as a solvent.

Incompatibilities: Reacts with oxidants and strong acids. Attacks copper and its alloys.

Permissible Exposure Limits in Air
Protective Action Criteria (PAC)
AIHA WEEL: 2 ppm, [skin] TWA; 5 ppm STEL (15 min) [skin], *as picolines.*
2-picoline
TEEL-0: 2 ppm
PAC-1: 5 ppm
PAC-2: 5 ppm
PAC-3: 300 ppm
3-picoline
TEEL-0: 2 ppm
PAC-1: 5 ppm
PAC-2: 125 ppm
PAC-3: 600 ppm
Russia: 5 mg/m^3.[43] (*o*-isomer)

Permissible Concentration in Water: There are no US criteria but the maximum allowable concentration in Class I waters for the production of drinking water has been set in Russia at 0.05 mg/L.[43] The EPA has suggested[32] a permissible ambient goal of 316 μg/L based on health effects.

Determination in Water: Octanol–water coefficient: Log K_{ow} = 1.2 (all isomers)

Routes of Entry: Inhalation, ingestion, skin and/or eye contact.

Harmful Effects and Symptoms
Short Term Exposure: Corrosive to the eyes, skin, and respiratory tract. Exposure can cause nausea, vomiting, diarrhea, and abdominal pain. *Inhalation:* May cause irritation to mouth, nose, and throat. Odor is very disagreeable above 30 ppm. However, people often become insensitive to the odor after a period of time, so odor detection cannot be relied upon as an indication of exposure. Exposure far above the OEL may cause you to pass out. Levels of 8000 ppm caused death in all exposed rats within 1.5 h. *Skin:* Readily absorbed and may contribute to symptoms. Irritation with rash or burning sensation on contact. May cause severe irritation if not promptly removed. *Eyes:* May cause severe irritation. *Ingestion:* Irritation and upset of

digestive system may occur. Muscle weakness, loss of coordination, diarrhea, and unconsciousness may result. Animal studies suggest that death may occur by ingestion of 1–2 fluid ounces for a 150-lb person.

Long Term Exposure: Repeated exposure to picoline can cause headache, dizziness, weakness, loss of coordination, double vision, and coma. Animal studies suggest that symptoms similar to those listed under ingestion would occur. Liver and kidney damage may also occur. Corrosive substances can irritate the lungs; bronchitis may develop.

Points of Attack: Skin, eyes, liver and kidneys, lungs.

Medical Surveillance: Liver and kidney function tests. Lung function tests. Consider chest X-ray following acute overexposure. Liver and kidney function tests.

First Aid: If this chemical gets into the eyes, remove any contact lenses at once and irrigate immediately for at least 15 min, occasionally lifting upper and lower lids. Seek medical attention immediately. If this chemical contacts the skin, remove contaminated clothing and wash immediately with soap and water. Speed in removing material from skin is of extreme importance. Shampoo hair promptly if contaminated. Seek medical attention immediately. If this chemical has been inhaled, remove from exposure, begin rescue breathing (using universal precautions, including resuscitation mask) if breathing has stopped and CPR if heart action has stopped. Transfer promptly to a medical facility. When this chemical has been swallowed, get medical attention. Give large quantities of water and induce vomiting. Do not make an unconscious person vomit.

Personal Protective Methods: Wear protective gloves and clothing to prevent any reasonable probability of skin contact. Safety equipment suppliers/manufacturers can provide recommendations on the most protective glove/clothing material for your operation. All protective clothing (suits, gloves, footwear, headgear) should be clean, available each day, and put on before work. Contact lenses should not be worn when working with this chemical. Wear splash-proof chemical goggles and face shield unless full face-piece respiratory protection is worn. Employees should wash immediately with soap when skin is wet or contaminated. Provide emergency showers and eyewash.

Respirator Selection: Follow regulations in OSHA 29CFR1910.134 or European Standard EN149. Use a NIOSH/MSHA- or European Standard EN149-approved respirator; or use an approved supplied-air respirator with a full face-piece operated in the positive-pressure mode, or with a full face-piece, hood, or helmet in the continuous-flow mode; or use a NIOSH/MSHA- or European Standard EN149-approved self-contained breathing apparatus with a full face-piece operated in a pressure-demand or other positive-pressure mode. A respirator providing protection against organic vapors may be of use to sensitive individuals and during entry or escape from a contaminated area.

Storage: Color Code—Red: Flammability Hazard: Store in a flammable liquid storage area or approved cabinet away from ignition sources and corrosive and reactive materials.

Prior to working with this chemical you should be trained on its proper handling and storage. Before entering confined space where this chemical may be present, check to make sure that an explosive concentration is not a danger. Store in tightly closed containers in a cool, well-ventilated area away from oxidizers, strong acids, acid chlorides, chloroformates, copper metals, and alloys. Where possible, automatically pump liquid from drums or other storage containers to process containers. Sources of ignition, such as smoking and open flames, are prohibited where this chemical is handled, used, or stored. Metal containers involving the transfer of 5 gallons or more of this chemical should be grounded and bonded. Drums must be equipped with self-closing valves, pressure vacuum bungs, and flame arresters. Use only nonsparking tools and equipment, especially when opening and closing containers of this chemical. Wherever this chemical is used, handled, manufactured, or stored, use explosion-proof electrical equipment and fittings.

Shipping: Picolines require a shipping label of "FLAMMABLE LIQUID." They fall in DOT Hazard Class 3 and Packing Group II.

Spill Handling: Evacuate and restrict persons not wearing protective equipment from area of spill or leak until cleanup is complete. Remove all ignition sources. Establish forced ventilation to keep levels below explosive limit. Absorb liquids in vermiculite, dry sand, earth, peat, carbon, or a similar material and deposit in sealed containers. Keep this chemical out of a confined space, such as a sewer, because of the possibility of an explosion, unless the sewer is designed to prevent the buildup of explosive concentrations. It may be necessary to contain and dispose of this chemical as a hazardous waste. If material or contaminated runoff enters waterways, notify downstream users of potentially contaminated waters. Contact your local or federal environmental protection agency for specific recommendations. If employees are required to clean up spills, they must be properly trained and equipped. OSHA 1910.120(q) may be applicable.

Fire Extinguishing: This chemical is a flammable liquid. Poisonous gases, including nitrogen oxides, are produced in fire. Use dry chemical, carbon dioxide, or alcohol foam extinguishers. Vapors are heavier than air and will collect in low areas. Vapors may travel long distances to ignition sources and flashback. Vapors in confined areas may explode when exposed to fire. Containers may explode in fire. Storage containers and parts of containers may rocket great distances, in many directions. If material or contaminated runoff enters waterways, notify downstream users of potentially contaminated waters. Notify local health and fire officials and pollution control agencies. From a secure, explosion-proof location, use water spray to cool exposed containers. If cooling streams are ineffective (venting sound increases in volume and pitch, tank discolors, or shows any signs of deforming), withdraw immediately to a secure position. If employees are expected to fight fires, they must be trained and equipped in OSHA 1910.156. The only respirators recommended for firefighting are self-contained breathing apparatuses that have full face-pieces and are operated in a pressure-demand or other positive-pressure mode.

References

US Environmental Protection Agency. (April 30, 1980). *2-Picoline: Health and Environmental Effects Profile No. 148*. Washington, DC: Office of Solid Waste

New York State Department of Health. (March 1986). *Chemical Fact Sheet: Picoline(s)*. Albany, NY: Bureau of Toxic Substance Assessment

New Jersey Department of Health and Senior Services. (August, 1999). *Hazardous Substances Fact Sheet: Picoline*. Trenton, NJ

Picric acid   P:0730

Molecular Formula: $C_6H_3N_3O_7$
Common Formula: $C_6H_2(NO_2)_3OH$
Synonyms: Acide picrique (French); Acido picrico (Spanish); Carbazotic acid; C.I. 10305; 2-Hydroxy-1,3,5-trinitrobenzene; Lyddite; Melinite; Nitroxanthic acid; PA; Pertite; Phenol trinitrate; Phenol, 2,4,6-trinitro-; Picral; Picronitric acid; Pikrinsaeure (German); Shimose; Trinitrophenol; Trinitrofenol (Spanish); 1,3,5-Trinitrophenol; 2,4,6-Trinitrophenol
CAS Registry Number: 88-89-1
RTECS® Number: TJ7875000
UN/NA & ERG Number: UN0154 (dry or wetted with <30% water, by mass)/112; UN3364 (wetted *with not* <10% water, by mass)/113; UN1344 (wetted with not <30% water, by mass)/113
EC Number: 201-865-9 [*Annex I Index No.:* 609-009-00-X]
Regulatory Authority and Advisory Bodies
Department of Homeland Security Screening Threshold Quantity (pounds): *Release hazard* 5000 (commercial grade); *Theft hazard* 400 (commercial grade).
US EPA Gene-Tox Program, Inconclusive: *D. melanogaster* sex-linked lethal.
An OSHA Class A Explosive (1910.109).
Air Pollutant Standard Set. See below, "Permissible Exposure Limits in Air" section.
EPCRA Section 313 Form R *de minimis* concentration reporting level: 1.0%.
Canada, WHMIS, Ingredients Disclosure List Concentration 1.0%.
European/International Regulations: Hazard Symbol: E, T; R3; R4; R23/24/25; Safety phrases: S1/2; S28; S35; R36/37; R45 (see Appendix 4).
WGK (German Aquatic Hazard Class): 2—Hazard to waters.
Description: Picric acid is a pale yellow, odorless solid. Usually found in solution with 10—20% water. *Must be kept wetted; the crystalline form is highly unstable.* The dry crystal form is explosive upon rapid heating or mechanical shock.

Molecular weight = 229.12; Specific gravity (H_2O:1) = 1.76 at 25°C; Boiling point = (explodes above 300°C); Freezing/Melting point = 122.2°C; Vapor pressure = 1 mmHg at 25°C; Flash point = 150°C; Autoignition temperature = (explodes) 300°C. Hazard Identification (based on NFPA-704 M Rating System): Health 3, Flammability 4, Reactivity 4; (*wet*): Health 2, Flammability 2, Reactivity 2. Slightly soluble in water; solubility = 1%.

Potential Exposure: Compound Description: Mutagen. Picric acid is used in the synthesis of dye intermediates and in manufacturing picrates; in the manufacture of explosives, rocket fuels, fireworks, colored glass, matches, electric batteries, and disinfectants. It is also used in the pharmaceutical and leather industries, in copper and steel etching, forensic chemistry, histology, textile printing, and photographic emulsions.

Incompatibilities: Violent reaction with oxidizers and reducing materials. Air or oxygen is not required for decomposition. Shock-sensitive compounds can be formed on contact with plaster, concrete. An explosive mixture results when the aqueous solution crystallizes. May explosively decompose from heat, shock, friction, or concussion. Copper, lead, zinc and other metals, or their salts can form other salts that are initiators and much more sensitive to shock than this chemical. Corrodes metals.

Permissible Exposure Limits in Air
Conversion factor: 1 ppm = 9.37 mg/m³ at 25°C & 1 atm.
OSHA PEL: 0.1 mg/m³ TWA [skin].
NIOSH REL: 0.1 mg/m³ TWA [skin]; 0.3 mg/m³ STEL.
ACGIH TLV®[1]: 0.1 mg/m³ TWA.
NIOSH IDLH: 75 mg/m³.
Protective Action Criteria (PAC)
TEEL-0: 0.1 mg/m³
PAC-1: 0.3 mg/m³
PAC-2: 15 mg/m³
PAC-3: 75 mg/m³
DFG MAK: 0.1 mg/m³, inhalable fraction; [skin] danger of skin sensitization; Carcinogen Category 3B.
Australia: TWA 0.1 mg/m³; STEL 0.3 mg/m³, [skin], 1993; Austria: MAK 0.1 mg/m³, [skin], 1999; Belgium: TWA 0.1 mg/m³; STEL 0.3 mg/m³, [skin], 1993; Denmark: TWA 0.1 mg/m³, [skin], 1999; Finland: TWA 0.1 mg/m³; STEL 0.3 mg/m³, [skin], 1999; France: VME 0.1 mg/m³, [skin], 1999; Hungary: TWA 0.1 mg/m³; STEL 0.2 mg/m³, [skin], 1993; the Netherlands: MAC-TGG 0.1 mg/m³, [skin], 2003; Norway: TWA 0.1 mg/m³, 1999; the Philippines: TWA 0.1 mg/m³, [skin], 1993; Poland: MAC (TWA) 0.1 mg/m³; MAC (STEL) 0.3 mg/m³, 1999; Switzerland: MAK-W 0.1 mg/m³, KZG-W 0.2 mg/m³, [skin], 1999; Turkey: TWA 0.1 mg/m³, [skin], 1993; United Kingdom: TWA 0.1 mg/m³; STEL 0.3 mg/m³, 2000; Argentina, Bulgaria, Columbia, Jordan, South Korea, New Zealand, Singapore, Vietnam: ACGIH TLV®: TWA 0.1 mg/m³. Several states have set guidelines or standards for picric acid in ambient air[60] ranging from 0.33 µg/m³ (New York) to 1.0 µg/m³ (Florida, South Carolina) to 1.0−3.0 µg/m³ (North Dakota) to 1.6 µg/m³ (Virginia) to 2.0 µg/m³ (Connecticut and Nevada). Russia set a MAC of 10 µg/m³ for ambient air on a once-daily basis.

Determination in Air: Collection on a mixed cellulose ester membrane filter, extraction with aqueous methanol measurement by high-performance liquid chromatography with UV detector. See NIOSH (II-4), Method #S-228.

Permissible Concentration in Water: To protect human health: no criteria set due to insufficient data.[6] Russia[35, 43] set a MAC of 0.5 mg/L in water bodies used for domestic purposes.

Determination in Water: Methylene chloride extraction followed by gas chromatography with flame ionization or electron capture detection (EPA Method 604) or gas chromatography plus mass spectrometry (EPA Method 625). Octanol−water coefficient: Log K_{ow} = 2.0.

Routes of Entry: Inhalation, ingestion, skin and/or eye contact. Absorbed by the skin.

Harmful Effects and Symptoms
Short Term Exposure: LD_{50} = (oral-rat) 200 mg/kg. Irritates the eyes, skin, and respiratory tract. Corneal injury may occur from exposure to picric acid dust and solutions. Dust or fume may cause eye irritation which may be aggravated by sensitization. Inhalation of high concentrations of dust by one worker caused temporary coma followed by weakness, myalgia, anuria, and later polyuria. Following ingestion of picric acid, there may be headache, vertigo, nausea, vomiting, diarrhea, yellow coloration of the skin, hematuria, and albuminuria. High doses may cause destruction of erythrocytes, hemorrhagic nephritis, and hepatitis. High doses which cause systemic intoxication will color all tissues yellow, including the conjunctive and aqueous humor, and cause yellow vision.

Long Term Exposure: Picric acid dust or solutions are potent skin sensitizers. The cutaneous lesions which appear usually on exposed areas of the upper extremities consist of dermatitis with erythema and vesicular eruptions. Desquamation may occur following repeated or prolonged contact. Skin usually turns yellow upon contact, and areas around nose and mouth as well as the hair are most often affected. May cause liver effects, hepatitis, hematuria (blood in the urine), albuminuria, kidney effects, nephritis.

Points of Attack: Eyes, skin, kidneys, liver, blood.

Medical Surveillance: NIOSH lists the following tests: liver function tests; urinalysis (routine). Preplacement and periodic medical examinations should focus on skin disorders (such as hypersensitivity, atopic dermatitis) and liver and kidney function. Examination by a qualified allergist. Complete blood count (CBC).

First Aid: If this chemical gets into the eyes, remove any contact lenses at once and irrigate immediately for at least 15 min, occasionally lifting upper and lower lids. Seek medical attention immediately. If this chemical contacts the skin, remove contaminated clothing and wash immediately with soap and water. Seek medical attention immediately. If this chemical has been inhaled, remove from exposure,

begin rescue breathing (using universal precautions, including resuscitation mask) if breathing has stopped and CPR if heart action has stopped. Transfer promptly to a medical facility. When this chemical has been swallowed, get medical attention. Give large quantities of water and induce vomiting. Do not make an unconscious person vomit.

Personal Protective Methods: Wear protective gloves and clothing to prevent any reasonable probability of skin contact. Safety equipment suppliers/manufacturers can provide recommendations on the most protective glove/clothing material for your operation. Natural rubber, Neoprene™, and Nitrile are among the recommended protective materials. All protective clothing (suits, gloves, footwear, headgear) should be clean, available each day, and put on before work. Contact lenses should not be worn when working with this chemical. Wear splash or dust-proof chemical goggles and face shield unless full face-piece respiratory protection is worn. Employees should wash immediately with soap when skin is wet or contaminated. Provide emergency showers and eyewash.

Respirator Selection: *Up to 0.5 mg/m³:* Qm (APF = 25) (any quarter-mask respirator). *Up to 1 mg/m³):* 95 XQ [any particulate respirator equipped with an N95, R95, or P95 filter (including N95, R95, and P95 filtering face-pieces) except quarter-mask respirators]. The following filters may also be used: N99, R99, P99, N100, R100, P100; or Sa (APF = 10) (any supplied-air respirator). *Up to 2.5 mg/m³:* Sa:Cf (APF = 25) (any supplied-air respirator operated in a continuous-flow mode) or PaprHie (APF = 25) (any powered, air-purifying respirator with a high-efficiency particulate filter). *Up to 5 mg/m³:* 100F (APF = 50) (any air-purifying, full-face-piece respirator with an N100, R100, or P100 filter) or SaT: Cf (APF = 50) (any supplied-air respirator that has a tight-fitting face-piece and is operated in a continuous-flow mode) or PaprTHie (APF = 50) (any powered, air-purifying respirator with a tight-fitting face-piece and a high-efficiency particulate filter) or SCBAF (APF = 50) (any self-contained breathing apparatus with a full face-piece) or SaF (APF = 50) (any supplied-air respirator with a full face-piece). *Up to 75 mg/m³:* SaF: Pd,Pp (APF = 2000) (any supplied-air respirator that has a full face-piece and is operated in a pressure-demand or other positive-pressure mode). *Emergency or planned entry into unknown concentrations or IDLH conditions:* SCBAF: Pd,Pp (APF = 10,000) (any NIOSH/MSHA- or European Standard EN 149-approved self-contained breathing apparatus that has a full face-piece and is operated in a pressure-demand or other positive-pressure mode) or Sa: Pd,Pp (APF = 1000): ASCBA (any supplied-air respirator that has a full face-piece and is operated in a pressure-demand or other positive-pressure mode in combination with an auxiliary, self-contained breathing apparatus operated in a pressure-demand or other positive-pressure mode). *Escape:* 100F (APF = 50) (any air-purifying, full-face-piece respirator with an N100, R100, or P100 filter) or SCBAE (any appropriate escape-type, self-contained breathing apparatus).

Storage: Explosive. Color Code—Red (*wet*): Flammability Hazard: Store in a flammable materials storage area. Prior to working with this chemical you should be trained on its proper handling and storage. Store in an explosion-proof refrigerator away from oxidizers, reducing agents, and metals. Where possible, automatically pump liquid from drums or other storage containers to process containers. Sources of ignition, such as smoking and open flames, are prohibited where this chemical is handled, used, or stored. Metal containers involving the transfer of 5 gallons or more of this chemical should be grounded and bonded. Drums must be equipped with self-closing valves, pressure vacuum bungs, and flame arresters. Use only nonsparking tools and equipment, especially when opening and closing containers of this chemical. Wherever this chemical is used, handled, manufactured, or stored, use explosion-proof electrical equipment and fittings.

Shipping: Trinitrophenol or picric acid, dry or wetted with <30% water, by mass, requires a shipping label of "EXPLOSIVES" It falls in Hazard Class 4.1.

Spill Handling: Evacuate persons not wearing protective equipment from area of spill or leak until cleanup is complete. Remove all ignition sources. Dampen spilled material with alcohol to avoid dust, then transfer material to a suitable container for eventual disposal. Collect powdered material in the most convenient and safe manner and deposit in sealed containers. Ventilate area after cleanup is complete. It may be necessary to contain and dispose of this chemical as a hazardous waste. If material or contaminated runoff enters waterways, notify downstream users of potentially contaminated waters. Contact your local or federal environmental protection agency for specific recommendations. If employees are required to clean up spills, they must be properly trained and equipped. OSHA 1910.120(q) may be applicable.

Fire Extinguishing: This chemical is a flammable solid. More powerful than TNT, picric acid explodes above 572°F/300°C. Use dry chemical, carbon dioxide, water spray, or alcohol foam extinguishers. Poisonous gases are produced in fire, including nitrogen oxides. If material or contaminated runoff enters waterways, notify downstream users of potentially contaminated waters. Notify local health and fire officials and pollution control agencies. Containers may explode. From a secure, explosion-proof location, use water spray to cool exposed containers. If cooling streams are ineffective (venting sound increases in volume and pitch, tank discolors, or shows any signs of deforming), withdraw immediately to a secure position. If employees are expected to fight fires, they must be trained and equipped in OSHA 1910.156. The only respirators recommended for firefighting are self-contained breathing apparatuses that have full face-pieces and are operated in a pressure-demand or other positive-pressure mode.

Disposal Method Suggested: Controlled incineration in a rotary kiln incinerator equipped with particulate abatement and wet scrubber devices.[22]

Reference
US Environmental Protection Agency. (1980). *Nitrophenols: Ambient Water Quality Criteria.* Washington, DC

Picrotoxin

P:0740

Molecular Formula: $C_{30}H_{34}O_{13}$

Synonyms: Cocculin; Cocculus; Coques du levant (French); Fish berry; Indian berry; Oriental berry; Picrotin, compounded with picrotoxinin (1:1); Picrotoxine

CAS Registry Number: 124-87-8

RTECS® Number: TJ9100000

UN/NA & ERG Number: UN3172/153

EC Number: 204-716-6

Regulatory Authority and Advisory Bodies

Superfund/EPCRA 40CFR355, Extremely Hazardous Substances: TPQ = 500/10,000 lb (227/4540 kg).

Reportable Quantity (RQ): 500 lb (227 kg).

European/International Regulations: not listed in Annex 1.

WGK (German Aquatic Hazard Class): No value assigned.

Description: Picrotoxin is an odorless crystalline solid with a very bitter taste. Molecular weight = 602.64; Freezing/Melting point = 203°C. Hazard Identification (based on NFPA-704 M Rating System): Health 3, Flammability 1, Reactivity 0. Practically insoluble in cold water; soluble in boiling water.

Potential Exposure: An alkaloid poison and convulsant. Used in medicine as a central nervous system stimulant and antidote for barbiturate poisoning. Reportedly, this material is not currently regarded as a useful therapeutic agent since it is not a selective respiratory stimulant.

Permissible Exposure Limits in Air

Protective Action Criteria (PAC)

TEEL-0: 3 mg/m^3

PAC-1: 7.5 mg/m^3

PAC-2: 15 mg/m^3

PAC-3: 15 mg/m^3

Routes of Entry: Inhalation, ingestion, skin and/or eye contact.

Harmful Effects and Symptoms

Short Term Exposure: Highly toxic and a dose of 20 mg may produce symptoms of severe poisoning. A human lethal dose of 1.5 mg/kg has been reported. It is an alkaloid convulsant poison. Picrotoxin is a powerful stimulant and affects all portions of the central nervous system. At doses approaching convulsant levels, signs and symptoms include salivation, elevated blood pressure, frequent vomiting, rapid breathing.

Points of Attack: Central nervous system.

First Aid: If this chemical gets into the eyes, remove any contact lenses at once and irrigate immediately for at least 15 min, occasionally lifting upper and lower lids. Seek medical attention immediately. If this chemical contacts the skin, remove contaminated clothing and wash immediately with soap and water. Seek medical attention immediately. If this chemical has been inhaled, remove from exposure, begin rescue breathing (using universal precautions, including resuscitation mask) if breathing has stopped and CPR if heart action has stopped. Transfer promptly to a medical facility. When this chemical has been swallowed, get medical attention. Give large quantities of water and induce vomiting. Do not make an unconscious person vomit. Keep victim quiet and maintain normal body temperature. Effects may be delayed; keep victim under observation.

Personal Protective Methods: Wear protective gloves and clothing to prevent any reasonable probability of skin contact. Safety equipment suppliers/manufacturers can provide recommendations on the most protective glove/clothing material for your operation. All protective clothing (suits, gloves, footwear, headgear) should be clean, available each day, and put on before work. Contact lenses should not be worn when working with this chemical. Wear dust-proof chemical goggles and face shield unless full face-piece respiratory protection is worn. Employees should wash immediately with soap when skin is wet or contaminated. Provide emergency showers and eyewash.

Respirator Selection: Follow regulations in OSHA 29CFR1910.134 or European Standard EN149. Use a NIOSH/MSHA- or European Standard EN149-approved respirator; or use an approved supplied-air respirator with a full face-piece operated in the positive-pressure mode, or with a full face-piece, hood, or helmet in the continuous-flow mode; or use a NIOSH/MSHA- or European Standard EN149-approved self-contained breathing apparatus with a full face-piece operated in a pressure-demand or other positive-pressure mode.

Storage: Color Code—Blue: Health Hazard/Poison: Store in a secure poison location. Prior to working with this chemical you should be trained on its proper handling and storage. Store in tightly closed containers in a cool, well-ventilated area away. Where possible, automatically transfer material from drums or other storage containers to process containers.

Shipping: Toxins, extracted from living sources, require a shipping label of "POISONOUS/TOXIC MATERIALS". It falls in Hazard Class 6.1 and Packing Group I.

Spill Handling: Evacuate persons not wearing protective equipment from area of spill or leak until cleanup is complete. Stay upwind; keep out of low areas. Ventilate closed spaces before entering them. Remove and isolate contaminated clothing at the site. If water pollution occurs, notify appropriate authorities. Do not touch spilled material; stop leak if you can do it without risk. Use water spray to reduce vapors. *Small spills*: take up with sand or other noncombustible absorbent material and place into containers for later disposal. *Small dry spills*: with clean shovel place material into clean, dry container and cover; move containers from spill area. *Large spills*: dike far ahead of spill for later disposal. Ventilate area after cleanup is complete. It may be necessary to contain and dispose of this chemical as a hazardous waste. If material or contaminated runoff enters waterways, notify downstream users of potentially contaminated waters. Contact your local or federal environmental protection agency for specific recommendations. If employees are required to clean up spills, they must be properly trained and equipped. OSHA 1910.120(q) may be applicable.

Fire Extinguishing: Use dry chemical, carbon dioxide, water spray, or foam for small fires. Use water spray, fog, or foam for large fires. Move container from fire area if this can be done without risk. Isolate hazard area and deny entry. Wear positive pressure breathing apparatus and special protective clothing. Poisonous gases are produced in fire. If material or contaminated runoff enters waterways, notify downstream users of potentially contaminated waters. Notify local health and fire officials and pollution control agencies. From a secure, explosion-proof location, use water spray to cool exposed containers. If cooling streams are ineffective (venting sound increases in volume and pitch, tank discolors, or shows any signs of deforming), withdraw immediately to a secure position. If employees are expected to fight fires, they must be trained and equipped in OSHA 1910.156. The only respirators recommended for firefighting are self-contained breathing apparatuses that have full face-pieces and are operated in a pressure-demand or other positive-pressure mode.

Reference

US Environmental Protection Agency. (November 30, 1987). *Chemical Hazard Information Profile: Picrotoxin.* Washington, DC: Chemical Emergency Preparedness Program

Pindone   P:0760

Molecular Formula: $C_{14}H_{14}O_3$
Common Formula: $C_6H_4(CO)_2CHCOC(CH_3)_3$
Synonyms: Chemrat; *tert*-Butyl valone; 1,3-Dioxo-2-pivaloy-lindane; Pival®; Pivalyl; 2-Pivalyl-1,3-indandione; Pivalyl Valone®; Pivaldione (French)
CAS Registry Number: 83-26-1
RTECS® Number: NK6300000
UN/NA & ERG Number: UN2588/151
EC Number: 201-462-8 [*Annex I Index No.:* 606-016-00-X]
Regulatory Authority and Advisory Bodies
US EPA, FIFRA 1998 Status of Pesticides: RED completed. Air Pollutant Standard Set. See below, "Permissible Exposure Limits in Air" section.
European/International Regulations: Hazard Symbol: T, N; Risk phrases: R25 R48/25;R50/53; Safety phrases: S1/2; S37; S45; S60; S61 (see Appendix 4).
WGK (German Aquatic Hazard Class): No value assigned.
Description: Pindone is a bright yellow crystalline solid. Almost odorless. Molecular weight = 230.28; Specific gravity (H_2O:1) = 1.06 at 25°C; Boiling point = decomposes; Freezing/Melting point = 110°C. Practically insoluble in water; solubility = 0.002% at 25°C.
Potential Exposure: Compound Description: Agricultural Chemical. Pindone is used as an anticoagulant and rodenticide. Those involved in manufacture, formulation, and application of this chemical.

Incompatibilities: None reported.
Permissible Exposure Limits in Air
OSHA PEL: 0.1 mg/m³ TWA.
NIOSH REL: 0.1 mg/m³ TWA.
ACGIH TLV®[1]: 0.1 mg/m³ TWA.
NIOSH IDLH: 100 mg/m³.
No TEEL available.
Australia: TWA 0.1 mg/m³, 1993; Belgium: TWA 0.1 mg/m³, 1993; Denmark: TWA 0.1 mg/m³, 1999; France: VME 0.1 mg/m³, 1999; Norway: TWA 0.1 mg/m³, 1999; Switzerland: MAK-W 0.1 mg/m³, 1999; the Netherlands: MAC-TGG 0.1 mg/m³, 2003; Argentina, Bulgaria, Columbia, Jordan, South Korea, New Zealand, Singapore, Vietnam: ACGIH TLV®: TWA 0.1 mg/m³. Several states have set guidelines or standards for pindone in ambient air[60] ranging from 1.0−3.0 μg/m³ (North Dakota) to 1.6 μg/m³ (Virginia) to 2.0 μg/m³ (Connecticut and Nevada).
Determination in Air: No method available.
Routes of Entry: Inhalation, ingestion.
Harmful Effects and Symptoms
Short Term Exposure: Nosebleeds (epistaxis), excessive bleeding of minor cuts and bruises, smoky urine, black tarry stools, abdominal and back pain. Reduced blood clotting which leads to hemorrhaging; symptoms resembling warfarin: depressed formation of prothrombin and capillary fragility; leading to hemorrhages.
Points of Attack: Blood prothrombin.
Medical Surveillance: NIOSH lists the following tests: blood plasma, Prothrombin Time; Complete blood count; urinalysis (routine), red blood cells/count.
First Aid: If this chemical gets into the eyes, remove any contact lenses at once and irrigate immediately for at least 15 min, occasionally lifting upper and lower lids. Seek medical attention immediately. If this chemical contacts the skin, remove contaminated clothing and wash immediately with soap and water. Seek medical attention immediately. If this chemical has been inhaled, remove from exposure, begin rescue breathing (using universal precautions, including resuscitation mask) if breathing has stopped and CPR if heart action has stopped. Transfer promptly to a medical facility. When this chemical has been swallowed, get medical attention. Give large quantities of water and induce vomiting. Do not make an unconscious person vomit.
Personal Protective Methods: Wear protective gloves and clothing to prevent any reasonable probability of skin contact. Safety equipment suppliers/manufacturers can provide recommendations on the most protective glove/clothing material for your operation. All protective clothing (suits, gloves, footwear, headgear) should be clean, available each day, and put on before work. Contact lenses should not be worn when working with this chemical. Wear dust-proof chemical goggles and face shield unless full face-piece respiratory protection is worn. Employees should wash immediately with soap when skin is wet or contaminated. Provide emergency showers and eyewash.

Respirator Selection: Up to 0.5 mg/m³: Qm (APF – 25) (any quarter-mask respirator). *Up to 1 mg/m³*): 95XQ (APF = 10) [any particulate respirator equipped with an N95, R95, or P95 filter (including N95, R95, and P95 filtering face-pieces) except quarter-mask respirators. The following filters may also be used: N99, R99, P99, N100, R100, P100] or Sa (APF = 10) (any supplied-air respirator). *Up to 2.5 mg/m³:* Sa:Cf (APF = 25) (any supplied-air respirator operated in a continuous-flow mode) or PaprHie (APF = 25) (any powered, air-purifying respirator with a high-efficiency particulate filter). *Up to 5 mg/m³:* 100F (APF = 50) (any air-purifying, full-face-piece respirator with an N100, R100, or P100 filter) or SaT: Cf (APF = 50) (any supplied-air respirator that has a tight-fitting face-piece and is operated in a continuous-flow mode) or PaprTHie (APF = 50) (any powered, air-purifying respirator with a tight-fitting face-piece and a high-efficiency particulate filter) or SCBAF (APF = 50) (any self-contained breathing apparatus with a full face-piece) or SaF (APF = 50) (any supplied-air respirator with a full face-piece). *Up to 100 mg/m³:* Sa: Pd,Pp (APF = 1000) (any supplied-air respirator that has a full face-piece and is operated in a pressure-demand or other positive-pressure mode). *Emergency or planned entry into unknown concentrations or IDLH conditions:* SCBAF: Pd, Pp (APF = 10,000) (any NIOSH/MSHA- or European Standard EN 149-approved self-contained breathing apparatus that has a full face-piece and is operated in a pressure-demand or other positive-pressure mode) or SaF: Pd, Pp: ASCBA (APF = 10,000) (any supplied-air respirator that has a full face-piece and is operated in a pressure-demand or other positive-pressure mode in combination with an auxiliary, self-contained breathing apparatus operated in a pressure-demand or other positive-pressure mode). *Escape:* 100F (APF = 50) (any air-purifying, full-face-piece respirator with an N100, R100, or P100 filter) or SCBAE (any appropriate escape-type, self-contained breathing apparatus).

Storage: Color Code—Blue: Health Hazard/Poison: Store in a secure poison location. Prior to working with this chemical you should be trained on its proper handling and storage. Store in a refrigerator or a cool, dry place.

Shipping: Pesticides, solid, toxic, n.o.s. require a shipping label of "POISONOUS/TOXIC MATERIALS." (Packing Group III): It falls in Hazard Class 6.1 and Packing Group III.

Spill Handling: Remove all sources of ignition and dampen spilled material with 60–70% ethanol to avoid airborne dust, then transfer material to a suitable container. Wash surfaces well with soap and water. Ventilate area after cleanup is complete. It may be necessary to contain and dispose of this chemical as a hazardous waste. If material or contaminated runoff enters waterways, notify downstream users of potentially contaminated waters. Contact your local or federal environmental protection agency for specific recommendations. If employees are required to clean up

spills, they must be properly trained and equipped. OSHA 1910.120(q) may be applicable.

Fire Extinguishing: Use dry chemical, carbon dioxide, water spray, or alcohol foam extinguishers. Poisonous gases are produced in fire. If material or contaminated runoff enters waterways, notify downstream users of potentially contaminated waters. Notify local health and fire officials and pollution control agencies. From a secure, explosion-proof location, use water spray to cool exposed containers. If cooling streams are ineffective (venting sound increases in volume and pitch, tank discolors, or shows any signs of deforming), withdraw immediately to a secure position. If employees are expected to fight fires, they must be trained and equipped in OSHA 1910.156. The only respirators recommended for firefighting are self-contained breathing apparatuses that have full face-pieces and are operated in a pressure-demand or other positive-pressure mode.

Disposal Method Suggested: Dissolve or mix the material with a combustible solvent and burn in a chemical incinerator equipped with an afterburner and scrubber. All federal, state, and local environmental regulations must be observed.[22] In accordance with 40CFR165, follow recommendations for the disposal of pesticides and pesticide containers. Must be disposed properly by following package label directions or by contacting your local or federal environmental control agency or by contacting your regional EPA office.

Reference

US Environmental Protection Agency, Special Review and Reregistration Division Office of Pesticide Programs. (1998). *Agency Status of Pesticides in Registration, Reregistration, and Special Review* (Rainbow Report). Washington, DC

Piperazine   P:0770

Molecular Formula: $C_4H_{10}N_2$; $C_4H_{12}Cl_2N_2$

Common Formula: $C_4H_{10}N_2 \cdot 2HCl$; $C_4H_{10}N_2$ HCl (monochloride)

Synonyms: Antiren; *N,N*-Diethylene diamine; 1,4-Diethylenediamine; Dihydrochloride salt of diethylenediamine; Dispermine; Dowzene; Hexahydro-1,4-diazine; Hexahydropyrazine; Lumbrical; Piperazidine; Piperazin (German); Piperazine dihydrochloride; Piperazine hydrochloride; Pyrazine hexahydride

CAS Registry Number: 110-85-0; 142-64-3 (hydrochloride)

RTECS® Number: TL4025000 (hydrochloride); TK7800000

UN/NA & ERG Number: UN2579/153

EC Number: 203-808-3 [*Annex I Index No.:* 612-057-00-4]; 205-551-2 [*Annex I Index No.:* 612-241-00-4] (hydrochloride)

Regulatory Authority and Advisory Bodies

Banned or Severely Restricted (several countries) (UN).[13]
Air Pollutant Standard Set. See below, "Permissible Exposure Limits in Air" section.

Canada, WHMIS, Ingredients Disclosure List Concentration) 0.1% (Piperazine and Piperazine dihydrochloride).

European/International Regulations: Hazard Symbol: Xn, C; Risk phrases: R34; R42/43; R62; R63; Safety phrases: S1/2; S22; S26; S36/37/39; S45; (*hydrochloride*): Hazard Symbol: Xn, C, N; Risk phrases: R36/38; R42/43; R62; R63; R52/53; Safety phrases: S1/2; S22; S36/37; S45; S61; S63 (see Appendix 4).

WGK (German Aquatic Hazard Class): 1—Low hazard to waters.

Description: Piperazine and Piperazine dihydrochloride, are white to cream-colored needles or powder with a characteristic ammonia-like odor. Combustible solids that do not easily ignite. Molecular weight = 86.16 (piperazine); 159.08 (dihydrochloride); Boiling point = 146°C; Freezing/Melting point = 106°C; 335°C (*dihydrochloride*); Flash point = 110°C. Hazard Identification (based on NFPA-704 M Rating System): (*piperazine*) Health 2, Flammability 2, Reactivity 0. Soluble in water.

Potential Exposure: Compound Description (Piperazine): Agricultural Chemical; Drug, Human Data; Primary Irritant; (dihydrochloride) Agricultural Chemical; Mutagen. Piperazine is used to manufacture anthelmintics, antifilarials, antihistamines, and tranquilizers; the dihydrochloride is used in the manufacture of fibers, pharmaceuticals, and insecticides. They are used as an intermediate in the manufacture of pesticides, rubber chemicals, and fibers. Also, piperazine is widely available, effective, and safe when used on an occasional basis against ascaride infections. It is also considerably cheaper than other anthelminthic drugs. In some countries where ascariasis is not endemic and where piperazine was used predominantly for the treatment of pinworn, it has been withdrawn from use on the grounds that other effective drugs are now available. Clinical dosages occasionally induce transient neurological signs and, in some circumstances, the drug may generate small amounts of *nitrosamine* in the stomach, which at considerably greater dosage in experimental animals has been demonstrated to have a carcinogenic potential.

Incompatibilities: Violent reaction with strong oxidizers and dicyanofurazan. Incompatible with nitrogen compounds, carbon tetrachloride. Attacks aluminum, copper, nickel, magnesium, and zinc.

Permissible Exposure Limits in Air

ACGIH TLV (*piperazine and piperazine dihydrochloride*): *2010 Notice of intended change:* 0.1 mg/m^3 measured as inhalable fraction and vapor TWA; Danger of sensitization; not classifiable as a human carcinogen.

NIOSH REL (*dihydrochloride*): 5 mg/m^3 TWA.

Piperazine

Protective Action Criteria (PAC)

TEEL-0: 2 mg/m^3

PAC-1: 6 mg/m^3

PAC-2: 40 mg/m^3

PAC-3: 500 mg/m^3

DFG MAK (*piperazine*): Danger of skin and airway sensitization.

Denmark: TWA 0.1 ppm (0.35 mg/m^3), 1999; Norway: TWA 0.1 ppm (0.3 mg/m^3), 1999; Sweden: NGV 0.1 ppm (0.3 mg/m^3), KTV 0.3 ppm (1 mg/m^3), 1999; the Netherlands: MAC-TGG 0.1 mg/m^3, 2003.

Dihydrochloride

OSHA PEL: None.

NIOSH REL: 5 mg/m^3 TWA.

ACGIH TLV$^{®[1]}$: 5 mg/m^3 TWA.

DFG MAK: Danger of skin and airway sensitization; No numerical value established. Data may be available.

Australia: TWA 5 mg/m^3, 1993; Belgium: TWA 5 mg/m^3, 1993; Denmark: TWA 5 mg/m^3, 1999; France: VME 5 mg/m^3, 1999; Switzerland: MAK-W 5 mg/m^3, 1999; United Kingdom: TWA 5 mg/m^3, 2000; Argentina, Bulgaria, Columbia, Jordan, South Korea, New Zealand, Singapore, Vietnam: ACGIH TLV$^®$: TWA 5 mg/m^3, 2003.

Determination in Air: No method available.

Routes of Entry: Inhalation, skin absorption, ingestion, skin and/or eye contact.

Harmful Effects and Symptoms

Short Term Exposure: Piperazine can affect you when breathed in and by passing through your skin. Piperazine is a corrosive chemical and eye contact can cause severe irritation and burns. Skin contact can cause irritation or a skin allergy, with rash at even very low exposure levels. Exposure can cause a lung allergy to develop, with cough and wheezing triggered by even low exposures. Higher exposures can cause pulmonary edema, a medical emergency that can be delayed for several hours. This can cause death. High exposures can cause weakness, tremors, visual changes, and trigger seizures. It can also interfere with the ability of the blood to carry oxygen, causing headaches, dizziness, and cyanosis, a bluish color to the skin and lips.

Long Term Exposure: Repeated exposure to piperazine dihydrochloride can cause skin sensitization and asthma-like allergy.

Points of Attack: Lungs.

Medical Surveillance: If symptoms develop or overexposure is suspected, the following may be useful: evaluation by a qualified allergist, including careful exposure history and special testing, may help diagnose skin allergy. Lung function tests. These may be normal if the person is not having an attack at the time of the test. Blood methemoglobin level.

First Aid: If this chemical gets into the eyes, remove any contact lenses at once and irrigate immediately for at least 15 min, occasionally lifting upper and lower lids. Seek medical attention immediately. If this chemical contacts the skin, remove contaminated clothing and wash immediately with soap and water. Seek medical attention immediately. If this chemical has been inhaled, remove from exposure, begin rescue breathing (using universal precautions, including resuscitation mask) if breathing has stopped and CPR if heart action has stopped. Transfer promptly to a medical

facility. When this chemical has been swallowed, get medical attention. If victim is *conscious*, administer water or milk. Do not induce vomiting. Medical observation is recommended for 24—48 h after breathing overexposure, as pulmonary edema may be delayed. As first aid for pulmonary edema, a doctor or authorized paramedic may consider administering a corticosteroid spray.

Note to physician: Treat for methemoglobinemia. Spectrophotometry may be required for precise determination of levels of methemoglobin in urine.

Personal Protective Methods: Wear protective gloves and clothing to prevent any reasonable probability of skin contact. Safety equipment suppliers/manufacturers can provide recommendations on the most protective glove/clothing material for your operation. All protective clothing (suits, gloves, footwear, headgear) should be clean, available each day, and put on before work. Contact lenses should not be worn when working with this chemical. Wear dust-proof chemical goggles and face shield unless full face-piece respiratory protection is worn. Employees should wash immediately with soap when skin is wet or contaminated. Provide emergency showers and eyewash.

Respirator Selection: Where there is potential for exposures *over 5 mg/m³*, use a NIOSH/MSHA- or European Standard EN149-approved full-face-piece respirator with a high-efficiency particulate filter. Greater protection is provided by a powered air-purifying respirator. *Where there is potential for high exposures*, use a NIOSH/MSHA- or European Standard EN149-approved supplied-air respirator with a full face-piece operated in the positive-pressure mode, or with a full face-piece, hood, or helmet in the continuous-flow mode; or use a NIOSH/MSHA- or European Standard EN149-approved self-contained breathing apparatus with a full face-piece operated in a pressure-demand or other positive-pressure mode.

Storage: Color Code—White: Corrosive or Contact Hazard; Store separately in a corrosion-resistant location. Prior to working with this chemical you should be trained on its proper handling and storage. Piperazine must be stored to avoid contact with oxidizers (such as perchlorates, peroxides, permanganates, chlorates, and nitrates) since violent reactions occur. Sources of ignition, such as smoking and open flames, are prohibited where piperazine is used, handled, or stored in a manner that could create a potential fire or explosion hazard. Store in tightly closed containers in a cool, well-ventilated area. Where possible, automatically transfer material from drums or other storage containers to process containers. Sources of ignition, such as smoking and open flames, are prohibited where this chemical is handled, used, or stored. Metal containers involving the transfer of this chemical should be grounded and bonded. Wherever this chemical is used, handled, manufactured, or stored, use explosion-proof electrical equipment and fittings.

Shipping: Piperazine requires a shipping label of "CORROSIVE." It falls in Hazard Class 8 and Packing Group III.

Spill Handling: Evacuate persons not wearing protective equipment from area of spill or leak until cleanup is complete. Remove all ignition sources. Collect powdered material in the most convenient and safe manner and deposit in sealed containers. Ventilate area after cleanup is complete. It may be necessary to contain and dispose of this chemical as a hazardous waste. If material or contaminated runoff enters waterways, notify downstream users of potentially contaminated waters. Contact your local or federal environmental protection agency for specific recommendations. If employees are required to clean up spills, they must be properly trained and equipped. OSHA 1910.120(q) may be applicable.

Fire Extinguishing: Piperazine is a combustible solid. Use dry chemical, water spray, or alcohol foam extinguishers. Piperazine dihydrochloride may burn but does not readily ignite. Extinguish fire using an agent suitable for type of surrounding fire. Poisonous gases, including nitrogen oxides and hydrogen chloride (hydrochloride), are produced in fire. If material or contaminated runoff enters waterways, notify downstream users of potentially contaminated waters. Notify local health and fire officials and pollution control agencies. From a secure, explosion-proof location, use water spray to cool exposed containers. If cooling streams are ineffective (venting sound increases in volume and pitch, tank discolors, or shows any signs of deforming), withdraw immediately to a secure position. If employees are expected to fight fires, they must be trained and equipped in OSHA 1910.156. The only respirators recommended for firefighting are self-contained breathing apparatuses that have full face-pieces and are operated in a pressure-demand or other positive-pressure mode.

References

New Jersey Department of Health and Senior Services. (September 2004). *Hazardous Substances Fact Sheet: Piperazine Dihydrochloride*. Trenton, NJ

New Jersey Department of Health and Senior Services. (April 2004). *Hazardous Substances Fact Sheet: Piperazine*. Trenton, NJ

Piperidine   P:0780

Molecular Formula: $C_5H_{11}N$

Synonyms: Azacyclohexane; Cyclopentimine; Cypentil; Hexahydropyridine; Hexazane; Pentamethyleneimine; Peperidin (German)

CAS Registry Number: 110-89-4

RTECS® Number: TM3500000

UN/NA & ERG Number: UN2401/132

EC Number: 203-813-0 [*Annex I Index No.:* 613-027-00-3]

Regulatory Authority and Advisory Bodies

Department of Homeland Security Screening Threshold Quantity (pounds): *Release hazard* 10,000 ($\geq 1.00\%$ concentration).

Air Pollutant Standard Set. See below, "Permissible Exposure Limits in Air" section.

Clean Air Act: Accidental Release Prevention/Flammable Substances, (Section 112[r], Table 3), TQ = 15,000 lb (6810 kg).

Superfund/EPCRA 40CFR355, Extremely Hazardous Substances: TPQ = 1000 lb (454 kg).

Reportable Quantity (RQ): 1000 lb (454 kg).

European/International Regulations: Hazard Symbol: F, T; Risk phrases: R11; R23/24; R34; Safety phrases: S1/2; S16; S26; S27; S45 (see Appendix 4).

WGK (German Aquatic Hazard Class): No value assigned.

Description: Piperidine is a clear, colorless liquid with an amine-like odor. Molecular weight = 85.17; Boiling point = 106°C; Freezing/Melting point = −7°C; Flash point = 16°C. Soluble in water.

Potential Exposure: Compound Description: Mutagen; Reproductive Effector; Primary Irritant. Piperidine is used in agriculture and pharmaceuticals; as an intermediate for rubber accelerators; as a solvent; as a curing agent for rubber and epoxy resins; catalyst for condensation reactions; as an ingredient in oils and fuels; as a complexing agent; in the manufacture of local anesthetics; in analgesics; pharmaceuticals, wetting agents; and germicides; in synthetic flavoring. Not registered as a pesticide in the United States.

Incompatibilities: Piperidine is a medium strong base. Reacts violently with oxidizers.

Permissible Exposure Limits in Air

AIHA WEEL: 1 ppm TWA [skin].

Protective Action Criteria (PAC)*

TEEL-0: 1 ppm

PAC-1: **6.6** ppm

PAC-2: **33** ppm

PAC-3: **110** ppm

*AEGLs (Acute Emergency Guideline Levels) & ERPGs (Emergency Response Planning Guideline) are in **bold face**.

United Kingdom: TWA 1 ppm/3.5 mg/m^3 [skin].

Russia[43] MAC (work-place air) 0.2 mg/m^3.

Permissible Concentration in Water: Russia[43] set a MAC of 0.06 mg/L in water bodies used for domestic purposes.

Routes of Entry: Inhalation, ingestion, skin and/or eye contact.

Harmful Effects and Symptoms

Short Term Exposure: Corrosive to the eyes, skin, and respiratory tract. Sore throat, coughing, labored breathing, and dizziness occur after inhalation. Higher exposures can cause pulmonary edema, a medical emergency that can be delayed for several hours. This can cause death. Exposure may cause increased blood pressure. May cause permanent injury after short exposure to small amounts. Ingestion may involve both irreversible and reversible changes. 30−60 mg/kg may cause symptoms in humans. Symptoms upon oral administration include weakness, nausea, vomiting, salivation, labored respiration, muscular paralysis, and asphyxiation. Redness, pain, and burns occur upon contact with skin.

Long Term Exposure: Irritating substances may cause lung irritation; bronchitis may develop. May affect the liver and kidneys.

Points of Attack: Lungs, blood, liver, kidneys.

Medical Surveillance: Monitor blood pressure. Lung function tests. Liver and kidney function tests. Consider chest X-ray following acute overexposure.

First Aid: If this chemical gets into the eyes, remove any contact lenses at once and irrigate immediately for at least 15 min, occasionally lifting upper and lower lids. Seek medical attention immediately. If this chemical contacts the skin, remove contaminated clothing and wash immediately with soap and water. Seek medical attention immediately. If this chemical has been inhaled, remove from exposure, begin rescue breathing (using universal precautions, including resuscitation mask) if breathing has stopped and CPR if heart action has stopped. Transfer promptly to a medical facility. When this chemical has been swallowed, get medical attention. Give large quantities of water and induce vomiting. Do not make an unconscious person vomit. Medical observation is recommended for 24−48 h after breathing overexposure, as pulmonary edema may be delayed. As first aid for pulmonary edema, a doctor or authorized paramedic may consider administering a corticosteroid spray.

Personal Protective Methods: Wear protective gloves and clothing to prevent any reasonable probability of skin contact. Safety equipment suppliers/manufacturers can provide recommendations on the most protective glove/clothing material for your operation. All protective clothing (suits, gloves, footwear, headgear) should be clean, available each day, and put on before work. Contact lenses should not be worn when working with this chemical. Wear splash-proof chemical goggles and face shield unless full face-piece respiratory protection is worn. Employees should wash immediately with soap when skin is wet or contaminated. Provide emergency showers and eyewash.

Respirator Selection: Follow regulations in OSHA 29CFR1910.134 or European Standard EN149. Use a NIOSH/MSHA- or European Standard EN149-approved respirator; or use an approved supplied-air respirator with a full face-piece operated in the positive-pressure mode, or with a full face-piece, hood, or helmet in the continuous-flow mode; or use a NIOSH/MSHA- or European Standard EN149-approved self-contained breathing apparatus with a full face-piece operated in a pressure-demand or other positive-pressure mode.

Storage: Color Code—Red: Flammability Hazard: Store in a flammable liquid storage area or approved cabinet away from ignition sources and corrosive and reactive materials. Prior to working with this chemical you should be trained on its proper handling and storage. Store in tightly closed containers in a cool, well-ventilated area away from oxidizers. Where possible, automatically pump liquid from drums or other storage containers to process containers. Sources of ignition, such as smoking and open flames, are prohibited where this chemical is handled, used, or stored. Metal containers involving the transfer of 5 gallons or more of this chemical should be grounded and bonded. Drums must be equipped with self-closing valves, pressure vacuum bungs, and flame arresters. Use only nonsparking tools and equipment, especially when opening and closing containers of this chemical. Wherever this chemical is used, handled, manufactured, or stored, use explosion-proof electrical equipment and fittings.

Shipping: Piperidine requires a shipping label of "FLAMMABLE LIQUID, CORROSIVE." It falls in Hazard Class 3 and Packing Group II.

Spill Handling: Evacuate and restrict persons not wearing protective equipment from area of spill or leak until cleanup is complete. Shut off ignition sources; no flares, smoking, or flames in hazard area. Do not touch spilled material; stop leak if you can do so without risk. Use water spray to reduce vapors; do not get water inside container. *Small spills:* absorb with sand or other noncombustible absorbent material and place into containers for later disposal. *Large spills:* dike far ahead of spill for later disposal. Ventilate area of spill or leak. Absorb liquids in vermiculite, dry sand, earth, peat, carbon, or a similar material and deposit in sealed containers. Keep this chemical out of a confined space, such as a sewer, because of the possibility of an explosion, unless the sewer is designed to prevent the buildup of explosive concentrations. It may be necessary to contain and dispose of this chemical as a hazardous waste. If material or contaminated runoff enters waterways, notify downstream users of potentially contaminated waters. Contact your local or federal environmental protection agency for specific recommendations. If employees are required to clean up spills, they must be properly trained and equipped. OSHA 1910.120(q) may be applicable.

Fire Extinguishing: This chemical is a flammable liquid. *Small fires:* dry chemical, carbon dioxide, water spray, or alcohol foam. *Large fires:* water spray, fog, or alcohol foam. Move container from fire area if you can do it without risk. Do not get water inside container. Cool containers that are exposed to flames with water from the side until well after fire is out. Withdraw immediately in case of rising sound from venting safety device or any discoloration of tank due to fire. Keep unnecessary people away; isolate hazard area and deny entry. Stay upwind; keep out of low area. Wear self-contained (positive pressure if available) breathing apparatus and full protective clothing. Isolate for ½ mile in all directions if tank car or truck is involved in fire. Poisonous gases, including nitrogen oxides, are produced in fire. Vapors are heavier than air and will collect in low areas. Vapors may travel long distances to ignition sources and flashback. Vapors in confined areas may explode when exposed to fire. Containers may explode in fire. Storage containers and parts of containers may rocket great distances, in many directions. If material or contaminated runoff enters waterways, notify downstream users of potentially contaminated waters. Notify local health and fire officials and pollution control agencies. From a secure, explosion-proof location, use water spray to cool exposed containers. If cooling streams are ineffective (venting sound increases in volume and pitch, tank discolors, or shows any signs of deforming), withdraw immediately to a secure position. If employees are expected to fight fires, they must be trained and equipped in OSHA 1910.156. The only respirators recommended for firefighting are self-contained breathing apparatuses that have full face-pieces and are operated in a pressure-demand or other positive-pressure mode.

References

US Environmental Protection Agency. (November 30, 1987). *Chemical Hazard Information Profile: Piperidine.* Washington, DC: Chemical Emergency Preparedness Program

New Jersey Department of Health and Senior Services. (September, 1999). *Hazardous Substances Fact Sheet: Piperidine.* Trenton, NJ

Pirimifos-ethyl   P:0790

Molecular Formula: $C_{13}H_{24}N_3O_3PS$

Synonyms: O-[2-(Diethylamino)-6-methyl-4-pyrimidinyl] O,O-diethyl phosphorothioate; 2-Diethylamino-6-methyl-pyrimidin-4-yl diethylphosphorothionate; O,O-Diethyl O-(2-diethylamino-6-methyl-4-pyrimidinyl) phosphorothioate; Diethyl O-(2-diethylamino-6-methyl-4-pyrimidinyl) phosphorothioate; Diethyl 2-dimethylamino-4-methylpyrimidin-6-yl phosphorothionate; Ethyl pirimiphos; Fernex; Phosphorothioic acid, O-[2-(diethylamino)-6-methyl-4-pyrimidinyl] O,O-diethyl ester; PP211; Primicid; Primifosethyl; Primotec; Prinicid; R 42211; Solgard

CAS Registry Number: 23505-41-1

RTECS® Number: TF1610000

UN/NA & ERG Number: UN3278 (organophosphorus compound, toxic n.o.s.)/151

EC Number: 245-704-0 [*Annex I Index No.:* 015-099-00-6]

Regulatory Authority and Advisory Bodies

Superfund/EPCRA 40CFR355, Extremely Hazardous Substances: TPQ = 1000 lb (454 kg).

Reportable Quantity (RQ): 1000 lb (454 kg).

US DOT Regulated Marine Pollutant (49CFR172.101, Appendix B), severe pollutant.

US DOT 49CFR172.101, Inhalation Hazard Chemical as organophosphates.

European/International Regulations: Hazard Symbol: T, N; Risk phrases: R21; R25; R50/53; Safety phrases: S1/2; S23; S36/37; S45S; S60; S61 (see Appendix 4).

WGK (German Aquatic Hazard Class): No value assigned.

Description: Pirimifos-ethyl is a straw-colored liquid. Molecular weight = 333.43; It decomposes at 130°C; no boiling point can be determined; Vapor pressure = 0.0003 mmHg. Hazard Identification (based on NFPA-704 M Rating System): Health 3, Flammability 1, Reactivity 1. Decomposes in water.

Potential Exposure: Those involved in the manufacture, formulation, and application of this organophosphate soil insecticide.

Permissible Exposure Limits in Air
Protective Action Criteria (PAC)
TEEL-0: 5 mg/m^3
PAC-1: 15 mg/m^3
PAC-2: 25 mg/m^3
PAC-3: 60 mg/m^3

Determination in Water: Fish Tox = 1.70268000 ppb MATC (HIGH)

Routes of Entry: Inhalation, ingestion, skin and/or eye contact. Absorbed through the skin.

Harmful Effects and Symptoms

Short Term Exposure: As with other organophosphorus pesticides, symptoms are secondary to cholinesterase inhibition: headache, giddiness, blurred vision, nervousness, weakness, nausea, cramps, diarrhea, and discomfort in the chest. Other signs include sweating, tearing, salivation, vomiting, cyanosis, convulsions, coma, loss of reflexes, and loss of sphincter control.

Long Term Exposure: Cholinesterase inhibitor; cumulative effect is possible. This chemical may damage the nervous system with repeated exposure, resulting in convulsions, respiratory failure. May cause liver damage. Human Tox; 1.40000 ppb (HIGH).

Points of Attack: Respiratory system, lungs, central nervous system, cardiovascular system, skin, eyes, plasma and red blood cell cholinesterase.

Medical Surveillance: Before employment and at regular times after that, the following are recommended: plasma and red blood cell cholinesterase levels (tests for the enzyme poisoned by this chemical). If exposure stops, plasma levels return to normal in 1−2 weeks while red blood cell levels may be reduced for 1−3 months.

When cholinesterase enzyme levels are reduced by 25% or more below preemployment levels, risk of poisoning is increased, even if results are in lower ranges of "normal." Reassignment to work not involving organophosphate or carbamate pesticides is recommended until enzyme levels recover. If symptoms develop or overexposure occurs, repeat the above tests as soon as possible and get an examination of the nervous system. Also, consider complete blood count. Consider chest X-ray following acute overexposure. Do not drink any alcoholic beverages before or during use. Alcohol promotes absorption of organic phosphates.

First Aid: If this chemical gets into the eyes, remove any contact lenses at once and irrigate immediately for at least 15 min, occasionally lifting upper and lower lids. Seek medical attention immediately. If this chemical contacts the skin, remove contaminated clothing and wash immediately with soap and water. Speed in removing material from skin is of extreme importance. Shampoo hair promptly if contaminated. Seek medical attention immediately. If this chemical has been inhaled, remove from exposure, begin rescue breathing (using universal precautions, including resuscitation mask) if breathing has stopped and CPR if heart action has stopped. Transfer promptly to a medical facility. When this chemical has been swallowed, get medical attention. Give large quantities of water and induce vomiting. Do not make an unconscious person vomit.

Personal Protective Methods: Wear protective gloves and clothing to prevent any reasonable probability of skin contact. Safety equipment suppliers/manufacturers can provide recommendations on the most protective glove/clothing material for your operation. All protective clothing (suits, gloves, footwear, headgear) should be clean, available each day, and put on before work. Contact lenses should not be worn when working with this chemical. Wear splash-proof chemical goggles and face shield unless full face-piece respiratory protection is worn. Employees should wash immediately with soap when skin is wet or contaminated. Provide emergency showers and eyewash.

Respirator Selection: Follow regulations in OSHA 29CFR1910.134 or European Standard EN149. Use a NIOSH/MSHA- or European Standard EN149-approved respirator; or use an approved supplied-air respirator with a full face-piece operated in the positive-pressure mode, or with a full face-piece, hood, or helmet in the continuous-flow mode; or use a NIOSH/MSHA- or European Standard EN149-approved self-contained breathing apparatus with a full face-piece operated in a pressure-demand or other positive-pressure mode.

Storage: Color Code—Blue: Health Hazard/Poison: Store in a secure poison location. Prior to working with this chemical you should be trained on its proper handling and storage. Store in tightly closed containers in a cool, well-ventilated area away from oxidizers. Where possible, automatically pump liquid from drums or other storage containers to process containers. Sources of ignition, such as smoking and open flames, are prohibited where this chemical is handled, used, or stored. Metal containers involving the transfer of 5 gallons or more of this chemical should be grounded and bonded. Drums must be equipped with self-closing valves, pressure vacuum bungs, and flame arresters. Use only nonsparking tools and equipment, especially when opening and closing containers of this chemical. Wherever this chemical is used, handled, manufactured, or stored, use explosion-proof electrical equipment and fittings.

Shipping: This compound requires a shipping label of "POISONOUS/TOXIC MATERIALS." It falls in Hazard Class 6.1 and Packing Group III.

Spill Handling: Evacuate and restrict persons not wearing protective equipment from area of spill or leak until cleanup is complete. As with other organophosphorus pesticides, stay upwind; keep out of low areas. Ventilate closed spaces before entering them. Do not touch spilled material; stop leak if you can do so without risk. Use water spray to reduce vapors. *Small spills:* absorb with sand or other non-combustible absorbent material and place into containers for later disposal. *Large spills:* dike far ahead of spill for later disposal. Remove all ignition sources. Ventilate area of spill or leak. Keep this chemical out of a confined space, such as a sewer, because of the possibility of an explosion, unless the sewer is designed to prevent the buildup of explosive concentrations. It may be necessary to contain and dispose of this chemical as a hazardous waste. If material or contaminated runoff enters waterways, notify downstream users of potentially contaminated waters. Contact your local or federal environmental protection agency for specific recommendations. If employees are required to clean up spills, they must be properly trained and equipped. OSHA 1910.120(q) may be applicable.

Initial isolation and protective action distances: Distances shown are likely to be affected during the first 30 min after materials are spilled and could increase with time. If more than one tank car, cargo tank, portable tank, or large cylinder involved in the incident is leaking, the protective action distance may need to be increased. You may need to seek emergency information from CHEMTREC at (800) 424-9300 or seek professional environmental engineering assistance from the US EPA Environmental Response Team at (908) 548-8730 (24-h response line).

Small spills (From a small package or a small leak from a large package)
First: Isolate in all directions (feet/meters)   100/30
Then: Protect persons downwind (miles/kilometers)
Day   0.3/0.4
Night   0.8/1.2

Large spills (From a large package or from many small packages)
First: Isolate in all directions (feet/meters)   600/200
Then: Protect persons downwind (miles/kilometers)
Day   1.6/2.6
Night   2.8/4.5

Fire Extinguishing: This material may burn but does not ignite readily. For small fires, use dry chemical, carbon dioxide, water spray, or foam. For large fires, use water spray, fog, or foam. Stay upwind; keep out of low areas. Move containers from fire area if you can do it without risk. Fight fire from maximum distance. Dike fire control water for later disposal; do not scatter the material. Poisonous gases, including nitrogen oxide, phosphorus oxides, and sulfur oxides, are produced in fire. Vapors are heavier than air and will collect in low areas. Vapors in confined areas may explode when exposed to fire. Containers may explode in fire. Storage containers and parts of containers may rocket great distances, in many directions. If material or contaminated runoff enters waterways, notify downstream users of potentially contaminated waters. Notify local health and fire officials and pollution control agencies. From a secure, explosion-proof location, use water spray to cool exposed containers. If cooling streams are ineffective (venting sound increases in volume and pitch, tank discolors, or shows any signs of deforming), withdraw immediately to a secure position. If employees are expected to fight fires, they must be trained and equipped in OSHA 1910.156. The only respirators recommended for firefighting are self-contained breathing apparatuses that have full face-pieces and are operated in a pressure-demand or other positive-pressure mode.

Disposal Method Suggested: In accordance with 40CFR165, follow recommendations for the disposal of pesticides and pesticide containers. Must be disposed properly by following package label directions or by contacting your local or federal environmental control agency or by contacting your regional EPA office.

References
US Environmental Protection Agency. (November 30, 1987). *Chemical Hazard Information Profile: Pirimiphos-Ethyl.* Washington, DC: Chemical Emergency Preparedness Program

Platinum and compounds   P:0800

Molecular Formula: Pt
Synonyms: Elemental platinum; Platin (German); Platinum black
CAS Registry Number: 7440-06-4 (Platinum metal); 16941-12-1 (Chloroplatinic acid); 10025-65-7 (Platinous chloride); 13454-96-1 [Platinum(IV) chloride]; 592-06-3 (Platinum cyanide)
RTECS® Number: TP2160000 (Platinum metal); FW7040000 (Chloroplatinic acid); TP2275000 (Platinous chloride); TP2275500 (Platinum(IV) chloride)
UN/NA & ERG Number: Metal powder, in bulk, may be pyrophoric: UN2545 (powder, dry)/135; UN2507 (*chloroplatinic acid*)/ 154
EC Number: 231-116-1 (platinum)
Regulatory Authority and Advisory Bodies
Air Pollutant Standard Set. See below, "Permissible Exposure Limits in Air" section.
Canada, WHMIS, Ingredients Disclosure List Concentration 1.0%.
Carcinogenicity (Platinum(IV) chloride): EPA Gene-Tox Program, Positive: *B. subtilis* rec assay; *D. melanogaster* sex-linked lethal.
WGK (German Aquatic Hazard Class): Nonwater polluting agent. (*metal*)
Description: Platinum is a soft, ductile, malleable, silver-white metal. It is found in the metallic form and as the arsenide, sperrylite. It forms complex soluble salts, such as Na_2PtCl_6. It also forms halides. Metallic platinum is insoluble in water. Platinum(IV) chloride is red-brown crystals or powder. Molecular weight = 195.09; 336.89 [Platinum(IV) chloride] Freezing/Melting point = (decomposes) 370°C. Soluble in water.
Potential Exposure: Compound Description (metal): Drug, Tumorigen, (platinum(IV) chloride) Mutagen; Reproductive Effector; Primary Irritant. Platinum and its alloys are utilized because of their resistance to corrosion and oxidation,

particularly at high temperatures; their high electrical conductivity; and their excellent catalytic properties. They are used in relays, contacts and tubes in electronic equipment, in spark plug electrodes for aircraft, and windings in high-temperature electrical furnaces. Platinum alloys are used for standards for weight, length, and temperature measurement. Platinum and platinum catalysts (e.g., hexachloroplatinic acid, H_2PtCl_6) are widely used in the chemical industry in persulfuric, nitric, and sulfuric acid production, in the synthesis of organic compounds and vitamins, and for producing higher octane gasoline. They are coming into use in catalyst systems for control of exhaust pollutants from automobiles. They are used in the equipment for handling molten glass and manufacturing fibrous glass; in laboratory, medical, and dental apparatus; in electroplating; in photography; in jewelry; and in X-ray fluorescent screens. Because platinum complexes are used as antitumor agents, the potential for carcinogenic activity is present; tests to clarify this aspect should be conducted. While low levels of emissions of platinum particulate have been observed from some catalyst-equipped automobiles, the major potential source of Pt is from the disposal of spent catalysts.

Incompatibilities: Platinum metal is incompatible with aluminum, acetone, arsenic, ethane, hydrazine, hydrogen peroxide, lithium, phosphorus, selenium, tellurium, various fluorides. Platinum(IV) chloride and finely divided powders are incompatible with oxidizers.

Permissible Exposure Limits in Air

OSHA PEL (*soluble salts, as Pt*) 0.002 mg[Pt]/m^3 TWA.

NIOSH REL (*metal*): 1 mg/m^3; (*soluble salts, as Pt*): 0.002 mg[Pt]/m^3 TWA.

ACGIH TLV®[1] (*soluble salts, as Pt*): 1 mg/m^3 (metal); 0.002 mg[Pt]/m^3 TWA.

NIOSH IDLH: 4 mg [Pt]/m^3.

Protective Action Criteria (PAC)

TEEL-0: 1 mg/m^3

PAC-1: 3 mg/m^3

PAC-2: 4 mg/m^3

PAC-3: 4 mg/m^3

DFG MAK (*chloroplatinates*): 0.002 mg[Pt]/m^3 Ceiling Concentration (peak should not be exceeded); danger of skin and airway sensitization.

Protective Action Criteria (PAC) 16941-12-1, Dihydrogen hexachloroplatinate.

TEEL-0: 0.0042 mg/m^3

PAC-1: 0.25 mg/m^3

PAC-2: 1.5 mg/m^3

PAC-3: 8.4 mg/m^3

Australia: TWA 1 mg/m^3, 1993; Austria: MAK 1 mg/m^3, 1999; Belgium: TWA 1 mg/m^3, 1993; Finland: TWA 1 mg/m^3, 1999; France: VME 1 mg/m^3, 1999; Hungary: TWA 0.001 mg/m^3; STEL 0.002 mg/m^3, 1993; Norway: TWA 0.002 mg/m^3, 1999; the Netherlands: MAC-TGG 1 mg/m^3, 2003; Switzerland: MAK-W 0.002 mg[Pt]/m^3, 1999; Argentina, Bulgaria, Columbia, Jordan, South Korea, New Zealand, Singapore, Vietnam: ACGIH TLV®: TWA

1 mg/m^3. Several states have set guidelines or standards for platinum in ambient air[60] ranging from 0.1 μg/m^3 (Nevada) to 0.4−20.0 μg/m^3 (Connecticut) to 10.0 μg/m^3 (North Dakota) to 330.0 μg/m^3 (Virginia).

Platinum(IV)chloride

Arab Republic of Egypt: TWA 0.002 mg[Pt]/m^3 (*dust*), 1993; Australia: TWA 0.002 mg[Pt]/m^3, 1993; Austria: MAK 0.002 mg[Pt]/m^3, 1999; Belgium: TWA 0.002 mg [Pt]/m^3, 1993; Hungary: STEL 0.002 mg[Pt]/m^3, 1993; Norway: TWA 0.002 mg[Pt]/m^3, 1999; the Philippines: TWA 0.002 mg[Pt]/m^3, 1993; Switzerland: MAK-W 0.002 mg[Pt]/m^3, 1999; United Kingdom: LTEL 0.002 mg [Pt]/m^3, 1993; Argentina, Bulgaria, Columbia, Jordan, South Korea, New Zealand, Singapore, Vietnam: ACGIH TLV®: TWA 0.002 mg[Pt]/m^3

Platinum cyanide

Protective Action Criteria (PAC)

TEEL-0: 0.00253 mg/m^3

PAC-1: 0.0076 mg/m^3

PAC-2: 0.0127 mg/m^3

PAC-3: 5.07 mg/m^3

Determination in Air: Use NIOSH Analytical Method (IV) #7300, #7303, Elements by ICP; #8310, Metals in urine; #8005, Elements in blood or tissue, OSHA Analytical Method ID-121; ID-130-SG; NIOSH II(7), Method S-19, Soluble salts.

Routes of Entry: Inhalation, ingestion, skin and/or eye contact.

Harmful Effects and Symptoms

Short Term Exposure: Metal dust and fume may cause irritation of the eyes, skin, and respiratory tract. Metal particles in the eye can cause scratching and possible damage. Hazards arise from the dust, droplets, spray, or mist of complex salts of platinum but not from the metal itself. These salts are sensitizers of the skin, nasal mucosa, and bronchi, and cause allergic phenomena. One case of contact dermatitis from wearing a ring made of platinum alloy is recorded.

Long Term Exposure: Characteristic symptoms of poisoning occur after 2−6 months of exposure and include pronounced irritation of the throat and nasal passages, which results in violent sneezing and coughing; bronchial irritation, which causes respiratory distress; and irritation of the skin, which produces cracking, bleeding, and pain. Respiratory symptoms can be so severe that exposed individuals may develop status asthmaticus. After recovery, most individuals develop allergic symptoms and experience further asthma attacks when exposed to even minimal amounts of platinum dust or mists. Mild cases of dermatitis involve only erythema and urticaria of the hands and forearms. More severe cases affect the face and neck. All pathology is limited to allergic manifestations. EPA research efforts indicate that platinum is more active biologically and toxicologically than previously believed. It methylates in aqueous media, establishing a previously unrecognized biotransformation and distribution mechanism.

Points of Attack: Respiratory system, skin, eyes.

Medical Surveillance: NIOSH lists the following tests: whole blood (chemical/metabolite); biologic tissue/biopsy; urine (chemical/metabolite). In preemployment and periodic physical examinations, the skin, eyes, and respiratory tract are most important. Any history of skin or pulmonary allergy should be noted, as well as exposure to other irritants or allergens, and smoking history. Periodic assessment of pulmonary function may be useful.

First Aid: If this chemical gets into the eyes, remove any contact lenses at once and irrigate immediately for at least 15 min, occasionally lifting upper and lower lids. Seek medical attention immediately. If this chemical contacts the skin, remove contaminated clothing and wash immediately with soap and water. Seek medical attention immediately. If this chemical has been inhaled, remove from exposure, begin rescue breathing (using universal precautions, including resuscitation mask) if breathing has stopped and CPR if heart action has stopped. Transfer promptly to a medical facility. When chloroplatinic acid has been swallowed, get medical attention. If victim is *conscious*, administer water or milk. Do not induce vomiting.

Personal Protective Methods: Wear protective gloves and clothing to prevent any reasonable probability of skin contact. Safety equipment suppliers/manufacturers can provide recommendations on the most protective glove/clothing material for your operation. All protective clothing (suits, gloves, footwear, headgear) should be clean, available each day, and put on before work. Contact lenses should not be worn when working with this chemical. Wear appropriate splash- or dust-proof chemical goggles and face shield unless full face-piece respiratory protection is worn. Employees should wash immediately with soap when skin is wet or contaminated. Provide emergency showers and eyewash.

Respirator Selection: (for soluble Pt salts): *Up to 0.05 mg/m³:* Sa:Cf* (APF = 25) (any supplied-air respirator operated in a continuous-flow mode). *Up to 0.1 mg/m³:* 100F (APF = 50) (any air-purifying, full-face-piece respirator with an N100, R100, or P100 filter) or SCBAF (APF = 50) (any self-contained breathing apparatus with a full face-piece) or SaF (APF = 50) (any supplied-air respirator with a full face-piece). *Up to 4 mg/m³:* SaF: Pd,Pp (APF = 2000) (any supplied-air respirator that has a full face-piece and is operated in a pressure-demand or other positive-pressure mode). *Emergency or planned entry into unknown concentrations or IDLH conditions:* SCBAF: Pd, Pp (APF = 10,000) (any NIOSH/MSHA- or European Standard EN 149-approved self-contained breathing apparatus that has a full face-piece and is operated in a pressure-demand or other positive-pressure mode) or SaF: Pd, Pp: ASCBA (APF = 10,000) (any supplied-air respirator that has a full face-piece and is operated in a pressure-demand or other positive-pressure mode in combination with an auxiliary, self-contained breathing apparatus operated in a pressure-demand or other positive-pressure mode). *Escape:* 100 F (APF = 50) (any air-purifying, full-face-piece respirator with an N100, R100, or P100 filter) or SCBAE (any appropriate escape-type, self-contained breathing apparatus).
*Substance causes eye irritation and damage; eye protection needed.

Storage: Color Code—White (*chloroplatinic acid*): Corrosive or Contact Hazard; Store separately in a corrosion-resistant location. Prior to working with this chemical you should be trained on its proper handling and storage.

Shipping: Chloroplatinic acid, solid, is the only soluble salt cited. This compound requires a shipping label of "CORROSIVE." It falls in Hazard Class 8 and Packing Group III. Fulminating platinum is also cited. Its Hazard class is "FORBIDDEN."
Dry powder, in bulk, requires a shipping label of "SPONTANEOUSLY COMBUSTIBLE." It falls in Hazard Class 4.2 and Packing Group II.

Spill Handling: Evacuate persons not wearing protective equipment from area of spill or leak until cleanup is complete. Remove all ignition sources. Collect powdered material in the most convenient and safe manner and deposit in sealed containers. Ventilate area after cleanup is complete. It may be necessary to contain and dispose of this chemical as a hazardous waste. If material or contaminated runoff enters waterways, notify downstream users of potentially contaminated waters. Contact your local or federal environmental protection agency for specific recommendations. If employees are required to clean up spills, they must be properly trained and equipped. OSHA 1910.120(q) may be applicable.

Fire Extinguishing: This chemical is a combustible solid. Use dry chemicals appropriate for metal fires. *Do not use water.* Poisonous gases are produced in fire, including Pt. If material or contaminated runoff enters waterways, notify downstream users of potentially contaminated waters. Notify local health and fire officials and pollution control agencies. From a secure, explosion-proof location, use water spray to cool exposed containers. If cooling streams are ineffective (venting sound increases in volume and pitch, tank discolors, or shows any signs of deforming), withdraw immediately to a secure position. If employees are expected to fight fires, they must be trained and equipped in OSHA 1910.156. The only respirators recommended for firefighting are self-contained breathing apparatuses that have full face-pieces and are operated in a pressure-demand or other positive-pressure mode.

Disposal Method Suggested: Catalyst disposal is expected to be the largest contributor of Pt to the environment. The value of the metal would help to offset the cost of reclaiming the Pt from discarded catalysts. If direct vehicular emissions of Pt are found to be significant, particulate taps, which are available at reasonable cost, may provide a technological solution. In any event, recovery and recycling is the preferred technique for both health and economic reasons. Details of platinum recovery and recycling from plating wastes, platinum metal refinery effluents, spent catalysts, and precious metals scrap have been published.

References

US Environmental Protection Agency. (April 1974). *A Literature Search and Analysis of Information Regarding Uses, Production, Consumption, Reported Medical Cases and Toxicology of Platinum and Palladium*, Report PB-238,546. Research Triangle Park, NC

National Academy of Sciences. (1977). Medical and Biologic Effects and Environmental Pollutants: Platinum Group Metals. Washington, DC

Sax, N. I. (Ed.). (1981). *Dangerous Properties of Industrial Materials Report*, 1, No. 3, 74–75

US Environmental Protection Agency. (October 31, 1985). *Chemical Hazard Information Profile: Platinous Chloride*. Washington, DC: Chemical Emergency Preparedness Program

US Environmental Protection Agency. (October 31, 1985). *Chemical Hazard Information Profile: Platinum Tetrachloride*. Washington, DC: Chemical Emergency Preparedness Program

New Jersey Department of Health and Senior Services. (September 2002). *Hazardous Substances Fact Sheet: Platinum*. Trenton, NJ

Polybrominated biphenyls (PBBs)
P:0810

Molecular Formula: $C_{12}H_4Br_6$
Common Formula: $Br_3C_6H_2–C_6H_2Br_3$
Synonyms: Decabromobiphenyl; Firemaster BP-6®; Firemaster FF-1®; HBB; NCI-C53634; 2,4,5,2′,4′,5′-Hexabromobiphenyl; PBBs; PBB (BP-6); PBB (FF-1); Polybrominated biphenyl (BP-6); Polybrominated biphenyl (FF-1); Tetrabromo(tetrabromophenyl)benzene
CAS Registry Number: 36355-01-8 (hexabromobiphenyl); 27858-07-7 (octabromobiphenyl); 59536-65-1 [Firemaster BP-6
Polybrominated biphenyls category include the following:

p-Bromodiphenyl ether	101-55-3
Decabromobiphenyl	13654-09-6
Decabromodiphenyl ether	1163-19-5
*p,p′*Dibromodiphenyl ether	2050-47-7
Hexabromobiphenyl	59080-40-9
Hexabromo-1,1′-biphenyl	36355-01-8
Hexabromodiphenyl ether	36483-60-0
Nonabromodiphenyl ether	63936-56-1
Octabromobiphenyl	27858-07-7
Octabromobiphenyl	61288-13-9
Octabromodiphenyl ether	32536-52-0
Pentabromodiphenyl ether	32534-81-9
Polybrominated biphenyl	59536-65-1
Polybrominated biphenyl mixture	67774-32-7
Tetrabromodiphenyl ether	40088-47-9
Tribromodiphenyl ether	49690-94-0

RTECS® Number: DV5330000 (hexabromobiphenyl); LK5060000 (Firemaster BP-6); LK5065000 (Firemaster FF-1)
UN/NA & ERG Number: UN3152 (Polyhalogenated biphenyls, solid)/171; UN3151 (Polyhalogenated biphenyls, liquid)/171
EC Number: 252-994-2 (hexabromobiphenyl); 248-696-7 (octabromobiphenyl or [tetrabromo(tetrabromophenyl)benzene]; 237-137-2 (decabromobiphenyl)
Regulatory Authority and Advisory Bodies
Carcinogenicity: IARC: Human Limited Evidence, animal Sufficient Evidence, *possibly carcinogenic to humans*, Group 2B; NTP: Reasonably anticipated to be a human carcinogen.
Chemicals Subject to TSCA 12(b) Export Notification Requirements, Section 5: Any combination of the following substances resulting from a chemical reaction (as well as the individual chemicals): Tetrabromodiphenyl ether (CAS:40088-47-9); Pentabromodiphenyl ether (CAS: 32534-81-9); Hexabromodiphenyl ether (CAS: 36483-60-0); Heptabromodiphenyl ether (CAS: 68928-80-3); Octabromodiphenyl ether (CAS: 32536-52-0); Nonabromodiphenyl ether (CAS: 63936-56-1). For details, see the Proposed Significant New Use Rule (69 FR 70404, December 6, 2004).
Banned or Severely Restricted (Canada, USA) (UN)[13] (Germany, EEC) (UN).[35]
RCRA, 40CFR261, Appendix 8 Hazardous Constituents, waste number not listed.
RCRA 40CFR264, Appendix 9; TSD Facilities Ground Water Monitoring List. Suggested test method(s) (PQL µg/L): 8080 (50); 8250 (100).
EPCRA Section 313 Form R *de minimis* concentration reporting level: 0.1%.
California Proposition 65 Chemical: Cancer 1/1/88; Reproductive toxin 10/1/94.
Canada, WHMIS, Ingredients Disclosure List Concentration 0.1% (67774-32-7).
Rotterdam Convention Annex III [Chemicals Subject to the Prior Informed Consent Procedure (PIC)] [(13654-09-6 (deca-); 36355-01-8 (hexa-); 27858-07-7 (octa-)].
List of Stockholm Convention POPs: Annex A (Elimination) included in the same category are the following: hexabromobiphenyl (CAS 59080-40-9); hexabromobiphenyl ether (CAS 59080-40-9) and heptabromodiphenyl ether (CAS 68928-80-3); Tetrabromodiphenyl ether CAS (40088-47-9); and Pentabromodiphenyl ether (CAS 32534-81-9).
European/International Regulations: not listed in Annex 1.
WGK (German Aquatic Hazard Class): No value assigned.
Description: PBBs do not occur as natural products. Hexabromobiphenyl is the predominant isomer. These materials are heavy, highly brominated compounds. Typical is hexabromobiphenyl, $Br_3C_6H_2–C_6H_2Br_3$: Molecular weight = 627.62; decomposes at 300°C to 400°C. It will be used as an illustrative example of such compounds. PBBs

are produced by direct bromination of biphenyl, and it could be anticipated that very complex mixtures of compounds differing from each other both in number of bromine atoms per molecule and by positional isomerism are formed. The possibility also exists (analogous to the PCBs) that halogenated dibenzofurans (e.g., Brominated dibenzofurans) may be trace contaminants in certain PBB formulations. Hazard Identification (based on NFPA-704 M Rating System): Health 4, Flammability 1, Reactivity 0.

Potential Exposure: The polybrominated biphenyls (PBBs) are inert substances and have been employed, primarily as fire retardants. For example, the PBBs were incorporated into thermoplastics at a concentration of about 15% to increase the heat stability of the plastic to which it is added. In 1973, 1—2 tons of PBBs, a highly toxic flame retardant, were accidentally mixed with an animal feed supplement and fed to cattle in Michigan. Contamination also resulted from traces of PBBs being discharged into the environment at the manufacturing site and at other facilities involved in handling PBBs. Approximately 250 dairy and 500 cattle farms were quarantined, tens of thousands of swine and cattle and more than one million chickens were destroyed, and lawsuits involving hundreds of millions of dollars were instituted. Before the nature of the contamination was recognized, many of the contaminated animals had been slaughtered, marketed, eaten; and eggs and milk of the contaminated animals were also consumed. Thus, large numbers of people have been exposed to PBBs; they are persistent in the environment and are concentrated in body fat. While commercial manufacture and distribution of PBBs have currently ceased, the full extent of the problem has not yet been assessed.

Permissible Exposure Limits in Air
No standards or TEEL available.

Routes of Entry: Inhalation, ingestion, skin and/or eye contact.

Harmful Effects and Symptoms
The accidental contamination, in 1973, of animal feed and livestock throughout Michigan by polybrominated biphenyl flame retardants (Firemaster BP-6) has stimulated extensive studies of the potential for water contamination, transport, bioaccumulation, biological and toxicological nature of this class of environmental agent. While no immediate adverse health effects were noted in several thousand Michigan farm families that consumed milk and dairy products contaminated with PBBs, it is not possible to determine at this date any chronic or delayed effects that might be attributed to the PBBs or the potential ability of this chemical to cause birth defects.

Firemaster FF-1 (Firemaster BP-6 containing 2% of calcium trisilicate)—a mixture of pentabromobiphenyl, hexabromobiphenyl, and heptabromobiphenyl, with hexabromobiphenyl being the major component—administered by gavage produced neoplastic nodules and hepatocellular carcinomas in female Sherman strain rats. In another bioassay, Firemaster FF-1, also administered by gavage, was carcinogenic to Fisher 344 rats and B6C3F1 mice of each sex,

inducing neoplastic nodules, hepatocellular carcinomas, and chloangiocarcinomas in rats and hepatocellular carcinomas in mice.

Long Term Exposure: Confirmed carcinogen. Experimental teratogenic and reproductive effects. Mutation data reported.

Points of Attack: Liver, kidneys, skin.

Medical Surveillance: Consider the points of attack in preplacement and periodic physical examinations. Liver and kidney function tests.

First Aid: If this chemical gets into the eyes, remove any contact lenses at once and irrigate immediately for at least 15 min, occasionally lifting upper and lower lids. Seek medical attention immediately. If this chemical contacts the skin, remove contaminated clothing and wash immediately with soap and water. Seek medical attention immediately. If this chemical has been inhaled, remove from exposure, begin rescue breathing (using universal precautions, including resuscitation mask) if breathing has stopped and CPR if heart action has stopped. Transfer promptly to a medical facility. When this chemical has been swallowed, get medical attention. Give large quantities of water and induce vomiting. Do not make an unconscious person vomit.

Personal Protective Methods: Wear protective gloves and clothing to prevent any reasonable probability of skin contact. Safety equipment suppliers/manufacturers can provide recommendations on the most protective glove/clothing material for your operation. All protective clothing (suits, gloves, footwear, headgear) should be clean, available each day, and put on before work. Contact lenses should not be worn when working with this chemical. Wear chemical goggles and face shield unless full face-piece respiratory protection is worn. Employees should wash immediately with soap when skin is wet or contaminated. Provide emergency showers and eyewash.

Respirator Selection: Follow regulations in OSHA 29CFR1910.134 or European Standard EN149. Use a NIOSH/MSHA- or European Standard EN149-approved respirator; or use an approved supplied-air respirator with a full face-piece operated in the positive-pressure mode, or with a full face-piece, hood, or helmet in the continuous-flow mode; or use a NIOSH/MSHA- or European Standard EN149-approved self-contained breathing apparatus with a full face-piece operated in a pressure-demand or other positive-pressure mode.

Storage: Color Code—Blue: Health Hazard/Poison: Store in a secure poison location. Prior to working with this chemical you should be trained on its proper handling and storage. A regulated, marked area should be established where this chemical is handled, used, or stored in compliance with OSHA Standard 1910.1045.

Shipping: Polyhalogenated biphenyls, liquid, require a shipping label of "CLASS 9." They fall in Hazard Class 9 and Packing Group II.

Spill Handling: Evacuate persons not wearing protective equipment from area of spill or leak until cleanup is

complete. Remove all ignition sources. It may be necessary to contain and dispose of this chemical as a hazardous waste. If material or contaminated runoff enters waterways, notify downstream users of potentially contaminated waters. Contact your local or federal environmental protection agency for specific recommendations. If employees are required to clean up spills, they must be properly trained and equipped. OSHA 1910.120(q) may be applicable.

Fire Extinguishing: Poisonous gases are produced in fire, including bromine. If material or contaminated runoff enters waterways, notify downstream users of potentially contaminated waters. Notify local health and fire officials and pollution control agencies. From a secure, explosion-proof location, use water spray to cool exposed containers. If cooling streams are ineffective (venting sound increases in volume and pitch, tank discolors, or shows any signs of deforming), withdraw immediately to a secure position. If employees are expected to fight fires, they must be trained and equipped in OSHA 1910.156. The only respirators recommended for firefighting are self-contained breathing apparatuses that have full face-pieces and are operated in a pressure-demand or other positive-pressure mode.

References

US Environmental Protection Agency. (December 1979). *Status Assessment of Toxic Chemicals: Polybrominated Biphenyls*, Report EPA-600/2-79-210k. Washington, DC

National Toxicology Program. (1982). *NTP: Technical Report on the Toxicology and Carcinogenesis Bioassay of Polybrominated Biphenyl Mixture (Firemaster FF-1)*, Technical Report Series No. 244, NIH Publication No. 82-1800. Research Triangle Park, NC

Polychlorinated biphenyls (PCBs)
P:0820

Molecular Formula: $C_{12}H_{10-x}Cl_x$

Synonyms: Aroclor; Aroclor 1221; Aroclor 1232; Aroclor 1242; Aroclor 1248; Aroclor 1254; Aroclor 1260; Aroclor 1262; Aroclor 1268; Aroclor 2565; Aroclor 4465; Biphenyl, Chlorinated; 1,1'-Biphenyl, chloro derivs.; Biphenyl, polychloro-; Chlophen; Chlorextol; Chlorinated biphenyl; Chlorinated diphenyl; Chlorinated diphenylene; Chloro biphenyl; Chloro 1,1-biphenyl; Clophen; Dykanol; Diphenyl, chlorinated; Fenclor; Inerteen; Kanechlor; Kanechlor 300; Kanechlor 400; Kanechlor 500; Montar; Noflamol; PCB; PCBS; Phenochlor; Phenoclor; Polychlorobiphenyl; Pyralene; Pyranol; Santotherm; Santotherm FR; Sovol; the rminol FR-1

CAS Registry Number: 1336-36-3 (Aroclor PCBs); 53469-21-9; *(alt.)* 11104-29-3 (Aroclor 1242) (42% Cl); 12672-29-6 (Aroclor 1248) (48% Cl); 11097-69-1 (Aroclor 1254) (54% Cl); 11096-82-5 (Aroclor 1260) (60% Cl); 37324-23-5 (Aroclor 1262) (62% Cl); 11100-14-4 (Aroclor 1268) (68% Cl); 55720-99-5 (PCB oxide)

RTECS® Number: TQ1350000; TQ1356000 (Aroclor 1242) (42% Cl); TQ1358000 (Aroclor 1248) (48% Cl); TQ1360000 (Aroclor 1254) (54% Cl); TQ1362000 (Aroclor 1260) (60% Cl); TQ1364000 (Aroclor 1262) (62% Cl); TQ1366000 (Aroclor 1268) (68% Cl)

UN/NA & ERG Number: UN2315/171

EC Number: 215-648-1 [*Annex I Index No.:* 602-039-00-4] (PCBs)

Regulatory Authority and Advisory Bodies

Carcinogenicity: (*1336-36-3*): Carcinogenicity: IARC: Animal Sufficient Evidence; Human Limited Evidence, Group 2A, 1998; EPA: Sufficient evidence from animal studies; inadequate evidence or no useful data from epidemiologic studies; NTP: Reasonably anticipated to be a human carcinogen; NIOSH: Potential occupational carcinogen.

Chemicals Subject to TSCA 12(b) Export Notification Requirements, Section 6.

Rotterdam Convention Annex III [Chemicals Subject to the Prior Informed Consent Procedure (PIC)] *as PCBs*.

Aroclor 1242 NTP: 11th Report on Carcinogens, 2002: Reasonably anticipated to be a human carcinogen; IARC *(PCB)*: Animal Sufficient Evidence; Human Limited Evidence, Group 2A, 1987; EPA *(PCB)*: Sufficient evidence from animal studies; inadequate evidence or no useful data from epidemiologic studies.

Aroclor 1254 NTP: 11th Report on Carcinogens, 2004: Reasonably anticipated to be a human carcinogen; NCI: Carcinogenesis Bioassay (feed); equivocal evidence: rat; IARC *(PCB)*: Animal Sufficient Evidence; Human Limited Evidence, Group 2A, 1987; EPA *(PCB)*: Sufficient evidence from animal studies; inadequate evidence or no useful data from epidemiologic studies.

US EPA Gene-Tox Program, Negative: Cytogenetics—male germ cell; Rodent dominant lethal (Aroclor 1242); Negative: SHE—clonal assay; Rodent dominant lethal; Negative: Sperm morphology—mouse; Inconclusive: Mammalian micronucleus (Aroclor 1254).

Banned or Severely Restricted (many countries) (UN).[13]

Persistent Organic Pollutants (UN).

Air Pollutant Standard Set. See below, "Permissible Exposure Limits in Air" section.

Clean Air Act: Hazardous Air Pollutants (Title I, Part A, Section 112).

Clean Water Act: Section 311 Hazardous Substances/RQ 40CFR117.3 (same as CERCLA, see below); 40CFR401.15 Section 307 Toxic Pollutants; 40CFR423, Appendix A.

Priority Pollutants; Section 313 Water Priority Chemicals (57FR41331, 9/9/92).

RCRA 40CFR268.48; 61FR15654, Universal Treatment Standards: Wastewater (mg/L), 0.10; Nonwastewater (mg/kg), 10, total PCBs, sum of all PCB isomers, or all AROCLORS.

RCRA 40CFR264, Appendix 9; TSD Facilities Ground Water Monitoring List. Suggested test method(s) (PQL µg/L): 8080 (50); 8250 (100).

Safe Drinking Water Act: MCL, 0.0005 mg/L; MCLG, zero; Regulated chemical (47 FR 9352).

Reportable Quantity (RQ): 1 lb (0.454 kg).

EPCRA Section 313 Form R *de minimis* concentration reporting level: 0.1%.

US DOT Regulated Marine Pollutant (49CFR172.101, Appendix B), severe pollutant.

California Proposition 65 Chemical: Cancer 10/1/89; Reproductive toxin 1/1/91; (containing ≥60% chlorine by molecular weight) 1/1/88.

Canada, WHMIS, Ingredients Disclosure List Concentration 0.1%.

Note: The EPA requires the following: All PCB transformer locations must be cleared of stored combustible materials (solvents, paints, paper, etc.). All PCB transformers must be registered with the local fire department. All PCB-containing equipment must be posted with a large yellow label and the exterior door of the vault, machinery room door and any other means of exit must also be marked with PCB yellow ID labels. All PCB-containing transformers must be inspected every 3[3] weeks; leaks must be repaired within 2[2] days and reported to the EPA within 5[5] days. In order to prevent fires, EPA recently required that high-voltage network transformers be removed and that enhanced electrical protection be added on many types of PCB transformers in commercial buildings.

List of Stockholm Convention POPs: Annex A (Elimination); Annex C (Unintentional production and release) *as PCBs*.

European/International Regulations: Hazard Symbol (*PCB*): Xn, N; Risk phrases: R33; R50/53; Safety phrases: S2; S35; S60; S61 (see Appendix 4).

WGK (German Aquatic Hazard Class): 3—Severe hazard to waters.

Description: Arochlor 1242 is a colorless to light yellow-colored viscous liquid with a mild, hydrocarbon odor. Molecular weight = 258 (approx.); Specific gravity (H_2O:1) = 1.39 at 25°C; Boiling point = 325–366°C; Freezing/Melting point = −18°C; Vapor pressure = 0.001 mmHg at 25°C. Insoluble in water. Arochlor 1254 is a colorless to pale-yellow, viscous liquid (resinous state) or solid (below 10°C) with a mild, hydrocarbon odor. Molecular weight = 326 (approx.); Specific gravity (H_2O:1) = 1.39 at 25°C; Boiling point = 365–390°C; Freezing/Melting point = −18°C; Vapor pressure = 0.00006 mmHg at 25°C. Insoluble in water. Hazard Identification (based on NFPA-704 M Rating System): Health 2, Flammability 1, Reactivity 0. $C_{12}H_{10-x}Cl_x$, diphenyl rings, in which one or more hydrogen atoms are replaced by a chlorine atom. Most widely used are chlorodiphenyl (42% chlorine), containing 3 chlorine atoms in unassigned positions, and chlorodiphenyl (54% chlorine) containing 5 chlorine atoms in unassigned positions. These compounds are light, straw-colored liquids with typical chlorinated aromatic odors; 42% chlorodiphenyl is a mobile liquid and 54% chlorodiphenyl is a viscous liquid. Insoluble in water. Polychlorinated biphenyls are prepared by the chlorination of biphenyl and hence are complex mixtures containing isomers of chlorobiphenyls with different chlorine contents. It should be noted that there are 209 possible compounds obtainable by substituting chlorine for hydrogen from 1–10 different positions on the biphenyl ring system. An estimated 40–70 different chlorinated biphenyl compounds can be present in each of the higher chlorinated commercial mixtures. For example, Aroclor 1254 contains 69 different molecules, which differ in the number and position of chlorine atoms. It should also be noted that certain PCB commercial mixtures (no longer produced in the United States) but produced in France, Germany, and Japan have been shown to contain other classes of chlorinated derivatives, e.g., chlorinated naphthalenes and chlorinated dibenzofurans. The possibility that naphthalene and dibenzofuran contaminate the technical biphenyl feedstock used in the preparation of the commercial PCB mixtures cannot be excluded.

Potential Exposure: Compound Description (Aroclor 1242): Agricultural Chemical; Tumorigen, Mutagen; Reproductive Effector; Human Data; (Aroclor 1254) Agricultural Chemical; Tumorigen, Mutagen; Reproductive Effector. PCBs are mixtures of individual chemicals which are no longer produced in the United States but are still found in the environment. Chlorinated diphenyls are used alone and in combination with chlorinated naphthalenes. They are stable, thermoplastic, and nonflammable; they are used in heat transfer and hydraulic fluids, lubricants, and insecticide formulations; they found use in insulation for electric cables and wires; in the production of electric condensers; as additives for extreme pressure lubricants; and as a coating in foundry use. Polychlorinated biphenyls (PCBs, first introduced into commercial use more than 45 years ago) are one member of a class of chlorinated aromatic organic compounds which are of increasing concern because of their apparent ubiquitous dispersal, persistence in the environment, and tendency to accumulate in food chains, with possible adverse effects on animals at the top of food webs, including man.

Incompatibilities: Strong oxidizers.

Permissible Exposure Limits in Air

Protective Action Criteria (PAC)

1336-36-3 (Aroclor PCBs); (42% Cl)
TEEL-0: 0.04 mg/m^3
PAC-1: 0.125 mg/m^3
PAC-2: 0.75 mg/m^3
PAC-3: 500 mg/m^3
12674-11-2 (Aroclor 1016)
TEEL-0: 12.5 mg/m^3
PAC-1: 40 mg/m^3
PAC-2: 300 mg/m^3
PAC-3: 500 mg/m^3
53469-21-9; (Aroclor 1242) (42% Cl)
TEEL-0: 1 mg/m^3
PAC-1: 1 mg/m^3
PAC-2: 1 mg/m^3

PAC-3: 500 mg/m^3
11104-28-2 (Aroclor 1221) (21% Cl); 11141-16-5 (Aroclor 1232) (32% Cl); 12672-29-6 (Aroclor 1248) (48% Cl)
TEEL-0: 3 mg/m^3
PAC-1: 7.5 mg/m^3
PAC-2: 60 mg/m^3
PAC-3: 500 mg/m^3
11097-69-1 (Aroclor 1254) (54% Cl)
TEEL-0: 0.5 mg/m^3
PAC-1: 35 mg/m^3
PAC-2: 250 mg/m^3
PAC-3: 400 mg/m^3
11096-82-5 (Aroclor 1260) (60% Cl); (Aroclor 1261/1262) (61% Cl)
TEEL-0: 0.4 mg/m^3
PAC-1: 1.25 mg/m^3
PAC-2: 7.5 mg/m^3
PAC-3: 500 mg/m^3
37324-23-5 (Aroclor 1262) (62% Cl)
TEEL-0: 0.6 mg/m^3
PAC-1: 1.5 mg/m^3
PAC-2: 12.5 mg/m^3
PAC-3: 500 mg/m^3
11100-14-4 (Aroclor 1268) (68% Cl)
TEEL-0: 0.0125 mg/m^3
PAC-1: 0.04 mg/m^3
PAC-2: 0.3 mg/m^3
PAC-3: 500 mg/m^3
OSHA PEL: 1 mg/m^3 TWA [skin]
NIOSH REL: 0.001 mg/m^3 TWA (The REL also applies to other PCBs); A potential occupational carcinogen. Limit exposure to lowest feasible concentration. See *NIOSH Pocket Guide*, Appendix A.
ACGIH TLV®[1]: 1 mg/m^3 TWA [skin].
DFG MAK: 0.1 ppm/1.1 mg/m^3 TWA; Peak Limitation Category II(8) [skin]; Carcinogen Category 3B; Pregnancy Risk Group B.
NIOSH IDLH: potential occupational carcinogen 5 mg/m^3.
Australia: TWA 1 mg/m^3; STEL 2 mg/m^3, [skin], carcinogen, 1993; Austria: MAK 0.1 ppm (1 mg/m^3), [skin], suspected carcinogen, 1999; Belgium: TWA 1 mg/m^3; STEL 2 mg/m^3, [skin], 1993; Finland: TWA 0.5 mg/m^3; STEL 1.5 mg/m^3, [skin], 1993; France: VME 1 mg/m^3, [skin], 1999; Japan: 0.1 mg/m^3, [skin], 2A carcinogen, 1999; the Netherlands: MAC-TGG 1 mg/m^3, [skin], 2003; the Philippines: TWA 1 mg/m^3, [skin], 1993; Poland: MAC (TWA) 1 mg/m^3, 1999; Sweden: NGV 0.01 mg/m^3, KTV 0.3 mg/m^3, [skin], carcinogen, 1999; Switzerland: MAK-W 0.1 ppm (1 mg/m^3), [skin], 1999; United Kingdom: TWA 0.1 mg/m^3, [skin], 2000; Argentina, Bulgaria, Columbia, Jordan, South Korea, New Zealand, Singapore, Vietnam: ACGIH TLV®: TWA 1 mg/m^3 [skin]
54% chlorine
OSHA PEL: 0.5 mg/m^3 TWA [skin].
NIOSH REL: 0.001 mg/m^3 (applies to all PCBs) TWA [skin]. See *NIOSH Pocket Guide*, Appendix A.

ACGIH TLV®[1]: 0.5 mg/m^3 TWA [skin]; confirmed animal carcinogen with unknown relevance to humans.
DFG MAK: 0.05 ppm/0.70 mg/m^3 TWA; Peak Limitation Category II(8) [skin]; Carcinogen Category 3B; Pregnancy Risk Group B.
NIOSH IDLH: 5 mg/m^3.
Australia: TWA 0.5 mg/m^3; STEL 1 mg/m^3, [skin], Carcinogen, 1993; Austria: MAK 0.05 ppm (0.5 mg/m^3), [skin], suspected carcinogen, 1999; Belgium: TWA 0.5 mg/m^3; STEL 1 mg/m^3, [skin], 1993; France: VME 0.5 mg/m^3, [skin], 1999; Japan: 0.1 mg/m^3, [skin], 2A carcinogen, 1999; Norway: TWA 0.01 mg/m^3, 1999; the Netherlands: MAC-TGG 0.5 mg/m^3, [skin], 2003; Poland: MAC (TWA) 1 mg/m^3, 1999; Sweden: NGV 0.1 mg/m^3, KTV 0.3 mg/m^3, [skin], carcinogen, 1999; Switzerland: MAK-W 0.05 ppm (0.5 mg/m^3), [skin], 1999; Turkey: TWA 1 mg/m^3, [skin], 1993; United Kingdom: TWA 0.1 mg/m^3, [skin], 2000; Argentina, Bulgaria, Columbia, Jordan, South Korea, New Zealand, Singapore, Vietnam: ACGIH TLV®: confirmed animal carcinogen with unknown relevance to humans. The Czech Republic has set a TWA of 0.5 mg/m^3 and a ceiling value of 1.0 mg/m^3.
PCB oxide
0.5 mg/m^3, inhalable fraction [skin].
Several states have set guidelines or standards for PCB's in ambient air[60] as follows (all in µg/m^3).
Determination in Air: Use NIOSH Analytical Method #5503, Polychlorobiphenyls; #8004, Polychlorobiphenyls in serum; #PV-2089, Chlorodiphenyl (42% Chlorine); #PV-2088, Chlorodiphenyl (54% Chlorine).
Permissible Concentration in Water: To protect freshwater aquatic life: 0.014 µg/L as a 24-h average. To protect saltwater aquatic life: 0.030 µg/L as a 24-h average. To protect human health: preferably zero. An additional lifetime cancer risk of 1 in 100,000 results at a level of 0.00079 µg/L.[6] The EPA has set a maximum contaminant level of 0.0005 milligrams PCBs per liter of drinking water (0.0005 mg/L). The EPA requires that spills of 1 lb or more of PBCs be reported to the EPA.
Determination in Water: Gas chromatography (EPA Method 608) or gas chromatography plus mass spectrometry (EPA Method 625). Octanol−water coefficient: Log K_{ow} = 6.25 (54% Chlorine) (estimated).
Routes of Entry: Inhalation, ingestion, skin and/or eye contact. Inhalation of fume or vapor and percutaneous absorption of liquid, ingestion, skin and/or eye contact.
Harmful Effects and Symptoms
Short Term Exposure: *Inhalation:* May produce irritation to nose, throat, and lungs. The vapors can cause coughing and/or difficulty in breathing. Levels above 10 mg/m^3 are reported to be unbearable. Inhalation may contribute significantly to all symptoms of long-term exposure. *Skin:* Absorption moderate. Contributes significantly to all symptoms of long-term exposure. Sensitized individuals may develop a rash after 2 days exposure by contact or inhalation. *Eyes:* May produce irritation and burns. Levels of

$10\ mg/m^3$ are severely irritating. *Ingestion:* Absorption in digestive system contributes significantly to all symptoms of long-term exposure. There are no reported deaths of humans due to a single ingestion. However, experiments in animals suggest that ingestion of 6–10 fluid ounces would cause death to a healthy a 150-lb adult.

Long Term Exposure: Repeated or prolonged contact with skin may cause acne-like skin rash (chloroacne). The substance may cause liver damage. High exposure can damage the nervous system. PCBs are readily absorbed into the body by all routes of exposure. They may persist in tissues for years after exposure stops. The symptoms below may be due to PCBs or to chemical contaminants. High levels of PCB vapor, $1–10\ mg/m^3$, may produce burning feeling in eyes, nose, and face; dry throat; lung and throat irritation; nausea, dizziness, and aggravation of acne. These may be felt immediately or be delayed weeks or months. Chemical acne, black heads, dark patches on skin, and unusual eye discharge have been reported by all routes of exposure. Although some sensitive individuals have reported these effects after 2 days, onset may not occur for months. These effects may last for months. Digestive disturbance have been reported in some individuals. PCBs may impair the function of the immune system. High levels of PCBs have been shown to produce cancer and birth defects in laboratory animals. Whether PCBs produce these effects in humans is not known.

Points of Attack: Liver, skin, nervous system. Cancer site in animals: liver.

Medical Surveillance: NIOSH lists the following tests: Adipose Tissue; whole blood (chemical/metabolite); blood serum; blood plasma; liver function tests. Before beginning employment and at regular times after that, the following are recommended: Serum Triglycerides level. Examination of the skin. Examination of the nervous system. Nerve conduction studies should be considered.

First Aid: If this chemical gets into the eyes, remove any contact lenses at once and irrigate immediately for at least 15 min, occasionally lifting upper and lower lids. Seek medical attention immediately. If this chemical contacts the skin, remove contaminated clothing and wash immediately with soap and water. Seek medical attention immediately. If this chemical has been inhaled, remove from exposure, begin rescue breathing (using universal precautions, including resuscitation mask) if breathing has stopped and CPR if heart action has stopped. Transfer promptly to a medical facility. When this chemical has been swallowed, get medical attention. Give large quantities of water and induce vomiting. Do not make an unconscious person vomit.

Personal Protective Methods: Wear protective gloves and clothing to prevent any reasonable probability of skin contact. *Chlorodiphenyl* (42% chlorine) 53469-21-9. Prevent skin contact. **8 h** (more than 8 h of resistance to breakthrough $>0.1\ \mu g/cm/min$): butyl rubber gloves, suits, boots; Neoprene™ rubber gloves, suits, boots; Teflon™ gloves, suits, boots; Viton™ gloves, suits; Saranex™ coated suits,

Barricade™ coated suits; Responder™ suits; **4 h** (At least 4 but <8 h of resistance to breakthrough $>0.1\ \mu g/cm^2/min$): 4H™ and Silver Shield™ gloves. *Chlorodiphenyl* (54% chlorine) 11097-69-1. Prevent skin contact. **8 h** (more than 8 h of resistance to breakthrough $>0.1\ \mu g/cm/min$): butyl rubber gloves, suits, boots; Neoprene™ rubber gloves, suits, boots; Teflon™ gloves, suits, boots; Viton™ gloves, suits, Saranex™ coated suits, Barricade™ coated suits; Responder™ suits; **4 h** (At least 4 but <8 h of resistance to breakthrough $>0.1\ \mu g/cm^2/min$): 4H™ and Silver Shield™ gloves. Safety equipment suppliers/manufacturers can provide recommendations on the most protective glove/clothing material for your operation. All protective clothing (suits, gloves, footwear, headgear) should be clean, available each day, and put on before work. Contact lenses should not be worn when working with this chemical. Wear splash-proof chemical goggles and face shield unless full face-piece respiratory protection is worn. Employees should wash immediately with soap when skin is wet or contaminated. Provide emergency showers and eyewash. Specific engineering controls are recommended in NIOSH Criteria Document: #77-225, *Occupational Exposure to Polychlorinated Biphenyls.*

Respirator Selection: NIOSH: *At any detectable concentration:* SCBAF: Pd,Pp (APF = 10,000) (any self-contained breathing apparatus that has a full face-piece and is operated in a pressure-demand or other positive-pressure mode) or SaF: Pd,Pp: ASCBA (APF = 10,000) (any supplied-air respirator that has a full face-piece and is operated in a pressure-demand or other positive-pressure mode in combination with an auxiliary, self-contained breathing apparatus operated in a pressure-demand or other positive-pressure mode). *Escape:* GmFOv100 (APF = 50) (any air-purifying, full-face-piece respirator (gas mask) with a chin-style, front- or back-mounted organic vapor canister having an N100, R100, or P100 filter) or SCBAE (any appropriate escape-type, self-contained breathing apparatus).

Storage: Color Code—Blue: Health Hazard/Poison: Store in a secure poison location. Prior to working with these chemicals you should be trained on its proper handling and storage. Store in tightly closed containers in a cool, well-ventilated area away from strong oxidizers (such as chlorine, bromine, and fluorine). A regulated, marked area should be established where this chemical is handled, used, or stored in compliance with OSHA Standard 1910.1045.

Shipping: Polychlorinated biphenyls require a shipping label of "CLASS 9." They fall in DOT Hazard Class 9 and Packing Group II.

Spill Handling: Evacuate and restrict persons not wearing protective equipment from area of spill or leak until cleanup is complete. Remove all ignition sources. Ventilate area of spill or leak. Absorb liquids in vermiculite, dry sand, earth, peat, carbon, or a similar material and deposit in sealed containers. Keep this chemical out of a confined space, such as a sewer, because of the possibility of an explosion, unless the sewer is designed to prevent the buildup of explosive concentrations. It may be necessary to contain and dispose of this chemical as a hazardous waste. If material or contaminated runoff enters waterways, notify downstream users of potentially contaminated waters. Contact your local or federal environmental protection agency for specific recommendations. If employees are required to clean up spills, they must be properly trained and equipped. OSHA 1910.120(q) may be applicable. Spills of 1[1] lb or more must be reported to the EPA.

Fire Extinguishing: PCBs may burn but do not easily ignite. Poisonous gases, including dioxin and chlorinated dibenzofurans, are produced in fire. Use dry chemical, carbon dioxide, or alcohol foam extinguishers. Vapors are heavier than air and will collect in low areas. Vapors in confined areas may explode when exposed to fire. Containers may explode in fire. Storage containers and parts of containers may rocket great distances, in many directions. If material or contaminated runoff enters waterways, notify downstream users of potentially contaminated waters. Notify local health and fire officials and pollution control agencies. From a secure, explosion-proof location, use water spray to cool exposed containers. If cooling streams are ineffective (venting sound increases in volume and pitch, tank discolors, or shows any signs of deforming), withdraw immediately to a secure position. If employees are expected to fight fires, they must be trained and equipped in OSHA 1910.156. The only respirators recommended for firefighting are self-contained breathing apparatuses that have full face-pieces and are operated in a pressure-demand or other positive-pressure mode.

Disposal Method Suggested: Incineration at 1648°C with scrubbing to remove any chlorine-containing products. In addition, some chemical waste landfills have been approved for PCB disposal. More recently, treatment with metallic sodium has been advocated which yields a low molecular weight polyphenylene and sodium chloride.

References

National Institute for Occupational Safety and Health. (1977). *Criteria for a Recommended Standard: Occupational Exposure to Polychlorinated Biphenyls*, NIOSH Document No. 77-225

US Environmental Protection Agency. (1980). *Polychlorinated Biphenyls: Ambient Water Quality Criteria*. Washington, DC

National Academy of Sciences. (1979). *Polychlorinated Biphenyls*. Washington, DC

World Health Organization. (1976). Polychlorinated Biphenyls and Triphenyls, Environmental Health Criteria No. 2. Geneva, Switzerland

Sax, N. I. (Ed.). *Dangerous Properties of Industrial Materials Report*, 3, No. 4, 95–100 (1983) and 6, No. 2, 28–34 (1986)

US Public Health Service. (November 1987). *Toxicological Profile for Selected PCB's*. Atlanta, GA: Agency for Toxic Substances and Disease Registry

New York State Department of Health. (March 1986). *Chemical Fact Sheet: PCB's*. Version 2. Albany, NY: Bureau of Toxic Substance Assessment

New Jersey Department of Health and Senior Services. (April 2002). *Hazardous Substances Fact Sheet: Polychlorinated Biphenyls*. Trenton, NJ

Portland cement   P:0830

Molecular Formula: Ca_2O_4Si; $Al_2Ca_3O_6$; Ca_3O_5Si; $Al_2Ca_4Fe_2O_{10}$

Common Formula: $2CaO \cdot SiO_2$; $3CaO \cdot Al_2O_3$; $3CaO \cdot SiO_2$; $4CaO \cdot Al_2O_3 \cdot Fe_2O_3$

Synonyms: Cement; Hydraulic cement; Portland cement silicate

CAS Registry Number: 65997-15-1; 68475-76-3 (flue dust, portland cement)

RTECS® Number: VV8770000

EC Number: 266-043-4; 270-659-9 (flue dust, portland cement)

Regulatory Authority and Advisory Bodies
Air Pollutant Standard Set. See below, "Permissible Exposure Limits in Air" section.

Description: Portland cement is a class of hydraulic cements containing tri- and dicalcium silicate in addition to varying amounts of alumina, tricalcium aluminate, and iron oxide. The quartz content of most is below 1%. The average composition of regular Portland cement is as follows: CaO (64.0%); SiO_2 (21.0%); Al_2O_3 (5.8%); Fe_2O_3 (2.9%); MgO (2.5%); Alkali oxides (1.4%); SO_3 (1.7%). Freezing/Melting point = >1050°C. Reacts with water.

Potential Exposure: Cement is used as a binding agent in mortar and concrete (a mixture of cement, gravel, and sand). Potentially hazardous exposure may occur during both the manufacture and use of cement.

Incompatibilities: None reported.

Permissible Exposure Limits in Air
OSHA PEL: 50 mppcf TWA or 15 mg/m^3, total dust TWA; 5 mg/m^3 TWA, respirable fraction.

NIOSH REL: 10 mg/m^3, total dust TWA; 5 mg/m^3 TWA, respirable fraction.

ACGIH TLV®[1]: 1 mg/m^3, respirable fraction TWA the value is for particulate matter containing no asbestos and <1% crystalline silica (free SiO_2), not classifiable as a human carcinogen.

NIOSH IDLH: 5000 mg/m^3.

No TEEL available.

Australia: TWA 10 mg/m^3, 1993; Belgium: TWA 10 mg/m^3, 1993; Sweden: NGV 10 mg/m^3 (total dust), 1999; Sweden: NGV 5 mg/m^3 (respirable dust), 1999; Switzerland: MAK-W 6 mg/m^3, 1999; United Kingdom: TWA 10 mg/m^3, total inhalable dust, 2000; the Netherlands: MAC-TGG 10 mg/m^3, 2003; Argentina, Bulgaria, Columbia, Jordan, South Korea, New Zealand, Singapore, Vietnam: ACGIH TLV®: TWA 10 mg/m^3 for particulate matter containing no asbestos and <1% crystalline silica (free SiO_2).

Determination in Air: NIOSH Analytical Method (IV) #0500, Particulates NOR, total dust; OSHA Analytical Method ID-207, Portland cement (total dust).

Routes of Entry: Inhalation, ingestion, skin and/or eye contact.

Harmful Effects and Symptoms

Short Term Exposure: Irritates the eyes, skin, and nose with cough, expectoration, exertional dyspnea (breathing difficulty), wheezing.

Long Term Exposure: May cause chronic bronchitis. Exposure may produce cement dermatitis which is usually due to primary irritation from the alkaline, hygroscopic, and abrasive properties of cement. Chronic irritation of the eyes and nose may occur. In some cases, cement workers have developed an allergic sensitivity to constituents of cement, such as hexavalent chromate. It is not unusual for cement dermatitis to be prolonged and to involve covered areas of the body. No documented cases of pneumoconiosis or other systemic manifestations attributed to finished Portland cement exposure have been reported. Conflicting reports of pneumoconiosis from cement dust appear related to exposures that occurred in mining, quarrying, or crushing silica-containing raw materials.

Points of Attack: Respiratory system, eyes, skin.

Medical Surveillance: NIOSH lists the following tests: chest X-ray; pulmonary function tests: forced vital capacity, forced expiratory volume (1 s). Preemployment and periodic medical examinations should stress significant respiratory problems, smoking history, and allergic skin sensitivities. The eyes should be examined. Patch test studies may be useful in dermatitis cases.

First Aid: If this chemical gets into the eyes, remove any contact lenses at once and irrigate immediately for at least 15 min, occasionally lifting upper and lower lids. Seek medical attention immediately. If this chemical contacts the skin, remove contaminated clothing and wash immediately with soap and water. Seek medical attention immediately. If this chemical has been inhaled, remove from exposure, begin rescue breathing (using universal precautions, including resuscitation mask) if breathing has stopped and CPR if heart action has stopped. Transfer promptly to a medical facility. When this chemical has been swallowed, get medical attention. Give large quantities of water and induce vomiting. Do not make an unconscious person vomit.

Personal Protective Methods: Wear protective gloves and clothing to prevent prolonged skin contact. Safety equipment suppliers/manufacturers can provide recommendations on the most protective glove/clothing material for your operation. All protective clothing (suits, gloves, footwear, headgear) should be clean, available each day, and put on before work. Contact lenses should not be worn when working with this chemical. Wear dust-proof chemical goggles and face shield unless full face-piece respiratory protection is worn. Employees should wash immediately with soap when skin is wet or contaminated. Provide emergency showers and eyewash.

Respirator Selection: NIOSH: *Up to 50 mg/m³:* Qm (APF = 25) (any quarter-mask respirator). *Up to 100 mg/m³:* 95XQ (APF = 10) (any particulate respirator equipped with an N95, R95, or P95 filter (including N95, R95, and P95 filtering face-pieces) except quarter-mask respirators. The following filters may also be used: N99, R99, P99, N100, R100, P100) or Sa (APF = 10) (any supplied-air respirator). *Up to 250 mg/m³:* Sa:Cf (APF = 25) (any supplied-air respirator operated in a continuous-flow mode) or PaprHie (APF = 25) (any powered, air-purifying respirator with a high-efficiency particulate filter). *Up to 500 mg/m³:* 100F (APF = 50) (any air-purifying, full-face-piece respirator with an N100, R100, or P100 filter) or SaT: Cf (APF = 50) (any supplied-air respirator that has a tight-fitting face-piece and is operated in a continuous-flow mode) or PaprTHie (APF = 50) (any powered, air-purifying respirator with a tight-fitting face-piece and a high-efficiency particulate filter) or SCBAF (APF = 50) (any self-contained breathing apparatus with a full face-piece) or SaF (APF = 50) (any supplied-air respirator with a full face-piece). *Up to 5000 mg/m³:* Sa: Pd,Pp (APF = 1000) (any supplied-air respirator operated in a pressure-demand or other positive-pressure mode). *Emergency or planned entry into unknown concentrations or IDLH conditions:* SCBAF: Pd,Pp (APF = 10,000) (any NIOSH/MSHA- or European Standard EN 149-approved self-contained breathing apparatus that has a full face-piece and is operated in a pressure-demand or other positive-pressure mode) or SaF: Pd,Pp: ASCBA (APF = 10,000) (any supplied-air respirator that has a full face-piece and is operated in a pressure-demand or other positive-pressure mode in combination with an auxiliary, self-contained breathing apparatus operated in a pressure-demand or other positive-pressure mode). *Escape:* 100F (APF = 50) (any air-purifying, full-face-piece respirator with an N100, R100, or P100 filter) or SCBAE (any appropriate escape-type, self-contained breathing apparatus).

Storage: Color Code—Green: General storage may be used. Prior to working with this chemical you should be trained on its proper handling and storage. Store in tightly closed containers in a cool, well-ventilated area.

Spill Handling: Evacuate persons not wearing protective equipment from area of spill or leak until cleanup is complete. Remove all ignition sources. Collect powdered material in the most convenient and safe manner and deposit in sealed containers. Ventilate area after cleanup is complete. It may be necessary to contain and dispose of this chemical as a hazardous waste. If material or contaminated runoff enters waterways, notify downstream users of potentially contaminated waters. Contact your local or federal environmental protection agency for specific recommendations. If employees are required to clean up spills, they must be properly trained and equipped. OSHA 1910.120(q) may be applicable.

Fire Extinguishing: This chemical is a noncombustible solid. Use any extinguishers suitable for surrounding fire. If employees are expected to fight fires, they must be trained and equipped in OSHA 1910.156. The only respirators recommended for firefighting are self-contained breathing apparatuses that have full face-pieces and are operated in a pressure-demand or other positive-pressure mode.

Disposal Method Suggested: Landfill. In some cases, recovery of cement from cement kiln dust or ready-mix concrete residues may be economic, and the technology is available.

Potassium metal   P:0840

Molecular Formula: K
Synonyms: Elemental potassium; Kalium
CAS Registry Number: 7440-09-7
RTECS® Number: TS6460000
UN/NA & ERG Number: UN2257/130; UN1420 (metal alloys)/138; metal powder, in bulk, may be pyrophoric: UN3089 (Metal powder, flammable, n.o.s.)/170
EC Number: 231-119-8 [*Annex I Index No.:* 019-001-00-2]
Regulatory Authority and Advisory Bodies
European/International Regulations: Hazard Symbol: F, C; Risk phrases: R14; R34; Safety phrases: S1/2; S5 (If appropriate); S8; S45 (see Appendix 4).
WGK (German Aquatic Hazard Class): No value assigned.
Description: Potassium is a soft silvery metal. Molecular weight = 39.10; Boiling point = 774°C; Freezing/Melting point = 64°C; Autoignition temperature = 441°C. Hazard Identification (based on NFPA-704 M Rating System) (*lump*): Health 3, Flammability 3, Reactivity 3₩. Reacts violently with water.
Potential Exposure: Used as a reagent and in sodium—potassium alloys which are used as high-temperature heat transfer media.
Incompatibilities: Air contact causes spontaneous ignition. Violent reaction with water, forming heat, spattering, corrosive potassium hydroxide, and explosive hydrogen. The heat from the reaction can ignite the hydrogen that is generated. A powerful reducing agent. Violent reaction with oxidizers, organic materials, carbon dioxide, heavy metal compounds, carbon tetrachloride, halogenated hydrocarbons, easily oxidized materials, and many other substances. Store under nitrogen, mineral oil, or kerosene. Oxidizes and forms unstable peroxides under storage conditions. Potassium metal containing an oxide coating is an extremely dangerous explosion hazard and should be removed and destroyed.
Permissible Exposure Limits in Air
Protective Action Criteria (PAC)
TEEL-0: 0.075 mg/m³
PAC-1: 0.2 mg/m³
PAC-2: 1.5 mg/m³
PAC-3: 300 mg/m³

Permissible Concentration in Water: No criteria set. (Reacts with water).
Routes of Entry: Inhalation, ingestion, skin and/or eye contact.
Harmful Effects and Symptoms
Short Term Exposure: Potassium can affect you when breathed in. Inhalation of dusts or mists can irritate the eyes, nose, throat, and lungs with sneezing, coughing, and sore throat. Higher exposures may cause a buildup of fluid in the lungs (pulmonary edema). This can cause death. Skin and eye contact can cause severe burns leading to permanent damage.
Long Term Exposure: Prolonged exposure to fumes can cause sores of the inner nose and nasal septum. Fumes can irritate the lungs; bronchitis may develop.
Points of Attack: Lungs, skin
Medical Surveillance: Before beginning employment and at regular times after that, for those with frequent or potentially high exposures, the following are recommended: lung function tests. If symptoms develop or overexposure is suspected, the following may be useful: consider chest X-ray after acute overexposure.
First Aid: Eye Contact: Immediately remove any contact lenses and flush with large amounts of water. Continue without stopping for at least 30 min, occasionally lifting upper and lower lids. Seek medical attention immediately.
Skin Contact: Quickly remove contaminated clothing. Immediately wash area with large amounts of water. Seek medical attention immediately.
Breathing: Remove the person from exposure. Begin rescue breathing (using universal precautions, including resuscitation mask) if breathing has stopped and CPR if heart action has stopped. Transfer promptly to a medical facility. Medical observation is recommended for 24—48 h after breathing overexposure, as pulmonary edema may be delayed.
Personal Protective Methods: Wear protective gloves and clothing to prevent any reasonable probability of skin contact. Safety equipment suppliers/manufacturers can provide recommendations on the most protective glove/clothing material for your operation. All protective clothing (suits, gloves, footwear, headgear) should be clean, available each day, and put on before work. Contact lenses should not be worn when working with this chemical. Wear dust-proof chemical goggles and face shield unless full face-piece respiratory protection is worn. Employees should wash immediately with soap when skin is wet or contaminated. Provide emergency showers and eyewash.
Respirator Selection: Where there is potential for exposure to potassium, use a NIOSH/MSHA- or European Standard EN149-approved respirator equipped with particulate (dust/fume/mist) filters. Particulate filters must be checked every day before work for physical damage, such as rips or tears, and replaced as needed. *Where there is potential for high exposures*, use a NIOSH/MSHA- or European Standard EN149-approved supplied-air respirator with a full

face-piece operated in the positive mode or with a full face-piece, hood, or helmet in the continuous-flow mode; or use a NIOSH/MSHA- or European Standard EN149-approved self-contained breathing apparatus with a full face-piece operated in a pressure-demand or other positive-pressure mode.

Storage: Color Code—Red Stripe (*lump*): Flammability Hazard: Do not store in the same area as other flammable materials. Prior to working with this chemical you should be trained on its proper handling and storage. Potassium must be stored to avoid contact with carbon monoxide and moisture, compounds of heavy metals (such as silver oxide and silver chloride) and carbon tetrachloride since violent reactions occur. Store under nitrogen, mineral oil, or kerosene. Sources of ignition, such as smoking and open flames, are prohibited where potassium metal is used, handled, or stored in a manner that could create a potential fire or explosion hazard.

Shipping: Potassium requires a shipping label of "DANGEROUS WHEN WET." It falls in Hazard Class 4.3 and Packing Group II. Potassium, metal alloys, requires a shipping label of "DANGEROUS WHEN WET." It falls in Hazard Class 4.3 and Packing Group II. For Metal powder, flammable, n.o.s. the required label is "SPONTANEOUSLY COMBUSTIBLE." They fall in Hazard Class 4.2 and Packing Group II.

Spill Handling: Evacuate persons not wearing protective equipment from area of spill or leak until cleanup is complete. Remove all ignition sources. Collect powdered material in the most convenient and safe manner and deposit in sealed containers. Ventilate area after cleanup is complete. Keep potassium out of a confined space, such as a sewer, because of the possibility of an explosion, unless the sewer is designed to prevent the buildup of explosive concentrations. It may be necessary to contain and dispose of this chemical as a hazardous waste. If material or contaminated runoff enters waterways, notify downstream users of potentially contaminated waters. Contact your local or federal environmental protection agency for specific recommendations. If employees are required to clean up spills, they must be properly trained and equipped. OSHA 1910.120(q) may be applicable.

Fire Extinguishing: Potassium is spontaneously combustible. Use dry chemical, dry graphite, soda ash, or lime extinguishers. *Do not use water*, carbon dioxide, or foam. Potassium may ignite in the presence of moisture or itself if exposed to air. Poisonous gases are produced in fire. If material or contaminated runoff enters waterways, notify downstream users of potentially contaminated waters. Notify local health and fire officials and pollution control agencies. From a secure, explosion-proof location, use water spray to cool exposed containers. If cooling streams are ineffective (venting sound increases in volume and pitch, tank discolors, or shows any signs of deforming), withdraw immediately to a secure position. If employees are expected to fight fires, they must be

trained and equipped in OSHA 1910.156. The only respirators recommended for firefighting are self-contained breathing apparatuses that have full face-pieces and are operated in a pressure-demand or other positive-pressure mode.

Reference
New Jersey Department of Health and Senior Services. (May 2003). *Hazardous Substances Fact Sheet: Potassium.* Trenton, NJ

Potassium arsenate   P:0850

Molecular Formula: AsH_2KO_4
Common Formula: KH_2AsO_4
Synonyms: Arsenic acid, Monopotassium salt; Arseniato potasico (Spanish); Macquer's salt; Monopotassium arsenate; Monopotassium dihydrogen arsenate; Potassium acid arsenate; Potassium arsenate, monobasic; Potassium dihydrogen arsenate; Potassium hydrogen arsenate
CAS Registry Number: 7784-41-0
RTECS® Number: CG1100000
UN/NA & ERG Number: UN1677/151
EC Number: 232-065-8
Regulatory Authority and Advisory Bodies
Carcinogenicity: NTP: 11th Report on Carcinogens, 2004: Known to be a human carcinogen; IARC: Human Sufficient Evidence, 1980; Animal Limited Evidence, *carcinogenic to humans*, Group 1, 1987.
Air Pollutant Standard Set. See below, "Permissible Exposure Limits in Air" section.
Clean Water Act: Section 311 Hazardous Substances/RQ 40CFR117.3 (same as CERCLA, see below); Section 313 Water Priority Chemicals (57FR41331, 9/9/92).
Reportable Quantity (RQ): 1 lb (0.454 kg).
California Proposition 65 Chemical: Cancer 2/27/87.
Canada, WHMIS, Ingredients Disclosure List Concentration 1.0%.
As arsenic compounds:
Clean Air Act: Hazardous Air Pollutants (Title I, Part A, Section 112) as arsenic compounds.
Clean Water Act: Toxic Pollutant (Section 401.15) as arsenic and compounds.
RCRA, 40CFR261, Appendix 8 Hazardous Constituents, waste number not listed.
Reportable Quantity (RQ): 1 lb (0.454 kg).
EPCRA (Section 313): Includes any unique chemical substance that contains arsenic as part of that chemical's infrastructure. Form R *de minimis* concentration reporting level: inorganics 0.1%; organics 1.0%.
US DOT Regulated Marine Pollutant (49CFR172.101, Appendix B) as arsenates, liquid, n.o.s.; arsenates, solid, n.o.s.; arsenical pesticides liquid, toxic, flammable, n.o.s.
California Proposition 65 Chemical: (*inorganic arsenic*) Cancer 2/27/87.

Canada: Priority Substance List & Restricted Substances/ Ocean Dumping FORBIDDEN (CEPA), National Pollutant Release Inventory (NPRI) (arsenic compounds).

European/International Regulations: Hazard Symbol: T, N; Risk phrases: R45; R23/25; R50/53; Safety phrases: S53; S45; S60; S61 (see Appendix 4).

WGK (German Aquatic Hazard Class): 3—Severe hazard to waters.

Description: Potassium arsenate is a colorless to white crystalline solid. Molecular weight = 180.04; Specific gravity = 3.74; Boiling point = 463°C; Freezing/Melting point = 288°C. Hazard Identification (based on NFPA-704 M Rating System): Health 3, Flammability 0, Reactivity 1.

Potential Exposure: Compound Description: Tumorigen. Potassium arsenate is used in the textile, tanning, preserving hides; in the textile printing and paper industries; and as an insecticide in fly baits, especially for fly paper; as laboratory reagent.

Incompatibilities: A weak base. Reacts with strong oxidizers, bromine azide, acids and decomposes on contact with strong acids producing acetic acid fumes. Arsine, a very deadly gas, can be released in the presence of acid, acid mists, or hydrogen gas.

Permissible Exposure Limits in Air

OSHA PEL: 0.010 mg[As]/m^3 TWA; cancer hazard that can be inhaled. See [1910.1018].

NIOSH REL: 0.002 mg[As]/m^3 15 min Ceiling Concentration. A potential occupational carcinogen. Limit exposure to lowest feasible concentration; See Appendix A.

ACGIH TLV®[1]: 0.01 mg[As]/m^3 TWA; Confirmed Human Carcinogen.

NIOSH IDLH: potential occupational carcinogen 5 mg [As]/m^3.

Protective Action Criteria (PAC)

TEEL-0: 0.024 mg/m^3

PAC-1: 0.35 mg/m^3

PAC-2: 2.5 mg/m^3

PAC-3: 12 mg/m^3

DFG TRK: 0.10 mg[As]/m^3; BAT: 1.30 μg[As]/L in urine/ end-of-shift; Carcinogen Category 1.

Arab Republic of Egypt: TWA 0.2 mg/m^3, 1993; Australia: TWA 0.05 mg/m^3, carcinogen, 1993; Belgium: TWA 0.2 mg/m^3, 1993; Denmark: TWA 0.05 mg/m^3, 1999; Finland: carcinogen, 1993; France: VME 0.2 mg/m^3, 1993; Hungary: STEL 0.5 mg/m^3, carcinogen, 1993; India: TWA 0.2 mg/m^3, 1993; Norway: TWA 0.02 mg/m^3, 1999; the Philippines: TWA 0.5 mg/m^3, 1993; Poland: MAC (TWA) 0.01 mg/m^3, 1999; Sweden: NGV 0.03 mg/m^3, carcinogen, 1999; Switzerland: TWA 0.1 mg/m^3, carcinogen, 1999; Thailand: TWA 0.5 mg/m^3, 1993; Turkey: TWA 0.5 mg (As)/m^3, 1993; Turkey: TWA 0.5 mg/m^3, 1993; United Kingdom: TWA 0.1 mg/m^3, carcinogen, 2000; Argentina, Bulgaria, Columbia, Jordan, South Korea, New Zealand, Singapore, Vietnam: ACGIH: TLV: Confirmed Human Carcinogen. Russia[43] set a MAC of 0.003 mg/m^3 on a daily average basis for residential areas. Several states have set guidelines or standards for arsenic in ambient air[60]: 0.06 mg/m^3 (California Prop. 65), 0.0002 μg/m^3 (Rhode Island), 0.00023 μg/m^3 (North Carolina), 0.024 μg/m^3 (Pennsylvania), 0.05 μg/m^3 (Connecticut), 0.07−0.39 μg/m^3 (Montana), 0.67 μg/m^3 (New York), 1.0 μg/m^3 (South Carolina), 2.0 μg/m^3 (North Dakota), 3.3 μg/m^3 (Virginia), 5 μg/m^3 (Nevada).

Determination in Air: Use NIOSH Analytical Methods (inorganic arsenic): #7300, #7301,#7303, #7900, #9102; OSHA Analytical Methods ID-105.

Permissible Concentration in Water: EPA[6] recommends a zero concentration of arsenic for human health reasons but has set a guideline of 50 μg/L[61] for drinking water.

Determination in Water: See OSHA Analytical Method ID-105 for arsenic.

Routes of Entry: Inhalation, ingestion, skin and/or eye contact. Absorbed through the skin.

Harmful Effects and Symptoms

Short Term Exposure: Potassium arsenate can affect you when breathed, and may enter through skin. Potassium arsenate is a carcinogen; handle with extreme caution. Eye contact causes irritation, burns and red, watery eyes. Skin contact can cause burning, itching, and rash. Breathing can cause irritation with sneezing and coughing. High or repeated exposures can cause disturbed sleep, with numbness and weakness of arms and legs; and can cause poor appetite; nausea, cramps, and if severe, vomiting and diarrhea.

Long Term Exposure: Long-term exposure can cause ulcer or hole in the nasal septum; hoarseness and sore eyes also occur. Repeated exposure can cause nervous system damage. Repeated skin contact can cause thickened skin and/or patchy area of darkening and loss of pigment.

Points of Attack: Nervous system, skin.

Medical Surveillance: Before beginning employment and at regular times after that, the following are recommended: examination of the nose, skin, eyes, nails, nervous system. Test for urine arsenic (may not be accurate within 2 days of eating shellfish or fish; most accurate at the end of a workday). At NIOSH recommended exposure levels, urine arsenic should not be greater than 50−100 μg/L of urine. After suspected overexposure, repeat these tests. Also, examine your skin periodically for abnormal growths. Skin cancer from arsenic is easily cured when detected early.

First Aid: If this chemical gets into the eyes, remove any contact lenses at once and irrigate immediately for at least 15 min, occasionally lifting upper and lower lids. Seek medical attention immediately. If this chemical contacts the skin, remove contaminated clothing and wash immediately with soap and water. Seek medical attention immediately. If this chemical has been inhaled, remove from exposure, begin rescue breathing (using universal precautions, including resuscitation mask) if breathing has stopped and CPR if heart action has stopped. Transfer promptly to a medical facility. When this chemical has been swallowed, get medical attention. Give large quantities of water and induce vomiting. Do not make an unconscious person vomit.

Note to physician: For severe poisoning BAL has been used. For milder poisoning penicillamine (*not penicillin*) has been used, both with mixed success. Side effects occur with such treatment and it is never a substitute for controlling exposures. It can only be done under strict medical care.

Personal Protective Methods: Wear protective gloves and clothing to prevent any reasonable probability of skin contact. Safety equipment suppliers/manufacturers can provide recommendations on the most protective glove/clothing material for your operation. All protective clothing (suits, gloves, footwear, headgear) should be clean, available each day, and put on before work. Contact lenses should not be worn when working with this chemical. Eye protection is included in the recommended respiratory protection. Employees should wash immediately with soap when skin is wet or contaminated. Provide emergency showers and eyewash.

Respirator Selection: *At concentrations above the NIOSH REL, or where there is no REL, at any detectable concentration:* Sa (APF = 10) (any supplied-air respirator) or SCBAF (APF = 50) (any self-contained breathing apparatus with a full face-piece). *Emergency or planned entry into unknown concentrations or IDLH conditions:* SCBAF: Pd,Pp (APF = 10,000) (any self-contained breathing apparatus that has a full face-piece and is operated in a pressure-demand or other positive-pressure mode) or SaF: Pd,Pp: ASCBA (APF = 10,000) (any supplied-air respirator that has a full face-piece and is operated in a pressure-demand or other positive-pressure mode in combination with an auxiliary, self-contained breathing apparatus operated in a pressure-demand or other positive-pressure mode). *Escape:* GmFAg100 (APF = 50) [any air-purifying, full-face-piece respirator (gas mask) with a chin-style, front- or back-mounted acid gas canister having an N100, R100, or P100 filter] or SCBAE (any appropriate escape-type, self-contained breathing apparatus).

Storage: Color Code—Blue: Health Hazard/Poison: Store in a secure poison location. Prior to working with this chemical you should be trained on its proper handling and storage. A regulated, marked area should be established where potassium arsenate is handled, used, or stored as required by OSHA Standard 29 CFR 1910.1018 for inorganic arsenic. Potassium arsenate must be stored to avoid contact with acids. Store in tightly closed containers in a cool, well-ventilated area.

Shipping: This compound requires a shipping label of "POISONOUS/TOXIC MATERIALS." It falls in Hazard Class 6.1 and Packing Group II.

Spill Handling: Evacuate persons not wearing protective equipment from area of spill or leak until cleanup is complete. Remove all ignition sources. Use HEPA vacuum or wet method to reduce dust during cleanup. Do not dry sweep. Deposit in sealed containers. Ventilate area after cleanup is complete. It may be necessary to contain and dispose of this chemical as a hazardous waste. If material or contaminated runoff enters waterways, notify downstream

users of potentially contaminated waters. Contact your local or federal environmental protection agency for specific recommendations. If employees are required to clean up spills, they must be properly trained and equipped. OSHA 1910.120(q) may be applicable.

Fire Extinguishing: Extinguish fire using an agent suitable for type of surrounding fire. Potassium arsenate itself does not burn. Poisonous gases are produced in fire, including arsenic trioxide and potassium oxide. If material or contaminated runoff enters waterways, notify downstream users of potentially contaminated waters. Notify local health and fire officials and pollution control agencies. From a secure, explosion-proof location, use water spray to cool exposed containers. If cooling streams are ineffective (venting sound increases in volume and pitch, tank discolors, or shows any signs of deforming), withdraw immediately to a secure position. If employees are expected to fight fires, they must be trained and equipped in OSHA 1910.156. The only respirators recommended for firefighting are self-contained breathing apparatuses that have full face-pieces and are operated in a pressure-demand or other positive-pressure mode.

References
Sax, N. I. (Ed.). (1983). *Dangerous Properties of Industrial Materials Report*, 3, No. 4, 101–103
New Jersey Department of Health and Senior Services. (October 2004). *Hazardous Substances Fact Sheet: Potassium Arsenate*. Trenton, NJ

Potassium arsenate   P:0860

Molecular Formula: As_2HKO_4
Common Formula: $KAsO_2 \cdot HAsO_2$; $AsH_3O_3 \cdot xK$
Synonyms: Arsenito potasico (Spanish); Arsenous acid, potassium salt; Arsenite de potassium (French); Arsonic acid, potassium salt; Fowler's solution (liquid); Kaliumarsenit (German); NSC 3060; Potassium metaarsenite
CAS Registry Number: 13464-35-2; 1332-10-1 (solution)
RTECS® Number: CG3800000
UN/NA & ERG Number: UN1678/154
EC Number: 236-680-2 [*Annex I Index No.:* 033-002-00-5] (potassium arsenite); 233-337-9 (potassium arsonate)

Regulatory Authority and Advisory Bodies
Carcinogenicity: NTP: 11th Report on Carcinogens, 2004: Known to be a human carcinogen; IARC: Human Sufficient Evidence, Animal Inadequate Evidence, 1980.
Air Pollutant Standard Set. See below, "Permissible Exposure Limits in Air" section.
Clean Water Act: Section 311 Hazardous Substances/RQ 40CFR117.3 (same as CERCLA, see below); Section 313 Water Priority Chemicals (57FR41331, 9/9/92).
Superfund/EPCRA 40CFR355, Extremely Hazardous Substances: TPQ = 500/10,000 lb (227/4540 kg) (10124-50-2).

Reportable Quantity (RQ): 1 lb (0.454 kg) (10124-50-2).
Arsenic compounds:
Clean Air Act: Hazardous Air Pollutants (Title I, Part A, Section 112) as arsenic compounds.
Clean Water Act: Toxic Pollutant (Section 401.15) as arsenic and compounds.
RCRA, 40CFR261, Appendix 8 Hazardous Constituents, waste number not listed.
Reportable Quantity (RQ): 1 lb (0.454 kg).
EPCRA (Section 313): Includes any unique chemical substance that contains arsenic as part of that chemical's infrastructure. Form R *de minimis* concentration reporting level: inorganics 0.1%; organics 1.0%.
US DOT Regulated Marine Pollutant (49CFR172.101, Appendix B) as arsenates, liquid, n.o.s.; arsenates, solid, n.o.s.; arsenical pesticides liquid, toxic, flammable, n.o.s.
California Proposition 65 Chemical: Cancer 2/27/87.
Canada, WHMIS, Ingredients Disclosure List Concentration 1.0% as arsenic, water-soluble compounds, n.o.s.
Canada: Priority Substance List & Restricted Substances/ Ocean Dumping FORBIDDEN (CEPA), National Pollutant Release Inventory (NPRI) (arsenic compounds).
European/International Regulations: Hazard Symbol: T, N; Risk phrases: (*as arsenic compound*) R45; R23/25; R50/53; R53; Safety phrases: S1/2; S20/21; S28; S45; S60; S61 (see Appendix 4).
WGK (German Aquatic Hazard Class): 3—Highly water polluting.
Description: Potassium arsenite is a white crystalline solid. Decomposes below Freezing/Melting point at 300°C. Hazard Identification (based on NFPA-704 M Rating System): Health 3, Flammability 0, Reactivity 0. Soluble in water.
Potential Exposure: Compound Description: Agricultural Chemical; Tumorigen, Drug, Mutagen; Human Data. Potassium metaarsenite is used in veterinary medicine; and for chronic dermatitis in man. Potassium arsenite reduces silver salt to metallic silver during mirror silvering. Currently, it is probably not being used for this purpose.
Incompatibilities: A weak base. Reacts with acids and decomposes on contact with strong acids producing acetic acid fumes. Arsine, a very deadly gas, can be released in the presence of acid or acid mist. Incompatible with alkaloidal salts, strong oxidizers, bromine azide, hypophosphites, sulfites in acid solution, iron salts, heavy metals and heavy metal compounds. Hydrogen gas can react with inorganic arsenic to form the highly toxic gas, arsine.
Permissible Exposure Limits in Air
OSHA PEL: 0.010 mg[As]/m^3 TWA; cancer hazard that can be inhaled. See [1910.1018].
NIOSH REL: 0.002 mg[As]/m^3 15 min Ceiling Concentration. A potential occupational carcinogen. Limit exposure to lowest feasible concentration; See Appendix A.
ACGIH TLV®[1]: 0.01 mg[As]/m^3 TWA; Confirmed Human Carcinogen.
NIOSH IDLH: potential occupational carcinogen 5 mg [As]/m^3.

Protective Action Criteria (PAC)
TEEL-0: 0.0533 mg/m^3
PAC-1: 2 mg/m^3
PAC-2: 14 mg/m^3
PAC-3: 26.7 mg/m^3
DFG TRK: 0.10 mg[As]/m^3; BAT: 1.30 μg[As]/L in urine/ end-of-shift; Carcinogen Category 1.
Arab Republic of Egypt: TWA 0.2 mg/m^3, 1993; Australia: TWA 0.05 mg/m^3, carcinogen, 1993; Belgium: TWA 0.2 mg/ m^3, 1993; Denmark: TWA 0.05 mg/m^3, 1999; Finland: carcinogen, 1993; France: VME 0.2 mg/m^3, 1993; Hungary: STEL 0.5 mg/m^3, carcinogen, 1993; India: TWA 0.2 mg/m^3, 1993; Norway: TWA 0.02 mg/m^3, 1999; the Philippines: TWA 0.5 mg/m^3, 1993; Poland: MAC (TWA) 0.01 mg/m^3, 1999; Sweden: NGV 0.03 mg/m^3, carcinogen, 1999; Switzerland: TWA 0.1 mg/m^3, carcinogen, 1999; Thailand: TWA 0.5 mg/m^3, 1993; Turkey: TWA 0.5 mg(As)/m^3, 1993; Turkey: TWA 0.5 mg/m^3, 1993; United Kingdom: TWA 0.1 mg/m^3, carcinogen, 2000; Argentina, Bulgaria, Columbia, Jordan, South Korea, New Zealand, Singapore, Vietnam: ACGIH: TLV: Confirmed Human Carcinogen. Russia[43] set a MAC of 0.003 mg/m^3 on a daily average basis for residential areas. Several states have set guidelines or standards for arsenic in ambient air: [60] 0.06 mg/m^3 (California Prop. 65), 0.0002 μg/m^3 (Rhode Island), 0.00023 μg/m^3 (North Carolina), 0.024 μg/m^3 (Pennsylvania), 0.05 μg/m^3 (Connecticut), 0.07−0.39 μg/m^3 (Montana), 0.67 μg/m^3 (New York), 1.0 μg/m^3 (South Carolina), 2.0 μg/m^3 (North Dakota), 3.3 μg/m^3 (Virginia), 5 μg/m^3 (Nevada).
Determination in Air: Use NIOSH Analytical Methods (inorganic arsenic): #7300, #7301, #7303, #7900, #9102; OSHA Analytical Methods ID-105.
Permissible Concentration in Water: EPA[6] recommends a zero concentration of arsenic on a human health basis. EPA[61] has set a guideline for drinking water of 50 μg/L of arsenic.
Determination in Water: See OSHA Analytical Method ID-105 for arsenic.
Routes of Entry: Inhalation, ingestion, skin and/or eye contact. Absorbed through the skin.
Harmful Effects and Symptoms
Short Term Exposure: Irritates the eyes, skin, and respiratory tract. Skin contact can cause irritation with a burning sensation; itching. High exposure can cause loss of appetite; garlic or metallic taste; nausea, vomiting, and muscle cramps. May affect the central nervous system, digestive tract, circulatory system, causing loss of fluids and electrolytes; may cause collapse, shock, and death. Very toxic: probable oral lethal dose in humans is 50−500 mg/kg, or between 1 teaspoonful and 1 oz for a 150-lb adult. Nausea, vomiting, and diarrhea result from arsenic ingestion. Patient may go into collapse and shock with weak, rapid pulse, cold sweat, coma, and death. Exposure at low level may result in death.
Long Term Exposure: May cause liver damage, cirrhosis, jaundice. A skin allergen. Repeated or prolonged contact

may cause skin sensitization. Itching or skin pigmentation changes may occur. May affect the peripheral nervous system, causing weakness in the hands and feet. May cause an ulcer in the inner nose, perforation of the nasal septum, and cirrhosis. This substance is carcinogenic to humans. Chronic arsenic poisoning may manifest itself by loss of appetite; cramps, nausea, constipation, or diarrhea.

Points of Attack: Liver, skin.

Medical Surveillance: Liver function testing. Examination of the nervous system. Examination of the nose, eyes, nails, and skin. Test for urine arsenic. Arsenic should not be greater than 100 μg/g of creatinine in the urine. Tests are most accurate at the end of a workday and may be inaccurate within 2 days of eating shellfish. If abnormal growths are detected on the skin, they can be easily cured when detected early.

First Aid: If this chemical gets into the eyes, remove any contact lenses at once and irrigate immediately for at least 15 min, occasionally lifting upper and lower lids. Seek medical attention immediately. If this chemical contacts the skin, remove contaminated clothing and wash immediately with soap and water. Seek medical attention immediately. If this chemical has been inhaled, remove from exposure, begin rescue breathing (using universal precautions, including resuscitation mask) if breathing has stopped and CPR if heart action has stopped. Transfer promptly to a medical facility. When this chemical has been swallowed, get medical attention. Give large quantities of water and induce vomiting. Do not make an unconscious person vomit.

Note to physician: For severe poisoning BAL [British Anti-Lewisite, Dimercaprol, dithiopropanol $(C_3H_8OS_2)$] has been used to treat toxic symptoms of certain heavy metals poisoning—including arsenic. Although BAL is reported to have a large margin of safety, caution must be exercised because toxic effects may be caused by excessive dosage. Most can be prevented by premedication with 1-ephedrine sulfate (CAS: 134-72-5). For milder poisoning *penicillamine (not penicillin)* has been used, both with mixed success. Side effects occur with such treatment and it is never a substitute for controlling exposure. It can only be done under strict medical care.

Personal Protective Methods: Wear protective gloves and clothing to prevent any reasonable probability of skin contact. Safety equipment suppliers/manufacturers can provide recommendations on the most protective glove/clothing material for your operation. All protective clothing (suits, gloves, footwear, headgear) should be clean, available each day, and put on before work. Contact lenses should not be worn when working with this chemical. Wear dust-proof chemical goggles and face shield unless full face-piece respiratory protection is worn. Employees should wash immediately with soap when skin is wet or contaminated. Provide emergency showers and eyewash.

Respirator Selection: At concentrations above the NIOSH REL, or where there is no REL, at any detectable concentration: Sa (APF = 10) (any supplied-air respirator) or SCBAF (APF = 50) (any self-contained breathing apparatus with a full face-piece). *Emergency or planned entry into unknown concentrations or IDLH conditions:* SCBAF: Pd,Pp (APF = 10,000) (any self-contained breathing apparatus that has a full face-piece and is operated in a pressure-demand or other positive-pressure mode) or SaF: Pd,Pp: ASCBA (APF = 10,000) (any supplied-air respirator that has a full face-piece and is operated in a pressure-demand or other positive-pressure mode in combination with an auxiliary, self-contained breathing apparatus operated in a pressure-demand or other positive-pressure mode). *Escape:* GmFAg100 (APF = 50) [any air-purifying, full-face-piece respirator (gas mask) with a chin-style, front- or back-mounted acid gas canister having an N100, R100, or P100 filter] or SCBAE (any appropriate escape-type, self-contained breathing apparatus).

Storage: Color Code—Blue: Health Hazard/Poison: Store in a secure poison location. Prior to working with this chemical you should be trained on its proper handling and storage. A regulated, marked area should be established where potassium arsenite is handled, used, or stored as required by OSHA Standard 29 CFR 1910.1018 for inorganic arsenic. Store in tightly closed containers in a cool, well-ventilated area away from acids, acid fumes, alkaloidal salts, strong oxidizers, bromine azide, hypophosphites, sulfites in acid solution, iron salts, heavy metals and heavy metal compounds. Where possible, automatically transfer material from drums or other storage containers to process containers. Sources of ignition, such as smoking and open flames, are prohibited where this chemical is handled, used, or stored. Metal containers involving the transfer of this chemical should be grounded and bonded. Wherever this chemical is used, handled, manufactured, or stored, use explosion-proof electrical equipment and fittings.

Shipping: Potassium arsenite (solid) requires a shipping label of "POISONOUS/TOXIC MATERIALS." It falls in Hazard Class 6.1 and Packing Group II.

Spill Handling: Evacuate persons not wearing protective equipment from area of spill or leak until cleanup is complete. Remove all ignition sources. Avoid inhalation and skin contact. Dike far ahead of spill for later disposal. Do not touch spilled material; stop leak if you can do it without risk. Collect powdered material in the most convenient and safe manner and deposit in sealed containers. Ventilate area after cleanup is complete. It may be necessary to contain and dispose of this chemical as a hazardous waste. If material or contaminated runoff enters waterways, notify downstream users of potentially contaminated waters. Contact your local or federal environmental protection agency for specific recommendations. If employees are required to clean up spills, they must be properly trained and equipped. OSHA 1910.120(q) may be applicable.

Fire Extinguishing: Extinguish fire using agents suitable for surrounding fire. Potassium arsenite itself does not burn. Keep unnecessary people away; stay upwind; keep out of low areas. Wear full protective clothing and self-contained breathing apparatus. Poisonous gases are produced in fire, including arsenic, phosgene, and potassium oxide. If material or contaminated runoff enters waterways, notify downstream users of potentially contaminated waters. Notify local health and fire officials and pollution control agencies. From a secure, explosion-proof location, use water spray to cool exposed containers. If cooling streams are ineffective (venting sound increases in volume and pitch, tank discolors, or shows any signs of deforming), withdraw immediately to a secure position. If employees are expected to fight fires, they must be trained and equipped in OSHA 1910.156. The only respirators recommended for firefighting are self-contained breathing apparatuses that have full face-pieces and are operated in a pressure-demand or other positive-pressure mode.

References
US Environmental Protection Agency. (November 30, 1987). *Chemical Hazard Information Profile: Potassium Arsenite*. Washington, DC: Chemical Emergency Preparedness Program
New Jersey Department of Health and Senior Services. (August 1999). *Hazardous Substances Fact Sheet: Potassium Arsenite*. Trenton, NJ

Potassium bromate   P:0870

Molecular Formula: $BrKO_3$
Common Formula: $KBrO_3$
Synonyms: Bromato potasico (Spanish); Bromic acid, potassium salt
CAS Registry Number: 7758-01-2
RTECS® Number: EF8725000
UN/NA & ERG Number: UN1484/140
EC Number: 231-829-8 [*Annex I Index No.:* 035-003-00-6]
Regulatory Authority and Advisory Bodies
Carcinogenicity: IARC: Human Inadequate Evidence, animal Sufficient Evidence, *possibly carcinogenic to humans*, Group 2B, 1999; EPA: Likely to produce cancer in humans (inhalation, as bromates); Available data are inadequate for an assessment of human carcinogenic potential (oral route, as bromates); Limited evidence of carcinogenicity based on epidemiologic studies.
Air Pollutant Standard Set. See below, "Permissible Exposure Limits in Air" section.
California Proposition 65 Chemical: Cancer 1/1/90.
EPCRA Section 313 Form R *de minimis* concentration reporting level: 0.1%.
European/International Regulations: Hazard Symbol: T, O; Risk phrases: R45; R9; R25; Safety phrases: S53; S45 (see Appendix 4).
WGK (German Aquatic Hazard Class): No value assigned.
Description: Potassium bromate is a white crystalline solid. Molecular weight = 167.01; Decomposes at 370°C; Freezing/Melting point = 350°C; also reported at 435°C. Hazard Identification (based on NFPA-704 M Rating

System): Health 2, Flammability 0, Reactivity 3 (Oxidizer). Soluble in water.
Potential Exposure: Compound Description: Tumorigen, Mutagen; Human Data. Potassium bromate is used as animal feed additive, food additive; flavor and packaging material; as a laboratory reagent; and oxidizing agent.
Incompatibilities: A strong oxidizer. Reacts violently with combustibles, organics, reducing agents, powdered metals, metal sulfides, carbon, sulfur, phosphorus, ammonium salts, oxidizers. Incompatible with aluminum, copper.
Permissible Exposure Limits in Air
AIHA WEEL: 0.1 mg/m^3 TWA.
Protective Action Criteria (PAC)
TEEL-0: 0.1 mg/m^3
PAC-1: 0.3 mg/m^3
PAC-2: 60 mg/m^3
PAC-3: 60 mg/m^3
United Kingdom: carcinogen, 2000.
Routes of Entry: Inhalation, ingestion, skin and/or eye contact.
Harmful Effects and Symptoms
Short Term Exposure: Potassium bromate can affect you when breathed in and may enter the body through the skin. Skin and eye contact can cause irritation, and burns may occur with prolonged contact. Breathing the mist or dust can irritate the nose, throat, and bronchial tubes, causing sneezing and coughing. Overexposure can cause kidney damage. Ingestion may affect the gastrointestinal tract and central nervous system.
Long Term Exposure: This substance is possibly carcinogenic to humans. May affect the nervous system causing headache, irritability, impaired thinking, and personality changes. May cause kidney damage. May cause lung irritation; bronchitis may develop.
Points of Attack: Lungs, kidneys, nervous system.
Medical Surveillance: Before beginning employment and at regular times after that, for those with frequent or potentially high exposures, the following are recommended: lung function tests. If symptoms develop or overexposure is suspected, the following may also be useful: kidney function tests. Examination of the nervous system. Interview for brain damage.
First Aid: If this chemical gets into the eyes, remove any contact lenses at once and irrigate immediately for at least 15 min, occasionally lifting upper and lower lids. Seek medical attention immediately. If this chemical contacts the skin, remove contaminated clothing and wash immediately with soap and water. Speed in removing material from skin is of extreme importance. Shampoo hair promptly if contaminated. Seek medical attention immediately. If this chemical has been inhaled, remove from exposure, begin rescue breathing (using universal precautions, including resuscitation mask) if breathing has stopped and CPR if heart action has stopped. Transfer promptly to a medical facility. When this chemical has been swallowed, get medical attention. Give large

quantities of water and induce vomiting. Do not make an unconscious person vomit.

Personal Protective Methods: Wear protective gloves and clothing to prevent any reasonable probability of skin contact. Safety equipment suppliers/manufacturers can provide recommendations on the most protective glove/clothing material for your operation. All protective clothing (suits, gloves, footwear, headgear) should be clean, available each day, and put on before work. Contact lenses should not be worn when working with this chemical. Wear dust-proof chemical goggles and face shield unless full face-piece respiratory protection is worn. Employees should wash immediately with soap when skin is wet or contaminated. Provide emergency showers and eyewash.

Respirator Selection: Where there is potential for exposure to potassium bromate, use a NIOSH/MSHA- or European Standard EN149-approved full-face-piece respirator equipped with particulate (dust/fume/mist) filters. Particulate filters must be checked every day before work for physical damage, such as rips or tears, and replaced as needed. *Where there is potential for high exposures,* use a NIOSH/MSHA- or European Standard EN149-approved supplied-air respirator with a full face-piece operated in the positive-pressure mode, or with a full face-piece, hood, or helmet in the continuous-flow mode; or use a NIOSH/MSHA- or European Standard EN149-approved self-contained breathing apparatus with a full face-piece operated in a pressure-demand or other positive-pressure mode.

Storage: Color Code—Yellow: Reactive Hazard; Store in a location separate from other materials, especially flammables and combustibles. Prior to working with this chemical you should be trained on its proper handling and storage. Potassium bromate must be stored to avoid contact with oxidizers (such as perchlorates, peroxides, permanganates, chlorates, and nitrates) and other incompatible materials listed above since violent reactions can occur. Protect storage containers against physical damage. Avoid storage on wood floors. Where possible, automatically transfer material from drums or other storage containers to process containers. Sources of ignition, such as smoking and open flames, are prohibited where this chemical is handled, used, or stored. Metal containers involving the transfer of this chemical should be grounded and bonded. Wherever this chemical is used, handled, manufactured, or stored, use explosion-proof electrical equipment and fittings. See OSHA Standard 1910.104 and NFPA 43A *Code for the Storage of Liquid and Solid Oxidizers* for detailed handling and storage regulations. A regulated, marked area should be established where this chemical is handled, used, or stored in compliance with OSHA Standard 1910.1045.

Shipping: This compound requires a shipping label of "OXIDIZER." It falls in Hazard Class 5.1 and Packing Group II.

Spill Handling: Evacuate persons not wearing protective equipment from area of spill or leak until cleanup is complete. Remove all ignition sources. Collect powdered material in the most convenient and safe manner and deposit in sealed containers. Ventilate area after cleanup is complete. Keep potassium bromate out of a confined space, such as a sewer, because of the possibility of an explosion, unless the sewer is designed to prevent the buildup of explosive concentrations. It may be necessary to contain and dispose of this chemical as a hazardous waste. If material or contaminated runoff enters waterways, notify downstream users of potentially contaminated waters. Contact your local or federal environmental protection agency for specific recommendations. If employees are required to clean up spills, they must be properly trained and equipped. OSHA 1910.120(q) may be applicable.

Fire Extinguishing: This chemical is a combustible solid. Use only water. Do not use chemical or carbon dioxide extinguishers. Poisonous gases are produced in fire, including bromine, potassium oxide, and oxygen, which increases fire hazard. If material or contaminated runoff enters waterways, notify downstream users of potentially contaminated waters. Notify local health and fire officials and pollution control agencies. Containers may explode in fire. From a secure, explosion-proof location, use water spray to cool exposed containers. If cooling streams are ineffective (venting sound increases in volume and pitch, tank discolors, or shows any signs of deforming), withdraw immediately to a secure position. If employees are expected to fight fires, they must be trained and equipped in OSHA 1910.156. The only respirators recommended for firefighting are self-contained breathing apparatuses that have full face-pieces and are operated in a pressure-demand or other positive-pressure mode.

References

Sax, N. I. (Ed.). (1981). *Dangerous Properties of Industrial Materials Report*, 1, No. 7, 70—71

New Jersey Department of Health and Senior Services. (July 2005). *Hazardous Substances Fact Sheet: Potassium Bromate*. Trenton, NJ

Potassium chlorate   P:0880

Molecular Formula: $ClKO_3$
Common Formula: $KClO_3$
Synonyms: Berthollet's salt; Chlorate de potassium (French); Chlorate of potash; Chloric acid, Potassium salt; Fekabit; Kaliumchlorat (German); Oxymuriate of potash; Pearl ash; Potash chlorate; Potassium (chlorate de) (French); Potassium oxymuriate; Potcrate; Salt of tartar
CAS Registry Number: 3811-04-9
RTECS® Number: FO0350000
UN/NA & ERG Number: UN2427 (solution)/140; UN1485 (solid)/140
EC Number: 223-289-7[017-004-00-3]
Regulatory Authority and Advisory Bodies
Department of Homeland Security Screening Threshold Quantity (pounds): *Theft hazard* 400 (Commercial grade).

FDA—over-the-counter drug.

WGK (German Aquatic Hazard Class): 2—Hazard to waters.

Description: Potassium chlorate is a white crystalline solid. Molecular weight = 122.55; Boiling point = decomposes below BP at 400°C; Freezing/Melting point = 360°C. Hazard Identification (based on NFPA-704 M Rating System): Health 2, Flammability 0, Reactivity 1 (Oxidizer). Soluble in water.

Potential Exposure: Compound Description: Agricultural Chemical; Human Data. Potassium chlorate is used in the manufacture of soap, glass, pottery, and many potassium salts; as an oxidizing agent, in explosives, matches, textile printing; disinfectants, and bleaching.

Incompatibilities: A strong oxidizer. Decomposes on heating above 400°C, on contact with strong acids producing toxic fumes including chlorine dioxide, chlorine fumes, and oxygen. Violent reaction with combustibles, oxidizers, strong acids, and reducing materials. Attacks many metals in the presence of water.

Permissible Exposure Limits in Air

Protective Action Criteria (PAC)

TEEL-0: 12.5 mg/m^3

PAC-1: 40 mg/m^3

PAC-2: 300 mg/m^3

PAC-3: 350 mg/m^3

Routes of Entry: Inhalation, ingestion, skin and/or eye contact.

Harmful Effects and Symptoms

Short Term Exposure: Potassium chlorate can affect you when breathed in. Irritates the eyes, skin, and respiratory tract. Prolonged contact can cause eye and skin burns and possible permanent damage. Exposure lowers the ability of the blood to carry oxygen (methemoglobinemia). This can result in a bluish color to skin and lips, headache, dizziness, collapse, and even death. Breathing the dust or mist can cause nose and throat irritation with sneezing, coughing, and sore throat. High exposure may cause kidney damage.

Long Term Exposure: May cause lung damage; bronchitis may develop. Repeated exposures may affect the kidneys and central nervous system.

Points of Attack: Lungs, kidneys, nervous system.

Medical Surveillance: Before beginning employment and at regular times after that, for those with frequent or potentially high exposures, the following are recommended: lung function tests. If symptoms develop or overexposure is suspected, the following may be useful: A blood test for methemoglobin level. Kidney function tests.

First Aid: If this chemical gets into the eyes, remove any contact lenses at once and irrigate immediately for at least 15 min, occasionally lifting upper and lower lids. Seek medical attention immediately. If this chemical contacts the skin, remove contaminated clothing and wash immediately with soap and water. Speed in removing material from skin is of extreme importance. Shampoo hair promptly if contaminated. Seek medical attention immediately. If this chemical has been inhaled, remove from exposure, begin rescue breathing (using universal precautions, including resuscitation mask) if breathing has stopped and CPR if heart action has stopped. Transfer promptly to a medical facility. When this chemical has been swallowed, get medical attention. Give large quantities of water and induce vomiting. Do not make an unconscious person vomit.

Note to physician: Treat for methemoglobinemia. Spectrophotometry may be required for precise determination of levels of methemoglobin in urine.

Personal Protective Methods: Wear protective gloves and clothing to prevent any reasonable probability of skin contact. Safety equipment suppliers/manufacturers can provide recommendations on the most protective glove/clothing material for your operation. All protective clothing (suits, gloves, footwear, headgear) should be clean, available each day, and put on before work. Contact lenses should not be worn when working with this chemical. Wear dust-proof chemical goggles and face shield unless full face-piece respiratory protection is worn. Employees should wash immediately with soap when skin is wet or contaminated. Provide emergency showers and eyewash.

Respirator Selection: Where there is potential for exposures to potassium chlorate, use a NIOSH/MSHA- or European Standard EN149-approved full-face-piece respirator equipped with particulate (dust/fume/mist) filters. Particulate filters must be checked every day before work for physical damage, such as rips or tears, and replaced as needed. *Where there is potential for high exposures*, use a NIOSH/MSHA- or European Standard EN149-approved supplied-air respirator with a full face-piece operated in the positive-pressure mode, or with a full face-piece, hood, or helmet in the continuous-flow mode; or use a NIOSH/MSHA- or European Standard EN149-approved self-contained breathing apparatus with a full face-piece operated in a pressure-demand or other positive-pressure mode.

Storage: Color Code—Yellow: Reactive Hazard; Store in a location separate from other materials, especially flammables and combustibles. Prior to working with this chemical you should be trained on its proper handling and storage. Potassium chlorate must be stored to avoid contact with oxidizers (such as perchlorates, peroxides, permanganates, chlorates, and nitrates) and strong acids (such as hydrochloric, sulfuric, and nitric); since violent reactions occur. Protect storage containers from physical damage. Avoid storage on wood floors. Where possible, automatically transfer material from drums or other storage containers to process containers. Sources of ignition, such as smoking and open flames, are prohibited where this chemical is handled, used, or stored. Metal containers involving the transfer of this chemical should be grounded and bonded. Wherever this chemical is used, handled, manufactured, or stored, use explosion-proof electrical equipment and fittings. See OSHA Standard 1910.104 and NFPA 43A *Code for the Storage of Liquid and Solid Oxidizers* for detailed handling and storage regulations.

Shipping: Potassium chlorate requires a shipping label of "OXIDIZER." It falls in Hazard Class 5.1 and Packing Group II. Potassium chlorate, *aqueous solution,* requires a shipping label of "OXIDIZER." It falls in Hazard Class 5.1 and Packing Group II or III.

Spill Handling: Evacuate persons not wearing protective equipment from area of spill or leak until cleanup is complete. Remove all ignition sources. Collect powdered material in the most convenient and safe manner and deposit in sealed containers. Ventilate area after cleanup is complete. It may be necessary to contain and dispose of this chemical as a hazardous waste. If material or contaminated runoff enters waterways, notify downstream users of potentially contaminated waters. Contact your local or federal environmental protection agency for specific recommendations. If employees are required to clean up spills, they must be properly trained and equipped. OSHA 1910.120(q) may be applicable.

Fire Extinguishing: This chemical is a combustible solid. Use water spray only. Do not use dry chemical, halon, foam, or carbon dioxide extinguishers. Poisonous gases are produced in fire. If material or contaminated runoff enters waterways, notify downstream users of potentially contaminated waters. Notify local health and fire officials and pollution control agencies. From a secure, explosion-proof location, use water spray to cool exposed containers. If cooling streams are ineffective (venting sound increases in volume and pitch, tank discolors, or shows any signs of deforming), withdraw immediately to a secure position. If employees are expected to fight fires, they must be trained and equipped in OSHA 1910.156. The only respirators recommended for firefighting are self-contained breathing apparatuses that have full face-pieces and are operated in a pressure-demand or other positive-pressure mode.

Reference

New Jersey Department of Health and Senior Services. (October 2004). *Hazardous Substances Fact Sheet: Potassium Chlorate.* Trenton, NJ

Potassium chloroplatinates   P:0890

Molecular Formula: Cl_4K_2Pt (tetra-); Cl_6K_2Pt (hexa-)
Common Formula: K_2PtCl_4; K_2PtCl_6
Synonyms: tetra-: Dipotassium tetrachloroplatinate; Potassium tetrachloroplatinate(II); Tetrachlorodipotassium platinate
hexa-: Dipotassium hexachloroplatinate; Hexachlorodipotassium platinate; Hexachloroplatinate(2-) dipotassium; Platinic potassium chloride; Potassium chloroplatinate; Potassium hexachloroplatinate(IV); Potassium hexachloroplatinate(4 +); Potassium platinic chloride
CAS Registry Number: 16921-30-5 (hexa-); 10025-99-7 (tetra-)
RTECS® Number: PT1650000 (hexa-); TP1850000 (tetra-)

UN/NA & ERG Number: UN3290 [Toxic solid, corrosive, inorganic, n.o.s., (*Platinium (IV) potassium chloride*)]/154
EC Number: 240-979-3 [*Annex I Index No.:* 078-007-00-3] (dipotassium hexachloroplatinate); 233-050-9 [*Annex I Index No.:* 078-004-00-7] (dipotassium tetrachloroplatinate)
Regulatory Authority and Advisory Bodies
Air Pollutant Standard Set. See below, "Permissible Exposure Limits in Air" section.
Canada, WHMIS, Ingredients Disclosure List Concentration 0.1%.
European/International Regulations (hexa-): Hazard Symbol: T; Risk phrases: R25; R41; R42/43; Safety phrases: S1/2; S22; S26; S36/37/39; S53; S45; (tetra-): Hazard Symbol: T; Risk phrases: R25; R38; R41; R42/43; Safety phrases: S2; S22; S26; S36/37/39; S45 (see Appendix 4).
WGK (German Aquatic Hazard Class): No value assigned.
Description: Potassium tetrachloroplatinate(II), K_2PtCl_4 is a ruby-red crystalline solid and Potassium hexachloroplatinate (IV), K_2PtCl_6 is an orange-yellow solid. Molecular weight = 485.89 (hexa-); 415.09 (tetra-); Freezing/Melting point = (decomposes) at 250°C (hexa-). Hazard Identification (based on NFPA-704 M Rating System): (hexa-) Health 3, Flammability 0, Reactivity 0. Soluble in water.
Potential Exposure: These materials are used in photography.
Permissible Exposure Limits in Air
OSHA PEL: 0.002 mg[Pt]/m³ TWA
NIOSH REL: 0.002 mg[Pt]/m³ TWA
ACGIH TLV®[1]: 1 0.002 mg[Pt]/m³ TWA
NIOSH IDLH: 4 mg [Pt]/m³
No TEEL available
DFG MAK: No numerical value established. Data may be available; however, 2 µg[Pt]/m³ peak should not be exceeded; danger of skin and airway sensitization, as chloroplatinates.
Several states have set guidelines or standards for platinum in ambient air[60] ranging from 0.1 µg/m³ (Nevada) to 0.4-20.0 µg/m³ (Connecticut) to 10.0 µg/m³ (North Dakota) to 330.0 µg/m³ (Virginia).
Determination in Air: Use NIOSH II(7) Method #S-19 (soluble salts)
Routes of Entry: Inhalation, ingestion, skin and/or eye contact.
Harmful Effects and Symptoms
Short Term Exposure: Potassium chloroplatinates can affect you when breathed in. Severe allergy can develop to potassium chloroplatinates. Symptoms include asthma (with cough, wheezing, and/or shortness of breath), runny nose, and/or skin rash, sometimes with hives. If allergy develops, even small future exposure can trigger significant symptoms. Some persons exposed to this type of chemical have developed lung scarring. Family members can develop allergy to dust carried home on work clothing. It may irritate the eyes, nose, and throat. High exposure may cause irritability and even seizures.

Long Term Exposure: May cause skin sensitization and dermatitis and/or asthma-like allergy. Tetrachloroplatinates are mutagens.

Points of Attack: Skin, lungs.

Medical Surveillance: Before beginning employment and at regular times after that, the following are recommended. Lung function tests. These may be normal if the person is not having an attack at the time of the test. If symptoms develop or overexposure is suspected, the following may be useful: chest X-ray every 3 years should be considered if above tests are not normal. Evaluation by a qualified allergist, including careful exposure history and special testing, may help diagnose skin allergy.

First Aid: If this chemical gets into the eyes, remove any contact lenses at once and irrigate immediately for at least 15 min, occasionally lifting upper and lower lids. Seek medical attention immediately. If this chemical contacts the skin, remove contaminated clothing and wash immediately with soap and water. Seek medical attention immediately. If this chemical has been inhaled, remove from exposure, begin rescue breathing (using universal precautions, including resuscitation mask) if breathing has stopped and CPR if heart action has stopped. Transfer promptly to a medical facility. When this chemical has been swallowed, get medical attention. Give large quantities of water and induce vomiting. Do not make an unconscious person vomit.

Personal Protective Methods: Wear protective gloves and clothing to prevent any reasonable probability of skin contact. Safety equipment suppliers/manufacturers can provide recommendations on the most protective glove/clothing material for your operation. All protective clothing (suits, gloves, footwear, headgear) should be clean, available each day, and put on before work. Contact lenses should not be worn when working with this chemical. Wear dust-proof chemical goggles and face shield unless full face-piece respiratory protection is worn. Employees should wash immediately with soap when skin is wet or contaminated. Provide emergency showers and eyewash.

Respirator Selection: (for soluble Pt salts): *Up to 0.05 mg/m³:* Sa:Cf* (APF = 25) (any supplied-air respirator operated in a continuous-flow mode). *Up to 0.1 mg/m³:* 100F (APF = 50) (any air-purifying, full-face-piece respirator with an N100, R100, or P100 filter) or SCBAF (APF = 50) (any self-contained breathing apparatus with a full face-piece) or SaF (APF = 50) (any supplied-air respirator with a full face-piece). *Up to 4 mg/m³:* SaF: Pd,Pp (APF = 2000) (any supplied-air respirator that has a full face-piece and is operated in a pressure-demand or other positive-pressure mode). *Emergency or planned entry into unknown concentrations or IDLH conditions:* SCBAF: Pd, Pp (APF = 10,000) (any NIOSH/MSHA- or European Standard EN 149-approved self-contained breathing apparatus that has a full face-piece and is operated in a pressure-demand or other positive-pressure mode) or SaF: Pd,Pp: ASCBA (APF = 10,000) (any supplied-air respirator that

has a full face-piece and is operated in a pressure-demand or other positive-pressure mode in combination with an auxiliary, self-contained breathing apparatus operated in a pressure-demand or other positive-pressure mode). *Escape:* 100 F (APF = 50) (any air-purifying, full-face-piece respirator with an N100, R100, or P100 filter) or SCBAE (any appropriate escape-type, self-contained breathing apparatus).

* Substance cause eye irritation and damage; eye protection needed.

Storage: Color Code—Blue: Health Hazard/Poison: Store in a secure poison location. Prior to working with this chemical you should be trained on its proper handling and storage. Store in tightly closed containers in a cool, well-ventilated area.

Shipping: Toxic solid, corrosive, inorganic, n.o.s., (*Platinum (IV) potassium chloride*) requires a shipping label of "POISONOUS/TOXIC MATERIALS, CORROSIVE." It falls in DOT/UN Hazard Class 6.1 and Packing Group II.

Spill Handling: Evacuate persons not wearing protective equipment from area of spill or leak until cleanup is complete. Remove all ignition sources. Collect powdered material in the most convenient and safe manner and deposit in sealed containers. Ventilate area after cleanup is complete. It may be necessary to contain and dispose of this chemical as a hazardous waste. If material or contaminated runoff enters waterways, notify downstream users of potentially contaminated waters. Contact your local or federal environmental protection agency for specific recommendations. If employees are required to clean up spills, they must be properly trained and equipped. OSHA 1910.120(q) may be applicable.

Fire Extinguishing: These chemicals may burn but do not readily ignite. Use dry chemical, carbon dioxide, water spray, or foam extinguishers. Poisonous gases are produced in fire, including chlorides and potassium oxide. If material or contaminated runoff enters waterways, notify downstream users of potentially contaminated waters. Notify local health and fire officials and pollution control agencies. From a secure, explosion-proof location, use water spray to cool exposed containers. If cooling streams are ineffective (venting sound increases in volume and pitch, tank discolors, or shows any signs of deforming), withdraw immediately to a secure position. If employees are expected to fight fires, they must be trained and equipped in OSHA 1910.156. The only respirators recommended for firefighting are self-contained breathing apparatuses that have full face-pieces and are operated in a pressure-demand or other positive-pressure mode.

References

New Jersey Department of Health and Senior Services. (March 2001). *Hazardous Substances Fact Sheet: Potassium Tetrachloroplatinate.* Trenton, NJ

New Jersey Department of Health and Senior Services. (June 2002). *Hazardous Substances Fact Sheet: Potassium Hexachloroplatinate.* Trenton, NJ

Potassium chromate   P:0900

Molecular Formula: CrK_2O_4; $Cr_2K_2O_7$

Common Formula: K_2CrO_4; $K_2Cr_2O_7$

Synonyms: *chromate:* Bipotassium chromate; Chromate of potassium; Dipotassium chromate; Dipotassium monochromate; Neutral potassium chromate; Potassium chromate (VI); Tarapacaite

dichromate: Bichromate of potash; Chromic acid, dipotassium salt; Dipotassium dichromate; Iopezite; Kaliumdichromat (German); Potassium bichromate; Potassium dichromate(VI)

CAS Registry Number: 7789-00-6; 7778-50-9 (dichromate)

RTECS® Number: GB2940000 (chromate); HX7680000 (dichromate)

UN/NA & ERG Number: UN1479/140

EC Number: 232-140-5 [*Annex I Index No.:* 024-006-00-8]; 231-906-6 [*Annex I Index No.:* 024-002-00-6] (dichromate)

Regulatory Authority and Advisory Bodies

Carcinogenicity: IARC: Human Sufficient Evidence; Animal Sufficient Evidence, *carcinogenic to humans*, Group 1, 1997; NTP: 11th Report on Carcinogens, 2004: Known to be a human carcinogen.

US EPA Gene-Tox Program, Positive: SHE—clonal assay; *In vitro* cytogenetics—nonhuman; Positive: *B. subtilis* rec assay; Positive: CHO gene mutation (dichromate).

US EPA, FIFRA, 1998 Status of Pesticides: Supported.

Air Pollutant Standard Set. See below, "Permissible Exposure Limits in Air" section.

Canada, WHMIS, Ingredients Disclosure List Concentration 1.0% as chromate; 0.1% as dichromate.

Water Pollution Standard Proposed (EPA).[6]

Hazardous Substance (EPA) (RQ = 1000/454).[4]

Priority Toxic Pollutant (EPA).[6]

As chromium compounds:

Clean Air Act: Hazardous Air Pollutants (Title I, Part A, Section 112).

Clean Water Act: Toxic Pollutant (Section 401.15); 40CFR401.15 Section 307 Toxic Pollutants as chromium and compounds.

RCRA, 40CFR261, Appendix 8 Hazardous Constituents, waste number not listed.

Canada, WHMIS, Ingredients Disclosure List Concentration 0.1%.

European/International Regulations (*chromate*): Hazard Symbol: T+, N; Risk phrases: R49; R46; R36/37/38; R43; R50/53; Safety phrases: S53; S45; S60; S61; (*dichromate*). Hazard Symbol: T+, N; Risk phrases: R45; R46; R60; R61; R8; R21; R25; R26; R34; R42/43;R48/23; R50/53; Safety phrases: S53; S45; S60; S61 (see Appendix 4).

WGK (German Aquatic Hazard Class): No value assigned.

Description: Potassium chromate(VI) is a yellow crystalline solid. Molecular weight = 194.20; Freezing/Melting point = 971°C. Soluble in water. Potassium dichromate(VI) is a yellowish-red crystalline solid. Molecular weight = 294.20; Boiling point = decomposition at 500°C; Freezing/Melting point = 398°C. Hazard Identification (based on NFPA-704 M Rating System): Health 3, Flammability 0, Reactivity 3 (Oxidizer). Soluble in water.

Potential Exposure: Compound Description: Tumorigen, Mutagen; Reproductive Effector; Human Data; Natural Product; Primary Irritant. Potassium chromate is used in printing, photomechanical processing, chrome-pigment production, and wool preservative methods; to make dyes, pigments, inks, and enamels; as an oxidizing agent, analytical reagent; in electroplating, explosives.

Incompatibilities: A powerful oxidizer. Violent reactions with combustibles, organics, powdered metals, or easily oxidizable substances. Contact with hydrazine causes explosion.

Permissible Exposure Limits in Air

Protective Action Criteria (PAC)

TEEL-0: 0.0187 mg/m³

PAC-1: 2 mg/m³

PAC-2: 12.5 mg/m³

PAC-3: 56 mg/m³

OSHA PEL: 0.1 mg[CrO_3]/m³ Ceiling Concentration.

NIOSH REL: 0.001mg[CrO_3]/m³ TWA, potential carcinogen, limit exposure to lowest feasible level. NIOSH considers all Cr(VI) compounds (including chromic acid; *tert*-butyl chromate; zinc chromate; and chromyl chloride) to be potential occupational carcinogens.

ACGIH TLV®[1]: 0.05 mg[Cr]/m³ TWA, Confirmed Human Carcinogen; BEI issued.

DFG MAK: Danger of skin sensitization; Carcinogen Category 2; TRK: 0.05 mg[Cr]/m³; 20 μg/L [Cr] in urine at end-of-shift.

NIOSH IDLH: 15 mg[Cr(VI)]/m³.

United Kingdom: carcinogen, 2000; the former USSR-UNEP/IRPTC joint project[43] give a MAC of 0.01 mg/m³ in work-place air. Connecticut[60] has set a guideline of 0.25 μg/m³ for chromium trioxide in ambient air; Sweden[35] for potassium chromate at 0.02 mg/m³ as a TWA in work-place air. North Carolina[60] has set a guideline of zero for ambient air for both compounds.

7778-50-9 (dichromate)

Protective Action Criteria (PAC)

TEEL-0: 0.0141 mg/m³

PAC-1: 4 mg/m³

PAC-2: 30 mg/m³

PAC-3: 42.4 mg/m³

Determination in Air: Use NIOSH Analytical Methods #7600, 7604, 7605, 7703, 9101 and OSHA Analytical Methods ID-103, ID-215, W-4001.

Permissible Concentration in Water: The EPA[6] has designated chromium as a priority toxic pollutant. To protect human health, the limits are trivalent chromium (in chromates): 170 μg/L and hexavalent chromium (in dichromates) 50 μg/L.

Determination in Water: Total chromium may be determined by digestion followed by atomic absorption, or by colorimetry (diphenylcarbazide), or by inductively coupled plasma (CP) optical emission spectrometry. Chromium (VI) may be determined by extraction and atomic absorption or colorimetry (using diphenylhydrazide). Dissolved total Cr or Cr(VI) may be determined by 0.45 μm filtration followed by the above-cited methods.[49]

Routes of Entry: Inhalation, ingestion, skin and/or eye contact.

Harmful Effects and Symptoms

Short Term Exposure: Potassium chromates can affect you when breathed in. It can also pass into inner layers of the skin. Eye contact can cause severe damage with possible loss of vision. Irritation of nose, throat, and bronchial tubes can occur, with cough and/or wheezing. Skin contact can cause severe irritation, deep ulcers, or an allergic skin rash.

Long Term Exposure: Potassium chromate is a human carcinogen. Potassium chromates can cause a sore or perforated nasal septum; with bleeding, discharge or crusting. May cause skin allergy. Can cause lung irritation; bronchitis may develop.

Points of Attack: Lungs, skin.

Medical Surveillance: NIOSH lists the following tests: Blood gas analysis, complete blood count; chest X-ray, electrocardiogram, liver function tests; pulmonary function tests; sputum cytology, urine (chemical/metabolite), urinalysis (routine), white blood cell count/differential. Before beginning employment and at regular times after that, for those with frequent or potentially high exposures, the following are recommended: examination of the nose and skin. If symptoms develop or overexposure is suspected, the following may be useful: evaluation by a qualified allergist, including careful exposure history and special testing, may help diagnose skin allergy. Also, check your skin daily for little bumps or blisters, the first sign of "chrome ulcers." If not treated early, these can last for years after exposure.

First Aid: If this chemical gets into the eyes, remove any contact lenses at once and irrigate immediately for at least 15 min, occasionally lifting upper and lower lids. Seek medical attention immediately. If this chemical contacts the skin, remove contaminated clothing and wash immediately with soap and water. Seek medical attention immediately. If this chemical has been inhaled, remove from exposure, begin rescue breathing (using universal precautions, including resuscitation mask) if breathing has stopped and CPR if heart action has stopped. Transfer promptly to a medical facility. When this chemical has been swallowed, get medical attention. Give large quantities of water and induce vomiting. Do not make an unconscious person vomit.

Personal Protective Methods: Wear protective gloves and clothing to prevent any reasonable probability of skin contact. Prevent skin contact. (As chromic acid and chromates) **8 h** (more than 8 h of resistance to breakthrough >0.1 μg/cm/min): polyethylene gloves, suits, boots; polyvinyl chloride gloves, suits, boots; Saranex™ coated suits; **4 h** (At least 4 but <8 h of resistance to breakthrough >0.1 μg/cm²/min): butyl rubber gloves, suits, boots; Viton™ gloves, suits.

Safety equipment suppliers/manufacturers can provide recommendations on the most protective glove/clothing material for your operation, For *potassium dichromate*, Neoprene™ and polyvinyl chloride are among the recommended protective materials. All protective clothing (suits, gloves, footwear, headgear) should be clean, available each day, and put on before work. Contact lenses should not be worn when working with this chemical. Wear dust-proof chemical goggles and face shield unless full face-piece respiratory protection is worn. Employees should wash immediately with soap when skin is wet or contaminated. Provide emergency showers and eyewash.

Respirator Selection: NIOSH, as chromates: *at any concentrations above the NIOSH REL, or where there is no REL, at any detectable concentration:* SCBAF: Pd,Pp (APF = 10,000) (any self-contained breathing apparatus that has a full face-piece and is operated in a pressure-demand or other positive-pressure mode) or SaF: Pd,Pp: ASCBA (APF = 10,000) (any supplied-air respirator that has a full face-piece and is operated in a pressure-demand or other positive-pressure mode in combination with an auxiliary, self-contained breathing apparatus operated in a pressure-demand or other positive-pressure mode). *Escape:* 100F (APF = 50) (any air-purifying, full-face-piece respirator with an N100, R100, or P100 filter) or SCBAE (any appropriate escape-type, self-contained breathing apparatus).

Storage: Color Code—Yellow: Reactive Hazard; Store in a location separate from other materials, especially flammables and combustibles. Color Code—Blue: Health Hazard/Poison: Store in a secure poison location. Prior to working with this chemical you should be trained on its proper handling and storage. Potassium chromates must be stored to avoid contact with combustible, organic, or other easily oxidized materials (such as paper, wood, sulfur, aluminum, hydrazine, and plastics) since violent reactions occur. A regulated, marked area should be established where potassium chromate is handled, used, or stored. Where possible, automatically transfer material from drums or other storage containers to process containers. Sources of ignition, such as smoking and open flames, are prohibited where this chemical is handled, used, or stored. Metal containers involving the transfer of this chemical should be grounded and bonded. Wherever this chemical is used, handled, manufactured, or stored, use explosion-proof electrical equipment and fittings. See OSHA Standard 1910.104 and NFPA 43A *Code for the Storage of Liquid and Solid Oxidizers* for detailed handling and storage regulations. A regulated, marked area should be established where this chemical is handled, used, or stored in compliance with OSHA Standard 1910.1045.

Shipping: Oxidizing solid, n.o.s. requires a shipping label of "OXIDIZER." They fall in DOT Hazard Class 5.1 and Packing Group II or III.

Spill Handling: Evacuate persons not wearing protective equipment from area of spill or leak until cleanup is complete. Remove all ignition sources. Collect powdered material in the most convenient and safe manner and deposit in

sealed containers. Ventilate area after cleanup is complete. Keep potassium dichromate out of a confined space, such as a sewer, because of the possibility of an explosion, unless the sewer is designed to prevent the buildup of explosive concentrations. It may be necessary to contain and dispose of this chemical as a hazardous waste. If material or contaminated runoff enters waterways, notify downstream users of potentially contaminated waters. Contact your local or federal environmental protection agency for specific recommendations. If employees are required to clean up spills, they must be properly trained and equipped. OSHA 1910.120(q) may be applicable.

Fire Extinguishing: This chemical may burn but does not easily ignite. Use dry chemical, carbon dioxide, water spray, or foam extinguishers. Poisonous gases are produced in fire, including chromium and potassium oxides. If material or contaminated runoff enters waterways, notify downstream users of potentially contaminated waters. Notify local health and fire officials and pollution control agencies. From a secure, explosion-proof location, use water spray to cool exposed containers. If cooling streams are ineffective (venting sound increases in volume and pitch, tank discolors, or shows any signs of deforming), withdraw immediately to a secure position. If employees are expected to fight fires, they must be trained and equipped in OSHA 1910.156. The only respirators recommended for firefighting are self-contained breathing apparatuses that have full face-pieces and are operated in a pressure-demand or other positive-pressure mode.

References

Sax, N. I. (Ed.). *Dangerous Properties of Industrial Materials Report*, 1, No. 7, 71–73 (1981) and 8, No. 5, 86–94 (1988) (Potassium Chromate)

New Jersey Department of Health and Senior Services. (September 1996). *Hazardous Substances Fact Sheet: Potassium Dichromate.* Trenton, NJ

New Jersey Department of Health and Senior Services. (June 2003). *Hazardous Substances Fact Sheet: Potassium Chromate.* Trenton, NJ

Potassium cyanide   P:0910

Molecular Formula: CNK
Common Formula: KCN
Synonyms: Cianuro potasico (Spanish); Cyanide of potassium; Cyanure de potassium (French); Hydrocyanic acid, Potassium salt; Kalium-cyanid (German)
CAS Registry Number: 151-50-8
RTECS® Number: TS8750000
UN/NA & ERG Number: UN1680/157
EC Number: 205-793-3
Regulatory Authority and Advisory Bodies
Department of Homeland Security Screening Threshold Quantity (pounds): Sabotage/Contamination Hazard: A placarded amount (commercial grade).

Clean Water Act: Section 311 Hazardous Substances/RQ 40CFR117.3 (same as CERCLA, see below); Section 313 Water Priority Chemicals (57FR41331, 9/9/92).
US EPA Hazardous Waste Number (RCRA No.): P098.
RCRA, 40CFR261, Appendix 8 Hazardous Constituents.
Superfund/EPCRA 40CFR355, Extremely Hazardous Substances: TPQ = 100 lb (45.4 kg).
Reportable Quantity (RQ): 10 lb (4.54 kg).
US DOT Regulated Marine Pollutant (49CFR172.101, Appendix B).
Canada, National Pollutant Release Inventory (NPRI); CEPA Priority Substance List, Ocean dumping prohibited.
As cyanide compounds:
Clean Air Act: Hazardous Air Pollutants (Title I, Part A, Section 112).
Clean Water Act: 40CFR423, Appendix A, Priority Pollutants, as cyanide, total.
US EPA Hazardous Waste Number (RCRA No.): P030 as cyanides soluble salts and complexes, n.o.s.
RCRA, 40CFR261, Appendix 8 Hazardous Constituents. as cyanides, soluble salts and complexes, n.o.s.
Canada, WHMIS, Ingredients Disclosure List Concentration 1.0% as cyanide compounds, inorganic, n.o.s.
WGK (German Aquatic Hazard Class): 3—Severe hazard to waters.

Description: Potassium cyanide is available as white lumps, granular powder, or colorless solution. It may be shipped as capsules, tablets, or pellets. Toxic hydrogen cyanide gas released by potassium cyanide has a distinctive, mild, bitter almond odor, but many people cannot detect it; the odor does not provide adequate warning of hazardous concentrations. Molecular weight = 65.12; Specific gravity (H$_2$O:1) = 1.55 at 20°C; Boiling point = 1625°C; Freezing/Melting point = 633.9°C. Hazard Identification (based on NFPA-704 M Rating System): Health 4, Flammability 0, Reactivity 0. Soluble in water; solubility = 72% at 25°C.

Potential Exposure: Compound Description: Organometallic, Mutagen; Reproductive Effector; Human Data. Used in electroplating, steel hardening, extraction of precious metals form ores, as a fumigant, in insecticides, a reagent in analytical chemistry.

Incompatibilities: Potassium cyanide decomposes on contact with water, humidity, carbon dioxide, strong acids (such as hydrochloric, sulfuric, and nitric acids), and acid salts, producing highly toxic and highly flammable hydrogen cyanide gas. Potassium cyanide absorbs water from air (is hygroscopic or deliquescent); the aqueous solution is a strong base. Incompatible with organic anhydrides, isocyanates, alkylene oxides, epichlorohydrin, aldehydes, alcohols, glycols, phenols, cresols, caprolactum, strong oxidizers, nitrogen trichloride, sodium chlorate. Attacks aluminum, copper, zinc in the presence of moisture.

Permissible Exposure Limits in Air
OSHA PEL: 5 mg[CN]/m^3/4.7 ppm TWA.
NIOSH REL: 5 mg[CN]/m^3/4.7 ppm/10 min Ceiling Concentration.

ACGIH TLV®[1]: 5 mg[CN]/m³ [skin] Ceiling Concentration.
Protective Action Criteria (PAC)*
TEEL-0: 5.3 mg/m³
PAC-1: **5.3** mg/m³
PAC-2: **19** mg/m³
PAC-3: **40** mg/m³
*AEGLs (Acute Emergency Guideline Levels) & ERPGs (Emergency Response Planning Guideline) are in **bold face**.
DFG MAK: 2 mg[CN]/m³, inhalable fraction TWA; Peak Limitation Category II(1) [skin]; Pregnancy Risk Group: C.
NIOSH IDLH: 25 mg[CN]/m³.
Skin contact may contribute significantly in overall exposure. Australia: TWA 5 mg/m³, [skin], 1993; Austria: MAK 5 mg [CN]/m³, [skin], 1999; Denmark: TWA 5 mg/m³, [skin], 1999; France: VME 5 mg[CN]/m³, [skin], 1999; Poland: TWA 0.3 mg[CN]/m³, ceiling 10 mg[CN]/m³, 1999; Switzerland: MAK-W 5 mg/m³, KZG-W 10 mg/m³, [skin], 1999; United Kingdom: TWA 5 mg[CN]/m³, [skin], 2000; Argentina, Bulgaria, Columbia, Jordan, South Korea, New Zealand, Singapore, Vietnam: ACGIH TLV®: Ceiling Concentration 5 mg/m³ [skin]. Russia[43] has set a MAC value of 0.009 mg/m³ for ambient air in residential areas on a momentary basis and 0.004 mg/m³ on a daily average basis. Several states have set guidelines or standards for cyanides in ambient air[60] ranging from 16.7 μg/m³ (New York) to 50.0 μg/m³ (Florida and North Dakota) to 80.0 μg/m³ (Virginia) to 100 μg/m³ (Connecticut and South Dakota) to 125 μg/m³ (South Carolina) to 119.0 μg/m³ (Nevada).
Determination in Air: Use NIOSH Analytical Method (IV) #7904, Cyanides. See also Method #6010, Hydrogen cyanide.[18]
Permissible Concentration in Water: In 1976 the EPA criterion was 5.0 μg/L for freshwater and marine aquatic life and wildlife. As of 1980, the criteria are: *To protect freshwater aquatic life:* 3.5 μg/L as a 24-h average, never to exceed 52.0 μg/L. *To protect saltwater aquatic life:* 30.0 μg/L on an acute toxicity basis; 2.0 μg/L on a chronic toxicity basis. *To protect human health:* 200 μg/L. The allowable daily intake for man is 8.4 mg/day.[6] On the international scene, the South African Bureau of Standards has set 10 μg/L, the World Health Organization (WHO) 10 μg/L, and Germany 50 μg/L as drinking water standards. Other international limits[35] include an EEC limit of 50 μg/L; Mexican limits of 200 μg/L in drinking water and 1.0 μg/L in coastal waters; and a Swedish limit of 100 μg/L. Russia[43] set a MAC of 100 μg/L in water bodies used for domestic purposes and 50 μg/L in water for fishery purposes. The US EPA[49] has determined a no-observed-adverse-effect-level (NOAEL) of 10.8 mg/kg/day which yields a lifetime health advisory of 154 μg/L. States which have set guidelines for cyanides in drinking water[61] include Arizona at 160 μg/L and Kansas at 220 μg/L.
Determination in Water: Distillation followed by silver nitrate titration or colorimetric analysis using pyridine pyrazolone (or barbituric acid).

Routes of Entry: Inhalation, ingestion, skin and/or eye contact. Absorbed through the skin.
Harmful Effects and Symptoms
Short Term Exposure: Potassium cyanide is corrosive to the eyes, skin, and the respiratory tract. Contact can cause skin and eye burns, and possible permanent eye damage. Inhalation can cause lung irritation with coughing, sneezing, and difficult breathing; slow gasping respiration. Corrosive if swallowed. These substances may affect the central nervous system. Symptoms include headaches, confusion, nausea, pounding heart, weakness, unconsciousness, and death.
Long Term Exposure: Repeated or prolonged contact with potassium cyanide may cause thyroid gland enlargement and interfere with thyroid function. May cause nosebleed and sores in the nose; changes in blood cell count. May cause central nervous system damage with headache, dizziness, confusion; nausea, vomiting, pounding heart, weakness in the arms and legs, unconsciousness, and death. May affect liver and kidney function.
Points of Attack: Liver, kidneys, skin, cardiovascular system, central nervous system, thyroid gland.
Medical Surveillance: Consider the points of attack in preplacement and periodic physical examinations. Urine thiocyanate levels. Blood cyanide levels. Complete blood count (CBC). Evaluation of thyroid function. Liver function tests. Kidney function tests. Central nervous system tests. EKG.
First Aid: If this chemical gets into the eyes, remove any contact lenses at once and irrigate immediately for at least 15 min, occasionally lifting upper and lower lids. Seek medical attention immediately. If this chemical contacts the skin, remove contaminated clothing and wash immediately with soap and water. Speed in removing material from skin is of extreme importance. Shampoo hair promptly if contaminated. Seek medical attention immediately. If this chemical has been inhaled, remove from exposure, begin rescue breathing (using universal precautions, including resuscitation mask) if breathing has stopped and CPR if heart action has stopped. Transfer promptly to a medical facility. When this chemical has been swallowed, get medical attention. Give large quantities of water and induce vomiting. Do not make an unconscious person vomit. Keep under observation for 24−48 h as symptoms may return. Use amyl nitrate capsules if symptoms develop. All area employees should be trained regularly in emergency measures for cyanide poisoning and in CPR. A cyanide antidote kit should be kept in the immediate work area and must be rapidly available. Kit ingredients should be replaced every 1−2 years to ensure freshness. Persons trained in the use of this kit, oxygen use, and CPR must be quickly available.
Personal Protective Methods: Wear protective gloves and clothing to prevent any reasonable probability of skin contact. Safety equipment suppliers/manufacturers can provide recommendations on the most protective glove/clothing material for your operation. All protective clothing (suits, gloves, footwear, headgear) should be clean, available each day, and put on before work. Contact lenses should not be

worn when working with this chemical. Wear splash-proof chemical goggles and face shield when working with liquid unless full face-piece respiratory protection is worn. Wear dust-proof goggles and face shield when working with powders or dust unless full face-piece respiratory protection is worn. Employees should wash immediately with soap when skin is wet or contaminated. Provide emergency showers and eyewash.

Respirator Selection: When used as a weapon, use SCBA Respirator Certified By NIOSH For CBRN Environments. Up to 25 mg/m³: Sa (APF = 10) (any supplied-air respirator) or SCBAF (APF = 50) (any self-contained breathing apparatus with full face-piece). *Emergency or planned entry into unknown concentrations or IDLH conditions:* SCBAF: Pd,Pp (APF = 10,000) (any self-contained breathing apparatus that has a full face-piece and is operated in a pressure-demand or other positive-pressure mode) or Sa: Pd,Pp (APF = 1000): ASCBA (any supplied-air respirator that has a full face-piece and is operated in a pressure-demand or other positive-pressure mode in combination with an auxiliary self-contained breathing apparatus operated in a pressure-demand or other positive-pressure mode). *Escape:* GmFS100 (APF = 50) [any air-purifying, full-face-piece respirator (gas mask) with a chin-style, front- or back-mounted canister providing protection against the compound of concern and having an N100, R100, or P100 filter] or SCBAE (any appropriate escape-type, self-contained breathing apparatus).

Storage: Color Code—Blue: Health Hazard/Poison: Store in a secure poison location. Prior to working with this chemical you should be trained on its proper handling and storage. Before entering confined space where this chemical may be present, check to make sure that an explosive concentration does not exist. *Note to physician:* For severe poisoning BAL [British Anti-Lewisite, Dimercaprol, dithiopropanol $(C_3H_8OS_2)$] has been used to treat toxic symptoms of certain heavy metals poisoning—including arsenic. Although BAL is reported to have a large margin of safety, caution must be exercised because toxic effects may be caused by excessive dosage. Most can be prevented by premedication with 1-ephedrine sulfate (CAS: 134-72-5). For milder poisoning *penicillamine (not penicillin)* has been used, both with mixed success. Side effects occur with such treatment and it is never a substitute for controlling exposure. It can only be done under strict medical care. *Note to physician:* For severe poisoning BAL [British Anti-Lewisite, Dimercaprol, dithiopropanol $(C_3H_8OS_2)$] has been used to treat toxic symptoms of certain heavy metals poisoning—including arsenic. Although BAL is reported to have a large margin of safety, caution must be exercised because toxic effects may be caused by excessive dosage. Most can be prevented by premedication with 1-ephedrine sulfate (CAS: 134-72-5). For milder poisoning *penicillamine (not penicillin)* has been used, both with mixed success. Side effects occur with such treatment and it is never a substitute for controlling exposure. It can only be done under strict medical care.

Store in tightly closed containers in a cool, well-ventilated area away from strong acids, acid salts, oxidizers, light, and moisture. Where possible, automatically transfer material from drums or other storage containers to process containers.

Shipping: Potassium cyanide require a shipping label of "POISONOUS/TOXIC MATERIALS." It falls in Hazard Class 6.1 and in Packing Group I.

Spill Handling: Evacuate persons not wearing protective equipment from area of spill or leak until cleanup is complete. Remove all ignition sources. This chemical has a lower explosive limit; ventilate closed spaces before entering them. Collect powdered material in the most convenient and safe manner and deposit in sealed containers. Ventilate area after cleanup is complete. It may be necessary to contain and dispose of this chemical as a hazardous waste. If material or contaminated runoff enters waterways, notify downstream users of potentially contaminated waters. Contact your local or federal environmental protection agency for specific recommendations. If employees are required to clean up spills, they must be properly trained and equipped. OSHA 1910.120(q) may be applicable.

Initial isolation and protective action distances: Distances shown are likely to be affected during the first 30 min after materials are spilled and could increase with time. If more than one tank car, cargo tank, portable tank, or large cylinder involved in the incident is leaking, the protective action distance may need to be increased. You may need to seek emergency information from CHEMTREC at (800) 424-9300 or seek professional environmental engineering assistance from the US EPA Environmental Response Team at (908) 548-8730 (24-h response line). UN 1680 (Potassium cyanide) is on the DOT's list of dangerous water-reactive materials which create large amounts of toxic vapor when *spilled in water:* Dangerous from 0.5 to 10 km (0.3−6.0 miles) downwind.

Fire Extinguishing: KCN is not combustible itself but it decomposes in the presence of moisture, damp air, or carbon dioxide, producing highly toxic and flammable hydrogen cyanide gas and oxides of nitrogen. NO acidic dry chemical extinguishers. NO hydrous agents. NO water. NO carbon dioxide. Use dry chemical and foam on surrounding fires. Vapors are heavier than air and may collect in low areas. Containers may explode in fire. Storage containers and parts of containers may rocket great distances, in many directions. If material or contaminated runoff enters waterways, notify downstream users of potentially contaminated waters. Notify local health and fire officials and pollution control agencies. From a secure, explosion-proof location, use water spray to cool exposed containers. Do not allow water to enter open containers. If cooling streams are ineffective (venting sound increases in volume and pitch, tank discolors, or shows any signs of deforming), withdraw immediately to a secure position. If employees are expected to fight fires, they must be trained and equipped in OSHA

1910.156. The only respirators recommended for firefighting are self-contained breathing apparatuses that have full face-pieces and are operated in a pressure-demand or other positive-pressure mode.

Disposal Method Suggested: Consult with environmental regulatory agencies for guidance on acceptable disposal practices. Generators of waste containing this contaminant ($\geq$100 kg/mo) must conform with EPA regulations governing storage, transportation, treatment, and waste disposal. In accordance with 40CFR165, follow recommendations for the disposal of pesticides and pesticide containers. Must be disposed properly by following package label directions or by contacting your local or federal environmental control agency or by contacting your regional EPA office. Add strong alkaline hypochlorite and react for 24 h. Then flush to sewer with large volumes of water.[22]

Reference

New Jersey Department of Health and Senior Services. (August 2005). *Hazardous Substances Fact Sheet: Potassium Cyanide.* Trenton NJ

Potassium dichloro-isocyanurate   P:0920

Molecular Formula: $C_3Cl_2KN_3O_3$
Common Formula: $KCl_2(NCO)_3$
Synonyms: ACL-59; Dichloroisocyanuric acid, potassium salt; Dichloro-*s*-triazine-2,4,6(1H,3H,5H)-trione potassium deriv; 1,3-Dichloro-*s*-triazine-2,4,6(1H,3H,5H)trione potassium salt; Dichlor-*s*-triazin-2,4,6(1H,3H,5H)trione potassium; Isocyanuric acid, dichloro-, potassium salt; Potassium dichloro-*s*-triazinetrione; Potassium troclosene; *s*-Triazine-2,4,6(1H,3H,5H)-trione, dichloro-, potassium deriv; 1,3,5-Triazine-2,4,6(1H,3H,5H)-trione, 1,3-dichloro-, potassium salt; Troclosene potassium
CAS Registry Number: 2244-21-5
RTECS® Number: XZ1850000
UN/NA & ERG Number: UN2465 (Dichloroisocyanuric acid salts)/140
EC Number: 218-828-8 [*Annex I Index No.:* 613-030-00-X]
Regulatory Authority and Advisory Bodies
Canada, WHMIS, Ingredients Disclosure List Concentration 1.0%.
European/International Regulations: Hazard Symbol: Xn, N; Risk phrases: R22; R31; R36/37; R50/53; Safety phrases: S2; S8; S26; S41; S60; S61 (see Appendix 4).
WGK (German Aquatic Hazard Class): No value assigned.
Description: Potassium dichloroisocyanurate is a white crystalline solid with a chlorine odor. Molecular weight = 237.07; Freezing/Melting point = (decomposes) 250°C. Hazard Identification (based on NFPA-704 M Rating System): Health 3, Flammability 0, Reactivity 2. Oxidizer. Reacts with water; Slightly soluble; solubility = 10−50 mg/mL at 20°C.[NTP]

Potential Exposure: Potassium dichloroisocyanurate is used in household bleaches, dishwashing compounds, and detergents.
Incompatibilities: A strong oxidizer; violent reaction with reducing agents, combustibles, organics, easily chlorinated or oxidized materials, ammonia, urea, other nitrogen compounds, calcium hypochloride, other alkalies, and moisture.
Permissible Exposure Limits in Air
No standards or TEEL available.
Routes of Entry: Inhalation, ingestion, skin and/or eye contact.
Harmful Effects and Symptoms
Short Term Exposure: Potassium dichloroisocyanurate can affect you when breathed in. Exposure can severely irritate the eyes. It can also irritate the nose, throat, and air passages. Contact can cause skin and eye irritation.
Long Term Exposure: May cause lung irritation and damage.
Points of Attack: Lungs, skin.
Medical Surveillance: Before beginning employment and at regular times after that, for those with frequent or potentially high exposures, the following is recommended: lung function tests. consider chest X-ray following acute overexposure.
First Aid: If this chemical gets into the eyes, remove any contact lenses at once and irrigate immediately for at least 15 min, occasionally lifting upper and lower lids. Seek medical attention immediately. If this chemical contacts the skin, remove contaminated clothing and wash immediately with soap and water. Seek medical attention immediately. If this chemical has been inhaled, remove from exposure, begin rescue breathing (using universal precautions, including resuscitation mask) if breathing has stopped and CPR if heart action has stopped. Transfer promptly to a medical facility. When this chemical has been swallowed, get medical attention. Give large quantities of water and induce vomiting. Do not make an unconscious person vomit.
Personal Protective Methods: Wear protective gloves and clothing to prevent any reasonable probability of skin contact. Safety equipment suppliers/manufacturers can provide recommendations on the most protective glove/clothing material for your operation. All protective clothing (suits, gloves, footwear, headgear) should be clean, available each day, and put on before work. Contact lenses should not be worn when working with this chemical. Wear dust-proof chemical goggles and face shield unless full face-piece respiratory protection is worn. Employees should wash immediately with soap when skin is wet or contaminated. Provide emergency showers and eyewash.
Respirator Selection: Where there is potential for exposure to potassium dichloroisocyanurate, use a NIOSH/MSHA- or European Standard EN149-approved supplied-air respirator with a full face-piece operated in the positive-pressure mode, or with a full face-piece, hood, or helmet in the continuous-flow mode; or use a NIOSH/MSHA- or European Standard EN149-approved self-contained breathing

apparatus with a full face-piece operated in a pressure-demand or other positive-pressure mode.

Storage: Color Code—Yellow: Reactive Hazard; Store in a location separate from other materials, especially flammables and combustibles. Prior to working with this chemical you should be trained on its proper handling and storage. Store in tightly closed containers in a cool, well-ventilated area away from combustibles (like wood, paper, and oil), ammonia, urea, other nitrogen compounds, calcium hypochlorite, other alkalis, and moisture. Do not store on wooden floors. Where possible, automatically transfer material from drums or other storage containers to process containers. Sources of ignition, such as smoking and open flames, are prohibited where this chemical is handled, used, or stored. Metal containers involving the transfer of this chemical should be grounded and bonded. Wherever this chemical is used, handled, manufactured, or stored, use explosion-proof electrical equipment and fittings. See OSHA Standard 1910.104 and NFPA 43A *Code for the Storage of Liquid and Solid Oxidizers* for detailed handling and storage regulations.

Shipping: Dichloroisocyanuric acid, dry requires a shipping label of "OXIDIZER." It falls in Hazard Class 5.1 and Packing Group II.

Spill Handling: Evacuate persons not wearing protective equipment from area of spill or leak until cleanup is complete. Remove all ignition sources. Collect powdered material in the most convenient and safe manner and deposit in sealed containers. Ventilate area after cleanup is complete. *Do not use water.* It may be necessary to contain and dispose of this chemical as a hazardous waste. If material or contaminated runoff enters waterways, notify downstream users of potentially contaminated waters. Contact your local or federal environmental protection agency for specific recommendations. If employees are required to clean up spills, they must be properly trained and equipped. OSHA 1910.120(q) may be applicable.

Fire Extinguishing: Potassium dichloroisocyanurate is not flammable. Potassium dichloroisocyanurate is a strong oxidizer and a dangerous fire hazard on contact with combustibles (such as wood, paper, and oil). Poisonous gases are produced in fire, including chlorine gas, potassium oxide, and oxides of nitrogen. Use dry chemical, or CO_2 extinguishers. Use water to keep fire-exposed containers cool. If material or contaminated runoff enters waterways, notify downstream users of potentially contaminated waters. Notify local health and fire officials and pollution control agencies. From a secure, explosion-proof location, use water spray to cool exposed containers. If cooling streams are ineffective (venting sound increases in volume and pitch, tank discolors, or shows any signs of deforming), withdraw immediately to a secure position. If employees are expected to fight fires, they must be trained and equipped in OSHA 1910.156. The only respirators recommended for firefighting are self-contained breathing apparatuses that have full face-pieces and are operated in a pressure-demand or other positive-pressure mode.

Reference
New Jersey Department of Health and Senior Services. (May 2002). *Hazardous Substances Fact Sheet: Potassium Dichloroisocyanurate.* Trenton, NJ

Potassium ferrocyanide   P:0930

Molecular Formula: $C_6K_4N_6Fe$
Common Formula: $K_4Fe(CN)_6$
Synonyms: Ferrate(4-), hexacyano-, tetrapotassium; Ferrate (4-), hexakis(cyano-C)-, tetrapotassium, (OC-6-11)-; Potassium ferrocyanate; Potassium ferrocyanide; Potassium hexacyanoferrate; Potassium hexacyanoferrate(II); Tetrapotassium ferrocyanide; Tetrapotassium hexacyanoferrate; Tetrapotassium hexacyanoferrate(II); Tetrapotassium hexacyanoferrate(4-)
CAS Registry Number: 13943-58-3
RTECS® Number: LI8219000
UN/NA & ERG Number: Not regulated.
EC Number: 237-722-2
Regulatory Authority and Advisory Bodies
Clean Air Act: Hazardous Air Pollutants (Title I, Part A, Section 112).
Clean Water Act: 40CFR423, Appendix A, Priority Pollutants as cyanide, total.
US EPA Hazardous Waste Number (RCRA No.): P030 as cyanides soluble salts and complexes, n.o.s.
RCRA, 40CFR261, Appendix 8 Hazardous Constituents. as cyanides, soluble salts and complexes, n.o.s.
EPCRA (Section 313): $X + CN^-$ where $X = H^+$ or any other group where a formal dissociation may occur. For example, KCN or Ca(CN)2. Form R *de minimis* concentration reporting level: 1.0%.
US DOT Regulated Marine Pollutant (49CFR172.101, Appendix B) as cyanide mixtures, cyanide solutions or cyanides, inorganic, n.o.s.
Canada, WHMIS, Ingredients Disclosure List Concentration 1.0%; National Pollutant Release Inventory (NPRI); CEPA Priority Substance List, Ocean dumping prohibited.
WGK (German Aquatic Hazard Class): 2—Hazard to waters.
Description: Tetrapotassium hexacyanoferrate is a lemon-yellow crystalline solid. Molecular weight = 368.37; Boiling point = (decomposes). Hazard Identification (based on NFPA-704 M Rating System): Health 1, Flammability 0, Reactivity 2.
Potential Exposure: Used in dyeing, tempering steel, explosives, process engraving, and lithography.
Incompatibilities: Acids or acid fumes may cause release of highly toxic cyanide fumes. Reacts violently with ammonia, copper nitrate; sodium nitrite or chromate.
Permissible Exposure Limits in Air
Protective Action Criteria (PAC)
TEEL-0: 11.8 mg/m^3

PAC-1: 35.4 mg/m^3
PAC-2: 59 mg/m^3
PAC-3: 59 mg/m^3
OSHA PEL: 5 mg[CN]/m^3/4.7 ppm TWA.
NIOSH REL: 5 mg[CN]/m^3/4.7 ppm/10 min Ceiling Concentration.
ACGIH TLV®[1]: 5 mg[CN]/m^3 [skin] Ceiling Concentration.
DFG MAK: 2 mg[CN]/m^3, inhalable fraction TWA; Peak Limitation Category II(1) [skin]; Pregnancy Risk Group: C.
NIOSH IDLH: 25 mg[CN]/m^3.

Determination in Air: Use NIOSH Analytical Method #7904, Cyanides.

Permissible Concentration in Water: As of 1980, the criteria are: *To protect freshwater aquatic life:* 3.5 µg/L as a 24-h average, never to exceed 52.0 µg/L. *To protect saltwater aquatic life:* 30.0 µg/L on an acute toxicity basis; 2.0 µg/L on a chronic toxicity basis. *To protect human health:* 200 µg/L. The allowable daily intake for man is 8.4 mg/day.[6] On the international scene, the South African Bureau of Standards has set 10 µg/L, the World Health Organization (WHO) 10 µg/L and Germany 50 µg/L as drinking water standards. Other international limits[35] include an EEC limit of 50 µg/L; Mexican limits of 200 µg/L in drinking water and 1.0 µg/L in coastal waters and a Swedish limit of 100 µg/L. Russia[43] set a MAC of 100 µg/L in water bodies used for domestic purposes and 50 µg/L in water for fishery purposes. The US EPA[49] has determined a no-observed-adverse-effect-level (NOAEL) of 10.8 mg/kg/day which yields a lifetime health advisory of 154 µg/L. States which have set guidelines for cyanides in drinking water[61] include Arizona at 160 µg/L and Kansas at 220 µg/L.

Routes of Entry: Inhalation, ingestion, skin and/or eye contact.

Harmful Effects and Symptoms

Short Term Exposure: Insufficient data are available on the effect of this substance on human health. However, this is a cyanide compound; therefore, utmost care must be taken. May cause eye irritation. May be harmful if swallowed. Animal studies suggest that the lethal dose is 4–8 oz for an adult.

Long Term Exposure: No information available.

Medical Surveillance: There is no special test for this chemical. However, if illness occurs or overexposure is suspected, medical attention is recommended.

First Aid: If this chemical gets into the eyes, remove any contact lenses at once and irrigate immediately for at least 15 min, occasionally lifting upper and lower lids. Seek medical attention immediately. If this chemical contacts the skin, remove contaminated clothing and wash immediately with soap and water. Seek medical attention immediately. If this chemical has been inhaled, remove from exposure, begin rescue breathing (using universal precautions, including resuscitation mask) if breathing has stopped and CPR if heart action has stopped. Transfer promptly to a medical facility. When this chemical has been swallowed, get medical attention. Give large quantities of water and induce vomiting. Do not make an unconscious person vomit. Use amyl nitrate capsules if symptoms of cyanide poisoning develop. All area employees should be trained regularly in emergency measures for cyanide poisoning and in CPR. A cyanide antidote kit should be kept in the immediate work area and must be rapidly available. Kit ingredients should be replaced every 1–2 years to ensure freshness. Persons trained in the use of this kit, oxygen use, and CPR must be quickly available.

Personal Protective Methods: Wear protective gloves and clothing to prevent any reasonable probability of skin contact. Safety equipment suppliers/manufacturers can provide recommendations on the most protective glove/clothing material for your operation. All protective clothing (suits, gloves, footwear, headgear) should be clean, available each day, and put on before work. Contact lenses should not be worn when working with this chemical. Wear dust-proof chemical goggles and face shield unless full face-piece respiratory protection is worn. Employees should wash immediately with soap when skin is wet or contaminated. Provide emergency showers and eyewash.

Respirator Selection: Follow regulations in OSHA 29CFR1910.134 or European Standard EN149. Use a NIOSH/MSHA- or European Standard EN149-approved respirator; or use an approved half-face, dust/mist respirator. *For emergencies or instances where the exposure levels are not known*, use a full-face positive-pressure, air-supplied respirator operated in the positive-pressure mode, or use a NIOSH/MSHA- or European Standard EN149-approved self-contained breathing apparatus with a full face-piece operated in a pressure-demand or other positive-pressure mode. *Note:* Air-purifying respirators do not protect workers in oxygen-deficient atmospheres.

Storage: Color Code—Green: General storage may be used. Prior to working with this chemical you should be trained on its proper handling and storage. Store away from incompatible materials listed above. Where possible, automatically transfer material from drums or other storage containers to process containers. Sources of ignition, such as smoking and open flames, are prohibited where this chemical is handled, used, or stored. Metal containers involving the transfer of this chemical should be grounded and bonded. Wherever this chemical is used, handled, manufactured, or stored, use explosion-proof electrical equipment and fittings.

Shipping: Not regulated.

Spill Handling: Evacuate persons not wearing protective equipment from area of spill or leak until cleanup is complete. Remove all ignition sources. Collect powdered material in the most convenient and safe manner and deposit in sealed containers. Ventilate area after cleanup is complete. It may be necessary to contain and dispose of this chemical as a hazardous waste. If material or contaminated runoff enters waterways, notify downstream users of potentially

contaminated waters. Contact your local or federal environmental protection agency for specific recommendations. If employees are required to clean up spills, they must be properly trained and equipped. OSHA 1910.120(q) may be applicable.

Fire Extinguishing: Use dry chemical, carbon dioxide, water spray, or alcohol foam extinguishers. Poisonous gases are produced in fire. If material or contaminated runoff enters waterways, notify downstream users of potentially contaminated waters. Notify local health and fire officials and pollution control agencies. From a secure, explosion-proof location, use water spray to cool exposed containers. If cooling streams are ineffective (venting sound increases in volume and pitch, tank discolors, or shows any signs of deforming), withdraw immediately to a secure position. If employees are expected to fight fires, they must be trained and equipped in OSHA 1910.156. The only respirators recommended for firefighting are self-contained breathing apparatuses that have full face-pieces and are operated in a pressure-demand or other positive-pressure mode.

Reference
New York State Department of Health. (February 1986). *Chemical Fact Sheet: Potassium Ferrocyanide.* Albany, NY: Bureau of Toxic Substance Assessment

Potassium fluoride   P:0940

Molecular Formula: FK
Common Formula: KF
Synonyms: Fluorure de potassium (French); Potassium fluorure (French)
CAS Registry Number: 7789-23-3
RTECS® Number: TT0700000
UN/NA & ERG Number: UN1812/154
EC Number: 232-151-5 [*Annex I Index No.:* 009-005-00-2]
Regulatory Authority and Advisory Bodies
Air Pollutant Standard Set. See below, "Permissible Exposure Limits in Air" section.
European/International Regulations: Hazard Symbol: T; Risk phrases: R23/24/25; Safety phrases: S1/2; S26; S45 (see Appendix 4).
WGK (German Aquatic Hazard Class): 1—Low hazard to waters.
Description: Potassium fluoride is a white crystalline solid. Molecular weight = 58.10; Boiling point = 1505°C; Freezing/Melting point = 860°C. Hazard Identification (based on NFPA-704 M Rating System): Health 3, Flammability 0, Reactivity 0. Soluble in water.
Potential Exposure: Potassium fluoride is used in etching glass, as a preservative, and as an insecticide.
Incompatibilities: Strong acids.
Permissible Exposure Limits in Air
OSHA PEL: 3 ppm/2.5 mg[F]/m^3 TWA.
NIOSH REL: 3 ppm/2.5 mg[F]/m^3 TWA; 6 ppm/5 mg[F]/m^3, 15 min Ceiling Concentration.

ACGIH TLV®[1]: 2.5 mg[F]/m^3 TWA; not classifiable as a human carcinogen; BEI: 3 mg[F]/g creatinine in urine *prior* to end-of-shift; 10 mg[F]/g creatinine in urine end-of-shift.
Protective Action Criteria (PAC)
TEEL-0: 7.65 mg/m^3
PAC-1: 7.65 mg/m^3
PAC-2: 30 mg/m^3
PAC-3: 500 mg/m^3
DFG MAK: 1 mg[F]/m^3, inhalable fraction [skin]; Peak Limitation Category II(4); Pregnancy Risk Group C; BAT: 7.0 mg[F]/g creatinine in urine at end-of-shift; 4.0 mg[F]/g creatinine in urine at the beginning of the next shift.
NIOSH IDLH: 250 mg[F]/m^3.
Russia[43] has set a MAC value of 0.03 mg/m^3 for soluble fluorides in ambient air in residential areas on a momentary basis and 0.01 mg/m^3 on a daily average basis. Several states have set limits for fluoride in ambient air[60] ranging from as low as 2.85 μg/m^3 (Iowa) to as high as 60,000 μg/m^3 (Kentucky). The reader is referred to the entry on "Fluorides" for more detail.
Permissible Concentration in Water: Fluoride is a safe drinking water act-regulated chemical (47FR9352, 56FR 3594); MCL = 4.0 mg/L; MCLG = 4.0 mg/L; SMCL = 2.0 mg/L. The state of Maine has set 2.4 mg/L as a guideline for drinking water. Arizona[61] has set 1.8 mg/L as a standard for drinking water.
Routes of Entry: Inhalation, ingestion, skin and/or eye contact.
Harmful Effects and Symptoms
Short Term Exposure: Potassium fluoride can affect you when breathed in. Inhalation of dust or mist can cause severe irritation and burns of the eyes and skin. May cause permanent eye damage. Inhalation can cause irritation of the nose and throat causing sneezing, coughing, and sore throat. High exposure can irritate the lungs, causing a buildup of fluid in the lungs. This can cause death.
Long Term Exposure: These effects do not occur at the levels of fluorides used in water to prevent cavities. Repeated exposure can cause fluoride to buildup in the body. Can irritate the lungs; bronchitis may develop. Repeated exposure can cause fluoride to build up in the body causing stiffness, brittle bones, and crippling. Prolonged contact can cause sores in the nose and perforated septum.
Points of Attack: Lungs, skin.
Medical Surveillance: NIOSH lists the following tests: chest X-ray, electrocardiogram, pulmonary function tests: forced vital capacity, forced expiratory volume (1 s); pelvic X-ray; sputum cytology; urine (chemical/metabolite); urine (chemical/metabolite) pre- and postshift; urinalysis (routine); complete blood count/differential.
First Aid: If this chemical gets into the eyes, remove any contact lenses at once and irrigate immediately for at least 30 min, occasionally lifting upper and lower lids. Seek medical attention immediately. If this chemical contacts the skin, remove contaminated clothing and wash immediately

with soap and water. Seek medical attention immediately. If this chemical has been inhaled, remove from exposure, begin rescue breathing (using universal precautions, including resuscitation mask) if breathing has stopped and CPR if heart action has stopped. Transfer promptly to a medical facility. If victim is *conscious*, administer water or milk. Do not induce vomiting. Medical observation is recommended for 24–48 h after breathing overexposure, as pulmonary edema may be delayed. As first aid for pulmonary edema, a doctor or authorized paramedic may consider administering a corticosteroid spray.

Personal Protective Methods: Wear protective gloves and clothing to prevent any reasonable probability of skin contact. Safety equipment suppliers/manufacturers can provide recommendations on the most protective glove/clothing material for your operation. All protective clothing (suits, gloves, footwear, headgear) should be clean, available each day, and put on before work. Contact lenses should not be worn when working with this chemical. Wear dust-proof chemical goggles and face shield unless full face-piece respiratory protection is worn. Employees should wash immediately with soap when skin is wet or contaminated. Provide emergency showers and eyewash.

Respirator Selection: NIOSH/OSHA *12.5 mg/m³:* Qm (APF = 25) (any quarter-mask respirator). *25 mg/m³:* 95XQ (APF = 10)* [any particulate respirator equipped with an N95, R95, or P95 filter (including N95, R95, and P95 filtering face-pieces) except quarter-mask respirators. The following filters may also be used: N99, R99, P99, N100, R100, P100] or SA* (any supplied-air respirator). *62.5 mg/m³:* Sa:Cf (APF = 25)*[†] (any supplied-air respirator operated in a continuous-flow mode) or PaprHie (APF = 25)* *if not present as a fume* (any powered, air-purifying respirator with a high-efficiency particulate filter). *125 mg/m³:* 100F (APF = 50)[†] [any particulate respirator equipped with an N95, R95, or P95 filter (including N95, R95, and P95 filtering face-pieces) except quarter-mask respirators. The following filters may also be used: N99, R99, P99, N100, R100, P100] or SCBAF (APF = 50) (any self-contained breathing apparatus with a full face-piece) or SaF (APF = 50) (any supplied-air respirator with a full face-piece). *250 mg/m³:* Sa: Pd,Pp (APF = 1000) (any supplied-air respirator operated in a pressure-demand or other positive-pressure mode). *Emergency or planned entry into unknown concentrations or IDLH conditions:* SCBAF: Pd, Pp (APF = 10,000) (any self-contained breathing apparatus that has a full faceplate and is operated in a pressure-demand or other positive-pressure mode) or SaF: Pd,Pp: ASCBA (APF = 10,000) (any supplied-air respirator that has a full face-piece and is operated in a pressure-demand or other positive-pressure mode in combination with an auxiliary, self-contained breathing apparatus operated in a pressure-demand or other positive-pressure mode). *Escape:* 100F (APF = 50)[†] [any particulate respirator equipped with an N95, R95, or P95 filter (including N95, R95, and P95 filtering face-pieces) except quarter-mask respirators. The

following filters may also be used: N99, R99, P99, N100, R100, P100] or SCBAE (any appropriate escape-type, self-contained breathing apparatus).

*Substance reported to cause eye irritation or damage; may require eye protection.

[†]May need acid gas sorbent.

Storage: Color Code—Blue: Health Hazard/Poison: Store in a secure poison location. Prior to working with this chemical you should be trained on its proper handling and storage. Potassium fluoride must be stored to avoid contact with strong acids (such as hydrochloric, sulfuric, and nitric) since violent reactions occur. Where possible, automatically transfer material from drums or other storage containers to process containers. Sources of ignition, such as smoking and open flames, are prohibited where this chemical is handled, used, or stored. Metal containers involving the transfer of this chemical should be grounded and bonded. Wherever this chemical is used, handled, manufactured, or stored, use explosion-proof electrical equipment and fittings.

Shipping: This compound requires a shipping label of "POISONOUS/TOXIC MATERIALS." It falls in Hazard Class 6.1 and Packing Group III.

Spill Handling: Evacuate persons not wearing protective equipment from area of spill or leak until cleanup is complete. Remove all ignition sources. Collect powdered material in the most convenient and safe manner and deposit in sealed containers. Ventilate area after cleanup is complete. It may be necessary to contain and dispose of this chemical as a hazardous waste. If material or contaminated runoff enters waterways, notify downstream users of potentially contaminated waters. Contact your local or federal environmental protection agency for specific recommendations. If employees are required to clean up spills, they must be properly trained and equipped. OSHA 1910.120(q) may be applicable.

Fire Extinguishing: Use dry chemical, carbon dioxide, water spray, or foam extinguishers. Poisonous gases are produced in fire, including potassium oxide and fluorine. If material or contaminated runoff enters waterways, notify downstream users of potentially contaminated waters. Notify local health and fire officials and pollution control agencies. From a secure, explosion-proof location, use water spray to cool exposed containers. If cooling streams are ineffective (venting sound increases in volume and pitch, tank discolors, or shows any signs of deforming), withdraw immediately to a secure position. If employees are expected to fight fires, they must be trained and equipped in OSHA 1910.156. The only respirators recommended for firefighting are self-contained breathing apparatuses that have full face-pieces and are operated in a pressure-demand or other positive-pressure mode.

Reference

New Jersey Department of Health and Senior Services. (November 2004). *Hazardous Substances Fact Sheet: Potassium Fluoride*. Trenton, NJ

Potassium hydroxide   P:0950

Molecular Formula: HKO
Common Formula: KOH
Synonyms: Caustic potash; Hidroxido potasico (Spanish); Hydroxide de potassium (French); Kaliumhydroxid (German); KOH; LYE; Potassa; Potasse caustique (French); Potassium hydrate; Potassium (hydrixyde de) (French)
CAS Registry Number: 1310-58-3
RTECS® Number: TT2100000
UN/NA & ERG Number: UN1813 (solid)/154; UN1814 (solution)/154
EC Number: 215-181-3 [*Annex I Index No.:* 019-002-00-8]
Regulatory Authority and Advisory Bodies
Air Pollutant Standard Set. See below, "Permissible Exposure Limits in Air" section.
US EPA, FIFRA 1998 Status of Pesticides: Canceled.
Clean Water Act: Section 311 Hazardous Substances/RQ 40CFR117.3 (same as CERCLA, see below).
Reportable Quantity (RQ): 1000 lb (454 kg).
Canada, WHMIS, Ingredients Disclosure List Concentration 1.0%.
European/International Regulations: not listed in Annex 1.
WGK (German Aquatic Hazard Class): 1—Low hazard to waters.
Description: Potassium hydroxide is a white deliquescent solid. Molecular weight = 56.11; Specific gravity $(H_2O:1)$ = 2.04 at 25°C; Boiling point = 1324°C; Freezing/Melting point = 380°C; also reported at 405°C (varies with water content). Hazard Identification (based on NFPA-704 M Rating System) (45%): Health 3, Flammability 0, Reactivity 1. Soluble in water; solubility = 107% at 15°C; reaction.
Potential Exposure: Compound Description: Agricultural Chemical; Mutagen, Primary Irritant. Used in the manufacture of other potassium compounds and in the general use of KOH as an alkali.
Incompatibilities: A strong base. Violent reaction with acids, water, metals (when wet), halogenated hydrocarbons, maleic anhydride. Heat is generated if KOH comes in contact with water and carbon dioxide from the air. Corrosive to zinc, aluminum, tin, and lead in the presence of moisture, forming a combustible/explosive hydrogen gas. Can absorb water from air and give off sufficient heat to ignite surrounding combustible materials.
Permissible Exposure Limits in Air
OSHA PEL: None.
NIOSH REL: 2 mg/m³ Ceiling Concentration.
ACGIH TLV®[1]: 2 mg/m³ Ceiling Concentration.
Protective Action Criteria (PAC)
TEEL-0: 0.1 mg/m³
PAC-1: 0.3 mg/m³
PAC-2: 2 mg/m³
PAC-3: 125 mg/m³
Australia: TWA 2 mg/m³, 1993; Austria: MAK 2 mg/m³, 1999; Belgium: STEL 2 mg/m³, 1993; Denmark: TWA 2 mg/m³, 1999; Finland: TWA 2 mg/m³, 1999; Japan: STEL 2 mg/m³, 1999; Norway: TWA 2 mg/m³, 1999; the Netherlands: MAC 2 mg/m³, 2003; Switzerland: MAK-W 2 mg/m³, 1999; United Kingdom: STEL 2 mg/m³, 2000; Argentina, Bulgaria, Columbia, Jordan, South Korea, New Zealand, Singapore, Vietnam: ACGIH TLV®: Ceiling Concentration 2 mg/m³. Several states have set guidelines or standards for KOH in ambient air[60] ranging from 16.0 μg/m³ (Virginia) to 20 μg/m³ (North Dakota) to 48 μg/m³ (Nevada).
Determination in Air: Use NIOSH Analytical Method (IV) #7401, alkaline Dusts.
Routes of Entry: Inhalation, ingestion, skin and/or eye contact.
Harmful Effects and Symptoms
Short Term Exposure: Potassium hydroxide can affect you when breathed in. Potassium hydroxide is highly corrosive. Eye contact causes immediate severe burns and can lead to blindness. Skin contact causes severe skin burns. Exposure can irritate the nose, throat, and airways, causing sneezing, coughing, and sores in the nose. Higher levels can irritate the lungs and cause a buildup of fluid (pulmonary edema). This can cause death. Ingestion may be fatal; causes epigastrium, hermatemesis, collapse, and stricture of esophagus.
Long Term Exposure: May cause sores in the nose and perforation of the nasal septum. May cause lung damage.
Points of Attack: Eyes, skin, respiratory system.
Medical Surveillance: Before beginning employment and at regular times after that, for those with frequent or potentially high exposures, the following is recommended: lung function tests. If symptoms develop or overexposure is suspected, the following may be useful: consider chest X-ray after acute overexposure.
First Aid: If this chemical gets into the eyes, remove any contact lenses at once and irrigate immediately for at least 45 min, occasionally lifting upper and lower lids. Seek medical attention immediately. If this chemical contacts the skin, remove contaminated clothing and wash immediately with soap and water. Seek medical attention immediately. If this chemical has been inhaled, remove from exposure, begin rescue breathing (using universal precautions, including resuscitation mask) if breathing has stopped and CPR if heart action has stopped. Transfer promptly to a medical facility. When this chemical has been swallowed, get medical attention. If victim is *conscious*, administer water or milk. Do not induce vomiting. Medical observation is recommended for 24−48 h after breathing overexposure, as pulmonary edema may be delayed. As first aid for pulmonary edema, a doctor or authorized paramedic may consider administering a corticosteroid spray.
Personal Protective Methods: Wear protective gloves and clothing to prevent any reasonable probability of skin contact. Safety equipment suppliers/manufacturers can provide recommendations on the most protective glove/clothing material for your operation. For solutions of 30−70%,

natural rubber; Neoprene™, nitrile, nitrile + PVC, Neoprene™ + natural rubber, and polyethylene are recommended. All protective clothing (suits, gloves, footwear, headgear) should be clean, available each day, and put on before work. Contact lenses should not be worn when working with this chemical. Wear splash-proof chemical goggles and face shield when working with liquid, or wear dust-proof goggles and face shield when working with powders or dusts unless full face-piece respiratory protection is worn. Employees should wash immediately with soap when skin is wet or contaminated. Provide emergency showers and eyewash.

Respirator Selection: Where there is potential for exposures over 2 mg/m³, use a NIOSH/MSHA- or European Standard EN 149-approved full-face-piece respirator with a high-efficiency particulate filter. More protection is provided by a powered air-purifying respirator. Particulate filters must be checked every day before work for physical damage, such as rips or tears, and replaced as needed. *Where there is potential for high exposures*, use a NIOSH/MSHA- or European Standard EN 149-approved supplied-air respirator with a full face-piece operated in the positive-pressure mode, or with a full face-piece, hood, or helmet in the continuous-flow mode; or use a NIOSH/MSHA- or European Standard EN 149-approved self-contained breathing apparatus with a full face-piece operated in a pressure-demand or other positive-pressure mode.

Storage: Color Code—White Stripe: Contact Hazard; Store separately; not compatible with materials in solid white category. Prior to working with this chemical you should be trained on its proper handling and storage. Potassium hydroxide must be stored to avoid contact with water or moisture and metals since violent reactions occur. Store in tightly closed containers in a cool, well-ventilated area away from acids, explosives, combustible materials, and organic peroxides. Where possible, automatically pump liquid from drums or other storage containers to process containers. Sources of ignition, such as smoking and open flames, are prohibited where this chemical is handled, used, or stored.

Shipping: Potassium hydroxide, solid or solution, requires a shipping label of "CORROSIVE." The solid falls in DOT Hazard Class 8 and Packing Group II. The solution falls in DOT Hazard Class 8 and Packing Group II or III.

Spill Handling: Evacuate persons not wearing protective equipment from area of spill or leak until cleanup is complete. Remove all ignition sources. Ventilate the area of spill or leak. Absorb liquids in vermiculite, dry sand, earth, or a similar material and deposit in sealed containers. Collect powdered material in the most convenient and safe manner and deposit in sealed containers. It may be necessary to contain and dispose of this chemical as a hazardous waste. If material or contaminated runoff enters waterways, notify downstream users of potentially contaminated waters. Contact your local or federal environmental protection agency for specific recommendations. If employees are required to clean up spills, they must be properly trained and equipped. OSHA 1910.120(q) may be applicable.

Fire Extinguishing: Potassium hydroxide may ignite surrounding material if it absorbs water. Extinguish fire using an agent suitable for type of surrounding fire. Potassium hydroxide itself does not burn. Poisonous gases are produced in fire, including potassium oxide. If material or contaminated runoff enters waterways, notify downstream users of potentially contaminated waters. Notify local health and fire officials and pollution control agencies. From a secure, explosion-proof location, use water spray to cool exposed containers. If cooling streams are ineffective (venting sound increases in volume and pitch, tank discolors, or shows any signs of deforming), withdraw immediately to a secure position. If employees are expected to fight fires, they must be trained and equipped in OSHA 1910.156. The only respirators recommended for firefighting are self-contained breathing apparatuses that have full face-pieces and are operated in a pressure-demand or other positive-pressure mode.

Disposal Method Suggested: Dilute with large volume of water, neutralize and flush to sewer.[22]

Reference
New Jersey Department of Health and Senior Services. (May 2001). *Hazardous Substances Fact Sheet: Potassium Hydroxide*. Trenton, NJ

Potassium nitrate   P:0960

Molecular Formula: KNO_3
Synonyms: Kaliumnitrat (German); Niter; Nitre; Nitric acid, potassium salt; Saltpeter; Vicknite
CAS Registry Number: 7757-79-1
RTECS® Number: TT3700000
UN/NA & ERG Number: UN1486/140
EC Number: 231-818-8
Regulatory Authority and Advisory Bodies
Department of Homeland Security Screening Threshold Quantity (pounds): *Theft hazard* 400 (Commercial grade).
US EPA, FIFRA 1998 Status of Pesticides: RED completed.
FDA—over-the-counter drug.
Air Pollutant Standard Set. See below, "Permissible Exposure Limits in Air" section.
WGK (German Aquatic Hazard Class): 1—Low hazard to waters.
Description: Potassium nitrate is an odorless, white or colorless crystalline powder with a salty taste. Molecular weight = 179.31; Boiling point = (decomposes) 400°C; Freezing/Melting point = 334°C. Hazard Identification (based on NFPA-704 M Rating System): Health 2, Flammability 0, Reactivity 3 (Oxidizer). Highly soluble in water; solubility = 36% at 25°C.
Potential Exposure: Compound Description: Drug, Mutagen; Reproductive Effector. Used to make explosives,

gunpowder, fireworks, rocket fuel, matches, fertilizer, fluxes, glass manufacture; and as a diuretic.

Incompatibilities: A powerful oxidizer. Dangerously reactive and friction- and shock-sensitive when mixed with organic materials and many materials. Violent reactions with reducing agents, chemically active metals, charcoal, trichloroethylene.

Permissible Exposure Limits in Air

Protective Action Criteria (PAC)

TEEL-0: 0.4 mg/m^3

PAC-1: 1.25 mg/m^3

PAC-2: 7.5 mg/m^3

PAC-3: 500 mg/m^3

Russia: STEL 5 mg/m^3, 1993.

Routes of Entry: Inhalation, ingestion, skin and/or eye contact.

Harmful Effects and Symptoms

Short Term Exposure: Contact can cause eye and skin irritation. Inhalation can cause respiratory tract irritation, coughing, and wheezing. High levels of exposure can interfere with the blood's ability to carry oxygen, causing headache, dizziness, cyanosis, methemoglobinemia, with blue color to the skin and lips. Higher levels can cause breathing difficulty, collapse, and death.

Long Term Exposure: There is limited evidence that this chemical can damage the developing fetus.

Points of Attack: Blood.

Medical Surveillance: Blood test for methemoglobin.

First Aid: If this chemical gets into the eyes, remove any contact lenses at once and irrigate immediately for at least 15 min, occasionally lifting upper and lower lids. Seek medical attention immediately. If this chemical contacts the skin, remove contaminated clothing and wash immediately with soap and water. Seek medical attention immediately. If this chemical has been inhaled, remove from exposure, begin rescue breathing (using universal precautions, including resuscitation mask) if breathing has stopped and CPR if heart action has stopped. Transfer promptly to a medical facility. When this chemical has been swallowed, get medical attention. Give large quantities of water and induce vomiting. Do not make an unconscious person vomit.

Note to physician: Treat for methemoglobinemia. Spectrophotometry may be required for precise determination of levels of methemoglobin in urine.

Personal Protective Methods: Wear protective gloves and clothing to prevent any reasonable probability of skin contact. Safety equipment suppliers/manufacturers can provide recommendations on the most protective glove/clothing material for your operation. All protective clothing (suits, gloves, footwear, headgear) should be clean, available each day, and put on before work. Contact lenses should not be worn when working with this chemical. Wear dust-proof chemical goggles and face shield unless full face-piece respiratory protection is worn. Employees should wash immediately with soap when skin is wet or contaminated. Provide emergency showers and eyewash.

Respirator Selection: Follow regulations in OSHA 29CFR1910.134 or European Standard EN149. Use a NIOSH/MSHA- or European Standard EN149-approved respirator; or use an approved supplied-air respirator with a full face-piece operated in the positive-pressure mode, or with a full face-piece, hood, or helmet in the continuous-flow mode; or use a NIOSH/MSHA- or European Standard EN149-approved self-contained breathing apparatus with a full face-piece operated in a pressure-demand or other positive-pressure mode.

Storage: Color Code—Yellow: Reactive Hazard; Store in a location separate from other materials, especially flammables and combustibles. Prior to working with this chemical you should be trained on its proper handling and storage. Store in tightly closed containers in a cool, well-ventilated area away from all other materials. Do not store on wooden floors. Where possible, automatically transfer material from drums or other storage containers to process containers. Sources of ignition, such as smoking and open flames, are prohibited where this chemical is handled, used, or stored. Metal containers involving the transfer of this chemical should be grounded and bonded. Wherever this chemical is used, handled, manufactured, or stored, use explosion-proof electrical equipment and fittings. See OSHA Standard 1910.104 and NFPA 43A *Code for the Storage of Liquid and Solid Oxidizers* for detailed handling and storage regulations. See also 29 CFR 1910.101 for specific regulations on storage of compressed gas cylinders.

Shipping: Potassium nitrate requires a shipping label of "OXIDIZER." It falls in Hazard Class 5.1 and Packing Group III.

Spill Handling: Evacuate persons not wearing protective equipment from area of spill or leak until cleanup is complete. Remove all ignition sources. Collect powdered material in the most convenient and safe manner and deposit in sealed containers. Ventilate area after cleanup is complete. It may be necessary to contain and dispose of this chemical as a hazardous waste. If material or contaminated runoff enters waterways, notify downstream users of potentially contaminated waters. Contact your local or federal environmental protection agency for specific recommendations. If employees are required to clean up spills, they must be properly trained and equipped. OSHA 1910.120(q) may be applicable.

Fire Extinguishing: This chemical is a combustible solid. Use dry chemical, carbon dioxide, water spray, or alcohol foam extinguishers. Poisonous gases are produced in fire. If material or contaminated runoff enters waterways, notify downstream users of potentially contaminated waters. Notify local health and fire officials and pollution control agencies. From a secure, explosion-proof location, use water spray to cool exposed containers. If cooling streams are ineffective (venting sound increases in volume and pitch, tank discolors, or shows any signs of deforming), withdraw immediately to a secure position. If employees are expected to fight fires, they must be trained and equipped in OSHA 1910.156. The only respirators recommended for firefighting are self-contained breathing apparatuses that have full face-pieces and are operated in a pressure-demand or other positive-pressure mode.

Reference

New Jersey Department of Health and Senior Services. (November 2004). *Hazardous Substances Fact Sheet: Potassium Nitrate*. Trenton, NJ

Potassium nitrite   P:0970

Molecular Formula: KNO_2

Synonyms: Kaliumnitrat (German); Niter; Nitre; Nitrous acid, Potassium salt; Saltpeter; Vicknite

CAS Registry Number: 7758-09-0

RTECS® Number: TT3750000

UN/NA & ERG Number: UN1488/140

EC Number: 231-832-4 [*Annex I Index No.:* 007-011-00-X]

Regulatory Authority and Advisory Bodies

Air Pollutant Standard Set. See below, "Permissible Exposure Limits in Air" section.

Canada, WHMIS, Ingredients Disclosure List Concentration 0.1%.

European/International Regulations: Hazard Symbol: C; Risk phrases: R22; R35; Safety phrases: S1/2; S26; S36/37/39; S45 (see Appendix 4).

WGK (German Aquatic Hazard Class): 2—Hazard to waters.

Description: Potassium nitrite is a white to yellowish crystalline solid. Molecular weight = 85.11; Boiling point = decomposition starts at 350°C; explosion at 535°C; Freezing/Melting point = 441°C. Hazard Identification (based on NFPA-704 M Rating System): Health 0, Flammability 1, Reactivity 2 (Oxidizer). Soluble in water; solubility = 280 g/100 mL.

Potential Exposure: Compound Description: Drug, Mutagen; Reproductive Effector; Human Data. Potassium nitrite is used in chemical analysis, as a food additive; in fertilizers; in medications as a vasodilator and as antidote for cyanide poisoning.

Incompatibilities: A strong oxidizer. Reacts violently with combustible and reducing materials. Heat above 530°C may cause explosion. Incompatible with cyanide salts; boron, ammonium sulfate, potassium amide, and acids. Decomposes on contact with even weak acids, producing toxic nitrogen oxide fumes.

Permissible Exposure Limits in Air

Protective Action Criteria (PAC)

TEEL-0: 0.04 mg/m^3

PAC-1: 0.1 mg/m^3

PAC-2: 0.75 mg/m^3

PAC-3: 500 mg/m^3

Routes of Entry: Inhalation, ingestion, skin and/or eye contact.

Harmful Effects and Symptoms

Short Term Exposure: Potassium nitrite can affect you when breathed in. Contact can cause eye and skin burns. Breathing the dust or mist can irritate the nose, throat, and lungs, and may cause cough with phlegm. Higher exposures can cause pulmonary edema, a medical emergency that can be delayed for several hours. This can cause death. High levels can affect the vascular system and interfere with the ability of the blood to carry oxygen (methemoglobinemia), causing headaches, weakness, dizziness, and cyanosis, a bluish color to the skin and lips. Higher levels can cause troubled breathing, collapse, and even death.

Long Term Exposure: Repeated skin contact causes dermatitis, drying, and cracking. May cause lung irritation; bronchitis may develop. There is limited evidence that potassium nitrite may damage the developing fetus.

Points of Attack: Eyes, skin, blood, lungs.

Medical Surveillance: If symptoms develop or overexposure is suspected, the following may be useful: blood test for methemoglobin. Lung function tests. Consider chest X-ray after acute overexposure.

First Aid: If this chemical gets into the eyes, remove any contact lenses at once and irrigate immediately for at least 15 min, occasionally lifting upper and lower lids. Seek medical attention immediately. If this chemical contacts the skin, remove contaminated clothing and wash immediately with soap and water. Seek medical attention immediately. If this chemical has been inhaled, remove from exposure, begin rescue breathing (using universal precautions, including resuscitation mask) if breathing has stopped and CPR if heart action has stopped. Transfer promptly to a medical facility. When this chemical has been swallowed, get medical attention. Give large quantities of water and induce vomiting. Do not make an unconscious person vomit. Medical observation is recommended for 24−48 h after breathing overexposure, as pulmonary edema may be delayed. As first aid for pulmonary edema, a doctor or authorized paramedic may consider administering a corticosteroid spray.

Note to physician: Treat for methemoglobinemia. Spectrophotometry may be required for precise determination of levels of methemoglobin in urine.

Personal Protective Methods: Wear protective gloves and clothing to prevent any reasonable probability of skin contact. Safety equipment suppliers/manufacturers can provide recommendations on the most protective glove/clothing material for your operation. All protective clothing (suits, gloves, footwear, headgear) should be clean, available each day, and put on before work. Contact lenses should not be worn when working with this chemical. Wear dust-proof chemical goggles and face shield unless full face-piece respiratory protection is worn. Employees should wash immediately with soap when skin is wet or contaminated. Provide emergency showers and eyewash.

Respirator Selection: Where there is potential for exposure to potassium nitrite, use a NIOSH/MSHA- or European

Standard EN149-approved full-face-piece respirator with a high-efficiency particulate filter. Greater protection is provided by a powered air-purifying respirator. *Where there is potential for high exposures*, use a NIOSH/MSHA- or European Standard EN149-approved supplied-air respirator with a full face-piece operated in the positive-pressure mode, or with a full face-piece, hood, or helmet in the continuous-flow mode; or use a NIOSH/MSHA- or European Standard EN149-approved self-contained breathing apparatus with a full face-piece operated in a pressure-demand or other positive-pressure mode.

Storage: Color Code—Yellow: Reactive Hazard; Store in a location separate from other materials, especially flammables and combustibles. Prior to working with this chemical you should be trained on its proper handling and storage. Store in tightly closed containers in a cool, well-ventilated area away from acids, cyanide salts, boron, ammonium sulfate, and potassium amide. Where possible, automatically transfer material from drums or other storage containers to process containers. Sources of ignition, such as smoking and open flames, are prohibited where this chemical is handled, used, or stored. Metal containers involving the transfer of this chemical should be grounded and bonded. Wherever this chemical is used, handled, manufactured, or stored, use explosion-proof electrical equipment and fittings. See OSHA Standard 1910.104 and NFPA 43A *Code for the Storage of Liquid and Solid Oxidizers* for detailed handling and storage regulations.

Shipping: Potassium nitrite requires a shipping label of "OXIDIZER." It falls in Hazard Class 5.1 and Packing Group II.

Spill Handling: Evacuate persons not wearing protective equipment from area of spill or leak until cleanup is complete. Remove all ignition sources. Collect powdered material in the most convenient and safe manner and deposit in sealed containers. Ventilate area after cleanup is complete. Keep potassium nitrite out of a confined space, such as a sewer, because of the possibility of an explosion, unless the sewer is designed to prevent the buildup of explosive concentrations. It may be necessary to contain and dispose of this chemical as a hazardous waste. If material or contaminated runoff enters waterways, notify downstream users of potentially contaminated waters. Contact your local or federal environmental protection agency for specific recommendations. If employees are required to clean up spills, they must be properly trained and equipped. OSHA 1910.120(q) may be applicable.

Fire Extinguishing: Does not burn but may ignite other combustible materials. Use dry chemical, CO_2, water spray, or foam extinguishers. Poisonous gases are produced in fire, including oxides of nitrogen. If material or contaminated runoff enters waterways, notify downstream users of potentially contaminated waters. Notify local health and fire officials and pollution control agencies. From a secure, explosion-proof location, use water spray to cool exposed containers. If cooling streams are ineffective (venting sound increases in volume and pitch, tank discolors, or shows any signs of deforming), withdraw immediately to a secure position. If employees are expected to fight fires, they must be trained and equipped in OSHA 1910.156. The only respirators recommended for firefighting are self-contained breathing apparatuses that have full face-pieces and are operated in a pressure-demand or other positive-pressure mode.

Reference
New Jersey Department of Health and Senior Services. (November 2004). *Hazardous Substances Fact Sheet: Potassium Nitrite.* Trenton, NJ

Potassium permanganate   P:0980

Molecular Formula: $KMnO_4$
Synonyms: Cairox; Chameleon mineral; C.I. 77755; Condy's crystals; Kaliumpermanganat (German); Permanganic acid, potassium salt; Permanganate de potassium (French); Permanganate of potash; Permanganato potasico (Spanish); Potassium (permanganate de) (French); Purple salt
CAS Registry Number: 7722-64-7
RTECS® Number: SD6475000
UN/NA & ERG Number: UN1490/140
EC Number: 231-760-3 [*Annex I Index No.:* 025-002-00-9]
Regulatory Authority and Advisory Bodies
Department of Homeland Security Screening Threshold Quantity (pounds): *Theft hazard* 400 (Commercial grade).
US EPA Gene-Tox Program, Negative: *In vitro* cytogenetics—nonhuman; *N. crassa*—reversion; Negative: *B. subtilis* rec assay.
US EPA, FIFRA 1998 Status of Pesticides: Pesticide subject to registration or re-registration.
Clean Water Act: Section 311 Hazardous Substances/RQ 40CFR117.3 (same as CERCLA, see below).
Reportable Quantity (RQ): 100 lb (45.4 kg).
European/International Regulations: Hazard Symbol: O, Xn, N; Risk phrases: R8; R22; R36/37/38; R42/43; Safety phrases: S2; S22; S24; S26; S37 (see Appendix 4).
WGK (German Aquatic Hazard Class): 3—Severe hazard to waters.

Description: Potassium permanganate is a dark purple crystalline solid. Molecular weight = 158.04; Freezing/Melting point = (decomposition, with evolution of oxygen) <240°C. Hazard Identification (based on NFPA-704 M Rating System): Health 3, Flammability 0, Reactivity 3 (Oxidizer). Highly soluble in water.

Potential Exposure: Compound Description: Agricultural Chemical; Mutagen; Reproductive Effector; Human Data. Potassium permanganate is used in solutions as a disinfectant, topical antibacterial agent; deodorizer, bleaching agent; and in air and water purification.

Incompatibilities: Potassium permanganate is a powerful oxidizing agent, that is, it will initiate a fire or explosion if

2226 Potassium permanganate

brought into contact with reducing materials, combustibles, organic materials; strong acids; or oxidizable solid, liquid, or gas; glycerine, ethylene glycol, polypropylene, hydroxylamine, hydrogen trisulfide, antimony, arsenic, sulfuric acid, hydrogen peroxide, phosphorus, and any finely divided combustible material. It will decompose, and release oxygen if brought into contact with heat, alcohol, acids, ferrous salts, iodides and oxalates.

Permissible Exposure Limits in Air

OSHA PEL: 5 mg[Mn]/m^3 Ceiling Concentration.

NIOSH: 1 mg[Mn]/m^3 TWA; 3 mg[Mn]/m^3 STEL.

ACGIH TLV®[1]: TWA 0.2 mg[Mn]/m^3, inorganic compounds.

NIOSH IDLH: 500 mg[Mn]/m^3.

Protective Action Criteria (PAC)

TEEL-0: 0.575 mg/m^3

PAC-1: 8.63 mg/m^3

PAC-2: 14.4 mg/m^3

PAC-3: 500 mg/m^3

DFG MAK: 0.5 mg[Mn]/m^3 inhalable fraction (Mn and its inorganic compounds); Pregnancy Risk Group C.

Australia: TWA 5 mg[Mn]/m^3, 1993; Belgium: TWA 5 mg [Mn]/m^3, 1993; Denmark: TWA 2.5 mg[Mn]/m^3, 1999; Finland: TWA 0.5 mg[Mn]/m^3, 1999; Hungary: TWA 0.3 mg[Mn]/m^3, short-term exposure limit 0.6 mg[Mn]/m^3, 1993; Japan: 0.3 mg[Mn]/m^3, respirable dust, 1999; Poland: MAC (TWA) 0.3 mg[Mn]/m^3; MAC 5 mg[Mn]/m^3, 1999; Sweden: NGV 1 mg[Mn]/m^3, KTV 2.5 mg[Mn]/m^3 (respirable dust), 1999; Sweden: NGV 2.5 mg[Mn]/m^3, KTV 5 mg[Mn]/m^3 (total dust), 1999; United Kingdom: LTEL 5 mg[Mn]/m^3, 1993; Argentina, Bulgaria, Columbia, Jordan, South Korea, New Zealand, Singapore, Vietnam: ACGIH TLV®: TWA 0.2 mg[Mn]/m^3.

Routes of Entry: Inhalation, ingestion, skin and/or eye contact.

Harmful Effects and Symptoms

Short Term Exposure: Inhalation: Irritates the respiratory tract, causing coughing and chest tightness. Higher exposures can cause pulmonary edema, a medical emergency that can be delayed for several hours. This can cause death. *Skin:* Concentrated solutions may cause severe irritation and burns. Dilute solutions can cause brown staining of the skin and hardening of outer skin layer. Penetration is poor. *Eyes:* Concentrated solution or crystalline material can cause severe irritation and damage that may be permanent. *Ingestion:* Dilute solutions (1%) may cause burning of the throat, nausea, vomiting, and stomach pain. Concentrations of 2–3% may cause anemia and swelling of the throat with a possibility of suffocation. More concentrated solutions may result in above symptoms plus the onset of kidney damage and circulatory collapse. The probable lethal dose is 1½ teaspoons (10 g) for a 150-lb (70 kg) adult.

Long Term Exposure: May cause mutations and might pose a cancer risk or reproduction hazard. May cause lung effects.

Points of Attack: Lungs, skin.

Medical Surveillance: NIOSH lists the following tests (for manganese and fume): whole blood (chemical/metabolite); biologic tissue/biopsy; Complete blood count; chest X-ray; pulmonary function tests; urine (chemical/metabolite); urinalysis (routine). For those with frequent or potentially high exposure the following are recommended before beginning work and at regular times after that: lung function tests. If symptoms develop or overexposure is suspected, the following may be useful: consider chest X-ray after acute overexposure.

First Aid: If this chemical gets into the eyes, remove any contact lenses at once and irrigate immediately for at least 15 min, occasionally lifting upper and lower lids. Seek medical attention immediately. If this chemical contacts the skin, remove contaminated clothing and wash immediately with soap and water. Seek medical attention immediately. If this chemical has been inhaled, remove from exposure, begin rescue breathing (using universal precautions, including resuscitation mask) if breathing has stopped and CPR if heart action has stopped. Transfer promptly to a medical facility. When this chemical has been swallowed, get medical attention. Give egg whites and milk. A tracheotomy may be required if swelling in throat blocks air. Medical observation is recommended for 24–48 h after breathing overexposure, as pulmonary edema may be delayed. As first aid for pulmonary edema, a doctor or authorized paramedic may consider administering a corticosteroid spray.

Personal Protective Methods: Wear protective gloves and clothing to prevent any reasonable probability of skin contact. Safety equipment suppliers/manufacturers can provide recommendations on the most protective glove/clothing material for your operation. All protective clothing (suits, gloves, footwear, headgear) should be clean, available each day, and put on before work. Contact lenses should not be worn when working with this chemical. Wear splash-proof chemical goggles and face shield when working with liquid unless full face-piece respiratory protection is worn. Wear dust-proof goggles when working with powders or dust unless full face-piece respiratory protection is worn. Employees should wash immediately with soap when skin is wet or contaminated. Provide emergency showers and eyewash.

Respirator Selection: Up to 10 mg/m³: 95XQ (APF = 10) [any particulate respirator equipped with an N95, R95, or P95 filter (including N95, R95, and P95 filtering face-pieces) except quarter-mask respirators. The following filters may also be used: N99, R99, P99, N100, R100, P100] or Sa (APF = 10) (any supplied-air respirator). *Up to 25 mg/m³:* Sa:Cf (APF = 25) (any supplied-air respirator operated in a continuous-flow mode); PaprHie (APF = 25) (any powered air-purifying respirator with a high-efficiency particulate filter). *Up to 50 mg/m³:* 100F (APF = 50) (any air-purifying, full-face-piece respirator with an N100, R100, or P100 filter) or SaT: Cf (APF = 50) (any supplied-air respirator that has a tight-fitting face-piece and is operated in a continuous-flow mode) or PaprTHie (APF = 50) (any powered, air-purifying respirator with a tight-fitting face-piece and a high-efficiency particulate filter) or SCBAF (APF = 50) (any self-contained breathing apparatus with a full face-piece) or SaF (APF = 50) (any supplied-air respirator with a full face-piece). *Up to 500 mg/m³:* Sa: Pd,Pp (APF = 1000) (any supplied-air respirator operated in a pressure-demand or other positive-pressure mode). *Emergency or planned entry into unknown concentrations or IDLH conditions:* SCBAF: Pd,Pp (APF = 10,000) (any self-contained breathing apparatus that has a full face-piece and is operated in a pressure-demand or other positive-pressure mode); ASCBA (any supplied-air respirator that has a full face-piece and is operated in a pressure-demand or other positive-pressure mode in combination with an auxiliary self-contained positive-pressure breathing apparatus). *Escape:* 100F (APF = 50) (any air-purifying, full-face-piece respirator with an N100, R100, or P100 filter) or SCBAE (any appropriate escape-type, self-contained breathing apparatus).

Storage: Color Code—Yellow: Reactive Hazard; Store in a location separate from other materials, especially flammables and combustibles. Prior to working with this chemical you should be trained on its proper handling and storage. Potassium permanganate must be stored to avoid contact with strong acids (such as hydrochloric, sulfuric, and nitric), any organic material, or any other combustible or oxidizable solid, liquid, or gas since violent reactions occur. Store in tightly closed containers in a cool, well-ventilated area. Protect containers from physical damage. Where possible, automatically transfer material from drums or other storage containers to process containers. Sources of ignition, such as smoking and open flames, are prohibited where this chemical is handled, used, or stored. Metal containers involving the transfer of this chemical should be grounded and bonded. Wherever this chemical is used, handled, manufactured, or stored, use explosion-proof electrical equipment and fittings. See OSHA Standard 1910.104 and NFPA 43A *Code for the Storage of Liquid and Solid Oxidizers* for detailed handling and storage regulations.

Shipping: Potassium permanganate requires a shipping label of "OXIDIZER." It falls in Hazard Class 5.1 and Packing Group II.

Spill Handling: Evacuate persons not wearing protective equipment from area of spill or leak until cleanup is complete. Remove all ignition sources. *Liquid:* Ventilate area of spill or leak. Absorb liquids in vermiculite, dry sand, earth, peat, carbon, or a similar material and deposit in sealed containers. *Dry material:* Collect powdered material in the most convenient and safe manner and deposit in sealed containers. Ventilate area after cleanup is complete. *Large spills:* Clean up should be performed by trained personnel. Cover the weak reducing agents, such as sodium thiosulfate, bisulfites, or ferrous salts. Bisulfites or ferrous salts need an additional promoter of three molar sulfuric acids to accelerate reaction. Transfer slurry or sludge to large container of water and neutralize with soda ash. Keep potassium permanganate out of a confined space, such as a sewer, because of the possibility of an explosion, unless the sewer is designed to prevent the buildup of explosive concentrations. It may be necessary to contain and dispose of this chemical as a hazardous waste. If material or contaminated runoff enters waterways, notify downstream users of potentially contaminated waters. Contact your local or federal environmental protection agency for specific recommendations. If employees are required to clean up spills, they must be properly trained and equipped. OSHA 1910.120(q) may be applicable.

Fire Extinguishing: Potassium permanganate does not burn. However, it is a powerful oxidizer. Potassium permanganate decomposes at 240°C/464°F and releases oxygen which will greatly intensify an ongoing fire. Use dry chemical or carbon dioxide extinguishers. Poisonous gases are produced in fire, including potassium oxide. If material or contaminated runoff enters waterways, notify downstream users of potentially contaminated waters. Notify local health and fire officials and pollution control agencies. From a secure, explosion-proof location, use water spray to cool exposed containers. If cooling streams are ineffective (venting sound increases in volume and pitch, tank discolors, or shows any signs of deforming), withdraw immediately to a secure position. If employees are expected to fight fires, they must be trained and equipped in OSHA 1910.156. The only respirators recommended for firefighting are self-contained breathing apparatuses that have full face-pieces and are operated in a pressure-demand or other positive-pressure mode.

Disposal Method Suggested: React with reducing agent, neutralize and flush to sewer.[22]

References

Sax, N. I. (Ed.). (1988). *Dangerous Properties of Industrial Materials Report*, 8, No. 4, 2–12

New Jersey Department of Health and Senior Services. (February 1986). *Hazardous Substances Fact Sheet: Potassium Permanganate*. Version 2 and 3. Albany, NY: Bureau of Toxic Substance Assessment

New Jersey Department of Health and Senior Services. (May 2002). *Hazardous Substances Fact Sheet: Potassium Permanganate*. Trenton, NJ

Potassium persulfate   P:0990

Molecular Formula: $K_2O_8S_2$
Common Formula: $K_2S_2O_8$
Synonyms: Anthion; Dipotassium peroxodisulphate; Dipotassium persulfate; Potassium persulphate; Potassium peroxydisulphate; Potassium peroxydisulfate; Peroxydisulfuric acid, Disodium salt
CAS Registry Number: 7727-21-1
RTECS® Number: SE0400000
UN/NA & ERG Number: UN1492/140

EC Number: 231-781-8 [*Annex I Index No.:* 016-061-00-1] [dipotassium peroxodisulphate]

Regulatory Authority and Advisory Bodies

European/International Regulations: Hazard Symbol: O, Xn; Risk phrases: R8; R22; R36/37/38; R42/43; Safety phrases: S2; S22; S24; S26; S37 (see Appendix 4).

WGK (German Aquatic Hazard Class): 1—Low hazard to waters.

Description: Potassium persulfate is a colorless or white, odorless crystalline material. Molecular weight = 272.34; Boiling point = 109.5°C; Freezing/Melting point = (decomposes) <100°C. Hazard Identification (based on NFPA-704 M Rating System): Health 3, Flammability 0, Reactivity 3. Oxidizer. Soluble in water.

Potential Exposure: Potassium persulfate is used as a bleaching and oxidizing agent; it is used in redox polymerization catalysts; in the defiberizing of wet strength paper and in the desizing of textiles. Soluble in water.

Incompatibilities: A strong oxidizer. Combustible, organic or other readily oxidizable materials, sulfur, metallic dusts, such as aluminum dust, chlorates, and perchlorates. Attacks chemically active metals. Keep away from moisture.

Permissible Exposure Limits in Air

ACGIH: 0.1 mg [S_2O_8]/m^3 TWA *as persulfates.*

Protective Action Criteria (PAC)

TEEL-0: 0.1 mg/m^3

PAC-1: 10 mg/m^3

PAC-2: 60 mg/m^3

PAC-3: 350 mg/m^3

Permissible Concentration in Water: No criteria set. (Aqueous solution decomposes even at room temperature).

Routes of Entry: Inhalation, ingestion, skin and/or eye contact.

Harmful Effects and Symptoms

Short Term Exposure: Potassium persulfate can affect you when breathed in. Breathing the dust or mist can cause eye, nose, and throat irritation with sneezing, coughing, and sore throat. Contact with skin and eyes can cause burns and permanent damage. Prolonged or repeated exposures may lead to sores of the inner nose. Higher exposures can cause pulmonary edema, a medical emergency that can be delayed for several hours. This can cause death.

Long Term Exposure: May cause sores of the inner nose. Skin contact may cause skin rash, with dryness and cracking. May cause lung irritation.

Points of Attack: Lungs.

Medical Surveillance: For those with frequent or potentially high exposure the following are recommended before beginning work and at regular times after that: lung function tests. If symptoms develop or overexposure is suspected, the following may also be useful: consider chest X-ray after acute overexposure.

First Aid: If this chemical gets into the eyes, remove any contact lenses at once and irrigate immediately for at least 15 min, occasionally lifting upper and lower lids. Seek medical attention immediately. If this chemical contacts the skin, remove contaminated clothing and wash immediately with soap and water. Seek medical attention immediately. If this chemical has been inhaled, remove from exposure, begin rescue breathing (using universal precautions, including resuscitation mask) if breathing has stopped and CPR if heart action has stopped. Transfer promptly to a medical facility. When this chemical has been swallowed, get medical attention. Give large quantities of water and induce vomiting. Do not make an unconscious person vomit. Medical observation is recommended for 24—48 h after breathing overexposure, as pulmonary edema may be delayed. As first aid for pulmonary edema, a doctor or authorized paramedic may consider administering a corticosteroid spray.

Personal Protective Methods: Wear protective gloves and clothing to prevent any reasonable probability of skin contact. Safety equipment suppliers/manufacturers can provide recommendations on the most protective glove/clothing material for your operation. All protective clothing (suits, gloves, footwear, headgear) should be clean, available each day, and put on before work. Contact lenses should not be worn when working with this chemical. Wear dust-proof chemical goggles and face shield unless full face-piece respiratory protection is worn. Employees should wash immediately with soap when skin is wet or contaminated. Provide emergency showers and eyewash.

Respirator Selection: Use a NIOSH/MSHA- or European Standard EN 149-approved dust mask when dust is encountered. Where there is potential for exposures *over 5 mg/m^3,* use a NIOSH/MSHA- or European Standard EN149-approved full-face-piece respirator with a high-efficiency particulate filter. Greater protection is provided by a powered air-purifying respirator. *Where there is potential for high exposures,* use a NIOSH/MSHA- or European Standard EN149-approved supplied-air respirator with a full face-piece operated in the positive-pressure mode, or with a full face-piece, hood, or helmet in the continuous-flow mode; or use a NIOSH/MSHA- or European Standard EN149-approved self-contained breathing apparatus with a full face-piece operated in a pressure-demand or other positive-pressure mode.

Storage: Color Code—Yellow: Reactive Hazard; Store in a location separate from other materials, especially flammables and combustibles. Prior to working with this chemical you should be trained on its proper handling and storage. Potassium persulfate must be stored to avoid contact with oxidizers (such as perchlorates, peroxides, permanganates, chlorates, and nitrates), strong oxidizers (such as chlorine, bromine, and fluorine), and chemically active metals (such as potassium, sodium, magnesium, and zinc) since violent reactions occur. Protect storage against physical damage. Store in tightly closed containers in a cool, well-ventilated area away from moisture. See OSHA Standard 1910.104 and NFPA 43A *Code for the Storage of Liquid and Solid Oxidizers* for detailed handling and storage regulations.

Shipping: Potassium persulfate requires a shipping label of "OXIDIZER." It falls in Hazard Class 5.1 and Packing Group III.

Spill Handling: Evacuate persons not wearing protective equipment from area of spill or leak until cleanup is complete. Remove all ignition sources. Collect powdered material in the most convenient and safe manner and deposit in sealed containers. Ventilate area after cleanup is complete. Keep potassium persulfate out of a confined space, such as a sewer, because of the possibility of an explosion, unless the sewer is designed to prevent the buildup of explosive concentrations. It may be necessary to contain and dispose of this chemical as a hazardous waste. If material or contaminated runoff enters waterways, notify downstream users of potentially contaminated waters. Contact your local or federal environmental protection agency for specific recommendations. If employees are required to clean up spills, they must be properly trained and equipped. OSHA 1910.120(q) may be applicable.

Fire Extinguishing: This chemical is a combustible solid. Use dry chemical, carbon dioxide, water spray, or alcohol foam extinguishers. Poisonous gases are produced in fire, including oxides of sulfur. If material or contaminated runoff enters waterways, notify downstream users of potentially contaminated waters. Notify local health and fire officials and pollution control agencies. From a secure, explosion-proof location, use water spray to cool exposed containers. If cooling streams are ineffective (venting sound increases in volume and pitch, tank discolors, or shows any signs of deforming), withdraw immediately to a secure position. If employees are expected to fight fires, they must be trained and equipped in OSHA 1910.156. The only respirators recommended for firefighting are self-contained breathing apparatuses that have full face-pieces and are operated in a pressure-demand or other positive-pressure mode.

Disposal Method Suggested: Use large volumes of reducing agents (e.g., bisulfites). Neutralize with soda ash and drain into sewer with abundant water.

Reference

New Jersey Department of Health and Senior Services. (May 1986). *Hazardous Substances Fact Sheet: Potassium Persulfate.* Trenton, NJ

Potassium silver cyanide   P:1000

Molecular Formula: AgC_2KN_2
Common Formula: $KAg(CN)_2$
Synonyms: Cianuro de plata y potasio (Spanish); Dicyano potassium argentate; Potassium dicyanoargentate; Silver potassium cyanide
CAS Registry Number: 506-61-6
RTECS® Number: TT5775000
UN/NA & ERG Number: UN1588/157
EC Number: 208-047-0 (potassium dicyanoargentate)

Regulatory Authority and Advisory Bodies
US EPA Hazardous Waste Number (RCRA No.): P099.
RCRA, 40CFR261, Appendix 8 Hazardous Constituents.
Superfund/EPCRA 40CFR355, Extremely Hazardous Substances: TPQ = 500 lb (227 kg).
Reportable Quantity (RQ): 1 lb (0.454 kg).
Canada, WHMIS, Ingredients Disclosure List Concentration 1.0%.
Cyanide compounds:
Clean Air Act: Hazardous Air Pollutants (Title I, Part A, Section 112).
Clean Water Act: 40CFR423, Appendix A, Priority Pollutants, as cyanide, total.
US EPA Hazardous Waste Number (RCRA No.): P030 as cyanides soluble salts and complexes, n.o.s.
RCRA, 40CFR261, Appendix 8 Hazardous Constituents. as cyanides, soluble salts and complexes, n.o.s.
EPCRA (Section 313): $X + CN^-$ where $X = H^+$ or any other group where a formal dissociation may occur. For example, KCN or Ca(CN)2. Form R *de minimis* concentration reporting level: 1.0%.
Canada, WHMIS, Ingredients Disclosure List Concentration 1.0%, Cyanide compounds, inorganic, n.o.s.
US DOT Regulated Marine Pollutant (49CFR172.101, Appendix B) as cyanide mixtures, cyanide solutions or cyanides, inorganic, n.o.s.
Silver compounds:
Clean Water Act: Section 307 Toxic Pollutants as silver and compounds.
RCRA Section 261 Hazardous Constituents, as silver compounds, n.o.s., waste number not listed.
EPCRA (Section 313): Includes any unique chemical substance that contains silver as part of that chemical's infrastructure. Form R *de minimis* concentration reporting level: 1.0%.
WGK (German Aquatic Hazard Class): 3—Severe hazard to waters.

Description: Potassium silver cyanide is a white crystalline solid. Molecular weight = 199.01; Hazard Identification (based on NFPA-704 M Rating System): Health 3, Flammability 0, Reactivity 0. Soluble in water.

Potential Exposure: Potassium silver cyanide is used in silver plating; as a bactericide; and in the manufacture of antiseptics. Not registered as a pesticide in the United States.

Incompatibilities: Contact with acid, acid fumes, water, steam, or when heated to decomposition emits toxic and flammable cyanide vapors. May be light sensitive.

Permissible Exposure Limits in Air
OSHA PEL: 5 mg[CN]/m^3/4.7 ppm TWA.
NIOSH REL: 5 mg[CN]/m^3/4.7 ppm/10 min Ceiling Concentration.
ACGIH TLV®[1]: 5 mg[CN]/m^3 [skin] Ceiling Concentration.
NIOSH IDLH: 25 mg[CN]/m^3.
Protective Action Criteria (PAC)
TEEL-0: 0.0184 mg/m^3

PAC-1: 2.5 mg/m^3
PAC-2: 18.4 mg/m^3
PAC-3: 18.4 mg/m^3
DFG MAK: 2 mg[CN]/m^3, inhalable fraction TWA; Peak Limitation Category II(1) [skin]; Pregnancy Risk Group: C.
Determination in Air: Use NIOSH Analytical Method (IV) #7904, Cyanides. See also Method #6010, Hydrogen cyanide.[18]

Permissible Concentration in Water: In 1976 the EPA criterion was 5.0 µg/L for freshwater and marine aquatic life and wildlife. As of 1980, the criteria are: *To protect freshwater aquatic life:* 3.5 µg/L as a 24-h average, never to exceed 52.0 µg/L. *To protect saltwater aquatic life:* 30.0 µg/L on an acute toxicity basis; 2.0 µg/L on a chronic toxicity basis. *To protect human health:* 200 µg/L. The allowable daily intake for man is 8.4 mg/day.[6]
On the international scene, the South African Bureau of Standards has set 10 µg/L, the World Health Organization (WHO) 10 µg/L, and Germany 50 µg/L as drinking water standards. Other international limits[35] include an EEC limit of 50 µg/L; Mexican limits of 200 µg/L in drinking water and 1.0 µg/L in coastal waters; and a Swedish limit of 100 µg/L. Russia[43] set a MAC of 100 µg/L in water bodies used for domestic purposes and 50 µg/L in water for fishery purposes.
The US EPA[49] has determined a no-observed-adverse-effect-level (NOAEL) of 10.8 mg/kg/day which yields a lifetime health advisory of 154 µg/L. States which have set guidelines for cyanides in drinking water[61] include Arizona at 160 µg/L and Kansas at 220 µg/L.
Determination in Water: Distillation followed by silver nitrate titration or colorimetric analysis using pyridine pyrazolone (or barbituric acid).
Routes of Entry: Inhalation, ingestion, skin and/or eye contact. Absorbed through the skin.

Harmful Effects and Symptoms
Short Term Exposure: The primary health hazard is as a cyanide. It is poisonous and may be fatal if inhaled, swallowed, or absorbed through the skin. Fire may produce irritating or poisonous gases. As a cyanide, massive doses may produce, without warning, sudden loss of consciousness and prompt death from respiratory arrest. Smaller but still lethal doses result in illness that may be prolonged for one or more hours. Other symptoms may include numbness in throat, salivation, nausea, anxiety, dizziness, irregular breathing, odor of bitter almonds may be noted on breath, blood pressure may rise, slowing of the heartbeat, sensation of constriction in the chest, unconsciousness followed by violent convulsions and paralysis. LD$_{50}$ = (oral-rat) 21 mg/kg.
First Aid: If this chemical gets into the eyes, remove any contact lenses at once and irrigate immediately for at least 15 min, occasionally lifting upper and lower lids. Seek medical attention immediately. If this chemical contacts the skin, remove contaminated clothing and wash immediately with soap and water. Seek medical attention immediately. If

this chemical has been inhaled, remove from exposure, begin rescue breathing (using universal precautions, including resuscitation mask) if breathing has stopped and CPR if heart action has stopped. Transfer promptly to a medical facility. When this chemical has been swallowed, get medical attention. Give large quantities of water and induce vomiting. Do not make an unconscious person vomit. Keep victim quiet and maintain normal body temperature. Effects may be delayed; keep victim under observation.
Use amyl nitrate capsules if symptoms develop. All area employees should be trained regularly in emergency measures for cyanide poisoning and in CPR. A cyanide antidote kit should be kept in the immediate work area and must be rapidly available. Kit ingredients should be replaced every 1–2 years to ensure freshness. Persons trained in the use of this kit, oxygen use, and CPR must be quickly available.
Personal Protective Methods: Wear protective gloves and clothing to prevent any reasonable probability of skin contact. Safety equipment suppliers/manufacturers can provide recommendations on the most protective glove/clothing material for your operation. All protective clothing (suits, gloves, footwear, headgear) should be clean, available each day, and put on before work. Contact lenses should not be worn when working with this chemical. Wear dust-proof chemical goggles and face shield unless full face-piece respiratory protection is worn. Employees should wash immediately with soap when skin is wet or contaminated. Provide emergency showers and eyewash.
Respirator Selection: *Up to 25 mg/m^3:* Sa (APF = 10) (any supplied-air respirator) or SCBAF (APF = 50) (any self-contained breathing apparatus with full face-piece). *Emergency or planned entry into unknown concentrations or IDLH conditions:* SCBAF: Pd,Pp (APF = 10,000) (any self-contained breathing apparatus that has a full face-piece and is operated in a pressure-demand or other positive-pressure mode) or SaF: Pd,Pp: ASCBA (APF = 10,000) (any supplied-air respirator that has a full face-piece and is operated in a pressure-demand or other positive-pressure mode in combination with an auxiliary self-contained breathing apparatus operated in a pressure-demand or other positive-pressure mode). *Escape:* GmFS100 (APF = 50) [any air-purifying, full-face-piece respirator (gas mask) with a chin-style, front- or back-mounted canister providing protection against the compound of concern and having an N100, R100, or P100 filter] or SCBAE (any appropriate escape-type, self-contained breathing apparatus).
Storage: Color Code—Blue: Health Hazard/Poison: Store in a secure poison location. Prior to working with this chemical you should be trained on its proper handling and storage.
Shipping: Cyanides, inorganic, solid, n.o.s. require a shipping label of "POISONOUS/TOXIC MATERIALS." They fall in Hazard Class 6.1.
Spill Handling: Evacuate persons not wearing protective equipment from area of spill or leak until cleanup is complete. Remove all ignition sources. Collect powdered

material in the most convenient and safe manner and deposit in sealed containers. Ventilate area after cleanup is complete. It may be necessary to contain and dispose of this chemical as a hazardous waste. If material or contaminated runoff enters waterways, notify downstream users of potentially contaminated waters. Contact your local or federal environmental protection agency for specific recommendations. If employees are required to clean up spills, they must be properly trained and equipped. OSHA 1910.120(q) may be applicable.

Fire Extinguishing: Use dry chemical, carbon dioxide, water spray, or foam for *small fires*; and water spray, fog, or foam for *large fires*. Move containers of this material away from fire area if this can be done without risk. Isolate hazard area and deny entry. Stay upwind; keep out of low areas. Ventilate closed spaces before entering them. Wear positive pressure breathing apparatus and special protective clothing. Fight fire from maximum distance. Dike fire control water for later disposal. Do not scatter the material. Poisonous gases are produced in fire. If material or contaminated runoff enters waterways, notify downstream users of potentially contaminated waters. Notify local health and fire officials and pollution control agencies. From a secure, explosion-proof location, use water spray to cool exposed containers. If cooling streams are ineffective (venting sound increases in volume and pitch, tank discolors, or shows any signs of deforming), withdraw immediately to a secure position. If employees are expected to fight fires, they must be trained and equipped in OSHA 1910.156. The only respirators recommended for firefighting are self-contained breathing apparatuses that have full face-pieces and are operated in a pressure-demand or other positive-pressure mode.

Disposal Method Suggested: Consult with environmental regulatory agencies for guidance on acceptable disposal practices. Generators of waste containing this contaminant ($\geq$100 kg/mo) must conform with EPA regulations governing storage, transportation, treatment, and waste disposal.

Reference

US Environmental Protection Agency. (November 30, 1987). *Chemical Hazard Information Profile: Potassium Silver Cyanide*. Washington, DC: Chemical Emergency Preparedness Program

Potassium sulfide   P:1010

Molecular Formula: K_2S
Synonyms: Dipotassium monosulfide; Dipotassium sulfide; Hepar sulfurous; Potassium monosulfide
CAS Registry Number: 1312-73-8
RTECS® Number: TT6000000 (anhydrous); TT6008000 (hydrated)
UN/NA & ERG Number: UN1382 (anhydrous)/135; UN1847 (hydrated)/153
EC Number: 215-197-0 [*Annex I Index No.:* 016-006-00-1]

Regulatory Authority and Advisory Bodies
European/International Regulations: Hazard Symbol: C; Risk phrases: R31; R34; R50; Safety phrases: S1/2; S26; S45; S61 (see Appendix 4).
WGK (German Aquatic Hazard Class): 2—Hazard to waters.
Description: Potassium sulfide is a brownish-red crystalline solid. Molecular weight = 110.26; Freezing/Melting point = 840°C. Soluble in water.
Potential Exposure: Potassium sulfide is used as a reagent in analytical chemistry; and in pharmaceutical preparations.
Incompatibilities: May explosively decompose from shock, friction, or concussion. May spontaneously ignite on contact with air. The aqueous solution is a strong base; reacts violently with strong acids and acid fumes. The solid material decomposes on contact with acids producing hydrogen sulfide, and oxidizers producing sulfur dioxide.
Permissible Exposure Limits in Air
No standards or TEEL available.
Routes of Entry: Inhalation, ingestion, skin and/or eye contact.
Harmful Effects and Symptoms
Short Term Exposure: Potassium sulfide can affect you when breathed in. Breathing the dust or mist can irritate the eyes, nose, and throat with sneezing, coughing, and sore throat. Potassium sulfide is a corrosive chemical and contact with skin and eyes can cause burns. High exposures can cause pulmonary edema, a medical emergency that can be delayed for several hours. This can cause death.
Long Term Exposure: Corrosive materials can cause lung problems; bronchitis may develop. Prolonged exposure can lead to sores or ulcers of the inner lining of the nose.
Points of Attack: Lungs, skin.
Medical Surveillance: Before beginning employment and at regular times after that, for those with frequent or potentially high exposures, the following are recommended: lung function tests. If symptoms develop or overexposure is suspected, the following may be useful: consider chest X-ray after acute overexposure.
First Aid: If this chemical gets into the eyes, remove any contact lenses at once and irrigate immediately for at least 15 min, occasionally lifting upper and lower lids. Seek medical attention immediately. If this chemical contacts the skin, remove contaminated clothing and wash immediately with soap and water. Seek medical attention immediately. If this chemical has been inhaled, remove from exposure, begin rescue breathing (using universal precautions, including resuscitation mask) if breathing has stopped and CPR if heart action has stopped. Transfer promptly to a medical facility. When this chemical has been swallowed, get medical attention. If victim is *conscious*, administer water or milk. Do not induce vomiting. Medical observation is recommended for 24—48 h after breathing overexposure, as pulmonary edema may be delayed. As first aid for pulmonary edema, a doctor or authorized paramedic may consider administering a corticosteroid spray.

Personal Protective Methods: Wear protective gloves and clothing to prevent any reasonable probability of skin contact. Safety equipment suppliers/manufacturers can provide recommendations on the most protective glove/clothing material for your operation. All protective clothing (suits, gloves, footwear, headgear) should be clean, available each day, and put on before work. Contact lenses should not be worn when working with this chemical. Wear dust-proof chemical goggles and face shield unless full face-piece respiratory protection is worn. Employees should wash immediately with soap when skin is wet or contaminated. Provide emergency showers and eyewash.

Respirator Selection: Where there is potential for exposures to potassium sulfide use a NIOSH/MSHA- or European Standard EN149-approved full-face-piece respirator equipped with particulate (dust/fume/mist) filters. Particulate filters must be checked every day before work for physical damage, such as rips or tears, and replaced as needed. Where there is potential for high exposures and liquid potassium sulfide, use a NIOSH/MSHA- or European Standard EN149-approved supplied-air respirator with a full face-piece operated in the positive-pressure mode, or with a full face-piece, hood, or helmet in the continuous-flow mode; or use a NIOSH/MSHA- or European Standard EN149-approved self-contained breathing apparatus with a full face-piece operated in a pressure-demand or other positive-pressure mode.

Storage: Color Code—White: Corrosive or Contact Hazard; Store separately in a corrosion-resistant location. Prior to working with this chemical you should be trained on its proper handling and storage. Potassium sulfide must be stored to avoid contact with oxidizers (such as perchlorates, peroxides, permanganates, chlorates, and nitrates) and strong acids (such as hydrochloric, sulfuric, and nitric) since violent reactions occur. Sources of ignition, such as smoking and open flames, are prohibited where potassium sulfide is used, handled, or stored in a manner that could create a potential fire or explosion hazard. Protect storage against physical damage.

Shipping: Anhydrous potassium sulfide requires a shipping label of "SPONTANEOUSLY COMBUSTIBLE." It falls in Hazard Class 4.2 and Packing Group II. Potassium sulfide, hydrated with not <30% water of crystallization requires a shipping label of "CORROSIVE." It falls in Hazard Class 8 and Packing Group II.

Spill Handling: Evacuate persons not wearing protective equipment from area of spill or leak until cleanup is complete. Remove all ignition sources. Collect powdered material in the most convenient and safe manner and deposit in sealed containers. Ventilate area after cleanup is complete. It may be necessary to contain and dispose of this chemical as a hazardous waste. If material or contaminated runoff enters waterways, notify downstream users of potentially contaminated waters. Contact your local or federal environmental protection agency for specific recommendations. If employees are required to clean up spills, they must be properly trained and equipped. OSHA 1910.120(q) may be applicable.

Fire Extinguishing: This chemical is a combustible solid. Use dry chemical, carbon dioxide, water spray, or alcohol foam extinguishers. Poisonous gases are produced in fire, including hydrogen sulfide and sulfur oxides. If material or contaminated runoff enters waterways, notify downstream users of potentially contaminated waters. Notify local health and fire officials and pollution control agencies. From a secure, explosion-proof location, use water spray to cool exposed containers. If cooling streams are ineffective (venting sound increases in volume and pitch, tank discolors, or shows any signs of deforming), withdraw immediately to a secure position. If employees are expected to fight fires, they must be trained and equipped in OSHA 1910.156. The only respirators recommended for firefighting are self-contained breathing apparatuses that have full face-pieces and are operated in a pressure-demand or other positive-pressure mode.

Reference
New Jersey Department of Health and Senior Services. (March 2001). *Hazardous Substances Fact Sheet: Potassium Sulfide.* Trenton, NJ

Procarbazine & procarbazine hydrochloride   P:1020

Molecular Formula: $C_{12}H_{19}N_3O$
Common Formula: $CH_3NHNHCH_2C_6H_4CONHCH(CH_3)_2$
Synonyms: Ibenzmethyzine; 2-(p-Isopropyl carbamoyl benzyl)-1-methylhydrazine; N-Isopropyl-α-(2-methylhydrazino)-p-toluamide N isopropyl; Matulane; 4-[(2-Methylhydrazino)methyl]-N-isopropylbenzamide; 1-[Methyl-2-(-isopropylcarbamoyl)benzyl]hydrazine; MIH; Natulan; NSC-77213; PCB; RO 4-6467
hydrochloride: Ibenzmethyzine hydrochloride; Ibenzmethyzin hydrochloride; IBZ; 1-(p-Isopropylcarbamoylbenzyl)-2-methylhydrazine hydrochloride; 2-[p-(Isopropylcarbamoyl)benzyl]-1-methylhydrazine hydrochloride; N-Isopropyl-p-(2-methylhydrazinomethyl)benzamidehydrochloride; N-Isopropyl-α-(2-methylhydrazino)-p-toluamide hydrochloride; Matulane; MBH; N-(1-Methylethyl)-4-[(2-methylhydrazino)methyl]benzamide monohydrochloride; p-(N'-Methylhydrazinomethyl)-N-isopropylbenzamide hydrochloride; 1-Methyl-2-p-(isopropylcarbamoyl)-benzohydrazine hydrochloride; 1-Methyl-2-(p-isopropylcarbamoylbenzyl)hydrazine hydrochloride; MIH hydrochloride; Nathulane; Natulan; Natulanar; Natulan hydrochloride; NCI-C01810; NSC-77213; PCBhydrochloride; Procarbazin (German); RO 4-6467
CAS Registry Number: 671-16-9; 366-70-1 (hydrochloride)
RTECS® Number: XS4550000; XS472000 (hydrochloride)
UN/NA & ERG Number: UN2811 (toxic solid, organic, n.o.s.)/154

EC Number: 211-582-2; 206-678-6 (hydrochloride)
Regulatory Authority and Advisory Bodies
Carcinogenicity: IARC: (hydrochloride) Animal Sufficient
Evidence; Human Limited Evidence, *probably carcinogenic
to humans*, Group 2A, 1998; NTP: Reasonably anticipated
to be a human carcinogen; NTP: Report on Carcinogens,
2004; NCI: Carcinogenesis Studies (ipr); clear evidence:
mouse, rat 1979.
California Proposition 65 Chemical: Cancer (methylhydrazine and its salts) 7/1/92; (procarbazine) 1/1/88; (procarbazine hydrochloride) cancer 1/1/88; Developmental/
Reproductive toxin 7/1/90.
WGK (German Aquatic Hazard Class): No value assigned.
Description: Procarbazine is a white to pale yellow crystalline powder with a slight odor. Molecular weight = 221.34.
The hydrochloride has a similar description. Molecular
weight = 257.80; Freezing/Melting point = 223−236°C
(hydrochloride). Soluble in water.
Potential Exposure: Procarbazine is available in capsule
form. The primary use of this drug is as an antineoplastic
agent in the treatment of advanced Hodgkin's disease and
oat-cell carcinoma of the lung. The hydrochloride compound is used in treatment. The FDA approved use of procarbazine hydrochloride in 1969 and indicated that the drug
should be used as an adjunct to standard therapy. Possible
exposure occurs during manufacture of the drug and direct
exposure during its subsequent administration to patients.
Some of the metabolites of procarbazine hydrochloride are
both carcinostatic and carcinogenic.
Incompatibilities: When heated to decomposition, it produces hydrogen chloride and nitrogen oxides. Incompatible
with strong acids, strong alkalies.
Permissible Exposure Limits in Air: No standard set.
Routes of Entry: Inhalation, ingestion, skin and/or eye
contact.
Harmful Effects and Symptoms
Short Term Exposure: Laboratory exposure of animals to
procarbazine was studied by IP (intraperitoneal) injection.
In rats, malignant lymphoma, adenocarcinoma of the mammary gland, and olfactory neuroblastomas were induced in
statistically significant numbers. In mice, malignant lymphoma or leukemia, olfactory neuroblastomas, alveolar/
bronchiolar adenoma, and adenocarcinoma of the uterus
were induced in statistically significant numbers. Can cause
nausea, vomiting, diarrhea, stomach pain, loss of appetite,
and weight loss. Symptoms of exposure include[52]: nausea,
vomiting, anorexia, dry mouth, dysphagia, diarrhea, constipation, chills and fever, sweating, weakness, edema, cough,
dermatitis, jaundice, headache, insomnia, coma.
Long Term Exposure: A probable human carcinogen.
There is some evidence that it causes cancer of the nervous
system, blood-forming organs, breast, lung, uterus, and
blood or bone marrow in humans. It has been shown to
cause cancer in the same sites in animals. May damage the
testes. The hydrochloride is a teratogen in animals and may
decrease the body's ability to produce blood cells, causing

reduced white blood cells with increased infection and general weakness, reduced platelets, causing bleeding when cut
or bruised, and/or reduced blood cells (anemia).
Points of Attack: Blood.
Medical Surveillance: Complete blood count (CBC).
First Aid: Skin Contact[52]: Flood all areas of body that
have contacted the substance with water. Do not wait to
remove contaminated clothing; do it under the water stream.
Use soap to help assure removal. Isolate contaminated
clothing when removed to prevent contact by others. *Eye
Contact:* Remove any contact lenses at once. Flush eyes
well with copious quantities of water or normal saline for at
least 20−30 min. Seek medical attention.
Inhalation: Leave contaminated area immediately; breathe
fresh air. Proper respiratory protection must be supplied to
any rescuers. If coughing, difficult breathing, or any other
symptoms develop, seek medical attention at once, even if
symptoms develop many hours after exposure. *Ingestion:* If
convulsions are not present, give a glass or two of water or
mild to dilute the substance. Assure that the person's airway
is unobstructed and contact a hospital or poison center immediately for advice on whether or not to induce vomiting.
Personal Protective Methods: Wear protective gloves and
clothing to prevent any reasonable probability of skin contact. Safety equipment suppliers/manufacturers can provide
recommendations on the most protective glove/clothing
material for your operation. All protective clothing (suits,
gloves, footwear, headgear) should be clean, available each
day, and put on before work. Contact lenses should not be
worn when working with this chemical. Wear dust-proof
chemical goggles and face shield unless full face-piece
respiratory protection is worn. Employees should wash
immediately with soap when skin is wet or contaminated.
Provide emergency showers and eyewash.
Respirator Selection: Follow regulations in OSHA
29CFR1910.134 or European Standard EN149. Use a
NIOSH/MSHA- or European Standard EN149-approved
respirator; or use an approved supplied-air respirator with a
full face-piece operated in the positive-pressure mode, or
with a full face-piece, hood, or helmet in the continuous-
flow mode; or use a NIOSH/MSHA- or European Standard
EN149-approved self-contained breathing apparatus with a
full face-piece operated in a pressure-demand or other positive-pressure mode.
Storage: Color Code—Blue: Health Hazard/Poison: Store
in a secure poison location. Prior to working with this
chemical you should be trained on its proper handling and
storage. Store in a refrigerator or in a cool, dry place. A regulated, marked area should be established where this chemical is handled, used, or stored in compliance with OSHA
Standard 1910.1045.
Shipping: Toxic solids, organic, n.o.s. require a shipping
label of "POISONOUS/TOXIC MATERIALS." They fall
in Hazard Class 6.1.
Spill Handling: Evacuate persons not wearing protective
equipment from area of spill or leak until cleanup is

complete. Remove all ignition sources. Dampen spilled material with water to avoid dust, than transfer material to a suitable container. Use absorbent dampened with water to pick up remaining material. Wash surfaces well with soap and water. Ventilate area after cleanup is complete. It may be necessary to contain and dispose of this chemical as a hazardous waste. If material or contaminated runoff enters waterways, notify downstream users of potentially contaminated waters. Contact your local or federal environmental protection agency for specific recommendations. If employees are required to clean up spills, they must be properly trained and equipped. OSHA 1910.120(q) may be applicable.

Fire Extinguishing: This chemical is a combustible solid. Use dry chemical, carbon dioxide, water spray, or alcohol foam extinguishers. Poisonous gases are produced in fire. If material or contaminated runoff enters waterways, notify downstream users of potentially contaminated waters. Notify local health and fire officials and pollution control agencies. From a secure, explosion-proof location, use water spray to cool exposed containers. If cooling streams are ineffective (venting sound increases in volume and pitch, tank discolors, or shows any signs of deforming), withdraw immediately to a secure position. If employees are expected to fight fires, they must be trained and equipped in OSHA 1910.156. The only respirators recommended for firefighting are self-contained breathing apparatuses that have full face-pieces and are operated in a pressure-demand or other positive-pressure mode.

References

National Cancer Institute. (1979). *Bioassay of Procarbazine for Possible Carcinogenicity*, DHHS Publication No. (NIH) 79-819. Springfield, VA: National Technical Information Service

New Jersey Department of Health and Senior Services. (August 2002). *Hazardous Substances Fact Sheet: Procarbazine Hydrochloride.* Trenton, NJ

Promecarb   P:1030

Molecular Formula: $C_{12}H_{17}NO_2$
Common Formula: $C_6H_3(CH_3)(OCONHCH_3)CH(CH_3)_2$
Synonyms: Carbamic acid, methyl-, *m-cym*-5-yl ester; Carbamic acid, 3-methyl-5-(1-methylethyl)phenyl-, methyl ester; Carbamic acid, *N*-methyl-, 3-methyl-5-isopropylphenyl ester; Carbamult; Carbanilic acid, 3-isopropyl-5-methyl-, methyl ester; *m-cym*-5-yl-methylcarbamate; ENT27,300; ENT 27,300-A; EP316; 3-Isopropyl-5-methylcarbamic acid methyl ester; 3-Isopropyl-5-methylphenyl *N*-methylcarbamate; 5-Isopropyl-*m*-tolyl methyl-carbamate; Methylcarbamic acid *m-cym*-5-yl ester; *N*-Methylcarbamic acid 3-methyl-5-isopropylphenyl ester; 5-Methyl *m*-cumenyl methylcarbamate; 3-Methyl-5-isopropyl-*N*-methyl carbamate; (3-Methyl-5-isopropylphenyl)-*N*-methylcarbamat (German); 3-Methyl-5-isopropylphenyl-*N*-methyl carbamate;

3-Methyl-5-(1-methylethyl)phenol methylcarbamate; 3-Methyl-5-(1-methylethyl)phenyl-carbamic acid methyl ester; Minacide; Morton EP-316; Phenol, 3-methyl-5-(1-methylethyl)-, methylcarbamate; Schering 34615; UC 9880; Union Carbide UC-9880
CAS Registry Number: 2631-37-0
RTECS® Number: FB8050000
UN/NA & ERG Number: UN2757/151
EC Number: 220-113-0 [*Annex I Index No.:* 006-037-00-9]
Regulatory Authority and Advisory Bodies
US EPA Hazardous Waste Number (RCRA No.): P201.
Superfund/EPCRA [40CFR 302 and 355, F R: 8/16/06,Vol 71, No. 158] Reportable Quantity (RQ): 1000 lb (454 kg).
RCRA, 40CFR261, Appendix 8 Hazardous Constituents.
RCRA 40CFR268.48; 61FR15654, Universal Treatment Standards: Wastewater (mg/L), 0.056; Nonwastewater (mg/kg), 1.4
Superfund/EPCRA 40CFR355, Extremely Hazardous Substances: TPQ = 500/10,000 lb (227/4540 kg).
Reportable Quantity (RQ): 1000 lb (454 kg).
US DOT Regulated Marine Pollutant (49CFR172.101, Appendix B).
European/International Regulations: Hazard Symbol: T, N; Risk phrases: R25; R50/53; Safety phrases: S1/2; S24; S37; 45; S60; S61 (see Appendix 4).
WGK (German Aquatic Hazard Class): 3—Severe hazard to waters.
Description: Promecarb is a colorless, odorless, crystalline solid. Molecular weight = 207.30; Freezing/Melting point = 87−88°C. Hazard Identification (based on NFPA-704 M Rating System): Health 3, Flammability 1, Reactivity 1. Slightly soluble in water.
Potential Exposure: Those involved in the manufacture, formulation, or application of this nonsystemic contact insecticide.
Incompatibilities: Alkalis.
Permissible Exposure Limits in Air
Protective Action Criteria (PAC)
TEEL-0: 3 mg/m³
PAC-1: 10 mg/m³
PAC-2: 16 mg/m³
PAC-3: 25 mg/m³
Determination in Water: Fish Tox = 33.48370000 ppb (INTERMEDIATE).
Routes of Entry: Inhalation, ingestion, skin and/or eye contact.
Harmful Effects and Symptoms
Short Term Exposure: Promecarb is highly toxic by ingestion and is absorbed through the intact skin. It is a reversible cholinesterase inhibitor and its effects are related to action on the nervous system. Diarrhea, nausea, vomiting, excessive salivation, headache, pinpoint pupils, and uncoordinated muscle movements are all common symptoms of exposure to carbamate insecticides.
Long Term Exposure: Cholinesterase inhibitor; cumulative effect is possible. This chemical may damage the nervous

system with repeated exposure, resulting in convulsions, respiratory failure. May cause liver damage. Human Tox; 350.00000 ppb (VERY LOW).

Points of Attack: Respiratory system, lungs, central nervous system, cardiovascular system, skin, eyes, plasma and red blood cell cholinesterase.

Medical Surveillance: Before employment and at regular times after that, the following are recommended: plasma and red blood cell cholinesterase levels (tests for the enzyme poisoned by this chemical). If exposure stops, plasma levels return to normal in 1–2 weeks while red blood cell levels may be reduced for 1–3 months.

When cholinesterase enzyme levels are reduced by 25% or more below preemployment levels, risk of poisoning is increased, even if results are in lower ranges of "normal." Reassignment to work not involving carbamate pesticides is recommended until enzyme levels recover. If symptoms develop or overexposure occurs, repeat the above tests as soon as possible and get an examination of the nervous system. Also, consider complete blood count. Consider chest X-ray following acute overexposure.

First Aid: If this chemical gets into the eyes, remove any contact lenses at once and irrigate immediately for at least 15 min, occasionally lifting upper and lower lids. Seek medical attention immediately. If this chemical contacts the skin, remove contaminated clothing and wash immediately with soap and water. Seek medical attention immediately. If this chemical has been inhaled, remove from exposure, begin rescue breathing (using universal precautions, including resuscitation mask) if breathing has stopped and CPR if heart action has stopped. Transfer promptly to a medical facility. When this chemical has been swallowed, get medical attention. Give large quantities of water and induce vomiting. Do not make an unconscious person vomit.

Personal Protective Methods: Wear protective gloves and clothing to prevent any reasonable probability of skin contact. Safety equipment suppliers/manufacturers can provide recommendations on the most protective glove/clothing material for your operation. All protective clothing (suits, gloves, footwear, headgear) should be clean, available each day, and put on before work. Contact lenses should not be worn when working with this chemical. Wear dust-proof chemical goggles and face shield unless full face-piece respiratory protection is worn. Employees should wash immediately with soap when skin is wet or contaminated. Provide emergency showers and eyewash.

Respirator Selection: Follow regulations in OSHA 29CFR1910.134 or European Standard EN149. Use a NIOSH/MSHA- or European Standard EN149-approved respirator; or use an approved supplied-air respirator with a full face-piece operated in the positive-pressure mode, or with a full face-piece, hood, or helmet in the continuous-flow mode; or use a NIOSH/MSHA- or European Standard EN149-approved self-contained breathing apparatus with a full face-piece operated in a pressure-demand or other positive-pressure mode.

Storage: Color Code—Blue: Health Hazard/Poison: Store in a secure poison location. Prior to working with this chemical you should be trained on its proper handling and storage. Store in tightly closed containers in a cool, well-ventilated area away from alkaline materials.

Shipping: Promecarb falls into the class of Carbamate pesticides, solid, toxic, n.o.s. This compound requires a shipping label of "POISONOUS/TOXIC MATERIALS." It falls in Hazard Class 6.1 and Packing Group III.

Spill Handling: Keep unnecessary people away; isolate hazard areas and deny entry. Stay upwind and keep out of low areas. Do not touch spilled material or breathe the dusts, vapors, or fumes from burning materials. Use water spray to reduce vapors. Do not handle broken packages without protective equipment. Wash away any material that may have contacted the body with soap and water. Take up *small spills* with sand or other noncombustible absorbent material and place in containers for later disposal. *Small dry spills:* with clean shovel place material into clean, dry container and cover; move containers from spill area. Dike far ahead of *large spills* for later disposal. Ventilate area after cleanup is complete. It may be necessary to contain and dispose of this chemical as a hazardous waste. If material or contaminated runoff enters waterways, notify downstream users of potentially contaminated waters. Contact your local or federal environmental protection agency for specific recommendations. If employees are required to clean up spills, they must be properly trained and equipped. OSHA 1910.120(q) may be applicable.

Soil Adsorption Index (K_{oc}) = 200 (estimate).

Fire Extinguishing: Extinguish fire using an agent suitable for the surrounding fire, as the material itself burns with difficulty. Use water in flooding quantities as a fog. Use alcohol foam, carbon dioxide, or dry chemical. Poisonous gases are produced in fire. If material or contaminated runoff enters waterways, notify downstream users of potentially contaminated waters. Notify local health and fire officials and pollution control agencies. From a secure, explosion-proof location, use water spray to cool exposed containers. If cooling streams are ineffective (venting sound increases in volume and pitch, tank discolors, or shows any signs of deforming), withdraw immediately to a secure position. If employees are expected to fight fires, they must be trained and equipped in OSHA 1910.156. The only respirators recommended for firefighting are self-contained breathing apparatuses that have full face-pieces and are operated in a pressure-demand or other positive-pressure mode.

Disposal Method Suggested: Small quantities: treat with alkali and then bury. *Large quantities:* incineration.[22] In accordance with 40CFR165, follow recommendations for the disposal of pesticides and pesticide containers. Must be disposed properly by following package label directions or by contacting your local or federal environmental control agency or by contacting your regional EPA office.

Reference
US Environmental Protection Agency. (November 30, 1987). *Chemical Hazard Information Profile: Promecarb.* Washington, DC: Chemical Emergency Preparedness Program

Pronamide   P:1040

Molecular Formula: $C_{12}H_{11}Cl_2NO$
Synonyms: Benzamide, 3,5-dichloro-*N*-(1,1-dimethyl-2-propynyl); Campbell's Rapier; 3,5-Dichloro-*N*-(1,1-dimethyl-2-propynyl)benzamide; 3,5-Dichloro-*N*-(1,1-dimethylprop-2-ynyl)benzamide; 3,5-Dichloro-*N*-(1,1-dimethyl-propynyl) benzamide; *N*-(1,1-Dimethylpropynyl)-3,5-dichlorobenzamide; Kerb; Kerb50W; Kerb propyzamide 50; Propyzamide; Rapier
CAS Registry Number: 23950-58-5
RTECS® Number: CV3460000
UN/NA & ERG Number: UN3077/171
EC Number: 245-951-4 [*Annex I Index No.:* 616-055-00-4]
Regulatory Authority and Advisory Bodies
Banned or Severely Restricted (USA) (UN).[13]
US EPA Hazardous Waste Number (RCRA No.): U192.
RCRA, 40CFR261, Appendix 8 Hazardous Constituents.
RCRA 40CFR268.48; 61FR15654, Universal Treatment Standards: Wastewater (mg/L), 0.093; Nonwastewater (mg/kg), 1.5.
RCRA 40CFR264, Appendix 9; TSD Facilities Ground Water Monitoring List. Suggested test method(s) (PQL µg/L): 8270 (10).
EPCRA Section 313 Form R *de minimis* concentration reporting level: 1.0%.
California Proposition 65 Chemical: Cancer 5/1/96.
Hazard Symbols: Xn, N, Risk phrases: R40; R50/53, Safety phrases: S36/37; S60; S61 (see Appendix 4).
WGK (German Aquatic Hazard Class): No value assigned.
Description: Pronamide is a colorless crystalline solid. Molecular weight = 256.14; Freezing/Melting point = 155−156°C; Vapor pressure = 0.0001 mmHg at 25°C. Insoluble in water; solubility = 15 mg/L.
Potential Exposure: Those involved in the manufacture, formulation, and application of this selective herbicide.
Permissible Exposure Limits in Air
No standards or TEEL available.
Permissible Concentration in Water: The EPA has derived a lifetime health advisory of 0.052 mg/L (52 µg/L).
Determination in Water: Extraction with methylene chloride, separation by capillary-column gas chromatography; then measurement using a nitrogen−phosphorus detector. Fish Tox = 13901.63079000 ppb (VERY LOW).
Routes of Entry: Inhalation, ingestion, skin and/or eye contact.
Harmful Effects and Symptoms
Short Term Exposure: Eye contact can cause irritation. Inhalation can cause irritation of the respiratory tract with cough, phlegm, and/or chest tightness. The acute oral LD_{50} for male rats is 8350 mg/kg and for female rats is 5620 mg/kg (insignificantly toxic in both cases).
Long Term Exposure: Applying the criteria described in EPA's final guidelines for assessment of carcinogenic risk, pronamide has tentatively been classified in Group C: possible human carcinogen. This category is for substances with limited evidence of carcinogenicity in animals in the absence of human data. There is limited animal evidence of liver cancer. Human Tox = 22.72727 ppb (INTERMEDIATE).
Points of Attack: Cancer site in animals: liver.
First Aid: If this chemical gets into the eyes, remove any contact lenses at once and irrigate immediately for at least 15 min, occasionally lifting upper and lower lids. Seek medical attention immediately. If this chemical contacts the skin, remove contaminated clothing and wash immediately with soap and water. Seek medical attention immediately. If this chemical has been inhaled, remove from exposure, begin rescue breathing (using universal precautions, including resuscitation mask) if breathing has stopped and CPR if heart action has stopped. Transfer promptly to a medical facility. When this chemical has been swallowed, get medical attention. Give large quantities of water and induce vomiting. Do not make an unconscious person vomit.
Personal Protective Methods: Wear protective gloves and clothing to prevent any reasonable probability of skin contact. Safety equipment suppliers/manufacturers can provide recommendations on the most protective glove/clothing material for your operation. All protective clothing (suits, gloves, footwear, headgear) should be clean, available each day, and put on before work. Contact lenses should not be worn when working with this chemical. Wear dust-proof chemical goggles and face shield unless full face-piece respiratory protection is worn. Employees should wash immediately with soap when skin is wet or contaminated. Provide emergency showers and eyewash.
Respirator Selection: Follow regulations in OSHA 29CFR1910.134 or European Standard EN149. Use a NIOSH/MSHA- or European Standard EN149-approved respirator; or use an approved supplied-air respirator with a full face-piece operated in the positive-pressure mode, or with a full face-piece, hood, or helmet in the continuous-flow mode; or use a NIOSH/MSHA- or European Standard EN149-approved self-contained breathing apparatus with a full face-piece operated in a pressure-demand or other positive-pressure mode.
Storage: Color Code—Green: General storage may be used. Prior to working with this chemical you should be trained on its proper handling and storage. Store in tightly closed containers in a cool, well-ventilated area away from strong oxidizers. Where possible, automatically transfer material from drums or other storage containers to process containers. A regulated, marked area should be established where this chemical is handled, used, or stored in compliance with OSHA Standard 1910.1045.

Shipping: The name of this material is not on the DOT list of materials[19] for label and packaging standards. However, based on regulations, it may be classified[52] as an Environmentally hazardous substances, solid, n.o.s. This chemical requires a shipping label of "CLASS 9." It falls in Hazard Class 9 and Packing Group III.[20, 21]

Spill Handling: Evacuate persons not wearing protective equipment from area of spill or leak until cleanup is complete. Remove all ignition sources. Use HEPA vacuum or wet method to reduce dust during cleanup. Do not dry sweep. Collect powdered material in the most convenient and safe manner and deposit in sealed containers. Ventilate area after cleanup is complete. It may be necessary to contain and dispose of this chemical as a hazardous waste. If material or contaminated runoff enters waterways, notify downstream users of potentially contaminated waters. Contact your local or federal environmental protection agency for specific recommendations. If employees are required to clean up spills, they must be properly trained and equipped. OSHA 1910.120(q) may be applicable. Soil Adsorption Index (K_{oc}) = 200.

Fire Extinguishing: This chemical is a combustible solid. Use dry chemical, carbon dioxide, or water spray extinguishers. Poisonous gases are produced in fire, including hydrogen chloride and oxides of nitrogen. If material or contaminated runoff enters waterways, notify downstream users of potentially contaminated waters. Notify local health and fire officials and pollution control agencies. From a secure, explosion-proof location, use water spray to cool exposed containers. If cooling streams are ineffective (venting sound increases in volume and pitch, tank discolors, or shows any signs of deforming), withdraw immediately to a secure position. If employees are expected to fight fires, they must be trained and equipped in OSHA 1910.156. The only respirators recommended for firefighting are self-contained breathing apparatuses that have full face-pieces and are operated in a pressure-demand or other positive-pressure mode.

Disposal Method Suggested: Consult with environmental regulatory agencies for guidance on acceptable disposal practices. Generators of waste containing this contaminant (≥100 kg/mo) must conform with EPA regulations governing storage, transportation, treatment, and waste disposal. In accordance with 40CFR165, follow recommendations for the disposal of pesticides and pesticide containers. Must be disposed properly by following package label directions or by contacting your local or federal environmental control agency or by contacting your regional EPA office.

References
US Environmental Protection Agency. (August 1987). *Health Advisory: Pronamide.* Washington, DC: Office of Drinking Water
New Jersey Department of Health and Senior Services. (July 2005). *Hazardous Substances Fact Sheet: Pronamide.* Trenton, NJ

Propachlor P:1045

Molecular Formula: $C_{11}H_{14}ClNO$
Common Formula: $(ClCH_2CO)N(C_6H_5)CH(CH_3)_2$
Synonyms: Acetamide, 2-chloro-N-isopropyl-; Acetamide, 2-chloro-N-(1-methylethyl)-N-phenyl-; Aclid; AI3-51503; Albrass; Bexton; Bexton 4L; Chloressigsaeure-N-isopropyl-anilid (German); α-Chloro-N-isopropylacetanilide; 2-Chloro-N-isopropylacetanilide; 2-Chloro-N-isopropyl-N-phenylacetamide; 2-Chloro-N-(1-methylethyl)-N-phenylacetamide; CIPA; CP31393; N-Isopropyl-α-chloroacetanilide; N-Isopropyl-2-chloroacetanilide; Niticid; Propachlore; Propacloro (Spanish); Ramrod; Ramrod 65; Satecid
CAS Registry Number: 1918-16-7
RTECS® Number: AE1575000
UN/NA & ERG Number: UN2588/151
EC Number: 217-638-2 [*Annex I Index No.:* 616-008-00-8]
Regulatory Authority and Advisory Bodies
Air Pollutant Standard Set. See below, "Permissible Exposure Limits in Air" section.
EPCRA Section 313 Form R *de minimis* concentration reporting level: 1.0%.
California Proposition 65 Chemical: Cancer 2/27/01.
European/International Regulations: Hazard Symbol: Xn, N; Risk phrases: R22; R36; R43; R50/53; Safety phrases: S2; S24; S37; S60; S61 (see Appendix 4).
WGK (German Aquatic Hazard Class): No value assigned.
Description: Propachlor is a light tan solid. Molecular weight = 211.71; Boiling point = 110°C at 0.03 mm; Freezing/Melting point = 67−76°C; Flash point = about 316°C. Slightly soluble in water.
Potential Exposure: Those engaged in the manufacture, formulation, and application of this preemergence herbicide, which is used to combat annual grasses and broad-leaved weeds in corn, soybeans, cotton, sugar cane, and vegetable crops.
Incompatibilities: Incompatible with alkaline materials, strong acids; strong oxidizers. Attacks carbon steel.
Permissible Exposure Limits in Air Russia[35, 43] set a MAC of 0.5 mg/m³ in work-place air.
Permissible Concentration in Water: Russia[35, 43] set a MAC of 0.01 mg/L in water bodies used for domestic purposes and of zero in water for fishery purposes. The EPA has set a lifetime health advisory of 0.092 mg/L (92 μg/L). States which have set guidelines for propachlor in drinking water include Kansas at 700 μg/L and Maine at 200 μg/L.
Determination in Water: Fish Tox = 17.92643000 ppb MATC (INTERMEDIATE).
Routes of Entry: Inhalation, ingestion, skin and/or eye contact.
Harmful Effects and Symptoms
Short Term Exposure: The maximal tolerated dosage of propachlor without adverse effect is reported as 133.3 mg/kg/day in both rats and dogs. Other workers reported slight organ pathology in rats, mice, and rabbits at 100 mg/kg/day

or higher; this agrees approximately with the former data. Apparently, no long-term toxicity studies have been completed that would contribute information on reproductive effects or carcinogenic potential of propachlor or its degradation products, which include aniline derivatives. These studies are needed.

Long Term Exposure: May be a mutagen. Human Tox = 10.93750 ppb; Chronic Human Carcinogen Level (INTERMEDIATE).

First Aid: If this chemical gets into the eyes, remove any contact lenses at once and irrigate immediately for at least 15 min, occasionally lifting upper and lower lids. Seek medical attention immediately. If this chemical contacts the skin, remove contaminated clothing and wash immediately with soap and water. Seek medical attention immediately. If this chemical has been inhaled, remove from exposure, begin rescue breathing (using universal precautions, including resuscitation mask) if breathing has stopped and CPR if heart action has stopped. Transfer promptly to a medical facility. When this chemical has been swallowed, get medical attention. Give large quantities of water and induce vomiting. Do not make an unconscious person vomit.

Personal Protective Methods: Wear protective gloves and clothing to prevent any reasonable probability of skin contact. Safety equipment suppliers/manufacturers can provide recommendations on the most protective glove/clothing material for your operation. All protective clothing (suits, gloves, footwear, headgear) should be clean, available each day, and put on before work. Contact lenses should not be worn when working with this chemical. Wear dust-proof chemical goggles and face shield unless full face-piece respiratory protection is worn. Employees should wash immediately with soap when skin is wet or contaminated. Provide emergency showers and eyewash.

Respirator Selection: Follow regulations in OSHA 29CFR1910.134 or European Standard EN149. Use a NIOSH/MSHA- or European Standard EN149-approved respirator; or use an approved supplied-air respirator with a full face-piece operated in the positive-pressure mode, or with a full face-piece, hood, or helmet in the continuous-flow mode; or use a NIOSH/MSHA- or European Standard EN149-approved self-contained breathing apparatus with a full face-piece operated in a pressure-demand or other positive-pressure mode.

Storage: Color Code—Blue: Health Hazard/Poison: Store in a secure poison location. Prior to working with this chemical you should be trained on its proper handling and storage. Store in tightly closed containers in a cool, well-ventilated area away from strong oxidizers. Where possible, automatically transfer material from drums or other storage containers to process containers. Sources of ignition, such as smoking and open flames, are prohibited where this chemical is handled, used, or stored. Metal containers involving the transfer of this chemical should be grounded and bonded. Wherever this chemical is used, handled, manufactured, or stored, use explosion-proof electrical equipment and fittings.

Shipping: Pesticides, solid, toxic, n.o.s. require a shipping label of "POISONOUS/TOXIC MATERIALS." It falls in Hazard Class 6.1 and Packing Group III.

Spill Handling: Evacuate persons not wearing protective equipment from area of spill or leak until cleanup is complete. Remove all ignition sources. Collect powdered material in the most convenient and safe manner and deposit in sealed containers. Ventilate area after cleanup is complete. It may be necessary to contain and dispose of this chemical as a hazardous waste. If material or contaminated runoff enters waterways, notify downstream users of potentially contaminated waters. Contact your local or federal environmental protection agency for specific recommendations. If employees are required to clean up spills, they must be properly trained and equipped. OSHA 1910.120(q) may be applicable. Soil Adsorption Index (K_{oc}) = 80.

Fire Extinguishing: This chemical is a combustible solid. Use dry chemical, carbon dioxide, water spray, or alcohol foam extinguishers. Poisonous gases are produced in fire, including nitrogen oxides and chlorine. If material or contaminated runoff enters waterways, notify downstream users of potentially contaminated waters. Notify local health and fire officials and pollution control agencies. From a secure, explosion-proof location, use water spray to cool exposed containers. If cooling streams are ineffective (venting sound increases in volume and pitch, tank discolors, or shows any signs of deforming), withdraw immediately to a secure position. If employees are expected to fight fires, they must be trained and equipped in OSHA 1910.156. The only respirators recommended for firefighting are self-contained breathing apparatuses that have full face-pieces and are operated in a pressure-demand or other positive-pressure mode.

Disposal Method Suggested: Alkaline hydrolysis would yield N-isopropylaniline. However, incineration at 850°C together with flue gas scrubbing is the preferred disposal method.[22]

References

US Environmental Protection Agency. (August 1987). *Health Advisory: Propachlor*. Washington, DC: Office of Drinking Water

US Environmental Protection Agency. *Integrated Risk Information System (IRIS) Propachlor*. Washington, DC. Various dates. <http://www.epa.gov/IRIS/subst/0096.htm>

Propadiene   P:1050

Molecular Formula: C_3H_4
Common Formula: $H_2C{=}C{=}CH_2$
Synonyms: Allene; Allylenel; Dimethylenemethane; 1,2-Propadiene; Propadieno (Spanish)
CAS Registry Number: 463-49-0
RTECS® Number: BA040000
UN/NA & ERG Number: UN2200/116
EC Number: 207-335-3 (allene)

Regulatory Authority and Advisory Bodies
Department of Homeland Security Screening Threshold Quantity (pounds): *Release hazard* 10,000 (≥1.00% concentration).
Clean Air Act: Accidental Release Prevention/Flammable Substances, (Section 112[r], Table 3), TQ = 10,000 lb (4540 kg).
European/International Regulations: not listed in Annex 1.
WGK (German Aquatic Hazard Class): No value assigned.
Description: Propadiene is a colorless, flammable gas or liquid with a sweet odor. Molecular weight = 40.07; Boiling point = −32°C; Freezing/Melting point = −137°C. Explosive limits: LEL = 2.1%; UEL—unknown. Hazard Identification (based on NFPA-704 M Rating System): Health 2, Flammability 4, Reactivity 3. Insoluble in water.
Potential Exposure: Used in chemical synthesis and as a component in mixtures with methyl acetylene, which make up specialty welding gases.
Incompatibilities: Violent reaction with strong oxidizers, strong acids, nitrogen oxides.
Permissible Exposure Limits in Air
Protective Action Criteria (PAC)
TEEL-0: 40 ppm
PAC-1: 125 ppm
PAC-2: 750 ppm
PAC-3: 4000 ppm
Determination in Air: Use NIOSH Analytical Method (IV) #1500, Hydrocarbons.
Routes of Entry: Inhalation, ingestion, skin and/or eye contact.
Harmful Effects and Symptoms
Short Term Exposure: Propadiene can affect you when breathed in. Exposure can cause irritation of the eyes, nose, and throat. Very high levels can cause you to feel dizzy, lightheaded, and to pass out. Extremely high levels could cause death. Contact with the liquid may cause frostbite.
Long Term Exposure: Unknown at this time.
First Aid: If this chemical gets into the eyes, remove any contact lenses at once and irrigate immediately for at least 15 min, occasionally lifting upper and lower lids. Seek medical attention immediately. If this chemical contacts the skin, remove contaminated clothing and wash immediately with soap and water. Seek medical attention immediately. If this chemical has been inhaled, remove from exposure, begin rescue breathing (using universal precautions, including resuscitation mask) if breathing has stopped and CPR if heart action has stopped. Transfer promptly to a medical facility. When this chemical has been swallowed, get medical attention. Give large quantities of water and induce vomiting. Do not make an unconscious person vomit. If frostbite has occurred, seek medical attention immediately; do *NOT* rub the affected areas or flush them with water. In order to prevent further tissue damage, do *NOT* attempt to remove frozen clothing from frostbitten areas. If frostbite has *NOT* occurred, immediately and thoroughly wash contaminated skin with soap and water.

Personal Protective Methods: Wear appropriate personal protective clothing to prevent the skin from becoming frozen from contact with the evaporating liquid or from contact with vessels containing the liquid. Safety equipment suppliers/manufacturers can provide recommendations on the most protective glove/clothing material for your operation. All protective clothing (suits, gloves, footwear, headgear) should be clean, available each day, and put on before work. Contact lenses should not be worn when working with this chemical. Wear gas-proof goggles unless full face-piece respiratory protection is worn. Wear splash-proof chemical goggles and face shield when working with liquid unless full face-piece respiratory protection is worn. Employees should wash immediately with soap when skin is wet or contaminated. Provide emergency showers and eyewash.
Respirator Selection: Where there is potential for exposures to propadiene, use a NIOSH/MSHA- or European Standard EN149-approved supplied-air respirator with a full face-piece operated in the positive-pressure mode, or with a full face-piece, hood, or helmet in the continuous-flow mode; or use a NIOSH/MSHA- or European Standard EN149-approved self-contained breathing apparatus with a full face-piece operated in a pressure-demand or other positive-pressure mode.
Storage: Color Code—Red Stripe: Flammability Hazard: Store separately from all other flammable materials. Prior to working with this chemical you should be trained on its proper handling and storage. Before entering confined space where this chemical may be present, check to make sure that an explosive concentration does not exist. Propadiene must be stored to avoid contact with strong oxidizers (such as chlorine, bromine, and fluorine), strong acids (such as hydrochloric, sulfuric, and nitric), and nitrogen oxides since violent reactions occur. Sources of ignition, such as smoking and open flames, are prohibited where propadiene is handled, used, or stored. Wherever propadiene is used, handled, manufactured, or stored, use explosion-proof electrical equipment and fittings. Store cylinders in well-ventilated areas away from potential heat sources. Protect cylinders from physical damage. Procedures for the handling, use, and storage of cylinders should be in compliance with OSHA 1910.101 and 1910.169, as with the recommendations of the Compressed Gas Association.
Shipping: Propadiene, inhibited, requires a shipping label of "FLAMMABLE GAS." It falls in Hazard Class 2.1.
Spill Handling: Evacuate and restrict persons not wearing protective equipment from area of spill or leak until cleanup is complete. Remove all ignition sources. Establish forced ventilation to keep levels below explosive limit. Ventilate area of leak to disperse the gas. Stop the flow of gas if it can be done safely. If source of leak is a cylinder and the leak cannot be stopped in place, remove leaking cylinder to a safe place in the open air, and repair leak or allow cylinder to empty. Keep this chemical out of confined space, such as a sewer, because of the possibility of an explosion,

unless the sewer is designed to prevent the buildup of explosive concentrations. It may be necessary to contain and dispose of this chemical as a hazardous waste. Contact your local or federal environmental protection agency for specific recommendations. If employees are required to clean up spills, they must be properly trained and equipped. OSHA 1910.120(q) may be applicable.

Fire Extinguishing: This chemical is a flammable gas. Poisonous gases are produced in fire. Do not extinguish the fire unless the flow of gas can be stopped and any remaining gas is out of the line. Specially trained personnel may use fog lines to cool exposures and let the fire burn itself out. Vapors are heavier than air and will collect in low areas. Vapors may travel long distances to ignition sources and flashback. Vapors in confined areas may explode when exposed to fire. Containers may explode in fire. Storage containers and parts of containers may rocket great distances, in many directions. If material or contaminated runoff enters waterways, notify downstream users of potentially contaminated waters. Notify local health and fire officials and pollution control agencies. From a secure, explosion-proof location, use water spray to cool exposed containers. If cooling streams are ineffective (venting sound increases in volume and pitch, tank discolors, or shows any signs of deforming), withdraw immediately to a secure position. If cylinders are exposed to excessive heat from fire or flame contact, withdraw immediately to a secure location. If employees are expected to fight fires, they must be trained and equipped in OSHA 1910.156. The only respirators recommended for firefighting are self-contained breathing apparatuses that have full face-pieces and are operated in a pressure-demand or other positive-pressure mode.

Reference

New Jersey Department of Health and Senior Services. (February 2001). *Hazardous Substances Fact Sheet: Propadiene.* Trenton, NJ

Propane   P:1060

Molecular Formula: C_3H_8
Common Formula: $CH_3CH_2CH_3$
Synonyms: A-108; Dimethylmethane; Hydrocarbon propellent A-108; *n*-Propane; Propano (Spanish); Propyl hydride
CAS Registry Number: 74-98-6
RTECS® Number: TX2275000
UN/NA & ERG Number: UN1978/115; UN1075 (liquefied)/115
EC Number: 200-827-9 [*Annex I Index No.:* 601-003-00-5]
Regulatory Authority and Advisory Bodies
Department of Homeland Security Screening Threshold Quantity (pounds): *Release hazard* 60,000. *Note*: Facilities are not required to count propane in tanks of 10,000 lb or less. The higher threshold of 60,000 lb is set to focus on the screening of high-volume propane users; not on nonindustrial propane users.

Air Pollutant Standard Set. See below, "Permissible Exposure Limits in Air" section.
Clean Air Act: Accidental Release Prevention/Flammable Substances, (Section 112[r], Table 3), TQ = 10,000 lb (4540 kg).
European/International Regulations: Hazard Symbol: F + ; Risk phrases: R12; Safety phrases: S2; S9; S16 (see Appendix 4).
WGK (German Aquatic Hazard Class): Nonwater polluting agent.

Description: Propane is a colorless gas that is odorless when pure (a foul-smelling odorant is often added). The odor threshold is 2700 ppm. Molecular weight = 44.11; Boiling point = −42.2°C; Freezing/Melting point = −187.8°C; Relative vapor density (air = 1) = 1.55; Vapor pressure = 8.4 atm at 21; Flash point = flammable gas (−104°C); Autoignition temperature = 450°C. Explosive limits: LEL = 2.1%°C; UEL = 9.5%. Hazard Identification (based on NFPA-704 M Rating System): Health 1, Flammability 4, Reactivity 0. Slightly soluble in water; solubility = 0.01%.

Potential Exposure: Compound Description: Agricultural Chemical. Propane is used as a household, industrial, and vehicle fuel; it is used as a refrigerant and aerosol propellant; it is used as an intermediate in petrochemical manufacture.

Incompatibilities: Flammable gas. Strong oxidizers may cause fire and explosions. Liquid attacks some plastics, rubber, and coatings.

Permissible Exposure Limits in Air
Conversion factor: 1 ppm = 1.80 mg/m^3 at 25°C & 1 atm.
OSHA PEL: 1000 ppm/1800 mg/m^3 TWA.
NIOSH REL: 1000 ppm/1800 mg/m^3 TWA.
ACGIH TLV®[1]: 1000 ppm TWA as aliphatic hydrocarbon gas (C_1−C_4).
NIOSH IDLH: 2100 ppm [LEL].
Protective Action Criteria (PAC)*
TEEL-0: 1000 ppm
PAC-1: **5500** ppm
PAC-2: **17,000** ppm
PAC-3: **33,000** ppm
*AEGLs (Acute Emergency Guideline Levels) & ERPGs (Emergency Response Planning Guideline) are in **bold face**.
DFG MAK: 1000 ppm/1800 mg/m^3 TWA; Peak Limitation Category II(4); Pregnancy Risk Group D.
Australia: asphyxiant, 1993; Austria: MAK 1000 ppm (1800 mg/m^3), 1999; Belgium: asphyxiant, 1993; Denmark: TWA 1000 ppm (1800 mg/m^3), 1999; Finland: TWA 800 ppm (1100 mg/m^3), 1999; Hungary: asphyxiant, 1993; the Philippines: TWA 1000 ppm (1800 mg/m^3), 1993; Switzerland: MAK-W 1000 ppm (1800 mg/m^3), 1999; United Kingdom: asphyxiant, 2000; Argentina, Bulgaria, Columbia, Jordan, South Korea, New Zealand, Singapore, Vietnam: ACGIH TLV®: TWA 2500 ppm.
Determination in Air: By combustible gas meter NIOSH (II-2) Method #S-87; OSHA Analytical Methods PV-2077.

Permissible Concentration in Water: No criteria set, but EPA[32] has suggested a permissible ambient goal of 120,000 µg/L based on health effects.

Routes of Entry: Inhalation, ingestion, skin and/or eye contact.

Harmful Effects and Symptoms

Short Term Exposure: Very high levels may produce the following symptoms, primarily due to lack of oxygen: dizziness, lightheadedness, disorientation, headache, numbness, vomiting, unconsciousness, and death from suffocation. Narcotic at high levels. Contact with the liquid can cause frostbite.

Long Term Exposure: No effects reported.

First Aid: If this chemical gets into the eyes, remove any contact lenses at once and irrigate immediately for at least 15 min, occasionally lifting upper and lower lids. Seek medical attention immediately. If this chemical contacts the skin, remove contaminated clothing and wash immediately with soap and water. Seek medical attention immediately. If this chemical has been inhaled, remove from exposure, begin rescue breathing (using universal precautions, including resuscitation mask) if breathing has stopped and CPR if heart action has stopped. Transfer promptly to a medical facility. When this chemical has been swallowed, get medical attention. Give large quantities of water and induce vomiting. Do not make an unconscious person vomit. If frostbite has occurred, seek medical attention immediately; do NOT rub the affected areas or flush them with water. In order to prevent further tissue damage, do NOT attempt to remove frozen clothing from frostbitten areas. If frostbite has NOT occurred, immediately and thoroughly wash contaminated skin with soap and water.

Personal Protective Methods: Wear protective gloves and clothing to prevent any reasonable probability of skin contact. Safety equipment suppliers/manufacturers can provide recommendations on the most protective glove/clothing material for your operation. Neoprene™, nitrile + PVC, Polyurethane, and polyethylene are among the recommended protective materials. All protective clothing (suits, gloves, footwear, headgear) should be clean, available each day, and put on before work. Contact lenses should not be worn when working with this chemical. Wear gas-proof chemical goggles and face shield unless full face-piece respiratory protection is worn. Employees should wash immediately with soap when skin is wet or contaminated. Provide emergency showers and eyewash.

Respirator Selection: 2100 ppm: Sa (APF = 10) (any supplied-air respirator) or SCBAF (APF = 50) (any self-contained breathing apparatus with a full face-piece). *Emergency or planned entry into unknown concentrations or IDLH conditions:* SCBAF: Pd,Pp (APF = 10,000) (any self-contained breathing apparatus that has a full face-piece and is operated in a pressure-demand or other positive-pressure mode) or SaF: Pd,Pp: ASCBA (APF = 10,000) (any supplied-air respirator that has a full face-piece and is operated in a pressure-demand or other positive-pressure

mode in combination with an auxiliary self-contained breathing apparatus operated in a pressure-demand or other positive-pressure mode). *Escape:* SCBAE (any appropriate escape-type, self-contained breathing apparatus).

Storage: Color Code—Red Stripe: Flammability Hazard: Store separately from all other flammable materials. Prior to working with this chemical you should be trained on its proper handling and storage. Before entering confined space where this chemical may be present, check to make sure that an explosive concentration is not a danger. Propane must be stored to avoid contact with strong oxidizers (such as chlorine, bromine, and fluorine) since violent reactions occur. Sources of ignition, such as smoking and open flames, are prohibited where propane is handled, or stored. Use only nonsparking tools and equipment, especially when opening and closing containers of propane. Wherever propane is used, handled, manufactured, or stored, use explosion-proof electrical equipment and fittings. Procedures for the handling, use, and storage of cylinders should be in compliance with OSHA 1910.101 and 1910.169, as with the recommendations of the Compressed Gas Association.

Shipping: Propane; or Petroleum gases, liquefied; or Liquefied petroleum gas Propane, requires a shipping label of "FLAMMABLE GAS."LD$_{50}$ = (oral-rat). They fall in DOT Hazard Class 2.1.

Spill Handling: Evacuate and restrict persons not wearing protective equipment from area of spill or leak until cleanup is complete. Remove all ignition sources. Establish forced ventilation to keep levels below explosive limit. Stop the flow of gas if it can be done safely. If source of leak is a cylinder and the leak cannot be stopped in place, remove leaking cylinder to a safe place in the open air, and repair leak or allow cylinder to empty. Keep this chemical out of confined space, such as a sewer, because of the possibility of an explosion, unless the sewer is designed to prevent the buildup of explosive concentrations. It may be necessary to contain and dispose of this chemical as a hazardous waste. Contact your local or federal environmental protection agency for specific recommendations. If employees are required to clean up spills, they must be properly trained and equipped. OSHA 1910.120(q) may be applicable.

Fire Extinguishing: This chemical is a flammable gas. Poisonous gases are produced in fire. Do not extinguish the fire unless the flow of gas can be stopped and any remaining gas is out of the line. Specially trained personnel may use fog lines to cool exposures and let the fire burn itself out. Vapors are heavier than air and will collect in low areas. Vapors may travel long distances to ignition sources and flashback. Vapors in confined areas may explode when exposed to fire. Containers may explode in fire. Storage containers and parts of containers may rocket great distances, in many directions. If material or contaminated run-off enters waterways, notify downstream users of potentially contaminated waters. Notify local health and fire officials and pollution control agencies. From a secure, explosion-proof location, use water spray to cool exposed containers. If cooling streams are ineffective (venting sound increases in volume and pitch, tank discolors, or shows any signs of deforming), withdraw immediately to a secure position. If cylinders are exposed to excessive heat from fire or flame contact, withdraw immediately to a secure location. If employees are expected to fight fires, they must be trained and equipped in OSHA 1910.156. The only respirators recommended for firefighting are self-contained breathing apparatuses that have full face-pieces and are operated in a pressure-demand or other positive-pressure mode.

Disposal Method Suggested: Dissolve or mix the material with a combustible solvent and burn in a chemical incinerator equipped with an afterburner and scrubber. All federal, state, and local environmental regulations must be observed.

References

New York State Department of Health. (February 1986). *Chemical Fact Sheet: Propane.* Albany, NY: Bureau of Toxic Substance Assessment

New Jersey Department of Health and Senior Services. (May 2004). *Hazardous Substances Fact Sheet: Propane.* Trenton, NJ

Propane sultone   P:1070

Molecular Formula: $C_3H_6O_3S$

Synonyms: 3-Hydroxy-1-propanesulphonic acid sultone; 3-Hydroxy-1-propanesulphonic acid γ-sultone; 3-Hydroxy-1-propanesulphonic acid sulfone; 1,2-Oxathrolane 2,2-dioxide; 1-propanesulfonic acid-3-hydroxy-g-sultone; 1,3-Propanesultone; 1,2-Oxathiolane 2,2-dioxide; 1-Propanesulfonic acid-3-hydroxy-g-sulfone; Propane sultone

CAS Registry Number: 1120-71-4

RTECS® Number: RP5425000

UN/NA & ERG Number: UN2811 (toxic solid, organic, n.o.s.)/154

EC Number: 214-317-9 [*Annex I Index No.:* 016-032-00-3]

Regulatory Authority and Advisory Bodies

Carcinogenicity: NTP: 11th Report on Carcinogens, 2004: Reasonably anticipated to be a human carcinogen; IARC: Human No Adequate Data, Animal Sufficient Evidence, *possibly carcinogenic to humans,* Group 2B, 1999; NIOSH: Potential occupational carcinogen.

US EPA Gene-Tox Program, Positive: Carcinogenicity—mouse/rat EPA; Positive: Cell transform.—RLV F344 rat embryo EPA; Positive: Cell transform.—SA7/SHE; Host-mediated assay EPA; Positive: *E. coli* polA without S9; Histidine reversion—Ames test EPA; Positive: *S. cerevisiae* gene conversion; *S. cerevisiae*—homozygosis EPA; Positive: *S. pombe*—reversion EPA; Positive/dose response: *In vitro* SCE—nonhuman EPA; Positive/dose response: *In vitro* UDS—human fibroblast EPA; Inconclusive: SHE—clonal assay.

Banned or Severely Restricted (Sweden) (UN).[13]

Very Toxic Substance (World Bank).[15]

Air Pollutant Standard Set. See below, "Permissible Exposure Limits in Air" section.

Clean Air Act: Hazardous Air Pollutants (Title I, Part A, Section 112).

US EPA Hazardous Waste Number (RCRA No.): U193.

RCRA, 40CFR261, Appendix 8 Hazardous Constituents.

Reportable Quantity (RQ): 10 lb (4.54 kg).

EPCRA Section 313 Form R *de minimis* concentration reporting level: 0.1%.

California Proposition 65 Chemical: Cancer 1/1/88.

Canada, WHMIS, Ingredients Disclosure List Concentration 0.1%.

European/International Regulations: Hazard Symbol: T; Risk phrases: R45; R21/22; Safety phrases: S53; S45 (see Appendix 4).

WGK (German Aquatic Hazard Class): 3—Severe hazard to waters.

Description: Propane sultone is a white crystalline solid or a colorless liquid above 30°C. It releases a foul odor as it melts. Molecular weight = 122.15; Specific gravity $(H_2O:1) = 1.39$ at 25°C; Boiling point = 180°C at 30 mm; 155−157°C at 14 mm; Freezing/Melting point = 31°C; Flash point >113°C.[50] Hazard Identification (based on NFPA-704 M Rating System): Health 2, Flammability 1, Reactivity 0. Soluble in water; solubility = 10%.

Potential Exposure: Compound Description: Tumorigen, Mutagen; Reproductive Effector; Primary Irritant. Those involved in use of this chemical intermediate to introduce the sulfopropyl group ($-CH_2CH_2CH_2SO_3-$) into molecules of other products.

Incompatibilities: Strong oxidizers.

Permissible Exposure Limits in Air

OSHA PEL: None.

NIOSH REL: A potential occupational carcinogen. [skin]; Limit exposure to lowest feasible concentration. See *NIOSH Pocket Guide*, Appendix A.

ACGIH TLV®[1]: Exposures by all routes should be carefully controlled to levels as low as possible. Confirmed animal carcinogen with unknown relevance to humans.

Protective Action Criteria (PAC)

TEEL-0: 0.15 mg/m^3

PAC-1: 0.5 mg/m^3

PAC-2: 3.5 mg/m^3

PAC-3: 250 mg/m^3

DFG MAK: [skin] Carcinogen Category 2.

NIOSH IDLH: Not determined. Potential occupational carcinogen.

Austria [skin], carcinogen, 1999; Switzerland: carcinogen, 1999; United Kingdom: carcinogen, 2000; Argentina, Bulgaria, Columbia, Jordan, South Korea, New Zealand, Singapore, Vietnam: ACGIH TLV®: confirmed animal carcinogen with unknown relevance to humans. Several states have set guidelines or standards for propane sultone in ambient air[60] ranging from zero (North Dakota) to 0.03 μg/m^3 (New York) to 3.0 μg/m^3 (Virginia).

Routes of Entry: Inhalation, skin absorption, ingestion, skin and/or eye contact.

Harmful Effects and Symptoms

Short Term Exposure: Irritates the eyes, skin, and respiratory system. More irritating if heated.

Long Term Exposure: Potential occupational carcinogen. Other long-term effects are unknown at this time.

Points of Attack: Eyes, skin, respiratory system. Cancer site in animals: skin, blood or bone marrow, brain or spine.

First Aid: If this chemical gets into the eyes, remove any contact lenses at once and irrigate immediately for at least 15 min, occasionally lifting upper and lower lids. Seek medical attention immediately. If this chemical contacts the skin, remove contaminated clothing and wash immediately with soap and water. Seek medical attention immediately. If this chemical has been inhaled, remove from exposure, begin rescue breathing (using universal precautions, including resuscitation mask) if breathing has stopped and CPR if heart action has stopped. Transfer promptly to a medical facility. When this chemical has been swallowed, get medical attention. Give large quantities of water and induce vomiting. Do not make an unconscious person vomit.

Personal Protective Methods: Wear protective gloves and clothing to prevent any reasonable probability of skin contact. Safety equipment suppliers/manufacturers can provide recommendations on the most protective glove/clothing material for your operation. All protective clothing (suits, gloves, footwear, headgear) should be clean, available each day, and put on before work. Contact lenses should not be worn when working with this chemical. Wear splash-proof chemical goggles and face shield when working with liquid unless full face-piece respiratory protection is worn. Wear dust-proof goggles and face shield when working with powders or dust unless full face-piece respiratory protection is worn. Employees should wash immediately with soap when skin is wet or contaminated. Provide emergency showers and eyewash.

Respirator Selection: *At any detectable concentration:* SCBAF: Pd,Pp (APF = 10,000) (any NIOSH/MSHA- or European Standard EN 149-approved self-contained breathing apparatus that has a full face-piece and is operated in a pressure-demand or other positive-pressure mode) or SaF: Pd,Pp: ASCBA (APF = 10,000) (any supplied-air respirator that has a full face-piece and is operated in a pressure-demand or other positive-pressure mode in combination with an auxiliary, self-contained breathing apparatus operated in a pressure-demand or other positive-pressure mode). *Escape:* GmFOv100 (APF = 50) [any air-purifying, full-face-piece respirator (gas mask) with a chin-style, front- or back-mounted organic vapor canister having an N100, R100, or P100 filter] or SCBAE (any appropriate escape-type, self-contained breathing apparatus).

Storage: Color Code—Blue: Health Hazard/Poison: Store in a secure poison location. Prior to working with this chemical you should be trained on its proper handling and storage. Store in tightly closed containers in a cool,

well-ventilated area away from strong oxidizers. Where possible, automatically transfer material from drums or other storage containers to process containers. Sources of ignition, such as smoking and open flames, are prohibited where this chemical is handled, used, or stored. Metal containers involving the transfer of this chemical should be grounded and bonded. Wherever this chemical is used, handled, manufactured, or stored, use explosion-proof electrical equipment and fittings. A regulated, marked area should be established where this chemical is handled, used, or stored in compliance with OSHA Standard 1910.1045.

Shipping: Toxic solids, organic, n.o.s. require a shipping label of "POISONOUS/TOXIC MATERIALS." They fall in Hazard Class 6.1.

Spill Handling: Evacuate persons not wearing protective equipment from area of spill or leak until cleanup is complete. Remove all ignition sources. Collect powdered material in the most convenient and safe manner and deposit in sealed containers. Ventilate area after cleanup is complete. It may be necessary to contain and dispose of this chemical as a hazardous waste. If material or contaminated runoff enters waterways, notify downstream users of potentially contaminated waters. Contact your local or federal environmental protection agency for specific recommendations. If employees are required to clean up spills, they must be properly trained and equipped. OSHA 1910.120(q) may be applicable.

Fire Extinguishing: This chemical is a combustible solid. Use dry chemical, carbon dioxide, water spray, or alcohol foam extinguishers. Poisonous gases are produced in fire, including sulfur oxides. If material or contaminated runoff enters waterways, notify downstream users of potentially contaminated waters. Notify local health and fire officials and pollution control agencies. From a secure, explosion-proof location, use water spray to cool exposed containers. If cooling streams are ineffective (venting sound increases in volume and pitch, tank discolors, or shows any signs of deforming), withdraw immediately to a secure position. If employees are expected to fight fires, they must be trained and equipped in OSHA 1910.156. The only respirators recommended for firefighting are self-contained breathing apparatuses that have full face-pieces and are operated in a pressure-demand or other positive-pressure mode.

References

Sax, N. I. (Ed.). (1984). *Dangerous Properties of Industrial Materials Report*, 4, No. 3, 82–85

New Jersey Department of Health and Senior Services. (June 2000). *Hazardous Substances Fact Sheet: 1,2-Oxathiolane-2,2-Dioxide.* Trenton, NJ

Propanil   P:1080

Molecular Formula: C$_9$H$_9$Cl$_2$NO
Common Formula: Cl$_2$C$_6$H$_3$NHCOC$_2$H$_5$
Synonyms: AI3-31382; BAY 30130; Chem rice; Crystal propanil-4; DCPA; *N*-(3,4-Dichlorophenyl)propanamide;

3′,4′-Dichlorophenylpropionanilide; 3′,4′-Dichloropropion-anilide; 3,4-Dichloropropionanilide; Dichloropropionani-lide; Dipram; DPA; Farmco propanil; FW-734; Herbax technical; Montrose propanil; NSC31312; Propanamide, *N*-(3,4-dichlorophenyl)-; Propanide; Propionanilide, 3′,4′-dichloro-; Propionic acid 3,4-dichloroanilide; Rogue; Stam; Stam LV10; Stam F-34; Stampede 3E; Stam supernox; Strel; Surpur; Synpran N; Vertac

CAS Registry Number: 709-98-8; *(alt.)* 11096-32-5
RTECS® Number: UE4900000
DOT ID and ERG Number: UN3077/171
EC Number: 211-914-6 [*Annex I Index No.:* 616-009-00-3]
Regulatory Authority and Advisory Bodies
US EPA Gene-Tox Program, Positive: *B. subtilis* rec assay; 1988, Negative: *Aspergillus*—forward mutation; *E. coli* pol A without S9; 1988, Negative: Histidine reversion—Ames test; 1988, Negative: *In vitro* UDS—human fibroblast; TRP reversion; 1988, Negative: *S. cerevisiae*—homozygosis.
US EPA, FIFRA, 1998 Status of Pesticides: Supported.
Air Pollutant Standard Set. See below, "Permissible Exposure Limits in Air" section.
Safe Drinking Water Act: Priority List (55 FR 1470) as DCPA (and its acid metabolites).
EPCRA Section 313 Form R *de minimis* concentration reporting level: 1.0%.
European/International Regulations: Hazard Symbol: Xn, N; Risk phrases: R22; R50; Safety phrases: S2; S22; S61 (see Appendix 4).
WGK (German Aquatic Hazard Class): 3—Severe hazard to waters.
Description: Propanil is a colorless solid. The technical product is a brown crystalline solid. Molecular weight = 218.09; Specific gravity (H_2O:1) = 1.22 at 25°C; Freezing/Melting point = 92−93°C (pure); 88−91°C (technical grade); Freezing/Melting point (pure) = 89−92°C (pure); 85−89°C (technical grade). Hazard Identification (based on NFPA-704 M Rating System): Health 2, Flammability 0, Reactivity 0. Insoluble in water. Commercial formulations use carrier solvents that may change the physical properties shown.
Potential Exposure: Compound Description: Agricultural Chemical; Mutagen. Propanil is used as a postemergent herbicide for rice and spring wheat. Those involved in the manufacture, formulation, and application of this contact herbicide.
Permissible Exposure Limits in Air
No TEEL available.
Russia[43] set a MAC of 0.1 mg/m³ in work-place air and has set a MAC value of 0.005 mg/m³ for ambient air in residential areas on a once-daily basis and 0.001 mg/m³ on a daily average basis.
Permissible Concentration in Water: Russia[43] set a MAC of 0.1 mg/L in water bodies used for domestic purposes.
Determination in Water: Fish Tox = 0.48990000 ppb (EXTRA HIGH). Octanol−water coefficient: Log K_{ow} = 3.1.

Routes of Entry: Inhalation, ingestion, skin and/or eye contact.
Harmful Effects and Symptoms
Short Term Exposure: Propanil is well tolerated by experimental animal on a chronic basis, and there is little or no indication of mutagenic or oncogenic properties of the compound. The highest no-adverse-effect concentration of propanil based on reproduction in the rat and acute, subchronic, and chronic studies in rats and dogs is 400 ppm in the diet. Based on this data, an ADI was calculated at 0.02 mg/kg/day. LD50 (oral-rat) 2756 mg/kg (male); 2343 mg/kg (female).
Long Term Exposure: Human Tox = 35.00000 ppb (INTERMEDIATE).
First Aid: If this chemical gets into the eyes, remove any contact lenses at once and irrigate immediately for at least 15 min, occasionally lifting upper and lower lids. Seek medical attention immediately. If this chemical contacts the skin, remove contaminated clothing and wash immediately with soap and water. Seek medical attention immediately. If this chemical has been inhaled, remove from exposure, begin rescue breathing (using universal precautions, including resuscitation mask) if breathing has stopped and CPR if heart action has stopped. Transfer promptly to a medical facility. When this chemical has been swallowed, get medical attention. Give large quantities of water and induce vomiting. Do not make an unconscious person vomit.
Personal Protective Methods: Wear protective gloves and clothing to prevent any reasonable probability of skin contact. Safety equipment suppliers/manufacturers can provide recommendations on the most protective glove/clothing material for your operation. All protective clothing (suits, gloves, footwear, headgear) should be clean, available each day, and put on before work. Contact lenses should not be worn when working with this chemical. Wear dust-proof chemical goggles and face shield unless full face-piece respiratory protection is worn. Employees should wash immediately with soap when skin is wet or contaminated. Provide emergency showers and eyewash.
Respirator Selection: Follow regulations in OSHA 29CFR1910.134 or European Standard EN149. Use a NIOSH/MSHA- or European Standard EN149-approved respirator; or use an approved supplied-air respirator with a full face-piece operated in the positive-pressure mode, or with a full face-piece, hood, or helmet in the continuous-flow mode; or use a NIOSH/MSHA- or European Standard EN149-approved self-contained breathing apparatus with a full face-piece operated in a pressure-demand or other positive-pressure mode.
Storage: Color Code—Green: General storage may be used. Prior to working with this chemical you should be trained on its proper handling and storage. Store in tightly closed containers in a cool, well-ventilated area.
Shipping: This chemical requires a shipping label of "CLASS 9." It falls in Hazard Class 9 and Packing Group III.[20, 21]

Spill Handling: Evacuate persons not wearing protective equipment from area of spill or leak until cleanup is complete. Remove all ignition sources. Collect powdered material in the most convenient and safe manner and deposit in sealed containers. Ventilate area after cleanup is complete. It may be necessary to contain and dispose of this chemical as a hazardous waste. If material or contaminated runoff enters waterways, notify downstream users of potentially contaminated waters. Contact your local or federal environmental protection agency for specific recommendations. If employees are required to clean up spills, they must be properly trained and equipped. OSHA 1910.120(q) may be applicable. Soil Adsorption Index (K_{oc}) = 149.

Fire Extinguishing: This chemical is a combustible solid. Use dry chemical, carbon dioxide, water spray, or alcohol foam extinguishers. Poisonous gases are produced in fire. If material or contaminated runoff enters waterways, notify downstream users of potentially contaminated waters. Notify local health and fire officials and pollution control agencies. From a secure, explosion-proof location, use water spray to cool exposed containers. If cooling streams are ineffective (venting sound increases in volume and pitch, tank discolors, or shows any signs of deforming), withdraw immediately to a secure position. If employees are expected to fight fires, they must be trained and equipped in OSHA 1910.156. The only respirators recommended for firefighting are self-contained breathing apparatuses that have full face-pieces and are operated in a pressure-demand or other positive-pressure mode.

Disposal Method Suggested: Hydrolysis in acidic or basic media yields the more toxic substance, 3,4-dichloraniline, and is not recommended.

References

US Environmental Protection Agency, Special Review and Reregistration Division Office of Pesticide Programs. (1998). *Agency Status of Pesticides in Registration, Reregistration, and Special Review* (Rainbow Report). Washington, DC

Propargyl alcohol   P:1090

Molecular Formula: C_3H_4O
Common Formula: $HC{\equiv}CCH_2OH$
Synonyms: AI3-24359; Alcohol propargilico (Spanish); Ethynylcarbinol; Ethynyl methanol; 1-Hydroxy-2-propyne; 3-Hydroxy-1-propyne; Methanol, ethynyl-; Propiolic alcohol; 1-Propyne-3-ol; 3-Propynol; 2-Propynol; 2-Propyn-1-ol; 1-Propyn-3-ol; Prop-2-yn-1-ol; 2-Propynyl alcohol; Propynyl alcohol
CAS Registry Number: 107-19-7
RTECS® Number: UK5075000
UN/NA & ERG Number: UN2929/131
EC Number: 203-471-2 [*Annex I Index No.:* 603-078-00-X]

Regulatory Authority and Advisory Bodies

Air Pollutant Standard Set. See below, "Permissible Exposure Limits in Air" section.
US EPA Hazardous Waste Number (RCRA No.): P102.
RCRA, 40CFR261, Appendix 8 Hazardous Constituents.
Reportable Quantity (RQ): 1000 lb (454 kg).
EPCRA Section 313 Form R *de minimis* concentration reporting level: 1.0%.
Canada, WHMIS, Ingredients Disclosure List Concentration 1.0%.
European/International Regulations: Hazard Symbol: T, N; Risk phrases: R10; R23/24/25; R34; R51/53; Safety phrases: S1/2; S26; S28; S36; S45; S61 (see Appendix 4).
WGK (German Aquatic Hazard Class): 2—Hazard to waters.

Description: Propargyl alcohol is a colorless liquid with a geranium-like odor. Molecular weight = 56.07; Specific gravity (H_2O:1) = 0.97 at 25°C; Boiling point = 119.9°C; Freezing/Melting point = −52°C; Flash point = 36°C. Explosive limits: LEL = 3.4%; UEL = 70%. Hazard Identification (based on NFPA-704 M Rating System): Health 4, Flammability 3, Reactivity 3. Soluble in water.

Potential Exposure: Compound Description: Agricultural Chemical; Tumorigen, Mutagen. Propargyl alcohol is used as a corrosion inhibitor, soil fumigant, solvent, stabilizer, and chemical intermediate.

Incompatibilities: Violent reaction with phosphorus pentoxide, oxidizers. May polymerize under the influence of heat, oxidizers, peroxides, light. Attacks many plastics.

Permissible Exposure Limits in Air

Conversion factor: 1 ppm = 2.29 mg/m^3 at 25°C & 1 atm.
OSHA PEL: None.
NIOSH REL: 1 ppm/2 mg/m^3 TWA [skin].
ACGIH TLV®[1]: 1 ppm/2.3 mg/m^3 TWA [skin].
Protective Action Criteria (PAC)*
TEEL-0: 1 ppm
PAC-1: **2.5** ppm
PAC-2: **16** ppm
PAC-3: **74** ppm
*AEGLs (Acute Emergency Guideline Levels) & ERPGs (Emergency Response Planning Guideline) are in **bold face**.
DFG MAK: 2 ppm; 4.7 mg/m^3 TWA; Peak Limitation Category I(2); [skin]; Pregnancy Risk Group D.
Australia: TWA 1 ppm (2 mg/m^3), [skin], 1993; Austria: MAK 2 ppm (9 mg/m^3), [skin], 1999; Belgium: TWA 1 ppm (2.3 mg/m^3), [skin], 1993; Denmark: TWA 1 ppm (2.5 mg/m^3), [skin], 1999; Finland: TWA 1 ppm (2 mg/m^3); STEL 3 ppm (6 mg/m^3), [skin], 1999; France: VME 1 ppm (2 mg/m^3), [skin], 1999; the Netherlands: MAC-TGG 2 mg/m^3, [skin], 2003; the Philippines: TWA 1 ppm (1 mg/m^3), [skin], 1993; Russia: STEL 1 mg/m^3, 1993; Switzerland: MAK-W 1 ppm (2 mg/m^3), [skin], 1999; United Kingdom: TWA 1 ppm (2.3 mg/m^3); STEL 3 ppm, [skin], 2000: Argentina, Bulgaria, Columbia, Jordan, South Korea, New Zealand, Singapore, Vietnam: ACGIH TLV®: TWA 1 ppm [skin].

Determination in Air: Use OSHA Analytical Method 97.

Routes of Entry: Inhalation, skin absorption, ingestion, skin and/or eye contact.

Harmful Effects and Symptoms

Short Term Exposure: Propargyl alcohol can affect you when breathed and by passing through skin. Irritates the eyes, skin, and respiratory tract. Contact can severely burn the eyes, causing permanent damage. Skin contact can irritate the skin and allow dangerous amounts to enter the body. Affects the central nervous system. Exposure can cause you to feel dizzy, lightheaded, and to have trouble concentrating. High exposures can cause liver and kidney damage, coma, and death.

Long Term Exposure: Propargyl alcohol may damage the liver and kidneys.

Points of Attack: Skin, respiratory system, central nervous system, liver, kidneys.

Medical Surveillance: If symptoms develop or overexposure is suspected, the following may be useful: liver and kidney function tests.

First Aid: If this chemical gets into the eyes, remove any contact lenses at once and irrigate immediately for at least 15 min, occasionally lifting upper and lower lids. Seek medical attention immediately. If this chemical contacts the skin, remove contaminated clothing and wash immediately with soap and water. Seek medical attention immediately. If this chemical has been inhaled, remove from exposure, begin rescue breathing (using universal precautions, including resuscitation mask) if breathing has stopped and CPR if heart action has stopped. Transfer promptly to a medical facility. When this chemical has been swallowed, get medical attention. Give large quantities of water and induce vomiting. Do not make an unconscious person vomit.

Personal Protective Methods: Wear protective gloves and clothing to prevent any reasonable probability of skin contact. Safety equipment suppliers/manufacturers can provide recommendations on the most protective glove/clothing material for your operation. All protective clothing (suits, gloves, footwear, headgear) should be clean, available each day, and put on before work. Contact lenses should not be worn when working with this chemical. Wear splash-proof chemical goggles and face shield unless full face-piece respiratory protection is worn. Employees should wash immediately with soap when skin is wet or contaminated. Provide emergency showers and eyewash.

Respirator Selection: Follow regulations in OSHA 29CFR1910.134 or European Standard EN149. Use a NIOSH/MSHA- or European Standard EN149-approved respirator; or use an approved supplied-air respirator with a full face-piece operated in the positive-pressure mode, or with a full face-piece, hood, or helmet in the continuous-flow mode; or use a NIOSH/MSHA- or European Standard EN149-approved self-contained breathing apparatus with a full face-piece operated in a pressure-demand or other positive-pressure mode.

Storage: (1) Color Code—Red: Flammability Hazard: Store in a flammable liquid storage area or approved cabinet away from ignition sources and corrosive and reactive materials. (2) Color Code—Blue: Health Hazard/Poison: Store in a secure poison location. Prior to working with this chemical you should be trained on its proper handling and storage. Before entering confined space where this chemical may be present, check to make sure that an explosive concentration is not a danger. Propargyl alcohol must be stored to avoid contact with alkalis, mercury(II) sulfate; oxidizing materials and phosphonic anhydride since violent reactions occur. Sources of ignition, such as smoking and open flames, are prohibited where propargyl alcohol is used, handled, or stored in a manner that could create a potential fire or explosion hazard. Metal containers involving the transfer of 5 gallons or more of propargyl alcohol should be grounded and bonded. Drums must be equipped with self-closing valves, pressure vacuum bungs, and flame arresters. Use only nonsparking tools and equipment, especially when opening and closing containers of Propargyl alcohol. Wherever propargyl alcohol is used, handled, manufactured, or stored, use explosion-proof electrical equipment and fittings.

Shipping: This compound requires a shipping label of "POISONOUS/TOXIC MATERIALS, FLAMMABLE LIQUID." It falls in Hazard Class 6.1 and Packing Group II.

Spill Handling: Evacuate and restrict persons not wearing protective equipment from area of spill or leak until cleanup is complete. Remove all ignition sources. Establish forced ventilation to keep levels below explosive limit. Absorb liquids in vermiculite, dry sand, earth, peat, carbon, or a similar material and deposit in sealed containers. Keep this chemical out of a confined space, such as a sewer, because of the possibility of an explosion, unless the sewer is designed to prevent the buildup of explosive concentrations. It may be necessary to contain and dispose of this chemical as a hazardous waste. If material or contaminated runoff enters waterways, notify downstream users of potentially contaminated waters. Contact your local or federal environmental protection agency for specific recommendations. If employees are required to clean up spills, they must be properly trained and equipped. OSHA 1910.120(q) may be applicable.

Fire Extinguishing: This chemical is a flammable liquid. Poisonous gases, including nitrogen oxides and chlorine, are produced in fire. Use dry chemical, carbon dioxide, or alcohol foam extinguishers. Vapors are heavier than air and will collect in low areas. Vapors may travel long distances to ignition sources and flashback. Vapors in confined areas may explode when exposed to fire. Containers may explode in fire. Storage containers and parts of containers may rocket great distances, in many directions. If material or contaminated runoff enters waterways, notify downstream users of potentially contaminated waters. Notify local health and fire officials and pollution control agencies. From a secure, explosion-proof location, use water spray to cool exposed containers. If cooling streams are ineffective (venting sound increases in volume and pitch, tank discolors, or shows any signs of deforming), withdraw immediately to a secure position. If employees are expected to fight fires, they must be trained and equipped in OSHA 1910.156. The only respirators recommended for firefighting are self-contained breathing apparatuses that have full face-pieces and are operated in a pressure-demand or other positive-pressure mode.

Disposal Method Suggested: Dissolve or mix the material with a combustible solvent and burn in a chemical incinerator equipped with an afterburner and scrubber. All federal, state, and local environmental regulations must be observed. Consult with environmental regulatory agencies for guidance on acceptable disposal practices. Generators of waste containing this contaminant ($\geq$100 kg/mo) must conform with EPA regulations governing storage, transportation, treatment, and waste disposal.

Reference

New Jersey Department of Health and Senior Services. (November 2004). *Hazardous Substances Fact Sheet: Propargyl Alcohol.* Trenton, NJ

Propargyl bromide   P:1100

Molecular Formula: C_3H_3Br
Common Formula: $BrCH_2C\equiv CH$
Synonyms: γ-Bromoallylene; 3-Bromopropyne; 3-Bromo-1-propyne; Bromuro de propargilo (Spanish)
CAS Registry Number: 106-96-7
RTECS® Number: UK4375000
UN/NA & ERG Number: UN2345/130
EC Number: 203-447-1
Regulatory Authority and Advisory Bodies
Superfund/EPCRA 40CFR355, Extremely Hazardous Substances: TPQ = 10 lb (4.54 kg).
Reportable Quantity (RQ): 10 lb (4.54 kg).
European/International Regulations: not listed in Annex 1.
WGK (German Aquatic Hazard Class): 3—Severe hazard to waters.
Description: Propargyl bromide is a colorless liquid with a sharp odor. Molecular weight = 118.97; Boiling point = 89–90°C; Freezing/Melting point = −61.07°C; Flash point = 10°C; Autoignition temperature = 324°C. Explosive limits: LEL = 3.0%; UEL—unknown. Hazard Identification (based on NFPA-704 M Rating System): Health 3, Flammability 3, Reactivity 4. Insoluble in water.
Potential Exposure: This material is used as a soil fumigant. Not registered as a pesticide in the United States.
Incompatibilities: Violent reaction with oxidizers. Becomes shock- or heat-sensitive when mixed with trichloronitromethane or chloropicrin. Detonates when heated to 220°C,

or when heated while confined. May explode on contact with copper, copper alloys, mercury, silver.
Permissible Exposure Limits in Air
AIHA WEEL: 0.1 ppm TWA [skin]
Protective Action Criteria (PAC)
TEEL-0: 0.03 mg/m^3
PAC-1: 0.03 mg/m^3
PAC-2: 0.03 mg/m^3
PAC-3: 20 mg/m^3
Routes of Entry: Inhalation, ingestion, skin and/or eye contact.
Harmful Effects and Symptoms
Short Term Exposure: This material is very poisonous if swallowed. If inhaled, may be harmful. Skin and eye contact may cause burns. Symptoms of exposure include skin irritation and tearing of the eyes.
First Aid: If this chemical gets into the eyes, remove any contact lenses at once and irrigate immediately for at least 15 min, occasionally lifting upper and lower lids. Seek medical attention immediately. If this chemical contacts the skin, remove contaminated clothing and wash immediately with soap and water. Seek medical attention immediately. If this chemical has been inhaled, remove from exposure, begin rescue breathing (using universal precautions, including resuscitation mask) if breathing has stopped and CPR if heart action has stopped. Transfer promptly to a medical facility. When this chemical has been swallowed, get medical attention. Give large quantities of water and induce vomiting. Do not make an unconscious person vomit. Keep victim quiet and maintain normal body temperature.
Personal Protective Methods: Wear protective gloves and clothing to prevent any reasonable probability of skin contact. Safety equipment suppliers/manufacturers can provide recommendations on the most protective glove/clothing material for your operation. All protective clothing (suits, gloves, footwear, headgear) should be clean, available each day, and put on before work. Contact lenses should not be worn when working with this chemical. Wear splash-proof chemical goggles and face shield unless full face-piece respiratory protection is worn. Employees should wash immediately with soap when skin is wet or contaminated. Provide emergency showers and eyewash.
Respirator Selection: Follow regulations in OSHA 29CFR1910.134 or European Standard EN149. Use a NIOSH/MSHA- or European Standard EN149-approved respirator; or use an approved supplied-air respirator with a full face-piece operated in the positive-pressure mode, or with a full face-piece, hood, or helmet in the continuous-flow mode; or use a NIOSH/MSHA- or European Standard EN149-approved self-contained breathing apparatus with a full face-piece operated in a pressure-demand or other positive-pressure mode.
Storage: Color Code—Red: Flammability Hazard: Store in a flammable liquid storage area or approved cabinet away from ignition sources and corrosive and reactive materials. Prior to working with this chemical you should be trained

on its proper handling and storage. Before entering confined space where this chemical may be present, check to make sure that an explosive concentration is not a danger. Store in tightly closed containers in a cool, well-ventilated area away from oxidizers and heat. Where possible, automatically pump liquid from drums or other storage containers to process containers. Sources of ignition, such as smoking and open flames, are prohibited where this chemical is handled, used, or stored. Metal containers involving the transfer of 5 gallons or more of this chemical should be grounded and bonded. Drums must be equipped with self-closing valves, pressure vacuum bungs, and flame arresters. Use only nonsparking tools and equipment, especially when opening and closing containers of this chemical. Wherever this chemical is used, handled, manufactured, or stored, use explosion-proof electrical equipment and fittings.

Shipping: This compound requires a shipping label of "FLAMMABLE LIQUID." It falls in Hazard Class 3 and Packing Group II.

Spill Handling: Evacuate and restrict persons not wearing protective equipment from area of spill or leak until cleanup is complete. Remove all ignition sources. Establish forced ventilation to keep levels below explosive limit. Absorb liquids in vermiculite, dry sand, earth, peat, carbon, or a similar material and deposit in sealed containers. Keep this chemical out of a confined space, such as a sewer, because of the possibility of an explosion, unless the sewer is designed to prevent the buildup of explosive concentrations. It may be necessary to contain and dispose of this chemical as a hazardous waste. If material or contaminated runoff enters waterways, notify downstream users of potentially contaminated waters. Contact your local or federal environmental protection agency for specific recommendations. If employees are required to clean up spills, they must be properly trained and equipped. OSHA 1910.120(q) may be applicable.

Fire Extinguishing: This chemical is a flammable liquid. Poisonous gases, including bromine, are produced in fire. Wear self-contained breathing apparatus and full protective clothing. Move container from fire area if you can do it without risk. Use dry chemical, carbon dioxide, or foam extinguishers. Vapors are heavier than air and will collect in low areas. Vapors may travel long distances to ignition sources and flashback. Vapors in confined areas may explode when exposed to fire. Containers may explode in fire. Storage containers and parts of containers may rocket great distances, in many directions. If material or contaminated runoff enters waterways, notify downstream users of potentially contaminated waters. Notify local health and fire officials and pollution control agencies. From a secure, explosion-proof location, use water spray to cool exposed containers. Do not get water inside container. If cooling streams are ineffective (venting sound increases in volume and pitch, tank discolors, or shows any signs of deforming), withdraw immediately to a secure position. If employees are expected to fight fires, they must be trained and equipped in OSHA 1910.156. The only respirators recommended for firefighting are self-contained breathing apparatuses that have full face-pieces and are operated in a pressure-demand or other positive-pressure mode.

Reference
US Environmental Protection Agency. (November 30, 1987). *Chemical Hazard Information Profile: Propargyl Bromide*. Washington, DC: Chemical Emergency Preparedness Program

Propazine   P:1110

Molecular Formula: $C_9H_{16}ClN_5$
Synonyms: 2,4-bis(Isopropylamino)-6-chloro-*s*-triazine; 2,4-Bis(propylamino)-6-chlor-1,3,5-triazin (German); 2-Chloro-4,6-bis(isopropylamino)-*s*-triazine; Gesamil®; MAXX-90; Milogard®; Plantulin; Primatol P; Propasin; Propazin; Prozinex
CAS Registry Number: 139-40-2
RTECS® Number: XY5300000
UN/NA & ERG Number: UN2753/151
EC Number: 205-359-9 [*Annex I Index No.:* 613-067-00-1]
Regulatory Authority and Advisory Bodies
US EPA, FIFRA 1998 Status of Pesticides: Canceled.
Air Pollutant Standard Set. See below, "Permissible Exposure Limits in Air" section.
European/International Regulations: Hazard Symbol: Xn, N; Risk phrases: R40; R50/53; Safety phrases: S2; S36/37; S60; S61 (see Appendix 4).
WGK (German Aquatic Hazard Class): No value assigned.
Description: Propazine is a colorless crystalline solid. Molecular weight = 229.75; Freezing/Melting point = 214°C; Vapor pressure = 1.3×10^{-7} mmHg at 20°C. Slightly soluble in water.
Potential Exposure: Compound Description: Agricultural Chemical; Tumorigen; Reproductive Effector. Those involved in the manufacture, formulation, and application of this preemergence selective herbicide used to control annual broadleaf weeds and grasses.
Permissible Exposure Limits in Air
Russia[43] set a MAC of 5.0 mg/m³ in work-place air and a MAC of 0.04 mg/m³ in ambient air in residential areas both on a momentary and a daily average basis.
Permissible Concentration in Water: Russia[43] set a MAC of 1.0 mg/L in water bodies used for domestic purposes. The US EPA has determined a lifetime health advisory of 0.014 mg/L (14 µg/L) (see "References" below). States which have set guidelines for propazine in drinking water[61] include Kansas at 325 µg/L and Maine at 93 µg/L.
Determination in Water: Analysis of propazine is by a gas chromatographic (GC) method applicable to the determination of certain nitrogen- and phosphorus-containing pesticides in water samples. In this method, approximately 1 L of sample is extracted with methylene chloride. The extract

is concentrated and the compounds are separated using capillary column GC. Measurement is made using a nitrogen—phosphorus detector. The method detection limit has not been determined for propazine, but it is estimated that the detection limits for analytes included in this method are in the range of 0.1−2 μg/L. Fish Tox = 938.12580000 ppb (VERY LOW). Octanol−water coefficient: Log K_{ow} = 2.9.

Routes of Entry: Inhalation, ingestion, skin and/or eye contact. May be absorbed by the skin.

Harmful Effects and Symptoms

Short Term Exposure: May cause eye irritation. Contact dermatitis was reported in workers involved in propazine manufacturing. Poisonous if ingested. No other information on the health effects of propazine in humans was found in the available literature.

Long Term Exposure: May cause skin allergy. Human Tox = 10.00000 ppb (INTERMEDIATE).

Points of Attack: Skin.

Medical Surveillance: Examination by a qualified allergist.

First Aid: If this chemical gets into the eyes, remove any contact lenses at once and irrigate immediately for at least 15 min, occasionally lifting upper and lower lids. Seek medical attention immediately. If this chemical contacts the skin, remove contaminated clothing and wash immediately with soap and water. Seek medical attention immediately. If this chemical has been inhaled, remove from exposure, begin rescue breathing (using universal precautions, including resuscitation mask) if breathing has stopped and CPR if heart action has stopped. Transfer promptly to a medical facility. When this chemical has been swallowed, get medical attention. Give large quantities of water and induce vomiting. Do not make an unconscious person vomit.

Personal Protective Methods: Wear protective gloves and clothing to prevent any reasonable probability of skin contact. Safety equipment suppliers/manufacturers can provide recommendations on the most protective glove/clothing material for your operation. All protective clothing (suits, gloves, footwear, headgear) should be clean, available each day, and put on before work. Contact lenses should not be worn when working with this chemical. Wear dust-proof chemical goggles and face shield unless full face-piece respiratory protection is worn. Employees should wash immediately with soap when skin is wet or contaminated. Provide emergency showers and eyewash.

Respirator Selection: Follow regulations in OSHA 29CFR1910.134 or European Standard EN149. Use a NIOSH/MSHA- or European Standard EN149-approved respirator; or use an approved supplied-air respirator with a full face-piece operated in the positive-pressure mode, or with a full face-piece, hood, or helmet in the continuous-flow mode; or use a NIOSH/MSHA- or European Standard EN149-approved self-contained breathing apparatus with a full face-piece operated in a pressure-demand or other positive-pressure mode.

Storage: Color Code—Blue: Health Hazard/Poison: Store in a secure poison location. Prior to working with this chemical you should be trained on its proper handling and storage.

Shipping: Triazine pesticides, solid, toxic, require a shipping label of "POISONOUS/TOXIC MATERIALS." The Hazard Class is 6.1 and Packing Group is III.

Spill Handling: Evacuate persons not wearing protective equipment from area of spill or leak until cleanup is complete. Remove all ignition sources. Collect powdered material in the most convenient and safe manner and deposit in sealed containers. Ventilate area after cleanup is complete. It may be necessary to contain and dispose of this chemical as a hazardous waste. If material or contaminated runoff enters waterways, notify downstream users of potentially contaminated waters. Contact your local or federal environmental protection agency for specific recommendations. If employees are required to clean up spills, they must be properly trained and equipped. OSHA 1910.120(q) may be applicable. Soil Adsorption Index (K_{oc}) = 154.

Fire Extinguishing: This chemical is a combustible solid. Use dry chemical, carbon dioxide, water spray, or alcohol foam extinguishers. Poisonous gases are produced in fire, including nitrogen oxides and chlorine. If material or contaminated runoff enters waterways, notify downstream users of potentially contaminated waters. Notify local health and fire officials and pollution control agencies. From a secure, explosion-proof location, use water spray to cool exposed containers. If cooling streams are ineffective (venting sound increases in volume and pitch, tank discolors, or shows any signs of deforming), withdraw immediately to a secure position. If employees are expected to fight fires, they must be trained and equipped in OSHA 1910.156. The only respirators recommended for firefighting are self-contained breathing apparatuses that have full face-pieces and are operated in a pressure-demand or other positive-pressure mode.

Reference

US Environmental Protection Agency. (August 1987). *Health Advisory: Propazine.* Washington, DC: Office of Drinking Water

Propham   P:1120

Molecular Formula: $C_{10}H_{13}NO_2$
Common Formula: $C_6H_5NHCOOCH(CH_3)_2$
Synonyms: Ban-hoe; Beet-kleen; Carbanilic acid, isopropyl ester; Chem-hoe; IPPC; Isopropyl carbanilate; Isopropyl carbanilic acid ester; Isopropyl-N-phenyl-carbamat (German); o-Isopropyl-N-phenyl carbamate; Isopropyl phenylcarbamate; Isopropyl-N-phenyl carbamate; Isopropyl-N-phenyurethan (German) Ortho grass killer; N-Phenylcarbamate d'isopropyle (French); Phenylcarbamic acid 1-methylethyl ester; N-Phenyl isopropyl carbamate; Premalox; Profam; Propham; Triherbide; Triherbide-IPC; Tuberit; Tuberite
CAS Registry Number: 122-42-9

RTECS® Number: ED9100000
UN/NA & ERG Number: UN2757/151
EC Number: 204-542-0

Regulatory Authority and Advisory Bodies
Carcinogenicity: IARC: Animal Inadequate Evidence; Human No Available Data, *not classifiable as carcinogenic to humans*, Group 3, 1987.
Air Pollutant Standard Set. See below, "Permissible Exposure Limits in Air" section.
US EPA Hazardous Waste Number (RCRA No.): U363.
Superfund/EPCRA [40CFR 302 and 355, F R: 8/16/06, Vol 71, No. 158] Reportable Quantity (RQ): 1000 lb (454 kg).
RCRA, 40CFR261, Appendix 8 Hazardous Constituents. Reportable Quantity (RQ): 1 lb (0.454 kg).
European/International Regulations: not listed in Annex 1.
WGK (German Aquatic Hazard Class): No value assigned.

Description: Propham is a colorless crystalline solid. Molecular weight = 179.24; Freezing/Melting point = 87−88°C; 84°C (technical grade). Decomposition temperature = 150°C. Insoluble in water.

Potential Exposure: Those involved in the manufacture, formulation, and application of this grass-control herbicide.

Permissible Exposure Limits in Air: Russia[43] set a MAC of 2.0 mg/m^3 in work-place air and a MAC of 0.02 mg/m^3 in ambient air of residential areas both on a momentary and a daily average basis.

Permissible Concentration in Water: Russia[43] set a MAC of 0.2 mg/L in water bodies used for domestic purposes. The EPA has set a lifetime health advisory of 0.12 mg/L.

Determination in Water: Analysis of propham is by a high-performance liquid chromatographic (HPLC) method applicable to the determination of certain carbamate and urea pesticides in water samples. This method requires a solvent extraction of approximately 1 L of sample with methylene chloride using a separatory funnel. The methylene chloride extract dried and concentrated to a volume of 10 mL or less. Compounds are separated by HPLC, and measurement is conducted with a UV detector. The method detection limit has not been determined for propham, but it is estimated that the detection limits for analytes included in this method are in the range of 1−5 μg/L. Fish Tox = 5112.55273000 ppb (VERY LOW).

Routes of Entry: Inhalation, ingestion, skin and/or eye contact.

Harmful Effects and Symptoms
Short Term Exposure: Doses of 2000 mg/kg to rats produced loss of righting reflex, ptosis, piloerection, decreased locomotor activity, chronic pulmonary disease, rugation and irregular thickening of the stomach. The acute oral LD$_{50}$ values in male and female rats were reported to be 3000 ± 232 mg/kg and 2360 ± 118 mg/kg, respectively. Carbamates are cholinesterase inhibitors. Symptoms of exposure include headache, giddiness, blurred vision, nervousness, weakness, nausea, cramps, diarrhea, and discomfort in the chest. Signs include sweating, tearing, salivation, vomiting, cyanosis, convulsions, coma, loss of reflexes, and loss of sphincter control.

Long Term Exposure: Cholinesterase inhibitor; cumulative effect is possible. This chemical may damage the nervous system with repeated exposure, resulting in convulsions, respiratory failure. May cause liver damage. Human Tox = 100.00000 ppb (VERY LOW).

Points of Attack: Respiratory system, lungs, central nervous system, cardiovascular system, skin, eyes, plasma and red blood cell cholinesterase.

Medical Surveillance: Before employment and at regular times after that, the following are recommended: plasma and red blood cell cholinesterase levels (tests for the enzyme poisoned by this chemical). If exposure stops, plasma levels return to normal in 1−2 weeks while red blood cell levels may be reduced for 1−3 months.
When cholinesterase enzyme levels are reduced by 25% or more below preemployment levels, risk of poisoning is increased, even if results are in lower ranges of "normal." Reassignment to work not involving carbamate pesticides is recommended until enzyme levels recover. If symptoms develop or overexposure occurs, repeat the above tests as soon as possible and get an examination of the nervous system. Also, consider complete blood count. Consider chest X-ray following acute overexposure.

First Aid: If this chemical gets into the eyes, remove any contact lenses at once and irrigate immediately for at least 15 min, occasionally lifting upper and lower lids. Seek medical attention immediately. If this chemical contacts the skin, remove contaminated clothing and wash immediately with soap and water. Speed in removing material from skin is of extreme importance. Shampoo hair promptly if contaminated. Seek medical attention immediately. If this chemical has been inhaled, remove from exposure, begin rescue breathing (using universal precautions, including resuscitation mask) if breathing has stopped and CPR if heart action has stopped. Transfer promptly to a medical facility. When this chemical has been swallowed, get medical attention. Give large quantities of water and induce vomiting. Do not make an unconscious person vomit.

Personal Protective Methods: Wear protective gloves and clothing to prevent any reasonable probability of skin contact. Safety equipment suppliers/manufacturers can provide recommendations on the most protective glove/clothing material for your operation. All protective clothing (suits, gloves, footwear, headgear) should be clean, available each day, and put on before work. Contact lenses should not be worn when working with this chemical. Wear dust-proof chemical goggles and face shield unless full face-piece respiratory protection is worn. Employees should wash immediately with soap when skin is wet or contaminated. Provide emergency showers and eyewash.

Respirator Selection: Follow regulations in OSHA 29CFR1910.134 or European Standard EN149. Use a NIOSH/MSHA- or European Standard EN149-approved respirator; or use an approved supplied-air respirator with a

full face-piece operated in the positive-pressure mode, or with a full face-piece, hood, or helmet in the continuous-flow mode; or use a NIOSH/MSHA- or European Standard EN149-approved self-contained breathing apparatus with a full face-piece operated in a pressure-demand or other positive-pressure mode.

Storage: Color Code—Blue: Health Hazard/Poison: Store in a secure poison location. Prior to working with this chemical you should be trained on its proper handling and storage. Store in tightly closed containers in a cool, well-ventilated area away from strong oxidizers. Where possible, automatically transfer material from drums or other storage containers to process containers. Sources of ignition, such as smoking and open flames, are prohibited where this chemical is handled, used, or stored. Metal containers involving the transfer of this chemical should be grounded and bonded. Wherever this chemical is used, handled, manufactured, or stored, use explosion-proof electrical equipment and fittings.

Shipping: Carbamate pesticides, solid, toxic, requires a shipping label of "POISONOUS/TOXIC MATERIALS." It falls in Hazard Class 6.1 and Packing Group III.

Spill Handling: Evacuate persons not wearing protective equipment from area of spill or leak until cleanup is complete. Remove all ignition sources. Collect powdered material in the most convenient and safe manner and deposit in sealed containers. Ventilate area after cleanup is complete. It may be necessary to contain and dispose of this chemical as a hazardous waste. If material or contaminated runoff enters waterways, notify downstream users of potentially contaminated waters. Contact your local or federal environmental protection agency for specific recommendations. If employees are required to clean up spills, they must be properly trained and equipped. OSHA 1910.120(q) may be applicable. Soil Adsorption Index (K_{oc}) = 200 (estimate).

Fire Extinguishing: This chemical is a combustible solid. Use dry chemical, carbon dioxide, water spray, or alcohol foam extinguishers. Poisonous gases are produced in fire, including nitrogen oxides. If material or contaminated run-off enters waterways, notify downstream users of potentially contaminated waters. Notify local health and fire officials and pollution control agencies. From a secure, explosion-proof location, use water spray to cool exposed containers. If cooling streams are ineffective (venting sound increases in volume and pitch, tank discolors, or shows any signs of deforming), withdraw immediately to a secure position. If employees are expected to fight fires, they must be trained and equipped in OSHA 1910.156. The only respirators recommended for firefighting are self-contained breathing apparatuses that have full face-pieces and are operated in a pressure-demand or other positive-pressure mode.

Disposal Method Suggested: Consult with environmental regulatory agencies for guidance on acceptable disposal practices. Generators of waste containing this contaminant (≥100 kg/mo) must conform with EPA regulations governing storage, transportation, treatment, and waste disposal. In accordance with 40CFR165, follow recommendations for the disposal of pesticides and pesticide containers. Must be disposed properly by following package label directions or by contacting your local or federal environmental control agency or by contacting your regional EPA office.

Reference
US Environmental Protection Agency. (August 1987). *Health Advisory: Propham.* Washington, DC: Office of Drinking Water

β-Propiolactone   P:1130

Molecular Formula: $C_3H_4O_2$

Synonyms: Betaprone; BPL; Hydracrylic acid, β-lactone; 3-Hydroxypropionic acid lactone; NSC21626; 2-Oxetanone; Propanoic acid, 3-hydroxy-, β-lactone; 3-Propanolide; Propanolide; 1,3-Propiolactone; 3-Propiolactone; Propiolactone; β-Propionolactone; Propionolactone, b

CAS Registry Number: 57-57-8

RTECS® Number: RQ7350000

UN/NA & ERG Number: UN2810/153

EC Number: 200-340-1[*Annex I Index No.:* 606-031-00-1]

Regulatory Authority and Advisory Bodies
Carcinogenicity: NTP: 11th Report on Carcinogens, 2004: Reasonably anticipated to be a human carcinogen; IARC: Human No Adequate Data, Animal Sufficient Evidence, *possibly carcinogenic to humans*, Group 2B, 1999; NIOSH: Potential occupational carcinogen; OSHA: Potential human carcinogen.

US EPA Gene-Tox Program, Positive: Carcinogenicity—mouse/rat; SHE—focus assay; Positive: Cell transform.—SA7/SHE; Positive: *D. melanogaster*—reciprocal translocation; Positive: Host-mediated assay; L5178Y cells *In vitro*—TK test; Positive: *N. crassa*—forward mutation; *N. crassa*—reversion; Positive: *E. coli* polA without S9; Histidine reversion—Ames test; Positive: *D. melanogaster* sex-linked lethal; Positive: *In vitro* UDS—human fibroblast; Positive: *S. cerevisiae* gene conversion; *S. cerevisiae*—homozygosis; Positive: *S. cerevisiae*—reversion; Positive/dose response: *In vitro* SCE—nonhuman; Negative: Sperm morphology—mouse; Inconclusive: Mammalian micronucleus; Positive: CHO gene mutation.

OSHA, 29CFR1910 Specifically Regulated Chemicals (See CFR 1910.1013).

Air Pollutant Standard Set. See below, "Permissible Exposure Limits in Air" section.

Clean Air Act: Hazardous Air Pollutants (Title I, Part A, Section 112).

Superfund/EPCRA 40CFR355, Extremely Hazardous Substances: TPQ = 500 lb (227 kg).

Reportable Quantity (RQ): 10 lb (4.54 kg).

EPCRA Section 313 Form R *de minimis* concentration reporting level: 0.1%.

California Proposition 65 Chemical: Cancer 1/1/88.

European/International Regulations: Hazard Symbol: T+; Risk phrases: R45; R26; R36/38; Safety phrases: S53; S45. (see Appendix 4).

WGK (German Aquatic Hazard Class): No value assigned.

Description: β-Propiolactone is a colorless liquid which slowly hydrolyzes to hydracrylic acid and must be cooled to remain stable. Molecular weight = 72.07; Specific gravity (H_2O:1) = 1.15 at 25°C; Boiling point = (decomposes) 161.7°C; Freezing/Melting point = −33.3°C; Vapor pressure = 3 mmHg at 25°C; Flash point = 75°C. Explosive limits: LEL = 2.9%; UEL—unknown. Hazard Identification (based on NFPA-704 M Rating System): Health 0, Flammability 2, Reactivity 0. Soluble in water; solubility = 37%.

Potential Exposure: Compound Description: Agricultural Chemical; Tumorigen, Mutagen. β-Propiolactone is used as a chemical intermediate in synthesis of acrylic acid and esters, acrylate plastics; as a vapor sterilizing agent; phase disinfectant; and a viricidal agent.

Incompatibilities: Acetates, halogens, thiocyanates, thiosulfates, strong oxidizers, strong bases. May polymerize upon storage or due to warming. Stable if kept under refrigeration at 40−50°F/5−10°C.

Permissible Exposure Limits in Air

OSHA PEL: Cancer suspect agent. Exposures of workers to this chemical is to be controlled through the required use of engineering controls, work practices, and personal protective equipment, including respirators. See 29CFR 1910.1003-1910.1016 for specific details of these requirements.

NIOSH REL: A potential occupational carcinogen. [skin]; Limit exposure to lowest feasible concentration. See *NIOSH Pocket Guide*, Appendix A.

ACGIH TLV®[1]: 0.5 ppm/1.5 mg/m³ TWA; confirmed animal carcinogen with unknown relevance to humans.

NIOSH IDLH: Not determined. Potential occupational carcinogen.

Protective Action Criteria (PAC)

TEEL-0: 0.5 ppm

PAC-1: 0.509 ppm

PAC-2: 5.09 ppm

PAC-3: 150 ppm

DFG MAK: [skin] Carcinogen Category 2.

Australia: TWA 0.5 ppm (1.5 mg/m³), carcinogen, 1993; Austria: carcinogen, 1999; Belgium: TWA 0.5 ppm (1.5 mg/m³), carcinogen, 1993; Denmark: TWA 0.1 ppm (1.5 mg/m³), 1999; Finland: carcinogen, 1999; France: carcinogen, 1993; Poland: MAC (TWA) 1 mg/m³, 1999; Sweden: carcinogen, 1999; Switzerland: MAK-W 0.5 ppm (1.5 mg/m³), carcinogen, 1999; United Kingdom: carcinogen, 2000; Argentina, Bulgaria, Columbia, Jordan, South Korea, New Zealand, Singapore, Vietnam: ACGIH TLV®: confirmed animal carcinogen with unknown relevance to humans. Several states have set guidelines or standards for propiolactone in ambient air[60] ranging from zero (North Dakota) to 5.0 μg/m³ (New York) to 7.5 μg/m³ (South Carolina) to 15 μg/m³ (Florida and Virginia) to 22.5 μg/m³ (Connecticut) to 36.0 μg/m³ (Nevada).

Routes of Entry: Inhalation, ingestion, skin absorption, skin and/or eye contact.

Harmful Effects and Symptoms

Short Term Exposure: Corrosive to the eyes. May cause corneal opacity and blindness. Irritates the respiratory tract. Contact with skin causes irritation, burns, and blistering; fluid from blisters may cause additional blistering of adjacent skin. Ingestion causes burns of mouth and stomach. The toxicity potential of this material via inhalation or ingestion is high; may cause death or permanent injury after very short exposures to small quantities.

Long Term Exposure: A potential occupational carcinogen. May cause frequent urination, dysuria, hematuria (blood in the urine). May affect the liver and kidneys.

Points of Attack: Kidneys, skin, lungs, eyes, liver. Cancer site in animals: liver, skin, and stomach.

Medical Surveillance: Based on its high toxicity and carcinogenic effects in animals, preplacement and periodic examinations should include a history of exposure to other carcinogens, alcohol and smoking habits, medication, and family history. The skin, eye, lung, liver, and kidney should be evaluated. Sputum cytology may be helpful in evaluating the presence or absence of carcinogenic effects. Kidney and liver function tests. Periodic lung function tests.

First Aid: If this chemical gets into the eyes, remove any contact lenses at once and irrigate immediately for at least 15 min, occasionally lifting upper and lower lids. Seek medical attention immediately. If this chemical contacts the skin, remove contaminated clothing and wash immediately with soap and water. Seek medical attention immediately. If this chemical has been inhaled, remove from exposure, begin rescue breathing (using universal precautions, including resuscitation mask) if breathing has stopped and CPR if heart action has stopped. Transfer promptly to a medical facility. When this chemical has been swallowed, get medical attention. Give large quantities of water and induce vomiting. Do not make an unconscious person vomit.

Personal Protective Methods: Wear protective gloves and clothing to prevent any reasonable probability of skin contact. Safety equipment suppliers/manufacturers can provide recommendations on the most protective glove/clothing material for your operation. Butyl rubber is among the recommended protective materials. All protective clothing (suits, gloves, footwear, headgear) should be clean, available each day, and put on before work. Contact lenses should not be worn when working with this chemical. Wear splash-proof chemical goggles and face shield unless full face-piece respiratory protection is worn. Employees should wash immediately with soap when skin is wet or contaminated. Provide emergency showers and eyewash. Specific engineering controls are required for this chemical. Refer to OSHA Standard: *beta-Propiolactone, 29 CFR 1910.1013.*

Respirator Selection: Employees engaged in handling operations involving this chemical must be provided with, and required to wear and use, a half-mask filter-type respirator for dusts, mists, and fumes. A respirator affording higher levels of protection than this respirator may be substituted. *At any detectable concentration:* SCBAF: Pd, Pp (APF = 10,000) (any NIOSH/MSHA- or European Standard EN 149-approved self-contained breathing apparatus that has a full face-piece and is operated in a pressure-demand or other positive-pressure mode) or SaF: Pd,Pp: ASCBA (APF = 10,000) (any supplied-air respirator that has a full face-piece and is operated in a pressure-demand or other positive-pressure mode in combination with an auxiliary, self-contained breathing apparatus operated in a pressure-demand or other positive-pressure mode). *Escape:* GmFOv (APF = 50) [any air-purifying, full-face-piece respirator (gas mask) with a chin-style, front- or back-mounted organic vapor canister] or SCBAE (any appropriate escape-type, self-contained breathing apparatus).

Storage: Color Code—Blue: Health Hazard/Poison: Store in a secure poison location. Prior to working with this chemical you should be trained on its proper handling and storage. Before entering confined space where this chemical may be present, check to make sure that an explosive concentration is not a danger. Store in a refrigerator or a freezer in glass containers and protect from air and light. Where possible, automatically pump liquid from drums or other storage containers to process containers. Sources of ignition, such as smoking and open flames, are prohibited where this chemical is handled, used, or stored. Metal containers involving the transfer of 5 gallons or more of this chemical should be grounded and bonded. Drums must be equipped with self-closing valves, pressure vacuum bungs, and flame arresters. Use only nonsparking tools and equipment, especially when opening and closing containers of this chemical. Wherever this chemical is used, handled, manufactured, or stored, use explosion-proof electrical equipment and fittings. A regulated, marked area should be established where this chemical is handled, used, or stored in compliance with OSHA Standard 1910.1045.

Shipping: Toxic liquids, organic, n.o.s. require a label of "POISONOUS/TOXIC MATERIALS." It falls in Hazard Class 6.1 and Packing Group I.

Spill Handling: Evacuate and restrict persons not wearing protective equipment from area of spill or leak until cleanup is complete. Remove all ignition sources. Establish forced ventilation to keep levels below explosive limit. Absorb liquids in vermiculite, dry sand, earth, peat, carbon, or a similar material and deposit in sealed containers. Keep this chemical out of a confined space, such as a sewer, because of the possibility of an explosion, unless the sewer is designed to prevent the buildup of explosive concentrations. It may be necessary to contain and dispose of this chemical as a hazardous waste. If material or contaminated runoff enters waterways, notify downstream users of potentially contaminated waters. Contact your local or federal environmental protection agency for specific recommendations. If employees are required to clean up spills, they must be properly trained and equipped. OSHA 1910.120(q) may be applicable.

Fire Extinguishing: This chemical is a combustible liquid. Poisonous gases are produced in fire. Use dry chemical, carbon dioxide, or alcohol foam extinguishers. Vapors are heavier than air and will collect in low areas. Vapors may travel long distances to ignition sources and flashback. Vapors in confined areas may explode when exposed to fire. Containers may explode in fire. Storage containers and parts of containers may rocket great distances, in many directions. If material or contaminated runoff enters waterways, notify downstream users of potentially contaminated waters. Notify local health and fire officials and pollution control agencies. From a secure, explosion-proof location, use water spray to cool exposed containers. If cooling streams are ineffective (venting sound increases in volume and pitch, tank discolors, or shows any signs of deforming), withdraw immediately to a secure position. If employees are expected to fight fires, they must be trained and equipped in OSHA 1910.156. The only respirators recommended for firefighting are self-contained breathing apparatuses that have full face-pieces and are operated in a pressure-demand or other positive-pressure mode.

Disposal Method Suggested: Dissolve or mix the material with a combustible solvent and burn in a chemical incinerator equipped with an afterburner and scrubber. All federal, state, and local environmental regulations must be observed.

References

Sax, N. I. (Ed.). (1983). *Dangerous Properties of Industrial Materials Report*, 3, No. 2, 57–60

US Environmental Protection Agency. (November 30, 1987). *Chemical Hazard Information Profile: Propiolactone, Beta-*. Washington, DC: Chemical Emergency Preparedness Program

New Jersey Department of Health and Senior Services. (August 2002). *Hazardous Substances Fact Sheet: Beta-Propiolactone*. Trenton, NJ

Propionaldehyde   P:1140

Molecular Formula: C_3H_6O
Common Formula: CH_3CH_2COH
Synonyms: Aldehyde propionique (French); Methylacetaldehyde; NCI-C61029; Propaldehyde; Propanal; *n*-Propanal; 1-Propanal; Propanaldehyde; 1-Propanone; Propional; Propionic aldehyde; Propyl aldehyde; Propylic aldehyde
CAS Registry Number: 123-38-6
RTECS® Number: UE0350000
UN/NA & ERG Number: UN1275/129
EC Number: 204-623-0 [*Annex I Index No.:* 605-018-00-8]
Regulatory Authority and Advisory Bodies
Carcinogenicity: EPA: Inadequate Information to assess carcinogenic potential.

Clean Air Act: Hazardous Air Pollutants (Title I, Part A, Section 112).

Reportable Quantity (RQ): 1 lb (0.454 kg).

EPCRA Section 313 Form R *de minimis* concentration reporting level: 1.0%.

European/International Regulations: Hazard Symbol: F, Xi; Risk phrases: R11; R36/37/38; Safety phrases: S2; S9; S16; S29 (see Appendix 4).

WGK (German Aquatic Hazard Class): 1—Low hazard to waters.

Description: Propionaldehyde is a colorless liquid with a strong fruity odor. Molecular weight = 58.09; Boiling point = 49°C; Freezing/Melting point = −81°C; Flash point = −30°C; −9°C (oc); Autoignition temperature = 207°C. Explosive limits: LEL = 2.6%; UEL = 17%. Hazard Identification (based on NFPA-704 M Rating System): Health 2, Flammability 3, Reactivity 2. Slightly soluble in water.

Potential Exposure: Compound Description: Mutagen, Primary Irritant. Used as a synthetic flavoring; as a disinfectant and preservative; to make propionic acid; in plastic and rubber manufacturing; to make alkyl resins and plasticizers.

Incompatibilities: Incompatible with strong acids, amines. Violent reaction with strong oxidizers. Strong caustics; reducing agents can cause explosive polymerization. Can self-ignite if finely dispersed on porous or combustible material, such as fabric. Heat or ultraviolet light can cause decomposition.

Permissible Exposure Limits in Air

AIHA WEEL: 20 ppm TWA.

ACGIH TLV®[1]: 20 ppm/48 mg/m^3 TWA.

Protective Action Criteria (PAC)*

TEEL-0: 20 ppm

PAC-1: **45** ppm

PAC-2: **260** ppm

PAC-3: **840** ppm

*AEGLs (Acute Emergency Guideline Levels) & ERPGs (Emergency Response Planning Guideline) are in **bold face**.

Determination in Air: NIOSH Analytical Method #2539, aldehydes, screening.

Routes of Entry: Inhalation, ingestion, skin and/or eye contact.

Harmful Effects and Symptoms

Short Term Exposure: Irritates the skin causing a burning sensation and rash on contact. Inhalation can irritate the respiratory tract and may cause nosebleeds, sore throat, cough, and phlegm. Higher exposures can cause pulmonary edema, a medical emergency that can be delayed for several hours. This can cause death.

Long Term Exposure: Can irritate the lungs; bronchitis may develop. Testing has not been completed to determine the carcinogenicity of propionaldehyde. However, the limited studies to date indicate that these substances have chemical reactivity and mutagenicity similar to acetaldehyde and malonaldehyde. Therefore, NIOSH recommends that careful consideration should be given to reducing exposures to this aldehyde. Further information can be found in the *NIOSH Current Intelligence Bulletin 55: Carcinogenicity of Acetaldehyde and Malonaldehyde, and Mutagenicity of Related Low-Molecular-Weight Aldehydes* [DHHS (NIOSH), Publication No. 91-112].

Points of Attack: Lungs, skin.

Medical Surveillance: Lung function test. Consider chest X-ray following acute overexposure.

First Aid: If this chemical gets into the eyes, remove any contact lenses at once and irrigate immediately for at least 15 min, occasionally lifting upper and lower lids. Seek medical attention immediately. If this chemical contacts the skin, remove contaminated clothing and wash immediately with soap and water. Seek medical attention immediately. If this chemical has been inhaled, remove from exposure, begin rescue breathing (using universal precautions, including resuscitation mask) if breathing has stopped and CPR if heart action has stopped. Transfer promptly to a medical facility. When this chemical has been swallowed, get medical attention. Give large quantities of water and induce vomiting. Do not make an unconscious person vomit. Medical observation is recommended for 24−48 h after breathing overexposure, as pulmonary edema may be delayed. As first aid for pulmonary edema, a doctor or authorized paramedic may consider administering a corticosteroid spray.

Personal Protective Methods: Wear protective gloves and clothing to prevent any reasonable probability of skin contact. Safety equipment suppliers/manufacturers can provide recommendations on the most protective glove/clothing material for your operation. All protective clothing (suits, gloves, footwear, headgear) should be clean, available each day, and put on before work. Contact lenses should not be worn when working with this chemical. Wear splash-proof chemical goggles and face shield unless full face-piece respiratory protection is worn. Employees should wash immediately with soap when skin is wet or contaminated. Provide emergency showers and eyewash.

Respirator Selection: Follow regulations in OSHA 29CFR1910.134 or European Standard EN149. Use a NIOSH/MSHA- or European Standard EN149-approved respirator; or use an approved supplied-air respirator with a full face-piece operated in the positive-pressure mode, or with a full face-piece, hood, or helmet in the continuous-flow mode; or use a NIOSH/MSHA- or European Standard EN149-approved self-contained breathing apparatus with a full face-piece operated in a pressure-demand or other positive-pressure mode.

Storage: Color Code—Red: Flammability Hazard: Store in a flammable liquid storage area or approved cabinet away from ignition sources and corrosive and reactive materials. Prior to working with this chemical you should be trained on its proper handling and storage. Before entering confined space where this chemical may be present, check to make sure that an explosive concentration does not exist. Store in tightly closed containers in a cool, well-ventilated area away from oxidizers. Where possible, automatically pump liquid from drums or other storage containers to process containers. Sources of ignition, such as smoking and open flames, are prohibited where this chemical is handled, used, or stored. Metal containers involving the transfer of 5 gallons or more of this chemical should be grounded and bonded. Drums must be equipped with self-closing valves, pressure vacuum bungs, and flame arresters. Use only non-sparking tools and equipment, especially when opening and closing containers of this chemical. Wherever this chemical is used, handled, manufactured, or stored, use explosion-proof electrical equipment and fittings.

Shipping: Propionaldehyde requires a label of "FLAMMABLE LIQUID." It falls in Hazard Class 3 and Packing Group II.

Spill Handling: Evacuate and restrict persons not wearing protective equipment from area of spill or leak until cleanup is complete. Remove all ignition sources. Establish forced ventilation to keep levels below explosive limit. Ventilate area of spill or leak. Absorb liquids in vermiculite, dry sand, earth, peat, carbon, or a similar material and deposit in sealed containers. Keep this chemical out of a confined space, such as a sewer, because of the possibility of an explosion, unless the sewer is designed to prevent the buildup of explosive concentrations. It may be necessary to contain and dispose of this chemical as a hazardous waste. If material or contaminated runoff enters waterways, notify downstream users of potentially contaminated waters. Contact your local or federal environmental protection agency for specific recommendations. If employees are required to clean up spills, they must be properly trained and equipped. OSHA 1910.120(q) may be applicable.

Fire Extinguishing: This chemical is a combustible liquid. Poisonous gases are produced in fire. *Do not use water.* Use dry chemical, carbon dioxide extinguishers. Vapors are heavier than air and will collect in low areas. Vapors may travel long distances to ignition sources and flashback. Vapors in confined areas may explode when exposed to fire. Containers may explode in fire. Storage containers and parts of containers may rocket great distances, in many directions. If material or contaminated runoff enters waterways, notify downstream users of potentially contaminated waters. Notify local health and fire officials and pollution control agencies. From a secure, explosion-proof location, use water spray to cool exposed containers. If cooling streams are ineffective (venting sound increases in volume and pitch, tank discolors, or shows any signs of deforming), withdraw immediately to a secure position. If employees are expected to fight fires, they must be trained and equipped in OSHA 1910.156. The only respirators recommended for firefighting are self-contained breathing apparatuses that have full face-pieces and are operated in a pressure-demand or other positive-pressure mode.

Reference
New Jersey Department of Health and Senior Services. (May 2006). *Hazardous Substances Fact Sheet: Propionaldehyde.* Trenton, NJ

Propionic acid   P:1150

Molecular Formula: $C_3H_6O_2$
Common Formula: CH_3CH_2COOH
Synonyms: Acide propionique (French); Carbonyethane; Carboxyethane; Ethanecarboxylic acid; Ethylformic acid; Metacetonic acid; Methylacetic acid; Propanoic acid; Propionic acid Grain preserver; Prozoin; Pseudoacetic acid; Sentry grain preserver; Tenox P grain preservative
CAS Registry Number: 79-09-4
RTECS® Number: UE5950000
UN/NA & ERG Number: UN1848/132
EC Number: 201-176-3 [*Annex I Index No.:* 607-089-00-0]
Regulatory Authority and Advisory Bodies
Air Pollutant Standard Set. See below, "Permissible Exposure Limits in Air" section.
US EPA, FIFRA 1998 Status of Pesticides: Pesticide subject to registration or re-registration.
FDA—over-the-counter drug.
Clean Water Act: Section 311 Hazardous Substances/RQ 40CFR117.3 (same as CERCLA, see below).
Reportable Quantity (RQ): 5000 lb (2270 kg).
Canada, WHMIS, Ingredients Disclosure List Concentration 1.0%.
European/International Regulations: Hazard Symbol: C; Risk phrases: R34; Safety phrases: S1/2; S23; S36; S45 (see Appendix 4).
WGK (German Aquatic Hazard Class): 1—Low hazard to waters.

Description: Propionic acid is a colorless liquid with a pungent odor. The odor threshold is 0.16 ppm. Molecular weight = 74.09; Specific gravity (H_2O:1) = 0.99 at 25°C; Boiling point = 141°C; Freezing/Melting point = −21°C; Vapor pressure = 3 mmHg at 25°C; Flash point = 52.2°C (cc); 57°C (oc); Autoignition temperature = 465°C. Explosive limits: LEL = 2.9%; UEL = 12.1%.[17] Hazard Identification (based on NFPA-704 M Rating System): Health 3, Flammability 2, Reactivity 1. Soluble in water.

Potential Exposure: Compound Description: Agricultural Chemical; Mutagen, Primary Irritant. Propionic acid is used in the manufacture of inorganic propionates and propionate esters which are used as mold inhibitors, electroplating additives, emulsifying agents, flavors, and perfumes. It is an intermediate in pesticide manufacture, pharmaceutic manufacture; and in the production of cellulose propionate plastics. Also, used as grain preservative.

Incompatibilities: The substance is a medium strong acid. Incompatible with sulfuric acid, strong bases, ammonia, isocyanates, alkylene oxides, epichlorohydrin. Reacts with bases, strong oxidizers, and amines, causing fire and explosion hazard. Attacks many metals forming flammable/explosive hydrogen gas.

Permissible Exposure Limits in Air
Conversion factor: 1 ppm = 3.03 mg/m³ at 25°C & 1 atm.
OSHA PEL: None.

NIOSH REL: 10 ppm/30 mg/m^3 TWA; 15 ppm/45 mg/m^3 STEL.

ACGIH TLV$^{®[1]}$: 10 ppm/30 mg/m^3 TWA.

Protective Action Criteria (PAC)

TEEL-0: 10 ppm

PAC-1: 15 ppm

PAC-2: 15 ppm

PAC-3: 350 ppm

DFG MAK: 10 ppm/31 mg/m^3 TWA; Peak Limitation Category I(2); [skin]; Pregnancy Risk Group C.

Australia: TWA 10 ppm (30 mg/m^3); STEL 15 ppm, 1993; Austria: MAK 10 ppm (30 mg/m^3), 1999; Belgium: TWA 10 ppm (30 mg/m^3); STEL 15 ppm, 1993; Denmark: TWA 10 ppm (30 mg/m^3), 1999; Finland: TWA 10 ppm, [skin], 1999; France: VME 10 ppm (30 mg/m^3), 1999; the Netherlands: MAC-TGG 31 mg/m^3, 2003; Russia: STEL 20 mg/m^3, 1993; Sweden: NGV 10 ppm (30 mg/m^3), KTV 15 ppm (45 mg/m^3), 1999; Switzerland: MAK-W 10 ppm (30 mg/m^3), KZG-W 20 ppm (60 mg/m^3), 1999; United Kingdom: TWA 10 ppm (31 mg/m^3); STEL 15 ppm (46 mg/m^3), 2000; Argentina, Bulgaria, Columbia, Jordan, South Korea, New Zealand, Singapore, Vietnam: ACGIH TLV$^®$: TWA 10 ppm. Russia[43] set a MAC of 0.015 mg/m^3 for ambient air in residential areas (15 μg/m^3) on a momentary basis. Several states have set guidelines or standards for propionic acid in ambient air[60] ranging from 300−450 μg/m^3 (North Dakota) to 500 μg/m^3 (Virginia) to 600 μg/m^3 (Connecticut) to 714 μg/m^3 (Nevada).

Determination in Air: No method available.

Determination in Water: Octanol−water coefficient: Log $K_{ow} = 0.31$.

Routes of Entry: Inhalation, ingestion, skin and/or eye contact.

Harmful Effects and Symptoms

Short Term Exposure: Corrosive to the eyes, skin, and respiratory tract. Contact can cause severe eye burns, leading to permanent damage. Skin contact causes skin burns. Inhalation can cause irritation of the respiratory tract with mild cough; asthmatic response was found in medical reports of acute exposure of workers.

Long Term Exposure: May cause an asthma-like allergy. May irritate the lungs; may lead to lung damage.

Points of Attack: Skin, eyes, respiratory system.

Medical Surveillance: For those with frequent or potentially high exposure (half the TLV or greater), the following are recommended before beginning work and at regular times after that: lung function tests. These may be normal if person is not having an attack at the time of the test.

First Aid: If this chemical gets into the eyes, remove any contact lenses at once and irrigate immediately for at least 15 min, occasionally lifting upper and lower lids. Seek medical attention immediately. If this chemical contacts the skin, remove contaminated clothing and wash immediately with soap and water. Seek medical attention immediately. If this chemical has been inhaled, remove from exposure, begin rescue breathing (using universal precautions, including resuscitation mask) if breathing has stopped and CPR if heart action has stopped. Transfer promptly to a medical facility. When this chemical has been swallowed, get medical attention. If victim is *conscious*, administer water or milk. Do not induce vomiting.

Personal Protective Methods: Wear protective gloves and clothing to prevent any reasonable probability of skin contact. Safety equipment suppliers/manufacturers can provide recommendations on the most protective glove/clothing material for your operation. ACGIH recommends Neoprene$^{™}$, nitrile rubber and polyvinyl chloride as protective materials. All protective clothing (suits, gloves, footwear, headgear) should be clean, available each day, and put on before work. Contact lenses should not be worn when working with this chemical. Wear splash-proof chemical goggles and face shield unless full face-piece respiratory protection is worn. Employees should wash immediately with soap when skin is wet or contaminated. Provide emergency showers and eyewash.

Respirator Selection: Where there is potential for exposures *over 10 ppm*, use a NIOSH/MSHA- or European Standard EN149-approved full face-piece powered air-purifying respirators. *Where there is potential for high exposures*, use a NIOSH/MSHA- or European Standard EN149-approved supplied air-purifying respirators. *Where there is potential for high exposures*, use a NIOSH/MSHA- or European Standard EN149-approved supplied-air respirator with a full face-piece operated in the positive-pressure mode, or with a full face-piece, hood, or helmet in the continuous-flow mode; or use a NIOSH/MSHA- or European Standard EN149-approved self-contained breathing apparatus with a full face-piece operated in a pressure-demand or other positive-pressure mode.

Storage: Color Code—White: Corrosive or Contact Hazard; Store separately in a corrosion-resistant location. Prior to working with this chemical you should be trained on its proper handling and storage. Before entering confined space where this chemical may be present, check to make sure that an explosive concentration is not a danger. Store in containers made of aluminum or stainless steel. Propionic acid will corrode steel. Store in tightly closed containers in a cool, well-ventilated area away from oxidizers (such as perchlorates, peroxides, permanganates, chlorates, and nitrates), ignition sources, or heat. Outside or detached storage is preferred. Where possible, automatically pump liquid from drums or other storage containers to process containers. Sources of ignition, such as smoking and open flames, are prohibited where this chemical is handled, used, or stored. Metal containers involving the transfer of 5 gallons or more of this chemical should be grounded and bonded. Drums must be equipped with self-closing valves, pressure vacuum bungs, and flame arresters. Use only nonsparking tools and equipment, especially when opening and closing containers of this chemical. Wherever this chemical is used, handled, manufactured, or stored, use explosion-proof electrical equipment and fittings.

Shipping: This compound requires a shipping label of "CORROSIVE." It falls in Hazard Class 8 and Packing Group III.

Spill Handling: Evacuate and restrict persons not wearing protective equipment from area of spill or leak until cleanup is complete. Remove all ignition sources. Establish forced ventilation to keep levels below explosive limit. Cover spilled material with soda ash or sodium bicarbonate. Mix and add water. Neutralize and drain to sewer with plenty of water. Use water spray to dilute spill and disperse vapor. Absorb liquids in vermiculite, dry sand, earth, peat, carbon, or a similar material and deposit in sealed containers. Keep this chemical out of a confined space, such as a sewer, because of the possibility of an explosion, unless the sewer is designed to prevent the buildup of explosive concentrations. It may be necessary to contain and dispose of this chemical as a hazardous waste. If material or contaminated runoff enters waterways, notify downstream users of potentially contaminated waters. Contact your local or federal environmental protection agency for specific recommendations. If employees are required to clean up spills, they must be properly trained and equipped. OSHA 1910.120(q) may be applicable.

Fire Extinguishing: This chemical is a flammable liquid. Poisonous gases are produced in fire. Use dry chemical, carbon dioxide, or alcohol foam extinguishers. Vapors are heavier than air and will collect in low areas. Vapors may travel long distances to ignition sources and flashback. Vapors in confined areas may explode when exposed to fire. Containers may explode in fire. Storage containers and parts of containers may rocket great distances, in many directions. If material or contaminated runoff enters waterways, notify downstream users of potentially contaminated waters. Notify local health and fire officials and pollution control agencies. From a secure, explosion-proof location, use water spray to cool exposed containers. If cooling streams are ineffective (venting sound increases in volume and pitch, tank discolors, or shows any signs of deforming), withdraw immediately to a secure position. If employees are expected to fight fires, they must be trained and equipped in OSHA 1910.156. The only respirators recommended for firefighting are self-contained breathing apparatuses that have full face-pieces and are operated in a pressure-demand or other positive-pressure mode.

Disposal Method Suggested: Incineration in admixture with flammable solvent.

Reference

New Jersey Department of Health and Senior Services. (April 2001). *Hazardous Substances Fact Sheet: Propionic Acid.* Trenton, NJ

Propionic anhydride   P:1160

Molecular Formula: $C_6H_{10}O_3$
Synonyms: Methylacetic anhydride; Propanoic anhydride; Propionic acid anhydride; Propionyl oxide

CAS Registry Number: 123-62-6
RTECS® Number: UF9100000
UN/NA & ERG Number: UN2496/156
EC Number: 204-638-2 [*Annex I Index No.:* 607-010-00-X]
Regulatory Authority and Advisory Bodies
Clean Water Act: Section 311 Hazardous Substances/RQ 40CFR117.3 (same as CERCLA, see below).
Reportable Quantity (RQ): 5000 lb (2270 kg).
European/International Regulations: Hazard Symbol: C; Risk phrases: R34; Safety phrases: S1/2; S26; S45 (see Appendix 4).
WGK (German Aquatic Hazard Class): 1—Low hazard to waters.

Description: Propionic anhydride is a colorless liquid with a strong unpleasant odor. Molecular weight = 130.16; Boiling point = 167−169°C; Freezing/Melting point = −45°C; Flash point = 63°C; Autoignition temperature = 285°C. Explosive limits: LEL = 1.3%; UEL = 9.5%. Hazard Identification (based on NFPA-704 M Rating System): Health 3, Flammability 2, Reactivity 1. Decomposes in water.

Potential Exposure: Used in the manufacture of perfumes, flavorings, alkyd resins; dyestuffs, pharmaceuticals; as an esterifying agent for fats, oils, and cellulose; dehydrating medium for nitrations and sulfonations.

Incompatibilities: Oxidizers, strong acids, strong bases, reducing agents, alcohols, and metals. Contact water forms propionic acid.

Permissible Exposure Limits in Air
Protective Action Criteria (PAC)
TEEL-0: 10 mg/m^3
PAC-1: 30 mg/m^3
PAC-2: 200 mg/m^3
PAC-3: 500 mg/m^3

Permissible Concentration in Water: High concentrations are dangerous to aquatic life. May be dangerous if it enters water intakes.

Routes of Entry: Inhalation, ingestion, skin and/or eye contact.

Harmful Effects and Symptoms

Short Term Exposure: Irritates the eyes, skin, and respiratory tract. Eye contact can cause burns and permanent damage. Contact with liquid causes burns of skin. Ingestion causes burns of the mouth and stomach. High exposures can cause pulmonary edema, a medical emergency that can be delayed for several hours. This can cause death.

Long Term Exposure: May cause skin allergy. Can irritate the lungs; bronchitis may develop.

First Aid: If this chemical gets into the eyes, remove any contact lenses at once and irrigate immediately for at least 15 min, occasionally lifting upper and lower lids. Seek medical attention immediately. If this chemical contacts the skin, remove contaminated clothing and wash immediately with soap and water. Seek medical attention immediately. If this chemical has been inhaled, remove from exposure, begin rescue breathing (using universal precautions,

including resuscitation mask) if breathing has stopped and CPR if heart action has stopped. Transfer promptly to a medical facility. When this chemical has been swallowed, get medical attention. If victim is *conscious*, administer water or milk. Do not induce vomiting. Medical observation is recommended for 24–48 h after breathing overexposure, as pulmonary edema may be delayed. As first aid for pulmonary edema, a doctor or authorized paramedic may consider administering a corticosteroid spray.

Personal Protective Methods: Wear protective gloves and clothing to prevent any reasonable probability of skin contact. Safety equipment suppliers/manufacturers can provide recommendations on the most protective glove/clothing material for your operation. All protective clothing (suits, gloves, footwear, headgear) should be clean, available each day, and put on before work. Contact lenses should not be worn when working with this chemical. Wear splash-proof chemical goggles and face shield unless full face-piece respiratory protection is worn. Employees should wash immediately with soap when skin is wet or contaminated. Provide emergency showers and eyewash.

Respirator Selection: Follow regulations in OSHA 29CFR1910.134 or European Standard EN149. Use a NIOSH/MSHA- or European Standard EN149-approved respirator; or use an approved supplied-air respirator with a full face-piece operated in the positive-pressure mode, or with a full face-piece, hood, or helmet in the continuous-flow mode; or use a NIOSH/MSHA- or European Standard EN149-approved self-contained breathing apparatus with a full face-piece operated in a pressure-demand or other positive-pressure mode.

Storage: Color Code—White: Corrosive or Contact Hazard; Store separately in a corrosion-resistant location. Prior to working with this chemical you should be trained on its proper handling and storage. Before entering confined space where this chemical may be present, check to make sure that an explosive concentration does not exist. Store in tightly closed containers in a cool, well-ventilated area away from oxidizers. Where possible, automatically pump liquid from drums or other storage containers to process containers. Sources of ignition, such as smoking and open flames, are prohibited where this chemical is handled, used, or stored. Metal containers involving the transfer of 5 gallons or more of this chemical should be grounded and bonded. Drums must be equipped with self-closing valves, pressure vacuum bungs, and flame arresters. Use only non-sparking tools and equipment, especially when opening and closing containers of this chemical. Wherever this chemical is used, handled, manufactured, or stored, use explosion-proof electrical equipment and fittings.

Shipping: Propionic anhydride requires a shipping label of "CORROSIVE" It falls in Hazard Class 8 and Packing Group III.

Spill Handling: Evacuate and restrict persons not wearing protective equipment from area of spill or leak until cleanup is complete. Remove all ignition sources. Establish forced ventilation to keep levels below explosive limit. Absorb liquids in vermiculite, dry sand, earth, peat, carbon, or a similar material and deposit in sealed containers. Keep this chemical out of a confined space, such as a sewer, because of the possibility of an explosion, unless the sewer is designed to prevent the buildup of explosive concentrations. It may be necessary to contain and dispose of this chemical as a hazardous waste. If material or contaminated runoff enters waterways, notify downstream users of potentially contaminated waters. Contact your local or federal environmental protection agency for specific recommendations. If employees are required to clean up spills, they must be properly trained and equipped. OSHA 1910.120(q) may be applicable.

Fire Extinguishing: This chemical is a combustible liquid. Poisonous gases are produced in fire. Use dry chemical, carbon dioxide, or polymer alcohol foam extinguishers. Vapors are heavier than air and will collect in low areas. Vapors may travel long distances to ignition sources and flashback. Vapors in confined areas may explode when exposed to fire. Containers may explode in fire. Storage containers and parts of containers may rocket great distances, in many directions. If material or contaminated runoff enters waterways, notify downstream users of potentially contaminated waters. Notify local health and fire officials and pollution control agencies. From a secure, explosion-proof location, use water spray to cool exposed containers. If cooling streams are ineffective (venting sound increases in volume and pitch, tank discolors, or shows any signs of deforming), withdraw immediately to a secure position. If employees are expected to fight fires, they must be trained and equipped in OSHA 1910.156. The only respirators recommended for firefighting are self-contained breathing apparatuses that have full face-pieces and are operated in a pressure-demand or other positive-pressure mode.

Disposal Method Suggested: Incinerator with afterburner.

Reference

New Jersey Department of Health and Senior Services. (July 1999). *Hazardous Substances Fact Sheet: Propionic Anhydride*. Trenton, NJ

Propionitrile   P:1170

Molecular Formula: C_3H_5N
Common Formula: CH_3CH_2CN
Synonyms: Cianuro de etilo (Spanish); Cyanoethane; Ether cyanatus; Ethyl cyanide; Hydrocyanic ether; Propanenitrile; Propionic nitrile propylnitrile
CAS Registry Number: 107-12-0
RTECS® Number: UF9625000
UN/NA & ERG Number: UN2404/131
EC Number: 203-464-4
Regulatory Authority and Advisory Bodies
Department of Homeland Security Screening Threshold Quantity (pounds): *Release hazard* 10,000 (≥1.00% concentration).

Air Pollutant Standard Set. See below, "Permissible Exposure Limits in Air" section.

Clean Air Act: Accidental Release Prevention/Flammable Substances, (Section 112[r], Table 3), TQ = 10,000 lb (4540 kg).

US EPA Hazardous Waste Number (RCRA No.): P101.

RCRA, 40CFR261, Appendix 8 Hazardous Constituents.

RCRA 40CFR264, Appendix 9; TSD Facilities Ground. Water Monitoring List. Suggested test method(s) (PQL μg/L): 8015 (60); 8240 (5).

Superfund/EPCRA 40CFR355, Extremely Hazardous Substances: TPQ = 500 lb (227 kg).

Reportable Quantity (RQ): 10 lb (4.54 kg).

Canada, WHMIS, Ingredients Disclosure List Concentration 1.0%.

European/International Regulations: not listed in Annex 1.

WGK (German Aquatic Hazard Class): 1—Low hazard to waters.

Description: Propionitrile is a colorless liquid with a pleasant, sweetish, ethereal odor. Molecular weight = 55.09; Specific gravity (H_2O:1) = 0.78 at 25°C; Boiling point = 97.2°C; Freezing/Melting point = − 91.7°C; Flash point = 2.2°C (cc). Explosive limits: LEL = 3.1%; UEL—unknown. Hazard Identification (based on NFPA-704 M Rating System): Health 4, Flammability 3, Reactivity 1. Soluble in water; solubility = 12%.

Potential Exposure: Compound Description: Mutagen; Reproductive Effector; Primary Irritant. Used as a solvent in petroleum refining, as a chemical intermediate, a raw material for drug manufacture, and a setting agent.

Incompatibilities: Strong oxidizers and reducing agents; strong acids and bases. Hydrogen cyanide is produced when propionitrile is heated to decomposition. Reacts with acids, steam, warm water, producing toxic and flammable hydrogen cyanide fumes.

Permissible Exposure Limits in Air

Conversion factor: 1 ppm = 2.25 mg/m³ at 25°C & 1 atm.

OSHA PEL: None.

NIOSH REL: TWA 6 ppm/14 mg/m³.

ACGIH TLV®[1]: None.

Protective Action Criteria (PAC)*

TEEL-0: 6 ppm

PAC-1: 6 ppm

PAC-2: **7** ppm

PAC-3: **37** ppm

Determination in Air: Use NIOSH Analytical Method (IV). *AEGLs (Acute Emergency Guideline Levels) & ERPGs (Emergency Response Planning Guideline) are in **bold face**. #1606.

Routes of Entry: Inhalation, skin absorption, ingestion, skin and/or eye contact.

Harmful Effects and Symptoms

Short Term Exposure: Contact may cause burns to skin and eyes. May affect the iron metabolism, causing asphyxia. It is highly toxic. Forms cyanide in the body. This super toxic compound has a probable oral lethal dose in humans of less than 5 mg/kg or a taste (less than 7 drops) for a 70-kg (150 lb) person. Exposure results in headache, dizziness, rapid pulse, deep, rapid breathing, nausea, vomiting, unconsciousness, convulsions, and possible death. May cause cyanosis (blue coloration of skin and lips caused by lack of oxygen).

Long Term Exposure: Chronic exposure over long periods may cause fatigue and weakness. Can cause same general symptoms as hydrogen cyanide but onset of symptoms is likely to be slower. May cause liver and kidney damage.

Points of Attack: In animals: liver, kidney damage.

Medical Surveillance: Liver and kidney function tests.

First Aid: If this chemical gets into the eyes, remove any contact lenses at once and irrigate immediately for at least 15 min, occasionally lifting upper and lower lids. Seek medical attention immediately. If this chemical contacts the skin, remove contaminated clothing and wash immediately with soap and water. Seek medical attention immediately. If this chemical has been inhaled, remove from exposure, begin rescue breathing (using universal precautions, including resuscitation mask) if breathing has stopped and CPR if heart action has stopped. Transfer promptly to a medical facility. When this chemical has been swallowed, get medical attention. Give large quantities of water and induce vomiting. Do not make an unconscious person vomit.

Use amyl nitrate capsules if symptoms develop. All area employees should be trained regularly in emergency measures for cyanide poisoning and in CPR. A cyanide antidote kit should be kept in the immediate work area and must be rapidly available. Kit ingredients should be replaced every 1−2 years to ensure freshness. Persons trained in the use of this kit, oxygen use, and CPR must be quickly available.

Personal Protective Methods: Wear protective gloves and clothing to prevent any reasonable probability of skin contact. Safety equipment suppliers/manufacturers can provide recommendations on the most protective glove/clothing material for your operation. All protective clothing (suits, gloves, footwear, headgear) should be clean, available each day, and put on before work. Contact lenses should not be worn when working with this chemical. Wear splash-proof chemical goggles and face shield unless full face-piece respiratory protection is worn. For engineering controls, see NIOSH Criteria Document 212 *Nitriles*.

Respirator Selection: 60 ppm: CcrOv (APF = 10) [any chemical cartridge respirator with organic vapor cartridge (s)] or Sa (APF = 10) (any supplied-air respirator). *150 ppm:* Sa:Cf (APF = 25) (any supplied-air respirator operated in a continuous-flow mode) or PaprOv (APF = 25) [any powered, air-purifying respirator with organic vapor cartridge(s)]. *300 ppm:* CcrFOv (APF = 50) [any air-purifying, full-face-piece respirator (gas mask) with a chin-style, front- or back-mounted acid gas canister] or GmFOv (APF = 50) [any air-purifying, full-face-piece respirator (gas mask) with a chin-style, front- or back-mounted acid gas canister] or PaprTOv (APF = 50) [any powered, air-purifying respirator with a tight-fitting face-piece and organic vapor cartridge(s)] or SCBAF (APF = 50) (any self-contained breathing apparatus with a full face-piece) or SaF (APF = 50) (any supplied-air respirator with a full face-piece). *1000 ppm:* SaF: Pd,Pp (APF = 2000) (any supplied-air respirator that has a full face-piece and is operated in a pressure-demand or other positive-pressure mode). *Emergency or planned entry into unknown concentrations or IDLH conditions:* SCBAF: Pd,Pp (APF = 10,000) (any self-contained breathing apparatus that has a full face-piece and is operated in a pressure-demand or other positive-pressure mode) or SaF: Pd,Pp: ASCBA (APF = 10,000) (any supplied-air respirator that has a full face-piece and is operated in a pressure-demand or other positive-pressure mode in combination with an auxiliary self-contained breathing apparatus operated in a pressure-demand or other positive-pressure mode). *Escape:* GmFOv (APF = 50) [any air-purifying, full-face-piece respirator (gas mask) with a chin-style, front- or back-mounted organic vapor canister] or SCBAE (any appropriate escape-type, self-contained breathing apparatus).

Storage: Color Code—Red: Flammability Hazard: Store in a flammable liquid storage area or approved cabinet away from ignition sources and corrosive and reactive materials. Prior to working with this chemical you should be trained on its proper handling and storage. Before entering confined space where this chemical may be present, check to make sure that an explosive concentration is not a danger. Store in an explosion-proof refrigerator. Where possible, automatically pump liquid from drums or other storage containers to process containers. Sources of ignition, such as smoking and open flames, are prohibited where this chemical is handled, used, or stored. Metal containers involving the transfer of 5 gallons or more of this chemical should be grounded and bonded. Drums must be equipped with self-closing valves, pressure vacuum bungs, and flame arresters. Use only nonsparking tools and equipment, especially when opening and closing containers of this chemical. Wherever this chemical is used, handled, manufactured, or stored, use explosion-proof electrical equipment and fittings.

Shipping: This compound requires a shipping label of ""FLAMMABLE LIQUID, POISONOUS/TOXIC MATERIALS." It falls in Hazard Class 3 and Packing Group II.

Spill Handling: Evacuate and restrict persons not wearing protective equipment from area of spill or leak until cleanup is complete. Remove all ignition sources. Do not touch spilled material; stop leak if you can do it without risk. Use water spray to reduce vapors. *Small spills:* take up with sand or other noncombustible absorbent material and place into containers for later disposal. *Large spills:* dike far ahead of spill for later disposal. Establish forced ventilation to keep levels below explosive limit. Absorb liquids in vermiculite, dry sand, earth, peat, carbon, or a similar material and deposit in sealed containers. Keep this chemical out of a confined space, such as a sewer, because of the possibility of an explosion, unless the sewer is designed to prevent the buildup of explosive concentrations. It may be necessary to contain and dispose of this chemical as a hazardous waste. If material or contaminated runoff enters waterways, notify downstream users of potentially contaminated waters. Contact your local or federal environmental protection agency for specific recommendations. If employees are required to clean up spills, they must be properly trained and equipped. OSHA 1910.120(q) may be applicable.

Fire Extinguishing: This chemical is a flammable liquid. Poisonous gases, including hydrogen cyanide, are produced in fire. Use dry chemical, carbon dioxide, or alcohol foam extinguishers. Vapors are heavier than air and will collect in low areas. Vapors may travel long distances to ignition sources and flashback. Vapors in confined areas may explode when exposed to fire. Containers may explode in fire. Storage containers and parts of containers may rocket great distances, in many directions. If material or contaminated runoff enters waterways, notify downstream users of potentially contaminated waters. Notify local health and fire officials and pollution control agencies. From a secure, explosion-proof location, use water spray to cool exposed containers. If cooling streams are ineffective (venting sound increases in volume and pitch, tank discolors, or shows any signs of deforming), withdraw immediately to a secure position. If employees are expected to fight fires, they must be trained and equipped in OSHA 1910.156. The only respirators recommended for firefighting are self-contained breathing apparatuses that have full face-pieces and are operated in a pressure-demand or other positive-pressure mode.

Disposal Method Suggested: Alcoholic NaOH followed by calcium hypochlorite may be used. Incineration is also recommended.[22] Consult with environmental regulatory agencies for guidance on acceptable disposal practices. Generators of waste containing this contaminant (≥100 kg/mo) must conform with EPA regulations governing storage, transportation, treatment, and waste disposal.

References

US Environmental Protection Agency. (September 2, 1983). *Chemical Hazard Information Profile Draft Report: Propionitrile.* Washington, DC

US Environmental Protection Agency. (November 30, 1987). *Chemical Hazard Information Profile: Propionitrile.* Washington, DC: Chemical Emergency Preparedness Program

Propoxur   P:1180

Molecular Formula: $C_{11}H_{15}NO_3$

Synonyms: 58-12-315; Arprocarb; BAY39007; BAY 5122; Bayer 39007; Bayer B 5122; Baygon; Blattanex; Blattosep; Bolfo; Boruho; Boruho 50; Brygou; Carbamic acid, methyl-, o-isopropoxyphenyl ester; Dalf dust; ENT25,671; Invisi-gard; IPMC; o-(2-Isopropoxyphenyl) N-methylcarbamate; o-Isopropoxyphenyl N-methylcarbamate; o-Isopropoxyphenyl methylcarbamate; 2-Isopropoxyphenyl

N-methylcarbamate; 2-Isopropoxyphenyl methylcarbamate; 2-(1-Methylethoxy)phenyl N-methylcarbamate; OMS 33; PHC; Phenol, 2-(1-methylethoxy)-, methylcarbamate; Propotox; Propoxylor; Sendran; Suncide; Tendex; Unden

CAS Registry Number: 114-26-1

RTECS® Number: FC3150000

UN/NA & ERG Number: UN2757/151

EC Number: 204-043-8 [*Annex I Index No.:* 006-16-00-4]

Regulatory Authority and Advisory Bodies

US EPA Gene-Tox Program, Negative: *B. subtilis* rec assay; TRP reversion; Negative: *S. cerevisiae* gene conversion; Inconclusive: *B. subtilis* rec assay.

US EPA, FIFRA 1998 Status of Pesticides: RED completed.

Air Pollutant Standard Set. See below, "Permissible Exposure Limits in Air" section.

Clean Air Act: Hazardous Air Pollutants (Title I, Part A, Section 112).

US EPA Hazardous Waste Number (RCRA No.): U411.

RCRA, 40CFR261, Appendix 8 Hazardous Constituents.

RCRA 40CFR268.48; 61FR15654, Universal Treatment Standards: Wastewater (mg/L), 0.056; Nonwastewater (mg/kg), 1.4.

Reportable Quantity (RQ): 1 lb (0.454 kg).

EPCRA Section 313 Form R *de minimis* concentration reporting level: 1.0%.

US DOT Regulated Marine Pollutant (49CFR172.101, Appendix B).

California Proposition 65 Chemical: Cancer 8/11/06.

European/International Regulations: Hazard Symbol: T, N; Risk phrases: R25; R50/53; Safety phrases: S1/2; S37; S45; S60; S61 (see Appendix 4).

WGK (German Aquatic Hazard Class): 3—Severe hazard to waters.

Description: Propoxur is a colorless crystalline powder with a faint characteristic odor. Molecular weight = 209.27; Boiling point = N/A (decomposes); Freezing/Melting point = 91°C; Flash point ≥ 149°C. Hazard Identification (based on NFPA-704 M Rating System): Health 2, Flammability 1, Reactivity 0. Poor solubility in water.

Potential Exposure: Compound Description: Agricultural Chemical; Mutagen; Reproductive Effector; Human Data. Personnel engaged in the manufacture, formulation, and application of this carbamate agricultural chemical and pesticide.

Incompatibilities: Strong oxidizers, alkalis, heat, and moisture. Emits highly toxic methyl isocyanate fumes when heated to decomposition.

Permissible Exposure Limits in Air

OSHA PEL: None.

NIOSH REL: 0.5 mg/m^3 TWA.

ACGIH TLV®[1]: 0.5 mg/m^3 TWA; confirmed animal carcinogen with unknown relevance to humans. BEI$_A$ issued for Acetylcholinesterase-inhibiting pesticides.

Protective Action Criteria (PAC)

TEEL-0: 0.5 mg/m^3

PAC-1: 1.5 mg/m^3

PAC-2: 2.5 mg/m^3

PAC-3: 20 mg/m^3

DFG MAK: 2 mg/m^3 measured as the, inhalable fraction TWA; Peak Limitation Category II(8).

Australia: TWA 0.5 mg/m^3, 1993; Austria: MAK 0.5 mg/m^3, 1999; Belgium: TWA 0.5 mg/m^3, 1993; Denmark: TWA 0.5 mg/m^3, 1999; Finland: TWA 0.5 mg/m^3; STEL 1.5 mg/m^3, 1993; France: VME 0.5 mg/m^3, 1999; the Netherlands: MAC-TGG 0.5 mg/m^3, 2003; Poland: MAC (TWA) 0.5 mg/m^3; MAC (STEL) 2 mg/m^3, 1999; Switzerland: MAK-W 0.5 mg/m^3, 1999; Argentina, Bulgaria, Columbia, Jordan, South Korea, New Zealand, Singapore, Vietnam: ACGIH TLV®: confirmed animal carcinogen with unknown relevance to humans. Several states have set guidelines or standards for Baygon in ambient air[60] ranging from 5−20 μg/m^3 (North Dakota) to 8 μg/m^3 (Virginia) to 10 μg/m^3 (Connecticut) to 12 μg/m^3 (Nevada).

Determination in Air: No method available.

Determination in Water: Fish Tox = 168.01099000 MATC (LOW). Octanol−water coefficient: Log K_{ow} = 1.5.

Routes of Entry: Inhalation, ingestion, skin and/or eye contact.

Harmful Effects and Symptoms

Short Term Exposure: Propoxur can affect you when breathed in and quickly enters the body by passing through the skin. Severe poisoning can occur from skin contact. It is a moderately toxic carbamate chemical. Exposure can cause severe carbamate poisoning, with symptoms of headaches, sweating, nausea and vomiting, diarrhea, muscle twitching, loss of coordination, and even death. May affect the nervous system, liver, kidneys. A cholinesterase inhibitor.

Long Term Exposure: Propoxur may cause mutations. Handle with extreme caution. It may damage the developing fetus. Cholinesterase inhibitor; cumulative effect is possible. This chemical may damage the nervous system with repeated exposure, resulting in convulsions, respiratory failure. May cause liver damage. Human Tox = 94.85095 ppb CHCL (Chronic Human Carcinogen Level) (LOW).

Points of Attack: Central nervous system, liver, kidneys, gastrointestinal tract, blood cholinesterase.

Medical Surveillance: If symptoms develop or overexposure is suspected, the following may be useful: serum and RBC cholinesterase levels (a test for the enzyme in the body affected by propoxur). These tests are useful only if done 1−2 hours after exposure and can return to normal before the person feels well. Before employment and at regular times after that, the following are recommended: plasma and red blood cell cholinesterase levels (tests for the enzyme poisoned by this chemical). If exposure stops, plasma levels return to normal in 1−2 weeks while red blood cell levels may be reduced for 1−3 months. When cholinesterase enzyme levels are reduced by 25% or more below preemployment levels, risk of poisoning is increased, even if results are in lower ranges of "normal." Reassignment to work not involving carbamate pesticides is recommended until enzyme levels recover. If symptoms develop or overexposure occurs, repeat the above tests as soon as possible and get an examination of the nervous system. Also, consider complete blood count. Consider chest X-ray following acute overexposure.

First Aid: If this chemical gets into the eyes, remove any contact lenses at once and irrigate immediately for at least 15 min, occasionally lifting upper and lower lids. Seek medical attention immediately. If this chemical contacts the skin, remove contaminated clothing and wash immediately with soap and water. Seek medical attention immediately. If this chemical has been inhaled, remove from exposure, begin rescue breathing (using universal precautions, including resuscitation mask) if breathing has stopped and CPR if heart action has stopped. Transfer promptly to a medical facility. When this chemical has been swallowed, get medical attention. Give large quantities of water and induce vomiting. Do not make an unconscious person vomit.

Personal Protective Methods: Wear protective gloves and clothing to prevent any reasonable probability of skin contact. Safety equipment suppliers/manufacturers can provide recommendations on the most protective glove/clothing material for your operation. All protective clothing (suits, gloves, footwear, headgear) should be clean, available each day, and put on before work. Contact lenses should not be worn when working with this chemical. Wear dust-proof chemical goggles and face shield unless full face-piece respiratory protection is worn. Employees should wash immediately with soap when skin is wet or contaminated. Provide emergency showers and eyewash.

Respirator Selection: Where there is potential for exposures over 0.5 mg/m^3, use a NIOSH/MSHA- or European Standard EN149-approved supplied-air respirator with a full face-piece operated in the positive-pressure mode, or with a full face-piece, hood, or helmet in the continuous-flow mode; or use a NIOSH/MSHA- or European Standard EN149-approved self-contained breathing apparatus with a full face-piece operated in a pressure-demand or other positive-pressure mode.

Storage: Color Code—Blue: Health Hazard/Poison: Store in a secure poison location. Prior to working with this chemical you should be trained on its proper handling and storage. Store in tightly closed containers in a cool, well-ventilated area away from heat, moisture, oxidizers, alkaline environments. Where possible, automatically transfer material from drums or other storage containers to process containers. Sources of ignition, such as smoking and open flames, are prohibited where this chemical is handled, used, or stored. Metal containers involving the transfer of this chemical should be grounded and bonded. Wherever this chemical is used, handled, manufactured, or stored, use explosion-proof electrical equipment and fittings.

Shipping: Carbamate pesticides, solid, toxic, require a shipping label of "POISONOUS/TOXIC MATERIALS." This material falls in DOT Hazard Class 6.1 and Packing Group III.

Spill Handling: Evacuate persons not wearing protective equipment from area of spill or leak until cleanup is complete. Remove all ignition sources. Collect powdered material in the most convenient and safe manner and deposit in sealed containers. Ventilate area after cleanup is complete. It may be necessary to contain and dispose of this chemical as a hazardous waste. If material or contaminated runoff enters waterways, notify downstream users of potentially contaminated waters. Contact your local or federal environmental protection agency for specific recommendations. If employees are required to clean up spills, they must be properly trained and equipped. OSHA 1910.120(q) may be applicable. Soil Adsorption Index (K_{oc}) = 30.

Fire Extinguishing: This chemical is a combustible solid. Use dry chemical, carbon dioxide, water spray, or alcohol foam extinguishers. Highly toxic gases are produced in fire, including methyl isocyanate. If material or contaminated runoff enters waterways, notify downstream users of potentially contaminated waters. Notify local health and fire officials and pollution control agencies. From a secure, explosion-proof location, use water spray to cool exposed containers. If cooling streams are ineffective (venting sound increases in volume and pitch, tank discolors, or shows any signs of deforming), withdraw immediately to a secure position. If employees are expected to fight fires, they must be trained and equipped in OSHA 1910.156. The only respirators recommended for firefighting are self-contained breathing apparatuses that have full face-pieces and are operated in a pressure-demand or other positive-pressure mode.

Disposal Method Suggested: In accordance with 40CFR165, follow recommendations for the disposal of pesticides and pesticide containers. Must be disposed properly by following package label directions or by contacting your local or federal environmental control agency or by contacting your regional EPA office. Consult with environmental regulatory agencies for guidance on acceptable disposal practices. Generators of waste containing this contaminant ($\geq$100 kg/mo) must conform with EPA regulations governing storage, transportation, treatment, and waste disposal.

Reference

New Jersey Department of Health and Senior Services. (July 2005). *Hazardous Substances Fact Sheet: Propoxur.* Trenton, NJ

n-Propyl acetate   P:1190

Molecular Formula: $C_5H_{10}O_2$
Common Formula: $CH_3COOCH_2CH_2CH_3$
Synonyms: Acetate de propyle normal (French); Acetic acid, propyl ester; Acetic acid, *n*-propyl ester; 1-Acetoxypropane; Propyl acetate; *n*-Propyl acetate; 1-Propyl acetate; Propylacetate

Isopropyl Acetate—see separate record

CAS Registry Number: 109-60-4
RTECS® Number: AJ3675000
UN/NA & ERG Number: UN1276/129
EC Number: 203-686-1 [*Annex I Index No.:* 607-024-00-6]
Regulatory Authority and Advisory Bodies
Air Pollutant Standard Set. See below, "Permissible Exposure Limits in Air" section.
Canada, WHMIS, Ingredients Disclosure List Concentration 1.0%.
European/International Regulations: Hazard Symbol: F; Risk phrases: R1; R 36; R66; R67; Safety phrases: S2; S16; S26; S29; S33 (see Appendix 4).
WGK (German Aquatic Hazard Class): 1—Low hazard to waters.

Description: *n*-Propyl acetate is a colorless liquid with a mild, fruity odor. The odor threshold is 70 mg/m$^{3[41]}$ and 2.8 mg/m^3 (New Jersey Fact Sheet). Molecular weight = 102.15 (both isomers); Specific gravity (H$_2$O:1) = 0.84 at 25°C; Boiling point = 101.7°C; Freezing/Melting point = −92.2°C; Vapor pressure = 25 mmHg at 25°C; Flash point = 13°C; Autoignition temperature = 450°C. Explosive limits: LEL = 1.7% at 38°C; UEL = 8.0%. Hazard Identification (based on NFPA-704 M Rating System): Health 1, Flammability 3, Reactivity 0. Slightly soluble in water; solubility = 2%.

Potential Exposure: Compound Description (both isomers): Human Data; Primary Irritant. Propyl acetate is a used as a solvent for plastics and cellulose ester resins; perfume ingredient; component of food flavoring. It is also used as a chemical intermediate.

Incompatibilities: Contact with nitrates, strong oxidizers, strong alkalis, strong acids; may pose risk of fire and explosions. Attacks plastic.

Permissible Exposure Limits in Air
Conversion factor: 1 ppm = 4.18 mg/m^3 at 25°C & 1 atm.
OSHA PEL: 200 ppm/840 mg/m^3 TWA.
NIOSH REL: 200 ppm/840 mg/m^3 TWA; 250 ppm/1050 mg/m^3 STEL.
ACGIH TLV®[1]: 200 ppm/835 mg/m^3 TWA; 250 ppm/1040 mg/m^3 STEL.
NIOSH IDLH: 1700 ppm.
Protective Action Criteria (PAC)
TEEL-0: 200 ppm
PAC-1: 250 ppm
PAC-2: 250 ppm
PAC-3: 1700 ppm
DFG MAK: 100 ppm/420 mg/m^3 TWA; Peak Limitation Category I(2); Pregnancy Risk Group D.
Australia: TWA 200 ppm (840 mg/m^3); STEL 250 ppm, 1993; Austria: MAK 200 ppm (840 mg/m^3), 1999; Belgium: TWA 200 ppm (835 mg/m^3); STEL 250 ppm (1040 mg/m^3), 1993; Finland: TWA 200 ppm (840 mg/m^3); STEL 250 ppm (1050 mg/m^3), 1993; France: VME 200 ppm (840 mg/m^3), 1999; Hungary: TWA 200 mg/m^3; STEL 600 mg/m^3, [skin], 1993; the Netherlands: MAC-TGG 420 mg/m^3, 2003; Norway: TWA 100 ppm

(420 mg/m^3), 1999; the Philippines: TWA 200 ppm (840 mg/m^3), 1993; Poland: MAC (TWA) 200 mg/m^3; MAC (STEL) 1000 mg/m^3, 1999; Russia: TWA 200 ppm; STEL 200 mg/m^3, 1993; Switzerland: MAK-W 200 ppm (840 mg/m^3), KZG-W 400 ppm, 1999; Turkey: TWA 200 ppm (840 mg/m^3), 1993; United Kingdom: TWA 200 ppm (849 mg/m^3); STEL 250 ppm (1060 mg/m^3); Argentina, Bulgaria, Columbia, Jordan, South Korea, New Zealand, Singapore, Vietnam: ACGIH TLV®: STEL 250 ppm.
Several states have set guidelines or standards for propyl acetate in ambient air[60] ranging from 8.4−10.5 mg/m^3 (North Dakota) to 14 mg/m^3 (Virginia) to 16.8 mg/m^3 (Connecticut) to 20 mg/m^3 (Nevada).

Determination in Air: Use NIOSH Analytical Method (IV) #1401, alcohols II; #1405, alcohols, Combined; OSHA Analytical Method 7, Organic Vapors.

Determination in Water: Octanol−water coefficient: Log K_{ow} = 1.23.

Routes of Entry: Inhalation, ingestion, skin and/or eye contact.

Harmful Effects and Symptoms
Short Term Exposure: *n*-Propyl acetate can affect you when breathed in. Exposure can irritate the eyes, nose, and throat. Very high levels are narcotic and may affect the nervous system and cause you to feel dizzy, lightheaded, and to pass out.

Long Term Exposure: Prolonged or repeated contact can cause drying and cracking of the skin.

Points of Attack: Eyes, skin, respiratory system, central nervous system.

Medical Surveillance: There is no special test for this substance. However, if illness occurs or overexposure is suspected, medical attention is recommended.

First Aid: If this chemical gets into the eyes, remove any contact lenses at once and irrigate immediately for at least 15 min, occasionally lifting upper and lower lids. Seek medical attention immediately. If this chemical contacts the skin, remove contaminated clothing and wash immediately with soap and water. Seek medical attention immediately. If this chemical has been inhaled, remove from exposure, begin rescue breathing (using universal precautions, including resuscitation mask) if breathing has stopped and CPR if heart action has stopped. Transfer promptly to a medical facility. When this chemical has been swallowed, get medical attention. Give large quantities of water and induce vomiting. Do not make an unconscious person vomit.

Personal Protective Methods: Wear protective gloves and clothing to prevent any reasonable probability of skin contact. Safety equipment suppliers/manufacturers can provide recommendations on the most protective glove/clothing material for your operation. Butyl rubber, polyvinyl alcohol, and Silvershield™ are recommended protective materials. All protective clothing (suits, gloves, footwear, headgear) should be clean, available each day, and put on before work. Contact lenses should not be worn when working with this chemical. Wear splash-proof chemical goggles and face shield unless full face-piece respiratory protection is worn. Employees should wash immediately with soap when skin is wet or contaminated. Provide emergency showers and eyewash.

Respirator Selection: Up to 1700 ppm: Sa:Cf (APF = 25) (any supplied-air respirator operated in a continuous-flow mode) or CcrFOv (APF = 50) [any air-purifying, full-face-piece respirator (gas mask) with a chin-style, front- or back-mounted acid gas canister] or GmFOv (APF = 50) [any air-purifying, full-face-piece respirator (gas mask) with a chin-style, front- or back-mounted organic vapor canister] or PaprOv (APF = 25) [any powered, air-purifying respirator with organic vapor cartridge(s)] or SCBAF (APF = 50) (any self-contained breathing apparatus with a full face-piece) or SaF (APF = 50) (any supplied-air respirator with a full face-piece). *Emergency or planned entry into unknown concentrations or IDLH conditions:* SCBAF: Pd,Pp (APF = 10,000) (any self-contained breathing apparatus that has a full face-piece and is operated in a pressure-demand or other positive-pressure mode) or SaF: Pd,Pp: ASCBA (APF = 10,000) (any supplied-air respirator that has a full face-piece and is operated in a pressure-demand or other positive-pressure mode in combination with an auxiliary self-contained breathing apparatus operated in a pressure-demand or other positive-pressure mode). *Escape:* GmFOv (APF = 50) [any air-purifying, full-face-piece respirator (gas mask) with a chin-style, front- or back-mounted organic vapor canister] or SCBAE (any appropriate escape-type, self-contained breathing apparatus).

Storage: Color Code—Red: Flammability Hazard: Store in a flammable liquid storage area or approved cabinet away from ignition sources and corrosive and reactive materials. Prior to working with this chemical you should be trained on its proper handling and storage. Before entering confined space where this chemical may be present, check to make sure that an explosive concentration is not a danger. *n*-Propyl acetate must be stored to avoid contact with nitrates, strong oxidizers (such as chlorine, bromine, and fluorine), and strong acids (such as hydrochloric, sulfuric, and nitric) since violent reactions occur. Store in tightly closed containers in a cool, well-ventilated area away from heat. *n*-Propyl acetate will dissolve some plastics and resins. Sources of ignition, such as smoking and open flames, are prohibited where *n*-propyl acetate is handled, used, or stored. Metal containers involving the transfer of 5 gallons or more of *n*-propyl acetate should be grounded and bonded. Drums must be equipped with self-closing valves, pressure vacuum bungs, and flame arresters. Use only non-sparking tools and equipment, especially when opening and closing containers of *n*-propyl acetate.

Shipping: n-Propyl acetate requires a shipping label of "FLAMMABLE LIQUID." It falls in Hazard Class 3 and Packing Group II.

Spill Handling: Evacuate and restrict persons not wearing protective equipment from area of spill or leak until cleanup is complete. Remove all ignition sources. Establish forced ventilation to keep levels below explosive limit. Absorb liquids in vermiculite, dry sand, earth, peat, carbon, or a similar material and deposit in sealed containers. Keep this chemical out of a confined space, such as a sewer, because of the possibility of an explosion, unless the sewer is designed to prevent the buildup of explosive concentrations. It may be necessary to contain and dispose of this chemical as a hazardous waste. If material or contaminated runoff enters waterways, notify downstream users of potentially contaminated waters. Contact your local or federal environmental protection agency for specific recommendations. If employees are required to clean up spills, they must be properly trained and equipped. OSHA 1910.120(q) may be applicable.

Fire Extinguishing: This chemical is a flammable liquid. Poisonous gases are produced in fire. Use dry chemical, carbon dioxide, or alcohol foam extinguishers. Vapors are heavier than air and will collect in low areas. Vapors may travel long distances to ignition sources and flashback. Vapors in confined areas may explode when exposed to fire. Containers may explode in fire. Storage containers and parts of containers may rocket great distances, in many directions. If material or contaminated runoff enters waterways, notify downstream users of potentially contaminated waters. Notify local health and fire officials and pollution control agencies. From a secure, explosion-proof location, use water spray to cool exposed containers. If cooling streams are ineffective (venting sound increases in volume and pitch, tank discolors, or shows any signs of deforming), withdraw immediately to a secure position. If employees are expected to fight fires, they must be trained and equipped in OSHA 1910.156. The only respirators recommended for firefighting are self-contained breathing apparatuses that have full face-pieces and are operated in a pressure-demand or other positive-pressure mode.

Disposal Method Suggested: Dissolve or mix the material with a combustible solvent and burn in a chemical incinerator equipped with an afterburner and scrubber. All federal, state, and local environmental regulations must be observed.

Reference

New Jersey Department of Health and Senior Services. (March 2001). *Hazardous Substances Fact Sheet: n-Propyl Acetate.* Trenton, NJ

Propyl alcohol   P:1200

Molecular Formula: C_3H_8O (n-); $C_6H_{14}O$ (iso-)
Synonyms: Alcohol C-3; Alcool propylique (French); Ethyl carbinol; 1-Hydroxypropane; Optal; Osmosol extra; Propanol-1; 1-Propanol; *n*-Propanol; Propanole (German); Propyl alcohol; Propyl alcohol, *normal*; 1-Propyl alcohol; *n*-Propyl alkohol (German); Propylic alcohol

(*iso-*) 2-Methylpentan-1-ol

CAS Registry Number: 71-23-8; 105-30-6 (iso-)

RTECS® Number: UH8225000 (n-)

UN/NA & ERG Number: UN1274 (*n*-propanol)/129

EC Number: 200-746-9 [*Annex I Index No.:* 603-003-00-0 (*n*-)]; 203-285-1 [2-methylpentan-1-ol]

Regulatory Authority and Advisory Bodies

US EPA Gene-Tox Program, Negative: *In vitro* SCE—nonhuman.

US EPA, FIFRA 1998 Status of Pesticides: Canceled.

Air Pollutant Standard Set. See below, "Permissible Exposure Limits in Air" section.

Canada, WHMIS, Ingredients Disclosure List Concentration 1.0%.

European/International Regulations: Hazard Symbol: F, Xi; Risk phrases: R11; R47; R67; Safety phrases: S2; S7; S16; S26; S39 (see Appendix 4).

WGK (German Aquatic Hazard Class): 1—Low hazard to waters.

Description: The two isomers of propyl alcohol are *n*-propyl alcohol and isopropyl alcohol. The odor threshold is 5.3 ppm[41] and 2.6 ppm (NJ). Both are colorless, volatile liquids. Isopropyl alcohol is discussed in a separate entry in this volume. *n*-Propanol: Molecular weight = 60.11; Specific gravity (H_2O:1) = 0.81 at 25°C; Boiling point = 97°C; Freezing/Melting point = −127°C; Vapor pressure = 15 mmHg at 25°C; Flash point = 22.2°C; Autoignition temperature = 412°C. Explosive limits: LEL = 2.2%; UEL = 13.7%. Hazard Identification (based on NFPA-704 M Rating System): Health 1, Flammability 3, Reactivity 0. Soluble in water.

Potential Exposure: Compound Description (*n*-): Agricultural Chemical; Tumorigen, Mutagen; Reproductive Effector; Human Data; Primary Irritant. *n*-Propyl alcohol is used as a solvent in lacquers, dopes; to make cosmetics, dental lotions, cleaners, polishes, and pharmaceuticals; as a surgical antiseptic. It is a solvent for vegetable oils, natural gums and resins; rosin, shellac, certain synthetic resins; ethyl cellulose and butyral; as a degreasing agent; as a chemical intermediate.

Incompatibilities: Strong oxidizers.

Permissible Exposure Limits in Air

OSHA PEL: 200 ppm/500 mg/m³ TWA.

NIOSH REL: 200 ppm/500 mg/m³ TWA; 250 ppm/625 mg/m³ STEL [skin].

ACGIH TLV®[1]: 100 ppm/246 mg/m³ TWA; Not Classifiable as a Human Carcinogen.

NIOSH IDLH: 800 ppm.

Protective Action Criteria (PAC)

TEEL-0: 200 ppm

PAC-1: 250 ppm

PAC-2: 250 ppm

PAC-3: 800 ppm

isopropanol

OSHA PEL: 400 ppm/980 mg/m³ TWA.

NIOSH REL: 400 ppm/980 mg/m³ TWA; 500 ppm/1225 mg/m³ STEL

ACGIH TLV®[1]: 200 ppm/492 mg/m³ TWA; 400 ppm/984 mg/m³ STEL, Not Classifiable as a Human Carcinogen; BEI issued.

TEEL-0: 0.25 ppm

PAC-1: 0.75 ppm

PAC-2: 5 ppm

PAC-3: 150 ppm

DFG MAK: 200 ppm/500 mg/m³ [skin]; BAT: 50 mg [Acetone]/L in blood or urine/end-of-shift as isopropyl alcohol.

Australia: TWA 200 ppm (500 mg/m³); 200 ppm (492 mg/m³) STEL [skin], 1993; Austria: MAK 200 ppm (500 mg/m³), 1999; Belgium: TWA 200 ppm (492 mg/m³); STEL 250 ppm, [skin], 1993; Denmark: TWA 200 ppm (500 mg/m³), [skin], 1999; Finland: TWA 200 ppm (500 mg/m³); STEL 250 ppm, [skin], 1999; France: VME 200 ppm (500 mg/m³), 1999; Hungary: TWA 100 mg/m³; STEL 200 mg/m³, 1993; Norway: TWA 100 ppm (245 mg/m³), 1999; Poland: MAC (TWA) 200 mg/m³; MAC (STEL) 600 mg/m³, 1999; Russia: STEL 10 mg/m³, 1993; Sweden: NGV 150 ppm (350 mg/m³), KTV 250 ppm (600 mg/m³), 1999; Switzerland: MAK-W 200 ppm (500 mg/m³), [skin], 1999; Turkey: TWA 200 ppm (500 mg/m³), 1993; United Kingdom: TWA 200 ppm (500 mg/m³); STEL 250 ppm, [skin], 2000; Argentina, Bulgaria, Columbia, Jordan, South Korea, New Zealand, Singapore, Vietnam: ACGIH TLV®: STEL 250 ppm [skin]. Russia[35, 43] set a MAC value of 0.3 mg/m³ for ambient air in residential areas[60] both on a momentary and a daily average basis. Several states have set guidelines or standards for propanol in ambient air[60] ranging from 0.67 mg/m³ (Massachusetts) to 5.0−6.25 mg/m³ (North Dakota) to 8.0 mg/m³ (Virginia) to 10.0 mg/m³ (Connecticut) to 11.905 mg/m³ (Nevada).

Determination in Air: Use NIOSH Analytical Method #1401, alcohols.[18]

Permissible Concentration in Water: EPA[32] has suggested a permissible ambient goal of 6900 μg/L based on health effects. Russia[35, 43] set a MAC of 0.25 mg/L in water bodies used for domestic purposes.

Routes of Entry: Inhalation of vapor, percutaneous absorption, ingestion, skin and/or eye contact.

Harmful Effects and Symptoms

Short Term Exposure: Propyl alcohol can affect you when breathed in and by passing through your skin. Irritates the eyes, skin, and respiratory tract. Prolonged skin contact can cause a burning sensation and rash. Exposure to high concentrations can affect the central nervous system and cause headaches, drowsiness, dizziness, and confusion; may cause ataxia, gastrointestinal pain, abdominal cramps, nausea, vomiting, diarrhea. High levels can cause unconsciousness.

Long Term Exposure: Repeated skin exposure may cause drying and cracking of the skin. Propyl alcohol may cause mutations. Handle with extreme caution. This chemical is listed by the state of New Jersey as a Special Health Hazard and a data sheet (listed below) states that this chemical may be a carcinogen in humans: It has been shown to cause liver carcinomas and sarcoma, spleen sarcoma, and leukemia in animals. It should be treated with caution.

Points of Attack: Eyes, skin, respiratory system, gastrointestinal tract, central nervous system. May cause liver damage. It may cause brain or nerve damage.

Medical Surveillance: Liver function tests. Evaluate for brain and nerve effects including cerebellar, autonomic and peripheral nervous systems, changes in memory, concentration, sleeping pattern, mood, headaches, and fatigue. Positive and borderline individuals should be referred for neuropsychological testing.

First Aid: If this chemical gets into the eyes, remove any contact lenses at once and irrigate immediately for at least 15 min, occasionally lifting upper and lower lids. Seek medical attention immediately. If this chemical contacts the skin, remove contaminated clothing and wash immediately with soap and water. Seek medical attention immediately. If this chemical has been inhaled, remove from exposure, begin rescue breathing (using universal precautions, including resuscitation mask) if breathing has stopped and CPR if heart action has stopped. Transfer promptly to a medical facility. When this chemical has been swallowed, get medical attention. Give large quantities of water and induce vomiting. Do not make an unconscious person vomit.

Personal Protective Methods: Wear protective gloves and clothing to prevent any reasonable probability of skin contact. Safety equipment suppliers/manufacturers can provide recommendations on the most protective glove/clothing material for your operation. Neoprene™, Teflon™, nitrile, and polyvinyl acetate are recommended protective materials. All protective clothing (suits, gloves, footwear, headgear) should be clean, available each day, and put on before work. Contact lenses should not be worn when working with this chemical. Wear splash-proof chemical goggles and face shield unless full face-piece respiratory protection is worn. Employees should wash immediately with soap when skin is wet or contaminated. Provide emergency showers and eyewash.

Respirator Selection: Up to 800 ppm: CcrOv (APF = 10) [any chemical cartridge respirator with organic vapor cartridge(s)] or PaprOv (APF = 25) [any powered, air-purifying respirator with organic vapor cartridge(s)] or GmFOv (APF = 50) [any air-purifying, full-face-piece respirator (gas mask) with a chin-style, front- or back-mounted organic vapor canister]; Sa (APF = 10) (any supplied-air respirator) or SCBAF (APF = 50) (any self-contained breathing apparatus with a full face-piece). *Emergency or planned entry into unknown concentrations or IDLH conditions:* SCBAF: Pd,Pp (APF = 10,000) (any self-contained breathing apparatus that has a full face-piece and is operated in a pressure-demand or other positive-pressure mode) or SaF: Pd,Pp: ASCBA (APF = 10,000) (any supplied-air respirator that has a full face-piece and is operated in a pressure-demand or other positive-pressure mode in combination with an auxiliary self-contained breathing apparatus operated in a pressure-demand or other positive-pressure mode). *Escape:* GmFOv (APF = 50) [any air-purifying, full-face-piece respirator (gas mask) with a chin-style, front- or back-mounted organic vapor canister] or SCBAE (any appropriate escape-type, self-contained breathing apparatus).

Note: Substance reported to cause eye irritation or damage; may require eye protection.

Storage: Color Code—Red: Flammability Hazard: Store in a flammable liquid storage area or approved cabinet away from ignition sources and corrosive and reactive materials. Prior to working with this chemical you should be trained on its proper handling and storage. Before entering confined space where this chemical may be present, check to make sure that an explosive concentration is not a danger. Propyl alcohol must be stored to avoid contact with strong oxidizers (such as chlorine and bromine) since violent reactions occur. Store in tightly closed containers in a cool, well-ventilated area away from heat. Sources of ignition, such as smoking and open flames, are prohibited where propyl alcohol is used, handled, or stored in a manner that could create a potential fire or explosion hazard. Metal containers involving the transfer of 5 gallons or more of propyl alcohol should be grounded and bonded. Drums must be equipped with self-closing valves, pressure vacuum bungs, and flame arresters. A regulated, marked area should be established where this chemical is handled, used, or stored in compliance with OSHA Standard 1910.1045.

Shipping: This compound requires a shipping label of "FLAMMABLE LIQUID." It falls in Hazard Class 3 and Packing Group II or III.

Spill Handling: Evacuate and restrict persons not wearing protective equipment from area of spill or leak until cleanup is complete. Remove all ignition sources. Establish forced ventilation to keep levels below explosive limit. Absorb liquids in vermiculite, dry sand, earth, peat, carbon, or a similar material and deposit in sealed containers. Keep this chemical out of a confined space, such as a sewer, because of the possibility of an explosion, unless the sewer is designed to prevent the buildup of explosive concentrations. It may be necessary to contain and dispose of this chemical as a hazardous waste. If material or contaminated runoff enters waterways, notify downstream users of potentially contaminated waters. Contact your local or federal environmental protection agency for specific recommendations. If employees are required to clean up spills, they must be properly trained and equipped. OSHA 1910.120(q) may be applicable.

Fire Extinguishing: This chemical is a flammable liquid. Poisonous gases are produced in fire. Use dry chemical, carbon dioxide, or alcohol foam extinguishers. Vapors are heavier than air and will collect in low areas. Vapors may travel long distances to ignition sources and flashback. Vapors in confined areas may explode when exposed to fire. Containers may explode in fire. Storage containers and parts of containers may rocket great distances, in many directions. If material or contaminated runoff enters waterways, notify downstream users of potentially contaminated waters. Notify local health and fire officials and pollution control agencies. From a secure, explosion-proof location, use water spray to cool exposed containers. If cooling streams are ineffective (venting sound increases in volume and pitch, tank discolors, or shows any signs of deforming), withdraw immediately to a secure position. If employees are expected to fight fires, they must be trained and equipped in OSHA 1910.156. The only respirators recommended for firefighting are self-contained breathing apparatuses that have full face-pieces and are operated in a pressure-demand or other positive-pressure mode.

Disposal Method Suggested: Dissolve or mix the material with a combustible solvent and burn in a chemical incinerator equipped with an afterburner and scrubber. All federal, state, and local environmental regulations must be observed.

References

US Environmental Protection Agency. (March 31, 1983). *Chemical Hazard Information Profile Draft Report: n-Propanol.* Washington, DC

New Jersey Department of Health and Senior Services. (June 2005). *Hazardous Substances Fact Sheet: Propyl Alcohol.* Trenton, NJ

Propylamine   P:1210

Molecular Formula: C_3H_9N
Common Formula: $CH_3CH_2CH_2NH_2$
Synonyms: 1-Aminopropane; 1-Iodopropane; Mono-*n*-propylamine; Monopropylamine; Propanamine; *n*-Propilamina (Spanish); Propylamine
CAS Registry Number: 107-10-8
RTECS® Number: UN9100000
UN/NA & ERG Number: UN1277/132
EC Number: 203-462-3
Regulatory Authority and Advisory Bodies
Air Pollutant Standard Set. See below, "Permissible Exposure Limits in Air" section.
US EPA Hazardous Waste Number (RCRA No.): U194.
RCRA, 40CFR261, Appendix 8 Hazardous Constituents.
Reportable Quantity (RQ): 5000 lb (2270 kg).
Canada, WHMIS, Ingredients Disclosure List Concentration 1.0%.
European/International Regulations: not listed in Annex 1.
WGK (German Aquatic Hazard Class): 1—Low hazard to waters.
Description: *n*-Propylamine is a water-white liquid with a strong irritating odor similar to that of ammonia. Molecular weight = 59.13; Specific gravity (H_2O:1) = 0.71 at 25°C; Boiling point = 49°C; Freezing/Melting point = −83°C; Flash point = < −37°C; Autoignition temperature = 318°C. Explosive limits: LEL = 2.0%; UEL = 10.4%. Hazard Identification (based on NFPA-704 M Rating System): Health 3, Flammability 3, Reactivity 0. Soluble in water.

Potential Exposure: Compound Description: Primary Irritant. Propylamine is used to make textile resins, drugs, pesticides, and other chemicals.
Incompatibilities: Violent reaction with oxidizers and mercury, strong acids, organic anhydrides, isocyanates, aldehydes, nitroparaffins, halogenated hydrocarbons, alcohols, and many other compounds. Attacks many metals and alloys, especially copper. Aqueous solutions may attack glass.
Permissible Exposure Limits in Air
Protective Action Criteria (PAC)
TEEL-0: 15 ppm
PAC-1: 50 ppm
PAC-2: 250 ppm
PAC-3: 250 ppm
Finland: STEL 5 ppm (12 mg/m^3), [skin], 1999; Russia[43] set a MAC of 5 mg/m^3 in work-place air. It should be recognized that propylamine can be absorbed through your skin, thereby increasing your exposure.
Permissible Concentration in Water: Russia[43] set a MAC of 0.5 mg/L in water bodies used for domestic purposes.
Determination in Water: Octanol−water coefficient: Log $K_{ow} = 0.2$.
Routes of Entry: Inhalation, ingestion, skin and/or eye contact. Absorbed through the skin.
Harmful Effects and Symptoms
Short Term Exposure: Propylamine can affect you when breathed in and by passing through your skin. Corrosive to the eyes, skin, and respiratory tract. Propylamine can cause severe eye burns leading to permanent damage and blindness. Contact can cause severe skin burns. Breathing Propylamine can irritate the lungs, causing coughing and/or shortness of breath. Higher exposures can cause a buildup of fluid in the lungs (pulmonary edema). This can cause death. Overexposure may damage lungs, liver, kidneys, and heart muscle.
Long Term Exposure: Repeated lower exposure may damage the lungs, liver, kidneys, and/or heart muscle. Some amines cause skin or lung sensitization and allergy; however, it is not known if this chemical causes these allergies.
Points of Attack: Lungs, liver, kidneys, heart.
Medical Surveillance: Before beginning employment and at regular times after that, the following are recommended: lung function tests. If symptoms develop or overexposure is suspected, the following may also be useful: tests for kidney and liver function. Consider chest X-ray after acute overexposure. Evaluation by a qualified allergist, including careful exposure history and special testing, may help diagnose skin allergy.
First Aid: If this chemical gets into the eyes, remove any contact lenses at once and irrigate immediately for at least 15 min, occasionally lifting upper and lower lids. Seek medical attention immediately. If this chemical contacts the skin, remove contaminated clothing and wash immediately with soap and water. Seek medical attention immediately. If this chemical has been inhaled, remove from exposure, begin rescue breathing (using universal precautions, including

resuscitation mask) if breathing has stopped and CPR if heart action has stopped. Transfer promptly to a medical facility. When this chemical has been swallowed, get medical attention. Give large quantities of water and induce vomiting. Do not make an unconscious person vomit. Medical observation is recommended for 24–48 h after breathing overexposure, as pulmonary edema may be delayed. As first aid for pulmonary edema, a doctor or authorized paramedic may consider administering a corticosteroid spray.

Personal Protective Methods: Wear protective gloves and clothing to prevent any reasonable probability of skin contact. Safety equipment suppliers/manufacturers can provide recommendations on the most protective glove/clothing material for your operation. Teflon™ and polyvinyl acetate are among the recommended protective materials. All protective clothing (suits, gloves, footwear, headgear) should be clean, available each day, and put on before work. Contact lenses should not be worn when working with this chemical. Wear splash-proof chemical goggles and face shield unless full face-piece respiratory protection is worn. Employees should wash immediately with soap when skin is wet or contaminated. Provide emergency showers and eyewash.

Respirator Selection: Where there is potential for exposure to propylamine, use a NIOSH/MSHA- or European Standard EN149-approved supplied-air respirator with a full face-piece operated in the positive-pressure mode, or with a full face-piece, hood, or helmet in the continuous-flow mode; or use a MSHA/NOSH-approved self-contained breathing apparatus with a full face-piece operated in a pressure-demand or other positive-pressure mode.

Storage: Color Code—Red: Flammability Hazard: Store in a flammable liquid storage area or approved cabinet away from ignition sources and corrosive and reactive materials. Prior to working with this chemical you should be trained on its proper handling and storage. Before entering confined space where this chemical may be present, check to make sure that an explosive concentration is not a danger. Propylamine is incompatible with strong acids (such as hydrochloric, sulfuric, and nitric), acid anhydrides, acid chlorides, strong oxidizers (such as chlorine, bromine, and fluorine), carbon dioxide, and triethynyl aluminum. Store in tightly closed containers in a cool, well-ventilated area away from oxidizers. Sources of ignition, such as smoking and open flames, are prohibited where Propylamine is used, handled, or stored in a manner that could create a potential fire or explosion hazard. Metal containers involving the transfer of 5 gallons or more of propylamine should be grounded and bonded. Drums must be equipped with self-closing valves, pressure vacuum bungs, and flame arresters. Use only nonsparking tools and equipment, especially when opening and closing containers of propylamine. Wherever propylamine is used, handled, manufactured, or stored, use explosion-proof electrical equipment and fittings.

Shipping: This compound requires a shipping label of "FLAMMABLE LIQUID, CORROSIVE." It falls in Hazard Class 3 and Packing Group II.

Spill Handling: Evacuate and restrict persons not wearing protective equipment from area of spill or leak until cleanup is complete. Remove all ignition sources. Establish forced ventilation to keep levels below explosive limit. Absorb liquids in vermiculite, dry sand, earth, peat, carbon, or a similar material and deposit in sealed containers. Keep this chemical out of a confined space, such as a sewer, because of the possibility of an explosion, unless the sewer is designed to prevent the buildup of explosive concentrations. It may be necessary to contain and dispose of this chemical as a hazardous waste. If material or contaminated runoff enters waterways, notify downstream users of potentially contaminated waters. Contact your local or federal environmental protection agency for specific recommendations. If employees are required to clean up spills, they must be properly trained and equipped. OSHA 1910.120(q) may be applicable.

Fire Extinguishing: This chemical is a flammable liquid. Poisonous gases are produced in fire. Use dry chemical, carbon dioxide, or alcohol foam extinguishers. Water may be ineffective for fighting fires. Vapors are heavier than air and will collect in low areas. Vapors may travel long distances to ignition sources and flashback. Vapors in confined areas may explode when exposed to fire. Containers may explode in fire. Storage containers and parts of containers may rocket great distances, in many directions. If material or contaminated runoff enters waterways, notify downstream users of potentially contaminated waters. Notify local health and fire officials and pollution control agencies. From a secure, explosion-proof location, use water spray to cool exposed containers. If cooling streams are ineffective (venting sound increases in volume and pitch, tank discolors, or shows any signs of deforming), withdraw immediately to a secure position. If employees are expected to fight fires, they must be trained and equipped in OSHA 1910.156. The only respirators recommended for firefighting are self-contained breathing apparatuses that have full face-pieces and are operated in a pressure-demand or other positive-pressure mode.

Disposal Method Suggested: Consult with environmental regulatory agencies for guidance on acceptable disposal practices. Generators of waste containing this contaminant ($\geq$100 kg/mo) must conform with EPA regulations governing storage, transportation, treatment, and waste disposal.

Reference
New Jersey Department of Health and Senior Services. (March 2001). *Hazardous Substances Fact Sheet: Propylamine*. Trenton, NJ

Propyl chloroformate   P:1220

Molecular Formula: $C_4H_7ClO_2$
Common Formula: $ClCOOC_3H_7$
Synonyms: Carbonochloridic acid, Propyl ester; Chloroformic acid propyl ester; Propyl chlorocarbonate; *n*-Propyl chloroformate

CAS Registry Number: 109-61-5
RTECS® Number: LQ6830000
UN/NA & ERG Number: UN2740/155
EC Number: 203-687-7 [*Annex I Index No.:* 607-142-00-8]
Regulatory Authority and Advisory Bodies
Department of Homeland Security Screening Threshold Quantity (pounds): *Release hazard* 10,000 (≥1.00% concentration).
Clean Air Act: Accidental Release Prevention/Flammable Substances, (Section 112[r], Table 3), TQ = 15,000 lb (6810 kg).
Superfund/EPCRA 40CFR355, Extremely Hazardous Substances: TPQ = 500 lb (227 kg).
Reportable Quantity (RQ): 500 lb (227 kg).
US DOT 49CFR172.101, Inhalation Hazardous Chemical.
Canada, WHMIS, Ingredients Disclosure List Concentration 1.0%.
European/International Regulations: Hazard Symbol: F, T; Risk phrases: R11; R23; R34; Safety phrases: S1/2; S16; S26; S36; S45 (see Appendix 4).
WGK (German Aquatic Hazard Class): No value assigned.
Description: Propyl chloroformate is a colorless liquid. Molecular weight = 122.56; Boiling point = 114−116°C; Flash point = −50°C. Hazard Identification (based on NFPA-704 M Rating System): Health 3, Flammability 4, Reactivity 2W. Insoluble in water; slowly decomposes.
Potential Exposure: Propyl chloroformate is used in organic synthesis; as an intermediate for polymerization initiators; military poison gas.
Incompatibilities: Oxidizers, water, and alcohol
Permissible Exposure Limits in Air
Protective Action Criteria (PAC)*
TEEL-0: 0.75 ppm
PAC-1: 2 ppm
PAC-2: **3.7** ppm
PAC-3: **11** ppm
*AEGLs (Acute Emergency Guideline Levels) & ERPGs (Emergency Response Planning Guideline) are in **bold face**.
Routes of Entry: Inhalation, ingestion, skin and/or eye contact. Absorbed through the skin.
Harmful Effects and Symptoms
Short Term Exposure: A lacrimator; vapors cause tearing. Corrosive to eyes, skin, and mucous membranes. May cause burns and permanent eye damage. Poisonous: may be fatal if inhaled, swallowed, or absorbed through skin.
Long Term Exposure: May cause lung damage.
Points of Attack: Lungs.
Medical Surveillance: Lung function tests. Consider chest X-ray following acute overexposure.
First Aid: If this chemical gets into the eyes, remove any contact lenses at once and irrigate immediately for at least 15 min, occasionally lifting upper and lower lids. Seek medical attention immediately. If this chemical contacts the skin, remove contaminated clothing and wash immediately with soap and water. Seek medical attention immediately. If

this chemical has been inhaled, remove from exposure, begin rescue breathing (using universal precautions, including resuscitation mask) if breathing has stopped and CPR if heart action has stopped. Transfer promptly to a medical facility. When this chemical has been swallowed, get medical attention. If victim is *conscious*, administer water or milk. Do not induce vomiting. Keep victim quiet and maintain normal body temperature. Effects may be delayed; keep victim under observation.
Personal Protective Methods: Wear protective gloves and clothing to prevent any reasonable probability of skin contact. Safety equipment suppliers/manufacturers can provide recommendations on the most protective glove/clothing material for your operation. All protective clothing (suits, gloves, footwear, headgear) should be clean, available each day, and put on before work. Contact lenses should not be worn when working with this chemical. Wear splash-proof chemical goggles and face shield unless full face-piece respiratory protection is worn. Employees should wash immediately with soap when skin is wet or contaminated. Provide emergency showers and eyewash.
Respirator Selection: Follow regulations in OSHA 29CFR1910.134 or European Standard EN149. Use a NIOSH/MSHA- or European Standard EN149-approved respirator; or use an approved supplied-air respirator with a full face-piece operated in the positive-pressure mode, or with a full face-piece, hood, or helmet in the continuous-flow mode; or use a NIOSH/MSHA- or European Standard EN149-approved self-contained breathing apparatus with a full face-piece operated in a pressure-demand or other positive-pressure mode.
Storage: Color Code—Red: Flammability Hazard: Store in a flammable liquid storage area or approved cabinet away from ignition sources and corrosive and reactive materials. Prior to working with this chemical you should be trained on its proper handling and storage. Store in tightly closed containers in a cool, well-ventilated area away from oxidizers. Where possible, automatically pump liquid from drums or other storage containers to process containers. Sources of ignition, such as smoking and open flames, are prohibited where this chemical is handled, used, or stored. Metal containers involving the transfer of 5 gallons or more of this chemical should be grounded and bonded. Drums must be equipped with self-closing valves, pressure vacuum bungs, and flame arresters. Use only nonsparking tools and equipment, especially when opening and closing containers of this chemical. Wherever this chemical is used, handled, manufactured, or stored, use explosion-proof electrical equipment and fittings.
Shipping: This compound requires a shipping label of "FLAMMABLE LIQUID, POISON, CORROSIVE," It falls in Hazard Class 6.1 and Packing Group I.
Spill Handling: Evacuate and restrict persons not wearing protective equipment from area of spill or leak until cleanup is complete. Remove all ignition sources. Ventilate area of spill or leak. Absorb liquids in vermiculite, dry sand, earth,

peat, carbon, or a similar material and deposit in sealed containers. Keep this chemical out of a confined space, such as a sewer, because of the possibility of an explosion, unless the sewer is designed to prevent the buildup of explosive concentrations. It may be necessary to contain and dispose of this chemical as a hazardous waste. If material or contaminated runoff enters waterways, notify downstream users of potentially contaminated waters. Contact your local or federal environmental protection agency for specific recommendations. If employees are required to clean up spills, they must be properly trained and equipped. OSHA 1910.120(q) may be applicable.

Initial isolation and protective action distances
Distances shown are likely to be affected during the first 30 min after materials are spilled and could increase with time. If more than one tank car, cargo tank, portable tank, or large cylinder involved in the incident is leaking, the protective action distance may need to be increased. You may need to seek emergency information from CHEMTREC at (800) 424-9300 or seek professional environmental engineering assistance from the US EPA Environmental Response Team at (908) 548-8730 (24-h response line).

Small spills (From a small package or a small leak from a large package)
First: Isolate in all directions (feet/meters)   100/30
Then: Protect persons downwind (miles/kilometers)
Day   0.1/0.2
Night   0.2/0.3

Large spills (From a large package or from many small packages)
First: Isolate in all directions (feet/meters)   200/60
Then: Protect persons downwind (miles/kilometers)
Day   0.5/0.8
Night   0.8/1.3

Fire Extinguishing: This chemical is a flammable liquid. Poisonous gases, including hydrogen chloride and chlorine, are produced in fire. *Small fires:* Dry chemical, carbon dioxide extinguishers. Move container from fire area if you can do so without risk. Dike fire control water for later disposal; do not scatter the material. Keep unnecessary people away; isolate hazard area and deny entry. Stay upwind; keep out of low areas. Wear positive pressure breathing apparatus and special protective clothing. Isolate for ½ mile in all directions if tank car or truck is involved in fire. Vapors are heavier than air and will collect in low areas. Vapors may travel long distances to ignition sources and flashback. Vapors in confined areas may explode when exposed to fire. Containers may explode in fire. Storage containers and parts of containers may rocket great distances, in many directions. If material or contaminated runoff enters waterways, notify downstream users of potentially contaminated waters. Notify local health and fire officials and pollution control agencies. From a secure, explosion-proof location, use water spray to cool exposed containers. If cooling streams are ineffective (venting sound increases in volume and pitch, tank discolors, or shows any signs of deforming), withdraw immediately to a secure position. Spray cooling water on containers that are exposed to flames until after fire is out. If employees are expected to fight fires, they must be trained and equipped in OSHA 1910.156. The only respirators recommended for firefighting are self-contained breathing apparatuses that have full face-pieces and are operated in a pressure-demand or other positive-pressure mode.

References
US Environmental Protection Agency. (November 30, 1987). *Chemical Hazard Information Profile: Propyl Chloroformate*. Washington, DC: Chemical Emergency Preparedness Program
New Jersey Department of Health and Senior Services. (September 2001). *Hazardous Substances Fact Sheet: Propyl Chloroformate*. Trenton, NJ

Propylene   P:1230

Molecular Formula: C_3H_6
Common Formula: $CH_2{=}CHCH_3$
Synonyms: Isobutylene; Methylethene; Methylethylene; NCI-C50077; Propene; 1-Propene; Propileno (Spanish)
CAS Registry Number: 115-07-1; *(alt.)* 676-63-1; *(alt.)* 33004-01-2
RTECS® Number: UC6470000
UN/NA & ERG Number: UN1077/115
EC Number: 204-062-1 [*Annex I Index No.:* 601-011-00-9]
Regulatory Authority and Advisory Bodies
Department of Homeland Security Screening Threshold Quantity (pounds): *Release hazard* 10,000 ($\geq$1.00% concentration).
Carcinogenicity: IARC: Animal Inadequate Evidence; Human Inadequate Evidence, *not classifiable as carcinogenic to humans*, Group 3, 1994; NCI: Carcinogenesis Studies (inhalation); no evidence: mouse, rat.
Air Pollutant Standard Set. See below, "Permissible Exposure Limits in Air" section.
Clean Air Act: Accidental Release Prevention/Flammable Substances, (Section 112[r], Table 3), TQ = 10,000 lb (4540 kg).
EPCRA Section 313 Form R *de minimis* concentration reporting level: 1.0%.
European/International Regulations: Hazard Symbol: F+; Risk phrases: R12; Safety phrases: S2; S9; S16; S33 (see Appendix 4).
WGK (German Aquatic Hazard Class): Nonwater polluting agent.
Description: Propylene is a colorless gas with a slight odor. The odor threshold is 23 ppm. Molecular weight = 42.09; Specific gravity (H_2O:1) = 0.51 at 25°C; Boiling point = −47°C; Freezing/Melting point = −185°C; Flash point = Flammable gas (−72°C); Autoignition temperature = 455°C. Explosive limits: LEL = 2.0%; UEL = 11.1%. Hazard Identification (based on NFPA-704

M Rating System): Health 1, Flammability 4, Reactivity 1. Insoluble in water.

Potential Exposure: Compound Description: Tumorigen. Propylene is used in the production of fabricated polymers, fibers, polypropylene resins, solvents, isopropyl alcohol, propylene dimer and trimer as gasoline components and detergent raw materials, propylene oxide, cumene, synthetic glycerol, isoprene, and oxo-alcohols.

Incompatibilities: Violent reaction with oxidizers and many other compounds. Able to form unstable peroxides; can polymerize, especially in heat, direct sunlight, oxidizers and other chemicals.

Permissible Exposure Limits in Air
ACGIH TLV®[1]: 500 ppm/860 mg/m^3 TWA, not classifiable as a human carcinogen.
Protective Action Criteria (PAC)
TEEL-0: 500 ppm
PAC-1: 1500 ppm
PAC-2: 10,000 ppm
PAC-3: 20,000 ppm
Australia: asphyxiant, 1993; Belgium: asphyxiant, 1993; Hungary: asphyxiant, 1993; Russia: STEL 100 mg/m^3, 1993; Switzerland: MAK-W 10,000 ppm (17,500 mg/m^3), 1999; United Kingdom: asphyxiant, 2000; the Netherlands: MAC-TGG 900 mg/m^3, 2003; Argentina, Bulgaria, Columbia, Jordan, South Korea, New Zealand, Singapore, Vietnam: ACGIH TLV®: not classifiable as a human carcinogen. Russia[43] set a MAC of 3.0 mg/m^3 in ambient air in residential areas both on a momentary and a daily average basis.
The oxygen content should be tested regularly to ensure that it is at least 19% by volume.

Permissible Concentration in Water: Russia[43] set a MAC of 0.5 mg/L in water bodies used for domestic purposes.

Determination in Water: Octanol–water coefficient: Log $K_{ow} = 1.8$.

Routes of Entry: Inhalation, ingestion, skin and/or eye contact.

Harmful Effects and Symptoms
Short Term Exposure: Propylene can affect you when breathed in. Exposure to high levels can cause you to feel dizzy and lightheaded. Very high levels can cause you to pass out from lack of oxygen. Death can result. Contact with liquefied propylene can cause frostbite.

Long Term Exposure: Exposure may cause an irregular heartbeat. It may also damage the liver.

Points of Attack: Liver, heart, liver.

Medical Surveillance: If symptoms develop or overexposure is suspected, the following may be useful: liver function tests. Holter monitor (a special 24-h EKG to look for irregular heartbeat).

First Aid: If this chemical gets into the eyes, remove any contact lenses at once and irrigate immediately for at least 15 min, occasionally lifting upper and lower lids. Seek medical attention immediately. If this chemical contacts the skin, remove contaminated clothing and wash immediately with soap and water. Seek medical attention immediately. If this chemical has been inhaled, remove from exposure, begin rescue breathing (using universal precautions, including resuscitation mask) if breathing has stopped and CPR if heart action has stopped. Transfer promptly to a medical facility. When this chemical has been swallowed, get medical attention. Give large quantities of water and induce vomiting. Do not make an unconscious person vomit. If frostbite has occurred, seek medical attention immediately; do NOT rub the affected areas or flush them with water. In order to prevent further tissue damage, do NOT attempt to remove frozen clothing from frostbitten areas. If frostbite has NOT occurred, immediately and thoroughly wash contaminated skin with soap and water.

Personal Protective Methods: Wear protective gloves and clothing to prevent any reasonable probability of skin contact. Safety equipment suppliers/manufacturers can provide recommendations on the most protective glove/clothing material for your operation. All protective clothing (suits, gloves, footwear, headgear) should be clean, available each day, and put on before work. Contact lenses should not be worn when working with this chemical. Wear splash-proof chemical goggles and face shield when working with liquid unless full face-piece respiratory protection is worn. Employees should wash immediately with soap when skin is wet or contaminated. Provide emergency showers and eyewash.

Respirator Selection: Exposure to propylene is dangerous because it can replace oxygen and lead to suffocation. Only NIOSH/MSHA- or European Standard EN 149-approved self-contained breathing apparatus with a full face-piece operated in positive-pressure mode should be used in oxygen-deficient environments.

Storage: Color Code—Red Stripe: Flammability Hazard: Store separately from all other flammable materials. Prior to working with this chemical you should be trained on its proper handling and storage. Before entering confined space where this chemical may be present, check to make sure that an explosive concentration is not a danger. Propylene must be stored to avoid contact with oxidizers (such as perchlorates, peroxides, permanganates, chlorates, and nitrates) since violent reactions occur. Store in tightly closed containers in a cool, well-ventilated area away from heat or direct sunlight. Sources of ignition, such as smoking and open flames, are prohibited where propylene is handled, used, or stored. Use only nonsparking tools and equipment, especially when opening and closing containers of propylene. Wherever propylene is used, handled, manufactured, or stored, use explosion-proof electrical equipment and fittings. Piping should be electrically bonded and grounded. Procedures for the handling, use, and storage of propylene cylinders should be in compliance with OSHA 1910.101 and 1910.169, as with the recommendations of the Compressed Gas Association.

Shipping: This compound requires a shipping label of "FLAMMABLE GAS." It falls in Hazard Class 2.1.

Spill Handling: Evacuate and restrict persons not wearing protective equipment from area of spill or leak until cleanup is complete. Remove all ignition sources. Establish forced ventilation to keep levels below explosive limit. Put on protective clothing and equipment. Remove tank or cylinder to an open area. Allow to bleed off slowly into atmosphere. Absorb liquids in vermiculite, dry sand, earth, peat, carbon, or a similar material and deposit in sealed containers. Keep this chemical out of a confined space, such as a sewer, because of the possibility of an explosion, unless the sewer is designed to prevent the buildup of explosive concentrations. It may be necessary to contain and dispose of this chemical as a hazardous waste. If material or contaminated runoff enters waterways, notify downstream users of potentially contaminated waters. Contact your local or federal environmental protection agency for specific recommendations. If employees are required to clean up spills, they must be properly trained and equipped. OSHA 1910.120(q) may be applicable.

Fire Extinguishing: This chemical is a flammable gas or liquid. Poisonous gases, including carbon dioxide and carbon monoxide, are produced in fire. Stop flow of gas. Use carbon dioxide or dry chemical. Use water spray, fog, or foam extinguishers. Vapors are heavier than air and will collect in low areas. Vapors may travel long distances to ignition sources and flashback. Vapors in confined areas may explode when exposed to fire. Containers may explode in fire. Storage containers and parts of containers may rocket great distances, in many directions. If material or contaminated runoff enters waterways, notify downstream users of potentially contaminated waters. Notify local health and fire officials and pollution control agencies. From a secure, explosion-proof location, use water spray to cool exposed containers and protect firefighters. If cooling streams are ineffective (venting sound increases in volume and pitch, tank discolors, or shows any signs of deforming), withdraw immediately to a secure position. If employees are expected to fight fires, they must be trained and equipped in OSHA 1910.156. The only respirators recommended for firefighting are self-contained breathing apparatuses that have full face-pieces and are operated in a pressure-demand or other positive-pressure mode.

Disposal Method Suggested: Controlled incineration.

References

New York State Department of Health. (April 1986). *Chemical Fact Sheet: Propylene.* Albany, NY: Bureau of Toxic Substance Assessment

New Jersey Department of Health and Senior Services. (May 2004). *Hazardous Substances Fact Sheet: Propylene.* Trenton, NJ

Propylene chlorohydrins   P:1240

Molecular Formula: C_3H_7ClO
Common Formula: $CH_3CHClCH_2OH$

Synonyms: 2-Chloro-1-propanol;   2-Chloropropanol; β-Chloropropyl   alcohol;   2-Chloropropyl   alcohol;   1-Propanol, 2-chloro-; 1-Chloro-2-propanol with 2-chloro-1-propanol (127-00-4)

CAS Registry Number: 78-89-7 (2-chloropropan-1-ol); 127-00-4 (1-chloropropan-2-ol)

RTECS® Number: UA8925000

UN/NA & ERG Number: UN2611/131

EC Number: 201-154-3 (2-chloropropan-1-ol); 204-819-6 (1-chloropropan-2-ol).

Regulatory Authority and Advisory Bodies
European/International Regulations: not listed in Annex 1.
WGK (German Aquatic Hazard Class): No value assigned.

Description: Propylene chlorohydrin is a colorless liquid. Molecular weight = 94.55; Boiling point = 133−134°C; Flash point = 52°C. Hazard Identification (based on NFPA-704 M Rating System): Health 2, Flammability 2, Reactivity 0. Soluble in water.

Potential Exposure: This material is used in organic synthesis.

Incompatibilities: Strong oxidizers.

Permissible Exposure Limits in Air
78-89-7 & 204-819-6
ACGIH TLV®[1]: 1 ppm/4 mg/m^3 [skin], not classifiable as a human carcinogen.

Routes of Entry: Inhalation, ingestion, skin and/or eye contact. Absorbed through the skin.

Harmful Effects and Symptoms
Short Term Exposure: Propylene chlorohydrin can affect you when breathed in and by passing through your skin. Contact can severely irritate and may burn the eyes. Breathing propylene chlorohydrin can irritate the nose and throat. Contact can irritate the skin. Overexposure may cause you to be lightheaded, unsteady, and drowsy.

Long Term Exposure: No chronic health effects are known at this time.

Points of Attack: Skin, eyes.

Medical Surveillance: There is no special test for this chemical. However, if illness occurs or overexposure is suspected, medical attention is recommended.

First Aid: If this chemical gets into the eyes, remove any contact lenses at once and irrigate immediately for at least 15 min, occasionally lifting upper and lower lids. Seek medical attention immediately. If this chemical contacts the skin, remove contaminated clothing and wash immediately with soap and water. Seek medical attention immediately. If this chemical has been inhaled, remove from exposure, begin rescue breathing (using universal precautions, including resuscitation mask) if breathing has stopped and CPR if heart action has stopped. Transfer promptly to a medical facility. When this chemical has been swallowed, get medical attention. Give large quantities of water and induce vomiting. Do not make an unconscious person vomit.

Personal Protective Methods: Wear protective gloves and clothing to prevent any reasonable probability of skin contact. Safety equipment suppliers/manufacturers can provide

recommendations on the most protective glove/clothing material for your operation. All protective clothing (suits, gloves, footwear, headgear) should be clean, available each day, and put on before work. Contact lenses should not be worn when working with this chemical. Wear splash-proof chemical goggles and face shield unless full face-piece respiratory protection is worn. Employees should wash immediately with soap when skin is wet or contaminated. Provide emergency showers and eyewash.

Respirator Selection: Where there is potential for exposure to propylene chlorohydrin, use a NIOSH/MSHA- or European Standard EN149-approved supplied-air respirator with a full face-piece operated in the positive-pressure mode, or with a full face-piece, hood, or helmet in the continuous-flow mode; or use a NIOSH/MSHA- or European Standard EN149-approved self-contained breathing apparatus with a full face-piece operated in a pressure-demand or other positive-pressure mode.

Storage: Color Code—Blue: Health Hazard/Poison: Store in a secure poison location. Prior to working with this chemical you should be trained on its proper handling and storage. Store in tightly closed containers in a cool, well-ventilated area away from strong oxidizers (such as chlorine, bromine, and fluorine). Sources of ignition, such as smoking and open flames, are prohibited where propylene chlorohydrin is used, handled, or stored in a manner that could create a potential fire or explosion hazard. Where possible, automatically transfer material from drums or other storage containers to process containers. Sources of ignition, such as smoking and open flames, are prohibited where this chemical is handled, used, or stored. Metal containers involving the transfer of this chemical should be grounded and bonded. Wherever this chemical is used, handled, manufactured, or stored, use explosion-proof electrical equipment and fittings.

Shipping: This compound requires a shipping label of "POISONOUS/TOXIC MATERIALS." It falls in Hazard Class 6.1 and Packing Group II.

Spill Handling: Evacuate and restrict persons not wearing protective equipment from area of spill or leak until cleanup is complete. Remove all ignition sources. Ventilate area of spill or leak. Absorb liquids in vermiculite, dry sand, earth, peat, carbon, or a similar material and deposit in sealed containers. Keep this chemical out of a confined space, such as a sewer, because of the possibility of an explosion, unless the sewer is designed to prevent the buildup of explosive concentrations. It may be necessary to contain and dispose of this chemical as a hazardous waste. If material or contaminated runoff enters waterways, notify downstream users of potentially contaminated waters. Contact your local or federal environmental protection agency for specific recommendations. If employees are required to clean up spills, they must be properly trained and equipped. OSHA 1910.120(q) may be applicable.

Fire Extinguishing: This chemical is a combustible liquid. Poisonous gases, including hydrogen chloride and carbon monoxide, are produced in fire. Use dry chemical, carbon dioxide, or foam extinguishers. Vapors are heavier than air and will collect in low areas. Vapors may travel long distances to ignition sources and flashback. Vapors in confined areas may explode when exposed to fire. Containers may explode in fire. Storage containers and parts of containers may rocket great distances, in many directions. If material or contaminated runoff enters waterways, notify downstream users of potentially contaminated waters. Notify local health and fire officials and pollution control agencies. From a secure, explosion-proof location, use water spray to cool exposed containers. If cooling streams are ineffective (venting sound increases in volume and pitch, tank discolors, or shows any signs of deforming), withdraw immediately to a secure position. If employees are expected to fight fires, they must be trained and equipped in OSHA 1910.156. The only respirators recommended for firefighting are self-contained breathing apparatuses that have full face-pieces and are operated in a pressure-demand or other positive-pressure mode.

Disposal Method Suggested: Dissolve or mix the material with a combustible solvent and burn in a chemical incinerator equipped with an afterburner and scrubber. All federal, state, and local environmental regulations must be observed.

Reference

New Jersey Department of Health and Senior Services. (March 1998). *Hazardous Substances Fact Sheet: Propylene Chlorohydrin.* Trenton, NJ

Propylene glycol   P:1250

Molecular Formula: $C_3H_8O_2$
Common Formula: $CH_3CHOHCH_2OH$
Synonyms: 1,2-Dihydroxypropane; Methyl ethylene glycol; Methyl glycol; 1,2-Propanediol
CAS Registry Number: 57-55-6
RTECS® Number: TY2000000
UN/NA & ERG Number: None assigned; UN1913 (combustible liquid; *Note:* must be preheated)/128
EC Number: 200-338-0
Regulatory Authority and Advisory Bodies
US EPA Gene-Tox Program, Negative: SHE—clonal assay.
US EPA, FIFRA, 1998 Status of Pesticides: Supported.
Air Pollutant Standard Set. See below, "Permissible Exposure Limits in Air" section.
FDA—over-the-counter drug.
Canada, WHMIS, Ingredients Disclosure List Concentration 1.0%.
European/International Regulations: not listed in Annex 1.
WGK (German Aquatic Hazard Class): 1—Low hazard to waters.
Description: Propylene glycol is a colorless, odorless, syrupy liquid. Molecular weight = 76.11; Boiling point = 188°C; Specific gravity (H_2O:1) = 1.04 at 25°C;

Freezing/Melting point $= -59°C$; Flash point $= 99°C$ (cc); Autoignition temperature $= 371°C$. Explosive limits: LEL $= 2.6\%$; UEL $= 12.5\%$. Hazard Identification (based on NFPA-704 M Rating System): Health 0, Flammability 1 (must be preheated), Reactivity 0; NJDHSS: Health 1, Flammability 2, Reactivity 0. Soluble in water.

Potential Exposure: Compound Description: Agricultural Chemical; Mutagen; Reproductive Effector; Human Data; Hormone, Primary Irritant. Propylene glycol is used as a solvent, emulsifying agent, food and feed additive, flavoring agent; in the manufacturing of plastics; as a plasticizer, surface-active agent, antifreeze, solvent, disinfectant, hydroscopic agent; coolant in refrigeration systems; pharmaceutical; brake fluid; and many others.

Incompatibilities: Contact with oxidizing agents, such as permanganates or dichromates, can cause a violent reaction.

Permissible Exposure Limits in Air
AIHA WEEL: 10 mg/m^3 TWA.
Protective Action Criteria (PAC)
TEEL-0: 10 mg/m^3
PAC-1: 10 mg/m^3
PAC-2: 10 mg/m^3
PAC-3: 500 mg/m^3
United Kingdom: TWA 150 ppm, total vapor and particulates; STEL 10 mg/m^3, particulates, 2000.

Determination in Air: NIOSH Analytical Method #5523, Glycols.

Determination in Water: Octanol−water coefficient: Log $K_{ow} = -0.9$.

Routes of Entry: Inhalation, ingestion, skin and/or eye contact. Absorbed through the skin.

Harmful Effects and Symptoms
Short Term Exposure: Irritates the eyes, skin, and respiratory tract. Two fluid ounces (60 mL) has caused stupor which lasted for a few hours which was followed by complete recovery.

Long Term Exposure: A mild allergen. Repeated or prolonged contact may cause skin sensitization and allergy. Therapeutic doses given for over a year have been associated with seizures; no further seizures occurred upon withdrawal of medication.

Points of Attack: Skin.

Medical Surveillance: Examination by a qualified allergist.

First Aid: If this chemical gets into the eyes, remove any contact lenses at once and irrigate immediately for at least 15 min, occasionally lifting upper and lower lids. Seek medical attention immediately. If this chemical contacts the skin, remove contaminated clothing and wash immediately with soap and water. Seek medical attention immediately. If this chemical has been inhaled, remove from exposure, begin rescue breathing (using universal precautions, including resuscitation mask) if breathing has stopped and CPR if heart action has stopped. Transfer promptly to a medical facility. When this chemical has been swallowed, get medical attention. Give large quantities of water and induce vomiting. Do not make an unconscious person vomit.

Personal Protective Methods: Wear protective gloves and clothing to prevent any reasonable probability of skin contact. Safety equipment suppliers/manufacturers can provide recommendations on the most protective glove/clothing material for your operation. Polyethylene, Nitrile + PVC, butyl rubber; polyvinyl chloride, Tychem® (from E.I. du Pont de Nemours & Company), Trellchem™, and Responder™ are among the recommended protective materials. Butyl rubber, Nitrile, Neoprene, and Neoprene + natural rubber are among the glove materials. All protective clothing (suits, gloves, footwear, headgear) should be clean, available each day, and put on before work. Contact lenses should not be worn when working with this chemical. Wear splash-proof chemical goggles and face shield unless full face-piece respiratory protection is worn. Employees should wash immediately with soap when skin is wet or contaminated. Provide emergency showers and eyewash.

Respirator Selection: Follow regulations in OSHA 29CFR1910.134 or European Standard EN149. Use a NIOSH/MSHA- or European Standard EN149-approved respirator; or use an approved supplied-air respirator with a full face-piece operated in the positive-pressure mode, or with a full face-piece, hood, or helmet in the continuous-flow mode; or use a NIOSH/MSHA- or European Standard EN149-approved self-contained breathing apparatus with a full face-piece operated in a pressure-demand or other positive-pressure mode.

Storage: Color Code—Green: General storage may be used. Prior to working with this chemical you should be trained on its proper handling and storage. Before entering confined space where this chemical may be present, check to make sure that an explosive concentration is not a danger. Store in a well-ventilated area away from ignition sources. Store in tightly closed containers in a cool, well-ventilated area away from oxidizers. Where possible, automatically pump liquid from drums or other storage containers to process containers. Sources of ignition, such as smoking and open flames, are prohibited where this chemical is handled, used, or stored. Metal containers involving the transfer of 5 gallons or more of this chemical should be grounded and bonded. Drums must be equipped with self-closing valves, pressure vacuum bungs, and flame arresters. Use only non-sparking tools and equipment, especially when opening and closing containers of this chemical. Wherever this chemical is used, handled, manufactured, or stored, use explosion-proof electrical equipment and fittings.

Spill Handling: Evacuate and restrict persons not wearing protective equipment from area of spill or leak until cleanup is complete. Remove all ignition sources. Establish forced ventilation to keep levels below explosive limit. Absorb liquids in vermiculite, dry sand, earth, peat, carbon, or a similar material and deposit in sealed containers. Keep this chemical out of a confined space, such as a sewer, because of the possibility of an explosion, unless the sewer is designed to prevent the buildup of explosive concentrations.

It may be necessary to contain and dispose of this chemical as a hazardous waste. If material or contaminated runoff enters waterways, notify downstream users of potentially contaminated waters. Contact your local or federal environmental protection agency for specific recommendations. If employees are required to clean up spills, they must be properly trained and equipped. OSHA 1910.120(q) may be applicable.

Fire Extinguishing: This chemical is a combustible liquid. Poisonous gases are produced in fire. Use dry chemical, carbon dioxide, or alcohol foam extinguishers. Vapors are heavier than air and will collect in low areas. Vapors may travel long distances to ignition sources and flashback. Vapors in confined areas may explode when exposed to fire. Containers may explode in fire. Storage containers and parts of containers may rocket great distances, in many directions. If material or contaminated runoff enters waterways, notify downstream users of potentially contaminated waters. Notify local health and fire officials and pollution control agencies. From a secure, explosion-proof location, use water spray to cool exposed containers. If cooling streams are ineffective (venting sound increases in volume and pitch, tank discolors, or shows any signs of deforming), withdraw immediately to a secure position. If employees are expected to fight fires, they must be trained and equipped in OSHA 1910.156. The only respirators recommended for firefighting are self-contained breathing apparatuses that have full face-pieces and are operated in a pressure-demand or other positive-pressure mode.

Disposal Method Suggested: Dissolve or mix the material with a combustible solvent and burn in a chemical incinerator equipped with an afterburner and scrubber. All federal, state, and local environmental regulations must be observed.

References
New York State Department of Health. (April 1986). *Chemical Fact Sheet: Propylene Glycol.* Albany, NY: Bureau of Toxic Substance Assessment
New Jersey Department of Health and Senior Services. (September 2009). *Hazardous Substances Fact Sheet: Propylene Glycol.* Trenton, NJ

Propylene glycol dinitrate   P:1260

Molecular Formula: $C_3H_6N_2O_6$
Common Formula: $CH_3CHONO_2CH_2ONO_2$
Synonyms: PGDN; Propylene glycol-1,2-dinitrate; 1,2-Propanediol, dinitrate; 1,2-Propylene glycol dinitrate
Note: Otto fuel is primarily propylene glycol dinitrate.
CAS Registry Number: 6423-43-4; 106602-80-6 (otto fuel)
RTECS® Number: TY6300000
UN/NA & ERG Number: FORBIDDEN to be transported
EC Number: 229-180-0

Regulatory Authority and Advisory Bodies
Air Pollutant Standard Set. See below, "Permissible Exposure Limits in Air" section.
Canada, WHMIS, Ingredients Disclosure List Concentration 1.0%.
European/International Regulations: not listed in Annex 1.
WGK (German Aquatic Hazard Class): No value assigned.
Description: Propylene glycol dinitrate is an explosive, colorless, high-boiling liquid (solid below $-8°C$) with a disagreeable odor. Molecular weight = 166.11; Specific gravity (H_2O:1) = 1.23 at 25°C; Boiling point = (decomposes) 207°C at 760 mmHg; Freezing/Melting point = $-7.8°C$; Vapor pressure = 0.07 mmHg at 22°C. Flash point = 99°C. Strong oxidizer. Slightly soluble in water; solubility = 0.1%.
Potential Exposure: Compound Description: Human Data; Primary Irritant. Propylene glycol dinitrate has been used as a torpedo propellant. The explosion potential is similar to ethylene glycol dinitrate.
Incompatibilities: A strong oxidizer. Contact with ammonia compounds, amines, strong acids, reducing agents, combustible materials may result in fire and explosion. It is similar to ethylene glycol dinitrate in explosion potential. Propylene glycol dinitrate may explode if strongly shocked or heated.
Permissible Exposure Limits in Air
Conversion factor: 6.79 mg/m^3 at 25°C & 1 atm.
OSHA PEL: None.
NIOSH REL: 0.05 ppm/0.3 mg/m^3 TWA [skin].
ACGIH TLV®[1]: 0.05 ppm/0.34 mg/m^3 TWA [skin]; BEI$_M$ issued for Methemoglobin inducers.
Protective Action Criteria (PAC)*
6423-43-4 (propylene glycol dinitrate) & 106602-80-6 (otto fuel which is primarily propylene glycol dinitrate)
TEEL-0: 0.05 ppm
PAC-1: **0.17** ppm
PAC-2: **1.0** ppm
PAC-3: **13** ppm
*AEGLs (Acute Emergency Guideline Levels) & ERPGs (Emergency Response Planning Guideline) are in **bold face**.
DFG MAK: 0.05 ppm/0.34 mg/m^3 TWA; Peak Limitation Category I(1) [skin].
Australia: TWA 0.05 ppm (0.3 mg/m^3), [skin], 1993; Austria: MAK 0.05 ppm (0.3 mg/m^3), [skin], 1999; Belgium: TWA 0.5 ppm (0.35 mg/m^3), [skin], 1993; Denmark: TWA 0.02 ppm (0.2 mg/m^3), [skin], 1999; Finland: TWA 0.02 ppm (0.2 mg/m^3); STEL 0.06 ppm, [skin], 1999; France: VME 0.05 ppm (0.3 mg/m^3), [skin], 1999; the Netherlands: MAC-TGG 0.3 mg/m^3, [skin], 2003; Sweden: NGV 0.1 ppm (0.7 mg/m^3), KTV 0.3 ppm (2 mg/m^3), [skin], 1999; Switzerland: MAK-W 0.05 ppm (0.35 mg/m^3), [skin], 1999; United Kingdom: TWA 0.2 ppm (1.4 mg/m^3); STEL 0.2 ppm, [skin], 2000; Argentina, Bulgaria, Columbia, Jordan, South Korea, New Zealand, Singapore, Vietnam: ACGIH TLV®: TWA 0.05 ppm [skin]. Several states have set guidelines or

standards for propylene glycol dinitrate in ambient air[60] ranging from 3 μg/m³ (North Dakota) to 5 μg/m³ (Virginia) to 6 μg/m³ (Connecticut) to 7 μg/m³ (Nevada).

Determination in Air: See OSHA Analytical Method 43.

Routes of Entry: Inhalation, ingestion, skin and/or eye contact. Absorbed through the skin.

Harmful Effects and Symptoms

Short Term Exposure: Propylene glycol dinitrate can affect you when breathed in and by passing through your skin. Exposure can cause headaches, stuffy nose, eye irritation, a "drunken" feeling with impaired balance, visual disturbance. Higher exposures can interfere with the ability of the blood to carry oxygen (methemoglobinemia). This causes cyanosis, a bluish color to the skin and can lead to death. Propylene glycol dinitrate can cause the blood pressure to drop rapidly. Human volunteers at 0.2 ppm exposure exhibited headaches and disruption in visually evoked response. At 0.5 ppm, marked impairment in balance was noted. At 1.5 ppm eye irritation occurred, in addition.[53]

Long Term Exposure: May cause liver and kidney damage.

Points of Attack: Eyes, central nervous system, blood, liver, kidneys.

Medical Surveillance: If symptoms develop or overexposure is suspected, the following may be useful: blood tests for methemoglobin level. Liver and kidney function tests.

First Aid: If this chemical gets into the eyes, remove any contact lenses at once and irrigate immediately for at least 15 min, occasionally lifting upper and lower lids. Seek medical attention immediately. If this chemical contacts the skin, remove contaminated clothing and wash immediately with soap and water. Seek medical attention immediately. If this chemical has been inhaled, remove from exposure, begin rescue breathing (using universal precautions, including resuscitation mask) if breathing has stopped and CPR if heart action has stopped. Transfer promptly to a medical facility. When this chemical has been swallowed, get medical attention. Give large quantities of water and induce vomiting. Do not make an unconscious person vomit. Medical observation is recommended for 24−48 h after breathing overexposure, as pulmonary edema may be delayed. As first aid for pulmonary edema, a doctor or authorized paramedic may consider administering a corticosteroid spray.

Note to physician: Treat for methemoglobinemia. Spectrophotometry may be required for precise determination of levels of methemoglobin in urine.

Personal Protective Methods: Wear protective gloves and clothing to prevent any reasonable probability of skin contact. Safety equipment suppliers/manufacturers can provide recommendations on the most protective glove/clothing material for your operation. All protective clothing (suits, gloves, footwear, headgear) should be clean, available each day, and put on before work. Contact lenses should not be worn when working with this chemical. Wear splash-proof chemical goggles and face shield unless full face-piece respiratory protection is worn. Employees should wash immediately with soap when skin is wet or contaminated. Provide emergency showers and eyewash.

Respirator Selection: Where there is potential for exposures *over 0.05* ppm, use a NIOSH/MSHA- or European Standard EN149-approved supplied-air respirator with a full face-piece operated in the positive-pressure mode, or with a full face-piece, hood, or helmet in the continuous-flow mode; or use a NIOSH/MSHA- or European Standard EN149-approved self-contained breathing apparatus with a full face-piece operated in a pressure-demand or other positive-pressure mode.

Storage: Explosive. Prior to working with this chemical you should be trained on its proper handling and storage. Propylene glycol dinitrate must be stored to avoid contact with ammonia compounds, amines, oxidizers, reducers, or combustible material since violent reactions occur. Store in tightly closed containers in a cool, well-ventilated area away from heat and avoid shocking or jolting containers. Where possible, automatically pump liquid from drums or other storage containers to process containers. Sources of ignition, such as smoking and open flames, are prohibited where this chemical is handled, used, or stored. Metal containers involving the transfer of 5 gallons or more of this chemical should be grounded and bonded. Drums must be equipped with self-closing valves, pressure vacuum bungs, and flame arresters. Use only nonsparking tools and equipment, especially when opening and closing containers of this chemical. Wherever this chemical is used, handled, manufactured, or stored, use explosion-proof electrical equipment and fittings.

Shipping: Dinitropropylene glycol is FORBIDDEN.

Spill Handling: Evacuate and restrict persons not wearing protective equipment from area of spill or leak until cleanup is complete. Remove all ignition sources. Ventilate area of spill or leak. Absorb liquids in vermiculite, dry sand, earth, peat, carbon, or a similar material and deposit in sealed containers. Keep this chemical out of a confined space, such as a sewer, because of the possibility of an explosion, unless the sewer is designed to prevent the buildup of explosive concentrations. It may be necessary to contain and dispose of this chemical as a hazardous waste. If material or contaminated runoff enters waterways, notify downstream users of potentially contaminated waters. Contact your local or federal environmental protection agency for specific recommendations. If employees are required to clean up spills, they must be properly trained and equipped. OSHA 1910.120(q) may be applicable.

Fire Extinguishing: This chemical is a combustible liquid. Poisonous gases, including nitrogen oxides, are produced in fire. Use dry chemical, carbon dioxide, or alcohol foam extinguishers. Vapors are heavier than air and will collect in low areas. Vapors may travel long distances to ignition sources and flashback. Vapors in confined areas may explode when exposed to fire. Containers may explode in fire. Storage containers and parts of containers may rocket great distances, in many directions. If material or contaminated runoff enters waterways, notify downstream users of potentially contaminated waters. Notify local health and fire officials and pollution control agencies. From a secure, explosion-proof location, use water spray to cool exposed containers. If cooling streams are ineffective (venting sound increases in volume and pitch, tank discolors, or shows any signs of deforming), withdraw immediately to a secure position. If employees are expected to fight fires, they must be trained and equipped in OSHA 1910.156. The only respirators recommended for firefighting are self-contained breathing apparatuses that have full face-pieces and are operated in a pressure-demand or other positive-pressure mode.

Disposal Method Suggested: Dissolve or mix the material with a combustible solvent and burn in a chemical incinerator equipped with an afterburner and scrubber. All federal, state, and local environmental regulations must be observed.

Reference

New Jersey Department of Health and Senior Services. (August 2005). *Hazardous Substances Fact Sheet: Propylene Glycol Dinitrate*. Trenton, NJ

Propylene glycol monomethyl ether   P:1270

Molecular Formula: $C_4H_{10}O_2$
Common Formula: $CH_3OCH_2CHOHCH_3$
Synonyms: Dowtherm® 209; 1-Methoxy-2-hydroxypropane; 1-Methoxy-2-propanol; 2-Methoxy-1-methylethanol; PGME; Propylene glycol methyl ether
CAS Registry Number: 107-98-2
RTECS® Number: UB7700000
UN/NA & ERG Number: UN3092/129
EC Number: 203-539-1
Regulatory Authority and Advisory Bodies
Air Pollutant Standard Set. See below, "Permissible Exposure Limits in Air" section.
Glycol ethers:
Clean Air Act: Hazardous Air Pollutants (Title I, Part A, Section 112) includes mono- and di-ethers of ethylene glycol, diethyl glycol, and triethylene glycol R—$(OCH_2CH_2)_n$—OR′, where $n = 1, 2,$ or 3; R = alkyl or aryl groups; R′ = R, H, or groups which when removed, yield glycol ethers with the structure: R—$(OCH_2CH)_n$—OH. Polymers are excluded from the glycol category.
EPCRA Section 313: Certain glycol ethers are covered. R—$(OCH_2CH_2)_n$—OR′; Where $n = 1, 2,$ or 3; R = alkyl C7 or less; or R = phenyl or alkyl substituted phenyl; R′ + H, or alkyl C7 or less; or OR′ consisting of carboxylic ester, sulfate, phosphate, nitrate, or sulfonate. Form R *de minimis* concentration reporting level: 1.0%.
Canada, WHMIS, Ingredients Disclosure List Concentration 1.0%.

European/International Regulations: Hazard Symbol: None assigned; Risk phrases: 10; Safety phrases: S2; S24 (see Appendix 4).
WGK (German Aquatic Hazard Class): 1—Low hazard to waters.
Description: Propylene glycol monomethyl ether is a colorless liquid with an ethereal odor. The odor threshold is 10 ppm. Molecular weight = 90.14; Specific gravity (H_2O:1) = 0.96 at 25°C; Boiling point = 120°C; Freezing/Melting point = (sets to glass) −96°C; Vapor pressure = 12 mmHg at 25°C; Flash point = 36.1°C; Autoignition temperature = 272°C. Explosive limits: LEL = 1.6%; UEL = 13.8%. Hazard Identification (based on NFPA-704 M Rating System): Health 0, Flammability 3, Reactivity 0. Soluble in water.
Potential Exposure: Compound Description: Reproductive Effector; Human Data; Primary Irritant. Propylene glycol monomethyl ether is used as a solvent for coatings; cellulose esters and acrylics; acrylics dyes; inks; and stains. It may also be used as a heat-transfer fluid.
Incompatibilities: Reacts with oxidizers, strong acids, acid chlorides, acid anhydrides, isocyanates, aluminum, and copper. Hygroscopic (i.e., absorbs moisture from the air). May slowly form reactive peroxides during prolonged storage.
Permissible Exposure Limits in Air
Conversion factor: 1 ppm = 3.69 mg/m³ at 25°C & 1 atm.
OSHA PEL: None.
NIOSH REL: 100 ppm/360 mg/m³ TWA: 150 ppm/540 mg/m³ STEL.
ACGIH TLV®[1]: 100 ppm TWA; 150 ppm STEL.
Protective Action Criteria (PAC)
TEEL-0: 100 ppm
PAC-1: 150 ppm
PAC-2: 150 ppm
PAC-3: 1250 ppm
DFG MAK: 100 ppm/370 mg/m³ TWA; Peak Limitation Category I(2); Pregnancy Risk Group C.
Austria: MAK 100 ppm (375 mg/m³), 1999; Belgium: TWA 100 ppm (369 mg/m³); STEL 150 ppm (553 mg/m³), 1993; Denmark: TWA 50 ppm (185 mg/m³), 1999; Finland: TWA 100 ppm (360 mg/m³); STEL 150 ppm, [skin], 1999; France: VME 100 ppm (360 mg/m³), 1999; Norway: TWA 50 ppm (180 mg/m³), 1999; the Netherlands: MAC-TGG 375 mg/m³, 2003; Switzerland: MAK-W 100 ppm (360 mg/m³), KZG-W 200 ppm (720 mg/m³), 1999; United Kingdom: TWA 100 ppm (375 mg/m³); STEL 300 ppm, [skin], 2000; Argentina, Bulgaria, Columbia, Jordan, South Korea, New Zealand, Singapore, Vietnam: ACGIH TLV®: STEL 150 ppm. Several states have set guidelines or standards for PGME in ambient air[60] ranging from 3.6−5.4 mg/m³ (North Dakota) to 6.0 mg/m³ (Virginia) to 7.2 mg/m³ (Connecticut).
Determination in Air: Use NIOSH Analytical Method #2554; OSHA Analytical Method 99.
Routes of Entry: Inhalation, ingestion, skin and/or eye contact. Absorbed through the skin.

Harmful Effects and Symptoms

Short Term Exposure: Propylene glycol monomethyl ether can affect you when breathed in and by passing through your skin. Contact can irritate the eyes and skin. Exposure can irritate the nose and throat. Very high levels may cause lung, liver, and kidney damage. Very high levels of propylene glycol monomethyl ether may cause central nervous system depression, dizziness and lightheadedness, and unconsciousness.

Long Term Exposure: Causes skin dryness, dermatitis. May cause liver and kidney damage. Can irritate the lungs; bronchitis may develop.

Points of Attack: Eyes, skin, respiratory system, central nervous system.

Medical Surveillance: If symptoms develop or overexposure is suspected, the following may be useful: liver and kidney function tests; lung function tests.

First Aid: If this chemical gets into the eyes, remove any contact lenses at once and irrigate immediately for at least 15 min, occasionally lifting upper and lower lids. Seek medical attention immediately. If this chemical contacts the skin, remove contaminated clothing and wash immediately with soap and water. Seek medical attention immediately. If this chemical has been inhaled, remove from exposure, begin rescue breathing (using universal precautions, including resuscitation mask) if breathing has stopped and CPR if heart action has stopped. Transfer promptly to a medical facility. When this chemical has been swallowed, get medical attention. Give large quantities of water and induce vomiting. Do not make an unconscious person vomit.

Personal Protective Methods: Wear protective gloves and clothing to prevent any reasonable probability of skin contact. Safety equipment suppliers/manufacturers can provide recommendations on the most protective glove/clothing material for your operation. All protective clothing (suits, gloves, footwear, headgear) should be clean, available each day, and put on before work. Contact lenses should not be worn when working with this chemical. Wear splash-proof chemical goggles and face shield unless full face-piece respiratory protection is worn. Employees should wash immediately with soap when skin is wet or contaminated. Provide emergency showers and eyewash.

Respirator Selection: Where there is potential for exposures over 100 ppm, use a NIOSH/MSHA- or European Standard EN 149-approved respirator with an organic vapor cartridge/canister. More protection is provided by a full-face-piece respirator than by a half-mask respirator, and even greater protection is provided by a powered air-purifying respirator. *Where there is potential for high exposures*, use a NIOSH/MSHA- or European Standard EN 149-approved supplied-air respirator with a full face-piece operated in the positive-pressure mode, or with a full face-piece, hood, or helmet in the continuous-flow mode; or use a NIOSH/MSHA- or European Standard EN 149-approved self-contained breathing apparatus with a full face-piece operated in a pressure-demand or other positive-pressure mode.

Storage: Color Code—Red: Flammability Hazard: Store in a flammable liquid storage area or approved cabinet away from ignition sources and corrosive and reactive materials. Prior to working with this chemical you should be trained on its proper handling and storage. Before entering confined space where this chemical may be present, check to make sure that an explosive concentration is not a danger. Store in tightly closed containers in a cool, well-ventilated area away from strong oxidizers (such as chlorine, bromine, and fluorine). Where possible, automatically transfer material from drums or other storage containers to process containers. Sources of ignition, such as smoking and open flames, are prohibited where this chemical is handled, used, or stored. Metal containers involving the transfer of this chemical should be grounded and bonded. Wherever this chemical is used, handled, manufactured, or stored, use explosion-proof electrical equipment and fittings.

Shipping: 1-Methoxy-2-propanol requires a shipping label of "FLAMMABLE LIQUID." It falls in Hazard Class 3. Packing Group III.

Spill Handling: Evacuate and restrict persons not wearing protective equipment from area of spill or leak until cleanup is complete. Remove all ignition sources. Establish forced ventilation to keep levels below explosive limit. Absorb liquids in vermiculite, dry sand, earth, peat, carbon, or a similar material and deposit in sealed containers. Keep this chemical out of a confined space, such as a sewer, because of the possibility of an explosion, unless the sewer is designed to prevent the buildup of explosive concentrations. It may be necessary to contain and dispose of this chemical as a hazardous waste. If material or contaminated runoff enters waterways, notify downstream users of potentially contaminated waters. Contact your local or federal environmental protection agency for specific recommendations. If employees are required to clean up spills, they must be properly trained and equipped. OSHA 1910.120(q) may be applicable.

Fire Extinguishing: This chemical is a flammable liquid. Poisonous gases, including nitrogen oxides, are produced in fire. Use dry chemical, carbon dioxide, or foam extinguishers. Water may be ineffective. Vapors are heavier than air and will collect in low areas. Vapors may travel long distances to ignition sources and flashback. Vapors in confined areas may explode when exposed to fire. Containers may explode in fire. Storage containers and parts of containers may rocket great distances, in many directions. If material or contaminated runoff enters waterways, notify downstream users of potentially contaminated waters. Notify local health and fire officials and pollution control agencies. From a secure, explosion-proof location, use water spray to cool exposed containers. If cooling streams are ineffective (venting sound increases in volume and pitch, tank discolors, or shows any signs of deforming), withdraw immediately to a secure position. If employees are expected to fight fires, they must be trained and equipped in OSHA 1910.156. The only respirators recommended for firefighting are self-contained breathing apparatuses that have full face-pieces and are operated in a pressure-demand or other positive-pressure mode.

Disposal Method Suggested: Dissolve or mix the material with a combustible solvent and burn in a chemical incinerator equipped with an afterburner and scrubber. All federal, state, and local environmental regulations must be observed.

Reference

New Jersey Department of Health and Senior Services. (August 2005). *Hazardous Substances Fact Sheet: Propylene Glycol Monomethyl Ether.* Trenton, NJ

Propyleneimine   P:1280

Molecular Formula: C_3H_7N

Synonyms: Aziridina, 2-metil (Spanish); Aziridine, 2-methyl-; 2-Methylazacyclopropane; 2-Methylaziridine; 2-Methylethylenimine; 2-Methylethylen imine; Propilenimina (Spanish); Propylene imine; 1,2-Propyleneimine

CAS Registry Number: 75-55-8

RTECS® Number: CM8050000

UN/NA & ERG Number: UN1921 (stabilized)/131

EC Number: 200-878-7 [*Annex I Index No.:* 613-033-00-6]

Regulatory Authority and Advisory Bodies

Department of Homeland Security Screening Threshold Quantity (pounds): *Release hazard* 10,000 (≥1.00% concentration).

Carcinogenicity: IARC: Human No Adequate Data, Animal Sufficient Evidence, *possibly carcinogenic to humans*, Group 2B, 1999; NTP: 11th Report on Carcinogens, 2004: Reasonably anticipated to be a human carcinogen; NIOSH: Potential occupational carcinogen.

US EPA Gene-Tox Program, Positive: Carcinogenicity—mouse/rat; SHE—clonal assay; Positive: Cell transform.—RLV F344 rat embryo; Host-mediated assay; Positive: *E. coli* polA without S9; Histidine reversion—Ames test; Positive: *In vitro* UDS—human fibroblast; *S. cerevisiae*—homozygosis; Positive: *S. pombe*—reversion; Positive/dose response: Cell transform.—BALB/c-3T3.

Banned or Severely Restricted (Belgium, Sweden) (UN).[13]

Toxic Substance (World Bank).[15]

Air Pollutant Standard Set. See below, "Permissible Exposure Limits in Air" section.

Clean Air Act: Hazardous Air Pollutants (Title I, Part A, Section 112); Accidental Release Prevention/Flammable Substances, (Section 112[r], Table 3), TQ = 10,000 lb (4540 kg).

US EPA Hazardous Waste Number (RCRA No.): P067.

RCRA, 40CFR261, Appendix 8 Hazardous Constituents.

Superfund/EPCRA 40CFR355, Extremely Hazardous Substances: TPQ = 10,000 lb (4540 kg).

Reportable Quantity (RQ): 1 lb (0.454 kg).

EPCRA Section 313 Form R *de minimis* concentration reporting level: 0.1%.

California Proposition 65 Chemical: Cancer 1/1/88.

Canada, WHMIS, Ingredients Disclosure List Concentration 0.1%.

European/International Regulations: Hazard Symbol: F, T +, N; Risk phrases: R45; R11; R26/27/28; R41; R51/53; Safety phrases: S53; S45; S61 (see Appendix 4).

WGK (German Aquatic Hazard Class): 3—Severe hazard to waters.

Description: Propyleneimine is a fuming, colorless, oily liquid with a strong ammonia-like odor. Molecular weight = 57.11; Specific gravity (H_2O:1) = 0.81 at 25°C; Boiling point = 66.7°C; Freezing/Melting point = −63°C; Vapor pressure = 112 mmHg at 25°C; Flash point = − 10°C (cc). Explosive limits: LEL = 2.3%; UEL = 15.1%. Hazard Identification (based on NFPA-704 M Rating System): Health 3, Flammability 4, Reactivity 0. Soluble in water.

Potential Exposure: Compound Description: Tumorigen, Mutagen. Propyleneimine is used in the modification of latex surface coating resins; in the production of polymers for use in the paper and textile industries as coatings and adhesives.

Incompatibilities: Reacts with acids, strong oxidizers, water, carbonyl compounds, quinones, sulfonyl halides. May explode in heat. Subject to violent polymerization in contact with acids. Hydrolyzes in water to form methylethanolamine.

Permissible Exposure Limits in Air

OSHA PEL: 2 ppm/5 mg/m³ TWA [skin].

NIOSH REL: 2 ppm/5 mg/m³ TWA [skin]; A potential occupational carcinogen; Limit exposure to lowest feasible concentration. See *NIOSH Pocket Guide*, Appendix A.

ACGIH TLV®[1]: 2 ppm/5 mg/m³ TWA; 0.4 ppm/1 mg/m³ STEL [skin]; confirmed animal carcinogen with unknown relevance to humans.

NIOSH IDLH: (potential occupational carcinogen) 100 ppm.

Protective Action Criteria (PAC)*

TEEL-0: 0.2 ppm

PAC-1: 0.4 ppm

PAC-2: **12 ppm**

PAC-3: **23 ppm**

*AEGLs (Acute Emergency Guideline Levels) & ERPGs (Emergency Response Planning Guideline) are in **bold face**.

DFG MAK: [skin] Carcinogen Category 2; Germ Cell Mutation Category 3B.

Australia: TWA 2 ppm (5 mg/m³), [skin], carcinogen, 1993; Austria: [skin], carcinogen, 1999; Denmark: TWA 2 ppm (5 mg/m³), [skin], 1999; Finland: STEL 2 ppm (5 mg/m³), [skin], carcinogen, 1999; France: carcinogen, 1993; Japan: 2 ppm (4.7 mg/m³), [skin], 1999; the Netherlands: MAC-TGG 0.6 μg/m³, 2003; the Philippines: TWA 2 ppm (5 mg/m³), [skin], 1993; Russia: TWA 2 ppm, 1993; Switzerland: MAK-W 2 ppm (5 mg/m³), [skin], carcinogen, 1999; Turkey: TWA 2 ppm (5 mg/m³), [skin], 1993; United

Kingdom: carcinogen, 2000; Argentina, Bulgaria, Columbia, Jordan, South Korea, New Zealand, Singapore, Vietnam: ACGIH TLV®: confirmed animal carcinogen with unknown relevance to humans. Several states have set guidelines or standards for propyleneimine in ambient air[60] ranging from zero (North Dakota) to $12 \, \mu g/m^3$ (Pennsylvania) to $50 \, \mu g/m^3$ (Connecticut and Virginia) to $119 \, \mu g/m^3$ (Nevada).

Determination in Air: No method available.

Routes of Entry: Inhalation, skin absorption, ingestion, eye and/or skin contact.

Harmful Effects and Symptoms

Short Term Exposure: Severely irritates the eyes, skin, and respiratory tract. Contact may cause skin burns and permanent eye damage. Symptoms of exposure include inflammation and blistering of the skin, eye, and upper respiratory tract; irritation, nausea, itching, and periodic vomiting. Headache, dizziness, and pain in the temple. Shortness of breath; and increased nasal and laryngeal secretion are seen. It is toxic after acute exposure. LD_{50} = (oral-rat) 19 mg/kg (highly toxic).

Long Term Exposure: Caused skin drying and dermatitis. A potential occupational carcinogen. May affect the kidneys.

Points of Attack: Eyes, skin. Cancer site in animals: nasal cavity, breast, brain, blood or bone marrow.

Medical Surveillance: NIOSH lists the following tests: complete blood count; granulocytic leukemia. Before beginning employment and at regular times after that, for those with frequent or potentially high exposures, the following are recommended: lung function tests.

First Aid: If this chemical gets into the eyes, remove any contact lenses at once and irrigate immediately for at least 15 min, occasionally lifting upper and lower lids. Seek medical attention immediately. If this chemical contacts the skin, remove contaminated clothing and wash immediately with soap and water. Seek medical attention immediately. If this chemical has been inhaled, remove from exposure, begin rescue breathing (using universal precautions, including resuscitation mask) if breathing has stopped and CPR if heart action has stopped. Transfer promptly to a medical facility. When this chemical has been swallowed, get medical attention. Give large quantities of water and induce vomiting. Do not make an unconscious person vomit.

Personal Protective Methods: Wear protective gloves and clothing to prevent any reasonable probability of skin contact. Safety equipment suppliers/manufacturers can provide recommendations on the most protective glove/clothing material for your operation. All protective clothing (suits, gloves, footwear, headgear) should be clean, available each day, and put on before work. Contact lenses should not be worn when working with this chemical. Wear splash-proof chemical goggles and face shield unless full face-piece respiratory protection is worn. Employees should wash immediately with soap when skin is wet or contaminated. Provide emergency showers and eyewash.

Respirator Selection: NIOSH: *At concentrations above the NIOSH REL, or where there is no REL, at any detectable concentration:* SCBAF: Pd,Pp (APF = 10,000) (any self-contained breathing apparatus that has a full face-piece and is operated in a pressure-demand or other positive-pressure mode) or SaF: Pd,Pp: ASCBA (APF = 10,000) (any supplied-air respirator that has a full face-piece and is operated in a pressure-demand or other positive-pressure mode in combination with an auxiliary self-contained breathing apparatus operated in a pressure-demand or other positive-pressure mode). *Escape:* GmFS (APF = 50) [any air-purifying, full-face-piece respirator (gas mask) with a chin-style, front- or back-mounted canister providing protection against the compound of concern] or SCBAE (any appropriate escape-type, self-contained breathing apparatus).

Storage: Color Code—Red: Flammability Hazard: Store in a flammable liquid storage area or approved cabinet away from ignition sources and corrosive and reactive materials. Prior to working with this chemical you should be trained on its proper handling and storage. Propyleneimine must be stored to avoid contact with acids and strong oxidizers (such as chlorine, bromine, and fluorine) because violent reactions occur. Store in tightly closed containers in a cool, well-ventilated area. Sources of ignition, such as smoking and open flames, are prohibited where propyleneimine is used, handled, or stored in a manner that could create a potential fire or explosion hazard. Metal containers involving the transfer of 5 gallons or more of Propyleneimine should be grounded and bonded. Drums must be equipped with self-closing valves, pressure vacuum bungs, and flame arresters. Use only nonsparking tools and equipment, especially when opening and closing containers of propyleneimine. A regulated, marked area should be established where this chemical is handled, used, or stored in compliance with OSHA Standard 1910.1045.

Shipping: This compound requires a shipping label of "FLAMMABLE LIQUID." It falls in Hazard Class 3 and Packing Group I.

Spill Handling: Evacuate and restrict persons not wearing protective equipment from area of spill or leak until cleanup is complete. Remove all ignition sources. Avoid breathing vapors; avoid bodily contact with the material. Toxic gases, including oxides of nitrogen and carbon monoxide, are released in a fire. Do not handle broken packages without protective equipment. Wash away any material which may have contacted the body with copious amounts of water or soap with water. Use water spray to disperse vapors and dilute standing pools of liquid. Keep sparks and flames away. Attempt to stop leak if it can be done without hazard. Ventilate area of spill or leak. Absorb liquids in vermiculite, dry sand, earth, peat, carbon, or a similar material and deposit in sealed containers. Keep this chemical out of a confined space, such as a sewer, because of the possibility of an explosion, unless the sewer is designed to prevent the buildup of explosive

concentrations. It may be necessary to contain and dispose of this chemical as a hazardous waste. If material or contaminated runoff enters waterways, notify downstream users of potentially contaminated waters. Contact your local or federal environmental protection agency for specific recommendations. If employees are required to clean up spills, they must be properly trained and equipped. OSHA 1910.120(q) may be applicable.

Fire Extinguishing: This chemical is a flammable liquid. Poisonous gases are produced in fire. Use dry chemical, carbon dioxide, or alcohol foam extinguishers. Vapors are heavier than air and will collect in low areas. Vapors may travel long distances to ignition sources and flashback. Vapors in confined areas may explode when exposed to fire. Containers may explode in fire. Storage containers and parts of containers may rocket great distances, in many directions. If material or contaminated runoff enters waterways, notify downstream users of potentially contaminated waters. Notify local health and fire officials and pollution control agencies. From a secure, explosion-proof location, use water spray to cool exposed containers. If cooling streams are ineffective (venting sound increases in volume and pitch, tank discolors, or shows any signs of deforming), withdraw immediately to a secure position. If employees are expected to fight fires, they must be trained and equipped in OSHA 1910.156. The only respirators recommended for firefighting are self-contained breathing apparatuses that have full face-pieces and are operated in a pressure-demand or other positive-pressure mode.

Disposal Method Suggested: Consult with environmental regulatory agencies for guidance on acceptable disposal practices. Generators of waste containing this contaminant ($\geq$100 kg/mo) must conform with EPA regulations governing storage, transportation, treatment, and waste disposal. Controlled incineration (incinerator equipped with a scrubber or thermal unit to reduce nitrogen oxides emissions).

References

Dermer, O. C., & Ham, G. E. (1969). *Ethyleneimine and Other Aziridines*. New York: Academic Press

US Environmental Protection Agency. (November 30, 1987). *Chemical Hazard Information Profile: Propyleneimine*. Washington, DC: Chemical Emergency Preparedness Program

New Jersey Department of Health and Senior Services. (August 2005). *Hazardous Substances Fact Sheet: Propylene Imine*. Trenton, NJ

Propylene oxide   P:1290

Molecular Formula: C$_3$H$_6$O

Synonyms: AD 6; Epoxypropane; 1,2-Epoxypropane; 2,3-Epoxypropane; Methyl ethylene oxide; Methyl oxirane; NCI-C50099; Oxido de propileno (Spanish); Oxyde de propylene (French); Oxirane, methyl-; Propane, 1,2-epoxy-; Propene oxide; Propylene epoxide; 1,2-Propylene oxide

CAS Registry Number: 75-56-9
RTECS® Number: TZ2975000
UN/NA & ERG Number: UN1280/127
EC Number: 200-879-2 [*Annex I Index No.:* 603-055-00-4]
Regulatory Authority and Advisory Bodies
Department of Homeland Security Screening Threshold Quantity (pounds): *Release hazard* 10,000 ($\geq$1.00% concentration).

Carcinogenicity: IARC: Human Inadequate Evidence, Animal Sufficient Evidence, *possibly carcinogenic to humans*, Group 2B, 1997; NTP: 11th Report on Carcinogens, 2004: Reasonably anticipated to be a human carcinogen; EPA: Sufficient evidence from animal studies; inadequate evidence or no useful data from epidemiologic studies; NIOSH: Potential occupational carcinogen; NCI: Carcinogenesis Studies (inhalation); clear evidence: mouse; some evidence: rat; equivocal evidence: rat

US EPA Gene-Tox Program, Positive: Carcinogenicity—mouse/rat; Cell transform.—SA7/SHE EPA; Positive: *N. crassa*—reversion; *D. melanogaster* sex-linked lethal EPA; Positive: *S. pombe*—reversion EPA; Negative: Rodent dominant lethal EPA; Positive: CHO gene mutation.

Highly Reactive Substance and Explosive (World Bank).[15]

Air Pollutant Standard Set. See below, "Permissible Exposure Limits in Air" section.

Clean Air Act: Hazardous Air Pollutants (Title I, Part A, Section 112); Accidental Release Prevention/Flammable Substances, (Section 112[r], Table 3), TQ = 10,000 lb (4540 kg).

Clean Water Act: Section 311 Hazardous Substances/RQ 40CFR117.3 (same as CERCLA, see below); Section 313 Water Priority Chemicals (57FR41331, 9/9/92).

Superfund/EPCRA 40CFR355, Extremely Hazardous Substances: TPQ = 10,000 lb (4540 kg).

Reportable Quantity (RQ): 100 lb (45.4 kg).

EPCRA Section 313 Form R *de minimis* concentration reporting level: 0.1%.

California Proposition 65 Chemical: Cancer 1/1/88.

Canada, WHMIS, Ingredients Disclosure List Concentration 1.0%.

European/International Regulations: Hazard Symbol: F +, T; R45; R46; R12; R20/21/22; R36/37/38; Safety phrases: S53; S45 (see Appendix 4).

WGK (German Aquatic Hazard Class): 3—Severe hazard to waters.

Description: Propylene oxide is a colorless liquid with an ether-like odor. Molecular weight = 58.09; Boiling point = 34.4°C; Specific gravity (H$_2$O:1) = 0.83 at 25°C; Freezing/Melting point = −112.2°C; Vapor pressure = 445 mmHg at 25°C; Flash point = −37.2°C; Autoignition temperature = 450°C. Explosive limits: LEL = 2.3%; UEL = 36−38.5%. Hazard Identification (based on NFPA-704 M Rating System): Health 3, Flammability 4, Reactivity 2. Soluble in water; solubility = 41%.

Potential Exposure: Compound Description: Agricultural Chemical; Tumorigen, Mutagen; Reproductive Effector;

Human Data; Hormone, Primary Irritant. Propylene oxide is used as an intermediate in the production of polyether polyols and propylene glycol; as a fumigant; in the production of adducts as urethane foam ingredients; in detergent manufacture; as a component in brake fluids.

Incompatibilities: Reacts with strong oxidizers, anhydrous metal chlorides, chlorine, iron, strong acids, caustics and peroxides. Polymerization may occur due to high temperatures or contamination with alkalis, aqueous acids, amines, metal chlorides, and acidic alcohols. Attacks some plastics, coatings, and rubber.

Permissible Exposure Limits in Air

Conversion factor: 1 ppm = 2.38 mg/m^3 at 25°C & 1 atm.
OSHA PEL: 100 ppm/240 mg/m^3 TWA.
NIOSH REL: A potential occupational carcinogen. [skin]; Limit exposure to lowest feasible concentration. See *NIOSH Pocket Guide*, Appendix A.
ACGIH TLV®[1]: 2 ppm/4.8 mg/m^3 TWA, sensitizer; confirmed animal carcinogen with unknown relevance to humans. (2000)
NIOSH IDLH: 400 ppm (potential occupational carcinogen).
Protective Action Criteria (PAC)*
TEEL-0: 2 ppm
PAC-1: **73** ppm
PAC-2: **290** ppm
PAC-3: **870** ppm
*AEGLs (Acute Emergency Guideline Levels) & ERPGs (Emergency Response Planning Guideline) are in **bold face**.
DFG MAK: [skin] Carcinogen Category 2.
Australia: TWA 20 ppm (50 mg/m^3), carcinogen, 1993; Austria: carcinogen, 1999; Belgium: TWA 20 ppm (48 mg/m^3), 1993; Denmark: TWA 5 ppm (12 mg/m^3), [skin], 1999; Finland: TWA 5 ppm (12 mg/m^3), carcinogen, 1999; France: VME 20 ppm (50 mg/m^3), carcinogen, 1999; the Netherlands: MAC-TGG 6 mg/m^3, 2003; the Philippines: TWA 100 ppm (240 mg/m^3), 1993; Russia: STEL 1 mg/m^3, [skin], 1993; Sweden: NGV 5 ppm (12 mg/m^3), KTV 10 ppm (25 mg/m^3), 1999; Switzerland: MAK-W 2.5 ppm (6 mg/m^3), carcinogen, 1999; Turkey: TWA 100 ppm (240 mg/m^3), 1993; United Kingdom: TWA 5 ppm (12 mg/m^3), carcinogen, 2000; Argentina, Bulgaria, Columbia, Jordan, South Korea, New Zealand, Singapore, Vietnam: ACGIH TLV®: confirmed animal carcinogen with unknown relevance to humans. Russia set a MAC of 0.08 mg/m^3 for ambient air in residential areas. Several states have set guidelines or standards for propylene oxide in ambient air[60] ranging from 0.0833 μg/m^3 (Kansas) to 1.0 μg/m^3 (Connecticut) to 15.0 μg/m^3 (Massachusetts) to 167.0 μg/m^3 (New York) to 250.0 μg/m^3 (South Carolina) to 500.0 μg/m^3 (North Dakota) to 625.0 μg/m^3 (Pennsylvania) to 850.0 μg/m^3 (Virginia) to 1190.0 μg/m^3 (Nevada).

Determination in Air: Use NIOSH Analytical Method #1612. Propylene oxide; OSHA Analytical Method 88.
Permissible Concentration in Water: Russia has set[35] a MAC value of 0.01 mg/L in water bodies used for domestic purposes and 0.005 mg/L in water bodies used for fishery purposes.

Determination in Water: Octanol−water coefficient: Log K_{ow} = <0.02.
Routes of Entry: Inhalation, ingestion, skin and/or eye contact.

Harmful Effects and Symptoms
Short Term Exposure: Propylene oxide may irritate or burn the skin, eyes, and respiratory tract. Contact with the liquid can cause blindness and death. This material is moderately toxic by inhalation and ingestion. It may cause irreversible and reversible changes. Skin contact with the material or solutions of the material cause irritation; diluted solutions are more irritating than undiluted materials. Dermatitis (red, inflamed skin) is common. Exposure may cause mild depression of the central nervous system, and eye, nasal, and lung irritation. Other signs and symptoms of acute exposure may include headache, nausea, vomiting, and unconsciousness. Victims may appear as if they are in a drunken stupor. Pulmonary edema may occur.

Long Term Exposure: Repeated or prolonged contact may cause skin sensitization. This substance is possibly carcinogenic to humans. May decrease fertility in males and females. May cause lung damage; pneumonia may develop.
Points of Attack: Eyes, skin, respiratory system. Cancer site in animals: nasal cavity, stomach.
Medical Surveillance: Before beginning employment and at regular times after that, the following are recommended: lung function tests. If symptoms develop or overexposure is suspected, the following may be useful: liver function tests. Examination of the eyes and vision.
First Aid: If this chemical gets into the eyes, remove any contact lenses at once and irrigate immediately for at least 15 min, occasionally lifting upper and lower lids. Seek medical attention immediately. If this chemical contacts the skin, remove contaminated clothing and wash immediately with soap and water. Speed in removing material from skin is of extreme importance. Shampoo hair promptly if contaminated. Seek medical attention immediately. If this chemical has been inhaled, remove from exposure, begin rescue breathing (using universal precautions, including resuscitation mask) if breathing has stopped and CPR if heart action has stopped. Transfer promptly to a medical facility. When this chemical has been swallowed, get medical attention. Give large quantities of water and induce vomiting. Do not make an unconscious person vomit. Medical observation is recommended for 24−48 h after breathing overexposure, as pulmonary edema may be delayed. As first aid for pulmonary edema, a doctor or authorized paramedic may consider administering a corticosteroid spray.
Personal Protective Methods: Wear protective gloves and clothing to prevent any reasonable probability of skin contact. Safety equipment suppliers/manufacturers can provide recommendations on the most protective glove/clothing material for your operation. Teflon™ and butyl rubber are

among the recommended protective materials. All protective clothing (suits, gloves, footwear, headgear) should be clean, available each day, and put on before work. Contact lenses should not be worn when working with this chemical. Wear splash-proof chemical goggles and face shield unless full face-piece respiratory protection is worn. Employees should wash immediately with soap when skin is wet or contaminated. Provide emergency showers and eyewash.

Respirator Selection: NIOSH: *At concentrations above the NIOSH REL, or where there is no REL, at any detectable concentration:* SCBAF: Pd,Pp (APF = 10,000) (any self-contained breathing apparatus that has a full face-piece and is operated in a pressure-demand or other positive-pressure mode) or SaF: Pd,Pp: ASCBA (APF = 10,000) (any supplied-air respirator that has a full face-piece and is operated in a pressure-demand or other positive-pressure mode in combination with an auxiliary self-contained breathing apparatus operated in a pressure-demand or other positive-pressure mode). *Escape:* GmFS (APF = 50) [any air-purifying, full-face-piece respirator (gas mask) with a chin-style, front- or back-mounted canister providing protection against the compound of concern] or SCBAE (any appropriate escape-type, self-contained breathing apparatus).

Storage: Color Code—Red: Flammability Hazard: Store in a flammable liquid storage area or approved cabinet away from ignition sources and corrosive and reactive materials. Prior to working with this chemical you should be trained on its proper handling and storage. Before entering confined space where this chemical may be present, check to make sure that an explosive concentration is not a danger. Propylene oxide must be stored to avoid contact with anhydrous metal chlorides, iron, strong acids (such as hydrochloric, sulfuric, and nitric), strong bases and peroxides, copper or copper alloys since violent reactions occur. Propylene oxide will attack some forms of plastics, rubber, and coatings. Store in tightly closed containers in a cool, well-ventilated area. Sources of ignition, such as smoking and open flames, are prohibited where propylene oxide is used, handled, or stored in a manner that could create a potential fire or explosion hazard. Metal containers involving the transfer of 5 gallons or more of propylene oxide should be grounded and bonded. Drums must be equipped with self-closing valves, pressure vacuum bungs, and flame arresters. Use only nonsparking tools and equipment, especially when opening and closing containers of propylene oxide. Wherever propylene oxide is used, handled, manufactured, or stored, use explosion-proof electrical equipment and fittings. A regulated, marked area should be established where this chemical is handled, used, or stored in compliance with OSHA Standard 1910.1045.

Shipping: This compound requires a shipping label of "FLAMMABLE LIQUID." It falls in Hazard Class 3 and Packing Group I.

Spill Handling: Evacuate and restrict persons not wearing protective equipment from area of spill or leak until cleanup is complete. Remove all ignition sources. Establish forced ventilation to keep levels below explosive limit. Absorb liquids in vermiculite, dry sand, earth, peat, carbon, or a similar material and deposit in sealed containers. Keep this chemical out of a confined space, such as a sewer, because of the possibility of an explosion, unless the sewer is designed to prevent the buildup of explosive concentrations. It may be necessary to contain and dispose of this chemical as a hazardous waste. If material or contaminated runoff enters waterways, notify downstream users of potentially contaminated waters. Contact your local or federal environmental protection agency for specific recommendations. If employees are required to clean up spills, they must be properly trained and equipped. OSHA 1910.120(q) may be applicable.

Fire Extinguishing: This chemical is a flammable liquid. Poisonous gases are produced in fire. Extinguish with dry chemical, carbon dioxide, water spray, fog, or foam. Firefighting should be done from a safe distance or from a protected location. Wear self-contained (positive pressure if available) breathing apparatus and full protective clothing. Isolate for ½ mile in all directions if tank car or truck is involved in fire. Move container from area if you can do so without risk. Spray cooling water on containers that are exposed to flames until well after fire is out. For massive fire in cargo area, use unmanned hose holder or monitor nozzles; if this is impossible, withdraw from area and let fire burn. Withdraw immediately in case of rising sound from venting safety device or any discoloration of tank due to fire. Vapors are heavier than air and will collect in low areas. Vapors may travel long distances to ignition sources and flashback. Vapors in confined areas may explode when exposed to fire. Containers may explode in fire. Storage containers and parts of containers may rocket great distances, in many directions. If material or contaminated run-off enters waterways, notify downstream users of potentially contaminated waters. Notify local health and fire officials and pollution control agencies. From a secure, explosion-proof location, use water spray to cool exposed containers. If cooling streams are ineffective (venting sound increases in volume and pitch, tank discolors, or shows any signs of deforming), withdraw immediately to a secure position. If employees are expected to fight fires, they must be trained and equipped in OSHA 1910.156. The only respirators recommended for firefighting are self-contained breathing apparatuses that have full face-pieces and are operated in a pressure-demand or other positive-pressure mode.

Disposal Method Suggested: Concentrated waste containing no peroxides—discharge liquid at a controlled rate near a pilot flame. Concentrated waste containing peroxides—perforation of a container of the waste from a safe distance followed by open burning.[22]

References
Bogyo, D. A., Lande, S. S., Meylan, W. M., Howard, P. H., & Santodonato, J. (March 1980). Syracuse Research Corp. Center for Chemical Hazard Assessment, *Investigation of*

Selected Potential Environmental Contaminants: Epoxides, Report EPA-560/11-08-005. Washington, DC: US Environmental Protection Agency

US Environmental Protection Agency. (November 30, 1987). *Chemical Hazard Information Profile: Propylene Oxide*. Washington, DC: Chemical Emergency Preparedness Program

New York State Department of Health. (April 1986). *Chemical Fact Sheet: Propylene Oxide*. Version 2. Albany, NY: Bureau of Toxic Substance Assessment

New Jersey Department of Health and Senior Services. (April 2002). *Hazardous Substances Fact Sheet: Propylene Oxide*. Trenton, NJ

Propyl isocyanate   P:1300

Molecular Formula: C_4H_7NO
Common Formula: C_3H_7NCO
Synonyms: 1-Isocyanatopropane; Isocyanic acid, propyl ester; *m*-Propyl isocyanate; 1-Propyl isocyanate
CAS Registry Number: 110-78-1
RTECS® Number: NR190000
UN/NA & ERG Number: UN2482/155
EC Number: 203-803-6
Regulatory Authority and Advisory Bodies
Canada, WHMIS, Ingredients Disclosure List Concentration 0.1%.
European/International Regulations: not listed in Annex 1.
WGK (German Aquatic Hazard Class): 1—Low hazard to waters.
Description: Propyl isocyanate is a colorless to light yellow liquid with a sharp odor. Molecular weight = 85.12; Flash point = 23°C. Insoluble; reacts with water.
Potential Exposure: This material is used in making other chemicals and insecticides.
Incompatibilities: Violent reactions with strong oxidizers, water, amines, strong bases, and alcohols.
Permissible Exposure Limits in Air
Protective Action Criteria (PAC) Isocyanate-bearing waste (as CNs, n.o.s.).
TEEL-0: 5 mg/m^3
PAC-1: 15 mg/m^3
PAC-2: 25 mg/m^3
PAC-3: 25 mg/m^3
Routes of Entry: Inhalation, ingestion, skin and/or eye contact. May be absorbed by the skin.
Harmful Effects and Symptoms
Short Term Exposure: Propyl isocyanate can affect you when breathed in and possibly by passing through your skin. Irritates the eyes, skin, nose, throat, and lungs. Prolonged contact may cause severe irritation and permanent eye damage. Very high exposures can cause pulmonary edema, a medical emergency that can be delayed for several hours. This can cause death.

Long Term Exposure: Very irritating substances may affect the lungs. Many isocyanates cause an asthma-like lung allergy. It is not known for certain if propyl isocyanate does.
Points of Attack: Lungs.
Medical Surveillance: For those with frequent or potentially high exposure the following are recommended before beginning work and at regular times after that: lung function tests. These may be normal if the person is not having an attack at the time of the test.
First Aid: If this chemical gets into the eyes, remove any contact lenses at once and irrigate immediately for at least 15 min, occasionally lifting upper and lower lids. Seek medical attention immediately. If this chemical contacts the skin, remove contaminated clothing and wash immediately with soap and water. Seek medical attention immediately. If this chemical has been inhaled, remove from exposure, begin rescue breathing (using universal precautions, including resuscitation mask) if breathing has stopped and CPR if heart action has stopped. Transfer promptly to a medical facility. When this chemical has been swallowed, get medical attention. Give large quantities of water and induce vomiting. Do not make an unconscious person vomit. Medical observation is recommended for 24−48 h after breathing overexposure, as pulmonary edema may be delayed. As first aid for pulmonary edema, a doctor or authorized paramedic may consider administering a corticosteroid spray.
Personal Protective Methods: Wear protective gloves and clothing to prevent any reasonable probability of skin contact. Safety equipment suppliers/manufacturers can provide recommendations on the most protective glove/clothing material for your operation. All protective clothing (suits, gloves, footwear, headgear) should be clean, available each day, and put on before work. Contact lenses should not be worn when working with this chemical. Wear splash-proof chemical goggles and face shield unless full face-piece respiratory protection is worn. Employees should wash immediately with soap when skin is wet or contaminated. Provide emergency showers and eyewash.
Respirator Selection: Where there is potential for exposures to propyl isocyanate, use a NIOSH/MSHA- or European Standard EN149-approved supplied-air respirator with a full face-piece operated in the positive-pressure mode, or with a full face-piece, hood, or helmet in the continuous-flow mode; or use a NIOSH/MSHA- or European Standard EN149-approved self-contained breathing apparatus with a full face-piece operated in a pressure-demand or other positive-pressure mode.
Storage: Color Code—Blue: Health Hazard/Poison: Store in a secure poison location. Prior to working with this chemical you should be trained on its proper handling and storage. Propyl isocyanate must be stored to avoid contact with strong oxidizers, water, amines, strong bases (such as sodium hydroxide or potassium hydroxide) and alcohols since violent reactions occur. Store in tightly closed

containers in a cool, well-ventilated area away from heat, light, and moisture. Sources of ignition, such as smoking and open flames, are prohibited where propyl isocyanate is handled, used, or stored. Metal containers involving the transfer of 5 gallons or more of propyl isocyanate should be grounded and bonded. Drums must be equipped with self-closing valves, pressure vacuum bungs, and flame arresters. Use only nonsparking tools and equipment, especially when opening and closing containers of propyl isocyanate.

Shipping: This compound requires a shipping label of "POISONOUS/TOXIC MATERIALS, FLAMMABLE LIQUID." It falls in Hazard Class 6.1 and Packing Group I.

Spill Handling: Evacuate and restrict persons not wearing protective equipment from area of spill or leak until cleanup is complete. Remove all ignition sources. Ventilate area of spill or leak. Absorb liquids in vermiculite, dry sand, earth, peat, carbon, or a similar material and deposit in sealed containers. Keep this chemical out of a confined space, such as a sewer, because of the possibility of an explosion, unless the sewer is designed to prevent the buildup of explosive concentrations. It may be necessary to contain and dispose of this chemical as a hazardous waste. If material or contaminated runoff enters waterways, notify downstream users of potentially contaminated waters. Contact your local or federal environmental protection agency for specific recommendations. If employees are required to clean up spills, they must be properly trained and equipped. OSHA 1910.120(q) may be applicable.

Initial isolation and protective action distances

Distances shown are likely to be affected during the first 30 min after materials are spilled and could increase with time. If more than one tank car, cargo tank, portable tank, or large cylinder involved in the incident is leaking, the protective action distance may need to be increased. You may need to seek emergency information from CHEMTREC at (800) 424-9300 or seek professional environmental engineering assistance from the US EPA Environmental Response Team at (908) 548-8730 (24-h response line).

Small spills (From a small package or a small leak from a large package)

First: Isolate in all directions (feet/meters)   300/100
Then: Protect persons downwind (miles/kilometers)
Day   0.8/1.2
Night   1.7/2.7

Large spills (From a large package or from many small packages)

First: Isolate in all directions (feet/meters)   2500/800
Then: Protect persons downwind (miles/kilometers)
Day   6.0/9.6
Night   7.0+/11.0+ distance can be larger in certain atmospheric conditions

Fire Extinguishing: Propyl isocyanate is a flammable liquid. Poisonous gases are produced in fire, including hydrogen cyanide and oxides of nitrogen. Use dry chemical, CO_2, or foam extinguishers. *Do not use water.* Vapors are heavier than air and will collect in low areas. Vapors may travel long distances to ignition sources and flashback. Vapors in confined areas may explode when exposed to fire. Containers may explode in fire. Storage containers and parts of containers may rocket great distances, in many directions. If material or contaminated runoff enters waterways, notify downstream users of potentially contaminated waters. Notify local health and fire officials and pollution control agencies. From a secure, explosion-proof location, use water spray to cool exposed containers. If cooling streams are ineffective (venting sound increases in volume and pitch, tank discolors, or shows any signs of deforming), withdraw immediately to a secure position. If employees are expected to fight fires, they must be trained and equipped in OSHA 1910.156. The only respirators recommended for firefighting are self-contained breathing apparatuses that have full face-pieces and are operated in a pressure-demand or other positive-pressure mode.

Reference

New Jersey Department of Health and Senior Services. (April 2001). *Hazardous Substances Fact Sheet: Propyl Isocyanate.* Trenton, NJ

n-Propyl nitrate   P:1310

Molecular Formula: $C_3H_7NO_2$
Common Formula: $CH_3CH_2CH_2NO_2$
Synonyms: Nitrate de propyle *normal* (French); Nitric acid, propyl ester; Propyl nitrate; Nitric acid, *n*-propyl ester; Propyl ester of nitric acid
CAS Registry Number: 627-13-4
RTECS® Number: UK0350000
UN/NA & ERG Number: UN1865/131
EC Number: 210-985-0
Regulatory Authority and Advisory Bodies
Air Pollutant Standard Set. See below, "Permissible Exposure Limits in Air" section.
Canada, WHMIS, Ingredients Disclosure List Concentration 1.0%.
European/International Regulations: not listed in Annex 1.
WGK (German Aquatic Hazard Class): No value assigned.

Description: *n*-Propyl nitrate is a colorless to pale yellow liquid with an ether-like odor. Molecular weight = 105.11; Specific gravity (H_2O:1) = 1.07 at 25°C; Boiling point = 110.6°C; Freezing/Melting point = −100°C; Vapor pressure = 18 mmHg at 25°C; Flash point = 20°C. Explosive limits: LEL = 2%; UEL = 100%. Hazard Identification (based on NFPA-704 M Rating System): Health 2, Flammability 3, Reactivity 3 (Oxidizer). Slightly soluble in water.

Potential Exposure: Propyl nitrate has been used as an intermediate as a rocket propellant and as an ignition improver in diesel fuels.

Incompatibilities: Reacts with strong oxidizers, combustible materials. A shock-sensitive explosive. May explode on heating. Forms explosive mixtures with combustible materials.

Permissible Exposure Limits in Air
Conversion factor: 1 ppm = 4.30 mg/m^3 at 25°C & 1 atm.
OSHA PEL: 25 ppm/110 mg/m^3 TWA.
NIOSH REL: 25 ppm/105 mg/m^3 TWA; 40 ppm/170 mg/m^3 STEL.
ACGIH TLV®[1]: 25 ppm/107 mg/m^3 TWA; 40 ppm/172 mg/m^3 STEL; BEI$_M$ issued for methemoglobin inducers.
NIOSH IDLH: 500 ppm.
Protective Action Criteria (PAC)
TEEL-0: 25 ppm
PAC-1: 40 ppm
PAC-2: 100 ppm
PAC-3: 500 ppm
DFG MAK: 25 ppm/110 mg/m^3 TWA; Peak Limitation Category II(2).
Australia: TWA 25 ppm (110 mg/m^3); STEL 40 ppm, 1993; Austria: MAK 25 ppm (110 mg/m^3), 1999; Belgium: TWA 25 ppm (107 mg/m^3); STEL 40 ppm (172 mg/m^3), 1993; Denmark: TWA 25 ppm (110 mg/m^3), 1999; Finland: TWA 25 ppm (105 mg/m^3), 1999; Norway: TWA 20 ppm (90 mg/m^3), 1999; the Philippines: TWA 25 ppm (110 mg/m^3), 1993; the Netherlands: MAC-TGG 110 mg/m^3, 2003; Switzerland: MAK-W 25 ppm (110 mg/m^3), 1999; Turkey: TWA 25 ppm (110 mg/m^3), 1993; Argentina, Bulgaria, Columbia, Jordan, South Korea, New Zealand, Singapore, Vietnam: ACGIH TLV®: STEL 40 ppm. Several states have set guidelines or standards for propyl nitrate in ambient air[60] ranging from 1.05–1.70 mg/m^3 (North Dakota) to 1.75 mg/m^3 (Virginia) to 2.1 mg/m^3 (Connecticut) to 2.5 mg/m^3 (Nevada).

Determination in Air: Use NIOSH II(3), Method #S-227; OSHA Analytical Method 7.
Routes of Entry: Inhalation, ingestion, skin and/or eye contact.
Harmful Effects and Symptoms
Short Term Exposure: None listed for humans according to NIOSH[2] but other sources[24] state that vapor inhalation causes low blood pressure, hypotony, hemoglobin defect, anoxia, and cyanosis. In animals: irritation of the eyes, skin; methemoglobinemia, anoxia, cyanosis, dyspnea (breathing difficulty), weakness, dizziness, headache.
Points of Attack: May attack the blood.
Medical Surveillance: NIOSH lists the following tests: whole blood (chemical/metabolite), methemoglobin; Complete blood count; urinalysis.
First Aid: If this chemical gets into the eyes, remove any contact lenses at once and irrigate immediately for at least 15 min, occasionally lifting upper and lower lids. Seek medical attention immediately. If this chemical contacts the skin, remove contaminated clothing and wash immediately with soap and water. Seek medical attention immediately. If this chemical has been inhaled, remove from exposure, begin rescue breathing (using universal precautions,

including resuscitation mask) if breathing has stopped and CPR if heart action has stopped. Transfer promptly to a medical facility. When this chemical has been swallowed, get medical attention. Give large quantities of water and induce vomiting. Do not make an unconscious person vomit.
Note to physician: Consider treatment for methemoglobinemia. Spectrophotometry may be required for precise determination of levels of methemoglobin in urine.
Personal Protective Methods: Wear protective gloves and clothing to prevent any reasonable probability of skin contact. Safety equipment suppliers/manufacturers can provide recommendations on the most protective glove/clothing material for your operation. All protective clothing (suits, gloves, footwear, headgear) should be clean, available each day, and put on before work. Contact lenses should not be worn when working with this chemical. Wear splash-proof chemical goggles and face shield unless full face-piece respiratory protection is worn. Employees should wash immediately with soap when skin is wet or contaminated. Provide emergency showers and eyewash.
Respirator Selection: *Up to 250 ppm:* Sa (APF = 10) (any supplied-air respirator). *Up to 500 ppm:* Sa:Cf (APF = 25) (any supplied-air respirator operated in a continuous-flow mode) or SCBAF (APF = 50) (any self-contained breathing apparatus with a full face-piece) or SaF (APF = 50) (any supplied-air respirator with a full face-piece). *Emergency or planned entry into unknown concentrations or IDLH conditions:* SCBAF: Pd,Pp (APF = 10,000) (any self-contained breathing apparatus that has a full face-piece and is operated in a pressure-demand or other positive-pressure mode) or SaF: Pd,Pp: ASCBA (APF = 10,000) (any supplied-air respirator that has a full face-piece and is operated in a pressure-demand or other positive-pressure mode in combination with an auxiliary self-contained breathing apparatus operated in a pressure-demand or other positive-pressure mode). *Escape:* GmFS* [any air-purifying, full-face-piece respirator (gas mask) with a chin-style, front- or back-mounted canister providing protection against the compound of concern] or SCBAE (any appropriate escape-type, self-contained breathing apparatus).
*End-of-service life indicator (ESLI) required.
Storage: (1) Color Code—Red: Flammability Hazard: Store in a flammable liquid storage area or approved cabinet away from ignition sources and corrosive and reactive materials. (2) Color Code—Yellow: Reactive Hazard; Store in a location separate from other materials, especially flammables and combustibles. Prior to working with this chemical you should be trained on its proper handling and storage. Before entering confined space where this chemical may be present, check to make sure that an explosive concentration is not a danger. Protect containers against physical damage. Outdoor or detached storage is preferred. Indoor storage should be in a flammable liquid storage room. Propyl nitrate should be isolated from combustible materials and oxidizing agents. Wear Neoprene™ gloves,

plastic coverall, and self-contained breathing apparatus. Where possible, automatically pump liquid from drums or other storage containers to process containers. Sources of ignition, such as smoking and open flames, are prohibited where this chemical is handled, used, or stored. Metal containers involving the transfer of 5 gallons or more of this chemical should be grounded and bonded. Drums must be equipped with self-closing valves, pressure vacuum bungs, and flame arresters. Use only nonsparking tools and equipment, especially when opening and closing containers of this chemical. Wherever this chemical is used, handled, manufactured, or stored, use explosion-proof electrical equipment and fittings. See OSHA Standard 1910.104 and NFPA 43A *Code for the Storage of Liquid and Solid Oxidizers* for detailed handling and storage regulations.

Shipping: This compound requires a shipping label of "FLAMMABLE LIQUID." It falls in Hazard Class 3 and Packing Group II.

Spill Handling: Evacuate and restrict persons not wearing protective equipment from area of spill or leak until cleanup is complete. Remove all ignition sources. Establish forced ventilation to keep levels below explosive limit. After covering the spills with soda ash, mix and spray with water. Scoop into a bucket of water and let it stand for 2 h. Neutralize with 6M-HCl and pass into the drain with sufficient water. Absorb liquids in vermiculite, dry sand, earth, peat, carbon, or a similar material and deposit in sealed containers. Keep this chemical out of a confined space, such as a sewer, because of the possibility of an explosion, unless the sewer is designed to prevent the buildup of explosive concentrations. It may be necessary to contain and dispose of this chemical as a hazardous waste. If material or contaminated runoff enters waterways, notify downstream users of potentially contaminated waters. Contact your local or federal environmental protection agency for specific recommendations. If employees are required to clean up spills, they must be properly trained and equipped. OSHA 1910.120(q) may be applicable.

Fire Extinguishing: This chemical is a flammable liquid. Poisonous gases are produced in fire. Use dry chemical, carbon dioxide, or alcohol foam extinguishers. On fires in which containers are not exposed, use dry chemical, foam, or carbon dioxide. Water spray may be ineffective as an extinguishing agent. Vapors are heavier than air and will collect in low areas. Vapors may travel long distances to ignition sources and flashback. Vapors in confined areas may explode when exposed to fire. Containers may explode in fire. Storage containers and parts of containers may rocket great distances, in many directions. If material or contaminated runoff enters waterways, notify downstream users of potentially contaminated waters. Notify local health and fire officials and pollution control agencies. From a secure, explosion-proof location, use water spray to cool exposed containers. Use water from unmanned monitors or householders to keep fire-exposed containers cool. If cooling streams are ineffective (venting sound increases in volume and pitch, tank discolors, or shows any signs of deforming), withdraw immediately to a secure position. If employees are expected to fight fires, they must be trained and equipped in OSHA 1910.156. The only respirators recommended for firefighting are self-contained breathing apparatuses that have full face-pieces and are operated in a pressure-demand or other positive-pressure mode.

Disposal Method Suggested: Incineration: large quantities of material may require nitrogen oxide removal by catalytic or scrubbing processes.[22] An alternative route suggested involves pouring over soda ash, neutralizing with HCl, and flushing to the drain with water.

Reference
New Jersey Department of Health and Senior Services. (July 2001). *Hazardous Substances Fact Sheet: n-Propyl Nitrate*. Trenton, NJ

Prothoate   P:1320

Molecular Formula: $C_9H_{20}NO_3PS_2$
Synonyms: AC18682; American Cyanamid 18682; *O,O*-Diethyldithiophosphorylacetic acid *N*-monoisopropylamide; *O,O*-Diethyl *S*-(*N*-isopropylcarbamoylmethyl) dithiophosphate; *O,O*-Diethyl *S*-(*N*-isopropylcarbamoylmethyl) phosphorodithioate; *O,O*-Diethyl *S*-isopropylcarbamoylmethyl phosphorodithioate; ENT 24,652; FAC; FAC20; Fostion; Isopropyl diethyldithiophosphorylacetamide; *N*-Isopropyl-2-mercaptoacetamide *S*-ester with *O,O*-diethyl phosphorodithioate; L 343; *N*-Monoisopropylamide of *O,O*-diethyl-dithiophosphorylacetic acid; Oleofac; Phosphorodithioic acid *O,O*-diethyl esters, ester with *n*-isopropyl-2-mercapto-acetamide; Phosphorodithioic acid, *O,O*-diethyl *S*-(2-[(1-methylethyl)amino]-2-oxoethyl) ester; Telefos; Trimethoate
CAS Registry Number: 2275-18-5
RTECS® Number: TD8225000
UN/NA & ERG Number: UN2783 (organophosphorus pesticides, solid, toxic)/152
EC Number: 218-893-2 [*Annex I Index No.:* 015-032-00-0]
Regulatory Authority and Advisory Bodies
Banned or Severely Restricted (Malaysia) (UN).[13]
Superfund/EPCRA 40CFR355, Extremely Hazardous Substances: TPQ = 100/10,000 lb (45.4/4540 kg).
Reportable Quantity (RQ): 100 lb (45.4 kg).
US DOT Regulated Marine Pollutant (49CFR172.101, Appendix B).
US DOT 49CFR172.101, Inhalation Hazard Chemical as organophosphates.
European/International Regulations: Hazard Symbol: T+, N; Risk phrases: R27/28; R52/53; Safety phrases: S1/2; S28; S36/37; S45; S61 (see Appendix 4).
WGK (German Aquatic Hazard Class): No value assigned.
Description: Prothoate is an amber to yellow crystalline solid with a camphor-like odor. Molecular weight = 285.39; Boiling point = 135°C at 0.1 mm; Freezing/Melting point = 29°C. Hazard Identification (based on NFPA-704 M

Rating System): Health 4, Flammability 1, Reactivity 0. Slightly soluble in water.

Potential Exposure: Those involved in the manufacture, formulation, and application of the systemic acaricide and insecticide.

Permissible Exposure Limits in Air
Protective Action Criteria (PAC)
TEEL-0: 0.35 mg/m^3
PAC-1: 1 mg/m^3
PAC-2: 1.7 mg/m^3
PAC-3: 7.5 mg/m^3

Determination in Air: OSHA versatile sampler-2; Toluene/Acetone; Gas chromatography/Flame photometric detection for sulfur, nitrogen, or phosphorus; NIOSH Analytical Method (IV) Method #5600, Organophosphorus pesticides.

Routes of Entry: Inhalation, ingestion, skin and/or eye contact.

Harmful Effects and Symptoms

Short Term Exposure: This is a highly toxic material capable of causing death or permanent injury due to exposures during normal use. Small doses at frequent intervals are additive. Similar to parathion. Symptoms may include nausea, vomiting, abdominal cramps, diarrhea, excessive salivation, headache, giddiness, dizziness, weakness, muscle twitching, difficult breathing, blurring or dimness of vision, and loss of muscle coordination. Death may occur from failure of the respiratory center, paralysis of the respiratory muscles, intense bronchoconstriction, or all the three.

Long Term Exposure: Cholinesterase inhibitor; cumulative effect is possible. This chemical may damage the nervous system with repeated exposure, resulting in convulsions, respiratory failure. May cause liver damage.

Points of Attack: Respiratory system, lungs, central nervous system, cardiovascular system, skin, eyes, plasma and red blood cell cholinesterase.

Medical Surveillance: Before employment and at regular times after that, the following are recommended: plasma and red blood cell cholinesterase levels (tests for the enzyme poisoned by this chemical). If exposure stops, plasma levels return to normal in 1−2 weeks while red blood cell levels may be reduced for 1−3 months.

When cholinesterase enzyme levels are reduced by 25% or more below preemployment levels, risk of poisoning is increased, even if results are in lower ranges of "normal." Reassignment to work not involving organophosphate or carbamate pesticides is recommended until enzyme levels recover. If symptoms develop or overexposure occurs, repeat the above tests as soon as possible and get an examination of the nervous system. Also, consider complete blood count. Consider chest X-ray following acute overexposure. Do not drink any alcoholic beverages before or during use. Alcohol promotes absorption of organic phosphates.

First Aid: If this chemical gets into the eyes, remove any contact lenses at once and irrigate immediately for at least 15 min, occasionally lifting upper and lower lids. Seek medical attention immediately. If this chemical contacts the skin, remove contaminated clothing and wash immediately with soap and water. Speed in removing material from skin is of extreme importance. Shampoo hair promptly if contaminated. Seek medical attention immediately. If this chemical has been inhaled, remove from exposure, begin rescue breathing (using universal precautions, including resuscitation mask) if breathing has stopped and CPR if heart action has stopped. Transfer promptly to a medical facility. When this chemical has been swallowed, get medical attention. Give large quantities of water and induce vomiting. Do not make an unconscious person vomit. Keep victim quiet and maintain normal body temperature. Effects may be delayed; keep victim under observation. If material has been ingested, induce vomiting with ipecac.

Personal Protective Methods: Wear protective gloves and clothing to prevent any reasonable probability of skin contact. Safety equipment suppliers/manufacturers can provide recommendations on the most protective glove/clothing material for your operation. All protective clothing (suits, gloves, footwear, headgear) should be clean, available each day, and put on before work. Contact lenses should not be worn when working with this chemical. Wear splash-proof chemical goggles and face shield unless full face-piece respiratory protection is worn. Employees should wash immediately with soap when skin is wet or contaminated. Provide emergency showers and eyewash.

Respirator Selection: Follow regulations in OSHA 29CFR1910.134 or European Standard EN149. Use a NIOSH/MSHA- or European Standard EN149-approved respirator; or use an approved supplied-air respirator with a full face-piece operated in the positive-pressure mode, or with a full face-piece, hood, or helmet in the continuous-flow mode; or use a NIOSH/MSHA- or European Standard EN149-approved self-contained breathing apparatus with a full face-piece operated in a pressure-demand or other positive-pressure mode.

Storage: Color Code—Blue: Health Hazard/Poison: Store in a secure poison location. Prior to working with this chemical you should be trained on its proper handling and storage. Store in tightly closed containers in a cool, well-ventilated area. Where possible, automatically transfer material from drums or other storage containers to process containers. Sources of ignition, such as smoking and open flames, are prohibited where this chemical is handled, used, or stored. Metal containers involving the transfer of this chemical should be grounded and bonded. Wherever this chemical is used, handled, manufactured, or stored, use explosion-proof electrical equipment and fittings.

Shipping: This compound requires a shipping label of "POISONOUS/TOXIC MATERIALS." It falls in Hazard Class 6.1 and Packing Group I.

Spill Handling: Evacuate and restrict persons not wearing protective equipment from area of spill or leak until cleanup is complete. Remove all ignition sources. Ventilate area of spill or leak. Stay upwind; keep out of low areas. Ventilate closed spaces before entering them. Do not touch spilled material; stop leak if you can do it without risk. Use water spray to reduce vapors. *Small spills:* take up with sand or other noncombustible absorbent material and place into containers for later disposal. *Small dry spills:* with clean shovel place material into clean, dry container and cover; move containers from spill area. *Large spills:* dike far ahead of spill for later disposal. Keep this chemical out of a confined space, such as a sewer, because of the possibility of an explosion, unless the sewer is designed to prevent the buildup of explosive concentrations. It may be necessary to contain and dispose of this chemical as a hazardous waste. If material or contaminated runoff enters waterways, notify downstream users of potentially contaminated waters. Contact your local or federal environmental protection agency for specific recommendations. If employees are required to clean up spills, they must be properly trained and equipped. OSHA 1910.120(q) may be applicable.

Fire Extinguishing: Poisonous gases, including nitrogen oxides, sulfur oxides, phosphorus oxides, are produced in fire. This material may burn but does not readily ignite. For small fires, use dry chemical, carbon dioxide, water spray, or foam. For large fires, use water spray, fog, or foam. In fighting fires, stay upwind; keep out of low areas. Move containers from fire area if you can do it without risk. Fight fire from maximum distance. Dike fire control water for later disposal; do not scatter the material. Wear positive pressure breathing apparatus and special protective clothing. Vapors are heavier than air and will collect in low areas. Vapors may travel long distances to ignition sources and flashback. Vapors in confined areas may explode when exposed to fire. Containers may explode in fire. Storage containers and parts of containers may rocket great distances, in many directions. If material or contaminated runoff enters waterways, notify downstream users of potentially contaminated waters. Notify local health and fire officials and pollution control agencies. From a secure, explosion-proof location, use water spray to cool exposed containers. If cooling streams are ineffective (venting sound increases in volume and pitch, tank discolors, or shows any signs of deforming), withdraw immediately to a secure position. If employees are expected to fight fires, they must be trained and equipped in OSHA 1910.156. The only respirators recommended for firefighting are self-contained breathing apparatuses that have full face-pieces and are operated in a pressure-demand or other positive-pressure mode.

Disposal Method Suggested: In accordance with 40CFR165, follow recommendations for the disposal of pesticides and pesticide containers. Must be disposed properly by following package label directions or by contacting your local or federal environmental control agency or by contacting your regional EPA office.

Reference

US Environmental Protection Agency. (November 30, 1987). *Chemical Hazard Information Profile: Prothoate.* Washington, DC: Chemical Emergency Preparedness Program

Pyrene   P:1330

Molecular Formula: $C_{16}H_{10}$
Synonyms: Benzo(def)phenanthrene; Pireno (Spanish); Pyren (German); β-Pyrene
CAS Registry Number: 129-00-0
UN/NA & ERG Number: UN1325 (flammable solid)/133
RTECS® Number: UR2450000
EC Number: 204-927-3
Regulatory Authority and Advisory Bodies
Carcinogenicity: IARC: Animal Inadequate Evidence; Human No Adequate Data, *not classifiable as carcinogenic to humans*, Group 3, 1987; EPA: Not Classifiable as to human carcinogenicity.
US EPA Gene-Tox Program, Negative: Cell transform.—BALB/c-3T3; SHE—clonal assay; Negative: Cell transform.—mouse embryo; Negative: Cell transform.—RLV F344 rat embryo; Negative: *In vitro* cytogenetics—nonhuman; Host-mediated assay; Negative: Histidine reversion—Ames test; *In vivo* SCE—nonhuman; Negative: Sperm morphology—mouse; *In vitro* UDS—human fibroblast; Negative: V79 cell culture—gene mutation; Inconclusive: Carcinogenicity—mouse/rat; Mammalian micronucleus; Inconclusive: *In vitro* SCE—nonhuman.
Air Pollutant Standard Set. See below, "Permissible Exposure Limits in Air" section.
OSHA, 29CFR1910 Specifically Regulated Chemicals (See CFR 1910.1002) as coal tar pitch volatiles.
RCRA 40CFR268.48; 61FR15654, Universal Treatment Standards: Wastewater (mg/L), 0.067; Nonwastewater (mg/kg), 8.2.
RCRA 40CFR264, Appendix 9; TSD Facilities Ground Water Monitoring List. Suggested test method(s) (PQL μg/L): 8100 (200); 8270 (10).
Superfund/EPCRA 40CFR355, Extremely Hazardous Substances: TPQ = 1000/10,000 lb (454/4540 kg).
Reportable Quantity (RQ): 5000 lb (2270 kg).
Canada, WHMIS, Ingredients Disclosure List Concentration 1.0%.
European/International Regulations: not listed in Annex 1.
WGK (German Aquatic Hazard Class): No value assigned.
Description: Pyrene is a colorless crystalline solid when pure or pale yellow plates (impure). Polynuclear aromatic hydrocarbons (PAHs) are compounds containing multiple benzene rings and are also called polycyclic aromatic hydrocarbons. Solids and solutions have a blue fluorescence (Merck Index). Molecular weight = 202.26; Boiling point 404°C at 760 mmHg[2]; Freezing/Melting point = 151.2°C.[2] Flash point = 199°C. Explosive limits: LEL = 0.6%; UEL = 3.9%. Hazard Identification (based on NFPA-704 M Rating System): Health 1, Flammability 2, Reactivity 0. Practically insoluble in water.
Potential Exposure: Compound Description (Toxicity evaluation)[77]: Tumorigen, Mutagen. Primary Irritant. Pyrene is used as an industrial chemical and in biochemical research.

Permissible Exposure Limits in Air
NIOSH IDLH: 80 mg/m^3.
Protective Action Criteria (PAC)
TEEL-0: 0.75 mg/m^3
PAC-1: 2.5 mg/m^3
PAC-2: 1.7 mg/m^3
PAC-3: 15 mg/m^3
DFG MAK: [skin].
Russia set a MAC of 0.3 mg/m^3 in work-place air.

Determination in Air: Use NIOSH Analytical Method #5506 polynuclear aromatic hydrocarbons by HPLC; NIOSH Analytical Method #5515, Polynuclear aromatic hydrocarbons by GC; OSHA Analytical Method ID-58.

Permissible Concentration in Water: Zero is recommended for maximum protection of human health.[6] Kansas[61] has set a guideline for pyrene in drinking water of 0.029 μg/L. The World Health Organization (WHO) recommends a maximum of 2 μg/L for specific PAHs, but this list does not include pyrene.

Determination in Water: Octanol–water coefficient: Log $K_{ow} = 4.88-5.32$.

Routes of Entry: Inhalation, ingestion, skin and/or eye contact. Can be absorbed through the skin.

Harmful Effects and Symptoms

Short Term Exposure: Pyrene is a skin irritant, a suspected mutagen; and an equivocal tumor-causing agent. Workers exposed to 3–5 mg/m^3 of pyrene exhibited some teratogenic effects. Pyrene is a polycyclic aromatic hydrocarbon (PAH). The acute toxicity of pure PAHs appears low when administered orally or dermally to rats or mice. Human exposure to PACs or PAHs is almost exclusively via the gastrointestinal and respiratory tracts, and approximately 99% is ingested in the diet. Despite the high concentrations of pyrene to which humans may be exposed through food, there is currently little information available to implicate diet-derived PAHs as the cause of serious health effects.

Long Term Exposure: The DFG[3] states that PAHs are present at particularly high levels in coal tar oils and related pyrolysis products of organic materials and are carcinogenic (category 1) in animal studies.

Points of Attack: Skin, respiratory system, bladder, liver, kidneys.

Medical Surveillance: Preplacement and regular physical examinations are indicated for workers having contact with acenaphthene in the workplace. NIOSH lists: complete blood count; chest X-ray; pulmonary function tests: Forced Vital Capacity; Forced Expiratory Volume (1 s); photopatch testing; sputum cytology; urinalysis (routine); cytology, hematuria.[2]

First Aid: Skin Contact[52]: Flood all areas of body that have contacted the substance with water. Do not wait to remove contaminated clothing; do it under the water stream. Use soap to help assure removal. Isolate contaminated clothing when removed to prevent contact by others. Eye Contact: Remove any contact lenses at once. Immediately flush eyes well with copious quantities of water or normal saline for at least 20–30 min. Seek medical attention. Inhalation: Leave contaminated area immediately; breathe fresh air. Proper respiratory protection must be supplied to any rescuers. If coughing, difficult breathing, or any other symptoms develop, seek medical attention at once, even if symptoms develop many hours after exposure. Ingestion: Contact a physician, hospital, or poison center at once. If the victim is unconscious or convulsing, do not induce vomiting or give anything by mouth. Assure that this airway is open and lay him on his side with his head lower than his body and transport immediately to a medical facility. If conscious and not convulsing, give a glass of water to dilute the substance. Vomiting should not be induced without a physician's advice.

Personal Protective Methods: Wear protective gloves and clothing to prevent any reasonable probability of skin contact. Safety equipment suppliers/manufacturers can provide recommendations on the most protective glove/clothing material for your operation. All protective clothing (suits, gloves, footwear, headgear) should be clean, available each day, and put on before work. Contact lenses should not be worn when working with this chemical. Wear dust-proof chemical goggles and face shield unless full face-piece respiratory protection is worn. Employees should wash immediately with soap when skin is wet or contaminated. Provide emergency showers and eyewash.

Respirator Selection: NIOSH: *At any detectable concentration over 0.1 mg/m^3*: SCBAF: Pd,Pp (APF = 10,000) (any NIOSH/MSHA- or European Standard EN 149-approved self-contained breathing apparatus that has a full face-piece and is operated in a pressure-demand or other positive-pressure mode) or SaF: Pd,Pp: ASCBA (APF = 10,000) (any supplied-air respirator that has a full face-piece and is operated in a pressure-demand or other positive-pressure mode in combination with an auxiliary, self-contained breathing apparatus operated in a pressure-demand or other positive-pressure mode). Escape: GmFOv100 (APF = 50) [any air-purifying, full-face-piece respirator (gas mask) with a chin-style, front- or back-mounted organic vapor canister having an N100, R100, or P100 filter] or SCBAE (any appropriate escape-type, self-contained breathing apparatus).

Storage: Color Code—Red: Flammability Hazard: Store in a flammable materials storage area. Prior to working with this chemical you should be trained on its proper handling and storage. Store in a cool, dry place.

Shipping: Flammable solids, organic, n.o.s. require a shipping label of "FLAMMABLE MATERIALS." It falls in Hazard Class 4.1 and Packing Group II.

Spill Handling: Evacuate persons not wearing protective equipment from area of spill or leak until cleanup is complete. Remove all ignition sources. Remove all sources of ignition and dampen spilled material with toluene to avoid airborne dust, then transfer material to a suitable container. Ventilate the spill area and use absorbent dampened with toluene to pick up remaining material. Wash surfaces well with soap and water. Seal all wastes in vapor-tight plastic bags for eventual disposal. Collect powdered material in the most convenient and safe manner and deposit in sealed containers. Ventilate area after cleanup is complete. It may be necessary to contain and dispose of this chemical as a hazardous waste. If material or contaminated runoff enters waterways, notify downstream users of potentially contaminated waters. Contact your local or federal environmental protection agency for specific recommendations. If employees are required to clean up spills, they must be properly trained and equipped. OSHA 1910.120(q) may be applicable.

Fire Extinguishing: This chemical is a combustible solid. Use dry chemical, carbon dioxide, water spray, or alcohol foam extinguishers. Poisonous gases are produced in fire. If material or contaminated runoff enters waterways, notify downstream users of potentially contaminated waters. Notify local health and fire officials and pollution control agencies. From a secure, explosion-proof location, use water spray to cool exposed containers. If cooling streams are ineffective (venting sound increases in volume and pitch, tank discolors, or shows any signs of deforming), withdraw immediately to a secure position. If employees are expected to fight fires, they must be trained and equipped in OSHA 1910.156. The only respirators recommended for firefighting are self-contained breathing apparatuses that have full face-pieces and are operated in a pressure-demand or other positive-pressure mode.

Disposal Method Suggested: Dissolve or mix the material with a combustible solvent and burn in a chemical incinerator equipped with an afterburner and scrubber. All federal, state, and local environmental regulations must be observed.[22]

References

US Environmental Protection Agency. (November 30, 1987). *Chemical Hazard Information Profile: Pyrene.* Washington, DC: Chemical Emergency Preparedness Program

US EPA. (April 1975). Identification of Organic Compounds in Effluents from Industrial Sources, EPA-560/3-75-002

Eller, P. M., Cassinelli, M. E. (Eds.). (1998). *NIOSH Manual of Analytical Methods (NMAM®)* (4th ed.). 2nd Supplement. National Institute for Occupational Safety and Health, DHHS (NIOSH), Publication No. 98-119. Cincinnati, OH

Pyrethrins or pyrethrum   P:1340

Molecular Formula: $C_{20-21}H_{28-30}O_{3-5}$; $C_{20}H_{28}O_3$/$C_{21}H_{28}O_5$/$C_{21}H_{30}O_3$/$C_{22}H_{30}O_5$/$C_{21}H_{28}O_3$/$C_{22}H_{28}O_5$
Synonyms: Buhach; Chrysanthemum cinerareaefolium; Cinerin I; Cinerin II; Dalmation insect flowers; Firmotox; Insect powder; Jasmolin I; Jasmolin II; Piretrina (Spanish);

Pyrethrin I; Pyrethrin II; Pyrethrum; Pyrethrum insecticide; Trieste flowers
CAS Registry Number: 8003-34-7; 121-21-1 (I); 121-29-9 (II)
RTECS® Number: UR4200000
UN/NA & ERG Number: UN2902 (Pesticides, liquid, toxic, n.o.s.)/151
EC Number: 232-319-8; 204-455-8 [*Annex I Index No.:* 613-023-00-1] (I); 204-462-6 [*Annex I Index No.:* 613-024-00-7] (II)
Regulatory Authority and Advisory Bodies
US EPA, FIFRA 1998 Status of Pesticides: Canceled.
Air Pollutant Standard Set. See below, "Permissible Exposure Limits in Air" section.
Type I:
Clean Water Act: Section 311 Hazardous Substances/RQ 40CFR117.3 (same as CERCLA, see below).
US EPA Hazardous Waste Number (RCRA No.): P008.
RCRA, 40CFR261, Appendix 8 Hazardous Constituents.
Reportable Quantity (RQ): 1 lb (0.454 kg).
Type II:
Clean Water Act: Section 311 Hazardous Substances/RQ 40CFR117.3 (same as CERCLA, see below).
Reportable Quantity (RQ): 1 lb (0.454 kg).
European/International Regulations (*I &II*): Hazard Symbol: Xn, N; Risk phrases: R0/21/22; R50/53; Safety phrases: S2; S13; S60; S61 (see Appendix 4).
WGK (German Aquatic Hazard Class): 3—Severe hazard to waters.

Description: The pyrethrins are a variable mixture of compounds which are found in pyrethrum flowers: cinerin, jasmolin, and pyrethrin. Pyrethrins are a brown, viscous oil or solid. Molecular weight = 316−378; 372.45; Specific gravity (H_2O:1) = 1.0 (approx.) at 25°C; Boiling point = 170°C at 0.1 mm (decomposition); Flash point = 82−88°C. Hazard Identification (based on NFPA-704 M Rating System): Health 2, Flammability 1, Reactivity 0. Insoluble in water.
Potential Exposure: Compound Description: Agricultural Chemical; Reproductive Effector; Human Data; Natural Product. Pyrethrins are used as an ingredient of various contact insecticides. Those engaged in the isolation, formulation, or application of these materials.
Incompatibilities: Violent reaction with strong oxidizers, alkaline materials.
Permissible Exposure Limits in Air
OSHA PEL: 5 mg/m³ TWA.
NIOSH REL: 5 mg/m³ TWA.
ACGIH TLV®[1]: 5 mg/m³ TWA; not classifiable as a human carcinogen.
NIOSH IDLH: 5000 mg/m³.
Protective Action Criteria (PAC)
TEEL-0: 1 mg/m³
PAC-1: 3 mg/m³
PAC-2: 20 mg/m³
PAC-3: 100 mg/m³

DFG MAK: 5 mg/m^3, inhalable fraction, danger of skin sensitization; Peak Limitation Category II(2).

Arab Republic of Egypt: TWA 5 mg/m^3, 1993; Australia: TWA 5 mg/m^3, 1993; Austria: MAK 5 mg/m^3, 1999; Belgium: TWA 5 mg/m^3, 1993; Denmark: TWA 5 mg/m^3, 1999; Finland: TWA 5 mg/m^3; STEL 10 mg/m^3, 1999; France: VME 5 mg/m^3, 1999; the Netherlands: MAC-TGG 5 mg/m^3, 2003; Norway: TWA 5 mg/m^3, 1999; Thailand: TWA 5 mg/m^3, 1993; Turkey: TWA 5 mg/m^3, 1993; United Kingdom: TWA 5 mg/m^3; STEL 10 mg/m^3, 2000; Argentina, Bulgaria, Columbia, Jordan, South Korea, New Zealand, Singapore, Vietnam: ACGIH TLV®: not classifiable as a human carcinogen. Several states have set guidelines or standards for pyrethrum in ambient air[60] ranging from 16.7 μg/m^3 (New York) to 50.0 μg/m^3 (Florida and South Carolina) to 50.0–100.0 μg/m^3 (North Dakota) to 80.0 μg/m^3 (Virginia) to 100.0 μg/m^3 (Connecticut) to 119.0 μg/m^3 (Nevada).

Determination in Air: Use NIOSH Analytical Method (IV) #5008, Pyrethrum; OSHA Analytical Method 7.[18]

Determination in Water: Fish Tox: 2.38748000 ppb (HIGH).

Routes of Entry: Inhalation, ingestion, skin and/or eye contact.

Harmful Effects and Symptoms

Short Term Exposure: Pyrethrum can affect you when breathed in and by passing through your skin. Irritates the eyes and respiratory tract. High exposure can affect the nervous system causing headache, nausea, vomiting, fatigue, and restlessness and rhinorrhea (discharge of thin nasal mucus).

Long Term Exposure: High or repeated exposure can cause lung allergy (with cough, wheezing, and/or shortness of breath) or hay fever symptoms (sneezing, runny or stuffy nose). Allergic "pneumonia" can also occur with cough, chest pain, breathing difficulty, and abnormal chest X-ray. Repeated attacks may lead to permanent scarring. Skin allergy may also develop with rash and itching, even with lower exposures. Skin contact can cause rash with redness, blisters, and intense itching. A severe generalized allergy can occur with weakness and collapse. Human Tox = 68.09339 ppb. Chronic Human Carcinogen Level (CHCL) (LOW).

Points of Attack: Respiratory system, skin, central nervous system.

Medical Surveillance: Before beginning employment and at regular times after that, the following are recommended: lung function tests. These may be normal if the person is not having an attack at the time of the test. Consider chest X-ray if lung symptoms are present. Evaluation by a qualified allergist, including careful exposure history and special testing, may help diagnose skin allergy.

First Aid: If this chemical gets into the eyes, remove any contact lenses at once and irrigate immediately for at least 15 min, occasionally lifting upper and lower lids. Seek medical attention immediately. If this chemical contacts the skin, remove contaminated clothing and wash immediately with soap and water. Seek medical attention immediately. If this chemical has been inhaled, remove from exposure, begin rescue breathing (using universal precautions, including resuscitation mask) if breathing has stopped and CPR if heart action has stopped. Transfer promptly to a medical facility. When this chemical has been swallowed, get medical attention. Give large quantities of water and induce vomiting. Do not make an unconscious person vomit.

Personal Protective Methods: Wear protective gloves and clothing to prevent any reasonable probability of skin contact. Safety equipment suppliers/manufacturers can provide recommendations on the most protective glove/clothing material for your operation. All protective clothing (suits, gloves, footwear, headgear) should be clean, available each day, and put on before work. Contact lenses should not be worn when working with this chemical. Wear eye protection to prevent any possibility of eye contact. Employees should wash immediately with soap when skin is wet or contaminated. Provide emergency showers and eyewash.

Respirator Selection: *50 mg/m^3:* CcrOv95 (APF = 10) [Any air-purifying half-mask respirator with organic vapor cartridge(s) in combination with an N95, R95, or P95 filter. The following filters may also be used: N99, R99, P99, N100, R100, P100] or Sa (APF = 10) (any supplied-air respirator). *125 mg/m^3:* Sa:Cf (APF = 25) (any supplied-air respirator operated in a continuous-flow mode) or PaprOvHie (APF = 25) (any powered air-purifying respirator with an organic vapor cartridge in combination with a high-efficiency particulate filter). *250 mg/m^3:* CcrFOv100 (APF = 50) [Any air-purifying full-face-piece respirator equipped with organic vapor cartridge(s) in combination with an N100, R100, or P100 filter] or PaprTOvHie (APF = 50) [any powered, air-purifying respirator with a tight-fitting face-piece and organic vapor cartridge(s) in combination with a high-efficiency particulate filter] or SCBAF (APF = 50) (any self-contained breathing apparatus with full face-piece) or SaF (APF = 50) (any supplied-air respirator with a full face-piece). *5000 mg/m^3:* SaF: Pd,Pp (APF = 2000) (any supplied-air respirator that has a full face-piece and is operated in a pressure-demand or other positive-pressure mode). *Emergency or planned entry into unknown concentrations or IDLH conditions:* SCBAF: Pd, Pp (APF = 10,000) (any self-contained breathing apparatus that has a full face-piece and is operated in a pressure-demand or other positive-pressure mode) or SaF: Pd,Pp: ASCBA (APF = 10,000) (any supplied-air respirator that has a full face-piece and is operated in a pressure-demand or other positive-pressure mode in combination with an auxiliary self-contained breathing apparatus operated in a pressure-demand or other positive-pressure mode). *Escape:* GmFOv100 (APF = 50) [any air-purifying, full-face-piece respirator (gas mask) with a chin-style, front- or back-mounted organic vapor canister having an N100, R100, or P100 filter] or SCBAE (any appropriate escape-type, self-contained breathing apparatus).

Note: Substance reported to cause eye irritation or damage; may require eye protection.

Storage: Color Code—Blue: Health Hazard/Poison: Store in a secure poison location. Prior to working with this chemical you should be trained on its proper handling and storage. Pyrethrum must be stored to avoid contact with oxidizers (such as perchlorates, peroxides, permanganates, chlorates, and nitrates) and alkalines since violent reactions occur. Store in tightly closed containers in a cool, well-ventilated area. Where possible, automatically transfer material from drums or other storage containers to process containers. Sources of ignition, such as smoking and open flames, are prohibited where this chemical is handled, used, or stored. Metal containers involving the transfer of this chemical should be grounded and bonded. Wherever this chemical is used, handled, manufactured, or stored, use explosion-proof electrical equipment and fittings.

Shipping: Pesticides, liquid, toxic, n.o.s. require a shipping label of "POISONOUS/TOXIC MATERIALS." They fall in Hazard Class 6.1. and Packing Group I to III.

Spill Handling: Evacuate and restrict persons not wearing protective equipment from area of spill or leak until cleanup is complete. Remove all ignition sources. Ventilate area of spill or leak. Absorb liquids in vermiculite, dry sand, earth, peat, carbon, or a similar material and deposit in sealed containers. Keep this chemical out of a confined space, such as a sewer, because of the possibility of an explosion, unless the sewer is designed to prevent the buildup of explosive concentrations. It may be necessary to contain and dispose of this chemical as a hazardous waste. If material or contaminated runoff enters waterways, notify downstream users of potentially contaminated waters. Contact your local or federal environmental protection agency for specific recommendations. If employees are required to clean up spills, they must be properly trained and equipped. OSHA 1910.120(q) may be applicable. Soil Adsorption Index (K_{oc}): 100,000 (estimate).

Fire Extinguishing: This chemical is a combustible liquid. Poisonous gases, including acid fumes, are produced in fire. Use dry chemical, carbon dioxide, or alcohol foam extinguishers. Vapors are heavier than air and will collect in low areas. Vapors may travel long distances to ignition sources and flashback. Vapors in confined areas may explode when exposed to fire. Containers may explode in fire. Storage containers and parts of containers may rocket great distances, in many directions. If material or contaminated runoff enters waterways, notify downstream users of potentially contaminated waters. Notify local health and fire officials and pollution control agencies. From a secure, explosion-proof location, use water spray to cool exposed containers. If cooling streams are ineffective (venting sound increases in volume and pitch, tank discolors, or shows any signs of deforming), withdraw immediately to a secure position. If employees are expected to fight fires, they must be trained and equipped in OSHA 1910.156. The only respirators recommended for firefighting are self-contained breathing apparatuses that have full face-pieces and are operated in a pressure-demand or other positive-pressure mode.

Reference

New Jersey Department of Health and Senior Services. (August 2002). *Hazardous Substances Fact Sheet: Pyrethrum.* Trenton, NJ

Pyridine   P:1345

Molecular Formula: C_5H_5N
Synonyms: Azabenzene; Azine; CP 32; NCI-C55301; Pyridin (German)
CAS Registry Number: 110-86-1
RTECS® Number: UR8400000
UN/NA & ERG Number: UN1282/129
EC Number: 203-809-9 [*Annex I Index No.:* 613-002-00-7]
Regulatory Authority and Advisory Bodies
Carcinogenicity: IARC: Animal Limited Evidence; Human Inadequate Evidence, *not classifiable as carcinogenic to humans,* Group 3, 2000; NCI: Carcinogenesis studies (water); clear evidence: mouse; equivocal evidence: rat; NTP: Carcinogenesis studies (drinking water); some evidence: rat.

US EPA Gene-Tox Program, Positive/dose response: *In vitro* SCE—nonhuman.

Air Pollutant Standard Set. See below, "Permissible Exposure Limits in Air" section.

US EPA Hazardous Waste Number (RCRA No.): U196, DO38.

RCRA, 40CFR261, Appendix 8 Hazardous Constituents.

RCRA Toxicity Characteristic (Section 261.24), Maximum Concentration of Contaminants, regulatory level, 5.0 mg/L.

RCRA 40CFR268.48; 61FR15654, Universal Treatment Standards: Wastewater (mg/L), 0.014; Nonwastewater (mg/kg), 16.

RCRA 40CFR264, Appendix 9; TSD Facilities Ground Water Monitoring List. Suggested test method(s) (PQL µg/L): 8240 (5); 8270 (10).

Reportable Quantity (RQ): 1000 lb (454 kg).

EPCRA Section 313 Form R *de minimis* concentration reporting level: 1.0%.

California Proposition 65 Chemical: Cancer 5/17/02.

Canada, WHMIS, Ingredients Disclosure List Concentration 1.0%.

European/International Regulations: Hazard Symbol: F, Xn; Risk phrases: R11; R20/21/22; Safety phrases: S2; S26; S28 (see Appendix 4).

WGK (German Aquatic Hazard Class): 2—Hazard to waters.

Description: Pyridine is a colorless liquid with a nauseating, fish-like odor. The odor threshold is 0.17 ppm. Molecular weight = 79.11; Specific gravity (H_2O:1) = 0.98 at 25°C; Boiling point = 115.6°C. Melting/Freezing point = −42.2°C; Vapor pressure = 16 mmHg at 25°C;

Flash point = 20°C; Autoignition temperature = 482°C. Explosive limits: LEL = 1.8%; UEL = 12.4%. Hazard Identification (based on NFPA-704 M Rating System): Health 3, Flammability 3, Reactivity 0. Soluble in water.

Potential Exposure: Compound Description: Tumorigen, Mutagen. Primary Irritant. Pyridine is used as a solvent in the chemical industry and as a denaturant for ethyl alcohol; as an intermediate in the production of pesticides; in pharmaceuticals; in the manufacture of paints, explosives, dyestuffs, rubber, vitamins, sulfa drugs, and disinfectants.

Incompatibilities: Violent reaction with strong oxidizers, strong acids, chlorosulfonic acid, maleic anhydride, oleum iodine.

Permissible Exposure Limits in Air

Conversion factor: 1 ppm = 3.24 mg/m^3 at 25°C & 1 atm.
OSHA PEL: 5 ppm/15 mg/m^3 TWA.
NIOSH REL: 5 ppm/15 mg/m^3 TWA.
ACGIH TLV®[1]: 1 ppm/3.1 mg/m^3 TWA.
NIOSH IDLH: 1000 ppm.
Protective Action Criteria (PAC)
TEEL-0: 1 ppm
PAC-1: 3 ppm
PAC-2: 5 ppm
PAC-3: 1000 ppm
DFG MAK: [skin] 5 ppm/16 mg/m^3 TWA; Peak Limitation Category II(2).
Arab Republic of Egypt: TWA 5 ppm (15 mg/m^3), 1993; Australia: TWA 5 ppm (15 mg/m^3), 1993; Austria: MAK 5 ppm (15 mg/m^3), 1999; Belgium: TWA 5 ppm (16 mg/m^3), 1993; Denmark: TWA 5 ppm (15 mg/m^3), 1999; Finland: TWA 5 ppm (15 mg/m^3); STEL 10 ppm (30 mg/m^3), [skin], 1999; France: VME 5 ppm (15 mg/m^3), VLE 10 ppm, 1999; the Netherlands: MAC-TGG 0.9 mg/m^3, 2003; Norway: TWA 5 ppm (15 mg/m^3), 1999; the Philippines: TWA 5 mg/m^3, 1993; Poland: MAC (TWA) 5 mg/m^3; MAC (STEL) 30 mg/m^3, 1999; Russia: STEL 5 mg/m^3, 1993; Sweden: NGV 5 ppm (16 mg/m^3), KTV 10 ppm (35 mg/m^3), 1999; Switzerland: MAK-W 5 ppm (15 mg/m^3), KZG-W 10 ppm (30 mg/m^3), 1999; Turkey: TWA 5 ppm (15 mg/m^3), 1993; United Kingdom: TWA 5 ppm (16 mg/m^3); STEL 10 ppm (33 mg/m^3), 2000; Argentina, Bulgaria, Columbia, Jordan, South Korea, New Zealand, Singapore, Vietnam: ACGIH TLV®: TWA 5 ppm. The Czech Republic: MAC 5.0 mg/m^3.[35] Russia[35, 43] has also set a MAC of 0.08 mg/m^3 for the ambient air in residential areas both on a momentary and a daily average basis. Several states have set guidelines or standards for pyridine in ambient air[60] ranging from 2.0 μg/m^3 (New York) to 150 μg/m^3 (Indiana) to 150.0−300.0 μg/m^3 (North Dakota) to 250.0 μg/m^3 (Virginia) to 300.0 μg/m^3 (Connecticut, Florida) to 357.0 μg/m^3 (Nevada).

Determination in Air: Use NIOSH Analytical Method #1613[18]; OSHA Analytical Method 7.

Permissible Concentration in Water: EPA[32] has suggested a permissible ambient goal of 207 μg/L. Russia[43] set a MAC of 200 μg/L in water bodies used for domestic purposes and 10 μg/L in water bodies used for fishery purposes.

Determination in Water: Octanol−water coefficient: Log K_{ow} = 0.65.

Routes of Entry: Inhalation of vapor, percutaneous absorption of liquids, ingestion, skin and/or eye contact.

Harmful Effects and Symptoms

Short Term Exposure: Irritates the eyes, skin, and respiratory tract. May affect the central nervous system; can cause stomach upset, headache, dizziness, lightheadedness, confusion, coma, and death. *Inhalation:* May cause nose and throat irritation at low levels. Above 5 ppm, the odor may no longer be detected but a characteristic taste may remain. Exposures of 6−12 ppm have caused headache, dizziness, nervousness, trouble sleeping, nausea, and vomiting. *Skin:* Contact with liquid may cause painful irritation and first-degree burns. Longer contact may cause second-degree burns. *Eyes:* Vapors may cause irritation. Contact with liquid can cause irritation, burns, and permanent eye damage. *Ingestion:* 2−3 mL (1/15−1/10 fl oz) may cause loss of appetite, nausea, fatigue, and depression. Swallowing ½ cup has resulted in vomiting, diarrhea, fever, abdominal pain, bluish tint to the skin, confusion, hallucinations, severe lung congestion, and death.

Long Term Exposure: Affects the central nervous system, brain, liver, and kidneys. Can cause skin allergy. Ingestion of about 2 mL (1/10 fl oz) three times/day for a prolonged period resulted in lever and kidney damage and death. Long-term inhalation of levels of 125 ppm in addition to the symptoms listed above may result in damage to the nervous system.

Points of Attack: Eyes, skin, central nervous system, liver, kidneys, gastrointestinal tract.

Medical Surveillance: NIOSH lists the following tests: Expired Air; liver function tests; urinalysis. If symptoms develop or overexposure is suspected, the following may be useful: examination of the nervous system; interview for brain effects, kidney function tests. Evaluation by a qualified allergist, including careful exposure history and special testing, may help diagnose skin allergy.

First Aid: If this chemical gets into the eyes, remove any contact lenses at once and irrigate immediately for at least 15 min, occasionally lifting upper and lower lids. Seek medical attention immediately. If this chemical contacts the skin, remove contaminated clothing and wash immediately with soap and water. Seek medical attention immediately. If this chemical has been inhaled, remove from exposure, begin rescue breathing (using universal precautions, including resuscitation mask) if breathing has stopped and CPR if heart action has stopped. Transfer promptly to a medical facility. When this chemical has been swallowed, get medical attention. Give large quantities of water and induce vomiting. Do not make an unconscious person vomit.

Personal Protective Methods: Wear protective gloves and clothing to prevent any reasonable probability of skin contact. Safety equipment suppliers/manufacturers can provide

recommendations on the most protective glove/clothing material for your operation. Polyethylene is among the recommended protective materials. All protective clothing (suits, gloves, footwear, headgear) should be clean, available each day, and put on before work. Contact lenses should not be worn when working with this chemical. Wear splash-proof chemical goggles and face shield unless full face-piece respiratory protection is worn. Employees should wash immediately with soap when skin is wet or contaminated. Provide emergency showers and eyewash.

Respirator Selection: 125 ppm: Sa:Cf (APF = 25) (any supplied-air respirator operated in a continuous-flow mode) or PaprOv (APF = 25) [any powered, air-purifying respirator with organic vapor cartridge(s)]. *250 ppm:* CcrFOv (APF = 50) [any air-purifying, full-face-piece respirator (gas mask) with a chin-style, front- or back-mounted acid gas canister] or GmFOv (APF = 50) [any air-purifying, full-face-piece respirator (gas mask) with a chin-style, front- or back-mounted organic vapor canister] or PaprTOv (APF = 50) [any powered, air-purifying respirator with a tight-fitting face-piece and organic vapor cartridge(s)] or SCBAF (APF = 50) (any self-contained breathing apparatus with a full face-piece) or SaF (APF = 50) (any supplied-air respirator with a full face-piece). *1000 ppm:* SaF: Pd,Pp (APF = 2000) (any supplied-air respirator that has a full face-piece and is operated in a pressure-demand or other positive-pressure mode). *Emergency or planned entry into unknown concentrations or IDLH conditions:* SCBAF: Pd, Pp (APF = 10,000) (any self-contained breathing apparatus that has a full face-piece and is operated in a pressure-demand or other positive-pressure mode) or SaF: Pd,Pp: ASCBA (APF = 10,000) (any supplied-air respirator that has a full face-piece and is operated in a pressure-demand or other positive-pressure mode in combination with an auxiliary self-contained breathing apparatus operated in a pressure-demand or other positive-pressure mode). *Escape:* GmFOv (APF = 50) [any air-purifying, full-face-piece respirator (gas mask) with a chin-style, front- or back-mounted organic vapor canister] or SCBAE (any appropriate escape-type, self-contained breathing apparatus).

Note: Substance causes eye irritation or damage; eye protection needed.

Storage: Color Code—Red: Flammability Hazard: Store in a flammable liquid storage area or approved cabinet away from ignition sources and corrosive and reactive materials. Prior to working with this chemical you should be trained on its proper handling and storage. Before entering confined space where this chemical may be present, check to make sure that an explosive concentration is not a danger. Pyridine must be stored to avoid contact with strong oxidizers (such as chloride, bromine, and fluorine), strong acids (such as hydrochloride, sulfuric, and nitric), chlorosulfonic acid, maleic anhydride, and oleum iodine because violent reactions occur. Sources of ignition, such as smoking and open flames, are prohibited where pyridine is used, handles,

or stored in a manner that could create a potential fire or explosion hazard. Metal containers used in the transfer of 5 gallons or more of pyridine should be grounded and bonded. Drums must be equipped with self-closing valves, pressure vacuum bungs, and flame arresters. Use only non-sparking tools and equipment, especially when opening and closing containers of pyridine. Wherever pyridine is used, handled, manufactured, or stored, use explosion-proof electrical equipment and fittings.

Shipping: This compound requires a shipping label of "FLAMMABLE LIQUID, POISONOUS/TOXIC MATERIALS." It falls in Hazard Class 3 and Packing Group II.

Spill Handling: Evacuate and restrict persons not wearing protective equipment from area of spill or leak until cleanup is complete. Remove all ignition sources. Establish forced ventilation to keep levels below explosive limit. Absorb liquids in vermiculite, dry sand, earth, peat, carbon, or a similar material and deposit in sealed containers. Keep this chemical out of a confined space, such as a sewer, because of the possibility of an explosion, unless the sewer is designed to prevent the buildup of explosive concentrations. It may be necessary to contain and dispose of this chemical as a hazardous waste. If material or contaminated runoff enters waterways, notify downstream users of potentially contaminated waters. Contact your local or federal environmental protection agency for specific recommendations. If employees are required to clean up spills, they must be properly trained and equipped. OSHA 1910.120(q) may be applicable.

Fire Extinguishing: This chemical is a flammable liquid. Poisonous gases, including nitrogen oxides and cyanides, are produced in fire. Use dry chemical, carbon dioxide, or alcohol foam extinguishers. Vapors are heavier than air and will collect in low areas. Vapors may travel long distances to ignition sources and flashback. Vapors in confined areas may explode when exposed to fire. Containers may explode in fire. Storage containers and parts of containers may rocket great distances, in many directions. If material or contaminated runoff enters waterways, notify downstream users of potentially contaminated waters. Notify local health and fire officials and pollution control agencies. From a secure, explosion-proof location, use water spray to cool exposed containers. If cooling streams are ineffective (venting sound increases in volume and pitch, tank discolors, or shows any signs of deforming), withdraw immediately to a secure position. If employees are expected to fight fires, they must be trained and equipped in OSHA 1910.156. The only respirators recommended for firefighting are self-contained breathing apparatuses that have full face-pieces and are operated in a pressure-demand or other positive-pressure mode.

Disposal Method Suggested: Controlled incineration whereby nitrogen oxides are removed from the effluent gas by scrubber, catalytic or thermal devices.[22]

References

US Environmental Protection Agency. (April 30, 1980). *Pyridine: Health and Environmental Effects Profile No. 150*. Washington, DC: Office of Solid Waste

New York State Department of Health. (March 1986). *Chemical Fact Sheet: Pyridine*. Version 2. Albany, NY: Bureau of Toxic Substance Assessment

New Jersey Department of Health and Senior Services. (March 2002). *Hazardous Substances Fact Sheet: Pyridine*. Trenton, NJ

Pyriminil   P:1350

Molecular Formula: $C_{13}H_{12}N_4O_3$

Synonyms: DLP787; DPL-87; *N*-(4-Nitrophenyl)-*N'*-(3-pyridinylmethyl)urea; 1-(4-Nitrophenyl)-3-(3-pyridinylmethyl) urea; Priminil; *N*-3-Pyridylmethyl-*N'*,*p*-nitrophenylurea; 1-(3-Pyridylmethyl)-3-(4-nitrophenyl)urea; Pyridylmethyl-*N'*, *p*-nitrophenylurea; Pyrinuron; RH-787; Urea, *N*-(4-nitrophenyl)-*N'*-(3-pyridinylmethyl)-; Vacor

CAS Registry Number: 53558-25-1

RTECS® Number: YI9690000

UN/NA & ERG Number: UN2767/151

EC Number: 258-626-7

Regulatory Authority and Advisory Bodies

Superfund/EPCRA 40CFR355, Extremely Hazardous Substances: TPQ = 100/10,000 lb (45.4/4540 kg).

Reportable Quantity (RQ): 100 lb (45.4 kg).

European/International Regulations: not listed in Annex 1.

WGK (German Aquatic Hazard Class): No value assigned.

Description: Pyriminil is a yellow crystalline solid resembling corn meal. Molecular weight = 272.29; Freezing/Melting point = 223°C (decomposition). Hazard Identification (based on NFPA-704 M Rating System): Health 3, Flammability 1, Reactivity 0.

Potential Exposure: Those involved in the manufacture, formulation, or application of this single-dose, acute rodenticide. Not registered as a pesticide in the United States.

Permissible Exposure Limits in Air

Protective Action Criteria (PAC)

TEEL-0: 1.25 mg/m^3

PAC-1: 3.5 mg/m^3

PAC-2: 6.2 mg/m^3

PAC-3: 20 mg/m^3

Routes of Entry: Inhalation, ingestion, skin and/or eye contact.

Harmful Effects and Symptoms

Short Term Exposure: This chemical may cause death by cardiovascular collapse and respiratory failure. Symptoms include nausea, vomiting, abdominal pain, chills, mental confusion, anorexia, aching, dilated pupils, dehydration, chest pain, urinary retention, irregular heartbeat, and muscular weakness. Exposure may also result in visual disturbances, central nervous system depression, and tremors.

Long Term Exposure: It may damage the pancreas, causing diabetes. Human survivors regularly develop an insulin-deficient, ketosis-prone form of diabetes mellitus. Also, it affects the central nervous system.

Points of Attack: Central nervous system.

Medical Surveillance: Blood sugar. Examination of the nervous system.

First Aid: If this chemical gets into the eyes, remove any contact lenses at once and irrigate immediately for at least 15 min, occasionally lifting upper and lower lids. Seek medical attention immediately. If this chemical contacts the skin, remove contaminated clothing and wash immediately with soap and water. Seek medical attention immediately. If this chemical has been inhaled, remove from exposure, begin rescue breathing (using universal precautions, including resuscitation mask) if breathing has stopped and CPR if heart action has stopped. Transfer promptly to a medical facility. When this chemical has been swallowed, get medical attention. Give large quantities of water and induce vomiting. Do not make an unconscious person vomit. Keep victim quiet and maintain normal body temperature. Effects may be delayed; keep victim under observation.

Personal Protective Methods: Wear protective gloves and clothing to prevent any reasonable probability of skin contact. Safety equipment suppliers/manufacturers can provide recommendations on the most protective glove/clothing material for your operation. All protective clothing (suits, gloves, footwear, headgear) should be clean, available each day, and put on before work. Contact lenses should not be worn when working with this chemical. Wear dust-proof chemical goggles and face shield unless full face-piece respiratory protection is worn. Employees should wash immediately with soap when skin is wet or contaminated. Provide emergency showers and eyewash.

Respirator Selection: Follow regulations in OSHA 29CFR1910.134 or European Standard EN149. Use a NIOSH/MSHA- or European Standard EN149-approved respirator; or use an approved supplied-air respirator with a full face-piece operated in the positive-pressure mode, or with a full face-piece, hood, or helmet in the continuous-flow mode; or use a NIOSH/MSHA- or European Standard EN149-approved self-contained breathing apparatus with a full face-piece operated in a pressure-demand or other positive-pressure mode.

Storage: Color Code—Blue: Health Hazard/Poison: Store in a secure poison location. Prior to working with this chemical you should be trained on its proper handling and storage. Store in tightly closed containers in a cool, well-ventilated area. Where possible, automatically transfer material from drums or other storage containers to process containers.

Shipping: Phenylurea pesticide, solid, toxic, n.o.s. This compound requires a shipping label of "POISONOUS/TOXIC MATERIALS." It falls in Hazard Class 6.1 and Packing Group I.

Spill Handling: Evacuate persons not wearing protective equipment from area of spill or leak until cleanup is complete. Remove all ignition sources. Stay upwind; keep out of low areas. Ventilate closed spaces before entering them. Remove and isolate contaminated clothing at the site. Do not touch spilled material; stop leak if you can do so without risk. Use water spray to reduce vapors. *Small spills:* absorb with sand or other noncombustible absorbent material and place into containers for later disposal. *Small dry spills:* with clean shovel place material into clean, dry container and cover; move containers from spill area. *Large spills:* dike far ahead of spill for later disposal. It may be necessary to contain and dispose of this chemical as a hazardous waste. If material or contaminated runoff enters waterways, notify downstream users of potentially contaminated waters. Contact your local or federal environmental protection agency for specific recommendations. If employees are required to clean up spills, they must be properly trained and equipped. OSHA 1910.120(q) may be applicable.

Fire Extinguishing: Poisonous gases are produced in fire, including nitrogen oxides. *Small fires:* dry chemical, carbon dioxide, water spray, or foam. *Large fires:* water spray, fog, or foam. Move container from fire area if you can do so without risk. Fight fire from maximum distance. Dike fire control water for later disposal; do not scatter the material. Keep unnecessary people away; isolate hazard area and deny entry. Stay upwind; keep out of low areas. Ventilate closed spaces before entering them. Wear positive pressure breathing apparatus and special protective clothing. Remove and isolate contaminated clothing at the site. If material or contaminated runoff enters waterways, notify downstream users of potentially contaminated waters. Notify local health and fire officials and pollution control agencies. From a secure, explosion-proof location, use water spray to cool exposed containers. If cooling streams are ineffective (venting sound increases in volume and pitch, tank discolors, or shows any signs of deforming), withdraw immediately to a secure position. If employees are expected to fight fires, they must be trained and equipped in OSHA 1910.156. The only respirators recommended for firefighting are self-contained breathing apparatuses that have full face-pieces and are operated in a pressure-demand or other positive-pressure mode.

Disposal Method Suggested: In accordance with 40CFR165, follow recommendations for the disposal of pesticides and pesticide containers. Must be disposed properly by following package label directions or by contacting your local or federal environmental control agency or by contacting your regional EPA office.

References

US Environmental Protection Agency. (November 30, 1987). *Chemical Hazard Information Profile: Pyriminil.* Washington, DC: Chemical Emergency Preparedness Program

New Jersey Department of Health and Senior Services. (May 2002). *Hazardous Substances Fact Sheet: Pyriminil.* Trenton, NJ

Quartz

See entry under "Silica."

Quinoline   Q:0050

Molecular Formula: C_9H_7N
Synonyms: 1-Azanaphthalene; B-500; 1-Benzazine; 1-Benzine; Benzo(b)pyridine; Benzopyridine; Chinoleine; Leucol; Leucoline; Leukol; Quinoleina (Spanish); Quinolin
CAS Registry Number: 91-22-5
RTECS® Number: VA9275000
UN/NA & ERG Number: UN2656/154
EC Number: 202-051-6 [*Annex I Index No.:* 613-281-00-5]
Regulatory Authority and Advisory Bodies
Carcinogenicity: EPA: Likely to produce cancer in humans; EPA: Sufficient evidence from animal studies; inadequate evidence or no useful data from epidemiologic studies.
Clean Air Act: Hazardous Air Pollutants (Title I, Part A, Section 112).
Clean Water Act: Section 311 Hazardous Substances/RQ 40CFR117.3 (same as CERCLA, see below); Section 313 Water Priority Chemicals (57FR41331, 9/9/92).
Reportable Quantity (RQ): 5000 lb (2270 kg).
EPCRA Section 313 Form R *de minimis* concentration reporting level: 1.0%.
California Proposition 65 Chemical: (*Quinoline and its strong acid salts*) Cancer 10/24/97.
Canada, WHMIS, Ingredients Disclosure List Concentration 1.0%.
European/International Regulations: Hazard Symbol: T, N; Risk phrases: R45; R21/22; R36/38; R68; R51/53; Safety phrases: S1/2; S53; S45; S61 (see Appendix 4).
WGK (German Aquatic Hazard Class): 2—Hazard to waters.
Description: Quinoline is a colorless liquid with a penetrating amine odor. Turns brown on exposure to light. Molecular weight = 129.17; Boiling point = 238°C; Freezing/Melting point = −15°C; Flash point = 101−107°C; Autoignition temperature = 480°C. Explosive limits: LEL = 1.2%; UEL = 7%. Hazard Identification (based on NFPA-704 M Rating System): Health 2, Flammability 1, Reactivity 0. Insoluble in cold water; highly soluble in hot.
Potential Exposure: In manufacture of quinoline derivatives (dyes and pesticides); in synthetic fuel manufacture. Occurs in cigarette smoke.
Incompatibilities: Reacts, possibly violently, with strong oxidants, strong acids, perchromates, nitrogen tetroxide, and maleic anhydride. Keep away from moisture, steam, and light. Contact with hydrogen peroxide may cause explosion.

Unpredictably violent, this substance has been the source of various plant accidents.
Permissible Exposure Limits in Air
AIHA WEEL: 0.001 ppm TWA [skin].
Protective Action Criteria (PAC)
TEEL-0: 0.001 ppm
PAC-1: 0.6 ppm
PAC-2: 5 ppm
PAC-3: 25 ppm
The state of New York has set a guideline for quinoline in ambient air[60] of 0.03 µg/m³.
Determination in Air: No methods available.
Routes of Entry: Inhalation, ingestion, skin and/or eye contact.
Harmful Effects and Symptoms
Short Term Exposure: Skin or eye contact can cause burns. Vapors can irritate the eyes and respiratory tract and may cause sore throat, nosebleeds, hoarseness, cough, phlegm, and/or tightness in the chest. High exposures can cause pulmonary edema, a medical emergency that can be delayed for several hours. This can cause death. High exposure can cause nausea, vomiting, gastrointestinal cramping, fever, dizziness, fatigue, rapid and irregular pulse, troubled breathing, collapse, and even death from paralysis of muscles needed for breathing. Liver damage may also occur.
Long Term Exposure: May affect the liver and retina of the eyes. May cause liver damage. May lead to development of skin allergy with rash and itching. Very irritating substances may cause lung damage. This substance caused mutations and is possibly carcinogenic to humans.
Points of Attack: Nervous system, liver and kidneys, eyes.
Medical Surveillance: Eye examination. Liver and kidney function tests. Examination of the nervous system.
First Aid: Skin Contact[52]*:* Flood all areas of body that have contacted the substance with water. Do not wait to remove contaminated clothing; do it under the water stream. Use soap to help assure removal. Isolate contaminated clothing when removed to prevent contact by others.
Eye Contact: Remove any contact lenses at once. Flush eyes well with copious quantities of water or normal saline for at least 20−30 min. Seek medical attention.
Inhalation: Leave contaminated area immediately; breathe fresh air. Proper respiratory protection must be supplied to any rescuers. If coughing, difficult breathing, or any other symptoms develop, seek medical attention at once, even if symptoms develop many hours after exposure.
Ingestion: If convulsions are not present, give a glass or two of water or milk to dilute the substance. Assure that the person's airway is unobstructed and contact a hospital or poison center immediately for advice on whether or not to induce vomiting. Medical observation is recommended for 24−48 h after breathing overexposure, as pulmonary edema may be delayed. As first aid for pulmonary edema, a doctor

or authorized paramedic may consider administering a corticosteroid spray.

Personal Protective Methods: Wear protective butyl rubber gloves and protective clothing to prevent any reasonable probability of skin contact. Safety equipment suppliers/ manufacturers can provide recommendations on the most protective glove/clothing material for your operation. All protective clothing (suits, gloves, footwear, headgear) should be clean, available each day, and put on before work. Contact lenses should not be worn when working with this chemical. Wear splash-proof chemical goggles and face shield unless full face-piece respiratory protection is worn. Employees should wash immediately with soap when skin is wet or contaminated. Provide emergency showers and eyewash.

Respirator Selection: Follow regulations in OSHA 29CFR1910.134 or European Standard EN149. Use a NIOSH/MSHA- or European Standard EN149-approved respirator; or use an approved supplied-air respirator with a full face-piece operated in the positive-pressure mode, or with a full face-piece, hood, or helmet in the continuous-flow mode; or use a NIOSH/MSHA- or European Standard EN149-approved self-contained breathing apparatus with a full face-piece operated in pressure-demand or other positive-pressure mode.

Storage: Color Code—Blue: Health Hazard/Poison: Store in a secure poison location. Prior to working with this chemical you should be trained on its proper handling and storage. Before entering confined space where this chemical may be present, check to make sure that an explosive concentration is not a danger. Store in a refrigerator under an inert atmosphere and protect from exposure to light, moisture, strong oxidants, strong acids, perchromates, nitrogen tetroxide, maleic anhydride. Where possible, automatically pump liquid from drums or other storage containers to process containers. Sources of ignition, such as smoking and open flames, are prohibited where this chemical is handled, used, or stored. Metal containers involving the transfer of 5 gallons or more of this chemical should be grounded and bonded. Drums must be equipped with self-closing valves, pressure vacuum bungs, and flame arresters. Use only non-sparking tools and equipment, especially when opening and closing containers of this chemical. Wherever this chemical is used, handled, manufactured, or stored, use explosion-proof electrical equipment and fittings.

Shipping: Quinoline requires a shipping label of "POISONOUS/TOXIC MATERIALS." It falls in Hazard Class 6.1 and Packing Group III.

Spill Handling: Evacuate and restrict persons not wearing protective equipment from area of spill or leak until cleanup is complete. Remove all ignition sources. Establish forced ventilation to keep levels below explosive limit. Use absorbent substance to pick up spilled material. Follow by washing surface well first with alcohol, then with soap and water. Absorb liquids in vermiculite, dry sand, earth, peat, carbon, or a similar material and deposit in sealed containers. Keep this chemical out of a confined space, such as a sewer, because of the possibility of an explosion, unless the sewer is designed to prevent the buildup of explosive concentrations. It may be necessary to contain and dispose of this chemical as a hazardous waste. If material or contaminated runoff enters waterways, notify downstream users of potentially contaminated waters. Contact your local or federal environmental protection agency for specific recommendations. If employees are required to clean up spills, they must be properly trained and equipped. OSHA 1910.120(q) may be applicable.

Fire Extinguishing: This chemical is a combustible liquid. Poisonous gases, including nitrogen oxides, carbon monoxide, and carbon dioxide, are produced in fire. Use dry chemical, carbon dioxide, water spray, or foam extinguishers. Vapors are heavier than air and will collect in low areas. Vapors may travel long distances to ignition sources and flashback. Vapors in confined areas may explode when exposed to fire. Containers may explode in fire. Storage containers and parts of containers may rocket great distances, in many directions. If material or contaminated runoff enters waterways, notify downstream users of potentially contaminated waters. Notify local health and fire officials and pollution control agencies. From a secure, explosion-proof location, use water spray to cool exposed containers. If cooling streams are ineffective (venting sound increases in volume and pitch, tank discolors, or shows any signs of deforming), withdraw immediately to a secure position. If employees are expected to fight fires, they must be trained and equipped in OSHA 1910.156. The only respirators recommended for firefighting are self-contained breathing apparatuses that have full face-pieces and are operated in a pressure-demand or other positive-pressure mode.

Disposal Method Suggested: Dissolve or mix the material with a combustible solvent and burn in a chemical incinerator equipped with an afterburner and scrubber. All federal, state, and local environmental regulations must be observed.

References

US Environmental Protection Agency. (December 29, 1983). *Chemical Hazard Information Profile Draft Report: Quinoline*. Washington, DC

New Jersey Department of Health and Senior Services. (June 2000). *Hazardous Substances Fact Sheet: Quinoline*. Trenton, NJ

Quinone   Q:0100

Molecular Formula: $C_6H_4O_2$
Synonyms: Benzo-chinon (German); 1,4-Benzoquine; *p*-Benzoquinona (Spanish); *p*-Benzoquinone; 1,4-Benzoquinone; Benzoquinone; *p*-Chinon (German); Chinon

(German); Chinone; Cyclohexadeinedione; 1,4-Cyclohexadienedione; 2,5-Cyclohexadiene-1,4-dione; 1,4-Cyclohexadiene dioxide; 1,4-Dioxybenzene; 1,4-Dioxybenzol; NCI-C55845; Quinona (Spanish); *p*-Quinone

CAS Registry Number: 106-51-4

RTECS® Number: DK2625000

UN/NA & ERG Number: UN2587/153

EC Number: 203-405-2 [*Annex I Index No.:* 606-013-00-3]

Regulatory Authority and Advisory Bodies

Carcinogenicity: IARC: Animal Inadequate Evidence; Human No Adequate Data, *not classifiable as carcinogenic to humans*, Group 3, 1999.

US EPA Gene-Tox Program, Inconclusive: *D. melanogaster* sex-linked lethal.

Air Pollutant Standard Set. See below, "Permissible Exposure Limits in Air" section.

Clean Air Act: Hazardous Air Pollutants (Title I, Part A, Section 112).

US EPA Hazardous Waste Number (RCRA No.): U197.

RCRA, 40CFR261, Appendix 8 Hazardous Constituents.

Reportable Quantity (RQ): 10 lb (4.54 kg).

EPCRA Section 313 Form R *de minimis* concentration reporting level: 1.0%.

Canada, WHMIS, Ingredients Disclosure List Concentration 1.0%.

European/International Regulations: Hazard Symbol: T; Risk phrases: R23/25; R36/37/38; R50; Safety phrases: S1/2; S26; S28-S45; S61 (see Appendix 4).

WGK (German Aquatic Hazard Class): 3—Severe hazard to waters.

Description: Quinone is a yellow, crystalline material or large yellow, monoclinic prisms with a pungent, irritating odor. Molecular weight = 108.10; Specific gravity $(H_2O:1) = 1.32$ at 25°C; Boiling point = (sublimes) about 180°C; Freezing/Melting point = 115.6°C; 224°C (decomposes); Vapor pressure = 0.1 mmHg at 25°C; Flash point (depending on humidity) = 38–93°C; Autoignition temperature = 560°C. Hazard Identification (based on NFPA-704 M Rating System): Health 3, Flammability 2, Reactivity 1. Slightly soluble in water.

Potential Exposure: Compound Description: Agricultural Chemical; Tumorigen, Mutagen. Due to this compound's ability to react with certain nitrogen compounds to form colored substances, quinone is widely used in the dye, textile, chemical, tanning, and cosmetic industries. It is used as an industrial chemical, laboratory reagent, and as an intermediate in chemical synthesis for hydroquinone and other chemicals.

Incompatibilities: Strong oxidizers, some combustible substances, reducing agents, and strong bases. Decomposes exothermically on warming above 60°C, when moist, producing carbon monoxide.

Permissible Exposure Limits in Air

Conversion factor: 1 ppm = 4.42 mg/m^3 at 25°C & 1 atm.

OSHA PEL: 0.1 ppm/0.4 mg/m^3 TWA.

NIOSH REL: 0.1 ppm/0.4 mg/m^3 TWA.

ACGIH TLV®[1]: 0.1 ppm/0.44 mg/m^3 TWA.

NIOSH IDLH: 100 mg/m^3.

Protective Action Criteria (PAC)

TEEL-0: 0.1 ppm

PAC-1: 0.35 ppm

PAC-2: 2.5 ppm

PAC-3: 22.6 ppm

DFG MAK: Danger of skin sensitization; Carcinogen Category 3B; Germ Cell Mutagen Group 3B.

Arab Republic of Egypt: TWA 0.1 ppm (0.4 mg/m^3), 1993; Australia: TWA 0.1 ppm (0.4 mg/m^3), 1993; Austria: MAK 0.1 ppm (0.4 mg/m^3), 1999; Belgium: TWA 0.1 ppm (0.44 mg/m^3), 1993; Denmark: TWA 0.1 ppm (0.4 mg/m^3), 1999; Finland: TWA 0.1 ppm (0.4 mg/m^3); STEL 0.3 ppm, [skin], 1999; France: VME 0.1 ppm (0.4 mg/m^3), VLE 0.3 ppm (1.5 mg/m^3), 1999; the Netherlands: MAC-TGG 0.4 mg/m^3, 2003; the Philippines: TWA 5 ppm (15 mg/m^3), 1993; Poland: TWA 0.1 mg/m^3; STEL 0.4 mg/m^3, 1999; Russia: STEL 0.05 mg/m^3, 1993; Sweden: NGV 0.1 ppm (0.4 mg/m^3), KTV 0.3 ppm (1.3 mg/m^3), 1999; Switzerland: MAK-W 0.1 ppm (0.4 mg/m^3), KZG-W 0.2 ppm (0.8 mg/m^3), 1999; Turkey: TWA 0.1 ppm (0.4 mg/m^3), 1993; United Kingdom: TWA 0.1 ppm (0.45 mg/m^3); STEL 0.3 ppm (1.3 mg/m^3), 2000; New Zealand, Singapore, Vietnam: ACGIH TLV®: TWA 0.1 ppm.

Determination in Air: Use NIOSH II(4), Method #S-181, Quinone.

Determination in Water: Octanol−water coefficient: Log $K_{ow} = 0.21$.

Routes of Entry: Inhalation, ingestion, skin and/or eye contact.

Harmful Effects and Symptoms

Short Term Exposure: The vapor irritates the eyes, skin, and respiratory tract; can cause nosebleeds, hoarseness, cough, phlegm, and/or tightness in the chest. Skin contact can cause severe irritation. Solid quinone in contact with skin or the lining of the nose and throat may produce discoloration, severe irritation, swelling, and the formation of ulcers, papules, and vesicles. Prolonged contact with the skin may cause ulceration. Quinone vapor is highly irritating to the eyes. Following prolonged exposure to vapor, brownish conjunctival stains may appear. These may be followed by corneal opacities and structural changes in the cornea and loss of visual acuity. The early pigmentary stains are reversible, while the corneal dystrophy tends to be progressive. Further effects reported[52] include vomiting, collapse, and coma.

Long Term Exposure: Repeated or prolonged contact with skin may cause dermatitis. The substance may have effects on the skin and eyes, resulting in discoloration, inflammation and injury of the corneal epithelium, keratitis (inflammation of the cornea), reduced vision. Can cause lung irritation; bronchitis may develop. May cause kidney damage.

Points of Attack: Eyes, skin, lungs, kidneys.

Medical Surveillance: Ophthalmic examination. Careful examination of the eyes, including visual acuity and slit lamp examinations, should be done during placement and periodic examinations. Lung function tests. Kidney function tests. Also evaluate skin.

First Aid: If this chemical gets into the eyes, remove any contact lenses at once and irrigate immediately for at least 15 min, occasionally lifting upper and lower lids. Seek medical attention immediately. If this chemical contacts the skin, remove contaminated clothing and wash immediately with soap and water. Seek medical attention immediately. If this chemical has been inhaled, remove from exposure, begin rescue breathing (using universal precautions, including resuscitation mask) if breathing has stopped and CPR if heart action has stopped. Transfer promptly to a medical facility. When this chemical has been swallowed, get medical attention. Give large quantities of water and induce vomiting. Do not make an unconscious person vomit.

Personal Protective Methods: Wear protective gloves and clothing to prevent any reasonable probability of skin contact. Safety equipment suppliers/manufacturers can provide recommendations on the most protective glove/clothing material for your operation. All protective clothing (suits, gloves, footwear, headgear) should be clean, available each day, and put on before work. Contact lenses should not be worn when working with this chemical. Wear splash-proof chemical goggles and face shield unless full face-piece respiratory protection is worn. Employees should wash immediately with soap when skin is wet or contaminated. Provide emergency showers and eyewash.

Respirator Selection: Up to 10 mg/m³: Sa:Cf (APF = 25)* (any supplied-air respirator operated in a continuous-flow mode). *Up to 20 mg/m³:* SCBAF (APF = 50) (any self-contained breathing apparatus with a full face-piece) or SaF (APF = 50) (any supplied-air respirator with a full face-piece). *Up to 100 mg/m³:* SaF: Pd,Pp (APF = 2000) (any supplied-air respirator that has a full face-piece and is operated in a pressure-demand or other positive-pressure mode). *Emergency or planned entry into unknown concentrations or IDLH conditions:* SCBAF: Pd,Pp (APF = 10,000) (any NIOSH/MSHA- or European Standard EN 149-approved self-contained breathing apparatus that has a full face-piece and is operated in a pressure-demand or other positive-pressure mode) or SaF: Pd,Pp: ASCBA (APF = 10,000) (any supplied-air respirator that has a full face-piece and is operated in a pressure-demand or other positive-pressure mode in combination with an auxiliary, self-contained breathing apparatus operated in a pressure-demand or other positive-pressure mode). *Escape:* GmFOv100 (APF = 50) [any air-purifying, full-face-piece respirator (gas mask) with a chin-style, front- or back-mounted organic vapor canister having an N100, R100, or P100 filter] or SCBAE (any appropriate escape-type, self-contained breathing apparatus).

*Substance causes eye irritation or damage; eye protection needed.

Storage: Color Code—Blue: Health Hazard/Poison: Store in a secure poison location. Prior to working with this chemical you should be trained on its proper handling and storage. Store in a refrigerator or a cool, dry place. Where possible, automatically transfer material from drums or other storage containers to process containers. Sources of ignition, such as smoking and open flames, are prohibited where this chemical is handled, used, or stored. Metal containers involving the transfer of this chemical should be grounded and bonded. Wherever this chemical is used, handled, manufactured, or stored, use explosion-proof electrical equipment and fittings.

Shipping: This compound requires a shipping label of "POISONOUS/TOXIC MATERIALS." It falls in Hazard Class 6.1 and Packing Group II.

Spill Handling: Evacuate persons not wearing protective equipment from area of spill or leak until cleanup is complete. Remove all ignition sources. Dampen spilled material with alcohol to avoid dust, then transfer material to a suitable container. Use absorbent dampened with alcohol to pick up remaining material. Wash surfaces well with soap and water. Collect powdered material in the most convenient and safe manner and deposit in sealed containers. Ventilate area after cleanup is complete. It may be necessary to contain and dispose of this chemical as a hazardous waste. If material or contaminated runoff enters waterways, notify downstream users of potentially contaminated waters. Contact your local or federal environmental protection agency for specific recommendations. If employees are required to clean up spills, they must be properly trained and equipped. OSHA 1910.120(q) may be applicable.

Fire Extinguishing: This chemical is a combustible solid. Use dry chemical, carbon dioxide, water spray, or alcohol foam extinguishers. Poisonous gases are produced in fire, including carbon monoxide. If material or contaminated runoff enters waterways, notify downstream users of potentially contaminated waters. Notify local health and fire officials and pollution control agencies. Containers may explode in fire. From a secure, explosion-proof location, use water spray to cool exposed containers. If cooling streams are ineffective (venting sound increases in volume and pitch, tank discolors, or shows any signs of deforming), withdraw immediately to a secure position. If employees are expected to fight fires, they must be trained and equipped in OSHA 1910.156. The only respirators recommended for firefighting are self-contained breathing apparatuses that have full face-pieces and are operated in a pressure-demand or other positive-pressure mode.

Disposal Method Suggested: Controlled incineration (982°C, 2.0 s minimum).

References

US Environmental Protection Agency, Quinone. (April 30, 1980). *Health and Environmental Effects Profile No. 157.* Washington, DC: Office of Solid Waste

New Jersey Department of Health and Senior Services. (June 2004). *Hazardous Substances Fact Sheet: p-Benzoquinone*. Trenton, NJ

Quintozene   Q:0110

Molecular Formula: $C_6Cl_5NO_2$
Synonyms: Avicol (pesticide); Bartilex; Batrilex; Benzene, pentachloronitro-; Botrilex; Brassicol; Brassicol 75; Brassicol earthcide; Brassicol super; Chinozan; Fartox; Folosan; Fomac 2; Fungichlor; GC 3944-3-4; Kobu; Kobutol; KP2; Marisan forte; NCI-C00419; Nitropentachlorobenzene; Olipsan; Olpisan; PCNB; Pentachlornirtobenzol (German); Pentachloronitrobenzene; Pentagen; Phomasan; PKHNB; Quinosan; Quintocene; Quintoceno (Spanish); Quintozen; RTU 1010; Saniclor 30; Terrachlor; Terraclor; Terraclor 30 G; Terrafun; Tilcarex; Tri-PCNB; Tritisan
CAS Registry Number: 82-68-8; *(alt.)* 39378-26-2
RTECS® Number: DA6550000
UN/NA & ERG Number: UN2588/151
EC Number: 201-435-0 [*Annex I Index No.:* 609-043-00-5]
Regulatory Authority and Advisory Bodies
Carcinogenicity: NCI: Carcinogenesis Studies (feed); no evidence: mouse; NTP: Carcinogenesis Studies (feed); no evidence: mouse; IARC: Animal Limited Evidence; Human No Adequate Data, *not classifiable as carcinogenic to humans*, Group 3, 1987.
US EPA Gene-Tox Program, Negative: Host-mediated assay; *In vitro* UDS—human fibroblast; Negative: TRP reversion; *S. cerevisiae*—homozygosis; Negative/limited: Carcinogenicity—mouse/rat; Inconclusive: *B. subtilis* rec assay; *E. coli* polA without S9; Inconclusive: Histidine reversion—Ames test; Inconclusive: *D. melanogaster* sex-linked lethal.
US EPA, FIFRA, 1998 Status of Pesticides: Supported.
US EPA Hazardous Waste Number (RCRA No.): U185.
RCRA, 40CFR261, Appendix 8 Hazardous Constituents.
RCRA 40CFR268.48; 61FR15654, Universal Treatment Standards: Wastewater (mg/L), 0.055; Nonwastewater (mg/kg), 4.8.
RCRA 40CFR264, Appendix 9; TSD Facilities Ground Water Monitoring List. Suggested test method(s) (PQL μg/L): 8270 (10).
Reportable Quantity (RQ): 100 lb (45.4 kg).
EPCRA Section 313 Form R *de minimis* concentration reporting level: 1.0%.
European/International Regulations: Hazard Symbol: T+, N; Risk phrases: R24/25; R26; R36/37/38; R40; R50/53; Safety phrases: S1/2; S22; 36/37; S45; S52; S60; S61 (see Appendix 4).
WGK (German Aquatic Hazard Class): No value assigned.

Description: Quintozene is a colorless to cream-colored crystalline material with a musty odor. Molecular weight = 295.32; Boiling point = 328°C; Freezing/Melting point = 142−146°C; Vapor pressure = 1×10^{-4} mbar at 25°C. Insoluble in water.
Potential Exposure: Compound Description: Agricultural Chemical; Tumorigen, Mutagen; Reproductive Effector. Quintozene is used as a chemical intermediate, soil and seed fungicide, slime control in industrial waters, herbicide. A rebuttal presumption against registration was issued on October 13, 1977 by EPA on the basis of oncogenicity.
Incompatibilities: Oxidizers, strong bases. Decomposition products include nitrogen oxides and hydrogen chloride.
Permissible Exposure Limits in Air
ACGIH TLV®[1]: 0.5 mg/m³ TWA; not classifiable as a human carcinogen.
Protective Action Criteria (PAC)
TEEL-0: 0.5 mg/m³
PAC-1: 1.5 mg/m³
PAC-2: 300 mg/m³
PAC-3: 500 mg/m³
Denmark: TWA 0.5 mg/m³, 1999; the Netherlands: MAC-TGG 0.5 mg/m³, 2003.
Determination in Water: Fish Tox = 64.49856000 ppb (INTERMEDIATE).
Routes of Entry: Inhalation, ingestion, skin and/or eye contact.
Harmful Effects and Symptoms
Short Term Exposure: Irritates the eyes, skin, and respiratory tract. Eye contact can cause corneal damage and permanent injury. High levels can interfere with the blood's ability to carry oxygen causing methemoglobinemia; cyanosis, with blue color to the skin and lips. Higher levels can cause trouble breathing, collapse, and even death. A chloro-mononitrobenzene, this chemical may have similar toxic effects as nitrobenzene.
Long Term Exposure: May cause sensitization and skin allergy. May cause damage of the liver and kidneys. Human Tox = 2.10000 ppb (HIGH).
Points of Attack: Blood, skin, liver, kidneys.
Medical Surveillance: Blood methemoglobin levels. Evaluation by a qualified allergist. Liver and kidney function tests.
First Aid: If this chemical gets into the eyes, remove any contact lenses at once and irrigate immediately for at least 15 min, occasionally lifting upper and lower lids. Seek medical attention immediately. If this chemical contacts the skin, remove contaminated clothing and wash immediately with soap and water. Seek medical attention immediately. If this chemical has been inhaled, remove from exposure, begin rescue breathing (using universal precautions, including resuscitation mask) if breathing has stopped and CPR if heart action has stopped. Transfer promptly to a medical facility. When this chemical has been swallowed, get medical attention. Give large quantities of water and

induce vomiting. Do not make an unconscious person vomit.

Note to physician: Treat for methemoglobinemia. Spectrophotometry may be required for precise determination of levels of methemoglobin in urine.

Personal Protective Methods: Wear protective gloves and clothing to prevent any reasonable probability of skin contact. Safety equipment suppliers/manufacturers can provide recommendations on the most protective glove/clothing material for your operation. All protective clothing (suits, gloves, footwear, headgear) should be clean, available each day, and put on before work. Contact lenses should not be worn when working with this chemical. Wear dust-proof chemical goggles and face shield unless full face-piece respiratory protection is worn. Employees should wash immediately with soap when skin is wet or contaminated. Provide emergency showers and eyewash.

Respirator Selection: Where there is potential for exposure to this chemical above 0.5 mg/m^3, use a NIOSH/MSHA- or European Standard EN149-approved supplied-air respirator with a full face-piece operated in the positive-pressure mode, or with a full face-piece, hood, or helmet in the continuous-flow mode; or use a NIOSH/MSHA- or European Standard EN149-approved self-contained breathing apparatus with a full face-piece operated in pressure-demand or other positive-pressure mode.

Storage: Color Code—Blue: Health Hazard/Poison: Store in a secure poison location. Prior to working with this chemical you should be trained on its proper handling and storage. Store in tightly closed containers in a cool, well-ventilated area away from strong oxidizers. Where possible, automatically transfer material from drums or other storage containers to process containers. Sources of ignition, such as smoking and open flames, are prohibited where this chemical is handled, used, or stored. Metal containers involving the transfer of this chemical should be grounded and bonded. Wherever this chemical is used, handled, manufactured, or stored, use explosion-proof electrical equipment and fittings.

Shipping: Pesticides, solid, toxic, n.o.s. require a shipping label of "POISONOUS/TOXIC MATERIALS." This material falls in Hazard Class 6.1 and Packing Group II or III.

Spill Handling: Evacuate persons not wearing protective equipment from area of spill or leak until cleanup is complete. Remove all ignition sources. Collect powdered material in the most convenient and safe manner and deposit in sealed containers. Ventilate area after cleanup is complete. It may be necessary to contain and dispose of this chemical as a hazardous waste. If material or contaminated runoff enters waterways, notify downstream users of potentially contaminated waters. Contact your local or federal environmental protection agency for specific recommendations. If employees are required to clean up spills, they must be properly trained and equipped. OSHA 1910.120(q) may be applicable. Soil Adsorption Index (K_{oc}) = 5000 (estimate).

Fire Extinguishing: This chemical is a combustible solid. Use dry chemical, carbon dioxide, water spray, or alcohol foam extinguishers. Poisonous gases are produced in fire, including nitrogen oxides and hydrogen chloride. If material or contaminated runoff enters waterways, notify downstream users of potentially contaminated waters. Notify local health and fire officials and pollution control agencies. From a secure, explosion-proof location, use water spray to cool exposed containers. If cooling streams are ineffective (venting sound increases in volume and pitch, tank discolors, or shows any signs of deforming), withdraw immediately to a secure position. If employees are expected to fight fires, they must be trained and equipped in OSHA 1910.156. The only respirators recommended for firefighting are self-contained breathing apparatuses that have full face-pieces and are operated in a pressure-demand or other positive-pressure mode.

Disposal Method Suggested: Consult with environmental regulatory agencies for guidance on acceptable disposal practices. Generators of waste containing this contaminant ($\geq$100 kg/mo) must conform with EPA regulations governing storage, transportation, treatment, and waste disposal. In accordance with 40CFR165, follow recommendations for the disposal.

References

US Environmental Protection Agency, Special Review and Reregistration Division Office of Pesticide Programs. (1998). *Agency Status of Pesticides in Registration, Reregistration, and Special Review* (Rainbow Report). Washington, DC

New Jersey Department of Health and Senior Services. (January 2007). *Hazardous Substances Fact Sheet: Chloronitrobenzenes (Mixed Isomers)*. Trenton, NJ

Quinuclidinyl benzilate (Agent BZ, WMD)   Q:0120

Molecular Formula: $C_{21}H_{23}NO_3$
Synonyms: Agent 15; Agent-Buzz; 1-Azabicyclo(2,2,2)octan-3-ol, benzilate (ester); Benzeneacetic acid, α-hydroxy-α-phenyl-, 1-azabicyclo(2.2.2)oct-3-yl ester; Benzilic acid, 3-quinuclidinyl ester; BUZZ; BZ; 3-Chinuclidylbenzilate; 3-(2,2-Diphenyl-2-hydroxyethanoyloxy)-quinuclidine; Oksilidin; QNB; 3-Quinuclidinol benzilate; 3-Quinuclidinyl benzilate; Quinuclidinyl benzilate
Hydrochloride:
1-Azabicyclo(2.2.2)octan-3-ol, benzylate (ester), hydrochloride; Benzeneacetic acid, alpha-hydroxy-alpha-phenyl-, 1-azabicyclo(2.2.2)oct-3-yl ester, HCL; Benzilic acid, 3-quinuclidinyl ester, hydrochloride; CHEKB; HNB 3; 3-Quinuclidinyl benzilate hydrochloride; Quinuclidyl benzylate hydrochloride; Ro 2-3308

CAS Registry Number: 6581-06-2; 62869-69-6; 13004-56-3 (hydrochloride)
RTECS®Number: DD4639000; VD6300000 (hydrochloride)
UN/NA & ERG Number: UN 2810 (Toxic liquids, organic, n.o.s.)/153
EC Number: Not established
Regulatory Authority and Advisory Bodies
Department of Homeland Security Screening Threshold Quantity (pounds): Sabotage/Contamination Hazard: A placarded amount (commercial grade).
Report any release of WMD to National Response Center 1-800-424-8802.
While not a mandated "Federally listed" waste, BZ is more toxic than most RCRA listed chemicals. However, BZ is a "listed" hazardous waste in some states where it may have been stockpiled by the military.
WGK (German Aquatic Class): No value assigned.
Description: Quinuclidinol benzilate (QNB) is a white crystalline solid. Odorless. Molecular weight = 337.41; Freezing/Melting point = 167.5°C; also 164−165°C; Boiling point = 320°C; Solid density = 0.51 g/cm³ (bulk); 1.33 g/cm³ (crystal); Vapor density = 11.6; Vapor pressure = negligible; about 0.5 mg/m³ at 70°C; Flash point = 246°C; Hazard Identification (based on NFPA-704 M Rating System): Health 3; Flammability 0; Reactivity 0. Slightly soluble in water.
History of the agent: QNB is an incapacitating agent and has been considered historically for use as a chemical warfare agent.[CDC] As a chemical weapon, QNB was mixed with a pyrotechnic mixture in small bombs designed for aerial delivery. Once exploded, the mixture produced a potentially dangerous aerosol of QNB. In the early 1960s at Pine Bluff Arsenal, Arkansas, the United States produced munitions containing QNB. Production ceased in the late 1960s and all stockpiles at every military installation were eventually destroyed by 1990. Iraq's "Agent 15" is believed to be identical to, or closely related to QNB. In general, the effectiveness of QNB as a WMD has proven both highly undependable and unpredictable.
Potential Exposure: Mutagen, drug; incapacitating agent. QNB is a glycolate anticholinergic compound that affects the central nervous system (CNS) and peripheral nervous system (PNS) and is related to the drugs atropine, scopolamine, and hyoscyamine. QNB is nonirritating; symptoms are delayed for several hours. QNB can be used to contaminate water, food, and agricultural products. A highly potent drug and central nervous system (CNS) depressant, QNB is a delayed-action incapacitating agent, usually dispersed as an aerosol, but it can also be used to penetrate skin when mixed with a solvent (such as DMSO), and to contaminate food and water. QNB appears to be widely used in pharmacologic research. The key to protection from QNB is prevention from entering the body with good quality aerosol filter and impermeable gloves and clothing. QMB is stable in most solvents, with a half-life of 3−4 weeks in

moist air; it can be dispersed even with heat-producing munitions.
Incompatibilities: May form explosive mixture with air. Decomposes at about 170°C in air under prolonged heating. After 1 or 2 h at 200°C, it is completely decomposed. Rate of decomposition is both temperature and purity dependent. No effect on steel or stainless steel after 3 months at 71°C. Aluminum and anodized aluminum are mildly attacked after 3 months at 71°C. Contact with metals may evolve flammable and potentially explosive hydrogen gas.
Persistence of Chemical Agent: QNB is very persistent in soil and water and on surfaces.
Permissible Exposure Limits in Air:
Protective Action Criteria (PAC)*
6581-06-2
TEEL-0: 0.0075 mg/m³
PAC-1: 0.02 mg/m³
PAC-2: **0.037** mg/m³
PAC-3: **0.69** mg/m³
*AEGLs (Acute Emergency Guideline Levels) & ERPGs (Emergency Response Planning Guideline) are in **bold face.**
Determination in Air: No method is available for detecting QNB in environmental samples.[CDC]
Determination in Water: No method is available for detecting QNB in environmental samples.[CDC]
Routes of Entry: Respiratory system, ingestion, skin, or eye contact. Skin and eye absorption is possible, depending on liquid solvents that might be used to enhance absorption. Inhalation and ingestion of QNB are important routes of exposure for the solid.[CDC]
Harmful Effects and Symptoms
QNB is nonirritating stunning agent. The onset of incapacitation is dose dependent. Symptoms may occur as early as 1 h following exposure and continue for 48 h. The onset of incapacitation is dose dependent. It might occur as early as 1 h after exposure and continue from 48 h to 4 days depending on level of exposure. An untreated casualty may require from 3 to 4 days to achieve full recovery from the effects of QNB intoxication. QNB affects a victim's ability to remember, solve problems, make sound decisions and judgments, pay attention to assigned tasks, and understand instructions. High levels of exposure can completely destroy a victim's ability to perform any tasks, military and/or otherwise.
Short Term Exposure: A very potent drug, QNB, can cause confusion, dream-like state, hallucinations, and severe delirium; it also affects circulation, digestion, salivation, sweating, and vision. QNB causes dilation of the pupils, which is extremely uncomfortable to most people. QNB produces profound mental disturbances at a dose of 0.1−0.2 mg. Signs and symptoms of exposure include agitation, restlessness, dizziness, giddiness; failure to obey orders, confusion, erratic behavior; stumbling or staggering; vomiting; hallucinations; blurred vision; dry, flushed skin; urinary retention;

ileus; tachycardia; hypertension; and elevated temperature ($>101°F/38°C$). QNB may cause short-term memory loss, and variable levels of side effects are experienced in different people. Impairments caused by QNB are generally temporary and unlikely to be fatal; however, they can be severe with high exposures.[NIOSH] QNB is an anticholinergic. LCt_{50} = High; estimated to be 200,000 mg-min/m^3; ICt_{50} = 112 mg-min/m^3. LD50 = (iv-mouse) 18 mg/kg[US Army]; 25 mg/kg; TD_{Lo} (lowest published toxic dose) = (subcutaneous-human) 3 μg/kg.[NIOSH]

Delayed effects of exposure: Widespread formation of clots in the blood vessels (disseminated intravascular coagulation) is a potential complication in a victim with marked agitation and/or exceptionally high body temperature (hyperthermia).[NIOSH]

Long Term Exposure: Information is unavailable about the carcinogenicity, developmental toxicity, or reproductive toxicity from chronic or repeated exposure to QNB.

Points of Attack: CNS (central nervous system); PNS (peripheral nervous system).

Medical Surveillance: In a victim with marked agitation and/or exceptionally high body temperature (hyperthermia), clotting studies (e.g., prothrombin time, activated partial thromboplastin time, and international normalized ratio) are recommended. Urine tests.

Decontamination: Establish the decontamination corridor upwind and uphill of the hot zone. The warm zone should include two decontamination corridors. One decontamination corridor is used to enter the warm zone and the other for exiting the warm zone into the cold zone. The decontamination zone for exiting should be upwind and uphill from the zone used to enter. Decontamination area workers should wear appropriate PPE. A solution of detergent and water (which should have a pH value of at least 8 but should not exceed a pH value of 10.5) should be available for use in decontamination procedures. Soft brushes should be available to remove contamination from the PPE. Labeled, durable 6-mil polyethylene bags should be available for disposal of contaminated PPE.

INDIVIDUAL DECONTAMINATION: *Decontamination of First Responder:* Begin washing PPE of the first responder using soap and water solution and a soft brush. Always move in a downward motion (from head to toe). Make sure to get into all areas, especially folds in the clothing. Wash and rinse (using cold or warm water) until the contaminant is thoroughly removed. Remove PPE by rolling downward (from head to toe) and avoid pulling PPE off over the head. Remove the SCBA after other PPE has been removed. Place all PPE in labeled durable 6-mil polyethylene bags.

Decontamination of Patient/Victim: Remove the patient/victim from the contaminated area and into the decontamination corridor. Remove all clothing (at least down to their undergarments) and place the clothing in a labeled durable 6-mil polyethylene bag. Thoroughly wash and rinse (using cold or warm water) the contaminated skin of the patient/victim using a soap and water solution. Be careful not to break the patient/victim's skin during the decontamination process, and cover all open wounds. Cover the patient/victim to prevent shock and loss of body heat. Move the patient/victim to an area where emergency medical treatment can be provided.

First Aid: Eyes: After removing patient/victim from the source of exposure, immediately wash eyes with large amounts of tepid water for at least 15 min. Do not allow the patient/victim to rub eyes. Monitor the patient/victim for signs of whole-body (systemic) effects; if signs appear, see the *Inhalation* section for treatment recommendations. Seek medical attention immediately.

Ingestion: After removing patient/victim from the source of exposure, immediately ensure that the patient/victim has an unobstructed airway. *Do not induce vomiting.* See the *Inhalation* section for first aid recommendations. Seek medical attention immediately.

Inhalation: After removing patient/victim from the source of exposure, immediately, evaluate respiratory function and pulse. Ensure that the patient/victim has an unobstructed airway. If shortness of breath occurs or breathing is difficult, administer oxygen. Assist ventilation as required. Always use a barrier or bag-valve-mask device. If breathing has ceased, provide artificial respiration. Monitor for exceptionally high body temperature. If body temperature is elevated above $102°F/39°C$, initiate immediate and vigorous cooling (as for heatstroke), using 72° to 75°F/22° to 24°C water and air circulation (fanning), wet cloths and air circulation, or maximum exposure to air in a shaded area with maximum air circulation. Do not use ice for skin cooling. Give fluids sparingly. Manage dryness and coating of the mouth and lips using moist swabs and small amounts of petroleum jelly. Monitor for skin abrasions caused by repetitive movements. Remove potentially harmful items, including cigarettes, matches, medications, and small items that could be accidentally ingested, from the patient/victim's possession. Consider loose restraint of disoriented or agitated patient/victims. Consider separation of affected individuals into small groups to minimize potential crowd control problems. Seek medical attention immediately. *Skin:* After removing patient/victim from the source of exposure, immediately, see the *Decontamination* section (above). See the *Inhalation* section for first aid recommendations. Seek medical attention immediately.

Note to health professionals: Treatment of QNB intoxication largely consists of supportive care: use of intravenous fluids and appropriate cooling measures to address elevated core body temperature and judicious use of sedation using benzodiazepines. Severe toxicity may require the use of physostigmine in a hospital setting.*[CDC]

Antidote: *Some military references suggest the use of physostigmine to temporarily increase synaptic acetylcholine concentrations. Physostigmine poses its own risks of side effects and interactions with other drugs and should be used only by physician or other medical personnel familiar with its safe use. Suggested dosages for physostigmine in the treatment of QNB (BZ) poisoning follow: *Test dose:* If the diagnosis is in doubt, a dose of 1 mg might be given. If a slight improvement occurs, routine dosing should be given.

Routine dosing: Doses of about 45 mcg/kg for adults have been recommended. This might be modified by the response. A mental status examination should be done every hour and the dose and time interval of dosing should be modified according to whether the mental status is improved or not. As the patient improves, the dosage requirement will decrease.

Routes of administration: IM: 45 mcg/kg in adults (20 mg/kg in children); IV: 30 mcg/kg slowly (1 mg/min); PO: 60 mcg/kg if patient is cooperative (because of bitter taste, consider diluting in juice). For each route, titrate about every 60 min to mental status.[USAMRICD]

Personal Protective Methods: First Responders should use a NIOSH- or European Standard EN 149-certified Chemical, Biological, Radiological, Nuclear (CBRN) Self Contained Breathing Apparatus (SCBA) with a Level A protective suit when entering an area with an unknown contaminant or when entering an area where the concentration of the contaminant is unknown. Level A protection should be used until monitoring results confirm the contaminant and the concentration of the contaminant. *Note:* Safe use of protective clothing and equipment requires specific skills developed through training and experience. *LEVEL A: (RED ZONE):* Used when the greatest level of skin, respiratory, and eye protection is required. This is the maximum protection for workers in danger of exposure to unknown chemical hazards or levels above the IDLH or greater than the AEGL-2 (not established). A Totally-Encapsulating Chemical Protective (TECP) suit that provides protection against CBRN agents. Chemical-resistant gloves (outer). Chemical-resistant gloves (inner). Chemical-resistant boots with a steel toe and shank. Coveralls, long underwear, and a hard hat worn under the TECP suit are optional items.

Respirator Selection: Where there is potential for exposure to QNB, use a NIOSH- or European Standard EN 149-certified CBRN full-face-piece SCBA operated in a pressure-demand mode or a pressure-demand supplied air hose respirator with an auxiliary escape bottle.

Storage: QNB is stable in most solvents, with a half-life of 3–4 weeks in moist air. Store at 2–8°C.

Shipping: Quinuclidinol benzilate (QNB) shipping name is "TOXIC, LIQUIDS, ORGANIC, N.O.S." Subsidiary Hazardous Class or Division Label "Poison (Toxic)." Hazardous Class or Division 6.1. Packing Group III.

Spill Handling: QNB aerosol is heavier than air. It will spread along the ground and collect and stay in poorly ventilated, low-lying, or confined areas (e.g., sewers, basements, and tanks). Hazardous concentrations may develop quickly in enclosed, poorly ventilated, or low-lying areas. Keep out of these areas. Stay upwind. *Environment/spillage disposal:* Do not touch or walk through the spilled agent if at all possible. However, if you must, personnel should wear the appropriate PPE during environmental decontamination. See the PPE section of this record for detailed information. Keep combustibles (e.g., wood, paper, and oil) away from the spilled agent. Use water spray to reduce aerosols or divert aerosol cloud drift. Avoid allowing water runoff to contact the spilled agent. Do not direct water at the spill or the source of the leak. Stop the leak if it is possible to do so without risk to personnel. Prevent entry into waterways, sewers, basements, or confined areas. Isolate the area until aerosol has dispersed. Ventilate the area.[NIOSH] Evacuate and restrict persons not wearing protective equipment from area of spill or leak until cleanup is complete. Remove all ignition sources. Ventilate area of spill or leak. Spills must be contained by covering with vermiculite, diatomaceous earth, clay, fine sand, sponges, and paper or cloth towels. This containment is followed by treatment with copious amounts of aqueous sodium hydroxide solution (a minimum 10 wt. %). Scoop up all material and place in a fully removable head drum with a high-density polyethylene liner. The decontamination solution must be treated with excess bleach to destroy the CN formed during hydrolysis. Cover the contents with additional bleach before affixing the drum head. After sealing the head, the exterior of the drum shall be decontaminated and then labeled in accordance with IAW, EPA, and DOT regulations. All leaking containers shall be overpacked with vermiculite placed between the interior and exterior containers. Decontaminate and label per IAW, EPA, and DOT regulations. Dispose of the material per IAW waste disposal methods provided below. Conduct general area monitoring with an approved monitor to confirm that the atmospheric concentrations do not exceed the airborne exposure limit. If 10 wt. % sodium hydroxide is not available, then the following decontaminants may be used instead and are listed in order of preference: Decontaminating Solution No. 2 [DS2: (2% NaOH, 70% diethylenetriamine, 28% ethylene glycol monomethyl ether)], sodium carbonate, and Supertropical Bleach Slurry (STB). Keep this chemical out of a confined space, such as a sewer, because of the possibility of an explosion, unless the sewer is designed to prevent the buildup of explosive concentrations. It may be necessary to contain and dispose of this chemical as a hazardous waste. If material or contaminated runoff enters waterways, notify downstream users of potentially contaminated waters. Contact your local or federal environmental protection agency for specific recommendations. If employees are required to clean up spills, they must be properly trained and equipped. OSHA 1910.120(q) may be applicable.

Initial isolation and protective action distances

Distances shown are likely to be affected during the first 30 min after materials are spilled and could increase with time. If more than one tank car, cargo tank, portable tank, or large cylinder involved in the incident is leaking, the protective action distance may need to be increased. You may need to seek emergency information from CHEMTREC at

(800) 424-9300 or seek professional environmental engineering assistance from the US EPA Environmental Response Team at (908) 548-8730 (24-h response line).

QNB, when used as a weapon

Small spills (From a small package or a small leak from a large package)

First: Isolate in all directions (feet/meters)   100/30

Then: Protect persons downwind (miles/kilometers)

Day   0.1/0.2

Night   0.1/0.2

Large spills (From a large package or from many small packages)

First: Isolate in all directions (feet/meters)   100/30

Then: Protect persons downwind (miles/kilometers)

Day   0.1/0.2

Night   0.3/0.5

Fire Extinguishing: QNB (BZ) is combustible; it may burn but does not ignite readily. QNB (BZ) might be transported in a molten form. QNB (BZ) may decompose upon heating producing corrosive and/or toxic gases. Containers may explode when heated. In case of fire, evacuate the area, including yourself. If there is some reason that you have to put out the fire—for example, there are things nearby that cannot be allowed to burn—use unattended equipment, then evacuate everyone immediately, including yourself. When heated, aerosols may form explosive mixtures with air, presenting an explosion hazard indoors, outdoors, and in sewers. Containers may explode when heated. QNB (BZ) aerosol is heavier than air. It will spread along the ground and collect and stay in poorly ventilated, low-lying, or confined areas (e.g., sewers, basements, and tanks). Hazardous concentrations may develop quickly in enclosed, poorly ventilated, or low-lying areas. Keep out of these areas. Stay upwind. Fire may produce irritating, corrosive, and/or toxic gases.

If you must extinguish a QNB (BZ) fire—Small fires: Use dry chemical, carbon dioxide, or water spray. *Large fires*: Use dry chemical, carbon dioxide, alcohol-resistant foam, or water spray. Move containers from the fire area if it is possible to do so without risk to personnel. Dike fire control water for later disposal; do not scatter the material. For fire involving tanks or car/trailer loads, fight the fire from maximum distance or use unmanned hose holders or monitor nozzles. Do not get water inside containers. Cool containers with flooding quantities of water until well after the fire is out. Withdraw immediately in case of rising sound from venting safety devices or discoloration of tanks. Always stay away from tanks engulfed in fire. Run-off from fire control or dilution water may be corrosive and/or toxic, and it may cause pollution. If the situation allows, control and properly dispose of run-off (effluent).

Disposal Method Suggested:

See "Spill Handling," above.

References

Ketchum, J. S., & Sidell, F. R. (1997). Incapacitating agents. In: R. Zajtchuk & R. F. Bellamy (Eds.), *Textbook of Military Medicine: Medical Aspects of Chemical and Biologic Warfare* (pp. 287–305). Washington, DC: Office of the Surgeon General at TMM Publications, Borden Institute, Walter Reed Army Medical Center

Byrd, G. D., Paule, R. C., Sander, L. C., Sniegoski, L. T., White, E. V., & Bausum, H. T. (1992). Determination of 3-Quinuclidinyl Benzilate (QNB) and its major metabolites in urine by isotope dilution gas chromatography/mass spectrometry. *Journal of Analytical Toxicology, 16*(3), 182–187; US Army Field Manual (DA FM) 3-9 (PCN 320 008457 00); US Navy Publication No. P-467; US Air Force Manual No. 355-7; *Potential Military Chemical/Biological Agents and Compounds*; Headquarters, Department of the Army, Headquarters, Department of the Navy, Headquarters, Department of the Air Force; Washington, DC, December, 1990

CDC/NIOSH, *The Emergency Response Safety and Health Database*, <http://www.cdc.gov/NIOSH/ershdb/Emergency ResponseCard_29750015.html>

R

Radium compounds   R:0050

Molecular Formula: Ra; Cl_2Ra (radium chloride); Br_2Ra (radium bromate)

Synonyms: Actinium-X (^{226}Ra); ^{224}Ra; ^{226}Ra; ^{228}Ra; Radio (Spanish); Radium-224; Radium 226; Radium 228; Radium, isotope of mass 226; Radium, isotope of mass 228; Thorium-X (^{226}Ra)

Radium bromide: Radium (II) bromide; Radium dibromide
radium chloride: Radium (II) chloride; Radium dichloride
CAS Registry Number: 7440-14-4 (Radium); 10031-23-9 (Radium bromide); 10025-66-8 (Radium chloride)

Other radium compounds:
Radium carbonate ($RaCO_3$) 7116-98-5
Radium fluoride (RaF_2) 20610-49-5
Radium hydroxide [$Ra(OH)_2$] 98966-86-0
Radium nitrate (Ra_3NO_3) 10213-12-4
Radium sulfate ($RaSO_4$) 7446-16-4
RTECS®Number: Not available
UN/NA & ERG Number: UN2915 (Radioactive material, Type A package nonspecial form, nonfissile or fissile-excepted)/163
EC Number: 231-122-4 (Radium); 233-035-7 (Radium chloride); 233-086-5 (Radium bromide)

Regulatory Authority and Advisory Bodies
Carcinogenicity: IARC: Known carcinogen; EPA and the National Academy of Sciences, Committee on Biological Effects of Ionizing Radiation: Radium is a known human carcinogen. Radium-224; Radium-226; Radium-228; Radon-222 and their decay products Category 1.
Radionuclides and Reportable Quantities (RQ) listed under CERCLA (see 40CFR Part 302, Table 302.4, Appendix B, for more information): Radium-223 RQ = 1 curies; Radium-224 RQ = 10 curies; Radium-225 RQ = 1 curies; Radium-226 RQ = 0.1 curies; Radium-227 RQ = 1000 curies; Radium-228 RQ = 0.1 curies; Radon-220 RQ = 1 curies; Radon-222 RQ = 1 curies.
Listed in the TSCA inventory.
The Clean Air Act authorizes EPA to establish annual limits, known as National Emission Standards for Hazardous Air Pollutants (NESHAP) for the maximum amount of radium and other radionuclides that may be released to the air. Radium = 10 mRem.
The Safe Drinking Water Act authorizes EPA to limit the Maximum Contaminant Levels (MCL) of radium and other radionuclides in publicly supplied drinking water.
Description: Radium (Ra) is a radioactive element, found naturally in the environment. Ra is a silvery-white-metallic solid at 25°C; it tarnishes black when exposed to air. It is an alkaline earth metal; there are 33 isotopes, all of them are unstable. Radium is commonly available as radium bromide ($RaBr_2$) or radium chloride ($RaCl_2$). Atomic weight = 226.025; Density = 5.5 g/cm^3 at 28°C; Freezing/

Melting point = 700°C; Boiling point = 1140°C; 1536°C; 1737°C [Merck]; Density = 5.3 g/cm^3 at 26.85°C; Vapor pressure = 327 Pa at 700°C. *Radium bromide* is a white to brownish crystalline solid. Radioactive. Molecular weight = 395.83; Density = 5.79 at 20°C; Specific gravity = (H_2O:1) 3.34; Freezing/Melting point = 728°C (decomposes); Boiling point = 900°C (sublimes). Soluble in cold water. *Radium chloride* is a yellowish-white or brownish crystalline solid. Radioactive. Molecular weight = 296.9 g/mol; Density = 4.91; Freezing/Melting point = 1000°C.

Potential Exposure: Radium is not available as a pure metal but is found in very small quantities in uranium and thorium ores. Uranium and thorium are found in small amounts in most rocks and soil; radium is formed when these elements break down in the environment. One ton of uranium ore yields only slightly more than 0.1 g of radium. Radium is formed from the radioactive decay; and as a by-product of refining these ores. Radium exists in several isotope forms. Two of the principal radium isotopes found in the environment are ^{226}Ra (radium-226) and ^{228}Ra (radium-228). Radium compounds, due to their geologically short half-life and intense radioactivity, are quite rare. A single gram of ^{226}Ra produces 10^{-4} mm of radon (Rn) a day. Radium-226, radium's most stable isotope has a half-life of about 1603−1620 years, and remains in the body for life. Radium, when used to produce radon gas, is used for treating various types of cancer; in radiography of metals; and combined with other metals, such as beryllium, as a neutron emitting source used in research and for calibrating radiation instruments. Until the 1960s, radium was a component in self-luminous paints used for watch, compass, and aircraft instrument dials and other aircraft and military instrumentation. A less dangerous radioactive source, ^{60}Co (cobalt-60), replaced radium in luminous paint. The greatest health risk from radium comes from exposure to its radioactive decay product, radon (Rn). Radon is common in many soils and can collect in buildings, including homes.

Incompatibilities: Metallic radium is highly chemically reactive. It forms compounds that are very similar to barium compounds, making separation of the two elements difficult.[EPA] On contact with water, radium forms flammable hydrogen gas. Radium bromate is a strong oxidizer; avoid contact with a combustible material (clothing, wood, paper, oil, etc.).

Permissible Exposure Limits:
Recommended Occupational Annual Dose Limits (NCRP and NRC):
Whole body: 5 Rem (0.05 Sv)/1 year.
Lens of eye: 15 Rem (0.15 Sv).
Skin: 50 Rem (0.5 Sv).
Hands or feet: 50 Rem (0.5 Sv).
Cumulative occupational limit: 1 Rem (0.01 Sv) × age.

The limit for radiation exposure for a declared pregnant radiation worker is 500 mRem (0.005 Sv) for the entire gestation period. The separate limit for the fetus is 500 mRem (0.005 Sv) for the entire gestation; exposures to the fetus must be uniform and must be maintained at or below 50 mRem (0.0005 Sv)/mo.

Radiological Dispersal Device (RDD) Incidents

Particulate sampling can be performed to measure the radioactivity of dust in the air and to further characterize exposures. Based on the sampling data, the respirator protection factor needed to meet the relevant exposure limits can be determined (see Table below), and the specific type of respirator needed can be identified. According to the data, it may be possible to downgrade or remove requirements for respiratory protection. However, until exposures have been characterized, responders and support personnel should continue to utilize full-face air-purifying P100 respirators, or higher respiratory protection (e.g., SCBAs, PAPRs).

Guidance Matrix for Radiological Dispersal Device (RDD) Incidents

OSHA Occupational Exposure Limits for Ionizing Radiation 29 CFR 1910.1096	Rem (Sv)/calendar qtr.
Whole body, head, and trunk	—
Active blood-forming organs	—
Lens of eye, or Gonads	1.25 (0.0125 Sv)
Hands and forearms	—
Feet and ankles	18.75 (0.1875 Sv)
Skin of whole body	7.5 (0.075 Sv)
Minors (workers under age 18)	10% of above limits
Workers over age 18	3 Rem (0.3 Sv) may be permitted under conditions specified in 29CFR1910.1096(b)(2)

Note: The Department of Homeland Security is currently chairing an interagency workgroup which is in the process of assessing the protective action guidance for response to an RDD event.

EPA Limits and Advisory Levels: United States, annual radiation exposure limits are found in Title 10, Part 20 of the Code of Federal Regulations, and in equivalent state regulations.

Public dose limits due to licensed activities (NRC) 100 mRem (0.01 Sv)/year.
Air: Radium NESHAP = 10 mRem (0.1 Sv).*
Water: ^{226}Ra & ^{228}Ra (combined radium) MCL = 5 pCi/L; ^{224}Ra = 15 picocuries/liter (5 pCi/L).*
Indoor Air (advisory "action level"): 4 pCi/L radon (Rn).
*Both the air and water standards limit the increased lifetime cancer risk to about 2 in 10,000.

Permissible Concentration in Water: The US EPA has set an MCL of 5 pCi/L in drinking water for Radium-226 and Radium-228 (combined) and 15 pCi/L for Radium-224.

Permissible Concentration in Soil: The US EPA has set a soil concentration limit of 5 pCi/g for Radium-226 in uranium and thorium mill tailings in the first 15 cm of soil and 15 picocuries per gram (5 pCi/g) in deeper soil.

Routes of Entry: Inhalation, ingestion, external exposure to alpha-, beta-, and gamma-rays.

Harmful Effects and Symptoms

Radium is highly radioactive and its decay product, radon (Rn) gas, is also radioactive. Radium is over one million times more radioactive than the same mass of uranium. Since radium is chemically similar to calcium, it has the potential to cause great harm by replacing it in bones and teeth. The later effects include an increase in cavities and broken teeth.

Short Term Exposure: Inhalation, injection, ingestion or body exposure to radium can cause cancer and other disorders. Exposure to high levels can increase the risk of bone, liver, and breast cancer. Radium emits several kinds of radiations, in particular, alpha particles and gamma rays. Alpha particles are generally harmful only if emitted inside the body. However, both internal and external exposure to gamma radiation is harmful. Gamma rays can penetrate the body, so gamma emitters like radium can result in exposures even when the source is a distance away. Radium bromate: Highly hazardous in case of skin contact (irritant), of eye contact (irritant), of ingestion, of inhalation. Slightly hazardous in case of skin contact; prolonged contact may result in skin burns, inflammation (including reddening, itching, and scaling), blistering, and ulcerations. Inhalation overexposure may cause respiratory irritation. Inflammation of the eye is characterized by redness, watering, and itching. Exposure from gamma rays is measured in units of roentgens (Rads); 100 Rads = 1 Gray (Gy) = 1 J/kg. For example: Radiation sickness = total body exposure of 100 Rads (or 1 Gy). Radiation sickness and death in half the population = total body exposure of 400 Rads (or 4 Gy). Without medical treatment, nearly everyone who receives more than 400 Rads will die within 30 days. 100,000 Rads causes almost immediate unconsciousness and death within an hour.[NYT]

Long Term Exposure: Long-term exposure to radium increases the risk of developing several diseases. Inhaled or ingested radium increases the risk of developing lymphoma, bone cancer, and diseases that affect the formation of blood, such as leukemia and aplastic anemia. These effects usually take years to develop. External exposure to radium's gamma radiation increases the risk of cancer to varying degrees in all tissues and organs. Historically, Marie Curie's death due to aplastic anemia has been blamed on exposure and handling of radium. Patients who were injected with radium in Germany, from 1946 to 1950, for the treatment of certain diseases including tuberculosis were significantly shorter as adults than people who were not treated.[CDC]

Points of Attack: Liver, blood (anemia), eyes (cataracts) and other organs, bones and teeth.

Medical Surveillance: There are tests that are used to determine exposure to radium or other radioactive substances. For example, a whole body count can measure the total amount of radioactivity in the body, and urine and feces can be tested for the presence of radionuclides. There is no test that can detect external exposure to radium's gamma radiation, unless the doses were very high, and cellular damage is detectable. Another test measures radon (Rn) in exhaled air [radon (Rn) is a breakdown product of radium]. None of these tests are routinely performed in a doctor's office; they require specialized laboratory equipment. These tests are unable to quantify exposure, nor can they be used to predict whether a patient will develop long-term or harmful health effects.

First Aid: Unless you are dressed in appropriate protective gear to prevent self-contaminating, do not provide medical attention. Evacuate the victim from area of exposure to a safe area as soon as possible. To stop ongoing contamination, have the victim remove clothing, if possible, and place clothing in a sealed garbage bag or container. Check the victim's breathing and pulse; start CPR, if necessary.

Skin: If skin contamination has occurred, measure levels of contamination with a survey meter, record results, and begin decontamination by gentle washing with plenty of water (warm if possible) and nonabrasive and disinfecting soap, washing downwards towards extremities, not upwards. Dry body and cover the irritated skin with an emollient. If burns are obvious, do not use ointments. Wrap victim in a clean, soft blanket. Seek immediate medical attention; evacuate the victim to nearest emergency medical facility.

Eyes: Check for and remove any contact lenses. Immediately flush eyes with cold water. Avoid the use of an eye ointment. Seek immediate medical attention; evacuate the victim to nearest emergency medical facility.

Inhalation: Allow the victim to rest in a well-ventilated area. If breathing is difficult, administer oxygen. Seek immediate medical attention; evacuate the victim to nearest emergency medical facility.

Ingestion: Do not induce vomiting. Loosen tight clothing such as a collar, tie, belt. Seek immediate medical attention; evacuate the victim to nearest emergency medical facility.

Personal Protective Methods: *Radium bromate:* Engineering Controls: Use process enclosures, local exhaust ventilation, or other engineering controls to keep airborne levels below recommended exposure limits. If user operations generate dust, fume or mist, use ventilation to keep exposure to airborne contaminants below the exposure limit. Emergency responders to an explosion or the resulting fires will generally not know they are being exposed to radiation unless they utilize a radiation-detecting device. There is no practical personal protective equipment (PPE) to protect First Responders against externally penetrating *gamma* radiation. Monitoring devices are the only means to ensure that responders do not enter an area where exposures

to external *gamma* radiation is excessive. Personal protective equipment (PPE) to prevent skin contamination of particulates is very effective against particulate-borne radiation hazards (i.e., *alpha* and *beta* particles). Typical firefighter "turnout" gear, including a SCBA and radiation detection device or dosimeter, may be adequate for this purpose (depending on radiation level). The use of turnout gear or any disposable protective clothing suitable for particulate exposure should be followed by appropriate decontamination of personnel and equipment. *Personal Protection in Case of a Large Spill:* Splash goggles; full suit; dust respirator; boots; gloves. A self-contained breathing apparatus should be used to avoid inhalation of the product. Suggested protective clothing might not be sufficient; consult a specialist BEFORE handling radioactive material.

Respirator Selection: Protection of internal organs from inhalation of radioactive particulates can be provided by wearing an appropriate particulate respirator. The SCBAs will provide the highest level of protection. Responders should utilize at least a full-face air-purifying respirator with a P-100 or HEPA filter, as appropriate. Respiratory protection specifically approved by NIOSH for CBRN exposures is desirable. However, where specific CBRN-approved respirators are not available, the incident commander may allow alternative NIOSH-approved respirators, such as SCBAs, or full-face powered or nonpowered air-purifying respirators with P-100 or HEPA filters, as appropriate. It should be noted that these recommendations for respiratory protection are designed ONLY for protection against inhalation of radioactive particulates, and do not consider protection that may be necessary for other contaminants, such as chemical or biological agents.

At any concentrations above the NIOSH REL, or where there is no REL, at any detectable concentration: SCBAF: Pd,Pp (APF = 10,000) (any NIOSH/MSHA- or European Standard EN 149-approved self-contained breathing apparatus that has a full face-piece and is operated in a pressure-demand or other positive-pressure mode) or SaF: Pd,Pp: ASCBA (APF = 10,000) (any supplied-air respirator that has a full face-piece and is operated in a pressure-demand or other positive-pressure mode in combination with an auxiliary, self-contained breathing apparatus operated in a pressure-demand or other positive-pressure mode).

Storage: Store in a ventilated area to prevent accumulation of radon (Rn).

Shipping: Radioactive material, Type A package nonspecial form, nonfissile or fissile-excepted, requires a label of "RADIOACTIVE MATERIAL." It falls in Hazard Class 7. *A1 and A2 values for Radium-226 taken from §173.435 (see also Table A-1 in 10CFR71 (Appendix A): A1(Special Form*) 0.3 TBq (8.11 Ci); A2 (Normal Form) 0.02TBq (0.541Ci))*

*Special form materials are limited to those materials which, if released from a package, would present a hazard only due to direct *external radiation*. Usually, due to the high physical integrity of a special form material, radioactive material contamination is not expected even under severe accident conditions. This high physical integrity is occasionally the result of *inherent natural* properties of the material, such as its being in nondispersible solid form. Most often, however, it is an *acquired characteristic*, resulting from being welded (encapsulated) into an extremely durable metal capsule.

The A1 and A2 quantities for each radionuclide are basically the maximum activity that can be transported in a Type A package. For many radionuclides, the regulations allow substantially larger quantities of special form material to be placed in a Type A packaging than when the material is in "normal form," i.e., "nonspecial form." Special form radioactive material is defined in 49 CFR 173.403. Special form sources must have at least one external physical dimension which exceeds 5 mm (0.197″). The minimum dimension requirement makes the capsule easier to see and recover in the event of its release from the package during an accident. *Special form encapsulations are required to be constructed in a manner that they can only be opened by destroying the capsule.* This requirement prevents the inadvertent loosening or opening of the capsule either during transport or following an accident. The testing requirements for determination of whether radioactive materials qualify as "special form" are found in 49CFR173.469, which describes tests for high temperature, impact, percussion, bending, and leakage.

Spill Handling: Notify authorities that spill has occurred. Do not touch spilled material. Cover with absorbent paper or dike with absorbent. Isolate the area to prevent unnecessary spread of the material and personnel exposures. Prevent entry into sewers, basements, or confined areas; dike if needed. Using appropriate monitoring equipment, survey the spill site; evaluate the presence of contamination on an individual's skin and clothing and on lab equipment. If skin or clothing contamination is present, assume a MAJOR spill has occurred. Do not attempt to clean it up. Confine all potentially contaminated individuals in order to prevent the further spread of contamination. If possible, shield the source, but only if it can be done without significantly increasing radiation exposure to yourself or other personnel. Leave the affected room and lock the doors in order to prevent entry. Do what you can to prevent further spreading of contamination to unrestricted areas. Contact local and state authorities and the Department of Energy (DOE) Radiological Response Team. Call for assistance on disposal.

Fire Extinguishing: If "Radioactive material, Type A package nonspecial form" is involved in a fire situation, first contact the Department of Energy (DOE) Radiological Response Team, as well as state and local authorities. Do not use water; use suitable dry powder, graphite, soda ash, powdered sodium chloride, or sand.

Disposal Method Suggested: Radioactive material considered waste and must be retained in containers for disposition by the authorizing institution. Drain disposal is prohibited. It is the responsibility of the operating institution to arrange for the proper disposal of all forms of any radioisotopes. The use, storage, transportation, labeling, and disposal of radioactive material are regulated through the Nuclear Regulatory Commission (NRC) using 10 CFR (Code of Federal Regulations) as the regulatory basis and 49 CFR (Transportation). (The CFRs are available at no cost on the internet at http://www.access.gpo.gov/nara/cfr/index.html)

References

Agency for Toxic Substances and Disease Registry (ATSDR). Division of Toxicology, US Department of Health and Human Services, Public Health Service. (1990). *Toxicological Profile for Radium.* Atlanta, GA

Bentor, Y. (November 2, 2010). *Chemical Element.com— Radium.* <http://www.chemicalelements.com/elements/ra.html>

29 CFR 1910.1096. *Ionizing Radiation.* OSHA Standard.

Environmental Protection Agency (EPA), Office of Radiation Programs. (May 1992). *Manual of Protective Action Guides and Protective Actions for Nuclear Incidents.* 16 MB PDF, 274 p

US Department of Energy (DOE). (May 2001). *Radiological Emergency Response Health and Safety Manual,* Report DOE/NV/11718-440. 1 MB PDF, 103 p

US Army Medical Research Institute of Chemical Defense (USAMRICD). Chemical Casualty Care Division (USAMRICD) MCMR-UV-ZM. (July 2000). *Field Management of Chemical Casualties* (2nd ed.). Aberdeen Proving Grounds, MD

Reserpine   R:0100

Molecular Formula: $C_{33}H_{40}N_2O_9$

Synonyms: Abesta; Abicol; Adelfan; Adelphane; Adelphin; Adelphin-esidrex-K; Alkarau; Alkaserp; Alserin; Anquil; Apoplon; Apsical; Arcum R-S; Ascoserp; Ascoserpina; Austrapine; Banasil; Banisil; Benazyl; Bendigon; Bioserpine; Brinderdin; Briserine; Broserpine; Butiserpazide-25; Butiserpazide-50; Butiserpine; Cardioserpin; Carditivo; Carrserp; Crystoserpine; Daerbon; Deserpine; Diupres; Diutensen-R; Drenusil-R; Dypertane compound; Eberpine; Eberspine; Ebserpine; Elerpine; Elfanex; Enipresser; ENT 50,146; Escaspere; Eserpine; Eskaserp; Gamaserpin; Gammaserpine; Gilucard; H 520; Helfoserpin; Hexaplin; Hiposerpil; Hiserpia; Hydromox R; Hydropres; Hydropreska; Hygroton-resperine; Hypercal B; Hypertane Forte; Hypertensan; Idoserp; Idsoserp; Interpina; Key-Serpine; Kitene; Klimanosid; "L," Carpserp; Lemiserp; Loweserp; Marnitension simple; Maviserpin; Mayserpine; Mephaserpin; Methylreserpate 3,4,5,-trimethoxybenzoic acid; Methylreserpate 3,4,5,-trimethoxybenzoic acid ester; Mio-pressin; Modenol; Naquival; NCI-C50157; Nembuserpin; Neo-antitensol; Neoserfin; Rau-sed; Rauwoleaf;

Reserpine

Recipin; Regroton; Renese R; R-E-S; Resaltex; Resedin; Resedrex; Resedril; Reserpex; Reserpoid; Serpasil; Serpasil Apresoline; Serpine; 3,4,5-Trimethoxybenzoyl methyl reserpate; Yohimban-16-carboxylic acid derivative of benz(g)indolo(2,3-a)quinolizine; Yohimban-16-carboxylic acid, 11,17-Dimethoxy-18-(3,4,5-trimethoxybenxoyl)oxy-, methyl ester; Yohimban-16-carboxylic acid, 11,17-dimethoxy-18-(3,4,5-trimethoxybenxoyl)oxy-, methyl ester, (3b,16b,17a,18b,20a)

CAS Registry Number: 50-55-5
RTECS® Number: ZG0350000
UN/NA & ERG Number: UN1544/151
EC Number: 200-047-9
Regulatory Authority and Advisory Bodies
Carcinogenicity: IARC: Animal Inadequate Evidence; Human Limited Evidence, *not classifiable as carcinogenic to humans*, Group 3, 1987; NTP: Reasonably anticipated to be a human carcinogen; NCI: Carcinogenesis studies (feed); clear evidence: mouse, rat, 1980.
US EPA Hazardous Waste Number (RCRA No.): U200.
RCRA, 40CFR261, Appendix 8 Hazardous Constituents.
Reportable Quantity (RQ): 5000 lb (2270 kg).
European/International Regulations: not listed in Annex 1.
WGK (German Aquatic Hazard Class): No value assigned.
Description: Reserpine is a white to pale buff to slightly yellow crystalline substance that darkens on exposure to light. Molecular weight = 608.75; Freezing/Melting point = 264–265°C (decomposes). Hazard Identification (based on NFPA-704 M Rating System): Health 3, Flammability 0, Reactivity 0. May be combustible.
Potential Exposure: Compound Description: Drug, Mutagen; Reproductive Effector; Human Data; Natural Product. An alkaloid. Reserpine, a pharmaceutical, is a naturally occurring substance that is isolated from the roots of the plant *Rauwolfia serpentina*. Insoluble in water. Reserpine is used as a hypertensive for humans and animals, tranquilizer, and sedative. Permitted for use as an additive in food for human consumption, and the feed and drinking water of food-producing animals.
Incompatibilities: A weak acid; keep away from bases. Incompatible with strong oxidizers and strong reducing agents.
Permissible Exposure Limits in Air
No standards or TEEL available.
Routes of Entry: Ingestion.
Harmful Effects and Symptoms
Short Term Exposure: Massive overexposure may cause decreased blood pressure, convulsions, and coma. Reserpine is highly toxic to man. In humans, 0.014 mg/kg produces psychotropic effects. Symptoms include nausea, diarrhea, excessive salivation, nasal stuffiness, drowsiness, nightmares, emotional depression and other psychotropic effects, extra systoles, angina pain, edema and weight gain sometimes associated with frank congestive heart failure, thrombocytopenia, tremor, muscular stiffness, severe hypotension in conjunction with general anesthetic administration.

Long Term Exposure: May be a human carcinogen. There is limited evidence that this chemical causes breast cancer in humans, and breast and testes cancer in animals. Reproductive activity: may also be a mutagen and teratogen. Reserpine may damage the developing fetus. People taking reserpine daily for medical purposes have developed nightmares, severe depression, cramps, diarrhea, and weight gain. It is unknown if these symptoms occur from repeated occupational exposure.
Points of Attack: Nervous system. *Cancer site:* breast in humans; breast and testes in animals.
Medical Surveillance: There is no special test for this chemical. If overexposure or illness is suspected, medical attention is recommended.
First Aid: Skin Contact: Flood all areas of body that have contacted the substance with water. Do not wait to remove contaminated clothing; do it under the water stream. Use soap to help assure removal. Isolate contaminated clothing when removed to prevent contact by others.[52]
Eye Contact: Remove any contact lenses at once. Immediately flush eyes well with copious quantities of water or normal saline for at least 20–30 min. Seek medical attention.
Inhalation: Leave contaminated area immediately; breathe fresh air. Proper respiratory protection must be supplied to any rescuers. If coughing, difficult breathing, or any other symptoms develop, seek medical attention at once, even if symptoms develop many hours after exposure.
Ingestion: Contact a physician, hospital, or poison center at once. If the victim is unconscious or convulsing, do not induce vomiting or give anything by mouth. Assure that the patient's airway is open and lay him on his side with his head lower than this body and transport immediately to a medical facility. If conscious and not convulsing, give a glass of water to dilute the substance. Vomiting should not be induced without a physician's advice.
Personal Protective Methods: Wear protective gloves and clothing to prevent any reasonable probability of skin contact. Safety equipment suppliers/manufacturers can provide recommendations on the most protective glove/clothing material for your operation. All protective clothing (suits, gloves, footwear, headgear) should be clean, available each day, and put on before work. Contact lenses should not be worn when working with this chemical. Wear dust-proof chemical goggles and face shield unless full face-piece respiratory protection is worn. Employees should wash immediately with soap when skin is wet or contaminated. Provide emergency showers and eyewash. Specific engineering controls are required for drug manufacture by the Food and Drug Administration. Refer to 21CFR210, *FDA Regulation for Good Manufacturing Practices.*
Respirator Selection: Follow regulations in OSHA 29CFR1910.134 or European Standard EN149. Use a NIOSH/MSHA- or European Standard EN149-approved respirator; or use an approved supplied-air respirator with a full face-piece operated in the positive-pressure mode, or with a

full face-piece, hood, or helmet in the continuous-flow mode; or use a NIOSH/MSHA- or European Standard EN149-approved self-contained breathing apparatus with a full face-piece operated in a pressure-demand or other positive-pressure mode.

Storage: Color Code—Blue: Health Hazard/Poison: Store in a secure poison location. Prior to working with this chemical you should be trained on its proper handling and storage. Store in a refrigerator or in a cool, dry place. Protect from exposure to light and acids. Where possible, automatically transfer material from drums or other storage containers to process containers. A regulated, marked area should be established, or stored in compliance with OSHA Standard 1910.1045.

Shipping: Alkaloids, solid, n.o.s. require a shipping label of "POISONOUS/TOXIC MATERIALS." Respirine falls in Hazard Class 3.1 and Packing Group II or III.

Spill Handling: Evacuate persons not wearing protective equipment from area of spill or leak until cleanup is complete. Remove all ignition sources:[52] Remove all sources of ignition and dampen spilled material with toluene to avoid airborne dust, then transfer material to a suitable container. Ventilate the spill area and use absorbent dampened with toluene to pick up remaining material. Wash surfaces well with soap and water. Collect powdered material in the most convenient and safe manner and deposit in sealed containers. Ventilate area after cleanup is complete. It may be necessary to contain and dispose of this chemical as a hazardous waste. If material or contaminated runoff enters waterways, notify downstream users of potentially contaminated waters. Contact your local or federal environmental protection agency for specific recommendations. If employees are required to clean up spills, they must be properly trained and equipped. OSHA 1910.120(q) may be applicable.

Fire Extinguishing: This chemical may burn but does not easily ignite. Use dry chemical, carbon dioxide, water spray, or foam extinguishers. Poisonous gases are produced in fire, including nitrogen oxides. If material or contaminated runoff enters waterways, notify downstream users of potentially contaminated waters. Notify local health and fire officials and pollution control agencies. From a secure, explosion-proof location, use water spray to cool exposed containers. If cooling streams are ineffective (venting sound increases in volume and pitch, tank discolors, or shows any signs of deforming), withdraw immediately to a secure position. If employees are expected to fight fires, they must be trained and equipped in OSHA 1910.156. The only respirators recommended for firefighting are self-contained breathing apparatuses that have full face-pieces and are operated in a pressure-demand or other positive-pressure mode.

Disposal Method Suggested: Consult with environmental regulatory agencies for guidance on acceptable disposal practices. Generators of waste containing this contaminant (≥100kg/mo) must conform with EPA regulations governing storage, transportation, treatment, and waste disposal.

References

Sax, N. I. (Ed.). (1981). *Dangerous Properties of Industrial Materials Report*, 1, No. 4, 90—92

New Jersey Department of Health and Senior Services. (October 2001). *Hazardous Substances Fact Sheet: Reserpine*. Trenton, NJ

Resorcinol   R:0110

Molecular Formula: $C_6H_6O_2$

Common Formula: $1,3\text{-}C_6H_4(OH)_2$

Synonyms: Benzene, *m*-dihydroxy-; *m*-Benzenediol; 1,3-Benzenediol; C.I. 76505; C.I. Developer 4; C.I. Oxidation base 31; Developer O; Developer R; Developer RS; *m*-Dihydroxybenzene; 1,3-Dihydroxybenzene; *m*-Dioxybenzene; Durafur developer G; Fouramine RS; Fourrine 79; Fourrine EW; *m*-Hydroquinone; 3-Hydroxycyclohexadien-1-one; *m*-Hydroxyphenol; 3-Hydroxyphenol; Nako TGG; NCI-C05970; Pelagol grey RS; Pelagol RS; Phenol, *m*-hydroxy-; Resorcin; Resorcina (Spanish); Resorcine

CAS Registry Number: 108-46-3

RTECS® Number: VG9625000

UN/NA & ERG Number: UN2876/153

EC Number: 203-585-2 [*Annex I Index No.*: 604-010-00-1]

Regulatory Authority and Advisory Bodies

Carcinogenicity: IARC: Animal Inadequate Evidence; Human No Adequate Data, *not classifiable as carcinogenic to humans*, Group 3, 1999; NCI: Carcinogenesis Studies (gavage); no evidence: mouse, rat; NTP: Carcinogenesis Studies (gavagae); no evidence: mouse, rat.

US EPA Gene-Tox Program, Negative: *N. crassa—aneuploidy*; Histidine reversion—Ames test; Inconclusive: Mammalian micronucleus.

Air Pollutant Standard Set. See below, "Permissible Exposure Limits in Air" section.

FDA—over-the-counter drug.

Clean Water Act: Section 311 Hazardous Substances/RQ 40CFR117.3 (same as CERCLA, see below).

US EPA Hazardous Waste Number (RCRA No.): U201.

RCRA, 40CFR261, Appendix 8 Hazardous Constituents.

Reportable Quantity (RQ): 5000 lb (2270 kg).

Canada, WHMIS, Ingredients Disclosure List Concentration 1.0%.

European/International Regulations: Hazard Symbol: Xn, N; Risk phrases: R22; R36/38; R50; Safety phrases: S2; S26; S61 (see Appendix 4).

WGK (German Aquatic Hazard Class): 1—Low hazard to waters.

Description: Resorcinol is a white crystalline solid with a characteristic odor and a sweetish taste. Turns pink on exposure to air or light, or contact with iron. Molecular weight = 110.12; Boiling point = 277.2°C; Freezing/Melting point = 108.9°C; Vapor pressure = 0.0002 mmHg at 25°C; Flash point = 127.2°C (cc); Autoignition temperature = 607°C. Explosive limits: LEL = 1.4% at 200°C. UEL—unknown. Hazard Identification (based on NFPA-704 M Rating System): Health 3, Flammability 1, Reactivity 0. Highly soluble in water; solubility = 110%.

Potential Exposure: Compound Description: Drug, Tumorigen, Mutagen, Human Data; Primary Irritant. Resorcinol is weakly antiseptic; resorcinol compounds are used in the production of resorcinol−formaldehyde adhesives; or as an intermediate; in pharmaceuticals and hair dyes for human use. Major industrial uses are as adhesives in rubber products and tires, wood adhesive resins, and as ultraviolet absorbers in polyolefin plastics. Resorcinol is also a by-product of coal conversion and is a component of cigarette smoke. Thus, substantial opportunity exists for human exposure.

Incompatibilities: Reacts with oxidizers, nitric acid, oil, ferric salts, methanol, acetanilide, albumin, antipyrene, alkalies, urethane, ammonia, amino compounds. Hygroscopic; absorbs moisture from the air.

Permissible Exposure Limits in Air

Conversion factor: 1 ppm = 4.50 mg/m^3 at 25°C & 1 atm.
OSHA PEL: None.
NIOSH REL: 10 ppm/45 mg/m^3 TWA; 20 ppm/90 mg/m^3 STEL.
ACGIH TLV®[1]: 10 ppm/45 mg/m^3 TWA; 20 ppm/90 mg/m^3 STEL; not classifiable as a human carcinogen.
Protective Action Criteria (PAC)
TEEL-0: 10 ppm
PAC-1: 20 ppm
PAC-2: 20 ppm
PAC-3: 20 ppm
DFG MAK: Danger of skin sensitization.
Australia: TWA 10 ppm (45 mg/m^3); STEL 20 ppm, 1993; Austria: MAK 10 ppm (45 mg/m^3), 1999; Belgium: TWA 10 ppm (45 mg/m^3); STEL 20 ppm, 1993; Denmark: TWA 10 ppm (45 mg/m^3), 1999; EU OEL: TWA 10 ppm, 45 mg/m^3; 2000; Finland: TWA 10 ppm (45 mg/m^3); STEL 20 ppm (90 mg/m^3), 1999; France: VME 10 ppm (45 mg/m^3), 1999; Hungary: TWA 45 mg/m^3; STEL 90 mg/m^3, [skin], 1993; the Netherlands: MAC-TGG 45 mg/m^3, 2003; Poland: MAC (TWA) 45 mg/m^3; MAC (STEL) 90 mg/m^3, 1999; Sweden: NGV 10 ppm (45 mg/m^3), [skin], 1999; Switzerland: MAK-W 10 ppm (45 mg/m^3), 1999; United Kingdom: TWA 10 ppm (46 mg/m^3); STEL 20 ppm (92 mg/m^3), 2000; Argentina, Bulgaria, Columbia, Jordan, South Korea, New Zealand, Singapore, Vietnam: ACGIH TLV®: STEL 20 ppm. Several states have set guidelines or standards for resorcinol in ambient air[60] ranging from 3.1 µg/m^3 (Massachusetts) to 450.0−900.0 µg/m^3 (North Dakota) to 750.0 µg/m^3 (Virginia) to 900.0 µg/m^3 (Connecticut, Florida, New York) to 1071.0 µg/m^3 (Nevada).

Determination in Air: OSHA versatile sampler-7; Methanol; Gas chromatography/Flame ionization detection; NIOSH Analytical Method (IV) #5701; OSHA Analytical Method: PV-2053.

Permissible Concentration in Water: Russia[35,43] set a MAC of 0.1 mg/L in water bodies used for domestic purposes and 0.004 mg/L in water bodies used for fishery purposes. The state of Maine has set a guideline of 140 µg/L for drinking water.[61]

Determination in Water: Octanol−water coefficient: Log K_{ow} = 0.8−0.9.

Routes of Entry: Inhalation, ingestion, eye and/or skin contact. Absorbed through the skin.

Harmful Effects and Symptoms

Short Term Exposure: Irritates the eyes, skin, and respiratory tract. Skin contact causes a severe rash and blistering. Eye contact can cause burns and permanent damage. May affect the blood, causing formation of methemoglobin, cyanosis, convulsions, restlessness, increased heart rate, dyspnea (breathing difficulty), dizziness, drowsiness, hypothermia, hemoglobinuria. High exposure can cause kidney and/or liver damage. *Inhalation:* Humans occupationally exposed to 10 ppm showed no effects. Experiments with animals showed no toxic effects after 8-h exposure to 625 ppm. *Skin:* Itching, irritation, redness, swelling, and chemical burns have been reported from contact with solutions of 3−25% strength. Skin absorption is significant and leads to loss of oxygen-carrying capacity of blood, convulsions, and death. *Eyes:* Animal tests suggest that 10% solution may cause irritation, pain, and corneal damage. Powdered resorcinol may cause chemical burns to the cornea resulting in blindness. *Ingestion:* May cause excessive sweating, low blood pressure, slowed breathing, tremors, breakdown of blood cells and death. One teaspoonful may cause death.

Long Term Exposure: Repeated or prolonged contact may cause skin sensitization and allergy. Repeated exposure can cause lung damage. May cause kidney, liver, spleen damage.

Points of Attack: Eyes, skin, respiratory system, cardiovascular system, central nervous system, blood, spleen, liver, kidneys.

Medical Surveillance: For those with frequent or potentially high exposure (half the TLV or greater, or significant skin contact), the following are recommended before beginning work and at regular times after that: lung function tests. If symptoms develop or overexposure is suspected, the following may be useful: urine test for resorcinol. Tests for kidney and liver function. Methemoglobin level. Evaluation by a qualified allergist, including careful exposure history and special testing, may help diagnose skin allergy.

First Aid: If this chemical gets into the eyes, remove any contact lenses at once and irrigate immediately for at least 15 min, occasionally lifting upper and lower lids. Seek medical attention immediately. If this chemical contacts the skin, remove contaminated clothing and wash immediately with soap and water. Seek medical attention immediately. If this chemical has been inhaled, remove from exposure, begin rescue breathing (using universal precautions, including resuscitation mask) if breathing has stopped and CPR if

heart action has stopped. Transfer promptly to a medical facility. When this chemical has been swallowed, get medical attention. Give large quantities of water and induce vomiting. Do not make an unconscious person vomit.

Note to physician: Treat for methemoglobinemia. Spectrophotometry may be required for precise determination of levels of methemoglobin in urine. Emergency treatment and management is similar to phenol.

Personal Protective Methods: Wear protective gloves and clothing to prevent any reasonable probability of skin contact. Safety equipment suppliers/manufacturers can provide recommendations on the most protective glove/clothing material for your operation. All protective clothing (suits, gloves, footwear, headgear) should be clean, available each day, and put on before work. Contact lenses should not be worn when working with this chemical. Wear dust-proof chemical goggles and face shield unless full face-piece respiratory protection is worn. Employees should wash immediately with soap when skin is wet or contaminated. Provide emergency showers and eyewash.

Respirator Selection: Where there is potential for exposures *over 10 ppm*, use a NIOSH/MSHA- or European Standard EN149-approved full-face-piece respirator equipped with particulate (dust/fume/mist) filters. Particulate filters must be checked every day before work for physical damage, such as rips or tears, and replaced as needed. Where there is potential for high exposure, use a NIOSH/MSHA- or European Standard EN149-approved supplied-air respirator with a full face-piece operated in the positive-pressure mode, or with a full face-piece, hood, or helmet in the continuous-flow mode; or use NIOSH/MSHA- or European Standard EN 149-approved self-contained breathing apparatus with a full face-piece operated in a pressure-demand or other positive-pressure mode.

Storage: Color Code—Blue: Health Hazard/Poison: Store in a secure poison location. Prior to working with this chemical you should be trained on its proper handling and storage. Before entering confined space where this chemical may be present, check to make sure that an explosive concentration does not exist. Store in tightly closed containers in a cool, well-ventilated area away from oxidizers, oil, ferric salts, methanol, acetanilide, albumin, antipyrine, and urethane. Where possible, automatically transfer material from drums or other storage containers to process containers. Sources of ignition, such as smoking and open flames, are prohibited where this chemical is handled, used, or stored. Metal containers involving the transfer of this chemical should be grounded and bonded. Wherever this chemical is used, handled, manufactured, or stored, use explosion-proof electrical equipment and fittings.

Shipping: Resorcinol requires a shipping label of "POISONOUS/TOXIC MATERIALS." It falls in Hazard Class 6.1 and Packing Group III.

Spill Handling: Evacuate persons not wearing protective equipment from area of spill or leak until cleanup is complete. Establish forced ventilation to keep levels below explosive limit. Remove all ignition sources. Collect powdered material in the most convenient and safe manner and deposit in sealed containers. Liquid solutions should be absorbed on sand or vermiculite and shoveled into suitable containers. Scrub spill area with soap and water. It may be necessary to contain and dispose of this chemical as a hazardous waste. If material or contaminated runoff enters waterways, notify downstream users of potentially contaminated waters. Contact your local or federal environmental protection agency for specific recommendations. If employees are required to clean up spills, they must be properly trained and equipped. OSHA 1910.120(q) may be applicable.

Fire Extinguishing: This chemical is a combustible solid. Use dry chemical, carbon dioxide, water spray, or alcohol foam extinguishers. Poisonous gases are produced in fire. If material or contaminated runoff enters waterways, notify downstream users of potentially contaminated waters. Notify local health and fire officials and pollution control agencies. From a secure, explosion-proof location, use water spray to cool exposed containers. If cooling streams are ineffective (venting sound increases in volume and pitch, tank discolors, or shows any signs of deforming), withdraw immediately to a secure position. If employees are expected to fight fires, they must be trained and equipped in OSHA 1910.156. The only respirators recommended for firefighting are self-contained breathing apparatuses that have full face-pieces and are operated in a pressure-demand or other positive-pressure mode.

Disposal Method Suggested: Consult with environmental regulatory agencies for guidance on acceptable disposal practices. Generators of waste containing this contaminant ($\geq$100kg/mo) must conform with EPA regulations governing storage, transportation, treatment, and waste disposal. Dissolve in a combustible solvent and incinerate.

References

US Environmental Protection Agency. (April 30, 1980). *Resorcinol: Health and Environmental Effects Profile No. 152.* Washington, DC: Office of Solid Waste

Sax, N. I. (Ed.). (1980). *Dangerous Properties of Industrial Materials Report,* 1, No. 2, 58–59

New York State Department of Health. (March 1986). *Chemical Fact Sheet: Resorcinol.* Albany, NY: Bureau of Toxic Substance Assessment

New Jersey Department of Health and Senior Services. (June 2001). *Hazardous Substances Fact Sheet: Resorcinol.* Trenton, NJ

Rhodium   R:0120

Molecular Formula: Rh; $O_3Rh_2 \cdot xH_2O$ [Rhodium(III) oxide, hydrate]

Synonyms: Elemental rhodium; Rhodium black; Rhodium-103

CAS Registry Number: 7440-16-6; 123542-79-0 [Rhodium (III) oxide, hydrate]; 12137-27-8 (Rhodium(IV) oxide); 12036-35-0 (Rhodium oxide, solids)
RTECS® Number: VI9069000 (metal)
UN/NA & ERG Number: UN3089 (metal powder, flammable, n.o.s.)/170
EC Number: 231-125-0; 234-846-9 [Rhodium(III) oxide, hydrate]
Regulatory Authority and Advisory Bodies
Listed on the TSCA inventory.
Air Pollutant Standard Set. See below, "Permissible Exposure Limits in Air" section.
Canada, WHMIS, Ingredients Disclosure List Concentration 1.0%, elemental.
WGK (German Aquatic Hazard Class): Nonwater polluting agent (metal).
Description: Rhodium, together with platinum, palladium, iridium, ruthenium, and osmium, is one of the platinum-group metals in Group VIII of the Periodic Table. Rhodium metal is a white, hard, ductile, malleable solid with a bluish-gray luster. Molecular weight = 102.91; Specific gravity [metal] (H_2O:1) = 12.41 at 25°C; Boiling point = 3727°C; Freezing/Melting point = 1966°C. Insoluble in water.
Potential Exposure: Rhodium has few applications by itself, as in rhodium plating of white gold jewelry or plating of electrical parts, such as commutator slip rings, but, mainly, rhodium is used as a component of platinum alloys. Rhodium-containing catalysts have been proposed for use in automotive catalytic converters for exhaust gas cleanup.
Incompatibilities: Flammable as a dust or powder. Violent reaction with strong oxidizers, bromine pentafluoride, bromine trifluoride, chlorine trifluoride, oxygen difluoride.
Permissible Exposure Limits in Air
Metal fume and insoluble compounds
OSHA PEL: 0.1 mg[Rh]/m³ TWA.
NIOSH REL: 0.1 mg[Rh]/m³ TWA.
ACGIH TLV®[1] (*elemental*): 1 mg[Rh]/m³; not classifiable as a human carcinogen.
Protective Action Criteria (PAC)
Metal
TEEL-0: 0.1 mg/m³
PAC-1: 3 mg/m³
PAC-2: 5 mg/m³
PAC-3: 100 mg/m³
DFG MAK: Carcinogen Category 3B.
NIOSH IDLH: 100 mg[Rh]/m³.
Australia: TWA 1 mg[Rh]/m³, 1993; Belgium: TWA 1 mg [Rh]/m³, 1993; Finland: TWA 0.1 mg[Rh]/m³, 1999; France: VME 1 mg[Rh]/m³, 1999; Norway: TWA 0.1 mg [Rh]/m³, 1999; the Philippines: TWA 0.1 mg[Rh]/m³, 1993; Switzerland: MAK-W 0.1 mg[Rh]/m³, 1999; the Netherlands: MAC-TGG 0.1 mg[Rh]/m³, 2003; Argentina, Bulgaria, Columbia, Jordan, South Korea, New Zealand, Singapore, Vietnam: ACGIH TLV®: not classifiable as a human carcinogen.

Soluble compounds
OSHA PEL: 0.001 mg[Rh]/m³ TWA.
NIOSH REL: 0.001 mg[Rh]/m³ TWA.
ACGIH TLV®[1]: 0.01 mg[Rh]/m³; not classifiable as a human carcinogen.
NIOSH IDLH: 2 mg[Rh]/m³.
DFG MAK (*inorganic only*): Carcinogen Category 3B.
Several states have guidelines or standards for rhodium in ambient air[60] ranging from 0.16 μg/m³ (Virginia) to 2.0 μg/m³ (Connecticut), to 10.0 μg/m³ (North Dakota) to 24.0 μg/m³ (Nevada).
Determination in Air: Use NIOSH II(3) Method #S-188.
Routes of Entry: Inhalation.
Harmful Effects and Symptoms
Short Term Exposure: May cause metal fume fever. Rhodium(III) chloride and its hydrate are extremely toxic. May cause serious eye damage.
Long Term Exposure: Possible respiratory sensitization.
Points of Attack: Respiratory system. Ingestion [Rhodium (III) chloride].
Medical Surveillance: Examination by a qualified allergist. Lung function tests. Consider chest X-ray following acute overexposure to metal fume or dust.
First Aid: If this chemical gets into the eyes, remove any contact lenses at once and irrigate immediately for at least 15 min, occasionally lifting upper and lower lids. Seek medical attention immediately. If this chemical contacts the skin, remove contaminated clothing and wash immediately with soap and water. Seek medical attention immediately. If this chemical has been inhaled, remove from exposure, begin rescue breathing (using universal precautions, including resuscitation mask) if breathing has stopped and CPR if heart action has stopped. Transfer promptly to a medical facility. When this chemical has been swallowed, get medical attention. Give large quantities of water and induce vomiting. Do not make an unconscious person vomit. If metal fume fever develops, it may last less than 36 h.
Note to physician: In case of fume inhalation, treat for pulmonary edema. Give prednisone or other corticosteroid orally to reduce tissue response to fume. Positive-pressure ventilation may be necessary. Treat metal fume fever with bed rest, analgesics, and antipyretics.
Personal Protective Methods: Wear protective gloves and clothing to prevent any reasonable probability of skin contact. Safety equipment suppliers/manufacturers can provide recommendations on the most protective glove/clothing material for your operation. All protective clothing (suits, gloves, footwear, headgear) should be clean, available each day, and put on before work. Contact lenses should not be worn when working with this chemical. Wear dust-proof chemical goggles and face shield unless full face-piece respiratory protection is worn. Employees should wash immediately with soap when skin is wet or contaminated. Provide emergency showers and eyewash.
Respirator Selection: *Up to 0.5 mg/m³:* Qm (APF = 25) (any quarter-mask respirator). *Up to 1 mg/m³:* 95XQ

(APF – 10) [any particulate respirator equipped with an N95, R95, or P95 filter (including N95, R95, and P95 filtering face-pieces) except quarter-mask respirators. The following filters may also be used: N99, R99, P99, N100, R100, P100] or Sa (APF = 10) (any supplied-air respirator). *Up to 2.5 mg/m³:* Sa:Cf (APF = 25) (any supplied-air respirator operated in a continuous-flow mode) or PaprHie (APF = 25) (any powered, air-purifying respirator with a high-efficiency particulate filter). *Up to 5 mg/m³:* 100F (APF = 50) (any air-purifying, full-face-piece respirator with an N100, R100, or P100 filter) or SaT: Cf (APF = 50) (any supplied-air respirator that has a tight-fitting face-piece and is operated in a continuous-flow mode) or PaprTHie (APF = 50) (any powered, air-purifying respirator with a tight-fitting face-piece and a high-efficiency particulate filter) or SCBAF (APF = 50) (any self-contained breathing apparatus with a full face-piece) or SaF (APF = 50) (any supplied-air respirator with a full face-piece). *Up to 100 mg/m³:* Sa: Pd,Pp (APF = 1000) (any supplied-air respirator operated in a pressure-demand or other positive-pressure mode). *Emergency or planned entry into unknown concentrations or IDLH conditions:* SCBAF: Pd,Pp (APF = 10,000) (any NIOSH/MSHA- or European Standard EN 149-approved self-contained breathing apparatus that has a full face-piece and is operated in a pressure-demand or other positive-pressure mode) or SaF: Pd,Pp: ASCBA (APF = 10,000) (any supplied-air respirator that has a full face-piece and is operated in a pressure-demand or other positive-pressure mode in combination with an auxiliary, self-contained breathing apparatus operated in a pressure-demand or other positive-pressure mode). *Escape:* 100F (APF = 50) (any air-purifying, full-face-piece respirator with an N100, R100, or P100 filter) or SCBAE (any appropriate escape-type, self-contained breathing apparatus).

Storage: Color Code—Red (*powder*): Flammability Hazard: Store in a flammable materials storage area. Prior to working with this material you should be trained on its proper handling and storage. Store in tightly closed containers in a cool, well-ventilated area away from strong oxidizers and sources of ignition. Where possible, automatically transfer material from drums or other storage containers to process containers. Sources of ignition, such as smoking and open flames, are prohibited where this chemical is handled, used, or stored. Metal containers involving the transfer of this chemical should be grounded and bonded. Wherever this chemical is used, handled, manufactured, or stored, use explosion-proof electrical equipment and fittings.

Sipping: Flammable powder requires a shipping label of "FLAMMABLE SOLID." It falls in Hazard Class 4.1 and Packing Group III.

Spill Handling: Evacuate persons not wearing protective equipment from area of spill or leak until cleanup is complete. Remove all ignition sources. Collect powdered material in the most convenient and safe manner and deposit in sealed containers. Ventilate area after cleanup is complete. It may be necessary to contain and dispose of this chemical as a hazardous waste. If material or contaminated runoff enters waterways, notify downstream users of potentially contaminated waters. Contact your local or federal environmental protection agency for specific recommendations. If employees are required to clean up spills, they must be properly trained and equipped. OSHA 1910.120(q) may be applicable.

Fire Extinguishing: This chemical is a combustible solid. Use dry chemical, carbon dioxide, water spray, or alcohol foam extinguishers. Poisonous gases are produced in fire. If material or contaminated runoff enters waterways, notify downstream users of potentially contaminated waters. Notify local health and fire officials and pollution control agencies. From a secure, explosion-proof location, use water spray to cool exposed containers. If cooling streams are ineffective (venting sound increases in volume and pitch, tank discolors, or shows any signs of deforming), withdraw immediately to a secure position. If employees are expected to fight fires, they must be trained and equipped in OSHA 1910.156. The only respirators recommended for firefighting are self-contained breathing apparatuses that have full face-pieces and are operated in a pressure-demand or other positive-pressure mode.

Disposal Method Suggested: Recovery in view of the high economic value. Recovery techniques for recycling of rhodium in plating wastes and spent catalysts have been described in the literature.

Rhodium trichloride   R:0130

Molecular Formula: Cl_3Rh

Common Formula: $RhCl_3 \cdot xH_2O$ (trihydrate)

Synonyms: Hydrated rhodium trichloride; Rhodium chloride; Rhodium(III) chloride (1:3); Rhodium chloride, trihydrate

CAS Registry Number: 10049-07-7; 13569-65-8 (trihydrate)

RTECS® Number: VI9275000

UN/NA & ERG Number: UN3260 (Corrosive solid, acidic, inorganic, n.o.s.)/154

EC Number: 233-165-4 (trichloride)

Regulatory Authority and Advisory Bodies

Air Pollutant Standard Set. See below, "Permissible Exposure Limits in Air" section.

Canada, WHMIS, Ingredients Disclosure List Concentration 1.0%.

Superfund/EPCRA 40CFR355, Extremely Hazardous Substances: dropped from listing in 1988.

WGK (German Aquatic Hazard Class): No value assigned.

Description: Rhodium trichloride is a red-brown or black, odorless solid or liquid. Molecular weight = 209.26; 263.32 (trihydrate); Boiling point = 775°C (sublimes); Freezing/Melting point = 400−500°C (decomposition). Hazard Identification (based on NFPA-704 M Rating System): Health 3, Flammability 0, Reactivity 1. Insoluble in water.

Rhodium trichloride trihydrate, $RhCl_3 \cdot xH_2O$, is a deep-red crystalline solid. Soluble in water.

Potential Exposure: Compound Description: Mutagen. Rhodium trichloride is used in hydrosilylation, hydrogenation, carbonylation, oxidation, arylation. See also "Rhodium Metal." In plating operations and in catalyst preparation, the metal will be used as the trichloride.

Incompatibilities: Strong oxidizers.

Permissible Exposure Limits in Air

OSHA PEL: 0.001 mg[Rh]/m^3 TWA.

NIOSH REL: 0.001 mg[Rh]/m^3 TWA.

ACGIH TLV®[1]: 0.01 mg[Rh]/m^3; not classifiable as a human carcinogen.

No TEEL available.

DFG MAK: Carcinogen Category 3B.

NIOSH IDLH: 2 mg[Rh]/m^3.

Australia: TWA 1 mg[Rh]/m^3, 1993; Belgium: TWA 1 mg [Rh]/m^3, 1993; Denmark: TWA 0.001 mg[Rh]/m^3, 1999; Finland: TWA 0.001 mg[Rh]/m^3, [skin], 1999; France: VME 1 mg[Rh]/m^3, 1993; Norway: TWA 0.1 mg[Rh]/m^3, 1999; the Philippines: TWA 0.1 mg[Rh]/m^3, 1993; Switzerland: MAK-W 0.1 mg[Rh]/m^3, 1999; United Kingdom: TWA 0.001 mg[Rh]/m^3; STEL 0.003 mg[Rh]/m^3, 2000; Argentina, Bulgaria, Columbia, Jordan, South Korea, New Zealand, Singapore, Vietnam: ACGIH TLV®: Not classifiable as a human carcinogen. Several states have guidelines or standards for rhodium in ambient air[60] ranging from 0.16 μg/m^3 (Virginia) to 2.0 μg/m^3 (Connecticut) to 10.0 μg/m^3 (North Dakota) to 24.0 μg/m^3 (Nevada).

Determination in Air: Use NIOSH II(3), Method #S-189.

Routes of Entry: Inhalation, ingestion, skin and/or eye contact.

Harmful Effects and Symptoms

Short Term Exposure: May cause eye irritation or serious damage. In animals: irritation of the eyes; central nervous system damage. LD_{50} = (oral-rat) 1302 mg/kg.

Long Term Exposure: May cause central nervous system damage. Tumorigenic and mutagenic effects have been reported in experimental lab animals.

Points of Attack: Eyes, central nervous system.

Medical Surveillance: Consider the point of attack in pre-placement and periodic physical examinations.

First Aid: If this chemical gets into the eyes, remove any contact lenses at once and irrigate immediately for at least 15 min, occasionally lifting upper and lower lids. Seek medical attention immediately. If this chemical contacts the skin, remove contaminated clothing and wash immediately with soap and water. Seek medical attention immediately. If this chemical has been inhaled, remove from exposure, begin rescue breathing (using universal precautions, including resuscitation mask) if breathing has stopped and CPR if heart action has stopped. Transfer promptly to a medical facility. When this chemical has been swallowed, get medical attention. Give large quantities of water and induce vomiting. Do not make an unconscious person vomit.

Personal Protective Methods: Wear protective gloves and clothing to prevent any reasonable probability of skin contact. Safety equipment suppliers/manufacturers can provide recommendations on the most protective glove/clothing material for your operation. All protective clothing (suits, gloves, footwear, headgear) should be clean, available each day, and put on before work. Contact lenses should not be worn when working with this chemical. Wear dust-proof chemical goggles and face shield unless full face-piece respiratory protection is worn. Employees should wash immediately with soap when skin is wet or contaminated. Provide emergency showers and eyewash.

Respirator Selection: *Up to 0.5 mg/m^3:* Qm (APF = 25) (any quarter-mask respirator). *Up to 1 mg/m^3:* 95XQ (APF = 10) [any particulate respirator equipped with an N95, R95, or P95 filter (including N95, R95, and P95 filtering face-pieces) except quarter-mask respirators. The following filters may also be used: N99, R99, P99, N100, R100, P100] or Sa (APF = 10) (any supplied-air respirator). *Up to 2.5 mg/m^3:* Sa:Cf (APF = 25) (any supplied-air respirator operated in a continuous-flow mode) or PaprHie (APF = 25) (any powered, air-purifying respirator with a high-efficiency particulate filter). *Up to 5 mg/m^3:* 100F (APF = 50) (any air-purifying, full-face-piece respirator with an N100, R100, or P100 filter) or SaT: Cf (APF = 50) (any supplied-air respirator that has a tight-fitting face-piece and is operated in a continuous-flow mode) or PaprTHie (APF = 50) (any powered, air-purifying respirator with a tight-fitting face-piece and a high-efficiency particulate filter) or SCBAF (APF = 50) (any self-contained breathing apparatus with a full face-piece) or SaF (APF = 50) (any supplied-air respirator with a full face-piece). *Up to 100 mg/m^3:* Sa: Pd,Pp (APF = 1000) (any supplied-air respirator operated in a pressure-demand or other positive-pressure mode). *Emergency or planned entry into unknown concentrations or IDLH conditions:* SCBAF: Pd,Pp (APF = 10,000) (any NIOSH/MSHA- or European Standard EN 149-approved self-contained breathing apparatus that has a full face-piece and is operated in a pressure-demand or other positive-pressure mode) or SaF: Pd,Pp: ASCBA (APF = 10,000) (any supplied-air respirator that has a full face-piece and is operated in a pressure-demand or other positive-pressure mode in combination with an auxiliary, self-contained breathing apparatus operated in a pressure-demand or other positive-pressure mode). *Escape:* 100F (APF = 50) (any air-purifying, full-face-piece respirator with an N100, R100, or P100 filter) or SCBAE (any appropriate escape-type, self-contained breathing apparatus).

Storage: Color Code—White: Corrosive or Contact Hazard; Store separately in a corrosion-resistant location. Prior to working with this chemical you should be trained on its proper handling and storage. Store in tightly closed containers in a cool, well-ventilated area away from strong oxidizers. Where possible, automatically transfer material from drums or other storage containers to process containers.

Shipping: Rhodium trichloride is not specifically cited in DOT regulations. However it is covered under "CORROSIVE, ACIDIC, INORGANIC, N.O.S."[19] and labeled "CORROSIVE." It falls in Hazard Class 8 and Packing Group III.

Spill Handling: Evacuate persons not wearing protective equipment from area of spill or leak until cleanup is complete. Remove all ignition sources. Collect powdered material in the most convenient and safe manner and deposit in sealed containers. Ventilate area after cleanup is complete. It may be necessary to contain and dispose of this chemical as a hazardous waste. If material or contaminated runoff enters waterways, notify downstream users of potentially contaminated waters. Contact your local or federal environmental protection agency for specific recommendations. If employees are required to clean up spills, they must be properly trained and equipped. OSHA 1910.120(q) may be applicable.

Fire Extinguishing: This chemical is a combustible solid. Use dry chemical, carbon dioxide, water spray, or alcohol foam extinguishers. Poisonous gases are produced in fire, including chlorides. If material or contaminated runoff enters waterways, notify downstream users of potentially contaminated waters. Notify local health and fire officials and pollution control agencies. From a secure, explosion-proof location, use water spray to cool exposed containers. If cooling streams are ineffective (venting sound increases in volume and pitch, tank discolors, or shows any signs of deforming), withdraw immediately to a secure position. If employees are expected to fight fires, they must be trained and equipped in OSHA 1910.156. The only respirators recommended for firefighting are self-contained breathing apparatuses that have full face-pieces and are operated in a pressure-demand or other positive-pressure mode.

Disposal Method Suggested: Recovery and reclaiming wherever possible in view of high economic value. See "Rhodium Metal."

References

US Environmental Protection Agency. (October 31, 1985). *Chemical Hazard Information Profile: Rhodium Trichloride.* Washington, DC: Chemical Emergency Preparedness Program

Riot Control Agents/Tear Gas

Highly irritating agents generally used by law enforcement for crowd control and by individuals for protection. Not designed to kill or cause serious injury, these agents are meant to weaken a victim, and render them helpless and compliant.

2-Chloroacetophenone (CN) see C:0750

o-Chlorobenzylidenemalononitrile (CS)* see C:0810

Chloropicrin (PS) see C:0980

CNB (a mixture of 10% 2-Chloroacetophenone (CN), 45% Benzene, 45% Carbon tetrachloride)

CNC (a mixture of 30% 2-Chloroacetophenone (CN) in Chloroform) see C:0810

CNS (a mixture of 23% 2-Chloroacetophenone (CN), 38% Chloropicrin (PS), 38.4% Chloroform) see C:0810 and C:0980

*CS1, CS2, CSX, all contain *o*-Chlorobenzylidenemalononitrile (CS) as agent, see C:0810

Ricin   R:0135

Molecular Formula: None

Synonyms: African coffee tree; Castor; Castor bean; Castor oil; Ricine; Ricino (Spanish); Ricinus lectin; Ricinus agglutinin; Ricinus toxin; Lectin isolated from seeds of the castor bean; Ricinus communis protein/polypeptide; Steadfast

CAS Registry Number: 9009-86-3; (*alt.*) 9067-26-9; 96638-28-7 (Ricin, A chain)

RTECS Number: VJ2625000

UN/NA & ERG Number: UN3462 (Toxin, extracted from living sources, n.o.s.)/153

EC Number: None assigned.

Regulatory Authority and Advisory Bodies

Report any release of WMD to National Response Center 1-800-424-8802.

European/International Regulations: Hazard Symbol: T; Risk phrases: R26/27/28; R42/43; Safety phrases: S23; S28; S36/37; S45 (see Appendix 4).

WGK (German Aquatic Class) No value assigned to these CAS numbers. *However, based on toxicity the following may apply:* 3—Severe hazard to waters.

Description: Ricin is a lethal, delayed-action cytotoxin; it is persistent in the environment. Ricin is a white powder. Molecular weight = (approx.) 64,000−66,000 Da; Volatility = negligible; Vapor pressure = negligible at 20°C; Boiling point = decomposes; Ricin is detoxified in 10 min at 80°C/176°F or in 1 h at 50°C/122°F at pH 7.8. Hazard Identification (based on NFPA-704 M Rating System): Health 4; Flammability 1; Reactivity 0.[CDC] Ricin is stable under ambient conditions and destroyed by heat and contact with solution household bleach. Michaelis constant $(K_M) = 0.1$ imol/L for ribosomes; Enzymatic constant $(K_{cat}) = 1500$/min. The toxin is soluble in water.

Detection: Various tests for the detection of Ricin are available, including "Pro-Strips™" from Avant Technologies, the BIO-911™ test strip developed by Osborn Scientific Group, Lakeside, AZ. This one-step test is reported to detect, within minutes, the presence of minute quantities of the toxin (less than 50 ng); the freeze-dried assay kit to detect crude Ricin called Ruggedized Advanced Pathogen Identification Device (Rapid)™, Rapid LT™, and Razor™ instrument developed by Idaho Technologies, Inc. Ricin is detectable in urine, plasma, and environmental samples. *Instant Check Abrin/Ricin Detection Kit*, Catalog Number: IC-RA-003-10. Kit Contents: 10 test units each test contains 2 tests (1 Abrin and 1 Ricin) per unit, lyophilized reagent, reconstitution buffer, wash buffer, transfer pipettes, instruction booklet. EY Laboratories, Inc., 107 N. Amphlett Blvd, San Mateo, CA 94401, USA, Toll Free (North America) 1-800-821-0044; Telephone: 1-650-342-3296, Option 2 or 3 Fax: +1 (650) 342 2648. Also, in 2009, researchers at Yeshiva University's Albert Einstein College of Medicine have developed an accurate test. Polymerase Chain Reaction (PCR) can detect castor bean DNA in most Ricin preparations. See also "Water Detection," below.

History of the material: Ricin is a lectin—a plant glycoprotein that binds and agglutinates animal cells. Ricin comes from the seeds (or beans) of the castor bean plant, *Ricinus communis.* The beans are processed to form castor oil, formerly used as a laxative and now used in industry as a lubricant. Annual worldwide processing of castor beans is approximately 1 million tons to produce castor oil; waste mash from this production is 3–5% Ricin by weight. Large scale production and use of Ricin by an enemy is fairly difficult. It is estimated that a ton of Ricin would be required to produce a mass casualty biological weapon; therefore an enemy would have to produce a very large quantity in order to cover a large area of a battlefield. For centuries farmers have known the dangers to farm animals that have eaten castor beans. In fact, people found castor beans in 6000-year-old Egyptian tombs. Castor oil has been used for centuries as a medicine, and scientists have known about Ricin's effect on stopping the body from making proteins since 1800. During WWII the United States and Britain worked together on building bombs carrying Ricin, but the only reported military use was by Iraq in their war with Iran. The US Army discovered large stores of Ricin in Al Qaeda caves in Afghanistan. This relatively inexpensive, accessible, natural source allows easy preparation of large quantities of Ricin; therefore, there is little motivation to produce it synthetically. Large-scale production of Ricin by recombinant DNA techniques is probably possible. Certain governments have successfully used Ricin to murder political enemies. In 1978, Georgi Markov, a Bulgarian journalist living in London, was assassinated when a Ricin pellet was injected into his leg using a specialized gun disguised as an umbrella. In February 2004, Ricin was found in the Senate mail room on Capitol Hill in Washington, DC. No one became ill from that exposure. Since 1978, worldwide, there have been more than a dozen other known incidents involving Ricin. Although Ricin has the potential for use as a terrorist weapon, it is also being investigated as a treatment for cancer and AIDS.

Potential Exposure: Potentially Fatal. Mutagen, Teratogen. Ricin, a protein found in castor beans, is one of the most incredibly potent and lethal substances known to humankind—500 μg (an amount capable of fitting on the head of a pin)—is capable of killing an adult within a few days. The castor plant and castor beans are important industrial plants and vast amounts of castor plants and beans are grown commercially. The castor plant, originally native to Africa, is an attractive plant that can be found in people's yards as ornamentals. If the flowers are removed before they produce seeds, the risk of accidents can be averted. Ricin poisoning can occur following accidental ingestion of castor beans. In some countries castor beans are used as beads on necklaces; both children and adults have eaten just a few beans and died from the poison. This invasive 8–10 foot tall plant can be found in more than 25 US states and other places, including Australia. The plants die in freezing weather. The beans are turned into important and valuable industrial products. After the castor oil is squeezed from the castor beans, about 5% of what's left is Ricin. Although people have used weak castor oil as a laxative, purgative, and general folk-medicine "cure all," pure Ricin kills in a few days. Ricin get into the body's cells and inhibits protein synthesis. Once this essential mechanism is halted, within hours the body's system begin to shut down and the body begins to die. Ricin poisoning is not contagious and does not spread from person to person. Because Ricin is not a living thing but rather a chemical made by living things, it probably would NOT make as effective a military weapon as viruses or bacteria like the ebola virus or anthrax. Nevertheless, it would make an effective terrorist weapon because it is easy to make and easy to use. It is estimated that a ton of Ricin would be required to produce a mass casualty biological weapon. With no known antidote*, vaccine, or other effective therapy available for Ricin poisoning, the threat of this agent being released into the environment as an aerosol, or added to the food or water supply, could be catastrophic. Ricin is extremely hazardous when freeze-dried, forming a light, easily-dispersed powder. For this reason, research laboratory personnel are usually directed to avoid freeze-drying Ricin.

*Various public and private sector organizations are working on the development of antibody therapy that shows promise in neutralizing Ricin in animals. If successful, this therapy could result in a vaccine for humans and potentially reduce the threat of Ricin being used as a terrorist weapon.

Incompatibilities: Product is considered stable at ambient temperature. Ricin and other protein toxins can be destroyed by exposure to 0.1% sodium hypochlorite solution (household bleach) for 10 min. The toxic portion of Ricin is heat stable at ambient temperatures and begins to decompose at 60°C/140°F. At 80°C/176°F, most of the toxicity is lost in about 30 min. When heated to decomposition, Ricin emits toxic nitrogen oxides. Avoid contact with strong oxidizers.

Permissible Exposure Limits in Air:
Protective Action Criteria (PAC)[CDC]
TEEL-0: 0.025 mg/m^3
PAC-1: 0.075 mg/m^3
PAC-2: 0.5 mg/m^3
PAC-3: 1.5 mg/m^3
ChemWatch®:   0.00006 mg/m^3 Ceiling   limit (sensitizer).[CDC]

Determination in Air: No NIOSH or OSHA methods available.

Permissible Concentration in Water: Unknown.

Determination in Water:

Detecting Ricin in Water; Summary of Results

Technology	Contaminant Presence/ Absence	False Pos (+)/Neg(−) Responses	Consistency	Lowest Detectable Conc.
BADD™ Test Strips	9/21	0/0	100%	20 mg/L
BioVerify Test Kits	15/22	0/3	97%	0.0005 mg/L
RAMP® Test	12/15	0/0	100%	5 mg/L
BioThreat Alert® Test Strips*	15/15	2/1	100%	0.035 mg/L
Enzyme Linked Immunosorbent Assay (ELISA)	12/15	0/0	100%	0.0075 mg/L
QTL Biosensor	12/15	2/2	90%	0.25 mg/L

*The most accurate results for Ricin using the Bio-Threat® Alert test strips.

Source: EPA/Homeland Security (2004−2006): see full report at http://www.epa.gov/nhsrc/tte_immtestkitpandb.html

(Tetracore, Inc.) detected 15/15 for Ricin (2004−2006).

Routes of Entry: Can be absorbed by all routes of exposure. Ricin normally enters the body by ingestion. Aerosolized Ricin would enter the body by inhalation. The toxin attaches to cell surfaces of a variety of tissues, particularly the stomach lining if ingested or the moist, upper respiratory tissues if inhaled.

Harmful Effects and Symptoms

The symptoms depend on purity of the Ricin, the level of exposure, and route of exposure (inhalation, ingestion, or injection), the dose received. The LD_{50} = (oral-ingestion) 1−20 mg/kg (the equivalent of 8 castor beans); (human-injection) 1−1.75 µg/kg; (human-inhalation) 21−42 µg/kg. As little as 500 µg of Ricin (about what might fit on the head of a pin; 1/228th of an aspirin tablet) may be enough to kill an adult. When Ricin toxins get inside the victim's body, they block protein synthesis (the body's ability to regenerate protein). Initial symptoms usually appear between 6−10 h and 3 days. Clinical signs may appear as early as 45 min after ingestion if the victim has an empty stomach. Death can occur within 36−72 h of exposure, depending on the route of exposure and upon the dose received. Death from Ricin poisoning could take place within 36−48 h of exposure, whether by injection, ingestion, or inhalation. Victims may linger for 10−12 days before death or recovery, also depending upon the level of exposure. However, if a person lives longer than 5 days after Ricin poisoning, without complications, they are less likely to die.

Short Term Exposure: The following lists do not convey prioritization or indicate specificity.

Inhalation: Within a few hours of inhaling significant amounts of Ricin, the likely symptoms would be respiratory distress, difficult or labored breathing, shortness of breath, chest tightness, hypoxemia, fever, cough, nausea, sweating, aching muscles. Heavy sweating may follow as noncardiogenic pulmonary edema develops.* This would make breathing even more difficult, and the skin might turn blue.

Excess fluid in the lungs would be diagnosed by X-ray or by listening to the chest with a stethoscope. Finally, low blood pressure, blue skin, multisystem organ failure, respiratory failure may occur, and possible death.

Skin: It is uncertain if Ricin can be absorbed through the skin. It is generally believed that the risk of toxicity is low from contact with unbroken skin; however, Ricin may be absorbed through irritated, damaged, abraded, wounded, or injured skin; or through normal skin when Ricin is dissolved in a solvent carrier. If enough ricin gets through the skin, the symptoms will be similar to those described for ingestion. If Ricin is injected, the muscles and lymph nodes near the injection site would die. This could lead to liver and kidney failure and death. *Eyes:* Tearing, swelling of the eyelids, pain, redness, corneal injury. Urinary excretion of Ricin is probably slow and limited with the estimated half-life being about 8 days. LD_{50} = (humans) 1 mg/kg; a single seed can be fatal. LD_{50} = (oral-mice) about 3 µg/kg by injection or aerosol.

Ingestion: Generally within a few hours symptoms will appear, including nausea, vomiting, abdominal pain and cramping, diarrhea (possibly bloody), lowered blood pressure, hallucinations, and bloody urine, gastrointestinal bleeding, low or no urinary output, dilation of the pupils, fever, thirst, sore throat, headache, vascular collapse, and shock. Severe vomiting and diarrhea may result from severe dehydration and multisystem organ failure. In a few days, organs (liver, spleen, and kidneys) might stop working, and possible death.

*Ricin-induced pulmonary edema would be expected to occur much later (1−3 days postexposure) compared to other agents such as phosgene (about 6 h postexposure).

Late phase complications: Ricin has cell-killing (cytotoxic) effects on the liver, central nervous system, kidneys, and adrenal glands, typically 2−5 days after exposure. The patient may be asymptomatic (showing no symptoms of exposure) during the preceding 1−5 days.[CDC]

Pulmonary edema would be expected to occur much later (1−3 days postexposure) compared to that induced by other agents such as phosgene (about 6 h postexposure).

Long Term Exposure: Repeated exposures to Ricin may cause allergic/asthma-like symptoms with congestion of nose and throat; itchy, watery eyes; hives; tightness of the chest; and in acute cases, wheezing. May affect kidneys, liver, pancreas. Based on animal studies, Ricin may be capable of causing severe lung damage in humans. Information about carcinogenicity, developmental toxicity, or reproductive toxicity from chronic or repeated exposure to Ricin is unknown at this time.

Points of Attack: Lungs, eyes, skin (primarily through cuts and bruises). The risk of toxicity from unbroken skin exposure to Ricin is low but there is still a danger of allergic skin reactions.

Medical Surveillance: Unfortunately Ricin symptoms resemble those that are similar to other diseases, especially the common flu. Laboratory testing might include metabolic acidosis, increased liver function tests, increased renal function tests, hematuria, leukocytosis (2- to 5-fold higher than normal).

First Aid: No antidote exists for Ricin. Make all exposed people go to the nearest hospital emergency department. Make all potentially exposed people shower and change clothes. In case you have gotten Ricin on your skin or your clothing, immediately shower and change clothes. See a doctor as soon as possible: since it can take only a day and a half for symptoms to appear, all exposed persons should get to a doctor that same day. If exposure includes contact with Ricin, remove it *off or out* of the body as quickly as possible. Treatment needs to be provided in a hospital setting. Make the doctors and nurses aware of the potential for exposure so that they can protect themselves; and provide the most appropriate treatments and therapies. If you have to wait for medical assistance, make the victim comfortable. If Ricin has been ingested, the airway must be secured and stomach pumping (gastric lavage) may be considered. Vigorous gastric lavage is recommended only if it can be done shortly after ingestion (generally within 1 h). Risk of aggravating injury to the lining of the gastrointestinal (GI) tract must be considered.

To minimize the effects of the poisoning and to keep the patient stable, Ricin poisoning is treated through supportive therapy. The types of supportive medical care would depend on several factors, such as the route by which victim(s) were poisoned (i.e., whether poisoning was by inhalation, ingestion, eye, or skin exposure). Do not induce vomiting. Rinse mouth, flush stomach with activated charcoal* (if the Ricin has been very recently ingested), wash out the victim's eyes with water, especially if the eyes are red and irritated. Supportive care could include intravenous fluid input and support of circulation and respiration; giving medications to treat conditions such as seizure and low blood pressure. Fluid input is critical, as fluid losses of up to 2−1/2 L are probable. If individual is drowsy or unconscious, do not give anything by mouth. In the event of vomiting, lean patient forward or place on left side (head-down position, if possible) to maintain open airway and prevent aspiration. Fluid and electrolyte balance should be monitored and restored if abnormal. Early and aggressive IV fluid and electrolyte replacement. If the victim's lungs fill with fluid, you administer oxygen if his breathing is difficult or labored.

*Superactivated charcoal may be of little value for large molecules such as Ricin.[USAMRICD]

Decontamination: When Ricin symptoms appear and you know the location of the attack, the area should be decontaminated with responders wearing Level A (fully encapsulated suit with SCBA). Otherwise, stay away. Move as fast as possible; extra minutes before decontamination might make a big difference. If the Ricin attack was aerosolized (by small particles floating in the air), and you have the equipment (*this is very important*), then you have to decontaminate as soon as possible. If you do not have the equipment and training, do not enter the hot zone to rescue and decontaminate victims. Even if you think you are not contaminated, be sure to thoroughly shower and change clothes as soon as you can after the incident. If possible, place all clothing in a labeled durable 6-mil polyethylene bag. If the victim cannot move, decontaminate, if possible, without touching and without entering the hot zone. To prevent spreading the agent, be certain that victims are decontaminated as much as possible before allowing them to leave the area. During the decontamination process, be careful not to break the patient's/victim's skin, and take care to cover all open wounds. The CDC recommends a decontamination procedure solution of detergent and water (with a pH value in the alkaline range of more than 8 but less than 10.5). Soft brushes should be available to remove contamination from the personal protective equipment. Also, recommended for cleanup of people and surfaces is household bleach. Use a fresh (made daily) solution of 0.5% sodium hypochlorite [diluted household bleach (10%, or one part bleach to nine parts water)]. Contact time: 15 min. Do not get bleach solution in the victim's eyes, open wounds (especially of the abdomen, spinal cord, or brain), or mouth. Wash off the diluted bleach solution after 15 min. Labeled, durable 6-mil polyethylene bags should be available for disposal of clothing and contaminated personal protective equipment. In the absence of pH adjusted solutions, wash the victim with lots of warm water with soap. Use clean water from any source; if possible, use a hose (spray or fog to prevent injury to the victim) or other system so that you would not have to touch the victim; do not even wait for soap or for the victim to remove clothing (at least down to undergarments), begin washing immediately. Immediate flush the eyes with water for at least 15 min. Wash−strip−wash−evacuate upwind and uphill: the idea is to immediately wash with water, then have the victim (not the responder) remove all the victim's clothing, then wash again (with soap if available), and then move away from the hot zone in an upwind and uphill direction.

Personal Protective Methods: Level A (Fully encapsulated suit with SCBA) protective suit when entering an area with an unknown contaminant or when entering an area where the concentration of the contaminant is unknown. Level A (Fully encapsulated suit with SCBA) protection should be used until monitoring results confirm the contaminant and the concentration of the contaminant. Recommended protective clothing and materials include Tychem® BR or Responder® CSM. *Note:* Safe use of protective clothing and equipment requires specific skills developed through training and experience.

Respirator Selection: Pressure demand, self-contained breathing apparatus [SCBA; Chemical, Biological, Radiological, Nuclear (CBRN), if available] is recommended in response to nonroutine emergency situations.

Storage: Color Code—Blue: Health Hazard/Poison (Toxic): Store in a secure, locked poison location. Store in a cool, dry, well-ventilated area and away from food stuff containers.

Shipping: Toxin, extracted from living sources, n.o.s. requires a label of "POISON (TOXIC)." This falls in Hazard Class: 6.1(a) Packing Group: III,[NIOSH] but you may want to treat it as Packing Group I.

Spill Handling: You must be careful! Avoid generating dust. Since a Ricin attack may be by small particles of Ricin floating in the air (aerosol), do not breathe it or get it on you. Remember that the victim's clothes or body may have Ricin; touch them and you can become a victim. If you think Ricin particles might be in the area, stay out until you are equipped with Level A (Fully encapsulated suit with SCBA), proper personal protective equipment (PPE), including protective clothing and respiratory protection. Shut off heating, ventilation, and air conditioning (HVAC) systems to prevent the tiny aerosolized (floating in the air) Ricin droplets from spreading throughout the building. Keep the public away. Immediately call for medical assistance. For those contaminated by Ricin, do not allow anyone to leave the hot zone. The Ricin chemical does not dissipate quickly, and "survives" in clothing and bedding for long periods of time. Consequently, people as well as "things" must be quarantined and decontaminated. Do not allow anyone to leave until medical people examine them. Do not breathe, touch, or eat anything that might be contaminated with Ricin. Notify the federal authorities, local health, and pollution/environmental agencies. See Decontamination.

Fire Extinguishing: Foam, dry chemical powder, BCF [bromochlorodifluoromethane (Halon 1211)] (where regulations permit), carbon dioxide, water spray or fog—large fires only. When heated to decomposition, Ricin emits toxic nitrogen oxides.

Disposal Method Suggested: Do not clean up or dispose of Ricin, except under supervision of a specialist.

References
CDC/NIOSH. *The Emergency Response Safety and Health Database.* <http://www.cdc.gov/NIOSH/ershdb/EmergencyResponseCard_29750015.html>
Defence Research and Development Canada. (2010). *Unique Partnership Provides Promising Lead on Medical Countermeasures against Ricin.* Suffield, AB, Canada. <http://www.css.drdc-rddc.gc.ca/crti/invest/stories-exemplaires/02_0007ta-eng.asp>
Franz, D. R. (1997). *Defense against Toxin Weapons, Revised Edition.* Fort Detrick, MD: US Army Medical Research and Material Command, US Army Medical Research Institute of Infectious Diseases
American Medical Association's. (2002). *Quick Reference Guide to Biological Weapons.* Chicago, IL
US Army Medical Research Institute of Infectious Diseases. (2001). *Medical Management of Biological Casualties Handbook* (4th ed.). Frederick, MD: Ft. Detrick
Tebbett, I., & Grundmann, O. *Forensics Magazine: Ricin on the Rise: Are We Prepared?* <http://www.forensicmag.com/article/ricin-rise-are-we-prepared?page=0,1>

Ronnel   R:0140

Molecular Formula: $C_8H_8Cl_3O_3PS$
Common Formula: $Cl_3C_6H_2OP(S)(OCH_3)_2$
Synonyms: Dermaphos; *O,O*-Dimethyl *O*-2,4,5-trichlorophenyl phosphorothioate; *O,O*-Dimethyl *O*-(2,4,5-trichlorophenyl) thiophosphate; Dimethyl trichlorophenyl thiophosphate; *O,O*-Dimethyl-*O*-(2,4,5-trichlorphenyl)-thionophosphat (German); Dow ET 14; Dow ET 57; Ectoral; ENT 23,284; ET 14; ET 57; Etrolene; Fenchlorfos; Fenchlorophos; Fenchlorphos; Karlan; Korlan; Korlane; Nanchor; Nanker; Nankor; Phosphorothioic acid, *O,O*-dimethyl *O*-(2,4,5-trichlorophenyl) ester; Thiophosphate de *O,O*-dimethyle et de *O*-(2,4,5-trichlorophenyle) (French); Trichlorometafos; 2,4,5-Trichlorophenol, *O*-ester with *O,O*-dimethyl phosphorothioate; *O*-(2,4,5-Trichlor-phenyl)-*O,O*-dimethyl-monothiophosphat (German); Trolen; Trolene; Viozene
CAS Registry Number: 299-84-3
RTECS® Number: TG0525000
UN/NA & ERG Number: UN2783 (organophosphorus pesticides, solid, toxic)/152
EC Number: 206-082-6 [*Annex I Index No.:* 015-052-00-X]
Regulatory Authority and Advisory Bodies
US EPA, FIFRA 1998 Status of Pesticides: Canceled.
Air Pollutant Standard Set. See below, "Permissible Exposure Limits in Air" section.
US DOT 49CFR172.101, Inhalation Hazard Chemical as organophosphates.
European/International Regulations: Hazard Symbol: Xn, N; Risk phrases: 21/22; R50/53; Safety phrases: S2; S25; S36/37; S60; S61 (see Appendix 4).
WGK (German Aquatic Hazard Class): No value assigned.
Description: Ronnel is a white to light tan crystalline solid. Molecular weight = 321.54; Specific gravity (H_2O:1) = 1.48 at 25°C; Boiling point = (decomposes); Freezing/Melting point = 41°C. Practically insoluble in water; solubility = 0.004% at 25°C.
Potential Exposure: Compound Description: Agricultural Chemical; Mutagen; Reproductive Effector. Those involved in manufacture, formulation, and application of this insecticide for farm (livestock) and household uses.
Incompatibilities: Strong oxidizers. Store at temperatures not exceeding 25–30°C.
Permissible Exposure Limits in Air
OSHA PEL: 15 mg/m³ TWA.
NIOSH REL: 10 mg/m³ TWA.
ACGIH TLV®[1]: 5 mg/m³ TWA measured as inhalable fraction and vapor; not classifiable as a human carcinogen; BEI_A issued for Acetylcholinesterase-inhibiting pesticides.

NIOSH IDLH: 300 mg/m^3.
No TEEL available.
Australia: TWA 10 mg/m^3, 1993; Belgium: TWA 10 mg/m^3, 1993; Denmark: TWA 5 mg/m^3, 1999; France: VME 10 mg/m^3, 1999; Norway: TWA 5 mg/m^3, 1999; the Philippines: TWA 10 mg/m^3, 1993; Russia: STEL 0.3 mg/m^3, 1993; the Netherlands: MAC-TGG 10 mg/m^3, 2003; United Kingdom: TWA 10 mg/m^3, 2000; Argentina, Bulgaria, Columbia, Jordan, South Korea, New Zealand, Singapore, Vietnam: ACGIH TLV®: not classifiable as a human carcinogen. Several states have set guidelines or standards for ronnel in ambient air[60] ranging from 100 μ/m^3 (North Dakota) to 160 μg/m^3 (Virginia) to 200 μg/m^3 (Connecticut) to 238 μg/m^3 (Virginia).

Determination in Air: Use NIOSH Analytical Method (IV) Method #5600, Organophosphorus Pesticides; OSHA Analytical Method PV-2054.

Permissible Concentration in Water: Mexico[35] has set maximum permissible concentration of 50 μg/L in estuaries and 5 μg/L in coastal waters.

Determination in Water: Octanol−water coefficient: Log $K_{ow} = 4.9$.

Routes of Entry: Inhalation, ingestion, skin and/or eye contact. Absorbed through the skin.

Harmful Effects and Symptoms

Short Term Exposure: Irritates the eyes. Organic phosphorus insecticides are absorbed by the skin, as well as by the respiratory and gastrointestinal tracts. They are cholinesterase inhibitors. Symptoms of exposure include headache, giddiness, blurred vision, nervousness, weakness, nausea, cramps, diarrhea, and discomfort in the chest. Signs include sweating, tearing, salivation, vomiting, cyanosis, convulsions, coma, loss of reflexes, and loss of sphincter control.

Long Term Exposure: Cholinesterase inhibitor; cumulative effect is possible. This chemical may damage the nervous system with repeated exposure, resulting in convulsions, respiratory failure. There is limited evidence that ronnel may damage the developing fetus. May cause personality changes, such as depression, anxiety, irritability. High or repeated exposure may cause nerve damage causing weakness, a feeling of "pins and needles" in the arms and legs; and poor coordination.

Points of Attack: Skin, central nervous system, blood plasma.

Medical Surveillance: Before employment and at regular times after that, the following are recommended: plasma and red blood cell cholinesterase levels (tests for the enzyme poisoned by this chemical). If exposure stops, plasma levels return to normal in 1−2 weeks while red blood cell levels may be reduced for 1−3 months. When cholinesterase enzyme levels are reduced by 25% or more below preemployment levels, risk of poisoning is increased, even if results are in lower ranges of "normal." Reassignment to work not involving organophosphate or carbamate pesticides is recommended until enzyme levels recover. If symptoms develop or overexposure occurs, repeat the above tests as soon as possible and get an examination of the nervous system. Also, consider complete blood count. Consider chest X-ray following acute overexposure. Do not drink any alcoholic beverages before or during use. Alcohol promotes absorption of organic phosphates. Examination of the nervous system.

First Aid: If this chemical gets into the eyes, remove any contact lenses at once and irrigate immediately for at least 15 min, occasionally lifting upper and lower lids. Seek medical attention immediately. If this chemical contacts the skin, remove contaminated clothing and wash immediately with soap and water. Speed in removing material from skin is of extreme importance. Shampoo hair promptly if contaminated. Seek medical attention immediately. If this chemical has been inhaled, remove from exposure, begin rescue breathing (using universal precautions, including resuscitation mask) if breathing has stopped and CPR if heart action has stopped. Transfer promptly to a medical facility. When this chemical has been swallowed, get medical attention. Give large quantities of water and induce vomiting. Do not make an unconscious person vomit.

Personal Protective Methods: Wear protective gloves and clothing to prevent any reasonable probability of skin contact. Safety equipment suppliers/manufacturers can provide recommendations on the most protective glove/clothing material for your operation. All protective clothing (suits, gloves, footwear, headgear) should be clean, available each day, and put on before work. Contact lenses should not be worn when working with this chemical. Wear dust-proof chemical goggles and face shield unless full face-piece respiratory protection is worn. Employees should wash immediately with soap when skin is wet or contaminated. Provide emergency showers and eyewash.

Respirator Selection: NIOSH: *Up to 100 mg/m^3:* CcrOv95 (APF = 10) [any air-purifying half-mask respirator with organic vapor cartridge(s) in combination with an N95, R95, or P95 filter. The following filters may also be used: N99, R99, P99, N100, R100, P100] or Sa (APF = 10) (any supplied-air respirator). *Up to 250 mg/m^3:* Sa:Cf (APF = 25) (any supplied-air respirator operated in a continuous-flow mode) or PaprOvHie (APF = 25) (any powered air-purifying respirator with an organic vapor cartridge in combination with a high-efficiency particulate filter). *Up to 300 mg/m^3:* CcrFOv100 (APF = 50) [any air-purifying full-face-piece respirator equipped with organic vapor cartridge(s) in combination with an N100, R100, or P100 filter] or GmFOv100 (APF = 50) [any air-purifying, full-face-piece respirator (gas mask) with a chin-style, front- or back-mounted organic vapor canister having an N100, R100, or P100 filter] or PaprTOvHie* (APF = 50) [any powered, air-purifying respirator with a tight-fitting face-piece and organic vapor cartridge(s) in combination with a high-efficiency particulate filter]; or SCBAF (APF = 50) (any self-contained breathing apparatus with a full face-piece) or SaF (APF = 50) (any supplied-air respirator with a full face-piece). *Emergency or planned entry into unknown concentrations or IDLH conditions:* SCBAF: Pd,Pp (APF = 10,000) (any NIOSH/MSHA- or European Standard EN 149-approved self-contained breathing apparatus that has a full face-piece and is operated in a pressure-demand or other positive-pressure mode) or SaF: Pd,Pp: ASCBA (APF = 10,000) (any supplied-air respirator that has a full face-piece and is operated in a pressure-demand or other positive-pressure mode in combination with an auxiliary, self-contained breathing apparatus operated in a pressure-demand or other positive-pressure mode). *Escape:* GmFOv100 (APF = 50) [any air-purifying, full-face-piece respirator (gas mask) with a chin-style, front- or back-mounted organic vapor canister having an N100, R100, or P100 filter] or SCBAE (any appropriate escape-type, self-contained breathing apparatus).

*Substance reported to cause eye irritation or damage; may require eye protection.

Storage: Color Code—Blue: Health Hazard/Poison: Store in a secure poison location. Prior to working with this chemical you should be trained on its proper handling and storage. Store in tightly closed containers in a cool, well-ventilated area away from strong oxidizers. Where possible, automatically transfer material from drums or other storage containers to process containers. Sources of ignition, such as smoking and open flames, are prohibited where this chemical is handled, used, or stored. Metal containers involving the transfer of this chemical should be grounded and bonded. Wherever this chemical is used, handled, manufactured, or stored, use explosion-proof electrical equipment and fittings.

Shipping: Organophosphorus pesticides, solid, toxic, require a shipping label of "POISONOUS/TOXIC MATERIALS." It falls in Hazard Class 6.1 and Packing Group III.

Spill Handling: Evacuate persons not wearing protective equipment from area of spill or leak until cleanup is complete. Remove all ignition sources. Collect powdered material in the most convenient and safe manner and deposit in sealed containers. Ventilate area after cleanup is complete. It may be necessary to contain and dispose of this chemical as a hazardous waste. If material or contaminated runoff enters waterways, notify downstream users of potentially contaminated waters. Contact your local or federal environmental protection agency for specific recommendations. If employees are required to clean up spills, they must be properly trained and equipped. OSHA 1910.120(q) may be applicable.

Fire Extinguishing: This chemical is a combustible solid. Use dry chemical, carbon dioxide, water spray, or alcohol foam extinguishers. Poisonous gases are produced in fire, including oxides of phosphorus, sulfur dioxide, dimethyl sulfide, and trichlorophenol. If material or contaminated runoff enters waterways, notify downstream users of potentially contaminated waters. Notify local health and fire officials and pollution control agencies. From a secure, explosion-proof location, use water spray to cool exposed containers. If cooling streams are ineffective (venting sound increases in volume and pitch, tank discolors, or shows any signs of deforming), withdraw immediately to a secure position. If employees are expected to fight fires, they must be trained and equipped in OSHA 1910.156. The only respirators recommended for firefighting are self-contained breathing apparatuses that have full face-pieces and are operated in a pressure-demand or other positive-pressure mode.

Disposal Method Suggested: Incineration with added flammable solvent in furnace equipped with afterburner and alkali scrubber.[22] In accordance with 40CFR165, follow recommendations for the disposal of pesticides and pesticide containers. Must be disposed properly by following package label directions or by contacting your local or federal environmental control agency or by contacting your regional EPA office.

References

New Jersey Department of Health and Senior Services. (August 2005). *Hazardous Substances Fact Sheet: Ronnel.* Trenton, NJ

Rotenone   R:0150

Molecular Formula: $C_{23}H_{22}O_6$

Synonyms: Arol gordon dust; Barbasco; (1)Benzopyrano (3,4-b)furo(2,3-H)(1)benzopyran-6(6aH)-one, 1,2,12,12a-tetrahydro-8,9-dimethoxy-2-(1-methylethenyl), [2R-(2a, 6 (a)a,12(a)a)]; Cenol garden dust; Chem fish; Chem-mite; Cube; Cube extract; Cube-pulver; Cube root; Cubor; Curex flea duster; Dactinol; Deril; Derrin; Derris; Dri-kil; ENT 133; Extrax; Fish-tox; Green cross warble powder; Haiari; Liquid derris; Mexide; NCI-C55210; Nicouline; Noxfish; Paraderil; Powder and root; Prentox; Pro-nox fish; Ro-Ko; Ronone; Rotefive; Rotefour; Rotenon; Rotenona (Spanish); Rotessenol; Rotocide; [2R-(2a,6aa,12aa)]-1,2,12,12a-Tetrahydro-8,9-dimethoxy-2-(1-methylethenyl) (1)-benzo-pyrano(3,4-b)furo(2,3-H) (1)-benzopyran-6(6aH)one; Tubatoxin

CAS Registry Number: 83-79-4; *(alt.)* 12679-58-2

RTECS® Number: DJ2800000

UN/NA & ERG Number: UN2588/151

EC Number: 201-501-9 [*Annex I Index No.:* 650-005-00-2]

Regulatory Authority and Advisory Bodies

Carcinogenicity: NCI: Carcinogenesis Studies (feed); equivocal evidence: rat; no evidence: mouse.

US EPA, FIFRA, 1998 Status of Pesticides: Supported.

Air Pollutant Standard Set. See below, "Permissible Exposure Limits in Air" section.

European/International Regulations: Hazard Symbol: T, N; Risk phrases: R25; R36/37/38; R50/53; Safety phrases: S1/2; S22; S24/25; S36; S45; S60; S61 (see Appendix 4).

WGK (German Aquatic Hazard Class): No value assigned.

Description: Rotenone is a colorless to red odorless crystalline solid. Molecular weight = 394.45; Boiling point =

(decomposes below BP); 210–220°C at 0.5 mm; Freezing/Melting point = 165.6°C; Vapor pressure = 1×10^{-6} mmHg at 20°C. Hazard Identification (based on NFPA-704 M Rating System): Health 3, Flammability 1, Reactivity 0. Insoluble in water.

Potential Exposure: Compound Description: Agricultural Chemical; Tumorigen, Mutagen; Reproductive Effector; Human Data; Natural Product; Primary Irritant. Those involved in extraction from derris root, formulation, or application of this insecticide. Rotenone is used as a pharmaceutical and veterinary drug.

Incompatibilities: Strong oxidizers, alkalies.

Permissible Exposure Limits in Air

OSHA PEL: 5 mg/m³ TWA.

NIOSH REL: 5 mg/m³ TWA.

ACGIH TLV®[1]: 5 mg/m³ TWA; not classifiable as a human carcinogen.

NIOSH IDLH: 2500 mg/m³.

Protective Action Criteria (PAC)

TEEL-0: 5 mg/m³

PAC-1: 15 mg/m³

PAC-2: 25 mg/m³

PAC-3: 125 mg/m³

DFG MAK: [skin].

Australia: TWA 5 mg/m³, 1993; Austria: MAK 5 mg/m³, 1999; Belgium: TWA 5 mg/m³, 1993; Denmark: TWA 5 mg/m³, 1999; Finland: TWA 5 mg/m³; STEL 10 mg/m³, 1993; the Philippines: TWA 5 mg/m³, 1993; the Netherlands: MAC-TGG 5 mg/m³, 2003; Turkey: TWA 5 mg/m³, 1993; United Kingdom: TWA 5 mg/m³; STEL 10 mg/m³, 2000; Argentina, Bulgaria, Columbia, Jordan, South Korea, New Zealand, Singapore, Vietnam: ACGIH TLV®: not classifiable as a human carcinogen. Several states have set guidelines or standards for rotenone in ambient air[60] ranging from 1.67 μg/m³ (New York) to 50.0 μg/m³ (Florida and South Carolina) to 50.0–100.0 μg/m³ (North Dakota) to 80.0 μg/m³ (Virginia) to 100.0 μg/m³ (Connecticut) to 119.0 μg/m³ (Nevada).

Determination in Air: Use NIOSH Analytical Method (IV) #5007, Rotenone.

Permissible Concentration in Water: The state of Maine has set a guideline of 4.0 μg/L for rotenone in drinking water.[61]

Determination in Water: Fish Tox = 1.49403000 ppb (HIGH); Octanol–water coefficient: Log K_{ow} = 4.1.

Routes of Entry: Inhalation, ingestion, skin and/or eye contact.

Harmful Effects and Symptoms

Short Term Exposure: Irritates the eyes, skin, and respiratory tract. Eye contact can cause severe irritation and permanent damage. Exposure can cause numbness of the mucous membrane, nausea, vomiting, abdominal pain, muscular tremors, incoherence, colonic convulsions, stupor. May cause a severe drop in blood sugar. Higher exposures can cause pulmonary edema, a medical emergency that can be delayed for several hours. This can cause death.

Long Term Exposure: May affect the liver and kidneys. There is limited evidence that rotenone causes cancer of the liver and breast in animals, and damage to the developing fetus. There is limited evidence that this chemical is stored in breast milk and passed on to nursing infants. Repeated skin contact can cause severe rash. Human Tox = 28.00000 ppb (INTERMEDIATE).

Points of Attack: Central nervous system, eyes, respiratory system, liver, and kidneys.

Medical Surveillance: Consider the points of attack in preplacement and periodic physical examinations. Lung function tests. Blood sugar. Liver function tests.

First Aid: If this chemical gets into the eyes, remove any contact lenses at once and irrigate immediately for at least 15 min, occasionally lifting upper and lower lids. Seek medical attention immediately. If this chemical contacts the skin, remove contaminated clothing and wash immediately with soap and water. Speed in removing material from skin is of extreme importance. Seek medical attention immediately. If this chemical has been inhaled, remove from exposure, begin rescue breathing (using universal precautions, including resuscitation mask) if breathing has stopped and CPR if heart action has stopped. Transfer promptly to a medical facility. When this chemical has been swallowed, get medical attention. Give large quantities of water and induce vomiting. Do not make an unconscious person vomit. Medical observation is recommended for 24–48 h after breathing overexposure, as pulmonary edema may be delayed. As first aid for pulmonary edema, a doctor or authorized paramedic may consider administering a corticosteroid spray.

Personal Protective Methods: Wear protective gloves and clothing to prevent any reasonable probability of skin contact. Safety equipment suppliers/manufacturers can provide recommendations on the most protective glove/clothing material for your operation. All protective clothing (suits, gloves, footwear, headgear) should be clean, available each day, and put on before work. Contact lenses should not be worn when working with this chemical. Wear dust-proof chemical goggles and face shield unless full face-piece respiratory protection is worn. Employees should wash immediately with soap when skin is wet or contaminated. Provide emergency showers and eyewash.

Respirator Selection: *Up to 50 mg/m³:* CcrOv95 (APF = 10) [any air-purifying half-mask respirator with organic vapor cartridge(s) in combination with an N95, R95, or P95 filter]. The following filters may also be used: N99, R99, P99, N100, R100, P100] or Sa (APF = 10) (any supplied-air respirator). *Up to 125 mg/m³:* Sa:Cf (APF = 25) (any supplied-air respirator operated in a continuous-flow mode) or PaprOvHie (APF = 25) (any powered air-purifying respirator with an organic vapor cartridge in combination with a high-efficiency particulate filter). *Up to 250 mg/m³:* CcrFOv100 (APF = 50) [any air-purifying full-face-piece respirator equipped with organic vapor cartridge(s) in combination with an N100, R100, or P100 filter] or GmFOv100 (APF = 50) [any air-purifying, full-face-piece respirator (gas mask) with a chin-style, front- or back-mounted organic vapor canister having an N100, R100, or P100 filter] or PaprTOvHie (APF = 50) [any powered, air-purifying respirator with a tight-fitting face-piece and organic vapor cartridge(s) in combination with a high-efficiency particulate filter] or SaT: Cf (APF = 50) (any supplied-air respirator that has a tight-fitting face-piece and is operated in a continuous-flow mode) or SCBAF (APF = 50) (any self-contained breathing apparatus with a full face-piece) or SaF (APF = 50) (any supplied-air respirator with a full face-piece). *Up to 2500 mg/m³:* Sa: Pd,Pp (APF = 1000) (any supplied-air respirator operated in a pressure-demand or other positive-pressure mode). *Emergency or planned entry into unknown concentrations or IDLH conditions:* SCBAF: Pd,Pp (APF = 10,000) (any NIOSH/MSHA- or European Standard EN 149-approved self-contained breathing apparatus that has a full face-piece and is operated in a pressure-demand or other positive-pressure mode) or SaF: Pd,Pp: ASCBA (APF = 10,000) (any supplied-air respirator that has a full face-piece and is operated in a pressure-demand or other positive-pressure mode in combination with an auxiliary, self-contained breathing apparatus operated in a pressure-demand or other positive-pressure mode). *Escape:* GmFOv100 (APF = 50) [any air-purifying, full-face-piece respirator (gas mask) with a chin-style, front- or back-mounted organic vapor canister having an N100, R100, or P100 filter] or SCBAE (any appropriate escape-type, self-contained breathing apparatus).

Storage: Color Code—Blue: Health Hazard/Poison: Store in a secure poison location. Prior to working with this chemical you should be trained on its proper handling and storage. Store in a refrigerator under an inert atmosphere and protect from prolonged exposure to light. Keep away from oxidizers and alkalis. Where possible, automatically transfer material from drums or other storage containers to process containers. Sources of ignition, such as smoking and open flames, are prohibited where this chemical is handled, used, or stored.

Shipping: Rotenone requires a shipping label of "POISONOUS/TOXIC MATERIALS." It falls in Hazard Class 6.1 and Packing Group II.

Spill Handling: Evacuate persons not wearing protective equipment from area of spill or leak until cleanup is complete. Remove all ignition sources. Dampen spilled material with alcohol to avoid dust, then transfer material to a suitable container. Use absorbent dampened with alcohol to pick up remaining material. Wash surfaces well with soap and water. Collect powdered material in the most convenient and safe manner and deposit in sealed containers. Ventilate area after cleanup is complete. It may be necessary to contain and dispose of this chemical as a hazardous waste. If material or contaminated runoff enters waterways, notify downstream users of potentially contaminated waters. Contact your local or federal environmental protection agency for specific recommendations. If employees are required to clean up spills, they must be properly trained and equipped. OSHA 1910.120(q) may be applicable. Soil Adsorption Index (K_{oc}) = 10,000 (estimate).

Fire Extinguishing: This chemical is a combustible solid. Use dry chemical, carbon dioxide, water spray, or foam extinguishers. Poisonous gases are produced in fire. If material or contaminated runoff enters waterways, notify downstream users of potentially contaminated waters. Notify local health and fire officials and pollution control agencies. From a secure, explosion-proof location, use water spray to cool exposed containers. If cooling streams are ineffective (venting sound increases in volume and pitch, tank discolors, or shows any signs of deforming), withdraw immediately to a secure position. If employees are expected to fight fires, they must be trained and equipped in OSHA 1910.156. The only respirators recommended for firefighting are self-contained breathing apparatuses that have full face-pieces and are operated in a pressure-demand or other positive-pressure mode.

Disposal Method Suggested: Rotenone is decomposed by light and alkali to less insecticidal products. It is readily detoxified by the action of light and air. It is also detoxified by heating; 2 h at 100°C results in 76% decomposition. Oxidation products are probably nontoxic. Incineration has been recommended as a disposal procedure. Burial with lime would also present minimal danger to the environment.[22] In accordance with 40CFR165, follow recommendations for the disposal of pesticides and pesticide containers. Must be disposed properly by following package label directions or by contacting your local or federal environmental control agency or by contacting your regional EPA office.

References

New Jersey Department of Health and Senior Services. (October 2000). *Hazardous Substances Fact Sheet: Rotenone.* Trenton, NJ

S

Saccharin   S:0100

Molecular Formula: $C_7H_5NO_3S$

Synonyms: 550 Saccharine; Anhydro-*o*-sulfaminebenzoic acid; 3-Benzisothiazolinone 1,1-dioxide; 1,2-Benzisothiazolin-3-one, 1,1-dioxide, and salts; 1,2-Benzisothiazol-3(2H)-one 1,1-dioxide; *o*-Benzoic acid sulfimide; *o*-Benzoic sulfimide; Benzoic sulfimide; *o*-Benzoic sulphimide; Benzoicsulphimide; Benzoic sulphinide; *o*-Benzosulfimide; Benzo-2-sulphimide; Benzosulphimide; Benzo-sulphimide; *o*-Benzoyl sulfimide; 1,2-Dihydro-2-keto-benzisosulfonazole; 1,2-Dihydro-2-ketobenzisosulphonazole; 2,3-Dihydro-3-oxobenzisosulfonazole; 2,3-Dihydro-3-oxo-benzisosulphonazole; 1,1-Dioxide-1,2-benzoisothiazol-3(2H)-one; Garantose; Glucid; Gluside; Glycophenol; Glycosin; Hermesetas; 3-Hydroxybenzisothiazole-*S*,*S*-dioxide; Kandiset; Natreen; Neosaccarin; Sacarina (Spanish); Saccharimide; Saccharina; Saccharin acid; Saccharinol; Saccharinose; Saccharol; Saccharose; Saxin; Sucre Edulcor; Sucrette; *o*-Sulfobenzimide; *o*-Sulfobenzoic acid imide; 2-Sulphobenzoic imide; Sweeta; Sycorin; Sykose; Syncal; Zaharina

CAS Registry Number: 81-07-2

RTECS® Number: DE4200000

EC Number: 201-321-0 [1,2-benzisothiazol-3(2H)-one 1,1-dioxide]

Regulatory Authority and Advisory Bodies

Carcinogenicity: IARC: Animal Inadequate Evidence; Human No Adequate Data, *not classifiable as carcinogenic to humans*, Group 3.

US EPA Hazardous Waste Number (RCRA No.): U202.

RCRA, 40CFR261, Appendix 8 Hazardous Constituents.

Reportable Quantity (RQ): 100 lb (45.4 kg).

California Proposition 65 Chemical: Delisted 4/6/01.

EPCRA Section 313 Form R *de minimis* concentration reporting level: 0.1%.

European/International Regulations: not listed in Annex 1.

WGK (German Aquatic Hazard Class): 2—Hazard to waters.

Description: Saccharin is a crystalline solid with a sweet taste (500 times sweeter than sugar). Molecular weight = 183.19; Freezing/Melting point = 229°C (decomposes). Hazard Identification (based on NFPA-704 M Rating System): Health 1, Flammability 1, Reactivity 0. Slightly soluble in water.

Potential Exposure: The information provided has to do, primarily, with the manufacturing of saccharin. Saccharin has been used as a nonnutritive sweetening agent. At one point the US consumption pattern for all forms of saccharin has been estimated as 45% in soft drinks; 18% in tabletop sweeteners; 14% in fruits, juices, sweets, chewing gum, and jellies; 10% in cosmetics and oral hygiene products; 7% in drugs, such as coating on pills; 2% in tobacco; 2% in electroplating; and 2% for miscellaneous uses. Human exposure to saccharin occurs primarily through ingestion because of its use in many dietetic foods and drinks and some personal hygiene products, including toothpastes and mouthwashes. The general public is exposed to saccharin, especially by persons required to reduce sugar intake.

Incompatibilities: Keep away from strong oxidizers.

Permissible Exposure Limits in Air

No standards or TEEL available.

Routes of Entry: Inhalation, ingestion, skin and/or eye contact.

Harmful Effects and Symptoms

Short Term Exposure: May cause skin irritation. Exposure to very high levels can cause loss of appetite, nausea, vomiting, and diarrhea.

Long Term Exposure: Saccharin in very high doses has been shown to cause bladder cancer in male rats. Other animal species have not shown this effect. High exposures in certain susceptible individuals may cause skin allergy.

Points of Attack: Bladder, skin.

Medical Surveillance: Before beginning employment and at regular times after that, for those with frequent or potentially high exposures, the following is recommended: Urine cytology (a special test for abnormal cells in the urine). Evaluation by a qualified allergist.

First Aid: If this chemical gets into the eyes, remove any contact lenses at once and irrigate immediately for at least 15 min, occasionally lifting upper and lower lids. Seek medical attention immediately. If this chemical contacts the skin, remove contaminated clothing and wash immediately with soap and water. Seek medical attention immediately. If this chemical has been inhaled, remove from exposure, begin rescue breathing (using universal precautions, including resuscitation mask) if breathing has stopped and CPR if heart action has stopped. Transfer promptly to a medical facility.

Personal Protective Methods: Wear protective gloves and clothing to prevent any reasonable probability of skin contact. Safety equipment suppliers/manufacturers can provide recommendations on the most protective glove/clothing material for your operation. All protective clothing (suits, gloves, footwear, headgear) should be clean, available each day, and put on before work. Contact lenses should not be worn when working with this chemical. Wear dust-proof chemical goggles and face shield unless full face-piece respiratory protection is worn. Employees should wash immediately with soap when skin is wet or contaminated. Provide emergency showers and eyewash.

Respirator Selection: At any exposure level, use a NIOSH/MSHA- or European Standard EN149-approved supplied-air respirator with a full face-piece operated in the positive-pressure mode, or with a full face-piece, hood, or helmet in the continuous-flow mode; or use a NIOSH/MSHA- or European Standard EN149-approved self-contained breathing apparatus with a full face piece operated in pressure-demand or other positive-pressure mode.

Storage: Color Code—Green: General storage may be used. Prior to working with this chemical you should be trained on its proper handling and storage. Store in tightly closed containers in a cool, well-ventilated area. A regulated, marked area should be established, or stored in compliance with OSHA Standard 1910.1045.

Spill Handling: Evacuate persons not wearing protective equipment from area of spill or leak until cleanup is complete. Remove all ignition sources. Use HEPA vacuum or wet method to reduce dust during cleanup. Do not dry sweep. Collect powdered material in the most convenient and safe manner and deposit in sealed containers. Ventilate area after cleanup is complete. It may be necessary to contain and dispose of this chemical as a hazardous waste. If material or contaminated runoff enters waterways, notify downstream users of potentially contaminated waters. Contact your local or federal environmental protection agency for specific recommendations. If employees are required to clean up spills, they must be properly trained and equipped. OSHA 1910.120(q) may be applicable.

Fire Extinguishing: This chemical may burn but does not easily ignite. Use extinguishers suitable for surrounding fires. Poisonous gases are produced in fire, including oxides of nitrogen and sulfur. If material or contaminated runoff enters waterways, notify downstream users of potentially contaminated waters. Notify local health and fire officials and pollution control agencies. From a secure, explosion-proof location, use water spray to cool exposed containers. If cooling streams are ineffective (venting sound increases in volume and pitch, tank discolors, or shows any signs of deforming), withdraw immediately to a secure position. If employees are expected to fight fires, they must be trained and equipped in OSHA 1910.156. The only respirators recommended for firefighting are self-contained breathing apparatuses that have full face-pieces and are operated in a pressure-demand or other positive-pressure mode.

Disposal Method Suggested: Consult with environmental regulatory agencies for guidance on acceptable disposal practices. Generators of waste containing this contaminant ($\geq$100 kg/mo) must conform with EPA regulations governing storage, transportation, treatment, and waste disposal.

References

Sax, N. I. (Ed.). (1982). *Dangerous Properties of Industrial Materials Report*, 2, No. 6, 18—21

New Jersey Department of Health and Senior Services. (June 2006). *Hazardous Substances Fact Sheet: Saccharin.* Trenton, NJ

Safrole   S:0110

Molecular Formula: $C_{10}H_{10}O_2$
Synonyms: 5-Allyl-1,3-benzodioxole;   Allylcatechol methylene ether; Allyldioxybenzene methylene ether;

1-Allyl-3,4-methylenedioxybenzene; 4-Allyl-1,2-(methylenedioxy)benzene; m-Allylpyrocatechinmethylene ether; Allylpyrocatechol methylene ether; Benzene, 4-allyl-1,2-(methylenedioxy)-; 1,3-Benzodioxole, 5-(2-propenyl)-; 3,4-Methylenedioxy-allylbenzene; 3-(3,4-Methylenedioxyphenyl)prop-1-ene; Methylene ester of allyl-pyrocatechol; 5-(2-Propenyl)-1,3-benzodioxole; Rhyuno oil; Safrene; Safrol (Spanish); Safrole MF; Shikimole; Shikomol

CAS Registry Number: 94-59-7
RTECS® Number: CY2800000
UN/NA & ERG Number: UN3077/171
EC Number: 202-345-4 [*Annex I Index No.:* 605-020-00-9]
Regulatory Authority and Advisory Bodies
Carcinogenicity: IARC: Human No Adequate Data, animal Sufficient Evidence, *possibly carcinogenic to humans,* Group 2B, 1987; NTP: Reasonably anticipated to be a human carcinogen.
Banned or Severely Restricted (US).[13]
US EPA Hazardous Waste Number (RCRA No.): U203.
RCRA, 40CFR261, Appendix 8 Hazardous Constituents.
RCRA 40CFR268.48; 61FR15654, Universal Treatment Standards: Wastewater (mg/L), 0.081; Nonwastewater (mg/kg), 22.
RCRA 40CFR264, Appendix 9; TSD Facilities Ground Water Monitoring List. Suggested test method(s) (PQL µg/L): 8270 (10).
Reportable Quantity (RQ): 100 lb (45.4 kg).
EPCRA Section 313 Form R *de minimis* concentration reporting level: 0.1%.
California Proposition 65 Chemical: Cancer 1/1/88.
Canada, WHMIS, Ingredients Disclosure List Concentration 1.0%.
European/International Regulations: Hazard Symbol: T; Risk phrases: R45; R22; R68; Safety phrases: S53; S45 (see Appendix 4).
WGK (German Aquatic Hazard Class): No value assigned.
Description: Safrole is a colorless to yellow liquid with an odor of camphor or sassafras. Molecular weight = 162.20; Boiling point = 232—234°C; Freezing/Melting point = 11°C; Flash point = 100°C. Hazard Identification (based on NFPA-704 M Rating System): Health 1, Flammability 1, Reactivity 0. Insoluble in water.
Potential Exposure: This compound has been used to flavor beverages and foods. It is also reported to be used in soap manufacture, perfumery, sleep aids, sedatives, and pesticides. The FDA estimated exposure to safrole of the general public through food consumption was extremely low since the Agency prohibited its use in food. Derived from oil of sassafras or camphor. Minimal exposure may occur through the use of edible spices, including nutmeg and mace, which contain low levels of naturally occurring safrole.
Incompatibilities: Oxidizing agents.
Permissible Exposure Limits in Air
Protective Action Criteria (PAC)
TEEL-0: 5 mg/m³
PAC-1: 15 mg/m³

PAC-2: 100 mg/m^3
PAC-3: 500 mg/m^3

Permissible Concentration in Water: The compound does not pose a hazard to the general population through consumption of drinking water because safrole is insoluble in water.

Routes of Entry: Inhalation, ingestion, skin and/or eye contact.

Harmful Effects and Symptoms

Short Term Exposure: Irritates the eyes, skin, and respiratory tract. Contact can cause severe skin and eye irritation. Ingestion may cause abdominal burning, nausea, and vomiting; diarrhea, dysuria, hematuria, unconsciousness, shallow respiration, and convulsions. Inhalation may cause dizziness, rapid and shallow breathing, tachycardia, bronchial irritation, and unconsciousness or convulsions. Other effects may include cyanosis, delirium, and circulatory collapse.

Long Term Exposure: May be a carcinogen in humans.

First Aid: *Skin Contact:* Flood all areas of body that have contacted the substance with water. Do not wait to remove contaminated clothing; do it under the water stream. Use soap to help assure removal. Isolate contaminated clothing when removed to prevent contact by others.[52]

Eye Contact: Remove any contact lenses at once. Immediately flush eyes well with copious quantities of water or normal saline for at least 20–30 min. Seek medical attention.

Inhalation: Leave contaminated area immediately; breathe fresh air. Proper respiratory protection must be supplied to any rescuers. If coughing, difficult breathing, or any other symptoms develop, seek medical attention at once, even if symptoms develop many hours after exposure.

Ingestion: Contact a physician, hospital, or poison center at once. If the victim is unconscious or convulsing, do not induce vomiting or give anything by mouth. Assure that the patient's airway is open and lay him on his side with his head lower than his body and transport immediately to a medical facility. If conscious and not convulsing, give a glass of water to dilute the substance. Vomiting should not be induced without a physician's advice.

Personal Protective Methods: Wear protective gloves and clothing to prevent any reasonable probability of skin contact. Safety equipment suppliers/manufacturers can provide recommendations on the most protective glove/clothing material for your operation. All protective clothing (suits, gloves, footwear, headgear) should be clean, available each day, and put on before work. Contact lenses should not be worn when working with this chemical. Wear dust-proof chemical goggles and face shield unless full face-piece respiratory protection is worn. Employees should wash immediately with soap when skin is wet or contaminated. Provide emergency showers and eyewash.

Respirator Selection: Follow regulations in OSHA 29CFR1910.134 or European Standard EN149. Use a NIOSH/MSHA- or European Standard EN149-approved respirator; or use an approved supplied-air respirator with a full face-piece operated in the positive-pressure mode, or with a full face-piece, hood, or helmet in the continuous-flow mode; or use a NIOSH/MSHA- or European Standard EN149-approved self-contained breathing apparatus with a full face-piece operated in pressure-demand or other positive-pressure mode.

Storage: Color Code—Green: General storage may be used. Prior to working with this chemical you should be trained on its proper handling and storage. Store in a refrigerator or a cool, dry place. Where possible, automatically transfer material from drums or other storage containers to process containers. Sources of ignition, such as smoking and open flames, are prohibited where this chemical is handled, used, or stored. Metal containers involving the transfer of this chemical should be grounded and bonded. Wherever this chemical is used, handled, manufactured, or stored, use explosion-proof electrical equipment and fittings. A regulated, marked area should be established, or stored in compliance with OSHA Standard 1910.1045.

Shipping: Environmentally hazardous substances, solid, n.o.s. require a shipping label of "CLASS 9." It falls in Hazard Class 9 and Packing Group III.[20, 21]

Spill Handling: Evacuate persons not wearing protective equipment from area of spill or leak until cleanup is complete. Remove all ignition sources[52]: Remove all sources of ignition, ventilate the spill area, and use absorbent to pick up spilled material. Follow by washing surfaces well, first with 60–70% ethanol and water, then with soap and water. Collect powdered material in the most convenient and safe manner and deposit in sealed containers. Ventilate area after cleanup is complete. It may be necessary to contain and dispose of this chemical as a hazardous waste. If material or contaminated runoff enters waterways, notify downstream users of potentially contaminated waters. Contact your local or federal environmental protection agency for specific recommendations. If employees are required to clean up spills, they must be properly trained and equipped. OSHA 1910.120(q) may be applicable.

Fire Extinguishing: This chemical is a combustible solid. Use dry chemical, carbon dioxide, water spray, or alcohol foam extinguishers. Poisonous gases are produced in fire. If material or contaminated runoff enters waterways, notify downstream users of potentially contaminated waters. Notify local health and fire officials and pollution control agencies. From a secure, explosion-proof location, use water spray to cool exposed containers. If cooling streams are ineffective (venting sound increases in volume and pitch, tank discolors, or shows any signs of deforming), withdraw immediately to a secure position. If employees are expected to fight fires, they must be trained and equipped in OSHA 1910.156. The only respirators recommended for firefighting are self-contained breathing apparatuses that have full face-pieces and are operated in a pressure-demand or other positive-pressure mode.

Disposal Method Suggested: Consult with environmental regulatory agencies for guidance on acceptable disposal practices. Generators of waste containing this contaminant ($\geq$100 kg/mo) must conform with EPA regulations governing storage, transportation, treatment, and waste disposal. In accordance with 40CFR165, follow recommendations for the disposal of pesticides and pesticide containers. Must be disposed properly by following package label directions or by contacting your local or federal environmental control agency or by contacting your regional EPA office.

Salicylic acid   S:0120

Molecular Formula: $C_7H_6O_3$
Common Formula: $C_6H_4(OH)COOH$
Synonyms: *o*-Hydroxybenzoic acid; 2-Hydroxybenzoic acid; Keralyt; Orthohydroxybenzoic acid; Retarder W; SA; SAx
CAS Registry Number: 69-72-7
RTECS® Number: VO0525000
UN/NA & ERG Number: Not regulated.
EC Number: 200-712-3
Regulatory Authority and Advisory Bodies
US EPA Gene-Tox Program, Positive: *S. cerevisiae*—reversion; Negative: Histidine reversion—Ames test.
US EPA, FIFRA 1998 Status of Pesticides: Canceled.
FDA—over-the-counter drug.
Canada, WHMIS, Ingredients Disclosure List Concentration 0.1%.
European/International Regulations: not listed in Annex 1.
WGK (German Aquatic Hazard Class): 1—Low hazard to waters.
Description: Salicylic acid is a white crystalline solid. Molecular weight = 138.13; Boiling point = 76°C (sublimes); Freezing/Melting point = 157°C; Flash point = 157°C; Autoignition temperature = 540°C. Explosive limits: LEL = 1.1% at 200°C; UEL—unknown. Hazard Identification (based on NFPA-704 M Rating System): Health 2, Flammability 1, Reactivity 1. Insoluble in water.
Potential Exposure: Compound Description: Drug, Mutagen; Reproductive Effector; Human Data; Primary Irritant. Used as a topical keratolytic agent; in the manufacture of aspirin, salicylates, resins, as a dyestuff intermediate; prevulcanization inhibitor; analytical reagent; fungicide, antiseptic, and food preservative.
Incompatibilities: Iron salts; lead acetate; iodine. Forms an explosive mixture in air.
Permissible Exposure Limits in Air
Protective Action Criteria (PAC)
TEEL-0: 0.3 mg/m^3
PAC-1: 0.75 mg/m^3
PAC-2: 6 mg/m^3
PAC-3: 400 mg/m^3

Routes of Entry: Ingestion, inhalation, eyes and/or skin contact. Absorbed through the skin.
Harmful Effects and Symptoms
Short Term Exposure: Overexposure may affect the central nervous system and the body's acid—base balance, causing delirium and tremors. *Inhalation:* May cause ringing in the ears, confusion, rapid pulse and breathing; headache, dizziness, nausea, and vomiting. *Skin:* May be very irritating and cause skin sores. May act as a systemic poison if applied to large areas of the skin. *Eyes:* Causes irritation; may be severe. *Ingestion:* 10 g may cause headache, dizziness, nausea, and vomiting. Ingestion of about 1 oz may be fatal.
Long Term Exposure: Repeated large doses may cause, in addition to the symptoms listed above, abdominal pain; loss of appetite; heartburn, poor digestion; stomach ulcers; bleeding of the stomach; iron-deficiency anemia; restlessness, incoherent speech; tremor, kidney damage; coma, convulsions, and death. Repeated or prolonged contact with skin may cause acne-like skin sores.
Points of Attack: Skin, blood, kidneys.
Medical Surveillance: Complete blood count (CBC). Kidney function tests. Evaluation by a qualified allergist.
First Aid: If this chemical gets into the eyes, remove any contact lenses at once and irrigate immediately for at least 15 min, occasionally lifting upper and lower lids. Seek medical attention immediately.
Skin contact: remove contaminated clothing and wash immediately with soap and water. Seek medical attention immediately. If this chemical has been inhaled, remove from exposure, begin rescue breathing (using universal precautions, including resuscitation mask) if breathing has stopped and CPR if heart action has stopped. Transfer promptly to a medical facility. When this chemical has been swallowed, get medical attention. Give large quantities of water and induce vomiting. Do not make an unconscious person vomit.
Note to physician: Induced emesis—gastric lavage, activated charcoal, or a combination of these—may be necessary to clear the gastrointestinal tract. Sodium bicarbonate (IV) with added potassium may be necessary for blood acidosis.
Personal Protective Methods: Wear protective gloves and clothing to prevent skin contact. Safety equipment suppliers/manufacturers can provide recommendations on the most protective glove/clothing material for your operation. All protective clothing (suits, gloves, footwear, headgear) should be clean, available each day, and put on before work. Contact lenses should not be worn when working with this chemical. Wear dust-proof chemical goggles and face shield unless full face-piece respiratory protection is worn. Employees should wash immediately with soap when skin is wet or contaminated. Provide emergency showers and eyewash.
Respirator Selection: Follow regulations in OSHA 29CFR1910.134 or European Standard EN149. Use a NIOSH/MSHA- or European Standard EN149-approved respirator; or use an approved supplied-air respirator with a

full face-piece operated in the positive-pressure mode, or with a full face-piece, hood, or helmet in the continuous-flow mode; or use a NIOSH/MSHA- or European Standard EN149-approved self-contained breathing apparatus with a full face-piece operated in pressure-demand or other positive-pressure mode.

Storage: Color Code—Green: General storage may be used. Prior to working with this chemical you should be trained in its proper handling and storage. Before entering confined space where this chemical may be present, check to make sure that an explosive concentration does not exist. Store in tightly closed containers in a cool, well-ventilated area away from light. Where possible, automatically pump liquid from drums or other storage containers to process containers.

Shipping: Not regulated.

Spill Handling: Evacuate persons not wearing protective equipment from area of spill or leak until cleanup is complete. Remove all ignition sources. Establish forced ventilation to keep levels below explosive limit. Collect powdered material in the most convenient and safe manner and deposit in sealed containers. Ventilate area after cleanup is complete. It may be necessary to contain and dispose of this chemical as a hazardous waste. If material or contaminated runoff enters waterways, notify downstream users of potentially contaminated waters. Contact your local or federal environmental protection agency for specific recommendations. If employees are required to clean up spills, they must be properly trained and equipped. OSHA 1910.120(q) may be applicable.

Fire Extinguishing: This chemical is a combustible solid. Use dry chemical, carbon dioxide, water spray, or alcohol foam extinguishers. Poisonous gases are produced in fire, including nitrogen oxides. If material or contaminated runoff enters waterways, notify downstream users of potentially contaminated waters. Notify local health and fire officials and pollution control agencies. From a secure, explosion-proof location, use water spray to cool exposed containers. If cooling streams are ineffective (venting sound increases in volume and pitch, tank discolors, or shows any signs of deforming), withdraw immediately to a secure position. If employees are expected to fight fires, they must be trained and equipped in OSHA 1910.156. The only respirators recommended for firefighting are self-contained breathing apparatuses that have full face-pieces and are operated in a pressure-demand or other positive-pressure mode.

References

New York State Department of Health. (April 1986). *Chemical Fact Sheet: Salicylic Acid.* Albany, NY: Bureau of Toxic Substance Assessment

Sarin (Agent GB, WMD)   S:0130

Molecular Formula: $C_4H_{10}FO_2P$
Common Formula: $(CH_3)_2CHOP(CH_3)OF$

Synonyms: Fluoroisopropoxymethyl oxide; GB (military designation); IMPF; Isopropoxymethylphoryl, fluoride; Isopropyhyl methylphosphonofluoridate; Isopropyl methylfluorophosphate; *O*-Isopropyl methylphosphonofluoridate; Isopropyl methylphosphonofluoridate; Isopropyl-methyl-phosphoryl fluoride; Methylfluorophosphoric acid isopropyl ester; Methylfluorphosphorsaeureisopropylester (German); Methylphosphonofluoridic acid isopropyl ester; Methylphosphonofluoridic acid 1-methylethyl ester; MFI; Sarina (Spanish); Sarin II; T-144; T-2106; Tl 1618; Trilone 46

CAS Registry Number: 107-44-8; 50642-23-4
RTECS® Number: TA84000000
UN/NA & ERG Number: UN2810/153
EC Number: None assigned.

Regulatory Authority and Advisory Bodies

Department of Homeland Security Screening Threshold Quantity: *Theft hazard* CUM 100 g (*107-44-8; 1445-76-7 Chlorosarin, a precursor*).

Carcinogenicity: GB is not listed by the International Agency for Research on Cancer (IARC); American Conference of Governmental Industrial Hygienists (ACGIH); Occupational Safety and Health Administration (OSHA); or National Toxicology Program (NTP) as a carcinogen.

OSHA 29CFR1910.119, Appendix A, Process Safety List of Highly Hazardous Chemicals, TQ = 100 lb (45 kg).

Superfund/EPCRA 40CFR355, Extremely Hazardous Substances: TPQ = 10 lb (4.54 kg).

Reportable Quantity (RQ): 1 lb (0.454 kg).

US DOT 49CFR172.101, Inhalation Hazard Chemical.

Note: Army Regulation, AR 50-6, deals specifically with the shipment of chemical agents; must be escorted in accordance with Army Regulation, AR 740-32.

WGK (German Aquatic Hazard Class): No value assigned.

Description: Sarin (GB), a nerve agent, is one of the most toxic of the known chemical warfare agents. Exposure to sarin can cause death in minutes. A fraction of an ounce (1−10 mL) of GB on the skin can be fatal. GB is an odorless, colorless, tasteless, nonflammable liquid at 15°C and 1 atm. *GB has no warning properties, especially when pure, and it can take away your sense of smell.* Molecular weight = 140.09; Specific gravity (H_2O:1) = 1.080 at 25°C; Boiling point = 147°C; Freezing/Melting point = −57°C; Vapor pressure = 2.9 mmHg at 25°C; 2.1 mmHg at 20°C; Liquid density = 1.10 g/mL at 20°C; Relative vapor density (air = 1) = 4.86; Vapor density = 4.9; Volatility = 22,000 mg/m^3 at 25°C. Flash point = >280°C; 78°C (cc).[NIOSH] Hazard Identification (based on NFPA-704 M Rating System): Health 4, Flammability 1, Reactivity 0. Sinks in water; soluble in water and hydrolyzes. The hydrolysis products are considerably less toxic than the material itself. *Note:* If it is used as a weapon, notify US Department of Defense: Army. If a means of detection is available, use M-8 paper (detection: yellow) or M256-A1 Detector Kit (detection limit: 0.005 mg/m^3). *Warning: A single drop on* *the skin can be fatal.* Damage and/or death may occur before chemical detection can take place.

Potential Exposure: GB is used as a quick-acting chemical warfare nerve agent; nerve gas. Both the liquid and the vapor can kill you. Very small amounts can hurt you in 1 min or less, and can quickly lead to death. A single drop, if vaporized, can kill everyone in a room![USCG] Sarin is 26 times more deadly than cyanide gas and 20 times more deadly than potassium cyanide.

Persistence of Chemical Agent: Sarin (GB): Summer: 10 min to 24 h; Winter: 2 h to 3 days.

Incompatibilities: Attacks tin, magnesium, cadmium-plated steel, and some aluminums. GB decomposes tin, magnesium, cadmium-plated steel, and aluminum. Slightly corrosive to brass, copper, and lead. No attack on 1020 steel, Inconel®, and K-Monel®. Hydrolyzed by water. In acid conditions, GB hydrolyzes, forming hydrofluoric acid (HF). Rapidly hydrolyzed by dilute aqueous sodium hydroxide, or sodium carbonate, forming relatively nontoxic products of polymers and isopropyl alcohol. Contact with metals may evolve flammable hydrogen gas.
STEL: 0.0001 mg/m^3
IDLH: 0.1 mg/m^3

Permissible Exposure Limits in Air
Protective Action Criteria (PAC) GB*
TEEL-0: 0.00015 ppm
PAC-1: **0.00048** ppm
PAC-2: **0.006** ppm
PAC-3: **0.022** ppm
*AEGLs (Acute Emergency Guideline Levels) & ERPGs (Emergency Response Planning Guideline) are in **bold face**.
WPL (Worker population limit): 0.00003 mg/m^3.

The suggested permissible airborne exposure concentration of sarin (GB) for an 8-h workday or a 40-h workweek is an 8-h time-weighted average (TWA) of 0.00003 mg/m^3 (2×10^{-5} ppm). This value is based on the TWA of GB as proposed in the USAEHA Technical Guide No. 169, *Occupational Health Guidelines for the Evaluation and Control of Occupational Exposure to Nerve Agents GA, GB, GD, and VX.*
GPL (General population limit): 0.000001 mg/m^3.
Also, the general population limits (as recommended by the Surgeon General's Working Group, US Department of Health): 0.000003 mg/m^3. *Human Toxicity:* The human lethal dose (man) is approximately 0.01 mg/kg. LCt$_{50}$ = 100 mg-min/m^3. LD$_{50}$ [skin] = 1.7 g/70 kg [man] (*Medical Aspects of Chemical and Biological Warfare, Part I*, Walter Reed Medical Center, 1997).

Determination in Air: Available monitoring equipment for agent GB is the M8/M9 detector paper, detector ticket, blue band tube, M256/M256A1 kits, bubbler, Depot Area Air Monitoring System (DAAMS); Automatic Continuous Air Monitoring System (ACAMS); Real Time Monitoring (RTM); Demilitarization Chemical Agent Concentrator (DCAC); M8/M43, M8A1/M43A2, Hydrogen Flame Photometric Emission Detector (HYPED); CAM-M1, Miniature Chemical Agent Monitor (MINICAM); and the Real Time Analytical Platform (RTAP). Real-time, low-level monitors (with alarm) are required for GB operations. In their absence, an IDLH atmosphere must be presumed. Laboratory operations conducted in appropriately maintained and alarmed engineering controls require only periodic low-level monitoring.

Determination in Water: Use M-272 Chemical Agent Water Testing Kit. Detection limit for nerve agents is 0.02 mg/L. Also, for cyanides, distillation followed by silver nitrate titration or colorimetric analysis using pyridine pyrazolone (or barbituric acid). *Food chain concentration potential:* sarin is soluble in water; bioconcentration in aquatic organisms is not expected to be an important fate process [US Army Corps of Engineers. Special Report 86–38, Britton, K. B., *Low Temperature Effects on Sorption, Hydrolysis, and Photolysis of Organophosphates: A Literature Review, p. 23.* Washington DC, 1986]. Octanol–water coefficient: Log K_{ow} = (estimated) −1.4.

Water pollution: If used as a weapon, utilize an M272 Water Detection Kit (Detection limit: 0.02 mg/L). Dangerous to aquatic life in high concentrations. May be dangerous if it enters water intakes. Notify local health and pollution control officials. Notify operators of nearby water intakes. This material will be broken down in water quickly, but small amounts may evaporate. This material will be broken down in moist soil quickly. Small amounts may evaporate into the air or travel below the soil surface and contaminated groundwater.

Routes of Entry: Inhalation, ingestion, skin and/or eye contact. Absorbed through the skin.

Harmful Effects and Symptoms
Short Term Exposure: Extremely toxic; vapor LCt$_{50}$ = 100 mg-min/m^3. Extremely active cholinesterase inhibitor. The toxic effects of this chemical are similar to parathion, but more severe and dangerous. Liquid, LD$_{50}$ (skin) = 1.7 g/70 kg man; sarin: (mice) (200 μg/kg). A single drop on the skin can cause death. Death may occur within 15 min after fatal dose is absorbed. Symptoms of overexposure may occur within minutes or hours—depending upon dose. They include miosis (constriction of pupils) and visual effects; headache and pressure sensation; runny nose and nasal congestion; salivation, tightness in the chest; nausea, vomiting, giddiness, anxiety, difficulty in thinking; difficulty sleeping, nightmares, muscle twitches, tremors, weakness, abdominal cramps; diarrhea, involuntary urination, and defecation.

Long Term Exposure: Cholinesterase inhibitor; cumulative effect is possible. This chemical may damage the nervous system with repeated exposure, resulting in convulsions and respiratory failure. Limited data suggest that chronic or repeated exposure to GB may result in a delayed postural sway and/or impaired psychomotor performance (neuropathy).

Points of Attack: Respiratory system; central nervous system; skin, eyes, plasma and red blood cell cholinesterase. Liver and kidney damage.

Medical Surveillance: Patients/victims who have severe exposure should be evaluated for persistent central nervous system (CNS) effects. Consider the points of attack in pre-placement and periodic physical examinations. Complete blood count (CBC). Evaluation of thyroid function. Lung function tests. Central nervous system tests. Liver and kidney function tests.

First Aid: Administration of antidotes is a critical step in managing a patient/victim. However, this may be difficult to achieve in the Red Zone, because the antidotes may not be readily available, and procedures or policies for their administration in the Red Zone may be lacking. Do not administer antidotes preventatively; there is no benefit to doing so. Diazepam (or other benzodiazepines) should be administered when there is evidence of seizures, usually seen in cases of moderate to severe exposure to a nerve agent. Remember, physical findings of localized exposure often precede systemic exposure and physical findings.[NIOSH]

Inhalation: Hold breath until respiratory protective mask is donned. If severe signs of agent exposure appear (chest tightens, pupil constriction, a lack of coordination, etc.), immediately administer, in rapid succession, all three Nerve Agent Antidote Kit(s), Mark I injectors (or atropine if directed by the local physician). Injections using the Mark I kit injectors may be repeated at 5–20 min intervals if signs and symptoms are progressing until three series of injections have been administered. No more injections will be given unless directed by medical personnel. In addition, a record will be maintained of all injections given. If breathing has stopped, give artificial respiration. Mouth-to-mouth resuscitation should be used when approved mask-bag or oxygen delivery systems are not available. Do not use mouth-to-mouth resuscitation when facial contamination exists. If breathing is difficult, administer oxygen. *Seek medical attention immediately.*

Eye contact: Immediately flush eyes with water for 10–15 min, then don respiratory protective mask. Although miosis (pinpointing of the pupils) may be an early sign of agent exposure, an injection will not be administered when miosis is the only sign present. Instead, the individual will be taken *immediately* to the medical treatment facility for observation.

Skin contact: Don respiratory protective mask and remove contaminated clothing. Immediately wash contaminated skin with copious amounts of soap and water, 10% sodium carbonate solution, or 5% liquid household bleach. Rinse well with water to remove decontaminant. Administer an intramuscular injection with the *Mark I* Kit injectors only if local sweating and muscular twitching symptoms are observed. *Seek medical attention immediately.*

Ingestion: Do not induce vomiting. First symptoms are likely to be gastrointestinal. Immediately administer an intramuscular injection of the *Mark I* kit autoinjectors.

Medical Treatment: Electrocardiogram (ECG), and adequacy of respiration and ventilation, should be monitored. Supplemental oxygenation, frequent suctioning of secretions, insertion of a tube into the trachea (endotracheal intubation), and assisted ventilation may be required. Diazepam (5–10 mg in adults and 0.2–0.5 mg/kg in children) may be used to control convulsions. Lorazepam or other benzodiazepines may be used, but barbiturates, phenytoin, and other anticonvulsants are not effective. Administration of atropine (if not already given) should precede the administration of benzodiazepines in order to best control seizures. Patients/victims who have inhalation exposure and who complain of chest pain, chest tightness, or cough should be observed and examined periodically for 6–12 h to detect delayed-onset inflammation of the large airways (bronchitis), inflammatory lung disease (pneumonia), accumulation of fluid in the lungs (pulmonary edema), or respiratory failure.[NIOSH]

Decontamination: This is very important. The rapid physical removal of a chemical agent is essential. If you do not have the equipment and training, do not enter the hot or the warm zone to rescue and/or decontaminate victims. Medical personnel should wear the proper PPE. If the victim cannot move, decontaminate without touching and without entering the hot or the warm zone. Nerve gases stay in clothing; *do not* touch with bare skin—if possible, seal contaminated clothes and personal belongings in a double bag. Use clean water from any source; if possible, use a hose (spray or fog to prevent injury to the victim) or other system to avoid touching the victim. Do not wait for soap or for the victim to remove clothing, begin washing immediately. Do not delay decontamination to obtain warm water; time is of the essence; use cold water instead. Immediately flush the eyes with water for at least 15 min. Use caution to avoid hypothermia in children and the elderly. Wash–strip–wash–evacuate upwind and uphill: Patients exposed to nerve agent by vapor only should be decontaminated by removing all clothing in a clean-air environment and shampooing or rinsing the hair to prevent vapor-off gassing; Patients exposed to liquid nerve agent should be decontaminated by washing in available clean water at least three times. Use liquid soap (dispose of container after use and replace), large amounts of water, and mild to moderate friction with a single-use sponge or washcloth in the first and second washes. Scrubbing of exposed skin with a brush is discouraged; skin damage may occur and may increase absorption. The third wash should be to rinse with large amounts of warm or hot water. Shampoo can be used to wash the hair. Decontaminate with diluted household bleach* (0.5%, or one part bleach to 200 parts water), but do not let any get in the victim's eyes, open wounds, or mouth. Wash off the diluted bleach solution after 15 min. Remember that the water you use to decontaminate the victims is dangerous. Be sure you have decontaminated the victims as much as you can before they are released from the area, so they do not spread the nerve gas. Rinse the eyes, mucous membranes, or open wounds with sterile saline or water and then move away from the hot zone in an upwind and uphill direction.

*The following can be used in addition to household bleach: (1) solids, powders, and solutions containing various types of bleach (NaOCl or Ca(OCl)$_2$); (2) DS2 (2% NaOH,

70% diethylenetriamine, 28% ethylene glycol monomethyl ether); (3) towelettes moistened with sodium hydroxide (NaOH) dissolved in water, phenol, ethanol, and ammonia. *Note:* Use 5% solution of common bleach (sodium hypochlorite) or calcium hypochlorite solution (48 oz per 5 gallons of water) to decontaminate scissors used in clothing removal, clothes and other items.

Personal Protective Methods: Wear protective "A-Level" PPE: rubber gloves, protective clothing; goggles, respirators, butyl rubber gloves M3 and M4, Norton chemical protective glove set; and Tyvek® "F" decontamination suit provides barrier protection against chemical warfare agents. Airtight, impermeable clothing was developed for personnel who must enter heavily contaminated areas. This clothing is made of butyl rubber or a coated fabric, such as Tyvek "F," and provides barrier protection against liquid chemical warfare agents. Although resistant to liquid chemical agents, impermeable protective clothing may be penetrated after a few hours of exposure to heavy concentration of agent. Consequently, liquid contamination on the clothing must be neutralized or removed as soon as possible. If the proper equipment is not available, or if the rescuers have not been trained in its use, call for assistance from the US Soldier and Biological Chemical Command—Edgewood Research Development and Engineering Center (from 0700-1630 EST call 410-671-4411, and from 1630-0700 EST call 410-278-5201; ask for the Staff Duty Officer). All protective clothing (suits, gloves, footwear, headgear) should be clean, available each day, and put on before work. Contact lenses should not be worn when working with this chemical. Wear splash-proof chemical goggles and face shield unless full face-piece respiratory protection is worn. Employees should wash immediately with soap when skin is wet or contaminated. Provide emergency showers and eyewash.

Swatch Test Results for Level A Suits and Chemical Protective Gloves for GB (Sarin)

Item	Breakthrough
25-mil Chemical protective gloves	>480 min
Kappler Suit Model 42483	350 min
TYCHEM 10,000 Pkg Style No. 12645	>480 min
Trellchem HPS suit	>480 min
Ready 1 Limited Use Suit: Model 91	>480 min
First Team XE HazMat suit	>480 min
Commander Ultrapro Suit, Style 79102	>480 min
Kappler Suit Model 50660	>480 min
TYCHEM Style No. 11645	>480 min
Trellchem TLU suit	>480 min
Chemturion Suit: Model 13	>480 min
Chempruf II BETEX Suit	225 min
Commander Brigade: F91	>480 min

Respirator Selection: When used as a weapon, use SCBA Respirator Certified By NIOSH For CBRN Environments. Emergency or planned entry into unknown concentrations or IDLH conditions: SCBAF: Pd,Pp (APF = 10,000) (any self-contained breathing apparatus that has a full face-piece and is operated in a pressure-demand or other positive-

pressure mode) or SaF: Pd,Pp: ASCBA (APF = 10,000) (any supplied-air respirator that has a full face-piece and is operated in a pressure-demand or other positive-pressure mode in combination with an auxiliary self-contained breathing apparatus operated in a pressure-demand or other positive-pressure mode). *Escape:* GmFOvHie [any air-purifying, full-face-piece respirator (gas mask) with a chin-style, front- or back-mounted canister having a high-efficiency particulate filter] or SCBAE (any appropriate escape-type, self-contained breathing apparatus). The US Army standard M40 Series (which replaced the M17A1 protective mask) provides complete respiratory protection against all known military toxic chemical agents, but it cannot be used in an oxygen-deficient environment and it does not afford protection against industrial toxics, such as ammonia and carbon monoxide. *It is not approved for civilian use.*

The following is taken from the Riegle Report:
Less than 0.0001 mg/m³: A full face-piece, chemical canister, air-purifying protective mask will be on hand for escape. (The M9- or M40-series masks are acceptable for this purpose). *0.0001–0.2 mg/m³:* A NIOSH/MSHA- or European Standard EN149-approved pressure demand full face-piece SCBA or supplied-air respirator with escape air cylinder may be used. Alternatively, a full face-piece, chemical canister air-purifying protective mask is acceptable for this purpose (for example, M9-, M17-, or M40-series mask or other mask certified as equivalent) is acceptable. *Greater than 0.2 mg/m³ or unknown concentration:* NIOSH/MSHA- or European Standard EN 149-approved pressure demand full face-piece SCBA suitable for use in high agent concentrations with protective ensemble.

Shipping: Toxic, liquids, organic, n.o.s. [Inhalation hazard, Packing Group I, Zone A] requires a shipping label of "POISONOUS/TOXIC MATERIALS." Inhalation Hazard. It falls in Hazard Class 6.1 and Packing Group III.[NIOSH]

Storage: In handling, the buddy system will be incorporated. No smoking, eating, and drinking in areas containing agent is permitted. Containers should be periodically inspected for leaks (either visually or by a detector kit). Stringent control over all personnel practices must be exercised. Decontamination equipment shall be conveniently located. Exits must be designed to permit rapid evacuation. Chemical showers, eyewash stations, and personal cleanliness facilities must be provided. Wash hands before meals and each worker will shower thoroughly with special attention given to hair, face, neck, and hands, using plenty of soap before leaving at the end of the workday.

Spill Handling: Spills must be contained by covering with vermiculite, diatomaceous earth, clay, fine sand, sponges, and paper or cloth towels. Decontaminate with copious amounts of aqueous sodium hydroxide solution (a minimum 10 wt. %). Scoop up all material and place in a fully removable head drum with a high-density polyethylene liner. Cover the contents of the drum with decontaminating solution as above before affixing the drum head. After sealing the head, the exterior of the drum shall be decontaminated and then labeled in accordance with IAW, EPA, and DOT regulations. All leaking containers shall be overpacked with vermiculite placed between the interior and exterior containers. Decontaminate and label in accordance with IAW, EPA, and DOT regulations. Dispose of material used to decontaminate exterior of drum. Conduct general area monitoring with an approved monitor to confirm that the atmospheric concentrations do not exceed the airborne exposure limit.

If 10 wt. % aqueous sodium hydroxide solution is not available, then the following decontaminants may be used instead and are listed in the order of preference: Decontaminating Solution No. 2 [DS2: (2% NaOH, 70% diethylenetriamine, 28% ethylene glycol monomethyl ether)], sodium carbonate, and Super-tropical Bleach Slurry (STB).

GB when used as a weapon

Small spills (From a small package or a small leak from a large package)

First: Isolate in all directions (feet/meters)   200/60
Then: Protect persons downwind (miles/kilometers)
Day   0.3/0.5
Night 0.8/1.3

Large spills (From a large package or from many small packages)

First: Isolate in all directions (feet/meters)   2500/800
Then: Protect persons downwind (miles/kilometers)
Day   1.4/2.3
Night 2.8/4.5

Fire: If tank, rail car, or tank truck is involved in fire, isolate for at least 800 m (½ mile) in all directions; also, consider initial evacuation for 800 m (½ mile) in all directions. Evacuate and restrict persons not wearing protective equipment from area of spill or leak until cleanup is complete. Remove all ignition sources. Ventilate area of spill or leak. Rapidly hydrolyzed by dilute aqueous sodium hydroxide or sodium carbonate forming relatively nontoxic products. Water alone removes the fluorine atom producing a nontoxic acid. Decontaminants include bleach slurry, dilute alkali, hot soapy water, steam, and ammonia. Absorb liquids in vermiculite, dry sand, earth, peat, carbon, or a similar material and deposit in sealed containers. Keep this chemical out of a confined space, such as a sewer, because of the possibility of an explosion, unless the sewer is designed to prevent the buildup of explosive concentrations. It may be necessary to contain and dispose of this chemical as a hazardous waste. If material or contaminated runoff enters waterways, notify downstream users of potentially contaminated waters. Contact your local or federal environmental protection agency for specific recommendations. If employees are required to clean up spills, they must be properly trained and equipped. OSHA 1910.120(q) may be applicable. See also above "Determination in Water." Soil: 2−24 h at 5°−25°C.

Fire Extinguishing: Highly volatile poison! (organophosphate). Breathing the vapor, skin [a single drop on the skin can be fatal] or eye contact, or swallowing the material can kill you; symptoms may be delayed for several hours. If exposure occurs, remove and isolate gear immediately and thoroughly decontaminate personnel. Storage containers and parts of containers may rocket great distances, in many directions. Vapors are heavier than air and will collect and stay in low areas. *Combustion products are less deadly than the material itself.* Toxic products of combustion may include carbon monoxide, fluorine, and phosphorus oxide. GB will react with steam or water to produce toxic and corrosive vapors. All persons not engaged in extinguishing the fire should be evacuated. Fires involving GB should be contained to prevent contamination to uncontrolled areas. When responding to a fire alarm in buildings or areas containing agents, firefighting personnel clothing (without TAP clothing) during chemical agent firefighting and fire rescue operations. Respiratory protection is required. Positive pressure, full face-piece, NIOSH-approved self-contained breathing apparatus (SCBA) will be worn where there is danger of oxygen deficiency and when directed by the fire chief or chemical accident/incident (CAI) operations officer. In cases where firefighters are responding to a chemical accident/incident for rescue/reconnaissance purposes, they will wear appropriate levels of protective clothing. Complete protection required; have available decontaminants (bleach, alkali) and atropine. Use dry chemical, carbon dioxide, or foam extinguishers. Vapors are heavier than air and will collect in low areas. If material or contaminated runoff enters waterways, notify downstream users of potentially contaminated waters. Notify local health and fire officials and pollution control agencies. From a secure, explosion-proof location, use water spray to cool exposed containers. If cooling streams are ineffective (venting sound increases in volume and pitch, tank discolors, or shows any signs of deforming), withdraw immediately to a secure position. If employees are expected to fight fires, they must be trained and equipped in OSHA 1910.156.

Disposal Method Suggested: Open pit burning or burying of GB or items containing or contaminated with GB in any quantity is prohibited. The detoxified GB using procedures above can be thermally destroyed by incineration in an EPA-approved incinerator in accordance with appropriate provisions of federal, state, and local RCRA regulations.

A minimum of 56 g of decon solution is required for each gram of GB. Decontaminant/agent solution is allowed to agitate for a minimum of 1 h. Agitation is not necessary following the first hour. At the end of 1 h, the resulting solution should be adjusted to a pH greater than 11.5. If the pH is below 11.5, sodium hydroxide (NaOH) should be added until a pH above 11.5 can be maintained for 60 min. An alternate solution for the decontamination of GB is 10 wt. % sodium carbonate in place of the 10% sodium hydroxide solution above. Continue with 56 g of decon to 1 g of agent. Agitate for 1 h but allow 3 h for the reaction. The final pH should be adjusted to above 10. It is also permitted to substitute 5.25% sodium hypochlorite or 25 wt. % monoethylamine (MEA) for the 10% sodium hydroxide solution above. MEA must be completely dissolved in water prior to addition of the agent. Continue with 56 g of decon for each gram of GB and provide agitation for 1 h. Continue with same ratios and time stipulations. Scoop up all material and place in a fully removable head drum with a high-density polyethylene liner. Cover the contents of the drum with decontaminating solution as above before affixing the drum head. After sealing the head, the exterior of the drum shall be decontaminated and then labeled per IAW, EPA, and DOT regulations. All leaking containers shall be overpacked with vermiculite placed between the interior and exterior containers. Decontaminate and label per IAW, EPA, and DOT regulations. Dispose of the material per IAW waste disposal methods provided below. Dispose of material used to decontaminate exterior of drum in accordance with IAW, federal, state, and local regulations. Conduct general area monitoring with an approved monitor to confirm that the atmospheric concentrations do not exceed the airborne exposure limit.

References

US Environmental Protection Agency. (November 30, 1987). *Chemical Hazard Information Profile: Sarin.* Washington, DC: Chemical Emergency Preparedness Program

The Riegle Report: A Report of Chairman Donald W. Riegle, Jr. and Ranking Member Alfonse M. D'Amato of the Committee on Banking, Housing and Urban Affairs with Respect to Export Administration, United States Senate, 103rd Congress, 2d Session (May 25, 1994)

New Jersey Department of Health and Senior Services. (April 2004). *Hazardous Substances Fact Sheet: Sarin.* Trenton, NJ

Schneider, A. L., et al. (2007). *CHRIS + CD-ROM Version 2.0, United Coast Guard Chemical Hazard Response Information System (COMDTINST 16465.12C).* Washington, DC: United States Coast Guard and the Department of Homeland Security

Belmont, R. B. (June 1998). *TR Tests of Level A Suits— Protection Against Chemical and Biological Warfare Agents and Simulants: Executive Summary.* Aberdeen Proving Ground, MD 21010-5423: CBRD-EN (US Army Chemical and Biological Defense Command)

Selenium   S:0140

Molecular Formula: Se
Synonyms: C.I. 77805; Colloidal selenium; Elemental selenium; Selenate; Selenio (Spanish); Selenium (colloidal); Selenium alloy; Selenium base; Selenium dust; Selenium element; Selenium homopolymer; Selenium powder
CAS Registry Number: 7782-49-2
RTECS® Number: VS7700000
UN/NA & ERG Number: UN2658 (powder)/152
EC Number: 231-957-4 [*Annex I Index No.:* 034-001-00-2]
Regulatory Authority and Advisory Bodies
Carcinogenicity: IARC: Animal Inadequate Evidence; Human Inadequate Evidence, *not classifiable as carcinogenic to humans*, Group 3, 1987; EPA: Not classifiable as to human carcinogenicity.
Banned or Severely Restricted (in agriculture) (Germany, UK) (UN).[13]
Air Pollutant Standard Set. See below, "Permissible Exposure Limits in Air" section.
Clean Water Act: 40CFR423, Appendix A, Priority Pollutants; Section 313 Water Priority Chemicals (57FR41331, 9/9/92); 40CFR401.15 Section 307 Toxic Pollutants.
US EPA Hazardous Waste Number (RCRA No.): D010.
RCRA, 40CFR261, Appendix 8 Hazardous Constituents, waste number not listed.
RCRA Toxicity Characteristic (Section 261.24), Maximum. Concentration of Contaminants, regulatory level, 1.0 mg/L.
RCRA 40CFR268.48; 61FR15654, Universal Treatment Standards: Wastewater (mg/L), 0.82; Nonwastewater (mg/L), 0.16 TCLP.
RCRA 40CFR264, Appendix 9; TSD Facilities Ground Water Monitoring List. Suggested test method(s) (PQL μg/L), total dust 6010 (750); 7740 (20); 7741 (20).
Safe Drinking Water Act: MCL, 0.05 mg/L; MCLG, 0.05 mg/L; Regulated chemical (47 FR 9352).
Reportable Quantity (RQ): 100 lb (45.4 kg).
EPCRA Section 313 Form R *de minimis* concentration reporting level: 1.0%.
Canada, WHMIS, Ingredients Disclosure List Concentration 0.1%.
European/International Regulations: Hazard Symbol: T; Risk phrases: R23/25-33; Safety phrases: S1/2; S20/21; S28; S45 (see Appendix 4).
WGK (German Aquatic Hazard Class): 2—Hazard to waters.
Description: Selenium exists in three forms: a red amorphous powder, a gray form, and red crystals. Occurs as an impurity in most sulfide ores. Selenium, along with tellurium, is found in the sludges and sediments from electrolytic copper refining. It may also be recovered in flue dust from burning pyrites in sulfuric acid manufacture. Molecular weight = 78.96; Specific gravity (H_2O:1) = 4.28 at 25°C; Boiling point = 685°C; Freezing/Melting point = 144°C (red crystalline form); 217°C (gray crystalline form). Hazard Identification (based on NFPA-704 M Rating System): (powder) Health 3, Flammability 1, Reactivity 1. Insoluble in water.
Potential Exposure: Compound Description: Tumorigen; Reproductive Effector; Human Data. Most of the selenium

produced is used in the manufacture of selenium rectifiers. It is also utilized as a pigment for ruby glass, paints, and dyes; as a vulcanizing agent for rubber; a decolorizing agent for green glass; a chemical catalyst in the Kjeldahl test; as an insecticide; in the manufacture of electrodes, selenium photocells, selenium cells, and semiconductor fusion mixtures; in photographic toning bathes; and for dehydrogenation of organic compounds. It is also used in veterinary medicine and in antidandruff shampoos. Se is used in radioactive scanning for the pancreas and for photostatic and X-ray xerography. It may be alloyed with stainless steel, copper, and cast steel. Selenium is a contaminant in most sulfide ores of copper, gold, nickel, and silver; and exposure may occur while removing selenium from these ores.

Incompatibilities: Reacts violently with strong acids and strong oxidizers, chromium trioxide, potassium bromate, cadmium. Reacts with incandescence on gentle heating with phosphorous and metals, such as nickel, zinc, sodium, potassium, platinum. Reacts with water at 50°C forming flammable hydrogen and selenious acids.

Permissible Exposure Limits in Air

The PEL and REL also apply to other selenium compounds (as Se) except selenium hexafluoride; the MAK applies to Se and its inorganic compounds.

OSHA PEL: 0.2 mg[Se]/m³ TWA.

NIOSH REL (*except selenium hexafluoride*): 0.2 mg[Se]/m³ TWA.

ACGIH TLV®[1]: 0.2 mg[Se]/m³ TWA.

NIOSH IDLH: 1 mg[Se]/m³.

Protective Action Criteria (PAC)

TEEL-0: 0.2 mg/m³

PAC-1: 0.6 mg/m³

PAC-2: 1 mg/m³

PAC-3: 1 mg/m³

DFG MAK (*metal and inorganic compounds*): 0.05 mg[Se]/m³, inhalable fraction TWA; Peak Limitation Category II (4); Carcinogen Category 3B; Pregnancy Risk Group C.

Arab Republic of Egypt: TWA 0.2 mg/m³, 1993; Australia: TWA 0.2 mg[Se]/m³, 1993; Austria: MAK 0.1 mg/m³, 1999; Belgium: TWA 0.2 mg/m³, 1993; Denmark: TWA 0.1 mg/m³, 1999; Finland: TWA 0.1 mg/m³; STEL 0.3 mg/m³, 1999; Hungary: STEL 0.1 mg/m³, 1993; the Netherlands: MAC-TGG 0.1 mg[Se]/m³, 2003; the Philippines: TWA 0.2 mg/m³, 1993; Poland: MAC (TWA) 0.1 mg/m³; MAC (STEL) 0.3 mg/m³, 1999; Sweden: NGV 0.1 mg/m³, 1999; Switzerland: MAK-W 0.1 mg/m³, 1999; Turkey: TWA 0.2 mg/m³, 1993; United Kingdom: TWA 0.1 mg/m³, 2000; Argentina, Bulgaria, Columbia, Jordan, South Korea, New Zealand, Singapore, Vietnam: ACGIH TLV®: TWA 0.2 mg [Se]/m³. Russia[43] set a MAC of 2.0 mg/m³ in work-place air. Several states have set guidelines or standards for selenium compounds as selenium in ambient air[60] ranging from 0.26−1.57 μg/m³ (Montana) to 0.27 μg/m³ (Massachusetts) to 0.66 μg/m³ (New York) to 2.0 μg/m³ (Florida and North Dakota) to 3.5 μg/m³ (Virginia) to 4.0 μg/m³ (Connecticut) to 5.0 μg/m³ (Nevada).

Determination in Air: NIOSH(IV) Methods #7300, Elements; #7301; #7303; #9102; NIOSH II(7), Method #S190; OSHA Analytical Method ID-121 and ID-125G.

Permissible Concentration in Water: To protect freshwater aquatic life: 35 μg/L as a 24-h average, never to exceed 260 μg/L for recoverable inorganic selenite. *To protect saltwater aquatic life:* 54 μg/L as a 24-h average, never to exceed 410 μg/L. To protect human health: 10 μg/L.[6] There are a variety of foreign standards for selenium in water.[35] The Czech Republic has set 0.1 mg/L in surface water, 0.05 mg/L in drinking water reserve, and 0.01 mg/L in drinking water. Germany[35] has set 0.008 mg/L as a maximum permissible concentration in drinking water. The EEC has set 0.01 mg/L as a MAC for drinking water. Mexico has set 0.01 mg/L for receiving waters used for drinking purposes. Russia has set 0.001 mg/L as a MAC for drinking water. WHO has set 0.01 mg/L as a limit for drinking water. Maine has set[61] a guideline of 0.01 mg/L and Minnesota 0.045 mg/L for drinking water. EPA[62] has proposed 0.05 mg/L as a limit for drinking water.

Determination in Water: Digestion followed by atomic absorption gives total selenium. Dissolved selenium is determined by 0.45 μm filtration prior to the above analysis.

Routes of Entry: Inhalation of dust or vapor, percutaneous absorption of liquid, ingestion, eye and/or skin contact.

Harmful Effects and Symptoms

The normal dietary intake of selenium, about 50−150 μg/day, is sufficient to meet the daily nutritional requirement for this essential nutrient. Selenium compounds can be toxic. However, at levels of daily intake that are only moderately higher than the nutritional requirement. The severity of the toxic effects of selenium would depend upon how much selenium was eaten and how often. Swallowing a quantity of concentrated sodium selenate or sodium selenite (for example, part of a bottle of sodium selenate designed to be administered to a flock of sheep, or large numbers of selenium supplement pills) would be life threatening without immediate treatment. If amounts of selenium that were moderately higher than the daily nutritional requirement were eaten over long periods of time, several health effects could occur, including brittle hair, deformed nails; and in extreme cases, numbness and a lack of coordination in arms and legs. These health effects have been observed in people living in several villages in the People's Republic of China who were exposed to foods that were high in selenium for months to years. There have been no reports of populations in the United States exhibiting these symptoms of extreme chronic selenium poisoning.

Short Term Exposure: Selenium can affect you when breathed in. Fumes can irritate the eyes and the respiratory tract. Contact can burn the eyes and skin. Inhalation of fume may cause symptoms of asphyxiation, chills and fever, and bronchitis. High levels can cause troubled breathing, lung irritation (pneumonitis), and headaches. Higher exposures can cause pulmonary edema, a medical emergency that can be delayed for several hours. This can cause death. Symptoms of exposure can cause visual disturbance, headache, chills, fever, dyspnea (breathing difficulty), bronchitis, metallic taste, garlic breath, gastrointestinal disturbance.

Long Term Exposure: Long-term exposure to selenium compounds may be a cause of amyotrophic lateral sclerosis in humans. Repeated or prolonged contact may cause skin dermatitis. May affect the central nervous system, blood, teeth, and bones. May cause nervousness, depression, pallor, digestive disturbances. Kidney and liver damage may occur. Daily ingestion of 25 mg of sodium selenite, containing 4−5 selenium, after 11 days showed signs of hair and fingernail loss, fatigue, nausea, vomiting, and garlic breath.

Points of Attack: Eyes, skin, respiratory system, liver, kidneys, blood, spleen. In animals: anemia, liver necrosis, cirrhosis, kidney, spleen damage.

Medical Surveillance: NIOSH lists the following tests: Body Hair/Nail; whole blood (chemical/metabolite); blood plasma; blood serum; liver function tests; urine (chemical/metabolite); urine (chemical/metabolite), 24-h collection; urinalysis; white blood cell count/differential. Urine test for selenium (normal is less than 100 μg/L of urine). If symptoms develop or overexposure is suspected, the following may be useful: consider chest X-ray after acute overexposure. Liver function tests.

First Aid: If this chemical gets into the eyes, remove any contact lenses at once and irrigate immediately for at least 15 min, occasionally lifting upper and lower lids. Seek medical attention immediately. If this chemical contacts the skin, remove contaminated clothing and wash immediately with soap and water. Seek medical attention immediately. If this chemical has been inhaled, remove from exposure, begin rescue breathing (using universal precautions, including resuscitation mask) if breathing has stopped and CPR if heart action has stopped. Transfer promptly to a medical facility. When this chemical has been swallowed, get medical attention. Give large quantities of water and induce vomiting. Do not make an unconscious person vomit. Medical observation is recommended for 24−48 h after breathing overexposure, as pulmonary edema may be delayed. As first aid for pneumonitis or pulmonary edema, a doctor or authorized paramedic may consider administering a corticosteroid spray.

Note to physician: For severe poisoning *do not* use BAL [British Anti-Lewisite, dimercaprol, dithiopropanol ($C_3H_8OS_2$)] as it is contraindicated or ineffective in poisoning from selenium.

Personal Protective Methods: Wear protective gloves and clothing to prevent any reasonable probability of skin contact. Safety equipment suppliers/manufacturers can provide recommendations on the most protective glove/clothing material for your operation. All protective clothing (suits, gloves, footwear, headgear) should be clean, available each day, and put on before work. Contact lenses should not be worn when working with this chemical. Wear dust-proof chemical goggles and face shield unless full face-piece respiratory protection is worn. Employees should wash immediately with soap when skin is wet or contaminated. Provide emergency showers and eyewash.

Respirator Selection: 1 mg/m³: Qm (APF = 25) (any quarter-mask respirator) or 95XQ (APF = 10) [any particulate respirator equipped with an N95, R95, or P95 filter (including N95, R95, and P95 filtering face-pieces) except quarter-mask respirators. The following filters may also be used: N99, R99, P99, N100, R100, P100] or 100F (APF = 50) (any air-purifying, full-face-piece respirator with an N100, R100, or P100 filter) or PAPRDM if not present as a fume (any powered, air-purifying respirator with a dust and mist filter) or PaprHie (APF = 25) (any powered, air-purifying respirator with a high-efficiency particulate filter) or Sa (APF = 10) (any supplied-air respirator) or SCBAF (APF = 50) (any self-contained breathing apparatus with a full face-piece) or SaF (APF = 50) (any supplied-air respirator with a full face-piece). *Emergency or planned entry into unknown concentrations or IDLH conditions:* SCBAF: Pd, Pp (APF = 10,000) (any self-contained breathing apparatus that has a full face-piece and is operated in a pressure-demand or other positive-pressure mode) or SaF: Pd,Pp: ASCBA (APF = 10,000) (any supplied-air respirator that has a full face-piece and is operated in a pressure-demand or other positive-pressure mode in combination with an auxiliary self-contained breathing apparatus operated in a pressure-demand or other positive-pressure mode). *Escape:* 100F (APF = 50) (any air-purifying, full-face-piece respirator with an N100, R100, or P100 filter) or SCBAE (any appropriate escape-type, self-contained breathing apparatus).

Note: Substance reported to cause eye irritation or damage; may require eye protection.

Storage: Color Code—Blue: Health Hazard/Poison: Store in a secure poison location. Prior to working with this chemical you should be trained on its proper handling and storage. Selenium must be stored to avoid contact with strong oxidizers (such as chlorine, bromine, and fluorine) and strong acids (such as hydrochloric, sulfuric, and nitric), since violent reactions occur. Store in tightly closed containers in a cool, well-ventilated area away from water.

Shipping: Selenium requires a shipping label of "POISONOUS/TOXIC MATERIALS." It falls in Hazard Class 6.1 and Packing Group III.

Spill Handling: Evacuate persons not wearing protective equipment from area of spill or leak until cleanup is complete. Remove all ignition sources. Collect powdered material in the most convenient and safe manner and deposit in sealed containers. Ventilate area after cleanup is complete. It may be necessary to contain and dispose of this chemical as a hazardous waste. If material or contaminated runoff enters waterways, notify downstream users of potentially contaminated waters. Contact your local or federal environmental protection agency for specific recommendations. If employees are required to clean up spills, they must be properly trained and equipped. OSHA 1910.120(q) may be applicable.

Fire Extinguishing: This chemical is a combustible solid. Use dry chemical, carbon dioxide, water spray, or foam extinguishers. Poisonous gases are produced in fire, including selenium. If material or contaminated runoff enters waterways, notify downstream users of potentially contaminated waters. Notify local health and fire officials and pollution control agencies. From a secure, explosion-proof location, use water spray to cool exposed containers. If cooling streams are ineffective (venting sound increases in volume and pitch, tank discolors, or shows any signs of deforming), withdraw immediately to a secure position. If employees are expected to fight fires, they must be trained and equipped in OSHA 1910.156. The only respirators recommended for firefighting are self-contained breathing apparatuses that have full face-pieces and are operated in a pressure-demand or other positive-pressure mode.

Disposal Method Suggested: Powdered selenium: dispose in a chemical waste landfill. When possible, recover selenium and return to suppliers.[22]

References

National Academy of Sciences. (1976). *Selenium.* Washington, DC (Also issued by EPA Health Effects Res. Lab. as Report EPA-600/1-76-014, Research Triangle Park, NC)

US Environmental Protection Agency. (1980). *Selenium: Ambient Water Quality Criteria.* Washington, DC

US Environmental Protection Agency. (April 30, 1980). *Selenium: Health and Environmental Effects Profile No. 153.* Washington, DC: Office of Solid Waste

US Public Health Service. (December 1988). *Toxicological Profile for Selenium.* Atlanta, GA: Agency for Toxic Substances and Disease Registry

New Jersey Department of Health and Senior Services. (September 2002). *Hazardous Substances Fact Sheet: Selenium.* Trenton, NJ

Sax, N. I. (Ed.). (1981). *Dangerous Properties of Industrial Materials Report*, 1, No. 3, 75–78

Selenium dioxide   S:0150

Molecular Formula: O_2Se
Common Formula: SeO_2
Synonyms: Dioxido de selenio (Spanish); Selenious acid anhydride; Selenious anhydride; Selenium(IV) dioxide; Selenium oxide; Selenium(IV) oxide
CAS Registry Number: 7446-08-4
RTECS® Number: VS8575000
UN/NA & ERG Number: UN3283/151; UN2811/154 (selenium oxide)
EC Number: 231-194-7
Regulatory Authority and Advisory Bodies
Air Pollutant Standard Set. See below, "Permissible Exposure Limits in Air" section.
Clean Water Act: Section 311 Hazardous Substances/RQ same as CERCLA; Section 313 Water Priority Chemicals

(57FR41331, 9/9/92) as selenium oxide; 40CFR 401.15 Section 307 Toxic Pollutants as selenium and compounds.
US EPA Hazardous Waste Number (RCRA No.): U204.
RCRA, 40CFR261, Appendix 8 Hazardous Constituents.
Superfund/EPCRA 40CFR355, Extremely Hazardous Substances: TPQ = 500 lb (227 kg).
Reportable Quantity (RQ): 10 lb (4.54 kg).
EPCRA Section 313 Form R *de minimis* concentration reporting level: 1.0%.
Canada, WHMIS, Ingredients Disclosure List Concentration 1.0%.
European/International Regulations: not listed in Annex 1.
WGK (German Aquatic Hazard Class): 2—Hazard to waters.

Description: Selenium dioxide is a white to slightly reddish crystalline solid or yellow liquid which forms a yellow-green vapor that has a sour and pungent odor. Odor threshold in air = 0.0002 mg/m^3. Molecular weight = 110.96; Sublimation point = 315°C. Hazard Identification (based on NFPA-704 M Rating System): Health 1, Flammability 0, Reactivity 0. Soluble in water; solubility = 40%.

Potential Exposure: Compound Description: Mutagen; Reproductive Effector. Selenium dioxide is used in the manufacture of selenium compounds, a reagent for alkaloids; an oxidizing agent; in paint and ink pigments; in metal "blueing" and etching; as a chemical catalyst; in photographic toners; in electric and photoelectric components; and others.

Incompatibilities: Strong acids may cause release of toxic hydrogen selenide gas. Water solution is a medium strong acid (selenious acid). Reacts with many substances producing toxic selenium vapors. Attacks many metals in the presence of water.

Permissible Exposure Limits in Air
OSHA PEL: 0.2 mg[Se]/m^3 TWA.
NIOSH REL: 0.2 mg[Se]/m^3 TWA.
ACGIH TLV®[1]: 0.2 mg[Se]/m^3 TWA.
NIOSH IDLH: 1 mg [Se]/m^3.
Protective Action Criteria (PAC)
TEEL-0: 0.281 mg/m^3
PAC-1: 3.5 mg/m^3
PAC-2: 6 mg/m^3
PAC-3: 6 mg/m^3
DFG MAK: 0.05 mg[Se]/m^3, inhalable fraction TWA; Peak Limitation Category II(4); Carcinogen Category 3B; Pregnancy Risk Group C.
Arab Republic of Egypt: TWA 0.2 mg[Se]/m^3, 1993; Australia: TWA 0.2 mg[Se]/m^3, 1993; Austria: MAK 0.1 mg[Se]/m^3, 1999; Belgium: TWA 0.2 mg[Se]/m^3, 1993; Denmark: TWA 0.1 mg[Se]/m^3, 1999; Finland: TWA 0.1 mg[Se]/m^3; STEL 0.3 mg[Se]/m^3, 1999; Hungary: STEL 0.1 mg[Se]/m^3, 1993; Japan: 0.1 mg[Se]/m^3, 1999; Norway: TWA 0.1 mg[Se]/m^3, 1999; the Philippines: TWA 0.2 mg[Se]/m^3, 1993; Poland: MAC (TWA) 0.1 mg[Se]/m^3; MAC (STEL) 0.3 mg[Se]/m^3, 1999; Sweden: NGV 0.1 mg[Se]/m^3, 1999; Switzerland: MAK-W 0.1 mg[Se]/m^3,

1999; Turkey. TWA 0.2 mg[Se]/m^3, 1993; United Kingdom: TWA 0.1 mg[Se]/m^3, 2000; Argentina, Bulgaria, Columbia, Jordan, South Korea, New Zealand, Singapore, Vietnam: ACGIH TLV®: TWA 0.2 mg[Se]/m^3. Russia[43] set a MAC of 0.1 μg/m^3 in ambient air in residential areas on a momentary basis and 0.05 μg/m^3 on a daily average basis.

Determination in Air: Use NIOSH(IV) Methods #7300, Elements; #7301; #7303; #9102; NIOSH II(7), Method #S190; OSHA Analytical Method ID-121 and ID-125G.

Permissible Concentration in Water: To protect freshwater aquatic life: 35 μg/L as a 24-h average, never to exceed 260 μg/L for recoverable inorganic selenite. *To protect saltwater aquatic life:* 54 μg/L as a 24-h average, never to exceed 410 μg/L. To protect human health: 10 μg/L.[6] There are a variety of foreign standards for selenium in water.[35] The Czech Republic has set 0.1 mg/L in surface water, 0.05 mg/L in drinking water reserve, and 0.01 mg/L in drinking water. Germany[35] has set 0.008 mg/L as a maximum permissible concentration in drinking water. The EEC has set 0.01 mg/L as a MAC for drinking water. Mexico has set 0.01 mg/L for receiving waters used for drinking purposes. Russia has set 0.001 mg/L as a MAC for drinking water. WHO has set 0.01 mg/L as a limit for drinking water. Maine has set[61] a guideline of 0.01 mg/L and Minnesota 0.045 mg/L for drinking water. EPA[62] has proposed 0.05 mg/L as a limit for drinking water.

Determination in Water: Digestion followed by atomic absorption gives total selenium. Dissolved selenium is determined by 0.45 μm filtration prior to the above analysis.

Routes of Entry: Inhalation, ingestion, skin, and/or eyes. Absorbed through the skin.

Harmful Effects and Symptoms

Short Term Exposure: Corrosive to the eyes, skin, and respiratory tract. *Inhalation:* Low-level exposures may cause garlic breath and metallic taste in the mouth. Tiredness, pallor, and indigestion have been reported. Unknown high levels have caused intense irritation of nose, throat, and lungs, with coughing, sneezing, congestion, dizziness, and headache. Inhalation may cause chemical pneumonitis. Higher exposures can cause pulmonary edema, a medical emergency that can be delayed for several hours. This can cause death. *Skin:* May cause dermatitis with itchy red bumps forming. Chemical burns may also occur. Intense pain will occur if selenium oxide penetrates under the fingernail. *Eyes:* May cause pain and irritation. Eyelids may become puffy, an allergic reaction known as "rose eye." *Ingestion:* No information specific to selenium oxide. Based on other selenium compounds, may cause nausea, vomiting, abdominal pain, diarrhea, metallic taste, and garlic odor on breath. Animal studies suggest that the lethal dose for an adult may be as low as 1/200 oz.

Long Term Exposure: Allergic sensitization may occur resulting in onset of skin rash and rose eye at very low levels. Discoloration of hair and nails may also affect many organ systems. Animal feeding studies suggest that liver,

spleen, and kidney damage is possible. Nasopharyngeal irritation, gastrointestinal distress, and persistent garlic breath may also occur. Long-term exposure to selenium compounds may be a cause of amyotrophic lateral sclerosis in humans. Repeated or prolonged contact may cause skin dermatitis. May affect the central nervous system, blood, teeth, and bones. May cause nervousness, depression, pallor, digestive disturbances. Kidney and liver damage may occur. Daily ingestion of 25 mg of sodium selenite, containing 4–5 selenium, after 11 days showed signs of hair and fingernail loss, fatigue, nausea, vomiting, and garlic breath.

Points of Attack: Skin, eyes, liver, kidneys, spleen.

First Aid: If this chemical gets into the eyes, remove any contact lenses at once and irrigate immediately for at least 15 min, occasionally lifting upper and lower lids. Seek medical attention immediately. If this chemical contacts the skin, remove contaminated clothing and wash immediately with soap and water. Seek medical attention immediately. If this chemical has been inhaled, remove from exposure, begin rescue breathing (using universal precautions, including resuscitation mask) if breathing has stopped and CPR if heart action has stopped. Transfer promptly to a medical facility. When this chemical has been swallowed, get medical attention. If victim is *conscious*, administer water or milk. Do not induce vomiting.

Note to physician: Inhalation: bronchodilators, decongestants, and oxygen may be used if necessary. Corticosteroids are useful for treating pneumonitis.

Note to physician: 10% sodium thiosulfate solutions, creams, and eye ointments will rapidly convert selenium oxide to less harmful red selenium. Urinary selenium may be useful indicator of degree of exposure. Liver function tests are suggested.

Note to physician: For severe poisoning *do not* use BAL [British Anti-Lewisite, dimercaprol, dithiopropanol ($C_3H_8OS_2$)] as it is contraindicated or ineffective in poisoning from selenium.

Personal Protective Methods: Wear protective gloves and clothing to prevent any reasonable probability of skin contact. Safety equipment suppliers/manufacturers can provide recommendations on the most protective glove/clothing material for your operation. All protective clothing (suits, gloves, footwear, headgear) should be clean, available each day, and put on before work. Contact lenses should not be worn when working with this chemical. Wear splash-proof chemical goggles and face shield when working with liquid, unless full face-piece respiratory protection is worn. Wear dust-proof goggles and face shield when working with powders or dust, unless full face-piece respiratory protection is worn. Employees should wash immediately with soap when skin is wet or contaminated. Provide emergency showers and eyewash.

Respirator Selection: 1 mg/m^3: Qm (APF = 25) (any quarter-mask respirator) or 95XQ (APF = 10) [any particulate respirator equipped with an N95, R95, or P95 filter (including N95, R95, and P95 filtering face-pieces) except quarter-mask respirators. The following filters may also be used: N99, R99, P99, N100, R100, P100] or 100F (APF = 50) (any air-purifying, full-face-piece respirator with an N100, R100, or P100 filter) or PAPRDM if not present as a fume (any powered, air-purifying respirator with a dust and mist filter) or PaprHie (APF = 25) (any powered, air-purifying respirator with a high-efficiency particulate filter) or Sa (APF = 10) (any supplied-air respirator) or SCBAF (APF = 50) (any self-contained breathing apparatus with a full face-piece) or SaF (APF = 50) (any supplied-air respirator with a full face-piece). *Emergency or planned entry into unknown concentrations or IDLH conditions:* SCBAF: Pd,Pp (APF = 10,000) (any self-contained breathing apparatus that has a full face-piece and is operated in a pressure-demand or other positive-pressure mode) or SaF: Pd,Pp: ASCBA (APF = 10,000) (any supplied-air respirator that has a full face-piece and is operated in a pressure-demand or other positive-pressure mode in combination with an auxiliary self-contained breathing apparatus operated in a pressure-demand or other positive-pressure mode). *Escape:* 100F (APF = 50) (any air-purifying, full-face-piece respirator with an N100, R100, or P100 filter) or SCBAE (any appropriate escape-type, self-contained breathing apparatus).

Note: Substance reported to cause eye irritation or damage; may require eye protection.

Storage: Color Code—Blue: Health Hazard/Poison: Store in a secure poison location. Prior to working with this chemical you should be trained on its proper handling and storage. Store away from acids and water. Where possible, automatically transfer material from drums or other storage containers to process containers. Sources of ignition, such as smoking and open flames, are prohibited where this chemical is handled, used, or stored. Metal containers involving the transfer of this chemical should be grounded and bonded. Wherever this chemical is used, handled, manufactured, or stored, use explosion-proof electrical equipment and fittings.

Shipping: This compound requires a shipping label of "POISONOUS/TOXIC MATERIALS." It falls in Hazard Class 6.1 and Packing Group II.

Spill Handling: Evacuate persons not wearing protective equipment from area of spill or leak until cleanup is complete. Remove all ignition sources. Use HEPA vacuum or wet method to reduce dust during cleanup. Do not dry sweep. Collect powdered material in the most convenient and safe manner and deposit in sealed containers. Ventilate area after cleanup is complete. For liquids, absorb on sand or vermiculite and shovel slurry into sealed container. It may be necessary to contain and dispose of this chemical as a hazardous waste. If material or contaminated runoff enters waterways, notify downstream users of potentially contaminated waters. Contact your local or federal environmental protection agency for specific recommendations. If employees are required to clean up spills, they must be properly trained and equipped. OSHA 1910.120(q) may be applicable.

Fire Extinguishing: This chemical is a combustible solid. Use dry chemical, carbon dioxide, water spray, or alcohol foam extinguishers. Poisonous gases are produced in fire, including selenium. If material or contaminated runoff enters waterways, notify downstream users of potentially contaminated waters. Notify local health and fire officials and pollution control agencies. From a secure, explosion-proof location, use water spray to cool exposed containers. If cooling streams are ineffective (venting sound increases in volume and pitch, tank discolors, or shows any signs of deforming), withdraw immediately to a secure position. If employees are expected to fight fires, they must be trained and equipped in OSHA 1910.156. The only respirators recommended for firefighting are self-contained breathing apparatuses that have full face-pieces and are operated in a pressure-demand or other positive-pressure mode.

Disposal Method Suggested: Consult with environmental regulatory agencies for guidance on acceptable disposal practices. Generators of waste containing this contaminant (≥100 kg/mo) must conform with EPA regulations governing storage, transportation, treatment, and waste disposal.

References
New York State Department of Health. (April 1986). *Chemical Fact Sheet: Selenium Oxide* (Version 2). Albany, NY: Bureau of Toxic Substance Assessment

Selenium hexafluoride   S:0160

Molecular Formula: F_6Se
Common Formula: SeF_6
Synonyms: Selenium fluoride
CAS Registry Number: 7783-79-1
RTECS® Number: VS9450000
UN/NA & ERG Number: UN2194/125
EC Number: None assigned
Regulatory Authority and Advisory Bodies
Department of Homeland Security Screening Threshold Quantity (pounds): *Theft hazard* 15 (≥1.67% concentration).
Carcinogenicity: IARC: Animal Inadequate Evidence; Human Inadequate Evidence, *not classifiable as carcinogenic to humans*, Group 3.
Very Toxic Substance (World Bank).[15]
Air Pollutant Standard Set. See below, "Permissible Exposure Limits in Air" section.
OSHA 29CFR1910.119, Appendix A. Process Safety List of Highly Hazardous Chemicals, TQ = 1000 lb (450 kg).
US DOT 49CFR172.101, Inhalation Hazardous Chemical.
EPCRA Section 313 Form R *de minimis* concentration reporting level: 1.0%.
Canada, WHMIS, Ingredients Disclosure List Concentration 1.0%.
European/International Regulations: not listed in Annex 1.
WGK (German Aquatic Hazard Class): No value assigned.

Description: Selenium hexafluoride is a nonflammable, colorless gas. Repulsive odor. Molecular weight = 192.96; Boiling point = −34.4°C (Sublimation point = −46°C); Freezing/Melting point = −50.6°C (also −39°C); Relative vapor density (air = 1) = 6.66. Hazard Identification (based on NFPA-704 M Rating System): Health 3, Flammability 0, Reactivity 1. Insoluble in water; slow reaction.

Potential Exposure: Selenium hexafluoride, a gas, is utilized as a gaseous electric insulator.

Incompatibilities: Hydrolyzes very slowly in cold water. Decomposes on heating, producing hydrogen fluoride, fluoride, and selenium. Contact with metal or acid will produce toxic gaseous hydrogen selenide.

Permissible Exposure Limits in Air
OSHA PEL: 0.05 ppm/0.4 mg[Se]/m^3 TWA.
NIOSH REL: 0.05 ppm TWA.
ACGIH TLV®[1]: 0.05 ppm/0.4 mg[Se]/m^3 TWA.
NIOSH IDLH: 2 ppm.
Protective Action Criteria (PAC)*
TEEL-0: 0.053 ppm
PAC-1: **0.053** ppm
PAC-2: **0.087** ppm
PAC-3: **0.26** ppm
*AEGLs (Acute Emergency Guideline Levels) & ERPGs (Emergency Response Planning Guideline) are in **bold face**.

Australia: TWA 0.05 ppm (0.2 mg/m^3), 1993; Austria: MAK 2.5 mg[F]/m^3, 1999; MAK 0.1 mg[Se]/m^3, 1999; Belgium: TWA 0.05 ppm (0.16 mg/m^3), 1993; Denmark: TWA 0.05 ppm (0.4 mg/m^3), 1999; Finland: TWA 0.05 ppm (0.4 mg/m^3); STEL 0.15 ppm (1.2 mg/m^3), 1999; France: VME 0.05 ppm (0.2 mg/m^3), 1999; Japan 0.1 mg [Se]/m^3, 1999; the Netherlands: MAC-TGG 0.2 mg[Se]/m^3, 2003; Norway: TWA 0.05 ppm (0.4 mg/m^3), 1999; the Philippines: TWA 0.05 ppm (0.4 mg/m^3), 1993; Poland: MAC (TWA) 1 mg[HF]/m^3, 1999; Sweden: NGV 2 mg[F]/m^3, 1999; Switzerland: MAK-W 0.05 ppm (0.4 mg/m^3), 1999; United Kingdom: TWA 0.1 mg[Se]/m^3, 2000; TWA 2.5 mg[F]/m^3, 2000; Argentina, Bulgaria, Columbia, Jordan, South Korea, New Zealand, Singapore, Vietnam: ACGIH TLV®: TWA 0.05 ppm. Several states have set guidelines or standards for SeF$_6$ in ambient air[60] ranging from 2 µg/m^3 (North Dakota) to 3.5 µg/m^3 (Virginia) to 4.0 µg/m^3 (Connecticut) to 5.0 µg/m^3 (Nevada).

Determination in Air: Collection by impinger or fritted bubbler, analysis by atomic absorption spectrometry.

Routes of Entry: Inhalation, skin and/or eye contact.

Harmful Effects and Symptoms
Short Term Exposure: Corrosive to the eyes and respiratory tract. Inhalation can cause pulmonary edema, a medical emergency that can be delayed for several hours. This can cause death. Contact with the liquid can cause frostbite.

Long Term Exposure: Repeated or prolonged contact may cause skin sensitization. May affect the central nervous system, liver, and kidneys.

Points of Attack: Respiratory system, liver, kidneys, skin. In animals: pulmonary irritation, edema.

Medical Surveillance: NIOSH lists the following tests: Blood Gas Analysis; chest X-ray, electrocardiogram, pulmonary function tests: forced vital capacity, forced expiratory volume (1 s); sputum cytology; urinalysis (routine); white blood cell count/differential. Urine test for selenium (should not exceed 100 µg/L of urine). Lung function tests. Consider chest X-ray following acute overexposure. Liver and kidney function tests. Examination by a qualified allergist. Examination of the nervous system.

First Aid: If this chemical gets into the eyes, remove any contact lenses at once and irrigate immediately for at least 15 min, occasionally lifting upper and lower lids. Seek medical attention immediately. If this chemical contacts the skin, remove contaminated clothing and wash immediately with soap and water. Seek medical attention immediately. If this chemical has been inhaled, remove from exposure, begin rescue breathing (using universal precautions, including resuscitation mask) if breathing has stopped and CPR if heart action has stopped. Transfer promptly to a medical facility. When this chemical has been swallowed, get medical attention. If victim is *conscious*, administer water or milk. Do not induce vomiting. Medical observation is recommended for 24−48 h after breathing overexposure, as pulmonary edema may be delayed. As first aid for pulmonary edema, a doctor or authorized paramedic may consider administering a corticosteroid spray. If frostbite has occurred, seek medical attention immediately; do *NOT* rub the affected areas or flush them with water. In order to prevent further tissue damage, do *NOT* attempt to remove frozen clothing from frostbitten areas. If frostbite has *NOT* occurred, immediately and thoroughly wash contaminated skin with soap and water.

Note to physician: For severe poisoning *do not* use BAL [British Anti-Lewisite, dimercaprol, dithiopropanol (C$_3$H$_8$OS$_2$)] as it is contraindicated or ineffective in poisoning from selenium.

Personal Protective Methods: Wear appropriate personal protective clothing to prevent the skin from becoming frozen from contact with the evaporating liquid or from contact with vessels containing the liquid. Safety equipment suppliers/manufacturers can provide recommendations on the most protective glove/clothing material for your operation. All protective clothing (suits, gloves, footwear, headgear) should be clean, available each day, and put on before work. Contact lenses should not be worn when working with this chemical. Wear gas-proof chemical goggles and face shield unless full face-piece respiratory protection is worn. Employees should wash immediately with soap when skin is wet or contaminated. Provide emergency showers and eyewash.

Respirator Selection: *Up to 0.5 ppm:* Sa (APF = 10) (any supplied-air respirator). *Up to 1.25 ppm:* Sa:Cf (APF = 25) (any supplied-air respirator operated in a continuous-flow mode). *Up to 2 ppm:* SaT: Cf (APF = 50) (any supplied-air respirator that has a tight-fitting face-piece and is operated in a continuous-flow mode) or SCBAF (APF = 50) (any self-contained breathing apparatus with a full face-piece) or SaF (APF = 50) (any supplied-air respirator with a full face-piece). *Emergency or planned entry into unknown concentrations or IDLH conditions:* SCBAF: Pd,Pp (APF = 10,000) (any self-contained breathing apparatus that has a full face-piece and is operated in a pressure-demand or other positive-pressure mode) or SaF: Pd,Pp: ASCBA (APF = 10,000) (any supplied-air respirator that has a full face-piece and is operated in a pressure-demand or other positive-pressure mode in combination with an auxiliary self-contained breathing apparatus operated in a pressure-demand or other positive-pressure mode). *Escape:* GmFS (APF = 50) [any air-purifying, full-face-piece respirator (gas mask) with a chin-style, front- or back-mounted canister providing protection against the compound of concern] or SCBAE (any appropriate escape-type, self-contained breathing apparatus).

Storage: Color Code—White stripe: Contact Hazard; Store separately; not compatible with materials in solid white category. Color Code—Blue: Health Hazard/Poison: Store in a secure poison location. Prior to working with this chemical you should be trained on its proper handling and storage. High concentrations cause a deficiency of oxygen with the risk of unconsciousness or death. Check that oxygen content is at least 19% before entering storage or spill area. Procedures for the handling, use, and storage of cylinders should be in compliance with OSHA 1910.101 and 1910.169, as with the recommendations of the Compressed Gas Association.

Shipping: Selenium hexafluoride requires a shipping label of "POISON GAS, CORROSIVE." It falls in Hazard Class 2.3.

Special precautions: Cylinders must be transported in a secure upright position, in a well-ventilated truck. It is a violation of transportation regulations to refill compressed gas cylinders without the express written permission of the owner.

Spill Handling: If in a building, evacuate building and confine vapors by closing doors and shutting down HVAC systems. Restrict persons not wearing protective equipment from area of spill or leak until cleanup is complete. Remove all ignition sources. Ventilate area of spill or leak to disperse the gas. Wear chemical protective suit with self-contained breathing apparatus to combat spills. Stay upwind and use water spray to "knock down" vapor; contain runoff. Stop the flow of gas, if it can be done safely from a distance. If source is a cylinder and the leak cannot be stopped in place, remove the leaking cylinder to a safe place; and repair leak or allow cylinder to empty. Keep this chemical out of confined spaces, such as a sewer, because of the possibility of explosion, unless the sewer is designed to prevent the buildup of explosive concentrations. If employees are required to clean up spills, they must be properly trained and equipped. OSHA 1910.120(q) may be applicable.

Initial isolation and protective action distances
Distances shown are likely to be affected during the first 30 min after materials are spilled and could increase with time. If more than one tank car, cargo tank, portable tank, or large cylinder involved in the incident is leaking, the protective action distance may need to be increased. You may need to seek emergency information from CHEMTREC at (800) 424-9300 or seek professional environmental engineering assistance from the US EPA Environmental Response Team at (908) 548-8730 (24-h response line).

Small spills (From a small package or a small leak from a large package)
First: Isolate in all directions (feet/meters)   200/60
Then: Protect persons downwind (miles/kilometers)
Day   0.3/0.5
Night   1.2/1.9

Large spills (From a large package or from many small packages)
First: Isolate in all directions (feet/meters)   1500/500
Then: Protect persons downwind (miles/kilometers)
Day   1.8/2.9
Night   4.0/6.5

Fire Extinguishing: Extinguish fire using agent suitable for type of surrounding fire. The material itself does not burn or burns with difficulty. Poisonous gases are produced in fire, including hydrogen fluoride, fluoride, and selenium. Do not extinguish the fire unless the flow of gas can be stopped and any remaining gas is out of the line. Specially trained personnel may use fog lines to cool exposures and let the fire burn itself out. Vapors are heavier than air and will collect in low areas. Containers may explode in fire. Storage containers and parts of containers may rocket great distances, in many directions. If material or contaminated runoff enters waterways, notify downstream users of potentially contaminated waters. Notify local health and fire officials and pollution control agencies. From a secure, explosion-proof location, use water spray to cool exposed containers. If cooling streams are ineffective (venting sound increases in volume and pitch, tank discolors, or shows any signs of deforming), withdraw immediately to a secure position. If cylinders are exposed to excessive heat from fire or flame contact, withdraw immediately to a secure location. If employees are expected to fight fires, they must be trained and equipped in OSHA 1910.156. The only respirators recommended for firefighting are self-contained breathing apparatuses that have full face-pieces and are operated in a pressure-demand or other positive-pressure mode.

Disposal Method Suggested: If possible, convert selenium compounds to an insoluble form with SO_2 before landfill or solidification.

References

New Jersey Department of Health and Senior Services. (September 2001). *Hazardous Substances Fact Sheet: Selenium Hexafluoride*. Trenton, NJ

Selenium oxychloride   S:0170

Molecular Formula: Cl_2OSe
Common Formula: $SeOCl_2$
Synonyms: Seleninyl chloride; Selenium chloride oxide; 9Seleninyl dichloride
CAS Registry Number: 7791-23-3
RTECS® Number: VS7000000
UN/NA & ERG Number: UN2879/157
EC Number: 232-244-0
Regulatory Authority and Advisory Bodies
Air Pollutant Standard Set. See below, "Permissible Exposure Limits in Air" section.
Clean Water Act: 40CFR401.15 Section 307 Toxic Pollutants as selenium and compounds.
RCRA, 40CFR261, Appendix 8 Hazardous Constituents, as selenium compounds, n.o.s., waste number not listed.
Superfund/EPCRA 40CFR355, Extremely Hazardous Substances: TPQ = 500 lb (227 kg).
Reportable Quantity (RQ): 500 lb (227 kg).
EPCRA Section 313 Form R *de minimis* concentration reporting level: 1.0%.
Canada, WHMIS, Ingredients Disclosure List Concentration 1.0%.
European/International Regulations: not listed in Annex 1.
WGK (German Aquatic Hazard Class): No value assigned.
Description: Selenium oxychloride is a colorless to yellowish liquid. Fumes in air. Molecular weight = 165.86; Boiling point = 176.3°C; Specific gravity (H_2O:1) = 2.42 at 25°C; Freezing/Melting point = 8.5−11°C. Hazard Identification (based on NFPA-704 M Rating System): Health 2, Flammability 0, Reactivity 1₩. Reacts with water.
Potential Exposure: Compound Description: Human Data. This material is used as a solvent for many substances, including metals and as a chlorinating agent; and resin plasticizer; as an ionizing solvent; for monochlorination of ketones.
Incompatibilities: Water and air reactive. The aqueous solution is a strong acid and oxidizer. Reacts violently with bases, reducing agents, powdered antimony, red and white phosphorus, disilver oxide, lead oxide, and potassium.
Note: Never pour water into this substance; when dissolving or diluting always add it slowly to the water.
Permissible Exposure Limits in Air
OSHA PEL: 0.2 mg[Se]/m³ TWA.
NIOSH REL: 0.2 mg[Se]/m³ TWA.
ACGIH TLV®[1]: 0.2 mg[Se]/m³ TWA.
NIOSH IDLH: 1 mg [Se]/m³.
Protective Action Criteria (PAC)
TEEL-0: 0.42 mg/m³
PAC-1: 1.26 mg/m³
PAC-2: 10 mg/m³
PAC-3: 10 mg/m³
DFG MAK: 0.05 mg[Se]/m³, inhalable fraction TWA; Peak Limitation Category II(4); Carcinogen Category 3B; Pregnancy Risk Group C.

Arab Republic of Egypt: TWA 0.2 mg[Se]/m³, 1993; Australia: TWA 0.2 mg[Se]/m³, 1993; Belgium: TWA 0.2 mg[Se]/m³, 1993; Denmark: TWA 0.1 mg[Se]/m³, 1999; Finland: TWA 0.1 mg[Se]/m³; STEL 0.3 mg[Se]/m³, 1999; Hungary: STEL 0.1 mg[Se]/m³, 1993; Norway: TWA 0.1 mg[Se]/m³, 1999; the Philippines: TWA 0.2 mg[Se]/m³, 1993; Poland: MAC (TWA) 0.1 mg[Se]/m³, 1993; Sweden: NGV 0.1 mg[Se]/m³, 1999; Switzerland: MAK-W 0.1 mg [Se]/m³, 1999; Turkey: TWA 0.2 mg[Se]/m³, 1993; United Kingdom: TWA 0.1 mg[Se]/m³, 2000; Argentina, Bulgaria, Columbia, Jordan, South Korea, New Zealand, Singapore, Vietnam: ACGIH TLV®: TWA 0.2 mg[Se]/m³. Several states have set guidelines or standards for selenium compounds as selenium in ambient air[60] ranging from 0.26−1.57 μg/m³ (Montana) to 0.27 μg/m³ (Massachusetts) to 0.66 μg/m³ (New York) to 2.0 μg/m³ (Florida and North Dakota) to 3.5 μg/m³ (Virginia) to 4.0 μg/m³ (Connecticut) to 5.0 μg/m³ (Nevada).
Determination in Air: NIOSH(IV) Methods #7300, Elements; #7301; #7303; #9102; NIOSH II(7), Method #S190; OSHA Analytical Method ID-121 and ID-125G.
Permissible Concentration in Water: To protect freshwater aquatic life: 35 μg/L as a 24-h average, never to exceed 260 μg/L for recoverable inorganic selenite. *To protect saltwater aquatic life:* 54 μg/L as a 24-h average, never to exceed 410 μg/L. To protect human health: 10 μg/L.[6] There are a variety of foreign standards for selenium in water.[35] The Czech Republic has set 0.1 mg/L in surface water, 0.05 mg/L in drinking water reserve, and 0.01 mg/L in drinking water. Germany[35] has set 0.008 mg/L as a maximum permissible concentration in drinking water. The EEC has set 0.01 mg/L as a MAC for drinking water. Mexico has set 0.01 mg/L for receiving waters used for drinking purposes. Russia has set 0.001 mg/L as a MAC for drinking water. WHO has set 0.01 mg/L as a limit for drinking water. Maine has set[61] a guideline of 0.01 mg/L and Minnesota 0.045 mg/L for drinking water. EPA[62] has proposed 0.05 mg/L as a limit for drinking water.
Determination in Water: Digestion followed by atomic absorption gives total selenium. Dissolved selenium is determined by 0.45 μm filtration prior to the above analysis.
Routes of Entry: Inhalation, ingestion, skin and/or eye contact.
Harmful Effects and Symptoms
Short Term Exposure: This material is very toxic and may cause death or permanent injury after very short exposures to small quantities. Inhalation of small quantities may be corrosive and irritating to the respiratory tract. It can burn and severely irritate the skin and eyes and cause burns to the mouth, esophagus, and stomach when ingested. Inhalation of this material may cause garlic breath, odor, nausea, vomiting, indigestion, fever, headache, lassitude, irritability, and unstable blood pressure.
Long Term Exposure: Long-term exposure to selenium compounds may be a cause of amyotrophic lateral sclerosis in humans. Repeated or prolonged contact may cause skin

dermatitis. May affect the central nervous system, blood, teeth, and bones. May cause nervousness, depression, pallor, digestive disturbances. Kidney and liver damage may occur. Daily ingestion of 25 mg of sodium selenite, containing 4–5 selenium, after 11 days showed signs of hair and fingernail loss, fatigue, nausea, vomiting, and garlic breath.

Points of Attack: Skin, lungs, liver.

Medical Surveillance: NIOSH lists the following tests: Blood Gas Analysis; chest X-ray, electrocardiogram, pulmonary function tests: forced vital capacity, forced expiratory volume (1 s); sputum cytology; urinalysis (routine); white blood cell count/differential. Urine test for selenium (should not exceed 100 µg/L of urine).

First Aid: If this chemical gets into the eyes, remove any contact lenses at once and irrigate immediately for at least 15 min, occasionally lifting upper and lower lids. Seek medical attention immediately. If this chemical contacts the skin, remove contaminated clothing and wash immediately with soap and water. Seek medical attention immediately. If this chemical has been inhaled, remove from exposure, begin rescue breathing (using universal precautions, including resuscitation mask) if breathing has stopped and CPR if heart action has stopped. Transfer promptly to a medical facility. When this chemical has been swallowed, get medical attention. If victim is *conscious*, administer water or milk. Do not induce vomiting. Medical observation is recommended for 24–48 h after breathing overexposure, as pulmonary edema may be delayed. As first aid for pulmonary edema, a doctor or authorized paramedic may consider administering a corticosteroid spray.

Note to physician: For severe poisoning *do not* use BAL [British Anti-Lewisite, dimercaprol, dithiopropanol $(C_3H_8OS_2)$] as it is contraindicated or ineffective in poisoning from selenium.

Personal Protective Methods: Wear protective gloves and clothing to prevent any reasonable probability of skin contact. Safety equipment suppliers/manufacturers can provide recommendations on the most protective glove/clothing material for your operation. All protective clothing (suits, gloves, footwear, headgear) should be clean, available each day, and put on before work. Contact lenses should not be worn when working with this chemical. Wear splash-proof chemical goggles and face shield unless full face-piece respiratory protection is worn. Employees should wash immediately with soap when skin is wet or contaminated. Provide emergency showers and eyewash.

Respirator Selection: 1 mg/m³: DM if not present as a fume (any dust and mist respirator) or any particulate respirator equipped with an N95, R95, or P95 filter (including N95, R95, and P95 filtering face-pieces) except quarter-mask respirators. The following filters may also be used: N99, R99, P99, N100, R100, P100; or 100F (APF = 50) (any air-purifying, full-face-piece respirator with an N100, R100, or P100 filter) or PAPRDM if not present as a fume (any powered, air-purifying respirator with a dust and mist filter) or PaprHie (APF = 25) (any powered, air-purifying respirator

with a high-efficiency particulate filter) or Sa (APF = 10) (any supplied-air respirator) or SCBAF (APF = 50) (any self-contained breathing apparatus with a full face-piece) or SaF (APF = 50) (any supplied-air respirator with a full face-piece). *Emergency or planned entry into unknown concentrations or IDLH conditions:* SCBAF: Pd,Pp (APF = 10,000) (any self-contained breathing apparatus that has a full face-piece and is operated in a pressure-demand or other positive-pressure mode) or SaF: Pd,Pp: ASCBA (APF = 10,000) (any supplied-air respirator that has a full face-piece and is operated in a pressure-demand or other positive-pressure mode in combination with an auxiliary self-contained breathing apparatus operated in a pressure-demand or other positive-pressure mode). *Escape:* 100F (APF = 50) (any air-purifying, full-face-piece respirator with an N100, R100, or P100 filter) or SCBAE (any appropriate escape-type, self-contained breathing apparatus).

Note: Substance reported to cause eye irritation or damage; may require eye protection.

Storage: Color Code—White: Corrosive or Contact Hazard; Store separately in a corrosion-resistant location. Prior to working with this chemical you should be trained on its proper handling and storage. Where possible, automatically pump liquid from drums or other storage containers to process containers. Sources of ignition, such as smoking and open flames, are prohibited where this chemical is handled, used, or stored. Metal containers involving the transfer of 5 gallons or more of this chemical should be grounded and bonded. Drums must be equipped with self-closing valves, pressure vacuum bungs, and flame arresters. Use only non-sparking tools and equipment, especially when opening and closing containers of this chemical. Wherever this chemical is used, handled, manufactured, or stored, use explosion-proof electrical equipment and fittings.

Shipping: Selenium oxychloride requires a shipping label of "CORROSIVE, POISONOUS/TOXIC MATERIALS." It falls in Hazard Class 8 and Packing Group I.

Spill Handling: Evacuate and restrict persons not wearing protective equipment from area of spill or leak until cleanup is complete. Remove all ignition sources. Ventilate area of spill or leak. Absorb liquids in vermiculite, dry sand, earth, peat, carbon, or a similar material and deposit in sealed containers. Keep this chemical out of a confined space, such as a sewer, because of the possibility of an explosion, unless the sewer is designed to prevent the buildup of explosive concentrations. It may be necessary to contain and dispose of this chemical as a hazardous waste. If material or contaminated runoff enters waterways, notify downstream users of potentially contaminated waters. Contact your local or federal environmental protection agency for specific recommendations. If employees are required to clean up spills, they must be properly trained and equipped. OSHA 1910.120(q) may be applicable.

Fire Extinguishing: This chemical is a combustible liquid. Poisonous gases, including selenium and chloride, are produced in fire. Stay upwind; keep out of low areas. Wear positive pressure breathing apparatus and special protective clothing. Use dry chemical, carbon dioxide, or alcohol foam extinguishers. Vapors are heavier than air and will collect in low areas. Vapors may travel long distances to ignition sources and flashback. Vapors in confined areas may explode when exposed to fire. Containers may explode in fire. Storage containers and parts of containers may rocket great distances, in many directions. If material or contaminated runoff enters waterways, notify downstream users of potentially contaminated waters. Notify local health and fire officials and pollution control agencies. From a secure, explosion-proof location, use water spray to cool exposed containers. If cooling streams are ineffective (venting sound increases in volume and pitch, tank discolors, or shows any signs of deforming), withdraw immediately to a secure position. If employees are expected to fight fires, they must be trained and equipped in OSHA 1910.156. The only respirators recommended for firefighting are self-contained breathing apparatuses that have full face-pieces and are operated in a pressure-demand or other positive-pressure mode.

References

US Environmental Protection Agency. (November 30, 1987). *Chemical Hazard Information Profile: Selenium Oxychloride*. Washington, DC: Chemical Emergency Preparedness Program

Selenium sulfides   S:0180

Molecular Formula: SSe; Se_4S_4; SeS_2; Se_2S_6
Common Formula: SeS; Se_4S_4; SeS_2; Se_2S_6
Synonyms: 7446-34-6 *(monosulfide):* NCI-C50033; Selenium monosulfide; Selenium sulfide; Selenium sulphide; Selensulfid (German); Sulfur selenide
7788-56-4 *(disulfide):* Exsel; Selenium(IV) disulfide (1:2); Selenium disulphide; Selenium sulfide; Sulfuro de selenio; Selsun
CAS Registry Number: 7446-34-6 (SeS); 7488-56-4 (SeS_2)
RTECS® Number: VT0525000 (SeS); VS8925000 (SeS_2)
UN/NA & ERG Number: UN2657 (SeS_2)/153
EC Number: 231-303-8 (selenium disulphide)
Regulatory Authority and Advisory Bodies
Carcinogenicity: IARC: Animal Inadequate Evidence; Human Inadequate Evidence, *not classifiable as carcinogenic to humans*, Group 3; EPA: Sufficient evidence from animal studies; inadequate evidence or no useful data from epidemiologic studies; NTP: Reasonably anticipated to be a human carcinogen; NCI: Carcinogenesis Studies (gavage); clear evidence: rat, mouse; NCI: Carcinogenesis Bioassay (dermal); no evidence: mouse.
Air Pollutant Standard Set. See below, "Permissible Exposure Limits in Air" section.
Clean Water Act: 40CFR401.15 Section 307 Toxic Pollutants as selenium and compounds.
US EPA Hazardous Waste Number (RCRA No.): U205.
RCRA, 40CFR261, Appendix 8 Hazardous Constituents.

Reportable Quantity (RQ): 10 lb (4.54 kg).
EPCRA Section 313 Form R *de minimis* concentration reporting level: 1.0%.
California Proposition 65 Chemical: Cancer 10/1/89.
Canada, WHMIS, Ingredients Disclosure List Concentration 0.1% (SeS_2).
European/International Regulations: not listed in Annex 1.
WGK (German Aquatic Hazard Class): No value assigned.
Description: There are various selenium sulfides: Selenium monosulfide, SeS, is orange-yellow powder or tablets, Molecular weight = 111.02; Boiling point = (decomposes) 118−119°C; Freezing/Melting point = 111°C. Selenium disulfide, SeS_2 which is a bright red to yellow material; Freezing/Melting point ≤100°C. Se_2S_6 which is a light orange crystalline solid. Freezing/Melting point = 121.5°C. Se_4S_4 is a red crystalline solid. Molecular weight = 143.08; Freezing/Melting point = 113°C. Hazard Identification (based on NFPA-704 M Rating System): Health 3, Flammability 1, Reactivity 0. Slightly soluble in water.
Potential Exposure: Selenium sulfide is used for the treatment of seborrhea, especially in shampoos. The chemical is available over the counter as Selsun®, a stabilized buffered suspension. FDA reports that selenium sulfide is an active ingredient in some drug products used for the treatment of dandruff and certain types of dermatitis. A dandruff shampoo containing 1% selenium sulfide is available without a prescription and is recommended for use once or twice a week. By prescription, selenium sulfide is available in a 2.5% shampoo or lotion, with the recommended application limited to 10 min for 7 days to avoid the possibility of acute toxic effects. Selenium sulfide is also used topically in veterinary medicine for eczemas and dermatomycoses.
Incompatibilities: Oxidizers, strong acids. Violent reaction with silver oxide.
Permissible Exposure Limits in Air
OSHA PEL: 0.2 mg[Se]/m³ TWA.
NIOSH REL: 0.2 mg[Se]/m³ TWA.
ACGIH TLV®[1]: 0.2 mg[Se]/m³ TWA.
Protective Action Criteria (PAC)
NIOSH IDLH: 1 mg [Se]/m³.
Monosulfide
TEEL-0: 0.281 mg/m³
PAC-1: 0.844 mg/m³
PAC-2: 12.5 mg/m³
PAC-3: 15 mg/m³
Disulfide
TEEL-0: 0.362 mg/m³
PAC-1: 1.09 mg/m³
PAC-2: 1.81 mg/m³
PAC-3: 60 mg/m³
DFG MAK: 0.05 mg[Se]/m³, inhalable fraction TWA; Peak Limitation Category II(4); Carcinogen Category 3B; Pregnancy Risk Group C.
Several states have set guidelines or standards for selenium compounds as selenium in ambient air[60] ranging from 0.26−1.57 μg/m³ (Montana) to 0.27 μg/m³ (Massachusetts)

to 0.66 µg/m^3 (New York) to 2.0 µg/m^3 (Florida and North Dakota) to 3.5 µg/m^3 (Virginia) to 4.0 µg/m^3 (Connecticut) to 5.0 µg/m^3 (Nevada).

Determination in Air: NIOSH(IV) Methods #7300, Elements; #7301; #7303; #9102; NIOSH II(7), Method #S190; OSHA Analytical Method ID-121 and ID-125G.

Permissible Concentration in Water: *To protect freshwater aquatic life:* 35 µg/L as a 24-h average, never to exceed 260 µg/L for recoverable inorganic selenite. *To protect saltwater aquatic life:* 54 µg/L as a 24-h average, never to exceed 410 µg/L. To protect human health: 10 µg/L.[6] There are a variety of foreign standards for selenium in water.[35] The Czech Republic has set 0.1 mg/L in surface water, 0.05 mg/L in drinking water reserve, and 0.01 mg/L in drinking water. Germany[35] has set 0.008 mg/L as a maximum permissible concentration in drinking water. The EEC has set 0.01 mg/L as a MAC for drinking water. Mexico has set 0.01 mg/L for receiving waters used for drinking purposes. Russia has set 0.001 mg/L as a MAC for drinking water. WHO has set 0.01 mg/L as a limit for drinking water. Maine has set[61] a guideline of 0.01 mg/L and Minnesota 0.045 mg/L for drinking water. EPA[62] has proposed 0.05 mg/L as a limit for drinking water.

Determination in Water: Digestion followed by atomic absorption gives total selenium. Dissolved selenium is determined by 0.45 µm filtration prior to the above analysis.

Routes of Entry: Inhalation, ingestion, skin and/or eye contact. Absorbed by the skin.

Harmful Effects and Symptoms

Short Term Exposure: Eye contact can cause irritation. The dust irritates the respiratory tract. Higher exposure can cause headaches, troubled breathing, and pneumonitis. Symptoms of exposure include irritation of the skin (dermatitis), eyes, and mucous membranes; eye injury; hair loss; discoloration of hair; garlic odor of the breath; depression, languor, giddiness, emotional instability; excess dental caries; pallor, nervousness; digestive disturbances; and nausea.[52]

Long Term Exposure: A probable carcinogen in humans, it has been shown to cause liver and lung cancer in animals. Long-term exposure to selenium compounds may be a cause of amyotrophic lateral sclerosis in humans. Repeated or prolonged contact may cause skin dermatitis. May affect the central nervous system, blood, teeth, and bones. May cause nervousness, depression, pallor, digestive disturbances. Kidney and liver damage may occur. Daily ingestion of 25 mg of sodium selenite, containing 4–5 selenium, after 11 days showed signs of hair and fingernail loss, fatigue, nausea, vomiting, and garlic breath.

Points of Attack: Liver.

Medical Surveillance: NIOSH lists the following tests: Blood Gas Analysis; chest X-ray, electrocardiogram, pulmonary function tests: forced vital capacity, forced expiratory volume (1 s); sputum cytology; urinalysis (routine); white blood cell count/differential. Urine test for selenium (should not exceed 100 µg/L of urine).

First Aid: Skin Contact: Flood all areas of body that have contacted the substance with water. Do not wait to remove contaminated clothing; do it under the water stream. Use soap to help assure removal. Isolate contaminated clothing when removed to prevent contact by others.[52]

Eye Contact: Remove any contact lenses at once. Immediately flush eyes well with copious quantities of water or normal saline for at least 20–30 min. Seek medical attention.

Inhalation: Leave contaminated area immediately; breathe fresh air. Proper respiratory protection must be supplied to any rescuers. If coughing, difficult breathing, or any other symptoms develop, seek medical attention at once, even if symptoms develop many hours after exposure.

Ingestion: Contact a physician, hospital, or poison center at once. If the victim is unconscious or convulsing, do not induce vomiting or give anything by mouth. Assure that the patient's airway is open and lay him on his side with his head lower than his body and transport immediately to a medical facility. If conscious and not convulsing, give a glass of water to dilute the substance. Vomiting should not be induced without a physician's advice.

Note to physician: Inhalation: Bronchodilators, decongestants, and oxygen may be used if necessary. Corticosteroids are useful for treating pneumonitis.

Note to physician: For severe poisoning *do not* use BAL [British Anti-Lewisite, dimercaprol, dithiopropanol ($C_3H_8OS_2$)] as it is contraindicated or ineffective in poisoning from selenium.

Personal Protective Methods: Wear protective gloves and clothing to prevent any reasonable probability of skin contact. Safety equipment suppliers/manufacturers can provide recommendations on the most protective glove/clothing material for your operation. All protective clothing (suits, gloves, footwear, headgear) should be clean, available each day, and put on before work. Contact lenses should not be worn when working with this chemical. Wear dust-proof chemical goggles and face shield unless full face-piece respiratory protection is worn. Employees should wash immediately with soap when skin is wet or contaminated. Provide emergency showers and eyewash.

Respirator Selection: Follow regulations in OSHA 29CFR1910.134 or European Standard EN149. Use a NIOSH/MSHA- or European Standard EN149-approved respirator; or use an approved supplied-air respirator with a full face-piece operated in the positive-pressure mode, or with a full face-piece, hood, or helmet in the continuous-flow mode; or use a NIOSH/MSHA- or European Standard EN149-approved self-contained breathing apparatus with a full face-piece operated in pressure-demand or other positive-pressure mode.

Storage: Color Code—Blue: Health Hazard/Poison: Store in a secure poison location. Prior to working with this chemical you should be trained on its proper handling and storage. Store in a refrigerator or a cool dry place. Where possible, automatically transfer material from drums or other storage containers to process containers. Sources of ignition, such as smoking and open flames, are prohibited where this chemical is handled, used, or stored. Metal containers involving the transfer of this chemical should be grounded and bonded. Wherever this chemical is used, handled, manufactured, or stored, use explosion-proof electrical equipment and fittings. A regulated, marked area should be established, or stored in compliance with OSHA Standard 1910.1045.

Shipping: Selenium disulfide requires a shipping label of "POISONOUS/TOXIC MATERIALS." It falls in Hazard Class 6.1 and Packing Group II.

Spill Handling: Evacuate persons not wearing protective equipment from area of spill or leak until cleanup is complete. Remove all ignition sources. Collect powdered material in the most convenient and safe manner and deposit in sealed containers. Ventilate area after cleanup is complete. It may be necessary to contain and dispose of this chemical as a hazardous waste. If material or contaminated runoff enters waterways, notify downstream users of potentially contaminated waters. Contact your local or federal environmental protection agency for specific recommendations. If employees are required to clean up spills, they must be properly trained and equipped. OSHA 1910.120(q) may be applicable.

Fire Extinguishing: This chemical is a combustible solid. Use dry chemical, carbon dioxide, water spray, or alcohol foam extinguishers. Poisonous gases are produced in fire, including sulfur oxides and selenium. If material or contaminated runoff enters waterways, notify downstream users of potentially contaminated waters. Notify local health and fire officials and pollution control agencies. From a secure, explosion-proof location, use water spray to cool exposed containers. If cooling streams are ineffective (venting sound increases in volume and pitch, tank discolors, or shows any signs of deforming), withdraw immediately to a secure position. If employees are expected to fight fires, they must be trained and equipped in OSHA 1910.156. The only respirators recommended for firefighting are self-contained breathing apparatuses that have full face-pieces and are operated in a pressure-demand or other positive-pressure mode.

References

New Jersey Department of Health and Senior Services. (October 2001). *Hazardous Substances Fact Sheet: Selenium Sulfide.* Trenton, NJ

Selenous acid   S:0190

Molecular Formula: H_2O_3Se
Common Formula: H_2SeO_3
Synonyms: Acide selenieux (French); Acido selenioso (Spanish); Hydrogen selenite; Monohydrated selenium dioxide; Selenious acid; Selenium dioxide
CAS Registry Number: 7783-00-8; *(alt.)* 11140-60-6
RTECS® Number: VS7175000
UN/NA & ERG Number: UN3283/151; UN2811/154

EC Number: 231-974-7
Regulatory Authority and Advisory Bodies
Carcinogenicity: EPA: Not Classifiable as to human carcinogenicity.
Air Pollutant Standard Set. See below, "Permissible Exposure Limits in Air" section.
Clean Water Act: 40CFR401.15 Section 307 Toxic Pollutants as selenium and compounds.
US EPA Hazardous Waste Number (RCRA No.): U204.
RCRA, 40CFR261, Appendix 8 Hazardous Constituents, as selenium compounds, n.o.s., waste number not listed.
Superfund/EPCRA 40CFR355, Extremely Hazardous Substances: TPQ = 1000/10,000 lb (454/4540 kg).
Reportable Quantity (RQ): 10 lb (4.54 kg).
EPCRA Section 313 Form R *de minimis* concentration reporting level: 1.0%.
Canada, WHMIS, Ingredients Disclosure List Concentration 1.0%.
European/International Regulations: not listed in Annex 1.
WGK (German Aquatic Hazard Class): No value assigned.
Description: Selenous acid is a colorless, crystalline solid. Molecular weight = 128.98; Freezing/Melting point = 70°C (decomposition). Hazard Identification (based on NFPA-704 M Rating System): Health 4, Flammability 0, Reactivity 1 (Oxidizer). Soluble in water.
Potential Exposure: Compound Description: Agricultural Chemical; Mutagen. It is used as a reagent for alkaloids and as an oxidizing agent. Isotope is used in labeling radiopharmaceuticals.
Incompatibilities: A strong oxidizer. Incompatible with reducing agents, combustibles, acids. Contact with acids produce toxic and gaseous hydrogen selenide. Attacks metals.
Permissible Exposure Limits in Air
OSHA PEL: 0.2 mg[Se]/m³ TWA.
NIOSH REL: 0.2 mg[Se]/m³ TWA.
ACGIH TLV®[1]: 0.2 mg[Se]/m³ TWA.
NIOSH IDLH: 1 mg [Se]/m³.
Protective Action Criteria (PAC)
TEEL-0: 0.327 mg/m³
PAC-1: 0.98 mg/m³
PAC-2: 250 mg/m³
PAC-3: 250 mg/m³
DFG MAK: 0.05 mg[Se]/m³, inhalable fraction TWA; Peak Limitation Category II(4); Carcinogen Category 3B; Pregnancy Risk Group C.
Arab Republic of Egypt: TWA 0.2 mg[Se]/m³, 1993; Australia: TWA 0.2 mg[Se]/m³, 1993; Austria: MAK 0.1 mg [Se]/m³, 1999; Belgium: TWA 0.2 mg[Se]/m³, 1993; Denmark: TWA 0.1 mg[Se]/m³, 1999; Finland: TWA 0.1 mg [Se]/m³; STEL 0.3 mg[Se]/m³, 1999; Hungary: STEL 0.1 mg [Se]/m³, 1993; Norway: TWA 0.1 mg[Se]/m³, 1999; the Philippines: TWA 0.2 mg[Se]/m³, 1993; Poland: MAC (TWA) 0.1 mg[Se]/m³, 1993; Sweden: NGV 0.1 mg[Se]/m³, 1999; Switzerland: MAK week 0.1 mg[Se]/m³, 1999; Turkey: TWA 0.2 mg[Se]/m³, 1993; United Kingdom:

TWA 0.1 mg[Se]/m^3, 2000; Argentina, Bulgaria, Columbia, Jordan, South Korea, New Zealand, Singapore, Vietnam: ACGIH TLV®: TWA 0.2 mg[Se]/m^3. Several states have set guidelines or standards for selenium compounds as selenium in ambient air[60] ranging from 0.26−1.57 μg/m^3 (Montana) to 0.27 μg/m^3 (Massachusetts) to 0.66 μg/m^3 (New York) to 2.0 μg/m^3 (Florida and North Dakota) to 3.5 μg/m^3 (Virginia) to 4.0 μg/m^3 (Connecticut) to 5.0 μg/m^3 (Nevada).

Determination in Air: NIOSH(IV) Methods #7300, Elements; #7301; #7303; #9102; NIOSH II(7), Method #S190; OSHA Analytical Method ID-121 and ID-125G.

Permissible Concentration in Water: To protect freshwater aquatic life: 35 μg/L as a 24-h average, never to exceed 260 μg/L for recoverable inorganic selenite. *To protect saltwater aquatic life:* 54 μg/L as a 24-h average, never to exceed 410 μg/L. To protect human health: 10 μg/L.[6] There are a variety of foreign standards for selenium in water.[35] The Czech Republic has set 0.1 mg/L in surface water, 0.05 mg/L in drinking water reserve, and 0.01 mg/L in drinking water. Germany[35] has set 0.008 mg/L as a maximum permissible concentration in drinking water. The EEC has set 0.01 mg/L as a MAC for drinking water. Mexico has set 0.01 mg/L for receiving waters used for drinking purposes. Russia has set 0.001 mg/L as a MAC for drinking water. WHO has set 0.01 mg/L as a limit for drinking water. Maine has set[61] a guideline of 0.01 mg/L and Minnesota 0.045 mg/L for drinking water. EPA[62] has proposed 0.05 mg/L as a limit for drinking water.

Determination in Water: Digestion followed by atomic absorption gives total selenium. Dissolved selenium is determined by 0.45 μm filtration prior to the above analysis.

Routes of Entry: Inhalation, ingestion, skin and/or eye contact. Absorbed through the skin.

Harmful Effects and Symptoms

Short Term Exposure: Selenous acid and its salts are capable of penetrating the skin and can produce acute poisonings. Corrosive to the eyes, skin, and respiratory tract. May affect the eyes, resulting in "rose-eye," an allergic-type reaction of the eyelids. It is highly toxic orally. Higher exposures can cause pulmonary edema, a medical emergency that can be delayed for several hours. This can cause death. Inorganic selenium compounds may cause dermatitis. Toxic effects are similar to those of selenium and other selenium compounds. Garlic odor of breath is a common symptom. Pallor, nervousness, depression, and digestive disturbances have been reported in cases of chronic exposure. The most common industrial injuries are irritations and burns of the skin.

Long Term Exposure: Repeated or prolonged contact may cause skin dermatitis. May affect the central nervous system, blood, teeth, and bones. May cause nervousness, depression, pallor, digestive disturbances. Kidney and liver damage may occur. Daily ingestion of 25 mg of sodium selenite, containing 4−5 selenium, after 11 days showed signs of hair and fingernail loss, fatigue, nausea, vomiting, and garlic breath.

Points of Attack: Skin, eyes, liver, kidneys, nervous system.

Medical Surveillance: NIOSH lists the following tests: Blood Gas Analysis; chest X-ray, electrocardiogram, pulmonary function tests: forced vital capacity, forced expiratory volume (1 s); sputum cytology; urinalysis (routine); white blood cell count/differential. Urine test for selenium (should not exceed 100 μg/L of urine). Examination of the nervous system. Examination by a qualified allergist.

First Aid: If this chemical gets into the eyes, remove any contact lenses at once and irrigate immediately for at least 15 min, occasionally lifting upper and lower lids. Seek medical attention immediately. If this chemical contacts the skin, remove contaminated clothing and wash immediately with soap and water. Seek medical attention immediately. If this chemical has been inhaled, remove from exposure, begin rescue breathing (using universal precautions, including resuscitation mask) if breathing has stopped and CPR if heart action has stopped. Transfer promptly to a medical facility. When this chemical has been swallowed, get medical attention. If victim is *conscious*, administer water or milk. Do not induce vomiting.

Note to physician: For severe poisoning *do not* use BAL [British Anti-Lewisite, dimercaprol, dithiopropanol (C$_3$H$_8$OS$_2$)] as it is contraindicated or ineffective in poisoning from selenium.

Personal Protective Methods: Wear protective gloves and clothing to prevent any reasonable probability of skin contact. Safety equipment suppliers/manufacturers can provide recommendations on the most protective glove/clothing material for your operation. All protective clothing (suits, gloves, footwear, headgear) should be clean, available each day, and put on before work. Contact lenses should not be worn when working with this chemical. Wear dust-proof chemical goggles and face shield unless full face-piece respiratory protection is worn. Employees should wash immediately with soap when skin is wet or contaminated. Provide emergency showers and eyewash.

Respirator Selection: Follow regulations in OSHA 29CFR1910.134 or European Standard EN149. Use a NIOSH/MSHA- or European Standard EN149-approved respirator; or use an approved supplied-air respirator with a full face-piece operated in the positive-pressure mode, or with a full face-piece, hood, or helmet in the continuous-flow mode; or use a NIOSH/MSHA- or European Standard EN149-approved self-contained breathing apparatus with a full face-piece operated in pressure-demand or other positive-pressure mode.

Storage: Color Code—Blue: Health Hazard/Poison: Store in a secure poison location. Prior to working with this chemical you should be trained in its proper handling and storage. Store in tightly closed containers in a cool, well-ventilated area away from reducing agents, acids, combustibles. Where possible, automatically transfer material from drums or other storage containers to process containers.

Shipping: This compound requires a shipping label of "POISONOUS/TOXIC MATERIALS." It falls in Hazard Class 6.1 and Packing Group II.

Spill Handling: Evacuate persons not wearing protective equipment from area of spill or leak until cleanup is complete. Remove all ignition sources. Collect powdered material in the most convenient and safe manner and deposit in sealed containers. Liquid containing selenium inorganic compounds should be absorbed in vermiculite, dry sand, earth, or similar material. Ventilate area after cleanup is complete. It may be necessary to contain and dispose of this chemical as a hazardous waste. If material or contaminated runoff enters waterways, notify downstream users of potentially contaminated waters. Contact your local or federal environmental protection agency for specific recommendations. If employees are required to clean up spills, they must be properly trained and equipped. OSHA 1910.120(q) may be applicable.

Fire Extinguishing: This chemical is a combustible solid. Use dry chemical, carbon dioxide, water spray, or alcohol foam extinguishers. Poisonous gases are produced in fire, including selenium oxides and selenium. If material or contaminated runoff enters waterways, notify downstream users of potentially contaminated waters. Notify local health and fire officials and pollution control agencies. From a secure, explosion-proof location, use water spray to cool exposed containers. If cooling streams are ineffective (venting sound increases in volume and pitch, tank discolors, or shows any signs of deforming), withdraw immediately to a secure position. If employees are expected to fight fires, they must be trained and equipped in OSHA 1910.156. The only respirators recommended for firefighting are self-contained breathing apparatuses that have full face-pieces and are operated in a pressure-demand or other positive-pressure mode.

Disposal Method Suggested: Consult with environmental regulatory agencies for guidance on acceptable disposal practices. Generators of waste containing this contaminant (≥ 100 kg/mo) must conform with EPA regulations governing storage, transportation, treatment, and waste disposal.

References

US Environmental Protection Agency. (November 30, 1987). *Chemical Hazard Information Profile: Selenious Acid.* Washington, DC: Chemical Emergency Preparedness Program

New Jersey Department of Health and Senior Services. (September 1999). *Hazardous Substances Fact Sheet: Selenious Acid.* Trenton, NJ

Semicarbazide hydrochloride S:0200

Molecular Formula: CH_6ClN_3O
Common Formula: $H_2NNHCONH_2 \cdot HCl$
Synonyms: Amidourea hydrochloride; Aminourea hydrochloride; Carbamylhydrazine hydrochloride; CH; Clorhidrato de semicarbazide (Spanish); Hydrazinecarboxamide monohydrochloride

CAS Registry Number: 563-41-7; 18396-65-1
RTECS® Number: VT3500000
UN/NA & ERG Number: Not regulated.
EC Number: 209-247-0

Regulatory Authority and Advisory Bodies
Carcinogenicity: IARC: Animal Sufficient Evidence; Human Inadequate Evidence, *not classifiable as carcinogenic to humans*, Group 3, 1987.
Superfund/EPCRA 40CFR355, Extremely Hazardous Substances: TPQ = 1000/10,000 lb (454/4540 kg).
Reportable Quantity (RQ): 1000 lb (454 kg).
European/International Regulations: not listed in Annex 1.
WGK (German Aquatic Hazard Class): No value assigned.

Description: Semicarbazide hydrochloride is a white crystalline solid. Molecular weight = 111.55; Freezing/Melting point = $172-175°C$ (decomposes). Hazard Identification (based on NFPA-704 M Rating System): Health 2, Flammability 1, Reactivity 0. Soluble in water.

Potential Exposure: This material is used as a reagent for ketones and aldehydes with which it affords crystalline compounds having characteristic freezing/melting points. Also used in isolation of hormones and certain fractions from essential oils.

Incompatibilities: Oxidizers. May ignite combustible materials (wood, oil, paper).

Permissible Exposure Limits in Air
Protective Action Criteria (PAC)
TEEL-0: 20 mg/m^3
PAC-1: 60 mg/m^3
PAC-2: 100 mg/m^3
PAC-3: 100 mg/m^3

Routes of Entry: Inhalation, ingestion, skin and/or eye contact.

Harmful Effects and Symptoms
Short Term Exposure: Contact can irritate the eyes and skin. Toxic by ingestion. Symptoms of exposure include convulsions; irritation of skin, eyes, and mucous membranes; gastroenteric disturbances; and anemia. High level of exposure may affect the nervous system.

Long Term Exposure: There is limited evidence that this chemical causes lung cancer in animals. High or repeated exposure may cause nerve damage with weakness, a feeling of "pins and needles," and loss of coordination in the limbs.

Points of Attack: Nervous system.

Medical Surveillance: Examination of the nervous system.

First Aid: If this chemical gets into the eyes, remove any contact lenses at once and irrigate immediately for at least 15 min, occasionally lifting upper and lower lids. Seek medical attention immediately. If this chemical contacts the skin, remove contaminated clothing and wash immediately with soap and water. Seek medical attention immediately. If this chemical has been inhaled, remove from exposure, begin rescue breathing (using universal precautions, including resuscitation mask) if breathing has stopped and CPR if

heart action has stopped. Transfer promptly to a medical facility. When this chemical has been swallowed, get medical attention. Give large quantities of water and induce vomiting. Do not make an unconscious person vomit.

Personal Protective Methods: Wear protective gloves and clothing to prevent any reasonable probability of skin contact. Safety equipment suppliers/manufacturers can provide recommendations on the most protective glove/clothing material for your operation. All protective clothing (suits, gloves, footwear, headgear) should be clean, available each day, and put on before work. Contact lenses should not be worn when working with this chemical. Wear dust-proof chemical goggles and face shield unless full face-piece respiratory protection is worn. Employees should wash immediately with soap when skin is wet or contaminated. Provide emergency showers and eyewash.

Respirator Selection: Follow regulations in OSHA 29CFR1910.134 or European Standard EN149. Use a NIOSH/MSHA- or European Standard EN149-approved respirator; or use an approved supplied-air respirator with a full face-piece operated in the positive-pressure mode, or with a full face-piece, hood, or helmet in the continuous-flow mode; or use a NIOSH/MSHA- or European Standard EN149-approved self-contained breathing apparatus with a full face-piece operated in pressure-demand or other positive-pressure mode.

Storage: Color Code—Blue: Health Hazard/Poison: Store in a secure poison location. Prior to working with this chemical you should be trained in its proper handling and storage. Store in a refrigerator and protect from oxidizers, combustible materials, moisture. Where possible, automatically transfer material from drums or other storage containers to process containers. Sources of ignition, such as smoking and open flames, are prohibited where this chemical is handled, used, or stored. Metal containers involving the transfer of this chemical should be grounded and bonded. Wherever this chemical is used, handled, manufactured, or stored, use explosion-proof electrical equipment and fittings. A regulated, marked area should be established, or stored in compliance with OSHA Standard 1910.1045.

Shipping: Not regulated.

Spill Handling: Evacuate persons not wearing protective equipment from area of spill or leak until cleanup is complete. Remove all ignition sources. Collect powdered material in the most convenient and safe manner and deposit in sealed containers. Ventilate area after cleanup is complete. It may be necessary to contain and dispose of this chemical as a hazardous waste. If material or contaminated runoff enters waterways, notify downstream users of potentially contaminated waters. Contact your local or federal environmental protection agency for specific recommendations. If employees are required to clean up spills, they must be properly trained and equipped. OSHA 1910.120(q) may be applicable.

Fire Extinguishing: This material may burn, but does not ignite readily. For small fires, use dry chemical, carbon dioxide, water spray, or alcohol-resistant foam. For large fires, use water spray, fog, or foam. Poisonous gases are produced in fire, including nitrogen oxides and hydrogen chloride. If material or contaminated runoff enters waterways, notify downstream users of potentially contaminated waters. Notify local health and fire officials and pollution control agencies. From a secure, explosion-proof location, use water spray to cool exposed containers. If cooling streams are ineffective (venting sound increases in volume and pitch, tank discolors, or shows any signs of deforming), withdraw immediately to a secure position. If employees are expected to fight fires, they must be trained and equipped in OSHA 1910.156. The only respirators recommended for firefighting are self-contained breathing apparatuses that have full face-pieces and are operated in a pressure-demand or other positive-pressure mode.

References

US Environmental Protection Agency. (November 30, 1987). *Chemical Hazard Information Profile: Semicarbazide Hydrochloride.* Washington, DC: Chemical Emergency Preparedness Program

Silane   S:0210

Molecular Formula: H_4Si
Common Formula: SiH_4
Synonyms: Monosilane; Silano (Spanish); Silicane; Silicon tetrahydride
CAS Registry Number: 7803-62-5
RTECS® Number: VV1400000
UN/NA & ERG Number: UN2203/116
EC Number: 232-263-4

Regulatory Authority and Advisory Bodies

Department of Homeland Security Screening Threshold Quantity (pounds): *Release hazard* 10,000 ($\geq 1.00\%$ concentration).

Air Pollutant Standard Set. See below, "Permissible Exposure Limits in Air" section.

Clean Air Act: Accidental Release Prevention/Flammable Substances (Section 112[r], Table 3), TQ = 10,000 lb (4540 kg).

Canada, WHMIS, Ingredients Disclosure List Concentration 1.0%.

Regulatory Authority and Advisory Bodies

Department of Homeland Security Screening Threshold Quantity (pounds): *Release hazard* 10,000 ($\geq 1.00\%$ concentration).

European/International Regulations: not listed in Annex 1.

WGK (German Aquatic Hazard Class): 1—Low hazard to waters.

Description: Silane is a colorless, pyrophoric gas with a repulsive odor. Molecular weight = 32.13; Boiling point = −169°C; Specific gravity (H_2O:1) = 1.31 at 25°C; Specific gravity (H_2O:1) = 1.31 at 25°C; Freezing/Melting point = −111.7°C; Flash point = −236°C. Relative vapor

mining of diatomaceous earth or fabrication of products therefrom.

Incompatibilities: Fluorine, oxygen difluoride, chlorine trifluoride.

Permissible Exposure Limits in Air

Silica, amorphous silica, fused

OSHA PEL: 30 mg/m^3 total dust/divided by the value "%SiO$_2$ + 2" TWA; *either one* of the following methods: 250 mppcf respirable dust/divided by the value "%SiO$_2$ + 5" *or* 10 mg/m^3 respirable dust/divided by the value "%SiO$_2$ + 2."

NIOSH REL: 0.05 mg/m^3, respirable dust TWA; See *NIOSH Pocket Guide*, Appendix A.

ACGIH TLV®[1]: withdrawn.

DFG MAK (*CAS: 60676-86-0 & 7699-41-4*): 0.3 mg/m^3 respirable fraction; Pregnancy Risk Group C.

Silica, amorphous precipitated and gel and diatomaceous earth, uncalcined

OSHA PEL: *either one of the methods:* 20 mppcf [millions of particles per cubic foot of air, based on impinger samples counted by light-field techniques] *or* 80 mg/m^3 divided by the value "%SiO$_2$," TWA.

NIOSH REL: 6 mg/m^3 TWA.

ACGIH TLV®[1]: withdrawn for the following: silica amorphous, precipitated and gel; silica fume; silica fused; diatomaceous earth, calcined.

NIOSH IDLH: 3000 mg/m^3.

Protective Action Criteria (PAC)

Amorphous hydrated

TEEL-0: 6 mg/m^3
PAC-1: 18 mg/m^3
PAC-2: 100 mg/m^3
PAC-3: 500 mg/m^3

Amorphous fumed

TEEL-0: 6 mg/m^3
PAC-1: 18 mg/m^3
PAC-2: 30 mg/m^3
PAC-3: 500 mg/m^3

DFG MAK (*diatomaceous earth, uncalcined*): 4 mg/m^3, inhalable fraction; Pregnancy Risk Group C.

Austria: MAK 4 mg/m^3, 1999; Norway: TWA 1.5 mg/m^3 (respirable dust), 1999; Switzerland: MAK-W 4 mg/m^3, 1999; Thailand: TWA 80 mg/m^3, 1993; United Kingdom: TWA 6 mg/m^3, total dust, 2000; United Kingdom: TWA 1.2 mg/m^3 (respirable dust), 2000; Argentina, Bulgaria, Columbia, Jordan, South Korea, New Zealand, Singapore, Vietnam: ACGIH TLV®: TWA 3 mg/m^3 (respirable fraction, for particulate matter containing no asbestos and <1% crystalline silica). Russia[43] set a MAC of 2 mg/m^3 in work-place air.

Determination in Air: Use NIOSH(IV) Method #7501 (any form); OSHA Analytical method ID-125G (fumed).

Routes of Entry: Inhalation, skin and/or eye contact.

Harmful Effects and Symptoms

Short Term Exposure: Amorphous fused silica can affect you when breathed in. Exposure can cause a very serious lung disease called silicosis, with cough and shortness of breath. Very high exposures can cause this problem to develop in a few weeks, or with lower exposures it may occur over many years. Silicosis can cause death. If silicosis develops, chances of getting tuberculosis are increased. The disease may progress, with or without continued exposure. If it does, this can be crippling or even fatal.

Points of Attack: Eyes, respiratory system.

Medical Surveillance: For those with frequent or potentially high exposure (half the TLV or greater), the following are recommended before beginning work and at regular times after that: lung function tests. Chest X-ray every 1−3 years. If abnormal chest X-ray develops, the following should be done periodically: skin test for TB.

First Aid: If this chemical gets into the eyes, remove any contact lenses at once and irrigate immediately for at least 15 min, occasionally lifting upper and lower lids. If this chemical has been inhaled, remove from exposure. Transfer promptly to a medical facility.

Personal Protective Methods: All protective clothing (suits, gloves, footwear, headgear) should be clean, available each day, and put on before work. Contact lenses should not be worn when working with this chemical. Wear dust-proof chemical goggles and face shield unless full face-piece respiratory protection is worn. Employees should wash immediately with soap when skin is wet or contaminated. Provide emergency showers and eyewash.

Respirator Selection: NIOSH: *Up to 30 mg/m^3:* Qm (APF = 25) (any quarter-mask respirator). *Up to 60 mg/m^3:* 95XQ (APF = 10) [Any particulate respirator equipped with an N95, R95, or P95 filter (including N95, R95, and P95 filtering face-pieces) except quarter-mask respirators. The following filters may also be used: N99, R99, P99, N100, R100, P100] or Sa (APF = 10) (any supplied-air respirator). *Up to 150 mg/m^3:* Sa:Cf (APF = 25) (any supplied-air respirator operated in a continuous-flow mode) or PaprHie (APF = 25) (any powered, air-purifying respirator with a high-efficiency particulate filter). *Up to 300 mg/m^3:* 100F (APF = 50) (any air-purifying, full-face-piece respirator with an N100, R100, or P100 filter) or SaT: Cf (APF = 50) (any supplied-air respirator that has a tight-fitting face-piece and is operated in a continuous-flow mode) or PaprTHie (APF = 50) (any powered, air-purifying respirator with a tight-fitting face-piece and a high-efficiency particulate filter) or SCBAF (APF = 50) (any self-contained breathing apparatus with a full face-piece) or SaF (APF = 50) (any supplied-air respirator with a full face-piece). *Up to 3000 mg/m^3:* Sa: Pd,Pp (APF = 1000) (any supplied-air respirator operated in a pressure-demand or other positive-pressure mode). *Emergency or planned entry into unknown concentrations or IDLH conditions:* SCBAF: Pd,Pp (APF = 10,000) (any self-contained breathing apparatus that has a full faceplate and is operated in a pressure-demand or other positive-pressure mode) or SaF: Pd,Pp: ASCBA (APF = 10,000) (any supplied-air respirator that has a full face-piece and is operated in a pressure-demand or other

positive-pressure mode in combination with an auxiliary, self-contained breathing apparatus operated in a pressure-demand or other positive-pressure mode). *Escape:* 100F (APF = 50) (any air-purifying, full-face-piece respirator with an N100, R100, or P100 filter) or SCBAE (any appropriate escape-type, self-contained breathing apparatus).
Storage: Color Code—Green: General storage may be used. Prior to working with this chemical you should be trained on its proper handling and storage. Amorphous fused silica must be stored to avoid contact with powerful oxidizers including fluorine, oxygen difluoride, chrome trifluoride, and manganese trioxide, since violent reactions occur.
Spill Handling: Evacuate persons not wearing protective equipment from area of spill or leak until cleanup is complete. Remove all ignition sources. Collect powdered material in the most convenient and safe manner and deposit in sealed containers. Ventilate area after cleanup is complete. It may be necessary to contain and dispose of this chemical as a hazardous waste. If material or contaminated runoff enters waterways, notify downstream users of potentially contaminated waters. Contact your local or federal environmental protection agency for specific recommendations. If employees are required to clean up spills, they must be properly trained and equipped. OSHA 1910.120(q) may be applicable.
Fire Extinguishing: Extinguish fire using an agent suitable for type of surrounding fire. Amorphous fused silica itself does not burn. If material or contaminated runoff enters waterways, notify downstream users of potentially contaminated waters. Notify local health and fire officials and pollution control agencies. If employees are expected to fight fires, they must be trained and equipped in OSHA 1910.156. The only respirators recommended for firefighting are self-contained breathing apparatuses that have full face-pieces and are operated in a pressure-demand or other positive-pressure mode.
Disposal Method Suggested: Sanitary landfill.
References
Sax, N. I. (Ed.). (1981). *Dangerous Properties of Industrial Materials Report*, 1, No. 6, 94
New Jersey Department of Health and Senior Services. (May 1999). *Hazardous Substances Fact Sheet: Silica, Amorphous-Dimataceous Earth.* Trenton, NJ
New Jersey Department of Health and Senior Services. (April 2003). *Hazardous Substances Fact Sheet: Silica, Amorphous (Fume).* Trenton, NJ
New Jersey Department of Health and Senior Services. (April 2002). *Hazardous Substances Fact Sheet: Silica, Amorphous Fused.* Trenton, NJ

Silica, crystalline   S:0230

Molecular Formula: O_2Si
Common Formula: SiO_2
Synonyms: Cristobalite: Calcined diatomite; Silica, cristobalite; Silica, crystalline-cristobalite

Silica, quartz: Agate; Amethyst; Chalcedony; Cherts; Flint; Onyx; Pure quartz; Quartz; Rose quartz; Sand; Silica flour (powdered crystalline silica); Silicic anhydride
Tridymite: Christensenite; Silica, crystalline-tridymite; Tridimite (French); α-Tridymite; Tridymite; Tridymite 118
Tripoli: Finely ground silica; Silica flour
CAS Registry Number: 14808-60-7 (crystalline quartz); 14464-46-1 (cristobalite); *(alt.)* 1317-48-2; 105269-70-3 (cristobalite); 15468-32-3; *(alt.)* 1317-94-8; 12414-70-9 (tridymite); 1317-95-9 (tripoli)
RTECS® Number: VV7325000 (cristobalite); VV7330000 (quartz); VV7335000 (tridymite); VV7336000 (tripoli)
EC Number: 238-878-4; 238-455-4 (cristobalite); 239-487-1 (tridymite)
Regulatory Authority and Advisory Bodies
Carcinogenicity: IARC (*cristobalite, tridymite, tripoli, and quartz*): Human Sufficient Evidence; Animal Sufficient Evidence, *carcinogenic to humans*, Group 1, 1997; NTP (*respirable cristobalite, tridymite, tripoli, and quartz*): 11th Report on Carcinogens, 2004: Known to be a human carcinogen; NIOSH (*cristobalite, tridymite, tripoli, and quartz*): Potential occupational carcinogen.
Air Pollutant Standard Set. See below, "Permissible Exposure Limits in Air" section.
California Proposition 65 Chemical: Cancer [Silica, crystalline (airborne particles of respirable size)] 10/1/88.
Canada, WHMIS, Ingredients Disclosure List Concentration 1.0%, all forms.
WGK (German Aquatic Hazard Class): Nonwater polluting agent.
Description: Crystalline silica is a component of many mineral dusts and materials which melts to a glass at very high temperature. Molecular weight = 60.09; Specific gravity (H_2O:1) = 2.66 at 25°C; Boiling point (quartz) = 2230°C; Freezing/Melting point (quartz) = 1610°C. Hazard Identification (based on NFPA-704 M Rating System): Health 1, Flammability 0, Reactivity 0. Insoluble in water.
Potential Exposure: Compound Description (cristobalite): Tumorigen, Human Data; (quartz): Tumorigen, Mutagen, Organometallic, Human Data; (tridymite): Tumorigen, Mutagen; Human Data. Cristobalite is used in the manufacture of water glass, refractories, abrasives, ceramics, and enamels. Quartz is used as a mineral, natural or synthetic fiber. Tridymite is used as a filtering and insulating media and as a refractory material for furnace linings. Workers are potentially exposed to crystalline silica in industries such as granite quarrying and cutting, foundry operations; metal, coal, dentistry, painting, and nonmetallic mining; and manufacture of clay and glass products.
Incompatibilities: Violent reactions with powerful oxidizers: fluorine, chlorine trifluoride; manganese trioxide; oxygen difluoride, hydrogen peroxide, etc.; acetylene; ammonia.
Permissible Exposure Limits in Air
OSHA PEL: (*silica, crystalline α-quartz, CAS: 14808-60-7 & Silica, crystalline tripoli CAS: 1317-95-9*): 30 mg/m³ total

dust/divided by the value "%SiO$_2$ + 2" TWA; *either one* of the following methods: 250 mppcf respirable dust/divided by the value "%SiO$_2$ + 5" *or* 10 mg/m^3 respirable dust/divided by the value "%SiO$_2$ + 2"; For Silica crystalline, tridymite, CAS: 15468-32-3 and Silica, crystalline cristobalite, CAS 14464-46-1 use ½ the values calculated above using the count or mass formula for quartz. See also Table Z-3 of 29 CFR 1910.1000.
NIOSH REL (*silica, crystalline, all forms*): 0.05 mg/m^3, respirable dust TWA; Limit exposure to lowest feasible concentration. See *NIOSH Pocket Guide*, Appendix A.
ACGIH TLV®[1] (*Silica, crystalline α-quartz, CAS: 14808-60-7 & Crystalline cristobalite, CAS 14464-46-1*): 0.025 mg/m^3 TWA, Respirable fraction of dust; Suspected Human Carcinogen.
NIOSH IDLH: (*cristobalite and tridymite*) 25 mg/m^3; (*quartz and tripoli*) 50 mg/m^3.
Protective Action Criteria (PAC)
14808-60-7 Silica, crystalline-quartz; (Silicon dioxide)
TEEL-0: 0.025 mg/m^3
PAC-1: 0.075 mg/m^3
PAC-2: 2 mg/m^3
PAC-3: 50 mg/m^3
14464-46-1 (cristobalite)
TEEL-0: 0.025 mg/m^3
PAC-1: 0.075 mg/m^3
PAC-2: 2 mg/m^3
PAC-3: 25 mg/m^3
DFG MAK (*silica, crystalline, all forms*): respirable; Carcinogen Category 1.
Austria: MAK 0.15 mg/m^3, 1999; Belgium: TWA 0.05 mg/m^3, 1993; Finland: TWA 0.1 mg/m^3, 1999; France: VME 10/(2X + 2), 1999; Norway: TWA 0.05 mg/m^3 (respirable dust), 1999; Russia: STEL 1 mg/m^3 (total dust), 1993; the Netherlands: MAC-TGG 0.075 mg/m^3, 2003; Switzerland: MAK-W 0.15 mg/m^3, 1999; Thailand: TWA 10 mg/m^3, 1993; United Kingdom: TWA0.3 mg/m^3, 2000; Argentina, Bulgaria, Columbia, Jordan, South Korea, New Zealand, Singapore, Vietnam: ACGIH TLV®: TWA 0.05 mg/m^3 (respirable dust).
Determination in Air: Use: Use NIOSH Analytical Method (cristobalite), #7500, Silica, crystalline, by XRD; #7601, by VIS; #7602, by IR; #7603, Silica, crystalline in coal mine dust, by IR; OSHA Analytical Method ID-142; 125-G.
Routes of Entry: Inhalation of dust, skin and/or eye contact.
Harmful Effects and Symptoms
Short Term Exposure: Irritates the eyes. Inhalation can cause cough, dyspnea (breathing difficulty), wheezing.
Long Term Exposure: Can cause decreased pulmonary function, progressive respiratory symptoms, fibrosis (silicosis). A potential occupational carcinogen. Silicosis is a very serious lung disease that can cause cough and shortness of breath. Silicosis can develop in a few weeks at very high exposures, or it may occur over many years with lower exposures. Silicosis can cause death. If silicosis develops,

risk of developing tuberculosis is increased. The disease may progress with or without continued exposure. If it does, this can be crippling or even fatal. Very fine silica or "silica flour" is even more hazardous.
Points of Attack: Eyes, respiratory system. Cancer site in animals: lungs.
Medical Surveillance: For those with frequent or potentially high exposure (half the TLV or greater), the following are recommended before beginning work and at regular times after that: lung function tests. Chest X-ray every 1−3 years. Chest X-rays should be read by a "B-reader," a doctor with special training for reading X-rays of the lungs. If abnormal chest X-ray develops, the following should be done periodically, skin test for tuberculosis.
First Aid: If this chemical gets into the eyes, remove any contact lenses at once and irrigate immediately for at least 15 min, occasionally lifting upper and lower lids. Seek medical attention immediately. If this chemical contacts the skin, remove contaminated clothing and wash immediately with soap and water. Seek medical attention immediately. If this chemical has been inhaled, remove from exposure, begin rescue breathing (using universal precautions, including resuscitation mask) if breathing has stopped and CPR if heart action has stopped. Transfer promptly to a medical facility. When this chemical has been swallowed, get medical attention. Give large quantities of water and induce vomiting. Do not make an unconscious person vomit.
Personal Protective Methods: Wear protective gloves and clothing to prevent any reasonable probability of skin contact. Safety equipment suppliers/manufacturers can provide recommendations on the most protective glove/clothing material for your operation. All protective clothing (suits, gloves, footwear, headgear) should be clean, available each day, and put on before work. Contact lenses should not be worn when working with this chemical. Wear dust-proof chemical goggles and face shield unless full face-piece respiratory protection is worn. Employees should wash immediately with soap when skin is wet or contaminated. Provide emergency showers and eyewash. Specific engineering controls are recommended in the NIOSH document CT-112-23b, *Control Technology for Bag Filling Operations at Manville Products Corporation, Lompoc, CA.* NIOSH recommends that *silica sand* or materials containing *more than 1% free silica* be prohibited as abrasive substances in abrasive blasting and cleaning operations. The New Jersey Department of Health and Senior Services document *Stop Silicosis in Sandblasters; Use Silica Substitutes* recommends the substitution of less toxic *alumina* in pottery, and *quartz free grit* in abrasive blasting for silica, cristobalite. Follow OSHA regulations (OSHA Standard 1910.94) for abrasive blasting operations.
Respirator Selection: NIOSH: *Up to 0.5 mg/m^3:* 95XQ (APF = 10) (any particulate respirator equipped with an N95, R95, or P95 filter (including N95, R95, and P95 filtering face-pieces) except quarter-mask respirators. The following filters may also be used: N99, R99, P99, N100,

R100, P100). *Up to 1.25 mg/m³:* PaprHie (APF = 25) (any powered, air-purifying respirator with a high-efficiency particulate filter) or Sa:Cf (APF = 25) (any supplied-air respirator operated in a continuous-flow mode). *Up to 2.5 mg/m³:* 100F (APF = 50) (any air-purifying, full-face-piece respirator with an N100, R100, or P100 filter) or PaprTHie (APF = 50) (any powered, air-purifying respirator with a tight-fitting face-piece and a high-efficiency particulate filter). *Up to 25 mg/m³:* Sa: Pd,Pp (APF = 1000) (any supplied-air respirator operated in a pressure-demand or other positive-pressure mode). *Emergency or planned entry into unknown concentrations or IDLH conditions:* SCBAF: Pd, Pp (APF = 10,000) (any self-contained breathing apparatus that has a full faceplate and is operated in a pressure-demand or other positive-pressure mode) or SaF: Pd,Pp: ASCBA (APF = 10,000) (any supplied-air respirator that has a full face-piece and is operated in a pressure-demand or other positive-pressure mode in combination with an auxiliary, self-contained breathing apparatus operated in a pressure-demand or other positive-pressure mode). *Escape:* 100F (APF = 50) (any air-purifying, full-face-piece respirator with an N100, R100, or P100 filter) or SCBAE (any appropriate escape-type, self-contained breathing apparatus).

Storage: Color Code—Green: General storage may be used. Prior to working with this chemical you should be trained on its proper handling and storage. Silica must be stored to avoid contact with strong oxidizers (such as chlorine, bromine, and fluorine), since violent reactions occur. A regulated, marked area should be established, or stored in compliance with OSHA Standard 1910.1045.

Spill Handling: Evacuate persons not wearing protective equipment from area of spill or leak until cleanup is complete. Remove all ignition sources. Use HEPA vacuum or wet method to reduce dust during cleanup. Do not dry sweep. Collect powdered material in the most convenient and safe manner and deposit in sealed containers. Ventilate area after cleanup is complete. It may be necessary to contain and dispose of this chemical as a hazardous waste. If material or contaminated runoff enters waterways, notify downstream users of potentially contaminated waters. Contact your local or federal environmental protection agency for specific recommendations. If employees are required to clean up spills, they must be properly trained and equipped. OSHA 1910.120(q) may be applicable.

Fire Extinguishing: Extinguish fire using an agent suitable for type of surrounding fire. Silica itself does not burn. If employees are expected to fight fires, they must be trained and equipped in OSHA 1910.156. The only respirators recommended for firefighting are self-contained breathing apparatuses that have full face-pieces and are operated in a pressure-demand or other positive-pressure mode.

Disposal Method Suggested: Sanitary landfill.

References

National Institute for Occupational Safety and Health. (1974). *Criteria for a Recommended Standard:*
Occupational Exposure to Crystalline Silica, NIOSH Document No. 75-120. Washington, DC

New Jersey Department of Health and Senior Services. (April 2002). *Hazardous Substances Fact Sheet: Silica, Quartz.* Trenton, NJ

New Jersey Department of Health and Senior Services. (April 2002). *Hazardous Substances Fact Sheet: Silica, Tripoli.* Trenton, NJ

New Jersey Department of Health and Senior Services. (April 2002). *Hazardous Substances Fact Sheet: Silica Cristobalite.* Trenton, NJ

Silicon   S:0240

Molecular Formula: Si
Synonyms: Defoamer S-10; Elemental silicon; Silicon, amorphous powder
CAS Registry Number: 7440-21-3
RTECS® Number: VW0400000
UN/NA & ERG Number: UN1346 (powder, amorphous)/170
EC Number: 231-130-8
Regulatory Authority and Advisory Bodies
Air Pollutant Standard Set. See below, "Permissible Exposure Limits in Air" section.
WGK (German Aquatic Hazard Class): No value assigned.
Description: Silicon is a nonmetallic element which is known as silicon metal. Does not occur free in nature, but is found in silicon dioxide (silica) and in various silicates.[NIOSH] It is a steel-gray crystalline solid or a black-brown amorphous material. Molecular weight = 28.09; Specific gravity (H₂O:1) = 2.33 at 25°C; Boiling point = 2345°C; Freezing/Melting point = 1410°C. Minimum Explosive Concentration = 160 g/m³. Ignition temperature of dust cloud = Unknown; Minimum Explosive concentration = 0.11 oz/ft³.[USBM] Relative explosion hazard of dust: Strong. Hazard Identification (based on NFPA-704 M Rating System): [powder] Health 1, Flammability 2, Reactivity 1. Nearly insoluble in water.
Potential Exposure: Compound Description: Primary Irritant. Silicon may be used in the manufacture of silanes, silicon tetrachloride, ferrosilicon, silicones. It is used in purified elemental form in transistors and photovoltaic cells.
Incompatibilities: The powder is highly flammable. Keep this material away from calcium, carbonates, chlorine, fluorine, oxidizers, cesium carbide, alkaline carbonates.
Permissible Exposure Limits in Air
OSHA PEL: 15 mg/m³ TWA, total dust; 5 mg/m³ TWA, respirable fraction.
NIOSH REL: 10 mg/m³ TWA, total dust; 5 mg/m³ TWA, respirable fraction.
ACGIH TLV®[1]: withdrawn.
Protective Action Criteria (PAC)
TEEL-0: 15 mg/m³

PAC-1: 45 mg/m^3
PAC-2: 75 mg/m^3
PAC-3: 500 mg/m^3
Australia: TWA 10 mg/m^3, 1993; Belgium: TWA 10 mg/m^3, 1993; Denmark: TWA 10 mg/m^3, 1999; France: VME 10 mg/m^3, 1999; Norway: TWA 10 mg/m^3, 1999; Switzerland: MAK-W 4 mg/m^3, 1999; United Kingdom: TWA 10 mg/m^2, total inhalable dust, 2000; the Netherlands: MAC-TGG 10 mg/m^3, 2003; Argentina, Bulgaria, Columbia, Jordan, South Korea, New Zealand, Singapore, Vietnam: ACGIH TLV®: TWA 10 mg/m^3.

Determination in Air: Filter; none; Gravimetric; NIOSH IV, Particulates NOR: Method #0500, total dust; Method #0600 (respirable dust).

Routes of Entry: Inhalation, ingestion, skin and/or eye contact.

Harmful Effects and Symptoms

Short Term Exposure: Causes irritation of the eyes, skin, and upper respiratory system; cough. Silicon dust does not produce significant organic disease or toxic effect when exposures are kept under reasonable control. Unpleasant deposits may be caused in eyes, ears, and nasal passages, and injury to the skin and mucous membranes may be caused by the dust itself or by cleansing procedures used for its removal.

Points of Attack: Eyes, skin, respiratory system.

Medical Surveillance: Lung function test. Consider chest X-ray following acute overexposure.

First Aid: If this chemical gets into the eyes, remove any contact lenses at once and irrigate immediately for at least 15 min, occasionally lifting upper and lower lids. Seek medical attention immediately. If this chemical has been inhaled, remove from exposure. Transfer promptly to a medical facility.

Personal Protective Methods: Wear protective gloves and clothing to prevent any reasonable probability of skin contact. Safety equipment suppliers/manufacturers can provide recommendations on the most protective glove/clothing material for your operation. All protective clothing (suits, gloves, footwear, headgear) should be clean, available each day, and put on before work. Contact lenses should not be worn when working with this chemical. Wear dust-proof chemical goggles and face shield unless full face-piece respiratory protection is worn. Employees should wash immediately with soap when skin is wet or contaminated. Provide emergency showers and eyewash.

Respirator Selection: Follow regulations in OSHA 29CFR1910.134 or European Standard EN149. Use a NIOSH/MSHA- or European Standard EN149-approved respirator; or use an approved supplied-air respirator with a full face-piece operated in the positive-pressure mode, or with a full face-piece, hood, or helmet in the continuous-flow mode; or use a NIOSH/MSHA- or European Standard EN149-approved self-contained breathing apparatus with a full face-piece operated in pressure-demand or other positive-pressure mode.

Storage: Color Code—Red: Flammability Hazard (*amorphous powder*): Store in a flammable materials storage area. Color Code—Blue: Health Hazard/Poison: Store in a secure poison location. Prior to working with this chemical you should be trained on its proper handling and storage. Store in tightly closed containers in a cool, well-ventilated area away from strong oxidizers and other incompatible materials.

Shipping: Silicon powder, amorphous requires a shipping label of "FLAMMABLE SOLID." It falls in Hazard Class 4.1 and Packing Group III.

Spill Handling: Evacuate persons not wearing protective equipment from area of spill or leak until cleanup is complete. Remove all ignition sources. Collect powdered material in the most convenient and safe manner and deposit in sealed containers. Ventilate area after cleanup is complete. It may be necessary to contain and dispose of this chemical as a hazardous waste. If material or contaminated runoff enters waterways, notify downstream users of potentially contaminated waters. Contact your local or federal environmental protection agency for specific recommendations. If employees are required to clean up spills, they must be properly trained and equipped. OSHA 1910.120(q) may be applicable.

Fire Extinguishing: This chemical is a combustible solid. Use dry chemical, carbon dioxide, water spray, or alcohol foam extinguishers. Poisonous gases are produced in fire. If material or contaminated runoff enters waterways, notify downstream users of potentially contaminated waters. Notify local health and fire officials and pollution control agencies. From a secure, explosion-proof location, use water spray to cool exposed containers. If cooling streams are ineffective (venting sound increases in volume and pitch, tank discolors, or shows any signs of deforming), withdraw immediately to a secure position. If employees are expected to fight fires, they must be trained and equipped in OSHA 1910.156. The only respirators recommended for firefighting are self-contained breathing apparatuses that have full face-pieces and are operated in a pressure-demand or other positive-pressure mode.

Silicon carbide   S:0250

Molecular Formula: CSi
Common Formula: SiC
Synonyms: Carbon silicide; Carborundum®, Crystolon®, Carbonite®, Electrolon®; Silicon monocarbide
CAS Registry Number: 409-21-2; *(alt.)* 12504-67-5; *(alt.)* 66039-27-8
RTECS® Number: VW0450000
EC Number: 206-991-8
Regulatory Authority and Advisory Bodies
Air Pollutant Standard Set. See below, "Permissible Exposure Limits in Air" section.

WGK (German Aquatic Hazard Class): No value assigned.

Description: Silicon carbide is a yellow to green to bluish-black, iridescent crystalline substance. Colorless when pure. Molecular weight = 40.10; Specific gravity (H_2O:1) = 3.23 at 25°C; Boiling point = (sublimes); Freezing/Melting point = (sublimes with decomposition) 2700°C. Hazard Identification (based on NFPA-704 M Rating System): Health 2, Flammability 1, Reactivity 0. Insoluble in water.

Potential Exposure: Compound Description: Tumorigen. Those involved in the manufacture of silicon carbide abrasives, refractories, and semiconductors. Silicon carbide fibers are also produced in fibrous form as reinforcing fibers for composite materials.

Incompatibilities: None listed. Sublimes with decomposition at 2700°C.

Permissible Exposure Limits in Air
OSHA PEL: 15 mg/m^3 TWA, total dust; 5 mg/m^3 TWA, respirable fraction.
NIOSH REL: 10 mg/m^3 TWA, total dust; 5 mg/m^3 TWA, respirable fraction.
ACGIH TLV®[1]: *Nonfibrous:* 10 mg/m^3 TWA, inhalable fraction; particulate matter containing no asbestos and <1% crystalline silica; 3 mg/m^3 TWA, respirable fraction, for particulate matter containing no asbestos and <1% crystalline silica; *fibrous (including whiskers):* 0.1 f/cc, TWA, respirable fibers; length >5 μm; aspect ratio ≥ 3:1, as determined by the membrane filter method at 400−450 × magnification (4-mm objective), using phase-contrast illumination; Suspected Human Carcinogen.
Protective Action Criteria (PAC)
TEEL-0: 15 mg/m^3
PAC-1: 45 mg/m^3
PAC-2: 250 mg/m^3
PAC-3: 500 mg/m^3
DFG MAK: 1.5 mg/m^3, respirable fraction (previously "fine dust"); 4 mg/m^3, inhalable fraction; Pregnancy Risk Group C.
Australia: TWA 10 mg/m^3, 1993; Austria: MAK 4 mg/m^3, 1999; Belgium: TWA 10 mg/m^3, 1993; France: VME 10 mg/m^3, 1999; Switzerland: MAK-W 4 mg/m^3, 1999; the Netherlands: MAC-TGG 10 mg/m^3, 2003; United Kingdom: TWA 4 mg/m^3 (respirable dust), 2000; Argentina, Bulgaria, Columbia, Jordan, South Korea, New Zealand, Singapore, Vietnam: ACGIH TLV®: Suspected Human Carcinogen.

Determination in Air: Use NIOSH IV Method #0500, total dust, Method #0600 (respirable dust), Particulates NOR.

Routes of Entry: Inhalation, ingestion, skin and/or eye contact.

Harmful Effects and Symptoms
Short Term Exposure: Irritates the eyes, skin, upper respiratory system; cough.
Long Term Exposure: Silicon carbide can alter the course of inhalation tuberculosis leading to extensive fibrosis and progressive disease.
Points of Attack: Respiratory system.

Medical Surveillance: Lung function tests. Consider chest X-ray following acute overexposure.

First Aid: If this chemical gets into the eyes, remove any contact lenses at once and irrigate immediately for at least 15 min, occasionally lifting upper and lower lids. Seek medical attention immediately. If this chemical contacts the skin, remove contaminated clothing and wash immediately with soap and water. Seek medical attention immediately. If this chemical has been inhaled, remove from exposure. Transfer promptly to a medical facility.

Personal Protective Methods: Wear protective gloves and clothing to prevent any reasonable probability of skin contact. Safety equipment suppliers/manufacturers can provide recommendations on the most protective glove/clothing material for your operation. All protective clothing (suits, gloves, footwear, headgear) should be clean, available each day, and put on before work. Contact lenses should not be worn when working with this chemical. Wear dust-proof chemical goggles and face shield unless full face-piece respiratory protection is worn. Employees should wash immediately with soap when skin is wet or contaminated. Provide emergency showers and eyewash.

Respirator Selection: Follow regulations in OSHA 29CFR1910.134 or European Standard EN149. Use a NIOSH/MSHA- or European Standard EN149-approved respirator; or use an approved supplied-air respirator with a full face-piece operated in the positive-pressure mode, or with a full face-piece, hood, or helmet in the continuous-flow mode; or use a NIOSH/MSHA- or European Standard EN149-approved self-contained breathing apparatus with a full face-piece operated in pressure-demand or other positive-pressure mode.

Storage: Color Code—Green: General storage may be used. Prior to working with this chemical you should be trained on its proper handling and storage.

Spill Handling: Evacuate persons not wearing protective equipment from area of spill or leak until cleanup is complete. Remove all ignition sources. Collect powdered material in the most convenient and safe manner and deposit in sealed containers. Ventilate area after cleanup is complete. It may be necessary to contain and dispose of this chemical as a hazardous waste. If material or contaminated runoff enters waterways, notify downstream users of potentially contaminated waters. Contact your local or federal environmental protection agency for specific recommendations. If employees are required to clean up spills, they must be properly trained and equipped. OSHA 1910.120(q) may be applicable.

Fire Extinguishing: This chemical is a noncombustible solid. Use any extinguishing agent suitable for surrounding fire. If material or contaminated runoff enters waterways, notify downstream users of potentially contaminated waters. Notify local health and fire officials and pollution control agencies. If employees are expected to fight fires, they must be trained and equipped in OSHA 1910.156. The only respirators recommended for firefighting are self-contained breathing apparatuses that have full face-pieces and are operated in a pressure-demand or other positive-pressure mode.

Disposal Method Suggested: Landfill.

References
New Jersey Department of Health and Senior Services. (February 2007). *Hazardous Substances Fact Sheet: Silicon Carbide.* Trenton, NJ

Silver   S:0260

Molecular Formula: Ag

Synonyms: Argentum; C.I. 77820; Elemental silver; Shell silver; Silber (German); Silver metal

Soluble compounds: Silver acetate; Silver bromate; Silver bromide; Silver carbonate; Silver chloride; Silver chromate; Silver cyanide; Silver cichromate; Silver hydroxide; Silver iodate; Silver iodide; Silver nitrite; Silver oxalate; Silver sulfate; Silver sulfide; Silver thiocyanate

CAS Registry Number: 7440-22-4

RTECS® Number: VW3500000

EC Number: 231-131-3

Regulatory Authority and Advisory Bodies
Carcinogenicity (*metal*): EPA: Not Classifiable as to human carcinogenicity.

Air Pollutant Standard Set. See below, "Permissible Exposure Limits in Air" section.

Clean Water Act: 40CFR423, Appendix A, Priority Pollutants; Section 313 Water Priority Chemicals (57FR41331, 9/9/92); 40CFR401.15 Section 307 Toxic Pollutants as *silver and compounds*.

US EPA Hazardous Waste Number (RCRA No.): D011.

RCRA, 40CFR261, Appendix 8 Hazardous Constituents, waste number not listed.

RCRA Toxicity Characteristic (Section 261.24), Maximum. Concentration of Contaminants, regulatory level, 5.0 mg/L. Land Ban chemical.

RCRA 40CFR268.48; 61FR15654, Universal Treatment Standards: Wastewater (mg/L), 0.43; Nonwastewater (mg/L), 0.30 TCLP.

RCRA 40CFR264, Appendix 9; TSD Facilities Ground Water Monitoring List. Suggested test method(s) (PQL μg/L): 6010 (70); 7760 (100).

Safe Drinking Water Act: SMCL, 0.1 mg/L.

Reportable Quantity (RQ): 1000 lb (454 kg).

EPCRA Section 313 Form R *de minimis* concentration reporting level: 1.0% (includes any unique chemical substance that contains silver as part of that chemical's infrastructure).

Canada, WHMIS, Ingredients Disclosure List Concentration 1.0%.

European/International Regulations: not listed in Annex 1.

WGK (German Aquatic Hazard Class): 1—Slightly water polluting (*metal*); 3—Highly water polluting (*colloidal*).

Description: Silver is a white lustrous metal that is extremely ductile and malleable. Molecular weight = 107.87; Specific gravity (H_2O:1) = 10.49 at 25°C (metal); Boiling point = 2212°C; Freezing/Melting point = 961°C. Hazard Identification (based on NFPA-704 M Rating System) (*powder*): Health 0, Flammability 0, Reactivity 1. Insoluble in water but soluble in hot sulfuric and nitric acids. Perhaps the most common soluble silver compounds are silver nitrate ($AgNO_3$) and silver cyanide (AgCN).

Potential Exposure: Silver may be alloyed with copper, aluminum, cadmium, lead, or antimony. The alloys are used in the manufacture of silverware, jewelry, coins, ornaments, plates, commutators, scientific instruments, automobile bearing, and grids in storage batteries. Silver is used in chrome-nickel steels, in solders and brazing alloys; in the application of metallic films on glass and ceramics, to increase corrosion resistance to sulfuric acid, in photographic films, plates, and paper; as an electroplated undercoating for nickel and chrome; as a bactericide for sterilizing water; fruit juices; vinegar, etc.; in bus bars and windings in electrical plants; in dental amalgams; and as a chemical catalyst in the synthesis of aldehydes. Because of its resistance to acetic and other food acids, it is utilized in the manufacture of pipes, valves, vats, pasteurizing coils and nozzles for the milk, vinegar, cider, brewing, and acetate rayon silk industries.

Incompatibilities: Acetylene, ammonia, hydrogen peroxide, bromoazide, chlorine trifluoride, ethyleneimine, oxalic acid, tartaric acid.

Permissible Exposure Limits in Air
Silver metal dust and fume
OSHA PEL: 0.01 mg[Ag]/m^3 TWA.
NIOSH REL: 0.01 mg[Ag]/m^3 TWA.
ACGIH TLV®[1]: *silver metal dust and fume*: 0.1 mg[Ag]/m^3 TWA; *silver salts*: 0.01 mg[Ag]/m^3 TWA.
NIOSH IDLH: 10 mg[Ag]/m^3.
Protective Action Criteria (PAC)
TEEL-0: 0.01 mg/m^3
PAC-1: 0.3 mg/m^3
PAC-2: 0.5 mg/m^3
PAC-3: 10 mg/m^3
DFG MAK: *silver metal dust and fume*: 0.1 mg[Ag]/m^3, inhalable fraction TWA; Peak Limitation Category II(8); Pregnancy Risk Group D. *Silver salts*: 0.01 mg[Ag]/m^3, inhalable fraction TWA; Peak Limitation Category I(2); Pregnancy Risk Group D.
Australia: TWA 0.1 mg/m^3, 1993; Austria: MAK 0.01 mg/m^3, 1999; Belgium: TWA 0.1 mg/m^3, 1993; Denmark: TWA 0.01 mg/m^3, 1999; Finland: TWA 0.1 mg/m^3, 1999; France: VME 0.1 mg/m^3, 1999; Japan: 0.01 mg/m^3, 1999; the Netherlands: MAC-TGG 0.1 mg/m^3, 2003; Norway: TWA 0.01 mg/m^3, 1999; Russia: STEL 1 mg/m^3, 1993; Sweden: NGV 0.1 mg/m^3, 1999; Switzerland: MAK-W 0.01 mg/m^3, 1999; United Kingdom: TWA 0.1 mg/m^3, 2000; Argentina, Bulgaria, Columbia, Jordan, South Korea,

New Zealand, Singapore, Vietnam: ACGIH TLV®: TWA 0.1 mg/m³ *silver metal dust*. Several states have set guidelines or standards for silver in ambient air[60] ranging from 0.01–0.08 µg/m³ (Montana) to 0.2 µg/m³ (Connecticut and Nevada) to 1.0 µg/m³ (North Dakota) to 16,000 µg/m³ (Virginia).

As silver soluble compounds
OSHA PEL: 0.01 mg/m³ TWA.
NIOSH REL: 0.01 mg/m³ TWA.
ACGIH TLV®[1]: 0.01 mg/m³ TWA.

Determination in Air: NIOSH Analytical Method #7300 Elements by ICP; #8005, Elements in blood or tissue, #8310 Metals in urine; OSHA Analytical Method ID-121.

Permissible Concentration in Water: *To protect freshwater aquatic life:* should not exceed e[1.72 in(hardness)—6.52] µg/L at any time. *To protect saltwater aquatic life:* never to exceed 2.3 µg/L. To protect human health: 50 µg/L.[6] The state of Maine recommends that silver in drinking water should not exceed 50 µg/L.[61]

Determination in Water: Digestion followed by atomic absorption or colorimetric determination (with Dithizone) or by inductively coupled plasma (ICP) optical emission spectrometry. This gives total silver. Dissolved silver may be determined by these same methods preceded by 0.45 µm filtration.

Routes of Entry: Inhalation of fumes or dust, ingestion of solutions or dust, eye and/or skin contact.

Harmful Effects and Symptoms

Short Term Exposure: Causes irritation of the eyes, skin, and respiratory tract. Ingestion of dust can cause gastrointestinal disturbance.

Long Term Exposure: Silver can affect you when breathed in. Repeated exposure to fine silver dust or fumes can cause blue-gray staining of the eyes, mouth, throat, internal organs, and skin. This occurs slowly and may take years to develop. Once present, it does not go away. It can be very disfiguring. Skin contact can cause silver to become imbedded in small cuts in the skin, forming a permanent tattoo. Can cause sores on the inner lining of the nose and may cause perforation of the nasal septum.

Points of Attack: Nasal septum, skin, eyes.

Medical Surveillance: NIOSH lists the following tests: whole blood (chemical/metabolite); blood serum; biologic tissue/biopsy; Pigmentation Evidence; urine (chemical/metabolite). Before beginning employment and at regular times after that, for those with frequent or potentially high exposures, the following are recommended: Slit lamp examination by an eye doctor. Examination of the skin, nose, and throat. If symptoms develop or overexposure is suspected, the following may be useful: kidney function tests.

First Aid: If this chemical gets into the eyes, remove any contact lenses at once and irrigate immediately for at least 15 min, occasionally lifting upper and lower lids. Seek medical attention immediately. If this chemical contacts the skin, remove contaminated clothing and wash immediately with soap and water. Seek medical attention immediately. If

this chemical has been inhaled, remove from exposure, begin rescue breathing (using universal precautions, including resuscitation mask) if breathing has stopped and CPR if heart action has stopped. Transfer promptly to a medical facility. When this chemical has been swallowed, get medical attention. Give large quantities of water and induce vomiting. Do not make an unconscious person vomit.

Personal Protective Methods: Wear protective gloves and clothing to prevent any reasonable probability of skin contact. Safety equipment suppliers/manufacturers can provide recommendations on the most protective glove/clothing material for your operation. All protective clothing (suits, gloves, footwear, headgear) should be clean, available each day, and put on before work. Contact lenses should not be worn when working with this chemical. Wear dust-proof chemical goggles and face shield unless full face-piece respiratory protection is worn. Employees should wash immediately with soap when skin is wet or contaminated. Provide emergency showers and eyewash.

Respirator Selection: *0.25 mg/m³:* Sa:Cf (APF = 25) (any supplied-air respirator operated in a continuous-flow mode) or PaprHie (APF = 25) (any powered, air-purifying respirator with a high-efficiency particulate filter). *0.5 mg/m³:* 100F (APF = 50) (any air-purifying, full-face-piece respirator with an N100, R100, or P100 filter) or SCBAF (APF = 50) (any self-contained breathing apparatus with a full face-piece) or SaF (APF = 50) (any supplied-air respirator with a full face-piece). *10 mg/m³:* SaF: Pd,Pp (APF = 2000) (any supplied-air respirator that has a full face-piece and is operated in a pressure-demand or other positive-pressure mode). *Emergency or planned entry into unknown concentrations or IDLH conditions:* SCBAF: Pd, Pp (APF = 10,000) (any self-contained breathing apparatus that has a full face-piece and is operated in a pressure-demand or other positive-pressure mode) or SaF: Pd,Pp: ASCBA (any supplied-air respirator that has a full face-piece and is operated in a pressure-demand or other positive-pressure mode in combination with an auxiliary self-contained breathing apparatus operated in a pressure-demand or other positive-pressure mode). *Escape:* 100F (APF = 50) (any air-purifying, full-face-piece respirator with an N100, R100, or P100 filter) or SCBAE (any appropriate escape-type, self-contained breathing apparatus).

Note: Substance causes eye irritation or damage; eye protection needed.

Storage: Color Code—Green: General storage may be used. Prior to working with this chemical you should be trained on its proper handling and storage. Silver must be stored to avoid contact with acetylene, ammonia, hydrogen peroxide, or ethyleneimine, since violent reactions occur.

Spill Handling: Evacuate persons not wearing protective equipment from area of spill or leak until cleanup is complete. Remove all ignition sources. Collect powdered material in the most convenient and safe manner and deposit in sealed containers. Ventilate area after cleanup is complete. It may be necessary to contain and dispose of this chemical

as a hazardous waste. If material or contaminated runoff enters waterways, notify downstream users of potentially contaminated waters. Contact your local or federal environmental protection agency for specific recommendations. If employees are required to clean up spills, they must be properly trained and equipped. OSHA 1910.120(q) may be applicable.

Fire Extinguishing: This chemical is a noncombustible solid, but flammable in the form of dust or powder. Use dry chemical, carbon dioxide, water spray, or alcohol foam extinguishers. Poisonous gases are produced in fire. If material or contaminated runoff enters waterways, notify downstream users of potentially contaminated waters. Notify local health and fire officials and pollution control agencies. From a secure, explosion-proof location, use water spray to cool exposed containers. If cooling streams are ineffective (venting sound increases in volume and pitch, tank discolors, or shows any signs of deforming), withdraw immediately to a secure position. If employees are expected to fight fires, they must be trained and equipped in OSHA 1910.156. The only respirators recommended for firefighting are self-contained breathing apparatuses that have full face-pieces and are operated in a pressure-demand or other positive-pressure mode.

Disposal Method Suggested: Recovery, wherever possible, in view of economic value of silver. Techniques for silver recovery from photoprocessing and electroplating wastewaters have been developed and patented.

References

US Environmental Protection Agency. (1980). *Silver: Ambient Water Quality Criteria.* Washington, DC
US Environmental Protection Agency. (April 30, 1980). *Silver: Health and Environmental Effects Profile No. 154.* Washington, DC: Office of Solid Waste
Sax, N. I. (Ed.). (1980). *Dangerous Properties of Industrial Materials Report,* 1, No. 1, 52–55
New Jersey Department of Health and Senior Services. (October 2002). *Hazardous Substances Fact Sheet: Silver.* Trenton, NJ

Silver cyanide   S:0270

Molecular Formula: CAgN
Synonyms: Cianuro de plata (Spanish); Cyanure d'argent (French)
CAS Registry Number: 506-64-9
RTECS® Number: VW3850000
UN/NA & ERG Number: UN1684/151
EC Number: 208-048-6
Regulatory Authority and Advisory Bodies
Air Pollutant Standard Set. See below, "Permissible Exposure Limits in Air" section.
Clean Water Act: 40CFR401.15 Section 307 Toxic Pollutants as silver and compounds.

US EPA Hazardous Waste Number (RCRA No.): P104.
RCRA, 40CFR261, Appendix 8 Hazardous Constituents.
Reportable Quantity (RQ): 1 lb (0.454 kg).
EPCRA Section 313 Form R *de minimis* concentration reporting level: 1.0%.
US DOT Regulated Marine Pollutant (49CFR172.101, Appendix B).
Canada, WHMIS, Ingredients Disclosure List Concentration 1.0%; National Pollutant Release Inventory (NPRI); CEPA Priority Substance List, Ocean dumping prohibited, as cyanides.
WGK (German Aquatic Hazard Class): No value assigned.
Description: Silver cyanide is a white or grayish odorless powder which darkens when exposed to light. Molecular weight = 133.89; Freezing/Melting point = 320°C (decomposition). Hazard Identification (based on NFPA-704 M Rating System): Health 3, Flammability 1, Reactivity 0. Insoluble in water.
Potential Exposure: It is primarily used in silver plating.
Incompatibilities: Acetylene, ammonia, and hydrogen peroxide (H_2O_2). Acid and acid fumes produce hydrogen cyanide. Contact with fluorine is violently explosive at normal temperatures. Incompatible with phosphorus cyanide.
Permissible Exposure Limits in Air
OSHA PEL: 5 mg[CN]/m³/4.7 ppm TWA.
NIOSH REL: 5 mg[CN]/m³/4.7 ppm/10 min, Ceiling Concentration.
ACGIH TLV®[1]: 5 mg[CN]/m³ [skin] Ceiling Concentration.
Protective Action Criteria (PAC)
TEEL-0: 0.0124 mg/m³
PAC-1: 0.35 mg/m³
PAC-2: 2.5 mg/m³
PAC-3: 12.4 mg/m³
DFG MAK: 2 mg[CN]/m³, inhalable fraction TWA; Peak Limitation Category II(1) [skin]; Pregnancy Risk Group: C.
NIOSH IDLH: 25 mg[CN]/m³.
Silver salts: 0.01 mg[Ag]/m³, respirable fraction TWA; Peak Limitation Category I(2); Pregnancy Risk Group D.
NIOSH IDLH: 10 mg[Ag]/m³
Determination in Air: Use NIOSH Analytical Method #7904, Cyanides. See also entry for Silver.
Permissible Concentration in Water: Again, both silver and cyanide can be considered. The EPA limit for silver in drinking water is 50 μg/L and that for cyanides is 154 μg/L.[61]
Routes of Entry: Inhalation, ingestion, skin and/or eye contact. Absorbed through the skin.
Harmful Effects and Symptoms
Short Term Exposure: Silver cyanide can affect you when breathed in and by passing through your skin. Heating releases deadly cyanide gas. Skin or eye contact can cause irritation. A deadly poison if ingested.
Long Term Exposure: Repeated exposure can slowly cause the eyes, inner nose, throat, skin, and body organs to turn a blue-gray color. This may take years to develop but is

permanent. High or repeated exposure may cause kidney damage.

Points of Attack: Kidneys, skin.

Medical Surveillance: To detect early changes in body color, careful periodic examinations of the eyes, inner nose, throat, and skin are useful. Persons with high or frequent exposure should consider periodic tests for kidney function. If cyanide exposure is suspected, immediate medical attention is recommended. Consider urine test for thiocyanates to detect cyanide exposure.

First Aid: If this chemical gets into the eyes, remove any contact lenses at once and irrigate immediately for at least 15 min, occasionally lifting upper and lower lids. Seek medical attention immediately. If this chemical contacts the skin, remove contaminated clothing and wash immediately with soap and water. Seek medical attention immediately. If this chemical has been inhaled, remove from exposure, begin rescue breathing (using universal precautions, including resuscitation mask) if breathing has stopped and CPR if heart action has stopped. Transfer promptly to a medical facility. When this chemical has been swallowed, get medical attention. Give large quantities of water and induce vomiting. Do not make an unconscious person vomit.

Note: Use amyl nitrate capsules if symptoms develop. All area employees should be trained regularly in emergency measures for cyanide poisoning and in CPR. A cyanide antidote kit should be kept in the immediate work area and must be rapidly available. Kit ingredients should be replaced every 1–2 years to ensure freshness. Persons trained in the use of this kit, oxygen use, and CPR must be quickly available.

Personal Protective Methods: Wear protective gloves and clothing to prevent any reasonable probability of skin contact. Safety equipment suppliers/manufacturers can provide recommendations on the most protective glove/clothing material for your operation. All protective clothing (suits, gloves, footwear, headgear) should be clean, available each day, and put on before work. Contact lenses should not be worn when working with this chemical. Wear dust-proof chemical goggles and face shield unless full face-piece respiratory protection is worn. Employees should wash immediately with soap when skin is wet or contaminated. Provide emergency showers and eyewash.

Respirator Selection: *0.25 mg/m³:* Sa:Cf (APF = 25) (any supplied-air respirator operated in a continuous-flow mode) or PaprHie (APF = 25) (any powered, air-purifying respirator with a high-efficiency particulate filter). *0.5 mg/m³:* 100F (APF = 50) (any air-purifying, full-face-piece respirator with an N100, R100, or P100 filter) or SCBAF (APF = 50) (any self-contained breathing apparatus with a full face-piece) or SaF (APF = 50) (any supplied-air respirator with a full face-piece). *10 mg/m³:* SaF: Pd,Pp (APF = 2000) (any supplied-air respirator that has a full face-piece and is operated in a pressure-demand or other positive-pressure mode). *Emergency or planned entry into unknown concentrations or IDLH conditions:* SCBAF: Pd, Pp (APF = 10,000) (any self-contained breathing apparatus that has a full face-piece and is operated in a pressure-demand or other positive-pressure mode) or SaF: Pd,Pp: ASCBA (any supplied-air respirator that has a full face-piece and is operated in a pressure-demand or other positive-pressure mode in combination with an auxiliary self-contained breathing apparatus operated in a pressure-demand or other positive-pressure mode). *Escape:* 100F (APF = 50) (any air-purifying, full-face-piece respirator with an N100, R100, or P100 filter) or SCBAE (any appropriate escape-type, self-contained breathing apparatus).

Note: Substance causes eye irritation or damage; eye protection needed.

Storage: Color Code—Blue: Health Hazard/Poison: Store in a secure poison location. Prior to working with this chemical you should be trained on its proper handling and storage. Store in tightly closed containers in a cool, well-ventilated area away from acetylene, ammonia, and hydrogen peroxide. Protect from light. Silver cyanide and fluorine are violently explosive at normal temperatures.

Shipping: This compound requires a shipping label of "POISONOUS/TOXIC MATERIALS." It falls in Hazard Class 6.1 and Packing Group II.

Spill Handling: Evacuate persons not wearing protective equipment from area of spill or leak until cleanup is complete. Remove all ignition sources. Collect powdered material in the most convenient and safe manner and deposit in sealed containers. Ventilate area after cleanup is complete. It may be necessary to contain and dispose of this chemical as a hazardous waste. If material or contaminated runoff enters waterways, notify downstream users of potentially contaminated waters. Contact your local or federal environmental protection agency for specific recommendations. If employees are required to clean up spills, they must be properly trained and equipped. OSHA 1910.120(q) may be applicable.

Fire Extinguishing: This chemical is a combustible solid. Use dry chemical, carbon dioxide, water spray, or foam extinguishers. Poisonous gases are produced in fire, including cyanide and nitrogen oxides. If material or contaminated runoff enters waterways, notify downstream users of potentially contaminated waters. Notify local health and fire officials and pollution control agencies. From a secure, explosion-proof location, use water spray to cool exposed containers. If cooling streams are ineffective (venting sound increases in volume and pitch, tank discolors, or shows any signs of deforming), withdraw immediately to a secure position. If employees are expected to fight fires, they must be trained and equipped in OSHA 1910.156. The only respirators recommended for firefighting are self-contained breathing apparatuses that have full face-pieces and are operated in a pressure-demand or other positive-pressure mode.

Disposal Method Suggested: Consult with environmental regulatory agencies for guidance on acceptable disposal practices. Generators of waste containing this contaminant (≥100 kg/mo) must conform with EPA regulations governing storage, transportation, treatment, and waste disposal.

References

New Jersey Department of Health and Senior Services. (June 2000). *Hazardous Substances Fact Sheet: Silver Cyanide*. Trenton, NJ

Silver nitrate   S:0280

Molecular Formula: $AgNO_3$

Synonyms: Lunar caustic; Nitrate d'argent (French); Nitrato de plata (Spanish); Nitric acid, silver(1+) salt; Nitric acid, silver(I) salt; Silbernitrat (German); Silver(1+) nitrate; Silver(I) nitrate

CAS Registry Number: 7761-88-8

RTECS® Number: VW4725000

UN/NA & ERG Number: UN1493/140

EC Number: 231-853-9 [*Annex I Index No.:* 047-001-00-2]

Regulatory Authority and Advisory Bodies

US EPA Gene-Tox Program, Positive: Cell transform.—SA7/SHE; Negative: *E. coli* polA without S9.

US EPA, FIFRA 1998 Status of Pesticides: Canceled.

Air Pollutant Standard Set. See below, "Permissible Exposure Limits in Air" section.

FDA—over-the-counter drug.

Clean Water Act: Section 311 Hazardous Substances/RQ 40CFR117.3 (same as CERCLA, see below); Section 313 Water Priority Chemicals (57FR 41331, 9/9/92).

40CFR401.15 Section 307 Toxic Pollutants as silver and compounds.

RCRA, 40CFR261, Appendix 8 Hazardous Constituents, as silver compounds, n.o.s., waste number not listed.

Land Ban chemical.

Reportable Quantity (RQ): 1 lb (0.454 kg).

EPCRA Section 313 Form R *de minimis* concentration reporting level: 1.0%.

Canada, WHMIS, Ingredients Disclosure List Concentration 1.0%.

European/International Regulations: Hazard Symbol: O, C, N; Risk phrases: R8; R34; R50/53; Safety phrases: S1/2; S26; S36/37/39; S45; S60; S61 (see Appendix 4).

WGK (German Aquatic Hazard Class): 3—Severe hazard to waters.

Description: Silver nitrate is a colorless to dark gray, odorless, crystalline solid. Molecular weight = 169.88; Boiling point = 444°C (decomposes); Freezing/Melting point = 212°C. Hazard Identification (based on NFPA-704 M Rating System): Health 2, Flammability 0, Reactivity 3 (Oxidizer). Soluble in water.

Potential Exposure: Compound Description: Tumorigen; Mutagen; Reproductive Effector; Human Data; Primary Irritant. Silver nitrate is used in photography, silver plating; as an antiseptic; in chemical reactions; and mirror manufacturing; as starting material in production of other silver compounds.

Incompatibilities: A strong oxidizer. Reacts violently with combustible and reducing materials. Reacts with acetylene forming a shock-sensitive explosive. Reacts with alkalis, antimony salts, ammonia, arsenites, bromides, carbonates, chlorides, iodides, hydrogen peroxide, thiocyanates, ferrous salts, oils, hypophosphites, morphine salts, creosote, phosphates, tannic acid, tartarates, halides, vegetable extracts, and others. Attacks some forms of plastics, rubber, and coatings.

Permissible Exposure Limits in Air

As silver soluble compounds

OSHA PEL: 0.01 mg/m³ TWA.

NIOSH REL: 0.01 mg/m³ TWA.

ACGIH TLV®[1]: 0.01 mg/m³ TWA.

NIOSH IDLH: 10 mg[Ag]/m³.

Protective Action Criteria (PAC)

TEEL-0: 0.0157 mg/m³

PAC-1: 15.7 mg/m³

PAC-2: 15.7 mg/m³

PAC-3: 15.7 mg/m³

DFG MAK: *Silver salts:* 0.01 mg[Ag]/m³, respirable fraction TWA; Peak Limitation Category I(2); Pregnancy Risk Group D.

Arab Republic of Egypt: TWA 0.01 mg[Ag]/m³, 1993; Australia: TWA 0.01 mg[Ag]/m³, 1993; Belgium: TWA 0.01 mg[Ag]/m³, 1993; Denmark: TWA 0.01 mg[Ag]/m³, 1999; Finland: TWA 0.01 mg[Ag]/m³; STEL 0.03 mg[Ag]/m³, [skin], 1999; France: VME 0.01 mg[Ag]/m³, 1999; Japan: 0.01 mg[Ag]/m³, 1999; Norway: TWA 0.01 mg[Ag]/m³, 1999; the Philippines: TWA 0.01 mg[Ag]/m³, 1993; Poland: MAC (TWA) 0.01 mg[Ag]/m³, 1999; Sweden: NGV 0.01 mg[Ag]/m³, 1999; Switzerland: TWA 0.01 mg[Ag]/m³, 1999; United Kingdom: TWA 0.01 mg[Ag]/m³, 2000; Argentina, Bulgaria, Columbia, Jordan, South Korea, New Zealand, Singapore, Vietnam: ACGIH TLV®: TWA 0.01 mg[Ag]/m³.

Determination in Air: See entry for Silver.

Permissible Concentration in Water: EPA[6] and the state of Maine[61] have set guidelines of 50 μg/L for silver in drinking water.

Routes of Entry: Inhalation, ingestion, skin and/or eye contact.

Harmful Effects and Symptoms

Short Term Exposure: Corrosive to the eyes, skin, and respiratory tract. May affect the blood, causing formation of methemoglobin. *Inhalation:* May cause irritation of the nose, throat, and lungs. *Skin:* May cause irritation. Concentrated solutions may cause burns, sores, and discoloration of skin. Solid materials will cause chemical burns especially if skin is wet. *Eyes:* May cause irritation which can be severe. Solid material may cause chemical burns and permanent damage.

Ingestion: May cause burns to mouth and throat, abdominal pain, diarrhea, and dizziness. Shock and convulsions may develop. Estimated lethal dose is 2 g or 1/14 oz for a 150-lb adult.

Long Term Exposure: All forms of silver accumulate and are excreted very slowly. Blue-gray discoloration (argyreia)

of eyes, nose, throat, and skin may occur. This discoloration is likely to be permanent. May affect the blood causing formation of methemoglobin. Very irritating substances can cause lung effects. High or repeated exposure may cause kidney damage.

Points of Attack: Skin, kidneys, blood.

Medical Surveillance: Before beginning employment and at regular times after that, for those with frequent or potentially high exposures, the following are recommended: lung function tests. Examination of the eyes, nose, throat, and skin for changes in color. If symptoms develop or overexposure is suspected, the following may be useful: kidney function tests. Test for blood methemoglobin level.

First Aid: If this chemical gets into the eyes, remove any contact lenses at once and irrigate immediately for at least 15 min, occasionally lifting upper and lower lids. Seek medical attention immediately. If this chemical contacts the skin, remove contaminated clothing and wash immediately with soap and water. Seek medical attention immediately. If this chemical has been inhaled, remove from exposure, begin rescue breathing (using universal precautions, including resuscitation mask) if breathing has stopped and CPR if heart action has stopped. Transfer promptly to a medical facility. When this chemical has been swallowed, get medical attention. If victim is *conscious*, administer water or milk. Do not induce vomiting.

Note to physician: If swallowed, perform gastric lavage, using 1−2% sodium chloride water every 15 min, followed by saline catharsis. Treat for methemoglobinemia. Spectrophotometry may be required for precise determination of levels of methemoglobin in urine.

Personal Protective Methods: Wear protective gloves and clothing to prevent any reasonable probability of skin contact. Safety equipment suppliers/manufacturers can provide recommendations on the most protective glove/clothing material for your operation. All protective clothing (suits, gloves, footwear, headgear) should be clean, available each day, and put on before work. Contact lenses should not be worn when working with this chemical. Wear splash-proof chemical goggles and face shield when working with solutions containing silver nitrate. Wear dust-proof goggles and face shield when working with powders or dust, unless full face-piece respiratory protection is worn. Employees should wash immediately with soap when skin is wet or contaminated. Provide emergency showers and eyewash.

Respirator Selection: *0.25 mg/m³:* Sa:Cf (APF = 25) (any supplied-air respirator operated in a continuous-flow mode) or PaprHie (APF = 25) (any powered, air-purifying respirator with a high-efficiency particulate filter). *0.5 mg/m³:* 100F (APF = 50) (any air-purifying, full-face-piece respirator with an N100, R100, or P100 filter) or SCBAF (APF = 50) (any self-contained breathing apparatus with a full face-piece) or SaF (APF = 50) (any supplied-air respirator with a full face-piece). *10 mg/m³:* SaF: Pd,Pp (APF = 2000) (any supplied-air respirator that has a full face-piece and is operated in a pressure-demand or other positive-

pressure mode). *Emergency or planned entry into unknown concentrations or IDLH conditions:* SCBAF: Pd,Pp (APF = 10,000) (any self-contained breathing apparatus that has a full face-piece and is operated in a pressure-demand or other positive-pressure mode) or SaF: Pd,Pp: ASCBA (any supplied-air respirator that has a full face-piece and is operated in a pressure-demand or other positive-pressure mode in combination with an auxiliary self-contained breathing apparatus operated in a pressure-demand or other positive-pressure mode). *Escape:* 100F (APF = 50) (any air-purifying, full-face-piece respirator with an N100, R100, or P100 filter) or SCBAE (any appropriate escape-type, self-contained breathing apparatus).

Note: Substance causes eye irritation or damage; eye protection needed.

Storage: Color Code—Yellow: Reactive Hazard; Store in a location separate from other materials, especially flammables and combustibles. Prior to working with this chemical you should be trained on its proper handling and storage. Silver nitrate must be stored to avoid contact with oils and fuels, since violent reactions occur. Store in tightly closed containers in a cool, well-ventilated area away from oxidizers and alkalis. Silver nitrate should not be stored in plastic containers. See OSHA Standard 1910.104 and NFPA 43A *Code for the Storage of Liquid and Solid Oxidizers* for detailed handling and storage regulations.

Shipping: This compound requires a shipping label of "OXIDIZER." It falls in Hazard Class 5.1 and Packing Group II.

Spill Handling: Solid material: evacuate persons not wearing protective equipment from area of spill or leak until cleanup is complete. Remove all ignition sources. Collect powdered material in the most convenient and safe manner and deposit in sealed containers. Ventilate area after cleanup is complete. It may be necessary to contain and dispose of this chemical as a hazardous waste. If material or contaminated runoff enters waterways, notify downstream users of potentially contaminated waters. Contact your local or federal environmental protection agency for specific recommendations. If employees are required to clean up spills, they must be properly trained and equipped. OSHA 1910.120(q) may be applicable.

Liquid: Evacuate and restrict persons not wearing protective equipment from area of spill or leak until cleanup is complete. Remove all ignition sources. Ventilate area of spill or leak. Absorb liquids in vermiculite, dry sand, earth, peat, carbon, or a similar material and deposit in sealed containers. Keep this chemical out of a confined space, such as a sewer, because of the possibility of an explosion, unless the sewer is designed to prevent the buildup of explosive concentrations. It may be necessary to contain and dispose of this chemical as a hazardous waste. If material or contaminated runoff enters waterways, notify downstream users of potentially contaminated waters. Contact your local or federal environmental protection agency for specific recommendations. If employees are required to clean up spills, they must be properly trained and equipped. OSHA 1910.120(q) may be applicable.

Fire Extinguishing: Nonflammable but is a strong oxidizer capable of increasing the intensity of an existing fire and the flammability of combustible, organic, or other oxidizable materials. Use type of extinguisher appropriate to other burning materials. Poisonous gases are produced in fire. If material or contaminated runoff enters waterways, notify downstream users of potentially contaminated waters. Notify local health and fire officials and pollution control agencies. From a secure, explosion-proof location, use water spray to cool exposed containers. If cooling streams are ineffective (venting sound increases in volume and pitch, tank discolors, or shows any signs of deforming), withdraw immediately to a secure position. If employees are expected to fight fires, they must be trained and equipped in OSHA 1910.156. The only respirators recommended for firefighting are self-contained breathing apparatuses that have full face-pieces and are operated in a pressure-demand or other positive-pressure mode.

References

Sax, N. I. (Ed.). (1980). *Dangerous Properties of Industrial Materials Report*, 1, No. 1, 52–53

New York State Department of Health. (April 1986). *Chemical Fact Sheet: Silver Nitrate* (Version 2). Albany, NY: Bureau of Toxic Substance Assessment

US Environmental Protection Agency, Special Review and Reregistration Division Office of Pesticide Programs. (1998). *Agency Status of Pesticides in Registration, Reregistration, and Special Review* (Rainbow Report). Washington, DC

New Jersey Department of Health and Senior Services. (May 2000). *Hazardous Substances Fact Sheet: Silver Nitrate*. Trenton, NJ

Silver picrate   S:0290

Molecular Formula: $C_6H_2AgN_3O_7$
Common Formula: $C_6H_2(NO_2)_3OAg$
Synonyms: Picragol; Picrotol; 2,4,6-Trinitro-phenol silver (1+) salt
CAS Registry Number: 146-84-9
RTECS® Number: TJ7891000
UN/NA & ERG Number: UN1347 (wetted with not <30% water, by mass)/113
EC Number: 205-682-5
Regulatory Authority and Advisory Bodies
Air Pollutant Standard Set. See below, "Permissible Exposure Limits in Air" section.
Silver compounds
Clean Water Act: 40CFR401.15 Section 307 Toxic Pollutants as silver and compounds.
RCRA, 40CFR261, Appendix 8 Hazardous Constituents, as silver compounds, n.o.s., waste number not listed.

Land Ban chemical.
EPCRA Section 313: Includes any unique chemical substance that contains silver as part of that chemical's infrastructure. Form R *de minimis* concentration reporting level: 1.0%.
Canada, WHMIS, Ingredients Disclosure List Concentration 1.0% as silver, soluble compounds.
WGK (German Aquatic Hazard Class): No value assigned.
Description: Silver picrate is a yellow powder or crystalline material which turns brown when heated or exposed to light. Molecular weight = 335.98. Soluble in water.
Potential Exposure: Used in antibacterial medicines.
Incompatibilities: Organics or other oxidizable materials. Dried out material is unstable and a severe explosion risk; protect from shock, light, and vibration.
Permissible Exposure Limits in Air
As silver soluble compounds
OSHA PEL: 0.01 mg/m^3 TWA.
NIOSH REL: 0.01 mg/m^3 TWA.
ACGIH TLV®[1]: 0.01 mg/m^3 TWA.
No TEEL available.
DFG MAK: *silver salts:* 0.01 mg[Ag]/m^3, respirable fraction TWA; Peak Limitation Category I(2); Pregnancy Risk Group D.
NIOSH IDLH: 10 mg[Ag]/m^3.
Determination in Air: See entry for Silver.
Permissible Concentration in Water: The EPA[6] and the state of Maine[61] have set limits of 50 µg/L for silver in drinking water.
Routes of Entry: Inhalation, ingestion, skin and/or eye contact. Absorbed by the skin.
Harmful Effects and Symptoms
Short Term Exposure: Silver picrate can affect you when breathed in and by passing through your skin. Eye contact can cause burns with possible damage. Skin contact can cause severe irritation or burns.
Long Term Exposure: Repeated exposure can slowly cause the eyes, inner nose, throat, skin, and body organs to turn a blue-gray color. This usually takes years to develop but is permanent. High or repeated exposure can cause kidney damage.
Points of Attack: Skin, eyes, kidneys.
Medical Surveillance: To detect early changes in body color, careful periodic examinations of the eyes, inner nose, throat, and skin are useful. Persons with high or frequent exposure should consider periodic tests for kidney function.
First Aid: If this chemical gets into the eyes, remove any contact lenses at once and irrigate immediately for at least 15 min, occasionally lifting upper and lower lids. Seek medical attention immediately. If this chemical contacts the skin, remove contaminated clothing and wash immediately with soap and water. Seek medical attention immediately. If this chemical has been inhaled, remove from exposure, begin rescue breathing (using universal precautions, including resuscitation mask) if breathing has stopped and CPR if heart action has stopped. Transfer promptly to a medical

facility. When this chemical has been swallowed, get medical attention. Give large quantities of water and induce vomiting. Do not make an unconscious person vomit.

Personal Protective Methods: Wear protective gloves and clothing to prevent any reasonable probability of skin contact. Safety equipment suppliers/manufacturers can provide recommendations on the most protective glove/clothing material for your operation. All protective clothing (suits, gloves, footwear, headgear) should be clean, available each day, and put on before work. Contact lenses should not be worn when working with this chemical. Wear dust-proof chemical goggles and face shield unless full face-piece respiratory protection is worn. Employees should wash immediately with soap when skin is wet or contaminated. Provide emergency showers and eyewash.

Respirator Selection: $0.25 \, mg/m^3$: Sa:Cf (APF = 25) (any supplied-air respirator operated in a continuous-flow mode) or PaprHie (APF = 25) (any powered, air-purifying respirator with a high-efficiency particulate filter). $0.5 \, mg/m^3$: 100F (APF = 50) (any air-purifying, full-face-piece respirator with an N100, R100, or P100 filter) or SCBAF (APF = 50) (any self-contained breathing apparatus with a full face-piece) or SaF (APF = 50) (any supplied-air respirator with a full face-piece). $10 \, mg/m^3$: SaF: Pd,Pp (APF = 2000) (any supplied-air respirator that has a full face-piece and is operated in a pressure-demand or other positive-pressure mode). *Emergency or planned entry into unknown concentrations or IDLH conditions:* SCBAF: Pd, Pp (APF = 10,000) (any self-contained breathing apparatus that has a full face-piece and is operated in a pressure-demand or other positive-pressure mode) or SaF: Pd,Pp: ASCBA (any supplied-air respirator that has a full face-piece and is operated in a pressure-demand or other positive-pressure mode in combination with an auxiliary self-contained breathing apparatus operated in a pressure-demand or other positive-pressure mode). *Escape:* 100F (APF = 50) (any air-purifying, full-face-piece respirator with an N100, R100, or P100 filter) or SCBAE (any appropriate escape-type, self-contained breathing apparatus).

Note: Substance causes eye irritation or damage; eye protection needed.

Storage: Treat as an explosive material. Color Code—Red Stripe: Flammability Hazard: Store separately from all other flammable materials. Prior to working with this chemical you should be trained on its proper handling and storage. Store in tightly closed containers in a cool, well-ventilated area away from organics or other readily oxidizable materials. Protect from shock, light, and vibration. Sources of ignition, such as smoking and open flames, are prohibited where silver picrate is used, handled, or stored in a manner that could create a potential fire or explosion hazard. Wherever silver picrate is used, handled, manufactured, or stored, use explosion-proof electrical equipment and fittings.

Shipping: Silver picrate (dry) is FORBIDDEN. Silver picrate, wetted with not <30% water, by mass, requires a shipping label of "FLAMMABLE SOLID." It falls in Hazard Class 4.1 and Packing Group I.

Spill Handling: Evacuate persons not wearing protective equipment from area of spill or leak until cleanup is complete. Remove all ignition sources. Collect powdered material in the most convenient and safe manner and deposit in sealed containers. Ventilate area after cleanup is complete. Keep silver picrate out of a confined space, such as a sewer, because of the possibility of an explosion, unless the sewer is designed to prevent the buildup of explosive concentrations. It may be necessary to contain and dispose of this chemical as a hazardous waste. If material or contaminated runoff enters waterways, notify downstream users of potentially contaminated waters. Contact your local or federal environmental protection agency for specific recommendations. If employees are required to clean up spills, they must be properly trained and equipped. OSHA 1910.120(q) may be applicable.

Fire Extinguishing: Silver picrate is a flammable material which may be ignited by heat, sparks, or flames. Containers may explode if exposed to heat, flame, or shock. Cover with sand, earth, or water spray, and keep it wet. Poisonous gases are produced in fire. If material or contaminated runoff enters waterways, notify downstream users of potentially contaminated waters. Notify local health and fire officials and pollution control agencies. From a secure, explosion-proof location, use water spray to cool exposed containers. If cooling streams are ineffective (venting sound increases in volume and pitch, tank discolors, or shows any signs of deforming), withdraw immediately to a secure position. If employees are expected to fight fires, they must be trained and equipped in OSHA 1910.156. The only respirators recommended for firefighting are self-contained breathing apparatuses that have full face-pieces and are operated in a pressure-demand or other positive-pressure mode.

References

New Jersey Department of Health and Senior Services. (May 2000). *Hazardous Substances Fact Sheet: Silver Picrate.* Trenton, NJ

Silvex   S:0300

Molecular Formula: $C_9H_7Cl_3O_3$
Common Formula: $Cl_3C_6H_2OCH(CH_3)COOH$
Synonyms: Acide, 2-(2,4,5-trichloro-phenoxy) propionique (French); Amchen 2,4,5-TP; Aqua-Vex; Color-set; Dedweed; Double strength; Fenomore; Fenoprop; Fruitone T; Fruit-O-Net; Kuran; Kuron; Kurosal; Kurosalg; Miller NU set; Propon; Silvex herbicide; Silvi-RHAP; STA-fast; 2,4,5-TC; 2,4,5-TCPPA; 2,4,5-TP; 2,4,5-Trichlorophenoxy-α-; α-(2,4,5-Trichlorophenoxy)propanoic acid; 2-(2,4,5-Trichlorophenoxy)propanoic acid; 2-(2,4,5-Trichlorphenoxy)propionsaeure (German)
CAS Registry Number: 93-72-1
RTECS® Number: UF8225000

UN/NA & ERG Number: UN2765 (Phenoxy pesticides, solid, toxic)/152

EC Number: 202-271-2

Regulatory Authority and Advisory Bodies

Carcinogenicity: EPA: Not Classifiable as to human carcinogenicity.

Banned or Severely Restricted (several countries) (UN).[13]

Clean Water Act: Section 311 Hazardous Substances/RQ 40CFR117.3 (same as CERCLA, see below) as 2,4,5-tp acid.

US EPA Hazardous Waste Number (RCRA No.): U233.

RCRA, 40CFR261, Appendix 8 Hazardous Constituents.

RCRA 40CFR264, Appendix 9; TSD Facilities Ground Water Monitoring List. Suggested test method(s) (PQL μg/L): 8150 (2).

Safe Drinking Water Act: MCL, 0.05 mg/L; MCLG, 0.05 mg/L; Regulated chemical (47 FR 9352) as 2,4,5-TP.

RCRA 40CFR268.48; 61FR15654, Universal Treatment Standards: Wastewater (mg/L), 0.72; Nonwastewater (mg/kg), 7.9.

Reportable Quantity (RQ): 100 lb (45.4 kg).

European/International Regulations: not listed in Annex 1.

WGK (German Aquatic Hazard Class): No value assigned.

Description: Silvex is a colorless powder; Freezing/Melting point = 179–181°C; Vapor pressure = 1×10^{-7} mmHg at 20°C. Hazard Identification (based on NFPA-704 M Rating System): Health 2, Flammability 1, Reactivity 0. Slightly water soluble.

Potential Exposure: Those engaged in the manufacture, formulation, and application of this herbicide.

Incompatibilities: None listed.

Permissible Exposure Limits in Air

Protective Action Criteria (PAC)

TEEL-0: 1.5 mg/m^3

PAC-1: 5 mg/m^3

PAC-2: 35 mg/m^3

PAC-3: 250 mg/m^3

Permissible Concentration in Water: Surface water levels should never exceed 2.5 ppb silvex (butoxyethyl ester); 2.0 ppb (propyleneglycolbutylether ester) (EPA).

Determination in Water: Fish Tox = 1515.29908000 ppb (VERY LOW).

Routes of Entry: Inhalation, ingestion, skin and/or eye contact.

Harmful Effects and Symptoms

Short Term Exposure: May cause skin and eye irritation. Poisonous by ingestion. Approximate lethal dose = 2.4 tablespoonful/150 lb man.

Long Term Exposure: Silvex has caused liver and kidney damage in experimental animals. Human Tox = 56.00000 ppb (LOW).

Points of Attack: Liver, kidneys, skin, eyes.

Medical Surveillance: Liver and kidney function tests.

First Aid: If this chemical gets into the eyes, remove any contact lenses at once and irrigate immediately for at least 15 min, occasionally lifting upper and lower lids. Seek medical attention immediately. If this chemical contacts the skin, remove contaminated clothing and wash immediately with soap and water. Seek medical attention immediately. If this chemical has been inhaled, remove from exposure, begin rescue breathing (using universal precautions, including resuscitation mask) if breathing has stopped and CPR if heart action has stopped. Transfer promptly to a medical facility. When this chemical has been swallowed, get medical attention. Give large quantities of water and induce vomiting. Do not make an unconscious person vomit.

Personal Protective Methods: Wear protective gloves and clothing to prevent any reasonable probability of skin contact. Safety equipment suppliers/manufacturers can provide recommendations on the most protective glove/clothing material for your operation. All protective clothing (suits, gloves, footwear, headgear) should be clean, available each day, and put on before work. Contact lenses should not be worn when working with this chemical. Wear dust-proof chemical goggles and face shield unless full face-piece respiratory protection is worn. Employees should wash immediately with soap when skin is wet or contaminated. Provide emergency showers and eyewash.

Respirator Selection: Follow regulations in OSHA 29CFR1910.134 or European Standard EN149. Use a NIOSH/MSHA- or European Standard EN149-approved respirator; or use an approved supplied-air respirator with a full face-piece operated in the positive-pressure mode, or with a full face-piece, hood, or helmet in the continuous-flow mode; or use a NIOSH/MSHA- or European Standard EN149-approved self-contained breathing apparatus with a full face-piece operated in pressure-demand or other positive-pressure mode.

Storage: Color Code—Blue: Health Hazard/Poison: Store in a secure poison location. Prior to working with this chemical you should be trained on its proper handling and storage. A regulated, marked area should be established, or stored in compliance with OSHA Standard 1910.1045.

Shipping: Phenoxy pesticides, solid, toxic and "POISONOUS/TOXIC MATERIALS." This material falls in Hazard Class 6.1 and Packing Group III.

Spill Handling: Evacuate persons not wearing protective equipment from area of spill or leak until cleanup is complete. Remove all ignition sources. Collect powdered material in the most convenient and safe manner and deposit in sealed containers. Ventilate area after cleanup is complete. It may be necessary to contain and dispose of this chemical as a hazardous waste. If material or contaminated runoff enters waterways, notify downstream users of potentially contaminated waters. Contact your local or federal environmental protection agency for specific recommendations. If employees are required to clean up spills, they must be properly trained and equipped. OSHA 1910.120(q) may be applicable. Soil Adsorption Index (K_{oc}) = 300.

Fire Extinguishing: This chemical is a combustible solid. Use dry chemical, carbon dioxide, water spray, or alcohol foam extinguishers. Poisonous gases are produced in fire, including chlorine. If material or contaminated runoff enters waterways, notify downstream users of potentially contaminated waters. Notify local health and fire officials and pollution control agencies. From a secure, explosion-proof location, use water spray to cool exposed containers. If cooling streams are ineffective (venting sound increases in volume and pitch, tank discolors, or shows any signs of deforming), withdraw immediately to a secure position. If employees are expected to fight fires, they must be trained and equipped in OSHA 1910.156. The only respirators recommended for firefighting are self-contained breathing apparatuses that have full face-pieces and are operated in a pressure-demand or other positive-pressure mode.

Disposal Method Suggested: Consult with environmental regulatory agencies for guidance on acceptable disposal practices. Generators of waste containing this contaminant ($\geq$100 kg/mo) must conform with EPA regulations governing storage, transportation, treatment, and waste disposal.[1] Mix with excess sodium carbonate, add water and let stand for 24 h before flushing down the drain with excess water; or[2] pour onto vermiculite and incinerate with wood, paper, and waste alcohol. EPA compared disposal procedures. They concluded that incineration was difficult and unreasonably expensive. They concluded that land-spreading permitted exposure to silvex and the contaminant TCDD. The preferred method was disposal in a secure hazardous waste landfill.

Simazine   S:0310

Molecular Formula: $C_7H_{12}ClN_5$

Synonyms: A 2079; AI3-51142; Aktinit S; Aquazine; Batazina; 2,4-Bis(aethylamino)-6-chlor-1,3,5-triazin (German); 2,4-Bis(ethylamino)-6-chloro-s-triazine; Bitemol; Bitemol S-50; Cat (Japan); CDT; Cekusan; Cekuzina-S; Cet; 1-Chloro-3,5-bis(ethylamino)-2,4,6-triazine; 2-Chloro-4,6-bis(ethylamino)-s-triazine; 2-Chloro-4,6-bis(ethylamino)-1,3,5-triazine; 6-Chloro-N,N'-diethyl-1,3,5-triazine-2,4-diamine; 6-Chloro-N^2,N^4-diethyl-1,3,5-triazine-2,4-diamine; 6-Chloro-N,N'-diethyl-1,3,5-triazine-2,4-diyldiamine; Framed; G 27692; Geigy 27692; Gesaran; Gesatop; Gesatop-50; H 1803; Herbazin 50; Herbex; Herboxy; Hungazin DT; NSC 25999; Premazine; Primatel S; Primatol S; Princep 80W; Simadex; Simanex; Simazina (Spanish); Simazine 80W; Tafazine; Tafazine 50-W; Taphazine; Triazine A 384; s-Triazine, 2-chloro-4,6-bis(ethylamino)-; 1,3,5-Triazine-2,4-diamine, 6-chloro-N,N'-diethyl-; W 6658; Weedex; Zeapur

CAS Registry Number: 122-34-9

RTECS® Number: XY5250000

UN/NA & ERG Number: UN2753/151

EC Number: 204-535-2 [*Annex I Index No.:* 612-088-00-3]

Regulatory Authority and Advisory Bodies

Air Pollutant Standard Set. See below, "Permissible Exposure Limits in Air" section.

Safe Drinking Water Act: MCL, 0.004 mg/L; MCLG, 0.004 mg/L; Regulated chemical (47 FR 9352).

EPCRA Section 313 Form R *de minimis* concentration reporting level: 1.0%.

European/International Regulations: Hazard Symbol: Xn, N; Risk phrases: R40; R50/53; Safety phrases: S2; S36/37; S46; S60; S61 (see Appendix 4).

WGK (German Aquatic Hazard Class): 3—Severe hazard to waters.

Description: Simazine is a combustible, white crystalline solid with little or no odor; Freezing/Melting point = (decomposes) 225°C; Vapor pressure = 2.2×10^{-8} mmHg at 20°C. Hazard Identification (based on NFPA-704 M Rating System): Health 1, Flammability 1, Reactivity 0. Insoluble in water.

Potential Exposure: Those involved in the manufacture, formulation, and application of this preemergence herbicide.

Incompatibilities: Powder forms explosive mixture with air. Incompatible with strong oxidizers.

Permissible Exposure Limits in Air: Russia set a MAC of 2.0 mg/m^3 for simazine in work-place air[35] and a MAC of 0.02 mg/m^3 for ambient air in residential areas both on a momentary and a daily average basis.

Permissible Concentration in Water: Russia set a MAC of 2.4 µg/L in water bodies used for fishery purposes and a limit of zero in water bodies used for domestic purposes. A lifetime health advisory of 35 µg/L has been developed by EPA (see "References" below). Various states have developed guidelines for simazine in drinking water[61] ranging from 150 µg/L (California) to 430 µg/L (Maine) to 2150 µg/L (Wisconsin).

Determination in Water: Analysis of simazine is by a gas chromatographic (GC) method applicable to the determination of certain nitrogen−phosphorus-containing pesticides in water samples. In this method, approximately 1 L of sample is extracted with methylene chloride. The extract is concentrated and the compounds are separated using capillary column GC. Measurement is made using a nitrogen−phosphorus detector. The method detection limit has not been determined for this compound but it is estimated that the detection limits for the method analytes are in the range of 0.1−2 µg/L. Fish Tox = 1732.04831000 ppb (VERY LOW). Octanol−water coefficient: Log K_{ow} = 2.08.

Routes of Entry: Inhalation, ingestion, skin and/or eye contact.

Harmful Effects and Symptoms

Short Term Exposure: May cause skin or eye irritation. Moderately poisonous if ingested. Approximate lethal dose = 1.5 cupful/150 lb man. No case of poisoning in humans from simazine has been reported, although exposure to simazine has caused acute and subacute dermatitis in Russia, characterized by erythema, slight edema, moderate pruritus, and burning lasting 4−5 days.

Long Term Exposure: Repeated exposure may cause weight loss and reduced red blood cell count. *Chronic*

Toxicity—simazine fed to rats for 2 years at 1.0, 10, and 100 ppm produced no difference between treated and control animals in gross appearance or behavior. The rats fed 100 ppm had approximately twice as many thyroid and mammary tumors as the control animals, but it was stated that these were not attributable to simazine. A 2-year chronic-feeding study of simazine in dogs with simazine 80W fed at 15, 150, and 1500 ppm showed only a slight thyroid hyperplasia at 1500 ppm and slight increases in serum alkaline phosphatase and serum glutamic oxaloacetic transaminase in several of the dogs fed 1500 ppm. Human Tox = 4.00000 ppb (HIGH).

Points of Attack: Blood.

Medical Surveillance: Complete blood count.

First Aid: If this chemical gets into the eyes, remove any contact lenses at once and irrigate immediately for at least 15 min, occasionally lifting upper and lower lids. Seek medical attention immediately. If this chemical contacts the skin, remove contaminated clothing and wash immediately with soap and water. Seek medical attention immediately. If this chemical has been inhaled, remove from exposure, begin rescue breathing (using universal precautions, including resuscitation mask) if breathing has stopped and CPR if heart action has stopped. Transfer promptly to a medical facility. When this chemical has been swallowed, get medical attention. Give large quantities of water and induce vomiting. Do not make an unconscious person vomit.

Personal Protective Methods: Wear protective gloves and clothing to prevent any reasonable probability of skin contact. Safety equipment suppliers/manufacturers can provide recommendations on the most protective glove/clothing material for your operation. All protective clothing (suits, gloves, footwear, headgear) should be clean, available each day, and put on before work. Contact lenses should not be worn when working with this chemical. Wear dust-proof chemical goggles and face shield unless full face-piece respiratory protection is worn. Employees should wash immediately with soap when skin is wet or contaminated. Provide emergency showers and eyewash.

Respirator Selection: Follow regulations in OSHA 29CFR1910.134 or European Standard EN149. Use a NIOSH/MSHA- or European Standard EN149-approved respirator; or use an approved supplied-air respirator with a full face-piece operated in the positive-pressure mode, or with a full face-piece, hood, or helmet in the continuous-flow mode; or use a NIOSH/MSHA- or European Standard EN149-approved self-contained breathing apparatus with a full face-piece operated in pressure-demand or other positive-pressure mode.

Storage: Color Code—Blue: Health Hazard/Poison: Store in a secure poison location. Prior to working with this chemical you should be trained on its proper handling and storage.

Shipping: Triazine pesticides, solid, toxic, require a shipping label of "POISONOUS/TOXIC MATERIALS." They fall in Hazard Class 6.1 and Packing Group I to III.

Spill Handling: Evacuate persons not wearing protective equipment from area of spill or leak until cleanup is complete. Remove all ignition sources. Collect powdered material in the most convenient and safe manner and deposit in sealed containers. Ventilate area after cleanup is complete. It may be necessary to contain and dispose of this chemical as a hazardous waste. If material or contaminated runoff enters waterways, notify downstream users of potentially contaminated waters. Contact your local or federal environmental protection agency for specific recommendations. If employees are required to clean up spills, they must be properly trained and equipped. OSHA 1910.120(q) may be applicable. Soil Adsorption Index (K_{oc}) = 130.

Fire Extinguishing: This chemical is a combustible solid. Use dry chemical, carbon dioxide, water spray, or alcohol foam extinguishers. Poisonous gases are produced in fire. If material or contaminated runoff enters waterways, notify downstream users of potentially contaminated waters. Notify local health and fire officials and pollution control agencies. From a secure, explosion-proof location, use water spray to cool exposed containers. If cooling streams are ineffective (venting sound increases in volume and pitch, tank discolors, or shows any signs of deforming), withdraw immediately to a secure position. If employees are expected to fight fires, they must be trained and equipped in OSHA 1910.156. The only respirators recommended for firefighting are self-contained breathing apparatuses that have full face-pieces and are operated in a pressure-demand or other positive-pressure mode.

Disposal Method Suggested: Strong acid or alkaline hydrolysis leads to complete degradation of simazine. However, large quantities of simazine should be incinerated in a unit operating at 850°C equipped with off-gas scrubbing equipment.[22] In accordance with 40CFR165, follow recommendations for the disposal of pesticides and pesticide containers. Must be disposed properly by following package label directions or by contacting your local or federal environmental control agency or by contacting your regional EPA office.

References

Sax, N. I. (Ed.). (1987). *Dangerous Properties of Industrial Materials Report*, 7, No. 4, 109–113

US Environmental Protection Agency. (August 1987). *Health Advisory: Simazine*. Washington, DC: Office of Drinking Water

Soapstone   S:0320

Molecular Formula: $H_2Mg_3O_{12}Si_4$

Common Formula: $3MgO \cdot 4SiO_2 \cdot H_2O$

Synonyms: Massive talc; Silicate soapstone; Soapstone silicate; Steatite; Talc

CAS Registry Number: None listed.

RTECS® Number: VV8780000

Regulatory Authority and Advisory Bodies

Carcinogenicity: IARC: Human Inadequate Evidence (talc not containing asbestiform fibers); Sufficient (for talc containing asbestiform fibers).

Air Pollutant Standard Set. See below, "Permissible Exposure Limits in Air" section.

Canada, WHMIS, Ingredients Disclosure List Concentration 1.0%.

Description: Soapstone is an odorless gray-white crystalline solid. Molecular weight = 351.31; Specific gravity (H_2O:1) = 2.7–2.8 at 25°C; Freezing/Melting point = 900–1000°C. Negligible solubility in water.

Potential Exposure: Soapstone is used as a pigment in paints, varnishes, rubber, and soap. It is used in lubricating molds and machinery. In massive form, it is used as a heat insulator.

Incompatibilities: Silicates can react violently with lithium.

Permissible Exposure Limits in Air

OSHA PEL: 20 mppcf TWA.

NIOSH REL: 6 mg/m^3, total dust TWA; 3 mg/m^3 (respirable dust) TWA.

ACGIH TLV®[1]: 6 mg/m^3 (for particulate matter containing no asbestos and <1% crystalline silica) TWA; 3 mg/m^3 (respirable fraction, for particulate matter containing no asbestos and <1% crystalline silica) TWA.

NIOSH IDLH: 3000 mg/m^3.

Protective Action Criteria (PAC)

Talc

TEEL-0: 2 mg/m^3

PAC-1: 2 mg/m^3

PAC-2: 10 mg/m^3

PAC-3: 500 mg/m^3

DFG MAK: *for talcum* suspended 2004; Carcinogen Category 3B.

The Netherlands: MAC-TGG 5 mg/m^3 (total dust); 2.5 mg/m^3 (respirable dust), 2003.

Determination in Air: Filter; none; Gravimetric; NIOSH Analytical Method (IV) #0500, Particulates NOR, total dust.

Routes of Entry: Inhalation, eye and/or skin contact.

Harmful Effects and Symptoms

Short Term Exposure: Irritates the eyes and respiratory tract causing coughing, wheezing.

Long Term Exposure: Pneumoconiosis: cough, dyspnea (breathing difficulty); digital clubbing; cyanosis, basal crackles; or pulmonale.

Points of Attack: Respiratory system, cardiovascular system.

Medical Surveillance: NIOSH lists the following tests: chest X-ray; pulmonary function tests: forced vital capacity, forced expiratory volume (1 s). Consider the points of attack in preplacement and periodic physical examinations. Lung function tests.

First Aid: If this chemical gets into the eyes, remove any contact lenses at once and irrigate immediately for at least 15 min, occasionally lifting upper and lower lids. Seek medical attention immediately. If this chemical contacts the skin, remove contaminated clothing and wash immediately with soap and water. Seek medical attention immediately. If this chemical has been inhaled, remove from exposure, begin rescue breathing (using universal precautions, including resuscitation mask) if breathing has stopped and CPR if heart action has stopped. Transfer promptly to a medical facility. When this chemical has been swallowed, get medical attention. Give large quantities of water and induce vomiting. Do not make an unconscious person vomit.

Personal Protective Methods: Wear dust-proof chemical goggles and face shield unless full face-piece respiratory protection is worn. Employees should wash immediately with soap when skin is wet or contaminated. Provide emergency showers and eyewash.

Respirator Selection: NIOSH: *Up to 30 mg/m^3:* Qm (APF = 25) (any quarter-mask respirator). *Up to 60 mg/m^3:* 95 XQ [any particulate respirator equipped with an N95, R95, or P95 filter (including N95, R95, and P95 filtering face-pieces) except quarter-mask respirators. The following filters may also be used: N99, R99, P99, N100, R100, P100] or Sa (APF = 10) (any supplied-air respirator). *Up to 150 mg/m^3:* PaprHie (APF = 25) (any powered, air-purifying respirator with a high-efficiency particulate filter). *Up to 300 mg/m^3:* 100F (APF = 50) (any air-purifying, full-face-piece respirator with an N100, R100, or P100 filter) or SaT: Cf (APF = 50) (any supplied-air respirator that has a tight-fitting face-piece and is operated in a continuous-flow mode) or PaprTHie (APF = 50)* (any powered, air-purifying respirator with a tight-fitting face-piece and a high-efficiency particulate filter) or SCBAF (APF = 50) (any self-contained breathing apparatus with a full face-piece) or SaF (APF = 50) (any supplied-air respirator with a full face-piece). *Up to 3000 mg/m^3:* SaF: Pd,Pp (APF = 2000) (any supplied-air respirator that has a full face-piece and is operated in a pressure-demand or other positive-pressure mode). *Emergency or planned entry into unknown concentrations or IDLH conditions:* SCBAF: Pd,Pp (APF = 10,000) (any self-contained breathing apparatus that has a full face-piece and is operated in a pressure-demand or other positive-pressure mode) or SaF: Pd,Pp: ASCBA (APF = 10,000) (any supplied-air respirator that has a full face-piece and is operated in a pressure-demand or other positive-pressure mode in combination with an auxiliary self-contained breathing apparatus operated in a pressure-demand or other positive-pressure mode). *Escape:* 100F (APF = 50) (any air-purifying, full-face-piece respirator with an N100, R100, or P100 filter) or SCBAE (any appropriate escape-type, self-contained breathing apparatus).

*Substance reported to cause eye irritation or damage; may require eye protection.

Storage: Color Code—Green: General storage may be used. Prior to working with this chemical you should be trained on its proper handling and storage. Store in a cool, dry place.

Spill Handling: Evacuate persons not wearing protective equipment from area of spill or leak until cleanup is complete. Remove all ignition sources. Dampen spilled material with water to avoid airborne dust, then transfer material to a suitable container for disposal. Ventilate area after cleanup is complete. It may be necessary to contain and dispose of this chemical as a hazardous waste. If material or contaminated runoff enters waterways, notify downstream users of potentially contaminated waters. Contact your local or federal environmental protection agency for specific recommendations. If employees are required to clean up spills, they must be properly trained and equipped. OSHA 1910.120(q) may be applicable.

Fire Extinguishing: This chemical is a combustible solid. Use dry chemical, carbon dioxide, water spray, or alcohol foam extinguishers. Poisonous gases are produced in fire. If material or contaminated runoff enters waterways, notify downstream users of potentially contaminated waters. Notify local health and fire officials and pollution control agencies. From a secure, explosion-proof location, use water spray to cool exposed containers. If cooling streams are ineffective (venting sound increases in volume and pitch, tank discolors, or shows any signs of deforming), withdraw immediately to a secure position. If employees are expected to fight fires, they must be trained and equipped in OSHA 1910.156. The only respirators recommended for firefighting are self-contained breathing apparatuses that have full face-pieces and are operated in a pressure-demand or other positive-pressure mode.

Disposal Method Suggested: Sanitary landfill.

Sodium   S:0330

Molecular Formula: Na
Synonyms: Elemental sodium; Natrium; Sodio (Spanish); Sodium element; Sodium, metal liquid alloy; Sodium metal
CAS Registry Number: 7440-23-5
RTECS® Number: VY0686000
UN/NA & ERG Number: UN1428/138
EC Number: 231-132-9 [Annex I Index No.: 011-001-00-0]
Regulatory Authority and Advisory Bodies
Clean Water Act: Section 311 Hazardous Substances/RQ 40CFR117.3 (same as CERCLA, see below).
Reportable Quantity (RQ): 10 lb (4.54 kg).
European/International Regulations: Hazard Symbol: F, C; Risk phrases: R14/15; R34; Safety phrases: S1/2; S5 (If appropriate); S8; S43; S45 (see Appendix 4).
WGK (German Aquatic Hazard Class): 1—Low hazard to waters.
Description: Sodium is a soft silvery white metallic element. Pyrophoric solid or molten liquid. Odorless. Molecular weight = 22.99; Boiling point = 881.4°C; Freezing/Melting point = 98°C; Autoignition temperature ≥115°C (in dry air). Hazard Identification (based on NFPA-704 M Rating System): Health 3, Flammability 3, Reactivity 3W. Violent reaction with water.

Potential Exposure: Those involved in tetra-alkyl lead manufacture using lead-sodium alloy as a reactant; those using sodium as a liquid metal coolant, as a catalyst, or in the manufacture of sodium hydride, borohydride, or peroxide.
Incompatibilities: A strong reducing agent. A dangerous fire hazard when exposed to heat and moisture. Violent reaction with water, forming sodium hydroxide. Violent reaction with oxidizers, acids, halogenated hydrocarbons, phosphorus and phosphorus compounds, sulfur and sulfur compounds, and many other chemicals.
Permissible Exposure Limits in Air
Protective Action Criteria (PAC)
TEEL-0: 0.75 mg/m^3
PAC-1: 2 mg/m^3
PAC-2: 15 mg/m^3
PAC-3: 500 mg/m^3
Permissible Concentration in Water: The metal reacts with water. Sodium ion limit is 10 mg/L as desirable in drinking water; 200 mg/L may be injurious to humans. Russia[43] set a MAC of 120 mg/L in water bodies used for domestic purposes. Several states have set guidelines for sodium in drinking water[61] ranging from 20 mg/L (Massachusetts) to 100 mg/L (Kansas).
Routes of Entry: Inhalation, ingestion, skin and/or eye contact.
Harmful Effects and Symptoms
Short Term Exposure: Inhalation: Contact with water, including perspiration, causes the formation of sodium hydroxide fumes which are highly irritating to skin, eyes, nose, and throat. May cause sneezing and coughing. Very severe exposures may result in difficult breathing, coughing, and chemical bronchitis. High exposures can cause pulmonary edema, a medical emergency that can be delayed for several hours. This can cause death. *Skin:* Contact may cause itching, tingling, thermal and caustic burns, and may cause permanent damage. *Eyes:* May cause tearing, very painful irritation, and burns. Contact with eyes may result in permanent damage and loss of sight. *Ingestion:* Causes immediate intense burning sensation in mouth, throat, and stomach, followed by salivation, vomiting, rapid breathing, symptoms of shock, diarrhea, loss of consciousness, and death.
Long Term Exposure: Very irritating substances may affect the lungs. It is not known, however, if sodium causes lung damage.
Points of Attack: Lungs.
Medical Surveillance: Before beginning employment and at regular times after that, for those with frequent or potentially high exposure to sodium metal, the following are recommended: lung function tests. If symptoms develop or overexposure is suspected, the following may be useful: consider chest X-ray after acute overexposure.
First Aid: If this chemical gets into the eyes, remove any contact lenses at once and irrigate immediately for at least 30 min, occasionally lifting upper and lower lids. Seek

medical attention immediately. Wipe the chemical off the skin with a dry cloth. Then quickly remove contaminated clothing. Immediately wash area with large amounts of water. Seek medical attention immediately. If this chemical has been inhaled, remove from exposure, begin rescue breathing (using universal precautions, including resuscitation mask) if breathing has stopped and CPR if heart action has stopped. Transfer promptly to a medical facility. When this chemical has been swallowed, get medical attention. Give large quantities of water and induce vomiting. Do not make an unconscious person vomit. Medical observation is recommended for 24–48 h after breathing overexposure, as pulmonary edema may be delayed. As first aid for pulmonary edema, a doctor or authorized paramedic may consider administering a corticosteroid spray.

Personal Protective Methods: Wear protective gloves and clothing to prevent any reasonable probability of skin contact. Safety equipment suppliers/manufacturers can provide recommendations on the most protective glove/clothing material for your operation. All protective clothing (suits, gloves, footwear, headgear) should be clean, available each day, and put on before work. Contact lenses should not be worn when working with this chemical. Wear dust-proof chemical goggles and face shield unless full face-piece respiratory protection is worn. Employees should wash immediately with soap when skin is wet or contaminated. Provide emergency showers and eyewash.

Respirator Selection: Where there is potential for exposure to sodium, use a NIOSH/MSHA- or European Standard EN149-approved full face-piece respirator equipped with particulate (dust/fume/mist) filters. Particulate filters must be checked every day before work for physical damage, such as rips or tears, and replaced as needed. *Where there is potential for high exposures*, use a NIOSH/MSHA- or European Standard EN149-approved supplied-air respirator with a full face-piece operated in the positive-pressure mode, or with a full face-piece, hood, or helmet in the continuous-flow mode; or use a NIOSH/MSHA- or European Standard EN149-approved self-contained breathing apparatus with a full face-piece operated in pressure-demand or other positive-pressure mode.

Storage: (1) Color Code—Red Stripe (*lump*): Flammability Hazard: Do not store in the same area as other flammable materials. (2) Color Code—Yellow Stripe (*strong reducing agent*): Reactivity Hazard; Store separately in an area isolated from flammables, combustibles, or other yellow coded materials. Prior to working with this chemical you should be trained on its proper handling and storage. Sodium must be stored to avoid contact with water, halogenated hydrocarbons, phosphorus and phosphorus compounds, and sulfur and sulfur compounds, since violent reactions occur. Protect storage containers from physical damage. Keep in an inert atmosphere or under oil. Requires special precautions to avoid contact with moisture, including condensation from other objects and perspiration. Store in a dry, fire resistive room exclusively for sodium storage. Sprinkler systems

should not be used, so keep combustibles away. Remove only the amount of sodium needed for immediate use and place dry in a friction top metal can under a layer of oil. Transport in a special container with a tight-fitting cover. Where sodium is used, handled, manufactured, or stored, use explosion-proof electrical equipment and fittings.

Shipping: Sodium metal requires a shipping label of "DANGEROUS WHEN WET." It falls in Hazard Class 4.3 and Packing Group I. *Note:* Finely divided sodium is pyrophoric.

Spill Handling: Evacuate persons not wearing protective equipment from area of spill or leak until cleanup is complete. Remove all ignition sources. Put on proper protective equipment. Blanket with appropriate inert material. Scoop up and place in a suitable, sealed containers. Ventilate area after cleanup is complete. It may be necessary to contain and dispose of this chemical as a hazardous waste. If material or contaminated runoff enters waterways, notify downstream users of potentially contaminated waters. Contact your local or federal environmental protection agency for specific recommendations. If employees are required to clean up spills, they must be properly trained and equipped. OSHA 1910.120(q) may be applicable.

Fire Extinguishing: This chemical is a combustible solid. A severe fire risk on contact with water or moisture. Blanket with dry soda ash, graphite, salt, dry limestone, or other approved dry powder. *Do not use water*, carbon dioxide, or halogenated extinguishers. Poisonous gases are produced in fire, including sodium oxide. If material or contaminated runoff enters waterways, notify downstream users of potentially contaminated waters. Notify local health and fire officials and pollution control agencies. From a secure, explosion-proof location, use water spray to cool exposed containers. If cooling streams are ineffective (venting sound increases in volume and pitch, tank discolors, or shows any signs of deforming), withdraw immediately to a secure position. If employees are expected to fight fires, they must be trained and equipped in OSHA 1910.156. The only respirators recommended for firefighting are self-contained breathing apparatuses that have full face-pieces and are operated in a pressure-demand or other positive-pressure mode.

Disposal Method Suggested: Incineration with absorption of oxide fumes.

References
Sax, N. I. (Ed.). (1981). *Dangerous Properties of Industrial Materials Report*, 1, No. 8, 85–88
Sittig, M. (1956). *Sodium, Its Manufacture, Properties and Uses.* American Chemical Monograph No. 133. New York: Reinhold Publishing Corp.
New York State Department of Health. (May 1986). *Chemical Fact Sheet: Sodium.* Albany, NY: Bureau of Toxic Substance Assessment
New Jersey Department of Health and Senior Services. (April 2001). *Hazardous Substances Fact Sheet: Sodium.* Trenton, NJ

Sodium aluminate   S:0340

Molecular Formula: $AlNaO_2$

Common Formula: $NaAlO_2$ ($Na_2O \cdot Al_2O_3$)

Synonyms: β-Alumina; *beta*-Alumina; Aluminum sodium oxide; J 242; Maxifloc 8010; Monofrax H; Nalco 680; Sodium aluminum oxide; Sodium polyaluminate; VSA 45

CAS Registry Number: 11138-49-1

RTECS® Number: BD1600000

EC Number: 234-391-6

Regulatory Authority and Advisory Bodies

Air Pollutant Standard Set. See below, "Permissible Exposure Limits in Air" section.

Canada, WHMIS, Ingredients Disclosure List Concentration 1.0%.

UN/NA & ERG Number: UN2812 (solid)/154; UN1819 (solution)/154.

WGK (German Aquatic Hazard Class): No value assigned.

Description: Sodium aluminate is a white crystalline solid or solution. Molecular weight = 82.12; Freezing/Melting point = 1650°C. Hazard Identification (based on NFPA-704 M Rating System): Health 3, Flammability 0, Reactivity 1. Soluble in water.

Potential Exposure: Used in water and waste treatment; papermaking industry; in printing on fabrics; in the manufacture of pigments, milk glass, and soap; hardening building stone; sizing paper; as a water softener.

Incompatibilities: The aqueous solution is a strong base. Reacts violently with acid. Incompatible with organic anhydrides, isocyanates, alkylene oxides, epichlorohydrin, aldehydes, alcohols, glycols, caprolactum, chlorocarbons. Attacks copper, tin, aluminum, and zinc.

Permissible Exposure Limits in Air

OSHA PEL: None.

NIOSH REL: 2 mg[Al]/m^3 TWA.

ACGIH TLV®[1]: 2 mg[Al]/m^3 TWA as soluble salts and alkyls, n.o.s.

Protective Action Criteria (PAC)

TEEL-0: 6.08 mg/m^3

PAC-1: 18.2 mg/m^3

PAC-2: 30.4 mg/m^3

PAC-3: 150 mg/m^3

Australia: TWA 2 mg[Al]/m^3, 1993; Belgium: TWA 2 mg [Al]/m^3, 1993; Denmark: TWA 2 mg[Al]/m^3, 1999; France: VME 2 mg[Al]/m^3, 1999; Norway: TWA 2 mg[Al]/m^3, 1999; Russia: TWA 2 mg[Al]/m^3, 1993; Sweden: NGV 2 mg[Al]/m^3, 1999; Switzerland: MAK-W 2 mg[Al]/m^3, 1999; United Kingdom: TWA 10 mg/m^3, total dust inhalable dust; TWA 4 mg/m^3, respirable dust, 2000; Argentina, Bulgaria, Columbia, Jordan, South Korea, New Zealand, Singapore, Vietnam: ACGIH TLV®: TWA 2 mg[Al]/m^3.

Determination in Air: Use NIOSH #7013 Aluminum; #7300, Elements; #7303; OSHA Analytical Method ID-121.

Routes of Entry: Inhalation, ingestion, skin and/or eye contact.

Harmful Effects and Symptoms

Short Term Exposure: Sodium aluminate can affect you when breathed in. Sodium aluminate solution is a corrosive chemical. Skin or eye contact can cause severe irritation or burns, with possible damage. Breathing sodium aluminate dust can irritate the air passages. This may cause cough with phlegm and/or shortness of breath. Inhalation of dust may cause pulmonary edema, a medical emergency that can be delayed for several hours. This can cause death. Corrosive on ingestion.

Long Term Exposure: Irritating substances may affect the lungs. It is not known whether sodium aluminate causes lung damage.

Points of Attack: Lungs.

Medical Surveillance: Before beginning employment and at regular times after that, for those with frequent or potentially high exposures, the following are recommended: lung function tests. Consider chest X-ray following acute overexposure.

First Aid: If this chemical gets into the eyes, remove any contact lenses at once and irrigate immediately for at least 30 min, occasionally lifting upper and lower lids. Seek medical attention immediately. If this chemical contacts the skin, remove contaminated clothing and wash immediately with soap and water. Seek medical attention immediately. If this chemical has been inhaled, remove from exposure, begin rescue breathing (using universal precautions, including resuscitation mask) if breathing has stopped and CPR if heart action has stopped. Transfer promptly to a medical facility. When this chemical has been swallowed, get medical attention. If victim is *conscious*, administer water or milk. Do not induce vomiting. Medical observation is recommended for 24−48 h after breathing overexposure, as pulmonary edema may be delayed. As first aid for pulmonary edema, a doctor or authorized paramedic may consider administering a corticosteroid spray.

Personal Protective Methods: Wear protective gloves and clothing to prevent any reasonable probability of skin contact. Safety equipment suppliers/manufacturers can provide recommendations on the most protective glove/clothing material for your operation. All protective clothing (suits, gloves, footwear, headgear) should be clean, available each day, and put on before work. Contact lenses should not be worn when working with this chemical. Wear splash-proof chemical goggles and face shield when working with liquid, unless full face-piece respiratory protection is worn. Wear dust-proof goggles and face shield when working with powders or dust, unless full face-piece respiratory protection is worn. Employees should wash immediately with soap when skin is wet or contaminated. Provide emergency showers and eyewash. Medical observation is recommended for 24−48 h after breathing overexposure, as pulmonary edema may be delayed. As first aid for pulmonary edema, a doctor or authorized paramedic may consider administering a corticosteroid spray.

Respirator Selection: Where there is potential for exposure to solid sodium aluminate, use a NIOSH/MSHA- or

European Standard EN149-approved full face-piece respirator with a high-efficiency particulate filter. Greater protection is provided by a powered air-purifying respirator. Where there is potential for high exposures or liquid sodium aluminate, use a NIOSH/MSHA- or European Standard EN149-approved supplied-air respirator with a full face-piece operated in the positive-pressure mode, or with a full face-piece, hood, or helmet in the continuous-flow mode; or use a NIOSH/MSHA- or European Standard EN149-approved self-contained breathing apparatus with a full face-piece operated in pressure-demand or other positive-pressure mode.

Storage: Color Code—White: Corrosive or Contact Hazard; Store separately in a corrosion-resistant location. Prior to working with this chemical you should be trained on its proper handling and storage. Store in tightly closed containers in a cool, well-ventilated area away from incompatible materials. Where possible, automatically transfer material from drums or other storage containers to process containers.

Shipping: Sodium aluminate, solid, requires a shipping label of "CORROSIVE." It falls in Hazard Class 8 and Packing Group III. Sodium aluminate, solution, requires a shipping label of "CORROSIVE." It falls in Hazard Class 8 and Packing Group II.

Spill Handling: Evacuate persons not wearing protective equipment from area of spill or leak until cleanup is complete. Remove all ignition sources. Absorb liquids in vermiculite, dry sand, earth, or a similar material and deposit in sealed containers. Collect powdered material in the most convenient and safe manner and deposit in sealed containers. Ventilate area after cleanup is complete. It may be necessary to contain and dispose of this chemical as a hazardous waste. If material or contaminated runoff enters waterways, notify downstream users of potentially contaminated waters. Contact your local or federal environmental protection agency for specific recommendations. If employees are required to clean up spills, they must be properly trained and equipped. OSHA 1910.120(q) may be applicable.

Fire Extinguishing: This material is nonflammable. Use agent suitable for surrounding fire. Poisonous gases are produced in fire. If material or contaminated runoff enters waterways, notify downstream users of potentially contaminated waters. Notify local health and fire officials and pollution control agencies. From a secure, explosion-proof location, use water spray to cool exposed containers. If cooling streams are ineffective (venting sound increases in volume and pitch, tank discolors, or shows any signs of deforming), withdraw immediately to a secure position. If employees are expected to fight fires, they must be trained and equipped in OSHA 1910.156. The only respirators recommended for firefighting are self-contained breathing apparatuses that have full face-pieces and are operated in a pressure-demand or other positive-pressure mode.

References

New Jersey Department of Health and Senior Services. (October 1986). *Hazardous Substances Fact Sheet: Sodium Aluminate.* Trenton, NJ
New York State Department of Health. (April 2000). *Chemical Fact Sheet: Sodium Aluminate.* Albany, NY: Bureau of Toxic Substance Assessment

Sodium aluminum fluoride   S:0350

Molecular Formula: AlF_6Na_3
Common Formula: Na_3AlF_6
Synonyms: Aluminum sodium fluoride; Cryolite; ENT 24,984; Kryolith (German); Natriumaluminumfluorid (German); Natriumhexafluoroaluminate (German); Sodium aluminofluoride; Sodium fluoaluminate; Sodium hexafluoroaluminate; Trisodium hexafluoroaluminate; Villiaumite
CAS Registry Number: 15096-52-3; 13775-53-6 (synthetic)
RTECS® Number: WA9625000
UN/NA & ERG Number: UN3077/171
EC Number: 239-148-8 [*Annex I Index No.:* 009-016-00-2]
Regulatory Authority and Advisory Bodies
US EPA, FIFRA 1998 Status of Pesticides: RED completed.
Air Pollutant Standard Set. See below, "Permissible Exposure Limits in Air" section.
Canada, WHMIS, Ingredients Disclosure List Concentration 1.0%.
European/International Regulations: Hazard Symbol (*natural and synthetic*): T, N; Risk phrases: R20/22; R48/23/25; R51/53; Safety phrases: S1/2; S22; S37; S45; S61 (see Appendix 4).
WGK (German Aquatic Hazard Class): 3—Severe hazard to waters.

Description: Sodium aluminum fluoride is a colorless to dark red or black, crystalline solid. Odorless. Loses color on heating. Molecular weight = 209.95; Boiling point = (decomposes); Freezing/Melting point = 1000°C. Very slightly soluble in water.

Potential Exposure: Compound Description: Agricultural Chemical; Mutagen. Sodium aluminum fluoride is used in making pesticides, ceramics, glass, and polishes; in refining reduction of aluminum, flux, glass, and enamel.

Incompatibilities: Strong acids, strong oxidizers.

Permissible Exposure Limits in Air
OSHA PEL: 3 ppm/2.5 mg[F]/m³ TWA.
NIOSH REL: 3 ppm/2.5 mg[F]/m³ TWA.
NIOSH IDLH: 250 mg[F]/m³.
Australia: TWA 2.5 mg[F]/m³, 1993; Austria: MAK 2.5 mg [F]/m³, 1999; Belgium: TWA 2.5 mg[F]/m³, 1993; Finland: TWA 2.5 mg[F]/m³, 1999; France: VME 2.5 mg[F]/m³, 1999; Hungary: TWA 1 mg[F]/m³; STEL 2 mg[F]/m³, 1993; Norway: TWA 0.6 mg[F]/m³, 1999; the Philippines: TWA 2.5 mg[F]/m³, 1993; Poland: MAC (TWA) 1 mg [HF]/m³, MAC (STEL) 3 mg[HF]/m³, 1999; Russia: STEL

0.5 ppm (2.5 mg/m^3), 1993; Sweden: NGV 2 mg[F]/m^3, 1999; Switzerland: MAK-W 1.8 ppm (1.5 mg[F]/m^3), KZG-W 3.6 ppm (3.0 mg[F]/m^3), 1999; Thailand: TWA 2.5 mg [F]/m^3, 1993; United Kingdom: TWA 2.5 mg[F]/m^3, 2000; LTEL 2.5 mg[F]/m^3, 1993; Argentina, Bulgaria, Columbia, Jordan, South Korea, New Zealand, Singapore, Vietnam: ACGIH TLV®: not classifiable as a human carcinogen.

As aluminum, soluble salts
OSHA PEL: None.
NIOSH REL: 2 mg[Al]/m^3 TWA.
No TEEL available.
Australia: TWA 2 mg[Al]/m^3, 1993; Belgium: TWA 2 mg [Al]/m^3, 1993; Denmark: TWA 2 mg[Al]/m^3, 1999; France: VME 2 mg[Al]/m^3, 1993; Russia: TWA 2 mg[Al]/m^3, 1993; Sweden: NGV 2 mg[Al]/m^3, 1999; Switzerland: MAK-W 2 mg[Al]/m^3, 1999; United Kingdom: LTEL 2 mg [Al]/m^3, 1993; Argentina, Bulgaria, Columbia, Jordan, South Korea, New Zealand, Singapore, Vietnam: ACGIH TLV®: TWA 2 mg[Al]/m^3.

Determination in Air: Use NIOSH Analytical Method (IV) #7902, Fluoride; for aluminum, soluble salts, use NIOSH #7013 Aluminum; #7300, Elements; #7303; OSHA Analytical Method ID-121.

Permissible Concentration in Water: Drinking water, *aluminum* guidelines[61] are 1.43 mg/L in Massachusetts and 5.0 mg/L in Kansas. *Fluoride* guidelines are 1.8 mg/L in Arizona, 2.4 mg/L in Maine, and 4.0 mg/L according to EPA. Toxic to aquatic organisms.

Routes of Entry: Inhalation, ingestion, skin and/or eye contact.

Harmful Effects and Symptoms

Short Term Exposure: Sodium aluminum fluoride can affect you when breathed in. Eye contact can cause severe irritation, burns with possible loss of vision. Skin contact can cause irritation and even burns, especially if prolonged. Breathing can irritate the nose, throat, and air passages. Higher exposures can cause pulmonary edema, a medical emergency that can be delayed for several hours. This can cause death. Exposure can cause nausea, abdominal pain, diarrhea, salivation, thirst, sweating.

Long Term Exposure: Repeated exposure can cause stiff spine; calcification of ligaments of ribs, pelvis. Repeated or high exposures may cause permanent lung damage.

Points of Attack: Eyes, skin, lungs, skeletal system.

Medical Surveillance: NIOSH lists the following tests: chest X-ray, electrocardiogram, pulmonary function tests: forced vital capacity, forced expiratory volume (1 s); pelvic X-ray; sputum cytology; urine (chemical/metabolite); urine (chemical/metabolite) pre- and postshift; urinalysis (routine); complete blood count/differential. Urine fluoride test (levels above 3–4 mg/L at the end of exposure represent increased exposure).

First Aid: If this chemical gets into the eyes, remove any contact lenses at once and irrigate immediately for at least 15 min, occasionally lifting upper and lower lids. Seek medical attention immediately. If this chemical contacts the skin, remove contaminated clothing and wash immediately with soap and water. Speed in removing material from skin is of extreme importance. Seek medical attention immediately. If this chemical has been inhaled, remove from exposure, begin rescue breathing (using universal precautions, including resuscitation mask) if breathing has stopped and CPR if heart action has stopped. Transfer promptly to a medical facility. When this chemical has been swallowed, get medical attention. If victim is *conscious*, administer water or milk. Do not induce vomiting. Medical observation is recommended for 24–48 h after breathing overexposure, as pulmonary edema may be delayed. As first aid for pulmonary edema, a doctor or authorized paramedic may consider administering a corticosteroid spray.

Personal Protective Methods: Wear protective gloves and clothing to prevent any reasonable probability of skin contact. Safety equipment suppliers/manufacturers can provide recommendations on the most protective glove/clothing material for your operation. All protective clothing (suits, gloves, footwear, headgear) should be clean, available each day, and put on before work. Contact lenses should not be worn when working with this chemical. Wear dust-proof chemical goggles and face shield unless full face-piece respiratory protection is worn. Employees should wash immediately with soap when skin is wet or contaminated. Provide emergency showers and eyewash. Specific engineering controls are recommended in NIOSH Criteria Document #76-103: *Inorganic fluorides.*

Respirator Selection: NIOSH/OSHA *12.5 mg/m^3:* Qm (APF = 25) (any quarter-mask respirator). *25 mg/m^3:* 95XQ (APF = 10)* [any particulate respirator equipped with an N95, R95, or P95 filter (including N95, R95, and P95 filtering face-pieces) except quarter-mask respirators. The following filters may also be used: N99, R99, P99, N100, R100, P100] or SA* (any supplied-air respirator). *62.5 mg/m^3:* Sa:Cf (APF = 25)*† (any supplied-air respirator operated in a continuous-flow mode) or PaprHie (APF = 25)* *if not present as a fume* (any powered, air-purifying respirator with a high-efficiency particulate filter). *125 mg/m^3:* 100F (APF = 50)† [any particulate respirator equipped with an N95, R95, or P95 filter (including N95, R95, and P95 filtering face-pieces) except quarter-mask respirators. The following filters may also be used: N99, R99, P99, N100, R100, P100] or SCBAF (APF = 50) (any self-contained breathing apparatus with a full face-piece) or SaF (APF = 50) (any supplied-air respirator with a full face-piece). *250 mg/m^3:* Sa: Pd,Pp (APF = 1000) (any supplied-air respirator operated in a pressure-demand or other positive-pressure mode). *Emergency or planned entry into unknown concentrations or IDLH conditions:* SCBAF: Pd, Pp (APF = 10,000) (any self-contained breathing apparatus that has a full faceplate and is operated in a pressure-demand or other positive-pressure mode) or SaF: Pd,Pp: ASCBA (APF = 10,000) (any supplied-air respirator that has a full face-piece and is operated in a pressure-demand or other positive-pressure mode in combination with an auxiliary, self-contained breathing apparatus operated in a pressure-demand or other positive-pressure mode). *Escape:* 100F (APF = 50)[†] [any particulate respirator equipped with an N95, R95, or P95 filter (including N95, R95, and P95 filtering face-pieces) except quarter-mask respirators. The following filters may also be used: N99, R99, P99, N100, R100, P100] or SCBAE (any appropriate escape-type, self-contained breathing apparatus).

*Substance reported to cause eye irritation or damage; may require eye protection.

[†]May need acid gas sorbent.

Storage: Color Code—Green: General storage may be used. Prior to working with this chemical you should be trained on its proper handling and storage. Sodium aluminum fluoride must be stored to avoid contact with strong acids (such as hydrochloric, sulfuric, and nitric), since violent reactions occur. Store in tightly closed containers in a cool, well-ventilated area.

Shipping: Environmentally hazardous substances, solid, n.o.s. require a shipping label of "CLASS 9." It falls in Hazard Class 9 and Packing Group III.[20, 21]

Spill Handling: Evacuate persons not wearing protective equipment from area of spill or leak until cleanup is complete. Remove all ignition sources. Collect powdered material in the most convenient and safe manner and deposit in sealed containers. Ventilate area after cleanup is complete. It may be necessary to contain and dispose of this chemical as a hazardous waste. If material or contaminated runoff enters waterways, notify downstream users of potentially contaminated waters. Contact your local or federal environmental protection agency for specific recommendations. If employees are required to clean up spills, they must be properly trained and equipped. OSHA 1910.120(q) may be applicable.

Fire Extinguishing: This material is not flammable. Use dry chemicals appropriate for metal fires. Poisonous gases are produced in fire, including sodium oxides and fluorine. If material or contaminated runoff enters waterways, notify downstream users of potentially contaminated waters. Notify local health and fire officials and pollution control agencies. From a secure, explosion-proof location, use water spray to cool exposed containers. If cooling streams are ineffective (venting sound increases in volume and pitch, tank discolors, or shows any signs of deforming), withdraw immediately to a secure position. If employees are expected to fight fires, they must be trained and equipped in OSHA 1910.156. The only respirators recommended for firefighting are self-contained breathing apparatuses that have full face-pieces and are operated in a pressure-demand or other positive-pressure mode.

Disposal Method Suggested: In accordance with 40CFR165, follow recommendations for the disposal of pesticides and pesticide containers. Must be disposed properly by following package label directions or by contacting your local or federal environmental control agency or by contacting your regional EPA office.

References
US Environmental Protection Agency, Special Review and Reregistration Division Office of Pesticide Programs. (1998). *Agency Status of Pesticides in Registration, Reregistration, and Special Review* (Rainbow Report). Washington, DC
New Jersey Department of Health and Senior Services. (April 2000). *Hazardous Substances Fact Sheet: Sodium Aluminum Fluoride.* Trenton, NJ

Sodium aluminum hydride   S:0360

Molecular Formula: AlH_4Na
Common Formula: $NaAlH_4$
Synonyms: Aluminate(1-), tetrahydro-, sodium, (*t*-4)-; Aluminum sodium hydride; Sah 22; Sodium aluminum tetrahydride; Sodium tetrahydroaluminate(1-)
CAS Registry Number: 13770-96-2
RTECS® Number: BD01800000
UN/NA & ERG Number: UN2835/138
EC Number: 237-400-1
Regulatory Authority and Advisory Bodies
Air Pollutant Standard Set. See below, "Permissible Exposure Limits in Air" section.
Canada, WHMIS, Ingredients Disclosure List Concentration 1.0%.
European/International Regulations: not listed in Annex 1.
WGK (German Aquatic Hazard Class): 1—Low hazard to waters.
Description: Sodium aluminum hydride is a white crystalline solid. Molecular weight = 54.01; Freezing/Melting point = 183°C (decomposes). Hazard Identification (based on NFPA-704 M Rating System): Health 2, Flammability 1, Reactivity ₩. Reaction with water may be dangerous.
Potential Exposure: Used in chemical synthesis.
Incompatibilities: A strong reducing agent. Violent reaction with water, air, oxidizers, acids, alcohols, and ethers. Reaction with water may cause fire or explosion. May ignite spontaneously in moist air.
Permissible Exposure Limits in Air
Protective Action Criteria (PAC)
TEEL-0: 3 mg/m^3
PAC-1: 7.5 mg/m^3
PAC-2: 60 mg/m^3
PAC-3: 300 mg/m^3
The limits for "aluminum pyro powders" may be applicable (NJ).
OSHA PEL: None.
NIOSH REL: 5 mg[Al]/m^3 TWA.
ACGIH TLV®[1]: 5 mg[Al]/m^3 TWA.
The limits for aluminum soluble salts, for reference
OSHA PEL: None.
NIOSH REL: 2 mg[Al]/m^3 TWA.
ACGIH TLV®[1]: 2 mg[Al]/m^3 TWA as soluble salts and alkyls, n.o.s.

Determination in Air: Use NIOSH #7300; #7301; #7303.

Permissible Concentration in Water: This material reacts vigorously with water so this category is not pertinent.

Routes of Entry: Inhalation, ingestion, skin and/or eye contact.

Harmful Effects and Symptoms

Short Term Exposure: Sodium aluminum hydride can affect you when breathed in. Exposure can irritate the eyes, skin, and respiratory tract. If wet, sodium aluminum hydride may cause burns and permanent damage.

Long Term Exposure: May cause lung injury, pulmonary fibrosis.

Points of Attack: Skin, respiratory system.

Medical Surveillance: For those with frequent or potentially high exposure (half the TLV or greater), the following are recommended before beginning work and at regular times after that: lung function tests. Consider chest X-ray following acute overexposure.

First Aid: If this chemical gets into the eyes, remove any contact lenses at once and irrigate immediately for at least 15 min, occasionally lifting upper and lower lids. Seek medical attention immediately. If this chemical contacts the skin, remove contaminated clothing and wash immediately with soap and water. Seek medical attention immediately. If this chemical has been inhaled, remove from exposure, begin rescue breathing (using universal precautions, including resuscitation mask) if breathing has stopped and CPR if heart action has stopped. Transfer promptly to a medical facility. When this chemical has been swallowed, get medical attention. Give large quantities of water and induce vomiting. Do not make an unconscious person vomit.

Personal Protective Methods: Wear protective gloves and clothing to prevent any reasonable probability of skin contact. Safety equipment suppliers/manufacturers can provide recommendations on the most protective glove/clothing material for your operation. All protective clothing (suits, gloves, footwear, headgear) should be clean, available each day, and put on before work. Contact lenses should not be worn when working with this chemical. Wear dust-proof chemical goggles and face shield unless full face-piece respiratory protection is worn. Employees should wash immediately with soap when skin is wet or contaminated. Provide emergency showers and eyewash.

Respirator Selection: Where there is potential for exposures *over 5 mg/m³*, use a NIOSH/MSHA- or European Standard EN149-approved full face-piece respirator with a high-efficiency particulate filter. Greater protection is provided by a powered air-purifying respirator. *Where there is potential for high exposures*, use a NIOSH/MSHA- or European Standard EN149-approved supplied-air respirator with a full face-piece operated in the positive-pressure mode, or with a full face-piece, hood, or helmet in the continuous-flow mode; or use a NIOSH/MSHA- or European Standard EN149-approved self-contained breathing apparatus with a full face-piece operated in pressure-demand or other positive-pressure mode.

Storage: Dangerous if any form of moisture is present. Color Code—Yellow Stripe (*strong reducing agent*): Reactivity Hazard; Store separately in an area isolated from flammables, combustibles, or other yellow coded materials. Prior to working with this chemical you should be trained on its proper handling and storage. Sodium aluminum hydride must be stored to avoid contact with water, air, oxidizers, acids, alcohols, and ethers, since violent reactions occur. Sources of ignition, such as smoking and open flames, are prohibited where sodium aluminum hydride is handled, used, or stored. Use only nonsparking tools and equipment, especially when opening and closing containers of sodium aluminum hydride. Wherever sodium aluminum hydride is used, handled, manufactured, or stored, use explosion-proof electrical equipment and fittings.

Shipping: This compound requires a shipping label of "DANGEROUS WHEN WET." It falls in Hazard Class 4.3 and Packing Group II.

Spill Handling: Evacuate persons not wearing protective equipment from area of spill or leak until cleanup is complete. Remove all ignition sources. Collect powdered material in the most convenient and safe manner and deposit in sealed containers. Ventilate area after cleanup is complete. It may be necessary to contain and dispose of this chemical as a hazardous waste. If material or contaminated runoff enters waterways, notify downstream users of potentially contaminated waters. Contact your local or federal environmental protection agency for specific recommendations. If employees are required to clean up spills, they must be properly trained and equipped. OSHA 1910.120(q) may be applicable.

Fire Extinguishing: Sodium aluminum hydride is a flammable solid. It can ignite spontaneously in moist air. The flame may be invisible. *Do not use water.* Use dry chemical, soda ash, or lime extinguishers. Fires may restart after it has been extinguished. Poisonous and explosive gases are produced in fire, including sodium oxides and hydrogen. If material or contaminated runoff enters waterways, notify downstream users of potentially contaminated waters. Notify local health and fire officials and pollution control agencies. From a secure, explosion-proof location, use water spray to cool exposed containers. If cooling streams are ineffective (venting sound increases in volume and pitch, tank discolors, or shows any signs of deforming), withdraw immediately to a secure position. If employees are expected to fight fires, they must be trained and equipped in OSHA 1910.156. The only respirators recommended for firefighting are self-contained breathing apparatuses that have full face-pieces and are operated in a pressure-demand or other positive-pressure mode.

References
New Jersey Department of Health and Senior Services. (May 2002). *Hazardous Substances Fact Sheet: Sodium Aluminum Hydride*. Trenton, NJ

Sodium arsenate S:0370

Molecular Formula: $AsH_3Na_xO_4$; $AsHNa_2O_4$
Common Formula: $Na_xH_3AsO_4$; HNa_2AsO_4
Synonyms: Arsenite de sodium (French); Arsenito sodico (Spanish); Arsenous acid, sodium salt; Atlas A; Chem Pels C; Chem-Sen 56; Disodium arsenate heptahydrate; Kill-All; Penite; Prodalumnol; Sodanit; Sodium metaarsenite
Dibasic: Arsenic acid disodium salt; Disodium arsenate; Disodium hydrogen arsenate; Sodium arsenate, dibasic
Heptahydrate: Dibasic sodium arsenate heptahydrate; Disodium arsenate, heptahydrate; Sodium acid arsenate, heptahydrate; Sodium arsenate, dibasic, heptahydrate; Sodium arsenate heptahydrate; Sodium arseniate heptahydrate
CAS Registry Number: 7631-89-2; 7778-43-0 (dibasic; disodium hydrogenarsenate); *(alt.)* 55957-14-7; 10048-95-0 (heptahydrate)
RTECS® Number: CG1225000; CG0875000 (dibasic); CG0900000 (heptahydrate)
UN/NA & ERG Number: UN1685/151
EC Number: 231-547-5; 231-902-4 [*Annex I Index No.:* 033-005-00-1] *(dibasic; disodium hydrogenarsenate)*
Regulatory Authority and Advisory Bodies
Department of Homeland Security Screening Threshold Quantity (pounds): *Release hazard* 15,000 ($\geq$1.00% concentration).
Carcinogenicity: NTP: 11th Report on Carcinogens, 2004: Known to be a human carcinogen; IARC: Human Sufficient Evidence, 1980; Animal Limited Evidence, *carcinogenic to humans*, Group 1, 1987.
Air Pollutant Standard Set. See below, "Permissible Exposure Limits in Air" section.
Clean Water Act: Section 311 Hazardous Substances/RQ (same as CERCLA); Section 313 Water Priority Chemicals (57FR41331, 9/9/92).
Superfund/EPCRA 40CFR355, Extremely Hazardous Substances: TPQ = 1000/10,000 lb (454/4540 kg).
Reportable Quantity (RQ): 1 lb (0.454 kg).
California Proposition 65 Chemical: Cancer 2/27/87.
Canada, WHMIS, Ingredients Disclosure List Concentration 0.1%; 1% [disodium hydrogen arsenate (dibasic)].
Canada: Priority Substance List & Restricted Substances/Ocean Dumping FORBIDDEN (CEPA), National Pollutant Release Inventory (NPRI) arsenic compounds.
European/International Regulations: Hazard Symbol: T, N; Risk phrases: R23/25; R53; Safety phrases: S1/2; S20/21; S28; S45; S60; S61 (see Appendix 4).
WGK (German Aquatic Hazard Class): 3—Severe hazard to waters.
Description: Sodium arsenate and sodium arsenate, dibasic are both white crystalline, odorless solids. Molecular weight = 202.94; 185.91 (dibasic); 427.00 (heptahydrate); Boiling point = 180°C (decomposition); Freezing/Melting point = 85°C; 57°C (dibasic). Hazard Identification (based

on NFPA-704 M Rating System): Health 3, Flammability 0, Reactivity 0. Soluble in water.
Potential Exposure: Compound Description (dibasic): Mutagen, (heptahydrate) Tumorigen, Mutagen; Reproductive Effector. Sodium arsenate is used in dyeing and printing; making other arsenates; as a germicide; in dyeing with turkey-red oil.
Incompatibilities: Acids, chemically active metals. Arsine, a very deadly gas, can be released in the presence of acid, acid mists, or hydrogen gas. Attacks many metals (such as aluminum, iron, and zinc) in the presence of moisture producing arsenic and arsine fumes.
Permissible Exposure Limits in Air
Inorganic arsenic compounds
OSHA PEL: 0.010 mg[As]/m³ TWA; cancer hazard that can be inhaled. See [1910.1018].
NIOSH REL: 0.002 mg[As]/m³ 15 min Ceiling Concentration. A potential occupational carcinogen. Limit exposure to lowest feasible concentration; See Appendix A.
ACGIH TLV®[1]: 0.01 mg[As]/m³ TWA; Confirmed Human Carcinogen.
Protective Action Criteria (PAC)
TEEL-0: 0.0271 mg/m³
PAC-1: 4 mg/m³
PAC-2: 13.5 mg/m³
PAC-3: 13.5 mg/m³
DFG TRK: 0.10 mg[As]/m³; BAT: 1.30 μg[As]/L in urine/end-of-shift; Carcinogen Category 1.
NIOSH IDLH: potential occupational carcinogen 5 mg[As]/m³.
Arab Republic of Egypt: TWA 0.2 mg/m³, 1993; Australia: TWA 0.05 mg/m³, carcinogen, 1993; Belgium: TWA 0.2 mg/m³, 1993; Denmark: TWA 0.05 mg/m³, 1999; Finland: carcinogen, 1993; France: VME 0.2 mg/m³, 1993; Hungary: STEL 0.5 mg/m³, carcinogen, 1993; India: TWA 0.2 mg/m³, 1993; Norway: TWA 0.02 mg/m³, 1999; the Philippines: TWA 0.5 mg/m³, 1993; Poland: MAC (TWA) 0.01 mg/m³, 1999; Sweden: NGV 0.03 mg/m³, carcinogen, 1999; Switzerland: TWA 0.1 mg/m³, carcinogen, 1999; Thailand: TWA 0.5 mg/m³, 1993; Turkey: TWA 0.5 mg (As)/m³, 1993; Turkey: TWA 0.5 mg/m³, 1993; United Kingdom: TWA 0.1 mg/m³, carcinogen, 2000; Argentina, Bulgaria, Columbia, Jordan, South Korea, New Zealand, Singapore, Vietnam: ACGIH: TLV: Confirmed Human Carcinogen. Russia[43] set a MAC of 0.003 mg/m³ on an average daily basis for residential areas. Several states have set guidelines or standards for arsenic in ambient air[60]: 0.06 mg/m³ (California Prop. 65), 0.0002 μg/m³ (Rhode Island), 0.00023 μg/m³ (North Carolina), 0.024 μg/m³ (Pennsylvania), 0.05 μg/m³ (Connecticut), 0.07−0.39 μg/m³ (Montana), 0.67 μg/m³ (New York), 1.0 μg/m³ (South Carolina), 2.0 μg/m³ (North Dakota), 3.3 μg/m³ (Virginia), 5 μg/m³ (Nevada).
Determination in Air: NIOSH Analytical Methods (inorganic arsenic): #7300, #7301,#7303, #7900, #9102; OSHA Analytical Methods ID-105. The American Conference of

Government Industrial Hygienists (ACGIH) Method 803 measures total particulate arsenic in air.

Permissible Concentration in Water: EPA[61] has set 50 µg/L of arsenic as a standard for drinking water. This standard has been in effect for 50 years. The Clinton administration reduced this level to 10 µg/L, and the G.W. Bush administration rescinded this change in 2001.

Routes of Entry: Inhalation, ingestion, skin and/or eye contact. Absorbed through the skin.

Harmful Effects and Symptoms

Short Term Exposure: Irritates the eyes, skin, and respiratory tract. May affect the central nervous system, digestive tract, circulatory system, causing loss of fluids and electrolytes, collapse, shock, and death. Exposure at low level may result in death. Death can occur due to a severe decrease in blood pressure. Thickening of skin on palms and soles following chronic low-level exposures. Symptoms of exposure include a feeling of constriction of throat, difficulty in swallowing; abdominal discomfort with pain, vomiting, watery diarrhea; sweetish metallic taste; garlicky odor of breath and stools; and dizziness with frontal headaches. Symptoms usually appear one-half to 1 h after ingestion but may be delayed many hours.

Long Term Exposure: Repeated or prolonged contact may cause skin sensitization and dermatitis. May affect the peripheral nervous system, skin, mucous membranes, causing neuropathy, skin thickening and pigmentation disorders, ulcers and perforation of nasal septum, and liver cirrhosis. Nerve damage may include "pins and needles," burning, numbness, and later weakness in the limbs. This substance is carcinogenic to humans. A probable teratogen in humans.

Points of Attack: Skin, nervous system, liver.

Medical Surveillance: Before first exposure and every 6−12 months thereafter, a medical history and examination are recommended, including: examination of the nose, skin eyes, nails, and nervous system. Test for urine arsenic (may not be accurate within 2 days of eating shellfish or fish; most accurate at the end of a workday). At NIOSH recommended exposure levels, urine arsenic should not be greater than 100 µg/g creatinine in the urine. Liver function tests.

First Aid: If this chemical gets into the eyes, remove any contact lenses at once and irrigate immediately for at least 15 min, occasionally lifting upper and lower lids. Seek medical attention immediately. If this chemical contacts the skin, remove contaminated clothing and wash immediately with soap and water. Seek medical attention immediately. If this chemical has been inhaled, remove from exposure, begin rescue breathing (using universal precautions, including resuscitation mask) if breathing has stopped and CPR if heart action has stopped. Transfer promptly to a medical facility. When this chemical has been swallowed, get medical attention. Give a slurry of activated charcoal in water to drink and induce vomiting. Do not make an unconscious person vomit.

Antidotes and Special Procedures: For severe poisoning BAL have been used. For milder poisoning, penicillamine (*not penicillin*) has been used, both with mixed success. Side effects occur with such treatment and it is never a substitute for controlling exposure. It can only be done under strict medical care.

Personal Protective Methods: Wear protective gloves and clothing to prevent any reasonable probability of skin contact. Safety equipment suppliers/manufacturers can provide recommendations on the most protective glove/clothing material for your operation. All protective clothing (suits, gloves, footwear, headgear) should be clean, available each day, and put on before work. Contact lenses should not be worn when working with this chemical. Eye protection is included in the recommended respiratory protection. Employees should wash immediately with soap when skin is wet or contaminated. Provide emergency showers and eyewash. Specific engineering controls are required under OSHA 1910.1018, *Inorganic Arsenic.* See also NIOSH Criteria Document #75-149, *Inorganic Arsenic.*

Respirator Selection: *At concentrations above the NIOSH REL, or where there is no REL, at any detectable concentration:* Sa (APF = 10) (any supplied-air respirator) or SCBAF (APF = 50) (any self-contained breathing apparatus with a full face-piece). *Emergency or planned entry into unknown concentrations or IDLH conditions:* SCBAF: Pd,Pp (APF = 10,000) (any self-contained breathing apparatus that has a full face-piece and is operated in a pressure-demand or other positive-pressure mode) or SaF: Pd,Pp: ASCBA (APF = 10,000) (any supplied-air respirator that has a full face-piece and is operated in a pressure-demand or other positive-pressure mode in combination with an auxiliary, self-contained breathing apparatus operated in a pressure-demand or other positive-pressure mode). *Escape:* GmFAg100 (APF = 50) [any air-purifying, full-face-piece respirator (gas mask) with a chin-style, front- or back-mounted acid gas canister having an N100, R100, or P100 filter] or SCBAE (any appropriate escape-type, self-contained breathing apparatus).

Storage: Color Code—Blue: Health Hazard/Poison: Store in a secure poison location. Prior to working with this chemical you should be trained on its proper handling and storage. Sodium arsenate must be stored to avoid contact with chemically active metals (such as potassium, sodium, magnesium, and zinc), since violent reactions occur. Sodium arsenate must be stored to avoid contact with acids. Store in tightly closed containers in a cool, well-ventilated area away from heat. A regulated, marked area should be established, or stored in compliance with OSHA Standard 1910.1045.

Shipping: This compound requires a shipping label of "POISONOUS/TOXIC MATERIALS." It falls in Hazard Class 6.1 and Packing Group II.

Spill Handling: Evacuate persons not wearing protective equipment from area of spill or leak until cleanup is complete. Remove all ignition sources. Do not touch spilled material; stop leak if you can do so without risk. *Small spills:* absorb with sand or other noncombustible absorbent material and place into containers for later disposal. *Small dry spills:* with clean shovel place material into clean, dry container and cover; move containers from spill area. *Large spills:* dike far ahead of spill for later disposal. It may be necessary to contain and dispose of this chemical as a hazardous waste. If material or contaminated runoff enters waterways, notify downstream users of potentially contaminated waters. Contact your local or federal environmental protection agency for specific recommendations. If employees are required to clean up spills, they must be properly trained and equipped. OSHA 1910.120(q) may be applicable.

Fire Extinguishing: Extinguish fire using an agent suitable for type of surrounding fire. Sodium arsenate itself does not burn. Poisonous gases are produced in fire. If material or contaminated runoff enters waterways, notify downstream users of potentially contaminated waters. Notify local health and fire officials and pollution control agencies. From a secure, explosion-proof location, use water spray to cool exposed containers. If cooling streams are ineffective (venting sound increases in volume and pitch, tank discolors, or shows any signs of deforming), withdraw immediately to a secure position. If employees are expected to fight fires, they must be trained and equipped in OSHA 1910.156. The only respirators recommended for firefighting are self-contained breathing apparatuses that have full face-pieces and are operated in a pressure-demand or other positive-pressure mode.

Disposal Method Suggested: Dissolve in minimum quantity of concentrated, reagent hydrochloric acid. Filter if necessary. Dilute with water until white precipitate forms. Add just enough 6 M HCl to redissolve. Saturate with hydrogen sulfide. Filter, wash the precipitate; dry, package, and ship to the supplier. (Sax, DPIMR see below).

References

Sax, N. I. (Ed.). (1982). *Dangerous Properties of Industrial Materials Report*, 2, No. 6, 71−73

US Environmental Protection Agency. (November 30, 1987). *Chemical Hazard Information Profile: Sodium Arsenate*. Washington, DC: Chemical Emergency Preparedness Program

New Jersey Department of Health and Senior Services. (April 2000). *Hazardous Substances Fact Sheet: Sodium Arsenate*. Trenton, NJ

Sodium arsenite   S:0380

Molecular Formula: AsO_2Na
Common Formula: $NaAsO_2$
Synonyms: Arsenite de sodium (French); Arsenito sodico (Spanish); Arsenous acid, sodium salt; Atlas A; Chem Pels C; Chem-Sen 56; Disodium arsenate heptahydrate; Kill-All; Penite; Prodalumnol; Sodanit; Sodium metaarsenite
CAS Registry Number: 7784-46-5

RTECS® Number: CG3675000
UN/NA & ERG Number: UN2027 (solid)/151; UN1686 (solution)/154
EC Number: 232-070-5
Regulatory Authority and Advisory Bodies
Carcinogenicity: IARC: Human Sufficient Evidence, 1980; Animal Limited Evidence, *carcinogenic to humans*, Group 1, 1987; EPA: Human Carcinogen; NTP: Known to be a human carcinogen; Known to be a human carcinogen; NIOSH: Potential occupational carcinogen; OSHA: Potential human carcinogen.
Banned or Severely Restricted (several countries) (UN).[13]
Clean Water Act: Section 311 Hazardous Substances/RQ 40CFR117.3 (same as CERCLA, see below); Section 313 Water Priority Chemicals (57FR41331, 9/9/92).
Superfund/EPCRA 40CFR355, Extremely Hazardous Substances: TPQ = 500/10,000 lb (227/4540 kg).
Reportable Quantity (RQ): 1 lb (0.454 kg).
Air Pollutant Standard Set. See below, "Permissible Exposure Limits in Air" section.
California Proposition 65 Chemical: Cancer 2/27/87.
Canada: Priority Substance List & Restricted Substances/ Ocean Dumping FORBIDDEN (CEPA), National Pollutant Release Inventory (NPRI) (arsenic compounds).
European/International Regulations: not listed in Annex 1.
WGK (German Aquatic Hazard Class): 3—Severe hazard to waters.

Description: Sodium arsenite is a white or grayish-white powder or flakes. Molecular weight = 129.91; Freezing/Melting point = 615°C; Boiling point = decomposes. Hazard Identification (based on NFPA-704 M Rating System): Health 3, Flammability 0, Reactivity 0. Highly soluble in water.

Potential Exposure: This material is used in manufacturing of arsenical soap for use on skin, treating vines against certain scale diseases, wood preservation, as a reagent in preparation of methylene iodide, corrosion inhibitor, and for herbicidal and pesticidal purposes.

Incompatibilities: Chemically active metals. Arsine, a very deadly gas, can be released in the presence of acid, acid mists, or hydrogen gas.

Permissible Exposure Limits in Air
OSHA PEL: 0.010 mg[As]/m^3 TWA; cancer hazard that can be inhaled. See [1910.1018].
NIOSH REL: 0.002 mg[As]/m^3 [15 min] Ceiling Concentration. A potential occupational carcinogen. Limit exposure to lowest feasible concentration; See *NIOSH Pocket Guide*, Appendix A.
ACGIH TLV®[1]: 0.01 mg[As]/m^3 TWA; Confirmed Human Carcinogen.
NIOSH IDLH: potential occupational carcinogen 5 mg[As]/m^3.
Protective Action Criteria (PAC)
TEEL-0: 0.0173 mg/m^3
PAC-1: 1.25 mg/m^3
PAC-2: 8.67 mg/m^3
PAC-3: 8.67 mg/m^3

DFG TRK: 0.10 mg[As]/m^3; BAT: 1.30 µg[As]/L in urine/end-of-shift; Carcinogen Category 1.

Arab Republic of Egypt: TWA 0.2 mg/m^3, 1993; Australia: TWA 0.05 mg/m^3, carcinogen, 1993; Belgium: TWA 0.2 mg/m^3, 1993; Denmark: TWA 0.05 mg/m^3, 1999; Finland: carcinogen, 1993; France: VME 0.2 mg/m^3, 1993; Hungary: STEL 0.5 mg/m^3, carcinogen, 1993; India: TWA 0.2 mg/m^3, 1993; Norway: TWA 0.02 mg/m^3, 1999; the Philippines: TWA 0.5 mg/m^3, 1993; Poland: MAC (TWA) 0.01 mg/m^3, 1999; Sweden: NGV 0.03 mg/m^3, carcinogen, 1999; Switzerland: TWA 0.1 mg/m^3, carcinogen, 1999; Thailand: TWA 0.5 mg/m^3, 1993; Turkey: TWA 0.5 mg (As)/m^3, 1993; Turkey: TWA 0.5 mg/m^3, 1993; United Kingdom: TWA 0.1 mg/m^3, carcinogen, 2000; Argentina, Bulgaria, Columbia, Jordan, South Korea, New Zealand, Singapore, Vietnam: ACGIH: TLV: Confirmed Human Carcinogen. Russia[43] set a MAC of 0.003 mg/m^3 on an average daily basis for residential areas. Several states have set guidelines or standards for arsenic in ambient air[60]: 0.06 mg/m^3 (California Prop. 65), 0.0002 µg/m^3 (Rhode Island), 0.00023 µg/m^3 (North Carolina), 0.024 µg/m^3 (Pennsylvania), 0.05 µg/m^3 (Connecticut), 0.07–0.39 µg/m^3 (Montana), 0.67 µg/m^3 (New York), 1.0 µg/m^3 (South Carolina), 2.0 µg/m^3 (North Dakota), 3.3 µg/m^3 (Virginia), 5 µg/m^3 (Nevada).

Determination in Air: NIOSH Analytical Methods (inorganic arsenic): #7300, #7301, #7303, #7900, #9102; OSHA Analytical Methods ID-105. The American Conference of Government Industrial Hygienists (ACGIH) Method 803 measures total particulate arsenic in air.

Permissible Concentration in Water: The EPA[61] has set a limit of 0.05 µg/L in drinking water.

Routes of Entry: Inhalation, ingestion, skin and/or eye contact. Absorbed by the skin.

Harmful Effects and Symptoms

Short Term Exposure: Sodium arsenite may irritate or burn the skin, eyes, and mucous membranes. Skin contact can cause burning sensation, itching, and rash. Extremely toxic: probable oral lethal dose (human) 5–50 mg/kg, between 7 drops and one teaspoon for a 70-kg person (150 lb). Poisonous if swallowed or inhaled. Signs and symptoms of acute exposure to sodium arsenite may be severe and include headache, vomiting, stomach pain, vomiting, cough, dyspnea (shortness of breath), hypotension (low blood pressure), and chest pain. Gastrointestinal effects include difficulty in swallowing, intense thirst, generalized abdominal pain, vomiting, and painful diarrhea; blood may be noted in the vomitus and feces. A weak pulse, cyanosis (blue tint to the skin and mucous membranes), and cold extremities may also be observed. Neurological effects include giddiness, delirium, mania, stupor, weakness, headache, dizziness, and fainting. Convulsions, paralysis, and coma may occur.

Long Term Exposure: Repeated or prolonged contact may cause skin sensitization and dermatitis. May affect the peripheral nervous system, skin, mucous membranes, causing neuropathy, skin thickening and pigmentation disorders, ulcers and perforation of nasal septum, and liver cirrhosis. Nerve damage may include "pins and needles," burning, numbness, and later weakness in the limbs. This substance is carcinogenic to humans. A probable teratogen in humans.

Points of Attack: Liver, kidneys, skin, lungs, lymphatic system.

Medical Surveillance: For those with frequent or potentially high exposure (half the TLV or greater or significant skin contact), the following are recommended before beginning work and at regular times after that: examination of the nose, skin, eyes, nails, and nervous system. Test for urine arsenic (may not be accurate within 2 days of eating shellfish or fish; most accurate at the end of a workday). At NIOSH recommended exposure levels, urine arsenic should not be greater than 100 µg/g creatinine in the urine.

First Aid: If this chemical gets into the eyes, remove any contact lenses at once and irrigate immediately for at least 15 min, occasionally lifting upper and lower lids. Seek medical attention immediately. If this chemical contacts the skin, remove contaminated clothing and wash immediately with soap and water. Seek medical attention immediately. If this chemical has been inhaled, remove from exposure, begin rescue breathing (using universal precautions, including resuscitation mask) if breathing has stopped and CPR if heart action has stopped. Transfer promptly to a medical facility. When this chemical has been swallowed, get medical attention. Give a slurry of activated charcoal in water to drink and induce vomiting. Do not make an unconscious person vomit.

Antidotes and Special Procedures: For severe poisoning BAL has been used. For milder poisoning penicillamine (*not penicillin*) has been used, both with mixed success. Side effects occur with such treatment and it is never a substitute for controlling exposure. It can only be done under strict medical care.

Personal Protective Methods: Wear protective gloves and clothing to prevent any reasonable probability of skin contact. Safety equipment suppliers/manufacturers can provide recommendations on the most protective glove/clothing material for your operation. All protective clothing (suits, gloves, footwear, headgear) should be clean, available each day, and put on before work. Contact lenses should not be worn when working with this chemical. Eye protection is included in the recommended respiratory protection. Employees should wash immediately with soap when skin is wet or contaminated. Provide emergency showers and eyewash. Specific engineering controls are required under OSHA 1910.1018, *Inorganic Arsenic.* See also NIOSH Criteria Document #75-149, *Inorganic Arsenic.*

Respirator Selection: At concentrations above the NIOSH REL, or where there is no REL, at any detectable concentration: Sa (APF = 10) (any supplied-air respirator) or SCBAF (APF = 50) (any self-contained breathing apparatus with a full face-piece). *Emergency or planned entry into unknown concentrations or IDLH conditions:* SCBAF: Pd,Pp (APF = 10,000) (any self-contained breathing apparatus that has a full face-piece and is operated in a pressure-demand or other positive-pressure mode) or SaF: Pd,Pp: ASCBA (APF = 10,000) (any supplied-air respirator that has a full face-piece and is operated in a pressure-demand or other positive-pressure mode in combination with an auxiliary, self-contained breathing apparatus operated in a pressure-demand or other positive-pressure mode). *Escape:* GmFAg100 (APF = 50) [any air-purifying, full-face-piece respirator (gas mask) with a chin-style, front- or back-mounted acid gas canister having an N100, R100, or P100 filter] or SCBAE (any appropriate escape-type, self-contained breathing apparatus).

Storage: Color Code—Blue: Health Hazard/Poison: Store in a secure poison location. Prior to working with this chemical you should be trained on its proper handling and storage. Sodium arsenite must be stored to avoid contact with chemically active metals (such as potassium, sodium, magnesium, and zinc), since violent reactions occur. Store in tightly closed containers in a cool, well-ventilated area away from heat. A regulated, marked area should be established, or stored in compliance with OSHA Standard 1910.1045.

Shipping: Sodium arsenite, solid, or sodium arsenite, aqueous solutions, require a shipping label of "POISONOUS/ TOXIC MATERIALS." It falls in Hazard Class 6.1 and Packing Group II. Aqueous solutions fall in DOT Hazard Class 6.1 and Packing Group II or III.

Spill Handling: Evacuate persons not wearing protective equipment from area of spill or leak until cleanup is complete. Remove all ignition sources. Collect powdered material in the most convenient and safe manner and deposit in sealed containers. Ventilate area after cleanup is complete. It may be necessary to contain and dispose of this chemical as a hazardous waste. If material or contaminated runoff enters waterways, notify downstream users of potentially contaminated waters. Contact your local or federal environmental protection agency for specific recommendations. If employees are required to clean up spills, they must be properly trained and equipped. OSHA 1910.120(q) may be applicable.

Fire Extinguishing: Extinguish fire using an agent suitable for type of surrounding fire. Sodium arsenite itself does not burn. Poisonous gases are produced in fire, including arsenic and sodium monoxide. If material or contaminated runoff enters waterways, notify downstream users of potentially contaminated waters. Notify local health and fire officials and pollution control agencies. From a secure, explosion-proof location, use water spray to cool exposed containers. If cooling streams are ineffective (venting sound increases in volume and pitch, tank discolors, or shows any signs of deforming), withdraw immediately to a secure position. If employees are expected to fight fires, they must be trained and equipped in OSHA 1910.156. The only respirators recommended for firefighting are self-contained breathing apparatuses that have full face-pieces and are operated in a pressure-demand or other positive-pressure mode.

Disposal Method Suggested: The arsenic may be precipitated as calcium arsenite. It should be stored until recycled.[22] In accordance with 40CFR165, follow recommendations for the disposal of pesticides and pesticide containers. Must be disposed properly by following package label directions or by contacting your local or federal environmental control agency or by contacting your regional EPA office.

References

US Environmental Protection Agency. (November 30, 1987). *Chemical Hazard Information Profile: Sodium Arsenite.* Washington, DC: Chemical Emergency Preparedness Program

New Jersey Department of Health and Senior Services. (March 2002). *Hazardous Substances Fact Sheet: Sodium Arsenite.* Trenton, NJ

Sodium azide   S:0390

Molecular Formula: N_3Na
Common Formula: NaN_3
Synonyms: AI3-50436; Axium; Azida sodico (Spanish); Azide; Azium; Azoture de sodium (French); Dazoe; Hydrazoic acid, Sodium salt; Kazoe; Natriumazid (German); NCI-C06462; Smite; Sodium salt of hydrazoic acid
CAS Registry Number: 26628-22-8; *(alt.)* 12136-89-9
RTECS® Number: VY8050000
UN/NA & ERG Number: UN1687/153
EC Number: 247-852-1 [*Annex I Index No.:* 011-004-00-7]
Regulatory Authority and Advisory Bodies

Department of Homeland Security Screening Threshold Quantity (pounds): *Theft hazard* 400 (Commercial grade).

Carcinogenicity: NCI: Carcinogenesis Studies (gavage); no evidence: rat; (EPA) Gene-Tox Program: Inconclusive: Carcinogenicity: mouse/rat.

US EPA Gene-Tox Program, Inconclusive: Carcinogenicity—mouse/rat; Positive: L5178Y cells *In vitro*—TK test; Positive: *D. melanogaster* sex-linked lethal; Positive: *S. cerevisiae* gene conversion; Positive: *S. cerevisiae*—forward mutation; *S. cerevisiae*—reversion; Negative: *In vitro* cytogenetics—human lymphocyte; Negative: Sperm morphology—mouse; *In vitro* UDS—human fibroblast; Negative: TRP reversion; TRP reversion.

Air Pollutant Standard Set. See below, "Permissible Exposure Limits in Air" section.

US EPA Hazardous Waste Number (RCRA No.): P105.

RCRA, 40CFR261, Appendix 8 Hazardous Constituents.

Superfund/EPCRA 40CFR355, Extremely Hazardous Substances: TPQ = 500 lb (227 kg).

Reportable Quantity (RQ): 1000 lb (454 kg).

EPCRA Section 313 Form R *de minimis* concentration reporting level: 1.0%.

Canada, WHMIS, Ingredients Disclosure List Concentration 1.0%.

European/International Regulations: Hazard Symbol: T+, N; Risk phrases: R28; R32; R50/53; Safety phrases: S1/2; S28; S45; S60; S61 (see Appendix 4).
WGK (German Aquatic Hazard Class): 2—Hazard to waters.

Description: Sodium azide is a colorless to white, odorless, crystalline solid. Combustible solid above 300°C. Molecular weight = 65.02; Specific gravity (H$_2$O:1) = 1.85 at 25°C; Boiling point = (decomposes); Freezing/Melting point = (the solid crystals decompose with the evolution of nitrogen gas, leaving a residue of sodium oxide) 275°C. Hazard Identification (based on NFPA-704 M Rating System): Health 4, Flammability 0, Reactivity 2. Soluble in water; reaction; solubility = 42% at 17°C.

Potential Exposure: Compound Description: Agricultural Chemical; Tumorigen, Drug, Mutagen; Human Data. Sodium azide is used as preservative and diluent. It has been used for a wide variety of military, laboratory, medicine and commercial purposes. It is used extensively as an intermediate in the production of lead azide, commonly used in detonators and other explosives. Reported to be used in automobile air-bag inflation. One of the largest potential exposures is that to automotive workers, repairmen, and wreckers, if sodium azide is used as the inflation chemical. Commercial applications include use as a fungicide, nematocide, and soil sterilizing agent and as a preservative for seeds and wine. The lumber industry has used sodium azide to limit the growth of enzymes responsible for formation of brown stain on sugar pine, while the Japanese beer industry used it to prevent the growth of a fungus which darkens its product. The chemical industry has used sodium azide as a retarder in the manufacture of sponge rubber, to prevent coagulation of styrene and butadiene latexes stored in contact with metals; and to decompose nitrites in the presence of nitrates.

Incompatibilities: Reacts explosively and/or forms explosive and/or shock-sensitive compounds with acids and many metals. Contact with water forms hydrazoic acid. Combustible solid (if heated above 300°C). May explode when heated above its melting point, especially if heating is rapid. Reacts with acids, producing toxic, shock-sensitive, and explosive hydrogen azide. It forms explosive compounds with phosgene, brass, zinc, trifluoroacrylol fluoride, and nitrogen-diluted bromine vapor. Reacts with benzoyl chloride and potassium hydroxide, bromine, carbon disulfide, copper, lead, nitric acid, barium carbonate, sulfuric acid, chromium(II) hypochlorite, dimethyl sulfate, dibromomalononitrile, silver, mercury. Over a period of time, sodium azide may react with copper, lead, brass, or solder in plumbing systems to form an accumulation of the *highly explosive* and shock-sensitive compounds of lead azide and copper azide.

Permissible Exposure Limits in Air
OSHA PEL: None.
NIOSH REL: 0.1 ppm (as HN$_3$) [skin] Ceiling Concentration; 0.3 mg/m^3 (as NaN$_3$) [skin] Ceiling Concentration.

ACGIH TLV®[1]: 0.11 ppm (as HN$_3$) [skin] Ceiling Concentration; 0.29 mg/m^3 (as NaN$_3$) [skin] Ceiling Concentration; not classifiable as a human carcinogen.
Protective Action Criteria (PAC)
TEEL-0: 0.2 mg/m^3
PAC-1: 3 mg/m^3
PAC-2: 20 mg/m^3
PAC-3: 25 mg/m^3
DFG MAK: 0.2 mg/m^3, inhalable fraction TWA; Peak Limitation Category I(2); Pregnancy Risk Group D.
Australia: TWA 0.1 ppm (0.3 mg/m^3), 1993; Austria: MAK 0.07 ppm (0.2 mg/m^3), 1999; Belgium: STEL 0.11 ppm (0.3 mg/m^3), 1993; Denmark: TWA 0.3 mg/m^3, 1999; Finland: TWA 0.1 ppm (0.3 mg/m^3); STEL 0.3 ppm (0.9 mg/m^3), 1999; France: VLE 0.1 ppm (0.3 mg/m^3), 1999; the Netherlands: MAC-TGG 0.1 mg/m^3, [skin], 2003; Switzerland: MAK-W 0.07 ppm (0.2 mg/m^3), 1999; United Kingdom: STEL 0.3 mg[NaN$_3$], 2000; Argentina, Bulgaria, Columbia, Jordan, South Korea, New Zealand, Singapore, Vietnam: ACGIH TLV®: Ceiling Concentration 0.29 mg/m^3. Several states have set guidelines or standards for sodium azide in ambient air[60] ranging from 0.7 μg/m^3 (Nevada) to 2.5 μg/m^3 (Virginia) to 3.0 μg/m^3 (North Dakota).

Determination in Air: Use OSHA Analytical Method ID-121.

Permissible Concentration in Water: No criteria set. (Sodium azide reacts with water to produce hydrazoic acid).

Routes of Entry: Inhalation, ingestion, skin contact. Absorbed through the skin.

Harmful Effects and Symptoms
Short Term Exposure: Severely irritates the eyes, skin, and respiratory tract. Contact of skin causes redness and pain. Contact with eyes causes redness, pain, and blurred vision; may cause loss of vision. Inhalation or ingestion causes dizziness, weakness, blurred vision, slight shortness of breath, hypotension, slow heart rate, abdominal pain, and spasms. Serious cases of exposure may result in convulsions, unconsciousness, and death. Exposure slightly above the exposure limits in air can cause death by affecting the central nervous system. Azides can cause blood pressure to drop and thus have action similar to cyanides and nitrites. Higher exposures can cause pulmonary edema, a medical emergency that can be delayed for several hours. This can cause death. Sodium azide is a broad-spectrum metabolic poison that interferes with oxidation enzymes and inhibits nuclear phosphorylation. Although the effects in these systems are complex, there is general agreement that azide causes a dissociation of phosphorylation and cellular respiration. For this reason parallels have been drawn to other metabolic inhibitors, such as cyanide, malonitrile, and fluoride.

Long Term Exposure: May cause kidney damage. Sodium azide is a potent mutagen in barley, peas, rice, and soybeans. It is also a very effective mutagen in bacteria. Its potency as a mutagen is comparable to the nitrosamines as a class. For these reasons sodium azide has been suspected of being a carcinogen. However, several studies have been performed to determine whether it is a carcinogen. In each instance the results were negative.

Points of Attack: Eyes, skin, central nervous system, cardiovascular system, kidneys.

Medical Surveillance: If symptoms develop or overexposure is suspected, the following may be useful: examination of the nervous system and vision (including visual fields). Lung function tests. Consider chest X-ray after acute overexposure.

First Aid: If this chemical gets into the eyes, remove any contact lenses at once and irrigate immediately for at least 15 min, occasionally lifting upper and lower lids. Seek medical attention immediately. If this chemical contacts the skin, remove contaminated clothing and wash immediately with soap and water. Seek medical attention immediately. If this chemical has been inhaled, remove from exposure, begin rescue breathing (using universal precautions, including resuscitation mask) if breathing has stopped and CPR if heart action has stopped. Transfer promptly to a medical facility. When this chemical has been swallowed, get medical attention. If victim is *conscious*, administer water or milk. Do not induce vomiting. Medical observation is recommended for 24–48 h after breathing overexposure, as pulmonary edema may be delayed. As first aid for pulmonary edema, a doctor or authorized paramedic may consider administering a corticosteroid spray.

Personal Protective Methods: Wear protective gloves and clothing to prevent any reasonable probability of skin contact. Safety equipment suppliers/manufacturers can provide recommendations on the most protective glove/clothing material for your operation. All protective clothing (suits, gloves, footwear, headgear) should be clean, available each day, and put on before work. Contact lenses should not be worn when working with this chemical. Wear dust-proof goggles and face shield when working with powders or dust, unless full face-piece respiratory protection is worn. Where sodium azide may be present in solution, wear splash-proof chemical goggles and face shield, unless full face-piece respiratory protection is worn. Employees should wash immediately with soap when skin is wet or contaminated. Provide emergency showers and eyewash.

Respirator Selection: Where there is potential for exposures over 0.1 ppm as a dust, fume, or mist, use a NIOSH/MSHA- or European Standard EN149-approved full face-piece respirator with a high-efficiency particulate filter. Greater protection is provided by a powered air-purifying respirator. *Where there is potential for high exposures*, or for exposures to hydrazoic acid vapor, use a NIOSH/MSHA- or European Standard EN149-approved supplied-air respirator with a full face-piece operated in the positive-pressure mode, or with a full face-piece, hood, or helmet in the continuous-flow mode; or use a NIOSH/MSHA- or European Standard EN149-approved self-contained breathing apparatus with a full face-piece operated in pressure-demand or other positive-pressure mode.

Storage: Color Code—Blue: Health Hazard/Poison: Store in a secure poison location. Prior to working with this chemical you should be trained on its proper handling and storage. Sodium azide must be stored to avoid contact with benzoyl chloride, potassium hydroxide, bromine, copper, and lead, since violent reactions occur. Store in tightly closed containers in a cool, well-ventilated area away from water, heat, and acids. On contact with water it forms hydrazoic acid which is explosive. Danger of explosion exists from friction, heat, or contamination. Where possible, automatically transfer material from drums or other storage containers to process containers. Sources of ignition, such as smoking and open flames, are prohibited where this chemical is handled, used, or stored. Metal containers involving the transfer of this chemical should be grounded and bonded. Wherever this chemical is used, handled, manufactured, or stored, use explosion-proof electrical equipment and fittings.

Shipping: This compound requires a shipping label of "POISONOUS/TOXIC MATERIALS." It falls in Hazard Class 6.1 and Packing Group II.

Spill Handling: Evacuate persons not wearing protective equipment from area of spill or leak until cleanup is complete. Remove all ignition sources. Do not touch spilled material; stop leak if you can do so without risk. Use water spray to reduce vapors. *Small spills:* absorb with sand or other noncombustible absorbent material and place into containers for later disposal. *Small dry spills:* with clean shovel place material into clean, dry container and cover; move containers from spill area. *Large spills:* dike far ahead of spill for later disposal. Ventilate area after cleanup is complete. It may be necessary to contain and dispose of this chemical as a hazardous waste. If material or contaminated runoff enters waterways, notify downstream users of potentially contaminated waters. Contact your local or federal environmental protection agency for specific recommendations. If employees are required to clean up spills, they must be properly trained and equipped. OSHA 1910.120(q) may be applicable.

Fire Extinguishing: This chemical is a combustible solid. Rapid heating above 275°C can cause rapid decomposition and explosion. If material is on fire or involved in fire, use water in flooding quantities as fog. Use foam, carbon dioxide, or dry chemical. *Small fires:* dry chemical, carbon dioxide, water spray, or foam. *Large fires:* water spray, fog, or foam. Move container from fire area if you can do so without risk. For massive fire in cargo area, use unmanned hose holder or monitor nozzles; if this is impossible, withdraw from area and let fire burn. Poisonous gases are produced in fire, including nitrogen oxides. If material or contaminated runoff enters waterways, notify downstream users of potentially contaminated waters. Notify local health and fire officials and pollution control agencies. From a secure, explosion-proof location, use water spray to cool exposed containers. If cooling streams are ineffective (venting sound increases in volume and pitch, tank discolors, or shows any signs of deforming), withdraw immediately to a secure position. If employees are expected to fight fires, they must be trained and equipped in OSHA 1910.156. The only respirators recommended for firefighting are self-contained breathing apparatuses that have full face-pieces and are operated in a pressure-demand or other positive-pressure mode.

Disposal Method Suggested: Consult with environmental regulatory agencies for guidance on acceptable disposal practices. Generators of waste containing this contaminant (≥ 100 kg/mo) must conform with EPA regulations governing storage, transportation, treatment, and waste disposal. Disposal may be accomplished by reaction with sulfuric acid solution and sodium nitrate in a hard rubber vessel. Nitrogen dioxide is generated by this reaction and the gas is run through a scrubber before it is released to the atmosphere. Controlled incineration is also acceptable (after mixing with other combustible wastes) with adequate scrubbing and ash disposal facilities.

References

National Institute for Occupational Safety and Health. (1977). *Profiles on Occupational Hazards for Criteria Document Priorities: Sodium Azide*, Report PB-274,073. Cincinnati, OH, pp. 306–308

US Environmental Protection Agency. (August 1, 1977). *Chemical Hazard Information Profile: Sodium Azide.* Washington, DC

Sax, N. I. (Ed.). (1982). *Dangerous Properties of Industrial Materials Report*, 2, No. 6, 74–78

US Environmental Protection Agency. (November 30, 1987). *Chemical Hazard Information Profile: Sodium Azide.* Washington, DC: Chemical Emergency Preparedness Program

New Jersey Department of Health and Senior Services. (October 1998). *Hazardous Substances Fact Sheet: Sodium Azide.* Trenton, NJ

Sodium benzoate   S:0400

Molecular Formula: $C_7H_5NaO_2$
Common Formula: C_6H_5COONa
Synonyms: Antimol; Benzoate of soda; Benzoate sodium; Benzoesaeure (Na-salz) (German); Benzoic acid, sodium salt; Sobenate; Sodium benzoic acid
CAS Registry Number: 532-32-1
RTECS® Number: DH6650000
UN/NA & ERG Number: Not regulated.
EC Number: 208-534-8
Regulatory Authority and Advisory Bodies
US EPA Gene-Tox Program, Negative: TRP reversion.
US EPA, FIFRA 1998 Status of Pesticides: Canceled.
WGK (German Aquatic Hazard Class): 1—Low hazard to waters.
Description: Sodium benzoate is a white crystalline solid. It is odorless and nonflammable. Molecular weight = 144.11; Decomposes at 120°C; Melting point = >300°C. Hazard

Identification (based on NFPA-704 M Rating System): Health 3, Flammability 2, Reactivity 1. Soluble in water.
Potential Exposure: Compound Description: Agricultural Chemical; Drug, Mutagen; Reproductive Effector; Human Data; Primary Irritant. Sodium benzoate is used as a food and feed additive, flavor, packaging material; pharmaceutical; preservative for food products and tobacco; antifungal agent; antiseptic, rust and mildew inhibitor; intermediate in the manufacture of dyes. Used as a human hygiene biocidal product.
Incompatibilities: Strong oxidizers.
Permissible Exposure Limits in Air
Protective Action Criteria (PAC)
TEEL-0: 4 mg/m^3
PAC-1: 12.5 mg/m^3
PAC-2: 75 mg/m^3
PAC-3: 400 mg/m^3
Sodium salt
TEEL-0: 100 mg/m^3
PAC-1: 350 mg/m^3
PAC-2: 500 mg/m^3
PAC-3: 500 mg/m^3
Routes of Entry: Inhalation, ingestion, skin and/or eye contact.
Harmful Effects and Symptoms
Short Term Exposure: An eye and skin irritant. The accumulation of dust in the eyes, ears, nose, throat, and lungs may be sufficiently unpleasant and distracting to make work near machinery hazardous. Irritation may result from abrasion or chemical action. Sodium benzoate can cause allergic reactions. Sodium benzoate has been given GRAS (Generally Recognized As Safe) status by the Food and Drug Administration at the levels currently being used as a food preservative. Ingestion of 8–10 g (1/3 oz) may cause nausea and vomiting. Twelve grams has caused gastric pain and loss of appetite. These symptoms disappear when exposure stops. LD$_{50}$ = (oral-rat) 27 mg/kg.
Long Term Exposure: Sodium benzoate may produce an allergic reaction and, in addition, may intensify the symptoms of allergies to other substances.
Points of Attack: Skin, lungs.
Medical Surveillance: Examination by a qualified allergist. Lung function tests.
First Aid: If this chemical gets into the eyes, remove any contact lenses at once and irrigate immediately for at least 15 min, occasionally lifting upper and lower lids. Seek medical attention immediately. If this chemical contacts the skin, remove contaminated clothing and wash immediately with soap and water. Seek medical attention immediately. If this chemical has been inhaled, remove from exposure, begin rescue breathing (using universal precautions, including resuscitation mask) if breathing has stopped and CPR if heart action has stopped. Transfer promptly to a medical facility. When this chemical has been swallowed, get medical attention. Give large quantities of water and induce vomiting. Do not make an unconscious person vomit.

Personal Protective Methods: Avoid prolonged skin contact. Safety equipment suppliers/manufacturers can provide recommendations on the most protective glove/clothing material for your operation. All protective clothing (suits, gloves, footwear, headgear) should be clean, available each day, and put on before work. Contact lenses should not be worn when working with this chemical. Wear dust-proof chemical goggles and face shield unless full face-piece respiratory protection is worn. Employees should wash immediately with soap when skin is wet or contaminated. Provide emergency showers and eyewash.

Respirator Selection: Follow regulations in OSHA 29CFR1910.134 or European Standard EN149. Use a NIOSH/MSHA- or European Standard EN149-approved respirator; or use an approved supplied-air respirator with a full face-piece operated in the positive-pressure mode, or with a full face-piece, hood, or helmet in the continuous-flow mode; or use a NIOSH/MSHA- or European Standard EN149-approved self-contained breathing apparatus with a full face-piece operated in pressure-demand or other positive-pressure mode.

Storage: Color Code—Blue: Health Hazard/Poison: Store in a secure poison location. Prior to working with this chemical you should be trained on its proper handling and storage. Store in tightly closed containers in a cool, well-ventilated area away from strong oxidizers. Where possible, automatically transfer material from drums or other storage containers to process containers. Sources of ignition, such as smoking and open flames, are prohibited where this chemical is handled, used, or stored. Metal containers involving the transfer of this chemical should be grounded and bonded. Wherever this chemical is used, handled, manufactured, or stored, use explosion-proof electrical equipment and fittings.

Spill Handling: Evacuate persons not wearing protective equipment from area of spill or leak until cleanup is complete. Remove all ignition sources. Collect powdered material in the most convenient and safe manner and deposit in sealed containers. Ventilate area after cleanup is complete. It may be necessary to contain and dispose of this chemical as a hazardous waste. If material or contaminated runoff enters waterways, notify downstream users of potentially contaminated waters. Contact your local or federal environmental protection agency for specific recommendations. If employees are required to clean up spills, they must be properly trained and equipped. OSHA 1910.120(q) may be applicable.

Fire Extinguishing: Sodium benzoate is combustible. It may emit acrid fumes when heated to decomposition (at 120°C/248°F). Use dry chemical, carbon dioxide, water spray, or foam extinguishers. Poisonous gases are produced in fire, including sodium oxide. If material or contaminated runoff enters waterways, notify downstream users of potentially contaminated waters. Notify local health and fire officials and pollution control agencies. From a secure, explosion-proof location, use water spray to cool exposed containers. If cooling streams are ineffective (venting sound increases in volume and pitch, tank discolors, or shows any signs of deforming), withdraw immediately to a secure position. If employees are expected to fight fires, they must be trained and equipped in OSHA 1910.156. The only respirators recommended for firefighting are self-contained breathing apparatuses that have full face-pieces and are operated in a pressure-demand or other positive-pressure mode.

References

New York State Department of Health. (March 1986). *Chemical Fact Sheet: Sodium Benzoate.* Albany, NY: Bureau of Toxic Substance Assessment

US Environmental Protection Agency, Special Review and Reregistration Division Office of Pesticide Programs. (1998). *Agency Status of Pesticides in Registration, Reregistration, and Special Review* (Rainbow Report). Washington, DC

Sodium bisulfate   S:0410

Molecular Formula: $HNaO_3S$
Common Formula: $NaHSO_3$
Synonyms: Amersite 2; Bisulfite de sodium (French); Bisulfito sodico (Spanish); Hydrogen sulfite sodium; Sodium acid sulfite; Sodium hydrogen sulfite; Sodium hydrogen sulfite; Sodium pyrosulfite; Sodium metabisulfite; Sodium sulhydrate; Sulfurous acid, monosodium salt
CAS Registry Number: 7631-90-5; *(alt.)* 57414-01-4; *(alt.)* 69098-86-8; *(alt.)* 89830-27-3; *(alt.)* 91829-63-9
RTECS® Number: VZ2000000
UN/NA & ERG Number: UN2693 (solution)/154
EC Number: 231-548-0 [*Annex I Index No.:* 016-064-00-8]
Regulatory Authority and Advisory Bodies
Carcinogenicity: IARC: Animal Inadequate Evidence; Human Inadequate Evidence, *not classifiable as carcinogenic to humans*, Group 3, 1992.
US EPA Gene-Tox Program, Negative: Rodent dominant lethal; Rodent heritable translocation; Negative: Mouse specific locus; TRP reversion.
US EPA, FIFRA, 1998 Status of Pesticides: Supported.
Air Pollutant Standard Set. See below, "Permissible Exposure Limits in Air" section.
Clean Water Act: Section 311 Hazardous Substances/RQ 40CFR117.3 (same as CERCLA, see below).
Reportable Quantity (RQ): 5000 lb (2270 kg).
Canada, WHMIS, Ingredients Disclosure List Concentration 1.0%.
European/International Regulations: Hazard Symbol: Xn; Risk phrases: R22; R31; Safety phrases: S2; S25; S46 (see Appendix 4).
WGK (German Aquatic Hazard Class): 1—Low hazard to waters.
Description: Sodium bisulfite is a white crystalline solid with a slight odor of sulfur dioxide and a disagreeable taste. Slowly oxidized to the sulfate on exposure to air. Molecular

weight = 104.06; Specific gravity (H_2O:1) = 1.48 at 25°C; Boiling point = (decomposes). Hazard Identification (based on NFPA-704 M Rating System): Health 1, Flammability 0, Reactivity 0. Soluble in water.

Potential Exposure: Compound Description: Agricultural Chemical; Tumorigen, Mutagen. Sodium bisulfite is used in the digestion of wood pulp; in the tanning of leather; in the dyeing of textiles; as a photographic reducing agent; as a food preservative; as an additive in electroplating; as disinfectant, bleach, antioxidant; and as inhibitor of yeast and bacteria in winemaking.

Incompatibilities: Aqueous solution is a weak acid. Incompatible with strong acids, such as hydrochloric and nitric, and oxidizers, such as perchlorates, peroxides, permanganates, chlorates, and nitrates. Reacts with bases forming sulfate. Slowly oxidizes to sulfate in air. Heat causes decomposition. Slowly oxidized to the sulfate on exposure to air. Contact with oxidizers or acids forms sulfur dioxide gas. Attacks some metals in the presence of moisture.

Permissible Exposure Limits in Air
OSHA PEL: None.
NIOSH REL: 5 mg/m³ TWA.
ACGIH TLV®[1]: 5 mg/m³ TWA; not classifiable as a human carcinogen.
Protective Action Criteria (PAC)
TEEL-0: 5 mg/m³
PAC-1: 25 mg/m³
PAC-2: 150 mg/m³
PAC-3: 500 mg/m³
DFG MAK: Sensitizing substances as sulfites.
Australia: TWA 5 mg/m³, 1993; Belgium: TWA 5 mg/m³, 1993; Denmark: TWA 5 mg/m³, 1999; France: VME 5 mg/m³, 1999; Norway: TWA 5 mg/m³, 1999; Switzerland: MAK-W 5 mg/m³, 1999; United Kingdom: TWA 5 mg/m³, 2000; the Netherlands: MAC-TGG 5 mg/m³, 2003; Argentina, Bulgaria, Columbia, Jordan, South Korea, New Zealand, Singapore, Vietnam: ACGIH TLV®: not classifiable as a human carcinogen. Several states have set guidelines or standards for sodium bisulfite in ambient air[60] ranging from 5.0 μg/m³ (North Dakota) to 80.0 μg/m³ (Virginia) to 100.0 μg/m³ (Connecticut) to 119.0 μg/m³ (Nevada).

Determination in Air: Use NIOSH Analytical Method (IV) #0500, Particulates NOR, total dust.

Permissible Concentration in Water: Maryland[61] has set a guideline of 70.0 μg/L for sodium bisulfite in drinking water.

Routes of Entry: Inhalation, ingestion, skin and/or eye contact.

Harmful Effects and Symptoms
Short Term Exposure: Corrosive and can cause severe skin and eye irritation and burns. Irritates the respiratory tract, causing cough, wheezing, and shortness of breath. Ingestion may cause irritation to mouth, throat, and stomach. Allergic response may occur. This could include itching of ears and legs, nausea, cough, tightening of throat, and reddening of the skin.

Long Term Exposure: An asthma-like allergy may develop after repeated exposure. Skin allergy may develop. May irritate the lungs, causing bronchitis to develop with cough, phlegm, and/or shortness of breath.

Points of Attack: Lungs, skin, eyes.

Medical Surveillance: For those with frequent or potentially high exposure (half the TLV or greater), the following are recommended before beginning work and at regular times after that: lung function tests. Evaluation by a qualified allergist.

First Aid: If this chemical gets into the eyes, remove any contact lenses at once and irrigate immediately for at least 15 min, occasionally lifting upper and lower lids. Seek medical attention immediately. If this chemical contacts the skin, remove contaminated clothing and wash immediately with soap and water. Seek medical attention immediately. If this chemical has been inhaled, remove from exposure, begin rescue breathing (using universal precautions, including resuscitation mask) if breathing has stopped and CPR if heart action has stopped. Transfer promptly to a medical facility. When this chemical has been swallowed, get medical attention. If victim is *conscious*, administer water or milk. Do not induce vomiting.

Note to physician: Converted to sulfuric acid in stomach. Acute obstruction of alimentary canal may occur up to 3 weeks following ingestion.

Personal Protective Methods: Wear protective gloves and clothing to prevent any reasonable probability of skin contact. Safety equipment suppliers/manufacturers can provide recommendations on the most protective glove/clothing material for your operation. All protective clothing (suits, gloves, footwear, headgear) should be clean, available each day, and put on before work. Contact lenses should not be worn when working with this chemical. Wear dust-proof chemical goggles and face shield unless full face-piece respiratory protection is worn. Employees should wash immediately with soap when skin is wet or contaminated. Provide emergency showers and eyewash.

Respirator Selection: Where there is potential for exposures *over 5 mg/m³*, use a NIOSH/MSHA- or European Standard EN149-approved full face-piece respirator equipped with particulate (dust/fume/mist) filters. Particulate filters must be checked every day before work for physical damage, such as rips or tears, and replaced as needed. Where there is potential for high exposure, use a NIOSH/MSHA- or European Standard EN149-approved supplied-air respirator with a full face-piece operated in the positive-pressure mode, or with a full face-piece, hood, or helmet in the continuous-flow mode; or use a NIOSH/MSHA- or European Standard EN149-approved self-contained breathing apparatus with a full operated in pressure-demand or other positive-pressure mode.

Storage: Color Code—White: Corrosive or Contact Hazard; Store separately in a corrosion-resistant location. Prior to working with this chemical you should be trained on its proper handling and storage. Store in tightly closed containers in a dark, cool, well-ventilated area away from strong acids, such as hydrochloric and nitric, and oxidizers, such as perchlorates, peroxides, permanganates, chlorates, and nitrates.

Shipping: Bisulfites, inorganic, aqueous solutions, n.o.s. require a shipping label of "CORROSIVE." It falls in Hazard Class 8 and Packing Group II.

Spill Handling: Evacuate persons not wearing protective equipment from area of spill or leak until cleanup is complete. Remove all ignition sources. Use HEPA vacuum or wet method to reduce dust during cleanup. Do not dry sweep. Collect powdered material in the most convenient and safe manner and deposit in sealed containers. Cover spilled liquid with soda ash, absorb on vermiculite or other inert material. Ventilate area after cleanup is complete. It may be necessary to contain and dispose of this chemical as a hazardous waste. If material or contaminated runoff enters waterways, notify downstream users of potentially contaminated waters. Contact your local or federal environmental protection agency for specific recommendations. If employees are required to clean up spills, they must be properly trained and equipped. OSHA 1910.120(q) may be applicable.

Fire Extinguishing: This chemical is noncombustible. Extinguish fire using an agent suitable for type of surrounding fire. Sodium bisulfite itself does not burn. Poisonous gases are produced in fire, including sulfur oxide. If material or contaminated runoff enters waterways, notify downstream users of potentially contaminated waters. Notify local health and fire officials and pollution control agencies. From a secure, explosion-proof location, use water spray to cool exposed containers. If cooling streams are ineffective (venting sound increases in volume and pitch, tank discolors, or shows any signs of deforming), withdraw immediately to a secure position. If employees are expected to fight fires, they must be trained and equipped in OSHA 1910.156. The only respirators recommended for firefighting are self-contained breathing apparatuses that have full face-pieces and are operated in a pressure-demand or other positive-pressure mode.

Disposal Method Suggested: Dump into water, add soda ash, then neutralize with HCl; flush to sewer with large volumes of water.

References

New Jersey Department of Health and Senior Services. (August 1998). *Hazardous Substances Fact Sheet: Sodium Bisulfite.* Trenton, NJ

New York State Department of Health. (April 1986). *Chemical Fact Sheet: Sodium Bisulfite.* Albany, NY: Bureau of Toxic Substance Assessment

US Environmental Protection Agency, Special Review and Reregistration Division Office of Pesticide Programs. (1998). *Agency Status of Pesticides in Registration, Reregistration, and Special Review* (Rainbow Report). Washington, DC

Sodium cacodylate   S:0420

Molecular Formula: $C_2H_6AsNaO_2$
Common Formula: $(CH_3)_2AsOONa$
Synonyms: Alkarsodyl; Ansar 160; Arsecodile; Arsycodile; Bolls-Eye; Cacodilato sodico (Spanish); Cacodylate de sodium (French); Cacodylic acid sodium salt; Chemaid; Dimethylarsinic acid, sodium salt; [(Dimethylarsino)oxy] sodium-arsenic-oxide; [(Dimethylarsino)oxy]sodium As-oxide; Dutch-treat; Hydrodimethylarsine oxide, sodium salt; Kakodylan Dodny; Phytar 560; Rad-E-Cate; Rad-E-Cate 16; Rad-E-Cate-25; Rad-E-Cate-35; Silvisar; Sodium dimethylarsinate; Sodium dimethyl arsonate; Sodium salt of cacodylic acid
CAS Registry Number: 124-65-2; 6131-99-3 (trihydrate)
RTECS® Number: CH7890000
UN/NA & ERG Number: UN1688/152
EC Number: 204-708-2
Regulatory Authority and Advisory Bodies
Carcinogenicity: NTP: 11th Report on Carcinogens, 2004: Known to be a human carcinogen; IARC: Human Sufficient Evidence, 1980; Animal Limited Evidence, *carcinogenic to humans*, Group 1, 1987.
Banned or Severely Restricted (Portugal) (UN).[13]
Air Pollutant Standard Set. See below, "Permissible Exposure Limits in Air" section.
Superfund/EPCRA 40CFR355, Extremely Hazardous Substances: TPQ = 100/10,000 lb (45.4/4540 kg).
Reportable Quantity (RQ): 100 lb (45.4 kg).
As arsenic compounds
Clean Air Act: Hazardous Air Pollutants (Title I, Part A, Section 112) as arsenic compounds.
Clean Water Act: Toxic Pollutant (Section 401.15) as arsenic and compounds.
RCRA, 40CFR261, Appendix 8 Hazardous Constituents, waste number not listed.
Reportable Quantity (RQ): 1 lb (0.454 kg).
EPCRA (Section 313): Includes any unique chemical substance that contains arsenic as part of that chemical's infrastructure. Form R *de minimis* concentration reporting level: organics 1.0%.
US DOT Regulated Marine Pollutant (49CFR172.101, Appendix B) as arsenates, liquid, n.o.s.; arsenates, solid, n.o.s.; arsenical pesticides liquid, toxic, flammable, n.o.s.
Canada: Priority Substance List & Restricted Substances/ Ocean Dumping FORBIDDEN (CEPA), National Pollutant Release Inventory (NPRI) (arsenic compounds).
European/International Regulations: Hazard Symbol: T, N; Risk phrases: R45; R23/25; R50/53; Safety phrases: S53; S45; S60; S61 (see Appendix 4).
WGK (German Aquatic Hazard Class): No value assigned.
Description: Sodium Cacodylate is a white crystalline solid which occurs as the trihydrate. It liquefies in the water of hydration at 60°C and becomes anhydrous at 120°C. Molecular weight = 159.99. Hazard Identification (based on

NFPA-704 M Rating System): Health 3, Flammability 0, Reactivity 0. Soluble in water.

Potential Exposure: This material has been used as a non-selective herbicide and for general weed control.

Incompatibilities: Incompatible with oxidizers, strong bases, acids, active metals (iron, aluminum, zinc). Contact with acids forms highly toxic dimethylarsine gas. Attacks some metals.

Permissible Exposure Limits in Air
Arsenic, organic compounds
OSHA PEL: 0.5 mg[As]/m^3 TWA.
NIOSH REL: Not established. See NIOSH Pocket Guide, Appendix A.
ACGIH TLV®[1]: 0.01 mg[As]/m^3 TWA; Confirmed Human Carcinogen; BEI established.
Protective Action Criteria (PAC)
TEEL-0: 1.07 mg/m^3
PAC-1: 3.2 mg/m^3
PAC-2: 40 mg/m^3
PAC-3: 500 mg/m^3

Determination in Air: Use NIOSH Analytical Method (IV) #5022, arsenic, Organo-.

Permissible Concentration in Water: The EPA[61] has set a limit of 0.05 µg/L in drinking water.

Routes of Entry: Inhalation, ingestion, skin and/or eye contact.

Harmful Effects and Symptoms
Short Term Exposure: Sodium cacodylate is corrosive to the skin, eyes, and mucous membranes. Moderately toxic; probable oral lethal dose in humans is 0.5−5 g/kg or between 1 oz and 1 pint (or 1 lb) for a 70-kg (150-lb) person. It may cause disturbances of the blood, kidneys, and nervous system. Acute exposure to sodium cacodylate may be fatal. Headache, red-stained eye, and a garlicky odor of the breath may be the first effects noticed. Other signs and symptoms include generalized weakness, intense thirst, muscle cramping, seizures, toxic delirium, and shock. Nausea, vomiting, anorexia, abdominal pain, and diarrhea may occur. Hypotension (low blood pressure), tachycardia (rapid heart rate), pulmonary edema, ventricular fibrillation, and other cardiac abnormalities are usually found following severe exposure.

Long Term Exposure: Repeated exposure may cause ulcers and hole in the nasal septum. Hoarseness and sore eyes also occur. Repeated contact may cause thickened skin, pigmentation changes. May cause liver damage and nerve damage, causing sensation of "pins and needles," weakness, and loss of coordination in the limbs. May cause gastrointestinal tract and reproductive effects. Repeated exposures can cause metallic taste; poor appetite; nausea, vomiting, diarrhea, and stomach pain; seizures and death.

Points of Attack: Skin, respiratory system, kidneys, central nervous system, liver, gastrointestinal tract, reproductive system.

Medical Surveillance: Examination of the nose, skin, eyes, and nails. Examination of the nervous system. Liver function tests. Test for urine arsenic. NIOSH recommended that exposure levels should not exceed 100 µg/L of creatinine in the urine. Results may be accurate within 2 days of eating shellfish or fish (which may increase arsenic levels); they are most accurate at the end of a workday.

First Aid: If this chemical gets into the eyes, remove any contact lenses at once and irrigate immediately for at least 15 min, occasionally lifting upper and lower lids. Seek medical attention immediately. If this chemical contacts the skin, remove contaminated clothing and wash immediately with soap and water. Seek medical attention immediately. If this chemical has been inhaled, remove from exposure, begin rescue breathing (using universal precautions, including resuscitation mask) if breathing has stopped and CPR if heart action has stopped. Transfer promptly to a medical facility. When this chemical has been swallowed, get medical attention. Give a slurry of activated charcoal in water to drink and induce vomiting. Do not make an unconscious person vomit. Obtain authorization and/or further instructions from the local hospital for administration of an antidote or performance of other invasive procedures. Rush to health-care facility.

Personal Protective Methods: Wear protective gloves and clothing to prevent any reasonable probability of skin contact. Safety equipment suppliers/manufacturers can provide recommendations on the most protective glove/clothing material for your operation. All protective clothing (suits, gloves, footwear, headgear) should be clean, available each day, and put on before work. Contact lenses should not be worn when working with this chemical. Wear dust-proof chemical goggles and face shield unless full face-piece respiratory protection is worn. Employees should wash immediately with soap when skin is wet or contaminated. Provide emergency showers and eyewash.

Note to physician: For severe poisoning BAL [British Anti-Lewisite, dimercaprol, dithiopropanol (C$_3$H$_8$OS$_2$)] has been used to treat toxic symptoms of certain heavy metals poisoning—including arsenic. Although BAL is reported to have a large margin of safety, caution must be exercised, because toxic effects may be caused by excessive dosage. Most can be prevented by premedication with 1-ephedrine sulfate (CAS: 134-72-5). For milder poisoning *penicillamine (not penicillin)* has been used, both with mixed success. Side effects occur with such treatment and it is never a substitute for controlling exposure. It can only be done under strict medical care.

Respirator Selection: *At any concentrations above the NIOSH REL, or where there is no REL, at any detectable concentration:* SCBAF: Pd,Pp (APF = 10,000) (any self-contained breathing apparatus that has a full faceplate and is operated in a pressure-demand or other positive-pressure mode) or SaF: Pd,Pp: ASCBA (APF = 10,000) (any supplied-air respirator that has a full face-piece and is operated in a pressure-demand or other positive-pressure mode in combination with an auxiliary, self-contained breathing apparatus operated in a pressure-demand or other positive-pressure mode). *Escape:* GmFAg100 (APF = 50) [any air-purifying, full-face-piece respirator (gas mask) with a chin-style, front- or back-mounted acid gas canister having an N100, R100, or P100 filter] or SCBAE (any appropriate escape-type, self-contained breathing apparatus).

Storage: Color Code—Blue: Health Hazard/Poison: Store in a secure poison location. Prior to working with this chemical you should be trained on its proper handling and storage.

Shipping: Sodium cacodylate requires a shipping label of "POISONOUS/TOXIC MATERIALS." It falls in Hazard Class 6.1 and Packing Group II.

Spill Handling: Evacuate persons not wearing protective equipment from area of spill or leak until cleanup is complete. Remove all ignition sources. Collect powdered material in the most convenient and safe manner and deposit in sealed containers. Ventilate area after cleanup is complete. It may be necessary to contain and dispose of this chemical as a hazardous waste. If material or contaminated runoff enters waterways, notify downstream users of potentially contaminated waters. Contact your local or federal environmental protection agency for specific recommendations. If employees are required to clean up spills, they must be properly trained and equipped. OSHA 1910.120(q) may be applicable.

Fire Extinguishing: Sodium cacodylate itself does not burn. Use any agent suitable for surrounding fire. Stay upwind; keep out of low areas. Wear self-contained breathing apparatus and full protective clothing. Move container from fire area, if you can do so without risk. Poisonous gases are produced in fire, including arsenic oxides. If material or contaminated runoff enters waterways, notify downstream users of potentially contaminated waters. Notify local health and fire officials and pollution control agencies. From a secure, explosion-proof location, use water spray to cool exposed containers. If cooling streams are ineffective (venting sound increases in volume and pitch, tank discolors, or shows any signs of deforming), withdraw immediately to a secure position. If employees are expected to fight fires, they must be trained and equipped in OSHA 1910.156. The only respirators recommended for firefighting are self-contained breathing apparatuses that have full face-pieces and are operated in a pressure-demand or other positive-pressure mode.

Disposal Method Suggested: For cacodylic acid, precipitate as calcium arsenate and calcium arsenite by treatment with excess lime water. Recycle if possible. If not, put in secure storage for possible disposal in leach-proof dumps.[22] In accordance with 40CFR165, follow recommendations for the disposal of pesticides and pesticide containers. Must be disposed properly by following package label directions or by contacting your local or federal environmental control agency or by contacting your regional EPA office.

References

US Environmental Protection Agency. (November 30, 1987). *Chemical Hazard Information Profile: Sodium Cacodylate.* Washington, DC: Chemical Emergency Preparedness Program

New Jersey Department of Health and Senior Services. (August 1999). *Hazardous Substances Fact Sheet: Sodium Cacodylate.* Trenton, NJ

Sodium chlorate   S:0430

Molecular Formula: $ClNaO_3$
Common Formula: $NaClO_3$
Synonyms: Asex; Atlacide; Atratol B-herbatox; Chlorate of soda; Chlorate salt of sodium; Chlorax; Chloric acid, Sodium salt; Chlorsaure (German); De-Fol-Ate; Desolet; Drexel defol; Drop leaf; Evau-superfall; Grain sorghum harvest aid; Granex OK; Harvest-aid; Klorex; Kusa-tohrukusa-tol; Lorex; Natriumchlorat (German); Ortho C-1 defoliant & weed killer; Oxycil; Rasikal; Shed-α-leaf; Shed-α-leaf "L"; Soda chlorate; Sodium (chlorate de) (French); Travex; Tumbleaf; United Chemical defoliant No. 1; Val-Drop
CAS Registry Number: 7775-09-9; *(alt.)* 11096-45-0
RTECS® Number: FO0525000
UN/NA & ERG Number: UN1495 (solid)/140; UN2428 (solution)/140
EC Number: 017-005-00-9
Regulatory Authority and Advisory Bodies
Department of Homeland Security Screening Threshold Quantity (pounds): *Theft hazard* 400 (Commercial grade).
US EPA, FIFRA, 1998 Status of Pesticides: Supported.
Highly Reactive Substance and Explosive (World Bank).[15]
European/International Regulations: not listed in Annex 1.
WGK (German Aquatic Hazard Class): 2—Hazard to waters.

Description: Sodium chlorate is a white crystalline solid. Molecular weight = 106.44; Decomposes below boiling point at ca. 300°C; Freezing/Melting point = 248°C (also listed at 255°C and 264°C). Hazard Identification (based on NFPA-704 M Rating System): Health 1, Flammability 0, Reactivity 2 (Oxidizer). Soluble in water.

Potential Exposure: Compound Description: Agricultural Chemical; Tumorigen, Mutagen, Human Data; Primary Irritant. Sodium chlorate is used to manufacture dyes, explosives, in paper pulp processing and as a weed killer; used as a constituent of atratol and pramitol.

Incompatibilities: A strong oxidizer. Reacts violently with combustibles, sulfuric acid, and reducing materials. Reacts with strong acids giving off carbon dioxide. Explosions may be caused by contact with ammonia salts, ammonium thiosulfate, antimony sulfide, arsenic, carbon, charcoal, organic matter, organic acids, thiocyanates, chemically active metals, oils, metal sulfides, nitrobenzene, powdered metals, sugar. Reacts with some organic contaminants forming shock-sensitive mixtures. Decomposes on heating above 300°C or on burning, producing oxygen and toxic chlorine fumes. Attacks zinc, magnesium, and steel.

Permissible Exposure Limits in Air
Protective Action Criteria (PAC)
TEEL-0: 0.15 mg/m^3
PAC-1: 0.4 mg/m^3
PAC-2: 3 mg/m^3
PAC-3: 75 mg/m^3

Permissible Concentration in Water: Russia[43] set a MAC of 20 mg/L in water bodies used for domestic purposes.

Routes of Entry: Inhalation, ingestion, skin and/or eye contact.

Harmful Effects and Symptoms
Short Term Exposure: Sodium chlorate can affect you when breathed in. Eye or skin contact can cause severe irritation and even burns. Breathing sodium chlorate, especially dust or mist, can irritate the nose and throat. It can also cause cyanosis, causing the skin to turn blue (methemoglobinemia), because it interferes with the blood's ability to carry oxygen. Damage to red blood cells (hemolytic anemia) can also occur. If severe or repeated, this can cause kidney damage. Ingestion can cause kidney damage. The effects may be delayed.

Long Term Exposure: Repeated or prolonged contact with skin may cause dermatitis. Kidney damage can occur from severe or repeated damage to red blood cells resulting from exposure. Very irritating substances may cause lung damage.

Points of Attack: Kidneys, lungs, skin.

Medical Surveillance: Before beginning employment and at regular times after that, for those with frequent or potentially high exposures, the following are recommended: lung function tests. If symptoms develop or overexposure is suspected, the following may be useful: complete blood count (CBC). Test for methemoglobin if skin is blue.

First Aid: If this chemical gets into the eyes, remove any contact lenses at once and irrigate immediately for at least 15 min, occasionally lifting upper and lower lids. Seek medical attention immediately. If this chemical contacts the skin, remove contaminated clothing and wash immediately with soap and water. Seek medical attention immediately. If this chemical has been inhaled, remove from exposure, begin rescue breathing (using universal precautions, including resuscitation mask) if breathing has stopped and CPR if heart action has stopped. Transfer promptly to a medical facility. When this chemical has been swallowed, get medical attention. Give large quantities of water and induce vomiting. Do not make an unconscious person vomit. Keep victim under medical observation.

Antidotes and Special Procedures: Do not use methylene blue to treat methemoglobinemia from sodium chlorate as it can cause increased toxicity.

Note to physician: Treat for methemoglobinemia. Spectrophotometry may be required for precise determination of levels of methemoglobin in urine.

Personal Protective Methods: Wear protective gloves and clothing to prevent any reasonable probability of skin contact. Safety equipment suppliers/manufacturers can provide recommendations on the most protective glove/clothing material for your operation. All protective clothing (suits, gloves, footwear, headgear) should be clean, available each day, and put on before work. Contact lenses should not be worn when working with this chemical. Wear dust-proof chemical goggles and face shield unless full face-piece respiratory protection is worn. Employees should wash immediately with soap when skin is wet or contaminated. Provide emergency showers and eyewash.

Respirator Selection: Where there is potential for exposure to sodium chlorate, use a NIOSH/MSHA- or European Standard EN149-approved full face-piece respirator equipped with particulate (dust/fume/mist) filters. Particulate filters must be checked every day before work for physical damage, such as rips or tears, and replaced as needed. *Where there is potential for high exposures,* use a NIOSH/MSHA- or European Standard EN149-approved supplied-air respirator with a full face-piece operated in the positive-pressure mode, or with a full face-piece, hood, or helmet in the continuous-flow mode; or use a NIOSH/MSHA- or European Standard EN149-approved self-contained breathing apparatus with a full face-piece operated in pressure-demand or other positive-pressure mode.

Storage: Color Code—Yellow: Reactive Hazard; Store in a location separate from other materials, especially flammables and combustibles. Prior to working with this chemical you should be trained on its proper handling and storage. Sodium chlorate must be stored to avoid contact with ammonium thiosulfate, antimony sulfide, arsenic, carbon, charcoal, organic matter, organic acids, sulfuric acid, thiocyanates, and chemically active metals (such as potassium, sodium, magnesium, and zinc), since violent reactions occur. Protect storage containers against physical damage. Store in tightly closed containers in a cool, well-ventilated area away from sources of heat. See OSHA Standard 1910.104 and NFPA 43A *Code for the Storage of Liquid and Solid Oxidizers* for detailed handling and storage regulations.

Shipping: This compound requires a shipping label of "OXIDIZER." It falls in Hazard Class 5.1 and Packing Group II.

Spill Handling: Evacuate persons not wearing protective equipment from area of spill or leak until cleanup is complete. Remove all ignition sources. Collect powdered material in the most convenient and safe manner and deposit in sealed containers. Ventilate area after cleanup is complete. Keep sodium chlorate out of a confined space, such as a sewer, because of the possibility of an explosion, unless the sewer is designed to prevent the buildup of explosive concentrations. It may be necessary to contain and dispose of this chemical as a hazardous waste. If material or contaminated runoff enters waterways, notify downstream users of potentially contaminated waters. Contact your local or federal environmental protection agency for specific recommendations. If employees are required to clean up spills, they must be properly trained and equipped. OSHA 1910.120(q) may be applicable.

Fire Extinguishing: Sodium chlorate may burn, but does not readily ignite. Heat above 300°C produces oxygen which can increase the intensity of fire and may ignite other combustible materials. Flood areas with water unless that is incompatible with other materials in the fire area. Poisonous gases are produced in fire, including chlorine. If material or contaminated runoff enters waterways, notify downstream users of potentially contaminated waters. Notify local health and fire officials and pollution control agencies. Containers may explode in fire. From a secure, explosion-proof location, use water spray to cool exposed containers. If cooling streams are ineffective (venting sound increases in volume and pitch, tank discolors, or shows any signs of deforming), withdraw immediately to a secure position. If employees are expected to fight fires, they must be trained and equipped in OSHA 1910.156. The only respirators recommended for firefighting are self-contained breathing apparatuses that have full face-pieces and are operated in a pressure-demand or other positive-pressure mode.

Disposal Method Suggested: In accordance with 40CFR165, follow recommendations for the disposal of pesticides and pesticide containers. Must be disposed properly by following package label directions or by contacting your local or federal environmental control agency or by contacting your regional EPA office.

References

Sax, N. I. (Ed.). (1983). *Dangerous Properties of Industrial Materials Report*, 3, No. 1, 28–32

US Environmental Protection Agency, Special Review and Reregistration Division Office of Pesticide Programs. (1998). *Agency Status of Pesticides in Registration, Reregistration, and Special Review* (Rainbow Report). Washington, DC

New Jersey Department of Health and Senior Services. (February 2001). *Hazardous Substances Fact Sheet: Sodium Chlorate.* Trenton, NJ

Sodium chloroplatinate   S:0440

Molecular Formula: $Cl_6Pt \cdot 2Na \cdot 4H_2O$
Synonyms: Platinate(2-), hexachlorodisodium, tetrahydrate; Platinic sodium chloride; Sodium hexachloroplatinate(IV); Sodium platinic chloride; Sodium platinic chloride
CAS Registry Number: 1307-82-0
EC Number: None assigned
Regulatory Authority and Advisory Bodies
Air Pollutant Standard Set. See below, "Permissible Exposure Limits in Air" section.
Canada, WHMIS, Ingredients Disclosure List Concentration 0.1% as platinum, water-soluble salts.

European/International Regulations: not listed in Annex 1.
WGK (German Aquatic Hazard Class): No value assigned.
Description: Sodium chloroplatinate is a yellow-orange or red-brown crystalline solid. Odorless. Molecular weight = 561.9; Melting point = loses water at 100°C; Vapor pressure = essentially zero. Soluble in water; solubility = 50 g/100 g at 20°C.
Potential Exposure: Used as a catalyst.
Incompatibilities: Oxidizers.
Permissible Exposure Limits in Air
OSHA PEL: 0.002 mg[Pt]/m³ TWA.
NIOSH REL: 0.002 mg[Pt]/m³ TWA.
ACGIH TLV®[1]: 1 0.002 mg[Pt]/m³ TWA.
No TEEL available.
DFG MAK: No numerical value established. Data may be available; however, 2 μg[Pt]/m³ peak should not be exceeded; danger of skin and airway sensitization, as chloroplatinates.
NIOSH IDLH: 4 mg [Pt]/m³.
Several states have set limits on platinum in ambient air[60] as well. See the entry on platinum for details.
Determination in Air: Use NIOSH II(7) Method #S-19 (soluble salts).
Routes of Entry: Inhalation, ingestion, skin and/or eye contact.
Harmful Effects and Symptoms
Short Term Exposure: Chloroplatinates can affect you when breathed in. Severe allergy can develop to potassium chloroplatinates. Symptoms include asthma (with cough, wheezing, and/or shortness of breath), runny nose, and/or skin rash, sometimes with hives. If allergy develops, even small future exposure can trigger significant symptoms. Some persons exposed to this type of chemical have developed lung scarring. Family members can develop allergy to dust carried home on work clothing. It may irritate the eyes, nose, and throat. High exposure may cause irritability and even seizures.
Long Term Exposure: May cause skin sensitization and dermatitis and/or asthma-like allergy. Repeated exposures can cause sores or ulcers in the lining of the nose. Tetrachloroplatinates are mutagens.
Points of Attack: Skin, lungs.
Medical Surveillance: Before beginning employment and at regular times after that, the following are recommended. Lung function tests. These may be normal if the person is not having an attack at the time of the test. If symptoms develop or overexposure is suspected, the following may be useful: chest X-ray every 3 years should be considered if above tests are not normal. Evaluation by a qualified allergist, including careful exposure history and special testing, may help diagnose skin allergy.
First Aid: If this chemical gets into the eyes, remove any contact lenses at once and irrigate immediately for at least 15 min, occasionally lifting upper and lower lids. Seek medical attention immediately. If this chemical contacts the skin, remove contaminated clothing and wash immediately

with soap and water. Seek medical attention immediately. If this chemical has been inhaled, remove from exposure, begin rescue breathing (using universal precautions, including resuscitation mask) if breathing has stopped and CPR if heart action has stopped. Transfer promptly to a medical facility. When this chemical has been swallowed, get medical attention. Give large quantities of water and induce vomiting. Do not make an unconscious person vomit.

Personal Protective Methods: Wear protective gloves and clothing to prevent any reasonable probability of skin contact. Safety equipment suppliers/manufacturers can provide recommendations on the most protective glove/clothing material for your operation. All protective clothing (suits, gloves, footwear, headgear) should be clean, available each day, and put on before work. Contact lenses should not be worn when working with this chemical. Wear dust-proof chemical goggles and face shield unless full face-piece respiratory protection is worn. Employees should wash immediately with soap when skin is wet or contaminated. Provide emergency showers and eyewash.

Respirator Selection: (for soluble Pt salts): *Up to 0.05 mg/m³:* Sa:Cf (APF = 25) (any supplied-air respirator operated in a continuous-flow mode). *Up to 0.1 mg/m³:* 100F (APF = 50) (any air-purifying, full-face-piece respirator with an N100, R100, or P100 filter) or SCBAF (APF = 50) (any self-contained breathing apparatus with a full face-piece) or SaF (APF = 50) (any supplied-air respirator with a full face-piece). *Up to 4 mg/m³:* SaF: Pd,Pp (APF = 2000) (any supplied-air respirator that has a full face-piece and is operated in a pressure-demand or other positive-pressure mode). *Emergency or planned entry into unknown concentrations or IDLH conditions:* SCBAF: Pd,Pp (APF = 10,000) (any NIOSH/MSHA- or European Standard EN 149-approved self-contained breathing apparatus that has a full face-piece and is operated in a pressure-demand or other positive-pressure mode) or SaF: Pd,Pp: ASCBA (APF = 10,000) (any supplied-air respirator that has a full face-piece and is operated in a pressure-demand or other positive-pressure mode in combination with an auxiliary, self-contained breathing apparatus operated in a pressure-demand or other positive-pressure mode). *Escape:* 100 F (APF = 50) (any air-purifying, full-face-piece respirator with an N100, R100, or P100 filter) or SCBAE (any appropriate escape-type, self-contained breathing apparatus).

Note: Substance cause eye irritation and damage; eye protection needed.

Storage: Color Code—Blue: Health Hazard/Poison: Store in a secure poison location. Prior to working with this chemical you should be trained on its proper handling and storage. Store in tightly closed containers in a cool, well-ventilated area away from strong oxidizers. Where possible, automatically transfer material from drums or other storage containers to process containers.

Spill Handling: Evacuate persons not wearing protective equipment from area of spill or leak until cleanup is complete. Remove all ignition sources. Collect powdered material in the most convenient and safe manner and deposit in sealed containers. Ventilate area after cleanup is complete. It may be necessary to contain and dispose of this chemical as a hazardous waste. If material or contaminated runoff enters waterways, notify downstream users of potentially contaminated waters. Contact your local or federal environmental protection agency for specific recommendations. If employees are required to clean up spills, they must be properly trained and equipped. OSHA 1910.120(q) may be applicable.

Fire Extinguishing: This chemical may burn but does not easily ignite. Use dry chemical, carbon dioxide, water spray, or foam extinguishers. Poisonous gases are produced in fire, including chlorides and sodium oxides. If material or contaminated runoff enters waterways, notify downstream users of potentially contaminated waters. Notify local health and fire officials and pollution control agencies. From a secure, explosion-proof location, use water spray to cool exposed containers. If cooling streams are ineffective (venting sound increases in volume and pitch, tank discolors, or shows any signs of deforming), withdraw immediately to a secure position. If employees are expected to fight fires, they must be trained and equipped in OSHA 1910.156. The only respirators recommended for firefighting are self-contained breathing apparatuses that have full face-pieces and are operated in a pressure-demand or other positive-pressure mode.

References

New Jersey Department of Health and Senior Services. (April 2000). *Hazardous Substances Fact Sheet: Sodium Chloroplatinate.* Trenton, NJ

Sodium chromates   S:0445

Common Formula: $CrO_4 \cdot 2Na$; $Cr_2O_7 \cdot 2Na$ (dihydrate); $CrO_4 \cdot _2Na \cdot 10H_2O$ (decahydrate)

Synonyms: Chromate of soda; Chromium disodium oxide; Chromium sodium oxide; Cromato sodico (Spanish); Disodium chromate; Neutral sodium chromate; Sodium chromate(VI)

Bichromate: Bichromate of soda; Bichromate de sodium (French); Chromic acid, disodium salt; Chromium sodium oxide; Disodium dichromate; Natriumdichromat (German); Sodium bichromate; Sodium chromate; Sodium dichromate de (French); Sodium dichromate (dihydrate); Sodium dichromate(VI)

Decahydrate: Chromic acid, disodium salt, decahydrate; Sodium chromate decahydrate

CAS Registry Number: 10588-01-9; 7775-11-3 (disodium chromate); 7789-12-0 (dihydrate); 13517-17-4 (decahydrate); 10034-82-9 (tetrahydrate)

RTECS® Number: GB2955000; GB2957000 (decahydrate)

UN/NA & ERG Number: UN3288 (Toxic solid, inorganic, n.o.s.)/151

EC Number: 231-889-5 [*Annex I Index No.:* 024-018-00-3]

Regulatory Authority and Advisory Bodies
Carcinogenicity: IARC: Human Sufficient Evidence; Animal Sufficient Evidence, *carcinogenic to humans*, Group 1, 1997; NTP: 11th Report on Carcinogens, 2004: Known to be a human carcinogen; EPA *(inhalation)*: Known human carcinogen; EPA *(oral)*: Not Classifiable as to human carcinogenicity; NTP: Known to be a human carcinogen.

US EPA Gene-Tox Program, Positive/dose response: TRP reversion.

US EPA, FIFRA 1998 Status of Pesticides: Canceled.

Air Pollutant Standard Set. See below, "Permissible Exposure Limits in Air" section.

Clean Water Act: Section 311 Hazardous Substances/RQ 40CFR117.3 (same as CERCLA, see below); Section 313 Water Priority Chemicals (57FR41331, 9/9/92).

Reportable Quantity (RQ): 10 lb (4.54 kg).

As chromium compounds

Clean Air Act: Hazardous Air Pollutants (Title I, Part A, Section 112).

Clean Water Act: Toxic Pollutant (Section 401.15); 40CFR401.15 Section 307 Toxic Pollutants as chromium and compounds.

RCRA, 40CFR261, Appendix 8 Hazardous Constituents, waste number not listed.

California Proposition 65 Chemical: *(hexavalent chromium)* Cancer 2/27/87; Developmental/Reproductive toxin (male, female) 12/19/08.

Canada, WHMIS, Ingredients Disclosure List Concentration 0.1%.

European/International Regulations: Hazard Symbol: O, T+, N; Risk phrases: R45; R46; R60; R61; R8; R21; R25; R26; R34; R42/43; R48/23; R50/53; Safety phrases: S53; S45; S60; S61.

European/International Regulations: Hazard Symbol: E, T+, N; Risk phrases: R45; R46; R60; R61; R2; R8; R21; R25; R26; R34; R42/43; R48/23; R50/53; Safety phrases: S53; S45; S60; S61 (see Appendix 4).

WGK (German Aquatic Hazard Class): 3—Severe hazard to waters.

Description: Sodium chromate, including the hexahydrate, is a yellow crystalline solid that can also be used in solution. Molecular weight = 161.98; Freezing/Melting point = 760–792°C. Soluble in water. The dichromate is a red or red-orange crystalline solid. Molecular weight = 161.98; Freezing/Melting point = 357°C (decomposes at 400°C). Hazard Identification (based on NFPA-704 M Rating System): Health 2, Flammability 0, Reactivity 0. Soluble in water.

Potential Exposure: Compound Description: Tumorigen, Mutagen; Reproductive Effector. Used to make dyes, inks, pigments, and other chromates; in leather tanning, a corrosion inhibitor in circulating water systems; metal treatment; a drilling mud additive; chemical intermediate for chromium catalysts; colorimetry, oxidizing agent; bleaching agent; an algicide, fungicide, insecticide; in wood preservation.

Incompatibilities: Aqueous solution in a base. A strong oxidizer. Violent reaction with reducing agents, combustibles, strong acids, organic materials.

Permissible Exposure Limits in Air
Protective Action Criteria (PAC)
Disodium chromate
TEEL-0: 0.0156 mg/m^3
PAC-1: 1.25 mg/m^3
PAC-2: 7.5 mg/m^3
PAC-3: 46.7 mg/m^3
Dichromate
TEEL-0: 0.0126 mg/m^3
PAC-1: 20 mg/m^3
PAC-2: 37.8 mg/m^3
PAC-3: 37.8 mg/m^3
Decahydrate
TEEL-0: 0.0329 mg/m^3
PAC-1: 2.5 mg/m^3
PAC-2: 20 mg/m^3
PAC-3: 98.7 mg/m^3
Tetrahydrate
TEEL-0: 0.0291 mg/m^3
PAC-1: 2.5 mg/m^3
PAC-2: 15 mg/m^3
PAC-3: 87.4 mg/m^3
As chromium(VI) inorganic soluble compounds
OSHA PEL: 0.005 mg[Cr(VI)]/m^3 TWA Concentration. See 29CFR1910.1026.

NIOSH REL: 0.001 mg[Cr]/m^3 TWA, potential carcinogen, limit exposure to lowest feasible level. NIOSH considers all Cr(VI) compounds (including chromic acid, *tert*-butyl chromate, zinc chromate, and chromyl chloride) to be potential occupational carcinogens. See *NIOSH Pocket Guide*, Appendix A & C.

ACGIH TLV®[1]: 0.05 mg[Cr]/m^3 TWA, Confirmed Human Carcinogen; BEI issued.

DFG MAK: Danger of skin sensitization; Carcinogen Category 1; TRK: 0.05 mg[Cr]/m^3; 20 μg/L [Cr] in urine at end-of-shift.

NIOSH IDLH: 15 mg[Cr(VI)]/m^3.

United Kingdom: carcinogen, 2000. The former USSR-UNEP/IRPTC joint project[43] set a MAC of 0.01 mg/m^3 in work-place air. Connecticut[60] has set a guideline of 0.25 μg/m^3 for chromium trioxide in ambient air.

Determination in Air: Use NIOSH Analytical Methods #7600, #7604, #7605, #7703, #9101; OSHA Analytical Methods ID-103, ID-215, W-4001.

Permissible Concentration in Water: The EPA[6] has designated chromium as a priority toxic pollutant. To protect human health, the limits are trivalent chromium (in chromates): 170 μg/L and hexavalent chromium (in dichromates) 50 μg/L.

Determination in Water: Total chromium may be determined by digestion followed by atomic absorption or by colorimetry (diphenylcarbazide) or by inductively coupled plasma (CP) optical emission spectrometry. Chromium(VI) may be determined by extraction and atomic absorption or colorimetry (using diphenylhydrazide). Dissolved total Cr or Cr(VI) may be determined by 0.45 μm filtration followed by the above-cited methods.[49]

Routes of Entry: Inhalation, ingestion, skin and/or eye contact.

Harmful Effects and Symptoms

Short Term Exposure: Sodium chromate can affect you when breathed in. It can also pass into inner layers of the skin. A corrosive. Eye contact can cause severe damage with possible loss of vision. Irritation of nose, throat, and bronchial tubes can occur, with cough and/or wheezing. Skin contact can cause severe irritation, deep ulcers, or an allergic skin rash.

Long Term Exposure: Sodium chromate is a human carcinogen. Sodium chromates can cause a sore or perforated nasal septum, with bleeding, discharge, or crusting. May cause skin allergy with eczema-like rash. Can cause lung irritation or allergy; bronchitis may develop. May cause kidney damage.

Points of Attack: Skin, lungs, kidneys. *Cancer site:* throat and lungs.

Medical Surveillance: NIOSH lists the following tests: Blood gas analysis, complete blood count; chest X-ray, electrocardiogram, liver function tests; pulmonary function tests; sputum cytology, urine (chemical/metabolite), urinalysis (routine), white blood cell count/differential. Examination of the nose and skin. If symptoms develop or overexposure is suspected, the following may be useful: evaluation by a qualified allergist, including careful exposure history and special testing, may help diagnose skin allergy. Also check your skin daily for little bumps or blisters, the first sign of "chrome ulcers." If not treated early, these can last for years after exposure. Urine test for chromates. This test is most accurate shortly after exposure.

First Aid: If this chemical gets into the eyes, remove any contact lenses at once and irrigate immediately for at least 15 min, occasionally lifting upper and lower lids. Seek medical attention immediately. If this chemical contacts the skin, remove contaminated clothing and wash immediately with soap and water. Seek medical attention immediately. If this chemical has been inhaled, remove from exposure, begin rescue breathing (using universal precautions, including resuscitation mask) if breathing has stopped and CPR if heart action has stopped. Transfer promptly to a medical facility. When this chemical has been swallowed, get medical attention. If victim is *conscious*, administer water or milk. Do not induce vomiting.

Personal Protective Methods: Wear protective gloves and clothing to prevent any reasonable probability of skin contact. Prevent skin contact. (As chromic acid and chromates) **8 h** (more than 8 h of resistance to breakthrough >0.1 μg/cm/min): polyethylene gloves, suits, boots; polyvinyl chloride gloves, suits, boots; Saranex™ coated suits; **4 h** (At least 4 but <8 h of resistance to breakthrough >0.1 μg/cm²/min): butyl rubber gloves, suits, boots; Viton™ gloves, suits.

Safety equipment suppliers/manufacturers can provide recommendations on the most protective glove/clothing material for your operation. All protective clothing (suits, gloves, footwear, headgear) should be clean, available each day, and put on before work. Contact lenses should not be worn when working with this chemical. Wear splash-proof chemical goggles and face shield when working with liquid, unless full face-piece respiratory protection is worn. Wear dust-proof goggles and face shield when working with powders or dust, unless full face-piece respiratory protection is worn. Employees should wash immediately with soap when skin is wet or contaminated. Provide emergency showers and eyewash. Specific engineering controls are recommended in NIOSH criteria document #76-129 [Chromium(VI)].

Respirator Selection: NIOSH: *At any concentrations above the NIOSH REL, or where there is no REL, at any detectable concentration:* SCBAF: Pd,Pp (APF = 10,000) (any self-contained breathing apparatus that has a full face-piece and is operated in a pressure-demand or other positive-pressure mode) or SaF: Pd,Pp: ASCBA (APF = 10,000) (any supplied-air respirator that has a full face-piece and is operated in a pressure-demand or other positive-pressure mode in combination with an auxiliary, self-contained breathing apparatus operated in a pressure-demand or other positive-pressure mode). *Escape:* 100F (APF = 50) (any air-purifying, full-face-piece respirator with an N100, R100, or P100 filter) or SCBAE (any appropriate escape-type, self-contained breathing apparatus).

Storage: Color Code—Blue: Health Hazard/Poison: Store in a secure poison location. Prior to working with this chemical you should be trained on its proper handling and storage. Store in tightly closed containers in a cool, well-ventilated area away from combustibles, organics, or other easily oxidized materials. Where possible, automatically transfer material from drums or other storage containers to process containers. A regulated, marked area should be established where this chemical is handled, used, or stored in compliance with OSHA Standard 1910.1045.

Shipping: Toxic solid, inorganic, n.o.s. requires a shipping label of "POISONOUS/TOXIC MATERIALS." It falls in Hazard Class 6.1 and Packing Group III.

Spill Handling: Evacuate persons not wearing protective equipment from area of spill or leak until cleanup is complete. Remove all ignition sources. Collect powdered material in the most convenient and safe manner and deposit in sealed containers. Ventilate area after cleanup is complete. It may be necessary to contain and dispose of this chemical as a hazardous waste. If material or contaminated runoff enters waterways, notify downstream users of potentially contaminated waters. Contact your local or federal environmental protection agency for specific recommendations. If employees are required to clean up spills, they must be properly trained and equipped. OSHA 1910.120(q) may be applicable.

Fire Extinguishing: This chemical may burn but does not easily ignite. Use dry chemical, carbon dioxide, water spray, or foam extinguishers. Poisonous gases are produced in fire, including chromium oxide and sodium oxide. If material or contaminated runoff enters waterways, notify downstream users of potentially contaminated waters. Notify local health and fire officials and pollution control agencies. From a secure, explosion-proof location, use water spray to cool exposed containers. If cooling streams are ineffective (venting sound increases in volume and pitch, tank discolors, or shows any signs of deforming), withdraw immediately to a secure position. If employees are expected to fight fires, they must be trained and equipped in OSHA 1910.156. The only respirators recommended for firefighting are self-contained breathing apparatuses that have full face-pieces and are operated in a pressure-demand or other positive-pressure mode.

References

US Environmental Protection Agency, Special Review and Reregistration Division Office of Pesticide Programs. (1998). *Agency Status of Pesticides in Registration, Reregistration, and Special Review* (Rainbow Report). Washington, DC

New Jersey Department of Health and Senior Services. (April 2000). *Hazardous Substances Fact Sheet: Sodium Dichromate*. Trenton, NJ

New Jersey Department of Health and Senior Services. (October 2001). *Hazardous Substances Fact Sheet: Sodium Chromate*. Trenton, NJ

Sodium cyanide   S:0450

Molecular Formula: NaCN
Synonyms: Cianuro sodico (Spanish); Cyanide of sodium; Cyanobrik; Cyanogran; Cyanure de sodium (French); Cymag; Hydrocyanic acid, sodium salt; Prussiate of soda; Sodium cyanide, solid; Sodium cyanide, solution
CAS Registry Number: 143-33-9; 10034-82-9 (tetrahydrate)
RTECS® Number: VZ7530000
UN/NA & ERG Number: UN1689/157
EC Number: 205-599-4
Regulatory Authority and Advisory Bodies
Department of Homeland Security Screening Threshold Quantity (pounds): Sabotage/Contamination Hazard: A placarded amount (commercial grade).
Air Pollutant Standard Set. See below, "Permissible Exposure Limits in Air" section.
US EPA, FIFRA 1998 Status of Pesticides: RED completed.
Clean Water Act: Section 311 Hazardous Substances/RQ 40CFR117.3 (same as CERCLA, see below); Section 313 Water Priority Chemicals (57FR41331, 9/9/92).
US EPA Hazardous Waste Number (RCRA No.): P106.
RCRA, 40CFR261, Appendix 8 Hazardous Constituents.
Superfund/EPCRA 40CFR355, Extremely Hazardous Substances: TPQ = 100 lb (45.4 kg).

Reportable Quantity (RQ): 10 lb (4.54 kg).
EPCRA Section 313: See Cyanide Compounds.
US DOT Regulated Marine Pollutant (49CFR172.101, Appendix B).
Canada, WHMIS, Ingredients Disclosure List Concentration 0.1%; National Pollutant Release Inventory (NPRI); CEPA Priority Substance List, Ocean dumping prohibited.
As cyanide compounds
Clean Air Act: Hazardous Air Pollutants (Title I, Part A, Section 112).
Clean Water Act: 40CFR423, Appendix A, Priority Pollutants, as cyanide, total.
US EPA Hazardous Waste Number (RCRA No.): P030 as cyanides soluble salts and complexes, n.o.s.
RCRA, 40CFR261, Appendix 8 Hazardous Constituents, as cyanides, soluble salts and complexes, n.o.s.
European/International Regulations: Hazard Symbol: T+; Risk phrases: R26/27/28; R32; Safety phrases: S1/2; S7; S28; S29; S45 (see Appendix 4).
WGK (German Aquatic Hazard Class): 3—Severe hazard to waters.
Description: Sodium cyanide is found as white granules, flakes, or lumps. Sodium cyanide is shipped as pellets or briquettes. Odorless when dry. It absorbs water from air (is hygroscopic or deliquescent). Hydrogen cyanide gas released by sodium cyanide has a distinctive mild, bitter almond odor, but a large proportion of people cannot detect it; the odor does not provide adequate warning of hazardous concentrations. Molecular weight = 49.01; Boiling point = 1496°C; Freezing/Melting point = 564°C. Hazard Identification (based on NFPA-704 M Rating System): Health 3, Flammability 0, Reactivity 0. Soluble in water (reactive).
Potential Exposure: Compound Description: Agricultural Chemical; Tumorigen, Mutagen; Reproductive Effector; Human Data. Sodium cyanide is used as a solid or in solution to extract metal ores, in electroplating and metal cleaning baths; in metal hardening; in treatment of rabbit and rat burrows and holes and termite nests; in insecticides.
Incompatibilities: Strong oxidizers, such as acids, acid salts, chlorates, and nitrates. Sodium cyanide decomposes on contact with acids, acid salts, water, moisture, and carbon dioxide, releasing highly toxic and flammable hydrogen cyanide gas. Aqueous solution is a strong base; it reacts violently with acid and is corrosive. Reacts violently with acid, strong oxidizers, such as nitrates and chlorates. Decomposes in the presence of air, moisture, or carbon dioxide producing highly toxic and flammable hydrogen cyanide gas. Absorbs moisture from the air forming a corrosive syrup. Corrosive to active metals, such as aluminum, copper, and zinc.
Permissible Exposure Limits in Air
OSHA PEL: 5 mg[CN]/m^3 TWA.
NIOSH REL: 4.7 ppm/5 mg[CN]/m^3 [10 min] Ceiling Concentration.
ACGIH TLV®[1]: 5 mg[CN]/m^3 [skin] Ceiling Concentration.

NIOSH IDLH: 25 mg[CN]/m^3.
Protective Action Criteria (PAC)*
TEEL-0: **4** mg/m^3
PAC-1: **4.0** mg/m^3
PAC-2: **14** mg/m^3
PAC-3: **30** mg/m^3
*AEGLs (Acute Emergency Guideline Levels) & ERPGs (Emergency Response Planning Guideline) are in **bold face**.
DFG MAK: 3.8 mg[CN]/m^3, inhalable fraction TWA; Peak Limitation Category II(1) [skin]; Pregnancy Risk Group: C. Skin contact may contribute significantly in overall exposure.
Australia: TWA 5 mg/m^3, [skin], 1993; Austria: MAK 5 mg [CN]/m^3, [skin], 1999; Denmark: TWA 5 mg/m^3, [skin], 1999; France: VME 5 mg[CN]/m^3, [skin], 1999; Poland: TWA 0.3 mg[CN]/m^3, ceiling 10 mg[CN]/m^3, 1999; Switzerland: MAK-W 5 mg/m^3, KZG-W 10 mg/m^3, [skin], 1999; United Kingdom: TWA 5 mg[CN]/m^3, [skin], 2000; Argentina, Bulgaria, Columbia, Jordan, South Korea, New Zealand, Singapore, Vietnam: ACGIH TLV®: Ceiling Concentration 5 mg/m^3 [skin]. Russia[43] has set a MAC value of 0.009 mg/m^3 for ambient air in residential areas on a momentary basis and 0.004 mg/m^3 on an average daily basis. Several states have set guidelines or standards for cyanides in ambient air[60] ranging from 16.7 µg/m^3 (New York) to 50.0 µg/m^3 (Florida and North Dakota) to 80.0 µg/m^3 (Virginia) to 100 µg/m^3 (Connecticut and South Dakota) to 125 µg/m^3 (South Carolina) to 119.0 µg/m^3 (Nevada).
Determination in Air: Use NIOSH Analytical Method (IV) #7904, Cyanides. See also Method #6010, Hydrogen cyanide.[18]

Permissible Concentration in Water: In 1976 the EPA criterion was 5.0 µg/L for freshwater and marine aquatic life and wildlife. As of 1980, the criteria are: *To protect freshwater aquatic life:* 3.5 µg/L as a 24-h average, never to exceed 52.0 µg/L. *To protect saltwater aquatic life:* 30.0 µg/L on an acute toxicity basis; 2.0 µg/L on a chronic toxicity basis. *To protect human health:* 200 µg/L. The allowable daily intake for man is 8.4 mg/day.[6]
On the international scene, the South African Bureau of Standards has set 10 µg/L, the World Health Organization (WHO) 10 µg/L, and Germany 50 µg/L as drinking water standards.
Other international limits[35] include an EEC limit of 50 µg/L; Mexican limits of 200 µg/L in drinking water and 1.0 µg/L in coastal waters and a Swedish limit of 100 µg/L. Russia[43] set a MAC of 100 µg/L in water bodies used for domestic purposes and 50 µg/L in water for fishery purposes.
The US EPA[49] has determined a no-observed-adverse-effect-level (NOAEL) of 10.8 mg/kg/day which yields a lifetime health advisory of 154 µg/L. States which have set guidelines for cyanides in drinking water[61] include Arizona at 160 µg/L and Kansas at 220 µg/L.
Determination in Water: Distillation followed by silver nitrate titration or colorimetric analysis using pyridine

pyrazolone (or barbituric acid). Octanol–water coefficient: Log K_{ow} (estimated) = −1.69.
Routes of Entry: Inhalation, skin absorption, ingestion, skin and/or eye contact.
Harmful Effects and Symptoms
Sodium cyanide releases hydrogen cyanide gas, a highly toxic chemical asphyxiant that interferes with the body's ability to use oxygen. Exposure to sodium cyanide can be rapidly fatal. It has whole-body (systemic) effects, particularly affecting those organ systems most sensitive to low oxygen levels: the central nervous system (brain), the cardiovascular system (heart and blood vessels), and the pulmonary system (lungs).[NIOSH]
Short Term Exposure: Sodium cyanide can be absorbed through the skin, thereby increasing exposure. Sodium cyanide is corrosive to the eyes, skin, and respiratory tract. Contact can cause skin and eye burns, and possible permanent eye damage. Inhalation can cause lung irritation with coughing, sneezing, and difficult breathing; slow gasping respiration. Corrosive if swallowed. These substances may affect the central nervous system. Symptoms include headaches, confusion, nausea, pounding heart, weakness, and unconsciousness.
Long Term Exposure: Repeated or prolonged contact with sodium cyanide may cause thyroid gland enlargement and interfere with thyroid function. May cause nosebleed and sores in the nose; changes in blood cell count. May cause central nervous system damage with headache, dizziness, confusion, nausea, vomiting, pounding heart, weakness in the arms and legs, unconsciousness, and death. May affect liver and kidney function. Repeated lower exposures can cause sores in the nose with nosebleeds.
Points of Attack: Liver, kidneys, skin, cardiovascular system, central nervous system, thyroid gland.
Medical Surveillance: Consider the points of attack in preplacement and periodic physical examinations. Urine thiocyanate levels. Blood cyanide levels. Complete blood count (CBC). Evaluation of thyroid function. Liver function tests. Kidney function tests. Central nervous system tests. EKG. Smokers may have somewhat higher blood cyanide and urine thiocyanate levels.
First Aid: If this chemical gets into the eyes, remove any contact lenses at once and irrigate immediately for at least 15 min, occasionally lifting upper and lower lids. Seek medical attention immediately. If this chemical contacts the skin, remove contaminated clothing and wash immediately with soap and water. Seek medical attention immediately. If this chemical has been inhaled, remove from exposure, begin rescue breathing (using universal precautions, including resuscitation mask) if breathing has stopped and CPR if heart action has stopped. Transfer promptly to a medical facility. When this chemical has been swallowed, get medical attention. Give large quantities of water and induce vomiting. Do not make an unconscious person vomit.
Use amyl nitrate capsules if symptoms develop. All area employees should be trained regularly in emergency

measures for cyanide poisoning and in CPR. A cyanide antidote kit should be kept in the immediate work area and must be rapidly available. Kit ingredients should be replaced every 1–2 years to ensure freshness. Persons trained in the use of this kit, oxygen use, and CPR must be quickly available.

Personal Protective Methods: Wear protective gloves and clothing to prevent any reasonable probability of skin contact. Safety equipment suppliers/manufacturers can provide recommendations on the most protective glove/clothing material for your operation. *Polyethylene (for sodium cyanide, less than 30%, only) is among the recommended protective materials.* All protective clothing (suits, gloves, footwear, headgear) should be clean, available each day, and put on before work. Contact lenses should not be worn when working with this chemical. Wear splash-proof chemical goggles and face shield when working with liquid, unless full face-piece respiratory protection is worn. Wear dust-proof goggles and face shield when working with powders or dust, unless full face-piece respiratory protection is worn. Employees should wash immediately with soap when skin is wet or contaminated. Provide emergency showers and eyewash.

Respirator Selection: *When used as a weapon, use SCBA Respirator Certified By NIOSH For CBRN Environments. Up to 25 mg/m³:* Sa (APF = 10) (any supplied-air respirator) or SCBAF (APF = 50) (any self-contained breathing apparatus with full face-piece). *Emergency or planned entry into unknown concentrations or IDLH conditions:* SCBAF: Pd,Pp (APF = 10,000) (any self-contained breathing apparatus that has a full face-piece and is operated in a pressure-demand or other positive-pressure mode) or SaF: Pd,Pp: ASCBA (APF = 10,000) (any supplied-air respirator that has a full face-piece and is operated in a pressure-demand or other positive-pressure mode in combination with an auxiliary, self-contained breathing apparatus operated in a pressure-demand or other positive-pressure mode). *Escape:* GmFS100 (APF = 50) [any air-purifying, full-face-piece respirator (gas mask) with a chin-style, front- or back-mounted canister providing protection against the compound of concern and having an N100, R100, or P100 filter] or SCBAE (any appropriate escape-type, self-contained breathing apparatus).

Storage: Color Code—Blue: Health Hazard/Poison: Store in a secure poison location. Prior to working with this chemical you should be trained on its proper handling and storage. Store in tightly closed containers in a cool, well-ventilated area away from strong acids, acid salts, oxidizers, light, and moisture. Where possible, automatically transfer material from drums or other storage containers to process containers.

Shipping: Sodium cyanide requires a "POISONOUS/TOXIC MATERIALS" label. It falls in Hazard Class 6.1 and in Packing Group I.

Spill Handling: Evacuate persons not wearing protective equipment from area of spill or leak until cleanup is complete. Keep water away from release. Remove all ignition sources. Collect powdered material in the most convenient and safe manner and deposit in sealed containers. Ventilate area after cleanup is complete. It may be necessary to contain and dispose of this chemical as a hazardous waste. If material or contaminated runoff enters waterways, notify downstream users of potentially contaminated waters. Contact your local or federal environmental protection agency for specific recommendations. If employees are required to clean up spills, they must be properly trained and equipped. OSHA 1910.120(q) may be applicable.

Initial isolation and protective action distances
Distances shown are likely to be affected during the first 30 min after materials are spilled and could increase with time. If more than one tank car, cargo tank, portable tank, or large cylinder involved in the incident is leaking, the protective action distance may need to be increased. You may need to seek emergency information from CHEMTREC at (800) 424-9300 or seek professional environmental engineering assistance from the US EPA Environmental Response Team at (908) 548-8730 (24-h response line).
UN 1689 (Sodium cyanide) is on the DOT's list of dangerous water-reactive materials which create large amounts of toxic vapor when *spilled in water:* Dangerous from 0.5 to 10 km (0.3–6.0 miles) downwind.

Fire Extinguishing: Material does not burn; fight surrounding fire with an agent appropriate for the burning material. NaCN decomposes in the presence of moisture, damp air, or carbon dioxide, producing highly toxic and flammable hydrogen cyanide gas and oxides of nitrogen. *NO* acidic dry chemical extinguishers; *NO* hydrous agents; *NO* water; *NO* carbon dioxide directly on material. Vapors are heavier than air and may collect in low areas. If material or contaminated runoff enters waterways, notify downstream users of potentially contaminated waters. Notify local health and fire officials and pollution control agencies. From a secure, explosion-proof location, use water spray to cool exposed containers. Do not allow water to enter open containers. If cooling streams are ineffective (venting sound increases in volume and pitch, tank discolors, or shows any signs of deforming), withdraw immediately to a secure position. If employees are expected to fight fires, they must be trained and equipped in OSHA 1910.156. The only respirators recommended for firefighting are self-contained breathing apparatuses that have full face-pieces and are operated in a pressure-demand or other positive-pressure mode.

Disposal Method Suggested: Consult with environmental regulatory agencies for guidance on acceptable disposal practices. Generators of waste containing this contaminant (≥100 kg/mo) must conform with EPA regulations governing storage, transportation, treatment, and waste disposal. In accordance with 40CFR165, follow recommendations for the disposal of pesticides and pesticide containers. Must be disposed properly by following package label directions or by contacting your local or federal environmental control agency or by contacting your regional EPA office. Add strong alkaline hypochlorite and allow to react for 24 h. Then flush to sewer with large volumes of water.[22]

References
US Environmental Protection Agency, Special Review and Reregistration Division Office of Pesticide Programs. (1998). *Agency Status of Pesticides in Registration, Reregistration, and Special Review* (Rainbow Report). Washington, DC
New Jersey Department of Health and Senior Services. (August 2006). *Hazardous Substances Fact Sheet: Sodium Cyanide*. Trenton, NJ

Sodium dichloroisocyanurate S:0460

Molecular Formula: $C_3HCl_2N_3O_3 \cdot Na$
Synonyms: ACL 60; CDB 63; Dichloroisocyanuric acid sodium salt; Dikonit; Dimanin C; FI Clor 60S; OCI 56; SDIC; Simpla; Sodium dichlorisocyanurate; Sodium dichlorocyanurate; Sodium dichloroisocyanurate; Sodium-1,3-dichloro-1,3,5-triazine-2,4-dione-6-oxide; 1-Sodium-3,5-dichloro-*s*-triazine-2,4,6-trione; 1-Sodium-3,5-dichloro-1,3,5-triazine-2,4,6-trione; Sodium dichloro-*s*-triazinetrione, dry, containing more than 39% available chlorine; Sodium salt of dichloro-*s*-triazinetrione
CAS Registry Number: 2893-78-9
RTECS® Number: XZ1900000
UN/NA & ERG Number: UN2465/140
EC Number: 220-767-7 [*Annex I Index No.:* 613-030-00-X]
Regulatory Authority and Advisory Bodies
Canada, WHMIS, Ingredients Disclosure List Concentration 0.1%.
US EPA, FIFRA 1998 Status of Pesticides: RED completed.
European/International Regulations: Hazard Symbol: Xn, N; Risk phrases: R22; R31; R36/37; R50/53; Safety phrases: S2; S8; S26; S41; S60; S61 (see Appendix 4).
WGK (German Aquatic Hazard Class): 2—Hazard to waters.
Description: Dichloroisocyanuric acid, sodium salt, is a white crystalline powder having a chlorine odor. Thermally unstable. Molecular weight = 220.96; Specific gravity (H_2O:1) = 1.10 at 25°C; Freezing/Melting point = (decomposes) 230°C. NFPA 704 M Hazard Identification (anhydrous): Health 2, Flammability 0, Reactivity 2 (Oxidizer). Soluble in water; solubility = 25%.
Potential Exposure: Compound Description: Agricultural Chemical; Reproductive Effector; Human Data; Primary Irritant. It is used in cleaning; making dry bleaches, detergents, sanitizers, and disinfectants; in swimming pool and sewage treatment.
Incompatibilities: Contact with water, acids, or acid fumes releases toxic chlorine gas. A powerful oxidizer. Violent reaction with reducing agents, organic matter, easily chlorinated or oxidized materials. Incompatible with ammonium salts; amines forming nitrogen trichloride.

Permissible Exposure Limits in Air
No standards or TEEL available.
Routes of Entry: Inhalation, ingestion, skin and/or eye contact.
Harmful Effects and Symptoms
Short Term Exposure: Contact can cause severe eye and permanent damage. Skin contact can cause severe irritation especially if skin is moist or material is in solution. Inhalation can cause irritation of the respiratory tract. The dry material is less irritating to the skin. A strong poison and corrosive if ingested; may cause liver damage.
Long Term Exposure: Highly irritating substances may affect the lungs.
Points of Attack: Lungs, eyes, skin, liver.
Medical Surveillance: Lung function tests. Liver function tests.
First Aid: If this chemical gets into the eyes, remove any contact lenses at once and irrigate immediately for at least 15 min, occasionally lifting upper and lower lids. Seek medical attention immediately. If this chemical contacts the skin, remove contaminated clothing and wash immediately with soap and water. Seek medical attention immediately. If this chemical has been inhaled, remove from exposure, begin rescue breathing (using universal precautions, including resuscitation mask) if breathing has stopped and CPR if heart action has stopped. Transfer promptly to a medical facility. When this chemical has been swallowed, get medical attention. If victim is *conscious*, administer water or milk. Do not induce vomiting.
Personal Protective Methods: Wear protective gloves and clothing to prevent any reasonable probability of skin contact. Safety equipment suppliers/manufacturers can provide recommendations on the most protective glove/clothing material for your operation. All protective clothing (suits, gloves, footwear, headgear) should be clean, available each day, and put on before work. Contact lenses should not be worn when working with this chemical. Wear splash-proof chemical goggles and face shield when working with liquid, unless full face-piece respiratory protection is worn. Wear dust-proof goggles and face shield when working with powders or dust, unless full face-piece respiratory protection is worn. Employees should wash immediately with soap when skin is wet or contaminated. Provide emergency showers and eyewash.
Respirator Selection: Follow regulations in OSHA 29CFR1910.134 or European Standard EN149. Use a NIOSH/MSHA- or European Standard EN149-approved respirator; or use an approved supplied-air respirator with a full face-piece operated in the positive-pressure mode, or with a full face-piece, hood, or helmet in the continuous-flow mode; or use a NIOSH/MSHA- or European Standard EN149-approved self-contained breathing apparatus with a full face-piece operated in pressure-demand or other positive-pressure mode.
Storage: Color Code—Yellow: Reactive Hazard; Store in a location separate from other materials, especially flammables and combustibles. Prior to working with this chemical you

should be trained on its proper handling and storage. Store in tightly closed containers in a cool, well-ventilated area away from ammonium compounds, hydrated salts, moisture, and combustible materials. Where possible, automatically transfer material from drums or other storage containers to process containers. Wherever this chemical is used, handled, manufactured, or stored, use explosion-proof electrical equipment and fittings. See OSHA Standard 1910.104 and NFPA 43A *Code for the Storage of Liquid and Solid Oxidizers* for detailed handling and storage regulations.

Shipping: Dichloroisocyanuric acid salts require a shipping label of "OXIDIZER." They fall in Hazard Class 5.1 and Packing Group II.

Spill Handling: Evacuate persons not wearing protective equipment from area of spill or leak until cleanup is complete. Keep water away from release. Remove all ignition sources. Collect powdered material in the most convenient and safe manner and deposit in sealed containers. Ventilate area after cleanup is complete. It may be necessary to contain and dispose of this chemical as a hazardous waste. If material or contaminated runoff enters waterways, notify downstream users of potentially contaminated waters. Contact your local or federal environmental protection agency for specific recommendations. If employees are required to clean up spills, they must be properly trained and equipped. OSHA 1910.120(q) may be applicable.

Fire Extinguishing: This chemical is a noncombustible solid but it is a powerful oxidizer and may increase the activity of an existing fire. May burn rapidly. Use flooding quantities of water, from a distance. Do *NOT* use extinguishers containing ammonia. Poisonous gases are produced in fire, including sodium oxide, chlorine and nitrogen oxides. If material or contaminated runoff enters waterways, notify downstream users of potentially contaminated waters. Notify local health and fire officials and pollution control agencies. Containers may explode in fire. From a secure, explosion-proof location, use water spray to cool exposed containers. If cooling streams are ineffective (venting sound increases in volume and pitch, tank discolors, or shows any signs of deforming), withdraw immediately to a secure position. If employees are expected to fight fires, they must be trained and equipped in OSHA 1910.156. The only respirators recommended for firefighting are self-contained breathing apparatuses that have full face-pieces and are operated in a pressure-demand or other positive-pressure mode.

References
New Jersey Department of Health and Senior Services. (May 2002). *Hazardous Substances Fact Sheet: Sodium Dichloro-Isocyanate*. Trenton, NJ

Sodium fluoride   S:0470

Molecular Formula: NaF
Synonyms: Alcoa sodium fluoride; Antibulit; Cav-trol; Checkmate; Chemiflour; Credo; Disodium difluoride; F1-Tabs; FDA 0101; Floridine; Florocid; Flour-O-kote; Flozenges; Fluoral; Fluorident; Fluorigard; Fluorineed; Fluorinse; Fluoritab; Fluorure de sodium (French); Fluoruro sodico (Spanish); Flura-gel; Flurcare; Fungol B; Gel II; Gelution; Iradicav; Karidium; Karigel; Kari-rinse; Lea-cov; Lemoflur; Luride; Nafeen; Natrium fluoride; NCI: C55221; Nuflour; Ossalin; Ossin; Pediaflor; Pedident; Pennwhite; Pergantene; Phos-flur; Point two; Pro-portion; Rafluor; Rescue squad; Roach salt; Sodium hydrofluoride; Sodium monofluoride; So-Flo; Stay-Flo; Studaflour; Super-Dent; T-Fluoride; the ra-flur-N; Trisodium trifluoride; Villiaumite; Zendium

CAS Registry Number: 7681-49-4; *(alt.)* 39287-69-9
RTECS® Number: WB0350000
UN/NA & ERG Number: UN1690/154
EC Number: 231-667-8 [*Annex I Index No.:* 009-004-00-7]

Regulatory Authority and Advisory Bodies
Carcinogenicity: NCI: Carcinogenesis Studies (water); equivocal evidence: rat; no evidence: mouse; IARC: Animal Inadequate Evidence; Human Inadequate Evidence, *not classifiable as carcinogenic to humans*, Group 3, 1987.
US EPA Gene-Tox Program, Positive: *D. melanogaster*—whole sex chrom. loss; Negative: *D. melanogaster*—nondisjunction; *N. crassa*—aneuploidy; Negative: *In vivo* SCE—nonhuman; *S. cerevisiae* gene conversion; Inconclusive: *D. melanogaster*—partial sex chrom. loss; Inconclusive: Histidine reversion—Ames test; Inconclusive: *D. melanogaster* sex-linked lethal.
US EPA, FIFRA, 1998 Status of Pesticides: Supported.
FDA—over-the-counter drug.
Clean Water Act: Section 311 Hazardous Substances/RQ 40CFR117.3 (same as CERCLA, see below).
Reportable Quantity (RQ): 1000 lb (454 kg).
European/International Regulations: Hazard Symbol: T; Risk phrases: R25; R32; R36/38; Safety phrases: S1/2; S22; S36; S45 (see Appendix 4).
WGK (German Aquatic Hazard Class): 1—Low hazard to waters.

Description: Sodium fluoride is a white powder or colorless crystals. Often used in a solution. Odorless. Pesticide grade is often dyed blue. Molecular weight = 41.99; Specific gravity (H_2O:1) = 2.78 at 25°C; Boiling point = 1703.9°C; Freezing/Melting point = 992.8°C. Hazard Identification (based on NFPA-704 M Rating System): Health 3, Flammability 0, Reactivity 0. Slightly soluble in water; solubility = 4%.

Potential Exposure: Compound Description: Agricultural Chemical; Tumorigen, Drug, Mutagen; Reproductive Effector; Human Data; Primary Irritant. Widely used in the chemical industry; in water treatment and fluoridation of drinking water; as an insecticide, fungicide, and rodenticide; chemical cleaning; electroplating, glass manufacture; vitreous enamels; preservative for adhesives; toothpastes, disinfectant, dental prophylaxis; also used orally in the treatment of various bone diseases to increase bone density and to relieve bone pain.

Incompatibilities: Strong oxidizers, acids.

Permissible Exposure Limits in Air

OSHA PEL: 3 ppm/2.5 mg[F]/m^3 TWA.

NIOSH REL: 3 ppm/2.5 mg[F]/m^3 TWA; 6 ppm/5 mg[F]/m^3, 15 min Ceiling Concentration.

ACGIH TLV®[1]: 2.5 mg[F]/m^3 TWA; not classifiable as a human carcinogen; BEI: 3 mg[F]/g creatinine in urine *prior* to end-of-shift; 10 mg[F]/g creatinine in urine end-of-shift.

Protective Action Criteria (PAC)

TEEL-0: 5.53 mg/m^3

PAC-1: 5.53 mg/m^3

PAC-2: 5.53 mg/m^3

PAC-3: 500 mg/m^3

DFG MAK: 1 mg[F]/m^3, inhalable fraction [skin]; Peak Limitation Category II(4); Pregnancy Risk Group C; BAT: 7.0 mg[F]/g creatinine in urine at end-of-shift; 4.0 mg[F]/g creatinine in urine at the beginning of the next shift.

NIOSH IDLH: 250 mg[F]/m^3.

Australia: TWA 2.5 mg[F]/m^3, 1993; Austria: MAK 2.5 mg[F]/m^3, 1999; Belgium: TWA 2.5 mg[F]/m^3, 1993; Finland: TWA 2.5 mg[F]/m^3, 1999; France: VME 2.5 mg[F]/m^3, 1999; Hungary: TWA 1 mg[F]/m^3; STEL 2 mg[F]/m^3, 1993; Norway: TWA 0.6 mg[F]/m^3, 1999; the Philippines: TWA 2.5 mg[F]/m^3, 1993; Poland: MAC (TWA) 1 mg[HF]/m^3, MAC (STEL) 3 mg[HF]/m^3, 1999; Russia: STEL 0.5 ppm (2.5 mg/m^3), 1993; Sweden: NGV 2 mg[F]/m^3, 1999; Switzerland: MAK-W 1.8 ppm (1.5 mg[F]/m^3), KZG-W 3.6 ppm (3.0 mg[F]/m^3), 1999; Thailand: TWA 2.5 mg[F]/m^3, 1993; United Kingdom: TWA 2.5 mg[F]/m^3, 2000; LTEL 2.5 mg[F]/m^3, 1993; Argentina, Bulgaria, Columbia, Jordan, South Korea, New Zealand, Singapore, Vietnam: ACGIH TLV®: not classifiable as a human carcinogen. Several states have set limits for fluoride in ambient air[60] ranging from as low as 2.85 μg/m^3 (Iowa) to as high as 60,000 μg/m^3 (Kentucky).

Determination in Air: Use NIOSH Analytical Method (IV) #7902, Fluorides; #7906; OSHA Analytical Method ID-110.

Permissible Concentration in Water: As with air, the applicable regulations are those for the fluoride ion. The values which have been set for drinking water[61] are a standard of 4.0 mg/L set by EPA and a guideline of 2.4 mg/L set by the state of Maine.

Routes of Entry: Inhalation, ingestion, skin and/or eye contact. Liquid can be absorbed through the skin.

Harmful Effects and Symptoms

Short Term Exposure: Sodium fluoride can affect you when breathed in. Inhalation of dust or mist can cause severe irritation and burns of the eyes and skin. Irritates the eyes and respiratory system. May cause permanent eye damage. Exposure can cause nausea, abdominal pain, diarrhea, salivation, thirst, sweating.

Long Term Exposure: Repeated or prolonged industrial contact can cause dermatitis. Repeated exposure can cause fluoride to build up in the body. Can irritate the lungs; bronchitis may develop. Repeated exposure can cause fluoride to build up in the body causing stiffness, brittle bones; stiff

spine; calcification of ligaments of ribs, pelvis; and crippling. Repeated exposures can cause weakness and muscle twitching; tremors, convulsions, coma, and even death. May cause kidney damage. Prolonged contact can cause sores in the nose and perforated septum. High concentrations can damage the developing fetus. These effects *DO NOT* occur when sodium fluoride is used in drinking water for dental cavity prevention.

Points of Attack: Eyes, skin, respiratory system, central nervous system, skeleton, kidneys.

Medical Surveillance: NIOSH lists the following tests: chest X-ray, electrocardiogram, pulmonary function tests: forced vital capacity, forced expiratory volume (1 s); pelvic X-ray; sputum cytology; urine (chemical/metabolite); urine (chemical/metabolite) pre- and postshift; urinalysis (routine); complete blood count/differential.

First Aid: If this chemical gets into the eyes, remove any contact lenses at once and irrigate immediately for at least 15 min, occasionally lifting upper and lower lids. Seek medical attention immediately. If this chemical contacts the skin, remove contaminated clothing and wash immediately with soap and water. Seek medical attention immediately. If this chemical has been inhaled, remove from exposure, begin rescue breathing (using universal precautions, including resuscitation mask) if breathing has stopped and CPR if heart action has stopped. Transfer promptly to a medical facility. When this chemical has been swallowed, get medical attention. Give large quantities of water and induce vomiting. Do not make an unconscious person vomit.

Personal Protective Methods: Wear protective gloves and clothing to prevent any reasonable probability of skin contact. Safety equipment suppliers/manufacturers can provide recommendations on the most protective glove/clothing material for your operation. All protective clothing (suits, gloves, footwear, headgear) should be clean, available each day, and put on before work. Contact lenses should not be worn when working with this chemical. Wear dust-proof chemical goggles and face shield unless full face-piece respiratory protection is worn. Employees should wash immediately with soap when skin is wet or contaminated. Provide emergency showers and eyewash.

Respirator Selection: NIOSH/OSHA *12.5 mg/m^3:* Qm (APF = 25) (any quarter-mask respirator). *25 mg/m^3:* 95XQ (APF = 10)* [any particulate respirator equipped with an N95, R95, or P95 filter (including N95, R95, and P95 filtering face-pieces) except quarter-mask respirators. The following filters may also be used: N99, R99, P99, N100, R100, P100] or SA* (any supplied-air respirator). *62.5 mg/m^3:* Sa:Cf (APF = 25)*† (any supplied-air respirator operated in a continuous-flow mode) or PaprHie (APF = 25)* *if not present as a fume* (any powered, air-purifying respirator with a high-efficiency particulate filter). *125 mg/m^3:* 100F (APF = 50)† [any particulate respirator equipped with an N95, R95, or P95 filter (including N95, R95, and P95 filtering face-pieces) except quarter-mask respirators. The following filters may also be used: N99,

R99, P99, N100, R100, P100] or SCBAF (APF = 50) (any self-contained breathing apparatus with a full face-piece) or SaF (APF = 50) (any supplied-air respirator with a full face-piece). *250 mg/m³:* Sa: Pd,Pp (APF = 1000) (any supplied-air respirator operated in a pressure-demand or other positive-pressure mode). *Emergency or planned entry into unknown concentrations or IDLH conditions:* SCBAF: Pd, Pp (APF = 10,000) (any self-contained breathing apparatus that has a full faceplate and is operated in a pressure-demand or other positive-pressure mode) or SaF: Pd,Pp: ASCBA (APF = 10,000) (any supplied-air respirator that has a full face-piece and is operated in a pressure-demand or other positive-pressure mode in combination with an auxiliary, self-contained breathing apparatus operated in a pressure-demand or other positive-pressure mode). *Escape:* 100F (APF = 50)† [any particulate respirator equipped with an N95, R95, or P95 filter (including N95, R95, and P95 filtering face-pieces) except quarter-mask respirators. The following filters may also be used: N99, R99, P99, N100, R100, P100] or SCBAE (any appropriate escape-type, self-contained breathing apparatus).

*Substance reported to cause eye irritation or damage; may require eye protection.

†May need acid gas sorbent.

Storage: Color Code—Blue: Health Hazard/Poison: Store in a secure poison location. Prior to working with this chemical you should be trained on its proper handling and storage.

Shipping: Sodium fluoride requires a shipping label of "POISONOUS/TOXIC MATERIALS." It falls in Hazard Class 6.1 and Packing Group III.

Spill Handling: Evacuate persons not wearing protective equipment from area of spill or leak until cleanup is complete. Remove all ignition sources. Collect powdered material in the most convenient and safe manner and deposit in sealed containers. Ventilate area after cleanup is complete. Absorb liquids in vermiculite, dry sand, earth, peat, carbon, or a similar material and deposit in sealed containers. It may be necessary to contain and dispose of this chemical as a hazardous waste. If material or contaminated runoff enters waterways, notify downstream users of potentially contaminated waters. Contact your local or federal environmental protection agency for specific recommendations. If employees are required to clean up spills, they must be properly trained and equipped. OSHA 1910.120(q) may be applicable.

Fire Extinguishing: This chemical is a combustible solid. Use dry chemical, carbon dioxide, water spray, or alcohol foam extinguishers. Poisonous gases are produced in fire, including fluorine and oxides of sodium. If material or contaminated runoff enters waterways, notify downstream users of potentially contaminated waters. Notify local health and fire officials and pollution control agencies. From a secure, explosion-proof location, use water spray to cool exposed containers. If cooling streams are ineffective (venting sound increases in volume and pitch, tank discolors, or shows any signs of deforming), withdraw immediately to a secure position. If employees are expected to fight fires, they must be trained and equipped in OSHA 1910.156. The only respirators recommended for firefighting are self-contained breathing apparatuses that have full face-pieces and are operated in a pressure-demand or other positive-pressure mode.

Disposal Method Suggested: In accordance with 40CFR165, follow recommendations for the disposal of pesticides and pesticide containers. Must be disposed properly by following package label directions or by contacting your local or federal environmental control agency or by contacting your regional EPA office.

References

US Environmental Protection Agency, Special Review and Reregistration Division Office of Pesticide Programs. (1998). *Agency Status of Pesticides in Registration, Reregistration, and Special Review* (Rainbow Report). Washington, DC

New Jersey Department of Health and Senior Services. (November 2004). *Hazardous Substances Fact Sheet: Sodium Fluoride.* Trenton, NJ

Sodium fluoroacetate   S:0480

Molecular Formula: $C_2H_2FNaO_2$
Common Formula: FCH_2COONa
Synonyms: 1080; Acetic acid, fluoro-, sodium salt; AI3-08434; Compound 1080; Fluoacetato sodico (Spanish); Fluorakil 3; Fluoressigsaeure (German); Fluoroacetic acid, sodium salt; Fratol; Furatol; Monofluoressigsaure, natrium (German); Natriumfluoracetat (German); NSC 77690; Ratbane 1080; SMFA; Sodium fluoacetate; Sodium fluoacetic acid; Sodium fluoracetate; Sodium fluoracetate de (French); Sodium monofluoroacetate; Ten-Eighty; TL 869; Yasoknock

CAS Registry Number: 62-74-8
RTECS® Number: AH9100000
UN/NA & ERG Number: UN2629/151
EC Number: 200-548-2 [*Annex I Index No.:* 607-169-00-5]

Regulatory Authority and Advisory Bodies

Banned or Severely Restricted (several countries) (UN).[13]
Air Pollutant Standard Set. See below, "Permissible Exposure Limits in Air" section.
US EPA PESTICIDE CHEMICAL CODE 075003.
US EPA Hazardous Waste Number (RCRA No.): P058.
RCRA, 40CFR261, Appendix 8 Hazardous Constituents.
Superfund/EPCRA 40CFR355, Extremely Hazardous Substances: TPQ = 10/10,000 lb (4.54/4540 kg).
Reportable Quantity (RQ): 10 lb (4.54 kg).
EPCRA Section 313 Form R *de minimis* concentration reporting level: 1.0%.
California Proposition 65 Chemical: Cancer; Developmental/Reproductive toxin 11/6/98.
European/International Regulations: Hazard Symbol: T+, N; Risk phrases: R26/27/28; R50; Safety phrases: S1/2; S13; S22; S36/37; S45; S61 (see Appendix 4).

WGK (German Aquatic Hazard Class): 3—Severe hazard to waters.

Description: Sodium fluoroacetate is a fluffy, colorless, odorless, hygroscopic solid (sometimes dyed black). Molecular weight = 100.03; Boiling point = (decomposes); Freezing/Melting point = 200°C (decomposes below MP). Hazard Identification (based on NFPA-704 M Rating System): Health 4, Flammability 0, Reactivity 0. Soluble in water.

Potential Exposure: Compound Description: Agricultural Chemical; Mutagen; Reproductive Effector; Human Data. Those involved in the manufacture, formulation, and application of this highly toxic, immediate-action rodenticide.

Incompatibilities: Alkaline metals and carbon disulfide.[24]

Permissible Exposure Limits in Air
OSHA PEL: 0.05 mg/m^3 TWA [skin].
NIOSH REL: 0.05 mg/m^3 TWA; 0.15 mg/m^3 STEL [skin].
ACGIH TLV®[1]: 0.05 mg/m^3 TWA [skin].
NIOSH IDLH: 2.5 mg/m^3.
Protective Action Criteria (PAC)
TEEL-0: 0.05 mg/m^3
PAC-1: 0.15 mg/m^3
PAC-2: 0.5 mg/m^3
PAC-3: 2.5 mg/m^3
DFG MAK: 0.05 mg/m^3, inhalable fraction TWA; Peak Limitation Category II(4) [skin]; Pregnancy Risk Group C.
Arab Republic of Egypt: TWA 0.05 mg/m^3, 1993; Australia: TWA 0.05 mg/m^3; STEL 0.15 mg/m^3, [skin], 1993; Austria: MAK 0.05 mg/m^3, [skin], 1999; Belgium: TWA 0.05 mg/m^3; STEL 0.15 mg/m^3, [skin], 1993; Denmark: TWA 0.05 mg/m^3, [skin], 1999; Finland: TWA 0.05 mg/m^3, STEL 0.15 mg/m^3, [skin], 1993; France: VME 0.05 mg/m^3, [skin], 1999; the Netherlands: MAC-TGG 0.05 mg/m^3, [skin], 2003; Norway: TWA 0.05 mg/m^3, 1999; the Philippines: TWA 0.05 mg/m^3, [skin], 1993; Switzerland: MAK-W 0.05 mg/m^3, KZG-W 1 mg/m^3, [skin], 1999; Turkey: TWA 0.05 mg/m^3, [skin], 1993; United Kingdom: TWA 0.05 mg/m^3; STEL 0.15 mg/m^3, [skin], 2000; Argentina, Bulgaria, Columbia, Jordan, South Korea, New Zealand, Singapore, Vietnam: ACGIH TLV®: TWA 0.05 mg/m^3 [skin].

Determination in Air: Use NIOSH II(5) Method #S-301.

Routes of Entry: Inhalation, skin absorption, ingestion, eye and/or skin contact.

Harmful Effects and Symptoms
Short Term Exposure: May affect the cardiovascular system and central nervous system, causing cardiac disorders and respiratory failure. Exposure may result in death. This material is super toxic. Higher exposures can cause pulmonary edema, a medical emergency that can be delayed for several hours. This can cause death. The probable oral lethal dose in humans is less than 5 mg/kg, or a taste (less than 7 drops) for a 150-lb person. Symptoms include nausea, vomiting, apprehension, auditory hallucinations, facial paresthesia, twitching face muscle, pulsus altenans, ectopic heartbeat, ventricular fibrillation. Symptoms are usually

seen within one-half hour of exposure, but severe effects may be delayed as long as 20 h. A rebuttable presumption against registration of sodium fluoroacetate for pesticidal uses was issued on December 1, 1976 by the US EPA on the basis of reductions in nontarget and endangered species and because there is no human antidote.

Long Term Exposure: May cause liver and kidney damage. Affects the central nervous system, causing epileptiform convulsive seizures that may be followed by severe depression.

Points of Attack: Cardiovascular system, lungs, kidneys, liver, central nervous system.

Medical Surveillance: Consider the points of attack in preplacement and periodic physical examinations. Liver and kidney function tests. Lung function tests. Consider chest X-ray following acute overexposure. Examination of the nervous system.

First Aid: If this chemical gets into the eyes, remove any contact lenses at once and irrigate immediately for at least 15 min, occasionally lifting upper and lower lids. Seek medical attention immediately. If this chemical contacts the skin, remove contaminated clothing and wash immediately with soap and water. Speed in removing material from skin is of extreme importance. Seek medical attention immediately. If this chemical has been inhaled, remove from exposure, begin rescue breathing (using universal precautions, including resuscitation mask) if breathing has stopped and CPR if heart action has stopped. Transfer promptly to a medical facility. When this chemical has been swallowed, get medical attention. Give large quantities of water and induce vomiting. Do not make an unconscious person vomit.

Personal Protective Methods: Wear protective gloves and clothing to prevent any reasonable probability of skin contact. Safety equipment suppliers/manufacturers can provide recommendations on the most protective glove/clothing material for your operation. All protective clothing (suits, gloves, footwear, headgear) should be clean, available each day, and put on before work. Contact lenses should not be worn when working with this chemical. Wear dust-proof chemical goggles and face shield unless full face-piece respiratory protection is worn. Employees should wash immediately with soap when skin is wet or contaminated. Provide emergency showers and eyewash.

Respirator Selection: *When used as a weapon, use SCBA Respirator Certified By NIOSH For CBRN Environments. Up to 0.25 mg/m^3:* Qm (APF = 25) (any quarter-mask respirator). *Up to 0.5 mg/m^3:* 95QX [any particulate respirator equipped with an N95, R95, or P95 filter (including N95, R95, and P95 filtering face-pieces) except quarter-mask respirators. The following filters may also be used: N99, R99, P99, N100, R100, P100] or Sa (APF = 10) (any supplied-air respirator). *Up to 1.25 mg/m^3:* Sa:Cf (APF = 25) (any supplied-air respirator operated in a continuous-flow mode) or PaprHie (APF = 25) (any powered, air-purifying respirator with a high-efficiency particulate filter). *Up to*

2.5 mg/m³: 100SaT (100F) (APF = 50) (any air-purifying, full-face-piece respirator with an N100, R100, or P100 filter) or SaT: Cf (APF = 50) (any supplied-air respirator that has a tight-fitting face-piece and is operated in a continuous-flow mode) or PaprTHie (APF = 50) (any powered, air-purifying respirator with a tight-fitting face-piece and a high-efficiency particulate filter) or SCBAF (APF = 50) (any self-contained breathing apparatus with a full face-piece) or SaF (APF = 50) (any supplied-air respirator with a full face-piece). *Emergency or planned entry into unknown concentrations or IDLH conditions:* SCBAF: Pd,Pp (APF = 10,000) (any NIOSH/MSHA- or European Standard EN 149-approved self-contained breathing apparatus that has a full face-piece and is operated in a pressure-demand or other positive-pressure mode) or SaF: Pd,Pp: ASCBA (APF = 10,000) (any supplied-air respirator that has a full face-piece and is operated in a pressure-demand or other positive-pressure mode in combination with an auxiliary, self-contained breathing apparatus operated in a pressure-demand or other positive-pressure mode). *Escape:* 100F (APF = 50) (any air-purifying, full-face-piece respirator with an N100, R100, or P100 filter) or SCBAE (any appropriate escape-type, self-contained breathing apparatus).

Storage: Color Code—Blue: Health Hazard/Poison: Store in a secure poison location. Prior to working with this chemical you should be trained on its proper handling and storage. Where possible, automatically transfer material from drums or other storage containers to process containers. Sources of ignition, such as smoking and open flames, are prohibited where this chemical is handled, used, or stored. Metal containers involving the transfer of this chemical should be grounded and bonded. Wherever this chemical is used, handled, manufactured, or stored, use explosion-proof electrical equipment and fittings.

Shipping: Sodium fluorosilicate requires a shipping label of "POISONOUS/TOXIC MATERIALS." It falls in Hazard Class 6.1 and Packing Group I.

Spill Handling: Evacuate persons not wearing protective equipment from area of spill or leak until cleanup is complete. Remove all ignition sources. Collect powdered material in the most convenient and safe manner and deposit in sealed containers. Ventilate area after cleanup is complete. It may be necessary to contain and dispose of this chemical as a hazardous waste. If material or contaminated runoff enters waterways, notify downstream users of potentially contaminated waters. Contact your local or federal environmental protection agency for specific recommendations. If employees are required to clean up spills, they must be properly trained and equipped. OSHA 1910.120(q) may be applicable.

Fire Extinguishing: This chemical is a noncombustible solid. Use dry chemical, carbon dioxide, water spray, or foam extinguishers. Poisonous gases are produced in fire, including fluorine and sodium oxide. If material or contaminated runoff enters waterways, notify downstream users of potentially contaminated waters. Notify local health and fire officials and pollution control agencies. From a secure, explosion-proof location, use water spray to cool exposed containers. If cooling streams are ineffective (venting sound increases in volume and pitch, tank discolors, or shows any signs of deforming), withdraw immediately to a secure position. If employees are expected to fight fires, they must be trained and equipped in OSHA 1910.156. The only respirators recommended for firefighting are self-contained breathing apparatuses that have full face-pieces and are operated in a pressure-demand or other positive-pressure mode.

Disposal Method Suggested: Consult with environmental regulatory agencies for guidance on acceptable disposal practices. Generators of waste containing this contaminant (≥100 kg/mo) must conform with EPA regulations governing storage, transportation, treatment, and waste disposal. In accordance with 40CFR165, follow recommendations for the disposal of pesticides and pesticide containers. Must be disposed properly by following package label directions or by contacting your local or federal environmental control agency or by contacting your regional EPA office. This compound is unstable at temperatures above 110°C and decomposes at 200°C. Thus, careful incineration has been suggested as a disposal procedure. According to their procedure, the product should be mixed with large amounts of vermiculite, sodium bicarbonate, and sand-soda ash. Slaked lime should also be added to the mixture. Two incineration procedures for this mixture are suggested. The better of these procedures is to burn the mixture in a closed incinerator equipped with an afterburner and an alkali scrubber. The other procedure suggests that the mixture be covered with scrap wood and paper in an open incinerator. (The incinerator should be lighted by means of an excelsior train).[22]

References

US Environmental Protection Agency. (November 30, 1987). *Chemical Hazard Information Profile: Sodium Fluoroacetate*. Washington, DC: Chemical Emergency Preparedness Program

Sodium hexafluorosilicate   S:0490

Molecular Formula: F_6Na_2Si
Common Formula: Na_2SiF_6
Synonyms: Destruxol applex; (2-)-Disodium hexafluorosilicate; Disodiumsilicofluoride; Ens-zem weevil bait; ENT 1,501; Fluosilicate de sodium; Natriumsilicofluorid (German); Ortho earwig bait; Ortho weevil bait; Prodan; PSC Co-Op weevil bait; Safsan; Salufer; Silicon sodium fluoride; Sodium fluorosilicate; Sodium fluosilicate; Sodium silicofluoride; Super prodan
CAS Registry Number: 16893-85-9; *(alt.)* 1310-02-7; *(alt.)* 39413-34-8
RTECS® Number: VV8410000
UN/NA & ERG Number: UN2674/154
EC Number: 240-934-8 [Annex I Index No.: 009-012-00-0]

Regulatory Authority and Advisory Bodies

Carcinogenicity: IARC: Animal Inadequate Evidence; Human Inadequate Evidence, *not classifiable as carcinogenic to humans*, Group 3, 1987.

US EPA, FIFRA 1998 Status of Pesticides: Canceled.

Banned or Severely Restricted (in UK) (UN).[13]

Air Pollutant Standard Set. See below, "Permissible Exposure Limits in Air" section.

European/International Regulations: Hazard Symbol: T; Risk phrases: R23/24/25; Safety phrases: S1/2; S26; S45 (see Appendix 4).

WGK (German Aquatic Hazard Class): 2—Hazard to waters.

Description: Sodium hexafluorosilicate is a white crystalline solid. Molecular weight = 188.06. Hazard Identification (based on NFPA-704 M Rating System): Health 2, Flammability 0, Reactivity 0. Practically insoluble in water.

Potential Exposure: Compound Description: Agricultural Chemical; Tumorigen, Primary Irritant. This material is used as an intermediate in the production of synthetic cyrolite; as a drinking-water additive; as an insecticide in delousing and in mothproofing of woolens.

Incompatibilities: Reacts with acids to produce hydrogen fluoride, which is a highly corrosive and poisonous gas.

Permissible Exposure Limits in Air

OSHA PEL: 3 ppm/2.5 mg[F]/m^3 TWA.

NIOSH REL: 3 ppm/2.5 mg[F]/m^3 TWA; 6 ppm/5 mg[F]/m^3, 15 min Ceiling Concentration.

ACGIH TLV®[1]: 2.5 mg[F]/m^3 TWA; not classifiable as a human carcinogen; BEI: 3 mg[F]/g creatinine in urine *prior* to end-of-shift; 10 mg[F]/g creatinine in urine end-of-shift.

Protective Action Criteria (PAC)

TEEL-0: 4.12 mg/m^3

PAC-1: 4.12 mg/m^3

PAC-2: 7.5 mg/m^3

PAC-3: 412 mg/m^3

DFG MAK: 1 mg[F]/m^3, inhalable fraction [skin]; Peak Limitation Category II(4); Pregnancy Risk Group C; BAT: 7.0 mg[F]/g creatinine in urine at end-of-shift; 4.0 mg[F]/g creatinine in urine at the beginning of the next shift.

NIOSH IDLH: 250 mg[F]/m^3.

Australia: TWA 2.5 mg[F]/m^3, 1993; Austria: MAK 2.5 mg [F]/m^3, 1999; Belgium: TWA 2.5 mg[F]/m^3, 1993; Finland: TWA 2.5 mg[F]/m^3, 1999; France: VME 2.5 mg[F]/m^3, 1999; Hungary: TWA 1 mg[F]/m^3; STEL 2 mg[F]/m^3, 1993; Norway: TWA 0.6 mg[F]/m^3, 1999; the Philippines: TWA 2.5 mg[F]/m^3, 1993; Poland: MAC (TWA) 1 mg [HF]/m^3, MAC (STEL) 3 mg[HF]/m^3, 1999; Russia: STEL 0.5 ppm (2.5 mg/m^3), 1993; Sweden: NGV 2 mg[F]/m^3, 1999; Switzerland: MAK-W 1.8 ppm (1.5 mg[F]/m^3), KZG-W 3.6 ppm (3.0 mg/m^3), 1999; Thailand: TWA 2.5 mg [F]/m^3, 1993; United Kingdom: TWA 2.5 mg[F]/m^3, 2000; LTEL 2.5 mg[F]/m^3, 1993; Argentina, Bulgaria, Columbia, Jordan, South Korea, New Zealand, Singapore, Vietnam: ACGIH TLV®: not classifiable as a human carcinogen.

Determination in Air: Use NIOSH Analytical Method (IV) #7902, Fluorides; #7906.

Permissible Concentration in Water: As with air, the applicable regulations are those for the fluoride ion. The values which have been set for drinking water[61] are a standard of 4.0 mg/L set by EPA and a guideline of 2.4 mg/L set by the state of Maine.

Routes of Entry: Inhalation, ingestion, skin and/or eye contact.

Harmful Effects and Symptoms

Short Term Exposure: Inhalation: May cause difficult breathing and burning of the mouth, throat, and nose which may result in bleeding. These may be felt at 7.5 mg/m^3. Nausea, vomiting, profuse sweating, and excess thirst may occur at higher levels. *Skin:* May cause rash, itching, and burning of skin. Solutions of 1% strength may cause sores if not removed promptly. *Eyes:* May cause severe irritation. *Ingestion:* Most reported instances of fluoride toxicity are due to accidental ingestion and it is difficult to associate symptoms with dose. Five to forty milligrams may cause diarrhea and vomiting. More severe symptoms of burning and painful abdomen; sores in mouth, throat, and digestive tract; tremors, convulsions, and shock will occur around a dose of 1 g. Death may result by ingestion of 2−5 g.

Long Term Exposure: Fluoride may increase bone density, stimulate new bone growth, or cause calcium deposits in ligaments. This may become a problem at levels of 20−50 mg/m^3 or higher. May cause mottling of the bones or teeth at this level, resulting in fluorosis. May cause lung damage.

Points of Attack: Bones, lungs.

Medical Surveillance: Lung function tests. DEXA bone densitometry scan.

First Aid: Inhalation: Move person to fresh air. Give artificial respiration, if necessary. If the nose is bleeding put absorbent material (like cotton) into the nasal openings. Do not pack the nostrils. Change the material often. Seek medical attention. *Skin:* Remove soiled clothing. Wash skin with soap and water for at least 5 min. Seek medical attention, if necessary. *Eyes:* Wash eyes with slow, steady stream of water for at least 15 min. Seek medical attention immediately. *Ingestion:* Seek medical attention. Give aluminum hydroxide gel, if conscious.

Note to physician: Inject intravenously 10 mL of 10% calcium gluconate solution. Gastric lavage with lime water of 1% calcium chloride.

Personal Protective Methods: Wear protective gloves and clothing to prevent any reasonable probability of skin contact. Safety equipment suppliers/manufacturers can provide recommendations on the most protective glove/clothing material for your operation. All protective clothing (suits, gloves, footwear, headgear) should be clean, available each day, and put on before work. Contact lenses should not be worn when working with this chemical. Wear dust-proof chemical goggles and face shield unless full face-piece respiratory protection is worn. Employees should wash

immediately with soap when skin is wet or contaminated. Provide emergency showers and eyewash.

Respirator Selection: NIOSH/OSHA *12.5 mg/m³:* Qm (APF = 25) (any quarter-mask respirator). *25 mg/m³:* 95XQ (APF = 10)* [any particulate respirator equipped with an N95, R95, or P95 filter (including N95, R95, and P95 filtering face-pieces) except quarter-mask respirators. The following filters may also be used: N99, R99, P99, N100, R100, P100] or SA* (any supplied-air respirator). *62.5 mg/m³:* Sa:Cf (APF = 25)*† (any supplied-air respirator operated in a continuous-flow mode) or PaprHie (APF = 25)* *if not present as a fume* (any powered, air-purifying respirator with a high-efficiency particulate filter). *125 mg/m³:* 100F (APF = 50)† [any particulate respirator equipped with an N95, R95, or P95 filter (including N95, R95, and P95 filtering face-pieces) except quarter-mask respirators. The following filters may also be used: N99, R99, P99, N100, R100, P100] or SCBAF (APF = 50) (any self-contained breathing apparatus with a full face-piece) or SaF (APF = 50) (any supplied-air respirator with a full face-piece). *250 mg/m³:* Sa: Pd,Pp (APF = 1000) (any supplied-air respirator operated in a pressure-demand or other positive-pressure mode). *Emergency or planned entry into unknown concentrations or IDLH conditions:* SCBAF: Pd, Pp (APF = 10,000) (any self-contained breathing apparatus that has a full faceplate and is operated in a pressure-demand or other positive-pressure mode) or SaF: Pd,Pp: ASCBA (APF = 10,000) (any supplied-air respirator that has a full face-piece and is operated in a pressure-demand or other positive-pressure mode in combination with an auxiliary, self-contained breathing apparatus operated in a pressure-demand or other positive-pressure mode). *Escape:* 100F (APF = 50)† [any particulate respirator equipped with an N95, R95, or P95 filter (including N95, R95, and P95 filtering face-pieces) except quarter-mask respirators. The following filters may also be used: N99, R99, P99, N100, R100, P100] or SCBAE (any appropriate escape-type, self-contained breathing apparatus).

*Substance reported to cause eye irritation or damage; may require eye protection.

†May need acid gas sorbent.

Storage: Color Code—Blue: Health Hazard/Poison: Store in a secure poison location. Prior to working with this chemical you should be trained on its proper handling and storage. Store in a cool, dry area that is well ventilated. Protect from damage and acids.

Shipping: Sodium fluorosilicate requires a shipping label of "POISONOUS/TOXIC MATERIALS." It falls in Hazard Class 6.1 and Packing Group III.

Spill Handling: Evacuate persons not wearing protective equipment from area of spill or leak until cleanup is complete. Remove all ignition sources. Enter spill area only with protective clothing and devices. Treat with soda ash or slaked lime. Use an industrial vacuum cleaner to remove the spill. Clean up with soap and water is allowed only if exposure and contamination are not increased to above the recommended levels. Ventilate area after cleanup is complete. It may be necessary to contain and dispose of this chemical as a hazardous waste. If material or contaminated runoff enters waterways, notify downstream users of potentially contaminated waters. Contact your local or federal environmental protection agency for specific recommendations. If employees are required to clean up spills, they must be properly trained and equipped. OSHA 1910.120(q) may be applicable.

Fire Extinguishing: This material is nonflammable. Use agents suitable to surrounding fire. Poisonous gases are produced in fire, including fluorine and sodium oxide. If material or contaminated runoff enters waterways, notify downstream users of potentially contaminated waters. Notify local health and fire officials and pollution control agencies. From a secure, explosion-proof location, use water spray to cool exposed containers. If cooling streams are ineffective (venting sound increases in volume and pitch, tank discolors, or shows any signs of deforming), withdraw immediately to a secure position. If employees are expected to fight fires, they must be trained and equipped in OSHA 1910.156. The only respirators recommended for firefighting are self-contained breathing apparatuses that have full face-pieces and are operated in a pressure-demand or other positive-pressure mode.

References
New York State Department of Health. (February 1986). *Chemical Fact Sheet: Sodium Hexafluorosilicate* (Version 2). Albany, NY: Bureau of Toxic Substance Assessment

Sodium hydroxide   S:0500

Molecular Formula: HNaO
Common Formula: NaOH
Synonyms: Caustic soda; Caustic soda, bead; Caustic soda, dry; Caustic soda, flake; Caustic soda, granular; Caustic soda, solid; Hidroxido sodico (Spanish); Hydroxyde of sodium (French); Lewis Red Devil Lye; Lye; Lye solution; Natriumhydroxid (German); Pels® soda lye; Sodium hydrate; Sodium hydrate solution; Sodium hydroxide, bead; Sodium hydroxide caustic soda solution; Sodium hydroxide, dry; Sodium hydroxide, flake; Sodium hydroxide, granular; Sodium hydroxide liquid; Sodium hydroxide, solid; Sodium hydroxide solution; Sodium (hydroxyde de) (French); White caustic; White caustic, solution
CAS Registry Number: 1310-73-2; *(alt.)* 8012-01-9
RTECS® Number: WB4900000
UN/NA & ERG Number: UN1823 (dry, solid)/154; UN1824 (solution)/154
EC Number: 215-185-5 [*Annex I Index No.:* 011-002-00-6]
Regulatory Authority and Advisory Bodies
US EPA Gene-Tox Program, Negative: Cell transform.—SA7/SHE.
US EPA, FIFRA 1998 Status of Pesticides: RED completed.

Air Pollutant Standard Set. See below, "Permissible Exposure Limits in Air" section.

Clean Water Act: Section 311 Hazardous Substances/RQ 40CFR117.3 (same as CERCLA, see below).

Reportable Quantity (RQ): 1000 lb (454 kg).

Canada, WHMIS, Ingredients Disclosure List Concentration 1.0%.

European/International Regulations: Hazard Symbol: T, N; Risk phrases: R45; R22; R50/53; Safety phrases: S53; S45; S60; S61 (see Appendix 4).

WGK (German Aquatic Hazard Class): 1—Low hazard to waters.

Description: Sodium hydroxide is a white, odorless, deliquescent material sold as pellets, flakes, lumps, or sticks. Aqueous solutions are known as soda lye. Molecular weight = 40.00; Specific gravity (H_2O:1) = 2.13 at 25°C; Boiling point = 1390°C; Freezing/Melting point = 318.3°C. Hazard Identification (based on NFPA-704 M Rating System): Health 3, Flammability 0, Reactivity 1. Highly soluble in water; solubility = 111%.

Potential Exposure: Compound Description: Agricultural Chemical; Mutagen, Human Data; Primary Irritant. Sodium hydroxide is utilized to neutralize acids and make sodium salts in petroleum refining, viscose rayon, cellophane, plastic production; and in the reclamation of solutions of their salts. It is used in the manufacture of mercerized cotton, paper, explosives, and dyestuffs in metal cleaning; electrolytic extraction of zinc; tin plating; oxide coating; laundering, bleaching, dishwashing; and it is used in the chemical industries.

Incompatibilities: A strong base and a strong oxidizer. Violent reaction with acid. Incompatible with water, flammable liquids, organic halogens, nitromethane, nitrocompounds, and combustibles. Rapidly absorbs carbon dioxide and water from air. Contact with moisture or water may generate heat. Corrosive to metals. Contact with zinc, aluminum, tin, and lead in the presence of moisture, forming explosive hydrogen gas. Attacks some forms of plastics, rubber, or coatings.

Permissible Exposure Limits in Air

OSHA PEL: 2 mg/m³.

NIOSH REL: 2 mg/m³ Ceiling Concentration.

ACGIH TLV®[1]: 2 mg/m³ Ceiling Concentration.

Protective Action Criteria (PAC)*

TEEL-0: 0.5 mg/m³

PAC-1: 0.5 mg/m³

PAC-2: **5** mg/m³

PAC-3: **50** mg/m³

*AEGLs (Acute Emergency Guideline Levels) & ERPGs (Emergency Response Planning Guideline) are in **bold face**.

DFG MAK: No numerical value established.

NIOSH IDLH: 10 mg/m³.

Australia: TWA 2 mg/m³, 1993; Austria: MAK 2 mg/m³, 1999; Belgium: STEL 2 mg/m³, 1993; Denmark: TWA 2 mg/m³, 1999; Finland: TWA 2 mg/m³, 1999; France:

VME 2 mg/m³, 1999; Japan: 2 mg/m³, 1999; the Netherlands: MAC-TGG 2 mg/m³, 2003; Norway: TWA 2 mg/m³, 1999; the Philippines: TWA 2 mg/m³, 1993; Poland: MAC (TWA) 0.5 mg/m³; MAC (STEL) 1 mg/m³, 1999; Sweden: TGV 2 mg/m³, 1999; Switzerland: MAK-W 2 mg/m³, KZG-W 4 mg/m³, 1999; Thailand: TWA 2 mg/m³, 1993; Turkey: TWA 2 mg/m³, 1993; United Kingdom: STEL 2 mg/m³, 2000; Argentina, Bulgaria, Columbia, Jordan, South Korea, New Zealand, Singapore, Vietnam: ACGIH TLV®: Ceiling Concentration 2 mg/m³. Russia has set 10 μg/m³ as a MAC for ambient air in residential areas on a once-daily basis. Several states have set guidelines or standards for sodium hydroxide in ambient air[60] ranging from 16.0 μg/m³ (Virginia) to 20.0 μg/m³ (North Dakota and South Carolina) to 40.0 μg/m³ (Connecticut and South Dakota) to 48.0 μg/m³ (Nevada).

Determination in Air: Use NIOSH Analytical Method (IV) #7401, alkaline Dusts; OSHA Analytical Method ID-121.

Permissible Concentration in Water: There are no criteria for NaOH as such. The EPA has, however, recommended criteria for pH as follows: to protect freshwater aquatic life: pH 6.5–9.0; to protect saltwater aquatic life: pH 6.5–8.5; and to protect humans' drinking water: pH 5–9.

Routes of Entry: Inhalation of dust or mist, ingestion, skin and/or eye contact.

Harmful Effects and Symptoms

Highly corrosive to the eyes, skin, and the respiratory tract. *Inhalation:* Can cause severe irritation of the nose and throat and inflammation of the lungs. Higher exposures can cause pulmonary edema, a medical emergency that can be delayed for several hours. This can cause death. *Skin:* Can cause severe irritation and deep burns. *Eyes:* Can cause severe irritation, corneal burns, and blindness. *Ingestion:* Can cause burning of the mouth and throat, nausea, vomiting, abdominal pains, and diarrhea (occasionally with blood). Can also cause swelling of the larynx and subsequent suffocation, holes in stomach and intestines, heart failure, coma. Death has resulted from swallowing less than 1/3 oz of the solid.

Long Term Exposure: Skin irritation may develop from repeated exposure to the solid or low concentrations of the liquid. Irritation to the lungs, nose, throat, and mouth may occur if exposed to low levels for long periods of time. May cause temporary loss of hair.

Points of Attack: Eyes, skin, respiratory system.

Medical Surveillance: NIOSH lists the following tests: chest X-ray; pulmonary function tests.

First Aid: If this chemical gets into the eyes, remove any contact lenses at once and irrigate immediately for at least 30 min, occasionally lifting upper and lower lids. Seek medical attention immediately. If this chemical contacts the skin, remove contaminated clothing and wash immediately with large amounts of water. Seek medical attention immediately. If this chemical has been inhaled, remove from exposure, begin rescue breathing (using universal precautions, including resuscitation mask) if breathing has stopped and CPR if heart action has stopped. Transfer promptly to a medical facility. When this chemical has been swallowed, get medical attention. If victim is *conscious*, administer water or milk. Do not induce vomiting. Medical observation is recommended for 24–48 h after breathing overexposure, as pulmonary edema may be delayed. As first aid for pulmonary edema, a doctor or authorized paramedic may consider administering a corticosteroid spray.

Personal Protective Methods: Wear protective gloves and clothing to prevent any reasonable probability of skin contact. Safety equipment suppliers/manufacturers can provide recommendations on the most protective glove/clothing material for your operation.

For 30–70% solution: Sealed chemical materials with good to excellent resistance: butyl rubber; natural rubber; Neoprene™, nitrile rubber; polyethylene, PVC, Teflon™, Viton™, Viton™/chlorobutyl rubber; Saranex™, Silvershield™, Sol-vex® nitrile.

For less than 30% solution: Viton + chlorobutyl rubber, Silvershield™, polyvinyl chloride; polyethylene, Nitrile + polyvinyl chloride; and chlorinated polyethylene are among the recommended protective materials. Neoprene™ + Natural rubber and polyethylene.

All protective clothing (suits, gloves, footwear, headgear) should be clean, available each day, and put on before work. Contact lenses should not be worn when working with this chemical. Wear splash-proof chemical goggles and face shield when working with liquid, unless full face-piece respiratory protection is worn. Wear dust-proof goggles and face shield when working with powders or dust, unless full face-piece respiratory protection is worn. Employees should wash immediately with soap when skin is wet or contaminated. Provide emergency showers and eyewash.

Respirator Selection: Up to 10 mg/m³: Sa:Cf* (APF = 25) (any supplied-air respirator operated in a continuous-flow mode) or 100F (APF = 50) (any air-purifying, full-face-piece respirator with an N100, R100, or P100 filter) or PaprHie* (APF = 25) (any powered air-purifying respirator with a high-efficiency particulate filter) or SCBAF (APF = 50) (any self-contained breathing apparatus with a full face-piece) or SaF (APF = 50) (any supplied-air respirator with a full face-piece). *Emergency or planned entry into unknown concentrations or IDLH conditions:* SCBAF: Pd,Pp (APF = 10,000) (any self-contained breathing apparatus that has a full face-piece and is operated in a pressure-demand or other positive-pressure mode) or SaF: Pd,Pp: ASCBA (APF = 10,000) (any supplied-air respirator that has a full face-piece and is operated in a pressure-demand or other positive-pressure mode in combination with an auxiliary self-contained breathing apparatus operated in a pressure-demand or other positive-pressure mode). *Escape:* 100F (APF = 50) (any air-purifying, full-face-piece respirator with an N100, R100, or P100 filter) or SCBAE (any appropriate escape-type, self-contained breathing apparatus).

*Substance causes eye irritation or damage; eye protection needed.

Storage: Color Code—White Stripe: Contact Hazard; Store separately; not compatible with materials in solid white category. Prior to working with this chemical you should be trained on its proper handling and storage. Sodium hydroxide must be stored to avoid contact with water, acids, flammable liquids, organic halogen compounds, metals, or nitro compounds, because violent reactions occur. Store in tightly closed containers in a cool, well-ventilated area away from water. Where possible, automatically transfer material from drums or other storage containers to process containers. Sources of ignition, such as smoking and open flames, are prohibited where this chemical is handled, used, or stored. Metal containers involving the transfer of this chemical should be grounded and bonded. Wherever this chemical is used, handled, manufactured, or stored, use explosion-proof electrical equipment and fittings.

Shipping: Sodium hydroxide, solid, or sodium hydroxide solution requires a shipping label of "CORROSIVE." It falls in Hazard Class 8 and Packing Group II. The solution falls in Hazard Class 8 and Packing Groups II or III.

Spill Handling: Evacuate persons not wearing protective equipment from area of spill or leak until cleanup is complete. Remove all ignition sources. Wear protective clothing. For the solid, sweep into large vessel containing a large amount of water. Neutralize with weak hydrochloric acid. For solution, neutralize with weak hydrochloric acid. Pick up with mop or water vacuum. Ventilate area after cleanup is complete. It may be necessary to contain and dispose of this chemical as a hazardous waste. If material or contaminated runoff enters waterways, notify downstream users of potentially contaminated waters. Contact your local or federal environmental protection agency for specific recommendations. If employees are required to clean up spills, they must be properly trained and equipped. OSHA 1910.120(q) may be applicable.

Fire Extinguishing: Extinguish fire using an agent suitable for type of surrounding fire. Sodium hydroxide itself does not burn, but it is a strong oxidizer and may ignite combustibles, such as wood, paper, oil, etc. Poisonous gases are produced in fire, including sodium oxide. If material or contaminated runoff enters waterways, notify downstream users of potentially contaminated waters. Notify local health and fire officials and pollution control agencies. From a secure, explosion-proof location, use water spray to cool exposed containers. If cooling streams are ineffective (venting sound increases in volume and pitch, tank discolors, or shows any signs of deforming), withdraw immediately to a secure position. If employees are expected to fight fires, they must be trained and equipped in OSHA 1910.156. The only respirators recommended for firefighting are self-contained breathing apparatuses that have full face-pieces and are operated in a pressure-demand or other positive-pressure mode.

Disposal Method Suggested: Discharge into tank containing water, neutralize, then flush to sewer with water.

References

National Institute for Occupational Safety and Health. (1976). *Criteria for a Recommended Standard: Occupational Exposure to Sodium Hydroxide*, NIOSH Document No. 76-105. Washington, DC

Sax, N. I. (Ed.). (1984). *Dangerous Properties of Industrial Materials Report*, 4, No. 3, 85–89

New York State Department of Health. (February 1986). *Chemical Fact Sheet: Sodium Hydroxide* (Version 3). Albany, NY: Bureau of Toxic Substance Assessment

US Environmental Protection Agency, Special Review and Reregistration Division Office of Pesticide Programs. (1998). *Agency Status of Pesticides in Registration, Reregistration, and Special Review* (Rainbow Report). Washington, DC

New Jersey Department of Health and Senior Services. (May 2001). *Hazardous Substances Fact Sheet: Sodium Hydroxide*. Trenton, NJ

Sodium metabisulfite   S:0510

Molecular Formula: $O_5S_2 \cdot 2Na$
Common Formula: $Na_2S_2O_5$
Synonyms: Disodium disulfite; Disodium disulphite; Disodium pyrosulfite; Disulfurous acid, disodium salt; Pyrosulfurous acid, sodium salt; Sodium disulfite; Sodium metabisulfite; Sodium metabisulphite; Sodium pyrosulfite
CAS Registry Number: 7681-57-4; *(alt.)* 7757-74-6; *(alt.)* 15771-29-6
RTECS® Number: UX8225000
UN/NA & ERG Number: UN2693/154
EC Number: 231-673-0 [*Annex I Index No.:* 016-063-00-2] (disodium disulphite)
Regulatory Authority and Advisory Bodies
Carcinogenicity: IARC: Animal Inadequate Evidence; Human Inadequate Evidence, *not classifiable as carcinogenic to humans*, Group 3, 1992.
US EPA Gene-Tox Program, Negative: TRP reversion.
Air Pollutant Standard Set. See below, "Permissible Exposure Limits in Air" section.
Canada, WHMIS, Ingredients Disclosure List Concentration 1.0%.
European/International Regulations: Hazard Symbol: Xn; Risk phrases: R22; R31; R41; Safety phrases: S2; S26; S39; 46 (see Appendix 4).
WGK (German Aquatic Hazard Class): 1—Low hazard to waters.
Description: Sodium metabisulfite is a white crystalline powder with a sulfur dioxide odor. It may be considered the anhydride of 2 molecules of sodium disulfite. Molecular weight = 190.10; Specific gravity (H_2O:1) = 1.40 at 25°C; Boiling point = (decomposes); Freezing/Melting point = (decomposes) ≥ 150°C. Hazard Identification (based on NFPA-704 M Rating System): Health 2, Flammability 0, Reactivity 0. Soluble in water; solubility = 54%.

Potential Exposure: Compound Description: Agricultural Chemical; Tumorigen, Mutagen; Reproductive Effector; Primary Irritant. Sodium metabisulfite is used as an antioxidant in pharmaceutical preparations and as a preservative in foods. People with asthma have a greater chance of having an allergic reaction with this chemical. Individuals allergic to *sodium bisulfite* (a food preservative found in some wine, fresh shrimp, packaged foods, and restaurant salads and potatoes) may have a severe reaction when exposed to sodium metabisulfite.
Incompatibilities: A strong reducing agent. Mixtures with water form a strong corrosive. Incompatible with reducing agents and combustibles. Heat causes decomposition. Slowly oxidized to the sulfate on exposure to air and moisture. Attacks metals.
Permissible Exposure Limits in Air
OSHA PEL: None.
NIOSH REL: 5 mg/m³ TWA.
ACGIH TLV®[1]: 5 mg/m³ TWA; not classifiable as a human carcinogen.
Protective Action Criteria (PAC)
TEEL-0: 5 mg/m³
PAC-1: 15 mg/m³
PAC-2: 100 mg/m³
PAC-3: 500 mg/m³
DFG MAK: Sensitizing substances as sulfites.
Australia: TWA 5 mg/m³, 1993; Belgium: TWA 5 mg/m³, 1993; Denmark: TWA 5 mg/m³, 1999; France: VME 5 mg/m³, 1999; Norway: TWA 5 mg/m³, 1999; Switzerland: MAK-W 5 mg/m³, 1999; United Kingdom: TWA 5 mg/m³, 2000; the Netherlands: MAC-TGG 5 mg/m³, 2003; Argentina, Bulgaria, Columbia, Jordan, South Korea, New Zealand, Singapore, Vietnam: ACGIH TLV®: not classifiable as a human carcinogen. Several states have set guidelines or standards for sodium metabisulfite in ambient air[60] ranging from 50.0 μg/m³ (North Dakota) to 80.0 μg/m³ (Virginia) to 100.0 μg/m³ (Connecticut) to 119.0 μg/m³ (Nevada).
Determination in Air: Filter; none; Gravimetric; NIOSH Analytical Method (IV) #0500, Particulates NOR, total dust.
Routes of Entry: Inhalation, skin and/or eye contact, ingestion.
Harmful Effects and Symptoms
Short Term Exposure: Sodium metabisulfite can affect you when breathed in. Exposure can irritate the nose, throat, and sinuses. It can also irritate the lungs, causing coughing, wheezing, and shortness of breath. Very severe general (anaphylactic) reactions can also occur in those allergic to sodium metabisulfite that can be fatal. Contact can irritate the skin.
Long Term Exposure: Sodium metabisulfite may cause an asthma-like allergy. Future exposures can cause asthma attacks with cough, shortness of breath, wheezing, and/or chest tightness. Can cause lung irritation; bronchitis may develop.

Points of Attack: Eyes, skin, respiratory system.

Medical Surveillance: For those with frequent or potentially high exposure (half the TLV or greater), the following are recommended before beginning work and at regular times after that: lung function tests. If symptoms develop or overexposure is suspected, the following may be useful: evaluation by a qualified allergist, including careful exposure history and special testing, may help diagnose skin allergy.

First Aid: If this chemical gets into the eyes, remove any contact lenses at once and irrigate immediately for at least 15 min, occasionally lifting upper and lower lids. Seek medical attention immediately. If this chemical contacts the skin, remove contaminated clothing and wash immediately with soap and water. Seek medical attention immediately. If this chemical has been inhaled, remove from exposure, begin rescue breathing (using universal precautions, including resuscitation mask) if breathing has stopped and CPR if heart action has stopped. Transfer promptly to a medical facility. When this chemical has been swallowed, get medical attention. If victim is *conscious,* administer water or milk. Do not induce vomiting.

Personal Protective Methods: Wear protective gloves and clothing to prevent any reasonable probability of skin contact. Safety equipment suppliers/manufacturers can provide recommendations on the most protective glove/clothing material for your operation. All protective clothing (suits, gloves, footwear, headgear) should be clean, available each day, and put on before work. Contact lenses should not be worn when working with this chemical. Wear dust-proof chemical goggles and face shield unless full face-piece respiratory protection is worn. Employees should wash immediately with soap when skin is wet or contaminated. Provide emergency showers and eyewash.

Respirator Selection: Where there is potential for exposures *over 5 mg/m³,* use a NIOSH/MSHA- or European Standard EN149-approved respirator equipped with particulate (dust/fume/mist) filters. Particulate filters must be checked every day before work for physical damage, such as rips or tears, and replaced as needed. *Where there is potential for high exposures,* use a NIOSH/MSHA- or European Standard EN149-approved supplied-air respirator with a full face-piece operated in the positive-pressure mode, or with a full face-piece, hood, or helmet in the continuous-flow mode; or use a NIOSH/MSHA- or European Standard EN149-approved self-contained breathing apparatus with a full face-piece operated in pressure-demand or other positive-pressure mode.

Storage: (1) Color Code—White: Corrosive or Contact Hazard; Store separately in a corrosion-resistant location. (2) Color Code—Yellow Stripe (*strong reducing agent*): Reactivity Hazard; Store separately in an area isolated from flammables, combustibles, or other yellow coded materials. Prior to working with this chemical you should be trained on its proper handling and storage. Store in tightly closed containers in a cool, well-ventilated area away from combustible materials, such as wood and paper. DOT requires sodium metabisulfite to be packed in earthenware, glass, metal, or plastic during transport.

Shipping: Bisulfites, inorganic, aqueous solutions, n.o.s. require a shipping label of "CORROSIVE." It falls in Hazard Class 8 and Packing Group II.

Spill Handling: Evacuate persons not wearing protective equipment from area of spill or leak until cleanup is complete. Remove all ignition sources. Collect powdered material in the most convenient and safe manner and deposit in sealed containers. Ventilate area after cleanup is complete. It may be necessary to contain and dispose of this chemical as a hazardous waste. If material or contaminated runoff enters waterways, notify downstream users of potentially contaminated waters. Contact your local or federal environmental protection agency for specific recommendations. If employees are required to clean up spills, they must be properly trained and equipped. OSHA 1910.120(q) may be applicable.

Fire Extinguishing: Sodium metabisulfite may burn, but does not readily ignite. Use dry chemical, CO_2, water spray, or foam extinguishers. Sodium metabisulfite may ignite nearby combustible materials. Poisonous gases are produced in fire, including sulfur oxides. If material or contaminated runoff enters waterways, notify downstream users of potentially contaminated waters. Notify local health and fire officials and pollution control agencies. From a secure, explosion-proof location, use water spray to cool exposed containers. If cooling streams are ineffective (venting sound increases in volume and pitch, tank discolors, or shows any signs of deforming), withdraw immediately to a secure position. If employees are expected to fight fires, they must be trained and equipped in OSHA 1910.156. The only respirators recommended for firefighting are self-contained breathing apparatuses that have full face-pieces and are operated in a pressure-demand or other positive-pressure mode.

References

New Jersey Department of Health and Senior Services. (August 2005). *Hazardous Substances Fact Sheet: Sodium Metabisulfite.* Trenton, NJ

Sodium pentachlorophenate   S:0520

Molecular Formula: C₆Cl₅NaO

Common Formula: C₆Cl₅ONa

Synonyms: AI3-16418; Dow dormant fungicide; Dowicide G; Dowicide G-ST; GR 48-11PS; GR 48-32S; Napclor-G; PCP-sodium; PCP sodium salt; Pentachlorophenate sodium; Pentachlorophenol, sodium salt; Pentachlorophenoxy sodium; Pentaclorofenato sodico (Spanish); Pentaphenate; Phenol, pentachloro-, sodium salt; Phenol, pentachloro-, sodium salt, monohydrate; PKHFN; Santobrite; Santobrite D; Sodium PCP; Sodium pentachlorophenol; Sodium

pentachlorophenolate; Sodium pentachlorophenoxide; Sodium, (pentachlorophenoxy)-; Sodium pentachlorphenate; Weedbeads
CAS Registry Number: 131-52-2
RTECS® Number: SM6490000
UN/NA & ERG Number: UN2567/154
EC Number: 205-025-2 [*Annex I Index No.:* 604-003-00-3]
Regulatory Authority and Advisory Bodies
US EPA Gene-Tox Program, Positive: *B. subtilis* rec assay; Inconclusive: *D. melanogaster* sex-linked lethal.
US EPA, FIFRA 1998 Status of Pesticides: Canceled.
Air Pollutant Standard Set. See below, "Permissible Exposure Limits in Air" section.
US EPA Hazardous Waste Number (RCRA No.): listed as "None."
RCRA, 40CFR261, Appendix 8 Hazardous Constituents.
EPCRA Section 313 Form R *de minimis* concentration reporting level: 1.0%.
US DOT Regulated Marine Pollutant (49CFR172.101, Appendix B).
European/International Regulations: Hazard Symbol: T+, N; Risk phrases: R24/24; R26; R36/37/38; R40; R51/53; Safety phrases: S1/2; S28; S52; S60; S61 (see Appendix 4).
WGK (German Aquatic Hazard Class): 3—Severe hazard to waters.
Description: Sodium pentachlorophenate is a crystalline solid with a phenolic odor. Molecular weight = 288.30. Soluble in water; solubility = 33% at 25°C.
Potential Exposure: Compound Description: Agricultural Chemical; Mutagen; Reproductive Effector. Uses include: wood preservative; as a fungicide in water-based latex paints; preservation of cellulose products, textiles, adhesives, leather, pulp, paper and industrial waste systems; a contact and preemergence herbicide; general disinfectant and control of the intermediate snail host of schistosomiasis. The technical grade of sodium pentachlorophenate usually contain toxic microcontaminants including polychlorinated dibenzodioxins and dibenzofurans (132-64-9 and others).
Incompatibilities: Oxidizers.
Permissible Exposure Limits in Air
Protective Action Criteria (PAC)
TEEL-0: 1 mg/m^3
PAC-1: 3.5 mg/m^3
PAC-2: 24 mg/m^3
PAC-3: 75 mg/m^3
Russia[43] set a MAC of 0.1 mg/m^3 in work-place air and MAC values for ambient air in residential areas as follows: 0.005 mg/m^3 on a momentary basis and 0.001 mg/m^3 on a daily average basis.
Determination in Air: No method available.
Permissible Concentration in Water: Russia[43] set a MAC of 5 mg/m^3 in water bodies used for domestic purposes and 0.0005 mg/L in water bodies used for fishery purposes.
Routes of Entry: Inhalation, ingestion, skin and/or eye contact.
Harmful Effects and Symptoms

Short Term Exposure: Exposure to fine dusts or sprays cause burning in eyes and painful irritation in upper respiratory tract. If inhaled, it will induce violent coughing and sneezing. Skin irritation results from brief exposures, causing a burning sensation or rash. Symptoms of severe systemic intoxication include loss of appetite, respiratory difficulties, anesthesia, fever, sweating, difficulty in breathing, and rapidly progressive coma. Severe intoxications, including fatalities, have been reported from uncontrolled use. This compound causes inflamed gastric mucosa, congestion of the lungs, edema in the brain, cardiac dilatation, degeneration of the liver and kidneys. Individuals suffering from kidney and liver diseases have a lowered resistance and should not be exposed.
Long Term Exposure: May cause skin allergy. May cause anemia. May damage the liver and kidneys. Repeated exposure can cause headache, weakness, sweating, fever, muscle twitching, dizziness, confusion, and death.
Points of Attack: Skin, blood, liver, kidneys.
Medical Surveillance: Complete blood count (CBC). Liver and kidney function tests. Evaluation by a qualified allergist.
First Aid: If this chemical gets into the eyes, remove any contact lenses at once and irrigate immediately for at least 15 min, occasionally lifting upper and lower lids. Seek medical attention immediately. If this chemical contacts the skin, remove contaminated clothing and wash immediately with soap and water. Seek medical attention immediately. If this chemical has been inhaled, remove from exposure, begin rescue breathing (using universal precautions, including resuscitation mask) if breathing has stopped and CPR if heart action has stopped. Transfer promptly to a medical facility. When this chemical has been swallowed, get medical attention. Give large quantities of water and induce vomiting. Do not make an unconscious person vomit.
Personal Protective Methods: Wear protective gloves and clothing to prevent any reasonable probability of skin contact. Safety equipment suppliers/manufacturers can provide recommendations on the most protective glove/clothing material for your operation. All protective clothing (suits, gloves, footwear, headgear) should be clean, available each day, and put on before work. Contact lenses should not be worn when working with this chemical. Wear dust-proof chemical goggles and face shield unless full face-piece respiratory protection is worn. Employees should wash immediately with soap when skin is wet or contaminated. Provide emergency showers and eyewash.
Respirator Selection: Follow regulations in OSHA 29CFR1910.134 or European Standard EN149. Use a NIOSH/MSHA- or European Standard EN149-approved respirator; or use an approved supplied-air respirator with a full face-piece operated in the positive-pressure mode, or with a full face-piece, hood, or helmet in the continuous-flow mode; or use a NIOSH/MSHA- or European Standard EN149-approved self-contained breathing apparatus with a full face-piece operated in pressure-demand or other positive-pressure mode.

Storage: Color Code—Blue: Health Hazard/Poison: Store in a secure poison location. Prior to working with this chemical you should be trained on its proper handling and storage. Where possible, automatically transfer material from drums or other storage containers to process containers. Sources of ignition, such as smoking and open flames, are prohibited where this chemical is handled, used, or stored. Metal containers involving the transfer of this chemical should be grounded and bonded. Wherever this chemical is used, handled, manufactured, or stored, use explosion-proof electrical equipment and fittings.

Shipping: Sodium pentachlorophenate requires a shipping label of "POISONOUS/TOXIC MATERIALS." It falls in Hazard Class 6.1 and Packing Group II.

Spill Handling: Evacuate persons not wearing protective equipment from area of spill or leak until cleanup is complete. Remove all ignition sources. Ventilate area of spill. Collect spilled material in most convenient manner and deposit in sealed containers for later disposal. Liquids should be absorbed in vermiculite, dry sand, earth, or similar material. Keep unnecessary people away; isolate hazard area and deny entry. Stay upwind; keep out of low areas. Do not touch spilled material; stop leak if you can do so without risk. *Small spills:* absorb with sand or other noncombustible absorbent material and place into containers for later disposal. *Small dry spills:* with clean shovel, place material into clean, dry container and cover; move containers from spill area. *Large spills:* dike far ahead of spill for later disposal. Ventilate area after cleanup is complete. It may be necessary to contain and dispose of this chemical as a hazardous waste. If material or contaminated runoff enters waterways, notify downstream users of potentially contaminated waters. Contact your local or federal environmental protection agency for specific recommendations. If employees are required to clean up spills, they must be properly trained and equipped. OSHA 1910.120(q) may be applicable.

Fire Extinguishing: This chemical is a combustible solid. It is noncombustible. For small fires, use dry chemical, carbon dioxide, water spray, or foam. For large fires, use water spray, fog, or foam. Wear self-contained (positive pressure) breathing apparatus and full protective clothing. Move container from fire area if possible. Runoff from fire control or dilution water may cause pollution. Poisonous gases are produced in fire, including hydrogen chloride, chlorides, and sodium oxides. If material or contaminated runoff enters waterways, notify downstream users of potentially contaminated waters. Notify local health and fire officials and pollution control agencies. From a secure, explosion-proof location, use water spray to cool exposed containers. If cooling streams are ineffective (venting sound increases in volume and pitch, tank discolors, or shows any signs of deforming), withdraw immediately to a secure position. If employees are expected to fight fires, they must be trained and equipped in OSHA 1910.156. The only respirators recommended for firefighting are self-contained breathing apparatuses that have full face-pieces and are operated in a pressure-demand or other positive-pressure mode.

Disposal Method Suggested: Consult with environmental regulatory agencies for guidance on acceptable disposal practices. Generators of waste containing this contaminant ($\geq$100 kg/mo) must conform with EPA regulations governing storage, transportation, treatment, and waste disposal.

References

Sax, N. I. (Ed.). (1986). *Dangerous Properties of Industrial Materials Report*, 6, No. 2, 5–30

US Environmental Protection Agency. (November 30, 1987). *Chemical Hazard Information Profile: Sodium Pentachlorophenate*. Washington, DC: Chemical Emergency Preparedness Program

New Jersey Department of Health and Senior Services. (August 1999). *Hazardous Substances Fact Sheet: Sodium Pentachlorophenate*. Trenton, NJ

US Environmental Protection Agency, Special Review and Reregistration Division Office of Pesticide Programs. (1998). *Agency Status of Pesticides in Registration, Reregistration, and Special Review* (Rainbow Report). Washington, DC

Sodium selenite   S:0530

Molecular Formula: Na_2O_3Se

Common Formula: Na_2SeO_3

Synonyms: Disodium selenite; Natriumselenit (German); Selenious acid, Disodium salt; Selenito sodico (Spanish)

CAS Registry Number: 10102-18-8

RTECS® Number: VS73500000

UN/NA & ERG Number: UN2630/151

EC Number: 233-267-9 [*Annex I Index No.:* 034-003-00-3]

Regulatory Authority and Advisory Bodies

Carcinogenicity: IARC: Animal Inadequate Evidence Group 3, 1987.

US EPA Gene-Tox Program, Positive: Histidine reversion—Ames test; Positive/dose response: *In vitro* SCE—human lymphocytes; Positive/dose response: *In vitro* SCE—human; Positive/dose response: *In vitro* UDS—human fibroblast.

Very Toxic Substance (World Bank).[15]

NTP: Toxicity studies, Report TOX-38, October, 2000.

Air Pollutant Standard Set. See below, "Permissible Exposure Limits in Air" section.

Water Pollution Standard Proposed (EPA) (6, 62).

Superfund/EPCRA 40CFR355, Extremely Hazardous Substances: TPQ = 100 lb (45.4 kg).[7]

Hazardous Substance RQ = 100 lb (45.4 kg).[4]

Priority Toxic Pollutant (EPA).[6]

Canada, WHMIS, Ingredients Disclosure List Concentration 1.0%.

European/International Regulations: Hazard Symbol: T+, N; Risk phrases: R23; R28; R31; R43; R51/53; Safety phrases: S1/2; S28; S36/37; S45; S61 (see Appendix 4).

WGK (German Aquatic Hazard Class): 2—Hazard to waters.

Description: Sodium selenite is a white crystalline substance. Molecular weight = 172.94; Freezing/Melting point = (decomposes) 320°C. Hazard Identification (based on NFPA-704 M Rating System): Health 3, Flammability 0, Reactivity 0. Highly soluble in water; solubility = >80% at 20°C.

Potential Exposure: Compound Description: Tumorigen, Mutagen; Reproductive Effector; Human Data. Sodium selenite is used in glass manufacturing and as an alkaloidal reagent; for removing green color from glass during its manufacture; alkaloidal reagent; reagent in bacteriology; testing germination of seeds; decorating porcelain; as a livestock feed additive.

Incompatibilities: The aqueous solution is a medium strong base. Reacts with water, strong acids, hot surfaces, causing decomposition and a toxic hazard.

Permissible Exposure Limits in Air
OSHA PEL: 0.2 mg[Se]/m^3 TWA.
NIOSH REL: 0.2 mg[Se]/m^3 TWA.
ACGIH TLV®[1]: 0.2 mg[Se]/m^3 TWA.
Protective Action Criteria (PAC)
TEEL-0: 0.438 mg/m^3
PAC-1: 1.31 mg/m^3
PAC-2: 2.19 mg/m^3
PAC-3: 2.19 mg/m^3
DFG MAK: 0.05 mg[Se]/m^3 inhalable fraction; Carcinogen Category 3; Pregnancy Risk Group C.
NIOSH IDLH: 1 mg [Se]/m^3.
Arab Republic of Egypt: TWA 0.2 mg[Se]/m^3, 1993; Australia: TWA 0.2 mg[Se]/m^3, 1993; Austria: MAK 0.1 mg[Se]/m^3, 1999; Belgium: TWA 0.2 mg[Se]/m^3, 1993; Denmark: TWA 0.1 mg[Se]/m^3, 1999; Finland: TWA 0.1 mg[Se]/m^3; STEL 0.3 mg[Se]/m^3, 1999; Hungary: STEL 0.1 mg[Se]/m^3, 1993; Norway: TWA 0.1 mg[Se]/m^3, 1999; the Philippines: TWA 0.2 mg[Se]/m^3, 1993; Poland: MAC (TWA) 0.1 mg[Se]/m^3, 1993; Sweden: NGV 0.1 mg [Se]/m^3, 1999; Switzerland: MAK-W 0.1 mg[Se]/m^3, 1999; Turkey: TWA 0.2 mg[Se]/m^3, 1993; United Kingdom: TWA 0.1 mg[Se]/m^3, 2000; Argentina, Bulgaria, Columbia, Jordan, South Korea, New Zealand, Singapore, Vietnam: ACGIH TLV®: TWA 0.2 mg[Se]/m^3. Russia[43] set a MAC of 0.1 μg/m^3 in ambient air in residential areas on a momentary basis and 0.05 μg/m^3 on a daily average basis.

Permissible Concentration in Water: To protect freshwater aquatic life: 35 μg/L as a 24-h average, never to exceed 260 μg/L for recoverable inorganic selenite. *To protect saltwater aquatic life:* 54 μg/L as a 24-h average, never to exceed 410 μg/L. To protect human health: 10 μg/L.[6] There are a variety of foreign standards for selenium in water.[35] The Czech Republic has set 0.1 mg/L in surface water, 0.05 mg/L in drinking water reserve, and 0.01 mg/L in drinking water. Germany[35] has set 0.008 mg/L as a maximum permissible concentration in drinking water. The EEC has set 0.01 mg/L as a MAC for drinking water.

Mexico has set 0.01 mg/L for receiving waters used for drinking purposes. Russia has set 0.001 mg/L as a MAC for drinking water. WHO has set 0.01 mg/L as a limit for drinking water. Maine has set[61] a guideline of 0.01 mg/L and Minnesota 0.045 mg/L for drinking water. EPA[62] has proposed 0.05 mg/L as a limit for drinking water.

Routes of Entry: Inhalation, ingestion, eye and/or skin contact.

Harmful Effects and Symptoms

Short Term Exposure: Irritates the eyes, skin, and respiratory tract. Inhalation of the dust can cause pulmonary edema, a medical emergency that can be delayed for several hours. This can cause death. May affect the liver, kidneys, heart, nervous system, and gastrointestinal tract. *Inhalation:* Dust or fumes can cause irritation of the nose, throat, and lungs; nausea, vomiting, intestinal disturbance; garlic odor on breath; headache, fatigue, and irritability. *Skin:* Can cause irritation, burning, and red or yellow discoloration. *Eyes:* Can cause irritation and injury. *Ingestion:* Can cause nausea, vomiting, abdominal pain, diarrhea, metallic taste, and garlic odor on breath. Elemental selenium has low acute systemic toxicity, but dust or fumes can cause serious irritation of the respiratory tract. In humans, a concentration of 5 ppm in food or 0.5 ppm in milk or water has been estimated to be dangerous (selenium compounds). Animal studies suggest that the lethal dose for an adult may be as low as 1/200 oz.

Long Term Exposure: Long-term exposure to selenium compounds may be a cause of amyotrophic lateral sclerosis in humans. Repeated or prolonged contact may cause skin dermatitis. May affect the central nervous system, blood, teeth, and bones. May cause nervousness, depression, pallor, digestive disturbances. Kidney and liver damage may occur. Daily ingestion of 25 mg of sodium selenite, containing 4—5 selenium, after 11 days showed signs of hair and fingernail loss, fatigue, nausea, vomiting, and garlic breath.

Points of Attack: Liver, kidneys, heart, nervous system, gastrointestinal tract.

Medical Surveillance: Urine test for selenium (should be less than 100 μg/L of urine). Liver and kidney function tests. Examination of the nervous system. EKG.

First Aid: If this chemical gets into the eyes, remove any contact lenses at once and irrigate immediately for at least 15 min, occasionally lifting upper and lower lids. Seek medical attention immediately. If this chemical contacts the skin, remove contaminated clothing and wash immediately with soap and water. Seek medical attention immediately. If this chemical has been inhaled, remove from exposure, begin rescue breathing (using universal precautions, including resuscitation mask) if breathing has stopped and CPR if heart action has stopped. Transfer promptly to a medical facility. When this chemical has been swallowed, get medical attention. Give large quantities of water and induce vomiting. Do not make an unconscious person vomit. Wear protective gloves when inducing vomiting. Medical observation is recommended for 24—48 h after breathing overexposure, as pulmonary edema may be delayed. As first aid for pulmonary edema, a doctor or authorized paramedic may consider administering a corticosteroid spray.

Note to physician: For severe poisoning *do not* use BAL [British Anti-Lewisite, dimercaprol, dithiopropanol $(C_3H_8OS_2)$] as it is contraindicated or ineffective in poisoning from selenium.

Personal Protective Methods: Wear protective gloves and clothing to prevent any reasonable probability of skin contact. Change gloves frequently. Safety equipment suppliers/manufacturers can provide recommendations on the most protective glove/clothing material for your operation. All protective clothing (suits, gloves, footwear, headgear) should be clean, available each day, and put on before work. Contact lenses should not be worn when working with this chemical. Wear dust-proof chemical goggles and face shield unless full face-piece respiratory protection is worn. Employees should wash immediately with soap when skin is wet or contaminated. Provide emergency showers and eyewash.

Respirator Selection: For levels up to 2 mg/m³ use a respirator with full face-piece and dust and mist filters. For levels up to 7.5 mg/m³ use a powered air-purifying respirator with dust and mist filters or supplied-air respirator operated in continuous-flow mode. For levels up to 10 mg/m³ use a high-efficiency particulate filter respirator with a full face-piece. For levels up to 100 mg/m³ use a Type C supplied-air respirator with a full face-piece operated in a positive-pressure mode. For levels above 100 mg/m³ or use in areas of unknown concentrations use a self-contained Type C supplied-air respirator with an auxiliary self-contained breathing apparatus, both with full face-piece and operated in a positive-pressure mode. For firefighting use a self-contained breathing apparatus with a full face-piece operated in a positive-pressure mode. For escape from a contaminated area use a high-efficiency particulate filter respirator with a full face-piece or an escape self-contained breathing apparatus with a full face-piece.

Storage: Color Code—Blue: Health Hazard/Poison: Store in a secure poison location. Prior to working with this chemical you should be trained on its proper handling and storage. Store at room temperature in watertight containers. A regulated, marked area should be established where this chemical is handled, used, or stored in compliance with OSHA Standard 1910.1045.

Shipping: This compound requires a shipping label of "POISONOUS/TOXIC MATERIALS." It falls in Hazard Class 6.1 and Packing Group I.

Spill Handling: Evacuate persons not wearing protective equipment from area of spill or leak until cleanup is complete. Remove all ignition sources. Collect powdered material in the most convenient and safe manner and deposit in sealed containers. Ventilate area after cleanup is complete. It may be necessary to contain and dispose of this chemical as a hazardous waste. If material or contaminated runoff enters waterways, notify downstream users of potentially contaminated waters. Contact your local or federal environmental protection agency for specific recommendations. If employees are required to clean up spills, they must be properly trained and equipped. OSHA 1910.120(q) may be applicable.

Fire Extinguishing: May burn but will not ignite readily. When heated to decomposition, may emit toxic fumes of selenium and sodium oxide. *Small fires:* dry chemical, carbon dioxide, water spray, or foam. *Large fires:* water spray, fog, or foam. Poisonous gases are produced in fire, including selenium and sodium oxides. If material or contaminated runoff enters waterways, notify downstream users of potentially contaminated waters. Notify local health and fire officials and pollution control agencies. From a secure, explosion-proof location, use water spray to cool exposed containers. If cooling streams are ineffective (venting sound increases in volume and pitch, tank discolors, or shows any signs of deforming), withdraw immediately to a secure position. If employees are expected to fight fires, they must be trained and equipped in OSHA 1910.156. The only respirators recommended for firefighting are self-contained breathing apparatuses that have full face-pieces and are operated in a pressure-demand or other positive-pressure mode.

Disposal Method Suggested: Liquid or solid: make a strongly acidic solution using hydrochloric acid. Slowly add sodium sulfite to the cold solution. Stir mixture producing sulfur dioxide. Heat, forming dark-gray selenium and black tellurium. Let stand overnight. Filter and dry. Ship to supplier.

References

Sax, N. I. (Ed.). (1983). *Dangerous Properties of Industrial Materials Report*, 3, No. 6, 75–77

US Environmental Protection Agency. (November 30, 1987). *Chemical Hazard Information Profile: Sodium Selenite*. Washington, DC: Chemical Emergency Preparedness Program

New York State Department of Health. (April 1986). *Chemical Fact Sheet: Sodium Selenite* (Version 2). Albany, NY: Bureau of Toxic Substance Assessment

Sodium sulfate   S:0540

Molecular Formula: Na_2O_4S
Common Formula: Na_2SO_4
Synonyms: Disodium sulfate; Glaubers salt; Natriumsulfat (German); Salt cake; Sodium sulphate; Sulfuric acid, Disodium salt; the nardite; Trona
CAS Registry Number: 7757-82-6
RTECS® Number: WE1650000
EC Number: 231-820-9
Regulatory Authority and Advisory Bodies
US EPA Gene-Tox Program, Negative: Cell transform.—SA7/SHE
European/International Regulations: not listed in Annex 1.

WGK (German Aquatic Hazard Class): 1—Low hazard to waters.

Description: Sodium sulfate is a white crystalline solid. It is frequently found as the decahydrate. Molecular weight = 142.04; Specific gravity (H_2O:1) = 2.71 at 25°C; Freezing/Melting point = 888°C. Hazard Identification (based on NFPA-704 M Rating System): Health 1, Flammability 0, Reactivity 1. Soluble in water.

Potential Exposure: Compound Description: Tumorigen, Mutagen; Reproductive Effector. Sodium sulfate is used in the manufacture of glass; as a precipitating agent in the manufacture of silver emulsions; as an analytical reagent; in making ultramarine and paper pulp; in ceramic glazes and pharmaceuticals; as a food additive; and as a filler in synthetic detergents.

Incompatibilities: Violent reaction with aluminum, magnesium. Attacks metals in the presence of moisture.

Permissible Exposure Limits in Air
Protective Action Criteria (PAC)
TEEL-0: 75 mg/m^3
PAC-1: 200 mg/m^3
PAC-2: 500 mg/m^3
PAC-3: 500 mg/m^3

Permissible Concentration in Water: Maryland[61] has set a guideline of 210 µg/L for sodium sulfate in drinking water.

Routes of Entry: Inhalation, ingestion, skin and/or eye contact.

Harmful Effects and Symptoms
Short Term Exposure: Inhalation: May cause irritation of nose and throat at high dust levels. *Skin:* No information found. *Eyes:* May cause irritation of nose and throat at high dust levels. *Ingestion:* Irritation of the digestive tract, vomiting, and diarrhea may result from ingestion of 1/2 oz.

Long Term Exposure: No information found.

First Aid: If this chemical gets into the eyes, remove any contact lenses at once and irrigate immediately for at least 15 min, occasionally lifting upper and lower lids. Seek medical attention immediately. If this chemical contacts the skin, remove contaminated clothing and wash immediately with soap and water. Seek medical attention immediately. If this chemical has been inhaled, remove from exposure, begin rescue breathing (using universal precautions, including resuscitation mask) if breathing has stopped and CPR if heart action has stopped. Transfer promptly to a medical facility. When this chemical has been swallowed, get medical attention. Give large quantities of water and induce vomiting. Do not make an unconscious person vomit.

Personal Protective Methods: Wear protective gloves and clothing to prevent any reasonable probability of skin contact. Safety equipment suppliers/manufacturers can provide recommendations on the most protective glove/clothing material for your operation. All protective clothing (suits, gloves, footwear, headgear) should be clean, available each day, and put on before work. Contact lenses should not be worn when working with this chemical. Wear dust-proof chemical goggles and face shield unless full face-piece respiratory protection is worn. Employees should wash immediately with soap when skin is wet or contaminated. Provide emergency showers and eyewash.

Respirator Selection: Wear a dust mask if necessary.

Storage: Color Code—Green: General storage may be used. Prior to working with this chemical you should be trained on its proper handling and storage. Keep tightly closed in a cool place.

Shipping: Prior to working with this chemical you should be trained on its proper handling and storage. Store in tightly closed containers in a cool, well-ventilated area away from strong oxidizers. Where possible, automatically transfer material from drums or other storage containers to process containers. Sources of ignition, such as smoking and open flames, are prohibited where this chemical is handled, used, or stored. Metal containers involving the transfer of this chemical should be grounded and bonded. Wherever this chemical is used, handled, manufactured, or stored, use explosion-proof electrical equipment and fittings.

Spill Handling: Evacuate persons not wearing protective equipment from area of spill or leak until cleanup is complete. Remove all ignition sources. Collect powdered material in the most convenient and safe manner and deposit in sealed containers. Ventilate area after cleanup is complete. It may be necessary to contain and dispose of this chemical as a hazardous waste. If material or contaminated runoff enters waterways, notify downstream users of potentially contaminated waters. Contact your local or federal environmental protection agency for specific recommendations. If employees are required to clean up spills, they must be properly trained and equipped. OSHA 1910.120(q) may be applicable.

Fire Extinguishing: This chemical is a noncombustible solid. Use dry chemical, carbon dioxide, water spray, or foam extinguishers. Poisonous gases are produced in fire, including sulfur oxides. If material or contaminated runoff enters waterways, notify downstream users of potentially contaminated waters. Notify local health and fire officials and pollution control agencies. From a secure, explosion-proof location, use water spray to cool exposed containers. If cooling streams are ineffective (venting sound increases in volume and pitch, tank discolors, or shows any signs of deforming), withdraw immediately to a secure position. If employees are expected to fight fires, they must be trained and equipped in OSHA 1910.156. The only respirators recommended for firefighting are self-contained breathing apparatuses that have full face-pieces and are operated in a pressure-demand or other positive-pressure mode.

Disposal Method Suggested: Do not discharge waste sodium sulfate directly into sewers or surface waters. Recovered sodium sulfate may be disposed of by burial in a landfill.

References
New York State Department of Health. (March 1986). *Chemical Fact Sheet: Sodium Sulfate.* Albany, NY: Bureau of Toxic Substance Assessment

US Environmental Protection Agency, Special Review and Reregistration Division Office of Pesticide Programs. (1998). *Agency Status of Pesticides in Registration, Reregistration, and Special Review* (Rainbow Report). Washington, DC

Sodium sulfite   S:0550

Molecular Formula: Na_2O_3S
Common Formula: Na_2SO_3
Synonyms: Disodium sulfite; Sodium sulfite(2:1); Sodium sulfite, anhydrous; Sulftech; Sulfurous acid, Sodium salt(1:2)
CAS Registry Number: 7757-83-7; *(alt.)* 10579-83-6; 68135-69-3
RTECS® Number: WE2150000
UN/NA & ERG Number: UN3260/154
EC Number: 231-821-4
Regulatory Authority and Advisory Bodies
Carcinogenicity: IARC: Human Inadequate Evidence; Animal Inadequate Evidence Group 3, 1992.
US EPA Gene-Tox Program, Positive: *S. cerevisiae*—reversion.
US EPA, FIFRA 1998 Status of Pesticides: Canceled.
European/International Regulations: not listed in Annex 1.
WGK (German Aquatic Hazard Class): 1—Low hazard to waters.
Description: Sodium sulfite is a white crystalline solid. Molecular weight = 126.04; Specific gravity (H_2O:1) = 2.63 at 25°C; Freezing/Melting point = (decomposes) 538°C. Hazard Identification (based on NFPA-704 M Rating System): Health 1, Flammability 0, Reactivity 0. Soluble in water.
Potential Exposure: Compound Description: Tumorigen, Mutagen. Sodium sulfite is used as a reducing agent; in boiler water treatment; food applications; in photographic developers and fixers; in bleaching of wool, paper, textiles, straw, and silk; manufacture of dyes; dechlorination; preservation of meat, fruit, and egg products; silvering of glass.
Incompatibilities: A strong reducing agent. Incompatible with strong oxidizers, combustibles, organic materials. Reacts with strong acids, producing toxic sulfur dioxide.
Permissible Exposure Limits in Air
Protective Action Criteria (PAC)
TEEL-0: 1 mg/m^3
PAC-1: 3 mg/m^3
PAC-2: 20 mg/m^3
PAC-3: 100 mg/m^3
Permissible Concentration in Water: Maryland[61] has set a guideline of 100 μg/L for sodium sulfite in drinking water.
Routes of Entry: Inhalation, ingestion, skin and/or eye contact.
Harmful Effects and Symptoms
Short Term Exposure: Inhalation: May cause asthmatic reactions. Animal studies indicate that levels of 1 mg/m^3

may cause slight swelling of lung tissue and irritation. *Skin:* Corrosive. Causes irritation and burns. *Eyes:* Corrosive. Causes severe irritation and burns. *Ingestion:* Solutions cause gastric irritation by the liberation of sulfurous acid. Because of rapid oxidation to sulfate, sulfites are well tolerated until large doses are reached; then violent colic and diarrhea, circulatory disturbances; central nervous depression; and death can occur. Increased dosage will result in sudden, violent colic and diarrhea, circulatory disturbances; slowed breathing; fainting, rapid pulse; and death. The lethal dose may be about 10 g (about 1/3 oz).
Long Term Exposure: Repeated or prolonged contact may cause skin sensitization. Repeated or prolonged inhalation exposure may cause asthma.
Points of Attack: Skin, lungs.
Medical Surveillance: Examination by a qualified allergist.
First Aid: If this chemical gets into the eyes, remove any contact lenses at once and irrigate immediately for at least 15 min, occasionally lifting upper and lower lids. Seek medical attention immediately. If this chemical contacts the skin, remove contaminated clothing and wash immediately with soap and water. Seek medical attention immediately. If this chemical has been inhaled, remove from exposure, begin rescue breathing (using universal precautions, including resuscitation mask) if breathing has stopped and CPR if heart action has stopped. Transfer promptly to a medical facility. When this chemical has been swallowed, get medical attention. If victim is *conscious*, administer water or milk. Do not induce vomiting.
Personal Protective Methods: Wear protective gloves and clothing to prevent any reasonable probability of skin contact. Safety equipment suppliers/manufacturers can provide recommendations on the most protective glove/clothing material for your operation. All protective clothing (suits, gloves, footwear, headgear) should be clean, available each day, and put on before work. Contact lenses should not be worn when working with this chemical. Wear dust-proof chemical goggles and face shield unless full face-piece respiratory protection is worn. Employees should wash immediately with soap when skin is wet or contaminated. Provide emergency showers and eyewash.
Respirator Selection: Follow regulations in OSHA 29CFR1910.134 or European Standard EN149. Use a NIOSH/MSHA- or European Standard EN149-approved respirator; or use an approved supplied-air respirator with a full face-piece operated in the positive-pressure mode, or with a full face-piece, hood, or helmet in the continuous-flow mode; or use a NIOSH/MSHA- or European Standard EN149-approved self-contained breathing apparatus with a full face-piece operated in pressure-demand or other positive-pressure mode.
Storage: (1) Color Code—White: Corrosive or Contact Hazard; Store separately in a corrosion-resistant location. (2) Color Code—Yellow Stripe (*strong reducing agent*): Reactivity Hazard; Store separately in an area isolated from flammables, combustibles, or other yellow coded materials.

Prior to working with this chemical you should be trained on its proper handling and storage. Store in a cool, dry location in closed containers.

Shipping: Corrosive solid, acidic, inorganic, n.o.s. requires a shipping label of "CORROSIVE." It falls in Hazard Class 8 and Packing Group III.

Spill Handling: Evacuate persons not wearing protective equipment from area of spill or leak until cleanup is complete. Remove all ignition sources. Collect powdered material in the most convenient and safe manner and deposit in sealed containers. Ventilate area after cleanup is complete. It may be necessary to contain and dispose of this chemical as a hazardous waste. If material or contaminated runoff enters waterways, notify downstream users of potentially contaminated waters. Contact your local or federal environmental protection agency for specific recommendations. If employees are required to clean up spills, they must be properly trained and equipped. OSHA 1910.120(q) may be applicable.

Fire Extinguishing: Nonflammable. Use agents suitable for surrounding fire. Decomposes above 538°C; Toxic sulfur dioxide gases formed in fire. If material or contaminated runoff enters waterways, notify downstream users of potentially contaminated waters. Notify local health and fire officials and pollution control agencies. From a secure, explosion-proof location, use water spray to cool exposed containers. If cooling streams are ineffective (venting sound increases in volume and pitch, tank discolors, or shows any signs of deforming), withdraw immediately to a secure position. If employees are expected to fight fires, they must be trained and equipped in OSHA 1910.156. The only respirators recommended for firefighting are self-contained breathing apparatuses that have full face-pieces and are operated in a pressure-demand or other positive-pressure mode.

References

New York State Department of Health. (March 1986). *Chemical Fact Sheet: Sodium Sulfite.* Albany, NY: Bureau of Toxic Substance Assessment

US Environmental Protection Agency, Special Review and Reregistration Division Office of Pesticide Programs. (1998). *Agency Status of Pesticides in Registration, Reregistration, and Special Review* (Rainbow Report). Washington, DC

Sodium tellurite   S:0560

Molecular Formula: Na_2O_3Te
Common Formula: Na_2TeO_3
Synonyms: Sodium tellurate(IV); Telluric acid, disodium salt; Tellurous acid, disodium salt; Telurito sodico (Spanish)
CAS Registry Number: 10102-20-2
RTECS® Number: WY24500000
UN/NA & ERG Number: UN3284 (tellurium compound, n.o.s.)/151

EC Number: 233-268-4
Regulatory Authority and Advisory Bodies
Air Pollutant Standard Set. See below, "Permissible Exposure Limits in Air" section.
Superfund/EPCRA 40CFR355, Extremely Hazardous Substances: TPQ = 500/10,000 lb (227/4540 kg).
Reportable Quantity (RQ): 500 lb (227 kg).
Canada, WHMIS, Ingredients Disclosure List Concentration 1.0%.
European/International Regulations: not listed in Annex 1.
WGK (German Aquatic Hazard Class): No value assigned.

Description: Sodium tellurite is a white crystalline solid. Molecular weight = 221.58; Hazard Identification (based on NFPA-704 M Rating System): Health 2, Flammability 0, Reactivity 0. Slightly soluble in water.

Potential Exposure: Used in bacteriology and medicine. Formerly used as pesticide.

Incompatibilities: Cadmium, nitric acid, halogens, oxidizers.

Permissible Exposure Limits in Air
OSHA PEL: 0.1 mg[Te]/m³ TWA.
NIOSH REL: 0.1 mg[Te]/m³ TWA.
ACGIH TLV®[1]: 0.1 mg[Te]/m³ TWA.
NIOSH IDLH: 25 mg[Te]/m³.
Protective Action Criteria (PAC)
TEEL-0: 0.174 mg/m³
PAC-1: 0.521 mg/m³
PAC-2: 20 mg/m³
PAC-3: 43.4 mg/m³
DFG MAK: 0.1 mg[Te]/m³, inhalable fraction.
Australia: TWA 0.1 mg/m³, 1993; Austria: MAK 0.1 mg/m³, 1999; Belgium: TWA 0.1 mg/m³, 1993; Denmark: TWA 0.1 mg/m³, 1999; Finland: TWA 0.1 mg/m³, STEL 0.3 mg/m³, 1999; France: VME 0.1 mg/m³, 1999; the Netherlands: MAC-TGG 0.1 mg[Te]/m³, 2003; the Philippines: TWA 0.1 mg/m³, 1993; Poland: MAC (TWA) 0.01 mg/m³; MAC (STEL) 0.03 mg/m³, 1999; Sweden: NGV 0.1 mg/m³, 1999; Switzerland: MAK-W 0.1 mg/m³, KZG-W 0.5 mg/m³, 1999; Turkey: TWA 0.1 mg/m³, 1993; United Kingdom: TWA 0.1 mg/m³, 2000; Argentina, Bulgaria, Columbia, Jordan, South Korea, New Zealand, Singapore, Vietnam: ACGIH TLV®: TWA 0.1 mg[Te]/m³. Several states have set guidelines or standards for tellurium in ambient air[60] ranging from 1.0 µg/m³ (North Dakota) to 1.6 µg/m³ (Virginia) to 2.0 µg/m³ (Connecticut and Nevada).

Determination in Air: Use NIOSH Analytical Method (IV) #7900. See also #7300, Elements; #7301; #7303; #9102; OSHA Analytical Method ID-121.

Permissible Concentration in Water: EPA[32] has suggested a permissible ambient goal of 1.4 µg/L based on health effects. Russia[43] set a MAC of 0.01 mg/L for tellurium in water bodies used for domestic purposes.

Routes of Entry: Inhalation, ingestion, skin and/or eye contact. Absorbed through the skin.

Harmful Effects and Symptoms

Short Term Exposure: Irritates the skin and eyes. Irritates the respiratory tract, causing cough and wheezing. The material is both an oral and dermal toxic hazard. The material is toxic by ingestion. Oral ingestion of tellurium compounds is generally regarded as extremely toxic. The probable oral lethal dose is 5–50 mg/kg or between 7 drops and 1 teaspoonful for a 70-kg (150 lb) person. Tellurium compounds are regarded as super toxic for skin exposures. Symptoms of exposure are as follows: sleepiness, fatigue, stupor; loss of appetite; nausea, vomiting, stomach pain; metallic taste; garlic odor of the breath and sweat; dryness of the mouth or excessive salivation; renal pain; bronchitis, irregular breathing; cyanosis, fatty degeneration of the liver; and unconsciousness.

Long Term Exposure: Repeated or prolonged exposure to tellurium compounds may cause reproductive damage, liver and kidney damage, lung irritation. Bronchitis may develop with cough, phlegm, and/or shortness of breath.

Points of Attack: Liver, kidneys, lungs.

Medical Surveillance: Lung function tests, liver and kidney function tests.

First Aid: Skin Contact: Flood all areas of body that have contacted the substance with water. Do not wait to remove contaminated clothing; do it under the water stream. Use soap to help assure removal. Isolate contaminated clothing when removed to prevent contact by others.[52]

Eye Contact: Remove any contact lenses at once. Immediately flush eyes well with copious quantities of water or normal saline for at least 20–30 min. Seek medical attention.

Inhalation: Leave contaminated area immediately; breathe fresh air. Proper respiratory protection must be supplied to any rescuers. If coughing, difficult breathing, or any other symptoms develop, seek medical attention at once, even if symptoms develop many hours after exposure.

Ingestion: If unconscious or convulsing, do not induce vomiting or give anything by mouth. Assure that victim's airway is open and lay him on his side with his head lower than his body and transport at once to a medical facility. If conscious and not convulsing, give a glass of water to dilute the substance. If medical advice is not readily available, consider inducing vomiting of this toxic material. Transport at once to a medical facility.

Personal Protective Methods: Wear protective gloves and clothing to prevent any reasonable probability of skin contact. Safety equipment suppliers/manufacturers can provide recommendations on the most protective glove/clothing material for your operation. All protective clothing (suits, gloves, footwear, headgear) should be clean, available each day, and put on before work. Contact lenses should not be worn when working with this chemical. Wear dust-proof chemical goggles and face shield unless full face-piece respiratory protection is worn. Employees should wash immediately with soap when skin is wet or contaminated. Provide emergency showers and eyewash.

Respirator Selection: *Up to 0.5 mg/m³:* Qm (APF = 25) (any quarter-mask respirator). *Up to 1 mg/m³:* 95QX [any particulate respirator equipped with an N95, R95, or P95 filter (including N95, R95, and P95 filtering face-pieces) except quarter-mask respirators. The following filters may also be used: N99, R99, P99, N100, R100, P100] or Sa (APF = 10) (any supplied-air respirator). *Up to 2.5 mg/m³:* Sa:Cf (APF = 25) (any supplied-air respirator operated in a continuous-flow mode) or PaprHie (APF = 25) (any powered, air-purifying respirator with a high-efficiency particulate filter). *Up to 5 mg/m³:* 100F (APF = 50) (any air-purifying, full face-piece respirator with an N100, R100, or P100 filter) or SaT: Cf (APF = 50) (any supplied-air respirator that has a tight-fitting face-piece and is operated in a continuous-flow mode) or PaprTHie (APF = 50) (any powered, air-purifying respirator with a tight-fitting face-piece and a high-efficiency particulate filter) or SCBAF (APF = 50) (any self-contained breathing apparatus with a full face-piece) or SaF (APF = 50) (any supplied-air respirator with a full-face-piece). *Up to 25 mg/m³:* Sa: Pd,Pp (APF = 1000) (any supplied-air respirator operated in a pressure-demand or other positive-pressure mode). *Emergency or planned entry into unknown concentrations or IDLH conditions:* SCBAF: Pd,Pp (APF = 10,000) (any self-contained breathing apparatus that has a full-face-piece and is operated in a pressure-demand or other positive-pressure mode) or SaF: Pd,Pp: ASCBA (APF = 10,000) (any supplied-air respirator that has a full-face-piece and is operated in a pressure-demand or other positive-pressure mode in combination with an auxiliary self-contained breathing apparatus operated in a pressure-demand or other positive-pressure mode). *Escape:* 100F (APF = 50) (any air-purifying, full-face-piece respirator with an N100, R100, or P100 filter) or SCBAE (any appropriate escape-type, self-contained breathing apparatus).

Note: Substance reported to cause eye irritation or damage; may require eye protection.

Storage: Color Code—Blue: Health Hazard/Poison: Store in a secure poison location. Prior to working with this chemical you should be trained on its proper handling and storage. Store in a refrigerator or a cool, dry place.

Shipping: Tellurium compound, n.o.s. requires a shipping label of "POISONOUS/TOXIC MATERIALS." They fall in Hazard Class 6.1 and Packing Group I to III.

Spill Handling: Evacuate persons not wearing protective equipment from area of spill or leak until cleanup is complete. Remove all ignition sources. Use HEPA vacuum or wet method to reduce dust during cleanup. Do not dry sweep. Collect powdered material in the most convenient and safe manner and deposit in sealed containers. Ventilate area after cleanup is complete. It may be necessary to contain and dispose of this chemical as a hazardous waste. If material or contaminated runoff enters waterways, notify downstream users of potentially contaminated waters. Contact your local or federal environmental protection agency for specific recommendations. If employees are required to clean up spills, they must be properly trained and equipped. OSHA 1910.120(q) may be applicable.

Fire Extinguishing: This chemical is a combustible solid. Use dry chemical, carbon dioxide, water spray, or foam extinguishers. Do not use halogens. Poisonous gases are produced in fire, including tellurium and sodium oxides. If material or contaminated runoff enters waterways, notify downstream users of potentially contaminated waters. Notify local health and fire officials and pollution control agencies. Containers may explode in fire. From a secure, explosion-proof location, use water spray to cool exposed containers. If cooling streams are ineffective (venting sound increases in volume and pitch, tank discolors, or shows any signs of deforming), withdraw immediately to a secure position. If employees are expected to fight fires, they must be trained and equipped in OSHA 1910.156. The only respirators recommended for firefighting are self-contained breathing apparatuses that have full face-pieces and are operated in a pressure-demand or other positive-pressure mode.

References

US Environmental Protection Agency. (November 30, 1987). *Chemical Hazard Information Profile: Sodium Tellurite.* Washington, DC: Chemical Emergency Preparedness Program

New Jersey Department of Health and Senior Services. (August 1999). *Hazardous Substances Fact Sheet: Sodium Tellurite.* Trenton, NJ

Soman (Agent GD, WMD)   S:0565

Molecular Formula: $C_7H_{16}FO_2P$
Common Formula: $CH_3P(O)(F)OCH(CH_3)C(CH_3)3$
Synonyms: EA 1210; Fluoromethylpinacolyloxyphosphine Oxide; GD (military designation); Methyl-fluoropinacolyl-phosphonate; Methyl-pinacolyloxyfluorophosphine oxide; Methyl-pinacolyloxyphosphonyl flouride; PFMP; Phosphonofluoridic acid, methyl-, 1, 2, 2-trimethylpropyl ester; Pinacolyl methylphosphonofluoridate; Pinacolyl methanefluorophosphonate; Pinacolyl methylfluorophosphonate; Pinacolyloxymethylphosphonyl flouride; Somain; Thickened GD; TGD; 1,2,2-Trimethylpropylmethylphosphonofluoridate; 1,2,2,-Trimethylpropoxyfluoromethylphosphine oxide; Zoman
Chlorosoman
O-Pinacolyl methylphosphonochloridate; méthylphosphono-chloridate de O-pinacolyle (French)
CAS Registry Number: 96-64-0; 50642-24-5; 7040-57-5 (Chlorosoman)
RTECS® Number: None found
UN/NA & ERG Number: UN2810/153
EC Number: Not assigned
Regulatory Authority and Advisory Bodies
Department of Homeland Security Screening Threshold Quantity: *Theft hazard* CUM 100 g (*96-64-0; 7040-57-5 chlorosoman*).
Carcinogenicity: GD is not listed by the International Agency for Research on Cancer (IARC), American

Conference of Governmental Industrial Hygienists (ACGIH), Occupational Safety and Health Administration (OSHA), or National Toxicology Program (NTP) as a carcinogen.

US DOT 49CFR172.101, Inhalation Hazard Chemical.
European/International Regulations: not listed in Annex 1.
WGK (German Aquatic Hazard Class): No value assigned.
Description: Soman (GD) is a fluorinated organophosphorus compound. Exposure to soman can cause death in minutes. A fraction of an ounce (1−10 mL) of soman on the skin can be fatal.[NIOSH] When pure, GD is a colorless liquid with fruity odor. With impurities, or upon aging, GD is an amber to dark-brown oily liquid with an odor of rotten fruit or camphor (like Vicks® Vapo-Rub). Do not rely on odor for detection; not everyone can smell low concentrations of this chemical. Molecular weight = 182.17; Boiling point = 198°C; 167−200°C; Specific gravity (H_2O = 1) = 1.022 at 25°C; Melting point: −42°C; Vapor pressure = 0.40 mmHg at 25°C; Vapor density (air = 1) = 6.3; Volatility = 3900 mg/m^3 at 25°C; Flash point = 121°C (oc). Flammability limits: Unknown. Hazard Identification (based on NFPA-704 M Rating System): Health 4, Flammability 1, Reactivity 1. Slightly soluble in water. Thickened agent GD [TGD] is essentially the same as GD except for viscosity. Viscosity = (approx.) 1180 centistokes uses "K125" (acryloid copolymer, 5%) to create TGD. K125 is not known to be a hazardous material except in a finely divided, powder form.[CDC]
Potential Exposure: Agent GD is a quick-acting chemical warfare nerve agent (nerve gas). Medical treatment of soman is difficult because it permanently binds to receptors in the body in minutes. Large amounts of the vapor or liquid can hurt you in minutes, and can quickly lead to death.
Persistence of Chemical Agent: Soman (GD): Summer: 10 min to 24 h; Winter: 2 h to 3 days.
Incompatibilities: Stable after storage in steel for 3 months at 65°C. GD corrodes steel at the rate of 1×10^{-5} inch/month. Raising the pH increases the rate of decomposition significantly. GD decomposes slowly in water; will hydrolyze to form HF−H−H−O−CH$_3$ and $(CH_3)_3$−C−C−O−P−OH. GD reacts readily with bases and weak acids. Under acid conditions, GD hydrolyzes, forming hydrofluoric acid (HF). Flammable hydrogen gas produced by the corrosive vapors reacting with metals, concrete, etc. may be present.
Permissible Exposure Limits in Air
IDLH: 0.05 mg/m^3.[US Army]
Acute Exposure Guideline Levels (AEGLs)
Level 1—potential minor discomfort or noticeable effects; reversible
10 min—0.00049 ppm;   30 min—0.00028 ppm;   1 h—0.0002 ppm; 4 h—0.0001 ppm; 8 h—0.00050 ppm
Level 2—potentially impacting functional abilities or ability to escape; potential delayed recovery
10 min—0.0062 ppm;   30 min—0.0035 ppm;   1 h—0.0024 ppm; 4 h—0.0013 ppm; 8 h—0.00091 ppm

Level 3—Life threatening; level of potential initial fatalities 10 min—0.053 ppm; 30 min—0.027 ppm; 1 h—0.018 ppm; 4 h—0.0098 ppm; 8 h—0.0071 ppm

Protective Action Criteria (PAC) GD*

TEEL-0: 0.00003 ppm

PAC-1: **0.00018** ppm

PAC-2: **0.0022** ppm

PAC-3: **0.017** ppm

*AEGLs (Acute Emergency Guideline Levels) & ERPGs (Emergency Response Planning Guideline) are in **bold face**.

The suggested permissible airborne exposure concentration of soman (GD) for an 8-h workday or a 40-hour workweek is an 8-h time-weighted average (TWA) of 0.00003 mg/m^3 (2×10^{-5} ppm). This value is based on the TWA of GB as proposed in the USAEHA Technical Guide No. 169, *Occupational Health Guidelines for the Evaluation and Control of Occupational Exposure to Nerve Agents GA, GB, GD, and VX*. Also, the general population limits (as recommended by the Surgeon General's Working Group, US Department of Health): 0.000003 mg/m^3.

Determination in Air: Available monitoring equipment for soman (GD) is the Automatic Chemical Agent Detector Alarm (ACADA), bubblers (GC method), and Chemical Agent Monitor (CAM).

Determination in Water: Log K_{ow} (estimated) = 1.02:Log $K_{benzene-water}$ = 1.61. Soman dissolves in water and remains very dangerous. To prevent anyone from drinking water mixed with soman, notify local health and pollution control officials. Also, notify operators of nearby water intakes and advise shutting water intakes. Hydrolysis by acidic, neutral, and basic mechanisms, giving fluoride and pinacolyl methylphosphonate. Rapidly hydrolyzed in basic solutions (e.g., Na_2CO_3, NaOH, or KOH) with a half-life of approximately 1 min at pH 11 at 25°C. Soman and its hydrolysis products exhibit no significant phototransformations in sunlight. Soman and its hydrolysis products are thermally stable at temperatures less than 49°C. Use M272 Chemical Agent Water Testing Kit. Detection limit for nerve agents is 0.02 mg/L. Bleaching powder (chlorinated lime) destroys soman but gives rise to cyanogen chloride (CAS: 506-77-4). See table of contents or name index for location of entry for: Cyanogen chloride.

Routes of Entry: Skin absorption, absorption through eyes, and inhalation.

Harmful Effects and Symptoms

Short Term Exposure: Soman (GD) is a lethal anticholinesterase agent with the median lethal dose in humans being: LCt$_{50}$ (inhalation) = 70 mg-min/m^3 (t = 10 min); LD$_{50}$ (PC, bare skin) = 0.35 g/man (70 kg). *One to several minutes following overexposure to airborne soman (GD)*, the following acute symptoms appear: *Local effects* (lasting 1−15 days, increase with dose): *Eyes:* Miosis (constriction of pupils); redness, pressure sensation on eyes. *Inhalation:* Rhinorrhea (runny nose), nasal congestion, tightness in chest, wheezing, salivation, nausea, vomiting. *Systemic effects* (increases with dose): When inhaled soman (GD) will cause excessive secretion causing coughing/breathing difficulty; salivation and sweating; vomiting, diarrhea, stomach cramps; involuntary urination/defecation; generalized muscle twitching/muscle cramps; CNS depression including anxiety, restlessness, giddiness, insomnia, excessive dreaming, and nightmares. With more severe exposure, also headache, tremor, drowsiness, concentration difficulty; memory impairment; confusion, unsteadiness on standing or walking; and progressing to death. After exposure to liquid soman (GD), the following acute symptoms appear: *Local effects*: *Eyes:* Miosis (constriction of pupils); redness, pressure sensation on eyes. *Ingestion:* salivation, anorexia, nausea, vomiting, abdominal cramps, diarrhea, involuntary defecation, heartburn. *Skin:* Sweating, muscle twitching. If recovery from nerve agent poisoning occurs, it will be complete unless anoxia or convulsions have gone unchecked so long that irreversible central nervous system changes due to anoxemia have occurred.

Long Term Exposure: Limited data suggest delayed neuropathy (postural sway, psychomotor performance). Miosis has been noted up to 62 days.[CDC] Chronic exposure to soman (GD) causes forgetfulness, thinking difficulty, vision disturbances, muscular aches/pains. Mild or moderately exposed people usually recover completely. Severely exposed people are not likely to survive. Unlike some organophosphate pesticides, nerve agents have not been associated with neurological problems lasting more than 1−2 weeks after the exposure. Although certain organophosphate pesticides have been shown to be teratogenic in animals, these effects have not been documented in carefully controlled toxicological evaluations for soman (GD). The inhibition of cholinesterase enzymes throughout the body by nerve agents is more or less irreversible so that their effects are prolonged. Until the tissue cholinesterase enzymes are restored to normal activity, probably by very slow regeneration over a period of weeks or 2−3 months if damage is severe, there is a period of increased susceptibility to the effects of another exposure to any nerve agent. During this period the effects of repeated exposures are cumulative; after a single exposure, daily exposure to concentrations of a nerve agent insufficient to produce symptoms may result in the onset of symptoms after several days. Continued daily exposure may be followed by increasingly severe effects. After symptoms subside, increased susceptibility persists for one to several days. The degree of exposure required to produce recurrence of symptoms, and the severity of these symptoms, depend on duration of exposure and time intervals between exposures. Increased susceptibility is not limited to the particular nerve agent initially absorbed. Agent GD is not listed by the International Agency for Research on Cancer (IARC); American Conference of Governmental Industrial Hygienists (ACGIH); Occupational Safety and Health Administration (OSHA); or National Toxicology Program (NTP) as a carcinogen.

Points of Attack: Respiratory system, lungs, central nervous system, cardiovascular system, skin, eyes, plasma and red blood cell cholinesterase, liver, kidneys.

Medical Surveillance: Consider the points of attack in pre-placement and periodic physical examinations. Urine thio-cyanate levels. Complete blood count (CBC). Evaluation of thyroid function. Liver function tests. Kidney function tests. Central nervous system tests. EKG.

First Aid: Inhalation: Hold breath until respiratory protective mask is donned. If severe signs of agent exposure appear (chest tightens, pupil constriction, a lack of coordination, etc.), immediately administer, in rapid succession, all three Nerve Agent Antidote Kit(s); Mark I injectors (or atropine if directed by the local physician). Injections using the Mark I kit injectors may be repeated at 5–20 min intervals if signs and symptoms are progressing until three series of injections have been administered. No more injections will be given unless directed by medical personnel. In addition, a record will be maintained of all injections given. If breathing has stopped, give artificial respiration. Mouth-to-mouth resuscitation should be used when approved mask-bag of oxygen delivery systems are not available. Do not use mouth-to-mouth resuscitation when facial contamination exists. If breathing is difficult, administer oxygen. Seek medical attention *immediately.*

Eye contact: Immediately flush eyes with water for 10–15 min, then don respiratory protective mask. Although miosis (pinpointing of the pupils) may be an early sign of agent exposure, an injection will not be administered when miosis is the only sign present. Instead, the individual will be taken *immediately* to the medical treatment facility for observation.

Skin contact: Don respiratory protective mask and remove contaminated clothing. Immediately wash contaminated skin with copious amounts of soap and water, 10% sodium carbonate solution, or 5% liquid household bleach. Rinse well with water to remove decontaminant. Administer nerve agent antidote kit, Mark I, only if local sweating and muscular twitching symptoms are present. Seek medical attention *immediately.*

Ingestion: Do not induce vomiting. First symptoms are likely to be gastrointestinal. *Immediately* administer Nerve Agent Antidote kit, Mark I. Seek medical attention immediately. If there is no apparent breathing, artificial resuscitation will be started immediately, but *do not* use mouth-to-mouth resuscitation when facial contamination exists; in this case, use mechanical resuscitator. The situation will dictate method of choice. When appropriate and trained personnel are available, cardiopulmonary resuscitation (CPR) may be necessary.

Notes for physician and qualified medical personnel: An individual who has received a known agent exposure or who exhibits definite signs or symptoms of agent exposure shall be given an intramuscular injection immediately with the MARK I kit autoinjectors. Some of the early symptoms of a *vapor exposure* may be rhinorrhea (runny nose) and/or tightness in the chest with shortness of breath (bronchial constriction). Some of the early symptoms of a *percutaneous exposure* may be local muscular twitching or sweating at the area of exposure followed by nausea or vomiting. Although miosis (pinpointing of the pupils) may be an early sign of agent exposure, an injection shall not be administered when myosis is the only sign present. Instead, the individual shall be taken immediately to the medical facility for observation. Injections using the MARK I kit injectors (or atropine only if directed by the local physician) may be repeated at 5–20 min intervals if signs and symptoms are progressing until three series of injections have been administered. No more injections will be given unless directed by medical personnel. In addition, a record will be maintained of all injections given. Administer, in rapid succession, all three MARK I kit injectors (or atropine if directed by the local physician) in the case of SEVERE signs of agent exposure. If indicated, CPR should be started immediately. Mouth-to-mouth resuscitation should be used when approved mask-bag or oxygen delivery systems are not available. Do not use mouth-to-mouth resuscitation when facial contamination exists. *CAUTION:* atropine does not act as a prophylactic and shall not be administered until an agent exposure has been ascertained.

Medical Treatment: Electrocardiogram (ECG), and adequacy of respiration and ventilation, should be monitored. Supplemental oxygenation, frequent suctioning of secretions, insertion of a tube into the trachea (endotracheal intubation), and assisted ventilation may be required. Diazepam (5–10 mg in adults and 0.2–0.5 mg/kg in children) may be used to control convulsions. Lorazepam or other benzodiazepines may be used, but barbiturates, phenytoin, and other anticonvulsants are not effective. Administration of atropine (if not already given) should precede the administration of benzodiazepines in order to best control seizures. Patients/victims who have inhalation exposure and who complain of chest pain, chest tightness, or cough should be observed and examined periodically for 6–12 h to detect delayed-onset inflammation of the large airways (bronchitis), inflammatory lung disease (pneumonia), accumulation of fluid in the lungs (pulmonary edema), or respiratory failure.

Decontamination: This is very important. The rapid physical removal of a chemical agent is essential. If you do not have the equipment and training, do not enter the hot or the warm zone to rescue and/or decontaminate victims. Medical personnel should wear the proper PPE. If the victim cannot move, decontaminate without touching and without entering the hot or the warm zone. Nerve gases stay in clothing; *do not* touch with bare skin—if possible, seal contaminated clothes and personal belongings in a sealed double bag. Use clean water from any source; if possible, use a hose (spray or fog to prevent injury to the victim) or other system to avoid touching the victim. Do not wait for soap or for the victim to remove clothing; begin washing immediately. Do not delay decontamination to obtain warm water; time is of the essence; use cold water instead. Immediately flush the eyes with water for at least 15 min. Use caution to avoid hypothermia in children and the elderly. Wash–strip–wash–evacuate upwind and uphill: Patients exposed to nerve agent by vapor only should be decontaminated by removing all clothing in a clean-air environment; and shampooing or rinsing the hair to prevent vapor-off gassing; patients exposed to liquid nerve agent should be decontaminated by washing in available clean water at least three times. Use liquid soap (dispose of container after use and replace), large amounts of water, and mild to moderate friction with a single-use sponge or washcloth in the first and second washes. Scrubbing of exposed skin with a brush is discouraged; skin damage may occur and may increase absorption. The third wash should be to rinse with large amounts of warm or hot water. Shampoo can be used to wash the hair. Decontaminate with diluted household bleach* (0.5%, or one part bleach to 200 parts water), but do not let any get in the victim's eyes, open wounds, or mouth. Wash off the diluted bleach solution after 15 min. Remember that the water you use to decontaminate the victims is dangerous. Be sure you have decontaminated the victims as much as you can before they are released from the area, so they do not spread the nerve gas. Rinse the eyes, mucous membranes, or open wounds with sterile saline or water and then move away from the hot zone in an upwind and uphill direction.

*The following can be used in addition to household bleach: (1) solids, powders, and solutions containing various types of bleach (NaOCl or Ca(OCl)$_2$); (2) DS2 (2% NaOH, 70% diethylenetriamine, 28% ethylene glycol monomethyl ether); (3) towelettes moistened with sodium hydroxide (NaOH) dissolved in water, phenol, ethanol, and ammonia.

Note: Use 5% solution of common bleach (sodium hypochlorite) or calcium hypochlorite solution (48 oz per 5 gallons of water) to decontaminate scissors used in clothing removal, clothes and other items.

Personal Protective Methods: *Protective gloves*: Butyl Glove M3 and M4; Norton chemical protective glove set. *Eye protection*: chemical goggles. For splash hazards use goggles and face shield. *Other protective equipment*: Full protective clothing will consist of M9 mask and hood; butyl rubber suit (M3), M2A1 butyl boots; M3 and M4 gloves; unimpregnated underwear; or demilitarization protective ensemble (DPE). For laboratory operations, wear lab coats and have a protective mask readily available.

Respirator Selection: *When used as a weapon, use SCBA Respirator Certified By NIOSH For CBRN Environments.* Positive pressure, full face-piece, NIOSH/MSHA- or European Standard EN 149-approved self contained breathing apparatus (SCBA) will be worn where there is danger of oxygen deficiency and when directed by the fire chief or chemical accident/incident (CAI) operations officer. The M9 or M17 series mask may be worn in lieu of SCBA when there is no danger of oxygen deficiency. In cases where firefighters are responding to a chemical accident/ incident for rescue/reconnaissance purposes, they will wear appropriate levels of protective clothing.

From the Riegle Report
Less than 0.00003 mg/m^3: M9, M17, or M40 series mask shall be available for escape as necessary. *0.00003 mg/m^3 to*

0.06 mg/m^3: M9 or M40 series mask with Level A or Level B ensemble. *Greater than 0.06 mg/m^3:* or DPE or TAPES used with prior approval from AMC Field Safety Activity. *Note:* When DPE or TAPES is not available, the M9 or M40 series mask with Level A protective ensemble can be used. However, use time shall be restricted to the extent operationally feasible, and may not exceed 1 h. As an additional precaution, the cuffs of the sleeves and the legs of the M3 suit shall be taped to the gloves and boots respectively to reduce aspiration.

Local Exhaust: Mandatory. Must be filtered or scrubbed to limit exit concentration to <0.00001 mg/m^3 (averaged over 8 h/day, indefinitely).

Special: Chemical laboratory hoods shall have an average inward face velocity of 100 linear feet per minute (lfpm) + 10% with the velocity at any point not deviating from the average face velocity by more than 20%. Laboratory hoods shall be located such that cross-drafts do not exceed 20% of the inward face velocity. A visual performance test utilizing smoke-producing devices shall be performed in assessing the ability of the hood to contain agent GD. Emergency back-up power necessary: Hoods should be tested semiannually or after modification or maintenance operations. Operations should be performed 20 cm inside hood face. *Other:* Recirculation of exhaust air from agent areas is prohibited. No connection between agent areas and other areas through ventilation system is permitted.

Storage: In handling soman (GD), the buddy system will be incorporated. No smoking, eating, or drinking is permitted in areas containing agent GD. Containers should be periodically inspected for leaks (either visually or by a detector kit) and prior to transferring the containers from storage to work areas. Stringent control over all personnel practices must be exercised. Decontamination equipment shall be conveniently located. Exits must be designed to permit rapid evacuation. Chemical showers, eyewash stations, and personal cleanliness facilities shall be provided. Wash hands before meals and each worker will shower thoroughly with special attention given to hair, face, neck, and hands, using plenty of soap before leaving at the end of the workday. *Other precautions:* Agent GD must be double-contained in liquid and vapor-tight containers when in storage or when outside of the ventilation hood.

Shipping: Toxic, liquids, organic, n.o.s. [Inhalation hazard, Packing Group I, Zone B] requires a shipping label of "POISONOUS/TOXIC MATERIALS." (Pinacolyl methyl phosphonofluoridate), Inhalation Hazard. It falls in Hazard Class 6.1 and Packing Group I, Hazard Zone B.

Spill Handling: Warn pollution control authorities and advise shutting water intakes. Spills must be contained by covering with vermiculite, diatomaceous earth, clay, fine sand, sponges, and paper or cloth towels. This containment is followed by treatment with copious amounts of aqueous sodium hydroxide solution (a minimum of 10%). Scoop up all material and place in a fully removable head drum with a high-density polyethylene liner. Cover the contents of the drum with decontaminating solution as above before affixing the drum head. After sealing the head, the exterior of the drum shall be decontaminated and then labeled in accordance with IAW, EPA, and DOT regulations. All leaking containers shall be overpacked with vermiculite placed between the interior and exterior containers. Decontaminate and label in accordance with IAW, EPA, state, and DOT regulations. Dispose of material used to decontaminate exterior of drum in accordance with IAW, federal, state, and local regulations. Contaminated clothing will be placed in a fully removable head drum with a high-density polyethylene liner and the contents shall be covered with decontaminating solution as above before affixing the drum head. Conduct general area monitoring to confirm that the atmospheric concentrations do not exceed the exposure limits (see Section 8). If 10 wt. % aqueous sodium hydroxide solution is not available, then the following decontaminants may be used instead and are listed in the order of preference: Decontaminating Solution No. 2 [DS2: (2% NaOH, 70% diethylenetriamine, 28% ethylene glycol monomethyl ether)], sodium carbonate, and Supertropical Tropical Bleach Slurry (STB).

GD, when used as a weapon

Initial isolation and protective action distances

Distances shown are likely to be affected during the first 30 min after materials are spilled and could increase with time. If more than one tank car, cargo tank, portable tank, or large cylinder involved in the incident is leaking, the protective action distance may need to be increased. You may need to seek emergency information from CHEMTREC at (800) 424-9300 or seek professional environmental engineering assistance from the US EPA Environmental Response Team at (908) 548-8730 (24-h response line).

Small spills (From a small package or a small leak from a large package)

First: Isolate in all directions (feet/meters)   200/60
Then: Protect persons downwind (miles/kilometers)
Day   0.3/0.5
Night   0.5/0.8

Large spills (From a large package or from many small packages)

First: Isolate in all directions (feet/meters)   1500/500
Then: Protect persons downwind (miles/kilometers)
Day   1.1/1.8
Night   1.5/2.4

Fire Extinguishing: When heated, vapors may form explosive mixtures with air, presenting an explosion hazard indoors, outdoors, and in sewers. Containers may explode when heated. Fires involving soman (GD) should be contained to prevent contamination of uncontrolled areas. All persons not engaged in extinguishing the fire should be evacuated immediately. Contact with soman (GD) or its vapors can be fatal. When responding to a fire alarm in buildings or areas containing nerve agents, firefighting personnel should wear full firefighter protective clothing (without TAP clothing) during chemical agent firefighting and fire rescue operations. Respiratory protection is required. Positive pressure, full face-piece, NIOSH/MSHA- or European Standard EN 149-approved self-contained breathing apparatus (SCBA) will be worn where there is danger of oxygen deficiency and when directed by the fire chief or chemical accident/incident (CAI) operations officer. The M9 or M17 series mask may be worn in lieu of SCBA when there is no danger of oxygen deficiency. In cases where firefighters are responding to a chemical accident/incident for rescue/reconnaissance purposes, they will wear appropriate levels of protective clothing.

Disposal Method Suggested: A minimum of 55 g of decon solution is required per gram of soman (GD). Decontaminant/agent solution is allowed to agitate for a minimum of 1 h. Agitation is not necessary following the first hour provided a single phase is obtained. At the end of the first hour the pH should be checked and adjusted up to 11.5 with additional sodium hydroxide as required. An alternate solution for the decontamination of soman (GD) is 10% sodium carbonate in place of the 10% sodium hydroxide solution above. Continue with 55 g of decon per gram of GD. Agitate for 1 h and allow to react for 3 h. At the end of the third hour, adjust the pH to above 10. It is also permitted to substitute 5.25% sodium hypochlorite for the 10% sodium hydroxide solution above. Continue with 55 g of decon per gram of GD. Agitate for 1 h and allow to react for 3 h, then adjust the pH to above 10. Scoop up all material and place in a fully removable head and a high-density polyethylene liner. Cover the contents with additional decontaminating solution before affixing the drum head. After sealing the head, the exterior of the drum shall be decontaminated and then labeled in accordance with IAW, EPA, and DOT regulations. All contaminated clothing will be placed in a fully removable head drum with a high-density polyethylene liner. Cover the contents of the drum with decontaminating solution as above before affixing the drum head. After sealing the head, the exterior of the drum shall be decontaminated and then labeled per IAW, EPA, and DOT regulations. All leaking containers shall be overpacked with vermiculite placed between the interior and exterior containers. Decontaminate and label in accordance with IAW, EPA, and DOT regulations. Conduct general area monitoring to confirm that the atmospheric concentrations do not exceed the exposure limits. *Waste disposal method*: Open pit burning or burying of GD or items containing or contaminated with GD in any quantity is prohibited. The detoxified GD (using procedures above) can be thermally destroyed by incineration in an EPA approved incinerator in accordance with appropriate provisions of federal, state, and local RCRA regulations. *NOTE:* Several states define decontaminated surety material as a RCRA Hazardous Waste.

References

The Riegle Report: A Report of Chairman Riegle, D. W., Jr. and Ranking Member Alfonse M. D'Amato of the Committee on Banking, Housing and Urban Affairs with Respect to Export Administration, United States Senate, 103rd Congress, 2d Session, (May 25, 1994)

Schneider, A. L., et al. (2007). *CHRIS + CD-ROM Version 2.0, United Coast Guard Chemical Hazard Response Information System (COMDTINST 16465.12C).* Washington, DC: United States Coast Guard and the Department of Homeland Security

Stannic chloride, hydrated   S:0570

Molecular Formula: $Cl_4H_{10}O_5Sn$

Common Formula: $SnCl_4 \cdot 5H_2O$

Synonyms: Stannic chloride pentahydrate; Tetrachlorostannane pentahydrate

CAS Registry Number: 10026-06-9 (hydrated)

RTECS® Number: XP8870000

UN/NA & ERG Number: UN2440/154

EC Number: 050-001-00-5

Regulatory Authority and Advisory Bodies

Air Pollutant Standard Set. See below, "Permissible Exposure Limits in Air" section.

Canada, WHMIS, Ingredients Disclosure List Concentration 1.0%.

European/International Regulations: not listed in Annex 1.

WGK (German Aquatic Hazard Class): No value assigned.

Description: Stannic chloride is a white to yellow powder with a faint odor of HCl. Molecular weight = 350.59; Boiling point = 114°C; Freezing/Melting point = −33°C. Hazard Identification (based on NFPA-704 M Rating System): Health 3, Flammability 0, Reactivity 1. Soluble in water.

Potential Exposure: Hydrated stannic chloride is used for fixing certain textile dyes, and for treating silk to give weight to the fabric.

Incompatibilities: Reacts violently with water, forming corrosive hydrochloric acid and tin oxide fumes. Reacts with turpentine, alcohols, and amines, causing fire and explosion hazard. Attacks many metals; some forms of plastics, rubber, and coatings. Reacts with moist air to form hydrochloric acid.

Permissible Exposure Limits in Air

OSHA PEL: 2 mg[Sn]/m³ TWA.

NIOSH REL: 2 mg[Sn]/m³ TWA.

ACGIH TLV®[1]: 2 mg[Sn]/m³ TWA.

No TEEL available.

DFG MAK: *tin, inorganic compounds:* No numerical value established. Data may be available.

NIOSH IDLH: 100 mg[Sn]/m³.

Routes of Entry: Inhalation, ingestion, skin and/or eye contact.

Harmful Effects and Symptoms

Short Term Exposure: Hydrated stannic chloride can affect you when breathed in. Corrosive to the eyes, skin, and respiratory tract. Eye damage may be permanent. Breathing hydrated stannic chloride can irritate the throat and bronchial tubes, causing cough and/or difficulty in breathing. Higher exposures can cause pulmonary edema, a medical emergency that can be delayed for several hours. This can cause death.

Long Term Exposure: Changes may occur on chest X-ray with repeated exposure, resulting in impaired lung functions. Tin may interfere with the body's ability to absorb iron from the diet or from vitamins, contributing to iron deficiency anemia.

Points of Attack: Lungs, blood.

Medical Surveillance: Before beginning employment and at regular times after that, the following are recommended: lung function tests. Complete blood count (CBC). Hemoglobin or hematocrit. If acute breathing exposure to heated hydrated stannic chloride with chlorine gas is suspected, also consider chest X-ray.

First Aid: If this chemical gets into the eyes, remove any contact lenses at once and irrigate immediately for at least 15 min, occasionally lifting upper and lower lids. Seek medical attention immediately. If this chemical contacts the skin, remove contaminated clothing and wash immediately with soap and water. Seek medical attention immediately. If this chemical has been inhaled, remove from exposure, begin rescue breathing (using universal precautions, including resuscitation mask) if breathing has stopped and CPR if heart action has stopped. Transfer promptly to a medical facility. When this chemical has been swallowed, get medical attention. If victim is *conscious*, administer water or milk. Do not induce vomiting. Medical observation is recommended for 24−48 h after breathing overexposure, as pulmonary edema may be delayed. As first aid for pulmonary edema, a doctor or authorized paramedic may consider administering a corticosteroid spray.

Personal Protective Methods: Wear protective gloves and clothing to prevent any reasonable probability of skin contact. Safety equipment suppliers/manufacturers can provide recommendations on the most protective glove/clothing material for your operation. All protective clothing (suits, gloves, footwear, headgear) should be clean, available each day, and put on before work. Contact lenses should not be worn when working with this chemical. Wear dust-proof chemical goggles and face shield unless full face-piece respiratory protection is worn. Employees should wash immediately with soap when skin is wet or contaminated. Provide emergency showers and eyewash.

Respirator Selection: NIOSH/OSHA: *Up to 10 mg/m³:* Qm (APF = 25) (any quarter-mask respirator). *Up to 20 mg/m³:* 95XQ (APF = 10) [any particulate respirator equipped with an N95, R95, or P95 filter (including N95, R95, and P95 filtering face-pieces) except quarter-mask respirators. The following filters may also be used: N99, R99, P99, N100, R100, P100] or Sa* (APF = 10) (any supplied-air respirator). *Up to 50 mg/m³:* Sa:Cf (APF = 25)* (any supplied-air respirator operated in a continuous-flow mode) or PaprHie (APF = 25)* (any powered air-purifying respirator with a high-efficiency particulate filter). *Up to 100 mg/m³:* 100F (APF = 50) (any air-purifying, full-face-piece respirator with an N100, R100, or P100 filter. Click here for information on selection of N, R, or P filters) or SCBAF (APF = 50) (any self-contained breathing apparatus with a full face-piece) or SaF (APF = 50) (any supplied-air respirator with a full face-piece). *Emergency or planned entry into unknown concentrations or IDLH conditions:* SCBAF: Pd, Pp (APF = 10,000) (any self-contained breathing apparatus that has a full face-piece and is operated in a pressure-demand or other positive-pressure mode) or SAF; PD,PP: ASCBA (any supplied-air respirator that has a full face-piece and is operated in a pressure-demand or other positive-pressure mode in combination with an auxiliary self-contained positive-pressure breathing apparatus). *Escape:* 100F (APF = 50) (any air-purifying, full-face-piece respirator with an N100, R100, or P100 filter) or SCBAE (any appropriate escape-type, self-contained breathing apparatus).

*Substance reported to cause eye irritation or damage; may require eye protection.

Storage: Color Code—White: Corrosive or Contact Hazard; Store separately in a corrosion-resistant location. Prior to working with this chemical you should be trained on its proper handling and storage. Hydrated stannic chloride must be stored to avoid contact with water, moisture, chlorine, turpentine, ethylene oxide, and alkyl nitrates, since violent reactions occur. Store in tightly closed containers in a cool, well-ventilated area away from flammable and combustible materials. Keep hydrated stannic chloride dry and out of direct sunlight. Hydrated stannic chloride can attack some plastics, rubbers, and coatings.

Shipping: Stannic chloride, pentahydrate, requires a shipping label of "CORROSIVE." It falls in Hazard Class 8 and Packing Group III.

Spill Handling: Evacuate persons not wearing protective equipment from area of spill or leak until cleanup is complete. Remove all ignition sources. Collect powdered material in the most convenient and safe manner and deposit in sealed containers. Ventilate area after cleanup is complete. It may be necessary to contain and dispose of this chemical as a hazardous waste. If material or contaminated runoff enters waterways, notify downstream users of potentially contaminated waters. Contact your local or federal environmental protection agency for specific recommendations. If employees are required to clean up spills, they must be properly trained and equipped. OSHA 1910.120(q) may be applicable.

Fire Extinguishing: Extinguish fire using an agent suitable for type of surrounding fire. Hydrated stannic chloride itself does not burn. Poisonous gases are produced in fire, including hydrogen chloride, chlorine, and tin. If material or contaminated runoff enters waterways, notify downstream users of potentially contaminated waters. Notify local health and fire officials and pollution control agencies. From a secure, explosion-proof location, use water spray to cool exposed containers. If cooling streams are ineffective (venting sound increases in volume and pitch, tank discolors, or shows any signs of deforming), withdraw immediately to a secure position. If employees are expected to fight fires, they must be trained and equipped in OSHA 1910.156. The only respirators recommended for firefighting are self-contained breathing apparatuses that have full face-pieces and are operated in a pressure-demand or other positive-pressure mode.

References
New Jersey Department of Health and Senior Services. (September 2002). *Hazardous Substances Fact Sheet: Stannic Chloride, Hydrated.* Trenton, NJ

Stannous chloride   S:0580

Molecular Formula: Cl_2Sn; $Cl_2Sn \cdot 2H_2O$ (dihydrate)
Common Formula: $SnCl_2$
Synonyms: C.I. 77864; NCI-C02722; Stannochlor; Tin(II) chloride; Tin dichloride; Tin protochloride
Dihydrate: Stannous dichloride dihydrate; Stannochlor; Stannous chloride dihydrate; Tin(II) chloride, dihydrate (1:2:2)
CAS Registry Number: 7772-99-8; 10025-69-1 (dihydrate)
RTECS® Number: XP8700000; XP8850000 (dihydrate)
UN/NA & ERG Number: UN3260/154
EC Number: 231-868-0
Regulatory Authority and Advisory Bodies
Carcinogenicity: NCI: Carcinogenesis Studies (feed); equivocal evidence: rat; no evidence: mouse.
US EPA Gene-Tox Program, Negative: *B. subtilis* rec assay; TRP reversion; Negative/limited: Carcinogenicity—mouse/rat.
Air Pollutant Standard Set: ACGIH[1]; DFG[3]; OSHA.[58]
Canada, WHMIS, Ingredients Disclosure List Concentration 1.0%.
European/International Regulations: not listed in Annex 1.
WGK (German Aquatic Hazard Class): 1—Low hazard to waters.
Description: Stannous chloride is a white crystalline solid. Molecular weight = 189.59; 225.63 (dihydrate); Boiling point = (decomposes) 652°C; Freezing/Melting point = 246°C. Hazard Identification (based on NFPA-704 M Rating System): Health 2, Flammability 0, Reactivity 0. Soluble in water; solubility = 90% at 20°C.
Potential Exposure: Compound Description: Tumorigen, Mutagen; Reproductive Effector; (dihydrate) Mutagen; Reproductive Effector. Stannous chloride is used as a dye, pigment, and printing ink; in making chemicals; chemical preservatives; food additives; polymers, textiles, glass, silvering mirrors.
Incompatibilities: A strong reducing agent. Reacts violently with oxidants. Reacts violently with bromine trifluoride,

potassium, hydrazine hydrate, sodium, sodium peroxide, ethylene oxide, and nitrates. Keep away from moisture, sources of oxygen, and combustible materials.

Permissible Exposure Limits in Air
Includes dihydrate
OSHA PEL: 2 mg[Sn]/m^3 TWA.
NIOSH REL: 2 mg[Sn]/m^3 TWA.
ACGIH TLV®[1]: 2 mg[Sn]/m^3 TWA.
NIOSH IDLH: 100 mg[Sn]/m^3.
Protective Action Criteria (PAC)
TEEL-0: 3.19 mg/m^3
PAC-1: 9.58 mg/m^3
PAC-2: 50 mg/m^3
PAC-3: 160 mg/m^3
EUR OEL: 2 mg[Sn]/m^3 as TWA.
DFG MAK: *tin, inorganic compounds:* No numerical value established. Data may be available.
Australia: TWA 2 mg[Sn]/m^3, 1993; Austria: MAK 2 mg/m^3, 1999; Belgium: TWA 2 mg[Sn]/m^3, 1993; Denmark: TWA 2 mg[Sn]/m^3, 1999; Finland: TWA 2 mg [Sn]/m^3, 1999; Hungary: TWA 1 mg[Sn]/m^3; STEL 2 mg [Sn]/m^3, [skin], 1993; Norway: TWA 2 mg[Sn]/m^3, 1999; the Philippines: TWA 2 mg[Sn]/m^3, 1993; Poland: TWA 2 mg[Sn]/m^3, 1999; Switzerland: MAK-W 2 mg[Sn]/m^3, KZG-W 4 mg[Sn]/m^3, 1999; Thailand: TWA 2 mg[Sn]/m^3, 1993; United Kingdom: TWA 2 mg[Sn]/m^3; STEL 4 mg [Sn]/m^3, 2000; Argentina, Bulgaria, Columbia, Jordan, South Korea, New Zealand, Singapore, Vietnam: ACGIH TLV®: TWA 2 mg[Sn]/m^3.

Routes of Entry: Inhalation, ingestion, skin and/or eye contact.

Harmful Effects and Symptoms
Short Term Exposure: Stannous chloride can affect you when breathed in. Contact can cause skin and eye burns. Breathing stannous chloride dust or mist can irritate the throat and bronchial tubes, causing cough and/or difficulty in breathing. May affect the central nervous system and blood when ingested.

Long Term Exposure: Effects on the liver following ingestion. Changes may also occur on chest X-ray with repeated exposures; reduced lung function has been reported. Tin may interfere with the body's ability to absorb iron from the diet or from vitamins, contributing to iron deficiency anemia.

Points of Attack: Lungs, blood, liver.

Medical Surveillance: For those with frequent or potentially high exposure (half the TLV or greater), the following are recommended before beginning work and at regular times after that: lung function tests. Complete blood count (CBC); hemoglobin or hematocrit. Liver function tests.

First Aid: If this chemical gets into the eyes, remove any contact lenses at once and irrigate immediately for at least 20 min, occasionally lifting upper and lower lids. Seek medical attention immediately. If this chemical contacts the skin, remove contaminated clothing and wash immediately with soap and water. Seek medical attention immediately. If this chemical has been inhaled, remove from exposure,

begin rescue breathing (using universal precautions, including resuscitation mask) if breathing has stopped and CPR if heart action has stopped. Transfer promptly to a medical facility. When this chemical has been swallowed, get medical attention. If victim is *conscious*, administer water or milk. Do not induce vomiting.

Personal Protective Methods: Wear protective gloves and clothing to prevent any reasonable probability of skin contact. Safety equipment suppliers/manufacturers can provide recommendations on the most protective glove/clothing material for your operation. Neoprene™ and polyvinyl chloride are among the recommended protective materials. All protective clothing (suits, gloves, footwear, headgear) should be clean, available each day, and put on before work. Contact lenses should not be worn when working with this chemical. Wear dust-proof chemical goggles and face shield unless full face-piece respiratory protection is worn. Employees should wash immediately with soap when skin is wet or contaminated. Provide emergency showers and eyewash.

Respirator Selection: NIOSH/OSHA: *Up to 10 mg/m^3:* Qm (APF = 25) (any quarter-mask respirator). *Up to 20 mg/m^3:* 95XQ (APF = 10) [any particulate respirator equipped with an N95, R95, or P95 filter (including N95, R95, and P95 filtering face-pieces) except quarter-mask respirators. The following filters may also be used: N99, R99, P99, N100, R100, P100] or Sa* (APF = 10) (any supplied-air respirator). *Up to 50 mg/m^3:* Sa:Cf (APF = 25)* (any supplied-air respirator operated in a continuous-flow mode) or PaprHie (APF = 25)* (any powered air-purifying respirator with a high-efficiency particulate filter). *Up to 100 mg/m^3:* 100F (APF = 50) (any air-purifying, full-face-piece respirator with an N100, R100, or P100 filter. Click here for information on selection of N, R, or P filters) or SCBAF (APF = 50) (any self-contained breathing apparatus with a full face-piece) or SaF (APF = 50) (any supplied-air respirator with a full face-piece). *Emergency or planned entry into unknown concentrations or IDLH conditions:* SCBAF: Pd,Pp (APF = 10,000) (any self-contained breathing apparatus that has a full face-piece and is operated in a pressure-demand or other positive-pressure mode) or SAF; PD,PP: ASCBA (any supplied-air respirator that has a full face-piece and is operated in a pressure-demand or other positive-pressure mode in combination with an auxiliary self-contained positive-pressure breathing apparatus). *Escape:* 100F (APF = 50) (any air-purifying, full-face-piece respirator with an N100, R100, or P100 filter) or SCBAE (any appropriate escape-type, self-contained breathing apparatus).

*Substance reported to cause eye irritation or damage; may require eye protection.

Storage: (1) Color Code—White: Corrosive or Contact Hazard; Store separately in a corrosion-resistant location. (2) Color Code—Yellow Stripe (*strong reducing agent*): Reactivity Hazard; Store separately in an area isolated from flammables, combustibles, or other yellow coded materials. Prior to working with this chemical you should be trained on its proper handling and storage. Stannous chloride must be stored to avoid contact with bromine trifluoride, nitrates, potassium hydrazine hydrate, sodium peroxide, ethylene oxide, and hydrogen peroxide, since violent reactions occur. Store in tightly closed containers in a cool, well-ventilated area away from moisture, sources of oxygen, and combustible materials. Sources of ignition, such as smoking and open flames, are prohibited where stannous chloride is used, handled, or stored in a manner that could create a potential fire or explosion hazard.

Shipping: Corrosive solid, acidic, inorganic, n.o.s. requires a shipping label of "CORROSIVE." It falls in Hazard Class 8 and Packing Group III.

Spill Handling: Evacuate persons not wearing protective equipment from area of spill or leak until cleanup is complete. Remove all ignition sources. Collect powdered material in the most convenient and safe manner and deposit in sealed containers. Ventilate area after cleanup is complete. It may be necessary to contain and dispose of this chemical as a hazardous waste. If material or contaminated runoff enters waterways, notify downstream users of potentially contaminated waters. Contact your local or federal environmental protection agency for specific recommendations. If employees are required to clean up spills, they must be properly trained and equipped. OSHA 1910.120(q) may be applicable.

Fire Extinguishing: Extinguish fire using an agent suitable for type of surrounding fire. Stannous chloride itself does not burn. Stannous chloride may ignite combustible materials. Poisonous gases are produced in fire, including chlorine and stannous sulfate. If material or contaminated runoff enters waterways, notify downstream users of potentially contaminated waters. Notify local health and fire officials and pollution control agencies. From a secure, explosion-proof location, use water spray to cool exposed containers. If cooling streams are ineffective (venting sound increases in volume and pitch, tank discolors, or shows any signs of deforming), withdraw immediately to a secure position. If employees are expected to fight fires, they must be trained and equipped in OSHA 1910.156. The only respirators recommended for firefighting are self-contained breathing apparatuses that have full face-pieces and are operated in a pressure-demand or other positive-pressure mode.

References

New Jersey Department of Health and Senior Services. (August 2001). *Hazardous Substances Fact Sheet: Stannous Chloride.* Trenton, NJ

Stannous fluoride   S:0590

Molecular Formula: F_2Sn
Common Formula: SnF_2
Synonyms: Fluoristan®; Stannous fluoride; Tin bifluoride; Tin difluoride

CAS Registry Number: 7783-47-3
RTECS® Number: XQ3450000
UN/NA & ERG Number: UN3288 (Toxic solid, inorganic, n.o.s.)/151
EC Number: 231-999-3
Regulatory Authority and Advisory Bodies
Carcinogenicity: IARC: Animal Inadequate Evidence; Human Inadequate Evidence, *not classifiable as carcinogenic to humans*, Group 3, 1987.
US EPA Gene-Tox Program, Negative: *N. crassa—aneuploidy.*
FDA—over-the-counter drug.
Air Pollutant Standard Set. See below, "Permissible Exposure Limits in Air" section.
Canada, WHMIS, Ingredients Disclosure List Concentration 1.0%.
European/International Regulations: not listed in Annex 1.
WGK (German Aquatic Hazard Class): No value assigned.
Description: Stannous fluoride is a white crystalline solid with a bitter, salty taste. Molecular weight = 156.69; Boiling point = 849°C; Freezing/Melting point = 213°C. Hazard Identification (based on NFPA-704 M Rating System): Health 2, Flammability 0, Reactivity 0. Soluble in water; 30% at 20°C.
Potential Exposure: Compound Description: Drug, Tumorigen, Mutagen. Stannous fluoride is used in caries prophylactic; as an ingredient of cavity-preventing toothpastes.
Incompatibilities: Reacts with acids, forming hydrogen fluoride fumes. Reacts violently with chlorine.
Permissible Exposure Limits in Air
No TEEL available.
As tin, inorganic compounds
OSHA PEL: 2 mg[Sn]/m³ TWA.
NIOSH REL: 2 mg[Sn]/m³ TWA.
ACGIH TLV®[1]: 2 mg[Sn]/m³ TWA.
DFG MAK: *tin, inorganic compounds:* No numerical value established. Data may be available.
NIOSH IDLH: 100 mg[Sn]/m³.
As fluorides
OSHA PEL: 3 ppm/2.5 mg[F]/m³ TWA.
NIOSH REL: 3 ppm/2.5 mg[F]/m³ TWA.
ACGIH TLV®[1]: 2.5 mg[F]/m³ TWA; not classifiable as a human carcinogen; BEI: 3 mg[F]/g creatinine in urine *prior* to end-of-shift; 10 mg[F]/g creatinine in urine end-of-shift.
DFG MAK: 1 mg[F]/m³, inhalable fraction [skin]; Peak Limitation Category II(4); Pregnancy Risk Group C; BAT: 7.0 mg[F]/g creatinine in urine at end-of-shift; 4.0 mg[F]/g creatinine in urine at the beginning of the next shift.
NIOSH IDLH: 250 mg[F]/m³.
Australia: TWA 2 mg[Sn]/m³, 1993; Australia: TWA 2.5 mg[F]/m³, 1993; Austria: MAK 2.5 mg[F]/m³, 1999; Austria: MAK 2 mg/m³, 1999; Belgium: TWA 2 mg[Sn]/m³, 1993; Belgium: TWA 2.5 mg[F]/m³, 1993; Denmark: TWA 2.5 mg[F]/m³, 1999; Denmark: TWA 2 mg[Sn]/m³, 1999; Finland: TWA 2 mg[Sn]/m³, 1999; Finland: TWA

2.5 mg[F]/m³, 1999; France: VME 2.5 mg[F]/m³, 1999; Hungary: TWA 1 mg[F]/m³; STEL 2 mg[F]/m³, 1993; Hungary: TWA 1 mg[Sn]/m³; STEL 2 mg[Sn]/m³, [skin], 1993; Norway: TWA 0.6 mg[F]/m³, 1999; Norway: TWA 2 mg[Sn]/m³, 1999; the Philippines: TWA 2 mg[Sn]/m³, 1993; the Philippines: TWA 2.5 mg[F]/m³, 1993; Poland: MAC (TWA) 1 mg[HF]/m³; MAC (STEL) 3 mg[HF]/m³, 1999; Poland: TWA 2 mg[Sn]/m³, 1999; Sweden: NGV 2 mg[F]/m³, 1999; Switzerland: MAK-W 1.8 ppm (1.5 mg [F]/m³), KZG-W 3.6 ppm (3.0 mg[F]/m³), 1999; Switzerland: MAK-W 2 mg[Sn]/m³, KZG-W 4 mg[Sn]/m³, 1999; Thailand: TWA 2 mg[Sn]/m³, 1993; Thailand: TWA 2.5 mg[F]/m³, 1993; Turkey: TWA 2.5 mg[F]/m³, 1993; United Kingdom: TWA 2.5 mg[F]/m³, 2000; United Kingdom: TWA 2 mg[Sn]/m³; STEL 4 mg[Sn]/m³, 2000; Argentina, Bulgaria, Columbia, Jordan, South Korea, New Zealand, Singapore, Vietnam: ACGIH TLV®: Not classifiable as a human carcinogen.

Determination in Air: Filter; Acid; Hydride generation atomic absorption spectrometry; NIOSH Analytical Method (IV) #7900. See also #7300, Elements.

Permissible Concentration in Water: The EPA has set 4 mg/L as a standard for fluoride[61] but has no level set for tin.

Routes of Entry: Inhalation, ingestion, skin and/or eye contact.

Harmful Effects and Symptoms

Short Term Exposure: Stannous fluoride can affect you when breathed in. Corrosive to the eyes. Irritates the skin and respiratory tract. Prolonged skin contact can cause burns. Breathing stannous fluoride can irritate the throat and bronchial tubes, causing cough and/or difficulty in breathing. Changes may also occur on chest X-ray with repeated exposures. When heated, toxic fluorine gas can be released, causing pulmonary edema (fluid in the lungs), a medical emergency that can be delayed for several hours. This can cause death. If swallowed, may affect the gastrointestinal tract, causing shock.

Long Term Exposure: May affect the teeth and bones. Repeated exposure, usually after years, may cause changes in the chest X-ray; reduced lung function has been reported. Tin released by stannous fluoride may interfere with the body's ability to absorb iron, contributing to iron deficiency anemia.

Points of Attack: Lungs, blood.

Medical Surveillance: Before beginning employment and at regular times after that, the following is recommended: lung function tests. For persons exposed to levels half the TLV or greater, the following is recommended: hemoglobin or hematocrits. If acute breathing overexposure to heated stannous fluoride with fluorine gas is suspected, also consider chest X-ray.

First Aid: If this chemical gets into the eyes, remove any contact lenses at once and irrigate immediately for at least 15 min, occasionally lifting upper and lower lids. Seek medical attention immediately. If this chemical contacts the skin, remove contaminated clothing and wash immediately with soap and water. Seek medical attention immediately. If this chemical has been inhaled, remove from exposure, begin rescue breathing (using universal precautions, including resuscitation mask) if breathing has stopped and CPR if heart action has stopped. Transfer promptly to a medical facility. When this chemical has been swallowed, get medical attention. Give large quantities of water and induce vomiting. Do not make an unconscious person vomit. For heated stannous fluoride exposure, medical observation is recommended for 24−48 h after breathing overexposure, as pulmonary edema may be delayed. As first aid for pulmonary edema, a doctor or authorized paramedic may consider administering a corticosteroid spray. If swallowed, watch for symptoms of shock.

Personal Protective Methods: Wear protective gloves and clothing to prevent any reasonable probability of skin contact. Safety equipment suppliers/manufacturers can provide recommendations on the most protective glove/clothing material for your operation. All protective clothing (suits, gloves, footwear, headgear) should be clean, available each day, and put on before work. Contact lenses should not be worn when working with this chemical. Wear dust-proof chemical goggles and face shield unless full face-piece respiratory protection is worn. Employees should wash immediately with soap when skin is wet or contaminated. Provide emergency showers and eyewash.

Respirator Selection: NIOSH/OSHA: *Up to 10 mg/m³:* Qm (APF = 25) (any quarter-mask respirator). *Up to 20 mg/m³:* 95XQ (APF = 10) [any particulate respirator equipped with an N95, R95, or P95 filter (including N95, R95, and P95 filtering face-pieces) except quarter-mask respirators. The following filters may also be used: N99, R99, P99, N100, R100, P100] or Sa* (APF = 10) (any supplied-air respirator). *Up to 50 mg/m³:* Sa:Cf* (APF = 25) (any supplied-air respirator operated in a continuous-flow mode) or PaprHie (APF = 25) * (any powered air-purifying respirator with a high-efficiency particulate filter). *Up to 100 mg/m³:* 100F (APF = 50) (any air-purifying, full-face-piece respirator with an N100, R100, or P100 filter. Click here for information on selection of N, R, or P filters) or SCBAF (APF = 50) (any self-contained breathing apparatus with a full face-piece) or SaF (APF = 50) (any supplied-air respirator with a full face-piece). *Emergency or planned entry into unknown concentrations or IDLH conditions:* SCBAF: Pd,Pp (APF = 10,000) (any self-contained breathing apparatus that has a full face-piece and is operated in a pressure-demand or other positive-pressure mode) or SAF; PD,PP: ASCBA (any supplied-air respirator that has a full face-piece and is operated in a pressure-demand or other positive-pressure mode in combination with an auxiliary self-contained positive-pressure breathing apparatus). *Escape:* 100F (APF = 50) (any air-purifying, full-face-piece respirator with an N100, R100, or P100 filter) or SCBAE (any appropriate escape-type, self-contained breathing apparatus).

*Substance reported to cause eye irritation or damage; may require eye protection.

Storage: Color Code—Blue: Health Hazard/Poison: Store in a secure poison location. Prior to working with this chemical you should be trained on its proper handling and storage. Stannous fluoride must be stored to avoid contact with chlorine and turpentine, since violent reactions occur.

Shipping: Toxic solid, inorganic, n.o.s. require a shipping label of "POISONOUS/TOXIC MATERIALS." This material falls in Hazard Class 6.1 and Packing Group III.

Spill Handling: Evacuate persons not wearing protective equipment from area of spill or leak until cleanup is complete. Remove all ignition sources. Collect powdered material in the most convenient and safe manner and deposit in sealed containers. Ventilate area after cleanup is complete. It may be necessary to contain and dispose of this chemical as a hazardous waste. If material or contaminated runoff enters waterways, notify downstream users of potentially contaminated waters. Contact your local or federal environmental protection agency for specific recommendations. If employees are required to clean up spills, they must be properly trained and equipped. OSHA 1910.120(q) may be applicable.

Fire Extinguishing: This chemical is a noncombustible solid. Use dry chemical or CO_2 extinguishers. Poisonous fluorine gas is produced in fire. If material or contaminated runoff enters waterways, notify downstream users of potentially contaminated waters. Notify local health and fire officials and pollution control agencies. From a secure, explosion-proof location, use water spray to cool exposed containers. If cooling streams are ineffective (venting sound increases in volume and pitch, tank discolors, or shows any signs of deforming), withdraw immediately to a secure position. If employees are expected to fight fires, they must be trained and equipped in OSHA 1910.156. The only respirators recommended for firefighting are self-contained breathing apparatuses that have full face-pieces and are operated in a pressure-demand or other positive-pressure mode.

References

New Jersey Department of Health and Senior Services. (August 2002). *Hazardous Substances Fact Sheet: Stannous Fluoride.* Trenton, NJ

Stibine   S:0600

Molecular Formula: H_3Sb

Common Formula: SbH_3

Synonyms: Antimonwasserstoffes (German); Antimony hydride; Antimony trihydride; Hydrogen antimonide

CAS Registry Number: 7803-52-3

RTECS® Number: WJ0700000

UN/NA & ERG Number: UN2676/119

EC Number: 051-003-00-9

Regulatory Authority and Advisory Bodies

Department of Homeland Security Screening Threshold Quantity (pounds): *Theft hazard* 15 (≥0.67% concentration).

OSHA 29CFR1910.119, Appendix A. Process Safety List of Highly Hazardous Chemicals, TQ = 500 lb (227 kg).

Air Pollutant Standard Set. See below, "Permissible Exposure Limits in Air" section.

As antimony compounds

Clean Air Act: Hazardous Air Pollutants (Title I, Part A, Section 112).

Clean Water Act: Toxic Pollutant (Section 401.15).

RCRA, 40CFR261, Appendix 8 Hazardous Constituents, waste number not listed, as antimony compounds, n.o.s.

Safe Drinking Water Act: MCL 0.006 mg/L; MCLG, 0.006 mg/L.

EPCRA Section 313: Includes any unique chemical substance that contains antimony as part of that chemical's infrastructure. Form R *de minimis* concentration reporting level: 0.1%.

US DOT 49CFR172.101, Inhalation Hazardous Chemical.

Canada, WHMIS, Ingredients Disclosure List Concentration 1.0%.

European/International Regulations: not listed in Annex 1.

WGK (German Aquatic Hazard Class): No value assigned.

Description: Stibine is a colorless gas with a characteristic disagreeable odor. It is produced by dissolving zinc/antimony or magnesium-antimony in hydrochloride acid. Molecular weight = 124.78; Boiling point = −18.3°C; Freezing/Melting point = −87.8°C; Flash point = Flammable gas; Relative vapor density (air = 1) = 4.31; Vapor pressure = >1 mmHg at 25°C. Hazard Identification (based on NFPA-704 M Rating System): Health 4, Flammability 4, Reactivity 2. Poor solubility in water.

Potential Exposure: Stibine is used as a fumigating agent. Exposure to stibine usually occurs when stibine is released from antimony-containing alloys during the charging of storage batteries, when certain antimonial drosses are treated with water or acid, or when antimony-containing metals come in contact with acid. Operations generally involved are metallurgy, welding, or cutting with blow torches; soldering, filling of hydrogen balloons; etching of zinc; and chemical processes.

Incompatibilities: A flammable gas. Incompatible with acids, halogenated hydrocarbons, oxidizers, moisture, chlorine, ammonia. Reacts violently with chlorine, concentrated nitric acid, and ozone. Decomposes in air. Thermally unstable: quick decomposition at 200°C producing metallic antimony and explosive hydrogen gas.

Permissible Exposure Limits in Air

OSHA PEL: 0.1 ppm/0.5 mg/m³ TWA.

NIOSH REL: 0.1 ppm/0.5 mg/m³ TWA.

ACGIH TLV®[1]: 0.1 ppm/0.5 mg/m³ TWA.

NIOSH IDLH: 5 ppm

Protective Action Criteria (PAC)

TEEL-0: 0.1 ppm

PAC-1: 0.2 ppm

PAC-2: 1.5 ppm

PAC-3: 9.6 ppm

Austria: MAK 0.1 ppm (0.5 mg/m³), 1999; Denmark: TWA 0.05 ppm (0.25 mg/m³), 1999; France: VME 0.1 ppm (0.5 mg/m³), 1999; Norway: TWA 0.05 ppm (0.25 mg/m³), 1999; Poland: MAC (TWA) 0.2 mg/m³; MAC (STEL) 1.5 mg/m³, 1999; Sweden: TWA 0.05 ppm (0.3 mg/m³), 1999; United Kingdom: TWA 0.1 ppm (0.52 mg/m³); STEL 0.3 ppm, 2000; the Netherlands: MAC-TGG 0.5 mg[Sb]/m³. Several states have set guidelines or standards for stibine in ambient air[60] ranging from 5.0 µg/m³ (North Dakota) to 8.0 µg/m³ (Virginia) to 10.0 µg/m³ (Connecticut) to 12.0 µg/m³ (Nevada).

Determination in Air: Use NIOSH Analytical Method (IV) #6008, Stibine.

Permissible Concentration in Water: See regulatory section.

Routes of Entry: Inhalation of gas.

Harmful Effects and Symptoms

Short Term Exposure: Stibine can affect you when breathed in. May be fatal if absorbed through the skin or inhaled. A strong sensitizer. May cause severe allergic respiratory reaction. Exposure can cause rapid, fatal poisoning, with symptoms of headaches, nausea, dark or bloody urine, pain in the back and abdomen, slowed breathing, and death. Exposure can also irritate the lungs and may lead to a buildup of fluid (pulmonary edema), a medical emergency that can be delayed for several hours. This can cause death. Stibine destroys red blood cells and can also cause liver and kidney damage.

Long Term Exposure: Stibine destroys red blood cells (hemolysis). May affect the kidneys, liver; hemoglobinuria, hematuria (blood in the urine), hemolytic anemia; jaundice. May affect the central nervous system.

Points of Attack: Blood, liver, kidneys, respiratory system, CNS.

Medical Surveillance: NIOSH lists the following tests: liver function tests; red blood cells/count, RBC hemolysis; urine (chemical/metabolite), hemoglobin; urinalysis (routine); white blood cell count/differential. For those with frequent or potentially high exposure (half the TLV or greater), the following are recommended before beginning work and at regular times after that: lung function tests. If symptoms develop or overexposure is suspected, the following may be useful: consider chest X-ray after acute overexposure. Examination of the central nervous system. Examination by a qualified allergist.

First Aid: If this chemical gets into the eyes, remove any contact lenses at once and irrigate immediately for at least 15 min, occasionally lifting upper and lower lids. Seek medical attention immediately. If this chemical contacts the skin, remove contaminated clothing and wash immediately with soap and water. Seek medical attention immediately. If this chemical has been inhaled, remove from exposure, begin rescue breathing (using universal precautions, including resuscitation mask) if breathing has stopped and CPR if heart action has stopped. Transfer promptly to a medical facility. When this chemical has been swallowed, get medical attention. Give large quantities of water and induce vomiting. Do not make an unconscious person vomit. Medical observation is recommended for 24–48 h after breathing overexposure, as pulmonary edema may be delayed. As first aid for pulmonary edema, a doctor or authorized paramedic may consider administering a corticosteroid spray.

Personal Protective Methods: Wear protective gloves and clothing to prevent any reasonable probability of skin contact. Safety equipment suppliers/manufacturers can provide recommendations on the most protective glove/clothing material for your operation. All protective clothing (suits, gloves, footwear, headgear) should be clean, available each day, and put on before work. Contact lenses should not be worn when working with this chemical. Wear gas-proof chemical goggles and face shield unless full face-piece respiratory protection is worn. Employees should wash immediately with soap when skin is wet or contaminated. Provide emergency showers and eyewash.

Respirator Selection: *Up to 1 ppm:* Sa (APF = 10) (any supplied-air respirator). *Up to 2.5 ppm:* Sa:Cf (APF = 25) (any supplied-air respirator operated in a continuous-flow mode). *Up to 5 ppm:* SaT: Cf (APF = 50) (any supplied-air respirator that has a tight-fitting face-piece and is operated in a continuous-flow mode) or SCBAF (APF = 50) (any self-contained breathing apparatus with a full face-piece) or SaF (APF = 50) (any supplied-air respirator with a full face-piece). *Emergency or planned entry into unknown concentrations or IDLH conditions:* SCBAF: Pd,Pp (APF = 10,000) (any self-contained breathing apparatus that has a full face-piece and is operated in a pressure-demand or other positive-pressure mode) or SaF: Pd,Pp: ASCBA (APF = 10,000) (any supplied-air respirator that has a full face-piece and is operated in a pressure-demand or other positive-pressure mode in combination with an auxiliary self-contained positive-pressure breathing apparatus). *Escape:* GmFS (APF = 50) [any air-purifying, full-face-piece respirator (gas mask) with a chin-style, front- or back-mounted canister providing protection against the compound of concern] or SCBAE (any appropriate escape-type, self-contained breathing apparatus).

Storage: Color Code—Red Stripe: Flammability Hazard: Store separately from all other flammable materials. Color Code—Blue: Health Hazard/Poison: Store in a secure poison location. Prior to working with this chemical you should be trained on its proper handling and storage. Stibine must be stored to avoid contact with oxidizers (such as perchlorates, peroxides, permanganates, chlorates, and nitrates), strong acids (such as hydrochloric, sulfuric, and nitric), and halogenated hydrocarbons, since violent reactions occur. Store in tightly closed containers in a cool, well-ventilated area away from heat, sparks, and flames.

Shipping: Stibine requires a shipping label of "POISON GAS, FLAMMABLE GAS." It falls in Hazard Class 2.3. It is a violation of transportation regulations to refill compressed gas cylinders without the express written permission of the owner.

Spill Handling: Evacuate and restrict persons not wearing protective equipment from area of spill or leak until cleanup is complete. Remove all ignition sources. Ventilate area of leak to disperse the gas. Stop the flow of gas if it can be done safely. If source of leak is a cylinder and the leak cannot be stopped in place, remove leaking cylinder to a safe place in the open air, and repair leak or allow cylinder to empty. Keep this chemical out of confined space, such as a sewer, because of the possibility of explosion, unless the sewer is designed to prevent the buildup of explosive concentrations. It may be necessary to contain and dispose of this chemical as a hazardous waste. Contact your local or federal environmental protection agency for specific recommendations. If employees are required to clean up spills, they must be properly trained and equipped. OSHA 1910.120(q) may be applicable.

Initial isolation and protective action distances
Distances shown are likely to be affected during the first 30 min after materials are spilled and could increase with time. If more than one tank car, cargo tank, portable tank, or large cylinder involved in the incident is leaking, the protective action distance may need to be increased. You may need to seek emergency information from CHEMTREC at (800) 424-9300 or seek professional environmental engineering assistance from the US EPA Environmental Response Team at (908) 548-8730 (24-h response line).

Small spills (From a small package or a small leak from a large package)
First: Isolate in all directions (feet/meters) 200/60
Then: Protect persons downwind (miles/kilometers)
Day 0.2/0.3
Night 1.1/1.8

Large spills (From a large package or from many small packages)
First: Isolate in all directions (feet/meters) 1500/500
Then: Protect persons downwind (miles/kilometers)
Day 1.7/2.8
Night 4.5/7.2

Fire Extinguishing: This chemical is a flammable gas. Evolves hydrogen and poisonous gases, including antimony, are produced in fire. Do not extinguish the fire unless the flow of gas can be stopped and any remaining gas is out of the line. Specially trained personnel may use fog lines to cool exposures and let the fire burn itself out. Use flooding quantities of water. Do not use halogenated extinguishers. Vapors are heavier than air and will collect in low areas. Vapors may travel long distances to ignition sources and flashback. Vapors in confined areas may explode when exposed to fire. Containers may explode in fire. Storage containers and parts of containers may rocket great distances, in many directions. If material or contaminated runoff enters waterways, notify downstream users of potentially contaminated waters. Notify local health and fire officials and pollution control agencies. From a secure, explosion-proof location, use water spray to cool exposed containers. If cooling streams are ineffective (venting sound increases in volume and pitch, tank discolors, or shows any signs of deforming), withdraw immediately to a secure position. If cylinders are exposed to excessive heat from fire or flame contact, withdraw immediately to a secure location. If employees are expected to fight fires, they must be trained and equipped in OSHA 1910.156. The only respirators recommended for firefighting are self-contained breathing apparatuses that have full face-pieces and are operated in a pressure-demand or other positive-pressure mode.

Disposal Method Suggested: Dissolve in hydrochloric acid; add water to produce precipitate; add acid to dissolve again; precipitate with H_2S; filter and dry precipitate and return to suppliers.

References
Sax, N. I. (Ed.). (1982). *Dangerous Properties of Industrial Materials Report*, 2, No. 4, 17—18
New Jersey Department of Health and Senior Services. (August 2001). *Hazardous Substances Fact Sheet: Stibine.* Trenton, NJ

Stoddard solvent   S:0610

Molecular Formula: C_9H_{20}
Synonyms: Cleaning solvent; Dry cleaner naphtha; Mineral spirits; Naphtha safety solvent; Petroleum solvent; Spotting solvent; Varnoline; White spirits
CAS Registry Number: 8052-41-3
RTECS® Number: WJ8925000
UN/NA & ERG Number: UN1268/128
EC Number: 232-489-3 [Annex I Index No.: 649-345-00-4]
Regulatory Authority and Advisory Bodies
Air Pollutant Standard Set. See below, "Permissible Exposure Limits in Air" section.
Canada, WHMIS, Ingredients Disclosure List Concentration 1.0%.
European/International Regulations: Hazard Symbol: T; Risk phrases: R45; R46; R65; Safety phrases: S53; S45 (see Appendix 4).
WGK (German Aquatic Hazard Class): 2—Hazard to waters.

Description: Stoddard solvent is a refined petroleum solvent containing >65% C_{10} or higher hydrocarbons. Stoddard solvent is a colorless liquid with a kerosene-like odor. Odor threshold = 1 ppm[NY]–30 ppm. Molecular weight can vary = 140–160; Specific gravity (H_2O:1) = 0.765–0.795 at 25°C; Boiling point = 154–202°C[1]; 130–230°C; Flash point = 39–60°C; but, may vary from 21°C to 60°C based on composition; Autoignition temperature = 229–240°C; but may vary based on composition. Explosive limits: LEL = 0.6%; UEL = 8.0%. Hazard Identification (based on NFPA-704 M Rating System): Health 20, Flammability 2, Reactivity 0. Insoluble in water. See also "Naphthas" for related materials.

Potential Exposure: Compound Description: Primary Irritant. Stoddard solvent is used as a diluent in paints, coatings, and waxes; as a paint thinner; as a dry-cleaning agent; as a degreaser for metal parts; as an herbicide.

Incompatibilities: Forms explosive mixture with air. Keep away from strong oxidizers. Attacks some forms of plastics, rubber, and coatings.

Permissible Exposure Limits in Air

OSHA PEL: 500 ppm/2900 mg/m^3 TWA.

NIOSH REL: 350 mg/m^3 TWA; 1800 mg/m^3 [15 min] Ceiling Concentration.

ACGIH TLV®[1]: 100 ppm/525 mg/m^3 TWA.

NIOSH IDLH: 20,000 mg/m^3.

Protective Action Criteria (PAC)

TEEL-0: 100 ppm

PAC-1: 100 ppm

PAC-2: 350 ppm

PAC-3: 3850 ppm

Australia: TWA 790 mg/m^3, 1993; Belgium: TWA 100 ppm (525 mg/m^3), 1993; Denmark: TWA 25 ppm (140 mg/m^3), 1999; the Philippines: TWA 200 ppm (1150 mg/m^3), 1993; the Netherlands: MAC-TGG 575 mg/m^3, 2003; Argentina, Bulgaria, Columbia, Jordan, South Korea, New Zealand, Singapore, Vietnam: ACGIH TLV®: TWA 100 ppm. Several states have set guidelines or standards for Stoddard solvent in ambient air[60] ranging from 5.25−10.50 mg/m^3 (North Dakota) to 7.0 mg/m^3 (Connecticut) to 12.5 mg/m^3 (Nevada).

Determination in Air: Use NIOSH Analytical Method 1550, Naphthas; OSHA Analytical Method 48.

Determination in Water: Octanol−water coefficient: Log K_{ow} = 3.2−7.1.

Routes of Entry: Inhalation, ingestion, skin and/or eye contact.

Harmful Effects and Symptoms

Short Term Exposure: Inhalation: Causes irritation of the eyes and respiratory tract. Exposure to levels above 2400 mg/m^3 may cause headache, dizziness, nose and throat irritation. More severe exposures may cause nausea and vomiting, a feeling of intoxication, weakness, muscle twitches; and in extreme cases, convulsions, unconsciousness, and death. *Skin:* Contact with liquid may cause irritation and drying of skin. This can result in dermatitis. *Eyes:* Contact with liquid or vapor levels of 900−2400 mg/m^3 may cause irritation and tearing. *Ingestion:* Small amounts may cause headache, dizziness, nausea, vomiting, intoxication, weakness, muscle twitches, convulsions, and unconsciousness. May cause aspiration into the lungs and chemical pneumonia. As little as 3 oz may be fatal. If liquid is breathed into the lungs, as little as 1 oz may cause death due to respiratory failure.

Long Term Exposure: Prolonged or repeated contact with liquid may cause defatting of the skin with drying, irritation, and skin ulcers. Exposure to vapor may cause eye, nose and throat irritation, fatigue, headaches, anemia, jaundice, and damage to the liver and bone marrow. In animals: kidney damage. Repeated exposure may cause a rare reaction in some people that destroys blood cells (aplastic anemia). This can be fatal. Many petroleum-based solvents have been shown to cause brain and/or nerve damage. Effects may include reduced memory and concentration, personality changes, fatigue, sleep disturbances, reduced coordination, effects on the autonomic nerves and/or nerves to the limbs.

Points of Attack: Eyes, skin, respiratory system, central nervous system, liver, kidneys.

Medical Surveillance: If symptoms develop or overexposure is suspected, the following may be useful: complete blood count. Evaluation for brain effects. Liver and kidney function tests.

First Aid: If this chemical gets into the eyes, remove any contact lenses at once and irrigate immediately for at least 15 min, occasionally lifting upper and lower lids. Seek medical attention immediately. If this chemical contacts the skin, remove contaminated clothing and wash immediately with soap and water. Seek medical attention immediately. If this chemical has been inhaled, remove from exposure, begin rescue breathing (using universal precautions, including resuscitation mask) if breathing has stopped and CPR if heart action has stopped. Transfer promptly to a medical facility. When this chemical has been swallowed, get medical attention. Do *NOT* induce vomiting.

Note to physician: Treat symptomatically for central nervous system depression. Supportive treatment for pulmonary edema using oxygen may be needed when aspiration of liquids or massive exposure to vapors has occurred.

Personal Protective Methods: Wear protective gloves and clothing to prevent any reasonable probability of skin contact. Safety equipment suppliers/manufacturers can provide recommendations on the most protective glove/clothing material for your operation. Some of these manufacturers recommend *nitrile* or *polyvinyl alcohol* as a protective material. All protective clothing (suits, gloves, footwear, headgear) should be clean, available each day, and put on before work. Contact lenses should not be worn when working with this chemical. Wear splash-proof chemical goggles and face shield unless full face-piece respiratory protection is worn. Employees should wash immediately with soap when skin is wet or contaminated. Provide emergency showers and eyewash. For engineering controls see *Criteria for a Recommended Standard: Occupational Exposure to Refined Petroleum*, NIOSH Doc. No. 77-192.

Respirator Selection: NIOSH: *Up to 3500 mg/m^3:* CcrOv* (APF = 10) [any chemical cartridge respirator with organic vapor cartridge(s)] or Sa* (APF = 10) (any supplied-air respirator). *Up to 8750 mg/m^3:* Sa:Cf* (APF = 25) (any supplied-air respirator operated in a continuous-flow mode); or PaprOv* (APF = 25) [any powered, air-purifying respirator with organic vapor cartridge(s)]. *Up to 17,500 mg/m^3:* CcrFOv (APF = 50) [any chemical cartridge respirator with a full face-piece and organic vapor cartridge(s)] or GmFOv (APF = 50) [any air-purifying, full-face-piece respirator (gas mask) with a chin-style, front- or back-mounted organic vapor canister] or PaprTOv* (APF = 50) [any powered, air-purifying respirator with a tight-fitting face-piece and organic vapor cartridge(s)]; or SCBAF (APF = 50) (any self-contained breathing apparatus with a full face-piece) or SaF (APF = 50) (any supplied-air respirator with a full face-piece). *Up to 20,000 mg/m³:* SaF: Pd,Pp (APF = 2000) (any supplied-air respirator that has a full face-piece and is operated in a pressure-demand or other positive-pressure mode). *Emergency or planned entry into unknown concentrations or IDLH conditions:* SCBAF: Pd, Pp (APF = 10,000) (any self-contained breathing apparatus that has a full face-piece and is operated in a pressure-demand or other positive-pressure mode) or SaF: Pd,Pp: ASCBA (APF = 10,000) (any supplied-air respirator that has a full face-piece and is operated in a pressure-demand or other positive-pressure mode in combination with an auxiliary, self-contained breathing apparatus operated in a pressure-demand or other positive-pressure mode). *Escape:* GmFOv (APF = 50) [any air-purifying, full-face-piece respirator (gas mask) with a chin-style, front- or back-mounted organic vapor canister] or SCBAE (any appropriate escape-type, self-contained breathing apparatus).

*Substance reported to cause eye irritation or damage; may require eye protection.

Storage: Color Code—Red: Flammability Hazard: Store in a flammable liquid storage area or approved cabinet away from ignition sources and corrosive and reactive materials. Prior to working with this chemical you should be trained on its proper handling and storage. Before entering confined space where this chemical may be present, check to make sure that an explosive concentration does not exist. Stoddard solvent must be stored to avoid contact with strong oxidizers (such as chlorine, bromine, and fluorine), since violent reactions occur. Store in tightly closed containers in a cool, well-ventilated area. Sources of ignition, such as smoking and open flames, are prohibited where Stoddard solvent is used, handled, or stored in a manner that could create a potential fire or explosion hazard. Metal containers involving the transfer of 5 gallons or more of this chemical should be grounded and bonded. Drums must be equipped with self-closing valves, pressure vacuum bungs, and flame arresters. Use only nonsparking tools and equipment, especially when opening and closing containers of this chemical. Wherever this chemical is used, handled, manufactured, or stored, use explosion-proof electrical equipment and fittings.

Shipping: Petroleum distillates, n.o.s. require a shipping label of "FLAMMABLE LIQUID." It falls in Hazard Class 3 and Packing Group II.

Spill Handling: Evacuate and restrict persons not wearing protective equipment from area of spill or leak until cleanup is complete. Remove all ignition sources. Establish forced ventilation to keep levels below explosive limit. Absorb liquids in vermiculite, dry sand, earth, peat, carbon, or a similar material and deposit in sealed containers. Keep this chemical out of a confined space, such as a sewer, because of the possibility of an explosion, unless the sewer is designed to prevent the buildup of explosive concentrations. It may be necessary to contain and dispose of this chemical as a hazardous waste. If material or contaminated runoff enters waterways, notify downstream users of potentially contaminated waters. Contact your local or federal environmental protection agency for specific recommendations. If employees are required to clean up spills, they must be properly trained and equipped. OSHA 1910.120(q) may be applicable.

Fire Extinguishing: This chemical is a flammable liquid. Poisonous gases are produced in fire. Use dry chemical, carbon dioxide, or foam extinguishers. Vapors are heavier than air and will collect in low areas. Vapors may travel long distances to ignition sources and flashback. Vapors in confined areas may explode when exposed to fire. Containers may explode in fire. Storage containers and parts of containers may rocket great distances, in many directions. If material or contaminated runoff enters waterways, notify downstream users of potentially contaminated waters. Notify local health and fire officials and pollution control agencies. From a secure, explosion-proof location, use water spray to cool exposed containers. If cooling streams are ineffective (venting sound increases in volume and pitch, tank discolors, or shows any signs of deforming), withdraw immediately to a secure position. If employees are expected to fight fires, they must be trained and equipped in OSHA 1910.156. The only respirators recommended for firefighting are self-contained breathing apparatuses that have full face-pieces and are operated in a pressure-demand or other positive-pressure mode.

Disposal Method Suggested: Dissolve or mix the material with a combustible solvent and burn in a chemical incinerator equipped with an afterburner and scrubber. All federal, state, and local environmental regulations must be observed.

References

National Institute for Occupational Safety and Health. (1977). *Criteria for a Recommended Standard: Occupational Exposure to Refined Petroleum*, NIOSH Document No. 77-192. Washington, DC

New York State Department of Health. (March 1986). *Chemical Fact Sheet: Stoddard Solvent* (Version 2). Albany, NY: Bureau of Toxic Substance Assessment

New Jersey Department of Health and Senior Services. (August 1998). *Hazardous Substances Fact Sheet: Stoddard Solvent*. Trenton, NJ

Streptozotocin   S:0620

Molecular Formula: $C_8H_{15}N_3O_7$
Synonyms: 2-Deoxy-2-([(methylnitrosoamino)carbonyl]amino)-D-glucopyranose; 2-Deoxy-2-(3-methyl-3-nitrosoureido)-a (and b)-D-glucopyranose; 2-Deoxy-2-(3-methyl-3-nitrosoureido)-D-glucopyranose; D-Glucose, 2-deoxy-2-([(methylnitrosoamino)carbonyl]amino)-; *N*-D-Glucosyl

(2)-N'-nitrosomethylharstoff (German); N-D-Glucosyl-(2)-N'-nitro-somethylurea; NCI-C03167; NSC 85598; NSC-85998; STR; Streptozoticin; STRZ; STZ; U-9889; Zanosar

CAS Registry Number: 18883-66-4

RTECS® Number: LX5775000

UN/NA & ERG Number: UN3249 (Medicines, toxic, solid, n.o.s.)/151

EC Number: 242-646-8

Regulatory Authority and Advisory Bodies
Carcinogenicity: IARC: Human No Adequate Data, animal Sufficient Evidence, *possibly carcinogenic to humans*, Group 2B, 1978, 1999; NTP: Reasonably anticipated to be a human carcinogen.
US EPA Hazardous Waste Number (RCRA No.): U206.
RCRA, 40CFR261, Appendix 8 Hazardous Constituents.
Reportable Quantity (RQ): 1 lb (0.454 kg).
California Proposition 65 Chemical: Cancer 1/1/88; Developmental/Reproductive toxin (male, female) 8/20/99.
Hazard Symbols: T; Risk phrases: R45; Safety phrases: S45; S53.
European/International Regulations: not listed in Annex 1.
WGK (German Aquatic Hazard Class): No value assigned.

Description: Streptozotocin is a crystalline substance. Molecular weight = 265.6; Freezing/Melting point = 115°C. Soluble in water.

Potential Exposure: Used as a medicine to treat cancer. Steptozotocin (STR), a water-soluble antibiotic, has been of interest as a potential antineoplastic agent. STR is used in research for studies on diabetes because of its specific toxic action on B-cells of the pancreas, including hyperglycemia. STR is produced by the soil microorganism *Streptomyces achromogenes*. It has also been synthesized by laboratory procedures. Occupational exposure to STR is believed to be limited to pharmaceutical and research workers.

Incompatibilities: Alkalis.

Permissible Exposure Limits in Air
No standards or TEEL available.

Routes of Entry: Inhalation, ingestion, skin and/or eye contact.

Harmful Effects and Symptoms

Short Term Exposure: Can damage the kidneys and pancreas; may damage the nervous system. These effects may be permanent. High exposure can damage the bone marrow causing anemia, reduced white blood cells, and reduced platelets, causing a tendency to bleeding. Symptoms of exposure include abdominal pain; nausea, vomiting; central nervous system depression; renal or hepatic toxicity; proximal tubular damage; hematological toxicity, consisting of anemia, leukopenia, or thrombocytopenia; diabetes mellitus with severe toxic peripheral neuropathy.

Long Term Exposure: This chemical is a probable human carcinogen. There is evidence of liver, kidney, and pancreatic cancer in humans. There is evidence that it is a teratogen in animals. Exposure can cause kidney damage and pancreas damage causing diabetes. May damage the nervous system. This chemical is closely related to pyriminyl, a rat poison.

Points of Attack: Nervous system, kidneys, pancreas, liver

Medical Surveillance: Liver and kidney function tests. Test for blood sugar. Complete examination of the nervous system, including the autonomic nervous system.

First Aid: *Skin Contact:* Flood all areas of body that have contacted the substance with water. Do not wait to remove contaminated clothing; do it under the water stream. Use soap to help assure removal. Isolate contaminated clothing when removed to prevent contact by others.[52]

Eye Contact: Remove any contact lenses at once. Immediately flush eyes well with copious quantities of water or normal saline for at least 20–30 min. Seek medical attention.

Inhalation: Leave contaminated area immediately; breathe fresh air. Proper respiratory protection must be supplied to any rescuers. If coughing, difficult breathing, or any other symptoms develop, seek medical attention at once, even if symptoms develop many hours after exposure.

Personal Protective Methods: Wear protective gloves and clothing to prevent any reasonable probability of skin contact. Safety equipment suppliers/manufacturers can provide recommendations on the most protective glove/clothing material for your operation. All protective clothing (suits, gloves, footwear, headgear) should be clean, available each day, and put on before work. Contact lenses should not be worn when working with this chemical. Wear dust-proof chemical goggles and face shield unless full face-piece respiratory protection is worn. Employees should wash immediately with soap when skin is wet or contaminated. Provide emergency showers and eyewash.

Respirator Selection: Follow regulations in OSHA 29CFR1910.134 or European Standard EN149. Use a NIOSH/MSHA- or European Standard EN149-approved respirator; or use an approved supplied-air respirator with a full face-piece operated in the positive-pressure mode, or with a full face-piece, hood, or helmet in the continuous-flow mode; or use a NIOSH/MSHA- or European Standard EN149-approved self-contained breathing apparatus with a full face-piece operated in pressure-demand or other positive-pressure mode.

Storage: Color Code—Blue: Health Hazard/Poison: Store in a secure poison location. Prior to working with this chemical you should be trained on its proper handling and storage. Store in a refrigerator or in a cool, dry place. A regulated, marked area should be established where this chemical is handled, used, or stored in compliance with OSHA Standard 1910.1045.

Shipping: Medicine, solid, toxic, n.o.s. requires a shipping label of "POISONOUS/TOXIC MATERIALS." They fall in Hazard Class 6.1 and Packing Groups II or III.

Spill Handling: Evacuate persons not wearing protective equipment from area of spill or leak until cleanup is complete. Remove all ignition sources. Collect powdered material in the most convenient and safe manner and deposit in

sealed containers. Ventilate area after cleanup is complete. It may be necessary to contain and dispose of this chemical as a hazardous waste. If material or contaminated runoff enters waterways, notify downstream users of potentially contaminated waters. Contact your local or federal environmental protection agency for specific recommendations. If employees are required to clean up spills, they must be properly trained and equipped. OSHA 1910.120(q) may be applicable.

Fire Extinguishing: This chemical is a combustible solid. Use dry chemical, carbon dioxide, water spray, or alcohol foam extinguishers. Poisonous gases are produced in fire, including nitrogen oxides. If material or contaminated runoff enters waterways, notify downstream users of potentially contaminated waters. Notify local health and fire officials and pollution control agencies. From a secure, explosion-proof location, use water spray to cool exposed containers. If cooling streams are ineffective (venting sound increases in volume and pitch, tank discolors, or shows any signs of deforming), withdraw immediately to a secure position. If employees are expected to fight fires, they must be trained and equipped in OSHA 1910.156. The only respirators recommended for firefighting are self-contained breathing apparatuses that have full face-pieces and are operated in a pressure-demand or other positive-pressure mode.

References

Sax, N. I. (Ed.). (1981). *Dangerous Properties of Industrial Materials Report*, 1, No. 5, 80

Strontium chromate   S:0630

Molecular Formula: CrO$_4$Sr
Common Formula: SrCrO$_4$
Synonyms: Chromic acid, stontium salt (1:1); Chromic acid, strontium salt; C.I. Pigment yellow 32; Cromato de estronicio (Spanish); Deep lemon yellow; Strontium chromate; Strontium chromate(VI); Strontium chromate 12170; Strontium chromate A; Strontium chromate X-2396; Strontium yellow
CAS Registry Number: 7789-06-2; *(alt.)* 54322-60-0
RTECS® Number: BG3240000
UN/NA & ERG Number: UN3086/141
EC Number: 232-142-6 [*Annex I Index No.:* 024-009-00-4]
Regulatory Authority and Advisory Bodies
Carcinogenicity: IARC: Human Sufficient Evidence; Animal Sufficient Evidence, *carcinogenic to humans*, Group 1, 1990; NTP: 11th Report on Carcinogens, 2004: Known to be a human carcinogen; NIOSH: Potential occupational carcinogen.
US EPA Gene-Tox Program, Positive: Carcinogenicity—mouse/rat.
Air Pollutant Standard Set. See below, "Permissible Exposure Limits in Air" section.
Clean Water Act: Section 311 Hazardous Substances/RQ 40CFR117.3 (same as CERCLA, see below).

Reportable Quantity (RQ): 10 lb (4.54 kg).
California Proposition 65 Chemical: (*hexavalent chromium*) Cancer 2/27/87; Developmental/Reproductive toxin (male, female) 12/19/08.
Canada, WHMIS, Ingredients Disclosure List Concentration 0.1%.
European/International Regulations: Hazard Symbol: T, N; Risk phrases: R45; R22; R50/53; Safety phrases: S53; S60; S61 (see Appendix 4).
WGK (German Aquatic Hazard Class): No value assigned.
Description: Strontium chromate is a light yellow crystalline solid or powder. Molecular weight = 203.62. Hazard Identification (based on NFPA-704 M Rating System): Health 1, Flammability 0, Reactivity 1. Slightly soluble in water.
Potential Exposure: Compound Description: Tumorigen, Mutagen. Strontium chromate is used as a metal protective coating to prevent corrosion, in wash primers; and aluminum flake coatings; colorant in polyvinyl chloride resins and pyrotechnics.
Incompatibilities: Violent reaction with strong oxidizers, hydrazine. Incompatible with combustible, organic, or other readily oxidizable materials, such as paper, wood, sulfur, aluminum powder. Attacks plastics and coatings.
Permissible Exposure Limits in Air
OSHA PEL: 0.005 mg[Cr]/m^3 TWA.
NIOSH REL: 0.001 mg[Cr]/m^3 TWA, potential carcinogen, limit exposure to lowest feasible level. See *NIOSH Pocket Guide*, Appendix A & C.
ACGIH TLV®[1]: 0.0005 mg[Cr]/m^3 TWA, Suspected Human Carcinogen.
NIOSH IDLH: 15 mg[Cr(VI)]/m^3.
No TEEL available.
DFG MAK: [skin] Danger of skin sensitization; Carcinogen Category 1; TRK: 0.05 mg[Cr]/m^3; 20 μg/L [Cr] in urine at end-of-shift.
Arab Republic of Egypt: TWA 0.5 mg/m^3, 1993; Australia: TWA 0.5 mg/m^3, 1993; Austria: carcinogen, 1999; Belgium: TWA 0.5 mg/m^3, 1993; Denmark: TWA 0.001 mg [Cr]/m^3, 1999; Finland: TWA 0.5 mg/m^3; carcinogen, 1999; France: VME 0.05 mg[Cr]/m^3, 1999; [skin], 1993; India: TWA 0.5 mg/m^3, 1993; Japan: 0.01 mg[Cr]/m^3, 1999; Norway: TWA 0.02 mg[CrO$_3$]/m^3, 1999; the Philippines: TWA 0.5 mg/m^3, 1993; Poland: TWA 0.5 mg/m^3, 1999; Sweden: NGV 0.02 mg/m^3, carcinogen, 1999; Switzerland: MAK-W 0.05 mg[Cr]/m^3, 1999; Thailand: TWA 1 mg/m^3, 1993; United Kingdom: TWA 0.05 mg[Cr]/m^3, carcinogen, 2000; Argentina, Bulgaria, Columbia, Jordan, South Korea, New Zealand, Singapore, Vietnam: ACGIH TLV®: STEL 0.0005 mg[Cr]/m^3. The former USSR-UNEP/IRPTC joint project[43] set a MAC of 0.01 mg/m^3 in work-place air. Connecticut[60] has set a guideline for chromium trioxide in ambient air of 0.25 μg/m^3.
Determination in Air: Use NIOSH Analytical Methods #7600,#7604, #7605, #7703, #9101; OSHA Analytical Methods ID-103, ID-215, W-4001.

Permissible Concentration in Water: For the protection of freshwater aquatic life: *hexavalent chromium:* 0.29 μg/L as a 24-h average, never to exceed 21.0 μg/L. For the protection of saltwater aquatic life: *hexavalent chromium:* 18 μg/L as a 24-h average, never to exceed 1260 μg/L. To protect human health: *hexavalent chromium*; 50 μg/L according to EPA.[6] EPA[49] has set a long-term health advisory for adults of 0.84 mg/L and a lifetime health advisory of 0.12 mg/L (120 μg/L) for chromium. EPA's maximum drinking water level (MCL) is 0.1 mg/L.[62] Germany, Canada, EEC, and WHO[35] have set a limit of 0.05 mg/L in drinking water. The states of Maine and Minnesota have set guidelines of 50 μg/L and 120 μg/L, respectively, for chromium in drinking water.[61]

Determination in Water: Total chromium may be determined by digestion followed by atomic absorption or by colorimetry (diphenylcarbazide) or by inductively coupled plasma (CP) optical emission spectrometry. Chromium(VI) may be determined by extraction and atomic absorption or colorimetry (using diphenylhydrazide). Dissolved total Cr or Cr(VI) may be determined by 0.45 μm filtration followed by the above-cited methods.[49]

Routes of Entry: Inhalation, ingestion, skin and/or eye contact.

Harmful Effects and Symptoms

Short Term Exposure: Strontium chromate can affect you when breathed in. Strontium chromate is a carcinogen; handle with extreme caution. Contact may burn the skin and eyes. The vapor or dust may irritate the mouth, nose, and air passages.

Long Term Exposure: Repeated or prolonged exposure (at 20 mg/m^3) may damage the lungs, heart, liver, kidneys, and affect the nervous system. May affect the blood and brain, resulting in changes in white and red blood cells and neuro-vegetative disorders. Strontium chromate accumulates in the body with repeated exposure and effects can persist after exposure. Repeated or prolonged contact with skin may cause skin sensitization, dermatitis, or ulcers. Lungs may be affected or asthma may develop from repeated or prolonged exposure; and with some chromate exposures, nasal septim perforation has occurred. This substance is possibly carcinogenic to humans.

Points of Attack: Kidneys, liver, nervous system, heart, lungs.

Medical Surveillance: NIOSH lists the following tests: blood gas analysis, complete blood count; chest X-ray, electrocardiogram, liver function tests; pulmonary function tests; sputum cytology, urine (chemical/metabolite), urinalysis (routine), white blood cell count/differential. For those with frequent or potentially high exposure (half the TLV or greater, or significant skin contact), the following are recommended before beginning work and at regular times after that: tests for kidney and liver function. Examination of the nervous system.

First Aid: If this chemical gets into the eyes, remove any contact lenses at once and irrigate immediately for at least 15 min, occasionally lifting upper and lower lids. Seek medical attention immediately. If this chemical contacts the skin, remove contaminated clothing and wash immediately with soap and water. Seek medical attention immediately. If this chemical has been inhaled, remove from exposure, begin rescue breathing (using universal precautions, including resuscitation mask) if breathing has stopped and CPR if heart action has stopped. Transfer promptly to a medical facility. When this chemical has been swallowed, get medical attention. Give large quantities of water and demulcents and induce vomiting. Do not make an unconscious person vomit.

Personal Protective Methods: Wear protective gloves and clothing to prevent any reasonable probability of skin contact. Safety equipment suppliers/manufacturers can provide recommendations on the most protective glove/clothing material for your operation. All protective clothing (suits, gloves, footwear, headgear) should be clean, available each day, and put on before work. Contact lenses should not be worn when working with this chemical. Eye protection is included in the recommended respiratory protection. Employees should wash immediately with soap when skin is wet or contaminated. Provide emergency showers and eyewash.

Respirator Selection: NIOSH, as chromates: *At any concentrations above the NIOSH REL, or where there is no REL, at any detectable concentration:* SCBAF: Pd,Pp (APF = 10,000) (any self-contained breathing apparatus that has a full face-piece and is operated in a pressure-demand or other positive-pressure mode) or SaF: Pd,Pp: ASCBA (APF = 10,000) (any supplied-air respirator that has a full face-piece and is operated in a pressure-demand or other positive-pressure mode in combination with an auxiliary, self-contained breathing apparatus operated in a pressure-demand or other positive-pressure mode). *Escape:* 100F (APF = 50) (any air-purifying, full-face-piece respirator with an N100, R100, or P100 filter) or SCBAE (any appropriate escape-type, self-contained breathing apparatus).

Storage: Color Code—Blue: Health Hazard/Poison: Store in a secure poison location. Prior to working with this chemical you should be trained on its proper handling and storage. Strontium chromate must be stored to avoid contact with strong oxidizers (such as chlorine, bromine, and fluorine), since violent reactions occur. A regulated, marked area should be established where this chemical is handled, used, or stored in compliance with OSHA Standard 1910.1045.

Shipping: Toxic solids, oxidizing, n.o.s. require a shipping label of "POISONOUS/TOXIC MATERIALS, OXIDIZER." Strontium chloride falls in Hazard Class 6.1 and Packing Group I or II.

Spill Handling: Evacuate persons not wearing protective equipment from area of spill or leak until cleanup is complete. Remove all ignition sources. Collect powdered material in the most convenient and safe manner and deposit in sealed containers. Ventilate area after cleanup is complete.

It may be necessary to contain and dispose of this chemical as a hazardous waste. If material or contaminated runoff enters waterways, notify downstream users of potentially contaminated waters. Contact your local or federal environmental protection agency for specific recommendations. If employees are required to clean up spills, they must be properly trained and equipped. OSHA 1910.120(q) may be applicable.

Fire Extinguishing: This chemical is a noncombustible solid. Use dry chemical, carbon dioxide, water spray, or foam extinguishers. Poisonous gases are produced in fire. Inhaling the hot fumes may cause cyanosis. If material or contaminated runoff enters waterways, notify downstream users of potentially contaminated waters. Notify local health and fire officials and pollution control agencies. From a secure, explosion-proof location, use water spray to cool exposed containers. If cooling streams are ineffective (venting sound increases in volume and pitch, tank discolors, or shows any signs of deforming), withdraw immediately to a secure position. If employees are expected to fight fires, they must be trained and equipped in OSHA 1910.156. The only respirators recommended for firefighting are self-contained breathing apparatuses that have full face-pieces and are operated in a pressure-demand or other positive-pressure mode.

References

Sax, N. I. (Ed.). (1981). *Dangerous Properties of Industrial Materials Report*, 1, No. 7, 74—76

New Jersey Department of Health and Senior Services. (August 1998). *Hazardous Substances Fact Sheet: Strontium Chromate*. Trenton, NJ

Strontium nitrate   S:0640

Molecular Formula: N_2O_6Sr
Common Formula: $Sr(NO_3)_2$
Synonyms: Nitrate de strontium (French); Nitric acid, strontium salt; Strontium(II) nitrate (1:2)
CAS Registry Number: 10042-76-9
RTECS® Number: WK98000000
UN/NA & ERG Number: UN1507/140
EC Number: 233-131-9
Regulatory Authority and Advisory Bodies
Air Pollutant Standard Set. See below, "Permissible Exposure Limits in Air" section.
As nitrate compounds (water dissociable)
EPCRA Section 313: Reportable only when in aqueous solution. Form R *de minimis* concentration reporting level: 1.0%.
US DOT Regulated Marine Pollutant (49CFR172.101, Appendix B), as nitrates, inorganic, n.o.s.
European/International Regulations: not listed in Annex 1.
WGK (German Aquatic Hazard Class): 2—Hazard to waters.
Description: Strontium nitrate is a white crystalline solid. Molecular weight = 211.64; Boiling point = 645°C;

Freezing/Melting point = 570°C. Hazard Identification (based on NFPA-704 M Rating System): Health 3, Flammability 0, Reactivity 2 (Oxidizer). Soluble in water.
Potential Exposure: Strontium nitrate is used in matches, pyrotechnics, marine signals, and railroad flares.
Incompatibilities: A strong oxidizer. Violent reaction with reducing agents, combustibles, organics, or other readily oxidizable materials.
Permissible Exposure Limits in Air
Protective Action Criteria (PAC)
TEEL-0: 3.5 mg/m^3
PAC-1: 10 mg/m^3
PAC-2: 60 mg/m^3
PAC-3: 500 mg/m^3
Russia[43] set a MAC of 1.0 mg/m^3 in work-place air.
Routes of Entry: Inhalation, ingestion, skin and/or eye contact.
Harmful Effects and Symptoms
Short Term Exposure: Strontium nitrate can affect you when breathed in. Contact may burn the skin and eyes. The vapor or dust can irritate eyes, mouth, nose, and air passages. Higher levels may cause a chemical pneumonia.
Long Term Exposure: Repeated exposure (at 50 mg/m^3) has caused damage to the lungs, heart muscle, liver, kidneys, and blood forming organs; and affects the nervous system in animals. Strontium nitrate accumulates in the body with repeated exposure and effects can persist after exposure stops.
Points of Attack: Liver, kidneys, nervous system, blood.
Medical Surveillance: If symptoms develop or overexposure is suspected, the following may be useful: tests for kidney and liver function. Examination of the nervous system. Complete blood count (CBC). If respiratory symptoms are present, lung function tests are recommended. Persons on low calcium diets may be at greater risk of absorbing more strontium nitrate.
First Aid: If this chemical gets into the eyes, remove any contact lenses at once and irrigate immediately for at least 15 min, occasionally lifting upper and lower lids. Seek medical attention immediately. If this chemical contacts the skin, remove contaminated clothing and wash immediately with soap and water. Seek medical attention immediately. If this chemical has been inhaled, remove from exposure, begin rescue breathing (using universal precautions, including resuscitation mask) if breathing has stopped and CPR if heart action has stopped. Transfer promptly to a medical facility. When this chemical has been swallowed, get medical attention. Give large quantities of water and induce vomiting. Do not make an unconscious person vomit. Medical observation is recommended for 24 h in case chemical pneumonia or other respiratory symptoms are present.
Personal Protective Methods: Wear protective gloves and clothing to prevent any reasonable probability of skin contact. Safety equipment suppliers/manufacturers can provide recommendations on the most protective glove/clothing material for your operation. All protective clothing (suits,

gloves, footwear, headgear) should be clean, available each day, and put on before work. Contact lenses should not be worn when working with this chemical. Wear dust-proof chemical goggles and face shield unless full face-piece respiratory protection is worn. Employees should wash immediately with soap when skin is wet or contaminated. Provide emergency showers and eyewash.

Respirator Selection: Where there is potential for exposure to strontium nitrate, use a NIOSH/MSHA- or European Standard EN149-approved full face-piece respirator with a high-efficiency particulate filter. Greater protection is provided by a powered air-purifying respirator. *Where there is potential for high exposures,* use a NIOSH/MSHA- or European Standard EN149-approved supplied-air respirator with a full face-piece operated in the positive-pressure mode, or with a full face-piece, hood, or helmet in the continuous-flow mode; or use a NIOSH/MSHA- or European Standard EN149-approved self-contained breathing apparatus with a full face-piece operated in pressure-demand or other positive-pressure mode.

Storage: Color Code—Yellow: Reactive Hazard; Store in a location separate from other materials, especially flammables and combustibles. Prior to working with this chemical you should be trained on its proper handling and storage. Strontium nitrate must be stored to avoid contact with combustible, organic, or other readily oxidizable materials, since violent reactions occur. Store in tightly closed containers in a cool, well-ventilated area away from sources of heat. See OSHA Standard 1910.104 and NFPA 43A *Code for the Storage of Liquid and Solid Oxidizers* for detailed handling and storage regulations.

Shipping: Strontium nitrate requires a shipping label of "OXIDIZER." It falls in Hazard Class 5.1 and Packing Group III.

Spill Handling: Evacuate persons not wearing protective equipment from area of spill or leak until cleanup is complete. Remove all ignition sources. Collect powdered material in the most convenient and safe manner and deposit in sealed containers. Ventilate area after cleanup is complete. Keep strontium nitrate out of a confined space, such as a sewer, because of the possibility of an explosion, unless the sewer is designed to prevent the buildup of explosive concentrations. It may be necessary to contain and dispose of this chemical as a hazardous waste. If material or contaminated runoff enters waterways, notify downstream users of potentially contaminated waters. Contact your local or federal environmental protection agency for specific recommendations. If employees are required to clean up spills, they must be properly trained and equipped. OSHA 1910.120(q) may be applicable.

Fire Extinguishing: This chemical is a noncombustible solid. Use dry chemical, carbon dioxide, water spray, or foam extinguishers. Poisonous gases are produced in fire, including nitrogen oxides. If material or contaminated runoff enters waterways, notify downstream users of potentially contaminated waters. Notify local health and fire officials and pollution control agencies. From a secure, explosion-proof location, use water spray to cool exposed containers. If cooling streams are ineffective (venting sound increases in volume and pitch, tank discolors, or shows any signs of deforming), withdraw immediately to a secure position. If employees are expected to fight fires, they must be trained and equipped in OSHA 1910.156. The only respirators recommended for firefighting are self-contained breathing apparatuses that have full face-pieces and are operated in a pressure-demand or other positive-pressure mode.

References
New Jersey Department of Health and Senior Services. (March 2000). *Hazardous Substances Fact Sheet: Strontium Nitrate.* Trenton, NJ

Strychnine   S:0650

Molecular Formula: $C_{21}H_{22}N_2O_2$; $C_{21}H_{22}N_2O_2 \cdot \frac{1}{2}H_2O_4S$ (sulfate)

Synonyms: Boomer-Rid; Certox; Dolco mouse cereal; Estricnina (Spanish); Gopher bait; Gopher-gitter; Hare-rid; Kwik-kil; Mole death; Mouse-nots; Mouse-rid; Mouse-tox; Nux vomica; Pied piper mouse seed; Ro-Dex; Sanaseed; Strychnidin-10-one; Strychnin (German); Strychnos
Strychnine, sulfate(2:1): Antivampire; Strychinium sulfate; Strychnidin-10-one, sulfate (2:1); Strychnine hemisulfate; Strychnine sulfate; Vampirol

CAS Registry Number: 57-24-9; 60-41-3 (sulfate)
RTECS® Number: WL2275000; WL2550000 (sulfate)
UN/NA & ERG Number: UN1692/151
EC Number: 200-319-7 [*Annex I Index No.:* 614-003-00-5]; 200-477-7 (strychnine sulfate)

Regulatory Authority and Advisory Bodies
Banned or Severely Restricted (several countries) (UN).[13]
US EPA, FIFRA 1998 Status of Pesticides: RED completed.
US EPA, FIFRA 1998 Status of Pesticides: Canceled (sulfate).
FDA—over-the-counter drug.
Air Pollutant Standard Set. See below, "Permissible Exposure Limits in Air" section.
Clean Water Act: Section 311 Hazardous Substances/RQ 40CFR117.3 (same as CERCLA, see below).
US EPA Hazardous Waste Number (RCRA No.): P108.
RCRA, 40CFR261, Appendix 8 Hazardous Constituents.
Superfund/EPCRA 40CFR355, Extremely Hazardous Substances: TPQ = 100/10,000 lb (45.4/4540 kg).
Reportable Quantity (RQ): 10 lb (4.54 kg).
EPCRA Section 313 Form R *de minimis* concentration reporting level: 1.0% *(Strychnine and salts).*
US DOT Regulated Marine Pollutant (49CFR172.101, Appendix B).
European/International Regulations *(Strychnine and salts):* Hazard Symbol: T+, N; Risk phrases: R27/28; R50/53; Safety phrases: S1/2; S36/37; S45; S60; S61 (see Appendix 4).

WGK (German Aquatic Hazard Class): No value assigned.

Description: Strychnine is a colorless crystalline prisms or white powder. It is odorless; with a bitter taste. Molecular weight = 333.45; 383.49 (sulfate); Specific gravity (H_2O:1) = 1.36 at 25°C; Boiling point = (decomposes) 270°C at 5 mmHg; Freezing/Melting point = 267.8°C. Hazard Identification (based on NFPA-704 M Rating System): Health 3, Flammability 1, Reactivity 0. Slightly soluble in water; solubility = 0.02%.

Potential Exposure: Compound Description: Agricultural Chemical; Drug; Reproductive Effector; Human Data; Natural Product; (sulfate). Agricultural Chemical; Drug; Human Data. Strychnine is an alkaloid compound that has been widely used as a rodenticide/bait to kill rodents; a medicine, respiratory stimulant. Those involved in the extraction of the seeds of *Strychnos nux-vomica*, *Strychnos ignatii* (S. sancta Ingnatius), and *Strychnos tiente* (Upas tree); formulation or application of this rodent poison. The sulfate is used to kill gophers and moles. A common adulterant in illicit street drugs. Listed as a potential WMD biotoxin.[NIOSH]

Incompatibilities: Strong oxidizers. Dangerous when heated; forms toxic fumes, including nitrogen oxides. In the body, caffeine may increase the strychnine effect.

Permissible Exposure Limits in Air
OSHA PEL: 0.15 mg/m³ TWA.
NIOSH REL: 0.15 mg/m³ TWA.
ACGIH TLV®[1]: 0.15 mg/m³ TWA.
NIOSH IDLH: 3 mg/m³.
Protective Action Criteria (PAC)
TEEL-0: 0.15 mg/m³
PAC-1: 0.3 mg/m³
PAC-2: 3 mg/m³
PAC-3: 3 mg/m³
Australia: TWA 0.15 mg/m³, 1993; Austria: MAK 0.15 mg/m³, 1999; Belgium: TWA 0.15 mg/m³, 1993; Belgium: TWA 0.15 mg/m³, 1999; Finland: TWA 0.15 mg/m³; STEL 0.45 mg/m³, 1999; France: VME 0.15 mg/m³, 1999; the Netherlands: MAC-TGG 0.15 mg/m³, 2003; the Philippines: TWA 0.15 mg/m³, 1993; Poland: MAC (TWA) 0.15 mg/m³, 1999; Switzerland: MAK-W 0.15 mg/m³, KZG-W 0.3 mg/m³, 1999; Turkey: TWA 0.15 mg/m³, 1993; United Kingdom: TWA 0.15 mg/m³; STEL 0.45 mg/m³, 2000; Argentina, Bulgaria, Columbia, Jordan, South Korea, New Zealand, Singapore, Vietnam: ACGIH TLV®: TWA 0.15 mg/m³. Several states have set guidelines or standards for strychnine in ambient air[60] ranging from 1.5 μg/m³ (North Dakota) to 2.5 μg/m³ (Virginia) to 3.0 μg/m³ (Connecticut) to 4.0 μg/m³ (Nevada).

Determination in Air: Use NIOSH Analytical Method (IV) #5016.

Determination in Water: Octanol–water coefficient: Log K_{ow} = 1.68.

Routes of Entry: Inhalation of dust, ingestion, skin and/or eye contact.

Harmful Effects and Symptoms
Short Term Exposure: Affects the central nervous system, causing convulsions, muscle contractions, and respiratory failure. Super toxic; probable oral lethal dose in humans is less than 5 mg/kg, a taste (less than 7 drops) for a 70-kg (150 lb) person. It causes violent generalized convulsions. Death results from respiratory arrest as the respiratory muscles are in sustained spasm. The lowest lethal oral dose reported for humans if 30 mg/kg. Respiratory paralysis and arrest are likely to occur following severe exposure to strychnine. Signs and symptoms of acute exposure generally involve excitation of all portions of the central nervous system. Convulsions, bilateral horizontal nystagmus (rapid, synchronous, horizontal, oscillations of the eyeballs), agitation, restlessness, apprehension, and abrupt jerking movements of the extremities may occur. Victims may also experience stiffness, painful muscle cramping (especially in the legs), and opisthotonos (spasm in which the spine and extremities are bent with convexity forward, the body resting on the head and heels). Vomiting and renal failure, as well as cyanosis (blue tint to skin and mucous membranes) and rhabdomyolysis (destruction of skeletal muscle), may be found.

Long Term Exposure: Chronic allergen if inhaled or ingested.

Points of Attack: Central nervous system.

Medical Surveillance: Be aware of possible convulsions. Consider the points of attack in preplacement and periodic physical examinations. Examination by a qualified allergist.

First Aid: Remove victims from exposure. Emergency personnel should avoid self-exposure to strychnine. Warning—Any unnecessary sensory input may induce seizures. Isolate the victims from any avoidable distractions. Rush to a healthcare facility! Evaluate vital signs including pulse and respiratory rate, and note any trauma. If no pulse is detected, provide CPR. If not breathing, provide artificial respiration. If breathing is labored, administer oxygen or other respiratory support. Remove contaminated clothing as soon as possible. If eye exposure has occurred, remove any contact lenses at once; eyes must be flushed with lukewarm water for at least 15 min. Wash exposed skin areas thoroughly with soap and water. Obtain authorization and/or further instructions from the local hospital for administration of an antidote or performance of other invasive procedures.

Personal Protective Methods: Wear protective gloves and clothing to prevent any reasonable probability of skin contact. Safety equipment suppliers/manufacturers can provide recommendations on the most protective glove/clothing material for your operation. All protective clothing (suits, gloves, footwear, headgear) should be clean, available each day, and put on before work. Contact lenses should not be worn when working with this chemical. Wear dust-proof chemical goggles and face shield unless full face-piece respiratory protection is worn. Employees should wash immediately with soap when skin is wet or contaminated. Provide emergency showers and eyewash.

Respirator Selection: *Up to 0.75 mg/m³:* Qm (APF = 25) (any quarter-mask respirator). *Up to 1.5 mg/m³:* 95 XQ [any particulate respirator equipped with an N95, R95, or P95 filter (including N95, R95, and P95 filtering face-pieces) except quarter-mask respirators. The following filters may also be used: N99, R99, P99, N100, R100, P100] or Sa (APF = 10) (any supplied-air respirator). *Up to 3 mg/m³:* Sa:Cf (APF = 25) (any supplied-air respirator operated in a continuous-flow mode) or PaprHie (APF = 25) (any powered, air-purifying respirator with a dust and mist filter) or 100F (APF = 50) (any air-purifying, full-face-piece respirator with an N100, R100, or P100 filter) or SCBAF (APF = 50) (any self-contained breathing apparatus with a full face-piece) or SaF (APF = 50) (any supplied-air respirator with a full face-piece). *Emergency or planned entry into unknown concentrations or IDLH conditions:* SCBAF: Pd,Pp (APF = 10,000) (any self-contained breathing apparatus that has a full face-piece and is operated in a pressure-demand or other positive-pressure mode) or SaF: Pd,Pp: ASCBA (APF = 10,000) (any supplied-air respirator that has a full face-piece and is operated in a pressure-demand or other positive-pressure mode in combination with an auxiliary self-contained breathing apparatus operated in a pressure-demand or other positive-pressure mode). *Escape:* 100F (APF = 50) (any air-purifying, full-face-piece respirator with an N100, R100, or P100 filter) or SCBAE (any appropriate escape-type, self-contained breathing apparatus).

Storage: Color Code—Blue: Health Hazard/Poison: Store in a secure poison location. Prior to working with this chemical you should be trained on its proper handling and storage. Store in tightly closed containers in a cool, well-ventilated area away from strong oxidizers and heat. Where possible, automatically transfer material from drums or other storage containers to process containers. Sources of ignition, such as smoking and open flames, are prohibited where this chemical is handled, used, or stored. Metal containers involving the transfer of this chemical should be grounded and bonded. Wherever this chemical is used, handled, manufactured, or stored, use explosion-proof electrical equipment and fittings.

Shipping: Strychnine or strychnine salts requires a shipping label of "POISONOUS/TOXIC MATERIALS." They fall in DOT Hazard Class 6.1 and Packing Group I.

Spill Handling: Evacuate persons not wearing protective equipment from area of spill or leak until cleanup is complete. Remove all ignition sources. Do not touch spilled material; stop leak if you can do so without risk. *Small spills:* absorb with sand or other noncombustible absorbent material and place into containers for later disposal. *Small dry spills:* with clean shovel place material into clean, dry container and cover; move containers from spill area. *Large spills:* dike far ahead of spill for later disposal. Avoid breathing dusts, and fumes from burning materials. Keep upwind. Avoid bodily contact with the material. *Do Not* handle broken packages without protective equipment.

Wash away any material which may have contacted the body with copious amounts of water or soap and water. Ventilate area after cleanup is complete. It may be necessary to contain and dispose of this chemical as a hazardous waste. If material or contaminated runoff enters waterways, notify downstream users of potentially contaminated waters. Contact your local or federal environmental protection agency for specific recommendations. If employees are required to clean up spills, they must be properly trained and equipped. OSHA 1910.120(q) may be applicable.

Fire Extinguishing: Extinguish fire using agent suitable for type of surrounding fire; material itself does not burn or burns with difficulty. Use water in flooding quantities as fog. Use alcohol foam, carbon dioxide, or dry chemical. Poisonous gases are produced in fire, including nitrogen oxides. If material or contaminated runoff enters waterways, notify downstream users of potentially contaminated waters. Notify local health and fire officials and pollution control agencies. From a secure, explosion-proof location, use water spray to cool exposed containers. If cooling streams are ineffective (venting sound increases in volume and pitch, tank discolors, or shows any signs of deforming), withdraw immediately to a secure position. If employees are expected to fight fires, they must be trained and equipped in OSHA 1910.156. The only respirators recommended for firefighting are self-contained breathing apparatuses that have full face-pieces and are operated in a pressure-demand or other positive-pressure mode.

Disposal Method Suggested: Consult with environmental regulatory agencies for guidance on acceptable disposal practices. Generators of waste containing this contaminant (≥100 kg/mo) must conform with EPA regulations governing storage, transportation, treatment, and waste disposal. In accordance with 40CFR165, follow recommendations for the disposal of pesticides and pesticide containers. Must be disposed properly by following package label directions or by contacting your local or federal environmental control agency or by contacting your regional EPA office. Careful incineration has been recommended for disposal. Two procedures are suggested.[1] Pour or sift onto a thick layer of sand and soda ash mixture (90:10). Mix and shovel into a heavy paper box with much paper packing. Burn in incinerator. Fire may be augmented by adding excelsior and scrap wood. Stay on the upwind side.[2] Waste may be dissolved in flammable solvent (alcohols, benzene, etc.) and sprayed into fire box of an incinerator with afterburner and scrubber.[22]

References

Sax, N. I. (Ed.). *Dangerous Properties of Industrial Materials Report*, 2, No. 2, 63−65 (1982) and 8, No. 1, 78−83 (1988)

US Environmental Protection Agency. (November 30, 1987). *Chemical Hazard Information Profile: Strychnine.* Washington, DC: Chemical Emergency Preparedness Program

US Environmental Protection Agency. (November 30, 1987). *Chemical Hazard Information Profile: Strychnine Sulfate*. Washington, DC: Chemical Emergency Preparedness Program

US Environmental Protection Agency, Special Review and Reregistration Division Office of Pesticide Programs. (1998). *Agency Status of Pesticides in Registration, Reregistration, and Special Review* (Rainbow Report). Washington, DC

Styrene   S:0660

Molecular Formula: C_8H_8
Common Formula: $C_6H_5CH{=}CH_2$
Synonyms: Annamene; Benzene, ethenyl-; Benzene, vinyl-; Cinnamene; Cinnamenol; Cinnamol; Diarex HF 77; Estireno (Spanish); Ethylbenzene; Ethylene, phenyl-; NCI-C02200; Phenethylene; Phenylethene; Phenylethylene; Styrene monomer; Styrene monomer, inhibited; Styrol (German); Styrole; Styrolene; Styron; Styropol; Styropol SO; Styropor; Vinylbenzene; Vinylbenzol
CAS Registry Number: 100-42-5; *(alt.)* 79637-11-9
RTECS® Number: WL3675000
UN/NA & ERG Number: UN2055/128
EC Number: 202-851-5 [*Annex I Index No.:* 601-026-00-0]
Regulatory Authority and Advisory Bodies
Carcinogenicity: NCI: Carcinogenesis Bioassay (gavage); inadequate studies: mouse, rat; IARC: Human Limited Evidence, animal Sufficient Limited Evidence, *possibly carcinogenic to humans*, Group 2B, 2002.
Note: Do not confuse with *Styrene polymer* (CAS: 9003-53-6; RTECS: WL6475000): IARC: Animal Inadequate Evidence; Human No Adequate Data, *not classifiable as carcinogenic to humans*, Group 3, 1987.
US EPA Gene-Tox Program, Positive: *In vivo* cytogenetics—human lymphocyte; Host-mediated assay; Positive: *In vitro* human lymphocyte micronucleus; Positive: Histidine reversion—Ames test; Positive: *D. melanogaster* sex-linked lethal; Positive: *S. cerevisiae* gene conversion; Positive/limited: Carcinogenicity—mouse/rat; Negative: Cell transform.—SA7/SHE; *In vitro* UDS—human fibroblast; Negative: V79 cell culture-gene mutation; Inconclusive: *S. pombe*—forward mutation.
Air Pollutant Standard Set. See below, "Permissible Exposure Limits in Air" section.
Clean Air Act: Hazardous Air Pollutants (Title I, Part A, Section 112).
Clean Water Act: Section 311 Hazardous Substances/RQ 40CFR117.3 (same as CERCLA, see below); Section 313 Water Priority Chemicals (57FR41331, 9/9/92).
RCRA 40CFR264, Appendix 9; TSD Facilities Ground Water Monitoring List. Suggested test method(s) (PQL μg/L): 8020 (1); 8240 (5).
Safe Drinking Water Act: MCL, 0.1 mg/L; MCLG, 0.1 mg/L.
Reportable Quantity (RQ): 1000 lb (454 kg).

EPCRA Section 313 Form R *de minimis* concentration reporting level: 0.1%.
US DOT Regulated Marine Pollutant (49CFR172.101, Appendix B).
European/International Regulations: Hazard Symbol: Xn; Risk phrases: R10; R20; R36/38; Safety phrases: S2; S23.
Canada, WHMIS, Ingredients Disclosure List Concentration 0.1%.
European/International Regulations: not listed in Annex 1.
WGK (German Aquatic Hazard Class): 2—Hazard to waters.
Description: Styrene is a colorless to yellowish, very refractive, oily liquid with a penetrating odor. The odor threshold = 0.02−0.47 ppm. Molecular weight = 104.16; Specific gravity (H_2O:1) = 0.91 at 25°C; Boiling point = 145°C; Freezing/Melting point = −30.6°C; Vapor pressure = 5 mmHg; Flash point = 31°C; Autoignition temperature = 490°C. Explosive limits: LEL = 0.9%; UEL = 6.8%. Hazard Identification (based on NFPA-704 M Rating System): Health 2, Flammability 3, Reactivity 2. Practically insoluble in water; solubility = 0.03%.
Potential Exposure: Compound Description: Tumorigen, Mutagen; Reproductive Effector; Human Data; Primary Irritant. Styrene is used in the production of plastics and polystyrene resins. It is also used in combination with 1,3-butadiene or acrylonitrile to form copolymer elastomers, butadiene-styrene rubber, and acrylonitrile-butadiene-styrene (ABS). It is also used in the manufacture of protective coatings; resins, polyesters; in making insulators and in drug manufacture.
Incompatibilities: Styrene forms explosive mixture with air. A storage hazard above 31°C. Upon heating to 200°C, styrene polymerizes to form polystyrene, a plastic. Before entering confined space where this chemical may be present, check to make sure that an explosive concentration does not exist. Store in a cool, dry area away from oxidizers, catalysts for vinyl polymers, peroxides, strong acids, aluminum chloride. May polymerize if contaminated, subjected to heat; under the influence of light; and on contact with many compounds, such as oxygen, oxidizing agents, peroxides, and strong acids. Usually contains an inhibitor, such as *tert*-butylcatechol. Corrodes copper and copper alloys. Attacks some plastics, rubber, and coatings.
Permissible Exposure Limits in Air
Conversion factor: 1 ppm = 4.26 mg/m³ at 25°C & 1 atm.
OSHA PEL: 100 ppm TWA; 200 ppm Ceiling Concentration; 600 ppm [5 min max peak in any 3 h].
NIOSH REL: 50 ppm/215 mg/m³ TWA; 100 ppm/425 mg/m³ STEL.
ACGIH TLV®[1]: 20 ppm/85 mg/m³ TWA; 40 ppm/170 mg/m³ STEL, not classifiable as a human carcinogen. BEI: 800 mg[mendelic acid]/g creatinine in urine/end-of-shift; 300 mg[mendelic acid]/g creatinine in urine, prior to next shift; 240 mg[phenylglyoxylic acid]/g creatinine in urine/end-of-shift; 100 mg[phenylglyoxylic acid]/g creatinine in urine, prior to next shift; 0.55 mg[styrene]/L in blood/end-of-shift; 0.02 mg[styrene]/L in blood, prior to next shift.

Protective Action Criteria (PAC)*
TEEL-0: 20 ppm
PAC-1: **20** ppm
PAC-2: **130** ppm
PAC-3: **1100** ppm
*AEGLs (Acute Emergency Guideline Levels) & ERPGs (Emergency Response Planning Guideline) are in **bold face**.
DFG MAK: 20 ppm/86 mg/m³ TWA; Peak Limitation Category II(2); Carcinogen Category 5; Pregnancy Risk Category C; BAT: 600 mg[mendelic acid plus phenyl-glyoxylic acid]/g creatinine in urine/end-of-shift; for long term exposures: after several shifts.
NIOSH IDLH: 700 ppm.
Australia: TWA 50 ppm (215 mg/m³); STEL 100 ppm, 1993; Austria: MAK 20 ppm (85 mg/m³), 1999; Denmark: TWA 25 ppm (105 mg/m³), [skin], 1999; Finland: TWA 20 ppm (85 mg/m³); STEL 100 ppm (420 mg/m³), 1999; France: VME 50 ppm (215 mg/m³), 1999; Hungary: STEL 50 mg/m³, carcinogen, 1993; Japan: 50 ppm (210 mg/m³), 2B carcinogen, 1999; the Netherlands: MAC-TGG 107 mg/m³, 2003; Norway: TWA 25 ppm (105 mg/m³), 1999; the Philippines: TWA 100 ppm (420 mg/m³), 1993; Poland: MAC (TWA) 50 mg/m³; MAC (STEL) 200 mg/m³, 1999; Sweden: NGV 20 ppm (90 mg/m³), KTV 50 ppm (200 mg/m³), [skin], 1999; Switzerland: MAK-W 50 ppm (215 mg/m³), KZG-W 100 ppm (430 mg/m³), 1999; Thailand: TWA 100 ppm; STEL 200 ppm, 1993; Turkey: TWA 100 ppm (420 mg/m³), 1993; United Kingdom: TWA 100 ppm (430 mg/m³); STEL 250 ppm, 2000; Argentina, Bulgaria, Columbia, Jordan, South Korea, New Zealand, Singapore, Vietnam: ACGIH TLV®: STEL 40 ppm. Also, values for styrene in ambient air in residential areas vary. The Czech Republic[35] has set 0.015 mg/m³, both on a momentary and a daily average basis. Russia has set 0.003 mg/m³ both on a daily average and momentary basis[43] but values are also given[35] as 0.04 mg/m³ on a once-daily basis and 0.002 mg/m³ on a daily average basis. Several states have set guidelines or standards for styrene in ambient air[60] ranging from zero (North Carolina) to 3.45 μg/m³ (Indiana) to 30.0 μg/m³ (Rhode Island) to 34.48 μg/m³ (Kansas) to 39.0 μg/m³ (Massachusetts) to 716.0 μg/m³ (New York) to 2150.0−4250.0 μg/m³ (North Dakota) to 3600.0 μg/m³ (Virginia) to 4300.0 μg/m³ (Connecticut) to 5119.0 μg/m³ (Nevada).
Determination in Air: Use NIOSH Analytical Method (IV) #1501, Aromatic hydrocarbons; #3800. OSHA Analytical Method 9 or 89.
Permissible Concentration in Water: A lifetime health advisory of 140 μg/L has been determined by EPA.[48] Russia[43] set a MAC of 100 μg/L in water bodies used for domestic purposes. This same limit applies to water bodies used for fishery purposes. Several states have set guidelines for styrene in drinking water[61] ranging from 10.0 μg/L (Wisconsin) to 140 μg/L (Arizona and Minnesota) to 270 μg/L Maine.

Determination in Water: Styrene may be determined by a purge-and-trap gas chromatographic procedure.[48] Octanol−water coefficient: Log K_{ow} = 3.2.
Routes of Entry: Inhalation, ingestion, skin and/or eye contact. Absorbed through the skin.
Harmful Effects and Symptoms
Irritates the eyes, skin, and respiratory tract. *Inhalation:* At or above 100 ppm, styrene causes immediate eye and nose irritation; persistent metallic taste; headache, fatigue, slight muscular weakness; loss of appetite; drowsiness, feelings of drunkenness, decreased coordination; depression, unconsciousness; inflammation of the lung, kidney and liver damage, and death. *Skin:* Can cause drying, cracking, itching, burning, and sores. Absorption is moderate and can cause symptoms described above. *Eyes:* Can cause severe itching, tearing, and injury to the surface of the eye. *Ingestion:* Symptoms are same as inhalation. Additional symptoms may include severe irritation of the mouth, throat, and stomach. Swallowing the liquid may cause chemical pneumonitis.
Long Term Exposure: Repeated exposure to low levels can cause concentration problems, memory problems, learning disability, slowed reflexes; reduced coordination and manual dexterity; and trouble with balance; nausea, headache, fatigue, and a feeling of drunkenness. Continued exposures to levels near 400 ppm can cause eye and throat irritation, and slight impairment of coordination and balance. At higher air concentrations, nasal, eye, throat, and skin irritation becomes pronounced. Prolonged inhalation of vapors may cause respiratory tract obstruction. Very high levels may affect brain function and cause liver damage and death. Cases of liver damage have been found in workers employed for over 5 years in polystyrene plants and exposed to a concentration of 20−150 ppm. Styrene has been found to produce lung tumors in mice and cause changes in the genetic material of laboratory organisms. Whether it does so in humans is not known.
Points of Attack: Eyes, skin, respiratory system, central nervous system, liver, reproductive system.
Medical Surveillance: NIOSH lists the following tests: whole blood (chemical/metabolite); whole blood (chemical/metabolite), end-of-shift; whole blood (chemical/metabolite), end-of-workweek; whole blood (chemical/metabolite), prior to next shift; whole blood (chemical/metabolite), Prior to Shift, expired air, Expired Air, During Exposure. For those with frequent or potentially high exposure (half the TLV or greater, or significant skin contact), the following are recommended before beginning work and at regular times after that: examination of the nervous system. If symptoms develop or overexposure is suspected, the following may be useful: EEG (brain wave study).
First Aid: If this chemical gets into the eyes, remove any contact lenses at once and irrigate immediately for at least 15 min, occasionally lifting upper and lower lids. Seek medical attention immediately. If this chemical contacts the skin, remove contaminated clothing and wash immediately

with soap and water. Seek medical attention immediately. If this chemical has been inhaled, remove from exposure, begin rescue breathing (using universal precautions, including resuscitation mask) if breathing has stopped and CPR if heart action has stopped. Transfer promptly to a medical facility. When this chemical has been swallowed, get medical attention. Give large quantities of water and induce vomiting. Do not make an unconscious person vomit.

Note to physician: Inhalation: bronchodilators, decongestants, and oxygen may be used if necessary. Corticosteroids are useful for treating pneumonitis.

Personal Protective Methods: Wear protective gloves and clothing to prevent any reasonable probability of skin contact. Safety equipment suppliers/manufacturers can provide recommendations on the most protective glove/clothing material for your operation. Safety equipment manufacturers and styrene manufacturers recommend Teflon™, Viton™/chlorobutyl rubber, chlorinated polyethylene, and polyvinyl alcohol as the recommended protective materials. All protective clothing (suits, gloves, footwear, headgear) should be clean, available each day, and put on before work. Contact lenses should not be worn when working with this chemical. Wear splash-proof chemical goggles and face shield unless full face-piece respiratory protection is worn. Employees should wash immediately with soap when skin is wet or contaminated. Provide emergency showers and eyewash. Before entering a confined space where styrene monomer is present, check the oxygen level (at least 19% must be present) and that explosive concentration does not exist.

Respirator Selection: NIOSH: *500 ppm:* CcrOv* [any chemical cartridge respirator with organic vapor cartridge(s)] or SA* (any supplied-air respirator). *700 ppm:* Sa:Cf (APF = 25)* (any supplied-air respirator operated in a continuous-flow mode) or CcrFOv (APF = 50) [any chemical cartridge respirator with a full face-piece and organic vapor cartridge(s)] or GmFOv (APF = 50) [any air-purifying, full-face-piece respirator (gas mask) with a chin-style, front- or back-mounted organic vapor canister] or PaprOv* [any powered, air-purifying respirator with organic vapor cartridge(s)] or SCBAF (APF = 50) (any self-contained breathing apparatus with a full face-piece) or SaF (APF = 50) (any supplied-air respirator with a full face-piece). *Emergency or planned entry into unknown concentrations or IDLH conditions:* SCBAF: Pd,Pp (APF = 10,000) (any self-contained breathing apparatus that has a full face-piece and is operated in a pressure-demand or other positive-pressure mode) or SaF: Pd,Pp: ASCBA (APF = 10,000) (any supplied-air respirator that has a full face-piece and is operated in a pressure-demand or other positive-pressure mode in combination with an auxiliary, self-contained breathing apparatus operated in a pressure-demand or other positive-pressure mode). *Escape:* GmFOv (APF = 50) [any air-purifying, full-face-piece respirator (gas mask) with a chin-style, front- or back-mounted organic vapor canister] or SCBAE (any appropriate escape-type, self-contained breathing apparatus).

*Substance reported to cause eye irritation or damage; may require eye protection.

Storage: Color Code—Red: Flammability Hazard: Store in a flammable liquid storage area or approved cabinet away from ignition sources and corrosive and reactive materials. Prior to working with this chemical you should be trained on its proper handling and storage. A storage hazard above 31°C. Before entering a confined space where styrene monomer is present, check the oxygen level (at least 19% must be present) and that explosive concentration does not exist. Styrene monomer must be stored to avoid contact with oxidizing agents; and catalysts for vinyl polymerization, such as peroxides, strong acids (such as hydrochloric, sulfuric, and nitric); and aluminum chlorides, since violent reactions occur. Store in tightly closed containers in a cool, well-ventilated area. Sources of ignition, such as smoking and open flames, are prohibited where styrene monomer is handled, used, or stored. Metal containers involving the transfer of 5 gallons or more of styrene monomer should be grounded and bonded. Drums must be equipped with self-closing valves, pressure vacuum bungs, and flame arresters. Use only nonsparking tools and equipment, especially when opening and closing containers of styrene monomer. Wherever styrene monomer is used, handled, manufactured, or stored, use explosion-proof electrical equipment and fittings. Styrene monomer will corrode copper and copper alloys and dissolve rubber. Styrene monomer must be stored with an inhibitor to prevent explosive reactions. A regulated, marked area should be established where this chemical is handled, used, or stored in compliance with OSHA Standard 1910.1045.

Shipping: Styrene monomer, inhibited, requires a shipping label of "FLAMMABLE LIQUID." It falls in Hazard Class 3 and Packing Group II.

Spill Handling: Evacuate and restrict persons not wearing protective equipment from area of spill or leak until cleanup is complete. Remove all ignition sources. Establish forced ventilation to keep levels below explosive limit. Absorb liquids in vermiculite, dry sand, earth, peat, carbon, or a similar material and deposit in sealed containers. Keep this chemical out of a confined space, such as a sewer, because of the possibility of an explosion, unless the sewer is designed to prevent the buildup of explosive concentrations. It may be necessary to contain and dispose of this chemical as a hazardous waste. If material or contaminated runoff enters waterways, notify downstream users of potentially contaminated waters. Contact your local or federal environmental protection agency for specific recommendations. If employees are required to clean up spills, they must be properly trained and equipped. OSHA 1910.120(q) may be applicable.

Fire Extinguishing: This chemical is a flammable liquid. Poisonous gases, including styrene oxides, are produced in fire. Use dry chemical, carbon dioxide, or foam extinguishers. Vapors are heavier than air and will collect in low areas. Vapors may travel long distances to ignition sources and flashback. Vapors in confined areas may explode when exposed to fire. Containers may explode in fire. Storage containers and parts of containers may rocket great distances, in many directions. If material or contaminated runoff enters waterways, notify downstream users of potentially contaminated waters. Notify local health and fire officials and pollution control agencies. From a secure, explosion-proof location, use water spray to cool exposed containers. If cooling streams are ineffective (venting sound increases in volume and pitch, tank discolors, or shows any signs of deforming), withdraw immediately to a secure position. If employees are expected to fight fires, they must be trained and equipped in OSHA 1910.156. The only respirators recommended for firefighting are self-contained breathing apparatuses that have full face-pieces and are operated in a pressure-demand or other positive-pressure mode.

Disposal Method Suggested: Dissolve or mix the material with a combustible solvent and burn in a chemical incinerator equipped with an afterburner and scrubber. All federal, state, and local environmental regulations must be observed. In some cases, recovery and recycle of styrene monomer is economic and the technology is available.

References

Sax, N. I. (Ed.). *Dangerous Properties of Industrial Materials Report*, 1, No. 8, 92−95 (1981), 2, No. 6, 60−64 (1982), 6, No. 2, 110−115 (1986), 8, No. 3, 10−44 (1988)
New York State Department of Health. (June 1984). *Chemical Fact Sheet: Styrene* (Version 3). Albany, NY: Bureau of Toxic Substance Assessment
New Jersey Department of Health and Senior Services. (August 2006). *Hazardous Substances Fact Sheet: Styrene Monomer*. Trenton, NJ

Styrene oxide   S:0670

Molecular Formula: C_8H_8O
Synonyms: Benzene, (epoxyethyl)-; Epoxyethylbenzene; (Epoxyethyl)benzene; 1,2-Epoxyethylbenzene; Epoxystyrene; α,β-Epoxystyrene; Ethane, 1,2-epoxy-1-phenyl-; Oxido de estireno (Spanish); Oxirane, phenyl-; Phenethylene oxide; 1-Phenyl-1,2-epoxyethane; Phenylethylene oxide; 2-Phenyloxirane; Phenyloxirane; Styrene epoxide; Styrene 7,8-oxide; Styryl oxide
CAS Registry Number: 96-09-3; *(alt.)* 62497-63-6
RTECS® Number: CZ9625000
UN/NA & ERG Number: UN3082/171
EC Number: 202-476-7 [*Annex I Index No.:* 603-084-00-2]
Regulatory Authority and Advisory Bodies
Carcinogenicity: NTP: Carcinogenesis Studies (gavage); clear evidence: mouse, rat; NTP: 11th Report on Carcinogens, 2004: Reasonably anticipated to be a human carcinogen; IARC: Animal, Sufficient Evidence, 1997; Human, Inadequate Evidence, Group 2A, 1994.

US EPA Gene-Tox Program, Positive: Host-mediated assay; Positive: *In vitro* human lymphocyte micronucleus; Positive: Histidine reversion—Ames test; Positive: *D. melanogaster* sex-linked lethal; Positive: *In vitro* UDS—human fibroblast; Positive: V79 cell culture-gene mutation; Positive: *S. cerevisiae* gene conversion; Weakly Positive: *S. pombe*—forward mutation; Positive/dose response: TRP reversion; Positive/limited: Carcinogenicity—mouse/rat; Negative: Rodent dominant lethal; Inconclusive: Mammalian micronucleus.
Clean Air Act: Hazardous Air Pollutants (Title I, Part A, Section 112).
Reportable Quantity (RQ): 1 lb (0.454 kg).
EPCRA Section 313 Form R *de minimis* concentration reporting level: 0.1%.
California Proposition 65 Chemical: Cancer 10/1/88.
Canada, WHMIS, Ingredients Disclosure List Concentration 1.0%.
European/International Regulations: Hazard Symbol: T; Risk phrases: R45; R21; R36; Safety phrases: S53; S45 (see Appendix 4).
WGK (German Aquatic Hazard Class): No value assigned.
Description: Styrene oxide is a colorless to pale straw-colored liquid with a pleasant, sweet odor. Molecular weight = 120.16; Specific gravity $(H_2O:1)$ = 1.02 at 25°C; Boiling point = 194°C; Freezing/Melting point = −36.6°C; Flash point = 76°C; Autoignition temperature = 498°C. Explosive limits: LEL = 1%; UEL = 22%. Hazard Identification (based on NFPA-704 M Rating System): Health 2, Flammability 2, Reactivity 0. Slightly soluble in water.
Potential Exposure: Compound Description: Tumorigen, Mutagen; Reproductive Effector; Human Data. Styrene oxide is used as a reactive intermediate, especially to produce styrene glycol and its derivatives. Substantial amounts are also used in the epoxy resin industry as a diluent. It may also have applications in the preparation of agricultural and biological chemicals, cosmetics, and surface coatings and in the treatment of textiles and fibers. Styrene oxide is made in quantities in excess of a million pounds per year, and further, is a presumed metabolite of styrene which is produced in much greater quantities.
Incompatibilities: Forms an explosive mixture with air. May polymerize on heating above 200°C, under the influence of strong acids, strong bases, oxidizers, metal salts, such as aluminum chloride, catalysts for vinyl polymers.
Permissible Exposure Limits in Air
Protective Action Criteria (PAC)
TEEL-0: 4 ppm
PAC-1: 12.5 ppm
PAC-2: 50 ppm
PAC-3: 50 ppm
Austria: carcinogen, 1999; France: carcinogen, 1993.
Determination in Air: NIOSH Analytical Method (IV) #1501, aromatic Hydrocarbons.
Routes of Entry: Inhalation, skin absorption, ingestion, skin and/or eye contact.

Harmful Effects and Symptoms
Short Term Exposure: Toxic by ingestion, inhalation, and skin absorption. May be corrosive. Irritates the skin, eyes, and mucous membranes. If inhaled, causes irritation to mucosal surfaces; may cause headache, fatigue, weakness, nausea, vomiting, diarrhea, central nervous system depression; unsteadiness or feeling of drunkenness; peripheral and hepatic neuropathies. May cause you to pass out.
Long Term Exposure: Repeated or prolonged contact may cause skin sensitization and skin allergy, with drying and cracking. May cause liver damage. Probable carcinogen in humans; has been shown to cause stomach and liver cancer in animals. May cause decreased fertility in females.
Points of Attack: Skin, liver.
Medical Surveillance: Liver function tests. Evaluation by a qualified allergist. Drinking alcohol may increase the liver damage caused by styrene oxide.
First Aid: Skin Contact: Flood all areas of body that have contacted the substance with water. Do not wait to remove contaminated clothing; do it under the water stream. Use soap to help assure removal. Isolate contaminated clothing when removed to prevent contact by others.[52]
Eye Contact: Remove any contact lenses at once. Immediately flush eyes well with copious quantities of water or normal saline for at least 20–30 min. Seek medical attention.
Inhalation: Leave contaminated area immediately; breathe fresh air. Proper respiratory protection must be supplied to any rescuers. If coughing, difficult breathing, or any other symptoms develop, seek medical attention at once, even if symptoms develop many hours after exposure.
Ingestion: If unconscious or convulsing, do not induce vomiting or give anything by mouth. Assure that victim's airway is open and lay him on his side with his head lower than his body and transport at once to a medical facility. If conscious and not convulsing, give a glass of water to dilute the substance. If medical advice is not readily available, do not induce vomiting, and rush the victim to the nearest medical facility.
Personal Protective Methods: Wear protective gloves and clothing to prevent any reasonable probability of skin contact. Safety equipment suppliers/manufacturers can provide recommendations on the most protective glove/clothing material for your operation. Polyvinyl alcohol is recommended by some safety equipment manufacturers. All protective clothing (suits, gloves, footwear, headgear) should be clean, available each day, and put on before work. Contact lenses should not be worn when working with this chemical. Wear splash-proof chemical goggles and face shield unless full face-piece respiratory protection is worn. Employees should wash immediately with soap when skin is wet or contaminated. Provide emergency showers and eyewash.
Respirator Selection: NIOSH as styrene monomer: *500 ppm:* CcrOv* [any chemical cartridge respirator with organic vapor cartridge(s)] or SA* (any supplied-air respirator). *700 ppm:* Sa:Cf (APF = 25)* (any supplied-air respirator operated in a continuous-flow mode) or CcrFOv

(APF = 50) [any chemical cartridge respirator with a full face-piece and organic vapor cartridge(s)] or GmFOv (APF = 50) [any air-purifying, full-face-piece respirator (gas mask) with a chin-style, front- or back-mounted organic vapor canister] or PaprOv* [any powered, air-purifying respirator with organic vapor cartridge(s)] or SCBAF (APF = 50) (any self-contained breathing apparatus with a full face-piece) or SaF (APF = 50) (any supplied-air respirator with a full face-piece). *Emergency or planned entry into unknown concentrations or IDLH conditions:* SCBAF: Pd,Pp (APF = 10,000) (any self-contained breathing apparatus that has a full face-piece and is operated in a pressure-demand or other positive-pressure mode) or SaF: Pd,Pp: ASCBA (APF = 10,000) (any supplied-air respirator that has a full face-piece and is operated in a pressure-demand or other positive-pressure mode in combination with an auxiliary, self-contained breathing apparatus operated in a pressure-demand or other positive-pressure mode). *Escape:* GmFOv (APF = 50) [any air-purifying, full-face-piece respirator (gas mask) with a chin-style, front- or back-mounted organic vapor canister] or SCBAE (any appropriate escape-type, self-contained breathing apparatus).
*Substance reported to cause eye irritation or damage; may require eye protection.
Storage: Color Code—Green: General storage may be used. Prior to working with this chemical you should be trained on its proper handling and storage. Before entering confined space where this chemical may be present, check to make sure that an explosive concentration does not exist. Store in a cool, dry place. A fireproof area separated from strong acids is recommended.[57] Where possible, automatically pump liquid from drums or other storage containers to process containers. Sources of ignition, such as smoking and open flames, are prohibited where this chemical is handled, used, or stored. Metal containers involving the transfer of 5 gallons or more of this chemical should be grounded and bonded. Drums must be equipped with self-closing valves, pressure vacuum bungs, and flame arresters. Use only non-sparking tools and equipment, especially when opening and closing containers of this chemical. Wherever this chemical is used, handled, manufactured, or stored, use explosion-proof electrical equipment and fittings. A regulated, marked area should be established where this chemical is handled, used, or stored in compliance with OSHA Standard 1910.1045.
Shipping: The name of this material is not in the DOT list of materials[19] for label and packaging standards. However, based on regulations, it may be classified[52] as an Environmentally hazardous substance, liquid, n.o.s. It falls in Hazard Class 9 and Packing Group III.[20, 21]
Spill Handling: Evacuate and restrict persons not wearing protective equipment from area of spill or leak until cleanup is complete. Remove all ignition sources. Establish forced ventilation to keep levels below explosive limit. Absorb liquids in vermiculite, dry sand, earth, peat, carbon, or a similar material and deposit in sealed containers. Follow by

washing surfaces well, first with 60—70% ethanol, then with soap and with 60—70% ethanol. Keep this chemical out of a confined space, such as a sewer, because of the possibility of an explosion, unless the sewer is designed to prevent the buildup of explosive concentrations. It may be necessary to contain and dispose of this chemical as a hazardous waste. If material or contaminated runoff enters waterways, notify downstream users of potentially contaminated waters. Contact your local or federal environmental protection agency for specific recommendations. If employees are required to clean up spills, they must be properly trained and equipped. OSHA 1910.120(q) may be applicable.

Fire Extinguishing: This chemical is a flammable liquid. Poisonous gases, including styrene oxide, are produced in fire. Use dry chemical, carbon dioxide, alcohol foam, or polymer foam extinguishers. Vapors are heavier than air and will collect in low areas. Vapors may travel long distances to ignition sources and flashback. Vapors in confined areas may explode when exposed to fire. Containers may explode in fire. Storage containers and parts of containers may rocket great distances, in many directions. If material or contaminated runoff enters waterways, notify downstream users of potentially contaminated waters. Notify local health and fire officials and pollution control agencies. From a secure, explosion-proof location, use water spray to cool exposed containers. If cooling streams are ineffective (venting sound increases in volume and pitch, tank discolors, or shows any signs of deforming), withdraw immediately to a secure position. If employees are expected to fight fires, they must be trained and equipped in OSHA 1910.156. The only respirators recommended for firefighting are self-contained breathing apparatuses that have full face-pieces and are operated in a pressure-demand or other positive-pressure mode.

References

US Environmental Protection Agency. (March 9, 1978). *Chemical Hazard Information Profile: Styrene Oxide.* Washington, DC

New Jersey Department of Health and Senior Services. (December 1999). *Hazardous Substances Fact Sheet: Styrene Oxide.* Trenton, NJ

Subtilisins   S:0680

Molecular Formula: None listed.
Synonyms: Alcalase®; Alk®; Bacillus subtilis; Bacillus subtilis BPN; Bacillus subtilis Carlsburg; BPN; Fungosin; Maxatase®; Protease 150®; Proteolytic enzymes; Subtilisin BPN; Subtilisin Carlsburg
CAS Registry Number: 9014-01-1 (Subtilisin); 1395-21-7 (BPN)
RTECS® Number: CO9450000 (BPN); CO9550000 (Carlsburg)
EC Number: 232-752-2 [*Annex I Index No.:* 647-012-00-8] (subtilisin)

Regulatory Authority and Advisory Bodies
Air Pollutant Standard Set. See below, "Permissible Exposure Limits in Air" section.
Canada, WHMIS, Ingredients Disclosure List Concentration 0.1% (BPN).
European/International Regulations: Hazard Symbol: Xn; Risk phrases: R37/38; R41; R42; Safety phrases: S2; S22; S24; S26; S36/37/39 (see Appendix 4).
WGK (German Aquatic Hazard Class): No value assigned.
Description: These are proteolytic enzymes which take the form of light-colored, free-flowing powders. A protein containing numerous amino acids. Molecular weight = 28,000 (approx.).
Potential Exposure: Compound Description (BPN): Drug, Natural Product; Primary Irritant; (Carlsburg). Primary Irritant. These commercial proteolytic enzymes are used in laundry detergent formulations.
Incompatibilities: None reported.
Permissible Exposure Limits in Air
OSHA PEL: None.
NIOSH REL: 0.00006 mg/m^3 [60-min] STEL.
ACGIH TLV®[1] (*as crystalline active enzyme*): 0.00006 mg/m^3 Ceiling Concentration.
No TEEL available.
DFG MAK: Danger of airway sensitization.
(*BPN*) Australia: TWA 0.00006 mg/m^3, 1993; Denmark: TWA 0.00006 mg/m^3, 1999; Norway: TWA 0.00006 mg/m^3, 1999; Switzerland: MAK-W 0.00006 mg/m^3, 1999; United Kingdom: LTEL 0.00006 mg/m^3; STEL 0.00006 mg/m^3, 1993; Argentina, Bulgaria, Columbia, Jordan, South Korea, New Zealand, Singapore, Vietnam: ACGIH TLV®: Ceiling Concentration 0.00006 mg/m^3.
(*Carlsburg*) Sweden: TWA 1 glycine unit/m^3, ceiling 3 glycine units/m^3, 1999; United Kingdom: LTEL 0.00006 mg/m^3; STEL 0.00006 mg/m^3, 1993; Argentina, Bulgaria, Columbia, Jordan, South Korea, New Zealand, Singapore, Vietnam: ACGIH TLV®: Ceiling Concentration 0.00006 mg/m^3. Several states have set guidelines or standards for subtilisins in ambient air[60] ranging from zero (North Dakota) to 0.1 μg/m^3 (Nevada) to 1000.0 μg/m^3 (Virginia).
Determination in Air: No method available.
Routes of Entry: Inhalation of dust.
Harmful Effects and Symptoms
Short Term Exposure: Subtilisins can affect you when breathed in. Irritates the eyes, skin, and respiratory tract. Exposure can cause runny nose, congestion, sore throat, sweating, headache, chest pain, flu-like symptoms, cough, breathlessness, wheezing. Prolonged exposure may lead to chronic lung damage.
Long Term Exposure: Subtilisins may cause respiratory sensitization and enzyme asthma. Future exposures can cause asthma attacks with shortness of breath, wheezing, cough, and/or chest tightness. Contact can irritate the skin, especially in sweaty areas. Prolonged exposure may lead to chronic lung damage.

Points of Attack: Eyes, skin, respiratory system.

Medical Surveillance: Before beginning employment and at regular times after that, the following are recommended: lung function tests. These may be normal if the person is not having an attack at the time of the test. If symptoms develop or overexposure is suspected, the following may be useful: evaluation by a qualified allergist, including careful exposure history and special testing, may help diagnose allergy.

First Aid: If this chemical gets into the eyes, remove any contact lenses at once and irrigate immediately for at least 15 min, occasionally lifting upper and lower lids. Seek medical attention immediately. If this chemical contacts the skin, remove contaminated clothing and wash immediately with soap and water. Seek medical attention immediately. If this chemical has been inhaled, remove from exposure, begin rescue breathing (using universal precautions, including resuscitation mask) if breathing has stopped and CPR if heart action has stopped. Transfer promptly to a medical facility. When this chemical has been swallowed, get medical attention. Give large quantities of water and induce vomiting. Do not make an unconscious person vomit.

Personal Protective Methods: Wear protective gloves and clothing to prevent any reasonable probability of skin contact. Safety equipment suppliers/manufacturers can provide recommendations on the most protective glove/clothing material for your operation. All protective clothing (suits, gloves, footwear, headgear) should be clean, available each day, and put on before work. Contact lenses should not be worn when working with this chemical. Wear dust-proof chemical goggles and face shield unless full face-piece respiratory protection is worn. Employees should wash immediately with soap when skin is wet or contaminated. Provide emergency showers and eyewash.

Respirator Selection: Where there is potential for exposures *over 0.00006 mg/m³*, use a NIOSH/MSHA- or European Standard EN149-approved respirator with a high-efficiency particulate filter. More protection is provided by a full face-piece respirator than by a half-mask respirator, and even greater protection is provided by a powered air-purifying respirator. Particulate filters must be checked every day before work for physical damage, such as rips or tears, and replaced as needed.

Where there is potential for high exposures, use a NIOSH/MSHA- or European Standard EN149-approved supplied-air respirator with a full face-piece operated in the positive-pressure mode, or with a full face-piece, hood, or helmet in the continuous-flow mode; or use a NIOSH/MSHA- or European Standard EN149-approved self-contained breathing apparatus with a full face-piece operated in pressure-demand or other positive-pressure mode.

Storage: Color Code—Blue: Health Hazard/Poison: Store in a secure poison location. Prior to working with this chemical you should be trained on its proper handling and storage. Store in tightly closed containers in a cool, well-ventilated area.

Spill Handling: Evacuate persons not wearing protective equipment from area of spill or leak until cleanup is complete. Remove all ignition sources. Collect powdered material in the most convenient and safe manner and deposit in sealed containers. Ventilate area after cleanup is complete. It may be necessary to contain and dispose of this chemical as a hazardous waste. If material or contaminated runoff enters waterways, notify downstream users of potentially contaminated waters. Contact your local or federal environmental protection agency for specific recommendations. If employees are required to clean up spills, they must be properly trained and equipped. OSHA 1910.120(q) may be applicable.

Fire Extinguishing: Use dry chemical, carbon dioxide, water spray, or foam extinguishers. Poisonous gases are produced in fire. If material or contaminated runoff enters waterways, notify downstream users of potentially contaminated waters. Notify local health and fire officials and pollution control agencies. From a secure, explosion-proof location, use water spray to cool exposed containers. If cooling streams are ineffective (venting sound increases in volume and pitch, tank discolors, or shows any signs of deforming), withdraw immediately to a secure position. If employees are expected to fight fires, they must be trained and equipped in OSHA 1910.156. The only respirators recommended for firefighting are self-contained breathing apparatuses that have full face-pieces and are operated in a pressure-demand or other positive-pressure mode.

References

New Jersey Department of Health and Senior Services. (March 2003). *Hazardous Substances Fact Sheet: Subtilisins.* Trenton, NJ

Sucrose S:0690

Molecular Formula: $C_{12}H_{22}O_{11}$

Synonyms: Beet sugar; Cane sugar; Confectioner's sugar; α-δ-Glucopyranosyl β-δ-fructofuranoside; (α-δ-Glucosido)-β-δ-fructofuranoside; Granulated sugar; NCI-C56597; Rock candy; Saccharose; Saccharum; Sugar

CAS Registry Number: 57-50-1

RTECS® Number: WN6500000

EC Number: 200-334-9

Regulatory Authority and Advisory Bodies

US EPA Gene-Tox Program, Inconclusive: Mammalian micronucleus.

Air Pollutant Standard Set. See below, "Permissible Exposure Limits in Air" section.

European/International Regulations: not listed in Annex 1.

WGK (German Aquatic Hazard Class): No value assigned.

Description: Sucrose is a white crystalline solid; Molecular weight = 342.34; Specific gravity (H_2O:1) = 1.59 at 25°C; Boiling point = (decomposes); Freezing/Melting point = (decomposition) 160−186°C. Maximum Explosive concentration (MEC) = 45 g/m³. Hazard Identification (based on

NFPA-704 M Rating System): Health 0, Flammability 1, Reactivity 0. Highly soluble in water; solubility = 200%.

Potential Exposure: Compound Description: Mutagen; Reproductive Effector; Human Data. Widely used as a sweetener for foods in households and industry; in pharmacy as a preservative; in the manufacture of ink.

Incompatibilities: Oxidizers, sulfuric acid, nitric acid. A noncombustible solid, but fine airborne dust may explode.

Permissible Exposure Limits in Air

OSHA PEL: 15 mg/m³ TWA total dust; 5 mg/m³ respirable fraction TWA.

NIOSH REL: 10 mg/m³ TWA, total dust; 5 mg/m³ TWA respirable fraction.

ACGIH TLV®[1]: 10 mg/m³ TWA; not classifiable as a human carcinogen.

Protective Action Criteria (PAC)

TEEL-0: 15 mg/m³

PAC-1: 30 mg/m³

PAC-2: 50 mg/m³

PAC-3: 500 mg/m³

Australia: TWA 10 mg/m³, 1993; Belgium: TWA 10 mg/m³, 1993; France: VME 10 mg/m³, 1999; United Kingdom: TWA 10 mg/m³; STEL 20 mg/m³, 2000; the Netherlands: MAC-TGG 10 mg/m³, 2003; Argentina, Bulgaria, Columbia, Jordan, South Korea, New Zealand, Singapore, Vietnam: ACGIH TLV®: not classifiable as a human carcinogen. The state of Virginia has set 80.0 μg/m³ as a guideline for sucrose in ambient air.

Determination in Air: Filter; none; Gravimetric; NIOSH IV, Particulates NOR: Method #0500, total dust, Method #0600 (respirable).

Routes of Entry: Inhalation, skin and/or eye contact.

Harmful Effects and Symptoms

Short Term Exposure: Acute doses may cause irritation of the eyes, skin, and upper respiratory system; cough may develop.

Long Term Exposure: Chronic doses may cause obesity, peripheral vascular disease, possible exacerbation of diabetes mellitus, hypoinsulinemia, hyperlipemia, and dental cavities.

Points of Attack: Eyes, respiratory system.

Medical Surveillance: Blood sugar.

First Aid: Skin Contact: Use soap and water to clean.

Eye Contact: Remove any contact lenses at once. Immediately flush eyes well with copious quantities of water or normal saline. Seek medical attention if irritation or other symptoms develop.

Inhalation: Inhalation of powdered sucrose can cause coughing and irritation of the lungs. Leave the contaminated area and breathe fresh air. If persistent coughing or other symptoms develop, seek medical attention.

Ingestion: Ingestion of sucrose in moderate amounts poses no acute toxic problems unless the person is a diabetic.

Personal Protective Methods: Wear protective gloves and clothing. Safety equipment suppliers/manufacturers can provide recommendations on the most protective glove/clothing material for your operation. All protective clothing (suits, gloves, footwear, headgear) should be clean, available each day, and put on before work. Contact lenses should not be worn when working with this chemical. Wear dust-proof chemical goggles and face shield unless full face-piece respiratory protection is worn. Employees should wash immediately with soap when skin is wet or contaminated. Provide emergency showers and eyewash.

Respirator Selection: Dust mask.

Storage: Color Code—Green: General storage may be used. Prior to working with this chemical you should be trained on its proper handling and storage. Store in tightly closed containers in a cool, well-ventilated area away from oxidizers, strong acids, potassium hydroxide.

Spill Handling: Evacuate persons not wearing protective equipment from area of spill or leak until cleanup is complete. Remove all ignition sources. Dampen spilled material with water, then transfer material to a suitable container. Wash surfaces with soap and water. Use absorbent dampened with water to pick up remaining material. Ventilate area after cleanup is complete. It may be necessary to contain and dispose of this chemical as a hazardous waste. If material or contaminated runoff enters waterways, notify downstream users of potentially contaminated waters. Contact your local or federal environmental protection agency for specific recommendations. If employees are required to clean up spills, they must be properly trained and equipped. OSHA 1910.120(q) may be applicable.

Fire Extinguishing: This chemical is a noncombustible solid, but airborne particles may explode. Use dry chemical, carbon dioxide, water spray, or foam extinguishers. If employees are expected to fight fires, they must be trained and equipped in OSHA 1910.156. The only respirators recommended for firefighting are self-contained breathing apparatuses that have full face-pieces and are operated in a pressure-demand or other positive-pressure mode.

Sulfallate   S:0700

Molecular Formula: $C_8H_{14}ClNS_2$

Synonyms: CDEC; Chlorallyl diethyldithiocarbamate; 2-Chlorallyl diethyldithiocarbamate; 2-Chlorallyl-*N,N*-diethyldithiocarbamate; 2-Chloroallyl-*N,N*-diethyldithiocarbamate; 2-Chloro-2-propene-1-thiol diethyldithiocarbamate; 2-Chloro-2-propenyl-diethylcarbamodithioate; CP 4572; Diethylcarbamodithioic acid 2-chloro-2-propenyl ester; Diethyldithiocarbamic acid-2-chloroallyl ester; NCI-COO453; Thioallate; Vegadex; Vegadex super

CAS Registry Number: 95-06-7

RTECS® Number: EZ5075000

UN/NA & ERG Number: UN2992 (Carbamate pesticides, liquid, toxic)/151

EC Number: 202-388-9 [*Annex I Index No.:* 006-038-00-4]

Regulatory Authority and Advisory Bodies

Carcinogenicity: IARC: Animal Sufficient Evidence, *possibly carcinogenic to humans*, Group 2B, 1983; NTP: Reasonably anticipated to be a human carcinogen; NCI: Carcinogenesis Studies (feed); clear evidence: rat, mouse, 1978; NTP: Report on Carcinogens, 2004.

European/International Regulations: Hazard Symbol: T, N; Risk phrases: R45; R22; R50/53; Safety phrases: S53; S45; S60; S61.

California Proposition 65 Chemical: Cancer 1/1/88.

WGK (German Aquatic Hazard Class): No value assigned.

Description: Sulfallate is an amber liquid. Molecular weight = 223.80; Boiling point = 128–130°C under 1.0 mm pressure; Flash point = 88°C. Hazard Identification (based on NFPA-704 M Rating System): Health 1, Flammability 1, Reactivity 0. Slightly soluble in water.

Potential Exposure: The major use for sulfallate in the United States is as a preemergent selective herbicide to control certain annual grasses and broadleaf weeds around vegetable and fruit crops. Sulfallate has also been used for weed control among shrubbery and ornamental plants. Some dithiocarbamates have been used as rubber components.

Incompatibilities: Strong oxidizers.

Permissible Exposure Limits in Air

DFG MAK: Danger of skin sensitization (dithiocarbamates used as rubber components).

Routes of Entry: Inhalation, ingestion, skin and/or eye contact. May be absorbed by the skin.

Harmful Effects and Symptoms

Short Term Exposure: Irritates the eyes and skin. High exposure may cause fatigue, sleepiness, headache, dizziness, upset stomach; severe rash and personality changes, muscle weakness, and collapse may result. LD_{50} = (oral-rat) 850 mg/kg (slightly toxic).

Long Term Exposure: Repeated or prolonged skin contact may cause rash from irritation. A probable carcinogen in humans. Repeated exposure may cause kidney damage. Sulfallate, a chlorinated dithiocarbamate, administered in the feed, was carcinogenic to Osborne-Mendel rats and to B6C3F1 mice, inducing mammary gland tumors in females of both species, tumors of the fore stomach in male rats, and lung tumors in male mice. May be a cholinesterase inhibitor.

Points of Attack: Skin, eyes, kidneys, plasma and red blood cell cholinesterase. Cancer site in animals: stomach, lungs.

Medical Surveillance: Before employment and at regular times after that, the following are recommended: plasma and red blood cell cholinesterase levels (tests for the enzyme poisoned by this chemical). If exposure stops, plasma levels return to normal in 1–2 weeks while red blood cell levels may be reduced for 1–3 months. When cholinesterase enzyme levels are reduced by 25% or more below preemployment levels, risk of poisoning is increased, even if results are in lower ranges of "normal." Reassignment to work not involving carbamate pesticides is recommended until enzyme levels recover. If symptoms develop or overexposure occurs, repeat the above tests as soon as possible and get an examination of the nervous system. Also consider complete blood count. Kidney function tests. Consider chest X-ray following acute overexposure.

First Aid: *Skin Contact:* Flood all areas of body that have contacted the substance with water. Do not wait to remove contaminated clothing; do it under the water stream. Use soap to help assure removal. Isolate contaminated clothing when removed to prevent contact by others.[52]

Eye Contact: Remove any contact lenses at once. Flush eyes well with copious quantities of water or normal saline for at least 20–30 min. Seek medical attention.

Inhalation: Leave contaminated area immediately; breathe fresh air. Proper respiratory protection must be supplied to any rescuers. If coughing, difficult breathing, or any other symptoms develop, seek medical attention at once, even if symptoms develop many hours after exposure.

Ingestion: If convulsions are not present, give a glass or two of water or milk to dilute the substance. Assure that the person's airway is unobstructed and contact a hospital or poison center immediately for advice on whether or not to induce vomiting.

Personal Protective Methods: Wear protective gloves and clothing to prevent any reasonable probability of skin contact. Safety equipment suppliers/manufacturers can provide recommendations on the most protective glove/clothing material for your operation. All protective clothing (suits, gloves, footwear, headgear) should be clean, available each day, and put on before work. Contact lenses should not be worn when working with this chemical. Wear splash-proof chemical goggles and face shield unless full face-piece respiratory protection is worn. Employees should wash immediately with soap when skin is wet or contaminated. Provide emergency showers and eyewash.

Respirator Selection: Follow regulations in OSHA 29CFR1910.134 or European Standard EN149. Use a NIOSH/MSHA- or European Standard EN149-approved respirator; or use an approved supplied-air respirator with a full face-piece operated in the positive-pressure mode, or with a full face-piece, hood, or helmet in the continuous-flow mode; or use a NIOSH/MSHA- or European Standard EN149-approved self-contained breathing apparatus with a full face-piece operated in pressure-demand or other positive-pressure mode.

Storage: Color Code—Blue: Health Hazard/Poison: Store in a secure poison location. Prior to working with this chemical you should be trained on its proper handling and storage. Store in a refrigerator or in a cool, dry regulated area. Where possible, automatically transfer material from drums or other storage containers to process containers. Sources of ignition, such as smoking and open flames, are prohibited where this chemical is handled, used, or stored. Metal containers involving the transfer of this chemical should be grounded and bonded. Wherever this chemical is used, handled, manufactured, or stored, use explosion-proof electrical equipment and fittings. A regulated, marked area should be established where this chemical is handled, used, or stored in compliance with OSHA Standard 1910.1045.

Shipping: Carbamate pesticides, liquid, toxic, require a shipping label of "POISONOUS/TOXIC MATERIALS." Sulfallate falls in Hazard Class 6.1 and Packing Group II or III.

Spill Handling: Evacuate and restrict persons not wearing protective equipment from area of spill or leak until cleanup is complete. Remove all ignition sources. Ventilate area of spill or leak. Absorb liquids in vermiculite, dry sand, earth, peat, carbon, or a similar material and deposit in sealed containers. Keep this chemical out of a confined space, such as a sewer, because of the possibility of an explosion, unless the sewer is designed to prevent the buildup of explosive concentrations. It may be necessary to contain and dispose of this chemical as a hazardous waste. If material or contaminated runoff enters waterways, notify downstream users of potentially contaminated waters. Contact your local or federal environmental protection agency for specific recommendations. If employees are required to clean up spills, they must be properly trained and equipped. OSHA 1910.120(q) may be applicable.

Fire Extinguishing: This chemical is a combustible liquid. Poisonous gases, including nitrogen oxides, sulfur oxides, and chlorine, are produced in fire. Use dry chemical, carbon dioxide, or alcohol foam extinguishers. Vapors are heavier than air and will collect in low areas. Vapors may travel long distances to ignition sources and flashback. Vapors in confined areas may explode when exposed to fire. Containers may explode in fire. Storage containers and parts of containers may rocket great distances, in many directions. If material or contaminated runoff enters waterways, notify downstream users of potentially contaminated waters. Notify local health and fire officials and pollution control agencies. From a secure, explosion-proof location, use water spray to cool exposed containers. If cooling streams are ineffective (venting sound increases in volume and pitch, tank discolors, or shows any signs of deforming), withdraw immediately to a secure position. If employees are expected to fight fires, they must be trained and equipped in OSHA 1910.156. The only respirators recommended for firefighting are self-contained breathing apparatuses that have full face-pieces and are operated in a pressure-demand or other positive-pressure mode.

Disposal Method Suggested: Small amounts may be decomposed by strong oxidizing agent. Large amounts should be incinerated in a unit with effluent gas scrubbing.[22]

References

New Jersey Department of Health and Senior Services. (October 2001). *Hazardous Substances Fact Sheet: Sulfallate.* Trenton, NJ

Sulfolane   S:0710

Molecular Formula: $C_4H_8O_2S$
Common Formula: $C_4H_8SO_2$
Synonyms: Bondelane A; Bondolane A; Cyclic tetramethylene sulfone; Cyclotetramethylene; Dihydrobutadiene sulphone; 1,1-Dioxidetetrahydrothiofuran; 1,1-Dioxidetetrahydrothiophene; Dioxothiolan; 1,1-Dioxothiolan; Sulfalone; Sulfolan; Sulpholane; Sulphoxaline; Tetrahydrothiofen-1,1-dioxid; Tetrahydrothiophene dioxide; Tetrahydrothiophene 1,1-dioxide; 2,3,4,5-Tetrahydrothiophene-1,1-dioxide; Tetramethylene sulfone; Thiacyclopentane dioxide; Thiocyclopentane-1,1-dioxide; Thiolane-1,1-dioxide; Thiophane dioxide; Thiophan sulfone

CAS Registry Number: 126-33-0
RTECS® Number: XN0700000
EC Number: 204-783-1 [*Annex I Index No.:* 016-031-00-8]

Regulatory Authority and Advisory Bodies
European/International Regulations: Hazard Symbol: Xn; Risk phrases: R22; Safety phrases: S2; S25 (see Appendix 4). WGK (German Aquatic Hazard Class): 1—Low hazard to waters.

Description: Sulfolane is a colorless oily liquid. Molecular weight = 120.18; Boiling point = 285°C; Freezing/Melting point = 27°C; Flash point = 176°C. Hazard Identification (based on NFPA-704 M Rating System): Health 2, Flammability 1, Reactivity 0. Soluble in water.

Potential Exposure: Compound Description: Primary Irritant. Sulfolane is used primarily as a process solvent for extraction of aromatics and for purification of acid gases. Used as a curing agent for epoxy resins; in medicine as an antibacterial; in fractionation of wood tars, tall oil, and other fatty acids; a component of hydraulic fluid; in textile finishing.

Incompatibilities: Strong oxidizers. Contact with nitronium tetrafluoroborate(1-) is potentially explosive.

Permissible Exposure Limits in Air
Protective Action Criteria (PAC)
TEEL-0: 125 mg/m^3
PAC-1: 200 mg/m^3
PAC-2: 200 mg/m^3
PAC-3: 200 mg/m^3

Routes of Entry: Inhalation, ingestion, skin and/or eye contact. Absorbed through the skin.

Harmful Effects and Symptoms
Short Term Exposure: Irritating to the eyes. Lethality data indicate that sulfolane is not highly acutely toxic. Oral LD$_{50}$ values in the rat range from 1846 to 2500 mg/kg. Symptoms of neurotoxicity have been observed in rats, dogs, and monkeys; after ingestion, injection, inhalation, or dermal application of sulfolane. These effects include convulsions, hyperactivity, tremors, and ataxia. Other effects of acute exposure to sulfolane include alterations in stomach, intestines, lungs, and liver following ingestion; and lung and liver inflammation following inhalation.

Long Term Exposure: Sulfolane is neither a dermal sensitizer nor an irritant in the guinea pig, but it does cause eye injuries in rabbits. Subchronic inhalation exposure to sulfolane has resulted in leukopenia in rats and monkeys. No information was found on the carcinogenicity or teratogenicity of sulfolane. The compound was not mutagenic.

First Aid: If this chemical gets into the eyes, remove any contact lenses at once and irrigate immediately for at least 15 min, occasionally lifting upper and lower lids. Seek medical attention immediately. If this chemical contacts the skin, remove contaminated clothing and wash immediately with soap and water. Seek medical attention immediately. If this chemical has been inhaled, remove from exposure, begin rescue breathing (using universal precautions, including resuscitation mask) if breathing has stopped and CPR if heart action has stopped. Transfer promptly to a medical facility. When this chemical has been swallowed, get medical attention. Give large quantities of water and induce vomiting. Do not make an unconscious person vomit.

Personal Protective Methods: Wear protective gloves and clothing. Safety equipment suppliers/manufacturers can provide recommendations on the most protective glove/clothing material for your operation. All protective clothing (suits, gloves, footwear, headgear) should be clean, available each day, and put on before work. Contact lenses should not be worn when working with this chemical. Wear splash-proof chemical goggles and face shield unless full face-piece respiratory protection is worn. Employees should wash immediately with soap when skin is wet or contaminated. Provide emergency showers and eyewash.

Respirator Selection: Follow regulations in OSHA 29CFR1910.134 or European Standard EN149. Use a NIOSH/MSHA- or European Standard EN149-approved respirator; or use an approved supplied-air respirator with a full face-piece operated in the positive-pressure mode, or with a full face-piece, hood, or helmet in the continuous-flow mode; or use a NIOSH/MSHA- or European Standard EN149-approved self-contained breathing apparatus with a full face-piece operated in pressure-demand or other positive-pressure mode.

Storage: Color Code—Green: General storage may be used. Prior to working with this chemical you should be trained on its proper handling and storage. Store in tightly closed containers in a cool, well-ventilated area away from strong oxidizers. Where possible, automatically transfer material from drums or other storage containers to process containers. Sources of ignition, such as smoking and open flames, are prohibited where this chemical is handled, used, or stored. Metal containers involving the transfer of this chemical should be grounded and bonded. Wherever this chemical is used, handled, manufactured, or stored, use explosion-proof electrical equipment and fittings.

Spill Handling: Evacuate and restrict persons not wearing protective equipment from area of spill or leak until cleanup is complete. Remove all ignition sources. Ventilate area of spill or leak. Absorb liquids in vermiculite, dry sand, earth, peat, carbon, or a similar material and deposit in sealed containers. Keep this chemical out of a confined space, such as a sewer, because of the possibility of an explosion, unless the sewer is designed to prevent the buildup of explosive concentrations. It may be necessary to contain and dispose of this chemical as a hazardous waste. If material or contaminated runoff enters waterways, notify downstream users of potentially contaminated waters. Contact your local or federal environmental protection agency for specific recommendations. If employees are required to clean up spills, they must be properly trained and equipped. OSHA 1910.120(q) may be applicable.

Fire Extinguishing: Poisonous gases, including sulfur oxides, are produced in fire. Use dry chemical, carbon dioxide, or foam extinguishers. Vapors are heavier than air and will collect in low areas. Vapors may travel long distances to ignition sources and flashback. Vapors in confined areas may explode when exposed to fire. Containers may explode in fire. Storage containers and parts of containers may rocket great distances, in many directions. If material or contaminated runoff enters waterways, notify downstream users of potentially contaminated waters. Notify local health and fire officials and pollution control agencies. From a secure, explosion-proof location, use water spray to cool exposed containers. If cooling streams are ineffective (venting sound increases in volume and pitch, tank discolors, or shows any signs of deforming), withdraw immediately to a secure position. If employees are expected to fight fires, they must be trained and equipped in OSHA 1910.156. The only respirators recommended for firefighting are self-contained breathing apparatuses that have full face-pieces and are operated in a pressure-demand or other positive-pressure mode.

References

US Environmental Protection Agency. (June 21, 1984). *Chemical Hazard Information Profile: Sulfolane.* Washington, DC: Office of Toxic Substances

Sulfotep   S:0720

Molecular Formula: $C_8H_{20}O_5P_2S_2$

Synonyms: ASP 47; Bay E-393; Bayer-E-393; bis-*O,O*-Diethylphosphorothionic anhydride; Bladafum; Bladafume; Bladafun; Dithio; Dithiodiphosphoric acid, tetraethyl ester; Dithiofos; Dithion; Dithione; Dithiophos; Di(thiophosphoric) acid, tetraethyl ester; Dithiopyrophosphate de tetraethyle (French); Dithiotep; E393; ENT 16,273; Ethyl thiopyrophosphate; Lethalaire G-57; Pirofos; Plant dithio aerosol; Plantfume 103 smoke generator; Pyrophosphorodithioic acid, tetraethyl ester; Pyrophosphorodithioic acid, *O,O,O,O*-tetraethyl ester; Sulfatep; TEDP; TEDTP; Tetraethyldithiopyrophosphate; *O,O,O,O*-Tetraethyldithiopyrophosphate; Tetraethyl dithiopyrophosphate; Thiotepp

CAS Registry Number: 3689-24-5

RTECS® Number: XN4375000

UN/NA & ERG Number: UN1704/153

EC Number: 222-995-2 [*Annex I Index No.:* 015-027-00-3]

Regulatory Authority and Advisory Bodies

US EPA, FIFRA, 1998 Status of Pesticides: Supported.

Banned or Severely Restricted (former USSR) (UN).[13]

Very Toxic Substance (World Bank).[15]

Air Pollutant Standard Set. See below, "Permissible Exposure Limits in Air" section.

US EPA Hazardous Waste Number (RCRA No.): P109.

RCRA, 40CFR261, Appendix 8 Hazardous Constituents.

RCRA 40CFR264, Appendix 9; TSD Facilities Ground Water Monitoring List. Suggested test method(s) (PQL μg/L): 8270 (10).

Superfund/EPCRA 40CFR355, Extremely Hazardous Substances: TPQ = 500 lb (227 kg).

Reportable Quantity (RQ): 100 lb (45.4 kg).

US DOT Regulated Marine Pollutant (49CFR172.101, Appendix B).

US DOT 49CFR172.101, Inhalation Hazardous Chemical.

European/International Regulations: Hazard Symbol: T+, N; Risk phrases: R27/28; R50/53; Safety phrases: S1/2; S23; S28; S36/37; S45; S60; S61 (see Appendix 4).

WGK (German Aquatic Hazard Class): 3—Severe hazard to waters.

Description: Sulfotep is a yellow mobile liquid with a garlic-like odor. Molecular weight = 322.34; Specific gravity (H_2O:1) = 1.20 at 25°C; Boiling point = (decomposes) 131–135°C at 2 mmHg; Vapor pressure = 0.0002 mmHg. Hazard Identification (based on NFPA-704 M Rating System): Health 4, Flammability 1, Reactivity 1. Practically insoluble in water; solubility = 0.0007%. A pesticide that may be absorbed on a solid carrier or mixed in a more flammable liquid which will change the physical properties listed here.

Potential Exposure: Compound Description: Agricultural Chemical; Mutagen. Sulfotep is used in greenhouse fumigant formulations for control of aphids, spider mites, thrips, whiteflies, etc.

Incompatibilities: Strong oxidizers. Hydrolyzes very slowly in aqueous solution. Attacks some forms of plastic, rubber, and coatings. Corrosive to iron.

Permissible Exposure Limits in Air

Conversion factor: 1 ppm = 13.18 mg/m³ at 25°C & 1 atm.

OSHA PEL: 0.2 mg/m³ TWA [skin].

NIOSH REL: 0.2 mg/m³ TWA [skin].

ACGIH TLV®[1]: 0.2 mg/m³, measured as inhalable fraction and vapor TWA [skin] not classifiable as a human carcinogen.

NIOSH IDLH: 10 mg/m³.

Protective Action Criteria (PAC)

TEEL-0: 0.2 mg/m³

PAC-1: 0.5 mg/m³

PAC-2: 3.5 mg/m³

PAC-3: 10 mg/m³

DFG MAK: 0.0075 ppm/0.1 mg/m³ TWA; Peak Limitation Category II(2) [skin]; Pregnancy Risk Group C.

EEC OEL: 0.1 mg/m³ [skin].

Australia: TWA 0.2 mg/m³, [skin], 1993; Austria: MAK 0.015 ppm (0.2 mg/m³), [skin], 1999; Belgium: TWA 0.2 mg/m³, [skin], 1993; Denmark: TWA 0.015 ppm (0.2 mg/m³), [skin], 1999; France: VME 0.2 mg/m³, [skin], 1999; Norway: TWA 0.015 ppm (0.2 mg/m³), 1999; the Netherlands: MAC-TGG 0.1 mg/m³, [skin], 2003; Switzerland: MAK-W 0.015 ppm (0.2 mg/m³), [skin], 1999; Turkey: TWA 0.2 mg/m³, [skin], 1993; United Kingdom: TWA 0.2 mg/m³, [skin], 2000; Argentina, Bulgaria, Columbia, Jordan, South Korea, New Zealand, Singapore, Vietnam: ACGIH TLV®: not classifiable as a human carcinogen.

Several states have set guidelines or standards for sulfotep in ambient air ranging from 3.5 μg/m³ (Virginia) to 4.0 μg/m³ (Connecticut) to 5.0 μg/m³ (Nevada) to 20.0 μg/m³ (North Dakota).

Determination in Air: No test available.

Determination in Water: Octanol–water coefficient: Log K_{ow} = 3.99.

Routes of Entry: Inhalation, skin absorption, ingestion, skin and/or eye contact.

Harmful Effects and Symptoms

Short Term Exposure: Irritates the eyes and skin. Can cause rapid, fatal poisoning. Contact may cause eye pain, blurred vision. May affect the nervous system. Symptoms of exposure include lacrimation (discharge of tears); rhinorrhea (discharge of thin nasal mucous); headache, cyanosis, anorexia, nausea, vomiting, diarrhea, localized sweating; weakness, twitching, paralysis, Cheyne-Stokes respiration; convulsions, low blood pressure; cardiac irregular/irregularities; respiratory failure and death. Super toxic; probable oral lethal dose in humans is less than 5 mg/kg, or a taste (less than 7 drops) for a 70-kg (150-lb) person. It is a cholinesterase inhibitor. Material is similar to parathion in symptomatology, including nausea followed by vomiting; abdominal cramps; diarrhea, excessive salivation; headache, giddiness, dizziness, weakness, tightness in chest; blurring of vision; tearing, slurring of speech; confusion, troubled breathing; convulsions, coma, and even death.

Long Term Exposure: A cholinesterase inhibitor; cumulative effect is possible. This chemical may damage the nervous system with repeated exposure, resulting in convulsions, respiratory failure. Repeated exposure may cause personality changes of depression, anxiety, or irritability.

Points of Attack: Eyes, skin, respiratory system, central nervous system, cardiovascular system, blood cholinesterase.

Medical Surveillance: NIOSH lists the following tests: Blood serum; cholinesterase, red blood cells/count. Before employment and at regular times after that, the following are recommended: plasma and red blood cell cholinesterase levels (tests for the enzyme poisoned by this chemical). If exposure stops, plasma levels return to normal in 1–2 weeks while red blood cell levels may be reduced for 1–3 months. When cholinesterase enzyme levels are reduced by 25% or more below preemployment levels, risk of poisoning is increased, even if results are in lower ranges of "normal." Reassignment to work not involving organophosphate or carbamate pesticides is recommended until enzyme levels recover. If symptoms develop or overexposure occurs, repeat the above test as soon as possible and get an examination of the nervous system.

First Aid: If this chemical gets into the eyes, remove any contact lenses at once and irrigate immediately for at least 15 min, occasionally lifting upper and lower lids. Seek medical attention immediately. If this chemical contacts the skin, remove contaminated clothing and wash immediately with soap and water. Speed in removing material from skin is of extreme importance. Shampoo hair promptly if contaminated. Seek medical attention immediately. If this chemical has been inhaled, remove from exposure, begin rescue breathing (using universal precautions, including resuscitation mask) if breathing has stopped and CPR if heart action has stopped. Transfer promptly to a medical facility. When this chemical has been swallowed, get medical attention. Give large quantities of water and induce vomiting. Do not make an unconscious person vomit. Effects may be delayed. Keep under medical observation.

Personal Protective Methods: Wear protective gloves and clothing to prevent any reasonable probability of skin contact. Safety equipment suppliers/manufacturers can provide recommendations on the most protective glove/clothing material for your operation. All protective clothing (suits, gloves, footwear, headgear) should be clean, available each day, and put on before work. Contact lenses should not be worn when working with this chemical. Wear splash- or dust-proof chemical goggles and face shield unless full face-piece respiratory protection is worn. Employees should wash immediately with soap when skin is wet or contaminated. Provide emergency showers and eyewash.

Respirator Selection: Up to 2 mg/m³: Sa (APF = 10) (any supplied-air respirator). *Up to 5 mg/m³:* Sa:Cf (APF = 25) (any supplied-air respirator operated in a continuous-flow mode). *Up to 10 mg/m³:* SCBAF (APF = 50) (any self-contained breathing apparatus with a full face-piece) or SaF (APF = 50) (any supplied-air respirator with a full face-piece). *Emergency or planned entry into unknown concentrations or IDLH conditions:* SCBAF: Pd,Pp (APF = 10,000) (any NIOSH/MSHA- or European Standard EN 149-approved self-contained breathing apparatus that has a full face-piece and is operated in a pressure-demand or other positive-pressure mode) or SaF: Pd,Pp: ASCBA (APF = 10,000) (any supplied-air respirator that has a full face-piece and is operated in a pressure-demand or other positive-pressure mode in combination with an auxiliary, self-contained breathing apparatus operated in a pressure-demand or other positive-pressure mode). *Escape:* GmFOv100 (APF = 50) [any air-purifying, full-face-piece respirator (gas mask) with a chin-style, front- or back-mounted organic vapor canister having an N100, R100, or P100 filter] or SCBAE (any appropriate escape-type, self-contained breathing apparatus).

Storage: Color Code—Blue: Health Hazard/Poison: Store in a secure poison location. Prior to working with this chemical you should be trained on its proper handling and storage. Store in tightly closed containers in a cool,

well-ventilated area away from oxidizers and moisture. Where possible, automatically transfer material from drums or other storage containers to process containers. Sources of ignition, such as smoking and open flames, are prohibited where this chemical is handled, used, or stored. Metal containers involving the transfer of this chemical should be grounded and bonded. Wherever this chemical is used, handled, manufactured, or stored, use explosion-proof electrical equipment and fittings.

Shipping: Tetraethyl dithiopyrophosphate requires a shipping label of "POISONOUS/TOXIC MATERIALS." It falls in Hazard Class 6.1 and Packing Group II.

Spill Handling: Evacuate and restrict persons not wearing protective equipment from area of spill or leak until cleanup is complete. Remove all ignition sources. Ventilate area of spill or leak. Stay upwind; keep out of low areas. Ventilate closed spaces before entering them. Remove and isolate contaminated clothing at the site. Do not touch spilled material; stop leak if you can do it without risk. Use water spray to reduce vapors. *Small spills:* absorb liquids in vermiculite, dry sand, earth, peat, carbon, or a similar material and deposit in sealed containers. *Large spills:* dike far ahead of spill for later disposal. Keep this chemical out of a confined space, such as a sewer, because of the possibility of an explosion, unless the sewer is designed to prevent the buildup of explosive concentrations. It may be necessary to contain and dispose of this chemical as a hazardous waste. If material or contaminated runoff enters waterways, notify downstream users of potentially contaminated waters. Contact your local or federal environmental protection agency for specific recommendations. If employees are required to clean up spills, they must be properly trained and equipped. OSHA 1910.120(q) may be applicable.

Fire Extinguishing: Sulfotep may burn, but does not readily ignite. Use dry chemical, CO_2, water spray, or foam extinguishers. Poisonous gases, including oxides of phosphorus and sulfur and phosphoric acid mist, are produced in fire. Vapors are heavier than air and will collect in low areas. Vapors in confined areas may explode when exposed to fire. Containers may explode in fire. Storage containers and parts of containers may rocket great distances, in many directions. If material or contaminated runoff enters waterways, notify downstream users of potentially contaminated waters. Notify local health and fire officials and pollution control agencies. From a secure, explosion-proof location, use water spray to cool exposed containers. If cooling streams are ineffective (venting sound increases in volume and pitch, tank discolors, or shows any signs of deforming), withdraw immediately to a secure position. If employees are expected to fight fires, they must be trained and equipped in OSHA 1910.156. The only respirators recommended for firefighting are self-contained breathing apparatuses that have full face-pieces and are operated in a pressure-demand or other positive-pressure mode.

Disposal Method Suggested: Consult with environmental regulatory agencies for guidance on acceptable disposal practices. Generators of waste containing this contaminant (≥100 kg/mo) must conform with EPA regulations governing storage, transportation, treatment, and waste disposal. In accordance with 40CFR165, follow recommendations for the disposal of pesticides and pesticide containers. Must be disposed properly by following package label directions or by contacting your local or federal environmental control agency or by contacting your regional EPA office. Incineration with added flammable solvent in furnace equipped with afterburner and alkaline scrubber.[22]

References

US Environmental Protection Agency. (November 30, 1987). *Chemical Hazard Information Profile: Sulfotep.* Washington, DC: Chemical Emergency Preparedness Program

US Environmental Protection Agency, Special Review and Reregistration Division Office of Pesticide Programs. (1998). *Agency Status of Pesticides in Registration, Reregistration, and Special Review* (Rainbow Report). Washington, DC

New Jersey Department of Health and Senior Services. (December 2000). *Hazardous Substances Fact Sheet: Sulfotep.* Trenton, NJ

Sulfur   S:0730

Molecular Formula: S

Synonyms: Brimstone

CAS Registry Number: 7704-34-9; *(alt.)* 12673-82-4; *(alt.)* 12767-24-7; *(alt.)* 56591-09-4; *(alt.)* 56645-30-8; *(alt.)* 57035-13-9; *(alt.)* 63705-05-5

RTECS® Number: WS4250000

UN/NA & ERG Number: UN1350 (solid)/133; UN2448 (molten)/155

EC Number: 231-722-6 [*Annex I Index No.:* 016-094-00-1]

Regulatory Authority and Advisory Bodies

FDA—over-the-counter and proprietary drug.

US EPA, FIFRA 1998 Status of Pesticides: RED completed.

Air Pollutant Standard Set. See below, "Permissible Exposure Limits in Air" section.

European/International Regulations: Hazard Symbol: Xi; Risk phrases: R38; Safety phrases: S2; S46 (see Appendix 4). WGK (German Aquatic Hazard Class): Nonwater polluting agent (*metal*); 1—Slightly water polluting (*colloidal*).

Description: Sulfur is a yellow crystalline solid or powder. Often transported in the molten state. Molecular weight = 32.06; 256.5 (S_8); Boiling point = 445°C; Freezing/Melting point = 113−120°C (amorphous); 115°C (β-sulfur); 107°C (r-sulfur); Flash point = 207°C; 160°C (cc); Autoignition temperature = 232°C. Explosive limits: LEL = 35% (1400 g/m³). Hazard Identification (based on NFPA-704 M Rating System): Health 2, Flammability 2, Reactivity 0. Insoluble in water.

Potential Exposure: Compound Description: Agricultural Chemical; Human Data; Primary Irritant. Widely used in the manufacture of sulfuric acid, carbon bisulfide, drugs, fungicides, gunpowder, wood pulp, rubber, and other products.

Incompatibilities: Combustible solid. Liquid forms sulfur dioxide with air. Violent reaction with strong oxidizers, halogen compounds, phosphorus, sodium, tin, uranium, metal carbides, and other compounds. Forms explosive, shock-sensitive, or pyrophoric mixtures with ammonia, ammonium nitrate, bromates, calcium carbide, charcoal, chlorates, hydrocarbons, iodates, iron. Attacks steel when moist.

Permissible Exposure Limits in Air

Protective Action Criteria (PAC)

TEEL-0: 1.25 mg/m³

PAC-1: 4 mg/m³

PAC-2: 30 mg/m³

PAC-3: 150 mg/m³

Russia set a MAC of 6.0 mg/m³ for work-place air of elemental sulfur.

Determination in Air: Filter; Acid; Inductively coupled plasma; NIOSH Analytical Method (IV) #7300, Elements.

Permissible Concentration in Water: Mexico[35] set a MAC of 0.5 mg/L of sulfides in estuaries.

Routes of Entry: Inhalation, ingestion, skin and/or eye contact.

Harmful Effects and Symptoms

Short Term Exposure: Sulfur can affect you when breathed in. Irritates the eyes, skin, and respiratory tract. Exposure can cause inflammation of the nose and irritate the lungs.

Long Term Exposure: Repeated exposures may cause chronic bronchitis to develop with cough, phlegm, and/or shortness of breath. Contact can irritate the skin and may cause skin allergy. Repeated exposure to sulfur dust may cause permanent eye damage (clouding of the eye lens and chronic irritation).

Points of Attack: Skin, respiratory tract.

Medical Surveillance: Before beginning employment and at regular times after that, the following are recommended: lung function tests. Eye examination. If symptoms develop or overexposure is suspected, the following may be useful: evaluation by a qualified allergist, including careful exposure history and special testing, may help diagnose skin allergy.

First Aid: If this chemical gets into the eyes, remove any contact lenses at once and irrigate immediately for at least 15 min, occasionally lifting upper and lower lids. Seek medical attention immediately. If this chemical contacts the skin, remove contaminated clothing and wash immediately with soap and water. Seek medical attention immediately. If this chemical has been inhaled, remove from exposure, begin rescue breathing (using universal precautions, including resuscitation mask) if breathing has stopped and CPR if heart action has stopped. Transfer promptly to a medical facility. When this chemical has been swallowed, get medical attention. Give large quantities of water and induce vomiting. Do not make an unconscious person vomit.

Personal Protective Methods: Wear protective gloves and clothing to prevent any reasonable probability of skin contact. Safety equipment suppliers/manufacturers can provide recommendations on the most protective glove/clothing material for your operation. All protective clothing (suits, gloves, footwear, headgear) should be clean, available each day, and put on before work. Contact lenses should not be worn when working with this chemical. Wear splash or dust-proof chemical goggles and face shield unless full face-piece respiratory protection is worn. Employees should wash immediately with soap when skin is wet or contaminated. Provide emergency showers and eyewash.

Respirator Selection: Where there is potential for exposures to sulfur, use a NIOSH/MSHA- or European Standard EN149-approved full face-piece respirator equipped with particulate (dust/fume/mist) filters. Particulate filters must be checked every day before work for physical damage, such as rips or tears, and replaced as needed. *Where there is potential for high exposures*, use a NIOSH/MSHA- or European Standard EN149-approved supplied-air respirator with a full face-piece operated in the positive-pressure mode, or with a full face-piece, hood, or helmet in the continuous-flow mode; or use a NIOSH/MSHA- or European Standard EN149-approved self-contained breathing apparatus with a full face-piece operated in pressure-demand or other positive-pressure mode.

Storage: Color Code—Red: Flammability Hazard: Store in a flammable materials storage area. Prior to working with this chemical you should be trained on its proper handling and storage. Store solid sulfur in tightly closed containers in a cool, well-ventilated area away from oxidizers (such as perchlorates, peroxides, permanganates, chlorates, and nitrates); chemically active metals (such as potassium, sodium, lithium, and zinc); charcoal, phosphorus, metal nitrates (such as potassium nitrate), and metal halogenates (such as zinc bromate). Molten sulfur must be stored to avoid contact with oxidizers (such as perchlorates, peroxides, permanganates, chlorates, and nitrates), chemically active metals (such as potassium, sodium, lithium, and zinc), and metal carbides (such as calcium carbide), since violent reactions occur. Sources of ignition, such as smoking and open flames, are prohibited where sulfur is used, handled, or stored in a manner that could create a potential fire or explosion hazard.

Shipping: Sulfur requires a shipping label of "CLASS 9" (Domestic) or "FLAMMABLE SOLID" (International). It falls in Hazard Class 9 (Domestic) or 4.1 (International) and Packing Group III. Sulfur, molten, requires a shipping label of "FLAMMABLE SOLID." It falls in Hazard Class 4.1 and Packing Group III.

Spill Handling: Evacuate persons not wearing protective equipment from area of spill or leak until cleanup is complete. Remove all ignition sources. Use HEPA vacuum or wet method to reduce dust during cleanup. Do not dry sweep. Collect powdered material in the most convenient and safe manner and deposit in sealed containers.

Ventilate area after cleanup is complete. It may be necessary to contain and dispose of this chemical as a hazardous waste. If material or contaminated runoff enters waterways, notify downstream users of potentially contaminated waters. Contact your local or federal environmental protection agency for specific recommendations. If employees are required to clean up spills, they must be properly trained and equipped. OSHA 1910.120(q) may be applicable.

Fire Extinguishing: Sulfur is a combustible solid. Use dry chemical, water spray, or foam extinguishers. Fire may restart after it has been extinguished. Dust or vapor (from molten sulfur) forms explosive mixtures with air. Poisonous gases are produced in fire, including sulfur oxides. If material or contaminated runoff enters waterways, notify downstream users of potentially contaminated waters. Notify local health and fire officials and pollution control agencies. From a secure, explosion-proof location, use water spray to cool exposed containers. If cooling streams are ineffective (venting sound increases in volume and pitch, tank discolors, or shows any signs of deforming), withdraw immediately to a secure position. If employees are expected to fight fires, they must be trained and equipped in OSHA 1910.156. The only respirators recommended for firefighting are self-contained breathing apparatuses that have full face-pieces and are operated in a pressure-demand or other positive-pressure mode.

Disposal Method Suggested: Salvage for reprocessing or dump to landfill.

References

Sax, N. I. (Ed.). (1982). *Dangerous Properties of Industrial Materials Report*, 2, No. 2, 65–68. New York: Van Nostrand Reinhold Co.
US Environmental Protection Agency, Special Review and Reregistration Division Office of Pesticide Programs. (1998). *Agency Status of Pesticides in Registration, Reregistration, and Special Review* (Rainbow Report). Washington, DC
New Jersey Department of Health and Senior Services. (August 2002). *Hazardous Substances Fact Sheet: Sulfur*. Trenton, NJ

Sulfur chloride   S:0740

Molecular Formula: Cl_2S_2
Common Formula: S_2Cl_2
Synonyms: Chloride of sulfur; Cloruro de azufre (Spanish); Chlorosulfane; Disulfur dichloride; Monocloruro de azufe (Spanish); Sulfur monochloride (di-); Sulfur subchloride; Sulphur chloride (di-); Thiosulfurous dichloride
CAS Registry Number: 10025-67-9
RTECS® Number: WS4300000
UN/NA & ERG Number: UN1828/137
EC Number: 233-036-2 [*Annex I Index No.*: 016-012-00-4]

Regulatory Authority and Advisory Bodies

Department of Homeland Security Screening Threshold Quantity (pounds): Sabotage/Contamination Hazard: A placarded amount (commercial grade).

Air Pollutant Standard Set. See below, "Permissible Exposure Limits in Air" section.

Clean Water Act: Section 311 Hazardous Substances/RQ 40CFR117.3 (same as CERCLA, see below).

Reportable Quantity (RQ): 1000 lb (454 kg).

US DOT 49CFR172.101, Inhalation Hazardous Chemical.

Canada, WHMIS, Ingredients Disclosure List Concentration 1.0%.

European/International Regulations: Hazard Symbol: C; Risk phrases: R14; R20; R25; R29; R35; R50; Safety phrases: S1/2; S26; S36/37/39; S45; S61 (see Appendix 4).

WGK (German Aquatic Hazard Class): 3—Severe hazard to waters.

Description: Sulfur chloride is a fuming, oily liquid with a yellowish-red to amber color and a suffocating odor. It has an added hazard since it oxidizes and hydrolyzes to sulfur dioxide and hydrogen chloride. Molecular weight = 135.02; Specific gravity (H_2O:1) = 1.68 at 25°C; Boiling point = 137.8°C; Freezing/Melting point = −77.2°C; Flash point = 118°C (cc).[17] Autoignition temperature = 234°C. Hazard Identification (based on NFPA-704 M Rating System): Health 3, Flammability 1, Reactivity 1₩. Reacts with water (decomposes).

Potential Exposure: Compound Description: Human Data. Sulfur chloride finds use as a chlorinating agent, catalyst, and as an intermediate in the manufacture of organic chemicals, carbon tetrachloride, sulfur dyes, insecticides, synthetic rubber, and pharmaceuticals. Exposure may also occur during the extraction of gold, purification of sugar juice, finishing and dyeing textiles, processing vegetable oils, hardening wood, and vulcanization of rubber. Sulfur chloride has been used as a military poison.

Incompatibilities: Decomposes violently in water, forming hydrochloric acid, sulfur dioxide, sulfur, sulfite, thiosulfate, and hydrogen sulfide. Reacts with oxidizers, strong bases, peroxides, phosphorus oxides, organics, antimony, antimony sulfide, arsenic sulfide, mercury oxide, tin, alkenes, terpenes, unsaturated glycerides, chromyl chloride, methyl sulfoxide, dimethylformamide, acetone, and other compounds, causing fire and explosion hazard. Corrosive to many metals in the presence of water. Attacks some plastics, rubber, and coatings.

Permissible Exposure Limits in Air

Conversion factor 1 ppm = 5.52 mg/m^3.

OSHA PEL: 1 ppm/6 mg/m^3 TWA.

NIOSH REL: 1 ppm/6 mg/m^3 Ceiling Concentration.

ACGIH TLV®[1]: 1 ppm/5.5 mg/m^3 Ceiling Concentration.

NIOSH IDLH: 5 ppm.

Protective Action Criteria (PAC)*

TEEL-0: 0.53 ppm

PAC-1: **0.53** ppm

PAC-2: **6.4** ppm

PAC-3: **15** ppm

*AEGLs (Acute Emergency Guideline Levels) & ERPGs (Emergency Response Planning Guideline) are in **bold face**.

Australia: TWA 1 ppm (6 mg/m^3), 1993; Austria: MAK 1 ppm (6 mg/m^3), 1999; Belgium: STEL 1 ppm (5.5 mg/m^3), 1993; Denmark: TWA 1 ppm (6 mg/m^3), 1999; Finland: TWA 1 ppm (6 mg/m^3); STEL 2 ppm (12 mg/m^3), [skin], 1999; Hungary: STEL 5 mg/m^3, 1993; Japan: 1 ppm (5.5 mg/m^3), 1999; the Netherlands: MAC 6 mg/m^3, 2003; Norway: TWA 1 ppm (6 mg/m^3), 1999; the Philippines: TWA 6 mg/m^3, 1993; Poland: MAC (TWA) 5 mg/m^3; MAC (STEL) 15 mg/m^3, 1999; Russia: STEL 1 ppm (0.3 mg/m^3), [skin], 1993; Switzerland: MAK-W 1 ppm (6 mg/m^3), KZG-W 2 ppm (12 mg/m^3), 1999; United Kingdom: STEL 1 ppm (5.6 mg/m^3), 2000; Argentina, Bulgaria, Columbia, Jordan, South Korea, New Zealand, Singapore, Vietnam: ACGIH TLV®: Ceiling Concentration 1 ppm.

Several states have set guidelines or standards for sulfur monochloride in ambient air[60] ranging from 50.0 μg/m^3 (Virginia) to 60.0 μg/m^3 (North Dakota) to 120.0 μg/m^3 (Connecticut) to 143.0 μg/m^3 (Nevada).

Determination in Air: No method available.

Determination in Water: Violent reaction.

Routes of Entry: Inhalation of vapor, ingestion, skin and/or eye contact.

Harmful Effects and Symptoms

Short Term Exposure: A lacrimator and corrosive. Fumes can cause severe irritation to eyes, skin, and mucous membranes of the upper respiratory tract. Contact can cause severe irritation, burns, and permanent eye damage. Corrosive on ingestion. Higher exposures can cause pulmonary edema, a medical emergency that can be delayed for several hours. This can cause death. Exposure can cause headache, nausea, and dizziness. Although this compound is capable of producing severe pulmonary irritation, few serious cases of industrial exposure have been reported. This may be due to the pronounced irritant effects of sulfur chloride which serve as an immediate warning signal when concentration of the gas approaches a hazardous level.

Long Term Exposure: May cause lung damage; bronchitis may develop. Repeated exposure can cause drying and cracking of the skin.

Points of Attack: Respiratory system, skin, eyes, lungs.

Medical Surveillance: Preemployment and periodic examinations should give special emphasis to the skin, eyes, and respiratory system. Pulmonary function tests may be useful. Lung function tests. Consider chest X-ray following acute overexposure. Exposures may also include sulfur dioxide and hydrochloric acid. See also these compounds.

First Aid: If this chemical gets into the eyes, remove any contact lenses at once and irrigate immediately for at least 30 min, occasionally lifting upper and lower lids. Seek medical attention immediately. If this chemical contacts the skin, remove contaminated clothing and wash immediately with soap and water. Seek medical attention immediately. If this chemical has been inhaled, remove from exposure, begin rescue breathing (using universal precautions, including resuscitation mask) if breathing has stopped and CPR if heart action has stopped. Transfer promptly to a medical facility. When this chemical has been swallowed, get medical attention. If victim is *conscious*, administer water or milk. Do not induce vomiting. Medical observation is recommended for 24—48 h after breathing overexposure, as pulmonary edema may be delayed. As first aid for pulmonary edema, a doctor or authorized paramedic may consider administering a corticosteroid spray.

Personal Protective Methods: Wear protective gloves and clothing to prevent any reasonable probability of skin contact. Safety equipment suppliers/manufacturers can provide recommendations on the most protective glove/clothing material for your operation. All protective clothing (suits, gloves, footwear, headgear) should be clean, available each day, and put on before work. Contact lenses should not be worn when working with this chemical. Wear splash-proof chemical goggles and face shield unless full face-piece respiratory protection is worn. Employees should wash immediately with soap when skin is wet or contaminated. Provide emergency showers and eyewash.

Respirator Selection: 5 ppm: CcrFS (APF = 50) [any chemical cartridge respirator with a full face-piece and cartridge(s) providing protection against the compound of concern] or GmFS (APF = 50) [any air-purifying, full-face-piece respirator (gas mask) with a chin-style, front- or back-mounted canister providing protection against the compound of concern] or PaprS (APF = 25) [any powered, air-purifying respirator with cartridge(s) providing protection against the compound of concern] or SCBAF (APF = 50) (any self-contained breathing apparatus with a full face-piece) or SaF (APF = 50) (any supplied-air respirator with a full face-piece). *Emergency or planned entry into unknown concentrations or IDLH conditions:* SCBAF: Pd,Pp (APF = 10,000) (any self-contained breathing apparatus that has a full face-piece and is operated in a pressure-demand or other positive-pressure mode) or SaF: Pd,Pp: ASCBA (APF = 10,000) (any supplied-air respirator that has a full face-piece and is operated in a pressure-demand or other positive-pressure mode in combination with an auxiliary self-contained breathing apparatus operated in a pressure-demand or other positive-pressure mode). *Escape:* GmFS (APF = 50) [any air-purifying, full-face-piece respirator (gas mask) with a chin-style, front- or back-mounted canister providing protection against the compound of concern] or SCBAE (any appropriate escape-type, self-contained breathing apparatus).

Note: Substance reported to cause eye irritation or damage; may require eye protection.

Storage: Color Code—White: Corrosive or Contact Hazard; Store separately in a corrosion-resistant location. Prior to working with this chemical you should be trained on its proper handling and storage. Keep containers well closed and upright, away from heat. Do not allow water to get into container as this material reacts violently with water when in a closed vessel. Provisions should be made for washing down spills with large quantities of water. Vapor-tight electrical equipment is recommended to reduce corrosion. Separate from oxidizing materials. Where possible, automatically pump liquid from drums or other storage containers to process containers. Sources of ignition, such as smoking and open flames, are prohibited where this chemical is handled, used, or stored. Metal containers involving the transfer of 5 gallons or more of this chemical should be grounded and bonded. Drums must be equipped with self-closing valves, pressure vacuum bungs, and flame arresters. Use only nonsparking tools and equipment, especially when opening and closing containers of this chemical. Wherever this chemical is used, handled, manufactured, or stored, use explosion-proof electrical equipment and fittings.

Shipping: Sulfur chloride requires a shipping label of "CORROSIVE." It falls in Hazard Class 8 and Packing Group I.

Spill Handling: Evacuate and restrict persons not wearing protective equipment from area of spill or leak until cleanup is complete. Avoid contact with liquid and vapor. Keep people away. Wear goggles, self-contained breathing apparatus, and rubber over clothing (including gloves). Stop discharge if possible to do so without harm. Call fire department. Isolate and remove discharged material. Notify local health and pollution control agencies. Remove all ignition sources. Ventilate area of spill or leak. The aqueous solution is a strong acid that can be neutralized with lime or soda ash. Absorb liquids in vermiculite, dry sand, earth, peat, carbon, or a similar material and deposit in sealed containers. Keep this chemical out of a confined space, such as a sewer, because of the possibility of an explosion, unless the sewer is designed to prevent the buildup of explosive concentrations. It may be necessary to contain and dispose of this chemical as a hazardous waste. If material or contaminated runoff enters waterways, notify downstream users of potentially contaminated waters. Contact your local or federal environmental protection agency for specific recommendations. If employees are required to clean up spills, they must be properly trained and equipped. OSHA 1910.120(q) may be applicable.

Initial isolation and protective action distances
Distances shown are likely to be affected during the first 30 min after materials are spilled and could increase with time. If more than one tank car, cargo tank, portable tank, or large cylinder involved in the incident is leaking, the protective action distance may be increased. You may need to seek emergency information from CHEMTREC at (800) 424-9300 or seek professional environmental engineering assistance from the US EPA Environmental Response Team at (908) 548-8730 (24-h response line).

Small spills (From a small package or a small leak from a large package)

When spilled on land
First: Isolate in all directions (feet/meters)   100/30
Then: Protect persons downwind (miles/kilometers)
Day   0.1/0.2
Night   0.1/0.2

Large spills (From a large package or from many small packages)
First: Isolate in all directions (feet/meters)   200/60
Then: Protect persons downwind (miles/kilometers)
Day   0.5/0.8
Night   0.8/1.3

When spilled in Water
First: Isolate in all directions (feet/meters)   100/30
Then: Protect persons downwind (miles/kilometers)
Day   0.1/0.2
Night   0.1/0.2

Large spills (From a large package or from many small packages)
First: Isolate in all directions (feet/meters)   100/30
Then: Protect persons downwind (miles/kilometers)
Day   0.2/0.3
Night   0.8/1.3

Fire Extinguishing: This chemical is a combustible liquid. Poisonous gases, including hydrogen chloride, hydrogen sulfide, sulfur oxides, are produced in fire. *Do not use water.* Use chemical extinguishers. Decomposes on contact with water to produce heat and toxic and corrosive fumes. Vapors are heavier than air and will collect in low areas. Vapors may travel long distances to ignition sources and flashback. Vapors in confined areas may explode when exposed to fire. Containers may explode in fire. Storage containers and parts of containers may rocket great distances, in many directions. If material or contaminated runoff enters waterways, notify downstream users of potentially contaminated waters. Notify local health and fire officials and pollution control agencies. From a secure, explosion-proof location, use water spray to cool exposed containers. If cooling streams are ineffective (venting sound increases in volume and pitch, tank discolors, or shows any signs of deforming), withdraw immediately to a secure position. If employees are expected to fight fires, they must be trained and equipped in OSHA 1910.156. The only respirators recommended for firefighting are self-contained breathing apparatuses that have full face-pieces and are operated in a pressure-demand or other positive-pressure mode.

Disposal Method Suggested: Wearing protective equipment, spray carefully onto sodium ash/slaked lime mixture. Then spray with water, dilute, neutralize, and flush to drain.

References
Sax, N. I. (Ed.). (1985). *Dangerous Properties of Industrial Materials Report*, 5, No. 6, 90−92
New Jersey Department of Health and Senior Services. (October 1999). *Hazardous Substances Fact Sheet: Sulfur Chloride*. Trenton, NJ

Sulfur dioxide   S:0750

Molecular Formula: O$_2$S
Common Formula: SO$_2$

Synonyms: Bisulfite; Dioxido de azufe (Spanish); Fermenicide; Schwefelddioxyd (German); Sulfurous acid anhydride; Sulfurous anhydride; Sulfurous oxide; Sulphur dioxide; Sulphurous anhydride; Sulphurous oxide
CAS Registry Number: 7446-09-5; *(alt.)* 8014-94-6; *(alt.)* 12396-99-5; *(alt.)* 83008-56-4; *(alt.)* 89125-89-3
RTECS® Number: WS4550000
UN/NA & ERG Number: UN1079/125
EC Number: 231-195-2 [*Annex I Index No.:* 016-011-00-9]
Regulatory Authority and Advisory Bodies
Department of Homeland Security Screening Threshold Quantity (pounds): *Release hazard* 5000 (≥1.00% concentration); *Theft hazard* 500 (≥84.00% concentration).
Carcinogenicity: IARC: Human Inadequate Evidence; Animal Limited Evidence, *not classifiable as carcinogenic to humans*, Group 3, 1992.
US EPA, FIFRA, 1998 Status of Pesticides: Supported.
Toxic Chemical (World Bank).[15]
Air Pollutant Standard Set. See below, "Permissible Exposure Limits in Air" section.
OSHA 29CFR1910.119, Appendix A. Process Safety List of Highly Hazardous Chemicals, TQ = 1000 lb (450 kg).
Clean Air Act: Accidental Release Prevention/Flammable Substances (Section 112[r], Table 3), TQ = 5000 lb (2270 kg).
Superfund/EPCRA 40CFR355, Extremely Hazardous Substances: TPQ = 500 lb (227 kg).
Reportable Quantity (RQ): 500 lb (227 kg).
US DOT 49CFR172.101, Inhalation Hazardous Chemical.
Canada, WHMIS, Ingredients Disclosure List Concentration 1.0%.
European/International Regulations: Hazard Symbol: T; Risk phrases: R23; R34; Safety phrases: S1/2; S9; S26; S36/37/39; S45 (see Appendix 4).
WGK (German Aquatic Hazard Class): 1—Low hazard to waters.
Description: Sulfur dioxide is a noncombustible colorless gas at ambient temperatures with a characteristic strong suffocating odor. The odor threshold is 1.1 ppm. Shipped as a liquefied compressed gas. Molecular weight = 64.06; Boiling point = −10°C; Freezing/Melting point = −75.6°C; Relative vapor density (air = 1) = 2.26. Vapor pressure = 3.2 atm. Hazard Identification (based on NFPA-704 M Rating System): Health 2, Flammability 0, Reactivity 0. Soluble in water; solubility = 10%.
Potential Exposure: Compound Description: Agricultural Chemical; Tumorigen, Mutagen; Reproductive Effector; Human Data; Primary Irritant. Sulfur dioxide is used in the manufacture of sodium sulfite, sulfuric acid, sulfuryl chloride, thionyl chloride, organic sulfonates, disinfectants, fumigants, glass, wine, ice, industrial and edible protein, and vapor pressure thermometers. It is also used in the bleaching of beet sugar, flour, fruit, gelatin, glue, grain, oil, straw, textiles, wicker ware; wood pulp; wool; in the tanning of leather; in brewing and preserving; and in the refrigeration industry. Exposure may also occur in various other

industrial processes as it is a by-product of ore smelting, coal and fuel oil combustion, paper manufacturing, and petroleum refining.

Incompatibilities: Reacts with water to form sulfurous acid, a medium strong acid. Reacts violently with ammonia, acrolein, acetylene, alkali metals, such as sodium, potassium, magnesium, and zinc, chlorine, ethylene oxide, amines, butadiene. Attacks many metals including aluminum, iron, steel, brass, copper, nickel, especially in the presence of water or steam. Incompatible with halogens. Attacks plastics, rubber, and coatings.

Permissible Exposure Limits in Air

Conversion factor: 1 ppm = 2.62 mg/m^3 at 25°C & 1 atm.
OSHA PEL: 5 ppm/13 mg/m^3.
NIOSH REL: 2 ppm/5 mg/m^3 TWA; 5 ppm/13 mg/m^3 STEL.
ACGIH TLV®[1]: 0.25 ppm/0.65 mg/m^3 STEL; not classifiable as a human carcinogen.
NIOSH IDLH: 100 ppm.
Protective Action Criteria (PAC)
TEEL-0: 0.2 ppm
PAC-1: 0.2 ppm
PAC-2: 0.75 ppm
PAC-3: 30 ppm
DFG MAK: 0.5 ppm/1.3 mg/m^3 TWA; Peak Limitation Category I(1); a momentary Ceiling Concentration value of 1 mL/m^3/2.7 mg/m^3 should not be exceeded; Pregnancy Risk Group C.
Arab Republic of Egypt: TWA 5 ppm (13 mg/m^3), 1993; Australia: TWA 2 ppm (5 mg/m^3); STEL 5 ppm, 1993; Austria: MAK 2 ppm (5 mg/m^3), 1999; Belgium: TWA 2 ppm (5.2 mg/m^3); STEL 5 ppm (13 mg/m^3), 1993; Denmark: TWA 2 ppm (5 mg/m^3), 1999; Finland: TWA 2 ppm (5 mg/m^3); STEL 5 ppm (13 mg/m^3), 1999; France: VME 2 ppm (5 mg/m^3), VLE 5 ppm (10 mg/m^3), 1999; the Netherlands: MAC-TGG 5 mg/m^3, 2003; Japan: 1999; Norway: TWA 2 ppm (5 mg/m^3), 1999; the Philippines: TWA 5 ppm (13 mg/m^3), 1993; Poland: MAC (TWA) 2 mg/m^3; MAC (STEL) 5 mg/m^3, 1999; Russia: STEL 10 mg/m^3, [skin], 1993; Sweden: NGV 2 ppm (5 mg/m^3), TKV 5 ppm (13 mg/m^3), 1999; Switzerland: MAK-W 2 ppm (5 mg/m^3), KZG-W 4 ppm (10 mg/m^3), 1999; Thailand: TWA 5 ppm (13 mg/m^3), 1993; Turkey: TWA 5 ppm (13 mg/m^3), 1993; United Kingdom: TWA 2 ppm (5.3 mg/m^3); STEL 5 ppm (13 mg/m^3), 2000; Argentina, Bulgaria, Columbia, Jordan, South Korea, New Zealand, Singapore, Vietnam; ACGIH TLV®: STEL 5 ppm. Russia[43] set a MAC of 0.5 mg/m^3 for ambient air in residential areas on a momentary basis and 0.05 mg/m^3 on a daily average basis. Several states have set guidelines or standards for sulfur dioxide in ambient air[60] ranging from 1.2 μg/m^3 (Tennessee) to 80.0–1300.0 μg/m^3 (Arizona) to 119.0 μg/m^3 (Nevada) to 1300.0 μg/m^3 (Connecticut).

Determination in Air:
Use NIOSH Analytical Method (IV) #6004, Sulfur dioxide; OSHA Analytical Methods ID-200, ID-104 (for impinger).

Routes of Entry: Inhalation of gas, direct contact of gas or liquid phase on skin and mucous membranes.

Harmful Effects and Symptoms

Short Term Exposure: Inhalation: Causes irritation of the eyes, nose, throat. Nose and throat irritation may occur at 6–12 ppm. Between 10 and 50 ppm for 5–15 min, runny nose (rhinorrhea), difficult breathing, coughing and choking, reflex bronchoconstriction may occur. Levels of 50–100 ppm may only be tolerated for 30–60 min, and 400 ppm and above may immediately cause swelling and accumulation of fluid in throat and lungs (pulmonary edema), breathing stoppage, pneumonia, and death. Pulmonary edema is a medical emergency that can be delayed for several hours. This can cause death. *Skin:* Levels of 1% (10,000 ppm) may be irritating to moist skin. Liquid may cause frostbite and chemical burns. *Eyes:* Irritation may occur at 20 ppm and above. Liquid may cause frostbite, and permanent damage leading to blindness. *Ingestion:* Liquid may cause frostbite and chemical burns to mouth.

Long Term Exposure: Repeated or prolonged inhalation exposure may cause asthma. May cause chronic irritation of the eyes and respiratory tract; loss of the sense of smell; burning and dryness. Stomach problems may also occur. May cause permanent lung damage. Repeated exposure to 10 ppm may cause irritation to throat and lungs and an increased occurrence of nosebleeds. There is limited animal evidence that sulfur dioxide causes lung cancer in animals. There is limited evidence that sulfur dioxide may damage the developing fetus.

Note: Persons with asthma, subnormal pulmonary function, or cardiovascular disease are at greater risk. Sulfur dioxide may irritate the eyes and respiratory tract. Signs and symptoms of acute exposure to sulfur dioxide may be severe and include coughing, choking, dyspnea (shortness of breath), sneezing, wheezing, and chest discomfort. Upper airway edema (swelling) or obstruction, bronchoconstriction, pneumonia, pulmonary edema, and respiratory paralysis may occur. Fatigue may be noted. Gastrointestinal effects may include nausea, vomiting, and abdominal pain. Cyanosis (blue tint to skin and mucous membranes) may be noted following exposure to sulfur dioxide. It may cause death or permanent injury after very short exposure to small quantities. 1000 ppm causes death from 10 min to several hours by respiratory depression.

Points of Attack: Respiratory system, skin, eyes.

Medical Surveillance: NIOSH lists the following tests: chest X-ray, electrocardiogram, pulmonary function tests: forced vital capacity, forced expiratory volume (1 s); sputum cytology; white blood cell count/differential. Preplacement and periodic medical examinations should be concerned especially with the skin, eyes, and respiratory tract. Lung function tests. Examination of the eyes, nose, and throat. Consider chest X-ray following acute overexposure. Pulmonary function should be evaluated, as well as smoking habits, and exposure to other pulmonary irritants.

First Aid: If this chemical gets into the eyes, remove any contact lenses at once and irrigate immediately for at least 15 min, occasionally lifting upper and lower lids. Seek medical attention immediately. If this chemical contacts the skin, remove contaminated clothing and wash immediately with soap and water. Seek medical attention immediately. If this chemical has been inhaled, remove from exposure, begin rescue breathing (using universal precautions, including resuscitation mask) if breathing has stopped and CPR if heart action has stopped. Transfer promptly to a medical facility. When this chemical has been swallowed, get medical attention. Medical observation is recommended for 24–48 h after breathing overexposure, as pulmonary edema may be delayed. As first aid for pulmonary edema, a doctor or authorized paramedic may consider administering a corticosteroid spray. If frostbite has occurred, seek medical attention immediately; do *NOT* rub the affected areas or flush them with water. In order to prevent further tissue damage, do *NOT* attempt to remove frozen clothing from frostbitten areas. If frostbite has *NOT* occurred, immediately and thoroughly wash contaminated skin with soap and water.

Personal Protective Methods: Wear appropriate personal protective clothing to prevent the skin from becoming frozen from contact with the evaporating liquid or from contact with vessels containing the liquid. Safety equipment suppliers/manufacturers can provide recommendations on the most protective glove/clothing material for your operation. All protective clothing (suits, gloves, footwear, headgear) should be clean, available each day, and put on before work. Contact lenses should not be worn when working with this chemical. Wear gas-proof chemical goggles and face shield unless full face-piece respiratory protection is worn. Employees should wash immediately with soap when skin is wet or contaminated. Provide emergency showers and eyewash.

Respirator Selection: NIOSH: *20 ppm:* CcrS (APF = 10) [any chemical cartridge respirator with cartridge(s) providing protection against the compound of concern] or Sa (APF = 10) (any supplied-air respirator). *50 ppm:* Sa:Cf (APF = 25) (any supplied-air respirator operated in a continuous-flow mode) or PaprS (APF = 25) [any powered, air-purifying respirator with cartridge(s) providing protection against the compound of concern]. *100 ppm:* CcrFS (APF = 50) [any chemical cartridge respirator with a full face-piece and cartridge(s) providing protection against the compound of concern] or GmFS (APF = 50) [any air-purifying, full-face-piece respirator (gas mask) with a chin-style, front- or back-mounted canister providing protection against the compound of concern] or PaprTS (APF = 50) [any powered, air-purifying respirator with a tight-fitting face-piece and cartridge(s) providing protection against the compound of concern] or SaT: Cf (APF = 50) (any supplied-air respirator that has a tight-fitting face-piece and is operated in a continuous-flow mode) or SCBAF (APF = 50) (any self-contained breathing apparatus with a full face-piece) or SaF (APF = 50) (any supplied-air respirator with a full face-piece). *Emergency or planned entry into unknown concentrations or IDLH conditions:* SCBAF: Pd, Pp (APF = 10,000) (any self-contained breathing apparatus that has a full face-piece and is operated in a pressure-demand or other positive-pressure mode) or SaF: Pd,Pp: ASCBA (APF = 10,000) (any supplied-air respirator that has a full face-piece and is operated in a pressure-demand or other positive-pressure mode in combination with an auxiliary, self-contained breathing apparatus operated in a pressure-demand or other positive-pressure mode). *Escape:* GmFS (APF = 50) [any air-purifying, full-face-piece respirator (gas mask) with a chin-style, front- or back-mounted canister providing protection against the compound of concern] or SCBAE (any appropriate escape-type, self-contained breathing apparatus).

Note: Substance reported to cause eye irritation or damage; may require eye protection.

Storage: Color Code—White: Corrosive or Contact Hazard; Store separately in a corrosion-resistant location. Prior to working with this chemical you should be trained on its proper handling and storage. Keep below 130°F/54°C, protect containers from damage. Outdoor, ventilated, fireproof storage preferred. Procedures for the handling, use, and storage of cylinders should be in compliance with OSHA 1910.101 and 1910.169 with the recommendations of the Compressed Gas Association.

Shipping: Sulfur dioxide requires a shipping label of "POISON GAS, CORROSIVE." It falls in Hazard Class 2.3. Sulfurous acid (liquid SO_2) requires a shipping label of "CORROSIVE." It falls in Hazard Class 8 and Packing Group II. It is a violation of transportation regulations to refill compressed gas cylinders without the express written permission of the owner.

Special precautions: Cylinders must be transported in a secure upright position, in a well-ventilated truck.

Spill Handling: If in a building, evacuate building and confine vapors by closing doors and shutting down HVAC systems. Restrict persons not wearing protective equipment from area of spill or leak until cleanup is complete. Remove all ignition sources. Ventilate area of spill or leak to disperse the gas. Wear chemical protective suit with self-contained breathing apparatus to combat spills. Stay upwind and use water spray to "knock down" vapor; contain runoff. Stop the flow of gas, if it can be done safely from a distance. If source is a cylinder and the leak cannot be stopped in place, remove the leaking cylinder to a safe place; and repair leak or allow cylinder to empty. Keep this chemical out of confined spaces, such as a sewer, because of the possibility of explosion, unless the sewer is designed to prevent the buildup of explosive concentrations. If employees are required to clean up spills, they must be properly trained and equipped. OSHA 1910.120(q) may be applicable.

Initial isolation and protective action distances
Distances shown are likely to be affected during the first 30 min after materials are spilled and could increase with

time. If more than one tank car, cargo tank, portable tank, or large cylinder involved in the incident is leaking, the protective action distance may need to be increased. You may need to seek emergency information from CHEMTREC at (800) 424-9300 or seek professional environmental engineering assistance from the US EPA Environmental Response Team at (908) 548-8730 (24-h response line).

Small spills (From a small package or a small leak from a large package)
First: Isolate in all directions (feet/meters)   200/60
Then: Protect persons downwind (miles/kilometers)
Day   0.2/0.3
Night   0.7/1.2

Large spills (From a large package or from many small packages)
First: Isolate in all directions (feet/meters)   1250/400
Then: Protect persons downwind (miles/kilometers)
Day   1.3/2.0
Night   3.6/5.7

Fire Extinguishing: Not flammable. Extinguish fires with dry chemical, carbon dioxide, water spry, fog, or foam. Wear self-contained breathing apparatus and full protective clothing. Move container from fire area. Stay away from ends of tanks. Cool containers that are exposed to flames with water from the side until well after the fire is out. Isolate area until gas has dispersed. Keep unnecessary people away. Containers may explode in heat of fire, or they may rupture and release irritating toxic sulfur dioxide. Sulfur dioxide has explosive properties when it comes in contact with sodium hydride; potassium chlorate at elevated temperatures; ethanol; ether; at very cool temperatures ($-15°C$); fluorine; chlorine trifluoride; and chlorates. It will react with water or steam to produce toxic and corrosive fumes. Containers may explode in fire. Storage containers and parts of containers may rocket great distances, in many directions. Notify local health and fire officials and pollution control agencies. From a secure, explosion-proof location, use water spray to cool exposed containers. If cooling streams are ineffective (venting sound increases in volume and pitch, tank discolors, or shows any signs of deforming), withdraw immediately to a secure position. If cylinders are exposed to excessive heat from fire or flame contact, withdraw immediately to a secure location. If employees are expected to fight fires, they must be trained and equipped in OSHA 1910.156. The only respirators recommended for firefighting are self-contained breathing apparatuses that have full face-pieces and are operated in a pressure-demand or other positive-pressure mode.

Disposal Method Suggested: Pass into soda ash solution, then add calcium hypochlorite, neutralize, and flush to sewer with water (A-38).

References

National Institute for Occupational Safety and Health. (1974). *Criteria for a Recommended Standard: Occupational Exposure to Sulfur Dioxide*, NIOSH Document No. 74-111
World Health Organization. (1979). *Sulfur Oxides and Suspended Particulate Matter*, Environmental Health Criteria No. 8. Geneva, Switzerland
Sax, N. I. (Ed.). (1981). *Dangerous Properties of Industrial Materials Report*, 1, No. 3, 78-79
US Environmental Protection Agency. (November 30, 1987). *Chemical Hazard Information Profile: Sulfur Dioxide*. Washington, DC: Chemical Emergency Preparedness Program
New York State Department of Health. (March 1986). *Chemical Fact Sheet: Sulfur Dioxide* (Version 2). Albany, NY: Bureau of Toxic Substance Assessment
New Jersey Department of Health and Senior Services. (June 2000). *Hazardous Substances Fact Sheet: Sulfur Dioxide*. Trenton, NJ

Sulfur hexafluoride   S:0760

Molecular Formula: F_6S
Common Formula: SF_6
Synonyms: Hexafluorure de soufre (French); Sulfur fluoride
CAS Registry Number: 2551-62-4
RTECS® Number: WS4900000
UN/NA & ERG Number: UN1080/126
EC Number: Not assigned
Regulatory Authority and Advisory Bodies
Air Pollutant Standard Set. See below, "Permissible Exposure Limits in Air" section.
Canada, WHMIS, Ingredients Disclosure List Concentration 1.0%.
European/International Regulations: not listed in Annex 1.
WGK (German Aquatic Hazard Class): Nonwater polluting agent.

Description: Sulfur hexafluoride is a colorless, odorless gas. Molecular weight = 146.06; Boiling point = (sublimes) $-64°C$; Freezing/Melting point = (sublimes) $-51°C$; Relative vapor density (air = 1): 5.11; Vapor pressure = 21.5 atm. Shipped as a liquefied compressed gas. Condenses directly to a solid upon cooling. Practically insoluble in water; solubility = 0.003% at 25°C.

Potential Exposure: May contain highly toxic sulfur pentafluoride as an impurity. SF_6 is used in various electric power applications as a gaseous dielectric or insulator. The most extensive use is in high-voltage transformers. SF_6 is also used in waveguides, linear particle accelerators, Van de Graaff generators, chemically pumped continuous-wave lasers, transmission lines, and power distribution substations.

Nonelectrical applications include use as a protective atmosphere for casting of magnesium alloys and use as a leak detector or in tracing moving air masses. Several sources note that vitreous substitution of SF_6 in owl monkeys results in a greater ocular vascular permeability than that caused by saline. This implies that SF_6 could have an important use in retinal surgery.

Incompatibilities: May contain impurities that cause it to hydrolyze on contact with water, forming corrosive and toxic hydrogen fluoride. Vigorous reaction with disilane.

Permissible Exposure Limits in Air

Conversion factor: 1 ppm = 5.98 mg/m³ at 25°C & 1 atm.
OSHA PEL: 1000 ppm/6000 mg/m³ TWA.
NIOSH REL: 1000 ppm/6000 mg/m³ TWA.
ACGIH TLV®[1]: 1000 ppm/5970 mg/m³ TWA.
Protective Action Criteria (PAC)
TEEL-0: 1000 ppm
PAC-1: 3000 ppm
PAC-2: 5000 ppm
PAC-3: 5000 ppm
DFG MAK: 1000 ppm/6100 mg/m³ TWA; Peak Limitation Category II(8); Pregnancy Risk Group D.
Australia: TWA 1000 ppm (6000 mg/m³), 1993; Austria: MAK 1000 ppm (6000 mg/m³), 1999; Belgium: TWA 1000 ppm (5970 mg/m³), 1993; Denmark: TWA 1000 ppm (6000 mg/m³), 1999; Finland: TWA 1000 ppm (6000 mg/m³); STEL 1250 ppm (7500 mg/m³), 1999; France: VME 1000 ppm (6000 mg/m³), 1999; the Netherlands: MAC-TGG 6000 mg/m³, 2003; the Philippines: TWA 1000 ppm (6000 mg/m³), 1993; Poland: MAC (TWA) 1 mg[HF]/m³; MAC (STEL) 3 mg[HF]/m³, 1999; Russia: STEL 5000 mg/m³, 1993; Sweden: NGV 2 mg[F]/m³, 1999; Switzerland: MAK-W 1000 ppm (6000 mg/m³), 1999; Turkey: TWA 1000 ppm (6000 mg/m³), 1993; United Kingdom: TWA 1000 ppm (6070 mg/m³); STEL 1250 ppm, 2000; Argentina, Bulgaria, Columbia, Jordan, South Korea, New Zealand, Singapore, Vietnam: ACGIH TLV®: TWA 1000 ppm. Several states have set guidelines or standards for sulfur hexafluoride in ambient air[60] ranging from 60.0 mg/m³ (North Dakota) to 100.0 mg/m³ (Virginia) to 120.0 mg/m³ (Connecticut) to 142.857 mg/m³ (Nevada).

Determination in Air: Use NIOSH Analytical Method (IV) #6602.

Determination in Water: Octanol–water coefficient: Log K_{ow} = 1.68.

Routes of Entry: Inhalation, skin absorption, ingestion.

Harmful Effects and Symptoms

Short Term Exposure: Contact with the liquid may cause frostbite. Symptoms of exposure include asphyxia, increase (d) breathing rate, pulse rate, slight muscle incoordination, emotional upset, fatigue, nausea, vomiting. SF_6 is considered to be physiologically inert in the pure state. In high concentrations, however, pure SF_6 can act as a simple asphyxiant by displacing the necessary oxygen. Ordinarily, however, SF_6 does not exist in the pure state. It contains variable quantities of sulfur fluorides. In the presence of water, these sulfur fluorides can hydrolyze to yield hydrogen fluoride (HF) and oxyfluoride compounds, such as sulfuryl fluoride (SO_2F_2) and thionyl fluoride (SOF_2). These compounds have much more toxic health effects. Sulfur

hexafluoride may also be contaminated with more toxic sulfur compounds, such as S_2F_{10}.

Long Term Exposure: Repeated high exposures can cause deposits of fluorides in the bones (fluorosis) that may cause pain, disability, and mottling of the teeth. Repeated exposure can cause nausea, vomiting, loss of appetite, diarrhea, or constipation. Nosebleeds and sinus problems can also occur.

Points of Attack: Respiratory system, skeleton.

Medical Surveillance: Fluoride level in urine (use NIOSH #8308). Levels higher than 4 mg/L may indicate overexposure.

First Aid: If this chemical gets into the eyes, remove any contact lenses at once and irrigate immediately for at least 15 min, occasionally lifting upper and lower lids. Seek medical attention immediately. If this chemical contacts the skin, remove contaminated clothing and wash immediately with soap and water. Seek medical attention immediately. If this chemical has been inhaled, remove from exposure, begin rescue breathing (using universal precautions, including resuscitation mask) if breathing has stopped and CPR if heart action has stopped. Transfer promptly to a medical facility. If frostbite has occurred, seek medical attention immediately; do *NOT* rub the affected areas or flush them with water. In order to prevent further tissue damage, do *NOT* attempt to remove frozen clothing from frostbitten areas. If frostbite has *NOT* occurred, immediately and thoroughly wash contaminated skin with soap and water.

Personal Protective Methods: Wear appropriate personal protective clothing to prevent the skin from becoming frozen from contact with the evaporating liquid or from contact with vessels containing the liquid. Safety equipment suppliers/manufacturers can provide recommendations on the most protective glove/clothing material for your operation. All protective clothing (suits, gloves, footwear, headgear) should be clean, available each day, and put on before work. Contact lenses should not be worn when working with this chemical. Wear gas-proof chemical goggles and face shield unless full face-piece respiratory protection is worn. Employees should wash immediately with soap when skin is wet or contaminated. Provide emergency showers and eyewash.

Respirator Selection: NIOSH: *(fluorides) 12.5 mg/m³:* Qm (APF = 25) (any quarter-mask respirator). *25 mg/m³:* 95XQ (APF = 10) [any particulate respirator equipped with an N95, R95, or P95 filter (including N95, R95, and P95 filtering face-pieces) except quarter-mask respirators.] The following filters may also be used: N99, R99, P99, N100, R100, P100] or SA* (any supplied-air respirator). *62.5 mg/m³:* Sa:Cf (APF = 25)* (any supplied-air respirator operated in a continuous-flow mode) or PAPRDM*† *if not present as a fume* (any powered, air-purifying respirator with a dust and mist filter). *125 mg/m³:* HieF† (any air-purifying, full-face-piece respirator with a high-efficiency particulate filter) or SCBAF (APF = 50) (any self-contained breathing apparatus with a full face-piece) or SaF (APF = 50) (any supplied-air respirator with a full face-piece). *250 mg/m³:* Sa: Pd,Pp (APF = 1000) (any supplied-air respirator operated in a pressure-demand or other positive-pressure mode). *Emergency or planned entry into unknown concentrations or IDLH conditions:* SCBAF: Pd,Pp (APF = 10,000) (any self-contained breathing apparatus that has a full faceplate and is operated in a pressure-demand or other positive-pressure mode) or SaF: Pd,Pp: ASCBA (APF = 10,000) (any supplied-air respirator that has a full face-piece and is operated in a pressure-demand or other positive-pressure mode in combination with an auxiliary, self-contained breathing apparatus operated in a pressure-demand or other positive-pressure mode). *Escape:* HieF† (any air-purifying, full-face-piece respirator with a high-efficiency particulate filter) or SCBAE (any appropriate escape-type, self-contained breathing apparatus).

*Substance reported to cause eye irritation or damage; may require eye protection.

†May need acid gas sorbent.

Storage: Color Code—Green: General storage may be used. Prior to working with this chemical you should be trained on its proper handling and storage. Procedures for the handling, use, and storage of cylinders should be in compliance with OSHA 1910.101 and 1910.169, as with the recommendations of the Compressed Gas Association.

Shipping: Sulfur hexafluoride requires a shipping label of "NONFLAMMABLE GAS." It falls in Hazard Class 2.2.

Spill Handling: If in a building, evacuate building and confine vapors by closing doors and shutting down HVAC systems. Restrict persons not wearing protective equipment from area of spill or leak until cleanup is complete. Remove all ignition sources. Ventilate area of spill or leak to disperse the gas. Wear chemical protective suit with self-contained breathing apparatus to combat spills. Stay upwind and use water spray to "knock down" vapor; contain runoff. Stop the flow of gas, if it can be done safely from a distance. If source is a cylinder and the leak cannot be stopped in place, remove the leaking cylinder to a safe place; and repair leak or allow cylinder to empty. Keep this chemical out of confined spaces, such as a sewer, because of the possibility of explosion, unless the sewer is designed to prevent the buildup of explosive concentrations. If employees are required to clean up spills, they must be properly trained and equipped. OSHA 1910.120(q) may be applicable.

Fire Extinguishing: This chemical is a nonflammable gas. Poisonous gases are produced in fire, including sulfur oxides and fluorine compounds. Containers may explode in fire. Storage containers and parts of containers may rocket great distances, in many directions. If material or contaminated runoff enters waterways, notify downstream users of potentially contaminated waters. Notify local health and fire officials and pollution control agencies. From a secure, explosion-proof location, use water spray to cool exposed containers. If cooling streams are ineffective (venting sound increases in volume and pitch, tank discolors, or shows any signs of deforming), withdraw immediately to a secure position. If cylinders are exposed to excessive heat from fire or flame contact, withdraw immediately to a secure location. If employees are expected to fight fires, they must be trained and equipped in OSHA 1910.156. The only respirators recommended for firefighting are self-contained breathing apparatuses that have full face-pieces and are operated in a pressure-demand or other positive-pressure mode.

Disposal Method Suggested: Seal unused cylinders and return to suppliers.

References

US Environmental Protection Agency. (July 10, 1978). *Chemical Hazard Information Profile: Sulfur Hexafluoride.* Washington, DC

Sulfuric acid   S:0770

Molecular Formula: H_2O_4S
Common Formula: H_2SO_4
Synonyms: Acido sulfurico (Spanish); Acide sulfurique (French); BOV; Dihydrogen sulfate; Dipping acid; Hydrogen sulfate; Hydroot; Matting acid; Nordhausen acid; Oil of vitriol; Schwefelsaeureloesungen (German); Spirit of sulfur; Sulphuric acid; Vitriol brown oil; Vitriol, oil of-
CAS Registry Number: 7664-93-9; *(alt.)* 119540-51-1; *(alt.)* 127529-01-5; 8014-95-7 (fuming sulfuric acid; Oleum)
RTECS® Number: WS5600000; WS5605000 (fuming)
UN/NA & ERG Number: UN1830/137; UN1831 (fuming)/137; UN1832 (spent)/137
EC Number: 231-639-5
Regulatory Authority and Advisory Bodies
Department of Homeland Security Screening Threshold Quantity (pounds): *Release hazard* 10,000 (≥1.00% concentration). *(Oleum).*
Carcinogenicity (*strong inorganic acid mists of sulfuric acid*): IARC: Human Sufficient Evidence, *carcinogenic to humans,* Group 1, 1992; NTP: 11th Report on Carcinogens, 2004: Known to be a human carcinogen.
US EPA, FIFRA 1998 Status of Pesticides: RED completed.
Air Pollutant Standard Set. See below, "Permissible Exposure Limits in Air" section.
Clean Water Act: Section 311 Hazardous Substances/RQ 40CFR117.3 (same as CERCLA, see below); Section 313 Water Priority Chemicals (57FR41331, 9/9/92).
Superfund/EPCRA 40CFR355, Extremely Hazardous Substances: TPQ = 1000 lb (454 kg).
Reportable Quantity (RQ): 1000 lb (454 kg).
EPCRA Section 313 (acid aerosols including mists, vapors, gas, fog, and other airborne species of any particle size) Form R *de minimis* concentration reporting level: 0.1%.
California Proposition 65 Chemical: Cancer (Strong inorganic acid mists containing sulfuric acid) 4/14/03.

Canada, WHMIS, Ingredients Disclosure List Concentration 1.0%.

European/International Regulations: not listed in Annex 1. WGK (German Aquatic Hazard Class): 1—Low hazard to waters.

Description: Sulfuric acid is a colorless to dark-brown, odorless, oily liquid which is commercially sold at 93—98% H_2SO_4, the remainder being water. Molecular weight = 98.08; Specific gravity (H_2O:1) = 1.8 at 25°C; Boiling point = (decomposes) 340°C; Freezing/Melting point = 10.6°C; Vapor pressure = 0.001 mmHg. Fuming sulfuric acid (oleum) gives off free SO_2. Soluble in water; violent reaction.

Potential Exposure: Compound Description: Agricultural Chemical; Tumorigen, Mutagen; Reproductive Effector; Human Data; Primary Irritant. Used as a chemical feedstock in the manufacture of acetic acid, hydrochloric acid, citric acid, phosphoric acid, aluminum sulfate, ammonium sulfate, barium sulfate, copper sulfate, phenol, superphosphates, titanium dioxide, as well as synthetic fertilizers, nitrate explosives, artificial fibers, dyes, pharmaceuticals, detergents, glue, paint, and paper. It finds use as a dehydrating agent for esters and ethers due to its high affinity for water; as an electrolyte in storage batteries; for the hydrolysis of cellulose to obtain glucose; in the refining of mineral and vegetable oil; and in the leather industry. Other uses include fur and food processing; carbonization of wool fabrics; gas drying; uranium extraction from pitchblende; and laboratory analysis. Sulfuric acid is among the highest volume produced chemical in the United States.

Incompatibilities: A strong acid and oxidizer. Reacts violently with water with dangerous spattering and evolution of heat. Reacts violently with combustible and reducing materials, bases, organic materials, chlorates, carbides, picrates, fulminates, water, powdered metals. Corrosive to most common metals, forming explosive hydrogen gas.

Permissible Exposure Limits in Air
OSHA PEL: 1 mg/m³ TWA.
NIOSH REL: 1 mg/m³ TWA.
ACGIH TLV®[1]: 0.2 mg/m³ (measured as thoracic fraction of the aerosol) TWA; Suspected Human Carcinogen; H_2SO_4 contained in strong inorganic acid mists.
NIOSH IDLH: 15 mg/m³.
Protective Action Criteria (PAC)*
Includes oleum
TEEL-0: 0.2 mg/m³
PAC-1: **0.20** mg/m³
PAC-2: **8.7** mg/m³
PAC-3: **160** mg/m³
*AEGLs (Acute Emergency Guideline Levels) & ERPGs (Emergency Response Planning Guideline) are in **bold face**.
DFG MAK: 0.1 mg/m³, inhalable fraction TWA; Peak Limitation Category I(1), a momentary Ceiling value of 0.2 mg/m³ should not be exceeded; Carcinogen Category 4; Pregnancy Risk Group C.

Arab Republic of Egypt: TWA 1 mg/m³, 1993; Australia: TWA 1 mg/m³, 1993; Austria: MAK 1 mg/m³, 1999; Belgium: TWA 1 mg/m³; STEL 3 mg/m³, 1993; Denmark: TWA 1 mg/m³, 1999; Finland: TWA 1 mg/m³; STEL 3 mg/m³, [skin], 1999; France: VME 1 mg/m³, VLE 3 mg/m³, 1999; the Netherlands: MAC-TGG 1 mg/m³, 2003; Japan: 1 mg/m³, 1999; Norway: TWA 1 mg/m³, 1999; Poland: MAC (TWA) 1 mg/m³; MAC (STEL) 3 mg/m³, 1999; Russia: STEL 1 mg/m³, [skin], 1993; Sweden: NGV 1 mg/m³, TKV 3 mg/m³, 1999; Switzerland: MAK-W 1 mg/m³, KZG-W 2 mg/m³, 1999; Thailand: TWA 1 mg/m³, 1993; Turkey: TWA 1 mg/m³, 1993; United Kingdom: TWA 1 mg/m³, 2000; Argentina, Bulgaria, Columbia, Jordan, South Korea, New Zealand, Singapore, Vietnam: ACGIH TLV®: STEL 3 mg/m³. Several states have set guidelines or standards for sulfuric acid in ambient air[60] ranging from 2.381 μg/m³ (Kansas) to 10.0 μg/m³ (South Carolina) to 14.0 μg/m³ (Massachusetts) to 16.0 μg/m³ (Virginia) to 12.0—100.0 μg/m³ (North Carolina) to 20.0 μg/m³ (Connecticut) to 24.0 μg/m³ (Nevada).

Determination in Air: Use NIOSH Analytical Method (IV) #7903, Inorganic Acids; OSHA Analytical Method ID-113.

Permissible Concentration in Water: A guideline of 250 μg/L of sulfate has been recommended for drinking water by the EEC[35] and the state of Kansas.[61]

Routes of Entry: Inhalation, ingestion, eye and/or skin contact.

Harmful Effects and Symptoms
Short Term Exposure: Inhalation: May cause irritation of nose and throat at levels of 5 mg/m³. Swelling of the throat and lungs and inflammation of the bronchial membranes may occur at levels of 12—35 mg/m³. A few drops in the lung air passages may be fatal. Higher exposures can cause pulmonary edema, a medical emergency that can be delayed for several hours. This can cause death. *Skin:* May cause severe irritation, burns, and ulceration. *Eyes:* May cause severe irritation, damage to the cornea, and blindness. *Ingestion:* May cause damage to teeth; burning of the mouth, throat, and stomach; nausea, vomiting of blood and eroded tissue; holes in the stomach and intestines; shock; and kidney damage. Death may occur from as little as 1 oz. Signs and symptoms of acute ingestion of sulfuric acid may be severe and include salivation, intense thirst, difficulty in swallowing, pain, and shock. Oral, esophageal, and stomach burns are common. Vomitus generally has a coffee-ground appearance. The potential for circulatory collapse is high following ingestion of sulfuric acid. Acute inhalation exposure may result in sneezing, hoarseness, choking, laryngitis, dyspnea (shortness of breath), respiratory tract irritation, and chest pain. Bleeding of nose and gums, ulceration of the nasal and oral mucosa, pulmonary edema, chronic bronchitis, and pneumonia may also occur. If the eyes have come in contact with sulfuric acid, irritation, pain, swelling, corneal erosion, and blindness may result. Dermal exposure may result in severe burns, pain, and dermatitis (red, inflamed skin).

Long Term Exposure: Can cause chronic runny nose, tearing of the eyes, nose bleeding, and stomach upset. Risk of tooth erosion and pitting from repeated or prolonged exposure to the aerosol. Lungs may be affected by repeated or prolonged exposure to the aerosol; bronchitis may develop. Exposure to amounts greater than 3.0 mg/m^3 may cause all of the above symptoms in greater severity.

Points of Attack: Eyes, skin, respiratory system, teeth.

Medical Surveillance: NIOSH lists the following tests: chest X-ray; pulmonary function tests: forced vital capacity, forced expiratory volume (1 s). For those with frequent or potentially high exposure (half the TLV or greater), the following are recommended before beginning work and at regular times after that: lung function tests. Examination of the teeth. If symptoms develop or overexposure is suspected, the following may be useful: consider chest X-ray after acute overexposure.

First Aid: If this chemical gets into the eyes, remove any contact lenses at once and irrigate immediately for at least 15 min, occasionally lifting upper and lower lids. Seek medical attention immediately. If this chemical contacts the skin, remove contaminated clothing and wash immediately with soap and water. Seek medical attention immediately. If this chemical has been inhaled, remove from exposure, begin rescue breathing (using universal precautions, including resuscitation mask) if breathing has stopped and CPR if heart action has stopped. Transfer promptly to a medical facility. When this chemical has been swallowed, get medical attention. If victim is *conscious*, administer water or milk. Do not induce vomiting. Medical observation is recommended for 24–48 h after breathing overexposure, as pulmonary edema may be delayed. As first aid for pulmonary edema, a doctor or authorized paramedic may consider administering a corticosteroid spray.

Personal Protective Methods: Wear protective gloves and clothing to prevent any reasonable probability of skin contact. Safety equipment suppliers/manufacturers can provide recommendations on the most protective glove/clothing material for your operation. Polyethylene, Teflon™, Saranex®, Neoprene™/natural rubber, Viton are among the recommended protective materials for *sulfuric acid solution of more than 70%*. All protective clothing (suits, gloves, footwear, headgear) should be clean, available each day, and put on before work. Contact lenses should not be worn when working with this chemical. Wear splash-proof chemical goggles and face shield unless full face-piece respiratory protection is worn. Employees should wash immediately with soap when skin is wet or contaminated. Provide emergency showers and eyewash. For engineering controls see NIOSH Criteria Document #74-128, *Exposure to Sulfuric Acid.*

Respirator Selection: *Up to 15 mg/m^3:* Sa:Cf (APF = 25)* (any supplied-air respirator operated in a continuous-flow mode) or PaprAgHie (APF = 25) (APF = 25)* (any powered, air-purifying respirator with acid gas cartridge(s) in combination with a high-efficiency particulate filter) or CcrFAg100 (APF = 50) [any chemical cartridge respirator with a full face-piece and acid gas cartridge(s) in combination with an N100, R100, or P100 filter] or GmFAg100 (APF = 50) [any air-purifying, full-face-piece respirator (gas mask) with a chin-style, front- or back-mounted acid gas canister having an N100, R100, or P100 filter] or SCBAF (APF = 50) (any self-contained breathing apparatus with a full face-piece) or SaF (APF = 50) (any supplied-air respirator with a full face-piece). Emergency or planned entry into unknown concentrations or IDLH conditions: SCBAF; Pd,Pp (any self-contained breathing apparatus that has a full face-piece and is operated in a pressure-demand or other positive-pressure mode) or SaF: Pd,Pp: ASCBA (any supplied-air respirator that has a full face-piece and is operated in a pressure-demand or other positive-pressure mode in combination with an auxiliary self-contained positive-pressure breathing apparatus). *Escape:* GmFAg100 (APF = 50) [any air-purifying, full-face-piece respirator (gas mask) with a chin-style, front- or back-mounted acid gas canister having an N100, R100, or P100 filter] or SCBAE (any appropriate escape-type, self-contained breathing apparatus).

*Substance causes eye irritation or damage; eye protection needed.

Storage: Color Code—White: Corrosive or Contact Hazard; Store separately in a corrosion-resistant location. Prior to working with this chemical you should be trained on its proper handling and storage. Sulfuric acid must be stored to avoid contact with water, chlorates, chromates, carbides, fulminates, nitrates, picrates, and powdered metals, since violent reactions occur. Store in tightly closed containers in a cool, dry well-ventilated area away from sunlight and in an area with an acid-resistant cement floor. Sources of ignition, such as smoking and open flames, are prohibited where sulfuric acid is used, handled, or stored in a manner that could create a potential fire or explosion hazard. Always add acid to water, never the reverse. Sulfuric acid is extremely corrosive; handle with care and use proper equipment and practices. Wherever sulfuric acid is used, handled, manufactured, or stored, use explosion-proof electrical equipment and fittings. Contact of sulfuric acid with metal drums may cause the release of flammable, explosive hydrogen gas; therefore, storage drums should be coated with acid-resistant material.

Shipping: Sulfuric acid with >51% acid or sulfuric acid with not >51% acid requires a shipping label of "CORROSIVE." They fall in Hazard Class 8 and Packing Group II.

Sulfuric acid, fuming, with 30% or more free sulfur trioxide requires a shipping label of "CORROSIVE, POISONOUS/TOXIC MATERIALS." It falls in Hazard Class 8 and Packing Group I.

Sulfuric acid, fuming, with <30% free sulfur trioxide, requires a shipping label of "CORROSIVE." It falls in Hazard Class 8 and Packing Group I.

Sulfuric acid, spent, requires a shipping label of "CORROSIVE." It falls in Hazard Class 8 and Packing Group II.

Spill Handling: Extremely hazardous to health; areas may be entered with extreme care. Evacuate and restrict persons not wearing protective equipment from area of spill or leak until cleanup is complete. No skin surface should be exposed. Keep all sources of ignition away from containers because explosive mixtures of hydrogen may be produced during storage. Ventilate area of spill or leak. *Small spills:* cover area with sodium bicarbonate or soda ash/slaked lime. Shovel neutralized residues into containers for disposal, or (if not available) cover area with sand or earth and shovel into disposal containers. Other neutralizing agents are calcinated dolomite, calcium oxide and hydroxide, sodium carbonate. Place sulfuric acid absorbed in vermiculite in sealed containers. Keep unnecessary people away; isolate hazard area and deny entry. Stay upwind; keep out of low areas. Ventilate closed spaces before entering them. Notify proper authorities in case of water pollution. Do not touch spilled material. Use water spray to reduce vapor; do not get water inside container. Dike for later disposal. Keep this chemical out of a confined space, such as a sewer, because of the possibility of an explosion, unless the sewer is designed to prevent the buildup of explosive concentrations. It may be necessary to contain and dispose of this chemical as a hazardous waste. If material or contaminated runoff enters waterways, notify downstream users of potentially contaminated waters. Contact your local or federal environmental protection agency for specific recommendations. If employees are required to clean up spills, they must be properly trained and equipped. OSHA 1910.120(q) may be applicable.

Initial isolation and protective action distances

Distances shown are likely to be affected during the first 30 min after materials are spilled and could increase with time. If more than one tank car, cargo tank, portable tank, or large cylinder involved in the incident is leaking, the protective action distance may need to be increased. You may need to seek emergency information from CHEMTREC at (800) 424-9300 or seek professional environmental engineering assistance from the US EPA Environmental Response Team at (908) 548-8730 (24-h response line).

Sulfuric acid, fuming; Sulfuric acid, fuming, with not less than 30% free Sulfur trioxide

Small spills (*From a small package or a small leak from a large package*)

First: Isolate in all directions (feet/meters)   200/60
Then: Protect persons downwind (miles/kilometers)
Day   0.2/0.3
Night   0.6/1.4

Large spills (From a large package or from many small packages)

First: Isolate in all directions (feet/meters)   1000/300
Then: Protect persons downwind (miles/kilometers)

Day   1.8/2.9
Night   3.6/5.7

Fire Extinguishing: This chemical is a combustible liquid. Poisonous gases, including sulfur oxides, are produced in fire or on contact with water. Contact with metal releases flammable and explosive hydrogen gas. Use carbon dioxide or dry chemical. Use water on combustibles burning in vicinity of this material, but use extreme care as water applied directly to this acid results in generation of heat and causes splattering. Vapors are heavier than air and will collect in low areas. Vapors may travel long distances to ignition sources and flashback. Vapors in confined areas may explode when exposed to fire. Containers may explode in fire. Storage containers and parts of containers may rocket great distances, in many directions. If material or contaminated runoff enters waterways, notify downstream users of potentially contaminated waters. Notify local health and fire officials and pollution control agencies. From a secure, explosion-proof location, use water spray to cool exposed containers. If cooling streams are ineffective (venting sound increases in volume and pitch, tank discolors, or shows any signs of deforming), withdraw immediately to a secure position. If employees are expected to fight fires, they must be trained and equipped in OSHA 1910.156. The only respirators recommended for firefighting are self-contained breathing apparatuses that have full face-pieces and are operated in a pressure-demand or other positive-pressure mode.

Disposal Method Suggested: Add slowly to solution of soda ash and slaked lime with stirring; flush to drain with large volumes of water. Recovery and reuse of spent sulfuric acid may be a viable alternative to disposal, and processes are available.

References

National Institute for Occupational Safety and Health. (1974). *Criteria for a Recommended Standard: Occupational Exposure to Sulfuric Acid*, NIOSH Document No. 74-128

Sax, N. I. (Ed.). *Dangerous Properties of Industrial Materials Report*, 1, No. 5, 80–83 (1981) and 5, No. 3, 70–74 (1985)

US Environmental Protection Agency. (November 30, 1987). *Chemical Hazard Information Profile: Sulfuric Acid*. Washington, DC: Chemical Emergency Preparedness Program

New York State Department of Health. (March 1988). *Chemical Fact Sheet: Sulfuric Acid* (Version 2). Albany, NY: Bureau of Toxic Substance Assessment

Fuming sulfuric acid or oleum

New Jersey Department of Health and Senior Services. (January 1986). *Hazardous Substances Fact Sheet: Sulfuric Acid Fuming*. Trenton, NJ

US Environmental Protection Agency, Special Review and Reregistration Division Office of Pesticide Programs. (1998). *Agency Status of Pesticides in Registration, Reregistration, and Special Review* (Rainbow Report). Washington, DC

New Jersey Department of Health and Senior Services. (June 2002). *Hazardous Substances Fact Sheet: Sulfuric Acid Fuming.* Trenton, NJ

New Jersey Department of Health and Senior Services. (June 2002). *Hazardous Substances Fact Sheet: Sulfuric Acid.* Trenton, NJ

Sulfurous acid 2-(*p-tert*-butylphenoxy)-1-methylethyl-2-chloroethyl ester   S:0780

Molecular Formula: $C_{15}H_{23}ClO_4S$

Common Formula: $(CH_3)_3C-C_6H_4-OCH_2CH(CH_3)OSO_2-(CH_2)_2Cl$

Synonyms: 88-R; Acaracide; Aracide; Aramite®; Araramite-15W®; Aratron®; 2-(*p*-Butylphenoxy)isopropyl 2-chloroethyl sulfite; 2-(4-*tert*-Butylphenoxy)isopropyl-2-chloroethyl sulfite; Butylphenoxyisopropyl chloroethyl sulfite; 2-(*p,tert*-Butylphenoxy)isopropyl 2′-chloroethyl sulphite; 2-(*p,tert*-Butylphenoxy)-1-methylethyl 2-chloroethyl ester of sulphurous acid; 2-(*p*-Butylphenoxy)-1-methylethyl 2-chloroethyl sulfite; 2-(*p,tert*-Butylphenoxy)-1-methylethyl-2-chloroethyl sulfite ester; 2-(*p,tert*-Butylphenoxy)-1-methylethyl 2′-chloroethyl sulphite; 2-(*p,tert*-Butylphenoxy)-1-methylethyl sulphite of 2-chloroethanol; 1-(*p,tert*-Butylphenoxy)-2-propanol-2-chloroethyl sulfite; CES; 2-Chloroethanol-2-(*p,tert*-butylphenoxy)-1-methylethyl sulfite; 2-Chloroethanol ester with 2-(*p,tert*-butylphenoxy)-1-methylethyl sulfite; β-Chloroethyl-b′-(*p,tert*-butylphenoxy)-a′-methylethyl sulfite; β-Chloroethyl-β-(*p,tert*-butylphenoxy)-α-methylethyl sulphite; 2-Chloroethyl 1-methyl-2-(*p,tert*-butylphenoxy)ethyl sulphite; 2-Chloroethylsulfurous acid 2-[4-(1,1-dimethylethyl)phenoxy]-1-methylethyl ester; 2-Chloroethyl sulphite of 1-(*p,tert*-Butylphenoxy)-2-propanol; Compound 88R; ENT 16,519; *o*-Mite; Niagaramite

CAS Registry Number: 140-57-8

RTECS® Number: WT2975000

UN/NA & ERG Number: UN2902 (Pesticides, liquid, toxic, n.o.s.)/151

EC Number: None assigned.

Regulatory Authority and Advisory Bodies
Carcinogenicity: IARC: Human No Sufficient Data; Animal Sufficient Evidence, *possibly carcinogenic to humans,* Group 2B, 1987.

Air Pollutant Standard Set. See below, "Permissible Exposure Limits in Air" section.

Hazardous Waste Constituent (EPA/RCRA).[5]

California Proposition 65 Chemical: Cancer 7/1/87.

European/International Regulations: not listed in Annex 1.

WGK (German Aquatic Hazard Class): No value assigned.

Description: Aramite® is a heavy, dark-amber liquid. Molecular weight = 334.89; Boiling point = 175°C at 0.1 mmHg; Freezing/Melting point = −32°C; Vapor pressure = 9 mmHg at 25°C. Practically insoluble in water.

Potential Exposure: Aramite® is a miticide and antimicrobial agent. Aramite is regulated by EPA under the Federal Insecticide, Fungicide, and Rodenticide Act and the Resource Conservation and Recovery Act. The significant regulatory action was a voluntary cancellation of the active ingredient registration by the sole producer in 1975.

Incompatibilities: Incompatible with alkaline material, such as lime or Bordeaux mixture (slaked lime and copper sulfate solution).

Permissible Exposure Limits in Air
No TEEL available.

A limit on Aramite® in ambient air has been set in Pennsylvania[60] at 18.07 μg/m³.

Routes of Entry: Inhalation, ingestion, skin and/or eye contact.

Harmful Effects and Symptoms
Short Term Exposure: This material is slightly toxic (LD_{50} value for rats is 3900 mg/kg) but it is carcinogenic to animals.

Long Term Exposure: Aramite® is carcinogenic in the rat and dog following its oral administration. It produced liver tumors in the rat and carcinomas of the gall bladder and biliary ducts in the dog. Aramite® was tested in two strains of mice by the oral route and produced a significant increase of hepatomas in males of one strain.

Points of Attack: See above.

First Aid: If this chemical gets into the eyes, remove any contact lenses at once and irrigate immediately for at least 15 min, occasionally lifting upper and lower lids. Seek medical attention immediately. If this chemical contacts the skin, remove contaminated clothing and wash immediately with soap and water. Seek medical attention immediately. If this chemical has been inhaled, remove from exposure, begin rescue breathing (using universal precautions, including resuscitation mask) if breathing has stopped and CPR if heart action has stopped. Transfer promptly to a medical facility. When this chemical has been swallowed, get medical attention. Give large quantities of water and induce vomiting. Do not make an unconscious person vomit.

Personal Protective Methods: Wear protective gloves and clothing to prevent any reasonable probability of skin contact. Safety equipment suppliers/manufacturers can provide recommendations on the most protective glove/clothing material for your operation. All protective clothing (suits, gloves, footwear, headgear) should be clean, available each day, and put on before work. Contact lenses should not be worn when working with this chemical. Wear splash-proof chemical goggles and face shield unless full face-piece respiratory protection is worn. Employees should wash immediately with soap when skin is wet or contaminated. Provide emergency showers and eyewash.

Respirator Selection: Follow regulations in OSHA 29CFR1910.134 or European Standard EN149. Use a NIOSH/MSHA- or European Standard EN149-approved respirator; or use an approved supplied-air respirator with a full face-piece operated in the positive-pressure mode, or with a full face-piece, hood, or helmet in the continuous-flow mode; or use a NIOSH/MSHA- or European Standard EN149-approved self-contained breathing apparatus with a full face-piece operated in pressure-demand or other positive-pressure mode.

Storage: Color Code—Blue: Health Hazard/Poison: Store in a secure poison location. Prior to working with this chemical you should be trained on its proper handling and storage. Store in tightly closed containers in a cool, well-ventilated area away from alkalis, nitrates, and heat. This material may be thermally unstable (Verschueren). Where possible, automatically transfer material from drums or other storage containers to process containers. Sources of ignition, such as smoking and open flames, are prohibited where this chemical is handled, used, or stored. Metal containers involving the transfer of this chemical should be grounded and bonded. Wherever this chemical is used, handled, manufactured, or stored, use explosion-proof electrical equipment and fittings. A regulated, marked area should be established where this chemical is handled, used, or stored in compliance with OSHA Standard 1910.1045.

Shipping: Pesticides, liquid, toxic, n.o.s. require a shipping label of "POISONOUS/TOXIC MATERIALS." They fall in Hazard Class 6.1 and Packing Group I to III.

Spill Handling: Evacuate and restrict persons not wearing protective equipment from area of spill or leak until cleanup is complete. Remove all ignition sources. Ventilate area of spill or leak. Absorb liquids in vermiculite, dry sand, earth, peat, carbon, or a similar material and deposit in sealed containers. Keep this chemical out of a confined space, such as a sewer, because of the possibility of an explosion, unless the sewer is designed to prevent the buildup of explosive concentrations. It may be necessary to contain and dispose of this chemical as a hazardous waste. If material or contaminated runoff enters waterways, notify downstream users of potentially contaminated waters. Contact your local or federal environmental protection agency for specific recommendations. If employees are required to clean up spills, they must be properly trained and equipped. OSHA 1910.120(q) may be applicable.

Fire Extinguishing: This chemical is a combustible liquid. Poisonous gases, including chlorine and sulfur oxide, are produced in fire. Use dry chemical, carbon dioxide, or alcohol foam extinguishers. Vapors are heavier than air and will collect in low areas. Vapors may travel long distances to ignition sources and flashback. Vapors in confined areas may explode when exposed to fire. Containers may explode in fire. Storage containers and parts of containers may rocket great distances, in many directions. If material or contaminated runoff enters waterways, notify downstream users of potentially contaminated waters. Notify local health and fire officials and pollution control agencies. From a secure, explosion-proof location, use water spray to cool exposed containers. If cooling streams are ineffective

(venting sound increases in volume and pitch, tank discolors, or shows any signs of deforming), withdraw immediately to a secure position. If employees are expected to fight fires, they must be trained and equipped in OSHA 1910.156. The only respirators recommended for firefighting are self-contained breathing apparatuses that have full face-pieces and are operated in a pressure-demand or other positive-pressure mode.

Disposal Method Suggested: Acid or alkaline hydrolysis followed by flushing to sewer.

References

Sax, N. I. (Ed.). (1981). *Dangerous Properties of Industrial Materials Report*, 1, No. 3, 79—80

Sulfur pentafluoride   S:0790

Molecular Formula: $F_{10}S_2$
Common Formula: S_2F_{10}
Synonyms: Suflur decafluoride; Disulfur decafluoride; Disulphur decafluoride
CAS Registry Number: 5714-22-7
RTECS® Number: WS4480000
UN/NA & ERG Number: Not regulated
EC Number: 227-204-4 (disulphur decafluoride)
Regulatory Authority and Advisory Bodies
Air Pollutant Standard Set. See below, "Permissible Exposure Limits in Air" section.
OSHA 29CFR1910.119, Appendix A. Process Safety List of Highly Hazardous Chemicals, TQ = 250 lb (114 kg).
Canada, WHMIS, Ingredients Disclosure List Concentration 1.0%.
European/International Regulations: not listed in Annex 1.
WGK (German Aquatic Hazard Class): No value assigned.
Description: Sulfur pentafluoride is a colorless liquid or gas (above 29°C) with an odor like sulfur dioxide. Noncombustible liquid and nonflammable gas. Molecular weight = 254.12; Boiling point = 28.9°C; Freezing/Melting point = −92.2°C; Relative vapor density (air = 1) = 8.77; Specific gravity (H_2O:1) = 2.08 at 0°C. Hazard Identification (based on NFPA-704 M Rating System): Health 2, Flammability 0, Reactivity 0. Insoluble in water.
Potential Exposure: Sulfur pentafluoride is encountered as a by-product in the manufacture of sulfur hexafluoride, which is made by the direct fluorination of sulfur or sulfur dioxide.
Incompatibilities: Fluorides form explosive and toxic gases on contact with strong acids and acid fumes. Reacts with strong caustics. Decomposed in temperatures above 400°C forming toxic and corrosive fumes of sulfur oxides and sulfur fluorides; and as this chemical decomposes, it acts as both a strong oxidizer and a fluorinating agent.
Permissible Exposure Limits in Air
Conversion factor: 1 ppm = 10.39 mg/m³ at 25°C & 1 atm.
OSHA PEL: 0.025 ppm/0.25 mg/m³ TWA.
NIOSH REL: 0.01 ppm/0.1 mg/m³ Ceiling Concentration.

ACGIH TLV®[1]: 0.01 ppm/0.10 mg/m³ Ceiling Concentration.
Protective Action Criteria (PAC)
TEEL-0: 0.01 ppm
PAC-1: 0.01 ppm
PAC-2: 0.01 ppm
PAC-3: 1 ppm
DFG MAK: No numerical value established. Data may be available.
NIOSH IDLH: 1 ppm.
Australia: TWA 0.01 ppm (0.1 mg/m³), 1993; Austria: MAK 0.025 ppm (0.25 mg/m³), 1999; Belgium: STEL 0.01 ppm (0.1 mg/m³), 1993; Denmark: TWA 0.01 ppm (0.1 mg/m³), 1999; Finland: TWA 0.025 ppm, STEL 0.075 ppm, [skin], 1999; Norway: TWA 0.01 ppm (0.1 mg/m³), 1999; the Philippines: TWA 0.025 ppm (0.25 mg/m³), 1993; the Netherlands: MAC-TGG 0.1 mg/m³, 2003; Sweden: NGV 2 mg[F]/m³, 1999; Switzerland: MAK-W 0.01 ppm (0.1 mg/m³), KZG-W 0.02 ppm (0.2 mg/m³), 1999; Turkey: TWA 0.025 ppm (0.25 mg/m³), 1993; United Kingdom: TWA 0.025 ppm (0.26 mg/m³), STEL 0.075 ppm, 2000: Argentina, Bulgaria, Columbia, Jordan, South Korea, New Zealand, Singapore, Vietnam: ACGIH TLV®: Ceiling Concentration 0.01 ppm. Several states have set guidelines or standards for sulfur pentafluoride in ambient air[60] ranging from 0.8 μg/m³ (Virginia) to 1.0 μg/m³ (North Dakota) to 5.0 μg/m³ (Connecticut) to 6.0 μg/m³ (Nevada).
Determination in Air: Gaseous fluorides collected by impinger using caustic; particulates by filter. Analysis is by ion-specific electrode per NIOSH Analytical Method 7902.[18]
Permissible Concentration in Water: The EPA has set a standard of 4 mg/L for fluoride[61] and the state of Maine has set 2.4 mg/L as a guideline for drinking water. Arizona[61] has set 1.8 mg/L as a standard for drinking water.
Routes of Entry: Inhalation, ingestion, skin and/or eye contact.
Harmful Effects and Symptoms
Short Term Exposure: Irritates the eyes, skin, and respiratory tract. Fluorides can irritate and may damage the eyes. Skin contact can cause irritation, rash, or burning sensation. High repeated exposure can cause nausea, vomiting, loss of appetite, and bone and teeth changes. Extremely high levels could be fatal. Breathing can irritate the nose and throat, and cause nausea, headaches, and nosebleeds. Higher exposures can cause pulmonary edema, a medical emergency that can be delayed for several hours. This can cause death. Very high exposure can cause fluoride poisoning with stomach pain, weakness, convulsions, collapse, and death. These effects do not occur at the level of fluorides used in water for preventing cavities in teeth. In animals: pulmonary edema, hemorrhage.
Long Term Exposure: Repeated high exposures can cause deposits of fluorides in the bones (fluorosis), which may

cause pain, disability, and mottling of the teeth. May cause kidney damage.
Points of Attack: Respiratory system, central nervous system.
Medical Surveillance: NIOSH lists the following tests: chest X-ray; electrocardiogram; pulmonary function tests: forced vital capacity, forced expiratory volume (1 s); sputum cytology; white blood cell count/differential. Also consider kidney function tests. Fluoride level in urine (use NIOSH #8308). Levels higher than 4 mg/L may indicate overexposure.
First Aid: If this chemical gets into the eyes, remove any contact lenses at once and irrigate immediately for at least 15 min, occasionally lifting upper and lower lids. Seek medical attention immediately. If this chemical contacts the skin, remove contaminated clothing and wash immediately with soap and water. Seek medical attention immediately. If this chemical has been inhaled, remove from exposure, begin rescue breathing (using universal precautions, including resuscitation mask) if breathing has stopped and CPR if heart action has stopped. Transfer promptly to a medical facility. When this chemical has been swallowed, get medical attention. Give large quantities of water and do not induce vomiting. Do not make an unconscious person vomit. Medical observation is recommended for 24−48 h after breathing overexposure, as pulmonary edema may be delayed. As first aid for pulmonary edema, a doctor or authorized paramedic may consider administering a corticosteroid spray.
Personal Protective Methods: Wear protective gloves and clothing to prevent any reasonable probability of skin contact. Safety equipment suppliers/manufacturers can provide recommendations on the most protective glove/clothing material for your operation. All protective clothing (suits, gloves, footwear, headgear) should be clean, available each day, and put on before work. Contact lenses should not be worn when working with this chemical. Wear splash- or gas-proof chemical goggles and face shield unless full face-piece respiratory protection is worn. Employees should wash immediately with soap when skin is wet or contaminated. Provide emergency showers and eyewash.
Respirator Selection: NIOSH: *Up to 0.1 ppm:* Sa (APF = 10) (any supplied-air respirator). *Up to 0.25 ppm:* Sa:Cf (APF = 25) (any supplied-air respirator operated in a continuous-flow mode). *Up to 0.5 ppm:* SaT: Cf (APF = 50) (any supplied-air respirator that has a tight-fitting face-piece and is operated in a continuous-flow mode) or SCBAF (APF = 50) (any self-contained breathing apparatus with a full face-piece) or SaF (APF = 50) (any supplied-air respirator with a full face-piece). *Up to 1 ppm:* Sa: Pd,Pp (APF = 1000) (any supplied-air respirator operated in a pressure-demand or other positive-pressure mode). *Emergency or planned entry into unknown concentrations or IDLH conditions:* SCBAF: Pd,Pp (APF = 10,000) (any self-contained breathing apparatus that has a full faceplate

and is operated in a pressure-demand or other positive-pressure mode) or SaF: Pd,Pp: ASCBA (APF = 10,000) (any supplied-air respirator that has a full face-piece and is operated in a pressure-demand or other positive-pressure mode in combination with an auxiliary self-contained breathing apparatus operated in a pressure-demand or other positive-pressure mode). *Escape:* GmFAg (APF = 50) [any air-purifying, full-face-piece respirator (gas mask) with a chin-style, front- or back-mounted acid gas canister] or SCBAE (any appropriate escape-type, self-contained breathing apparatus).

Storage: Color Code—Blue: Health Hazard/Poison: Store in a secure poison location. Prior to working with this chemical you should be trained on its proper handling and storage. Fluorides must be stored to avoid contact with strong acids (such as hydrochloric, sulfuric, and nitric), since violent reactions occur. Fluorides form explosive gases on contact with nitric acid. Store in tightly closed containers in a cool, well-ventilated area away from water. Where possible, automatically transfer material from drums or other storage containers to process containers. Sources of ignition, such as smoking and open flames, are prohibited where this chemical is handled, used, or stored. Metal containers involving the transfer of this chemical should be grounded and bonded. Wherever this chemical is used, handled, manufactured, or stored, use explosion-proof electrical equipment and fittings. Procedures for the handling, use, and storage of cylinders should be in compliance with OSHA 1910.101 and 1910.169, as with the recommendations of the Compressed Gas Association.

Spill Handling: Liquid: Evacuate and restrict persons not wearing protective equipment from area of spill or leak until cleanup is complete. Remove all ignition sources. Ventilate area of spill or leak. Absorb liquids in vermiculite, dry sand, earth, peat, carbon, or a similar material and deposit in sealed containers. *Gas:* If in a building, evacuate building and confine vapors by closing doors and shutting down HVAC systems. Restrict persons not wearing protective equipment from area of spill or leak until cleanup is complete. Remove all ignition sources. Ventilate area of spill or leak to disperse the gas. Wear chemical protective suit with self-contained breathing apparatus to combat spills. Stay upwind and use water spray to "knock down" vapor; contain runoff. Stop the flow of gas, if it can be done safely from a distance. If source is a cylinder and the leak cannot be stopped in place, remove the leaking cylinder to a safe place; and repair leak or allow cylinder to empty. Keep this chemical out of confined spaces, such as a sewer, because of the possibility of explosion, unless the sewer is designed to prevent the buildup of explosive concentrations. If employees are required to clean up spills, they must be properly trained and equipped. OSHA 1910.120(q) may be applicable.

Fire Extinguishing: This chemical is a nonflammable gas. Poisonous gases are produced in fire. Vapors are heavier than air and will collect in low areas. Containers may explode in fire. Storage containers and parts of containers may rocket great distances, in many directions. If material or contaminated runoff enters waterways, notify downstream users of potentially contaminated waters. Notify local health and fire officials and pollution control agencies. From a secure, explosion-proof location, use water spray to cool exposed containers. If cooling streams are ineffective (venting sound increases in volume and pitch, tank discolors, or shows any signs of deforming), withdraw immediately to a secure position. If cylinders are exposed to excessive heat from fire or flame contact, withdraw immediately to a secure location. If employees are expected to fight fires, they must be trained and equipped in OSHA 1910.156

Sulfur tetrafluoride   S:0800

Molecular Formula: F_4S
Common Formula: SF_4
Synonyms: Sulfur fluoride (SF_4),(*t*-4)-; Sulphur fluoride; Sulphur tetrafluoride; Tetraflurosulfurane; Tetrafluoruro de azufre (Spanish); Tétrafluorure de soufre (French)
CAS Registry Number: 7783-60-0
RTECS® Number: WT4800000
UN/NA & ERG Number: UN2418/125
EC Number: 232-013-4
Regulatory Authority and Advisory Bodies
Department of Homeland Security Screening Threshold Quantity (pounds): *Release hazard* 2500 ($\geq$1.00% concentration); *Theft hazard* 15 ($\geq$1.33% concentration).
Air Pollutant Standard Set. See below, "Permissible Exposure Limits in Air" section.
OSHA 29CFR1910.119, Appendix A. Process Safety List of Highly Hazardous Chemicals, TQ = 250 lb (114 kg).
Clean Air Act: Accidental Release Prevention/Flammable Substances (Section 112[r], Table 3), TQ = 2500 lb (1135 kg).
Superfund/EPCRA 40CFR355, Extremely Hazardous Substances: TPQ = 100 lb (45.4 kg).
Reportable Quantity (RQ): 100 lb (45.4 kg).
OSHA 29CFR1910.119, Appendix A. Process Safety List of Highly Hazardous Chemicals, TQ = 250 lb (114 kg).
US DOT 49CFR172.101, Inhalation Hazardous Chemical.
Canada, WHMIS, Ingredients Disclosure List Concentration 1.0%.
European/International Regulations: not listed in Annex 1.
WGK (German Aquatic Hazard Class): No value assigned.
Description: Sulfur tetrafluoride is a colorless gas with an odor like sulfur dioxide. Shipped as a liquefied compressed gas. Molecular weight = 108.06; Specific gravity (H_2O:1) = 10.5 at 21°C; Boiling point = −40°C; Freezing/Melting point = −120°C; Relative vapor density (air = 1) = 3.78; Vapor pressure = 10.5 atm at 25°C. Reacts with water; highly soluble.

Potential Exposure: Sulfur tetrafluoride is used as a selective fluorinating agent in making water-repellent and oil-repellent materials and lubricity improvers. It is also used as a pesticide intermediate.

Incompatibilities: Moisture, concentrated sulfuric acid, dioxygen difluoride. Readily hydrolyzed by moisture, forming hydrofluoric acid and thionyl fluoride.

Permissible Exposure Limits in Air

Conversion factor: 1 ppm = 4.42 mg/m^3 at 25°C & 1 atm.
OSHA PEL: None.
NIOSH REL: 0.1 ppm/0.4 mg/m^3 Ceiling Concentration.
ACGIH TLV®[1]: 0.1 ppm/0.44 mg/m^3 Ceiling Concentration.
Protective Action Criteria (PAC)
TEEL-0: 0.1 ppm
PAC-1: 0.3 ppm
PAC-2: 2.08 ppm
PAC-3: 2.08 ppm
Australia: TWA 0.1 ppm (0.4 mg/m^3), 1993; Austria: MAK 2.5 mg[F]/m^3, 1999; Belgium: STEL 0.1 ppm (0.44 mg/m^3), 1993; Denmark: TWA 0.1 ppm (0.4 mg/m^3), 1999; Finland: TWA 0.1 ppm (0.4 mg/m^3), STEL 0.3 ppm, [skin], 1999; Norway: TWA 0.1 ppm (0.4 mg/m^3), 1999; Poland: MAC (TWA) 1 mg[HF]/m^3; MAC (STEL) 2 mg[HF]/m^3, 1999; the Netherlands: MAC 0.4 mg/m^3, 2003; United Kingdom: TWA 0.1 ppm (0.45 mg/m^3), STEL 0.3 ppm (1.3 mg/m^3), 2000; Argentina, Bulgaria, Columbia, Jordan, South Korea, New Zealand, Singapore, Vietnam; ACGIH TLV®: Ceiling Concentration 0.1 ppm. Several states have set guidelines or standards for sulfur tetrafluoride in ambient air[60] ranging from 3.5 μg/m^3 (Virginia) to 4.0 μg/m^3 (North Dakota) to 8.0 μg/m^3 (Connecticut) to 10.0 μg/m^3 (Nevada).

Determination in Air: Bubbler; Sodium hydroxide; Ion-specific electrode; OSHA Analytical Method. #ID11.

Permissible Concentration in Water: No criteria set. (SF$_4$ reacts violently with water to give SO$_2$ and HF).

Routes of Entry: Inhalation, ingestion, skin and/or eye contact.

Harmful Effects and Symptoms

Short Term Exposure: Exposure can severely irritate the nose, throat, and lungs. May cause skin burns (from SF$_4$ releasing hydrofluoric acid on exposure to moisture). High levels can cause a buildup of fluid in the lungs (pulmonary edema) with cough and shortness of breath. This can lead to death. Contact with liquid may cause frostbite. Sulfur tetrafluoride is about as toxic as phosgene. It is a strong irritant. The toxic effects are attributed largely to fluorine which is released upon hydrolysis. In animals: dyspnea (breathing difficulty), weakness, rhinorrhea (discharge of thin nasal mucous).

Long Term Exposure: May cause lung damage. Repeated high exposures can cause deposits of fluorides in the bones (fluorosis), which may cause pain, disability, and mottling of the teeth. May cause kidney damage.

Points of Attack: Eyes, skin, respiratory system.

Medical Surveillance: For those with frequent or potentially high exposure (half the TLV or greater), the following are recommended before beginning work and at regular times after that: lung function tests. Urine test for fluoride level (should not be above 4 mg/L). If symptoms develop or overexposure is suspected, the following may be useful: consider chest X-ray after acute overexposure.

First Aid: If this chemical gets into the eyes, remove any contact lenses at once and irrigate immediately for at least 15 min, occasionally lifting upper and lower lids. Seek medical attention immediately. If this chemical contacts the skin, remove contaminated clothing and wash immediately with soap and water. Seek medical attention immediately. If this chemical has been inhaled, remove from exposure, begin rescue breathing (using universal precautions, including resuscitation mask) if breathing has stopped and CPR if heart action has stopped. Transfer promptly to a medical facility. When this chemical has been swallowed, get medical attention. Give large quantities of water and induce vomiting. Do not make an unconscious person vomit. Medical observation is recommended for 24–48 h after breathing overexposure, as pulmonary edema may be delayed. As first aid for pulmonary edema, a doctor or authorized paramedic may consider administering a corticosteroid spray. If frostbite has occurred, seek medical attention immediately; do *NOT* rub the affected areas or flush them with water. In order to prevent further tissue damage, do *NOT* attempt to remove frozen clothing from frostbitten areas. If frostbite has *NOT* occurred, immediately and thoroughly wash contaminated skin with soap and water.

Personal Protective Methods: Wear appropriate personal protective clothing to prevent the skin from becoming frozen from contact with the evaporating liquid or from contact with vessels containing the liquid. Safety equipment suppliers/manufacturers can provide recommendations on the most protective glove/clothing material for your operation. All protective clothing (suits, gloves, footwear, headgear) should be clean, available each day, and put on before work. Contact lenses should not be worn when working with this chemical. Wear gas-proof chemical goggles and face shield unless full face-piece respiratory protection is worn. Employees should wash immediately with soap when skin is wet or contaminated. Provide emergency showers and eyewash.

Respirator Selection: NIOSH: *Fluorides: 12.5 mg/m^3:* Qm (APF = 25) (any quarter-mask respirator). *25 mg/m^3:* Any particulate respirator equipped with an N95, R95, or P95 filter (including N95, R95, and P95 filtering face-pieces) except quarter-mask respirators. The following filters may also be used: N99, R99, P99, N100, R100, P100; or SA* (any supplied-air respirator). *62.5 mg/m^3:* Sa:Cf (APF = 25)* (any supplied-air respirator operated in a continuous-flow mode) or PAPRDM*† *if not present as a fume* (any powered, air-purifying respirator with a dust and mist filter). *125 mg/m^3:* HieF† (any air-purifying, full-face-piece respirator with a high-efficiency particulate filter) or SCBAF

(APF = 50) (any self-contained breathing apparatus with a full face-piece) or SaF (APF = 50) (any supplied-air respirator with a full face-piece). *250 mg/m³:* Sa: Pd,Pp (APF = 1000) (any supplied-air respirator operated in a pressure-demand or other positive-pressure mode). *Emergency or planned entry into unknown concentrations or IDLH conditions:* SCBAF: Pd,Pp (APF = 10,000) (any self-contained breathing apparatus that has a full faceplate and is operated in a pressure-demand or other positive-pressure mode) or SaF: Pd,Pp: ASCBA (APF = 10,000) (any supplied-air respirator that has a full face-piece and is operated in a pressure-demand or other positive-pressure mode in combination with an auxiliary, self-contained breathing apparatus operated in a pressure-demand or other positive-pressure mode). *Escape:* HieF† (any air-purifying, full-face-piece respirator with a high-efficiency particulate filter) or SCBAE (any appropriate escape-type, self-contained breathing apparatus).

*Substance reported to cause eye irritation or damage; may require eye protection.

†May need acid gas sorbent.

Storage: Color Code—Yellow Stripe: Reactivity Hazard; Store separately in an area isolated from flammables, combustibles, or other yellow coded materials. Color Code—Blue: Health Hazard/Poison: Store in a secure poison location. Prior to working with this chemical you should be trained on its proper handling and storage. Sulfur tetrafluoride must be stored to avoid contact with water, since violent reactions occur. Store in tightly closed containers in a cool, well-ventilated area away from water, steam, or acids to avoid the production of toxic fumes. Procedures for the handling, use, and storage of cylinders should be in compliance with OSHA 1910.101 and 1910.169, as with the recommendations of the Compressed Gas Association.

Shipping: Sulfur tetrafluoride requires a shipping label of "POISON GAS, CORROSIVE." It falls in Hazard Class 2.3.

Special precautions: Cylinders must be transported in a secure upright position, in a well-ventilated truck. It is a violation of transportation regulations to refill compressed gas cylinders without the express written permission of the owner.

Spill Handling: If in a building, evacuate building and confine vapors by closing doors and shutting down HVAC systems. Restrict persons not wearing protective equipment from area of spill or leak until cleanup is complete. Remove all ignition sources. Ventilate area of spill or leak to disperse the gas. Wear chemical protective suit with self-contained breathing apparatus to combat spills. Stay upwind and use water spray to "knock down" vapor; contain runoff. Stop the flow of gas, if it can be done safely from a distance. If source is a cylinder and the leak cannot be stopped in place, remove the leaking cylinder to a safe place; and repair leak or allow cylinder to empty. Keep this chemical out of confined spaces, such as a sewer, because of the possibility of explosion, unless the sewer is designed to prevent the buildup of explosive concentrations. If employees are required to clean up spills, they must be properly trained and equipped. OSHA 1910.120(q) may be applicable.

Initial isolation and protective action distances
Distances shown are likely to be affected during the first 30 min after materials are spilled and could increase with time. If more than one tank car, cargo tank, portable tank, or large cylinder involved in the incident is leaking, the protective action distance may need to be increased. You may need to seek emergency information from CHEMTREC at (800) 424-9300 or seek professional environmental engineering assistance from the US EPA Environmental Response Team at (908) 548-8730 (24-h response line).

Small spills (From a small package or a small leak from a large package)
First: Isolate in all directions (feet/meters)   300/100
Then: Protect persons downwind (miles/kilometers)
Day   0.4/0.6
Night   1.6/2.5

Large spills (From a large package or from many small packages)
First: Isolate in all directions (feet/meters)   2500/800
Then: Protect persons downwind (miles/kilometers)
Day   2.9/4.7
Night   6.4/10.3

Fire Extinguishing: This chemical is a nonflammable gas. Poisonous gases, including sulfur dioxide and hydrogen fluoride, are produced in fire. *Do not use water*. For small fires, use dry chemical or carbon dioxide. Vapors are heavier than air and will collect in low areas. Containers may explode in fire. Storage containers and parts of containers may rocket great distances, in many directions. If material or contaminated runoff enters waterways, notify downstream users of potentially contaminated waters. Notify local health and fire officials and pollution control agencies. From a secure, explosion-proof location, use water spray to cool exposed containers. If cooling streams are ineffective (venting sound increases in volume and pitch, tank discolors, or shows any signs of deforming), withdraw immediately to a secure position. If cylinders are exposed to excessive heat from fire or flame contact, withdraw immediately to a secure location. If employees are expected to fight fires, they must be trained and equipped in OSHA 1910.156. The only respirators recommended for firefighting are self-contained breathing apparatuses that have full face-pieces and are operated in a pressure-demand or other positive-pressure mode.

References
US Environmental Protection Agency. (November 30, 1987). *Chemical Hazard Information Profile: Sulfur Tetrafluoride*. Washington, DC: Chemical Emergency Preparedness Program
New Jersey Department of Health and Senior Services. (February 2000). *Hazardous Substances Fact Sheet: Sulfur Tetrafluoride*. Trenton, NJ

Sulfuryl fluoride   S:0820

Molecular Formula: F_2O_2S
Common Formula: SO_2F_2
Synonyms: Fluorure de sulfuryle (French); Fluoruro de sulfurilo (Spanish); Sulfonyl fluoride; Sulfur difluoride dioxide; Sulfuric oxyfluoride; Sulphuryl fluoride; Sulphuryl difluoride; Vikane; Vikane fumigant
CAS Registry Number: 2699-79-8
RTECS® Number: WT5075000
UN/NA & ERG Number: UN2191/123
EC Number: 220-281-5 [*Annex I Index No.:* 009-015-00-7]
Regulatory Authority and Advisory Bodies
Department of Homeland Security Screening Threshold Quantity (pounds): Sabotage/Contamination Hazard: A placarded amount (commercial grade).
US EPA, FIFRA 1998 Status of Pesticides: RED completed.
Air Pollutant Standard Set. See below, "Permissible Exposure Limits in Air" section.
EPCRA Section 313 Form R *de minimis* concentration reporting level: 1.0%.
US DOT 49CFR172.101, Inhalation Hazardous Chemical.
European/International Regulations: Hazard Symbol: T, N; Risk phrases: R23; R48/20; R50; Safety phrases: S1/2; S45; S630; S60; S61 (see Appendix 4).
WGK (German Aquatic Hazard Class): No value assigned.
Description: Sulfuryl fluoride is a colorless, poisonous gas. Odorless. Molecular weight = 102.06; Boiling point = $-55.6°C$; Freezing/Melting point = $-135.6°C$; Relative vapor density (air = 1) = 3.72; Vapor pressure = 15.2 atm at 25°C. Slightly soluble in water; solubility = 0.2 at 0°C. Water and air reactive.
Potential Exposure: Compound Description: Agricultural Chemical; Reproductive Effector. Sulfuryl fluoride is used an insecticidal fumigant. It is also used in organic synthesis of drugs and dyes.
Incompatibilities: Can react with water, steam. Fluorides form explosive gases on contact with strong acids or acid fumes.
Permissible Exposure Limits in Air
Conversion factor: 1 ppm = $4.18 mg/m^3$ at 25°C & 1 atm.
OSHA PEL: 5 ppm/20 mg/m³ TWA.
NIOSH REL: 5 ppm/20 mg/m³ TWA; 10 ppm/42 mg/m³ STEL.
ACGIH TLV®[1]: 5 ppm/21 mg/m³ TWA; 10 ppm/40 mg/m³ STEL.
NIOSH IDLH: 200 ppm.
Protective Action Criteria (PAC)
TEEL-0: 5 ppm
PAC-1: 10 ppm
PAC-2: 21 ppm
PAC-3: 64 ppm
Australia: TWA 5 ppm (20 mg/m³), STEL 10 ppm, 1993; Austria: MAK 2.5 mg[F]/m³, 1999; Belgium: TWA 5 ppm (21 mg/m³), STEL 10 ppm (42 mg/m³), 1993; Denmark: TWA 5 ppm (20 mg/m³), 1999; Finland: TWA 5 ppm (20 mg/m³), STEL 10 ppm (40 mg/m³), 1999; France: VME 5 ppm (20 mg/m³), 1999; Norway: TWA 5 ppm (20 mg/m³), 1999; the Netherlands: MAC-TGG 20 mg/m³, 2003; Poland: MAC (TWA) 1 mg[HF]/m³; MAC (STEL) 3 mg[HF]/m³, 1999; Sweden: NGV 2 mg[F]/m³, 1999; Switzerland: MAK-W 5 ppm (20 mg/m³), 1999; United Kingdom: TWA 5 ppm (21 mg/m³), STEL 10 ppm (42 mg/m³), 2000; Argentina, Bulgaria, Columbia, Jordan, South Korea, New Zealand, Singapore, Vietnam: ACGIH TLV®: STEL 10 ppm. Several states have set guidelines or standards for sulfuryl fluoride in ambient air[60] ranging from 200–400 μg/m³ (North Dakota) to 350 μg/m³ (Virginia) to 400 μg/m³ (Connecticut) to 476 μg/m³ (Nevada).
Determination in Air: Use NIOSH Analytical Method (IV) #6012, Sulfuryl fluoride.
Routes of Entry: Inhalation, eye and/or skin contact (liquid).
Harmful Effects and Symptoms
Short Term Exposure: May cause conjunctivitis, rhinitis, pharyngitis, paresthesia. Contact with the liquid may cause frostbite. High exposures can cause pulmonary edema, a medical emergency that can be delayed for several hours. This can cause death. Overexposure can cause nausea, vomiting, itching, muscle twitching, tremors, and seizures.
Long Term Exposure: May cause kidney damage. Repeated high exposures can cause deposits of fluorides in the bones (fluorosis), which may cause pain, disability, and mottling of the teeth.
Points of Attack: Eyes, skin, respiratory system, central nervous system, kidneys.
Medical Surveillance: Consider the points of attack in preplacement and periodic physical examinations. The fluoride level in urine (for fluoride in urine use NIOSH #8308). Levels higher than 4 mg/L may indicate overexposure. If symptoms develop or overexposure is suspected, the following may be useful: consider chest X-ray after acute overexposure, kidney function tests, examination of the nervous system.
First Aid: If this chemical gets into the eyes, remove any contact lenses at once and irrigate immediately for at least 15 min, occasionally lifting upper and lower lids. Seek medical attention immediately. If this chemical contacts the skin, remove contaminated clothing and wash immediately with soap and water. Seek medical attention immediately. If this chemical has been inhaled, remove from exposure, begin rescue breathing (using universal precautions, including resuscitation mask) if breathing has stopped and CPR if heart action has stopped. Transfer promptly to a medical facility. Medical observation is recommended for 24–48 h after breathing overexposure, as pulmonary edema may be delayed. As first aid for pulmonary edema, a doctor or authorized paramedic may consider administering a corticosteroid spray. If frostbite has occurred, seek medical attention immediately; do *NOT* rub the affected areas or flush

them with water. In order to prevent further tissue damage, do *NOT* attempt to remove frozen clothing from frostbitten areas. If frostbite has *NOT* occurred, immediately and thoroughly wash contaminated skin with soap and water.

Personal Protective Methods: Safety equipment suppliers/manufacturers can provide recommendations on the most protective glove/clothing material for your operation. All protective clothing (suits, gloves, footwear, headgear) should be clean, available each day, and put on before work. Contact lenses should not be worn when working with this chemical. Wear gas-proof chemical goggles and face shield unless full face-piece respiratory protection is worn. Employees should wash immediately with soap when skin is wet or contaminated. Provide emergency showers and eyewash.

Respirator Selection: *Up to 50 ppm:* Sa* (APF = 10) (any supplied-air respirator). *Up to 125 ppm:* Sa:Cf* (APF = 25) (any supplied-air respirator operated in a continuous-flow mode). *Up to 200 ppm:* SCBAF (APF = 50) (any self-contained breathing apparatus with a full face-piece) or SaF (APF = 50) (any supplied-air respirator with a full face-piece). *Emergency or planned entry into unknown concentrations or IDLH conditions:* SCBAF: Pd,Pp (APF = 10,000) (any self-contained breathing apparatus that has a full face-piece and is operated in a pressure-demand or other positive-pressure mode) or SaF: Pd,Pp: ASCBA (APF = 10,000) (any supplied-air respirator that has a full face-piece and is operated in a pressure-demand or other positive-pressure mode in combination with an auxiliary self-contained breathing apparatus operated in a pressure-demand or other positive-pressure mode). *Escape:* GmFS100 (APF = 50) [any air-purifying, full-face-piece respirator (gas mask) with a chin-style, front- or back-mounted canister providing protection against the compound of concern and having an N100, R100, or P100 filter] or SCBAE (any appropriate escape-type, self-contained breathing apparatus).

*Substance reported to cause eye irritation or damage; may require eye protection.

Storage: Poison gas. Color Code—Blue: Health Hazard/Poison: Store in a secure poison location. Prior to working with this chemical you should be trained on its proper handling and storage. Store in tightly closed containers in a cool, well-ventilated area away from water, steam, and strong acids. Procedures for the handling, use, and storage of cylinders should be in compliance with OSHA 1910.101 and 1910.169, as with the recommendations of the Compressed Gas Association.

Shipping: Sulfuryl fluoride requires a shipping label of "POISON GAS." It falls in Hazard Class 2.3. It is a violation of transportation regulations to refill compressed gas cylinders without the express written permission of the owner.

Spill Handling: If in a building, evacuate building and confine vapors by closing doors and shutting down HVAC systems. Restrict persons not wearing protective equipment from area of spill or leak until cleanup is complete. Remove all ignition sources. Ventilate area of spill or leak to disperse the gas. Wear chemical protective suit with self-contained breathing apparatus to combat spills. Stay upwind and use water spray to "knock down" vapor; contain runoff. Stop the flow of gas, if it can be done safely from a distance. If source is a cylinder and the leak cannot be stopped in place, remove the leaking cylinder to a safe place; and repair leak or allow cylinder to empty. Keep this chemical out of confined spaces, such as a sewer, because of the possibility of explosion, unless the sewer is designed to prevent the buildup of explosive concentrations. If employees are required to clean up spills, they must be properly trained and equipped. OSHA 1910.120(q) may be applicable.

Initial isolation and protective action distances
Distances shown are likely to be affected during the first 30 min after materials are spilled and could increase with time. If more than one tank car, cargo tank, portable tank, or large cylinder involved in the incident is leaking, the protective action distance may need to be increased. You may need to seek emergency information from CHEMTREC at (800) 424-9300 or seek professional environmental engineering assistance from the US EPA Environmental Response Team at (908) 548-8730 (24-h response line).

Small spills (From a small package or a small leak from a large package)
First: Isolate in all directions (feet/meters)   100/30
Then: Protect persons downwind (miles/kilometers)
Day   0.1/0.2
Night   0.3/0.5

Large spills (From a large package or from many small packages)
First: Isolate in all directions (feet/meters)   1000/300
Then: Protect persons downwind (miles/kilometers)
Day   1.1/1.8
Night   3.1/4.9

Fire Extinguishing: This chemical is a nonflammable gas. Poisonous gases, including sulfur dioxide and hydrogen fluoride, are produced in fire. For small fires, use dry chemical or carbon dioxide extinguishers. Gas is heavier than air and will collect in low areas. Containers may explode in fire. Storage containers and parts of containers may rocket great distances, in many directions. Notify local health and fire officials and pollution control agencies. From a secure, explosion-proof location, use water spray to cool exposed containers. If cooling streams are ineffective (venting sound increases in volume and pitch, tank discolors, or shows any signs of deforming), withdraw immediately to a secure position. If cylinders are exposed to excessive heat from fire or flame contact, withdraw immediately to a secure location. If employees are expected to fight fires, they must be trained and equipped in OSHA 1910.156.

Disposal Method Suggested: Addition of soda ash—slaked lime solution to form the corresponding sodium and calcium salt solution. This solution can be safely discharged after dilution. The precipitated calcium fluoride may be buried or added to a landfill. Small amounts could also be released directly to the atmosphere without serious harm.

References

New Jersey Department of Health and Senior Services. (May 2000). *Hazardous Substances Fact Sheet: Sulfuryl Fluoride*. Trenton, NJ

US Environmental Protection Agency, Special Review and Reregistration Division Office of Pesticide Programs. (1998). *Agency Status of Pesticides in Registration, Reregistration, and Special Review* (Rainbow Report). Washington, DC

Sulphamic acid   S:0830

Molecular Formula: H_3NO_3S

Common Formula: H_2NSO_3H

Synonyms: Amidosulfonic acid; Amidosulfuric acid; Aminosulfonic acid; Sulfamic acid; Sulfamidic acid

CAS Registry Number: 5329-14-6

RTECS® Number: WO5950000

UN/NA & ERG Number: UN2967/154

EC Number: 226-218-8 [*Annex I Index No.*: 016-026-00-0]

Regulatory Authority and Advisory Bodies

US EPA, FIFRA, 1998 Status of Pesticides: Unsupported.

Canada, WHMIS, Ingredients Disclosure List Concentration 1%.

European/International Regulations: Hazard Symbol: Xi, N; Risk phrases: R36/38; R52/53; Safety phrases: S2; S26; S28; S61 (see Appendix 4).

WGK (German Aquatic Hazard Class): 1—Low hazard to waters.

Description: Sulphamic acid is a white crystalline solid. Molecular weight = 97.10; Freezing/Melting point = about 205°C (decomposition). Soluble in water; slowly reactive.

Potential Exposure: Compound Description: Agricultural Chemical; Primary Irritant. Sulphamic acid is used in metal and ceramic cleaning, bleaching paper pulp, and textiles; in acid cleaning; as a stabilizing agent for chlorine and hypochlorite in swimming pools; cooling towers; and paper mills.

Incompatibilities: The aqueous solution is a strong acid. Reacts violently with strong acids (especially fuming nitric acid), bases, chlorine. Reacts slowly with water, forming ammonium bisulfate. Incompatible with ammonia, amines, isocyanates, alkylene oxides, epichlorohydrin, oxidizers.

Permissible Exposure Limits in Air

Protective Action Criteria (PAC)

TEEL-0: 12.5 mg/m^3

PAC-1: 40 mg/m^3

PAC-2: 250 mg/m^3

PAC-3: 500 mg/m^3

Routes of Entry: Inhalation, ingestion, skin and/or eye contact.

Harmful Effects and Symptoms

Short Term Exposure: Sulphamic acid can affect you when breathed in. Contact can burn the skin and eyes. Breathing the dust (crystals) or vapor can irritate the nose, mouth, and lower airways; and may cause cough with phlegm. High exposures can cause pulmonary edema, a medical emergency that can be delayed for several hours. This can cause death.

Long Term Exposure: May cause dermatitis and skin allergy. Some closely related sulfonates cause skin allergy. May affect the lungs.

Points of Attack: Lungs, skin.

Medical Surveillance: Before beginning employment and at regular times after that, for those with frequent or potentially high exposures, the following are recommended: lung function tests. If symptoms develop or overexposure is suspected, the following may be useful: consider chest X-ray after acute overexposure. Evaluation by a qualified allergist, including careful exposure history and special testing, may help diagnose skin allergy.

First Aid: If this chemical gets into the eyes, remove any contact lenses at once and irrigate immediately for at least 15 min, occasionally lifting upper and lower lids. Seek medical attention immediately. If this chemical contacts the skin, remove contaminated clothing and wash immediately with soap and water. Seek medical attention immediately. If this chemical has been inhaled, remove from exposure, begin rescue breathing (using universal precautions, including resuscitation mask) if breathing has stopped and CPR if heart action has stopped. Transfer promptly to a medical facility. When this chemical has been swallowed, get medical attention. If victim is *conscious*, administer water or milk. Do not induce vomiting. Medical observation is recommended for 24−48 h after breathing overexposure, as pulmonary edema may be delayed. As first aid for pulmonary edema, a doctor or authorized paramedic may consider administering a corticosteroid spray.

Personal Protective Methods: Wear protective gloves and clothing to prevent any reasonable probability of skin contact. Safety equipment suppliers/manufacturers can provide recommendations on the most protective glove/clothing material for your operation. All protective clothing (suits, gloves, footwear, headgear) should be clean, available each day, and put on before work. Contact lenses should not be worn when working with this chemical. Wear splash-proof chemical goggles and face shield when working with liquid, unless full face-piece respiratory protection is worn. Wear dust-proof goggles and face shield when working with powders or dust, unless full face-piece respiratory protection is worn. Employees should wash immediately with soap when skin is wet or contaminated. Provide emergency showers and eyewash.

Respirator Selection: Where there is potential for exposures to sulphamic acid, use a NIOSH/MSHA- or European Standard EN149-approved full face-piece respirator with a high-efficiency particulate filter. Greater protection is provided by a powered air-purifying respirator. *Where there is potential for high exposures*, use a NIOSH/MSHA- or European Standard EN149-approved supplied-air respirator with a full face-piece operated in the positive-pressure mode, or with a full face-piece, hood, or helmet in the continuous-flow mode; or use a NIOSH/MSHA- or European Standard EN149-approved self-contained breathing apparatus with a full face-piece operated in pressure-demand or other positive-pressure mode.

Storage: Color Code—White: Corrosive or Contact Hazard; Store separately in a corrosion-resistant location. Prior to working with this chemical you should be trained on its proper handling and storage. Sulphamic acid must be stored to avoid contact with strong acids (such as hydrochloric, sulfuric, and nitric), since violent reactions occur. Store in tightly closed containers in a cool, well-ventilated area away from chlorine and nitric acid.

Shipping: Sulfamic acid requires a shipping label of "CORROSIVE." It falls in Hazard Class 8 and Packing Group III.

Spill Handling: Evacuate persons not wearing protective equipment from area of spill or leak until cleanup is complete. Remove all ignition sources. Collect powdered material in the most convenient and safe manner and deposit in sealed containers. Ventilate area after cleanup is complete. It may be necessary to contain and dispose of this chemical as a hazardous waste. If material or contaminated runoff enters waterways, notify downstream users of potentially contaminated waters. Contact your local or federal environmental protection agency for specific recommendations. If employees are required to clean up spills, they must be properly trained and equipped. OSHA 1910.120(q) may be applicable.

Fire Extinguishing: Sulphamic acid may burn, but does not readily ignite. Use dry chemical, CO_2, dry sand. *Do not use water* directly on material. If large quantities of combustibles are involved, use water in flooding quantities as spray and fog. Poisonous gases, including oxides of sulfur and nitrogen, are produced in fire. If material or contaminated runoff enters waterways, notify downstream users of potentially contaminated waters. Notify local health and fire officials and pollution control agencies. From a secure, explosion-proof location, use water spray to cool exposed containers. If cooling streams are ineffective (venting sound increases in volume and pitch, tank discolors, or shows any signs of deforming), withdraw immediately to a secure position. If employees are expected to fight fires, they must be trained and equipped in OSHA 1910.156. The only respirators recommended for firefighting are self-contained breathing apparatuses that have full face-pieces and are operated in a pressure-demand or other positive-pressure mode.

References

New Jersey Department of Health and Senior Services. (May 2000). *Hazardous Substances Fact Sheet: Sulphamic Acid.* Trenton, NJ

Sulprofos   S:0840

Molecular Formula: $C_{12}H_{19}O_2PS_3$

Common Formula: $CH_3S-C_6H_4-OP(S)(SC_3H_7)(OCH_2CH_3)$

Synonyms: AI3-29149; Bayer NTN 9306; Bay-NTN-9306; Bolstar (Bayer); *o*-Ethyl *o*-[4-(methylmercapto)phenyl]-*S-n*-propylphosphorothionothiolate; *o*-Ethyl *o*-[4-(methylthio)phenyl]phosphorodithioic acid *S*-propyl ester; *o*-Ethyl *o*-[4-(methylthio)phenyl]phosphorodithioic acid *S*-propyl ester; *o*-Ethyl *o*-(4-methylthiophenyl) *S*-propyl dithiophosphate; *o*-Ethyl *o*-[4-(methylthio)phenyl] *S*-propyl phosphorodithioate; Helothion; Mercaprofos; Mercaprophos; Phosphorodithioic acid, *o*-ethyl *o*-[4-(methylthio)phenyl] *S*-propyl ester; Phosphorothioic acid, *o*-Ethyl *o*-[4-(methylthio)phenyl] *S*-propyl ester

CAS Registry Number: 35400-43-2

RTECS® Number: TE4165000

UN/NA & ERG Number: UN3018 (organophosphorus pesticide, liquid, toxic)/152

EC Number: 252-545-0

Regulatory Authority and Advisory Bodies

US EPA, FIFRA 1998 Status of Pesticides: Canceled.

Banned or Severely Restricted (Germany, Malaysia) (UN).[13]

Air Pollutant Standard Set. See below, "Permissible Exposure Limits in Air" section.

EPCRA Section 313 Form R *de minimis* concentration reporting level: 1.0%.

US DOT Regulated Marine Pollutant (49CFR172.101, Appendix B), severe pollutant.

US DOT 49CFR172.101, Inhalation Hazard Chemical as organophosphates.

European/International Regulations: not listed in Annex 1.

WGK (German Aquatic Hazard Class): 3—Severe hazard to waters.

Description: Sulprofos is a tan colored liquid. Molecular weight = 322.46; Specific gravity $(H_2O:1) = 1.20$ at 25°C; Boiling point = 155−158°C at 0.1 mmHg; Vapor pressure = 63 mmHg at 20°C. Poor solubility in water.

Potential Exposure: Compound Description: Agricultural Chemical; Reproductive Effector. Those involved in the manufacture, formulation, and application of this insecticide that is used for control of certain lepidopterous, dipterous, and hemipterous insects on cotton, etc.

Incompatibilities: Strong oxidizers may cause release of toxic phosphorus oxides. Organophosphates, in the presence of strong reducing agents such as hydrides, may form highly toxic and flammable phosphine gas. Keep away from alkaline materials.

Permissible Exposure Limits in Air

OSHA PEL: None.

NIOSH REL: 1 mg/m³ TWA.

ACGIH TLV®[1]: 0.008 ppm/0.1 mg/m³ measured as inhalable fraction and vapor TWA; not classifiable as a human carcinogen; BEI_A issued for acetylcholinesterase inhibiting pesticides.

No TEEL available.

Australia: TWA 1 mg/m³, 1993; Belgium: TWA 1 mg/m³, 1993; France: VME 1 mg/m³, 1999; Switzerland: MAK-W 1 mg/m³, 1999; the Netherlands: MAC-TGG 1 mg/m³, 2003; Argentina, Bulgaria, Columbia, Jordan, South Korea, New Zealand, Singapore, Vietnam: ACGIH TLV®: not classifiable as a human carcinogen. Several states have set guidelines of standards for sulprofos in ambient air[60] ranging from 16.0 μg/m³ (Virginia) to 20.0 μg/m³ (Connecticut) to 24.0 μg/m³ (Nevada) to 100.0 μg/m³ (North Dakota).

Determination in Air: Use NIOSH Analytical Method (IV) #5600, Organophosphorus pesticides; OSHA Analytical Method PV-2037, Sulprofos.

Determination in Water: Fish Tox = 63.99264000 ppb (INTERMEDIATE).

Routes of Entry: Inhalation, ingestion, skin and/or eye contact. Absorbed by the skin.

Harmful Effects and Symptoms

Short Term Exposure: May affect the nervous system, causing convulsions and respiratory failure. Shows typical anticholinesterase effects. Sulprofos can affect you when breathed in and quickly enters the body by passing through the skin. Severe poisoning can occur from skin contact. It is an organophosphate pesticide. Exposure can cause rapid severe poisoning with headaches, sweating, nausea and vomiting, diarrhea, loss of coordination, convulsions, respiratory failure, and death. This is considered a moderately toxic compound (LD_{50} for rats is 65 mg/kg).

Long Term Exposure: Cholinesterase inhibitor; cumulative effect is possible. This chemical may damage the nervous system with repeated exposure, resulting in convulsions, respiratory failure. Human Tox = 21.00000 ppb (INTERMEDIATE).

Points of Attack: Respiratory system, central nervous system, cardiovascular system, blood cholinesterase.

Medical Surveillance: Before employment and at regular times after that, the following are recommended: plasma and red blood cell cholinesterase levels (tests for the enzyme poisoned by this chemical). If exposure stops, plasma levels return to normal in 1−2 weeks while red blood cell levels may be reduced for 1−3 months. When cholinesterase enzyme levels are reduced by 25% or more below preemployment levels, risk of poisoning is increased, even if results are in lower ranges of "normal." Reassignment to work not involving organophosphate or carbamate pesticides is recommended until enzyme levels recover. If symptoms develop or overexposure occurs, repeat the above tests as soon as possible and get an examination of the nervous system.

First Aid: If this chemical gets into the eyes, remove any contact lenses at once and irrigate immediately for at least 15 min, occasionally lifting upper and lower lids. Seek medical attention immediately. If this chemical contacts the skin, remove contaminated clothing and wash immediately with soap and water. Speed in removing material from skin is of extreme importance. Shampoo hair promptly if contaminated. Seek medical attention immediately. If this chemical has been inhaled, remove from exposure, begin rescue breathing (using universal precautions, including resuscitation mask) if breathing has stopped and CPR if heart action has stopped. Transfer promptly to a medical facility. When this chemical has been swallowed, get medical attention. Give large quantities of water and induce vomiting. Do not make an unconscious person vomit.

Personal Protective Methods: Wear protective gloves and clothing to prevent any reasonable probability of skin contact. Safety equipment suppliers/manufacturers can provide recommendations on the most protective glove/clothing material for your operation. All protective clothing (suits, gloves, footwear, headgear) should be clean, available each day, and put on before work. Contact lenses should not be worn when working with this chemical. Wear splash-proof chemical goggles and face shield unless full face-piece respiratory protection is worn. Employees should wash immediately with soap when skin is wet or contaminated. Provide emergency showers and eyewash.

Respirator Selection: Where there is potential for exposures *over 1 mg/m³*, use a NIOSH/MSHA- or European Standard EN149-approved supplied-air respirator with a full face-piece operated in the positive-pressure mode, or with a full face-piece, hood, or helmet in the continuous-flow mode; or use a NIOSH/MSHA- or European Standard EN149-approved self-contained breathing apparatus with a full face-piece operated in pressure-demand or other positive-pressure mode.

Storage: Color Code—Blue: Health Hazard/Poison: Store in a secure poison location. Prior to working with this chemical you should be trained on its proper handling and storage. Store in tightly closed containers in a cool, well-ventilated area. Where possible, automatically transfer material from drums or other storage containers to process containers.

Shipping: Organophosphorus pesticides, liquid, toxic, require a shipping label of "POISONOUS/TOXIC MATERIALS." It falls in Hazard Class 6.1 and Packing Group III.

Spill Handling: Evacuate and restrict persons not wearing protective equipment from area of spill or leak until cleanup is complete. Remove all ignition sources. Ventilate area of spill or leak. Absorb liquids in vermiculite, dry sand, earth, peat, carbon, or a similar material and deposit in sealed containers. Keep this chemical out of a confined space, such as a sewer, because of the possibility of an explosion, unless the sewer is designed to prevent the buildup of explosive concentrations. It may be necessary to contain and dispose of this chemical as a hazardous waste. If material or contaminated runoff enters waterways, notify downstream users of potentially contaminated waters. Contact your local or federal environmental protection agency for specific recommendations. If employees are required to clean up spills, they must be properly trained and equipped. OSHA

1910.120(q) may be applicable. Soil Adsorption Index (K_{oc}) = 12,000 (Estimate).

Fire Extinguishing: Sulprofos may burn, but does not readily ignite. Use dry chemical, CO_2, water spray, or foam extinguishers. Poisonous gases, including oxides of sulfur and phosphorus, are produced in fire. Vapors are heavier than air and will collect in low areas. Vapors may travel long distances to ignition sources and flashback. Vapors in confined areas may explode when exposed to fire. Containers may explode in fire. Storage containers and parts of containers may rocket great distances, in many directions. If material or contaminated runoff enters waterways, notify downstream users of potentially contaminated waters. Notify local health and fire officials and pollution control agencies. From a secure, explosion-proof location, use water spray to cool exposed containers. If cooling streams are ineffective (venting sound increases in volume and pitch, tank discolors, or shows any signs of deforming), withdraw immediately to a secure position. If employees are expected to fight fires, they must be trained and equipped in OSHA 1910.156. The only respirators recommended for firefighting are self-contained breathing apparatuses that have full face-pieces and are operated in a pressure-demand or other positive-pressure mode.

Disposal Method Suggested: In accordance with 40CFR165, follow recommendations for the disposal of pesticides and pesticide containers. Must be disposed properly by following package label directions or by contacting your local or federal environmental control agency or by contacting your regional EPA office.

References

New Jersey Department of Health and Senior Services. (October 1998). *Hazardous Substances Fact Sheet: Sulprofos.* Trenton, NJ

2,4,5-T

T:0100

Molecular Formula: $C_8H_5Cl_3O_3$

Common Formula: $Cl_3C_6H_2OCH_2COOH$

Synonyms: Acetic acid, (2,4,5-T)-; Acetic acid, (2,4,5-trichlorophenoxy)-; Acide 2,4,5-trichlorophenoxyacetique (French); Acido 2,4,5-triclorofenoxiacetico (Spanish); Amine; BCF-Bushkiller; Brush-Off 445 low-volatile brush killer; Brush rhap; Brushtox; Dacamine; Debroussaillant concentre; Debroussaillant super concentre; Decamine 4T; Ded-weed brush killer; Ded-weed LV-6 brush kill; Dinoxol; Envert-T; Estercide T-2 and T-245; Esteron; Esteron 245; Esteron brush killer; Fence rider; Forron; Forst U 46; Fortex; Fruitone A; Inverton 245; Line rider; Phortox; Reddon; Reddox; Spontox; Super D weedone; 2,4,5-T; T-5 brush kil; Tippon; T-Nox; Tormona; Transamine; Tributon; 2,4,5-Trichlorophenoxyacetic acid; (2,4,5-Trichlor-phenoxy)-essigsaeure (German); Trinoxol; Trioxal; Trioxon; Trioxone; Veon; Veon 245; Verton 2T; Visko rhap low-volatile ester; Weedar; Weedone

CAS Registry Number: 93-76-5

RTECS® Number: AJ8400000

UN/NA & ERG Number: UN3345/153

EC Number: 202-273-3 [*Annex I Index No.:* 607-041-00-9]

Regulatory Authority and Advisory Bodies

Carcinogenicity: IARC: Human Limited Evidence, animal Inadequate Evidence, *possibly carcinogenic to humans*, Group 2B, 1987.

US EPA Gene-Tox Program, Positive: *D. melanogaster* sex-linked lethal; *S. cerevisiae*—reversion; Positive/dose response: *In vivo* cytogenetics—nonhuman bone marrow; Negative: *D. melanogaster*—whole sex chromosome loss; Negative: *D. melanogaster*—nondisjunction; Inconclusive: Host-mediated assay; Mammalian micronucleus.

Banned or Severely Restricted (many countries) (UN).[13]

Air Pollutant Standard Set. See below, "Permissible Exposure Limits in Air" section.

Clean Water Act: Section 311 Hazardous Substances/RQ 40CFR117.3 (same as CERCLA, see below).

US EPA Hazardous Waste Number (RCRA No.): U232.

RCRA, 40CFR261, Appendix 8 Hazardous Constituents.

RCRA 40CFR268.48; 61FR15654, Universal Treatment Standards: Wastewater (mg/L), 0.72; Nonwastewater (mg/kg), 7.9.

RCRA 40CFR264, Appendix 9; TSD Facilities Ground Water Monitoring List. Suggested test method(s) (PQL µg/L): 8150 (2).

Safe Drinking Water Act: Priority List (55FR1470).

Reportable Quantity (RQ): 1000 lb (454 kg).

Rotterdam Convention Annex III [Chemicals Subject to the Prior Informed Consent Procedure (PIC)] (as 2,4,5-T and its salts and esters).

European/International Regulations: Hazard Symbol: Xn, N; Risk phrases: R2; R36/37/38; R50/53; Safety phrases: S2; S24; S60; S61 (see Appendix 4).

WGK (German Aquatic Hazard Class): 3—Severe hazard to waters.

Description: 2,4,5-T acid is an odorless, colorless to tan crystalline solid. Molecular weight = 255.48; Specific gravity $(H_2O:1) = 1.80$ at 25°C; Boiling point = (decomposes); Freezing/Melting point = 152.8°C; Vapor pressure = 1×10^{-7} mmHg. Slightly soluble in water; solubility = 0.03% at 25°C.

Potential Exposure: Compound Description: Agricultural Chemical; Tumorigen, Mutagen; Reproductive Effector. Those engaged in the manufacture, formulation, and application of this herbicide used to control woody and herbaceous weeds. The EPA has issued a rebuttable presumption against registration of 2,4,5-T for pesticide uses, however. The Vietnam war era defoliant, Agent Orange, was a mixture of 2,4,5-T and 2,4-D.

Incompatibilities: The aqueous solution is a weak acid. Incompatible with sulfuric acid, bases, ammonia, aliphatic amines; alkanolamines, isocyanates, alkylene oxides; epichlorohydrin; strong oxidizers, such as chlorine, bromine, fluorine, and strong bases.

Permissible Exposure Limits in Air

OSHA PEL: 10 mg/m³ TWA

NIOSH REL: 10 mg/m³ TWA

ACGIH TLV®[1]: 10 mg/m³; Not classifiable as a human carcinogen

NIOSH IDLH: 250 mg/m³

Protective Action Criteria (PAC)

TEEL-0: 10 mg/m³

PAC-1: 10 mg/m³

PAC-2: 10 mg/m³

PAC-3: 250 mg/m³

DFG MAK: 10 mg/m³, inhalable fraction [skin]; Peak Limitation Category II(2); Pregnancy Risk Group C

Australia: TWA 10 mg/m³, 1993; Austria: MAK 10 mg/m³ [skin], 1999; Belgium: TWA 10 mg/m³, 1993; Denmark: TWA 5 mg/m³ [skin], 1999; France: VME 10 mg/m³, 1993; Hungary: TWA 1 mg/m³, STEL 2 mg/m³ [skin], 1993; the Netherlands: MAC-TGG 10 mg/m³, 2003; the Philippines: TWA 10 mg/m³, 1993; Switzerland: MAK-W 10 mg/m³, STEL 50 mg/m³ [skin], 1999; Thailand: TWA 10 mg/m³, 1993; United Kingdom: TWA 10 mg/m³, STEL 20 mg/m³, 2000; Argentina, Bulgaria, Columbia, Jordan, South Korea, New Zealand, Singapore, Vietnam: ACGIH TLV®: Not classifiable as a human carcinogen. Several states have set guidelines or standards for 2,4,5-T in ambient air[60] ranging from 1.0 µg/m³ (Pennsylvania) to 100.0 µg/m³ (North Dakota) to 160 µg/m³ (Virginia) to 238.0 µg/m³ (Nevada).

Determination in Air: Use NIOSH Analytical Method (IV) #5001, 2,4,5-T.

Permissible Concentration in Water: The EPA (see "References" below) has set a lifetime health advisory of 21.0 µg/L. Mexico[35] has set limits of 100 µg/L in estuaries and 10 µg/L in coastal waters. The state of Kansas[61] has set a guideline for drinking water of 700.0 µg/L.

Determination in Water: Liquid–liquid extraction and gas chromatography (see EPA reference below). Fish Tox = 79154.36150000 ppb (VERY LOW). Octanol–water coefficient: Log $K_{ow} = 3.99$.

Routes of Entry: Inhalation, ingestion; skin and/or eye contact. Absorbed through the skin.

Harmful Effects and Symptoms

Short Term Exposure: Irritates the eyes, skin, and respiratory tract. *Inhalation:* Increasingly severe symptoms may include nose and throat irritation; weakness, tiredness, metallic taste in mouth; loss of appetite; diarrhea, heart problems; heart failure and death. *Skin:* Reddening and itching may develop. Absorption is slow, but may contribute significantly to total exposure. *Eyes:* Irritation may develop. *Ingestion:* 350 mg (0.01 oz) produces only a metallic taste in the mouth, lasting about 2 h. Approximately 4 teaspoonfuls (150-lb man) may cause weakness, tiredness, loss of appetite; diarrhea, heart problems, heart failure; and death. *Note:* Reported effects of 2,4,5-T are due to accidental exposures, often at unknown levels or duration. In addition, 2,4,5-T may be contaminated with very small amounts of another more toxic compound 2,3,7,8-tetrachlorodibenzo-p-dioxin (TCDD, Dioxin). Therefore, some of the symptoms of exposure to 2,4,5,-T may be due to contaminants.

Long Term Exposure: Levels above the standard may produce skin irritation, acne-like skin sores; loss of skin coloration in small patches; GI tract ulcer; and nerve disorders resulting in difficulty controlling muscles. Animal studies also indicate the possibility of an increased susceptibility to infection. Changes in generic material and birth defects have been reported in laboratory studies and may be due to 2,4,5-T or its contaminant. Whether these effects are produced in humans is unknown. In animals: ataxia, skin irritation; acne-like rash; liver damage. Human Tox = 70.00000 ppb (LOW).

Points of Attack: Skin, liver, gastrointestinal tract.

Medical Surveillance: NIOSH lists the following tests: blood plasma; urine (chemical/metabolite); urine (chemical/metabolite), 24-h collection. If symptoms develop or overexposure is suspected, liver or kidney function tests may be useful.

First Aid: If this chemical gets into the eyes, remove any contact lenses at once and irrigate immediately for at least 15 min, occasionally lifting upper and lower lids. Seek medical attention immediately. If this chemical contacts the skin, remove contaminated clothing and wash immediately with soap and water. Seek medical attention immediately. If this chemical has been inhaled, remove from exposure, begin rescue breathing (using universal precautions, including resuscitation mask) if breathing has stopped and CPR if heart action has stopped. Transfer promptly to a medical facility. When this chemical has been swallowed, get medical attention. Give large quantities of water and induce vomiting. Do not make an unconscious person vomit.

Note to physician: If ingested, remove by lavage or emesis. Use general supportive measures for CNS depression. Use quinidine for myotonia.

Personal Protective Methods: Wear protective gloves and clothing to prevent any reasonable probability of skin contact. Safety equipment suppliers/manufacturers can provide recommendations on the most protective glove/clothing material for your operation. All protective clothing (suits, gloves, footwear, headgear) should be clean, available each day, and put on before work. Contact lenses should not be worn when working with this chemical. Wear dust-proof chemical goggles and face shield unless full-face-piece respiratory protection is worn. Employees should wash immediately with soap when skin is wet or contaminated. Provide emergency showers and eyewash.

Respirator Selection: 50 mg/m³: Qm (APF = 25) (any quarter-mask respirator). *100 mg/m³:* 95XQ (APF = 10) [any particulate respirator equipped with an N95, R95, or P95 filter (including N95, R95, and P95 filtering face-pieces) except quarter-mask respirators. The following filters may also be used: N99, R99, P99, N100, R100, P100]; or Sa (APF = 10) (any supplied-air respirator). *250 mg/m³:* Sa:Cf (APF = 25) (any supplied-air respirator operated in a continuous-flow mode); or 100F (APF = 50) (any air-purifying, full-face-piece respirator with an N100, R100, or P100 filter); or PAPRDM, if not present as a fume (any powered, air-purifying respirator with a dust and mist filter); or SCBAF (APF = 50) (any self-contained breathing apparatus with a full face-piece); or SaF (APF = 50) (any supplied-air respirator with a full face-piece). *Emergency or planned entry into unknown concentrations or IDLH conditions:* SCBAF: Pd,Pp (APF = 10,000) (any self-contained breathing apparatus that has a full face-piece and is operated in a pressure-demand or other positive-pressure mode); or SaF: Pd,Pp: ASCBA (APF = 10,000) (any supplied-air respirator that has a full face-piece and is operated in a pressure-demand or other positive-pressure mode in combination with an auxiliary self-contained breathing apparatus operated in a pressure-demand or other positive-pressure mode). *Escape:* 100F (APF = 50) (any air-purifying, full-face-piece respirator with an N100, R100, or P100 filter); or SCBAE (any appropriate escape-type, self-contained breathing apparatus).

Storage: Color Code—Blue: Health Hazard/Poison: Store in a secure poison location. Prior to working with this chemical you should be trained on its proper handling and storage. Store in tightly closed containers in a cool, well-ventilated area away from strong oxidizers, such as chlorine, bromine, fluorine, and strong bases. Where possible, automatically transfer material from drums or other storage containers to process containers. A regulated, marked area should be established where this chemical is handled, used, or stored in compliance with OSHA Standard 1910.1045.

Shipping: This material falls under "Phenoxy pesticides, solid, toxic, n.o.s." This compound requires a shipping label of "POISONOUS/TOXIC MATERIALS." It falls in Hazard Class 6.1 and Packing Group III.

Spill Handling: Evacuate persons not wearing protective equipment from area of spill or leak until cleanup is complete. Remove all ignition sources. Collect powdered material in the most convenient and safe manner and deposit in sealed containers. Ventilate area after cleanup is complete. It may be necessary to contain and dispose of this chemical as a hazardous waste. If material or contaminated runoff enters waterways, notify downstream users of potentially contaminated waters. Contact your local or federal environmental protection agency for specific recommendations. If employees are required to clean up spills, they must be properly trained and equipped. OSHA 1910.120(q) may be applicable. Soil Adsorption Index $(K_{oc}) = 80$.

Fire Extinguishing: 2,4,5-T Acid may burn, but does not readily ignite. Use dry chemical, carbon dioxide, water spray, or foam extinguishers. Poisonous gases are produced in fire, including phosgene, hydrogen chloride, and chlorine. If material or contaminated runoff enters waterways, notify downstream users of potentially contaminated waters. Notify local health and fire officials and pollution control agencies. From a secure, explosion-proof location, use water spray to cool exposed containers. If cooling streams are ineffective (venting sound increases in volume and pitch, tank discolors, or shows any signs of deforming), withdraw immediately to a secure position. If employees are expected to fight fires, they must be trained and equipped in OSHA 1910.156. The only respirators recommended for firefighting are self-contained breathing apparatuses that have full face-pieces and are operated in a pressure-demand or other positive-pressure mode.

Disposal Method Suggested: Two disposal procedures have been discussed for 2,4,5-T: (1) Mix with excess sodium carbonate, add water and let stand for 24 h before flushing down the drain with excess water; and (2) pour onto vermiculite and incinerate with wood, paper, and waste alcohol.[22] In accordance with 40CFR165, follow recommendations for the disposal of pesticides and pesticide containers. Must be disposed properly by following package label directions or by contacting your local or federal environmental control agency, or by contacting your regional EPA office. Consult with environmental regulatory agencies for guidance on acceptable disposal practices. Generators of waste containing this contaminant (≥ 100 kg/mo) must conform with EPA regulations governing storage, transportation, treatment, and waste disposal.

References

US Environmental Protection Agency. (August 1987). *Health Advisory: 2,4,5-Trichlorphenoxy-Acetic Acid.* Washington, DC: Office of Drinking Water

Sax, N. I. (Ed.). (1983). *Dangerous Properties of Industrial Materials Report*, 3, No. 5, 20–21

New York State Department of Health. (March 1980). *Chemical Fact Sheet: 2,4,5-T.* Albany, NY: Bureau of Toxic Substance Assessment

New Jersey Department of Health and Senior Services. (August 2001). *Hazardous Substances Fact Sheet: 2,4,5-(Trichlorophenoxy) Acetic Acid.* Trenton, NJ

Tabun (Agent GA, WMD)   T:0110

Molecular Formula: $C_5H_{11}N_2O_2P$

Common Formula: $(CH_3)_2NPO(OC_2H_5)CN$

Synonyms: Dimethylamidoethoxyphosphotyl cyanide; Dimethylaminocyanphosphorsaeureaethylester (German); Dimethylphosphoramidocyanidic acid, ethyl ester; Ethyl dimethylamidocyanophosphate; Ethyl *N,N*-dimethyl-aminocyanophosphate; Ethyl *N,N*-dimethyl-phosphoramidocyanidate; Ethyl dimethyl-phosphoramidocyanidate; GA (military designation); Gelan 1; LE-100; MCE; Phosphoramidocyanidic acid, dimethyl-, ethyl ester; T-2104; Taboon A; TL 1578

CAS Registry Number: 77-81-6

RTECS® Number: TB4550000

UN/NA & ERG Number: UN2810/153

EC Number: None assigned.

Regulatory Authority and Advisory Bodies

Department of Homeland Security Screening Threshold Quantity: *Theft hazard* CUM 100 g.

Carcinogenicty: GA is not listed by the International Agency for Research on Cancer (IARC), American Conference of Governmental Industrial Hygienists (ACGIH), Occupational Safety and Health Administration (OSHA), or National Toxicology Program (NTP) as a carcinogen.

Superfund/EPCRA 40CFR355, Extremely Hazardous Substances: TPQ = 10 lb (4.54 kg).

Reportable Quantity (RQ): 10 lb (4.54 kg).

US DOT 49CFR172.101, Inhalation Hazard Chemical.

Note: Army Regulation, AR 50-6, deals specifically with the shipment of chemical agents; must be escorted in accordance with Army Regulation, AR 740-32.

US DOT Regulated Marine Pollutant (49CFR172.101, Appendix B) as cyanide mixtures, cyanide solutions.

Canada, National Pollutant Release Inventory (NPRI); CEPA Priority Substance List, Ocean dumping prohibited.

European/International Regulations: Not listed in Annex 1.

WGK (German Aquatic Hazard Class): No value assigned.

Description: Tabun (GA), an organophosphorous compound, is a nerve agent, and among the most toxic of the known chemical warfare agents. Exposure to tabun can cause death in minutes. A fraction of an ounce (1–10 mL) of tabun on the skin can be fatal. GA is a clear, colorless to brownish, oily liquid, with a slight fruity odor, like almonds. No odor when pure. *Warning:* Odor is not a reliable indicator of the presence of toxic amounts of tabun. It is tasteless. Molecular weight = 162.12; Specific gravity (H_2O:1) = 1.07 at 25°C; Boiling point = 230°C (decomposition); Freezing/Melting point $= -50°C$; Vapor density (air $= 1$) $= 5.6$; Vapor pressure $= 0.057$ mmHg at $25°C/0.07$ at $24°C$; Volatility $= 490$ mg/m^3 at $25°C$; Flash point $= 78°C$. Hazard Identification (based on NFPA-704 M Rating System): Health 4, Flammability 2, Reactivity 0. Soluble in water; readily hydrolyzed forming hydrogen cyanide; solubility $= 9.8\%$ at $25°C/7.2$ at $20°C$. It is chemically similar to malathion or parathion, and other organophosphates.

Potential Exposure: GA is a highly persistent (may remain as liquid for more than 24 h) chemical warfare agent; military nerve gas. Nerve agents are more toxic and potent than insecticides. *Note:* If used as a weapon, notify US Department of Defense: Army. Damage and/or death may occur before chemical detection can take place. Use M8 paper if available (Detection: yellow) or M256-A1 Detector Kit (Detection limit: 0.005 mg/m^3).

Persistence of Chemical Agent: Tabun (GA): Summer: 10 min to 24 h; Winter: 2 h to 3 days.

Incompatibilities: Tabun (GA) decomposes slowly in water. Under acid conditions, GA hydrolyzes to form hydrofluoric acid (HF). Raising the pH increases the rate of decomposition significantly. Rapidly hydrolyzed in basic solutions (Na$_2$CO$_3$, NaOH, or KOH) with a half-life of 1.5 min at pH 11 at $25°C$. GA and its hydrolysis products exhibit no significant phototransformations in sunlight. Tabun and its hydrolysis products are thermally stable at temperatures less than $49°C$. Contact with metals may evolve flammable hydrogen gas. Reacts with oxidizing materials. Tabun is destroyed by bleaching powder, but the reaction produces cyanogen chloride (CNCl). Decomposes within 6 months at $60°C$. Complete decomposition in 3.5 h at $150°C$. May produce hydrogen cyanide, oxides of nitrogen; oxides of phosphorus; carbon monoxide; and hydrogen cyanide.

Permissible Exposure Limits in Air
Protective Action Criteria (PAC) GA*
TEEL-0: 0.000125 ppm
PAC-1: 0.00042 ppm
PAC-2: **0.0053** ppm
PAC-3: **0.039** ppm
*AEGLs (Acute Emergency Guideline Levels) & ERPGs (Emergency Response Planning Guideline) are in **bold face**.
STEL: **0.0001** mg/m^3
The suggested permissible airborne exposure concentration for GA for an 8-h workday or a 40-h workweek is an 8 h time weight average (TWA) of 0.0001 mg/m^3 (2×10^{-5} ppm). This value is based on the TWA of GA as proposed in the USAEHA Technical Guide 169, *Occupational Health Guidelines for the Evaluation and Control of Occupational Exposure to Nerve Agents, GA, GB, GD, and VX.*
WPL (Worker population limit): 0.00003 mg/m^3.
GPL (General population limit): 0.000001 mg/m^3.

Determination in Air: Military chemical agent detection papers and kits (M18A2 for liquid; M256A1, M8A1, M18A2, ICAD, CAM for vapor). M8, M9 paper will quickly detect the presence of a nerve agent but will not identify the type of agent being used. Also, the following may be helpful: "Accuro®" or "Accuro® 2000" (Draeger®) Air Sampling/Gas (Vapor) System; or use NIOSH Analytical Method (IV) #5600, Organophosphorus pesticides; NIOSH Analytical Method #7904, Cyanides.

Determination in Water: Log $K_{ow} =$ (estimated) 0.29; also listed at 1.4. Tabun dissolves in water and remains very dangerous. To prevent anyone from drinking water mixed with tabun, notify local health and pollution control officials. Also, notify operators of nearby water intakes and advise shutting water intakes. Use M272 Chemical Agent Water Testing Kit. Detection limit for nerve agents is 0.02 mg/L. Also, for cyanides, distillation followed by silver nitrate titration or colorimetric analysis using pyridine pyrazolone (or barbituric acid). If used as a weapon, utilize an M272 Water Detection Kit (Detection limit: 0.02 mg/L). Dangerous to aquatic life in high concentrations. May be dangerous if it enters water intakes. This material will be broken down in water quickly, but small amounts may evaporate. This material will be broken down in moist soil quickly. Small amounts may evaporate into the air or travel below the soil surface and contaminate groundwater. Persists 1½ to 2 days in soil. Bleaching powder (chlorinated line) destroys tabun but gives rise to cyanogen chloride (CAS: 506-77-4). See table of contents or name index for location of entry for cyanogen chloride.

Routes of Entry: Skin absorption, absorption through eyes; and inhalation.

Harmful Effects and Symptoms
Short Term Exposure: Tabun is a nerve agent; it acts as a cholinesterase inhibitor. The median lethal dosage (vapor/respiratory) LCt$_{50} = 400$ mg-min/m^3 for humans; the median incapacitating dosage is 300 mg-min/m^3. Respiratory lethal dosages kill in 1–10 min; liquid in the eye kills nearly as rapidly. The LD$_{50}$ (skin) $= 1.0$g/70 kg/156.8 lb [human] (*Medical Aspects of Chemical and Biological Warfare, Part I*, Walter Reed Medical Center, 1997). Skin absorption great enough to cause death may occur in 1–2 min, but may be delayed for 1–2 h. Nerve agent symptoms include difficulty in breathing; drooling, nausea, vomiting, cramps, involuntary defecation and urination; twitching, jerking, staggering, headache, confusion, drowsiness, coma, and convulsions. Inhalation causes dimness of vision and pinpointing of the pupils. GA is an anticholinesterase agent similar in action to GB (sarin). Although only about half as toxic as GB (sarin) by inhalation, GA in low concentrations, is more irritating to the eyes than GB (sarin). The number and severity of symptoms which appear are dependent on the quantity, and rate of entry of the nerve agent which is introduced into the body (very small skin dosages sometimes cause local sweating and tremors with few other effects). Individuals poisoned by GA display approximately the same sequence of symptoms regardless of the route by which the poison enters the body

(whether by inhalation, absorption, or ingestion). These symptoms, in normal order of appearance, runny nose; tightness of chest; dimness of vision and pinpointing of the eye pupils; difficulty in breathing; drooling and excessive sweating; nausea, vomiting, cramps; and involuntary defecation and urination; twitching, jerking, staggering, headaches, confusion, drowsiness, coma, and convulsions. These symptoms are followed by cessation of breathing and death. Onset Time of Symptoms: Symptoms appear much more slowly from skin dosage than from respiratory dosage. Although skin absorption great enough to cause death may occur in 1–2 min, death may be delayed for 1–2 h. Respiratory lethal dosages kill in 1–10 min, and liquid in the eye kills almost as rapidly. Median Lethal Dosage, animals: LD_{50} (monkey, percutaneous) = 9.3 mg/kg (shaved skin); LCt_{50} (monkey, inhalation) = 187 mg-min/m^3 ($t = 10$); Median Lethal Dosage, Man: LCt_{50} (man, inhalation) = 135 mg-min/m^3 ($t = 0.5$–2 min) at RMV (Respiratory Minute Volume) of 15 l/min; 200 mg-min/m^3 at RMV of 10 l/min.

GA is not listed by the International Agency for Research on Cancer (IARC); American Conference of Governmental Industrial Hygienists (ACGIH); Occupational Safety and Health Administration (OSHA); or National Toxicology Program (NTP) as a carcinogen.

Long Term Exposure: Cholinesterase inhibitor; cumulative effect is possible. This chemical may damage the nervous system with repeated exposure, resulting in convulsions, respiratory failure.

Points of Attack: Respiratory system, lungs, central nervous system; cardiovascular system, skin, eyes, plasma and red blood cell cholinesterase. Liver, kidneys.

Medical Surveillance: Consider the points of attack in preplacement and periodic physical examinations. Urine thiocyanate levels. Complete blood count (CBC). Evaluation of thyroid function. Liver function tests. Kidney function tests. Central nervous system tests. EKG.

First Aid: *Inhalation:* Hold breath until respiratory protective mask is donned. If severe signs of agent exposure appear (chest tightens, pupil construction, a lack of coordination, etc.), immediately administer, in rapid succession, all three Nerve Agent Antidote Kit(s), Mark I injectors (or atropine if directed by the local physician). Injections using the Mark I kit injectors may be repeated at 5- to 20-min intervals if signs and symptoms are progressing until three series of injections have been administered. No more injections will be given unless directed by medical personnel. In addition, a record will be maintained of all injections given. If breathing has stopped, give artificial respiration. Mouth-to-mouth resuscitation should be used when approved mask-bag or oxygen delivery systems are not available. Do not use mouth-to-mouth resuscitation when facial contamination exists. If breathing is difficult, administer oxygen. Seek medical attention *immediately*.

Eye contact: immediately flush eyes with water for 10–15 min then don respiratory protective mask. Although miosis (pinpointing of the pupils) may be an early sign of agent exposure, an injection will not be administered when miosis is the only sign present. Instead, the individual will be taken immediately to the medical treatment facility for observation.

Skin contact: Don respiratory protection mask and remove contaminated clothing. Immediately wash contaminated skin with copious amounts of soap and water, 10% sodium carbonate solution, or 5% liquid household bleach. Rinse well with water to remove decontaminate. Use M258A1 and/or M291 kit for skin decontamination. Speed in removing material from skin is of extreme importance. Administer an intramuscular injection with the MARK I kit injectors only if local sweating and muscular twitching symptoms are observed. Seek medical attention *immediately*.

Ingestion: Do not induce vomiting. First symptoms are likely to be gastrointestinal. *Immediately* administer 2 mg intramuscular injection of the MARK I kit auto-injectors. Seek medical attention *immediately*.

Medical treatment: Electrocardiogram (ECG), and adequacy of respiration and ventilation, should be monitored. Supplemental oxygenation, frequent suctioning of secretions, insertion of a tube into the trachea (endotracheal intubation), and assisted ventilation may be required. Diazepam (5–10 mg in adults and 0.2–0.5 mg/kg in children) may be used to control convulsions. Lorazepam or other benzodiazepines may be used, but barbiturates, phenytoin, and other anticonvulsants are not effective. Administration of atropine (if not already given) should precede the administration of benzodiazepines in order to best control seizures. Patients/victims who have inhalation exposure and who complain of chest pain, chest tightness, or cough should be observed and examined periodically for 6–12 h to detect delayed-onset inflammation of the large airways (bronchitis), inflammatory lung disease (pneumonia), accumulation of fluid in the lungs (pulmonary edema), or respiratory failure.

Decontamination: This is very important. The rapid physical removal of a chemical agent is essential. If you do not have the equipment and training, do not enter the hot or the warm zone to rescue and/or decontaminate victims. Medical personnel should wear the proper PPE. If the victim cannot move, decontaminate without touching and without entering the hot or the warm zone. Nerve gases stay in clothing; *do not* touch with bare skin—if possible, seal contaminated clothes and personal belongings in a sealed double bag. Use clean water from any source; if possible, use a hose (spray or fog to prevent injury to the victim) or other system to avoid touching the victim. Do not wait for soap or for the victim to remove clothing, begin washing immediately. Do not delay decontamination to obtain warm water; time is of the essence; use cold water instead. Immediately flush the eyes with water for at least 15 min. Use caution to avoid hypothermia in children and the elderly. Wash–strip–wash–evacuate upwind and uphill: Patients exposed to nerve agent by vapor only should be decontaminated by removing all clothing in a clean-air environment and shampooing or rinsing the hair to prevent vapor-off gassing; Patients exposed to liquid nerve agent should be decontaminated by washing in available clean water at least three times. Use liquid soap (dispose of container after use and replace), large amounts of water, and mild-to-moderate friction with a single-use sponge or washcloth in the first and second washes. Scrubbing of exposed skin with a brush is discouraged; skin damage may occur and may increase absorption. The third wash should be to rinse with large amounts of warm or hot water. Shampoo can be used to wash the hair. Decontaminate with diluted household bleach* (0.5%, or one part bleach to 200 parts water), but do not let any get in the victim's eyes, open wounds, or mouth. Wash off the diluted bleach solution after 15 min. Remember that the water you use to decontaminate the victims is dangerous. Be sure you've decontaminated the victims as much as you can before they are released from the area, so they do not spread the nerve gas. Rinse the eyes, mucous membranes; or open wounds with sterile saline or water and then move away from the hot zone in an upwind and uphill direction.

*The following can be used in addition to household bleach: (1) solids, powders and solutions containing various types of bleach (NaOCl or Ca(OCl)$_2$); (2) DS2 (2% NaOH, 70% diethylenetriamine, 28% ethylene glycol monomethyl ether); (3) towelettes moistened with sodium hydroxide (NaOH) dissolved in water, phenol, ethanol, and ammonia.

Note: Use 5% solution of common bleach (sodium hypochlorite) or calcium hypochlorite solution (48 oz per 5 gallons of water) to decontaminate scissors used in clothing removal, clothes, and other items.

Personal Protective Methods: Wear protective gloves and clothing to prevent any reasonable probability of skin contact. Safety equipment suppliers/manufacturers can provide recommendations on the most protective glove/clothing material for your operation. Butyl rubber gloves and Tyvek® "F" decontamination suits provide barrier protection against chemical warfare agents. Although resistant to liquid chemical agents, impermeable protective clothing may be penetrated after a few hours of exposure to heavy concentration of agent. Consequently, liquid contamination on the clothing must be neutralized or removed as soon as possible. All protective clothing (suits, gloves, footwear, headgear) should be clean, available each day, and put on before work. Contact lenses should not be worn when working with this chemical. Wear splash-proof chemical goggles and face shield unless full-face-piece respiratory protection is worn. Employees should wash immediately with soap when skin is wet or contaminated. Provide emergency showers and eyewash.

Respirator Selection: *When used as a weapon, use SCBA Respirator Certified By NIOSH For CBRN Environments.* Follow regulations in OSHA 29CFR1910.134 or European Standard EN149. Use a NIOSH/MSHA- or European Standard EN149-approved respirator; or use an approved supplied-air respirator with a full face-piece operated in the positive-pressure mode, or with a full face-piece, hood, or helmet in the continuous-flow mode; or use a NIOSH/MSHA- or European Standard EN149-approved self-contained breathing apparatus with a full face-piece operated in pressure-demand or other positive-pressure mode. The M40 Series mask (which replaced the M17A1 protective mask) provides respiratory protection against all known military toxic chemical agents, but it cannot be used in an oxygen-deficient environment and is *not approved for civilian use*. It does not afford protection against industrial toxics, such as ammonia and carbon monoxide.

Storage: Color Code—Blue: Health Hazard/Poison: Store in a secure poison location. Prior to working with this chemical you should be trained on its proper handling and storage. Store in tightly closed containers in a cool, well-ventilated area away from strong oxidizers. Where possible, automatically transfer material from drums or other storage containers to process containers. Sources of ignition, such as smoking and open flames, are prohibited where this chemical is handled, used, or stored. Metal containers involving the transfer of this chemical should be grounded and bonded. Wherever this chemical is used, handled, manufactured, or stored, use explosion-proof electrical equipment and fittings.

Shipping: Toxic, liquids, organic, n.o.s. [Inhalation hazard, Packing Group I, Zone A] require a shipping label of "POISONOUS/TOXIC MATERIALS." It falls in Hazard Class 6.1 and Packing Group I.

Spill Handling: Evacuate and restrict persons not wearing protective equipment from area of spill or leak until cleanup is complete. Remove all ignition sources. Ventilate area of spill or leak. Spills must be contained by covering with vermiculite, diatomaceous earth; clay, fine sand; sponges, and paper or cloth towels. This containment is followed by treatment with copious amounts of aqueous sodium hydroxide solution (a minimum 10% wt.). Scoop up all material and place in a fully removable head drum with a high density polyethylene liner. The decontamination solution must be treated with excess bleach to destroy the CN formed during hydrolysis. Cover the contents with additional bleach before affixing the drum head. After sealing the head, the exterior of the drum shall be decontaminated and then labeled in accordance with IAW, EPA, and DOT regulations. All leaking containers shall be over-packed with vermiculite placed between the interior and exterior containers. Decontaminate and label per IAW, EPA, and DOT regulations. Dispose of the material per IAW waste disposal methods provided below. Conduct general area monitoring with an approved monitor to confirm that the atmospheric concentrations do not exceed the airborne exposure limit. If 10% wt. sodium hydroxide is not available then the following decontaminants may be used instead and are listed in order of preference: Decontaminating Solution No. 2 [DS2: (2% NaOH, 70% diethylenetriamine, 28% ethylene glycol monomethyl ether)], sodium carbonate and Supertropical

Bleach Slurry (STB). Keep this chemical out of a confined space, such as a sewer, because of the possibility of an explosion, unless the sewer is designed to prevent the buildup of explosive concentrations. It may be necessary to contain and dispose of this chemical as a hazardous waste. If material or contaminated runoff enters waterways, notify downstream users of potentially contaminated waters. Contact your local or federal environmental protection agency for specific recommendations. If employees are required to clean up spills, they must be properly trained and equipped. OSHA 1910.120(q) may be applicable.

Initial isolation and protective action distances

Distances shown are likely to be affected during the first 30 min after materials are spilled and could increase with time. If more than one tank car, cargo tank, portable tank, or large cylinder is involved in the incident is leaking, the protective action distance may need to be increased. You may need to seek emergency information from CHEMTREC at (800) 424-9300 or seek professional environmental engineering assistance from the US EPA Environmental Response Team at (908) 548-8730 (24-h response line).

GA, when used as a weapon

Small spills (From a small package or a small leak from a large package).

First: Isolate in all directions (feet/meters)   100/30.
Then: Protect persons downwind (miles/kilometers).
Day   0.1/0.2
Night   0.1/0.2

Large spills (From a large package or from many small packages).

First: Isolate in all directions (feet/meters)   300/100.
Then: Protect persons downwind (miles/kilometers).
Day   0.4/0.6
Night   0.4/0.6

Fire: If tank, rail car, or tank truck is involved in fire, isolate for at least 800 m (½ mile) in all directions; also, consider initial evacuation for 800 m (½ mile) in all directions.

Fire Extinguishing: Tabun is a combustible liquid. When heated, vapors may form explosive mixtures with air, presenting an explosion hazard indoors, outdoors, and in sewers. Containers may explode when heated. Water, fog, foam, carbon dioxide—avoid using extinguishing methods that will cause splashing or spreading of the GA. Poisonous gases, including hydrogen cyanide, oxides of nitrogen, oxides of phosphorus; carbon monoxide; and hydrogen cyanide may be produced in fire. Respiratory protection is required. Positive pressure, full face-piece, NIOSH-approved self-contained breathing apparatus (SCBA) will be worn where there is danger of oxygen deficiency and when directed by the fire chief or chemical accident/incident (CAI) operations officer. The M9 or M17 series mask may be worn in lieu of SCBA when there is no danger of oxygen deficiency. In cases where firefighters are responding to a chemical accident/incident for rescue/reconnaissance

purposes they will wear appropriate levels of protective clothing. Complete protection required; have decontaminants available (bleach, alkali) and atropine. Bleaching powder (chlorinated line) destroys tabun but gives rise to cyanogen chloride. Vapors are heavier than air and will collect in low areas. Vapors may travel long distances to ignition sources and flashback. Vapors in confined areas may explode when exposed to fire. Containers may explode in fire. Storage containers and parts of containers may rocket great distances, in many directions. If material or contaminated runoff enters waterways, notify downstream users of potentially contaminated waters. Notify local health and fire officials and pollution control agencies. From a secure, explosion-proof location, use water spray to cool exposed containers. If cooling streams are ineffective (venting sound increases in volume and pitch, tank discolors, or shows any signs of deforming), withdraw immediately to a secure position. If employees are expected to fight fires, they must be trained and equipped in OSHA 1910.156.

Disposal Method Suggested: A minimum of 56 g of decontamination solution is required for each gram of GA. The decontamination solution is agitated while GA is added and the agitation is maintained for at least 1 h. The resulting solution is allowed to react for 24 h. At the end of 24 h, the solution must be titrated to a pH between 10 and 12. After completion of the 24-h period, the decontamination solution must be treated with excess bleach (2.5 mole OCl-/mole GA) to destroy the CN formed during hydrolysis. Scoop up all material and place in a fully removable head drum with a high density polyethylene liner. Cover the contents with additional bleach before affixing the drum head. All contaminated clothing will be placed in a fully removable head drum with a high density polyethylene liner. Cover the contents of the drum with decontaminating solution as above before affixing the drum head. After sealing the head, the exterior of the drum shall be decontaminated and then labeled per IAW, state, EPA, and DOT regulations. All leaking containers shall be overpacked with vermiculite placed between the interior and exterior containers. Decontaminate and label in accordance with IAW, state, EPA, and DOT regulations. Conduct general area monitoring with an approved monitor to confirm that the atmospheric concentrations do not exceed the airborne exposure limit.

References

Sax, N. I. (Ed.). (1980). *Dangerous Properties of Industrial Materials Report*, 1, No. 2, 63

US Environmental Protection Agency. (November 30, 1987). *Chemical Hazard Information Profile: Tabun*. Washington, DC: Chemical Emergency Preparedness Program

The Riegle Report: A Report of Chairman Donald W. Riegle, Jr. and Ranking Member Alfonse M. D'Amato of the Committee on Banking, Housing and Urban Affairs with Respect to Export Administration, United States Senate, 103rd Congress, 2d Session, May 25, 1994

New Jersey Department of Health and Senior Services. (April 2004). *Hazardous Substances Fact Sheet: Tabun.* Trenton, NJ

Schneider, A. L., (Ed.) (2007). *CHRIS + CD-ROM Version 2.0 (United States Coast Guard Chemical Hazard Response Information System (COMDTINST 16465.12C).* Washington, DC: United States Coast Guard and the Department of Homeland Security

Talc (no asbestos and less than 1% quartz) T:0120

Molecular Formula: $H_2Mg_3O_{12}Si$
Common Formula: $Mg_3SiO_{10}(OH)_2$
Synonyms: Agalite; Alpine talc; Asbestine; C.I. 77718; Desertalc 57; Emtal 596; Fibrene C 400; French chalk; Hydrous magnesium silicate; Lo micron talc 1; Metro talc; Mistron; Mistron star; Mistron super frost; Mistron vapor; MP-12-50; MP 25-38; NCI-C06008; Nonasbestiform talc; Nonfibrous talc; Nytal; OOS; OXO; Puretalc USP; Seawhite; Sierra C-400; Snowgoose; Steatite talc; Supreme dense; Talc (nonasbestos form); Talcum
CAS Registry Number: 14807-96-6; *(alt.)* 11119-41-8; *(alt.)* 12420-12-1; *(alt.)* 37232-12-5; *(alt.)* 99638-63-8; *(alt.)* 110540-41-5
RTECS® Number: VV8790000
EC Number: 238-877-9
Regulatory Authority and Advisory Bodies
Carcinogenicty: IARC: Sufficient *(for talc containing asbestiform fibers)* Group 1; Human Inadequate Evidence *(talc not containing asbestiform fibers)* group 3; NTP *(for talc containing asbestiform fibers)*: Known to be a human carcinogen; NCI: Carcinogenesis Studies (inhalation); clear evidence: rat; no evidence: mouse.
FDA—over-the-counter drug.
Air Pollutant Standard Set. See below, "Permissible Exposure Limits in Air" section.
California Proposition 65 Chemical: Cancer *(talc containing asbestiform fibers)* 4/1/90
European/International Regulations: Not listed in Annex 1.
WGK (German Aquatic Hazard Class): Nonwater polluting agent.
Description: Talc is an odorless solid which exists in both a nonasbestos form and a fibrous form. This entry will be concerned with the nonfibrous form. Molecular weight = 96.33; Freezing/Melting point = 900°C. Hazard Identification (based on NFPA-704 M Rating System): Health 1, Flammability 0, Reactivity 0. Insoluble in water.
Potential Exposure: Compound Description: Tumorigen, Natural Product; Primary Irritant. Talc is used in the ceramics, paint, roofing, insecticide, paper, cosmetics, lubricant, pharmaceutical, and rubber industries; electrical insulation.
Incompatibilities: None reported.

Permissible Exposure Limits in Air
For talc containing asbestos fibers: See Asbestos.
The following is for talc containing NO asbestos fibers.
OSHA PEL *(containing less than 1% quartz)*: 20 mppcf TWA.
NIOSH REL: $2 mg/m^3$ TWA, respirable dust.
ACGIH TLV®[1]: $2 mg/m^3$ TWA (respirable fraction, for particulate matter containing no asbestos and <1% crystalline silica); not classifiable as a human carcinogen.
NIOSH IDLH: $1000 mg/m^3$.
Protective Action Criteria (PAC)
TEEL-0: $2 mg/m^3$
PAC-1: $6 mg/m^3$
PAC-2: $75 mg/m^3$
PAC-3: $500 mg/m^3$
Australia: TWA $2.5 mg/m^3$, 1993; Austria: MAK $5 mg/m^3$, 1999; Belgium: TWA $2 mg/m^3$, 1993; Finland: TWA $5 mg/m^3$, 1999; Switzerland: MAK-W $2 mg/m^3$ (respirable dust), 1999; United Kingdom: TWA $1 mg/m^3$, respirable dust, 2000; the Netherlands: MAC-TGG $1 mg/m^3$, 2003; Argentina, Bulgaria, Columbia, Jordan, South Korea, New Zealand, Singapore, Vietnam: ACGIH TLV®: not classifiable as a human carcinogen.
Determination in Air: Use NIOSH (III) P&CAM, Method #355.
Routes of Entry: Inhalation of dust, skin, and/or eye contact.
Harmful Effects and Symptoms
Short Term Exposure: Talc can affect you when breathed in. Can cause eye and lung irritation.
Long Term Exposure: May affect the lungs, causing talc fibrotic pneumoconiosis. Repeated high exposure can cause scarring of the lungs. Symptoms of shortness of breath and cough can develop. This disease can be disabling and fatal. Talc can cause the chest X-ray to become abnormal. Contact can cause eye irritation, and may lead to a reaction causing serious eye damage.
Points of Attack: Eyes, respiratory system; cardiovascular system.
Medical Surveillance: For those with frequent or potentially high exposure (half the TLV or greater), the following are recommended before beginning work and at regular times after that: lung function tests. Chest X-ray every 1−3 years, after five or more years of heavy exposure should be considered.
First Aid: If this chemical gets into the eyes, remove any contact lenses at once and irrigate immediately for at least 20 min, occasionally lifting upper and lower lids. Seek medical attention immediately. If this chemical contacts the skin, remove contaminated clothing and wash immediately with soap and water. Seek medical attention immediately. If this chemical has been inhaled, remove from exposure, begin rescue breathing (using universal precautions, including resuscitation mask) if breathing has stopped and CPR if heart action has stopped. Transfer promptly to a medical facility. When this chemical has been swallowed, get

medical attention. Give large quantities of water and induce vomiting. Do not make an unconscious person vomit.

Personal Protective Methods: Wear protective gloves and clothing to prevent any reasonable probability of skin contact. Safety equipment suppliers/manufacturers can provide recommendations on the most protective glove/clothing material for your operation. All protective clothing (suits, gloves, footwear, headgear) should be clean, available each day, and put on before work. Contact lenses should not be worn when working with this chemical. Wear dust-proof chemical goggles and face shield unless full-face-piece respiratory protection is worn. Employees should wash immediately with soap when skin is wet or contaminated. Provide emergency showers and eyewash.

Respirator Selection: NIOSH: *Up to 10 mg/m³:* Qm (APF = 25) (any quarter-mask respirator). *Up to 20 mg/m³:* 95XQ (APF = 10) [any particulate respirator equipped with an N95, R95, or P95 filter (including N95, R95, and P95 filtering face-pieces) except quarter-mask respirators. The following filters may also be used: N99, R99, P99, N100, R100, P100]; or Sa (APF = 10) (any supplied-air respirator). *Up to 50 mg/m³:* PaprHie (APF = 25) (any powered, air-purifying respirator with a high-efficiency particulate filter); or Sa:Cf (APF = 25) (any supplied-air respirator operated in a continuous-flow mode). *Up to 100 mg/m³:* 100F (APF = 50) (any air-purifying, full-face-piece respirator with an N100, R100, or P100 filter); or SaT:Cf (APF = 50) (any supplied-air respirator that has a tight-fitting face-piece and is operated in a continuous-flow mode); or PaprTHie (APF = 50) (any powered, air-purifying respirator with a tight-fitting face-piece and a high-efficiency particulate filter); or SCBAF (APF = 50) (any self-contained breathing apparatus with a full face-piece); SaF (any supplied-air respirator with a full face-piece). *Up to 1000 mg/m³:* Sa: Pd,Pp (APF = 1000) (any supplied-air respirator operated in a pressure-demand or other positive-pressure mode. *Emergency or planned entry into unknown concentrations or IDLH conditions:* SCBAF: Pd,Pp (APF = 10,000) (any self-contained breathing apparatus that has a full face-piece and is operated in a pressure-demand or other positive-pressure mode); or SaF: Pd,Pp: ASCBA (any supplied-air respirator that has a full face-piece and is operated in a pressure-demand or other positive-pressure mode in combination with an auxiliary self-contained breathing apparatus operated in a pressure-demand or other positive-pressure mode). *Escape:* 100F (APF = 50) (any air-purifying, full-face-piece respirator with an N100, R100, or P100 filter); or SCBAE (any appropriate escape-type, self-contained breathing apparatus).

Storage: Color Code—Blue (*only for talc containing asbestiform fibers*): Health Hazard/Poison: Store in a secure poison location. Color Code—Green (*talc powder or tablets*): General storage may be used. Prior to working with this chemical you should be trained on its proper handling and storage. Store in tightly closed containers in a cool, well-ventilated area.

Spill Handling: Evacuate persons not wearing protective equipment from area of spill or leak until cleanup is complete. Remove all ignition sources. Collect powdered material in the most convenient and safe manner and deposit in sealed containers. Ventilate area after cleanup is complete. It may be necessary to contain and dispose of this chemical as a hazardous waste. If material or contaminated runoff enters waterways, notify downstream users of potentially contaminated waters. Contact your local or federal environmental protection agency for specific recommendations. If employees are required to clean up spills, they must be properly trained and equipped. OSHA 1910.120(q) may be applicable.

Fire Extinguishing: Extinguish fire using an agent suitable for type of surrounding fire. Talc itself does not burn. Poisonous gases are produced in fire. If material or contaminated runoff enters waterways, notify downstream users of potentially contaminated waters. Notify local health and fire officials and pollution control agencies. From a secure, explosion-proof location, use water spray to cool exposed containers. If cooling streams are ineffective (venting sound increases in volume and pitch, tank discolors, or shows any signs of deforming), withdraw immediately to a secure position. If employees are expected to fight fires, they must be trained and equipped in OSHA 1910.156. The only respirators recommended for firefighting are self-contained breathing apparatuses that have full face-pieces and are operated in a pressure-demand or other positive-pressure mode.

Disposal Method Suggested: Landfill.

References

National Institute for Occupational Safety and Health. (October 1977). *Information Profiles on Potential Occupational Hazards: Talc,* Report PB-276, 678. Rockville, MD, pp. 51–53

New Jersey Department of Health and Senior Services. (December 2000). *Hazardous Substances Fact Sheet: Talc.* Trenton, NJ

Tantalum & tantalum oxide dusts T:0130

Molecular Formula: Ta; O_5Ta_2 (oxide)
Synonyms: Metal: Elemental tantalum; Tantalum 181; Tantalum metal
Oxide: Tantalic acid anhydride; Tantalum(V) oxide; Tantalum pentaoxide; Tantalum pentoxide
CAS Registry Number: 7440-25-7 (elemental); 1314-61-0 (oxide)
RTECS® Number: WW5505000 (elemental)
UN/NA & ERG Number: Metal powder, in bulk, may be pyrophoric. UN3089 (metal powder, flammable, n.o.s.)/170
EC Number: 231-135-5; 215-238-2 (pentoxide)
Regulatory Authority and Advisory Bodies

Air Pollutant Standard Set. See below, "Permissible Exposure Limits in Air" section.

Canada, WHMIS, Ingredients Disclosure List Concentration 1.0%.

European/International Regulations: Not listed in Annex 1.

WGK (German Aquatic Hazard Class): Nonwater polluting agent (*metal and oxide*).

Description: Tantalum is a refractory metal in Group V-B of the periodic table. The pure metal is ductile, steel-blue to gray solid or black, odorless powder. Molecular weight = 180.95; Specific gravity (H_2O:1) = 16.65 (*metal*); 14.40 (powder) at 25°C; Boiling point = 5425°C; Freezing/Melting point = 2996°C; 1872 (oxide); Maximum Explosive concentration (MEC) = <200 g/m^3. Hazard Identification (based on NFPA-704 M Rating System): Health 1, Flammability 0, Reactivity 0. Insoluble in water.

Potential Exposure: Compound Description: Tumorigen. Tantalum metal is used in electronic components: electric capacitors; rectifiers, in chemical equipment; in nuclear reactor components; in chemical equipment; body implants. Tantalum carbide is used in metal cutting tools and wear-resistant parts. Some tantalum salts are used in catalysts.

Incompatibilities: A combustible solid; the dry powder ignites spontaneously in air. Incompatible with lead chromate, strong oxidizers; bromine trifluoride; fluorine. Tantalum metal is attacked by hydrogen fluoride, fused alkalis, fuming sulfuric acid.

Permissible Exposure Limits in Air

(*metal, oxides, and dusts*)

OSHA PEL: 5 mg/m^3 TWA.

NIOSH REL: 5 mg/m^3 TWA; 10 mg[Ta]/m^3 STEL.

ACGIH TLV®[1]: Withdrawn due to insufficient data.

NIOSH IDLH: 2500 mg/m^3.

Protective Action Criteria (PAC)

TEEL-0: 5 mg/m^3

PAC-1: 10 mg/m^3

PAC-2: 200 mg/m^3

PAC-3: 500 mg/m^3

DFG MAK (*metal*): 1.5 mg/m^3, respirable fraction (previously "fine dust"); 4 mg[Ta]/m^3, inhalable fraction (previously "total dust"); Pregnancy Risk Group C.

Arab Republic of Egypt: TWA 0.1 mg/m^3, 1993; Australia: TWA 5 mg/m^3, 1993; Austria: MAK 5 mg/m^3, 1999; Belgium: TWA 5 mg/m^3, 1993; Denmark: TWA 5 mg/m^3, 1999; Finland: TWA 5 mg/m^3, 1999; France: VME 5 mg/m^3, 1993; the Netherlands: MAC-TGG 5 mg/m^3, 2003; the Philippines: TWA 5 mg/m^3, 1993; Poland: MAC (TWA) 5 mg/m^3, 1999; Russia: STEL 10 mg/m^3, 1993; Switzerland: MAK-W 5 mg/m^3, 1999; United Kingdom: TWA 5 mg/m^3, STEL 10 mg/m^3, 2000; Argentina, Bulgaria, Columbia, Jordan, South Korea, New Zealand, Singapore, Vietnam: ACGIH TLV®: TWA 5 mg. Several states have set guidelines or standards for tantalum in ambient air[60] ranging from 50−100 μg/m^3 (North Dakota) to 80 μg/m^3 (Virginia) to 100 μg/m^3 (Connecticut) to 119 μg/m^3 (Nevada).

Determination in Air: Use NIOSH Analytical Method (IV) #0500, Particulates NOR, total dust.

Permissible Concentration in Water: No criteria set. Insoluble.

Routes of Entry: Inhalation, skin, and/or eye contact.

Harmful Effects and Symptoms

Short Term Exposure: Irritates the eyes and skin. In animals: pulmonary irritation.

Long Term Exposure: May be a systemic poison.

Points of Attack: Eyes, skin, respiratory system.

Medical Surveillance: Lung function tests. Consider chest X-ray following acute overexposure.

First Aid: If this chemical gets into the eyes, remove any contact lenses at once and irrigate immediately for at least 15 min, occasionally lifting upper and lower lids. Seek medical attention immediately. If this chemical contacts the skin, remove contaminated clothing and wash immediately with soap and water. Seek medical attention immediately. If this chemical has been inhaled, remove from exposure, begin rescue breathing (using universal precautions, including resuscitation mask) if breathing has stopped and CPR if heart action has stopped. Transfer promptly to a medical facility. When this chemical has been swallowed, get medical attention. Give large quantities of water and induce vomiting. Do not make an unconscious person vomit.

Personal Protective Methods: Wear protective gloves and clothing to prevent any reasonable probability of skin contact. Safety equipment suppliers/manufacturers can provide recommendations on the most protective glove/clothing material for your operation. All protective clothing (suits, gloves, footwear, headgear) should be clean, available each day, and put on before work. Contact lenses should not be worn when working with this chemical. Wear dust-proof chemical goggles and face shield unless full-face-piece respiratory protection is worn. Employees should wash immediately with soap when skin is wet or contaminated. Provide emergency showers and eyewash.

Respirator Selection: Up to 25 mg/m^3: Qm (APF = 25) (any quarter-mask respirator). Up to 50 mg/m^3: 95XQ (APF = 10) [any particulate respirator equipped with an N95, R95, or P95 filter (including N95, R95, and P95 filtering face-pieces) except quarter-mask respirators. The following filters may also be used: N99, R99, P99, N100, R100, P100]; or any particulate respirator equipped with an N95, R95, or P95 filter (including N95, R95, and P95 filtering face-pieces) except quarter-mask respirators. The following filters may also be used: N99, R99, P99, N100, R100, P100; or Sa (APF = 10) (any supplied-air respirator). Up to 125 mg/m^3: Sa:Cf (APF = 25) (any supplied-air respirator operated in a continuous-flow mode); or PaprHie (APF = 25) (any powered, air-purifying respirator with a high-efficiency particulate filter). Up to 250 mg/m^3: 100F (APF = 50) (any air-purifying, full-face-piece respirator with an N100, R100, or P100 filter); or SaT: Cf (APF = 50) (any supplied-air respirator that has a tight-fitting face-piece and is operated in a continuous-flow mode); or PaprTHie

(APF = 50) (any powered, air-purifying respirator with a tight-fitting face-piece and a high-efficiency particulate filter); or SCBAF (APF = 50) (any self-contained breathing apparatus with a full face-piece); or SaF (APF = 50) (any supplied-air respirator with a full face-piece). *Up to 2,500 mg/m³:* Sa: Pd,Pp (APF = 1000) (any supplied-air respirator operated in a pressure-demand or other positive-pressure mode). *Emergency or planned entry into unknown concentrations or IDLH conditions:* SCBAF: Pd,Pp (APF = 10,000) (any self-contained breathing apparatus that has a full face-piece and is operated in a pressure-demand or other positive-pressure mode); or SaF: Pd,Pp: ASCBA (APF = 10,000) (any supplied-air respirator that has a full face-piece and is operated in a pressure-demand or other positive-pressure mode in combination with an auxiliary self-contained breathing apparatus operated in a pressure-demand or other positive-pressure mode). *Escape:* GmFOv100 (APF = 50) (100F (APF = 50) (any air-purifying, full-face-piece respirator with an N100, R100, or P100 filter); or SCBAE (any appropriate escape-type, self-contained breathing apparatus).

Note: Substance reported to cause eye irritation or damage; may require eye protection.

Storage: Color Code—Red (*pyrophoric powder*): Flammability Hazard: Store in a flammable materials storage area. Prior to working with this chemical you should be trained on its proper handling and storage.

Sipping: Metal powder, in bulk, may be pyrophoric. Flammable powder requires a shipping label of "FLAMMABLE SOLID." It falls in Hazard Class 4.1 and Packing Group III.

Spill Handling: Evacuate persons not wearing protective equipment from area of spill or leak until cleanup is complete. Remove all ignition sources. Collect powdered material in the most convenient and safe manner and deposit in sealed containers. Ventilate area after cleanup is complete. It may be necessary to contain and dispose of this chemical as a hazardous waste. If material or contaminated runoff enters waterways, notify downstream users of potentially contaminated waters. Contact your local or federal environmental protection agency for specific recommendations. If employees are required to clean up spills, they must be properly trained and equipped. OSHA 1910.120(q) may be applicable.

Fire Extinguishing: This chemical is a combustible solid. Use dry chemical, carbon dioxide; water spray; or alcohol foam extinguishers. Poisonous gases are produced in fire. If material or contaminated runoff enters waterways, notify downstream users of potentially contaminated waters. Notify local health and fire officials and pollution control agencies. From a secure, explosion-proof location, use water spray to cool exposed containers. If cooling streams are ineffective (venting sound increases in volume and pitch, tank discolors, or shows any signs of deforming), withdraw immediately to a secure position. If employees are expected to fight fires, they must be trained and equipped in OSHA 1910.156. The only respirators recommended for firefighting are self-contained breathing apparatuses that have full face-pieces and are operated in a pressure-demand or other positive-pressure mode.

Disposal Method Suggested: Sanitary landfill if necessary; recover if possible because of economic value. Technology exists for tantalum recovery from spent catalysts, for example.

TDE   T:0140

Molecular Formula: $C_{14}H_{10}Cl_4$
Common Formula: $ClC_6H_4CH(CHCl_2)C_6H_4Cl$
Synonyms: Benzene, 1,1'-(2,2-dichloroethylidene)bis(4-chloro-); 1,1-Bis(*p*-chlorophenyl)-2,2-dichloroethane; 1,1-Bis(4-chlorophenyl)-2,2-dichloroethane; 2,2-Bis(*p*-chlorophenyl)-1,1-dichloroethane; 2,2-Bis(4-chlorophenyl)-1,1-dichloroethane; DDD; *p,p'*-DDD (EPA); 1,1-Dichlor-2,2-bis(4-chlor-phenyl)-aethan (German); 1,1-Dichloro-2,2-bis(*p*-chlorophenyl)ethane; 1,1-Dichloro-2,2-bis(4-chlorophenyl)ethane; 1,1-Dichloro-2,2-bis(*p*-chlorophenyl)ethane; 1,1-Dichloro-2,2-di(4-chlorophenyl)ethane; *p,p*-'-Dichlorodiphenyldichloroethane; Dichlorodiphenyldichloroethane; Diclorodifeniltricloroetano (Spanish); Dilene; ENT 4,225; ME-1700; NCI-C00475; Rhothane; Rhothane D-3; Rothane; *p,p-'*-TDE; Tetrachlorodiphenylethane
CAS Registry Number: 72-54-8
RTECS® Number: KI070000
UN/NA & ERG Number: UN2761/151
EC Number: 200-783-0
Regulatory Authority and Advisory Bodies
Carcinogenicity: IARC: Animal Sufficient Evidence, *possibly carcinogenic to humans*, Group 2B, 1987; NCI: Carcinogenesis Studies (feed); clear evidence: rat; no evidence: mouse.
US EPA, FIFRA 1998 Status of Pesticides: Canceled.
Clean Water Act: Section 311 Hazardous Substances/RQ 40CFR117.3 (same as CERCLA, see below); 40CFR423, Appendix A, Priority Pollutants.
US EPA Hazardous Waste Number (RCRA No.): U060.
RCRA, 40CFR261, Appendix 8 Hazardous Constituents.
RCRA 40CFR268.48; 61FR15654, Universal Treatment Standards: Wastewater (mg/L), 0.023; Nonwastewater (mg/kg), 0.087.
RCRA 40CFR264, Appendix 9; TSD Facilities Ground Water Monitoring List. Suggested test method(s) (PQL µg/L): 8080 (0.1); 8270 (10).
Reportable Quantity (RQ): 1 lb (0.454 kg).
California Proposition 65 Chemical: Cancer 1/1/89.
European/International Regulations: Not listed in Annex 1.
WGK (German Aquatic Hazard Class): 3—Severe hazard to waters.

Description: TDE is a colorless, combustible, crystalline compound. Molecular weight = 320.04; Vapor pressure = 1×10^{-6} mmHg at 20°C; Freezing/Melting point = 109−110°C. Insoluble in water.

Potential Exposure: Those involved in the manufacture, formulation, and application of this insecticide. In an action of March 18, 1971, EPA canceled all pesticide uses of this product which is a metabolite of DDT. Hence it is no longer manufactured commercially.

Incompatibilities: Incompatible with alkalis, strong oxidizers.

Permissible Exposure Limits in Air
Protective Action Criteria (PAC)
TEEL-0: 1.5 mg/m^3
PAC-1: 5 mg/m^3
PAC-2: 35 mg/m^3
PAC-3: 500 mg/m^3

Permissible Concentration in Water: For the protection of freshwater aquatic life, the value is 0.6 μg/L, based on acute toxicity. For saltwater aquatic life, the value is 3.6 μg/L, based on acute toxicity. For the protection of human health, with respect to TDE, see criteria proposed for DDT, since TDE is a metabolite of DDT. Mexico[35] has set a limit of 30 μg/L in estuaries and 3 μg/L in coastal waters. Russia set a MAC of zero in surface water used for fishery purposes.

Determination in Water: Methylene chloride extraction followed by gas chromatography with electron capture or halogen-specific detection (EPA Method #608) or gas chromatography plus mass spectrometry (EPA Method #625). Fish Tox = 0.59514000 ppb (EXTRA HIGH).

Routes of Entry: Inhalation, skin absorption; ingestion; skin and/or eye contact.

Harmful Effects and Symptoms
Short Term Exposure: Since DDD is a metabolite of DDT, as well as a contaminant of commercial preparations of DDT, many of the effects of DDT could be mediated through DDD. Irritates the eyes, skin, and respiratory tract. May affect the central nervous system, causing convulsions and respiratory failure. Exposure may result in death. Symptoms of exposure include lethargy, anorexia, nausea, vomiting, diarrhea, paresthesia of tongue, lips, face; tremor, apprehension, dizziness, confusion, malaise (vague feeling of discomfort), headache, fatigue; convulsions; paresis of hands.

Long Term Exposure: Based on DDT, the chemical may affect the central nervous system. May damage the liver. A Suspected Human Carcinogen; there is some evidence that DDD is carcinogenic in mice; however, in other species, it appears to be noncarcinogenic. DDD has been shown to be mutagenic in drosophila, but not in yeast or bacteria. In cell culture, DDD causes chromosomal breaks.

Points of Attack: Eyes, skin, central nervous system; kidneys, liver, peripheral nervous system.

Medical Surveillance: Kidney and liver function tests. Examination of the nervous system.

First Aid: Skin Contact[52]: Flood all areas of the body that have contacted the substance with water. Do not wait to remove contaminated clothing; do it under the water stream. Use soap to help assure removal. Isolate contaminated clothing when removed to prevent contact by others.

Eye Contact: Remove any contact lenses at once. Flush eyes well with copious quantities of water or normal saline for at least 20–30 min. Seek medical attention.

Inhalation: Leave contaminated area immediately; breathe fresh air. Proper respiratory protection must be supplied to any rescuers. If coughing, difficult breathing or any other symptoms develop, seek medical attention at once, even if symptoms develop many hours after exposure.

Ingestion: If convulsions are not present, give a glass or two of water or milk to dilute the substance. Assure that the person's airway is unobstructed and contact a hospital or poison center immediately for advice on whether or not to induce vomiting. Medical observation is recommended following acute overexposure.

Personal Protective Methods: Wear protective gloves and clothing to prevent any reasonable probability of skin contact. Safety equipment suppliers/manufacturers can provide recommendations on the most protective glove/clothing material for your operation. All protective clothing (suits, gloves, footwear, headgear) should be clean, available each day, and put on before work. Contact lenses should not be worn when working with this chemical. Wear dust-proof chemical goggles and face shield unless full-face-piece respiratory protection is worn. Employees should wash immediately with soap when skin is wet or contaminated. Provide emergency showers and eyewash.

Respirator Selection: NIOSH: *at any concentrations above the NIOSH REL, or where there is no REL, at any detectable concentration:* SCBAF: Pd,Pp (APF = 10,000) (any self-contained breathing apparatus that has a full face-piece and is operated in a pressure-demand or other positive-pressure mode); or SaF: Pd,Pp: ASCBA (APF = 10,000) (any supplied-air respirator that has a full face-piece and is operated in a pressure-demand or other positive-pressure mode in combination with an auxiliary, self-contained breathing apparatus operated in a pressure-demand or other positive-pressure mode). *Escape:* GmFOv100 (APF = 50) [Any air-purifying, full-face-piece respirator (gas mask) with a chin-style, front- or back-mounted organic vapor canister having an N100, R100, or P100 filter]; or SCBAE (any appropriate escape-type, self-contained breathing apparatus).

Storage: Color Code—Blue: Health Hazard/Poison: Store in a secure poison location. Prior to working with this chemical you should be trained on its proper handling and storage. Store in a cool, dry place. Where possible, automatically transfer material from drums or other storage containers to process containers. Sources of ignition, such as smoking and open flames, are prohibited where this chemical is handled, used, or stored. Metal containers involving the transfer of this chemical should be grounded and bonded. Wherever this chemical is used, handled, manufactured, or stored, use explosion-proof electrical equipment and fittings. A regulated, marked area should be established where this chemical is handled, used, or stored in compliance with OSHA Standard 1910.1045.

Shipping: Organochlorine pesticides, solid, toxic, n.o.s. require a shipping label of "POISONOUS/TOXIC

MATERIALS." This material falls in DOT Hazard Class 6.1 and Packing Group III.

Spill Handling: Evacuate persons not wearing protective equipment from area of spill or leak until cleanup is complete. Remove all ignition sources. Use HEPA vacuum or wet method to reduce dust during cleanup. Do not dry sweep. Dampen spilled material with acetone to avoid dust. Collect powdered material in the most convenient and safe manner and deposit in sealed containers. Ventilate area after cleanup is complete. It may be necessary to contain and dispose of this chemical as a hazardous waste. If material or contaminated runoff enters waterways, notify downstream users of potentially contaminated waters. Contact your local or federal environmental protection agency for specific recommendations. If employees are required to clean up spills, they must be properly trained and equipped. OSHA 1910.120(q) may be applicable. Soil Adsorption Index $(K_{oc}) = 10,000$.

Fire Extinguishing: This chemical is a combustible solid. Use dry chemical, carbon dioxide; water spray; or alcohol foam extinguishers. Poisonous gases are produced in fire, including hydrogen chloride and chlorine. If material or contaminated runoff enters waterways, notify downstream users of potentially contaminated waters. Notify local health and fire officials and pollution control agencies. From a secure, explosion-proof location, use water spray to cool exposed containers. If cooling streams are ineffective (venting sound increases in volume and pitch, tank discolors, or shows any signs of deforming), withdraw immediately to a secure position. If employees are expected to fight fires, they must be trained and equipped in OSHA 1910.156. The only respirators recommended for firefighting are self-contained breathing apparatuses that have full face-pieces and are operated in a pressure-demand or other positive-pressure mode.

Disposal Method Suggested: Incineration in a unit operating above 850°C equipped with HCl scrubber. Incineration above 1200°C for 1−2 s is recommended. In accordance with 40CFR165, follow recommendations for the disposal of pesticides and pesticide containers. Must be disposed properly by following package label directions or by contacting your local or federal environmental control agency, or by contacting your regional EPA office.

References

US Environmental Protection Agency. (April 30, 1980). *DDD, Health and Environmental Effects Profile No. 58.* Washington, DC: Office of Solid Waste

US Environmental Protection Agency. (1980). *DDT— Ambient Water Quality Criteria.* Washington, DC

Tellurium   T:0150

Molecular Formula: Te

Synonyms: Aurum paradoxum; Elemental tellurium; Metallum problematum; Telloy; Telurio (Spanish); Tellurium elemental

CAS Registry Number: 13494-80-9; *(alt.)* 137322-20-4

UN/NA & ERG Number: UN3288 (Toxic solid, inorganic, n.o.s.)/151

RTECS® Number: WY2625000 (elemental)

UN/NA & ERG Number: Available for Te compounds only: 3284 (tellurium compound, n.o.s.)/151

EC Number: 236-813-4

Regulatory Authority and Advisory Bodies

Air Pollutant Standard Set. See below, "Permissible Exposure Limits in Air" section.

Superfund/EPCRA 40CFR355, Extremely Hazardous Substances: TPQ = 500/10,000 lb (227/4540 kg).

Reportable Quantity (RQ): 1 lb (0.454 kg).

Canada, WHMIS, Ingredients Disclosure List Concentration 1.0%.

European/International Regulations: Not listed in Annex 1.

WGK (German Aquatic Hazard Class): No value assigned.

Description: Tellurium is a grayish or silvery white, lustrous, crystalline, semimetallic element. It may exist in a hexagonal crystalline form or an amorphous powder. It is found in sulfide ores and is produced as a by-product of copper or bismuth refining. Molecular weight = 127.60; Specific gravity (H_2O:1) = 6.24 at 25°C; Boiling point = 990°C; Freezing/Melting point = 450°C; Autoignition temperature = 340°C. Insoluble in water.

Potential Exposure: Compound Description: Reproductive Effector; Human Data. The primary use of tellurium is in the vulcanization of rubber and as an additive in ferritic steel production. It is also used as a carbide stabilizer in cast iron, a chemical catalyst; a coloring agent in glazes and glass; a thermocoupling material in refrigerating equipment; as an additive to selenium rectifiers; in alloys of lead, copper, steel, and tin for increased resistance to corrosion and stress, workability, machinability, and creep strength; and in certain culture media in bacteriology. Since tellurium is present in silver, copper, lead, and bismuth ores, exposure may occur during purification of these ores.

Incompatibilities: Finely divided powder or dust may be flammable and explosive. Violent reaction with halogens, interhalogens, zinc and lithium silicide; with incandescence. Incompatible with oxidizers, cadmium; strong bases; chemically active metals; silver bromate; nitric acid.

Permissible Exposure Limits in Air

OSHA PEL: 0.1 mg[Te]/m³ TWA.

NIOSH REL: 0.1 mg[Te]/m³ TWA, *except tellurium hexafluoride and bismuth telluride.*

ACGIH TLV®[1]: 0.1 mg[Te]/m³ TWA, *except hydrogen telluride.*

NIOSH IDLH: 25 mg[Te]/m³.

Protective Action Criteria (PAC)

TEEL-0: 0.1 mg/m³

PAC-1: 0.3 mg/m³

PAC-2: 20 mg/m³

PAC-3: 25 mg/m³

Australia: TWA 0.1 mg/m³, 1993; Austria: MAK 0.1 mg/m³, 1999; Belgium: TWA 0.1 mg/m³, 1993; Denmark: TWA

0.1 mg/m^3, 1999; Finland: TWA 0.1 mg/m^3, STEL 0.3 mg/m^3, 1999; France: VME 0.1 mg/m^3, 1999; the Netherlands: MAC-TGG 0.1 mg[Te]/m^3, 2003; the Philippines: TWA 0.1 mg/m^3, 1993; Poland: MAC (TWA) 0.01 mg/m^3; MAC (STEL) 0.03 mg/m^3, 1999; Sweden: NGV 0.1 mg/m^3, 1999; Switzerland: MAK-W 0.1 mg/m^3, KZG-W 0.5 mg/m^3, 1999; Turkey: TWA 0.1 mg/m^3, 1993; United Kingdom: TWA 0.1 mg/m^3, 2000; Argentina, Bulgaria, Columbia, Jordan, South Korea, New Zealand, Singapore, Vietnam: ACGIH TLV®: TWA 0.1 mg[Te]/m^3. Several states have set guidelines or standards for tellurium in ambient air[60] ranging from 1.0 µg/m^3 (North Dakota) to 1.6 µg/m^3 (Virginia) to 2.0 µg/m^3 (Connecticut and Nevada).

Determination in Air: Use NIOSH Analytical Method (IV) #7900. See also #7300, Elements: #7301; #7303; #9102; OSHA Analytical Method ID-121.

Permissible Concentration in Water: EPA[32] has suggested a permissible ambient goal of 1.4 µg/L based on health effects. Russia[43] set a MAC for tellurium in water bodies used for domestic purposes of 0.01 mg/L.

Routes of Entry: Inhalation of dust or fume; percutaneous absorption from dust, ingestion; skin and/or eye contact.

Harmful Effects and Symptoms

Short Term Exposure: Irritates the eyes, skin, and respiratory tract. High exposures can cause pulmonary edema, a medical emergency that can be delayed for several hours. This can cause death. Causes central nervous system depression. Moderate skin and eye irritant. Tellurium is capable of doing harm within the body by replacing the essential element sulfur. Signs and symptoms of acute exposure to tellurium may include garlicky breath, metallic taste; sweating; dry mouth; drowsiness, no sweating headache; drowsiness, malaise, fatigue, lassitude, weakness, and dizziness. Gastrointestinal symptoms may include nausea, vomiting, anorexia, and constipation. High levels of the dust or fume may cause bronchitis or pneumonia to develop. In animals: central nervous system; red blood cell changes.

Long Term Exposure: Repeated exposure can cause garlic odor on the breath, nausea, vomiting, anorexia, metallic taste, and irritability. Kidney damage, liver injury, and pulmonary effects may also occur. Dermal exposure may result in dermatitis; red, inflamed skin; drying and cracking. At doses which are severely toxic to the mother, tellurium produces teratogenic effects. May damage the unborn fetus. High exposures may cause damage to the nervous system.

Points of Attack: Skin, central nervous system; kidneys, liver.

Medical Surveillance: Oral hygiene and the respiratory tract should receive special attention in preplacement or periodic examinations. Liver and kidney function tests. Blood tests for liver and kidney function. Examination of the nervous system. Consider chest X-ray following acute overexposure.

First Aid: If this chemical gets into the eyes, remove any contact lenses at once and irrigate immediately for at least 15 min, occasionally lifting upper and lower lids. Seek medical attention immediately. If this chemical contacts the skin, remove contaminated clothing and wash immediately with soap and water. Seek medical attention immediately. If this chemical has been inhaled, remove from exposure, begin rescue breathing (using universal precautions, including resuscitation mask) if breathing has stopped and CPR if heart action has stopped. Transfer promptly to a medical facility. When this chemical has been swallowed, get medical attention. Obtain authorization and/or further instructions from the local hospital for administration of an antidote or performance of other invasive procedures. Give a slurry of activated charcoal in water to drink. Seek medical attention. Give large quantities of water and induce vomiting. Do not make an unconscious person vomit. Medical observation is recommended for 24–48 h after breathing overexposure, as pulmonary edema may be delayed. As first aid for pulmonary edema, a doctor or authorized paramedic may consider administering a corticosteroid spray.

Note to physician: For severe poisoning *do not* use BAL [British Anti-Lewisite, dimercaprol, dithiopropanol ($C_3H_8OS_2$)] as it is contraindicated or ineffective in poisoning from tellurium.

Personal Protective Methods: Wear protective gloves and clothing to prevent any reasonable probability of skin contact. Safety equipment suppliers/manufacturers can provide recommendations on the most protective glove/clothing material for your operation. All protective clothing (suits, gloves, footwear, headgear) should be clean, available each day, and put on before work. Contact lenses should not be worn when working with this chemical. Wear dust-proof chemical goggles and face shield unless full-face-piece respiratory protection is worn. Employees should wash immediately with soap when skin is wet or contaminated. Provide emergency showers and eyewash.

Respirator Selection: Up to 0.5 mg/m^3: Qm (APF = 25) (any quarter-mask respirator). *Up to 1 mg/m^3:* 95QX [any particulate respirator equipped with an N95, R95, or P95 filter (including N95, R95, and P95 filtering face-pieces) except quarter-mask respirators. The following filters may also be used: N99, R99, P99, N100, R100, P100]; or Sa (APF = 10) (any supplied-air respirator). *Up to 2.5 mg/m^3:* Sa:Cf (APF = 25) (any supplied-air respirator operated in a continuous-flow mode); or PaprHie (APF = 25) (any powered, air-purifying respirator with a high-efficiency particulate filter). *Up to 5 mg/m^3:* 100F (APF = 50) (any air-purifying, full-face-piece respirator with an N100, R100, or P100 filter); or SaT: Cf (APF = 50) (any supplied-air respirator that has a tight-fitting face-piece and is operated in a continuous-flow mode); or PaprTHie (APF = 50) (any powered, air-purifying respirator with a tight-fitting face-piece and a high-efficiency particulate filter); or SCBAF (APF = 50) (any self-contained breathing apparatus with a full face-piece); or SaF (APF = 50) (any supplied-air respirator with a full face-piece). *Up to 25 mg/m^3:* Sa: Pd,Pp (APF = 1000) (any supplied-air respirator operated in a pressure-demand or other positive-pressure mode). *Emergency or planned entry into unknown concentrations or IDLH conditions:* SCBAF: Pd,Pp (APF = 10,000) (any self-contained breathing apparatus that has a full face-piece and is operated in a pressure-demand or other positive-pressure mode); or SaF: Pd,Pp: ASCBA (APF = 10,000) (any supplied-air respirator that has a full face-piece and is operated in a pressure-demand or other positive-pressure mode in combination with an auxiliary self-contained breathing apparatus operated in a pressure-demand or other positive-pressure mode). *Escape:* 100F (APF = 50) (any air-purifying, full-face-piece respirator with an N100, R100, or P100 filter); or SCBAE (any appropriate escape-type, self-contained breathing apparatus).

Note: Substance reported to cause eye irritation or damage; may require eye protection.

Storage: Color Code—Blue: Health Hazard/Poison: Store in a secure poison location. Prior to working with this chemical you should be trained on its proper handling and storage. Store in tightly closed containers in a cool, well-ventilated area away from strong oxidizers, chlorine, cadmium. Where possible, automatically transfer material from drums or other storage containers to process containers. Sources of ignition, such as smoking and open flames, are prohibited where this chemical is handled, used, or stored. Metal containers involving the transfer of this chemical should be grounded and bonded. Wherever this chemical is used, handled, manufactured, or stored, use explosion-proof electrical equipment and fittings.

Shipping: Toxic solid, inorganic, n.o.s. require a shipping label of "POISONOUS/TOXIC MATERIALS." This material falls in Hazard Class 6.1 and Packing Group III.

Spill Handling: Evacuate persons not wearing protective equipment from area of spill or leak until cleanup is complete. Remove all ignition sources. Collect powdered material in the most convenient and safe manner and deposit in sealed containers. Ventilate area after cleanup is complete. It may be necessary to contain and dispose of this chemical as a hazardous waste. If material or contaminated runoff enters waterways, notify downstream users of potentially contaminated waters. Contact your local or federal environmental protection agency for specific recommendations. If employees are required to clean up spills, they must be properly trained and equipped. OSHA 1910.120(q) may be applicable.

Fire Extinguishing: This chemical is a flammable solid. Use water spray. Tellurium will burn only slowly in air. Straight water streams will scatter molten tellurium oxide. Wear goggles, rubber gloves, and proper respirator with filter. Poisonous gases, including tellurium, are produced in fire. If material or contaminated runoff enters waterways, notify downstream users of potentially contaminated waters. Notify local health and fire officials and pollution control agencies. From a secure, explosion-proof location, use water spray to cool exposed containers. If cooling streams are ineffective (venting sound increases in volume and

pitch, tank discolors, or shows any signs of deforming), withdraw immediately to a secure position. If employees are expected to fight fires, they must be trained and equipped in OSHA 1910.156. The only respirators recommended for firefighting are self-contained breathing apparatuses that have full face-pieces and are operated in a pressure-demand or other positive-pressure mode.

References
US Environmental Protection Agency. (1979). *Chemical Hazard Information Profile: Tellurium.* Washington, DC
US Environmental Protection Agency. (November 30, 1987). *Chemical Hazard Information Profile: Tellurium.* Washington, DC: Chemical Emergency Preparedness Program

Tellurium hexafluoride   T:0160

Molecular Formula: F_6Te
Common Formula: TeF_6
Synonyms: Tellurium fluoride
CAS Registry Number: 7783-80-4
RTECS® Number: WY2800000
UN/NA & ERG Number: UN2195/125
EC Number: 232-027-0
Regulatory Authority and Advisory Bodies
Department of Homeland Security Screening Threshold Quantity (pounds): *Theft hazard* 15 (≥0.83.00% concentration).
Department of Homeland Security Screening Threshold Quantity (pounds): Sabotage/Contamination Hazard: A placarded amount (commercial grade).
Air Pollutant Standard Set. See below, "Permissible Exposure Limits in Air" section.
OSHA 29CFR1910.119, Appendix A. Process Safety List of Highly Hazardous Chemicals, TQ = 250 lb (114 kg).
Superfund/EPCRA 40CFR355, Extremely Hazardous Substances: TPQ = 100 lb (454 kg).
Reportable Quantity (RQ): 100 lb (45.4 kg).
US DOT 49CFR172.101, Inhalation Hazardous Chemical.
Canada, WHMIS, Ingredients Disclosure List Concentration 1.0%.
European/International Regulations: Not listed in Annex 1.
WGK (German Aquatic Hazard Class): No value assigned.
Description: Tellurium hexafluoride is a colorless gas with a repulsive odor. Molecular weight = 241.60; Boiling point = (sublimes before reaching its BP) −39°C; Freezing/Melting point = (sublimes) −38°C; Relative vapor density (air = 1) = 8.34. Decomposes in water.
Potential Exposure: Tellurium hexafluoride is stated to be a by-product of ore refining.
Incompatibilities: Hydrolyzes slowly in water to telluric acid. Emits highly toxic fumes when heated or on contact with acid or acid fumes.
Permissible Exposure Limits in Air
Conversion factor: 1 ppm = 9.88 mg/m^3 at 25°C & 1 atm.

OSHA PEL: 0.02 ppm/0.2 mg/m^3 TWA.
NIOSH REL: 0.02 ppm/0.2 mg/m^3 TWA.
ACGIH TLV$^{®[1]}$: 0.02 ppm/0.2 mg/m^3 TWA.
NIOSH IDLH: 1 ppm.
Protective Action Criteria (PAC)*
TEEL-0: 0.018 ppm
PAC-1: 0.018 ppm
PAC-2: **0.018 ppm**
PAC-3: **0.053 ppm**
*AEGLs (Acute Emergency Guideline Levels) & ERPGs (Emergency Response Planning Guideline) are in **bold face**.
DFG MAK: 0.1 mg[Te]/m^3, inhalable fraction, as Te and its compounds.
The following is for reference and consideration in that this compound is also a fluoride compound and some foreign countries use the fluoride standard.
OSHA PEL: 3 ppm/2.5 mg[F]/m^3 TWA.
NIOSH REL: 3 ppm/2.5 mg[F]/m^3 TWA; 6 ppm/5 mg[F]/m^3, 15 min. Ceiling Concentration.
ACGIH TLV$^{®[1]}$: 2.5 mg[F]/m^3 TWA; not classifiable as a human carcinogen; BEI: 3 mg[F]/g creatinine in urine *prior* to end-of-shift; 10 mg[F]/g creatinine in urine end-of-shift.
DFG MAK: 1 mg[F]/m^3, inhalable fraction [skin]; Peak Limitation Category II(4); Pregnancy Risk Group C; BAT: 7.0 mg[F]/g creatinine in urine at end-of-shift; 4.0 mg[F]/g creatinine in urine at the beginning of the next shift.
NIOSH IDLH: 250 mg[F]/m^3.
Australia: TWA 0.02 ppm (0.2 mg/m^3), 1993; Austria: MAK 2.5 mg[F]/m^3, 1999; Austria: MAK 0.1 mg[Te]/m^3, 1999; Belgium: TWA 0.02 ppm (0.1 mg/m^3), 1993; Denmark: TWA 0.02 ppm (0.2 mg/m^3), 1999; Finland: TWA 0.02 ppm (0.2 mg/m^3), STEL 0.06 ppm (0.6 mg/m^3), 1999; France: VME 0.02 ppm (0.2 mg/m^3), 1999; the Netherlands: MAC-TGG 0.2 mg/m^3, 2003; the Philippines: TWA 0.02 ppm (0.2 mg/m^3), 1993; Poland: MAC (TWA) 1 mg[HF]/m^3; MAC (STEL) 3 mg[HF]/m^3, 1999; Poland: MAC (TWA) 0.01 mg[Te]/m^3; MAC (STEL) 0.03 mg[Te]/m^3, 1999; Sweden: NGV 2 mg[F]/m^3, 1999; Sweden: NGV 0.1 mg[Te]/m^3, 1999; Switzerland: MAK-W 0.02 ppm (0.2 mg/m^3), 1999; United Kingdom: TWA 2.5 mg[F]/m^3; TWA 0.1 mg[Te]/m^3, 2000; Argentina, Bulgaria, Columbia, Jordan, South Korea, New Zealand, Singapore, Vietnam: ACGIH TLV$^®$: TWA 0.02 ppm. Several states have set guidelines or standards for TeF$_6$ in ambient air[60] ranging from 2.0 μg/m^3 (North Dakota) to 3.5 μg/m^3 (Virginia) to 4.0 μg/m^3 (Connecticut) to 5.0 μg/m^3 (Nevada).
Determination in Air: Use NIOSH II (3), Method #S-187.
Routes of Entry: Inhalation.
Harmful Effects and Symptoms
Short Term Exposure: Capable of causing death or permanent injury. *Acute*: the material is highly toxic by inhalation exposure and a strong irritant to skin, eyes, lungs, throat, and stomach. Death can occur from filling of the lungs with fluid (pulmonary edema) or from asphyxiation due to spasm of the throat (larynx), or bronchi. Signs and symptoms of acute exposure to tellurium hexafluoride may include drowsiness, malaise, lassitude, weakness, and dizziness.
Long Term Exposure: Dermal exposure may result in dermatitis; red, inflamed skin; drying and cracking. May cause kidney and liver injury. A metallic taste; garlicky breath; and profuse sweating may develop. Gastrointestinal effects may include nausea, vomiting, anorexia, and constipation.
Points of Attack: Respiratory system, liver, and kidneys.
Medical Surveillance: NIOSH lists the following tests: chest X-ray; pulmonary function tests: forced vital capacity, forced expiratory volume (1 s). Liver and kidney function tests.
First Aid: If this chemical gets into the eyes, remove any contact lenses at once and irrigate immediately for at least 15 min, occasionally lifting upper and lower lids. Seek medical attention immediately. If this chemical contacts the skin, remove contaminated clothing and wash immediately with soap and water. Seek medical attention immediately. If this chemical has been inhaled, remove from exposure, begin rescue breathing (using universal precautions, including resuscitation mask) if breathing has stopped and CPR if heart action has stopped. Transfer promptly to a medical facility. Medical observation is recommended for 24−48 h after breathing overexposure, as pulmonary edema may be delayed. As first aid for pulmonary edema, a doctor or authorized paramedic may consider administering a corticosteroid spray.
Note to physician: For severe poisoning *do not* use BAL [British Anti-Lewisite, dimercaprol, dithiopropanol (C$_3$H$_8$OS$_2$)] as it is contraindicated or ineffective in poisoning from tellurium.
Personal Protective Methods: Wear protective gloves and clothing to prevent any reasonable probability of skin contact. Safety equipment suppliers/manufacturers can provide recommendations on the most protective glove/clothing material for your operation. All protective clothing (suits, gloves, footwear, headgear) should be clean, available each day, and put on before work. Contact lenses should not be worn when working with this chemical. Wear gas-proof chemical goggles and face shield unless full-face-piece respiratory protection is worn. Employees should wash immediately with soap when skin is wet or contaminated. Provide emergency showers and eyewash.
Respirator Selection: *Up to 0.2 ppm:* Sa (APF = 10) (any supplied-air respirator). *Up to 0.5 ppm:* Sa:Cf (APF = 25) (any supplied-air respirator operated in a continuous-flow mode). *Up to 1 ppm:* SaT: Cf (APF = 50) (any supplied-air respirator that has a tight-fitting face-piece and is operated in a continuous-flow mode); or SCBAF (APF = 50) (any self-contained breathing apparatus with a full face-piece); or SaF (APF = 50) (any supplied-air respirator with a full face-piece). Emergency or planned entry into unknown concentrations or IDLH conditions: SCBAF: Pd,Pp (APF = 10,000) (any self-contained breathing apparatus that has a full face-piece and is operated in a pressure-demand or other positive-pressure mode); or SaF: Pd,Pp: ASCBA

(APF = 10,000) (any supplied-air respirator that has a full face-piece and is operated in a pressure-demand or other positive-pressure mode in combination with an auxiliary self-contained breathing apparatus operated in a pressure-demand or other positive-pressure mode). *Escape:* GmFS (APF = 50) [any air-purifying, full-face-piece respirator (gas mask) with a chin-style, front- or back-mounted canister providing protection against the compound of concern]; or SCBAE (any appropriate escape-type, self-contained breathing apparatus).

Storage: Poison gas. Color Code—White Stripe: Contact Hazard; Store separately; not compatible with materials in solid white category. Color Code—Blue: Health Hazard/ Poison: Store in a secure poison location. Prior to working with this chemical you should be trained on its proper handling and storage. Store in tightly closed containers in a cool, well-ventilated area away from water.

Procedures for the handling, use, and storage of cylinders should be in compliance with OSHA 1910.101 and 1910.169 with the recommendations of the Compressed Gas Association.

Shipping: Tellurium hexafluoride requires a shipping label of "POISON GAS, CORROSIVE." It falls in Hazard Class 2.3 and Packing Group I. It is a violation of transportation regulations to refill compressed gas cylinders without the express written permission of the owner.

Special precautions: Cylinders must be transported in a secure upright position, in a well-ventilated truck.

Spill Handling: Evacuate persons not wearing protective equipment from area of spill or leak until cleanup is complete. Remove all ignition sources. Ventilate area of leak to disperse vapors. Stop the flow of the leak. Remove the leaking container to a safe place in the open air and allow the leak to disperse. Use water spray to reduce vapor but do not put water on leak or spill area. *Small spills:* flush area with flooding amounts of water. *Large spills:* dike far ahead of spill for later disposal. Do not get water inside container. Isolate area until gas has dispersed. It may be necessary to contain and dispose of this chemical as a hazardous waste. If material or contaminated runoff enters waterways, notify downstream users of potentially contaminated waters. Contact your local or federal environmental protection agency for specific recommendations. If employees are required to clean up spills, they must be properly trained and equipped. OSHA 1910.120(q) may be applicable.

Initial isolation and protective action distances
Distances shown are likely to be affected during the first 30 min after materials are spilled and could increase with time. If more than one tank car, cargo tank, portable tank, or large cylinder is involved in the incident is leaking, the protective action distance may need to be increased. You may need to seek emergency information from CHEMTREC at (800) 424-9300 or seek professional environmental engineering assistance from the US EPA Environmental Response Team at (908) 548-8730 (24-h response line).

Small spills (From a small package or a small leak from a large package)
First: Isolate in all directions (feet/meters) 600/200.
Then: Protect persons downwind (miles/kilometers).
Day 0.8/1.2
Night 2.7/4.3/4.3
Large spills (From a large package or from many small packages)
First: Isolate in all directions (feet/meters) 3000/1000.
Then: Protect persons downwind (miles/kilometers).
Day 5.9/9.4
Night 7.0 + /11.0+

Fire Extinguishing: This chemical is a nonflammable gas. *Small fires:* dry chemical or carbon dioxide. *Large fires:* water spray, fog, or foam. Keep unnecessary people away; isolate hazard area and deny entry. Stay upwind; keep out of low areas. Ventilate closed spaces before entering them. Wear positive pressure breathing apparatus and full protective clothing. Do not get water inside container. Move container from fire area if you can do so without risk. Stay away from ends of tanks. Spray cooling water on containers that are exposed to flames until well after fire is out. Isolate area until gas has dispersed. Poisonous gases are produced in fire. If material or contaminated runoff enters waterways, notify downstream users of potentially contaminated waters. Notify local health and fire officials and pollution control agencies. From a secure, explosion-proof location, use water spray to cool exposed containers. If cooling streams are ineffective (venting sound increases in volume and pitch, tank discolors, or shows any signs of deforming), withdraw immediately to a secure position. If employees are expected to fight fires, they must be trained and equipped in OSHA 1910.156. The only respirators recommended for firefighting are self-contained breathing apparatuses that have full face-pieces and are operated in a pressure-demand or other positive-pressure mode.

References
US Environmental Protection Agency. (November 30, 1987). *Chemical Hazard Information Profile: Tellurium Hexafluoride*. Washington, DC: Chemical Emergency Preparedness Program

Temephos   T:0170

Molecular Formula: $C_{16}H_{20}O_6P_2S_3$
Synonyms: 27165; Abat; Abate; Abathion; AC 52160; AI3-27165; American Cyanamid AC-52,160; American Cyanamid CL-52160; American Cyanamid E.I. 52,160; Biothion; Bis- *p*-(*O*,*O*-dimethyl *O*-phenylphosphorothioate) sulfide; Bithion; Cl 52160; Difenphos; Difenthos; Difos; *O*, *O*-Dimethyl phosphorothioate *O*,*O*-diester with 4,4′-thiodiphenol; Diphos; Ecopro; Ecopro 1707; EI 52160; ENT 27,165; Nephis; Nephis 1G; Nimitex; Nimitox; Phenol,4,4′-thiodi-, *O*,*O*-diester with *O*,*O*-dimethyl phosphorothioate; Phosphorothioic acid, *O*,*O*′-dimethyl ester, *O*,*O*-diester with

4,4′-thiodiphenol; Phosphorothioic acid, O,O'-(thiodi-p-phenylene) O,O,O',O'-tetramethyl ester; Phosphorothioic acid, O,O'-(thiodi-4,1-phenylene) O,O,O',O'-tetramethyl ester; Swebate; Temefos (Spanish); Temophos; Tetrafenphos; O,O,O',O'-Tetramethyl O,O'-thiodi-p-phenylene bis(phosphorothioate); O,O,O',O'-Tetramethyl O,O'-thiodi-p-phenylene phosphorothioate; Tetramethyl O,O'-thiodi-p-phenylene phosphorothioate; O,O'-(Thiodi-4,1-phenylene) bis(O,O-dimethyl phosphorothioate); O,O'-(Thiodi-4,1-phenylene) phosphorothioic acid O,O,O',O'-tetramethyl ester; O,O'-(Thiodi-p-phenylene) O,O,O',O'-tetramethyl bis(phosphorothioate)

CAS Registry Number: 3383-96-8
RTECS® Number: TF6890000
UN/NA & ERG Number: UN2783 (organophosphorus pesticides, solid, toxic)/152
EC Number: 222-191-1 [015-025-00-2]

Regulatory Authority and Advisory Bodies
US EPA, FIFRA, 1998 Status of Pesticides: Supported.
Air Pollutant Standard Set. See below, "Permissible Exposure Limits in Air" section.
EPCRA Section 313 Form R *de minimis* concentration reporting level: 1.0%.
US DOT Regulated Marine Pollutant (49CFR172.101, Appendix B).
US DOT 49CFR172.101, Inhalation Hazard Chemical as organophosphates.
European/International Regulations: Not listed in Annex 1.
WGK (German Aquatic Hazard Class): No value assigned.

Description: Temephos is a crystalline solid. The technical product is a brown viscous liquid. Molecular weight = 466.48; Boiling point = 120−125°C; Freezing/Melting point = 30.6°C; Vapor pressure = 7×10^{-8} mmHg at 25°C. Practically insoluble in water.

Potential Exposure: Compound Description: Agricultural Chemical; Reproductive Effector; Primary Irritant. Those involved in the manufacture, formulation, and application of this insecticide which is used as a mosquito, black fly, and midge larvicide.

Incompatibilities: Strong acids; bases.

Permissible Exposure Limits in Air
OSHA PEL: 15 mg/m^3 TWA, total dust; 5 mg/m^3 TWA respirable fraction.
NIOSH REL: 10 mg/m^3 TWA, total dust; 5 mg/m^3 TWA respirable fraction.
ACGIH TLV®[1]: 1 mg/m^3 TWA, inhalable fraction and vapors [skin]; not classifiable as a human carcinogen; BEI$_A$ issued for acetylcholinesterase inhibiting pesticides.
No Teel available.
Australia: TWA 10 mg/m^3, 1993; Belgium: TWA 10 mg/m^3, 1993; France: VME 10 mg/m^3, 1999; Russia: STEL 0.5 mg/m^3 [skin], 1993; Switzerland: MAK-W 10 mg/m^3, 1999; the Netherlands: MAC-TGG 10 mg/m^3, 2003; Argentina, Bulgaria, Columbia, Jordan, South Korea, New Zealand, Singapore, Vietnam: ACGIH TLV®: TWA 10 mg/m^3. Several states have set guidelines or standards for

temephos in ambient air[60] ranging from 100 μg/m^3 (North Dakota) to 160 μg/m^3 (Virginia) to 200 μg/m^3 (Connecticut) to 238 μg/m^3 (Nevada).

Determination in Air: Use NIOSH Analytical Method #0500, #0600; NIOSH Analytical Method PV-2056.

Permissible Concentration in Water: No criteria set. Experience in the field for a period of more than 1 year has shown, however, that 1 mg/L in drinking water is without effect.

Determination in Water: Techniques used for residue determination include colorimetry and gas liquid chromatography, and may be applicable to water analysis. Fish Tox = 16.53952000 ppb (INTERMEDIATE). Octanol−water coefficient: Log K_{ow} = 5.96.

Routes of Entry: Inhalation, skin absorption; ingestion; skin and/or eye contact.

Harmful Effects and Symptoms
Short Term Exposure: Temephos can affect you when breathed in and quickly enters the body by passing through the skin. Severe poisoning can occur from skin contact. It is a moderately toxic organophosphate chemical. Exposure can cause rapid severe poisoning with headache, sweating, nausea, and vomiting; diarrhea, loss of coordination; and death.

Long Term Exposure: Cholinesterase inhibitor; cumulative effect is possible. This chemical may damage the nervous system with repeated exposure, resulting in convulsions, respiratory failure. May cause liver damage. Human Tox = 3.50000 ppb (HIGH).

Points of Attack: Respiratory system, lungs, central nervous system; cardiovascular system, skin, eyes, plasma, and red blood cell cholinesterase.

Medical Surveillance: Before employment and at regular times after that, the following are recommended: plasma and red blood cell cholinesterase levels (tests for the enzyme poisoned by this chemical). If exposure stops, plasma levels return to normal in 1−2 weeks while red blood cell levels may be reduced for 1−3 months.
When cholinesterase enzyme levels are reduced by 25% or more below preemployment levels, risk of poisoning is increased, even if results are in lower ranges of "normal." Reassignment to work not involving organophosphate or carbamate pesticides is recommended until enzyme levels recover. If symptoms develop or overexposure occurs, repeat the above tests as soon as possible and get an exam of the nervous system. Do not drink any alcoholic beverages before or during use. Alcohol promotes absorption of organic phosphates.

First Aid: If this chemical gets into the eyes, remove any contact lenses at once and irrigate immediately for at least 15 min, occasionally lifting upper and lower lids. Seek medical attention immediately. If this chemical contacts the skin, remove contaminated clothing and wash immediately with soap and water. Shampoo hair promptly if contaminated. Speed in removing material from skin is of extreme importance. Shampoo hair promptly if contaminated. Seek medical attention immediately. If this chemical has been inhaled, remove from exposure, begin rescue breathing (using universal precautions, including resuscitation mask) if breathing has stopped and CPR if heart action has stopped. Transfer promptly to a medical facility. When this chemical has been swallowed, get medical attention. Give large quantities of water and induce vomiting. Do not make an unconscious person vomit.

Personal Protective Methods: Wear protective gloves and clothing to prevent any reasonable probability of skin contact. Safety equipment suppliers/manufacturers can provide recommendations on the most protective glove/clothing material for your operation. All protective clothing (suits, gloves, footwear, headgear) should be clean, available each day, and put on before work. Contact lenses should not be worn when working with this chemical. Wear splash-proof chemical goggles and face shield when working with liquid, unless full-face-piece respiratory protection is worn. Wear dust-proof goggles and face shield when working with the crystals, unless full-face-piece respiratory protection is worn. Employees should wash immediately with soap when skin is wet or contaminated. Provide emergency showers and eyewash.

Respirator Selection: Where there is potential for exposures over *10 mg/m³*, use a NIOSH/MSHA- or European Standard EN149-approved full-face-piece respirator with a pesticide cartridge. Increased protection is obtained from full-face-piece air-purifying respirators. Where there is potential for exposure *over 10 mg/m³* as liquid temephos or for high exposures, use a NIOSH/MSHA- or European Standard EN149-approved supplied-air respirator with a full face-piece operated in the positive-pressure mode, or with a full face-piece, hood, or helmet in the continuous-flow mode; or use a NIOSH/MSHA- or European Standard EN149-approved self-contained breathing apparatus with a full face-piece operated in pressure-demand or other positive-pressure mode.

Storage: Color Code—Blue: Health Hazard/Poison: Store in a secure poison location. Prior to working with this chemical you should be trained on its proper handling and storage. Store in tightly closed containers in a cool, well-ventilated area away from strong acids and bases. Where possible, automatically transfer material from drums or other storage containers to process containers. Sources of ignition, such as smoking and open flames, are prohibited where this chemical is handled, used, or stored. Metal containers involving the transfer of this chemical should be grounded and bonded. Wherever this chemical is used, handled, manufactured, or stored, use explosion-proof electrical equipment and fittings.

Shipping: Organophosphorus pesticides, solid, toxic, require a shipping label of "POISONOUS/TOXIC MATERIALS." This material falls in Hazard Class 6.1 and Packing Group III.

Spill Handling: Evacuate persons not wearing protective equipment from area of spill or leak until cleanup is complete. Remove all ignition sources. Collect powdered material in the most convenient and safe manner and deposit in sealed containers. Ventilate area after cleanup is complete. It may be necessary to contain and dispose of this chemical as a hazardous waste. If material or contaminated runoff enters waterways, notify downstream users of potentially contaminated waters. Contact your local or federal environmental protection agency for specific recommendations. If employees are required to clean up spills, they must be properly trained and equipped. OSHA 1910.120(q) may be applicable.

Fire Extinguishing: This chemical is a combustible solid. Use dry chemical, carbon dioxide; water spray; or alcohol foam extinguishers. Poisonous gases are produced in fire, including sulfur and phosphorus oxides. If material or contaminated runoff enters waterways, notify downstream users of potentially contaminated waters. Notify local health and fire officials and pollution control agencies. From a secure, explosion-proof location, use water spray to cool exposed containers. If cooling streams are ineffective (venting sound increases in volume and pitch, tank discolors, or shows any signs of deforming), withdraw immediately to a secure position. If employees are expected to fight fires, they must be trained and equipped in OSHA 1910.156. The only respirators recommended for firefighting are self-contained breathing apparatuses that have full face-pieces and are operated in a pressure-demand or other positive-pressure mode.

Disposal Method Suggested: Essentially complete, hydrolysis occurs upon heating in concentrated KOH for 20 min.[22] Incineration is recommended for large quantities. In accordance with 40CFR165, follow recommendations for the disposal of pesticides and pesticide containers. Must be disposed of properly by following package label directions or by contacting your local or federal environmental control agency, or by contacting your regional EPA office.

References

US Environmental Protection Agency, Special Review and Reregistration Division Office of Pesticide Programs. (1998). *Agency Status of Pesticides in Registration, Reregistration, and Special Review* (Rainbow Report). Washington, DC

New Jersey Department of Health and Senior Services. (May 2000). *Hazardous Substances Fact Sheet: Temephos.* Trenton, NJ

TEPP   T:0180

Molecular Formula: $C_8H_{20}O_7P_2$

Synonyms: Bis-*O,O*-diethylphosphoric anhydride; Bladan; Bladon; Diphosphoric acid, Tetraethyl ester; ENT 18,771; Ethyl pyrophosphate, *tetra*-; Fosvex; Grisol; HEPT; Hexamite; Killax; Kilmite 40; Lethalaire G-52; Lirohex; Mortopal; Motopal; Nifos; Nifos T; Nifrost; Phosphoric acid, tetraethyl ester; Pyrophosphate de tetraethyle

(French); TEP; *O,O,O,O*-Tetraaethyl-diphosphat, bis(*O,O*-diaethylphosphorsaeure-anhydrid (German); Tetraethyl pyrophosphate; Tetraethyl pyrophosphate, liquid; Tetrastigmine; Tetron; Tetron-100; Vapotone

CAS Registry Number: 107-49-3

RTECS® Number: UX6825000

UN/NA & ERG Number: UN3018 (liquid)/152; UN2783 (organophosphorus pesticides, solid, toxic)/152

EC Number: 203-495-3 [*Annex I Index No.:* 015-025-00-2]

Regulatory Authority and Advisory Bodies: Very Toxic Substance (World Bank).[15]

Air Pollutant Standard Set. See below, "Permissible Exposure Limits in Air" section.

Clean Water Act: Section 311 Hazardous Substances/RQ 40CFR117.3 (same as CERCLA, see below).

US EPA Hazardous Waste Number (RCRA No.): P111.

Superfund/EPCRA 40CFR355, Extremely Hazardous Substances: TPQ = 100 lb (45.4 kg).

Reportable Quantity (RQ): 10 lb (4.54 kg).

US DOT Regulated Marine Pollutant (49CFR172.101, Appendix B).

US DOT 49CFR172.101, Inhalation Hazard Chemical as organophosphates.

European/International Regulations: Hazard Symbol: T+, N; Risk phrases: R27/28; R50; Safety phrases: S1/2; S36/37/39; S38; S45; S61 (see Appendix 4).

WGK (German Aquatic Hazard Class): No value assigned.

Description: TEPP is a colorless to amber liquid with a faint, fruity odor. Molecular weight = 290.22; Specific gravity (H_2O:1) = 1.19 at 25°C; Boiling point = Decomposes below BP at 170°C; 138°C under 2.3 mmHg; Freezing/Melting point = 0°C; Vapor pressure = 2×10^{-4} mmHg at 25°C. Soluble in water.

Potential Exposure: Compound Description: Agricultural Chemical; Drug; Human Data. Those engaged in the manufacture, formulation, and application of this aphicide and acaricide; used as an insecticide to control aphids, thrips, and mites; as an anticholinesterase.

Incompatibilities: Strong oxidizers, alkalis, water. Hydrolyzes quickly in water to form pyrophosphoric acid.

Permissible Exposure Limits in Air

Conversion factor: 1 ppm = 11.87 mg/m³ at 25°C & 1 atm.

OSHA PEL: 0.05 mg/m³ TWA [skin].

NIOSH REL: 0.05 mg/m³ TWA [skin].

ACGIH TLV®[1]: 0.01 mg/m³, measured as inhalable fraction and vapor TWA [skin]; BEI_A issued for acetylcholinesterase inhibiting pesticides.

NIOSH IDLH: 5 mg/m³.

Protective Action Criteria (PAC)

TEEL-0: 0.01 mg/m³

PAC-1: 0.15 mg/m³

PAC-2: 1 mg/m³

PAC-3: 5 mg/m³

DFG MAK: 0.005 ppm/0.06 mg/m³; Peak Limitation Category II(2) [skin].

Arab Republic of Egypt: TWA 0.004 ppm (0.05 mg/m³) [skin], 1993; Australia: TWA 0.004 ppm (0.05 mg/m³) [skin], 1993; Austria: MAK 0.005 ppm (0.05 mg/m³) [skin], 1999; Belgium: TWA 0.004 ppm (0.047 mg/m³) [skin], 1993; Denmark: TWA 0.004 ppm (0.05 mg/m³) [skin], 1999; France: VME 0.004 ppm (0.05 mg/m³) [skin], 1999; the Netherlands: MAC-TGG 0.05 mg/m³ [skin], 2003; the Philippines: TWA 0.05 mg/m³ [skin], 1993; Switzerland: MAK-W 0.005 ppm (0.05 mg/m³) [skin], 1999; Turkey: TWA 0.05 mg/m³ [skin], 1993 Occupational Exposure Limit United Kingdom: TWA 0.004 ppm (0.05 mg/m³), STEL 0.1 ppm, 2000; Argentina, Bulgaria, Columbia, Jordan, South Korea, New Zealand, Singapore, Vietnam: ACGIH TLV®: TWA 0.05 mg/m³ (skin).

Several states have set guidelines or standards for TEPP in ambient air[60] ranging from 1.0 μg/m³ (Connecticut) to 5.0 μg/m³ (North Dakota) to 80,000 μg/m³ (Virginia) to a much higher value for Nevada.

Determination in Air: Use NIOSH Analytical Method (IV) #2504, Tetraethyl Pyrophosphate; see also NIOSH Analytical Method (IV) Method #5600, Organophosphorus Pesticides.

Routes of Entry: Inhalation, skin absorption; ingestion; skin and/or eye contact.

Harmful Effects and Symptoms

Short Term Exposure: Symptoms of exposure include eye pain; blurred vision; lacrimation (discharge of tears); rhinorrhea (discharge of thin nasal mucous); headache, chest tightness; cyanosis, anorexia, nausea, vomiting, diarrhea, weakness, twitching, paralysis, Cheyne—Stokes respiration, convulsions, low blood pressure; cardiac irregular/irregularities; sweating. TEPP is classified as super toxic. Probable oral lethal dose in humans is less than 5 mg/kg (a taste) for a 150 lb person. A small drop in the eye may cause death. Small doses at frequent intervals are additive. Poisonings always develop at a rapid rate. It is a cholinesterase inhibitor.

Long Term Exposure: Cholinesterase inhibitor; cumulative effect is possible. This chemical may damage the nervous system with repeated exposure, resulting in convulsions, respiratory failure. May cause liver damage.

Points of Attack: Eyes, respiratory system; central nervous system; cardiovascular system, gastrointestinal tract; blood cholinesterase.

Medical Surveillance: Before employment and at regular times after that, the following are recommended: plasma and red blood cell cholinesterase levels (tests for the enzyme poisoned by this chemical). If exposure stops, plasma levels return to normal in 1—2 weeks while red blood cell levels may be reduced for 1—3 months.

When cholinesterase enzyme levels are reduced by 25% or more below preemployment levels, risk of poisoning is increased, even if results are in lower ranges of "normal." Reassignment to work not involving organophosphate or carbamate pesticides is recommended until enzyme levels recover. If symptoms develop or overexposure occurs, repeat the above tests as soon as possible and get an exam of the nervous system. Also consider complete blood count. Consider chest X-ray following acute overexposure. Do not drink any alcoholic beverages before or during use. Alcohol promotes absorption of organic phosphates.

First Aid: If this chemical gets into the eyes, remove any contact lenses at once and irrigate immediately for at least 15 min, occasionally lifting upper and lower lids. Seek medical attention immediately. If this chemical contacts the skin, remove contaminated clothing and wash immediately with soap and water. Speed in removing material from skin is of extreme importance. Shampoo hair promptly if contaminated. Seek medical attention immediately. If this chemical has been inhaled, remove from exposure, begin rescue breathing (using universal precautions, including resuscitation mask) if breathing has stopped and CPR if heart action has stopped. Transfer promptly to a medical facility. When this chemical has been swallowed, get medical attention. Give a slurry of activated charcoal in water to drink. *Do NOT* induce vomiting.

Personal Protective Methods: Wear protective gloves and clothing to prevent any reasonable probability of skin contact. Safety equipment suppliers/manufacturers can provide recommendations on the most protective glove/clothing material for your operation. All protective clothing (suits, gloves, footwear, headgear) should be clean, available each day, and put on before work. Contact lenses should not be worn when working with this chemical. Wear splash-proof chemical goggles and face shield unless full-face-piece respiratory protection is worn. Employees should wash immediately with soap when skin is wet or contaminated. Provide emergency showers and eyewash.

Respirator Selection: 0.5 mg/m^3: Sa (APF = 10) (any supplied-air respirator). 1.25 mg/m^3: Sa:Cf (APF = 25) (any supplied-air respirator operated in a continuous-flow mode). 2.5 mg/m^3: SaT: Cf (APF = 50) (any supplied-air respirator that has a tight-fitting face-piece and is operated in a continuous-flow mode); or SCBAF (APF = 50) (any self-contained breathing apparatus with a full face-piece); or SaF (APF = 50) (any supplied-air respirator with a full face-piece). 5 mg/m^3: Sa: Pd,Pp (APF = 1000) (any supplied-air respirator operated in a pressure-demand or other positive-pressure mode). *Emergency or planned entry into unknown concentrations or IDLH conditions:* SCBAF: Pd, Pp (APF = 10,000) (any self-contained breathing apparatus that has a full face-piece and is operated in a pressure-demand or other positive-pressure mode); or SaF: Pd,Pp: ASCBA (APF = 10,000) (any supplied-air respirator that has a full face-piece and is operated in a pressure-demand or other positive-pressure mode in combination with an auxiliary self-contained breathing apparatus operated in a pressure-demand or other positive-pressure mode). *Escape:* GmFOv100 (APF = 50) [Any air-purifying, full-face-piece respirator (gas mask) with a chin-style, front- or back-mounted organic vapor canister having an N100, R100, or P100 filter]; or SCBAE (any appropriate escape-type, self-contained breathing apparatus).

Storage: Color Code—Blue: Health Hazard/Poison: Store in a secure poison location. Prior to working with this chemical you should be trained on its proper handling and storage. Store in tightly closed containers in a cool, well-ventilated area away from strong oxidizers (such as chlorine, bromine, and fluorine). Where possible, automatically transfer material from drums or other storage containers to process containers. Sources of ignition, such as smoking and open flames, are prohibited where this chemical is handled, used, or stored. Metal containers involving the transfer of this chemical should be grounded and bonded. Wherever this chemical is used, handled, manufactured, or stored, use explosion-proof electrical equipment and fittings.

Shipping: Organophosphorus pesticides, liquid, toxic, require a shipping label of "POISONOUS/TOXIC MATERIALS." It falls in Hazard Class 6.1 and Packing Group I.
Organophosphorus pesticides, solid, toxic, require a shipping label of "POISONOUS/TOXIC MATERIALS." It falls in Hazard Class 6.1 and Packing Group I.

Spill Handling: Evacuate and restrict persons not wearing protective equipment from area of spill or leak until cleanup is complete. Remove all ignition sources. Ventilate area of spill or leak. Absorb liquids in vermiculite, dry sand; earth, peat, carbon, or a similar material and deposit in sealed containers. Keep this chemical out of a confined space, such as a sewer, because of the possibility of an explosion, unless the sewer is designed to prevent the buildup of explosive concentrations. It may be necessary to contain and dispose of this chemical as a hazardous waste. If material or contaminated runoff enters waterways, notify downstream users of potentially contaminated waters. Contact your local or federal environmental protection agency for specific recommendations. If employees are required to clean up spills, they must be properly trained and equipped. OSHA 1910.120(q) may be applicable.

Fire Extinguishing: This chemical is a noncombustible liquid that may be formulated with a flammable substance. Poisonous gases, including oxides of sulfur; oxides of phosphorus; and phosphoric acid, are produced in fire. Use dry chemical, carbon dioxide; or foam extinguishers if formulated with a flammable substance. Vapors are heavier than air and will collect in low areas. Vapors from a flammable carrier may travel long distances to ignition sources and flashback. Vapors in confined areas may explode when exposed to fire. Containers may explode in fire. Storage containers and parts of containers may rocket great distances, in many directions. If material or contaminated runoff enters waterways, notify downstream users of potentially contaminated waters. Notify local health and fire officials and pollution control agencies. From a secure, explosion-proof location, use water spray to cool exposed containers. If cooling streams are ineffective (venting sound increases in volume and pitch, tank discolors, or shows any signs of deforming), withdraw immediately to a secure position. If employees are expected to fight fires, they must be trained and equipped in OSHA 1910.156. The only respirators recommended for firefighting are self-contained breathing apparatuses that have full face-pieces and are operated in a pressure-demand or other positive-pressure mode.

Disposal Method Suggested: Consult with environmental regulatory agencies for guidance on acceptable disposal practices. Generators of waste containing this contaminant ($\geq$100 kg/mo) must conform with EPA regulations governing storage, transportation, treatment, and waste disposal. TEPP is 50% hydrolyzed in water in 6.8 h at 25°C, and 3.3 h at 38°C; 99% hydrolysis requires 45.2 h at 25°C, or 21.9 h at 38°C. Hydrolysis of TEPP yields nontoxic products. Incineration is, however, an option for TEPP disposal. In accordance with 40CFR165, follow recommendations for the disposal of pesticides and pesticide containers. Must be disposed properly by following package label directions or by contacting your local or federal environmental control agency, or by contacting your regional EPA office.

References

US Environmental Protection Agency. (November 30, 1987). *Chemical Hazard Information Profile: TEPP.* Washington, DC: Chemical Emergency Preparedness Program

New Jersey Department of Health and Senior Services. (January 2001). *Hazardous Substances Fact Sheet: TEPP.* Trenton, NJ

Terbufos   T:0190

Molecular Formula: $C_9H_{21}O_2PS_3$
Common Formula: $(C_2H_5O)_2PSSCH_2SC(CH_3)_3$
Synonyms: AC 921000; Counter; Counter 15G soil insecticide; Counter 15G soil insecticide-nematicide; *S*-([(1,1-Dimethylethyl)thio]methyl) *O,O*-diethyl phosphorodithioate; Phosphorodithioic acid, *S*-([(1,1-dimethylethyl)thio] methyl), *O,O*-diethyl ester; Phosphorodithioic acid, *S*-[(*tert*-butylthio)methyl], *O,O*-diethyl ester; *S*-[(*tert*-Butylthio) methyl] *O,O*-diethyl phosphorodithioate
CAS Registry Number: 13071-79-9
RTECS® Number: TD7200000
UN/NA & ERG Number: UN3018 (organophosphorus pesticide, liquid, toxic)/152
EC Number: 235-963-8 [*Annex I Index No.:* 015-139-00-2]
Regulatory Authority and Advisory Bodies
Superfund/EPCRA 40CFR355, Extremely Hazardous Substances: TPQ = 100 lb (45.4 kg).
Reportable Quantity (RQ): 100 lb (45.4 kg).
US DOT Regulated Marine Pollutant (49CFR172.101, Appendix B), severe pollutant.
US DOT 49CFR172.101, Inhalation Hazard Chemical as organophosphates.
European/International Regulations: Hazard Symbol: T+, N; Risk phrases: R27/28; R50/53; Safety phrases: S1/2; S36/37; S45; S60. S61 (see Appendix 4).

WGK (German Aquatic Hazard Class): 3—Severe hazard to waters.
Description: Terbufos is a colorless to pale yellow liquid. Molecular weight = 288.45; Boiling point = 70°C at 0.01 mm; Freezing/Melting point = −29°C; Vapor pressure = 3×10^{-4} mmHg at 20°C; Flash point = 88°C (oc). Hazard Identification (based on NFPA-704 M Rating System): Health 4, Flammability 3, Reactivity 0. Slightly soluble in water.
Potential Exposure: Those involved in the manufacture, formulation, or application of this soil insecticide.
Incompatibilities: Strong oxidizers may cause release of toxic phosphorus oxides. Organophosphates, in the presence of strong reducing agents such as hydrides, may form highly toxic and flammable phosphine gas. Keep away from alkaline materials.
Permissible Exposure Limits in Air
ACGIH TLV®[1]: 0.01 mg/m³ TWA, inhalable fraction and vapor [skin]; not classifiable as a human carcinogen; BEI issued (1999).
Protective Action Criteria (PAC)
TEEL-0: 0.01 mg/m³
PAC-1: 0.03 mg/m³
PAC-2: 1 mg/m³
PAC-3: 1 mg/m³
Determination in Air: Use NIOSH Analytical Method (IV) Method #5600, Organophosphorus Pesticides.
Permissible Concentration in Water: The EPA (see "References" below) has developed a lifetime health advisory of 0.18 μg/L.
Determination in Water: Analysis of terbufos is by a gas chromatographic (GC) method applicable to the determination of certain nitrogen–phosphorus containing pesticides in water samples. In this method, approximately 1 L of sample is extracted with methylene chloride. The extract is concentrated and the compounds are separated using capillary column GC. Measurement is made using a nitrogen–phosphorus detector. Fish Tox = 0.04733000 ppb (EXTRA HIGH).
Routes of Entry: Inhalation, ingestion, skin contact.
Harmful Effects and Symptoms
Short Term Exposure: This material may be fatal if swallowed, inhaled, or absorbed through the skin. Repeated inhalation or skin contact may progressively increase susceptibility to poisoning. Acute exposure to terbufos may produce the following signs and symptoms: pinpoint pupils; blurred vision; headache, dizziness, muscle spasms; and profound weakness. Vomiting, diarrhea, abdominal pain; seizures, and coma may also occur. The heart rate may decrease following oral exposure or increase following dermal exposure. Chest pain may be noted. Hypotension (low blood pressure) may be noted, although hypertension (high blood pressure) is not uncommon. Respiratory symptoms include dyspnea (shortness of breath), respiratory depression; and respiratory paralysis. Psychosis may occur.
Long Term Exposure: Cholinesterase inhibitor; cumulative effect is possible. This chemical may damage the nervous

system with repeated exposure, resulting in convulsions, respiratory failure. May cause liver damage. Human Tox: 0.90000 ppb (EXTRA HIGH).

Points of Attack: Respiratory system, lungs, central nervous system; cardiovascular system, skin, eyes, plasma, and red blood cell cholinesterase.

Medical Surveillance: Before employment and at regular times after that, the following are recommended: plasma and red blood cell cholinesterase levels (tests for the enzyme poisoned by this chemical). If exposure stops, plasma levels return to normal in 1–2 weeks while red blood cell levels may be reduced for 1–3 months.

When cholinesterase enzyme levels are reduced by 25% or more below preemployment levels, risk of poisoning is increased, even if results are in lower ranges of "normal." Reassignment to work not involving organophosphate or carbamate pesticides is recommended until enzyme levels recover. If symptoms develop or overexposure occurs, repeat the above tests as soon as possible and get an exam of the nervous system. Also consider complete blood count. Consider chest X-ray following acute overexposure. Do not drink any alcoholic beverages before or during use. Alcohol promotes absorption of organic phosphates.

First Aid: If this chemical gets into the eyes, remove any contact lenses at once and irrigate immediately for at least 15 min, occasionally lifting upper and lower lids. Seek medical attention immediately. If this chemical contacts the skin, remove contaminated clothing and wash immediately with soap and water. Seek medical attention immediately. If this chemical has been inhaled, remove from exposure, begin rescue breathing (using universal precautions, including resuscitation mask) if breathing has stopped and CPR if heart action has stopped. Transfer promptly to a medical facility. Obtain authorization and/or further instructions from the local hospital for administration of an antidote or performance of other invasive procedures. Transport to a health-care facility.

Personal Protective Methods: Wear protective gloves and clothing to prevent any reasonable probability of skin contact. Safety equipment suppliers/manufacturers can provide recommendations on the most protective glove/clothing material for your operation. All protective clothing (suits, gloves, footwear, headgear) should be clean, available each day, and put on before work. Contact lenses should not be worn when working with this chemical. Wear splash-proof chemical goggles and face shield unless full-face-piece respiratory protection is worn. Employees should wash immediately with soap when skin is wet or contaminated. Provide emergency showers and eyewash.

Respirator Selection: Follow regulations in OSHA 29CFR1910.134 or European Standard EN149. Use a NIOSH/MSHA- or European Standard EN149-approved respirator; or use an approved supplied-air respirator with a full face-piece operated in the positive-pressure mode, or with a full face-piece, hood, or helmet in the continuous-flow mode; or use a NIOSH/MSHA- or European Standard EN149-approved self-contained breathing apparatus with a full face-piece operated in pressure-demand or other positive-pressure mode.

Storage: Color Code—Blue: Health Hazard/Poison: Store in a secure poison location. Prior to working with this chemical you should be trained on its proper handling and storage. Store in tightly closed containers in a cool, well-ventilated area away from strong oxidizers. Where possible, automatically transfer material from drums or other storage containers to process containers. Sources of ignition, such as smoking and open flames, are prohibited where this chemical is handled, used, or stored. Metal containers involving the transfer of this chemical should be grounded and bonded. Wherever this chemical is used, handled, manufactured, or stored, use explosion-proof electrical equipment and fittings.

Shipping: Organophosphorus pesticides, liquid, toxic, require a shipping label of "POISONOUS/TOXIC MATERIALS." Terbufos falls in DOT Hazard Class 6.1 and Packing Group II.

Spill Handling: This is a liquid organophosphorus pesticide. Keep unnecessary people away; isolate hazard area and deny entry. Stay upwind; keep out of low areas. Ventilate closed spaces before entering them. Remove and isolate contaminated clothing at the site. Do not touch spilled material; stop leak if you can do so without risk. Use water spray to reduce vapors. *Small spills:* absorb with sand or other noncombustible absorbent material and place into containers for later disposal. *Large spills:* dike far ahead of spill for later disposal. Keep this chemical out of a confined space, such as a sewer, because of the possibility of an explosion, unless the sewer is designed to prevent the buildup of explosive concentrations. It may be necessary to contain and dispose of this chemical as a hazardous waste. If material or contaminated runoff enters waterways, notify downstream users of potentially contaminated waters. Contact your local or federal environmental protection agency for specific recommendations. If employees are required to clean up spills, they must be properly trained and equipped. OSHA 1910.120(q) may be applicable. Soil Adsorption Index (K_{oc}) = 500.

Fire Extinguishing: This chemical is a combustible liquid. Poisonous gases, including sulfur and phosphorus oxides, are produced in fire. Use dry chemical, carbon dioxide, or alcohol foam extinguishers. Vapors are heavier than air and will collect in low areas. Vapors may travel long distances to ignition sources and flashback. Vapors in confined areas may explode when exposed to fire. Containers may explode in fire. Storage containers and parts of containers may rocket great distances, in many directions. If material or contaminated runoff enters waterways, notify downstream users of potentially contaminated waters. Notify local health and fire officials and pollution control agencies. From a secure, explosion-proof location, use water spray to cool exposed containers. If cooling streams are ineffective (venting sound increases in volume and pitch, tank discolors, or shows any signs of deforming), withdraw immediately to a secure position. If employees are expected to fight fires, they must be trained and equipped in OSHA 1910.156. The only respirators recommended for firefighting are self-contained breathing apparatuses that have full face-pieces and are operated in a pressure-demand or other positive-pressure mode.

Disposal Method Suggested: In accordance with 40CFR 165 recommendations for the disposal of pesticides and pesticide containers. Must be disposed properly by following package label directions or by contacting your local or federal environmental control agency, or by contacting your regional EPA office.

References

US Environmental Protection Agency, Office of Drinking Water. (August 1987). *Health Advisory: Terbufos.* Washington, DC

US Environmental Protection Agency. (November 30, 1987). *Chemical Hazard Information Profile: Terbufos.* Washington, DC: Chemical Emergency Preparedness Program

Terephthalic acid   T:0200

Molecular Formula: $C_8H_6O_4$
Common Formula: $HOOCC_6H_4COOH$
Synonyms: Acide terephthalique (French); *p*-Benzenedicarboxylic acid; 1,4-Benzenedicarboxylic acid; *p*-Phthalic acid; Phthalic acid, *p*-isomer; TA 12; TA
CAS Registry Number: 100-21-0
RTECS® Number: WZ0875000
EC Number: 202-830-0
Regulatory Authority and Advisory Bodies
Air Pollutant Standard Set. See below, "Permissible Exposure Limits in Air" section.
Canada, WHMIS, Ingredients Disclosure List Concentration 1.0%.
European/International Regulations: Not listed in Annex 1.
WGK (German Aquatic Hazard Class): No value assigned.
Description: TPA is a white crystalline solid. Molecular weight = 166.14; Specific gravity (H_2O:1) = 1.51 at 25°C; Sublimation point ≥400°C; Flash point = 260°C; Autoignition temperature = 495°C. Hazard Identification (based on NFPA-704 M Rating System): Health 2, Flammability 1, Reactivity 0. Slightly soluble in water.
Potential Exposure: Compound Description: Drug, Mutagen, Primary Irritant. TPA is used primarily in the production of polyethylene terephthalate polymer for the fabrication of polyester fibers and films. A high-volume production chemical in the US.
Incompatibilities: Dust may form an explosive mixture with air. May react with strong oxidizers, such as chlorine or permanganate, and may form explosive compounds when exposed to nitric acid.

Permissible Exposure Limits in Air
ACGIH TLV®[1]: 10 mg/m³ TWA.
Protective Action Criteria (PAC)
TEEL-0: 10 mg/m³
PAC-1: 125 mg/m³
PAC-2: 500 mg/m³
PAC-3: 500 mg/m³
DFG MAK: 0.1 mg/m³, inhalable fraction TWA; Peak Limitation Category I(2); Pregnancy Risk Group C.
The Netherlands: MAC-TGG 10 mg/m³, 2003. Russia[43] set a MAC in work-place air of 0.1 mg/m³. Kansas[60] has set a guideline for ambient air of 556 μg/m³.
Permissible Concentration in Water: Russia[43] set a MAC in water bodies used for domestic purposes of 0.1 mg/L.
Determination in Water: Octanol−water coefficient: Log K_{ow} = 1.96.
Routes of Entry: Inhalation, ingestion, eye, and/or skin contact.
Harmful Effects and Symptoms
Short Term Exposure: *Inhalation:* May cause irritation to mouth, nose, or throat. *Skin:* May cause irritation, especially in open cuts or sores. *Eyes:* Can cause irritation. *Ingestion:* Mildly toxic.
Long Term Exposure: No information found.
First Aid: If this chemical gets into the eyes, remove any contact lenses at once and irrigate immediately for at least 15 min, occasionally lifting upper and lower lids. Seek medical attention immediately. If this chemical contacts the skin, remove contaminated clothing and wash immediately with soap and water. Seek medical attention immediately. If this chemical has been inhaled, remove from exposure, begin rescue breathing (using universal precautions, including resuscitation mask) if breathing has stopped and CPR if heart action has stopped. Transfer promptly to a medical facility. When this chemical has been swallowed, get medical attention. Give large quantities of water and induce vomiting. Do not make an unconscious person vomit.
Note to physician: May require supportive measures for allergic reaction. Urinary excretion is rapid.
Personal Protective Methods: Wear protective gloves and clothing to prevent any reasonable probability of skin contact. Safety equipment suppliers/manufacturers can provide recommendations on the most protective glove/clothing material for your operation. All protective clothing (suits, gloves, footwear, headgear) should be clean, available each day, and put on before work. Contact lenses should not be worn when working with this chemical. Wear dust-proof chemical goggles and face shield unless full-face-piece respiratory protection is worn. Employees should wash immediately with soap when skin is wet or contaminated. Provide emergency showers and eyewash.
Respirator Selection: Wear a dust mask. Where there is potential for overexposure to this chemical, use a NIOSH/MSHA- or European Standard EN149-approved supplied-air respirator with a full face-piece operated in the positive-pressure mode, or with a full face-piece, hood, or helmet in

the continuous-flow mode; or use a NIOSH/MSHA- or European Standard EN149-approved self-contained breathing apparatus with a full face-piece operated in pressure-demand or other positive-pressure mode.

Storage: Color Code—Green: General storage may be used. Prior to working with this chemical you should be trained on its proper handling and storage. Store in detached units of noncombustible construction. As far as possible use dust-tight equipment and vacuum cleaning.

Spill Handling: Evacuate persons not wearing protective equipment from area of spill or leak until cleanup is complete. Remove all ignition sources. Collect powdered material in the most convenient and safe manner and deposit in sealed containers. Ventilate area after cleanup is complete. It may be necessary to contain and dispose of this chemical as a hazardous waste. If material or contaminated runoff enters waterways, notify downstream users of potentially contaminated waters. Contact your local or federal environmental protection agency for specific recommendations. If employees are required to clean up spills, they must be properly trained and equipped. OSHA 1910.120(q) may be applicable.

Fire Extinguishing: This chemical is a combustible solid. Fine dust may form explosive mixture with air and produce a severe hazard. Use carbon dioxide, dry chemical, or water. Wear a self-contained breathing apparatus when fighting a fire. Poisonous gases are produced in fire. If material or contaminated runoff enters waterways, notify downstream users of potentially contaminated waters. Notify local health and fire officials and pollution control agencies. From a secure, explosion-proof location, use water spray to cool exposed containers. If cooling streams are ineffective (venting sound increases in volume and pitch, tank discolors, or shows any signs of deforming), withdraw immediately to a secure position. If employees are expected to fight fires, they must be trained and equipped in OSHA 1910.156. The only respirators recommended for firefighting are self-contained breathing apparatuses that have full face-pieces and are operated in a pressure-demand or other positive-pressure mode.

Disposal Method Suggested: Dissolve or mix the material with a combustible solvent and burn in a chemical incinerator equipped with an afterburner and scrubber. All federal, state, and local environmental regulations must be observed.

References

National Institute for Occupational Safety and Health. (December 1979). *Information Profile on Potential Occupational Hazards—Single Chemicals: Terephthalic Acid,* Report TR 79-607. Rockville, MD, pp. 115−119

New York State Department of Health. (April 1986). *Chemical Fact Sheet: Terephthalic Acid.* Albany, NY: Bureau of Toxic Substance Assessment

New Jersey Department of Health and Senior Services. (April 2000). *Hazardous Substances Fact Sheet: Terephthalic Acid.* Trenton, NJ

Terphenyls   T:0210

Molecular Formula: $C_{18}H_{14}$
Common Formula: $C_6H_5-C_6H_4-C_6H_5$
Synonyms: *ortho-isomer:* *o*-Diphenylbenzene; 1,2-Diphenylbenzene; 2-Phenylbiphenyl; *o*-Terphenyl; 1,2-Terphenyl; *o*-Triphenyl
para-isomer: *p*-Diphenylbenzene; 1,4-Diphenylbenzene; 4-Phenylbiphenyl; *p*-Terphenyl; 1,4-Terphenyl; *p*-Triphenyl
meta-isomer: *m*-Diphenylbenzene; 1,3-Diphenylbenzene; Isodiphenylbenzene; 3-Phenylbiphenyl; *m*-Terphenyl; 1,3-Terphenyl; *m*-Triphenyl
Hydrogenated: Hydrogenated terphenyls
Mixed isomers: Delowas S; Delowax OM; Diphenylbenzene; Gilotherm OM 2; Terbenzene; Triphenyl
CAS Registry Number: 92-06-8 (*m*-isomer); 84-15-1 (*o*-isomer); 92-94-4 (*p*-isomer); 26140-60-3 (*mixed isomers*); 61788-32-7 (hydrogenated)
RTECS® Number: WZ6470000 (*m*-isomer); WZ6472000 (*o*-isomer); WX6475000 (*p*-isomer); WZ6535000 (hydrogenated)
EC Number: 202-122-1 (*m*-); 201-517-6 (*o*-); 202-205-2 (*p*-); 247-477-3 (*mixed isomers*); 262-967-7 (hydrogenated)
Regulatory Authority and Advisory Bodies
Air Pollutant Standard Set. See below, "Permissible Exposure Limits in Air" section.
Canada, WHMIS, Ingredients Disclosure List Concentration 1.0% (*m*-, *o*-, *p*-isomers).
European/International Regulations: Not listed in Annex 1.
WGK (German Aquatic Hazard Class): No value assigned.
Description: There are three isomeric terphenyls, all having the formula $C_6H_5-C_6H_4-C_6H_5$. Pure terphenyl is a white, crystalline solid. The commercial grades are light yellow. All three isomers are unusually stable toward heat. Molecular weight = 230.32 (*m-, o-, p-isomers*). Hazard Identification (based on NFPA-704 M Rating System): Health 1, Flammability 1, Reactivity 0. All are insoluble in water.
The properties of the terphenyls are:
m-isomer
Specific gravity (H_2O:1) = 1.23 at 25°C; Boiling point = 365°C; Freezing/Melting point = 89°C; Vapor pressure = 0.01 mmHg at 93°C; Flash point = 191°C.
o-isomer
Specific gravity (H_2O:1) = 1.10 at 25°C; Boiling point = 332°C; Freezing/Melting point = 58°C; Vapor pressure = 0.09 mmHg at 93°C; Flash point = 163°C.
p-isomer
Specific gravity (H_2O:1) = 1.23 at 25°C; Boiling point = 405°C; Freezing/Melting point = 212.8°C; Vapor pressure = 1.23 mmHg at 93°C; Flash point = 207°C.
Potential Exposure: Compound Description (*p*-isomer): Agricultural Chemical. Terphenyl is used primarily as a heat storage and heat transfer agent. It is also used as a high-temperature lubricant; a constituent of waxes and

polishes; and as a plasticizer for resin-bodied paints; as a coolant and heat storage agent.

Incompatibilities: Materials are combustible. Avoid contact with strong oxidizers.

Permissible Exposure Limits in Air

Conversion factor: 1 ppm = 9.57 mg/m³ (*m-* and *p-*isomers); 1 ppm = 9.42 mg/m³ at 25°C & 1 atm (*o-*isomer).

All isomers

OSHA PEL: 1 ppm/9 mg/m³ Ceiling Concentration.
NIOSH REL: 0.5 ppm/5 mg/m³ Ceiling Concentration.
ACGIH TLV®[1]: 0.53 ppm/5 mg/m³ Ceiling Concentration.
NIOSH IDLH: 500 mg/m³.
Protective Action Criteria (PAC)
European/International Regulations: Not listed in Annex I.

p- and *mixed isomers*

TEEL-0: 0.4 mg/m³
PAC-1: 1.25 mg/m³
PAC-2: 9 mg/m³
PAC-3: 500 mg/m³
Denmark: Ceiling Concentration 0.5 ppm (5 mg/m³), 1999; Switzerland: MAK-W 0.5 ppm (5 mg/m³), 1999; United Kingdom: STEL 0.5 ppm (4.8 mg/m³), 2000.

Terphenyls, hydrogenated

Australia: Ceiling Concentration 0.5 ppm (5 mg/m³), 1993; Belgium: Ceiling Concentration 0.5 ppm (4.9 mg/m³), 1993; Denmark: Ceiling Concentration 0.4 ppm (4.4 mg/m³), 1999; France: VME 0.5 ppm (5 mg/m³), 1999; Russia: STEL 5 mg/m³, 1993; Switzerland: MAK-W 0.5 ppm (5 mg/m³), 1999; the Netherlands: MAC-TGG 5 mg/m³, 2003; Argentina, Bulgaria, Columbia, Jordan, South Korea, New Zealand, Singapore, Vietnam: ACGIH TLV®: Ceiling Concentration 0.5 ppm.

Russia: MAC 5 mg/m³. Several states have set guidelines or standards for terphenyls in ambient air[60] ranging from 40 μg/m³ (Virginia) to 50 μg/m³ (North Dakota) to 119 μg/m³ (Nevada).

Determination in Air: Use NIOSH Analytical Method #5021, *o-*terphenyl.

Routes of Entry: Inhalation of dusts and mists, ingestion; skin and/or eye contact.

Harmful Effects and Symptoms

Short Term Exposure: Irritation of eyes, skin, and respiratory tract. Skin or eye contact may cause thermal burns. Inhalation can cause lung irritation with coughing, wheezing, and/or shortness of breath. Symptoms of exposure include headache, sore throat.

Long Term Exposure: May affect the liver and kidneys. Repeated skin contact can cause dermatitis; drying and cracking. In animals: liver, kidney damage.

Points of Attack: Eyes, skin, respiratory system; liver, kidneys.

Medical Surveillance: Consider the points of attack in preplacement and periodic physical examinations. Lung function tests. Liver and kidney function tests.

First Aid: If this chemical gets into the eyes, remove any contact lenses at once and irrigate immediately for at least 15 min, occasionally lifting upper and lower lids. Seek medical attention immediately. If this chemical contacts the skin, remove contaminated clothing and wash immediately with soap and water. Seek medical attention immediately. If this chemical has been inhaled, remove from exposure, begin rescue breathing (using universal precautions, including resuscitation mask) if breathing has stopped and CPR if heart action has stopped. Transfer promptly to a medical facility. When this chemical has been swallowed, get medical attention. Give large quantities of water and induce vomiting. Do not make an unconscious person vomit.

Personal Protective Methods: Wear protective gloves and clothing to prevent any reasonable probability of skin contact. Safety equipment suppliers/manufacturers can provide recommendations on the most protective glove/clothing material for your operation. All protective clothing (suits, gloves, footwear, headgear) should be clean, available each day, and put on before work. Contact lenses should not be worn when working with this chemical. Wear dust-proof chemical goggles and face shield unless full-face-piece respiratory protection is worn. Employees should wash immediately with soap when skin is wet or contaminated. Provide emergency showers and eyewash.

Respirator Selection: NIOSH: *Up to 25 mg/m³:* Qm (APF = 25) (any quarter-mask respirator). *Up to 50 mg/m³:* 95XQ (APF = 10)* [any particulate respirator equipped with an N95, R95, or P95 filter (including N95, R95, and P95 filtering face-pieces) except quarter-mask respirators. The following filters may also be used: N99, R99, P99, N100, R100, P100]; or Sa (APF = 10) (any supplied-air respirator). *Up to 125 mg/m³:* Sa:Cf (APF = 25) (any supplied-air respirator operated in a continuous-flow mode); or PaprHie (APF = 25) (any powered, air-purifying respirator with a high-efficiency particulate filter). *Up to 250 mg/m³:* 100F (APF = 50) (any air-purifying, full-face-piece respirator with an N100, R100, or P100 filter); or SCBAF (APF = 50) (any self-contained breathing apparatus with a full face-piece); or SaF (APF = 50) (any supplied-air respirator with a full face-piece). *Up to 500 mg/m³:* SaF: Pd,Pp (APF = 2000) (any supplied-air respirator that has a full face-piece and is operated in a pressure-demand or other positive-pressure mode). *Emergency or planned entry into unknown concentrations or IDLH conditions:* SCBAF: Pd,Pp (APF = 10,000) (any self-contained breathing apparatus that has a full face-piece and is operated in a pressure-demand or other positive-pressure mode); or SaF: Pd,Pp: ASCBA (APF = 10,000) (any supplied-air respirator that has a full face-piece and is operated in a pressure-demand or other positive-pressure mode in combination with an auxiliary self-contained breathing apparatus operated in a pressure-demand or other positive-pressure mode). *Escape:* 100F (APF = 50) (any air-purifying, full-face-piece respirator with an N100, R100, or P100 filter); or SCBAE (any appropriate escape-type, self-contained breathing apparatus).

*Substance causes eye irritation or damage; eye protection needed.

Storage: Color Code—Green: General storage may be used. Prior to working with this chemical you should be trained on its proper handling and storage. Store in tightly closed containers in a cool, well-ventilated area away from strong oxidizers. Where possible, automatically transfer material from drums or other storage containers to process containers. Sources of ignition, such as smoking and open flames, are prohibited where this chemical is handled, used, or stored. Metal containers involving the transfer of this chemical should be grounded and bonded. Wherever this chemical is used, handled, manufactured, or stored, use explosion-proof electrical equipment and fittings.

Spill Handling: Evacuate persons not wearing protective equipment from area of spill or leak until cleanup is complete. Remove all ignition sources. Collect powdered material in the most convenient and safe manner and deposit in sealed containers. Ventilate area after cleanup is complete. It may be necessary to contain and dispose of this chemical as a hazardous waste. If material or contaminated runoff enters waterways, notify downstream users of potentially contaminated waters. Contact your local or federal environmental protection agency for specific recommendations. If employees are required to clean up spills, they must be properly trained and equipped. OSHA 1910.120(q) may be applicable.

Fire Extinguishing: Terphenyls are combustible solid, but do not easily ignite. Use dry chemical, carbon dioxide; water spray; alcohol foam or polymer foam extinguishers. Poisonous gases are produced in fire. If material or contaminated runoff enters waterways, notify downstream users of potentially contaminated waters. Notify local health and fire officials and pollution control agencies. From a secure, explosion-proof location, use water spray to cool exposed containers. If cooling streams are ineffective (venting sound increases in volume and pitch, tank discolors, or shows any signs of deforming), withdraw immediately to a secure position. If employees are expected to fight fires, they must be trained and equipped in OSHA 1910.156. The only respirators recommended for firefighting are self-contained breathing apparatuses that have full face-pieces and are operated in a pressure-demand or other positive-pressure mode.

Disposal Method Suggested: Dissolve or mix the material with a combustible solvent and burn in a chemical incinerator equipped with an afterburner and scrubber. All federal, state, and local environmental regulations must be observed.

References

New Jersey Department of Health and Senior Services. (November 1999). *Hazardous Substances Fact Sheet: Terphenyls.* Trenton, NJ

Testosterone   T:0220

Molecular Formula: $C_{19}H_{28}O_2$
Synonyms: Androlin; Andronaq; Androst-4-en-17(β)-ol-3-one; Androst-4-en-3-one, 17-β-hydroxy-; Androst-4-en-3-one,

17-hydroxy-, (17-β)-; Andrusol; Cristerone T; Geno-cristaux gremy; Homosteron; Homosterone; 17-β-Hydroxy-4-androsten-3-one; 17-β-Hydroxyandrost-4-en-3-one; 7-β-Hydroxy-androst-4-en-3-one; Malestrone (AMPS); Mertestate; Neo-testis; Oreton; Oreton-F; Orquisteron; Perandren; Percutacrine androgenique; Primotest; Primoteston; Sustanone; Synandrol F; Teslen; Testandrone; Testiculosterone; Testobase; Testopropon; Testosteroid; Testosteron; Testosterone hydrate; Testostosterone; Testoviron schering; Testoviron T; Testrone; Testryl; Virormone; Virosterone

CAS Registry Number: 58-22-0; 57-85-2 (testosterone propionate)
RTECS® Number: XA3030000
UN/NA & ERG Number: UN3249 (Medicine, solid, toxic, n.o.s.)/151
EC Number: 200-370-5; 200-351-1 (testosterone propionate)
Regulatory Authority and Advisory Bodies
Carcinogenicity: IARC: Animal Sufficient Evidence; Human Limited Evidence, 1979.
Air Pollutant Standard Set. See below, "Permissible Exposure Limits in Air" section.
California Proposition 65 Chemical: Cancer (testosterone and its esters) 1/1/88.
European/International Regulations: Not listed in Annex 1.
WGK (German Aquatic Hazard Class): No value assigned.
Description: Testosterone is an odorless, white or slight cream-colored crystals or crystalline powder. Molecular weight = 288.47; Freezing/Melting point = 155°C. Hazard Identification (based on NFPA-704 M Rating System): Health 2, Flammability 0, Reactivity 0. Insoluble in water.
Potential Exposure: Used as an androgenic, anabolic, and estrogenic hormone for both males and females.
Incompatibilities: None listed.
Permissible Exposure Limits in Air
No TEEL available.
Russia[43] set a MAC in work-place air of 0.005 mg/m^3 (5 μg/m^3).
Harmful Effects and Symptoms
Short Term Exposure: Testosterone can affect you when breathed in and by passing through your skin. Testosterone is a teratogen. Handle with extreme caution. Exposure can cause nausea, upset stomach; acne, and fluid and salt retention in body tissues.
Long Term Exposure: Women can also develop male features, baldness, increased body hair; deep voice and menstrual changes. Men can have lower sperm production, painful enlarged breasts; enlarged prostate; and excess of red blood cells. Repeated exposure can cause women to develop male characteristics including face and body hair, baldness, deepened voice, and enlarged female genitalia (clitoris). Higher levels can cause irregular menstrual cycles, lighter periods, and smaller breasts. If carried home (clothing, etc.) children in the home may have stunted growth, premature puberty in boys, and masculine appearance and abnormal genitals in girls.
Points of Attack: Blood cells, sex organs.

Medical Surveillance: Frequent exams (at least monthly) are recommended to evaluate for any signs or symptoms of exposure. If exposure is suspected, the following are recommended: red blood cell count (in men). Urine test for androsterone and/or etiocholanolone. (The more common 17-ketosteroid test is not sufficiently sensitive for such exposures).

First Aid: If this chemical gets into the eyes, remove any contact lenses at once and irrigate immediately for at least 15 min, occasionally lifting upper and lower lids. Seek medical attention immediately. If this chemical contacts the skin, remove contaminated clothing and wash immediately with soap and water. Seek medical attention immediately. If this chemical has been inhaled, remove from exposure, begin rescue breathing (using universal precautions, including resuscitation mask) if breathing has stopped and CPR if heart action has stopped. Transfer promptly to a medical facility. When this chemical has been swallowed, get medical attention. Give large quantities of water and induce vomiting. Do not make an unconscious person vomit.

Personal Protective Methods: Wear protective gloves and clothing to prevent any reasonable probability of skin contact. Safety equipment suppliers/manufacturers can provide recommendations on the most protective glove/clothing material for your operation. All protective clothing (suits, gloves, footwear, headgear) should be clean, available each day, and put on before work. Contact lenses should not be worn when working with this chemical. Wear dust-proof chemical goggles and face shield unless full-face-piece respiratory protection is worn. Employees should wash immediately with soap when skin is wet or contaminated. Provide emergency showers and eyewash.

Respirator Selection: Where there is potential for exposure to testosterone, use a NIOSH/MSHA- or European Standard EN149-approved supplied-air respirator with a full face-piece operated in the positive-pressure mode, or with a full face-piece, hood, or helmet in the continuous-flow mode; or use a NIOSH/MSHA- or European Standard EN149-approved self-contained breathing apparatus with a full face-piece operated in pressure-demand or other positive-pressure mode.

Storage: Color Code—Blue: Health Hazard/Poison: Store in a secure poison location. Prior to working with this chemical you should be trained on its proper handling and storage. Store in tightly closed containers in a cool, well-ventilated area. A regulated, marked area should be established where this chemical is handled, used, or stored in compliance with OSHA Standard 1910.1045.

Shipping: Medicine, solid, toxic, n.o.s., requires a shipping label of "POISONOUS/TOXIC MATERIALS." It falls in Hazard Class 6.1 and Packing Group II.

Spill Handling: Evacuate persons not wearing protective equipment from area of spill or leak until cleanup is complete. Remove all ignition sources. Collect powdered material in the most convenient and safe manner and deposit in sealed containers. Ventilate area after cleanup is complete.

It may be necessary to contain and dispose of this chemical as a hazardous waste. If material or contaminated runoff enters waterways, notify downstream users of potentially contaminated waters. Contact your local or federal environmental protection agency for specific recommendations. If employees are required to clean up spills, they must be properly trained and equipped. OSHA 1910.120(q) may be applicable.

Fire Extinguishing: Testosterone may burn, but does not readily ignite. Use dry chemical, carbon dioxide, water spray; or foam extinguishers. Poisonous gases are produced in fire. If material or contaminated runoff enters waterways, notify downstream users of potentially contaminated waters. Notify local health and fire officials and pollution control agencies. From a secure, explosion-proof location, use water spray to cool exposed containers. If cooling streams are ineffective (venting sound increases in volume and pitch, tank discolors, or shows any signs of deforming), withdraw immediately to a secure position. If employees are expected to fight fires, they must be trained and equipped in OSHA 1910.156. The only respirators recommended for firefighting are self-contained breathing apparatuses that have full face-pieces and are operated in a pressure-demand or other positive-pressure mode.

References

Sax, N. I. (Ed.). (1981). *Dangerous Properties of Industrial Materials Report*, 1, No. 3, 81−82
New Jersey Department of Health and Senior Services. (February 17, 1987). *Hazardous Substances Fact Sheet: Testosterone*. Trenton, NJ

Tetrachloro-dibenzo-*p*-dioxin   T:0230

Molecular Formula: $C_{12}H_4Cl_4O_2$

Synonyms: Dibenzo(b,e)-1,4-dioxin, 2,3,7,8-tetrachloro-; Dibenzo-*p*-dioxin, 2,3,7,8-tetrachloro-; Dioxine; Dioxin (herbicide contaminant); NCI-CO3714; TCDBD; 2,3,7,8-TCDD; TCDD; 2,3,7,8-Tetrachlorodibenzo(b,e)(1,4)dioxan; 2,3,6,7-Tetrachlorodibenzo-*p*-dioxin; 2,3,7,8-Tetrachloro-dibenzo-*p*-dioxin; 2,3,7,8-Tetrachlorodibenzo-1,4-dioxin; Tetrachlorodibenzodioxin; Tetradioxin

CAS Registry Number: 1746-01-6; *(alt.)* 56795-67-6

RTECS® Number: HP3500000

UN/NA & ERG Number: UN2811/154

EC Number: 217-122-7

Regulatory Authority and Advisory Bodies

Carcinogenicity: IARC: Human Limited Evidence; Animal Sufficient Evidence, *carcinogenic to humans*, Group 1, 1997; NTP: 11th Report on Carcinogens, 2004: Known to be a human carcinogen; Carcinogenesis studies; on test (2-year study), October, 2000; Carcinogenesis studies; on test (prechronic studies), October, 2000; Carcinogenesis studies; test completed (peer review), October, 2000; NCI: Carcinogenesis Studies (gavage); clear evidence: mouse, rat;

(dermal); clear evidence: mouse; NIOSH: Potential occupational carciogen.

US EPA Gene-Tox Program, Positive: Carcinogenicity—mouse/rat; Negative: Rodent dominant lethal.

Banned or Severely Restricted (UN).[13]

Persistent Organic Pollutants (UN) *as polychlorinated dibenzo-p-dioxins*.

Very Toxic Substance (World Bank).[15]

Air Pollutant Standard Set. See below, "Permissible Exposure Limits in Air" section.

Clean Air Act: Hazardous Air Pollutants (Title I, Part A, Section 112).

Clean Water Act: 40CFR423, Appendix A, Priority Pollutants; 40CFR401.15 Section 307 Toxic Pollutants.

RCRA, 40CFR261, Appendix 8 Hazardous Constituents, waste number not listed.

Safe Drinking Water Act: MCL, 0.000000003 mg/L; MCLG, zero; Regulated chemical (47 FR 9352).

RCRA 40CFR268.48; 61FR15654, Universal Treatment Standards: Wastewater (mg/L), 0.000063; Nonwastewater (mg/kg), 0.001.

RCRA 40CFR264, Appendix 9; TSD Facilities Ground Water Monitoring List. Suggested test method(s) (PQL μg/L): 8250 (0.005).

Reportable Quantity (RQ): 1 lb (0.454 kg).

California Proposition 65 Chemical: Cancer 1/1/88; Developmental/Reproductive toxin 4/1/91.

EPCRA Section 313 Form R *de minimis* concentration reporting level: 1.0%.

Canada, WHMIS, Ingredients Disclosure List Concentration 0.1%.

List of Stockholm Convention POPs: Annex C (Unintentional production and release) as *polychlorinated dibenzo-p-dioxins (PCDD)*.

European/International Regulations: Not listed in Annex 1.

WGK (German Aquatic Hazard Class): No value assigned.

Description: Tetrachlorodibenzo-*p*-dioxin is a white, needle-shaped, crystalline solid. Molecular weight = 321.96; Boiling point = (decomposition); Freezing/Melting point = 305°C; Vapor pressure = 2×10^{-6} mmHg at 25°C. Decomposition begins at 500°C and is virtually complete within 21 s at 800°C. Very slightly soluble in water; solubility = 0.00000002%. Polychlorinated dibenzo-*p*-dioxins are formed in the manufacturing process of all chlorophenols. However, the amount formed is dependent on the degree to which the temperature and pressure are controlled during production. An especially toxic dioxin, 2,3,7,8-tetrachlorodibenzo-*p*-dioxin (TCDD), is formed during the production of 2,4,5,-TCP (trichlorophenol) by the alkaline hydrolysis of 1,2,4,5-tetrachlorobenzene.

Potential Exposure: Compound Description: Agricultural Chemical; Tumorigen, Mutagen; Reproductive Effector; Human Data; Primary Irritant. TCDD has no uses except as a research chemical. As noted above, TCDD is an inadvertent contaminant in herbicide precursors and thus in the herbicides themselves. It is also formed during various combustion processes including the incineration of chemical wastes (chlorophenols, chlorinated benzenes, and biphenyl ethers). It may be found in flue gases, fly ash, and soot particles. It is highly persistent in soil, and contamination may be retained for years. TCDD is the most toxic of all the dioxins, and has the potential for bioaccumulation in animals. Thus, it is applied in herbicide formulations, but is not used *per se*. It has been estimated that approximately 2 million acres in the United States have been treated for weed control on one or more occasions with approximately 15 million pounds of TCDD contaminated 2,4,5,-T, 2,4,-D, or combinations of the two.

Incompatibilities: Decomposes in UV light.

Permissible Exposure Limits in Air

OSHA PEL: None.

NIOSH REL: Carcinogen; Limit exposure to lowest feasible concentration. See *NIOSH Pocket Guide*, Appendix A.

Protective Action Criteria (PAC)

TEEL-0: 0.00000001 mg/m^3

PAC-1: 0.0015 mg/m^3

PAC-2: 0.0075 mg/m^3

PAC-3: 0.0075 mg/m^3

DFG MAK: 1.0×10^{-8}, inhalable fraction Peak Limitation Category II(8) [skin]; Carcinogen Category 4; Pregnancy Risk Group C.

Austria: carcinogen, 1999; Finland: carcinogen, 1999; France: carcinogen, 1993; Switzerland: carcinogen, 1999. Several states have set guidelines or standards for TCDD in ambient air[60] ranging from zero (North Dakota and South Carolina) to 1.1 pg/m^3 (Massachusetts) to 0.003 ng/m^3 (North Carolina) to 0.0001 μg/m^3 (Pennsylvania) to 3.0 μg/m^3 (Virginia) to 450.0 μg/m^3 (Indiana).

Determination in Air: No method available.

Permissible Concentration in Water: There is insufficient data to permit the development of criteria for the protection of freshwater or saltwater aquatic life. For the protection of human health, the concentration is preferably zero. An additional lifetime cancer risk of 1 in 100,000 is posed at a concentration of 4.55×10^{-7} μg/L as of 1979. A concentration of 0.0039 μg/L was estimated to limit cancer risk to one in a million by EPA in 1980. States which have set guidelines for TCDD in drinking water include Maine at 2×10^{-4} μg/L and Minnesota at 2×10^{-6} μg/L.[61]

Determination in Water: Methylene chloride extraction followed by transfer to benzene and capillary column gas chromatography/mass spectrometry with electron impact ionization (EPA Method #613); or gas chromatography plus mass spectrometry (EPA Method #625).

Routes of Entry: Inhalation, skin absorption; ingestion; skin and/or eye contact.

Harmful Effects and Symptoms

Short Term Exposure: Note: 2,3,7,8-TCDD is one of the most toxic synthetic chemicals, and the most toxic of the 75 dioxins. LD$_{50}$ = (oral-rat) 0.05 mg/kg. Exposure can cause headache, weakness, and digestive disturbances. Most symptoms develop slowly, over many days. Inhalation can cause burning sensation in nose and throat, headache, dizziness, nausea, vomiting, pain in the joints; tiredness, emotional disorders; blurred vision; muscle pain; nervousness, irritability, and intolerance to cold. Itching, swelling, and redness, followed by acne-like eruptions of the skin known as chloracne commonly occur. Symptoms of chloracne may appear weeks or months after initial exposure and may last a few months or up to 15 years. Can cause abnormalities of liver, pancreas, circulatory system, respiratory system, and death. Skin contact with very small amounts can cause chloracne. Eye contact can cause burning and irritation. Animal studies suggest that daily exposure to amounts smaller than one grain of salt may cause severe symptoms and death within a few weeks.

Long Term Exposure: Can cause effects under inhalation, especially chloracne (an acne-like skin rash), as well as numbness and tingling in arms and legs. Can cause allergic dermatitis porphyria; gastrointestinal disturbance; possible reproductive problems. A blood abnormality may occur which may include light sensitive skin, blisters, dark skin coloration; excessive hair growth and dark red urine. Reproductive problems and an increased susceptibility to infection may occur. TCDD is considered a potential occupational carcinogen because extremely low levels cause cancer and birth defects in animals. May cause liver, kidney damage. May be a teratogen. May decrease fertility in males and females.

Points of Attack: Eyes, skin, liver, kidneys, reproductive system, liver, and kidneys. Cancer site in animals: tumors at many sites; liver, lung, mouth, tongue, skin cancer. Causes lymphomas in humans.

Medical Surveillance: In short, contact with TCDD should be avoided but obviously careful preplacement and regular physical exams should be carried out in those cases where worker exposure cannot be avoided, with emphasis on liver and kidney function studies. Examination of the nervous system.

First Aid: If this chemical gets into the eyes, remove any contact lenses at once and irrigate immediately for at least 15 min, occasionally lifting upper and lower lids. Seek medical attention immediately. If this chemical contacts the skin, remove contaminated clothing and wash immediately with soap and water. Seek medical attention immediately. If this chemical has been inhaled, remove from exposure, begin rescue breathing (using universal precautions, including resuscitation mask) if breathing has stopped and CPR if heart action has stopped. Transfer promptly to a medical facility. When this chemical has been swallowed, get medical attention. Give large quantities of water and induce vomiting. Do not make an unconscious person vomit.

Note to physician: Liver and nerve function screening tests should be performed. Also monitor serum triglycerides and cholesterol.

Personal Protective Methods: Wear protective gloves and clothing to prevent any reasonable probability of skin contact. Safety equipment suppliers/manufacturers can provide recommendations on the most protective glove/clothing material for your operation. All protective clothing (suits, gloves, footwear, headgear) should be clean, available each day, and put on before work. Contact lenses should not be worn when working with this chemical. Wear dust-proof chemical goggles and face shield unless full-face-piece respiratory protection is worn. Employees should wash immediately with soap when skin is wet or contaminated. Provide emergency showers and eyewash.

Respirator Selection: At any detectable concentration: SCBAF: Pd,Pp (APF = 10,000) (any NIOSH/MSHA- or European Standard EN 149-approved self-contained breathing apparatus that has a full face-piece and is operated in a pressure-demand or other positive-pressure mode); or SaF: Pd,Pp: ASCBA (APF = 10,000) (any supplied-air respirator that has a full face-piece and is operated in a pressure-demand or other positive-pressure mode in combination with an auxiliary, self-contained breathing apparatus operated in a pressure-demand or other positive-pressure mode). *Escape:* GmFOv100 (APF = 50) [any air-purifying, full-face-piece respirator (gas mask) with a chin-style, front- or back-mounted organic vapor canister having an N100, R100, or P100 filter]; or SCBAE (any appropriate escape-type, self-contained breathing apparatus).

Storage: Color Code—Blue: Health Hazard/Poison: Store in a secure poison location. Prior to working with this chemical you should be trained on its proper handling and storage. A regulated, marked area should be established where this chemical is handled, used, or stored in compliance with OSHA Standard 1910.1045. Use in isolated area with adequate ventilation, preferably a hood; segregated glassware and tools; and plastic-backed absorbent. Thoroughly wash hands and forearms after each manipulation and before leaving work area. Use the same precautions required for radioactive work.

Spill Handling: Evacuate persons not wearing protective equipment from area of spill or leak until cleanup is complete. Remove all ignition sources. Warn other workers of spill. Wear protective clothing, collect powdered material in the most convenient and safe manner and deposit in sealed containers. Rinse area with 1,1,1-trichloroethane, then wash with detergent and water. Ventilate area after cleanup is complete. It may be necessary to contain and dispose of this chemical as a hazardous waste. If material or contaminated runoff enters waterways, notify downstream users of potentially contaminated waters. Contact your local or federal environmental protection agency for specific recommendations. If employees are required to clean up spills, they must be properly trained and equipped. OSHA 1910.120(q) may be applicable.

Fire Extinguishing: Use dry chemical, carbon dioxide; water spray; or foam extinguishers. Poisonous gases are produced in fire. If material or contaminated runoff enters waterways, notify downstream users of potentially contaminated waters. Notify local health and fire officials and pollution control agencies. From a secure, explosion-proof location, use water spray to cool exposed containers. If cooling streams are ineffective (venting sound increases in volume and pitch, tank discolors, or shows any signs of deforming), withdraw immediately to a secure position. If employees are expected to fight fires, they must be trained and equipped in OSHA 1910.156. The only respirators recommended for firefighting are self-contained breathing apparatuses that have full face-pieces and are operated in a pressure-demand or other positive-pressure mode.

References

US Environmental Protection Agency. (1979). *2,3,7,8-Tetrachlorodibenzo-p-Dioxin: Ambient Water Quality Criteria*. Washington, DC

US Public Health Service. (November 1987). *Toxicological Profile for 2,3,7,8-Tetrachloro-Dibenzo-p-Dioxin*. Atlanta, GA: Agency for Toxic Substance & Disease Registry

US Environmental Protection Agency. (April 30, 1980). *TCDD, Health and Environmental Effects Profile No. 155*. Washington, DC: Office of Solid Waste

Sax, N. I. (Ed.). (1980). *Dangerous Properties of Industrial Materials Report*, 1, No. 2, 63–64

National Institute for Occupational Safety and Health. (January 23, 1984). *2,3,7,8-Tetrachlorodiabenzop-dioxin*. Current Intelligence Bulletin 40, DHHS (NIOSH), Publication No. 84-104. Cincinnati, OH

New York State Department of Health. (March 1986). *Chemical Fact Sheet: 2,3,7,8-TCDD (Dioxin)*. Albany, NY: Bureau of Toxic Substance Assessment

New Jersey Department of Health and Senior Services. (September, 2002). *Hazardous Substances Fact Sheet: 2,3,7,8-Tetrachlorodibenzo-p-dioxin*. Trenton, NJ.

Tetrachlorodifluoroethanes   T:0240

Molecular Formula: $C_2Cl_4F_2$
Common Formula: Cl_2FCCl_2F; CCl_3CF_2Cl
Synonyms: *1,1,1,2-:* 1,2-Difluoro-1,1,2,2-tetrachloroethane; Ethane, 1,1,2,2-tetrachloro-1,2-difluoro-; F-112; Freon 112; Genetron 112; Halocarbon 112; Refrigerant 111; *sym*-Tetrachloro-1,2-difluoroethane; 1,1,2,2-Tetrachloro-1,2-difluoroethane; Tetrachloro-1,2-difluoroethane
1,1,2,2-: 1,1-Difluoroperchloroethane; 2,2-Difluoro-1,1,1,2-tetrachlorethane; Ethane, 1,1,1,2-Tetrachloro-2,2-difluoro-; Halocarbon 112A; Refrigerant 112A; 1,1,1,2-Tetrachloro-2,2-difluoroethane
CAS Registry Number: 76-11-9 (1,1,1,2-T-2,2,-D); 76-12-0 (1,1,2,2-T-1,2,-D)
RTECS® Number: KI1420000 (1,1,2,2-T-1,2,-D); KI1425000 (1,1,1,2-T-2,2,-D)
UN/NA & ERG Number: Not regulated.
EC Number: 200-934-0 (1,1,1,2-T-2,2,-D); 200-935-6 (1,1,2,2-T-1,2,-D)
Regulatory Authority and Advisory Bodies
Air Pollutant Standard Set. See below, "Permissible Exposure Limits in Air" section.

Canada, WHMIS, Ingredients Disclosure List Concentration 1.0%.
European/International Regulations: Not listed in Annex 1.
WGK (German Aquatic Hazard Class): No value assigned (both CAS numbers).
Description: 1,1,2,2-tetrachloro-1,2-difluoroethane, CCl_2FCCl_2F, and 1,1,1,2-tetrachloro-2,2-difluoroethane, CCl_3CF_2Cl, are both colorless liquids or solids with slight ether-like odor. Molecular weight = 203.82 (either compound); Specific gravity (H_2O:1) = 1.65 at 25°C (both); Boiling point 93°C (1,1,2,2-) and 91.7°C (1,1,1,2-); Freezing/Melting point = 25°C (1,1,2,2-) and 41°C (1,1,1,2-); Vapor pressure = 40 mmHg at 25°C (both); 40 mmHg at 20°C[NIOSH]. Hazard Identification (based on NFPA-704 M Rating System): Health 2, Flammability 1, Reactivity 0. Both are practically insoluble in water; solubility = 0.01%.
Potential Exposure: Compound Description (1,1,2,2-): Tumorigen, Primary Irritant. 1,1,1,2- is used as a refrigerant, corrosion inhibitor, and blowing agent; making plastics. 1,1,2,2- is used as a refrigerant, a solvent extractant; and in the dry cleaning industry.
Incompatibilities: Acids and chemically active metals, such as potassium, beryllium, powdered aluminum; zinc, calcium, magnesium, and sodium.
Permissible Exposure Limits in Air
Conversion factor: 1 ppm = 8.34 mg/m^3 (both) at 25°C & 1 atm.
(1,1,1,2-)
OSHA PEL: 500 ppm/4170 mg/m^3 TWA.
NIOSH REL: 500 ppm/4170 mg/m^3 TWA.
ACGIH TLV®[1]: 100 ppm/834 mg/m^3 TWA.
NIOSH IDLH: 2000 ppm.
DFG MAK: 200 ppm/1700 mg/m^3 TWA; Peak Limitation Category II(2); Pregnancy Risk Group D.
Australia: TWA 500 ppm (4170 mg/m^3), 1993; Austria: MAK 500 ppm (4179 mg/m^3), 1999; Belgium: TWA 500 ppm (4170 mg/m^3), 1993; Denmark: TWA 500 ppm (4170 mg/m^3), 1999; Finland: TWA 500 ppm (4170 mg/m^3), STEL 625 ppm (5215 mg/m^3), 1999; France: VME 500 ppm (4170 mg/m^3), 1999; the Netherlands: MAC-TGG 4170 mg/m^3, 2003; Switzerland: MAK-W 500 ppm (4200 mg/m^3), 1999; United Kingdom: TWA 100 ppm (847 mg/m^3), STEL 100 ppm, 2000; Argentina, Bulgaria, Columbia, Jordan, South Korea, New Zealand, Singapore, Vietnam: ACGIH TLV®: TWA 500 ppm.
(1,1,2,2-)
OSHA PEL: 500 ppm/4170 mg/m^3 TWA.
NIOSH REL: 500 ppm/4170 mg/m^3 TWA.
ACGIH TLV®[1]: 500 TWA.
NIOSH IDLH: 2000 ppm
Protective Action Criteria (PAC)
1,1,2,2-T-1,2,-D (76-12-0)
TEEL-0: 50 ppm
PAC-1: 150 ppm
PAC-2: 2000 ppm
PAC-3: 2000 ppm

DFG MAK: 200 ppm/1700 mg/m³; Pregnancy Risk Group D.

Australia: TWA 500 ppm (4170 mg/m³), 1993; Austria: MAK 200 ppm (1690 mg/m³), 1999; Belgium: TWA 500 ppm (4170 mg/m³), 1993; Denmark: TWA 200 ppm (1665 mg/m³), 1999; Finland: TWA 500 ppm (4170 mg/m³), STEL 625 ppm (5215 mg/m³), 1999; France: VME 500 ppm (4170 mg/m³), 1999; the Netherlands: MAC-TGG 850 mg/m³, 2003; the Philippines: TWA 500 ppm (4170 mg/m³), 1993; Switzerland: MAK-W 500 ppm (4200 mg/m³), STEL 1000 ppm, 1999; United Kingdom: TWA 100 ppm (847 mg/m³), STEL 100 ppm, 2000; Argentina, Bulgaria, Columbia, Jordan, South Korea, New Zealand, Singapore, Vietnam: ACGIH TLV®: TWA 500 ppm. Several states have set guidelines or standards for these compounds in ambient air[60] ranging from 0.57 mg/m³ (Massachusetts) to 41.7 mg/m³ (North Dakota) to 51.7 mg/m³ (North Carolina) to 70.0 mg/m³ (Virginia) to 83.4 mg/m³ (Connecticut) to 99.296 mg/m³ (Nevada).

Determination in Air: NIOSH Analytical Method (IV) #1016; OSHA Analytical Method #7.

Routes of Entry: Inhalation, ingestion; skin and/or eye contact.

Harmful Effects and Symptoms

Short Term Exposure: Contact with either chemical can irritate the eyes and the skin, causing a rash or a burning feeling. 1,1,1,2-Tetrachloro-2,2-difluoroethane can affect you when breathed in. May cause central nervous system depression. High levels may irritate the lungs, causing coughing, shortness of breath; drowsiness, dyspnea (breathing difficulty). Higher exposures can cause pulmonary edema, a medical emergency that can be delayed for several hours. This can cause death. 1,1,2,2,-Tetrachloro-1,2-difluoroethane can affect you when breathed in. Exposure can cause you to become dizzy, lightheaded, and to pass out. Very high exposures could cause death. May affect the heart, causing an irregular rhythm. Exposure can cause you to become dizzy and lightheaded. Extremely high levels could cause you to pass out and even die. It can irritate the lungs, causing coughing and/or shortness of breath. Higher exposures can cause pulmonary edema, a medical emergency that can be delayed for several hours. This can cause death.

Long Term Exposure: Either compound may cause liver damage. Repeated exposure to either compound may reduce the number of white blood cells.

Points of Attack: Eyes, skin, respiratory system; central nervous system; heart.

Medical Surveillance: For the 1,1,1,2-compound: For those with frequent or potentially high exposure (half the TLV or greater), the following are recommended before beginning work and at regular times after that: lung function tests. If symptoms develop or overexposure is suspected, the following may be useful: Holter monitor (a special 24-h EKG to look for irregular heart rhythms). Consider chest X-ray after acute overexposure. Liver function tests. Complete blood count (CBC). For the *1,1,2,2-compound:* If symptoms develop or overexposure has occurred, blood tests for liver function, a complete blood count (CBC) and lung function tests may be useful. If heart rhythm symptoms are occurring, special 24-h EKG test might be indicated.

First Aid: If this chemical gets into the eyes, remove any contact lenses at once and irrigate immediately for at least 15 min, occasionally lifting upper and lower lids. Seek medical attention immediately. If this chemical contacts the skin, remove contaminated clothing and wash immediately with soap and water. Seek medical attention immediately. If this chemical has been inhaled, remove from exposure, begin rescue breathing (using universal precautions, including resuscitation mask) if breathing has stopped and CPR if heart action has stopped. Transfer promptly to a medical facility. When this chemical has been swallowed, get medical attention. Give large quantities of water and induce vomiting. Do not make an unconscious person vomit. Medical observation is recommended for 24–48 h after breathing overexposure, as pulmonary edema may be delayed. As first aid for pulmonary edema, a doctor or authorized paramedic may consider administering a corticosteroid spray. (This is particularly true of the 1,1,1,2-compound).

Personal Protective Methods: Wear protective gloves and clothing to prevent any reasonable probability of skin contact. Barrier® and Viton gloves; Tychem® BR, Tychem® TK, Responder®, and Trellchem suits. Safety equipment suppliers/manufacturers can provide recommendations on the most protective glove/clothing material for your operation. All protective clothing (suits, gloves, footwear, headgear) should be clean, available each day, and put on before work. Contact lenses should not be worn when working with this chemical. Wear splash-proof chemical goggles and face shield when working with liquid or dust-proof goggles and face shield when working with powders or dusts, unless full-face-piece respiratory protection is worn. Employees should wash immediately with soap when skin is wet or contaminated. Provide emergency showers and eyewash.

Respirator Selection: Up to 2000 ppm: Sa (APF = 10) (any supplied-air respirator); or SCBAF (APF = 50) (any self-contained breathing apparatus with a full face-piece). *Emergency or planned entry into unknown concentrations or IDLH conditions:* SCBAF: Pd,Pp (APF = 10,000) (any NIOSH/MSHA- or European Standard EN 149-approved self-contained breathing apparatus that has a full face-piece and is operated in a pressure-demand or other positive-pressure mode); or SaF: Pd,Pp: ASCBA (APF = 10,000) (any supplied-air respirator that has a full face-piece and is operated in a pressure-demand or other positive-pressure mode in combination with an auxiliary, self-contained breathing apparatus operated in a pressure-demand or other positive-pressure mode). *Escape:* GmFOv (APF = 50) [any air-purifying, full-face-piece respirator (gas mask) with a chin-style, front- or back-mounted organic vapor canister]; or

SCBAE (any appropriate escape-type, self-contained breathing apparatus).

Storage: Color Code—Green: General storage may be used. Prior to working with this chemical you should be trained on its proper handling and storage. Tetrachloro-difluoroethanes must be stored to avoid contact with chemically active metals (such as sodium, potassium, beryllium, zinc, powdered aluminum, or magnesium), since violent reactions occur. Store in tightly closed containers in a cool, well-ventilated area away from heat.

Shipping: Not regulated.

Spill Handling: Evacuate persons not wearing protective equipment from area of spill or leak until cleanup is complete. Remove all ignition sources. Absorb liquid in vermiculite, dry sand; earth, or similar material and deposit in sealed containers. Collect powdered material in the most convenient and safe manner and deposit in sealed containers. Ventilate area after cleanup is complete. It may be necessary to contain and dispose of this chemical as a hazardous waste. If material or contaminated runoff enters waterways, notify downstream users of potentially contaminated waters. Contact your local or federal environmental protection agency for specific recommendations. If employees are required to clean up spills, they must be properly trained and equipped. OSHA 1910.120(q) may be applicable.

Fire Extinguishing: Extinguish fire using an agent suitable for type of surrounding fire. Tetrachloro-difluoroethanes do not burn. Poisonous gases are produced in fire, including hydrogen chloride, hydrogen fluoride, and phosgene. If material or contaminated runoff enters waterways, notify downstream users of potentially contaminated waters. Notify local health and fire officials and pollution control agencies. From a secure, explosion-proof location, use water spray to cool exposed containers. If cooling streams are ineffective (venting sound increases in volume and pitch, tank discolors, or shows any signs of deforming), withdraw immediately to a secure position. If employees are expected to fight fires, they must be trained and equipped in OSHA 1910.156. The only respirators recommended for firefighting are self-contained breathing apparatuses that have full face-pieces and are operated in a pressure-demand or other positive-pressure mode.

References

New Jersey Department of Health and Senior Services. (September 2002). *Hazardous Substances Fact Sheet: 1,1,1,2-Tetrachloro-2,2-Difluoroethane.* Trenton, NJ

New Jersey Department of Health and Senior Services. (August 2002). *Hazardous Substances Fact Sheet: 1,1,2,2-Tetrachloro-1,2-Difluoroethane.* Trenton, NJ

1,1,1,2-Tetrachloroethane   T:0250

Molecular Formula: $C_2H_2Cl_4$
Common Formula: Cl_3CCH_2Cl

Synonyms: Ethane, 1,1,1,2-tetrachloro-; NCI-C52459; 1,1,1,2-Tetracloroetano (Spanish)

CAS Registry Number: 630-20-6
RTECS® Number: KI8450000
UN/NA & ERG Number: UN1702/151
EC Number: 211-135-1

Regulatory Authority and Advisory Bodies

Carcinogenicity: IARC: Animal Limited Evidence; Human No Adequate Data, *not classifiable as carcinogenic to humans*, Group 3, 1999; EPA: Possible Human Carcinogen; NCI: Carcinogenesis Studies (gavage); clear evidence: mouse; equivocal evidence: rat.

NTP: Toxicity studies, RPT#TOX-45, October 2000.

Air Pollutant Standard Set. See below, "Permissible Exposure Limits in Air" section.

US EPA Hazardous Waste Number (RCRA No.): U208.

RCRA, 40CFR261, Appendix 8 Hazardous Constituents.

RCRA 40CFR268.48; 61FR15654, Universal Treatment Standards: Wastewater (mg/L), 0.057; Nonwastewater (mg/kg), 6.0.

RCRA 40CFR264, Appendix 9; TSD Facilities Ground Water Monitoring List. Suggested test method(s) (PQL μg/L): 8010 (5); 8240 (5).

Safe Drinking Water Act: Priority List (55 FR 1470).

Reportable Quantity (RQ): 100 lb (45.4 kg).

EPCRA Section 313 Form R *de minimis* concentration reporting level: 1.0%.

US DOT Regulated Marine Pollutant (49CFR172.101, Appendix B)

Canada, WHMIS, Ingredients Disclosure List Concentration 1.0%.

European/International Regulations: Not listed in Annex 1.

WGK (German Aquatic Hazard Class): No value assigned.

Description: 1,1,1,2-Tetrachloroethane is a colorless to yellowish-red liquid. Molecular weight = 167.84; Specific gravity (H_2O:1) = 1.54 at 25°C; Boiling point = 135.6°C; Freezing/Melting point = −70°C; Vapor pressure = 14 mmHg at 25°C. Hazard Identification (based on NFPA-704 M Rating System): Health 3, Flammability 0, Reactivity 0. Slightly soluble in water; solubility = 0.1%.

Potential Exposure: Compound Description: Tumorigen, Mutagen. Primary Irritant. 1,1,1,2-Tetrachloroethane is used as a solvent and in manufacture of insecticides, herbicides, soil fumigants; blanches, paints, and a number of widely used products; as are the other chloroethanes.

Incompatibilities: Oxidizers, strong bases; dinitrogen tetraoxide, metal caustics, hot iron, potassium hydroxide, nitrogen tetroxide; 2,4-dinitrophenyl disulfide, chemically active metals: potassium, sodium, magnesium, and zinc.

Permissible Exposure Limits in Air

OSHA PEL: None

NIOSH REL: No numerical value but users are cautioned to handle with caution in the workplace. See *NIOSH Pocket Guide*, Appendix C.

Protective Action Criteria (PAC)

TEEL-0: 3 ppm

PAC-1: 7.5 ppm
PAC-2: 30 ppm
PAC-3: 30 ppm

Brazil[35] 4 ppm (27 mg/m^3) for work-place air over a 48-h week. Russia[43] set a MAC for tetrachloroethanes (no isomer indicated) of 5 mg/m^3 in work-place air.

Determination in Air: No method available.

Permissible Concentration in Water: For 1,1,1,2-tetrachoroethane the criterion to protect freshwater aquatic life is 9320 µg/L based on acute toxicity data. For saltwater aquatic life, no criterion for 1,1,1,2-tetrachloroethane can be derived using the guidelines, and there is insufficient data to estimate a criterion using other procedures. For the protection of human health, there is insufficient data to derive criteria for 1,1,1,2-tetrachloroethane.

Determination in Water: Octanol—water coefficient: Log K_{ow} = 2.4.

Routes of Entry: Inhalation, ingestion; skin and/or eye contact.

Harmful Effects and Symptoms

Short Term Exposure: Contact irritates the eyes and skin. Inhalation can irritate the respiratory tract causing coughing, wheezing, and/or shortness of breath. Can cause headache, nausea and vomiting. May cause central nervous system depression. Symptoms of exposure include weakness, restlessness, tremor, dizziness, drowsiness, irregular/irregularities of respiration; decreased muscle coordination; coma.

Long Term Exposure: There is limited evidence that this chemical causes liver cancer in animals. Can cause dermatitis; drying and cracking. May affect the liver and nervous system.

Points of Attack: Eyes, skin, central nervous system; liver.

Medical Surveillance: Liver function tests. Examination of the nervous system.

First Aid: If this chemical gets into the eyes, remove any contact lenses at once and irrigate immediately for at least 15 min, occasionally lifting upper and lower lids. Seek medical attention immediately. If this chemical contacts the skin, remove contaminated clothing and wash immediately with soap and water. Seek medical attention immediately. If this chemical has been inhaled, remove from exposure, begin rescue breathing (using universal precautions, including resuscitation mask) if breathing has stopped and CPR if heart action has stopped. Transfer promptly to a medical facility. When this chemical has been swallowed, get medical attention. Give large quantities of water and induce vomiting. Do not make an unconscious person vomit.

Personal Protective Methods: Wear protective gloves and clothing to prevent any reasonable probability of skin contact. Safety equipment suppliers/manufacturers can provide recommendations on the most protective glove/clothing material for your operation. Viton and polyvinyl alcohol are among the recommended protective materials. All protective clothing (suits, gloves, footwear, headgear) should be clean, available each day, and put on before work. Contact lenses should not be worn when working with this chemical. Wear splash-proof chemical goggles and face shield unless full-face-piece respiratory protection is worn. Employees should wash immediately with soap when skin is wet or contaminated. Provide emergency showers and eyewash.

Respirator Selection: Follow regulations in OSHA 29CFR1910.134 or European Standard EN149. Use a NIOSH/MSHA- or European Standard EN149-approved respirator; or use an approved supplied-air respirator with a full face-piece operated in the positive-pressure mode, or with a full face-piece, hood, or helmet in the continuous-flow mode; or use a NIOSH/MSHA- or European Standard EN149-approved self-contained breathing apparatus with a full face-piece operated in pressure-demand or other positive-pressure mode.

Storage: Color Code—Blue: Health Hazard/Poison: Store in a secure poison location. Prior to working with this chemical you should be trained on its proper handling and storage. A regulated, marked area should be established where this chemical is handled, used, or stored in compliance with OSHA Standard 1910.1045. Store in tightly closed containers in a cool, well-ventilated area away from potassium, sodium, dinitrogen tetraoxide; potassium hydroxide; nitrogen tetroxide; sodium potassium alloy; 2,4-dinitrophenyl disulfide. Metal containers involving the transfer of 5 gallons or more of this chemical should be grounded and bonded.

Shipping: Tetrachloroethane requires a shipping label of "POISONOUS/TOXIC MATERIALS." It falls in Hazard Class 6.1 and Packing Group II.

Spill Handling: Evacuate and restrict persons not wearing protective equipment from area of spill or leak until cleanup is complete. Remove all ignition sources. Ventilate area of spill or leak. Absorb liquids in vermiculite, dry sand; earth, peat, carbon, or a similar material and deposit in sealed containers. Keep this chemical out of a confined space, such as a sewer, because of the possibility of an explosion, unless the sewer is designed to prevent the buildup of explosive concentrations. It may be necessary to contain and dispose of this chemical as a hazardous waste. If material or contaminated runoff enters waterways, notify downstream users of potentially contaminated waters. Contact your local or federal environmental protection agency for specific recommendations. If employees are required to clean up spills, they must be properly trained and equipped. OSHA 1910.120(q) may be applicable.

Fire Extinguishing: This chemical is a noncombustible liquid. Poisonous gases, including chlorine and hydrogen chloride, are produced in fire. Use dry chemical, carbon dioxide; or alcohol foam extinguishers. Vapors are heavier than air and will collect in low areas. Storage containers and parts of containers may rocket great distances, in many directions. If material or contaminated runoff enters waterways, notify downstream users of potentially contaminated waters. Notify local health and fire officials and pollution control agencies. From a secure, explosion-proof location, use water spray to cool exposed containers. If cooling streams are ineffective (venting sound increases in volume and pitch, tank discolors, or shows any signs of deforming), withdraw immediately to a secure position. If employees are expected to fight fires, they must be trained and equipped in OSHA 1910.156. The only respirators recommended for firefighting are self-contained breathing apparatuses that have full face-pieces and are operated in a pressure-demand or other positive-pressure mode.

Disposal Method Suggested: Consult with environmental regulatory agencies for guidance on acceptable disposal practices. Generators of waste containing this contaminant ($\geq$100 kg/mo) must conform with EPA regulations governing storage, transportation, treatment, and waste disposal.

References

US Environmental Protection Agency. (April 30, 1980). *1,1,1,2-Tetrachloroethane, Health and Environmental Effects Profile No. 156.* Washington, DC: Office of Solid Waste

US Environmental Protection Agency. (1980). *Chlorinated Ethanes: Ambient Water Quality Criteria.* Washington, DC

Sax, N. I. (Ed.). (1984). *Dangerous Properties of Industrial Materials Report*, 4, No. 3, 93–95

New Jersey Department of Health and Senior Services. (October 1999). *Hazardous Substances Fact Sheet: 1,1,1,2-Tetrachloroethane.* Trenton, NJ

1,1,2,2-Tetrachloroethane   T:0260

Molecular Formula: $C_2H_2Cl_4$

Common Formula: $CHCl_2CHCl_2$

Synonyms: Acetylene tetrachloride; Bonoform; Cellon; 1,1-Dichloro-2,2-dichloroethane;   Ethane,1,1,2,2-tetrachloro-; NCI-C03554;   *sym*-Tetrachloroethane;   TCE;   1,1,2,2-Tetrachloraethan   (German);   1,1,2,2-Tetrachlorethane (French);   Tetrachlorethane;   1,1,2,2-Tetrachloro-; Tetrachlorure d'acetylene (French); Westron

CAS Registry Number: 79-34-5

RTECS® Number: KI8575000

UN/NA & ERG Number: UN1702/151

EC Number: 201-197-8 [*Annex I Index No.:* 602-015-00-3]

Regulatory Authority and Advisory Bodies

Carcinogenicity: IARC: Animal Limited Evidence; Human Inadequate Evidence, *not classifiable as carcinogenic to humans,* Group 3, 1999; EPA: Possible Human Carcinogen; NCI: Carcinogenesis Bioassay (gavage); clear evidence: mouse; equivocal evidence: rat; NIOSH: Potential occupational carcinogen.

US EPA Gene-Tox Program, Positive: *E. coli* polA without S9; *S. cerevisiae* gene conversion; Positive: *S. cerevisiae*—homozygosis; *S. cerevisiae*—reversion; Positive/limited: Carcinogenicity—mouse/rat.

NTP: Toxicity studies, RPT-TOX-45, October 2000.

Banned or Severely Restricted (UK, Belgium) (UN).[13]

Air Pollutant Standard Set. See below, "Permissible Exposure Limits in Air" section.

Clean Air Act: Hazardous Air Pollutants (Title I, Part A, Section 112).

US EPA Hazardous Waste Number (RCRA No.): U209.

RCRA 40CFR264, Appendix 9; TSD Facilities Ground Water Monitoring List.

Suggested methods (PQL μg/L): 8010 (0.5); 8240 (5).

RCRA 40CFR268.48; 61FR15654, Universal Treatment Standards: Wastewater (mg/L), 0.057; Nonwastewater (mg/kg), 6.0.

Safe Drinking Water Act: Priority List (55 FR 1470).

Reportable Quantity (RQ): 100 lb (45.4 kg).

EPCRA Section 313 Form R *de minimis* concentration reporting level: 1.0%.

US DOT Regulated Marine Pollutant (49CFR172.101, Appendix B).

California Proposition 65 Chemical: Cancer 1/1/90.

Canada, WHMIS, Ingredients Disclosure List Concentration 1.0%.

European/International Regulations: Hazard Symbol: T+, N; Risk phrases: R26/27; R51/53; Safety phrases: S1/2; S38; S45; S61 (see Appendix 4).

WGK (German Aquatic Hazard Class): 3—Severe hazard to waters.

Description: Tetrachloroethane is a heavy, volatile colorless to light yellow liquid. It has a sweetish, chloroform-like odor. The odor threshold is 0.5 ppm in water and 1.5 ppm in air. Molecular weight = 146.3; Specific gravity (H_2O:1) = 1.59 at 25°C; Boiling point = 146.7°C; Freezing/Melting point = -36.1°C; Vapor pressure = 5.8 mmHg at 25°C. Hazard Identification (based on NFPA-704 M Rating System): Health 3, Flammability 0, Reactivity 0. Slightly soluble in water; solubility = 0.29% at 20°C.

Potential Exposure: Compound Description: Agricultural Chemical; Tumorigen, Drug, Mutagen; Human Data. Tetrachloroethane is used as an intermediate in the trichloroethylene production from acetylene and as a solvent; as a dry cleaning agent; as a fumigant; in cement; and in lacquers. It is used in the manufacture of artificial silk, artificial leather, and artificial pearls. Recently, its use as a solvent has declined due to replacement by less toxic compounds. It is also used in the estimation of water content in tobacco and many drugs, and as a solvent for chromium chloride impregnation of furs.

Incompatibilities: Violent reaction with chemically active metals; strong caustics; strong acids; especially fuming sulfuric acid. Degrades slowly when exposed to air. Attacks plastic and rubber.

Permissible Exposure Limits in Air

OSHA PEL: 5 ppm/35 mg/m^3 TWA [skin].

NIOSH REL: 1 ppm/7 mg/m^3 TWA [skin]; potential occupational carcinogen; See *NIOSH Pocket Guide*, Appendix A and C (chloroethanes).

ACGIH TLV®[1]: 1 ppm/6.9 mg/m^3 TWA [skin]; confirmed animal carcinogen with unknown relevance to humans.

2518   1,1,2,2-Tetrachloroethane

NIOSH IDLH: potential occupational carcinogen 100 ppm.
Protective Action Criteria (PAC)
TEEL-0: 1 ppm
PAC-1: 3 ppm
PAC-2: 20 ppm
PAC-3: 100 ppm
DFG MAK: 1 ppm/7.0 mg/m^3 TWA; Peak Limitation Category II(2) [skin]; Carcinogen Category 3B; Pregnancy Risk Group D.
Australia: TWA 1 ppm (7 mg/m^3) [skin], 1993; Austria: MAK 1 ppm (7 mg/m^3) [skin], Suspected: carcinogen, 1993; Belgium: TWA 1 ppm (6.9 mg/m^3) [skin], 1993; Denmark: TWA 1 ppm (7 mg/m^3) [skin], 1999; Finland: TWA 1 ppm (7 mg/m^3), STEL 3 ppm (21 mg/m^3) [skin], 1999; France: VME 1 ppm (7 mg/m^3), VLE 5 ppm (35 mg/m^3), 1999; Japan: 1 ppm (6.9 mg/m^3) [skin], 1999; the Netherlands: MAC-TGG 7 mg/m^3 [skin], 2003; Norway: TWA 1 ppm (7 mg/m^3), 1999; the Philippines: TWA 5 ppm (35 mg/m^3) [skin], 1993; Poland: MAC (TWA) 5 mg/m^3; MAC (STEL) 35 mg/m^3, 1999; Russia: TWA 1 ppm, 1993; Switzerland: MAK-W 1 ppm (7 mg/m^3) [skin], 1999; Turkey: TWA 5 ppm (35 mg/m^3), 1993; Argentina, Bulgaria, Columbia, Jordan, South Korea, New Zealand, Singapore, Vietnam: ACGIH TLV®: confirmed animal carcinogen with unknown relevance to humans. Russia set a MAC for ambient air in residential areas[35] of 60 μg/m^3 on a once-daily basis. Several states have set guidelines or standards for 1,1,2,2-tetrachloroethane in ambient air[60] ranging from zero (North Carolina) to 1.2 μg/m^3 (Massachusetts) to 16.667 μg/m^3 (Kansas) to 23.3 μg/m^3 (New York) to 34.4 μg/m^3 (Connecticut) to 167.0 μg/m^3 (Pennsylvania).

Determination in Air: Adsorption on charcoal, workup with CS$_2$, analysis by gas chromatography/flame ionization. Use NIOSH Analytical Method #1019.[18]

Permissible Concentration in Water: To protect freshwater aquatic life: 9320 μg/L on an acute toxicity basis and 2400 μg/L on a chronic basis. *To protect saltwater aquatic life:* 9020 μg/L on an acute toxicity basis. *To protect human health:* preferably zero. An additional lifetime cancer risk of 1 in 100,000 is posed by a concentration of 1.7 μg/L. Russia[43] set a MAC in water bodies used for domestic purposes of 200 μg/L for tetrachloroethanes in general. Several states have set guidelines for 1,1,2,2-tetrachloroethane in drinking water[61] ranging from 0.5 μg/L (Arizona) to 1.7 μg/L (Kansas and Vermont) to 1.75 μg/L (Minnesota). New Mexico has set a standards of 10.0 μg/L.

Determination in Water: Inert gas purge followed by gas chromatography with halide specific detection (EPA Method #601) or gas chromatography plus mass spectrometry (EPA Method #624).

Routes of Entry: Inhalation of vapor and absorption of liquid through the skin (there is some evidence that tetrachloroethane absorbed through the skin affects the central nervous system only), ingestion, and/or eye contact.

Harmful Effects and Symptoms
Short Term Exposure: A central nervous system depressant. Symptoms may include nervousness, loss of appetite; constipation, tremors, fatigue, dizziness, nausea, vomiting, and headache. May cause liver and kidney damage. These symptoms have been reported after prolonged exposure to 75 ppm. *Inhalation:* Exposure of 116 ppm for 20 min has caused dizziness and vomiting. At 260 ppm for 10 min, irritation of nose and throat were felt also. At 335 ppm for 10 min, rapid fatigue was also experienced. These symptoms generally disappear when exposure stops. Large accidental exposures have resulted in death. *Skin:* Causes skin irritation. Absorption through skin is possible. Significant skin absorption may occur to produce toxic effects. Earliest and most common symptom is tremors of hands, followed by skin irritation, numbness, and effects listed above. Death has occurred from a combination of inhalation and skin absorption. *Eyes:* Causes irritation and tearing. *Ingestion:* Abdominal pain; nausea, and vomiting followed by similar symptoms as *inhalation*. As little as 3 mL (1/10 liquid oz) may cause unconsciousness. LD$_{50}$ = (oral-rat) 25 mg/kg.

Long Term Exposure: Repeated exposure may cause dermatitis, drying and cracking. May damage the blood forming organs, causing monocytosis (increased blood monocytes). May affect the central nervous system causing tremors, weakness, dizziness, decreased coordination; and even unconsciousness and death. May cause kidney and liver damage; jaundice, hepatitis, liver tenderness. Tetrachloroethane is a mutagen and a potential occupational carcinogen. There is limited evidence that this chemical is a teratogen in animals.

Points of Attack: Skin, liver, kidneys, central nervous system; gastrointestinal tract. Cancer site in animals: liver tumors.

Medical Surveillance: NIOSH lists the following tests: Blood Urea Nitrogen; liver function tests; urinalysis (routine). Preplacement and periodic examination should be comprehensive because of the possible involvement of many systems. Special attention should be given to liver, kidney, and bone marrow function, as well as to the central and peripheral nervous system. Alcoholism may be a predisposing factor.

First Aid: If this chemical gets into the eyes, remove any contact lenses at once and irrigate immediately for at least 15 min, occasionally lifting upper and lower lids. Seek medical attention immediately. If this chemical contacts the skin, remove contaminated clothing and wash immediately with soap and water. Seek medical attention immediately. If this chemical has been inhaled, remove from exposure, begin rescue breathing (using universal precautions, including resuscitation mask) if breathing has stopped and CPR if heart action has stopped. Transfer promptly to a medical facility. When this chemical has been swallowed, get medical attention. Give large quantities of water and induce vomiting. Do not make an unconscious person vomit.

Personal Protective Methods: Wear protective gloves and clothing to prevent any reasonable probability of skin contact.

Safety equipment suppliers/manufacturers can provide recommendations on the most protective glove/clothing material for your operation. Teflon™, Viton™, and polyvinyl alcohol are among the recommended protective materials. All protective clothing (suits, gloves, footwear, headgear) should be clean, available each day, and put on before work. Contact lenses should not be worn when working with this chemical. Wear splash-proof chemical goggles and face shield unless full-face-piece respiratory protection is worn. Employees should wash immediately with soap when skin is wet or contaminated. Provide emergency showers and eyewash.

Respirator Selection: At concentrations above the NIOSH REL, or where there is no REL, at any detectable concentration: SCBAF: Pd,Pp (APF = 10,000) (any NIOSH/MSHA- or European Standard EN 149-approved self-contained breathing apparatus that has a full face-piece and is operated in a pressure-demand or other positive-pressure mode); or SaF: Pd,Pp: ASCBA (APF = 10,000) (any supplied-air respirator that has a full face-piece and is operated in a pressure-demand or other positive-pressure mode in combination with an auxiliary, self-contained breathing apparatus operated in a pressure-demand or other positive-pressure mode). *Escape:* GmFOv (APF = 50) [any air-purifying, full-face-piece respirator (gas mask) with a chin-style, front- or back-mounted organic vapor canister]; or SCBAE (any appropriate escape-type, self-contained breathing apparatus).

Storage: Color Code—Blue: Health Hazard/Poison: Store in a secure poison location. Prior to working with this chemical you should be trained on its proper handling and storage. A regulated, marked area should be established where this chemical is handled, used, or stored in compliance with OSHA Standard 1910.1045. 1,1,2,2-Tetrachloroethane must be stored to avoid contact with chemically active metals (such as potassium, powdered aluminum; sodium, magnesium, and zinc) or strong acids (such as hydrochloric, sulfuric, and nitric), since violent reactions occur. Where possible, automatically transfer material from other storage containers to process containers.

Shipping: Tetrachloroethane requires a shipping label of "POISONOUS/TOXIC MATERIALS." It falls in Hazard Class 6.1 and Packing Group II.

Spill Handling: Evacuate and restrict persons not wearing protective equipment from area of spill or leak until cleanup is complete. Remove all ignition sources. Ventilate area of spill or leak. Absorb liquids in vermiculite, dry sand; earth, peat, carbon, or a similar material and deposit in sealed containers. Keep this chemical out of a confined space, such as a sewer, because of the possibility of an explosion, unless the sewer is designed to prevent the buildup of explosive concentrations. It may be necessary to contain and dispose of this chemical as a hazardous waste. If material or contaminated runoff enters waterways, notify downstream users of potentially contaminated waters. Contact your local or federal environmental protection agency for specific recommendations. If employees are required to clean up spills, they must be properly trained and equipped. OSHA 1910.120(q) may be applicable.

Fire Extinguishing: Extinguish fire using an agent suitable for type of surrounding fire. 1,1,2,2-Tetrachloroethane itself does not burn. Poisonous gases, including phosgene, chlorine, and hydrogen chloride, are produced in fire. Vapors are heavier than air and will collect in low areas. If material or contaminated runoff enters waterways, notify downstream users of potentially contaminated waters. Notify local health and fire officials and pollution control agencies. From a secure, explosion-proof location, use water spray to cool exposed containers. If cooling streams are ineffective (venting sound increases in volume and pitch, tank discolors, or shows any signs of deforming), withdraw immediately to a secure position. If employees are expected to fight fires, they must be trained and equipped in OSHA 1910.156. The only respirators recommended for firefighting are self-contained breathing apparatuses that have full face-pieces and are operated in a pressure-demand or other positive-pressure mode.

Disposal Method Suggested: Consult with environmental regulatory agencies for guidance on acceptable disposal practices. Generators of waste containing this contaminant (≥100 kg/mo) must conform with EPA regulations governing storage, transportation, treatment, and waste disposal. Incineration, preferably after mixing with another combustible fuel. Care must be exercised to assure complete combustion to prevent the formation of phosgene. An acid scrubber is necessary to remove the halo acids produced.[22]

References
National Institute for Occupational Safety and Health. (1977). *Criteria for a Recommended Standard: Occupational Exposure to 1,1,2,2-Tetrachoroethane.* NIOSH Document No. 77-121. Washington, DC
US Environmental Protection Agency. (1979). *Chemical Hazard Information Profile: 1,1,2,2-Tetrachloroethane.* Washington, DC
US Environmental Protection Agency. (1980). *Chlorinated Ethanes: Ambient Water Quality Criteria.* Washington, DC
US Environmental Protection Agency. (April 30, 1980). *1,1,2,2-Tetrachloroethane, Health and Environmental Effects Profile No. 157.* Washington, DC: Office of Solid Waste
Sax, N. I. (Ed.). *Dangerous Properties of Industrial Materials Report,* 1, No. 5, 84–85 (1981); 2, No. 6, 79–83 (1982); and 3, No. 2, 60–64 (1983)
US Public Health Service. (December 1988). *Toxicological Profile for 1,1,2,2-Tetrachloroethane.* Atlanta, GA: Agency for Toxic Substances and Disease Registry
New Jersey Department of Health and Senior Services. (October 1999). *Hazardous Substances Fact Sheet: 1,1,2,2-Tetrachloroethane.* Trenton, NJ
New York State Department of Health. (March 1986). *Chemical Fact Sheet: 1,1,2,2-Tetrachloroethane.* Albany, NY: Bureau of Toxic Substance Assessment (Version 3)

Tetrachloroethylene   T:0270

Molecular Formula: C_2Cl_4

Common Formula: $Cl_2C{=}CCl_2$

Synonyms: Ankilostin; Antisal 1; Carbon bichloride; Carbon dichloride; Didakene; Dilatin PT; Dowper; ENT 1,860; Ethene, tetrachloro-; Ethylene tetrachloride; Ethylene tetrachloro-; Fedal-UN; Freon 1110; NCI-C04580; Nema; PER; Perawin; PERC; Perchlor; Perchloraethylen, per (German); Perchlorethylene; Perchlorethylene, per (French); Perclene; Percloroetileno (Spanish); Percosolve; PERK; Perklone; Persec; Tetlen; Tetracap; Tetrachlora-ethen (German); 1,1,2,2-Tetrachloroethene; Tetrachloro-ethene; 1,1,2,2,-Tetrachloroethylene; Tetrachloroethylene; Tetracloroetileno (Spanish); Tetraleno; Tetralex; Tetravec; Tetroguer; Tetropil

CAS Registry Number: 127-18-4

RTECS® Number: KX3850000

UN/NA & ERG Number: UN1897

EC Number: 204-825-9

Regulatory Authority and Advisory Bodies

Carcinogenicity: IARC: Animal, Sufficient Evidence; Human, Limited Evidence, Group 2A, 1995; NTP: 11th Report on Carcinogens, 2004: Reasonably anticipated to be a human carcinogen; NCI: Carcinogenesis Studies (inhalation); clear evidence: mouse, rat; Carcinogenesis Bioassay (gavage); inadequate studies: rat; NTP: Carcinogenesis Studies (inhalation); clear evidence: mouse, rat; NIOSH: Potential occupational carcinogen.

US EPA Gene-Tox Program, Positive: Cell transform.—RLV F344 rat embryo; Positive: *S. cerevisiae* gene conversion; *S. cerevisiae*—homozygosis; Positive: *S. cerevisiae*—reversion; Positive/limited: Carcinogenicity—mouse/rat; Negative: Cell transform.—SA7/SHE.

US EPA, FIFRA 1998 Status of Pesticides: Canceled.

Banned or Severely Restricted (Japan) (UN).[13]

Air Pollutant Standard Set. See below, "Permissible Exposure Limits in Air" section.

Clean Air Act: Hazardous Air Pollutants (Title I, Part A, Section 112); 40CFR401.15 Section 307 Toxic Pollutants.

Clean Water Act: 40CFR423, Appendix A, Priority Pollutants.

US EPA Hazardous Waste Number (RCRA No.): U210.

RCRA, 40CFR261, Appendix 8 Hazardous Constituents.

RCRA 40CFR268.48; 61FR15654, Universal Treatment Standards: Wastewater (mg/L), 0.056; Nonwastewater (mg/kg), 6.0.

RCRA 40CFR264, Appendix 9; TSD Facilities Ground Water Monitoring List. Suggested test method(s) (PQL μg/L): 8010 (0.5); 8240 (5).

Safe Drinking Water Act: MCL, 0.005 mg/L; MCLG, zero; Regulated chemical (47 FR 9352).

Reportable Quantity (RQ): 100 lb (45.4 kg).

EPCRA Section 313 Form R *de minimis* concentration reporting level: 0.1%.

US DOT Regulated Marine Pollutant (49CFR172.101, Appendix B).

California Proposition 65 Chemical: Cancer 4/1/88.

Canada, WHMIS, Ingredients Disclosure List Concentration 1.0%.

European/International Regulations: Not listed in Annex 1.

WGK (German Aquatic Hazard Class): 3—Severe hazard to waters.

Description: Tetrachloroethylene is a clear, colorless, non-flammable liquid with a characteristic odor. The odor is noticeable at 47 ppm, though after a short period it may become inconspicuous, thereby becoming an unreliable warning signal. The odor threshold is variously given as 5 ppm[41] to 6.17 (3M). Molecular weight = 165.82; Specific gravity (H_2O:1) = 1.62 at 25°C; Boiling point = 121.1°C; Freezing/Melting point = −18.9°C; Vapor pressure = 14 mmHg at 25°C. Slightly soluble in water; solubility = 0.02%.

Potential Exposure: Compound Description: Agricultural Chemical; Tumorigen, Drug, Mutagen; Reproductive Effector; Human Data; Primary Irritant. Tetrachloroethylene is used in the textile industry and as a chemical intermediate or a heat-exchange fluid; a widely used solvent with particular use as a dry cleaning agent; a degreaser; a fumigant, and medically as an anthelmintic.

Incompatibilities: Violent reaction with strong oxidizers; powdered, chemically active metals, such as aluminum, lithium, beryllium, and barium; caustic soda; sodium hydroxide; potash. Tetrachloroethylene is quite stable. However, it reacts violently with concentrated nitric acid to give carbon dioxide as a primary product. Slowly decomposes on contact with moisture producing trichloroacetic acid and hydrochloric acid. Decomposes in UV light and in temperatures above 150°C, forming hydrochloric acid and phosgene.

Permissible Exposure Limits in Air

Conversion factor: 1 ppm = 6.78 mg/m^3 at 25°C & 1 atm.

OSHA PEL: 100 ppm TWA; 200 ppm Ceiling Concentration; 300 ppm (5-min maximum peak in any 3 h).

NIOSH REL: Potential occupational carcinogen; limit exposure to lowest feasible concentration; See *NIOSH Pocket Guide*, Appendix A.

ACGIH TLV®[1]: 25 ppm/170 mg/m^3 TWA; 100 ppm/685 mg/m^3 STEL, animal carcinogen with unknown relevance to humans; BEI issued.

NIOSH IDLH: potential occupational carcinogen 150 ppm.

Protective Action Criteria (PAC)*

TEEL-0: 25 ppm

PAC-1: **35** ppm

PAC-2: **230** ppm

PAC-3: **1200** ppm

*AEGLs (Acute Emergency Guideline Levels) & ERPGs (Emergency Response Planning Guideline) are in **bold face**.

DFG MAK: [skin]; Carcinogen Category 3B; BAT: 1 mg [tetrachloroethylene]/L in blood at the beginning of the next shift.

Arab Republic of Egypt: TWA 5 ppm (35 mg/m^3) [skin], 1993; Australia: TWA 50 ppm (335 mg/m^3), STEL 150 ppm, carcinogen, 1993; Austria: MAK 50 ppm (345 mg/m^3), Suspected: carcinogen, 1999; Belgium: TWA 50 ppm (339 mg/m^3), STEL 200 ppm (1368 mg/m^3), 1993; Denmark: TWA 10 ppm (70 mg/m^3) [skin], 1999; Finland: TWA 50 ppm (335 mg/m^3), STEL 75 ppm (520 mg/m^3) [skin], 1999; France: VME 50 ppm (335 mg/m^3), carcinogen, 1999; the Netherlands: MAC-TGG 240 mg/m^3 [skin], 2003; Japan: 50 ppm (340 mg/m^3), 2B carcinogen, 1999; Norway: TWA 20 ppm (130 mg/m^3), 1999; the Philippines: TWA 100 ppm (670 mg/m^3), 1993; Poland: MAC (TWA) 60 mg/m^3; MAC (STEL) 480 mg/m^3, 1999; Russia: TWA 50 ppm, STEL 10 mg/m^3, 1993; Sweden: NGV 10 ppm (70 mg/m^3), KTV 25 ppm (170 mg/m^3), carcinogen, 1999; Switzerland: MAK-W 50 ppm (345 mg/m^3), STEL 100 ppm [skin], 1999; Thailand: TWA 100 ppm, STEL 200 ppm, 1993; United Kingdom: TWA 50 ppm (345 mg/m^3), STEL 199 ppm, 2000; New Zealand, Singapore, Vietnam: ACGIH TLV®: STEL 100 ppm. Limits in ambient air in residential areas have been set by The Czech Republic[35] at 1.0 mg/m^3 on a daily average basis and 4.0 mg/m^3 on a half-hour basis as well as by Russia at 0.06 mg/m^3 on a daily average basis and 0.5 mg/m^3 on a once-daily basis. Several states have set guidelines or standards for tetrachloroethylene in ambient air[60] ranging from 0.05 μg/m^3 (Rhode Island) to 0.18 μg/m^3 (Massachusetts) to 1.7 μg/m^3 (Michigan) to 5.882 μg/m^3 (Kansas) to 21.0 μg/m^3 (North Carolina) to 1116.0 μg/m^3 (New York) to 1700.0 μg/m^3 (Connecticut) to 3350.0 μg/m^3 (Indiana, South Carolina, South Dakota) to 3350−13,400 μg/m^3 (North Dakota) to 5600 μg/m^3 (Virginia) to 7976.0 μg/m^3 (Nevada) to 8040 μg/m^3 (Pennsylvania).

Determination in Air: Use NIOSH Analytical Method #1003, Hydrocarbons, halogenated[18]; #2549, Volatile organic compounds; OSHA Analytical Method #1001.

Permissible Concentration in Water: *To protect freshwater aquatic life:* 5,280 μg/L on an acute toxicity basis and 840 μg/L on a chronic toxicity basis. *To protect saltwater aquatic life:* 10,200 μg/L on an acute toxicity basis and 450 μg/L on a chronic toxicity basis. *To protect human health:* preferably 0. An additional lifetime cancer risk of 1 in 100,000 is posed by a concentration of 8.0 μg/L.[6] A lifetime health advisory of 10 μg/L has been suggested by EPA.[48] A drinking water maximum of 5 μg/L has more recently been suggested by EPA.[62] WHO[35] has set 10 μg/L as a guideline for drinking water as has Japan. Several states have set guidelines for PCE in drinking water[61] ranging from 0.67 μg/L (New Hampshire) to 1.0 μg/L (Arizona) to 4.0 μg/L (California) to 5.0 μg/L (Colorado) to 7.0 μg/L (Kansas and Vermont) to 20.0 μg/L (Connecticut, Massachusetts and Wisconsin) to 35.0 μg/L (Maine). Standards have been set at 1.0 μg/L (New Jersey) to 3.0 μg/L (Florida) to 20.0 μg/L (New Mexico).

Determination in Water: Inert gas purge followed by gas chromatography with halide-specific detection (EPA Method #601); or gas chromatography plus mass spectrometry (EPA Method #624). Octanol−water coefficient: Log K_{ow} = 2.9.

Routes of Entry: Inhalation of vapor, percutaneous absorption of liquid, ingestion; skin and/or eye contact.

Harmful Effects and Symptoms

Short Term Exposure: *Inhalation:* Irritates the eyes and respiratory tract causing coughing and/or shortness of breath. High exposure can cause headache, dizziness, light-headedness, nausea, vomiting, and unconsciousness. Higher exposures can cause pulmonary edema, a medical emergency that can be delayed for several hours. This can cause death. Exposures of 200 ppm for 1 h can cause irritation of the nose, mouth and throat; dizziness, headaches and light-headedness; exposures of 1000 ppm for 30 min can cause difficult breathing; weakness, loss of muscle control; irritability, tremors, convulsions, paralysis, coma, heart irregularities and death. *Skin:* Contact can cause irritation and burns. Can cause dry, scaly skin; a mild-to-moderate burning sensation; redness and inflammation. *Eyes:* Can cause burning and irritation. *Ingestion:* Can cause nausea, vomiting, diarrhea, bloody stool; a reddening of face and neck; weakness and loss of muscle control.

Long Term Exposure: May affect the liver, kidneys, and nervous system. Exposures *over 200 ppm* during weeks or months can cause irritation of the respiratory tract, nausea, headache, sleeplessness, abdominal pain; constipation, dizziness, increased perspiration; fatigue, skin infection; kidney and liver damage; fluid in the lungs and coma. Long term exposure can cause dermatitis; drying and cracking of the skin. Tetrachloroethylene has caused liver cancer and birth defects in mice. Whether it causes cancer in humans is unknown. May damage the developing fetus.

Points of Attack: Liver, kidneys, lungs.

Medical Surveillance: NIOSH lists the following tests: whole blood (chemical/metabolite). For those with frequent or potentially high exposure (half the TLV or greater, or significant skin contact) the following are recommended before beginning work and at regular times after that: urinalysis. Liver function tests. Lung function tests. If symptoms develop or overexposure is suspected, the following may be useful: consider chest X-ray after acute overexposure. Special 24-h EKG (Holter monitor) to look for irregular hearth beat.

First Aid: If this chemical gets into the eyes, remove any contact lenses at once and irrigate immediately for at least 15 min, occasionally lifting upper and lower lids. Seek medical attention immediately. If this chemical contacts the skin, remove contaminated clothing and wash immediately with soap and water. Seek medical attention immediately. If this chemical has been inhaled, remove from exposure, begin rescue breathing (using universal precautions, including resuscitation mask) if breathing has stopped and CPR if heart action has stopped. Transfer promptly to a medical facility. When this chemical has been swallowed, get medical attention. Give large quantities of water and induce vomiting. Do not make an unconscious person vomit. Medical observation is recommended for 24–48 h after breathing overexposure, as pulmonary edema may be delayed. As first aid for pulmonary edema, a doctor or authorized paramedic may consider administering a corticosteroid spray.

Personal Protective Methods: Wear protective gloves and clothing to prevent any reasonable probability of skin contact. Safety equipment suppliers/manufacturers can provide recommendations on the most protective glove/clothing material for your operation. All protective clothing (suits, gloves, footwear, headgear) should be clean, available each day, and put on before work. Contact lenses should not be worn when working with this chemical. Wear splash-proof chemical goggles and face shield unless full-face-piece respiratory protection is worn. Employees should wash immediately with soap when skin is wet or contaminated. Provide emergency showers and eyewash. Exposure to tetrachloroethylene should not be controlled with the use of respirators except: during the time period necessary to install or implement engineering or work practice controls; in work situations in which engineering and work practice controls are technically not feasible; to supplement engineering and work practice controls when such controls fail to adequately control exposure to tetrachloroethylene; for operations which require entry into tanks or closed vessels; or in emergencies.

Respirator Selection: NIOSH: *At concentrations above the NIOSH REL, or where there is no REL, at any detectable concentration:* SCBAF: Pd,Pp (APF = 10,000) (any NIOSH/MSHA- or European Standard EN 149-approved self-contained breathing apparatus that has a full face-piece and is operated in a pressure-demand or other positive-pressure mode); or SaF: Pd,Pp: ASCBA (APF = 10,000) (any supplied-air respirator that has a full face-piece and is operated in a pressure-demand or other positive-pressure mode in combination with an auxiliary, self-contained breathing apparatus operated in a pressure-demand or other positive-pressure mode). *Escape:* GmFOv (APF = 50) [any air-purifying, full-face-piece respirator (gas mask) with a chin-style, front- or back-mounted organic vapor canister]; or SCBAE (any appropriate escape-type, self-contained breathing apparatus).

Storage: Color Code—Blue: Health Hazard/Poison: Store in a secure poison location. Prior to working with this chemical you should be trained on its proper handling and storage. A regulated, marked area should be established where this chemical is handled, used, or stored in compliance with OSHA Standard 1910.1045. Tetrachloroethylene must be stored to avoid contact with strong oxidizers, such as chlorine, bromine, and chlorine dioxide; chemically active metals, such as barium, lithium, and beryllium; and nitric acid, since violent reactions occur. Store in tightly closed containers in a cool, well-ventilated area away from heat.

Shipping: Tetrachloroethylene requires a shipping label of "POISONOUS/TOXIC MATERIALS." It falls in Hazard Class 6.1 and Packing Group III.

Spill Handling: Evacuate and restrict persons not wearing protective equipment from area of spill or leak until cleanup is complete. Remove all ignition sources. Ventilate area of spill or leak. Absorb liquids in vermiculite, dry sand; earth, peat, carbon, or a similar material and deposit in sealed containers. Keep this chemical out of a confined space, such as a sewer, because of the possibility of an explosion, unless the sewer is designed to prevent the buildup of explosive concentrations. It may be necessary to contain and dispose of this chemical as a hazardous waste. If material or contaminated runoff enters waterways, notify downstream users of potentially contaminated waters. Contact your local or federal environmental protection agency for specific recommendations. If employees are required to clean up spills, they must be properly trained and equipped. OSHA 1910.120(q) may be applicable.

Fire Extinguishing: Extinguish fire using an agent suitable for type of surrounding fire. Tetrachloroethylene itself does not burn. Poisonous gases, including hydrogen chloride, phosgene, and chlorine are produced in fire or heat above 150°C. Vapors are heavier than air and will collect in low areas. Storage containers and parts of containers may rocket great distances, in many directions. If material or contaminated runoff enters waterways, notify downstream users of potentially contaminated waters. Notify local health and fire officials and pollution control agencies. From a secure, explosion-proof location, use water spray to cool exposed containers. If cooling streams are ineffective (venting sound increases in volume and pitch, tank discolors, or shows any signs of deforming), withdraw immediately to a secure position. If employees are expected to fight fires, they must be trained and equipped in OSHA 1910.156. The only respirators recommended for firefighting are self-contained breathing apparatuses that have full face-pieces and are operated in a pressure-demand or other positive-pressure mode.

Disposal Method Suggested: Consult with environmental regulatory agencies for guidance on acceptable disposal practices. Generators of waste containing this contaminant (≥100 kg/mo) must conform with EPA regulations governing storage, transportation, treatment, and waste disposal. Incineration, preferably after mixing with another combustible fuel. Care must be exercised to assure complete combustion to prevent the formation of phosgene. An acid scrubber is necessary to remove the halo acids produced. Alternatively, PCE may be recovered from waste gases and reused.

References

National Institute for Occupational Safety and Health. (1986). *Criteria for a Recommended Standard: Occupational Exposure to Tetrachloroethylene*. NIOSH Document 76-185

National Institute for Occupational Safety and Health. (January 20, 1978). *Current Intelligence Bulletin No. 20: Tetrachloroethylene*. Washington, DC

US Public Health Service. (December 1987). *Toxicological Profile for Tetrachloroethylene*. Atlanta, GA: Agency for Toxic Substances and Disease Registry

US Environmental Protection Agency. (1980). *Tetrachloroethylene: Ambient Water Quality Criteria.* Washington, DC

US Environmental Protection Agency. (January 1980). *Health Assessment Document for Tetrachloroethylene.* Research Triangle Park, NC: Environmental Criteria and Assessment Office

US Environmental Protection Agency. (April 30, 1980). *Tetrachloroethylene, Health and Environmental Effects Profile No. 158.* Washington, DC: Office of Solid Waste

New York State Department of Health. (March 1986). *Chemical Fact Sheet: Tetrachloroethylene.* Albany, NY: Bureau of Toxic Substance Assessment (Version 2)

US Environmental Protection Agency, Special Review and Reregistration Division Office of Pesticide Programs. (1998). *Agency Status of Pesticides in Registration, Reregistration, and Special Review* (Rainbow Report). Washington, DC

New Jersey Department of Health and Senior Services. (March 2002). *Hazardous Substances Fact Sheet: Tetrachloroethylene.* Trenton, NJ

Tetracycline   T:0280

Molecular Formula: $C_{22}H_{24}N_2O_8$

Synonyms: Abramycin; Abricycline; Achromycin; Agromicina; Ambramicina; Ambramycin; Bio-tetra; Bristaciclin; Bristacycline; Cefracycline suspension; Criseocicline; Cyclomycin; Democracin; Deschlorobiomycin; Hostacyclin; Liquamycin; 6-Methyl-1,11-dioxy-2-naphthacenecarboxamide; Neocycline; Oletetrin; Panmycin; Polycycline; Purocyclina; Robitet; Sanclomycine; Sigmamycin; SK-Tetracycline; Steclin; T-125; Tetrabon; Tetracycline I; Tetracyn; Tetradecin; Tetraverine; Tsiklomitsin; ω-Mycin

Hydrochloride: Achro; Achromycin; Achromycin hydrochloride; Achromycin V; AI3-50120; Amycin, hydrochloride; Artomycin; Bristacycline; Cefracycline tablets; Chlorhydrate de tetracycline (French); Diacycine; 4-(Dimethylamino)-1,4,4a,5,5a,6,11,12a-octahydro-3,6,10,12,12a-pentahydroxy-6-methyl-1,11-dioxo-2-naphthacenecarboxamide monohydrochloride; Dumocycin; Medamycin; Mephacyclin; 2-Naphthacenecarbo xamide, 4-(dimethylamino)-1,4,4a,5,5a,6,11,12a-octahydro-3,6,10,12,12a-pentahydroxy-6-methyl-1,11-dioxo-, monohydrochloride; 2-Naphthacenecarbo xamide,4-(dimethylamino)-1,4,4a,5,5a,6,11,12a-octahydro-3,6, 10,12,12a-pentahydroxy-6-methyl-1,11-dioxo-, monohydrochloride [4s-(4a,4a.a,5aa,6b,12aa)]-; NCI-C55561; Neocyclin; Paltet; Panmycin hydrochloride; Polycycline hydrochloride; Qidtet; Quadracycline; Remicyclin; Ricycline; RO-Cycline; SK-Tetracycline; Steclin hydrochloride; Stilciclina; Subamycin; Supramycin; Sustamycin; T-250 Capsules; TC Hydrochloride; Tefilin; Teline; Telotrex; TET-CY; Tetrabakat; Tetrablet; Tetracaps; Tetrachel; Tetracompren;

Tetracycline chloride; Tetracyn hydrochloride; Tetra-D; Tetralution; Tetramavan; Tetramycin; Tetrasure; Tetra-Wedel; Tetrosol; Topicycline; Totomycin; Triphacyclin; U-5965; Unicin; Unimycin; Vetquamycin-324

CAS Registry Number: 60-54-8; 64-75-5 (hydrochloride)

RTECS® Number: QI8750000

UN/NA & ERG Number: UN3249 (Medicines, toxic, solid, n.o.s.)/151

EC Number: 200-481-9; 200-593-8 (hydrochloride)

Regulatory Authority and Advisory Bodies

Banned or Severely Restricted (several countries) (UN).[13]

Air Pollutant Standard Set. See below, "Permissible Exposure Limits in Air" section.

Hydrochloride:

EPCRA Section 313 Form R *de minimis* concentration reporting level: 1.0%.

California Proposition 65 Developmental/Reproductive toxin [Tetracycline (internal use)] 10/1/92; [Tetracycline hydrochloride (internal use)] 1/1/91.

European/International Regulations: Not listed in Annex 1.

WGK (German Aquatic Hazard Class): No value assigned.

Description: Tetracycline trihydrate is a white crystalline substance. Molecular weight = 444.48; Freezing/Melting point = 170−175°C (decomposition). It is commonly employed as the *hydrochloride*; Molecular weight = 480.94; Freezing/Melting point = 214°C (decomposes). Hazard Identification (based on NFPA-704 M Rating System): Health 3, Flammability 0, Reactivity 1. Tetracycline is slightly soluble in water; the hydrochloride is highly soluble in water.

Potential Exposure: Tetracycline is an antibiotic medicine used as capsules, tablets, or intravenous injections against certain infections in humans and animals.

Incompatibilities: Although no dangerous incompatibilities are reported, the potency of this medicine is reduced by heat, sunlight, and solutions with pH <2; and destroyed by caustic hydroxide solutions.

Permissible Exposure Limits in Air

Protective Action Criteria (PAC)

TEEL-0: 0.75 mg/m^3

PAC-1: 2.5 mg/m^3

PAC-2: 15 mg/m^3

PAC-3: 500 mg/m^3

Russia[43] set a MAC in work-place air of 0.1 mg/m^3 and also MAC values for ambient air in residential areas of 0.01 mg/m^3 on a momentary basis and 0.006 mg/m^3 on a daily average basis.

Determination in Air: No method available.

Routes of Entry: Ingestion, inhalation.

Harmful Effects and Symptoms

Short Term Exposure: Tetracycline can affect you when breathed in. Exposure can cause sneezing, itching of the nose; stomach upset; vomiting and diarrhea.

Long Term Exposure: Tetracycline can cause an allergic skin rash to develop. If skin allergy develops, sunlight may exacerbate the reaction. May cause stomach upset with nausea, vomiting, and diarrhea. Tetracycline is used as a medical drug. Used that way, it may cause liver and kidney damage. It is not known if this can occur with occupational exposure. May cause mutations. Whether it poses a cancer risk requires more study. Handle with extreme caution. It may also damage the developing fetus.

Points of Attack: Liver, kidneys, skin.

Medical Surveillance: If symptoms develop or overexposure is suspected, the following may be useful: evaluation by a qualified allergist, including careful exposure history and special testing, may help diagnose skin allergy. Liver and kidney function tests.

First Aid: If this chemical gets into the eyes, remove any contact lenses at once and irrigate immediately for at least 15 min, occasionally lifting upper and lower lids. Seek medical attention immediately. If this chemical contacts the skin, remove contaminated clothing and wash immediately with soap and water. Seek medical attention immediately. If this chemical has been inhaled, remove from exposure, begin rescue breathing (using universal precautions, including resuscitation mask) if breathing has stopped and CPR if heart action has stopped. Transfer promptly to a medical facility. When this chemical has been swallowed, get medical attention. Give large quantities of water and induce vomiting. Do not make an unconscious person vomit.

Personal Protective Methods: Wear protective gloves and clothing to prevent any reasonable probability of skin contact. Safety equipment suppliers/manufacturers can provide recommendations on the most protective glove/clothing material for your operation. All protective clothing (suits, gloves, footwear, headgear) should be clean, available each day, and put on before work. Contact lenses should not be worn when working with this chemical. Wear dust-proof chemical goggles and face shield unless full-face-piece respiratory protection is worn. Employees should wash immediately with soap when skin is wet or contaminated. Provide emergency showers and eyewash.

Respirator Selection: Where there is potential for high exposures, use a NIOSH/MSHA- or European Standard EN149-approved supplied-air respirator with a full face-piece operated in the positive-pressure mode, or with a full face-piece, hood, or helmet in the continuous-flow mode; or use a NIOSH/MSHA- or European Standard EN149-approved self-contained breathing apparatus with a full face-piece operated in pressure-demand or other positive-pressure mode.

Storage: Color Code—Blue: Health Hazard/Poison: Store in a secure poison location. Prior to working with this chemical you should be trained on its proper handling and storage. Store in tightly closed containers in a cool, well-ventilated area away from sunlight, heat, and caustic materials.

Shipping: Medicine, solid, toxic, n.o.s., requires a shipping label of "POISONOUS/TOXIC MATERIALS." This material falls in Hazard Class 6.1 and Packing Group III.

Spill Handling: Evacuate persons not wearing protective equipment from area of spill or leak until cleanup is complete. Remove all ignition sources. Collect powdered material in the most convenient and safe manner and deposit in sealed containers. Ventilate area after cleanup is complete. It may be necessary to contain and dispose of this chemical as a hazardous waste. If material or contaminated runoff enters waterways, notify downstream users of potentially contaminated waters. Contact your local or federal environmental protection agency for specific recommendations. If employees are required to clean up spills, they must be properly trained and equipped. OSHA 1910.120(q) may be applicable.

Fire Extinguishing: This chemical is a noncombustible solid. Use any agent suitable for surrounding fires. Poisonous gases are produced in fire, including nitrogen oxides (and in the case of the *hydrochloride:* hydrogen chloride). If material or contaminated runoff enters waterways, notify downstream users of potentially contaminated waters. Notify local health and fire officials and pollution control agencies. From a secure, explosion-proof location, use water spray to cool exposed containers. If cooling streams are ineffective (venting sound increases in volume and pitch, tank discolors, or shows any signs of deforming), withdraw immediately to a secure position. If employees are expected to fight fires, they must be trained and equipped in OSHA 1910.156. The only respirators recommended for firefighting are self-contained breathing apparatuses that have full face-pieces and are operated in a pressure-demand or other positive-pressure mode.

References

New Jersey Department of Health and Senior Services. (February 1986). *Hazardous Substances Fact Sheet: Tetracycline.* Trenton, NJ

Tetraethylenepentamine   T:0290

Molecular Formula: $C_8H_{23}N_5$

Common Formula: $H_2N(C_2H_4NH)_3C_2H_4NH_2$

Synonyms: *N*-(2-Aminoethyl)-*N*-(2-[(2-aminoethyl) amino] ethyl-1,2-ethanediamine); Amino ethyl-1,2-ethanediamine; D.E.H. 26; 1,2-Ethanediamine, *N*-(2-aminoethyl)-*N'*-(2-aminoethyl) aminoethyl-; 1,2-Ethanediamine, *N*-(2-aminoethyl)-*N'*-[2-(2-aminoethyl)ethyl]-; 1,4,7,10,13-Pentaazatridecane

CAS Registry Number: 112-57-2

RTECS® Number: KH8585000

UN/NA & ERG Number: UN2320/153

EC Number: 203-986-2 [*Annex I Index No.:* 612-060-00-0]

Regulatory Authority and Advisory Bodies

Canada, WHMIS, Ingredients Disclosure List Concentration 1.0%.

European/International Regulations: Hazard Symbol: C, N; Risk phrases: R21/22; R34; R43; R51/53; Safety phrases: S1/2; S26; S36/37/39; S45; S61 (see Appendix 4).

WGK (German Aquatic Hazard Class): 2—Hazard to waters.

Description: Tetraethylenepentamine is a yellow, viscous liquid. Molecular weight = 189.36; Boiling point = 333–340°C; Freezing/Melting point = −40 to −30°C; Flash point = 163–185°C. Hazard Identification (based on NFPA-704 M Rating System): Health 2, Flammability 1, Reactivity 0. Soluble in water.

Potential Exposure: Tetraethylenepentamine is used as a solvent for resins and dyes, in manufacture of synthetic rubber; and intermediate for oil additives; in papermaking.

Incompatibilities: Violent reaction with strong oxidizers. This chemical is strongly alkaline; reacts with acids.

Permissible Exposure Limits in Air

AIHA WEEL: 5 mg/m^3 TWA [skin]; Potential for dermal sensitization.

Protective Action Criteria (PAC)

TEEL-0: 5 mg/m^3

PAC-1: 50 mg/m^3

PAC-2: 350 mg/m^3

PAC-3: 500 mg/m^3

Routes of Entry: Inhalation, ingestion, eye, and/or skin contact. Absorbed through the skin.

Harmful Effects and Symptoms

Short Term Exposure: Tetraethylenepentamine can affect you when breathed in and by passing through your skin. Tetraethylenepentamine is a corrosive chemical, and skin or eye contact can cause burns. The vapor is strongly irritating to the nose, throat, and bronchial tubes. Higher exposures can cause pulmonary edema, a medical emergency that can be delayed for several hours. This can cause death.

Long Term Exposure: Repeated exposure may damage the liver, kidneys, heart muscle, and/or brain; impairing thinking, reasoning, and concentration. Some related chemicals can cause a skin or lung allergy to develop. It is not known whether this chemical can cause the same allergies.

Points of Attack: Lungs, liver, kidneys, heart, brain.

Medical Surveillance: Before beginning employment and at regular times after that, for those with frequent or potentially high exposures, the following are recommended: lung function tests. Exam of central nervous system. If symptoms develop or overexposure is suspected, the following may be useful: consider chest X-ray after acute overexposure. Liver and kidney function tests.

First Aid: If this chemical gets into the eyes, remove any contact lenses at once and irrigate immediately for at least 15 min, occasionally lifting upper and lower lids. Seek medical attention immediately. If this chemical contacts the skin, remove contaminated clothing and wash immediately with soap and water. Seek medical attention immediately. If this chemical has been inhaled, remove from exposure, begin rescue breathing (using universal precautions, including resuscitation mask) if breathing has stopped and CPR if heart action has stopped. Transfer promptly to a medical facility. When this chemical has been swallowed, get medical attention. If victim is *conscious*, administer 50/50 solution of vinegar/water, water, or milk. Do not induce vomiting. Medical observation is recommended for 24–48 h after breathing overexposure, as pulmonary edema may be delayed. As first aid for pulmonary edema, a doctor or authorized paramedic may consider administering a corticosteroid spray.

Personal Protective Methods: Wear protective gloves and clothing to prevent any reasonable probability of skin contact. Safety equipment suppliers/manufacturers can provide recommendations on the most protective glove/clothing material for your operation. All protective clothing (suits, gloves, footwear, headgear) should be clean, available each day, and put on before work. Contact lenses should not be worn when working with this chemical. Wear splash-proof chemical goggles and face shield unless full-face-piece respiratory protection is worn. Employees should wash immediately with soap when skin is wet or contaminated. Provide emergency showers and eyewash.

Respirator Selection: Where there is potential for exposures to tetraethylenepentamine, use a NIOSH/MSHA- or European Standard EN149-approved supplied-air respirator with a full face-piece operated in the positive-pressure mode, or with a full face-piece, hood, or helmet in the continuous-flow mode; or use a NIOSH/MSHA- or European Standard EN149-approved self-contained breathing apparatus with a full face-piece operated in pressure-demand or other positive-pressure mode.

Storage: Color Code—White: Corrosive or Contact Hazard; Store separately in a corrosion-resistant location. Prior to working with this chemical you should be trained on its proper handling and storage. Tetraethylenepentamine must be stored to avoid contact with oxidizers (such as perchlorates, peroxides, permanganates, chlorates, and nitrates), since violent reactions occur. Store in tightly closed containers in a cool, well-ventilated area. Sources of ignition, such as smoking and open flames, are prohibited where tetraethylenepentamine is used, handled, or stored in a manner that could create a potential fire or explosion hazard.

Shipping: Tetraethylenepentamine requires a shipping label of "CORROSIVE." It falls in Hazard Class 8 and Packing Group III.

Spill Handling: Evacuate and restrict persons not wearing protective equipment from area of spill or leak until cleanup is complete. Remove all ignition sources. Ventilate area of spill or leak. Absorb liquids in vermiculite, dry sand; earth, peat, carbon, or a similar material and deposit in sealed containers. Keep this chemical out of a confined space, such as a sewer, because of the possibility of an explosion, unless the sewer is designed to prevent the buildup of explosive concentrations. It may be necessary to contain and dispose of this chemical as a hazardous waste. If material or contaminated runoff enters waterways, notify downstream users of potentially contaminated waters. Contact your local or federal environmental protection agency for specific recommendations. If employees are required to clean up spills, they must be properly trained and equipped. OSHA 1910.120(q) may be applicable.

Fire Extinguishing: This chemical is a combustible liquid. Poisonous gases, including nitrogen oxides; are produced in fire. Use dry chemical, carbon dioxide; or foam extinguishers. Vapors are heavier than air and will collect in low areas. Vapors may travel long distances to ignition sources and flashback. Vapors in confined areas may explode when exposed to fire. Containers may explode in fire. Storage containers and parts of containers may rocket great distances, in many directions. If material or contaminated runoff enters waterways, notify downstream users of potentially contaminated waters. Notify local health and fire officials and pollution control agencies. From a secure, explosion-proof location, use water spray to cool exposed containers. If cooling streams are ineffective (venting sound increases in volume and pitch, tank discolors, or shows any signs of deforming), withdraw immediately to a secure position. If employees are expected to fight fires, they must be trained and equipped in OSHA 1910.156. The only respirators recommended for firefighting are self-contained breathing apparatuses that have full face-pieces and are operated in a pressure-demand or other positive-pressure mode.

References

New Jersey Department of Health and Senior Services. (February 2000). *Hazardous Substances Fact Sheet: Tetraethylenepentamine.* Trenton, NJ

Tetraethyl lead   T:0300

Molecular Formula: $C_8H_{20}Pb$
Common Formula: $Pb(C_2H_5)_4$
Synonyms: Lead, tetraethyl-; Motor fuel antiknock compound; NCI-C54988; NSC-22314; Piombo tetra-etile; Plumbane, tetraethyl-; TEL; Tetraethylolovo; Tetraethylplumbane; Tetraethylplumbium
CAS Registry Number: 78-00-2
RTECS® Number: TP4550000
UN/NA & ERG Number: UN1649/131
EC Number: 201-075-4; Listed in part 3 of Annex I to Regulation (EC) 689/2008; HS 2931.00.

Regulatory Authority and Advisory Bodies

Carcinogenicity: IARC: Animal Inadequate Evidence; Human Inadequate Evidence, *not classifiable as carcinogenic to humans*, Group 3, 1987; NTP: 11th Report on Carcinogens, 2004: Reasonably anticipated to be a human carcinogen.
US EPA Gene-Tox Program, Positive: Cell transform.—SA7/SHE.
Toxic Substance (World Bank).[15]
Air Pollutant Standard Set. See below, "Permissible Exposure Limits in Air" section.

Clean Water Act: Section 311 Hazardous Substances/RQ 40CFR117.3 (same as CERCLA, see below).
US EPA Hazardous Waste Number (RCRA No.): P110.
RCRA, 40CFR261, Appendix 8 Hazardous Constituents.
Superfund/EPCRA 40CFR355, Extremely Hazardous Substances: TPQ = 100 lb (45.4 kg).
Reportable Quantity (RQ): 10 lb (4.54 kg).
US DOT Regulated Marine Pollutant (49CFR172.101, Appendix B) (liquid).
Canada, WHMIS, Ingredients Disclosure List Concentration 1.0%.
Rotterdam Convention Annex III [Chemicals Subject to the Prior Informed Consent Procedure (PIC)].
European/International Regulations: listed in Annex 1, Part 3 without Index Number.
WGK (German Aquatic Hazard Class): 3—Severe hazard to waters.

Description: Tetraethyl lead is a colorless oily liquid with a sweet, slight musty odor. In commerce it is usually dyed red, orange, or blue. Tetraethyl lead will decompose in bright sunlight yielding needle-like crystals of tri-, di-, and monoethyl lead compounds, which have a garlic odor. Molecular weight = 323.47; Specific gravity (H_2O:1) = 1.65 at 25°C; Boiling point = (decomposes) 118.9°C; Freezing/Melting point = −130°C; Vapor pressure = 0.2 mmHg at 25°C; Flash point = 93°C; Autoignition temperature ≥110°C. Explosive limits: LEL = 1.8%; UEL—unknown. Hazard Identification (based on NFPA-704 M Rating System): Health 3, Flammability 2, Reactivity 3. Practically insoluble in water; solubility = 0.00002%.

Potential Exposure: Compound Description: Tumorigen, Organometallic, Mutagen; Reproductive Effector; Human Data. Tetraethyl lead is used as a component of antiknock mixes for gas and as an intermediate in making fungicides; Tetraethyl lead (used as an antiknock compound in gasoline) can also contain impurities, such as *ethylene dibromide* and *ethylene dichloride*.

Incompatibilities: Violent reaction with strong oxidizers, sulfuryl chloride; halogens, oils and fats; rust, potassium permanganate. Decomposes slowly in light and at room temperature, and more rapidly at temperatures above 110°C. Attacks rubber and some plastics and coatings.

Permissible Exposure Limits in Air

OSHA PEL: 0.075 mg[Pb]/m³ TWA [skin].
NIOSH REL: 0.075 mg[Pb]/m³ TWA [skin].
ACGIH TLV®[1]: 0.1 mg[Pb]/m³ [skin]; not classifiable as a human carcinogen.
NIOSH IDLH: 40 mg[Pb]/m³.
Protective Action Criteria (PAC)
TEEL-0: 0.1170 mg/m³
PAC-1: 0.468 mg/m³
PAC-2: 4 mg/m³
PAC-3: 62.4 mg/m³
DFG MAK: 0.05 mg[Pb]/m³; Peak Limitation Category II (2) [skin]; Pregnancy Risk Group B.

Arab Republic of Egypt: TWA 0.1 mg/m^3, 1993; Australia: TWA 0.1 mg/m^3 [skin], 1993; Austria: MAK 0.1 ppm (0.075 mg/m^3) [skin], 1999; Belgium: TWA 0.1 mg/m^3 [skin], 1993; Denmark: TWA 0.007 ppm (0.05 mg/m^3) [skin], 1999; France: VME 0.10 mg/m^3 [skin], 1999; Hungary: TWA 0.005 mg/m^3, STEL 0.01 mg/m^3 [skin], 1993; the Netherlands: MAC-TGG 0.05 mg/m^3 [skin], 2003; Norway: TWA 0.001 ppm (0.075 mg/m^3), 1999; the Philippines: TWA 0.075 mg/m^3 [skin], 1993; Poland: MAC (TWA) 0.05 mg/m^3; MAC (STEL) 0.1 mg/m^3, 1999; Russia: STEL 0.005 mg/m^3 [skin], 1993; Sweden: NGV 0.05 mg/m^3, ktv 0.2 mg/m^3 [skin], 1999; Switzerland: MAK-W 0.01 ppm (0.075 mg/m^3), KZG-W 0.02 ppm [skin], 1999; Thailand: TWA 0.075 mg/m^3, 1993; United Kingdom: LTEL 0.10 mg/m^3, 1993; Argentina, Bulgaria, Columbia, Jordan, South Korea, New Zealand, Singapore, Vietnam: ACGIH TLV®: not classifiable as a human carcinogen. Several states have set guidelines or standards for TEL in ambient air[60] ranging from 1.0 µg/m^3 (North Dakota) to 1.5 µg/m^3 (Connecticut) to 1.6 µg/m^3 (Virginia) to 2.0 µg/m^3 (Nevada).

Determination in Air: Use NIOSH Analytical Method (IV) #2533, Tetraethyl lead (as Pb).

Permissible Concentration in Water: No criteria set, but EPA[32] has suggested a permissible ambient goal of 1.4 µg/L based on health effects. Russia[35,43] set a MAC of zero in water bodies used for domestic purposes.

Determination in Water: Octanol–water coefficient: Log K_{ow} = 4.15. See also "Lead" entry.

Routes of Entry: Inhalation, skin absorption; ingestion; skin and/or eye contact.

Harmful Effects and Symptoms

Short Term Exposure: Tetraethyl lead is extremely poisonous; may be fatal if inhaled, swallowed, or absorbed through the skin. Irritates the moist eyes, skin, and respiratory tract. Contact may cause burns to skin and eyes; causing permanent loss of vision. Most symptoms of poisoning are due to the effects of tetraethyl lead on the nervous system. Signs and symptoms of acute exposure to tetraethyl lead may be severe and include intoxication, anxiety, irritability, insomnia, violent/frightening, strange dreams; headache, disorientation, hyperexcitability, delusions, reduced memory; hallucinations, personality changes; tremors, convulsions, and death. Muscular weakness, ataxia, tremors, convulsions, cerebral edema; and coma may occur. A metallic taste may be noted. Sneezing, bronchitis, and pneumonia may be observed. Bradycardia (slow heart rate), hypotension (low blood pressure), hypothermia, and pallor may also occur. Gastrointestinal symptoms include vomiting and diarrhea.

Long Term Exposure: May be a reproductive toxin; may damage the developing fetus. There is limited evidence that this chemical causes cancer in animals. High levels can cause muscle and joint pains, weakness, muscle cramps; and easy fatigue. Repeated exposure may cause lead to accumulate in the body. May cause kidney and brain damage, and damage to the blood cells; causing anemia.

Points of Attack: Central nervous system; cardiovascular system, kidneys, eyes.

Medical Surveillance: NIOSH lists the following tests: whole blood (chemical/metabolite); whole blood (chemical/metabolite): blood urea nitrogen, calcium, carbon dioxide; Sugar/Glucose; Biologic Monitoring of Urine every 3 months; urine (chemical/metabolite); urine (chemical/metabolite), end-of-shift; urinalysis (routine). If symptoms develop or overexposure is suspected, the following may be useful: Urine test for lead (levels of 0.1 mg/L of urine indicate increased exposure. Such levels increase risk from further exposure). Blood lead tests are not usually accurate with exposure to tetraethyl lead. Kidney function tests. Examination of the nervous system.

First Aid: If this chemical gets into the eyes, remove any contact lenses at once and irrigate immediately for at least 15 min, occasionally lifting upper and lower lids. Seek medical attention immediately. If this chemical contacts the skin, remove contaminated clothing and wash immediately with soap and water. Seek medical attention immediately. If this chemical has been inhaled, remove from exposure, begin rescue breathing (using universal precautions, including resuscitation mask) if breathing has stopped and CPR if heart action has stopped. Transfer promptly to a medical facility. When this chemical has been swallowed, get medical attention. Give large quantities of water and induce vomiting. Do not make an unconscious person vomit.

Personal Protective Methods: Wear protective gloves and clothing to prevent any reasonable probability of skin contact. Safety equipment suppliers/manufacturers can provide recommendations on the most protective glove/clothing material for your operation. All protective clothing (suits, gloves, footwear, headgear) should be clean, available each day, and put on before work. Contact lenses should not be worn when working with this chemical. Wear dust-proof chemical goggles and face shield unless full-face-piece respiratory protection is worn. Employees should wash immediately with soap when skin is wet or contaminated. Provide emergency showers and eyewash.

Respirator Selection: *0.75 ppm:* Sa (APF = 10) (any supplied-air respirator). *1.875 ppm:* Sa:Cf (APF = 25) (any supplied-air respirator operated in a continuous-flow mode). *3.75 ppm:* SaT: Cf (APF = 50) (any supplied-air respirator that has a tight-fitting face-piece and is operated in a continuous-flow mode); or SCBAF (APF = 50) (any self-contained breathing apparatus with a full face-piece); or SaF (APF = 50) (any supplied-air respirator with a full face-piece). *40 ppm:* Sa: Pd,Pp (APF = 1000) (any supplied-air respirator operated in a pressure-demand or other positive-pressure mode). *Emergency or planned entry into unknown concentrations or IDLH conditions:* SCBAF: Pd,Pp (APF = 10,000) (any self-contained breathing apparatus that has a full face-piece and is operated in a pressure-demand or other positive-pressure mode); or SaF: Pd,Pp: ASCBA (APF = 10,000) (any supplied-air respirator that has a full face-piece and is operated in a pressure-demand or other positive-pressure mode in combination with an auxiliary self-contained breathing apparatus operated in a pressure-demand or other positive-pressure mode). *Escape:* GmFOv (APF = 50) [any air-purifying, full-face-piece respirator (gas mask) with a chin-style, front- or back-mounted organic vapor canister]; or SCBAE (any appropriate escape-type, self-contained breathing apparatus).

Storage: Color Code—Blue: Health Hazard/Poison: Store in a secure poison location. Prior to working with this chemical you should be trained on its proper handling and storage. Before entering confined space where this chemical may be present, check to make sure that an explosive concentration does not exist. Tetraethyl lead must be stored to avoid contact with oxidizers (such as perchlorates, peroxides, permanganates, chlorates, and nitrates) and chemically active metals (such as potassium, sodium, magnesium, and zinc), since violent reactions occur. Store in tightly closed containers in a cool, well-ventilated area away from heat. Protect storage containers from physical damage. Sources of ignition, such as smoking and open flames, are prohibited where tetraethyl lead is used, handled, or stored in a manner that could create a potential fire or explosion hazard. Use only nonsparking tools and equipment, especially when opening and closing containers of tetraethyl lead. Wherever tetraethyl lead is used, handled, manufactured, or stored, use explosion-proof electrical equipment and fittings.

Shipping: Motor fuel antiknock mixtures require a shipping label of "POISONOUS/TOXIC MATERIALS." It falls in Hazard Class 6.1 and Packing Group I.

Spill Handling: Evacuate and restrict persons not wearing protective equipment from area of spill or leak until cleanup is complete. Remove all ignition sources. Establish forced ventilation to keep levels below explosive limit. Absorb liquids in vermiculite, dry sand; earth, peat, carbon, or a similar material and deposit in sealed containers. Keep this chemical out of a confined space, such as a sewer, because of the possibility of an explosion, unless the sewer is designed to prevent the buildup of explosive concentrations. It may be necessary to contain and dispose of this chemical as a hazardous waste. If material or contaminated runoff enters waterways, notify downstream users of potentially contaminated waters. Contact your local or federal environmental protection agency for specific recommendations. If employees are required to clean up spills, they must be properly trained and equipped. OSHA 1910.120(q) may be applicable.

Fire Extinguishing: This chemical is a combustible liquid. Poisonous gases, including lead and carbon monoxide, are produced in fire. Use dry chemical, carbon dioxide; mist and foam. Vapors are heavier than air and will collect in low areas. Vapors may travel long distances to ignition sources and flashback. Vapors in confined areas may explode when exposed to fire. Containers may explode in fire. Storage containers and parts of containers may rocket great distances, in many directions. If material or contaminated runoff enters waterways, notify downstream users of potentially contaminated waters. Notify local health and fire officials and pollution control agencies. Use water from unmanned monitors and hose-holders to keep fire-exposed containers cool. When stopping leak, use water spray to protect firefighters from a secure, explosion-proof location, use water spray to cool exposed containers. If cooling streams are ineffective (venting sound increases in volume and pitch, tank discolors, or shows any signs of deforming), withdraw immediately to a secure position. If employees are expected to fight fires, they must be trained and equipped in OSHA 1910.156. The only respirators recommended for firefighting are self-contained breathing apparatuses that have full face-pieces and are operated in a pressure-demand or other positive-pressure mode.

Disposal Method Suggested: Consult with environmental regulatory agencies for guidance on acceptable disposal practices. Generators of waste containing this contaminant (≥100 kg/mo) must conform with EPA regulations governing storage, transportation, treatment, and waste disposal. Controlled incineration with scrubbing for collection of lead oxides which may be recycled or landfilled. It is also possible to recover alkyl lead compound from wastewaters as an alternative to disposal.

References

US Environmental Protection Agency. (November 30, 1987). *Chemical Hazard Information Profile: Tetraethyl Lead.* Washington, DC: Chemical Emergency Preparedness Program

New York State Department of Health. (March 1986). *Chemical Fact Sheet: Tetraethyl Lead.* Albany, NY: Bureau of Toxic Substance Assessment

New Jersey Department of Health and Senior Services. (March 2002). *Hazardous Substances Fact Sheet: Tetraethyl Lead.* Trenton, NJ

Tetraethyltin   T:0310

Molecular Formula: $C_8H_{20}Sn$
Synonyms: Stannane, tetraethyl-; TET; Tetraethylstannane; Tetraethyltin; Tin, tetraethyl-
CAS Registry Number: 597-64-8
RTECS® Number: WH8625000
UN/NA & ERG Number: UN2788/153
EC Number: 209-906-2
Regulatory Authority and Advisory Bodies
Air Pollutant Standard Set. See below, "Permissible Exposure Limits in Air" section.
Superfund/EPCRA 40CFR355, Extremely Hazardous Substances: TPQ = 100 lb (45.4 kg).
Reportable Quantity (RQ): 100 lb (45.4 kg).
Canada, WHMIS, Ingredients Disclosure List Concentration 1.0%.
European/International Regulations: Not listed in Annex 1.

WGK (German Aquatic Hazard Class): No value assigned.

Description: Tetraethyltin is a colorless liquid. Molecular weight = 234.97; Boiling point = 181°C; Freezing/Melting point = −112°C; Flash point = 53°C. Hazard Identification (based on NFPA-704 M Rating System): Health 3, Flammability 3, Reactivity 0. Insoluble in water.

Potential Exposure: Used as biocide, bactericide, fungicide and insecticide; preservative for wood, textile, paper, and leather. Not registered as a pesticide in the US.

Incompatibilities: Strong oxidizers.

Permissible Exposure Limits in Air

OSHA PEL: 0.1 mg[Sn]/m^3 TWA.

NIOSH REL: 0.1 mg[Sn]/m^3 TWA [skin].

ACGIH TLV$^{®[1]}$: 0.1 mg[Sn]/m^3 TWA; 0.2 mg[Sn]/m^3 STEL [skin].

NIOSH IDLH: 25 mg[Sn]/m^3 [skin].

Protective Action Criteria (PAC)

TEEL-0: 0.198 mg/m^3

PAC-1: 0.396 mg/m^3

PAC-2: 7 mg/m^3

PAC-3: 49.5 mg/m^3

DFG MAK: 0.1 mg[Sn]/m^3 inhalable fraction [skin]; Pregnancy Risk Group D.

Determination in Air: Use NIOSH Analytical Method (IV) #5504, Organotin.

Permissible Concentration in Water: A MAC in water bodies used for domestic purposes of 0.2 μg/L has been set by Russia joint project.[43]

Routes of Entry: Inhalation, skin and/or eye contact. Absorbed through the skin.

Harmful Effects and Symptoms

Short Term Exposure: Irritates the eyes, skin, and respiratory tract. Contact may cause skin burns. Inhalation can cause coughing, wheezing, and/or shortness of breath. Toxic hazard rating is high for oral, intravenous, intraperitoneal administration. This material causes swelling of the brain and spinal cord. Exposure may result in muscular weakness and paralysis, leading to respiratory failure; convulsive movements; closure of eyelids and sensitivity to light; headaches, EEG changes; dizziness, psychological and neurological disturbances; vertigo (an illusion of movement), sore throat; cough, abdominal pain; nausea, vomiting, diarrhea, urine retention; paresis, focal anesthesia; pruritus. Higher levels can cause unconsciousness, collapse, and death.

Long Term Exposure: Repeated or prolonged contact can cause dermatitis; dry and cracked skin. May cause brain damage, hepatic necrosis; kidney damage.

Points of Attack: Skin, brain, kidneys.

Medical Surveillance: NIOSH lists the following tests: Glaucoma; kidney function tests; liver function tests; urine (chemical/metabolite); urinalysis (routine). Also consider, psychological testing; examination of the nervous system; EEG.

First Aid: If this chemical gets into the eyes, remove any contact lenses at once and irrigate immediately for at least 15 min, occasionally lifting upper and lower lids. Seek medical attention immediately. If this chemical contacts the skin, remove contaminated clothing and wash immediately with soap and water. Seek medical attention immediately. If this chemical has been inhaled, remove from exposure, begin rescue breathing (using universal precautions, including resuscitation mask) if breathing has stopped and CPR if heart action has stopped. Transfer promptly to a medical facility. When this chemical has been swallowed, get medical attention. Give large quantities of water and induce vomiting. Do not make an unconscious person vomit.

Personal Protective Methods: Wear protective gloves and clothing to prevent any reasonable probability of skin contact. Safety equipment suppliers/manufacturers can provide recommendations on the most protective glove/clothing material for your operation. All protective clothing (suits, gloves, footwear, headgear) should be clean, available each day, and put on before work. Contact lenses should not be worn when working with this chemical. Wear splash-proof chemical goggles and face shield unless full-face-piece respiratory protection is worn. Employees should wash immediately with soap when skin is wet or contaminated. Provide emergency showers and eyewash.

Respirator Selection: *Up to 1 mg/m^3:* CcrOvDM [any chemical cartridge respirator with organic vapor cartridge(s) in combination with a dust and mist filter]; or Sa (APF = 10) (any supplied-air respirator). *Up to 2.5 mg/m^3:* Sa:Cf (APF = 25) (any supplied-air respirator operated in a continuous-flow mode); or PaprOvHie (APF = 25) (any air-purifying full-face-piece respirator equipped with an organic vapor cartridge in combination with a high-efficiency particulate filter). *Up to 5 mg/m^3:* CcrFOv100 (APF = 50) (any air-purifying full-face-piece respirator equipped with organic vapor cartridge(s) in combination with an N100, R100, or P100 filter); or GmFOv100 (APF = 50) [any air-purifying, full-face-piece respirator (gas mask) with a chin-style, front- or back-mounted organic vapor canister having an N100, R100, or P100 filter]; or PaprTOvHie (APF = 50) [any powered, air-purifying respirator with a tight-fitting face-piece and organic vapor cartridge(s) in combination with a high-efficiency particulate filter]; or SaT: Cf (APF = 50) (any supplied-air respirator that has a tight-fitting face-piece and is operated in a continuous-flow mode); or SCBAF (APF = 50) (any self-contained breathing apparatus with a full face-piece); or SaF (APF = 50) (any supplied-air respirator with a full face-piece). *Up to 25 mg/m^3:* SaF: Pd,Pp (APF = 2000) (any supplied-air respirator that has a full face-piece and is operated in a pressure-demand or other positive-pressure mode). Emergency or planned entry into unknown concentrations or IDLH conditions: SCBAF: Pd,Pp (APF = 10,000) (any NIOSH/MSHA- or European Standard EN 149-approved self-contained breathing apparatus that has a full face-piece and is operated in a pressure-demand or other positive-pressure mode); or SaF: Pd,Pp: ASCBA (APF = 10,000) (any supplied-air respirator that has a full face-piece and is operated in a pressure-demand or other positive-pressure mode in combination with an auxiliary, self-contained breathing apparatus operated in a pressure-demand or other positive-pressure mode). *Escape:* GmFOv100 (APF = 50) [any air-purifying, full-face-piece respirator (gas mask) with a chin-style, front- or back-mounted organic vapor canister having an N100, R100, or P100 filter]; or SCBAE (any appropriate escape-type, self-contained breathing apparatus).

Storage: Color Code—Blue: Health Hazard/Poison: Store in a secure poison location. Prior to working with this chemical you should be trained on its proper handling and storage. Store in tightly closed containers in a cool, well-ventilated area away from strong oxidizers. Where possible, automatically transfer material from drums or other storage containers to process containers. Sources of ignition, such as smoking and open flames, are prohibited where this chemical is handled, used, or stored. Metal containers involving the transfer of this chemical should be grounded and bonded. Wherever this chemical is used, handled, manufactured, or stored, use explosion-proof electrical equipment and fittings.

Shipping: Organotin compounds, liquid, n.o.s., require a shipping label of "POISONOUS/TOXIC MATERIALS." Hazard Class 6.1 and Packing Group II.

Spill Handling: Evacuate and restrict persons not wearing protective equipment from area of spill or leak until cleanup is complete. Remove all ignition sources. Wearing protective equipment, absorb liquids in vermiculite, dry sand; earth, peat, carbon, or a similar material and deposit in sealed containers. Ventilate and wash area after cleanup is complete. Keep this chemical out of a confined space, such as a sewer, because of the possibility of an explosion, unless the sewer is designed to prevent the buildup of explosive concentrations. It may be necessary to contain and dispose of this chemical as a hazardous waste. If material or contaminated runoff enters waterways, notify downstream users of potentially contaminated waters. Contact your local or federal environmental protection agency for specific recommendations. If employees are required to clean up spills, they must be properly trained and equipped. OSHA 1910.120(q) may be applicable.

Fire Extinguishing: This material is a flammable liquid. Extinguish small fires with dry chemical, carbon dioxide; water spray or foam. For large fires, use water spray, fog, or foam. Poisonous gases are produced in fire. Vapors are heavier than air and will collect in low areas. Vapors may travel long distances to ignition sources and flashback. Vapors in confined areas may explode when exposed to fire. Containers may explode in fire. Storage containers and parts of containers may rocket great distances, in many directions. If material or contaminated runoff enters waterways, notify downstream users of potentially contaminated waters. Notify local health and fire officials and pollution control agencies. From a secure, explosion-proof location, use water spray to cool exposed containers. If cooling streams are ineffective (venting sound increases in volume and pitch, tank discolors, or shows any signs of deforming), withdraw immediately to a secure position. If employees are expected to fight fires, they must be trained and equipped in OSHA 1910.156. The only respirators recommended for firefighting are self-contained breathing apparatuses that have full face-pieces and are operated in a pressure-demand or other positive-pressure mode.

Disposal Method Suggested: In accordance with 40CFR 165 recommendations for the disposal of pesticides and pesticide containers. Must be disposed properly by following package label directions or by contacting your local or federal environmental control agency, or by contacting your regional EPA office.

References

US Environmental Protection Agency. (November 30, 1987). *Chemical Hazard Information Profile: Tetraethyltin.* Washington, DC: Chemical Emergency Preparedness Program

New Jersey Department of Health and Senior Services. (November 1999). *Hazardous Substances Fact Sheet: Tetraethyltin.* Trenton, NJ

Tetrafluoroethylene   T:0320

Molecular Formula: C_2F_4

Synonyms: Ethene, tetrafluoro-; Ethylene, tetrafluoro-; Fluroplast 4; Perfluoroethene; Perfluoroethylene; Teflon 1,1,2,2-Tetrafluoroethylene; Tetrafluoroethylene, Inhibited; Tetrafluoroethene; TFE

CAS Registry Number: 116-14-3

RTECS® Number: KX4000000

UN/NA & ERG Number: UN1081 (stabilized)/116P

EC Number: 204-126-9

Regulatory Authority and Advisory Bodies

Department of Homeland Security Screening Threshold Quantity (pounds): *Release hazard* 10,000 (≥1.00% concentration).

Carcinogenicity: IARC: Animal Sufficient Evidence, *possibly carcinogenic to humans*, Group 2B; NTP: Reasonably anticipated to be a human carcinogen.

California Proposition 65 Chemical: Cancer 5/1/97.

European/International Regulations: Not listed in Annex 1.

WGK (German Aquatic Hazard Class): 2—Hazard to waters.

Description: TFE is a colorless, flammable gas. Molecular weight = 100.02; Melting point = −143°C; Boiling point = −76°C; Flash point ≤0°C; Autoignition temperature = 188°C. Its flammable limits in air are LEL = 10%; UEL = 50%. Hazard Identification (based on NFPA-704 M Rating System): Health 3, Flammability 4, Reactivity 3. Insoluble in water.

Potential Exposure: Those involved in the production of TFE and the manufacture of fluorocarbon polymers.

Incompatibilities: Reacts with air. Hazardous polymerization may occur unless inhibited. Will explode at pressures above 2.7 bar if terpene inhibitor is not added. Inhibited monomer can decompose explosively in fire, under pressure, or upon contact with materials with which it can react exothermically. Violent reaction with oxygen, oxidizers, sulfur trioxide; halogen compounds.

Permissible Exposure Limits in Air
ACGIH TLV®[1]: 2 ppm/8.2 mg/m^3 TWA; confirmed animal carcinogen with unknown relevance to humans.
Protective Action Criteria (PAC)*
TEEL-0: 2 ppm
PAC-1: **220** ppm
PAC-2: **550** ppm
PAC-3: **3300** ppm
*AEGLs (Acute Emergency Guideline Levels) & ERPGs (Emergency Response Planning Guideline) are in **bold face**.
DFG MAK: Carcinogen Category 2.

Permissible Concentration in Water: Russia[43] set a MAC in water bodies used for domestic purposes of 0.5 mg/L.

Routes of Entry: Inhalation, eye, and/or skin contact. Absorbed through the skin.

Harmful Effects and Symptoms
Short Term Exposure: Tetrafluoroethylene can affect you when breathed in. Irritates the eyes, skin, and respiratory tract. Very high exposures can displace the oxygen in the air, causing lightheadedness, dizziness, poor coordination, and unconsciousness. High levels may also damage the liver and/or kidneys and irritate the lungs. Contact with liquefied gas may cause frostbite.

Long Term Exposure: May cause lung irritation; bronchitis may develop. May cause kidney and liver damage.

Points of Attack: Lungs, kidneys, liver.

Medical Surveillance: NIOSH lists the following tests: chest X-ray, electrocardiogram, pulmonary function tests: forced vital capacity, forced expiratory volume (1 s); pelvic X-ray; sputum cytology; urine (chemical/metabolite); urine (chemical/metabolite) pre- and postshift; urinalysis (routine); complete blood count/differential.

First Aid: If this chemical gets into the eyes, remove any contact lenses at once and irrigate immediately for at least 15 min, occasionally lifting upper and lower lids. Seek medical attention immediately. If this chemical contacts the skin, remove contaminated clothing and wash immediately with soap and water. Seek medical attention immediately. If this chemical has been inhaled, remove from exposure, begin rescue breathing (using universal precautions, including resuscitation mask) if breathing has stopped and CPR if heart action has stopped. Transfer promptly to a medical facility. If frostbite has occurred, seek medical attention immediately; do *NOT* rub the affected areas or flush them with water. In order to prevent further tissue damage, do *NOT* attempt to remove frozen clothing from frostbitten areas. If frostbite has *NOT* occurred, immediately and thoroughly wash contaminated skin with soap and water.

Personal Protective Methods: Wear appropriate personal protective clothing to prevent the skin from becoming frozen from contact with the evaporating liquid or from contact with vessels containing the liquid. Safety equipment suppliers/manufacturers can provide recommendations on the most protective glove/clothing material for your operation. Neoprene™, Viton™, butyl rubber; and polyvinyl alcohol are among the recommended protective materials. All protective clothing (suits, gloves, footwear, headgear) should be clean, available each day, and put on before work. Contact lenses should not be worn when working with this chemical. Wear gas-proof chemical goggles and face shield unless full-face-piece respiratory protection is worn. Employees should wash immediately with soap when skin is wet or contaminated. Provide emergency showers and eyewash.

Respirator Selection: Combustion by-products include hydrogen fluoride and carbonyl fluoride. The respiratory recommendations for *fluorides* has been included for reference: NIOSH: *(fluorides) 12.5 mg/m^3*: Qm (APF = 25) (any quarter-mask respirator). *25 mg/m^3*: Any particulate respirator equipped with an N95, R95, or P95 filter (including N95, R95, and P95 filtering face-pieces) except quarter-mask respirators. The following filters may also be used: N99, R99, P99, N100, R100, P100; or SA* (any supplied-air respirator). *62.5 mg/m^3*: Sa:Cf (APF = 25)* (any supplied-air respirator operated in a continuous-flow mode); or PAPRDM*† *if not present as a fume* (any powered, air-purifying respirator with a dust and mist filter). *125 mg/m^3*: HieF† (any air-purifying, full-face-piece respirator with a high-efficiency particulate filter); or SCBAF (APF = 50) (any self-contained breathing apparatus with a full face-piece); or SaF (APF = 50) (any supplied-air respirator with a full face-piece). *250 mg/m^3*: Sa: Pd,Pp (APF = 1000) (any supplied-air respirator operated in a pressure-demand or other positive-pressure mode). *Emergency or planned entry into unknown concentrations or IDLH conditions:* SCBAF: Pd,Pp (APF = 10,000) (any self-contained breathing apparatus that has a full faceplate and is operated in a pressure-demand or other positive-pressure mode); or SaF: Pd,Pp: ASCBA (APF = 10,000) (any supplied-air respirator that has a full face-piece and is operated in a pressure-demand or other positive-pressure mode in combination with an auxiliary, self-contained breathing apparatus operated in a pressure-demand or other positive-pressure mode). *Escape:* HieF† (any air-purifying, full-face-piece respirator with a high-efficiency particulate filter); or SCBAE (any appropriate escape-type, self-contained breathing apparatus).
*Substance reported to cause eye irritation or damage; may require eye protection.
†May need acid gas sorbent.

Storage: Color Code—Red Stripe: Flammability Hazard: Store separately from all other flammable materials. High concentrations cause a deficiency of oxygen with the risk of unconsciousness or death. Before entering confined space where this chemical may be present, check to make sure that an explosive concentration does not exist. Check that oxygen content is at least 19% before entering storage or spill area. Prior to working with this chemical you should be trained on its proper handling and storage. Tetrafluoroethylene must be stored to avoid contact with oxidizers (such as perchlorates, peroxides, permanganates, chlorates, and nitrates), since violent reactions occur. Store in tightly closed containers in a cool, well-ventilate area. Protect storage against physical damage. Will explode at pressures above 2.7 bar if terpene inhibitor is not added. Procedures for the handling, use, and storage of cylinders should be in compliance with OSHA 1910.101 and 1910.169, as with the recommendations of the Compressed Gas Association.

Shipping: Tetrafluoroethylene, inhibited, requires a shipping label of "FLAMMABLE GAS." It falls in Hazard Class 2.1.

Spill Handling: Evacuate and restrict persons not wearing protective equipment from area of spill or leak until cleanup is complete. High concentrations cause a deficiency of oxygen with the risk of unconsciousness or death. Check that oxygen content is at least 19% before entering storage or spill area. Remove all ignition sources. Ventilate area of leak to disperse the gas. Stop the flow of gas if it can be done safely. If source of leak is a cylinder and the leak cannot be stopped in place, remove leaking cylinder to a safe place in the open air, and repair leak or allow cylinder to empty. Absorb liquids in vermiculite, dry sand; earth, or a similar material and deposit in sealed containers. Keep this chemical out of confined space, such as a sewer because of the possibility of explosion, unless the sewer is designed to prevent the buildup of explosive concentrations. It may be necessary to contain and dispose of this chemical as a hazardous waste. Contact your local or federal environmental protection agency for specific recommendations. If employees are required to clean up spills, they must be properly trained and equipped. OSHA 1910.120(q) may be applicable.

Fire Extinguishing: This chemical is a flammable gas. Poisonous gases, including hydrogen fluoride and carbonyl fluoride, are produced in fire. Do not extinguish the fire unless the flow of gas can be stopped and any remaining gas is out of the line. Specially trained personnel may use fog lines to cool exposures and let the fire burn itself out. Vapors are heavier than air and will collect in low areas. Vapors may travel long distances to ignition sources and flashback. Vapors in confined areas may explode when exposed to fire. Containers may explode in fire. Storage containers and parts of containers may rocket great distances, in many directions. If material or contaminated runoff enters waterways, notify downstream users of potentially contaminated waters. Notify local health and fire officials and pollution control agencies. From a secure, explosion-proof location, use water spray to cool exposed containers. If cooling streams are ineffective (venting sound increases in volume and pitch, tank discolors, or shows any

signs of deforming), withdraw immediately to a secure position. If cylinders are exposed to excessive heat from fire or flame contact, withdraw immediately to a secure location. If employees are expected to fight fires, they must be trained and equipped in OSHA 1910.156. The only respirators recommended for firefighting are self-contained breathing apparatuses that have full face-pieces and are operated in a pressure-demand or other positive-pressure mode.

References
New Jersey Department of Health and Senior Services. (May 2004). *Hazardous Substances Fact Sheet: Tetrafluoroethylene.* Trenton, NJ

Tetrafluoromethane   T:0330

Molecular Formula: CF_4
Synonyms: Arcton O; Carbon fluoride; Carbon tetrafluoride; F 14; FC 14; Freon 14; Halocarbon 14; Halon 14; Methane, tetrafluoro-; Perfluoromethane; R 14
CAS Registry Number: 75-73-0
RTECS® Number: FG4920000
UN/NA & ERG Number: UN1982/126
Ec Number: 200-896-5
Regulatory Authority and Advisory Bodies
European/International Regulations: Not listed in Annex 1.
WGK (German Aquatic Hazard Class): No value assigned.
Description: Tetrafluoromethane is a colorless, odorless gas. Molecular weight = 88.01; Boiling point = $-128°C$; Freezing/Melting point = $-183.6°C$; Autoignition temperature $\geq 1100°C$. Insoluble in water.
Potential Exposure: Tetrafluoromethane is used in fire extinguishers and as a low temperature refrigerant.
Incompatibilities: Forms hydrogen fluoride and fluorides on decomposition with hot surfaces above 125°F/52°C or open flame. Incompatible with powdered metals; including aluminum, zinc, and beryllium.
Permissible Exposure Limits in Air
Protective Action Criteria (PAC)
TEEL-0: 200 ppm
PAC-1: 600 ppm
PAC-2: 5000 ppm
PAC-3: 25,000 ppm
Routes of Entry: Inhalation, skin, and/or eye contact.
Harmful Effects and Symptoms
Short Term Exposure: Tetrafluoromethane can affect you when breathed in. High levels can cause you to feel dizzy, lightheadedness, and to pass out. Very high levels could cause death. Similar chemicals can affect the cardiovascular system, causing irregular heartbeat, which could lead to death. Contact with the liquefied gas could cause frostbite. Exposure at high levels can cause depletion of oxygen, causing unconsciousness and death by suffocation.
Long Term Exposure: Unknown at this time.
Points of Attack: Cardiovascular system.

Medical Surveillance: If symptoms develop or overexposure is suspected, the following may be useful: Special 24-h EKG (Holter monitor) to look for irregular heart beat.

First Aid: If this chemical gets into the eyes, remove any contact lenses at once and irrigate immediately for at least 15 min, occasionally lifting upper and lower lids. Seek medical attention immediately. If this chemical contacts the skin, remove contaminated clothing and wash immediately with soap and water. Seek medical attention immediately. If this chemical has been inhaled, remove from exposure, begin rescue breathing (using universal precautions, including resuscitation mask) if breathing has stopped and CPR if heart action has stopped. Transfer promptly to a medical facility. If frostbite has occurred, seek medical attention immediately; do *NOT* rub the affected areas or flush them with water. In order to prevent further tissue damage, do *NOT* attempt to remove frozen clothing from frostbitten areas. If frostbite has *NOT* occurred, immediately and thoroughly wash contaminated skin with soap and water.

Personal Protective Methods: Wear appropriate personal protective clothing to prevent the skin from becoming frozen from contact with the evaporating liquid or from contact with vessels containing the liquid. Safety equipment suppliers/manufacturers can provide recommendations on the most protective glove/clothing material for your operation. All protective clothing (suits, gloves, footwear, headgear) should be clean, available each day, and put on before work. Contact lenses should not be worn when working with this chemical. Wear gas-proof goggles unless full-face-piece respiratory protections is worn. Wear splash-proof chemical goggles and face shield when working with liquid, unless full-face-piece respiratory protection is worn. Employees should wash immediately with soap when skin is wet or contaminated. Provide emergency showers and eyewash.

Respirator Selection: Where there is potential for exposure to tetrafluoromethane, use a NIOSH/MSHA- or European Standard EN149-approved supplied-air respirator with a full face-piece operated in the positive-pressure mode, or with a full face-piece, hood, or helmet in the continuous-flow mode; or use a NIOSH/MSHA- or European Standard EN149-approved self-contained breathing apparatus with a full face-piece operated in pressure-demand or other positive-pressure mode.

Storage: Color Code—Green: General storage may be used. High concentrations cause a deficiency of oxygen with the risk of unconsciousness or death. Check that oxygen content is at least 19% before entering storage or spill area. Prior to working with this chemical you should be trained on its proper handling and storage. Store in tightly closed containers in a cool, well-ventilated area away from powdered metals; including aluminum, zinc, and beryllium; and from open flames or temperatures above 125°F/51.6°C. Procedures for the handling, use, and storage of cylinders should be in compliance with OSHA 1910.101 and 1910.169, as with the recommendations of the Compressed Gas Association.

Shipping: Tetrafluoromethane, compressed, requires a shipping label of "NONFLAMMABLE GAS." It falls in Hazard Class 2.2.

Spill Handling: High concentrations cause a deficiency of oxygen with the risk of unconsciousness or death. Check that oxygen content is at least 19% before entering storage or spill area. If in a building, evacuate building and confine vapors by closing doors and shutting down HVAC systems. Restrict persons not wearing protective equipment from area of spill or leak until cleanup is complete. Remove all ignition sources. Ventilate area of spill or leak to disperse the gas. Wear chemical protective suit with self-contained breathing apparatus to combat spills. Stay upwind and use water spray to "knock down" vapor; contain runoff. Stop the flow of gas, if it can be done safely from a distance. If source is a cylinder and the leak cannot be stopped in place, remove the leaking cylinder to a safe place; and repair leak or allow cylinder to empty. Absorb liquids in vermiculite, dry sand; earth, or a similar material and deposit in sealed containers. Keep this chemical out of confined spaces, such as a sewer, because of the possibility of explosion, unless the sewer is designed to prevent the buildup of explosive concentrations. If employees are required to clean up spills, they must be properly trained and equipped. OSHA 1910.120(q) may be applicable.

Fire Extinguishing: Tetrafluoromethane may burn, but does not readily ignite. Use dry chemical or carbon dioxide extinguishers. Poisonous gases are produced in fire, including fluorides, such as hydrogen fluoride. Containers may explode in fire. If liquid or contaminated runoff enters waterways, notify downstream users of potentially contaminated waters. Notify local health and fire officials and pollution control agencies. From a secure, explosion-proof location, use water spray to cool exposed containers. If cooling streams are ineffective (venting sound increases in volume and pitch, tank discolors, or shows any signs of deforming), withdraw immediately to a secure position. If employees are expected to fight fires, they must be trained and equipped in OSHA 1910.156. The only respirators recommended for firefighting are self-contained breathing apparatuses that have full face-pieces and are operated in a pressure-demand or other positive-pressure mode.

References
New Jersey Department of Health and Senior Services. (November 2004). *Hazardous Substances Fact Sheet: Tetrafluoromethane.* Trenton, NJ

Tetrahydrofuran   T:0340

Molecular Formula: C_4H_8O
Synonyms: Butane, 1,4-epoxy-; Butylene oxide; Cyclotetramethylene oxide; Diethylene oxide; 1,4-Epoxybutane; Furanidine; Furan, tetrahydro-; Hydrofuran; NCI-C60560; Oxacyclopentane; Oxolane; Tetrahydrofuranne (French); Tetramethylene oxide; THF

CAS Registry Number: 109-99-9
RTECS® Number: LU5950000
UN/NA & ERG Number: UN2056/127
EC Number: 203-726-8 [*Annex I Index No.:* 603-025-00-0]

Regulatory Authority and Advisory Bodies
Carcinogenicity: NCI: Carcinogenesis Studies (inhalation); clear evidence: mouse; equivocal evidence: rat; NTP: Carcinogenesis Studies (inhalation): some evidence: rat.
Air Pollutant Standard Set. See below, "Permissible Exposure Limits in Air" section.
US EPA Hazardous Waste Number (RCRA No.): U213.
RCRA, 40CFR261, Appendix 8 Hazardous Constituents.
Reportable Quantity (RQ): 1000 lb (454 kg).
Canada, WHMIS, Ingredients Disclosure List Concentration 1.0%.
European/International Regulations: Hazard Symbol: F, Xi; Risk phrases: R11; R19; R36/37; Safety phrases: S2; S16; S29; S33 (see Appendix 4).
WGK (German Aquatic Hazard Class): 1—Low hazard to waters.

Description: Tetrahydrofuran is a colorless liquid with an ether-like odor. The odor threshold is listed at 3.8 (3M), 20−50 ppm[41] and 31 ppm. Molecular weight = 72.12; Specific gravity (H_2O:1) = 0.89 at 25°C; Boiling point = 66.1°C; Freezing/Melting point: −108.5°C; Vapor pressure = 132 mmHg; Flash point = −15°C (cc); Autoignition temperature = 321°C. Explosive limits: LEL = 2%; UEL = 11.8%. Hazard Identification (based on NFPA-704 M Rating System): Health 2, Flammability 3, Reactivity 1. Soluble in water.

Potential Exposure: Compound Description: Tumorigen, Mutagen; Reproductive Effector; Human Data. The primary use of tetrahydrofuran is as a solvent to dissolve synthetic resins, particularly polyvinyl chloride and vinylidene chloride copolymers. It is also used to cast polyvinyl chloride films, to coat substrates with vinyl and vinylidene chloride; and to solubilize adhesives based on or containing polyvinyl chloride resins. A second large market for THF is as an electrolytic solvent in the Grignard reaction-based production of tetramethyl lead. THF is used as an intermediate in the production of polytetramethylene glycol.

Incompatibilities: Forms thermally explosive peroxides in air on standing (in absence of inhibitors). Peroxides can be detonated by heating, friction, or impact. Reacts violently with strong oxidizers, strong bases, and some metal halides. Attacks some forms of plastics, rubber, and coatings.

Permissible Exposure Limits in Air
Conversion factor: 1 ppm = 2.95 mg/m³ at 25°C & 1 atm.
OSHA PEL: 200 ppm/590 mg/m³ TWA.
NIOSH REL: 200 ppm/590 mg/m³ TWA; 250 ppm/735 mg/m³ STEL.
ACGIH TLV®[1]: 50 ppm/147 mg/m³ TWA; 100 ppm/295 mg/m³ STEL [skin] confirmed animal carcinogen with unknown relevance to humans.
NIOSH IDLH: 2000 ppm [LEL].
Protective Action Criteria (PAC)*

TEEL-0: 50 ppm
PAC-1: **100** ppm
PAC-2: **500** ppm
PAC-3: **5000** ppm
*AEGLs (Acute Emergency Guideline Levels) & ERPGs (Emergency Response Planning Guideline) are in **bold face**.
DFG MAK: 50 ppm/150 mg/m³ TWA; Peak Limitation Category I(2) [skin]; Carcinogen Category 4; Pregnancy Risk Group C; BAT: 8 mg/L in urine/end-of-shift.
Australia: TWA 200 ppm (590 mg/m³), STEL 250 ppm, 1993; Austria: MAK 200 ppm (590 mg/m³), 1999; Belgium: TWA 200 ppm (590 mg/m³), STEL 250 ppm (738 mg/m³), 1993; Denmark: TWA 100 ppm (295 mg/m³), 1999; Finland: TWA 100 ppm (290 mg/m³), STEL 150 ppm (440 mg/m³), 1999; France: VME 200 ppm (590 mg/m³), 1999; Hungary: TWA 200 mg/m³, STEL 400 mg/m³, 1999; Japan: 200 ppm (590 mg/m³), 1999; the Netherlands: MAC-TGG 300 mg/m³ [skin], 2003; Norway: TWA 50 ppm (150 mg/m³), 1999; the Philippines: TWA 200 ppm (590 mg/m³), 1993; Poland: MAC (TWA) 600 mg/m³, MAC (STEL) 750 mg/m³, 1999; Russia: TWA 200 ppm, STEL 100 mg/m³, 1993; Sweden: NGV 50 ppm (150 mg/m³), KTV 80 ppm (250 mg/m³), 1999; Switzerland: MAK-W 200 ppm (590 mg/m³), STEL 1000 ppm, 1999; Turkey: TWA 200 ppm (590 mg/m³), 1993; United Kingdom: TWA 100 ppm (300 mg/m³), STEL 200 ppm [skin], 2000; Argentina, Bulgaria, Columbia, Jordan, South Korea, New Zealand, Singapore, Vietnam: ACGIH TLV®: STEL 250 ppm; Russia[43] set a MAC values for ambient air in residential areas of 0.2 mg/m³ both on a momentary and a daily average basis. Several states have set guidelines or standards for tetrahydrofuran in ambient air[61] ranging from zero (North Carolina) to 0.8 mg/m³ (Massachusetts) to 5.9−7.35 mg/m³ (North Dakota) to 9.8 mg/m³ (Virginia) to 11.8 mg/m³ (Connecticut, Florida, New York) to 14.048 mg/m³ (Nevada).

Determination in Air: Use NIOSH Analytical Method (IV) #1609[18]; 3800; OSHA Analytical Method #7.

Permissible Concentration in Water: No criteria set but EPA[32] has suggested a permissible ambient goal of 8100 μg/L (based on health effects). Russia[43] set a MAC in water bodies used for domestic purposes of 0.5 mg/L (500 μg/L). States which have set guidelines for tetrahydrofuran in drinking water include Wisconsin at 50.0 μg/L and New Hampshire at 154 μg/L.

Routes of Entry: Inhalation, ingestion; skin and/or eye contact. Absorbed through the skin.

Harmful Effects and Symptoms
Short Term Exposure: Tetrahydrofuran can affect you when breathed in and may enter the body through the skin. Eye contact causes severe irritation and possible damage. Skin contact causes severe irritation. If covered by clothing or prolonged, blistering can occur. The vapors irritate the eyes, nose, throat, and lungs. Very high exposures can

affect the central nervous system causing narcosis, unconsciousness, and rapid death. High exposure can damage the liver and kidneys.

Long Term Exposure: May cause dermatitis; drying and cracking. Repeated exposure may cause liver and kidney damage. May cause lung irritation; bronchitis may develop.

Points of Attack: Eyes, respiratory system; central nervous system; liver and kidneys.

Medical Surveillance: Before beginning employment and at regular times after that, for those with frequent or potentially high exposures, the following are recommended: lung function tests. If symptoms develop or overexposure is suspected, the following may be useful: consider chest X-ray after acute overexposure. Liver and kidney function tests.

First Aid: If this chemical gets into the eyes, remove any contact lenses at once and irrigate immediately for at least 15 min, occasionally lifting upper and lower lids. Seek medical attention immediately. If this chemical contacts the skin, remove contaminated clothing and wash immediately with soap and water. Seek medical attention immediately. If this chemical has been inhaled, remove from exposure, begin rescue breathing (using universal precautions, including resuscitation mask) if breathing has stopped and CPR if heart action has stopped. Transfer promptly to a medical facility. When this chemical has been swallowed, get medical attention. Give large quantities of water and induce vomiting. Do not make an unconscious person vomit.

Personal Protective Methods: Wear protective gloves and clothing to prevent any reasonable probability of skin contact. Safety equipment suppliers/manufacturers can provide recommendations on the most protective glove/clothing material for your operation. Teflon™ is among the recommended protective materials. All protective clothing (suits, gloves, footwear, headgear) should be clean, available each day, and put on before work. Contact lenses should not be worn when working with this chemical. Wear splash-proof chemical goggles and face shield unless full-face-piece respiratory protection is worn. Employees should wash immediately with soap when skin is wet or contaminated. Provide emergency showers and eyewash.

Respirator Selection: 2000 ppm: Sa:Cf (APF = 25) (any supplied-air respirator operated in a continuous-flow mode); CcrFOv (APF = 50) [any chemical cartridge respirator with a full face-piece and organic vapor cartridge(s)]; or GmFOv (APF = 50) [any air-purifying, full-face-piece respirator (gas mask) with a chin-style, front- or back-mounted acid gas canister]; or PaprOv (APF = 25) [any powered, air-purifying respirator with organic vapor cartridge(s)]; or SCBAF (APF = 50) (any self-contained breathing apparatus with a full face-piece); or SaF (APF = 50) (any supplied-air respirator with a full face-piece). *Emergency or planned entry into unknown concentrations or IDLH conditions:* SCBAF: Pd,Pp (APF = 10,000) (any self-contained breathing apparatus that has a full face-piece and is operated in a pressure-demand or other positive-pressure mode); or SaF: Pd,Pp: ASCBA (APF = 10,000) (any supplied-air respirator that has a full face-piece and is operated in a pressure-demand or other positive-pressure mode in combination with an auxiliary self-contained breathing apparatus operated in a pressure-demand or other positive-pressure mode). *Escape:* GmFOv (APF = 50) [any air-purifying, full-face-piece respirator (gas mask) with a chin-style, front- or back-mounted organic vapor canister]; or SCBAE (any appropriate escape-type, self-contained breathing apparatus).

Note: Substance causes eye irritation or damage; eye protection needed.

Storage: Color Code—Red: Flammability Hazard: Store in a flammable liquid storage area or approved cabinet away from ignition sources and corrosive and reactive materials. May form peroxides in storage. Prior to working with this chemical you should be trained on its proper handling and storage. Before entering confined space where this chemical may be present, check to make sure that an explosive concentration does not exist. Tetrahydrofuran must be stored to avoid contact with strong oxidizers (such as chlorine, bromine, and fluorine), since violent reactions occur. Store in tightly closed containers in a cool, well-ventilated area. Protect storage containers from physical damage. Sources of ignition, such as smoking and open flames, are prohibited where tetrahydrofuran is handled, used, or stored. Metal containers involving the transfer of 5 gallons or more of tetrahydrofuran should be grounded and bonded. Drums must be equipped with self-closing valves, pressure vacuum bungs, and flame arresters. Use only nonsparking tools and equipment, especially when opening and closing containers of tetrahydrofuran.

Shipping: Tetrahydrofuran requires a shipping label of "FLAMMABLE LIQUID." It falls in Hazard Class 3 and Packing Group II.

Spill Handling: Evacuate and restrict persons not wearing protective equipment from area of spill or leak until cleanup is complete. Remove all ignition sources. Establish forced ventilation to keep levels below explosive limit. Absorb liquids in vermiculite, dry sand; earth, peat, carbon, or a similar material and deposit in sealed containers. Keep this chemical out of a confined space, such as a sewer, because of the possibility of an explosion, unless the sewer is designed to prevent the buildup of explosive concentrations. It may be necessary to contain and dispose of this chemical as a hazardous waste. If material or contaminated runoff enters waterways, notify downstream users of potentially contaminated waters. Contact your local or federal environmental protection agency for specific recommendations. If employees are required to clean up spills, they must be properly trained and equipped. OSHA 1910.120(q) may be applicable.

Fire Extinguishing: This chemical is a flammable liquid. Poisonous gases are produced in fire. Use dry chemical, carbon dioxide; or alcohol foam extinguishers. Vapors are heavier than air and will collect in low areas. Vapors may travel long distances to ignition sources and flashback. Vapors in confined areas may explode when exposed to fire. Containers may explode in fire. Storage containers and parts of containers may rocket great distances, in many directions. If material or contaminated runoff enters waterways, notify downstream users of potentially contaminated waters. Notify local health and fire officials and pollution control agencies. From a secure, explosion-proof location, use water spray to cool exposed containers. If cooling streams are ineffective (venting sound increases in volume and pitch, tank discolors, or shows any signs of deforming), withdraw immediately to a secure position. If employees are expected to fight fires, they must be trained and equipped in OSHA 1910.156. The only respirators recommended for firefighting are self-contained breathing apparatuses that have full face-pieces and are operated in a pressure-demand or other positive-pressure mode.

Disposal Method Suggested: Consult with environmental regulatory agencies for guidance on acceptable disposal practices. Generators of waste containing this contaminant (≥100 kg/mo) must conform with EPA regulations governing storage, transportation, treatment, and waste disposal. Concentrated waste containing peroxides—perforation of a container of the waste from a safe distance followed by open burning.

References
US Environmental Protection Agency. (October 21, 1977). *Chemical Hazard Information Profile: Tetrahydrofuran.* Washington, DC (Revised edition issued 1979)
National Institute for Occupational Safety and Health. (1977). *Profiles on Occupational Hazards for Criteria Document Priorities*, Report PB-274,073. Cincinnati, OH, pp. 314−316
Sax, N. I. (Ed.). *Dangerous Properties of Industrial Materials Report*, 1, No. 2, 64−65 (1980) and 5, No. 5, 83−87 (1985)
New Jersey Department of Health and Senior Services. (May 2004). *Hazardous Substances Fact Sheet: Tetrahydrofuran.* Trenton, NJ

Tetramethyl lead   T:0360

Molecular Formula: $C_4H_{12}Pb$
Common Formula: $Pb(CH_3)_4$
Synonyms: Lead, tetramethyl-; Plumbane, tetramethyl-; Tetramethylplumbane; TML
CAS Registry Number: 75-74-1
RTECS® Number: TP4725000
UN/NA & ERG Number: UN1649 (motor fuel antiknock mixture)/131
EC Number: 200-897-0; Listed in Title I, Part 3 with no Index Number. HS2931.00
Regulatory Authority and Advisory Bodies
Department of Homeland Security Screening Threshold Quantity (pounds): *Release hazard* 10,000 (≥1.00% concentration).

Carcinogenicity: IARC: (organolead) Animal Inadequate Evidence; Human Inadequate Evidence, *not classifiable as carcinogenic to humans*, Group 3, 1987.
Toxic Substance (World Bank).[15]
Air Pollutant Standard Set. See below, "Permissible Exposure Limits in Air" section.
OSHA 29CFR1910.119, Appendix A. Process Safety List of Highly Hazardous Chemicals, TQ = 1000 lb (450 kg).
Clean Air Act: Accidental Release Prevention/Flammable Substances, (Section 112[r], Table 3), TQ = 10,000 lb (4540 kg).
EPCRA Section 313 (as organic lead compound) Form R *de minimis* concentration reporting level: 1.0%.
Superfund/EPCRA 40CFR355, Extremely Hazardous Substances: TPQ = 100 lb (45.4 kg).
Reportable Quantity (RQ): 100 lb (45.4 kg).
US DOT Regulated Marine Pollutant (49CFR172.101, Appendix B).
Canada, WHMIS, Ingredients Disclosure List Concentration 1.0%.
Rotterdam Convention Annex III [Chemicals Subject to the Prior Informed Consent Procedure (PIC)].
European/International Regulations: Listed in Annex 1, Part 3 without Index Number.
WGK (German Aquatic Hazard Class): 3—Severe hazard to waters.

Description: Tetramethyl lead is a colorless liquid with a slight musty odor. In commerce it is usually dyed red, orange, or blue. Molecular weight = 267.35; Boiling point ≥100°C (decomposes); Freezing/Melting point = −27°C; Flash point = 37°C. Explosive limits: LEL = 1.8%; UEL—unknown. Hazard Identification (based on NFPA-704 M Rating System): Health 3, Flammability 3, Reactivity 3W. Insoluble in water.

Potential Exposure: Compound Description: Organometallic; Reproductive Effector. Those engaged in the manufacture, distribution, and blending into gasoline of this antiknock agent for aviation gasoline.

Incompatibilities: Violent reaction with oxidizers, such as sulfuryl chloride or potassium permanganate; strong acids; especially nitric acid; chemically active metals. Decomposes and may explode in heat above 90°C. Attacks rubber.

Permissible Exposure Limits in Air
OSHA PEL: 0.075 mg[Pb]/m³ TWA [skin].
NIOSH REL: 0.075 mg[Pb]/m³ TWA [skin].
ACGIH TLV®[1]: 0.15 mg[Pb]/m³ [skin].
NIOSH IDLH: 40 mg[Pb]/m³.
Protective Action Criteria (PAC)
TEEL-0: 0.0968 mg/m³
PAC-1: 0.581 mg/m³
PAC-2: 4 mg/m³
PAC-3: 51.6 mg/m³
DFG MAK: 0.05 mg[Pb]/m³; Peak Limitation Category II (2); Peak Limitation Category II(2) [skin]; Pregnancy Risk Group B.

Arab Republic of Egypt: TWA 0.05 mg[Pb]/m³, 1993; Australia: TWA 0.15 mg[Pb]/m³, 1993; Australia: TWA 0.15 mg/m³ [skin], 1993; Austria: MAK 0.01 ppm (0.075 mg/m³) [skin], 1993; Austria: MAK 0.01 ppm (0.075 mg/m³), 1999; Belgium: TWA 0.15 mg[Pb]/m³, 1993; Belgium: TWA 0.15 mg/m³ [skin], 1993; Denmark: TWA 0.007 ppm (0.05 mg/m³) [skin], 1999; the Netherlands: MAC-TGG 0.05 mg/m³ [skin], 2003; Finland: TWA 0.1 mg[Pb]/m³, 1993; France: VME 0.15 mg[Pb]/m³ [skin], 1999; VME 0.15 mg/m³ [skin], 1993; Hungary: STEL 0.04 mg[Pb]/m³, carcinogen, 1993; Hungary: TWA 0.005 mg/m³, STEL 0.01 mg/m³ [skin], 1993; Norway: TWA 0.01 ppm (0.075 mg/m³), 1999; the Philippines: TWA 0.07 mg/m³ [skin], 1993; TWA 0.15 mg[Pb]/m³, 1993; Russia: STEL 0.005 ppm (0.01 mg[Pb]/m³), 1993; Sweden: NGV 0.05 mg[Pb]/m³, KTV 0.2 mg[Pb]/m³ [skin], 1999; NGV 0.05 mg/m³, STEL 0.2 mg/m³ [skin], 1993; Switzerland: MAK-W 0.01 ppm (0.075 mg/m³), KZG-W 0.02 ppm [skin], 1999; TWA 0.01 ppm (0.075 mg/m³), STEL 0.02 ppm [skin], 1993; TWA 0.1 mg[Pb]/m³, 1993; Thailand: TWA 0.07 mg/m³, 1993; TWA 0.2 mg[Pb]/m³, 1993; Turkey: TWA 0.2 mg[Pb]/m³, 1993; United Kingdom: LTEL 0.15 mg[Pb]/m³, 1993; Argentina, Bulgaria, Columbia, Jordan, South Korea, New Zealand, Singapore, Vietnam: ACGIH TLV®: TWA 0.15 mg[Pb]/m³ [skin]. Several states have set guidelines or standards for TML in ambient air[60] ranging from 1.5 μg/m³ (Connecticut and North Dakota) to 2.5 μg/m³ (Virginia) to 4.0 μg/m³ (Nevada).

Determination in Air: Use NIOSH Analytical Method (IV) #2534.

Permissible Concentration in Water: No criteria set, but EPA[32] has suggested a permissible ambient goal of 2 μg/L based on health effects.

Routes of Entry: Inhalation, skin absorption; ingestion; skin and/or eye contact.

Harmful Effects and Symptoms

Short Term Exposure: Vapors are very toxic. Tetramethyl lead irritates the moist skin, eyes, and mucous membrane. May affect the central nervous system. Fatal lead poisoning may occur by ingestion, vapor inhalation, or skin absorption. Several cases of acute toxicity, usually in the form of degenerative brain disease, have been described following occupational exposure. Signs and symptoms of acute exposure to tetramethyl lead may be severe and include nausea, delirium, mania, anxiety, irritability, headache, insomnia, disorientation, violent/frightening dreams; hyperexcitability, delusions, and hallucinations. Muscular weakness, tremor, a lack of coordination; convulsions, cerebral edema; and coma may occur. A metallic taste may be noted. Sneezing, bronchitis, and pneumonia may be noted. Bradycardia (slow heart rate), hypotension (low blood pressure), hypothermia, and pallor may also occur. Gastrointestinal symptoms include vomiting and diarrhea.

Long Term Exposure: High levels can cause muscle and joint pains, weakness, muscle cramps; and fatigue. Lead

can accumulate in the body with repeated exposure. May affect the kidneys.

Points of Attack: Central nervous system; cardiovascular system, kidneys.

Medical Surveillance: NIOSH lists the following tests: Biologic monitoring of urine every 3 months; urine (chemical/metabolite); urine (chemical/metabolite), end-of-shift; urinalysis (routine). If symptoms develop or overexposure is suspected, the following may be useful: urine test for lead levels of 0.1 mg/L of urine indicate increased exposure. Such levels increase risk from further exposure. Blood lead tests are not usually accurate with exposure to tetramethyl lead. Complete blood count (CBC). Kidney function tests. Examination of the nervous system.

First Aid: If this chemical gets into the eyes, remove any contact lenses at once and irrigate immediately for at least 15 min, occasionally lifting upper and lower lids. Seek medical attention immediately. If this chemical contacts the skin, remove contaminated clothing and wash immediately with soap and water. Seek medical attention immediately. If this chemical has been inhaled, remove from exposure, begin rescue breathing (using universal precautions, including resuscitation mask) if breathing has stopped and CPR if heart action has stopped. Transfer promptly to a medical facility. When this chemical has been swallowed, get medical attention. Give large quantities of water and induce vomiting. Do not make an unconscious person vomit. Effects may be delayed; medical observation is recommended.

Personal Protective Methods: Wear protective gloves and clothing to prevent any reasonable probability of skin contact. Safety equipment suppliers/manufacturers can provide recommendations on the most protective glove/clothing material for your operation. All protective clothing (suits, gloves, footwear, headgear) should be clean, available each day, and put on before work. Contact lenses should not be worn when working with this chemical. Wear splash-proof chemical goggles and face shield unless full-face-piece respiratory protection is worn. Employees should wash immediately with soap when skin is wet or contaminated. Provide emergency showers and eyewash.

Respirator Selection: 0.75 mg/m³: Sa (APF = 10) (any supplied-air respirator). *1.875 mg/m³:* Sa:Cf (APF = 25) (any supplied-air respirator operated in a continuous-flow mode). *3.75 mg/m³:* SaT: Cf (APF = 50) (any supplied-air respirator that has a tight-fitting face-piece and is operated in a continuous-flow mode); or SCBAF (APF = 50) (any self-contained breathing apparatus with a full face-piece); or SaF (APF = 50) (any supplied-air respirator with a full face-piece). *40 mg/m³:* Sa: Pd,Pp (APF = 1000) (any supplied-air respirator operated in a pressure-demand or other positive-pressure mode). *Emergency or planned entry into unknown concentrations or IDLH conditions:* SCBAF: Pd,Pp (APF = 10,000) (any self-contained breathing apparatus that has a full face-piece and is operated in a pressure-demand or other positive-pressure mode); or SaF: Pd,Pp: ASCBA

Tetramethyl succinonitrile

(APF = 10,000) (any supplied-air respirator that has a full face-piece and is operated in a pressure-demand or other positive-pressure mode in combination with an auxiliary self-contained breathing apparatus operated in a pressure-demand or other positive-pressure mode). *Escape:* GmFOv (APF = 50) [any air-purifying, full-face-piece respirator (gas mask) with a chin-style, front- or back-mounted organic vapor canister]; or SCBAE (any appropriate escape-type, self-contained breathing apparatus).

Storage: Color Code—Red: Flammability Hazard: Store in a flammable liquid storage area or approved cabinet away from ignition sources and corrosive and reactive materials. Prior to working with this chemical you should be trained on its proper handling and storage. Before entering confined space where this chemical may be present, check to make sure that an explosive concentration does not exist. Tetramethyl lead must be stored to avoid contact with oxidizers (such as perchlorates, peroxides, permanganates, chlorates, and nitrates) and chemically active metals (such as potassium, sodium, magnesium, and zinc), since violent reactions occur. Store in tightly closed containers in a cool, well-ventilated area away from heat. Protect storage containers from physical damage. Sources of ignition, such as smoking and open flames, are prohibited where Tetramethyl lead is handled, used, or stored. Metal containers involving the transfer of 5 gallons or more of tetramethyl lead should be grounded and bonded. Drums must be equipped with self-closing valves, pressure vacuum bungs, and flame arresters. Use only nonsparking tools and equipment, especially when opening and closing containers of tetramethyl lead. Wherever tetramethyl lead is used, handled, manufactured, or stored, use explosion-proof electrical equipment and fittings.

Shipping: Motor fuel antiknock mixtures requires a shipping label of "POISONOUS/TOXIC MATERIALS." It falls in Hazard Class 6.1 and Packing Group I.

Spill Handling: Evacuate and restrict persons not wearing protective equipment from area of spill or leak until cleanup is complete. Remove all ignition sources. Establish forced ventilation to keep levels below explosive limit. Absorb liquids in vermiculite, dry sand; earth, peat, carbon, or a similar material and deposit in sealed containers. Keep this chemical out of a confined space, such as a sewer, because of the possibility of an explosion, unless the sewer is designed to prevent the buildup of explosive concentrations. It may be necessary to contain and dispose of this chemical as a hazardous waste. If material or contaminated runoff enters waterways, notify downstream users of potentially contaminated waters. Contact your local or federal environmental protection agency for specific recommendations. If employees are required to clean up spills, they must be properly trained and equipped. OSHA 1910.120(q) may be applicable.

Fire Extinguishing: Tetramethyl lead is a flammable and reactive liquid. Use dry chemical, carbon dioxide, water spray; or foam extinguishers. Poisonous gases, including lead, lead oxides and carbon monoxide, are produced in fire. Vapors are heavier than air and will collect in low areas. Vapors may travel long distances to ignition sources and flashback. Vapors in confined areas may explode when exposed to fire. Containers may explode in fire. Storage containers and parts of containers may rocket great distances, in many directions. If material or contaminated runoff enters waterways, notify downstream users of potentially contaminated waters. Notify local health and fire officials and pollution control agencies. From a secure, explosion-proof location, use water spray to cool exposed containers. If cooling streams are ineffective (venting sound increases in volume and pitch, tank discolors, or shows any signs of deforming), withdraw immediately to a secure position. If employees are expected to fight fires, they must be trained and equipped in OSHA 1910.156. The only respirators recommended for firefighting are self-contained breathing apparatuses that have full face-pieces and are operated in a pressure-demand or other positive-pressure mode.

Disposal Method Suggested: Controlled incineration with scrubbing for collection of lead oxides which may be recycled or landfilled. It is also possible to recover alkyl lead compounds from wastewaters (A-58) as an alternative to disposal.

References

US Environmental Protection Agency. (November 30, 1987). *Chemical Hazard Information Profile: Tetramethyl Lead.* Washington, DC: Chemical Emergency Preparedness Program

New Jersey Department of Health and Senior Services. (March 2002). *Hazardous Substances Fact Sheet: Tetramethyl Lead.* Trenton, NJ

Tetramethyl succinonitrile   T:0370

Molecular Formula: $C_8H_{12}N_2$
Common Formula: $(CH_3)_2C(CN)C(CN)(CH_3)_2$
Synonyms: Tetramethylbutane dinitrile; Tetramethyl-succinic acid dinitrile; TMSN
CAS Registry Number: 3333-52-6
RTECS® Number: WN4025000
UN/NA & ERG Number: UN2811 (toxic solid, organic, n.o.s.)/154
EC Number: None assigned.
Regulatory Authority and Advisory Bodies
Air Pollutant Standard Set. See below, "Permissible Exposure Limits in Air" section.
As cyanide compounds:
Clean Air Act: Hazardous Air Pollutants (Title I, Part A, Section 112) as cyanide compound.
Clean Water Act: 40CFR423, Appendix A, Priority Pollutants as cyanide, total.
US EPA Hazardous Waste Number (RCRA No.): P030 as cyanides soluble salts and complexes, n.o.s.

RCRA, 40CFR261, Appendix 8 Hazardous Constituents. as cyanides, soluble salts and complexes, n.o.s.

EPCRA (Section 313): Form R *de minimis* concentration reporting level: 1.0%.

US DOT Regulated Marine Pollutant (49CFR172.101, Appendix B) as cyanide mixtures, cyanide solutions.

Canada, WHMIS, Ingredients Disclosure List Concentration 1.0%; National Pollutant Release Inventory (NPRI); CEPA Priority Substance List, Ocean dumping prohibited.

European/International Regulations: Not listed in Annex 1.

WGK (German Aquatic Hazard Class): 3—Severe hazard to waters.

Description: Tetramethyl succinonitrile is a colorless, odorless solid. Molecular weight = 136.22; Specific gravity (H_2O:1) = 1.07 at 25°C; Boiling point = sublimes; Freezing/Melting point = 170°C (sublimes). Insoluble in water.

Potential Exposure: Compound Description: Reproductive Effector. Tetramethyl succinonitrile is reported to be a breakdown product of azobisisobutyronitrile, which is used as a blowing agent or propellant for the production of vinyl foams. Forms cyanide in the body.

Incompatibilities: Strong oxidizers.

Permissible Exposure Limits in Air
Conversion factor: 1 ppm = 5.57 mg/m^3 at 25°C & 1 atm.
OSHA PEL: 3 mg/m^3/0.5 ppm TWA [skin].
NIOSH REL: 3 mg/m^3/0.5 ppm TWA [skin].
ACGIH TLV®[1]: 0.5 ppm TWA [skin].
No Teel available.
DFG MAK: [skin] No numerical value established. Data may be available.
NIOSH IDLH: 5 ppm.
Australia: TWA 0.5 ppm (3 mg/m^3) [skin], 1993; Austria: MAK 0.5 ppm (3 mg/m^3) [skin], 1999; Belgium: TWA 0.5 ppm (28 mg/m^3) [skin], 1993; Denmark: TWA 0.5 ppm (3 mg/m^3) [skin], 1999; Finland: TWA 0.5 ppm (3 mg/m^3), STEL 1.5 ppm (8.4 mg/m^3) [skin], 1999; France: VME 0.5 ppm (3 mg/m^3) [skin], 1999; the Netherlands: MAC-TGG 3 mg/m^3 [skin], 2003; the Philippines: TWA 0.5 ppm (3 mg/m^3) [skin], 1993; Switzerland: MAK-W 0.3 ppm (5 mg/m^3), KZG-W 2 ppm [skin], 1999; United Kingdom: TWA 0.5 ppm (2.8 mg/m^3), STEL 2 ppm [skin], 2000; Argentina, Bulgaria, Columbia, Jordan, South Korea, New Zealand, Singapore, Vietnam: ACGIH TLV®: TWA 0.5 ppm [skin]. Several states have set guidelines or standards for TMSN in ambient air[60] ranging from 30 μg/m^3 (North Dakota) to 50 μg/m^3 (Virginia) to 60 μg/m^3 (Connecticut) to 71 μg/m^3 (Nevada).

Determination in Air: Charcoal tube; CS2; Gas chromatography/Flame ionization detection; NIOSH II (3), Method #S155.

Permissible Concentration in Water: No criteria set, but EPA[32] has suggested a permissible ambient goal of 41 μg/L based on health effects.

Routes of Entry: Inhalation, skin absorption; ingestion; skin and/or eye contact.

Harmful Effects and Symptoms
Short Term Exposure: May affect the nervous system. Symptoms of exposure are headaches, nausea, convulsions, coma. May affect the liver, kidneys, gastrointestinal tract. Exposure to high concentrations may result in death. The fatal dose for humans is about 25 mg/kg of body weight.

Long Term Exposure: May affect the kidneys, liver.

Points of Attack: Central nervous system; liver, kidneys, gastrointestinal tract.

Medical Surveillance: Consider the points of attack in pre-placement and periodic physical examinations. Blood cyanide level. Kidney and liver functions. G.I. series. Examination of the nervous system.

First Aid: If this chemical gets into the eyes, remove any contact lenses at once and irrigate immediately for at least 15 min, occasionally lifting upper and lower lids. Seek medical attention immediately. If this chemical contacts the skin, remove contaminated clothing and wash immediately with soap and water. Seek medical attention immediately. If this chemical has been inhaled, remove from exposure, begin rescue breathing (using universal precautions, including resuscitation mask) if breathing has stopped and CPR if heart action has stopped. Transfer promptly to a medical facility. When this chemical has been swallowed, get medical attention. Give large quantities of water and induce vomiting. Do not make an unconscious person vomit.

Use amyl nitrate capsules if symptoms of cyanide poisoning develop. All area employees should be trained regularly in emergency measures for cyanide poisoning and in CPR. A cyanide antidote kit should be kept in the immediate work area and must be rapidly available. Kit ingredients should be replaced every 1−2 years to ensure freshness. Persons trained in the use of this kit, oxygen use, and CPR must be quickly available.

Personal Protective Methods: Wear protective gloves and clothing to prevent any reasonable probability of skin contact. Safety equipment suppliers/manufacturers can provide recommendations on the most protective glove/clothing material for your operation. All protective clothing (suits, gloves, footwear, headgear) should be clean, available each day, and put on before work. Contact lenses should not be worn when working with this chemical. Wear dust-proof chemical goggles and face shield unless full-face-piece respiratory protection is worn. Employees should wash immediately with soap when skin is wet or contaminated. Provide emergency showers and eyewash. See NIOSH Criteria Document 212 *Nitriles.*

Respirator Selection: Up to 28 mg/m^3: Sa (APF = 10) (any supplied-air respirator); or SCBAF (APF = 50) (any self-contained breathing apparatus with a full face-piece). *Emergency or planned entry into unknown concentrations or IDLH conditions:* SCBAF: Pd,Pp (APF = 10,000) (any self-contained breathing apparatus that has a full face-piece and is operated in a pressure-demand or other positive-pressure mode); or SaF: Pd,Pp: ASCBA (APF = 10,000) (any supplied-air respirator that has a full face-piece and is operated in a pressure-demand or other positive-pressure mode in combination with an auxiliary, self-contained breathing apparatus operated in a pressure-demand or other positive-pressure mode). *Escape:* GmFOv100 (APF = 50) [any air-purifying, full-face-piece respirator (gas mask) with a chin-style, front- or back-mounted organic vapor canister having an N100, R100, or P100 filter]; or SCBAE (any appropriate escape-type, self-contained breathing apparatus).

Storage: Color Code—Blue: Health Hazard/Poison: Store in a secure poison location. Prior to working with this chemical you should be trained on its proper handling and storage. Store in tightly closed containers in a cool, well-ventilated area away from strong oxidizers. Where possible, automatically transfer material from drums or other storage containers to process containers. Sources of ignition, such as smoking and open flames, are prohibited where this chemical is handled, used, or stored. Metal containers involving the transfer of this chemical should be grounded and bonded. Wherever this chemical is used, handled, manufactured, or stored, use explosion-proof electrical equipment and fittings.

Shipping: Toxic solids, organic, n.o.s., require a shipping label of "POISONOUS/TOXIC MATERIALS." It falls in Hazard Class 6.1 and Packing Group II.

Spill Handling: Evacuate persons not wearing protective equipment from area of spill or leak until cleanup is complete. Remove all ignition sources. Collect powdered material in the most convenient and safe manner and deposit in sealed containers. Ventilate area after cleanup is complete. It may be necessary to contain and dispose of this chemical as a hazardous waste. If material or contaminated runoff enters waterways, notify downstream users of potentially contaminated waters. Contact your local or federal environmental protection agency for specific recommendations. If employees are required to clean up spills, they must be properly trained and equipped. OSHA 1910.120(q) may be applicable.

Fire Extinguishing: This chemical is a combustible solid. Use dry chemical, carbon dioxide; water spray; or alcohol foam extinguishers. Poisonous gases are produced in fire. If material or contaminated runoff enters waterways, notify downstream users of potentially contaminated waters. Notify local health and fire officials and pollution control agencies. From a secure, explosion-proof location, use water spray to cool exposed containers. If cooling streams are ineffective (venting sound increases in volume and pitch, tank discolors, or shows any signs of deforming), withdraw immediately to a secure position. If employees are expected to fight fires, they must be trained and equipped in OSHA 1910.156. The only respirators recommended for firefighting are self-contained breathing apparatuses that have full face-pieces and are operated in a pressure-demand or other positive-pressure mode.

Disposal Method Suggested: Incineration—incinerator equipped with a scrubber or thermal unit to reduce nitrogen oxides emissions.

Tetranitromethane   T:0380

Molecular Formula: CN_4O_8
Common Formula: $C(NO_2)_4$
Synonyms: Methane, tetranitro-; NCI-C55947; Tetan; Tetranitrometano (Spanish); TNM
CAS Registry Number: 509-14-8
RTECS® Number: PB4025000
UN/NA & ERG Number: UN1510/143
EC Number: 208-094-7
Regulatory Authority and Advisory Bodies
Department of Homeland Security Screening Threshold Quantity (pounds): *Release hazard* 10,000 ($\geq$1.00% concentration).
Carcinogenicity: IARC: Human Inadequate Evidence; Animal Sufficient Evidence, *possibly carcinogenic to humans*, Group 2B, 1996; NTP: 11th Report on Carcinogens, 2004: Reasonably anticipated to be a human carcinogen; NCI: Carcinogenesis Studies (inhalation); clear evidence: mouse, rat.
Air Pollutant Standard Set. See below, "Permissible Exposure Limits in Air" section.
Clean Air Act: Accidental Release Prevention/Flammable Substances, (Section 112[r], Table 3), TQ = 10,000 lb (4540 kg).
US EPA Hazardous Waste Number (RCRA No.): P112.
RCRA, 40CFR261, Appendix 8 Hazardous Constituents.
Superfund/EPCRA 40CFR355, Extremely Hazardous Substances: TPQ = 500 lb (227 kg).
Reportable Quantity (RQ): 10 lb (4.54 kg).
US DOT 49CFR172.101, Inhalation Hazardous Chemical.
California Proposition 65 Chemical: Cancer 7/1/90.
Canada, WHMIS, Ingredients Disclosure List Concentration 1.0%.
European/International Regulations: Not listed in Annex 1.
WGK (German Aquatic Hazard Class): No value assigned.
Description: Tetranitromethane, a nitroparaffin, is a colorless to pale yellow liquid or solid with a pungent odor. It causes tears. Molecular weight = 196.05; Specific gravity (H_2O:1) = 1.62 at 25°C; Boiling point = 126.1°C; Freezing/Melting point 14°C; Vapor pressure = 8 mmHg at 25°C. Hazard Identification (based on NFPA-704 M Rating System): Health 3, Flammability 1, Reactivity 3 (Oxidizer). Insoluble in water.
Potential Exposure: Compound Description: Tumorigen, Mutagen. Tetranitromethane is used as a solvent for polymers and as a stabilizer; as an oxidizer in rocket propellant combinations. It is also used as an explosive in admixture with toluene.
Incompatibilities: Tetranitromethane is a powerful oxidizer. It is more easily detonated than TNT. Contact with hydrocarbons, alkalis, or metals form explosive mixtures. Contact with toluene or cotton may cause fire and explosion. Combustible material wet with tetranitromethane may be highly explosive. The potential for explosion is severe,

especially when exposed to heat, powerful oxidizers, or reducing agents; or, when subject to mild shock. Impurities can also cause explosives. Attacks some plastics, rubber, and coatings.

Permissible Exposure Limits in Air
Conversion factor: 1 ppm = 8.02 mg/m^3 at 25°C & 1 atm.
OSHA PEL: 1 ppm/8 mg/m^3 TWA.
NIOSH REL: 1 ppm/8 mg/m^3 TWA.
ACGIH TLV®[1]: 0.005 ppm/0.04 mg/m^3 TWA; confirmed animal carcinogen with unknown relevance to humans.
NIOSH IDLH: 4 ppm.
Protective Action Criteria (PAC)*
TEEL-0: 0.52 ppm
PAC-1: 0.52 ppm
PAC-2: **0.52** ppm
PAC-3: **1.7** ppm
*AEGLs (Acute Emergency Guideline Levels) & ERPGs (Emergency Response Planning Guideline) are in **bold face**.
DFG MAK: [skin] Carcinogen Category 2.
Australia: TWA 1 ppm (8 mg/m^3), 1993; Austria: carcinogen, 1999; Belgium: TWA 1 ppm (8 mg/m^3), 1993; Denmark: TWA 1 ppm (8 mg/m^3), 1999; Finland: TWA 1 ppm (8 mg/m^3), STEL 3 ppm (24 mg/m^3), 1999; France: VME 1 ppm (8 mg/m^3), 1999; the Netherlands: MAC-TGG 0.04 mg/m^3, 2003; the Philippines: TWA 1 ppm (8 mg/m^3), 1993; Poland: MAC (TWA) 0.04 mg/m^3, 1999; Russia: STEL 0.3 mg/m^3, 1993; Sweden: NGV 0.05 ppm (0.4 mg/m^3), KTV 0.2 ppm (0.8 mg/m^3), 1999; Switzerland: MAK-W 1 ppm (8 mg/m^3), carcinogen, 1999; Turkey: TWA 1 ppm (8 mg/m^3), 1993; Argentina, Bulgaria, Columbia, Jordan, South Korea, New Zealand, Singapore, Vietnam: ACGIH TLV®: confirmed animal carcinogen with unknown relevance to humans. Several states have set guidelines or standards for TNM in ambient air[60] ranging from 80 µg/m^3 (North Dakota) to 130 µg/m^3 (Virginia) to 160 µg/m^3 (Connecticut) to 190 µg/m^3 (Nevada).
Determination in Air: Use NIOSH Analytical Method (IV) #3513.
Permissible Concentration in Water: Russia[43] set a MAC in water bodies used for domestic purposes of 0.5 mg/L.
Routes of Entry: Inhalation, ingestion; skin and/or eye contact.
Harmful Effects and Symptoms
Short Term Exposure: Irritates the eyes, skin, and respiratory tract. Skin contact causes burns. After more prolonged inhalation, headache, dizziness, chest pain; dyspnea, and respiratory distress may occur. After prolonged exposure, central nervous system; heart, liver, and kidney damage can occur as well as pulmonary edema, a medical emergency that can be delayed for several hours. This can cause death. Can cause methemoglobinemia and cyanosis.
Long Term Exposure: Chronic signs and symptoms include weariness and pneumonia. May cause CNS, kidney, and liver damage.
Points of Attack: Respiratory system, eyes, skin, blood, central nervous system.

Medical Surveillance: Consider the points of attack in pre-placement and periodic physical examinations. Kidney and liver function tests. Complete blood count (CBC). Examination of the nervous system.
First Aid: If this chemical gets into the eyes, remove any contact lenses at once and irrigate immediately for at least 15 min, occasionally lifting upper and lower lids. Seek medical attention immediately. If this chemical contacts the skin, remove contaminated clothing and wash immediately with soap and water. Seek medical attention immediately. If this chemical has been inhaled, remove from exposure, begin rescue breathing (using universal precautions, including resuscitation mask) if breathing has stopped and CPR if heart action has stopped. Transfer promptly to a medical facility. When this chemical has been swallowed, get medical attention. Give large quantities of water and induce vomiting. Do not make an unconscious person vomit. Medical observation is recommended for 24−48 h after breathing overexposure, as pulmonary edema may be delayed. As first aid for pulmonary edema, a doctor or authorized paramedic may consider administering a corticosteroid spray.
Note to physician: Treat for methemoglobinemia. Spectrophotometry may be required for precise determination of levels of methemoglobinemia in urine.
Personal Protective Methods: Wear protective gloves and clothing to prevent any reasonable probability of skin contact. Safety equipment suppliers/manufacturers can provide recommendations on the most protective glove/clothing material for your operation. All protective clothing (suits, gloves, footwear, headgear) should be clean, available each day, and put on before work. Contact lenses should not be worn when working with this chemical. Wear splash-proof chemical goggles and face shield unless full-face-piece respiratory protection is worn. Employees should wash immediately with soap when skin is wet or contaminated. Provide emergency showers and eyewash.
Respirator Selection: *Up to 4 ppm:* Sa:Cf (APF = 25) (any supplied-air respirator operated in a continuous-flow mode); ccrFS (APF = 50) [any chemical cartridge respirator with a full face-piece and cartridge(s) providing protection against the compound of concern]; or GmFS (APF = 50) [any air-purifying, full-face-piece respirator (gas mask) with a chin-style, front- or back-mounted canister providing protection against the compound of concern]; or PaprS (APF = 25) [any powered, air-purifying respirator with cartridge(s) providing protection against the compound of concern]; or SCBAF (APF = 50) (any self-contained breathing apparatus with a full face-piece); or SaF (APF = 50) (any supplied-air respirator with a full face-piece). *Emergency or planned entry into unknown concentrations or IDLH conditions:* SCBAF: Pd,Pp (APF = 10,000) (any self-contained breathing apparatus that has a full face-piece and is operated in a pressure-demand or other positive-pressure mode); or SaF: Pd,Pp: ASCBA (APF = 10,000) (any supplied-air respirator that has a full

face-piece and is operated in a pressure-demand or other positive-pressure mode in combination with an auxiliary self-contained breathing apparatus operated in a pressure-demand or other positive-pressure mode). *Escape:* GmFS (APF = 50) [any air-purifying, full-face-piece respirator (gas mask) with a chin-style, front- or back-mounted canister providing protection against the compound of concern]; or SCBAE (any appropriate escape-type, self-contained breathing apparatus). *Note:* Substance causes eye irritation or damage; requires eye protection. Only nonoxidizable sorbents allowed (not charcoal).

Storage: Color Code—Yellow: Reactive Hazard; Store in a location separate from other materials, especially flammables and combustibles. A regulated, marked area should be established where this chemical is handled, used, or stored in compliance with OSHA Standard 1910.1045. Store in an explosion-proof freezer. Keep away from metals and other organic and easily oxidized compounds. Where possible, automatically transfer material from drums or other storage containers to process containers. Sources of ignition, such as smoking and open flames, are prohibited where this chemical is handled, used, or stored. Metal containers involving the transfer of this chemical should be grounded and bonded. Wherever this chemical is used, handled, manufactured, or stored, use explosion-proof electrical equipment and fittings. See OSHA Standard 1910.104 and NFPA 43A *Code for the Storage of Liquid and Solid Oxidizers* for detailed handling and storage regulations.

Shipping: Tetranitromethane requires a shipping label of "OXIDIZER, POISONOUS/TOXIC MATERIALS." It falls in Hazard Class 5.1 and Packing Group I. A plus sign (+) symbol indicates that the designated proper shipping name and hazard class of the material must always be shown whether or not the material or its mixtures or solutions meet the definitions of the class.

Spill Handling: Evacuate and restrict persons not wearing protective equipment from area of spill or leak until cleanup is complete. Remove all ignition sources. Ventilate area of spill or leak. Avoid shock and friction if liquid spills on combustible material, such as wood or paper. Use water spray to reduce vapors. Absorb liquids in vermiculite, dry sand; earth, peat, carbon, or a similar material and deposit in sealed containers. Flush area with flooding amounts of water and dike spill for later disposal. Keep this chemical out of a confined space, such as a sewer, because of the possibility of an explosion, unless the sewer is designed to prevent the buildup of explosive concentrations. It may be necessary to contain and dispose of this chemical as a hazardous waste. If material or contaminated runoff enters waterways, notify downstream users of potentially contaminated waters. Contact your local or federal environmental protection agency for specific recommendations. If employees are required to clean up spills, they must be properly trained and equipped. OSHA 1910.120(q) may be applicable.

Initial isolation and protective action distances
Distances shown are likely to be affected during the first 30 min after materials are spilled and could increase with time. If more than one tank car, cargo tank, portable tank, or large cylinder is involved in the incident is leaking, the protective action distance may need to be increased. You may need to seek emergency information from CHEMTREC at (800) 424-9300 or seek professional environmental engineering assistance from the US EPA Environmental Response Team at (908) 548-8730 (24-h response line).
Small spills (From a small package or a small leak from a large package)
First: Isolate in all directions (feet/meters)   100/30.
Then: Protect persons downwind (miles/kilometers).
Day   0.2/0.3
Night   0.2/0.3
Large spills (From a large package or from many small packages)
First: Isolate in all directions (feet/meters)   200/60.
Then: Protect persons downwind (miles/kilometers)
Day   0.4/0.6
Night   0.6/1.4

Fire Extinguishing: Combustible liquid, but difficult to ignite. Poisonous gases, including nitrogen oxides, are produced in fire. Material is a strong oxidizer. The potential for explosion is severe, especially when exposed to heat or to powerful oxidizing or reducing agents; or when shocked or heated. It is more easily detonated than TNT. Impurities can also cause explosives. Extinguish small fires with water only, no dry chemicals or carbon dioxide. For large fires, flood the fire area with water. Do not move cargo or vehicle if cargo has been exposed to heat. Cool containers that are exposed to flames with water from the side until well after fire is out. For massive fire, use unmanned hose holder or monitor nozzles; if this is impossible, withdraw from area and let fire burn. Vapors are heavier than air and will collect in low areas. Vapors may travel long distances to ignition sources and flashback. Vapors in confined areas may explode when exposed to fire. Containers may explode in fire. Storage containers and parts of containers may rocket great distances, in many directions. If material or contaminated runoff enters waterways, notify downstream users of potentially contaminated waters. Notify local health and fire officials and pollution control agencies. From a secure, explosion-proof location, use water spray to cool exposed containers. If cooling streams are ineffective (venting sound increases in volume and pitch, tank discolors, or shows any signs of deforming), withdraw immediately to a secure position. If employees are expected to fight fires, they must be trained and equipped in OSHA 1910.156. The only respirators recommended for firefighting are self-contained breathing apparatuses that have full face-pieces and are operated in a pressure-demand or other positive-pressure mode.

Disposal Method Suggested: Consult with environmental regulatory agencies for guidance on acceptable disposal practices. Generators of waste containing this contaminant

(≥100 kg/mo) must conform with EPA regulations governing storage, transportation, treatment, and waste disposal. Open burning at remote burning sites is not entirely satisfactory since it makes no provision for the control of the toxic effluents, nitrogen oxides and HCN. Suggested procedures are to employ modified closed pit burning, using blowers for air supply and passing the effluent combustion gases through wet scrubbers.

References

Sax, N. I. (Ed.). (1985). *Dangerous Properties of Industrial Materials Report*, 5, No. 5, 87−91

US Environmental Protection Agency. (November 30, 1987). *Chemical Hazard Information Profile: Tetranitromethane*. Washington, DC: Chemical Emergency Preparedness Program

Tetrasodium EDTA   T:0390

Molecular Formula: $C_{10}H_{12}N_2Na_4O_8$
Common Formula: $[(NaOOCCH_2)_2NCH_2]_2$
Synonyms: *N,N'*-1,2-Ethanediylbis[*N*-(carboxymethyl)] glycine tetrasodium salt; Ethylene dinitrilotetra-acetic acid tetrasodium salt; Sodium EDTA; Sodium ethylene-diaminetetraacetate; Sodium ethylenediaminetetraacetic acid; Sodium salt of ethylene-diaminetetraacetic acid; Tetrasodium ethylenediaminetetraacetate; Tetrasodium ethylene-diaminetetracetate; Tetrasodium (ethylenedinitrilo)-tetraacetate; Tetrasodium salt of EDTA; Tetrasodium salt of ethylenediaminetetracetic acid
CAS Registry Number: 64-02-8
RTECS® Number: AH5075000
UN/NA & ERG Number: UN3077/171
EC Number: 200-573-9 [*Annex I Index No.:* 607-428-00-2]
Regulatory Authority and Advisory Bodies
This chemical is not specifically listed by the EPA, but ETDA is regulated.
Clean Water Act: Section 311 Hazardous Substances/RQ 40CFR117.3 (same as CERCLA, see below).
Reportable Quantity (RQ): 5000 lb (2270 kg).
European/International Regulations: Hazard Symbol: Xn; Risk phrases: R22; R41; Safety phrases: S2; S26; S39; S46 (see Appendix 4).
WGK (German Aquatic Hazard Class): 2—Hazard to waters.
Description: Tetrasodium EDTA is a white crystalline solid. Molecular weight = 380.20; Freezing/Melting point = 240°C (decomposition). Highly soluble in water.
Potential Exposure: Used therapeutically in treating arteriosclerosis. Used as a metal cleaner; in detergents, liquid soaps; shampoos; metal chelating agent; in textiles industry to improve dyeing, scouring, and detergent operations; antioxidant.
Incompatibilities: Incompatible with strong acids, nitrates, oxidizers. Can solubilize metals.

Permissible Exposure Limits in Air
Protective Action Criteria (PAC)
TEEL-0: 40 mg/m³
PAC-1: 125 mg/m³
PAC-2: 500 mg/m³
PAC-3: 500 mg/m³
Permissible Concentration in Water: Russia[43] set a MAC for the disodium salt in water bodies used for fishery purposes of 0.05 mg/L. Maryland[61] has set a guideline for drinking water of 180 μg/L (0.18 mg/L).[61]
Harmful Effects and Symptoms
Short Term Exposure: Inhalation: Dust may cause irritation of the nose and throat. *Skin:* May cause irritation. Prolonged skin contact to high concentrations may cause irritation, even a mild burn. *Eyes:* May cause irritation. Alkaline solution can burn the eyes. *Ingestion:* Doses of 200 mg/kg have caused muscle spasms. Kidney damage has been reported at doses of 600 mg/kg for 4 days.
Long Term Exposure: ETDA may cause kidney injury, which may be due to chelating action, not inherent nephrotoxicity. Symptoms of exposure include vomiting, depression, and bloody diarrhea. It is not known if this chemical has the same effects.
Points of Attack: Kidneys.
Medical Surveillance: Kidney function tests.
First Aid: If this chemical gets into the eyes, remove any contact lenses at once and irrigate immediately for at least 15 min, occasionally lifting upper and lower lids. Seek medical attention immediately. If this chemical contacts the skin, remove contaminated clothing and wash immediately with soap and water. Seek medical attention immediately. If this chemical has been inhaled, remove from exposure, begin rescue breathing (using universal precautions, including resuscitation mask) if breathing has stopped and CPR if heart action has stopped. Transfer promptly to a medical facility. When this chemical has been swallowed, get medical attention. Give large quantities of water and induce vomiting. Do not make an unconscious person vomit.
Note to physician: May cause a negative calcium imbalance.
Personal Protective Methods: Wear protective gloves and clothing to prevent any reasonable probability of skin contact. Safety equipment suppliers/manufacturers can provide recommendations on the most protective glove/clothing material for your operation. All protective clothing (suits, gloves, footwear, headgear) should be clean, available each day, and put on before work. Contact lenses should not be worn when working with this chemical. Wear dust-proof chemical goggles and face shield unless full-face-piece respiratory protection is worn. Employees should wash immediately with soap when skin is wet or contaminated. Provide emergency showers and eyewash.
Respirator Selection: Follow regulations in OSHA 29CFR1910.134 or European Standard EN149. Use a NIOSH/MSHA- or European Standard EN149-approved respirator; or use an approved supplied-air respirator with a

full face-piece operated in the positive-pressure mode, or with a full face-piece, hood, or helmet in the continuous-flow mode; or use a NIOSH/MSHA- or European Standard EN149-approved self-contained breathing apparatus with a full face-piece operated in pressure-demand or other positive-pressure mode.

Storage: Color Code—Green: General storage may be used. Prior to working with this chemical you should be trained on its proper handling and storage. Store in a well-ventilated area away from sources of heat. Where possible, automatically transfer material from drums or other storage containers to process containers. Sources of ignition, such as smoking and open flames, are prohibited where this chemical is handled, used, or stored. Metal containers involving the transfer of this chemical should be grounded and bonded. Wherever this chemical is used, handled, manufactured, or stored, use explosion-proof electrical equipment and fittings.

Shipping: The name of this material is not on the DOT list of materials[19] for label and packaging standards. However, based on regulations, it may be classified[52] as Environmentally hazardous substances, solid, n.o.s. This chemical requires a shipping label of "CLASS 9." It falls in Hazard Class 9 and Packing Group III.[20, 21]

Spill Handling: Evacuate persons not wearing protective equipment from area of spill or leak until cleanup is complete. Remove all ignition sources. Collect powdered material in the most convenient and safe manner and deposit in sealed containers. Ventilate area after cleanup is complete. It may be necessary to contain and dispose of this chemical as a hazardous waste. If material or contaminated runoff enters waterways, notify downstream users of potentially contaminated waters. Contact your local or federal environmental protection agency for specific recommendations. If employees are required to clean up spills, they must be properly trained and equipped. OSHA 1910.120(q) may be applicable.

Fire Extinguishing: Nonflammable. Use extinguisher appropriate to other burning material. Poisonous gases are produced in fire, including sodium and nitrogen oxides. If material or contaminated runoff enters waterways, notify downstream users of potentially contaminated waters. Notify local health and fire officials and pollution control agencies. From a secure, explosion-proof location, use water spray to cool exposed containers. If cooling streams are ineffective (venting sound increases in volume and pitch, tank discolors, or shows any signs of deforming), withdraw immediately to a secure position. If employees are expected to fight fires, they must be trained and equipped in OSHA 1910.156. The only respirators recommended for firefighting are self-contained breathing apparatuses that have full face-pieces and are operated in a pressure-demand or other positive-pressure mode.

References
Sax, N. I. (Ed.). (1987). *Dangerous Properties of Industrial Materials Report*, 7, No. 4, 76—80 (for EDTA)

New York State Department of Health. (July 1986). *Chemical Fact Sheet: Tetrasodium-EDTA.* Albany, NY: Bureau of Toxic Substance Assessment

Tetrasodium pyrophosphate   T:0400

Molecular Formula: $Na_4O_7P_2$
Common Formula: $Na_4P_2O_7$
Synonyms: Natriumpyrophosphat; Phosphotex; Pyrophosphate; Sodium pyrophosphate; Tetranatriumpyrophosphat (German); Tetrasodium diphosphate; Tetrasodium pyrophosphate, anhydrous; TSPP; Victor TSPP
CAS Registry Number: 7722-88-5
RTECS® Number: UX7350000
UN/NA & ERG Number: Not regulated.
EC Number: 231-767-1
Regulatory Authority and Advisory Bodies
US EPA, FIFRA 1998 Status of Pesticides: Canceled.
Air Pollutant Standard Set. See below, "Permissible Exposure Limits in Air" section.
Canada, WHMIS, Ingredients Disclosure List Concentration 1.0%.
European/International Regulations: Not listed in Annex 1.
WGK (German Aquatic Hazard Class): 1—Low hazard to waters.
Description: TSPP is a colorless or white crystalline powder or granules. Molecular weight = 265.90; Specific gravity (H_2O:1) = 2.45 at 25°C; Boiling point = (decomposes); Freezing/Melting point = 987.8°C. Hazard Identification (based on NFPA-704 M Rating System): Health 2, Flammability 0, Reactivity 0. Soluble in water; solubility = 7% at 25°C.
Potential Exposure: Compound Description: Drug TSPP is used as a water softener, as a builder in synthetic detergents; as a metal cleaner; in boiler water treatment; in viscosity control of drilling muds and in textile scouring and dyeing.
Incompatibilities: Strong acids; magnesium. The aqueous solution is a weak base.
Permissible Exposure Limits in Air
OSHA PEL: None.
NIOSH REL: 5 mg/m^3 TWA.
ACGIH TLV®[1]: withdrawn.
Protective Action Criteria (PAC)
TEEL-0: 5 mg/m^3
PAC-1: 15 mg/m^3
PAC-2: 25 mg/m^3
PAC-3: 500 mg/m^3
Australia: TWA 5 mg/m^3, 1993; Belgium: TWA 5 mg/m^3, 1993; Denmark: TWA 5 mg/m^3, 1999; Finland: TWA 5 mg/m^3, 1999; France: VME 5 mg/m^3, 1999; Norway: TWA 5 mg/m^3, 1999; Switzerland: MAK-W 5 mg/m^3, 1999; United Kingdom: TWA 5 mg/m^3, 2000; Argentina, Bulgaria, Columbia, Jordan, South Korea, New Zealand, Singapore, Vietnam: ACGIH TLV®: TWA 5 mg/m^3.

Several states have set guidelines or standards for TSPP in ambient air[60] ranging from 50 $\mu g/m^3$ (North Dakota) to 80 $\mu g/m^3$ (Virginia) to 100 $\mu g/m^3$ (Connecticut) to 119 $\mu g/m^3$ (Nevada).

Determination in Air: Use NIOSH Analytical Method (IV) #0500, Particulates NOR, total dust.

Routes of Entry: Inhalation, ingestion; skin and/or eye contact.

Harmful Effects and Symptoms

Short Term Exposure: TSPP is basically of low toxicity but it is an alkaline material. The dust is irritating to the eyes, skin, and respiratory tract. Contact irritates the skin.

Long Term Exposure: May cause dermatitis. May affect the lungs.

Points of Attack: Eyes, skin, respiratory system.

Medical Surveillance: For those with frequent or potentially high exposure (half the TLV or greater), the following are recommended before beginning work and at regular times after that: lung function tests.

First Aid: If this chemical gets into the eyes, remove any contact lenses at once and irrigate immediately for at least 15 min, occasionally lifting upper and lower lids. Seek medical attention immediately. If this chemical contacts the skin, remove contaminated clothing and wash immediately with soap and water. Seek medical attention immediately. If this chemical has been inhaled, remove from exposure, begin rescue breathing (using universal precautions, including resuscitation mask) if breathing has stopped and CPR if heart action has stopped. Transfer promptly to a medical facility. When this chemical has been swallowed, get medical attention. Give large quantities of water and induce vomiting. Do not make an unconscious person vomit.

Personal Protective Methods: Wear protective gloves and clothing to prevent any reasonable probability of skin contact. Safety equipment suppliers/manufacturers can provide recommendations on the most protective glove/clothing material for your operation. All protective clothing (suits, gloves, footwear, headgear) should be clean, available each day, and put on before work. Contact lenses should not be worn when working with this chemical. Wear dust-proof chemical goggles and face shield unless full-face-piece respiratory protection is worn. Employees should wash immediately with soap when skin is wet or contaminated. Provide emergency showers and eyewash.

Respirator Selection: Where there is potential for exposures *over 5 mg/m³*, use a NIOSH/MSHA- or European Standard EN149-approved respirator equipped with particulate (dust/fume/mist) filters. Particulate filters must be checked every day before work for physical damage, such as rips or tears, and replaced as needed. *Where there is potential for high exposures*, use a NIOSH/MSHA- or European Standard EN149-approved supplied-air respirator with a full face-piece operated in the positive-pressure mode, or with a full face-piece, hood, or helmet in the continuous-flow mode; or use a NIOSH/MSHA- or European Standard EN149-approved self-contained breathing apparatus with a full face-piece operated in pressure-demand or other positive-pressure mode.

Storage: Color Code—Green: General storage may be used. Prior to working with this chemical you should be trained on its proper handling and storage. Store in tightly closed containers in a cool, well-ventilated area away from strong acids and magnesium.

Spill Handling: Evacuate persons not wearing protective equipment from area of spill or leak until cleanup is complete. Remove all ignition sources. Collect powdered material in the most convenient and safe manner and deposit in sealed containers. Ventilate area after cleanup is complete. It may be necessary to contain and dispose of this chemical as a hazardous waste. If material or contaminated runoff enters waterways, notify downstream users of potentially contaminated waters. Contact your local or federal environmental protection agency for specific recommendations. If employees are required to clean up spills, they must be properly trained and equipped. OSHA 1910.120(q) may be applicable.

Fire Extinguishing: Extinguish fire using an agent suitable for type or surrounding fire. Tetrasodium pyrophosphate itself does not burn. Poisonous gases are produced in fire, including oxides of phosphorus and sodium. If material or contaminated runoff enters waterways, notify downstream users of potentially contaminated waters. Notify local health and fire officials and pollution control agencies. From a secure, explosion-proof location, use water spray to cool exposed containers. If cooling streams are ineffective (venting sound increases in volume and pitch, tank discolors, or shows any signs of deforming), withdraw immediately to a secure position. If employees are expected to fight fires, they must be trained and equipped in OSHA 1910.156. The only respirators recommended for firefighting are self-contained breathing apparatuses that have full face-pieces and are operated in a pressure-demand or other positive-pressure mode.

References

US Environmental Protection Agency, Special Review and Reregistration Division Office of Pesticide Programs. (1998). *Agency Status of Pesticides in Registration, Reregistration, and Special Review* (Rainbow Report), Washington, DC

New Jersey Department of Health and Senior Services. (January 2001). *Hazardous Substances Fact Sheet: Tetrasodium Pyrophosphate.* Trenton, NJ

Tetryl   T:0410

Molecular Formula: $C_7H_5N_5O_8$

Common Formula: $(NO_2)_3C_6H_2N(NO_2)CH_3$

Synonyms: *N*-Methyl-*N*-2,4,6-tetranitroaniline; Nitramine; Picrylnitromethylamine; Pyrenite; Tetralite; *N*-2,4,5-Tetranitro-*N*-methylaniline; 2,4,6-Tetryl; 2,4,6-Trinitrophenyl-*N*-methylnitramine; 2,4,6-Trinitro-phenylmethylnitramine; Trinitrophenylmethylnitramine

CAS Registry Number: 479-45-8

RTECS® Number: BY6300000

UN/NA & ERG Number: UN0208/112

EC Number: 207-531-9 [*Annex I Index No.:* 612-017-00-6]

Regulatory Authority and Advisory Bodies

Carcinogenicity: The carcinogenicity of tetryl in humans and animals has not been studied.

Air Pollutant Standard Set. See below, "Permissible Exposure Limits in Air" section.

Canada, WHMIS, Ingredients Disclosure List Concentration 1.0%.

European/International Regulations: Hazard Symbol: E, T; Risk phrases: R3; R23/24/25; R33; Safety phrases: S1/2; S35; S36/37; S45; R63 (see Appendix 4).

WGK (German Aquatic Hazard Class): No value assigned.

Description: Tetryl is a colorless to yellow, odorless crystalline solid. High explosive material. Molecular weight = 287.17; Specific gravity (H_2O:1) = 1.57 at 25°C; Boiling point = 180−190°C (explodes); Freezing/Melting point = 129−131°C; Flash point = 187°C (explodes); Vapor pressure = <1 mmHg at 25°C. Practically insoluble in water; solubility = 0.02%.

Potential Exposure: Compound Description: Mutagen, Primary Irritant. Tetryl is used in explosives; as an intermediary detonating agent; and as a booster charge for military devices; it is also used as a chemical indicator. No longer manufactured or used in the US.

Incompatibilities: Violent reaction with hydrazine; oxidizable materials. May explosively decompose from heat, shock, friction, or concussion. Explosive decomposition/detonation from heat takes: approximately 1000 s at 160°C; 0.1 s at 500°C.

Permissible Exposure Limits in Air

OSHA PEL: 1.5 mg/m³ TWA [skin].

NIOSH REL: 1.5 mg/m³ TWA [skin].

ACGIH TLV®[1]: 1.5 mg/m³ TWA [skin].

NIOSH IDLH: 750 mg/m³.

Protective Action Criteria (PAC)

TEEL-0: 1.5 mg/m³

PAC-1: 7.5 mg/m³

PAC-2: 50 mg/m³

PAC-3: 500 mg/m³

DFG MAK: [skin] danger of skin sensitization; Carcinogen Category 3B.

Australia: TWA 1.5 mg/m³, 1993; Austria: MAK 1.5 mg/m³ [skin], 1999; Belgium: TWA 1.5 mg/m³, 1993; Denmark: TWA 1.5 mg/m³ [skin], 1999; Finland: TWA 1.5 mg/m³, STEL 3 mg/m³ [skin], 1993; France: VME 1.5 mg/m³ [skin], 1999; the Netherlands: MAC-TGG 1.5 mg/m³ [skin], 2003; the Philippines: TWA 1.5 mg/m³ [skin], 1993; Switzerland: MAK-W 1.5 mg/m³ [skin], 1999; Turkey: TWA 1.5 mg/m³, 1993; United Kingdom: TWA 1.5 mg/m³, STEL 3 mg/m³, 2000; Argentina, Bulgaria, Columbia, Jordan, South Korea, New Zealand, Singapore, Vietnam: ACGIH TLV®: TWA 1.5 mg/m³.

Several states have set guidelines or standards for tetryl in ambient air[60] ranging from 15 µg/m³ (North Dakota) to 25 µg/m³ (Virginia) to 30 µg/m³ (Connecticut) to 36 µg/m³ (Nevada).

Determination in Air: Use NIOSH II(3), Method #S225.

Routes of Entry: Inhalation, skin absorption; ingestion; skin and/or eye contact.

Harmful Effects and Symptoms

Short Term Exposure: Irritates the eyes, skin, and respiratory tract. May affect the nervous system. Contact may stain skin and hair yellow or orange. Tetryl is acutely irritating to the mucous membranes of the respiratory tract and the eyes, causing coughing, sneezing, epistaxis, conjunctivitis, and palpebral and periorbital edema.

Long Term Exposure: Tetryl is a potent sensitizer, and allergic dermatitis is common. Dermatitis first appears on exposed skin areas, but can spread to other parts of the body in fair-skinned individuals or those with poor personal hygiene. Repeated or prolonged inhalation exposure may cause asthma. The severest forms show massive generalized edema with partial obstruction of the trachea due to swelling of the tongue, and these cases require hospitalization. Tetryl exposure may cause irritability, easy fatigability; malaise, headaches, lassitude, insomnia, nausea, and vomiting. Anemia of the marrow depression or deficiency type has been observed among tetryl workers. Tetryl exposure has produced liver and kidney damage in animals. The substance may have effects on the liver, kidneys, and blood.

Points of Attack: Eyes, skin, respiratory system; central nervous system; liver, kidneys.

Medical Surveillance: NIOSH lists the following tests: complete blood count; anemia; liver function tests. Preplacement physical examination should give special attention to those individuals with a history of allergy, blood dyscrasias, or skin, liver, or kidney disease. Periodic examinations should be directed primarily to the control of dermatitis and allergic reactions, plus any effects on the respiratory tract, eyes, central nervous system; blood, liver, and kidneys.

First Aid: If this chemical gets into the eyes, remove any contact lenses at once and irrigate immediately for at least 15 min, occasionally lifting upper and lower lids. Seek medical attention immediately. If this chemical contacts the skin, remove contaminated clothing and wash immediately with soap and water. Seek medical attention immediately. If this chemical has been inhaled, remove from exposure, begin rescue breathing (using universal precautions, including resuscitation mask) if breathing has stopped and CPR if heart action has stopped. Transfer promptly to a medical facility. When this chemical has been swallowed, get medical attention. Give large quantities of water and induce vomiting. Do not make an unconscious person vomit.

Personal Protective Methods: Wear protective gloves and clothing to prevent any reasonable probability of skin contact. Safety equipment suppliers/manufacturers can provide

recommendations on the most protective glove/clothing material for your operation. All protective clothing (suits, gloves, footwear, headgear) should be clean, available each day, and put on before work. Contact lenses should not be worn when working with this chemical. Wear dust-proof chemical goggles and face shield unless full-face-piece respiratory protection is worn. Employees should wash immediately with soap when skin is wet or contaminated. Provide emergency showers and eyewash.

Respirator Selection: Up to 7.5 mg/m^3: Qm (APF = 25) (any quarter-mask respirator). Up to 15 mg/m^3: 95XQ* (APF = 10) [any particulate respirator equipped with an N95, R95, or P95 filter (including N95, R95, and P95 filtering face-pieces) except quarter-mask respirators. The following filters may also be used: N99, R99, P99, N100, R100, P100]; or Sa* (APF = 10) (any supplied-air respirator). Up to 37.5 mg/m^3: Sa:Cf* (APF = 25) (any supplied-air respirator operated in a continuous-flow mode); or PaprHie* (APF = 25) (any powered, air-purifying respirator with a high-efficiency particulate filter). Up to 75 mg/m^3: 100F (APF = 50) (any air-purifying, full-face-piece respirator with an N100, R100, or P100 filter); or SCBAF (APF = 50) (any self-contained breathing apparatus with a full face-piece); or SaF (APF = 50) (any supplied-air respirator with a full face-piece). Up to 750 mg/m^3: SaF: Pd,Pp (APF = 2000) (any supplied-air respirator that has a full face-piece and is operated in a pressure-demand or other positive-pressure mode). Emergency or planned entry into unknown concentrations or IDLH conditions: SCBAF: Pd, Pp (APF = 10,000) (any self-contained breathing apparatus that has a full faceplate and is operated in a pressure-demand or other positive-pressure mode); or SaF: Pd,Pp: ASCBA (APF = 10,000) (any supplied-air respirator that has a full face-piece and is operated in a pressure-demand or other positive-pressure mode in combination with an auxiliary, self-contained breathing apparatus operated in a pressure-demand or other positive-pressure mode). *Escape:* 100F (APF = 50) (any air-purifying, full-face-piece respirator with an N100, R100, or P100 filter); or SCBAE (any appropriate escape-type, self-contained breathing apparatus).

*Substance reported to cause eye irritation or damage; may require eye protection.

Storage: Explosive. Color Code—Red Stripe: Flammability Hazard: Store separately from all other flammable materials. Prior to working with this chemical you should be trained on its proper handling and storage. Store in tightly closed containers in a cool, well-ventilated area away from strong oxidizers, hydrazine and sources of heat. Where possible, automatically transfer material from drums or other storage containers to process containers. Sources of ignition, such as smoking and open flames, are prohibited where this chemical is handled, used, or stored. Metal containers involving the transfer of this chemical should be grounded and bonded. Wherever this chemical is used, handled, manufactured, or stored, use explosion-proof electrical equipment and fittings.

Shipping: Tetryl requires a shipping label of "EXPLOSIVE." It falls in Hazard Class 1.1D and Packing Group II.

Spill Handling: Evacuate persons not wearing protective equipment from area of spill or leak until cleanup is complete. Remove all ignition sources. Collect powdered material in the most convenient and safe manner and deposit in sealed containers. Ventilate area after cleanup is complete. It may be necessary to contain and dispose of this chemical as a hazardous waste. If material or contaminated runoff enters waterways, notify downstream users of potentially contaminated waters. Contact your local or federal environmental protection agency for specific recommendations. If employees are required to clean up spills, they must be properly trained and equipped. OSHA 1910.120(q) may be applicable.

Fire Extinguishing: This chemical is an explosive solid. If material is on fire and conditions permit, do not extinguish. Cool exposures using unattended monitors. If fire must be extinguished, use any agent appropriate for the burning material. Poisonous gases are produced in fire, including nitrogen oxides. If material or contaminated runoff enters waterways, notify downstream users of potentially contaminated waters. Notify local health and fire officials and pollution control agencies. Cool exposed containers from unattended equipment or remove intact containers if it can be done safely. If cooling streams are ineffective (venting sound increases in volume and pitch, tank discolors, or shows any signs of deforming), withdraw immediately to a secure position. If employees are expected to fight fires, they must be trained and equipped in OSHA 1910.156. The only respirators recommended for firefighting are self-contained breathing apparatuses that have full face-pieces and are operated in a pressure-demand or other positive-pressure mode.

Disposal Method Suggested: Solution in acetone and incineration in furnace equipped with afterburner and caustic soda solution scrubber.

References

New Jersey Department of Health and Senior Services. (September 2000). *Hazardous Substances Fact Sheet: Tetryl.* Trenton, NJ

Thallium & compounds   T:0420

Molecular Formula: Tl

Synonyms: Elemental: Ramor; Talio (TL) (Spanish); Thallium elemental

Acetate: Acetato de talio (Spanish); Thallium(1+) acetate; Thallium(I) acetate; Thallium monoacetate; Thallous acetate

Carbonate: Carbonato de talio (Spanish); Carbonic acid, dithallium(1+) salt; Carbonic acid, dithallium(I) salt;

Dithallium carbonate; Thallium(1+) carbonate; Thallium(I) carbonate; Thallous carbonate; Thiochoman-4-one, oxime

Chloride: Cloruro de talio (Spanish); Thallium(1+) chloride; Thallium(I) chloride; Thallium monochloride; Thallous chloride

Malonite: Formomalenic thallium; Malonic acid, thallium salt (1:2); Propanedioic acid, dithallium salt; Thallium malonite; Thallous malonate (EPA)

Nitrate: Nitrato de talio (Spanish); Nitric acid, thallium(1+) salt; Nitric acid, thallium(I) salt; Thallium mononitrate; Thallium(1+) nitrate (1:1); Thallium(I) nitrate; Thallous nitrate

Oxide: Dithallium trioxide; Oxido talico (Spanish); Thallic oxide (EPA); Thallium oxide; Thallium(3+) oxide; Thallium(III) oxide; Thallium peroxide; Thallium sesquioxide

Selinide: Thallium monoselenide

Sulfate: C.F.S; CFS-giftweizen; Dithallium sulfate; Dithallium(1+) sulfate; Dithallium(I) sulfate; Eccothal; M7-giftkoerner; Ratox; Rattengiftkonserv; Sulfato de talio (Spanish); Sulfuric acid, dithallium(+1) salt; Sulfuric acid, dithallium(I) salt(8CI,9CI); Sulfuric acid, Thallium salt; Sulfuric acid, thallium(1+) salt(1:2); Sulfuric acid, thallium (I) salt(1:2); Thallium sulfate; Thallium(1+) sulfate (2:1); Thallium(I) sulfate (2:1); Thallium sulfate; Thallous sulfate; Zelio sulfate(I)

CAS Registry Number: 7440-28-0 (elemental); 563-68-8 (acetate); 6533-73-9 (carbonate); 7799-12-0 (chloride); 2757-18-8 (malonite); 10102-45-1 (nitrate); 1314-32-5 (oxide); 12039-52-0 (selinide); 10031-59-1 (sulfate); 7446-18-6 [sulfate(I)]

RTECS® Number: XG3425000 (elemental); AJ5425000 (acetate); XG4000000 (carbonate); XG4200000 (chloride); OO1770000 (malonite); XG5950000 (nitrate); XG2975000 (oxide); XG6300000 (selinide); XG6600000 (sulfate); XG7800000 [sulfate(I)]

UN/NA & ERG Number: UN1707 (thallium compounds, n.o.s.)/151

EC Number: 081-001-00-3 (elemental); 081-002-00-9 [sulfate(I)]

Regulatory Authority and Advisory Bodies

Banned or Severely Restricted (many countries) (UN)[13].

Air Pollutant Standard Set. See below, "Permissible Exposure Limits in Air" section.

Thallium, elemental:

Clean Water Act: 40CFR401.15 Section 307 Toxic Pollutants; 40CFR423, Appendix A, Priority Pollutants.

RCRA, 40CFR261, Appendix 8 Hazardous Constituents, waste number not listed.

RCRA, 40CFR261, Appendix 8 Hazardous Constituents.

RCRA 40CFR268.48; 61FR15654, Universal Treatment Standards: Wastewater (mg/L), 1.4; Nonwastewater (mg/L), 0.78 TCLP.

RCRA 40CFR264, Appendix 9; TSD Facilities Ground Water Monitoring List. Suggested test method(s) (PQL μg/L); total dust 6010 (400); 7840 (1000); 7841 (10).

Safe Drinking Water Act: MCL, 0.002 mg/L; MCLG, 0.005 mg/L; Regulated chemical (47 FR 9352).

Reportable Quantity (RQ): 1000 lb (454 kg).

EPCRA Section 313 Form R *de minimis* concentration reporting level: 1.0% (Thallium compounds).

Thallium acetate:

Clean Water Act: 40CFR401.15 Section 307 Toxic Pollutants.

US EPA Hazardous Waste Number (RCRA No.): U214.

RCRA, 40CFR261, Appendix 8 Hazardous Constituents.

Reportable Quantity (RQ): 100 lb (45.4 kg).

EPCRA Section 313 Form R *de minimis* concentration reporting level: 1.0%.

Thallium carbonate; thallous carbonate:

Clean Water Act: 40CFR401.15 Section 307 Toxic Pollutants.

US EPA Hazardous Waste Number (RCRA No.): U215.

RCRA, 40CFR261, Appendix 8 Hazardous Constituents.

Superfund/EPCRA 40CFR355, Extremely Hazardous Substances: TPQ = 100/10,000 lb (45.4/4540 kg).

Reportable Quantity (RQ): 100 lb (45.4 kg).

EPCRA Section 313 Form R *de minimis* concentration reporting level: 1.0%.

Thallium chloride:

Clean Water Act: 40CFR401.15 Section 307 Toxic Pollutants.

US EPA Hazardous Waste Number (RCRA No.): U216.

RCRA, 40CFR261, Appendix 8 Hazardous Constituents.

Superfund/EPCRA 40CFR355, Extremely Hazardous Substances: TPQ = 100/10,000 lb (45.4/4540 kg).

Reportable Quantity (RQ): 100 lb (45.4 kg).

EPCRA Section 313 Form R *de minimis* concentration reporting level: 1.0%.

Thallium compounds:

Clean Water Act: 40CFR401.15 Section 307 Toxic Pollutants as thallium and compounds.

RCRA, 40CFR261, Appendix 8 Hazardous Constituents, waste number not listed.

PCRA Section 313 Form R *de minimis* concentration reporting level: 1.0%.

US DOT Regulated Marine Pollutant (49CFR172.101, Appendix B) as thallium compounds, n.o.s.; thallium compounds (pesticides).

Canada, WHMIS, Ingredients Disclosure List Concentration 1.0%.

European/International Regulations (*thallium*; 7440-28-0): Hazard Symbol: T+, N; Risk phrases: R26/28; R33; R53; Safety phrases: S1/2; S13; S28; S45; S61; (*thallium compounds*): Hazard Symbol: T+, N; Risk phrases: R28; R38; R48/25; R51/53; Safety phrases: S1/2; S13; S36/37; S45; S61; (*thallic sulfate*; 7446-18-6) Hazard Symbol: T+, N; Risk phrases: R28; R38; R48/25; R51/53; Safety phrases: S1/2; S13; S36/37; S45; S61.

Thallium nitrate:

Clean Water Act: 40CFR401.15 Section 307 Toxic Pollutants.

US EPA Hazardous Waste Number (RCRA No.): U217.
RCRA, 40CFR261, Appendix 8 Hazardous Constituents.
Reportable Quantity (RQ): 100 lb (45.4 kg).
EPCRA Section 313 Form R *de minimis* concentration reporting level: 1.0%.

Thallium(I) sulfate (7446-18-6):
Clean Water Act: Section 311 Hazardous Substances/RQ 40CFR117.3 (same as CERCLA, see below); 40CFR 401.15 Section 307 Toxic Pollutants.
US EPA Hazardous Waste Number (RCRA No.): P115.
RCRA, 40CFR261, Appendix 8 Hazardous Constituents.
Reportable Quantity (RQ): 100 lb (45.4 kg).
EPCRA Section 313 Form R *de minimis* concentration reporting level: 1.0%.
Superfund/EPCRA 40CFR355, Extremely Hazardous Substances: TPQ = 100/10,000 lb (45.4/4540 kg).

Thallium sulfate (10031-59-1):
Clean Water Act: Section 311 Hazardous Substances/RQ 40CFR117.3 (same as CERCLA, see below); 40CFR 401.15 Section 307 Toxic Pollutants.
Superfund/EPCRA 40CFR355, Extremely Hazardous Substances: TPQ = 100/10,000 lb (45.4/4540 kg).
Reportable Quantity (RQ): 100 lb (45.4 kg).
EPCRA Section 313 Form R *de minimis* concentration reporting level: 1.0%.
US DOT Regulated Marine Pollutant (49CFR172.101, Appendix B).

Thallium malonite:
Clean Water Act: 40CFR401.15 Section 307 Toxic Pollutants.
Superfund/EPCRA 40CFR355, Extremely Hazardous Substances: TPQ = 100/10,000 lb (45.4/4540 kg).
Reportable Quantity (RQ): 1 lb (0.454 kg).
EPCRA Section 313 Form R *de minimis* concentration reporting level: 1.0%.
WGK (German Aquatic Hazard Class): 2—Water polluting (*thallium chlorate, nitrate(I), nitrate(II), and sulfate*).

Description: Thallium is a soft, bluish-white, heavy, very soft metal insoluble in water and organic solvents. It turns gray on exposure to air. Molecular weight = 204.38; Boiling point = 1457°C; Freezing/Melting point = 304°C; Vapor pressure = 1.41×10^{-23} mmHg at 25°C. The nitrate is a colorless, crystalline powder. Hazard Identification (based on NFPA-704 M Rating System): Health 3, Flammability 1, Reactivity 0. Soluble in hot water. Thallium(I)carbonate(2:1) has molecular weight = 468.75. Thallium(I)sulfate(2:1) (7446-18-6) is a white or colorless, odorless crystalline solid. Molecular weight = 504.80; Freezing/Melting point = 632°C. Soluble in water. Thallium is usually obtained as a by-product from the flue dust generated during the roasting of pyrite ores in the smelting and refining of lead and zinc.

Potential Exposure: Compound Description: [elemental] Mutagen, Human Data; [thallium(I) carbonate] Mutagen; Reproductive Effector; [sulfate(I)] Agricultural Chemical; Reproductive Effector; Human Data. Thallium has not been produced in the United States since 1984, but is imported for use in the manufacture of electronics, optical lenses, and imitation precious jewels. It also has use in some chemical reactions and medical procedures. Thallium and its compounds are used as a rodenticide* and fungicide; in the manufacture of plates and prisms, high-density liquids; as insecticides, catalysts; in certain organic reactions, in phosphor activators; in bromoiodide crystals for lenses, plates, and prisms in infrared optical instruments; in photoelectric cells; in mineralogical analysis; alloyed with mercury in low-temperature thermometers, switches, and closures; in high-density liquids; in dyes and pigments; in fireworks; and imitation precious jewelry. It forms a stainless alloy with silver and a corrosion-resistant alloy with lead. Its medicinal use for epilation has been almost discontinued. Highly persistent in the environment.

*Thallium was used in the past as a rodenticide, it has been banned in the US due to its toxicity from accidental exposure. In some countries, thallium(I)sulfate(2:1) is still used as a rat poison and ant bait.

Incompatibilities: Varies. Thallium metal reacts violently with strong acids (such as hydrochloric, sulfuric, and nitric) and strong oxidizers (such as chlorine, bromine, and fluorine). Cold thallium ignites on contact with fluorine. Reacts with other halogens at room temperature.

Permissible Exposure Limits in Air
OSHA PEL: 0.1 mg[Tl]/m^3 TWA [skin].
NIOSH REL: 0.1 mg[Tl]/m^3 TWA [skin].
ACGIH TLV®[1]: 0.1 mg[Tl]/m^3 TWA [skin].
NIOSH IDLH: 15 mg[Tl]/m^3.
Protective Action Criteria (PAC)
TEEL-0: 0.02 mg/m^3
PAC-1: 0.6 mg/m^3
PAC-2: 3 mg/m^3
PAC-3: 3 mg/m^3
Protective Action Criteria (PAC) *thallium sulfate (10031-59-1)*
TEEL-0: 0.0213 mg/m^3
PAC-1: 0.064 mg/m^3
PAC-2: 2 mg/m^3
PAC-3: 16 mg/m^3
Protective Action Criteria (PAC)
7446-18-6 [thallium(I) sulfate]
TEEL-0: 0.123 mg/m^3
PAC-1: 0.37 mg/m^3
PAC-2: 2 mg/m^3
PAC-3: 18.5 mg/m^3
Protective Action Criteria (PAC)
6533-73-9 [thallium(I) carbonate (2:1)]
TEEL-0: 0.115 mg/m^3
PAC-1: 0.344 mg/m^3
PAC-2: 2 mg/m^3
PAC-3: 17.2 mg/m^3
Protective Action Criteria (PAC)
7791-12-0 [thallium(I) chloride]
TEEL-0: 0.0235 mg/m^3

PAC-1: 0.0704 mg/m^3
PAC-2: 2 mg/m^3
PAC-3: 17.6 mg/m^3
Protective Action Criteria (PAC)
12026-06-1 (thallium hydroxide)
TEEL-0: 0.0217 mg/m^3
PAC-1: 0.065 mg/m^3
PAC-2: 0.108 mg/m^3
PAC-3: 16.2 mg/m^3
Protective Action Criteria (PAC)
2757-18-8 (thallous malonate)
TEEL-0: 0.025 mg/m^3
PAC-1: 0.075 mg/m^3
PAC-2: 2 mg/m^3
PAC-3: 18.7 mg/m^3
10102-45-1 (thallium nitrate)
TEEL-0: 0.0261 mg/m^3
PAC-1: 0.0782 mg/m^3
PAC-2: 19.6 mg/m^3
PAC-3: 19.6 mg/m^3
13826-63-6 (thallium nitrite)
TEEL-0: 0.0245 mg/m^3
PAC-1: 0.0735 mg/m^3
PAC-2: 0.125 mg/m^3
PAC-3: 18.4 mg/m^3
1314-12-1 (thallium oxide)
TEEL-0: 0.0208 mg/m^3
PAC-1: 0.0623 mg/m^3
PAC-2: 0.104 mg/m^3
PAC-3: 15.6 mg/m^3

DFG MAK: *soluble compounds:* No numerical value established. Data may be available.
Australia: TWA 0.1 mg/m^3 [skin], 1993; Austria: MAK 0.1 mg/m^3, 1999; Belgium: TWA 0.1 mg/m^3 [skin], 1993; Finland: TWA 0.1 mg/m^3 [skin], 1999; France: VME 0.1 mg/m^3, 1999; Norway: TWA 0.1 mg/m^3, 1999; the Netherlands: MAC-TGG 0.1 mg/m^3 [skin], 2003; Poland: MAC (TWA) 0.1 mg[Tl]/m^3; MAC (STEL) 0.3 mg[Tl]/m^3, 1999; Switzerland: MAK-W 0.1 mg/m^3 [skin], 1999; Thailand: TWA 0.1 mg/m^3, 1993; Turkey: TWA 0.1 mg/m^3 [skin], 1993; United Kingdom: LTEL 0.1 mg/m^3 [skin], 1993; Argentina, Bulgaria, Columbia, Jordan, South Korea, New Zealand, Singapore, Vietnam: ACGIH TLV®: TWA 0.1 mg[Tl]/m^3 [skin]. Several states have set guidelines or standards for thallium soluble compounds in ambient air[60] ranging from 0.238 μg/m^3 (Kansas) to 0.33 μg/m^3 (New York) to 1.0 μg/m^3 (Florida, North Dakota) to 1.6 μg/m^3 (Virginia) to 2.0 μg/m^3 (Connecticut and Nevada) to 2.47 μg/m^3 (Pennsylvania).

Determination in Air: Use NIOSH Analytical Method (IV) #7300; #7301; #7303; #9102; #8005 (Elements in blood or tissue); OSHA Analytical Method ID-121.

Permissible Concentration in Water: *To protect freshwater aquatic life:* 1400 μg/L on an acute toxicity basis and 40 μg/L on a chronic basis. *To protect saltwater aquatic life:* 2130 μg/L on an acute toxicity basis. For the protection of human health from the toxic properties of thallium ingested through water and contaminated aquatic organisms, the ambient water criterion is 13.0 μg/L.[6] Kansas[61] has set a guideline for thallium in drinking water of 13.0 μg/L. Russia set a MAC in water bodies used for domestic purposes of 0.1 μg/L.

Determination in Water: Digestion followed by atomic absorption measurement gives total thallium. Dissolved thallium may be determined by the same procedure preceded by 0.45 μm filtration. Octanol−water coefficient: Log K_{ow} = (estimated) 0.23.

Routes of Entry: Ingestion and percutaneous absorption of dust, eye/skin contact.

Harmful Effects and Symptoms

Short Term Exposure: Thallium salts may be eye and skin irritants and skin sensitizes. Exposure can cause fatigue, weakness, poor appetite; insomnia and mood changes. Acute poisoning rarely occurs in industry, and is usually due to ingestion of thallium. When it occurs, gastrointestinal symptoms, abdominal colic; loss of kidney function; peripheral neuritis; strabismus, disorientation, convulsions, joint pain; and alopecia develop rapidly. The symptoms of acute thallium poisoning (except for gastrointestinal symptoms) do not become manifest until 12 h to 4 days after exposure. Death is due to damage to the central nervous system. Thallium may affect the peripheral and the central nervous system; liver and kidneys; the gastrointestinal tract; skin (hair) and the cardiovascular system; resulting in polyneuritis; optic nerve atrophy; encephalopathy, cardiac disturbances; liver and kidney damage; alopecia. Exposure may result in death. The nitrate can irritate and burn the skin and eyes. The nitrate can damage the nervous system causing headache, weakness, irritability, pain, "pins and needles" in arms and leg; convulsions, coma, and death. The sulfate(I) irritates the eyes and the skin. May affect the nervous system; cardiovascular system; kidneys and gastrointestinal tract. Exposure may result in death. Exposure may result in hair loss.

Long Term Exposure: Thallium is an extremely toxic and cumulative poison. In nonfatal occupational cases of moderate or long-term exposure, early symptoms usually include fatigue, limb pain; metallic taste in the mouth and loss of hair; although loss of hair is not always present as an early symptom. Later, peripheral neuritis, proteinuria, and joint pains occur. Occasionally, neurological signs are the presenting factor, especially in more severe poisonings. Long-term exposure may produce optic atrophy, paresthesia, and changes in papillary and superficial tendon reflexes (slowed responses). Some thallium compounds are teratogens in animals.

Points of Attack: Eyes, central nervous system; lungs, liver, kidneys, gastrointestinal tract; body hair.

Medical Surveillance: NIOSH lists the following tests: whole blood (chemical/metabolite); biologic tissue/biopsy; nerve conduction studies; urine (chemical/metabolite); urinalysis (routine). Preplacement and periodic examinations should give special consideration to the eyes, central nervous system; gastrointestinal symptoms; and liver and kidney function. Hair loss may be a significant sign.

First Aid: If this chemical gets into the eyes, remove any contact lenses at once and irrigate immediately for at least 15 min, occasionally lifting upper and lower lids. Seek medical attention immediately. If this chemical contacts the skin, remove contaminated clothing and wash immediately with soap and water. Seek medical attention immediately. If this chemical has been inhaled, remove from exposure, begin rescue breathing (using universal precautions, including resuscitation mask) if breathing has stopped and CPR if heart action has stopped. Transfer promptly to a medical facility. When this chemical has been swallowed, get medical attention. Give a slurry of activated charcoal in water to drink and induce vomiting. Do not make an unconscious person vomit. The symptoms of acute thallium poisoning (except for gastrointestinal symptoms) do not become manifest until 12 h to 4 days after exposure.

For severe poisoning BAL [British Anti-Lewisite, dimercaprol, dithiopropanol ($C_3H_8OS_2$)] has been used to treat toxic symptoms of certain heavy metals poisoning. In the case of thallium it may have SOME value. Although BAL is reported to have a large margin of safety, caution must be exercised, because toxic effects may be caused by excessive dosage. Most can be prevented by premedication with 1-ephedrine sulfate (CAS: 134-72-5).

Personal Protective Methods: Wear protective gloves and clothing to prevent any reasonable probability of skin contact. Safety equipment suppliers/manufacturers can provide recommendations on the most protective glove/clothing material for your operation. All protective clothing (suits, gloves, footwear, headgear) should be clean, available each day, and put on before work. Contact lenses should not be worn when working with this chemical. Wear dust-proof chemical goggles and face shield unless full-face-piece respiratory protection is worn. Employees should wash immediately with soap when skin is wet or contaminated. Provide emergency showers and eyewash.

Respirator Selection: NIOSH/OSHA [as thallium (soluble compounds)]: *0.5 mg/m³:* Qm (APF = 25), if not present as a fume (any quarter-mask respirator). *1 mg/m³:* 95XQ (APF = 10) [any particulate respirator equipped with an N95, R95, or P95 filter (including N95, R95, and P95 filtering face-pieces) except quarter-mask respirators. The following filters may also be used: N99, R99, P99, N100, R100, P100]; or Sa (APF = 10) (any supplied-air respirator). *2.5 mg/m³:* Sa:Cf (APF = 25) (any supplied-air respirator operated in a continuous-flow mode); or PAPRDM, if not present as a fume (any powered, air-purifying respirator with a dust and mist filter). *5 mg/m³:* 100F (APF = 50) (any air-purifying, full-face-piece respirator with an N100, R100, or P100 filter); or SaT: Cf (APF = 50) (any supplied-air respirator that has a tight-fitting face-piece and is operated in a continuous-flow mode); or PaprTHie (APF = 50) (any powered, air-purifying respirator with a tight-fitting face-piece

and a high-efficiency particulate filter); or SCBAF (APF = 50) (any self-contained breathing apparatus with a full face-piece); or SaF (APF = 50) (any supplied-air respirator with a full face-piece). *15 mg/m³:* SaF: Pd,Pp (APF = 2000) (any supplied-air respirator that has a full face-piece and is operated in a pressure-demand or other positive-pressure mode). *Emergency or planned entry into unknown concentrations or IDLH conditions:* SCBAF: Pd, Pp (APF = 10,000) (any self-contained breathing apparatus that has a full face-piece and is operated in a pressure-demand or other positive-pressure mode); or SaF: Pd,Pp: ASCBA (APF = 10,000) (any supplied-air respirator that has a full face-piece and is operated in a pressure-demand or other positive-pressure mode in combination with an auxiliary self-contained breathing apparatus operated in a pressure-demand or other positive-pressure mode). *Escape:* 100F (APF = 50) (any air-purifying, full-face-piece respirator with an N100, R100, or P100 filter); or SCBAE (any appropriate escape-type, self-contained breathing apparatus).

Storage: Color Code—Blue: Health Hazard/Poison: Store in a secure poison location. Prior to working with this chemical you should be trained on its proper handling and storage. Thallium must be stored to avoid contact with strong acids (such as hydrochloric, sulfuric, and nitric) and strong oxidizers (such as chlorine, bromine, and fluorine), since violent reactions occur.

Shipping: Thallium compounds n.o.s. require a shipping label of "POISONOUS/TOXIC MATERIALS." They fall in DOT Hazard Class 6.1 and Packing Group II.

Spill Handling: Evacuate persons not wearing protective equipment from area of spill or leak until cleanup is complete. Remove all ignition sources. Collect powdered material in the most convenient and safe manner and deposit in sealed containers. Ventilate area after cleanup is complete. It may be necessary to contain and dispose of this chemical as a hazardous waste. If material or contaminated runoff enters waterways, notify downstream users of potentially contaminated waters. Contact your local or federal environmental protection agency for specific recommendations. If employees are required to clean up spills, they must be properly trained and equipped. OSHA 1910.120(q) may be applicable.

Fire Extinguishing: Thallium *metal* may burn, but does not readily ignite. Extinguish fire using an agent suitable for type of surrounding fire. For the *nitrate* use water only; *do not* use chemicals, foam, or carbon dioxide. Poisonous gases produced in fire will vary, including (*sulfide:* thallium and sulfur oxides; *nitrates:* nitrogen oxides and thallium). If material or contaminated runoff enters waterways, notify downstream users of potentially contaminated waters. Notify local health and fire officials and pollution control agencies. From a secure, explosion-proof location, use water spray to cool exposed containers. If cooling streams are ineffective (venting sound increases in volume and pitch, tank discolors, or shows any signs of deforming), withdraw immediately to a secure position. If employees

are expected to fight fires, they must be trained and equipped in OSHA 1910.156. The only respirators recommended for firefighting are self-contained breathing apparatuses that have full face-pieces and are operated in a pressure-demand or other positive-pressure mode.

Disposal Method Suggested: Consult with environmental regulatory agencies for guidance on acceptable disposal practices. Generators of waste containing this contaminant (≥100 kg/mo) must conform with EPA regulations governing storage, transportation, treatment, and waste disposal. Dilute thallium solutions may be disposed of in chemical waste landfills. When possible, thallium should be recovered and returned to the suppliers.

References
US Environmental Protection Agency. (1980). *Thallium: Ambient Water Quality Criteria.* Washington, DC

US Environmental Protection Agency. (April 30, 1980). *Thallium, Health and Environmental Effects Profile No. 159.* Washington, DC: Office of Solid Waste

Sax, N. I. (Ed.). *Dangerous Properties of Industrial Materials Report,* 4, No. 1, 94–97 (1984) (Sulfate); 7, No. 2, 92–94 (1987) (Acetate); and 8, No. 4, 13–22 (1988) (Nitrate)

US Environmental Protection Agency. (November 30, 1987). *Chemical Hazard Information Profile: Thallous Carbonate.* Washington, DC: Chemical Emergency Preparedness Program

US Environmental Protection Agency. (November 30, 1987). *Chemical Hazard Information Profile: Thallous Chloride.* Washington, DC: Chemical Emergency Preparedness Program

US Environmental Protection Agency. (November 30, 1987). *Chemical Hazard Information Profile: Thallous Malonate.* Washington, DC: Chemical Emergency Preparedness Program

US Environmental Protection Agency. (October 31, 1985). *Chemical Hazard Information Profile: Thallic Oxide.* Washington, DC: Chemical Emergency Preparedness Program

US Environmental Protection Agency. (November 30, 1987). *Chemical Hazard Information Profile: Thallium Sulfate.* Washington, DC: Chemical Emergency Preparedness Program

US Environmental Protection Agency. (November 30, 1987). *Chemical Hazard Information Profile: Thallous Sulfate.* Washington, DC: Chemical Emergency Preparedness Program

New Jersey Department of Health and Senior Services. (December 2000). *Hazardous Substances Fact Sheet: Thallium Acetate.* Trenton, NJ

New Jersey Department of Health and Senior Services. (November 2000). *Hazardous Substances Fact Sheet: Thallium Sulfate.* Trenton, NJ

New Jersey Department of Health and Senior Services. (November 2004). *Hazardous Substances Fact Sheet: Thallium Nitrate.* Trenton, NJ

New Jersey Department of Health and Senior Services. (November 2004). *Hazardous Substances Fact Sheet: Thallium.* Trenton, NJ

Thioacetamide   T:0430

Molecular Formula: C_2H_5NS
Common Formula: CH_3CSNH_2
Synonyms: Acetamide, thio-; Acetimidic acid, thio-; Acetothioamide; Ethanethioamide; TAA; Thiacetamide; Tioacetamida (Spanish)
CAS Registry Number: 62-55-5
RTECS® Number: AC8925000
UN/NA & ERG Number: UN2811 (toxic solid, organic, n.o.s.)/154
EC Number: 200-541-4 [Annex I Index No.: 616-026-00-6]
Regulatory Authority and Advisory Bodies
Carcinogenicity: IARC: Human No Adequate Data; Animal Sufficient Evidence, *possibly carcinogenic to humans,* Group 2B, 1987; NTP: 11th Report on Carcinogens, 2002: Reasonably anticipated to be a human carcinogen.
US EPA Gene-Tox Program, Positive: Carcinogenicity—mouse/rat; SHE—clonal assay; Positive: Cell transform.—mouse embryo; Positive: Cell transform.—RLV F344 rat embryo; Host-mediated assay; Positive: *D. melanogaster* sex-linked lethal; Weakly Positive: *S. cerevisiae*—homozygosis; Negative: *E. coli* polA with S9; Histidine reversion—Ames test; Negative: Sperm morphology—mouse; Inconclusive: *E. coli* polA without S9.
Banned or Severely Restricted (Sweden) (UN)[13].
US EPA Hazardous Waste Number (RCRA No.): U218.
RCRA, 40CFR261, Appendix 8 Hazardous Constituents.
Reportable Quantity (RQ): 10 lb (4.54 kg).
EPCRA Section 313 Form R *de minimis* concentration reporting level: 0.1%.
Air Pollutant Standard Set. See below, "Permissible Exposure Limits in Air" section.
California Proposition 65 Chemical: Cancer 1/1/88.
Canada, WHMIS, Ingredients Disclosure List Concentration 0.1%.
European/International Regulations: Hazard Symbol: T, N; Risk phrases: R45; R22; R36/38; R52/53; Safety phrases: S53; S45; S61 (see Appendix 4).
WGK (German Aquatic Hazard Class): No value assigned.
Description: Thioacetamide is combustible, crystalline compound with a slight mercaptan odor. Molecular weight = 75.14; Freezing/Melting point = 115°C. Hazard Identification (based on NFPA-704 M Rating System): Health 2, Flammability 1, Reactivity 0. Soluble in water.
Potential Exposure: Compound Description: Tumorigen, Mutagen; Reproductive Effector. Thioacetamide is used as a replacement for hydrogen sulfide in qualitative analyses. Thioacetamide has been used as an organic solvent in the leather, textile, and paper industries; as an accelerator in the

vulcanization of buna rubber; and as a stabilizer of motor fuel.

Incompatibilities: Strong oxidizers; strong acids; strong bases.

Permissible Exposure Limits in Air
Protective Action Criteria (PAC)
TEEL-0: 1.5 mg/m^3
PAC-1: 5 mg/m^3
PAC-2: 40 mg/m^3
PAC-3: 125 mg/m^3
United Kingdom: carcinogen, 2000. North Dakota[60] has set a guideline for thioacetamide in ambient air at zero concentration.

Determination in Water: Octanol—water coefficient: Log K_{ow} = low, about −0.40.

Routes of Entry: Inhalation, ingestion; skin and/or eye contact. Absorbed by the skin.

Harmful Effects and Symptoms
Short Term Exposure: Skin and eye irritation; conjunctivitis. Irritates the respiratory tract causing coughing. Symptoms of exposure include fatigue, nausea, vomiting, anorexia, liver damage; respiratory depression; central nervous system depression; acidosis, hypotension, tremors, convulsions, and unconsciousness.

Long Term Exposure: May cause liver damage severe enough to cause death. May cause lung damage. May cause dermatitis; eczema. A potential occupational carcinogen.

Points of Attack: Liver, lungs.

Medical Surveillance: Liver function tests. Lung function tests.

First Aid: Skin Contact:[52] Flood all areas of body that have contacted the substance with water. Do not wait to remove contaminated clothing; do it under the water stream. Use soap to help assure removal. Isolate contaminated clothing when removed to prevent contact by others.

Eye Contact: Remove any contact lenses at once. Immediately flush eyes well with copious quantities of water or normal saline for at least 20—30 min. Seek medical attention.

Inhalation: Leave contaminated area immediately; breathe fresh air. Proper respiratory protection must be supplied to any rescuers. If coughing, difficult breathing or any other symptoms develop, seek medical attention at once, even if symptoms develop many hours after exposure.

Ingestion: Contact a physician, hospital, or poison center at once. If the victim is unconscious or convulsing, do not induce vomiting or give anything by mouth. Assure that the patient's airway is open and lay him on his side with his head lower than his body and transport immediately to a medical facility. If conscious and not convulsing, give a glass of water or milk to dilute the substance. Vomiting should not be induced without a physician's advice.

Personal Protective Methods: Wear protective gloves and clothing to prevent any reasonable probability of skin contact. Safety equipment suppliers/manufacturers can provide recommendations on the most protective glove/clothing material for your operation. All protective clothing (suits, gloves, footwear, headgear) should be clean, available each day, and put on before work. Contact lenses should not be worn when working with this chemical. Wear dust-proof chemical goggles and face shield unless full-face-piece respiratory protection is worn. Employees should wash immediately with soap when skin is wet or contaminated. Provide emergency showers and eyewash.

Respirator Selection: Follow regulations in OSHA 29CFR1910.134 or European Standard EN149. Use a NIOSH/MSHA- or European Standard EN149-approved respirator; or use an approved supplied-air respirator with a full face-piece operated in the positive-pressure mode, or with a full face-piece, hood, or helmet in the continuous-flow mode; or use a NIOSH/MSHA- or European Standard EN149-approved self-contained breathing apparatus with a full face-piece operated in pressure-demand or other positive-pressure mode.

Storage: Color Code—Blue: Health Hazard/Poison: Store in a secure poison location. Prior to working with this chemical you should be trained on its proper handling and storage. Store in a refrigerator or a cool, dry place. Where possible, automatically transfer material from other storage containers to process containers. A regulated, marked area should be established where this chemical is handled, used, or stored in compliance with OSHA Standard 1910.1045.

Shipping: Thioacetamide is a potential occupational carcinogen. Toxic solids, organic, n.o.s., requires a shipping label of "POISONOUS/TOXIC MATERIALS." It falls in Hazard Class 6.1 and Packing Group III.

Spill Handling: Evacuate persons not wearing protective equipment from area of spill or leak until cleanup is complete. Remove all ignition sources. Collect powdered material in the most convenient and safe manner and deposit in sealed containers. Ventilate area after cleanup is complete. Use absorbent dampened with 60—70% acetone to pick up remaining material. Wash surfaces well with soap and water. It may be necessary to contain and dispose of this chemical as a hazardous waste. If material or contaminated runoff enters waterways, notify downstream users of potentially contaminated waters. Contact your local or federal environmental protection agency for specific recommendations. If employees are required to clean up spills, they must be properly trained and equipped. OSHA 1910.120(q) may be applicable.

Fire Extinguishing: Thioacetamide is a combustible solid. Use dry chemical, carbon dioxide; water spray; or alcohol foam extinguishers. Poisonous gases are produced in fire, including hydrogen sulfide and nitrogen oxides. If material or contaminated runoff enters waterways, notify downstream users of potentially contaminated waters. Notify local health and fire officials and pollution control agencies. From a secure, explosion-proof location, use water spray to cool exposed containers. If cooling streams are ineffective (venting sound increases in volume and pitch, tank discolors, or shows any signs of deforming), withdraw immediately to a secure position. If employees are expected to fight fires, they must be trained and equipped in OSHA 1910.156. The only respirators recommended for firefighting are self-contained breathing apparatuses that have full face-pieces and are operated in a pressure-demand or other positive-pressure mode.

Disposal Method Suggested: Consult with environmental regulatory agencies for guidance on acceptable disposal practices. Generators of waste containing this contaminant (≥100 kg/mo) must conform with EPA regulations governing storage, transportation, treatment, and waste disposal. Treatment in an incinerator, boiler, or cement kiln.

References

Sax, N. I. (Ed.). *Dangerous Properties of Industrial Materials Report*, 1, No. 2, 66−67 (1980) and 5, No. 5, 91−94 (1985)

New Jersey Department of Health and Senior Services. (March 2002). *Hazardous Substances Fact Sheet: Thioacetamide*. Trenton, NJ

4,4′-Thiobis(6-*tert*-butyl-*m*-cresol)
T:0440

Molecular Formula: $C_{12}H_{18}OS$

Common Formula: $[(CH_3)_3CC_6H_2(OH)(CH_3)_2]S$

Synonyms: Bis-3-*tert*-butyl-4-hydroxy-6-methylphenyl) sulfide; Bis(4-hydroxy-5-*tert*-butyl-2-methylphenyl) sulfide; *m*-Cresol, 4,4′-thiobis(6-*tert*-butyl-); Disperse MB-61; Santonox; Santowhite crystals; Santox; Thioalkofen BM4; 4,4′-Thiobis(3-methyl-6-*tert*-butylphenol); 1,1′-Thiobis(2-methyl-4-hydroxy-5-*tert*-butylbenzene)

CAS Registry Number: 96-69-5

RTECS® Number: GP3150000

EC Number: 202-525-2

Regulatory Authority and Advisory Bodies

Carcinogenicity: NCI: Carcinogenesis Studies (feed); no evidence: mouse, rat.

Air Pollutant Standard Set. See below, "Permissible Exposure Limits in Air" section.

Canada, WHMIS, Ingredients Disclosure List Concentration 1.0%.

European/International Regulations: Not listed in Annex 1.

WGK (German Aquatic Hazard Class): 2—Hazard to waters.

Description: 4,4′-Thiobis(6-*tert*-butyl-*m*-cresol) is a light gray to tan powder with an aromatic odor. Molecular weight = 358.58; Specific gravity ($H_2O:1$) = 1.10 at 25°C; Freezing/Melting point = 150°C; Vapor pressure = 6×10^{-7} mmHg at 25°C; Flash point = 207°C. Hazard Identification (based on NFPA-704 M Rating System): Health 2, Flammability 1, Reactivity 0. Practically insoluble in water; solubility 0.08%.

Potential Exposure: Compound Description: Tumorigen. This material is used as an antioxidant in the plastics and rubber industries; in Neoprene™ and other synthetic rubbers; in polyethylene and polypropylene.

Incompatibilities: None reported. However, it is combustible and may react with strong oxidizers. Many sulfides evolve explosive hydrogen sulfide upon contact with moisture or acids.

Permissible Exposure Limits in Air

OSHA PEL: 15 mg/m³, total dust TWA; 5 mg/m³, respirable fraction TWA.

NIOSH REL: 10 mg/m³, total dust TWA; 5 mg/m³, respirable fraction TWA.

ACGIH TLV®[1]: 10 mg/m³, total dust TWA, not classifiable as a human carcinogen; *Notice of intended change*: 1 mg/m³ inhalable fraction TWA.

Protective Action Criteria (PAC)

TEEL-0: 15 mg/m³

PAC-1: 30 mg/m³

PAC-2: 300 mg/m³

PAC-3: 500 mg/m³

Denmark: TWA 10 mg/m³, 1999; France: VME 10 mg/m³, 1999; Switzerland: MAK-W 10 mg/m³, 1999; United Kingdom: TWA 10 mg/m³, STEL 20 mg/m³, 2000; the Netherlands: MAC-TGG 10 mg/m³, 2003; Argentina, Bulgaria, Columbia, Jordan, South Korea, New Zealand, Singapore, Vietnam: ACGIH TLV®: not classifiable as a human carcinogen. Several states have set guidelines or standards for this material in ambient air[60] ranging from 100 μg/m³ (North Dakota) to 160 μg/m³ (Virginia) to 200 μg/m³ (Connecticut) to 238 μg/m³ (Nevada).

Determination in Air: Use NIOSH(IV), Particulates NOR: Method #0500, total dust; Method #0600 (respirable).

Routes of Entry: Inhalation, ingestion; skin and/or eye contact.

Harmful Effects and Symptoms

Short Term Exposure: Irritates the eyes, skin, and respiratory tract. Although reported to be poisonous, this compound is insignificantly toxic on the basis of acute oral toxicity to rats. However, gastroenteritis, retarded weight gain, and enlarged livers resulted in rat feeding studies.

Long Term Exposure: Repeated or high exposures may cause liver damage.

Points of Attack: Eyes, skin, respiratory system; liver.

Medical Surveillance: Liver function tests.

First Aid: *Skin Contact:*[52] Flood all areas of body that have contacted the substance with water. Do not wait to remove contaminated clothing; do it under the water stream. Use soap to help assure removal. Isolate contaminated clothing when removed to prevent contact by others.

Eye Contact: Remove any contact lenses at once. Flush eyes well with copious quantities of water or normal saline for at least 20−30 min. Seek medical attention.

Inhalation: Leave contaminated area immediately; breathe fresh air. Proper respiratory protection must be supplied to any rescuers. If coughing, difficult breathing, or any other symptoms develop, seek medical attention at once, even if symptoms develop many hours after exposure.

Ingestion: If convulsions are not present, give a glass or two of water or milk to dilute the substance. Assure that the person's airway is unobstructed and contact hospital or poison center immediately for advice on whether or not to induce vomiting.

Personal Protective Methods: Wear protective gloves and clothing to prevent any reasonable probability of skin contact. Safety equipment suppliers/manufacturers can provide recommendations on the most protective glove/clothing material for your operation. All protective clothing (suits, gloves, footwear, headgear) should be clean, available each day, and put on before work. Contact lenses should not be worn when working with this chemical. Wear dust-proof chemical goggles and face shield unless full-face-piece respiratory protection is worn. Employees should wash immediately with soap when skin is wet or contaminated. Provide emergency showers and eyewash.

Respirator Selection: Where there is potential for exposures over *10 mg/m³*, use a NIOSH/MSHA- or European Standard EN149-approved respirator equipped with particulate (dust/fume/mist) filters. Particulate filters must be checked every day before work for physical damage, such as rips or tears, and replaced as needed. *Where there is potential for high exposures*, use a NIOSH/MSHA- or European Standard EN149-approved supplied-air respirator with a full face-piece operated in the positive-pressure mode, or with a full face-piece, hood, or helmet in the continuous-flow mode; or use a NIOSH/MSHA- or European Standard EN149-approved self-contained breathing apparatus with a full face-piece operated in pressure-demand or other positive-pressure mode.

Storage: Color Code—Blue: Health Hazard/Poison: Store in a secure poison location. Prior to working with this chemical you should be trained on its proper handling and storage. Store in tightly closed containers in a cool, well-ventilated area. Sources of ignition, such as smoking and open flames, are prohibited where this chemical is used, handled, or stored in a manner that could create a potential fire or explosion hazard. Use only nonsparking tools and equipment, especially when opening and closing containers of this chemical.

Spill Handling: Evacuate persons not wearing protective equipment from area of spill or leak until cleanup is complete. Remove all ignition sources. The spilled material may be dampened with alcohol to avoid dust. Collect powdered material in the most convenient and safe manner and deposit in sealed containers. Ventilate area after cleanup is complete. It may be necessary to contain and dispose of this chemical as a hazardous waste. If material or contaminated runoff enters waterways, notify downstream users of potentially contaminated waters. Contact your local or federal environmental protection agency for specific recommendations. If employees are required to clean up spills, they must be properly trained and equipped. OSHA 1910.120(q) may be applicable.

Fire Extinguishing: This chemical is a combustible solid. Use dry chemical, carbon dioxide; water spray; or foam extinguishers. Poisonous gases are produced in fire, including oxides of sulfur. If material or contaminated run-off enters waterways, notify downstream users of potentially contaminated waters. Notify local health and fire officials and pollution control agencies. From a secure, explosion-proof location, use water spray to cool exposed containers. If cooling streams are ineffective (venting sound increases in volume and pitch, tank discolors, or shows any signs of deforming), withdraw immediately to a secure position. If employees are expected to fight fires, they must be trained and equipped in OSHA 1910.156. The only respirators recommended for firefighting are self-contained breathing apparatuses that have full face-pieces and are operated in a pressure-demand or other positive-pressure mode.

References

New Jersey Department of Health and Senior Services. (January 2001). *Hazardous Substances Fact Sheet: 4,4'-Thiobis(6-t-Butyl m-Cresol).* Trenton, NJ

Thiofanox   T:0450

Molecular Formula: $C_9H_{18}N_2O_2S$
Common Formula: $(CH_3)_3CC(CH_2SCH_3)=NOCONHCH_3$
Synonyms: Dacamox; Diamond Shamrock DS-15647; 3,3-Dimethyl-1-(methylthio)-2-butanone-*O*-[(methylamino) carbonyl] oxime; DS-15647; ENT 27,851; Thiofanocarb (South Africa)
CAS Registry Number: 39196-18-4
RTECS® Number: EL8200000
UN/NA & ERG Number: UN2757/151
EC Number: 254-346-4 [*Annex I Index No.:* 006-064-00-6]
Regulatory Authority and Advisory Bodies
US EPA Hazardous Waste Number (RCRA No.): P045.
RCRA, 40CFR261, Appendix 8 Hazardous Constituents.
Superfund/EPCRA 40CFR355, Extremely Hazardous Substances: TPQ = 100/10,000 lb (45.4/4540 kg).
Reportable Quantity (RQ): 100 lb (45.4 kg).
European/International Regulations: Hazard Symbol: T+, N; Risk phrases: R27/28; R50/53; Safety phrases: S1/2; S27; S36/37; S45; S60, 61 (see Appendix 4).
WGK (German Aquatic Hazard Class): No value assigned.
Description: Thiofanox is a colorless solid with a pungent odor. Molecular weight = 218.35; Freezing/Melting point = 57°C. Hazard Identification (based on NFPA-704 M Rating System): Health 3, Flammability 0, Reactivity 0.
Potential Exposure: Those involved in the manufacture, formulation, and application of this systemic insecticide and acaricide.
Permissible Exposure Limits in Air
Protective Action Criteria (PAC)
TEEL-0: 10 mg/m³
PAC-1: 35 mg/m³
PAC-2: 100 mg/m³
PAC-3: 100 mg/m³
Routes of Entry: Inhalation, ingestion, eye, and/or skin contact.

Harmful Effects and Symptoms

Short Term Exposure: This material is moderately to highly toxic. It is a cholinesterase inhibitor. Symptoms of exposure include nausea, vomiting, abdominal cramps; diarrhea, excessive salivation; sweating, weakness, runny nose; tightness of chest (inhalation exposure); blurred vision; tearing, muscle spasm; loss of eye coordination; ocular pain, extreme dilation of the pupil; loss of muscle coordination; slurring of speech; difficulty in breathing; excessive respiratory tract mucus; skin discoloration; and hypertension. High exposures can cause pulmonary edema, a medical emergency that can be delayed for several hours. This can cause death.

Long Term Exposure: Cholinesterase inhibitor; cumulative effect is possible. This chemical may damage the nervous system with repeated exposure, resulting in convulsions, respiratory failure.

Points of Attack: Blood, eyes, lungs.

Medical Surveillance: Before employment and at regular times after that, the following are recommended: plasma and red blood cell cholinesterase levels (tests for the enzyme poisoned by this chemical). If exposure stops, plasma levels return to normal in 1–2 weeks while red blood cell levels may be reduced for 1–3 months. When cholinesterase enzyme levels are reduced by 25% or more below preemployment levels, risk of poisoning is increased, even if results are in lower ranges of "normal." Reassignment to work not involving carbamate or organophosphate pesticides is recommended until enzyme levels recover. If symptoms develop or overexposure occurs, repeat the above tests as soon as possible and get an exam of the nervous system. Also consider complete blood count. Consider chest X-ray following acute overexposure. Do not drink any alcoholic beverages before or during use. Eye examination. Lung function tests. Consider chest X-ray following acute overexposure.

First Aid: If this chemical gets into the eyes, remove any contact lenses at once and irrigate immediately for at least 15 min, occasionally lifting upper and lower lids. Seek medical attention immediately. If this chemical contacts the skin, remove contaminated clothing and wash immediately with soap and water. Speed in removing material from skin is of extreme importance. Shampoo hair promptly if contaminated. Seek medical attention immediately. If this chemical has been inhaled, remove from exposure, begin rescue breathing (using universal precautions, including resuscitation mask) if breathing has stopped and CPR if heart action has stopped. Transfer promptly to a medical facility. When this chemical has been swallowed, get medical attention. Give large quantities of water and induce vomiting. Do not make an unconscious person vomit. Medical observation is recommended for 24–48 h after breathing overexposure, as pulmonary edema may be delayed. As first aid for pulmonary edema, a doctor or authorized paramedic may consider administering a corticosteroid spray.

Personal Protective Methods: Wear protective gloves and clothing to prevent any reasonable probability of skin contact. Safety equipment suppliers/manufacturers can provide recommendations on the most protective glove/clothing material for your operation. All protective clothing (suits, gloves, footwear, headgear) should be clean, available each day, and put on before work. Contact lenses should not be worn when working with this chemical. Wear dust-proof chemical goggles and face shield unless full-face-piece respiratory protection is worn. Employees should wash immediately with soap when skin is wet or contaminated. Provide emergency showers and eyewash.

Respirator Selection: Follow regulations in OSHA 29CFR1910.134 or European Standard EN149. Use a NIOSH/MSHA- or European Standard EN149-approved respirator; or use an approved supplied-air respirator with a full face-piece operated in the positive-pressure mode, or with a full face-piece, hood, or helmet in the continuous-flow mode; or use a NIOSH/MSHA- or European Standard EN149-approved self-contained breathing apparatus with a full face-piece operated in pressure-demand or other positive-pressure mode.

Storage: Color Code—Blue: Health Hazard/Poison: Store in a secure poison location. Prior to working with this chemical you should be trained on its proper handling and storage. Store in tightly closed containers in a cool, well-ventilated area away from strong oxidizers. Where possible, automatically transfer material from drums or other storage containers to process containers. Sources of ignition, such as smoking and open flames, are prohibited where this chemical is handled, used, or stored. Metal containers involving the transfer of this chemical should be grounded and bonded. Wherever this chemical is used, handled, manufactured, or stored, use explosion-proof electrical equipment and fittings.

Shipping: Carbamate pesticides, solid, toxic, require a shipping label of "POISONOUS/TOXIC MATERIALS." It falls in Hazard Class 6.1 and Packing Group I.

Spill Handling: Evacuate persons not wearing protective equipment from area of spill or leak until cleanup is complete. Remove all ignition sources. Collect powdered material in the most convenient and safe manner and deposit in sealed containers. Ventilate area after cleanup is complete. It may be necessary to contain and dispose of this chemical as a hazardous waste. If material or contaminated runoff enters waterways, notify downstream users of potentially contaminated waters. Contact your local or federal environmental protection agency for specific recommendations. If employees are required to clean up spills, they must be properly trained and equipped. OSHA 1910.120(q) may be applicable.

Fire Extinguishing: Stay upwind; keep out of low areas. Ventilate closed spaces before entering. Use water spray, fog or foam. Move container from fire area (only without risk). Fight fire from maximum distance. Dike fire control water for later disposal; do not scatter material. Poisonous gases are produced in fire, including oxides of sulfur and nitrogen. If material or contaminated runoff enters waterways, notify downstream users of potentially contaminated waters. Notify local health and fire officials and pollution control agencies. From a secure, explosion-proof location, use water spray to cool exposed containers. If cooling streams are ineffective (venting sound increases in volume and pitch, tank discolors, or shows any signs of deforming), withdraw immediately to a secure position. If employees are expected to fight fires, they must be trained and equipped in OSHA 1910.156. The only respirators recommended for firefighting are self-contained breathing apparatuses that have full face-pieces and are operated in a pressure-demand or other positive-pressure mode.

Disposal Method Suggested: In accordance with 40CFR 165 recommendations for the disposal of pesticides and pesticide containers. Must be disposed properly by following package label directions or by contacting your local or federal environmental control agency, or by contacting your regional EPA office. Consult with environmental regulatory agencies for guidance on acceptable disposal practices. Generators of waste containing this contaminant ($\geq$100 kg/mo) must conform with EPA regulations governing storage, transportation, treatment, and waste disposal.

References
US Environmental Protection Agency. (November 30, 1987). *Chemical Hazard Information Profile: Thiofanox.* Washington, DC: Chemical Emergency Preparedness Program

Thioglycolic acid   T:0460

Molecular Formula: $C_2H_4O_2S$
Common Formula: $HSCH_2COOH$
Synonyms: Acetic acid, mercapto-; Acetyl mercaptan; Acide thioglycolique (French); Glycolic acid, 2-thio-; Glycolic acid, thio-; Mercaptoacetate; 2-Mercaptoacetic acid; α-Mercaptoacetic acid; Mercaptoacetic acid; 2-Thioglycolic acid; Thioglycolic acid; Thioglycollic acid; Thiovanic acid
CAS Registry Number: 68-11-1
RTECS® Number: AI5950000
UN/NA & ERG Number: UN1940/153
EC Number: 200-677-4 [*Annex I Index No.:* 607-090-00-6]
Regulatory Authority and Advisory Bodies
Air Pollutant Standard Set. See below, "Permissible Exposure Limits in Air" section.
Canada, WHMIS, Ingredients Disclosure List Concentration 1.0%.
European/International Regulations: Hazard Symbol: T; Risk phrases: R23/24/25; R34; Safety phrases: S1/2; S25; S27; S28; S45 (see Appendix 4).
WGK (German Aquatic Hazard Class): 1—Low hazard to waters.

Description: Thioglycolic acid is a colorless liquid with a strong unpleasant odor like rotten eggs. Molecular weight = 92.12; Specific gravity (H_2O:1) = 1.32 at 25°C; Boiling point = 123°C at 29 mmHg; also reported at 120°C at 20 mmHg; 104–106°C at 11 mmHg; Freezing/Melting point = −16.5°C; Vapor pressure = 10 mmHg at 18°C; Flash point = >110°C; Autoignition temperature = 350°C. Explosive limits: LEL = 5.9%; UEL—unknown. Soluble in water.

Potential Exposure: Compound Description: Agricultural Chemical; Primary Irritant. Thioglycolic acid is used to make thioglycolates; in sensitivity tests for iron; in formulations of permanent wave solutions and depilatories; in pharmaceutical manufacture; as a stabilizer in vinyl plastics.

Incompatibilities: Air, strong oxidizers; bases, active metals (e.g., sodium potassium, magnesium, calcium). Readily oxidized by air. Decomposition can lead to release of H_2S.

Permissible Exposure Limits in Air
Conversion factor: 1 ppm = 3.77 mg/m^3 at 25°C & 1 atm.
OSHA PEL: None.
NIOSH REL: 1 ppm/4 mg/m^3 TWA [skin].
ACGIH TLV®[1]: 1 ppm/3.8 mg/m^3 TWA [skin].
Protective Action Criteria (PAC)
TEEL-0: 1 ppm
PAC-1: 1 ppm
PAC-2: 1.25 ppm
PAC-3: 6 ppm
Australia: TWA 1 ppm (4 mg/m^3), 1993; Belgium: TWA 1 ppm (3.8 mg/m^3), 1993; Denmark: TWA 1 ppm (5 mg/m^3), 1999; Finland: TWA 1 ppm (5 mg/m^3), STEL 3 ppm (15 mg/m^3), 1993; France: VME 1 ppm (5 mg/m^3) [skin], 1999; Hungary: TWA 0.5 mg/m^3, STEL 1 mg/m^3, 1993; Norway: TWA 1 ppm (5 mg/m^3), 1999; the Netherlands: MAC-TGG 4 mg/m^3 [skin], 2003; Russia: STEL 0.1 mg/m^3 [skin], 1993; United Kingdom: TWA 1 ppm, 3.8 mg/m^3, 2000; Argentina, Bulgaria, Columbia, Jordan, South Korea, New Zealand, Singapore, Vietnam: ACGIH TLV®: TWA 1 ppm [skin]. Several states have set guidelines or standards for thioglycolic acid in ambient air[60] ranging from 40 μg/m^3 (North Dakota) to 80 μg/m^3 (Virginia) to 100 μg/m^3 (Connecticut) to 119 μg/m^3 (Nevada).

Determination in Air: No method available.

Determination in Water: Octanol–water coefficient: Log K_{ow} = low, <0.1.

Routes of Entry: Inhalation, skin absorption; ingestion; skin and/or eye contact.

Harmful Effects and Symptoms
Short Term Exposure: Corrosive to the eyes, skin, and respiratory tract. Contact can cause eye and skin burns and blisters. It can also cause conjuctival edema and corneal damage. Inhalation can cause pulmonary edema, a medical emergency that can be delayed for several hours. This can cause death. Ingesting the liquid may cause chemical pneumonitis. Exposure can cause death. Irritation, blistering, and severe burns of the skin, eyes, and mucous membranes; corrosion of tissues.[52] If ingested, severe burning pain in the

mouth, pharynx, and abdomen; vomiting, bloody diarrhea; sharp drop in blood pressure: and asphyxia. If inhaled; coughing, choking, headache, dizziness, weakness, tightness in the chest; air hunger; and cyanosis. In animals: weakness, asping respirations; convulsions.

Long Term Exposure: Repeated contact may cause a skin rash.

Points of Attack: Eyes, skin, respiratory system.

Medical Surveillance: Consider chest X-ray following acute overexposure.

First Aid: If this chemical gets into the eyes, remove any contact lenses at once and irrigate immediately for at least 20 min, occasionally lifting upper and lower lids. Seek medical attention immediately. If this chemical contacts the skin, remove contaminated clothing and wash immediately with soap and water. Seek medical attention immediately. If this chemical has been inhaled, remove from exposure, begin rescue breathing (using universal precautions, including resuscitation mask) if breathing has stopped and CPR if heart action has stopped. Transfer promptly to a medical facility. When this chemical has been swallowed, contact a physician, hospital or poison center at once. If the victim is unconscious or convulsing, do not induce vomiting or give anything by mouth. If conscious, do not induce vomiting. For dilute acid, give water, milk, milk of magnesia; A (OH)$_3$, or Ca(OH)$_2$. Avoid carbonates or bicarbonates. For concentrated acid, it may be dangerous to administer water or antacids. Some authorities suggest ice water or a snow slurry. Medical observation is recommended for 24−48 h after breathing overexposure, as pulmonary edema may be delayed. As first aid for pulmonary edema or pneumonitis, a doctor or authorized paramedic may consider administering a corticosteroid spray.

Personal Protective Methods: Wear protective gloves and clothing to prevent any reasonable probability of skin contact. Safety equipment suppliers/manufacturers can provide recommendations on the most protective glove/clothing material for your operation. All protective clothing (suits, gloves, footwear, headgear) should be clean, available each day, and put on before work. Contact lenses should not be worn when working with this chemical. Wear splash-proof chemical goggles and face shield unless full-face-piece respiratory protection is worn. Employees should wash immediately with soap when skin is wet or contaminated. Provide emergency showers and eyewash.

Respirator Selection: Where there is potential for exposures *over 1 ppm*, use a NIOSH/MSHA- or European Standard EN149-approved supplied-air respirator with a full face-piece operated in the positive-pressure mode, or with a full face-piece, hood, or helmet in the continuous-flow mode; or use a NIOSH/MSHA- or European Standard EN149-approved self-contained breathing apparatus with a full face-piece operated in pressure-demand or other positive-pressure mode.

Storage: Color Code—White: Corrosive or Contact Hazard; Store separately in a corrosion-resistant location. Prior to working with this chemical you should be trained on its proper handling and storage. Store in tightly closed containers in a cool, well-ventilated area away from oxidizers. Where possible, automatically pump liquid from drums or other storage containers to process containers. Sources of ignition, such as smoking and open flames, are prohibited where this chemical is handled, used, or stored.

Shipping: This compound requires a shipping label of "CORROSIVE." It falls in Hazard Class 8 and Packing Group II.

Spill Handling: Evacuate and restrict persons not wearing protective equipment from area of spill or leak until cleanup is complete. Remove all ignition sources. Ventilate area of spill or leak. Absorb liquids in vermiculite, dry sand; earth, peat, carbon, or a similar material and deposit in sealed containers. Keep this chemical out of a confined space, such as a sewer, because of the possibility of an explosion, unless the sewer is designed to prevent the buildup of explosive concentrations. It may be necessary to contain and dispose of this chemical as a hazardous waste. If material or contaminated runoff enters waterways, notify downstream users of potentially contaminated waters. Contact your local or federal environmental protection agency for specific recommendations. If employees are required to clean up spills, they must be properly trained and equipped. OSHA 1910.120(q) may be applicable.

Fire Extinguishing: Thioglycolic acid may burn, but does not readily ignite. Use dry chemical, carbon dioxide, water spray; or foam extinguishers. Poisonous gases, including sulfur oxides and hydrogen sulfide, are produced in fire. Vapors are heavier than air and will collect in low areas. Vapors may travel long distances to ignition sources and flashback. Vapors in confined areas may explode when exposed to fire. Containers may explode in fire. Storage containers and parts of containers may rocket great distances, in many directions. If material or contaminated runoff enters waterways, notify downstream users of potentially contaminated waters. Notify local health and fire officials and pollution control agencies. From a secure, explosion-proof location, use water spray to cool exposed containers. If cooling streams are ineffective (venting sound increases in volume and pitch, tank discolors, or shows any signs of deforming), withdraw immediately to a secure position. If employees are expected to fight fires, they must be trained and equipped in OSHA 1910.156. The only respirators recommended for firefighting are self-contained breathing apparatuses that have full face-pieces and are operated in a pressure-demand or other positive-pressure mode.

Disposal Method Suggested: Dissolve in flammable solvent and burn in furnace equipped with afterburner and alkaline scrubber.

References

New Jersey Department of Health and Senior Services. (August 2005). *Hazardous Substances Fact Sheet: Thioglycolic Acid*. Trenton, NJ

Thionazin   T:0470

Molecular Formula: $C_8H_{13}N_2O_3PS$

Synonyms: AC 18133; American Cyanamid 18133; CL 18133; Cynem; *O,O*-Diaethyl-*O*-(pyrazin-2yl)-monothiophosphat (German); *O,O*-Diaethyl-*O*-(2-pyrazinyl)-thionophosphat (German); *O,O*-Diethyl *O*-2-pyrazinyl phosphorothioate; Diethyl *O*-2-pyrazinyl phosphorothionate; *O,O*-Diethyl *O*-2-pyrazinyl phosphothionate; *O,O*-Diethyl *O*-pyrazinyl thiophosphate; EN 18133; ENT 25 580; Experimental nematocide 18,133; Nemafos; Nemaphos; Nematocide; Phosphorothioic acid, *O,O*-diethyl *O*-2-pyrazinyl ester; Pyrazinol *O*-ester with *O,O*-diethyl phosphorothioate pyrazinol *O*-ester; Thionazin; Zinophos®

CAS Registry Number: 297-97-2

RTECS® Number: TF5775000

UN/NA & ERG Number: UN3018 (organophosphorus pesticide, liquid, toxic)/152

EC Number: 206-049-6 [*Annex I Index No.:* 015-112-00-5]

Regulatory Authority and Advisory Bodies

Very Toxic Substance (World Bank).[15]

US EPA Hazardous Waste Number (RCRA No.): P040.

RCRA, 40CFR261, Appendix 8 Hazardous Constituents.

RCRA 40CFR264, Appendix 9; TSD Facilities Ground Water Monitoring List. Suggested test method(s) (PQL μg/L): 8270 (10).

Superfund/EPCRA 40CFR355, Extremely Hazardous Substances: TPQ = 500 lb (227 kg).

Reportable Quantity (RQ): 100 lb (45.4 kg).

US DOT 49CFR172.101, Inhalation Hazard Chemical as organophosphates.

European/International Regulations: Hazard Symbol: T+; Risk phrases: R27/28; Safety phrases: S1/2; S36/37/39; S38; S45 (see Appendix 4).

WGK (German Aquatic Hazard Class): No value assigned.

Description: Thionazin is an amber to colorless liquid. Molecular weight = 248.46; Boiling point = 80°C at 0.001 mmHg; Freezing/Melting point = −1.7°C. Hazard Identification (based on NFPA-704 M Rating System): Health 4, Flammability 1, Reactivity 0. Slightly soluble in water.

Potential Exposure: Those involved in the manufacture, formulation, and application of this insecticide, fungicide, and nematocide.

Incompatibilities: Strong oxidizers may cause release of toxic phosphorus oxides. Organophosphates, in the presence of strong reducing agents such as hydrides, may form highly toxic and flammable phosphine gas. Keep away from alkaline materials.

Permissible Exposure Limits in Air

Protective Action Criteria (PAC)

TEEL-0: 0.6 mg/m³

PAC-1: 2 mg/m³

PAC-2: 3.5 mg/m³

PAC-3: 3.5 mg/m³

Determination in Air: OSHA versatile sampler-2; Toluene/Acetone; Gas chromatography/Flame photometric detection for sulfur, nitrogen, or phosphorus; NIOSH Analytical Method (IV) Method #5600, Organophosphorus pesticides.

Routes of Entry: Inhalation, ingestion, eye and/or skin contact. Absorbed through the skin.

Harmful Effects and Symptoms

Short Term Exposure: Irritates the eyes, skin, and respiratory tract. Contact may cause burns to skin and eyes. Poisonous; may be fatal if inhaled, swallowed, or absorbed through skin. Acute effects include loss of appetite; nausea, vomiting, diarrhea, excessive salivation; papillary constriction; bronchoconstriction, muscle twitching; convulsions, and coma. Organic phosphorus insecticides are absorbed by the skin, as well as by the respiratory and gastrointestinal tracts. They are cholinesterase inhibitors. Symptoms of exposure include headache, giddiness; blurred vision; nervousness, weakness, nausea, cramps, diarrhea, and discomfort in the chest. Signs include sweating, tearing, salivation, vomiting, cyanosis, convulsions, coma, loss of reflexes, and loss of sphincter control.

Long Term Exposure: Cholinesterase inhibitor; cumulative effect is possible. This chemical may damage the nervous system with repeated exposure, resulting in convulsions, respiratory failure. May cause liver damage. May cause dermatitis.

Points of Attack: Respiratory system, lungs, central nervous system; cardiovascular system, skin, eyes, plasma, and red blood cell cholinesterase.

Medical Surveillance: Before employment and at regular times after that, the following are recommended: plasma and red blood cell cholinesterase levels (tests for the enzyme poisoned by this chemical). If exposure stops, plasma levels return to normal in 1−2 weeks while red blood cell levels may be reduced for 1−3 months. When cholinesterase enzyme levels are reduced by 25% or more below preemployment levels, risk of poisoning is increased, even if results are in lower ranges of "normal." Reassignment to work not involving organophosphate or carbamate pesticides is recommended until enzyme levels recover. If symptoms develop or overexposure occurs, repeat the above tests as soon as possible and get an exam of the nervous system. Also consider complete blood count. Consider chest X-ray following acute overexposure. Do not drink any alcoholic beverages before or during use. Alcohol promotes absorption of organic phosphates.

First Aid: If this chemical gets into the eyes, remove any contact lenses at once and irrigate immediately for at least 15 min, occasionally lifting upper and lower lids. Seek medical attention immediately. If this chemical contacts the skin, remove contaminated clothing and wash immediately with soap and water. Shampoo hair. Speed in removing material from skin is of extreme importance. Seek medical attention immediately. If this chemical has been inhaled, remove from exposure, begin rescue breathing (using universal precautions, including resuscitation mask) if breathing has stopped and CPR if heart action has stopped. Transfer promptly to a medical facility. When this chemical has been swallowed, get medical attention. Give large quantities of water and induce vomiting. Do not make an unconscious person vomit.

Personal Protective Methods: Wear protective gloves and clothing to prevent any reasonable probability of skin contact. Safety equipment suppliers/manufacturers can provide recommendations on the most protective glove/clothing material for your operation. All protective clothing (suits, gloves, footwear, headgear) should be clean, available each day, and put on before work. Contact lenses should not be worn when working with this chemical. Wear splash-proof chemical goggles and face shield unless full-face-piece respiratory protection is worn. Employees should wash immediately with soap when skin is wet or contaminated. Provide emergency showers and eyewash.

Respirator Selection: Follow regulations in OSHA 29CFR1910.134 or European Standard EN149. Use a NIOSH/MSHA- or European Standard EN149-approved respirator; or use an approved supplied-air respirator with a full face-piece operated in the positive-pressure mode, or with a full face-piece, hood, or helmet in the continuous-flow mode; or use a NIOSH/MSHA- or European Standard EN149-approved self-contained breathing apparatus with a full face-piece operated in pressure-demand or other positive-pressure mode.

Storage: Color Code—Blue: Health Hazard/Poison: Store in a secure poison location. Prior to working with this chemical you should be trained on its proper handling and storage.

Shipping: This compound requires a shipping label of "POISONOUS/TOXIC MATERIALS." It falls in Hazard Class 6.1 and Packing Group I.

Spill Handling: Evacuate and restrict persons not wearing protective equipment from area of spill or leak until cleanup is complete. Remove all ignition sources. Ventilate area of spill or leak. Do not touch spilled material; stop leak if possible; use water spray to reduce vapors. *Small spill*: take up with sand or other noncombustible absorbent material and place into container for later disposal. *Large spills*: dike far ahead of spill for later disposal. Avoid breathing vapors. Avoid bodily contact with materials. Do not handle broken packages without protective equipment. Wash away any material which may have contacted the body with copious amounts of water or soap and water. Keep this chemical out of a confined space, such as a sewer, because of the possibility of an explosion, unless the sewer is designed to prevent the buildup of explosive concentrations. It may be necessary to contain and dispose of this chemical as a hazardous waste. If material or contaminated runoff enters waterways, notify downstream users of potentially contaminated waters. Contact your local or federal environmental protection agency for specific recommendations. If employees are required to clean up spills, they must be properly trained and equipped. OSHA 1910.120(q) may be applicable.

Fire Extinguishing: Fight fire from maximum distance. Dike fire control water for later disposal; do not scatter the material. Extinguish fire using agent suitable for type of surrounding fire. Use water in flooding quantities as fog. Use foam, carbon dioxide; or dry chemicals. Poisonous gases, including oxides of nitrogen, phosphorus, and sulfur, are produced in fire. Vapors are heavier than air and will collect in low areas. Containers may explode in fire. Storage containers and parts of containers may rocket great distances, in many directions. If material or contaminated runoff enters waterways, notify downstream users of potentially contaminated waters. Notify local health and fire officials and pollution control agencies. From a secure, explosion-proof location, use water spray to cool exposed containers. If cooling streams are ineffective (venting sound increases in volume and pitch, tank discolors, or shows any signs of deforming), withdraw immediately to a secure position. If employees are expected to fight fires, they must be trained and equipped in OSHA 1910.156. The only respirators recommended for firefighting are self-contained breathing apparatuses that have full face-pieces and are operated in a pressure-demand or other positive-pressure mode.

Disposal Method Suggested: Consult with environmental regulatory agencies for guidance on acceptable disposal practices. Generators of waste containing this contaminant ($\geq$100 kg/mo) must conform with EPA regulations governing storage, transportation, treatment, and waste disposal.

References
US Environmental Protection Agency. (November 30, 1987). *Chemical Hazard Information Profile: Thionazin.* Washington, DC: Chemical Emergency Preparedness Program

Thionyl chloride   T:0480

Molecular Formula: Cl_2OS
Common Formula: $SOCl_2$
Synonyms: Sulfinyl chloride; Sulfur chloride oxide; Sulfur oxychloride; Sulfurous dichloride; Sulfurous oxychloride; Thionyl dichloride
CAS Registry Number: 7719-09-7
RTECS® Number: XM5150000
UN/NA & ERG Number: UN1836/137
EC Number: 231-748-8 [*Annex I Index No.:* 016-015-00-0]
Regulatory Authority and Advisory Bodies
Department of Homeland Security Screening Threshold Quantity (pounds): Sabotage/Contamination Hazard: A placarded amount (commercial grade).
Air Pollutant Standard Set. See below, "Permissible Exposure Limits in Air" section.
OSHA 29CFR1910.119, Appendix A. Process Safety List of Highly Hazardous Chemicals, TQ = 250 lb (114 kg).
US DOT 49CFR172.101, Inhalation Hazardous Chemical.
Canada, WHMIS, Ingredients Disclosure List Concentration 1.0%.

European/International Regulations: Hazard Symbol: C; Risk phrases: R14; R20/22; R29; R35; Safety phrases: S1/2; S26; S36/37/39; S45 (see Appendix 4).

WGK (German Aquatic Hazard Class): 1—Low hazard to waters.

Description: Thionyl chloride is a pale yellow to reddish liquid with a suffocating odor like sulfur dioxide. Fumes form when exposed to moist air. Molecular weight = 118.96; Specific gravity (H_2O:1) = 1.64 at 25°C; Boiling point = 79°C (decomposes to SO_2 and S_2Sl_2 at 140°C); Freezing/Melting point = −104°C; Vapor pressure = 100 mmHg at 21°C. Hazard Identification (based on NFPA-704 M Rating System): Health 4, Flammability 0, Reactivity 2W. Reacts with water.

Potential Exposure: Thionyl chloride is used as specialty chlorinating agent, particularly in preparation of organic acid chlorides; in organic synthesis; as a catalyst.

Incompatibilities: Reacts violently with water to form sulfur dioxide and hydrogen chloride. Keep away from water, acids, alkalis, ammonia, chloryl perchlorate.

Permissible Exposure Limits in Air

Conversion factor: 1 ppm = 4.87 mg/m^3 at 25°C & 1 atm.
OSHA PEL: None.
NIOSH REL: 1 ppm/5 mg/m^3 Ceiling Concentration.
Protective Action Criteria (PAC)*
ACGIH TLV®[1]: 0.2 ppm TWA Ceiling Concentration
TEEL-0: 0.06 ppm
PAC-1: **0.2** ppm
PAC-2: **2.4** ppm
PAC-3: **14** ppm
*AEGLs (Acute Emergency Guideline Levels) & ERPGs (Emergency Response Planning Guideline) are in **bold face**.
Australia: TWA 1 ppm (5 mg/m^3), 1993; Belgium: STEL 1 ppm (4.9 mg/m^3), 1993; Denmark: TWA 1 ppm (5 mg/m^3, 1999; Switzerland: MAK-W 1 ppm (5 mg/m^3), 1999; United Kingdom: STEL 1 ppm (4.9 mg/m^3), 2000; the Netherlands: MAC 5 mg/m^3, 2003; Argentina, Bulgaria, Columbia, Jordan, South Korea, New Zealand, Singapore, Vietnam: ACGIH TLV®: Ceiling Concentration 1 ppm.
North Dakota[60] has set a guideline for thionyl chloride in ambient air of 50 μg/m^3 (5 mg/m^3).

Determination in Air: No method available.

Routes of Entry: Inhalation, ingestion; skin and/or eye contact.

Harmful Effects and Symptoms

Short Term Exposure: A corrosive irritant to the eyes, skin, and mucous membrane. Contact causes eye and skin burns. Can cause dermatitis, rhinitis (inflammation of the nose), and pneumonia. High exposures can cause pulmonary edema, a medical emergency that can be delayed for several hours. This can cause death. This chemical is more toxic than sulfur dioxide.

Long Term Exposure: Highly irritating substances can cause lung irritation; bronchitis may develop.

Points of Attack: Lungs. Consider chest X-ray following acute overexposure.

Medical Surveillance: Lung function tests.

First Aid: If this chemical gets into the eyes, remove any contact lenses at once and irrigate immediately for at least 15 min, occasionally lifting upper and lower lids. Seek medical attention immediately. If this chemical contacts the skin, remove contaminated clothing and wash immediately with soap and water. Seek medical attention immediately. If this chemical has been inhaled, remove from exposure, begin rescue breathing (using universal precautions, including resuscitation mask) if breathing has stopped and CPR if heart action has stopped. Transfer promptly to a medical facility. When this chemical has been swallowed, get medical attention. If victim is *conscious*, administer water or milk. Do not induce vomiting. Medical observation is recommended for 24−48 h after breathing overexposure, as pulmonary edema may be delayed. As first aid for pulmonary edema, a doctor or authorized paramedic may consider administering a corticosteroid spray.

Note to medical personnel: Administer oxygen, using intermittent positive-pressure breathing apparatus: 5% solution of sodium bicarbonate may be used, as well as bronchodilators and decongestants.

Personal Protective Methods: Wear protective gloves and clothing to prevent any reasonable probability of skin contact. Safety equipment suppliers/manufacturers can provide recommendations on the most protective glove/clothing material for your operation. All protective clothing (suits, gloves, footwear, headgear) should be clean, available each day, and put on before work. Contact lenses should not be worn when working with this chemical. Wear splash-proof chemical goggles and face shield unless full-face-piece respiratory protection is worn. Employees should wash immediately with soap when skin is wet or contaminated. Provide emergency showers and eyewash.

Respirator Selection: Follow regulations in OSHA 29CFR1910.134 or European Standard EN149. Use a NIOSH/MSHA- or European Standard EN149-approved respirator; or use an approved supplied-air respirator with a full face-piece operated in the positive-pressure mode, or with a full face-piece, hood, or helmet in the continuous-flow mode; or use a NIOSH/MSHA- or European Standard EN149-approved self-contained breathing apparatus with a full face-piece operated in pressure-demand or other positive-pressure mode.

Storage: Color Code—White: Corrosive or Contact Hazard; Store separately in a corrosion-resistant location. Color Code—White: Corrosive or Contact Hazard; Store separately in a corrosion-resistant location. Prior to working with thionyl chloride you should be trained on its proper handling and storage. Store in tightly closed containers in a cool, well-ventilated area away from water, acids, alkalis, ammonia, chloryl perchloride. Where possible, automatically pump liquid from drums or other storage containers to process containers.

Shipping: This compound requires a shipping label of "CORROSIVE." It falls in Hazard Class 8 and Packing Group I.

Spill Handling: Evacuate and restrict persons not wearing protective equipment from area of spill or leak until cleanup is complete. Remove all ignition sources. Ventilate area of spill or leak. Absorb liquids in vermiculite, dry sand; earth, peat, carbon, or a similar material and deposit in sealed containers. Cover with sodium bicarbonate or an equal mixture of soda ash and slaked lime. After mixing, spray water from an atomizer with great caution. Transfer slowly into a large container of water. Neutralize and drain into the sewer with sufficient water. It may be necessary to contain and dispose of this chemical as a hazardous waste. If material or contaminated runoff enters waterways, notify downstream users of potentially contaminated waters. Contact your local or federal environmental protection agency for specific recommendations. If employees are required to clean up spills, they must be properly trained and equipped. OSHA 1910.120(q) may be applicable.

Initial isolation and protective action distances

Distances shown are likely to be affected during the first 30 min after materials are spilled and could increase with time. If more than one tank car, cargo tank, portable tank, or large cylinder is involved in the incident is leaking, the protective action distance may need to be increased. You may need to seek emergency information from CHEMTREC at (800) 424-9300 or seek professional environmental engineering assistance from the US EPA Environmental Response Team at (908) 548-8730 (24-h response line).

When spilled on land

Small spills (From a small package or a small leak from a large package)

First: Isolate in all directions (feet/meters)   100/30.
Then: Protect persons downwind (miles/kilometers).
Day   0.2/0.3
Night   0.5/0.8

Large spills (From a large package or from many small packages)

First: Isolate in all directions (feet/meters)   300/100.
Then: Protect persons downwind (miles/kilometers).
Day   0.6/0.9
Night   1.2/1.9

When spilled in water

Small spills (From a small package or a small leak from a large package)

First: Isolate in all directions (feet/meters)   100/30.
Then: Protect persons downwind (miles/kilometers).
Day   0.2/0.3
Night   0.9/1.5

Large spills (From a large package or from many small packages)

First: Isolate in all directions (feet/meters)   1000/300.
Then: Protect persons downwind (miles/kilometers).
Day   2.1/3.3

Night   4.7/7.5

Fire Extinguishing: This chemical is a noncombustible liquid. Poisonous gases are produced in fire. Use dry chemical, carbon dioxide; or alcohol foam extinguishers. Notify local health and fire officials and pollution control agencies. From a secure, explosion-proof location, use water spray to cool exposed containers. Keep water out of open containers. If cooling streams are ineffective (venting sound increases in volume and pitch, tank discolors, or shows any signs of deforming), withdraw immediately to a secure position. If employees are expected to fight fires, they must be trained and equipped in OSHA 1910.156. The only respirators recommended for firefighting are self-contained breathing apparatuses that have full face-pieces and are operated in a pressure-demand or other positive-pressure mode.

Disposal Method Suggested: Spray on a thick layer of a (1:1) mixture of dry soda ash and slaked lime behind a shield. After mixing, spray water from an atomizer with great precaution. Transfer slowly into a large amount of water. Neutralize and drain into the sewer with sufficient water.

Thiosemicarbazide   T:0490

Molecular Formula: CH_5N_3S
Common Formula: $H_2NNHCSNH_2$
Synonyms: AI3-16319; *N*-Aminothiourea; 1-Amino-2-thiourea; 1-Aminothiourea; Aminothiourea; Hydrazinecarbothioamide; Isothiosemicarbazide; Semicarbazide, 3-thio-; Semicarbazide, thio-; Thiocarbamoylhydrazine; Thiocarbamylhydrazine; 2-Thiosemicarbazide; 3-Thiosemicarbazide; Tiosemicarbazida (Spanish); TSC; TSZ
CAS Registry Number: 79-19-6
RTECS® Number: VT4200000
UN/NA & ERG Number: UN2771/151
EC Number: 201-184-7
Regulatory Authority and Advisory Bodies
US EPA Hazardous Waste Number (RCRA No.): P116.
RCRA, 40CFR261, Appendix 8 Hazardous Constituents.
Superfund/EPCRA 40CFR355, Extremely Hazardous Substances: TPQ = 100/10,000 lb (45.4/4540 kg).
Reportable Quantity (RQ): 100 lb (45.4 kg).
EPCRA Section 313 Form R *de minimis* concentration reporting level: 1.0%.
European/International Regulations: Not listed in Annex 1.
WGK (German Aquatic Hazard Class): 3—Severe hazard to waters.

Description: Thiosemicarbazide is an odorless, white crystalline powder. Molecular weight = 91.15; Freezing/Melting point = 180−184°C. Hazard Identification (based on NFPA-704 M Rating System): Health 3, Flammability 1, Reactivity 0. Soluble in water.

Potential Exposure: This compound is used as an intermediate for pharmaceuticals and herbicides; as a reagent for ketones and certain metals; in certain photography and dye operations; as a rodenticide. It is also effective for control of bacterial leaf blight of rice.

Incompatibilities: None listed, but may react with nitrates.

Permissible Exposure Limits in Air

Protective Action Criteria (PAC)

TEEL-0: 1.5 mg/m^3

PAC-1: 5 mg/m^3

PAC-2: 9.2 mg/m^3

PAC-3: 9.2 mg/m^3

Permissible Concentration in Water: No criteria set.

Determination in Water: Octanol—water coefficient: Log $K_{ow} = -0.67$. See Feigel, *Spot Tests in Organic Analysis*, Elsevier Publishing Company, New York.

Routes of Entry: Inhalation, ingestion, eye, and/or skin contact. Absorbed through the skin.

Harmful Effects and Symptoms

Short Term Exposure: This material is highly toxic by ingestion. May cause delayed toxic effects in blood and skin. May be mutagenic in human cells. Thiosemicarbazide may induce goiter and has also been reported to cause bone marrow depression with accompanying decreases in white blood cells and platelets. It may also cause skin irritation.

Long Term Exposure: May cause delayed toxic effects in blood and skin. May be mutagenic in human cells. May be a cholinesterase inhibitor; cumulative effect is possible. This chemical may damage the nervous system with repeated exposure, resulting in convulsions, respiratory failure.

Points of Attack: Respiratory system, lungs, central nervous system; cardiovascular system, skin, eyes, plasma, and red blood cell cholinesterase.

Medical Surveillance: Before employment and at regular times after that, the following are recommended: plasma and red blood cell cholinesterase levels (tests for the enzyme poisoned by this chemical). If exposure stops, plasma levels return to normal in 1—2 weeks while red blood cell levels may be reduced for 1—3 months.

When cholinesterase enzyme levels are reduced by 25% or more below preemployment levels, risk of poisoning is increased, even if results are in lower ranges of "normal." Reassignment to work not involving carbamate pesticides (or organophosphates) is recommended until enzyme levels recover. If symptoms develop or overexposure occurs, repeat the above tests as soon as possible and get an exam of the nervous system. Also consider complete blood count. Consider chest X-ray following acute overexposure. Do not drink any alcoholic beverages before or during use.

First Aid: If this chemical gets into the eyes, remove any contact lenses at once and irrigate immediately for at least 15 min, occasionally lifting upper and lower lids. Seek medical attention immediately. If this chemical contacts the skin, remove contaminated clothing and wash immediately with soap and water. Speed in removing material from skin is of extreme importance. Shampoo hair promptly if contaminated. Seek medical attention immediately. If this

chemical has been inhaled, remove from exposure, begin rescue breathing (using universal precautions, including resuscitation mask) if breathing has stopped and CPR if heart action has stopped. Transfer promptly to a medical facility. When this chemical has been swallowed, get medical attention. Give large quantities of water and induce vomiting. Do not make an unconscious person vomit.

Personal Protective Methods: Wear protective gloves and clothing to prevent any reasonable probability of skin contact. Safety equipment suppliers/manufacturers can provide recommendations on the most protective glove/clothing material for your operation. All protective clothing (suits, gloves, footwear, headgear) should be clean, available each day, and put on before work. Contact lenses should not be worn when working with this chemical. Wear dust-proof chemical goggles and face shield unless full-face-piece respiratory protection is worn. Employees should wash immediately with soap when skin is wet or contaminated. Provide emergency showers and eyewash.

Respirator Selection: Follow regulations in OSHA 29CFR1910.134 or European Standard EN149. Use a NIOSH/MSHA- or European Standard EN149-approved respirator; or use an approved supplied-air respirator with a full face-piece operated in the positive-pressure mode, or with a full face-piece, hood, or helmet in the continuous-flow mode; or use a NIOSH/MSHA- or European Standard EN149-approved self-contained breathing apparatus with a full face-piece operated in pressure-demand or other positive-pressure mode.

Storage: Color Code—Blue: Health Hazard/Poison: Store in a secure poison location. Prior to working with this chemical you should be trained on its proper handling and storage. Store in tightly closed containers in a cool, well-ventilated area. Where possible, automatically transfer material from drums or other storage containers to process containers.

Shipping: Thiosemicarbazide, pesticides, solid, toxic, requires a shipping label of "POISONOUS/TOXIC MATERIALS." It falls in Hazard Class 6.1 and Packing Group III.

Spill Handling: Avoid skin contact, ingestion, or inhalation. Do not touch spilled material; stop leak if you can do it without risk. *Small spills:* take up with sand or other noncombustible absorbent material and place into containers for later disposal. *Small dry spills:* with clean shovel place material into clean, dry container and cover; move containers from spill area. *Large spills:* dike far ahead of spill for later disposal. Remove all ignition sources. Ventilate area after cleanup is complete. It may be necessary to contain and dispose of this chemical as a hazardous waste. If material or contaminated runoff enters waterways, notify downstream users of potentially contaminated waters. Contact your local or federal environmental protection agency for specific recommendations. If employees are required to clean up spills, they must be properly trained and equipped. OSHA 1910.120(q) may be applicable.

Fire Extinguishing

Use dry chemical, carbon dioxide; water spray; or foam extinguishers. Poisonous gases are produced in fire. If material or contaminated runoff enters waterways, notify downstream users of potentially contaminated waters. Notify local health and fire officials and pollution control agencies. From a secure, explosion-proof location, use water spray to cool exposed containers. If cooling streams are ineffective (venting sound increases in volume and pitch, tank discolors, or shows any signs of deforming), withdraw immediately to a secure position. If employees are expected to fight fires, they must be trained and equipped in OSHA 1910.156. The only respirators recommended for firefighting are self-contained breathing apparatuses that have full face-pieces and are operated in a pressure-demand or other positive-pressure mode.

References

US Environmental Protection Agency. (November 30, 1987). *Chemical Hazard Information Profile: Thiosemicarbazide.* Washington, DC: Chemical Emergency Preparedness Program

Thiotepa   T:0500

Molecular Formula: $C_6H_{12}N_3PS$
Synonyms: 1,1',1''-Phosphinothioylidynetrisaziridine; Phosphorothioic acid triethylenetriamide; SK 6882; Tespamine; Thiofozil; Thiophosphamide; Thio-TEP; Tiofosfamid; Tiofozil; Triaziridinylphosphine sulfide; *N,N', N''*-Tri-1,2-ethanediylphosphorothioic triamide; *N,N',N''*-Tri-1,2-ethanediylthiophosphoramide; Tri(ethyleneimino) thiophosphoramide; *N,N',N''*-Triethylenephosphorothioic triamide; *N,N',N''*-Triethylenethiophosphamide; *N,N',N''*-Triethylen-ethiophosphoramide; Triethylenethiophosphorotriamide; Tris(1-aziridinyl)phosphine sulfide; Tris(ethylenimino)thiophosphate
CAS Registry Number: 52-24-4
RTECS® Number: SZ2975000
UN/NA & ERG Number: UN3249 (Medicines, toxic, solid, n.o.s.)/151
EC Number: 200-135-7
Regulatory Authority and Advisory Bodies
Carcinogenicity: IARC: Human Sufficient Evidence, 1978; Animal Sufficient Evidence, *carcinogenic to humans*, Group 1, 1997; NTP: Reasonably anticipated to be a human carcinogen; NCI: Carcinogenesis Studies (ipr); clear evidence: mouse, rat.
Banned or Severely Restricted (In Household Products) (Japan).[13]
California Proposition 65 Chemical: Cancer 1/1/88.
European/International Regulations: Not listed in Annex 1.
WGK (German Aquatic Hazard Class): No value assigned.
Description: Thiotepa is a crystalline substance. Molecular weight = 189.24; Freezing/Melting point = 51.5°C. Insoluble in water.

Potential Exposure: Thiotepa has been prescribed for a wide variety of neoplastic diseases (adenocarcinomas of the breast and the ovary; superficial carcinoma of the urinary bladder; controlling intracavitary or localized neoplastic disease; lymphomas, such as lymphosarcomas and Hodgkin's disease; as well as bronchogenic carcinoma). It is now largely superseded by other treatments.
Permissible Exposure Limits in Air
No standards or TEEL available.
Routes of Entry: Inhalation, ingestion, eye, and/or skin contact.
Harmful Effects and Symptoms
Long Term Exposure: Tris(1-aziridinyl)phosphine sulfide (Thiotepa) is carcinogenic in mice and rats after administration by various routes, producing a variety of malignant tumors. There are several reports and epidemiological studies suggesting the development of acute nonlymphocytic leukemia in patients treated with thiotepa for ovarian and other malignant tumors.
First Aid: If this chemical gets into the eyes, remove any contact lenses at once and irrigate immediately for at least 15 min, occasionally lifting upper and lower lids. Seek medical attention immediately. If this chemical contacts the skin, remove contaminated clothing and wash immediately with soap and water. Seek medical attention immediately. If this chemical has been inhaled, remove from exposure, begin rescue breathing (using universal precautions, including resuscitation mask) if breathing has stopped and CPR if heart action has stopped. Transfer promptly to a medical facility. When this chemical has been swallowed, get medical attention. Give large quantities of water and induce vomiting. Do not make an unconscious person vomit.
Personal Protective Methods: Wear protective gloves and clothing to prevent any reasonable probability of skin contact. Safety equipment suppliers/manufacturers can provide recommendations on the most protective glove/clothing material for your operation. All protective clothing (suits, gloves, footwear, headgear) should be clean, available each day, and put on before work. Contact lenses should not be worn when working with this chemical. Wear dust-proof chemical goggles and face shield unless full-face-piece respiratory protection is worn. Employees should wash immediately with soap when skin is wet or contaminated. Provide emergency showers and eyewash.
Respirator Selection: Follow regulations in OSHA 29CFR1910.134 or European Standard EN149. Use a NIOSH/MSHA- or European Standard EN149-approved respirator; or use an approved supplied-air respirator with a full face-piece operated in the positive-pressure mode, or with a full face-piece, hood, or helmet in the continuous-flow mode; or use a NIOSH/MSHA- or European Standard EN149-approved self-contained breathing apparatus with a full face-piece operated in pressure-demand or other positive-pressure mode.
Storage: Color Code—Blue: Health Hazard/Poison: Store in a secure poison location. Prior to working with this

chemical you should be trained on its proper handling and storage. A regulated, marked area should be established where this chemical is handled, used, or stored in compliance with OSHA Standard 1910.1045. Store in tightly closed containers in a cool, well-ventilated area away from strong oxidizers. Where possible, automatically transfer material from drums or other storage containers to process containers.

Shipping: Medicine, solid, toxic, n.o.s., require a shipping label of "POISONOUS/TOXIC MATERIALS." This material falls in Hazard Class 6.1 and Packing Group III.

Spill Handling: Evacuate persons not wearing protective equipment from area of spill or leak until cleanup is complete. Remove all ignition sources. Collect powdered material in the most convenient and safe manner and deposit in sealed containers. Ventilate area after cleanup is complete. It may be necessary to contain and dispose of this chemical as a hazardous waste. If material or contaminated runoff enters waterways, notify downstream users of potentially contaminated waters. Contact your local or federal environmental protection agency for specific recommendations. If employees are required to clean up spills, they must be properly trained and equipped. OSHA 1910.120(q) may be applicable.

Fire Extinguishing: This chemical is a combustible solid. Use dry chemical, carbon dioxide; water spray; or alcohol foam extinguishers. Poisonous gases are produced in fire, including oxides of nitrogen, sulfur, and phosphorus. If material or contaminated runoff enters waterways, notify downstream users of potentially contaminated waters. Notify local health and fire officials and pollution control agencies. From a secure, explosion-proof location, use water spray to cool exposed containers. If cooling streams are ineffective (venting sound increases in volume and pitch, tank discolors, or shows any signs of deforming), withdraw immediately to a secure position. If employees are expected to fight fires, they must be trained and equipped in OSHA 1910.156. The only respirators recommended for firefighting are self-contained breathing apparatuses that have full face-pieces and are operated in a pressure-demand or other positive-pressure mode.

References

Sax, N. I. (Ed.). (1980). Dangerous Properties of Industrial Materials Report, 1, No. 2, 69–70

Thiourea   T:0510

Molecular Formula: CH_4N_2S
Common Formula: H_2NCSNH_2
Synonyms: Carbamide, thio-; Isothiourea; Pseudothiourea; Pseudourea, 2-thio-; Sulfourea; Suluourea; Thiocarbamate; Thiocarbamide; β-Thiopseudourea; 2-Thiourea; THU; Tiourea (Spanish); TSIZP 34; Urea, 2-thio-
CAS Registry Number: 62-56-6

RTECS® Number: YU2800000
UN/NA & ERG Number: UN2811 (toxic solid, organic, n.o.s.)/154
EC Number: 200-543-5 [*Annex I Index No.:* 612-082-00-0]
Regulatory Authority and Advisory Bodies
Carcinogenicity: IARC: Animal Limited Evidence; Human No Adequate Data, *not classifiable as carcinogenic to humans*, Group 3, 2000; NTP: 11th Report on Carcinogens, 2004: Reasonably anticipated to be a human carcinogen; US EPA Gene-Tox Program, Positive: Carcinogenicity—mouse/rat; Cell transform.—SA7/F344 rat EPA; Positive: SHE—clonal assay; Cell transform.—RLV F344 rat embryo EPA; Positive: Host-mediated assay EPA; Negative: Cell transform.—SA7/SHE; *E. coli* polA with S9 EPA; Negative: Histidine reversion—Ames test; Sperm morphology—mouse EPA; Negative: *S. cerevisiae*—homozygosis EPA; Inconclusive: *E. coli* polA without S9.
Banned or Severely Restricted (Sweden) (UN).[13]
US EPA Hazardous Waste Number (RCRA No.): U219.
RCRA, 40CFR261, Appendix 8 Hazardous Constituents.
Reportable Quantity (RQ): 10 lb (4.54 kg).
EPCRA Section 313 Form R *de minimis* concentration reporting level: 0.1%.
California Proposition 65 Chemical: Cancer 1/1/88.
European/International Regulations: Hazard Symbol: Xn, N; Risk phrases: R22; R40; R51/53; Safety phrases: S2; S36/37; S61 (see Appendix 4).
WGK (German Aquatic Hazard Class): 3—Severe hazard to waters.

Description: Thiourea consists of colorless, lustrous crystals or powder with a bitter taste. Molecular weight = 76.13; Boiling point = decomposes below BP; Freezing/Melting point = 180−182°C; Flash point = 55°C. Partially soluble in water.

Potential Exposure: Compound Description: Agricultural Chemical; Drug, Tumorigen, Mutagen; Reproductive Effector; Human Data; Primary Irritant. Thiourea is used as rubber antiozonant, toning agent; corrosion inhibitor; and in pharmaceutical manufacture; in the manufacture of photosensitive papers; flame-retardant textile sizes; boiler water treatment. It is also used in photography; pesticide manufacture; in textile chemicals.

Incompatibilities: Reacts violently with acrolein, strong acids (nitric acid), and strong oxidants.

Permissible Exposure Limits in Air
Protective Action Criteria (PAC)
TEEL-0: 1.25 mg/m^3
PAC-1: 4 mg/m^3
PAC-2: 25 mg/m^3
PAC-3: 125 mg/m^3
DFG MAK: [skin] Danger of skin sensitization; danger of photo-contact sensitization; Carcinogen Category 3B.
Austria Suspected: carcinogen, 1999; Finland: carcinogen, 1999; France: carcinogen, 1993; Russia: STEL 0.3 mg/m^3, 1993; Sweden: carcinogen, 1999; the Netherlands: MAC-TGG 0.5 mg/m^3 [skin], 2003. Several states have set

guidelines or standards for thiourea in ambient air[60] ranging from zero (North Dakota) to 0.03 μg/m^3 (New York).

Determination in Water: Octanol—water coefficient: Log K_{ow} = −2.38/−0.95.

Routes of Entry: Inhalation, ingestion, eye, and/or skin contact.

Harmful Effects and Symptoms

Short Term Exposure: Irritates the eyes. Thiourea may affect you when breathed in. A related chemical (naphthylthiourea) in higher exposures can cause fluid in the lungs, a medical emergency. It is not known if thiourea has this effect.

Long Term Exposure: Repeated exposure can cause goiter (enlarged thyroid gland). Exposure may damage the bone marrow, causing reduced red blood cells; white blood cells and platelets (reduced blood clotting ability). Thiourea has been identified as a sensitizer in people suffering from photosensitivity. A possible human carcinogen.

Points of Attack: Thyroid gland, blood, skin. Cancer site in animals: thyroid, liver.

Medical Surveillance: Before beginning employment and at regular times after that, for those with frequent or potentially high exposures, the following are recommended. Lung function tests. Complete blood count. Consider thyroid test for thyroxin. If symptoms develop or overexposure is suspected, the following may be useful. Consider chest X-ray after acute overexposure. Evaluation by a qualified allergist, including careful exposure history and special testing, may help diagnose skin allergy.

First Aid: If this chemical gets into the eyes, remove any contact lenses at once and irrigate immediately for at least 15 min, occasionally lifting upper and lower lids. Seek medical attention immediately. If this chemical contacts the skin, remove contaminated clothing and wash immediately with soap and water. Seek medical attention immediately. If this chemical has been inhaled, remove from exposure, begin rescue breathing (using universal precautions, including resuscitation mask) if breathing has stopped and CPR if heart action has stopped. Transfer promptly to a medical facility. When this chemical has been swallowed, contact a physician, hospital, or poison center at once. If the victim is unconscious or convulsing, do not induce vomiting or give anything by mouth. Assure that the patient's airway is open and lay him on his side with has head lower than his body and transport immediately to a medical facility. If conscious and not convulsing, give a glass of water to dilute the substance. Vomiting should not be induced without a physician's advice.

Personal Protective Methods: Wear protective gloves and clothing to prevent any reasonable probability of skin contact. Safety equipment suppliers/manufacturers can provide recommendations on the most protective glove/clothing material for your operation. All protective clothing (suits, gloves, footwear, headgear) should be clean, available each day, and put on before work. Contact lenses should not be worn when working with this chemical. Eye protection is included in the recommended respiratory protection. Employees should wash immediately with soap when skin is wet or contaminated. Provide emergency showers and eyewash.

Respirator Selection: At any exposure level, use a NIOSH/MSHA- or European Standard EN149-approved supplied-air respirator with a full face-piece operated in the positive-pressure mode, or with a full face-piece, hood, or helmet in the continuous-flow mode; or use a NIOSH/MSHA- or European Standard EN149-approved self-contained breathing apparatus with a full face-piece operated in pressure-demand or other positive-pressure mode.

Storage: Color Code—Blue: Health Hazard/Poison: Store in a secure poison location. Prior to working with this chemical you should be trained on its proper handling and storage. A regulated, marked area should be established where this chemical is handled, used, or stored in compliance with OSHA Standard 1910.1045. Thiourea must be stored to avoid contact with acrolein, hydrogen peroxide; and nitric acid, since violent reactions occur. Store in tightly closed containers in a cool, well-ventilated area.

Shipping: Toxic solids, organic, n.o.s., require a shipping label of "POISONOUS/TOXIC MATERIALS." It falls in Hazard Class 6.1 and Packing Group III.

Spill Handling: Evacuate persons not wearing protective equipment from area of spill or leak until cleanup is complete. Remove all ignition sources. Collect powdered material in the most convenient and safe manner and deposit in sealed containers. Ventilate area after cleanup is complete. It may be necessary to contain and dispose of this chemical as a hazardous waste. If material or contaminated runoff enters waterways, notify downstream users of potentially contaminated waters. Contact your local or federal environmental protection agency for specific recommendations. If employees are required to cleanup spills, they must be properly trained and equipped. OSHA 1910.120(q) may be applicable.

Fire Extinguishing: This chemical is a combustible solid. Use dry chemical, carbon dioxide; water spray; or foam extinguishers. Poisonous gases are produced in fire, including nitrogen and sulfur oxides. If material or contaminated runoff enters waterways, notify downstream users of potentially contaminated waters. Notify local health and fire officials and pollution control agencies. From a secure, explosion-proof location, use water spray to cool exposed containers. If cooling streams are ineffective (venting sound increases in volume and pitch, tank discolors, or shows any signs of deforming), withdraw immediately to a secure position. If employees are expected to fight fires, they must be trained and equipped in OSHA 1910.156. The only respirators recommended for firefighting are self-contained breathing apparatuses that have full face-pieces and are operated in a pressure-demand or other positive-pressure mode.

Disposal Method Suggested: Consult with environmental regulatory agencies for guidance on acceptable disposal practices. Generators of waste containing this contaminant

(≥100 kg/mo) must conform with EPA regulations governing storage, transportation, treatment, and waste disposal.

References

US Environmental Protection Agency. (1979). *Chemical Hazard Information Profile: Thiourea.* Washington, DC

New Jersey Department of Health and Senior Services. (July 2002). *Hazardous Substances Fact Sheet: Thiourea.* Trenton, NJ

Thiram   T:0520

Molecular Formula: $C_6H_{12}N_2S_4$

Synonyms: Aapirol; Aatack; Aatiram; Accelerator T; Accelerator thiuram; Accel TMT; Aceto TETD; AI3-00987; Anles; Arasan; Arasan 42S; Arasan 42-S; Arasan 70; Arasan 70-S Red; Arasan 75; Arasan-M; Arasan-SF; Arasan-SF-X; Atiram; Attack; Aules; Bis(diethylthiocarbamoyl) sulfide; Bis[(dimethylamino)carbonothioyl] disulfide; Bis[(dimethylamino)carbonothioyl] disulphide; Bis(dimethylthiocarbamoyl) disulfide; Bis(dimethylthiocarbamoyl) disulphide; Chipco Thiram 75; Cunitex; Cyuram DS; Delsan; Disulfide, bis(dimethylthiocarbamoyl); α,α'-Dithiobis(dimethyl-thio)formamide; N,N-(Dithio-dicarbonothioyl)bis(N-methylmethanamine); Ekagom TB; ENT 987; Falitiram; Fermide; Fermide 850; Fernacol; Fernasan; Fernasan A; Fernide; Flo Pro T seed protectant; FMC 2070; Formalsol; Formamide, 1,1'-dithiobis(N,N-dimethylthio-); Hermal; Hermat TMT; Heryl; Hexathir; HY-VIC; Kregasan; Mercuram; Methyl thiram; Methylthiuram disulfide; Methyl tuads; Metiurac; NA2771; Nobecutan; Nomersan; Normersan; NSC 1771; Panoram 75; Polyram ultra; Pomarsol; Pomarsol forte; Pomasol; Puralin; Rezifilm; Royal TMTD; Sadoplon; Spotrete; Spotrete-F; SQ 1489; Sranan-SF-X; STCC 4941187; Teramethylthiuram disulfide; Tersan; Tersan 75; Tersantetramethyl diurane sulfide; Tetramethyldiurane sulphite; Tetramethylenethiuram disulfide; Tetramethylenethiuram disulphide; Tetramethylthiocarbamoyldisulphide; Tetramethyl-thioperoxydicarbonic diamide; Tetramethylthiuram; Tetramethylthiuram bisulfide; Tetramethylthiuram bisulphide; N,N,N',N'-Tetramethylthiuram disulfide; N,N-Tetramethylthiuram disulfide; Tetramethylthiuram disulfide; Tetramethylthiuram disulphide; Tetramethylthiuran disulphide; Tetramethylthiurane disulfide; Tetramethyl thiurane disulphide; Tetramethylthiurum disulfide; Tetramethylthiurum disulphide; Tetrapom; Tetrasipton; Tetrathiuram disulfide; Tetrathiuram disulphide; Thianosan; Thillate; Thimar; Thimer; Thioknock; Thioperoxydicarbo NIC diamide, tetramethyl-; Thioperoxydicarbonic diamide, tetramethyl-; Thiosan; Thioscabin; Thiotex; Thiotox; Thiram 75; Thiram 80; Thiramad; Thiram B; Thirame (French); Thirampa; Thirasan; Thiulin; Thiulix; Thiurad; Thiuram; Thiuram D; Thiuramin; Thiuramyl; Thylate; Tiram (Spanish); Tirampa; Tiuramyl; TMTD; TMTDS; Trametan; Tridipam; Tripomol; TTD; Tuads; Tuex; Tulisan; Vancida TM-95; Vancide TM; Vuagt-1-4; Vulcafor

TMTD; Vulkacit MTIC; Vulkacit Thiuram; Vulkacit Thiuram/C

CAS Registry Number: 137-26-8; *(alt.)* 12680-07-8; *(alt.)* 12680-62-5; *(alt.)* 39456-80-9; *(alt.)* 66173-72-6; *(alt.)* 93196-73-7

RTECS® Number: JO14000000

UN/NA & ERG Number: UN2771/151

EC Number: 205-286-2 [*Annex I Index No.:* 006-005-00-4]

Regulatory Authority and Advisory Bodies

Carcinogenicity: IARC: Animal Inadequate Evidence; Human Inadequate Evidence, *not classifiable as carcinogenic to humans,* Group 3, 1991.

US EPA Gene-Tox Program, Positive: Mammalian micronucleus; *B. subtilis* rec assay.

Air Pollutant Standard Set. See below, "Permissible Exposure Limits in Air" section.

US EPA Hazardous Waste Number (RCRA No.): U244.

RCRA, 40CFR261, Appendix 8 Hazardous Constituents.

Reportable Quantity (RQ): 10 lb (4.54 kg).

EPCRA Section 313 Form R *de minimis* concentration reporting level: 1.0%.

Canada, WHMIS, Ingredients Disclosure List Concentration 1.0%.

Rotterdam Convention Annex III [Chemicals Subject to the Prior Informed Consent Procedure (PIC)] (as dustable powder formulations containing a combination of: Benomyl at or >7%; carbofuran at or >10%; thiram at or >15%).

European/International Regulations: Hazard Symbol: Xn, N; Risk phrases: R20/22; R36/38; R43; R48/22; R50/53; Safety phrases: S2; S26; S36/37; S60; S61 (see Appendix 4).

WGK (German Aquatic Hazard Class): 3—Severe hazard to waters.

Description: Thiram is a colorless to yellow, crystalline solid (which is sometimes dyed blue) with a characteristic odor. Commercial pesticide products may be dyed blue. Molecular weight = 240.44; Boiling point = 129°C; Freezing/Melting point = 156°C; Flash point = 148°C. Hazard Identification (based on NFPA-704 M Rating System): Health 2, Flammability 1, Reactivity 0. Insoluble in water.

Potential Exposure: Compound Description: Agricultural Chemical; Drug, Tumorigen, Mutagen; Reproductive Effector; Human Data; Hormone, Primary Irritant. Some thiurams have been used as rubber components: thiram is used as a rubber accelerator and vulcanizer; a seed, nut, fruit, and mushroom disinfectant; a bacteriostat for edible oils and fats; and as an ingredient in suntan and antiseptic sprays and soaps. It is also used as a fungicide, rodent repellent; wood preservative; and may be used in the blending of lubricant oils.

Incompatibilities: Strong oxidizers; strong acids; oxidizable materials.

Permissible Exposure Limits in Air

OSHA PEL: 5 mg/m³ TWA.

NIOSH REL: 5 mg/m^3 TWA.
ACGIH TLV®[1]: 0.05 mg/m^3, measured as inhalable fraction and vapor TWA; not classifiable as a human carcinogen.
NIOSH IDLH: 100 mg/m^3.
Protective Action Criteria (PAC)
TEEL-0: 0.05 mg/m^3
PAC-1: 10 mg/m^3
PAC-2: 75 mg/m^3
PAC-3: 100 mg/m^3
DFG MAK: MAK: 1 mg/m^3, inhalable fraction TWA; Peak Limitation Category II(2); danger of skin sensitization; Pregnancy Risk Group C.
Australia: TWA 5 mg/m^3, 1993; Austria: MAK 5 mg/m^3, 1999; Belgium: TWA 5 mg/m^3, 1993; Denmark: TWA 1 mg/m^3, 1999; Finland: TWA 5 mg/m^3, STEL 10 mg/m^3 [skin], 1999; France: VME 5 mg/m^3, 1999; the Netherlands: MAC-TGG 5 mg/m^3, 2003; the Philippines: TWA 5 mg/m^3, 1993; Poland: MAC (TWA) 0.5 mg/m^3; MAC (STEL) 2 mg/m^3, 1999; Russia: STEL 0.5 mg/m^3 [skin], 1993; Sweden: TWA 1 mg/m^3, STEL 2 mg/m^3, 1999; Switzerland: MAK-W 5 mg/m^3, STEL 25 mg/m^3, 1999; Thailand: TWA 5 mg/m^3, 1993; Turkey: TWA 5 mg/m^3, 1993; United Kingdom: TWA 5 mg/m^3, STEL 10 mg/m^3 2000 Occupational Exposure Limit; Argentina, Bulgaria, Columbia, Jordan, South Korea, New Zealand, Singapore, Vietnam: ACGIH TLV®: not classifiable as a human carcinogen. Russia[35, 43] has set MAC values for ambient air in residential areas of 0.01 mg/m^3 on a momentary basis and 0.006 mg/m^3 on a daily average basis. Several states have set guidelines or standards for Thiram in ambient air[60] ranging from 50 µg/m^3 (North Dakota) to 80 µg/m^3 (Virginia) to 100 µg/m^3 (Connecticut) to 119 µg/m^3 (Nevada).

Determination in Air: Use NIOSH Analytical Method (IV) #5005, Thiram.
Permissible Concentration in Water: Russia[43] set a MAC in water bodies used for domestic purposes of 1.0 µg/L. Further, it set a MAC in water bodies used for fishery purposes of zero. The state of Maine[61] has set a guideline for thiram in drinking water of 10 µg/L.
Routes of Entry: Inhalation, ingestion; skin and/or eye contact.
Harmful Effects and Symptoms
Short Term Exposure: Irritates the eyes, skin, and respiratory tract. Skin irritation can lead to rash, and allergy. High exposures can cause kidney and liver damage. Brain and nerve damage can also occur. Inhalation can cause irritation of the respiratory tract with stuffy nose; nosebleeds, hoarseness, cough, and/or phlegm. *Inhalation:* Animal studies indicate that irritation of the nose and throat may occur at levels above 5 mg/m^3. *Skin:* Exposure to spray containing 45% Thiram resulted in irritation and skin sensitization. *Eyes:* May cause irritation, tearing, and sensitivity to light. *Ingestion:* No information available on human exposure. In animal studies, 38 ppm in food caused nausea, vomiting, diarrhea, hyperexcitability, weakness, and loss of muscle

control. Death may occur from ingestion of approximately one teaspoonful.
Note: Can cause extreme illness when exposure is combined with alcohol ingestion.
Long Term Exposure: Repeated or prolonged contact may cause skin sensitization. Prolonged contact has caused eye irritation; tearing, increased sensitivity to light; reduced night vision and blurred vision. Occupational exposures to 0.03 mg/m^3 over a 5-year period has caused mild irritation of the nose and throat. Whether it has this effect in humans is not known. May affect the thyroid and liver. Thiram has caused birth defects in laboratory animals and has been shown to be a teratogen in animals.
Points of Attack: Eyes, skin, respiratory system; central nervous system.
Medical Surveillance: Preplacement and periodic medical examinations should give special attention to history of skin allergy, eye irritation; and significant respiratory, liver, or kidney disease. Workers should be aware of the potentiating action of alcoholic beverages when working with tetramethylthiuram disulfide.
First Aid: If this chemical gets into the eyes, remove any contact lenses at once and irrigate immediately for at least 15 min, occasionally lifting upper and lower lids. Seek medical attention immediately. If this chemical contacts the skin, remove contaminated clothing and wash immediately with soap and water. Seek medical attention immediately. If this chemical has been inhaled, remove from exposure, begin rescue breathing (using universal precautions, including resuscitation mask) if breathing has stopped and CPR if heart action has stopped. Transfer promptly to a medical facility. When this chemical has been swallowed, get medical attention. Give large quantities of water and induce vomiting. Do not make an unconscious person vomit.
Personal Protective Methods: Wear protective gloves and clothing to prevent any reasonable probability of skin contact. Safety equipment suppliers/manufacturers can provide recommendations on the most protective glove/clothing material for your operation. All protective clothing (suits, gloves, footwear, headgear) should be clean, available each day, and put on before work. Contact lenses should not be worn when working with this chemical. Wear dust-proof chemical goggles and face shield unless full-face-piece respiratory protection is worn. Employees should wash immediately with soap when skin is wet or contaminated. Provide emergency showers and eyewash.
Respirator Selection: 50 mg/m^3: CcrOv95 (APF = 10) [any air-purifying half-mask respirator equipped with an organic vapor cartridge(s) in combination with an N95, R95, or P95 filter. The following filters may also be used: N99, R99, P99, N100, R100, P100]; or Sa (APF = 10) (any supplied-air respirator). *100 mg/m^3:* Sa:Cf (APF = 25) (any supplied-air respirator operated in a continuous-flow mode); or CcrFOv100 (APF = 50) [any air-purifying full-face-piece respirator equipped with organic vapor cartridge(s) in combination with an N100, R100, or P100 filter]; or

GmFOv100 (APF = 50) [Any air-purifying, full-face-piece respirator (gas mask) with a chin-style, front- or back-mounted organic vapor canister having an N100, R100, or P100 filter]; or PaprOvHie (APF = 25) (any powered, air-purifying respirator with an organic vapor cartridge in combination with a high-efficiency particulate filter); or SCBAF (APF = 50) (any self-contained breathing apparatus with a full face-piece); or SaF (APF = 50) (any supplied-air respirator with a full face-piece). *Emergency or planned entry into unknown concentrations or IDLH conditions:* SCBAF: Pd,Pp (APF = 10,000) (any self-contained breathing apparatus that has a full face-piece and is operated in a pressure-demand or other positive-pressure mode); or SaF: Pd,Pp: ASCBA (APF = 10,000) (any supplied-air respirator that has a full face-piece and is operated in a pressure-demand or other positive-pressure mode in combination with an auxiliary self-contained breathing apparatus operated in a pressure-demand or other positive-pressure mode). *Escape:* GmFOv100 (APF = 50) [Any air-purifying, full-face-piece respirator (gas mask) with a chin-style, front- or back-mounted organic vapor canister having an N100, R100, or P100 filter]; or SCBAE (any appropriate escape-type, self-contained breathing apparatus).
Note: Substance reported to cause eye irritation or damage; may require eye protection.
Storage: Color Code—Blue: Health Hazard/Poison: Store in a secure poison location. Prior to working with this chemical you should be trained on its proper handling and storage. Store in a cool place away from strong oxidizers; strong acids; oxidizable materials.
Shipping: This compound requires a shipping label of "POISONOUS/TOXIC MATERIALS." It falls in Hazard Class 6.1 and Packing Group III.
Spill Handling: Evacuate persons not wearing protective equipment from area of spill or leak until cleanup is complete. Remove all ignition sources. Cover the spill with weak solution of calcium hypochlorite (up to 15%). Collect powdered material in the most convenient and safe manner and deposit in sealed containers. Ventilate area after cleanup is complete. It may be necessary to contain and dispose of this chemical as a hazardous waste. If material or contaminated runoff enters waterways, notify downstream users of potentially contaminated waters. Contact your local or federal environmental protection agency for specific recommendations. If employees are required to clean up spills, they must be properly trained and equipped. OSHA 1910.120(q) may be applicable.
Fire Extinguishing: This chemical is a combustible solid. Use dry chemical, carbon dioxide; water spray; or alcohol foam extinguishers. Poisonous gases are produced in fire, including sulfur dioxide, carbon disulfide, and nitrogen oxides. If material or contaminated runoff enters waterways, notify downstream users of potentially contaminated waters. Notify local health and fire officials and pollution control agencies. From a secure, explosion-proof location, use water spray to cool exposed containers. If cooling streams are ineffective (venting sound increases in volume and pitch, tank discolors, or shows any signs of deforming), withdraw immediately to a secure position. If employees are expected to fight fires, they must be trained and equipped in OSHA 1910.156. The only respirators recommended for firefighting are self-contained breathing apparatuses that have full face-pieces and are operated in a pressure-demand or other positive-pressure mode.
Disposal Method Suggested: Consult with environmental regulatory agencies for guidance on acceptable disposal practices. Generators of waste containing this contaminant (≥100 kg/mo) must conform with EPA regulations governing storage, transportation, treatment, and waste disposal. Thiram can be dissolved in alcohol or other flammable solvent and burned in an incinerator with an afterburner and scrubber.

References

Sax, N. I. (Ed.). (1981). *Dangerous Properties of Industrial Materials Report*, 1, No. 5, 41–42
New York State Department of Health. (April 1986). *Chemical Fact Sheet: Thiram.* Albany, NY: Bureau of Toxic Substance Assessment
New Jersey Department of Health and Senior Services. (June 2000). *Hazardous Substances Fact Sheet: Thiram.* Trenton, NJ

Thorium & compounds   T:0525

Common Formula: Th; Th(NO$_3$)$_4$; ThO$_2$
Synonyms: *Metal:* Thorium-232; Thorium metal, pyrophoric
Chloride: (Spanish) Tetrochlorothorium; Thorium tetrachloride
Dioxide: Dioxido de torio (Spanish); Thoria; Thorium oxide (tho2); thorium(IV) oxide; Thorotrast; Thortrast; Umbrathor
Nitrate: Nitric acid, thorium(4+) salt; Thorium(4+) nitrate; Thorium(IV) nitrate; Thorium tetranitrate
CAS Registry Number: 7440-29-1 (elemental); 10026-08-1 (chloride); 13823-29-5 (nitrate); 1314-20-1 (dioxide); 13825-36-0 (hydroxide); 2040-52-0 (oxalate); 16045-17-3 (perchlorate)
RTECS® Number: XO6400000 (elemental); XO6825000 (nitrate); XO6950000 (dioxide); 237-514-1(nitrate)
UN/NA & ERG Number: Metal powder, in bulk, may be pyrophoric. UN2975 (metal, pyrophoric)/162; UN2976 (nitrate); UN3077/171 (dioxide)
EC Number: 231-139-7 (elemental); 233-056-1 (chloride); 215-225-1(dioxide)
Regulatory Authority and Advisory Bodies
Carcinogenicity: IARC: Human Sufficient Evidence [diagnostic injection of thorium-232 as stabilized thorium-232 dioxide in colloidal form (Thorotrast)]; Humans Inadequate Evidence (inhalation thorium-232), generally in Group, 2001; RTECS: (thorium dioxide).[9]

Thorium dioxide:
EPCRA Section 313 Form R *de minimis* concentration reporting level: 1.0%.
California Proposition 65 Chemical: Cancer 2/27/87.
European/International Regulations: Not listed in Annex 1.
WGK (German Aquatic Hazard Class): No value assigned (all CAS numbers shown above).

Description: Thorium is a silvery-white, soft, ductile metal which is a natural radioactive element. Atomic weight = 232.00; Freezing/Melting point = $1750 \pm 30°C$. Insoluble in water. It occurs in the minerals monazite, thorite, and thorinite; usually mixed with its disintegration products. Thorium chloride is a crystalline solid. Molecular weight = 373.80; Boiling point = 927.5°C; Freezing/Melting point = 770°C. Soluble in water; Thorium nitrate, $Th(NO_3)_4$ is a crystalline solid which decomposes at 500°C. Thorium oxide, ThO_2 is a white crystalline powder or solid. Molecular weight = 264.00; Boiling point = 4400°C; Freezing/Melting point = 3320°C. Insoluble in water.

Potential Exposure: Metallic thorium is used in nuclear reactors to produce nuclear fuel; in the manufacture of incandescent mantles; as an alloying material, especially with some of the lighter metals, e.g., magnesium as a reducing agent in metallurgy; for filament coatings in incandescent lamps and vacuum tubes; as a catalyst in organic synthesis; in ceramics; and in welding electrodes. Exposure may occur during production and use of thorium-containing materials, in the casting and machining of alloy parts; and from the fume produced during welding with thorium electrodes.

Thorium nitrate is an oxidizer. Contact with combustibles and reducing agents will cause violent combustion or ignition.

Incompatibilities: Metal: The powder may ignite spontaneously in air. Heating may cause violent combustion or explosion. May explosively decompose from shock, friction, or concussion. Potentially hazardous reactions with strong oxidizers, chlorine, fluorine, bromine, oxygen; phosphorus, nitryl fluoride; peroxyformic acid; silver, sulfur.

Permissible Exposure Limits in Air
Protective Action Criteria (PAC)
TEEL-0: 0.35 mg/m³
PAC-1: 1 mg/m³
PAC-2: 6 mg/m³
PAC-3: 35 mg/m³
Nitrate
TEEL-0: 0.75 mg/m³
PAC-1: 2 mg/m³
PAC-2: 15 mg/m³
PAC-3: 25 mg/m³
Perchlorate
TEEL-0: 31.4 mg/m³
PAC-1: 94.3 mg/m³
PAC-2: 157 mg/m³
PAC-3: 500 mg/m³

Oxide
TEEL-0: 25 mg/m³
PAC-1: 75 mg/m³
PAC-2: 500 mg/m³
PAC-3: 500 mg/m³
Oxalate
TEEL-0: 42.1 mg/m³
PAC-1: 126 mg/m³
PAC-2: 210 mg/m³
PAC-3: 500 mg/m³
Hydroxide
TEEL-0: 0.75 mg/m³
PAC-1: 0.75 mg/m³
PAC-2: 2.5 mg/m³
PAC-3: 75 mg/m³
North Dakota[60] has set a guideline for ambient air of zero for thorium dioxide.

Routes of Entry
Ingestion of liquid, inhalation of dust or gas, and percutaneous absorption.

Harmful Effects and Symptoms
Short Term Exposure: Irritates the eyes, skin, and respiratory tract. Inhalation may cause damage to the bone marrow and lungs. May affect the blood-forming system, reducing the ability to produce white blood cells, resulting in pernicious anemia.

Long Term Exposure: Gas and aerosols can penetrate the body by way of the respiratory system, the digestive system, and the skin. Only 0.001% of an ingested dose is retained in the body. Thorium, once deposited in the body, remains in the bones, lymph system; lungs and other body organs; and parenchymatous tissues for long periods of time. Low repeated exposures may scar the lungs, and damage the liver and kidneys. Characteristic effects of the activity of thorium and its disintegration products are changes in blood-forming, nervous and reticuloendothelial systems; and functional and morphological damage to lung and bone tissue. Only much later do illness and symptoms characteristic of chronic radiation disease appear. After a considerable time, neoplasms may occur and the immunological activity of the body may be reduced. Repeated or prolonged contact with skin may cause dermatitis. A potential occupational carcinogen. May cause heritable genetic damage. May cause birth defects.

Points of Attack: Bone marrow, lungs. Liver, kidneys, lungs. Cancer site in humans (thorium dioxide): liver.

Medical Surveillance: Monitoring of personnel for early symptoms and changes, such as abnormal leukocytes in the blood smear, may be of value. In cases of chronic or acute exposure, the determination of thorium in the urine or the use of whole body radiation counts and breath radon are useful methods of monitoring the exposure dose and excretion rates. White blood cell count. Lung function test. Consider periodic chest X-ray for persons with potentially high or repeated lower exposure.

First Aid: If this chemical gets into the eyes, remove any contact lenses at once and irrigate immediately for at least 15 min, occasionally lifting upper and lower lids. Seek medical attention immediately. If this chemical contacts the skin, remove contaminated clothing and wash immediately with soap and water. Seek medical attention immediately. If this chemical has been inhaled, remove from exposure, begin rescue breathing (using universal precautions, including resuscitation mask) if breathing has stopped and CPR if heart action has stopped. Transfer promptly to a medical facility. When this chemical has been swallowed, get medical attention. Give large quantities of water and induce vomiting. Do not make an unconscious person vomit.

Personal Protective Methods: Wear protective gloves and clothing to prevent any reasonable probability of skin contact. Safety equipment suppliers/manufacturers can provide recommendations on the most protective glove/clothing material for your operation. All protective clothing (suits, gloves, footwear, headgear) should be clean, available each day, and put on before work. Contact lenses should not be worn when working with this chemical. Wear dust-proof chemical goggles and face shield unless full-face-piece respiratory protection is worn. Employees should wash immediately with soap when skin is wet or contaminated. Provide emergency showers and eyewash.

Respirator Selection: Where there is potential for exposure to thorium nitrate, use a NIOSH/MSHA- or European Standard EN149-approved full-face-piece respirator with a high-efficiency particulate filter. Greater protection is provided by a powered, air-purifying respirator. *Where there is potential for high exposures*, use a NIOSH/MSHA- or European Standard EN149-approved supplied-air respirator with a full face-piece operated in the positive-pressure mode, or with a full face-piece, hood, or helmet in the continuous-flow mode; or use a NIOSH/MSHA- or European Standard EN149-approved self-contained breathing apparatus with a full face-piece operated in pressure-demand or other positive-pressure mode.

Storage: Radioactive. Color Code—Yellow Stripe: Reactivity Hazard; Store separately in an area isolated from flammables, combustibles, or other yellow-coded materials. Prior to working with this chemical you should be trained on its proper handling and storage. A regulated, marked area should be established where this chemical is handled, used, or stored in compliance with OSHA Standard 1910.1045. Thorium nitrate must be stored to avoid contact with combustible, organic, or other readily oxidizable materials, since violent reactions occur.

Shipping: Thorium metal, pyrophoric, requires a shipping label of "RADIOACTIVE, SPONTANEOUSLY COMBUSTIBLE." It falls in Hazard Class 7. Thorium nitrate, solid, requires a shipping label of "RADIOACTIVE, OXIDIZER." It falls in Hazard Class 7.

Spill Handling: Evacuate persons not wearing protective equipment from area of spill or leak until cleanup is complete. Remove all ignition sources. Collect powdered material in the most convenient and safe manner and deposit in sealed containers. Ventilate area after cleanup is complete. It may be necessary to contain and dispose of this chemical as a hazardous waste. If material or contaminated runoff enters waterways, notify downstream users of potentially contaminated waters. Contact your local or federal environmental protection agency for specific recommendations. If employees are required to clean up spills, they must be properly trained and equipped. OSHA 1910.120(q) may be applicable.

Fire Extinguishing: Thorium dioxide may burn, but does not readily ignite. Thorium nitrate can cause violent combustion or ignition when in contact with readily combustible substances. On thorium nitrate, use dry chemical, carbon dioxide; water spray; or foam extinguishers. Poisonous gases are produced in fire, including nitrogen oxides in thorium nitrate. If material or contaminated runoff enters waterways, notify downstream users of potentially contaminated waters. Notify local health and fire officials and pollution control agencies. From a secure, explosion-proof location, use water spray to cool exposed containers. If cooling streams are ineffective (venting sound increases in volume and pitch, tank discolors, or shows any signs of deforming), withdraw immediately to a secure position. If employees are expected to fight fires, they must be trained and equipped in OSHA 1910.156. The only respirators recommended for firefighting are self-contained breathing apparatuses that have full face-pieces and are operated in a pressure-demand or other positive-pressure mode.

Disposal Method Suggested: Recovery and recycling is in the preferred route.

References

Sax, N. I. (Ed.). (1988). *Dangerous Properties of Industrial Materials Report: Thorium Chloride*, 8, No. 4, 72–74
New Jersey Department of Health and Senior Services. (February 2001). *Hazardous Substances Fact Sheet: Thorium Nitrate*. Trenton, NJ
New Jersey Department of Health and Senior Services. (June 2006). *Hazardous Substances Fact Sheet: Thorium Dioxide*. Trenton, NJ

Tin & inorganic compounds   T:0530

Molecular Formula: Sn

Synonyms: Alloy 510; Alloy 511; Alloy 521; Alloy 725; Estano (Spanish); Metallic tin; Prepared bath 2137; Tin, elemental; Tin flake; Tin metal; Tin powder; Zinn (German)

Other tin inorganic compounds for reference: Tin fluoroborate (B_2F_8Sn) (13814-97-6); Tin hydroxide (H_2O_2Sn) (12026-24-3); Tin nitrate (41480-79-9); Tin nitrite (100737-27-7); Tin(II) chloride dihydrate ($C_{12}H_4O_2Sn$) (10025-69-1); Tin(II) oxide (1332-29-2); Tin(IV) oxide (O_2Sn) (18282-10-5); Stannic tetrachloride ($C_{14}Sn$); (7646-78-8); Stannous chloride (Cl_2Sn) (7772-99-8)

CAS Registry Number: 7440-31-5 (metal); 18282-10-5 [Tin(IV) oxide]; 21651-19-4 (oxide)

RTECS® Number: XP7320000 (elemental)
UN/NA & ERG Number: No citation.
EC Number: 231-141-8
Regulatory Authority and Advisory Bodies
Air Pollutant Standard Set. See below, "Permissible Exposure Limits in Air" section.
Canada, WHMIS, Ingredients Disclosure List Concentration 1.0%, tin, elemental and tin compounds, n.o.s.
European/International Regulations: Not listed in Annex 1.
WGK (German Aquatic Hazard Class): Nonwater polluting agent (metal, oxide).
Description: Tin is a gray to almost silver-white, ductile, malleable, lustrous metal. Specific gravity (H_2O:1) = 7.28 at 25°C; Molecular weight = 118.69; Boiling point = 2625°C; Melting point = 232°C. Ignition temperature of dust cloud = 630°C; Minimum Explosive concentration = 0.19 oz/ft^3.[USBM] Relative explosion hazard of dust: Moderate (due to oxide coating). Hazard Identification (based on NFPA-704 M Rating System): Health 2, Flammability 2, Reactivity 1. Insoluble in water. The primary commercial source of tin is cassiterite (SnO_2, tinstone).
Potential Exposure: Compound Description: Tumorigen; Human Data. The most important use of tin is as a protective coating for other metals, such as in the food and beverage canning industry; in roofing tiles; silverware, coated wire; household utensils; electronic components; and pistons. Common tin alloys are phosphor bronze; light brass; gun metal; high tensile brass; manganese bronze; die-casting alloys; bearing metals; type metal; and pewter. These are used as soft solders, fillers in automobile bodies; and as coatings for hydraulic brake parts; aircraft landing gear and engine parts. Metallic tin is used in the manufacture of collapsible tubes and foil for packaging. Exposures to tin may occur in mining, smelting, and refining; and in the production and use of tin alloys and solders. Inorganic tin compounds are important industrially in the production of ceramics; porcelain, enamel, glass; and inks; in the production of fungicides; anthelmintics, insecticides; as a stabilizer it is used in polyvinyl plastics and chlorinated rubber paints; and it is used in plating baths.
Incompatibilities: Tin is a reducing agent; keep away from strong oxidizers; turpentine, acids, bases, alkalis. Powder quickly corrodes in air, especially if moist.
Permissible Exposure Limits in Air
Metal
OSHA PEL: 2 mg[Sn]/m^3 TWA [also applies to other inorganic tin compounds (as Sn), including tin oxides].
NIOSH REL: 2 mg[Sn]/m^3 TWA [also applies to other inorganic tin compounds (as Sn) *except* tin oxides].
ACGIH TLV®[1]: 2 mg[Sn]/m^3 TWA [also applies to other inorganic tin compounds (as Sn), including tin oxides].
NIOSH IDLH: 100 mg[Sn]/m^3.
Protective Action Criteria (PAC)
TEEL-0: 2 mg/m^3
PAC-1: 20 mg/m^3

PAC-2: 100 mg/m^3
PAC-3: 100 mg/m^3
EUR OEL: 2 mg[Sn]/m^3 as TWA.
DFG MAK: *inorganic compounds:* Not established (Section IIb); *organic compounds:* 0.1 mg/m^3, inhalable fraction TWA; Peak Limitation Category II(2) [skin]; Pregnancy Risk Group D.
Australia: TWA 0.1 mg[Sn]/m^3 [skin], 1993; Australia: TWA 2 mg/m^3, 1993; Austria: MAK 2 mg/m^3, 1999; Belgium: TWA 2 mg/m^3, 1993; Denmark: TWA 2 g/m^3, 1999; Finland: TWA 2 mg/m^3, 1999; Hungary: TWA 1 mg/m^3, STEL 2 mg/m^3 [skin], 1993; the Netherlands: MAC-TGG 2 mg/m^3, 2003; Norway: TWA 1 mg/m^3, 1999; the Philippines: TWA 2 mg/m^3, 1993; Poland: TWA 2 mg/m^3, 1999; Switzerland: MAK-W 2 mg/m^3, KZG-W 4 mg/m^3, 1999; Thailand: TWA 2 mg/m^3, 1993; United Kingdom: TWA 2 mg[Sn]/m^3, STEL 4 mg[Sn]/m^3, 2000; Argentina, Bulgaria, Columbia, Jordan, South Korea, New Zealand, Singapore, Vietnam: ACGIH TLV®: TWA 2 mg/m^3.
Several states have set guidelines or standards for tin in ambient air[60] ranging from 1.6 µg/m^3 (Virginia) to 20.0 µg/m^3 (North Dakota) to 40.0 µg/m^3 (Connecticut) to 48.0 µg/m^3 (Nevada).
1332-29-2 [tin(II) oxide] and 18282-10-5 [Tin(IV) oxide; (stannic oxide)]
Protective Action Criteria (PAC)
TEEL-0: 2.54 mg/m^3
PAC-1: 7.62 mg/m^3
PAC-2: 1.27 mg/m^3
PAC-3: 1270 mg/m^3
Determination in Air: Use NIOSH Analytical Method (IV) #7300, Element; #7301; #7303; OSHA Analytical Method ID-121; ID-206.
Routes of Entry: Inhalation of dust, eye, and/or skin contact.
Harmful Effects and Symptoms
Short Term Exposure: Dust irritates the eyes, skin, and respiratory tract. Tin may be contaminated with toxic lead or arsenic, causing exposure to these chemicals from dust or fumes. Inorganic tin salts are irritants to the skin and mucous membranes; they may be strongly acidic or basic depending on the cation or anion present. In animals: vomiting, diarrhea, paralysis with muscle twitching.
Long Term Exposure: Exposure to dust or fumes of inorganic tin is known to cause a benign pneumoconiosis (stannosis). Tin or dust fumes can cause "spots" to appear on chest X-ray and may represent reduced lung function. This form of pneumoconiosis produces distinctive progressive X-ray changes of the lungs as long as exposure persists, but there is no distinctive fibrosis; no evidence of disability; and no special complicating factors. Because tin is so radio-opaque, early diagnosis is possible. Fumes can also cause chronic cough and may cause reduced lung function. Tin may interfere with the body's ability to absorb iron from food or vitamin pills, contributing to iron-deficiency anemia.

Points of Attack: Eyes, skin, respiratory system.

Medical Surveillance: NIOSH lists the following tests: whole blood (chemical/metabolite), Chest X-ray; urine (chemical/metabolite). Also, see NIOSH Analytical Method #8310, Metals in urine.

First Aid: If this chemical gets into the eyes, remove any contact lenses at once and irrigate immediately for at least 15 min, occasionally lifting upper and lower lids. Seek medical attention immediately. If this chemical contacts the skin, remove contaminated clothing and wash immediately with soap and water. Seek medical attention immediately. If this chemical has been inhaled, remove from exposure, begin rescue breathing (using universal precautions, including resuscitation mask) if breathing has stopped and CPR if heart action has stopped. Transfer promptly to a medical facility. When this chemical has been swallowed, get medical attention. Give large quantities of water and induce vomiting. Do not make an unconscious person vomit.

Personal Protective Methods: Wear protective gloves and clothing to prevent any reasonable probability of skin contact. Safety equipment suppliers/manufacturers can provide recommendations on the most protective glove/clothing material for your operation. All protective clothing (suits, gloves, footwear, headgear) should be clean, available each day, and put on before work. Contact lenses should not be worn when working with this chemical. Wear dust-proof chemical goggles and face shield unless full-face-piece respiratory protection is worn. Employees should wash immediately with soap when skin is wet or contaminated. Provide emergency showers and eyewash.

Respirator Selection: NIOSH/OSHA: *Up to 10 mg/m³:* Qm (APF = 25) (any quarter-mask respirator). *Up to 20 mg/m³:* 95XQ (APF = 10) [any particulate respirator equipped with an N95, R95, or P95 filter (including N95, R95, and P95 filtering face-pieces) except quarter-mask respirators. The following filters may also be used: N99, R99, P99, N100, R100, P100]; or Sa* (APF = 10) (any supplied-air respirator). *Up to 50 mg/m³:* Sa:Cf* (APF = 25) (any supplied-air respirator operated in a continuous-flow mode); or PaprHie (APF = 25)* (any powered, air-purifying respirator with a high-efficiency particulate filter). *Up to 100 mg/m³:* 100F (APF = 50) (any air-purifying, full-face-piece respirator with an N100, R100, or P100 filter; or SCBAF (APF = 50) (any self-contained breathing apparatus with a full face-piece); or SaF (APF = 50) (any supplied-air respirator with a full face-piece). *Emergency or planned entry into unknown concentrations or IDLH conditions:* SCBAF: Pd, Pp (APF = 10,000) (any self-contained breathing apparatus that has a full face-piece and is operated in a pressure-demand or other positive-pressure mode); or SAF; PD,PP: ASCBA (any supplied-air respirator that has a full face-piece and is operated in a pressure-demand or other positive-pressure mode in combination with an auxiliary self-contained positive-pressure breathing apparatus). *Escape:* 100F (APF = 50) (any air-purifying, full-face-piece respirator with an N100, R100, or P100 filter); or SCBAE (any appropriate escape-type, self-contained breathing apparatus).

Storage: Powdered tin is a moderate explosion risk. Color Code—Yellow Stripe (*strong reducing agent*): Reactivity Hazard; Store separately in an area isolated from flammables, combustibles, or other yellow-coded materials. Color Code—Green (*tin metal*): General storage may be used. Prior to working with this chemical you should be trained on its proper handling and storage. Tin must be stored to avoid contact with chlorine, bromine, bromine trifluoride; chlorine monofluoride; copper nitrate; turpentine, and potassium dioxide, since violent reactions occur. Sources of ignition, such as smoking and open flames, are prohibited where tin is used, handled, or stored in a manner that could create a potential fire or explosion hazard.

Spill Handling: Evacuate persons not wearing protective equipment from area of spill or leak until cleanup is complete. Remove all ignition sources. Collect powdered material in the most convenient and safe manner and deposit in sealed containers. Ventilate area after cleanup is complete. It may be necessary to contain and dispose of this chemical as a hazardous waste. If material or contaminated runoff enters waterways, notify downstream users of potentially contaminated waters. Contact your local or federal environmental protection agency for specific recommendations. If employees are required to clean up spills, they must be properly trained and equipped. OSHA 1910.120(q) may be applicable.

Fire Extinguishing: Tin dust may be a fire hazard. Toxic metal fumes may be produced in a fire. Use dry chemicals appropriate for extinguishing metal fires. *Do not use water.* If material or contaminated runoff enters waterways, notify downstream users of potentially contaminated waters. Notify local health and fire officials and pollution control agencies. From a secure, explosion-proof location, use water spray to cool exposed containers. If cooling streams are ineffective (venting sound increases in volume and pitch, tank discolors, or shows any signs of deforming), withdraw immediately to a secure position. If employees are expected to fight fires, they must be trained and equipped in OSHA 1910.156. The only respirators recommended for firefighting are self-contained breathing apparatuses that have full face-pieces and are operated in a pressure-demand or other positive-pressure mode.

References

US Environmental Protection Agency. (May 1977). *Toxicology of Metals, Vol. II: Tin, Report EPA-600/1-77-022.* Research Triangle Park, NC, pp. 405–426

Sax, N. I. (Ed.). (1981). *Dangerous Properties of Industrial Materials Report*, 1, No. 3, 82–83. New York, NY: Van Nostrand Reinhold Co.

New Jersey Department of Health and Senior Services. (April 2001). *Hazardous Substances Fact Sheet: Tin.* Trenton, NJ

Tin organic compounds   T:0540

Molecular Formula: $Sn(R)_{4-x}$ (radical); $C_{18}H_{16}OSn$ *(used as an example)*

Synonyms: varies, this is used as an example: Dowco 186; Du-Ter; ENT 28,009; Fenolovo; Fentin hydroxide; Fintine hydroxyde (French); Fintin hydroxid (German); Haitin; Hydroxyde de triphenyl-etain (French); Hydroxytriphenyltin; NCI-C00260; Suzu H; TPTH; Triphenyltin hydroxide; Triphenyltin oxide; Triphenyl-zinn-hydroxid (German); Tubotin; Vancide KS

See also entries for "Tetraethyltin" (597-64-8); "Cyhexatin" (13121-70-5)

CAS Registry Number: (varies) Fentin acetate ($C_{20}H_{18}O_2Sn$) (900-95-8); Fentin hydroxide ($C_{18}H_{16}OSn$) (76-87-9); Tin(IV) isopropoxide ($C_{12}H_{28}O_4Sn$) (1184-61-8); Triphenyltin chloride ($C_{18}H_{15}ClSn$) 639-58-7; Fentin chloride (C_3H_9ClSn) (1066-45-1)

RTECS® Number: WH8750000 (varies; used as an example)

UN/NA & ERG Number: UN2777 (organotin compound, liquid, n.o.s.); UN2786 (organotin pesticide, solid, toxic)/153; 3146 (organotin compound, solid, n.o.s.)/153; UN2787 (organotin pesticide, liquid, flammable, toxic)/131; UN2788 (organotin compounds, liquid, n.o.s.)/153; UN3019 (organotin pesticide, liquid, toxic, flammable)/131; UN3020 (organotin pesticides, liquid, toxic)/153; UN3145 (organotin compound, solid, n.o.s.)/153

EC Number: 200-990-6 [*Annex I Index No.:* 050-004-00-1] (fentin hydroxide)

Regulatory Authority and Advisory Bodies

Banned or Severely Restricted (Japan) (UN).[13]

Air Pollutant Standard Set. See below, "Permissible Exposure Limits in Air" section.

Canada, WHMIS, Ingredients Disclosure List Concentration 1.0%.

European/International Regulations (*900-95-8; fentin acetate* and *76-87-9; fentin hydroxide*): European/International Regulations: Hazard Symbol: T+, N; Risk phrases: R24/25; R26; R37/38; R40; R41; R48/23; R63; R50/53; Safety phrases: S1/2; S26; S28; S36/37/39; S45; S60; S61 (see Appendix 4).

WGK (German Aquatic Hazard Class): 3—Highly water polluting (*fentin acetate; fentin hydroxide; Triphenyltin chloride; Trimethyltin chloride*).

Description: $Sn(R)_{4-x}$ (radical)$_x$. A typical compound is triphenyltin hydroxide, $C_{18}H_{16}OSn$, which is a white powder; Freezing/Melting point $= 122°C$. Slightly soluble in water.

Potential Exposure: Organotin compounds are used as additives in a variety of products and processes. Diorganotins find application as heat stabilizers in plastics, as catalysts in the production of urethane foams; in the cold curing of rubber; and as scavengers for halogen acids. Tri- and tetraorganotins are used as preservatives for wood, leather, paper, paints, and textiles; and as biocides.

Incompatibilities: Strong oxidizers.

Permissible Exposure Limits in Air

OSHA PEL: 0.1 mg[Sn]/m^3 TWA (*Note*: The PEL applies to all organic tin compounds).

NIOSH REL: 0.1 mg[Sn]/m^3 TWA [skin] (*Note:* The REL applies to all organic tin compounds, except Cyhexatin).

ACGIH TLV®[1]: 0.1 mg[Sn]/m^3 TWA; 0.2 mg[Sn]/m^3 STEL [skin].

NIOSH IDLH: 25 mg[Sn]/m^3.

DFG MAK: 0.1 mg[Sn]/m^3; inhalable fraction TWA; Peak Limitation Category II(2); [skin]; Pregnancy Risk Group D.

Cyhexatin

Australia: TWA 0.1 mg[Sn]/m^3 [skin], 1993; Austria: MAK 0.1 mg[Sn]/m^3 [skin], 1999; Australia: TWA 5 mg/m^3, 1993; Belgium: TWA 5 mg/m^3, 1993; Denmark: TWA 0.1 mg[Sn]/m^3 [skin], 1999; Denmark: TWA 5 mg/m^3, 1999; Finland: TWA 0.1 mg[Sn]/m^3; STEL 0.3 mg[Sn]/m^3, 1999; France: VME 5 mg/m^3, 1999; the Netherlands: MAC-TGG 5 mg/m^3, 2003; Hungary: STEL 0.1 mg[Sn]/m^3 [skin], 1993; Norway: TWA 0.1 mg[Sn]/m^3, 1999; the Philippines: TWA 0.1 mg[Sn]/m^3, 1993; Russia: STEL 0.02 mg/m^3 [skin], 1993; Switzerland: MAK-W 0.1 mg [Sn]/m^3, KZG-W 0.2 mg[Sn]/m^3 [skin], 1999; Thailand: TWA 0.1 mg[Sn]/m^3, 1993; United Kingdom: TWA 5 mg/m^3; STEL 10 mg/m^3, 2000; Argentina, Bulgaria, Columbia, Jordan, South Korea, New Zealand, Singapore, Vietnam: ACGIH TLV®: not classifiable as a human carcinogen.

Phenyltin compounds

DFG MAK: 0.002 mg[Sn]/m^3; inhalable fraction TWA; Peak Limitation Category II(2); Pregnancy Risk Group C.

Determination in Air: Use NIOSH Analytical Method (IV) #5504, Organotin.

Permissible Concentration in Water: No criteria set, but EPA[32] has suggested a permissible ambient goal of 1.4 μg/L based on health effects.

Routes of Entry: Inhalation, skin absorption; ingestion; skin and/or eye contact.

Harmful Effects and Symptoms

Short Term Exposure: Organic tin compounds, especially tributyl and dibutyl compounds, may cause acute burns to the skin. The burns produce little pain but may itch. They heal without scarring. Clothing contaminated by vapors or liquids may cause subacute lesions and diffuse erythermatoid dermatitis on the lower abdomen, thighs, and groin of workmen who handle these compounds. The lesions heal rapidly on removal from contact. The eyes are rarely involved, but accidental splashing with tributyltin has caused lacrimation and conjunctival edema which lasted several days; there was no permanent injury. Triphenyltin hydroxide has been subjected to a carcinogenesis bioassay by NCL and found to be not carcinogenic. Certain organic tin compounds, especially alkyltin compounds, are highly toxic when ingested. The trialkyl and tetraalkyl compounds cause damage to the central nervous system with symptoms of headaches, dizziness, photophobia, vomiting, and urinary retention; some weakness and flaccid paralysis of the limbs in the most severe cases. Percutaneous absorption of these compounds has been postulated, but to date, deaths and serious injury have resulted only from ill-advised attempts at therapeutic use by mouth. The mechanism of action of the organotins is not clearly understood, although triethyltin is an extremely potent inhibitor of oxidative phosphorylation. Occasionally, mild organotin intoxication is seen in chemical laboratories with headache, nausea, and EEG changes. Symptoms also include sore throat and cough, abdominal pain; skin burns and itching.

Points of Attack: Central nervous system; eyes, liver, urinary tract, skin, blood.

Medical Surveillance: NIOSH lists the following tests: complete blood count [RBC hemolysis; electrocardiogram, especially on workers over 40 years; glaucoma, liver function tests; urine (chemical/metabolite)]. For organotins, preplacement and periodic examinations should include the skin, eyes, blood, central nervous system; liver and kidney function.

First Aid: If this chemical gets into the eyes, remove any contact lenses at once and irrigate immediately for at least 15 min, occasionally lifting upper and lower lids. Seek medical attention immediately. If this chemical contacts the skin, remove contaminated clothing and wash immediately with soap and water. Speed in removing material from skin is of extreme importance. Shampoo hair promptly if contaminated. Seek medical attention immediately. If this chemical has been inhaled, remove from exposure, begin rescue breathing (using universal precautions, including resuscitation mask) if breathing has stopped and CPR if heart action has stopped. Transfer promptly to a medical facility. When this chemical has been swallowed, get medical attention. Give large quantities of water and induce vomiting. Do not make an unconscious person vomit.

Personal Protective Methods: It is important that employees be trained in the correct use of personal protective equipment. Wear protective gloves and clothing to prevent any reasonable probability of skin contact. Safety equipment suppliers/manufacturers can provide recommendations on the most protective glove/clothing material for your operation. All protective clothing (suits, gloves, footwear, headgear) should be clean, available each day, and put on before work. Contact lenses should not be worn when working with this chemical. Depending on physical properties, wear splash or dust-proof chemical goggles and face shield unless full-face-piece respiratory protection is worn. Employees should wash immediately with soap when skin is wet or contaminated. Provide emergency showers and eyewash. Skin contact should be prevented by protective clothing, and, especially in the case of organic tin compounds, clean work clothes should be supplied daily and the worker required to shower following the shift and prior to change to street clothes.

Respirator Selection: *(organotin) Up to 1 mg/m³:* CcrOvDM [any chemical cartridge respirator with organic vapor cartridge(s) in combination with a dust and mist

filter]; or Sa (APF = 10) (any supplied-air respirator). *Up to 2.5 mg/m³:* Sa:Cf (APF = 25) (any supplied-air respirator operated in a continuous-flow mode); or PaprOvHie (APF = 25) (any air-purifying full-face-piece respirator equipped with an organic vapor cartridge in combination with a high-efficiency particulate filter). *Up to 5 mg/m³:* CcrFOv100 (APF = 50) [any air-purifying full-face-piece respirator equipped with organic vapor cartridge(s) in combination with an N100, R100, or P100 filter]; or GmFOv100 (APF = 50) [Any air-purifying, full-face-piece respirator (gas mask) with a chin-style, front- or back-mounted organic vapor canister having an N100, R100, or P100 filter]; or PaprTOvHie (APF = 50) [any powered, air-purifying respirator with a tight-fitting face-piece and organic vapor cartridge(s) in combination with a high-efficiency particulate filter]; or SaT: Cf (APF = 50) (any supplied-air respirator that has a tight-fitting face-piece and is operated in a continuous-flow mode); or SCBAF (APF = 50) (any self-contained breathing apparatus with a full face-piece); or SaF (APF = 50) (any supplied-air respirator with a full face-piece). *Up to 25 mg/m³:* SaF: Pd,Pp (APF = 2000) (any supplied-air respirator that has a full face-piece and is operated in a pressure-demand or other positive-pressure mode). *Emergency or planned entry into unknown concentrations or IDLH conditions:* SCBAF: Pd, Pp (APF = 10,000) (any self-contained breathing apparatus that has a full-faceplate and is operated in a pressure-demand or other positive-pressure mode); or SaF: Pd,Pp: ASCBA (any supplied-air respirator that has a full face-piece and is operated in a pressure-demand or other positive-pressure mode in combination with an auxiliary, self-contained breathing apparatus operated in a pressure-demand or other positive-pressure mode). *Escape:* GmFOv100 (APF = 50) [Any air-purifying, full-face-piece respirator (gas mask) with a chin-style, front- or back-mounted organic vapor canister having an N100, R100, or P100 filter]; or SCBAE (any appropriate escape-type, self-contained breathing apparatus).

Storage: Color Code—Blue: Health Hazard/Poison: Store in a secure poison location. Prior to working with this chemical you should be trained on its proper handling and storage.

Shipping: Organotin compounds, liquid, n.o.s., require a shipping label of "POISONOUS/TOXIC MATERIALS." Hazard Class 6.1 and Packing Group II.

Spill Handling: *Dry material:* Evacuate persons not wearing protective equipment from area of spill or leak until cleanup is complete. Remove all ignition sources. Collect powdered material in the most convenient and safe manner and deposit in sealed containers. Ventilate area after cleanup is complete. It may be necessary to contain and dispose of this chemical as a hazardous waste. If material or contaminated runoff enters waterways, notify downstream users of potentially contaminated waters. Contact your local or federal environmental protection agency for specific recommendations. If employees are required to clean up spills, they must be properly trained and equipped. OSHA 1910.120(q) may be applicable.

Fire Extinguishing: Dry material: This chemical is a combustible solid. Use dry chemical, carbon dioxide; water spray; or alcohol foam extinguishers. Poisonous gases are produced in fire, including tin. If material or contaminated runoff enters waterways, notify downstream users of potentially contaminated waters. Notify local health and fire officials and pollution control agencies. From a secure, explosion-proof location, use water spray to cool exposed containers. If cooling streams are ineffective (venting sound increases in volume and pitch, tank discolors, or shows any signs of deforming), withdraw immediately to a secure position. If employees are expected to fight fires, they must be trained and equipped in OSHA 1910.156. The only respirators recommended for firefighting are self-contained breathing apparatuses that have full face-pieces and are operated in a pressure-demand or other positive-pressure mode.

References

National Institute for Occupational Safety and Health. (1977). *Criteria for a Recommended Standard: Occupational Exposure to Organotin Compounds*, NIOSH Document No. 77-115

Tin tetrachloride   T:0550

Molecular Formula: Cl_4Sn
Common Formula: $SnCl_4$
Synonyms: Libavius fuming spirit; Stannic chloride, anhydrous; Tetrachlorostannane; Tin perchloride; Tin chloride, fuming; Tin perchloride; Tin tetrachloride, anhydrous; Zinntetrachlorid (German)
CAS Registry Number: 7646-78-8
RTECS® Number: XP8750000
UN/NA & ERG Number: UN1827/137
EC Number: 231-588-9 [*Annex I Index No.:* 050-001-00-5]
Regulatory Authority and Advisory Bodies
US EPA Gene-Tox Program, Negative: *B. subtilis* rec assay.
Air Pollutant Standard Set. See below, "Permissible Exposure Limits in Air" section.
Canada, WHMIS, Ingredients Disclosure List Concentration 1.0%.
European/International Regulations: Hazard Symbol: C, N; Risk phrases: R34; R52/53; Safety phrases: S1/2; S7/8; S26; S45; 61 (see Appendix 4).
WGK (German Aquatic Hazard Class): 1—Low hazard to waters.
Description: Tin tetrachloride is a colorless fuming liquid. Molecular weight = 260.49; Boiling point = 114°C; Freezing/Melting point = −33°C. Hazard Identification (based on NFPA-704 M Rating System): Health 3, Flammability 0, Reactivity 1. Reacts with water; soluble.
Potential Exposure: Compound Description: Mutagen, Hormone. Tin tetrachloride is used in the production of

blueprints and electroconductive readings, as a bleaching agent for sugar and resin stabilizer.
Incompatibilities: Water, turpentine, potassium, sodium, ethylene oxide; nitrates, alcohols, amines, chlorine, strong acids; strong bases.
Permissible Exposure Limits in Air
OSHA PEL: 2 mg[Sn]/m³ TWA [also applies to other inorganic tin compounds (as Sn) except tin oxides].
NIOSH REL: 2 mg[Sn]/m³ TWA [also applies to other inorganic tin compounds (as Sn) except tin oxides].
ACGIH TLV®[1]: 2 mg[Sn]/m³ TWA.
EUR OEL: 2 mg[Sn]/m³ as TWA.
Protective Action Criteria (PAC)
TEEL-0: 4.39 mg/m³
PAC-1: 4.39 mg/m³
PAC-2: 5 mg/m³
PAC-3: 219 mg/m³
DFG MAK: *tin, inorganic compounds:* No numerical value established. Data may be available.
NIOSH IDLH: 100 mg[Sn]/m³.
Australia: TWA 2 mg[Sn]/m³, 1993; Austria: MAK 2 mg/m³, 1999; Belgium: TWA 2 mg[Sn]/m³, 1993; Denmark: TWA 2 mg[Sn]/m³, 1999; Finland: TWA 2 mg[Sn]/m³, 1999; Hungary: TWA 1 mg[Sn]/m³; STEL 2 mg[Sn]/m³ [skin], 1993; Norway: TWA 2 mg[Sn]/m³, 1999; the Philippines: TWA 2 mg[Sn]/m³, 1993; Poland: TWA 2 mg [Sn]/m³, 1999; Switzerland: MAK-W 2 mg[Sn]/m³, KZG-W 4 mg[Sn]/m³, 1999; Thailand: TWA 2 mg[Sn]/m³, 1993; United Kingdom: TWA 2 mg[Sn]/m³; STEL 4 mg[Sn]/m³, 2000; Argentina, Bulgaria, Columbia, Jordan, South Korea, New Zealand, Singapore, Vietnam: ACGIH TLV®: TWA 2 mg[Sn]/m³.
Determination in Air: Filter; Acid; Inductively coupled plasma; NIOSH Analytical Method (IV) #7300, Element.
Routes of Entry: Inhalation, eyes, and/or skin contact.
Harmful Effects and Symptoms
Short Term Exposure: Tin tetrachloride can affect you when breathed in. Tin tetrachloride is a corrosive chemical and eye or skin contact can cause severe burns. Breathing tin tetrachloride can irritate the throat and bronchial tubes, causing cough, and/or difficulty breathing. Higher exposures can cause pulmonary edema, a medical emergency that can be delayed for several hours. This can cause death.
Long Term Exposure: May affect the nervous system. May cause lung irritation; bronchitis may develop. Changes may occur on chest X-ray with repeated exposures.
Points of Attack: Eyes, skin, respiratory system; central nervous system; liver, kidneys, urinary tract; blood.
Medical Surveillance: NIOSH lists the following tests: whole blood (chemical/metabolite). Chest X-ray; urine (chemical/metabolite). Also, see NIOSH Analytical Method #8310, Metals in urine. Before beginning employment and at regular times after that, the following are recommended: lung function tests. If symptoms develop or overexposure is suspected, the following may be useful: consider chest X-ray after acute overexposure. Hemoglobin or hematocrit.

First Aid: If this chemical gets into the eyes, remove any contact lenses at once and irrigate immediately for at least 15 min, occasionally lifting upper and lower lids. Seek medical attention immediately. If this chemical contacts the skin, remove contaminated clothing and wash immediately with soap and water. Seek medical attention immediately. If this chemical has been inhaled, remove from exposure, begin rescue breathing (using universal precautions, including resuscitation mask) if breathing has stopped and CPR if heart action has stopped. Transfer promptly to a medical facility. When this chemical has been swallowed, get medical attention. If victim is *conscious*, administer water or milk. Do not induce vomiting. Medical observation is recommended for 24—48 h after breathing over-exposure, as pulmonary edema may be delayed. As first aid for pulmonary edema, a doctor or authorized paramedic may consider administering a corticosteroid spray.

Personal Protective Methods: Wear protective gloves and clothing to prevent any reasonable probability of skin contact. Safety equipment suppliers/manufacturers can provide recommendations on the most protective glove/clothing material for your operation. All protective clothing (suits, gloves, footwear, headgear) should be clean, available each day, and put on before work. Contact lenses should not be worn when working with this chemical. Wear splash-proof chemical goggles and face shield unless full-face-piece respiratory protection is worn. Employees should wash immediately with soap when skin is wet or contaminated. Provide emergency showers and eyewash.

Respirator Selection: NIOSH/OSHA: *Up to 10 mg/m³:* Qm (APF = 25) (any quarter-mask respirator). *Up to 20 mg/m³:* 95XQ (APF = 10) [any particulate respirator equipped with an N95, R95, or P95 filter (including N95, R95, and P95 filtering face-pieces) except quarter-mask respirators. The following filters may also be used: N99, R99, P99, N100, R100, P100]; or Sa* (APF = 10) (any supplied-air respirator). *Up to 50 mg/m³:* Sa:Cf* (APF = 25) (any supplied-air respirator operated in a continuous-flow mode); or PaprHie (APF = 25)* (any powered, air-purifying respirator with a high-efficiency particulate filter). *Up to 100 mg/m³:* 100F (APF = 50) (any air-purifying, full-face-piece respirator with an N100, R100, or P100 filter; or SCBAF (APF = 50) (any self-contained breathing apparatus with a full face-piece); or SaF (APF = 50) (any supplied-air respirator with a full face-piece). *Emergency or planned entry into unknown concentrations or IDLH conditions:* SCBAF: Pd, Pp (APF = 10,000) (any self-contained breathing apparatus that has a full face-piece and is operated in a pressure-demand or other positive-pressure mode); or SAF; PD,PP: ASCBA (any supplied-air respirator that has a full face-piece and is operated in a pressure-demand or other positive-pressure mode in combination with an auxiliary self-contained positive-pressure breathing apparatus). *Escape:* 100F (APF = 50) (any air-purifying, full-face-piece respirator with an N100, R100, or P100 filter); or SCBAE (any appropriate escape-type, self-contained breathing apparatus).

Storage: Color Code—White: Corrosive or Contact Hazard; Store separately in a corrosion-resistant location. Prior to working with this chemical you should be trained on its proper handling and storage. Tin tetrachloride must be stored to avoid contact with water, alcohols, amines, chlorine, turpentine, ethylene oxide; alkyl nitrates; potassium and sodium, since violent reactions occur. Store in tightly closed containers in a cool, well-ventilated area away from flammable and combustible materials. Keep tin tetrachloride dry and out of direct sunlight. If moisture enters containers, pressure may cause the containers to burst. Tin tetrachloride can attack some plastics, rubbers, and coatings.

Shipping: This compound requires a shipping label of "CORROSIVE." It falls in Hazard Class 8 and Packing Group II.

Spill Handling: Evacuate and restrict persons not wearing protective equipment from area of spill or leak until cleanup is complete. Remove all ignition sources. Ventilate area of spill or leak. Absorb liquids in vermiculite, dry sand; earth, peat, carbon, or a similar material and deposit in sealed containers. Keep this chemical out of a confined space, such as a sewer, because of the possibility of an explosion, unless the sewer is designed to prevent the buildup of explosive concentrations. It may be necessary to contain and dispose of this chemical as a hazardous waste. If material or contaminated runoff enters waterways, notify downstream users of potentially contaminated waters. Contact your local or federal environmental protection agency for specific recommendations. If employees are required to clean up spills, they must be properly trained and equipped. OSHA 1910.120(q) may be applicable.

Fire Extinguishing: Tin tetrachloride does not burn, but it does react violently with water. *Do not use water.* Use dry chemical or carbon dioxide extinguishers to extinguish surrounding fire. Poisonous gases, including hydrogen chloride, chlorine, and tin oxides, are produced in fire. Vapors are heavier than air and will collect in low areas. Vapors may travel long distances to ignition sources and flashback. Vapors in confined areas may explode when exposed to fire. Containers may explode in fire. Storage containers and parts of containers may rocket great distances, in many directions. If material or contaminated runoff enters waterways, notify downstream users of potentially contaminated waters. Notify local health and fire officials and pollution control agencies. From a secure, explosion-proof location, use water spray to cool exposed containers, but do not get water inside containers or on spilled tin tetrachloride as poisonous gases will be formed. If cooling streams are ineffective (venting sound increases in volume and pitch, tank discolors, or shows any signs of deforming), withdraw immediately to a secure position. If employees are expected to fight fires, they must be trained and equipped in OSHA 1910.156. The only respirators recommended for firefighting are self-contained breathing apparatuses that have full face-pieces and are operated in a pressure-demand or other positive-pressure mode.

Disposal Method Suggested: SnCl₄: Pour onto sodium bicarbonate; spray with ammonium hydroxide while adding crushed ice; when reaction subsides, flush down drain.

References

New Jersey Department of Health and Senior Services. (October 1998). *Hazardous Substances Fact Sheet: Tin Tetrachloride.* Trenton, NJ

Titanium & compounds   T:0560

Molecular Formula: Ti

Synonyms: C.P. Titanium; IMI 115; NCI-C04251; Ontimet 30; Oremet; T 40; Titanate; Titanium 50A; Titanium alloy; Titanium, elemental; VT 1

Other titanium compounds: titanium carbide (CTi) 12070-08-5; titanium disulfide (S₂Ti) (12039-13-3); titanium hydride (H₂Ti) (7704-98-5); titanium monoxide (Oti) 12137-20-1; titanium sulfate (O₅STi) (13825-74-6); titanium trichloride (Cl₂Ti) (7705-07-9)

CAS Registry Number: 7440-32-6 (metal)

RTECS® Number: XR1700000 (elemental)

UN/NA & ERG Number: Metal powder, in bulk, may be pyrophoric. UN2546 (Titanium powder, dry)/135; UN2878 (Titanium sponge powders)/170; UN1352 (Titanium powder, wetted with not less than 25% water)170

EC Number: 231-142-3

Regulatory Authority and Advisory Bodies

European/International Regulations: Not listed in Annex 1. WGK (German Aquatic Hazard Class): Nonwater polluting agent.

Description: Titanium is a silvery metal or dry, dark-gray amorphous, lustrous powder. Molecular weight = 47.90; Boiling point = 3260°C; Freezing/Melting point = 1675°C; Autoignition temperature = 1200°C (solid); 480°C (powder in air). Hazard Identification (based on NFPA-704 M Rating System) (powder): Health 1, Flammability 3, Reactivity 1; (solid metal) Health 1, Flammability 0, Reactivity 0. It is brittle when cold and malleable when hot. The dry powder is easily ignited, and burns with an intense flame. Ignition temperature of dust cloud = 460°C; Minimum Explosive concentration = 0.45 oz/ft³.[USBM] Relative explosion hazard of dust: Severe. Insoluble in water. The most important minerals containing titanium are ilmenite, rutile, perovskite, and titanite or sphene. Insoluble in water.

Potential Exposure: Titanium metal, because of its low weight, high strength, and heat resistance, is used in the aerospace and aircraft industry as tubing, fittings, fire walls; cowlings, skin sections; jet compressors; and it is also used in surgical appliances. It is also used as control-wire casings in nuclear reactors, as a protective coating for mixers in the pulp-paper industry, and in other situations in which

protection against chlorides or acids is required; in vacuum lamp bulbs and X-ray tubes; as an addition to carbon and tungsten in electrodes and lamp filaments; and to the powder in the pyrotechnics industry. It forms alloys with iron, aluminum, tin, and vanadium, of which ferrotitanium is especially important in the steel industry. Other titanium compounds are utilized in smoke screens, as mordants in dyeing; in the manufacture of cemented metal carbides; as thermal insulators; and in heat-resistant surface coatings in paints and plastics.

Incompatibilities: Dust may ignite spontaneously in air. Violent reactions occur on contact with water, steam, halocarbons, halogens, and aluminum. The dry powder is a strong reducing agent; violent reaction with strong oxidizers.

Permissible Exposure Limits in Air

Protective Action Criteria (PAC)

Metal

TEEL-0: 0.6 mg/m³

PAC-1: 2 mg/m³

PAC-2: 12.5 mg/m³

PAC-3: 60 mg/m³

Titanium compounds including titanium carbide

TEEL-0: 10 mg/m³

PAC-1: 30 mg/m³

PAC-2: 50 mg/m³

PAC-3: 250 mg/m³

Determination in Air: Filter; Acid; Hydride generation atomic absorption spectrometry; NIOSH Analytical Method (IV) #7900. See also #7300, Elements.

Permissible Concentration in Water: No criteria set, but EPA[32] has suggested a permissible ambient goal for titanium compounds (as Ti) of 83 μg/L, based on health effects.

Routes of Entry: Inhalation of dust or fume, eyes.

Harmful Effects and Symptoms

Short Term Exposure: Titanium can affect you when breathed in. Contact may irritate the eyes. Breathing titanium may irritate the throat and air passages with cough and phlegm.

Long Term Exposure: Repeated exposure may cause chronic bronchitis and possibly emphysema. There is limited evidence that titanium may damage the developing fetus.

Points of Attack: Lungs.

Medical Surveillance: Preemployment and periodic physical examinations should give special attention to lung disease, especially if irritant compounds are involved. Chest X-ray should be included in both examinations and pulmonary function evaluated periodically. Smoking history should be taken. Careful attention should be given to the eyes and the skin.

First Aid: If this chemical gets into the eyes, remove any contact lenses at once and irrigate immediately for at least 15 min, occasionally lifting upper and lower lids. Seek medical attention immediately. If this chemical contacts the skin, remove contaminated clothing and wash immediately with soap and water. Seek medical attention immediately. If this chemical has been inhaled, remove from exposure, begin rescue breathing (using universal precautions, including resuscitation mask) if breathing has stopped and CPR if heart action has stopped. Transfer promptly to a medical facility. When this chemical has been swallowed, get medical attention. Give large quantities of water and induce vomiting. Do not make an unconscious person vomit.

Personal Protective Methods: Wear protective gloves and clothing to prevent any reasonable probability of skin contact. Safety equipment suppliers/manufacturers can provide recommendations on the most protective glove/clothing material for your operation. All protective clothing (suits, gloves, footwear, headgear) should be clean, available each day, and put on before work. Contact lenses should not be worn when working with this chemical. Wear dust-proof chemical goggles and face shield unless full-face-piece respiratory protection is worn. Employees should wash immediately with soap when skin is wet or contaminated. Provide emergency showers and eyewash.

Respirator Selection: Where there is potential for exposures to titanium, use a NIOSH/MSHA- or European Standard EN149-approved respirator equipped with particulate (Dust/fume/mist) filters. More protection is provided by a full-face-piece respirator than by a half-mask respirator, and even greater protection is provided by a powered, air-purifying respirator. Particulate filters must be checked every day before work for physical damage, such as rips or tears, and replaced as needed. *Where there is potential for high exposures*, use a NIOSH/MSHA- or European Standard EN149-approved supplied-air respirator with a full face-piece operated in the positive-pressure mode, or with a full face-piece, hood, or helmet in the continuous-flow mode; or use a NIOSH/MSHA- or European Standard EN149-approved self-contained breathing apparatus with a full face-piece operated in pressure-demand or other positive-pressure mode.

Storage: Titanium powder is a severe explosion risk. (1) Color Code—Red Stripe: Flammability Hazard: Store separately from all other flammable materials. (2) Color Code—Yellow Stripe (*strong reducing agent*): Reactivity Hazard; Store separately in an area isolated from flammables, combustibles, or other yellow-coded materials. Prior to working with this chemical you should be trained on its proper handling and storage. Titanium must be stored to avoid contact with water, steam, halocarbons, halogens, and aluminum, since violent reactions occur. Store in tightly closed containers in a cool, well-ventilated area. Sources of ignition, such as smoking and open flames, are prohibited where titanium is used, handled, or stored in a manner that could create a potential fire or explosion hazard. Use only non-sparking tools and equipment, especially when opening and closing containers of titanium. Wherever titanium is used, handled, manufactured, or stored, use explosion-proof electrical equipment and fittings.

Shipping: Titanium powders require a shipping label of "SPONTANEOUSLY COMBUSTIBLE." They fall in Hazard Class 4.2 and Packing Group III. Titanium powder, wetted [with not $<25\%$ water (a visible excess of water must be present) (a) mechanically produced, particle size $<53\,\mu m$; (b) chemically produced, particle size <840 microns require a shipping label of "FLAMMABLE SOLID." It falls in Hazard Class 4.1 and Packing Group II.

Spill Handling: Evacuate persons not wearing protective equipment from area of spill or leak until cleanup is complete. Remove all ignition sources. Use HEPA vacuum or wet method to reduce dust during cleanup. Do not dry sweep. Collect powdered material in the most convenient and safe manner and deposit in sealed containers. Ventilate area after cleanup is complete. It may be necessary to contain and dispose of this chemical as a hazardous waste. If material or contaminated runoff enters waterways, notify downstream users of potentially contaminated waters. Contact your local or federal environmental protection agency for specific recommendations. If employees are required to clean up spills, they must be properly trained and equipped. OSHA 1910.120(q) may be applicable.

Fire Extinguishing: Titanium is a flammable solid; dust and powders are an explosion hazard. *Do not use water* on burning titanium; this can cause explosion. Ordinary extinguishers are often ineffective. Use special extinguishers designed for metal fires. Poisonous gases are produced in fire. If material or contaminated runoff enters waterways, notify downstream users of potentially contaminated waters. Notify local health and fire officials and pollution control agencies. From a secure, explosion-proof location, use water spray to cool exposed containers. If cooling streams are ineffective (venting sound increases in volume and pitch, tank discolors, or shows any signs of deforming), withdraw immediately to a secure position. If employees are expected to fight fires, they must be trained and equipped in OSHA 1910.156. The only respirators recommended for firefighting are self-contained breathing apparatuses that have full face-pieces and are operated in a pressure-demand or other positive-pressure mode.

References

US Environmental Protection Agency. (May 1977). *Toxicology of Metals, Vol. II: Titanium, Report EPA-600/1-77-022.* Research Triangle Park, NC, pp. 427—441

Sax, N. I. (Ed.). *Dangerous Properties of Industrial Materials Report,* 1, No. 3, 83 (1981) and 4, No. 3, 27—29 (1984)

New Jersey Department of Health and Senior Services. (December 2000). *Hazardous Substances Fact Sheet: Titanium.* Trenton, NJ

Titanium dioxide   T:0570

Molecular Formula: O_2Ti
Common Formula: TiO_2

Synonyms: A-FIL cream; Anatase; Atlas white titanium; Austiox; Bayeritian; Bayertitan; Brookite; Calcotone white T; C.I. 77891; C.I. Pigment white 6; Cosmetic white C47-5175; C-Weiss 7 (German); Flamenco; Hombitan; Horse head A-410; KH 360; Kronos titanium dioxide; Levanox white RKB; NCI-C04240; Rayox; Runa RH20; Rutile; Tiofine; Tioxide; Titanium oxide; Tronox Unitane 0-110; Zopaque 1700 white

CAS Registry Number: 13463-67-7; *(alt.)* 1309-63-3

RTECS® Number: XR2275000

EC Number: 236-675-5

Regulatory Authority and Advisory Bodies

Carcinogenicity: IARC: Human Inadequate Evidence; Animal Sufficient Evidence, *possibly carcinogenic to humans*, Group 2B; NCI: Carcinogenesis Bioassay (feed); no evidence: mouse, rat; NIOSH: Potential occupational carcinogen.

US EPA Gene-Tox Program, Negative: Carcinogenicity—mouse/rat; Cell transform.—SA7/SHE.

FDA—over-the-counter drug.

Air Pollutant Standard Set. See below, "Permissible Exposure Limits in Air" section.

European/International Regulations: Not listed in Annex 1.

WGK (German Aquatic Hazard Class): Nonwater polluting agent.

Description: Titanium dioxide is an odorless white powder. Molecular weight = 79.90; Specific gravity (H_2O:1) = 4.26 at 25°C; Boiling point = 2500–3000°C; Freezing/Melting point = 1830–1850°C (decomposition). Hazard Identification (based on NFPA-704 M Rating System): Health 1, Flammability 0, Reactivity 0. Insoluble in water.

Potential Exposure: Compound Description: Tumorigen, Mutagen. Primary Irritant. Titanium dioxide is a white pigment used as a pigment in paint; in the rubber, plastics, ceramics, paint, and varnish industries, in dermatological preparations; and is used as a starting material for other titanium compounds; as a gem; in curing concrete; and in coatings for welding rods. It is also used in paper and cardboard manufacture.

Incompatibilities: Strong acids.

Permissible Exposure Limits in Air

OSHA PEL: 15 mg/m³, total dust TWA.

NIOSH REL: Potential occupational carcinogen; Limit exposure to lowest feasible concentration. See *NIOSH Pocket Guide*, Appendix A.

ACGIH TLV®[1]: 10 mg/m³ TWA, not classifiable as a human carcinogen.

NIOSH IDLH: 5000 mg/m³.

Protective Action Criteria (PAC)

TEEL-0: 15 mg/m³

PAC-1: 30 mg/m³

PAC-2: 50 mg/m³

PAC-3: 500 mg/m³

DFG MAK: 1.5 mg/m³, respirable fraction (previously "fine dust"); Pregnancy Risk Group C.

Arab Republic of Egypt: TWA 15 mg/m³, 1993; Australia: TWA 10 mg/m³, 1993; Austria: MAK 6 mg/m³, 1999; Belgium: TWA 10 mg/m³, 1993; Denmark: TWA 6mg[Ti]/m³, 1999; France: VME 10 mg/m³, 1999; Norway: TWA 5 mg/m³, 1999; the Netherlands: MAC-TGG 10 mg/m³, 2003; Poland: MAC (TWA) 10mg[Ti]/m³; MAC (STEL) 30mg[Ti]/m³, 1999; Sweden: NGV 5 mg/m³, 1999; Turkey: TWA 15 mg/m³, 1993; United Kingdom: TWA 10 mg/m³ (total inhalable dust); TWA 4 mg/m³ (respirable dust), 2000; Argentina, Bulgaria, Columbia, Jordan, South Korea, New Zealand, Singapore, Vietnam: ACGIH TLV®: not classifiable as a human carcinogen. Russia have set limits for ambient air in residential areas at 0.5 mg/m³ on a momentary basis. Several states have set guidelines or standards for titanium dioxide in ambient air[60] ranging from 0.13–0.79 μg/m³ (Montana) to 17.86 μg/m³ (Kansas) to 80.0 μg/m³ (Virginia) to 300.0 μg/m³ (Connecticut).

Determination in Air: Use NIOSH II(3), Method #S385; #7300, Elements by ICP.

Permissible Concentration in Water: No criteria set, but EPA[32] has suggested a permissible ambient goal for titanium compounds (as Ti) of 83 μg/L, based on health effects.

Routes of Entry: Inhalation of dust.

Harmful Effects and Symptoms

Short Term Exposure: Inhalation can cause irritation of the eyes and respiratory tract, causing cough and phlegm. Irritates the skin.

Long Term Exposure: High exposures may cause lung irritation; bronchitis may develop. Continued exposure may result in emphysema, lung scarring; lung fibrosis; and tumors. A potential occupational carcinogen.

Points of Attack: Respiratory system. Cancer site in animals: lung tumors.

Medical Surveillance: Before beginning employment and at regular times after that, for those with frequent or potentially high exposures, the following are recommended: lung function tests. If symptoms develop or overexposure is suspected, the following may be useful: chest X-ray should be considered.

First Aid: If this chemical gets into the eyes, remove any contact lenses at once and irrigate immediately for at least 15 min, occasionally lifting upper and lower lids. Seek medical attention immediately. If this chemical contacts the skin, remove contaminated clothing and wash immediately with soap and water. Seek medical attention immediately. If this chemical has been inhaled, remove from exposure, begin rescue breathing (using universal precautions, including resuscitation mask) if breathing has stopped and CPR if heart action has stopped. Transfer promptly to a medical facility. When this chemical has been swallowed, get medical attention. Give large quantities of water and induce vomiting. Do not make an unconscious person vomit.

Personal Protective Methods: Wear protective gloves and clothing to prevent any reasonable probability of skin contact. Safety equipment suppliers/manufacturers can provide recommendations on the most protective glove/clothing material for your operation. All protective clothing (suits,

gloves, footwear, headgear) should be clean, available each day, and put on before work. Contact lenses should not be worn when working with this chemical. Wear dust-proof chemical goggles and face shield unless full-face-piece respiratory protection is worn. Employees should wash immediately with soap when skin is wet or contaminated. Provide emergency showers and eyewash.

Respirator Selection: At concentrations above the NIOSH REL, or where there is no REL, at any detectable concentration: SCBAF: Pd,Pp (APF = 10,000) (any NIOSH/MSHA- or European Standard EN 149-approved self-contained breathing apparatus that has a full face-piece and is operated in a pressure-demand or other positive-pressure mode); or SaF: Pd,Pp: ASCBA (APF = 10,000) (any supplied-air respirator that has a full face-piece and is operated in a pressure-demand or other positive-pressure mode in combination with an auxiliary, self-contained breathing apparatus operated in a pressure-demand or other positive-pressure mode). *Escape:* 100F (APF = 50) (any air-purifying, full-face-piece respirator with an N100, R100, or P100 filter); or SCBAE (any appropriate escape-type, self-contained breathing apparatus).

Storage: Color Code—Green: General storage may be used. Prior to working with this chemical you should be trained on its proper handling and storage. A regulated, marked area should be established where this chemical is handled, used, or stored in compliance with OSHA Standard 1910.1045. Store in tightly closed containers in a cool, well-ventilated area away from strong acids and other metals. Where possible, automatically transfer material from other storage containers to process containers.

Spill Handling: Evacuate persons not wearing protective equipment from area of spill or leak until cleanup is complete. Remove all ignition sources. Collect powdered material in the most convenient and safe manner and deposit in sealed containers. Ventilate area after cleanup is complete. It may be necessary to contain and dispose of this chemical as a hazardous waste. If material or contaminated runoff enters waterways, notify downstream users of potentially contaminated waters. Contact your local or federal environmental protection agency for specific recommendations. If employees are required to clean up spills, they must be properly trained and equipped. OSHA 1910.120(q) may be applicable.

Fire Extinguishing: This chemical is a noncombustible solid. Use any extinguishing agent suitable for surrounding fires. Poisonous gases are produced in fire. If material or contaminated runoff enters waterways, notify downstream users of potentially contaminated waters. Notify local health and fire officials and pollution control agencies. From a secure, explosion-proof location, use water spray to cool exposed containers. If cooling streams are ineffective (venting sound increases in volume and pitch, tank discolors, or shows any signs of deforming), withdraw immediately to a secure position. If employees are expected to fight fires, they must be trained and equipped in OSHA 1910.156. The only respirators recommended for firefighting are self-contained breathing apparatuses that have full face-pieces and are operated in a pressure-demand or other positive-pressure mode.

Disposal Method Suggested: Landfill.

References

Sax, N. I. (Ed.). *Dangerous Properties of Industrial Materials Report*, 1, No. 3, 84 (1981) and 3, No. 1, 85—89 (1983)

New York State Department of Health. (March 1986). *Chemical Fact Sheet: Titanium Dioxide.* Albany, NY: Bureau of Toxic Substance Assessment (Version 2)

New Jersey Department of Health and Senior Services. (May 2006). *Hazardous Substances Fact Sheet: Titanium Dioxide.* Trenton, NJ

Titanium tetrachloride   T:0580

Molecular Formula: Cl₄Ti
Common Formula: TiCl₄
Synonyms: FM (military designation); Tetrochlorure de titane (French); Tetrachlorotitanium; Titane (tetrachlorure de) (French); Tetracloruro de titanio (Spanish); Titanium chloride (TICL4) (T-4)-; Titantetrachlorid (German); Titanium(IV) chloride
CAS Registry Number: 7550-45-0
RTECS® Number: XR1925000
UN/NA & ERG Number: UN1838/137
EC Number: 231-441-9 [*Annex I Index No.:* 022-001-00-5]

Regulatory Authority and Advisory Bodies
Department of Homeland Security Screening Threshold Quantity (pounds): *Release hazard* 2500 (1.00% concentration); *Theft hazard* 45 (≥13.33% concentration); Sabotage/Contamination Hazard: A placarded amount (commercial grade).
Air Pollutant Standard Set. See below, "Permissible Exposure Limits in Air" section.
Clean Air Act: Hazardous Air Pollutants (Title I, Part A, Section 112); Accidental Release Prevention/Flammable Substances (Section 112[r], Table 3), TQ = 2500 lb (1135 kg).
Superfund/EPCRA 40CFR355, Extremely Hazardous Substances: TPQ = 100 lb (45.4 kg).
Reportable Quantity (RQ): 1000 lb (454 kg).
EPCRA Section 313 Form R *de minimis* concentration reporting level: 1.0%.
US DOT 49CFR172.101, Inhalation Hazardous Chemical.
Canada, WHMIS, Ingredients Disclosure List Concentration 1.0%.
European/International Regulations: Hazard Symbol: C; Risk phrases: R14; R34; Safety phrases: S1/2; S7/8; S26; S36/37/39; S45 (see Appendix 4).
WGK (German Aquatic Hazard Class): 1—Low hazard to waters.

Description: Titanium tetrachloride is a noncombustible, colorless to light yellow liquid with a penetrating acrid odor. Molecular weight = 189.70; Boiling point = 136°C; Freezing/Melting point = −24°C. Reacts with water. Hazard Identification (based on NFPA-704 M Rating System): Health 3, Flammability 0, Reactivity 1W. Decomposes in water.

Potential Exposure: Used in the manufacture of titanium salts; mordant dye; titanium pigments; and used as a chemical intermediate for titanium metal; titanium dioxide; as an agent in smoke screens; polymerization catalyst; and iridescent agent in glass and pearl manufacturing.

Incompatibilities: Violent reaction with water or steam, releasing heat and hydrogen chloride fumes. Contact with moist air releases hydrogen chloride. Attacks many metals in presence of moisture.

Permissible Exposure Limits in Air
AIHA WEEL: 0.5 mg/m³ TWA.
Protective Action Criteria (PAC)*
TEEL-0: 0.645 ppm
PAC-1: 0.645 ppm
PAC-2: **1.0** ppm
PAC-3: **5.7** ppm
*AEGLs (Acute Emergency Guideline Levels) & ERPGs (Emergency Response Planning Guideline) are in **bold face**.

Routes of Entry: Inhalation, ingestion, eye, and/or skin contact.

Harmful Effects and Symptoms

Short Term Exposure: This compound is a highly corrosive, acute irritant to the skin, eyes, mucous membranes and the respiratory tract. It is capable of causing death or permanent injury due to exposures encountered in normal use. Even short contact may lead to eye inflammation which may result in corneal opacities. Inhalation symptoms include congestion and irritation of upper respiratory tract, coughing, burning of the throat; headache and weakness. Prolonged exposure to low concentrations may cause cough and pneumonia. High exposures can cause pulmonary edema, a medical emergency that can be delayed for several hours. This can cause death. Ingestion causes mouth, throat, and GI tract irritation, nausea, vomiting, cramps, and diarrhea.

Long Term Exposure: May cause respiratory problems; lung damage; bronchitis may develop.

Points of Attack: Respiratory system, eyes.

Medical Surveillance: Lung function tests. Eye examination. Consider chest X-ray following acute overexposure.

First Aid: If this chemical gets into the eyes, remove any contact lenses at once and irrigate immediately for at least 15 min, occasionally lifting upper and lower lids. Seek medical attention immediately. If this chemical contacts the skin, remove contaminated clothing and wash immediately with soap and water. Seek medical attention immediately. If this chemical has been inhaled, remove from exposure, begin rescue breathing (using universal precautions, including resuscitation mask) if breathing has stopped and CPR if heart action has stopped. Transfer promptly to a medical facility. When this chemical has been swallowed, get medical attention. If victim is *conscious*, administer water or milk. Do not induce vomiting. Medical observation is recommended for 24−48 h after breathing overexposure, as pulmonary edema may be delayed. As first aid for pulmonary edema, a doctor or authorized paramedic may consider administering a corticosteroid spray.

Personal Protective Methods: Wear protective gloves and clothing to prevent any reasonable probability of skin contact. Safety equipment suppliers/manufacturers can provide recommendations on the most protective glove/clothing material for your operation. All protective clothing (suits, gloves, footwear, headgear) should be clean, available each day, and put on before work. Contact lenses should not be worn when working with this chemical. Wear splash-proof chemical goggles and face shield unless full-face-piece respiratory protection is worn. Employees should wash immediately with soap when skin is wet or contaminated. Provide emergency showers and eyewash.

Respirator Selection: Follow regulations in OSHA 29CFR1910.134 or European Standard EN149. Use a NIOSH/MSHA- or European Standard EN149-approved respirator; or use an approved supplied-air respirator with a full face-piece operated in the positive-pressure mode, or with a full face-piece, hood, or helmet in the continuous-flow mode; or use a NIOSH/MSHA- or European Standard EN149-approved self-contained breathing apparatus with a full face-piece operated in pressure-demand or other positive-pressure mode.

Storage: Color Code—White: Corrosive or Contact Hazard; Store separately in a corrosion-resistant location. Prior to working with this chemical you should be trained on its proper handling and storage. Store away from possible contact with all forms of moisture. Store in tightly closed containers in a cool, well-ventilated area. Metal containers involving the transfer of this chemical should be grounded and bonded. Where possible, automatically pump liquid from drums or other storage containers to process containers.

Shipping: This compound requires a shipping label of "CORROSIVE, POISONOUS/TOXIC MATERIALS." It falls in Hazard Class 8 and Packing Group II. A plus sign (+) symbol indicates that the designated proper shipping name and hazard class of the material must always be shown whether or not the material or its mixtures or solutions meet the definitions of the class.

Spill Handling: Evacuate and restrict persons not wearing protective equipment from area of spill or leak until cleanup is complete. Remove all ignition sources. Ventilate area of spill or leak. Absorb liquids in vermiculite, dry sand; earth, peat, carbon, or a similar material and deposit in sealed containers. It may be necessary to contain and dispose of this chemical as a hazardous waste. If material or contaminated runoff enters waterways, notify downstream users of

potentially contaminated waters. Contact your local or federal environmental protection agency for specific recommendations. If employees are required to clean up spills, they must be properly trained and equipped. OSHA 1910.120(q) may be applicable.

Initial isolation and protective action distances

Distances shown are likely to be affected during the first 30 min after materials are spilled and could increase with time. If more than one tank car, cargo tank, portable tank, or large cylinder involved in the incident is leaking, the protective action distance may need to be increased. You may need to seek emergency information from CHEMTREC at (800) 424-9300 or seek professional environmental engineering assistance from the US EPA Environmental Response Team at (908) 548-8730 (24-h response line).

FM, when used as a weapon—not listed in DOT tables

When spilled on land

Small spills (From a small package or a small leak from a large package)

First: Isolate in all directions (feet/meters)   100/30.
Then: Protect persons downwind (miles/kilometers).
Day   0.1/0.2
Night   0.1/0.2

Large spills (From a large package or from many small packages)

When spilled in water

First: Isolate in all directions (feet/meters)   200/60.
Then: Protect persons downwind (miles/kilometers).
Day   0.4/0.6
Night   1.2/1.9

Small spills (From a small package or a small leak from a large package)

First: Isolate in all directions (feet/meters)   100/30.
Then: Protect persons downwind (miles/kilometers).
Day   0.1/0.2
Night   0.1/0.2

Large spills (From a large package or from many small packages)

First: Isolate in all directions (feet/meters)   200/60.
Then: Protect persons downwind (miles/kilometers).
Day   0.4/0.6
Night   1.2/1.9

Fire Extinguishing: Not flammable. For small fires, use dry chemical or carbon dioxide. For large fires, flood fire area with water from a distance. Do not get solid streams of water on spilled material. Move container from fire area if this can be done without risk. Cool containers exposed to flames with water until well after fire is out. Poisonous gases are produced in fire, including hydrogen chloride. If material or contaminated runoff enters waterways, notify downstream users of potentially contaminated waters. Notify local health and fire officials and pollution control agencies. From a secure, explosion-proof location, use water spray to cool exposed containers. If cooling streams are ineffective (venting sound increases in volume and pitch, tank discolors, or shows any signs of deforming),

withdraw immediately to a secure position. If employees are expected to fight fires, they must be trained and equipped in OSHA 1910.156. The only respirators recommended for firefighting are self-contained breathing apparatuses that have full face-pieces and are operated in a pressure-demand or other positive-pressure mode.

References

US Environmental Protection Agency. (November 30, 1987). *Chemical Hazard Information Profile: Titanium Tetrachloride.* Washington, DC: Chemical Emergency Preparedness Program

o-Tolidine   T:0590

Molecular Formula: $C_{14}H_{16}N_2$

Common Formula: $H_2NC_6H_3(CH_3)\text{-}C_6H_3(CH_3)NH_2$

Synonyms: Benzidine, 3,3'-dimethyl-; Bianisidine; (1,1'-Biphenyl)-4,4'-diamine, 3,3'-dimethyl-; 4,4'-Bi-*o*-toluidine; C.I. 37230; C.I. azoic diazo component 113; (4,4'-Diamine)-3,3'-dimethyl(1,1'-biphenyl); 4,4'-Diamino-3,3'-dimethylbiphenyl; Diaminoditolyl; 3,3'-Dimethylbenzidin (German); 3,3'-Dimethylbenzidine; 3,3'-Dimethyl-(1,1'-biphenyl)-4,4'-diamine; 3,3'-Dimethyl-4,4'-biphenyldiamine; 3,3'-Dimethyl-4,4'-diaminobiphenyl; 3,3'-Dimethyl-4,4'-diphenyldiamine; 3,3'-Dimethyldiphenyl-4,4'-diamine; 4,4'-Di-*o*-toluidine; Fast dark blue base R; 3,3'-Methylphenyl-4,4'-diamine; *o*-Tolidin (German); 3-Tolidin (German); *o*-Tolidina (Spanish); 2-Tolidina (Spanish); *o,o'*-Tolidine; 2-Tolidine; 3,3'-Tolidine; Tolidine

CAS Registry Number: 119-93-7

Dyes based on *o*-tolidine for reference: 612-82-8 (3,3'-dimethylbenzidine dihydrochloride); 41766-75-0 (3,3'-Dimethylbenzidine dihydrofluoride); 992-59-6 [Disodium *o*-tolidinediazobis(1-naphthylamine-4-sulfonate)]

RTECS® Number: DD1225000

UN/NA & ERG Number: UN2811 (toxic solid, organic, n.o.s.)/154

EC Number: 210-322-5 [*Annex I Index No.:* 612-081-00-5] (*4,4'-bi-o-toluidine dihydrochloride*) [*Annex I Index No.:* 612-041-00-7]

Regulatory Authority and Advisory Bodies

Carcinogenicity: IARC: Human No Adequate Data; Animal Sufficient Evidence, *possibly carcinogenic to humans*, Group 2B, 1987; NTP: 11th Report on Carcinogens, 2004: Reasonably anticipated to be a human carcinogen; NIOSH: Potential occupational carcinogen.

US EPA Gene-Tox Program, Positive: Carcinogenicity—mouse/rat; Positive: Cell transform.—RLV F344 rat embryo; Positive: Mammalian micronucleus; Histidine reversion—Ames test; Inconclusive: *In vitro* UDS—human fibroblast.

US EPA Hazardous Waste Number (RCRA No.): U095.
RCRA, 40CFR261, Appendix 8 Hazardous Constituents.

Reportable Quantity (RQ): 10 lb (4.54 kg).

EPCRA Section 313 Form R *de minimis* concentration reporting level: 1.0%.

California Proposition 65 Chemical: Cancer 1/1/88; 4/1/92 (dihydrochloride).

Canada, WHMIS, Ingredients Disclosure List Concentration 0.1%.

European/International Regulations: Hazard Symbol: T, N; Risk phrases: R45; R22; R51/53; Safety phrases: S53; S45; S61 (see Appendix 4).

WGK (German Aquatic Hazard Class): No value assigned.

Description: *o*-Tolidine is a white to reddish crystal or powder. Darkens on exposure to air. Often used in paste or wet cake form. Used as a basis for many dyes. Molecular weight = 212.32; Boiling point = 300°C (also listed at 200°C); Freezing/Melting point = 129°C; Flash point = 244°C; Autoignition temperature = 526°C. Hazard Identification (based on NFPA-704 M Rating System): Health 3, Flammability 2, Reactivity 0. Slightly soluble in water; solubility = 0.1% at 25°C.

Potential Exposure: Compound Description: Tumorigen, Mutagen. Over 75% of *o*-tolidine is used as a dye and as an intermediate in the production of dyestuffs and pigments. Approximately 20% of *o*-tolidine is used in the production of polyurethane-based high-strength elastomers, coatings, and rigid plastics. *o*-Tolidine has also been used in small quantities in chlorine test kits by water companies and swimming pool owners. Used as a laboratory agent to detect blood.

Incompatibilities: Strong oxidizers.

Permissible Exposure Limits in Air

OSHA PEL: Cancer suspect agent. Exposures of workers to this chemical is to be controlled through the required use of engineering controls, work practices; and personal protective equipment, including respirators. See 29 CFR 1910.1003-1910.1016 for specific details of these requirements.

NIOSH REL: Carcinogen: 0.02 mg/m^3 [60 min.] Ceiling Concentration; [skin]; Limit exposure to lowest feasible concentration. See *NIOSH Pocket Guide*, Appendices A and C.

Note: OSHA and NIOSH concluded that benzidine and benzidine-based dyes were potential occupational carcinogens and recommended that worker exposure be reduced to the lowest feasible level. OSHA and NIOSH further concluded that *o*-tolidine and *o*-dianisidine [119-90-4] (and dyes based on these chemicals) may present a cancer risk to workers and should be handled with caution and exposure minimized.

ACGIH TLV®[1]: [skin]; confirmed animal carcinogen with unknown relevance to humans.

Protective Action Criteria (PAC)

TEEL-0: 0.1 mg/m^3

PAC-1: 0.3 mg/m^3

PAC-2: 2 mg/m^3

PAC-3: 100 mg/m^3

No TEEL available for dihydrochloride.

DFG MAK: Carcinogen Category 2

Australia [skin], carcinogen, 1993; Austria: carcinogen, 1999; Finland: carcinogen, 1999; France: carcinogen, 1993; Switzerland: carcinogen, 1999; United Kingdom: carcinogen, 2000; Argentina, Bulgaria, Columbia, Jordan, South Korea, New Zealand, Singapore, Vietnam: ACGIH TLV®: confirmed animal carcinogen with unknown relevance to humans.

Guidelines or standards which have been set for *o*-tolidine in ambient air[60] range from zero (North Dakota) to 20.0 µg/m^3 (Virginia).

Determination in Air: OSHA Analytical Method #ID-71; NIOSH Analytical Method (IV) #5013, Dyes.

Determination in Water: Octanol−water coefficient: Log K_{ow} = 2.34.

Routes of Entry: Inhalation, skin absorption; ingestion; skin and/or eye contact.

Harmful Effects and Symptoms

Short Term Exposure: *o*-Tolidine can affect you when breathed in and can rapidly enter the body through the skin. High exposure can irritate the nose and throat.

Long Term Exposure: *o*-Tolidine may affect the kidneys and bladder. A potential occupational carcinogen.

Points of Attack: Eyes, respiratory system; liver, kidneys. Cancer site in animals: liver, bladder, and mammary gland tumors.

Medical Surveillance: Before employment and every 6 months thereafter, the following are recommended to detect bladder cancer at an early stage: Kidney function tests. Urine cytology test (a test for abnormal cells in urine).

First Aid: If this chemical gets into the eyes, remove any contact lenses at once and irrigate immediately for at least 15 min, occasionally lifting upper and lower lids. Seek medical attention immediately. If this chemical contacts the skin, remove contaminated clothing and wash immediately with soap and water. Seek medical attention immediately. If this chemical has been inhaled, remove from exposure, begin rescue breathing (using universal precautions, including resuscitation mask) if breathing has stopped and CPR if heart action has stopped. Transfer promptly to a medical facility. When this chemical has been swallowed, get medical attention. Give large quantities of water and induce vomiting. Do not make an unconscious person vomit.

Personal Protective Methods: Wear protective gloves and clothing to prevent any reasonable probability of skin contact. Safety equipment suppliers/manufacturers can provide recommendations on the most protective glove/clothing material for your operation. All protective clothing (suits, gloves, footwear, headgear) should be clean, available each day, and put on before work. Contact lenses should not be worn when working with this chemical. Wear dust-proof chemical goggles and face shield unless full-face-piece respiratory protection is worn. Employees should wash immediately with soap when skin is wet or contaminated. Provide emergency showers and eyewash.

Respirator Selection: At concentrations above the NIOSH REL, or where there is no REL, at any detectable concentration: SCBAF: Pd,Pp (APF = 10,000) (any self-contained breathing apparatus that has a full face-piece and is operated in a pressure-demand or other positive-pressure mode); or SaF: Pd,Pp: ASCBA (APF = 10,000) (any supplied-air respirator that has a full face-piece and is operated in a pressure-demand or other positive-pressure mode in combination with an auxiliary self-contained positive-pressure breathing apparatus). *Escape:* GmFOv100 (APF = 50) [any air-purifying, full-face-piece respirator (gas mask) with a chin-style, front- or back-mounted organic vapor canister having an N100, R100, or P100 filter].

Storage: Color Code—Blue: Health Hazard/Poison: Store in a secure poison location. Prior to working with this chemical you should be trained on its proper handling and storage. Store in tightly closed containers in a cool, well-ventilated area away from direct light. Sources of ignition, such as smoking and open flames, are prohibited where 3,3′-dimethylbenzidine is used, handled, or stored in a manner that could create a potential fire or explosion hazard. A regulated, marked area should be established where this chemical is handled, used, or stored in compliance with OSHA Standard 1910.1045.

Shipping: Toxic solids, organic, n.o.s., must be labeled "POISONOUS/TOXIC MATERIALS." *o*-Tolidine falls in Hazard Class 6.1 and Packing Group III.

Spill Handling: Evacuate persons not wearing protective equipment from area of spill or leak until cleanup is complete. Remove all ignition sources. Collect powdered material in the most convenient and safe manner and deposit in sealed containers. Ventilate area after cleanup is complete. It may be necessary to contain and dispose of this chemical as a hazardous waste. If material or contaminated runoff enters waterways, notify downstream users of potentially contaminated waters. Contact your local or federal environmental protection agency for specific recommendations. If employees are required to clean up spills, they must be properly trained and equipped. OSHA 1910.120(q) may be applicable.

Fire Extinguishing: This chemical is a combustible solid. Use dry chemical, carbon dioxide; water spray; or extinguishers. Poisonous gases are produced in fire, including nitrogen oxides. If material or contaminated runoff enters waterways, notify downstream users of potentially contaminated waters. Notify local health and fire officials and pollution control agencies. From a secure, explosion-proof location, use water spray to cool exposed containers. If cooling streams are ineffective (venting sound increases in volume and pitch, tank discolors, or shows any signs of deforming), withdraw immediately to a secure position. If employees are expected to fight fires, they must be trained and equipped in OSHA 1910.156. The only respirators recommended for firefighting are self-contained breathing apparatuses that have full face-pieces and are operated in a pressure-demand or other positive-pressure mode.

Disposal Method Suggested: Dissolve in flammable solvent and spray into firebox of an incinerator equipped with afterburner and scrubber.[22] Consult with environmental regulatory agencies for guidance on acceptable disposal practices. Generators of waste containing this contaminant (≥100 kg/mo) must conform with EPA regulations governing storage, transportation, treatment, and waste disposal.

References

Sax, N. I. (Ed.). (1985). *Dangerous Properties of Industrial Materials Report*, 5, No. 3, 75–77
New Jersey Department of Health and Senior Services. (January 2001). *Hazardous Substances Fact Sheet: 3,3′-Dimethylbenzidine*. Trenton, NJ

Toluene   T:0600

Molecular Formula: C_7H_8
Common Formula: $C_6H_5CH_3$
Synonyms: Antisal 1A; Benzene, methyl-; Black out black; CP 25; Methacide; Methane, phenyl-; Methylbenzene; Methylbenzol; NCI-C07272; Phenylmethane; Tolueno (Spanish); Toluol; Tolu-sol
CAS Registry Number: 108-88-3
RTECS® Number: XS5250000
UN/NA & ERG Number: UN1294/130
EC Number: 203-625-9 [*Annex I Index No.:* 601-021-00-3]
Regulatory Authority and Advisory Bodies
Carcinogenicity: IARC: Animal No Evidence; Human Inadequate Evidence, *not classifiable as carcinogenic to humans*, Group 3, 1999; EPA: Inadequate Information to assess carcinogenic potential; NCI: Carcinogenesis Studies, inhalation; no evidence: mouse, rat; US EPA Gene-Tox Program, Negative: Cell transform.—SA7/SHE; *In vitro* SCE—human; Negative: Sperm morphology—mouse; Inconclusive: *E. coli* polA without S9.
Air Pollutant Standard Set. See below, "Permissible Exposure Limits in Air" section.
Clean Air Act: Hazardous Air Pollutants (Title I, Part A, Section 112).
Clean Water Act: Section 311 Hazardous Substances/RQ 40CFR117.3 (same as CERCLA, see below); 40CFR 401.15 Section 307 Toxic Pollutants; 40CFR423, Appendix A, Priority Pollutants.
US EPA Hazardous Waste Number (RCRA No.): U220.
RCRA, 40CFR261, Appendix 8 Hazardous Constituents.
RCRA 40CFR268.48; 61FR15654, Universal Treatment Standards: Wastewater (mg/L), 0.080; Nonwastewater (mg/kg), 10.
RCRA 40CFR264, Appendix 9; TSD Facilities Ground Water Monitoring List. Suggested test method(s) (PQL µg/L): 8020 (2); 8240 (5).
Safe Drinking Water Act: MCL, 1.0 mg/L; MCLG, 1.0 mg/L; Regulated chemical (47 FR 9352).
Reportable Quantity (RQ): 1000 lb (454 kg).

EPCRA Section 313 Form R *de minimis* concentration reporting level: 1.0%.
California Proposition 65 Developmental/Reproductive toxin 1/1/91; (female) 8/7/09.
Canada, WHMIS, Ingredients Disclosure List Concentration 1.0%.
European/International Regulations: Hazard Symbol: F, Xn; Risk phrases: R11; R38; R48/20; R63; R65; R67; Safety phrases: S2; S36/37; S46; S62 (see Appendix 4).
WGK (German Aquatic Hazard Class): 2—Hazard to waters.

Description: Toluene is a clear, colorless, noncorrosive liquid with a sweet, pungent, benzene-like odor. The odor threshold in air is variously given as 0.17 ppm,[41] 2.9 ppm (NJ) and 8 ppm (EPA). The odor threshold in water is 0.04−1.0 mg/L. Molecular weight = 92.15; Specific gravity (H_2O:1) = 0.87; Boiling point = 111.1°C; Freezing/Melting point = −95°C; Vapor pressure = 21 mmHg; Flash point = 4°C (cc); Autoignition temperature = 480°C. Explosive limits: LEL = 1.1%; UEL = 7.1%. Hazard Identification (based on NFPA-704 M Rating System): Health 2, Flammability 3, Reactivity 0. Insoluble in water.

Potential Exposure: Compound Description: Tumorigen, Mutagen; Reproductive Effector; Human Data; Primary Irritant. Toluene is used as an industrial chemical, chemical intermediate; solvent, and emulsifier; may be encountered in the manufacture of benzene. It is also used as a chemical feed for toluene diisocyanate, phenol, benzyl and benzoyl derivatives; benzoic acid; toluene sulfonates; nitrotoluenes, vinyltoluene, and saccharin; as a solvent for paints and coatings; or as a component of automobile and aviation fuels.

Incompatibilities: Strong oxidizers may cause fire and explosions. Violent reaction with mixtures of nitric and sulfuric acid.

Permissible Exposure Limits in Air
Conversion factor: 1 ppm = 3.77 mg/m³ at 25°C & 1 atm.
OSHA PEL: 200 ppm/754 mg/m³ Ceiling Concentration; 300 ppm/500 ppm [10-min maximum peak per 8-h shift].
NIOSH REL: 100 ppm/375 mg/m³ TWA; 150 ppm/560 mg/m³ STEL.
ACGIH TLV®[1]: 20 ppm/75 mg/m³ TWA [skin]. BEI; 0.5 mg[o-cresol]/L in urine/end-of-shift; 1.6 g[hippuric acid]g/g creatinine in urine/end-of-shift; 0.05 mg[toluene]/L in blood, prior to last shift of workweek, not classifiable as a human carcinogen.
NIOSH IDLH: 500 ppm.
Protective Action Criteria (PAC)*
TEEL-0: 20 ppm
PAC-1: **200** ppm
PAC-2: **1200** ppm
PAC-3: **4500** ppm
*AEGLs (Acute Emergency Guideline Levels) & ERPGs (Emergency Response Planning Guideline) are in **bold face**.
DFG MAK: 50 ppm/190 mg/m³ TWA; Peak Limitation Category II(4) [skin], Pregnancy Risk Group: C;

BAT:0.1 mg[toluene]/L in blood/end-of-shift; 0.3 mg[o-cresol]/L in urine/end-of-shift, for long-term exposure, after several shifts.
Australia: TWA 100 ppm (375 mg/m³), STEL 150 ppm, 1993; Austria: MAK 100 ppm (380 mg/m³), 1999; Belgium: TWA 100 ppm (377 mg/m³), STEL 150 ppm (565 mg/m³), 1993; Denmark: TWA 35 ppm (130 mg/m³) [skin], 1999; Finland: TWA 100 ppm (375 mg/m³), STEL 150 ppm [skin], 1999; France: VME 100 ppm (375 mg/m³), VLE 150 ppm (550 mg/m³), 1999; Hungary: TWA 100 mg/m³, STEL 300 mg/m³ [skin], 1993; the Netherlands: MAC-TGG 150 mg/m³, 2003; Norway: TWA 25 ppm (94 mg/m³), 1999; the Philippines: TWA 100 ppm (375 mg/m³), 1993; Poland: MAC (TWA) 100 mg/m³; MAC (STEL) 350 mg/m³, 1999; Russia: TWA 100 ppm, STEL 50 mg/m³, 1993; Sweden: NGV 50 ppm (200 mg/m³), TKV 100 ppm (400 mg/m³) [skin], 1999; Switzerland: MAK-W 50 ppm (190 mg/m³), KZG-W 250 ppm (950 mg/m³), 1999; Thailand: TWA 200 ppm, STEL 300 ppm, 1993; Turkey: TWA 200 ppm (750 mg/m³), 1993; United Kingdom: TWA 50 ppm (191 mg/m³), STEL 150 ppm [skin], 2000; Argentina, Bulgaria, Columbia, Jordan, South Korea, New Zealand, Singapore, Vietnam: ACGIH TLV®: not classifiable as a human carcinogen. Russia[35, 43] has also set a MAC for ambient air in residential areas of 0.6 mg/m³, both on a momentary and a daily average basis. Several states have set guidelines or standards for toluene in ambient air[60] ranging from 0.05 mg/m³ (Massachusetts) to 1.0 mg/m³ (Arizona) to 0.4−2.0 mg/m³ (Rhode Island) to 1.875 mg/m³ (Indiana) to 3.75−5.60 mg/m³ (North Dakota) to 6.0 mg/m³ (Virginia) to 7.5 mg/m³ (Connecticut, New York, and South Dakota) to 8.929 mg/m³ (Nevada) to 4.7−56.0 mg/m³ (North Carolina).

Determination in Air: Use NIOSH Analytical Method #8002, Toluene in blood; #1500 Hydrocarbons (BP 36−126°C), #1501, Hydrocarbons, aromatic, #4000, Toluene (passive); OSHA Analytical Method #111.

Permissible Concentration in Water: To protect freshwater aquatic life: 17,500 µg/L on an acute toxicity basis. *To protect saltwater aquatic life:* 6300 µg/L on an acute toxicity basis and 5000 µg/L on a chronic basis. To protect human health—14.3 mg/L.[6] EPA more recently[48] set a lifetime health advisory of 2.42 mg/L and even more recently[62] proposed a drinking water standard of 2.0 mg/L (2000 µg/L). Several states have set guidelines for toluene in drinking water[61] ranging from 100 µg/L (California and Maine) to 340 µg/L (Massachusetts) to 343 µg/L (Wisconsin) to 750 µg/L (New Mexico) to 1000 µg/L (Connecticut) to 2000 µg/L (Arizona, Kansas, Minnesota, and Vermont). Russia[35,43] set a MAC in water bodies used both for household purposes and fishery purposes of 0.5 mg/L (500 µg/L). Octanol−water coefficient: Log K_{ow} = 2.69.

Determination in Water: Inert gas purge followed by gas chromatography and photoionization detection (EPA Method #602) or gas chromatography plus mass spectrometry (EPA Method #624).

Routes of Entry: Inhalation of vapor, percutaneous absorption of liquid, ingestion; skin and/or eye contact.

Harmful Effects and Symptoms

Short Term Exposure: Irritates the eyes and respiratory tract. Causes central nervous system depression. High levels of exposure may cause fatigue, weakness, confusion, euphoria, dizziness, headache, dilated pupils; lacrimation (discharge of tears); nervousness, muscle fatigue, insomnia, paresthesia; cardiac dysrhythmia; unconsciousness and death may occur. *Inhalation:* 100 ppm exposure can cause dizziness, drowsiness, and hallucinations. 100–200 ppm can cause depression; 200–500 ppm can cause headache; nausea, loss of appetite; loss of energy; loss of coordination and coma. In addition to the above, death has resulted from exposure to 10,000 ppm for an unknown time. *Skin:* Can cause dryness and irritation. Absorption may cause or increase the severity of symptoms listed above. *Eyes:* Can cause irritation at 300 ppm. *Ingestion:* Can cause a burning sensation in the mouth and stomach, upper abdominal pain; cough, hoarseness, headache, nausea, loss of appetite; loss of energy; loss of coordination; and coma.

Long Term Exposure: Repeated or prolonged contact with skin may cause dermatitis; drying, cracking, itching, and skin rash. May cause liver, kidney, and brain damage; decreased learning ability; psychological disorders. Levels below 200 ppm may produce headache, tiredness, and nausea. From 200 to 750 ppm symptoms may include insomnia, irritability, dizziness, some loss of memory; cause heart palpitations and loss of coordination. Blood effects and anemia have been reported but are probably due to contamination by benzene.

Points of Attack: Eyes, skin, respiratory system; central nervous system; liver, kidneys.

Medical Surveillance: Whole blood (chemical/metabolite); whole blood (chemical/metabolite), end-of-shift; whole blood (chemical/metabolite), end-of-workweek; whole blood (chemical/metabolite), prior to last shift of workweek, Expired Air, During Exposure; urine (chemical/metabolite). For those with frequent or potentially high exposure (half the TLV or greater, or significant skin contact), the following is recommended before beginning work and at regular times after that: Urinary hippuric acid excretion (at the end-of-shift) as an index of overexposure. If symptoms develop or overexposure is suspected, the following may be useful: exam of the nervous system. Liver and kidney function tests, and evaluation for renal tubular acidosis. Complete blood count (CBC).

First Aid: If this chemical gets into the eyes, remove any contact lenses at once and irrigate immediately for at least 15 min, occasionally lifting upper and lower lids. Seek medical attention immediately. If this chemical contacts the skin, remove contaminated clothing and wash immediately with soap and water. Seek medical attention immediately. If this chemical has been inhaled, remove from exposure, begin rescue breathing (using universal precautions, including resuscitation mask) if breathing has stopped and CPR if heart action has stopped. Transfer promptly to a medical facility. When this chemical has been swallowed, get medical attention. Give large quantities of water and induce vomiting. Do not make an unconscious person vomit.

Note to physician: Exposure to toluene at levels greater than 200 ppm may result in hippuric acid levels above 5 g/L urine. After elevated exposure, toluene may also be detected in blood.

Personal Protective Methods: Wear protective gloves and clothing to prevent any reasonable probability of skin contact. Safety equipment suppliers/manufacturers can provide recommendations on the most protective glove/clothing material for your operation. Viton™/Neoprene™, Teflon™, Viton™/chlorobutyl, and Silvershield™ are recommended protective materials. All protective clothing (suits, gloves, footwear, headgear) should be clean, available each day, and put on before work. Contact lenses should not be worn when working with this chemical. Wear splash-proof chemical goggles and face shield unless full-face-piece respiratory protection is worn. Employees should wash immediately with soap when skin is wet or contaminated. Provide emergency showers and eyewash.

Respirator Selection: *500 ppm:* CcrFOv (APF = 50) [any air-purifying, full-face-piece respirator (gas mask) with a chin-style, front- or back-mounted acid gas canister]; or PaprOv (APF = 25) [any powered, air-purifying respirator with organic vapor cartridge(s)]; or GmFOv (APF = 50) [any air-purifying, full-face-piece respirator (gas mask) with a chin-style, front- or back-mounted organic vapor canister]; or Sa (APF = 10) (any supplied-air respirator); or SCBAF (APF = 50) (any self-contained breathing apparatus with a full face-piece). *Emergency or planned entry into unknown concentrations or IDLH conditions:* SCBAF: Pd, Pp (APF = 10,000) (any self-contained breathing apparatus that has a full face-piece and is operated in a pressure-demand or other positive-pressure mode); or SaF: Pd,Pp: ASCBA (APF = 10,000) (any supplied-air respirator that has a full face-piece and is operated in a pressure-demand or other positive-pressure mode in combination with an auxiliary self-contained breathing apparatus operated in a pressure-demand or other positive-pressure mode). *Escape:* GmFOv (APF = 50) [any air-purifying, full-face-piece respirator (gas mask) with a chin-style, front- or back-mounted organic vapor canister]; or SCBAE (any appropriate escape-type, self-contained breathing apparatus).

Storage: Color Code—Red: Flammability Hazard: Store in a flammable liquid storage area or approved cabinet away from ignition sources and corrosive and reactive materials. Prior to working with this chemical you should be trained on its proper handling and storage. Before entering confined space where this chemical may be present, check to make sure that an explosive concentration does not exist. Toluene must be stored to avoid contact with strong oxidizers (such as chlorine, bromine, and fluorine), since violent reactions occur. Protect storage containers from physical damage. Sources of ignition, such as smoking and open flames, are prohibited where toluene is used, handled, or stored in a manner that could create a potential fire or explosion hazard. Metal containers involving the transfer of 5 gallons or more of toluene should be grounded and bonded. Drums must be equipped with self-closing, valves, pressure vacuum bungs; and flame arresters. Use only nonsparking tools and equipment, especially when opening and closing containers of toluene.

Shipping: Toluene requires a shipping label of "FLAMMABLE LIQUID." It falls in Hazard Class 3 and Packing Group II.

Spill Handling: Evacuate and restrict persons not wearing protective equipment from area of spill or leak until cleanup is complete. Remove all ignition sources. Establish forced ventilation to keep levels below explosive limit. Absorb liquids in vermiculite, dry sand; earth, peat, carbon, or a similar material and deposit in sealed containers. Keep this chemical out of a confined space, such as a sewer, because of the possibility of an explosion, unless the sewer is designed to prevent the buildup of explosive concentrations. It may be necessary to contain and dispose of this chemical as a hazardous waste. If material or contaminated runoff enters waterways, notify downstream users of potentially contaminated waters. Contact your local or federal environmental protection agency for specific recommendations. If employees are required to clean up spills, they must be properly trained and equipped. OSHA 1910.120(q) may be applicable.

Fire Extinguishing: This chemical is a flammable liquid. Poisonous gases may be produced in fire. Use dry chemical, carbon dioxide; or alcohol foam extinguishers. Vapors are heavier than air and will collect in low areas. Vapors may travel long distances to ignition sources and flashback. Vapors in confined areas may explode when exposed to fire. Containers may explode in fire. Storage containers and parts of containers may rocket great distances, in many directions. If material or contaminated runoff enters waterways, notify downstream users of potentially contaminated waters. Notify local health and fire officials and pollution control agencies. From a secure, explosion-proof location, use water spray to cool exposed containers. If cooling streams are ineffective (venting sound increases in volume and pitch, tank discolors, or shows any signs of deforming), withdraw immediately to a secure position. If employees are expected to fight fires, they must be trained and equipped in OSHA 1910.156. The only respirators recommended for firefighting are self-contained breathing apparatuses that have full face-pieces and are operated in a pressure-demand or other positive-pressure mode.

Disposal Method Suggested: Consult with environmental regulatory agencies for guidance on acceptable disposal practices. Generators of waste containing this contaminant (≥100 kg/mo) must conform with EPA regulations governing storage, transportation, treatment, and waste disposal.

References
US Environmental Protection Agency. (1979). *Chemical Hazard Information Profile: Toluene.* Washington, DC
US Environmental Protection Agency. (1980). *Toluene: Ambient Water Quality Criteria.* Washington, DC
National Institute for Occupational Safety and Health. (1973). *Criteria for a Recommended Standard: Occupational Exposure to Toluene.* NIOSH Document No. 73-11023
US Environmental Protection Agency. (April 30, 1980). *Toluene, Health and Environmental Effects Profile No. 160.* Washington, DC: Office of Solid Waste
Sax, N. I. (Ed.). *Dangerous Properties of Industrial Materials Report,* 2, No. 6, 83−87 (1982); 5, No. 5, 94−99 (1985); and 7, No. 5, 2−14 (1987)
US Public Health Service. (December 1988). *Toxicological Profile for Toluene.* Atlanta, GA: Agency for Toxic Substance and Disease Registry
New York State Department of Health. (March 1986). *Chemical Fact Sheet: Toluene.* Albany, NY: Bureau of Toxic Substance Assessment (Version 2)
New Jersey Department of Health and Senior Services. (August 1998). *Hazardous Substances Fact Sheet: Toluene.* Trenton, NJ

Toluene-2,4-diamine other toluenediamine isomers   T:0610

Molecular Formula: $C_7H_{10}N_2$
Common Formula: $H_3CC_6H_3(NH_2)_2$
Synonyms: 3-Amino-*p*-toluidine; 5-Amino-*o*-toluidine; Azogen developer H; 1,3-Benzenediamine, 4-methyl; Benzofur MT; C.I. 76035; C.I. Oxidation base; C.I. Oxidation base 20; C.I. Oxidation base 200; C.I. Oxidation base 35; Developer B; Developer DB; Developer DBJ; Developer H; Developer MC; Developer MT; Developer MT-CF; Developer MTD; Developer T; 1,3-Diamino-4-methylbenzene; 2,4-Diamino-1-toluene; Diaminotoluene; 2,4-Diaminotolueno (Spanish); 2,4-Diaminotoluol; Eucanine GB; Fouramine; Fouramine J; Fourrine 94; Fourrine M; 4-Methyl-1,3-benzenediamine; 4-Methyl-*m*-phenylenediamine; MTD; Nako TMT; NCI-C02302; Pelagol grey J; Pelagol J; Pontamine developer TN; Renal MD; TDA; Tetral G; 2,4-Tolamine; *m*-Toluenediamine; Toluene-2,4-diamine; *m*-Toluylenediamine; *m*-Tolylenediamine; 2,4-Tolylenediamine; 4-*m*-Tolylenediamine; Zoba GKE; Zogen developer H
Mixed isomers: Benzenediamine, *ar*-methyl-; Diaminotoluene; Diaminotolueno (Spanish); Methylphenylenediamine; Toluen diamina (Spanish); Toluene-*ar,ar'*-diamine; Toluene-*ar,ar*-diamine; Toluenediamine; Tolylenediamine
CAS Registry Number: 95-80-7 (toluene-2,4-diamine); 25376-45-8 (toluenediamine); *(alt.)* 26764-44-3; *(alt.)* 30143-

13-6; 108-71-4 (toluene-3,5-diamine); 95-70-5 (toluene-2,5-diamine); 823-40-5 (toluene-2,6-diamine)

RTECS® Number: XS9625000; XS944500 (toluenediamine)

UN/NA & ERG Number: UN1709 (2,4-toluylenediamine, solid or 2,4- toluenediamine. solid)/151; UN3814 (2,4-toluylenediamine solution or 2,4-toluenediamine solution)/151

EC Number: 202-453-1 [*Annex I Index No.:* 612-099-00-3] (toluene-2,4-diamine); 246-910-3 (toluenediamine); 203-609-1 (toluene-3,5-diamine)

Regulatory Authority and Advisory Bodies

Carcinogenicity: (*toluene-2,4-diamine*) IARC: Human No Adequate Data; Animal Sufficient Evidence, *possibly carcinogenic to humans*, Group 2B, 1987; NCI: Carcinogenesis Bioassay (feed); clear evidence: mouse, rat; NTP: 11th Report on Carcinogens, 2004: Reasonably anticipated to be a human carcinogen; NTP: Carcinogenesis studies; test completed (peer review), October 2000.

US EPA Gene-Tox Program, Positive: Carcinogenicity—mouse/rat; Histidine reversion—Ames test; Positive: *D. melanogaster* sex-linked lethal; Negative: *N. crassa—aneuploidy*; Sperm morphology—mouse; Inconclusive: SHE—clonal assay.

Air Pollutant Standard Set. See below, "Permissible Exposure Limits in Air" section.

Reportable Quantity (RQ): 10 lb (4.54 kg).

US EPA Hazardous Waste Number (RCRA No.): U221.

RCRA, 40CFR261, Appendix 8 Hazardous Constituents.

EPCRA Section 313 Form R *de minimis* concentration reporting level: 1.0%.

European/International Regulations (*95-80-7, toluene-2,4-diamine*): Hazard Symbol: T, N; Risk phrases: R45; R21; R25; R36; R43; R48/22; R62; R68; Safety phrases: S51/53; S53; S45; S61; (*95-70-5; toluene-2,6-diamine*): Hazard Symbol: T,N; Risk phrases: R20/21; R25; R43; R51/53; Safety phrases: S1/2; S24; S37; S45; S61 (see Appendix 4). WGK (German Aquatic Hazard Class) (*95-80-7*): 3—Highly water polluting.

Mixed isomers:

US EPA Hazardous Waste Number (RCRA No.): U221.

RCRA, 40CFR261, Appendix 8 Hazardous Constituents.

Reportable Quantity (RQ): 10 lb (4.54 kg).

EPCRA Section 313 Form R *de minimis* concentration reporting level: 1.0%.

California Proposition 65 Chemical: Cancer (95-80-7) 1/1/88; [25376-45-8 (mixed isomers)]1/1/90.

Canada, WHMIS, Ingredients Disclosure List Concentration 0.1% *as Toluene-2,4-diamine, Toluene-ar,ar′-diamine, Toluene-3,4-diamine, Toluene-3,5-diamine.*

Description: Toluene-2,4-diamine takes the form of colorless needles. Molecular weight = 122.19; Specific gravity (H_2O:1) = 1.05 (liquid at 100°C); Freezing/Melting point = 98.9°C; Vapor pressure = 1 mmHg at 107°C; Flash point = 148°C. Soluble in water.

Potential Exposure: Compound Description (*toluene-2,4-diamine*): Agricultural Chemical; Tumorigen, Mutagen;

Reproductive Effector; Primary Irritant. Compound Description (*toluene-2,6-diamine*): Mutagen. Toluene-2,4-diamine is a chemical intermediate for toluene diisocyanate (used in the production of flexible and rigid polyurethane foams, polyurethane coatings; cast elastomers including fabric coatings and polyurethane and other adhesives), for dyes used on textiles; leather, furs; and in hair-dye formulations.

Incompatibilities: Strong acids; chloroformates, oxidizers.

Permissible Exposure Limits in Air

OSHA PEL: None.

NIOSH REL: (*all isomers*) Potential occupational carcinogen; Limit exposure to lowest feasible concentration. See *NIOSH Pocket Guide*, Appendix A.

Mixed isomers

AIHA WEEL: 0.005 ppm TWA [skin].

Toluene-2,4-diamine

Protective Action Criteria (PAC)

TEEL-0: 0.005 ppm

PAC-1: 1.5 ppm

PAC-2: 12.5 ppm

PAC-3: 50 ppm

DFG MAK (*toluene-2,4-diamine*): [skin] danger of skin sensitization; Carcinogen Category 2; DFG TRK: *Air*: toluene-2,4-diamine 0.100 mg/m^3; *Urine*: Sampling time: end-of-exposure or end-of-shift, Total toluene-2,4-diamine 100 μg[creatinine]/g in urine.

Austria: carcinogen, 1999; Finland: carcinogen, 1999; Poland: MAC (TWA) 0.04 mg/m^3; MAC (STEL) 0.1 mg/m^3, 1999; Russia: STEL 2 mg/m^3 [skin], 1993; Sweden: carcinogen, 1993; Switzerland: MAK-W 0.1 mg/m^3, carcinogen, 1999.

Toluene-2,5-diamine

DFG MAK (*toluene-2,4-diamine*): [skin] danger of skin sensitization.

Determination in Air: Use NIOSH Analytical Method (IV) #5516, OSHA Analytical Method #ID-65.

Routes of Entry: Inhalation, eye, and/or skin contact.

Harmful Effects and Symptoms

Short Term Exposure: Irritates the eyes and skin. Eye contact may cause permanent damage. Skin contact may cause burns and blistering. Methemoglobinemia, central nervous system depression, and degeneration of the liver typically result from exposure to toluene-2,4-diamine. Exposure can cause cyanosis, headache, fatigue, dizziness, nausea, vomiting.

Long Term Exposure: May cause liver damage. Jaundice and anemia are reported. Repeated exposure causes CNS depression causing headache, weakness, dizziness, fatigue, nausea, vomiting, and possible death. There is limited evidence that toluenediamines may reduce fertility in males and may affect the developing fetus. 2,4-Diaminotoluene is carcinogenic in rats and after its oral administration, producing hepatocellular carcinomas; and its subcutaneous injection, inducing local sarcomas. Diaminotoluenes are mutagens.

Points of Attack: Central nervous system; liver, blood. Cancer site in animals: liver, mammary glands.

Medical Surveillance: Consider the points of attack in preplacement and periodic physical examination. Examination of the nervous system. Methemoglobin level. Complete blood count (CBC). Liver function tests.

First Aid: Skin Contact: [52] Flood all areas of body that have contacted the substance with water. Do not wait to remove contaminated clothing; do it under the water stream. Use soap to help assure removal. Isolate contaminated clothing when removed to prevent contact by others.

Eye Contact: Remove any contact lenses at once. Flush eyes well with copious quantities of water or normal saline for at least 20−30 min. Seek medical attention.

Inhalation: Leave contaminated area immediately; breathe fresh air. Proper respiratory protection must be supplied to any rescuers. If coughing, difficult breathing, or any other symptoms develop, seek medical attention at once, even if symptoms develop many hours after exposure.

Ingestion: If convulsions are not present, give a glass or two of water or milk to dilute the substance. Assure that the person's airway is unobstructed and contact a hospital or poison center immediately for advice on whether or not to induce vomiting.

Note to physician: Treat for methemoglobinemia. Spectrophotometry may be required for precise determination of levels of methemoglobinemia in urine.

Personal Protective Methods: Wear protective gloves and clothing to prevent any reasonable probability of skin contact. Safety equipment suppliers/manufacturers can provide recommendations on the most protective glove/clothing material for your operation. Butyl rubber is recommended. All protective clothing (suits, gloves, footwear, headgear) should be clean, available each day, and put on before work. Contact lenses should not be worn when working with this chemical. Wear dust-proof chemical goggles and face shield unless full-face-piece respiratory protection is worn. Employees should wash immediately with soap when skin is wet or contaminated. Provide emergency showers and eyewash.

Respirator Selection: Use self-contained breathing apparatus.

Storage: Color Code—Blue: Health Hazard/Poison: Store in a secure poison location. Prior to working with this chemical you should be trained on its proper handling and storage. Store in a cool, dry place or a refrigerator. A regulated, marked area should be established where this chemical is handled, used, or stored in compliance with OSHA Standard 1910.1045.

Shipping: 2,4-Toluylenediamine requires a shipping label of "POISONOUS/TOXIC MATERIALS." It falls in Hazard Class 6.1 and Packing Group III.

Spill Handling: Evacuate persons not wearing protective equipment from area of spill or leak until cleanup is complete. Remove all ignition sources. Dampen spilled material with alcohol to avoid dust, then transfer material to a suitable container. Use absorbent dampened with alcohol to pick up remaining material. Wash surfaces well with soap and water. Collect powdered material in the most convenient and safe manner and deposit in sealed containers. Ventilate area after cleanup is complete. It may be necessary to contain and dispose of this chemical as a hazardous waste. If material or contaminated runoff enters waterways, notify downstream users of potentially contaminated waters. Contact your local or federal environmental protection agency for specific recommendations. If employees are required to clean up spills, they must be properly trained and equipped. OSHA 1910.120(q) may be applicable.

Fire Extinguishing: This chemical may burn but does not easily ignite. Use dry chemical, carbon dioxide; water spray; or foam extinguishers. Poisonous gases are produced in fire, including carbon monoxide and nitrogen oxides. If material or contaminated runoff enters waterways, notify downstream users of potentially contaminated waters. Notify local health and fire officials and pollution control agencies. From a secure, explosion-proof location, use water spray to cool exposed containers. If cooling streams are ineffective (venting sound increases in volume and pitch, tank discolors, or shows any signs of deforming), withdraw immediately to a secure position. If employees are expected to fight fires, they must be trained and equipped in OSHA 1910.156. The only respirators recommended for firefighting are self-contained breathing apparatuses that have full face-pieces and are operated in a pressure-demand or other positive-pressure mode.

Disposal Method Suggested: Consult with environmental regulatory agencies for guidance on acceptable disposal practices. Generators of waste containing this contaminant ($\geq$100 kg/mo) must conform with EPA regulations governing storage, transportation, treatment, and waste disposal. Controlled incineration (oxides of nitrogen are removed from the effluent gas by scrubbers and/or thermal devices).

References

US Environmental Protection Agency. (1979). *Chemical Hazard Information Profile: Toluene-2,4-Diamine.* Washington, DC
US Environmental Protection Agency. (April 30, 1980). *2,4-Toluenediamine, Health and Environmental Effects Profile No. 161.* Washington, DC: Office of Solid Waste
Sax, N. I. (Ed.). (1985). *Dangerous Properties of Industrial Materials Report,* 5, No. 5, 99−103
New Jersey Department of Health and Senior Services. (December 2005). *Hazardous Substances Fact Sheet: 2,4-Diaminotoluene.* Trenton. NJ
New Jersey Department of Health and Senior Services. (December 2005). *Hazardous Substances Fact Sheet: Diaminetoluenes (mixed isomers).* Trenton, NJ

Toluene 2,4-diisocyanate   T:0620

Molecular Formula: $C_9H_6N_2O_2$
Common Formula: 2,4-$CH_3C_6H_3(NCO)_2$

Synonyms: Benzene, 2,4-diisocyanato-1-methyl-; Benzene, 2,4-diisocyanato-1-methyl-; Benzene,2,4-diisocyanatomethyl-; Cresorcinol diisocyanate; Desmodur T80; Di-*iso*-cyanatoluene; Di-isocyanate de toluylene (French); 2,4-Diisocyanato-1-methylbenzene; 2,4-Diisocyanatotoluene; Diisocyanat-toluol (German); Hylene T; Hylene TCPA; Hylene TLC; Hylene TM; Hylene TM-65; Hylene TRF; Isocyanic acid, 4-methyl-*m*-phenylene ester; Isocyanic acid, methylphenylene ester; 4-Methyl-phenylene diisocyanate; 4-Methyl-phenylene isocyanate; Mondur TDS; Nacconate 1OO; NCI-C50533; Niax TDI; Niax TDI-P; Scuranate; 2,4-TDI; TDI; TDI-80; Toluen-2,4-diisociato (Spanish); 2,4-Toluene diisocyanate; Toluene diisocyanate; Toluene di-isocyanate; Toluylene 2,4-diisocyanate; Tolyene 2,4-diisocyanate; 2,4-Tolylene diisocyanate; 2,4-Tolylenediisocyanat E; Tolylene 2,4-diisocyanate; Tuluylen diisocyanat (German); Tuluylene 2,4-diisocyanate; Voranate T-80; Voranate T-80, type I; Voranate T-80, type II

CAS Registry Number: 584-84-9

RTECS® Number: CZ6300000

UN/NA & ERG Number: UN2078/156

EC Number: 209-544-5 [*Annex I Index No.:* 615-006-00-4]

Regulatory Authority and Advisory Bodies

Carcinogenicity: IARC: Human Inadequate Evidence; Animal Sufficient Evidence, *possibly carcinogenic to humans*, Group 2B, 1999; NTP: 11th Report on Carcinogens, 2004: Reasonably anticipated to be a human carcinogen; NIOSH: Potential occupational carcinogen.

Air Pollutant Standard Set. See below, "Permissible Exposure Limits in Air" section.

Clean Air Act: Hazardous Air Pollutants (Title I, Part A, Section 112); Accidental Release Prevention/Flammable Substances (Section 112[r], Table 3), TQ = 10,000 lb (4540 kg).

Superfund/EPCRA 40CFR355, Extremely Hazardous Substances: TPQ = 500 lb (127.5 kg).

Reportable Quantity (RQ): 100 lb (45.4 kg).

EPCRA Section 313 Form R *de minimis* concentration reporting level: 0.1%.

Canada, WHMIS, Ingredients Disclosure List Concentration 0.1%.

European/International Regulations: Hazard Symbol: T, N; Risk phrases: R26; R36/37/38; R42/43; R52/53; Safety phrases: S1/2; S23; S36/37; S45; S61 (see Appendix 4).

WGK (German Aquatic Hazard Class): 2—Hazard to waters.

Description: Toluene diisocyanate is a colorless, yellow, or dark liquid or solid with a sweet, fruity, pungent odor. Toluene diisocyanate (technical) is an 80:20 mixture of 2,4- and 2,6-isomers. A solid above 71°F/22°C. The odor threshold is 0.4−2.14 ppm. Molecular weight = 174.17; Specific gravity (H_2O:1) = 1.22; Boiling point = 251.1°C; Freezing/Melting point = 21.7°C; Vapor pressure = 0.01 mmHg at 25°C; Relative vapor density (air = 1) = 6.1; Flash point = 127°C; Autoignition temperature = 620°C. Explosive limits: LEL = 0.9%; UEL = 9.5%. Hazard Identification (based on NFPA-704 M Rating System): Health 3, Flammability 1, Reactivity 3. Insoluble; reacts exothermically with water.

Potential Exposure: Compound Description: Tumorigen, Mutagen, Human Data; Primary Irritant. Toluene diisocyanate is used in the production of polyurethane flexible foams, coatings, and elastomers. It is more widely used than MDI (diphenylmethane diisocyanate). Polyurethanes are formed by the reaction of isocyanates with polyhydroxy compounds. Since the reaction proceeds rapidly at room temperature, the reactants must be mixed in pots or spray guns just before use. These resins can be produced with various physical properties, e.g., hard, flexible, semirigid foams; and have found many uses, e.g., upholstery padding; thermal insulation; molds, surface coatings; shoe inner soles; and in rubbers, adhesives, paints, and textile finishes. Because of TDI's high volatility, exposure can occur in all phases of its manufacture and use. MDI has a much lower volatility, and problems generally arise only in spray applications.

Incompatibilities: Strong oxidizers, water, acids, bases, and amines (may cause foam and spatter); alcohols. Reacts slowly with water to form carbon dioxide and polyureas (NIOSH). Contact with bases, tertiary amines, and acyl-chlorides may cause explosive polymerization.

Permissible Exposure Limits in Air

Conversion factor: 1 ppm = 7.13 mg/m^3 at 25°C & 1 atm.

OSHA PEL: 0.02 ppm/0.14 mg/m^3 Ceiling Concentration.

NIOSH REL: Potential occupational carcinogen; Limit exposure to lowest feasible concentration. See *NIOSH Pocket Guide*, Appendix A.

ACGIH TLV®[1]: 0.005 ppm/0.036 mg/m^3 TWA; 0.02 ppm/0.14 mg/m^3, STEL, danger of sensitization of respiratory tract; not classifiable as a human carcinogen.

Notice of intended change: 0.001 ppm/0.007 mg/m^3, measured as inhalable fraction and vapor TWA; 0.003 ppm/0.021 mg/m^3, measured as inhalable fraction and vapor STEL, danger of sensitization of respiratory tract; not classifiable as a human carcinogen.

NIOSH IDLH: Potential occupational carcinogen, 2.5 ppm

Protective Action Criteria (PAC)*

TEEL-0: 0.005 ppm

PAC-1: **0.020** ppm

PAC-2: **0.083** ppm

PAC-3: **0.51** ppm

*AEGLs (Acute Emergency Guideline Levels) & ERPGs (Emergency Response Planning Guideline) are in **bold face**.

DFG MAK: Danger of sensitization of the airways; Carcinogen Carcinogen Category 3A.

Arab Republic of Egypt: TWA 0.02 ppm (0.14 mg/m^3), 1993; Austria: MAK 0.01 ppm (0.07 mg/m^3), 1999; Belgium: TWA 0.005 ppm (0.036 mg/m^3), STEL 0.02 ppm, 1993; Denmark: TWA 0.005 ppm (0.035 mg/m^3), 1999; France: VME 0.01 ppm (0.08 mg/m^3), STEL 0.02 ppm (0.16 mg/m^3), 1993; Hungary: STEL 0.04 mg/m^3, 1993;

Japan; 0.02 ppm (0.14 mg/m^3), 1993; the Netherlands: MAC-TGG 0.04 mg/m^3, 2003; Norway: TWA 0.005 ppm (0.035 mg/m^3), 1999; the Philippines: TWA 0.02 ppm (0.14 mg/m^3), 1993; Poland: MAC (TWA) 0.035 mg/m^3; MAC (STEL) 0.070 mg/m^3, 1999; Russia: STEL 0.05 mg/m^3, 1993; Sweden: NGV 0.005 ppm (0.04 mg/m^3), KTV 0.01 ppm (0.07 mg/m^3), 1999; Switzerland: MAK-W 0.005 ppm (0.04 mg/m^3), KZG-W 0.01 ppm (0.08 mg/m^3), 1999; Thailand: TWA 0.02 ppm (0.14 mg/m^3), 1993; Turkey: TWA 0.02 ppm (0.14 mg/m^3), 1993; United Kingdom: TWA 0.02 [NCO]mg/m^3, STEL 0.07 mg[NCO]/m^3, 2000; Argentina, Bulgaria, Columbia, Jordan, South Korea, New Zealand, Singapore, Vietnam: ACGIH TLV®: STEL 0.02 ppm.

Russia has also set MAC values for ambient air in residential areas of 0.05 mg/m^3 on a momentary basis and 0.02 mg/m^3 on a daily average basis. Several states have set guidelines or standards for TDI in ambient air[60] ranging from 0.03–0.20 μg/m^3 (Rhode Island) to 0.13 μg/m^3 (New York) to 0.4 μg/m^3 (South Carolina) to 0.48 μg/m^3 (Massachusetts) to 0.70 μg/m^3 (Virginia) to 0.72 μg/m^3 (Connecticut and South Dakota) to 0.4–1.5 μg/m^3 (North Dakota) to 0.9 μg/m^3 (Nevada) to 4.8–16.0 μg/m^3 (North Carolina).

Determination in Air: Use NIOSH Analytical Method (IV) #5521, Isocyanates, monomeric, #5522 Isocyanates, #2535, Toluene-2,4-diisocyanate, and OSHA Analytical Method #18, superseded by #42 and #33.

Determination in Water: Octanol–water coefficient: Log $K_{ow} = 0.22$.

Routes of Entry: Inhalation of vapor, ingestion and skin, and/or eye contact.

Harmful Effects and Symptoms

Short Term Exposure: Irritate the eyes, skin, and respiratory tract. Inhalation of the vapor may cause asthmatic reactions; chemical bronchitis, pneumonitis, and pulmonary edema. Pulmonary edema is a medical emergency that can be delayed for several hours. This can cause death. *Note:* TDI is a strong sensitizer. Allergic individuals may experience symptoms at very low concentrations. *Inhalation:* Causes irritation of nose, throat, and lungs; insomnia, euphoria, difficulty in walking; loss of consciousness; poor memory; personality changes; irritability, and depression. Allergic response is also possible. Sensitive individuals may react to 0.007 ppm or less. Exposure to high levels leads to chemical pneumonia. Levels of 0.01–0.03 ppm reportedly caused no symptoms. At 0.03–0.07 ppm, respiratory illness with continuous coughing, sore throat; difficulty in breathing; fatigue and nocturnal sweating were reported. In another study, 10 min at 0.5 ppm caused nose and throat irritation. Lung damage may be permanent. *Skin:* If not removed promptly, may cause redness, pain, swelling, and blistering. TDI is corrosive and may be absorbed through the skin. Repeated contact has caused skin sensitization in humans and allergic eczema. *Eyes:* May cause redness, pain; blurred vision;

severe irritation, tears and damage to the cornea. Prolonged contact may cause permanent damage. *Ingestion:* May cause sore throat, abdominal pain; diarrhea, and irritation of mouth and stomach.

Long Term Exposure: May produce asthma-like allergy and chronic lungs disease; chronic obstructive bronchitis; emphysema, chemical bronchitis; asthmatic syndrome. May cause memory loss and concentration problems; psychological effects; central nervous system effects. May cause chest tightness (sometimes very severe), sneezing, cyanosis (blue coloration), blood changes; and collapse. Sensitization may occur after exposure to spills or other unusually high concentrations. Decreased lung function has been reported from estimated exposure to 0.02 ppm for 2 years. It has also been reported that excessive loss of pulmonary function occurs at 0.0035 ppm possibly to 0.002 ppm. Sensitization has been reported on the first exposure at concentration below 0.05 ppm and as late as 14 years (0.06 ppm) after first exposure. This substance is a probable carcinogen in humans. There is limited evidence that TDI may cause temporary impotence in males.

Points of Attack: Eyes, skin, respiratory system. Cancer site in animals: pancreas, liver, mammary gland; circulatory system; and skin tumors.

Medical Surveillance: NIOSH lists the following tests: Blood Gas Analysis; blood plasma; chest X-ray, electrocardiogram, pulmonary function tests: forced vital capacity, forced expiratory volume (1 s); pulmonary function tests; pre- and postshift; sputum cytology; urine (chemical/metabolite); white blood cell count/differential. Preplacement and periodic medical examinations should include chest roentgenography, pulmonary function tests; and an evaluation of any respiratory disease or history of allergy. Periodic pulmonary function tests may be useful in detecting the onset of pulmonary sensitization. See also "References."[4]

First Aid: If this chemical gets into the eyes, remove any contact lenses at once and irrigate immediately for at least 15 min, occasionally lifting upper and lower lids. Seek medical attention immediately. If this chemical contacts the skin, remove contaminated clothing and wash immediately with soap and water. Seek medical attention immediately. If this chemical has been inhaled, remove from exposure, begin rescue breathing (using universal precautions, including resuscitation mask) if breathing has stopped and CPR if heart action has stopped. Transfer promptly to a medical facility. When this chemical has been swallowed, get medical attention. Give large quantities of water and induce vomiting. Do not make an unconscious person vomit. Medical observation is recommended for 24–48 h after breathing overexposure, as pulmonary edema may be delayed. As first aid for pneumonitis or pulmonary edema, a doctor or authorized paramedic may consider administering a corticosteroid spray.

Personal Protective Methods: Wear protective gloves and clothing to prevent any reasonable probability of skin contact. Safety equipment suppliers/manufacturers can provide recommendations on the most protective glove/clothing material for your operation. Teflon™, Silvershield™, Viton™, polyethylene, nitrile, chlorinated polyethylene; and polyvinyl alcohol are among the recommended protective materials. All protective clothing (suits, gloves, footwear, headgear) should be clean, available each day, and put on before work. Contact lenses should not be worn when working with this chemical. Wear splash-proof chemical goggles and face shield when working with liquid, unless full-face-piece respiratory protection is worn. Wear dust-proof goggles and face shield when working with powders or dust, unless full-face-piece respiratory protection is worn. Employees should wash immediately with soap when skin is wet or contaminated. Provide emergency showers and eyewash.

Respirator Selection: NIOSH: *At any concentrations above the NIOSH REL, or where there is no REL, at any detectable concentration:* SCBAF: Pd,Pp (APF = 10,000) (any self-contained breathing apparatus that has a full face-piece and is operated in a pressure-demand or other positive-pressure mode); or SaF: Pd,Pp: ASCBA (APF = 10,000) (any supplied-air respirator that has a full face-piece and is operated in a pressure-demand or other positive-pressure mode in combination with an auxiliary self-contained breathing apparatus operated in a pressure-demand or other positive-pressure mode). *Escape:* GmFOv (APF = 50) [any air-purifying, full face-piece respirator (gas mask) with a chin-style, front- or back-mounted organic vapor canister]; or SCBAE (any appropriate escape-type, self-contained breathing apparatus).

Storage: Color Code—Blue: Health Hazard/Poison: Store in a secure poison location. Prior to working with this chemical you should be trained on its proper handling and storage. Before entering confined space where this chemical may be present, check to make sure that an explosive concentration does not exist. Store in tightly closed containers in a cool well-ventilated area away from amines, strong bases (such as sodium hydroxide), and alcohols. Toluene 2,4-diisocyanate should not be stored in contact with water, because they react and release carbon dioxide gas. Toluene 2,4-diisocyanate will polymerize and rupture containers at temperatures over 177°C/350°F. At normal temperatures (21°C/70°F) Toluene 2,4-diisocyanate levels quickly exceed the PEL and therefore proper ventilation must be in practice or personal protective equipment must be worn. A regulated, marked area should be established where this chemical is handled, used, or stored in compliance with OSHA Standard 1910.1045.

Shipping: This compound requires a shipping label of "POISONOUS/TOXIC MATERIALS." It falls in Hazard Class 6.1 and Packing Group II.

Spill Handling: Evacuate and restrict persons not wearing protective equipment from area of spill or leak until cleanup is complete. Remove all ignition sources. Establish forced ventilation to keep levels below explosive limit. *Solid:* Collect powdered material in the most convenient and safe manner and deposit in sealed containers. Ventilate area after cleanup is complete. *Liquid:* Ventilate area of spill or leak. Absorb liquids in vermiculite, dry sand; earth, peat, carbon, or a similar material and deposit in sealed containers. Keep this chemical out of a confined space, such as a sewer, because of the possibility of an explosion, unless the sewer is designed to prevent the buildup of explosive concentrations. It may be necessary to contain and dispose of this chemical as a hazardous waste. If material or contaminated runoff enters waterways, notify downstream users of potentially contaminated waters. Contact your local or federal environmental protection agency for specific recommendations. If employees are required to clean up spills, they must be properly trained and equipped. OSHA 1910.120(q) may be applicable.

Fire Extinguishing: *Solid:* Water gently applied to surface or foam may cause frothing which will extinguish the fire (NFPA). If material is on fire or involved in fire, do not extinguish fire unless flow can be stopped. Use water in flooding quantities as fog. Solid streams of water may be ineffective. Cool all affected containers with flooding quantities of water. Apply water from as far a distance as possible. Use "alcohol" foam, carbon dioxide, or dry chemical. Use water spray to absorb vapor. Poisonous gases are produced in fire, including cyanide and oxides of nitrogen. If material or contaminated runoff enters waterways, notify downstream users of potentially contaminated waters. Notify local health and fire officials and pollution control agencies. From a secure, explosion-proof location, use water spray to cool exposed containers. If cooling streams are ineffective (venting sound increases in volume and pitch, tank discolors, or shows any signs of deforming), withdraw immediately to a secure position. If employees are expected to fight fires, they must be trained and equipped in OSHA 1910.156. The only respirators recommended for firefighting are self-contained breathing apparatuses that have full face-pieces and are operated in a pressure-demand or other positive-pressure mode.

Liquid: This chemical is a combustible liquid. Poisonous gases, including cyanide and oxides of nitrogen, are produced in fire. Use dry chemical, carbon dioxide; or alcohol foam extinguishers. Vapors are heavier than air and will collect in low areas. Vapors may travel long distances to ignition sources and flashback. Vapors in confined areas may explode when exposed to fire. Containers may explode in fire. Storage containers and parts of containers may rocket great distances, in many directions. If material or contaminated runoff enters waterways, notify downstream users of potentially contaminated waters. Notify local health and fire officials and pollution control agencies. From a secure, explosion-proof location, use water spray to cool exposed containers. If cooling streams are ineffective (venting sound increases in volume and pitch, tank discolors, or shows any signs of deforming), withdraw immediately to a secure position. If employees are expected to fight fires, they must be trained and equipped in OSHA 1910.156.

The only respirators recommended for firefighting are self-contained breathing apparatuses that have full face-pieces and are operated in a pressure-demand or other positive-pressure mode.

Disposal Method Suggested: Consult with environmental regulatory agencies for guidance on acceptable disposal practices. Generators of waste containing this contaminant (≥100 kg/mo) must conform with EPA regulations governing storage, transportation, treatment, and waste disposal. Controlled incineration (oxides of nitrogen are removed from the effluent gas by scrubbers and/or thermal devices). In accordance with 40CFR165, follow recommendations for the disposal of pesticides and pesticide containers. Must be disposed properly by following package label directions or by contacting your local or federal environmental control agency, or by contacting your regional EPA office.

References

National Institute for Occupational Safety and Health. (1973). *Criteria for a Recommended Standard: Occupational Exposure to Toluene Diisocyanate.* NIOSH Document No. 73-11022

National Institute for Occupational Safety and Health. (October 1977). *Information Profiles on Potential Occupational Hazards: Organoisocyanates,* Report PB-276,678. Rockville, MD, pp. 265−275

US Environmental Protection Agency. (April 30, 1980). *Toluene Diisocyanate, Health and Environmental Effects Profile No. 162.* Washington, DC: Office of Solid Waste

US Environmental Protection Agency. (November 30, 1987). *Chemical Hazard Information Profile: Toluene 2,4-Diisocyanate.* Washington, DC: Chemical Emergency Preparedness Program

US Environmental Protection Agency. (July 25, 1984). *Chemical Hazard Information Profile: Toluene Diisocyanate.* Washington, DC: Office of Toxic Substances

New York State Department of Health. (May 1986). *Chemical Fact Sheet: Toluene Diisocyanate.* Albany, NY: Bureau of Toxic Substance Assessment (Version 2)

New Jersey Department of Health and Senior Services. (April 2002). *Hazardous Substances Fact Sheet: Toluene 2,4-Diisocyanate.* Trenton, NJ

Toluene sulfonic acid T:0630

Molecular Formula: $C_7H_8O_3S$

Common Formula: $C_6H_4(CH_3)SO_3H$

Synonyms: Benzenesulfonic acid, methyl-; Benzenesulfonic acid, methyl ester; Manro PTSA 65 E; Manro PTSA 65 H; Manro PTSA 65 LS; Methylbenzene sulfonic acid; *p*-Methylbenzenesulfonic acid; 4-Methylbenzenesulfonic acid; *p*-Methylphenylsulfonic acid; Toluenesulfonic acid; 4-Toluenesulfonic acid; *p*-Toluenesulphonic acid; *p*-Tolylsulfonic acid; Tosic acid; TSA-HP; TSA-MH

CAS Registry Number: 104-15-4; 25231-46-3 (NJ and WHMIS); *(alt.)* 402-47-1; *(alt.)* 100901-72-2; *(alt.)* 114213-96-6; *(alt.)* 126033-27-0; *(alt.)* 128739-80-0; *(alt.)* 144647-92-7; *(alt.)* 156627-46-2; 80-48-8 (Toluenesulfonic acid, methyl ester, *p*-)

RTECS® Number: XT1630000

UN/NA & ERG Number: UN2583 (solid, with >5% free H_2SO_4)/153; UN2584 (liquid, with >5% free H_2SO_4)/153; UN2585 (solid, with >5% free H_2SO_4)/153; UN2586 (liquid, with >5% free H_2SO_4)/153

EC Number: 203-180-0 [*Annex I Index No.:* 016-030-00-2] (*toluene-4-sulfonic acid*)

Regulatory Authority and Advisory Bodies

US EPA, FIFRA 1998 Status of Pesticides: Canceled.

Canada, WHMIS, Ingredients Disclosure List Concentration 1.0% as toluenesulfonic acid (25231-46-3).

European/International Regulations: Hazard Symbol (*104-15-4*): Xi; Risk phrases: R36/37/38; Safety phrases: S2; S26; S37 (see Appendix 4).

WGK (German Aquatic Hazard Class) (*104-15-4*): 1— Slightly water polluting.

Description: Toluene sulfonic acid is a colorless liquid or colorless crystalline solid. Odorless when pure; the technical grade has a slightly aromatic odor; Freezing/Melting point = 105°C. Molecular weight = 172.21; Boiling point = 140°C at 20 mm; Flash point = 184°C. Hazard Identification (based on NFPA-704 M Rating System): Health 3, Flammability 1, Reactivity 0. Soluble in water.

Potential Exposure: Toluene sulfonic acid is used as catalyst, stabilizer for monomers and polymers; cleaning agents; a plating additive; in synthesis of dyes; drugs; and other chemicals.

Incompatibilities: The solution is a strong acid. Incompatible with sulfuric acid, caustics, ammonia, amines, amides, organic anhydrides; isocyanates, vinyl acetate; alkylene oxides; epichlorohydrin, and combustible materials. Attacks metals in the presence of moisture forming hydrogen gas.

Permissible Exposure Limits in Air

Protective Action Criteria (PAC)

80-48-8 (Toluenesulfonic acid, methyl ester, p-)

TEEL-0: 0.75 mg/m³

PAC-1: 2.5 mg/m³

PAC-2: 15 mg/m³

PAC-3: 150 mg/m³

Routes of Entry: Inhalation, ingestion, eye, and/or skin contact.

Harmful Effects and Symptoms

Short Term Exposure: Toluene sulfonic acid can affect you when breathed in. Corrosive to the eyes and skin; can cause severe irritation and burns. Irritates the respiratory tract causing burning sensation; dryness, and coughing. Overexposure can cause poisoning, which can lead to death. High levels can burn the lungs, causing a buildup of fluid, which can also cause death.

Long Term Exposure: Can irritate the lungs; bronchitis may develop with cough, phlegm, and/or shortness of breath.

Points of Attack: Lungs.

Medical Surveillance: Before beginning employment and at regular times after that, for those with frequent or potentially high exposures, the following are recommended: lung function tests. Exam of the eyes, nose and throat. If symptoms develop or overexposure is suspected, the following may be useful: consider chest X-ray after acute overexposure.

First Aid: If this chemical gets into the eyes, remove any contact lenses at once and irrigate immediately for at least 15 min, occasionally lifting upper and lower lids. Seek medical attention immediately. If this chemical contacts the skin, remove contaminated clothing and wash immediately with soap and water. Seek medical attention immediately. If this chemical has been inhaled, remove from exposure, begin rescue breathing (using universal precautions, including resuscitation mask) if breathing has stopped and CPR if heart action has stopped. Transfer promptly to a medical facility. When this chemical has been swallowed, get medical attention. If victim is *conscious*, administer water or milk. Do not induce vomiting. Medical observation is recommended for 24–48 h after breathing overexposure, as pulmonary edema may be delayed. As first aid for pulmonary edema, a doctor or authorized paramedic may consider administering a corticosteroid spray.

Personal Protective Methods: Wear protective gloves and clothing to prevent any reasonable probability of skin contact. Safety equipment suppliers/manufacturers can provide recommendations on the most protective glove/clothing material for your operation. Neoprene™, chlorinated polyethylene, and polyvinyl chloride are among the recommended protective materials. All protective clothing (suits, gloves, footwear, headgear) should be clean, available each day, and put on before work. Contact lenses should not be worn when working with this chemical. Wear splash-proof chemical goggles and face shield when working with liquid, unless full-face-piece respiratory protection is worn. Wear dust-proof goggles and face shield when working with powders or dust, unless full-face-piece respiratory protection is worn. Employees should wash immediately with soap when skin is wet or contaminated. Provide emergency showers and eyewash.

Respirator Selection: Where there is potential for exposures to solid toluene sulfonic acid, use a NIOSH/MSHA- or European Standard EN149-approved full-face-piece respirator with a high-efficiency particulate filter. Greater protection is provided by a powered, air-purifying respirator. Where there is potential for high exposures to toluene sulfonic acid in liquid form, use a NIOSH/MSHA- or European Standard EN149-approved supplied-air respirator with a full face-piece operated in the positive-pressure mode, or with a full face-piece, hood, or helmet in the continuous-flow mode; or use a NIOSH/MSHA- or European Standard EN149-approved self-contained breathing apparatus with a full face-piece operated in pressure-demand or other positive mode.

Storage: Color Code—White: Corrosive or Contact Hazard; Store separately in a corrosion-resistant location. Prior to working with this chemical you should be trained on its proper handling and storage. Store in tightly closed containers in a cool, well-ventilated area away from bases, combustibles, and other incompatible materials listed above. Where possible, automatically pump liquid or transfer dry material from drums or other storage containers to process containers. Sources of ignition, such as smoking and open flames, are prohibited where this chemical is handled, used, or stored.

Shipping: Alkyl sulfonic acids, liquid, or aryl sulfonic acids, liquid (with >5% free sulfuric acid), require a shipping label of "CORROSIVE." It falls in Hazard Class 8 and Packing Group II.

Alkyl sulfonic acids, liquid, or aryl sulfonic acids, liquid (with not >5% free sulfuric acid) requires a shipping label of "CORROSIVE." It falls in Hazard Class 8 and Packing Group III.

Alkyl sulfonic acids, solid, or aryl sulfonic acids, solid (with not >5% free sulfuric acid) falls in DOT Hazard Class 8 and Packing Group III.

Alkyl sulfonic acids, solid, or aryl sulfonic acids, solid (with >5% free sulfuric acid) requires a shipping label of "CORROSIVE." It falls in Hazard Class 8 and Packing Group II.

Spill Handling: Dry material: Evacuate persons not wearing protective equipment from area of spill or leak until cleanup is complete. Remove all ignition sources. For *small spills,* flush with water, rinse with dilute sodium bicarbonate or lime solution. Collect material in the most convenient and safe manner and deposit in sealed containers. Ventilate area after cleanup is complete. It may be necessary to contain and dispose of this chemical as a hazardous waste. If material or contaminated runoff enters waterways, notify downstream users of potentially contaminated waters. Contact your local or federal environmental protection agency for specific recommendations. If employees are required to clean up spills, they must be properly trained and equipped. OSHA 1910.120(q) may be applicable.

Liquid: Evacuate and restrict persons not wearing protective equipment from area of spill or leak until cleanup is complete. Remove all ignition sources. Ventilate area of spill or leak. Absorb liquids in vermiculite, dry sand; earth, peat, carbon, or a similar material and deposit in sealed containers. It may be necessary to contain and dispose of this chemical as a hazardous waste. If material or contaminated runoff enters waterways, notify downstream users of potentially contaminated waters. Contact your local or federal environmental protection agency for specific recommendations. If employees are required to clean up spills, they must be properly trained and equipped. OSHA 1910.120(q) may be applicable.

Fire Extinguishing: Toluene sulfonic acid may burn, but does not readily ignite. Use dry chemical, carbon dioxide, water spray; or foam extinguishers. However, water may cause frothing. Poisonous gases are produced in fire, including sulfur oxides. If material or contaminated runoff enters waterways, notify downstream users of potentially contaminated waters. Notify local health and fire officials and pollution control agencies. From a secure, explosion-proof location, use water spray to cool exposed containers. If cooling streams are ineffective (venting sound increases in volume and pitch, tank discolors, or shows any signs of deforming), withdraw immediately to a secure position. If employees are expected to fight fires, they must be trained and equipped in OSHA 1910.156. The only respirators recommended for firefighting are self-contained breathing apparatuses that have full face-pieces and are operated in a pressure-demand or other positive-pressure mode.

References

US Environmental Protection Agency, Special Review and Reregistration Division Office of Pesticide Programs. (1998). *Agency Status of Pesticides in Registration, Reregistration, and Special Review* (Rainbow Report). Washington, DC

New Jersey Department of Health and Senior Services. (May 2003). *Hazardous Substances Fact Sheet: p-Toluene Sulfonic Acid.* Trenton, NJ

o-Toluidine   T:0640

Molecular Formula: C_7H_9N
Common Formula: $CH_3C_6H_4NH_2$
Synonyms: 1-Amino-2-methylbenzene; 2-Amino-1-methylbenzene; *o*-Aminotoluene; 2-Aminotoluene; Aniline, 2-methyl-; Benzenamine, 2-methyl-; C.I. 37077; 1-Methyl-1,2-amino-benzene; 1-Methyl-2-aminobenzene; 2-Methyl-1-aminobenzene; *o*-Methylaniline; 2-Methylaniline; *o*-Methylbenzenamine; 2-Methylbenzenamine; *o*-Toluidina (Spanish); 2-Toluidine; Toluidine, *o*-; *o*-Tolylamine
CAS Registry Number: 95-53-4; 636-21-5 (hydrochloride)
RTECS® Number: XU2975000
UN/NA & ERG Number: UN1708/153
EC Number: 202-429-0 [*Annex I Index No.:* 612-091-00-X]; 211-252-8 (hydrochloride)

Regulatory Authority and Advisory Bodies

Carcinogenicity: IARC *(95-53-4)*: Human Sufficient Evidence, 1978; Animal Sufficient Evidence, *carcinogenic to humans*, Group 1; NTP *(95-53-4 & 636-21-5)*: 11th Report on Carcinogens, 2004: Reasonably anticipated to be a human carcinogen; NIOSH *(95-53-4)*: Potential occupational carcinogen.
US EPA Gene-Tox Program, Positive: Carcinogenicity—mouse/rat; Positive: Cell transform.—RLV F344 rat embryo; Positive: *E. coli* polA without S9; Negative: Sperm morphology—mouse; *S. cerevisiae*—homozygosis; Inconclusive: Histidine reversion—Ames test.
Clean Air Act: Hazardous Air Pollutants (Title I, Part A, Section 112).
US EPA Hazardous Waste Number (RCRA No.): U328.

RCRA, 40CFR261, Appendix 8 Hazardous Constituents.
RCRA 40CFR264, Appendix 9; TSD Facilities Ground Water Monitoring List. Suggested test method(s) (PQL μg/L): 8270 (10).
Reportable Quantity (RQ): 100 lb (45.4 kg).
EPCRA Section 313 Form R *de minimis* concentration reporting level: 0.1%.
Air Pollutant Standard Set. See below, "Permissible Exposure Limits in Air" section.
California Proposition 65 Chemical: Cancer 1/1/88; *o*-Toluidine hydrochloride 1/1/88.
Canada, WHMIS, Ingredients Disclosure List Concentration 0.1%.
European/International Regulations: Hazard Symbol *(95-53-4)*: T, N; Risk phrases: R45; R23/25; R36; R50; Safety phrases: S53; S45; S61 (see Appendix 4).
WGK (German Aquatic Hazard Class): 3—Highly water polluting (*CAS: 95-53-4*).

Description: *o*-Toluidine is a colorless to pale yellow liquid with a weak, pleasant, aromatic odor. The odor threshold = 0.25 ppm. Molecular weight = 107.17; Relative density (H_2O:1): 1.01; Boiling point = 200°C; Freezing/Melting point = −16°C; Vapor pressure = 0.3 mmHg at 21°C; Flash point = 85°C (cc); Autoignition temperature = 482°C. Explosive limits: LEL = 1.5%; UEL—unknown. Slightly soluble in water; solubility = 2%.
Potential Exposure: Compound Description: Tumorigen, Mutagen; Reproductive Effector; Human Data; Primary Irritant. *o*-Toluidine is used as an intermediate in the manufacture of dyes; as an intermediate in pharmaceutical manufacture; in textile printing; in rubber accelerators; in production of *o*-aminoazotoluene.
Incompatibilities: Strong oxidizers, especially nitric acid; bases.

Permissible Exposure Limits in Air

Conversion factor: 1 ppm = 4.38 mg/m^3 at 25°C & 1 atm.
OSHA PEL: 5 ppm/22 mg/m^3 TWA [skin].
NIOSH REL: [skin] Potential occupational carcinogen; Limit exposure to lowest feasible concentration. See *NIOSH Pocket Guide*, Appendix A.
ACGIH TLV®[1]: 2 ppm/8.8 mg/m^3 TWA [skin], confirmed animal carcinogen with unknown relevance to humans; BEI_M issued as methemoglobin inducers.
NIOSH IDLH: Potential occupational carcinogen, 50 ppm.
Protective Action Criteria (PAC)
TEEL-0: 5 ppm
PAC-1: 5 ppm
PAC-2: 5 ppm
PAC-3: 50 ppm
DFG MAK: [skin] Carcinogen Category 1; Germ Cell Mutation Category 3A.
Australia: TWA 2 ppm (9 mg/m^3) [skin], carcinogen, 1993; Austria [skin], carcinogen, 1999; Belgium: TWA 2 ppm (8.8 mg/m^3) [skin], Carcinogen 1993; Denmark: TWA 2 ppm (9 mg/m^3) [skin], 1999; Finland: TWA 5 ppm (22 mg/m^3), STEL 10 ppm (44 mg/m^3) [skin], carcinogen,

1999; France: VME 2 ppm (9 mg/m^3), carcinogen, 1999; Carcinogen (salts), 1993; Japan; 1 ppm (4.4 mg/m^3) [skin], 2B carcinogen, 1999; Norway: TWA 1 ppm (4.5 mg/m^3), 1999; the Philippines: TWA 5 ppm (22 mg/m^3) [skin], 1993; Poland: MAC (TWA) 3 mg/m^3; MAC (STEL) 9 mg/m^3, 1999; Russia: TWA 0.5 mg/m^3, STEL 1 mg/m^3 [skin], carcinogen, 1993; Sweden: carcinogen, 1999; Switzerland: MAK-W 0.1 ppm (0.5 mg/m^3) [skin], carcinogen, 1999; Turkey: TWA 5 ppm (22 mg/m^3) [skin], 1993; United Kingdom: carcinogen, 2000; Argentina, Bulgaria, Columbia, Jordan, South Korea, New Zealand, Singapore, Vietnam: ACGIH TLV®: confirmed animal carcinogen with unknown relevance to humans.

Also, many states have set guidelines for ambient air.

Determination in Air: Use NIOSH Analytical Method #2002, amines, aromatic, #2017, #8317, OSHA Analytical Method ID-73.

Permissible Concentration in Water: No criteria set, but EPA[32] has suggested an ambient water goal of 304 µg/L based on health effects.

Determination in Water: Octanol—water coefficient: Log $K_{ow} = 1.32$.

Routes of Entry: Inhalation, skin absorption; ingestion; skin and/or eye contact.

Harmful Effects and Symptoms

Short Term Exposure: Exposure may affect the blood causing cyanosis and formation of methemoglobin. Exposure to high concentrations may result in damage to kidneys and bladder.

Inhalation: Combines with blood cells to prevent binding of oxygen. Early symptoms are headache, nausea, vomiting, diarrhea, low blood pressure; increased salivation and loss of appetite. Causes cyanosis; lips, fingernails, and tongue may turn blue. Symptoms may begin to appear at 6 ppm after several hours, or at 100 ppm after 1 h. Continued exposure may lead to difficult breathing; dizziness, stupor, unconsciousness, and death. *Skin:* Causes irritation. May cause excessive drying of skin and irritation. Absorption is significant and may increase severity of symptoms listed under inhalation. *Eyes:* Causes irritation, redness, and chemical burns. *Ingestion:* Animal studies suggest that symptoms as listed under inhalation would occur, and that death may result from ingestion of about 2 oz by a 150 lb person.

Long Term Exposure: Skin and inhalation exposures may cause the formation of methemoglobin and cyanosis, mild blue coloration of the skin due to lack of oxygen in the blood. Loss of appetite and weight, headache, dizziness may occur. Irritation of the kidneys and bladder may occur, with decreased functions and damage. *o*-Toluidine is a potential occupational carcinogen.

Points of Attack: Eyes, skin, blood, kidneys, liver, cardio-vascular system. *Cancer site:* bladder cancer.

Medical Surveillance: NIOSH lists the following tests: whole blood (chemical/metabolite), Methemoglobin; Complete blood count; red blood cells/count; urine (chemical/metabolite); urine (chemical/metabolite) [Whole Blood (chemical/metabolite)]; urinalysis (routine)]. Before beginning employment and at regular times after that, the following is recommended: Urine exam for blood and abnormal cells (urine cytology) and blood. If symptoms develop or overexposure is suspected, the following may be useful: methemoglobin level every 3—6 h for 18—24 h. Liver function tests.

First Aid: If this chemical gets into the eyes, remove any contact lenses at once and irrigate immediately for at least 15 min, occasionally lifting upper and lower lids. Seek medical attention immediately. If this chemical contacts the skin, remove contaminated clothing and wash immediately with soap and water. Seek medical attention immediately. If this chemical has been inhaled, remove from exposure, begin rescue breathing (using universal precautions, including resuscitation mask) if breathing has stopped and CPR if heart action has stopped. Transfer promptly to a medical facility. When this chemical has been swallowed, get medical attention. Give large quantities of water and induce vomiting. Do not make an unconscious person vomit. Effects may be delayed. Medical observation is recommended.

Note to physician: Treat for methemoglobinemia. Spectrophotometry may be required for precise determination of levels of methemoglobinemia in urine.

Personal Protective Methods: Wear protective gloves and clothing to prevent any reasonable probability of skin contact. Safety equipment suppliers/manufacturers can provide recommendations on the most protective glove/clothing material for your operation. Teflon™ is among the recommended protective materials. All protective clothing (suits, gloves, footwear, headgear) should be clean, available each day, and put on before work. Contact lenses should not be worn when working with this chemical. Wear splash-proof chemical goggles and face shield unless full-face-piece respiratory protection is worn. Employees should wash immediately with soap when skin is wet or contaminated. Provide emergency showers and eyewash.

Respirator Selection: *At any concentrations above the NIOSH REL, or where there is no REL, at any detectable concentration:* SCBAF: Pd,Pp (APF = 10,000) (any NIOSH/MSHA- or European Standard EN 149-approved self-contained breathing apparatus that has a full face-piece and is operated in a pressure-demand or other positive-pressure mode); or SaF: Pd,Pp: ASCBA (APF = 10,000) (any supplied-air respirator that has a full face-piece and is operated in a pressure-demand or other positive-pressure mode in combination with an auxiliary self-contained breathing apparatus operated in a pressure-demand or other positive-pressure mode). *Escape:* GmFOv (APF = 50) [any air-purifying, full-face-piece respirator (gas mask) with a chin-style, front- or back-mounted organic vapor canister]; or SCBAE (any appropriate escape-type, self-contained breathing apparatus).

Storage: Color Code—Blue: Health Hazard/Poison: Store in a secure poison location. Prior to working with this chemical you should be trained on its proper handling and storage. Before entering confined space where this chemical may be present, check to make sure that an explosive concentration does not exist. o-Toluidine must be stored to avoid contact with strong oxidizers (such as chlorine, bromine, and fluorine) because violent reactions occur. Store in tightly closed containers in a cool, well-ventilated area away from heat. Sources of ignition, such as smoking and open flames, are prohibited where o-toluidine is used, handled, or stored in a manner that could create a potential fire or explosion hazard. A regulated, marked area should be established where this chemical is handled, used, or stored in compliance with OSHA Standard 1910.1045.

Shipping: This compound requires a shipping label of "POISONOUS/TOXIC MATERIALS." It falls in Hazard Class 6.1 and Packing Group II.

Spill Handling: Evacuate and restrict persons not wearing protective equipment from area of spill or leak until cleanup is complete. Remove all ignition sources. Establish forced ventilation to keep levels below explosive limit. Absorb liquids in vermiculite, dry sand; earth, peat, carbon, or a similar material and deposit in sealed containers. Keep this chemical out of a confined space, such as a sewer, because of the possibility of an explosion, unless the sewer is designed to prevent the buildup of explosive concentrations. It may be necessary to contain and dispose of this chemical as a hazardous waste. If material or contaminated runoff enters waterways, notify downstream users of potentially contaminated waters. Contact your local or federal environmental protection agency for specific recommendations. If employees are required to clean up spills, they must be properly trained and equipped. OSHA 1910.120(q) may be applicable.

Fire Extinguishing: This chemical is a combustible liquid. Poisonous gases, including nitrogen oxides, are produced in fire. Use dry chemical, carbon dioxide; or alcohol foam extinguishers. Vapors are heavier than air and will collect in low areas. Vapors may travel long distances to ignition sources and flashback. Vapors in confined areas may explode when exposed to fire. Containers may explode in fire. Storage containers and parts of containers may rocket great distances, in many directions. If material or contaminated runoff enters waterways, notify downstream users of potentially contaminated waters. Notify local health and fire officials and pollution control agencies. From a secure, explosion-proof location, use water spray to cool exposed containers. If cooling streams are ineffective (venting sound increases in volume and pitch, tank discolors, or shows any signs of deforming), withdraw immediately to a secure position. If employees are expected to fight fires, they must be trained and equipped in OSHA 1910.156. The only respirators recommended for firefighting are self-contained breathing apparatuses that have full face-pieces and are operated in a pressure-demand or other positive-pressure mode.

Disposal Method Suggested: Consult with environmental regulatory agencies for guidance on acceptable disposal practices. Generators of waste containing this contaminant (≥100 kg/mo) must conform with EPA regulations governing storage, transportation, treatment, and waste disposal. Controlled incineration (oxides of nitrogen are removed from the effluent gas by scrubbers and/or thermal devices).

References
US Environmental Protection Agency. (February 23, 1984). *Chemical Hazard Information Profile Draft Report: o-Toluidine; o-Toluidine Hydrochloride.* Washington, DC
New York State Department of Health. (March 1986). *Chemical Fact Sheet: o-Toluidine.* Albany, NY: Bureau of Toxic Substance Assessment (Version 2)
Sax, N. I. (Ed.). (1982). *Dangerous Properties of Industrial Materials Report*, 2, No. 1, 121–123
New Jersey Department of Health and Senior Services. (January 2001). *Hazardous Substances Fact Sheet: o-Toluidine.* Trenton, NJ

Toxaphene   T:0650

Molecular Formula: $C_{10}H_{10}Cl_8$
Synonyms: 8001-35-2; Agricide Maggot killer (F); Alltex; Alltox; Anatox; Attac-2; Attac 6; Attac 6-3; Camphechlor; Camphene, octachloro-; Camphochlor; Camphoclor; Camphofene huileux; Canfeclor; Chem-Phene; Chlorinated camphene; Chlorocamphene; Clor Chem T-590; Compound 3956; Crestoxo; Cristoxo 90; ENT 9,735; Estonox; Fascoterpene; Geniphene; GY-Phene; Hercules 3956; Hercules Toxaphene; Kamfochlor; M 5055; Melipax; Motox; NCI-C00259; Octachlorocamphene; PCC; PCHK; Penphene; Phenacide; Phenatox; Polychlorcamphene; Polychlorinated camphene; Polychlorocamphene; Strobane T; Strobane T 90; Synthetic 3956; Technical chlorinated camphene, 67–69% chlorine; Toxadust; Toxafeno (Spanish); Toxakil; Toxaphen (German); Toxaspray; Toxon 63; Toxyphen; Vertac 90%; Vertac toxaphene 90
CAS Registry Number: 8001-35-2; *(alt.)* 8022-04-6
RTECS® Number: XW5250000
UN/NA & ERG Number: UN2761/151
EC Number: 232-283-3 [*Annex I Index No.:* 602-044-00-1]
Regulatory Authority and Advisory Bodies
Carcinogenicity: IARC: Human Inadequate Evidence; Animal Sufficient Evidence, *possibly carcinogenic to humans*, Group 2B; EPA: Sufficient evidence from animal studies; inadequate evidence or no useful data from epidemiologic studies; NTP: 11th Report on Carcinogens, 2004: Reasonably anticipated to be a human carcinogen; NCI: Carcinogenesis Bioassay (feed); clear evidence: mouse; equivocal evidence: rat.
US EPA Gene-Tox Program, Positive: Carcinogenicity—mouse/rat.
US EPA, FIFRA 1998 Status of Pesticides: Canceled.
Banned or Severely Restricted (many countries) (UN).[13]
Persistent Organic Pollutants (UN).

Air Pollutant Standard Set. See below, "Permissible Exposure Limits in Air" section.

Clean Air Act: Hazardous Air Pollutants (Title I, Part A, Section 112).

Clean Water Act: Section 311 Hazardous Substances/RQ 40CFR117.3 (same as CERCLA, see below); Toxic Pollutant (Section 401.15); 40CFR423, Appendix A, Priority Pollutants; Section 313 Water Priority Chemicals (57FR41331, 9/9/92).

US EPA Hazardous Waste Number (RCRA No.): P123.

RCRA, 40CFR261, Appendix 8 Hazardous Constituents.

RCRA Toxicity Characteristic (Section 261.24), Maximum Concentration of Contaminants, regulatory level, 0.5 mg/L.

RCRA 40CFR268.48; 61FR15654, Universal Treatment Standards: Wastewater (mg/L), 0.0095; Nonwastewater (mg/kg), 2.6.

RCRA 40CFR264, Appendix 9; TSD Facilities Ground Water Monitoring List. Suggested test method(s) (PQL μg/L): 8080 (2); 8250 (10).

Safe Drinking Water Act: MCL, 0.003 mg/L; MCLG, zero; Regulated chemical (47 FR 9352).

Superfund/EPCRA 40CFR355, Extremely Hazardous Substances: TPQ = 500/10,000 lb (227/4540 kg).

Reportable Quantity (RQ): 1 lb (0.454 kg).

EPCRA Section 313 Form R de minimis concentration reporting level: 0.1%.

US DOT Regulated Marine Pollutant (49CFR172.101, Appendix B).

Rotterdam Convention Annex III [Chemicals Subject to the Prior Informed Consent Procedure (PIC)].

California Proposition 65 Chemical: Cancer 1/1/88.

List of Stockholm Convention POPs: Annex A (Elimination).

European/International Regulations: Hazard Symbol: T, N; Risk phrases: R21; R25; R37/38; R40; R50/53; Safety phrases: S1/2; S36/37; S45; S60; S61. (see Appendix 4).

WGK (German Aquatic Hazard Class): No value assigned.

Description: Toxaphene is an amber, waxy solid with a mild, piney, chlorine- and camphor-like odor. Usually dissolved in a flammable solvent. Flammability depends on the solvent used. Molecular weight = 413.80; Specific gravity (H_2O:1) = 1.65 at 25°C; Boiling point = (decomposes); Freezing/Melting point = 65−90°C; Vapor pressure = 4×10^{-6} mmHg at 20°C; Vapor pressure = 0.4 mmHg at 25°C. Flash point = 135°C (solid); 29°C (solution); Autoignition temperature = 530°C. Explosive limits: LEL = 1.1%; UEL = 6.4% (solvents only). Hazard Identification (based on NFPA-704 M Rating System): Health 3, Flammability 0, Reactivity 0. Practically insoluble in water; solubility = 0.0003%.

Potential Exposure: Compound Description: Agricultural Chemical; Tumorigen, Mutagen; Reproductive Effector; Human Data; Natural Product; Primary Irritant. Toxaphene is a broad spectrum, chlorinated hydrocarbon pesticide that is used as an insecticide for the control of grasshoppers, army-worms. and all major cotton pests.

Incompatibilities: Contact with strong oxidizers may cause fire and explosion hazard. Decomposes, producing fumes of hydrogen chloride and chlorine in heat above 155°C, on contact with strong bases; strong sunlight; and catalysts, such as iron. Slightly corrosive to metals in the presence of moisture.

Permissible Exposure Limits in Air

OSHA PEL: 0.5 mg/m³ TWA [skin].

NIOSH REL: A potential occupational carcinogen. Limit exposure to lowest feasible concentration. See *NIOSH Pocket Guide*, Appendix A.

ACGIH TLV®[1]: 0.5 mg/m³ TWA [skin]; 1 mg/m³ STEL [skin]; confirmed animal carcinogen with unknown relevance to humans.

NIOSH IDLH: 200 mg/m³ Potential occupational carcinogen.

Protective Action Criteria (PAC)

TEEL-0: 0.5 mg/m³

PAC-1: 1 mg/m³

PAC-2: 20 mg/m³

PAC-3: 200 mg/m³

DFG MAK: [skin]; Carcinogen Category 2.

Arab Republic of Egypt: TWA 0.5 mg/m³ [skin], 1993; Australia: TWA 0.5 mg/m³, STEL 1 mg/m³ [skin], 1993; Austria: MAK 0.5 mg/m³ [skin], 1999; Belgium: TWA 0.5 mg/m³, STEL 1 mg/m³ [skin], 1993; Denmark: TWA 0.5 mg/m³ [skin], 1999; Finland: TWA 0.5 mg/m³, STEL 1.5 mg/m³ [skin], 1999; France: VME 0.5 mg/m³ [skin], carcinogen, 1999; the Netherlands: MAC-TGG 0.5 mg/m³ [skin], 2003; Norway: TWA 0.5 mg/m³, 1999; the Philippines: TWA 0.5 mg/m³ [skin], 1993; Switzerland: MAK-W 0.5 mg/m³ [skin], 1999; Thailand: TWA 0.5 mg/m³, 1993; Turkey: TWA 0.5 mg/m³ [skin], 1993; Argentina, Bulgaria, Columbia, Jordan, South Korea, New Zealand, Singapore, Vietnam: ACGIH TLV®: STEL 1 mg/m³ [skin].

Russia set a MAC for ambient air in residential areas of 7.0 μg/m³ on a once-daily basis. Several states have set guidelines or standards for toxaphene in ambient air[60] ranging from 1.19 μg/m³ (Kansas) to 1.2 μg/m³ (Pennsylvania) to 1.67 μg/m³ (New York) to 2.5 μg/m³ (Connecticut and South Carolina) to 5.0 μg/m³ (Florida) to 5.0−10.0 μg/m³ (North Dakota) to 8.0 μg/m³ (Virginia) to 12.0 μg/m³ (Nevada).

Determination in Air: Use NIOSH Analytical Method #5039, Chlorinated camphene.

Permissible Concentration in Water: *To protect freshwater aquatic life:* 0.013 μg/L as a 24-h average, never to exceed 1.6 μg/L. *To protect saltwater aquatic life:* never to exceed 0.07 μg/L. *To protect human health:* preferably zero. An additional lifetime cancer risk of 1 in 100,000 is presented by a concentration of 0.0071 μ/l.[6] More recently, EPA[62] has proposed a drinking water standard of 5.0 μg/L. States which have set drinking water guidelines include Maine at 5.0 μg/L and Minnesota at 0.3 μg/L.[61] Mexico[35] has set MPC values of 30 μg/L for estuaries, 3.0 μg/L for coastal waters, and 5.0 μg/L for receiving waters used for drinking water supply. Fish Tox = 0.03900000 ppb MATC (EXTRA HIGH).

Determination in Water: Gas chromatography (EPA Method #608) or gas chromatography plus mass spectrometry (EPA Method #625). Octanol−water coefficient: Log $K_{ow} = 3.32$.

Routes of Entry: Inhalation, skin absorption; ingestion, eye, and/or skin contact.

Harmful Effects and Symptoms

Short Term Exposure: High concentrations can irritate the skin and eyes. A nervous system depressant causing tremors, weakness, dizziness, increased saliva; convulsions, unconsciousness, and possible death. High exposures can cause pulmonary edema, a medical emergency that can be delayed for several hours. This can cause death. Inhalation of spray at unknown levels has caused bronchitis and inflammation of the lungs. If spilled on clothing and allowed to remain, can cause pain and reddening. May be absorbed through skin contributing to symptoms described under ingestion. Absorbed through the stomach and intestines. 0.6 g (about 1/50 oz) has caused convulsions. Other symptoms include nausea, vomiting, bluish coloration of the skin; and coma. Estimated lethal dose for an adult is between 2 and 7 g (1/15−1/4 oz).

Long Term Exposure: Aplastic anemia (low blood count) is an uncommon but serious reaction to toxaphene. High or repeated exposure may cause liver and kidney damage. Toxaphene is suspected of causing brain damage. Changes in genetic material have been observed in workers exposed to toxaphene. Toxaphene is potential occupational carcinogen.

Human Tox = 3.00000 ppb (HIGH).

Points of Attack: Central nervous system; skin. Cancer site in animals: liver cancer; thyroid tumors.

Medical Surveillance: Consider the points of attack in preplacement and periodic physical examinations. Complete blood count (CBC). Complete examination of the nervous system. Liver and kidney function tests. Consider chest X-ray following acute overexposure.

First Aid: If this chemical gets into the eyes, remove any contact lenses at once and irrigate immediately for at least 15 min, occasionally lifting upper and lower lids. Seek medical attention immediately. If this chemical contacts the skin, remove contaminated clothing and wash immediately with soap and water. Seek medical attention immediately. If this chemical has been inhaled, remove from exposure, begin rescue breathing (using universal precautions, including resuscitation mask) if breathing has stopped and CPR if heart action has stopped. Transfer promptly to a medical facility. When this chemical has been swallowed, get medical attention. Give large quantities of water and induce vomiting. Do not make an unconscious person vomit. Medical observation is recommended for 24−48 h after breathing overexposure, as pulmonary edema may be delayed. As first aid for pulmonary edema, a doctor or authorized paramedic may consider administering a corticosteroid spray.

Personal Protective Methods: Wear protective gloves and clothing to prevent any reasonable probability of skin contact. Safety equipment suppliers/manufacturers can provide recommendations on the most protective glove/clothing material for your operation. All protective clothing (suits, gloves, footwear, headgear) should be clean, available each day, and put on before work. Contact lenses should not be worn when working with this chemical. Wear splash-proof chemical goggles and face shield when working with liquid, unless full-face-piece respiratory protection is worn. Employees should wash immediately with soap when skin is wet or contaminated. Provide emergency showers and eyewash.

Respirator Selection: NIOSH: *At any concentrations above the NIOSH REL, or where there is no REL, at any detectable concentration:* SCBAF: Pd,Pp (APF = 10,000) (any self-contained breathing apparatus that has a full face-piece and is operated in a pressure-demand or other positive-pressure mode); or SaF: Pd,Pp: ASCBA (APF = 10,000) (any supplied-air respirator that has a full face-piece and is operated in a pressure-demand or other positive-pressure mode in combination with an auxiliary self-contained breathing apparatus operated in a pressure-demand or other positive-pressure mode). *Escape:* GmFOv100 (APF = 50) [Any air-purifying, full-face-piece respirator (gas mask) with a chin-style, front- or back-mounted organic vapor canister having an N100, R100, or P100 filter]; or SCBAE (any appropriate escape-type, self-contained breathing apparatus).

Storage: Color Code—Blue: Health Hazard/Poison: Store in a secure poison location. Prior to working with this chemical you should be trained on its proper handling and storage. A regulated, marked area should be established where this chemical is handled, used, or stored in compliance with OSHA Standard 1910.1045. Before entering confined space where this chemical may be present, check to make sure that an explosive concentration does not exist. Store in sealed containers in a well-ventilated area away from oxidizers. Store in tightly closed containers in a cool, well-ventilated area away from heat. Sources of ignition, such as smoking and open flames, are prohibited where this chemical is used, handled, or stored in a manner that could create A potential fire or explosion hazard. Metal containers involving the transfer of 5 gallons or more of this chemical should be grounded and bonded. Drums must be equipped with self-closing valves, pressure vacuum bungs; and flame arresters. Use only nonsparking tools and equipment, especially when opening and closing containers of this chemical.

Shipping: Organochlorine pesticides, solid toxic, require a shipping label of "POISONOUS/TOXIC MATERIALS." It falls in Hazard Class 6.1.

Spill Handling: Evacuate persons not wearing protective equipment from area of spill or leak until cleanup is

complete. Remove all ignition sources. Stay upwind; keep out of low areas. Establish forced ventilation to keep levels below explosive limit. Wear boots, protective gloves, goggles, and positive-pressure breathing apparatus. Wash away any material which may have contacted the body with copious amounts of water or soap and water. In case of land spill, dig a pit, pond, lagoon, or holding area to contain the liquid or solid material. Cover solids with a plastic sheet to prevent dissolving in rain or firefighting water. In case of water spill, if camphechlor is dissolved, apply activated carbon at 10 times the spilled amounts in the region of 10 ppm or greater concentration. Remove trapped material with suction hoses. Use mechanical dredges or lifts to remove immobilized masses of pollutants and precipitates. Collect powdered material in the most convenient and safe manner and deposit in sealed containers. Ventilate area after cleanup is complete. It may be necessary to contain and dispose of this chemical as a hazardous waste. If material or contaminated runoff enters waterways, notify downstream users of potentially contaminated waters. Contact your local or federal environmental protection agency for specific recommendations. If employees are required to clean up spills, they must be properly trained and equipped. OSHA 1910.120(q) may be applicable.

Fire Extinguishing: This chemical is a combustible solid that is not easy to ignite, but is usually dissolved in a flammable carrier. Flammability depends on the type of solvent being used. Use dry chemical, carbon dioxide; water spray; or foam extinguishers. Poisonous gases are produced in fire, including chlorine and hydrogen chloride. If material or contaminated runoff enters waterways, notify downstream users of potentially contaminated waters. Notify local health and fire officials and pollution control agencies. From a secure, explosion-proof location, use water spray to cool exposed containers. If cooling streams are ineffective (venting sound increases in volume and pitch, tank discolors, or shows any signs of deforming), withdraw immediately to a secure position. If employees are expected to fight fires, they must be trained and equipped in OSHA 1910.156. The only respirators recommended for firefighting are self-contained breathing apparatuses that have full face-pieces and are operated in a pressure-demand or other positive-pressure mode.

Disposal Method Suggested: Consult with environmental regulatory agencies for guidance on acceptable disposal practices. Generators of waste containing this contaminant (≥100 kg/mo) must conform with EPA regulations governing storage, transportation, treatment, and waste disposal. Incineration of flammable solvent mixture in furnace equipped with afterburner and alkali scrubber.

References
US Environmental Protection Agency. (1980). *Toxaphene: Ambient Water Quality Criteria*. Washington, DC
US Environmental Protection Agency. (1979). *Reviews of the Environmental Effects of Pollutants: X. Toxaphene.* Report No. EPA-600/1-79-044

US Environmental Protection Agency. (April 30, 1980). *Toxaphene, Health and Environmental Effects Profile No. 163*. Washington, DC: Office of Solid Waste
US Environmental Protection Agency. (November 30, 1987). *Chemical Hazard Information Profile: Camphechlor.* Washington, DC: Chemical Emergency Preparedness Program
Sax, N. I. (Ed.). *Dangerous Properties of Industrial Materials Report*, 2, No. 2, 68–70 (1982), 4, No. 1, 27–28 (1984), and 7, No. 5, 100–107 (1987).
New York State Department of Health. (April 1986). *Chemical Fact Sheet: Toxaphene.* Albany, NY: Bureau of Toxic Substance Assessment
US Environmental Protection Agency, Special Review and Reregistration Division Office of Pesticide Programs. (1998). *Agency Status of Pesticides in Registration, Reregistration, and Special Review* (Rainbow Report). Washington, DC
New Jersey Department of Health and Senior Services. (May 2001). *Hazardous Substances Fact Sheet: Toxaphene.* Trenton, NJ

Tributyl phosphate   T:0660

Molecular Formula: $C_{12}H_{27}O_4P$
Common Formula: $(C_4H_9O)_3PO$
Synonyms: Butyl phosphate, tri-; Celluphos 4; Phosphoric acid, tributyl ester; TBP; Tributyle (phosphate de) (French); Tributylphosphat (German); Tri-*n*-butyl Phosphate
CAS Registry Number: 126-73-8
RTECS® Number: IC7700000
UN/NA & ERG Number: No citation.
EC Number: 204-800-2 [*Annex I Index No.:* 015-014-00-2]
Regulatory Authority and Advisory Bodies
Air Pollutant Standard Set. See below, "Permissible Exposure Limits in Air" section.
Canada, WHMIS, Ingredients Disclosure List Concentration 1.0%.
European/International Regulations: Hazard Symbol: Xn; Risk phrases: R22; R38; R40; Safety phrases: S2; S36/37; S45 (see Appendix 4).
WGK (German Aquatic Hazard Class): 2—Hazard to waters.
Description: Tributyl phosphate is a colorless to pale yellow odorless liquid. Molecular weight = 266.36; Specific gravity $(H_2O:1) = 0.98$; Boiling point = 289°C (decomposes); Freezing/Melting point ≤−80°C; Vapor pressure = 0.004 mmHg at 25°C; Autoignition temperature = 410°C; Flash point = 146°C (oc); Autoignition temperature = 410°C. Hazard Identification (based on NFPA-704 M Rating System): Health 2, Flammability 1, Reactivity 0. Poor solubility in water.
Potential Exposure: Tributyl phosphate is used as plasticizer, solvent, and grinding assistant; in inks and as a

dielectric material; as an antifoaming agent. It is also used as a solvent in uranium extraction and as a solvent for cellulose esters. It may be used as a heat exchange medium.

Incompatibilities: Alkalis, strong oxidizers; water, moist air. Reacts with warm water producing phosphoric acid and butanol. Attacks some forms of plastics, rubber, and coatings.

Permissible Exposure Limits in Air

Conversion factor: 1 ppm = 10.89 mg/m³ at 25°C & 1 atm.
OSHA PEL: 5 mg/m³ TWA.
NIOSH REL: 0.2 ppm/2.5 mg/m³ TWA
ACGIH TLV®[1]: 0.2 ppm/2.2 mg/m³ TWA.
NIOSH IDLH: 30 ppm.
Protective Action Criteria (PAC)
TEEL-0: 0.2 ppm
PAC-1: 6 ppm
PAC-2: 30 ppm
PAC-3: 30 ppm
DFG MAK: 1 ppm/11 mg/m³ TWA; Peak Limitation Category II(4) [skin]; Carcinogen Category 4; Pregnancy Risk Group C.
Compound Description: Tumorigen, Mutagen; Reproductive Effector; Primary Irritant.
Australia: TWA 0.2 ppm (2.5 mg/m³), 1993; Belgium: TWA 0.2 ppm (2.2 mg/m³), 1993; Denmark: TWA 0.2 ppm (2.5 mg/m³), 1999; Finland: TWA 5 mg/m³, STEL 10 mg/m³ [skin], 1999; France: VME 0.2 ppm (2.5 mg/m³), 1999; Norway: TWA 0.2 ppm (2.5 mg/m³), 1999; the Philippines: TWA 5 mg/m³, 1993; the Netherlands: MAC-TGG 5 mg/m³, 2003; Switzerland: MAK-W 0.2 ppm (2.5 mg/m³), 1999; United Kingdom: TWA 5 mg/m³, STEL 5 mg/m³, 2000; Argentina, Bulgaria, Columbia, Jordan, South Korea, New Zealand, Singapore, Vietnam: ACGIH TLV®: TWA 0.2 ppm.
Several states have set guidelines or standards for TBP in ambient air[60] ranging from 25 μg/m³ (North Dakota) to 40 μg/m³ (Virginia) to 50 μg/m³ (Connecticut) to 59 μg/m³ (Nevada).

Determination in Air: Use NIOSH Analytical Method (IV) #5034.

Determination in Water: Octanol−water coefficient: Log K_{ow} = 4.02.

Permissible Concentration in Water: Russia[43] set a MAC in water bodies used for domestic purposes of 0.01 mg/L.

Routes of Entry: Inhalation of mist, eye and/or skin contact, ingestion.

Harmful Effects and Symptoms

Short Term Exposure: Tributyl phosphate can affect you when breathed in. Skin or eye contact can cause severe irritation or even burns. Breathing overexposure can irritate the nose, throat, and bronchial tubes and cause headaches, weakness, muscle twitching; nausea, collapse, and even death. Inhalation exposures can cause pulmonary edema, a medical emergency that can be delayed for several hours. This can cause death. Animal data have shown weak anticholinesterase activity.

Long Term Exposure: Similar to other highly irritating substances, Tributyl phosphate may be able to cause lung damage. Animal data have shown weak anticholinesterase activity.

Points of Attack: Eyes, skin, respiratory system.

Medical Surveillance: For those with frequent or potentially high exposure (half the TLV or greater), the following are recommended before beginning work and at regular times after that: Consider lung function tests. If symptoms develop or overexposure is suspected, the following may be useful: consider chest X-ray after acute overexposure.

First Aid: If this chemical gets into the eyes, remove any contact lenses at once and irrigate immediately for at least 15 min, occasionally lifting upper and lower lids. Seek medical attention immediately. If this chemical contacts the skin, remove contaminated clothing and wash immediately with soap and water. Seek medical attention immediately. If this chemical has been inhaled, remove from exposure, begin rescue breathing (using universal precautions, including resuscitation mask) if breathing has stopped and CPR if heart action has stopped. Transfer promptly to a medical facility. When this chemical has been swallowed, get medical attention. Give large quantities of water and induce vomiting. Do not make an unconscious person vomit. Medical observation is recommended for 24−48 h after breathing overexposure, as pulmonary edema may be delayed. As first aid for pulmonary edema, a doctor or authorized paramedic may consider administering a corticosteroid spray.

Personal Protective Methods: Wear protective gloves and clothing to prevent any reasonable probability of skin contact. Safety equipment suppliers/manufacturers can provide recommendations on the most protective glove/clothing material for your operation. All protective clothing (suits, gloves, footwear, headgear) should be clean, available each day, and put on before work. Contact lenses should not be worn when working with this chemical. Wear splash-proof chemical goggles and face shield unless full-face-piece respiratory protection is worn. Employees should wash immediately with soap when skin is wet or contaminated. Provide emergency showers and eyewash.

Respirator Selection: NIOSH: *2 ppm:* Sa (APF = 10) (any supplied-air respirator). *5 ppm:* Sa:Cf (APF = 25) (any supplied-air respirator operated in a continuous-flow mode). *10 ppm:* SCBAF (APF = 50) (any self-contained breathing apparatus with a full face-piece); or SaF (APF = 50) (any supplied-air respirator with a full face-piece). *50 ppm:* SaF: Pd,Pp (APF = 2000) (any supplied-air respirator that has a full face-piece and is operated in a pressure-demand or other positive-pressure mode). *Emergency or planned entry into unknown concentrations or IDLH conditions:* SCBAF: Pd,Pp (APF = 10,000) (any self-contained breathing apparatus that has a full face-piece and is operated in a pressure-demand or other positive-pressure mode); or SaF: Pd,Pp: ASCBA (APF = 10,000) (any supplied-air respirator that has a full face-piece and is operated in a pressure-demand or other positive-pressure mode in combination with an auxiliary self-contained breathing apparatus operated in a pressure-demand or other positive-pressure mode). *Escape:* GmFOv 100 [any air-purifying, full-face-piece respirator (gas mask) with a chin-style, front- or back-mounted organic vapor canister having an N100, R100, or P100 filter]; or SCBAE (any appropriate escape-type, self-contained breathing apparatus).

Storage: Color Code—Blue: Health Hazard/Poison: Store in a secure poison location. Prior to working with this chemical you should be trained on its proper handling and storage. Tributyl phosphate must be stored to avoid contact with strong oxidizers, such as chlorine, bromine, and fluorine, since violent reactions occur. Store in tightly closed containers in a cool, well-ventilated area. Sources of ignition, such as smoking and open flames, are prohibited where tributyl phosphate is used, handled, or stored in a manner that could create a potential fire or explosion hazard.

Spill Handling: Evacuate and restrict persons not wearing protective equipment from area of spill or leak until cleanup is complete. Remove all ignition sources. Ventilate area of spill or leak. Absorb liquids in vermiculite, dry sand; earth, peat, carbon, or a similar material and deposit in sealed containers. Keep this chemical out of a confined space, such as a sewer, because of the possibility of an explosion, unless the sewer is designed to prevent the buildup of explosive concentrations. It may be necessary to contain and dispose of this chemical as a hazardous waste. If material or contaminated runoff enters waterways, notify downstream users of potentially contaminated waters. Contact your local or federal environmental protection agency for specific recommendations. If employees are required to clean up spills, they must be properly trained and equipped. OSHA 1910.120(q) may be applicable.

Fire Extinguishing: This chemical is a combustible liquid. Poisonous gases, including phosphorus oxides; phosphine, and phosphoric acid; are produced in fire. Use dry chemical, carbon dioxide; or foam extinguishers. Vapors are heavier than air and will collect in low areas. Vapors may travel long distances to ignition sources and flashback. Vapors in confined areas may explode when exposed to fire. Containers may explode in fire. Storage containers and parts of containers may rocket great distances, in many directions. If material or contaminated runoff enters waterways, notify downstream users of potentially contaminated waters. Notify local health and fire officials and pollution control agencies. From a secure, explosion-proof location, use water spray to cool exposed containers. If cooling streams are ineffective (venting sound increases in volume and pitch, tank discolors, or shows any signs of deforming), withdraw immediately to a secure position. If employees are expected to fight fires, they must be trained and equipped in OSHA 1910.156. The only respirators recommended for firefighting are self-contained breathing apparatuses that have full face-pieces and are operated in a pressure-demand or other positive-pressure mode.

Disposal Method Suggested: Tributyl phosphate may be recovered from nuclear fuel processing operations.

References

New Jersey Department of Health and Senior Services. (May 2000). *Hazardous Substances Fact Sheet: Tributyl Phosphate.* Trenton, NJ

Trichlorfon   T:0670

Molecular Formula: $C_4H_8Cl_3O_4P$

Common Formula: $(CH_3O)_2POCHOHCCl_3$

Synonyms: Aerol 1 (pesticide); Agroforotox; Anthon; Bay 15922; Bayer 15922; Bayer L 13/59; Bilarcil; Bovinox; Briton; Britten; Cekufon; Chlorak; Chlorofos; Chloroftalm; Chlorophos; Chlorophthalm; Chloroxyphos; Ciclo-som; Combot; Combot equine; Danex; DEP (pesticide); Depthon; DETF; Dimethoxy-2,2,2-trichloro-1-hydroxy-ethylphosphine oxide; *O,O*-Dimethyl (1-hydroxy-2,2,2-trichloraethyl)phosphat (German); *O,O*-Dimethyl (1-hydroxy-2,2,2-trichloraethyl)phosphonsaeure ester (German); *O,O*-Dimethyl (1-hydroxy-2,2,2-trichloroethyl) phosphonate; Dimethyl 1-hydroxy-2,2,2-trichloroethylphosphonate; *O,O*-Dimethyl (2,2,2-trichloro-1-hydroxyethyl) phosphonate; Dimethyl (2,2,2-trichloro-1-hydroxyethyl)phosphonate; Dimetox; Dipterex; Dipterex 50; Diptevur; Ditrifon; Dylox; Dylox-Metasystox-R; Dyrex; Dyvon; ENT 19,763; Equino-acid; Equino-aid; Flibol E; Fliegenteller; Forotox; Foschlor; Foschlor R; Foschlor R-50; 1-Hydroxy-2,2,2-trichloroethyl-phosphonic acid dimethyl ester; Hypodermacid; Leivasom; Loisol; Masoten; Mazoten; Methyl chlorophos; Metifonate; Metrifonate; Metriphonate; NCI-C54831; Neguvon; Neguvon A; Phoschlor; Phoschlor R50; Phosphonic acid (2,2,2-trichloro-1-hydroxyethyl)-, dimethyl ester; Polfoschlor; Proxol; Ricifon; Ritsifon; Satox 20WSC; Soldep; Sotipox; 2,2,2-Trichloro-1-hydroxyethyl-phosphonate, dimethyl ester; (2,2,2-Trichloro-1-hydroxyethyl)phosphonic acid dimethyl ester; Trichlorophene; Trichlorphon; Trichlorphon FN; Trinex; Tugon; Tugon fly bait; Tugon stable spray; Vermicide bayer 2349; Volfartol; Votexit; WEC 50; Wotexit

CAS Registry Number: 52-68-6

RTECS® Number: AO700000

UN/NA & ERG Number: UN2783 (organophosphorus pesticides, solid, toxic)/152

EC Number: 200-149-3 [*Annex I Index No.:* 015-021-00-0]

Regulatory Authority and Advisory Bodies

Carcinogenicity: IARC: Animal, Inadequate Evidence; Human No Adequate Data, *not classifiable as carcinogenic to humans*, Group 3, 1987.

US EPA Gene-Tox Program, Positive: Body fluid assay; Host-mediated assay; Positive: Histidine reversion—Ames test; *S. cerevisiae*—homozygosis; Weakly Positive: *In vitro* UDS—human fibroblast; Positive/dose response: TRP reversion; Negative: *D. melanogaster* sex-linked lethal;

Inconclusive: Mammalian micronucleus; *B. subtilis* rec assay; Inconclusive: *E. coli* polA without S9.

US EPA, FIFRA 1998 Status of Pesticides: RED completed. Air Pollutant Standard Set. See below, "Permissible Exposure Limits in Air" section.

Clean Water Act: Section 311 Hazardous Substances/RQ 40CFR117.3 (same as CERCLA, see below).

Reportable Quantity (RQ): 100 lb (45.4 kg).

Dropped from Extremely Hazardous Substance (EPA-SARA) in 1988.

EPCRA Section 313 Form R *de minimis* concentration reporting level: 1.0%.

US DOT 49CFR172.101, Inhalation Hazard Chemical as organophosphates.

US DOT Regulated Marine Pollutant (49CFR172.101, Appendix B).

European/International Regulations: Hazard Symbol: Xn, N; Risk phrases: R22; R43; R50/53; Safety phrases: S2; S24; S37; S60; S61 (see Appendix 4).

WGK (German Aquatic Hazard Class): 3—Severe hazard to waters.

Description: Trichlorfon is a white to pale yellow crystalline solid. Molecular weight $= 257.44$; Specific gravity $(H_2O:1) = 1.73$ at 25°C; Boiling point $= 100°C$ at 1 mmHg; Freezing/Melting point $= 83-84°C$; Vapor pressure $= 2 \times 10^{-6}$ mmHg at 20°C. Hazard Identification (based on NFPA-704 M Rating System): Health 2, Flammability 1, Reactivity 0. Soluble in water; solubility $= 15.4$ g/100 mL at 25°C.

Potential Exposure: Compound Description: Agricultural Chemical; Tumorigen, Drug, Mutagen; Reproductive Effector; Human Data; Primary Irritant. Trichlorfon is used as an agricultural and forest insecticide.

Incompatibilities: Alkaline materials: lime, lime sulfur, etc. Corrosive to iron and steel.

Permissible Exposure Limits in Air

OSHA PEL: None.

NIOSH REL: None.

ACGIH TLV®[1]: 1 mg/m^3 (intermittent) TWA, sensitizer, not classifiable as a human carcinogen; BEI$_A$ issued for acetylcholinesterase inhibiting pesticides.

Protective Action Criteria (PAC)

TEEL-0: 1 mg/m^3

PAC-1: 3 mg/m^3

PAC-2: 13 mg/m^3

PAC-3: 500 mg/m^3

Poland: MAC (TWA) 0.5 mg/m^3, MAC (STEL) 2 mg/m^3, 1999; Russia: STEL 0.5 mg/m^3 [skin] 1993. Russia[43] set a MAC in residential areas of 0.04 mg/m^3 on a momentary basis and 0.02 mg/m^3 on a daily average basis.

Determination in Air: OSHA versatile sampler-2; Toluene/Acetone; Gas chromatography/Flame photometric detection for sulfur, nitrogen, or phosphorus; NIOSH Analytical Method (IV) Method #5600, Organophosphorus Pesticides.

Permissible Concentration in Water: Russia[35,43] set a MAC in water bodies used for domestic purposes of 0.05 mg/L and in water bodies used for fishery purposes of zero.

Determination in Water: Fish Tox $= 24.99773000$ (ppb) (INTERMEDIATE). Octanol—water coefficient: Log $K_{ow} = 0.48$.

Routes of Entry: Inhalation, ingestion, skin contact.

Harmful Effects and Symptoms

Short Term Exposure: Very toxic: probable oral lethal dose (human) 50—500 mg/kg, between 1 teaspoon and 1 oz for 150 lb (70 kg) person. Toxicity is relatively low among organic phosphate insecticides, although it is a potent cholinesterase inhibitor. Skin sensitivity has been reported. Symptoms of exposure: muscle weakness; twitching, respiratory depression; sweating, vomiting, diarrhea, chest and abdominal distress; sometimes pulmonary edema; excessive salivation; headache, giddiness, vertigo, and weakness; runny nose and sensation of tightness in chest (inhalation), blurring of vision; tearing, ocular pain; loss of muscle coordination; and slurring of speech.

Long Term Exposure: Cholinesterase inhibitor; cumulative effect is possible. This chemical may damage the nervous system with repeated exposure, resulting in convulsions, respiratory failure. May cause liver damage. Human Tox; 14.00000 ppb (INTERMEDIATE).

Points of Attack: Respiratory system, lungs, central nervous system; cardiovascular system, skin, eyes, plasma, and red blood cell cholinesterase.

Medical Surveillance: Before employment and at regular times after that, the following are recommended: plasma and red blood cell cholinesterase levels (tests for the enzyme poisoned by this chemical). If exposure stops, plasma levels return to normal in 1—2 weeks while red blood cell levels may be reduced for 1—3 months. When cholinesterase enzyme levels are reduced by 25% or more below preemployment levels, risk of poisoning is increased, even if results are in lower ranges of "normal." Reassignment to work not involving organophosphate or carbamate pesticides is recommended until enzyme levels recover. If symptoms develop or overexposure occurs, repeat the above tests as soon as possible and get an exam of the nervous system. Also consider complete blood count. Consider chest X-ray following acute overexposure. Do not drink any alcoholic beverages before or during use. Alcohol promotes absorption of organic phosphates.

First Aid: If this chemical gets into the eyes, remove any contact lenses at once and irrigate immediately for at least 15 min, occasionally lifting upper and lower lids. Seek medical attention immediately. If this chemical contacts the skin, remove contaminated clothing and wash immediately with soap and water. Speed in removing material from skin is of extreme importance. Shampoo hair promptly if contaminated. Seek medical attention immediately. If this chemical has been inhaled, remove from exposure, begin rescue breathing (using universal precautions, including resuscitation mask) if breathing has stopped and CPR if heart action has stopped. Transfer promptly to a medical facility. When this chemical has been swallowed, get medical attention. Give large quantities of water and induce vomiting. Do not make an unconscious person vomit. Medical observation is recommended for 24–48 h after breathing overexposure, as pulmonary edema may be delayed. As first aid for pulmonary edema, a doctor or authorized paramedic may consider administering a corticosteroid spray.

Personal Protective Methods: Wear protective gloves and clothing to prevent any reasonable probability of skin contact. Safety equipment suppliers/manufacturers can provide recommendations on the most protective glove/clothing material for your operation. All protective clothing (suits, gloves, footwear, headgear) should be clean, available each day, and put on before work. Contact lenses should not be worn when working with this chemical. Wear dust-proof chemical goggles and face shield unless full-face-piece respiratory protection is worn. Employees should wash immediately with soap when skin is wet or contaminated. Provide emergency showers and eyewash.

Respirator Selection: Follow regulations in OSHA 29CFR1910.134 or European Standard EN149. Use a NIOSH/MSHA- or European Standard EN149-approved respirator; or use an approved supplied-air respirator with a full face-piece operated in the positive-pressure mode, or with a full face-piece, hood, or helmet in the continuous-flow mode; or use a NIOSH/MSHA- or European Standard EN149-approved self-contained breathing apparatus with a full face-piece operated in pressure-demand or other positive-pressure mode. A/P2 filter respirator for organic vapor and harmful dust. *Escape:* GmFOv 100 [any air-purifying, full-face-piece respirator (gas mask) with a chin-style, front- or back-mounted organic vapor canister having an N100, R100, or P100 filter]; or SCBAE (any appropriate escape-type, self-contained breathing apparatus).

Storage: Color Code—Blue: Health Hazard/Poison: Store in a secure poison location. Prior to working with this chemical you should be trained on its proper handling and storage. Store in tightly closed containers in a cool, well-ventilated area away from alkaline materials. Where possible, automatically transfer material from other storage containers to process containers.

Shipping: This compound requires a shipping label of "POISONOUS/TOXIC MATERIALS." It falls in Hazard Class 6.1 and Parking Group III.

Spill Handling: As for other organophosphorus pesticides, stay upwind; keep out of low areas. Ventilate closed spaces before entering them. Wear positive-pressure breathing apparatus and special protective clothing. Do not touch spilled material; stop leak if you can do so without risk. Use water spray to reduce vapors. *Small spills:* absorb with sand or other noncombustible absorbent material and place into containers for later disposal. *Small dry spills:* with clean shovel place material into clean dry containers and cover; move containers from spill area. *Large spills:* dike far ahead of spill for later disposal. Ventilate area after cleanup is complete. It may be necessary to contain and dispose of this chemical as a hazardous waste. If material or contaminated runoff enters waterways, notify downstream users of potentially contaminated waters. Contact your local or federal environmental protection agency for specific recommendations. If employees are required to clean up spills, they must be properly trained and equipped. OSHA 1910.120(q) may be applicable. Soil Adsorption Index $(K_{oc}) = 10$.

Fire Extinguishing: This material may burn, but does not ignite readily. For small fires, use dry chemical, carbon dioxide; water spray; or foam. For large fires, use water spray, fog, or foam. Poisonous gases are produced in fire, including chlorine and phosphorus oxides. If material or contaminated runoff enters waterways, notify downstream users of potentially contaminated waters. Notify local health and fire officials and pollution control agencies. From a secure, explosion-proof location, use water spray to cool exposed containers. If cooling streams are ineffective (venting sound increases in volume and pitch, tank discolors, or shows any signs of deforming), withdraw immediately to a secure position. If employees are expected to fight fires, they must be trained and equipped in OSHA 1910.156. The only respirators recommended for firefighting are self-contained breathing apparatuses that have full face-pieces and are operated in a pressure-demand or other positive-pressure mode.

Disposal Method Suggested: Add a combustible solvent and burn in a furnace equipped with an afterburner and an alkali scrubber.[24] In accordance with 40CFR165, follow recommendations for the disposal of pesticides and pesticide containers. Must be disposed properly by following package label directions or by contacting your local or federal environmental control agency, or by contacting your regional EPA office.

References

Sax, N. I. (Ed.). (1987). *Dangerous Properties of Industrial Materials Report*, 7, No. 2, 95–101

US Environmental Protection Agency. (October 31, 1985). *Chemical Hazard Information Profile: Trichlorophon.* Washington, DC: Chemical Emergency Preparedness Program

US Environmental Protection Agency, Special Review and Reregistration Division Office of Pesticide Programs. (1998). *Agency Status of Pesticides in Registration, Reregistration, and Special Review* (Rainbow Report). Washington, DC

New Jersey Department of Health and Senior Services. (November 2004). *Hazardous Substances Fact Sheet: Trichlorofon.* Trenton, NJ

Trichloroacetic acid   T:0680

Molecular Formula: $C_2HCl_3O_2$
Common Formula: Cl_3CCOOH

Synonyms: Acetic acid, trichloro-; Aceto-caustin; Acide trichloracetique (French); Amchem grass killer; Dow sodium TCA solution; Konesta; Sodium TCA solution; TCA; Trichloressigsaeure (German); Trichloroethanoic acid; Varitox

CAS Registry Number: 76-03-9

RTECS® Number: AJ7875000

UN/NA & ERG Number: UN1839 (solid)/153; UN2564 (solution)/153

EC Number: 200-927-2 [*Annex I Index No.:* 607-004-00-7]

Regulatory Authority and Advisory Bodies

Carcinogenicity: IARC: Animal, Limited Evidence; Human Inadequate Evidence, *not classifiable as carcinogenic to humans*, Group 3, 1999; EPA: Possible Human Carcinogen.

US EPA Gene-Tox Program, Inconclusive: Histidine reversion—Ames test.

US EPA, FIFRA 1998 Status of Pesticides: Canceled.

Air Pollutant Standard Set. See below, "Permissible Exposure Limits in Air" section.

Canada, WHMIS, Ingredients Disclosure List Concentration 1.0%.

European/International Regulations: Hazard Symbol: C, N; Risk phrases: R35; R50/53; Safety phrases: S1/2; S26; S36/37/39; S45; S60; S61 (see Appendix 4).

WGK (German Aquatic Hazard Class): 2—Hazard to waters.

Description: Trichloroacetic acid is a colorless crystalline solid which is used in liquid solutions. Molecular weight = 163.38; Specific gravity (H_2O:1) = 1.62 at 25°C; Boiling point = 197.8°C; Freezing/Melting point = 57.8°C; Vapor pressure = 1 mmHg at 51°C. Hazard Identification (based on NFPA-704 M Rating System): Health 2, Flammability 0, Reactivity 0. Soluble in water.

Potential Exposure: Compound Description: Agricultural Chemical; Drug, Tumorigen, Mutagen; Reproductive Effector; Primary Irritant. Trichloroacetic acid is used as medication, in organic syntheses; and as a reagent for albumin detection; as an intermediate in pesticide manufacture and in the production of sodium trichloroacetate which is itself an herbicide.

Incompatibilities: Moisture, iron, zinc, aluminum, strong oxidizers. Corrosive to metals.

Permissible Exposure Limits in Air

Conversion factor: 1 ppm = 6.68 mg/m^3 at 25°C & 1 atm.

OSHA PEL: None.

NIOSH REL: 1 ppm/7 mg/m^3 TWA.

ACGIH TLV®[1]: 1 ppm/6.7 mg/m^3 TWA, confirmed animal carcinogen with unknown relevance to humans.

Protective Action Criteria (PAC)

TEEL-0: 1 ppm

PAC-1: 1 ppm

PAC-2: 2 ppm

PAC-3: 25 ppm

Austria: MAK 1 ppm (5 mg/m^3), 1999; Belgium: TWA 1 ppm (6.7 mg/m^3), 1993; Denmark: TWA 1 mg/m^3, 1999; France: VME 1 ppm (5 mg/m^3), 1999; Norway: TWA 0.75 ppm (5 mg/m^3), 1999; Russia: STEL 5 mg/m^3 [skin], 1993; Switzerland: MAK-W 1 ppm (7 mg/m^3), 1999; the Netherlands: MAC-TGG 1 mg/m^3, 2003; Argentina, Bulgaria, Columbia, Jordan, South Korea: ACGIH TLV®: confirmed animal carcinogen with unknown relevance to humans. Several states have set guidelines or standards for TCA in ambient air[60] ranging from 70 μg/m^3 (North Dakota) to 80 μg/m^3 (Virginia) to 100 μg/m^3 (Connecticut) to 119 μg/m^3 (Nevada).

Determination in Air: No method available.

Determination in Water: Fish Tox: 538434.64955000 ppb (VERY LOW).

Routes of Entry: Inhalation, ingestion; skin and/or eye contact.

Harmful Effects and Symptoms

Short Term Exposure: Inhalation: Causes irritation to respiratory tract with choking, coughing, dizziness, and weakness. Swelling of throat and lungs can occur. Higher exposures can cause pulmonary edema, a medical emergency that can be delayed for several hours. This can cause death. *Skin:* Corrosive to the skin. Burns and blisters may result if not removed promptly. May cause thickening of skin. *Eyes:* Corrosive to the eyes. Can cause extremely painful burns and sores on eyes, which can result in blindness. *Ingestion:* May cause intense burning of mouth, throat, and stomach; vomiting, diarrhea, and fatigue. Throat may swell to block airway. The estimated lethal dose is about 1 g (1/30 oz).

Long Term Exposure: Fumes may produce irritation of throat and lungs with persistent cough. Disturbances of the digestive tract may also be noticed. These should only be significant at levels above the recommended occupational exposure limit. Human Tox = 60.00000 ppb (LOW).

Points of Attack: Eyes, skin, respiratory system; gastrointestinal tract.

Medical Surveillance: For those with frequent or potentially high exposure (half the TLV or greater) the following are recommended before beginning work and at regular times after that: lung function tests. If symptoms develop or overexposure is suspected, the following may be useful: consider chest X-ray after acute overexposure.

First Aid: If this chemical gets into the eyes, remove any contact lenses at once and irrigate immediately for at least 15 min, occasionally lifting upper and lower lids. Seek medical attention immediately. If this chemical contacts the skin, remove contaminated clothing and wash immediately with soap and water. Seek medical attention immediately. If this chemical has been inhaled, remove from exposure, begin rescue breathing (using universal precautions, including resuscitation mask) if breathing has stopped and CPR if heart action has stopped. Transfer promptly to a medical facility. When this chemical has been swallowed, get medical attention. If victim is *conscious*, administer water or milk. Do not induce vomiting. Medical observation is recommended for 24—48 h after breathing overexposure, as pulmonary edema may be delayed. As first aid for pulmonary edema, a doctor or authorized paramedic may consider administering a corticosteroid spray.

Personal Protective Methods: Wear protective gloves and clothing to prevent any reasonable probability of skin contact. Safety equipment suppliers/manufacturers can provide recommendations on the most protective glove/clothing material for your operation. All protective clothing (suits, gloves, footwear, headgear) should be clean, available each day, and put on before work. Contact lenses should not be worn when working with this chemical. Wear dust-proof goggles and face shield when working with powders or dust, unless full-face-piece respiratory protection is worn. Wear splash-proof chemical goggles and face shield when working with liquid, unless full-face-piece respiratory protection is worn. Employees should wash immediately with soap when skin is wet or contaminated. Provide emergency showers and eyewash.

Respirator Selection: Where there is potential for exposures *over 1 ppm*, use a NIOSH/MSHA- or European Standard EN149-approved respirator with a high-efficiency particulate filter. More protection is provided by a full-face-piece respirator than by a half-mask respirator, and even greater protection is provided by a powered, air-purifying respirator. Particulate filters must be checked every day before work for physical damage, such as rips or tears, and replaced as needed. If trichloroacetic acid is a liquid, where the potential exists for exposures *over 1 ppm*, use a NIOSH/MSHA- or European Standard EN149-approved full-face-piece respirator with an acid gas canister. Increased protection is obtained from full-face-piece powered, air-purifying respirators.

Storage: Color Code—White: Corrosive or Contact Hazard; Store separately in a corrosion-resistant location. Prior to working with this chemical you should be trained on its proper handling and storage. Store in a cool, dry, well-ventilated place away from strong oxidizers, strong bases. Where possible, automatically transfer material from other storage containers to process containers.

Shipping: Trichloroacetic acid, solid, requires a shipping label of "CORROSIVE." Trichloroacetic acid, solution. Both fall in DOT Hazard Class 8 and Packing Group II.

Spill Handling: Evacuate persons not wearing protective equipment from area of spill or leak until cleanup is complete. Remove all ignition sources. Wearing protective equipment and clothing, spread soda ash on spill and mop up with water. Shovel slurry into appropriate container. Ventilate area after cleanup is complete. It may be necessary to contain and dispose of this chemical as a hazardous waste. If material or contaminated runoff enters waterways, notify downstream users of potentially contaminated waters. Contact your local or federal environmental protection agency for specific recommendations. If employees are required to clean up spills, they must be properly trained and equipped. OSHA 1910.120(q) may be applicable. Soil Adsorption Index (K_{oc}) = 3.

Fire Extinguishing: Extinguish fire using an agent suitable for type of surrounding fire. Trichloroacetic acid itself does not burn. Poisonous gases are produced in fire, including chlorine, chloroform, phosgene, and hydrogen chloride. If material or contaminated runoff enters waterways, notify downstream users of potentially contaminated waters. Notify local health and fire officials and pollution control agencies. From a secure, explosion-proof location, use water spray to cool exposed containers. If cooling streams are ineffective (venting sound increases in volume and pitch, tank discolors, or shows any signs of deforming), withdraw immediately to a secure position. If employees are expected to fight fires, they must be trained and equipped in OSHA 1910.156. The only respirators recommended for firefighting are self-contained breathing apparatuses that have full face-pieces and are operated in a pressure-demand or other positive-pressure mode.

References

New York State Department of Health. (March 1986). *Chemical Fact Sheet: Trichloroacetic Acid.* Albany, NY: Bureau of Toxic Substance Assessment

US Environmental Protection Agency, Special Review and Reregistration Division Office of Pesticide Programs. (1998). *Agency Status of Pesticides in Registration, Reregistration, and Special Review* (Rainbow Report). Washington, DC

New Jersey Department of Health and Senior Services. (May 2004). *Hazardous Substances Fact Sheet: Trichloroacetic Acid.* Trenton, NJ

Trichloroacetyl chloride   T:0690

Molecular Formula: C_2Cl_4O

Common Formula: Cl_3CCOCl

Synonyms: Acetyl chloride, trichloro-; Cloruro de tricloroacetilo (Spanish); NSC 190466; Superpalite; Trichloroacetic acid chloride; Trichloroacetochloride

CAS Registry Number: 76-02-8

RTECS® Number: AO7140000

UN/NA & ERG Number: UN2442/156

EC Number: 200-926-7

Regulatory Authority and Advisory Bodies

Air Pollutant Standard Set. See below, "Permissible Exposure Limits in Air" section.

Superfund/EPCRA 40CFR355, Extremely Hazardous Substances: TPQ = 500 lb (127.5 kg).

Reportable Quantity (RQ): 500 lb (127.5 kg).

EPCRA Section 313 Form R *de minimis* concentration reporting level: 1.0%.

US DOT 49CFR172.101, Inhalation Hazardous Chemical.

Canada, WHMIS, Ingredients Disclosure List Concentration 1.0%.

European/International Regulations: Not listed in Annex 1.

WGK (German Aquatic Hazard Class): No value assigned.

Description: Trichloroacetyl chloride is a clear liquid. Molecular weight = 181.82; Boiling point = 118°C; Freezing/Melting point = −146°C. Hazard Identification (based on NFPA-704 M Rating System): Health 3, Flammability 0, Reactivity 1W. Decomposes in water.

Incompatibilities: Violent reaction with water, forming hydrochloric acid and trichloroacetic acid. Keep away from strong oxidizers.

Potential Exposure: Used in chemical syntheses.

Permissible Exposure Limits in Air

Protective Action Criteria (PAC)

TEEL-0: 0.03 ppm

PAC-1: 0.075 ppm

PAC-2: 0.606 ppm

PAC-3: 6 ppm

Russia[43] set a MAC in work-place air[43] of 0.1 mg/m^3.

Routes of Entry: Inhalation, ingestion, eye, and/or skin contact.

Harmful Effects and Symptoms

Short Term Exposure: Corrosive. Severe irritation to the eyes, skin, and respiratory tract. Skin or eye contact may cause burns. Moderately toxic by ingestion and inhalation. High exposures can cause pulmonary edema, a medical emergency that can be delayed for several hours. This can cause death.

Long Term Exposure: Highly irritating substances may affect the lungs; bronchitis may develop.

Points of Attack: Lungs.

Medical Surveillance: Lung function tests. Consider chest X-ray following acute overexposure.

First Aid: If this chemical gets into the eyes, remove any contact lenses at once and irrigate immediately for at least 15 min, occasionally lifting upper and lower lids. Seek medical attention immediately. If this chemical contacts the skin, remove contaminated clothing and wash immediately with soap and water. Seek medical attention immediately. If this chemical has been inhaled, remove from exposure, begin rescue breathing (using universal precautions, including resuscitation mask) if breathing has stopped and CPR if heart action has stopped. Transfer promptly to a medical facility. When this chemical has been swallowed, get medical attention. If victim is *conscious*, administer water or milk. Do not induce vomiting. Medical observation is recommended for 24−48 h after breathing overexposure, as pulmonary edema may be delayed. As first aid for pulmonary edema, a doctor or authorized paramedic may consider administering a corticosteroid spray.

Personal Protective Methods: Wear protective gloves and clothing to prevent any reasonable probability of skin contact. Safety equipment suppliers/manufacturers can provide recommendations on the most protective glove/clothing material for your operation. All protective clothing (suits, gloves, footwear, headgear) should be clean, available each day, and put on before work. Contact lenses should not be worn when working with this chemical. Wear splash-proof chemical goggles and face shield unless full-face-piece respiratory protection is worn. Employees should wash immediately with soap when skin is wet or contaminated. Provide emergency showers and eyewash.

Respirator Selection: Follow regulations in OSHA 29CFR1910.134 or European Standard EN149. Use a NIOSH/MSHA- or European Standard EN149-approved respirator; or use an approved supplied-air respirator with a full face-piece operated in the positive-pressure mode, or with a full face-piece, hood, or helmet in the continuous-flow mode; or use a NIOSH/MSHA- or European Standard EN149-approved self-contained breathing apparatus with a full face-piece operated in pressure-demand or other positive-pressure mode.

Storage: Color Code—White: Corrosive or Contact Hazard; Store separately in a corrosion-resistant location. Prior to working with this chemical you should be trained on its proper handling and storage. Store in tightly closed containers in a cool, well-ventilated area. Where possible, automatically transfer material from other storage containers to process containers.

Shipping: This compound requires a shipping label of "CORROSIVE, POISONOUS/TOXIC MATERIALS." It falls in Hazard Class 8 and Packing Group II.

Spill Handling: Evacuate and restrict persons not wearing protective equipment from area of spill or leak until cleanup is complete. Remove all ignition sources. Ventilate area of spill or leak. Absorb liquids in vermiculite, dry sand; earth, peat, carbon, or a similar material and deposit in sealed containers. Keep this chemical out of a confined space, such as a sewer, because of the possibility of an explosion, unless the sewer is designed to prevent the buildup of explosive concentrations. It may be necessary to contain and dispose of this chemical as a hazardous waste. If material or contaminated runoff enters waterways, notify downstream users of potentially contaminated waters. Contact your local or federal environmental protection agency for specific recommendations. If employees are required to clean up spills, they must be properly trained and equipped. OSHA 1910.120(q) may be applicable.

Fire Extinguishing: Material may burn but does not ignite readily. Material may react violently with water. Extinguish with dry chemical, carbon dioxide; water spray, fog, or foam. Move container from fire area if you can do so without risk. Spray cooling water on containers that are exposed to flames until well after fire is out. Keep unnecessary people away; isolate hazard area and deny entry. Stay upwind; keep out of low areas. Wear positive-pressure breathing apparatus and special protective clothing. Poisonous gases are produced in fire. Vapors are heavier than air and will collect in low areas. Containers may explode in fire. Storage containers and parts of containers may rocket great distances, in many directions. If material or contaminated runoff enters waterways, notify downstream users of potentially contaminated waters. Notify local health and fire officials and pollution control agencies. From a secure, explosion-proof location, use water spray to cool exposed containers. If cooling streams are ineffective (venting sound increases in volume and pitch, tank discolors, or shows any signs of deforming), withdraw immediately to a secure position. If employees are expected to fight fires, they must be trained and equipped in OSHA 1910.156. The only respirators recommended for firefighting are self-contained breathing apparatuses that have full face-pieces and are operated in a pressure-demand or other positive-pressure mode.

References
US Environmental Protection Agency. (November 30, 1987). *Chemical Hazard Information Profile: Trichloroacetyl Chloride*. Washington, DC: Chemical Emergency Preparedness Program

1,2,4-Trichlorobenzene   T:0700

Molecular Formula: $C_6H_3Cl_3$
Synonyms: Benzene, 1,2,4-trichloro-; Hostetex L-PEC; 1,2,5-Trichlorobenzene; 1,3,4-Trichlorobenzene; *asym*-Trichlorobenzene; 1,2,4-Trichlorobenzol; 1,2,4-Triclorobenceno (Spanish)
CAS Registry Number: 120-82-1
RTECS® Number: DC2100000
UN/NA & ERG Number: UN2321/153
EC Number: 204-428-0 [*Annex I Index No.:* 602-087-00-6]
Regulatory Authority and Advisory Bodies
Carcinogenicity: EPA: Not Classifiable as to human carcinogenicity.
Air Pollutant Standard Set. See below, "Permissible Exposure Limits in Air" section.
RCRA, 40CFR261, Appendix 8 Hazardous Constituents, waste number not listed.
RCRA 40CFR268.48; 61FR15654, Universal Treatment Standards: Wastewater (mg/L), 0.055; Nonwastewater (mg/kg), 19.
RCRA 40CFR264, Appendix 9; TSD Facilities Ground Water Monitoring List. Suggested test method(s) (PQL µg/L): 8270 (10).
Safe Drinking Water Act: Regulated chemical (47 FR 9352) as trichlorobenzene.
Safe Drinking Water Act: MCL, 0.07 mg/L; MCLG, 0.7 mg/L.
Reportable Quantity (RQ): 100 lb (45.4 kg).
EPCRA Section 313 Form R *de minimis* concentration reporting level: 1.0%.
US DOT Regulated Marine Pollutant (49CFR172.101, Appendix B) as trichlorobenzenes, liquid.
Canada, WHMIS, Ingredients Disclosure List Concentration 1.0%.
European/International Regulations: Hazard Symbol: Xn, N; Risk phrases: R22; R38; R50/53; Safety phrases: S2; S37/39; S60; S61 (see Appendix 4).
WGK (German Aquatic Hazard Class): 3—Severe hazard to waters.
Description: 1,2,4-Trichlorobenzene is a low-melting solid or liquid with a pleasant, aromatic odor. The odor threshold

is 1.4 ppm. Molecular weight = 181.44; Specific gravity $(H_2O:1)$ = 1.45 at 25°C; Boiling point = 213.3°C; Freezing/Melting point = 17.2°C; Vapor pressure = 1 mmHg; Flash point = 105°C; Autoignition temperature = 571°C. Explosive limits: LEL = 2.5%; UEL = 6.6%, both at 150°C. Hazard Identification (based on NFPA-704 M Rating System): Health 2, Flammability 1, Reactivity 0. Insoluble in water; solubility = 0.003%.
Potential Exposure: Compound Description: Tumorigen, Mutagen; Reproductive Effector; Human Data; Primary Irritant. 1,2,4-Trichlorobenzene is used as a dye carrier, herbicide intermediate; a heat transfer medium; a dielectric fluid in transformers; a degreaser; a lubricant; as an industrial chemical; solvent, emulsifier, and as a potential insecticide against termites. The other trichlorobenzene isomers are not used in any quantity.
Incompatibilities: Reacts violently with oxidants, acids, acid fumes; steam.
Permissible Exposure Limits in Air
Conversion factor: 1 ppm = 7.42 mg/m³ at 25°C & 1 atm.
OSHA PEL: None.
NIOSH REL: 5 ppm/40 mg/m³ Ceiling Concentration.
ACGIH TLV®[1]: 5 ppm/37 mg/m³ Ceiling Concentration.
European OEL: 2 ppm/15.1 mg/m³ TWA; 5 ppm, 37.8 mg/m³ STEL [skin].[2003]
Protective Action Criteria (PAC)
TEEL-0: 0.25 ppm
PAC-1: 0.75 ppm
PAC-2: 5 ppm
PAC-3: 40 ppm
DFG MAK: [skin] Carcinogen Category 3B.
Australia: TWA 5 ppm (40 mg/m³), 1993; Austria: MAK 5 ppm (40 mg/m³), 1999; Belgium: STEL 5 ppm (37 mg/m³), 1993; Denmark: TWA 5 ppm (40 mg/m³), 1999; Finland: TWA 5 ppm (40 mg/m³), STEL 10 ppm (74 mg/m³) [skin], 1999; France: VME 5 ppm (40 mg/m³), 1999; the Netherlands: MAC-TGG 15.1 mg/m³ [skin], 2003; Switzerland: MAK-W 5 ppm (40 mg/m³), 1999; United Kingdom: TWA 1 ppm (7.6 mg/m³) [skin], 2000; Argentina, Bulgaria, Columbia, Jordan, South Korea, New Zealand, Singapore, Vietnam: ACGIH TLV®: Ceiling Concentration 5 ppm.
Several states have set guidelines or standards for 1,2,4-trichlorobenzene in ambient air[60] ranging from 133 µg/m³ (New York) to 350 µg/m³ (Virginia) to 400 µg/m³ (North Dakota and South Carolina) to 800 µg/m³ (Connecticut) to 952 µg/m³ (Nevada).
Determination in Air: Use NIOSH Analytical Method (IV) #5517, Polychlorobenzenes.
Determination in Water: Octanol−water coefficient: Log K_{ow} = 3.98.
Permissible Concentration in Water: To protect human health—no criterion developed due to insufficient data.[6]
Several states have set guidelines or standards for 1,2,4-trichlorobenzene in drinking water.[61] These include a guideline of 13 µg/L (Kansas) and a standard of 8 µg/L

(New Jersey). Russia[43] set a MAC for water bodies used for domestic purposes of 30 μg/L.

Determination in Water: Methylene chloride extraction followed by concentration, gas chromatography with electron capture detection (EPA Method #612) or gas chromatography plus mass spectrometry (EPA Method #625).

Routes of Entry: Inhalation, skin absorption; ingestion; skin and/or eye contact.

Harmful Effects and Symptoms

Short Term Exposure: Inhalation: May cause irritation to the nose and throat, nervousness, restlessness, tremors, increased heart rate and blood pressure; weakness, digestive disturbances; weight loss; and headache.

Skin: May cause severe irritation. Prolonged contact may cause skin burns.

Eyes: Causes irritation. Levels greater than 5 ppm may cause severe irritation.

Ingestion: Animal studies suggest that a dose of 2 oz may cause liver damage and death.

Long Term Exposure: Removes the skin's natural oils, causing drying and cracking. Possible teratogenic effects. May cause liver and kidney damage.

Points of Attack: Eyes, skin, respiratory system; liver, reproductive system.

Medical Surveillance: If symptoms develop or overexposure is suspected, the following may be useful: liver and kidney function test. Urinary 2,5-dichlorophenol excretion test.

First Aid: If this chemical gets into the eyes, remove any contact lenses at once and irrigate immediately for at least 15 min, occasionally lifting upper and lower lids. Seek medical attention immediately. If this chemical contacts the skin, remove contaminated clothing and wash immediately with soap and water. Seek medical attention immediately. If this chemical has been inhaled, remove from exposure, begin rescue breathing (using universal precautions, including resuscitation mask) if breathing has stopped and CPR if heart action has stopped. Transfer promptly to a medical facility. When this chemical has been swallowed, get medical attention. Give large quantities of water. Do not make an unconscious person vomit.

Personal Protective Methods: Wear protective gloves and clothing to prevent any reasonable probability of skin contact. Safety equipment suppliers/manufacturers can provide recommendations on the most protective glove/clothing material for your operation. Teflon™, Neoprene™, and polyvinyl alcohol are among the recommended protective materials. All protective clothing (suits, gloves, footwear, headgear) should be clean, available each day, and put on before work. Contact lenses should not be worn when working with this chemical. Wear splash-proof chemical goggles and face shield unless full-face-piece respiratory protection is worn. Employees should wash immediately with soap when skin is wet or contaminated. Provide emergency showers and eyewash.

Respirator Selection: Where there is potential for exposures *over 5 ppm*, use a NIOSH/MSHA- or European Standard EN149-approved supplied-air respirator with a full face-piece operated in the positive-pressure mode, or with a full face-piece, hood, or helmet in the continuous-flow mode; or use a NIOSH/MSHA- or European Standard EN149-approved self-contained breathing apparatus with a full face-piece operated in pressure-demand or other positive-pressure mode.

Storage: Color Code—Blue: Health Hazard/Poison: Store in a secure poison location. Prior to working with this chemical you should be trained on its proper handling and storage. Before entering confined space where this chemical may be present, check to make sure that an explosive concentration does not exist. 1,2,4-Trichlorobenzene must be stored to avoid contact with oxidizers (such as perchlorates, peroxides, permanganates, chlorates, and nitrates), since violent reactions occur. Sources of ignition, such as smoking and open flames, are prohibited where 1,2,4-trichlorobenzene is used, handled, or stored in a manner that could create a potential fire or explosion hazard.

Shipping: This compound requires a shipping label of "POISONOUS/TOXIC MATERIALS." It falls in Hazard Class 6.1 and Packing Group III.

Spill Handling: Evacuate and restrict persons not wearing protective equipment from area of spill or leak until cleanup is complete. Remove all ignition sources. Establish forced ventilation to keep levels below explosive limit. Absorb liquids in vermiculite, dry sand; earth, peat, carbon, or a similar material and deposit in sealed containers. Keep this chemical out of a confined space, such as a sewer, because of the possibility of an explosion, unless the sewer is designed to prevent the buildup of explosive concentrations. It may be necessary to contain and dispose of this chemical as a hazardous waste. If material or contaminated runoff enters waterways, notify downstream users of potentially contaminated waters. Contact your local or federal environmental protection agency for specific recommendations. If employees are required to clean up spills, they must be properly trained and equipped. OSHA 1910.120(q) may be applicable.

Fire Extinguishing: This chemical is a combustible liquid. Poisonous gases, including phosgene, chlorine, and hydrogen chloride, are produced in fire. Use dry chemical, carbon dioxide; or alcohol foam extinguishers. Vapors are heavier than air and will collect in low areas. Vapors may travel long distances to ignition sources and flashback. Vapors in confined areas may explode when exposed to fire. Containers may explode in fire. Storage containers and parts of containers may rocket great distances, in many directions. If material or contaminated runoff enters waterways, notify downstream users of potentially contaminated waters. Notify local health and fire officials and pollution control agencies. From a secure, explosion-proof location, use water spray to cool exposed containers. If cooling streams are ineffective (venting sound increases in volume and pitch, tank discolors, or shows any signs of deforming), withdraw immediately to a secure position. If employees

are expected to fight fires, they must be trained and equipped in OSHA 1910.156. The only respirators recommended for firefighting are self-contained breathing apparatuses that have full face-pieces and are operated in a pressure-demand or other positive-pressure mode.

Disposal Method Suggested: Incineration, preferably after mixing with another combustible fuel. Care must be exercised to assure complete combustion to prevent the formation of phosgene. An acid scrubber is necessary to remove the halo acids produced.[22]

References

US Environmental Protection Agency. (1980). *Chlorinated Benzenes: Ambient Water Quality Criteria.* Washington, DC

Sax, N. I. (Ed.). (1984). *Dangerous Properties of Industrial Materials Report*, 4, No. 3, 96−99

New York State Department of Health. (March 1986). *Chemical Fact Sheet: 1,2,4-Trichlorobenzene.* Albany, NY: Bureau of Toxic Substance Assessment (Version 2)

New Jersey Department of Health and Senior Services. (August 2005). *Hazardous Substances Fact Sheet: 1,2,4-Trichlorobenzene.* Trenton, NJ

Trichloro(chloromethyl)-silane T:0710

Molecular Formula: CH_2Cl_4Si
Common Formula: $Si(CH_2Cl)Cl_3$
Synonyms: (Chloromethyl)trichlorosilane; Chloromethyl (trichloro)silane; Trichloro(chloromethyl)silane
CAS Registry Number: 1558-25-4
RTECS® Number: VV2200000
UN/NA & ERG Number: UN1295(trichlorosilane)/139
EC Number: 216-316-9
Regulatory Authority and Advisory Bodies
Superfund/EPCRA 40CFR355, Extremely Hazardous Substances: TPQ = 100 lb (45.4 kg).
OSHA 29CFR1910.119, Appendix A, Process Safety List of Highly Hazardous Chemicals, TQ = 100 lb (45 kg).
Reportable Quantity (RQ): 100 lb (45 kg).
European/International Regulations: Not listed in Annex 1.
WGK (German Aquatic Hazard Class): No value assigned.
Description: Trichloro(chloromethyl)-silane is a colorless liquid with a sharp, biting odor. Molecular weight = 183.92; Boiling point = 117°C; Freezing/Melting point = 111°C. Hazard Identification (based on NFPA-704 M Rating System): Health 3, Flammability 2, Reactivity 1W. Violent reaction with water.
Potential Exposure: Used in the synthesis of polysiloxane (silicone polymers).
Incompatibilities: Contact with water forms hydrochloric acid. Incompatible with strong oxidizers, acids. Corrodes metals in the presence of moisture. Contact with ammonia can cause a self-igniting compound. Some chlorosilanes self-ignite in air.

Permissible Exposure Limits in Air
Protective Action Criteria (PAC)*
TEEL-0: 0.2 ppm
PAC-1: **0.6** ppm
PAC-2: **7.3** ppm
PAC-3: **33** ppm
*AEGLs (Acute Emergency Guideline Levels) & ERPGs (Emergency Response Planning Guideline) are in **bold face**.
Routes of Entry: Inhalation, ingestion; skin, and/or eye contact.
Harmful Effects and Symptoms
Short Term Exposure: Irritates the eyes, skin, and respiratory tract. Contact with the eyes causes irritation, pain, swelling; corneal erosion and blindness may result. Contact with the skin causes dermatitis (red, inflamed skin), severe burns; pain, and shock generally follow dermal exposure. Acute inhalation exposure may result in sneezing, choking, laryngitis, dyspnea (shortness of breath), respiratory tract irritation; and chest pain. Bleeding of nose and gums, ulceration of the nasal and oral mucosa; pulmonary edema; chronic bronchitis; and pneumonia may also occur. If ingested, symptoms include increased salivation; intense thirst; difficulty in swallowing; chills, pain, and shock. Oral, esophageal, and stomach burns are common. Vomitus generally has a coffee-ground appearance. The potential for circulatory collapse is high following ingestion.
Long Term Exposure: Highly irritating substances may cause lung effects.
Points of Attack: Lungs.
Medical Surveillance: Before beginning employment and at regular times after that, for those with frequent or potentially high exposures, the following are recommended: lung function tests. If symptoms develop or overexposure is suspected, the following may be useful: consider chest X-ray after acute overexposure.
First Aid: If this chemical gets into the eyes, remove any contact lenses at once and irrigate immediately for at least 15 min, occasionally lifting upper and lower lids. Seek medical attention immediately. If this chemical contacts the skin, remove contaminated clothing and wash immediately with soap and water. Seek medical attention immediately. If this chemical has been inhaled, remove from exposure, begin rescue breathing (using universal precautions, including resuscitation mask) if breathing has stopped and CPR if heart action has stopped. Transfer promptly to a medical facility. When this chemical has been swallowed, get medical attention. Obtain authorization and/or further instructions from the local hospital for administration of an antidote or performance of other invasive procedures. Rush to a health care facility. Medical observation is recommended for 24−48 h after breathing overexposure, as pulmonary edema may be delayed. As first aid for pulmonary edema, a doctor or authorized paramedic may consider administering a corticosteroid spray.

Personal Protective Methods: Wear protective gloves and clothing to prevent any reasonable probability of skin contact. Safety equipment suppliers/manufacturers can provide recommendations on the most protective glove/clothing material for your operation. All protective clothing (suits, gloves, footwear, headgear) should be clean, available each day, and put on before work. Contact lenses should not be worn when working with this chemical. Wear splash-proof chemical goggles and face shield unless full-face-piece respiratory protection is worn. Employees should wash immediately with soap when skin is wet or contaminated. Provide emergency showers and eyewash.

Respirator Selection: Prior to working with this chemical you should be trained on its proper handling and storage. Store in tightly closed containers in a cool, well-ventilated area away from oxidizers. Where possible, automatically pump liquid from drums or other storage containers to process containers. Sources of ignition, such as smoking and open flames, are prohibited where this chemical is handled, used, or stored.

Storage: Color Code—Red: Flammability Hazard: Store in a flammable liquid storage area or approved cabinet away from ignition sources and corrosive and reactive materials. Prior to working with this chemical you should be trained on its proper handling and storage. Store in tightly closed containers in a cool, well-ventilated area away from strong oxidizers. Where possible, automatically transfer material from drums or other storage containers to process containers. Sources of ignition, such as smoking and open flames, are prohibited where this chemical is handled, used, or stored. Metal containers involving the transfer of this chemical should be grounded and bonded. Wherever this chemical is used, handled, manufactured, or stored, use explosion-proof electrical equipment and fittings.

Shipping: Trichlorosilanes require a shipping label of "DANGEROUS WHEN WET, FLAMMABLE LIQUID." They fall in Hazard Class 4.3 and Packing Group I.

Spill Handling: Evacuate and restrict persons not wearing protective equipment from area of spill or leak until cleanup is complete. Remove all ignition sources. Ventilate area of spill or leak. Stay upwind; keep out of low areas. If water pollution occurs, notify appropriate authorities. Shut off ignition sources; no flares, smoking, or flames in hazard area. Do not touch spilled material; stop leak if you can do so without risk. Use water spray to reduce vapors do not get water inside container. *Small spills:* flush area with flooding amounts of water. *Large spills:* dike far ahead of spill for later disposal. Absorb liquids in vermiculite, dry sand; earth, peat, carbon, or a similar material and deposit in sealed containers. Keep this chemical out of a confined space, such as a sewer, because of the possibility of an explosion, unless the sewer is designed to prevent the buildup of explosive concentrations. It may be necessary to contain and dispose of this chemical as a hazardous waste. If material or contaminated runoff enters waterways, notify downstream users of potentially contaminated waters.

Contact your local or federal environmental protection agency for specific recommendations. If employees are required to clean up spills, they must be properly trained and equipped. OSHA 1910.120(q) may be applicable.

Initial isolation and protective action distances as Chlorosilanes, corrosive, n.o.s.

Distances shown are likely to be affected during the first 30 min after materials are spilled and could increase with time. If more than one tank car, cargo tank, portable tank, or large cylinder is involved in the incident is leaking, the protective action distance may need to be increased. You may need to seek emergency information from CHEMTREC at (800) 424-9300 or seek professional environmental engineering assistance from the US EPA Environmental Response Team at (908) 548-8730 (24-h response line).

When spilled in water as *trichlorosilane*

Small spills (From a small package or a small leak from a large package)

First: Isolate in all directions (feet/meters)   100/30.
Then: Protect persons downwind (miles/kilometers).
Day   0.1/0.2
Night   0.2/0.3

Large spills (From a large package or from many small packages)

First: Isolate in all directions (feet/meters)   200/60.
Then: Protect persons downwind (miles/kilometers).
Day   0.5/0.8
Night   1.4/2.3

Fire Extinguishing: This chemical is a flammable liquid. Poisonous gases, including chlorine, are produced in fire. Use dry chemical, carbon dioxide or foam extinguishers. *Do not use water.* Vapor explosion hazard indoors, outdoors, or in sewers. Runoff to sewer may create fire or explosion hazard. Move container from fire area if you can do it without risk. Vapors are heavier than air and will collect in low areas. Vapors may travel long distances to ignition sources and flashback. Vapors in confined areas may explode when exposed to fire. Containers may explode in fire. Storage containers and parts of containers may rocket great distances, in many directions. If material or contaminated runoff enters waterways, notify downstream users of potentially contaminated waters. Notify local health and fire officials and pollution control agencies. From a secure, explosion-proof location, use water spray to cool exposed containers. Do not get water inside container. Cool containers that are exposed to flames with water from the side until well after fire is out. Stay away from ends of tanks. If cooling streams are ineffective (venting sound increases in volume and pitch, tank discolors, or shows any signs of deforming), withdraw immediately to a secure position. If employees are expected to fight fires, they must be trained and equipped in OSHA 1910.156. The only respirators recommended for firefighting are self-contained breathing apparatuses that have full face-pieces and are operated in a pressure-demand or other positive-pressure mode.

References

US Environmental Protection Agency. (November 30, 1987). *Chemical Hazard Information Profile: Trichloro (Chloromethyl)Silane.* Washington, DC: Chemical Emergency Preparedness Program

1,1,1-Trichloroethane   T:0720

Molecular Formula: $C_2H_3Cl_3$
Common Formula: CH_3CCl_3
Synonyms: Aerothene TT; CF 2; Chlorotene; Chlorothene NU; Chlorothene; Chlorothene NU; Chlorothene SM; Chlorothene VG; Chlorten; Ethana NU; Ethane, 1,1,1-trichloro-; ICI-CF 2; Inhibisol; Methyl chloroform; Methyltrichloromethane; NCI-C04626; Solvent 111; Strobane; α-T; Tafclean; 1,1,1-TCE; 1,1,1-Trichloraethan (German); 1,1,1-Trichlorethane; α-Trichloroethane; Trichloro-1,1,1-ethane (French); Trichloroethane; Trichloromethylmethane; 1,1,1-Tricloroetano (Spanish); Tri-ethane
CAS Registry Number: 71-55-6; (*alt.*) 74552-83-3
RTECS® Number: KJ2975000
UN/NA & ERG Number: UN2831/160
EC Number: 200-756-3 [*Annex I Index No.:* 602-013-00-2]
Regulatory Authority and Advisory Bodies
Carcinogenicity: IARC: Animal, Inadequate Evidence; Human Inadequate Evidence, *not classifiable as carcinogenic to humans*, Group 3, 1999; EPA: Inadequate Information to assess carcinogenic potential; NIOSH: Potential occupational carcinogen.
US EPA Gene-Tox Program, Positive: Cell transform.— RLV F344 rat embryo; Negative: Sperm morphology—mouse; Inconclusive: Carcinogenicity—mouse/rat; Mammalian micronucleus.
US EPA, FIFRA 1998 Status of Pesticides: Canceled.
Air Pollutant Standard Set. See below, "Permissible Exposure Limits in Air" section.
Clean AIR Act: Hazardous Air Pollutants (Title I, Part A, Section 112); Stratospheric ozone protection (Title VI, Subpart A, Appendix A), Class I, Ozone Depletion Potential = 0.1, all isomers except 1,1,2-trichlorethane.
Clean Water Act: Toxic Pollutant (Section 401.15) as chlorinated ethanes.
US EPA Hazardous Waste Number (RCRA No.): U226.
RCRA 40CFR261, Appendix 8 Hazardous Constituents.
RCRA 40CFR268.48; 61FR15654, Universal Treatment Standards: Wastewater (mg/L), 0.054; Nonwastewater (mg/kg), 6.0.
RCRA 40CFR264, Appendix 9; TSD Facilities Ground Water Monitoring List. Suggested test method(s) (PQL µg/L): 8240 (5).
Safe Drinking Water Act: MCL, 0.2 mg/L; MCLG, 0.20 mg/L; Regulated chemical (47 FR 9352).
Reportable Quantity (RQ): 1000 lb (454 kg).
EPCRA Section 313 Form R *de minimis* concentration reporting level: 0.1%.
Canada, WHMIS, Ingredients Disclosure List Concentration 1.0%.
Hazard Symbol: Xn, N; Risk phrases: R20; R59; Safety phrases: S2; S24/25; S59; S61 (see Appendix 4).
WGK (German Aquatic Hazard Class): 3—Severe hazard to waters.
Description: 1,1,1-Trichloroethane is a colorless, nonflammable liquid with an odor similar to chloroform. The odor threshold is 120 ppm (NJ) or 400 ppm (NY). Molecular weight = 133.40; Specific gravity (H_2O:1) = 1.34 at 25°C; Boiling point = 73.9°C; Freezing/Melting point = −30.6°C; Vapor pressure = 100 mmHg at 25°C; Flash point = none; Autoignition temperature = 537°C. Explosive limits: LEL = 7.5%; UEL = 12.5%. Hazard Identification (based on NFPA-704 M Rating System): Health 2, Flammability 1, Reactivity 0. Practically insoluble in water; solubility = 0.4%.
Potential Exposure: Compound Description: Agricultural Chemical; Tumorigen, Mutagen; Reproductive Effector; Human Data; Primary Irritant. 1,1,1-Trichloroethane is used as a cleaning solvent, chemical intermediate for vinylidene chloride. In liquid form it is used as a degreaser and for cold cleaning, dip-cleaning; and bucket cleaning of metals. Other industrial applications of 1,1,1-trichloroethane's solvent properties include its use as a dry-cleaning agent; a vapor degreasing agent; and a propellant. In recent years, 1,1,1-trichloroethane has found wide use as a substitute for carbon tetrachloride.
Incompatibilities: Strong caustics; strong oxidizers; chemically active metals, such as aluminum, magnesium powder; sodium, potassium. Reacts slowly with water to form hydrochloric acid. Upon contact with hot metal or exposure to ultraviolet radiation, it will decompose to form the irritant gases hydrochloric acid, phosgene, and dichloroacetylene. Forms shock-sensitive mixtures with potassium or its alloys. Attacks natural rubber.
Permissible Exposure Limits in Air
Conversion factor: 1 ppm = 5.46 mg/m³ at 25°C & 1 atm.
OSHA PEL: 350 ppm/1900 mg/m³ TWA.
NIOSH REL: 350 ppm/1900 mg/m³ [15 min] Ceiling Concentration. NIOSH considers methyl chloroform to be a potential occupational carcinogen. Limit exposure to lowest feasible concentration. See *NIOSH Pocket Guide*, Appendix C.
ACGIH TLV®[1]: 350 ppm/1910 mg/m³ TWA, 450 ppm/2460 mg/m³ STEL, not classifiable as a human carcinogen; BEI: 40 ppm methyl chloroform in end-exhaled air prior to last shift of workweek; 10 mg/L trichloroacetic acid in urine at end-of-workweek; 30 mg/L total trichloroethanol in urine, end-of-shift at end-of-workweek; 1 mg/L total trichloroethanol in blood, end-of-shift at end-of-workweek.
NIOSH IDLH: 700 ppm.
Protective Action Criteria (PAC)*
TEEL-0: 230 ppm
PAC-1: **230** ppm

PAC-2: **600** ppm

PAC-3: **4200** ppm

*AEGLs (Acute Emergency Guideline Levels) & ERPGs (Emergency Response Planning Guideline) are in **bold face**.

DFG MAK: 200 ppm/1100 mg/m³ TWA; Peak Limitation Category II(1) [skin], Pregnancy Risk Group: C; BAT: 550 µg/L in blood after several shifts [for long-time exposure]; at the beginning of next shift.

Australia: TWA 125 ppm (680 mg/m³), 1993; Austria: MAK 200 ppm (1080 mg/m³), 1999; Belgium: TWA 350 ppm (1910 mg/m³), STEL 450 ppm (2460 mg/m³), 1993; Denmark: TWA 50 ppm (275 mg/m³), 1999; Finland: TWA 100 ppm (540 mg/m³), STEL 250 ppm (1400 mg/m³), 1999; France: VME 300 ppm (1650 mg/m³), VLE 450 ppm (2500 mg/m³), 1999; Hungary: TWA 100 mg/m³, STEL 300 mg/m³ [skin], 1993; the Netherlands: MAC-TGG 555 mg/m³, 2003; Norway: TWA 50 ppm (270 mg/m³), 1999; the Philippines: TWA 350 ppm (1900 mg/m³), 1993; Poland: MAC (TWA) 300 mg/m³; MAC (STEL) 1400 mg/m³, 1999; Russia: TWA 200 ppm, STEL 20 mg/m³, 1993; Sweden: NGV 50 ppm (300 mg/m³), KTV 90 ppm (500 mg/m³), 1999; Switzerland: MAK-W 200 ppm (1080 mg/m³), KZG-W 1000 ppm, 1999; Turkey: TWA 350 ppm (1900 mg/m³), 1993; United Kingdom: TWA 200 ppm (1110 mg/m³), STEL 400 ppm, 2000; Argentina, Bulgaria, Columbia, Jordan, South Korea, New Zealand, Singapore, Vietnam: ACGIH TLV®: STEL 450 ppm. Several states have set guidelines or standards for 1,1,1-Trichloroethane in ambient air[60] ranging from 1.3 mg/m³ (Massachusetts) to 12.0−245.0 mg/m³ (North Carolina) to 19.0 mg/m³ (Indiana) to 19.0−24.5 mg/m³ (North Dakota) to 32.0 mg/m³ (Virginia) to 38.0 mg/m³ (Connecticut, New York, South Dakota) to 45.238 mg/m³ (Nevada). Russia set a MAC for ambient air[43] in residential areas of 2.0 mg/m³ on a once-daily basis and 0.2 mg/m³ on a daily average basis.

Determination in Air: Use NIOSH Analytical Method #1003, Hydrocarbons, halogenated[18]; OSHA Analytical Method 14. See also NIOSH Analytical Method #2549, Volatile organic compounds.

Determination in Water: Octanol−water coefficient: Log K_{ow} = 2.49.

Permissible Concentration in Water: To protect freshwater aquatic life: 18,000 µg/L on an acute toxicity basis. *To protect saltwater aquatic life:* 31,200 µg/L on an acute toxicity basis. To protect human health, on the basis of fish consumption alone—1,030,000 µg/L.[6] The EPA has recently set a lifetime health advisory for 1,1,1-trichloroethane of 0.2 mg/L (200 µg/L).[48] Several states have set guidelines or standards for drinking water ranging from 26 µg/L (New Jersey) to 60 µg/L (New Mexico) to 140 µg/L (Massachusetts) to 200 µg/L (California, Maine, Minnesota) to 300 µg/L (Connecticut).

Determination in Water: Charcoal tube; CS2; Gas chromatography/Flame ionization detection; NIOSH Analytical Method (IV) #1003, Halogenated hydrocarbons.

Routes of Entry: Inhalation of vapor, moderate skin adsorption, ingestion; skin and/or eye contact.

Harmful Effects and Symptoms

Short Term Exposure: Irritates the eyes, skin, and respiratory tract. *Inhalation:* Exposure can cause headache, lassitude (weakness, exhaustion), central nervous system depressant/depression; poor equilibrium; cardiac arrhythmias; liver damage. May affect the heart and central nervous system; kidneys and liver; causing cardiac disorders and respiratory failure. Levels above 900 ppm can cause dizziness, mental confusion; drowsiness, loss of coordination, and unconsciousness. Death may result. *Skin:* Contact can cause irritation and rash. Absorption is moderate; may contribute significantly to health hazard. *Eyes:* Contact causes irritation. The vapor has caused irritation at levels of 450 ppm. *Ingestion:* May cause symptoms similar to inhalation. In addition, may cause mouth, throat, and stomach irritation.

Long Term Exposure: High exposures may damage the liver and kidneys. Prolonged contact can cause thickening and cracking of the skin. Repeated or prolonged contact at levels of 450 ppm or above may result in irritation and dry, scaly, fractured skin. Additionally, NIOSH recommends that this chemical be treated in the workplace with caution because of its structural similarity to other chloroethanes shown to be carcinogenic in animals.

Points of Attack: Eyes, skin, central nervous system; cardiovascular system, liver.

Medical Surveillance: NIOSH lists the following tests: whole blood (chemical/metabolite); whole blood (chemical/metabolite), During Exposure; whole blood (chemical/metabolite), end-of-shift; whole blood (chemical/metabolite), end-of-workweek; whole blood (chemical/metabolite), end-of-workweek, expired air, expired air, 16 h following end-of-exposure; expired air, end-of-workweek; expired air, prior to next shift; expired air prior to last shift of workweek; urine (chemical/metabolite); urine (chemical/metabolite), end-of-shift; urine (chemical/metabolite), end-of-shift at end-of-workweek; urine (chemical/metabolite), end-of-workweek; urine (chemical/metabolite), prior to next shift. Consider the skin, liver function; cardiac status, especially arrythmias; in preplacement or periodic examinations. Expired air analyses may be useful in monitoring exposure. Persons with heart disease may be at an increased risk of irregular heartbeat from very high exposures.

First Aid: If this chemical gets into the eyes, remove any contact lenses at once and irrigate immediately for at least 15 min, occasionally lifting upper and lower lids. Seek medical attention immediately. If this chemical contacts the skin, remove contaminated clothing and wash immediately with soap and water. Seek medical attention immediately. If this chemical has been inhaled, remove from exposure, begin rescue breathing (using universal precautions, including resuscitation mask) if breathing has stopped and CPR if heart action has stopped. Transfer promptly to a medical facility. When this chemical has been swallowed, get

medical attention. Give large quantities of water and induce vomiting. Do not make an unconscious person vomit.

Personal Protective Methods: Wear protective gloves and clothing to prevent any reasonable probability of skin contact: **8 h**: polyvinyl alcohol gloves; Viton™ gloves, suits; 4H™ and Silver Shield™ gloves; Barricade™ coated suits; CPF3™ suits; Responder™ suits; Trychem 1000™ suits; **4 h**: Teflon™ gloves, suits, boots. Also, protective clothing of leather or Neoprene™ may offer some protection. Safety equipment suppliers/manufacturers can provide recommendations on the most protective glove/clothing material for your operation. All protective clothing (suits, gloves, footwear, headgear) should be clean, available each day, and put on before work. Contact lenses should not be worn when working with this chemical. Wear splash-proof chemical goggles and face shield unless full-face-piece respiratory protection is worn. Employees should wash immediately with soap when skin is wet or contaminated. Provide emergency showers and eyewash.

Respirator Selection: Up to 700 ppm: Sa (APF = 10) (any supplied-air respirator);* or SCBA (any self-contained breathing apparatus with a full face-piece). *Emergency or planned entry into unknown concentrations or IDLH conditions:* SCBAF: Pd,Pp (APF = 10,000) (any self-contained breathing apparatus that has a full face-piece and is operated in a pressure-demand or other positive-pressure mode); or SaF: Pd,Pp: ASCBA (any supplied-air respirator that has a full face-piece and is operated in a pressure-demand or other positive-pressure mode in combination with an auxiliary self-contained positive-pressure breathing apparatus). *Escape:* GmFOv (APF = 50) [any air-purifying, full-face-piece respirator (gas mask) with a chin-style, front- or back-mounted organic vapor canister]; or SCBAE (any appropriate escape-type, self-contained breathing apparatus).

*Substance reported to cause eye irritation or damage; may require eye protection.

Storage: Color Code—Blue: Health Hazard/Poison: Store in a secure poison location. Prior to working with this chemical you should be trained on its proper handling and storage. Before entering confined space where this chemical may be present, check to make sure that an explosive concentration does not exist. Methyl chloroform must be stored to avoid contact with strong caustics (such as sodium and potassium hydroxide); acetone, strong oxidizers (such as chlorine, chlorine dioxide, and bromine); and chemically active metals (such as potassium, aluminum, zinc, and magnesium); since violent reactions occur. Do not allow vapor near sources of ultraviolet light, such as arc welding, because poisonous gases may be produced. Store in tightly closed containers in a cool, well-ventilated area away from heat and moisture. Do not use aluminum containers.

Shipping: This compound requires a shipping label of "POISONOUS/TOXIC MATERIALS." It falls in Hazard Class 6.1 and Packing Group III.

Spill Handling: Evacuate and restrict persons not wearing protective equipment from area of spill or leak until cleanup is complete. Remove all ignition sources. Establish forced ventilation to keep levels below explosive limit. Absorb liquids in vermiculite, dry sand; earth, peat, carbon, or a similar material and deposit in sealed containers. Keep this chemical out of a confined space, such as a sewer, because of the possibility of an explosion, unless the sewer is designed to prevent the buildup of explosive concentrations. It may be necessary to contain and dispose of this chemical as a hazardous waste. If material or contaminated runoff enters waterways, notify downstream users of potentially contaminated waters. Contact your local or federal environmental protection agency for specific recommendations. If employees are required to clean up spills, they must be properly trained and equipped. OSHA 1910.120(q) may be applicable.

Fire Extinguishing: 1,1,1-Trichloroethane's combustible mixtures with air do not readily ignite, but ignition may occur in conditions of excess oxygen or in the presence of a high-energy ignition sources, such as a furnace; or welding. Use dry chemical, carbon dioxide, or foam extinguishers. Poisonous gases, including hydrogen chloride and phosgene, are produced in fire. Vapors are heavier than air and will collect in low areas. Vapors in confined areas may explode when exposed to fire. Containers may explode in fire. Storage containers and parts of containers may rocket great distances, in many directions. If material or contaminated runoff enters waterways, notify downstream users of potentially contaminated waters. Notify local health and fire officials and pollution control agencies. From a secure, explosion-proof location, use water spray to cool exposed containers. If cooling streams are ineffective (venting sound increases in volume and pitch, tank discolors, or shows any signs of deforming), withdraw immediately to a secure position. If employees are expected to fight fires, they must be trained and equipped in OSHA 1910.156. The only respirators recommended for firefighting are self-contained breathing apparatuses that have full face-pieces and are operated in a pressure-demand or other positive-pressure mode.

Disposal Method Suggested: Consult with environmental regulatory agencies for guidance on acceptable disposal practices. Generators of waste containing this contaminant (≥100 kg/mo) must conform with EPA regulations governing storage, transportation, treatment, and waste disposal. Incineration, preferably after mixing with another combustible fuel. Care must be exercised to assure complete combustion to prevent the formation of phosgene. An acid scrubber is necessary to remove the halo acids produced. As an alternative to disposal, trichloroethane may be recovered from waste gases and liquids from various processes and recycled.

References

National Institute for Occupational Safety and Health. (1976). *Criteria for a Recommended Standard: Occupational Exposure to 1,1,1-Trichloroethane (Methyl Chloroform)*. NIOSH Document No. 76-184. Washington, DC

US Environmental Protection Agency. (1980). *Chlorinated Ethanes: Ambient Water Quality Criteria*. Washington, DC
US Environmental Protection Agency. (April 30, 1980). *1,1,1-Trichloroethane, Health and Environmental Effects Profile No. 164*. Washington, DC: Office of Solid Waste
Sax, N. I. (Ed.). *Dangerous Properties of Industrial Materials Report*, 1, No. 1, 124–126 (1982) and 5, No. 6, 28–30 (1985)
New York State Department of Health. (April 1986). *Chemical Fact Sheet: 1,1,1-Trichloroethane*. Albany, NY: Bureau of Toxic Substance Assessment
US Environmental Protection Agency, Special Review and Reregistration Division Office of Pesticide Programs. (1998). *Agency Status of Pesticides in Registration, Reregistration, and Special Review* (Rainbow Report). Washington, DC
New Jersey Department of Health and Senior Services. (February 2001). *Hazardous Substances Fact Sheet: Methyl Chloroform*. Trenton, NJ

1,1,2-Trichloroethane   T:0730

Molecular Formula: $C_2H_3Cl_3$
Common Formula: $CH_2ClCHCl_2$
Synonyms: Cement-339; Ethane trichloride; Ethane, 1,1,2-trichloro-; NCI-C04579; β-T; 1,2,2-Trichloroethane; β-Trichloroethane; 1,1,2-Tricloroetano (Spanish); Vinyl trichloride
CAS Registry Number: 79-00-5
RTECS®Number: KJ3150000
UN/NA & ERG Number: UN3082/171
EC Number: 201-166-9 [*Annex I Index No.:* 602-014-00-8]
Regulatory Authority and Advisory Bodies
Carcinogenicity: IARC: Animal, Inadequate Evidence; Human No Adequate Data, *not classifiable as carcinogenic to humans*, Group 3, 1999; EPA: Possible Human Carcinogen; NCI: Carcinogenesis Bioassay (gavage); clear evidence: mouse; no evidence: rat; NIOSH: Potential occupational carcinogen.
Air Pollutant Standard Set. See below, "Permissible Exposure Limits in Air" section.
Clean AIR Act: Hazardous Air Pollutants (Title I, Part A, Section 112).
US EPA Hazardous Waste Number (RCRA No.): U227.
RCRA 40CFR261, Appendix 8 Hazardous Constituents.
RCRA 40CFR268.48; 61FR15654, Universal Treatment Standards: Wastewater (mg/L), 0.054; Nonwastewater (mg/kg), 6.0.
RCRA 40CFR264, Appendix 9; TSD Facilities Ground.
Water Monitoring List. Suggested test method(s) (PQL μg/L): 8010 (0.2); 8240 (5).
Safe Drinking Water Act: MCL, 0.005 mg/L; MCLG, 0.003 mg/L; Regulated chemical (47 FR 9352).
Reportable Quantity (RQ): 100 lb (45.4 kg).

EPCRA Section 313 Form R *de minimis* concentration reporting level: 1.0%.
California Proposition 65 Chemical: Cancer 10/1/90.
Canada, WHMIS, Ingredients Disclosure List Concentration 1.0%.
European/International Regulations: Hazard Symbol: Xn; Risk phrases: R20/21/22; R40; R66; Safety phrases: S2; S9; S36; S37; S46 (see Appendix 4).
WGK (German Aquatic Hazard Class): 3—Severe hazard to waters.
Description: 1,1,2-Trichloroethane is a colorless, nonflammable liquid with a sweet, chloroform-like odor. Molecular weight = 133.40; Specific gravity (H_2O:1) = 1.44 at 25°C; Boiling point = 113.9°C; Freezing/Melting point = −36°C. Vapor pressure = 19 mmHg at 25°C; Explosive limits are: LEL = 6.0%; UEL = 15.5%. Hazard Identification (based on NFPA-704 M Rating System): Health 3, Flammability 1, Reactivity 0. Practically insoluble in water; solubility = 0.4%.
Potential Exposure: Compound Description: Agricultural Chemical; Tumorigen, Drug, Mutagen, Primary Irritant. 1,1,2-Trichloroethane is used as an intermediate in the production of vinylidine chloride, and a component of adhesives; as a solvent; but is not as widely used as is its isomer 1,1,1-trichloroethane; it is an isomer of 1,1,1-Trichloroethane *but should not be confused with it toxicologically.* 1,1,2-Trichloroethane is comparable to carbon tetrachloride and tetrachloroethane in toxicity.
Incompatibilities: Strong oxidizers, strong caustics; chemically active metals, such as aluminum, magnesium powders, sodium, potassium. Attacks many plastics, rubber, coatings, steel, and zinc.
Permissible Exposure Limits in Air
Conversion factor: 1 ppm = 5.46 mg/m³ at 25°C & 1 atm.
OSHA PEL: 10 ppm/45 mg/m³ TWA [skin].
NIOSH REL: 10 ppm/45 mg/m³ TWA [skin], Potential occupational carcinogen; limit exposure to lowest feasible concentration. See *NIOSH Pocket Guide* Appendices A & C.
ACGIH TLV®[1]: 10 ppm TWA [skin], animal carcinogen with Unknown Relevance to Humans.
Protective Action Criteria (PAC)
TEEL-0: 10 ppm
PAC-1: 100 ppm
PAC-2: 100 ppm
PAC-3: 100 ppm
DFG MAK: 10 ppm/55 mg/m³ TWA; Peak Limitation Category II(2) [skin]; Carcinogen Category 3B.
NIOSH IDLH: 100 ppm.
Arab Republic of Egypt: TWA 10 ppm (40 mg/m³) [skin], 1993; Australia: TWA 10 ppm (45 mg/m³) [skin], 1993; Austria: MAK 10 ppm (55 mg/m³) [skin], Suspected: carcinogen, 1999; Belgium: TWA 10 ppm (55 mg/m³) [skin], 1993; Denmark: TWA 10 ppm (54 mg/m³) [skin], 1999; Finland: TWA 10 ppm (54 mg/m³), STEL 20 ppm (110 mg/m³) [skin], 1999; Hungary: TWA 10 mg/m³, STEL 20 mg/m³ [skin],

1993; the Netherlands: MAC-TGG 45 mg/m^3 [skin], 2003; Norway: TWA 10 ppm (54 mg/m^3), 1999; Russia: TWA 10 ppm, 1993; Switzerland: MAK-W 10 ppm (55 mg/m^3), KZG-W 50 ppm (275 mg/m^3) [skin], 1999; United Kingdom: LTEL 10 ppm (45 mg/m^3), STEL 20 ppm [skin], 1993; Argentina, Bulgaria, Columbia, Jordan, South Korea, New Zealand, Singapore, Vietnam: ACGIH TLV®: confirmed animal carcinogen with unknown relevance to humans. Several states have set guidelines or standards for 1,1,2-trichloroethane in ambient air[60] ranging from 0.00 μg/m^3 (Massachusetts) to 7.0 μg/m^3 (Rhode Island) to 27.0 μg/m^3 (Pennsylvania) to 107.143 μg/m^3 (Kansas) to 150.0 μg/m^3 (New York) to 225.0 μg/m^3 (Connecticut and Indiana) to 450.0 μg/m^3 (Florida and North Dakota) to 900.0 μg/m^3 (Virginia) to 1071.0 μg/m^3 (Nevada).

Determination in Air: Use NIOSH Analytical Method (IV) #1003, Halogenated hydrocarbons, OSHA Analytical Method 11.

Determination in Water: Octanol–water coefficient: Log K_{ow} = 2.35.

Permissible Concentration in Water: *To protect freshwater aquatic life:* 18,000 μg/L on an acute toxicity basis and 9400 μg/L on a chronic basis. *To protect saltwater aquatic life:* no criteria developed due to insufficient data. *To protect human health:* preferably zero. An additional lifetime cancer risk of 1 in 100,000 is posed by a concentration of 6.0 μg/L.[6] Several states have set guidelines or standards for 1,1,2-trichloroethane in drinking water[61] ranging from 1.0 μg/L (Arizona) to 6.11 μg/L (Minnesota) to 10.0 μg/L (New Mexico) to 100.0 μg/L (California).

Determination in Water: Inert gas purge followed by gas chromatography with halide-specific detection (EPA Method 601) or gas chromatography plus mass spectrometry (EPA Method 624).

Routes of Entry: Inhalation of vapor, absorption through the skin, ingestion; skin and/or eye contact.

Harmful Effects and Symptoms

Short Term Exposure: Exposure may cause central nervous depression. May affect the kidneys and liver. *Inhalation:* Inhalation may produce headache, lassitude, dizziness, a lack of coordination; low blood pressure; irregular heartbeat; coma and death from respiratory arrest. Exposure to vapor concentrations near 2000 ppm for 5 min causes central nervous system depression and anesthetic effects. Symptoms are nasal irritation, drowsiness an equilibrium disturbances. Death may result from 13,600 ppm for 2 h. *Skin:* Can cause irritation and chemical burns if allowed to remain on the skin for a prolonged period. May be absorbed through the skin to cause or increase the severity of symptoms listed above. *Eyes:* Can cause irritation. *Ingestion:* May cause effects similar to those listed under inhalation. Laboratory studies with animals suggest that the probable lethal dose for humans is about 12 oz. Liver and kidney damage have occurred in animals.

Long Term Exposure: Can destroy the skin's natural oils, causing drying and cracking. Inhalation may cause liver and kidney damage. NIOSH recommends this chemical be treated as a potential occupational carcinogen. Has caused cancer in laboratory animals. Whether it does so in humans is unknown.

Points of Attack: Eyes, respiratory system; central nervous system; liver, kidneys. Cancer site in animals: liver cancer.

Medical Surveillance: NIOSH lists the following tests: Expired Air. For those with frequent or potentially high exposure (half the TLV or greater, or significant skin contact) the following are recommended before beginning work and at regular times after that: Liver and kidney function tests. If symptoms develop or overexposure is suspected, the following may be useful: special 24-h EKG (Holter monitor), to look for irregular heartbeat.

First Aid: If this chemical gets into the eyes, remove any contact lenses at once and irrigate immediately for at least 15 min, occasionally lifting upper and lower lids. Seek medical attention immediately. If this chemical contacts the skin, remove contaminated clothing and wash immediately with soap and water. Seek medical attention immediately. If this chemical has been inhaled, remove from exposure, begin rescue breathing (using universal precautions, including resuscitation mask) if breathing has stopped and CPR if heart action has stopped. Transfer promptly to a medical facility. When this chemical has been swallowed, get medical attention. Give large quantities of water and induce vomiting. Do not make an unconscious person vomit.

Personal Protective Methods: Wear protective gloves and clothing to prevent any reasonable probability of skin contact. Safety equipment suppliers/manufacturers can provide recommendations on the most protective glove/clothing material for your operation. Viton and Teflon™ are recommended as protective materials. All protective clothing (suits, gloves, footwear, headgear) should be clean, available each day, and put on before work. Contact lenses should not be worn when working with this chemical. Wear splash-proof chemical goggles and face shield unless full-face-piece respiratory protection is worn. Employees should wash immediately with soap when skin is wet or contaminated. Provide emergency showers and eyewash.

Respirator Selection: *At concentrations above the NIOSH REL, or where there is no REL, at any detectable concentration:* SCBAF: Pd,Pp (APF = 10,000) (any NIOSH/MSHA- or European Standard EN 149-approved self-contained breathing apparatus that has a full face-piece and is operated in a pressure-demand or other positive-pressure mode); or SaF: Pd,Pp: ASCBA (APF = 10,000) (any supplied-air respirator that has a full face-piece and is operated in a pressure-demand or other positive-pressure mode in combination with an auxiliary, self-contained breathing apparatus operated in a pressure-demand or other positive-pressure mode). *Escape:* GmFOv (APF = 50) [any air-purifying, full-face-piece respirator (gas mask) with a chin-style, front- or back-mounted organic vapor canister]; or SCBAE (any appropriate escape-type, self-contained breathing apparatus).

Storage: Color Code—Blue: Health Hazard/Poison: Store in a secure poison location. Prior to working with this chemical you should be trained on its proper handling and storage. Before entering confined space where this chemical may be present, check to make sure that an explosive concentration does not exist. 1,1,2-Trichloroethane must be stored to avoid contact with strong oxidizers (such as chlorates, nitrates, peroxides, chlorine, and bromine); strong caustics; and chemically active metals (such as potassium, magnesium, zinc and sodium); because violent reactions occur. Store in tightly closed containers in a cool, well-ventilated area away from heat. Sources of ignition, such as smoking and open flames, are prohibited where 1,1,2-trichloroethane is used, handled, or stored in a manner that could create A potential fire or explosion hazard. A regulated, marked area should be established where this chemical is handled, used, or stored in compliance with OSHA Standard 1910.1045.

Shipping: Environmentally hazardous substances, liquid, n.o.s., require a shipping label of "CLASS 9." 1,1,2-Trichloroethane falls in Hazard Class 9 and Packing Group III.[20, 21]

Spill Handling: Evacuate and restrict persons not wearing protective equipment from area of spill or leak until cleanup is complete. Remove all ignition sources. Establish forced ventilation to keep levels below explosive limit. Absorb liquids in vermiculite, dry sand; earth, peat, carbon, or a similar material and deposit in sealed containers. It may be necessary to contain and dispose of this chemical as a hazardous waste. If material or contaminated runoff enters waterways, notify downstream users of potentially contaminated waters. Contact your local or federal environmental protection agency for specific recommendations. If employees are required to clean up spills, they must be properly trained and equipped. OSHA 1910.120(q) may be applicable.

Fire Extinguishing: Not flammable under normal conditions but can be ignited by high-energy ignition sources, such as a furnace or welding. Use dry chemical, carbon dioxide, or foam extinguishers and water to keep fire-exposed containers cool. Poisonous gases, including hydrogen chloride and phosgene, are produced in fire. Vapors are heavier than air and will collect in low areas. Containers may explode in fire. Storage containers and parts of containers may rocket great distances, in many directions. If material or contaminated runoff enters waterways, notify downstream users of potentially contaminated waters. Notify local health and fire officials and pollution control agencies. From a secure, explosion-proof location, use water spray to cool exposed containers. If cooling streams are ineffective (venting sound increases in volume and pitch, tank discolors, or shows any signs of deforming), withdraw immediately to a secure position. If employees are expected to fight fires, they must be trained and equipped in OSHA 1910.156. The only respirators recommended for firefighting are self-contained breathing apparatuses that have full face-pieces and are operated in a pressure-demand or other positive-pressure mode.

Disposal Method Suggested: Consult with environmental regulatory agencies for guidance on acceptable disposal practices. Generators of waste containing this contaminant (≥100 kg/mo) must conform with EPA regulations governing storage, transportation, treatment, and waste disposal. Incineration, preferably after mixing with another combustible fuel. Care must be exercised to assure complete combustion to prevent the formation of phosgene. An acid scrubber is necessary to remove the halo acids produced.

References

US Environmental Protection Agency. (August 1, 1978). *Chemical Hazard Information Profile: 1,1,2-Trichloroethane.* Washington, DC (Revised issue 1979)

US Environmental Protection Agency. (1980). *Chlorinated Ethanes: Ambient Water Quality Criteria.* Washington, DC

US Environmental Protection Agency. (April 30, 1980). *1,1,2-Trichloroethane, Health and Environmental Effects Profile No. 165.* Washington, DC: Office of Solid Waste

Sax, N. I. (Ed.). *Dangerous Properties of Industrial Materials Report*, 2, No. 6, 88–90 (1982) and 3, No. 2, 66–69 (1983)

US Public Health Service. (December 1988). *Toxicological Profile for 1,1,2-Trichloroethane.* Atlanta, GA: Agency for Toxic Substances and Disease Registry

New York State Department of Health. (May 1986). *Chemical Fact Sheet: 1,1,2-Trichloroethane.* Albany, NY: Bureau of Toxic Substance Assessment (Version 2)

New Jersey Department of Health and Senior Services. (March 2002). *Hazardous Substances Fact Sheet: 1,1,2-Trichloroethane.* Trenton, NJ

Trichloroethylene   T:0740

Molecular Formula: C_2HCl_3
Common Formula: $ClCH{=}CCl_2$
Synonyms: Acetylene trichloride; Algylen; Anamenth; Benzinol; Blacosolv; Cecolene; Chlorilen; Chlorylea; Chorylen; Circosolv; Crawhaspol; Densinfluat; Dow-tri; Dukeron; Ethene, trichloro-; Ethinyl trichloride; Ethylene trichloride; Ethylene, trichloro-; Fleck-flip; Fluate; Germalgene; Halocarbon 113; Lanadin; Lethurin; Narcogen; Narkosoid; NCI-C04546; Nialk; Perm-α-chlor; Petzinol; TCE; Threthylen; Threthylene; Trethylene; Tri; Triad; Triasol; Trichloraethen (German); Trichloran; Trichloren; Trichlorethene (French); Trichloroethene; 1,1,2-Trichloroethylene; Trichloroethylene tri (French); Trichlororan; 1,1,2-Trichloro-1,2,2-trifluoroethane; Triclene; Tricloroetileno (Spanish); Trielene; Trielin; Trieline; Trilentrilene; Trimar; Tri-plus; TTE; Vestrol; Vitran; Westrosol
CAS Registry Number: 79-01-6; (*alt.*) 52037-46-4
RTECS®Number: KX4550000

UN/NA & ERG Number: UN1710/160
EC Number: 201-167-4 [*Annex I Index No.:* 602-027-00-9]
Regulatory Authority and Advisory Bodies
IARC: Animal, Sufficient Evidence; Human, Limited Evidence, Group 2A, 1995; NCI: Carcinogenesis Studies (gavage); clear evidence: mouse; no evidence: rat; NTP: 11th Report on Carcinogens, 2004: Reasonably anticipated to be a human carcinogen; NIOSH: Potential occupational carcinogen.
US EPA Gene-Tox Program, Positive: Cell transform.—RLV F344 rat embryo; Host-mediated assay; Positive: Mouse spot test; Sperm morphology—mouse; Positive: *S. cerevisiae* gene conversion; *S. cerevisiae*—homozygosis; Positive: *S. cerevisiae*—reversion; Positive/limited: Carcinogenicity—mouse/rat; Negative: *D. melanogaster* sex-linked lethal; Inconclusive: Histidine reversion—Ames test.
Air Pollutant Standard Set. See below, "Permissible Exposure Limits in Air" section.
Clean AIR Act: Hazardous Air Pollutants (Title I, Part A, Section 112).
Clean Water Act: Section 311 Hazardous Substances/RQ 40CFR117.3 (same as CERCLA, see below); 40CFR 401.15 Section 307 Toxic Pollutants; 40CFR423, Appendix A, Priority Pollutants.
US EPA Hazardous Waste Number (RCRA No.): U228; D040.
RCRA Toxicity Characteristic (Section 261.24), Maximum Concentration of Contaminants, regulatory level, 0.5 mg/L.
RCRA 40CFR261, Appendix 8 Hazardous Constituents.
RCRA 40CFR268.48; 61FR15654, Universal Treatment Standards: Wastewater (mg/L), 0.054; Nonwastewater (mg/kg), 6.0.
RCRA 40CFR264, Appendix 9; TSD Facilities Ground Water Monitoring List. Suggested test method(s) (PQL µg/L): 8010 (1); 8240 (5).
Safe Drinking Water Act: MCL, 0.005 mg/L; MCLG, zero; Regulated chemical (47 FR 9352).
Reportable Quantity (RQ): 100 lb (45.4 kg).
EPCRA Section 313 Form R *de minimis* concentration reporting level: 1.0%.
California Proposition 65 Chemical: Cancer 4/1/88.
Canada, WHMIS, Ingredients Disclosure List Concentration 1.0%.
European/International Regulations: Hazard Symbol: T, N; Risk phrases: R45; R36/38; R52/53; R67; Safety phrases: S53; S45; S61 (see Appendix 4).
WGK (German Aquatic Hazard Class): 3—Severe hazard to waters.
Description: Trichloroethylene, a colorless (often dyed blue), nonflammable, noncorrosive liquid that has the "sweet" odor characteristic of some chlorinated hydrocarbons. The odor threshold is 25−50 ppm. Molecular weight = 131.38; Specific gravity (H_2O:1) = 1.46 at 25°C; Boiling point = 87.2°C; Freezing/Melting point = −72.8°C; Vapor pressure = 58 mmHg at 25°C;

Autoignition temperature = 410°C. Explosive limits: LEL = 8.0%; UEL = 10.5% at 25°C; LEL = 7.8%; UEL = 52% at 100°C. Hazard Identification (based on NFPA-704 M Rating System): Health 3, Flammability 1, Reactivity 0. Practically insoluble in water; solubility = 0.1%.
Potential Exposure: Compound Description: Agricultural Chemical; Tumorigen, Drug, Mutagen; Reproductive Effector; Human Data; Primary Irritant. Trichloroethylene is used as a vapor degreaser of metal parts, as a solvent; and as a drug. It is also used for extracting caffeine from coffee, as a dry-cleaning agent; and as a chemical intermediate in the production of pesticides; in making waxes, gums, resins, tars, paints, varnishes, and specific chemicals, such as chloroacetic acid.
Incompatibilities: Contact with strong caustics causes decomposition and the production of highly toxic and flammable dichloroacetylene. Violent reaction with chemically active metals; powders, or shavings, such as aluminum, barium, lithium, sodium, magnesium, titanium. Violent reaction with aluminum in the presence of dilute hydrochloric acid. Decomposition of trichloroethylene, due to contact with hot metal or ultraviolet radiation, forms hazardous products including chlorine gas, hydrogen chloride; and phosgene. Keep this chemical away from high temperatures, such as arc welding or cutting, unshielded resistance heating; open flames; and high-intensity ultraviolet light. Slowly decomposed by light in presence of moisture, with formulation of hydrochloric acid.
Permissible Exposure Limits in Air
Conversion factor: 1 ppm = 5.37 mg/m^3 at 25°C & 1 atm.
OSHA PEL: 100 ppm TWA; 200 ppm; 300 ppm [5-min maximum peak in any 2 h] Ceiling Concentration.
NIOSH REL: 25 ppm TWA; 2 ppm [60 min when used as a waste anesthetic gas] STEL; Potential occupational carcinogen. Limit exposure to lowest feasible concentration. See the *NIOSH Pocket Guide*, Appendix A & C.
ACGIH TLV®[1]: 10 ppm TWA; 25 ppm STEL, Suspected Human Carcinogen. BEI: 100 mg[trichloroacetic acid]/g creatinine in urine, end-of-workweek; 300 mg[trichloroacetic acid and trichloroethanol]/g creatinine in urine, end-of-shift, end-of-workweek; 4 mg[trichloroethylene]/g in blood, end-of-shift, end-of-workweek.
NIOSH IDLH: 1000 ppm.
Protective Action Criteria (PAC)*
TEEL-0: 10 ppm
PAC-1: **130** ppm
PAC-2: **450** ppm
PAC-3: **3800** ppm
*AEGLs (Acute Emergency Guideline Levels) & ERPGs (Emergency Response Planning Guideline) are in **bold face**.
DFG MAK: Carcinogen Category: 1; Germ Cell Mutagen Group: 3B; TRK: 5 mg[trichloroethanol]/L in blood, end-of-exposure, end-of-shift; for long-term exposure, after several shifts; 100 mg[trichloroacetic acid]/L in urine, end-of-

exposure, end-of-shift; for long-term exposure, after several shifts.

Australia: TWA 50 ppm (270 mg/m^3), STEL 200 ppm, 1993; Austria: MAK 50 ppm (270 mg/m^3), Suspected: carcinogen, 1999; Belgium: TWA 50 ppm (269 mg/m^3), STEL 200 ppm (1070 mg/m^3), 1993; Denmark: TWA 10 ppm (55 mg/m^3), 1999; Finland: TWA 30 ppm (160 mg/m^3), STEL 45 ppm (240 mg/m^3) [skin], 1999; France: VME 75 ppm (405 mg/m^3), VLE 200 ppm (1080 mg/m^3), carcinogen, 1999; Hungary: TWA 10 mg/m^3, STEL 40 mg/m^3, 1993; the Netherlands: MAC-TGG 190 mg/m^3, 2003; Norway: TWA 20 ppm (110 mg/m^3), 1999; the Philippines: TWA 100 ppm (535 mg/m^3), 1993; Poland: MAC (TWA) 50 mg/m^3; MAC (STEL) 400 mg/m^3, 1999; Russia: TWA 50 ppm, STEL 10 mg/m^3, 1993; Sweden: NGV 10 ppm (50 mg/m^3), KTV 25 ppm (140 mg/m^3), carcinogen, 1999; Switzerland: MAK-W 50 ppm (260 mg/m^3), KZG-W 250 ppm (1300 mg/m^3), 1999; Thailand: TWA 100 ppm, STEL 200 ppm, 1993; Turkey: TWA 100 ppm (535 mg/m^3), 1993; United Kingdom: TWA 100 ppm (550 mg/m^3), STEL 150 ppm [skin], 2000; Argentina, Bulgaria, Columbia, Jordan, South Korea, New Zealand, Singapore, Vietnam: ACGIH TLV®: STEL 100 ppm. Russia has set MAC values for ambient air in residential areas of 4.0 mg/m^3 on a momentary basis and 1.0 mg/m^3 on a daily average basis. Several states have set guidelines or standards for trichloroethylene in ambient air[60] ranging from zero (North Carolina) to 0.25 μg/m^3 (Arizona) to 0.8 μg/m^3 (Michigan) to 2.43 μg/m^3 (Kansas) to 6.1 μg/m^3 (Massachusetts) to 900.0 μg/m^3 (New York) to 1350.0 μg/m^3 (Connecticut) to 2675.0 μg/m^3 (Indiana) to 2700.0 μg/m^3 (Florida and South Dakota) to 4500.0 μg/m^3 (Virginia) to 2700.0−10,800.0 μg/m^3 (North Dakota) to 6429 μg/m^3 (Nevada) to 6750 μg/m^3 (South Carolina) to 6840 μg/m^3 (Pennsylvania).

Determination in Air: Use NIOSH Analytical Method, #1022, Trichloroethylene, by portable GC. #3800; OSHA Analytical Method 1001.

Determination in Water: Octanol−water coefficient: Log K_{ow} = 2.42.

Permissible Concentration in Water: *To protect freshwater aquatic life:* 45,000 μg/L on an acute toxicity basis. *To protect saltwater aquatic life:* 2000 μg/L on an acute toxicity basis. *To protect human health:* preferably zero. An additional lifetime cancer risk of 1 in 100,000 is posed by a concentration of 27 μg/L.[6] EPA more recently has proposed a maximum contaminant level of 5 μg/L for trichloroethylene.[48] Japan has set[35] a maximum permissible concentration in drinking water of 30 μg/L. Russia has set a limit of 60 μg/L and WHO has set 30 μg/L as a guideline.[35] Several states have set standards and guidelines for trichloroethylene in drinking water[61] ranging from 1.0 μg/L (New Jersey) to 2.8 μg/L (New Hampshire) to 3.0 μg/L (Florida) to 5.0 μg/L (California, Maine, and Colorado) to 25 μg/L (Connecticut) to 31.2 μg/L (Minnesota).

Determination in Water: Inert gas purge followed by gas chromatography with halide-specific detection (EPA Method #601) or gas chromatography plus mass spectrometry (EPA Method #624).

Routes of Entry: Inhalation percutaneous absorption, ingestion; skin and/or eye contact.

Harmful Effects and Symptoms

Short Term Exposure: Exposure to the vapor irritates the eyes, skin, and respiratory tract. High exposures can cause pulmonary edema, a medical emergency that can be delayed for several hours. This can cause death. Inhalation causes headache, sleepiness, nausea, vomiting, dizziness, and coughing have been felt around 100 ppm. Unconsciousness can result at 3000 ppm. Exposure to 8000 ppm can cause death. Can be absorbed through skin. Can cause skin irritation, burning, or redness; blistering can occur. Can cause eye irritation; burning sensation; and/or watering, and can cause permanent damage. Ingestion can cause chemical pneumonitis and diminish kidney action. It can cause drunkenness, vomiting, diarrhea, or abdominal pain. Unconsciousness, liver or kidney damage, vision distortion, and death have been reported at large doses. Exposure to TCE may affect the central nervous system causing lightheadedness, dizziness, visual disturbances; feeling of excitement; nausea, and vomiting. High levels can cause irregular heartbeat; unconsciousness, and death.

Long Term Exposure: Contact with vapor levels near 100 ppm can cause giddiness, nervous exhaustion, increased sensitivity to alcohol including redness in the face (trichloroethylene blush); the ability to become addicted to the vapor; as well as effects of acute exposure listed above. Higher levels can cause irregular heartbeat. Repeated contact with hands can cause excessive dryness, cracking, burning, loss of sense of touch, or temporary paralysis of fingers. Most of these effects seem to go away after exposure has stopped. Trichloroethylene is considered a cancer suspect agent because high levels cause liver cancer in mice. Whether it causes cancer in humans is unknown. May affect the liver and kidney.

Points of Attack: Eyes, skin, respiratory system; heart, liver, kidneys, central nervous system. Cancer site in animals: liver and kidneys.

Medical Surveillance: NIOSH lists the following tests: whole blood (chemical/metabolite); whole blood (chemical/metabolite), end-of-shift; whole blood (chemical/metabolite), end-of-shift-, end-of-workweek; whole blood (chemical/metabolite), end-of-workweek, expired air, Expired Air, end-of-workweek; Expired Air, prior to next shift; urine (chemical/metabolite); urine (chemical/metabolite), end-of-shift; urine (chemical/metabolite), end-of-shift at end-of-workweek; urine (chemical/metabolite), end-of-workweek; urine (chemical/metabolite), prior to next shift. For those with frequent or potentially high exposure (half the TLV or greater, or significant skin contact), the following are recommended before beginning work and at regular times after that: Liver function tests. If symptoms develop or overexposure is suspected, the following may be useful: exam of the nervous system. Consider nerve conduction tests. Urinary trichloroacetic acid level (for repeated exposures) or blood trichloroethylene levels (for acute exposure). Consider chest X-ray after acute overexposure. Evaluation by a qualified allergist, including careful exposure history and special testing, may help diagnose skin allergy. Kidney function tests.

First Aid: If this chemical gets into the eyes, remove any contact lenses at once and irrigate immediately for at least 15 min, occasionally lifting upper and lower lids. Seek medical attention immediately. If this chemical contacts the skin, remove contaminated clothing and wash immediately with soap and water. Seek medical attention immediately. If this chemical has been inhaled, remove from exposure, begin rescue breathing (using universal precautions, including resuscitation mask) if breathing has stopped and CPR if heart action has stopped. Transfer promptly to a medical facility. When this chemical has been swallowed, get medical attention. Give large quantities of water and induce vomiting. Do not make an unconscious person vomit.

Note to physician: Inhalation: Bronchodilators, decongestants, and oxygen may be used if necessary. Corticosteroids are useful for treating pneumonitis.

Personal Protective Methods: Wear protective gloves and clothing to prevent any reasonable probability of skin contact. Safety equipment suppliers/manufacturers can provide recommendations on the most protective glove/clothing material for your operation. Teflon™ and Silvershield™ are among the recommended materials. All protective clothing (suits, gloves, footwear, headgear) should be clean, available each day, and put on before work. Contact lenses should not be worn when working with this chemical. Wear splash-proof chemical goggles and face shield unless full-face-piece respiratory protection is worn. Employees should wash immediately with soap when skin is wet or contaminated. Provide emergency showers and eyewash.

Respirator Selection: NIOSH: *At concentrations above the NIOSH REL, or where there is no REL, at any detectable concentration:* SCBAF: Pd,Pp (APF = 10,000) (any NIOSH/MSHA- or European Standard EN 149-approved self-contained breathing apparatus that has a full face-piece and is operated in a pressure-demand or other positive-pressure mode); or SaF: Pd,Pp: ASCBA (APF = 10,000) (any supplied-air respirator that has a full face-piece and is operated in a pressure-demand or other positive-pressure mode in combination with an auxiliary, self-contained breathing apparatus operated in a pressure-demand or other positive-pressure mode). *Escape:* GmFOv (APF = 50) [any air-purifying, full-face-piece respirator (gas mask) with a chin-style, front- or back-mounted organic vapor canister]; or SCBAE (any appropriate escape-type, self-contained breathing apparatus).

Storage: Color Code—Blue: Health Hazard/Poison: Store in a secure poison location. Prior to working with this chemical you should be trained on its proper handling and storage. Before entering confined space where this chemical may be present, check to make sure that an explosive concentration does not exist. Trichloroethylene must be handled and stored away from operations which generate high temperatures, such as arc welding or cutting; unshielded resistance heating; open flames; and high-intensity ultraviolet light. It must also be handled to avoid contact with hot metals. Poisonous gases, such as phosgene, and hydrogen chloride are formed. Prevent contact of trichloroethylene with strong alkalis, such as sodium hydroxide or potassium hydroxide, because a highly flammable, toxic liquid is produced. Also prevent contact with aluminum in the presence of dilute hydrochloric acid, because a violent reaction will occur. Prevent contact with chemically active metals; powders, or shavings, such as barium, lithium, sodium, or magnesium; and titanium powders or shavings, since an explosion can occur. A regulated, marked area should be established where this chemical is handled, used, or stored in compliance with OSHA Standard 1910.1045.

Shipping: Trichloroethylene requires a shipping label of "POISONOUS/TOXIC MATERIALS." It falls in Hazard Class 6.1 and Packing Group III.

Spill Handling: Evacuate and restrict persons not wearing protective equipment from area of spill or leak until cleanup is complete. Remove all ignition sources. Establish forced ventilation to keep levels below explosive limit. Absorb liquids in vermiculite, dry sand; earth, peat, carbon, or a similar material and deposit in sealed containers. Keep this chemical out of a confined space, such as a sewer, because of the possibility of an explosion, unless the sewer is designed to prevent the buildup of explosive concentrations. It may be necessary to contain and dispose of this chemical as a hazardous waste. If material or contaminated runoff enters waterways, notify downstream users of potentially contaminated waters. Contact your local or federal environmental protection agency for specific recommendations. If employees are required to clean up spills, they must be properly trained and equipped. OSHA 1910.120(q) may be applicable.

Fire Extinguishing: This chemical is a combustible liquid. Poisonous gases, including hydrogen chloride, chlorine gas, and phosgene, are produced in fire. Use dry chemical, carbon dioxide, or alcohol foam extinguishers. Vapors are heavier than air and will collect in low areas. Vapors may travel long distances to ignition sources and flashback. Vapors in confined areas may explode when exposed to fire. Containers may explode in fire. Storage containers and parts of containers may rocket great distances, in many directions. If material or contaminated runoff enters waterways, notify downstream users of potentially contaminated waters. Notify local health and fire officials and pollution control agencies. From a secure, explosion-proof location, use water spray to cool exposed containers. If cooling streams are ineffective (venting sound increases in volume and pitch, tank discolors, or shows any signs of deforming), withdraw immediately to a secure position. If employees are expected to fight fires, they must be trained and equipped in OSHA 1910.156. The only respirators recommended for firefighting are self-contained breathing apparatuses that have full face-pieces and are operated in a pressure-demand or other positive-pressure mode.

Disposal Method Suggested: Consult with environmental regulatory agencies for guidance on acceptable disposal practices. Generators of waste containing this contaminant (≥100 kg/mo) must conform with EPA regulations governing storage, transportation, treatment, and waste disposal. Incineration, preferably after mixing with another combustible fuel. Care must be exercised to assure complete combustion to prevent the formation of phosgene. An acid scrubber is necessary to remove the halo acids produced. An alternative to disposal for TCE is recovery and recycling.

References

National Institute for Occupational Safety and Health. (1973). *Criteria for a Recommended Standard: Occupational Exposure to Trichloroethylene.* NIOSH Document No. 73-11025

National Institute for Occupational Safety and Health. (January 1978). *Special Occupational Hazard Review with Control Recommendations: Trichloroethylene.* NIOSH Document No. 78-130. Washington, DC

US Environmental Protection Agency. (1980). *Trichloroethylene: Ambient Water Quality Criteria.* Washington, DC

US Environmental Protection Agency. (December 1979). *Status Assessment of Toxic Chemicals: Trichloroethylene,* Report EPA-600/2-79-210m. Cincinnati, OH

US Environmental Protection Agency. (April 30, 1980). *Trichloroethylene, Health and Environmental Effects Profile No. 166.* Washington, DC: Office of Solid Waste

US Public Health Service. (January 1988). *Toxicological Profile for Trichloroethylene.* Atlanta, GA: Agency for Toxic Substances and Disease Registry

Sax, N. I. (Ed.). *Dangerous Properties of Industrial Materials Report,* 1, No. 2, 67–69 (1980), 3, No. 1, 89–94 (1953), 4, No. 3, 30–32 (1984), and 7, No. 1, 83–92 (1987)

New York State Department of Health. (March 1986). *Chemical Fact Sheet: Trichloroethylene.* Albany, NY: Bureau of Toxic Substance Assessment (Version 3)

New Jersey Department of Health and Senior Services. (January 2000). *Hazardous Substances Fact Sheet: Trichloroethylene.* Trenton, NJ

Trichloroisocyanuric acid   T:0750

Molecular Formula: $C_3Cl_3N_3O_3$
Common Formula: $(ClNCO)_3$
Synonyms: ACL 85; CBD 90; Fichlor 91; FI Clor 91; Isocyanuric chloride; NSC-405124; Symclosen; Symclosene; Trichlorinated isocyanuric acid; Trichlorocyanuric acid; Trichloroisocyanic acid, 1,3,5-Trichloroisocyanuric acid; Trichloroisocyanuric acid; 1,3,5-Trichloro-1,3,5-triazinetrione; Trichloro-*s*-triazine-2,4,6 (1H,3H,5H)-trione; Trichloro-*s*-triazinetrione; 1,3,5-Trichloro-2,4,6-trioxohexahydro-*s*-triazine

CAS Registry Number: 87-90-1
RTECS® Number: XZ1925000
UN/NA & ERG Number: UN2468 (dry)/140
EC Number: 201-782-8 [*Annex I Index No.:* 613-031-00-5]

Regulatory Authority and Advisory Bodies
Cyanide compounds:
Clean Air Act: Hazardous Air Pollutants (Title I, Part A, Section 112) as cyanide compound.
Clean Water Act: 40CFR423, Appendix A, Priority Pollutants as cyanide, total.
US EPA Hazardous Waste Number (RCRA No.): P030 as cyanides soluble salts and complexes, n.o.s.
RCRA 40CFR261, Appendix 8 Hazardous Constituents as cyanides, soluble salts and complexes, n.o.s.
EPCRA (Section 313): X + CN- where X = H + or any other group where a formal dissociation may occur. For example, KCN or Ca(CN)₂; Form R *de minimis* concentration reporting level: 1.0%.
US DOT Regulated Marine Pollutant (49CFR172.101, Appendix B) as cyanide mixtures, cyanide solutions or cyanides, inorganic, n.o.s.
Canada, WHMIS, Ingredients Disclosure List Concentration 1.0%; National Pollutant Release Inventory (NPRI); CEPA Priority Substance List, Ocean dumping prohibited.
TRICHLOROISOCYANURIC ACID 87-90-1 European/ International Regulations: Hazard Symbol: O, Xn, N; Risk phrases: R8; R22; R31; R36/37; R50/53; Safety phrases: S2;8; S26; S41; S60; S61 (see Appendix 4).
WGK (German Aquatic Hazard Class): 2—Hazard to waters.

Description: Trichloroisocyanuric acid is a white crystalline solid with a chlorine odor. Molecular weight = 232.41; Freezing/Melting point = 225–230°C (with decomposition). Hazard Identification (based on NFPA-704 M Rating System): Health 2, Flammability 1, Reactivity 2 OX.

Potential Exposure: This material is used in household bleaches and detergents.

Incompatibilities: A powerful oxidizer. Violent reaction with reducing agents; combustible materials. Trichloroisocyanuric acid can release poisonous chlorine, nitrogen oxides, and cyanides when heated to high temperatures. Contact with water may also release toxic chemicals.

Permissible Exposure Limits in Air
Protective Action Criteria (PAC)
TEEL-0: 30 mg/m³
PAC-1: 75 mg/m³
PAC-2: 500 mg/m³
PAC-3: 500 mg/m³

Determination in Air: Use NIOSH Analytical Method (IV) #7904, Cyanides, OSHA Analytical Method ID-120. See also Method #6010, Hydrogen Cyanide.[18]

Permissible Concentration in Water: In 1976 the EPA criterion was 5.0 µg/L for freshwater and marine aquatic life and wildlife. As of 1980, the criteria are: *To protect freshwater aquatic life:* 3.5 µg/L as a 24-h average, never to exceed 52.0 µg/L. *To protect saltwater aquatic life:* 30.0 µg/L on an acute toxicity basis; 2.0 µg/L on a chronic toxicity basis. *To protect human health:* 200 µg/L. The allowable daily intake for man is 8.4 mg/day.[6] On the international scene, the South African Bureau of Standards has set 10 µg/L, the World Health Organization (WHO) 10 µg/L, and Germany 50 µg/L as drinking water standards. Other international limits[35] include an EEC limit of 50 µg/L; Mexican limits of 200 µg/L in drinking water and 1.0 µg/L in coastal waters and a Swedish limit of 100 µg/L. Russia[43] set a MAC of 100 µg/L in water bodies used for domestic purposes and 50 µg/L in water for fishery purposes. The US EPA[49] has determined a no-observed-adverse-effect-level (NOAEL) of 10.8 mg/kg/day which yields a lifetime health advisory of 154 µg/L. States which have set guidelines for cyanides in drinking water[61] include Arizona at 160 µg/L and Kansas at 220 µg/L.

Determination in Water: Distillation followed by silver nitrate titration or colorimetric analysis using pyridine pyrazolone (or barbituric acid).

Routes of Entry: Inhalation, ingestion, eyes, and/or skin contact. Absorbed through the skin.

Harmful Effects and Symptoms

Short Term Exposure: Trichloroisocyanuric acid can affect you when breathed in and by passing through your skin. Exposure can irritate the eyes, skin, nose, throat, and air passages. Contact can cause skin or eye irritation.

Long Term Exposure: Repeated exposure may cause skin irritation. Highly irritating substances may affect the lungs, although it is not known for certain that this chemical causes lung damage.

Points of Attack: Lungs.

Medical Surveillance: For those with frequent or potentially high exposure the following are recommended before beginning work and at regular times after that: lung function tests.

First Aid: If this chemical gets into the eyes, remove any contact lenses at once and irrigate immediately for at least 15 min, occasionally lifting upper and lower lids. Seek medical attention immediately. If this chemical contacts the skin, remove contaminated clothing and wash immediately with soap and water. Seek medical attention immediately. If this chemical has been inhaled, remove from exposure, begin rescue breathing (using universal precautions, including resuscitation mask) if breathing has stopped and CPR if heart action has stopped. Transfer promptly to a medical facility. When this chemical has been swallowed, get medical attention. Give large quantities of water and induce vomiting. Do not make an unconscious person vomit. Use amyl nitrate capsules if symptoms of cyanide poisoning develop. All area employees should be trained regularly in emergency measures for cyanide poisoning and in CPR.

A cyanide antidote kit should be kept in the immediate work area and must be rapidly available. Kit ingredients should be replaced every 1–2 years to ensure freshness. Persons trained in the use of this kit; oxygen use, and CPR must be quickly available.

Personal Protective Methods: Wear protective gloves and clothing to prevent any reasonable probability of skin contact. Safety equipment suppliers/manufacturers can provide recommendations on the most protective glove/clothing material for your operation. All protective clothing (suits, gloves, footwear, headgear) should be clean, available each day, and put on before work. Contact lenses should not be worn when working with this chemical. Wear dust-proof chemical goggles and face shield unless full-face-piece respiratory protection is worn. Employees should wash immediately with soap when skin is wet or contaminated. Provide emergency showers and eyewash.

Respirator Selection: None listed. Following is the information for *cyanide compounds*: *Up to 25 mg/m³:* Sa (APF = 10) (any supplied-air respirator); or SCBAF (APF = 50) (any self-contained breathing apparatus with full face-piece). *Emergency or planned entry into unknown concentrations or IDLH conditions:* SCBAF: Pd,Pp (APF = 10,000) (any self-contained breathing apparatus that has a full face-piece and is operated in a pressure-demand or other positive-pressure mode); or SaF: Pd,Pp: ASCBA (any supplied-air respirator that has a full face-piece and is operated in a pressure-demand or other positive-pressure mode in combination with an auxiliary self-contained breathing apparatus operated in a pressure-demand or other positive-pressure mode). *Escape:* GmFS100 (APF = 50) [any air-purifying, full-face-piece respirator (gas mask) with a chin-style, front- or back-mounted canister providing protection against the compound of concern and having an N100, R100, or P100 filter]; or SCBAE (any appropriate escape-type, self-contained breathing apparatus).

Storage: Color Code—Yellow: Reactive Hazard; Store in a location separate from other materials, especially flammables and combustibles. Prior to working with this chemical you should be trained on its proper handling and storage. Store in tightly closed containers in a cool well-ventilated area away from combustibles (such as wood, paper, and oil). Sources of ignition, such as smoking and open flames, are prohibited where trichloroisocyanuric acid is used, handled, or stored in a manner that could create a potential fire or explosion hazard. See OSHA Standard 1910.104 and NFPA 43A *Code for the Storage of Liquid and Solid Oxidizers* for detailed handling and storage regulations.

Shipping: This compound requires a shipping label of "OXIDIZER." It falls in Hazard Class 5.1 and Packing Group II.

Spill Handling: Evacuate persons not wearing protective equipment from area of spill or leak until cleanup is complete. Remove all ignition sources. Collect powdered material in the most convenient and safe manner and deposit in

sealed containers. Ventilate area after cleanup is complete. It may be necessary to contain and dispose of this chemical as a hazardous waste. If material or contaminated runoff enters waterways, notify downstream users of potentially contaminated waters. Contact your local or federal environmental protection agency for specific recommendations. If employees are required to clean up spills, they must be properly trained and equipped. OSHA 1910.120(q) may be applicable.

Fire Extinguishing: Trichloroisocyanuric acid may burn, but does not readily ignite. Trichloroisocyanuric acid is a strong oxidizer and a dangerous fire risk on contact with combustibles (like paper, wood, and oil). Poisonous gases are produced in fire, including chorine, nitrogen oxides; and cyanides. Use dry chemical, carbon dioxide, or water spray extinguishers. If material or contaminated runoff enters waterways, notify downstream users of potentially contaminated waters. Notify local health and fire officials and pollution control agencies. From a secure, explosion-proof location, use water spray to cool exposed containers. If cooling streams are ineffective (venting sound increases in volume and pitch, tank discolors, or shows any signs of deforming), withdraw immediately to a secure position. If employees are expected to fight fires, they must be trained and equipped in OSHA 1910.156. The only respirators recommended for firefighting are self-contained breathing apparatuses that have full face-pieces and are operated in a pressure-demand or other positive-pressure mode.

References
New Jersey Department of Health and Senior Services. (December 2000). *Hazardous Substances Fact Sheet: Trichloroisocyanuric Acid.* Trenton, NJ

Trichloronate   T:0760

Molecular Formula: $C_{10}H_{12}Cl_3O_2PS$
Common Formula: $C_6H_2(Cl_3)-O-P(S)(CH_2CH_3)OCH_2CH_3$
Synonyms: o-Aethyl-o-(2,4,5-trichlorphenyl)-aethylthiono-phosphonat (German); Agrisil; Agritox; Bay 37289; Bayer 37289; Bayer S 4400; Chemagro 37289; ENT 25,712; o-Ethyl o-2,4,5-trichlorophenyl ethylphosphonothioate; Ethyl trichlorophenylethylphosphonothioate; Fenophosphon; Phytosol; Stauffer N-3049; Trichloronat; 2,4,5-Trichlorophenol o-ester with o-ethyl ethylphosphonothioate; Wirkstoff 37289
CAS Registry Number: 327-98-0
RTECS® Number: TB0700000
UN/NA & ERG Number: UN3018 (organophosphorus pesticide, liquid, toxic)/152
EC Number: 206-326-1 [*Annex I Index No.:* 015-098-00-0]
Regulatory Authority and Advisory Bodies
Superfund/EPCRA 40CFR355, Extremely Hazardous Substances: TPQ = 500 lb (227 kg).

Reportable Quantity (RQ): 500 lb (227 kg).
US DOT Regulated Marine Pollutant (49CFR172.101, Appendix B).
US DOT 49CFR172.101, Inhalation Hazard Chemical as organic phosphate.
European/International Regulations: Hazard Symbol: T+, N; Risk phrases: R24; R28; R50/53; Safety phrases: S1/2; S23; S28; S36/37; S45; S60; S61 (see Appendix 4).
WGK (German Aquatic Hazard Class): No value assigned.
Description: Trichloronate is an amber-colored liquid. Molecular weight = 333.60; Boiling point = 108°C at 0.01 mmHg. Hazard Identification (based on NFPA-704 M Rating System): Health 3, Flammability 1, Reactivity 1. Soluble in water; solubility = 50 mg/L at 20°C.
Potential Exposure: Those involved in the manufacture, formulation, or application of this nonsystemic, organophosphate insecticide which is used for the control of soil insects.
Incompatibilities: Strong oxidizers may cause release of toxic phosphorus oxides. Organophosphates, in the presence of strong reducing agents such as hydrides, may form highly toxic and flammable phosphine gas. Keep away from alkaline materials.
Permissible Exposure Limits in Air
Protective Action Criteria (PAC)
TEEL-0: 2 mg/m³
PAC-1: 6 mg/m³
PAC-2: 10 mg/m³
PAC-3: 300 mg/m³
Determination in Air: OSHA versatile sampler-2; Toluene/Acetone; Gas chromatography/Flame photometric detection for sulfur, nitrogen, or phosphorus; NIOSH Analytical Method (IV) Method #5600, Organophosphorus Pesticides.
Determination in Water: Fish Tox = 14.47908000 (ppb) (INTERMEDIATE).
Routes of Entry: Inhalation, skin absorption, ingestion, skin and/or eye contact.
Harmful Effects and Symptoms
Short Term Exposure: Toxic effects are due to action on the nervous system. It has high oral toxicity and death can occur in acute poisonings. Delayed neurotoxicity has been reported. Symptoms of exposure include headache, dizziness, nausea, salivation, vomiting, abdominal pain, diarrhea, chest pain, decreased heart rate, excessive discharge of mucous from the air passages, difficult breathing, contraction of the pupil, blurred vision, profuse perspiration, muscle twitching and spasms, profound weakness, psychotic behavior, uncoordination, unconsciousness, rarely, convulsions. Low level absorption syndrome is similar to influenza. High dosage may cause toxic psychosis similar to alcoholism. Exposures may be misdiagnosed as asthma and heart failure. Organic phosphorus insecticides are absorbed by the skin as well as by the respiratory and gastrointestinal tracts. They are cholinesterase inhibitors. Symptoms of exposure include headache, giddiness, blurred vision, nervousness, weakness, nausea, cramps, diarrhea,

and discomfort in the chest. Signs include sweating, tearing, salivation, vomiting, cyanosis, convulsions, coma, loss of reflexes, and loss of sphincter control.

Long Term Exposure: Cholinesterase inhibitor; cumulative effect is possible. This chemical may damage the nervous system with repeated exposure, resulting in convulsions, respiratory failure. May cause liver damage.

Points of Attack: Respiratory system, lungs, central nervous system, cardiovascular system, skin, eyes, plasma and red blood cell cholinesterase.

Medical Surveillance: Before employment and at regular times after that, the following are recommended: plasma and red blood cell cholinesterase levels (tests for the enzyme poisoned by this chemical). If exposure stops, plasma levels return to normal in 1–2 weeks while red blood cell levels may be reduced for 1–3 months. When cholinesterase enzyme levels are reduced by 25% or more below preemployment levels, risk of poisoning is increased, even if results are in lower ranges of "normal." Reassignment to work not involving organophosphate or carbamate pesticides is recommended until enzyme levels recover. If symptoms develop or overexposure occurs, repeat the above tests as soon as possible and get an examination of the nervous system. Also, consider complete blood count. Consider chest X-ray following acute overexposure. Do not drink any alcoholic beverages before or during use. Alcohol promotes absorption of organic phosphates.

First Aid: If this chemical gets into the eyes, remove any contact lenses at once and irrigate immediately for at least 15 min, occasionally lifting upper and lower lids. Seek medical attention immediately. If this chemical contacts the skin, remove contaminated clothing and wash immediately with soap and water. Seek medical attention immediately. If this chemical has been inhaled, remove from exposure, begin rescue breathing (using universal precautions, including resuscitation mask) if breathing has stopped and CPR if heart action has stopped. Transfer promptly to a medical facility. When this chemical has been swallowed, get medical attention. Give large quantities of water and induce vomiting. Do not make an unconscious person vomit. Keep victim quiet and maintain normal body temperature. Effects may be delayed; keep victim under observation.

Personal Protective Methods: Wear protective gloves and clothing to prevent any reasonable probability of skin contact. Safety equipment suppliers/manufacturers can provide recommendations on the most protective glove/clothing material for your operation. All protective clothing (suits, gloves, footwear, and headgear) should be clean, available each day, and put on before work. Contact lenses should not be worn when working with this chemical. Wear splash-proof chemical goggles and face shield unless full-face-piece respiratory protection is worn. Employees should wash immediately with soap when skin is wet or contaminated. Provide emergency showers and eyewash.

Respirator Selection: Follow regulations in OSHA 29CFR1910.134 or European Standard EN149. Use a NIOSH/MSHA- or European Standard EN149-approved respirator; or use an approved supplied-air respirator with a full face-piece operated in the positive-pressure mode, or with a full face-piece, hood, or helmet in the continuous-flow mode; or use a NIOSH/MSHA- or European Standard EN149-approved self-contained breathing apparatus with a full face-piece operated in pressure-demand or other positive-pressure mode.

Storage: Color Code—Blue: Health Hazard/Poison: Store in a secure poison location. Prior to working with this chemical you should be trained on its proper handling and storage. Store in tightly closed containers in a cool, well-ventilated area away from strong bases. Where possible, automatically transfer material from drums or other storage containers to process containers. Sources of ignition, such as smoking and open flames, are prohibited where this chemical is handled, used, or stored. Metal containers involving the transfer of this chemical should be grounded and bonded. Wherever this chemical is used, handled, manufactured, or stored, use explosion-proof electrical equipment and fittings. This compound requires a shipping label of "POISONOUS/TOXIC MATERIALS." It falls in Hazard Class 6.1 and Packing Group I.

Spill Handling: Evacuate and restrict persons not wearing protective equipment from area of spill or leak until cleanup is complete. Remove all ignition sources. Ventilate area of spill or leak. Do not touch spilled material; stop leak if you can do so without risk. Use water spray to reduce vapors. *Small spills:* Absorb liquids in vermiculite, dry sand, earth, peat, carbon, or a similar material and deposit in sealed containers. *Large spills:* dike far ahead of spill for later disposal. Stay upwind; keep out of low areas. Ventilate closed spaces before entering them. Remove and isolate contaminated clothing at the site. Keep this chemical out of a confined space, such as a sewer, because of the possibility of an explosion, unless the sewer is designed to prevent the buildup of explosive concentrations. It may be necessary to contain and dispose of this chemical as a hazardous waste. If material or contaminated runoff enters waterways, notify downstream users of potentially contaminated waters. Contact your local or federal environmental protection agency for specific recommendations. If employees are required to clean up spills, they must be properly trained and equipped. OSHA 1910.120(q) may be applicable.

Fire Extinguishing: Poisonous gases, including phosphorus oxides, are produced in fire. *Small fires:* dry chemical, carbon dioxide, water spray, or foam. *Large fires:* water spray, fog, or foam. Move container from fire area if you can do it without risk. Dike fire control water for later disposal; do not scatter the material. Wear positive pressure breathing apparatus and special protective clothing. Vapors are heavier than air and will collect in low areas. Containers may explode in fire. Storage containers and parts of containers may rocket great distances, in many directions. If material or contaminated runoff enters waterways, notify downstream users of potentially contaminated waters.

Notify local health and fire officials and pollution control agencies. From a secure, explosion-proof location, use water spray to cool exposed containers. If cooling streams are ineffective (venting sound increases in volume and pitch, tank discolors, or shows any signs of deforming), withdraw immediately to a secure position. If employees are expected to fight fires, they must be trained and equipped in OSHA 1910.156. The only respirators recommended for firefighting are self-contained breathing apparatuses that have full face-pieces and are operated in a pressure-demand or other positive-pressure mode.

References

US Environmental Protection Agency. (November 30, 1987). *Chemical Hazard Information Profile: Trichloronate*. Washington, DC: Chemical Emergency Preparedness Program

Trichlorophenols   T:0770

Molecular Formula: $C_6H_3Cl_3O$
Common Formula: $HOC_6H_2Cl_3$
Synonyms: 25167-82-2: Omal; Phenachlor; Phenol, trichloro-; Trichlorofenol (Spanish); Triclorofenol
15950-66-0: Phenol, 2,3,4-trichloro-; 2,3,4-Trichlorofenol (Spanish); 2,3,4-Trichlorophenol; Trichlorophenol, 2,3,4-
933-78-8: Phenol, 2,3,5-trichloro-; 2,3,5-Trichlorofenol (Spanish); 2,3,5-Trichlorophenol trichlorophenol, 2,3,5-
933-75-5: Phenol, 2,3,6-trichloro-; 2,3,6-Trichlorofenol (Spanish); 2,3,6-Trichlorophenol; Trichlorophenol, 2,3,6-
88-06-2: Dowicide 2S; NCI-CO2904; Omal; Phenachlor; Phenol, 2,4,6-trichloro-; 2,4,6-Trichlorfenol (Spanish); 1,3,5-Trichloro-2-hydroxybenzene; Trichlorophenol, 2,4,6-; 2,4,6-Trichlorophenos
609-19-8: Phenol, 3,4,5-trichloro-; 3,4,5-Trichlorofenol (Spanish); 3,4,5-Trichlorophenol; Trichlorophenol, 3,4,5-
CAS Registry Number: 25167-82-2 (mixed isomers); 15950-66-0 (2,3,4-); 933-78-8 (2,3,5-); 933-75-5 (2,3,6-); 95-95-4 (2,4,5-); 88-06-2 (2,4,6-); 609-19-8 (3,4,5-)
RTECS® Number: SN1400000 (2,4,5-); SN1575000 (2,4,6-); SN1650000 (3,4,5-); SN1300000 (2,3,6-)
UN/NA & ERG Number: UN2020 (solid)/153
EC Number: 246-694-0 (mixed isomers); 240-083-2 (2,3,4-); 213-272-2 (2,3,5-); 213-271-7 (2,3,6-); 202-467-8 [*Annex I Index No.:* 604-017-00-X] (2,4,5-); 201-795-9 [*Annex I Index No.:* 604-018-00-5] (2,4,6-); 210-183-0 (3,4,5-)

Regulatory Authority and Advisory Bodies
Carcinogenicity: IARC[9]: Human Inadequate Evidence, animal Sufficient Evidence, 1982, *possibly carcinogenic to humans*, Group B2, 1987; NTP: Reasonably anticipated to be a human carcinogen.
Air Pollutant Standard Set. See below, "Permissible Exposure Limits in Air" section.
Mixed isomers-isomer:
Clean Water Act: Section 311 Hazardous Substances/RQ 40CFR117.3 (same as CERCLA, see below).

Reportable Quantity (RQ): 10 lb (4.54 kg).
2,3,4-; 2,3,5-; 2,3,6-; 3,4,5-isomers:
Reportable Quantity (RQ): 10 lb (4.54 kg).
2,4,5-isomer:
Carcinogenicity: IARC: Animal Limited Evidence, 1999; Human Limited Evidence, 1986.
US EPA, FIFRA 1998 Status of Pesticides: Canceled.
Clean Air Act: Hazardous Air Pollutants (Title I, Part A, Section 112).
US EPA Hazardous Waste Number (RCRA No.): U230.
RCRA Toxicity Characteristic (Section 261.24), Maximum. Concentration of Contaminants, regulatory level, 400.0 mg/L.
RCRA, 40CFR261, Appendix 8 Hazardous Constituents.
RCRA 40CFR268.48; 61FR15654, Universal Treatment Standards: Wastewater (mg/L), 0.18; Nonwastewater (mg/kg), 7.4.
RCRA 40CFR264, Appendix 9; TSD Facilities Ground Water Monitoring List. Suggested test method(s) (PQL µg/L): 8270 (10).
Reportable Quantity (RQ): 10 lb (4.54 kg).
EPCRA Section 313 Form R *de minimis* concentration reporting level: 1.0%.
European/International Regulations (*2,4,5-isomer*): Hazard Symbol: Xn, N; Risk phrases: R22; R36/38; R50/53; Safety phrases: S2; S26; S28; S60; S61 (see Appendix 4).
WGK (German Aquatic Hazard Class) (*2,4,5*): 3—Highly water polluting.
2,4,6-isomer:
Carcinogenicity: NCI: Carcinogenesis Bioassay (feed); clear evidence: mouse, rat; NTP: 11th Report on Carcinogens, 2004: Reasonably anticipated to be a human carcinogen; IARC: Animal Limited Evidence, 1999; Human Limited Evidence, 1986.
US EPA Gene-Tox Program, Positive: Carcinogenicity—mouse/rat; *S. cerevisiae*—forward mutation; Negative: Mouse spot test; Histidine reversion—Ames test; Negative: *S. cerevisiae* gene conversion; *S. cerevisiae*—homozygosis.
Clean Air Act: Hazardous Air Pollutants (Title I, Part A, Section 112).
US EPA Hazardous Waste Number (RCRA No.): U231.
RCRA Toxicity Characteristic (Section 261.24), Maximum. Concentration of Contaminants, regulatory level, 2.0 mg/L.
RCRA, 40CFR261, Appendix 8 Hazardous Constituents.
RCRA 40CFR268.48; 61FR15654, Universal Treatment Standards: Wastewater (mg/L), 0.035; Nonwastewater (mg/kg), 7.4.
RCRA 40CFR264, Appendix 9; TSD Facilities Ground Water Monitoring List. Suggested test method(s) (PQL µg/L): 8040 (5); 8270 (10).
Reportable Quantity (RQ): 10 lb (4.54 kg).
EPCRA Section 313 Form R *de minimis* concentration reporting level: 0.1%.
California Proposition 65 Chemical: Cancer (2,4,6-) 1/1/88.
European/International Regulations (*2,4,6-*): Hazard Symbol: Xn; Risk phrases: R22; R36/38; R40; Safety phrases: S2; S36/37 (see Appendix 4).

WGK (German Aquatic Hazard Class) (*2,4,6*-): No value assigned.

Description: Trichlorophenols exists as 6 isomers (2,4,5-; 3,4,5-; 2,4,6-; 2,3,4-; 2,3,5-; and 2,3,6-). The most important (heavily regulated) are the 2,4,5- and 2,4,6-isomers. The 2,4,5-isomer is white powder or needles; 2,3,5- and 2,3,6- are colorless crystals; 2,4,5- is a gray crystalline solid or flakes; 2,4,6- is a colorless to light yellow crystalline solid. They have a phenolic odor. Molecular weight = 197.44; Boiling point = 248−253°C (*2,3,5*-); 253°C (*2,3,6*-); 253°C (*2,4,5*-); 246°C (*2,4,6*-); Freezing/Melting point = 84°C (*2,3,4*-); 62°C (*2,3,5*-); 58°C (*2,3,6*-); 67°C (*2,4,5*-); 70°C (*2,4,6*-); Flash point = 78°C (*2,3,6*-); 61°C (*2,4,6*-). Hazard Identification (based on NFPA-704 M Rating System): (*2,4,5*-; *2,4,6*-) Health 2, Flammability 1, Reactivity 0. All isomers are slightly soluble or practically insoluble in water.

Potential Exposure: Compound Description (2,4,5-): Agricultural Chemical; Tumorigen, Mutagen; Reproductive Effector; (2,4,6-): Agricultural Chemical; Tumorigen, Mutagen; Reproductive Effector; Human Data; Primary Irritant (2,3,6-) Mutagen; Human Data. 2,4,5-TCP is used as antifungal agent in adhesives and as preservative in polyvinyl acetate emulsions. 2,4,6-T is used in manufacturing slime-control agents and as an effective germicide and preservative; to produce defoliant 2,4,5-T and related products. Also used directly as a fungicide, anti-mildew and preservative agent; algicide, bactericide. 2,4,6-TCP is used to produce 2,3,4,6-TCP and PCP. Used directly as germicide, bactericide, glue and wood preservative; and anti-mildew treatment. 2,3,6-TCP is used as intermediate in production of fungicides and plant growth regulators.

Incompatibilities: Perhaps the most notable incompatibility is the reaction of 2,4,5-trichlorophenol in alkaline medium at high temperatures to produce dioxin. (2,3,4-isomer) reacts with oxidizers, acid anhydrides, and acid chlorides. (2,3,5-isomer) Decomposes on heating, on burning, and on contact with strong oxidants, producing toxic and corrosive fumes of hydrogen chloride. The substance is a weak acid. (2,3,6-isomer); the substance is a weak acid. pH = 4.8/4.2-; (2,4,6-isomer); reacts violently with strong oxidants and is incompatible with acid chlorides and acid anhydrides.

Permissible Exposure Limits in Air
Protective Action Criteria (PAC)
2,3,6-
TEEL-0: 1.25 mg/m^3
PAC-1: 4 mg/m^3
PAC-2: 25 mg/m^3
PAC-3: 125 mg/m^3
2,4,5-
TEEL-0: 40 mg/m^3
PAC-1: 125 mg/m^3
PAC-2: 350 mg/m^3
PAC-3: 350 mg/m^3
2,4,6-
TEEL-0: 10 mg/m^3

PAC-1: 30 mg/m^3
PAC-2: 200 mg/m^3
PAC-3: 350 mg/m^3
DFG MAK: No numerical value established. Data may be available.
Denmark: TWA 0.5 mg/m^3, [skin] 1999 (2,4,5- and 2,3,6- isomers).
Sweden: MAC 0.5 mg/m^3; STEL 1.5 mg/m^3. Russia set a MAC of 3.0 μg/m^3 in ambient air in residential areas for the 2,4,6-isomer on a once-daily basis. Several states have set guidelines or standards for the trichlorophenols in ambient air.[60] Massachusetts has set zero for the 2,4,6-isomer and 1.6 μg/m^3 for the 2,4,5-isomer. Pennsylvania has set 3500 μg/m^3 for the 2,4,5-isomer on a 1-year exposure basis.

Determination in Air: Use NIOSH: (*o*-chlorophenol) P&CAM Method #337 (chlorophenols).

Permissible Concentration in Water: For 2,4,6-trichlorophenol, to protect freshwater aquatic life: 970 μg/L on a chronic toxicity basis. *To protect saltwater aquatic life:* no criteria developed due to insufficient data. To protect human health—for 2,4,5-TCP, 2600 μg/L; for 2,4,6-TCP, preferably zero. An additional lifetime cancer risk of 1 in 100,000 occurs at a level of 12 μg/L. These are based on organoleptic effects. A limit based on toxicological effects for 2,4,5-TCP would be 1600 μg/L.[6] Kansas[61] has set a guideline of 1.0 μg/L for the 2,4,5-isomer in drinking water. Values for the 2,4,6-isomer have been set by Kansas at 17.0 μg/L, by Minnesota at 17.5 μg/L, and by Maine at 700.0 μg/L. The WHO[35] has set a limit of 10.0 μg/L for the 2,4,6-isomer in drinking water.

Determination in Water: Methylene chloride extraction followed by gas chromatography with flame ionization or electron capture detection (EPA Method 604); or gas chromatography plus mass spectrometry (EPA Method 625).

Routes of Entry: Inhalation, ingestion, skin and/or eye contact may be absorbed through the skin.

Harmful Effects and Symptoms
Short Term Exposure: Trichlorophenols irritates the eyes, skin, and the respiratory tract. A central nervous system depressant. High exposures can cause weakness, difficulty in breathing, tremors, convulsions, coma, and possible death. See also *chlorophenols.*

2,3,5-isomer: A mixture of trichlorophenols may cause irritation of the skin, eyes, and respiratory tract. These substances may cause acute metabolic effects resulting in damage in several organs, notably the CNS. Some technical products may contain highly toxic impurities including polychlorinated dibenzo-*p*-dioxins and -furans.

For the 2,4,5-isomer[52]: irritation of the skin, eyes, nose, and pharynx; redness and edema of the skin; dermatitis, corneal injury; iritis; sweating, thirst, nausea, vomiting, diarrhea, abdominal pain; cyanosis, hyperactivity, stupor, decreased activity and motor weakness; increase followed by decrease in respiratory rate and urinary output; fever; increased bowel action; lung, liver, or kidney damage; convulsions, collapse, and coma.

Long Term Exposure: Repeated or prolonged contact with skin may cause dermatitis, drying, and cracking. May affect the liver and kidneys. A related chemical, *phenol*, can cause liver and kidney damage. May be carcinogenic to humans. If any of the trichlorophenols is contaminated with 2,3,7,8-tetra-chlorodibenzo-p-dioxin, the following effects may occur: acne-like skin rash; liver damage; nervous system damage with symptoms of weakness, pain in the legs, and numbness.

Points of Attack: Inhalation: Human (2,4,5-): lung, thorax, or respiration; structural or functional change in trachea or bronchi; lung, thorax, or respiration: other changes. Animal tests: change in liver weight, changes in spleen weight; other changes. Cancer site in animals (2,4,6-): liver and leukemia.

Medical Surveillance: Liver and kidney function tests. Complete blood count (CBC).

First Aid: Skin Contact[52]: Flood all areas of body that have contacted the substance with water. Do not wait to remove contaminated clothing; do it under the water stream. Use soap to help assure removal. Isolate contaminated clothing when removed to prevent contact by others.

Eye Contact: Remove any contact lenses at once. Immediately flush eyes well with copious quantities of water or normal saline for at least 20–30 min. Seek medical attention.

Inhalation: Leave contaminated area immediately; breathe fresh air. Proper respiratory protection must be supplied to any rescuers. If coughing, difficult breathing, or any other symptoms develop, seek medical attention at once, even if symptoms develop many hours after exposure.

Ingestion: If unconscious or convulsing, do not induce vomiting or give anything by mouth. Assure that victim's airway is open and lay him on his side with his head lower than his body and transport at once to a medical facility. If conscious and not convulsing, give a slurry of activated charcoal in water. If medical advice is not readily available, do not induce vomiting, and rush the victim to the nearest medical facility.

Personal Protective Methods: Wear protective gloves and clothing to prevent any reasonable probability of skin contact. Safety equipment suppliers/manufacturers can provide recommendations on the most protective glove/clothing material for your operation. All protective clothing (suits, gloves, footwear, headgear) should be clean, available each day, and put on before work. Contact lenses should not be worn when working with this chemical. Wear dust-proof chemical goggles and face shield unless full-face-piece respiratory protection is worn. Employees should wash immediately with soap when skin is wet or contaminated. Provide emergency showers and eyewash.

Respirator Selection: Follow regulations in OSHA 29CFR1910.134 or European Standard EN149. Use a NIOSH/MSHA- or European Standard EN149-approved respirator; or use an approved supplied-air respirator with a full face-piece operated in the positive-pressure mode, or with a full face-piece, hood, or helmet in the continuous-flow mode; or use a NIOSH/MSHA- or European Standard EN149-approved self-contained breathing apparatus with a full face-piece operated in pressure-demand or other positive-pressure mode.

Storage: Color Code—Blue: Health Hazard/Poison: Store in a secure poison location. Prior to working with this chemical you should be trained on its proper handling and storage. Store in a cool dry place or a refrigerator away from oxidizing agents and other incompatible materials listed above. Where possible, automatically transfer material from drums or other storage containers to process containers. Sources of ignition, such as smoking and open flames, are prohibited where this chemical is handled, used, or stored. Metal containers involving the transfer of this chemical should be grounded and bonded. Wherever this chemical is used, handled, manufactured, or stored, use explosion-proof electrical equipment and fittings. A regulated, marked area should be established where this chemical is handled, used, or stored in compliance with OSHA Standard 1910.1045.

Shipping: Chlorophenols, solid, require a shipping label of "POISONOUS/TOXIC MATERIALS." It falls in Hazard Class 6.1 and Packing Group III.

Spill Handling: Evacuate persons not wearing protective equipment from area of spill or leak until cleanup is complete. Remove all ignition sources. Remove all sources of ignition and dampen spilled material with 60–70% ethanol to avoid airborne dust, then transfer material to a sealed container. Ventilate the spill area and use absorbent dampened with 60–70% ethanol to pick up remaining material. Wash surfaces well with soap and water. Ventilate area after cleanup is complete. It may be necessary to contain and dispose of this chemical as a hazardous waste. If material or contaminated runoff enters waterways, notify downstream users of potentially contaminated waters. Contact your local or federal environmental protection agency for specific recommendations. If employees are required to clean up spills, they must be properly trained and equipped. OSHA 1910.120(q) may be applicable.

Fire Extinguishing: Trichlorophenols are combustible, but are not easy to ignite. Use dry chemical, carbon dioxide, water spray, or alcohol foam extinguishers. Poisonous gases are produced in fire, including carbon monoxide, carbon dioxide, hydrogen chloride, and chlorine. If material or contaminated runoff enters waterways, notify downstream users of potentially contaminated waters. Notify local health and fire officials and pollution control agencies. From a secure, explosion-proof location, use water spray to cool exposed containers. If cooling streams are ineffective (venting sound increases in volume and pitch, tank discolors, or shows any signs of deforming), withdraw immediately to a secure position. If employees are expected to fight fires, they must be trained and equipped in OSHA 1910.156. The only respirators recommended for firefighting are self-contained breathing apparatuses that have full face-pieces and are operated in a pressure-demand or other positive-pressure mode.

Disposal Method Suggested: Consult with environmental regulatory agencies for guidance on acceptable disposal practices. Generators of waste containing this contaminant (≥100 kg/mo) must conform with EPA regulations governing storage, transportation, treatment, and waste disposal. Incineration, preferably after mixing with another combustible fuel. Care must be exercised to assure complete combustion to prevent the formation of phosgene. An acid scrubber is necessary to remove the halo acids produced.[22]

References

US Environmental Protection Agency. (1980). *Chlorinated Phenols: Ambient Water Quality Criteria.* Washington, DC
US Environmental Protection Agency. (April 30, 1980). *2,4,6-Trichlorophenol: Health and Environmental Effects Profile No. 168.* Washington, DC: Office of Solid Waste
US Environmental Protection Agency, Special Review and Reregistration Division Office of Pesticide Programs. (1998). *Agency Status of Pesticides in Registration, Reregistration, and Special Review* (Rainbow Report). Washington, DC
New Jersey Department of Health and Senior Services. (August 2002). *Hazardous Substances Fact Sheet: 2,4,6-Trichlorophenol.* Trenton, NJ

1,2,3-Trichloropropane   T:0780

Molecular Formula: $C_3H_5Cl_3$
Common Formula: $CH_2ClCHClCH_2Cl$
Synonyms: AI3-26040; Allyl trichloride; Glycerol trichlorohydrin; Glyceryl trichlorohydrin; NCI-C60220; NSC 35403; Propane, 1,2,3-trichloro-; Trichlorohydrin; Trichloropropane; 1,2,3-Tricloropropano (Spanish)
CAS Registry Number: 96-18-4
RTECS® Number: TZ9275000
UN/NA & ERG Number: UN2810/153
EC Number: 202-486-1 [*Annex I Index No.:* 602-062-00-X]
Regulatory Authority and Advisory Bodies
Carcinogenicity: IARC: Animal, Sufficient Evidence; Human, Insufficient Evidence, Group 2A, 1995; NCI: Carcinogenesis Studies (gavage); clear evidence: mouse, rat; NTP: 11th Report on Carcinogens, 2004: Reasonably anticipated to be a human carcinogen; EPA: Likely to produce cancer in humans; NIOSH: Potential occupational carcinogen.
Air Pollutant Standard Set. See below, "Permissible Exposure Limits in Air" section.
RCRA, 40CFR261, Appendix 8 Hazardous Constituents, waste number not listed.
RCRA 40CFR268.48; 61FR15654, Universal Treatment Standards: Wastewater (mg/L), 0.85; Nonwastewater (mg/kg), 30.
RCRA 40CFR264, Appendix 9; TSD Facilities Ground Water Monitoring List. Suggested test method(s) (PQL μg/L): 8010 (10); 8240 (5).
Safe Drinking Water Act: Priority List (55 FR 1470).

EPCRA Section 313 Form R *de minimis* concentration reporting level: 1.0%.
California Proposition 65 Chemical: Cancer 10/1/92.
Canada, WHMIS, Ingredients Disclosure List Concentration 1.0%.
European/International Regulations: Hazard Symbol: T; Risk phrases: R45; R60; R20/21/22; Safety phrases: S53; S45 (see Appendix 4).
WGK (German Aquatic Hazard Class): No value assigned.
Description: 1,2,3-Trichloropropane is a colorless liquid with a strong acid odor. Molecular weight = 147.43; Specific gravity $(H_2O:1) = 1.39$ at 25°C; Boiling point = 156°C; Freezing/Melting point = −14°C; Flash point = 71°C (cc); Autoignition temperature = 304°C. Explosive limits: LEL = 3.2% at 120°C; UEL = 12.6% at 150°C. Hazard Identification (based on NFPA-704 M Rating System): Health 3, Flammability 2, Reactivity 0. Poor solubility in water; solubility = 0.1%.
Potential Exposure: Compound Description: Tumorigen, Mutagen; Reproductive Effector; Primary Irritant. Trichloropropane dissolves oils, fats, waxes, chlorinated rubber, and numerous resins; it is used as a paint and varnish remover, a solvent, and a degreasing agent.
Incompatibilities: Violent decomposition with chemically active metals; strong bases. Vigorous reaction with strong oxidizers. Keep away from chlorinated rubber, resins and waxes, and sunlight.
Permissible Exposure Limits in Air
Conversion factor: 1 ppm = 6.03 mg/m³ at 25°C & 1 atm.
OSHA PEL: 50 ppm/300 mg/m³ TWA.
NIOSH REL: 10 ppm/60 mg/m³ TWA [skin], Potential occupational carcinogen. Limit exposure to lowest feasible concentration. See *NIOSH Pocket Guide*, Appendix A.
ACGIH TLV®[1]: 10 ppm/60 mg/m³ TWA [skin] confirmed animal carcinogen with unknown relevance to humans.
NIOSH IDLH: 1000 ppm.
Protective Action Criteria (PAC)
TEEL-0: 10 mg/m³
PAC-1: 30 mg/m³
PAC-2: 50 mg/m³
PAC-3: 100 mg/m³
DFG MAK: [skin] Carcinogen Category 2.
Australia: TWA 10 ppm (60 mg/m³), [skin], 1993; Austria: MAK 50 ppm (300 mg/m³), 1999; Belgium: TWA 10 ppm (60 mg/m³), [skin], 1993; Denmark: TWA 10 ppm (60 mg/m³), [skin], 1999; Finland: TWA 50 ppm (300 mg/m³), STEL 75 ppm (450 mg/m³), 1999; Norway: TWA 10 ppm (60 mg/m³), 1999; the Netherlands: MAC-TGG 0.108 mg/m³, [skin], 2003; Russia: STEL 2 mg/m³, 1993; Switzerland: carcinogen, 1999; United Kingdom: TWA 50 ppm (306 mg/m³), STEL 75 ppm, 2000; Argentina, Bulgaria, Columbia, Jordan, South Korea; New Zealand, Singapore, Vietnam: ACGIH TLV®: confirmed animal carcinogen with unknown relevance to humans.
Russia[43] set a MAC of 0.05 mg/m³ for ambient air in residential areas on a daily average basis. Several states have

set guidelines or standards for 1,2,3-trichloropropane in ambient air[60] ranging from 3.0—4.5 mg/m^3 (North Dakota) to 5.0 mg/m^3 (Virginia) to 6.0 mg/m^3 (Connecticut) to 7.143 mg/m^3 (Nevada).

Determination in Air: Use NIOSH Analytical Method #1003, Hydrocarbons, halogenated; OSHA Analytical Method 7.[18]

Permissible Concentration in Water: Russia[43] has set 0.07 mg/L as a MAC in water bodies used for domestic purposes.

Determination in Water: Octanol—water coefficient: Log $K_{ow} = 2.27$.

Routes of Entry: Inhalation, skin absorption, ingestion, skin and/or eye contact.

Harmful Effects and Symptoms

Short Term Exposure: Trichloropropane is highly toxic by inhalation and moderately toxic by skin absorption. It is a local irritant and produces a number of unpleasant sensory effects; irritates the eyes, skin, and respiratory tract. Humans exposed to trichloropropane at 100 ppm found this to be an objectionable level of exposure, and all reported eye and throat irritation as well as an unpleasant odor. Also, according to NIOSH, skin irritation, central nervous system depression, and liver injury may result. Exposure to high concentrations may result in unconsciousness.

Long Term Exposure: Repeated contact can cause dermatitis; drying and cracking of the skin. May affect the heart and damage the liver. Although this chemical has not been adequately evaluated for brain and disturbed sleep, many solvents and petroleum-based products can cause these effects. Symptoms of exposure can include reduced memory, reduced ability to concentrate, personality changes, fatigue, sleep disturbances, reduced coordination, weakness, and/or feeling of "pins and needles" in extremities. A potential occupational carcinogen.

Points of Attack: Eyes, skin, respiratory system, central nervous system, liver, kidneys. Cancer site in animals: forestomach, liver, and mammary glands.

Medical Surveillance: NIOSH lists the following tests: Expired Air. Liver function tests; EKG, evaluation for brain effects and psychological changes.

First Aid: If this chemical gets into the eyes, remove any contact lenses at once and irrigate immediately for at least 15 min, occasionally lifting upper and lower lids. Seek medical attention immediately. If this chemical contacts the skin, remove contaminated clothing and wash immediately with soap and water. Seek medical attention immediately. If this chemical has been inhaled, remove from exposure, begin rescue breathing (using universal precautions, including resuscitation mask) if breathing has stopped and CPR if heart action has stopped. Transfer promptly to a medical facility. When this chemical has been swallowed, get medical attention. Give large quantities of water and induce vomiting. Do not make an unconscious person vomit.

Personal Protective Methods: Wear protective gloves and clothing to prevent any reasonable probability of skin contact. Safety equipment suppliers/manufacturers can provide recommendations on the most protective glove/clothing material for your operation. Viton and PVC are among the recommended materials. All protective clothing (suits, gloves, footwear, headgear) should be clean, available each day, and put on before work. Contact lenses should not be worn when working with this chemical. Wear splash-proof chemical goggles and face shield unless full-face-piece respiratory protection is worn. Employees should wash immediately with soap when skin is wet or contaminated. Provide emergency showers and eyewash.

Respirator Selection: At concentrations above the NIOSH REL, or where there is no REL, at any detectable concentration: SCBAF: Pd,Pp (APF = 10,000) (any NIOSH/MSHA- or European Standard EN 149-approved self-contained breathing apparatus that has a full face-piece and is operated in a pressure-demand or other positive-pressure mode) or SaF: Pd,Pp: ASCBA (APF = 10,000) (any supplied-air respirator that has a full face-piece and is operated in a pressure-demand or other positive-pressure mode in combination with an auxiliary, self-contained breathing apparatus operated in a pressure-demand or other positive-pressure mode). *Escape:* GmFOv (APF = 50) [any air-purifying, full-face-piece respirator (gas mask) with a chin-style, front- or back-mounted organic vapor canister] or SCBAE (any appropriate escape-type, self-contained breathing apparatus).

Storage: Color Code—Blue: Health Hazard/Poison: Store in a secure poison location. Prior to working with this chemical you should be trained on its proper handling and storage. Before entering confined space where this chemical may be present, check to make sure that an explosive concentration does not exist. Store in a refrigerator or a cool, dry place and keep away from chemically active metals, oxidizers, strong caustics. Where possible, automatically pump liquid from drums or other storage containers to process containers. Sources of ignition, such as smoking and open flames, are prohibited where this chemical is handled, used, or stored.

Shipping: Toxic, liquids, organic, n.o.s. require a shipping label of "POISONOUS/TOXIC MATERIALS." 1,2,3-Trichloropropane falls in DOT Hazard Class 6.1 and Parking Group III.

Spill Handling: Evacuate and restrict persons not wearing protective equipment from area of spill or leak until cleanup is complete. Remove all ignition sources. Establish forced ventilation to keep levels below explosive limit. Absorb liquids in vermiculite, dry sand, earth, peat, carbon, or a similar material and deposit in sealed containers. Keep this chemical out of a confined space, such as a sewer, because of the possibility of an explosion, unless the sewer is designed to prevent the buildup of explosive concentrations. It may be necessary to contain and dispose of this chemical as a hazardous waste. If material or contaminated runoff enters waterways, notify downstream users of potentially contaminated waters. Contact your local or federal environmental protection agency for specific recommendations. If employees are required to clean up spills, they must be properly trained and equipped. OSHA 1910.120(q) may be applicable.

Fire Extinguishing: This chemical is a combustible liquid. Poisonous gases, including phosgene, carbon monoxide, chlorine, and hydrogen chloride, are produced in fire. Use dry chemical, carbon dioxide, alcohol foam, or polymer foam extinguishers. Vapors are heavier than air and will collect in low areas. Vapors may travel long distances to ignition sources and flashback. Vapors in confined areas may explode when exposed to fire. Containers may explode in fire. Storage containers and parts of containers may rocket great distances, in many directions. If material or contaminated runoff enters waterways, notify downstream users of potentially contaminated waters. Notify local health and fire officials and pollution control agencies. From a secure, explosion-proof location, use water spray to cool exposed containers. If cooling streams are ineffective (venting sound increases in volume and pitch, tank discolors, or shows any signs of deforming), withdraw immediately to a secure position. If employees are expected to fight fires, they must be trained and equipped in OSHA 1910.156. The only respirators recommended for firefighting are self-contained breathing apparatuses that have full face-pieces and are operated in a pressure-demand or other positive-pressure mode.

Disposal Method Suggested: Incineration, preferably after mixing with another combustible fuel. Care must be exercised to assure complete combustion to prevent the formation of phosgene. An acid scrubber is necessary to remove the halo acids produced.

References

National Institute for Occupational Safety and Health. (1977). *Profiles on Occupational Hazards for Criteria Document Priorities*, Report PB-274,073. Cincinnati, OH, pp. 289—291

US Environmental Protection Agency. (April 30, 1980). *1,2,3-Trichloropropane: Health and Environmental Effects Profile No. 169*. Washington, DC: Office of Solid Waste New Jersey Department of Health and Senior Services. (May 1999). *Hazardous Substances Fact Sheet: 1,2,3-Trichloropropane*. Trenton, NJ

1,1,2-Trichloro-1,2,2-trifluoroethane   T:0790

Molecular Formula: $C_2Cl_3F_3$
Common Formula: CCl_2FCClF_2
Synonyms: Arcton 63; Arklone P; Asahifron 113; Daiflon S 3; Distillex DS5; Ethane, 1,1,2-trichloro-1,2,2,-trifluoro-; F 113; FC 113; Fluorocarbon 113; Forane 113; Freon 113TR-T; Freon TF; Frigen 113; Frigen 113A; Frigen 113TR; Frigen 113TR-N; Frigen 113TR-T; Genesolv D solvent; Genetron 113; Isceon 113; Kaiser chemicals 11; Khladon 113; Ledon 113; MS-180 freon TF solvent; R 113; Refrigerant 113; Refrigerant R 113; 1,1,2-Trichloro-1,2,2-trifluoroethane; 1,1,2-Trichlorotrifluoroethane; 1,1,2-Tricloro-fluoetano (Spanish); 1,1,2-Trifluoro-1,2,2-trichloroethane; 1,1,2-Trifluorotrichloro ethane

CAS Registry Number: 76-13-1
RTECS® Number: KJ4000000
UN/NA & ERG Number: UN3082/171
EC Number: 200-936-1
Regulatory Authority and Advisory Bodies
Air Pollutant Standard Set. See below, "Permissible Exposure Limits in Air" section.
Clean Air Act: Stratospheric ozone protection (Title VI, Subpart A, Appendix A), Class I, Ozone Depletion Potential = 0.8.
RCRA 40CFR268.48; 61FR15654, Universal Treatment Standards: Wastewater (mg/L), 0.057; Nonwastewater (mg/kg), 30.
EPCRA Section 313 Form R *de minimis* concentration reporting level: 1.0%.
Canada, WHMIS, Ingredients Disclosure List Concentration 1.0%.
European/International Regulations: not listed in Annex 1.
WGK (German Aquatic Hazard Class): 2—Hazard to waters.

Description: TTE is a colorless liquid with an odor like carbon tetrachloride at high concentrations. A gas above 48°C. The odor threshold is 45—68 ppm. Molecular weight = 187.37; Specific gravity (H_2O:1) = 1.56 at 25°C; Boiling point = 48°C; Freezing/Melting point = −36°C; Vapor pressure = 285 mmHg at 25°C. Noncombustible liquid at ordinary temperatures, but the gas will ignite and burn weakly at 680°C (autoignition temperature). Hazard Identification (based on NFPA-704 M Rating System): Health 2, Flammability 0, Reactivity 2. Practically insoluble in water; solubility = 0.02%.

Potential Exposure: Compound Description: Drug; Reproductive Effector; Human Data; Primary Irritant. TTE is used as a solvent and refrigerant; it is used in fire extinguishers; as a blowing agent and as an intermediate in the production of chlorotrifluoroethylene monomer by reaction with zinc.

Incompatibilities: Violent reaction with chemically active metals (such as powdered aluminum, beryllium, magnesium, and zinc); calcium. Contact with alloys containing more than 2% Mg causes decomposition releasing hydrogen chloride, hydrogen fluoride, and carbon monoxide.

Permissible Exposure Limits in Air
Conversion factor: 1 ppm = 7.67 mg/m³ at 25°C & 1 atm.
OSHA PEL: 1000 ppm/7600 mg/m³ TWA.
NIOSH REL: 1000 ppm/7600 mg/m³ TWA; 1250 ppm/9500 mg/m³ STEL.
ACGIH TLV®[1]: 1000 ppm/7670 mg/m³ TWA; 1250 ppm/9590 mg/m³ STEL, not classifiable as a human carcinogen.
NIOSH IDLH: 2000 ppm.

Protective Action Criteria (PAC)

TEEL-0: 1000 ppm

PAC-1: 1250 ppm

PAC-2: 1500 ppm

PAC-3: 2000 ppm

DFG MAK: 500 ppm/3900 mg/m^3 TWA; Peak Limitation Category II(2); Pregnancy Risk Group D.

Australia: TWA 1000 ppm (7600 mg/m^3), STEL 1250 ppm, 1993; Austria: MAK 500 ppm (3800 mg/m^3), 1999; Belgium: TWA 1000 ppm (7670 mg/m^3), STEL 1250 ppm, 1993; Denmark: TWA 500 ppm (3800 mg/m^3), 1999; Finland: TWA 1000 ppm (7600 mg/m^3), STEL 1250 ppm, 1999; France: VME 1000 ppm (7600 mg/m^3), VLE 1250 ppm, 1999; Hungary: STEL 40 mg/m^3, 1993; the Netherlands: MAC-TGG 1170 mg/m^3, 2003; Norway: TWA 500 ppm (3800 mg/m^3), 1999; the Philippines: TWA 1000 ppm (7600 mg/m^3), 1993; Russia: TWA 500 ppm, STEL 5000 mg/m^3, 1993; Sweden: NGV 500 ppm (4000 mg/m^3), KTV 750 ppm (6000 mg/m^3), 1999; Switzerland: MAK-W 500 ppm (3800 mg/m^3), 1999; Turkey: TWA 1000 ppm (7600 mg/m^3), 1993; United Kingdom: TWA 1000 ppm (7790 mg/m^3), STEL 1250 ppm, 2000; Argentina, Bulgaria, Columbia, Jordan, South Korea, New Zealand, Singapore, Vietnam: ACGIH TLV®: STEL 1250 ppm. Several states have set guidelines or standards for TTE in ambient air[60] ranging from 0.152 mg/m^3 (Connecticut) to 0.225 mg/m^3 (Indiana) to 76.0-95.0 mg/m^3 (North Dakota) to 180.952 mg/m^3 (Nevada) to 950.0 mg/m^3 (North Carolina).

Determination in Air: Use NIOSH Analytical Method #1020, 1,1,2-Trichloro-1,2,2-trifluoroethane, #2549, Volatile organic compound; OSHA Analytical Method 113.

Determination in Water: Octanol−water coefficient: Log $K_{ow} = 3.30$.

Routes of Entry: Inhalation, ingestion, skin and/or eye contact.

Harmful Effects and Symptoms

Short Term Exposure: Irritates the eyes and respiratory tract. May be a central nervous system depressant in high concentrations, causing headache, dizziness, loss of coordination, and unconsciousness; asphyxiation can result. *Inhalation:* No effects may be felt below 1000 ppm. Irregular heartbeat (arrhythmia) can occur at levels above 2000 ppm; may cause unconsciousness and possible death. Levels above 2500 ppm may cause loss of concentration, tiredness, and a feeling of heaviness of the head. Levels above 200,000 ppm may cause irritation of the nose and lungs, tremors, and coma. *Ingestion:* No information is available on human ingestion. However, animal studies show that Freon 113 causes unresponsiveness, facial swelling, diarrhea, and bleeding lungs. *Skin:* May cause irritation and frostbite if TTE is cold from refrigeration.

Long Term Exposure: Skin contact can cause dermatitis, drying, and cracking. Can accumulate in brain and kidneys from exposure below 2000 ppm. Will generally not produce noticeable effects at this level and may pass from the body within a week after exposure stops.

Points of Attack: Skin, heart, central nervous system, cardiovascular system.

Medical Surveillance: If symptoms develop or overexposure is suspected, the following may be useful: Special 24-h EKG (Holter monitor) to look for irregular heartbeat.

First Aid: If this chemical gets into the eyes, remove any contact lenses at once and irrigate immediately for at least 15 min, occasionally lifting upper and lower lids. Seek medical attention immediately. If this chemical contacts the skin, remove contaminated clothing and wash immediately with soap and water. Seek medical attention immediately. If this chemical has been inhaled, remove from exposure, begin rescue breathing (using universal precautions, including resuscitation mask) if breathing has stopped and CPR if heart action has stopped. Transfer promptly to a medical facility. When this chemical has been swallowed, get medical attention. Give large quantities of water and induce vomiting. Do not make an unconscious person vomit. If frostbite has occurred, seek medical attention immediately; do *NOT* rub the affected areas or flush them with water. In order to prevent further tissue damage, do *NOT* attempt to remove frozen clothing from frostbitten areas. If frostbite has *NOT* occurred, immediately and thoroughly wash contaminated skin with soap and water.

Personal Protective Methods: Wear appropriate personal protective clothing to prevent the skin from becoming frozen from contact with the evaporating liquid or from contact with vessels containing the liquid. Safety equipment suppliers/manufacturers can provide recommendations on the most protective glove/clothing material for your operation. All protective clothing (suits, gloves, footwear, headgear) should be clean, available each day, and put on before work. Contact lenses should not be worn when working with this chemical. Wear splash-proof chemical goggles and face shield unless full-face-piece respiratory protection is worn. Employees should wash immediately with soap when skin is wet or contaminated. Provide emergency showers and eyewash.

Respirator Selection: Up to 2000 ppm: Sa (APF = 10) (any supplied-air respirator) or SCBAF (APF = 50) (any self-contained breathing apparatus with a full face-piece). *Emergency or planned entry into unknown concentrations or IDLH conditions:* SCBAF: Pd,Pp (APF = 10,000) (any NIOSH/MSHA- or European Standard EN 149-approved self-contained breathing apparatus that has a full face-piece and is operated in a pressure-demand or other positive-pressure mode) or SaF: Pd,Pp: ASCBA (APF = 10,000) (any supplied-air respirator that has a full face-piece and is operated in a pressure-demand or other positive-pressure mode in combination with an auxiliary, self-contained breathing apparatus operated in a pressure-demand or other positive-pressure mode). *Escape:* GmFOv (APF = 50) [any air-purifying, full-face-piece respirator (gas mask) with a chin-style, front- or back-mounted organic vapor canister] or SCBAE (any appropriate escape-type, self-contained breathing apparatus).

Storage: Color Code—Green: General storage may be used. Prior to working with this chemical you should be trained on its proper handling and storage. 1,1,2-Trichloro-1,2,2-trifluoroethane must be stored to avoid contact with chemically active metals, such as calcium, powdered aluminum, zinc, magnesium, and beryllium because violent reactions occur. Store in tightly closed containers in a cool, well-ventilated area away from oxidizers. Where possible, automatically pump liquid from drums or other storage containers to process containers. Sources of ignition, such as smoking and open flames, are prohibited where this chemical is handled, used, or stored.

Shipping: The name of this material is not in the DOT list of materials[19] for label and packaging standards. However, based on regulations, it may be classified[52] as an Environmentally hazardous substances, liquid, n.o.s. It falls in Hazard Class 9 and Packing Group III.[20, 21]

Spill Handling: Evacuate and restrict persons not wearing protective equipment from area of spill or leak until cleanup is complete. Remove all ignition sources. Ventilate area of spill or leak. Absorb liquids in vermiculite, dry sand, earth, peat, carbon, or a similar material and deposit in sealed containers. Keep this chemical out of a confined space, such as a sewer, because of the possibility of an explosion, unless the sewer is designed to prevent the buildup of explosive concentrations. It may be necessary to contain and dispose of this chemical as a hazardous waste. If material or contaminated runoff enters waterways, notify downstream users of potentially contaminated waters. Contact your local or federal environmental protection agency for specific recommendations. If employees are required to clean up spills, they must be properly trained and equipped. OSHA 1910.120(q) may be applicable.

Fire Extinguishing: 1,1,2-Trichloro-1,2,2-trifluoroethane itself does not easily burn. Extinguish fire using agent suitable for type of surrounding fire. The gas will ignite and burn weakly when exposed to high heat or flame. Poisonous gases, including carbonyl fluoride, chlorine, hydrogen chloride, hydrogen fluoride, and phosgene, are produced in fire. Containers may explode in fire. Storage containers and parts of containers may rocket great distances, in many directions. If material or contaminated runoff enters waterways, notify downstream users of potentially contaminated waters. Notify local health and fire officials and pollution control agencies. From a secure, explosion-proof location, use water spray to cool exposed containers. If cooling streams are ineffective (venting sound increases in volume and pitch, tank discolors, or shows any signs of deforming), withdraw immediately to a secure position. If employees are expected to fight fires, they must be trained and equipped in OSHA 1910.156. The only respirators recommended for firefighting are self-contained breathing apparatuses that have full face-pieces and are operated in a pressure-demand or other positive-pressure mode.

Disposal Method Suggested: Incineration, preferably after mixing with another combustible fuel. Care must be exercised to assure complete combustion to prevent the formation of phosgene. An acid scrubber is necessary to remove the halo acids produced.

References

New York State Department of Health. (March 1986). *Chemical Fact Sheet: Trichlorotrifluoroethane.* Albany, NY: Bureau of Toxic Substance Assessment

Sax, N. I. (Ed.). (1986). *Dangerous Properties of Industrial Materials Report,* 6, No. 3, 91–93

New Jersey Department of Health and Senior Services. (June 2000). *Hazardous Substances Fact Sheet: 1,1,2-Trichloro-1,2,2-Trifluoroethane.* Trenton, NJ

Tricresyl phosphates   T:0800

Molecular Formula: $C_{21}H_{21}O_4P$
Common Formula: $(CH_3C_6H_4O)_3PO$
Synonyms: o-Cresyl phosphate; Fosfito de tricresilo (Spanish); Phosflex 179-C; Phosphoric acid, tri-o-cresyl ester; Phosphoric acid, tris(methyl phenyl) ester; TCP; TOCP; TOFK; o-Tolyl phosphate; TPTP; Tri-o-cresyl ester of phosphoric acid; o-Tricresyl phosphate; Tri-o-cresyl phosphate; Tricresyl phosphate, o-; o-Trikresylphosphat (German); Tri-2-methylphenyl phosphate; Tris(o-cresyl) phosphate; Tris(o-methylphenyl) phosphate; Tri-o-tolyl phosphate; Tri-2-tolyl phosphate
CAS Registry Number: 78-30-8 (o-isomer); 563-04-2 (m-isomer); 78-32-0 (p-isomer); 1330-78-5 (mixed isomers)
RTECS® Number: TD0350000
UN/NA & ERG Number: UN2574/151
EC Number: 201-103-5 [*Annex I Index No.:* 015-015-00-8] (o-isomer); 209-241-8 (m-isomer); 201-105-6 [*Annex I Index No.:* 015-016-00-3] (p-isomer); 215-548-8 (mixed isomers or tris(methylphenyl) phosphate)

Regulatory Authority and Advisory Bodies
Air Pollutant Standard Set. See below, "Permissible Exposure Limits in Air" section.
European/International Regulations (o-o-o; o-o-m; o-o-p; o-m-m; o-m-p; o-p-p): Hazard Symbol: T, N; Risk phrases: R39; R23/24/25; R51/53; Safety phrases: S1/2; S20/21; S28; S45; S6; (m-m-m; m-m-p; m-p-p; p-p-p) Hazard Symbol: Xn, N; Risk phrases: R21/2251/53; Safety phrases: S2; S28; S61 (see Appendix 4).
WGK (German Aquatic Hazard Class): No value assigned (*all isomers*).

Description: Tricresyl phosphates are available as the o-isomer (TOCP), the m-isomer (TMCP), and p-isomer (TPCP). The ortho-isomer is the most toxic of the three; the meta- and para-isomers are relatively inactive. The commercial product may contain the *ortho*-isomer as a contaminant unless special precautions are taken during manufacture. Pure tri-*para*-cresyl phosphate is a solid, and ortho- and meta- are liquids (see below). The *tri-o-cresyl phosphate* will be discussed here as the specific example of these compounds because it is the most toxic of the tricresyl

Tricresyl phosphates

phosphates and specifically regulated by OSHA. TOCP is a colorless to pale yellow, odorless liquid or solid (below 52°F/11°C). Molecular weight = 368.39; Specific gravity (H$_2$O:1) = 1.20 at 25°C; Boiling point = 410°C (with decomposition); Freezing/Melting point = 11°C; Vapor pressure = 0.00002 mmHg at 25°C; Flash point = 110−225°C; Autoignition temperature = 385°C. Hazard Identification (based on NFPA-704 M Rating System): Health 2, Flammability 1, Reactivity 0. Practically insoluble in water.

Potential Exposure: Compound Description: Agricultural Chemical; Drug, Mutagen; Reproductive Effector. Tricresyl phosphate is used as an additive in hydraulic fluids; as a plasticizer, pigment dispersant, flame retardant; as a plasticizer for chlorinated rubber, vinyl plastics, polystyrene, polyacrylic and polymethacrylic esters; as an adjuvant in milling of pigment pastes; as a solvent and as a binder in nitrocellulose and various natural resins; and as an additive to synthetic lubricants and gasoline. It is also used in the recovery of phenol in coke-oven wastewaters.

Incompatibilities: Contact with magnesium may cause explosion. Contact with strong oxidizers may cause fire and explosions.

Permissible Exposure Limits in Air
OSHA PEL: 0.1 mg/m^3 TWA.
NIOSH REL: 0.1 mg/m^3 TWA [skin].
ACGIH TLV$^{®[1]}$: 0.1 mg/m^3 TWA [skin], not classifiable as a human carcinogen; BEI$_A$ issued for acetylcholinesterase-inhibiting pesticides.
NIOSH IDLH: 40 mg/m^3.
Protective Action Criteria (PAC)
TEEL-0: 0.1 ppm
PAC-1: 0.3 ppm
PAC-2: 0.6 ppm
PAC-3: 40 ppm
Australia: TWA 0.1 mg/m^3, [skin], 1993; Austria: MAK 0.1 mg/m^3, 1999; Belgium: TWA 0.1 mg/m^3, [skin], 1993; Denmark: TWA 0.1 mg/m^3, 1999; Finland: TWA 0.1 mg/m^3, STEL 0.3 mg/m^3, [skin], 1999; France: VME 0.1 mg/m^3, [skin], 1999; Hungary: TWA 0.1 mg/m^3, STEL 0.2 mg/m^3, [skin], 1993; the Netherlands: MAC-TGG 0.1 mg/m^3, 2003; the Philippines: TWA 0.1 mg/m^3, 1993; Poland: MAC (TWA) 0.1 mg/m^3; MAC (STEL) 0.3 mg/m^3, 1999; Switzerland: MAK-W 0.1 mg/m^3, 1999; United Kingdom: TWA 0.1 mg/m^3, STEL 0.3 mg/m^3, 2000; Argentina, Bulgaria, Columbia, Jordan, South Korea, New Zealand, Singapore, Vietnam: ACGIH TLV$^®$: not classifiable as a human carcinogen. Several states have set guidelines or standards for the compound in ambient air[60] ranging from 1.0 μg/m^3 (North Dakota) to 2.0 μg/m^3 (Connecticut and Nevada) to 160.0 μg/m^3 (Virginia).

Determination in Air: Use NIOSH Analytical Method (IV) #5037, Triorthocresyl phosphate.

Routes of Entry: Inhalation, skin absorption, ingestion, skin and/or eye contact. The widespread epidemics of poisoning that have occurred have been due to ingested ortho-isomer as a contaminant of foodstuffs. Experimental human studies with labeled phosphorus derivatives show that only 0.4% of the applied dose was absorbed.

Harmful Effects and Symptoms
Short Term Exposure: May affect the central and peripheral nervous systems, causing impaired functions (paralysis). Exposure above exposure limits may cause permanent paralysis. The major effects from inhaling, swallowing, or absorbing tricresyl phosphate through the skin are on the spinal cord and peripheral nervous system; the poison attacking the anterior horn cells and pyramidal tract as well as the peripheral nerves. Gastrointestinal symptoms on acute exposure (nausea, vomiting, diarrhea, and abdominal pain) are followed by a latent period of 3−30 days with the progressive development of muscle soreness and numbness of fingers, calf muscles, and toes; with foot and wrist drop. In chronic intoxication, the gastrointestinal symptoms pass unnoticed, and after a long latent period, flaccid paralysis of limb and leg muscles appears. There are minor sensory changes and no loss of sphincter control.

Long Term Exposure: May affect the nervous system, causing peripheral neuropathy, cramps in calves, paresthesia in feet or hands, weakness in the feet, wrist drop, muscular paralysis.

Points of Attack: Peripheral nervous system, central nervous system.

Medical Surveillance: Preplacement and periodic examinations should include evaluation of spinal cord and neuromuscular function, especially in the extremities; and a history of exposure to other organophosphate esters, pesticides, or neurotoxic agents. Periodic cholinesterase determination may relate to exposure, but not necessarily to neuromuscular effect.

First Aid: If this chemical gets into the eyes, remove any contact lenses at once and irrigate immediately for at least 15 min, occasionally lifting upper and lower lids. Seek medical attention immediately. If this chemical contacts the skin, remove contaminated clothing and wash immediately with soap and water. Speed in removing material from skin is of extreme importance. Shampoo hair promptly if contaminated. Seek medical attention immediately. If this chemical has been inhaled, remove from exposure, begin rescue breathing (using universal precautions, including resuscitation mask) if breathing has stopped and CPR if heart action has stopped. Transfer promptly to a medical facility. When this chemical has been swallowed, get medical attention. Give large quantities of water and induce vomiting. Do not make an unconscious person vomit. Effects may be delayed. Medical observation is recommended.

Personal Protective Methods: Wear protective gloves and clothing to prevent any reasonable probability of skin contact. Safety equipment suppliers/manufacturers can provide recommendations on the most protective glove/clothing material for your operation. Viton$^{™}$, polyvinyl chloride, Nitrile + PVC, butyl rubber, and polyethylene are among the recommended protective materials. All protective clothing (suits, gloves, footwear, headgear) should be clean, available each day, and put on before work. Contact lenses should not be worn when working with this chemical. Wear splash-proof chemical goggles and face shield unless full-face-piece respiratory protection is worn. Employees should wash immediately with soap when skin is wet or contaminated. Provide emergency showers and eyewash.

Respirator Selection: NIOSH: *0.5 mg/m³:* Qm (APF = 25) (any quarter-mask respirator). *1 mg/m³:* 95XQ (APF = 10) [any particulate respirator equipped with an N95, R95, or P95 filter (including N95, R95, and P95 filtering face-pieces) except quarter-mask respirators. The following filters may also be used: N99, R99, P99, N100, R100, P100] or Sa (APF = 10) (any supplied-air respirator). *2.5 mg/m³:* Sa:Cf (APF = 25) (any supplied-air respirator operated in a continuous-flow mode); PaprHie (APF = 25) (any powered, air-purifying respirator with a high-efficiency particulate filter). *5 mg/m³:* 100F (APF = 50) (any air-purifying, full-face-piece respirator with an N100, R100, or P100 filter) or SaT: Cf (APF = 50) (any supplied-air respirator that has a tight-fitting face-piece and is operated in a continuous-flow mode) or PaprTHie (APF = 50) (any powered, air-purifying respirator with a tight-fitting face-piece and a high-efficiency particulate filter) or SCBAF (APF = 50) (any self-contained breathing apparatus with a full face-piece) or SaF (APF = 50) (any supplied-air respirator with a full face-piece). *40 mg/m³:* Sa: Pd,Pp (APF = 1000) (any supplied-air respirator operated in a pressure-demand or other positive-pressure mode). *Emergency or planned entry into unknown concentrations or IDLH conditions:* SCBAF: Pd,Pp (APF = 10,000) (any self-contained breathing apparatus that has a full-faceplate and is operated in a pressure-demand or other positive-pressure mode) or SaF: Pd,Pp: ASCBA (APF = 10,000) (any supplied-air respirator that has a full face-piece and is operated in a pressure-demand or other positive-pressure mode in combination with an auxiliary, self-contained breathing apparatus operated in a pressure-demand or other positive-pressure mode). *Escape:* 100F (APF = 50) (any air-purifying, full-face-piece respirator with an N100, R100, or P100 filter) or SCBAE (any appropriate escape-type, self-contained breathing apparatus).

Storage: Color Code—Blue: Health Hazard/Poison: Store in a secure poison location. Prior to working with this chemical you should be trained on its proper handling and storage. Store in a refrigerator or a cool, dry place away from oxidizing materials. Where possible, automatically pump liquid from drums or other storage containers to process containers. Sources of ignition, such as smoking and open flames, are prohibited where this chemical is handled, used, or stored.

Shipping: Tricresyl phosphate with >3% ortho-isomer requires a shipping label of "POISONOUS/TOXIC MATERIALS." It falls in Hazard Class 6.1 and Packing Group II.

Spill Handling: *Solid* (below 52°F/11°C): Evacuate persons not wearing protective equipment from area of spill or leak until cleanup is complete. Remove all ignition sources. Collect powdered material in the most convenient and safe manner and deposit in sealed containers. Ventilate area after cleanup is complete. It may be necessary to contain and dispose of this chemical as a hazardous waste. If material or contaminated runoff enters waterways, notify downstream users of potentially contaminated waters. Contact your local or federal environmental protection agency for specific recommendations. If employees are required to clean up spills, they must be properly trained and equipped. OSHA 1910.120(q) may be applicable.

Liquid: Evacuate and restrict persons not wearing protective equipment from area of spill or leak until cleanup is complete. Remove all ignition sources. Ventilate area of spill or leak. Absorb liquids in vermiculite, dry sand, earth, peat, carbon, or a similar material and deposit in sealed containers. Keep this chemical out of a confined space, such as a sewer, because of the possibility of an explosion, unless the sewer is designed to prevent the buildup of explosive concentrations. It may be necessary to contain and dispose of this chemical as a hazardous waste. If material or contaminated runoff enters waterways, notify downstream users of potentially contaminated waters. Contact your local or federal environmental protection agency for specific recommendations. If employees are required to clean up spills, they must be properly trained and equipped. OSHA 1910.120(q) may be applicable.

Fire Extinguishing: This chemical is a combustible liquid. Poisonous gases, including phosphorus oxides, are produced in fire. Use dry chemical, carbon dioxide, or alcohol foam extinguishers. Vapors are heavier than air and will collect in low areas. Containers may explode in fire. Storage containers and parts of containers may rocket great distances, in many directions. If material or contaminated runoff enters waterways, notify downstream users of potentially contaminated waters. Notify local health and fire officials and pollution control agencies. From a secure, explosion-proof location, use water spray to cool exposed containers. If cooling streams are ineffective (venting sound increases in volume and pitch, tank discolors, or shows any signs of deforming), withdraw immediately to a secure position. If employees are expected to fight fires, they must be trained and equipped in OSHA 1910.156. The only respirators recommended for firefighting are self-contained breathing apparatuses that have full face-pieces and are operated in a pressure-demand or other positive-pressure mode.

References
Sax, N. I. (Ed.). *Dangerous Properties of Industrial Materials Report*, 2, No. 2, 73–75 (1982) and 2, No. 3, 83–84 (1982)
New Jersey Department of Health and Senior Services. (February 2007). *Hazardous Substances Fact Sheet: Tricresylphosphate (mixed isomers)*. Trenton, NJ

Triethylamine   T:0810

Molecular Formula: $C_6H_{15}N$

Common Formula: $(C_2H_5)_3N$

Synonyms: (Diethylamino)ethane;   *N,N*-Diethyl-ethaneamine; Ethanamine, *N,N*-diethyl-; TEA; TEN; Triaethylamin (German); Trietilamina (Spanish)

CAS Registry Number: 121-44-8

RTECS® Number: YE0175000

UN/NA & ERG Number: UN1296/132

EC Number: 204-469-4 [*Annex I Index No.:* 612-004-00-5]

Regulatory Authority and Advisory Bodies

Air Pollutant Standard Set. See below, "Permissible Exposure Limits in Air" section.

Clean Air Act: Hazardous Air Pollutants (Title I, Part A, Section 112).

Clean Water Act: Section 311 Hazardous Substances/RQ 40CFR117.3 (same as CERCLA, see below).

US EPA Hazardous Waste Number (RCRA No.): U404.

RCRA, 40CFR261, Appendix 8 Hazardous Constituents.

Reportable Quantity (RQ): 5000 lb (2270 kg).

EPCRA Section 313 Form R *de minimis* concentration reporting level: 1.0%.

European/International Regulations: Hazard Symbol: F, C; Risk phrases: R11; R20/21/22; R35; Safety phrases: S1/2; S3; S16; S26; S29; S36/37/39; S45 (see Appendix 4).

WGK (German Aquatic Hazard Class): 1—Low hazard to waters.

Description: Triethylamine is a colorless liquid with a strong ammonia-like odor. The odor threshold is 0.48 ppm. Molecular weight = 101.22; Specific gravity (H_2O:1) = 0.73 at 25°C; Boiling point = 89°C; Freezing/Melting point = −115°C; Vapor pressure = 54 mmHg at 25°C; Flash point = −7°C; Autoignition temperature = 249°C; also listed at 230°C. Explosive limits: LEL = 1.2%; UEL = 8.0%. Hazard Identification (based on NFPA-704 M Rating System): Health 3, Flammability 3, Reactivity 0. Slightly soluble in water; solubility = 2%.

Potential Exposure: Compound Description: Tumorigen, Mutagen, Human Data; Primary Irritant. Triethylamine is used as a solvent; corrosion inhibitor; in chemical synthesis; and accelerator activators; paint remover; base in methylene chloride or other chlorinated solvents. TEA is used to solubilize 2,4,5-T in water and serves as a selective extractant in the purification of antibiotics. It is used to manufacture quaternary ammonia compounds and octadecyloxymethyl-triethylammonium chloride; an agent used in textile treatment.

Incompatibilities: A strong base. Violent reaction with strong acids, halogenated compounds, and strong oxidizers. Attacks some forms of plastics, rubber, and coatings. Corrosive to aluminum, zinc, copper, and their alloys in the presence of moisture.

Permissible Exposure Limits in Air

Conversion factor: 1 ppm = 4.14 mg/m³ at 25°C & 1 atm.

OSHA PEL: 25 ppm/100 mg/m³ TWA.

NIOSH REL: No established REL; See *NIOSH Pocket Guide*, Appendix D.

ACGIH TLV®[1]: 1 ppm/4.1 mg/m³ TWA; 3 ppm/12.4 mg/m³ STEL, not classifiable as a human carcinogen [skin].

NIOSH IDLH: 200 ppm.

Protective Action Criteria (PAC)

TEEL-0: 1 ppm

PAC-1: 3 ppm

PAC-2: 3 ppm

PAC-3: 200 ppm

DFG MAK: 1 ppm/4.2 mg/m³ TWA; Peak Limitation Category I(2); Pregnancy Risk Group D.

Europe OEL: 2 ppm/8.4 mg/m³ TWA; 3 ppm/12.6 mg/m³ STEL [skin] 2002.

Australia: TWA 10 ppm (40 mg/m³), STEL 15 ppm, 1993; Austria: MAK 2.5 ppm (10 mg/m³), 1999; Belgium: TWA 10 ppm (41 mg/m³), STEL 15 ppm, 1993; Denmark: TWA 10 ppm (40 mg/m³), 1999; Finland: TWA 10 mg/m³, STEL 20 mg/m³, [skin], 1999; France: VLE 10 ppm (40 mg/m³), 1999; Hungary: TWA 20 mg/m³, STEL 40 mg/m³, 1993; the Netherlands: MAC-TGG 20 mg/m³, [skin], 2003; Norway: TWA 10 ppm (40 mg/m³), 1999; the Philippines: TWA 25 ppm (100 mg/m³), 1993; Russia: STEL 10 mg/m³, [skin], 1993; Sweden: NGV 2 ppm (8 mg/m³), KTV 10 ppm (40 mg/m³), 1999; Switzerland: MAK-W 10 ppm (40 mg/m³), KZG-W 20 ppm (80 mg/m³), 1999; Turkey: TWA 25 ppm (100 mg/m³), 1993; United Kingdom: TWA 10 ppm (42 mg/m³), STEL 15 ppm (63 mg/m³), 2000; Argentina, Bulgaria, Columbia, Jordan, South Korea, New Zealand, Singapore, Vietnam: ACGIH TLV®: STEL 3 ppm [skin]. Russia[43] set a MAC of 0.14 mg/m³ in ambient air of residential areas both on a momentary and a daily average basis. Several states have set guidelines or standards for triethylamine in ambient air[60] ranging from 5.6 μg/m³ (Massachusetts) to 20.0−200.0 μg/m³ (Rhode Island) to 400.0−600.0 μg/m³ (North Dakota) to 660.0 μg/m³ (Virginia) to 800.0 μg/m³ (Connecticut) to 952.0 μg/m³ (Nevada).

Determination in Air: Use NIOSH Analytical Method (II-3) #S-152, OSHA Analytical Method PV2060.

Determination in Water: Octanol−water coefficient: Log K_{ow} = 1.45.

Permissible Concentration in Water: Russia[43] set a MAC of 2.0 mg/L in water bodies used for domestic purposes.

Routes of Entry: Inhalation, ingestion, skin absorption, skin and/or eye contact.

Harmful Effects and Symptoms

Short Term Exposure: Triethylamine can affect you when breathed in and by passing through your skin. Corrosive to the eyes, skin, and respiratory tract. Contact can cause severe eye damage. Breathing the vapor can irritate the lungs. Higher exposures can cause pulmonary edema, a medical emergency that can be delayed for several hours. This can cause death. A central nervous system depressant.

Long Term Exposure: Can irritate the lungs; repeated exposures may cause bronchitis to develop. May cause skin allergy. May damage the kidneys and liver. Similar compounds can cause an asthma-like allergy to develop. Once allergy develops, even very small future exposures can cause wheezing, chest tightness, and shortness of breath. In animals: myocardial, kidney, liver damage.

Points of Attack: Eyes, skin, respiratory system, cardiovascular system, liver, kidneys.

Medical Surveillance: For those with frequent or potentially high exposure (half the TLV or greater, or significant skin contact), the following are recommended before beginning work and at regular times after that: lung function tests; liver function tests. If symptoms develop or overexposure is suspected, the following may be useful: liver and kidney function tests; examination of the eyes and vision; evaluation by a qualified allergist, including careful exposure history and special testing, may help diagnose allergy.

First Aid: If this chemical gets into the eyes, remove any contact lenses at once and irrigate immediately for at least 15 min, occasionally lifting upper and lower lids. Seek medical attention immediately. If this chemical contacts the skin, remove contaminated clothing and wash immediately with soap and water. Seek medical attention immediately. If this chemical has been inhaled, remove from exposure, begin rescue breathing (using universal precautions, including resuscitation mask) if breathing has stopped and CPR if heart action has stopped. Transfer promptly to a medical facility. When this chemical has been swallowed, get medical attention. If victim is *conscious*, administer water or milk. Do not induce vomiting. Give large quantities of water and induce vomiting. Do not make an unconscious person vomit. Medical observation is recommended for 24−48 h after breathing overexposure, as pulmonary edema may be delayed. As first aid for pulmonary edema, a doctor or authorized paramedic may consider administering a corticosteroid spray.

Personal Protective Methods: Wear protective gloves and clothing to prevent any reasonable probability of skin contact. Safety equipment suppliers/manufacturers can provide recommendations on the most protective glove/clothing material for your operation. Nitrile, Viton™, and chlorinated polyethylene are among the recommended protective materials. All protective clothing (suits, gloves, footwear, headgear) should be clean, available each day, and put on before work. Contact lenses should not be worn when working with this chemical. Wear splash-proof chemical goggles and face shield unless full-face-piece respiratory protection is worn. Employees should wash immediately with soap when skin is wet or contaminated. Provide emergency showers and eyewash. Provide: eyewash (>1%), quick drench (>1%).

Respirator Selection: OSHA: *200 ppm:* Sa:Cf (APF = 25) (any supplied-air respirator operated in a continuous-flow mode) or SCBAF (APF = 50) (any self-contained breathing apparatus with a full face-piece) or SaF (APF = 50) (any supplied-air respirator with a full face-piece). *Emergency or planned entry into unknown concentrations or IDLH*

conditions: SCBAF: Pd,Pp (APF = 10,000) (any self-contained breathing apparatus that has a full face-piece and is operated in a pressure-demand or other positive-pressure mode) or SaF: Pd,Pp: ASCBA (APF = 10,000) (any supplied-air respirator that has a full face-piece and is operated in a pressure-demand or other positive-pressure mode in combination with an auxiliary self-contained breathing apparatus operated in a pressure-demand or other positive-pressure mode). *Escape:* GmFS (APF = 50) [any air-purifying, full-face-piece respirator (gas mask) with a chin-style, front- or back-mounted canister providing protection against the compound of concern] or SCBAE (any appropriate escape-type, self-contained breathing apparatus).

Storage: Color Code—Red: Flammability Hazard: Store in a flammable liquid storage area or approved cabinet away from ignition sources and corrosive and reactive materials. Prior to working with this chemical you should be trained on its proper handling and storage. Before entering confined space where this chemical may be present, check to make sure that an explosive concentration does not exist. Triethylamine must be stored to avoid contact with strong acids (such as hydrochloric, sulfuric, and nitric) or oxidizers (such as perchlorates, peroxides, permanganates, chlorates, and nitrates) because violent reactions occur. Store in tightly closed containers in a cool, well-ventilated area away from heat. Sources of ignition, such as smoking and open flames, are prohibited where triethylamine is used, handled, or stored in a manner that could create a potential fire or explosion hazard. Metal containers involving the transfer of 5 gallons or more of triethylamine should be grounded and bonded. Drums must be equipped with self-closing valves, pressure vacuum bungs, and flame arresters. Use only nonsparking tools and equipment, especially when opening and closing containers of triethylamine.

Shipping: This compound requires a shipping label of "FLAMMABLE LIQUID." It falls in Hazard Class 3 and Packing Group II.

Spill Handling: Evacuate and restrict persons not wearing protective equipment from area of spill or leak until cleanup is complete. Remove all ignition sources. Establish forced ventilation to keep levels below explosive limit. Absorb liquids in vermiculite, dry sand, earth, peat, carbon, or a similar material and deposit in sealed containers. Keep this chemical out of a confined space, such as a sewer, because of the possibility of an explosion, unless the sewer is designed to prevent the buildup of explosive concentrations. It may be necessary to contain and dispose of this chemical as a hazardous waste. If material or contaminated runoff enters waterways, notify downstream users of potentially contaminated waters. Contact your local or federal environmental protection agency for specific recommendations. If employees are required to clean up spills, they must be properly trained and equipped. OSHA 1910.120(q) may be applicable.

Fire Extinguishing: This chemical is a flammable liquid. Poisonous gases, including nitrogen oxides and carbon monoxide, are produced in fire. Use dry chemical, carbon dioxide, or alcohol foam extinguishers. Water may be ineffective. Vapors are heavier than air and will collect in low areas. Vapors may travel long distances to ignition sources and flashback. Vapors in confined areas may explode when exposed to fire. Containers may explode in fire. Storage containers and parts of containers may rocket great distances, in many directions. If material or contaminated runoff enters waterways, notify downstream users of potentially contaminated waters. Notify local health and fire officials and pollution control agencies. From a secure, explosion-proof location, use water spray to cool exposed containers. If cooling streams are ineffective (venting sound increases in volume and pitch, tank discolors, or shows any signs of deforming), withdraw immediately to a secure position. If employees are expected to fight fires, they must be trained and equipped in OSHA 1910.156. The only respirators recommended for firefighting are self-contained breathing apparatuses that have full face-pieces and are operated in a pressure-demand or other positive-pressure mode.

Disposal Method Suggested: Controlled incineration (incinerator equipped with a scrubber or thermal unit to reduce nitrogen oxides emissions).

References

US Environmental Protection Agency. (April 1, 1978). *Chemical Hazard Information Profile: Ethylamines.* Washington, DC

Sax, N. I. (Ed.). (1983). *Dangerous Properties of Industrial Materials Report*, 3, No. 6, 81–83

New Jersey Department of Health and Senior Services. (June 2003). *Hazardous Substances Fact Sheet: Triethylamine.* Trenton, NJ

Trifluorobromomethane   T:0820

Molecular Formula: CBrF$_3$

Synonyms: Bromofluroform; Bromotrifluormetano (Spanish); Bromotrifluoromethane; Carbon monobromide trifluoride; F 13 B1; FC 13 B1; Flugex 13 B1; Fluorocarbon 1301; Freon 13 B1; Halon 1301; Methane, bromotrifluoro-; R 13 B1; Refrigerant 1301; Trifluoromethyl bromide; Trifluoromonobromomethane

CAS Registry Number: 75-63-8

RTECS® Number: PA5425000

UN/NA & ERG Number: UN1009/126

EC Number: 200-887-6

Regulatory Authority and Advisory Bodies

Air Pollutant Standard Set. See below, "Permissible Exposure Limits in Air" section.

Clean Air Act: Stratospheric ozone protection (Title VI, Subpart A, Appendix A), Class I, Ozone Depletion Potential = 10.0.

EPCRA Section 313 Form R *de minimis* concentration reporting level: 1.0%.

Canada, WHMIS, Ingredients Disclosure List Concentration 1.0%.

European/International Regulations: not listed in Annex 1.

WGK (German Aquatic Hazard Class): 1—Low hazard to waters.

Description: Trifluorobromomethane is a colorless gas with a slight ethereal odor. Shipped as a liquefied compressed gas. Molecular weight = 148.92; Specific gravity (H$_2$O:1) = 1.50 at 25°C; Boiling point = −58°C; Freezing/Melting point = −166°C. Slightly soluble in water; solubility = 0.03%.

Potential Exposure: Compound Description: Human Data. This material is used as a fire extinguishing agent, a chemical intermediate, and as a refrigerant.

Incompatibilities: Chemically active metals, such as calcium, powdered aluminum, zinc, magnesium. Attacks some plastics, rubber, and coatings.

Permissible Exposure Limits in Air

Conversion factor: 1 ppm = 6.09 mg/m^3 at 25°C & 1 atm.

OSHA PEL: 1000 ppm/6100 mg/m^3 TWA.

NIOSH REL: 1000 ppm/6100 mg/m^3 TWA.

ACGIH TLV®[1]: 1000 ppm/6090 mg/m^3 TWA.

Protective Action Criteria (PAC)

TEEL-0: 1000 ppm

PAC-1: 3500 ppm

PAC-2: 25,000 ppm

PAC-3: 40,000 ppm

DFG MAK: 1000 ppm/6200 mg/m^3 TWA; Peak limitation Category II(8); Pregnancy Group C.

NIOSH IDLH: 40,000 ppm.

Australia: TWA 1000 ppm (6100 mg/m^3), 1993; Austria: MAK 1000 ppm (6100 mg/m^3), 1999; Belgium: TWA 1000 ppm (6090 mg/m^3), 1993; Denmark: TWA 1000 ppm (6100 mg/m^3), 1999; Finland: TWA 1000 ppm (6100 mg/m^3), STEL 1250 ppm (7625 mg/m^3), 1999; France: VME 1000 ppm (6100 mg/m^3), 1999; the Netherlands: MAC-TGG 6100 mg/m^3, 2003; the Philippines: TWA 1000 ppm (6100 mg/m^3), 1993; Russia: STEL 3000 mg/m^3, 1993; Switzerland: MAK-W 1000 ppm (6100 mg/m^3), 1999; Turkey: TWA 1000 ppm (6100 mg/m^3), 1993; United Kingdom: TWA 1000 ppm (6190 mg/m^3), STEL 1200 ppm, 2000; Argentina, Bulgaria, Columbia, Jordan, South Korea, New Zealand, Singapore, Vietnam: ACGIH TLV®: TWA 1000 ppm. Several states have set guidelines or standards for this compound in ambient air[60] ranging from 61.0 mg/m^3 (North Dakota) to 100 mg/m^3 (Virginia) to 122 mg/m^3 (Connecticut) to 145.238 mg/m^3 (Nevada).

Determination in Air: Absorption on charcoal, workup with methylene chloride; analysis by gas chromatography.

Determination in Water: Octanol−water coefficient: Log $K_{ow} = 1.86$.

Routes of Entry: Inhalation, eye and/or skin contact with the liquid.

Harmful Effects and Symptoms

Short Term Exposure: Trifluorobromomethane can affect you when breathed in. Irritates the eyes. Contact with the liquid may cause frostbite. Exposure can cause CNS depression, causing lightheadedness and trouble concentrating. Breathing high concentrations of the vapor may cause the heart to beat irregularly or to stop.

Long Term Exposure: Unknown at this time. However, some related chemicals are known to cause liver damage with high or repeated exposure.

Points of Attack: Central nervous system, heart.

Medical Surveillance: If symptoms develop or overexposure has occurred, the following may be useful: A special 24-h EKG (Holter monitor) to look for irregular heartbeat; consider liver function tests. Examination of the nervous system.

First Aid: If this chemical gets into the eyes, remove any contact lenses at once and irrigate immediately for at least 15 min, occasionally lifting upper and lower lids. Seek medical attention immediately. If this chemical contacts the skin, remove contaminated clothing and wash immediately with soap and water. Seek medical attention immediately. If this chemical has been inhaled, remove from exposure, begin rescue breathing (using universal precautions, including resuscitation mask) if breathing has stopped and CPR if heart action has stopped. Transfer promptly to a medical facility. When this chemical has been swallowed, get medical attention. Give large quantities of water and induce vomiting. Do not make an unconscious person vomit. If frostbite has occurred, seek medical attention immediately; do *NOT* rub the affected areas or flush them with water. In order to prevent further tissue damage, do *NOT* attempt to remove frozen clothing from frostbitten areas. If frostbite has *NOT* occurred, immediately and thoroughly wash contaminated skin with soap and water.

Personal Protective Methods: Wear appropriate personal protective clothing to prevent the skin from becoming frozen from contact with the evaporating liquid or from contact with vessels containing the liquid. Safety equipment suppliers/manufacturers can provide recommendations on the most protective glove/clothing material for your operation. All protective clothing (suits, gloves, footwear, headgear) should be clean, available each day, and put on before work. Contact lenses should not be worn when working with this chemical. Wear gas-proof chemical goggles and face shield unless full-face-piece respiratory protection is worn. Employees should wash immediately with soap when skin is wet or contaminated. Provide emergency showers and eyewash.

Respirator Selection: *Up to 10,000 ppm:* Sa (APF = 10) (any supplied-air respirator). *Up to 25,000 ppm:* Sa:Cf (APF = 25) (any supplied-air respirator operated in a continuous-flow mode). *Up to 40,000 ppm:* SaT: Cf (APF = 50) (any supplied-air respirator that has a tight-fitting face-piece and is operated in a continuous-flow mode) or SCBAF (APF = 50) (any self-contained breathing apparatus with a full face-piece) or SaF (APF = 50) (any supplied-air respirator with a full face-piece). *Emergency or planned entry into unknown concentrations or IDLH*

conditions: SCBAF: Pd,Pp (APF = 10,000) (any NIOSH/MSHA- or European Standard EN 149-approved self-contained breathing apparatus that has a full face-piece and is operated in a pressure-demand or other positive-pressure mode) or SaF: Pd,Pp: ASCBA (APF = 10,000) (any supplied-air respirator that has a full face-piece and is operated in a pressure-demand or other positive-pressure mode in combination with an auxiliary, self-contained breathing apparatus operated in a pressure-demand or other positive-pressure mode). *Escape:* GmFOv (APF = 50) [any air-purifying, full-face-piece respirator (gas mask) with a chin-style, front- or back-mounted organic vapor canister] or SCBAE (any appropriate escape-type, self-contained breathing apparatus).

Storage: Color Code—Green: General storage may be used. Prior to working with this chemical you should be trained on its proper handling and storage. Trifluorobromomethane must be stored to avoid contact with chemically active metals (such as calcium, powdered aluminum, zinc, and magnesium) since violent reactions occur. Store in tightly closed containers in a cool, well-ventilated area away from heat.

Shipping: This compound requires a shipping label of "NONFLAMMABLE GAS." It falls in Hazard Class 2.2.

Spill Handling: If in a building, evacuate building and confine vapors by closing doors and shutting down HVAC systems. Restrict persons not wearing protective equipment from area of spill or leak until cleanup is complete. Remove all ignition sources. Ventilate area of spill or leak to disperse the gas. Wear chemical protective suit with self-contained breathing apparatus to combat spills. Stay upwind and use water spray to "knock down" vapor; contain runoff. Stop the flow of gas, if it can be done safely from a distance. If source is a cylinder and the leak cannot be stopped in place, remove the leaking cylinder to a safe place; and repair leak or allow cylinder to empty. Absorb liquids in vermiculite, dry sand, earth, or a similar material and deposit in sealed containers. Keep this chemical out of confined spaces, such as a sewer, because of the possibility of explosion, unless the sewer is designed to prevent the buildup of explosive concentrations. If employees are required to clean up spills, they must be properly trained and equipped. OSHA 1910.120(q) may be applicable.

Fire Extinguishing: This chemical is a nonflammable gas. Poisonous gases, including phosgene, hydrogen bromide, and hydrogen fluoride, are produced in fire. Extinguish fire using an agent suitable for type of surrounding fire. Vapors are heavier than air and will collect in low areas. Containers may explode in fire. Storage containers and parts of containers may rocket great distances, in many directions. If material or contaminated runoff enters waterways, notify downstream users of potentially contaminated waters. Notify local health and fire officials and pollution control agencies. From a secure, explosion-proof location, use water spray to cool exposed containers. If cooling streams are ineffective (venting sound increases in volume and pitch, tank discolors, or shows any signs of deforming), withdraw immediately to a secure position. If employees are expected to fight fires, they must be trained and equipped in OSHA 1910.156. The only respirators recommended for firefighting are self-contained breathing apparatuses that have full face-pieces and are operated in a pressure-demand or other positive-pressure mode.

Disposal Method Suggested: Incineration, preferably after mixing with another combustible fuel. Care must be exercised to assure complete combustion to prevent the formation of phosgene. An acid scrubber is necessary to remove the halo acids produced.

References

New Jersey Department of Health and Senior Services. (June 2006). *Hazardous Substances Fact Sheet: Trifluorobromomethane.* Trenton, NJ

Trifluoroethane   T:0830

Molecular Formula: $C_2H_3F_3$
Synonyms: *1,1,1-isomer:* Ethane, 1,1,1-trifluoro-; FC143A; Fluorocarbon FC143A; Methylfluoroform; R 143A; 1,1,1-Trifluoroethane; 1,1,1-Trifluoroform
1,1,2-isomer: Ethane, 1,1,2-trifluoro-; R 143
CAS Registry Number: 420-46-2 (1,1,1-isomer); 430-66-0 (1,1,2-isomer); 27987-06-0 (mixed isomers)
RTECS® Number: KJ4100000 (trifluoroethane); KJ4110000 (1,1,1-isomer)
UN/NA & ERG Number: UN2035 (1,1,1-Trifluoroethane, compressed or Refrigerant gas, R 143)/115
EC Number: 206-996-5 (1,1,1-isomer); 207-066-1 (1,1,2-isomer); 248-764-6 (mixed isomers)
Regulatory Authority and Advisory Bodies
Air Pollutant Standard Set. See below, "Permissible Exposure Limits in Air" section.
European/International Regulations: not listed in Annex 1.
WGK (German Aquatic Hazard Class): 1—Slightly water polluting (*CAS: 420-46-2*).
Description: 1,1,1-Trifluoroethane is a colorless gas or a liquid under pressure. Molecular weight = 84.05; Boiling point = −111°C.
Potential Exposure: The 1,1,1-isomer is used to make other chemicals and may be used as a refrigerant.
Incompatibilities: None listed.
Permissible Exposure Limits in Air
AIHA WEEL: 1000 ppm TWA, as 1,1,1-Trifluoroethane.
Russia: The 1,1,1-isomer has a MAC of 3000 mg/m³ in work-place air.[43]
No TEEL available.
Routes of Entry: Inhalation.
Harmful Effects and Symptoms
Short Term Exposure: Trifluoroethane can affect you when breathed in. Inhalation may cause central nervous system depression. Higher levels may cause dizziness, lightheadedness,

unconsciousness, and asphyxiation. Contact with the liquid may cause frostbite. Exposure may cause cardiac arrhythmia (irregular heartbeat).
Long Term Exposure: No chronic health effects are known at this time.
Points of Attack: Heart, central nervous system.
Medical Surveillance: If symptoms develop or overexposure is suspected, the following may be useful: Holter monitor (a special 24-h EKG to look for irregular heartbeats). Examination of the nervous system.
First Aid: If this chemical gets into the eyes, remove any contact lenses at once and irrigate immediately for at least 15 min, occasionally lifting upper and lower lids. If this chemical has been inhaled, remove from exposure, begin rescue breathing (using universal precautions, including resuscitation mask) if breathing has stopped and CPR if heart action has stopped. Transfer promptly to a medical facility. If frostbite has occurred, seek medical attention immediately; do *NOT* rub the affected areas or flush them with water. In order to prevent further tissue damage, do *NOT* attempt to remove frozen clothing from frostbitten areas. If frostbite has *NOT* occurred, immediately and thoroughly wash contaminated skin with soap and water.
Personal Protective Methods: Wear protective gloves and clothing to prevent any reasonable probability of skin contact. Safety equipment suppliers/manufacturers can provide recommendations on the most protective glove/clothing material for your operation. All protective clothing (suits, gloves, footwear, headgear) should be clean, available each day, and put on before work. Contact lenses should not be worn when working with this chemical. Wear splash-proof chemical goggles and face shield when working with liquids, unless full-face-piece respiratory protection is worn. Employees should wash immediately with soap when skin is wet or contaminated. Provide emergency showers and eyewash. Where exposure to cold equipment, vapors, or liquid may occur, employees should be provided with special clothing designed to prevent the freezing of body tissues.
Respirator Selection: Where there is potential for exposure to trifluoroethane, use a NIOSH/MSHA- or European Standard EN149-approved supplied-air respirator with a full face-piece operated in the positive-pressure mode, or with a full-face-piece, hood, or helmet in the continuous-flow mode; or use a NIOSH/MSHA- or European Standard EN149-approved self-contained breathing apparatus with a full face-piece operated in pressure-demand or other positive-pressure mode.
Storage: Color Code—Red Stripe: Flammability Hazard: Store separately from all other flammable materials. Prior to working with this chemical you should be trained on its proper handling and storage. Store in tightly closed containers in a cool, well-ventilated area. Outdoor or detached storage is recommended. Sources of ignition, such as smoking and open flames, are prohibited where trifluoroethane is used, handled, or stored in a manner that could create a potential fire or explosion hazard. Procedures for the

handling, use, and storage of cylinders should be in compliance with OSHA 1910.101 and 1910.169, as with the recommendations of the Compressed Gas Association.

Shipping: This compound requires a shipping label of "FLAMMABLE GAS." It falls in Hazard Class 2.1.

Spill Handling: Evacuate and restrict persons not wearing protective equipment from area of spill or leak until cleanup is complete. Remove all ignition sources. Ventilate area of leak to disperse the gas. Stop the flow of gas if it can be done safely. If source of leak is a cylinder and the leak cannot be stopped in place, remove leaking cylinder to a safe place in the open air, and repair leak or allow cylinder to empty. Keep this chemical out of confined space, such as a sewer, because of the possibility of explosion, unless the sewer is designed to prevent the buildup of explosive concentrations. It may be necessary to contain and dispose of this chemical as a hazardous waste. Contact your local or federal environmental protection agency for specific recommendations. If employees are required to clean up spills, they must be properly trained and equipped. OSHA 1910.120(q) may be applicable.

Fire Extinguishing: Trifluoroethane is a flammable gas. Use dry chemical, CO_2, water spray, or foam extinguishers. Use water spray to keep fire-exposed containers cool. Poisonous gases, including hydrogen fluoride, are produced in fire. Vapors are heavier than air and will collect in low areas. Vapors may travel long distances to ignition sources and flashback. Vapors in confined areas may explode when exposed to fire. Containers may explode in fire. Storage containers and parts of containers may rocket great distances, in many directions. If material or contaminated runoff enters waterways, notify downstream users of potentially contaminated waters. Notify local health and fire officials and pollution control agencies. From a secure, explosion-proof location, use water spray to cool exposed containers. If cooling streams are ineffective (venting sound increases in volume and pitch, tank discolors, or shows any signs of deforming), withdraw immediately to a secure position. If employees are expected to fight fires, they must be trained and equipped in OSHA 1910.156. The only respirators recommended for firefighting are self-contained breathing apparatuses that have full face-pieces and are operated in a pressure-demand or other positive-pressure mode.

References

New Jersey Department of Health and Senior Services. (March 1987). *Hazardous Substances Fact Sheet: Trifluoroethane.* Trenton, NJ

Trifluralin   T:0840

Molecular Formula: $C_{10}H_9F_3N_3O_4$

Common Formula: $C_3H_7N—C_6H_2(NO_2)_2CF_3$

Synonyms: Agreflan; Agriflan 24; Autumn kite; Benzenamine, 2,6-dinitro-*N,N*-dipropyl-4-(trifluoromethyl-); Benzeneamine, 2,6-dinitro-*N,N*-dipropyl-4-(trifluoromethylaniline);

Campbell's trifluron; Chandor; Crisalin; Devrinol T; Digermin; 2,6-Dinitro-*N,N*-dipropyl-4-(trifluoromethyl)aniline; 2,6-Dinitro-*N,N*-dipropyl-4-(trifluoromethyl)benzenamine; 2,6-Dinitro-*N,N*-di-*N*-propyl-α,α,α-trifluro-*p*-toluidine; 2,6-Dinitro-4-trifluormethyl-*N,N*-dipropylanilin (German); 4-(Di-*N*-propylamino)-3,5-dinitro-1-trifluoromethylbenzene; *N,N*-Di-*N*-propyl-2,6-dinitro-4-trifluoromethylaniline; *N,N*-Dipropyl-4-trifluoromethyl-2,6-dinitroaniline; Elancolan; Ethane, trifluoro-; Flint; Ipersan; Janus; L-36352; Lilly 36,352; Linnet; Marksman; Marksman 2, Trigard; M.T.F; NCI-C00442; Nitran; Olitref; Onslaught; Sinflowan; Solo; Su seguro carpidor; Synfloran; *p*-Toluidine, α,α,α-trifluoro-2,6-dinitro-*N,N*-dipropyl-; Trefanocide; Treficon; Treflan®; Treflanocide elancolan; Trifarmon; α,α,α-Trifluoro-2,6-dinitro-*N,N*-dipropyl-*p*-toluidine; Trifluralina (Spanish); Trifluralina 600; Trifluraline; Triflurex; Trifurex; Trigard; Trikepin; Trilin; Trilin 10G; Trim; Trimaran; Tripart Trifluralin 48 EC; Tristar®

CAS Registry Number: 1582-09-8; *(alt.)* 39300-53-3; *(alt.)* 52627-52-8; *(alt.)* 61373-95-3; *(alt.)* 75635-23-3

RTECS® Number: XU9275000

UN/NA & ERG Number: UN2588/151

EC Number: 216-428-8 [*Annex I Index No.:* 609-046-00-1]

Regulatory Authority and Advisory Bodies

Carcinogenicity: IARC: Animal, Limited Evidence; Human Inadequate Evidence, *not classifiable as carcinogenic to humans*, Group 3, 1991; NCI: Carcinogenesis Bioassay (feed); clear evidence: mouse; no evidence: rat; EPA: Possible Human Carcinogen.

US EPA, FIFRA 1998 Status of Pesticides: RED completed.

US EPA Gene-Tox Program, Positive: *N. crassa*—aneuploidy; Weakly Positive: *S. cerevisiae*—homozygosis; Positive/limited: Carcinogenicity—mouse/rat; Negative: *D. melanogaster* sex-linked lethal; Negative: *In vitro* UDS—human fibroblast; TRP reversion; Inconclusive: *B. subtilis* rec assay; *E. coli* polA without S9.

Banned or Severely Restricted (in US) (UN).[13]

Clean Air Act: Hazardous Air Pollutants (Title I, Part A, Section 112).

Safe Drinking Water Act: Priority List (55 FR 1470).

EPCRA Section 313 Form R *de minimis* concentration reporting level: 1.0%.

European/International Regulations (*containing <0.5 ppm NPDA*): Hazard Symbol: Xi; Risk phrases: R36; R43; Safety phrases: S2; S24; S37 (see Appendix 4).

WGK (German Aquatic Hazard Class): 2—Hazard to waters.

Description: Trifluralin is an orange crystalline solid. Molecular weight = 335.32; Boiling point = 139°C; Freezing/Melting point = 49°C; Vapor pressure = 0.0001 mmHg. Hazard Identification (based on NFPA-704 M Rating System): Health 1, Flammability 1, Reactivity 0. Practically insoluble in water.

Potential Exposure: Compound Description: Agricultural Chemical; Tumorigen, Mutagen; Reproductive Effector.

Those involved in the manufacture, formulation, and application of this selective preemergence herbicide.

Incompatibilities: Store in temperatures above 4.4°C. Fluorocarbons can react violently with barium, potassium, sodium.

Permissible Exposure Limits in Air
Protective Action Criteria (PAC)
TEEL-0: 40 mg/m^3
PAC-1: 125 mg/m^3
PAC-2: 150 mg/m^3
PAC-3: 150 mg/m^3
The state of Pennsylvania has set a guideline of 1150 μg/m^3 for trifluralin in ambient air.[60]

Permissible Concentration in Water: EPA has set a lifetime health advisory of 2.0 μg/L. The state of Maine has set a guideline of 200.0 μg/L for drinking water.[61]

Determination in Water: Octanol—water coefficient: Log K_{ow} = 5.07. Fish Tox = 1.57645000 ppb MATC (HIGH).

Routes of Entry: Inhalation, skin and/or eye contact.

Harmful Effects and Symptoms

Short Term Exposure: Inhalation can cause irritation of the respiratory tract with cough, phlegm, and/or tightness in the chest. The vapor can cause eye and skin irritation. Skin contact can cause irritation and rash which can be exacerbated by sunlight. The majority of reported trifluralin exposure cases were occupational in nature. Other reported symptoms include respiratory involvement, abdominal cramps, nausea, diarrhea, headache, lethargy, and paresthesia following dermal or inhalation exposure.

Long Term Exposure: May cause skin sensitization. High or repeated exposure may affect the liver and kidneys and/or cause anemia. There is some dispute about the actual carcinogenic effect of trifluralin. NCI[9] reports clear evidence of carcinogenicity in mice but not in rats. Some authorities feel that dipropyl nitrosamine formed in trifluralin manufacture and contained in the technical material might be the actual culprit, and the purified trifluralin might not have this problem. Human Tox = 5.00000 ppb (HIGH).

Points of Attack: Skin, eyes, liver, kidneys, blood.

Medical Surveillance: Kidney and liver function tests. Complete blood count (CBC).

First Aid: Skin Contact[52]*:* Flood all areas of body that have contacted the substance with water. Do not wait to remove contaminated clothing; do it under the water stream. Use soap to help assure removal. Isolate contaminated clothing when removed to prevent contact by others.

Eye Contact: Remove any contact lenses at once. Flush eyes well with copious quantities of water or normal saline for at least 20—30 min. Seek medical attention.

Inhalation: Leave contaminated area immediately; breathe fresh air. Proper respiratory protection must be supplied to any rescuers. If coughing, difficult breathing, or any other symptoms develop, seek medical attention at once, even if symptoms develop many hours after exposure.

Ingestion: If convulsions are not present, give a glass or two of water or milk to dilute the substance. Assure that the

person's airway is unobstructed and contact a hospital or poison center immediately for advice on whether or not to induce vomiting.

Personal Protective Methods: Wear protective gloves and clothing to prevent any reasonable probability of skin contact. Safety equipment suppliers/manufacturers can provide recommendations on the most protective glove/clothing material for your operation. All protective clothing (suits, gloves, footwear, headgear) should be clean, available each day, and put on before work. Contact lenses should not be worn when working with this chemical. Wear dust-proof chemical goggles and face shield unless full-face-piece respiratory protection is worn. Employees should wash immediately with soap when skin is wet or contaminated. Provide emergency showers and eyewash.

Respirator Selection: Follow regulations in OSHA 29CFR1910.134 or European Standard EN149. Use a NIOSH/MSHA- or European Standard EN149-approved respirator; or use an approved supplied-air respirator with a full face-piece operated in the positive-pressure mode, or with a full face-piece, hood, or helmet in the continuous-flow mode; or use a NIOSH/MSHA- or European Standard EN149-approved self-contained breathing apparatus with a full face-piece operated in pressure-demand or other positive-pressure mode.

Storage: Color Code—Blue: Health Hazard/Poison: Store in a secure poison location. Prior to working with this chemical you should be trained on its proper handling and storage. Store in a cool, dry place and protect from prolonged exposure to light.

Shipping: Trifluralin, an organofluorine, can be classed as Pesticides, solid, toxic, n.o.s., which require a shipping label of "POISONOUS/TOXIC MATERIALS." It falls in Hazard Class 6.1.

Spill Handling: Evacuate persons not wearing protective equipment from area of spill or leak until cleanup is complete. Remove all ignition sources. Dampen spilled material with alcohol to avoid dust, then transfer material to a suitable container. Use absorbent dampened with alcohol to pick up remaining material. Wash surfaces well with soap and water. Collect waste material in the most convenient and safe manner and deposit in sealed containers. Ventilate area after cleanup is complete. It may be necessary to contain and dispose of this chemical as a hazardous waste. If material or contaminated runoff enters waterways, notify downstream users of potentially contaminated waters. Contact your local or federal environmental protection agency for specific recommendations. If employees are required to clean up spills, they must be properly trained and equipped. OSHA 1910.120(q) may be applicable. Soil Adsorption Index (K_{oc}) = 8000.

Fire Extinguishing: This chemical is a combustible solid. Use dry chemical, carbon dioxide, water spray, or alcohol foam extinguishers. Poisonous gases are produced in fire, including nitrogen oxide and hydrogen fluoride. If material or contaminated runoff enters waterways, notify downstream users of potentially contaminated waters. Notify local health and fire officials and pollution control agencies. From a secure, explosion-proof location, use water spray to cool exposed containers. If cooling streams are ineffective (venting sound increases in volume and pitch, tank discolors, or shows any signs of deforming), withdraw immediately to a secure position. If employees are expected to fight fires, they must be trained and equipped in OSHA 1910.156. The only respirators recommended for firefighting are self-contained breathing apparatuses that have full face-pieces and are operated in a pressure-demand or other positive-pressure mode.

Disposal Method Suggested: Dissolve or mix the material with a combustible solvent and burn in a chemical incinerator equipped with an afterburner and scrubber. All federal, state, and local environmental regulations must be observed. Trifluralin does contain fluorine, and therefore incineration presents the increased hazard of HF in the off-gases. Prior to incineration, fluorine-containing compounds should be mixed with slaked lime plus vermiculite, sodium carbonate, or sand-soda ash mixture (90-10).

References
Sax, N. I. (Ed.). (1980). *Dangerous Properties of Industrial Materials Report*, 1, No. 2, 70–71
US Environmental Protection Agency. (August 1987). *Health Advisory: Trifluralin.* Washington, DC: Office of Drinking Water

Triisobutyl aluminum

See entry under "Aluminum alkyls."

Trimellitic anhydride   T:0850

Molecular Formula: $C_9H_4O_5$
Synonyms: Anhydrotrimellic acid; 1,2,4-Benzenetricarboxylic acid anhydride; 1,2,4-Benzenetricarboxylic acid, cyclic 1,2-anhydride; 1,2,4-Benzenetricarboxylic anhydride; 4-Carboxyphthalic anhydride; 1,3-Dihydro-1,3-dioxo-5-isobenzofurancarboxylic acid; 1,3-Dioxo-5-phthalancarboxylic acid; Diphenylmethane-4,4'-diisocyanate-trimellic anhydride-ethomid HT polymer; NCI-C56633; TMA; TMAN; Trimellic acid anhydride; Trimellitic acid cyclic-1,2-anhydride
CAS Registry Number: 552-30-7
RTECS® Number: DC2050000
EC Number: 209-008-0 [*Annex I Index No.:* 607-097-00-4]
Regulatory Authority and Advisory Bodies
Air Pollutant Standard Set. See below, "Permissible Exposure Limits in Air" section.
Canada, WHMIS, Ingredients Disclosure List Concentration 0.1%.
European/International Regulations: Hazard Symbol: Xn; Risk phrases: R37; R41; R42/43; Safety phrases: S2; S22; S26; S36/37/39 (see Appendix 4).

WGK (German Aquatic Hazard Class): No value assigned.
Description: Trimellitic anhydride is a crystalline solid. It is the anhydride of trimellitic acid (1,2,4-benzenetricarboxylic acid). Molecular weight = 192.13; Boiling point = 245°C; Freezing/Melting point = 166.7°C; Vapor pressure = 0.000004 mmHg; Flash point = 227°C. Hazard Identification (based on NFPA-704 M Rating System): Health 3, Flammability 1, Reactivity 1W. Highly soluble; reacts with water.
Potential Exposure: Compound Description: Drug. TMA is used to produce trimellitate plasticizers, poly (amide-imide) polymers; in paints, enamels, and coatings; polymers, polyesters; as a curing agent for epoxy and other resins; in vinyl plasticizers; agricultural chemicals; dyes and pigments; pharmaceuticals, surface active agents; modifiers, intermediates, and specialty chemicals.
Incompatibilities: Dust can cause an explosion. Violent reaction with strong oxidizers. Reacts slowly with water, forming trimellitic acid.
Permissible Exposure Limits in Air
Conversion factor: 1 ppm = 7.86 mg/m^3 at 25°C & 1 atm.
OSHA PEL: None.
NIOSH REL: 0.005 ppm/0.04 mg/m^3 TWA, Should be handled in the workplace as an extremely toxic substance.
ACGIH TLV®[1]: 0.0005 mg/m^3 measured as inhalable fraction and vapor TWA; 0.002 mg/m^3 measured as inhalable fraction and vapor Ceiling Concentration; [skin]; danger of sensitization.
DFG MAK: 0.04 mg/m^3 TWA; Peak Limitation Category I (1); danger of sensitization of the airways.
Australia: TWA 0.005 ppm (0.04 mg/m^3), 1993; Austria: MAK 0.005 ppm (0.04 mg/m^3), 1999; Belgium: TWA 0.005 ppm (0.059 mg/m^3), 1993; Denmark: TWA 0.1 mg/m^3, 1999; Finland: TWA 0.005 ppm, 1999; Norway: TWA 0.005 ppm (0.04 mg/m^3), 1999; Russia: STEL 0.1 mg/m^3, 1993; Sweden: NGV 0.04 mg/m^3, TGV 0.08 mg/m^3, 1999; Switzerland: MAK-W 0.005 ppm (0.04 mg/m^3), KZG-W 0.01 ppm (0.08 mg/m^3), 1999; United Kingdom: TWA 0.04 mg/m^3, STEL 0.12 mg/m^3, 2000; Argentina, Bulgaria, Columbia, Jordan, South Korea, New Zealand, Singapore, Vietnam: ACGIH TLV®: Ceiling Concentration 0.04 mg/m^3. Several states have set guidelines or standards for trimellitic anhydride in ambient air[60] ranging from 0.4 µg/m^3 (North Dakota) to 0.7 µg/m^3 (Virginia) to 0.8 µg/m^3 (Connecticut) to 1.0 µg/m^3 (Nevada).
Determination in Air: Use NIOSH Analytical Method (IV) #5036, Trimellitic anhydride, OSHA Analytical Method 98.
Routes of Entry: Inhalation, ingestion, skin and/or eye contact.
Harmful Effects and Symptoms
Short Term Exposure: Trimellitic anhydride can affect you when breathed in. Irritates the eyes, skin, and respiratory tract. High exposure may cause runny nose, cough, wheezing and shortness of breath, malaise (vague feeling of discomfort), fever, muscle aches. "TMA-flu" with symptoms of cough, chills, shortness of breath, body aches, weakness,

and coughing up blood may be delayed for 4—12 h following overexposure. Higher exposures can cause pulmonary edema, a medical emergency that can be delayed for several hours. This can cause death.

Long Term Exposure: Repeated inhalation exposure may cause asthma which may be delayed and is aggravated by physical activity. Allergic asthma occurs sometimes weeks or months after exposure. Trimellitic anhydride can also cause "Pulmonary Disease-Anemia Syndrome" with low blood count, and different lung changes.

Points of Attack: Eyes, skin, respiratory system, blood.

Medical Surveillance: Before beginning employment and at regular times after that, the following are recommended: lung function test—these may be normal if the person is not having an attack at the time of the test. If symptoms develop or overexposure is suspected, the following may be useful: special tests for trimellitic anhydride allergy (IgE antibodies against TM-HAS). Evaluation by a qualified allergist, including careful exposure history and special testing, may help diagnose allergy. Consider chest X-ray following acute overexposure.

First Aid: If this chemical gets into the eyes, remove any contact lenses at once and irrigate immediately for at least 15 min, occasionally lifting upper and lower lids. Seek medical attention immediately. If this chemical contacts the skin, remove contaminated clothing and wash immediately with soap and water. Seek medical attention immediately. If this chemical has been inhaled, remove from exposure, begin rescue breathing (using universal precautions, including resuscitation mask) if breathing has stopped and CPR if heart action has stopped. Transfer promptly to a medical facility. When this chemical has been swallowed, get medical attention. Give large quantities of water and induce vomiting. Do not make an unconscious person vomit. Medical observation is recommended for 24—48 h after breathing overexposure, as pulmonary edema may be delayed. As first aid for pulmonary edema, a doctor or authorized paramedic may consider administering a corticosteroid spray.

Personal Protective Methods: Wear protective gloves and clothing to prevent any reasonable probability of skin contact. Safety equipment suppliers/manufacturers can provide recommendations on the most protective glove/clothing material for your operation. All protective clothing (suits, gloves, footwear, headgear) should be clean, available each day, and put on before work. Contact lenses should not be worn when working with this chemical. Wear dust-proof chemical goggles and face shield unless full-face-piece respiratory protection is worn. Employees should wash immediately with soap when skin is wet or contaminated. Provide emergency showers and eyewash.

Respirator Selection: Where there is potential for exposures *over 0.005 ppm,* use a NIOSH/MSHA- or European Standard EN149-approved supplied-air respirator with a full face-piece operated in the positive-pressure mode, or with a full face-piece, hood, or helmet in the continuous-flow mode; or use a NIOSH/MSHA- or European Standard EN149-approved self-contained breathing apparatus with a full face-piece operated in pressure-demand or other positive-pressure mode.

Storage: Color Code—Blue: Health Hazard/Poison: Store in a secure poison location. Prior to working with this chemical you should be trained on its proper handling and storage. Store in tightly closed containers in a cool well-ventilated area.

Spill Handling: Evacuate persons not wearing protective equipment from area of spill or leak until cleanup is complete. Remove all ignition sources. Collect powdered material in the most convenient and safe manner and deposit in sealed containers. Ventilate area after cleanup is complete. It may be necessary to contain and dispose of this chemical as a hazardous waste. If material or contaminated runoff enters waterways, notify downstream users of potentially contaminated waters. Contact your local or federal environmental protection agency for specific recommendations. If employees are required to clean up spills, they must be properly trained and equipped. OSHA 1910.120(q) may be applicable.

Fire Extinguishing: This chemical is a combustible solid (no flash point can be found in the literature). Use dry chemical, carbon dioxide, water spray, or alcohol foam extinguishers. Poisonous gases are produced in fire. If material or contaminated runoff enters waterways, notify downstream users of potentially contaminated waters. Notify local health and fire officials and pollution control agencies. From a secure, explosion-proof location, use water spray to cool exposed containers. If cooling streams are ineffective (venting sound increases in volume and pitch, tank discolors, or shows any signs of deforming), withdraw immediately to a secure position. If employees are expected to fight fires, they must be trained and equipped in OSHA 1910.156. The only respirators recommended for firefighting are self-contained breathing apparatuses that have full face-pieces and are operated in a pressure-demand or other positive-pressure mode.

Disposal Method Suggested: Dissolve or mix the material with a combustible solvent and burn in a chemical incinerator equipped with an afterburner and scrubber. All federal, state, and local environmental regulations must be observed.

References

National Institute for Occupational Safety and Health. (February 3, 1978). *Trimellitic Anhydride (TMA),* Current Intelligence Bulletin No. 21. Washington, DC

US Environmental Protection Agency. (March 3, 1978). *Chemical Hazard Information Profile: Trimellitic Anhydride.* Washington, DC

Sax, N. I. (Ed.). (1985). *Dangerous Properties of Industrial Materials Report,* 5, No. 6, 30—31

New Jersey Department of Health and Senior Services. (May 2000). *Hazardous Substances Fact Sheet: Trimellitic Anhydride.* Trenton, NJ

Trimethylamine   T:0860

Molecular Formula: C_3H_9N
Common Formula: $(CH_3)_3N$
Synonyms: *N,N*-Dimethylmethanamine; Methamine, *N,N*-dimethyl-
CAS Registry Number: 75-50-3
RTECS® Number: YH2880000
UN/NA & ERG Number: UN1083 (anhydrous)/118; UN1297 (aqueous solutions)/132
EC Number: 200-875-0 [*Annex I Index No.:* 612-001-00-9]
Regulatory Authority and Advisory Bodies
Department of Homeland Security Screening Threshold Quantity (pounds): *Release hazard* 10,000 ($\geq$1.00% concentration).
Air Pollutant Standard Set. See below, "Permissible Exposure Limits in Air" section.
Clean Air Act: Accidental Release Prevention/Flammable Substances, (Section 112[r], Table 3), TQ = 10,000 lb (4540 kg).
Clean Water Act: Section 311 Hazardous Substances/RQ 40CFR117.3 (same as CERCLA, see below).
Reportable Quantity (RQ): 100 lb (45.4 kg).
Canada, WHMIS, Ingredients Disclosure List Concentration 1.0%.
European/International Regulations: Hazard Symbol: F+, Xn; Risk phrases: R12; R20; R37/38; R41; Safety phrases: S2; S16; S26; S39 (see Appendix 4).
WGK (German Aquatic Hazard Class): 2—Hazard to waters.
Description: Trimethylamine is a strong, fishy, ammoniacal-smelling gas. Flammable gas. Shipped as a compressed gas. It may be present in an aqueous solution. The odor threshold is 0.00011–0.87 ppm. *Warning:* The odor threshold range is so broad that odor alone should not be used as a warning of potentially hazardous exposures. Molecular weight = 59.13; Boiling point = 3°C; Specific gravity (H_2O:1) = 0.58 at 25°C; Freezing/Melting point = −117°C; Vapor pressure = 1454 mmHg at 21°C; Flash point = flammable gas; Autoignition temperature = 190°C. Explosive limits: LEL = 2.0%; UEL = 11.6%. Hazard Identification (based on NFPA-704 M Rating System): Health 3, Flammability 4, Reactivity 0. Soluble in water; solubility = 48% at 30°C.
Potential Exposure: Compound Description: Reproductive Effector. Trimethylamine is used as a chemical intermediate in organic synthesis of quaternary ammonium compounds, as an insect attractant, as a warning agent in natural gas, as a flotation agent.
Incompatibilities: A medium strong base. Violent reaction with strong oxidizers (such as chlorine, bromine, fluorine), ethylene oxide, nitrosating agents (e.g., sodium nitrite), mercury, strong acids. Corrosive to many metals (e.g., zinc, brass, aluminum, copper, tin, and their alloys).

Permissible Exposure Limits in Air
Conversion factor: 1 ppm = 2.42 mg/m^3 at 25°C & 1 atm.
OSHA PEL: None.
NIOSH REL: 10 ppm/24 mg/m^3 TWA; 15 ppm/36 mg/m^3 STEL.
ACGIH TLV®[1]: 5 ppm/12 mg/m^3 TWA; 15 ppm/36 mg/m^3 STEL.
AIHA WEEL: 1 ppm TWA.
Protective Action Criteria (PAC)*
TEEL-0: 5 mg/m^3
PAC-1: **8.0** mg/m^3
PAC-2: **120** mg/m^3
PAC-3: **380** mg/m^3
*AEGLs (Acute Emergency Guideline Levels) & ERPGs (Emergency Response Planning Guideline) are in **bold face**.
DFG MAK: 2 ppm/4.9 mg/m^3 TWA; Peak Limitation Category I(2); Pregnancy Risk Group C.
Australia: TWA 10 ppm (24 mg/m^3), STEL 15 ppm, 1993; Belgium: TWA 10 ppm (24 mg/m^3), STEL 15 ppm, 1993; Denmark: TWA 10 ppm (24 mg/m^3), 1999; France: VLE 10 ppm (25 mg/m^3), 1999; Hungary: TWA 5 mg/m^3, STEL 10 mg/m^3, 1993; Norway: TWA 10 ppm (24 mg/m^3), 1999; the Netherlands: MAC-TGG 1 mg/m^3, 2003; Argentina, Bulgaria, Columbia, Jordan, South Korea, New Zealand, Singapore, Vietnam: ACGIH TLV®: STEL 15 ppm. Several states have set guidelines or standards for trimethylamine in ambient air[60] ranging from 57.143 µg/m^3 (Kansas) to 240–360.0 µg/m^3 (North Dakota) to 400.0 µg/m^3 (Virginia) to 480.0 µg/m^3 (Connecticut) to 952.0 µg/m^3 (Nevada).
Determination in Air: Use OSHA Analytical Method PV2060.
Determination in Water: Octanol–water coefficient: Log K_{ow} = 0.2.
Routes of Entry: Inhalation, ingestion (solution), skin and/or eye contact.
Harmful Effects and Symptoms
Short Term Exposure: Trimethylamine can affect you when breathed in. Exposure can irritate the eyes, skin, and respiratory tract. Very high levels may cause a buildup of fluid in the lungs (pulmonary edema). This can cause death. Contact can cause severe irritation and burns to the eyes and skin. May cause blurred vision, corneal necrosis. Contact may cause frostbite.
Long Term Exposure: Long-term effects are not known at this time.
Points of Attack: Eyes, skin, respiratory system.
Medical Surveillance: For those with frequent or potentially high exposure (half the TLV or greater), the following are recommended before beginning work and at regular times after that: lung function tests. If symptoms develop or overexposure is suspected, the following may be useful: consider chest X-ray following acute overexposure. Evaluation by a qualified allergist, including careful exposure history and special testing, may help diagnose skin allergy.

First Aid: If this chemical gets into the eyes, remove any contact lenses at once and irrigate immediately for at least 15 min, occasionally lifting upper and lower lids. Seek medical attention immediately. If this chemical contacts the skin, remove contaminated clothing and wash immediately with soap and water. Seek medical attention immediately. If this chemical has been inhaled, remove from exposure, begin rescue breathing (using universal precautions, including resuscitation mask) if breathing has stopped and CPR if heart action has stopped. Transfer promptly to a medical facility. When this chemical has been swallowed, get medical attention. Give large quantities of water and induce vomiting. Do not make an unconscious person vomit. Medical observation is recommended for 24–48 h after breathing overexposure, as pulmonary edema may be delayed. As first aid for pulmonary edema, a doctor or authorized paramedic may consider administering a corticosteroid spray.

Personal Protective Methods: Wear protective gloves and clothing to prevent any reasonable probability of skin contact. Safety equipment suppliers/manufacturers can provide recommendations on the most protective glove/clothing material for your operation. All protective clothing (suits, gloves, footwear, headgear) should be clean, available each day, and put on before work. Contact lenses should not be worn when working with this chemical. Wear splash-proof chemical goggles and face shield when working with liquid, unless full-face-piece respiratory protection is worn. Wear gas-proof goggles, unless full-face-piece respiratory protection is worn. Employees should wash immediately with soap when skin is wet or contaminated. Provide emergency showers and eyewash.

Respirator Selection: Where there is potential for exposures *over 10 ppm*, use a NIOSH/MSHA- or European Standard EN149-approved supplied-air respirator with a full face-piece operated in the positive-pressure mode, or with a full face-piece, hood, or helmet in the continuous-flow mode; or use a NIOSH/MSHA- or European Standard EN149-approved self-contained breathing apparatus with a full face-piece operated in pressure-demand or other positive-pressure mode.

Storage: Color Code—Red Stripe: Flammability Hazard: Store separately from all other flammable materials. Prior to working with this chemical you should be trained on its proper handling and storage. Before entering confined space where this chemical may be present, check to make sure that an explosive concentration does not exist. Trimethylamine must be stored to avoid contact with strong oxidizers (such as chlorine, bromine, and fluorine) and mercury since violent reactions occur. Sources of ignition, such as smoking and open flames, are prohibited where trimethylamine is used, handled, or stored in a manner that could create a potential fire or explosion hazard. Use only nonsparking tools and equipment, especially when opening and closing containers of trimethylamine. Wherever trimethylamine is used, handled, manufactured, or stored, use explosion-proof electrical equipment and fittings. Procedures for the handling, use, and storage of cylinders should be in compliance with OSHA 1910.101 and 1910.169, as with the recommendations of the Compressed Gas Association.

Shipping: Trimethylamine, anhydrous, requires a shipping label of "FLAMMABLE GAS." It falls in Hazard Class 2.1. Trimethylamine, aqueous solutions, requires a shipping label of "FLAMMABLE LIQUID, CORROSIVE." Trimethylamine, aqueous solutions, with not >50% trimethylamine by mass falls in DOT Hazard Class 3 and Packing Group I, II, or III.

Spill Handling: Evacuate and restrict persons not wearing protective equipment from area of spill or leak until cleanup is complete. Remove all ignition sources. Establish forced ventilation to keep levels below explosive limit. Stop the flow of gas if it can be done safely. If source of leak is a cylinder and the leak cannot be stopped in place, remove leaking cylinder to a safe place in the open air, and repair leak or allow cylinder to empty. For a spill of a liquid solution of trimethylamine, absorb liquid with sodium bisulfate, sand, vermiculite, earth, or a similar material. Keep this chemical out of confined space, such as a sewer, because of the possibility of explosion, unless the sewer is designed to prevent the buildup of explosive concentrations. It may be necessary to contain and dispose of this chemical as a hazardous waste. Contact your local or federal environmental protection agency for specific recommendations. If employees are required to clean up spills, they must be properly trained and equipped. OSHA 1910.120(q) may be applicable.

Fire Extinguishing: Trimethylamine is a flammable gas or liquid. For a small fire involving trimethylamine gas, use dry chemical or CO_2 extinguishers. For small water solution fires, use dry chemical, CO_2, water spray, or foam extinguishers. Poisonous gases are produced in fire, including oxides of nitrogen. Do not extinguish the fire unless the flow of gas can be stopped and any remaining gas is out of the line. Specially trained personnel may use fog lines to cool exposures and let the fire burn itself out. Vapors are heavier than air and will collect in low areas. Vapors may travel long distances to ignition sources and flashback. Vapors in confined areas may explode when exposed to fire. Containers may explode in fire. Storage containers and parts of containers may rocket great distances, in many directions. If liquid or contaminated runoff enters waterways, notify downstream users of potentially contaminated waters. Notify local health and fire officials and pollution control agencies. From a secure, explosion-proof location, use water spray to cool exposed containers. If cooling streams are ineffective (venting sound increases in volume and pitch, tank discolors, or shows any signs of deforming), withdraw immediately to a secure position. If cylinders are exposed to excessive heat from fire or flame contact, withdraw immediately to a secure location. If employees are expected to fight fires, they must be trained and equipped in OSHA 1910.156. The only respirators recommended for firefighting are self-contained breathing apparatuses that have full face-pieces and are operated in a pressure-demand or other positive-pressure mode.

References

Sax, N. I. (Ed.). *Dangerous Properties of Industrial Materials Report*, 2, No. 2, 70–73 (1982) and 5, No. 6, 96–98 (1985)

New Jersey Department of Health and Senior Services. (June 2003). *Hazardous Substances Fact Sheet: Trimethylamine*. Trenton, NJ

2,4,6-Trimethylaniline   T:0870

Molecular Formula: $C_9H_{13}N$
Common Formula: $C_6H_2(CH_3)_3NH_2$
Synonyms: Aminomesitylene; 2-Amino-1,3,5-trimethylbenzene; Aniline, 2,4,6-trimethyl-; Mesidine; Mesitylamine; Mezidine; 2,4,6-Trimethylbenzenamine
CAS Registry Number: 88-05-1
RTECS® Number: BZ0700000
UN/NA & ERG Number: UN2810/153
EC Number: 201-794-3
Regulatory Authority and Advisory Bodies
Carcinogenicity: IARC: Animal Inadequate Evidence; Human Inadequate Evidence, *not classifiable as carcinogenic to humans*, Group 3, 1982.
CERCLA/SARA 40CFR302 Extremely Hazardous Substances: TPQ = 500 lb (227 kg).
Reportable Quantity (RQ): 500 lb (227 kg).
Canada, WHMIS, Classification D2B. On NDSL list.
European/International Regulations: not listed in Annex 1.
WGK (German Aquatic Hazard Class): No value assigned.
Description: Aniline, 2,4,6-trimethyl- is a clear liquid. Molecular weight = 135.21; Boiling point = 233°C; Freezing/Melting point = −5°C. Insoluble in water.
Potential Exposure: Used on small scale in organic synthesis.
Permissible Exposure Limits in Air
Protective Action Criteria (PAC)
TEEL-0: 0.125 mg/m³
PAC-1: 0.4 mg/m³
PAC-2: 2.9 mg/m³
PAC-3: 40 mg/m³
Routes of Entry: Ingestion.
Harmful Effects and Symptoms
Short Term Exposure: This material is moderately toxic orally. It is also considered highly toxic by unspecified routes. It is a skin and eye irritant. The danger of acute poisoning is represented by methemoglobinemia leading to adverse effects on the red cells. A number of the amines may act as skin sensitizers. Repeated exposure results in narrowing of peripheral vision, increase in size of blind spot, and decrease in photosensitivity. The LC_{50} inhalation (mouse) is 0.29 mg/L/2 h.
Long Term Exposure: Suspect occupational carcinogen. This material is a suspect carcinogen on the basis of being an aromatic amine but, unlike the 2,4,5-trimethyl-isomer, is not positive for animals.

Points of Attack: Blood, eyes, respiratory system.
First Aid: If this chemical gets into the eyes, remove any contact lenses at once and irrigate immediately for at least 15 min, occasionally lifting upper and lower lids. Seek medical attention immediately. If this chemical contacts the skin, remove contaminated clothing and wash immediately with soap and water. Seek medical attention immediately. If this chemical has been inhaled, remove from exposure, begin rescue breathing (using universal precautions, including resuscitation mask) if breathing has stopped and CPR if heart action has stopped. Transfer promptly to a medical facility. When this chemical has been swallowed, get medical attention. Give large quantities of water and induce vomiting. Do not make an unconscious person vomit.
Note to physician: Treat for methemoglobinemia. Spectrophotometry may be required for precise determination of levels of methemoglobin in urine.
Personal Protective Methods: Wear protective gloves and clothing to prevent any reasonable probability of skin contact. Safety equipment suppliers/manufacturers can provide recommendations on the most protective glove/clothing material for your operation. All protective clothing (suits, gloves, footwear, headgear) should be clean, available each day, and put on before work. Contact lenses should not be worn when working with this chemical. Wear splash-proof chemical goggles and face shield unless full-face-piece respiratory protection is worn. Employees should wash immediately with soap when skin is wet or contaminated. Provide emergency showers and eyewash.
Respirator Selection: Follow regulations in OSHA 29CFR1910.134 or European Standard EN149. Use a NIOSH/MSHA- or European Standard EN149-approved respirator; or use an approved supplied-air respirator with a full face-piece operated in the positive-pressure mode, or with a full face-piece, hood, or helmet in the continuous-flow mode; or use a NIOSH/MSHA- or European Standard EN149-approved self-contained breathing apparatus with a full face-piece operated in pressure-demand or other positive-pressure mode.
Storage: Color Code—Blue: Health Hazard/Poison: Store in a secure poison location. Prior to working with this chemical you should be trained on its proper handling and storage. Store in tightly closed containers in a cool, well-ventilated area. Where possible, automatically pump liquid from drums or other storage containers to process containers. Sources of ignition, such as smoking and open flames, are prohibited where this chemical is handled, used, or stored. Metal containers involving the transfer of 5 gallons or more of this chemical should be grounded and bonded. Drums must be equipped with self-closing valves, pressure vacuum bungs, and flame arresters. Use only nonsparking tools and equipment, especially when opening and closing containers of this chemical. Wherever this chemical is used, handled, manufactured, or stored, use explosion-proof electrical equipment and fittings. A regulated, marked area

should be established where this chemical is handled, used, or stored in compliance with OSHA Standard 1910.1045.

Shipping: Toxic, liquids, organic, n.o.s. require a shipping label of "POISONOUS/TOXIC MATERIALS." 1,2,3-Trichloropropane falls in DOT Hazard Class 6.1 and Parking Group III.

Spill Handling: Evacuate persons not wearing protective equipment from area of spill or leak until cleanup is complete. Remove all ignition sources. Do not touch spilled material; stop leak if you can do so without risk. Use water spray to reduce vapors. *Small spills*: absorb with sand or other noncombustible absorbent material and place into containers for later disposal. *Large spills*: dike far ahead of spill for later disposal. Wear positive pressure breathing apparatus and special protective clothing. It may be necessary to contain and dispose of this chemical as a hazardous waste. If material or contaminated runoff enters waterways, notify downstream users of potentially contaminated waters. Contact your local or federal environmental protection agency for specific recommendations. If employees are required to clean up spills, they must be properly trained and equipped. OSHA 1910.120(q) may be applicable.

Fire Extinguishing: This chemical is a combustible liquid. Poisonous gases, including nitrogen oxides, are produced in fire. *Small fires:* dry chemical, carbon dioxide, water spray, or foam. *Large fires:* water spray, fog, or foam. Move container from fire area if you can do so without risk. Fight fire from maximum distance. Dike fire control water for later disposal; do not scatter the material. Keep unnecessary people away; isolate hazard area and deny entry. Stay upwind; keep out of low areas. Ventilate closed spaces before entering them. Wear positive pressure breathing apparatus and special protective clothing. If water pollution occurs, notify appropriate authorities. Vapors are heavier than air and will collect in low areas. Vapors may travel long distances to ignition sources and flashback. Vapors in confined areas may explode when exposed to fire. Storage containers and parts of containers may rocket great distances, in many directions. If material or contaminated runoff enters waterways, notify downstream users of potentially contaminated waters. Notify local health and fire officials and pollution control agencies. From a secure, explosion-proof location, use water spray to cool exposed containers. If cooling streams are ineffective (venting sound increases in volume and pitch, tank discolors, or shows any signs of deforming), withdraw immediately to a secure position. If employees are expected to fight fires, they must be trained and equipped in OSHA 1910.156. The only respirators recommended for firefighting are self-contained breathing apparatuses that have full face-pieces and are operated in a pressure-demand or other positive-pressure mode.

References

US Environmental Protection Agency. (November 30, 1987). *Chemical Hazard Information Profile: Aniline, 2,4,6-Trimethyl.* Washington, DC: Chemical Emergency Preparedness Program.

Trimethylbenzenes   T:0880

Molecular Formula: C_9H_{12}
Common Formula: $C_6H_3(CH_3)_3$
*See also separate **Mesitylene** 108-67-8 (1,3,5-Trimethylbenzene) record*
Synonyms: *1,2,3-isomer:* Hemellitol; Hemimellite; 1,2,3-Trimethylbenzene; 1,2,3-Trimetilbenceno (Spanish)
1,2,4-isomer: Pseudocumene; PSI-Cumene; *asym*-Trimethylbenzene; 1,2,4-Trimethylbenzene; 1,2,4-Trimetilbenceno (Spanish)
1,3,5-isomer: Mesitylene, *sym*-trimethylbenzene; *sym*-Trimethylbenzene; 1,3,5-Trimethylbenzene; 1,3,5-Trimetilbenceno (Spanish)
Mixed isomers: Trimethylbenzene, mixed isomers; Trimetilbenceno (Spanish)
Note: Hemimellitene is a mixture of the 1,2,3-isomer with up to 10% of related aromatics, such as the 1,2,4-isomer.
CAS Registry Number: 526-73-8 (1,2,3-isomer); 95-63-6 (1,2,4-isomer); 108-67-8 (1,3,5-isomer); 25551-13-7 (mixed isomers)
RTECS® Number: OX6825000 (1,3,5-isomer); DC3325000 (1,2,4-isomer); DC3300000 (1,2,3-isomer); DC3225000 (mixed isomers)
UN/NA & ERG Number: UN2325 (1,3,5-isomer)/129; UN1993 (flammable liquids, n.o.s.)/128
EC Number: 208-394-8 (1,2,3-isomer); 202-436-9 [*Annex I Index No.:* 601-043-00-3] (1,2,4-isomer); 203-604-4 [*Annex I Index No.:* 601-025-00-5] (1,3,5-isomer); 247-099-9 (mixed isomers)

Regulatory Authority and Advisory Bodies

Air Pollutant Standard Set. See below, "Permissible Exposure Limits in Air" section.
1,2,4-isomer:
EPCRA Section 313 Form R *de minimis* concentration reporting level: 1.0%.
Canada, WHMIS, Ingredients Disclosure List Concentration 0.1% (1,2,4- and 1,3,5-isomers); 1% 1,2,3- and mixed-isomers).
Note: Only the 1,2,4-isomer is regulated by the US EPA. All three isomers are "regulated" by NIOSH. The 1,3,5-isomer was dropped from the EPA's Extremely Hazardous Substance List (EPA-SARA) in 1988.
European/International Regulations (*95-63-6*): Hazard Symbol: Xn; Risk phrases: R10; R20; R36/37/38; Safety phrases S2; S26 (see Appendix 4).
WGK (German Aquatic Hazard Class): 2—Water polluting *95-63-6 (1,2,4-isomer).*
Description: Trimethylbenzenes exists in three isomeric forms. All isomers are clear, colorless liquids with a distinctive, aromatic odor. The odor threshold for the class is 0.55 ppm. Molecular weight = 120.21; Specific gravity (H_2O:1) = 0.89 at 25°C; Boiling point = 176°C (1,2,3-isomer); 169°C (1,2,4-isomer); 164°C (1,3,5-isomer); Freezing/Melting point = −25°C (1,2,3-isomer); Vapor

pressure = 1 mmHg at 17°C (1,2,3-); 1 mmHg at 13°C (1,2,4-); Flash points = 44°C (1,2,3- & 1,2,4-isomers); 50°C (1,3,5-isomer); 53°C [1,2,3-isomer (90.5%)]; Autoignition temperature = 470°C (1,2,3-isomer); 479°C [1,2,3-isomer (90.5%)]; 500°C (1,2,4-isomer); 559°C (1,3,5-isomer). Explosive limits: LEL = 0.9%; UEL = 6.4% (1,2,4-isomer); LEL = 0.8%; UEL = 6.6% (1,2,3-isomer). NFPA-704 M Hazard Identification (all isomers): Health 1, Flammability 2, Reactivity 0. Practically insoluble in water; solubility = 0.005% (1,2,3-isomer); 0.006% (1,2,4-isomer).

Potential Exposure: Compound Description (1,2,3- and 1,2,4-isomers): Mutagen, (mixed isomers) Primary Irritant. These materials are used as solvents and in dye and perfume manufacture. The 1,2,3-isomer is used as raw material in chemical synthesis and as an ultraviolet stabilizer. The 1,2,4-isomer is used as the raw material for trimellitic anhydride manufacture. These compounds are found in diesel engine exhaust fumes.

Incompatibilities: Oxidizers (perchlorates, peroxides, permanganates, chlorates, nitrates), strong oxidizers (chlorine, bromine, fluorine), and nitric acid.

Permissible Exposure Limits in Air

All isomers

Conversion factor: 1 ppm = 4.92 mg/m^3 at 25°C & 1 atm.
OSHA PEL: None.
NIOSH REL: 25 ppm/125 mg/m^3 TWA.
ACGIH TLV®[1] (*mixed isomers*): 25 ppm/123 mg/m^3 TWA.
DFG MAK: (*all isomers*) 20 ppm/100 mg/m^3 TWA; Peak Limitation Category II(2); Pregnancy Group C.

1,2,3-isomer

Austria: MAK 25 ppm (125 mg/m^3), 1999; Denmark: TWA 25 ppm (120 mg/m^3), 1999; Japan; 25 ppm (120 mg/m^3), 1999; Norway: TWA 20 ppm (100 mg/m^3), 1999; United Kingdom: TWA 25 ppm (125 mg/m^3), 2000; the Netherlands: MAC-TGG 100 mg/m^3, 2003.

1,2,4-isomer

Protective Action Criteria (PAC)*
TEEL-0: 25 ppm
PAC-1: **140** ppm
PAC-2: **360** ppm
PAC-3: 400 ppm
*AEGLs (Acute Emergency Guideline Levels) & ERPGs (Emergency Response Planning Guideline) are in **bold face**.

Denmark: TWA 25 ppm (120 mg/m^3), 1999; Japan; 25 ppm (120 mg/m^3), 1999; Norway: TWA 20 ppm (100 mg/m^3), 1999; United Kingdom: TWA 25 ppm (125 mg/m^3), 2000; the Netherlands: MAC-TGG 100 mg/m^3, 2003.

Mixed isomers

Australia: TWA 25 ppm (125 mg/m^3), 1993; Austria: MAK 25 ppm (125 mg/m^3), 1999; Belgium: TWA 25 ppm (123 mg/m^3), 1993; Denmark: TWA 25 ppm (120 mg/m^3), 1999; Finland: TWA 25 ppm (120 mg/m^3), STEL 40 ppm (200 mg/m^3), [skin], 1999; France: VME 25 ppm (125 mg/m^3), 1999; Norway: TWA 20 ppm (100 mg/m^3), 1999; the

Netherlands: MAC-TGG 100 mg/m^3, 2003; Sweden: NGV 25 ppm (120 mg/m^3), KTV 35 ppm (170 mg/m^3), 1999; Switzerland: MAK-W 25 ppm (125 mg/m^3), 1999; United Kingdom: LTEL 25 ppm (125 mg/m^3), STEL 35 ppm, 1993; Argentina, Bulgaria, Columbia, Jordan, South Korea, New Zealand, Singapore, Vietnam: ACGIH TLV®: TWA 25 ppm. Several states have set guidelines or standard for Trimethyl benzenes in ambient air[60] ranging from 1.25−1.70 mg/m^3 (North Dakota) to 2.1 mg/m^3 (Virginia) to 2.5 mg/m^3 (Connecticut) to 2.976 mg/m^3 (Nevada).

Determination in Air: Use OSHA Analytical Method PV-2091.

Determination in Water: Octanol−water coefficient: Log K_{ow} = 3.7 (1,2,3-).

Routes of Entry: Inhalation, percutaneous absorption, ingestion, skin and/or eye contact.

Harmful Effects and Symptoms

Short Term Exposure: Trimethyl benzene can affect you when breathed in. Irritates the eyes, skin, and respiratory tract. Exposure can cause you to feel dizzy, lightheaded, and to pass out. Symptoms of exposure can also include headache, drowsiness, fatigue, dizziness, nausea, a lack of coordination, vomiting, nervousness, tenseness, confusion. Liquid deposition in lungs causes bronchitis or chemical pneumonitis.

Long Term Exposure: Repeated exposures can cause headaches, tiredness, and a feeling of nervous tension. Can affect the blood cells and the blood's clotting ability; hypochromic anemia. Delayed or chronic health hazard is possible including asthmatic bronchitis with coughing and/or shortness of breath. The use of alcoholic beverages enhances the effect. May cause liver damage. The liquid destroys the skin's natural oils, causing drying and cracking.

Points of Attack: Eyes, skin, respiratory system, central nervous system, blood, liver.

Medical Surveillance: Before beginning employment and at regular times after that, the following are recommended: lung function tests. Complete blood count and platelet count. If symptoms develop or overexposure is suspected, the following may be useful: liver function tests. Complete blood count (CBC) and platelet count.

First Aid: If this chemical gets into the eyes, remove any contact lenses at once and irrigate immediately for at least 15 min, occasionally lifting upper and lower lids. Seek medical attention immediately. If this chemical contacts the skin, remove contaminated clothing and wash immediately with soap and water. Seek medical attention immediately. If this chemical has been inhaled, remove from exposure, begin rescue breathing (using universal precautions, including resuscitation mask) if breathing has stopped and CPR if heart action has stopped. Transfer promptly to a medical facility. When this chemical has been swallowed, get medical attention. Give large quantities of water and induce vomiting. Do not make an unconscious person vomit.

Note to physician: Inhalation: Bronchodilators, decongestants, and oxygen may be used if necessary. Corticosteroids are useful for treating pneumonitis.

Personal Protective Methods: Wear protective gloves and clothing to prevent any reasonable probability of skin contact. Safety equipment suppliers/manufacturers can provide recommendations on the most protective glove/clothing material for your operation. All protective clothing (suits, gloves, footwear, headgear) should be clean, available each day, and put on before work. Contact lenses should not be worn when working with this chemical. Wear splash-proof chemical goggles and face shield unless full-face-piece respiratory protection is worn. Employees should wash immediately with soap when skin is wet or contaminated. Provide emergency showers and eyewash.

Respirator Selection: Where there is potential for exposures *over 25 ppm*, use a NIOSH/MSHA- or European Standard EN149-approved respirator with an organic vapor cartridge/canister. More protection is provided by a full-face-piece respirator than by a half-mask respirator, and even greater protection is provided by a powered air-purifying respirator. *Where there is potential for high exposures*, use a NIOSH/MSHA- or European Standard EN149-approved supplied-air respirator with a full face-piece operated the positive-pressure mode, or with a full face-piece, hood, or helmet in the continuous-flow mode; or use a NIOSH/MSHA- or European Standard EN149-approved self-contained breathing apparatus with a full face-piece operated in pressure-demand or other positive-pressure mode.

Storage: Color Code—Red: Flammability Hazard: Store in a flammable liquid storage area or approved cabinet away from ignition sources and corrosive and reactive materials. Prior to working with this chemical you should be trained on its proper handling and storage. Before entering confined space where this chemical may be present, check to make sure that an explosive concentration does not exist. Trimethylbenzene must be stored to avoid contact with oxidizers (such as perchlorates, peroxides, permanganates, chlorates, and nitrates) and strong oxidizers (such as chlorine, bromine, and fluorine) since violent reactions occur. Store in tightly closed containers in a cool, well-ventilated area away from heat. Sources of ignition, such as smoking and open flames, are prohibited where this chemical is used, handled, or stored in a manner that could create a potential fire or explosion hazard. Metal containers involving the transfer of 5 gallons or more of this chemical should be grounded and bonded. Drums must be equipped with self-closing valves, pressure vacuum bungs, and flame arresters. Use only nonsparking tools and equipment, especially when opening and closing containers of this chemical.

Shipping: Flammable liquids, toxic, n.o.s., require a shipping label of "FLAMMABLE LIQUID, POISONOUS/TOXIC MATERIALS." The 1,2,3-, 1,2,4-, and mixed isomers fall in DOT Hazard Class 3 and Packing Group II. 1,3,5-Trimethylbenzene requires a shipping label of "FLAMMABLE LIQUID." It falls in Hazard Class 3 and Packing Group II.

Spill Handling: Evacuate and restrict persons not wearing protective equipment from area of spill or leak until cleanup is complete. Remove all ignition sources. Establish forced ventilation to keep levels below explosive limit. Absorb liquids in vermiculite, dry sand, earth, peat, carbon, or a similar material and deposit in sealed containers. Keep this chemical out of a confined space, such as a sewer, because of the possibility of an explosion, unless the sewer is designed to prevent the buildup of explosive concentrations. It may be necessary to contain and dispose of this chemical as a hazardous waste. If material or contaminated runoff enters waterways, notify downstream users of potentially contaminated waters. Contact your local or federal environmental protection agency for specific recommendations. If employees are required to clean up spills, they must be properly trained and equipped. OSHA 1910.120(q) may be applicable.

Fire Extinguishing: This chemical is a flammable liquid. Poisonous gases are produced in fire. *Small fires:* dry chemical, carbon dioxide, water spray, or alcohol-resistant foam. *Large fires:* water spray, fog, or alcohol foam. Move container from fire area if you can do so without risk. Spray cooling water on containers that are exposed to flames until well after fire is out. For massive fire in cargo area, use unmanned hose holder or monitor nozzles; if this is impossible, withdraw from area and let fire burn. Isolate for one-half mile in all directions if tank car or truck is involved in fire. Vapors are heavier than air and will collect in low areas. Vapors may travel long distances to ignition sources and flashback. Vapors in confined areas may explode when exposed to fire. Containers may explode in fire. Storage containers and parts of containers may rocket great distances, in many directions. If material or contaminated runoff enters waterways, notify downstream users of potentially contaminated waters. Notify local health and fire officials and pollution control agencies. From a secure, explosion-proof location, use water spray to cool exposed containers. If cooling streams are ineffective (venting sound increases in volume and pitch, tank discolors, or shows any signs of deforming), withdraw immediately to a secure position. If employees are expected to fight fires, they must be trained and equipped in OSHA 1910.156. The only respirators recommended for firefighting are self-contained breathing apparatuses that have full face-pieces and are operated in a pressure-demand or other positive-pressure mode.

Disposal Method Suggested: Dissolve or mix the material with a combustible solvent and burn in a chemical incinerator equipped with an afterburner and scrubber. All federal, state, and local environmental regulations must be observed.

References

US Environmental Protection Agency. (October 31, 1985). *Chemical Hazard Information Profile: Mesitylene.* Washington, DC: Chemical Emergency Preparedness Program

New Jersey Department of Health and Senior Services. (May 2003). *Hazardous Substances Fact Sheet: Trimethylbenzene (mixed isomers).* Trenton, NJ

Trimethylchlorosilane T:0890

Molecular Formula: C_3H_9ClSi
Synonyms: Chlorotrimethylsilane; Monochloro-trimethyl-silicon; NSC 15750; Silane, chlorotrimethyl-; Silane, trimethylchloro-; Silicane, chlorotrimethyl-; Silylium, trimethyl-, chloride; STCC 4907680; TL 1163; Trimethylsilyl chloride; Trimetilclorosilano (Spanish)
CAS Registry Number: 75-77-4
RTECS® Number: VV2710000
UN/NA & ERG Number: UN1298/132
EC Number: 200-900-5
Regulatory Authority and Advisory Bodies
Department of Homeland Security Screening Threshold Quantity (pounds): *Release hazard* 10,000 ($\geq$1.00% concentration).
Clean Air Act: Accidental Release Prevention/Flammable Substances, (Section 112[r], Table 3), TQ = 10,000 lb (4540 kg).
Superfund/EPCRA 40CFR355, Extremely Hazardous Substances: TPQ = 1000 lb (454 kg).
Reportable Quantity (RQ): 1000 lb (454 kg).
EPCRA Section 313 Form R *de minimis* concentration reporting level: 1.0%.
European/International Regulations: not listed in Annex 1.
WGK (German Aquatic Hazard Class): 1—Low hazard to waters.
Description: Trimethylchlorosilane is a colorless, fuming liquid with an irritating odor. Molecular weight = 108.66; Specific gravity (H_2O:1) = 0.85 at 25°C; Boiling point = 57°C; Freezing/Melting point = −58°C; Flash point = −15°C (cc); Vapor pressure = 200 mmHg at 18°C; Autoignition temperature = 395°C. Hazard Identification (based on NFPA-704 M Rating System): Health 3, Flammability 3, Reactivity 2₩. Reacts with water.
Potential Exposure: Compound Description: Tumorigen, Mutagen. Primary Irritant. Trimethylchlorosilane is used as an intermediate to make silicone products, including lubricants.
Incompatibilities: Moisture and air contact forms hydrochloric acid. Violent reaction with moisture forming corrosive chloride gases, including hydrogen chloride. Vigorous reaction with aluminum. Store in temperatures below 21°C.
Permissible Exposure Limits in Air
AIHA WEEL: 5 ppm, ceiling.
Protective Action Criteria (PAC)*
TEEL-0: 0.6 ppm
PAC-1: **1.8** ppm
PAC-2: **22** ppm

PAC-3: **100** ppm
*AEGLs (Acute Emergency Guideline Levels) & ERPGs (Emergency Response Planning Guideline) are in **bold face**.
Determination in Air: No method available.
Routes of Entry: Inhalation, ingestion, eye and/or skin contact.
Harmful Effects and Symptoms
Short Term Exposure: Irritates the respiratory tract causing coughing and wheezing. Contact can cause severe skin and eye irritation and burns with possible permanent damage. Higher exposures can cause pulmonary edema, a medical emergency that can be delayed for several hours. This can cause death.
Long Term Exposure: Can irritate the lungs; bronchitis may develop with coughing, phlegm, and/or shortness of breath. There is limited evidence that trimethylchlorosilane causes cancer in animals.
Points of Attack: Cancer site in animals: lungs.
Medical Surveillance: Lung function tests. Consider chest X-ray following acute overexposure.
First Aid: If this chemical gets into the eyes, remove any contact lenses at once and irrigate immediately for at least 30 min, occasionally lifting upper and lower lids. Seek medical attention immediately. If this chemical contacts the skin, remove contaminated clothing and wash immediately with soap and water. Seek medical attention immediately. If this chemical has been inhaled, remove from exposure, begin rescue breathing (using universal precautions, including resuscitation mask) if breathing has stopped and CPR if heart action has stopped. Transfer promptly to a medical facility. When this chemical has been swallowed, get medical attention. If victim is *conscious*, administer water or milk. Do not induce vomiting. Medical observation is recommended for 24−48 h after breathing overexposure, as pulmonary edema may be delayed. As first aid for pulmonary edema, a doctor or authorized paramedic may consider administering a corticosteroid spray.
Personal Protective Methods: Wear protective gloves and clothing to prevent any reasonable probability of skin contact. Safety equipment suppliers/manufacturers can provide recommendations on the most protective glove/clothing material for your operation. All protective clothing (suits, gloves, footwear, headgear) should be clean, available each day, and put on before work. Contact lenses should not be worn when working with this chemical. Wear-splash-proof chemical goggles and face shield unless full face-piece respiratory protection is worn. Employees should wash immediately with soap when skin is wet or contaminated. Provide emergency showers and eyewash.
Respirator Selection: Follow regulations in OSHA 29CFR1910.134 or European Standard EN149. Use a NIOSH/MSHA- or European Standard EN149-approved respirator; or use an approved supplied-air respirator with a full face-piece operated in the positive-pressure mode, or with a full face-piece, hood, or helmet in the continuous-flow mode; or use

a NIOSH/MSHA- or European Standard EN149-approved self-contained breathing apparatus with a full face-piece operated in a pressure-demand or other positive-pressure mode.

Storage: Color Code—Red: Flammability Hazard: Store in a flammable liquid storage area or approved cabinet away from ignition sources and corrosive and reactive materials. Prior to working with this chemical you should be trained on its proper handling and storage. Store in tightly closed containers in a cool (below 70°F/21°C), well-ventilated area away from moisture of any form and strong oxidizers. Sources of ignition, such as smoking and open flames, are prohibited where this chemical is used, handled, or stored in a manner that could create a potential fire or explosion hazard. Metal containers involving the transfer of 5 gallons or more of this chemical should be grounded and bonded. Drums must be equipped with self-closing valves, pressure vacuum bungs, and flame arresters. Use only nonsparking tools and equipment, especially when opening and closing containers of this chemical.

Shipping: Trimethylchlorosilane requires a shipping label of "FLAMMABLE LIQUID, CORROSIVE." It falls in Hazard Class 3 and Packing Group II.

Spill Handling: Evacuate and restrict persons not wearing protective equipment from area of spill or leak until cleanup is complete. Remove all ignition sources. Ventilate area of spill or leak. Absorb liquids in vermiculite, dry sand, earth, peat, carbon, or a similar material and deposit in sealed containers. Keep this chemical out of a confined space, such as a sewer, because of the possibility of an explosion, unless the sewer is designed to prevent the buildup of explosive concentrations. It may be necessary to contain and dispose of this chemical as a hazardous waste. If material or contaminated runoff enters waterways, notify downstream users of potentially contaminated waters. Contact your local or federal environmental protection agency for specific recommendations. If employees are required to clean up spills, they must be properly trained and equipped. OSHA 1910.120(q) may be applicable.

Initial isolation and protective action distances

Distances shown are likely to be affected during the first 30 min after materials are spilled and could increase with time. If more than one tank car, cargo tank, portable tank, or large cylinder involved in the incident is leaking, the protective action distance may need to be increased. You may need to seek emergency information from CHEMTREC at (800) 424-9300 or seek professional environmental engineering assistance from the US EPA Environmental Response Team at (908) 548-8730 (24-h response line).

Small spills (From a small package or a small leak from a large package)

When spilled in water
First: Isolate in all directions (feet/meters)   100/30
Then: Protect persons downwind (miles/kilometers)
Day   0.1/0.2
Night   0.1/0.2

Large spills (From a large package or from many small packages)
First: Isolate in all directions (feet/meters)   100/30
Then: Protect persons downwind (miles/kilometers)
Day   0.3/0.5
Night   0.7/1.2

Fire Extinguishing: This chemical is a flammable liquid. Poisonous gases, including hydrogen chloride, are produced in fire. Use dry chemical or carbon dioxide. Fire may restart after it has been extinguished. Vapors are heavier than air and will collect in low areas. Vapors may travel long distances to ignition sources and flashback. Vapors in confined areas may explode when exposed to fire. Containers may explode in fire. Storage containers and parts of containers may rocket great distances, in many directions. If material or contaminated runoff enters waterways, notify downstream users of potentially contaminated waters. Notify local health and fire officials and pollution control agencies. From a secure, explosion-proof location, use water spray to cool exposed containers. If cooling streams are ineffective (venting sound increases in volume and pitch, tank discolors, or shows any signs of deforming), withdraw immediately to a secure position. If employees are expected to fight fires, they must be trained and equipped in OSHA 1910.156. The only respirators recommended for firefighting are self-contained breathing apparatuses that have full face-pieces and are operated in a pressure-demand or other positive-pressure mode.

References

New Jersey Department of Health and Senior Services. (June 2003). *Hazardous Substances Fact Sheet: Trimethylchlorosilane.* Trenton, NJ

Trimethyl phosphite   T:0900

Molecular Formula: $C_3H_9O_3P$
Common Formula: $(CH_3O)_3P$
Synonyms: Methyl phosphite; Phosphorus acid, trimethyl ester; TMP; Trimethoxyphosphine; Trimethyl ester of phosphorous acid
CAS Registry Number: 121-45-9
RTECS® Number: TH1400000
UN/NA & ERG Number: UN2329/130
EC Number: 204-471-5
Regulatory Authority and Advisory Bodies
Department of Homeland Security Screening Threshold Quantity (pounds): *Theft hazard* 220 (≥80.00% concentration).
Air Pollutant Standard Set. See below, "Permissible Exposure Limits in Air" section.
Canada, WHMIS, Ingredients Disclosure List Concentration 1.0%.
European/International Regulations: not listed in Annex 1.
WGK (German Aquatic Hazard Class): 1—Low hazard to waters.

Description: Trimethyl phosphite is a colorless liquid with a distinctive pungent, pyridine-like odor. The odor threshold is 0.0001 ppm. Molecular weight = 124.09; Specific gravity (H_2O:1) = 1.05 at 25°C; Boiling point = 111°C; Freezing/Melting point = −78°C; Vapor pressure = 24 mmHg at 25°C; Flash point = 28°C. Hazard Identification (based on NFPA-704 M Rating System): Health 1, Flammability 2, Reactivity 0. Insoluble in water.

Potential Exposure: Compound Description: Reproductive Effector; Primary Irritant. Trimethyl phosphite is a flame retardant and used as an intermediate in the manufacture of a number of pesticides and organophosphorus additives.

Incompatibilities: Reacts (hydrolyzes) with water. Violent reaction with magnesium perchlorate. Incompatible with air, moisture, oxidizers, strong bases. Store and handle under a nitrogen blanket.

Permissible Exposure Limits in Air

Conversion factor: 1 ppm = 5.08 mg/m^3 at 25°C & 1 atm.
OSHA PEL: None.
NIOSH REL: 2 ppm/10 mg/m^3 TWA.
ACGIH TLV®[1]: 2 ppm TWA.
Protective Action Criteria (PAC)*
TEEL-0: 2 ppm
PAC-1: **6.1** ppm
PAC-2: **61** ppm
PAC-3: **310** ppm
*AEGLs (Acute Emergency Guideline Levels) & ERPGs (Emergency Response Planning Guideline) are in **bold face**.
Australia: TWA 2 ppm (10 mg/m^3), 1993; Belgium: TWA 2 ppm (10 mg/m^3), 1993; Denmark: TWA 0.5 ppm (2.6 mg/m^3), 1999; Finland: TWA 0.5 ppm (2.6 mg/m^3), STEL 10 ppm (52 mg/m^3), [skin], 1999; France: VME 2 ppm (10 mg/m^3), 1999; the Netherlands: MAC-TGG 10 mg/m^3, 2003; United Kingdom: TWA 2 ppm (10 mg/m^3), 2000; Argentina, Bulgaria, Columbia, Jordan, South Korea, New Zealand, Singapore, Vietnam: ACGIH TLV®: TWA 2 ppm. Several states have set guidelines or standards for Trimethyl phosphite in ambient air[60] ranging from 100 μg/m^3 (North Dakota) to 160 μg/m^3 (Virginia) to 200 μg/m^3 (Connecticut) to 238 μg/m^3 (Nevada).

Determination in Air

DFG MAK: No numerical value established. Data may be available.

Routes of Entry: Inhalation, ingestion, skin and/or eye contact.

Harmful Effects and Symptoms

Short Term Exposure: Trimethyl phosphite can affect you when breathed in and by passing through your skin. The vapor irritates the eyes, skin, and upper respiratory tract. Contact can severely irritate and may permanently damage the eyes. High exposures can cause pulmonary edema, a medical emergency that can be delayed for several hours. This can cause death. Trimethyl phosphite can severely irritate the skin and can cause a rash and skin allergy. Exposure may damage the liver and kidneys.

Long Term Exposure: Can cause lung irritation; bronchitis may develop. Exposure may cause emphysema. May damage the liver and kidneys. In animals: teratogenic effects.

Points of Attack: Eyes, skin, respiratory system, reproductive system.

Medical Surveillance: Before beginning employment and at regular times after that, the following are recommended: lung function tests. If symptoms develop or overexposure is suspected, the following may be useful: evaluation by a qualified allergist, including careful exposure history and special testing, may help diagnose skin allergy; and liver and kidney function tests. Consider chest X-ray following acute overexposure.

First Aid: If this chemical gets into the eyes, remove any contact lenses at once and irrigate immediately for at least 15 min, occasionally lifting upper and lower lids. Seek medical attention immediately. If this chemical contacts the skin, remove contaminated clothing and wash immediately with soap and water. Seek medical attention immediately. If this chemical has been inhaled, remove from exposure, begin rescue breathing (using universal precautions, including resuscitation mask) if breathing has stopped and CPR if heart action has stopped. Transfer promptly to a medical facility. When this chemical has been swallowed, get medical attention. Give large quantities of water and induce vomiting. Do not make an unconscious person vomit. Medical observation is recommended for 24−48 h after breathing overexposure, as pulmonary edema may be delayed. As first aid for pulmonary edema, a doctor or authorized paramedic may consider administering a corticosteroid spray.

Personal Protective Methods: Wear protective gloves and clothing to prevent any reasonable probability of skin contact. Safety equipment suppliers/manufacturers can provide recommendations on the most protective glove/clothing material for your operation. All protective clothing (suits, gloves, footwear, headgear) should be clean, available each day, and put on before work. Contact lenses should not be worn when working with this chemical. Wear splash-proof chemical goggles and face shield unless full face-piece respiratory protection is worn. Employees should wash immediately with soap when skin is wet or contaminated. Provide emergency showers and eyewash.

Respirator Selection: Where there is potential for exposures *over 2 ppm*, use a NIOSH/MSHA- or European Standard EN149-approved full-face-piece respirator with an organic vapor cartridge/canister. Increased protection is obtained from full-face-piece powered air-purifying respirators. *Where there is potential for high exposures*, use a NIOSH/MSHA- or European Standard EN149-approved supplied-air respirator with a full face-piece operated in the positive-pressure mode, or with a full-face-piece, hood, or helmet in the continuous-flow mode; or use a NIOSH/MSHA- or European Standard EN149-approved self-contained breathing apparatus with a full face-piece operated in a pressure-demand or other positive-pressure mode.

Storage: Color Code—Red: Flammability Hazard: Store in a flammable liquid storage area or approved cabinet away from ignition sources and corrosive and reactive materials. Prior to working with this chemical you should be trained on its proper handling and storage. Store in tightly closed containers in a cool well-ventilated area away from magnesium diperchlorate. Sources of ignition, such as smoking and open flames, are prohibited where trimethyl phosphite in handled, used, or stored. Metal containers involving the transfer of 5 gallons or more of trimethyl phosphite should be grounded and bonded. Drums must be equipped with self-closing valves, pressure vacuum bungs, and flame arresters.

Shipping: Trimethyl phosphite requires a shipping label of "FLAMMABLE LIQUID." It falls in Hazard Class 3 and Packing Group II.

Spill Handling: Evacuate and restrict persons not wearing protective equipment from area of spill or leak until cleanup is complete. Remove all ignition sources. Ventilate area of spill or leak. Absorb liquids in vermiculite, dry sand, earth, peat, carbon, or a similar material and deposit in sealed containers. Keep this chemical out of a confined space, such as a sewer, because of the possibility of an explosion, unless the sewer is designed to prevent the buildup of explosive concentrations. It may be necessary to contain and dispose of this chemical as a hazardous waste. If material or contaminated runoff enters waterways, notify downstream users of potentially contaminated waters. Contact your local or federal environmental protection agency for specific recommendations. If employees are required to clean up spills, they must be properly trained and equipped. OSHA 1910.120(q) may be applicable.

Fire Extinguishing: This chemical is a flammable liquid. Poisonous gases, including carbon monoxide, phosphine, and phosphorus oxides, are produced in fire. Use dry chemical, carbon dioxide, alcohol foam, or polymer foam extinguishers. Vapors are heavier than air and will collect in low areas. Vapors may travel long distances to ignition sources and flashback. Vapors in confined areas may explode when exposed to fire. Containers may explode in fire. Storage containers and parts of containers may rocket great distances, in many directions. If material or contaminated runoff enters waterways, notify downstream users of potentially contaminated waters. Notify local health and fire officials and pollution control agencies. From a secure, explosion-proof location, use water spray to cool exposed containers. If cooling streams are ineffective (venting sound increases in volume and pitch, tank discolors, or shows any signs of deforming), withdraw immediately to a secure position. If employees are expected to fight fires, they must be trained and equipped in OSHA 1910.156. The only respirators recommended for firefighting are self-contained breathing apparatuses that have full face-pieces and are operated in a pressure-demand or other positive-pressure mode.

References
New Jersey Department of Health and Senior Services. (October 1998). *Hazardous Substances Fact Sheet: Trimethyl Phosphite.* Trenton, NJ

Trinitrobenzene   T:0910

Molecular Formula: $C_6H_3N_3O_6$
Common Formula: $1,3,5-C_6H_3(NO_2)_3$
Synonyms: Benzene, 1,3,5-trinitro-; TNB; 1,3,5-Trinitrobenceno (Spanish); *sym*-Trinitrobenzene; *symmetrical*-Trinitrobenzene; Trinitrobenzene; Trinitrobenzene, dry; Trinitrobenzol (German)
CAS Registry Number: 99-35-4
RTECS® Number: DC3850000
UN/NA & ERG Number: UN0214 (dry or wetted with <30% water, by mass)/112; UN1354 (wetted with not <30% water, by mass)/113
EC Number: 202-752-7 [*Annex I Index No.:* 609-005-00-8]
Regulatory Authority and Advisory Bodies
Department of Homeland Security Screening Threshold Quantity (pounds): *Release hazard* 5000 (commercial grade); *Theft hazard* 400 (commercial grade).
Explosive Substance (World Bank).[15]
Reportable Quantity (RQ): 10 lb (4.54 kg).
US EPA Hazardous Waste Number (RCRA No.): U234.
RCRA, 40CFR261, Appendix 8 Hazardous Constituents.
RCRA 40CFR264, Appendix 9; TSD Facilities Ground Water Monitoring List. Suggested test method(s) (PQL µg/L): 8270 (10).
European/International Regulations: not listed in Annex 1.
WGK (German Aquatic Hazard Class): No value assigned.
Description: Trinitrobenzene is a yellow crystalline solid. Molecular weight = 213.12; Freezing/Melting point = $122-123°C$. Hazard Identification (based on NFPA-704 M Rating System): (*wetted or with <30% water*) Health 2, Flammability 4, Reactivity 4; (wetted with >30% water) Health 2, Flammability 4, Reactivity 2. Insoluble in water.
Potential Exposure: Trinitrobenzene is used as an explosive, and as a vulcanizing agent for natural rubber. Trinitrobenzene may be more powerful than TNT; and it is reported to be less sensitive to impact than TNT. However it is difficult to produce and is not used as widely as TNT.
Incompatibilities: Sensitive to shock and heat. Incompatible with initiating explosives, combustible materials, oxidizers.
Permissible Exposure Limits in Air
Protective Action Criteria (PAC)
TEEL-0: 3 mg/m^3
PAC-1: 7.5 mg/m^3
PAC-2: 60 mg/m^3
PAC-3: 125 mg/m^3
Permissible Concentration in Water: Russia[43] has set a MAC of 0.4 mg/L in water bodies used for domestic purposes.

Routes of Entry: Inhalation, ingestion, eye and/or skin contact.

Harmful Effects and Symptoms

Short Term Exposure: Irritates the eyes, skin, and respiratory tract.

Long Term Exposure: Information on the carcinogenicity, mutagenicity, teratogenicity, or adverse reproductive effects of trinitrobenzene was not found in the available literature. Trinitrobenzene has been reported to produce liver damage, central nervous system damage, and methemoglobin formation in animals. Breathing difficulties have also been reported.

Points of Attack: Liver, central nervous system, respiratory system.

Medical Surveillance: Lung function tests. Liver function tests. Examination of the nervous system.

First Aid: If this chemical gets into the eyes, remove any contact lenses at once and irrigate immediately for at least 15 min, occasionally lifting upper and lower lids. Seek medical attention immediately. If this chemical contacts the skin, remove contaminated clothing and wash immediately with soap and water. Seek medical attention immediately. If this chemical has been inhaled, remove from exposure, begin rescue breathing (using universal precautions, including resuscitation mask) if breathing has stopped and CPR if heart action has stopped. Transfer promptly to a medical facility. When this chemical has been swallowed, get medical attention. Give large quantities of water and induce vomiting. Do not make an unconscious person vomit.

Note to physician: Treat for methemoglobinemia. Spectrophotometry may be required for precise determination of levels of methemoglobin in urine.

Personal Protective Methods: Wear protective gloves and clothing to prevent any reasonable probability of skin contact. Safety equipment suppliers/manufacturers can provide recommendations on the most protective glove/clothing material for your operation. All protective clothing (suits, gloves, footwear, headgear) should be clean, available each day, and put on before work. Contact lenses should not be worn when working with this chemical. Wear dust-proof chemical goggles and face shield unless full face-piece respiratory protection is worn. Employees should wash immediately with soap when skin is wet or contaminated. Provide emergency showers and eyewash.

Respirator Selection: Follow regulations in OSHA 29CFR1910.134 or European Standard EN149. Use a NIOSH/MSHA- or European Standard EN149-approved respirator; or use an approved supplied-air respirator with a full face-piece operated in the positive-pressure mode, or with a full-face-piece, hood, or helmet in the continuous-flow mode; or use a NIOSH/MSHA- or European Standard EN149-approved self-contained breathing apparatus with a full face-piece operated in a pressure-demand or other positive-pressure mode.

Storage: Treat as an explosive. Color Code—Red Stripe: Flammability Hazard: Store separately from all other flammable materials. Prior to working with this chemical you should be trained on its proper handling and storage. Store in tightly closed containers in a cool, well-ventilated area away from heat, explosives, oxidizable materials. Sources of ignition, such as smoking and open flames, are prohibited where this chemical is used, handled, or stored in a manner that could create a potential fire or explosion hazard. Use only nonsparking tools and equipment, especially when opening and closing containers of this chemical.

Shipping: Trinitrobenzene, dry or wetted with <30% water, by mass, requires a shipping label of "EXPLOSIVE." It falls in Hazard Class 1.1D and Packing Group II. Trinitrobenzene, wetted with not <30% water, by mass, requires a shipping label of "FLAMMABLE SOLID." It falls in Hazard Class 4.1 and Packing Group I.

Spill Handling: Evacuate and restrict persons not wearing protective equipment from area of spill or leak until cleanup is complete. Remove all ignition sources. Ventilate area of spill or leak. Absorb liquids in vermiculite, dry sand, earth, peat, carbon, or a similar material and deposit in sealed containers. Keep this chemical out of a confined space, such as a sewer, because of the possibility of an explosion, unless the sewer is designed to prevent the buildup of explosive concentrations. It may be necessary to contain and dispose of this chemical as a hazardous waste. If material or contaminated runoff enters waterways, notify downstream users of potentially contaminated waters. Contact your local or federal environmental protection agency for specific recommendations. If employees are required to clean up spills, they must be properly trained and equipped. OSHA 1910.120(q) may be applicable.

Fire Extinguishing: This chemical is an explosive solid. If material is on fire and conditions permit, do not extinguish. Cool exposures using unattended monitors. If fire must be extinguished, use any agent appropriate for the burning material. Poisonous gases are produced in fire, including nitrogen oxides. If material or contaminated runoff enters waterways, notify downstream users of potentially contaminated waters. Notify local health and fire officials and pollution control agencies. Cool exposed containers from unattended equipment or remove intact containers if it can be done safely. If cooling streams are ineffective (venting sound increases in volume and pitch, tank discolors, or shows any signs of deforming), withdraw immediately to a secure position. If employees are expected to fight fires, they must be trained and equipped in OSHA 1910.156. The only respirators recommended for firefighting are self-contained breathing apparatuses that have full face-pieces and are operated in a pressure-demand or other positive-pressure mode.

Disposal Method Suggested: Consult with environmental regulatory agencies for guidance on acceptable disposal practices. Generators of waste containing this contaminant (≥100 kg/mo) must conform with EPA regulations governing storage, transportation, treatment, and waste disposal. Dissolve in a combustible solvent and spray into an incinerator equipped with afterburner and scrubber.

References

US Environmental Protection Agency. (April 30, 1980). *Trinitrobenzene, Health and Environmental Effects Profile No. 171.* Washington, DC: Office of Solid Waste

Trinitrotoluene   T:0920

Molecular Formula: $C_7H_5N_3O_6$

Synonyms: Entsufon; 1-Methyl-2,4,6-trinitrobenzene; NCI-C56155; α-TNT; TNT; TNT-tolite (French); Tolit; Tolite; Toluene, 2,4,6-trinitro,-(wet); *sym*-Trinitrotoluene; 2,4,6-Trinitrotoluene; Trinitrotoluene; Trinitrotoluene, wet; *s*-Trinitrotoluol; *sym*-Trinitrotoluol; 2,4,6-Trinitrotoluol (German); Trinitrotoluol; Tritol; Trotyl; Trotyl oil

CAS Registry Number: 118-96-7

RTECS® Number: XU0175000

UN/NA & ERG Number: UN1356 (wetted with not <30% water, by mass)/113; UN0209 (TNT, dry or wetted with <30% water, by mass)/112

EC Number: 204-289-6 [*Annex I Index No.:* 609-008-00-4]

Regulatory Authority and Advisory Bodies

Department of Homeland Security Screening Threshold Quantity (pounds): *Release hazard* 5000 (commercial grade); *Theft hazard* 400 (commercial grade).

Carcinogenicity: IARC: Animal, Inadequate Evidence; Human, Inadequate Evidence, *not classifiable as carcinogenic to humans*, Group 3, 1996.

US EPA Gene-Tox Program, Positive: Histidine reversion—Ames test.

Air Pollutant Standard Set. See below, "Permissible Exposure Limits in Air" section.

Canada, WHMIS, Ingredients Disclosure List Concentration 1.0%.

European/International Regulations: Hazard Symbol: E, T, N; R2; R23/24/25; R33; R51/53; Safety phrases: S1/2; S35; S45; S61 (see Appendix 4).

WGK (German Aquatic Hazard Class): No value assigned.

Description: TNT exists in five isomers; the most commonly used is 2,4,6-trinitrotoluene. It is a colorless to pale yellow odorless solid (pellets, cast blocks, and cast slabs) or crushed flakes. Molecular weight = 227.15; Specific gravity (H_2O:1) = 1.65 at 25°C; Boiling point = (explodes) 240°C; Freezing/Melting point = 80°C; Vapor pressure = 0.0002 mmHg at 25°C; Ionization potential = 10.59 eV. It explodes at 232°C (also reported at 240°C and 297°C) but burns at 295°C when not confined. TNT is a relatively stable high explosive. Practically insoluble in water; solubility = 0.01% at 25°C.

Potential Exposure: Compound Description: Agricultural Chemical; Tumorigen, Mutagen; Reproductive Effector; Human Data; Primary Irritant. TNT is used as an explosive, i.e., as a bursting charge in military explosive shells, bombs, grenades, and mines; and an intermediate in dyestuffs and photographic chemicals.

Incompatibilities: Strong oxidizers, ammonia, strong alkalis, combustible materials, heat. Violent reaction with reducing agents. Rapid heating will result in detonation. Explodes when heated to 232°C. May explosively decompose from shock, friction, or concussion. Reacts with heavy metals.

Permissible Exposure Limits in Air

OSHA PEL: 1.5 mg/m³ TWA [skin].

NIOSH REL: 0.5 mg/m³ TWA [skin].

ACGIH TLV®[1]: 0.1 mg/m³ TWA [skin].

DFG MAK: 0.011 ppm/0.1 mg/m³ TWA; Peak Limitation Category II(2) [skin]; Carcinogen Category 3B; Pregnancy Risk Group D.

NIOSH IDLH: 500 mg/m³.

Arab Republic of Egypt: TWA 0.5 mg/m³, 1993; Australia: TWA 0.5 mg/m³, [skin], 1993; Austria: MAK 0.01 ppm (0.1 mg/m³), [skin], Suspected: carcinogen, 1999; Belgium: TWA 0.5 mg/m³, [skin], 1993; Denmark: TWA 0.1 mg/m³, [skin], 1999; Finland: TWA 0.5 mg/m³, STEL 3 mg/m³, [skin], 1999; France: VME 0.5 mg/m³, [skin], 1999; the Netherlands: MAC-TGG 0.1 mg/m³, [skin], 2003; Norway: TWA 0.1 mg/m³, 1999; the Philippines: TWA 1.5 mg/m³, [skin], 1993; Poland: MAC (TWA) 1 mg/m³; MAC (STEL) 3 mg/m³, 1999; Russia: TWA 0.1 mg/m³, STEL 0.5 mg/m³, [skin], 1993; Sweden: NGV 0.1 mg/m³, KTV 0.2 mg/m³, [skin], 1999; Switzerland: MAK-W 0.1 mg/m³, [skin], 1999; Turkey: TWA 1.5 mg/m³, [skin], 1993; United Kingdom: TWA 0.3 mg/m³, [skin]; TWA 0.2 ppm (0.89 mg/m³), [skin], carcinogen, 2000; Argentina, Bulgaria, Columbia, Jordan, South Korea, New Zealand, Singapore, Vietnam: ACGIH TLV®: TWA 0.1 mg/m³ [skin]. Several states have set guidelines or standards for TNT in ambient air[60] ranging from 5.0 μg/m³ (North Dakota) to 8.0 μg/m³ (Virginia) to 10.0 μg/m³ (Connecticut) to 12.0 μg/m³ (Nevada).

Determination in Air: Use OSHA Analytical Method 44.

Determination in Water: Octanol—water coefficient: Log K_{ow} = 1.60

Permissible Concentration in Water: Russia[43] set a MAC of 0.5 mg/L in water bodies used for domestic purposes.

Routes of Entry: Inhalation of dust, fume, or vapor; ingestion of dust; percutaneous absorption from dust; skin and/or eye contact.

Harmful Effects and Symptoms

Short Term Exposure: Exposure to trinitrotoluene may cause irritation of the eyes, nose, and throat with sneezing, cough, and sore throat. It may cause skin irritation, dermatitis, and may give the skin, hair, and nails a yellowish color. Numerous fatalities have occurred in workers exposed to TNT from toxic hepatitis or aplastic anemia. TNT exposure may also cause methemoglobinemia with cyanosis, weakness, drowsiness, dyspnea, and unconsciousness. Ingestion may cause hallucinations or distorted perceptions, cyanosis (blue color to the skin, lips, and fingertips), and gastrointestinal changes.

Long Term Exposure: May affect the liver, causing hepatitis and jaundice. May affect the blood, causing hemolysis, formation of methemoglobin. Cyanosis may also occur. May affect vision, causing cataracts. In addition, it may cause muscular pains, heart irregularities, renal irritation, menstrual irregularities, and peripheral neuritis.

Points of Attack: Blood, liver, eyes, cardiovascular system, central nervous system, kidneys, skin.

Medical Surveillance: NIOSH lists the following tests: complete blood count; aplastic anemia; liver function tests; urine (chemical/metabolite); urinalysis (routine). Placement or periodic examinations should give special considerations to history of allergic reactions, blood dyscrasias, reactions to medications, and alcohol intake. The skin, eyes, blood, and liver and kidney function should be followed. Urine may be examined for TNT using the Webster test or for the urinary metabolite 2,6-dinitro-4-aminotoluene; however, both may be negative if there is liver injury.

First Aid: If this chemical gets into the eyes, remove any contact lenses at once and irrigate immediately for at least 15 min, occasionally lifting upper and lower lids. Seek medical attention immediately. If this chemical contacts the skin, remove contaminated clothing and wash immediately with soap and water. Seek medical attention immediately. If this chemical has been inhaled, remove from exposure, begin rescue breathing (using universal precautions, including resuscitation mask) if breathing has stopped and CPR if heart action has stopped. Transfer promptly to a medical facility. When this chemical has been swallowed, get medical attention. Give large quantities of water and induce vomiting. Do not make an unconscious person vomit.

Note to physician: Treat for methemoglobinemia. Spectrophotometry may be required for precise determination of levels of methemoglobin in urine.

Personal Protective Methods: Wear protective gloves and clothing to prevent any reasonable probability of skin contact. Safety equipment suppliers/manufacturers can provide recommendations on the most protective glove/clothing material for your operation. Polyvinyl chloride is among the recommended protective materials. All protective clothing (suits, gloves, footwear, headgear) should be clean, available each day, and put on before work. Contact lenses should not be worn when working with this chemical. Wear splash-proof chemical goggles and face shield unless full face-piece respiratory protection is worn. Employees should wash immediately with soap when skin is wet or contaminated. Provide emergency showers and eyewash. The Webster skin test (colorimetric tests with alcoholic sodium hydroxide) or indicator soap should be used to make sure workers have washed all TNT off their skins.

Respirator Selection: NIOSH: *Up to 5 mg/m³:* Sa* (APF = 10) (any supplied-air respirator). *Up to 12.5 mg/m³:* Sa:Cf* (APF = 25) (any supplied-air respirator operated in a continuous-flow mode). *Up to 25 mg/m³:* SCBAF (APF = 50) (any self-contained breathing apparatus with a full-face-piece; or SaF (APF = 50) (any supplied-air respirator with a full face-piece). *Up to 500 mg/m³:* SaF: Pd,Pp (APF = 2000) (any supplied-air respirator that has a

full face-piece and is operated in a pressure-demand or other positive-pressure mode). *Emergency or planned entry into unknown concentrations or IDLH conditions:* SCBAF: Pd,Pp (APF = 10,000) (any self-contained breathing apparatus that has a full face-piece and is operated in a pressure-demand or other positive-pressure mode) or SaF: Pd,Pp: ASCBA (APF = 10,000) (any supplied-air respirator that has a full face-piece and is operated in a pressure-demand or other positive-pressure mode in combination with an auxiliary, self-contained breathing apparatus operated in a pressure-demand or other positive-pressure mode). *Escape:* GmFOv100 (APF = 50) [any air-purifying, full-face-piece respirator (gas mask) with a chin-style, front- or back-mounted organic vapor canister having an N100, R100, or P100 filter] or SCBAE (any appropriate escape-type, self-contained breathing apparatus).

*Substance reported to cause eye irritation or damage; may require eye protection.

Storage: TNT is an explosive. Color Code—Red Stripe: Flammability Hazard: Store separately from all other flammable materials. Prior to working with this chemical you should be trained on its proper handling and storage. Store in an explosion-proof refrigerator and keep away from reducing agents.[52] Keep material wet with water or treat as an explosive. Keep away from heat, sources of ignition, metal, nitric acid, and reducing materials. Protect containers from shock. Use only nonsparking tools and equipment, especially when opening and closing containers of this chemical. A regulated, marked area should be established where this chemical is handled, used, or stored in compliance with OSHA Standard 1910.1045.

Shipping: Trinitrotoluene, wetted with not <30% water, by mass, requires a shipping label of "FLAMMABLE SOLID." It falls in Hazard Class 4.1 and Packing Group I. Trinitrotoluene or TNT, dry or wetted with <30% water, by mass, requires a shipping label of "EXPLOSIVE." It falls in Hazard Class 1.1D and Packing Group II.

Spill Handling: Evacuate persons not wearing protective equipment from area of spill or leak until cleanup is complete. Remove all ignition sources. Dampen spilled material with alcohol to avoid dust. Collect waste material in the most convenient and safe manner and deposit in sealed containers. Ventilate area after cleanup is complete. It may be necessary to contain and dispose of this chemical as a hazardous waste. If material or contaminated runoff enters waterways, notify downstream users of potentially contaminated waters. Contact your local or federal environmental protection agency for specific recommendations. If employees are required to clean up spills, they must be properly trained and equipped. OSHA 1910.120(q) may be applicable.

Fire Extinguishing: This chemical is a dangerously explosive solid. If material is on fire and conditions permit, do not extinguish. Evacuate area and let burn. Cool exposures using unattended monitors. If fire must be extinguished, use any agent appropriate for the burning material. Poisonous gases are produced in fire, including nitrogen oxides. If material or contaminated runoff enters waterways, notify downstream users of potentially contaminated waters. Notify local health and fire officials and pollution control agencies. Cool exposed containers from unattended equipment or remove intact containers if it can be done safely. If cooling streams are ineffective (venting sound increases in volume and pitch, tank discolors, or shows any signs of deforming), withdraw immediately to a secure position. If employees are expected to fight fires, they must be trained and equipped in OSHA 1910.156. The only respirators recommended for firefighting are self-contained breathing apparatuses that have full face-pieces and are operated in a pressure-demand or other positive-pressure mode.

Disposal Method Suggested: TNT is dissolved in acetone and incinerated. The incinerator should be equipped with an afterburner and a caustic soda solution scrubber.

References

US Environmental Protection Agency. (1979). *Chemical Hazard Information Profile: 2,4,6-Trinitrotoluene.* Washington, DC

Sax, N. I. (Ed.). *Dangerous Properties of Industrial Materials Report,* 2, No. 5, 93–96 (1982) and 8, No. 4, 75–80 (1988)

Triphenylamine   T:0930

Molecular Formula: $C_{18}H_{15}N$
Common Formula: $(C_6H_5)_3N$
Synonyms: N,N-Diphenylaniline; N,N-Diphenylbenzenamine
CAS Registry Number: 603-34-9
RTECS® Number: YK2680000
EC Number: 210-035-5
Regulatory Authority and Advisory Bodies
Air Pollutant Standard Set. See below, "Permissible Exposure Limits in Air" section.
Canada, WHMIS, Ingredients Disclosure List Concentration 1.0%.
European/International Regulations: not listed in Annex 1.
WGK (German Aquatic Hazard Class): No value assigned.
Description: Triphenylamine is a colorless crystalline solid. Molecular weight = 245.34; Specific gravity (H_2O:1) = 0.77 at 25°C; Boiling point = 365°C; also reported at 195°C at 10 mmHg; Freezing/Melting point = 127°C. Hazard Identification (based on NFPA-704 M Rating System): Health 2, Flammability 0, Reactivity 0. Insoluble in water.
Potential Exposure: Triphenylamine is used as a primary photoconductor and in making photographic film coated on photographic film bases.
Incompatibilities: Aldehydes, ketones, nitrates, oxidizers, oxygen, and peroxides.
Permissible Exposure Limits in Air
OSHA PEL: None.
NIOSH REL: 5 mg/m³ TWA.

ACGIH TLV®[1]: Withdrawn due to insufficient data.
Australia: TWA 5 mg/m³, 1993; Belgium: TWA 5 mg/m³, 1993; Denmark: TWA 5 mg/m³, 1999; Finland: TWA 5 mg/m³, STEL 10 mg/m³, [skin], 1999; France: VME 5 mg/m³, 1999; Norway: TWA 5 mg/m³, 1999; IN Argentina, Bulgaria, Columbia, Jordan, South Korea, New Zealand, Singapore, Vietnam: ACGIH TLV®: TWA 5 mg/m³. Several states have set guidelines or standards for triphenylamine in ambient air[60] ranging from 8.0 μg/m³ (Virginia) to 50.0 μg/m³ (North Dakota) to 100.0 μg/m³ (Connecticut).
Determination in Air: No method available.
Routes of Entry: Inhalation, ingestion, skin and/or eye contact.
Harmful Effects and Symptoms
Short Term Exposure: Triphenylamine can affect you when breathed in and by passing through your skin. Contact can irritate the skin.
Long Term Exposure: No known long-term effects are known although some related aromatic amines can cause skin and lung allergies and have been shown to be carcinogenic to the human bladder, ureter, prostate, intestines, lung, and liver.
Points of Attack: Skin.
First Aid: *Skin Contact*[52]: Flood all areas of body that have contacted the substance with water. Do not wait to remove contaminated clothing; do it under the water stream. Use soap to help assure removal. Isolate contaminated clothing when removed to prevent contact by others. *Eye Contact:* Remove any contact lenses at once. Immediately flush eyes well with copious quantities of water or normal saline for at least 20–30 min. Seek medical attention. *Inhalation:* Leave contaminated area immediately; breathe fresh air. Proper respiratory protection must be supplied to any rescuers. If coughing, difficult breathing, or any other symptoms develop, seek medical attention at once, even if symptoms develop many hours after exposure. *Ingestion:* Contact a physician, hospital, or poison center at once. If the victim is unconscious or convulsing, do not induce vomiting or give anything by mouth. Assure that the patient's airway is open and lay him on his side with his head lower than his body and transport immediately to a medical facility. If conscious and not convulsing, give a glass of water to dilute the substance. Vomiting should not be induced without a physician's advice.
Personal Protective Methods: Wear protective gloves and clothing to prevent any reasonable probability of skin contact. Safety equipment suppliers/manufacturers can provide recommendations on the most protective glove/clothing material for your operation. All protective clothing (suits, gloves, footwear, headgear) should be clean, available each day, and put on before work. Contact lenses should not be worn when working with this chemical. Wear dust-proof chemical goggles and face shield unless full face-piece respiratory protection is worn. Employees should wash immediately with soap when skin is wet or contaminated. Provide emergency showers and eyewash.

Respirator Selection: Where there is potential for exposures *over 5 mg/m³*, use a NIOSH/MSHA- or European Standard EN149-approved full-face-piece respirator equipped with particulate (dust/fume/mist) filters. Particulate filters must be checked every day before work for physical damage, such as rips or tears, and replaced as needed. *Where there is potential for high exposures*, use a NIOSH/MSHA- or European Standard EN149-approved supplied-air respirator with a full face-piece operated in the positive-pressure mode, or with a full-face-piece, hood, or helmet in the continuous-flow mode; or use a NIOSH/MSHA- or European Standard EN149-approved self-contained breathing apparatus with a full face-piece operated in a pressure-demand or other positive-pressure mode.

Storage: Color Code—Green: General storage may be used. Prior to working with this chemical you should be trained on its proper handling and storage. Store in tightly closed containers in a cool well-ventilated area away from aldehydes, ketones, nitrates, oxidizers, oxygen, and peroxides. Where possible, automatically transfer material from storage containers to process containers.

Spill Handling: Evacuate persons not wearing protective equipment from area of spill or leak until cleanup is complete. Remove all ignition sources. The powdered material may be dampened with 60−70% acetone to avoid airborne dust. Collect powdered material in the most convenient and safe manner and deposit in sealed containers. Ventilate area after cleanup is complete. It may be necessary to contain and dispose of this chemical as a hazardous waste. If material or contaminated runoff enters waterways, notify downstream users of potentially contaminated waters. Contact your local or federal environmental protection agency for specific recommendations. If employees are required to clean up spills, they must be properly trained and equipped. OSHA 1910.120(q) may be applicable.

Fire Extinguishing: Use dry chemical, carbon dioxide, water spray, or foam extinguishers. Poisonous gases are produced in fire, including nitrogen oxides. If material or contaminated runoff enters waterways, notify downstream users of potentially contaminated waters. Notify local health and fire officials and pollution control agencies. From a secure, explosion-proof location, use water spray to cool exposed containers. If cooling streams are ineffective (venting sound increases in volume and pitch, tank discolors, or shows any signs of deforming), withdraw immediately to a secure position. If employees are expected to fight fires, they must be trained and equipped in OSHA 1910.156. The only respirators recommended for firefighting are self-contained breathing apparatuses that have full face-pieces and are operated in a pressure-demand or other positive-pressure mode.

References

New Jersey Department of Health and Senior Services. (November 2000). *Hazardous Substances Fact Sheet: Triphenylamine.* Trenton, NJ

Triphenyl phosphate   T:0940

Molecular Formula: $C_{18}H_{15}O_4P$
Common Formula: $(C_6H_5O)_3PO$
Synonyms: Celluflex TPP; Disflamoll-TP; Fosfato de trifenilo (Spanish); Phenyl phosphate; Phiosflex-TPP; Phosphoric acid, triphenyl ester; TP; TPP; Triphenoxyphosphine oxide; Triphenyl ester of phosphoric acid
CAS Registry Number: 115-86-6
RTECS® Number: TC8400000
UN/NA & ERG Number: UN3077/171
EC Number: 204-112-2
Regulatory Authority and Advisory Bodies
US EPA TSCA Section 8(e) Risk Notification, 8EHQ-0892-9169; 8EHQ-0892-8839; 8EHQ-0892-9290.
Air Pollutant Standard Set. See below, "Permissible Exposure Limits in Air" section.
Canada, WHMIS, Ingredients Disclosure List Concentration 1.0%.
European/International Regulations: not listed in Annex 1.
WGK (German Aquatic Hazard Class): 2—Hazard to waters.

Description: Triphenyl phosphate is a colorless crystalline powder with a faint, phenol-like odor. Molecular weight = 326.30; Specific gravity (H_2O:1) = 1.29 at 25°C; Boiling point = 413.3°C; Freezing/Melting point = 49°C; Vapor pressure = 1 mmHg at 193°C; Flash point = 220°C (cc). Hazard Identification (based on NFPA-704 M Rating System): Health 2, Flammability 1, Reactivity 0. Practically insoluble in water; solubility 0.002% at 54°C.

Potential Exposure: Triphenyl phosphate is used to impregnate roofing paper and as a fire-resistant plasticizer in plastics; for cellulose esters in lacquers and varnishes. Used in making adhesives, gasoline additives, flotation agents, insecticides, surfactants, antioxidants, and stabilizers. A substitute for camphor.

Incompatibilities: Contact with strong oxidizers, strong acids, and nitrates may cause fire or explosions. Phosphates are incompatible with antimony pentachloride, magnesium, silver nitrate, zinc acetate.

Permissible Exposure Limits in Air
OSHA PEL: 3 mg/m³ TWA.
NIOSH REL: 3 mg/m³ TWA.
ACGIH TLV®[1]: 3 mg/m³ TWA, not classifiable as a human carcinogen.
NIOSH IDLH: 1000 mg/m³.
Australia: TWA 3 mg/m³, 1993; Belgium: TWA 3 mg/m³, 1993; Denmark: TWA 3 mg/m³, 1999; Finland: TWA 3 mg/m³, STEL 6 mg/m³, [skin], 1999; France: VME 3 mg/m³, 1999; Norway: TWA 3 mg/m³, 1999; the Netherlands: MAC-TGG 3 mg/m³, 2003; Switzerland: MAK-W 3 mg/m³, 1999; United Kingdom: LTEL 3 mg/m³, STEL 6 mg/m³, 1993; Argentina, Bulgaria, Columbia, Jordan, South Korea, New Zealand,

Singapore, Vietnam: ACGIH TLV®: not classifiable as a human carcinogen. Several states have set guidelines or standards for triphenyl phosphate in ambient air[60] ranging from 1.6 μg/m³ (Virginia) to 30.0 μg/m³ (North Dakota) to 60.0 μg/m³ (Connecticut) to 71.0 μg/m³ (Nevada).

Determination in Air: Use NIOSH Analytical Method (IV) #5038, Triphenyl phosphate.

Determination in Water: Octanol−water coefficient: Log K_{ow} = 4.59.

Routes of Entry: Inhalation, ingestion, skin and/or eye contact. Slowly absorbed by the skin.

Harmful Effects and Symptoms

Short Term Exposure: Slowly absorbed by the skin. May cause eye and skin irritation.

Long Term Exposure: Minor changes in blood enzymes. In animals: muscular weakness, paralysis. May be a cholinesterase inhibitor, but not a potent one.

Points of Attack: Blood, peripheral nervous system.

Medical Surveillance: Consider the blood in preplacement and periodic physical examinations. Examination of the peripheral nervous system.

First Aid: If this chemical gets into the eyes, remove any contact lenses at once and irrigate immediately for at least 15 min, occasionally lifting upper and lower lids. Seek medical attention immediately. If this chemical contacts the skin, remove contaminated clothing and wash immediately with soap and water. Seek medical attention immediately. If this chemical has been inhaled, remove from exposure, begin rescue breathing (using universal precautions, including resuscitation mask) if breathing has stopped and CPR if heart action has stopped. Transfer promptly to a medical facility. When this chemical has been swallowed, rinse mouth and get medical attention.

Personal Protective Methods: Safety equipment suppliers/manufacturers can provide recommendations on the most protective glove/clothing material for your operation. All protective clothing (suits, gloves, footwear, headgear) should be clean, available each day, and put on before work. Contact lenses should not be worn when working with this chemical. Wear dust-proof chemical goggles and face shield unless full face-piece respiratory protection is worn. Employees should wash immediately with soap when skin is wet or contaminated. Provide emergency showers and eyewash.

Respirator Selection: *Up to 15 mg/m³:* Qm (APF = 25) (any quarter-mask respirator). *Up to 30 mg/m³:* 95XQ (APF = 10) [any particulate respirator equipped with an N95, R95, or P95 filter (including N95, R95, and P95 filtering face-pieces) except quarter-mask respirators. The following filters may also be used: N99, R99, P99, N100, R100, P100] or Sa (APF = 10) (any supplied-air respirator). *Up to 75 mg/m³:* Sa:Cf (APF = 25) (any supplied-air respirator operated in a continuous-flow mode) or PaprHie (APF = 25) (any powered, air-purifying respirator with a high-efficiency particulate filter). *Up to 150 mg/m³:* 100F (APF = 50) (any air-purifying, full-face-piece respirator

with an N100, R100, or P100 filter) or SaT: Cf (APF = 50) (any supplied-air respirator that has a tight-fitting face-piece and is operated in a continuous-flow mode) or PaprTHie (APF = 50) (any powered, air-purifying respirator with a tight-fitting face-piece and a high-efficiency particulate filter) or SCBAF (APF = 50) (any self-contained breathing apparatus with a full-face-piece) or SaF (APF = 50) (any supplied-air respirator with a full-face-piece). *Up to 1000 mg/m³:* SA: PD,PP (any supplied-air respirator operated in a pressure-demand or other positive-pressure mode). *Emergency or planned entry into unknown concentrations or IDLH conditions:* SCBAF: Pd,Pp (APF = 10,000) (any NIOSH/MSHA- or European Standard EN 149-approved self-contained breathing apparatus that has a full face-piece and is operated in a pressure-demand or other positive-pressure mode) or SaF: Pd,Pp: ASCBA (APF = 10,000) (any supplied-air respirator that has a full-face-piece and is operated in a pressure-demand or other positive-pressure mode in combination with an auxiliary, self-contained breathing apparatus operated in a pressure-demand or other positive-pressure mode). *Escape:* 100F (APF = 50) (any air-purifying, full-face-piece respirator with an N100, R100, or P100 filter) or SCBAE (any appropriate escape-type, self-contained breathing apparatus).

Storage: Color Code—Blue: Health Hazard/Poison: Store in a secure poison location. Prior to working with this chemical you should be trained on its proper handling and storage. Store in a cool, dry place away from antimony pentachloride, magnesium, silver nitrate, zinc acetate.

Shipping: The name of this material is not on the DOT list of materials[19] for label and packaging standards. However, based on regulations, it may be classified[52] as Environmentally hazardous substances, solid, n.o.s. This chemical requires a shipping label of "CLASS 9." It falls in Hazard Class 9 and Packing Group III.[20, 21]

Spill Handling: Evacuate persons not wearing protective equipment from area of spill or leak until cleanup is complete. Remove all ignition sources. Dampen spilled material with alcohol to avoid dust. Collect powdered material in the most convenient and safe manner and deposit in sealed containers. Ventilate area after cleanup is complete. It may be necessary to contain and dispose of this chemical as a hazardous waste. If material or contaminated runoff enters waterways, notify downstream users of potentially contaminated waters. Contact your local or federal environmental protection agency for specific recommendations. If employees are required to clean up spills, they must be properly trained and equipped. OSHA 1910.120(q) may be applicable.

Fire Extinguishing: This chemical is a combustible solid. Use dry chemical, carbon dioxide, water spray, or alcohol foam extinguishers. Poisonous gases are produced in fire, including oxides of phosphorus. If material or contaminated runoff enters waterways, notify downstream users of potentially contaminated waters. Notify local health and fire officials and pollution control agencies. From a secure, explosion-proof location, use water spray to cool exposed containers. If cooling streams are ineffective (venting sound increases in volume and pitch, tank discolors, or shows any signs of deforming), withdraw immediately to a secure position. If employees are expected to fight fires, they must be trained and equipped in OSHA 1910.156. The only respirators recommended for firefighting are self-contained breathing apparatuses that have full face-pieces and are operated in a pressure-demand or other positive-pressure mode.

Disposal Method Suggested: Incinerate in furnace equipped with alkaline scrubber.

References

Sax, N. I. (Ed.). (1986). *Dangerous Properties of Industrial Materials Report*, 6, No. 4, 91−100

New Jersey Department of Health and Senior Services. (May 2001). *Hazardous Substances Fact Sheet: Triphenyl Phosphate*. Trenton, NJ

Triphenyltin compounds   T:0950

Molecular Formula: $C_{20}H_{18}O_2Sn$; $C_{18}H_{15}ClSn$; $C_{18}H_{16}OSn$

Common Formula: $(C_6H_5)_3SnOOCCH_3$; $(C_6H_5)_3SnCl$; $(C_6H_5)_3SnOH$

Synonyms: acetate: Acetate de triphenyl-etain (French); Acetotriphenylstannine; Acetoxy-triphenyl-stannan (German); Acetoxytriphenylstannane; Acetoxytriphenyltin; (Acetyloxy)triphenyl-stannane; Batasan; Brestan; ENT 25,208; Fenolovo acetate; Fentin acetat (German); Fentin acetate; Fentine acetate (French); GC 6936; HOE-2824; Liromatin; Lirostanol; Phentin acetate; Phentinoacetate; Stannane, acetoxytriphenyl-; Suzi; Tinestan; Tinestan 60 WP; Tin triphenyl acetate; TPTA; TPZA; Triphenylacetostannane; Triphenyltin acetate; Triphenyl-zinnacetat (German); Tubotin

Chloride: AI3-25207; Aquatin 20 EC; Brestanol; Chlorotriphenylstannane; Chlorotriphenyltin; Fentin chloride; GC 8993; General chemicals 8993; HOE 2872; LS 4442; NSC 43675; Phenostat-C; Stannane, chlorotriphenyl-; Tinmate; TPTC; Triphenylchlorostannane; Triphenyl-chlorotin; Triphenyltin chloride (EPA)

Hydroxide: AI3-28009; Brestan H 47.5 WP fungicide; Dowco 186; Duter; Du-Ter; Duter extra; Du-Ter fungicide; Du-Ter fungicide wettable powder; Du-Ter PB-47 fungicide; Du-Ter W-50; Du-Tur flowable-30; ENT 28,009; Fentin; Fentin hydroxide; Fintine hydroxyde (French); Fintin hydroxid (German); Flo tin 4L; Haitin; Haitin WP 20 (fentin hydroxide 20%); Haitin WP 60 (fentin hydroxide 60%); Hydroxyde de triphenyl-etain (French); Hydroxytriphenylstannane; Hydroxytriphenyltin; IDA, IMC Flo-tin 4L; K19; NCI-C00260; NSC 113243; Phenostat-H; Stannane, hydroxytriphenyl-; Stannol, triphenyl-; Super tin; Super tin 4l gardian flowable fungicide; Suzu H; Tin, hydroxytriphenyl-; TN IV; TPTH; TPTH technical;

TPTOH; Triphenylstannanol; Triphenylstannium hydroxide; Triphenyltin(IV) hydroxide; Triphenyltin hydroxide (EPA); Triphenyltin hydroxide organotin fungicide; Triphenyltin oxide; Triphenyl-zinnhydroxid (German); Triple-tin; Triple tin 4l; Tubotin; Vancide KS; Vito spot fungicide; Wesley technical triphenyltin hydroxide

CAS Registry Number: 900-95-8 (acetate); 639-58-7 (chloride); 76-87-9 (hydroxide); 752-74-9 [Tris(triphenylstannyl) isocyanurate]

RTECS® Number: WH6650000 (acetate); WH6860000 (chloride); WH8575000 (hydroxide)

UN/NA & ERG Number: UN2786 (organotin pesticides, solid, toxic)/153; UN3020 (organotin pesticides, liquid, toxic)/153

EC Number: 212-984-0 [050-003-00-6] (fentin acetate); 211-358-4 (fentin chloride); 200-990-6 [050-004-00-1] (fentin hydroxide)

Regulatory Authority and Advisory Bodies

Air Pollutant Standard Set. See below, "Permissible Exposure Limits in Air" section.

Acetate:

US DOT Regulated Marine Pollutant (49CFR172.101, Appendix B) as triphenyltin compounds.

European/International Regulations (*900-95-8*): Hazard Symbol: T + , N; Risk phrases: R24/25; R26; R36/38; R43; R50/53; Safety phrases: S1/2; S36/37; S45; S60; S61.

Chloride:

Superfund/EPCRA 40CFR355, Extremely Hazardous Substances: TPQ = 500/10,000 lb (227/4540 kg) (chloride). Reportable Quantity (RQ): 1 lb (0.454 kg) (chloride).

EPCRA Section 313 Form R *de minimis* concentration reporting level: 1.0%.

US DOT Regulated Marine Pollutant (49CFR172.101, Appendix B) as triphenyltin compounds.

European/International Regulations: not listed in Annex I.

Hydroxide:

Carcinogenicity: NCI: Carcinogenesis Bioassay (feed); no evidence: mouse, rat.

US EPA Gene-Tox Program, Negative: Carcinogenicity—mouse/rat.

US EPA, FIFRA, 1998 Status of Pesticides: Supported.

EPCRA Section 313 Form R *de minimis* concentration reporting level: 1.0%.

US DOT Regulated Marine Pollutant (49CFR172.101, Appendix B) as triphenyltin compounds.

California Proposition 65 Chemical: (*Triphenyltin hydroxide*) Cancer 1/1/92; Developmental/Reproductive toxin 3/18/02.

Canada, WHMIS, Ingredients Disclosure List Concentration 0.1% (acetate and hydroxide); 1% (chloride).

European/International Regulations (*76-87-9*): Hazard Symbol: T+, N; Risk phrases: R24/25; R26; R37/38; R40; R41; R48/23; R63; R50/53; Safety phrases: S1/2; S26; S28; S36/37/39; S45; S60; 61 (see Appendix 4).

WGK (German Aquatic Hazard Class): 3—Highly water polluting (*acetate, chloride, hydroxide*).

Description: Triphenyltin acetate is a white solid. Molecular weight = 409.07; Freezing/Melting point = 122°C. Practically insoluble in water. Triphenyltin chloride is a colorless to yellow crystalline solid with a characteristic odor. Molecular weight = 385.47; Boiling point = 240°C at 13.5 mmHg; Freezing/Melting point = 106°C. Hazard Identification (based on NFPA-704 M Rating System): Health 3, Flammability 2, Reactivity 0. Insoluble in water. *Triphenyltin hydroxide* is a white crystalline solid. Molecular weight = 367.03; Freezing/Melting point = 118−120°C (decomposes). Practically insoluble in water.

Potential Exposure: Compound Description (hydroxide): Agricultural Chemical; Tumorigen, Mutagen, Organometallic; Reproductive Effector; Human Data; Primary Irritant. The hydroxide is used in vinyl products to protect against mildew growth and stiffening by bacteria and fungi. At risk are those engaged in the manufacture, formulation, and application of insecticides used for fungus, algae, and mollusk control; as a chemosterilant.

Incompatibilities: Triphenyltin chloride: violent reaction with strong oxidizers. Keep away from moisture.

Permissible Exposure Limits in Air

OSHA PEL: 0.1 mg[Sn]/m³ TWA.

NIOSH REL: 0.1 mg[Sn]/m³ TWA [skin].

ACGIH TLV®[1]: 0.1 mg[Sn]/m³ TWA; 0.2 mg[Sn]/m³ STEL [skin].

NIOSH IDLH: 25 mg[Sn]/m³.

Protective Action Criteria (PAC) (chloride)

TEEL-0: 0.325 mg/m³

PAC-1: 0.65 mg/m³

PAC-2: 20 mg/m³

PAC-3: 81.2 mg/m³

DFG MAK: 0.1 mg[Sn]/m³, inhalable fraction Peak Limitation Category II(2); [skin]; Pregnancy Risk Group D.

Hydroxide

Australia: TWA 0.1 mg[Sn]/m³, [skin], 1993; Austria: MAK 0.1 mg[Sn]/m³, [skin], 1999; Belgium: TWA 0.1 mg [Sn]/m³, [skin], 1993; Denmark: TWA 0.1 mg[Sn]/m³, [skin], 1999; Finland: TWA 0.1 mg[Sn]/m³, STEL 0.3 mg [Sn]/m³, 1999; France: VME 0.1 mg[Sn]/m³, VLE 0.2 mg [Sn]/m³, 1999; Hungary: STEL 0.1 mg[Sn]/m³, [skin], 1993; Norway: TWA 0.1 mg[Sn]/m³, 1999; the Philippines: TWA 0.1 mg[Sn]/m³, 1993; Switzerland: MAK-W 0.1 mg [Sn]/m³, KZG-W 0.2 mg[Sn]/m³, [skin], 1999; Thailand: TWA 0.1 mg[Sn]/m³, 1993; United Kingdom: TWA 0.1 mg [Sn]/m³, STEL 0.2 mg[Sn]/m³, [skin], 2000; Argentina, Bulgaria, Columbia, Jordan, South Korea, New Zealand, Singapore, Vietnam: ACGIH TLV®: Not classifiable as a human carcinogen.

Determination in Air: Use NIOSH Analytical Method (IV) #5504, Organotin compounds.

Routes of Entry: Inhalation, skin absorption, ingestion, skin and/or eye contact.

Harmful Effects and Symptoms

Short Term Exposure: These chemicals are strong poisons. Toxic and irritating to the eyes, skin, and respiratory system. Dermal exposure may lead to severe skin burns as well as renal failure, and possible death in the case of the chloride. Symptom of exposure include headache, vertigo (an illusion of movement), psycho-neurologic disturbance, sore throat, cough, abdominal pain, vomiting, urine retention, paresis, focal anesthesia.

Long Term Exposure: Exposure may affect the nervous system causing headache, nausea, vomiting, dizziness, decreased coordination, muscle weakness, and visual changes. Triphenyltin chloride can irritate the lungs; bronchitis may develop. In animals: hemolysis; hepatic necrosis; kidney damage.

Points of Attack: Kidneys, liver.

Medical Surveillance: Kidney and liver function tests. Evaluation of the nervous system. Lung function tests.

First Aid: For triphenyltin hydroxide: Skin Contact[52]: Flood all areas of body that have contacted the substance with water. Do not wait to remove contaminated clothing; do it under the water stream. Use soap to help assure removal. Speed in removing material from skin is of extreme importance. Shampoo hair promptly if contaminated. Isolate contaminated clothing when removed to prevent contact by others.

Eye Contact: Remove any contact lenses at once. Immediately flush eyes well with copious quantities of water or normal saline for at least 20−30 min. Seek medical attention.

Inhalation: Leave contaminated area immediately; breathe fresh air. Proper respiratory protection must be supplied to any rescuers. If coughing, difficult breathing, or any other symptoms develop, seek medical attention at once, even if symptoms develop many hours after exposure.

Ingestion: Contact a physician, hospital, or poison center at once. If the victim is unconscious or convulsing, do not induce vomiting or give anything by mouth. Assure that the patient's airway is open and lay him on his side with his head lower than his body and transport immediately to a medical facility. If conscious and not convulsing, give a glass of water to dilute the substance. Vomiting should not be induced without a physician's advice.

Personal Protective Methods: Wear protective gloves and clothing to prevent any reasonable probability of skin contact. Safety equipment suppliers/manufacturers can provide recommendations on the most protective glove/clothing material for your operation. All protective clothing (suits, gloves, footwear, headgear) should be clean, available each day, and put on before work. Contact lenses should not be worn when working with this chemical. Wear splash-proof chemical goggles and face shield when working with liquid unless full face-piece respiratory protection is worn. Wear dust-proof goggles and face shield when working with powders or dust unless full face-piece respiratory protection is worn. Employees should wash immediately with soap when skin is wet or contaminated. Provide emergency showers and eyewash.

Respirator Selection: NIOSH/OSHA *[Tin, organic compounds as (Sn)]: Up to 1 mg/m³:* CcrOv95 (APF = 10) [any air-purifying half-mask respirator with organic vapor cartridge(s) in combination with an N95, R95, or P95 filter. The following filters may also be used: N99, R99, P99, N100, R100, P100] or Sa (APF = 10) (any supplied-air respirator). *Up to 2.5 mg/m³:* Sa:Cf (APF = 25) (any supplied-air respirator operated in a continuous-flow mode) or PaprOvHie (APF = 25) (any air-purifying full-face-piece respirator equipped with an organic vapor cartridge in combination with a high-efficiency particulate filter). *Up to 5 mg/m³:* CcrFOv100 (APF = 50) [any air-purifying full-face-piece respirator equipped with organic vapor cartridge(s) in combination with an N100, R100, or P100 filter] or GmFOv100 (APF = 50) [any air-purifying, full-face-piece respirator (gas mask) with a chin-style, front- or back-mounted organic vapor canister having an N100, R100, or P100 filter] or PaprTOvHie (APF = 50) [any powered, air-purifying respirator with a tight-fitting face-piece and organic vapor cartridge (s) in combination with a high-efficiency particulate filter] or SaT: Cf (APF = 50) (any supplied-air respirator that has a tight-fitting face-piece and is operated in a continuous-flow mode) or SCBAF (APF = 50) (any self-contained breathing apparatus with a full face-piece) or SaF (APF = 50) (any supplied-air respirator with a full face-piece). *Up to 25 mg/m³:* SaF: Pd,Pp (APF = 2000) (any supplied-air respirator that has a full face-piece and is operated in a pressure-demand or other positive-pressure mode). *Emergency or planned entry into unknown concentrations or IDLH conditions:* SCBAF: Pd,Pp (APF = 10,000) (any self-contained breathing apparatus that has a full faceplate and is operated in a pressure-demand or other positive-pressure mode) or SaF: Pd,Pp: ASCBA (any supplied-air respirator that has a full face-piece and is operated in a pressure-demand or other positive-pressure mode in combination with an auxiliary, self-contained breathing apparatus operated in a pressure-demand or other positive-pressure mode). *Escape:* GmFOv100 (APF = 50) [Any air-purifying, full-face-piece respirator (gas mask) with a chin-style, front- or back-mounted organic vapor canister having an N100, R100, or P100 filter] or SCBAE (any appropriate escape-type, self-contained breathing apparatus).

Storage: Color Code—Blue: Health Hazard/Poison: Store in a secure poison location. Prior to working with this chemical you should be trained on its proper handling and storage. Store in tightly closed containers in a cool, well-ventilated area away from incompatible materials. Where possible, automatically transfer material from storage containers to process containers. Sources of ignition, such as smoking and open flames, are prohibited where this chemical is handled, used, or stored.

Shipping: Organotin compounds, solid, n.o.s., and Organotin pesticides, liquid, toxic, require a shipping label of "POISONOUS/TOXIC MATERIALS." They fall in DOT Hazard Class 6.1.

Spill Handling: Evacuate persons not wearing protective equipment from area of spill or leak until cleanup is complete. Remove all ignition sources. Stay upwind; keep out of low areas. Ventilate closed spaces before entering them.

Remove and isolate contaminated clothing at the site. Do not touch spilled material; stop leak if you can do it without risk. Use water spray to reduce vapors. *Small spills:* absorb with sand or other noncombustible absorbent material and place into containers for later disposal. *Small dry spills:* with clean shovel place material into clean, dry container and cover; move containers from spill area. *Large spills:* dike far ahead of spill for later disposal. Ventilate area after cleanup is complete. It may be necessary to contain and dispose of this chemical as a hazardous waste. If material or contaminated runoff enters waterways, notify downstream users of potentially contaminated waters. Contact your local or federal environmental protection agency for specific recommendations. If employees are required to clean up spills, they must be properly trained and equipped. OSHA 1910.120(q) may be applicable.

Fire Extinguishing: *Small fires:* dry chemical, carbon dioxide, water spray, or foam. *Large fires:* water spray, fog, or foam. Stay upwind; keep out of low areas. Ventilate closed spaces before entering them. Wear positive pressure breathing apparatus and special protective clothing. Move container from fire area is you can do so without risk. Fight fire from maximum distance. Dike fire control water for later disposal; do not scatter the material. Poisonous gases are produced in fire, including tin oxides and hydrogen chloride (triphenyltin chloride). If material or contaminated runoff enters waterways, notify downstream users of potentially contaminated waters. Notify local health and fire officials and pollution control agencies. From a secure, explosion-proof location, use water spray to cool exposed containers. If cooling streams are ineffective (venting sound increases in volume and pitch, tank discolors, or shows any signs of deforming), withdraw immediately to a secure position. If employees are expected to fight fires, they must be trained and equipped in OSHA 1910.156. The only respirators recommended for firefighting are self-contained breathing apparatuses that have full face-pieces and are operated in a pressure-demand or other positive-pressure mode.

References

US Environmental Protection Agency. (November 30, 1987). *Chemical Hazard Information Profile: Acetoxytriphenyl Stannane.* Washington, DC: Chemical Emergency Preparedness Program

US Environmental Protection Agency. (November 30, 1987). *Chemical Hazard Information Profile: Triphenyltin Chloride.* Washington, DC: Chemical Emergency Preparedness Program

Sax, N. I. (Ed.). (1982). *Dangerous Properties of Industrial Materials Report, 2,* No. 4, 92–94

Tris(2-chloroethyl)amine (Agent HN-3, WMD)   T:0960

Molecular Formula: $C_6H_{12}Cl_3N$
Common Formula: $(ClCH_2CH_2)_3N$

Synonyms: AI3-16198; 2-Chloro-*N*,*N*-bis(2-chloroethyl)etha-namine; HN-3 (military designation); Nitrogen Mustard-3; TL 145; Trichlormethine; Tri-(2-chloroethyl)amine; 2,2′,2″-Trichlorotriethylamine; Tris(2-chloroethyl)amine; Tris (β-chloroethyl)amine; TS160

CAS Registry Number: 555-77-1

RTECS® Number: YE2625000

UN/NA & ERG Number: UN2810/153

EC Number: None assigned.

Regulatory Authority and Advisory Bodies

Department of Homeland Security Screening Threshold Quantity: *Theft hazard* CUM 100 g.

Superfund/EPCRA 40CFR355, Extremely Hazardous Substances: TPQ = 100 lb (45.4 kg).

Reportable Quantity (RQ): 100 lb (45.4 kg).

US DOT 49CFR172.101, Inhalation Hazardous Chemical.

HN-3 is a suspected carcinogen, developmental toxin, and reproductive toxin.

European/International Regulations: not listed in Annex 1.

WGK (German Aquatic Hazard Class): No value assigned.

Description: HN-3, a nitrogen mustard blister agent (vesicants), is a colorless to pale yellow liquid. Pure material is odorless; otherwise it has a faint fish- or soap-like odor. Density: 1.2347 at 25°C; Molecular weight = 204.53; Boiling point = 256°C (HN-3 decomposes before its boiling point is reached or condenses under all conditions; the reactions involved could generate enough heat to cause an explosion[NIOSH]); Freezing/Melting point = −3.9°C; Vapor pressure = 0.0106 mmHg at 20°C; 0.0109 mmHg at 25°C. Volatility: 0.120 mg/L at 25°C. Hazard Identification (based on NFPA-704 M Rating System): Health 4, Flammability 1, Reactivity 0. Sparingly soluble in water; solubility = 160 mg/L at 25°C.

Incompatibilities: HN-3 is not stable; it undergoes slow but steady polymerization. Avoid contamination with oxidizing agents, e.g., nitrates, oxidizing acids, chlorine bleaches, pool chlorine, which may result in ignition. Unstable in the presence of light and heat and forms dimers at temperatures above 50°C. HN-3 decomposes before its boiling point is reached or condenses under all conditions; the reactions involved could generate enough heat to cause an explosion. Polymerizes slowly, so munitions would be effective for several years. Heated to decomposition emits hydrogen chloride and nitrogen oxide.

Note: Chlorinating agents destroy nitrogen mustards. Dry chlorinated lime and chloramines with a high content of active chlorine vigorously chlorinate nitrogen mustards to the carbon chain giving low-toxicity products. In the presence of water, this interaction proceeds less actively. They are rapidly oxidized by peracids in aqueous solution at weakly alkaline pH. In acid solution the oxidation is much slower.

Potential Exposure: Sulfur mustards were formerly used as a gas warfare agent. Nitrogen mustards have not previously been used in warfare.[NIOSH] Exposure to nitrogen mustard damages the eyes, skin, and respiratory tract and suppresses the immune system. Although the nitrogen mustards cause cellular changes within minutes of contact, the onset of pain and other symptoms is delayed. Exposure to large amounts can be fatal.[NIOSH]

Permissible Exposure Limits in Air

Protective Action Criteria (PAC) HN-3*

TEEL-0: 0.001 mg/m^3

PAC-1: 0.003 mg/m^3

PAC-2: **0.022** mg/m^3

PAC-3: **0.37** mg/m^3

*AEGLs (Acute Emergency Guideline Levels) & ERPGs (Emergency Response Planning Guideline) are in **bold face**.

AEL (US Military): 0.003 mg/m^3.

Determination in Water: A water contaminant. Octanol−water coefficient: Log K_{ow} = (estimated) 2.27.[NIOSH]

Routes of Entry: Inhalation, ingestion, skin contact.

Harmful Effects and Symptoms

Nitrogen mustards are extremely toxic and may damage the eyes, skin, and respiratory tract and suppress the immune system. Although these agents cause cellular changes within minutes of contact, the onset of pain and other symptoms is delayed. Thus, patients/victims arriving immediately from the scene of nitrogen mustard exposure are not likely to have signs and symptoms. The sooner after exposure that symptoms occur, the more likely they are to progress and become severe.[NIOSH]

Short Term Exposure: Extremely toxic and may damage the eyes, skin, and respiratory tract and suppress the immune system. Although these agents cause cellular changes within minutes of contact, the onset of pain and other symptoms is delayed. Most toxic of the nitrogen mustards. The median lethal dose for inhalation is 1500 mg-min/m^3; for skin absorption (masked personnel) is 10,000 mg-min/m^3. The medium incapacitating dose for eye injury is 200 mg-min/m^3; for skin absorption is 2500 mg-min/m^3. Irritates the eyes in quantities which do not significantly damage the skin or respiratory tract, insofar as single exposures are concerned. After mild vapor exposure, there may be no skin lesions. After severe vapor exposures, or after exposure to the liquid, erythema may appear. Irritation and itching may occur. Later, blisters may appear in the erythematous areas. Effects on the respiratory tract include irritation of the nose and throat, hoarseness progressing to loss of voice, and a persistent cough. Fever, labored respiration, and moist rales develop. Bronchial pneumonia may appear after the first 24 h. Following ingestion or systemic absorption, material causes inhibition of cell mitosis, resulting in depression of the blood-forming mechanism and injury to other tissues. Severe diarrhea, which may be hemorrhagic, occurs. Lesions are most marked in the small intestine and consist of degenerative changes and nercosis in the mucous membranes. Ingestion of 2−6 mg causes nausea and vomiting.

Long Term Exposure: Chronic or repeated exposure to HN-3 may cause bone marrow suppression resulting in damage to the blood-forming (hematopoietic) system, lymph node damage, weakening of the immune system, kidney damage, and reproductive system damage.[NIOSH] Early signs of bone marrow suppression include: a low white blood cell count; an increased risk for developing infections; a tendency for easy bruising and bleeding. May cause lymph node damage and a weakened immune system. It also causes liver and kidney damage, damage to the reproductive systems of both men and women leading to decreased fertility. It is mutagenic, toxic to the developing embryo, and carcinogenic.

First Aid: There is no antidote for nitrogen mustard toxicity.

Because health effects due to nitrogen mustard may not occur until several hours after exposure, patients/victims should be observed in a hospital setting for at least 24 h. Gastric lavage is contraindicated following ingestion of this agent due to the risk of perforation of the esophagus or upper airway. If this chemical gets into the eyes, remove any contact lenses at once and irrigate immediately for at least 15 min, occasionally lifting upper and lower lids. Seek medical attention immediately. If this chemical contacts the skin, remove contaminated clothing and wash immediately with soap and water. Speed in removing material from skin is of extreme importance. Seek medical attention immediately. If this chemical has been inhaled, remove from exposure, begin rescue breathing (using universal precautions, including resuscitation mask) if breathing has stopped and CPR if heart action has stopped. Transfer promptly to a medical facility. When this chemical has been swallowed, get medical attention. Give large quantities of water and induce vomiting. Do not make an unconscious person vomit. Keep victim quiet and maintain normal body temperature. Effects may be delayed; keep victim under observation.

Personal Protective Methods: Wear Totally Encapsulating Chemical Protective (TECP) suit that provides protection against CBRN agents; Chemical-resistant inner and outer gloves; Chemical-resistant boots with a steel toe and shank; Coveralls, long underwear, and a hard hat worn under the TECP suit are optional items. Take all necessary precautions to prevent any reasonable probability of skin contact. Safety equipment suppliers/manufacturers can provide recommendations on the most protective glove/clothing material for your operation. All protective clothing (suits, gloves, footwear, headgear) should be clean, available each day, and put on before work. Contact lenses should not be worn when working with this chemical. Wear splash-proof chemical goggles and face shield unless full face-piece respiratory protection is worn. Employees should wash immediately with soap when skin is wet or contaminated. Provide emergency showers and eyewash.

Decontamination: Decontamination of all potentially exposed areas within minutes after exposure is the only effective means of decreasing tissue damage.[NIOSH] Remove clothes and place contaminated clothes and personal belongings in a sealed double bag. Decontamination

of mustard-exposed victims by either vapor or liquid should be performed within the first 2 min following the exposure to prevent tissue damage. If not accomplished within the first several minutes, decontamination should still be performed to ensure any residual liquid mustard is removed from the skin or clothes, or to ensure any trapped mustard vapor is removed with the clothing. Removing trapped mustard vapor will prevent vapor off-gassing or subsequent cross-contamination of other emergency responders/health-care providers or the health-care facility. Physical removal of the mustard agent, rather than detoxification or neutralization, is the most important principle in patient decontamination. Mustard is not detoxified by water alone and will remain in decontamination effluent (in dilute concentrations) if hydrolysis has not taken place.

(1) Patients exposed to vapor should be decontaminated by removing all clothing in a clean air environment and shampooing or rinsing the hair to prevent vapor off-gassing.

(2) Patients exposed to liquid should be decontaminated by (a) Washing in warm or hot water at least three times. Use liquid soap (dispose of container after use and replace), large volumes of water, and mild to moderate friction with a single-use sponge or washcloth in the first and second washes. Scrubbing of exposed skin with a brush is discouraged because skin damage may occur which may enhance absorption. The third wash should be to rinse with large amounts of warm or hot water. Shampoo can be used to wash the hair. The rapid physical removal of a chemical agent is essential. If warm or hot water is not available, but cold water is, use cold water. Do not delay decontamination to obtain warm water. (b) Rinse the eyes, mucous membranes, or open wounds with sterile saline or water.

(3) The health-care provider should (a) Check the victim after the three washes to verify adequate decontamination before allowing entry to the medical treatment facility. If the washes were inadequate, repeat the entire process. (b) Be prepared to stabilize conventional injuries during the decontamination process. Careful decontamination can be a time-consuming process. The health-care provider may have to enter the contaminated area to treat the casualty during this process. Medical personnel should wear the proper PPE and evaluate the exposed workers.

Respirator Selection: *When used as a weapon, use SCBA Respirator Certified By NIOSH For CBRN Environments.* Where a potential exposure to the chemical exists, use a NIOSH-certified CBRN full-face-piece SCBA operated in a pressure-demand mode or a pressure-demand supplied air-hose respirator with an auxiliary escape bottle; or use a NIOSH/MSHA- or European Standard EN149-approved supplied-air respirator with a full face-piece operated in the positive-pressure mode, or with a full face-piece, hood, or helmet in the continuous-flow mode; or use a NIOSH/MSHA- or European Standard EN149-approved self-contained breathing apparatus (SCBA) with a full face-piece operated in a pressure-demand or other positive-pressure mode.

Storage: Color Code—Blue: Health Hazard/Poison: Store in a secure poison location. Prior to working with this chemical you should be trained on its proper handling and storage. Store in tightly closed containers in a cool, well-ventilated area away from heat. Sources of ignition, such as smoking and open flames, are prohibited where this chemical is used, handled, or stored in a manner that could create a potential fire or explosion hazard. Metal containers involving the transfer of 5 gallons or more of this chemical should be grounded and bonded. Drums must be equipped with self-closing valves, pressure vacuum bungs, and flame arresters. Use only nonsparking tools and equipment, especially when opening and closing containers of this chemical.

Shipping: Toxic liquids, organic, n.o.s. [Inhalation hazard, Packing Group I, Zone B] require a shipping label of "POISONOUS/TOXIC MATERIALS." Inhalation Hazard. It falls in Hazard Class 6.1 and Packing Group 1.

Spill Handling: Evacuate and restrict persons not wearing protective equipment from area of spill or leak until cleanup is complete. Remove all ignition sources. Ventilate area of spill or leak. Avoid inhalation and skin contact. Do not touch spilled material; stop leak if you can do so without risk. Use water spray to reduce vapors. *Small spills:* absorb with sand or other noncombustible absorbent material and place into containers for later disposal. *Large spills:* dike far ahead of spill for later disposal. Keep this chemical out of a confined space, such as a sewer, because of the possibility of an explosion, unless the sewer is designed to prevent the buildup of explosive concentrations. It may be necessary to contain and dispose of this chemical as a hazardous waste. If material or contaminated runoff enters waterways, notify downstream users of potentially contaminated waters. Contact your local or federal environmental protection agency for specific recommendations. If employees are required to clean up spills, they must be properly trained and equipped. OSHA 1910.120(q) may be applicable.

Initial isolation and protective action distances

Distances shown are likely to be affected during the first 30 min after materials are spilled and could increase with time. If more than one tank car, cargo tank, portable tank, or large cylinder involved in the incident is leaking, the protective action distance may need to be increased. You may need to seek emergency information from CHEMTREC at (800) 424-9300 or seek professional environmental engineering assistance from the US EPA Environmental Response Team at (908) 548-8730 (24-h response line).

Small spills (From a small package or a small leak from a large package)

HN-3, when used as a weapon

First: Isolate in all directions (feet/meters)   100/30
Then: Protect persons downwind (miles/kilometers)
Day   0.1/0.2
Night   0.1/0.2

Large spills (From a large package or from many small packages)

First: Isolate in all directions (feet/meters)   100/30
Then: Protect persons downwind (miles/kilometers)
Day   0.1/0.2
Night   0.1/0.2

Fire Extinguishing: Poisonous gases, including nitrogen oxides and hydrogen chloride, are produced in fire. Use dry chemical, carbon dioxide, or foam extinguishers. Containers may explode in fire. Storage containers and parts of containers may rocket great distances, in many directions. If material or contaminated runoff enters waterways, notify downstream users of potentially contaminated waters. Notify local health and fire officials and pollution control agencies. From a secure, explosion-proof location, use water spray to cool exposed containers. If cooling streams are ineffective (venting sound increases in volume and pitch, tank discolors, or shows any signs of deforming), withdraw immediately to a secure position. If employees are expected to fight fires, they must be trained and equipped in OSHA 1910.156. The only respirators recommended for firefighting are self-contained breathing apparatuses that have full face-pieces and are operated in a pressure-demand or other positive-pressure mode.

References

US Environmental Protection Agency. (November 30, 1987). *Chemical Hazard Information Profile: Tris(2-Chloroethyl)Amine.* Washington, DC: Chemical Emergency Preparedness Program

Tris(2,3-dibromopropyl)-phosphate   T:0970

Molecular Formula: $C_9H_{15}Br_6O_4P$
Common Formula: $[BrCH_2CH(Br)CH_2O]_3P = O$
Synonyms: 3PBR; Anfram 3PB; Apex 462-5; Bromkal P 67-6HP; 2,3-Dibromo-1-propanol phosphate; ES 685; Firemaster LV-T 23P; Firemaster T 23; Firemaster T 23P; Firemaster T 23P-LV; Flacavon R; Flammex AP; Flammex LV-T 23P; Flammex T 23P; Fosfato de tris(2,3-dibromo-propilo) (Spanish); Fyrol HB 32; NCI-C03270; Phoscon PE 60; Phoscon UF-S; Phosphoric acid tris(2,3-dibromopropyl) ester; 1-Propanol, 2,3-dibromo-, phosphate (3:1); T 23P; TDBP; TDBPP; Tris; Tris BP; Tris(dibromopropyl) phosphate; Tris(2,3-dibromopropyl) phosphoric acid ester; Tris (flame retardant); Zetofex ZN
CAS Registry Number: 126-72-7
RTECS® Number: UB0350000
UN/NA & ERG Number: UN2811 (toxic solid, organic, n.o.s.)/154
EC Number: 204-799-9
Regulatory Authority and Advisory Bodies
Carcinogenicity: IARC: Animal, Sufficient Evidence; Human, Inadequate Evidence, Group 2A, 1999; NTP: 11th Report on Carcinogens, 2004: Reasonably anticipated to be a human carcinogen; NCI: Carcinogenesis Bioassay (feed); clear evidence: mouse, rat.

US EPA Gene-Tox Program, Positive: Carcinogenicity—mouse/rat; Positive: *D. melanogaster*—reciprocal translocation; Positive: Host-mediated assay; Histidine reversion—Ames test; Positive: Sperm morphology—mouse; *D. melanogaster* sex-linked lethal; Positive/dose response: *In vitro* SCE—nonhuman; *In vivo* SCE—nonhuman.

Banned or Severely Restricted (many countries) (UN).[13]

US EPA Hazardous Waste Number (RCRA No.): U235.

RCRA, 40CFR261, Appendix 8 Hazardous Constituents.

RCRA 40CFR268.48; 61FR15654, Universal Treatment Standards: Wastewater (mg/L), 0.11; Nonwastewater (mg/kg), 0.10.

Reportable Quantity (RQ): 10 lb (4.54 kg).

EPCRA Section 313 Form R *de minimis* concentration reporting level: 0.1%.

Rotterdam Convention Annex III [Chemicals Subject to the Prior Informed Consent Procedure (PIC)].

California Proposition 65 Chemical: Cancer 1/1/88.

European/International Regulations: not listed in Annex 1.

WGK (German Aquatic Hazard Class): No value assigned.

Description: TDBP is a crystalline solid. Molecular weight = 697.67; Flash point ≥112°C.

Potential Exposure: Compound Description: Tumorigen, Mutagen; Reproductive Effector; Primary Irritant. Tris-BP is used as a flame retardant additive for synthetic textiles and plastics. It was applied to fabrics used for children's clothes (sleepwear in particular) with some used as a flame retardant in other materials, such as urethane foams. Commercial preparations of tris-BP can be obtained in two grades, viz, HV (high in volatiles) and LV (low in volatiles). A typical LV sample has been reported to contain the following impurities[1]: 0.05% 1,2-dibromo-3-chloropropane (BrCH$_2$CHBrCH$_2$Cl); 0.05% 1,2,3-tribromopropane (BrCH$_2$CHBrCH$_2$Br); and 0.20% 2,3-dibromopropanol (BrCH$_2$CHBrCH$_2$OH). Use and exposure have greatly decreased after a ruling by the Consumer Product Safety Commission in April 1977.

Incompatibilities: Acids, bases.

Permissible Exposure Limits in Air

No numerical OELs have been set.

Finland: carcinogen, 1999; France: carcinogen, 1993; Sweden: carcinogen, 1999.

Routes of Entry: TDBP was added to fabrics used for children's garments to the extent of 5−10% by weight. A child wearing such garment and chewing on a sleeve or collar could easily ingest some TDBP, particularly if the garment had not been laundered before use.

Harmful Effects and Symptoms

Short Term Exposure: May cause nausea, vomiting, gastrointestinal irritation. May cause central nervous system depression, headaches, dizziness.

Long Term Exposure: May cause skin sanitization and allergy; chronic lung disease; kidney damage, such as renal tubular necrosis; liver damage; testicular atrophy; and sterility. Suspected mutagenesis and carcinogenesis.

Points of Attack: Lungs, kidneys, liver, central nervous system, reproductive system.

Medical Surveillance: Lung function tests. Liver and kidney function tests. Examination of the nervous system.

First Aid: Skin Contact[52]: Flood all areas of body that have contacted the substance with water. Do not wait to remove contaminated clothing; do it under the water stream. Use soap to help assure removal. Isolate contaminated clothing when removed to prevent contact by others.

Eye Contact: Remove any contact lenses at once. Immediately flush eye well with copious quantities of water or normal saline for at least 20−30 min. Seek medical attention.

Inhalation: Leave contaminated area immediately; breathe fresh air. Proper respiratory protection must be supplied to any rescuers. If coughing, difficult breathing, or any other symptoms develop, seek medical attention at once, even if symptoms develop many hours after exposure.

Ingestion: Contact a physician, hospital, or poison center at once. If the victim is unconscious or convulsing, do not induce vomiting or give anything by mouth. Assure that the patient's airway is open and lay him on his side with his head lower than his body and transport immediately to a medical facility. If conscious and not convulsing, give a glass of water to dilute the substance. Vomiting should not be induced without a physician's advice.

Personal Protective Methods: Wear protective gloves and clothing to prevent any reasonable probability of skin contact. Safety equipment suppliers/manufacturers can provide recommendations on the most protective glove/clothing material for your operation. All protective clothing (suits, gloves, footwear, headgear) should be clean, available each day, and put on before work. Contact lenses should not be worn when working with this chemical. Wear splash-proof chemical goggles and face shield when working with liquid unless full face-piece respiratory protection is worn. Wear dust-proof goggles and face shield when working with powders or dust unless full face-piece respiratory protection is worn. Employees should wash immediately with soap when skin is wet or contaminated. Provide emergency showers and eyewash.

Respirator Selection: Follow regulations in OSHA 29CFR1910.134 or European Standard EN149. Use a NIOSH/MSHA- or European Standard EN149-approved respirator; or use an approved supplied-air respirator with a full face-piece operated in the positive-pressure mode, or with a full face-piece, hood, or helmet in the continuous-flow mode; or use a NIOSH/MSHA- or European Standard EN149-approved self-contained breathing apparatus with a full face-piece operated in a pressure-demand or other positive-pressure mode.

Storage: Color Code—Blue: Health Hazard/Poison: Store in a secure poison location. Prior to working with this chemical you should be trained on its proper handling and storage. Store in a cool, dry place and keep away from acids and bases. Where possible, automatically transfer material from storage containers to process containers. Sources of ignition, such as smoking and open flames, are prohibited where this chemical is handled, used, or stored. A regulated, marked area should be established where this chemical is handled, used, or stored in compliance with OSHA Standard 1910.1045.

Shipping: Toxic solids, organic, n.o.s., require a shipping label of "POISONOUS/TOXIC MATERIALS." It falls in Hazard Class 6.1 and Packing Group III.

Spill Handling: Evacuate and restrict persons not wearing protective equipment from area of spill or leak until cleanup is complete. Remove all ignition sources. Ventilate area of spill or leak. Absorb liquids in vermiculite, dry sand, earth, peat, carbon, or a similar material and deposit in sealed containers. Follow by washing surfaces well, first with 60–70% acetone, then with soap and water. Keep this chemical out of a confined space, such as a sewer, because of the possibility of an explosion, unless the sewer is designed to prevent the buildup of explosive concentrations. It may be necessary to contain and dispose of this chemical as a hazardous waste. If material or contaminated runoff enters waterways, notify downstream users of potentially contaminated waters. Contact your local or federal environmental protection agency for specific recommendations. If employees are required to clean up spills, they must be properly trained and equipped. OSHA 1910.120(q) may be applicable.

Fire Extinguishing: This chemical is a combustible liquid. Poisonous gases, including phosphorus oxides and bromine, are produced in fire. Use dry chemical, carbon dioxide, or alcohol foam extinguishers. Containers may explode in fire. Storage containers and parts of containers may rocket great distances, in many directions. If material or contaminated runoff enters waterways, notify downstream users of potentially contaminated waters. Notify local health and fire officials and pollution control agencies. From a secure, explosion-proof location, use water spray to cool exposed containers. If cooling streams are ineffective (venting sound increases in volume and pitch, tank discolors, or shows any signs of deforming), withdraw immediately to a secure position. If employees are expected to fight fires, they must be trained and equipped in OSHA 1910.156. The only respirators recommended for firefighting are self-contained breathing apparatuses that have full face-pieces and are operated in a pressure-demand or other positive-pressure mode.

Disposal Method Suggested: Consult with environmental regulatory agencies for guidance on acceptable disposal practices. Generators of waste containing this contaminant (≥100 kg/mo) must conform with EPA regulations governing storage, transportation, treatment, and waste disposal.

References
US Environmental Protection Agency. (April 1976). *Summary Characterization of Selected Chemicals of Near-Term Interest*, Report EPA 560/4-76-004. Washington, DC: Office of toxic Substances
US Environmental Protection Agency. (December 1979). *Status Assessment of Toxic Chemicals: Tris(2,3-Dibromopropyl) Phosphate*, Report EPA-600/2-79-210n. Cincinnati, OH
US Environmental Protection Agency. (August 1976). *Investigation of Selected Potential Environmental Contaminants: Haloalkyl Phosphates*, Report EPA-560/2-76-007. Washington, DC

Trypan blue   T:0980

Molecular Formula: $C_{32}H_{24}N_6Na_4O_{14}S_4$
Common Formula: $C_{32}H_{24}N_6O_{14}S_4 \cdot 4Na$
Synonyms: AI3-26698; Amanil sky blue; Amanil sky blue R; Amidine blue 4B; Azidinblau 3B; Azidine blue 3B; Azirdinblau 3B; Azurro diretto 3B; Bencidal blue 3B; Benzaminblau 3B; Benzamine blue; Benzamine blue 3B; Benzanil blue 3BN; Benzanil blue R; Benzoblau 3B; Benzo blue; Benzo blue 3B; Benzo blue 3BS; Bleu diamine; Bleu diazole N 3B; Bleu directe 3B; Bleue diretto 3B; Bleu trypane N; Blue 3B; Blue EMB; Brasilamina blue 3B; Brasilazina blue 3B; Centraline blue 3B; Chloramiblau 3B; Chloramine blue; Chloramine blue 3B; Chlorazol blue 3B; Chrome leather blue 3B; C.I. 23850; C.I. Direct blue 14; C.I. Direct blue 14, tetrasodium salt; Congoblau 3B; Congo blue; Congo blue 3B; Cresotine blue 3B; Diaminblau 3B; Diamine blue; Diamine blue 3B; Diaminineblue; Dianilblau; Dianilblau H3G; Dianil blue; Dianil blue H3G; Diaphtamine blue TH; Diazine blue 3B; Diazol blue 3B; 3,3'-[(3,3'-Dimethyl(1,1'-biphenyl)-4,4'-diyl)bis(azo)bis(5-amino-4-hydroxynaphthalene-2,7-disulphonate); Diphenyl blue 3B; Directakol blue 3BL; Directblau 3B; Direct blue 14; Direct blue 3B; Direct blue D3B; Direct blue FFN; Direct blue H3G; Direct blue M3B; Hispamin blue 3B; Naphtamine blue 2B; Naphtamine blue 3B; 2,7-Naphthalenedisulfonic acid, 2-57-13,3'-([3,3'-dimethyl(1,1'-biphenyl)-4,4'-diyl]bis(azo))bis(5-amino-4-hydroxy-, tetrasodium salt; 2,7-Naphthalenedisulfonic acid, 3,3'-([3,3'-dimethyl(1,1'-biphenyl)-4,4'-diyl]bis(azo))bis(5-amino-4-hydroxy-), tetrasodium salt; 2,7-Naphthalenedisulfonic acid, 3,3'-[(3,3'-dimethyl-4,4'-biphenylylene)bis(azo)]bis(5-amino-4-hydroxy-), tetrasodium salt; Naphthaminblau 3B; Naphthamine blue 3B; Naphthylamine blue; NCI: C61289; Niagara blue; Niagara blue 3B; NSC 11247; Orion blue 3B; Paramine blue 3B; Parkibleu; Parkipan; Pontamine blue 3B; Pyrazol blue 3B; Pyrotropblau; Renolblau 3B; Sodium ditolyl-diazobis-8-amino-1-naphthol-3,6-disulfonate; Sodium ditolyldisazobis-8-amino-1-naphthol-3,6-disulfonate; Sodium ditolyldisazobis-8-amino-1-naphthol-3,6-disulphonate; TB; Tetrasodium; Trianol direct blue 3B; Triazolblau 3B; Tripan blue; Trypan blue BPC; Trypan blue sodium salt; Trypane blue
CAS Registry Number: 72-57-1
RTECS® Number: QJ6475000
UN/NA & ERG Number: UN3143 (Dyes, solid, toxic, n.o.s.)/151

EC Number: 200-786-7

Regulatory Authority and Advisory Bodies
Carcinogenicity: IARC: Human Inadequate Evidence; Animal Sufficient Evidence, *possibly carcinogenic to humans,* Group 2B, 1987.
US EPA Hazardous Waste Number (RCRA No.): U236.
RCRA, 40CFR261, Appendix 8 Hazardous Constituents.
Reportable Quantity (RQ): 10 lb (4.54 kg).
EPCRA Section 313 Form R *de minimis* concentration reporting level: 0.1%.
California Proposition 65 Chemical: Cancer 1/1/89.
European/International Regulations: not listed in Annex 1.
WGK (German Aquatic Hazard Class): No value assigned.

Description: Trypan blue is a dark blue crystalline solid or powder. Molecular weight = 964.88. Hazard Identification (based on NFPA-704 M Rating System): Health 0, Flammability 1, Reactivity 0. Soluble in water.

Potential Exposure: Used in dyeing textiles; leather and paper; as a biological stain.

Incompatibilities: Strong oxidizers.

Permissible Exposure Limits in Air
Protective Action Criteria (PAC)
TEEL-0: 2 mg/m^3
PAC-1: 6 mg/m^3
PAC-2: 40 mg/m^3
PAC-3: 500 mg/m^3

Harmful Effects and Symptoms
Short Term Exposure: Inhalation: No symptoms reported. *Skin:* may stain skin. *Eyes:* application of 0.2 mL (0.007 oz) of a 1% solution caused no eye irritation. *Ingestion:* moderately toxic. Probable lethal dose between 1 oz and 1 lb for a 150-lb person.

Long Term Exposure: A potential occupational carcinogen, mutagen, and teratogen. Has been shown to cause birth defects, cancer, and liver injury in laboratory animals.

First Aid: If this chemical gets into the eyes, remove any contact lenses at once and irrigate immediately for at least 15 min, occasionally lifting upper and lower lids. Seek medical attention immediately. If this chemical contacts the skin, remove contaminated clothing and wash immediately with soap and water. Seek medical attention immediately. If this chemical has been inhaled, remove from exposure, begin rescue breathing (using universal precautions, including resuscitation mask) if breathing has stopped and CPR if heart action has stopped. Transfer promptly to a medical facility. When this chemical has been swallowed, get medical attention. Give large quantities of water and induce vomiting. Do not make an unconscious person vomit.

Personal Protective Methods: Wear protective gloves and clothing to prevent any reasonable probability of skin contact. Safety equipment suppliers/manufacturers can provide recommendations on the most protective glove/clothing material for your operation. All protective clothing (suits, gloves, footwear, headgear) should be clean, available each day, and put on before work. Contact lenses should not be worn when working with this chemical. Wear dust-proof chemical goggles and face shield unless full face-piece respiratory protection is worn. Employees should wash immediately with soap when skin is wet or contaminated. Provide emergency showers and eyewash.

Respirator Selection: Follow regulations in OSHA 29CFR1910.134 or European Standard EN149. Use a NIOSH/MSHA- or European Standard EN149-approved respirator; or use an approved supplied-air respirator with a full face-piece operated in the positive-pressure mode, or with a full-face-piece, hood, or helmet in the continuous-flow mode; or use a NIOSH/MSHA- or European Standard EN149-approved self-contained breathing apparatus with a full face-piece operated in a pressure-demand or other positive-pressure mode.

Storage: Color Code—Blue: Health Hazard/Poison: Store in a secure poison location. Prior to working with this chemical you should be trained on its proper handling and storage. Avoid creating dust. Where possible, automatically pump material from storage containers to process containers. Sources of ignition, such as smoking and open flames, are prohibited where this chemical is handled, used, or stored. A regulated, marked area should be established where this chemical is handled, used, or stored in compliance with OSHA Standard 1910.1045.

Shipping: Dyes, solid, toxic, n.o.s. [or] Dye intermediates, solid, toxic, n.o.s. require a shipping label of "POISONOUS/TOXIC MATERIALS." Tris(2,3-dibromopropyl) phosphate falls in Hazard Class 6.1.

Spill Handling: Evacuate persons not wearing protective equipment from area of spill or leak until cleanup is complete. Remove all ignition sources. Collect powdered material in the most convenient and safe manner and deposit in sealed containers. Ventilate area after cleanup is complete. It may be necessary to contain and dispose of this chemical as a hazardous waste. If material or contaminated runoff enters waterways, notify downstream users of potentially contaminated waters. Contact your local or federal environmental protection agency for specific recommendations. If employees are required to clean up spills, they must be properly trained and equipped. OSHA 1910.120(q) may be applicable.

Fire Extinguishing: This chemical is a noncombustible solid. Use extinguisher appropriate for burning material. Poisonous gases are produced in fire, including oxides of sodium, nitrogen, and sulfur. If material or contaminated runoff enters waterways, notify downstream users of potentially contaminated waters. Notify local health and fire officials and pollution control agencies. From a secure, explosion-proof location, use water spray to cool exposed containers. If cooling streams are ineffective (venting sound increases in volume and pitch, tank discolors, or shows any signs of deforming), withdraw immediately to a secure position. If employees are expected to fight fires, they must be trained and equipped in OSHA 1910.156. The only respirators recommended for firefighting are self-contained breathing apparatuses that have full face-pieces and are operated in a pressure-demand or other positive-pressure mode.

Disposal Method Suggested: Consult with environmental regulatory agencies for guidance on acceptable disposal practices. Generators of waste containing this contaminant (≥100 kg/mo) must conform with EPA regulations governing storage, transportation, treatment, and waste disposal.

References

New York State Department of Health. (July 1986). *Chemical Fact Sheet: Trypan Blue.* Albany, NY: Bureau of Toxic Substance Assessment

Tungsten & insoluble compounds   T:0985

Common Formula: W; WC; WCCo; WCNi; WCTi

Synonyms: metal: Tungsten, elemental; Tungsten metal; Wolfram

Tungsten carbide (cemented): Cemented tungsten carbide; Cemented WC; Hard metal

Tungsten other insoluble compounds: Tungsten(IV) oxide 12036-22-5; Tungsten trioxide 1314-35-8; Tungstic acid (7783-03-1); 12070-12-1 (tungsten carbide)

CAS Registry Number: 7440-33-7 (elemental); 12718-69-3 (92% W, 8% Co); 11107-01-0 (85% W, 15% Co); 37329-49-0 (78% W: 14% Co: 8% Ti)

RTECS® Number: YO7175000 (elemental); YO7525000 (92% W: 8% Co); Y07350000 (85% W, 15% Co); YO7700000 (78% W: 14% Co: 8% Ti)

UN/NA & ERG Number: UN3189 (metal powder, self-heating, n.o.s.)/135

EC Number: 231-143-9 (tungsten)

Regulatory Authority and Advisory Bodies

Carcinogenicity: NIOSH (tungsten carbide containing >0.3% Ni) NIOSH: Potential occupational carcinogen.

Air in Pollutant Standard Set. See below, "Permissible Exposure Limits Air" section.

Canada, WHMIS, Ingredients Disclosure List Concentration 1.0%, for tungsten and its compounds.

European/International Regulations: not listed in Annex 1.

WGK (German Aquatic Hazard Class): Nonwater polluting agent.

Description: Tungsten is a hard, brittle, steel-gray to tin-white metal or fine powder. Molecular weight = 183.85; Boiling point = 5927°C; Freezing/Melting point = 3410°C. Insoluble in water. Tungsten carbide is a gray powder; Freezing/Melting point = 2780°C. Cemented tungsten carbide is a mixture, generally consisting of 85–95% tungsten carbide (WC) and 5–15% cobalt (Co). Physical properties vary depending upon the specific mixture. Insoluble in water. Tungsten carbide is a gray powder. Molecular weight = 195.86; Boiling point: 6000°C at 760 mmHg; Melting point = 2730–2830°C; solubility in water = <1 mg/mL at 18°C.[NTP] Hazard Identification (based on NFPA-704 M Rating System): Health 1, Flammability 2, Reactivity 0.

Potential Exposure: Compound Description (tungsten): Reproductive Effector; Primary Irritant. Tungsten is used in ferrous and nonferrous alloys, and for filaments in incandescent lamps. It has been stated that the principal health hazards from tungsten and its compounds arise from inhalation of aerosols during mining and milling operations. The principal compounds of tungsten to which workers are exposed are ammonium paratungstate, oxides of tungsten (WO_3, W_2O_5, WO_2), metallic tungsten, and tungsten carbide. In the production and use of tungsten carbide tools for machining, exposure to the cobalt used as a binder or cementing substance may be the most important hazard to the health of the employees. Since the cemented tungsten carbide industry uses such other metals as tantalum, titanium, niobium, nickel, chromium, and vanadium in the manufacturing process, the occupational exposures are generally to mixed dust.

Incompatibilities: Tungsten: The finely divided powder is combustible and may ignite spontaneously in air. Incompatible with bromine trifluoride, chlorine trifluoride, fluorine, iodine pentafluoride. *Tungsten carbide:* Incompatible with strong oxidizers: fluorine (may cause ignition), chlorine trifluoride, iodine pentafluoride, mercuric iodine, oxides of nitrogen, lead dioxide, strong acid mixtures (i.e., HNO_3/HCl mixture).

Permissible Exposure Limits in Air

Tungsten and insoluble compounds

OSHA PEL: None.

NIOSH REL: 5 mg[W]/m³ TWA; 10 mg[W]/m³ STEL [also applies to other insoluble compounds (as W)].

ACGIH TLV®[1]: 5 mg[W]/m³ TWA; 10 mg[W]/m³ STEL [also applies to other insoluble compounds (as W)].

Protective Action Criteria (PAC)

TEEL-0: 5 mg/m³

PAC-1: 10 mg/m³

PAC-2: 150 mg/m³

PAC-3: 500 mg/m³

Denmark: TWA 5 mg[W]/m³, 1999; Norway: TWA 5 mg[W]/m³, 1999; Poland: MAC (TWA) fume and dust 5 mg[W]/m³, 1999; Russia: STEL 2 mg[W]/m³, 1993; Sweden: NGV 5 mg[W]/m³, 1999; United Kingdom: TWA 5 mg[W]/m³, STEL 10 mg[W]/m³, 2000.

Tungsten soluble compounds

OSHA PEL: None.

NIOSH REL: 1 mg[W]/m³ TWA; 3 mg[W]/m³ STEL.

ACGIH TLV®[1]: 1 mg[W]/m³ TWA; 3 mg[W]/m³ STEL.

Tungsten carbide containing >2% Co

NIOSH REL: 0.05 mg[W]/m³ TWA; See *NIOSH Pocket Guide*, Appendix C.

Tungsten carbide containing >0.3% Ni

NIOSH REL: 0.015 mg[W]/m³ TWA; Potential occupational carcinogen; Reduce exposure to lowest feasible level; See *NIOSH Pocket Guide*, Appendix A & C.

Tungsten carbide

Protective Action Criteria (PAC)

For 12070-12-1 (92% W, 8% Co)

TEEL-0: 5.33 mg/m^3
PAC-1: 10.7 mg/m^3
PAC-2: 26.6 mg/m^3
PAC-3: 125 mg/m^3

Cemented tungsten carbide (example: 85%: 15%) or *tungsten carbide, mixed with cobalt and titanium* (example: 78% W: 14% Co: 8% Ti) also known as "hard metal" refers to a mixture of tungsten carbide, cobalt, and sometimes metal oxides or carbides and other metals (including nickel). When the *cobalt* (Co) content exceeds 2%, its contribution to the potential hazard is judged to exceed that of tungsten carbide. Therefore, the NIOSH REL (10-h TWA) for cemented tungsten carbide containing >2% Co is 0.05 mg [Co]/m^3; the applicable OSHA PEL is 0.1 mg[Co]/m^3 (8-h TWA). *Nickel* (Ni) may sometimes be used as a binder rather than cobalt. NIOSH considers cemented tungsten carbide containing nickel to be a potential occupational carcinogen and recommends a REL of 0.015 mg [Ni]/m^3 (10-h TWA). The OSHA PEL for *insoluble nickel*, 1 mg (Ni)/m^3 8-h TWA applies to mixtures of tungsten carbide and nickel.

Tungsten carbide
Australia: TWA 5 mg[W]/m^3, STEL 10 mg[W]/m^3, 1993; Austria: MAK 5 mg[W]/m^3, 1999; Belgium: TWA 5 mg [W]/m^3, STEL 10 mg[W]/m^3, 1993; Denmark: TWA 5 mg [W]/m^3, 1999; Finland: TWA 5 mg[W]/m^3, 1999; Norway: TWA 5 mg[W]/m^3, 1999; the Philippines: TWA 1 mg[W]/m^3, 1993; Poland: MAC (TWA) 5 mg[W]/m^3, 1999; Sweden: NGV 5 mg[W]/m^3, 1999; Switzerland: MAK 5 mg [W]/m^3, 1999; United Kingdom: TWA 5 mg[W]/m^3, STEL 10 mg[W]/m^3, 2000; Argentina, Bulgaria, Columbia, Jordan, South Korea, New Zealand, Singapore, Vietnam: ACGIH TLV®: TWA 5 mg[W]/m^3; STEL 10 mg[W]/m^3.
For both tungsten and tungsten carbide, Russia[43] set a MAC of 6.0 mg[W]/m^3 in work-place air. Several states have set guidelines and standards for tungsten in ambient air[60] ranging from 16.0 µg/m^3 (Virginia) to 20.0 µg/m^3 (Connecticut) to 24.0 µg/m^3 (Nevada).

Determination in Air: For tungsten metal and tungsten soluble compounds: Use NIOSH Analytical Method (IV) #7074, #7301, OSHA Analytical Method ID-213. See also Method #7300, Elements. There is no specific method for *cemented tungsten carbide*.

Permissible Concentration in Water: No criteria set, but EPA[32] has suggested a permissible ambient goal of 14 µg/L based on health effects. Russia[43] set a MAC of 0.1 mg [W]/L in water bodies used for domestic purposes.

Determination in Water: By neutron activation analysis.

Routes of Entry: Inhalation, ingestion, skin and/or eye contact.

Harmful Effects and Symptoms
Short Term Exposure: *Tungsten* can affect you when breathed in. Irritates the eyes, skin, and respiratory system. Some tungsten compounds can cause lung and skin problems. *Tungsten carbide* can affect you when breathed in. There is no health effects from exposure to pure tungsten

carbide alone. However, tungsten carbide is often combined with nickel or cobalt to make cemented tungsten carbide (hard metal). Exposure to *tungsten carbide combined with cobalt or nickel* can cause skin irritation.

Long Term Exposure: Long-term exposure to *tungsten metal* and *cemented tungsten carbide* may cause diffuse pulmonary fibrosis (lung scarring), loss of appetite, nausea, cough. May cause blood changes. Exposure to *tungsten carbide combined with cobalt or nickel* can cause skin sensitization, lung allergy, with wheezing, coughing, and shortness of breath. Repeated exposure can cause pulmonary fibrosis. Long-term exposure to *tungsten-soluble compounds:* in animals: central nervous system disturbances; diarrhea; respiratory failure; behavioral, body weight, and blood changes.

Points of Attack: *Metal* and *cemented tungsten carbides:* Eyes, skin, respiratory system, blood. *Soluble compounds:* Eyes, skin, respiratory system, central nervous system, gastrointestinal tract.

Medical Surveillance: If you are exposed to tungsten alone, no medical tests are necessary. If you are exposed to cemented tungsten carbide (hard metal), the following are recommended before beginning employment and at regular times after that: lung function tests; chest X-ray every 2–3 years after five or more years of exposure. If symptoms develop or overexposure is suspected, the following may be useful: evaluation by a qualified allergist, including careful exposure history and special testing, may help diagnose skin allergy.

First Aid: If this chemical gets into the eyes, remove any contact lenses at once and irrigate immediately for at least 15 min, occasionally lifting upper and lower lids. Seek medical attention immediately. If this chemical contacts the skin, remove contaminated clothing and wash immediately with soap and water. Seek medical attention immediately. If this chemical has been inhaled, remove from exposure, begin rescue breathing (using universal precautions, including resuscitation mask) if breathing has stopped and CPR if heart action has stopped. Transfer promptly to a medical facility. When this chemical has been swallowed, get medical attention. Give large quantities of water and induce vomiting. Do not make an unconscious person vomit.

Personal Protective Methods: Wear protective gloves and clothing to prevent any reasonable probability of skin contact. Safety equipment suppliers/manufacturers can provide recommendations on the most protective glove/clothing material for your operation. All protective clothing (suits, gloves, footwear, headgear) should be clean, available each day, and put on before work. Contact lenses should not be worn when working with this chemical. Wear dust-proof chemical goggles and face shield when working with powders or dust unless full face-piece respiratory protection is worn. Employees should wash immediately with soap when skin is wet or contaminated. Provide emergency showers and eyewash.

Respirator Selection: NIOSH *(tungsten metal and insoluble compounds):* Up to 50 mg[W]/m^3: 100XQ (APF = 10) [Any air-purifying respirator with an N100, R100, or P100 filter (including N100, R100, and P100 filtering face-pieces) except quarter-mask respirators] or Sa (APF = 10) (any supplied-air respirator) or SCBAF (APF = 50) (any self-contained breathing apparatus with a full face-piece). *Emergency or planned entry into unknown concentrations or IDLH conditions:* SCBAF: Pd,Pp (APF = 10,000) (any NIOSH/MSHA- or European Standard EN 149-approved self-contained breathing apparatus that has a full face-piece and is operated in a pressure-demand or other positive-pressure mode) or SaF: Pd,Pp: ASCBA (APF = 10,000) (any supplied-air respirator that has a full face-piece and is operated in a pressure-demand or other positive-pressure mode in combination with an auxiliary, self-contained breathing apparatus operated in a pressure-demand or other positive-pressure mode). *Escape:* 100F (APF = 50) (any air-purifying, full-face-piece respirator with an N100, R100, or P100 filter) or SCBAE (any appropriate escape-type, self-contained breathing apparatus).

Respirator for Tungsten carbide (cemented) containing Cobalt.

NIOSH/OSHA, for cobalt metal dust and fume: *0.25 mg/m^3: if not present as a fume* Qm* (APF = 25) (any quarter-mask respirator). *0.5 mg/m^3:* 95XQ* (APF = 10) [any particulate respirator equipped with an N95, R95, or P95 filter (including N95, R95, and P95 filtering face-pieces) except quarter-mask respirators. The following filters may also be used: N99, R99, P99, N100, R100, P100] or Sa* (APF = 10) (any supplied-air respirator). *1.25 mg/m^3:* Sa:Cf (APF = 25)* (any supplied-air respirator operated in a continuous-flow mode) or PaprHie* (APF = 25) (any powered, air-purifying respirator with a high-efficiency particulate filter). *2.5 mg/m^3:* 100F (APF = 50) (any air-purifying, full-face-piece respirator with an N100, R100, or P100 filter) or SCBAF (APF = 50) (any self-contained breathing apparatus with a full face-piece) or SaF (APF = 50) (any supplied-air respirator with a full face-piece). *20 mg/m^3:* SaF: Pd,Pp (APF = 2000) (any supplied-air respirator that has a full face-piece and is operated in a pressure-demand or other positive-pressure mode. *Emergency or planned entry into unknown concentrations or IDLH conditions:* SCBAF: Pd, Pp (APF = 10,000) (any self-contained breathing apparatus that has a full face-piece and is operated in a pressure-demand or other positive-pressure mode) or SaF: Pd,Pp: ASCBA (APF = 10,000) (any supplied-air respirator that has a full face-piece and is operated in a pressure-demand or other positive-pressure mode in combination with an auxiliary, self-contained breathing apparatus operated in a pressure-demand or other positive-pressure mode). *Escape:* 100F (APF = 50) (any air-purifying, full-face-piece respirator with an N100, R100, or P100 filter) or SCBAE (any appropriate escape-type, self-contained breathing apparatus). *Substance reported to cause eye irritation or damage; may require eye protection.

Respirator for Tungsten carbide (cemented) containing Nickel.

At concentrations above the NIOSH REL, or where there is no REL, at any detectable concentration: SCBAF: Pd,Pp (APF = 10,000) (any NIOSH/MSHA- or European Standard EN 149-approved self-contained breathing apparatus that has a full face-piece and is operated in a pressure-demand or other positive-pressure mode) or SaF: Pd,Pp: ASCBA (APF = 10,000) (any supplied-air respirator that has a full face-piece and is operated in a pressure-demand or other positive-pressure mode in combination with an auxiliary, self-contained breathing apparatus operated in a pressure-demand or other positive-pressure mode). *Escape:* GmFOv (APF = 50) [any air-purifying, full-face-piece respirator (gas mask) with a chin-style, front- or back-mounted organic vapor canister] or SCBAE (any appropriate escape-type, self-contained breathing apparatus).

Storage: Color Code—Red Stripe: Flammability Hazard: Store separately from all other flammable materials. Prior to working with this chemical you should be trained on its proper handling and storage. Tungsten must be stored to avoid contact with fluorine and chlorine compounds since violent reactions occur. Store in a tightly closed container in a cool, well-ventilated area. Sources of ignition, such as smoking and open flames, are prohibited where tungsten is used, handled, or stored in a manner that could create a potential fire or explosion hazard. Tungsten carbide must be stored to avoid contact with fluorine, chlorine trifluoride, iodine pentafluoride, lead dioxide, nitrous oxide, nitrogen dioxide, and mercurium iodine since violent reactions occur. Store in tightly closed containers in a cool well-ventilated area.

Shipping: Metal powder, self-heating, n.o.s. requires a label of "SPONTANEOUSLY COMBUSTIBLE." Tungsten powdered metal falls in Hazard Class 4.2 and Packing Group II.

Spill Handling: Evacuate persons not wearing protective equipment from area of spill or leak until cleanup is complete. Remove all ignition sources. Collect powdered material in the most convenient and safe manner and deposit in sealed containers; do not sweep in the case of tungsten carbide. Ventilate area after cleanup is complete. It may be necessary to contain and dispose of this chemical as a hazardous waste. If material or contaminated runoff enters waterways, notify downstream users of potentially contaminated waters. Contact your local or federal environmental protection agency for specific recommendations. If employees are required to clean up spills, they must be properly trained and equipped. OSHA 1910.120(q) may be applicable.

Fire Extinguishing: Tungsten is a flammable powder. Use dry chemicals appropriate for extinguishing metal fires. In the case of tungsten carbide, extinguish fire using an agent suitable for type of surrounding fire. Tungsten carbide itself does not burn. Poisonous gases are produced in fire. If material or contaminated runoff enters waterways, notify downstream users of potentially contaminated waters. Notify local health and fire officials and pollution control agencies. From a secure, explosion-proof location, use water spray to cool exposed containers. If cooling streams are ineffective (venting sound increases in volume and pitch, tank discolors, or shows any signs of deforming), withdraw immediately to a secure position. If employees are expected to fight fires, they must be trained and equipped in OSHA 1910.156. The only respirators recommended for firefighting are self-contained breathing apparatuses that have full face-pieces and are operated in a pressure-demand or other positive-pressure mode.

Disposal Method Suggested: Recovery of tungsten from sintered metal carbides, scrap, and spent catalysts has been described as an alternative to disposal.

References

US Environmental Protection Agency. (May 1977). *Toxicology of Metals, Vol. II: Tungsten*, Report EPA-600/1-77-022. Research Triangle Park, NC, pp. 442–453

National Institute for Occupational Safety and Health. (September 1977). *Criteria for a Recommended Standard: Occupational Exposure to Tungsten and Cemented Tungsten Carbide*, NIOSH Document No. 77-127

New Jersey Department of Health and Senior Services. (November 2000). *Hazardous Substances Fact Sheet: Tungsten*. Trenton, NJ

New Jersey Department of Health and Senior Services. (August 2005). *Hazardous Substances Fact Sheet: Tungsten Carbide*. Trenton, NJ

Tungsten hexafluoride   T:0990

Molecular Formula: F_6W
Common Formula: WF_6
Synonyms: Tungsten fluoride
CAS Registry Number: 7783-82-6
RTECS® Number: YO7720000
UN/NA & ERG Number: UN2196/125
EC Number: 232-029-1
Regulatory Authority and Advisory Bodies
Department of Homeland Security Screening Threshold Quantity (pounds): *Theft hazard* 45 (≥7.10% concentration).
Air Pollutant Standard Set. See below, "Permissible Exposure Limits in Air" section.
Canada, WHMIS, Ingredients Disclosure List Concentration 1.0%.
European/International Regulations: Not listed in Annex 1.
WGK (German Aquatic Hazard Class): No value assigned.
Description: Tungsten hexafluoride is a toxic, colorless gas or a light yellow liquid. Molecular weight = 297.85; Freezing/Melting point = 2.5°C. Boiling point = 17.5–19.5°C. Reacts with water (decomposes).
Potential Exposure: A strong halogenating agent. Used to apply tungsten coatings to other surfaces by vapor

deposition process; making electronics and components; in the manufacture of other chemicals.
Incompatibilities: Decomposes on contact with water and moist air, forming corrosive hydrofluoric acid. Violent reaction on contact with methyl silicate.
Permissible Exposure Limits in Air
Protective Action Criteria (PAC)
TEEL-0: 1.62 mg/m³
PAC-1: 4.86 mg/m³
PAC-2: 15 mg/m³
PAC-3: 150 mg/m³
Determination in Air: NIOSH Analytical Method (IV) #7074. See also Method #7300; NIOSH Analytical Method (IV) #7902, Fluorides.
Permissible Concentration in Water: The EPA has set a standard of 4 mg/L for fluoride[61] and the state of Maine has set 2.4 mg/L as a guideline for drinking water. Arizona[61] has set 1.8 mg/L as a standard for drinking water.
Routes of Entry: Inhalation, eye and/or skin contact.
Harmful Effects and Symptoms
Short Term Exposure: Tungsten hexafluoride is a corrosive chemical. Irritates the eyes, skin, and respiratory tract. Contact with liquid may cause frostbite. Tungsten hexafluoride can affect you when breathed in. Exposure to tungsten hexafluoride may expose you to both tungsten and fluorides. Exposure to very high levels of fluorides may cause symptoms of nausea, vomiting, abnormal pain, convulsions, and kidney damage. Higher exposures can cause pulmonary edema, a medical emergency that can be delayed for several hours. This can cause death.
Long Term Exposure: Repeated high exposures may affect kidneys. Repeated high exposures can cause deposits of fluorides in the bones (fluorosis) that may cause pain, disability, and mottling of the teeth. Repeated exposure can cause nausea, vomiting, loss of appetite, diarrhea, or constipation. Nosebleeds and sinus problems can also occur.
Points of Attack: Eyes, respiratory system, central nervous system, skeleton, kidneys, skin.
Medical Surveillance: For those with frequent or potentially high exposure (half the TLV or greater), the following are recommended before beginning work and at regular times after that: lung function tests. Fluoride level in urine (for fluoride in urine, use NIOSH #8308). Levels higher than 4 mg/L may indicate overexposure. If symptoms develop or overexposure is suspected, the following may be useful: consider chest X-ray after acute overexposure. Kidney function tests. Consider chest X-ray following acute overexposure.
First Aid: If this chemical gets into the eyes, remove any contact lenses at once and irrigate immediately for at least 15 min, occasionally lifting upper and lower lids. Seek medical attention immediately. If this chemical contacts the skin, remove contaminated clothing and wash immediately with soap and water. Seek medical attention immediately. If this chemical has been inhaled, remove from exposure,

begin rescue breathing (using universal precautions, including resuscitation mask) if breathing has stopped and CPR if heart action has stopped. Transfer promptly to a medical facility. When this chemical has been swallowed, get medical attention. Give large quantities of water and induce vomiting. Do not make an unconscious person vomit. Medical observation is recommended for 24—48 h after breathing overexposure, as pulmonary edema may be delayed. As first aid for pulmonary edema, a doctor or authorized paramedic may consider administering a corticosteroid spray. If frostbite has occurred, seek medical attention immediately; do *NOT* rub the affected areas or flush them with water. In order to prevent further tissue damage, do *NOT* attempt to remove frozen clothing from frostbitten areas. If frostbite has *NOT* occurred, immediately and thoroughly wash contaminated skin with soap and water.

Personal Protective Methods: Wear protective gloves and clothing to prevent any reasonable probability of skin contact. Safety equipment suppliers/manufacturers can provide recommendations on the most protective glove/clothing material for your operation. All protective clothing (suits, gloves, footwear, headgear) should be clean, available each day, and put on before work. Contact lenses should not be worn when working with this chemical. Wear splash-proof chemical goggles and face shield when working with liquid unless full face-piece respiratory protection is worn. Wear gas-proof goggles and face shield unless full face-piece respiratory protection is worn. Employees should wash immediately with soap when skin is wet or contaminated. Provide emergency showers and eyewash. Specific engineering controls are recommended in NIOSH Criteria Document #76-103: *Inorganic fluorides.*

Respirator Selection:
Fluorides
NIOSH/OSHA *12.5 mg/m³:* Qm (APF = 25) (any quarter-mask respirator). *25 mg/m³:* 95XQ (APF = 10)* [any particulate respirator equipped with an N95, R95, or P95 filter (including N95, R95, and P95 filtering face-pieces) except quarter-mask respirators. The following filters may also be used: N99, R99, P99, N100, R100, P100] or Sa* (any supplied-air respirator). *62.5 mg/m³:* Sa:Cf (APF = 25)*† (any supplied-air respirator operated in a continuous-flow mode) or PaprHie (APF = 25)* *if not present as a fume* (any powered, air-purifying respirator with a high-efficiency particulate filter). *125 mg/m³:* 100F (APF = 50)⁺ [any particulate respirator equipped with an N95, R95, or P95 filter (including N95, R95, and P95 filtering face-pieces) except quarter-mask respirators. The following filters may also be used: N99, R99, P99, N100, R100, P100] or SCBAF (APF = 50) (any self-contained breathing apparatus with a full face-piece) or SaF (APF = 50) (any supplied-air respirator with a full face-piece). *250 mg/m³:* Sa: Pd,Pp (APF = 1000) (any supplied-air respirator operated in a pressure-demand or other positive-pressure mode). *Emergency or planned entry into unknown concentrations or IDLH conditions:* SCBAF: Pd,Pp (APF = 10,000) (any self-contained

breathing apparatus that has a full faceplate and is operated in a pressure-demand or other positive-pressure mode) or SaF: Pd,Pp: ASCBA (APF = 10,000) (any supplied-air respirator that has a full face-piece and is operated in a pressure-demand or other positive-pressure mode in combination with an auxiliary, self-contained breathing apparatus operated in a pressure-demand or other positive-pressure mode). *Escape:* 100F (APF = 50)⁺ [any particulate respirator equipped with an N95, R95, or P95 filter (including N95, R95, and P95 filtering face-pieces) except quarter-mask respirators. The following filters may also be used: N99, R99, P99, N100, R100, P100] or SCBAE (any appropriate escape-type, self-contained breathing apparatus).
*Substance reported to cause eye irritation or damage; may require eye protection.
†May need acid gas sorbent.
Tungsten (insoluble compounds)
NIOSH *(tungsten metal):* Up to *50 mg/m³:* 100XQ (APF = 10) [Any air-purifying respirator with an N100, R100, or P100 filter (including N100, R100, and P100 filtering face-pieces) except quarter-mask respirators] or Sa (APF = 10) (any supplied-air respirator) or SCBAF (APF = 50) (any self-contained breathing apparatus with a full face-piece). *Emergency or planned entry into unknown concentrations or IDLH conditions:* SCBAF: Pd,Pp (APF = 10,000) (any NIOSH/MSHA- or European Standard EN 149-approved self-contained breathing apparatus that has a full face-piece and is operated in a pressure-demand or other positive-pressure mode) or SaF: Pd,Pp: ASCBA (APF = 10,000) (any supplied-air respirator that has a full face-piece and is operated in a pressure-demand or other positive-pressure mode in combination with an auxiliary, self-contained breathing apparatus operated in a pressure-demand or other positive-pressure mode). *Escape:* 100F (APF = 50) (any air-purifying, full face-piece respirator with an N100, R100, or P100 filter) or SCBAE (any appropriate escape-type, self-contained breathing apparatus).

Storage: Poison gas. Color Code—White Stripe: Contact Hazard; Store separately, not compatible with materials in solid white category. Prior to working with this chemical you should be trained on its proper handling and storage. Store in tightly closed container in a cool well-ventilated area away from water. Procedures for the handling, use, and storage of cylinders should be in compliance with OSHA 1910.101 and 1910.169 with the recommendations of the Compressed Gas Association.

Shipping: Tungsten hexafluoride requires a shipping label of "POISON GAS, CORROSIVE." It falls in Hazard Class 2.3.

Special precautions: Cylinders must be transported in a secure upright position, in a well-ventilated truck. It is a violation of transportation regulations to refill compressed gas cylinders without the express written permission of the owner.

Spill Handling: If in a building, evacuate building and confine vapors by closing doors and shutting down HVAC systems. Restrict persons not wearing protective equipment from area of spill or leak until cleanup is complete. Remove all ignition sources. Ventilate area of spill or leak to disperse the gas. Wear chemical protective suit with self-contained breathing apparatus to combat spills. Stay upwind and use water spray to "knock down" vapor; contain runoff. Stop the flow of gas, if it can be done safely from a distance. If source is a cylinder and the leak cannot be stopped in place, remove the leaking cylinder to a safe place; and repair leak or allow cylinder to empty. Absorb liquids in vermiculite, dry sand, earth, or a similar material and deposit in sealed containers. Keep this chemical out of confined spaces, such as a sewer, because of the possibility of explosion, unless the sewer is designed to prevent the buildup of explosive concentrations. If employees are required to clean up spills, they must be properly trained and equipped. OSHA 1910.120(q) may be applicable.

Initial isolation and protective action distances

Distances shown are likely to be affected during the first 30 min after materials are spilled and could increase with time. If more than one tank car, cargo tank, portable tank, or large cylinder involved in the incident is leaking, the protective action distance may need to be increased. You may need to seek emergency information from CHEMTREC at (800) 424-9300 or seek professional environmental engineering assistance from the US EPA Environmental Response Team at (908) 548-8730 (24-h response line).

Small spills (From a small package or a small leak from a large package)

First: Isolate in all directions (feet/meters)   100/30
Then: Protect persons downwind (miles/kilometers)
Day   0.1/0.2
Night   0.5/0.8

Large spills (From a large package or from many small packages)

First: Isolate in all directions (feet/meters)   500/150
Then: Protect persons downwind (miles/kilometers)
Day   0.6/0.9
Night   1.8/2.8

Fire Extinguishing: Tungsten hexafluoride may burn but does not readily ignite. For small fires, use dry chemical or CO_2 extinguishers. Poisonous gases, including fluorine, are produced in fire. Vapors are heavier than air and will collect in low areas. Containers may explode in fire. Storage containers and parts of containers may rocket great distances, in many directions. If material or contaminated runoff enters waterways, notify downstream users of potentially contaminated waters. Notify local health and fire officials and pollution control agencies. From a secure, explosion-proof location, use water spray to cool exposed containers. If cooling streams are ineffective (venting sound increases in volume and pitch, tank discolors, or shows any signs of deforming), withdraw immediately to a secure position. If cylinders are exposed to excessive heat from fire or flame contact, withdraw immediately to a secure location. If employees are expected to fight fires, they must be trained and equipped in OSHA 1910.156. The only respirators recommended for firefighting are self-contained breathing apparatuses that have full face-pieces and are operated in a pressure-demand or other positive-pressure mode.

Disposal Method Suggested: Nonrefillable cylinders should be disposed of in accordance with local, state, and federal regulations. Allow remaining gas to vent slowly into atmosphere in an unconfined area or exhaust hood. Refillable-type cylinders should be returned to original supplier with any valve caps and outlet plugs secured and valve protection caps in place.

References

New Jersey Department of Health and Senior Services. (November 2000). *Hazardous Substances Fact Sheet: Tungsten Hexafluoride*. Trenton, NJ

Turpentine   T:1000

Molecular Formula: $C_{10}H_{16}$ (approx.)

Synonyms: Gum spirits; Gum turpentine; Oil of turpentine; Spirits of turpentine; Steam distilled turpentine; Sulfate wood turpentine; Terebenthine (French); Terpentin oel (German); Turpentine steam distilled; Turps; Wood turpentine

CAS Registry Number: 8006-64-2; selected monoterpenes: 80-56-8 (α-Pinene); 127-91-3; 13466-78-9 (Carene); 498-15-7 (Carene)[CHRIS Manual]

RTECS® Number: YO8400000

UN/NA & ERG Number: UN1299/128

EC Number: 232-350-7 [*Annex I Index No.:* 650-002-00-6]

Regulatory Authority and Advisory Bodies

FDA—over-the-counter drug.

US EPA, FIFRA 1998 Status of Pesticides: Canceled/New AI. Air Pollutant Standard Set. See below, "Permissible Exposure Limits in Air" section.

Canada, WHMIS, Ingredients Disclosure List Concentration 1.0%.

European/International Regulations (*8006-64-2*): Hazard Symbol: Xn, N; Risk phrases: R10; R20/21/22; R36/38; R43; R51/53; R65; Safety phrases: S2; S36/37; S46; S61; S62 (see Appendix 4).

WGK (German Aquatic Hazard Class): 2—Water polluting (*CAS: 8006-64-2*).

Description: Turpentine is the oleoresin from species of *Pinus* (Pinaceae) trees. The crude oleoresin (gum turpentine) is a yellowish, sticky, opaque mass, and the distillate (oil of turpentine) is a colorless, volatile liquid with a characteristic odor. Chemically, it contains: *alpha*-pinene; *beta*-pinene; camphene, monocyclic terpene; and terpene alcohols. Molecular weight = 136 (approx.); Specific gravity (H_2O:1) = 0.86 at 25°C; Boiling point = 153.8–170°C; Freezing/Melting point = −50 to −60°C; Vapor pressure = 4 mmHg; Flash point = 35°C (cc), also listed at 30–46°C (cc); Autoignition temperature = 253°C, also listed at 220–255°C. Explosive limit: LEL = 0.8%; UEL = 6%. Hazard Identification (based on NFPA-704 M

Rating System): Health 1, Flammability 3, Reactivity 0. Insoluble in water.

Potential Exposure: Compound Description: Tumorigen, Human Data; Natural Product; Primary Irritant. Turpentines have found wide use as chemical feedstock for the manufacture of floor, furniture, shoe, and automobile polishes, camphor, cleaning materials, inks, putty, mastics, cutting and grinding fluids, paint thinners, resins, and degreasing solutions. Recently, *alpha-* and *beta-*pinenes, which can be extracted, have found use as volatile bases for various compounds. The components *d*-α-pinene and 3-carene, or their hydroperoxides, may be the cause of eczema and toxic effects of turpentine.

Incompatibilities: Forms an explosive mixture with air. Violent reaction with strong oxidizers, especially chlorine; chromic anhydride; stannic chloride; chromyl chloride.

Permissible Exposure Limits in Air

Conversion factor: 1 ppm = 5.56 mg/m^3 (approx.) at 25°C & 1 atm.

OSHA PEL (8006-64-2): 100 ppm/560 mg/m^3 TWA.

NIOSH REL (8006-64-2): 100 ppm/560 mg/m^3 TWA (Intended change).

ACGIH TLV[®][1] (*terpentine and selected monoterpenes*): 20 ppm/112 mg/m^3 TWA, sensitizer; not classifiable as a human carcinogen.

NIOSH IDLH: 200 ppm.

Protective Action Criteria (PAC)

8006-64-2

TEEL-0: 20 ppm

PAC-1: 20 ppm

PAC-2: 20 ppm

PAC-3: 800 ppm

DFG MAK (8006-64-2): 100 ppm/560 mg/m3; danger of skin sensitization.

Australia: TWA 100 ppm (560 mg/m^3), 1993; Austria: MAK 100 ppm (560 mg/m^3), 1999; Belgium: TWA 100 ppm (556 mg/m^3), 1993; Denmark: TWA 25 ppm (140 mg/m^3), 1999; Finland: TWA 100 ppm (560 mg/m^3), STEL 150 ppm, [skin], 1999; France: VME 100 ppm (560 mg/m^3), 1999; Hungary: TWA 300 mg/m^3, STEL 600 mg/m^3, 1993; the Netherlands: MAC-TGG 560 mg/m^3, 2003; the Philippines: TWA 100 ppm (560 mg/m^3), 1993; Poland: MAC (TWA) 300 mg/m^3; MAC (STEL) 840 mg/m^3, 1999; Russia: STEL 300 mg/m^3, 1993; Sweden: NGV 25 ppm (150 mg/m^3), KTV 50 ppm (300 mg/m^3), [skin], 1999; Turkey: TWA 100 ppm (560 mg/m^3), 1993; United Kingdom: TWA 100 ppm (566 mg/m^3), STEL 150 ppm, 2000; New Zealand, Singapore, Vietnam: ACGIH TLV®: not classifiable as a human carcinogen.

Russia[35, 43] set a MAC of 2.0 mg/m^3 for ambient air in residential areas on a momentary basis and 1.0 mg/m^3 on a daily average basis. Several states have set guidelines or standards for turpentine in ambient air[60] ranging from 5.6–8.4 mg/m^3 (North Dakota) to 9.3 mg/m^3 (Virginia) to 11.2 mg/m^3 (Connecticut and New York) to 13.333 mg/m^3 (Nevada).

α-*Pinene (80-56-8)*

Protective Action Criteria (PAC)

TEEL-0: 20 ppm

PAC-1: 20 ppm

PAC-2: 60 ppm

PAC-3: 300 ppm

Determination in Air: Use NIOSH Analytical Method #1551, Terpentine, #2549, Volatile organic compound.[18]

Permissible Concentration in Water: Russia[43] set a MAC of 0.2 mg/L in water bodies used for domestic purposes.

Routes of Entry: Inhalation of vapor and percutaneous absorption of liquid are the usual paths of occupational exposure. However, symptoms have been reported to develop from percutaneous absorption alone. Ingestion and/or skin and/or eye contact are also routes of entry.

Harmful Effects and Symptoms

Short Term Exposure: Turpentine can affect you when breathed in and by passing through your skin. Exposure can irritate the eyes, nose, and throat. Higher levels can affect the CNS, causing headache, vertigo, dizziness, abdominal pain, nausea, vomiting, diarrhea, confusion, and rapid pulse. Swallowing the liquid may cause chemical pneumonia to develop. High exposures can cause pulmonary edema, a medical emergency that can be delayed for several hours. This can cause death. Still higher levels can cause hematuria (blood in the urine), albuminuria, kidney damage, convulsions, and death.

Long Term Exposure: Repeated or prolonged contact may cause skin sensitization and allergy. The liquid destroys the skin's natural oils, causing dryness and cracking. May damage the kidneys, bladder, and the nervous system. Can irritate the lungs; bronchitis may develop. Various kinds of products are in use as turpentine oil. Their respective toxicities and tendency to cause eczema, which probably arise from their content of δ-α-pinene and 3-carene, vary considerably. However, systematic comparative investigations for these products are lacking.

Points of Attack: Eyes, skin, respiratory system, central nervous system, kidneys.

Medical Surveillance: If symptoms develop or overexposure is suspected, the following may be useful: kidney function tests. Evaluation by a qualified allergist, including careful exposure history and special testing, may help diagnose skin allergy. Examination of the nervous system. Consider chest X-ray following acute overexposure.

First Aid: If this chemical gets into the eyes, remove any contact lenses at once and irrigate immediately for at least 15 min, occasionally lifting upper and lower lids. Seek medical attention immediately. If this chemical contacts the skin, remove contaminated clothing and wash immediately with soap and water. Seek medical attention immediately. If this chemical has been inhaled, remove from exposure, begin rescue breathing (using universal precautions, including resuscitation mask) if breathing has stopped and CPR if heart action has stopped. Transfer promptly to a medical facility. When this chemical has been swallowed, give plenty of water to drink and get medical attention. Do not induce vomiting. Medical observation is recommended for 24–48 h after breathing overexposure, as pulmonary edema may be delayed. As first aid for pulmonary edema, a doctor or authorized paramedic may consider administering a corticosteroid spray.

Personal Protective Methods: Wear protective gloves and clothing to prevent any reasonable probability of skin contact. Safety equipment suppliers/manufacturers can provide recommendations on the most protective glove/clothing material for your operation. Teflon™ and polyvinyl alcohol are among the recommended protective materials. All protective clothing (suits, gloves, footwear, headgear) should be clean, available each day, and put on before work. Contact lenses should not be worn when working with this chemical. Wear splash-proof chemical goggles and face shield unless full face-piece respiratory protection is worn. Employees should wash immediately with soap when skin is wet or contaminated. Provide emergency showers and eyewash.

Respirator Selection: *800 ppm:* Sa:Cf (APF = 25) (any supplied-air respirator operated in a continuous-flow mode) or PaprOv (APF = 25) [any powered, air-purifying respirator with organic vapor cartridge(s)] or CcrFOv (APF = 50) [any air-purifying, full-face-piece respirator (gas mask) with a chin-style, front- or back-mounted acid gas canister] or GmFOv (APF = 50) [any air-purifying, full face-piece respirator (gas mask) with a chin-style, front- or back-mounted acid gas canister] or SCBAF (APF = 50) (any self-contained breathing apparatus with a full face-piece) or SaF (APF = 50) (any supplied-air respirator with a full face-piece). *Emergency or planned entry into unknown concentrations or IDLH conditions:* SCBAF: Pd,Pp (APF = 10,000) (any self-contained breathing apparatus that has a full face-piece and is operated in a pressure-demand or other positive-pressure mode) or SaF: Pd,Pp: ASCBA (APF = 10,000) (any supplied-air respirator that has a full face-piece and is operated in a pressure-demand or other positive-pressure mode in combination with an auxiliary self-contained breathing apparatus operated in a pressure-demand or other positive-pressure mode). *Escape:* GmFOv (APF = 50) [any air-purifying, full-face-piece respirator (gas mask) with a chin-style, front- or back-mounted organic vapor canister] or SCBAE (any appropriate escape-type, self-contained breathing apparatus).

Storage: Color Code—Red: Flammability Hazard: Store in a flammable liquid storage area or approved cabinet away from ignition sources and corrosive and reactive materials. Prior to working with this chemical you should be trained on its proper handling and storage. Before entering confined space where this chemical may be present, check to make sure that an explosive concentration does not exist. Store in tightly closed containers in a cool well-ventilated area away from oxidizers (such as perchlorates, peroxides, permanganates, chlorates, and nitrates). Sources of ignition, such as smoking and open flames, are prohibited where turpentine is used, handled, or stored in a manner that could create a potential fire or explosion hazard. Metal containers involving the transfer of 5 gallons or more of turpentine should be grounded and bonded. Drums must be equipped with self-closing values, pressure vacuum bungs, and flame arresters. Use only nonsparking tools and equipment, especially when opening and closing containers of turpentine.

Shipping: This compound requires a shipping label of "FLAMMABLE LIQUID." It falls in Hazard Class 3 and Packing Group II.

Spill Handling: Evacuate and restrict persons not wearing protective equipment from area of spill or leak until cleanup is complete. Remove all ignition sources. Establish forced ventilation to keep levels below explosive limit. Absorb liquids in vermiculite, dry sand, earth, peat, carbon, or a similar material and deposit in sealed containers. Keep this chemical out of a confined space, such as a sewer, because of the possibility of an explosion, unless the sewer is designed to prevent the buildup of explosive concentrations. It may be necessary to contain and dispose of this chemical as a hazardous waste. If material or contaminated runoff enters waterways, notify downstream users of potentially contaminated waters. Contact your local or federal environmental protection agency for specific recommendations. If employees are required to clean up spills, they must be properly trained and equipped. OSHA 1910.120(q) may be applicable.

Fire Extinguishing: This chemical is a flammable liquid. Toxic fumes and gases are produced in fire. Use dry chemical, carbon dioxide, or alcohol foam extinguishers. Vapors are heavier than air and will collect in low areas. Vapors may travel long distances to ignition sources and flashback. Vapors in confined areas may explode when exposed to fire. Containers may explode in fire. Storage containers and parts of containers may rocket great distances, in many directions. If material or contaminated runoff enters waterways, notify downstream users of potentially contaminated waters. Notify local health and fire officials and pollution control agencies. From a secure, explosion-proof location, use water spray to cool exposed containers. If cooling streams are ineffective (venting sound increases in volume and pitch, tank discolors, or shows any signs of deforming), withdraw immediately to a secure position. If employees are expected to fight fires, they must be trained and equipped in OSHA 1910.156. The only respirators recommended for firefighting are self-contained breathing apparatuses that have full face-pieces and are operated in a pressure-demand or other positive-pressure mode.

Disposal Method Suggested: Dissolve or mix the material with a combustible solvent and burn in a chemical incinerator equipped with an afterburner and scrubber. All federal, state, and local environmental regulations must be observed.

References

Sax, N. I. (Ed.). (1982). *Dangerous Properties of Industrial Materials Report*, 2, No. 2, 75–76

New Jersey Department of Health and Senior Services. (October 1996). *Hazardous Substances Fact Sheet: Turpentine*. Trenton, NJ

U

Uranium & compounds U:0100

Molecular Formula: U

Synonyms: Metal: Uranium 1; Uranium metal

Acetate: Acetato de uranilo (Spanish); Uranium acetate; Uranium bis(aceto-*o*)dioxo-; Uranium oxyacetate; Uranyl acetate

Uranium(IV)oxide: Black uranium oxide; Uranium dioxide; Uranous oxide (UO_2)

Uranyl nitrate, solid (10102-06-4): Bis(nitrato-*O,O'*)dioxo uranium; Nitrato de uranilo (Spanish); Uranium bis(nitrato-*o*)dioxo-, (T-4); Uranyl nitrate (EPA)

Uranyl nitrate (36478-76-9): Uranium, bis(nitrato-*O,O'*) dioxo, (OC-6-11)-

Nitrate hexahydrate: Bis(nitrato)dioxouranium hexahydrate; Dinitratodioxouranium, hexahydrate

Uranium sulfate (1314-64-3): Uranium (soluble compounds, as U); Uranium sulfate trihydrate; Uranyl sulfate trihydrate ($UO_2SO_4 \cdot 3H_2O$)

CAS Registry Number: 7440-61-1 (elemental); *(alt.)* 24678-82-8; 541-09-3 (acetate); 1344-57-6 (dioxide); 10102-06-4 (nitrate); 36478-76-9 (uranyl nitrate); 13520-83-7 (uranyl nitrate hexahydrate); 6159-44-0 (uranyl acetate dihydrate)

RTECS® Number: YR3490000 (elemental); YR3850000 (nitrate); YR4705000 [uranium(IV)oxide]

UN/NA & ERG Number: UN2979 (uranium metal, pyrophoric)/162; UN2979 [Radioactive Materials (Low to Moderate Level Radiation)]/162; UN2909 (Radioactive material, excepted package, articles manufactured from depleted Uranium)/161; UN2910 (Radioactive material, excepted package, articles manufactured from depleted Uranium)/161; UN2980 (uranyl nitrate hexahydrate solution)/162; UN2981 (uranyl nitrate, solid)/162

EC Number: 231-170-6 (elemental)

Regulatory Authority and Advisory Bodies

Carcinogenicity: NIOSH (*uranium, insoluble and soluble compounds*): Potential occupational carcinogen.

Air Pollutant Standard Set. See below, "Permissible Exposure Limits in Air" section.

Clean Water Act: Section 311 Hazardous Substances/RQ 40CFR117.3 (same as CERCLA, see below).

Reportable Quantity (RQ): 100 lb (45.4 kg).

Canada, WHMIS, Ingredients Disclosure List Concentration 1.0%.

European/International Regulations (*uranium and compounds*): Hazard Symbol: T+; Risk phrases: R26/28; R33; R53; Safety phrases: S1/2; S20/21; S45; S61 (see Appendix 4).

WGK (German Aquatic Hazard Class): 3—Highly water polluting (*uranyl acetate dihydrate*).

Description: Uranium is a silver-white, malleable, ductile, lustrous solid. Weakly radioactive but must be handled with caution. A combustible solid in the form of powder or turnings. Insoluble in water. Molecular weight = 238.00; Specific gravity ($H_2O:1$) = 19.05 at 25°C (metal); Boiling point = 3813°C; Freezing/Melting point = 1147°C; Autoignition temperature = 20°C (dust cloud); Ignition temperature of dust cloud = 20°C; Minimum Explosive concentration = 0.060 oz/ft^3.[USBM] Relative explosion hazard of dust: Severe. Hazard Identification (based on NFPA-704 M Rating System): Health 2, Flammability 0, Reactivity 0; Minimum explosive concentration (MEC) = 60 g/m^3.

In the natural state, uranium consists of three isotopes: ^{238}U (99.28%), ^{234}U (0.006%), and ^{235}U (0.714%).

There are over 100 uranium minerals; those of commercial importance are the oxides and oxygenous salts. The processing of uranium ore generally involves extraction then leaching either by an acid or by a carbonate method. The metal may be obtained from its halides by fused salt electrolysis.

Uranium(IV)oxide is a black to brown crystalline solid or powder. Molecular weight = 270.03; Freezing/Melting point = 2865°C. Insoluble in water.

Uranyl chloride (7791-26-6) is a bright yellow crystalline solid. Molecular weight = 349.90; Freezing/Melting point = 578°C. Highly soluble in water (unstable).

Uranyl nitrate, solid (10102-06-4) is a yellow crystalline solid. Molecular weight = 394.02; Boiling point = 118°C; Freezing/Melting point = 60°C. Hazard Identification (based on NFPA-704 M Rating System): Health 3, Flammability 0, Reactivity 2 (Oxidizer).

Uranyl nitrate, hexahydrate

Hazard Identification (based on NFPA-704 M Rating System): Health 3, Flammability 0, Reactivity 1.

Uranyl sulfate: Molecular weight = 420.2. Soluble in water.

Potential Exposure: The primary use of natural uranium is in nuclear energy as a fuel for nuclear reactors, in plutonium production, and as feeds for gaseous diffusion plants. It is also a source of radium salts. Uranium compounds are used in staining glass, glazing ceramics, and enameling; in photographic processes; for alloying steels; and as a catalyst for chemical reactions; radiation shielding; and aircraft counterweights. Uranium presents both chemical and radiation hazards, and exposures may occur during mining, processing of the ore, and production of uranium metal.

Incompatibilities: Uranium: Metal powder is radioactive, pyrophoric (ignites spontaneously in air), and a strong reducing agent. Keep away fromchlorine, fluorine, nitric acid, nitric oxide, selenium, sulfur, carbon dioxide, carbon tetrachloride. Complete coverage of uranium metal scrap or turnings with oil is essential for prevention of fire.

Uranium(IV)oxide: May spontaneously ignite on contact with air when heated above 700°C.

Uranium hydride: Keep away from strong oxidizers, water, halogenated hydrocarbons.

Uranyl chloride: Aqueous solutions are chemically unstable.

Uranyl nitrate(s): Keep away from combustible materials; reducing agents; Uranium hexafluoride: water.

Uranium hexafluoride: Water.

Permissible Exposure Limits in Air

OSHA PEL: (*natural & insoluble compounds*) 0.25 mg[U]/m^3 TWA; (*soluble*) 0.05 mg[U]/m^3 TWA.

NIOSH REL: (*natural & insoluble compounds*) 0.2 mg[U]/m^3 TWA; 0.6 mg[U]/m^3 STEL; (*soluble*) 0.05 mg[U]/m^3 TWA; Potential occupational carcinogen. Limit exposure to lowest feasible concentration. See *NIOSH Pocket Guide*, Appendix A.

ACGIH TLV®[1]: 0.2 mg[U]/m^3 TWA; 0.6 mg[U]/m^3 STEL, Confirmed Human Carcinogen; BEI issued.

NIOSH IDLH: 10 mg[U]/m^3.

Protective Action Criteria (PAC)*

Uranium metal and some insoluble compounds

TEEL-0: 0.25 mg/m^3

PAC-1: 0.6 mg/m^3

PAC-2: 2.5 mg/m^3

PAC-3: 10 mg/m^3

1344-57-6 (uranium oxide)

TEEL-0: 0.284 mg/m^3

PAC-1: 0.681 mg/m^3

PAC-2: **10** mg/m^3

PAC-3: **30** mg/m^3

*AEGLs (Acute Emergency Guideline Levels) & ERPGs (Emergency Response Planning Guideline) are in **bold face**.

36478-76-9 (uranyl nitrate)

TEEL-0: 0.0828 mg/m^3

PAC-1: 0.993 mg/m^3

PAC-2: 3 mg/m^3

PAC-3: 16.6 mg/m^3

Uranium, soluble compounds

TEEL-0: 0.05 mg/m^3

PAC-1: 0.6 mg/m^3

PAC-2: 2 mg/m^3

PAC-3: 10 mg/m^3

DFG MAK (uranium compounds): 0.25 mg[U]/m^3, measured as the inhalable fraction.

Australia: TWA 0.2 mg[U]/m^3, STEL 0.6 mg[U]/m^3, 1993; Belgium: TWA 0.2 mg[U]/m^3, STEL 0.5 mg[U]/m^3, 1993; Belgium: TWA 0.2 mg[U]/m^3, STEL 0.6 mg[U]/m^3, 1993; Denmark: TWA 0.2 mg[U]/m^3, 1999; Finland: TWA 0.2 mg[U]/m^3, 1999; Norway: TWA 0.2 mg[U]/m^3, 1999; the Netherlands: MAC-TGG 0.2 mg[U]/m^3, 2003; the Philippines: TWA 0.25 mg[U]/m^3, 1993; Poland: MAC (TWA) 0.015 mg[U]/m^3; MAC (STEL) 0.12 mg[U]/m^3, 1999; Poland: MAC (TWA) 0.075 mg[U]/m^3, 1993; Russia: STEL 0.015 mg[U]/m^3, 1993; Russia: STEL 0.075 mg[U]/m^3, 1993; Switzerland: MAK-W 0.2 mg[U]/m^3, 1999; Turkey: TWA 0.05 mg[U]/m^3, 1993; United Kingdom: LTEL 0.2 mg[U]/m^3, STEL 0.6 mg[U]/m^3, 1993.

Uranium, soluble compounds

OSHA PEL: 0.05 mg[U]/m^3 TWA.

NIOSH REL: TWA 0.05 mg[U]/m^3, Potential occupational carcinogen. Limit exposure to lowest feasible concentration. See *NIOSH Pocket Guide*, Appendix A.

ACGIH TLV®[1]: 0.2 mg[U]/m^3 TWA; 0.6 mg[U]/m^3 STEL, Confirmed Human Carcinogen.

DFG MAK (uranium compounds): 0.25 mg[U]/m^3, measured as the inhalable fraction.

NIOSH IDLH: 10 mg[U]/m^3.

Russia[43] sets a MAC of 0.015 mg[U]/m^3 in work-place air for soluble uranium compounds and 0.075 mg[U]/m^3 for insoluble compounds. Several states have set guidelines or standards for uranium in ambient air[60] ranging from 2.0 to 6.0 μg/m^3 (North Dakota) to 3.5 μg/m^3 (Virginia) to 4.0 μg/m^3 (Connecticut) to 5.0 μg/m^3 (Nevada).

EPA Limits and advisory levels:

United States, annual radiation exposure limits are found in Title 10, part 20 of the Code of Federal Regulations, and in equivalent state regulations.

Public dose limits due to licensed activities (NRC) 100 mRem (0.01 Sv)/year.

Air: Radium NESHAP = 10 mRem (0.1 Sv)*

Water: ^{226}Ra & ^{228}Ra (combined radium) MCL = 5 pCi/L; ^{224}Ra = 15 picocuries/liter (5 pCi/L)*

Indoor Air (advisory "action level"): 4 pCi/L radon (Rn)

*Both the air and water standards limit the increased lifetime cancer risk to about 2 in 10,000.

Guidance Matrix for Radiological Dispersal Device (RDD) Incidents

OSHA Occupational Exposure Limits for Ionizing Radiation 29 CFR 1910.1096	Rem (Sv)/calendar quarter
Whole body, head and trunk, active blood-forming organs, lens of eye, or gonads	1.25 (0.0125 Sv)
Hands and forearms, feet, and ankles	18.75 (0.1875 Sv)
Skin of whole body	7.5 (0.075 Sv)
Minors (workers under 18 yrs)	10% of above limits
Majors (workers over 18 yrs)	ADVANCE \d43 Rem (0.3 Sv) may be permitted under conditions specified in 29 CFR 1910.1096(b)(2)

Note: The Department of Homeland Security is currently chairing an interagency workgroup which is in the process of assessing the protective action guidance for response to an RDD event.

Determination in Air: No method is available.

Permissible Concentration in Water: No criteria set, but EPA[32] has suggested a permissible ambient goal of 3 μg/L based on health effects. Several states have set guidelines and standards for uranium in drinking water ranging from 10 μCi/L (Massachusetts) to 30 μCi/L (California) to 35 μCi/L (Arizona) to 15 μg/L (Colorado).

Routes of Entry: Inhalation of fume, dust, or gas; ingestion; skin and/or eye contact. The following uranium salts are reported to be capable of penetrating intact skin: uranyl nitrate, $UO_2(NO_3)_2 \cdot 6H_2H$; uranyl fluoride, UO_2F_2; uranium pentachloride, UCl_5; uranium trioxide (uranyl oxide), UO_3; sodium diuranate [sodium uranate (VI), $Na_2U_2O_7$]; uranium hexafluoride, UF_6.

Harmful Effects and Symptoms

Short Term Exposure: Prolonged contact with skin should be avoided to prevent radiation injury.

Uranium and its compounds are highly toxic substances. The compounds which are soluble in body fluids possess the highest toxicity. Poisoning has generally occurred as a result of accidents. Acute chemical toxicity produces damage primarily to the kidneys. Kidney changes precede in time and degree the effects on the liver. Chronic poisoning with prolonged exposure gives chest findings of pneumoconiosis, pronounced blood changes, and generalized injury.

It is difficult to separate the toxic chemical effects of uranium and its compounds from their radiation effects. The chronic radiation effects are similar to those produced by ionizing radiation. Reports now confirm that carcinogenicity is related to dose and exposure time. Cancer of the lung, osteosarcoma, and lymphoma have all been reported.

For soluble compounds: lacrimation, conjunctivitis, shortness of breath, coughing, chest rales, nausea, vomiting, skin burns, casts in urine, albuminuria, high blood urea nitrogen, lymphatic cancer. *For insoluble compounds:* dermatitis; cancer of lymphatic and blood-forming tissues.

Long Term Exposure: Can cause dermatitis, kidney damage, blood changes. A potential occupational carcinogen. Potential for cancer is a result of *alpha*-emitting properties and radioactive decay products (e.g., radon).

Points of Attack: For soluble compounds: respiratory system, blood, liver, kidneys, lymphatic system, skin, bone marrow. *Cancer site:* lungs. *For metal and insoluble compounds:* skin, kidneys, bone marrow, lymphatic system. *Cancer site:* lungs.

Medical Surveillance: NIOSH lists the following tests: whole blood (chemical/metabolite); complete blood count; chest X-ray; urine (chemical/metabolite); urinalysis (routine). Special attention should be given to the blood, lung, kidney, and liver in preemployment physical examinations. In periodic examinations, tests for blood changes, changes in chest X-rays, or for renal injury and liver damage are advisable. Uranium excretion in the urine has been used as an index of exposure. Whole body counting may also be useful. Blood Urea Nitrogen (BUN) for soluble compounds.

First Aid: If this chemical gets into the eyes, remove any contact lenses at once and irrigate immediately for at least 15 min, occasionally lifting upper and lower lids. Seek medical attention immediately. If this chemical contacts the skin, remove contaminated clothing and wash immediately with soap and water. Seek medical attention immediately. If this chemical has been inhaled, remove from exposure, begin rescue breathing (using universal precautions, including resuscitation mask) if breathing has stopped and CPR if heart action has stopped. Transfer promptly to a medical facility. When this chemical has been swallowed, get medical attention. Give large quantities of water and induce vomiting. Do not make an unconscious person vomit.

Personal Protective Methods: Soluble compounds, especially UF_6: Wear appropriate clothing to prevent any possibility of skin contact with UF_6. Wear eye protection to prevent any possibility of eye contact. Employees should wash immediately when skin is wet or contaminated with UF_6 and daily at the end of each work shift. Work clothing should be changed daily if it is possible that clothing is contaminated with UF_6. Remove nonimpervious clothing immediately if wet or contaminated UF_6. Provide emergency showers and eyewash if UF_6 is involved.

Insoluble compounds: Wear protective gloves and clothing to prevent any reasonable probability of skin contact. Safety equipment suppliers/manufacturers can provide recommendations on the most protective glove/clothing material for your operation. All protective clothing (suits, gloves, footwear, headgear) should be clean, available each day, and put on before work. Contact lenses should not be worn when working with this chemical. Wear dust-proof chemical goggles and face shield unless full face-piece respiratory protection is worn. Employees should wash immediately with soap when skin is wet or contaminated. Provide emergency showers and eyewash.

Respirator Selection:

Soluble Uranium Compounds: At concentrations above the NIOSH REL, or where there is no REL, at any detectable concentration: SCBAF: Pd,Pp (APF = 10,000) (any NIOSH/MSHA- or European Standard EN 149-approved self-contained breathing apparatus that has a full face-piece and is operated in a pressure-demand or other positive-pressure mode) or SaF: Pd,Pp: ASCBA (any supplied-air respirator that has a full face-piece and is operated in a pressure-demand or other positive-pressure mode in combination with an auxiliary, self-contained breathing apparatus operated in a pressure-demand or other positive-pressure mode). *Escape (halides):* 100F (APF = 50) (any air-purifying, full-face-piece respirator with an N100, R100, or P100 filter) or SCBAE (any appropriate escape-type, self-contained breathing apparatus). *Escape (nonhalides):* 100F (APF = 50) (any air-purifying, full-face-piece respirator with an N100, R100, or P100 filter) or SCBAE (any appropriate escape-type, self-contained breathing apparatus).

Insoluble Uranium Compounds: At concentrations above the NIOSH REL, or where there is no REL, at any detectable concentration: SCBAF: Pd,Pp (APF = 10,000) (any NIOSH/MSHA- or European Standard EN 149-approved self-contained breathing apparatus that has a full face-piece and is operated in a pressure-demand or other positive-pressure mode) or SaF: Pd,Pp: ASCBA (APF = 10,000) (any supplied-air respirator that has a full face-piece and is operated in a pressure-demand or other positive-pressure mode in combination with an auxiliary, self-contained breathing apparatus operated in a pressure-demand or other positive-pressure mode). *Escape:* 100F (APF = 50) (any air-purifying, full-face-piece respirator with an N100, R100, or P100 filter) or SCBAE (any appropriate escape-type, self-contained breathing apparatus) or SCBAE (any appropriate escape-type, self-contained breathing apparatus).

Storage: Radioactive. (1) Color Code—Red Stripe: Flammability Hazard: Store separately from all other flammable materials. (2) Color Code—Yellow Stripe (*strong reducing agent*): Reactivity Hazard; Store separately in an area isolated from flammables, combustibles, or other yellow-coded materials. Prior to working with this chemical you should be trained on its proper handling and storage. A regulated, marked area should be established where this chemical is handled, used, or stored in compliance with OSHA Standard 1910.1045.

Shipping: Uranium metal, pyrophoric, requires a shipping label of "RADIOACTIVE, SPONTANEOUSLY COMBUSTIBLE." It falls in Hazard Class 7.

Uranyl nitrate, solid, requires a shipping label of "RADIOACTIVE, OXIDIZER." It falls in Hazard Class 7.

Uranyl nitrate, hexahydrate solution, requires a shipping label of "CORROSIVE." It falls in Hazard Class 7.

Spill Handling: *Dry material:* Evacuate persons not wearing protective equipment from area of spill or leak until cleanup is complete. Remove all ignition sources. Collect powdered material in the most convenient and safe manner and deposit in sealed containers. Ventilate area after cleanup is complete. Uranium chips or turnings should be covered with oil to prevent fires, and collected in sealed containers for later disposal. *Liquid:* Evacuate and restrict persons not wearing protective equipment from area of spill or leak until cleanup is complete. Remove all ignition sources. Ventilate area of spill or leak. Absorb liquids in vermiculite, dry sand, earth, peat, carbon, or a similar material and deposit in sealed containers. Keep this chemical out of a confined space, such as a sewer, because of the possibility of an explosion, unless the sewer is designed to prevent the buildup of explosive concentrations. It may be necessary to contain and dispose of this chemical as a hazardous waste. If material or contaminated runoff enters waterways, notify downstream users of potentially contaminated waters. Contact your local or federal environmental protection agency for specific recommendations. If employees are required to clean up spills, they must be properly trained and equipped. OSHA 1910.120(q) may be applicable. Seek professional environmental engineering assistance from the US EPA's Environmental Response Team (ERT), Edison, NJ. Telephone 24-h hotline: 908-548-8730.

Fire Extinguishing: Uranium is an explosion hazard as dust or solid when exposed to flame. In case of fire, contact the local, state, or department of energy radiological response team. *Do not use water.* Use graphite, soda ash, powdered sodium chloride, or suitable dry powder.

Poisonous gases are produced in fire. If material or contaminated runoff enters waterways, notify downstream users of potentially contaminated waters. Notify local health and fire officials and pollution control agencies. From a secure, explosion-proof location, use water spray to cool exposed containers. If cooling streams are ineffective (venting sound increases in volume and pitch, tank discolors, or shows any signs of deforming), withdraw immediately to a secure position. If employees are expected to fight fires, they must be trained and equipped in OSHA 1910.156. The only respirators recommended for firefighting are self-contained breathing apparatuses that have full face-pieces and are operated in a pressure-demand or other positive-pressure mode.

Disposal Method Suggested: Disposal of wastes containing uranium (uranium and compounds) should follow guidelines set forth by the nuclear regulatory commission. Contact the nuclear regulatory commission regarding disposal notification. Recovery for reprocessing is the preferred method. Processes are available for uranium recovery from process wastewaters and process scrap. Burial at an authorized radioactive burial site.

References

US Environmental Protection Agency. (May 1977). *Toxicology of Metals, Vol. II: Uranium*, Report EPA-600/1-77-022. Research Triangle Park, NC, pp. 454–472

Sax, N. I. (Ed.). *Dangerous Properties of Industrial Materials Report*, 2, No. 2, 78–79 (1982), Uranyl Acetate and 4, No. 1, 99–102, Uranyl nitrate (1984)

Agency for Toxic Substances and Disease Registry (ATSDR). (1990). *Toxicological Profile for Radium*. Atlanta, GA: Division of Toxicology, US Department of Health and Human Services, Public Health Service

Bentor, Y. (November 2, 2010). *Chemical Element.com—Radium.* <http://www.chemicalelements.com/elements/ra.html>

29 CFR 1910.1096. *Ionizing Radiation.* OSHA Standard.

Environmental Protection Agency (EPA), Office of Radiation Programs. (May 1992). *Manual of Protective Action Guides and Protective Actions for Nuclear Incidents.* 16 MB PDF, 274 p.

US Department of Energy (DOE). (May 2001). *Radiological Emergency Response Health and Safety Manual*, Report DOE/NV/11718-440. 1 MB PDF, 103 p.

US Army Medical Research Institute of Chemical Defense (USAMRICD), Chemical Casualty Care Division (USAMRICD) MCMR-UV-ZM. (July 2000). *Field Management of Chemical Casualties* (2nd ed.). Aberdeen Proving Grounds, MD

Urea   U:0110

Molecular Formula: CH_4N_2O
Common Formula: H_2NCONH_2

Synonyms: Carbamide; Carbamide resin; Carbamimidic acid; Carbonyl diamide; Carbonyldiamine; Isourea; NCI-C02119; Prespersion, 75 Urea; Pseudourea; Supercel 3000; Ureaphil; Ureophil; Urevert; Varioform II

CAS Registry Number: 57-13-6

RTECS® Number: YR6250000

EC Number: 200-315-5

Regulatory Authority and Advisory Bodies

US EPA Gene-Tox Program, Positive: *In vitro* cytogenetics—human lymphocyte; Negative: Sperm morphology—mouse; Inconclusive: *E. coli* polA without S9.

US EPA, FIFRA 1998 Status of Pesticides: Canceled.

FDA—over-the-counter and proprietary drug.

Air Pollutant Standard Set. See below, "Permissible Exposure Limits in Air" section.

European/International Regulations: not listed in Annex 1.

WGK (German Aquatic Hazard Class): 1—Low hazard to waters.

Description: Urea is a white crystalline solid. Molecular weight = 60.07; Boiling point = (decomposes); Freezing/Melting point = 133°C. Hazard Identification (based on NFPA-704 M Rating System): Health 1, Flammability 0, Reactivity 0. Soluble in water.

Potential Exposure: Compound Description: Agricultural Chemical; Tumorigen, Drug, Mutagen; Reproductive Effector; Human Data; Primary Irritant. Urea is used in ceramics, cosmetics, paper processing, resins, adhesives, animal feeds; in the manufacture of isocyanurates, resins, and plastics; as a stabilizer in explosives; in medicines; and others.

Incompatibilities: Violent reaction with strong oxidizers, chlorine, permanganates, dichromates, nitrites, inorganic chlorides, chlorites, and perchlorates. Contact with hypochlorites can result in the formation of explosive compounds.

Permissible Exposure Limits in Air

AIHA WEEL: 10 mg/m^3 TWA.

Protective Action Criteria (PAC)

TEEL-0: 10 mg/m^3

PAC-1: 10 mg/m^3

PAC-2: 15 mg/m^3

PAC-3: 500 mg/m^3

The state of New York[61] has set a guidelines of 0.03 μg/m^3 for urea in ambient air.

Russia set a MAC of 0.2 mg/m^3 in ambient air in residential areas on a daily average basis.

Permissible Concentration in Water: Russia[43] set a MAC of 80.0 mg/L in water bodies used for domestic purposes.

Routes of Entry: Inhalation, ingestion, eye and/or skin contact.

Harmful Effects and Symptoms

Short Term Exposure: *Inhalation:* Causes irritation of the respiratory tract. Dust may cause difficult breathing especially if the person has asthma. *Skin:* May cause irritation, burning, or stinging. *Eyes:* Causes irritation. *Ingestion:* There have been no reported cases of human toxicity.

However, some toxic effects have been seen in sheep with impaired liver function.

Long Term Exposure: Prolonged skin contact may cause dermatitis.

Points of Attack: Skin.

First Aid: If this chemical gets into the eyes, remove any contact lenses at once and irrigate immediately for at least 15 min, occasionally lifting upper and lower lids. Seek medical attention immediately. If this chemical contacts the skin, remove contaminated clothing and wash immediately with soap and water. Seek medical attention immediately. If this chemical has been inhaled, remove from exposure, begin rescue breathing (using universal precautions, including resuscitation mask) if breathing has stopped and CPR if heart action has stopped. Transfer promptly to a medical facility. When this chemical has been swallowed, get medical attention. Give large quantities of water and induce vomiting. Do not make an unconscious person vomit.

Personal Protective Methods: Wear protective gloves and clothing to prevent any reasonable probability of skin contact. Safety equipment suppliers/manufacturers can provide recommendations on the most protective glove/clothing material for your operation. All protective clothing (suits, gloves, footwear, headgear) should be clean, available each day, and put on before work. Contact lenses should not be worn when working with this chemical. Wear dust-proof chemical goggles and face shield unless full face-piece respiratory protection is worn. Employees should wash immediately with soap when skin is wet or contaminated. Provide emergency showers and eyewash.

Respirator Selection: Follow regulations in OSHA 29CFR1910.134 or European Standard EN149. Use a NIOSH/MSHA- or European Standard EN149-approved respirator; or use an approved supplied-air respirator with a full face-piece operated in the positive-pressure mode, or with a full face-piece, hood, or helmet in the continuous-flow mode; or use a NIOSH/MSHA- or European Standard EN149-approved self-contained breathing apparatus with a full face-piece operated in a pressure-demand or other positive-pressure mode.

Storage: Color Code—Green: General storage may be used. Prior to working with this chemical you should be trained on its proper handling and storage. Store in tightly closed containers in a cool, well-ventilated area away from oxidizers. Where possible, automatically transfer material from storage containers to process containers.

Spill Handling: Evacuate persons not wearing protective equipment from area of spill or leak until cleanup is complete. Remove all ignition sources. Collect powdered material in the most convenient and safe manner and deposit in sealed containers. Ventilate area after cleanup is complete. It may be necessary to contain and dispose of this chemical as a hazardous waste. If material or contaminated runoff enters waterways, notify downstream users of potentially contaminated waters. Contact your local or federal environmental protection agency for specific recommendations. If employees are required to clean up spills, they must be properly trained and equipped. OSHA 1910.120(q) may be applicable.

Fire Extinguishing: This chemical is a combustible solid. Use dry chemical, carbon dioxide, water spray, or foam extinguishers. Poisonous gases are produced in fire, including nitrogen oxides. If material or contaminated runoff enters waterways, notify downstream users of potentially contaminated waters. Notify local health and fire officials and pollution control agencies. From a secure, explosion-proof location, use water spray to cool exposed containers. If cooling streams are ineffective (venting sound increases in volume and pitch, tank discolors, or shows any signs of deforming), withdraw immediately to a secure position. If employees are expected to fight fires, they must be trained and equipped in OSHA 1910.156. The only respirators recommended for firefighting are self-contained breathing apparatuses that have full face-pieces and are operated in a pressure-demand or other positive-pressure mode.

Disposal Method Suggested: Controlled incineration in equipment containing a scrubber or thermal unit to reduce nitrogen oxide emissions.

References

New York State Department of Health. (April 1986). *Chemical Fact Sheet: Urea.* Albany, NY: Bureau of Toxic Substance Assessment

US Environmental Protection Agency, Special Review and Reregistration Division Office of Pesticide Programs. (1998). *Agency Status of Pesticides in Registration, Reregistration, and Special Review* (Rainbow Report). Washington, DC

Urethane   U:0120

Molecular Formula: $C_3H_7NO_2$
Common Formula: $H_2NCOOC_2H_5$
Synonyms: A 11032; Aethylcarbamat (German); Aethylurethan (German); Carbamic acid, ethyl ester; Carbamidsaeure-aethylester (German); Estane 5703; Ethyl carbamate; Ethylurethan; *o*-Ethylurethane; Ethyl urethane; Leucethane; Leucothane; NSC 746; Pracarbamin; Pracarbamine; U-Compound; Uretano (Spanish); Urethan; Urethane
CAS Registry Number: 51-79-6
RTECS® Number: FA8400000
UN/NA & ERG Number: UN2811 (toxic solid, organic, n.o.s.)/154
EC Number: 200-123-1[*Annex I Index No.:* 607-149-00-6]
Regulatory Authority and Advisory Bodies
Carcinogenicity: IARC: Animal, Sufficient Evidence; Human, Inadequate Evidence, Group 2A; NTP: 11th Report on Carcinogens, 2004: Reasonably anticipated to be a human carcinogen; NTP: Carcinogenesis studies; on test (2-year study), October 2000; NTP: Toxicity studies, RPT#TOX-52, October 2000.

US EPA Gene-Tox Program, Positive: Carcinogenicity—mouse/rat; SHE—clonal assay; Positive: Cell transform.—mouse embryo; Positive: Cell transform.—RLV F344 rat embryo; Positive: *D. melanogaster*—whole sex chrom. loss; Positive: *D. melanogaster*—reciprocal translocation; Positive: *Mammalian micronucleus*; *N. crassa—reversion*; Positive: *D. melanogaster* sex-linked lethal; Positive: *S. cerevisiae* gene conversion; Positive/dose response: *In vitro* SCE—nonhuman; Negative: *D. melanogaster*—nondisjunction; Host-mediated assay; Negative: *E. coli* polA with S9; Histidine reversion—Ames test; Negative: Sperm morphology—mouse; TRP reversion; Negative: *S. cerevisiae*—homozygosis; Inconclusive: *E. coli* polA without S9; *In vitro* UDS—human fibroblast.
Banned or Severely Restricted (many countries) (UN).[13]
Air Pollutant Standard Set. See below, "Permissible Exposure Limits in Air" section.
Clean Air Act: Hazardous Air Pollutants (Title I, Part A, Section 112).
US EPA Hazardous Waste Number (RCRA No.): U238.
RCRA, 40CFR261, Appendix 8 Hazardous Constituents.
Reportable Quantity (RQ): 100 lb (45.4 kg).
EPCRA Section 313 Form R *de minimis* concentration reporting level: 0.1%.
California Proposition 65 Chemical: Cancer 1/1/88; Developmental/Reproductive toxin 10/1/94.
Canada, WHMIS, Ingredients Disclosure List Concentration 0.1%.
European/International Regulations: Hazard Symbol: T; Risk phrases: R45; Safety phrases: S53; S45 (see Appendix 4).
WGK (German Aquatic Hazard Class): No value assigned.
Description: Urethane is a colorless, almost odorless crystalline solid or powder. Molecular weight = 89.11; Specific gravity (H_2O:1) = 1.11 at 25°C; Boiling point = 183°C; Freezing/Melting point = 49°C; Flash point = 92°C (cc). Hazard Identification (based on NFPA-704 M Rating System): Health 3, Flammability 2, Reactivity 0. Slightly soluble in water.
Potential Exposure: Compound Description: Agricultural Chemical; Drug, Tumorigen, Mutagen; Reproductive Effector. Urethane is used as a chemical intermediate in the manufacture of pharmaceuticals, pesticides, and fungicides; in the preparation of amino resins. It may be reacted with formaldehyde to give cross-linking agents that impart wash-and-wear properties to fabrics. It has also been used as a solubilizer and cosolvent in the manufacture of pesticides, fumigants, and cosmetics. It was formerly used in the treatment of leukemia. It occurs when diethylpyrocarbonate, a preservative used in wines, fruit juices, and soft drinks, is added to aqueous solutions.
Incompatibilities: Gallium, perchlorate, and strong oxidizers.
Permissible Exposure Limits in Air
Protective Action Criteria (PAC)
TEEL-0: 500 mg/m^3

PAC-1. 500 mg/m^3
PAC-2: 500 mg/m^3
PAC-3: 500 mg/m^3

DFG MAK: [skin] Carcinogen Category 2; Germ Cell Mutation Category 3A, *as carbamic acid ethyl ester.*

Austria: carcinogen, 1999; Finland: carcinogen, 1993; Sweden: carcinogen, 1999; Switzerland: carcinogen, 1999. Several states have set guidelines or standards for urethane in ambient air[60] ranging from zero (North Dakota) to 0.03 μg/m^3 (New York) to 5000.0 μg/m^3 (South Carolina).

Determination in Water: Octanol−water coefficient: Log K_{ow} = −0.15

Routes of Entry: Inhalation, ingestion, skin and/or eye contact.

Harmful Effects and Symptoms

Short Term Exposure: Urethane can affect you when breathed in and by passing through your skin. High exposures may affect the CNS, causing dizziness, lightheadedness, and unconsciousness. Very high exposures can cause damage to the liver, brain, and blood-forming organs.

Long Term Exposure: Urethane is a carcinogen and may be a teratogen. Handle with extreme caution. Repeated exposures can damage the liver, brain, and blood-forming organs.

Points of Attack: Liver, brain, blood.

Medical Surveillance: Before beginning employment, and at regular times after that, for those with frequent or potentially high exposures, the following are recommended: liver function tests. Complete blood count. If symptoms develop or overexposure suspected, the following may be useful: examination of the nervous system.

First Aid: Skin Contact[52]*:* Flood all areas of body that have contacted the substance with water. Do not wait to remove contaminated clothing; do it under the water stream. Use soap to help assure removal. Isolate contaminated clothing when removed to prevent contact by others.

Eye Contact: Remove any contact lenses at once. Immediately flush eyes well with copious quantities of water or normal saline for at least 20−30 min. Seek medical attention.

Inhalation: Leave contaminated area immediately; breathe fresh air. Proper respiratory protection must be supplied to any rescuers. If coughing, difficult breathing, or any other symptoms develop, seek medical attention at once, even if symptoms develop many hours after exposure.

Ingestion: Contact a physician, hospital, or poison center at once. If the victim is unconscious or convulsing, do not induce vomiting or give anything by mouth. Assure that this airway is open and lay him on his side with his head lower than his body and transport immediately to a medical facility. If conscious and not convulsing, give a glass of water or milk to dilute the substance. Vomiting should not be induced without a physician's advice.

Personal Protective Methods: Wear protective gloves and clothing to prevent any reasonable probability of skin contact. Safety equipment suppliers/manufacturers can provide recommendations on the most protective glove/clothing material for your operation. All protective clothing (suits, gloves, footwear, headgear) should be clean, available each day, and put on before work. Contact lenses should not be worn when working with this chemical. Wear dust-proof chemical goggles and face shield unless full face-piece respiratory protection is worn. Employees should wash immediately with soap when skin is wet or contaminated. Provide emergency showers and eyewash.

Respirator Selection: Follow regulations in OSHA 29CFR1910.134 or European Standard EN149. Use a NIOSH/MSHA- or European Standard EN149-approved respirator; or use an approved supplied-air respirator with a full face-piece operated in the positive-pressure mode, or with a full face-piece, hood, or helmet in the continuous-flow mode; or use a NIOSH/MSHA- or European Standard EN149-approved self-contained breathing apparatus with a full face-piece operated in a pressure-demand or other positive-pressure mode.

Storage: Color Code—Blue: Health Hazard/Poison: Store in a secure poison location. Prior to working with this chemical you should be trained on its proper handling and storage. Urethane must be stored to avoid contact with strong oxidizers (such as chlorine, bromine, and fluorine), strong acids (such as hydrochloric, sulfuric, and nitric), strong bases, camphor, menthol, salol, or thymol since violent reactions occur. Store in tightly closed containers in a cool, well-ventilated area. A regulated, marked area should be established where this chemical is handled, used, or stored in compliance with OSHA Standard 1910.1045.

Shipping: Toxic solids, organic, n.o.s. require a label of "POISONOUS/TOXIC MATERIALS." Urethane falls in Hazard Class 6.1.

Spill Handling: Evacuate persons not wearing protective equipment from area of spill or leak until cleanup is complete. Remove all ignition sources. Collect powdered material in the most convenient and safe manner and deposit in sealed containers. Ventilate area after cleanup is complete. It may be necessary to contain and dispose of this chemical as a hazardous waste. If material or contaminated runoff enters waterways, notify downstream users of potentially contaminated waters. Contact your local or federal environmental protection agency for specific recommendations. If employees are required to clean up spills, they must be properly trained and equipped. OSHA 1910.120(q) may be applicable.

Fire Extinguishing: This chemical is a combustible solid. Use dry chemical, carbon dioxide, water spray, or alcohol foam extinguishers. Poisonous gases are produced in fire, including nitrogen oxides. If material or contaminated run-off enters waterways, notify downstream users of potentially contaminated waters. Notify local health and fire officials and pollution control agencies. From a secure, explosion-proof location, use water spray to cool exposed containers. If cooling streams are ineffective (venting sound increases in volume and pitch, tank discolors, or shows any signs of deforming), withdraw immediately to a secure position. If employees are expected to fight fires, they must be trained and equipped in OSHA 1910.156. The only respirators recommended for firefighting are self-contained breathing apparatuses that have full face-pieces and are operated in a pressure-demand or other positive-pressure mode.

Disposal Method Suggested: Consult with environmental regulatory agencies for guidance on acceptable disposal practices. Generators of waste containing this contaminant (≥100 kg/mo) must conform with EPA regulations governing storage, transportation, treatment, and waste disposal. Controlled incineration (incinerator equipped with a scrubber or thermal unit to reduce nitrogen oxides emissions).

References

US Environmental Protection Agency. (1979). *Chemical Hazard Information Profile: Urethane.* Washington, DC
New Jersey Department of Health and Senior Services. (May 2001). *Hazardous Substances Fact Sheet: Urethane.* Trenton, NJ

V

Valeraldehyde   V:0100

Molecular Formula: $C_5H_{10}O$
Common Formula: $CH_3(CH_2)_3CHO$
Synonyms: Amyl aldehyde; Butyl formal; *n*-Pentanal; Pentanal; Valeral; *n*-Valeraldehyde; Valerianic aldehyde; Valeric acid aldehyde; *n*-Valeric aldehyde; Valeric aldehyde
CAS Registry Number: 110-62-3
RTECS® Number: YV3600000
UN/NA & ERG Number: UN2058/129
EC Number: 203-784-4
Regulatory Authority and Advisory Bodies
Air Pollutant Standard Set. See below, "Permissible Exposure Limits in Air" section.
Canada, WHMIS, Ingredients Disclosure List Concentration 1.0%.
European/International Regulations: not listed in Annex 1.
WGK (German Aquatic Hazard Class): 1—Low hazard to waters.
Description: Valeraldehyde is a colorless liquid with a strong acrid, pungent odor. The odor threshold is 0.028 ppm. Molecular weight = 86.15; Specific gravity (H_2O:1) = 0.81 at 25°C; Boiling point = 103°C; Freezing/Melting point = −92°C; Vapor pressure = 26 mmHg; Flash point = 12°C (oc); Autoignition temperature = 222°C. Explosive limits in air = LEL = 1.4%; UEL = 7.2%. Hazard Identification (based on NFPA-704 M Rating System): Health 3, Flammability 3, Reactivity 0. Moderately soluble in water; solubility = 1.4% at 20°C.
Potential Exposure: Compound Description: Mutagen, Primary Irritant. Valeraldehyde is used in food flavorings and in resin chemistry. It is also used in the acceleration of rubber vulcanization.
Incompatibilities: Strong oxidizers, caustics, amines.
Permissible Exposure Limits in Air
Conversion factor: 1 ppm = 3.53 mg/m³ at 25°C & 1 atm.
OSHA PEL: None.
NIOSH REL: 50 ppm/175 mg/m³ TWA, See Appendix C (Aldehydes) of the *NIOSH Pocket Guide*. Limited studies to date indicate that these substances have chemical reactivity and mutagenicity similar to acetaldehyde and malonaldehyde.
ACGIH TLV®[1]: 50 ppm/176 mg/m³ TWA.
No TEEL available.
Australia: TWA 50 ppm (175 mg/m³), 1993; Belgium: TWA 50 ppm (176 mg/m³), 1993; Denmark: TWA 50 ppm (175 mg/m³), 1999; Finland: TWA 50 ppm (175 mg/m³), STEL 75 ppm (265 mg/m³), 1999; France: VME 50 ppm (175 mg/m³), 1999; Norway: TWA 0.3 mg/m³, 1999; Switzerland: MAK-W 50 ppm (175 mg/m³), 1999; the Netherlands: MAC-TGG 175 mg/m³, 2003; Argentina, Bulgaria, Columbia, Jordan, South Korea, New Zealand, Singapore, Vietnam; ACGIH TLV®: TWA 50 ppm. Several states have set guidelines or standards for valeraldehyde in ambient air[60] ranging from 1.75 mg/m³ (North Dakota) to 2.9 mg/m³ (Virginia) to 3.5 mg/m³ (Connecticut) to 4.167 mg/m³ (Nevada).
Determination in Air: Use NIOSH Analytical Method (IV), #2018, #2536, valeraldehyde, OSHA Analytical Method 85.
Determination in Water: Octanol−water coefficient: Log K_{ow} = 1.31. See also NIOSH (IV) #2539, aldehydes, screening.
Routes of Entry: Inhalation, ingestion, skin and/or eye contact.
Harmful Effects and Symptoms
Short Term Exposure: Valeraldehyde can affect you when breathed in. Contact can severely irritate the eyes, skin, nose, and throat. Exposure to very high levels can cause you to feel dizzy and lightheaded. Poisonous if swallowed.
Long Term Exposure: Testing has not been completed to determine the carcinogenicity of *n*-valeraldehyde. However, the limited studies to date indicate that these substances have chemical reactivity and mutagenicity similar to acetaldehyde and malonaldehyde. Therefore, NIOSH recommends that careful consideration should be given to reducing exposures to this aldehyde. Further information can be found in the *NIOSH Current Intelligence Bulletin 55: Carcinogenicity of Acetaldehyde and Malonaldehyde, and Mutagenicity of Related Low-Molecular-Weight Aldehydes* [DHHS (NIOSH), Publication No. 91-112].
Points of Attack: Eyes, skin, respiratory system.
Medical Surveillance: See Long Term Exposure.
First Aid: If this chemical gets into the eyes, remove any contact lenses at once and irrigate immediately for at least 15 min, occasionally lifting upper and lower lids. Seek medical attention immediately. If this chemical contacts the skin, remove contaminated clothing and wash immediately with soap and water. Seek medical attention immediately. If this chemical has been inhaled, remove from exposure, begin rescue breathing (using universal precautions, including resuscitation mask) if breathing has stopped and CPR if heart action has stopped. Transfer promptly to a medical facility. When this chemical has been swallowed, get medical attention. Give large quantities of water and induce vomiting. Do not make an unconscious person vomit.
Personal Protective Methods: Wear protective gloves and clothing to prevent any reasonable probability of skin contact. Safety equipment suppliers/manufacturers can provide recommendations on the most protective glove/clothing material for your operation. All protective clothing (suits, gloves, footwear, headgear) should be clean, available each day, and put on before work. Contact lenses should not be worn when working with this chemical. Wear splash-proof chemical goggles and face shield unless full-face-piece respiratory protection is worn. Employees should wash immediately with soap when skin is wet or contaminated. Provide emergency showers and eyewash.

Respirator Selection: Where there is potential for exposures *over 50 ppm*, use a NIOSH/MSHA- or European Standard EN149-approved full-face-piece respirator with an organic vapor cartridge/canister. Increased protection is obtained from full-face-piece powered air-purifying respirators. *Where there is potential for high exposures*, use a NIOSH/MSHA- or European Standard EN149-approved supplied-air respirator with a full-face-piece operated in the positive-pressure mode, or with a full face-piece, hood, or helmet in the continuous-flow mode; or use a NIOSH/MSHA- or European Standard EN149-approved self-contained breathing apparatus with a full face-piece operated in pressure-demand or other positive-pressure mode.

Storage: Color Code—Red: Flammability Hazard: Store in a flammable liquid storage area or approved cabinet away from ignition sources and corrosive and reactive materials. Prior to working with this chemical you should be trained on its proper handling and storage. Store in tightly closed containers in a cool, well-ventilated area. Sources of ignition, such as smoking and open flames, are prohibited where valeraldehyde is handled, used, or stored. Metal containers involving the transfer of 5 gallons or more of valeraldehyde should be grounded and bonded. Drums must be equipped with self-closing valves, pressure vacuum bungs, and flame arresters. Use only nonsparking tools and equipment, especially when opening and closing containers of valeraldehyde.

Shipping: This compound requires a shipping label of "FLAMMABLE LIQUID." It falls in Hazard Class 3 and Packing Group II.

Spill Handling: Evacuate and restrict persons not wearing protective equipment from area of spill or leak until cleanup is complete. Remove all ignition sources. Ventilate area of spill or leak. Absorb liquids in vermiculite, dry sand, earth, peat, carbon, or a similar material and deposit in sealed containers. Keep this chemical out of a confined space, such as a sewer, because of the possibility of an explosion, unless the sewer is designed to prevent the buildup of explosive concentrations. It may be necessary to contain and dispose of this chemical as a hazardous waste. If material or contaminated runoff enters waterways, notify downstream users of potentially contaminated waters. Contact your local or federal environmental protection agency for specific recommendations. If employees are required to clean up spills, they must be properly trained and equipped. OSHA 1910.120(q) may be applicable.

Fire Extinguishing: This chemical is a flammable liquid. Poisonous gases are produced in fire. Use dry chemical, carbon dioxide, or alcohol foam extinguishers. Vapors are heavier than air and will collect in low areas. Vapors may travel long distances to ignition sources and flashback. Vapors in confined areas may explode when exposed to fire. Containers may explode in fire. Storage containers and parts of containers may rocket great distances, in many directions. If material or contaminated runoff enters waterways, notify downstream users of potentially contaminated waters. Notify local health and fire officials and pollution control agencies. From a secure, explosion-proof location, use water spray to cool exposed containers. If cooling streams are ineffective (venting sound increases in volume and pitch, tank discolors, or shows any signs of deforming), withdraw immediately to a secure position. If employees are expected to fight fires, they must be trained and equipped in OSHA 1910.156. The only respirators recommended for firefighting are self-contained breathing apparatuses that have full face-pieces and are operated in a pressure-demand or other positive-pressure mode.

Disposal Method Suggested: Dissolve or mix the material with a combustible solvent and burn in a chemical incinerator equipped with an afterburner and scrubber. All federal, state, and local environmental regulations must be observed.

References

New Jersey Department of Health and Senior Services. (November 2000). *Hazardous Substances Fact Sheet: Valeraldehyde.* Trenton, NJ

Vanadium & inorganic compounds   V:0110

Molecular Formula: V

Synonyms: Elemental vanadium; Vanadio (Spanish); Vanadium-51; Vanadium, elemental

CAS Registry Number: 7440-62-2; 12070-10-9 (Vanadium carbide); *(alt.)* 11130-21-5

RTECS® Number: YW1355000

UN/NA & ERG Number: UN3285/151 (vanadium compound, n.o.s.)

EC Number: 231-171-1; 235-122-5 (vanadium carbide); 235-122-5 (vanadium carbide)

Regulatory Authority and Advisory Bodies

Carcinogenicity: DFG MAK: Carcinogen Category 2.

Air Pollutant Standard Set. See below, "Permissible Exposure Limits in Air" section.

Except when contained in an alloy:

Clean Water Act: Section 313 Water Priority Chemicals (57FR41331, 9/9/92).

RCRA 40CFR268.48; 61FR15654, Universal Treatment Standards: Wastewater (mg/L), 4.3; Nonwastewater (mg/L), 0.23.

RCRA 40CFR264, Appendix 9; TSD Facilities Ground Water Monitoring List. Suggested test method(s) (PQL µg/L): total dust 6010 (80); 7910 (2000); 7911 (40).

Safe Drinking Water Act: Priority List (55 FR 1470) as vanadium.

EPCRA Section 313 Form R *de minimis* concentration reporting level: 1.0%.

Canada, WHMIS, Ingredients Disclosure List Concentration 1.0%, elemental.

European/International Regulations: not listed in Annex 1.

WGK (German Aquatic Hazard Class): Nonwater polluting agent.

Description: Vanadium is a light-gray or silver-white ductile solid, lustrous powder, or fused hard lump. Molecular weight = 50.94; Boiling point = 3380°C; Freezing/Melting point = 1917°C. Hazard Identification (based on NFPA-704 M Rating System) (*Fume and dust*): Health 3, Flammability 0, Reactivity 0. Practically insoluble in water. It is produced by roasting the ores, thermal decomposition of the iodide; or from petroleum residues; slags from ferrovanadium production; or soot from oil burning.
Vanadium carbide (CV) is a dark-gray powder. Insoluble in water.

Potential Exposure: Most of the vanadium produced is used in ferrovanadium and of this the majority is used in high speed and other alloy steels with only small amounts in tool or structural steels. It is usually combined with chromium, nickel, manganese, boron, and tungsten in steel alloys. Vanadium carbide is used in cutting tool bits. Melting point 2750−2810°C.

Incompatibilities: Violent reaction with strong oxidizers: chlorine, bromine trifluoride, lithium, nitryl fluoride, chlorine trifluoride.

Permissible Exposure Limits in Air
NIOSH IDLH = 35 mg[V]/m^3.
Protective Action Criteria (PAC)
TEEL-0: 0.06 mg/m^3
PAC-1: 1.5 mg/m^3
PAC-2: 12.5 mg/m^3
PAC-3: 35 mg/m^3
DFG MAK: inhalable fraction, Carcinogen Category: 2; Germ cell mutagen group: 2.
Dust and fume: see vanadium pentoxide V:0120.

Determination in Air: Use NIOSH Analytical Method #7300, Elements, #7504, Vanadium oxides.

Permissible Concentration in Water: Russia[43] set a MAC of 0.1 mg/L in water bodies used for domestic purposes. There is no US standard for vanadium in drinking water, but EPA[32] has suggested a permissible ambient goal of 7 μg/L based on health effects. The lack of data on acute or chronic oral toxicity is not surprising because of the extremely low absorption of vanadium from the gastrointestinal tract. Inhaled vanadium can produce adverse health effects.

Determination in Water: Russia set a limit of 0.1 mg/L for vanadium as a maximum permissible limit for water basins.

Routes of Entry: *For dust and fume:* Inhalation, ingestion, skin and/or eye contact.

Harmful Effects and Symptoms

Short Term Exposure: Vanadium can affect you when breathed in. Exposure may irritate the eyes, nose, throat, and air passages, with cough and phlegm. Eye contact may cause irritation. See also "Vanadium Pentoxide." *Ingestion:* acute poisoning in animals by ingestion of vanadium compounds has been shown to cause nervous disturbances, leg paralysis, respiratory failure, convulsions, bloody diarrhea, coma, and death.

Long Term Exposure: Can cause greenish-black coloration of the tongue and metallic taste. May cause eczema. Dust and/or fume may cause respiratory irritation; bronchitis may develop, with cough, fine rales, wheezing, dyspnea (breathing difficulty). See also "Vanadium Pentoxide."

Points of Attack: Eyes, skin, respiratory system.

Medical Surveillance: For those with frequent or potentially high exposure (half the TLV or greater), the following are recommended before beginning work and at regular times after that: lung function tests. If symptoms develop or overexposure is suspected, the following may be useful: Urine test for vanadium. An EEC guideline recommends the following based on vanadium urine levels: an every-4-month control at the workplace if a 50 μg/L is reached and an annual control if the 5 μg/L urine level is reached. In case levels exceed 50 μg/L urine, workers should be temporarily removed from risk. See also "Vanadium Pentoxide."

First Aid: If this chemical gets into the eyes, remove any contact lenses at once and irrigate immediately for at least 15 min, occasionally lifting upper and lower lids. Seek medical attention immediately. If this chemical contacts the skin, remove contaminated clothing and wash immediately with soap and water. Seek medical attention immediately. If this chemical has been inhaled, remove from exposure, begin rescue breathing (using universal precautions, including resuscitation mask) if breathing has stopped and CPR if heart action has stopped. Transfer promptly to a medical facility. When this chemical has been swallowed, get medical attention. Give large quantities of water and induce vomiting. Do not make an unconscious person vomit.

Personal Protective Methods: Wear protective gloves and clothing to prevent any reasonable probability of skin contact. Safety equipment suppliers/manufacturers can provide recommendations on the most protective glove/clothing material for your operation. All protective clothing (suits, gloves, footwear, headgear) should be clean, available each day, and put on before work. Contact lenses should not be worn when working with this chemical. Wear dust-proof chemical goggles and face shield unless full-face-piece respiratory protection is worn. Employees should wash immediately with soap when skin is wet or contaminated. Provide emergency showers and eyewash.

Respirator Selection: NIOSH/OSHA (as V): *0.5 mg/m^3:* 100XQ* (APF = 10) (any air-purifying respirator with an N100, R100, or P100 filter (including N100, R100, and P100 filtering face-pieces) except quarter-mask respirators) or SA* (any supplied-air respirator). *1.25 mg/m^3:* Sa:Cf* (APF = 25) (any supplied-air respirator operated in a continuous-flow mode) or PaprHie* (APF = 25) (any powered, air-purifying respirator with a high-efficiency particulate filter). *2.5 mg/m^3:* 100F (APF = 50) (any air-purifying, full-face-piece respirator with an N100, R100, or P100 filter) or PaprHie* (APF = 25) (any powered, air-purifying respirator with a high-efficiency particulate filter) or SCBAF (APF = 50) (any self-contained breathing apparatus with a full face-piece) or SaF (APF = 50) (any supplied-air respirator with a full face-piece). *Emergency or planned entry into unknown concentrations or IDLH conditions:* SCBAF: Pd,Pp (APF = 10,000) (any self-contained breathing apparatus that has a full face-piece and is operated in a pressure-demand or other positive-pressure mode) or SaF: Pd,Pp: ASCBA (APF = 10,000) (any supplied-air respirator that has a full face-piece and is operated in a pressure-demand or other positive-pressure mode in combination with an auxiliary self-contained breathing apparatus operated in a pressure-demand or other positive-pressure mode). *Escape:* 100F (APF = 50) (any air-purifying, full-face-piece respirator with an N100, R100, or P100 filter) or SCBAE (any appropriate escape-type, self-contained breathing apparatus).

*Substance reported to cause eye irritation or damage; may require eye protection.

Storage: Color Code—Blue: Health Hazard/Poison: Store in a secure poison location. Prior to working with this chemical you should be trained on its proper handling and storage. Vanadium must be stored to avoid contact with oxidizers (such as chlorates, nitrates, chlorine, and bromine trifluoride), since violent reactions occur. Store the powder in tightly closed containers in a cool, well-ventilated area away from heat and sparks. Use only nonsparking tools and equipment, especially when opening and closing containers of the powder.

Shipping: Vanadium compound, n.o.s. requires a shipping label of "POISONOUS/TOXIC MATERIALS." It falls in Hazard Class 6.1 and can fall into Packing Groups I, II, or III.

Spill Handling: Evacuate persons not wearing protective equipment from area of spill or leak until cleanup is complete. Remove all ignition sources. Collect powdered material in the most convenient and safe manner and deposit in sealed containers. Ventilate area after cleanup is complete. It may be necessary to contain and dispose of this chemical as a hazardous waste. If material or contaminated runoff enters waterways, notify downstream users of potentially contaminated waters. Contact your local or federal environmental protection agency for specific recommendations. If employees are required to clean up spills, they must be properly trained and equipped. OSHA 1910.120(q) may be applicable.

Fire Extinguishing: Large pieces of vanadium are not combustible. The fine powder can burn. Use dry chemicals appropriate for extinguishing metal fires. *Do not use water.* Poisonous gases are produced in fire, including vanadium oxide. If material or contaminated runoff enters waterways, notify downstream users of potentially contaminated waters. Notify local health and fire officials and pollution control agencies. From a secure, explosion-proof location, use water spray to cool exposed containers. If cooling streams are ineffective (venting sound increases in volume and pitch, tank discolors, or shows any signs of deforming), withdraw immediately to a secure position. If employees are expected to fight fires, they must be trained and equipped in OSHA 1910.156. The only respirators recommended for firefighting are self-contained breathing apparatuses that have full face-pieces and are operated in a pressure-demand or other positive-pressure mode.

References

National Institute for Occupational Safety and Health. (1977). *Criteria for a Recommended Standard: Occupational Exposure to Vanadium*, NIOSH Document No. 77-222

National Academy of Sciences. (1974). *Medical and Biologic Effects of Environmental Pollutants: Vanadium.* Washington, DC

New Jersey Department of Health and Senior Services. (January 2007). *Hazardous Substances Fact Sheet: Vanadium.* Trenton, NJ

Vanadium pentoxide   V:0120

Molecular Formula: O_5V_2

Common Formula: V_2O_5

Synonyms: Anhydride vanadique (French); C.I. 77938; Divanadium pentoxide; Pentoxido de vanadilo (Spanish); UN2862; Vanadic acid anhydride; Vanadic anhydride; Vanadium oxide; Vanadium(5+) oxide; Vanadium(V) oxide; Vanadiumpentoxid (German); Vanadium pentoxide; Vanadium, pentoxyde de (French)

CAS Registry Number: 1314-62-1

RTECS® Number: YW2450000 (*dust*); YW2460000 (fume)

UN/NA & ERG Number: UN2862/151

EC Number: 215-239-8 [*Annex I Index No.:* 023-001-00-8]

Regulatory Authority and Advisory Bodies

Carcinogenicity: IARC: Human Inadequate Evidence; Animal Sufficient Evidence, *possibly carcinogenic to humans*, Group 2B; NCI: Carcinogenesis Studies (*dust*): (inhalation); clear evidence: mouse; rat; NTP: Carcinogenesis studies (fume); on test (2-year study), October 2000.

An OSHA specifically regulated substance.

US EPA Gene-Tox Program, Positive: *B. subtilis* rec assay.

Air Pollutant Standard Set. See below, "Permissible Exposure Limits in Air" section.

Clean Water Act: Section 311 Hazardous Substances/RQ 40CFR117.3 (same as CERCLA, see below).

US EPA Hazardous Waste Number (RCRA No.): P120.

RCRA, 40CFR261, Appendix 8 Hazardous Constituents.

RCRA 40CFR264, Appendix 9; TSD Facilities Ground Water Monitoring List. Suggested test method(s) (PQL μg/L): total dust 6010 (80); 7910 (2000); 7911 (40).

Superfund/EPCRA 40CFR355, Extremely Hazardous Substances: TPQ = 100/10,000 lb (45.4/4540 kg).

Reportable Quantity (RQ): 1000 lb (454 kg).

California Proposition 65 Chemical 2/11/05.

EPCRA Section 313 Form R *de minimis* concentration reporting level: 1.0%.

Canada, WHMIS, Ingredients Disclosure List Concentration 0.1% pentoxide.

European/International Regulations: Hazard Symbol: T, Xn; Risk phrases: R20/22; R37; R48/23; R51/53; R63; R68; Safety phrases: S1/2; S36/37; S38; S45; S61 (see Appendix 4). WGK (German Aquatic Hazard Class): 3—Severe hazard to waters.

Description: Vanadium pentoxide dust is an odorless, yellow to red crystal, or powder; or a fume when vanadium is heated. Vanadium pentoxide fume is a finely divided particulate dispersed in air. Molecular weight = 181.88; Specific gravity (H_2O:1) = 3.36 at 25°C; Boiling point = 1750°C (decomposes); Freezing/Melting point = 690°C. Hazard Identification (based on NFPA-704 M Rating System): Health 3, Flammability 0, Reactivity 0. Practically insoluble in water; solubility = 0.8%.

Potential Exposure: Compound Description (*dust*): Mutagen; Reproductive Effector; Human Data; Primary Irritant; (fume) Tumorigen. Vanadium pentoxide [V_2O_5] is an industrial catalyst in oxidation reactions, used in glass and ceramic glazes, a steel additive, and used in welding electrode coatings.

Incompatibilities: Strong acids, lithium, chlorine trifluoride, peroxyformic acid, combustible substances.

Permissible Exposure Limits in Air

OSHA PEL: 0.5 mg[V_sO_5]/m^3 (*dust*), respirable fraction; 0.1 mg[V_sO_5]/m^3 (fume) Ceiling Concentration.

NIOSH REL: 0.05 mg[V]/m^3 Ceiling Concentration [15 min, except vanadium metal and vanadium carbide].

ACGIH TLV®[1]: withdrawn.

NIOSH IDLH = 35 mg[V]/m^3.

Protective Action Criteria (PAC)

TEEL-0: 0.179 mg/m^3

PAC-1: 1 mg/m^3

PAC-2: 7 mg/m^3

PAC-3: 125 mg/m^3

DFG MAK: Carcinogen Category: 2; Germ cell mutagen group: 2; (DFG 2005).

Listed under "dust"

Austria: MAK 0.05 mg/m^3, 1999; Denmark: TWA 0.03 mg [V]/m^3, 1999; Japan; 0.5 mg[V_2O_5]/m^3, 1999; Poland: MAC (TWA) 0.05 mg/m^3; MAC (STEL) 0.5 mg/m^3, 1999; Switzerland: MAK-W 50 ppm (175 mg/m^3), 1999; United Kingdom: TWA 0.5 mg[V]/m^3, total inhalable (*dust*); TWA 0.04 mg[V]/m^3, (fume and respirable dust), 2000.

Listed under "fume"

Austria: MAK 0.05 mg/m^3, 1999; Japan; 0.1 mg [V_2O_5]/m^3, 1999; Poland: MAC (TWA) 0.05 mg/m^3; MAC (STEL) 0.1 mg/m^3, 1999; Sweden: NGV 0.2 mg[V]/m^3, 1999; Switzerland: MAK-W 50 ppm (175 mg/m^3), 1999; United Kingdom: TWA 0.04 mg[V]/m^3, (fume and respirable dust); TWA 5 mg/m^3, 2000. Russia[43] set a MAC of 0.002 mg/m^3 in ambient air in residential areas on a daily average basis. Several states have set guidelines or standards for vanadium pentoxide in ambient air[60] ranging from 0.14 μg/m^3 (Massachusetts) to 0.8 μg/m^3 (Virginia) to 1.0 μg/m^3 (Connecticut and Nevada) to 5.0 μg/m^3 (North Dakota).

Determination in Air: Use NIOSH Analytical Method (IV) #7300, Elements by ICP, #7301; #7303; #7504, Vanadium oxides; #9102, OSHA Analytical Method ID-185.

Determination in Water: Russia set a limit of 0.1 mg/L for vanadium as a maximum permissible limit for water basins.

Routes of Entry: Inhalation, ingestion, skin and/or eye contact.

Harmful Effects and Symptoms

Short Term Exposure: Vanadium pentoxide is irritating to the skin, eyes, and mucous membranes. Blindness and epistaxis (bloody nose) are further complications. Headache, dry mouth, dizziness, nervousness, insomnia, and tremor may be found. Probable oral lethal dose for humans is between 5 and 50 mg/kg or between 7 drops and 1 teaspoon for a 70-kg (150-lb) person. Toxicity is about the same magnitude as pentavalent arsenic. Acute exposure to vanadium pentoxide may result in pulmonary irritation, bronchospasm, hemoptysis (coughing up of blood), emphysema, anorexia, black stools, and pulmonary edema. Pulmonary edema is a medical emergency that can be delayed for several hours, and can cause death.

Long Term Exposure: May cause dermatitis, eczema, skin allergy. Lungs may be affected by repeated inhalation of dust or fumes; bronchitis may develop, with cough, fine rales, wheezing, dyspnea (breathing difficulty). May cause a green staining of the tongue and skin.

Points of Attack: Eyes, skin, respiratory system.

Medical Surveillance: Before beginning employment and at regular times after that, the following are recommended: lung function tests. If symptoms develop or overexposure is suspected, the following may be useful: evaluation by a qualified allergist, including careful exposure history and special testing, may help diagnose skin allergy. Consider chest X-ray after acute overexposure. Test for urine level of vanadium. An EEC guideline recommends the following based on vanadium urine levels: an every-4-month control at the workplace if a 50 μg/L is reached and an annual control if the 5 μg/L urine level is reached. In case levels exceed 50 μg/L urine, workers should be temporarily removed from risk.

An OSHA specifically regulated substance. See NIOSH Publication No. 2005-110, December 2004.

First Aid: If this chemical gets into the eyes, remove any contact lenses at once and irrigate immediately for at least 15 min, occasionally lifting upper and lower lids. Seek medical attention immediately. If this chemical contacts the skin, remove contaminated clothing and wash immediately with soap and water. Seek medical attention immediately. If this chemical has been inhaled, remove from exposure, begin rescue breathing (using universal precautions, including resuscitation mask) if breathing has stopped and CPR if heart action has stopped. Transfer promptly to a medical facility. When this chemical has been swallowed, get medical attention. Give large quantities of water and induce vomiting. Do not make an unconscious person vomit. Medical observation is recommended for 24—48 h after breathing overexposure, as pulmonary edema may be delayed. As first aid for pulmonary edema, a doctor or authorized paramedic may consider administering a corticosteroid spray.

Personal Protective Methods: Wear protective gloves and clothing to prevent any reasonable probability of skin contact. Safety equipment suppliers/manufacturers can provide recommendations on the most protective glove/clothing material for your operation. All protective clothing (suits, gloves, footwear, headgear) should be clean, available each day, and put on before work. Contact lenses should not be worn when working with this chemical. Wear dust-proof chemical goggles and face shield unless full-face-piece respiratory protection is worn. Employees should wash immediately with soap when skin is wet or contaminated. Provide emergency showers and eyewash.

Respirator Selection: NIOSH/OSHA (as V): *0.5 mg/m³:* 100XQ* (APF = 10) (any air-purifying respirator with an N100, R100, or P100 filter (including N100, R100, and P100 filtering face-pieces) except quarter-mask respirators) or SA* (any supplied-air respirator). *1.25 mg/m³:* Sa:Cf* (APF = 25) (any supplied-air respirator operated in a continuous-flow mode) or PaprHie* (APF = 25) (any powered, air-purifying respirator with a high-efficiency particulate filter). *2.5 mg/m³:* 100F (APF = 50) (any air-purifying, full-face-piece respirator with an N100, R100, or P100 filter) or PaprHie* (APF = 25) (any powered, air-purifying respirator with a high-efficiency particulate filter) or SCBAF (APF = 50) (any self-contained breathing apparatus with a full face-piece) or SaF (APF = 50) (any supplied-air respirator with a full face-piece). *Emergency or planned entry into unknown concentrations or IDLH conditions:* SCBAF: Pd, Pp (APF = 10,000) (any self-contained breathing apparatus that has a full face-piece and is operated in a pressure-demand or other positive-pressure mode) or SaF: Pd,Pp: ASCBA (APF = 10,000) (any supplied-air respirator that has a full face-piece and is operated in a pressure-demand or other positive-pressure mode in combination with an auxiliary self-contained breathing apparatus operated in a pressure-demand or other positive-pressure mode). *Escape:* 100F (APF = 50) (any air-purifying, full-face-piece respirator with an N100, R100, or P100 filter) or SCBAE (any appropriate escape-type, self-contained breathing apparatus).

*Substance reported to cause eye irritation or damage; may require eye protection.

Storage: Color Code—Blue: Health Hazard/Poison: Store in a secure poison location. Prior to working with this chemical you should be trained on its proper handling and storage. Vanadium pentoxide must be stored to avoid contact with chlorine trifluoride, lithium, and peroxyformic acid, since violent reactions occur. Store in tightly closed containers in a cool, well-ventilated area away from heat.

Shipping: Vanadium pentoxide, nonfused form, requires a shipping label of "POISONOUS/TOXIC MATERIALS." It falls in Hazard Class 6.1 and Packing Group II.

Spill Handling: Evacuate persons not wearing protective equipment from area of spill or leak until cleanup is complete. Remove all ignition sources. Do not touch spilled materials; stop leak if you can do so without risk. Use water spray to reduce vapors. *Small spills:* absorb with sand or other noncombustible absorbent material and place into containers for later disposal. *Small dry spills:* with clean shovel place material into clean, dry container and cover; move containers from spill area. *Large spills:* dike far ahead of spill for later disposal. Ventilate area after cleanup is complete. It may be necessary to contain and dispose of this chemical as a hazardous waste. If material or contaminated runoff enters waterways, notify downstream users of potentially contaminated waters. Contact your local or federal environmental protection agency for specific recommendations. If employees are required to clean up spills, they must be properly trained and equipped. OSHA 1910.120(q) may be applicable.

Fire Extinguishing: Not combustible but will increase the intensity of an existing fire. Extinguish fire using an agent suitable for type of surrounding fire. Vanadium pentoxide itself does not burn, but dust may increase the intensity of fire when in contact with combustible materials. Poisonous gases are produced in fire, including oxides of vanadium. If material or contaminated runoff enters waterways, notify downstream users of potentially contaminated waters. Notify local health and fire officials and pollution control agencies. From a secure, explosion-proof location, use water spray to cool exposed containers. If cooling streams are ineffective (venting sound increases in volume and pitch, tank discolors, or shows any signs of deforming), withdraw immediately to a secure position. If employees are expected to fight fires, they must be trained and equipped in OSHA 1910.156. The only respirators recommended for firefighting are self-contained breathing apparatuses that have full face-pieces and are operated in a pressure-demand or other positive-pressure mode.

Disposal Method Suggested: Consult with environmental regulatory agencies for guidance on acceptable disposal practices. Generators of waste containing this contaminant (≥100 kg/mo) must conform with EPA regulations governing storage, transportation, treatment, and waste disposal. Vanadium pentoxide may be salvaged or disposed of in a sanitary landfill.

References

US Environmental Protection Agency. (November 30, 1987). *Chemical Hazard Information Profile: Vanadium Pentoxide*. Washington, DC: Chemical Emergency Preparedness Program

Sax, N. I. (Ed.). *Dangerous Properties of Industrial Materials Report*, 2, No. 2, 83–84 (1982) and 8, No. 4, 81–92 (1988)

New Jersey Department of Health and Senior Services. (October 1998). *Hazardous Substances Fact Sheet: Vanadium Pentoxide*. Trenton, NJ

Vanadium tetrachloride   V:0130

Molecular Formula: Cl$_4$V
Common Formula: VCl$_4$
Synonyms: Vanadium chloride; Vanadium(IV) chloride
CAS Registry Number: 7632-51-1
RTECS® Number: YW2625000
UN/NA & ERG Number: UN2444/137
EC Number: 231-561-1
Regulatory Authority and Advisory Bodies
Carcinogenicity: DFG MAK: Carcinogen Category 2.
Air Pollutant Standard Set. See below, "Permissible Exposure Limits in Air" section.
EPCRA Section 313 Form R *de minimis* concentration reporting level: 1.0%.
Canada, WHMIS, Ingredients Disclosure List Concentration 1.0%.
European/International Regulations: not listed in Annex 1.
WGK (German Aquatic Hazard Class): No value assigned.
Description: Vanadium tetrachloride is a thick, reddish-brown liquid that gives off fumes on exposure to moist air. Molecular weight = 192.74; Boiling point = 148.5°C at 75 mmHg; Freezing/Melting point = −28 ± 2°C. Hazard Identification (based on NFPA-704 M Rating System) (*Fume and dust*): Health 3, Flammability 0, Reactivity 2W. Decomposes in water.
Potential Exposure: Vanadium tetrachloride is used as a fixative in textile dyeing and in the manufacture of other vanadium compounds.
Incompatibilities: Water, lithium, chlorine, trifluoride, combustible materials. Vanadium tetrachloride is a reactive chemical and is an explosion hazard. See storage and handling section.
Permissible Exposure Limits in Air
NIOSH IDLH = 35 mg[V]/m^3.
Protective Action Criteria (PAC)
TEEL-0: 0.6 mg/m^3
PAC-1: 2 mg/m^3
PAC-2: 12.5 mg/m^3
PAC-3: 60 mg/m^3
DFG MAK: inhalable fraction, Carcinogen Category: 2; Germ cell mutagen group: 2; (DFG 2005).
Determination in Air: Use Method #7300, Elements, #7504, Vanadium oxides.
Permissible Concentration in Water: Russia set a limit of 0.1 mg/L for vanadium as a maximum permissible limit for water basins.
Determination in Water: Russia set a limit of 0.1 mg/L for vanadium as a maximum permissible limit for water basins.
Routes of Entry: Inhalation, ingestion, skin and/or eye contact.
Harmful Effects and Symptoms
Short Term Exposure: Vanadium tetrachloride can affect you when breathed in and by passing through your skin. Vanadium tetrachloride is a corrosive chemical and eye contact can cause irritation and possible damage. Exposure can irritate the eyes, nose, throat, and lungs, with cough and shortness of breath. Higher exposures can cause pneumonia or pulmonary edema (fluid in the lungs). Pulmonary edema is a medical emergency that can be delayed for several hours. This can cause death.
Long Term Exposure: Repeated exposure can cause lung irritation; bronchitis may develop.
Points of Attack: Lungs.
Medical Surveillance: Before beginning employment and at regular times after that, for those with frequent or potentially high exposures, the following are recommended: lung function tests. If symptoms develop or overexposure is suspected, the following may be useful: Urine test for vanadium. Test for urine level of vanadium. An EEC guideline recommends the following based on vanadium urine levels: an every-4-month control at the workplace if a 50 µg/L is reached and an annual control if the 5 µg/L urine level is reached. In case levels exceed 50 µg/L urine, workers should be temporarily removed from risk.
First Aid: If this chemical gets into the eyes, remove any contact lenses at once and irrigate immediately for at least 30 min, occasionally lifting upper and lower lids. Seek medical attention immediately. If this chemical contacts the skin, remove contaminated clothing and wash immediately with soap and water. Seek medical attention immediately. If this chemical has been inhaled, remove from exposure, begin rescue breathing (using universal precautions, including resuscitation mask) if breathing has stopped and CPR if heart action has stopped. Transfer promptly to a medical facility. When this chemical has been swallowed, get medical attention. If victim is *conscious*, administer water or milk. Do not induce vomiting. Medical observation is recommended for 24−48 h after breathing overexposure, as pulmonary edema may be delayed. As first aid for pulmonary edema, a doctor or authorized paramedic may consider administering a corticosteroid spray.
Personal Protective Methods: Wear protective gloves and clothing to prevent any reasonable probability of skin contact. Safety equipment suppliers/manufacturers can provide recommendations on the most protective glove/clothing material for your operation. All protective clothing (suits, gloves, footwear, headgear) should be clean, available each day, and put on before work. Contact lenses should not be worn when working with this chemical. Wear splash-proof chemical goggles and face shield unless full-face-piece respiratory protection is worn. Employees should wash immediately with soap when skin is wet or contaminated. Provide emergency showers and eyewash.
Respirator Selection: NIOSH/OSHA (as V): *0.5 mg/m^3:* 100XQ* (APF = 10) (any air-purifying respirator with an N100, R100, or P100 filter (including N100, R100, and P100 filtering face-pieces) except quarter-mask respirators) or SA* (any supplied-air respirator). *1.25 mg/m^3:* Sa:Cf* (APF = 25) (any supplied-air respirator operated in a continuous-flow mode) or PaprHie* (APF = 25) (any

powered, air-purifying respirator with a high-efficiency particulate filter). *2.5 mg/m³:* 100F (APF = 50) (any air-purifying, full-face-piece respirator with an N100, R100, or P100 filter) or PaprHie* (APF = 25) (any powered, air-purifying respirator with a high-efficiency particulate filter) or SCBAF (APF = 50) (any self-contained breathing apparatus with a full face-piece) or SaF (APF = 50) (any supplied-air respirator with a full face-piece). *Emergency or planned entry into unknown concentrations or IDLH conditions:* SCBAF: Pd,Pp (APF = 10,000) (any self-contained breathing apparatus that has a full face-piece and is operated in a pressure-demand or other positive-pressure mode) or SaF: Pd,Pp: ASCBA (APF = 10,000) (any supplied-air respirator that has a full face-piece and is operated in a pressure-demand or other positive-pressure mode in combination with an auxiliary self-contained breathing apparatus operated in a pressure-demand or other positive-pressure mode). *Escape:* 100F (APF = 50) (any air-purifying, full-face-piece respirator with an N100, R100, or P100 filter) or SCBAE (any appropriate escape-type, self-contained breathing apparatus).

*Substance reported to cause eye irritation or damage; may require eye protection.

Storage: Color Code—White: Corrosive or Contact Hazard; Store separately in a corrosion-resistant location. Prior to working with this chemical you should be trained on its proper handling and storage. Vanadium tetrachloride must be stored to avoid contact with water, heat, and poisonous gases since violent reactions occur. Store in tightly closed containers in a cool, well-ventilated area away from radiant heat and flammable and combustible materials, lithium, chlorine, and trifluoride. Vanadium tetrachloride decomposes slowly to vanadium trichloride and chloride. Open containers in dry oxygen-free atmosphere or in inert gas, wearing appropriate personal protective equipment. Chill to below 20°C before opening.

Shipping: Vanadium tetrachloride requires a shipping label of "CORROSIVE." It falls in Hazard Class 8 and Packing Group I.

Spill Handling: Evacuate and restrict persons not wearing protective equipment from area of spill or leak until cleanup is complete. Remove all ignition sources. Ventilate area of spill or leak. *Do not use water* or wet method. Absorb liquids in vermiculite, dry sand, earth, peat, carbon, or a similar material and deposit in sealed containers. Keep out of sewers because of the possibility of fire and explosion. It may be necessary to contain and dispose of this chemical as a hazardous waste. If material or contaminated runoff enters waterways, notify downstream users of potentially contaminated waters. Contact your local or federal environmental protection agency for specific recommendations. If employees are required to clean up spills, they must be properly trained and equipped. OSHA 1910.120(q) may be applicable.

Fire Extinguishing: Vanadium tetrachloride itself does not burn, but it may increase the intensity of fire since it is an oxidizer. Use dry chemical or CO_2 extinguishers on surrounding fire. *Do not use water* on material. Poisonous gases, including hydrogen chloride, vanadium oxides, and vanadium, are produced in fire. Vapors are heavier than air and will collect in low areas. Containers may explode in fire. Storage containers and parts of containers may rocket great distances, in many directions. If material or contaminated runoff enters waterways, notify downstream users of potentially contaminated waters. Notify local health and fire officials and pollution control agencies. From a secure, explosion-proof location, use water spray to cool exposed containers. Use water spray to keep fire-exposed containers cool and to reduce vapors but do not get water inside containers of vanadium tetrachloride or on spilled material. If cooling streams are ineffective (venting sound increases in volume and pitch, tank discolors, or shows any signs of deforming), withdraw immediately to a secure position. If employees are expected to fight fires, they must be trained and equipped in OSHA 1910.156. The only respirators recommended for firefighting are self-contained breathing apparatuses that have full face-pieces and are operated in a pressure-demand or other positive-pressure mode.

References

New Jersey Department of Health and Senior Services. (August 2005). *Hazardous Substances Fact Sheet: Vanadium Tetrachloride.* Trenton, NJ

Vanadyl sulfate   V:0140

Molecular Formula: O_5SV
Common Formula: $VOSO_4$
Synonyms: C.I. 77940; Oxysulfatovanadium; Vanadium, oxysulfato (2-)-*o*-; Vanadium oxysulfide
CAS Registry Number: 27774-13-6
RTECS® Number: YW1925000
UN/NA & ERG Number: UN2931/151
EC Number: 248-652-7
Regulatory Authority and Advisory Bodies
Carcinogenicity: DFG MAK: Carcinogen Category 2.
Air Pollutant Standard Set. See below, "Permissible Exposure Limits in Air" section.
Clean Water Act: Section 311 Hazardous Substances/RQ 40CFR117.3 (same as CERCLA, see below).
Reportable Quantity (RQ): 1000 lb (454 kg).
EPCRA Section 313 Form R *de minimis* concentration reporting level: 1.0%.
Canada, WHMIS, Ingredients Disclosure List Concentration 1.0%.
European/International Regulations: not listed in Annex 1.
WGK (German Aquatic Hazard Class): No value assigned.
Description: Vanadyl sulfate is an odorless pale blue powder. Molecular weight = 163.1; Hazard Identification (based on NFPA-704 M Rating System): Health 2, Flammability 0, Reactivity 0. Highly soluble in water.

Potential Exposure: Vanadyl sulfate is used as a fixative for textile dyes, a colorant for glass and ceramics, a reducing agent, and a catalyst.

Incompatibilities: Sulfates react with aluminum, magnesium. Incompatible with strong oxidizers.

Permissible Exposure Limits in Air
NIOSH IDLH = 35 mg[V]/m^3.
Protective Action Criteria (PAC)
TEEL-0: 1 mg/m^3
PAC-1: 3 mg/m^3
PAC-2: 20 mg/m^3
PAC-3: 112 mg/m^3
DFG MAK: inhalable fraction, Carcinogen Category: 2; Germ cell mutagen group: 2; (DFG 2005).

Determination in Air: Use Method #7300, Elements, #7504, Vanadium oxides.

Permissible Concentration in Water: Russia[43] set a MAC of 0.1 mg/L in water bodies used for domestic purposes.

Routes of Entry: Inhalation, ingestion, skin and/or eye contact.

Harmful Effects and Symptoms

Short Term Exposure: Toxic when ingested. Vanadyl sulfate can affect you when breathed in. Contact can irritate the skin. Exposure can irritate the eyes, nose, throat, and lungs with cough and phlegm. Higher exposures can cause pneumonia and/or pulmonary edema, a medical emergency that can be delayed for several hours. This can cause death.

Long Term Exposure: Lungs may be affected by repeated inhalation; bronchitis may develop. May cause a greenish-black discoloration of the tongue and skin.

Points of Attack: Lungs.

Medical Surveillance: For those with frequent or potentially high exposure (half the TLV or grater), the following are recommended before beginning work and at regular times after that: lung function tests. If symptoms develop or overexposure is suspected, the following may be useful: Urine test for vanadium. An EEC guideline recommends the following based on vanadium urine levels: an every-4-month control at the workplace if a 50 μg/L is reached and an annual control if the 5 μg/L urine level is reached. In case levels exceed 50 μg/L urine, workers should be temporarily removed from risk. Consider chest X-ray after acute overexposure.

First Aid: If this chemical gets into the eyes, remove any contact lenses at once and irrigate immediately for at least 15 min, occasionally lifting upper and lower lids. Seek medical attention immediately. If this chemical contacts the skin, remove contaminated clothing and wash immediately with soap and water. Seek medical attention immediately. If this chemical has been inhaled, remove from exposure, begin rescue breathing (using universal precautions, including resuscitation mask) if breathing has stopped and CPR if heart action has stopped. Transfer promptly to a medical facility. When this chemical has been swallowed, get medical attention. Give large quantities of water and induce vomiting. Do not make an unconscious person vomit. Medical observation is recommended for 24–48 h after breathing overexposure, as pulmonary edema may be delayed. As first aid for pulmonary edema, a doctor or authorized paramedic may consider administering a corticosteroid spray.

Personal Protective Methods: Wear protective gloves and clothing to prevent any reasonable probability of skin contact. Safety equipment suppliers/manufacturers can provide recommendations on the most protective glove/clothing material for your operation. All protective clothing (suits, gloves, footwear, headgear) should be clean, available each day, and put on before work. Contact lenses should not be worn when working with this chemical. Wear dust-proof chemical goggles and face shield unless full-face-piece respiratory protection is worn. Employees should wash immediately with soap when skin is wet or contaminated. Provide emergency showers and eyewash.

Respirator Selection: NIOSH/OSHA (as V): *0.5 mg/m^3:* 100XQ* (APF = 10) (any air-purifying respirator with an N100, R100, or P100 filter (including N100, R100, and P100 filtering face-pieces) except quarter-mask respirators) or SA* (any supplied-air respirator). *1.25 mg/m^3:* Sa:Cf* (APF = 25) (any supplied-air respirator operated in a continuous-flow mode) or PaprHie* (APF = 25) (any powered, air-purifying respirator with a high-efficiency particulate filter). *2.5 mg/m^3:* 100F (APF = 50) (any air-purifying, full face-piece respirator with an N100, R100, or P100 filter) or PaprHie* (APF = 25) (any powered, air-purifying respirator with a high-efficiency particulate filter) or SCBAF (APF = 50) (any self-contained breathing apparatus with a full face piece) or SaF (APF = 50) (any supplied-air respirator with a full face-piece). *Emergency or planned entry into unknown concentrations or IDLH conditions:* SCBAF: Pd,Pp (APF = 10,000) (any self-contained breathing apparatus that has a full-face-piece and is operated in a pressure-demand or other positive-pressure mode) or SaF: Pd,Pp: ASCBA (APF = 10,000) (any supplied-air respirator that has a full face-piece and is operated in a pressure-demand or other positive-pressure mode in combination with an auxiliary self-contained breathing apparatus operated in a pressure-demand or other positive-pressure mode). *Escape:* 100F (APF = 50) (any air-purifying, full-face-piece respirator with an N100, R100, or P100 filter) or SCBAE (any appropriate escape-type, self-contained breathing apparatus).
*Substance reported to cause eye irritation or damage; may require eye protection.

Storage: (1) Color Code—Yellow Stripe (*strong reducing agent*): Reactivity Hazard; Store separately in an area isolated from flammables, combustibles, or other yellow-coded materials. (2) Color Code—Blue: Health Hazard/Poison: Store in a secure poison location. Prior to working with this chemical you should be trained on its proper handling and storage. Store in tightly closed containers in a cool, well-ventilated area.

Shipping: Vanadyl sulfate requires a shipping label of "POISONOUS/TOXIC MATERIALS." It falls in Hazard Class 6.1 and Packing Group II.

Spill Handling: Evacuate persons not wearing protective equipment from area of spill or leak until cleanup is complete. Remove all ignition sources. Collect powdered material in the most convenient and safe manner and deposit in sealed containers. Ventilate area after cleanup is complete. It may be necessary to contain and dispose of this chemical as a hazardous waste. If material or contaminated runoff enters waterways, notify downstream users of potentially contaminated waters. Contact your local or federal environmental protection agency for specific recommendations. If employees are required to clean up spills, they must be properly trained and equipped. OSHA 1910.120(q) may be applicable.

Fire Extinguishing: Extinguish fire using an agent suitable for type of surrounding fire. Use water spray to reduce vapors. Poisonous gases are produced in fire, including oxides of sulfur and vanadium. If material or contaminated runoff enters waterways, notify downstream users of potentially contaminated waters. Notify local health and fire officials and pollution control agencies. From a secure, explosion-proof location, use water spray to cool exposed containers. If cooling streams are ineffective (venting sound increases in volume and pitch, tank discolors, or shows any signs of deforming), withdraw immediately to a secure position. If employees are expected to fight fires, they must be trained and equipped in OSHA 1910.156. The only respirators recommended for firefighting are self-contained breathing apparatuses that have full face-pieces and are operated in a pressure-demand or other positive-pressure mode.

References

New Jersey Department of Health and Senior Services. (August 2005). *Hazardous Substances Fact Sheet: Vanadyl Sulfate.* Trenton, NJ

Vinyl acetate   V:0150

Molecular Formula: $C_4H_6O_2$

Common Formula: $CH_3COOCH=CH_2$

Synonyms: Acetate de vinyle (French); Acetic acid, ethenyl ester; Acetic acid, vinyl ester; Aceto de vinilo (Spanish); 1-Acetoxyethylene; Ethenyl acetate; Ethenyl ethanoate; Ethonic acid, ethenyl ester; Everflex 811; Plyamul 40305-00; Unocal 76 RES 6206; Unocal 76 RES S-55; VAC; VAM; Vinnapas A 50; Vinylacetat (German); Vinyl acetate H.Q.; Vinyl acetate monomer; Vinyl A monomer; Vinyle (acetate de) (French); Vinyl ethanoate; VYAC; Zeset T

CAS Registry Number: 108-05-4; *(alt.)* 61891-42-7; *(alt.)* 82041-23-4

RTECS® Number: AK0875000

UN/NA & ERG Number: UN1301 (stabilized)/129

EC Number: 203-545-4 [*Annex I Index No.:* 607-023-00-0]

Regulatory Authority and Advisory Bodies

Department of Homeland Security Screening Threshold Quantity (pounds): *Release hazard* 10,000 ($\geq$1.00% concentration).

Carcinogenicity: IARC: Human Inadequate Evidence; Animal Limited Evidence, *possibly carcinogenic to humans*, Group 2B, 1995.

US EPA Gene-Tox Program, Positive: Cell transform.—SA7/SHE; Positive/limited: Carcinogenicity—mouse/rat; Negative: Histidine reversion—Ames test.

Air Pollutant Standard Set. See below, "Permissible Exposure Limits in Air" section.

Clean Air Act: Hazardous Air Pollutants (Title I, Part A, Section 112); Accidental Release Prevention/Flammable Substances, (Section 112[r], Table 3), TQ = 15,000 lb (6825 kg).

Clean Water Act: Section 311 Hazardous Substances/RQ 40CFR117.3 (same as CERCLA, see below); Priority Pollutants (40CFR PART 423); Section 313 Water Priority Chemicals (57FR41331, 9/9/92).

RCRA 40CFR264, Appendix 9; TSD Facilities Ground Water Monitoring List. Suggested test method(s) (PQL µg/L): 8240 (5).

CERCLA Section 302, Extremely Hazardous Substances, TPQ = 1000 lb (455 kg).

Reportable Quantity (RQ): 5000 lb (2270 kg).

CERCLA Section 313 Form R *de minimis* concentration reporting level: 1.0%.

Canada, WHMIS, Ingredients Disclosure List Concentration 1.0%.

European/International Regulations: Hazard Symbol: F+; Risk phrases: R11; Safety phrases: S2; S16; S23; S29; S33 (see Appendix 4).

WGK (German Aquatic Hazard Class): 2—Hazard to waters.

Description: Vinyl acetate is a colorless, flammable liquid. The odor threshold is 0.12 ppm,[41] 0.3 ppm (NY, NJ). Molecular weight = 86.10; Specific gravity (H_2O:1) = 0.93 at 25°C; Boiling point = 72.2°C; Freezing/Melting point = −93.3°C; Vapor pressure = 83 mmHg at 25°C; Flash point = −8°C (cc); Autoignition temperature = 402°C. Explosive limits: LEL = 2.6%; UEL = 13.4%. Hazard Identification (based on NFPA-704 M Rating System): Health 2, Flammability 3, Reactivity 2. Slightly soluble (2%) in water.

Potential Exposure: Compound Description: Tumorigen, Mutagen; Reproductive Effector; Primary Irritant. Vinyl acetate is used primarily in polymerization processes to produce polyvinyl acetate, polyvinyl alcohol, and vinyl acetate copolymer. The polymers usually made as emulsions, suspensions, solutions, or resins are used to prepare adhesives, paints, paper coatings, and textile finishes. Low-molecular-weight vinyl acetate is used as a chewing gum base.

Incompatibilities: Violent reaction with strong oxidizers. The vapor may react vigorously with silica gel or aluminum. Acids, bases, silica gel, alumina, oxidizers, azo compounds. Ozone readily polymerizes in elevated temperatures, under the influence of light or peroxides. Usually contains a stabilizer [e.g., hydroquinone (limit to 2 months) or diphenylamine (for longer term)] to prevent polymerization.

Permissible Exposure Limits in Air
Conversion factor: 1 ppm = 3.52 mg/m^3 at 25°C & 1 atm.
OSHA PEL: None.
NIOSH REL: 4 ppm/15 mg/m^3 [15 min] Ceiling Concentration.
ACGIH TLV®[1]: 10 ppm/35 mg/m^3 TWA; 15 ppm/53 mg/m^3 STEL, confirmed animal carcinogen with unknown relevance to humans.
Protective Action Criteria (PAC)*
TEEL-0: 6.7 ppm
PAC-1: **6.7** ppm
PAC-2: **180** ppm
PAC-3: **610** ppm
*AEGLs (Acute Emergency Guideline Levels) & ERPGs (Emergency Response Planning Guideline) are in **bold face**.
DFG MAK: Carcinogen Category 3A.
Australia: TWA 10 ppm (30 mg/m^3), STEL 20 ppm, 1993; Austria: MAK 1 ppm (35 mg/m^3), Suspected: carcinogen, 1999; Belgium: TWA 10 ppm (35 mg/m^3), STEL 20 ppm, 1993; Denmark: TWA 10 ppm (30 mg/m^3), 1999; Finland: TWA 10 ppm (35 mg/m^3), STEL 20 ppm (70 mg/m^3), 1993; France: VME 10 ppm (30 mg/m^3), 1999; the Netherlands: MAC-TGG 18 mg/m^3, 2003; Poland: MAC (TWA) 10 mg/m^3; MAC (STEL) 30 mg/m^3, 1999; Sweden: NGV 5 ppm (18 mg/m^3), KTV 10 ppm (35 mg/m^3), 1999; Switzerland: MAK-W 10 ppm (35 mg/m^3), KZG-W 20 ppm (70 mg/m^3), 1999; Turkey: TWA 10 ppm (30 mg/m^3), 1993; United Kingdom: TWA 10 ppm (36 mg/m^3), STEL 20 ppm (72 mg/m^3), 2000; Argentina, Bulgaria, Columbia, Jordan, South Korea, New Zealand, Singapore, Vietnam: ACGIH TLV®: STEL 15 ppm.
Russia[43] set a MAC of 0.15 mg/m^3 in ambient air in residential areas both on a momentary and a daily average basis. Several states have set guidelines or standards for vinyl acetate in ambient air[60] ranging from 5.0 μg/m^3 (Virginia) to 9.6 μg/m^3 (Massachusetts) to 300.0–600.0 μg/m^3 (North Dakota) to 600.0 μg/m^3 (Connecticut) to 714.0 μg/m^3 (Nevada).
Determination in Air: Use NIOSH Analytical Method (IV) #1453, vinyl acetate; OSHA Analytical Method 51.
Determination in Water: Octanol–water coefficient: Log K_{ow} = <0.8.
Permissible Concentration in Water: Russia[43] set a MAC of 0.2 mg/L in water bodies used for domestic purposes.
Routes of Entry: Inhalation, ingestion, skin and/or eye contact.
Harmful Effects and Symptoms
Short Term Exposure: Irritates the eyes, skin, and respiratory tract, causing hoarseness, cough, loss of smell. A CNS depressant; high levels of exposure can cause fatigue, dizziness, lightheadedness, and disturbed sleep. *Inhalation:* Irritates the nose and throat, causing coughing and/or shortness of breath. Levels of 19 ppm for 4 h have caused slight throat irritation. The characteristic odor may not be recognized after about 2 h. Levels of 71 ppm for 1/2 h may cause

definite throat irritations. May cause lesions of the lung tissue. *Skin:* Contact can cause irritation, dryness. Prolonged contact can cause rash, burns, and blisters. *Eyes:* Levels of 22 ppm and above may produce eye reddening and irritation. Contact can cause burns.
Long Term Exposure: May affect the heart, central nervous system, and liver. Levels above 22 ppm for 1 year may cause reversible irritation to eyes, throat, and lungs, sometimes accompanied by skin irritation or rash. Vinyl acetate has been related to reproductive abnormalities. Exposure caused gradual deterioration of heart muscles.
Points of Attack: Liver, eyes, heart, skin.
Medical Surveillance: For those with frequent or potentially high exposure (half the TLV or greater), the following is recommended before beginning work and at regular times after that: lung function tests. If symptoms develop or overexposure is suspected, the following may be useful: liver function tests. Examination of the nervous system. EKG. Specific engineering controls are recommended for this chemical. Refer to NIOSH Criteria Document #78-205, Occupational Exposure to Vinyl Acetate.
First Aid: If this chemical gets into the eyes, remove any contact lenses at once and irrigate immediately for at least 15 min, occasionally lifting upper and lower lids. Seek medical attention immediately. If this chemical contacts the skin, remove contaminated clothing and wash immediately with soap and water. Seek medical attention immediately. If this chemical has been inhaled, remove from exposure, begin rescue breathing (using universal precautions, including resuscitation mask) if breathing has stopped and CPR if heart action has stopped. Transfer promptly to a medical facility. When this chemical has been swallowed, get medical attention. Give large quantities of water and induce vomiting. Do not make an unconscious person vomit.
Personal Protective Methods: Wear protective gloves and clothing to prevent any reasonable probability of skin contact. Safety equipment suppliers/manufacturers can provide recommendations on the most protective glove/clothing material for your operation. Teflon™, polyethylene and ethylene vinyl alcohol are recommended as protective materials. All protective clothing (suits, gloves, footwear, headgear) should be clean, available each day, and put on before work. Contact lenses should not be worn when working with this chemical. Wear splash-proof chemical goggles and face shield unless full-face-piece respiratory protection is worn. Employees should wash immediately with soap when skin is wet or contaminated. Provide emergency showers and eyewash.
Respirator Selection: NIOSH: *40 ppm:* CcrFOv (APF = 50) [any chemical cartridge respirator with a full face-piece and organic vapor cartridge(s)] or Sa (APF = 10) (any supplied-air respirator). *100 ppm:* Sa:Cf (APF = 25) (any supplied-air respirator operated in a continuous-flow mode) or PaprOv (APF = 25) [any powered, air-purifying respirator with organic vapor cartridge(s)]. *200 ppm:* CcrFOv (APF = 50) [any chemical cartridge respirator with a full face-piece and

organic vapor cartridge(s)] or GmFOv (APF = 50) [any air-purifying, full-face-piece respirator (gas mask) with a chin-style, front- or back-mounted acid gas canister] or PaprTOv (APF = 50) [any powered, air-purifying respirator with a tight-fitting face-piece and organic vapor cartridge (s)] or SCBAF (APF = 50) (any self-contained breathing apparatus with a full face-piece) or SaF (APF = 50) (any supplied-air respirator with a full face-piece). *400 ppm:* Sa: Pd,Pp (APF = 1000) (any supplied-air respirator operated in a pressure-demand or other positive-pressure mode). *Emergency or planned entry into unknown concentrations or IDLH conditions:* SCBAF: Pd,Pp (APF = 10,000) (any self-contained breathing apparatus that has a full face-piece and is operated in a pressure-demand or other positive-pressure mode) or SaF: Pd,Pp: ASCBA (APF = 10,000) (any supplied-air respirator that has a full face-piece and is operated in a pressure-demand or other positive-pressure mode in combination with an auxiliary, self-contained breathing apparatus operated in a pressure-demand or other positive-pressure mode). *Escape:* GmFOv (APF = 50) [any air-purifying, full-face-piece respirator (gas mask) with a chin-style, front- or back-mounted organic vapor canister] or SCBAE (any appropriate escape-type, self-contained breathing apparatus).

Note: Substance reported to cause eye irritation or damage; may require eye protection.

Storage: Color Code—Red: Flammability Hazard: Store in a flammable liquid storage area or approved cabinet away from ignition sources and corrosive and reactive materials. May form peroxides in storage. Prior to working with this chemical you should be trained on its proper handling and storage. Before entering confined space where this chemical may be present, check to make sure that an explosive concentration does not exist. Vinyl acetate must be stored to avoid contact with oxidizers, such as perchlorates, peroxides, permanganates, chlorates, and nitrates), since violent reactions occur. Store in tightly closed containers in a cool, well-ventilated area away from heat and direct sunlight. Sources of ignition, such as smoking and open flames, are prohibited where vinyl acetate is handled, used, or stored. Metal containers involving the transfer of 5 gallons or more of vinyl acetate should be grounded and bonded. Drums must be equipped with self-closing valves, pressure vacuum bungs, and flame arresters. Use only nonsparking tools and equipment, especially when opening and closing containers of vinyl acetate. Wherever vinyl acetate is used, handled, manufactured, or stored, use explosion-proof electrical equipment and fittings.

Shipping: This compound requires a shipping label of "FLAMMABLE LIQUID." It falls in Hazard Class 3 and Packing Group II.

Spill Handling: Evacuate and restrict persons not wearing protective equipment from area of spill or leak until cleanup is complete. Remove all ignition sources. Establish forced ventilation to keep levels below explosive limit. Absorb liquids in vermiculite, dry sand, earth, peat, carbon, or a similar material and deposit in sealed containers. Keep this chemical out of a confined space, such as a sewer, because of the possibility of an explosion, unless the sewer is designed to prevent the buildup of explosive concentrations. It may be necessary to contain and dispose of this chemical as a hazardous waste. If material or contaminated runoff enters waterways, notify downstream users of potentially contaminated waters. Contact your local or federal environmental protection agency for specific recommendations. If employees are required to clean up spills, they must be properly trained and equipped. OSHA 1910.120(q) may be applicable.

Fire Extinguishing: This chemical is a flammable liquid. Poisonous gases are produced in fire. Water may be ineffective. Use dry chemical, carbon dioxide, or foam extinguishers. Vapors are heavier than air and will collect in low areas. Vapors may travel long distances to ignition sources and flashback. Vapors in confined areas may explode when exposed to fire. Containers may explode in fire. Storage containers and parts of containers may rocket great distances, in many directions. If material or contaminated runoff enters waterways, notify downstream users of potentially contaminated waters. Notify local health and fire officials and pollution control agencies. From a secure, explosion-proof location, use water spray to cool exposed containers. If cooling streams are ineffective (venting sound increases in volume and pitch, tank discolors, or shows any signs of deforming), withdraw immediately to a secure position. If employees are expected to fight fires, they must be trained and equipped in OSHA 1910.156. The only respirators recommended for firefighting are self-contained breathing apparatuses that have full face-pieces and are operated in a pressure-demand or other positive-pressure mode.

Disposal Method Suggested: Dissolve or mix the material with a combustible solvent and burn in a chemical incinerator equipped with an afterburner and scrubber. All federal, state, and local environmental regulations must be observed.

References

National Institute for Occupational Safety and Health. (1978). *Criteria for a Recommended Standard: Occupational Exposure to Vinyl Acetate*, NIOSH Publication No. 78-205. Washington, DC

US Environmental Protection Agency. (April 23, 1984). *Chemical Hazard Information Profile Draft Report: Vinyl Acetate*. Washington, DC

US Environmental Protection Agency. (November 30, 1987). *Chemical Hazard Information Profile: Vinyl Acetate Monomer*. Washington, DC: Chemical Emergency Preparedness Program

New York State Department of Health. (March 1986). *Chemical Fact Sheet: Vinyl Acetate* (Version 2). Albany, NY: Bureau of Toxic Substance Assessment

Sax, N. I. (Ed.). (1989). *Dangerous Properties of Industrial Materials Report*, 9, No. 2, 89−106

New Jersey Department of Health and Senior Services. (April 2002). *Hazardous Substances Fact Sheet: Vinyl Acetate.* Trenton, NJ

Vinyl bromide   V:0160

Molecular Formula: C_2H_3Br
Common Formula: CH_2=$CHBr$
Synonyms: Bromoethene; Bromoethylene; Bromure de vinyle (French); Bromuro de vinilo (Spanish); Ethene, bromo-; Ethylene, bromo-; Monobromoethylene; Vinylbromid (German); Vinyle (bromure de) (French)
CAS Registry Number: 593-60-2
RTECS® Number: KU8400000
UN/NA & ERG Number: UN1085 (stabilized)/116
EC Number: 209-800-6 [*Annex I Index No.:* 602-024-00-2]
Regulatory Authority and Advisory Bodies
Carcinogenicity: IARC: Animal, Sufficient Evidence; Human, No Adequate Data, Group 2A, 1999 NTP: 11th Report on Carcinogens, 2004: Reasonably anticipated to be a human carcinogen; NIOSH: Potential Occupational carcinogen.
Air Pollutant Standard Set. See below, "Permissible Exposure Limits in Air" section.
Clean Air Act: Hazardous Air Pollutants (Title I, Part A, Section 112).
Reportable Quantity (RQ): 1 lb (0.454 kg).
EPCRA Section 313 Form R *de minimis* concentration reporting level: 0.1%.
California Proposition 65 Chemical 10/1/88.
Canada, WHMIS, Ingredients Disclosure List Concentration 0.1%.
European/International Regulations: Hazard Symbol: F+, T; Risk phrases: R45; R12 (Carc/Cat. 2); Safety phrases: S53; S45 (see Appendix 4).
WGK (German Aquatic Hazard Class): No value assigned.
Description: Vinyl bromide is a colorless gas or liquid with a pleasant odor. Shipped as a liquefied compressed gas with 0.1% phenol added to prevent polymerization. Molecular weight = 106.96; Specific gravity (H_2O:1) = 1.49 (liquid at 15.6°C); Boiling point = 15.6°C; Freezing/Melting point = −139.4°C; Vapor pressure = 1.4 atm; Relative vapor density (air = 1) = 3.97; Flash point = flammable gas; Autoignition temperature = 530°C. Explosive limits: LEL = 9.0%; UEL = 15.0%. Hazard Identification (based on NFPA-704 M Rating System): Health 2, Flammability 4, Reactivity 1. Insoluble in water.
Potential Exposure: Compound Description: Tumorigen, Mutagen. Vinyl bromide is used as an intermediate in organic synthesis and for the preparation of plastics by polymerization and copolymerization; as a comonomer with acrylonitrile and other vinyl monomers in modacrylic fibers; in the production of flame-retardant synthetic fibers.
Incompatibilities: Strong oxidizers (e.g., perchlorates, peroxides, chlorates, permanganates, and nitrates). May

polymerize in heat, light, especially sunlight. Add 0.1% phenol to gas to prevent polymerization.
Permissible Exposure Limits in Air
Conversion factor: 1 ppm = 4.38 mg/m³ at 25°C & 1 atm.
OSHA PEL: None.
NIOSH REL: Potential occupational carcinogen. Limit exposure to lowest feasible concentration. See *NIOSH Pocket Guide*, Appendix A.
ACGIH TLV®[11]: 0.5 ppm/2.2 mg/m³ TWA, Suspected Human Carcinogen.
Protective Action Criteria (PAC)
TEEL-0: 0.5 ppm
PAC-1: 3500 ppm
PAC-2: 6000 ppm
PAC-3: 6000 ppm
Australia: TWA 5 ppm (20 mg/m³), carcinogen, 1993; Belgium: TWA 5 ppm (22 mg/m³), carcinogen 1993; Denmark: TWA 5 ppm (20 mg/m³), 1999; Finland: TWA 5 ppm (20 mg/m³), STEL 10 ppm (40 mg/m³), carcinogen, 1999; Norway: TWA 1 ppm (4 mg/m³), 1999; Switzerland: MAK-W 5 ppm (22 mg/m³), 1999; United Kingdom: LTEL 5 ppm (20 mg/m³), 1993; the Netherlands: MAC-TGG 0.012 mg/m³, 2003; Argentina, Bulgaria, Columbia, Jordan, South Korea, New Zealand, Singapore, Vietnam: ACGIH TLV®: Suspected Human Carcinogen. Several states have set guidelines or standards for vinyl bromide in ambient air[60] ranging from zero (North Dakota) to 2.0 μg/m³ (Virginia) to 44.0 μg/m³ (Connecticut) to 48.07 μg/m³ (Pennsylvania) to 66.7 μg/m³ (New York) to 100.0 μg/m³ (South Carolina) to 200.0 μg/m³ (Florida) to 476.0 μg/m³ (Nevada).
Determination in Air: Charcoal tube; Ethanol; Gas chromatography/Flame ionization detection; NIOSH Analytical Method (IV) #1009, Vinyl bromide, OSHA Analytical Method 08.
Determination in Water: Octanol−water coefficient: Log K_{ow} = 1.6.
Routes of Entry: Inhalation, ingestion (liquid), skin and/or eye contact.
Harmful Effects and Symptoms
Short Term Exposure: Vinyl bromide can affect you when breathed in. Contact can irritate the eyes. A nervous system depressant; exposures may cause dizziness, lightheadedness, confusion, a lack of coordination, narcosis, nausea, vomiting. Contact with the liquid can cause frostbite.
Long Term Exposure: Potential occupational carcinogen.
Points of Attack: Eyes, skin, central nervous system, liver. Cancer site in animals: liver and lymph node.
Medical Surveillance: There are no special tests. However, medical attention is recommended if overexposure is suspected.
First Aid: If this chemical gets into the eyes, remove any contact lenses at once and irrigate immediately for at least 15 min, occasionally lifting upper and lower lids. Seek medical attention immediately. If this chemical contacts the skin, remove contaminated clothing and wash immediately with soap and water. Seek medical attention immediately.

If this chemical has been inhaled, remove from exposure, begin rescue breathing (using universal precautions, including resuscitation mask) if breathing has stopped and CPR if heart action has stopped. Transfer promptly to a medical facility. When this chemical has been swallowed, get medical attention. Give large quantities of water and induce vomiting. Do not make an unconscious person vomit.

Personal Protective Methods: Wear protective gloves and clothing to prevent any reasonable probability of skin contact. Safety equipment suppliers/manufacturers can provide recommendations on the most protective glove/clothing material for your operation. All protective clothing (suits, gloves, footwear, headgear) should be clean, available each day, and put on before work. Contact lenses should not be worn when working with this chemical. Wear gas-proof chemical goggles and face shield unless full-face-piece respiratory protection is worn. Employees should wash immediately with soap when skin is wet or contaminated. Provide emergency showers and eyewash.

Respirator Selection: *At concentrations above the NIOSH REL, or where there is NO REL, at any detectable concentration:* SCBAF: Pd,Pp (APF = 10,000) (any self-contained breathing apparatus that has a full face-piece and is operated in a pressure-demand or other positive-pressure mode) or SaF: Pd,Pp: ASCBA (APF = 10,000) (any supplied-air respirator that has a full face-piece and is operated in a pressure-demand or other positive-pressure mode in combination with an auxiliary self-contained breathing apparatus operated in a pressure-demand or other positive-pressure mode). *Escape:* GmFOv (APF = 50) [any air-purifying, full-face-piece respirator (gas mask) with a chin-style, front- or back-mounted organic vapor canister] or SCBAE (any appropriate escape-type, self-contained breathing apparatus).

Storage: Color Code—Red Stripe: Flammability Hazard: Store separately from all other flammable materials. Prior to working with this chemical you should be trained on its proper handling and storage. Before entering confined space where this chemical may be present, check to make sure that an explosive concentration does not exist. Vinyl bromide must be stored to avoid contact with oxidizers (such as perchlorates, peroxides, permanganates, chlorates, and nitrates) and heat or flame, since violent reactions occur. Sources of ignition, such as smoking and open flames, are prohibited where vinyl bromide is used, handled, or stored in a manner that could create a potential fire or explosion hazard. Procedures for the handling, use, and storage of cylinders should be in compliance with OSHA 1910.101 and 1910.169, as with the recommendations of the Compressed Gas Association. A regulated, marked area should be established where this chemical is handled, used, or stored in compliance with OSHA Standard 1910.1045.

Shipping: This compound requires a shipping label of "FLAMMABLE GAS." It falls in Hazard Class 2.1.

Spill Handling: Evacuate and restrict persons not wearing protective equipment from area of spill or leak until cleanup is complete. Remove all ignition sources. Establish forced ventilation to keep levels below explosive limit and to disperse the gas. Stop the flow of gas if it can be done safely. If source of leak is a cylinder and the leak cannot be stopped in place, remove leaking cylinder to a safe place in the open air, and repair leak or allow cylinder to empty. Keep this chemical out of confined space, such as a sewer, because of the possibility of explosion, unless the sewer is designed to prevent the buildup of explosive concentrations. It may be necessary to contain and dispose of this chemical as a hazardous waste. Contact your local or federal environmental protection agency for specific recommendations. If employees are required to clean up spills, they must be properly trained and equipped. OSHA 1910.120(q) may be applicable.

Fire Extinguishing: Vinyl bromide is a flammable gas. For small fires, use dry chemical, CO_2, water spray, or foam extinguishers. Poisonous gases are produced in fire. Do not extinguish the fire unless the flow of gas can be stopped and any remaining gas is out of the line. Specially trained personnel may use fog lines to cool exposures and let the fire burn itself out. Vapors are heavier than air and will collect in low areas. Vapors may travel long distances to ignition sources and flashback. Vapors in confined areas may explode when exposed to fire. Containers may explode in fire. Storage containers and parts of containers may rocket great distances, in many directions. If material or contaminated runoff enters waterways, notify downstream users of potentially contaminated waters. Notify local health and fire officials and pollution control agencies. From a secure, explosion-proof location, use water spray to cool exposed containers. If cooling streams are ineffective (venting sound increases in volume and pitch, tank discolors, or shows any signs of deforming), withdraw immediately to a secure position. If cylinders are exposed to excessive heat from fire or flame contact, withdraw immediately to a secure location. If employees are expected to fight fires, they must be trained and equipped in OSHA 1910.156. The only respirators recommended for firefighting are self-contained breathing apparatuses that have full face-pieces and are operated in a pressure-demand or other positive-pressure mode.

References

US Environmental Protection Agency. (January 30, 1978). *Chemical Hazard Information Profile: Vinyl Bromide.* Washington, DC

National Institute for Occupational Safety and Health. (September 21, 1978). *Vinyl Halides: Carcinogenicity,* Current Intelligence Bulletin No. 28. Washington, DC

Sax, N. I. (Ed.). *Dangerous Properties of Industrial Materials Report,* 2, No. 2, 87−88 (1982) and 4, No. 5, 58−63 (1984) and 9, No. 1, 80−88 (1989).

New Jersey Department of Health and Senior Services. (June 2000). *Hazardous Substances Fact Sheet: Vinyl Bromide.* Trenton, NJ

Vinyl chloride   V:0170

Molecular Formula: C_2H_3Cl
Common Formula: $CH_2{=}CHCl$
Synonyms: Chloroethene; Chloroethylene; Chlorure de vinyle (French); Cloruro de vinilo (Spanish); Ethene, chloro-; Ethylene, chloro-; Ethylene monochloride; Monochloroethene; Monochloroethylene; Troviduer; Trovidur; UN1086; VC; VCL; VCM; Vinylchlorid (German); Vinyl chloride monomer; Vinyl C monomer; Vinyle (chlorure de) (French)
CAS Registry Number: 75-01-4
RTECS® Number: KU9625000
UN/NA & ERG Number: UN1086 (stabilized)/116
EC Number: 200-831-0 [*Annex I Index No.:* 602-023-00-7]
Regulatory Authority and Advisory Bodies
Department of Homeland Security Screening Threshold Quantity (pounds): *Release hazard* 10,000 ($\geq$1.00% concentration).
Carcinogenicity: IARC: Human Sufficient Evidence; Animal Sufficient Evidence, *carcinogenic to humans*, Group 1, 1998; NTP: 11th Report on Carcinogens, 2004: Known to be a human carcinogen; EPA: Known human carcinogen; NIOSH: Potential occupational carcinogen; OSHA: Potential human carcinogen; US EPA Gene-Tox Program, Positive: Carcinogenicity—mouse/rat; Positive: *In vivo* cytogenetics—human lymphocyte; *E. coli* polA with S9; Positive: Histidine reversion—Ames test; Positive: *D. melanogaster* sex-linked lethal; Positive: *S. cerevisiae* gene conversion; *S. pombe*—forward mutation; Negative: *D. melanogaster*—reciprocal translocation; Negative: Rodent dominant lethal; Mouse spot test; Negative: *S. cerevisiae*—homozygosis.
Banned or Severely Restricted (several countries) (UN).[13]
OSHA, 29CFR1910 Specifically Regulated Chemicals (see CFR 1910.1017).
Air Pollutant Standard Set. See below, "Permissible Exposure Limits in Air" section.
Clean Air Act: Hazardous Air Pollutants (Title I, Part A, Section 112); Accidental Release Prevention/Flammable Substances, (Section 112[r], Table 3), TQ = 10,000 lb (4540 kg).
Clean Water Act: Section 313 Water Priority Chemicals (57FR41331, 9/9/92); 40CFR401.15 Section 307 Toxic Pollutants.
US EPA Hazardous Waste Number (RCRA No.): U043; D043.
RCRA Toxicity Characteristic (Section 261.24), Maximum. Concentration of Contaminants, regulatory level, 0.2 mg/L.
RCRA, 40CFR261, Appendix 8 Hazardous Constituents.
RCRA 40CFR268.48; 61FR15654, Universal Treatment Standards: Wastewater (mg/L), 0.27; Nonwastewater (mg/kg), 6.0.
RCRA 40CFR264, Appendix 9; TSD Facilities Ground Water Monitoring List. Suggested test method(s) (PQL µg/L): 8010 (2); 8240 (10).

Safe Drinking Water Act: MCL, 0.002 mg/L; MCLG, zero; Regulated chemical (47 FR 9352).
Reportable Quantity (RQ): 1 lb (0.454 kg).
EPCRA Section 313 Form R *de minimis* concentration reporting level: 0.1%.
California Proposition 65 Chemical 2/27/87.
Canada, WHMIS, Ingredients Disclosure List Concentration 0.1%.
European/International Regulations (as chloroethylene): Hazard Symbol: F+, T; Risk phrases: R45; R12; Safety phrases: S53; S45 (see Appendix 4).
WGK (German Aquatic Hazard Class): 2—Hazard to waters.
Description: Vinyl chloride is a flammable gas at room temperature, and is usually encountered as a cooled liquid. The colorless liquid forms vapors which have a pleasant, ethereal odor. The odor threshold is variously given as 260 ppm,[41] 3000 ppm (NJ fact sheet), 4000 ppm (NY fact sheet) in air and 3.4 ppm in water (EPA Toxicological profile). Shipped as a liquefied compressed gas. Molecular weight = 62.50; Specific gravity (H_2O:1) = 0.88 (Liquid) at 25°C; Boiling point = −14°C; Freezing/Melting point = −160°C; Relative vapor density (air = 1) = 2.21; Flash point = flammable gas at −75°C (cc); Autoignition temperature = 472°C. Explosive limits: LEL = 3.6%; UEL = 33.0%. Hazard Identification (based on NFPA-704 M Rating System): Health 2, Flammability 4, Reactivity 2. Insoluble in water; 0.1% at 25°C.
Potential Exposure: Compound Description: Agricultural Chemical; Tumorigen, Mutagen; Reproductive Effector; Human Data. Vinyl chloride is used as a vinyl monomer in the manufacture of polyvinyl chloride (vinyl chloride homopolymer) and other copolymer resins. It is also used as a chemical intermediate and as a solvent.
Incompatibilities: Copper, oxidizers, aluminum, peroxides, iron, steel. Polymerizes in air, sunlight, heat, and on contact with a catalyst, strong oxidizers, and metals, such as aluminum and copper, unless stabilized by inhibitors, such as phenol. Attacks iron and steel in the presence of moisture.
Permissible Exposure Limits in Air
Conversion factor: 1 ppm = 2.56 mg/m^3 at 25°C & 1 atm.
OSHA PEL: 1 ppm/2.56 mg/m^3 TWA; 5 ppm [Avg not exceeding any 15 min period STEL].
NIOSH REL: Potential occupational carcinogen. Limit exposure to lowest feasible concentration. See *NIOSH Pocket Guide*, Appendix A.
ACGIH TLV®[1]: 1 ppm/2.6 mg/m^3 TWA, Confirmed Human Carcinogen.
Protective Action Criteria (PAC)*
TEEL-0: 1 ppm
PAC-1: **250** ppm
PAC-2: **1200** ppm
PAC-3: **4800** ppm
*AEGLs (Acute Emergency Guideline Levels) & ERPGs (Emergency Response Planning Guideline) are in **bold face**.

DFG MAK: Carcinogen Category 1; TRK: 2 mL/m^3/ 5.2 mg/m^3; 2.4 mg[thioglycolic acid]/24 h, in urine; after several shifts.

Arab Republic of Egypt: TWA 2.5 mg/m^3, 1993; Australia: TWA 5 ppm (10 mg/m^3), carcinogen, 1993; Austria: carcinogen, 1999; Belgium: TWA 5 ppm (13 mg/m^3), carcinogen 1993; Denmark: TWA 1 ppm (3 mg/m^3), [skin], 1999; Finland: TWA 5 ppm (15 mg/m^3), STEL 10 ppm (30 mg/m^3), carcinogen, 1993; Hungary: STEL 10 mg/m^3, carcinogen, 1993; Japan 2.5 ppm (6.5 mg/m^3), carcinogen, 1999; the Netherlands: MAC-TGG 7.77 mg/m^3, 2003; Norway: TWA 1 ppm (3 mg/m^3), 1999; the Philippines: TWA 50 ppm (100 mg/m^3), 1993; Poland: MAC (TWA) 5 mg/m^3; MAC (STEL) 30 mg/m^3, 1999; Russia: TWA 1 mg/m^3, STEL 2.5 ppm (5 mg/m^3), 1993; Sweden: NGV 1 ppm (2.5 mg/m^3), KTV 5 ppm (13 mg/m^3), [skin], carcinogen, 1999; Switzerland: MAK-W 2 ppm (5.2 mg/m^3), carcinogen, 1999; Thailand: TWA 1 ppm (2.8 mg/m^3), 1993; Turkey: TWA 500 ppm (1300 mg/m^3), 1993; United Kingdom: TWA 7 ppm, carcubigen, 2000; Argentina, Bulgaria, Columbia, Jordan, South Korea, New Zealand, Singapore, Vietnam: ACGIH TLV$^®$: Confirmed Human Carcinogen. MAC values have been set for ambient air in residential areas[35] by Russia at 0.005 mg/m^3 on a momentary basis and by the Czech Republic at 0.3 mg/m^3 on a momentary basis and 0.1 mg/m^3 on a daily average basis. Several states have set guidelines or standards for vinyl chloride in ambient air[60] ranging from zero (North Dakota) to 0.038 μg/m^3 (North Carolina) to 0.4 μg/m^3 (Michigan and New York) to 1.0 μg/m^3 (Virginia) to 3.846 μg/m^3 (Kansas) to 3.9 μg/m^3 (Massachusetts) to 5.0 μg/m^3 (Pennsylvania) to 50.0 μg/m^3 (Connecticut, South Carolina, South Dakota) to 238.0 μg/m^3 (Nevada).

Determination in Air: Use NIOSH Analytical Method (IV) #1007, vinyl chloride, OSHA Analytical Method 04.

Permissible Concentration in Water: No criteria have been determined for the protection of freshwater or saltwater aquatic life due to insufficient data. For the protection of human health: preferably zero. An additional lifetime cancer risk of 1 in 100,000 is posed by a concentration of 20 μg/L.[6] A long-term health advisory for vinyl chloride of 46 μg/L has been set by EPA.[48] Several states have set guidelines or standards for vinyl chloride in drinking water[61] ranging from zero (Rhode Island) to 0.15 μg/L (Minnesota) to 1.0 μg/L (Arizona, Florida, Kansas, Massachusetts, New Mexico, Vermont) to 2.0 μg/L (California, Maine, New Jersey, Colorado) to 5.0 μg/L (New York).

Determination in Water: Inert gas purge followed by gas chromatography with halide-specific detection (EPA Method 601) or gas chromatography plus mass spectrometry (EPA Method 624). Octanol−water coefficient: Log K_{ow} = 0.62.

Routes of Entry: Inhalation, skin, and/or eye contact (liquid).

Harmful Effects and Symptoms

Short Term Exposure: *Inhalation:* May affect the central nervous system. High exposures can cause dizziness,
lightheadedness, sleepiness. Even higher levels can cause headache, nausea, weakness, unconsciousness, and possible death. Exposure at 8000 ppm for 5 min can cause a feeling of intoxication, tiredness, drowsiness, abdominal pain, numbness and tingling in fingers and toes, pains in joints, coughing, sneezing, irritability, loss of appetite and weight. *Skin:* Contact with liquid may cause frostbite; contact with vapor may cause irritation and rash. Absorption is possible through the skin. *Eyes:* Can cause severe and immediate irritation. Contact with the liquid may cause frostbite. *Ingestion:* Moderately toxic.

Long Term Exposure: May cause "scleroderma" a disease that causes the skin to become smooth, tight and shiny and causes the bones in the fingers to erode (resulting in club-like swelling and shortening of finger tips) and damage the blood vessels in the hands (Raynaud's syndrome). This causes the hands or feet to turn numb, pale or blue with even mild cold exposure. Connective tissue, bones, and joints of arms and legs may suffer damage. Repeated exposure can permanently damage the liver, spleen, kidneys, nervous system, and blood cells. Vinyl chloride can cause symptoms such as stomach ulcers and skin allergy. Not all symptoms disappear after exposure stops. Vinyl chloride has caused liver cancer in occupationally exposed individuals. It may damage the developing fetus and there is limited evidence that it is a teratogen in animals. An excess of spontaneous abortions has been reported among spouses of workers who have been exposed to vinyl chloride. Increased rates of birth defects have been reported in areas where vinyl chloride processing plants are located. Vinyl chloride's role in this increased risk is unknown at this time.

Points of Attack: Liver, central nervous system, blood, respiratory system, lymphatic system. Cancer site in humans: liver. Cancer site in animals: liver, brain, lung.

Medical Surveillance: OSHA mandates the following tests: blood serum; alkaline phosphatase; bilirubin, gamma glutamyl transpeptidase; serum glutamic oxaloacetic transaminase; serum glutamic pyruvic transaminase. NIOSH lists the following tests: alkaline phosphatase; blood serum: lactic dehydrogenase; chest X-ray; expired air; liver function tests: alkaline phosphatase, bilirubin, gamma glutamyl transpeptidase, serum glutamic oxaloacetic transaminase, serum glutamic pyruvic transaminase; pulmonary function tests; pulmonary function tests: forced vital capacity, forced expiratory volume (1 s); red blood cell count; urine (chemical/metabolite); urinalysis (routine): albumin, whole blood (chemical/metabolite), ultrasonography of the liver. Complete examination of the skin and nervous system. If symptoms develop or overexposure is suspected, the following may be useful: X-rays of the fingers. Test for "urinary thiodiglycolic acid" (normal level is usually less than 2 mg/L). Evaluation by a qualified allergist, including careful exposure history and special testing, may help diagnose skin allergy.

First Aid: If this chemical gets into the eyes, remove any contact lenses at once and irrigate immediately for at least

15 min, occasionally lifting upper and lower lids. Seek medical attention immediately. If this chemical contacts the skin, remove contaminated clothing and wash immediately with soap and water. Seek medical attention immediately. If this chemical has been inhaled, remove from exposure, begin rescue breathing (using universal precautions, including resuscitation mask) if breathing has stopped and CPR if heart action has stopped. Transfer promptly to a medical facility. When this chemical has been swallowed, get medical attention. Give large quantities of water and induce vomiting. Do not make an unconscious person vomit.

Personal Protective Methods: Wear protective gloves and clothing to prevent any reasonable probability of skin contact. Safety equipment suppliers/manufacturers can provide recommendations on the most protective glove/clothing material for your operation. Viton™, Silvershield™, and chlorinated polyethylene are among the recommended protective materials. All protective clothing (suits, gloves, footwear, headgear) should be clean, available each day, and put on before work. Contact lenses should not be worn when working with this chemical. Eye protection is included in the recommended respiratory protection. Employees should wash immediately with soap when skin is wet or contaminated. Provide emergency showers and eyewash.

Respirator Selection: *At concentrations above the NIOSH REL, or where there is NO REL, at any detectable concentration:* SCBAF: Pd,Pp (APF = 10,000) (any self-contained breathing apparatus that has a full face-piece and is operated in a pressure-demand or other positive-pressure mode) or SaF: Pd,Pp: ASCBA (APF = 10,000) (any supplied-air respirator that has a full face-piece and is operated in a pressure-demand or other positive-pressure mode in combination with an auxiliary self-contained breathing apparatus operated in a pressure-demand or other positive-pressure mode). *Escape:* GmFOv (APF = 50) [any air-purifying, full-face-piece respirator (gas mask) with a chin-style, front- or back-mounted organic vapor canister] or SCBAE (any appropriate escape-type, self-contained breathing apparatus).

Vinyl Chloride (1910.1017) ≤*10 ppm*: (1) Combination Type C supplied-air respirator, demand type, with half face-piece, and auxiliary self-contained air supply; (2) Type C supplied-air respirator, demand type, with half face-piece; or (3) Any chemical cartridge respirator with an organic vapor cartridge which provides a service life of at least 1 h for concentrations of vinyl chloride *Up to 10 ppm.* ≤*25 ppm*: (1) Powered air-purifying respirator with hood, helmet, full or half face-piece, and a canister which provides a service life of at least 4 h for concentrations of vinyl chloride up to 25 ppm; or (2) Gas mask with front- or back-mounted canister which provides a service life of at least 4 h for concentrations of vinyl chloride *Up to 25 ppm.* ≤*100 ppm*: (1) Combination Type C supplied-air respirator, demand type, with full face-piece, and auxiliary self-contained air supply; or (2) Open-circuit self-contained breathing apparatus with full face-piece, in demand mode; or (3) Type C supplied-air respirator, demand type, with full face-piece. ≤*1000 ppm*: Type C supplied-air respirator, continuous-flow type, with full or half face-piece, helmet, or hood. ≤*3600 ppm*: (1) Combination Type C supplied-air respirator, pressure demand type, with full or half face-piece, and auxiliary self-contained air supply; or (2) Combination type continuous-flow supplied-air respirator with full or half face-piece and auxiliary self-contained air supply. >*3600 ppm*: or unknown concentration Open-circuit self-contained breathing apparatus, pressure-demand type, with full face-piece.

Storage: Flammable gas. Color Code—Red Stripe: Flammability Hazard: Do not store in the same area as other flammable materials. Color Code—Blue: Health Hazard/ Poison: Store in a secure poison location. Prior to working with this chemical you should be trained on its proper handling and storage. Before entering confined space where this chemical may be present, check to make sure that an explosive concentration does not exist. Vinyl chloride must be stored to avoid contact with oxidizers (such as perchlorates, peroxides, permanganates, chlorates, and nitrates), since violent reactions occur. Sources of ignition, such as smoking and open flames, are prohibited where vinyl chloride is handled, used, or stored. Metal containers involving the transfer of 5 gallons of vinyl chloride should be grounded and bonded. Drums must be equipped with self-closing valves, pressure vacuum bungs, and flame arresters. Use only nonsparking tools and equipment, especially when opening and closing containers of vinyl chloride. Wherever vinyl chloride is used, handled, manufactured, or stored, use explosion-proof electrical equipment and fittings. Procedures for the handling, use, and storage of cylinders should be in compliance with OSHA 1910.101 and 1910.169, as with the recommendations of the Compressed Gas Association. A regulated, marked area should be established where this chemical is handled, used, or stored in compliance with OSHA Standard 1910.1045.

Shipping: Vinyl chloride requires a shipping label of "FLAMMABLE GAS." It falls in Hazard Class 2.1.

Spill Handling: Evacuate and restrict persons not wearing protective equipment from area of spill or leak until cleanup is complete. Remove all ignition sources. Establish forced ventilation to keep levels below explosive limit. Ventilate area of leak to disperse the gas. Establish forced ventilation to keep levels below explosive limit. Stop the flow of gas if it can be done safely. If source of leak is a cylinder and the leak cannot be stopped in place, remove leaking cylinder to a safe place in the open air, and repair leak or allow cylinder to empty. Keep this chemical out of confined space, such as a sewer, because of the possibility of explosion, unless the sewer is designed to prevent the buildup of explosive concentrations. It may be necessary to contain and dispose of this chemical as a hazardous waste. Contact your local or federal environmental protection agency for specific recommendations. If employees are required to clean up spills, they must be properly trained and equipped. OSHA 1910.120(q) may be applicable.

Fire Extinguishing: Vinyl chloride is a flammable gas. Use dry chemical or CO_2 extinguishers. Poisonous gases are produced in fire, including phosgene, hydrogen chloride, and carbon monoxide. Fire may restart after it has been extinguished. Do not extinguish the fire unless the flow of gas can be stopped and any remaining gas is out of the line. Specially trained personnel may use fog lines to cool exposures and let the fire burn itself out. Vapors are heavier than air and will collect in low areas. Vapors may travel long distances to ignition sources and flashback. Vapors in confined areas may explode when exposed to fire. Containers may explode in fire. Storage containers and parts of containers may rocket great distances, in many directions. If material or contaminated runoff enters waterways, notify downstream users of potentially contaminated waters. Notify local health and fire officials and pollution control agencies. From a secure, explosion-proof location, use water spray to cool exposed containers. If cooling streams are ineffective (venting sound increases in volume and pitch, tank discolors, or shows any signs of deforming), withdraw immediately to a secure position. If cylinders are exposed to excessive heat from fire or flame contact, withdraw immediately to a secure location. If employees are expected to fight fires, they must be trained and equipped in OSHA 1910.156. The only respirators recommended for firefighting are self-contained breathing apparatuses that have full face-pieces and are operated in a pressure-demand or other positive-pressure mode.

Disposal Method Suggested: Consult with environmental regulatory agencies for guidance on acceptable disposal practices. Generators of waste containing this contaminant ($\geq$100 kg/mo) must conform with EPA regulations governing storage, transportation, treatment, and waste disposal. Incineration, preferably after mixing with another combustible fuel. Care must be exercised to assure complete combustion to prevent the formation of phosgene. An acid scrubber is necessary to remove the halo acids produced.[22] A variety of techniques have been described for vinyl chloride recovery from PVC latexes.

References

US Environmental Protection Agency. (June 1975). *Scientific and Technical Assessment Report on Vinyl Chloride and Polyvinyl Chloride.* Washington, DC: Office of Research and Development
National Institute for Occupational Safety and Health. (September 21, 1978). *Vinyl Halides—Carcinogenicity,* Current Intelligence Bulletin No. 28. Washington, DC
US Environmental Protection Agency. (1980). *Vinyl Chloride: Ambient Water Quality Criteria.* Washington, DC
US Environmental Protection Agency. (April 30, 1980). *Chloroethene: Health and Environmental Effects Profile No. 45.* Washington, DC: Office of Solid Waste
Sittig, M. (1978). *Vinyl Chloride and PVC Manufacture: Process and Environmental Aspects.* Park Ridge, NJ: Noyes Data Corp.

Sax, N. I. (Ed.). *Dangerous Properties of Industrial Materials Report,* 1, No. 3, 85−87 (1981) and 6, No. 4, 13−43 (1986)
US Public Health Service. (January 1988). *Toxicological Profile for Vinyl Chloride.* Atlanta, GA: Agency for Toxic Substances and Disease Registry
New York State Department of Health. (March 1986). *Chemical Fact Sheet: Vinyl Chloride* (Version 2). Albany, NY: Bureau of Toxic Substance Assessment
New Jersey Department of Health and Senior Services. (June 2001). *Hazardous Substances Fact Sheet: Vinyl Chloride.* Trenton, NJ

4-Vinyl-1-cyclohexene   V:0180

Molecular Formula: C_8H_{12}
Common Formula: $C_6H_9CH{=}CH_2$
Synonyms: Butadiene dimer; Cyclohexenylethylene; 1-Ethenylcyclohexene; 4-Ethenyl-1-cyclohexene; NCI-C54999; 1,2,3,4-Tetrahydrostyrene; 1-Vinylcyclohex-3-ENE; 1-Vinylcyclohexene-3; 4-Vinyl-1-cyclohexene; 4-Vinylcyclohexene; 4-Vinylcyclohexene-1
CAS Registry Number: 100-40-3
RTECS® Number: GW6650000
UN/NA & ERG Number: UN1993/128
EC Number: 202-848-9
Regulatory Authority and Advisory Bodies
Carcinogenicity: IARC: Human Inadequate Evidence; Animal Sufficient Evidence, *possibly carcinogenic to humans,* Group 2B, 1994; NCI: Carcinogenesis Studies (gavage); clear evidence: mouse; inadequate study: rat.
California Proposition 65 Chemical: Cancer 2/27/87; Developmental/Reproductive toxin (female) 8/7/09.
European/International Regulations: not listed in Annex 1.
WGK (German Aquatic Hazard Class): 2—Hazard to waters.
Description: 4-Vinyl-1-cyclohexene, a cyclic alkene, is a flammable liquid. Molecular weight = 108.20; Specific gravity (H_2O:1) = 0.83 at 25°C; Boiling point = 130.3°C; Freezing/Melting point = −109°C; Flash point = 16−21°C; Autoignition temperature = 269°C. Hazard Identification (based on NFPA-704 M Rating System): Health 0, Flammability 3, Reactivity 2. Hydrolyzes in water.
Potential Exposure: Compound Description: Tumorigen; Reproductive Effector. 4-Vinyl-1-cyclohexene is used as an intermediate for the production of vinylcyclohexene dioxide, which is used as a reactive diluent in epoxy resins. Previous uses of 4-vinyl-1-cyclohexene include comonomer in the polymerization of other monomers and for halogenation to polyhalogenated derivatives which are used as flame retardants.
Incompatibilities: Hydrolyzes in water. Reacts with oxidizers, amines, alcohols.
Permissible Exposure Limits in Air
OSHA PEL: None.
NIOSH REL: None.

ACGIH TLV®[1]: 0.1 ppm/0.44 mg/m^3 TWA; confirmed animal carcinogen with unknown relevance to humans.

AIHA WEEL: 1 ppm/4.4 mg/m^3 TWA.

Protective Action Criteria (PAC)

TEEL-0: 0.1 ppm

PAC-1: 0.3 ppm

PAC-2: 3500 ppm

PAC-3: 3500 ppm

DFG MAK: [skin] Carcinogen Category: 2.

Poland: MAC (TWA) 10 mg/m^3, 1999; the Netherlands: MAC-TGG 0.4 mg/m^3, 2003.

Determination in Water: Octanol—water coefficient: Log K_{ow} = 3.93.

Routes of Entry: Inhalation, skin, and/or eye contact.

Harmful Effects and Symptoms

Short Term Exposure: Vapors cause irritation and smarting of the eyes and respiratory system if present in high concentrations. The effect is temporary.

Long Term Exposure: Workers exposed to 4-vinyl-1-cyclohexene experienced keratitis (inflammation of the cornea), rhinitis, headache, hypotonia, leucopenia (decrease in the number of white blood cells), neutrophilia, lymphocytosis, and impairment of pigment and carbohydrate metabolism. A confirmed carcinogen.

Points of Attack: Blood, eyes.

Medical Surveillance: Complete eye examination. Complete blood count and hematocrit.

First Aid: Skin Contact[52]: Flood all areas of body that have contacted the substance with water. Do not wait to remove contaminated clothing; do it under the water stream. Use soap to help assure removal. Isolate contaminated clothing when removed to prevent contact by others.

Eye Contact: Remove any contact lenses at once. Flush eyes well with copious quantities of water or normal saline for at least 20—30 min. Seek medical attention.

Inhalation: Leave contaminated area immediately; breathe fresh air. Proper respiratory protection must be supplied to any rescuers. If coughing, difficult breathing, or any other symptoms develop, seek medical attention at once, even if symptoms develop many hours after exposure.

Ingestion: If convulsions are not present, give a glass or two of water or milk to dilute the substance. Assure that the person's airway is unobstructed and contact a hospital or poison center immediately for advice on whether or not to induce vomiting.

Personal Protective Methods: Wear protective gloves and clothing to prevent any reasonable probability of skin contact. Safety equipment suppliers/manufacturers can provide recommendations on the most protective glove/clothing material for your operation. Viton is among the recommended protective materials. All protective clothing (suits, gloves, footwear, headgear) should be clean, available each day, and put on before work. Contact lenses should not be worn when working with this chemical. Wear splash-proof chemical goggles and face shield unless full-face-piece respiratory protection is worn. Employees should wash immediately with soap when skin is wet or contaminated. Provide emergency showers and eyewash.

Respirator Selection: Follow regulations in OSHA 29CFR1910.134 or European Standard EN149. Use a NIOSH/MSHA- or European Standard EN149-approved respirator; or use an approved supplied-air respirator with a full face-piece operated in the positive-pressure mode, or with a full-face-piece, hood, or helmet in the continuous-flow mode; or use a NIOSH/MSHA- or European Standard EN149-approved self-contained breathing apparatus with a full face-piece operated in pressure-demand or other positive-pressure mode.

Storage: Color Code—Red: Flammability Hazard: Store in a flammable liquid storage area or approved cabinet away from ignition sources and corrosive and reactive materials. Prior to working with this chemical you should be trained on its proper handling and storage. Store in an explosion-proof refrigerator. Protect from exposure to oxidizers, alcohols, amines, air, and light. Sources of ignition, such as smoking and open flames, are prohibited where this chemical is used, handled, or stored in a manner that could create a potential fire or explosion hazard. Metal containers involving the transfer of 5 gallons or more of this chemical should be grounded and bonded. Drums must be equipped with self-closing valves, pressure vacuum bungs, and flame arresters. Use only nonsparking tools and equipment, especially when opening and closing containers of this chemical. A regulated, marked area should be established where this chemical is handled, used, or stored in compliance with OSHA Standard 1910.1045.

Shipping: Flammable liquids, n.o.s. require a shipping label of "FLAMMABLE LIQUID." It falls in Hazard Class 3 and Packing Group II.

Spill Handling: Evacuate and restrict persons not wearing protective equipment from area of spill or leak until cleanup is complete. Remove all ignition sources. Ventilate area of spill or leak. Absorb liquids in vermiculite, dry sand, earth, peat, carbon, or a similar material and deposit in sealed containers. Keep this chemical out of a confined space, such as a sewer, because of the possibility of an explosion, unless the sewer is designed to prevent the buildup of explosive concentrations. It may be necessary to contain and dispose of this chemical as a hazardous waste. If material or contaminated runoff enters waterways, notify downstream users of potentially contaminated waters. Contact your local or federal environmental protection agency for specific recommendations. If employees are required to clean up spills, they must be properly trained and equipped. OSHA 1910.120(q) may be applicable.

Fire Extinguishing: This chemical is a flammable liquid. Poisonous gases are produced in fire. Use dry chemical, carbon dioxide, or alcohol foam extinguishers. Water may be ineffective due to the low flash point. Water spray may be used however. Vapors are heavier than air and will collect in low areas. Vapors may travel long distances to ignition sources and flashback. Vapors in confined areas may explode when exposed to fire. Containers may explode in fire. Storage containers and parts of containers may rocket great distances, in many directions. If material or contaminated runoff enters waterways, notify downstream users of potentially contaminated waters. Notify local health and fire officials and pollution control agencies. From a secure, explosion-proof location, use water spray to cool exposed containers. If cooling streams are ineffective (venting sound increases in volume and pitch, tank discolors, or shows any signs of deforming), withdraw immediately to a secure position. If employees are expected to fight fires, they must be trained and equipped in OSHA 1910.156. The only respirators recommended for firefighting are self-contained breathing apparatuses that have full face-pieces and are operated in a pressure-demand or other positive-pressure mode.

Disposal Method Suggested: Dissolve or mix the material with a combustible solvent and burn in a chemical incinerator equipped with an afterburner and scrubber. All federal, state, and local environmental regulations must be observed.

References

US Environmental Protection Agency. (September 19, 1985). *Chemical Hazard Information Profile: 4-Vinyl-1-Cyclohexene*. Washington, DC

Vinyl cyclohexene dioxide   V:0190

Molecular Formula: $C_8H_{12}O_2$

Synonyms: 1,2-Epoxy-4-(epoxyethyl)cyclohexane; 3-(Epoxyethyl)-7-oxabicyclo(4.1.0) heptane, vinyl cyclohexene diepoxide; NCI-C60139; 3-Oxiranyl-7-oxabicyclo(4.1.0) heptene; Ucet textile finish 11-74 (obs.); Unox epoxide 206; 4-Vinyl-1,2-cyclohexene diepoxide; 4-Vinyl-1-cyclohexene diepoxide; 4-Vinylcyclohexene diepoxide; Vinyl cyclohexene diepoxide; 1-Vinyl-3-cyclohexene dioxide; 4-Vinyl-1-cyclohexene dioxide; 4-Vinylcyclohexene dioxide

CAS Registry Number: 106-87-6; (*alt.*) 25550-49-6

RTECS® Number: RN8640000

UN/NA & ERG Number: UN2810/153

EC Number: 203-437-7 [*Annex I Index No.:* 603-066-00-4]

Regulatory Authority and Advisory Bodies

Carcinogenicity: IARC: Human Inadequate Evidence; Animal Sufficient Evidence, *possibly carcinogenic to humans*, Group 2B, 1994; NCI: Carcinogenesis Studies (gavage); clear evidence: mouse, rat; NTP: 11th Report on Carcinogens, 2004: Reasonably anticipated to be a human carcinogen; NIOSH: Potential occupational carcinogen.

Air Pollutant Standard Set. See below, "Permissible Exposure Limits in Air" section.

California Proposition 65 Chemical: Cancer 7/1/90; Developmental/Reproductive toxin 8/1/08.

Canada, WHMIS, Ingredients Disclosure List Concentration 0.1%.

European/International Regulations: Hazard Symbol: T; Risk phrases: R23/24/25; R40; Safety phrases: S1/2; S36/37; S45; S63 (see Appendix 4).

WGK (German Aquatic Hazard Class): No value assigned.

Description: Vinyl cyclohexene dioxide is a colorless liquid. Molecular weight = 140.20; Specific gravity (H_2O:1) = 1.10 at 25°C; Boiling point = 227.2°C; Freezing/Melting point = −109°C; Vapor pressure = 0.1 mmHg at 25°C; Flash point = 110°C; Autoignition temperature = 392°C. Hazard Identification (based on NFPA-704 M Rating System) (*Fume and dust*): Health 3, Flammability 1, Reactivity 0. Highly soluble in water; reacts slowly.

Potential Exposure: Compound Description: Tumorigen, Mutagen; Reproductive Effector; Primary Irritant. This material is used as a monomer in the production of epoxy resins for coatings and adhesives, as a chemical intermediate, and as a reactive diluent.

Incompatibilities: Reacts with alcohols, amines and other active hydrogen compounds, water. Slowly hydrolyzes in water.

Permissible Exposure Limits in Air

OSHA PEL: None.

NIOSH REL: 10 ppm/60 mg/m³ TWA [skin]; Potential occupational carcinogen. Limit exposure to lowest feasible concentration; See *NIOSH Pocket Guide*, Appendix A.

ACGIH TLV®[1]: 10 ppm/57 mg/m³ TWA [skin]; confirmed animal carcinogen with unknown relevance to humans.

Protective Action Criteria (PAC)

TEEL-0: 0.1 ppm

PAC-1: 0.35 ppm

PAC-2: 2.5 ppm

PAC-3: 75 ppm

DFG MAK: [skin], Carcinogen Category: 2.

Australia: TWA 10 ppm (60 mg/m³), [skin], carcinogen, 1993; Austria: carcinogen, 1999; Belgium: TWA 10 ppm (57 mg/m³), [skin], carcinogen, 1993; Denmark: TWA 10 ppm (60 mg/m³), 1999; Finland: TWA 10 ppm (60 mg/m³), STEL 20 ppm (120 mg/m³), carcinogen, 1999; Norway: TWA 10 ppm (60 mg/m³), 1999; the Netherlands: MAC-TGG 60 mg/m³, 2003; United Kingdom: LTEL 10 ppm (60 mg/m³), 1993; Argentina, Bulgaria, Columbia, Jordan, South Korea, New Zealand, Singapore, Vietnam: ACGIH TLV®: confirmed animal carcinogen with unknown relevance to humans. Several states have set guidelines or standards for vinyl cyclohexene dioxide in ambient air[60] ranging from zero (North Dakota) to 6.0 μg/m³ (Virginia) to 150.0 μg/m³ (Pennsylvania) to 600.0 μg/m³ (Connecticut) to 1429 μg/m³ (Nevada).

Determination in Air: No method available.

Determination in Water: Octanol−water coefficient: Log K_{ow} = 1.32.

Routes of Entry: Inhalation, skin absorption, ingestion, skin and/or eye contact.

Harmful Effects and Symptoms

Short Term Exposure: Vinyl cyclohexene dioxide can affect you when breathed in and by passing through your skin. Irritates the eyes, skin and severely irritates the nose, throat, and lungs, causing coughing and wheezing. Prolonged contact can cause severe burns and blisters.

Long Term Exposure: Vinyl cyclohexene dioxide may cause a skin allergy to develop. If this happens, very small future exposure can cause itching and a rash. It can cause lung irritation; bronchitis may develop. A potential occupational carcinogen and a mutagen; handle with extreme caution. It may also damage the testes (male reproductive system). In animals: irritation of the eyes, skin, respiratory system; testicular atrophy; leukopenia (reduced blood leukocytes), necrosis thymus; skin sensitization.

Points of Attack: Eyes, skin, respiratory system, blood, thymus, lungs, reproductive system. Cancer site in animals: skin.

Medical Surveillance: For those with frequent or potentially high exposure (half the TLV or grater), the following are recommended before beginning work and at regular times after that: lung function tests. If symptoms develop or overexposure is suspected, the following may be useful: White blood cell count. Evaluation by a qualified allergist, including careful exposure history and special testing, may help diagnose skin allergy.

First Aid: Eye Contact: Immediately remove any contact lenses and flush with large amounts of water for at least 15 min, occasionally lifting upper and lower lids. Seek medical attention immediately.

Skin Contact: Quickly remove contaminated clothing. Immediately wash contaminated skin with large amounts of water. Seek medical attention.

Breathing: Remove the person from exposure. Begin rescue breathing (using universal precautions, including resuscitation mask) if breathing has stopped and CPR if heart action has stopped. Transfer promptly to a medical facility.

Ingestion: If convulsions are not present, give a glass or two of water or milk to dilute the substance. Assure that the person's airway is unobstructed and contact a hospital or poison center immediately for advice on whether or not to induce vomiting.

Personal Protective Methods: Wear protective gloves and clothing to prevent any reasonable probability of skin contact. Safety equipment suppliers/manufacturers can provide recommendations on the most protective glove/clothing material for your operation. All protective clothing (suits, gloves, footwear, headgear) should be clean, available each day, and put on before work. Contact lenses should not be worn when working with this chemical. Wear splash-proof chemical goggles and face shield unless full-face-piece respiratory protection is worn. Employees should wash immediately with soap when skin is wet or contaminated. Provide emergency showers and eyewash.

Respirator Selection: At concentrations above the NIOSH REL, or where there is no REL, at any detectable concentration: SCBAF: Pd,Pp (APF = 10,000) (any self-contained breathing apparatus that has a full face-piece and is operated in a pressure-demand or other positive-pressure mode) or SaF: Pd,Pp: ASCBA (APF = 10,000) (any supplied-air respirator that has a full face-piece and is operated in a pressure-demand or other positive-pressure mode in combination with an auxiliary, self-contained breathing apparatus operated in a pressure-demand or other positive-pressure mode). *Escape:* GmFOv (APF = 50) [any air-purifying, full-face-piece respirator (gas mask) with a chin-style, front- or back-mounted organic vapor canister] or SCBAE (any appropriate escape-type, self-contained breathing apparatus).

Storage: Color Code—Blue: Health Hazard/Poison: Store in a secure poison location. Prior to working with this chemical you should be trained on its proper handling and storage. Store in tightly closed containers in a cool, well-ventilated area away from heat or flames, water, alcohols, amines. Sources of ignition, such as smoking and open flames, are prohibited where this chemical is used, handled, or stored in a manner that could create a potential fire or explosion hazard. Metal containers involving the transfer of 5 gallons or more of this chemical should be grounded and bonded. Drums must be equipped with self-closing valves, pressure vacuum bungs, and flame arresters. Use only non-sparking tools and equipment, especially when opening and closing containers of this chemical. A regulated, marked area should be established where this chemical is handled, used, or stored in compliance with OSHA Standard 1910.1045.

Shipping: Toxic, liquids, organic, n.o.s. require a shipping label of "POISONOUS/TOXIC MATERIALS." It falls in Hazard Class 6.1.

Spill Handling: Evacuate and restrict persons not wearing protective equipment from area of spill or leak until cleanup is complete. Remove all ignition sources. Ventilate area of spill or leak. Absorb liquids in vermiculite, dry sand, earth, peat, carbon, or a similar material and deposit in sealed containers. Keep this chemical out of a confined space, such as a sewer, because of the possibility of an explosion, unless the sewer is designed to prevent the buildup of explosive concentrations. It may be necessary to contain and dispose of this chemical as a hazardous waste. If material or contaminated runoff enters waterways, notify downstream users of potentially contaminated waters. Contact your local or federal environmental protection agency for specific recommendations. If employees are required to clean up spills, they must be properly trained and equipped. OSHA 1910.120(q) may be applicable.

Fire Extinguishing: This chemical is a combustible liquid. Poisonous gases are produced in fire. Use dry chemical, carbon dioxide, or alcohol foam extinguishers. Vapors are heavier than air and will collect in low areas. Containers may explode in fire. Storage containers and parts of containers may rocket great distances, in many directions. If material or contaminated runoff enters waterways, notify downstream users of potentially contaminated waters. Notify local health and fire officials and pollution control agencies. From a secure, explosion-proof location, use water spray to cool exposed containers. If cooling streams are ineffective (venting sound increases in volume and pitch, tank discolors, or shows any signs of deforming),

withdraw immediately to a secure position. If employees are expected to fight fires, they must be trained and equipped in OSHA 1910.156. The only respirators recommended for firefighting are self-contained breathing apparatuses that have full face-pieces and are operated in a pressure-demand or other positive-pressure mode.

Disposal Method Suggested: Concentrated waste containing no peroxides: discharge liquid at a controlled rate near a pilot flame. Concentrated waste containing peroxides: perforation of a container of the waste from a safe distance followed by open burning.

References

National Institute for Occupational Safety and Health. (October 1977). *Information Profiles on Potential Occupational Hazards: Vinyl Cyclohexene Dioxide*, Report PB-276,678. Rockville, MD, pp. 54—57

New Jersey Department of Health and Senior Services. (August 2005). *Hazardous Substances Fact Sheet: Vinyl Cyclohexene Dioxide*. Trenton, NJ

Vinyl ether   V:0200

Molecular Formula: C_4H_6O

Common Formula: $(CH_2=CH)_2O$

Synonyms: Divinyl ether; Divinyl oxide; Ethene, ethoxy-; Ethenyloxyethene; Ether, vinyl ethyl; Ethyl vinyl ether; Etoxyethene; EVE; 1,1-Oxybisethene; Vinamar; Vinesthene; Vinesthesin; Vinethen; Vinethene; Vinether; Vinidyl; Vinil etil eter (Spanish); Vinydan; Vinyl ethyl ether

CAS Registry Number: 109-93-3; 109-92-2

RTECS® Number: XZ6700000

UN/NA & ERG Number: UN1167(stabilized)/128

EC Number: 203-720-5

Regulatory Authority and Advisory Bodies

Department of Homeland Security Screening Threshold Quantity (pounds): *Release hazard* 10,000 ($\geq 1.00\%$ concentration).

Clean Air Act: Accidental Release Prevention/Flammable Substances, (Section 112[r], Table 3), TQ = 10,000 lb (4540 kg).

European/International Regulations: not listed in Annex 1.

WGK (German Aquatic Hazard Class): 1—Slightly water polluting (*CAS: 109-92-2*).

Description: Vinyl ether is a volatile liquid. Molecular weight = 70.10; Boiling point = 28°C; Freezing/Melting point = −101°C; Flash point ≤ −30°C; Autoignition temperature = 360°C. Explosive limits: LEL = 1.7%; UEL = 27%. Hazard Identification (based on NFPA-704 M Rating System): Health 2, Flammability 4, Reactivity 2. Insoluble in water.

Potential Exposure: It is used as an inhalation anesthetic; in formulation of copolymers with vinyl chloride.

Incompatibilities: Forms peroxides when exposed to air or oxygen; may polymerize explosively with evolution of acetylene gas. Store away from oxidizers, heat, and sunlight. Reacts with concentrated nitric acid. May accumulate static electricity.

Permissible Exposure Limits in Air

Protective Action Criteria (PAC)

109-92-2

TEEL-0: 15 ppm

PAC-1: 50 ppm

PAC-2: 350 ppm

PAC-3: 1500 ppm

Permissible Concentration in Water: No criteria set.

Routes of Entry: Inhalation, skin absorption, ingestion, skin and/or eye contact.

Harmful Effects and Symptoms

Short Term Exposure: A moderate health hazard. Irritates the eyes, skin, and respiratory tract. Affects the central nervous system, causing anesthesia, narcosis, and loss of consciousness.

Long Term Exposure: Prolonged exposure may cause liver damage.

Points of Attack: Central nervous system, liver.

Medical Surveillance: Liver function tests.

First Aid: If this chemical gets into the eyes, remove any contact lenses at once and irrigate immediately for at least 15 min, occasionally lifting upper and lower lids. Seek medical attention immediately. If this chemical contacts the skin, remove contaminated clothing and wash immediately with soap and water. Seek medical attention immediately. If this chemical has been inhaled, remove from exposure, begin rescue breathing (using universal precautions, including resuscitation mask) if breathing has stopped and CPR if heart action has stopped. Transfer promptly to a medical facility. When this chemical has been swallowed, get medical attention. Give large quantities of water and induce vomiting. Do not make an unconscious person vomit.

Personal Protective Methods: Wear skin protection. Wear protective gloves and clothing to prevent any reasonable probability of skin contact. Safety equipment suppliers/manufacturers can provide recommendations on the most protective glove/clothing material for your operation. All protective clothing (suits, gloves, footwear, headgear) should be clean, available each day, and put on before work. Contact lenses should not be worn when working with this chemical. Wear splash-proof chemical goggles and face shield unless full-face-piece respiratory protection is worn. Employees should wash immediately with soap when skin is wet or contaminated. Provide emergency showers and eyewash.

Respirator Selection: Follow regulations in OSHA 29CFR1910.134 or European Standard EN149. Use a NIOSH/MSHA- or European Standard EN149-approved respirator; or use an approved supplied-air respirator with a full-face-piece operated in the positive-pressure mode, or with a full face-piece, hood, or helmet in the continuous-flow mode; or use a NIOSH/MSHA- or European Standard EN149-approved self-contained breathing apparatus with a

full face-piece operated in pressure-demand or other positive-pressure mode.

Storage: Color Code—Red: Flammability Hazard: Store in a flammable liquid storage area or approved cabinet away from ignition sources and corrosive and reactive materials. Prior to working with this chemical you should be trained on its proper handling and storage. Before entering confined space where this chemical may be present, check to make sure that an explosive concentration does not exist. Store in tightly closed containers in a cool, well-ventilated area away from heat and oxidizers. Sources of ignition, such as smoking and open flames, are prohibited where this chemical is used, handled, or stored in a manner that could create a potential fire or explosion hazard. Metal containers involving the transfer of 5 gallons or more of this chemical should be grounded and bonded. Drums must be equipped with self-closing valves, pressure vacuum bungs, and flame arresters. Use only nonsparking tools and equipment, especially when opening and closing containers of this chemical. Protect against physical damage. Outside or detached storage is preferred. Inside storage should be in a standard flammable liquids storage room or cabinet. Protect against static electricity and lightning. For large-quantity storage rooms, protect with automatic sprinklers, total flooding carbon dioxide, or dry chemical systems.

Shipping: Divinyl ether, inhibited, requires a shipping label of "FLAMMABLE LIQUID." It falls in Hazard Class 3 and Packing Group II.

Spill Handling: Evacuate and restrict persons not wearing protective equipment from area of spill or leak until cleanup is complete. Remove all ignition sources. Establish forced ventilation to keep levels below explosive limit. Absorb liquids in vermiculite, dry sand, earth, peat, carbon, or a similar material and deposit in sealed containers. Keep this chemical out of a confined space, such as a sewer, because of the possibility of an explosion, unless the sewer is designed to prevent the buildup of explosive concentrations. It may be necessary to contain and dispose of this chemical as a hazardous waste. If material or contaminated runoff enters waterways, notify downstream users of potentially contaminated waters. Contact your local or federal environmental protection agency for specific recommendations. If employees are required to clean up spills, they must be properly trained and equipped. OSHA 1910.120(q) may be applicable.

Fire Extinguishing: This chemical is a flammable liquid. Poisonous gases are produced in fire. Use dry chemical, carbon dioxide, alcohol foam, or polymer foam extinguishers. Vapors are heavier than air and will collect in low areas. Vapors may travel long distances to ignition sources and flashback. Vapors in confined areas may explode when exposed to fire. Containers may explode in fire. Storage containers and parts of containers may rocket great distances, in many directions. If material or contaminated runoff enters waterways, notify downstream users of potentially contaminated waters. Notify local health and fire officials and pollution control agencies. From a secure, explosion-proof location, use water spray to cool exposed containers. If cooling streams are ineffective (venting sound increases in volume and pitch, tank discolors, or shows any signs of deforming), withdraw immediately to a secure position. If employees are expected to fight fires, they must be trained and equipped in OSHA 1910.156. The only respirators recommended for firefighting are self-contained breathing apparatuses that have full face-pieces and are operated in a pressure-demand or other positive-pressure mode.

Disposal Method Suggested: Allow to evaporate or incinerate. Beware of explosive peroxides in old containers.

References

Sax, N. I. (Ed.). *Dangerous Properties of Industrial Materials Report*, 1, No. 7, 78–79 (1981). New York: Van Nostrand Reinhold Co.

Vinyl fluoride   V:0210

Molecular Formula: C_2H_3F
Common Formula: $CHF=CH_2$
Synonyms: Ethene, fluoro-; Ethylene fluoro-; Fluorethylene; Fluoroethene; Fluoruro de vinilo (Spanish); Monofluroethylene; Vinyl fluoride monomer
CAS Registry Number: 75-02-5
RTECS® Number: YZ7351000
UN/NA & ERG Number: UN1860 (stabilized)/116
EC Number: 200-832-6
Regulatory Authority and Advisory Bodies
Department of Homeland Security Screening Threshhold Quantity (pounds): *Release hazard* 10,000 ($\geq$1.00% concentration).
Carcinogenicity: NTP: 11th Report on Carcinogens, 2004: Reasonably anticipated to be a human carcinogen; IARC: Animal, Sufficient Evidence; Human, Inadequate Evidence, Group 2A, 1995.
Air Pollutant Standard Set. See below, "Permissible Exposure Limits in Air" section.
Clean Air Act: Accidental Release Prevention/Flammable Substances, (Section 112[r], Table 3), TQ = 10,000 lb (4540 kg).
California Proposition 65 Chemical: Cancer 5/1/97.
European/International Regulations: not listed in Annex 1.
WGK (German Aquatic Hazard Class): No value assigned.
Description: Vinyl fluoride is a colorless gas. Molecular weight = 46.05; Boiling point = $-72.2°C$; Freezing/Melting point = $-160.6°C$; Relative vapor density (air = 1) = 1.60; Vapor pressure = 25.2 atm; Flash point = flammable gas; Autoignition temperature = $385°C$. Explosive limits: LEL = 2.6%; UEL = 21.7%. Hazard Identification (based on NFPA-704 M Rating System): Health 1, Flammability 4, Reactivity 2. Insoluble in water.
Potential Exposure: Compound Description: Tumorigen; Mutagen. Vinyl fluoride's primary use is as a chemical and

polymer intermediate; used to make polyvinyl fluoride (Tedlar®) film. Polyvinyl fluoride film is characterized by superior resistance to weather, high strength, and a high dielectric constant. It is used as a film laminate for building materials and in packaging electrical equipment. Polyvinyl fluoride film poses a hazard, so it is not recommended for food packaging. Polyvinyl fluoride evolves toxic fumes upon heating.

Incompatibilities: May polymerize. Inhibited with 0.2% terpenes to prevent polymerization. Violent reaction with oxidizers. May accumulate static electrical charges.

Permissible Exposure Limits in Air

Conversion factor: 1 ppm = 1.82 mg/m^3 at 25°C & 1 atm.
OSHA PEL: None.
NIOSH REL: 1 ppm TWA; 5 ppm Ceiling Concentration/15 min, CFR 1910.1017.
ACGIH TLV®[1]: 1 ppm, Suspected Human Carcinogen.
NIOSH IDLH: Not determined. Suspected occupational carcinogen.
Protective Action Criteria (PAC)
TEEL-0: 1 ppm
PAC-1: 7.5 ppm
PAC-2: 50 ppm
PAC-3: 75,000 ppm
The state of South Carolina has set a guideline of 19.0 μg/m^3 for vinyl fluoride in ambient air.[60]

Following are OELs for hydrogen fluoride which is produced as a decomposition product in the heat of fire.
OSHA PEL: 3 ppm/2.5 mg[F]/m^3 TWA.
NIOSH REL: 3 ppm/2.5 mg[F]/m^3 TWA; 6 ppm/5 mg[F]/m^3, 15 min Ceiling Concentration.
ACGIH TLV®[1]: 3 ppm/2.5 mg[F]/m^3 TWA; BEI: 3 mg [F]/g creatinine in urine *prior* to end-of-shift; 10 mg[F]/g creatinine in urine end-of-shift.
DFG MAK: 1 mg[F]/m^3, inhalable fraction [skin]; Peak Limitation Category II(4); Pregnancy Risk Group C; BAT: 7.0 mg[F]/g creatinine in urine at end-of-shift; 4.0 mg[F]/g creatinine in urine at the beginning of the next shift.
NIOSH IDLH: 30 ppm.

Determination in Air: There is no method listed by NIOSH. However, the following method is listed for gaseous fluorides: NIOSH Analytical Method 7902.[18]

Permissible Concentration in Water: The EPA has set a standard of 4 mg/L for fluoride[61] and the state of Maine has set 2.4 mg/L as a guideline for drinking water. Arizona[61] has set 1.8 mg/L as a standard for drinking water.

Routes of Entry: Inhalation, skin and/or eye contact (liquid).

Harmful Effects and Symptoms

Short Term Exposure: A CNS depressant, causing headache, dizziness, confusion, a lack of coordination, nausea, vomiting, narcosis, unconsciousness. In industrial use, inhalation of dusts of decomposed fluorocarbon polymers may cause polymer fume fever characterized by headache, aching joints, general feeling of discomfort, cough, shivering, chills, fever, rapid heartbeat, and chest discomfort.

Animal data indicate that concentrations at and above 30% (300,000 ppm) vinyl fluoride may cause symptoms of intoxication. Contact with the liquid may cause frostbite.

Long Term Exposure: May be a human carcinogen. Repeated high exposures may affect kidneys. Repeated high exposures can cause deposits of fluorides in the bones (fluorosis), which may cause pain, disability, and mottling of the teeth. Repeated exposure can cause nausea, vomiting, loss of appetite, diarrhea, or constipation. Nosebleeds and sinus problems can also occur.

Points of Attack: Central nervous system, skeleton, kidneys.

Medical Surveillance: Lung function tests. Fluoride level in urine (use NIOSH #8308). Levels higher than 4 mg/L may indicate overexposure. If symptoms develop or overexposure is suspected, the following may be useful: consider chest X-ray after acute overexposure. Kidney function tests. Consider chest X-ray following acute overexposure.

First Aid: If this chemical gets into the eyes, remove any contact lenses at once and irrigate immediately for at least 15 min, occasionally lifting upper and lower lids. Seek medical attention immediately. If this chemical contacts the skin, remove contaminated clothing and wash immediately with soap and water. Seek medical attention immediately. If this chemical has been inhaled, remove from exposure, begin rescue breathing (using universal precautions, including resuscitation mask) if breathing has stopped and CPR if heart action has stopped. Transfer promptly to a medical facility. When this chemical has been swallowed, get medical attention. Give large quantities of water and induce vomiting. Do not make an unconscious person vomit. If frostbite has occurred, seek medical attention immediately; do *NOT* rub the affected areas or flush them with water. In order to prevent further tissue damage, do *NOT* attempt to remove frozen clothing from frostbitten areas. If frostbite has *NOT* occurred, immediately and thoroughly wash contaminated skin with soap and water.

Personal Protective Methods: Wear appropriate personal protective clothing to prevent the skin from becoming frozen from contact with the evaporating liquid or from contact with vessels containing the liquid. Safety equipment suppliers/manufacturers can provide recommendations on the most protective glove/clothing material for your operation. All protective clothing (suits, gloves, footwear, headgear) should be clean, available each day, and put on before work. Contact lenses should not be worn when working with this chemical. Wear gas-proof or splash-proof chemical goggles and face shield unless full-face-piece respiratory protection is worn. Employees should wash immediately with soap when skin is wet or contaminated. Provide emergency showers and eyewash.

Respirator Selection: NIOSH: *10 ppm:* CcrOv (APF = 10) [any chemical cartridge respirator with organic vapor cartridge(s)] or Sa (APF = 10) (any supplied-air respirator). *25 ppm:* Sa:Cf (APF = 25) (any supplied-air respirator operated in a continuous-flow mode) or PaprOv (APF = 25) [any powered, air-purifying respirator with organic vapor cartridge(s)]. *50 ppm:* CcrFOv (APF = 50) [any chemical cartridge respirator with a full face-piece and organic vapor cartridge(s)] or GmFOv (APF = 50) [any air-purifying, full-face-piece respirator (gas mask) with a chin-style, front- or back-mounted acid gas canister] or PaprTOv (APF = 50) [any powered, air-purifying respirator with a tight-fitting face-piece and organic vapor cartridge(s)] or SCBAF (APF = 50) (any self-contained breathing apparatus with a full face-piece) or SaF (APF = 50) (any supplied-air respirator with a full face-piece). *200 ppm:* SaF: Pd,Pp (APF = 2000) (any supplied-air respirator that has a full face-piece and is operated in a pressure-demand or other positive-pressure mode). *Emergency or planned entry into unknown concentration or IDLH conditions:* SCBAF: Pd,Pp (APF = 10,000) (any self-contained breathing apparatus that has a full face-piece and is operated in a pressure-demand or other positive-pressure mode) or SaF: Pd,Pp: ASCBA (APF = 10,000) (any supplied-air respirator that has a full face-piece and is operated in a pressure-demand or other positive-pressure mode in combination with an auxiliary, self-contained breathing apparatus operated in a pressure-demand or other positive-pressure mode). *Escape:* GmFOv (APF = 50) [any air-purifying, full-face-piece respirator (gas mask) with a chin-style, front- or back-mounted organic vapor canister] or SCBAE (any appropriate escape-type, self-contained breathing apparatus).

Storage: Color Code—Red Stripe: Flammability Hazard: Store separately from all other flammable materials. Prior to working with this chemical you should be trained on its proper handling and storage. Before entering confined space where this chemical may be present, check to make sure that an explosive concentration does not exist. Store in a well-ventilated area away from sources of ignition. Use only nonsparking tools and equipment, especially when opening and closing containers of this chemical.

Shipping: Vinyl fluoride, inhibited, requires a shipping label of "FLAMMABLE GAS." It falls in Hazard Class 2.1.

Spill Handling: Evacuate and restrict persons not wearing protective equipment from area of spill or leak until cleanup is complete. Remove all ignition sources. Establish forced ventilation to keep levels below explosive limit. Stop the flow of gas if it can be done safely. If source of leak is a cylinder and the leak cannot be stopped in place, remove leaking cylinder to a safe place in the open air, and repair leak or allow cylinder to empty. Keep this chemical out of confined space, such as a sewer, because of the possibility of explosion, unless the sewer is designed to prevent the buildup of explosive concentrations. It may be necessary to contain and dispose of this chemical as a hazardous waste. Contact your local or federal environmental protection agency for specific recommendations. If employees are required to clean up spills, they must be properly trained and equipped. OSHA 1910.120(q) may be applicable.

Fire Extinguishing: This chemical is a flammable gas. Poisonous gases are produced in fire. Do not extinguish the fire unless the flow of gas can be stopped and any remaining gas is out of the line. Specially trained personnel may use fog lines to cool exposures and let the fire burn itself out. Vapors are heavier than air and will collect in low areas. Vapors may travel long distances to ignition sources and flashback. Vapors in confined areas may explode when exposed to fire. Containers may explode in fire. Storage containers and parts of containers may rocket great distances, in many directions. If material or contaminated runoff enters waterways, notify downstream users of potentially contaminated waters. Notify local health and fire officials and pollution control agencies. From a secure, explosion-proof location, use water spray to cool exposed containers. If cooling streams are ineffective (venting sound increases in volume and pitch, tank discolors, or shows any signs of deforming), withdraw immediately to a secure position. If cylinders are exposed to excessive heat from fire or flame contact, withdraw immediately to a secure location. If employees are expected to fight fires, they must be trained and equipped in OSHA 1910.156. The only respirators recommended for firefighting are self-contained breathing apparatuses that have full face-pieces and are operated in a pressure-demand or other positive-pressure mode.

References

US Environmental Protection Agency. (January 30, 1978). *Chemical Hazard Information Profile: Vinyl Fluoride.* Washington, DC

New York State Department of Health. (April 1986). *Chemical Fact Sheet: Vinyl Fluoride.* Albany, NY: Bureau of Toxic Substance Assessment

Vinylidene chloride   V:0220

Molecular Formula: $C_2H_2Cl_2$

Synonyms: Clorure de vinylidene (French); Cloruro de vinildeno (Spanish); 1,1-DCE; 1,1-Dichloroethene; 1,1-Dichloroethylene; Ethene, 1,1-dichloro-; Ethylene, 1,1-dichloro-; NCI-C54262; Sconatex; *asym*-Dichloroethylene; VDC; Vinylidene chloride(II); Vinylidene dichloride; Vinylidine chloride(II)

CAS Registry Number: 75-35-4

RTECS® Number: KV9275000

UN/NA & ERG Number: UN1303 (stabilized)/130

EC Number: 200-864-0 [Annex I Index No.: 602-025-00-8]

Regulatory Authority and Advisory Bodies

Department of Homeland Security Screening Threshold Quantity (pounds): *Release hazard* 10,000 (≥1.00% concentration).

Carcinogenicity: IARC: Animal, Limited Evidence; Human Inadequate Evidence, *not classifiable as carcinogenic to humans*, Group 3, 1999; EPA (*inhalation*) Suggestive evidence of carcinogenic potential; (*oral*): Available data are inadequate for an assessment of human carcinogenic potential; Possible Human Carcinogen; NCI: Carcinogenesis

Studies (gavage); no evidence: mouse, rat; NIOSH: Potential occupational carcinogen.

US EPA Gene-Tox Program, Positive: Histidine reversion— Ames test; Positive/limited: Carcinogenicity—mouse/rat; Negative: Rodent dominant lethal.

Air Pollutant Standard Set. See below, "Permissible Exposure Limits in Air" section.

Clean Air Act: Hazardous Air Pollutants (Title I, Part A, Section 112); Accidental Release Prevention/Flammable Substances, (Section 112[r], Table 3), TQ = 10,000 lb (4540 kg).

Clean Water Act: Section 311 Hazardous Substances/RQ 40CFR117.3 (same as CERCLA, see below); 40CFR423, Appendix A, Priority Pollutants; Section 313 Water Priority Chemicals (57FR41331, 9/9/92); Toxic Pollutant (Section 401.15).

US EPA Hazardous Waste Number (RCRA No.): U078; D029.

RCRA, 40CFR261, Appendix 8 Hazardous Constituents.

RCRA Toxicity Characteristic (Section 261.24), Maximum. Concentration of Contaminants, regulatory level, 0.7 mg/L.

RCRA 40CFR268.48; 61FR15654, Universal Treatment Standards: Wastewater (mg/L), 0.025; Nonwastewater (mg/kg), 6.0.

RCRA 40CFR264, Appendix 9; TSD Facilities Ground Water Monitoring List. Suggested test method(s) (PQL μg/L): 8010 (1); 8240 (5).

Safe Drinking Water Act: MCL, 0.007 mg/L; MCLG, 0.007 mg/L; Regulated chemical (47 FR 9352).

Reportable Quantity (RQ): 100 lb (45.4 kg).

EPCRA Section 313 Form R de minimis concentration reporting level: 1.0%.

US DOT Regulated Marine Pollutant (49CFR172.101, Appendix B).

Canada, WHMIS, Ingredients Disclosure List Concentration 1.0%.

European/International Regulations: Hazard Symbol: F+, Xn; Risk phrases: R12; R20; R40; Safety phrases: S2; S7; S16; S29; S36/36; S46 (see Appendix 4).

WGK (German Aquatic Hazard Class): 3—Severe hazard to waters.

Description: Vinylidene chloride is a volatile liquid or gas with a mild, sweet odor resembling that of chloroform. The odor threshold in air is 500 ppm. Molecular weight = 96.94; Specific gravity (H$_2$O:1) = 1.21 at 25°C; Boiling point = 31.7°C at 760 mm; Freezing/Melting point = −122.8°C; Vapor pressure = 500 mmHg at 25°C; Flash point = −18.9°C (cc); Autoignition temperature = 570°C. Explosive limits: LEL = 6.5%; UEL = 15.5%. Hazard Identification (based on NFPA-704 M Rating System): Health 2, Flammability 4, Reactivity 2. Practically insoluble in water; solubility = 0.04%.

Potential Exposure: Compound Description: Tumorigen, Mutagen; Reproductive Effector; Human Data. Vinylidene chloride is used in the manufacture of 1,1,1-trichloroethane (methyl chloroform). However, the manufacture of polyvinylidene copolymers is the major use of VDC. The extruded films of the copolymers are used in packaging and have excellent resistance to water vapor and most gases. The chief copolymer is Saran® (polyvinylidene chloride/ vinyl chloride), a transparent film used for food packaging. The films shrink when exposed to higher than normal temperatures. This characteristic is advantageous in the heat shrinking of overwraps on packaged goods and in the sealing of the wraps. Applications of VDC latexes include mixing in cement to produce high-strength mortars and concretes, and as binders for paints and nonwoven fabrics providing both water resistance and nonflammability. VDC polymer lacquers are also used in coating films and paper. VDC is also used to produce fibers. Monofilaments, made by extruding the copolymer, are used in the textile industry as furniture and automobile upholstery, drapery fabric, outdoor furniture, venetian-blind tape, and filter cloths.

Incompatibilities: Readily forms explosive peroxides; violent polymerization from heat or on contact with oxidizers, chlorosulfonic acid; nitric acid; or oleum; or under the influence of oxygen, sunlight, alkali metals; aluminum, copper. Explosive on heating or on contact with flames. Inhibitors, such as the monomethyl ether of hydroquinone, are added to prevent polymerization.

Permissible Exposure Limits in Air
OSHA PEL: None.
NIOSH PEL: None.
NIOSH REL: Potential occupational carcinogen. Limit exposure to lowest feasible concentration. See *NIOSH Pocket Guide*, Appendix A.
ACGIH TLV®[1]: 5 ppm/20 mg/m^3 TWA Not classifiable as a human carcinogen.
Protective Action Criteria (PAC)
TEEL-0: 5 ppm
PAC-1: 75 ppm
PAC-2: 500 ppm
PAC-3: 1000 ppm
DFG MAK: 2 ppm/8.0 mg/m^3; Peak Limitation Category II (2); Carcinogen Category 3B; Pregnancy Risk Group C.
Australia: TWA 5 ppm (20 mg/m^3), STEL 20 ppm, 1993; Austria: MAK 2 ppm (8 mg/m^3), Suspected: carcinogen, 1999; Belgium: TWA 5 ppm (20 mg/m^3), STEL 20 ppm (79 mg/m^3), 1993; Denmark: TWA 2 ppm (8 mg/m^3), 1999; Finland: TWA 10 ppm (40 mg/m^3), STEL 20 ppm (80 mg/m^3), carcinogen, 1999; France: VME 5 ppm (20 mg/m^3), 1999; the Netherlands: MAC-TGG 20 mg/m^3, 2003; Poland: MAC (TWA) 50 mg/m^3; MAC (STEL) 80 mg/m^3, 1999; Russia: STEL 50 mg/m^3, 1993; Sweden: NGV 5 ppm (20 mg/m^3), KTV 10 ppm (40 mg/m^3), 1999; Switzerland: MAK-W 2 ppm (8 mg/m^3), KZG-W 4 ppm (16 mg/m^3), 1999; United Kingdom: TWA 10 ppm (40 mg/m^3), 2000; Argentina, Bulgaria, Columbia, Jordan, South Korea, New Zealand, Singapore, Vietnam: ACGIH TLV®: not classifiable as a human carcinogen. Several states have set guidelines or standards for Vinylidene chloride in ambient air[60] ranging from 0.2 μg/m^3 (Massachusetts) to 0.238 μg/m^3

(Kansas) to 3.5 μg/m³ (Virginia) to 24.0 μg/m³ (Pennsylvania) to 66.7 μg/m³ (New York) to 120.0 μg/m³ (North Carolina) to 200.0 μg/m³ (Indiana) to 400.0 μg/m³ (Connecticut) to 476.0 μg/m³ (Nevada) to 200.0–10,000 μg/m³ (North Dakota).

Determination in Air: NIOSH Analytical Method (IV) #1015, Vinylidene chloride, OSHA Analytical Method 19.

Permissible Concentration in Water: *To protect freshwater aquatic life:* 11,600 μg/L on an acute toxicity basis for dichloroethylenes as a class. *To protect saltwater aquatic life:* 224,000 μg/L on an acute basis for dichloroethylenes as a class; 1700 μg/L as a 24-h average, never to exceed 3900 μg/L. *To protect human health:* preferably zero. An additional lifetime cancer risk of 1 in 100,000 is posed by a concentration of 0.33 μg/L.[6] Russia set a MAC of 0.6 μg/L in water bodies used for domestic purposes and WHO has set a limit of 0.3 μg/L for drinking water.

Determination in Water: Inert gas purge followed by gas chromatography with halide-specific detection (EPA Method 601) or gas chromatography plus mass spectrometry (EPA Method 624). Octanol–water coefficient: Log K_{ow} = 1.32.

Routes of Entry: Inhalation, skin absorption, ingestion, skin and/or eye contact.

Harmful Effects and Symptoms

Short Term Exposure: Vinylidene chloride can affect you when breathed and by passing through skin. Exposure can irritate the eyes, nose, and throat. Contact can irritate and burn the eyes and skin. High levels may affect the CNS, causing dizziness, headache, nausea, dyspnea (breathing difficulty), a "drunken" feeling, unconsciousness. Swallowing the liquid may cause chemical pneumonitis.

Long Term Exposure: Repeated or prolonged contact with skin may cause dermatitis with drying and cracking. Repeated exposure may damage the liver, kidneys, and lungs. A potential occupational carcinogen. Handle with extreme caution. It may damage the developing fetus and cause reproductive damage in males.

Points of Attack: Eyes, skin, respiratory system, central nervous system, liver, kidneys. Cancer site in animals: liver and kidney; skin.

Medical Surveillance: For those with frequent or potentially high exposure (half the TLV or greater, or significant skin contact), the following are recommended before beginning work and at regular times after that: lung function tests. Liver and kidney function tests.

First Aid: If this chemical gets into the eyes, remove any contact lenses at once and irrigate immediately for at least 15 min, occasionally lifting upper and lower lids. Seek medical attention immediately. If this chemical contacts the skin, remove contaminated clothing and wash immediately with soap and water. Seek medical attention immediately. If this chemical has been inhaled, remove from exposure, begin rescue breathing (using universal precautions, including resuscitation mask) if breathing has stopped and CPR if heart action has stopped. Transfer promptly to a medical facility. If victim is *conscious*, administer water or milk. Do not induce vomiting.

Note to physician: Inhalation: bronchodilators, decongestants, and oxygen may be used if necessary. Corticosteroids are useful for treating pneumonitis.

Personal Protective Methods: Wear protective gloves and clothing to prevent any reasonable probability of skin contact. Safety equipment suppliers/manufacturers can provide recommendations on the most protective glove/clothing material for your operation. All protective clothing (suits, gloves, footwear, headgear) should be clean, available each day, and put on before work. Contact lenses should not be worn when working with this chemical. Wear splash-proof chemical goggles and face shield unless full-face-piece respiratory protection is worn. Employees should wash immediately with soap when skin is wet or contaminated. Provide emergency showers and eyewash.

Respirator Selection: *At concentrations above the NIOSH REL, or where there is no REL, at any detectable concentration:* SCBAF: Pd,Pp (APF = 10,000) (any NIOSH/ MSHA- or European Standard EN 149-approved self-contained breathing apparatus that has a full face-piece and is operated in a pressure-demand or other positive-pressure mode) or SaF: Pd,Pp: ASCBA (APF = 10,000) (any supplied-air respirator that has a full face-piece and is operated in a pressure-demand or other positive-pressure mode in combination with an auxiliary, self-contained breathing apparatus operated in a pressure-demand or other positive-pressure mode). *Escape:* GmFOv100 (APF = 50) [any air-purifying full-face-piece respirator (gas mask) with a chin-style, front- or back-mounted organic vapor canister having an N100, R100, or P100 filter] or SCBAE (any appropriate escape-type, self-contained breathing apparatus).

Storage: Color Code—Red: Flammability Hazard: Store in a flammable liquid storage area or approved cabinet away from ignition sources and corrosive and reactive materials. May form peroxides in storage. High concentrations cause a deficiency of oxygen with the risk of unconsciousness or death. Check that oxygen content is at least 19% before entering storage or spill area. A regulated, marked area should be established where this chemical is handled, used, or stored in compliance with OSHA Standard 1910.1045. Color Code—Red: Flammability Hazard: Store in a flammable liquid storage area or approved cabinet away from ignition sources and corrosive and reactive materials. Prior to working with this chemical you should be trained on its proper handling and storage. Before entering confined space where this chemical may be present, check to make sure that an explosive concentration is not present. Vinylidene chloride must be stored to avoid contact with oxidizers, such as perchlorates, peroxides, permanganates, chlorates, and nitrates, and strong acids, such as hydrochloric, sulfuric, and nitric, since violent reactions occur. Store in tightly closed containers in a cool, well-ventilated area away from sources of heat. Protect storage containers from physical damage. Sources of ignition, such as smoking and open flames, are prohibited where vinylidene chloride is handled, used, or stored. Metal containers involving the transfer of 5 gallons or more of vinylidene chloride should be grounded and bonded. Drums must be equipped with self-closing valves, pressure vacuum bungs, and flame arresters. Use only nonsparking tools and equipment, especially when opening and closing containers of vinylidene chloride. Wherever vinylidene chloride is used, handled, manufactured, or stored, use explosion-proof electrical equipment and fittings. Procedures for the handling, use, and storage of cylinders should be in compliance with OSHA 1910.101 and 1910.169, as with the recommendations of the Compressed Gas Association.

Shipping: Vinylidene chloride, inhibited, requires a shipping label of "FLAMMABLE LIQUID." It falls in Hazard Class 3 and Packing Group I.

Spill Handling: Evacuate and restrict persons not wearing protective equipment from area of spill or leak until cleanup is complete. Remove all ignition sources. Establish forced ventilation to keep levels below explosive limit. Absorb liquids in vermiculite, dry sand, earth, peat, carbon, or a similar material and deposit in sealed containers. Keep this chemical out of a confined space, such as a sewer, because of the possibility of an explosion, unless the sewer is designed to prevent the buildup of explosive concentrations. It may be necessary to contain and dispose of this chemical as a hazardous waste. If material or contaminated runoff enters waterways, notify downstream users of potentially contaminated waters. Contact your local or federal environmental protection agency for specific recommendations. If employees are required to clean up spills, they must be properly trained and equipped. OSHA 1910.120(q) may be applicable.

Fire Extinguishing: This chemical is a flammable liquid. Poisonous gases, including phosgene, hydrogen chloride, and chlorine, are produced in fire. Do not extinguish fire unless flow can be stopped. Use dry chemical, carbon dioxide, or foam extinguishers. Vapors are heavier than air and will collect in low areas. Vapors may travel long distances to ignition sources and flashback. Vapors in confined areas may explode when exposed to fire. Containers may explode in fire. Storage containers and parts of containers may rocket great distances, in many directions. If material or contaminated runoff enters waterways, notify downstream users of potentially contaminated waters. Notify local health and fire officials and pollution control agencies. From a secure, explosion-proof location, use water spray to cool exposed containers. If cooling streams are ineffective (venting sound increases in volume and pitch, tank discolors, or shows any signs of deforming), withdraw immediately to a secure position. If employees are expected to fight fires, they must be trained and equipped in OSHA 1910.156. The only respirators recommended for firefighting are self-contained breathing apparatuses that have full face-pieces and are operated in a pressure-demand or other positive-pressure mode.

Disposal Method Suggested: Consult with environmental regulatory agencies for guidance on acceptable disposal practices. Generators of waste containing this contaminant (≥100 kg/mo) must conform with EPA regulations governing storage, transportation, treatment, and waste disposal. Incineration, preferably after mixing with another combustible fuel. Care must be exercised to assure complete combustion to prevent the formation of phosgene. An acid scrubber is necessary to remove the halo acids produced.[22]

References
The National Institute for Occupational Safety and Health. (September 21, 1978). *Vinyl Halides—Carcinogenicity*, Current Intelligence Bulletin No. 28. Washington, DC
US Environmental Protection Agency. (1980). *Dichloroethylenes: Ambient Water Quality Criteria*. Washington, DC
US Environmental Protection Agency. (December 1979). *Status Assessment of Toxic Chemicals: Vinylidene Chloride*, Report EPA-600/2-79-210a. Cincinnati, OH
US Environmental Protection Agency. (April 30, 1980). *1,1-Dichloroethylene: Health and Environmental Effects Profile No. 71*. Washington, DC: Office of Solid Waste
US Environmental Protection Agency. (April 30, 1980). *Dichloroethylenes: Health and Environmental Effects Profile No. 73*. Washington, DC: Office of Solid Waste
Sax, N. I. (Ed.). (1982). *Dangerous Properties of Industrial Materials Report*, 2, No. 6, 92−94
US Public Health Service. (December 1988). *Toxicological Profile for 1,1-Dichloroethene*. Atlanta, GA: Agency for Toxic Substances and Disease Registry
New Jersey Department of Health and Senior Services. (August 2002). *Hazardous Substances Fact Sheet: Vinylidene Chloride*. Trenton, NJ

Vinylidene fluoride   V:0230

Molecular Formula: $C_2H_2F_2$
Common Formula: $CH_2{=}CF_2$
Synonyms: 1,1-Difluoroethylene; Ethene, 1,1-difluoro-; Fluoruro de vinilideno (Spanish); Genetron®; Halocarbon 1132A; NCI-C60208; R1132A; Refrigerant gas, R 1132a; VDF; Vinylidene difluoride
CAS Registry Number: 75-38-7
RTECS® Number: KW0560000
UN/NA & ERG Number: UN1959/116
EC Number: 200-867-7
Regulatory Authority and Advisory Bodies
Department of Homeland Security Screening Threshold Quantity (pounds): *Release hazard* 10,000 (≥1.00% concentration).
Carcinogenicity: IARC: Animal, Inadequate Data; Human No Adequate Data, *not classifiable as carcinogenic to humans*, Group 3, 1999.
Clean Air Act: Accidental Release Prevention/Flammable Substances, (Section 112[r], Table 3), TQ = 10,000 lb (4540 kg).

European/International Regulations: not listed in Annex 1.
WGK (German Aquatic Hazard Class): No value assigned.
Description: Vinylidene fluoride is a colorless gas with a faint ethereal odor. Shipped as a liquefied compressed gas. Molecular weight = 64.04; Boiling point = −85.6°C; Freezing/Melting point = −143.9°C; Vapor pressure = 35.2 atm; Flash point (flammable gas) ≤ −65°C; Autoignition temperature = 640°C. Explosive limits: LEL = 5.5%; UEL = 21.3%. Hazard Identification (based on NFPA-704 M Rating System): Health 2, Flammability 4, Reactivity 0. Slightly soluble in water.
Potential Exposure: Compound Description: Tumorigen; Mutagen. Vinylidene fluoride is used in the formulation of many polymers and copolymers, such as chlorotrifluoro-ethylene-vinylidene fluoride (Kel F®), perfluoropropylene-vinylidene fluoride (Viton™, Fluorel®); polyvinylidene fluoride; and hexafluoropropylene-*tetra*-fluoroethylene-vinylidene fluoride; elastomeric copolymers. It is also used as a chemical intermediate in organic synthesis. NIOSH has estimated that 32,000 workers are exposed annually.
Incompatibilities: Contact with oxidizers can cause fire and explosions. Reacts with aluminum chloride. Capable of forming unstable peroxides, which can cause explosive polymerization. May accumulate static electricity, and cause ignition of its vapors.

Permissible Exposure Limits in Air
Conversion factor: 1 ppm = 2.62 mg/m³ at 25°C & 1 atm.
OSHA PEL: None.
NIOSH REL: 1 ppm TWA; 5 ppm Ceiling Concentration [15 min], use 29CFR1910.1017.
ACGIH TLV®[1]: 500 ppm/1310 mg/m³ TWA, Not classifiable as a human carcinogen.
NIOSH IDLH: Not determined. Suspected occupational carcinogen.
Protective Action Criteria (PAC)
TEEL-0: 500 ppm
PAC-1: 500 ppm
PAC-2: 1250 ppm
PAC-3: 25,000 ppm
DFG MAK: Carcinogen Category 3B.
Austria: Suspected: carcinogen, 1999; Finland: carcinogen, 1993.
Determination in Air: NIOSH Analytical Method #3800.[18]
Permissible Concentration in Water: The EPA has set a standard of 4 mg/L for fluoride[61] and the state of Maine has set 2.4 mg/L as a guideline for drinking water. Arizona[61] has set 1.8 mg/L as a standard for drinking water.
Routes of Entry: Inhalation, skin and/or eye contact (liquid).

Harmful Effects and Symptoms
Short Term Exposure: Vinylidene fluoride is considered toxic by inhalation. The lowest lethal concentration is 128,000 ppm for a 4-h exposure. Vinylidene fluoride has been reported to be nontoxic to rats at 800,000 ppm. Irritating to skin and pulmonary tract. May affect the CNS, causing dizziness, disorientation, headache, a lack of coordination, narcosis, dizziness, nausea, vomiting. Contact with the liquid can cause frostbite.
Long Term Exposure: Repeated high exposures may affect kidneys. Repeated high exposures may cause deposits of fluorides in the bones (fluorosis), which may cause pain, disability, and mottling of the teeth.
Points of Attack: Eyes, respiratory system, central nervous system, skeleton, kidneys, skin.
Medical Surveillance: Lung function tests. Fluoride level in urine (use NIOSH #8308). Levels higher than 4 mg/L may indicate overexposure. If symptoms develop or overexposure is suspected, the following may be useful: consider chest X-ray after acute overexposure. Kidney function tests. Consider chest X-ray following acute overexposure.
First Aid: If this chemical gets into the eyes, remove any contact lenses at once and irrigate immediately for at least 20 min, occasionally lifting upper and lower lids. Seek medical attention immediately. If this chemical contacts the skin, remove contaminated clothing and wash immediately with soap and water. Seek medical attention immediately. If this chemical has been inhaled, remove from exposure, begin rescue breathing (using universal precautions, including resuscitation mask) if breathing has stopped and CPR if heart action has stopped. Transfer promptly to a medical facility. If frostbite has occurred, seek medical attention immediately; do *NOT* rub the affected areas or flush them with water. In order to prevent further tissue damage, do *NOT* attempt to remove frozen clothing from frostbitten areas. If frostbite has *NOT* occurred, immediately and thoroughly wash contaminated skin with soap and water.
Personal Protective Methods: Wear protective gloves and clothing to prevent any reasonable probability of skin contact. Safety equipment suppliers/manufacturers can provide recommendations on the most protective glove/clothing material for your operation. Viton™ and butyl rubber are among the recommended protective materials. All protective clothing (suits, gloves, footwear, headgear) should be clean, available each day, and put on before work. Contact lenses should not be worn when working with this chemical. Wear gas-proof chemical goggles and face shield unless full-face-piece respiratory protection is worn. Employees should wash immediately with soap when skin is wet or contaminated. Provide emergency showers and eyewash.
Respirator Selection: NIOSH: *Up to 10 ppm:* CcrOv (APF = 10) [any chemical cartridge respirator with organic vapor cartridge(s)] or Sa (APF = 10) (any supplied-air respirator). *Up to 25 ppm:* Sa:Cf (APF = 25) (any supplied-air respirator operated in a continuous-flow mode); PaprOv (APF = 25) [any powered, air-purifying respirator with organic vapor cartridge(s)]. *Up to 50 ppm:* CcrFOv (APF = 50) [any chemical cartridge respirator with a full face-piece and organic vapor cartridge(s)] or GmFOv (APF = 50) [any air-purifying, full-face-piece respirator (gas mask) with a chin-style, front- or back-mounted organic vapor canister] or PaprTOv (APF = 50) [any

powered, air-purifying respirator with a tight-fitting face-piece and organic vapor cartridge(s)] or SCBAF (APF = 50) (any self-contained breathing apparatus with a full face-piece) or SaF (APF = 50) (any supplied-air respirator with a full face-piece); *Up to 200 ppm:* SaF: Pd,Pp (APF = 2000) (any supplied-air respirator that has a full face-piece and is operated in a pressure-demand or other positive-pressure mode). *Emergency or planned entry into unknown concentrations or IDLH conditions:* SCBAF: Pd,Pp (APF = 10,000) (any self-contained breathing apparatus that has a full face-piece and is operated in a pressure-demand or other positive-pressure mode) or SaF: Pd,Pp: ASCBA (APF = 10,000) (any supplied-air respirator that has a full face-piece and is operated in a pressure-demand or other positive-pressure mode in combination with an auxiliary, self-contained breathing apparatus operated in a pressure-demand or other positive-pressure mode). *Escape:* GmFOv (APF = 50) [any air-purifying, full-face-piece respirator (gas mask) with a chin-style, front- or back-mounted organic vapor canister] or SCBAE (any appropriate escape-type, self-contained breathing apparatus).

Storage: Color Code—Red Stripe: Flammability Hazard: Store separately from all other flammable materials. May form peroxides in storage. Prior to working with this chemical you should be trained on its proper handling and storage. Before entering confined space where this chemical may be present, check to make sure that an explosive concentration does not exist. Store gas cylinders in a cool, dry place and use the safety precautions necessary with all compressed gases. High concentrations cause a deficiency of oxygen with the risk of unconsciousness or death. Check that oxygen content is at least 19% before entering storage or spill area. Store in tightly closed containers in a cool, well-ventilated area away from heat. Sources of ignition, such as smoking and open flames, are prohibited where this chemical is used, handled, or stored in a manner that could create a potential fire or explosion hazard. Use only non-sparking tools and equipment, especially when opening and closing containers of this chemical. Procedures for the handling, use, and storage of cylinders should be in compliance with OSHA 1910.101 and 1910.169, as with the recommendations of the Compressed Gas Association. A regulated, marked area should be established where this chemical is handled, used, or stored in compliance with OSHA Standard 1910.1045.

Shipping: 1,1-Difluoroethylene [or] Refrigerant gas, R 1132a requires a shipping label of "FLAMMABLE GAS." It falls in Hazard Class 2.1.

Spill Handling: Evacuate and restrict persons not wearing protective equipment from area of spill or leak until cleanup is complete. Remove all ignition sources. Establish forced ventilation to keep levels below explosive limit and to disperse the gas. Stop the flow of gas if it can be done safely. If source of leak is a cylinder and the leak cannot be stopped in place, remove leaking cylinder to a safe place in the open air, and repair leak or allow cylinder to empty.

Keep this chemical out of confined space, such as a sewer, because of the possibility of explosion, unless the sewer is designed to prevent the buildup of explosive concentrations. It may be necessary to contain and dispose of this chemical as a hazardous waste. Contact your local or federal environmental protection agency for specific recommendations. If employees are required to clean up spills, they must be properly trained and equipped. OSHA 1910.120(q) may be applicable.

Fire Extinguishing: This chemical is a flammable gas. Poisonous gases are produced in fire, including hydrogen fluoride, fluorine, and fluorides. Extinguish with CO_2 or dry chemical to allow access to valves to shut off supply if necessary. Do not extinguish the fire unless the flow of gas can be stopped and any remaining gas is out of the line. Specially trained personnel may use fog lines to cool exposures and let the fire burn itself out. Vapors are heavier than air and will collect in low areas. Vapors may travel long distances to ignition sources and flashback. Vapors in confined areas may explode when exposed to fire. Containers may explode in fire. Storage containers and parts of containers may rocket great distances, in many directions. If material or contaminated runoff enters waterways, notify downstream users of potentially contaminated waters. Notify local health and fire officials and pollution control agencies. From a secure, explosion-proof location, use water spray to cool exposed containers. If cooling streams are ineffective (venting sound increases in volume and pitch, tank discolors, or shows any signs of deforming), withdraw immediately to a secure position. If cylinders are exposed to excessive heat from fire or flame contact, withdraw immediately to a secure location. If employees are expected to fight fires, they must be trained and equipped in OSHA 1910.156. The only respirators recommended for firefighting are self-contained breathing apparatuses that have full face-pieces and are operated in a pressure-demand or other positive-pressure mode.

References

US Environmental Protection Agency. (January 30, 1978). *Chemical Hazard Information Profile: Vinylidene Fluoride.* Washington, DC

Vinyl toluene   V:0240

Molecular Formula: C_9H_{10}
Common Formula: $CH_3C_6H_4CH\!=\!CH_2$
Synonyms: Ethenylmethylbenzene; *m*-Methyl styrene; *p*-Methyl styrene; 3- and 4-Methyl styrene; Methyl styrene; NCI-C56406; Tolyethylene; *m*-Vinyl toluene; *p*-Vinyl toluene; *p*-Vinyltoluene; Vinyl toluene, inhibited; Vinyl toluene, mixed isomers
a-isomer: see *a*-Methylstyrene
m-isomer: 1-Ethenyl-3-methylbenzene; Benzene, 1-ethenyl-3-methyl-; *m*-Methylstyrene; 3-Methylstyrene; 1-Methyl-3-vinylbenzene; *m*-Vinyltoluene; 3-Vinyltoluene

CAS Registry Number: 25013-15-4; *(alt.)* 1321-45-5; 100-80-1 (*m*-isomer)
RTECS® Number: WL5075000
UN/NA & ERG Number: UN2618 (stabilized)/130
EC Number: 246-562-2
Regulatory Authority and Advisory Bodies
Carcinogenicity: NCI: Carcinogenesis Studies (inhalation); no evidence: mouse, rat; IARC: Animal Lack Carcinogenicity; Human Inadequate Evidence, *not classifiable as carcinogenic to humans*, Group 3, 1994; NTP: Carcinogenesis studies (α-isomer); on test (prechronic studies), October 2000.
Air Pollutant Standard Set. See below, "Permissible Exposure Limits in Air" section.
Canada, WHMIS, Ingredients Disclosure List Concentration 1.0%.
European/International Regulations: not listed in Annex 1.
WGK (German Aquatic Hazard Class): No value assigned.
Description: Vinyl toluene is a colorless liquid with a strong, disagreeable odor. It consists of mixed *meta-* and *para-*isomers. The odor threshold is 50 ppm. Molecular weight = 118.19; Specific gravity (H$_2$O:1) = 0.89 at 25°C; Boiling point = 170.6°C; Melting point = −76.7°C; Flash point = 53°C; Autoignition temperature = 538°C; also listed at 489−515°C. Explosive limits: LEL = 0.8%; UEL = 11.0%. Hazard Identification (based on NFPA-704 M Rating System): Health 2, Flammability 2, Reactivity 2. Practically insoluble in water; solubility = 0.009%.
Potential Exposure: Compound Description (mixed isomers): Tumorigen, Mutagen; Reproductive Effector; Human Data; Primary Irritant; (*m*-isomer) Tumorigen, Mutagen. Vinyl toluene is used in copolymers and as specialty monomer for paint, varnish, and polyester preparations; as a solvent and an organic intermediate.
Incompatibilities: Violent reaction with oxidizers; catalysts for vinyl polymerization, such as peroxides, strong acids, iron or aluminum salts. Usually inhibited with *tert*-butylcatechol to prevent polymerization.
Permissible Exposure Limits in Air
Conversion factor: 1 ppm = 4.83 mg/m^3 at 25°C & 1 atm.
Methyl styrene, all isomers
OSHA PEL: 100 ppm/480 mg/m^3 TWA.
NIOSH REL: 100 ppm/480 mg/m^3 TWA.
ACGIH TLV®[1]: 50 ppm/242 mg/m^3 TWA; 100 ppm/483 mg/m^3 STEL, Not classifiable as a human carcinogen.
No TEEL available.
DFG MAK: 100 ppm/490 mg/m^3 TWA; Peak Limitation Category I(2).
NIOSH IDLH: 400 ppm.
Australia: TWA 50 ppm (240 mg/m^3), STEL 100 ppm, 1993; Austria: MAK 100 ppm (480 mg/m^3), 1999; Belgium: TWA 50 ppm (242 mg/m^3), STEL 100 ppm (483 mg/m^3), 1993; Denmark: TWA 25 ppm (120 mg/m^3), [skin], 1999; Finland: TWA 50 ppm (240 mg/m^3), STEL 100 ppm (480 mg/m^3), 1999; France: VME 50 ppm (240 mg/m^3), 1999; the Netherlands: MAC-TGG 50 mg/m^3,

[skin], 2003; the Philippines: TWA 100 ppm (480 mg/m^3), 1993; Poland: MAC (TWA) 100 mg/m^3; MAC (STEL) 300 mg/m^3, 1999; Russia: STEL 50 mg/m^3, 1993; Sweden: NGV 10 ppm (50 mg/m^3), KTV 75 ppm (150 mg/m^3), [skin], 1999; Switzerland: MAK-W 50 ppm (240 mg/m^3) KZG-W 100 ppm (480 mg/m^3), 1999; Turkey: TWA 100 ppm (480 mg/m^3), 1993; United Kingdom: TWA 100 ppm (491 mg/m^3), STEL 150 ppm, 2000; Argentina, Bulgaria, Columbia, Jordan, South Korea, New Zealand, Singapore, Vietnam: ACGIH TLV®: STEL 100 ppm. Several states have set guidelines or standards for vinyl toluene in ambient air[60] ranging from 40 μg/m^3 (Virginia) to 2400.0 to 4850.0 μg/m^3 (North Dakota) to 5714 μg/m^3 (Nevada) to 9600.0 μg/m^3 (Connecticut).
m-isomer
Denmark: TWA 25 ppm (120 mg/m^3), [skin], 1999; France: VME 50 ppm (240 mg/m^3), 1999; Switzerland: MAK-W 50 ppm (240 mg/m^3), KZG-W 100 ppm (480 mg/m^3), 1999; United Kingdom: TWA 100 ppm (491 mg/m^3), STEL 150 ppm, 2000.
Determination in Air: Use NIOSH Analytical Method (IV) #1501, aromatic hydrocarbons, OSHA Analytical Method 7.
Determination in water: Octanol−water coefficient: Log K_{ow} = 3.58.
Routes of Entry: Inhalation, ingestion, skin and/or eye contact.
Harmful Effects and Symptoms
Short Term Exposure: Vinyl toluene can affect you when breathed in. Exposure can irritate the eyes, nose, and upper respiratory system. A CNS depressant; very high levels can cause you to become dizzy, lightheaded, and drowsy. In animals: narcosis.
Long Term Exposure: Repeated exposure may affect the liver, kidneys, and nervous system. There is limited evidence that vinyl toluene may damage the developing fetus.
Points of Attack: Eyes, skin, respiratory system, central nervous system.
Medical Surveillance: Liver and kidney function tests. Evaluation of the nervous system.
First Aid: If this chemical gets into the eyes, remove any contact lenses at once and irrigate immediately for at least 15 min, occasionally lifting upper and lower lids. Seek medical attention immediately. If this chemical contacts the skin, remove contaminated clothing and wash immediately with soap and water. Seek medical attention immediately. If this chemical has been inhaled, remove from exposure, begin rescue breathing (using universal precautions, including resuscitation mask) if breathing has stopped and CPR if heart action has stopped. Transfer promptly to a medical facility. When this chemical has been swallowed, get medical attention. Do not induce vomiting.
Personal Protective Methods: Wear protective gloves and clothing to prevent any reasonable probability of skin contact. Safety equipment suppliers/manufacturers can provide recommendations on the most protective glove/clothing material for your operation. All protective clothing (suits,

gloves, footwear, headgear) should be clean, available each day, and put on before work. Contact lenses should not be worn when working with this chemical. Wear splash-proof chemical goggles and face shield unless full-face-piece respiratory protection is worn. Employees should wash immediately with soap when skin is wet or contaminated. Provide emergency showers and eyewash.

Respirator Selection: 400 ppm: CcrOv (APF = 10) [any chemical cartridge respirator with organic vapor cartridge (s)] or PaprOv (APF = 25) [any powered, air-purifying respirator with organic vapor cartridge(s)] or GmFOv (APF = 50) [any air-purifying, full-face-piece respirator (gas mask) with a chin-style, front- or back-mounted organic vapor canister] or Sa (APF = 10) (any supplied-air respirator) or SCBAF (APF = 50) (any self-contained breathing apparatus with a full face-piece). *Emergency or planned entry into unknown concentrations or IDLH conditions:* SCBAF: Pd,Pp (APF = 10,000) (any self-contained breathing apparatus that has a full face-piece and is operated in a pressure-demand or other positive-pressure mode) or SaF: Pd,Pp: ASCBA (APF = 10,000) (any supplied-air respirator that has a full face-piece and is operated in a pressure-demand or other positive-pressure mode in combination with an auxiliary, self-contained breathing apparatus operated in a pressure-demand or other positive-pressure mode). *Escape:* GmFOv (APF = 50) [any air-purifying, full-face-piece respirator (gas mask) with a chin-style, front- or back-mounted organic vapor canister] or SCBAE (any appropriate escape-type, self-contained breathing apparatus).

Storage: Color Code—Red: Flammability Hazard: Store in a flammable liquid storage area or approved cabinet away from ignition sources and corrosive and reactive materials. Prior to working with this chemical you should be trained on its proper handling and storage. Before entering confined space where this chemical may be present, check to make sure that an explosive concentration does not exist. Vinyl toluene must be stored to avoid contact with oxidizers (such as perchlorates, peroxides, permanganates, chlorates, and nitrates), strong acids, and aluminum chloride, since violent reactions occur. Store in tightly closed containers in a cool, well-ventilated area away from heat. Sources of ignition, such as smoking and open flames, are prohibited where this chemical is used, handled, or stored in a manner that could create a potential fire or explosion hazard. Metal containers involving the transfer of 5 gallons or more of this chemical should be grounded and bonded. Drums must be equipped with self-closing valves, pressure vacuum bungs, and flame arresters. Use only nonsparking tools and equipment, especially when opening and closing containers of this chemical.

Shipping: This compound requires a shipping label of "FLAMMABLE LIQUID." It falls into Hazard Class 3 and Packing Group II.

Spill Handling: Evacuate and restrict persons not wearing protective equipment from area of spill or leak until cleanup is complete. Remove all ignition sources. Establish forced ventilation to keep levels below explosive limit. Absorb liquids in vermiculite, dry sand, earth, peat, carbon, or a similar material and deposit in sealed containers. Oil-skimming equipment and sorbent foams can be applied to slick if done immediately. Keep this chemical out of a confined space, such as a sewer, because of the possibility of an explosion, unless the sewer is designed to prevent the buildup of explosive concentrations. It may be necessary to contain and dispose of this chemical as a hazardous waste. If material or contaminated runoff enters waterways, notify downstream users of potentially contaminated waters. Contact your local or federal environmental protection agency for specific recommendations. If employees are required to clean up spills, they must be properly trained and equipped. OSHA 1910.120(q) may be applicable.

Fire Extinguishing: This chemical is a flammable liquid. Poisonous gases are produced in fire. Use dry chemical, carbon dioxide, alcohol foam, or polymer foam extinguishers. Vapors are heavier than air and will collect in low areas. Containers may explode in fire. Storage containers and parts of containers may rocket great distances, in many directions. If material or contaminated runoff enters waterways, notify downstream users of potentially contaminated waters. Notify local health and fire officials and pollution control agencies. From a secure, explosion-proof location, use water spray to cool exposed containers. If cooling streams are ineffective (venting sound increases in volume and pitch, tank discolors, or shows any signs of deforming), withdraw immediately to a secure position. If employees are expected to fight fires, they must be trained and equipped in OSHA 1910.156. The only respirators recommended for firefighting are self-contained breathing apparatuses that have full face-pieces and are operated in a pressure-demand or other positive-pressure mode.

Disposal Method Suggested: Dissolve or mix the material with a combustible solvent and burn in a chemical incinerator equipped with an afterburner and scrubber. All federal, state, and local environmental regulations must be observed.

References
New Jersey Department of Health and Senior Services. (November 2000). *Hazardous Substances Fact Sheet: Vinyl Toluene.* Trenton, NJ

Vomiting agents

Vomiting agents are not designed to kill or cause serious injury. They have been formulated to cause nausea and vomiting, causing their victims to become helpless. The major chemical warfare agent in this category and its code name are listed below along with its record number for quick access.

Adamsite, agent DM see A:0435.

VX (Agent VX, WMD)   V:0250

Molecular Formula: $C_{11}H_{26}NO_2PS$

Synonyms: *S*-(2-Diisopropylaminoethyl) *O*-ethyl methylphosphonothiolate; *EA-1701*; *O*-Ethyl *S*-(2-[bis(1-methylethyl)amino] ethyl) methylphosphonothioate; Ethyl *S*-diiosopropylaminoethylmethyl thiophosphonate; *O*-Ethyl *S*-diisopropylaminoethyl methylphosphonothioate; Ethyl *S*-dimethylaminoethyl methyl phosphonothiolate; Methylphosphonothioic acid, *S*-(2-[bis(methylethyl)amino]ethyl) *O*-ethyl ester; Methylphosphonothioic acid, *S*-[2-bis(1-methylethyl)amino]ethyl) *O*-ethyl ester; Phosphonothioic acid, methyl-, *S*-[2-[bis(1-methylethyl)aminoethyl] *O*-ethyl] ester; Phosphonothioic acid, methyl-, S-[2-[bis(1-methylethyl)amino]ethyl] O-ethyl; ester; TX-60; VX (military designation)

CAS Registry Number: 50782-69-9
RTECS® Number: TB1090000
UN/NA & ERG Number: UN2810/123
EC Number: None assigned

Regulatory Authority and Advisory Bodies
Department of Homeland Security Screening Threshold Quantity: *Theft hazard* CUM 100 g.
Carcinogenicity: VX is not listed by the International Agency for Research on Cancer (IARC); American Conference of Governmental Industrial Hygienists (ACGIH); Occupational Safety and Health Administration (OSHA); or National Toxicology Program (NTP) as a carcinogen.
Superfund/EPCRA 40CFR355, Extremely Hazardous Substances: TPQ = 100 lb (45.4 kg).
Reportable Quantity (RQ): 100 lb (45.4 kg).
US DOT 49CFR172.101, Inhalation Hazardous Chemical.
Note: Army Regulation, AR 50-6, deals specifically with the shipment of chemical agents; must be escorted in accordance with Army Regulation, AR 740-32.
European/International Regulations: not listed in Annex 1.
WGK (German Aquatic Hazard Class): No value assigned.

Description: VX, a sulfinated organophosphorus compound, is a nerve agent, and the most toxic of all known chemical warfare agents. VX can cause death in minutes. As little as one drop of VX on the skin can be fatal. VX is a colorless, to straw to amber-colored, odorless liquid. Looks like motor oil. Molecular weight = 267.37; Boiling point = 298°C (decomposition); Freezing/Melting point = −51°C; Volatility = 8.9–10.5 mg/m³ at 25°C Vapor density (air = 1): 9.2; Vapor pressure = 0.0007 mmHg at 25°C; Vapor density (air = 1) = 9.2; Flash point = 159°C. Hazard Identification (based on NFPA-704 M Rating System): Health 4, Flammability 1, Reactivity 0. Slightly to moderately soluble in water below 9.4°C; 30 g/L at 25°C. See "Medical Surveillance" below for odor detection.

Potential Exposure: VX is a quick-acting, military chemical nerve agent. VX is the most potent of all chemical warfare agents. It attacks the nervous system, causing the muscles to convulse uncontrollably. The nerve agent works similarly to pesticide and was originally developed in the early 1950s. Highly persistent, it can be dangerous for weeks and remains a liquid for more than 24 h. It poses little vapor hazard. The least volatile of the nerve agents, VX, is very slow to evaporate; about as slowly as motor oil. VX is highly efficient at skin penetration, more than any other of the "G" agents. It is used in the M-23 land mine. VX was never used in combat by the United States and all stockpiles of approximately 4400 tons of the Agent were destroyed in 2008 by the US Army Chemical Materials Agency (CMA).

Persistence of Chemical Agent: Agent VX: Summer: 2 days to a week; Winter: 2 days to weeks.

Incompatibilities: Contact with metals may evolve flammable hydrogen gas. Relatively stable at room temperature. Unstabilized VX of 95% purity decomposed at a rate of 5% a month at 71°C. At pH 12, the toxic by-product has a half-life of about 14 days and in 90 days there is about a 64-fold reduction.

Permissible Exposure Limits in Air
Conversion factor: 0.09145 ppm = 1 mg/m³.
Protective Action Criteria (PAC) VX*
TEEL-0: 0.000005 ppm
PAC-1: 0.000016 ppm
PAC-2: **0.000016** ppm
PAC-3: **0.00027** ppm
*AEGLs (Acute Emergency Guideline Levels) & ERPGs (Emergency Response Planning Guideline) are in **bold face**.
STEL: 0.00001 mg/m³.
WPL (worker population limit): 0.000001 mg/m³.
The suggested permissible airborne exposure concentration of VX for an 8-h workday or a 40-h workweek is an 8-h time-weighted average (TWA) of 0.00001 mg/m³ (9×10^{-7} ppm). This value is based on the TWA of VX as proposed in the USAEHA Technical Guide No. 169, *Occupational Health Guidelines for the Evaluation and Control of Occupational Exposure to Nerve Agents GA, GB, GD, and VX.*
GPL (general population limit): 0.0000006 mg/m³.

Determination in Air: Available monitoring equipment for agent VX is the M8/M9 detector paper, (ACADA), detector ticket; M256/M256A1 kits; bubbler. Depot Area Air Monitoring System (DAMMS); automated Continuous Air Monitoring System (ACMS); Real-Time Monitor (RTM); Demilitarization Chemical Agent Concentrator (DCAC); M8/M43, M8A1/M43A1, CAM-M1, Hydrogen Flame Photometric Emission Detector (HYFED); and the Miniature Chemical Agent Monitor (MINICAM).

Determination in Water: Octanol−water coefficient: Log K_{ow} (estimated) = 2.06. VX is hydrolyzed only slowly, and the hydrolysis products include EA2192, which is nearly as toxic as VX and is hydrolyzed over 1000 times more slowly. Oxidation using common bleach Na(OCl) and superchlorinated bleach (Ca(OCl)$_2$)-calcium hypochlorite (HTH) will decontaminate.

Permissible Concentration in Water: No criteria set.

Routes of Entry: Inhalation, skin and/or eye contact.

Harmful Effects and Symptoms

Summary: Exposure can result in loss of consciousness, convulsions, paralysis, and respiratory failure resulting in death.

Short Term Exposure: VX. Acts as a cholinesterase inhibitor. Lowest toxic oral dose (TD$_{LO}$) to humans is 4 mg/kg; lowest lethal skin dose to humans (LD$_{LO}$) is 86 mg/kg. Death occurs within 15 min after fatal dose is absorbed.

Also reported: VX is a lethal anticholinergic agent with the median dose in humans being: LC$_{50}$ (skin) = 0.135 mg/kg; ID LC$_{50}$ (Skin)—0.07–0.71 mg/kg; LCt$_{50}$ (inhalation) = 30 mg-min/m^3; LCt$_{50}$ (inhalation)—30 mg-min/m^3; LCt$_{50}$ (inhalation)—24 mg-min/m^3.

One to several minutes after overexposure to airborne VX the following acute symptoms appear: Local effects (lasting 1–15 days, increases with dose). *Eyes:* Miosis (constriction of pupils); redness, pressure sensation on eyes. *Inhalation:* Rhinorrhea (runny nose), nasal congestion, tightness in chest, wheezing, salivation, nausea, vomiting. *Systemic effects* (increases with dose): *Inhalation*—excessive secretion causing coughing/breathing difficulty, salivation and sweating, vomiting, diarrhea, stomach cramps, involuntary urination/defecation, generalized muscle twitching/muscle cramps, CNS depression including anxiety, restlessness, giddiness, insomnia, excessive dreaming, and nightmares. With more severe exposure, headache, tremor, drowsiness, concentration difficulty, memory impairment, confusion, unsteadiness on standing or walking, and progression to death may also occur.

After exposure to liquid VX, the following acute symptoms appear: Local efffects: Eyes: Miosis, redness, pressure sensation on eyes. *Ingestion:* salivation, anorexia, nausea, vomiting, abdominal cramps, diarrhea, involuntary defecation, heartburn. *Skin:* Sweating, muscle twitching. *Systemic effects:* Similar to generalized effects from exposure to airborne material.

Long Term Exposure: Chronic overexposure to VX causes forgetfulness, thinking difficulty, vision disturbances, muscular aches/pains. Although *cer-*organophosphate pesticides have been shown to be teratogenic in animals, these effects have not been documented in carefully controlled toxicological evaluations for VX.[US Army]

Medical Surveillance: A chemical agent monitor (CAM) can detect VX at 0.1 mg/m^3. The US military also has the following detectors for VX and other chemical agents: *M256A1 Chemical Agent Detector Kit*; *M8A1 Automatic Chemical Agent Alarm.*

First Aid: Immediate decontamination of the smallest drop is essential. For decontamination, the *M291 Skin Decontamination Kit* should be used.

Inhalation: Hold breath until respiratory protective mask is donned. If severe signs of agent exposure appear (chest tightens, pupil constriction, lack of coordination, etc.), immediately administer, in rapid succession; all three Nerve Agent Antidote Kit(s), *Mark 1* injectors (or atropine if directed by the local physician). Injections using the Mark I kit injectors may be repeated at 5- to 20-min intervals if signs and symptoms are progressing until three series of injections have been administered. No more injections will be given unless directed by medical personnel. In addition, a record will be maintained of all injections given. If breathing has stopped, give artificial respiration. Mouth-to-mouth resuscitation should be used when approved mask-bag or oxygen delivery systems are not available. Do not use mouth-to-mouth resuscitation when facial contamination exists. If breathing is difficult, administer oxygen. Seek medical attention *immediately.*

Eye contact: Immediately flush eyes with water for 10–15 min, then don respiratory protective mask. Although miosis (pinpointing of the pupils) may be an early sign of agent exposure, an injection will not be administered when miosis is the only sign present. Instead, the individual will be taken immediately to the medical treatment facility for observation.

Skin contact: Don respirator with protective mask and remove contaminated clothing. Immediately wash contaminated skin with a solution of 5% household bleach. Rinse well with water to remove excess bleach followed by copious soap and water wash. Administer nerve agent antidote kit, Mark I, only if local sweating and muscular twitching symptoms are present. Seek medical attention *immediately.*

Ingestion: Do not induce vomiting. First symptoms are likely to be gastrointestinal. Immediately administer nerve agent antidote kit, Mark I. Seek medical attention *immediately.*

Medical observation recommended.

Medical treatment: Electrocardiogram (ECG), and adequacy of respiration and ventilation, should be monitored. Supplemental oxygenation, frequent suctioning of secretions, insertion of a tube into the trachea (endotracheal intubation), and assisted ventilation may be required. Diazepam (5–10 mg in adults and 0.2–0.5 mg/kg in children) may be used to control convulsions. Lorazepam or other benzodiazepines may be used, but barbiturates, phenytoin, and other anticonvulsants are not effective. Administration of atropine (if not already given) should precede the administration of benzodiazepines in order to best control seizures. Patients/victims who have inhalation exposure and who complain of chest pain, chest tightness, or cough should be observed and examined periodically for 6–12 h to detect delayed-onset inflammation of the large airways (bronchitis), inflammatory lung disease (pneumonia), accumulation of fluid in the lungs (pulmonary edema), or respiratory failure.

Decontamination: This is very important. The rapid physical removal of a chemical agent is essential. If you do not have the equipment and training, do not enter the hot or the warm zone to rescue and/or decontaminate victims. Medical personnel should wear the proper PPE. If the victim cannot move, decontaminate without touching and without entering the hot or the warm zone. Nerve gases stay in clothing; *do not* touch with bare skin—if possible, seal contaminated clothes and personal belongings in a sealed double bag. Use clean water from any source; if possible, use a hose (spray or fog to prevent injury to the victim) or other system to avoid touching the victim. Do not wait for soap or for the victim to remove clothing, begin washing immediately. Do not delay decontamination to obtain warm water; time is of the essence; use cold water instead. Immediately flush the eyes with water for at least 15 min. Use caution to avoid hypothermia in children and the elderly. Wash—strip—wash—evacuate upwind and uphill: Patients exposed to nerve agent by vapor only should be decontaminated by removing all clothing in a clean-air environment and shampooing or rinsing the hair to prevent vapor-off gassing; Patients exposed to liquid nerve agent should be decontaminated by washing in available clean water at least three times. Use liquid soap (dispose of container after use and replace), large amounts of water, and mild to moderate friction with a single-use sponge or washcloth in the first and second washes. Scrubbing of exposed skin with a brush is discouraged; skin damage may occur and may increase absorption. The third wash should be to rinse with large amounts of warm or hot water. Shampoo can be used to wash the hair. Decontaminate with diluted household bleach* (0.5%, or one part bleach to 200 parts water), but do not let any get in the victim's eyes, open wounds, or mouth. Wash off the diluted bleach solution after 15 min. Remember that the water you use to decontaminate the victims is dangerous. Be sure you have decontaminated the victims as much as you can before they are released from the area, so they do not spread the nerve gas. Rinse the eyes, mucous membranes, or open wounds with sterile saline or water and then move away from the hot zone in an upwind and uphill direction.

*The following can be used in addition to household bleach: (1) solids, powders, and solutions containing various types of bleach (NaOCl or Ca(OCl)$_2$); (2) DS2 (2% NaOH, 70% diethylenetriamine, 28% ethylene glycol monomethyl ether); (3) towelettes moistened with sodium hydroxide (NaOH) dissolved in water, phenol, ethanol, and ammonia. *Note:* Use 5% solution of common bleach (sodium hypochlorite) or calcium hypochlorite solution (48 oz per 5 gallons of water) to decontaminate scissors used in clothing removal, clothes and other items.

Personal Protective Methods: For emergency situations, wear a "moon suit" consisting of a positive pressure, pressure-demand, full-face-piece self-contained breathing apparatus (SCBA) or pressure-demand supplied-air respirator SCBA with escape cylinder in combination with a fully encapsulating, chemical-resistant suit capable of maintaining a positive air pressure within the suit. *Protective gloves:* butyl glove M3 and M4; Norton chemical protective glove set. *Eye protection:* Chemical goggles. For splash hazards use goggles and face shield.

Other protective equipment: Full protective clothing will consist of M9 mask and hood; M3 butyl rubber suit; M2A1 butyl boots; M3 and M4 gloves; unimpregnated underwear; or demilitarization protective ensemble (DPE). For laboratory operations, wear lab coats and have a protective mask readily available. In addition, daily clean smock, foot covers, and head covers will be required when handling contaminated lab animals.

Respirator Selection: *When used as a weapon, use SCBA Respirator Certified By NIOSH For CBRN Environments. Less than 0.00001 mg/m^3:* M9, M17, or M40 series mask shall be available for escape as necessary. 0.00001 mg/m^3 to 0.02 mg/m^3: M9 or M40 series mask with Level A or Level B ensemble. *Greater than 0.02 mg/m^3:* or DPE or TAPES used with prior approval from AMC Field Safety Activity. *Note:* When DPE or TAPES is not available the M9 or M40 series mask with Level A protective ensemble can be used. However, use time shall be restricted to the extent operationally feasible, and may not exceed 1 h. As an additional precaution, the cuffs of the sleeves and the legs of the M3 suit shall be taped to the gloves and boots respectively to reduce aspiration.

Local Exhaust: Mandatory. Must be filtered or scrubbed to limit exit conc. to <0.00001 mg/m^3 (averaged over 8 h/day, indefinitely).

Special: Chemical laboratory hoods shall have an average inward face velocity of 100 linear feet per minute (lf/m) + 10% with the velocity at any point not deviating from the average face velocity by more than 20%. Laboratory hoods shall be located such that cross-drafts do not exceed 20% of the inward face velocity. A visual performance test utilizing smoke-producing devices shall be performed in assessing the ability of the hood to contain agent VX. Emergency back-up power necessary: Hoods should be tested semiannually or after modification or maintenance operations. Operations should be performed 20 cm inside hood face. *Other:* Recirculation of exhaust air from agent areas is prohibited. No connection between agent areas and other areas through ventilation system is permitted.

Storage: Color Code—Blue: Health Hazard/Poison: Store in a secure poison location. Prior to working with this chemical you should be trained on its proper handling and storage. *Precautions to be taken in handling and storing:* in handling, the buddy system will be incorporated. No smoking, eating, or drinking in areas containing agent is permitted. Containers should be periodically inspected for leaks (either visually or by a detector kit). Stringent control over all personnel practices must be exercised. Decontamination equipment shall be conveniently located. Exits must be designed to permit rapid evacuation. Chemical showers, eyewash stations, and personal cleanliness facilities shall be provided. Wash hands before meals and each worker will shower thoroughly with special attention given to hair, face, neck, and hands, using plenty of soap before leaving at the end of the workday. *Other precautions:* agent must be double-contained in liquid and vapor-tight containers when in storage or when outside of the ventilation hood.

Store in tightly closed containers in a cool, well-ventilated area away from heat. Sources of ignition, such as smoking and open flames, are prohibited where this chemical is used, handled, or stored in a manner that could create a potential fire or explosion hazard.

Shipping: Toxic, liquids, organic, n.o.s. requires a shipping label of "POISONOUS/TOXIC MATERIALS." It falls in Hazard Class 6.1 and Packing Group III. Driver shall be given full and complete information regarding shipment and conditions in case of emergency. AR 50-6 deals specifically with the shipment of chemical agents. Shipments of agent will be escorted in accordance with AR 740-32.

Spill Handling: See also above. Evacuate and restrict persons not wearing protective equipment from area of spill or leak until cleanup is complete. Remove all ignition sources. Ventilate area of spill or leak. Absorb liquids in vermiculite, dry sand, earth, peat, carbon, or a similar material and deposit in sealed containers. Keep this chemical out of a confined space, such as a sewer, because of the possibility of an explosion, unless the sewer is designed to prevent the buildup of explosive concentrations. It may be necessary to contain and dispose of this chemical as a hazardous waste. If material or contaminated runoff enters waterways, notify downstream users of potentially contaminated waters. Contact your local or federal environmental protection agency for specific recommendations. If employees are required to clean up spills, they must be properly trained and equipped. OSHA 1910.120(q) may be applicable. See Medical Surveillance section for decontamination kit. If kit is not available when needed, decontaminants include bleach slurry and hot soapy water.

Initial isolation and protective action distances

Distances shown are likely to be affected during the first 30 min after materials are spilled and could increase with time. If more than one tank car, cargo tank, portable tank, or large cylinder involved in the incident is leaking, the protective action distance may need to be increased. You may need to seek emergency information from CHEMTREC at (800) 424-9300 or seek professional environmental engineering assistance from the US EPA Environmental Response Team at (908) 548-8730 (24-h response line).

VX, when used as a weapon

Small spills (From a small package or a small leak from a large package)

First: Isolate in all directions (feet/meters)   100/30
Then: Protect persons downwind (miles/kilometers)
Day   0.1/0.2
Night   0.1/0.2

Large spills (From a large package or from many small packages)

First: Isolate in all directions (feet/meters)   200/60
Then: Protect persons downwind (miles/kilometers)
Day   0.2/0.3
Night   0.3/0.5

Fire Extinguishing: When heated, vapors may form explosive mixtures with air, presenting an explosion hazard indoors, outdoors, and in sewers. Containers may explode when heated. Use moon suit/respirator protection. Poisonous gases, including oxides of nitrogen and sulfur, are produced in fire. *Extinguishing media*: Water mist, fog, foam, CO_2—Avoid using extinguishing methods that will cause splashing or spreading of the VX. *Special firefighting procedures:* All persons not engaged in extinguishing the fire should be immediately evacuated from the area. Fires involving VX should be contained to prevent contamination to uncontrolled areas. When responding to a fire alarm in buildings or areas containing nerve agents, firefighting personnel should wear full firefighter protective clothing (without TAP clothing) during chemical agent firefighting and fire rescue operations. Respiratory protection is required. Positive pressure, full face-piece, NIOSH-approved self contained breathing apparatus (SCBA) will be worn where there is danger of oxygen deficiency and when directed by the fire chief of chemical accident/incident (CAI) operations officer. The only respirators recommended for firefighting are self-contained breathing apparatuses that have full face-pieces and are operated in a pressure-demand or other positive-pressure mode. The M9 or M17 series mask may be worn in lieu of SCBA when there is no danger of oxygen deficiency. In cases where firefighters are responding to a chemical accident/incident for rescue/reconnaissance purposes they will wear appropriate levels of protective clothing. Do not breathe fumes. Skin contact with V-agents must be avoided at all times. Although the fire may destroy most of the agent, care must still be taken to assure the agent or contaminated liquids do not further contaminate other areas or sewers. Contact with VX or VX vapors can be fatal. Vapors are heavier than air and will collect in low areas. Containers may explode in fire. Storage containers and parts of containers may rocket great distances, in many directions. If material or contaminated runoff enters waterways, notify downstream users of potentially contaminated waters. Notify local health and fire officials and pollution control agencies. From a secure, explosion-proof location, use water spray to cool exposed containers. If cooling streams are ineffective (venting sound increases in volume and pitch, tank discolors, or shows any signs of deforming), withdraw immediately to a secure position. If employees are expected to fight fires, they must be trained and equipped in OSHA 1910.156.

Disposal Method Suggested: Recommended field procedures (for quantities greater than 50 g): *NOTE:* These procedures can only be used with the approval of a qualified expert or safety officer. An alcoholic calcium hypochlorite (HTH) mixture is prepared by adding 100 mL of denatured ethanol to 900 mL slurry of 10% calcium hypochlorite (HTH) in water. This mixture should be made just prior to use since the HTH can react with the ethanol. Fourteen grams of alcoholic calcium hypochlorite (HTH) solution is used for each gram of VX. Agitate the contamination mixture as the VX is added. Continue the agitation for a minimum of 1 h. This reaction is reasonably exothermic and evolves substantial off gassing. The evolved reaction gases should be routed through a decontaminate-filled scrubber prior to release through filtration systems. After completion of 1 h minimum agitation, 10% sodium hydroxide is added in a quantity equal to that necessary to assure that a pH of 12.5 is maintained for a period not less than 24 h. Hold the material at a pH between 10 and 12 for a period not less than 90 days to ensure that a hazardous intermediate material is not formed. After sealing the head, the exterior of the drum shall be decontaminated and then labeled in accordance with IAW, EPA, and DOT regulations. All leaking containers shall be overpacked with vermiculite placed between the interior and exterior containers. Decontaminate and label per IAW, EPA, and DOT regulations. Conduct general area monitoring to confirm that the atmospheric concentrations do not exceed the airborne exposure limit. If the alcoholic calcium hypochlorite (HTH) mixture is not available, then the following decontaminates may be used instead and are listed in the order of preference: Decontaminating Solution No. 2 [DS2: (2% NaOH, 70% diethylenetriamine, 28% ethylene glycol monomethyl ether)], Supertropical Bleach Slurry (STB), and sodium hypochlorite. Open pit burning or burying of VX or items containing or contaminated with VX in any quantity is prohibited. The detoxified VX (using procedures above) can be thermally destroyed by incineration in an EPA-approved incinerator in accordance with appropriate provisions of federal, state, and local RCRA regulations.

Note: Several states define decontaminated surety material as a RCRA Hazardous Waste.

Recommended laboratory procedures (for quantities less than 50 g): If the active chlorine of the calcium hypochlorite (HTH) is at least 55%, then 80 g of a 10% slurry is required for each gram of VX. Proportionally more HTH is required if the chlorine activity of the HTH is lower than 55%. The mixture is agitated as the VX is added and the agitation is maintained for a minimum of 1 h. If phasing of the VX/decon solution continues after 5 min, an amount of denatured ethanol equal to a 10 wt. % of the total agent/decon shall be added to assist miscibility.

Note: Ethanol should be minimized to prevent the formation of a hazardous waste. Upon completion of the 1-h agitation, the decon mixture shall be adjusted to a pH between 10 and 11. Conduct general area monitoring to confirm that the atmospheric concentrations do not exceed the airborne exposure limit.

References

US Environmental Protection Agency. (November 30, 1987). *Chemical Hazard Information Profile: Phosphonothioic Acid, S-(2-[(bis(1-Methylethyl)Amino]ethyl)-Ethyl Ester.* Washington, DC: Chemical Emergency Preparedness Program

New Jersey Department of Health and Senior Services. (April 2006). *Hazardous Substances Fact Sheet: VX.* Trenton, NJ

Schneider, A. L., et al. (2007). *CHRIS + CD-ROM Version 2.0, United Coast Guard Chemical Hazard Response Information System (COMDTINST 16465.12C).* Washington, DC: United States Coast Guard and the Department of Homeland Security

The Riegle Report: A Report of Chairman Donald W. Riegle, Jr. and Ranking Member Alfonse M. D'Amato of the Committee on Banking, Housing and Urban Affairs with Respect to Export Administration, United States Senate, 103rd Congress, 2d Session, (May 25, 1994)

Besch, T., Moss, C., the Battlebook Project Team, et al. (1991). *MEDICAL NBC Battlebook* (304 p), Tech Guide 244. Ft. Leonard Wood, MO: The US Army Center for Health Promotion and Preventive Medicine (USACHPPM)

US Army Chemical Materials Agency (CMA). (2008). *Last VX Nerve Agent Munition Eliminated from CMA's Destruction Stockpile.* Aberdeen Proving Ground, MD. <http://www.cma.army.mil/home.aspx>

W

Warfarin   W:0100

Molecular Formula: $C_{19}H_{16}O_4$

Synonyms: 3-(α-Acetonylbenzyl)-4-hydroxycoumarin; Arab Rat Deth; Atrombine-K; 2H-1-Benzopyran-2-one,4-hydroxy-3-(3-oxo-1- phenylbutyl)-; Brumin; Compound 42; CO-RAX; Coumadin; Coumafene; Coumarin, 3-(α-acetonylbenzyl)-4-hydroxy-; D-CON; Dethmore; Eastern states duocide; Grovex sewer bait; 4-Hydroxy-3-(3-oxo-1-phenylbutyl)coumarin; Killgerm sewarin P; Kilmol; Kumander; Kypfarin; Liqua-tox; Mouse PAK; (Phenyl-1acetyl-2-ethyl)-3-hydroxy-4 coumarine (French); 3-(1'-Phenyl-2'-acetylethyl)-4-hydroxycoumarin; 3-(α-Phenyl-β-acetylethyl)-4-hydroxycoumarin; Plusbait; Prothromadin; Rat-*A*-Way; Rat-B-Gon; Rat-Gard; Rat & mice bait; Rat-o-cide; Ratron; Rats-no-more; Rax; RCR Squirrel killer; Rentokil; Rentokil biotrol; Rodentex; RO-Deth; Rodex blox; Rough & ready mouse mix; Sakarat; Sewarin; Solfarin; Sorexa plus; Sorex Cr1; Spray-trol branch roden-trol; Toxic chemical category code, N874; Twin light rat away; Vampirinip; Warfarine (French); Warf compound; Zoocoumarin

sodium: 3-(α-Acetonylbenzyl)-4-hydroxy-coumarin sodium salt; Athrombin; Coumadin sodium; 4-Hydroxy-3-(3-oxo-1-phenylbutyl)-2H-1-benzopyran-2-one sodium salt; Marevan (sodium salt); Panwarfin; Prothrombin; Ratsul soluble; Sodium coumadin; Sodium warfarin; Tintorane; Varfine; Waran; Warcoumin; Warfilone

CAS Registry Number: 81-81-2; 129-06-6 (sodium)

RTECS® Number: GN4550000; GN4725000 (sodium)

UN/NA & ERG Number: UN3027 (coumarin derivative pesticide, solid, poisonous)

EC Number: 201-377-6 [*Annex I Index No.:* 607-056-00-0]; 204-929-4 (sodium)

Regulatory Authority and Advisory Bodies

US EPA, FIFRA 1998 Status of Pesticides: RED completed.

Very Toxic Substance (World Bank).[15]

Air Pollutant Standard Set. See below, "Permissible Exposure Limits in Air" section.

US EPA Hazardous Waste Number (RCRA No.): P001; U248.

RCRA, 40CFR261, Appendix 8 Hazardous Constituents.

Superfund/EPCRA 40CFR355, Extremely Hazardous Substances: TPQ = 500/10,000 lb (227/4540 kg).

Reportable Quantity (RQ): 100 lb (45.4 kg).

EPCRA Section 313 Form R *de minimis* concentration reporting level: 1.0% (Warfarin and salts).

US DOT Regulated Marine Pollutant (49CFR172.101, Appendix B).

California Proposition 65 Chemical: Developmental/Reproductive toxin 7/1/87.

Warfarin sodium:

Superfund/EPCRA 40CFR355, Extremely Hazardous Substances: TPQ = 100/10,000 lb (45.4/4540 kg).

Reportable Quantity (RQ): 100 lb (45.4 kg).

EPCRA Section 313 Form R *de minimis* concentration reporting level: 1.0%.

European/International Regulations (*81-81-2*): Hazard Symbol: T; Risk phrases: R61; R48/25; Safety phrases: S3; S45 (see Appendix 4).

WGK (German Aquatic Hazard Class): No value assigned.

Description: Warfarin is a colorless, odorless crystalline solid. Molecular weight = 308.35. Combustible. Although warfarin is usually available commercially as the sodium salt, the following physical properties refer to the pure substance: Freezing/Melting point = 161°C (decomposes below BP); Vapor pressure = 0.09 mmHg at 25°C. Hazard Identification (based on NFPA-704 M Rating System) (*Fume and dust*): Health 4, Flammability 0, Reactivity 0. Practically insoluble in water; solubility = 0.002%.

Potential Exposure: Compound Description: Agricultural Chemical; Reproductive Effector; Human Data; Natural Product. Warfarin is used as an oral anticoagulant and as a rodenticide or rat poison.

Incompatibilities: Strong oxidizers, strong acids, strong bases. Dust mixtures with air may cause explosion.

Permissible Exposure Limits in Air

Warfarin 81-81-2

OSHA PEL: 0.1 mg/m³ TWA.

NIOSH REL: 0.1 mg/m³ TWA.

ACGIH TLV®[1]: 0.1 mg/m³ TWA.

NIOSH IDLH: 100 mg/m³.

Protective Action Criteria (PAC)

TEEL-0: 0.1 mg/m³

PAC-1: 0.3 mg/m³

PAC-2: 20 mg/m³

PAC-3: 100 mg/m³

Protective Action Criteria (PAC) (sodium)

TEEL-0: 1.5 mg/m³

PAC-1: 5 mg/m³

PAC-2: 9 mg/m³

PAC-3: 9 mg/m³

DFG MAK 0.5 mg/m³, inhalable fraction TWA; Peak Limitation Category II(2).

Compound Description: Agricultural Chemical; Reproductive Effector; Human Data; Natural Product.

Australia: TWA 0.1 mg/m³, 1993; Austria: MAK 0.1 mg/m³, 1999; Belgium: TWA 0.1 mg/m³, 1993; Denmark: TWA 0.1 mg/m³, 1999; Finland: TWA 0.1 mg/m³, STEL 0.3 mg/m³, 1999; France: VME 0.1 mg/m³, 1999; the Netherlands: MAC-TGG 0.1 mg/m³, 2003; Norway: TWA 0.1 mg/m³, 1999; the Philippines: TWA

0.1 mg/m^3, 1993; Russia: STEL 0.001 mg/m^3, 1993; Switzerland: MAK-W 0.1 mg/m^3, KZG-W 0.5 mg/m^3, 1999; Thailand: TWA 0.1 mg/m^3, 1993; Turkey: TWA 0.1 mg/m^3, 1993; United Kingdom: TWA 0.1 mg/m^3, STEL 0.3 mg/m^3, 2000; Argentina, Bulgaria, Columbia, Jordan, South Korea, New Zealand, Singapore, Vietnam: ACGIH TLV$^®$: TWA 0.1 mg/m^3.

Several states have set guidelines or standards for warfarin in ambient air[60] ranging from 0.016 μg/m^3 (Virginia) to 1.0−3.0 μg/m^3 (North Dakota) to 2.0 μg/m^3 (Connecticut and Nevada).

Determination in Air: Use NIOSH Analytical Method (IV) #5002.

Determination in water: Octanol−water coefficient: Log $K_{ow} = 2.51$.

Routes of Entry: Skin absorption, ingestion, inhalation, skin and/or eye contact.

Harmful Effects and Symptoms

Short Term Exposure: Warfarin is classified as very toxic, and may cause hemorrhage at even low levels. Probable oral lethal dose in humans is 50−500 mg/kg, between 1 teaspoon and 1 oz for a 150-lb person. Material is an anticoagulant. Toxic effects other than hemorrhage are rarely seen in humans. Other symptoms of warfarin exposure begin a few days or weeks after ingestion. They include epistaxis (nose bleed); bleeding gums; pallor, and sometimes hematomas around joints and on buttocks; blood in urine and feces; hematoma arms, legs; bleeding lips; mucous membrane hemorrhage; petechial rash; abnormal/abnormalities hematologic indices. Later, paralysis due to cerebral hemorrhage and finally, hemorrhagic shock and death may occur. Warfarin sodium is an anticoagulant. Hemorrhage is the most common sign and may be manifested by hemorrhagic skin rashes and lip, nose, and upper airway bleeding. Upper airway pain, difficulty in speaking and swallowing, and dyspnea (shortness of breath) may occur. Back pain may be noted.

Long Term Exposure: Anemia can result from severe or repeated bleeding. Repeated exposure may affect the liver and kidneys. Material is believed to be teratogenic in humans. There is limited evidence that warfarin may decrease fertility in females. Animal tests indicates that warfarin may cause malformations in human babies.

Points of Attack: Blood, cardiovascular system.

Medical Surveillance: NIOSH lists the following tests: blood plasma: prothrombin time; complete blood count; urine (chemical/metabolite): whole blood (chemical/metabolite); urinalysis (routine): red blood cells/count. Persons taking "blood thinning" medications are at increased risk.

First Aid: If this chemical gets into the eyes, remove any contact lenses at once and irrigate immediately for at least 15 min, occasionally lifting upper and lower lids. Seek medical attention immediately. If this chemical contacts the skin, remove contaminated clothing and wash immediately with soap and water. Seek medical attention immediately. If this chemical has been inhaled, remove from exposure, begin rescue breathing (using universal precautions, including resuscitation mask) if breathing has stopped and CPR if heart action has stopped. Transfer promptly to a medical facility. When this chemical has been swallowed, get medical attention. Give large quantities of water and induce vomiting. Do not make an unconscious person vomit. Medical observation is recommended.

Personal Protective Methods: Wear protective gloves and clothing to prevent any reasonable probability of skin contact. Safety equipment suppliers/manufacturers can provide recommendations on the most protective glove/clothing material for your operation. All protective clothing (suits, gloves, footwear, headgear) should be clean, available each day, and put on before work. Contact lenses should not be worn when working with this chemical. Wear dust-proof chemical goggles and face shield unless full face-piece respiratory protection is worn. Employees should wash immediately with soap when skin is wet or contaminated.

Respirator Selection: Up to 0.5 mg/m^3: Qm (APF = 25) (any quarter-mask respirator). *Up to 1 mg/m^3:* 95XQ (APF = 10) [any particulate respirator equipped with an N95, R95, or P95 filter (including N95, R95, and P95 filtering face-pieces) except quarter-mask respirators. The following filters may also be used: N99, R99, P99, N100, R100, P100] or Sa (APF = 10) (any supplied-air respirator). *Up to 2.5 mg/m^3:* Sa:Cf (APF = 25) (any supplied-air respirator operated in a continuous-flow mode) or PaprHie (APF = 25) (any powered, air-purifying respirator with a high-efficiency particulate filter). *Up to 5 mg/m^3:* 100F (APF = 50) (any air-purifying, full-face-piece respirator with an N100, R100, or P100 filter) or SaT: Cf (APF = 50) (any supplied-air respirator that has a tight-fitting face-piece and is operated in a continuous-flow mode) or PaprTHie (APF = 50) (any powered, air-purifying respirator with a tight-fitting face-piece and a high-efficiency particulate filter) or SCBAF (APF = 50) (any self-contained breathing apparatus with a full face-piece) or SaF (APF = 50) (any supplied-air respirator with a full face-piece). *Up to 100 mg/m^3:* Sa: Pd,Pp (APF = 1000) (any supplied-air respirator operated in a pressure-demand or other positive-pressure mode). *Emergency or planned entry into unknown concentrations or IDLH conditions:* SCBAF: Pd,Pp (APF = 10,000) (any NIOSH/MSHA- or European Standard EN 149-approved self-contained breathing apparatus that has a full face-piece and is operated in a pressure-demand or other positive-pressure mode) or SaF: Pd,Pp: ASCBA (APF = 10,000) (any supplied-air respirator that has a full face-piece and is operated in a pressure-demand or other positive-pressure mode in combination with an auxiliary, self-contained breathing apparatus operated in a pressure-demand or other positive-pressure mode). *Escape:* 100F (APF = 50) (any air-purifying, full-face-piece respirator with an N100, R100, or P100 filter) or SCBAE (any appropriate escape-type, self-contained breathing apparatus).

Storage: Color Code—Blue: Health Hazard/Poison: Store in a secure poison location. Prior to working with this

chemical you should be trained on its proper handling and storage. Store in tightly closed containers in a cool, well-ventilated area away from oxidizers. Where possible, automatically transfer material from other storage containers to process containers.

Shipping: Coumarin derivative pesticides, solid, toxic, require a shipping label of "POISONOUS/TOXIC MATERIALS." It falls in Hazard Class 6.1 and Packing Group I.

Spill Handling: Evacuate persons not wearing protective equipment from area of spill or leak until cleanup is complete. Remove all ignition sources. Collect powdered material in the most convenient and safe manner and deposit in sealed containers. Ventilate area after cleanup is complete. It may be necessary to contain and dispose of this chemical as a hazardous waste. If material or contaminated runoff enters waterways, notify downstream users of potentially contaminated waters. Contact your local or federal environmental protection agency for specific recommendations. If employees are required to clean up spills, they must be properly trained and equipped. OSHA 1910.120(q) may be applicable. Do not touch spilled material. Eating and smoking should not be permitted in areas where it is handled, processed, or stored. *Small spills:* sweep onto paper or other suitable material. Place in an appropriate container and burn in a safe place. *Large quantities:* may be destroyed by dissolving in a flammable solvent (e.g., alcohol) and atomizing in a combustion chamber.

Fire Extinguishing: Warfarin is combustible; however, no flash point can be found. Extinguish fire using agent suitable for type of surrounding fire. Use alcohol foam, carbon dioxide, or dry chemical. Wear full protective clothing and self-contained breathing apparatus when engaged in firefighting. Poisonous gases are produced in fire, including oxides of sodium (warfarin sodium). If material or contaminated runoff enters waterways, notify downstream users of potentially contaminated waters. Notify local health and fire officials and pollution control agencies. From a secure, explosion-proof location, use water spray to cool exposed containers. If cooling streams are ineffective (venting sound increases in volume and pitch, tank discolors, or shows any signs of deforming), withdraw immediately to a secure position. If employees are expected to fight fires, they must be trained and equipped in OSHA 1910.156. The only respirators recommended for firefighting are self-contained breathing apparatuses that have full face-pieces and are operated in a pressure-demand or other positive-pressure mode.

Disposal Method Suggested: Consult with environmental regulatory agencies for guidance on acceptable disposal practices. Generators of waste containing this contaminant ($\geq$100 kg/mo) must conform with EPA regulations governing storage, transportation, treatment, and waste disposal. Incineration.

References

US Environmental Protection Agency. (November 30, 1987). *Chemical Hazard Information Profile: Warfarin.* Washington, DC: Chemical Emergency Preparedness Program

US Environmental Protection Agency. (November 30, 1987). *Chemical Hazard Information Profile: Warfarin Sodium.* Washington, DC: Chemical Emergency Preparedness Program

X

Xylenes X:0100

Molecular Formula: C_8H_{10}
Common Formula: $C_6H_4(CH_3)_2$
Synonyms: m-isomer: Benzene, *m*-dimethyl-; Benzene, 1,3-dimethyl-; *m*-Dimethylbenzene; 1,3-Dimethylbenzene; *m*-Methyltoluene; *m*-Xileno (Spanish); *m*-Xylene; 1,3-Xylene; Xylene, *m*-; *m*-Xylol
o-isomer: Benzene-*o*-dimethyl; Benzene-1,2-dimethyl-; *o*-Dimethylbenzene; 1,2-Dimethylbenzene; *o*-Methyltoluene; 1,2-Methyltoluene; *o*-Xileno (Spanish); *o*-Xylene; 1,2-Xylene; Xylene, *o*-; *o*-Xylol
p-isomer: Benzene-*p*-dimethyl; Benzene-1,4-dimethyl; Chromar; *p*-Dimethylbenzene; 1,4-Dimethylbenzene; *p*-Methyltoluene; 4-Methyltoluene; Scintillar; *p*-Xileno (Spanish); *p*-Xylene; 1,4-Xylene; Xylene, *p*-; *p*-Xylol
CAS Registry Number: 108-38-3 (*m*-isomer); 95-47-6 (*o*-isomer); 106-42-3 (*p*-isomer); 1330-20-7 (mixed isomers)
RTECS® Number: ZE2100000 (mixed isomers); ZE2275000 (*m*-isomer); ZE2450000 (*o*-isomer); ZE2625000 (*p*-isomer)
UN/NA & ERG Number: UN1307 (all isomers)/130
EC Number: 203-576-3 [*Annex I Index No.:* 601-022-00-9] (*m*-isomer); 202-422-2 [*Annex I Index No.:* 601-022-00-9] (*o*-isomer); 203-396-5 [*Annex I Index No.:* 601-022-00-9] (*p*-isomer); 215-535-7 [*Annex I Index No.:* 601-022-00-9] (mixed isomers)

Regulatory Authority and Advisory Bodies

Carcinogenicity (*m*-, *o*-, *and p*-isomers): IARC: Animal, Inadequate Evidence; Human, Inadequate Evidence, *not classifiable as carcinogenic to humans*, Group 3, 1999; EPA: Available data are inadequate for an assessment of human carcinogenic potential.
Air Pollutant Standard Set. See below, "Permissible Exposure Limits in Air" section.
All isomers and mixtures:
Clean Air Act: Hazardous Air Pollutants (Title I, Part A, Section 112).
Clean Water Act: Section 311 Hazardous Substances/RQ 100 lb (45.4 kg); Section 313 Water Priority Chemicals (57FR41331, 9/9/92).
US EPA Hazardous Waste Number (RCRA No.): U239.
RCRA, 40CFR261, Appendix 8 Hazardous Constituents.
RCRA 40CFR268.48; 61FR15654, Universal Treatment Standards: Wastewater (mg/L), 0.32; Nonwastewater (mg/kg), 30.
RCRA 40CFR264, Appendix 9; TSD Facilities Ground Water Monitoring List. Suggested methods (PQL µg/L): total dust 8020 (5); 8240 (5).
Safe Drinking Water Act: Xylenes, total dust, MCL, 10 mg/L; MCLG, 10 mg/L; Regulated chemical (47 FR 9352).
Reportable Quantity (RQ): 1000 lb (454.0 kg).
EPCRA Section 313 Form R *de minimis* concentration reporting level: 1.0%.

Canada, WHMIS, Ingredients Disclosure List Concentration 1.0% (*o*- and *m*-isomers); 0.1% (*p*-isomer).
European/International Regulations (*all isomers*): Hazard Symbol: Xn; Risk phrases: R10; R20/21; R38; Safety phrases: S2; S25 (see Appendix 4).
WGK (German Aquatic Hazard Class): 2—Water polluting (*all isomers*).
Description: Xylene exists in three isomeric forms, *ortho-*, *meta-*, and *para-*xylene. Commercial xylene is a mixture of these three isomers and may also contain ethylbenzene as well as small amounts of toluene, trimethylbenzene, phenol, thiophene, pyridine, and other nonaromatic hydrocarbons. *m*-Xylene is predominant in commercial xylene. The physical properties of the three isomers are as follows:

Isomer	Melting Point (°C)	Boiling Point (°C)	Flash Point (°C)	Lower Expl. (%)	Upper Expl. (%)	Auto-Temp. (°C)
ortho-	−25	144	32	0.9	6.7	463
meta-	−48	139	27	1.1	7.0	527
para-	13	138	27	1.1	7.0	528

Odor threshold = 0.081−40 ppm. The range of odor threshold values is quite broad and caution is advised in relying on odor alone as a warning of potentially hazardous exposures. NFPA 704 M Hazard Identification (all isomers): Health 2, Flammability 3, Reactivity 0. All isomers are practically insoluble in water.
Potential Exposure: Compound Description (*m*-isomer): Tumorigen; Reproductive Effector; Human Data; Primary Irritant; (*o*-isomer) Tumorigen; Reproductive Effector; Human Data; (*p*-isomer) Tumorigen; Reproductive Effector. Xylene is used as a solvent; as a constituent of paint, lacquers, varnishes, inks, dyes, adhesives, cements, cleaning fluids, and aviation fuels; and as a chemical feed-stock for xylidines, benzoic acid, phthalic anhydride, isophthalic, and terephthalic acids, as well as their esters (which are specifically used in the manufacture of plastic materials and synthetic textile fabrics). Xylene is also used in the manufacture of quartz crystal oscillators, hydrogen peroxide, perfumes, insect repellants, epoxy resins, pharmaceuticals, and in the leather industry. *m*-Xylene is used as an intermediate in the preparation of isophthalic acid; *o*-xylene is used in the manufacture of phthalic anhydride and in pharmaceutical and insecticide synthesis. *p*-Xylene is used in pharmaceutical and insecticide systhesis and in the production of polyester.
Incompatibilities: Strong oxidizers, strong acids. Electrostatic charges can be generated from agitation or flow.

Permissible Exposure Limits in Air

Conversion factor: 1 ppm = 4.34 mg/m³ (*all isomers*) at 25°C & 1 atm.
OSHA PEL: 100 ppm/435 mg/m³ TWA.
NIOSH REL: 100 ppm/435 mg/m³ TWA; 150 ppm/655 mg/m³ STEL.

ACGIH TLV®[1]: 100 ppm/434 mg/m³ TWA; 150 ppm/651 mg/m³ STEL, not classifiable as a human carcinogen; BEI: (technical grade) 1.5 g[methylhippuric acids]/g creatinine in urine, end-of-shift.
NIOSH IDLH: 900 ppm.
Protective Action Criteria (PAC)*
All isomers and mixed isomers
TEEL-0: 100 ppm
PAC-1: **130** ppm
PAC-2: **920** ppm
PAC-3: **2500** ppm
*AEGLs (Acute Emergency Guideline Levels) & ERPGs (Emergency Response Planning Guideline) are in **bold face**.
DFG MAK: 100 ppm/440 mg/m³ TWA; Peak Limitation Category II(2) [skin]; Pregnancy Risk Group D; BAT: 1.5 mg[xylene]/L in blood, end-of-shift; 2000 mg[methylhippuric (toluric)acid]/L in urine, end-of-shift.
European OEL: 50 ppm TWA; 100 ppm STEL [skin] (2000) Austria: MAK 100 ppm (440 mg/m³), 1999; Denmark: TWA 35 ppm (150 mg/m³), [skin], 1999; France: VME 100 ppm (435 mg/m³), VLE 150 ppm (650 mg/m³), 1999; Japan; 100 ppm (430 mg/m³), 1999; Norway: TWA 25 ppm (108 mg/m³), 1999; Switzerland: MAK-W 100 ppm (435 mg/m³), KZG-W 200 ppm (870 mg/m³), 1999; United Kingdom: TWA 100 ppm (441 mg/m³), STEL 150 ppm, [skin], 2000. Russia[43] set a MAC value of 0.2 mg/m³ (200 μg/m³) for ambient air in residential areas both on a momentary and a daily average basis. Several states have set guidelines or standards for xylenes in ambient air[60] ranging from zero (Colorado) to 0.059 mg/m³ (Massachusetts) to 0.073 mg/m³ (Virginia) to 0.700 mg/m³ (Rhode Island) to 1.45 mg/m³ (New York) to 8.68 mg/m³ (Connecticut) to 8.70 mg/m³ (South Dakota) to 10.357 mg/m³ (Nevada) to 2.6−65.5 mg/m³ (North Carolina).
Determination in Air: Use NIOSH Analytical Method #1501, Hydrocarbons, aromatic, #3800, OSHA Analytical Method 1002.
Permissible Concentration in Water: Russia[43] set a MAC of 0.05 mg/L (50 μg/L) in water both for household and fishery purposes. The EPA has proposed a limit of 10 μg/L (10 ppm)[62] and has determined a lifetime health advisory of 400 μg/L.[48] Several states have set guidelines and standards for xylenes in drinking water[61] ranging from 44.0 μg/L (New Jersey) to 50.0 μg/L (New York) to 440.0 μg/L (Arizona, Kansas, Minnesota) to 620.0 μg/L (Massachusetts, Maine, New Mexico, Vermont, and Wisconsin).
Determination in Water: Octanol−water coefficient: Log K_{ow} = 3.20.
Routes of Entry: Inhalation, skin absorption, ingestion, skin and/or eye contact.
Harmful Effects and Symptoms
Short Term Exposure: Inhalation: Exposure to vapor can be irritating to the nose and throat. Inhalation of vapor at concentrations *above 200 ppm* or 3−5 min can lead to

xylene intoxication. Symptoms include headache, dizziness, nausea, and vomiting. If exposure continues, central nervous system depression characterized by shallow breathing and weak pulse can occur. Levels of 230 ppm for 15 min may cause lightheadedness without loss of equilibrium. Reversible liver and kidney damage in humans has followed exposure to sudden high concentrations of vapor. Such high levels may also give rise to lung congestion. Exposure to extremely high concentrations (10,000 ppm or more) of xylene vapors can lead to a strong narcotic effect with symptoms of slurred speech, stupor fatigue, confusion, unconsciousness, coma, and possible death. *Skin:* Contact with vapor or liquid can cause defatting which may lead to irritation, drying, and cracking. *Eyes:* Vapor and liquid may be irritating to the eye and eyelids at levels of 100 ppm for 15 min. *Ingestion:* Swallowing liquid xylene will bring about an immediate burning sensation in the mouth and throat. Irritation of the stomach and intestine can give rise to sharp stomach pains. Symptoms are the same as inhalation, except that lung congestion will not usually develop.
Long Term Exposure: Inhalation of xylene vapor and skin contact with liquid are the two most probable routes of long-term exposure. Symptoms of inhalation are dizziness, headache, and nausea. Long-term exposure has been associated with liver and kidney damage, intestinal tract disturbances, and central nervous system depression. Prolonged contact with skin can lead to irritation, dryness, and cracking. Repeated exposure can cause poor memory, difficulty in concentration, and other brain effects. It can also cause damage to the eye surface.
Points of Attack: Eyes, skin, respiratory system, central nervous system, gastrointestinal tract, blood, liver, kidneys.
Medical Surveillance: whole blood (chemical/metabolite); whole blood (chemical/metabolite), end-of-shift; Complete blood count; Complete blood count; hematopoietic depression, expired air, urine (chemical/metabolite); urine (chemical/metabolite): end-of-shift; urine (chemical/metabolite): end-of-work-week; urine (chemical/metabolite): Last 4 h of 8-h exposure; urinalysis (routine). For those with frequent or potentially high exposure (half the TLV or greater, or significant skin contact), the following is recommended before beginning work and at regular times after that: examination of the eyes by slit lamp. If symptoms develop or overexposure is suspected, the following may be useful: liver and kidney function tests. Urine concentration of *m-methylhippuric acid* (at the end-of-shift) as an index of overexposure.
First Aid: If this chemical gets into the eyes, remove any contact lenses at once and irrigate immediately for at least 15 min, occasionally lifting upper and lower lids. Seek medical attention immediately. If this chemical contacts the skin, remove contaminated clothing and wash immediately with soap and water. Seek medical attention immediately. If this chemical has been inhaled, remove from exposure, begin rescue breathing (using universal precautions, including resuscitation mask) if breathing has stopped and CPR if

heart action has stopped. Transfer promptly to a medical facility. If victim is *conscious*, administer water or milk. Do not induce vomiting.

Note to physician: May require supportive measures for pulmonary edema.

Personal Protective Methods: Wear protective gloves and clothing to prevent any reasonable probability of skin contact. Safety equipment suppliers/manufacturers can provide recommendations on the most protective glove/clothing material for your operation. Viton™, polyvinyl alcohol, and Teflon™ are among the recommended protective materials. All protective clothing (suits, gloves, footwear, headgear) should be clean, available each day, and put on before work. Contact lenses should not be worn when working with this chemical. Wear splash-proof chemical goggles and face shield unless full face-piece respiratory protection is worn. Employees should wash immediately with soap when skin is wet or contaminated. Provide emergency showers and eyewash.

Respirator Selection: *900 ppm:* CcrOv (APF = 10) [any chemical cartridge respirator with organic vapor cartridge(s)] or PaprOv (APF = 25) [any powered, air-purifying respirator with organic vapor cartridge(s)] or Sa (APF = 10) (any supplied-air respirator) or SCBAF (APF = 50) (any self-contained breathing apparatus with a full face-piece). *Emergency or planned entry into unknown concentrations or IDLH conditions:* SCBAF: Pd,Pp (APF = 10,000) (any self-contained breathing apparatus that has a full face-piece and is operated in a pressure-demand or other positive-pressure mode) or SaF: Pd,Pp: ASCBA (APF = 10,000) (any supplied-air respirator that has a full face-piece and is operated in a pressure-demand or other positive-pressure mode in combination with an auxiliary self-contained breathing apparatus operated in a pressure-demand or other positive-pressure mode). *Escape:* GmFOv (APF = 50) [any air-purifying, full-face-piece respirator (gas mask) with a chin-style, front- or back-mounted organic vapor canister] or SCBAE (any appropriate escape-type, self-contained breathing apparatus). *Note:* Substance reported to cause eye irritation or damage; may require eye protection.

Storage: Color Code—Red: Flammability Hazard: Store in a flammable liquid storage area or approved cabinet away from ignition sources and corrosive and reactive materials. Prior to working with this chemical you should be trained on its proper handling and storage. Xylenes must be stored to avoid contact with strong oxidizers (such as chlorine, bromine, and fluorine) since violent reactions occur. Sources of ignition, such as smoking and open flames, are prohibited where xylenes are used, handled, or stored in a manner that could create a potential fire or explosion hazard. Use only nonsparking tools and equipment, especially when opening and closing containers of xylenes. Protect storage containers from physical damage.

Shipping: This compound requires a shipping label of "FLAMMABLE LIQUID." It falls in Hazard Class 3 and Packing Group II.

Spill Handling: Evacuate and restrict persons not wearing protective equipment from area of spill or leak until cleanup is complete. Remove all ignition sources. Ventilate area of spill or leak. Absorb liquids in vermiculite, dry sand, earth, peat, carbon, or a similar material and deposit in sealed containers. Oil-skimming equipment and sorbent foams can be applied to slick if done immediately. Keep this chemical out of a confined space, such as a sewer, because of the possibility of an explosion, unless the sewer is designed to prevent the buildup of explosive concentrations. It may be necessary to contain and dispose of this chemical as a hazardous waste. If material or contaminated runoff enters waterways, notify downstream users of potentially contaminated waters. Contact your local or federal environmental protection agency for specific recommendations. If employees are required to clean up spills, they must be properly trained and equipped. OSHA 1910.120(q) may be applicable.

Fire Extinguishing: This chemical is a flammable liquid. Poisonous gases are produced in fire. Use dry chemical, carbon dioxide, or alcohol foam extinguishers. Vapors are heavier than air and will collect in low areas. Vapors may travel long distances to ignition sources and flashback. Vapors in confined areas may explode when exposed to fire. Containers may explode in fire. Storage containers and parts of containers may rocket great distances, in many directions. If material or contaminated runoff enters waterways, notify downstream users of potentially contaminated waters. Notify local health and fire officials and pollution control agencies. From a secure, explosion-proof location, use water spray to cool exposed containers. If cooling streams are ineffective (venting sound increases in volume and pitch, tank discolors, or shows any signs of deforming), withdraw immediately to a secure position. If employees are expected to fight fires, they must be trained and equipped in OSHA 1910.156. The only respirators recommended for firefighting are self-contained breathing apparatuses that have full face-pieces and are operated in a pressure-demand or other positive-pressure mode.

Disposal Method Suggested: Consult with environmental regulatory agencies for guidance on acceptable disposal practices. Generators of waste containing this contaminant (≥100 kg/mo) must conform with EPA regulations governing storage, transportation, treatment, and waste disposal. Incineration.

References

National Institute for Occupational Safety and Health. (1975). *Criteria for a Recommended Standard: Occupational Exposure to Xylene*, NIOSH Document No. 75-168

Sax, N. I. (Ed.). *Dangerous Properties of Industrial Materials Report*, 1, No. 7, 79–81 (1981) (*meta*-); 3, No. 3, 88–92 (1983); 4 No. 5, 75–88 (*para*-); and 4, No. 5, 63–75 (1984) (*ortho*-)

New York State Department of Health. (March 1986). *Chemical Fact Sheet: Xylenes.* Version 3. Albany, NY: Bureau of Toxic Substance Assessment

New Jersey Department of Health and Senior Services. (August 2006). *Hazardous Substances Fact Sheet: Xylenes.* Trenton, NJ

m-Xylene-α,α′-diamine   X:0110

Molecular Formula: $C_8H_{12}N_2$
Common Formula: $H_2NCH_2C_6H_4CH_2NH_2$
Synonyms: 1,3-Benzenedimethanamine; 1,3-Bis(amino-methyl)benzene; Methylamine, *m*-phenylenebis-; MXDA; *m*-Phenylenebis(methylamine); *m*-Xylylenediamine
CAS Registry Number: 1477-55-0
RTECS® Number: PF8970000
UN/NA & ERG Number: Not regulated.
EC Number: 216-032-5
Regulatory Authority and Advisory Bodies
Air Pollutant Standard Set. See below, "Permissible Exposure Limits in Air" section.
Canada, WHMIS, Ingredients Disclosure List Concentration 1.0%.
European/International Regulations: not listed in Annex 1.
WGK (German Aquatic Hazard Class): 2—Hazard to waters.
Description: *m*-Xylene-α,α′-diamine is a colorless liquid. Molecular weight = 136.22; Specific gravity ($H_2O:1$) = 1.032 at 25°C; Boiling point = 247.2°C; Freezing/Melting point = 15°C; Vapor pressure = 0.03 mmHg at 25°C; Flash point = 117°C. Hazard Identification (based on NFPA-704 M Rating System): Health 3, Flammability 0, Reactivity 0. Soluble in water.
Potential Exposure: Compound Description: Primary Irritant. *m*-Xylene-α,α′-diamine is a source of *m*-xylene diisocyanate and used as an intermediate in the manufacture of epoxy and polyamide resins.
Incompatibilities: Oxidizers, strong acids.
Permissible Exposure Limits in Air
OSHA PEL: None.
NIOSH REL: 0.1 mg/m³ Ceiling Concentration [skin].
ACGIH TLV®[1]: 0.1 mg/m³ Ceiling Concentration [skin]. No TEEL available.
DFG MAK: [skin] Danger of skin sensitization.
Australia: TWA 1 mg/m³, 1993; Belgium: STEL 0.1 mg/m³, 1993; Denmark: TWA 0.02 ppm (0.1 mg/m³), [skin], 1999; Finland: TWA 0.1 mg/m³, [skin], 1999; France: VLE 0.1 mg/m³, 1999; Norway: TWA 0.1 mg/m³, 1999; Switzerland: MAK-W 0.1 mg/m³, [skin], 1999; the Netherlands: MAC-0.1 mg/m³, 2003; Argentina, Bulgaria, Columbia, Jordan, South Korea, New Zealand, Singapore, Vietnam: ACGIH TLV®: Ceiling Concentration 0.1 mg/m³ [skin].
Several states have set guidelines or standards for MXDA in ambient air[60] ranging from 0.008 μg/m³ (Virginia) to 1.0 μg/m³ (North Dakota) to 2.0 μg/m³ (Nevada) to 33.0 μg/m³ (New York).

Determination in Air: Use OSHA Analytical Method 105
Permissible Concentration in Water: No criteria set.
Routes of Entry: Inhalation, skin absorption, ingestion, skin and/or eye contact.
Harmful Effects and Symptoms
Short Term Exposure: *m*-Xylene-α,α′-diamine can affect you when breathed in and by passing through your skin. Contact can severely irritate and burn the skin and eyes. Exposure can irritate the nose and throat. High levels may cause respiratory depression, tiredness, and unconsciousness. The LC_{50} (rat) = 700 ppm/1 h.
Long Term Exposure: *m*-Xylene-α,α′-diamine may cause skin and lung allergies. Once this happens, even very small future exposures may cause a skin rash to develop and/or may cause an asthma-like reaction with wheezing, coughing, and shortness of breath. May cause liver and kidney damage. In animals: liver, kidney, lung damage.
Points of Attack: Eyes, skin, respiratory system, liver, kidneys.
Medical Surveillance: Before beginning employment and at regular times after that, the following are recommended: lung function tests. These may be normal if person is not having an attack at the time of test. If symptoms develop or overexposure is suspected, the following may be useful: liver and kidney function tests. Evaluation by a qualified allergist, including careful exposure history and special testing, may help diagnose skin allergy.
First Aid: If this chemical gets into the eyes, remove any contact lenses at once and irrigate immediately for at least 30 min, occasionally lifting upper and lower lids. Seek medical attention immediately. If this chemical contacts the skin, remove contaminated clothing and wash immediately with soap and water. Seek medical attention immediately. If this chemical has been inhaled, remove from exposure, begin rescue breathing (using universal precautions, including resuscitation mask) if breathing has stopped and CPR if heart action has stopped. Transfer promptly to a medical facility. When this chemical has been swallowed, get medical attention. Give large quantities of water and induce vomiting. Do not make an unconscious person vomit.
Personal Protective Methods: Wear protective gloves and clothing to prevent any reasonable probability of skin contact. Safety equipment suppliers/manufacturers can provide recommendations on the most protective glove/clothing material for your operation. All protective clothing (suits, gloves, footwear, headgear) should be clean, available each day, and put on before work. Contact lenses should not be worn when working with this chemical. Wear splash-proof chemical goggles and face shield unless full face-piece respiratory protection is worn. Employees should wash immediately with soap when skin is wet or contaminated. Provide emergency showers and eyewash.
Respirator Selection: Where there is potential for exposures over *0.1 mg/m³*, use a NIOSH/MSHA- or European Standard EN149-approved supplied-air respirator with a full face-piece operated in the positive-pressure mode, or with a

full face-piece, hood, or helmet in the continuous-flow mode; or use a NIOSH/MSHA- or European Standard EN149-approved self-contained breathing apparatus with a full face-piece operated in a pressure-demand or other positive-pressure mode.

Storage: Color Code—White: Corrosive or Contact Hazard; Color Code—Blue: Health Hazard/Poison: Store in a secure poison location. Store separately in a corrosion-resistant location. Prior to working with this chemical you should be trained on its proper handling and storage. Store in tightly closed containers in a cool, well-ventilated area away from oxidizers (such as perchlorates, peroxides, permanganates, chlorates, and nitrates) and strong acids (such as hydrochloride, sulfuric, and nitric). Store in tightly closed containers in a cool, well-ventilated area away from heat. Sources of ignition, such as smoking and open flames, are prohibited where this chemical is used, handled, or stored in a manner that could create a potential fire or explosion hazard. Metal containers involving the transfer of 5 gallons or more of this chemical should be grounded and bonded. Drums must be equipped with self-closing valves, pressure vacuum bungs, and flame arresters. Use only nonsparking tools and equipment, especially when opening and closing containers of this chemical.

Shipping: Not regulated.

Spill Handling: Evacuate and restrict persons not wearing protective equipment from area of spill or leak until cleanup is complete. Remove all ignition sources. Ventilate area of spill or leak. Absorb liquids in vermiculite, dry sand, earth, peat, carbon, or a similar material and deposit in sealed containers. Keep this chemical out of a confined space, such as a sewer, because of the possibility of an explosion, unless the sewer is designed to prevent the buildup of explosive concentrations. It may be necessary to contain and dispose of this chemical as a hazardous waste. If material or contaminated runoff enters waterways, notify downstream users of potentially contaminated waters. Contact your local or federal environmental protection agency for specific recommendations. If employees are required to clean up spills, they must be properly trained and equipped. OSHA 1910.120(q) may be applicable.

Fire Extinguishing: m-Xylene-α,α'-diamine may burn but does not readily ignite. Use dry chemical, CO_2, water spray, or foam extinguishers. Poisonous gases, including oxides of nitrogen, are produced in fire. Vapors are heavier than air and will collect in low areas. Containers may explode in fire. Storage containers and parts of containers may rocket great distances, in many directions. If material or contaminated runoff enters waterways, notify downstream users of potentially contaminated waters. Notify local health and fire officials and pollution control agencies. From a secure, explosion-proof location, use water spray to cool exposed containers. If cooling streams are ineffective (venting sound increases in volume and pitch, tank discolors, or shows any signs of deforming), withdraw immediately to a secure position. If employees are expected to fight fires, they must be trained and equipped in OSHA 1910.156. The only respirators recommended for firefighting are self-contained breathing apparatuses that have full face-pieces and are operated in a pressure-demand or other positive-pressure mode.

References

New Jersey Department of Health and Senior Services. (December 2006). *Hazardous Substances Fact Sheet: m-Xylene-2,2'-Diamine.* Trenton, NJ

3,5-Xylenol   X:0120

Molecular Formula: $C_8H_{10}O$

Common Formula: $C_6H_3(CH_3)_2OH$

Synonyms: AI3-01553; 3,5-Dimethylphenol; 3,5-DMP; 1-Hydroxy-3,5-dimethylbenzene; Phenol, dimethyl-; *sym, m*-Xylene; 1,3,5-Xylenol

CAS Registry Number: 108-68-9; 1300-71-6 (mixed isomers)

RTECS® Number: ZE6475000

UN/NA & ERG Number: UN2261/153

EC Number: 203-606-5 [*Annex I Index No.:* 604-037-00-9]; 215-089-3 [*Annex I Index No.:* 604-006-00-X] (mixed isomers)

Regulatory Authority and Advisory Bodies

Mixed isomers:

Clean Water Act: Section 311 Hazardous Substances/RQ 40CFR117.3 (same as CERCLA, see below).

Reportable Quantity (RQ): 1000 lb (454 kg).

US DOT Regulated Marine Pollutant (49CFR172.101, Appendix B) as xylenols.

WHMIS Classifications, E, D2B.

European/International Regulations (*xylenols*): Hazard Symbol: T, N; Risk phrases: R24/25; R34; R51/53; Safety phrases: S1/2; S26; S36/3739; S45; S61 (see Appendix 4).

WGK (German Aquatic Hazard Class): 2—Water polluting (*CAS: 108-68-9*).

Description: 3,5-Xylenol is a crystalline solid. Molecular weight = 122.17; Boiling point = 220°C (sublimes); Freezing/Melting point = 64°C; Flash point = 109°C. Soluble in water. Hazard Identification (based on NFPA-704 M Rating System): Health 3, Flammability 2, Reactivity 0. Slightly soluble in water; solubility = 0.5%.

Potential Exposure: Compound Description: Tumorigen; Reproductive Effector; Primary Irritant. 3,5-Xylenol is used as an antioxidant, solvent, plasticizer, wetting agent; and in pharmaceuticals. Xylenols are also used in pesticides, fuel and lubricant additives; as a rubber chemical; in dyestuff manufacture. There are actually six xylenol isomers.

Incompatibilities: A weak organic acid. Keep away from oxidizers.

Permissible Exposure Limits in Air

No standards or TEEL available.

Permissible Concentration in Water: No criteria set.

Determination in Water: Octanol—water coefficient: Log $K_{ow} = 2.35$.

Routes of Entry: Inhalation, skin absorption, ingestion, skin and/or eye contact.

Harmful Effects and Symptoms

Short Term Exposure: Xylenol can affect you when breathed in and by passing through your skin. Can severely irritate the eyes, skin, and respiratory tract. Contact can severely burn the eyes and skin. LD_{50} = (oral-rat) 608 mg/kg.

Long Term Exposure: Repeated exposure may lead to vomiting, headaches, dizziness, and fainting. Chronic poisoning can cause digestive disturbances, nervous disorders, and skin eruptions. Closely related compounds can cause liver and kidney damage, and even collapse and death. Highly irritating substances can affect the lungs although it is not known if xylenols cause lung damage.

Points of Attack: Eyes, skin, respiratory system, liver, kidneys.

Medical Surveillance: If symptoms develop or overexposure is suspected, the following may be useful: Urinary phenol test. Liver and kidney function tests. Lung function tests.

First Aid: Get medical attention at once following exposure to this compound.

Inhalation: Remove patient immediately to fresh air; irritation of nose or throat may be somewhat relieved by spraying or gargling with water until all odor is gone; 100% oxygen inhalation is indicated for cyanosis or respiratory distress; keep patient warm but not hot.

Eyes: Flood with running water for 15 min; if physician is not immediately available, continue irrigation for another 15 min; 2—3 drops of 0.5% pontocaine or equivalent may be instilled after first 15 min; do not use oils or oily ointments unless ordered by physician.

Skin: Wash affected areas with large quantities of water or soapy water until all odor is gone; then wash with alcohol or 20% glycerin solution and more water; keep patient warm but not hot; cover chemical burns continuously with compresses wet with saturated solution of sodium thiosulfate; apply no salves or ointments for 24 h after injury.

Ingestion: Give large quantities of liquid (saltwater, weak sodium bicarbonate solution, milk, or gruel) followed by demulcent, such as raw egg white or corn starch paste; if profuse vomiting does not follow immediately, give a mild emetic (such as 1 tablespoon of mustard in glass of water) or tickle back of throat. Repeat procedure until vomitus is free of the odor. Some demulcent should be left in stomach after vomiting. Keep patient comfortably warm.

Personal Protective Methods: Wear protective gloves and clothing to prevent any reasonable probability of skin contact. Safety equipment suppliers/manufacturers can provide recommendations on the most protective glove/clothing material for your operation. All protective clothing (suits, gloves, footwear, headgear) should be clean, available each day, and put on before work. Contact lenses should not be worn when working with this chemical. Wear dust-proof chemical goggles and face shield unless full face-piece respiratory protection is worn. Employees should wash immediately with soap when skin is wet or contaminated. Provide emergency showers and eyewash.

Respirator Selection: Where there is potential for exposures to xylenols, use a NIOSH/MSHA- or European Standard EN149-approved full-face-piece respirator with a high-efficiency particulate filter. Greater protection is provided by a powered air-purifying respirator. *Where there is potential for high exposures*, use a NIOSH/MSHA- or European Standard EN149-approved supplied-air respirator with a full face-piece operated in the positive-pressure mode, or with a full face-piece, hood, or helmet in the continuous-flow mode; or use a NIOSH/MSHA- or European Standard EN149-approved self-contained breathing apparatus with a full face-piece operated in a pressure-demand or other positive-pressure mode.

Storage: Color Code—Blue: Health Hazard/Poison: Store in a secure poison location. Prior to working with this chemical you should be trained on its proper handling and storage. Store in tightly closed containers in a cool, well-ventilated area away from chemical oxidizers. Sources of ignition, such as smoking and open flames, are prohibited where xylenols is used, handled, or stored in a manner that could create a potential fire or explosion hazard.

Shipping: Xylenols require a shipping label of "POISONOUS/TOXIC MATERIALS." It falls in Hazard Class 6.1 and Packing Group II.

Spill Handling: Evacuate persons not wearing protective equipment from area of spill or leak until cleanup is complete. Remove all ignition sources. Collect powdered material in the most convenient and safe manner and deposit in sealed containers. Ventilate area after cleanup is complete. It may be necessary to contain and dispose of this chemical as a hazardous waste. If material or contaminated runoff enters waterways, notify downstream users of potentially contaminated waters. Contact your local or federal environmental protection agency for specific recommendations. If employees are required to clean up spills, they must be properly trained and equipped. OSHA 1910. 120(q) may be applicable.

Fire Extinguishing: This chemical is a combustible solid. Use dry chemical, carbon dioxide, water spray, or alcohol foam extinguishers. Poisonous gases are produced in fire. If material or contaminated runoff enters waterways, notify downstream users of potentially contaminated waters. Notify local health and fire officials and pollution control agencies. From a secure, explosion-proof location, use water spray to cool exposed containers. If cooling streams are ineffective (venting sound increases in volume and pitch, tank discolors, or shows any signs of deforming), withdraw immediately to a secure position. If employees are expected to fight fires, they must be trained and equipped in OSHA 1910.156. The only respirators recommended for firefighting are self-contained breathing apparatuses that have full face-pieces and are operated in a pressure-demand or other positive-pressure mode.

Disposal Method Suggested: Dissolve or mix the material with a combustible solvent and burn in a chemical incinerator equipped with an afterburner and scrubber. All federal, state, and local environmental regulations must be observed.

References

Sax, N. I. (Ed.). *Dangerous Properties of Industrial Materials Report*, 1, No. 7, 81–82 (1981) and 4, No. 1, 102–106 (1984)

New Jersey Department of Health and Senior Services. (February 2001). *Hazardous Substances Fact Sheet: Xylenol*. Trenton, NJ

Xylidines   X:0130

Molecular Formula: $C_8H_{11}N$

Common Formula: $(CH_3)_2C_6H_3NH_2$

Synonyms: *mixed isomers:* Aminodimethylbenzene; Aminoxylene; Dimethylaminobenzene; Dimethylaniline; Xylidine isomers

2,6-isomer: 1-Amino-2,6-dimethylbenzene; 2-Amino-1,3-dimethylbenzene; 2-Amino-*m*-xylene; 2-Amino-1,3-xylene; Benzenamine, 2,6-dimethyl-; 2,6-Dimethylaniline; 2,6-Dimethylbenzenamine; 2,6-Dimethylphenylamine; 2,6-Xilidina (Spanish); *o*-Xylidine; 2,6-Xylidine; 2,6-Xylylamine

Note: Dimethylaniline is also used as a synonym for *N,N*-Dimethylaniline

CAS Registry Number: 1300-73-8 (mixed isomers); 87-62-7 (2,6-); 87-59-2 (2,3-); 95-68-1 (2,4-); 95-78-3 (2,5-); 95-64-7 (3,4-); 108-69-0 (3,5-)

RTECS® Number: ZE8575000 (mixed isomers); ZE9275000 (2,6-)

UN/NA & ERG Number: UN1711/153

EC Number: 215-091-4 (mixed isomers); 201-758-7 [*Annex I Index No.:* 612-161-00-X] (2,6-); 201-755-0 (2,3-); 202-440-0 (2,4-); 202-451-0 (2,5-); 202-437-4 (3,4-); 203-607-0 (3,5-)

EC Number: [*Annex I Index No.:* 612-027-00-0]

Regulatory Authority and Advisory Bodies

Air Pollutant Standard Set. See below, "Permissible Exposure Limits in Air" section.

2,6-isomer*:

Carcinogenicity: NCI: Carcinogenesis Studies (feed); clear evidence: rat; IARC: Human Inadequate Evidence; Animal Sufficient Evidence, *possibly carcinogenic to humans*, Group 2B, 1993.

EPCRA Section 313 Form R *de minimis* concentration reporting level: 1.0%.

California Proposition 65 Chemical: Cancer 1/1/91 (2,6-).

Canada, WHMIS, Ingredients Disclosure List Concentration 1.0%, xylidine (mixed isomers).

*The 2,6-isomer is the only xylidine specifically regulated by the US EPA.

European/International Regulations (*87-62-7*): Hazard Symbol: T, N; Risk phrases: R23/24/25; R33; R51/53; Safety phrases: S1/2; S28; S36/37; S45; S61 (see Appendix 4).

WGK (German Aquatic Hazard Class): 3—Highly water polluting (*mixed isomers, dimethylaniline, and 3,5-isomer*). 2 for the (2,6-, 2,3-, 2,4-, 2,5-, and 3,4-isomers).

Description: There are six xylidine isomers. Xylidine, mixed isomers (principally made up of 2,4-, 2,5-, and 2,6-isomers), is a pale yellow to brown liquid with a weak, aromatic amine odor. The odor threshold is 0.056 ppm. Molecular weight = 121.20; Specific gravity (H_2O:1) = 0.98 at 25°C; Boiling point (mixed isomers) = 213–226°C; Freezing/Melting point = −36°C; Vapor pressure = <1 mmHg at 25°C; Flash point (2,6-isomer) = 91°C; (mixed isomers) 96.7°C. Explosive limits (2,6-isomer): LEL = 1.0%; UEL—unknown. Hazard Identification (based on NFPA-704 M Rating System): Health 3, Flammability 1, Reactivity 0. Insoluble in water.

Potential Exposure: Compound Description (2,6-isomer): Tumorigen, Mutagen, Natural Product. Xylidines are used in dyestuff manufacture; as intermediates in the manufacture of pesticides, antioxidants, pharmaceuticals, and other organic compounds.

Incompatibilities: Strong oxidizers. Contact with hypochlorite salts and bleaches form explosive chloroamines.

Permissible Exposure Limits in Air

Conversion factor: 1 ppm = 4.96 mg/m³ at 25°C & 1 atm.

OSHA PEL: 5 ppm/25 mg/m³ TWA [skin].

NIOSH REL: 2 ppm/10 mg/m³ TWA [skin].

ACGIH TLV®[1]: 0.5 ppm/2.5 mg/m³, inhalable fraction and vapor [skin], confirmed animal carcinogen with unknown relevance to humans; BEI_M issued for methemoglobin inducers.

NIOSH IDLH: 50 ppm.

DFG MAK: (*2,3-, 2,5-, 3,4-, 3,5-isomers*): [skin]; Carcinogen Category 3A; (*2,4- & 2,6-isomers*): [skin]; Carcinogen Category 2.

2,3-isomer

Protective Action Criteria (PAC)

TEEL-0: 0.5 ppm

PAC-1: 10 ppm

PAC-2: 10 ppm

PAC-3: 50 ppm

2,6-isomer

Protective Action Criteria (PAC)

TEEL-0: 0.5 ppm

PAC-1: 10 ppm

PAC-2: 25 ppm

PAC-3: 50 ppm

Denmark: TWA 0.5 ppm (2.5 mg/m³), [skin], 1999; France: VME 2 ppm (10 mg/m³), [skin], 1999; Poland: MAC (TWA) 10 mg/m³, 1999; Switzerland: MAK-W 2 ppm (10 mg/m³), [skin], 1999; United Kingdom: TWA 2 ppm (10 mg/m³), STEL 10 ppm, [skin], 2000.

Mixed isomers

ACGIH TLV®[1]: 0.5 ppm, inhalable fraction and vapor [skin], confirmed animal carcinogen with unknown relevance to humans; BEI issued (1999).

Protective Action Criteria (PAC)
TEEL-0: 0.5 ppm
PAC-1: 10 ppm
PAC-2: 10 ppm
PAC-3: 50 ppm
DFG MAK: [skin] Carcinogen Category 3A.
Arab Republic of Egypt: TWA 5 ppm (25 mg/m^3), [skin], 1993; Australia: TWA 2 ppm (10 mg/m^3), [skin], 1993; Austria: MAK 5 ppm (25 mg/m^3), [skin], 1999; Belgium: TWA 2 ppm (9.9 mg/m^3), [skin], 1993; Denmark: TWA 0.5 ppm (2.5 mg/m^3), [skin], 1999; Finland: TWA 5 ppm (25 mg/m^3), STEL 10 ppm (50 mg/m^3), [skin], 1999; France: VME 2 ppm (10 mg/m^3), [skin], 1999; the Netherlands: MAC-TGG 2.5 mg/m^3, 2003; Norway: TWA 1 ppm (5 mg/m^3), 1999; the Philippines: TWA 100 ppm (435 mg/m^3), [skin], 1993; Poland: MAC (TWA) 10 mg/m^3, 1999; Russia: STEL 3 mg/m^3, [skin], 1993; Switzerland: MAK-W 2 ppm (10 mg/m^3), [skin], 1999; Turkey: TWA 5 ppm (25 mg/m^3), [skin], 1993; United Kingdom: TWA 2 ppm (10 mg/m^3), STEL 10 ppm, [skin], 2000; Argentina, Bulgaria, Columbia, Jordan, South Korea, New Zealand, Singapore, Vietnam: ACGIH TLV®: confirmed animal carcinogen with unknown relevance to humans. The Czech Republic has set a TWA of 5.0 mg/m^3. Several states have set guidelines or standards for xylidine in ambient air[60] ranging from 1.6 µg/m^3 (Virginia) to 33.3 µg/m^3 (New York) to 50.0 µg/m^3 (South Carolina) to 100.0 µg/m^3 (Connecticut, Florida, and North Dakota) to 238.0 µg/m^3 (Nevada).

Determination in Air: Use NIOSH Analytical Method #2002, amines, aromatic.

Permissible Concentration in Water: No criteria set, but EPA[32] has suggested a permissible ambient concentration of 345 µg/L.

Determination in Water: Octanol–water coefficient: Log K_{ow} = (estimated) <2.0 (mixed isomers).

Routes of Entry: Inhalation, skin absorption, ingestion, skin and/or eye contact.

Harmful Effects and Symptoms

Short Term Exposure: Xylidine can affect you when breathed in and by passing through your skin. The effects of exposure may be delayed in case of skin absorption. Irritates the eyes, skin, and respiratory tract. Exposure can cause the formation of methemoglobin causing interference with the ability of the blood to carry oxygen, causing headaches, dizziness, nausea, vomiting, and cyanosis (a bluish color to the skin and lips). Higher levels can cause trouble breathing, collapse, and even death. High or repeated exposure may damage the liver.

Long Term Exposure: May cause anemia. Affects the kidneys, liver, and blood. 2,6-Xylidine has been shown to cause cancer in animals.

Points of Attack: Respiratory system, blood, liver, kidneys, cardiovascular system. Cancer site in animals: nose.

Medical Surveillance: US DHHS PHS CDC NIOSH, and US DOL OSHA list the following tests: whole blood (chemical/metabolite), methemoglobin; complete blood count. Liver and kidney function tests.

First Aid: If this chemical gets into the eyes, remove any contact lenses at once and irrigate immediately for at least 15 min, occasionally lifting upper and lower lids. Seek medical attention immediately. If this chemical contacts the skin, remove contaminated clothing and wash immediately with soap and water. Seek medical attention immediately. If this chemical has been inhaled, remove from exposure, begin rescue breathing (using universal precautions, including resuscitation mask) if breathing has stopped and CPR if heart action has stopped. Transfer promptly to a medical facility. When this chemical has been swallowed, get medical attention. Give large quantities of water and induce vomiting. Do not make an unconscious person vomit.

Note to physician: Treat for methemoglobinemia. Spectrophotometry may be required for precise determination of levels of methemoglobin in urine. If symptoms of serious cyanosis develop, *methylene blue* may be given as an antidote (by a trained medical person only), over 5 min. Repeat in 1 h if not improving. 100% oxygen can be given only by a trained person.

Personal Protective Methods: Wear protective gloves and clothing to prevent any reasonable probability of skin contact. Safety equipment suppliers/manufacturers can provide recommendations on the most protective glove/clothing material for your operation. All protective clothing (suits, gloves, footwear, headgear) should be clean, available each day, and put on before work. Contact lenses should not be worn when working with this chemical. Wear splash-proof chemical goggles and face shield unless full face-piece respiratory protection is worn. Employees should wash immediately with soap when skin is wet or contaminated. Provide emergency showers and eyewash.

Respirator Selection: NIOSH: *20 ppm:* CcrOv (APF = 10) [any chemical cartridge respirator with organic vapor cartridge(s)]; Sa (APF = 10) (any supplied-air respirator). *50 ppm:* Sa:Cf (APF = 25) (any supplied-air respirator operated in a continuous-flow mode) or CcrFOv (APF = 50) [any chemical cartridge respirator with a full face-piece and organic vapor cartridge(s)] or GmFOv (APF = 50) [any air-purifying, full face-piece respirator (gas mask) with a chin-style, front- or back-mounted organic vapor canister] or PaprOv (APF = 25) [any powered, air-purifying respirator with an organic vapor cartridge(s)] or SCBAF (APF = 50) (any self-contained breathing apparatus with a full face-piece) or SaF (APF = 50) (any supplied-air respirator with a full face-piece). *Emergency or planned entry into unknown concentrations or IDLH conditions:* SCBAF: Pd,Pp (APF = 10,000) (any self-contained breathing apparatus that has a full face-piece and is operated in a pressure-demand or other positive-pressure mode) or SaF: Pd,Pp: ASCBA (APF = 10,000) (any supplied-air respirator that has a full face-piece and is operated in a pressure-demand or other positive-pressure mode in combination with an auxiliary, self-contained breathing apparatus operated in a pressure-demand or other positive-pressure mode). *Escape:* GmFOv (APF = 50) [any air-purifying, full-face-piece respirator (gas mask) with a chin-style, front- or back-mounted organic vapor canister] or SCBAE (any appropriate escape-type, self-contained breathing apparatus).

Storage: Color Code—Blue: Health Hazard/Poison: Store in a secure poison location. Prior to working with this chemical you should be trained on its proper handling and storage. Before entering confined space where this chemical may be present, check to make sure that an explosive concentration does not exist. Xylidine must be stored to avoid contact with strong oxidizers (such as bromine, chlorine, or fluorine) since violent reactions occur. Contact with hypochlorite bleaches may form explosive chloro-amines. Store in tightly closed containers in a cool, dry, well-ventilated area away from heat sources. Sources of ignition, such as smoking and open flames, are prohibited where this chemical is used, handled, or stored in a manner that could create a potential fire or explosion hazard. Metal containers involving the transfer of 5 gallons or more of this chemical should be grounded and bonded. Drums must be equipped with self-closing valves, pressure vacuum bungs, and flame arresters. Use only nonsparking tools and equipment, especially when opening and closing containers of this chemical. A regulated, marked area should be established where this chemical is handled, used, or stored in compliance with OSHA Standard 1910.1045.

Shipping: Xylidines, solid require a shipping label of "POISONOUS/TOXIC MATERIALS." It falls in Hazard Class 6.1 and Packing Group II.

Spill Handling: Evacuate and restrict persons not wearing protective equipment from area of spill or leak until cleanup is complete. Remove all ignition sources. Establish forced ventilation to keep levels below explosive limit. Absorb liquids in vermiculite, dry sand, earth, peat, carbon, or a similar material and deposit in sealed containers. Keep this chemical out of a confined space, such as a sewer, because of the possibility of an explosion, unless the sewer is designed to prevent the buildup of explosive concentrations.

It may be necessary to contain and dispose of this chemical as a hazardous waste. If material or contaminated runoff enters waterways, notify downstream users of potentially contaminated waters. Contact your local or federal environmental protection agency for specific recommendations. If employees are required to clean up spills, they must be properly trained and equipped. OSHA 1910.120(q) may be applicable.

Fire Extinguishing: This chemical is a combustible liquid. Poisonous gases, including nitrogen oxides, are produced in fire. Use dry chemical, carbon dioxide, or alcohol foam extinguishers. Vapors are heavier than air and will collect in low areas. Containers may explode in fire. Storage containers and parts of containers may rocket great distances, in many directions. If material or contaminated runoff enters waterways, notify downstream users of potentially contaminated waters. Notify local health and fire officials and pollution control agencies. From a secure, explosion-proof location, use water spray to cool exposed containers. If cooling streams are ineffective (venting sound increases in volume and pitch, tank discolors, or shows any signs of deforming), withdraw immediately to a secure position. If employees are expected to fight fires, they must be trained and equipped in OSHA 1910.156. The only respirators recommended for firefighting are self-contained breathing apparatuses that have full face-pieces and are operated in a pressure-demand or other positive-pressure mode.

Disposal Method Suggested: Incineration; oxides of nitrogen are removed from the effluent gas by scrubber, catalytic or thermal device.

References

New Jersey Department of Health and Senior Services. (October 1986). *Hazardous Substances Fact Sheet: Xylidine*. Trenton, NJ

New Jersey Department of Health and Senior Services. (May 2006). *Hazardous Substances Fact Sheet: 2,6-Xylidine*. Trenton, NJ

Y

Yttrium & compounds   Y:0100

Molecular Formula: Y

Synonyms: Yttria; Yttrium-89; Yttrium, elemental

other yttrium compounds: Yttrium chloride 10361-92-9; Yttrium chloride, hexahydrate 10025-94-2; Yttrium nitrate 10361-93-0; Yttrium oxide 1314-36-9

CAS Registry Number: 7440-65-5

RTECS®Number: ZG2980000; ZG3150000 (chloride); ZG3675000 (nitrate); ZG3850000 (oxide)

UN/NA & ERG Number: UN3089 (metal powder, flammable, n.o.s.)/170

EC Number: 231-174-8; 233-801-0 (chloride); 233-802-6 (nitrate); 215-233-5 (oxide)

Regulatory Authority and Advisory Bodies

Air Pollutant Standard Set. See below, "Permissible Exposure Limits in Air" section.

Canada, WHMIS, Ingredients Disclosure List Concentration 1.0%, elemental and compounds.

European/International Regulations: not listed in Annex 1.

WGK (German Aquatic Hazard Class): No value assigned (metal); 1 Slightly water polluting (oxide)

Description: Yttrium is a silvery-white to dark-gray, or black solid. Odorless. An element in Group III-B of the Periodic Table. It is very similar to the rare earth metals. Molecular weight = 88.91; Specific gravity (H_2O:1) = 4.47 @ 25°C; Boiling point = 2927°C; Freezing/Melting point = 1509°C. Soluble in hot water.

Potential Exposure: Yttrium is used in iron and other alloys, in incandescent gas mantles, and as a deoxidizer for metals. Yttrium metal has a low cross section for neutron capture and is very stable at high temperatures. Further, it is very inert toward liquid uranium and many liquid uranium alloys. Thus, it may well have applications in nuclear power generation. The metal is usually prepared by reduction of the halide with an active metal, such as calcium. To identify and analyze this element, x-ray fluorescence spectroscopy is commonly employed.

Incompatibilities: Flammable in the form of dust in air. Reacts with oxidizers, halogens. Yttrium nitrate: combustible materials.

Permissible Exposure Limits in Air

Note: applies to yttrium and compounds.

OSHA PEL: 1 mg[Y]/m^3 TWA

NIOSH REL: 1 mg[Y]/m^3 TWA

ACGIH TLV®[1]: 1 mg[Y]/m^3 TWA

Protective Action Criteria (PAC)

NIOSH IDLH: 500 mg[Y]/m^3

metal

TEEL-0: 1 mg/m^3

PAC-1: 3 mg/m^3

PAC-2: 5 mg/m^3

PAC-3: 500 mg/m^3

chloride, hexahydrate

TEEL-0: 3.41 mg/m^3

PAC-1: 10.2 mg/m^3

PAC-2: 500 mg/m^3

PAC-3: 500 mg/m^3

oxide

TEEL-0: 1.18 mg/m^3

PAC-1: 3.54 mg/m^3

PAC-2: 150 mg/m^3

PAC-3: 500 mg/m^3

trioxide

TEEL-0: 1.27 mg/m^3

PAC-1: 3.81 mg/m^3

PAC-2: 40 mg/m^3

PAC-3: 500 mg/m^3

Australia: TWA 1 mg/m^3, 1993; Austria: MAK 1 mg/m^3, 1999; Belgium: TWA 1 mg/m^3, 1993; Denmark: TWA 1 mg/m^3, 1999; Finland: TWA 1 mg/m^3, 1999; France: VME 1 mg/m^3, 1999; the Netherlands: MAC-TGG 1 mg/m^3, 2003; the Philippines: TWA 5 ppm (25 mg/m^3), 1993; Poland: MAC (TWA) 1 mg/m^3, 1999; Switzerland: MAK-W 1 mg/m^3, 1999; United Kingdom: TWA 1 mg/m^3, STEL 3 mg/m^3, 2000; Argentina, Bulgaria, Columbia, Jordan, South Korea, New Zealand, Singapore, Vietnam: ACGIH TLV®: TWA 1 mg[Y]/m^3. Several states have set guidelines or standards for yttrium in ambient air[60] ranging from 0.16 μg/m^3 (Virginia); to 10.0 μg/m^3 (North Dakota); to 20.0 μg/m^3 (Connecticut); to 24.0 μg/m^3 (Nevada).

Determination in Air: Use NIOSH Analytical Method (IV) #7300, #7301, #7303, #9102, OSHA Analytical Method ID-121.

Permissible Concentration in Water: No criteria set.

Routes of Entry: Inhalation of dusts, ingestion; skin and/or eye contact.

Harmful Effects and Symptoms

Short Term Exposure: Eye irritation in humans.

Long Term Exposure: In animals: pulmonary irritation; eye injury; possible liver damage.

Points of Attack: Eyes, respiratory system; liver

Medical Surveillance: Consider the points of attack in pre-placement and periodic physical examinations. Liver function tests.

First Aid: If this chemical gets into the eyes, remove any contact lenses at once and irrigate immediately for at least 15 minutes, occasionally lifting upper and lower lids. Seek medical attention immediately. If this chemical contacts the skin, remove contaminated clothing and wash immediately with soap and water. Seek medical attention immediately. If this chemical has been inhaled, remove from exposure, begin rescue breathing (using universal precautions, including resuscitation mask) if breathing has stopped and CPR if heart action has stopped. Transfer promptly to a medical facility. When this chemical has been swallowed, get

medical attention. Give large quantities of water and induce vomiting. Do not make an unconscious person vomit.

Personal Protective Methods: Wear protective gloves and clothing to prevent any reasonable probability of skin contact. Safety equipment suppliers/manufacturers can provide recommendations on the most protective glove/clothing material for your operation. All protective clothing (suits, gloves, footwear, headgear) should be clean, available each day, and put on before work. Contact lenses should not be worn when working with this chemical. Wear dust-proof chemical goggles and face shield unless full-face-piece respiratory protection is worn. Employees should wash immediately with soap when skin is wet or contaminated. Provide emergency showers and eyewash.

Respirator Selection: *Up to 5 mg/m³:* Qm (APF = 25) (any quarter-mask respirator). *Up to 10 mg/m³:* 95XQ (APF = 10) [any particulate respirator equipped with an N95, R95, or P95 filter (including N95, R95, and P95 filtering face-pieces) except quarter-mask respirators. The following filters may also be used: N99, R99, P99, N100, R100, P100]; or Sa (APF = 10) (any supplied-air respirator). *Up to 25 mg/m³:* Sa: Cf (APF = 25) (any supplied-air respirator operated in a continuous-flow mode); or PaprHie (APF = 25) (any powered, air-purifying respirator with a high-efficiency particulate filter). *Up to 50 mg/m³:* 100F (APF = 50) (any air purifying, full face-piece respirator with and N100, R100, or P100 filter); or SaT: Cf (APF = 50) (any supplied-air respirator that has a tight-fitting face-piece and is operated in a continuous-flow mode); or PaprTHie (APF = 50) (any powered, air-purifying respirator with a tight-fitting face-piece and a high-efficiency particulate filter); or SCBAF (APF = 50) (any self-contained breathing apparatus with a full face-piece); or SaF (APF = 50) (any supplied-air respirator with a full face-piece). *Up to 500 mg/m³:* Sa: Pd,Pp (APF = 1000) (any supplied-air respirator operated in a pressure-demand or other positive-pressure mode). *Emergency or planned entry into unknown concentrations or IDLH conditions:* SCBAF: Pd,Pp (APF = 10,000) (any NIOSH/MSHA- or European Standard EN 149-approved self-contained breathing apparatus that has a full face-piece and is operated in a pressure-demand or other positive-pressure mode); or SaF: Pd,Pp: ASCBA (APF = 10,000) (any supplied-air respirator that has a full face-piece and is operated in a pressure-demand or other positive-pressure mode in combination with an auxiliary, self-contained breathing apparatus operated in a pressure-demand or other positive pressure mode). *Escape:* 100F (APF = 50) (any air purifying, full-face-piece respirator with and N100, R100, or P100 filter); or SCBAE (any appropriate escape-type, self-contained breathing apparatus).

Storage: Color Code—Red (*powder*): Flammability Hazard: Store in a flammable materials storage area. Prior to working with this chemical you should be trained on its proper handling and storage. Store in tightly closed containers in a cool, well-ventilated area away from oxidizers. Where possible, automatically transfer material from storage containers to process containers.

Shipping: Flammable powder requires a shipping label of "FLAMMABLE SOLID." It falls in Hazard Class 4.1 and Packing Group III.

Spill Handling: Evacuate persons not wearing protective equipment from area of spill or leak until clean-up is complete. Remove all ignition sources. Collect powdered material in the most convenient and safe manner and deposit in sealed containers. Ventilate area after cleanup is complete. It may be necessary to contain and dispose of this chemical as a hazardous waste. If material or contaminated runoff enters waterways, notify downstream users of potentially contaminated waters. Contact your local or federal environmental protection agency for specific recommendations. If employees are required to cleanup spills, they must be properly trained and equipped. OSHA 1910.120(q) may be applicable.

Fire Extinguishing: This chemical is a noncombustible solid in bulk form. Use dry chemical, carbon dioxide; water spray; or foam extinguishers. Poisonous gases are produced in fire. If material or contaminated runoff enters waterways, notify downstream users of potentially contaminated waters. Notify local health and fire officials and pollution control agencies. From a secure, explosion-proof location, use water spray to cool exposed containers. If cooling streams are ineffective (venting sound increases in volume and pitch, tank discolors, or shows any signs of deforming), withdraw immediately to a secure position. If employees are expected to fight fires, they must be trained and equipped in OSHA 1910.156. The only respirators recommended for fire fighting are self-contained breathing apparatuses that have full face-pieces and are operated in a pressure-demand or other positive-pressure mode.

Disposal Method Suggested: Recovery is indicated wherever possible. Specifically, processes are available for yttrium oxysulfide recovery from color television tube manufacture.

Reference
New Jersey Department of Health and Senior Services. (October 2002). *Hazardous Substances Fact Sheet: Yttrium,* Trenton, NJ

Z

Zinc & compounds   Z:0100

Molecular Formula: Zn

Synonyms: Asareo L15; Blue powder; C.I. 77945; C.I. Pigment black 16; C.I. Pigment metal 6; Emanay zinc dust; Jasad; Merrillite (powder); Pasco; Zinc dust; Zinc powder

CAS Registry Number: 7440-66-6

RTECS® Number: ZG8600000

UN/NA & ERG Number: UN1436 (powder or dust)/138; UN1383 (pyrophoric powder)/135

EC Number: 231-175-3 [*Annex I Index No.:* 030-001-00-1]

Regulatory Authority and Advisory Bodies

Carcinogenicity: EPA: Inadequate Information to assess carcinogenic potential; Available data are inadequate for an assessment of human carcinogenic potential.

Air Pollutant Standard Set. See below, "Permissible Exposure Limits in Air" section.

Metal:

US EPA, FIFRA 1998 Status of Pesticides: RED completed.

US EPA Gene-Tox Program, Inconclusive: *In vivo* cytogenetics—human lymphocyte.

Clean Water Act: 40CFR401.15 Section 307 Toxic Pollutants; 40CFR423, Appendix A, Priority Pollutants.

RCRA 40CFR268.48; 61FR15654, Universal Treatment Standards: Wastewater (mg/L), 2.61; Nonwastewater (mg/L), 5.3 TCLP. *Note:* these constituents are not "underlying hazardous constituents" in characteristic wastes, according to the definition at Section 268.2(i).

RCRA 40CFR264, Appendix 9; TSD Facilities Ground Water Monitoring List. Suggested test method(s) (PQL μg/L), total dust 6010 (20); 7950 (50).

Safe Drinking Water Act: SMCL, 5 mg/L; Priority List (55 FR 1470).

Reportable Quantity (RQ): 1000 lb (454 kg).

EPCRA Section 313 Form R *de minimis* concentration reporting level: 1.0%.

Fume or dust:

Clean Water Act: 40CFR401.15 Section 307 Toxic Pollutants; Section 313 Water Priority Chemicals (57FR 41331, 9/9/92).

Reportable Quantity (RQ): 1000 lb (454 kg).

EPCRA Section 313 Form R *de minimis* concentration reporting level: 1.0%.

Canada, WHMIS, Ingredients Disclosure List Concentration zinc metal not listed.

European/International Regulations (dust, pyrophoric): Hazard Symbol: F; Risk phrases: R15; R17; Safety phrases: S2; S7/8; S43 (see Appendix 4).

WGK (German Aquatic Hazard Class): 2—Water polluting (*zinc, grain size < = 1 mm*); Nonwater polluting agent (*metal*).

Description: Zinc is a soft silver-colored metal; the dust is odorless and gray. Molecular weight = 65.37; Boiling point = 908°C; Freezing/Melting point = 420°C; Autoignition temperature = 460°C. Ignition temperature of dust cloud = 600°C; Minimum explosive concentration = 0.48 oz/ft^3.[USBM] Relative explosion hazard of dust: Moderate. Hazard Identification (based on NFPA-704 M Rating System): Health 0, Flammability 2, Reactivity 0; (*powder*) Health 1, Flammability 3, Reactivity 2. The metal is insoluble in water. The dust reacts with water.

Potential Exposure: Compound Description: Tumorigen, Mutagen, Human Data; Primary Irritant. Zinc is used most commonly as a protective coating of other metals. In addition, it is used in alloys, such as bronze and brass, for electrical apparatus in many common goods; and in organic chemical extractions and reductions. Zinc chloride is a primary ingredient in smoke bombs used by military for screening purposes, crowd dispersal, and occasionally in firefighting exercises by both military and civilian communities. In pharmaceuticals, salts of zinc are used as solubilizing agents in many drugs, including insulin.

Incompatibilities: Dust may self-ignite in air. A strong reducing agent. Violent reaction with oxidizers, chromic anhydride, manganese chloride, chlorates, chlorine, and magnesium. Reacts with water and reacts violently with acids, alkali hydroxides, and bases forming highly flammable hydrogen gas. Reacts violently with sulfur, halogenated hydrocarbons, and many other substances, causing fire and explosion hazard.

Permissible Exposure Limits in Air

Protective Action Criteria (PAC)

TEEL-0: 1 mg/m^3

PAC-1: 3 mg/m^3

PAC-2: 240 mg/m^3

PAC-3: 500 mg/m^3

Zinc nitrate (7779-88-6)

TEEL-0: 1 mg/m^3

PAC-1: 3 mg/m^3

PAC-2: 20 mg/m^3

PAC-3: 100 mg/m^3

DFG MAK (*zinc & zinc inorganic compounds*): 0.1 mg/m^3, respirable fraction; Peak Limitation Category I(4); 2 mg/m^3, inhalable fraction (excluding zinc chloride) TWA; Peak Limitation Category II(2).

Arab Republic of Egypt: 0.1 mg/m^3 TWA, 1993.

Several states have set a standard for zinc metal in ambient air[60] ranging from 0.03 μg/m^3 (New York) to 6.55−39.29 μg/m^3 (Montana).

Determination in Air: Use NIOSH Analytical Method (IV) #7300, Elements, #7030 Zinc and compounds (as Zn), #8005, Elements in blood or tissue, #8310 Metals in urine, OSHA Analytical Methods ID-121; ID-125G.

Permissible Concentration in Water: There are a number of standards for *zinc* in water set around the world[35]: EC: 100−500 μg/L, for drinking water; Germany: 2000 μg/L, tor drinking water; Mexico: 10,000 μg/L, for estuaries; Mexico: 10 μg/L, for coastal waters; Russia: 5000 μg/L, for drinking water; Russia: 1000 μg/L, for surface water; Russia: 10 μg/L, in water for fishery purposes; WHO: 5000 μg/L, in water for esthetic quality. The US EPA[6] has set 5 mg/L (5000 μg/L) for the prevention of adverse effects due to the organoleptic properties of zinc. The state of Kansas has also set a drinking water limit of 5 mg/L.[61]

Determination in Water: There are a number of standards for *zinc* in water set around the world[35]: EC: 100–500 μg/L, for drinking water; Germany: 2000 μg/L, for drinking water; Mexico: 10,000 μg/L, for estuaries; Mexico: 10 μg/L, for coastal waters; Russia: 5000 μg/L, for drinking water; Russia: 1000 μg/L, for surface water; Russia: 10 μg/L, in water for fishery purposes; WHO: 5000 μg/L, in water for esthetic quality. The US EPA recommends there be no more than 5 ppm of drinking water because of taste. The state of Kansas has set a drinking water limit of 5 mg/L.[61]

Routes of Entry: Inhalation, ingestion, eye and/or skin contact.

Harmful Effects and Symptoms

Short Term Exposure: Zinc can affect you when breathed in. Zinc dust particles can irritate the eyes. Exposure to solid zinc is not known to cause acute or chronic health effects, but heated zinc may give off zinc oxide fumes that can cause health effects. Metal fragments can scratch the eyes. When zinc is refined, cadmium is released. *Cadmium is a cancer causing agent.* Inhalation of the dust or fume may cause metal fume fever.

Long Term Exposure: Repeated contact with the dust or fume may cause dermatitis. Ingestion of high levels of zinc can cause anemia, pancreas damage, and lower levels of high-density lipoprotein cholesterol (HDL, the good form of cholesterol). It is not known if high levels of zinc affect human reproduction or cause birth defects. Rats that were fed large amounts of zinc became infertile or had small babies. Zinc is an essential element in our diet. Not enough zinc can cause a loss of appetite, a decrease in the sense of taste and smell, slow wound healing, and skin sores, or a damaged immune system. The recommended dietary allowance (RDA) for zinc is 15 mg/day for men; 12 mg/day for women; 10 mg/day for children; 5 mg/day for infants. Harmful health effects generally begin at levels from 10 to 15 times the RDA (in the 100–250 mg/day range).

Points of Attack: Skin, blood, pancreas.

Medical Surveillance: Zinc can be measured in the blood or feces.

First Aid: If this chemical gets into the eyes, remove any contact lenses at once and irrigate immediately for at least 15 min, occasionally lifting upper and lower lids. Seek medical attention immediately. If this chemical contacts the skin, remove contaminated clothing and wash immediately with soap and water. Seek medical attention immediately. If this chemical has been inhaled, remove from exposure, begin rescue breathing (using universal precautions, including resuscitation mask) if breathing has stopped and CPR if

heart action has stopped. Transfer promptly to a medical facility. When this chemical has been swallowed, get medical attention. Give large quantities of water and induce vomiting. Do not make an unconscious person vomit.

Note to physician: For severe poisoning do not use BAL [British Anti-Lewisite, dimercaprol, dithiopropanol (C₃H₈OS₂)] as it is contraindicated or ineffective in poisoning from zinc or cadmium.

Personal Protective Methods: Wear protective gloves and clothing to prevent any reasonable probability of skin contact with the dust. Safety equipment suppliers/manufacturers can provide recommendations on the most protective glove/clothing material for your operation. All protective clothing (suits, gloves, footwear, headgear) should be clean, available each day, and put on before work. Contact lenses should not be worn when working with this chemical. Wear dust-proof chemical goggles and face shield unless full face-piece respiratory protection is worn. Employees should wash immediately with soap when skin is wet or contaminated. Provide emergency showers and eyewash.

Respirator Selection: Where there is potential for exposures to zinc dusts, use a NIOSH/MSHA- or European Standard EN149-approved respirator equipped with particulate (dust/fume/mist) filters. Particulate filters must be checked every day before work for physical damage, such as rips or tears, and replaced as needed. Where there is potential for high exposure, use a NIOSH/MSHA- or European Standard EN149-approved supplied-air respirator with a full face-piece operated in the positive-pressure mode, or with a full face-piece, hood, or helmet in the continuous-flow mode; or use a NIOSH/MSHA- or European Standard EN149-approved self-contained breathing apparatus with a full face-piece operated in a pressure-demand or other positive-pressure mode.

Storage: Color Code—Yellow Stripe (*powder is a strong reducing agent*): Reactivity Hazard; Store separately in an area isolated from flammables, combustibles, or other yellow-coded materials. Prior to working with this chemical you should be trained on its proper handling and storage. Zinc must be stored to avoid contact with chromic anhydride, manganese chloride, chlorates, chlorine, and magnesium since violent reactions occur. Store in tightly closed containers in a cool, well-ventilated area away from water, acids, and alkali hydroxides because flammable hydrogen gas is produced. Sources of ignition, such as smoking and open flames, are prohibited where zinc is used, handled, or stored in a manner that could create a potential fire or explosion hazard.

Shipping: Zinc powder or zinc dust requires a shipping label of "DANGEROUS WHEN WET, SPONTANEOUSLY COMBUSTIBLE." They fall in DOT Hazard Class 4.3.

Pyrophoric metals, n.o.s., or Pyrophoric alloys, n.o.s. require a label of "SPONTANEOUSLY COMBUSTIBLE." They fall in DOT Hazard Class 4.3 and Packing Group I.

Spill Handling: Evacuate persons not wearing protective equipment from area of spill or leak until cleanup is complete. Remove all ignition sources. Collect powdered material in the most convenient and safe manner and deposit in sealed containers. Ventilate area after cleanup is complete. It may be necessary to contain and dispose of this chemical as a hazardous waste. If material or contaminated runoff enters waterways, notify downstream users of potentially contaminated waters. Contact your local or federal environmental protection agency for specific recommendations. If employees are required to clean up spills, they must be properly trained and equipped. OSHA 1910.120(q) may be applicable.

Fire Extinguishing: Zinc is a combustible solid. Use dry chemical, sand, or foam extinguishers. Poisonous gases are produced in fire, including zinc oxides. If material or contaminated runoff enters waterways, notify downstream users of potentially contaminated waters. Notify local health and fire officials and pollution control agencies. From a secure, explosion-proof location, use water spray to cool exposed containers. If cooling streams are ineffective (venting sound increases in volume and pitch, tank discolors, or shows any signs of deforming), withdraw immediately to a secure position. If employees are expected to fight fires, they must be trained and equipped in OSHA 1910.156. The only respirators recommended for firefighting are self-contained breathing apparatuses that have full face-pieces and are operated in a pressure-demand or other positive-pressure mode.

Disposal Method Suggested: Zinc powder should be reclaimed. Unsalvageable waste may be buried in an approved landfill. Leachate should be monitored for zinc content.[22]

References

Sax, N. I. (Ed.). (1981). *Dangerous Properties of Industrial Materials Report*, 1, No. 7, 82–85

US Public Health Service. (December 1988). *Toxicological Profile for Zinc*, Atlanta, GA: Agency for Toxic Substances and Disease Registry

New York State Department of Health. (March 1986). *Chemical Fact Sheet: Zinc*. Albany, NY: Bureau of Toxic Substance Assessment

ATSDR. (September 1995). *Toxicological Fact Sheet: Zinc*. Atlanta, GA: US Department of Health and Human Services, Public Health Service

Note: In the interest of compactness and utility of this work, a few zinc compounds have been selected for inclusion. Data are available on other zinc compounds as follows:

Compounds: Sax, N. I. (Ed.). *Dangerous Properties of Industrial Material Report*
Zinc Acetate(I), No. 7, 88–90 (1981)
Zinc Borate(IV), No. 2, 93–96 (1984)
Zinc Carbonate(IV), No. 2, 98–100 (1984)
Zinc Cyanide(IV), No. 2, 100–102 (1984)
Zinc Dithionite (Zinc Hydrosulfite) 4, No. 1, 108–110 (1984)
Zinc Fluoride(III), No. 6, 83–85 (1983)

Zinc Nitrate(VIII), No. 5, 101–110 (1988)
Zinc Sulfate(V), No. 5, 106–113 (1985)
US Environmental Protection Agency. Special Review and Reregistration Division Office of Pesticide Programs. (1998). *Agency Status of Pesticides in Registration, Reregistration, and Special Review* (Rainbow Report). Washington, DC
New Jersey Department of Health and Senior Services. (October 2005). *Hazardous Substances Fact Sheet: Zinc*. Trenton, NJ
Compounds: New Jersey Fact Sheet
Zinc Acetate (September 2002)
Zinc Borate (September 2002)
Zinc Carbonate (September 2002)
Zinc Chlorate (September 2002)
Zinc Cyanide (December 2000)
Zinc Dithionite (December 2000)
Zinc Fluoride (December 2000)
Zinc Nitrate (November 2000)
Zinc Permanganate (November 2000)
Zinc Peroxide (November 2000)
Zinc Potassium Chromate (September 1998)
Zinc Sulfate (September 1998)

Zinc bromide   Z:0110

Molecular Formula: Br₂Zn
Common Formula: ZnBr₂
Synonyms: Anhydrous zinc bromide; Bromuro de zinc (Spanish); Zinc bromide, anhydrous; Zinc dibromide
CAS Registry Number: 7699-45-8
RTECS® Number: ZH1150000
EC Number: 231-718-4
Regulatory Authority and Advisory Bodies
Clean Water Act: Section 311 Hazardous Substances/RQ 40CFR117.3 (same as CERCLA, see below); 40CFR 401.15 Section 307 Toxic Pollutants; Section 313 Water Priority Chemicals (57FR41331, 9/9/92).
Reportable Quantity (RQ): 1000 lb (454 kg).
EPCRA Section 313 Form R *de minimis* concentration reporting level: 1.0%.
US DOT Regulated Marine Pollutant (49CFR172.101, Appendix B).
European/International Regulations: not listed in Annex 1.
WGK (German Aquatic Hazard Class): No value assigned.
Description: Zinc bromide is an odorless white crystalline solid. Boiling point = 650°C; Freezing/Melting point = 394°C. Hazard Identification (based on NFPA-704 M Rating System): Health 1, Flammability 0, Reactivity 0. Soluble in water.
Potential Exposure: Zinc bromide is used in photography, rayon manufacturing, and medicine.
Incompatibilities: Alkali metals. Violent reaction with metallic sodium or potassium. Store above 32°F/0°C.

Permissible Exposure Limits in Air
Protective Action Criteria (PAC)
TEEL-0: 2 mg/m^3
PAC-1: 6 mg/m^3
PAC-2: 40 mg/m^3
PAC-3: 200 mg/m^3
Neither ACGIH nor OSHA have set standards but the state of New York has set a guideline for ambient air of 3.0 μg/m^3.[60]
DFG MAK (*zinc & zinc inorganic compounds*): 0.1 mg/m^3, respirable fraction; Peak Limitation Category I(4); 2 mg/m^3, inhalable fraction (excluding zinc chloride) TWA; Peak Limitation Category II(2).

Permissible Concentration in Water: There are a number of standards for *zinc* in water set around the world[35]: EC: 100–500 μg/L, for drinking water; Germany: 2000 μg/L, for drinking water; Mexico: 10,000 μg/L, for estuaries; Mexico: 10 μg/L, for coastal waters; Russia: 5000 μg/L, for drinking water; Russia: 1000 μg/L, for surface water; Russia: 10 μg/L, in water for fishery purposes; WHO: 5000 μg/L, in water for esthetic quality. The US EPA[6] has set 5 mg/L (5000 μg/L), for the prevention of adverse effects due to the organoleptic properties of zinc. The state of Kansas has also set a drinking water limit of 5 mg/L.[61]

Routes of Entry: Inhalation, ingestion, eye and/or skin contact.

Harmful Effects and Symptoms
Short Term Exposure: *Inhalation:* Dust levels of 80 mg/m^3 may cause irritation of the mouth, nose, and throat resulting in coughing, wheezing, and difficult breathing. *Skin:* Can cause irritation and burns. *Eyes:* Can cause irritation and burns. *Ingestion:* Dust or solution may cause irritation to the mouth, throat, and digestive tract. Large doses can cause violent vomiting, severe stomach pain, diarrhea, shock, and collapse. Scars may form in the throat and stomach. Long lasting kidney irritation may occur. Less than 1 oz may cause death. Repeated exposure can lead to bromine poisoning, with symptoms of personality changes (such as depression), poor appetite, and confusion.

Long Term Exposure: Can cause headache, personality changes, poor appetite, lethargy, and confusion. Skin rash can occur with repeated exposure. Can irritate the lungs; bronchitis may develop.

Points of Attack: Lungs.

Medical Surveillance: For those with frequent or potentially high exposure, the following are recommended before beginning work and at regular times after that: lung function tests. If symptoms develop or overexposure is suspected, the following may be useful: blood test for bromide.

First Aid: If this chemical gets into the eyes, remove any contact lenses at once and irrigate immediately for at least 15 min, occasionally lifting upper and lower lids. Seek medical attention immediately. If this chemical contacts the skin, remove contaminated clothing and wash immediately with soap and water. Seek medical attention immediately. If

this chemical has been inhaled, remove from exposure, begin rescue breathing (using universal precautions, including resuscitation mask) if breathing has stopped and CPR if heart action has stopped. Transfer promptly to a medical facility. When this chemical has been swallowed, get medical attention. Give large quantities of water and induce vomiting. Do not make an unconscious person vomit.
Note to physician: Administer prompt and complete gastric lavage and demulcents. Observe for gastric perforations and late complications, such as pyloric stenosis. For eye exposure, rinsing with 0.05 molar neutral sodium edetate may help prevent or reverse corneal opacification.
Note to physician: For severe poisoning, *do not* use BAL [British Anti-Lewisite, dimercaprol, dithiopropanol (C$_3$H$_8$OS$_2$)] as it is contraindicated or ineffective in poisoning from zinc.

Personal Protective Methods: Wear protective gloves and clothing to prevent any reasonable probability of skin contact. Safety equipment suppliers/manufacturers can provide recommendations on the most protective glove/clothing material for your operation. All protective clothing (suits, gloves, footwear, headgear) should be clean, available each day, and put on before work. Contact lenses should not be worn when working with this chemical. Wear dust-proof chemical goggles and face shield unless full face-piece respiratory protection is worn. Employees should wash immediately with soap when skin is wet or contaminated. Provide emergency showers and eyewash.

Respirator Selection: Where there is potential for exposures to zinc bromide, use a NIOSH/MSHA- or European Standard EN149-approved full-face-piece respirator with a high-efficiency particulate filter. Greater protection is provided by a powered air-purifying respirator. *Where there is potential for high exposures*, use a NIOSH/MSHA- or European Standard EN149-approved supplied-air respirator with a full face-piece operated in the positive-pressure mode, or with a full face-piece, hood, or helmet in the continuous-flow mode; or use a NIOSH/MSHA- or European Standard EN149-approved self-contained breathing apparatus with a full face-piece operated in a pressure-demand or other positive-pressure mode.

Storage: Color Code—Green: General storage may be used. Prior to working with this chemical you should be trained on its proper handling and storage. Zinc bromide must be stored to avoid contact with metallic sodium or potassium since violent reactions occur. Store in tightly closed containers in a cool, well-ventilated area. Where possible, automatically transfer material from drums or other storage containers to process containers.

Spill Handling: Evacuate persons not wearing protective equipment from area of spill or leak until cleanup is complete. Remove all ignition sources. Collect powdered material in the most convenient and safe manner and deposit in sealed containers. Ventilate area after cleanup is complete. It may be necessary to contain and dispose of this chemical as a hazardous waste. If material or contaminated runoff enters waterways, notify downstream users of potentially contaminated waters. Contact your local or federal environmental protection agency for specific recommendations. If employees are required to clean up spills, they must be properly trained and equipped. OSHA 1910.120(q) may be applicable.

Fire Extinguishing: Extinguish fire using an agent suitable for type of surrounding fire. Zinc bromide itself does not burn. Poisonous gases are produced in fire, including hydrogen bromide and zinc. If material or contaminated runoff enters waterways, notify downstream users of potentially contaminated waters. Notify local health and fire officials and pollution control agencies. From a secure, explosion-proof location, use water spray to cool exposed containers. If cooling streams are ineffective (venting sound increases in volume and pitch, tank discolors, or shows any signs of deforming), withdraw immediately to a secure position.

References

Sax, N. I. (Ed.). (1984). *Dangerous Properties of Industrial Materials Report*, 4, No. 2, 96–98

New York State Department of Health. (March 1986). *Chemical Fact Sheet: Zinc Bromide.* Albany, NY: Bureau of Toxic Substance Assessment

New Jersey Department of Health and Senior Services. (October 1998). *Hazardous Substances Fact Sheet: Zinc Bromide.* Trenton, NJ

Zinc chloride   Z:0120

Molecular Formula: Cl$_2$Zn
Common Formula: ZnCl$_2$
Synonyms: Butter of zinc; Chlorure de zinc (French); Cloruro de zinc (Spanish); Tinning glux; Zinc butter; Zinc chloride, anhydrous; Zinc chloride fume; Zinc (chlorure de) (French); Zinc dichloride; Zinc muriate solution; Zinkchlorid (German)
CAS Registry Number: 7646-85-7
RTECS® Number: ZH1400000
UN/NA & ERG Number: UN2331 (anhydrous)/154; UN1840 (solution)/154
EC Number: 231-592-0 [*Annex I Index No.:* 030-003-00-2]
Regulatory Authority and Advisory Bodies
Carcinogenicity: EPA (*fume*): Inadequate Information to assess carcinogenic potential; US EPA Gene-Tox Program, Positive: Cell transform.—SA7/SHE; Host-mediated assay; Positive: Histidine reversion—Ames test; Negative: *In vitro* cytogenetics—human lymphocyte; Negative: *B. subtilis* rec assay; Sperm morphology—mouse.
FDA—over-the-counter drug.
Air Pollutant Standard Set. See below, "Permissible Exposure Limits in Air" section.
Clean Water Act: Section 311 Hazardous Substances/RQ 40CFR117.3 (same as CERCLA, see below); Clean Water

Act: 40CFR401.15 Section 307 Toxic Pollutants; Section 313 Water Priority Chemicals (57FR41331, 9/9/92). Reportable Quantity (RQ): 1000 lb (454 kg).
EPCRA Section 313 Form R *de minimis* concentration reporting level: 1.0%.
Canada, WHMIS, Ingredients Disclosure List Concentration 1.0%.
European/International Regulations: Hazard Symbol: C, N; Risk phrases: R22; R34; R50/53; Safety phrases: S1/2; S26; S36/37/39; S45; S60; S61 (see Appendix 4).
WGK (German Aquatic Hazard Class): 3—Severe hazard to waters.
Description: Zinc chloride is a white hexagonal, deliquescent crystals or colorless solution. The fume is a white particulate dispersed in air. Molecular weight = 136.27; Specific gravity (H$_2$O:1) = 2.91 at 25°C; Boiling point = 732°C; Freezing/Melting point = 290°C. Hazard Identification (based on NFPA-704 M Rating System): Health 3, Flammability 0, Reactivity 1. Soluble in water; solubility = 435% at 21°C.
Potential Exposure: Compound Description: Agricultural Chemical; Tumorigen, Mutagen; Reproductive Effector; Human Data. Zinc chloride is used in iron galvanizing; as a wood preservative; for dry battery cells; as a soldering flux; in textile finishing; in vulcanized fiber; reclaiming rubber; in oil and gas well operations; oil refining; manufacturing of parchment paper; in dyes; activated carbon; in chemical synthesis; in adhesives; dentists' cement; deodorants, disinfecting, and embalming solutions; and taxidermy. It is also produced by military screening smoke.
Incompatibilities: Aqueous solutions are strongly acidic. Incompatible with bases and potassium. Corrosive to metals.
Permissible Exposure Limits in Air
Fume
OSHA PEL: 1 mg/m^3 TWA.
NIOSH REL: 1 mg/m^3 TWA; 2 mg/m^3 STEL.
ACGIH TLV®[1]: 1 mg/m^3 TWA; 2 mg/m^3 STEL.
NIOSH IDLH: 50 mg/m^3.
Protective Action Criteria (PAC)
TEEL-0: 1 mg/m^3
PAC-1: 2 mg/m^3
PAC-2: 50 mg/m^3
PAC-3: 50 mg/m^3
DFG MAK (*fume*): 0.1 mg/m^3, respirable fraction; Peak Limitation Category I(1); 2 mg/m^3, inhalable fraction TWA; Peak Limitation Category I(4).
Arab Republic of Egypt: TWA 1 mg/m^3 (fume), 1993; Australia: TWA 1 mg/m^3, STEL 2 mg/m^3 (fume), 1993; Belgium: TWA 1 mg/m^3, STEL 2 mg/m^3 (fume), 1993; Denmark: TWA 0.5 mg[Zn]/m^3, 1999; Finland: TWA 1 mg/m^3 (fume), 1999; France: VME 1 mg/m^3 (fume), 1999; Norway: TWA 1 mg/m^3, 1999; the Netherlands: MAC-TGG 1 mg/m^3, 2003; Poland: TWA 1 mg/m^3, STEL 2 mg/m^3, 1999; Sweden: NGV 1 mg/m^3 (resp. dust), 1999; Switzerland: MAK-W 1 mg/m^3 (fume), 1999; Thailand: TWA 1 mg/m^3

(fume), 1993; Turkey: TWA 1 mg/m³ (fume), 1993; United Kingdom: TWA 1 mg/m³, STEL 2 mg/m³, 2000; Argentina, Bulgaria, Columbia, Jordan, South Korea, New Zealand, Singapore, Vietnam: ACGIH TLV®: STEL 2 mg/m³. Several states have set guidelines or standards for zinc chloride in ambient air[60] ranging from 0.16 μg/m³ (Virginia) to 3.3 μg/m³ (New York) to 10.0 μg/m³ (Florida, North Dakota, South Dakota) to 20.0 μg/m³ (Connecticut) to 24.0 μg/m³ (Nevada).

Determination in Air: Use OSHA Analytical Method #ID-121, #ID-125G.

Permissible Concentration in Water: There are a number of standards for *zinc* in water set around the world[35]: EC: 100–500 μg/L, for drinking water; Germany: 2000 μg/L, for drinking water; Mexico: 10,000 μg/L, for estuaries; Mexico: 10 μg/L, for coastal waters; Russia: 5000 μg/L, for drinking water; Russia: 1000 μg/L, for surface water; Russia: 10 μg/L, in water for fishery purposes; WHO: 5000 μg/L, in water for esthetic quality. The US EPA[6] has set 5 mg/L (5000 μg/L) for the prevention of adverse effects due to the organoleptic properties of zinc. The state of Kansas has also set a drinking water limit of 5 mg/L.[61]

Routes of Entry: Inhalation, ingestion, eye and/or skin contact. Absorbed through the skin.

Harmful Effects and Symptoms

Short Term Exposure: Inhalation: Corrosive to the respiratory tract. Higher exposures can cause pulmonary edema, a medical emergency that can be delayed for several hours. This can cause death. Symptoms of exposure can include cough, copious sputum, dyspnea (breathing difficulty), chest pain, bronchopneumonia, pulmonary fibrosis, cor pulmonale, fever, cyanosis, tachypnea. No effects were reported from 30 min exposure to 0.07–0.04 mg/m³. Exposure to dust or fumes above 80 mg/m³ for 2 min may cause nose and throat irritation, cough, chest pain, cyanosis (bluish skin), fever, nausea and vomiting, shortness of breath, difficult breathing, and pneumonia. Breathing difficulties may not appear for several hours. Fume concentrations over 52 mg/m³ may produce symptoms listed above. Fatal accidental exposures have occurred. *Skin:* Corrosive. Dust or solution can cause irritation and chemical burns particularly on areas where skin is broken. *Eyes:* Corrosive. Dust can cause burning irritation. Concentrated solutions are very dangerous to the eyes, causing extreme pain, redness, and swelling. Eye damage may result. *Ingestion:* Dust or solution may be irritating and corrosive to the mouth, throat, and digestive tract. Other symptoms may include stomach pain, nausea, vomiting, bloody diarrhea, swelling of the throat, blood in the urine, and shock. Less than 1 oz has killed an adult although recovery has been reported after ingestion of 4 oz.

Long Term Exposure: Repeated or prolonged contact with skin may cause dermatitis. There is limited evidence that zinc chloride is a teratogen in animals. Repeated exposure can lead to delayed permanent lung damage. Prolonged contact can cause skin burns and ulcers.

Points of Attack: Eyes, skin, respiratory system, cardiovascular system.

Medical Surveillance: NIOSH lists the following tests: Blood gas analysis, whole blood (chemical/metabolite), biologic tissue/biopsy, chest X-ray, electrocardiogram, pulmonary function tests: forced vital capacity, forced expiratory volume (1 s), sputum cytology, urine (chemical/metabolite), white blood cell count/differential.

First Aid: If this chemical gets into the eyes, remove any contact lenses at once and irrigate immediately for at least 15 min, occasionally lifting upper and lower lids. Seek medical attention immediately. If this chemical contacts the skin, remove contaminated clothing and wash immediately with soap and water. Seek medical attention immediately. If this chemical has been inhaled, remove from exposure, begin rescue breathing (using universal precautions, including resuscitation mask) if breathing has stopped and CPR if heart action has stopped. Transfer promptly to a medical facility. When this chemical has been swallowed, get medical attention. If victim is *conscious*, administer water or milk. Do not induce vomiting. Medical observation is recommended for 24–48 h after breathing overexposure, as pulmonary edema may be delayed.

Note to physician: Inhalation: Bronchodilators, decongestants, and oxygen may be used if necessary. Corticosteroids are useful for treating pneumonitis. Ingestion—gastric lavage with 5% sodium bicarbonates; Dimercaprol has been suggested for treatment. *Eyes:* rinsing with 0.05 molar neutral sodium edetate may prevent or reverse corneal opacification.

Note to physician: For severe poisoning *do not* use BAL [British Anti-Lewisite, dimercaprol, dithiopropanol $(C_3H_8OS_2)$] as it is contraindicated or ineffective in poisoning from zinc.

Personal Protective Methods: Wear protective gloves and clothing to prevent any reasonable probability of skin contact. Safety equipment suppliers/manufacturers can provide recommendations on the most protective glove/clothing material for your operation. All protective clothing (suits, gloves, footwear, headgear) should be clean, available each day, and put on before work. Contact lenses should not be worn when working with this chemical. Wear dust-proof goggles and use a face shield when working with powders or dust unless full face-piece respiratory protection is worn. Wear splash-proof chemical goggles and use a face shield when working with liquid unless full-face-piece respiratory protection is worn. Where the fume is present, wear gas-proof goggles and use face shield unless full face-piece respiratory protection is worn. Employees should wash immediately with soap when skin is wet or contaminated. Provide emergency showers and eyewash.

Respirator Selection: 10 mg/m³: 95XQ (APF = 10) [any particulate respirator equipped with an N95, R95, or P95 filter (including N95, R95, and P95 filtering face-pieces) except quarter-mask respirators. The following filters may also be used: N99, R99, P99, N100, R100, P100] or Sa (APF = 10) (any supplied-air respirator). *25 mg/m³:* Sa:Cf

(APF = 25) (any supplied-air respirator operated in a continuous-flow mode) or PaprHie (APF = 25) (any powered, air-purifying respirator with a high-efficiency particulate filter). *50 mg/m³:* 100F (APF = 50) (any air-purifying, full-face-piece respirator with an N100, R100, or P100 filter) or PaprTHie (APF = 50)* (any powered, air-purifying respirator with a tight-fitting face-piece and a high-efficiency particulate filter) or SCBAF (APF = 50) (any self-contained breathing apparatus with a full face-piece) or SaF (APF = 50) (any supplied-air respirator with a full face-piece). *Emergency or planned entry into unknown concentrations or IDLH conditions:* SCBAF: Pd,Pp (APF = 10,000) (any self-contained breathing apparatus that has a full face-piece and is operated in a pressure-demand or other positive-pressure mode) or SaF: Pd,Pp: ASCBA (APF = 10,000) (any supplied-air respirator that has a full face-piece and is operated in a pressure-demand or other positive-pressure mode in combination with an auxiliary self-contained breathing apparatus operated in a pressure-demand or other positive-pressure mode). *Escape:* 100F (APF = 50) (any air-purifying, full-face-piece respirator with an N100, R100, or P100 filter) or SCBAE (any appropriate escape-type, self-contained breathing apparatus).

*Substance reported to cause eye irritation or damage; may require eye protection.

Storage: Color Code—White: Corrosive or Contact Hazard; Store separately in a corrosion-resistant location. Prior to working with this chemical you should be trained on its proper handling and storage. Store in tightly closed containers in a cool, well-ventilated area away from potassium.

Shipping: Zinc chloride, anhydrous or solution, requires a shipping label of "CORROSIVE." It falls in Hazard Class 8 and Packing Group III.

Spill Handling: *Solid:* Evacuate persons not wearing protective equipment from area of spill or leak until cleanup is complete. Remove all ignition sources. Collect powdered material in the most convenient and safe manner and deposit in sealed containers. Wash spill area with soap and water. Ventilate area after cleanup is complete. It may be necessary to contain and dispose of this chemical as a hazardous waste. If material or contaminated runoff enters waterways, notify downstream users of potentially contaminated waters. Contact your local or federal environmental protection agency for specific recommendations. If employees are required to clean up spills, they must be properly trained and equipped. OSHA 1910.120(q) may be applicable. *Liquid:* Evacuate and restrict persons not wearing protective equipment from area of spill or leak until cleanup is complete. Remove all ignition sources. Ventilate area of spill or leak. Absorb liquids in vermiculite, dry sand, earth, peat, carbon, or a similar material and deposit in sealed containers. It may be necessary to contain and dispose of this chemical as a hazardous waste. If material or contaminated runoff enters waterways, notify downstream users of potentially contaminated waters. Contact your local or federal environmental protection agency for specific

recommendations. If employees are required to clean up spills, they must be properly trained and equipped. OSHA 1910.120(q) may be applicable.

Fire Extinguishing: This chemical is noncombustible. Use agent suitable for surrounding fire. Poisonous gases are produced in fire. If material or contaminated runoff enters waterways, notify downstream users of potentially contaminated waters. Notify local health and fire officials and pollution control agencies. From a secure, explosion-proof location, use water spray to cool exposed containers. If cooling streams are ineffective (venting sound increases in volume and pitch, tank discolors, or shows any signs of deforming), withdraw immediately to a secure position. If employees are expected to fight fires, they must be trained and equipped in OSHA 1910.156. The only respirators recommended for firefighting are self-contained breathing apparatuses that have full face-pieces and are operated in a pressure-demand or other positive-pressure mode.

Disposal Method Suggested: Dump in water; add soda ash and stir, then neutralize and flush to sewer with water. Alternatively, zinc chloride may be recovered from spent catalysts and used in acrylic fiber spinning solutions.

References

Sax, N. I. (Ed.). *Dangerous Properties of Industrial Materials Report*, 1, No. 7, 90—92 (1981) and 5, No. 3, 77—82 (1985)

New York State Department of Health. (April 1986). *Chemical Fact Sheet: Zinc Chloride.* Albany, NY: Bureau of Toxic Substance Assessment

New Jersey Department of Health and Senior Services. (December 2000). *Hazardous Substances Fact Sheet: Zinc Chloride.* Trenton, NJ

Zinc chromate   Z:0130

Molecular Formula: CrO_4Zn
Common Formula: $ZnCrO_4$
Synonyms: Basic zinc chromate; Basic zinc chromate X-2259; Buttercup yellow; Chromic acid, zinc salt; Chromium zinc oxide; C.I. 77955; C.I. Pigment yellow 36; Citron yellow; C.P. zinc yellow X-883; Primrose yellow; Pure zinc chrome; Pure zinc yellow; Zinc chromate C; Zinc chromate (VI) hydroxide; Zinc chromate O; Zinc chromate T; Zinc chromate Z; Zinc chrome; Zinc chrome (anti-corrosion); Zinc chrome yellow; Zinc chromium oxide; Zinc hydroxychromate; Zinc tetraoxychromate; Zinc tetraoxychromate 76A; Zinc tetraoxychromate 780B; Zinc tetroxychromate; Zinc yellow; Zinc yellow 1; Zinc yellow 1425; Zinc yellow 386N; Zinc yellow 40-9015; Zinc yellow AZ-16; Zinc yellow AZ-18; Zinc yellow KSH

CAS Registry Number: 13530-65-9; *(alt.)* 1308-13-0; *(alt.)* 1328-67-2; *(alt.)* 14675-41-3; 11103-86-9; 37300-23-5; 12018-19-8; 14018-95-2; 37224-57-0 (zinc potassium chromate)

RTECS® Number: GB3290000
UN/NA & ERG Number: UN3288 (toxic solid, inorganic, n.o.s.)/151
EC Number: 236-878-9 [*Annex I Index No.:* 024-007-00-3]
Regulatory Authority and Advisory Bodies
Carcinogenicity: IARC: Human Sufficient Evidence; Animal Sufficient Evidence, *carcinogenic to humans*, Group 1, 1997; NTP: 11th Report on Carcinogens, 2004: Known to be a human carcinogen; EPA *(inhalation)*: Known human carcinogen; EPA *(oral)*: Cannot be Determined; Not Classifiable as to human carcinogenicity; NTP: Known to be a human carcinogen.
US EPA Gene-Tox Program, Positive: Carcinogenicity— mouse/rat; Cell transform.—SA7/SHE; NIOSH: Potential occupational carcinogen.
Air Pollutant Standard Set. See below, "Permissible Exposure Limits in Air" section.
Chromium compounds:
Clean Air Act: Hazardous Air Pollutants (Title I, Part A, Section 112).
Clean Water Act: Toxic Pollutant (Section 401.15); 40CFR401.15 Section 307 Toxic Pollutants as chromium and compounds.
RCRA, 40CFR261, Appendix 8 Hazardous Constituents, waste number not listed.
EPCRA (Section 313): Includes any unique chemical substance that contains chromium as part of that chemical's infrastructure. Form R *de minimis* concentration reporting level: Chromium(VI) compounds: 0.1%.
California Proposition 65 Chemical: (*hexavalent chromium*) Cancer 2/27/87; Developmental/Reproductive toxin (male, female) 12/19/08.
Canada, WHMIS, Ingredients Disclosure List Concentration 0.1%.
European/International Regulations: Hazard Symbol: E, T + , N; Risk phrases: R45; R46; R60; R61; R2; R8; R21; R25; R26; R34; R42/43; R48/23; R50/53; Safety phrases: S53; S45; S60; S61.
European/International Regulations (*zinc chromates including zinc potassium chromate*): Hazard Symbol: T, N; Risk phrases: R45; R22; R43; R50/53; Safety phrases: S53; S45; S60; S61 (see Appendix 4).
WGK (German Aquatic Hazard Class): 3—Severe hazard to waters.
Description: Zinc chromate is a yellow crystalline powder. Molecular weight = 181.37; Freezing/Melting point = 316°C. Hazard Identification (based on NFPA-704 M Rating System): Health 2, Flammability 0, Reactivity 0. Soluble in water.
Potential Exposure: Compound Description: Tumorigen, Mutagen; Human Data. Zinc chromate is used as an anticorrosion pigment in primers and as a coloring agent; as a pigment in surface coatings and linoleum; to impart corrosion-resistance to epoxy laminates.
Incompatibilities: An oxidizer; reacts with reducing agents, combustibles, organic materials.

Permissible Exposure Limits in Air
OSHA PEL: 0.005 mg[Cr(VI)]/m^3 (*13530-65-9 only*) TWA; 0.1 mg[CrO^3]/m^3 (*11103-86-9 and 37300-23-5*) Ceiling Concentration. See 29CFR1910.1026.
NIOSH REL: 0.001 mg[Cr]/m^3 TWA, potential carcinogen, limit exposure to lowest feasible level. NIOSH considers all Cr(VI) compounds (including chromic acid, *tert*-butyl chromate, zinc chromate, and chromyl chloride) to be potential occupational carcinogens. See *NIOSH Pocket Guide*, Appendix A & C.
ACGIH TLV®[1]: 0.015 mg[Cr]/m^3 TWA, Confirmed Human Carcinogen; BEI issued.
NIOSH IDLH: 15 mg[Cr(VI)]/m^3.
Protective Action Criteria (PAC)
TEEL-0: 0.0174 mg/m^3
PAC-1: 0.4 mg/m^3
PAC-2: 2.5 mg/m^3
PAC-3: 52.3 mg/m^3
DFG MAK: [skin] Danger of skin sensitization; Carcinogen Category 1; Pregnancy Risk Category 2; TRK: 0.05 mg[Cr]/m^3; 20 μg/L [Cr] in urine at end-of-shift.
Australia: TWA 0.01 mg/m^3, carcinogen, 1993; Austria: carcinogen, 1999; Belgium: TWA 0.01 mg/m^3, carcinogen, 1993; Denmark: TWA 0.02 mg[Cr]/m^3, 1999; France: VME 0.05 mg[Cr]/m^3, 1999; Japan; 0.05 mg[Cr]/m^3, 1999; Norway: TWA 0.02 mg[CrO$_3$]/m^3, 1999; Poland: TWA 0.1 mg/m^3, STEL 0.3 mg/m^3, 1999; Sweden: TWA 0.02 mg[Cr]/m^3, carcinogen, 1999; Switzerland: MAK-W 0.01 mg[Cr]/m^3, carcinogen, 1999; United Kingdom: TWA 0.5 mg[Cr]/m^3, carcinogen, 2000; Argentina, Bulgaria, Columbia, Jordan, South Korea, New Zealand, Singapore, Vietnam: ACGIH TLV®: Confirmed Human Carcinogen. Several states have set guidelines or standards for zinc chromate in ambient air[60] ranging from 0.008 μg/m^3 (Virginia) to 0.5 μg/m^3 (Connecticut) to 1.0 μg/m^3 (Nevada).
Determination in Air: Use NIOSH Analytical Methods #7600,#7604, #7605, #7703, #9101; OSHA Analytical Methods ID-103, ID-215, W-4001.
Permissible Concentration in Water: Since zinc chromate may consist of compounds with various ZnO/Cr$_2$O$_3$ rations, it is best simply to refer to the EPA water quality criteria cited in the sections of the volume dealing with "Chromium" and with "Zinc Chloride." There are a number of standards for *zinc* in water set around the world[35]: The US EPA[6] has set 5 mg/L (5000 μg/L) for the prevention of adverse effects due to the organoleptic properties of zinc. The state of Kansas has also set a drinking water limit of 5 mg/L.[61] EC: 100−500 μg/L (drinking water); Germany: 2000 μg/L (drinking water); Mexico: 10,000 μg/L (estuaries); 10 μg/L (coastal waters); Russia: 5000 μg/L (drinking water); 1000 μg/L (surface water); 10 μg/L (for fishery purposes); WHO: 5000 μg/L (for esthetic quality). *As a hexavalent chromium compound:* For the protection of freshwater aquatic life: *Hexavalent chromium:* 0.29 μg/L as a 24-h average, never to exceed 21.0 μg/L. For the protection of saltwater aquatic

life: *Hexavalent chromium:* 18 μg/L as a 24-h average, never to exceed 1260 μg/L. *To protect human health:* hexavalent chromium 50 μg/L according to EPA.[6] US EPA[49] has set a long-term health advisory for adults of 0.84 mg/L and a lifetime health advisory of 0.12 mg/L (120 μg/L) for chromium. EPA's maximum drinking water level (MCL) is 0.1 mg/L.[62] Germany, Canada, EEC, and WHO[35] have set a limit of 0.05 mg/L in drinking water. The states of Maine and Minnesota have set a guideline of 50 μg/L for chromium in drinking water[61] for Maine and 120 μg/L for Minnesota.

Determination in Water: Total chromium may be determined by digestion followed by atomic absorption, or by colorimetry (diphenylcarbazide), or by inductively coupled plasma (CP) optical emission spectrometry. Chromium(VI) may be determined by extraction and atomic absorption or colorimetry (using diphenylhydrazide). Dissolved total Cr or Cr(VI) may be determined by 0.45 μm filtration followed by the above-cited methods.[49]

Routes of Entry: Inhalation, ingestion, eye and/or skin contact.

Harmful Effects and Symptoms

Short Term Exposure: Zinc chromate can affect you when breathed in and may enter the body through the skin. Zinc chromate is a carcinogen; handle with extreme caution. Zinc chromate can irritate the skin, causing a rash or skin ulcers. It can also trigger a skin allergy.

Long Term Exposure: Repeated exposure can cause a hole in the nasal septum (bone dividing the inner nose). Nose bleeds and sores are earlier signs. Zinc chromate is a human carcinogen. Repeated exposure may cause skin allergy with rash and itching.

Points of Attack: Skin. Cancer site in humans: Lung and throat.

Medical Surveillance: NIOSH lists the following tests for chromates: Blood gas analysis, complete blood count, chest X-ray, electrocardiogram, liver function tests; pulmonary function tests; sputum cytology, urine (chemical/metabolite), urinalysis (routine), white blood cell count/differential. Before beginning employment and at regular times after that, the following is recommended: Urine test for *chromates.* This test is most accurate shortly after exposure. If symptoms develop or overexposure is suspected, the following may be useful: evaluation by a qualified allergist, including careful exposure history and special testing, may help diagnose skin allergy.

First Aid: If this chemical gets into the eyes, remove any contact lenses at once and irrigate immediately for at least 15 min, occasionally lifting upper and lower lids. Seek medical attention immediately. If this chemical contacts the skin, remove contaminated clothing and wash immediately with soap and water. Seek medical attention immediately. If this chemical has been inhaled, remove from exposure, begin rescue breathing (using universal precautions, including resuscitation mask) if breathing has stopped and CPR if heart action has stopped. Transfer promptly to a medical facility. When this chemical has been swallowed, get medical attention. Give large quantities of water and induce vomiting. Do not make an unconscious person vomit.

Note to physician: For severe poisoning *do not* use BAL [British Anti-Lewisite, dimercaprol, dithiopropanol ($C_3H_8OS_2$)] as it is contraindicated or ineffective in poisoning from zinc.

Personal Protective Methods: Wear protective gloves and clothing to prevent any reasonable probability of skin contact. Prevent skin contact. (As chromic acid and chromates) **8 h** (more than 8 h of resistance to breakthrough >0.1 μg/cm/min): polyethylene gloves, suits, boots; polyvinyl chloride gloves, suits, boots; Saranex™ coated suits; **4 h** (At least 4 but <8 h of resistance to breakthrough >0.1 μg/cm²/min): butyl rubber gloves, suits, boots; Viton™ gloves, suits. Safety equipment suppliers/manufacturers can provide recommendations on the most protective glove/clothing material for your operation. All protective clothing (suits, gloves, footwear, headgear) should be clean, available each day, and put on before work. Contact lenses should not be worn when working with this chemical. Eye protection is included in the recommended respiratory protection. Employees should wash immediately with soap when skin is wet or contaminated. Provide emergency showers and eyewash. Specific engineering controls are recommended in NIOSH Criteria Document #76-129 [Chromium(VI)].

Respirator Selection: NIOSH, as chromates: *at any concentrations above the NIOSH REL, or where there is no REL, at any detectable concentration:* SCBAF: Pd,Pp (APF = 10,000) (any self-contained breathing apparatus that has a full face-piece and is operated in a pressure-demand or other positive-pressure mode) or SaF: Pd,Pp: ASCBA (APF = 10,000) (any supplied-air respirator that has a full face-piece and is operated in a pressure-demand or other positive-pressure mode in combination with an auxiliary, self-contained breathing apparatus operated in a pressure-demand or other positive-pressure mode). *Escape:* 100F (APF = 50) (any air-purifying, full-face-piece respirator with an N100, R100, or P100 filter) or SCBAE (any appropriate escape-type, self-contained breathing apparatus).

Storage: Color Code—Blue: Health Hazard/Poison: Store in a secure poison location. Store in tightly closed containers in a cool, well-ventilated area away from reducing agents. Where possible, automatically transfer material from other storage containers to process containers. A regulated, marked area should be established where this chemical is handled, used, or stored in compliance with OSHA Standard 1910.1045.

Shipping: Toxic solid, inorganic, n.o.s. requires a shipping label of "POISONOUS/TOXIC MATERIALS." They fall in Hazard Class 6.1.

Spill Handling: Evacuate persons not wearing protective equipment from area of spill or leak until cleanup is complete. Remove all ignition sources. Collect powdered material in the most convenient and safe manner and deposit in sealed containers. Ventilate area after cleanup is complete. It may be necessary to contain and dispose of this chemical as a hazardous waste. If material or contaminated runoff enters waterways, notify downstream users of potentially contaminated waters. Contact your local or federal environmental protection agency for specific recommendations. If employees are required to clean up spills, they must be properly trained and equipped. OSHA 1910.120(q) may be applicable.

Fire Extinguishing: Nonflammable. Use agent suitable for surrounding fire. Poisonous gases are produced in fire, including zinc and chromium. If material or contaminated runoff enters waterways, notify downstream users of potentially contaminated waters. Notify local health and fire officials and pollution control agencies. From a secure, explosion-proof location, use water spray to cool exposed containers. If cooling streams are ineffective (venting sound increases in volume and pitch, tank discolors, or shows any signs of deforming), withdraw immediately to a secure position. If employees are expected to fight fires, they must be trained and equipped in OSHA 1910.156. The only respirators recommended for firefighting are self-contained breathing apparatuses that have full face-pieces and are operated in a pressure-demand or other positive-pressure mode.

References
National Institute for Occupational Safety and Health. (1976). *Criteria for a Recommended Standard: Occupational Exposure to Chromium*, NIOSH Document No. 76-129. Rockville, MD
Sax, N. I. (Ed.). (1981). *Dangerous Properties of Industrial Materials Report*, 1, No. 7, 92−94
New Jersey Department of Health and Senior Services. (August 2002). *Hazardous Substances Fact Sheet: Zinc Chromate*. Trenton, NJ

Zinc oxide   Z:0140

Molecular Formula: OZn
Common Formula: ZnO
Synonyms: Akro-zinc bar 85; Akro zinc bar 90; Amalox; Azo-33; Azo-55; Azo-66; Azo-77; Azodox-55; Calamine; Chinese white; C.I. 77947; C.I. Pigment white 4; Emanay zinc oxide; Emar; Felling zinc oxide; Flowers of zinc; Green seal-8; HC (military designation); Hubbuck's white; Kadox-25; K-zinc; Ozide; Ozlo; Pasco; Permanent white; Philosopher's wool; Protox type 166; Protox type 167; Protox type 168; Protox type 169; Protox type 267; Protox type 268; Red-seal-9; Snow white; White flower of zinc; White seal-7; Zincite; Zincoid; Zinc oxide fume; Zinc peroxide; Zinc white
CAS Registry Number: 1314-13-2; *(alt.)* 8051-03-4; *(alt.)* 78590-82-6
RTECS® Number: ZH4810000
UN/NA & ERG Number: UN3077/171
EC Number: 215-222-5 [*Annex I Index No.:* 030-013-00-7]

Regulatory Authority and Advisory Bodies
Carcinogenicity: EPA: Not Classifiable as to human carcinogenicity; Inadequate Information to assess carcinogenic potential; Available data are inadequate for an assessment of human carcinogenic potential.
Air Pollutant Standard Set. See below, "Permissible Exposure Limits in Air" section.
FDA—over-the-counter drug.
US EPA, FIFRA 1998 Status of Pesticides: RED completed.
EPCRA Section 313 Form R *de minimis* concentration reporting level: 1.0%.
Canada, WHMIS, Ingredients Disclosure List Concentration 1.0%.
European/International Regulations: Hazard Symbol: N; Risk phrases: R50/53; Safety phrases: S60; S61 (see Appendix 4).
WGK (German Aquatic Hazard Class): 2—Hazard to waters.
Description: Zinc oxide is an amorphous, white or yellowish-white powder. Odorless. Molecular weight = 81.37; Specific gravity (H_2O:1) = 5.61 at 25°C; Freezing/Melting point = 1975°C. Hazard Identification (based on NFPA-704 M Rating System) *(powder)*: Health 2, Flammability 2, Reactivity 0. Practically insoluble in water; zinc oxide undergoes slow decomposition; solubility = 0.0004% at 18°C.
Potential Exposure: Compound Description: Mutagen; Reproductive Effector; Human Data; Primary Irritant. Zinc oxide is primarily used as a white pigment in rubber formulations and as a vulcanizing aid. It is also used as an anti-inflammatory agent; in photocopying; paints, chemicals, ceramics, lacquers, and varnishes; as a filler for plastic; in cosmetics, pharmaceuticals, and calamine lotion. Exposure may occur in the manufacture and use of zinc oxide and products, or through its formation as a fume when zinc or its alloys are heated. HC may have been used as a Choking/Pulmonary Agent.
Incompatibilities: Incompatible with linseed oil, magnesium. Contact with chlorinated rubber (at 215°C) may cause a violent reaction. Slowly decomposed by water.
Permissible Exposure Limits in Air
OSHA PEL: 15 mg/m³ TWA (*total dust*); 5 mg/m³ TWA (*respirable fraction and fume*).
NIOSH REL: (*dust only*): 5 mg/m³ TWA; 15 mg/m³ Ceiling Concentration; (*fume*) 5 mg/m³ TWA; 10 mg/m³ Ceiling Concentration.
ACGIH TLV®[1]: 2 mg/m³, respirable fraction TWA; 10 mg/m³, respirable fraction STEL.
NIOSH IDLH: 500 mg/m³.
Protective Action Criteria (PAC)
TEEL-0: 5 mg/m³
PAC-1: 10 mg/m³
PAC-2: 15 mg/m³
PAC-3: 500 mg/m³
DFG *MAK* (*zinc oxide and fume*): 0.1 mg/m³, respirable fraction TWA; Peak Limitation Category I(4); 2 mg/m³, inhalable fraction TWA; Peak Limitation Category I(2).

Arab Republic of Egypt: TWA 5 mg/m^3, 1993; Australia: TWA 10 mg/m^3, 1993; Australia: TWA 5 mg/m^3, STEL 10 mg/m^3 (fume), 1993; Belgium: TWA 10 mg/m^3, 1993; TWA 5 mg/m^3, STEL 10 mg/m^3 (fume), 1993; Denmark: TWA 4 mg[Zn]/m^3, 1999; Finland: TWA 5 mg/m^3 (fume), 1999; the Netherlands: MAC-TGG 5 mg/m^3, 2003; France: VME (fume) 5 mg/m^3, 1999; Hungary: TWA 5 mg/m^3, 1993; Japan; 5 mg/m^3 (fume), 1999; Norway: TWA 5 mg/m^3, 1999; the Philippines: TWA 1 mg/m^3, 1993; Poland: MAC (TWA) fume 5 mg/m^3; MAC (STEL) fume 10 mg/m^3, 1999; Sweden: NGV 5 mg/m^3, 1999; Switzerland: MAK-W 5 mg/m^3, 1999; Thailand: TWA 5 mg/m^3 (fume), 1993; Turkey: TWA 5 mg/m^3, 1993; United Kingdom: TWA 5 mg/m^3, STEL 10 mg/m^3, fume, 2000; Argentina, Bulgaria, Columbia, Jordan, South Korea, New Zealand, Singapore, Vietnam: ACGIH TLV®: STEL 10 mg/m^3. Russia[43] set a MAC of 0.05 mg/m^3 in ambient air in residential areas on a daily average basis. Several states have set guidelines or standards for zinc oxide fume in ambient air[60] ranging from 0.8 µg/m^3 (Virginia) to 16.7 µg/m^3 (New York) to 50.0 µg/m^3 (Florida) to 50.0−100.0 µg/m^3 (North Dakota) to 100.0 µg/m^3 (Connecticut) to 119.0 µg/m^3 (Nevada).

Determination in Air: Use NIOSH Analytical Method (IV) #7303, Elements by ICP, #7502, Zinc and compounds, OSHA Analytical Method #ID-121, #ID1-43.

Permissible Concentration in Water: There are a number of standards for *zinc* in water set around the world.[35] The US EPA[6] has set 5 mg/L (5000 µg/L) for the prevention of adverse effects due to the organoleptic properties of zinc. EC: 100−500 µg/L (drinking water); Germany: 2000 µg/L (drinking water); Mexico: 10,000 µg/L (estuaries); 10 µg/L (coastal waters); Russia: 5000 µg/L (drinking water); 1000 µg/L (surface water); 10 µg/L (for fishery purposes); WHO: 5000 µg/L (for esthetic quality). The state of Kansas has also set a drinking water limit of 5 mg/L.[61]

Routes of Entry: Inhalation, ingestion, eye and/or skin contact.

Harmful Effects and Symptoms

Short Term Exposure: *Note:* Symptoms of metallic or sweet taste and/or throat irritation or dryness may indicate overexposure. *Inhalation:* Irritates the respiratory tract. Exposure to fumes over 52 mg/m^3 can cause "metal fume fever." Onset of symptoms may be delayed 4−12 h. Symptoms include irritation of the nose, mouth, and throat; chills, muscle ache, nausea, fever, dry throat, cough, weakness, lassitude (weakness, exhaustion), metallic taste, headache, blurred vision, low back pain, vomiting, fatigue, malaise (vague feeling of discomfort), tightness in chest, dyspnea (breathing difficulty), rales, decreased pulmonary function, stomach pain, chills, fever, pains in the muscles and joints, thirst, bronchitis or pneumonia, and bluish tint to the skin. Higher exposures can cause pulmonary edema, a medical emergency that can be delayed for several hours. This can cause death.

Skin: Dust may cause irritation which can result in rash. *Eyes:* No information available. *Ingestion:* May cause abdominal discomfort, watery diarrhea, and cramps.

Long Term Exposure: Repeated or prolonged contact with skin may cause dermatitis. Repeated or prolonged inhalation exposure may cause asthma. There is limited evidence that zinc oxide may damage the developing fetus. Repeated overexposure may cause ulcer symptoms and affect the liver.

Points of Attack: Respiratory system, liver, skin.

Medical Surveillance: There are no special tests for this chemical. However, if overexposure is suspected or if illness occurs medical attention is recommended. Lung function tests. Liver function tests. GI series. Consider chest X-ray following acute overexposure.

First Aid: If this chemical gets into the eyes, remove any contact lenses at once and irrigate immediately for at least 15 min, occasionally lifting upper and lower lids. Seek medical attention immediately. If this chemical contacts the skin, remove contaminated clothing and wash immediately with soap and water. Seek medical attention immediately. If this chemical has been inhaled, remove from exposure, begin rescue breathing (using universal precautions, including resuscitation mask) if breathing has stopped and CPR if heart action has stopped. Transfer promptly to a medical facility. When this chemical has been swallowed, get medical attention. Give large quantities of water and induce vomiting. Do not make an unconscious person vomit. Medical observation is recommended for 24−48 h after breathing overexposure, as pulmonary edema may be delayed. As first aid for pulmonary edema, a doctor or authorized paramedic may consider administering a corticosteroid spray. If metal fume fever develops, it may last less than 36 h.

Note to physician: In case of fume inhalation, treat pulmonary edema. Give prednisone or other corticosteroid orally to reduce tissue response to fume. Positive-pressure ventilation may be necessary. Treat metal fume fever with bed rest, analgesics, and antipyretics.

Note to physician: For severe poisoning *do not* use BAL [British Anti-Lewisite, dimercaprol, dithiopropanol ($C_3H_8OS_2$)] as it is contraindicated or ineffective in poisoning from zinc.

Personal Protective Methods: Wear protective gloves and clothing to prevent any reasonable probability of skin contact. Safety equipment suppliers/manufacturers can provide recommendations on the most protective glove/clothing material for your operation. All protective clothing (suits, gloves, footwear, headgear) should be clean, available each day, and put on before work. Contact lenses should not be worn when working with this chemical. Wear dust-proof chemical goggles and face shield unless full face-piece respiratory protection is worn. Employees should wash immediately with soap when skin is wet or contaminated. Provide emergency showers and eyewash.

Respirator Selection: *Up to 50 mg/m^3:* 95XQ (APF = 10) [any particulate respirator equipped with an N95, R95, or P95 filter (including N95, R95, and P95 filtering face-pieces) except quarter-mask respirators. The following filters may also be used: N99, R99, P99, N100, R100,

P100] or Sa (APF = 10) (any supplied-air respirator). *Up to 125 mg/m³:* Sa:Cf (APF = 25) (any supplied-air respirator operated in a continuous-flow mode) or PaprHie (APF = 25) (any powered, air-purifying respirator with a dust, mist, and fume filter). *Up to 250 mg/m³:* 100F (APF = 50) (any air-purifying, full-face-piece respirator with a high-efficiency particulate filter) or SaT: Cf (APF = 50) (any supplied-air respirator that has a tight-fitting face-piece and is operated in a continuous-flow mode) or PaprTHie (APF = 50) (any powered, air-purifying respirator with a tight-fitting face-piece and a high-efficiency particulate filter) or SCBAF (APF = 50) (any self-contained breathing apparatus with a full face-piece) or SaF (APF = 50) (any supplied-air respirator with a full face-piece). *Up to 500 mg/m³:* Sa: Pd,Pp (APF = 1000) (any supplied-air respirator operated in a pressure-demand or other positive-pressure mode). *Emergency or planned entry into unknown concentrations or IDLH conditions:* SCBAF: Pd,Pp (APF = 10,000) (any NIOSH/MSHA- or European Standard EN 149-approved self-contained breathing apparatus that has a full face-piece and is operated in a pressure-demand or other positive-pressure mode) or SaF: Pd,Pp: ASCBA (APF = 10,000) (any supplied-air respirator that has a full face-piece and is operated in a pressure-demand or other positive-pressure mode in combination with an auxiliary, self-contained breathing apparatus operated in a pressure-demand or other positive-pressure mode). *Escape:* 100F (APF = 50) (any air-purifying, full-face-piece respirator with a high-efficiency particulate filter) or SCBAE (any appropriate escape-type, self-contained breathing apparatus).
Storage: Color Code—Green: General storage may be used. Store in tightly closed containers in a cool, well-ventilated area away from chlorinated rubber, magnesium, and linseed oil.
Shipping: The name of this material is not on the DOT list of materials[19] for label and packaging standards. However, based on regulations, it may be classified[52] as an Environmentally hazardous substances, solid, n.o.s. This chemical requires a shipping label of "CLASS 9." It falls in Hazard Class 9 and Packing Group III.[20, 21]
Spill Handling: Evacuate persons not wearing protective equipment from area of spill or leak until cleanup is complete. Remove all ignition sources. Collect powdered material in the most convenient and safe manner and deposit in sealed containers. Ventilate area after cleanup is complete. It may be necessary to contain and dispose of this chemical as a hazardous waste. If material or contaminated runoff enters waterways, notify downstream users of potentially contaminated waters. Contact your local or federal environmental protection agency for specific recommendations. If employees are required to clean up spills, they must be properly trained and equipped. OSHA 1910.120(q) may be applicable.
Fire Extinguishing: Does not burn. Extinguish fire using an agent suitable for type of surrounding fire. Poisonous gases are produced in fire, including oxides of zinc. If material or contaminated runoff enters waterways, notify

downstream users of potentially contaminated waters. Notify local health and fire officials and pollution control agencies. From a secure, explosion-proof location, use water spray to cool exposed containers. If cooling streams are ineffective (venting sound increases in volume and pitch, tank discolors, or shows any signs of deforming), withdraw immediately to a secure position. If employees are expected to fight fires, they must be trained and equipped in OSHA 1910.156. The only respirators recommended for firefighting are self-contained breathing apparatuses that have full face-pieces and are operated in a pressure-demand or other positive-pressure mode.
References
National Institute for Occupational Safety and Health. (1976). *Criteria for a Recommended Standard: Occupational Exposure to Zinc Oxide*, NIOSH Document No. 76-104
US Environmental Protection Agency. (July 1987). *Summary Review of Health Effects Associated with Zinc and Zinc Oxide*, Report EPA/600/8-87/022F. Research Triangle Park, NC
New York State Department of Health. (March 1986). *Chemical Fact Sheet: Zinc Oxide*. Albany, NY: Bureau of Toxic Substance Assessment
US Environmental Protection Agency. Special Review and Reregistration Division Office of Pesticide Programs. (1998). *Agency Status of Pesticides in Registration, Reregistration, and Special Review* (Rainbow Report). Washington, DC
New Jersey Department of Health and Senior Services. (January 2007). *Hazardous Substances Fact Sheet: Zinc Oxide*. Trenton, NJ

Zinc phosphide   Z:0150

Molecular Formula: P_2Zn_3
Common Formula: Zn_3P_2
Synonyms: Blue-ox; Fosfuro de zinc (Spanish); Kilrat; Mous-con; Phosphure de zinc (French); Phosvin; Ratol; Zinc (phosphure de) (French); Zinc-tox; Zinkphosphid (German); ZP
CAS Registry Number: 1314-84-7; (*alt.*) 39342-49-9
RTECS® Number: ZH4900000
UN/NA & ERG Number: UN1714/139
EC Number: 215-244-5 [*Annex I Index No.:* 015-006-00-9]
Regulatory Authority and Advisory Bodies
US EPA, FIFRA 1998 Status of Pesticides: RED completed.
Clean Water Act: Section 311 Hazardous Substances/RQ 40CFR117.3 (same as CERCLA, see below); 40CFR 401.15 Section 307 Toxic Pollutants; Section 313 Water Priority Chemicals (57FR41331, 9/9/92).
US EPA Hazardous Waste Number (RCRA No.): P122.
RCRA, 40CFR261, Appendix 8 Hazardous Constituents.
Superfund/EPCRA 40CFR355, Extremely Hazardous Substances: TPQ = 500 lb (227 kg). This material is a

reactive solid. The TPQ does not default to 10,000 lb for nonpowder, nonmolten, nonsolution form.

Reportable Quantity (RQ): 100 lb (45.4 kg).

EPCRA Section 313 Form R *de minimis* concentration reporting level: 1.0%.

European/International Regulations: Hazard Symbol: T+, F, N; Risk phrases: R15/29; R28; R32; R50/53; Safety phrases: S1/2; S28; S30; S36/37; S43; S45; S60; S61 (see Appendix 4).

WGK (German Aquatic Hazard Class): 3—Severe hazard to waters.

Description: Zinc phosphide is a gray crystalline solid. Molecular weight = 258.05; Boiling point = 1100°C; Freezing/Melting point = 420°C. Hazard Identification (based on NFPA-704 M Rating System): Health 3, Flammability 3, Reactivity 2W. Insoluble in water; slowly decomposes.

Potential Exposure: Compound Description: Agricultural Chemical; Mutagen; Reproductive Effector; Human Data. It is used as an acute single-feeding rodenticide.

Incompatibilities: Heat and contact with water causes decomposition, producing toxic and flammable fumes of phosphorus, zinc oxides, and toxic and flammable phosphine gas. Reacts violently with concentrated sulfuric acid, nitric acid, and other oxidizers. Reacts with hydrochloric acid or sulfuric acid with the evolution of spontaneously flammable phosphine gas. Incompatible with carbon dioxide, halogenated agents.

Permissible Exposure Limits in Air

Protective Action Criteria (PAC)*

TEEL-0: 0.05 ppm

PAC-1: 0.15 ppm

PAC-2: **1** ppm

PAC-3: **1.8** ppm

*AEGLs (Acute Emergency Guideline Levels) & ERPGs (Emergency Response Planning Guidelines) are in **bold face**.

Russia: TWA 0.1 mg/m^3, 1993.

Determination in Air: Use #7502, Zinc and compounds, OSHA Analytical Method #ID-121, #ID1-43.

Permissible Concentration in Water: There are a number of standards for zinc in water set around the world[35]: The EC: 100−500 µg/L (drinking water); Germany: 2000 µg/L (drinking water); Mexico: 10,000 µg/L (estuaries); 10 µg/L (coastal waters; Russia: 5000 µg/L (drinking water); 1000 µg/L (surface water); 10 µg/L (for fishery purposes); WHO: 5000 µg/L (for esthetic quality). US EPA[6] has set 5 mg/L (5000 µg/L) for the prevention of adverse effects due to the organoleptic properties of zinc. The state of Kansas has also set a drinking water limit of 5 mg/L.[61]

Routes of Entry: Inhalation, ingestion, eye and/or skin contact. Absorbed through the skin.

Harmful Effects and Symptoms

Short Term Exposure: Irritates the respiratory tract. Contact with the eyes can cause severe irritation, burns, and permanent damage. Skin contact causes irritation. This chemical is a CNS depressant. Inhalation of zinc phosphide dust is followed in several hours by vomiting, diarrhea, cyanosis (bluing of skin), rapid pulse, fever, and shock. The breath smells of phosphine. The compound is very caustic and may cause closing of the esophagus. Inhalation of phosphine (formed when zinc phosphide is exposed to flame, water, or acids) can cause pulmonary edema, a medical emergency that can be delayed for several hours. This can cause death. Zinc phosphide is very caustic when ingested and forms phosphine. The probable oral lethal dose is 5−50 mg/kg or between 7 drops and 1 teaspoonful for a 70-kg (150−lb) person. Most patients die after about 30 h from peripheral vascular collapse secondary to the compound's direct effects. Extensive liver damage and kidney damage can also occur. Ingestion of 4−5 g has produced death in human adults, but also doses of 25−50 g have been survived. The lowest oral lethal dose reported for women is 80 mg/kg. Symptoms of oral ingestion include nausea, abdominal pain, vomiting, tightness in chest, excitement, agitation and chills, faintness, weakness, dyspnea, fall in blood pressure, change in pulse rate, diarrhea, intense thirst, convulsions, paralysis, and coma. Early labored breathing, shock, halted urinary output, metabolic acidosis, muscle cramps, and convulsions are grave prognostic signs.

Long Term Exposure: The substance may cause effects on the liver, kidneys, heart, and nervous system. Repeated exposures to low exposures cause chronic poisoning, anemia, bronchitis, and gastrointestinal, visual, speech, and motor disturbances.

Points of Attack: Lungs, liver, kidneys, heart, blood, nervous system.

Medical Surveillance: Liver and kidney function tests. EKG. Lung function tests. Consider chest X-ray following acute overexposure. Complete blood count (CBC).

First Aid: If this chemical gets into the eyes, remove any contact lenses at once and irrigate immediately for at least 15 min, occasionally lifting upper and lower lids. Seek medical attention immediately. If this chemical contacts the skin, remove contaminated clothing and wash immediately with soap and water. Seek medical attention immediately. If this chemical has been inhaled get medical attention for phosphine, remove from exposure, begin rescue breathing (using universal precautions, including resuscitation mask) if breathing has stopped and CPR if heart action has stopped. Transfer promptly to a medical facility. When this chemical has been swallowed, get medical attention for phosphine poisoning. Give one tablespoonful of mustard in a glass of warm water; repeat until vomit fluid is clear; avoid use of all oils. Do not make an unconscious person vomit. Medical observation is recommended for 24−48 h after breathing overexposure, as pulmonary edema may be delayed. As first aid for pulmonary edema, a doctor or authorized paramedic may consider administering a corticosteroid spray.

Note to physician: For severe poisoning *do not* use BAL [British Anti-Lewisite, dimercaprol, dithiopropanol

$(C_3H_8OS_2)]$ as it is contraindicated or ineffective in poisoning from zinc.

Personal Protective Methods: Wear protective gloves and clothing to prevent any reasonable probability of skin contact. Safety equipment suppliers/manufacturers can provide recommendations on the most protective glove/clothing material for your operation. All protective clothing (suits, gloves, footwear, headgear) should be clean, available each day, and put on before work. Contact lenses should not be worn when working with this chemical. Wear dust-proof chemical goggles and face shield unless full face-piece respiratory protection is worn. Employees should wash immediately with soap when skin is wet or contaminated. Provide emergency showers and eyewash.

Respirator Selection: Where there is potential for exposures to zinc dusts, use a NIOSH/MSHA- or European Standard EN149-approved respirator equipped with particulate (dust/fume/mist) filters. Particulate filters must be checked every day before work for physical damage, such as rips or tears, and replaced as needed. Where there is potential for high exposure, use a NIOSH/MSHA- or European Standard EN149-approved supplied-air respirator with a full face-piece operated in the positive-pressure mode, or with a full-face-piece, hood, or helmet in the continuous-flow mode; or use a NIOSH/MSHA- or European Standard EN149-approved self-contained breathing apparatus with a full face-piece operated in a pressure-demand or other positive-pressure mode.

Storage: Color Code—Blue: Health Hazard/Poison: Store in a secure poison location. Prior to working with this chemical you should be trained on its proper handling and storage. Store in tightly closed containers in a cool, well-ventilated area away from moisture and oxidizers.

Shipping: Zinc phosphide requires a shipping label of "DANGEROUS WHEN WET, POISONOUS/TOXIC MATERIALS." It falls in Hazard Class 4.3 and Packing Group I.

Spill Handling: Evacuate persons not wearing protective equipment from area of spill or leak until cleanup is complete. Remove all ignition sources. *Do not use water.* Keep sparks, flames, and other sources of ignition away. Keep material out of water sources and sewers. Keep material dry. Avoid breathing dusts and fumes from burning material. Keep upwind. Avoid bodily contact with the material. Do not handle broken packages without protective equipment. Wash away any material which may have contacted the body with copious amounts of water or soap and water. For a land spill, dig a pit, pond, lagoon, or holding area to contain liquid or solid material. For water spill, neutralize with agricultural lime (slaked lime), crushed limestone, or sodium bicarbonate. Use mechanical dredges or lifts to remove immobilized wastes of pollutants and precipitates. *Small dry spills:* with clean shovel place material into clean, dry container and cover; move containers from spill area. *Large spills:* dike spill for later disposal. Cover powder spill with plastic sheet or tarp to minimize spreading. Clean up

only under supervision of an expert. Collect powdered material in the most convenient and safe manner and deposit in sealed containers. Ventilate area after cleanup is complete. It may be necessary to contain and dispose of this chemical as a hazardous waste. If material or contaminated runoff enters waterways, notify downstream users of potentially contaminated waters. Contact your local or federal environmental protection agency for specific recommendations. If employees are required to clean up spills, they must be properly trained and equipped. OSHA 1910.120(q) may be applicable. UN1714, zinc phosphide is on the DOT's list of dangerous water-reactive materials which create large amounts of toxic vapor when *spilled in water.* Dangerous from 0.5 to 10 km (0.3−6.0 miles) downwind.

Fire Extinguishing: *Do not use water* or foam. For small fires, use dry chemical, soda ash, or lime. For large fires, withdraw from area and let burn. *Do not use water* or any agent with an acid reaction (i.e., carbon dioxide or halogenated agents) as phosphine will be liberated. Poisonous gases, including phosphine, zinc oxide, and phosphorus, are produced in fire. Wear boots, protective gloves, and goggles. Wear self-contained breathing apparatus when fighting fires involving this material. Keep unnecessary people away; isolate hazard area and deny entry. Stay upwind; keep out of low areas. Move container from fire area if you can do so without risk.

Disposal Method Suggested: Consult with environmental regulatory agencies for guidance on acceptable disposal practices. Generators of waste containing this contaminant ($\geq$100 kg/mo) must conform with EPA regulations governing storage, transportation, treatment, and waste disposal.

References

US Environmental Protection Agency. (November 30, 1987). *Chemical Hazard Information Profile: Zinc Phosphide.* Washington, DC: Chemical Emergency Preparedness Program
US Environmental Protection Agency. Special Review and Reregistration Division Office of Pesticide Programs. (1998). *Agency Status of Pesticides in Registration, Reregistration, and Special Review* (Rainbow Report). Washington, DC
New Jersey Department of Health and Senior Services. (May 2000). *Hazardous Substances Fact Sheet: Zinc Phosphide.* Trenton, NJ

Zirconium & compounds   Z:0160

Molecular Formula: Zr
Common Formula: Zr
Synonyms: Zirconium metal
other zirconium compounds: Zirconium boride (12045-64-6); Zirconium carbide (12070-14-3); Zirconium chloride (10026-11-6); Zirconium fluoride (7783-64-4); Zirconium hydride (7704-99-6); Zirconium hydroxide (14475-63-9);

Zirconium nitrate (13746-89-9); Zirconium nitride (25658-42-8); Zirconium oxide (1314-23-4); Zirconium phosphide (12037-80-8); (Zirconium oxychloride) 7699-43-6; Zirconyl chloride (10026-11-6)

CAS Registry Number: 7440-67-7 (metal)

RTECS® Number: ZH7070000 (elemental)

UN/NA & ERG Number: UN1358 [Zirconium powder, wetted with not <25% water (a visible excess of water must be present); (a) mechanically produced, particle size <53 μm; (b) chemically produced, particle size <840 μm]/ 170; UN1932 (metal scrap)/135; UN2008 (metal powder, dry)/135; UN2858 [dry, coiled wire, finished metal sheets, strip (thinner than 254 μm but not thinner than 18 μm)]/170; UN2009[dry, finished sheets, strip or coiled wire]/135

EC Number: 231-176-9 [Annex I Index No.: 040-001-00-3]

Regulatory Authority and Advisory Bodies

Air Pollutant Standard Set. See below, "Permissible Exposure Limits in Air" section.

The FDA controls zirconium-containing drugs and/or cosmetic products.

Canada, WHMIS, Ingredients Disclosure List Concentration 1.0%.

European/International Regulations: Hazard Symbol (*powder, pyrophoric*): F; Risk phrases: R15; R17; Safety phrases: S2; S7/8; S43; (*powder stabilized*) Risk phrases: R17; Safety phrases: S2; S7/8; S43 (see Appendix 4).

WGK (German Aquatic Hazard Class): Nonwater polluting agent (*metal*); 1-Slightly water polluting (*acetate, dichloride, dioxide, propionate*).

Description: Zirconium is a grayish-white, lustrous metal in the form of platelets, flakes, or a bluish-black, amorphous powder. Molecular weight = 91.22 (elemental); Specific gravity (H_2O:1) = 6.51 at 25°C; Boiling point = 3577°C; Freezing/Melting point = 1857°C. Hazard Identification (based on NFPA-704 M Rating System): Health 1, Flammability 4, Reactivity 1. Insoluble in water. The powdered metal is a fire and explosive hazard: it may ignite *spontaneously* and can continue burning under water.

Zirconium is never found in the free state; the most common sources are the ores zircon and baddeleyite. It is generally produced by reduction of the chloride or iodide. The metal is very reactive, and the process is carried out under an atmosphere of inert gas.

Potential Exposure: Zirconium metal is used as a "getter" in vacuum tubes, a deoxidizer in metallurgy, a substitute for platinum; it is used in priming of explosive mixtures; flashlight powders; lamp filaments; flash bulbs; and construction of rayon spinnerets. Zirconium or its alloys (with nickel, cobalt, niobium, tantalum) are used as lining materials for pumps and pipes, for chemical processes, and for reaction vessels. Pure zirconium is a structural material for atomic reactor; and alloyed, particularly with aluminum, it is a cladding material for fuel rods in water-moderated nuclear reactors. A zirconium—columbium alloy is an excellent superconductor. Zircon ($ZrSiO_4$) is utilized as a foundry sand, an abrasive, a refractory in combination with zirconia, a coating for casting molds, a catalyst in alkyl and alkenyl hydrocarbon manufacture, a stabilizer in silicone rubbers, and as a gem stone; in ceramics, it is used as an opacifier for glazes and enamels and in fritted glass filters. Both zircon and zirconia (zirconium oxide, ZrO_2) bricks are used as linings for glass furnaces. Zirconia itself is used in die extrusion of metals and in spout linings for pouring metals, as a substitute for lime in oxyhydrogen flam, as a pigment; and an abrasive; it is used, too, in incandescent lights, as well as in the manufacture of enamels, white glass, and refractory crucibles. Other zirconium compounds are used in metal cutting tools, thermocouple jackets, waterproofing textiles, ceramics, and in treating dermatitis and poison ivy.

Incompatibilities: Violent reactions with oxidizers, air, alkali hydroxides, alkali metal compounds (such as chromates, dichromates, molybdates, salts, sulfates, and tungstates), borax, carbon tetrachloride, lead, lead oxide, phosphorus, potassium compounds. Incompatible with boron, carbon, nitrogen, halogens, lead, platinum, potassium nitrate. Powder may ignite *spontaneously* and can continue burning under water. Explodes if mixed with hydrated borax when heated. Fine powder may be stored completely immersed in water.

Permissible Exposure Limits in Air

OSHA PEL (compounds): 5 mg[Zr]/m³ TWA.

NIOSH REL: 5 mg[Zr]/m³ TWA; 10 mg/m³ STEL [The REL applies to all zirconium compounds (as Zr) except Zirconium tetrachloride].

ACGIH TLV®[1] (*elemental & zirconium compounds*): 5 mg[Zr]/m³ TWA; 10 mg/m³ STEL; not classifiable as a human carcinogen.

NIOSH IDLH: 50 mg[Zr]/m³.

Protective Action Criteria (PAC)

Metal

TEEL-0: 5 mg/m³

PAC-1: 10 mg/m³

PAC-2: 10 mg/m³

PAC-3: 25 mg/m³

1314-23-4 (zirconium oxide)

TEEL-0: 6.75 mg/m³

PAC-1: 13.5 mg/m³

PAC-2: 13.5 mg/m³

PAC-3: 33.8 mg/m³

DFG MAK: (*elemental and insoluble compounds*): 1 mg/m³, inhalable fraction TWA; Peak Limitation Category I(1) danger of sensitization of the airways and skin; Pregnancy Risk Group; (*soluble compounds*): danger of sensitization of the airways and skin.

Arab Republic of Egypt: TWA 5 mg/m³, 1993; Austria: MAK 5 mg/m³, 1999; Denmark: TWA 5 mg[Zr]/m³, 1999; Finland: TWA 5 mg/m³, 1999; Hungary: STEL 5 mg/m³, 1993; Poland: MAC (TWA) 5 mg/m³; MAC (STEL) 10 mg/m³, 1999; Russia: STEL 6 mg/m³, 1993; the Netherlands: MAC-TGG 5 mg/m³, 2003; Argentina, Bulgaria, Columbia, Jordan, South Korea, New Zealand, Singapore, Vietnam: ACGIH TLV®: STEL 10 mg/m³. Several states have set guidelines or standards for zirconium compounds in ambient air[60] ranging from 0.8 μg/m³ (Virginia) to 50.0−100.0 μg/m³ (North Dakota) to 100.0 μg/m³ (Connecticut) to 119.0 μg/m³ (Nevada).

Determination in Air: Use NIOSH Analytical Method (IV) #7300, Elements by ICP, #8005, Elements in blood or tissue. See also OSHA Analytical Method ID-121.

Permissible Concentration in Water: No criteria set.

Routes of Entry: Inhalation of dust or fume, eye and/or skin contact.

Harmful Effects and Symptoms

Short Term Exposure: Zirconium can affect you when breathed in. The dust can irritate the lungs, causing coughing and/or shortness of breath.

Long Term Exposure: Contact can cause an allergic skin reaction to develop with small nodules (granulomas). May cause change in chest X-ray; lung granulomas. In animals: irritation of skin, mucous membrane; X-ray evidence of retention in lungs.

Points of Attack: Skin, respiratory system.

Medical Surveillance: If symptoms develop or overexposure is suspected, the following may be useful: evaluation by a qualified allergist, including careful exposure history and special testing, may help diagnose skin allergy. If breathing problems occur, lung function tests and a chest X-ray should be considered.

First Aid: If this chemical gets into the eyes, remove any contact lenses at once and irrigate immediately for at least 15 min, occasionally lifting upper and lower lids. Seek medical attention immediately. If this chemical contacts the skin, remove contaminated clothing and wash immediately with soap and water. Seek medical attention immediately. If this chemical has been inhaled, remove from exposure, begin rescue breathing (using universal precautions, including resuscitation mask) if breathing has stopped and CPR if heart action has stopped. Transfer promptly to a medical facility. When this chemical has been swallowed, get medical attention. Give large quantities of water and induce vomiting. Do not make an unconscious person vomit.

Personal Protective Methods: Skin protection is not generally necessary, but it is probably advisable especially where there is a history of zircon granuloma from deodorants. Safety equipment suppliers/manufacturers can provide recommendations on the most protective glove/clothing material for your operation. All protective clothing (suits, gloves, footwear, headgear) should be clean, available each day, and put on before work. Contact lenses should not be worn when working with this chemical. Wear dust-proof chemical goggles and face shield unless full face-piece respiratory protection is worn. Employees should wash immediately with soap when skin is wet or contaminated. Provide emergency showers and eyewash.

Respirator Selection: 25 mg/m³: Qm (APF = 25) (any quarter-mask respirator). 50 mg/m³: 95XQ (APF = 10) [any particulate respirator equipped with an N95, R95, or P95 filter (including N95, R95, and P95 filtering face-pieces) except quarter-mask respirators. The following filters may also be used: N99, R99, P99, N100, R100, P100] or PaprHie (APF = 25) (any powered, air-purifying respirator with a dust, mist, and fume filter); 100F (APF = 50) (any air-purifying, full face-piece respirator with an N100, R100, or P100 filter) or Sa (APF = 10) (any supplied-air respirator) or SCBAF (APF = 50) (any self-contained breathing apparatus with a full face-piece). *Emergency or planned entry into unknown concentrations or IDLH conditions:* SCBAF: Pd,Pp (APF = 10,000) (any self-contained breathing apparatus that has a full face-piece and is operated in a pressure-demand or other positive-pressure mode) or SaF: Pd,Pp: ASCBA (APF = 10,000) (any supplied-air respirator that has a full face-piece and is operated in a pressure-demand or other positive-pressure mode in combination with an auxiliary, self-contained breathing apparatus operated in a pressure-demand or other positive-pressure mode). *Escape:* 100F (APF = 50) (any air-purifying, full-face-piece respirator with an N100, R100, or P100 filter) or SCBAE (any appropriate escape-type, self-contained breathing apparatus).

Storage: Color Code—Red Stripe (*powder*): Flammability Hazard: Store separately from all other flammable materials. Zirconium must be stored to avoid contact with oxidizers (such as perchlorates, peroxides, permanganates, chlorates, and nitrates) since violent reactions occur. Store in tightly closed containers in a cool, well-ventilated area away from flammable materials. Sources of ignition, such as smoking and open flames, are prohibited where zirconium is handled, used, or stored. Metal containers involving the transfer of 5 gallons or more of zirconium should be grounded and bonded. Drums must be equipped with self-closing valves, pressure vacuum bungs, and flame arresters. Use only nonsparking tools and equipment, especially when opening and closing containers of zirconium. Wherever zirconium is used, handled, manufactured, or stored, use explosion-proof electrical equipment and fittings.

Shipping: Dry zirconium powder requires a shipping label of "SPONTANEOUSLY COMBUSTIBLE." It falls in Hazard Class 4.2 and Packing Group II.

Spill Handling: Evacuate persons not wearing protective equipment from area of spill or leak until cleanup is complete. Remove all ignition sources. Collect powdered material in the most convenient and safe manner and deposit in sealed containers. Ventilate area after cleanup is complete. Keep zirconium out of a confined space, such as a sewer, because of the possibility of an explosion, unless the sewer is designed to prevent the buildup of explosive concentrations. It may be necessary to contain and dispose of this chemical as a hazardous waste. If material or contaminated runoff enters waterways, notify downstream users of potentially contaminated waters. Contact your local or federal environmental protection agency for specific recommendations. If employees are required to clean up spills, they must be properly trained and equipped. OSHA 1910.120(q) may be applicable.

Fire Extinguishing: Zirconium is a flammable powder. Fire may restart after it has been extinguished. Containers may explode in fire. Use dry chemicals appropriate for extinguishing metal fires, salt, dry sand. *Do not use water.* Poisonous gases are produced in fire. If material or contaminated runoff enters waterways, notify downstream users of potentially contaminated waters. Notify local health and fire officials and pollution control agencies. From a secure, explosion-proof location, use water spray to cool exposed containers. If cooling streams are ineffective (venting sound increases in volume and pitch, tank discolors, or shows any signs of deforming), withdraw immediately to a secure position. If employees are expected to fight fires, they must be trained and equipped in OSHA 1910.156. The only respirators recommended for firefighting are self-contained breathing apparatuses that have full face-pieces and are operated in a pressure-demand or other positive-pressure mode.

References

New Jersey Department of Health and Senior Services. (October, 1998). *Hazardous Substances Fact Sheet: Zirconium.* Trenton, NJ

General Guide to Chemical Resistant Gloves

Material	Generally suitable for
Butyl rubber	Aldehydes Carboxylic acids Glycols and ethers Hydroxyl compounds and alcohol peroxides
Latex	Limit use (see note below) Acetone Alcohols Alkalies and caustics Ammonium fluoride Dimethyl sulfoxide (DMSO) Phenol
Natural rubber	Plating solutions Alcohols Alkalies and caustics Cellosolve Degreasing solvents Mineral acids Oils
Neoprene	Plating solutions
Nitrile rubber	Alcohols Ammonium fluoride Freons Hexane Hydrofluoric and hydrochloric acid Perchloric acid Perchloroethylene Phosphoric acid Potassium and sodium hydroxide Water-soluble materials, dilute acids and bases
Vinyl	General prevention of contamination Medical examination Nuisance materials

Important Note: Latex gloves present a risk of irritation, allergic reaction, or sensitization which, for susceptible individuals, can be significant. The latex protein can leach out of the gloves into the user's skin, or into the glove powdering, if powdered gloves are used. This can lead to allergic skin reaction or a potentially more serious reaction if latex protein-contaminated powder is released into the environment and breathed in. The use of any sort of latex gloves should be considered carefully, and they should only be used when no other glove is appropriate. You are strongly advised not to use powdered latex gloves. *Users at University of Oxford should be aware that University Safety Policy statement S3/02 prohibits the use of powdered latex gloves for any purpose and advises that, if use of latex is essential, only latex gloves with low levels of extractable latex protein may be used.*
Printed with permission from the Physical and Theoretical Chemistry Laboratory, Oxford University.

Bibliography

[1] American Conference of Governmental Industrial Hygienists, 2007 TLVs® and BEIs® Threshold Limit Values for Chemical Substances and Physical Agents & Biological Exposure Indices, ACGIH, Cincinnati, OH, 2007.

[2] Department of Health and Human Services (National Institute for Occupational Safety and Health), NIOSH Pocket Guide to Chemical Hazards, DHHS, Cincinnati, OH, August 2006 (DHHS (NIOSH) Publication No. 2005-149).

[3] Deutsche Forschungsgemeinschaft (DFG), List of MAK and BAT Values 2006: Maximum Concentrations and Biological Tolerance Values at the Workplace: Report 42, Wiley-VCH Publishers, Hoboken, NJ, 2006.

[4] US Environmental Protection Agency, Water programs: hazardous substances, Fed. Regist. 43 (49) (1978) 10474−10508.

[5] US Environmental Protection Agency, Identification and testing of hazardous waste, Fed. Regist. 45 (98) (1980) 33084−33133.

[6] US Environmental Protection Agency, Fed. Regist. 43 (31 January 1978) 4109; See also Fed. Regist. 44 (30 July 1979) 44501; and also Fed. Regist. 45 (28 November 1980) 79318−79379.

[7] US Environmental Protection Agency, Emergency planning and community right-to-know programs, Fed. Regist. 51 (221) (1986) 41570−41594.

[8] US Environmental Protection Agency, Toxic chemical release reporting: community right-to-know, Fed. Regist. 52 (107) (1987) 21152−21208.

[9] R.P. Pohanish, HazMat Data: For First Responders, Transportation, Storage, and Security, second ed., John Wiley & Sons, Hoboken, NJ, 2004.

[10] US Department of Health and Human Services, Public Health Service, National Toxicology Program (NTP), Report on carcinogens, eleventh ed., DHHS, Research Triangle Park, NC, 2005.

[11] K. Verscheuren, Handbook of Environmental Data on Organic Chemicals, third ed., Van Nostrand Reinhold, New York, NY, 1996.

[12] International Agency for Research on Cancer, IARC Monographs on the Carcinogenic Risks of Chemicals to Humans, Lyon, France (various years).

[13] United Nations, Consolidated List of Products Whose Consumption and/or Sale Have Been Banned, Withdrawn, Severely Restricted or Not Approved by Governments, United Nations, Geneva, Switzerland, 1987 (Second issue, UN Sales No. E. 87.IV.1)

[14] US Environmental Protection Agency, Report on the Status of Chemicals in the Special Review Program, Registration Program and Data Call-in Program, Office of Pesticide Programs, Washington, DC, 14 March 2002.

[15] World Bank, Manual of Industrial Hazard Assessment Techniques, Office of Environmental and Scientific Affairs, Washington, DC, 1985.

[16] National Research Council, Drinking Water and Health, National Academy Press, Washington, DC, 1980 (See also Ref. 45).

[17] NFPA (National Fire Protection Association), Fire Protection Guide to Hazardous Materials, NFPA, Quincy, MA, 2001.

[18] National Institute of Occupational Safety and Health, in: P.C. Schlecht, P.F. O'Connor (Eds.), NIOSH Manual of Analytical Methods (electronic), fifth ed., NIOSH, Cincinnati, OH, 2007 (available on NIOSH web site http://www.cdc.gov/niosh/nmam).

[19] US Department of Transportation, Performance-oriented packaging standards 49CFR17179 Fed. Regist. 52(215) (1987) 42772−43000.

[20] United Nations, Recommendations on the Transport of Dangerous Goods, fourth revised ed., United Nations, New York, NY, 1986 (UN Sales No. E.85. VIII.3)

[21] United Nations, Recommendations on the Transport of Dangerous Goods: Tests and Criteria, United Nations, New York, NY, 1986 (UN Sales No. E.85.VIII.2)

[22] International Register of Potentially Toxic Chemicals, Treatment and Disposal Methods for Waste Chemicals, UNEP, Geneva, Switzerland, 1985 (UN Sale No. E.85.111.2)

[23] C.R. Worthing, S.B. Walker (Eds.), The Pesticide Manual, eighth ed., The British Crop Protection Council, Thornton Heath, England, 1987.

[24] The International Technical Information Institute, Toxic and Hazardous Industrial Chemicals Safety Manual for Handling and Disposal with Toxicity and Hazard Data, The International Technical Information Institute, Tokyo, Japan, 1986.

[25] US Environmental Protection Agency, Guidelines establishing test procedures for the analysis of pollutants: proposed regulations, Fed. Regist. 44 (233) (3 December 1979) 69464−69575; and also a corrected version in Fed. Regist. 44 (244) (18 December 1979) 75028−75052.

[26] M. Sittig (Ed.), Priority Toxic Pollutants: Health Impacts and Allowable Limits, Noyes Data Corp., Park Ridge, NJ, 1980.

[27] National Institute for Occupational Safety and Health, Occupational Diseases: A Guide to Their Recognition, DHEW (NIOSH), Washington, DC, June 1977 (Publication No. 77-181).

[28] N.H. Proctor, J.P. Hughes, M.L. Fischman, Chemical Hazards of the Workplace, third ed., Van Nostrand Reinhold, New York, NY, 1991.

[29] E.R. Plunkett, Handbook of Industrial Toxicology, third ed., Chemical Publishing Co., Inc., New York, NY, 1987.

[30] L. Parmeggiani (Ed.), Encyclopedia of Occupational Health and Safety, third ed., International Labor Office (ILO), Geneva, 1983.

[31] US Department of Transportation, Office of Hazardous Materials Initiatives and Training, Research and Special Programs Administration, Transport Canada, and Secretariate of Communications and Transportation of Mexico, 2004 Emergency Response Guidebook, Washington, DC, 2004.

[32] US Environmental Protection Agency, Multimedia Environmental Goals for Environmental Assessment, USEPA, Research Triangle Park, NC, November 1977 (Report EPA-600/7-77-136).

[33] Health and Safety Executive, Monitoring Strategies for Toxic Substances, Guidance Note EH 42, London, England, November 1984.

[34] International Register of Potentially Toxic Chemicals (IRPTC), IRPTC Legal File (1986), UNEP, Geneva, Switzerland, 1987 (UN Sales No. E.87.III.D5)

[35] US Environmental Protection Agency, Drinking water: proposed substitution of contaminants and proposed list of additional substances which may require regulation under the Safe Drinking Water Act, Fed. Regist. 52(130) (1987) 25720−25734.

[36] US Environmental Protection Agency, Hazardous waste management system; identification and testing of hazardous waste; notification requirements; reportable quantity adjustments; proposed rule, Fed. Regist. 51(114) (1986) 21648−21693.

[37] US Environmental Protection Agency, Land disposal restrictions for certain California list of hazardous wastes and modifications to the framework: final rule, Fed. Regist. 52(130) (1987) 25760−25792.

[38] US Environmental Protection Agency, Notice of the first priority list of hazardous substances that will be the subject of toxicological profiles and guidelines for the development of toxicological profiles, Fed. Regist. 52(74) (1987) 12868−12874.

[39] US Environmental Protection Agency, National primary drinking water regulations—synthetic organic chemicals: monitoring for unregulated contaminants: final rule, Fed. Regist. 52(130) (1987) 25690−25717.

[40] G. Weiss (Ed.), Hazardous Chemicals Data Book, second ed., Noyes Data Corp., Park Ridge, NJ, 1986.

[41] US Environmental Protection Agency, Guidelines establishing test procedures for the analysis of pollutants: interim final rule and request for comments and proposed regulation, Fed. Regist. 52(171) (1987) 33542−33557.

[42] United Nations Environment Program, Maximum Allowable Concentrations and Tentative Safe Exposure Levels of Harmful Substances in the Environmental Media (Hygiene Standards Officially Approved in the USSR), Center of International Projects, Moscow, USSR, 1984.

[43] R.J. Lewis Sr., Hazardous Chemicals Desk Reference, fifth ed., John Wiley & Sons, Hoboken, NJ, 2002.

[44] National Institute for Occupational Safety and Health, NIOSH Recommendations for Occupational Safety and Health Standards, Centers for Disease Control, Atlanta, GA, 26 September 1986 (Supplement to Morbidity and Mortality Weekly Report)

[45] National Research Council, Drinking Water and Health, National Academy of Sciences, Washington, DC, 1977 (See also Ref. 16).

[46] US Environmental Protection Agency, Office of Drinking Water, Health Advisories for 16 Pesticides, USEPA, Washington, DC, March 1987 (Report PB-87-200176).

[47] US Environmental Protection Agency, Office of Drinking Water, Health Advisories for 25 Organics, USEPA, Washington, DC, March 1987 (Report PB 87-235578).

[48] US Environmental Protection Agency, Office of Drinking Water, Health Advisories for Legionella and Seven Inorganics, USEPA, Washington, DC, March 1987 (Report PB 87-235586).

[49] S.A. Greene, R.P. Pohanish (Eds.), Sittig's Handbook of Pesticides and Agricultural Chemicals, William Andrew Publishing, Norwich, NY, 2005.

[50] US Environmental Protection Agency, Organic chemicals and plastics and synthetic fibers category effluent limitations guidelines, pretreatment standards and new source performance standards, Fed. Regist. 52 (214) (1987) 42522−42584.

[51] L.H. Keith, D.B. Walters (Eds.), Compendium of Safety Data Sheets for Research and Industrial Chemicals, vols. I−III, VCH Publishers, Inc., New York, NY, 1985 and vols. IV−VI, New York, NY, 1987.

[52] R.P. Pohanish, Rapid Guide to Hazardous Chemicals in the Environment, Van Nostrand Reinhold, New York, NY, 1997.

[53] US Congress, Office of Technology Assessment, Identifying and Regulating Carcinogens, US Government Printing Office, Washington, DC, 1987 (Report OTA-BPH-42).

[54] US Environmental Protection Agency, Pesticide Fact Handbook, Noyes Data Corp., Park Ridge, NJ, 1988.

[55] Dutch Association of Safety Experts, Dutch Chemical Industry Association and Dutch Safety Institute, Handling Chemicals Safely, second ed., Dutch Association of Safety Experts, Amsterdam, Netherlands, 1980.

[56] P.M. Eller, M.E. Cassinelli (Eds.), NIOSH Manual of Analytical Methods (NMAM®), fourth ed., National Institute for Occupational Safety and Health, Cincinnati, OH, 1998 (Second supplement. DHHS (NIOSH) Publication No. 98-119).

[57] US Department of Labor, OSHA, Air contaminants—final rule, Fed. Regist. 54(12) (1989) 2332–2983 (29CFR1910—Occupational Safety and Health Administration).

[58] New Jersey Drinking Water Institute, Maximum Contaminant Level Recommendations for Hazardous Contaminants in Drinking Water, New Jersey Drinking Water Institute, Trenton, NJ, 26 March 1987 (Appendix B: Health-Based Maximum Contaminant Level Support Documents).

[59] US Environmental Protection Agency, NATICH Data Base Report on State, Local and EPA Air Toxics Activities, Office of Air Quality Planning and Standards, Research Triangle Park, NC, July 1988.

[60] US Environmental Protection Agency, Summary of State and Federal Drinking Water Standards and Guidelines, Federal-State Toxicology and Regulatory Alliance Committee (FSTRAC), Office of Drinking Water, Washington, DC, February 1990.

[61] US Environmental Protection Agency, National primary and secondary drinking water regulations, Fed. Regist. 54(97) (1989) 22062–22160.

[62] US Department of Labor, Occupational Safety and Health Standards, US Department of Labor, Washington, DC, 1 July 1988 (29CFR1910).

[63] US Department of Transportation, Chemical Data Guide for Bulk Shipment by Water, United States Coast Guard, Washington, DC, 1990.

[64] New York State Department of Health, Bureau of Toxic Substance Assessment, Chemical Fact Sheets, Albany, NY (various issues and dates).

[65] US Environmental Protection Agency, Consolidated List of Chemicals Subject to the Emergency Planning and Community Right-to-Know Act (EPCRA) and Section 112(r) of The Clean Air Act, as Amended, USEPA, Washington, DC, October 2001 (EPA 550-B-01-003).

[66] P.M. Bomgardner (Ed.), Handling Hazardous Materials, American Trucking Association, Alexandria, VA, 1997.

[67] R.J. Lewis Sr., Hawley's Condensed Chemical Dictionary, thirteenth ed., Van Nostrand Reinhold, New York, NY, 1998.

[68] R.P. Pohanish, S.A. Greene, Hazardous Substance Resource Guide, second ed., Gale Research, Detroit, MI, 1997.

[69] New Jersey Department of Health and Senior Services, Right-to-Know Project, Hazardous Substance Fact Sheets, New Jersey Department of Health and Senior Services, Trenton, NJ, various dates from 1985 to 2007.

[70] S.R. Stricoff, L.J. Partridge Jr. (Eds.), NIOSH/OSHA Occupational Health Guidelines for Chemical Hazards, United States Department of Health and Human Services, Cincinnati, OH, 1997.

[71] US Environmental Protection Agency, Pollution Prevention Fact Sheets: Chemical Production, FREG-1 (PPIC), Washington, DC (various years).

[72] US Environmental Protection Agency, Polychlorinated Biphenyl (PWB) Information Package, TSCA Information Service, Washington, DC, April 1993.

[73] R.P. Pohanish, A.G. Stanley, Wiley Guide to Chemical Incompatibilities, second ed., John Wiley & Sons, New York, NY, 2003.

[74] R.J. Lewis Sr., Sax's Dangerous Properties of Industrial Materials, eleventh ed., John Wiley & Sons, Hoboken, NJ, 2005.

[75] ATSDR (Agency for Toxic Substances and Disease Registry), US Department of Health and Human Services, Public Health Service, Toxicological Fact Sheets, Atlanta, GA (various dates).

[76] US Department of Health and Human Services/National Institute for Occupational Safety and Health, CD-ROM NIOSH/OSHA Pocket Guide to Chemical Hazards and other Databases, US Department of Health and Human Services, Cincinnati, OH, September 2005 (DHHS (NIOSH) Publication No. 2005-151).

[77] FEMA (Federal Emergency Management Agency), US Fire Administration, Hazardous Materials for First Responders, FEMA, Washington, DC, 1999.

[78] US Senate Committee on Banking, Housing, and Urban Affairs, Congressional Record, Riegel Report: United States Dual Use Exports to Iraq and Their Possible Impact on the Health Consequences of the Persian Gulf War, US Government Printing Office, Washington, DC, 25 May 1994.

[79] US Environmental Protection Agency, Special Review and Reregistration Division Office of Pesticide Programs, Agency Status of Pesticides in Registration, Reregistration, and Special Review, US Government Printing Office, Washington, DC, 1998 (also known as the Rainbow Report).

[80] F.R. Sidell, Riot control agents, in: R. Zajtchuk, R.F. Bellamy (Eds.), Textbook of Military Medicine: Medical Aspects of Chemical and Biological Warfare, Office of the Surgeon General, TMM Publications, Borden Institute, Walter Reed Army Medical Center, Washington, DC, 1997, pp. 307–324.

[81] US Environmental Protection Agency, Chemical Emergency and Preparedness and Prevention, Extremely Hazardous Substances (EHS) Chemical Profiles and Emergency First Aid Guides, USEPA, Washington, DC, last updated 2007. http://yosemite.epa.gov/osuker/CeppoeHs.nsf/firstaid/100-44-7?OpenDocument.

[82] S.E. Flynn, J.J. Kirkpatrick, Ending the post 9/11 security neglect of America's chemical facilities. Written testimony before a hearing of the Committee on Homeland Security and Governmental Affairs, United States Senate on The Security of America's Chemical Facilities, Washington, DC, 2005.

[83] Chemical Safety and Hazard Investigation Board (CSB), 600K Report, CSB, Washington, DC, February 1999.

[84] A.L. Schneider (Ed.), CHRIS + CD-ROM Version 2.0 (United Coast Guard Chemical Hazard Response Information System (COMDTINST 16465.12C), United States Coast Guard and the Department of Homeland Security, Washington, DC, 2007.

[85] R.B. Belmont, TR Tests of Level A Suits—Protection Against Chemical and Biological Warfare Agents and Stimulants: Executive Summary, CBRD-EN (US Army Chemical and Biological Defense Command), Aberdeen Proving Ground, MD, 1998.

[86] US Department of Energy, Office of Emergency Management and Policy, Protective Action Criteria for Chemicals—Including AEGLs, ERPGs, & TEELs (aka "PAC data set"), Revision 26, developed under contract number DE-AC05-06OR23100 between Oak Ridge Associated Universities and the US Department of Energy, Oak Ridge, TN, and Washington, DC, September 2010. http://orise.orau.gov/emi/scapa/chem-pacs-teels/default.htm.

[87] Hazardous Substances Data Bank (HSDB), US National Library of Medicine, TOXNET. http://toxnet.nlm.nih.gov/cgi-bin/sis/search/r?dbs + hsdb:@term + @rn + 64-17-5.

[88] R.E. Gosselin, R.P. Smith, H.C. Hodge, Clinical Toxicology of Commercial Products, fifth ed., Williams & Wilkins Inc., Baltimore, MD, 1984.

[89] US Environmental Protection Agency, Integrated Risk Information System (IRIS), Washington, DC (various dates). http://www.epa.gov/iris/; http://www.epa.gov/IRIS/subst/0096.htm.

[90] US National Response Team, Hazards, chemicals and other materials, Washington, DC. http://www.nrt.org/production/NRT/NRTWeb.nsf/PagesByLevelCat/Level2NRTPublications?Opendocument.

[91] Centers for Disease Control and Prevention, National Institute for Occupational Safety and Health (NIOSH), Databases, Atlanta, GA, 2011. http://www.cdc.gov/niosh/database.html.

[92] T. Besch, C. Moss, the Battlebook Project Team, et al., Medical NBC Battlebook, The US Army Center for Health Promotion and Preventive Medicine (USACHPPM), Fort Leonard Wood, MO, 1991 (Tech Guide 244, 304 p.).

[93] US Army Chemical Materials Agency (CMA), Various Announcements and Press Releases, CMA, Aberdeen Proving Ground, MD, 2008.

[94] US Department of Health and Human Services, The Medical Management Guidelines (MMGs) for Acute Chemical Exposures, US Department of Health and Human Services, Atlanta, GA, 2006.

[95] US Department of Health and Human Services, Division of Toxicology and Environmental Medicine, Managing Hazardous Materials Incidents, US Department of Health and Human Services, Atlanta, GA, 1990.

[96] European Commission Joint Research Centre, Institute for Health and Consumer Protection, ESIS, European Chemical Substances Information System. http://ecb.jrc.ec.europa.eu/esis/.

[97] Environmental Agency for Germany (Umweltbundesamt), Commission for the Evaluation of Substances Hazardous to Waters, Documentation and Information on Substances Hazardous to Waters, WGK-Documentation according to Annex 3 of the VwVwS, Verwaltungssvorschrift wassergefährdende Stoffe (VwVwS), Berlin, Germany (Current). http://webrigoletto.uba.de/rigoletto/public/search.do.

[98] US Department of Commerce, National Oceanic and Atmospheric Administration (NOAA) Office of Response and Restoration (OR&R) and US Environmental Agency, Computer-Aided Management of Emergency Operations (CAMEO) web site. http://www.noaa.gov/index.html.

[99] Office of Environmental Health Hazard Assessment, State of California, Prop 65 List (Chemical Listed as Known to the State of California to Cause Cancer), Sacramento, CA, 7 January 2011.

[100] D.R. Franz, Defense Against Toxin Weapons, revised ed., US Army Medical Research and Material Command, US Army Medical Research Institute of Infectious Diseases, Fort Detrick, MD, 1997.

[101] American Medical Association, Quick Reference Guide to Biological Weapons, AMA, Chicago, IL, 2002.

[102] US Army Medical Research Institute of Infectious Diseases, Medical Management of Biological Casualties Handbook, fourth ed., US Army Medical Research Institute of Infectious Diseases, Ft. Detrick, Frederick, MD, 2001.

Appendix 1: Oxidizing Materials

Notes: Best efforts have been made to ensure the information below is as accurate as possible. This list is not (and cannot be) comprehensive. Therefore, the absence of an oxidizing material from this list *does not* mean that it is not an oxidizer. The Publisher or Author cannot accept any responsibility for the use or misuse of this information. Generally, solid oxidizers include the following materials: bromates, chlorates, chlorites, chromates, dichromates, hypochlorites, iodates, nitrates, nitrites, perchlorates, permanganates, peroxides, persulfates, picrates, and chemical oxygen generators (such as potassium superoxide). Also, many fertilizers are oxidizing materials.

Oxidizing Materials by Name and CAS

Aluminum nitrate	13473-90-0
Ammonium chloride	7446-70-0
Ammonium chromate	7788-98-9
Ammonium dichromate	7789-09-5
Ammonium nitrate	6484-52-2
Ammonium nitrate—phosphate mixture	57608-40-9
Ammonium nitrate—sulfate mixture	6484-52-2
Ammonium perchlorate	7790-98-9
Ammonium permanganate	13446-10-1
Ammonium persulfate	7727-54-0
Ammonium picrate (wet)	131-74-8
Amyl nitrate	463-04-7
Barium bromate	13967-90-3
Barium chlorate	13477-00-4
Barium hypochlorite	13477-10-6
Barium nitrate	10022-31-8
Barium perchlorate trihydrate	10294-39-0
Barium permanganate	7787-36-2
Barium peroxide	1304-29-6
Benzoyl peroxide	94-36-0
Beryllium nitrate	13597-99-4
Bromine	7726-95-6
Bromine chloride	13863-41-7
Bromine pentafluoride	7789-30-2
Bromine trifluoride	7787-71-5
t-Butyl peroxybenzoate	614-45-9
Cadmium nitrate	10325-94-7
Cadmium nitrate tetrahydrate	10022-68-1
Calcium chlorate	10137-74-3
Calcium chlorite	14674-72-7
Calcium hypochlorite*	7778-54-3
Calcium nitrate	10124-37-5
Calcium perchlorate	13477-36-6
Calcium permanganate	10118-76-0
Calcium peroxide	1305-79-9
Cesium nitrate	7789-18-6
Chloric acid solution	7790-93-4
Chlorine	7787-50-5
Chlorine oxide	10049-04-4
Chlorine trifluoride	7790-91-2
Chromic acid*	7738-94-5
Chromic anhydride	11115-74-5
Chromium nitrate	7789-02-8
Chromium oxychloride	14977-61-8
Chromium trioxide, anhydrous	1333-82-0
Cobalt nitrate	10141-05-6
Cobalt(II) perchlorate, hexahydrate	13478-33-6
Cumene hydroperoxide	80-15-9
Cupric nitrate (copper nitrate)	3251-23-8
Cyclohexanone peroxide	12262-58-7
Diacetone alcohol peroxide	N/A
Dimethylhexane dihydroperoxide	3025-88-5
Di-*t*-butyl peroxide	110-05-4
Dichloroisocyanuric acid	2782-57-2
Ferric nitrate	10421-48-4
Gallium(III) nitrate	13494-90-1
Guanidine nitrate	506-93-4
Hydrogen peroxide (27.5—52% by weight)*	7722-84-1
Indium nitrate	13770-61-1
Iodine pentafluoride	7783-66-6
Iron(II) nitrate hexahydrate (1:2:6)	13520-68-8
Isopropyl nitrate	1712-64-7
Isopropyl peroxydicarbonate	105-64-6
Lanthanum nitrate	10099-59-9
Lead dioxide	1309-60-0
Lead nitrate	10099-74-8
Lead perchlorate	13637-76-8
Lithium bichromate	N/A
Lithium chromate	14307-35-8
Lithium hypochlorite	13840-33-0
Lithium nitrate	7790-69-4
Lithium peroxide	12031-80-0
Magnesium bromate	7789-36-8
Magnesium chlorate	7791-19-7
Magnesium nitrate	10377-60-3
Magnesium perchlorate	10034-81-8
Magnesium peroxide	14452-57-4
Manganese dioxide	1313-13-9
Manganese nitrate	10377-66-9
Mercuric nitrate	10045-94-0
Mercurous nitrate	10415-75-5
Nickel nitrate	14216-75-2
Nickel nitrite	17861-62-0
Nickel perchlorate	13637-71-3
Nickel(II) nitrate	13138-45-9
Nitrogen dioxide	10102-44-0
Nitrogen trifluoride	7783-54-2
Nitrosylsulfuric acid	7782-78-7

Oxidizing Materials by CAS and Name

7738-94-5... *Chromic acid
7757-79-1... Potassium nitrate
7758-01-2... *Potassium bromate
7758-09-0... Potassium nitrite
7758-19-2... *Sodium chlorite
7761-88-8... Silver nitrate
7775-11-3... Sodium chromate
7775-09-9... Sodium chlorate
7775-27-1... Sodium persulfate
7778-50-9... Potassium dichromate
7778-54-3... *Calcium hypochlorite
7778-74-7... Potassium perchlorate
7779-88-6... Zinc nitrate
7782-78-7... Nitrosyl sulfuric acid
7783-54-2... Nitrogen trifluoride
7783-66-6... Iodine pentafluoride
7787-36-2... Barium permanganate
7787-50-5... Chlorine
7787-71-5... Bromine trifluoride
7788-98-9... Ammonium chromate
7789-00-6... Potassium chromate(VI)
7789-02-8... Chromium nitrate
7789-09-5... Ammonium dichromate
7789-12-0... Sodium dichromate
7789-18-6... Cesium nitrate
7789-30-2... Bromine pentafluoride
7789-36-8... Magnesium bromate
7789-38-0... Sodium bromate
7790-69-4... Lithium nitrate
7790-91-2... Chlorine trifluoride
7790-93-4... Chloric acid solution
7790-98-9... Ammonium perchlorate
7791-10-8... Strontium chlorate
7791-19-7... Magnesium chlorate
7791-27-7... Pyrosulfuryl chloride
10022-31-8... Barium nitrate
10022-68-1... Cadmium nitrate tetrahydrate
10028-15-6... Ozone
10034-81-8... Magnesium perchlorate
10042-76-9... Strontium nitrate
10045-94-0... Mercuric nitrate
10049-04-4... Chlorine oxide
10099-59-9... Lanthanum nitrate
10099-74-8... Lead nitrate
10101-50-5... *Sodium permanganate
10102-05-3... Palladium dinitrate
10102-06-4... Uranyl nitrate
10102-44-0... Nitrogen dioxide
10118-76-0... Calcium permanganate
10124-37-5... Calcium nitrate

10137-74-3... Calcium chlorate
10141-05-6... Cobalt nitrate
10294-39-0... Barium perchlorate trihydrate
10325-94-7... Cadmium nitrate
10361-95-2... Zinc chlorate
10377-60-3... Magnesium nitrate
10377-66-9... Manganese nitrate
10415-75-5... Mercurous nitrate
10421-48-4... Ferric nitrate
11115-74-5... Chromic anhydride
12030-88-5... Potassium superoxide
12031-80-0... Lithium peroxide
12034-12-7... Sodium superoxide
12262-58-7... Cyclohexanone peroxide
13138-45-9... Nickel(II) nitrate
13446-10-1... Ammonium permanganate
13450-97-0... Strontium perchlorate
13473-90-0... Aluminum nitrate
13477-00-4... Barium chlorate
13477-10-6... Barium hypochlorite
13477-36-6... Calcium perchlorate
13478-33-6... Cobalt(II) perchlorate, hexahydrate
13494-90-1... Gallium(III) nitrate
13494-98-9... Yttrium(III) nitrate hexahydrate
13517-27-6... Zinc bromate
13520-68-8... Iron(II) nitrate hexahydrate (1:2:6)
13530-65-9... Zinc chromate
13637-71-3... Nickel perchlorate
13637-76-8... Lead perchlorate
13746-89-9... Zirconium nitrate
13768-67-7... Ytterbium nitrate
13770-61-1... Indium nitrate
13823-29-5... Thorium(IV) nitrate
13840-33-0... Lithium hypochlorite
13863-41-7... Bromine chloride
13967-90-3... Barium bromate
14216-75-2... Nickel nitrate
14307-35-8... Lithium chromate
14452-57-4... Magnesium peroxide
14674-72-7... Calcium chlorite
14977-61-8... Chromium oxychloride
15630-89-4... Sodium percarbonate
17014-71-0... Potassium peroxide
17861-62-0... Nickel nitrite
23414-72-4... Zinc permanganate
57608-40-9............. Ammonium nitrate−phosphate mixture
63885-01-8... Zinc ammonium nitrite

Note: The asterisk (*) denotes some of those materials with potential for spontaneous ignition.

Appendix 2: Carcinogens

Carcinogen Index by Name and CAS

A-α-C (2-Amino-9H-pyrido [2,3-b]indole) 26148-68-5
Acetaldehyde 75-07-0
Acetamide 60-35-5
Acetochlor...34256-82-1
2-Acetylaminofluorene 53-96-3
Acifluorfen sodium...62476-59-9
Acrylamide..79-06-1
Acrylonitrile..107-13-1
Actinomycin D..50-76-0
AF-2 [2-(2-furyl)-3-(5-nitro-2-furyl)]
acrylamide...3688-53-7
Aflatoxins...various CAS
Alachlor 15972-60-8
Alcoholic beverages, when associated
with alcohol abuse .... no CAS
Aldrin 309-00-2
Allyl chloride (*delisted 1999*)...........................107-05-1
2-Aminoanthraquinone.......................................117-79-3
p-Aminoazobenzene 60-09-3
o-Aminoazotoluene..97-56-3
4-Aminobiphenyl (4-amin *o*-diphenyl)......................92-67-1
1-Amino-2,4-dibromo-anthraquinone.......................81-49-2
3-Amino-9-ethylcarbazole hydrochloride 6109-97-3
2-Aminofluorene...153-78-6
1-Amino-2-methylanthraquinone 82-28-0
2-Amino-5-(5-nitro-2-furyl)-1,3,4-thiadiazole 712-68-5
4-Amino-2-nitrophenol.......................................119-34-6
Amitrole..61-82-5
Amsacrine 51264-14-3
Analgesic mixtures containing phenacetin.........various CAS
Aniline 62-53-3
Aniline hydrochloride...142-04-1
o-Anisidine...90-04-0
o-Anisidine hydrochloride..................................134-29-2
Antimony oxide (Antimony trioxide)....................1309-64-4
Anthraquinone 84-65-1
Aramite 140-57-8
Areca nut.. no CAS
Aristolochic acids ....various CAS
Arsenic (inorganic arsenic compounds)................7440-38-2
   List of arsenic compounds, inorganic*
   Arsanilic acid [arsonic acid,
   (4-aminophenyl)-]... 98-50-0
   Arsenic pentoxide
   [arsenic oxide (As$_2$O$_5$)] 1303-28-2
   Arsenic sulfide [arsenic sulfide (As$_2$S$_3$)] 1303-33-9
   Arsenic trioxide [arsenic oxide (As$_2$O$_3$)] 1327-53-3
   Arsine 7784-42-1
   Calcium arsenate [arsenic acid (H$_3$AsO$_4$),
   calcium salt (2:3)].. 7778-44-1

Dimethylarsinic acid (arsinic acid,
dimethyl-).. 75-60-5
Lead arsenate [arsenic acid (H$_3$AsO$_4$),
lead(2+) salt (1:1)] 7784-40-9
Methanearsonic acid, disodium salt
(arsonic acid, methyl-, disodium salt)............ 144-21-8
Methanearsonic acid, monosodium salt
(arsonic acid, methyl-, monosodium salt)..... 2163-80-6
Potassium arsenate [arsenic acid
(H$_3$AsO$_4$), monopotassium salt] 7784-41-0
Potassium arsenite (arsenenous acid,
potassium salt) 13464-35-2
Sodium arsenate, sodium salt 7631-89-2
Sodium arsenite 7784-46-5
Sodium cacodylate (arsinic acid,
dimethyl-, sodium salt)................................. 124-65-2
Asbestos 1332-21-4
Asbestos, amosite..12172-73-5
   Asbestos, actinolite 77536-66-4
   Asbestos, anthophylite................................ 77536-67-5
   Asbestos, anthophyllite............................... 17068-78-9
   Asbestos, chrysotile 12001-29-5
   Asbestos, crocidolite................................... 12001-28-4
   Asbestos, tremolite 77536-68-6
Auramine 492-80-8
Azacitidine..320-67-2
Azaserine..115-02-6
Azathioprine...446-86-6
Azobenzene..103-33-3
Benthiavalicarb-isopropyl..................................177406-68-7
Benz[a]anthracene..56-55-3
Benzene...71-43-2
Benzidine [and its salts] 92-87-5
Benzidine-based dyes ....various CAS
Benzo[b]fluoranthene...205-99-2
Benzo[j]fluoranthene..205-82-3
Benzo[k]fluoranthene...207-08-9
Benzofuran..271-89-6
Benzo[a]pyrene...50-32-8
Benzotrichloride...98-07-7
Benzyl chloride...100-44-7
Benzyl violet 4B..1694-09-3
Beryllium (and beryllium compounds) 7440-41-7
   List of beryllium compounds*
   Beryllium chloride...................................... 7787-47-5
   Beryllium fluoride 7787-49-7
   Beryllium nitrate.. 13597-99-4
   Beryllium sulfate 13510-49-1
Betel quid with tobacco................................ no CAS
Betel quid without tobacco............................ no CAS
2,2-Bis(bromomethyl)-1,3-propanediol.................3296-90-0
Bis(2-chloroethyl)ether......................................111-44-4

N,N-Bis(2-chloroethyl)-2-naphthylamine
(Chlornapazine) 494-03-1
Bischloroethyl nitrosourea (BCNU)
(Carmustine) 154-93-8
Bis(chloromethyl)ether 542-88-1
Bis(2-chloro-1-methylethyl)ether,
technical grade .... no CAS
Bitumens, extracts of steam-refined
and air refined... no CAS
Bracken fern... no CAS
Bromate...15541-45-4
Bromochloroacetic acid.............................5589-96-8
Bromodichloromethane 75-27-4
Bromoethane 74-96-4
Bromoform..75-25-2
1,3-Butadiene 106-99-0
1,4-Butanediol dimethanesulfonate (Busulfan)..........55-98-1
Butylated hydroxyanisole 25013-16-5
β-Butyrolactone 3068-88-0
Cacodylic acid..75-60-5
Cadmium (and cadmium compounds) 7440-43-9
   *List of cadmium compounds**
   Cadmium acetate 543-90-8
   Cadmium bromide 7789-42-6
   Cadmium chloride 10108-64-2
   Cadmium cyanide 542-83-6
   Cadmium fluoroborate...................... 14486-19-2
   Cadmium nitrate 10325-94-7
   Cadmium nitrate tetrahydrate.................... 10022-68-1
   Cadmium oxide............................... 1306-19-0
   Cadmium oxide fume 1306-19-0
   Cadmium stearate 2223-93-0
   Cadmium succinate.......................... 141-00-4
   Cadmium sulfate 10124-36-4
Caffeic acid..331-39-5
Captafol..2425-06-1
Captan 133-06-2
Carbaryl 63-25-2
Carbazole 86-74-8
Carbon black (airborne, unbound
particles of respirable size)......................1333-86-4
Carbon tetrachloride 56-23-5
Carbon-black extracts .... no CAS
N-Carboxymethyl-*N*-nitrosourea 60391-92-6
Catechol 120-80-9
Ceramic fibers (airborne particles
of respirable size)....................................... no CAS
Certain combined chemotherapy for lymphomas ..... no CAS
Chlorambucil 305-03-3
Chloramphenicol..56-75-7
Chlordane...57-74-9
Chlordecone (Kepone)...............................143-50-0
Chlordimeform..6164-98-3
Chlorendic acid...115-28-6
Chlorinated paraffins (Average Chain length, C12;
approximately 60% chlorine by weight) 108171-26-2

p-Chloroaniline 106-47-8
p-Chloroaniline hydrochloride 20265-96-7
Chlorodibromomethane (*delisted 1999*)..................124-48-1
Chloroethane (Ethyl chloride) 75-00-3
1-(2-Chloroethyl)-3-cyclohexylnitrosourea
(CCNU) (Lomustine)............................13010-47-4
1-(2-Chloroethyl)-3-(4-methylcyclohexyl)-
1-nitrosourea (Methyl-CCNU) 13909-09-6
Chloroform...67-66-3
Chloromethyl methyl ether
(technical grade) 107-30-2
3-Chloro-2-methylpropene 563-47-3
1-Chloro-4-nitrobenzene............................100-00-5
4-Chloro-*o*-phenylenediamine 95-83-0
Chloroprene 126-99-8
Chlorothalonil 1897-45-6
p-Chloro-*o*-toluidine 95-69-2
p-Chloro-*o*-toluidine, strong acid salts of ....various CAS
5-Chloro-*o*-toluidine and its strong
acid salts..various CAS
Chlorotrianisene..569-57-3
Chlorozotocin...54749-90-5
Chromium (hexavalent compounds) ....various CAS
Chrysene 218-01-9
C.I. Acid Red 114 6459-94-5
C.I. Basic Red 9 monohydrochloride 569-61-9
C.I. Direct Blue 15.....................................2429-74-5
C.I. Direct Blue 218....................................28407-37-6
C.I. Solvent Yellow 14 842-07-9
Ciclosporin (Cyclosporin A;
Cyclosporine)....................59865-13-3; 79217-60-0
Cidofovir..113852-37-2
Cinnamyl anthranilate..................................87-29-6
Cisplatin 15663-27-1
Citrus Red No. 2 6358-53-8
Clofibrate 637-07-0
Cobalt metal powder...................................7440-48-4
Cobalt[II] oxide 1307-96-6
Cobalt sulfate 10124-43-3
Cobalt sulfate heptahydrate 10026-24-1
Coke oven emissions .... no CAS
Conjugated estrogens.................................. no CAS
Creosotes...various CAS
p-Cresidine..120-71-8
Cumene 98-82-8
Cupferron 135-20-6
Cycasin...14901-08-7
Cyclophosphamide (anhydrous)...............................50-18-0
Cyclophosphamide (hydrated)....................6055-19-2
Cytembena 21739-91-3
D&C Orange No. 17 3468-63-1
D&C Red No. 8 2092-56-0
D&C Red No. 9 5160-02-1
D&C Red No. 19 81-88-9
Dacarbazine 4342-03-4
Daminozide 1596-84-5

Nickel hydroxide 12054-48-7; 12125-56-3
Nickelocene 1271-28-9
Nickel oxide.......................................1313-99-1
Nickel refinery dust from the
pyrometallurgical process...................... no CAS
Nickel subsulfide 12035-72-2
Niridazole...61-57-4
Nitrapyrin 1929-82-4
Nitrilotriacetic acid 139-13-9
Nitrilotriacetic acid, trisodium
salt monohydrate.................................18662-53-8
5-Nitroacenaphthene............................602-87-9
5-Nitro-o-anisidine (delisted 2006) 99-59-2
o-Nitroanisole.....................................91-23-6
Nitrobenzene.......................................98-95-3
4-Nitrobiphenyl...................................92-93-3
6-Nitrochrysene 7496-02-8
Nitrofen (technical grade)......................1836-75-5
2-Nitrofluorene 607-57-8
Nitrofurazone......................................59-87-0
1-[(5-Nitrofurfurylidene)-amino]-2-
imidazolidinone..................................555-84-0
N-[4-(5-Nitro-2-furyl)-2-thiazolyl]acetamide...........531-82-8
Nitrogen mustard (Mechlorethamine) 51-75-2
Nitrogen mustard hydrochloride (Mechlorethamine
hydrochloride)....................................55-86-7
Nitrogen mustard N-oxide 126-85-2
Nitrogen mustard N-oxide hydrochloride................302-70-5
Nitromethane......................................75-52-5
2-Nitropropane....................................79-46-9
1-Nitropyrene......................................5522-43-0
4-Nitropyrene......................................57835-92-4
N-Nitrosodi-n-butylamine......................924-16-3
N-Nitrosodiethanolamine.......................1116-54-7
N-Nitrosodiethylamine...........................55-18-5
N-Nitrosodimethylamine.........................62-75-9
p-Nitrosodiphenylamine 156-10-5
N-Nitrosodiphenylamine.........................86-30-6
N-Nitrosodi-n-propylamine 621-64-7
N-Nitroso-N-ethylurea 759-73-9
3-(N-Nitrosomethylamino)-propionitrile.............60153-49-3
4-(N-Nitrosomethylamino)-1-(3-pyridyl)1-
butanone...64091-91-4
N-Nitrosomethylethylamine 10595-95-6
N-Nitroso-N-methylurea 684-93-5
N-Nitroso-N-methylurethane 615-53-2
N-Nitrosomethylvinylamine 4549-40-0
N-Nitrosomorpholine............................59-89-2
N-Nitrosonornicotine 16543-55-8
N-Nitrosopiperidine 100-75-4
N-Nitrosopyrrolidine............................930-55-2
N-Nitrososarcosine...............................13256-22-9
o-Nitrotoluene.................................... 88722
Norethisterone (Norethindrone)................68-22-4
Norethynodrel......................................68-23-5
Ochratoxin A 303-47-9

Oil Orange SS......................................2646-17-5
Oral contraceptives, combined .... no CAS
Oral contraceptives, sequential................ no CAS
Oryzalin 19044-88-3
Oxadiazon 19666-30-9
Oxazepam...604-75-1
Oxymetholone......................................434-07-1
Oxythioquinox (Chinomethionat)..............2439-01-2
Palygorskite fibers (>5 μm in length)..................12174-11-7
Panfuran S..794-93-4
Pentachlorophenol 87-86-5
Phenacetin..62-44-2
Phenazopyridine...................................94-78-0
Phenazopyridine hydrochloride................136-40-3
Phenesterin...3546-10-9
Phenobarbital.......................................50-06-6
Phenolphthalein....................................77-09-8
Phenoxybenzamine 59-96-1
Phenoxybenzamine hydrochloride.............63-92-3
o-Phenylenediamine and its salts.............95-54-5
Phenyl glycidyl ether.............................122-60-1
Phenylhydrazine and its salts .... no CAS
o-Phenylphenate, sodium........................132-27-4
o-Phenylphenol....................................90-43-7
PhiP(2-Amino-1-methyl-6-phenylimidazol
[4,5-b]pyridine)...................................105650-23-5
Pirimicarb...23103-98-2
Polybrominated biphenyls (PBBs) ....various CAS
*List of polybrominated biphenyl (PBB) compounds**
  p-Bromodiphenyl ether.......................... 101-55-3
  p,p'-Didibromodiphenyl ether 2050-47-7
  Decabromobiphenyl............................. 13654-09-6
  Decabromodiphenyl ether....................... 1163-19-5
  Hexabromobiphenyl............................. 59080-40-9
  Hexabromo-1,1'-biphenyl....................... 36355-01-8
  Hexabromodiphenyl ether...................... 36483-60-0
  Nonabromodiphenyl ether...................... 63936-56-1
  Octabromobiphenyl 27858-07-7
  Octabromobiphenyl 61288-13-9
  Octabromodiphenyl ether...................... 32536-52-0
  Pentabromodiphenyl ether..................... 32534-81-9
  Polybrominated biphenyl....................... 59536-65-1
  Polybrominated biphenyl mixture 67774-32-7
  Tetrabromodiphenyl ether 40088-47-9
  Tribromodiphenyl ether 49690-94-0
Polychlorinated biphenyls (PCBs)..................... no CAS
*List of polychlorinated biphenyl (PCB) compounds**
  Polychlorinated biphenyls 1336-36-3
  Biphenyl....................................... 92-52-4
  2-Chlorobiphenyl................................ 2051-60-7
  4-Chlorobiphenyl................................ 2051-62-9
  2,2'-Dichlorobiphenyl.......................... 13029-08-8
  2,3'-Dichlorobiphenyl.......................... 25569-80-6
  2,4'-Dichlorobiphenyl.......................... 34883-43-7
  4,4'-Dichlorobiphenyl.......................... 2050-68-2
  2,2',3,3',4,6-Hexachlorobiphenyl 38380-05-1

2,2′, 3,3′,6,6′-Hexachlorobiphenyl 38411-22-2
2,2′,3,4,4′,5-Hexachlorobiphenyl 35694-06-5
2,2′,3,4,4′,5′-Hexachlorobiphenyl............... 35065-28-2
2,2′,3′,4,5,6′-Hexachlorobiphenyl............... 38380-04-0
2,2′,4,4′,5,5′-Hexachlorobiphenyl............... 35065-27-1
2,2′,3,3′,4,4′,5-Heptachlorobiphenyl 35065-30-6
2,2′,3,3′,4,5,6′-Heptachlorobiphenyl 38441-25-5
2,2′,3,4,4′,5,5′-Heptachlorobiphenyl 35065-29-3
2,2′,3,3′,6-Pentachlorobiphenyl 52663-60-2
2,2′,3,4,5′-Pentachlorobiphenyl 38380-02-8
2,2′,3′,4,5-Pentachlorobiphenyl 41464-51-1
2,2′,3,4′,6-Pentachlorobiphenyl .... no CAS
2,2′,3,5′,6-Pentachlorobiphenyl 38379-99-6
2,2′,4,4′,5-Pentachlorobiphenyl 38380-01-7
2,2′,4,5,5′-Pentachlorobiphenyl 37680-73-2
2,3,3′,4,4′-Pentachlorobiphenyl 32598-14-4
2,3,3′,4′,6-Pentachlorobiphenyl 38380-03-9
2,3′,4,4′,5-Pentachlorobiphenyl 31508-00-6
2,2′3,5′-Tetrachlorobiphenyl 41464-39-5
2,2′,4,5′-Tetrachlorobiphenyl 41464-40-8
2,2′,5,5′-Tetrachlorobiphenyl 35693-99-3
2,3,4,4′-Tetrachlorobiphenyl 33025-41-1
2,3′,4,4′-Tetrachlorobiphenyl 32598-10-0
2,3′,4′,5-Tetrachlorobiphenyl 32598-11-1
3,3′,4,4′-Tetrachlorobiphenyl 32598-13-3
2,2′,3-Trichlorobiphenyl.......................... 38444-78-9
2,2′,5-Trichlorobiphenyl.......................... 37680-65-2
2,3′,4-Trichlorobiphenyl .... no CAS
2′,3,4-Trichlorobiphenyl.......................... 38444-86-9
2,4,4′-Trichlorobiphenyl 7012-37-5
2,4′,5-Trichlorobiphenyl.......................... 16606-02-3
Polychlorinated biphenyls (containing
60 or more % Cl by molecular weight).............various CAS
Polychlorinated dibenzo-p-dioxins.....................various CAS
Polychlorinated dibenzofurans .... no CAS
Polygeenan.......................................53973-98-1
Ponceau MX 3761-53-3
Ponceau 3R 3564-09-8
Potassium bromate................................7758-01-2
Primidone 125-33-7
Procarbazine.....................................671-16-9
Procarbazine hydrochloride.......................366-70-1
Procymidone 32809-16-8
Progesterone.....................................57-83-0
Pronamide 23950-58-5
Propachlor.......................................1918-16-7
1,3-Propane sultone 1120-71-4
Propargite.......................................2312-35-8
β-Propiolactone..................................57-57-8
Propoxur...114-26-1
Propylene glycol mono-t-butyl ether...............57018-52-7
Propylene oxide..................................75-56-9
Propylthiouracil 51-52-5
Pyridine...110-86-1
Quinoline and its strong acid salts .... no CAS
Radionuclides....................................... no CAS

Reserpine 50-55-5
Residual (heavy) fuel oils........................ no CAS
Resmethrin.......................................10453-86-8
Riddelliine......................................23246-96-0
Saccharin (delisted 2001) 81-07-2
Saccharin, sodium (delisted 2003) 128-44-9
Safrole..94-59-7
Selenium sulfide 7446-34-6
Shale-oils.......................................68308-34-9
Silica, crystalline (airborne particles of
respirable size) .... no CAS
Soots, tars, and mineral oils (untreated and
mildly treated oils and used engine oils).................. no CAS
Spirodiclofen....................................148477-71-8
Spironolactone 52-01-7
Stanozolol 10418-03-8
Sterigmatocystin 10048-13-2
Streptozotocin (streptozocin)....................18883-66-4
Strong inorganic acid mists containing
sulfuric acid .... no CAS
Styrene oxide 96-09-3
Sulfallate.......................................95-06-7
Sulfasalazine (salicylazosulfapyridine) 599-79-1
Talc containing asbestiform fibers .... no CAS
Tamoxifen and its salts..........................10540-29-1
Terrazole 2593-15-9
Testosterone and its esters......................58-22-0
2,3,7,8-Tetrachlorodibenzo-p-dioxin (TCDD) 1746-01-6
1,1,2,2-Tetrachloroethane........................79-34-5
Tetrachloroethylene (Perchloroethylene) 127-18-4
p-a,a,a-Tetrachlorotoluene 5216-25-1
Tetrafluoroethylene..............................116-14-3
Tetranitromethane...............................509-14-8
Thioacetamide....................................62-55-5
4,4′-Thiodianiline..............................139-65-1
Thiodicarb......................................59669-26-0
Thiouracil......................................141-90-2
Thiourea..62-56-6
Thorium dioxide 1314-20-1
Tobacco, oral use of smokeless products.................. no CAS
Tobacco smoke .... no CAS
Toluene diisocyanate 26471-62-5
o-Toluidine.....................................95-53-4
o-Toluidine hydrochloride........................636-21-5
p-Toluidine (delisted 1999) 106-49-0
Toxaphene (Polychlorinated camphenes)..............8001-35-2
Toxins derived from Fusarium moniliforme
(Fusarium verticillioides) .... no CAS
Treosulfan 299-75-2
Trichlormethine (Trimustine hydrochloride) 817-09-4
Trichloroethylene................................79-01-6
2,4,6-Trichlorophenol 88-06-2
1,2,3-Trichloropropane 96-18-4
2,4,5-Trimethylaniline (and its strong
acid salts)......................................137-17-7
Trimethyl phosphate 512-56-1

Carcinogen Index by CAS and Name

75-09-2 .... Dichloromethane (Methylene chloride)
75-21-8 .... Ethylene oxide
75-25-2 .... Bromoform
75-27-4 .... Bromodichloromethane
75-34-3 .... 1,1-Dichloroethane
75-52-5 .... Nitromethane
75-55-8 .... 2-Methylaziridine (Propyleneimine)
75-56-9 .... Propylene oxide
75-60-5 .... Cacodylic acid
76-44-8 .... Heptachlor
76-87-9 .... Triphenyltin hydroxide
77-09-8 .... Phenolphthalein
77-78-1 .... Dimethyl sulfate
78-79-5 .... Isoprene
78-87-5 .... 1,2-Dichloropropane
79-00-5 .... Vinyl trichloride (1,1,2-Trichloroethane)
79-01-6 .... Trichloroethylene
79-06-1 .... Acrylamide
79-34-5 .... 1,1,2,2-Tetrachloroethane
79-43-6 .... Dichloroacetic acid
79-44-7 .... Dimethylcarbamoyl chloride
79-46-9 .... 2-Nitropropane
81-07-2 .... Saccharin (delisted 2001)
81-49-2 .... 1-Amino-2,4-dibromo-anthraquinone
81-88-9 .... D&C Red No. 19
82-28-0 .... 1-Amino-2-methylanthraquinone
84-17-3 .... Dienestrol
84-65-1 .... Anthraquinone
86-30-6 .... N-Nitrosodiphenylamine
86-74-8 .... Carbazole
87-29-6 .... Cinnamyl anthranilate
87-62-7 .... 2,6-Xylidine (2,6-Dimethylaniline)
87-86-5 .... Pentachlorophenol
88-06-2 .... 2,4,6-Trichlorophenol
88722 .... o-Nitrotoluene
90-04-0 .... o-Anisidine
90-43-7 .... o-Phenylphenol
90-94-8 .... Michler's ketone
91-20-3 .... Naphthalene
91-23-6 .... o-Nitroanisole
91-59-8 .... 2-Naphthylamine
91-94-1 .... 3,3'-Dichlorobenzidine
92-67-1 .... 4-Aminobiphenyl (4-amino-diphenyl)
92-87-5 .... Benzidine [and its salts]
92-93-3 .... 4-Nitrobiphenyl
93-15-2 .... Methyleugenol
94-58-6 .... Dihydrosafrole
94-59-7 .... Safrole
94-78-0 .... Phenazopyridine
95-06-7 .... Sulfallate
95-53-4 .... o-Toluidine
95-54-5 .... o-Phenylenediamine and its salts
95-69-2 .... p-Chloro-o-toluidine
95-80-7 .... 2,4-Diaminotoluene
95-83-0 .... 4-Chloro-o-phenylenediamine
96-09-3 .... Styrene oxide

96-12-8 .... 1,2-Dibromo-3-chloropropane (DBCP) (male)
96-13-9 .... 2,3-Dibromo-1-propanol
96-18-4 .... 1,2,3-Trichloropropane
96-23-1 .... 1,3-Dichloro-2-propanol (1,3-DCP)
96-24-2 .... 3-Monochloropropane-1,2- diol (3-MCPD)
96-45-7 .... Ethylene thiourea
97-56-3 .... o-Aminoazotoluene
98-07-7 .... Benzotrichloride
98-82-8 .... Cumene
98-95-3 .... Nitrobenzene
99-59-2 .... 5-Nitro-o-anisidine (delisted 2006)
100-00-5 .... 1-Chloro-4-nitrobenzene
100-40-3 ....4-Vinylcyclohexene
100-41-4 .... Ethylbenzene
100-44-7 .... Benzyl chloride
100-75-4 .... N-Nitrosopiperidine
101-14-4 .... 4,4'-Methylene bis(2-chloroaniline)
101-61-1 .... 4,4'-Methylene
                        bis(N,N-dimethyl)benzenamine
101-77-9 .... 4,4'-Methylenedianiline
101-80-4 .... 4,4'-Diaminodiphenyl
                        ether(4,4'-Oxydianiline)
101-90-6 .... Diglycidyl resorcinol ether (DGRE)
103-33-3 .... Azobenzene
106-46-7 .... p-Dichlorobenzene
106-47-8 .... p-Chloroaniline
106-49-0 .... p-Toluidine (delisted 1999)
106-87-6 .... 4-Vinyl-1-cyclohexene diepoxide (Vinyl
                        cyclohexene dioxide)
106-89-8 .... Epichlorohydrin
106-93-4 .... Ethylene dibromide
106-99-0 .... 1,3-Butadiene
107-05-1 .... Allyl chloride (delisted 1999)
107-06-2 .... Ethylene dichloride (1,2-Dichloroethane)
107-13-1 .... Acrylonitrile
107-30-2 .... Chloromethyl methyl ether (technical grade)
110-00-9 .... Furan
110-86-1 .... Pyridine
111-44-4 .... Bis(2-chloroethyl)ether
114-26-1 .... Propoxur
115-02-6 .... Azaserine
115-28-6 .... Chlorendic acid
115-96-8 .... Tris(2-chloroethyl) phosphate
116-14-3 .... Tetrafluoroethylene
117-10-2 .... Dantron (Chrysazin;
                        1,8-Dihydroxy-anthraquinone)
117-79-3 .... 2-Aminoanthraquinone
117-81-7 .... Di(2-ethylhexyl)phthalate (DEHP)
118-74-1 .... Hexachlorobenzene
118-96-7 .... 2,4,6-Trinitrotoluene
119-34-6 .... 4-Amino-2-nitrophenol
119-90-4 .... 3,3'-Dimethoxybenzidine (o-Dianisidine)
119-93-7 .... 3,3'-Dimethylbenzidine (o-Tolidine)
120-58-1 .... Isosafrole (delisted 2006)
120-71-8 .... p-Cresidine
120-80-9 .... Catechol

121-14-2.................................... 2,4-Dinitrotoluene
122-60-1..............................Phenyl glycidyl ether
122-66-7............. Hydrazobenzene (1,2-Diphenylhydrazine)
123-91-1.. 1,4-Dioxane
124-48-1......... Chlorodibromomethane (*delisted 1999*)
125-33-7.. Primidone
126-07-8.. Griseofulvin
126-72-7............. Tris(2,3-dibromopropyl)phosphate
126-85-2......................Nitrogen mustard *N*-oxide
126-99-8.. Chloroprene
127-18-4...............Tetrachloroethylene (Perchloroethylene)
128-44-9............. Saccharin, sodium (*delisted 2003*)
129-15-7...................... 2-Methyl-1-nitroanthraquinone
(of uncertain purity)
129-43-1..........................1-Hydroxyanthraquinone
132-27-4.......................... *o*-Phenylphenate, sodium
133-06-2...Captan
133-07-3..Folpet
134-29-2......................*o*-Anisidine hydrochloride
134-32-7.............................. 1-Naphthylamine
135-20-6..Cupferron
136-35-6........................... Diazoaminobenzene
136-40-3......................Phenazopyridine hydrochloride
136-45-8......................Di-*n*-propyl isocinchomeronate
(MGK Repellent 326)
137-17-7.... 2,4,5-Trimethylaniline (and its strong acid salts)
137-41-7........................... Metam potassium
137-42-8......................................Metam sodium
139-13-9.............................Nitrilotriacetic acid
139-65-1.......................... 4,4′-Thiodianiline
139-91-3...................... 5-(Morpholinomethyl)-3-
[(5-nitrofurfuryl-idene)-amino]-2-oxazolidinone
140-57-8... Aramite
140-67-0... Estragole
140-88-5...Ethyl acrylate
141-90-2.................................... Thiouracil
142-04-1.....................Aniline hydrochloride
143-50-0.................... Chlordecone (Kepone)
148-82-3.. Melphalan
151-56-4.....................................Ethyleneimine
153-78-6.................................. 2-Aminofluorene
154-93-8..................... Bischloroethyl nitrosourea (BCNU)
(Carmustine)
156-10-5.................... *p*-Nitrosodiphenylamine
189-55-9...................... Dibenzo[a,i]pyrene
189-64-0...................... Dibenzo[a,h]pyrene
191-30-0...................... Dibenzo[a,l]pyrene
192-65-4...................... Dibenzo[a,e]pyrene
193-39-5........................... Indeno[1,2,3-cd]pyrene
194-59-2.....................7H-Dibenzo[c,g]carbazole
205-82-3.....................Benzo[j]fluoranthene
205-99-2.....................Benzo[b]fluoranthene
207-08-9.....................Benzo[k]fluoranthene
218-01-9..Chrysene
224-42-0........................... Dibenz[a,j]acridine
226-36-8........................... Dibenz[a,h]acridine

271-89-6.. Benzofuran
298-81-7......................8-Methoxypsoralen with ultraviolet
A therapy
299-75-2......................................Treosulfan
301-04-2......................................Lead acetate
302-01-2.. Hydrazine
302-70-5........ Nitrogen mustard *N*-oxide hydrochloride
303-34-4.. Lasiocarpine
303-47-9...Ochratoxin A
305-03-3..Chlorambucil
309-00-2...Aldrin
315-22-0....................................Monocrotaline
320-67-2.......................................Azacitidine
330-54-1..Diuron
331-39-5..Caffeic acid
366-70-1..........................Procarbazine hydrochloride
373-02-4.. Nickel acetate
389-08-2.. Nalidixic acid
434-07-1... Oxymetholone
443-48-1.......................................Metronidazole
446-86-6... Azathioprine
484-20-8........5-Methoxypsoralen with ultraviolet A therapy
492-80-8... Auramine
494-03-1............... *N*,*N*-Bis(2-chloroethyl)-2-naphthylamine
(Chlornapazine)
505-60-2...Mustard Gas
509-14-8........................... Tetranitromethane
510-15-6................. Ethyl-4,4′-dichlorobenzilate
512-56-1....................Trimethyl phosphate
513-37-1....................Dimethylvinylchloride
531-76-0...Merphalan
531-82-8.......... *N*-[4-(5-Nitro-2-furyl)-2-thiazolyl]acetamide
540-73-8........................... 1,2-Dimethylhydrazine
542-56-3........................... Isobutyl nitrite
542-75-6............................1,3-Dichloropropene
542-88-1.....................Bis(chloromethyl)ether
555-84-0......................1-[(5-Nitrofurfurylidene)-amino]-2-
imidazolidinone
556-52-5... Glycidol
563-47-3..................... 3-Chloro-2-methylpropene
569-57-3............................. Chlorotrianisene
569-61-9...................C.I. Basic Red 9 monohydrochloride
590-96-5....................................Methylazoxymethanol
592-62-1...................Methylazoxymethanol acetate
593-60-2..Vinyl bromide
598-55-0.................................Methyl carbamate
599-79-1............Sulfasalazine (salicylazosulfapyridine)
602-87-9....................................5-Nitroacenaphthene
604-75-1..Oxazepam
606-20-2.................................. 2,6-Dinitrotoluene
607-57-8.................................. 2-Nitrofluorene
608-73-1............... Hexachlorocyclohexane (technical grade)
612-82-8................. 3,3′-Dimethylbenzidine dihydrochloride
612-83-9................. 3,3′-Dichlorobenzidine dihydrochloride
613-35-4.............................. *N*,*N*′-Diacetylbenzidine
615-05-4................................. 2,4-Diaminoanisole

615-53-2...................................*N*-Nitroso-*N*-methylurethane
621-64-7.....................................*N*-Nitrosodi-*n*-propylamine
630-93-3...........Diphenylhydantoin (Phenytoin), sodium salt
631-64-1.. Dibromoacetic acid
636-21-5......................................*o*-Toluidine hydrochloride
637-07-0...Clofibrate
671-16-9.. Procarbazine
680-31-9.................................Hexamethylphosphoramide
684-93-5.....................................*N*-Nitroso-*N*-methylurea
712-68-5.........2-Amino-5-(5-nitro-2-furyl)-1,3,4-thiadiazole
759-73-9.......................................*N*-Nitroso-*N*-ethylurea
764-41-0...................................... 1,4-Dichloro-2-butene
765-34-4...Glycidaldehyde
794-93-4.. Panfuran S
817-09-4...........Trichlormethine (Trimustine hydrochloride)
822-36-6.. 4-Methylimidazole
838-88-0......................4,4′-Methylene bis(2-methylaniline)
842-07-9.....................................C.I. Solvent Yellow 14
924-16-3................................. *N*-Nitrosodi-*n*-butylamine
924-42-5.....................................*N*-Methylolacrylamide
930-55-2.................................... *N*-Nitrosopyrrolidine
1024-57-3....................................Heptachlor epoxide
1116-54-7............................. *N*-Nitrosodiethanolamine
1120-71-4...................................... 1,3-Propane sultone
1271-28-9.. Nickelocene
1303-00-0...................................Gallium arsenide
1307-96-6...Cobalt[II] oxide
1309-64-4.................... Antimony oxide (Antimony trioxide)
1313-99-1.. Nickel oxide
1314-20-1.. Thorium dioxide
1314-62-1.... Vanadium pentoxide (orthorhombic crystalline
form)
1332-21-4..Asbestos
1333-86-4........................Carbon black (airborne, unbound
particles of respirable size)
1335-32-6.. Lead subacetate
1464-53-5..Diepoxybutane
1596-84-5..Daminozide
1615-80-1.. 1,2-Diethylhydrazine
1694-09-3..Benzyl violet 4B
1746-01-6.......2,3,7,8-Tetrachlorodibenzo-*p*-dioxin (TCDD)
1836-75-5........................... Nitrofen (technical grade)
1897-45-6...Chlorothalonil
1918-16-7...Propachlor
1929-82-4.. Nitrapyrin
1937-37-7...................... Direct Black 38 (technical grade)
2092-56-0...D&C Red No. 8
2312-35-8.. Propargite
2385-85-5.. Mirex
2425-06-1.. Captafol
2429-74-5... C.I. Direct Blue 15
2439-01-2.......................... Oxythioquinox (Chinomethionat)
2475-45-8.. Disperse Blue 1
2593-15-9..Terrazole
2602-46-2...................Direct Blue 6 (technical grade)
2646-17-5..Oil Orange SS

2784-94-3...HC Blue 1
2973-10-6.. Diisopropyl sulfate
3068-88-0..β-Butyrolactone
3296-90-0................ 2,2-Bis(bromomethyl)-1,3-propanediol
3333-67-3...Nickel carbonate
3468-63-1................................... D&C Orange No. 17
3546-10-9.. Phenesterin
3564-09-8...Ponceau 3R
3570-75-0................ 2-(2-Formylhydrazino)-4-(5-nitro-2-
furyl)thiazole
3688-53-7....................AF-2 [2-(2-furyl)-3-(5-nitro-2-
furyl)]acrylamide
3697-24-3... 5-Methylchrysene
3761-53-3... Ponceau MX
3771-19-5.. Nafenopin
4342-03-4... Dacarbazine
4549-40-0..............................*N*-Nitrosomethylvinylamine
5160-02-1...D&C Red No. 9
5216-25-1................................ *p-a,a*-Tetrachlorotoluene
5522-43-0.. 1-Nitropyrene
5589-96-8...................... Bromochloroacetic acid
6055-19-2...................... Cyclophosphamide (hydrated)
6109-97-3.............3-Amino-9-ethylcarbazole hydrochloride
6164-98-3.. Chlordimeform
6358-53-8...................................Citrus Red No. 2
6459-94-5...................................C.I. Acid Red 114
7280-37-7...Estropipate
7439-92-1............................. Lead (and lead compounds)
7440-02-0.. Nickel (Metallic)
7440-38-2.............Arsenic (inorganic arsenic compounds)
7440-41-7...................Beryllium (and beryllium compounds)
7440-43-9................... Cadmium (and cadmium compounds)
7440-48-4... Cobalt metal powder
7446-27-7...Lead phosphate
7446-34-6... Selenium sulfide
7481-89-2...Zalcitabine
7496-02-8.. 6-Nitrochrysene
7758-01-2.............................. Potassium bromate
8001-35-2.............. Toxaphene (Polychlorinated camphenes)
8006-61-9.................. Unleaded gasoline (wholly vaporized)
8018-01-7.. Mancozeb
9004-66-4.. Iron dextran complex
9006-42-2... Metiram
10026-24-1....................................Cobalt sulfate heptahydrate
10034-93-2....................................Hydrazine sulfate
10048-13-2.. Sterigmatocystin
10124-43-3..Cobalt sulfate
10418-03-8...Stanozolol
10453-86-8...Resmethrin
10540-29-1............................... Tamoxifen and its salts
10595-95-6.................... *N*-Nitrosomethylethylamine
12035-72-2...Nickel subsulfide
12054-48-7; 12125-56-3...........................Nickel hydroxide
12122-67-7............................. Zineb (*delisted 1999*)
12174-11-7............... Palygorskite fibers (>5 μm in length)
12427-38-2.. Maneb

12510-42-8/ 66733-21-9 .... Erionite

13010-47-4 .... 1-(2-Chloroethyl)-3-cyclohexylnitrosourea (CCNU) (Lomustine)

13194-48-4 .... Ethoprop

13256-22-9 .... N-Nitrososarcosine

13463-39-3 .... Nickel carbonyl

13552-44-8 .... 4,4′-Methylenedianiline dihydrochloride

13909-09-6 .... 1-(2-Chloroethyl)-3-(4-methylcyclohexyl)-1-nitrosourea (Methyl-CCNU)

14901-08-7 .... Cycasin

15541-45-4 .... Bromate

15663-27-1 .... Cisplatin

15972-60-8 .... Alachlor

16071-86-6 .... Direct Brown 95 (technical grade)

16543-55-8 .... N-Nitrosonornicotine

16568-02-8 .... Gyromitrin (Acetaldehyde methylformylhydrazone)

18662-53-8 . Nitrilotriacetic acid, trisodiumsalt monohydrate

18883-66-4 .... Streptozotocin (streptozocin)

19044-88-3 .... Oryzalin

19666-30-9 .... Oxadiazon

20265-96-7 .... p-Chloroaniline hydrochloride

20325-40-0 ...... 3,3′-Dimethoxybenzidine dihydrochloride(o-Dianisidine dihydrochloride)

20830-81-3 .... Daunomycin

21739-91-3 .... Cytembena

22398-80-7 .... Indium phosphide

22506-53-2 .... 3,9-Dinitrofluoranthene

23103-98-2 .... Pirimicarb

23246-96-0 .... Riddelliine

23950-58-5 .... Pronamide

25013-16-5 .... Butylated hydroxyanisole

25316-40-9 .... Doxorubicin hydrochloride (Adriamycin)

25321-14-6 .... Dinitrotoluene mixture, 2,4-/2,6-

25812-30-0 .... Gemfibrozil

26148-68-5 .... A-α-C (2-Amino-9H-pyrido [2,3-b]indole)

26471-62-5 .... Toluene diisocyanate

28407-37-6 .... C.I. Direct Blue 218

28434-86-8 .... 3,3′-Dichloro-4,4′-diaminodiphenyl ether

30516-87-1 .... Zidovudine (AZT)

32809-16-8 .... Procymidone

34256-82-1 .... Acetochlor

34465-46-8 .... Hexachlorodibenzodioxin

36734-19-7 .... Iprodione

39156-41-7 .... 2,4-Diaminoanisole sulfate

42397-64-8 .... 1,6-Dinitropyrene

42397-65-9 .... 1,8-Dinitropyrene

50471-44-8 .... Vinclozolin

51264-14-3 .... Amsacrine

51338-27-3 .... Diclofop-methyl

53973-98-1 .... Polygeenan

54749-90-5 .... Chlorozotocin

55738-54-0 .... trans-2-[(Dimethylamino)methyl-imino]-5-[2-(5-nitro-2-furyl)vinyl]-1,3,4-oxadiazole

57018-52-7 .... Propylene glycol mono-t-butyl ether

57835-92-4 .... 4-Nitropyrene

59669-26-0 .... Thiodicarb

59865-13-3; 79217-60-0 .... Ciclosporin (Cyclosporin A; Cyclosporine)

60153-49-3 .... 3-(N-Nitrosomethylamino)-propionitrile

60391-92-6 .... N-Carboxymethyl-N-nitrosourea

60568-05-0 .... Furmecyclox

62450-06-0 .... Trp-p-1 (Tryptophan-p-1)

62450-07-1 .... Trp-p-2 (Tryptophan-p-2)

62476-59-9 .... Aciclofen sodium

64091-91-4 .... 4-(N-Nitrosomethylamino)-1-(3-pyridyl)1-butanone

67730-10-3 .... Glu-p-2(2-Aminodipyrido[1,2-a:3′,2′-d] imidazole)

67730-11-4 .... Glu-p-1(2-Amino-6-methyldipyrido [1,2- a:3′,2′-d]imidazole)

68006-83-7 .... Me-A-α-C (2-Amino-3-methyl-9H-pyrido[2,3-b]indole)

68308-34-9 .... Shale-oils

72490-01-8 .... Fenoxycarb

76180-96-6 .... IQ (2-Amino-3-methylimidazo [4,5-f] quinoline)

77094-11-2 ....MeIQ(2-Amino-3,4-dimethyl-imidazo[4,5-f]quinoline)

77439-76-0 .... MX (3-chloro-4-(dichloromethyl) 5-hydroxy-2(5H)-furanone)

77500-04-0 .... MeIQx (2-Amino-3,8-dimethyl-imidazo[4,5-f]quinoxaline)

77501-63-4 .... Lactofen

79748-81-5 .... Fusarin C

82410-32-0 .... Ganciclovir

105650-23-5 .... PhiP(2-Amino-1-methyl-6-phenylimidazol[4,5-b]pyridine)

105735-71-5 .... 3,7-Dinitrofluoranthene

108171-26-2 .... Chlorinated paraffins (Average Chain length, C12; approximately 60% chlorine by weight)

110235-47-7 .... Mepanipyrim

111406-87-2 .... Zileuton

113852-37-2 .... Cidofovir

116355-83-0 .... Fumonisin B1

140923-17-7/140923-25-7 .... Iprovalicarb

141112-29-0 .... Isoxaflutole

148477-71-8 .... Spirodiclofen

177406-68-7 .... Benthiavalicarb-isopropyl

no CAS .... Silica, crystalline (airborne particles of respirable size)

no CAS .... Soots, tars, and mineral oils (untreated and mildly treated oils and used engine oils)

no CAS .... Strong inorganic acid mists containing sulfuric acid

no CAS .... Phenylhydrazine and its salts

no CAS .... Toxins derived from Fusarium moniliforme (Fusarium verticillioides)

no CAS .... Nickel refinery dust from thepyrometallurgical process

no CAS .... Oral contraceptives, combined

no CAS .... Diesel engine exhaust

no CAS...Wood dust
no CAS......................... Ceramic fibers (airborne particles of respirable size)
no CAS................................ Oral contraceptives, sequential
no CAS... Residual (heavy) fuel oils
no CAS.. Radionuclides
no CAS.................................Certain combined chemotherapy for lymphomas
no CAS............................Talc containing asbestiform fibers
no CAS.. Aflatoxins
no CAS.................................... Diaminotoluene (mixed)
no CAS..................................... Alcoholic beverages, when associated with alcohol abuse
no CAS.. Bracken fern
no CAS....................................Bitumens, extracts of steam-refined and air refined
no CAS...........................Bis(2-chloro-1-methylethyl) ether, technical grade
no CAS...Tobacco smoke
no CAS............................. 2,4-Hexadienal (89% *trans*, *trans* isomer; 11% *cis*, *trans* isomer)
no CAS..................Tobacco, oral use of smokeless products
no CAS.............................. Betel quid without tobacco
no CAS.............5-Chloro-*o*-toluidine and its strong acid salts
no CAS............................ Herbal remedies containing plant species of the genus *Aristolochia*
no CAS.. Creosotes

no CAS.................................Polychlorinated dibenzofurans
no CAS...Betel quid with tobacco
no CAS...Methylmercury compounds
no CAS..................................... Conjugated estrogens
no CAS.....................Chromium (hexavalent compounds)
no CAS... Glasswool fibers (airborne particlesof respirable size)
no CAS........Gasoline engine exhaust (condensates/extracts)
no CAS..Coke oven emissions
no CAS........................... Quinoline and its strong acid salts
no CAS.. Areca nut
no CAS... Aristolochic acids
no CAS...Carbon-black extracts
no CAS.. Benzidine-based dyes
no CAS..............................3,3'-Dimethoxybenzidine-based dyesmetabolized to 3,3'-dimethylbenzidine
no CAS..............................3,3';-Dimethoxybenzidine-based dyesmetabolized to 3,3'-dimethoxybenzidine
various CAS...Nickel compounds
various CAS.......... *p*-Chloro-*o*-toluidine, strong acid salts of
various CAS........ Analgesic mixtures containing phenacetin
various CAS....................Polybrominated biphenyls (PBBs)
various CAS................... Polychlorinated dibenzo-*p*-dioxins
various CAS..............Polychlorinated biphenyls (containing 60 or more percent Cl by molecular weight)
various CAS................................Polychlorinated biphenyls
Various CAS.. Estrogens, steroidal

Appendix 3: Glossary

A

absorbent material Commercially packaged clay, kitty litter, or other material used to soak up liquid hazardous materials.

absorption Penetration of a substance across a biologic barrier (such as the skin) and into either the lymphatic system or bloodstream.

acaricide A chemical substance used to kill ticks and mites.

accident An unplanned energy transfer causing property damage and/or human injury. See also "incident."

accumulative effect The effect of a chemical substance on a biologic system when the substance is being absorbed at a rate that exceeds the body's ability to eliminate it from the system. Excessive accumulation of the substance in the system can lead to toxicity.

acid Any compound containing hydrogen replaceable by metals, and having a pH of zero to 6. Strong acids in the pH range of zero to 2 are corrosive and will cause chemical burns to the skin, eyes, and mucous membranes. Acids turn litmus red. See also "pH" and "strong acids."

acid gas A gas that forms an acid when dissolved in water.

acidosis A pathologic condition resulting from accumulation of acid in, or loss of base from, the body.

acrid Having a biting taste; sharp; pungent.

active ingredient The component that actually performs the primary function of a product. Products generally contain both active and inert ingredients, and both may be harmful. For example, insecticides in spray cans contain both chemicals having pesticidal action (active) and propellent gas (inert). Active ingredients are listed on product labels as percentage by weight or as pounds per gallon of concentrate.

acute The clinical term for a disease having a short and relatively severe course, measured in seconds, minutes, hours, or days, following exposure to a health hazard. Also, in animal testing, pertains to administration of an agent in a single dose.

acute effect Refers to an adverse health effect that usually occurs rapidly, sometimes immediately, as a result of a single, short significant exposure to a health hazard, without implying a degree of severity. Acute effects may include irritation, corrosivity, narcosis, and death. See also "chronic."

acute exposure Refers to a single exposure to a toxic substance that results in death or severe biological harm. Acute exposures are characterized as lasting no longer than 1 day.

acidosis A pathologic condition resulting from accumulation of acid in, or loss of base from, the body.

adrenal gland A hormone-secreting organ located above each kidney.

AEGL Acronym for Acute Emergency Guideline Levels. AEGLs represent threshold airborne exposure limits for the general public and are applicable to emergency exposures ranging from 10 min to 8 h. Three levels—AEGL-1, AEGL-2, AEGL-3—are developed for each of five exposure periods (10 min, 30 min, 1 h, 4 h, and 8 h) and are distinguished by varying degrees of severity of toxic effects. DOE guidance is to use the 1-h AEGL values, which appear in this database.[DOE] AEGL values are developed and published by the US Environmental Protection Agency (EPA). See also "How to Use this Book" in the front matter of this book.

aerosol A dispersed suspension of fine particles suspended in air (dispersed in a gas), the particle size often being in the $0.01-100\,\mu m$ range. Natural aerosols include smoke (solid particles) and fog (liquid particles). Man-made aerosols are manufactured by filling a valved container, usually a can, with a suspension (e.g., paint, insecticides, cosmetics) in a gas under pressure.

aliphatic Pertaining to an open-chain hydrocarbon compound. Substances such as methane and ethane are typical aliphatic hydrocarbons.

alkali Any ACID destroying compound having a pH of $8-14$. Strong alkalis (or bases) in the pH range of $12-14$ are considered corrosive and will cause chemical burns to the skin, eyes, and mucous membranes. Alkalis turn litmus blue. Widely used industrial alkali substances include sodium hydroxide, sodium carbonate, potassium hydroxide, and potassium carbonate. Common household products include DRANO™ and lye. See also "acid," "base," and "pH."

alkaloid An organic nitrogen base, of vegetable origin, usually toxic.

allergen A substance that causes the body to produce an antibody and which results in an allergy in hypersensitive people.

allergy A hypersensitive reaction of body tissues to specific substances. In similar concentrations and circumstances these same substances do not affect other persons. Allergic reactions in the workplace tend to affect the skin (see dermatitis) and lungs (see asthma).

ambient air concentration The concentration of a material in environmental air outside of buildings, that is, air to which the general public is exposed.

amine An organic compound that may be derived from ammonia (NH_3) by the replacement of one or more hydrogen atoms (H) by hydrocarbon groups or other chemical moieties; replacing one, two, or three hydrogen atoms gives primary, secondary, or tertiary amines, respectively; if a fourth group is added to a tertiary amine (R_3N), the compound formed is called a quaternary amine (R_4N+) and the nitrogen carries a positive charge.

anaerobic conditions Refers to the absence of oxygen.

amnesia Total or partial memory loss.

analgesia A pain-relieving agent that causes insensibility to pain without loss of consciousness.

analogue A compound that resembles another in structure; may be an isomer, but not necessarily.

analytical grade The highest available purity of a chemical.

anemia A deficiency of the blood caused by reduced red blood cell count or a reduction in the amount of hemoglobin per unit volume of blood.

anesthesia Total or partial loss of sensation with or without loss of consciousness.

angina pectoris Pain in chest caused by inadequate supply of blood to the heart.

anhydrous Containing no water.

anorexia Loss or reduction of appetite for food.

anosmia Total or partial loss of the sense of smell.

anoxia A reduction in the quantity of oxygen supplied by blood to cells or tissues.

anticonvulsant A substance that lessens the severity of convulsions.

antidote A remedy to relieve, prevent, or counteract the effects of a poison; that which counteracts anything noxious.

antiepileptic A substance that lessens the severity of epileptic seizures.

anuria Absence of urine in the bladder caused by the failure of the kidneys to produce urine. This is a possible symptom of chlorate or inorganic mercury poisoning.

apathy Reduced emotions with lack of interest in outside stimuli.

apnea Temporary cessation of breathing. A possible symptom of poisoning.

aquatic toxicology A branch of toxicology that deals with water pollution and its ecological effects.

aqueous Watery or water like.

aromatic compound Pertaining to a molecular ring structure hydrocarbon compound, characterized by the presence of the benzene nucleus.

aromatic hydrocarbon An organic chemical compound formed primarily from carbon (C) and hydrogen (H) atoms with a structure based on benzene rings *and* resembling benzene in chemical behavior; substituents on the rings(s) may contain atoms other than C or H.

arrhythmia Disturbed heartbeat.

arsenical A compound containing arsenic.

arylamine An organic compound formed from an aromatic hydrocarbon that has at least one amine group joined to it.

ataxia Unsteady walk or shaky movements due to neurological problems. May be a symptom of poisoning.

asphyxia Difficulty in breathing or respiratory arrest; suffocation.

asphyxiant Refers to a substance, usually a vapor or gas, that can cause suffocation, unconsciousness, or death by preventing the blood from carrying oxygen. Most simple asphyxiants (which have no inherent toxicity) are harmful to the body only when they become so concentrated that oxygen in the air is reduced (normally about 21%) to dangerous levels (18% or lower).

asthenia Reduced physical and psychological strength.

asthma Respiratory problem characterized by attacks of wheezing, shortness of breath, and/or coughing and resulting in difficult breathing due to contraction of air passages.

ataxia Loss or failure of muscular coordination, voluntary movement, or muscle control.

atrophy A loss of weight, volume, and activity of an organ, tissue, or cell; shrinkage.

autoignition temperature The minimum temperature at which a substance will ignite spontaneously, or cause self-sustained combustion in the absence of any heated element, spark, or flame. The closer the autoignition temperature is to room temperature, the greater the risk of fire.

awareness level (trained) First responders at the awareness level are those persons who, in the course of their normal duties, may be the first on the scene of an emergency involving hazardous materials. First responders at the awareness level are expected to recognize hazardous materials presence, protect themselves, call for trained personnel, and secure the area (ANSI/NFPA 472).

azide A compound that contains the monovalent $-N_3$ group.

azo- A prefix denoting the presence in a molecule of the group $-N=N-$. See also "diazo-."

B

bactericidal Destructive to bacteria. An agent (e.g., heat, light, or osmotic pressure) or a chemical, such as a pesticide, that kills bacteria or inhibits their growth is called a bactericide.

barbiturate A drug used as a sedative or hypnotic.

base A substance that reacts with acids to form salts and water. All bases create solutions having a pH of more than 7.0, the neutral point, and may be corrosive to skin and other human tissue. The terms alkali and caustic are closely related in meaning. See also "acid," "alkali," and "pH."

BEI Biological Exposure Index. The maximum recommended value of a substance in blood, urine, or exhaled air, recommended by the ACGIH. See "Threshold Limit Values and Biological Exposure Indices," published by the ACGIH, for an explanation.

benign Not harmful.

bile A yellow-green, bitter fluid secreted by the liver. Also called gall.

bioaccumulation The process by which a material in an organism's environment progressively concentrates within the organism.

bioassay The determination of the potency or concentration of a compound by its effect upon animals, isolated tissues, or microorganisms, as compared with a chemical or physical assay.

biodegradation Biotransformation; the conversion within an organism of molecules from one form to another, a change often associated with change in pharmacologic activity.

bld Blood effects. A toxicology term describing the effect on all blood elements including oxygen carrying or releasing capacity, pH, protein, and electrolytes.

blepharospasm Abnormal contraction of eyelid muscles.

BOD Biological oxygen demand. A test that measures the dissolved oxygen consumed by microbial life while assimilating and oxidizing the organic matter present in organic waste discharges. This test permits calculation of the effect of the discharges on the oxygen resources of the receiving water.

boiling point The temperature at which a product changes from a liquid to a vapor at normal atmospheric pressure (760 mmHg). Mixtures may not have a specific boiling point. As a general rule, material safety data sheets (MSDS) carry the initial boiling point or a boiling range for a mixture.

bowel The intestine, or the part of the digestive tract extending from the stomach to the anus.

bradycardia A decrease in the heartbeat rate to less than 60 beats per minute.

breakthrough time The time from initial chemical contact to detection.

breathing zone sample An air sample collected from the area around the nose of a worker to assess exposure to airborne contaminants.

bronchoconstriction Contraction with narrowing of bronchia.

bronchospasm Spasmodic contraction of the muscles surrounding the bronchia.

bulk density Mass of powdered or granulated solid material per unit of volume.

by-product Any material, other than the principal product, generated as a consequence of an industrial process.

C

C Symbol for Celsius or Centigrade, a unit of temperature in which the interval between the freezing point of water and the boiling point is divided into 100 units, or degrees, with 0°C representing the freezing point and 100°C the boiling point.

cancer A general term used to indicate any of various types of malignant neoplasms.

canister A personal air cleaning device usually worn by the user. The canister contains sorbents, catalysts, or other filter materials designed to remove gases, vapors, and liquid and solid particles from air drawn through it.

carcinogen Any substance causing the promotion or initiation of malignant or benign neoplasia (cancer) in humans or animals. A material is considered carcinogenic if (a) it is found to be a carcinogen or potential carcinogen by the International Agency for Research on Cancer (IARC); or, (b) if it is listed in the latest edition of the "Annual Report on Carcinogens," published by the National Toxicology Program (NTP); or, (c) it is regulated by OSHA as a carcinogen.

carcinogenesis The process by which normal tissue becomes cancerous.

carcinogenicity The power, ability, or tendency to produce cancerous tissue from normal tissue.

carcinoma A malignant neoplasm of the epithelium.

carcinoma in situ Noninvasive cancer.

carcinoma A malignant tumor; a type of cancer.

cardio A medical prefix that refers to the heart.

cardiovascular A medical term that refers to the heart and blood vessel system.

catalyst A substance that affects the rate of a chemical reaction, but that is neither changed nor consumed by the reaction.

cataract A disease of the eye in which the lens becomes gray-white and loses its clearness.

cathartic Substance that aids bowel movement and stimulates evacuation of the intestine.

cation An ion that carries a positive charge, e.g., sodium (N^+), (calcium Ca^{2+}), and ammonium (NH_4^+); the corresponding hydroxide is formed when combined with hydroxyl (OH^-) ions.

caustic Any strongly alkaline substance that has a corrosive effect on living tissue. See also "alkali."

ceiling limit The concentration of a substance that should not be exceeded, even for an instant. Also called ceiling concentration.

central nervous system (CNS) Refers to the brain and spinal cord, the main network of coordination and control for the entire body. Chemicals acting on the brain may cause CNS depression with symptoms of dizziness, headache, and drowsiness; higher exposure may cause unconsciousness, coma, and death.

characteristic hazardous waste An RCRA regulated waste classified as "hazardous" because of its ignitability, corrosivity, reactivity, or toxicity as determined by the Toxicity Characteristic Leachate Procedure (TCLP). It has an EPA Waste Code ranging from "D001" to "D043."

chelation A complex formation involving a metal ion and two or more polar groupings of a single molecule; chelation can be used to remove an ion from participation in biological reactions, as in the chelation of Ca^{2+} in blood by EDTA.

chemical Any element, chemical compound, or mixture of elements and/or compounds.

chemical burn Similar to a thermal burn from heat or fire, but caused by contact with a chemical substance.

chemical family A group of single elements or groups of compounds having a common chemical structure and name. Also known as chemical class. As used on a Material Safety Data Sheet (MSDS), more than one chemical family may be used if applicable.

chemical formula The chemical makeup of a substance using accepted written symbols. Although several kinds of formulas are used to indicate chemical constitution and physical structure, the *molecular formula* showing the actual kinds and numbers of atoms that comprise a molecule of a chemical substance is most commonly used in Material Safety Data Sheet (MSDS).

chemical hygiene plan A written action plan required by OSHA's regulation 29CFR1910, Occupational Exposures to Hazardous Chemicals in Laboratories.

chemical intermediate A chemical formed or used during the process of producing another chemical.

chemically active metals Usually refers (but not restricted) to chemicals such as sodium, potassium, beryllium, calcium, powdered aluminum, zinc, and magnesium. These metals can cause violent reactions with certain other substances and materials.

chemical name The scientific designation of a chemical as outlined by the Chemical Abstract Service (CAS) rules of nomenclature and the nomenclature system developed by the International Union of Pure and Applied Chemistry. Also defined by the OSHA Hazard Communication Standard (HCS) as a name which will clearly identify the chemical for the purpose of conducting a hazard evaluation. By this definition, more than one valid name for many chemicals is permitted. However, when chemicals with more than one valid name appears on the OSHA (29CFR1910 Table Z-1) or CERCLA (40CFR302.4) lists, this name is generally used as the chemical name on material safety data sheets (MSDS). If a chemical does not appear on these lists, then the most common chemical name should be used. Trade names that adequately identify a chemical may be used.

chemical manufacturer Defined in the OSHA Hazard Communication Standard (HCS) as an employer with a workplace where chemical(s) are produced for use or distribution.

chemical reaction Any chemical change, regardless of rate, or whether it occurs naturally or induced by human. There are many types of chemical reactions including decomposition, explosion, combustion, condensation, polymerization, neutralization.

chemical resistance The ability of a material to resist chemical reaction.

chemosterilant A chemical compound that causes an organism to become sterile after exposure to it.

chlorofluorocarbons (CFCs) A group of chemicals that depletes the earth's protective ozone layer. Chemical substance often used as refrigerants, solvents, and propellants as propellants in spray cans. CFCs are not destroyed in the lower atmosphere; they drift into the upper atmosphere where their chlorine components are released and destroy the ozone layer.

chloracne A severe acne-like affliction of the skin resulting from excessive exposure to certain chlorinated or halogenated chemical compounds, such as carbon tetrachloride, chloroform, trichloroethylene, biphenyls, dioxins, naphthalenes, and DDT.

chronic Refers to a change to an organism over a long period of time, measured in weeks, months, or years following repeated exposure to a health hazard.

chronic effect of overexposure Refers to an adverse health effect that develops slowly over a long period of time or from prolonged exposure to a health hazard without implying a degree of severity.

chronic toxicity Refers to permanent and irreversible health effects resulting from prolonged exposure to a toxic substance.

circulatory system The system consisting of the blood, blood vessels, lymph vessels, and heart; involved in circulating blood and lymph throughout the body.

cirrhosis Chronic progressive illness affecting the structure and function of the liver. Replacement of normal liver tissue with bands of fibrous tissue surrounding nodules of regenerating liver tissue.

closed cup Test for the flash point of a substance.

coagulant An agent that causes, stimulates, or accelerates coagulation, especially with reference to blood.

co-carcinogen Any substance, not itself carcinogenic, capable of enhancing the carcinogenic effect of another substance.

Code of Federal Regulations (CFR) A publication of the regulations promulgated under United States federal law. Changes to CFR are published in the "Federal Register" (FR). The CFR is divided into titles as follows:

Title 29: OSHA regulations, including the Hazard Communication Standard (HCS).

Title 40: EPA regulations, including TSCA.

Title 49: DOT regulations.

colic A sharp, crampy, and possibly painful disorder of the abdomen resulting from blockage, twisting, or muscle spasm.

collapsus Rapid decrease in strength or collapse of an organ.

coma A state of deep unconsciousness from which a victim cannot be wakened by external stimulants.

combustible liquid A material having a flash point at or above 37.8°C/100°F, but below 93.3°C/200°F, except that this term does not include any liquid mixture that has one or more components with a flash point above 93.3°C/200°F, which make up 99% or more of the total volume of the mixture as determined by tests listed in 49CFR173.115(d). Exceptions to this are found in 49CFR173.115(b).

commercial grade Less than the purest available form of a chemical; the purity normally produced for and adequate for commercial uses.

common name Any designation or identification such as a code name, code number, trade name, or generic name used to identify a chemical other than by its chemical name (OSHA).

component An ingredient or constituent part.

compound A substance consisting of two or more elements that have united chemically.

compressed gas Any material or mixture having in the container a pressure exceeding 40 psi at 70°F/21.1°C, or a pressure exceeding 104 psi at 130°F/54.4°C, regardless of the pressure at 70°F/21.1°C; or any liquid flammable material having a vapor pressure exceeding 40 psi absolute pressure at 100°F/37.8°C as determined by the American National Standard Method of Tests for Vapor Pressure of Petroleum Products (Reid Method) Z11.44-1973 (ASTM—American Society for Testing Materials D 323-72).

congenital A condition that begins to develop in the uterus and is existing at birth.

congestion Abnormal accumulation of blood in the vessels of tissue, an organ or other part of the body.

conjugated Bound together; in organic chemistry, conjugated refers to a molecular structure or substructure containing alternating double and single bonds between pairs of adjacent atoms.

conjunctivitis Irritation and inflammation of the conjunctiva, a part of the inner lining of the eyelids.

container Defined by OSHA as any bag, barrel, bottle, box, can, cylinder, drum, reaction vessel, storage tank, or the like that contains a hazardous chemical. Pipes and piping systems and engines, fuel tanks or other operating systems in a vehicle, are not considered containers (29CFR).

contaminant An impurity; in the environment, a chemical that is not ordinarily present and that may have deleterious effects.

contraindication Any condition that renders some particular treatment of disease improper or undesirable.

convulsions Violent, involuntary spasms or muscle contractions.

copolymer A chemical (polymer) made up of repetitive subunits (monomers) that are not all alike.

corrosive material Any liquid or solid with pH ranges of 2–6 or 12–14, and that cause visible destruction or irreversible alteration of living tissue, or a liquid that has a severe corrosion rate on steel. To determine whether a material is destructive or to cause irreversible alteration of human skin, refer to the method described in Appendix A of 49CFR173. A liquid is considered to have a severe corrosion rate if its corrosion rate exceeds 0.250 inch per year (IPY) on steel [Society of Automotive Engineers (SAE) 1020] at a test temperature of 130°F/54.4°C. Also see test described in NACE Standard TM-01-69.

cryogenic liquid Defined by DOT as a refrigerated liquefied gas having a boiling point colder than −130°F/−90°C at one atmosphere, absolute.

cutaneous Pertaining to the skin. See also "dermal."

cyanosis A bluish discoloration of the skin, lips, and mucous membrane, resulting from lack of oxygen in the blood hemoglobin, or excessive concentration of reduced hemoglobin in the blood.

cytotoxic Having a poisonous effect on cells.

D

decomposition Breakdown of a material or substance into parts, elements, or simpler compounds that may be caused by heat, chemical reaction, electrolysis, decay, biodegradation, or other process.

decontamination The removal of hazardous substances from employees and their equipment to the extent necessary to preclude the occurrence of foreseeable adverse health effects.

degradation The destructive effect a chemical may have on chemical-protective clothing, reducing its strength and flexibility, and permitting a direct route to skin contact.

defoliant A chemical spray or dust that causes leaves to drop off plants prematurely.

dermal penetration The act of entering the body by penetrating the layers of the skin.

dermatitis An inflammation of the skin.

delayed hazard The potential to cause an adverse effect which may not appear until after a long period of time. Carcinogenicity, teratogenicity, and certain target organ/system effects are examples of delayed hazards (ANSI).

deliquescent Substance which absorbs moisture from the air to the point of becoming liquid.

delirium A mental state of great excitement or confusion marked by speech disorders, anxiety, and often hallucinations.

density The mass (weight) per unit volume of a substance. For example, lead has much greater density than aluminum.

dermal Pertaining to the skin. See also "cutaneous."

dermatitis Inflammation of the skin from any cause.

dermatosis Generic name for all skin disorders.

desquamation Abnormal elimination of surficial layers of skin in small flakes.

diabetes A disease in which the body's ability to use sugar is impaired and which usually involves the abnormal appearance of sugar in the urine; characterized by excessive urination.

diamine An organic compound containing two amine groups, e.g., ethylenediamine, $H_2NCH_2CH_2NH_2$.

diarrhea Abnormally frequent discharge of loose, watery feces from the large intestine (colon). May be a symptom of poisoning.

diazo- A prefix denoting a compound containing the $-N'N-$ or $-N,N^+$ group. See also "azo-."

dimer A compound or unit produced by the combination of two like molecules.

diuretic That which increases volume of urine.

DL- Used separately, prefixes of D- for dextrorotary (rotated to the right) and L- for levorotary (rotated to the left) before the same chemical name refer to designations for optically active isomers that are chemically identical but that rotate plane polarized light in opposite directions; the isomers are mirror images of each other; when used together, DL designates a racemic mixture of the two isomers, whose optical activities cancel each other.

dose The amount of a chemical substance or drug to which a person has been exposed or absorbed into the body.

dysplasia Malformation; abnormal development.

dyspnea Difficulty or labored breathing. Shortness of breath. A possible symptom of poisoning.

dysuria Painful or difficulty in urinating.

E

easily oxidized materials A broad range of materials that includes combustible materials, organic materials, paper, wood, sulfur, aluminum, acetic acid, alcohols, fuels, oils, plastics, hydrazine, acetic anhydride, sulfuric acid, and many other chemicals. Easily oxidized materials can cause violent reactions with other substances and materials. See also "oxidizer," "strong oxidizers."

eczema Not a distinct disease but a general medical description for swelling of the skin, or rash of unknown cause. An inflammation of the outer layer of skin, characterized by redness, itching, crusting, and scaling.

edema Swelling caused by infiltration of fluid into the tissues or intercellular spaces.

effluent Wastewater discharged from a treatment plant, sewer, or industrial outfall into the environment, usually to surface waters.

effluent guidelines (listed as a toxic pollutant) Under the Clean Water Act, pollutants that are subject to technology-based standards (application of best available technology) developed for selected groups of industries.

electrolyte a substance, such as sodium chloride (NaCl), that dissociates into ions when fused (melted) or in solution, thereby becoming capable of conducting an electric current.

EIS Environmental Impact Statement. The results of a study to determine the probable effects of a proposed activity on the surrounding environment.

element The simplest form of a pure substance that cannot be broken down into simpler substances by chemical means.

embryotoxic Toxic effect on the embryo.

emphysema A condition of the lungs characterized by dilation or destruction of the pulmonary areola. An illness characterized by plugged passageways and a difficulty in exhaling.

employer Defined by OSHA as a person (including contractor or subcontractor) engaged in business, where chemicals are either used, distributed, or produced for use or distribution.

encephalopathy Generic name given to illnesses affecting the brain in general.

environment Includes water, air, and land and their interrelationship which exists among and between water, air, and land and all living things (EPA).

EP Extraction Procedure toxicity characteristics. Toxicity test performed on RCRA wastes.

environmental fate The distribution and transformation of a chemical from its first release until its ultimate removal from or recycling through the environment.

enzyme A protein produced in organisms capable of accelerating a particular biochemical reaction; a biological catalyst.

epidemiology Science concerned with the study of disease in a general population. Determination of the incidence (rate of occurrence) and distribution of a particular disease (as by age, sex, or occupation), which may provide information about the cause of the disease.

ergonomics The activity dealing with interactions between workers and their total working environment and their stresses relating to the elements of the environment, their tools and equipment.

ERPG Acronym for Emergency Response Planning Guidelines. Exposure limit values produced by the American Industrial Hygiene Association (AIHA). See also "How to Use this Book" in the front matter of this book.

erythema Abnormal flushing or redness of the skin due to increase in blood flow.

etiologic agents Airborne microorganisms capable of causing disease. These agents are the only nonchemical materials regulated by the US Department of Transportation (DOT).

etiology The science and study of all the factors that contribute to the development of a disease.

euphoria Intense feeling of well-being or elation.

evaporation rate A measure of the time required for a given amount of a substance to evaporate, compared with the time required for an equal amount of butyl acetate or ether to evaporate.

exothermic Heat producing.

expectoration Expulsion from the mouth of secretions from the respiratory tract.

explosive Any chemical compound, mixture, or device that produces a sudden, almost instantaneous release of pressure, gas, and heat when subjected to sudden shock, ignition source, pressure, or high temperature.

explosive limits Defined by the NFPA as the boundary-line mixture of vapor or gas with air, which, if ignited will just propagate the flame. They are known as the "lower and upper explosive limits," and are usually expressed in terms of percentage by volume of gas or vapor in air. Same as flammable limits. See also "LEL" and "UEL."

exposure People, property, or the environment that are subjected to the harmful effects of a hazardous material. Defined by OSHA as meaning an employee subjected to a hazardous chemical in the course of employment through a ROUTE OF ENTRY (e.g., ingestion, inhalation, skin contact, or absorption) and includes potential (e.g., accidental or possible) exposure.

exposure limits Concentrations of substances and conditions under which it is believed that nearly all workers may be repeatedly exposed, day after day, without adverse effects. Standard limits are established by ACGIH and OSHA.

F

F Degree Fahrenheit, a unit for measuring temperature. On the Fahrenheit scale, water boils at 100°C or greater.

Federal Register (FR) A daily publication of all US government documents required by law. It is the daily supplement to the Code of Federal Regulations (CFR).

fibrillation Rapid and chaotic contractions of many individual muscle fibers of the heart in the area of the ventricles, capable of causing cardiac arrest.

fibrosis Chronic lung affliction or scarring of the lung caused by an unusual increase of fibrous tissue, causing progressive respiratory problems, and often occurring after exposure to certain chemical substances.

fire diamond A visual hazard rating system of the National Fire Protection Association (NFPA). Provides general information about inherent hazards of materials: Health, Flammability, Reactivity and Special.

flammable Catches on fire and burns rapidly. The National Fire Protection Association and the US Department of Transportation define a flammable liquid as having a flash point below 100°F/37.8°C.

flammable aerosol An aerosol that, when tested by the method described in 16CFR1500.45, yields a flame projection exceeding 18 in at full valve opening, or a flashback (a flame extending back to the valve) at any degree of valve opening (29CFR).

flammable limits Range of gas or vapor concentrations (percent by volume) in air which will burn or explode if an ignition source is present. See also "LEL" and "UEL."

flash back A phenomenon characterized by vapor ignition and flame traveling back to the source of the vapor.

flash point The minimum temperature at which a substance gives off flammable vapors that are in contact with spark or flame will easily ignite and burn rapidly. The flash point is established by one of the two following methods: in a closed cup (see also "Pensky-Martens"), or inside its container; or in an open cup, or near the surface of the liquid. The lower the flash point of a liquid, the higher the risk of fire.

flatulence Accumulation of gas in the digestive tract.

fluorosis Characteristic chronic intoxication caused by fluorine and its derivatives.

foreseeable emergency Any potential occurrence such as, but not limited to, equipment failure, rupture of containers, or failure of control equipment, that could result in an uncontrolled release of a hazardous chemical into the workplace (OSHA).

Form R Forms that must be completed annually for the EPA by reporting industrial facilities for chemicals used above the threshold amount.

fume A suspension of very fine solid particles in air, or vapors from a volatile liquid.

fumigant A pesticide in vapor or gaseous form used to kill pests or disinfect materials.

fungicide A pesticide used to control, prevent, or kill fungi.

fungus A lower plant that feeds on other organic matter and lacks the chlorophyll and tissue differentiation of higher plants.

G

gas An air-like, formless fluid having the property of uniformly distributing itself throughout a space in air. A state of matter.

gastric lavage Washing out the stomach.

gastrointestinal (GI) Refers to the organs of stomach, intestines, and/or other organs from mouth to anus.

gene Heredity-carrying material that is part of the chromosome.

genotoxic Causing genetic damage.

gestation The development of the fetus in the uterus from conception to birth; during pregnancy.

glycosuria Abnormal presence of glucose in the urine.

granulomatosis Pulmonary lesion characterized by the formation of small nodules.

granulometry Indicates the size of powder, and usually expressed in microns (1 micron is 0.000001 meters, or about 1,000 times smaller than a grain of sand). Particles of less than 10 microns are capable of deep penetration and becoming deposited in the respiratory tract. Larger particles can deposit in the upper respiratory area such as the bronchia, and although they may be expulsed, they can be dissolved by the organism. Particle size makes it possible to deduce the method to be adopted for corrective action (e.g., ventilation, respirator).

H

halogens Refers to inorganic compounds containing astatine, bromine, chlorine, fluorine, and iodine.

hazardous classes A collection of terms established the United Nations Committee of Experts to categorize hazardous materials. The specific categories are: flammable liquids, explosives, gases, oxidizers, radioactive materials, corrosives, flammable solids, poisonous and infectious substances, and dangerous substances.

hazardous constituent of waste A list of chemicals which is referenced under certain Conservation and Recovery Act (RCRA) provisions to determine if a solid waste is a hazardous waste.

hazardous materials Refers generally to hazardous substances, petroleum, natural gas, synthetic gas, acutely toxic chemicals, and other toxic chemicals. Substances or materials which have been determined to be capable of posing an unreasonable risk to health, safety, and property.

hazardous waste Defined by RCRA as any solid or combination of solid wastes, which because of its physical, chemical, or infectious properties, may pose a health hazard when improperly managed. It must possess at least one of four characteristics—ignitability, corrosivity, reactivity, toxicity—or appears on special EPA lists.

hazardous waste code The number assigned to every hazardous waste listed under the Resource Conservation and

Recovery Act (RCRA); the code is used for notification, recordkeeping, and reporting requirements.

hazard warning The OSHA definition means any words, pictures, symbols, or combination thereof appearing on a label or other appropriate form of warning that conveys the hazards of the chemical(s) in the container(s).

health hazard Evidence based on scientific data (human or animal) that acute or chronic effects may occur (29CGR1910.1200).

hematoma Localized bleeding into tissue.

hematuria Presence of blood in the urine.

heme The prosthetic, oxygen-carrying, color-furnishing constituent of hemoglobin.

hemoglobin The red, respiratory protein of erythrocytes; transports oxygen from the lungs to the tissues.

hemolysis Destruction of red blood cells, releasing hemoglobin.

hematopoietic system System responsible for formation of blood cells (includes bone marrow and lymphatic organs).

hemorrhage Loss of a significant amount blood. Can be external or internal bleeding.

hepatic Referring to the liver.

hepatitis An inflammation of the liver.

hepatoma Tumor of the liver.

hepatomegalia Increase in liver volume.

hepatotoxic Refers to a substance that is toxic to the liver.

hormone Any of various chemical substances that are produced by the endocrine glands and that have specific regulatory effects on the activity of certain organs.

hydrolysis A chemical process whereby a compound is cleaved into two or more simpler compounds with the uptake of the H and OH parts of a water molecule on either side of the chemical bond cleaved.

hydrolyze To subject to hydrolysis.

hydroxyl The atom group or radical, OH.

hypnotic Sleep inducing. Also, a drug that induces sleep.

hot zone The area immediately surrounding the incident site. Appropriate protective clothing and equipment must be worn by all personnel in the hot zone. AWARENESS LEVEL and OPERATIONAL LEVEL trained personel are not permitted in the hot zone.

hydrate Chemical substance combined with water in a specific proportion.

hydrocarbons Chemical compounds that consist entirely of carbon and hydrogen.

hydrochlorofluorocarbons (HCFCs) A group of chemicals having a weaker negative impact on the ozone layer and developed as temporary substitutes for ozone-depleting chemicals, such as CFCs.

hydrolysis Chemical change to a substance in an aqueous environment leading to the formation of new products.

hygroscopic Substances with a tendency to absorb moisture from the air.

hyperkeratosis Increased thickness of the cornified layer of the epidermis such as a corn.

hyperpigmentation Excessive pigmentation of the skin.

hyperplasia Abnormal growth of normal tissue.

hyperreflexia Excessive reflexes.

hypertension Increased blood pressure.

hypotension Decreased blood pressure.

hypothermia Lowering of body temperature to below normal.

I

icterus Yellow coloration of the skin and mucosa.

ignition temperature The minimum temperature required to initiate or cause self-sustained combustion independent of a heat source.

incident An unplanned event that could have resulted in an accident or which diminishes efficiency or production. See "accident."

inhibitor A chemical that is added to another substance to prevent unwanted chemical change from occurring.

immediate hazard A hazard with immediate effects. See "acute effect."

immediate use The hazardous chemical that will only be used by the individual who transfers it from a labeled container (29CFR1910.1200).

immunosuppression Decrease in the immune response. Artificial prevention or diminution of the (natural) immune response, e.g., by irradiation or by administration of substances such as pharmaceutical antimetabolites or specific antibodies to prevent sensitization; immunosuppression, or immunodeficiency may also be used to describe the condition of acquired or congenital lowered immune response.

incompatibility Indicates whether a material can be placed in contact with certain other products or materials. The direct contact of incompatible materials can cause dangerous reactions and give off heat and toxic vapors.

inflammation The response of the tissues of the body to injury, infection, or irritation. Its chief symptoms are redness, heat, swelling, and pain.

ingestion Taking in by the mouth; swallowing.

inhalation Breathing into the lungs of a (contaminated) substance in the form of a gas, vapor, fume, mist, or dust.

initiator A chemical that permanently alters a cell or group of cells and, in the case of carcinogens, is tumor producing.

inorganic chemical Those chemical substances not containing carbon.

insoluble Products that cannot be dissolved in each other.

interaction Modification of the toxic effects of one substance by another. Depending on the substances involved, the effects of interaction can be amplified or mitigated.

irritant A substance or material capable of causing irritation to organs and body parts.

irritating material Any substance that upon contact with fire or air produces irritating fumes.

invasive Spreading beyond specific body tissues.

isomer One of two or more variations of a chemical, each of which has the same chemical formula but a different structural arrangement.

J

jaundice Yellowing of the skin or eyes caused by too much bilirubin in the blood, and indication of liver diseases, biliary obstructions, and hemolysis.

K

keratitis Inflammation of the cornea of the eye.
keratoconjunctivitis Inflammation of the cornea and the conjunctiva of the eye.
K_{ow} See "partition coefficient."

L

lacrimation Production or discharge of excess tears from the eyes.
lacrimator A chemical substance that causes the secretion of excess tears from the eyes.
laryngitis Inflammation of the larynx.
latency period A seemingly inactive period, as that between the exposure of tissue to an injurious agent and a manifest response; often used to identify the period between exposure to a carcinogen and development of a tumor.
LEL Lower Explosive Limit of a vapor or gas; the lowest concentration (lowest percentage of the substance in air) that will produce a flash of fire when an ignition source (heat, arc, or flame) is present. At lower concentrations, the mixture is too "lean" to burn. See "UEL."
lethargy Deep and prolonged sleep, or extreme indifference.
leukemia A cancer of the blood-forming tissues that is characterized by a marked increase in the number of abnormal white blood cells (leukocytes).
listed hazardous waste under the Resource Conservation and Recovery Act (RCRA), wastes from generic industrial processes, wastes from certain sectors of industry, and unused pure chemical products that have been shown to generally contain toxic chemicals that could pose a threat to human health and the environment, or that generally exhibit one of the characteristics of ignitability, corrosivity, reactivity, or toxicity are considered hazardous and must adhere to RCRA provisions.
lymph A clear liquid that is collected from the tissues throughout the body and that flows in lymphatic vessels.
lymphatic A medical term describing a small sac or node in which lymph is stored; pertaining to the lymph, lymph nodes, or vascular channels that transport lymph to the lymph nodes.

M

malaise A feeling of general discomfort, distress, or uneasiness; an out-of-sorts feeling.
malignant Tending to become progressively worse; life-threatening.

manganism Chronic intoxication caused by manganese and its derivatives.
maximum contaminant levels (MCLs) The maximum level of a contaminant permissible in a public water system.
MCL The Maximum Contaminant Level; the maximum allowable of a contaminant in public drinking water supplies under the Safe Drinking Water Act.
melanoma A neoplasm derived from cells that are capable of forming the pigment melanin.
meta- In chemistry, a prefix denoting that a compound is formed by two substitutions in the benzene ring separated by one carbon atom, i.e., linked to the first and third, second and fourth, etc., carbon atoms of the ring; usually abbreviated *m-*.
metastasis The appearance of a neoplasm in a part of the body remote from the site of its origin.
methemoglobinemia Presence of an abnormal concentration and form of hemoglobin which will not carry oxygen to the blood, resulting in anemia or cyanosis.
miscible Products capable of being completely mixed and staying mixed with each other (without separation into distinct components) under normal conditions.
mist Liquid droplets suspended in air.
molecular weight Weight (mass) of a molecule based on the sum of the atomic weights of the atoms that make up the molecule.
mucous membranes Membranes lining body cavities and covered by a viscous substance called mucus.
myosis Reduction in diameter of pupils of the eye.
melting point The temperature at which a product changes from the solid to liquid state at normal atmospheric pressure (760 mmHg).
Montreal Protocol A 1987 agreement made by 25 countries to reduce the production and consumption of ozone depleting chemicals. Since 1987, additional countries have joined this agreement.
mutagen A chemical substance or physical effect capable of inducing transmissible changes in the genetic material of a living cell that results in physical and functional changes in the descendants. Depending on the type of cells affected, ova or spermazoids, both male and female can be affected. Mutations can lead to birth defects, miscarriage, or cancer.
mutagenic Capable of causing mutations.
mutagenicity The capability to induce mutation, or permanent change, in genetic material.

N

narcosis A stupor, drowsiness, arrested activity or unconsciousness produced by the influence of narcotics or other chemical substances. Artificially induced sleep.
narcotic Any substance that induces narcosis.
National Fire Protection Association (NFPA) This international membership organization promotes/improves fire protection and prevention and establishes safeguards against loss of life and property by fire. Best known in industry for

the National Fire Codes. Among these codes is the NFPA 704M, the code for classifying substances according to their fire and explosion hazard (as they might be encountered under fire or related emergency conditions) using the familiar diamond-shaped label or placard with appropriate color, numbers, or symbols.

nausea The urge to vomit; a feeling of sickness in the stomach.

necrosis Cellular or tissue death.

neonatal Relating to or affecting the first 4 weeks after birth.

neoplasm Presence of a new growth of abnormal cells. A tumor.

nephritis Inflammation of the kidney.

nephropathy Any affliction of the kidneys.

nephrotoxic Toxic to the kidneys.

NESHAP (National Emission Standards for Hazardous Air Pollutants) Under the Clean Air Act, regulations set for industries that emit one of more of the listed hazardous air pollutants in significant quantities.

neuropathy Any affliction of the nervous system.

neurotoxic Toxic to nerve cells and the nervous system. The effect may produce emotional or behavioral abnormalities.

neuritis Inflammation of a nerve.

neural Refers to a nerve or the nervous system.

neutralize To eliminate potential hazards by inactivating strong acids, caustics, and oxidizers. For example, acid spills can be neutralized by adding an appropriate amount of caustic or alkali substances.

NIOSH National Institute for Occupational Safety and Health. Tests equipment, evaluates and approves respirators, conducts studies of workplace hazards, and proposes standards to OSHA.

nitrogen mustard A substituted mustard gas in which sulfur is replaced by an amino nitrogen.

nitrogen oxides (NO_x) Gases associated with the breakdown of the earth's protective ozone layer that are released primarily from the burning of fossil fuels.

nitroparaffin An organic compound in which one or more hydrogen molecules of an alkane are replaced by a nitro group.

nitrosamine A class of organic compounds that contain a $-NH_2$ and a $-NO$ radical.

non-liquefied compressed gas DOT describes as a gas, other than gas in solution, which under the charged pressure is entirely gaseous at a temperature of 21°C. (Code of Federal Regulations, Title 49, Department of Transportation).

O

oncogenicity The capacity to induce tumors.

operations level (trained) First responders at the operational level are those persons who respond to releases or potential releases of hazardous materials as part of the initial response to the incident for the purpose of protecting nearby persons, the environment, or property from the effects of the release. They shall be trained to respond in a defensive fashion to control the release from a safe distance and keep it from spreading (ANSI/NFPA 472).

optic nerve Nerve running from the eye to the centers of vision in the brain.

oral Used or taken into the body through the mouth.

organic compound A class of chemical compounds containing mainly carbon atoms.

organic peroxide An organic compound containing the bivalent $-O-O-$ structure and which may be considered a structural derivative of hydrogen peroxide where one or more of the hydrogen atoms have been replaced by organic radicals (49CFR173.151).

ortho- In chemistry, denoting that a compound has two substitutions on adjacent carbon atoms in a benzene ring; usually abbreviated *o-*.

oxidant The substance that is reduced and that, therefore, oxidizes the other component of an oxidation–reduction system.

oxidase One of a group of enzymes now termed oxidoreductases that bring about oxidation by the addition of oxygen to a metabolite or by the removal of hydrogen or of one or more electrons.

oxidation The act of combining or causing to combine with oxygen or to lose electrons.

oxidize Oxygenize; to combine or cause an element or radical to combine with oxygen or to lose electrons.

oxidizing agent or oxidizer A substance other than a blasting agent or explosive such as chlorate, permanganate, inorganic peroxide, or a nitrate, that yields oxygen or other gases readily, thereby causing fire of other, usually organic, materials. Perchlorates, peroxides, permanganates, chlorates, and nitrates are examples of oxidizers.

ozone layer The protective layer of molecules surrounding the earth. The ozone layer reduces the amount of high-energy ultraviolet radiation that reaches the earth's surface.

P

PAC Acronym for Protective Action Criteria exposure limits developed by the US Department of Energy (DOE) for hazardous chemicals. These exposure limits include TEELs, AEGLs, and ERPGs and are used to estimate the consequence of uncontrolled releases of hazardous materials and to plan for emergency response. See also definitions in the glossary.

para- In chemistry, a prefix designating two substitutions in the benzene ring arranged symmetrically, i.e., linked to opposite carbon atoms in the ring; usually abbreviated *p-*.

particulate Fine liquid or solid particles such as dust, smoke, mist, fumes, or smog suspended in air or atmospheric emissions.

partition coefficient The ratio of a substance's distribution between oil and water when they are in contact. A value of less than 1 indicates better solubility of the substance in oils and greases. Such a product is therefore likely to be absorbed by the skin. A value greater than 1 indicates a better solubility in water, and, therefore, can be absorbed by the mucosa. This information can be useful in assessing first aid requirements, and can help in the selection of proper protective equipment. Also known as the coefficient of water/oil distribution. Usually abbreviated K_{ow}.

palpitation Perception of one's own heartbeat. A racing, irregular beat, or pounding of the heart.

paralysis The loss of function and/or feeling.

pathology The study of the nature of diseases, especially of the structural and functional changes in body tissues, organs, or fluids caused by disease, physical and biological agents, and toxic substances.

Pensky-Martens A closed cup method for determining flash point.

percutaneous absorption Absorption through the skin.

peritonitis Inflammation of the peritoneum or the membrane lining the abdominal cavity and the organs contained within it.

permissible exposure limit (PEL) A legal occupational limit of exposure established and defined by OSHA. The limit of allowable exposure to a chemical contaminant expressed as a Time-Weighted Average (TWA) concentration during a work-day of 8 h within, or as a maximum concentration never to be exceeded either instantaneously (ceiling) or a Short Term Exposure Limit (STEL) during any maximum period of 15 min. The exposure, inhalation, or dermal permissible exposure limit specified in 29CFR1910, Subparts G and Z. These concentrations are expressed in parts per million (ppm) and or milligrams of the product per cubic meter of air (mg/m^3) at 25°C. Exposure limits published in 29CFR1910.1000.

peroxide Compounds containing two oxygen atoms bound together. The oxide of any series that contains the greatest number of oxygen atoms; applied most correctly to compounds containing an −O−O−link, as in hydrogen peroxide (H−O−O−H).

peroxy- Prefix denoting the presence of an extra O atom, as in peroxides, peroxy acids (e.g., hydrogen peroxide, peroxyformic acid).

pesticide As defined by the Federal Insecticide, Fungicide and Rodenticide Act (FIFRA), a pesticide includes "any substance or mixture of substances intended for preventing, killing, repelling, or mitigating any pest, and any substance or mixture of substances intended for use as a plant regulator, defoliant, or desiccant."

petroleum distillate A material produced by a combination of vaporization and condensation of petroleum.

personal protection equipment (PPE) Safety equipment designed to protect parts or all of the body from workplace hazards. Such protective equipment includes chemical resistant clothing, gloves, respirators, and eye protection.

pH A symbol representing the concentration in hydrogen ions (H^+) in aqueous solution. This logarithmic scale is expressed as a numerical value usually between 0 and 14. A pH of 7 indicates a neutral or noncorrosive substance. A pH between 0 and 7 indicates greater acidity. A pH between 7 and 14 indicates greater alkalinity. A pH of 0 (very acid) or 14 (very basic) are highly corrosive. The symbol is useful in the identification of the appropriate type of protective equipment necessary for handling a chemical material.

pharyngitis Inflammation of the throat or pharynx.

phlegm Thick mucus produced in the breathing passages.

photophobia Unpleasant to painful feeling in the eyes, caused by light.

photosensitive Substances that change in the presence of light.

photosensitization Abnormal reaction of the skin to sunlight.

physical hazard A chemical for which there is scientifically valid evidence that it is a combustible liquid, a compressed gas, explosive, flammable, an organic peroxide, an oxidizer, pyrophoric, unstable (reactive) or water reactive.

pituitary gland A small gland at the base of the brain that secretes several important hormones.

plasma The fluid part of blood or lymph, as distinguished from suspended material.

platelet An irregularly shaped disk found in blood, containing granules in the central part and, peripherally, clear protoplasm, but no definite nucleus; it is about one-third to one-half the size of an erythrocyte, and contains no hemoglobin.

pneumoconiosis Chronic affliction of the lungs due to the inhaling of certain types of dust.

pneumonia Acute infection of the lung, characterized by inflammation.

pneumonitis (chemical) Inflammation of the lungs, resulting from chemical irritation.

pneumopathy Any pulmonary affliction.

point source contamination Contamination to the environment from a specific source such as a smokestack or sewer pipe.

polychlorinated biphenyl (PCBs) A pathogenic and teratogenic industrial compound used as a heat-transfer agent, primarily in transformers. PCBs may accumulate in human or animal tissue. PCBs are banned from production in the United States.

polymer A chemical formed by the joining together of similar chemical subunits.

polymerization A chemical reaction in which one or more small molecules combine to form larger molecules. A hazardous polymerization is a reaction that takes place at a rate that releases large amounts of energy. If hazardous polymerization can occur with a given material, the MSDS usually will list conditions that could start the reaction and—since the material usually will list conditions that could start the reaction, and since the material usually contains a polymerization inhibitor—the length of time during which the inhibitor will be effective. The heat given off or the expansion in volume, or both, caused by the polymerization reaction could cause the container holding the product to break and the unpolymerized residual product to be spilled.

polyneuritis Inflammation of several nerves.

polyuria Elimination of an abnormally large amount of urine over a given period.

pro-carcinogen Product that must be changed by the organism in order to become a carcinogen.

promoter (of carcinogenesis) Substance capable of enhancing the carcinogenic effect of another substance.

prostration Extreme exhaustion, physical or mental.

pulmonary Pertaining to the lungs.

pulmonary edema A buildup of fluid in the lungs caused by congestive heart failure, lung damage, side effects of drugs, infections, or kidney failure.

pyrophoric Any chemical substance or mixture that ignites spontaneously in dry or moist air at or below 130°F/54.4°C.

pyrolysis Decomposition of a substance by heat in the absence of air.

psychosis A group of mental illnesses characterized by a change in personality and loss of contact with reality.

Q

quaternary In chemistry, the term used to describe a substance with four chemical groups attached to a central atom; when the central atom is a trivalent nitrogen atom (N), adding the fourth group places a positive charge on N; compounds thus formed are called quaternary ammonium compounds.

R

racemate A mixture of equal parts of isomers of opposite rotation.

racemic Denoting a mixture that is optically inactive, being composed of an equal number of dextro- and levorotary substances (see DL-), which are separable.

radioactive Having the property of emitting radiation (such as alpha, beta, or gamma rays) from an atomic nucleus.

radioactive material Any material, or combination of materials, that spontaneously emits ionizing radiation, and having a specific activity greater than 0.002 microcuries per gram.

radionuclides Radioactive-decay particles emitted from natural and manufactured sources, including cosmic rays, X-rays, radon, and coal-fired utilities.

radon A colorless, naturally occurring, radioactive, inert gaseous element formed by radioactive decay of radium atoms in soil or rocks.

reactive Unstable.

reactivity Chemical reaction with the release of energy. Undesirable effects such as pressure buildup, temperature increase, formation of noxious, toxic, or corrosive by-product, may occur because of the reactivity of a substance to heating, burning, direct contact with other materials, or other conditions in use or in storage.

Recommended Exposure Limit (REL) A 10-h average exposure limit during a 40-h work-week recommended by the National Institute for Occupational Safety and Health (NIOSH) for occupational exposures.

reducing agent In a reducing reaction (which occurs simultaneously with an oxidation reaction) the reducing agent is the chemical or substance which (1) combines with oxygen or (2) loses electrons to the reaction. Reducing agents react violently with oxidizing agents or oxidizers.

Registry of Toxic Effects of Chemical Substances (RTECS) Published by NIOSH, RTECS is a compendium of the known toxic and biological effects of many chemical substances.

regulated material A substance or material that is subject to regulations promulgated by any government agency.

relative density The ratio of the density of a material to the density of a standard material, such as water at a specified temperature.

remedial action The actual construction or implementation phase of a Superfund site cleanup that follows remedial design.

renal Pertaining to the kidney.

reportable quantity (RQ) An amount of a hazardous substance or "extremely hazardous substance" that, if released, requires notification to the National Response Center (800/424-8802) under the emergency release reporting requirements of the Emergency Planning and Community Right-to-Know Act (EPCRA) or under those of CERCLA (Superfund). For the purposes of the Department of Transportation, RQ means the quantity specified for the substance in the Appendix to the Hazardous Materials Table. For the purposes of SARA Title III, RQ means, for any CERCLA hazardous substance the reportable quantity established therein for such a substance, for any other substance the RQ is one pound.

respirator A devise worn by a person to filter dust particles or gas out of surrounding air before inhalation of air.

respiratory tract The structures and organs involved in breathing; includes the nose, larynx, trachea, bronchi, bronchioli, and lungs.

respiratory tract filters Those anatomical structures that remove particles from inhaled air.

responsible party Someone who can provide additional information on the substance if needed (29CFR1910.1200).

rhinitis Inflammation of mucosa of the nasal passages.

rheumatoid arthritis A chronic disease of the joints, marked by inflammatory changes of joint structures.

rodenticide Chemical substances used to kill mice, rats, and other rodents.

route of entry The means or natural route by which hazardous chemicals or other contaminants can penetrate the body. For example, the skin (by cutaneous absorption), the digestive system (by ingestion), or the respiratory system (by inhalation).

routes of exposure Also known as routes of entry, any one of the ways by which substances enter the body, such as through the skin, ingestion, eye contact, or by breathing.

S

sensitization Defense reaction by the organism following exposure to a contaminant, resulting in an allergy.

senstitizer A chemical that causes a substantial proportion of exposed people or animals to develop an allergic reaction in normal tissue after repeated exposure to the chemical.

shock An abnormal condition resulting from not enough blood flowing through the body causing reduced blood pressure and interference with bodily functions.

Short Term Exposure Limit (STEL) A 15-min time-weighted-average exposure limit that should not be exceeded at any time during a work-day, recommended by the National Institute for Occupational Safety and Health (NIOSH) or a concentration that it is believed a worker can be exposed to continuously for a short period of time without suffering from irritation, chronic or irreversible tissue damage, or narcosis of sufficient degree to increase the likelihood of accidental injury, impair self-rescue, or materially reduce work efficiency, recommended by the American Conference of Governmental Industrial Hygienists (ACGIH).

silicosis Chronic disease of the lungs (fibrosis) provoked by inhaling dust of crystalline silica. See "pneumoconiosis."

solid waste Any garbage, refuse or sludge, including solid, liquid, semisolid, or contained gaseous material resulting from industrial, commercial, agricultural, and mining operations, and community activities (excluding material in domestic sewage); discharges subject to regulation as point sources under the Federal Water Pollution Control Act, or any nuclear material or byproduct regulated under the Atomic Energy Act of 1954.

solubility The property of a substance describing the degree to which one material may be completely mixed or dissolved in another material. The degree of solubility of most substances increases with the rise in temperature; however, in the case of organic salts of calcium the substance may be more soluble in cold than in hot solvents.

solution Any homogeneous liquid mixture of two or more chemical compounds or elements that will not undergo any segregation normal to transportation (CFR, Title 49, DOT).

solvent A chemical liquid, capable of dissolving another substance. A term generally used to describe organic solvents.

spasm Involuntary muscle contraction.

specific gravity The weight of a material compared to the weight of an equal volume of water is an expression of the density (or heaviness) of a material. Insoluble materials with specific gravity of less than 1.0 will float in (or on water). Insoluble materials with specific gravity greater than 1.0 will sink in water. Most (but not all) flammable liquids have specific gravity less than 1.0 and, if not soluble, will float on water—an important consideration for fire suppression.

spill Another name for a leak. The methods, equipment, and precautions that should be used to control or clean up a leak or spill.

stability Relates to a material's ability to resist change in form or chemical nature (e.g., xylene decomposes when strongly heated, and gives off toxic fumes). For MSDS purposes, a material is stable if it remains in the same form under expected and reasonable conditions of storage or use. Conditions that may cause instability (dangerous change) are stated; for example, temperature above 150°F/66°C; shock from dropping.

stabilized Containing a small amount of another substance included to keep the first material from changing form.

stool Discharge from the bowels.

stratosphere That part of the earth's atmosphere, located above the troposphere, containing the ozone layer.

strong acids Refers, but not restricted, to chemicals such as hydrochloric, sulfuric, and nitric acids. Strong acids can cause violent reactions with certain other substances and materials.

strong bases Refers, but not restricted, to chemicals such as sodium hydroxide and potassium hydroxide. Strong bases can cause violent reactions with certain other substances and materials.

strong oxidizers Refers, but not restricted, to chemicals such as chlorine, bromine, and fluorine and many of their compounds Strong oxidizers can cause violent reactions with certain other substances and materials. See also "oxidizers," "oxidizing agent," "easily oxidized materials."

sublimate To go directly from the solid to the gaseous state without passing through the liquid state.

Superfund Amendments and Reauthorization Act (SARA) Title III of SARA establishes the first national program of emergency planning for dealing with hazardous chemicals. Includes detailed provisions for community planning and annual submission of information about hazardous chemicals to the EPA, states and local communities.

suspected carcinogen A substance which is known to cause cancer in test animals, but is only suspected of causing cancer in humans. Also referred to as an experimental or potential carcinogen.

systemic toxicity Adverse effects caused by a substance that affects the body as a whole rather than local or individual parts or organs.

T

tachycardia Increased speed of heart beat.

tachypnea Abnormally rapid breathing.

target organ The body affected by a specific chemical in a specific species.

TCLP (toxicity characteristics leaching procedure) Under the Resource Conservation and Recovery Act (RCRA), wastes are subject to this laboratory procedure to determine if they can be disposed of in Subtitle D landfills or if they require Subtitle C disposal (i.e., considered a hazardous waste).

technical grade A purity standard applied to a chemical that may contain multiple impurities.

TEEL Acronym for Temporary Emergency Exposure Limit. Airborne exposure limits, TEELs are intended for use until AEGLs or ERPGs are adopted for chemicals. Temporary emergency exposure limit (TEEL) values developed by the Subcommittee on Consequence Assessment and Protective Actions (SCAPA) which provides the US Department of Energy (DOE)/National Nuclear Security Administration (NNSA) and its contractors with technical information and recommendations for emergency preparedness to assist in safeguarding the health and safety of workers and the public. See also "How to Use this Book" in the front matter of this book.

teratogen A substance that has been demonstrated to cause birth defects by causing malformations in the fetus. Teratogenic contaminants can be qualified as being "proven" when an effect has been shown in humans, "possible" when an effect has been shown in animals, or suspected in humans, and "suspected" when an effect is suspected in animals.

tetanic Refers to persistence in a muscle contraction.

Threshold Limit Value®–Time-Weighted Average limit *(TLV®–TWA)* An 8-h average exposure limit during a 40-h work-week recommended by the American Conference of Governmental Industrial Hygienists (ACGIH) for occupational exposures.

Threshold Planning Quantity (TPQ) Under the Emergency Planning and Community Right-to-Know-Act, the presence of a chemical at or above this level requires certain emergency planning activities to be conducted. The quantity of a SARA extremely hazardous substance present at a facility above which the facility's owner/operator must give emergency planning notification to the SERC and LEPC.

Tier I or Tier II Describes hazardous substances inventory forms required under SARA Title III. These forms report quantities and locations of hazardous substances to various state agencies and planning committees.

Time-Weighted Average (TWA) The average concentration of a substance in air over the total time of exposure, usually expressed as an 8-h day.

tinnitus Ringing in one or both ears. May be a sign of hearing injury.

Title III The third part of SARA, also known as EPCRA, the Emergency Planning and Community Right-to-Know Act of 1986. Specifies requirements for organizing the planning process at the state and local levels for extremely hazardous substances; minimum plan content; requirements for fixed facility owners and operators to inform officials about extremely hazardous substances present at the facilities; and mechanisms for making information about extremely hazardous substances available to citizens.

tolerance Under the Federal Insecticide, Fungicide, and Rodenticide Act (FIFRA), the amount of pesticide residues allowed to remain in or on each treated food commodity.

topical Designed for direct application to a specific part of the body.

toxic Any substance capable of causing human injury or damage to living body tissue, impairment to the central nervous system, severe illness, and, in severe cases, death. A poison.

toxicology The branch of chemistry that deals with poisons.

toxicologist One who studies the nature, effects, and detection of poisons and the treatment of poisoning.

toxic pollutant Pollutants, which after discharge and upon exposure, cause adverse health effects.

Toxic Release Inventory (TRI) (Listed substance subject to reporting requirements.) An EPA database of release quantities by facilities of a growing list of chemicals and chemical categories into the nation's air, water, and land, and transfers to off-site locations for treatment or disposal, or for recycling and energy recovery. Under the Emergency Planning and Community Right-to-Know Act, certain industries are required on an annual basis to complete a Toxic Chemical Release Inventory Form for these chemicals.

trade name The manufacturer's commercial name for a chemical substance or product. A registered trade name contains the symbol "®".

trade secret Any confidential formula or information used to give the manufacturer, etc. an (economic) advantage over others who do not have the information (29CFR 1910.1200).

tumor A neoplasm; a mass of new tissue that persists and grows independently of its surrounding structures and that has no physiological use; it may be benign or malignant.

tumorigenic That which causes or produces tumors.

U

UEL Upper Explosive Limit of a vapor or gas; the highest concentration (highest percentage of the substance in air) that will produce a flash of fire when an ignition source (heat, arc, or flame) is present. At higher concentrations, the mixture is too "rich" to burn. See also "LEL."

UN identification number An international four digit number assigned to all hazardous materials regulated by the United Nations.

unstable A chemical in the pure state which will become self-reactive under conditions of shocks, pressure, or temperature (29CFR1910.1200).

UST Underground Storage Tank regulated under RCRA. A tank used to store CERCLA-regulated chemicals or petroleum products, with 10% or more of its volume underground, having connected piping.

ulceration Creation of an ulcer.

urban air toxics Under the Clean Air Act, the 33 air toxics that have been identified as posing the greatest potential health threat in urban areas.

V

vapor density Indicator of the number of times that the vapors of a substance are heavier or lighter than air. Vapor density measurement is taken at the boiling point. If the vapor density is greater than 1, the vapor will tend to collect at floor level. If the vapor density is less than 1, the vapor will rise in air.

vapor pressure When a substance evaporates, its vapors create a pressure in the surrounding atmosphere; therefore, vapor pressure is a measurement of how readily a liquid or a solid mixes with air at its surface. This measurement is expressed in millimeters of mercury (mmHg), at 68°F/20°C and normal atmospheric pressure (760 mmHg). A vapor pressure above 760 mm indicates a substance in the gaseous state. The higher a product's vapor pressure, the more it tends to evaporate, resulting in a higher concentration of the substance in air and therefore increases the likelihood of breathing it in.

vascular Pertaining to vessels or ducts that convey fluids such as blood, lymph, or sap; in human or veterinary medicine, vascular pertains to blood vessels.

vascular constriction Constriction with narrowing of blood vessels.

vascular dilation Dilation of the blood vessels.

vertigo Dizziness; giddiness.

volatiles A substance, usually a liquid, that easily vaporizes or evaporates to form a gas or vapor.

W

waste code Identifier assigned by the EPA consisting of a single letter (D, F, P, U, K) and three numbers in the format "Dxxx." This code identifies the type of hazardous waste stream being reported.

waste management Describes activities undertaken by facilities to treat, recycle or otherwise manage generated waste, including disposal and energy recovery.

waste stream Under RCRA, solid or liquid materials containing hazardous materials and generated as waste.

water reactive A chemical which reacts with water to release flammable or hazardous gas (29CFR1910.1200).

work area A defined space where hazardous substances are produced or used when employees are present (29CFR 1910.1200).

Appendix 4: European/International Hazard Codes, Risk Phrases, and Safety Phrases

Note: Followed by the GLOBALLY HARMONIZED SYSTEM OF CLASSIFICATION AND LABELING OF CHEMICALS (GHS)

Hazard Codes

Explosive (E)
Oxidizing (O)
Highly flammable (F)
Extremely flammable (F+)
Toxic (T)
Very toxic (T+)
Harmful (Xn)
Irritant (Xi)
Corrosive (C)
Dangerous to the environment (N)

(List of R-phrases) Risk Precaution Codes and Statements

R1—Explosive when dry.
R2—Risk of explosion by shock, friction, fire, or other sources of ignition.
R3—Extreme risk of explosion by shock, friction, fire, or other sources of ignition.
R4—Forms very sensitive metallic compounds.
R5—Heating may cause an explosion.
R6—Explosive when mixed with combustible materials.
R7—May cause fire.
R8—Contact with combustible material may cause fire.
R9—Explosive when mixed with combustible material.
R10—Flammable.
R11—Highly flammable.
R12—Extremely flammable.
R13—Extremely flammable liquefied gas.
R14—Reacts violently with water.
R14/15—Reacts violently with water, liberating extremely flammable gases.
R14/15—Reacts violently with water, liberating highly flammable gases.
R14/15—Reacts violently with water, liberating extremely flammable gases.
R15—Contact with water liberates highly flammable gases.
R15/29—Contact with water liberates toxic, highly flammable gas.
R16—Explosive when mixed with oxidizing substances.
R17—Spontaneously flammable in air.
R18—In use, may form flammable/explosive vapor—air mixture.
R19—May form explosive peroxides.
R20—Harmful by inhalation.

R20/21—Harmful by inhalation and in contact with the skin.
R20/21/22—Harmful by inhalation, in contact with skin and if swallowed.
R20/22—Harmful by inhalation and if swallowed.
R21—Harmful in contact with skin.
R21/22—Harmful in contact with the skin and if swallowed.
R22—Harmful if swallowed.
R23—Toxic by inhalation.
R23/24—Toxic by inhalation and in contact with skin.
R23/24/25—Toxic by inhalation, in contact with skin and if swallowed.
R23/25—Toxic by inhalation and if swallowed.
R24—Toxic in contact with skin.
R24/25—Toxic in contact with skin and if swallowed.
R25—Toxic if swallowed.
R26—Very toxic by inhalation.
R26/27—Very toxic by inhalation and in contact with the skin.
R26/27/28—Very toxic by inhalation, in contact with the skin and if swallowed.
R26/28—Very toxic by inhalation and if swallowed.
R27—Very toxic in contact with skin.
R27/28—Very toxic in contact with skin and if swallowed.
R28—Very toxic if swallowed.
R29—Contact with water liberates toxic gas.
R30—Can become highly flammable in use.
R31—Contact with acids liberates toxic gas.
R32—Contact with acids liberates very toxic gas.
R33—Danger of cumulative effects.
R34—Causes burns.
R35—Causes severe burns.
R36—Irritating to eyes.
R36/37—Irritating to eyes and respiratory system.
R36/37/38—Irritating to eyes, respiratory system and skin.
R36/38—Irritating to eyes and skin.
R37—Irritating to respiratory system.
R37/38—Irritating to respiratory system and skin.
R38—Irritating to skin.
R39—Danger of very serious irreversible effects.
R39/23—Toxic: danger of very serious irreversible effects through inhalation.
R39/23/24—Toxic: danger of very serious irreversible effects through inhalation and in contact with skin.
R39/23/24/25—Toxic: danger of very serious irreversible effects through inhalation, in contact with skin and if swallowed.

R39/23/25—Toxic: danger of very serious irreversible effects through inhalation and if swallowed.

R39/24—Toxic: danger of very serious irreversible effects in contact with skin.

R39/24/25—Toxic: danger of very serious irreversible effects in contact with skin and if swallowed.

R39/25—Toxic: danger of very serious irreversible effects if swallowed.

R39/26—Very toxic: danger of very serious irreversible effects through inhalation.

R39/26/27—Very toxic: danger of very serious irreversible effects through inhalation and in contact with skin.

R39/26/27/28—Very toxic: danger of very serious irreversible effects through inhalation, in contact with skin and if swallowed.

R39/26/28—Very toxic: danger of very serious irreversible effects through inhalation and if swallowed.

R39/27—Very toxic: danger of very serious irreversible effects in contact with skin.

R39/27/28—Very toxic: danger of very serious irreversible effects in contact with skin and if swallowed.

R39/28—Very toxic danger of very serious irreversible effects if swallowed.

R40—Possible risks of irreversible effects.

R40/20—Harmful: possible risk of irreversible effects through inhalation.

R40/20/21—Harmful: possible risk of irreversible effects through inhalation and in contact with skin.

R40/20/21/22—Harmful: possible risk of irreversible effects through inhalation, in contact with skin and if swallowed.

R40/20/22—Harmful: possible risk of irreversible effects through inhalation and if swallowed.

R40/21—Harmful: possible risk of irreversible effects in contact with the skin.

R40/21/22—Harmful: possible risk of irreversible effects in contact with skin and if swallowed.

R40/22—Harmful: possible risk of irreversible effects if swallowed.

R41—Risk of serious danger to eyes.

R42—May cause sensitization by inhalation.

R42/43—May cause sensitization by inhalation and skin contact.

R43—May cause sensitization by skin contact.

R44—Risk of explosion if heated under confinement.

R45—May cause cancer.

R46—May cause heritable genetic damage.

R47—May cause birth defects.

R48—Danger of serious damage to health by prolonged exposure.

R48/20—Harmful: danger of serious damage to health by prolonged exposure through inhalation.

R48/20/21—Harmful: danger of serious damage to health by prolonged exposure through inhalation and in contact with skin.

R48/20/21/22—Harmful: danger of serious damage to health by prolonged through inhalation, in contact with skin and if swallowed.

R48/20/22—Harmful: danger of serious damage to health by prolonged exposure through inhalation and in contact with skin.

R48/21—Harmful: danger of serious damage to health by prolonged exposure in contact with skin.

R48/21/22—Harmful: danger of serious damage to health by prolonged exposure in contact with skin and if swallowed.

R48/22—Harmful: danger of serious damage to health by prolonged exposure if swallowed.

R48/23—Toxic: danger of serious damage to health by prolonged exposure through inhalation.

R48/23/24—Toxic: danger of serious damage to health by prolonged exposure through inhalation and in contact with skin.

R48/23/24/25—Toxic: danger of serious damage to health by prolonged exposure through inhalation, in contact with skin and if swallowed.

R48/23/25—Toxic: danger of serious damage to health by prolonged exposure through inhalation and if swallowed.

R48/24—Toxic: danger of serious damage to health by prolonged exposure in contact with skin.

R48/24/25—Toxic: danger of serious damage to health by prolonged exposure in contact with skin and if swallowed.

R48/25—Toxic: danger of serious damage to health by prolonged exposure if swallowed.

R49—May cause cancer by inhalation.

R50—Very toxic to aquatic organisms.

R50/53—Very toxic to aquatic organisms, may cause long-term adverse effects in the aquatic environment.

R51—Toxic to aquatic organisms.

R51/53—Toxic to aquatic organisms, may cause long-term adverse effects in the aquatic environment.

R52—Harmful to aquatic organisms.

R52/53—Harmful to aquatic organisms, may cause long-term adverse effects in the aquatic environment.

R53—May cause long-term adverse effects in the aquatic environment.

R54—Toxic to flora.

R55—Toxic to fauna.

R56—Toxic to soil organisms.

R57—Toxic to bees.

R58—May cause long-term adverse effects in the environment.

R59—Dangerous to the ozone layer.

R60—May impair fertility.

R61—May cause harm to the unborn child.

R62—Possible risk of impaired fertility.

R63—Possible risk of harm to the unborn child.

R64—May cause harm to breast-fed babies.

R65—Harmful: may cause lung damage if swallowed.

R66—Repeated exposure may cause skin dryness or cracking.

R67—Vapors may cause drowsiness and dizziness.

R68—Possible risk of irreversible effects.

R68/20/21—Harmful: possible risk of irreversible effects through inhalation and in contact with skin.

R68/20/22—Harmful: possible risk of irreversible effects through inhalation and if swallowed.

R68/20/21/22—Harmful: possible risk of irreversible effects through inhalation, in contact with skin and if swallowed.

R68/21/22—Harmful: possible risk of irreversible effects in contact with skin and if swallowed.

(List of S-phrases) Safety Precautions

S1—Keep locked up.

S1/2—Keep locked up and out of the reach of children.

S2—Keep out of the reach of children.

S3—Keep in a cool place.

S3/7—Keep container tightly closed in a cool place

S3/7/9—Keep container tightly closed in a cool, well-ventilated place.

S3/9—Keep in a cool, well-ventilated place.

S3/9/14—Keep in a cool well-ventilated place away from __?__(incompatible materials to be indicated by the manufacturer).

S3/9/14/49—Keep only in the original container in a cool, well-ventilated place away from __?__(incompatible materials to be indicated by the manufacturer).

S3/9/49—Keep only the original container in a cool, well-ventilated place.

S3/14—Keep in a cool place away from __?__ (incompatible materials to be indicated by the manufacturer).

S4—Keep away from living quarters.

S5—Keep contents under __?__(appropriate liquid to be specified by the manufacturer).

S6—Keep under __?__(inert gas to be specified by the manufacturer).

S7—Keep containers tightly closed.

S7/8—Keep container tightly closed and dry.

S7/9—Keep in a container tightly closed and in a well-ventilated place.

S7/47—Keep container tightly closed and at temperature not exceeding __?__°C (*to be specified by the manufacturer*).

S8—Keep containers dry.

S8/10—Keep container wet, but keep the contents dry.

S9—Keep containers in a well-ventilated place.

S12—Do not keep container sealed.

S13—Keep away from food, drink, and animal foodstuffs.

S14—Keep away from __?__(incompatible materials to be indicated by the manufacturer).

S15—Keep away from heat.

S16—Keep away from sources of ignition—No smoking.

S17—Keep away from combustible material.

S18—Handle and open containers with care.

S20—When using do not eat or drink.

S20/21—When using do not eat, drink, or smoke.

S21—When using do not smoke.

S22—Do not breathe dust.

S23—Do not breathe gas/fumes/vapor/spray (appropriate wording to be specified by the manufacturer).

S24—Avoid contact with skin.

S24/25—Avoid contact with the skin or eyes.

S25—Avoid contact with eyes.

S26—In case of contact with the eyes, rinse immediately with plenty of water and seek medical advice.

S27—Take off immediately all contaminated clothing.

S27/28—After contact with skin, take off immediately all contaminated clothing, and wash immediately with plenty of __?__ (*to be specified by the manufacturer*).

S28—After contact with skin, wash immediately with plenty of __?__(to be specified by the manufacturer).

S29—Do not empty into drains.

S29/35—Do not empty into drains; dispose of this material and its container in a safe way.

S29/56—Do not empty into drains, dispose of this material and its container at hazardous or special waste collection point.

S30—Never add water to this product.

S33—Take precautionary measures against static discharge.

S34—Avoid shock and friction.

S35—This material and its container must be disposed of in a safe way.

S36—Wear suitable protective clothing.

S36/37—Wear suitable protective clothing and gloves.

S36/37/39—Wear suitable protective clothing, gloves and eye/face protection.

S36/39—Wear suitable protective clothing and eye/face protection.

S37—Wear suitable gloves.

S37/39—Wear suitable gloves and eye/face protection.

S38—In case of insufficient ventilation, wear suitable respiratory equipment.

S39—Wear eye/face protection.

S40—To clean the floor and all objects contaminated by this material, use __?__(appropriate wording to be specified by the manufacturer).

S41—In case of fire and/or explosion, do not breathe fumes.

S42—During fumigation/spraying wear suitable respiratory equipment (appropriate wording to be specified by the manufacturer).

S43—In case of fire, use __?__(indicate in the space the precise type of fire-fighting equipment. If water increases the risk, add "Never use water").

S44—If you feel unwell, seek medical advice (show the label where possible).

S45—In case of accident or if you feel unwell, seek medical advice immediately (show the label where possible).

S46—If swallowed, seek medical advice immediately and show this container or label.

S47—Keep at temperature not exceeding __?__ °C (to be specified by the manufacturer).

S47/49—Keep only in the original container at temperatures not exceeding __?__ °C (to be specified by the manufacturer).

S48—Keep wetted with __?__(appropriate material to be specified by the manufacturer).

S49—Keep only in the original container.

S50—Do not mix with __?__(to be specified by the manufacturer).

S51—Use only in well-ventilated areas.

S52—Not recommended for interior use on large surface areas.

S53—Avoid exposure—obtain special instructions before use.

S54—Obtain the consent of pollution control authorities before discharging to wastewater treatment plants.

S55—Treat using the best available techniques before discharge into drains or the aquatic environment.

S56—Do not discharge into drains or the environment, dispose to an authorized waste collection point.

S57—Use the appropriate containment to avoid environmental contamination.

S58—To be disposed of as a hazardous waste.

S59—Refer to manufacturer/supplier for information on recovery/recycling.

S60—This material and/or its container must be disposed of as hazardous waste.

S61—Avoid release to the environment. Refer to special instructions/safety data sheet.

S62—If swallowed, do not induce vomiting; seek medical advice immediately and show this container or label.

S63—In case of accident by inhalation; remove casualty to fresh air and keep at rest.

S64 —If swallowed, rinse mouth with water (only if the person is conscious).

Globally Harmonized System of Classification and Labeling of Chemicals (GHS)

Hazard statements are an essential element under the GHS, and will eventually replace the currently used R-phrases. In addition to hazard statements, containers and Material Safety Data Sheets (MSDS) will include, where necessary, the following: one or multiple pictograms, a signal word such as "Warning" or "Danger," and precautionary statements. The precautionary statements will indicate proper handling procedures aimed at protecting the user and other people who might come in contact with the substance during an accident or in the environment. The container and MSDS will also contain the name of the supplier, manufacturer, or importer.

Each hazard statement contains a four digit code, starting with the letter H (Hxxx). Statements appear under various headings grouped together by code number. The purpose of the four digit code is for reference only; however, following the code is the exact *phrase* as it should appear on labels and MSDS.

Hazard Statements
Physical Hazards Codes and Statements

H200—Unstable explosive.
H201—Explosive; mass explosion hazard.
H202—Explosive; severe projection hazard.
H203—Explosive; fire, blast, or projection hazard.
H204—Fire or projection hazard.
H205—May mass explode in fire.
H220—Extremely flammable gas.
H221—Flammable gas.
H222—Extremely flammable material.
H223—Flammable material.
H224—Extremely flammable liquid and vapor.
H225—Highly flammable liquid and vapor.
H226—Flammable liquid and vapor.
H227—Combustible liquid.
H228—Flammable solid.
H240—Heating may cause an explosion.
H241—Heating may cause a fire or explosion.
H242—Heating may cause a fire.
H250—Catches fire spontaneously if exposed to air.
H251—Self-heating; may catch fire.
H252—Self-heating in large quantities; may catch fire.
H260—In contact with water releases flammable gases which may ignite spontaneously.
H261—In contact with water releases flammable gas.
H270—May cause or intensify fire; oxidizer.
H271—May cause fire or explosion; strong oxidizer.
H272—May intensify fire; oxidizer.
H280—Contains gas under pressure; may explode if heated.
H281—Contains refrigerated gas; may cause cryogenic burns or injury.
H290—May be corrosive to metals.

Health Hazards

H300—Fatal if swallowed.
H301—Toxic if swallowed.
H302—Harmful if swallowed.
H303—May be harmful if swallowed.
H304—May be fatal if swallowed and enters airways.
H305—May be harmful if swallowed and enters airways.
H310—Fatal in contact with skin.
H311—Toxic in contact with skin.
H312—Harmful in contact with skin.
H313—May be harmful in contact with skin.
H314—Causes severe skin burns and eye damage.
H315—Causes skin irritation.
H316—Causes mild skin irritation.
H317—May cause an allergic skin reaction.
H318—Causes serious eye damage.

H319—Causes serious eye irritation.
H320—Causes eye irritation.
H330—Fatal if inhaled.
H331—Toxic if inhaled.
H332—Harmful if inhaled.
H333—May be harmful if inhaled.
H334—May cause allergy or asthma symptoms or breathing difficulties if inhaled.
H335—May cause respiratory irritation.
H336—May cause drowsiness or dizziness.
H340—May cause genetic defects.
H341—Suspected of causing genetic defects.
H350—May cause cancer.
H351—Suspected of causing cancer.
H360—May damage fertility or the unborn child.
H361—Suspected of damaging fertility or the unborn child.
H362—May cause harm to breast-fed children.
H370—Causes damage to organs.
H371—May cause damage to organs.
H372—Causes damage to organs through prolonged or repeated exposure.
H373—May cause damage to organs through prolonged or repeated exposure.

Environmental Hazards

H400—Very toxic to aquatic life.
H401—Toxic to aquatic life.
H402—Harmful to aquatic life.
H410—Very toxic to aquatic life with long lasting effects.
H411—Toxic to aquatic life with long lasting effects.
H412—Harmful to aquatic life with long lasting effects.
H413—May cause long lasting harmful effects to aquatic life.

Prevention Precautionary Codes and Statements

P201—Obtain special instructions before use.
P202—Do not handle until all safety precautions have been read and understood.
P210—Keep away from heat/sparks/open flames/hot surfaces—NO SMOKING.
P211—Do not spray on an open flame or other ignition source.
P220—Keep/Store away from clothing/. . ./combustible materials.
P221—Take any precautions to avoid mixing with combustibles.
P222—Do not allow contact with air.
P223—Keep away from any possible contact with water, because of violent reaction and possible flash fire.
P230—Keep wetted with __?__.
P231—Handle under inert gas.
P232—Protect from moisture.
P233—Keep container tightly closed.
P234—Keep only in original container.

P235—Keep cool.
P240—Ground/bond container and receiving equipment.
P241—Use explosion-proof electrical/ventilating/light/ . . ./equipment.
P242—Use only non-sparking tools.
P243—Take precautionary measures against static discharge.
P244—Keep reduction valves free from grease and oil.
P250—Do not subject to grinding/shock/. . ./friction.
P251—Pressurized container—Do not pierce or burn, even after use.
P260—Do not breathe dust/fume/gas/mist/vapors/spray.
P261—Avoid breathing dust/fume/gas/mist/vapors/spray.
P262—Do not get in eyes, on skin, or on clothing.
P263—Avoid contact during pregnancy/while nursing.
P264—Wash __?__ thoroughly after handling.
P270—Do not eat, drink, or smoke when using this product.
P271—Use only outdoors or in a well-ventilated area.
P272—Contaminated work clothing should not be allowed out of the workplace.
P273—Avoid release to the environment.
P280—Wear protective gloves/protective clothing/eye protection/face protection.
P281—Use personal protective equipment as required.
P282—Wear cold insulating gloves/face shield/eye protection.
P283—Wear fire/flame resistant/retardant clothing.
P284—Wear respiratory protection.
P285—In case of inadequate ventilation wear respiratory protection.
P231 + 232—Handle under inert gas. Protect from moisture.
P235 + 410—Keep cool. Protect from sunlight.

Response Precautionary Codes and Statements

P301—IF SWALLOWED −.
P302—IF ON SKIN −.
P303—IF ON SKIN (or hair) −.
P304—IF INHALED −.
P305—IF IN EYES −.
P306—IF ON CLOTHING −.
P307—IF exposed −.
P308—IF exposed or concerned −.
P309—IF exposed or you feel unwell −.
P310—Immediately call a POISON CENTER or doctor/ physician.
P311—Call a POISON CENTER or doctor/physician.
P312—Call a POISON CENTER or doctor/physician if you feel unwell.
P313—Get medical advice/attention.
P314—Get medical advice/attention if you feel unwell.
P315—Get immediate medical advice/attention.
P320—Specific treatment is urgent (see __?__ on this label).

P321—Specific treatment (see __?__ on this label).
P322—Specific measures (see __?__ on this label).
P330—Rinse mouth.
P331—Do NOT induce vomiting.
P332—If skin irritation occurs −.
P333—If skin irritation or a rash occurs −.
P334—Immerse in cool water/wrap in wet bandages.
P335—Brush off loose particles from skin.
P336—Thaw frosted parts with lukewarm water. Do not rub affected areas.
P337—If eye irritation persists −.
P338—Remove contact lenses if present and easy to do. Continue rinsing.
P340—Remove victim to fresh air and keep at rest in a position comfortable for breathing.
P341—If breathing is difficult, remove victim to fresh air and keep at rest in a position comfortable for breathing.
P342—If experiencing respiratory symptoms −.
P350—Gently wash with soap and water.
P351—Rinse continuously with water for several minutes.
P352—Wash with soap and water.
P353—Rinse skin with water/shower.
P360—Rinse immediately contaminated clothing and skin with plenty of water before removing clothes.
P361—Remove/Take off immediately all contaminated clothing.
P362—Take off contaminated clothing and wash before reuse.
P363—Wash contaminated clothing before reuse.
P370—In case of fire −.
P371—In case of major fire and large quantities −.
P372—Explosion risk in case of fire.
P373—DO NOT fight fire when fire reaches explosives.
P374—Fight fire with normal precautions from a reasonable distance.
P375—Fight fire remotely due to the risk of explosion.
P376—Stop leak if safe to do so.
P377—Leaking gas fire—do not extinguish unless leak can be stopped safely.
P378—Use __?__ for extinction.
P380—Evacuate area.
P381—Eliminate all ignition sources if safe to do so.
P301 + 310—IF SWALLOWED—Immediately call a POISON CENTER or doctor/physician.
P301 + 312—IF SWALLOWED—Call a POISON CENTER or doctor/physician if you feel unwell.
P301 + 330 + 331—IF SWALLOWED—Rinse mouth. DO NOT induce vomiting.
P302 + 334—IF ON SKIN—Immerse in cool water/wrap in wet bandages.
P302 + 350—IF ON SKIN—Gently wash with soap and water.
P302 + 352—IF ON SKIN—Wash with soap and water.
P303 + 361 + 353—IF ON SKIN (or hair)—Remove/Take off immediately all contaminated clothing. Rinse skin with water/shower.

P304 + 312—IF INHALED—Call a POISON CENTER or doctor/physician if you feel unwell.
P304 + 340—IF INHALED—Remove victim to fresh air and keep at rest in a position comfortable for breathing.
P304 + 341—IF INHALED—If breathing is difficult, remove victim to fresh air and keep at rest in a position comfortable for breathing.
P305 + 351 + 338—IF IN EYES—Rinse continuously with water for several minutes. Remove contact lenses if present and easy to do—continue rinsing.
P306 + 360—IF ON CLOTHING—Rinse immediately contaminated clothing and skin with plenty of water before removing clothes.
P307 + 311—IF exposed—Call a POISON CENTER or doctor/physician.
P308 + 313—IF exposed or concerned—Get medical advice/attention.
P309 + 311—IF exposed or you feel unwell—Call a POISON CENTER or doctor/physician.
P332 + 313—If skin irritation occurs—Get medical advice/attention.
P333 + 313—If skin irritation or a rash occurs—Get medical advice/attention.
P335 + 334—Brush off loose particles from skin. Immerse in cool water/wrap in wet bandages.
P337 + 313—Get medical advice/attention.
P342 + 311—Call a POISON CENTER or doctor/physician.
P370 + 376—In case of fire—Stop leak if safe to do so.
P370 + 378—In case of fire—Use __?__ for extinction.
P370 + 380—In case of fire—Evacuate area.
P370 + 380 + 375—In case of fire—Evacuate area. Fight fire remotely due to the risk of explosion.
P371 + 380 + 375—In case of major fire and large quantities—Evacuate area. Fight fire remotely due to the risk of explosion.

Storage Precautionary Codes and Statements

P401—Store __?__.
P402—Store in a dry place.
P403—Store in a well-ventilated place.
P404—Store in a closed container.
P405—Store locked up.
P406—Store in a corrosive resistant/. . . container with a resistant inner liner.
P407—Maintain air gap between stacks/pallets.
P410—Protect from sunlight.
P411—Store at temperatures not exceeding __?__ °C/ __?__ °F.
P412—Do not expose to temperatures exceeding 50°C/ 122°F.
P420—Store away from other materials.
P422—Store contents under __?__.
P402 + 404—Store in a dry place. Store in a closed container.

P403 + 233—Store in a well-ventilated place. Keep container tightly closed.

P403 + 235—Store in a well-ventilated place. Keep cool.

P410 + 403—Protect from sunlight. Store in a well-ventilated place.

P410 + 412—Protect from sunlight. Do not expose to temperatures exceeding 50°C/122°F.

P411 + 235—Store at temperatures not exceeding __?__ °C/ __?__ °F. Keep cool.

Disposal Precautionary Codes and Statements

P501—Dispose of contents/container to __?__.

Appendix 5: Synonym and Trade Name—Cross Index

Acetanilide, 2-chloro-2',6'-diethyl-
N-methoxymethyl)-.................................... A:0480
Acetate d'amyle (French) A:1300
Acetate de butyle (French) B:0810
Acetate de butyle secondaire (French) B:0810
Acetate de cuivre (French) C:1530
Acetate de methyle (French) M:0620
Acetate de plomb (French) L:0110
Acetate de propyle normal (French)....................... P:1190
Acetate de triphenyl-etain (French)........................T:0950
Acetate de vinyle (French) V:0150
Acetate d'isoamyle (French) I:0230
Acetate d'isobutyle (French) B:0810
Acetate fast orange R A:0850
Acetate phenylmercurique (French) P:0450
Acetato cromico (Spanish) C:1090
Acetato de amilo (Spanish) A:1300
Acetato de butilo (Spanish) B:0810
Acetato de butilo-sec (Spanish)........................B:0810
n-Acetato de butilo (Spanish) B:0810
Acetato de terc-butilo (Spanish)........................B:0810
Acetato de cobre (Spanish) C:1530
Acetato de etilo (Spanish) E:0300
Acetato de 2-etoxietilo (Spanish) E:0290
Acetato de isoamilo (Spanish)...........................I:0230
Acetato de isobutilo (Spanish) B:0810
Acetato de plomo (Spanish) L:0110
Acetato de talio (Spanish) T:0420
Acetato de uranilo (Spanish) U:0100
Acetato fenilmercurio (Spanish)......................... P:0450
Acetato(2-methoxyethyl)mercury....................M:0600
Acetdimethylamide.................................. D:1060
Acetehyde A:0100
Acetehyde A:0110
Acetene E:0540
Acetic acid.. **A:0160**
Acetic acid, (2,4,5-T)- T:0100
Acetic acid, allyl acetate............................... A:0530
Acetic acid amide A:0140
Acetic acid, ammonium salt A:0960
Acetic acid n-amyl ester.............................. A:1300
Acetic anhydride..................................... A:0170
Acetic acid anilide A:0150
Acetic acid (aqueous solution) A:0160
Acetic acid, bichloro- D:0430
Acetic acid bromide.................................. A:0280
Acetic acid, bromo-, ethyl ester E:0420
Acetic acid, 2-butoxy ester..........................B:0810
Acetic acid, butyl ester B:0810
Acetic acid, n-butyl ester............................B:0810
Acetic acid t-butyl ester..............................B:0810

Acetic acid, tert-butyl ester B:0810
Acetic acid, cadmium salt C:0110
Acetic acid chloride A:0290
Acetic acid, chloro- C:0740
Acetic acid (4-chloro-2-methylphenoxy)-..............M:0290
Acetic acid [(4-chloro-o-tolyl)-oxy]- M:0290
Acetic acid, chromium(3+) salt C:1090
Acetic acid, copper(2+) salt C:1530
Acetic acid, copper(II) salt C:1530
Acetic acid, cupric salt C:1530
Acetic acid, cyano-, ethyl ester E:0510
Acetic acid, dichloro- D:0430
Acetic acid (2,4-dichlorophenoxy)-.................... D:0100
Acetic acid, dimethyl- I:0310
Acetic acid, dimethylamide D:1060
Acetic acid-1,3-dimethylbutyl ester H:0340
Acetic acid, O,O-dimethyldithiophosphoryl-,
n-monomethylamide salt D:1040
Acetic acid, 1,1-dimethylethyl ester.......................B:0810
Acetic acid, ethenyl ester.............................. V:0150
Acetic acid, 2-ethoxyethyl ester......................E:0290
Acetic acid (ethylenedinitrilo)tetra- E:0570
Acetic acid, ethyl ester E:0300
Acetic acid, fluoro-, sodium salt S:0480
Acetic acid, glacial A:0160
Acetic acid, isobutyl ester B:0810
Acetic acid, isopentyl ester........................... A:1300
Acetic acid, isopentyl ester...............................I:0230
Acetic acid, lead(2+) salt L:0110
Acetic acid, lead(II) salt L:0110
Acetic acid, mercapto-...............................T:0460
Acetic acid, mercury(2+) salt M:0340
Acetic acid, mercury(II) salt..........................M:0340
Acetic acid 3-methoxybutyl ester.......................B:0800
Acetic acid, methyl ester.............................M:0620
Acetic acid, 1-methylpropyl ester B:0810
Acetic acid, 2-methylpropyl ester B:0810
Acetic acid, nitrilotri- N:0360
Acetic acid pentyl ester A:1300
Acetic acid, 2-pentyl ester A:1300
Acetic acid, phenylmercury derivative.................. P:0450
Acetic acid, 2-propenyl ester A:0530
Acetic acid, propyl ester P:1190
Acetic acid, n-propyl ester............................ P:1190
Acetic acid, trichloro-................................T:0680
Acetic acid, (2,4,5-trichlorophenoxy)- T:0100
Acetic acid, vinyl ester V:0150
Acetic aldehyde A:0110
Acetic anhydride **A:0170**
Acetic bromide....................................... A:0280
Acetic chloride....................................... A:0290

Acido etilendiaminotetraacetico (Spanish) E:0570
Acido fenilarsonico (Spanish) B:0320
Acido fluoborico (Spanish) F:0260
Acido fluorhidrico (Spanish) H:0450
Acido fluoroacetico (Spanish) F:0330
Acido formico (Spanish) F:0450
Acido fosforico Spanish) P:0590
Acido isobutirico (Spanish) I:0310
Acido malico (Spanish) M:0210
Acido metacrilico (Spanish) M:0490
Acido α-metacrilico (Spanish) M:0490
Acido naftalico (Spanish) N:0130
Acido nitrico (Spanish) N:0340
Acido nitrilotriacetico (Spanish) N:0360
Acido peracetico (Spanish) P:0290
Acido picrico (Spanish) P:0730
Acido selenioso (Spanish) S:0190
Acido sulfurico (Spanish) S:0770
Acido 2,4,5-triclorofenoxiacetico (Spanish) T:0100
Acifloctin A:0440
Acifluorfen A:0360
Acifluorfene A:0360
Acigena H:0240
Acillin A:1290
Acinetten A:0440
Acisal® A:0340
ACL 59 .... P:0920
ACL 60 .... S:0460
ACL 85 .... T:0750
Aclid P:1045
ACME MCPA amine 4 .... M:0290
Acnegel B:0430
Acnestrol D:0910
Acocantherin O:0150
ACP-M-728 .... C:0600
Acquinite A:0380
Acquinite C:0980
Acrehyde A:0380
Acricid® B:0475
Acridine.. A:0370
Acrilamida (Spanish) A:0390
Acrilato de n-butilo (Spanish) B:0830
Acrilato de 2-etilhexilo (Spanish) E:0710
Acrilato de etilo (Spanish) E:0320
Acrilato de metilo (Spanish) M:0650
Acrilonitrilo (Spanish) A:0410
Acrinet® A:0410
Acroleic acid A:0400
Acrolein.. A:0380
Acrolein acetal D:0780
Acroleine (French) A:0380

Acromona M:1340
Acrylaldehyde A:0380
Acrylaldehyde diethyl D:0780
Acrylamide A:0390
Acrylamide monomer A:0390
Acrylate d'ethyle (French) E:0320
Acrylate de methyle (French) M:0650
Acrylehyd (German) A:0380
Acrylehyde A:0380
Acrylic acid.. A:0400
Acrylic acid amide (50%) A:0390
Acrylic acid, butyl ester B:0830
Acrylic acid n-butyl ester B:0830
Acrylic acid chloride A:0420
Acrylic acid, 2-cyano-, methyl ester M:0790
Acrylic acid, ethyl ester E:0320
Acrylic acid, 2-ethylhexyl ester E:0710
Acrylic acid 2-hydroxypropyl ester H:0510
Acrylic acid, inhibited A:0400
Acrylic acid isobutyl ester I:0270
Acrylic acid, 2-methyl- M:0490
Acrylic acid methyl ester M:0650
Acrylic acid, 2-methyl-, methyl ester M:1060
Acrylic aldehyde A:0380
Acrylic amide A:0390
Acrylic amide 50% A:0390
Acrylic acid, glacial A:0400
Acrylnitril (German) A:0410
Acrylonitrile A:0410
Acrylonitrile monomer A:0410
Acrylon® A:0410
Acryloyl chloride.. A:0420
Acrylsaeuraethylester (German) E:0320
Acrylsaeuremethylester (German) M:0650
Actamer B:0560
Actedron A:1280
Acti-Aid® C:1730
Actidione C:1730
Actidione TGF C:1730
Actidone C:1730
Actinite PK A:1610
Actinomicina D (Spanish) A:0430
Actinomycin D.. A:0430
Actinomycin I A:0430
Actinomycindioic D acid, dilactone A:0430
Actispray C:1730
Activated ergosterol E:0190
Actybaryte B:0210
Acylpyrin® A:0340
AD A:0430
AD 1 (aluminum)... A:0660

Antimony(V) fluoride A:1430
Antimony hydride S:0600
Antimony lactate **A:1410**
Antimonyl potassium tartrate A:1440
Antimony pentachloride **A:1420**
Antimony pentafluoride **A:1430**
Antimony(5+) pentafluoride A:1430
Antimony(V) pentafluoride A:1430
Antimony perchloride A:1420
Antimony peroxide A:1480
Antimony potassium tartrate **A:1440**
Antimony powder A:1400
Antimony, regulus A:1400
Antimony (3+) salt (3:1) A:1410
Antimony sesquioxide A:1480
Antimony tribromide **A:1450**
Antimony trichloride **A:1460**
Antimony trifluoride **A:1470**
Antimony trihydride S:0600
Antimony trioxide **A:1480**
Antimony, white A:1480
Antimucin WDR P:0450
Antimycin A **A:1490**
Antinonin D:1340
Antinonnin D:1340
Antio F:0460
Antipiricullin A:1490
Antiren P:0770
Antisacer P:0510
Antisal 1 .... T:0270
Antisal 1A T:0600
Antisal 2B H:0450
Antivampire S:0650
Antiverm P:0360
Antivitium D:1570
Antlak D:0280
Antol E:0420
Antraceno (Spanish) A:1380
Antraquinona (Spanish) A:1390
ANTU **A:1500**
Anturat A:1500
Antywylegacz C:0710
Anyvim A:1350
AO A1 (aluminum) A:0660
AP P:0330
Apachlor C:0650
Apadrin M:1430
Apamidon P:0570
Aparasin L:0260
Apaurin D:0270
Apavap D:0690

Apavinphos M:1350
APC A:0220
Apco 2330 .... P:0380
Apex 462-5 .... T:0970
APFO A:1160
APGA A:0880
Aphamite P:0170
Aphtiria L:0260
Aplidal L:0260
Apomine black GX D:1550
Apoplon R:0100
Apozepam D:0270
APPA P:0560
Apsical R:0100
Apyonine auramarine base A:1620
Aqua ammonia A:0950
Aqua ammonia A:1110
Aquacat C:1300
Aquacide D:1540
Aqua fortis N:0340
Aqua-Kleen D:0100
Aqualin A:0380
Aqualine A:0380
Aqua mephyton P:0690
Aquamycetin C:0620
Aqua regia N:0340
Aqua-Sol flux B:0840
Aquathol E:0110
Aquatin 20 EC T:0950
Aqua-Vex S:0300
Aquazine S:0310
Aqueous acrylic acid (technical grade is 94%) A:0400
Aqueous ammonia A:1110
Aqueous hydrogen chloride H:0430
AR2 (aluminum) A:0660
Arab rat deth W:0100
Arabic gum G:0210
Aracide S:0780
Aragonite C:0230
Araldite hardener 972 .... D:0250
Araldite HT 901 .... P:0670
Aralo P:0170
Aramite® S:0780
Araamite®-15W S:0780
Arasan T:0520
Arasan 42-S T:0520
Arasan 42S T:0520
Arasan 70 .... T:0520
Arasan 70-S red T:0520
Arasan 75 .... T:0520
Arasan-M T:0520

Bromopropanes	**B:0730**
1-Bromo-2-propene	A:0560
3-Bromopropeno (Spanish)	A:0560
3-Bromopropylene	A:0560
3-Bromopropyne	P:1100
3-Bromo-1-propyne	P:1100
α-Bromotoluene	B:0440
ω-Bromotoluene	B:0440
Bromotrifluormetano (Spanish)	T:0820
Bromotrifluoromethane	T:0820
Bromure d'ethyle (French)	E:0410
Bromure de vinyle (French)	V:0160
Bromuro de alilo (Spanish)	A:0560
Bromuro de bencilo (Spanish)	B:0440
Bromuro de n-butilo (Spanish)	B:0880
Bromuro de cadmio (Spanish)	C:0120
Bromuro de cianogeno (Spanish)	C:1610
Bromure de cyanogen (French)	C:1610
Bromuro de metileno (Spanish)	M:0890
Bromuro de propargilo (Spanish)	P:1100
Bromuro de vinilo (Spanish)	V:0160
Bromuro de zinc (Spanish)	Z:0110
Bromwasserstoff (German)	H:0420
Bronze powder	C:1360
Brookite	T:0570
Broserpine	R:0100
BRP	N:0100
Brucina (Spanish)	B:0740
Brucine	**B:0740**
(-) Brucine	B:0740
(-) Brucine dihydrate	B:0810
(-)Brucine hydrate	B:0740
Brumin	W:0100
Brush Buster	D:0420
Brushkiller	H:0320
Brush-Off 445 low volatile brush killer	T:0100
Brush-Rhap	D:0100
Brush rhap	T:0100
Brushtox	T:0100
Brygou	P:1180
BSC-refined D	B:0340
BTS 27,419	A:0940
B-Selektonon	D:0100
B-Selektonon M	M:0290
BUCS	B:0790
BUFA	D:0410
Buddhist rosary bead	A:0025
Bufen	P:0450
Bufton	D:0910
Bu-gas	B:0770
Buhach	P:1340
Bunt-Cure	H:0190
Bunt-No-More	H:0190
Burmar Lab Clean	A:1110
Burmar Lab Clean	B:0840
Burmar Nophenol-922 HB	C:0570
Burnt island red	I:0210
Burnt lime	C:0320
Burnt sienna	I:0210
Burnt umber	I:0210
Burtolin	M:0220
Bush killer	D:0100
Busulfan	**B:0750**
Buta-1,3-dien (German)	B:0760
Butadiendioxyd (German)	D:0760
Butadiene	B:0760
Buta-1,3-diene	B:0760
1,3-Butadiene	**B:0760**
α-γ-Butadiene	B:0760
1,3-Butadiene, 2-chloro-	C:1000
Butadiene diepoxide	D:0760
1,3-Butadiene diepoxide	D:0760
Butadiene dimer	V:0180
Butadiene dioxide	D:0760
Butadiene, hexachloro-	H:0200
1,3-Butadiene, 1,1,2,3,4,4-hexachloro-	H:0200
1,3-Butadiene, 2-methyl	I:0420
1,3-Butadieno (Spanish)	B:0760
Butadione	B:0780
Butafume	B:0850
Butal	B:1030
Butaldehyde	B:1030
Butalyde	B:1030
Butan-1-ol	B:0840
Butan-2-ol	B:0840
Butanal	B:1030
1-Butanamine	B:0850
2-Butanamine	B:0850
1-Butanamine, n-butyl	D:0370
1-Butanamine, n-butyl-N-nitroso-	N:0560
Butanes	**B:0770**
n-Butane	B:0770
Butane, 1-bromo-	B:0880
Butane, 1-chloro-	B:0890
1,4-Butanedicarboxylic acid	A:0440
Butane diepoxide	D:0760
Butane, 1,2:3,4-diepoxy-	D:0760
Butane, 2,3-dimethyl-	D:1120
Butanedioic acid, 2,3-dihydroxy-[R-(R*,R*)]-, diammonium salt	A:1240
Butanedioic acid, [(dimethoxyphosphinothioyl) thio]-, diethyl ester	M:0190

E

E-D-BEE	E:0580
EDC	E:0590
EDCO	M:0720
Edetic	E:0570
Edetic acid	E:0570
Edicol Supra Rose B	C:1250
Edicol Supra Rose BS	C:1250
EDTA	E:0570
EDTA acid	E:0570
EE acetate	E:0290
EEC No. E924	P:0870
Eerex	B:0640
Eerex granular weed killer	B:0640
Eerex water soluble granular weed killer	B:0640
EF 121	C:1070
Effluderm (free base)	F:0370
Effusan	D:1340
Effusan 3436	D:1340
Efloran	M:1340
Efudex	F:0370
Efudix	F:0370
Efurix	F:0370
EFV 250/400	I:0190
EG	E:0610
EGBE	B:0790
EGDN	E:0630
EGEEA	E:0290
Egitol	H:0230
EGM	E:0640
EGME	E:0640
EGMEA (109-86-4)	M:0590
EGEEA	E:0290
Ehhanol, 2-(2-aminoethioxy)-	A:0810
EI 38555	C:0710
EI 47031	P:0540
EI 47300	F:0100
EI 47470	M:0330
EI 52160	T:0170
Eisendextran (German)	I:0200
Eisendimethyldithiocarbamat (German)	F:0130
Eisen(III)-tris(N,N-dimethyldithiocarbamat) (German)	F:0130
Ekagom TB	T:0520
Ekagom teds	D:1570
Ekatin WF & WF ULV	P:0170
Ekatox	P:0170
EKKO	P:0510
EKKO Capsules	P:0510
Ektafos	D:0710
Ektasolve EB solvent	B:0790
Ektasolve EE	E:0280
Ektofos	D:0710
EL 3911	P:0520
EL 4049	M:0190
Elaldehyde	P:0130
Elancolan	T:0840
Elaol	D:0410
Elastonon	A:1280
Elayl	E:0540
Elcoril	C:0610
Eldopaque	H:0490
Eldoquin	H:0490
Electro-CF 11	F:0360
Electro-CF 12	D:0500
Electro-CF 22	C:0850
Electrolon®	S:0250
Electronic E-2	H:0470
Elemental cadmium	C:0100
Elemental calcium	C:0200
Elemental carbon	C:0450
Elemental chromium	C:1130
Elemental copper	C:1360
Elemental gallium	G:0050
Elemental germanium	G:0110
Elemental hafnium	H:0100
Elemental indium	I:0120
Elemental iron	I:0190
Elemental manganese	M:0250
Elemental molybdenum	M:1410
Elemental platinum	P:0800
Elemental potassium	P:0840
Elemental rhodium	R:0120
Elemental selenium	S:0140
Elemental silicon	S:0240
Elemental silver	S:0260
Elemental sodium	S:0330
Elemental tantalum	T:0130
Elemental tellurium	T:0150
Elemental vanadium	V:0110
Elepsindon	P:0510
Elerpine	R:0100
Elfan WA sulphonic acid	D:1630
Elfanex	R:0100
Elgetol	D:1340
Elgetol 30	D:1340
Elgetol 318	D:1380
Elipol	D:1340
Elmasil	A:0910
Eloxyl	B:0430
Elyzol	M:1340
Emanay atomized aluminum powder	A:0660
Emanay zinc dust	Z:0100

ENT 9,233	H:0210	ENT 24,984	S:0350
ENT 9,234	II:0210	ENT 24,988	N:0100
ENT 9,735	T:0650	ENT 25,208	T:0950
ENT 9,932	C:0630	ENT 25,294	M:0300
ENT 14,689	F:0130	ENT 25,445	A:0910
ENT 14,875	M:0240	ENT 25,500	P:0350
ENT 15,108	P:0170	ENT 25,515	P:0570
ENT 15,152	H:0140	ENT 25,540	F:0120
ENT 15,349	E:0580	ENT 25,543	P:0350
ENT 15,406	D:0640	ENT 25,545	I:0250
ENT 15,949	A:0510	ENT 25,545-X	I:0250
ENT 16,087	P:0140	ENT 25,552-X	C:0630
ENT 16,225	D:0750	ENT 25,554	M:1080
ENT 16,273	S:0720	ENT 25,580	T:0470
ENT 16,391	C:0640	ENT 25,584	H:0150
ENT 16,519	S:0780	ENT 25,584	H:0140
ENT 17,034	M:0190	ENT 25,595-X	D:1300
ENT 17,251	E:0140	ENT 25,602-X	C:1490
ENT 17,291	O:0110	ENT 25,671	P:1180
ENT 17,292	M:1070	ENT 25,675	C:1640
ENT 17,295	D:0170	ENT 25,705	P:0560
ENT 17,510	A:0520	ENT 25,712	T:0760
ENT 17,798	E:0170	ENT 25,715	F:0100
ENT 17,957	C:1420	ENT 25,719	M:1390
ENT 18,596	E:0520	ENT 25,726	M:0550
ENT 18,771	T:0180	ENT 25,766	M:1360
ENT 18,862	D:0180	ENT 25,793	B:0475
ENT 18,870	M:0220	ENT 25,796	F:0400
ENT 19,060	I:0360	ENT 25,830	P:0540
ENT 19,109	D:1030	ENT 25,922	D:1300
ENT 19,244	I:0340	ENT 25,991	M:0330
ENT 19,507	D:0280	ENT 26,263	E:0660
ENT 19,763	T:0670	ENT 26,396	E:0770
ENT 20,738	D:0690	ENT 26,538	C:0410
ENT 22,014	A:1640	ENT 26,592	D:0760
ENT 22,374	M:1350	ENT 27,129	M:1430
ENT 22,879	D:1420	ENT 27,164	C:0440
ENT 23,233	A:1650	ENT 27,165	T:0170
ENT 23,284	R:0140	ENT 27,193	M:0540
ENT 23,437	D:1580	ENT 27,257	F:0460
ENT 23,648	D:0700	ENT 27,300	P:1030
ENT 23,708	C:0530	ENT 27,300-A	P:1030
ENT 23,969	C:0430	ENT 27,311	C:1070
ENT 23,979	E:0100	ENT 27,318	E:0270
ENT 24,042	P:0520	ENT 27,320	D:0210
ENT 24,105	E:0260	ENT 27,341	M:0560
ENT 24,482	D:0710	ENT 27,396	M:0520
ENT 24,652	P:1320	ENT 27,566	F:0440
ENT 24,653	E:0120	ENT 27,572	F:0050
ENT 24,945	F:0110	ENT 27,635	C:1080
ENT 24,969	C:0650	ENT 27,851	T:0450

ENT 27,967.. A:0940
ENT 28,009...T:0540
ENT 28,009 ....T:0950
ENT 50,146...R:0100
ENT 50,324...E:0650
ENT 50,434.. A:1440
ENT 50,882.. H:0290
ENT 51,762.. N:0700
ENTAC 349 biocide B:0840
ENTEC 327 surfactant.....................................B:0840
Enteromycetin...C:0620
Entex F:0120
Entizol...M:1340
Entomoxan..L:0260
Entonite 2073...B:0250
Entozyme P:0050
Entphosphorothioate F:0110
Entsufon...T:0920
ENU N:0620
Envert 171 .... D:0100
Envert DT D:0100
Envert-T T:0100
EN-Viron D concentrated phenolic
disinfectant.. H:0240
E.O E:0660
EO 5A I:0190
EP-161E...M:1030
EP-332..F:0440
EP30 ....P:0240
EP316..P:1030
Epal-6.. H:0310
E-Pam.. D:0270
Epamin..P:0510
Epanutin..P:0510
EPA pesticide chemical 004401 .... A:0910
EPA pesticide chemical 015801........................M:1350
EPA pesticide chemical 029801.......................... D:0420
EPA pesticide chemical 034801..........................F:0130
EPA pesticide chemical 035505.......................... D:1610
EPA pesticide chemical 037505.......................... D:1380
EPA pesticide chemical 038501.......................... D:1470
EPA pesticide chemical 041101..........................E:0270
EPA pesticide chemical 053301..........................F:0120
EPA pesticide chemical 057801.......................... D:0280
EPA pesticide chemical 080801.......................... A:0740
EPA pesticide chemical 080803.......................... A:1610
EPA pesticide chemical 090201..........................C:0540
EPA pesticide chemical 090501.......................... A:0480
EPA pesticide chemical 098301.......................... A:0490
EPA pesticide chemical 100101..........................C:1580
EPA pesticide chemical 106201 .... A:0940

Epasmir 5 ....P:0510
Epdantoine simple P:0510
Epelin...P:0510
EPF B20 fixer A:0160
Ephorran... D:1570
Epichlorhydrin (German)..................................E:0160
Epichlorhydrine (French)..................................E:0160
α-Epichlorohydrin E:0160
(DL)-α-Epichlorohydrin....................................E:0160
Epichlorohydrin E:0160
EPI-chlorohydrin E:0160
Epiclear B:0430
Epiclorhidrina (Spanish)..................................E:0160
Epicure DDM D:0250
Epihydrin alcohol... G:0160
Epihydrinaldehyde G:0170
Epihydrine aldehyde G:0170
Epikure DDM D:0250
Epilan P:0510
Epilantin..P:0510
Epinat P:0510
EPI-REZ 508 .... D:0970
EPI-REZ 510 .... D:0970
Epised..P:0510
EPN.. E:0170
Epon 828 .... D:0970
Epoxide A D:0970
1,2-Epoxyaethan (German)................................E:0660
1,2-Epoxy-3-allyloxypropane A:0590
1,2-Epoxybutane B:0910
1,4-Epoxybutane T:0340
1,2-Epoxy-3-butoxy propane B:0930
6-6 Epoxy chem resin finish, clear curing agent B:0810
6-6 Epoxy chem resin finish, clear curing agent B:0840
1,2-Epoxy-3-chloropropane E:0160
Epoxy cure agent B:0840
3,6-Epoxycyclohexane-1,2-dicarboxylic acid E:0110
1,2-Epoxy-4-(epoxyethyl)cyclohexane V:0190
Epoxyethane..E:0660
1,2-Epoxyethane E:0660
1,2-Epoxy-3-ethoxy-propane (DOT).......................E:0180
Epoxyethylbenzene S:0670
(Epoxyethyl)benzene S:0670
1,2-Epoxyethylbenzene....................................S:0670
3-(Epoxyethyl)-7-oxabicyclo(4.1.0) heptane,
vinyl cyclohexene diepoxide V:0190
Epoxy ethyloxy propane........................... E:0180
Epoxyheptachlor... H:0140
Epoxyheptachlor... H:0150
4,7-Epoxyisobenzofuran-1,3-dione, hexahydro-
3a, 7a-dimethyl-, (3a a, 4 b, 7 b, 7a a)-....................C:0380

G

G2 (Oxide)	A:0660
G 11	H:0240
G 301	D:0280
G 338	E:0520
G 23992	E:0520
G 24480	D:0280
G 27692	S:0310
G 30027	A:1610
GA (military designation)	T:0110
Galena	L:0220
Gallium	**G:0050**
Gallium chloride	G:0075
Gallium(3+) chloride	G:0075
Gallium(III) chloride	G:0075
Gallium trichloride	**G:0075**
Gallogama	L:0260
Gallotox	P:0450
Gamacid	L:0260
Gamaphex	H:0210
Gamaphex	L:0260
Gamaserpin	R:0100
Gamasol-90	D:1280
Gamene	L:0260
Gamixel	P:0150
Gammabenzene hexachlorocyclohexane (g isomer)	L:0260
Gammahexa	L:0260
Gammahexane	L:0260
Gammalex	L:0260
Gammalin	L:0260
Gammalin 20	L:0260
Gammaphex	L:0260
Gammasan 30	L:0260
Gammaserpine	R:0100
Gammaterr	L:0260
Gammex	L:0260
Gammexane	H:0210
Gammexane	L:0260
Gammexene	L:0260
Gammopaz	L:0260
Gamonil	C:0430
Gamophen	H:0240
Gamophene	H:0240
Ganeake	C:1350
Garantose	S:0100
Gardentox	D:0280
Garox	B:0430
Garrathion®	C:0530
Garvox	B:0220

Garvox 3G	B:0220
Gas de petroleo licuado (Spanish)	L:0270
Gas mostaza (Spanish)	M:1460
Gasolina (Spanish)	G:0100
Gasoline	**G:0100**
Gastracid	P:0330
Gastrotest	P:0330
GB (military designation)	S:0130
GBL	D:1530
GC 1189	C:0640
GC 3944-3-4	P:0230
GC 3944-3-4	Q:0110
GC 4072	C:0650
GC 6936	T:0950
GC 8993	T:0950
Gearphos	P:0170
Gebutox	D:1380
Geigy 338	E:0520
Geigy 13005	M:0540
Geigy 22870	D:1300
Geigy 24480	D:0280
Geigy 27692	S:0310
Geigy 30,027	A:1610
Geigy 30494	M:1080
Geigy G-23611	I:0360
Gel II	S:0470
Gelan 1	T:0110
Gelber phosphor (German)	P:0610
Gelbin	C:0260
Geltabs	E:0190
Gelution	S:0470
General chemicals 1189	C:0640
General chemicals 8993	T:0950
Genesolv 404 azeotrope	H:0300
Genesolv D solvent	T:0790
Genetron®	V:0230
Genetron 11	F:0360
Genetron 12	D:0500
Genetron 21	D:0570
Genetron 22	C:0850
Genetron 101	C:0840
Genetron 112	T:0240
Genetron 113	T:0790
Genetron 114	D:0680
Genetron 115	C:0930
Genetron 142b	C:0840
Genetron 316	D:0680
Geniphene	T:0650
Genisis	C:1350
Genithion	P:0170
Genitox®	D:0140

Green chromic oxide C:1160
Green cinnabar C:1160
Green cross warble powder R:0150
Greenfly aerosol spray M:0190
Green GA C:1160
Greenmaster autumn F:0180
Green oil A:1380
Greenockite C:0170
Green rouge C:1160
Green seal-8 .... Z:0140
Green vitriol iron monosulfate F:0220
Grey arsenic A:1520
Griffex A:1610
Griffin manex M:0240
Griffin super Cu C:1390
Grisol T:0180
Grocolene G:0150
Groundhog soltair D:1540
Grouticide 75 .... C:1040
Grovex sewer bait W:0100
Grundier arbezol P:0240
GS 6 .... I:0190
GS-13005 .... M:0520
GS-13005 .... M:0540
G-Strophanthin O:0150
GT41 .... B:0750
GT 2041 .... B:0750
GTN N:0510
p-Guaicol M:0610
Guanidine, cyano-,
methylmercury deriv M:1050
Guesarol® D:0140
Guicitrina A:1290
Guicitrine A:1290
Gum acacia G:0210
Gum arabic G:0210
Gum camphor C:0370
Gum ovaline G:0210
Gum Senegal G:0210
Gum spirits T:1000
Gum turpentine T:1000
Gun cotton N:0420
Gusathion® A:1650
Gusathion A A:1640
Gusathion A insecticide A:1640
Gusathion ethyl A:1640
Gusathion® M A:1650
Guthion® A:1650
Guthion ethyl A:1640
Guthion insecticide A:1640
Gylcidy butyl ether B:0930

Gynergen E:0200
Gynopharm D:0910
GY-Phene T:0650
Gypsine L:0120
Gypsum C:0350
Gyron® D:0140
Gygon D 1040
Gyycolic nitrile F:0420

H

H (military designation) M:1460
H 35-F 87 (BVM) F:0100
H 321 .... M:0550
H 520 .... R:0100
H 1803 .... S:0310
H 5727 .... P:0350
H 8757 .... P:0350
Haematite H:0130
Hafnium H:0100
Hafnium metal H:0100
Haiari R:0150
Haitin T:0540
Haitin T:0950
Haitin WP 20 (fentin hydroxide 20%) T:0950
Haitin WP 60 (fentin hydroxide 60%) T:0950
Halane® D:0510
Half-myderan E:0770
Halizan M:0480
Halocarbon 11 .... F:0360
Halocarbon 12/Ucon 12 .... D:0500
Halocarbon 14 .... T:0330
Halocarbon 112 .... T:0240
Halocarbon 112A T:0240
Halocarbon 113 .... T:0740
Halocarbon 1132A V:0230
Halocarbon 114 .... D:0680
Halocarbon 115 .... C:0930
Halomycetin C:0620
Halon D:0500
Halon 14 .... T:0330
Halon 104 .... C:0510
Halon 112 .... D:0570
Halon 122 .... D:0500
Halon 242 .... D:0680
Halon 1001 .... M:0720
Halon 1011 .... C:0820
Halon® 1202 .... D:0940
Halon 1211 .... C:0830
Halon 1301 .... T:0820
Halon 2001 .... E:0410

J

K

Malamar 50 .... M:0190
Malasol M:0190
Malaspray M:0190
Malataf M:0190
Malathion M:0190
Malathion 60 .... M:0190
Malathion E50 .... M:0190
Malathion LB concentrate M:0190
Malathion organophosphorous insecticide M:0190
Malathon M:0190
Malathyl M:0190
Malation (Spanish) M:0190
Malazide M:0220
Maldison (in Australia, New Zealand) M:0190
Maleic acid M:0200
Maleic acid anhydride M:0210
Maleic acid hydrazide M:0220
Maleic anhydride M:0210
Maleic hydrazide M:0220
Maleic hydrazide fungicide M:0220
Maleic hydrazine M:0220
Malein 30 .... M:0220
Maleinic acid M:0200
Maleinsaurehydrazid (German) M:0220
Malenic acid M:0200
N,N-Maleoylhydrazine M:0220
Malestrone (AMPS) T:0220
Malipur C:0410
Malix E:0100
Mallofeen P:0330
Mallophene P:0330
Malmed M:0190
Malonic acid dinitrile M:0230
Malonic acid, ethyl ester nitrile E:0510
Malonic acid, thallium salt (1:2) T:0420
Malonic dinitrile M:0230
Malonodinitrile M:0230
Malononitrile M:0230
Malononitrilo (Spanish) M:0230
Malphos M:0190
Malzid M:0220
Manam M:0240
Maneb M:0240
Maneb 80 .... M:0240
Maneba M:0240
Manebe (French) M:0240
Manebe 80 .... M:0240
Manebgan M:0240
Manesan M:0240
Manex M:0240
Mangan(II)-[N,N'-aethylen-bis(dithiocarbamate)]

(German) M:0240
Mangandioxid (German) M:0260
Manganese (dust and fume) M:0250
Manganese-55 .... M:0250
Manganese binoxide M:0260
Manganese (bioxyd de) (French) M:0260
Manganese black M:0260
Manganese dinitrate M:0270
Manganese dioxide M:0260
Manganese (dioxyde de) (French) M:0260
Manganese element M:0250
Manganese ethylene-1,2-bisdithiocarbamate M:0240
Manganese(II) ethylene di(dithiocarbamate) M:0240
Manganese, (methylcyclopentadienyl)
tricarbonyl- M:0280
Manganese nitrate M:0270
Manganese peroxide M:0260
Manganese superoxide M:0260
**Manganese, tricarbonyl
methylcyclopentadienyl M:0280**
Manganeso (Spanish) M:0250
Manganous dinitrate M:0270
Manganous ethylenebis(dithiocarbamate) M:0240
Manganous nitrate M:0270
Manialith L:0290
Manmade mineral fibers F:0240
Mannitol mustard H:0370
Manoc M:0240
Manro PTSA 65 E T:0630
Manro PTSA 65 H T:0630
Manro PTSA 65 LS T:0630
Manufactured iron oxides I:0210
Manzate M:0240
Manzate D M:0240
Manzate maneb fungicide M:0240
Manzeb M:0240
Manzin M:0240
MAOH M:0990
MAPP gas M:0640
Marble C:0230
Marevan (sodium salt) W:0100
Margarite M:1370
Marisan forte P:0230
Marisan forte Q:0110
Marisilan A:1290
Markem 320 cleaner B:0840
Markem thinner XF E:0290
Marksman T:0840
Marksman 2, trigard T:0840
Marlate M:0580
Marlate 50 .... M:0580

Methylenebis(aniline) D:0250
4,4′-Methylenebis(aniline) D:0250
4,4′-Methylenebis(Benzeneamine) D:0250
Methylenebis(3-chloro-4-aminobenzene) M:0850
Methylenebis(o-chloroaniline) M:0850
Methylene-4,4′-bis(o-chloroaniline) M:0850
4,4′-Methylene(bis)-chloroaniline M:0850
4,4′-Methylenebis (2-chloroaniline) M:0850
4,4′-Methylenebis(o-chloroaniline) M:0850
p,p′-Methylenebis(α-chloroaniline) M:0850
p,p′-Methylenebis(o-chloroaniline) M:0850
4,4′-Methylenebis-2-chlorobenzenamine M:0850
4,4′-Methylenebis (2-chloro-benzeneamine) M:0850
Methylenebis(4-cyclohexyl isocyanate) M:0860
Methylene-S,S′-bis(O,O-diaethyl-
dithiophosphat) (German) E:0260
4,4′-Methylenebis(N,N-dimethyl) aniline M:0870
4,4′-Methylene-bis-(N,N-dimethylaniline) M:0700
4,4′-Methylene bis(N,N-dimethylaniline) M:0870
Methylenebis(4-isocyanatobenzene) M:0880
1,1-Methylenebis(4-isocyanatobenzene) M:0880
1,1′-Methylenebis(4-isocyanatobenzene) M:0880
1,1-[Methylenebis(oxy)]bis(2-chloroethane) B:0490
Methylenebis(4-phenylene isocyanate) M:0880
Methylenebis(p-phenylene isocyanate) M:0880
Methylenebis(phenylisocyanate) M:0880
Methylene bisphenylisocyanate M:0880
Methylene bis(4-phenylisocyanate) M:0880
Methylenebis(4-phenylisocyanate) M:0880
Methylenebis(4,4′-phenylisocyanate) M:0880
Methylenebis(p-phenylisocyanate) M:0880
4,4′-Methylenebis(phenylisocyanate) M:0880
p,p′-Methylenebis(phenylisocyanate) M:0880
2,2′-Methylenebis(3,4,6-trichlorophenol) H:0240
2,2′-Methylenebis(3,5,6-trichlorophenol) H:0240
Methylene bromide M:0890
Methylene chloride M:0900
Methylene chlorobromide C:0820
Methylene cyanide M:0230
Methylene cyanohydrine F:0420
Methylenedianiline D:0250
p,p′-Methylenedianiline D:0250
4,4′-Methylenedibenzenamine D:0250
Methylene dibromide M:0890
Methylene dichloride M:0900
Methylene dimethyl ether M:0660
Methylenedinitrile M:0230
3,4-Methylenedioxy-allylbenzene S:0110
3-(3,4-Methylenedioxyphenyl)prop-1-ene S:0110
[1,2-(Methylenedioxy)-4-propyl]benzene D:0990
4,4′-Methylenedi(phenyldiisocyanate) M:0880

Methylenedi-p-phenylene diisocyanate M:0880
Methylenedi(p-phenylene diisocyanate) M:0880
Methylene di(phenylene isocyanate) M:0880
Methylenedi-p-phenylene isocyanate M:0880
Methylenedi(p-phenylene isocyanate) M:0880
4,4′-Methylenedi(phenylene isocyanate) M:0880
4,4′-Methylenedi-p-phenylene diisocyanate M:0880
4,4′-Methylene diphenylisocyanate M:0880
Methylene ester of allyl-pyrocatechol S:0110
Methylene glycol F:0410
Methylene oxide F:0410
a,S′-Methylene O,O,O′,O′-tetraethyl ester
phosphorodithioic acid E:0260
S,S′-Methylene O,O,O′,O′-tetraethyl
phosphorodithioate E:0260
Methyl ester of isocyanic acid M:1010
Methyl ester of methacrylic acid M:1060
Methyle (sulfate de) (French) D:1260
Methyl ethanoate M:0620
Methylethene P:1230
Methyl ether D:1180
Methyl ethoxol E:0640
2-(1-Methylethoxy)ethanol I:0440
2-(1-Methylethoxy)phenyl
N-methylcarbamate P:1180
1-Methylethyl acetate I:0450
1-Methylethylamine I:0470
N-(1-Methylethyl)-benzenamine I:0480
1-Methylethyl benzene C:1500
1-(Methylethyl) benzene M:1240
3-(1-Methylethyl)-1H-2,1,3-benzothiazain-
4(3H)-one-2,2-dioxide B:0240
Methyl ethyl bromomethane B:0880
Methyl ethyl carbinol B:0840
Methylethylene P:1230
Methyl ethylene glycol P:1250
Methyl ethylene oxide P:1290
2-Methylethylen imine P:1280
2-Methylethylenimine P:1280
1-Methylethyl ester of acetic acid I:0450
Methyl ethyl ether M:0910
1-(Methylethyl)-ethyl 3-methyl-4-(methylthio)
phenylphosphoramidate F:0050
4,4′-(1-Methylethylidene)bisphenol B:0550
2,2′-[(1-Methylethylidene)bis
(4,1-phenyleneoxy-methylene)] bisoxirane D:0970
Methyl ethyl ketone M:0920
Methyl ethyl ketone hydroperoxide M:0930
Methyl ethyl ketone peroxide M:0930
Methyl ethyl methane B:0770
Methylethylmethane B:0770

Methylsilicochloroform M:1280
Methylsilicon trichloride M:1280
Methylsilyl trichloride M:1280
Methyl styrene V:0240
3-Methylstyrene V:0240
3- and 4-Methyl styrene V:0240
α-Methylstyrene M:1240
m-Methyl styrene V:0240
m-Methylstyrene V:0240
p-Methyl styrene V:0240
Methyl sulfate D:1260
Methyl sulfhydrate M:1040
Methyl sulfocyanate M:1260
Methylsulfonic acid, ethyl ester E:0770
Methyl sulfoxide D:1280
Methyl sulphide D:1270
Methyl systox D:0180
Methyltetrahydrofuran **M:1250**
2-Methyltetrahydrofuran M:1250
Methyltetrahydrofuran, 2- M:1250
2-Methyl-3-(3,7,11,15-tetramethyl-2-
hexadecenyl)-1,4-naphthalenedione P:0690
N-Methyl-*N*-2,4,6-tetranitroaniline T:0410
Methyl thiocyanate **M:1260**
4-Methylthio-3,5-dimethylphenyl
methylcarbamate M:0550
2-Methylthio-4-ethylamino-6-
isopropylamino-*s*-triazine A:0740
Methylthiokyanat M:1260
Methylthiomethane D:1270
Methylthiophos M:1070
2-Methylthio-propionaldehyd-*o*-
(methylcarbamoyl)oxim (German) M:0560
4-(Methylthio)-3,5-xylyl methylcarbamate M:0550
4-(Methylthio)-3,5-xylyl
N-methylcarbamate M:0550
Methyl thiram T:0520
Methylthiuram disulfide T:0520
1,2-Methyltoluene X:0100
4-Methyltoluene X:0100
α-Methyltoluene E:0380
m-Methyltoluene X:0100
o-Methyltoluene X:0100
p-Methyltoluene X:0100
Methyl tribromide B:0710
Methyl trichloride C:0870
Methyltrichloromethane T:0720
Methyl trichlorosilane **M:1280**
1-Methyl-2,4,6-trinitrobenzene T:0920
Methyl tuads T:0520
Methyl-vinyl-cetone (French) M:1290

Methylvinylketon (German) M:1290
Methyl vinyl ketone **M:1290**
Methylvinylnitrosamine N:0630
Methylvinylnitrosamine (German) N:0630
1-Methyl-3-vinylbenzene V:0240
2-Methyl-5-vinyl pyridine **M:1300**
Methyl viologen P:0150
Methyl viologen chloride P:0150
Methyl viologen dichloride P:0150
Methyl viologen (reduced) P:0150
Methyl yellow D:1080
1-Methypropyl alcohol B:0840
Metidation (Spanish) M:0540
Metifonate T:0670
Metilamino (Spanish) M:0680
Metil azinfos (Spanish) A:1650
Metilchlorpindol C:1270
p,p'-Metilenbis(*o*-cloroanilina) (Spanish) M:0850
Metilenbis(fenilisocianato) (Spanish) M:0880
4,4'-Metilendianilina (Spanish) D:0250
Metil etil cetona (Spanish) M:0920
Metil fenil eter (Spanish) A:1370
5-Metilheptano-3-ona (Spanish) E:0350
Metilhidrazina (Spanish) M:0960
Metil isobutil cetona (Spanish) M:1000
Metilmercaptano (Spanish) M:1040
Metilparationa (Spanish) M:1070
Metilpiridina (Spanish) P:0720
2-Metilpropeno (Spanish) I:0280
Metil vinil cetona (Spanish) M:1290
Metindol I:0130
Metiocarb (Spanish) M:0550
Metiurac T:0520
Metmercapturon M:0550
Metolachlor **M:1310**
Metolcarb **M:1320**
Metomilo (Spanish) M:0560
Metopryl M:1210
Metox M:0580
Metoxicloro (Spanish) M:0580
Metramac A:0920
Metramak A:0920
Metribuzin **M:1330**
Metribuzina (Spanish) M:1330
Metrifonate T:0670
Metriphonate T:0670
Metron M:1070
Metronidaz M:1340
Metronidazol M:1340
Metronidazole **M:1340**
Metronidazolo M:1340

Monobasic lead acetate L:0200
Monobromobenzene B:0690
Monobromodichloromethane B:0700
Monobromodiphenyl ether B:0720
Mono bromodiphenyl oxide B:0720
Monobromoethane E:0410
Monobromoethylene V:0160
Monobromomethane M:0720
Monobutylamine B:0850
Mono-*n*-butylamine B:0850
Monochloracetic acid C:0740
Monochlorbenzene C:0770
Monochlorbenzol (German) C:0770
Monochlorbenzol (German) C:0780
Monochloressigsaeure (German) C:0740
Monochlorethane E:0480
Monochlorhydrine du glycol (French) E:0550
Monochloroacetaldehyde C:0730
Monochloroacetic acid C:0740
Monochloroacetyl chloride C:0760
Monochlorobenzene C:0770
Monochlorobenzene C:0780
Monochlorodifluoromethane C:0850
Monochlorodimethyl ether C:0890
Monochloroethane E:0480
Monochloroethanoic acid C:0740
2-Monochloroethanol E:0550
Monochloroethene V:0170
Monochloroethylene V:0170
Monochloromethane M:0750
Monochloromethyl ether B:0510
Monochloromethyl methyl ether C:0890
Mono-chloro-mono-bromo-methane C:0820
Monochlorosulfuric acid C:1030
Monochlorotrimethylsilicon T:0890
Monochromium oxide C:1180
Monochromium trioxide C:1100
Monochromium trioxide C:1180
Monocide C:0050
Monocite methacrylate monomer M:1060
Monocloropentafluoetano (Spanish) C:0930
Monocloruro de azufe (Spanish) S:0740
Monocron M:1430
Monocrotofos (Spanish) M:1430
Monocrotophos M:1430
Monocyanogen C:1600
Monodion P:0690
Monodrin M:1430
Monoethanolamine E:0240
Monoethylamine E:0340
Monoethyldichlorosilane E:0530

Monoethylene glycol E:0610
Monofluoressigsaeure (German) F:0330
Monofluoressigsaeure, natrium (German) S:0480
Monofluoroacetamide F:0320
Monofluoroacetate F:0330
Monofluoroacetic acid F:0330
Monofluorobenzene F:0350
Monofluorodichlorometha NE D:0570
Monofluroethylene V:0210
Monoflurotrichloromethane F:0360
Monofrax H S:0340
Monogermane G:0120
Monoglycocoard D:0950
Monohydrated selenium dioxide S:0190
Monohydroxybenzene P:0340
Monohydroxymethane M:0670
Monoisobutylamine B:0850
N-Monoisopropylamide of
O,O-diethyldithiopho-sphorylacetic acid P:1320
Monoisopropyl ether of ethylene glycol I:0440
Monoisopropylamine I:0470
Monomethylamine M:0680
Monomethyl aniline M:0700
N-Monomethylaniline M:0700
Monomethyldichlorosilane M:0840
Mono methyl ether hydroquinone M:0610
Monomethylhydrazine M:0960
Monomethyl mercury chloride M:0440
Monopentaerythritol P:0250
Monoperacetic acid P:0290
Monophenol P:0340
Monophenylhydrazine P:0420
Monopotassium arsenate P:0850
Monopotassium dihydrogen arsenate P:0850
Monopropylamine P:1210
Mono-*n*-propylamine P:1210
Monosan D:0100
Monosilane S:0210
Monosodium sulfite A:1230
Monoxido barico (Spanish) B:0170
Monoxido de nitrogeno (Spanish) N:0350
Monsanto butyl benzyl phthalate B:0870
Monsanto CP 47114 .... F:0100
Montar C:0050
Montar P:0820
Montmorillonite B:0250
Montrel® C:1490
Montrose propanil P:1080
Moon G:0150
Mopari D:0690
Morbicid F:0410

N

O

Phosphorodithioic acid, *o*-ethyl
o-[4-(methylthio)phenyl] *S*-propyl ester S:0840
Phosphorodithionic acid, *O,O*-diethyl
S-2-[(ethylthio)ethyl] ester D:1580
Phosphorodithionic acid, *S*-(2-(ethylthio)ethyl
O,O-diethyl ester D:1580
Phosphorofluoridic acid, diisopropyl ester I:0350
Phosphorothioate E:0120
Phosphorothioate, *O,O*-diethyl *O*-6-
(2-isopropyl-4-methylpyrimidyl) D:0280
Phosphorothioc acid A:0930
Phosphorothioic acid, *O*-(3-chloro-4-methyl-
2-oxo-2H-1-benzopyran-7-yl) *O,O*-diethyl ester C:1420
Phosphorothioic acid, *O*-(4-cyanophenyl)-
9,9-dimethyl ester C:1640
Phosphorothioic acid, *O*-(4-cyanophenyl)-
O,O-dimethyl ester C:1640
Phosphorothioic acid, *O*-[2-(diethylamino)-
6-methyl-4-pyrimidinyl] *O,O*-diethyl ester P:0790
Phosphorothioic acid, *O,O*-diethyl ester, *O*-ester
with 3-chloro-7-hydroxy-4-methylcoumarin C:1420
Phosphorothioic acid, *O,O*-diethyl *O*-2-
(ethylthio)ethyl ester, mixed with *O,O*-diethyl
S-2-(ethylthio)ethyl phosphorothioate D:0170
Phosphorothioic acid, *O,O*-diethyl *O*-
(isopropylmethylpyrimidyl) ester D:0280
Phosphorothioic acid, *O,O*-diethyl *O*-
(2-isopropyl-6-methyl-4-pyrimidinyl) ester D:0280
Phosphorothioic acid, *O,O*-diethyl *O*-
[6-methyl-2-(1-methylethyl)-
4-pyrimidinyl] ester D:0280
Phosphorothioic acid, *O,O*-diethyl
O-[*p*-(methylsulfinyl)phenyl] F:0110
Phosphorothioic acid, *O,O*-diethyl
O-(4-nitrophenyl) ester P:0170
Phosphorothioic acid, *O,O*-diethyl
O-(*p*-nitrophenyl) ester P:0170
Phosphorothioic acid, *O,O*-diethyl
O-2-pyrazinyl ester T:0470
Phosphorothioic acid, *O,O*-diethyl
O-(3,5,6-trichloro-2-pyridinyl) ester C:1070
Phosphorothioic acid, *O,O'*-dimethyl ester,
O,O-diester with 4,4'-thiodiphenol T:0170
Phosphorothioic acid, *O,O*-dimethyl ester,
O-ester with *p*-hydroxybenzonitrile C:1640
Phosphorothioic acid, *O,O*-dimethyl
O-[3-methyl-4-(methylthio)phenyl] ester F:0120
Phosphorothioic acid, *O,O*-dimethyl
O-(3-methyl-4-nitrophenyl) ester F:0100
Phosphorothioic acid, *O,O*-dimethyl
O-[4-(methylthio)-*m*-tolyl] ester F:0120

Phosphorothioic acid, *O,O*-dimethyl
O-(4-nitrophenyl) ester M:1070
Phosphorothioic acid, *O,O*-dimethyl
O-(*p*-nitrophenyl) ester M:1070
Phosphorothioic acid, *O,O*-dimethyl
O-(4-nitro-*m*-tolyl) ester F:0100
Phosphorothioic acid, *O,O*-dimethyl *O*-(2,4,5-
trichlorophenyl) ester R:0140
Phosphorothioic acid, *o*-ethyl *o*-[4-
(methylthio)phenyl] *S*-propyl ester S:0840
Phosphorothioic acid, *O*-2-(ethylthio)ethyl
O,O-dimethyl ester mixed with
S-2-(ethylthio)ethyl *O,O*-dimethyl
phosphorothioate D:0180
Phosphorothioic acid, *O,O',O'*-(thiodi-4,1-phenylene)
O,O,O',O'-tetramethyl ester T:0170
Phosphorothioic acid, *O,O'*-(thiodi- *p*-phenylene)
O,O,O',O'-tetramethyl ester T:0170
Phosphorothioic acid triethylenetriamide T:0500
Phosphorous acid P:0600
Phosphorous acid P:0600
Phosphorous chloride P:0660
Phosphorous hydride P:0580
Phosphorous trihydride P:0580
Phosphorous yellow P:0610
Phosphorpentachlorid (German) P:0630
Phosphorsaeureloesungen (German) P:0590
Phosphortrichlorid (German) P:0660
Phosphorus P:0610
Phosphorus-31 .... P:0610
Phosphorus acid, trimethy ester T:0900
Phosphorus chloride P:0660
Phosphorus chloride oxide P:0620
Phosphorus elemental, white P:0610
Phosphorus oxide P:0650
Phosphorus(5+) oxide P:0650
Phosphorus(V) oxide P:0650
Phosphorus oxychloride P:0620
Phosphorus oxytrichloride P:0620
Phosphorus pentachloride....................................P:0630
Phosphorus pentaoxide P:0650
Phosphorus pentasulfide P:0640
Phosphorus pentoxide..P:0650
Phosphorus perchloride P:0630
Phosphorus persulfide P:0640
Phosphorus sulfide P:0640
Phosphorus trichloride P:0660
Phosphorus trihydroxide P:0600
Phosphorwasserstoff (German) P:0580
Phosphoryl chloride P:0620
Phosphoryl hexamethyltriamide H:0290

Sinafid M-48 .... M:1070
Sinflowan T:0840
Sinituho P:0240
Sinox D:1340
Sinox general D:1380
Sintestrol D:0910
Sintomicetina C:0620
Sintomicetine R C:0620
Sipcam UK Rover 5000 .... C:1040
Siptox I M:0190
Sixty-three special E.C. insecticide M:1070
SK 106N N:0510
SK 6882 .... T:0500
SK 15673 .... M:0320
SK 20501 .... E:0130
SK-Ampicillin A:1290
SK-Digoxin D:0980
Skekhg E:0160
SK-Estrogens C:1350
Skellysolve-A P:0260
Skellysolve B H:0300
Skelly-Solve C H:0160
SK-Tetracycline T:0280
Slago A:1030
Slag wool F:0240
Slaked lime C:0293
Slaymor B:0650
Slimicide A:0380
Slo-Gro M:0220
Slow-Fe F:0220
Slug-tox M:0480
Smeesana A:1500
SMFA S:0480
Smidan P:0560
Smite S:0390
SMT F:0100
Smut-Go H:0190
SN 36056 .... F:0440
SNG N:0510
Snieciotox H:0190
Snip D:1300
Snip fly D:1300
Snowfloss S:0220
Snowgoose T:0120
Snow white Z:0140
SNP P:0170
So-Flo S:0470
Soapstone **S:0320**
Soapstone silicate S:0320
Sobenate S:0400
Soda chlorate S:0430

Sodanit S:0370
Sodanit S:0380
Sodanthon P:0510
Sodantoin P:0510
Sodestrin-H C:1350
Sodio (Spanish) S:0330
Sodium **S:0330**
Sodium acid arsenate, heptahydrate S:0370
Sodium acid sulfite S:0410
Sodium aluminate **S:0340**
Sodium aluminofluoride S:0350
Sodium aluminum fluoride **S:0350**
Sodium aluminum hydride **S:0360**
Sodium aluminum oxide S:0340
Sodium aluminum tetrahydride S:0360
Sodium arsenate **S:0370**
Sodium arsenate, dibasic S:0370
Sodium arsenate, dibasic, heptahydrate .. S:0370
Sodium arsenate heptahydrate S:0370
Sodium arseniate heptahydrate S:0370
Sodium arsenite **S:0380**
Sodium azide **S:0390**
Sodium benzoate **S:0400**
Sodium beryllium oxide B:0470
Sodium benzoic acid S:0400
Sodium bichromate S:0445
Sodium bisulfite **S:0410**
Sodium bismuthate B:0530
Sodium borate B:0580
Sodium borate decahydrate B:0580
Sodium cacodylate **S:0420**
Sodium chlorate **S:0430**
Sodium (chlorate de) (French) S:0430
Sodium chloroplatinate **S:0440**
Sodium chromates **S:0445**
Sodium chromate S:0445
Sodium chromate(VI) S:0445
Sodium chromate decahydrate S:0445
Sodium coumadin W:0100
Sodium cyanide **S:0450**
Sodium cyanide, solid S:0450
Sodium cyanide, solution S:0450
Sodium dichlorisocyanurate S:0460
Sodium dichlorocyanurate S:0460
Sodium dichloro-isocyanate **S:0460**
Sodium dichloroisocyanurate S:0460
Sodium-2-(2,4-dichlorophenoxy)ethyl sulfate D:0190
Sodium-2,4-dichlorophenoxyethyl sulphate D:0190
Sodium-2,4-dichlorophenyl cellosolve sulfate D:0190
Sodium-1,3-dichloro-1,3,5-triazine-
2,4-dione-6-oxide S:0460

Strontium chromate X-2396	S:0630
Strontium nitrate	**S:0640**
Strontium(II) nitrate (1:2)	S:0640
Strontium yellow	S:0630
Strophanthin G	O:0150
Strophoperm	O:0150
Strychinium sulfate	S:0650
Strychnidin-10-one	S:0650
Strychnidin-10-one, 2,3-dimethoxy-	B:0740
Strychnidin-10-one, sulfate (2:1)	S:0650
Strychnin (German)	S:0650
Strychnine	**S:0650**
Strychnine, 2,3-dimethoxy-	B:0740
Strychnine hemisulfate	S:0650
Strychnine sulfate	S:0650
Strychnos	S:0650
STRZ	S:0620
Studaflour	S:0470
Stuntman	M:0220
Styrene	**S:0660**
Styrene epoxide	S:0670
Styrene, α-methyl-	M:1240
Styrene monomer	S:0660
Styrene monomer, inhibited	S:0660
Styrene oxide	**S:0670**
Styrene 7,8-oxide	S:0670
Styrol (German)	S:0660
Styrole	S:0660
Styrolene	S:0660
Styron	S:0660
Styropol	S:0660
Styropol SO	S:0660
Styropor	S:0660
Styryl oxide	S:0670
STZ	S:0620
Subacetate lead	L:0200
Subaceto de plomo (Spanish)	L:0200
Subamycin	T:0280
Suberane	C:1670
Suberylene	C:1670
Subitex	D:1380
Subtilisin BPN	S:0680
Subtilisin Carlsburg	S:0680
Subtilisins	**S:0680**
Succinic acid 2,2-dimethylhydrazide	D:0120
Succinic acid, mercapto-, diethyl ester, *S*-ester with *O,O*-dimethyl phosphorodithioate	M:0190
Succinic-1,1-dimethyl hydrazide	D:0120
Sucker-stuff	M:0220
Sucre Edulcor	S:0100
Sucrette	S:0100

Sucrose	**S:0690**
Sudan yellow GG	D:1080
Sudan yellow GGA	D:1080
Sudan yellow R	A:0760
Sudan Yellow RRA	A:0770
Suflur decafluoride	S:0790
Sugai Congo red	C:1240
Sugai fast scarlet G base	N:0670
Sugar	S:0690
Sugar of lead	L:0110
Suladyne	P:0330
Sulfallate	**S:0700**
Sulfalone	S:0710
Sulfamate	A:1210
Sulfamato amonico (Spanish)	A:1210
Sulfamic acid	S:0830
Sulfamic acid, monoammonium salt	A:1210
Sulfamidic acid	S:0830
Sulfaminsaure (German)	A:1210
Sulfan	S:0810
Sulfate d'aluminium (French)	A:0730
Sulfate de cuivre (French)	C:1390
Sulfate de methyle (French)	D:1260
Sulfate de nicotine (French)	N:0310
Sulfate de plomb (French)	L:0210
Sulfate dimethylique (French)	D:1260
Sulfate mercurique (French)	M:0420
Sulfate of copper	C:1390
Sulfatep	S:0720
Sulfate wood turpentine	T:1000
Sulfato aluminico (Spanish)	A:0730
Sulfato barico (Spanish)	B:0210
Sulfato cromico (Spanish)	C:1120
Sulfato de cobre (Spanish)	C:1390
Sulfato de 3,3-diclorobenzidina (Spanish)	D:0470
Sulfato de dietilo (Spanish)	D:0920
Sulfato de dimetilo (Spanish)	D:1260
Sulfato de nicotina (Spanish)	N:0310
Sulfato de niquel (Spanish)	N:0290
Sulfato de niquel y amonio (Spanish)	N:0230
Sulfato de plomo (Spanish)	L:0210
Sulfato de talio (Spanish)	T:0420
Sulfato ferrico (Spanish)	F:0180
Sulfato ferroso (Spanish)	F:0220
Sulfato ferroso amonico (Spanish)	F:0200
Sulfato mercurico (Spanish)	M:0420
Sulficyl bis(methane)	D:1280
Sulfide, bis(2-chloroethyl)	M:1460
Sulfinyl chloride	T:0480
Sulfito amonico (Spanish)	A:1230
o-Sulfobenzimide	S:0100

Synchemicals couch and grass killer D:0670
Synchemicals total weed killer A:0910
Syndiol E:0210
Synestrin D:0910
Synfloran T:0840
Synklor C:0630
Synpenin A:1290
Synpor N:0420
Synpran N P:1080
Syntexan D:1280
Synthetic 3956 .... T:0650
Synthetic glycerin G:0150
Synthetic iron oxide I:0210
Synthetic mustard oil A:0610
Synthetic pyrethrins A:0520
Synthetic vitreous fibers F:0240
Synthoestrin D:0910
Synthofolin D:0910
Synthomycetin C:0620
Synthomycetine C:0620
Synthomycine C:0620
Syntofolin D:0910
Syntox total weed killer A:0910
Systam O:0110
Systemox D:0170
Systophos O:0110
Systox D:0170
Sytam O:0110
Szklarniak D:0690

T

2,4,5-T **T:0100**
α-T T:0720
β-T T:0730
T-5 brush kil T:0100
T-23P T:0970
T-40 .... T:0560
T-47 .... P:0170
T-125 .... T:0280
T-144 .... S:0130
T-250 Capsules T:0280
T-1703 .... I:0350
T-2002 .... D:1030
T-2104 .... T:0110
T-2106 .... S:0130
TA 12 .... T:0200
TAA T:0430
Taboon A T:0110
Tabun (WMD) **T:0110**
Tackle® A:0360

Tacosal P:0510
Tafazine S:0310
Tafazine 50-W S:0310
Tafclean T:0720
Tag P:0450
Tag-39 .... C:1350
Tag 331 .... P:0450
Tag fungicide P:0450
Tag HL 331 .... P:0450
Tahmabon M:0520
TAK M:0190
Takaoka Rhodamine B C:1250
Takineocol I:0460
Taktic® A:0940
Tal T:0200
Talbot L:0120
Talc S:0320
Talc (no asbestos and less than 1% quartz) **T:0120**
Talc (non-asbestos form) T:0120
Talcum T:0120
Talio (TL) (Spanish) T:0420
Talon® C:1070
Tamaron M:0520
Tampovagan stilboestrol D:0910
Tannex I:0130
Tanol secondaire (French) B:0840
Tantalic acid anhydride T:0130
Tantalum and tantalum oxide dusts **T:0130**
Tantalum 181 .... T:0130
Tantalum metal T:0130
Tantalum(V) oxide T:0130
Tantalum pentoxide T:0130
Tantalum pentaoxide T:0130
Tap 9VP D:0690
TAP85 .... L:0260
Taphazine S:0310
Tarapacaite P:0890
Tar camphor N:0120
Tardex 100 .... D:0160
Tardigal D:0950
Tar oil C:1290
Tarsan® B:0230
Tartar emetic A:1440
Tartaric acid, ammonium salt A:1240
Tartaric acid, antimony potassium salt A:1440
Tartaric acid, diammonium salt A:1240
1-Tartaric acid, diammonium salt A:1240
Tartarized antimony A:1440
Tartrated antimony A:1440
Tartrato amonico (Spanish) A:1240
Tartrato de antimonio y potasio (Spanish) A:1440

Tartrato de ergosterol (Spanish) E:0200
Task D:0690
Task Tabs D:0690
Tastox A:1440
TAT C:0630
TAT Chlor 4 .... C:0630
TATD D:1570
Tattoo B:0220
TB T:0980
TBA B:0840
TBE A:0320
TBP T:0660
TBP B:0560
TBT B:1000
2,4,5-TC S:0300
TCA T:0680
TCDBD T:0230
TCDD T:0230
2,3,7,8-TCDD T:0230
TCE T:0260
TCE T:0740
1,1,1-TCE T:0720
TC Hydrochloride T:0280
TCM C:0870
TCP T:0800
2,4,5-TCPPA S:0300
TDA T:0610
TDBP T:0970
TDBPP T:0970
TDE T:0140
p-,p-'-TDE T:0140
TDI T:0620
TDI-80 .... T:0620
2,4-TDI T:0620
TEA A:0650
TEA E:0240
TEA T:0810
Tear gas C:0750
Tebol-88 .... B:0840
Tebol-99 .... B:0840
Technical chlorinated camphene, 67-69% chlorine . T:0650
90 Technical glycerin G:0150
Tecquinol H:0490
TEDP S:0720
TEDTP S:0720
Tefilin T:0280
Teflon T:0320
T-Fluoride S:0470
T-gas E:0660
Tekresol C:1450
TEL T:0300

Telefos P:1320
Teline T:0280
Telloy T:0150
Telluric acid, disodium salt S:0560
Tellurium T:0150
Tellurium elemental T:0150
Tellurium fluoride T:0160
Tellurium hexafluoride T:0160
Tellurous acid, disodium salt S:0560
Telmicid D:1590
Telmid D:1590
Telmide D:1590
Telodrin I:0250
Telon fast black E D:1550
Telone D:0660
Telone II D:0660
Telotrex T:0280
Telurio (Spanish) T:0150
Telurito sodico (Spanish) S:0560
Telvar D:1610
Telvar diuron weed killer D:1610
Temefos (Spanish) T:0170
Temephos T:0170
Temic® A:0490
Temik® A:0490
Temik 10 G® A:0490
Temik G10® A:0490
Temophos T:0170
Temponitrin N:0510
Temus B:0650
TEN T:0810
Tenac D:0690
Tendex P:1180
Tendust N:0300
Ten-Eighty S:0480
Tennplas B:0370
Tenoran® C:1060
Tenox HQ H:0490
Tenox P grain preservative P:1150
Tensopam D:0270
Tentachlorure d'antimoine (French) A:1420
Tentos da America (Spanish) A:0025
Tentos dos mundos (Spanish) A:0025
Tenurid D:1570
Tenutex D:1570
TEOS E:0820
TEP T:0180
TEPP T:0180
Tequinol H:0490
Terabol M:0720
Teramethylthiuram disulfide T:0520

Thiophosphate de *O,O*-dimethyle et de
O-(3-methyl-4-methylthiophenyle) (French) F:0120
Thiophosphate de *O,O*-dimethyle et de
O-(3-methyl-4-nitrophenyle) (French) F:0100
Thiophosphate de *O,O*-dimethyle et de
O-(2,4,5-trichlorophenyle) (French) R:0140
Thiophosphoric acid 2-isopropyl-4-methyl-
6-pyrimidyl diethyl ester D:0280
Thiophosphoric anhydride P:0640
Thiophosphorsaeure-*O,S*-
dimethylesteramid (German) M:0520
2-Thiopropane D:1270
β-Thiopseudourea T:0510
Thiosan D:1570
Thiosan T:0520
Thioscabin D:1570
Thioscabin T:0520
Thiosemicarbazide **T:0490**
2-Thiosemicarbazide T:0490
3-Thiosemicarbazide T:0490
Thiosemicarbazone acetone A:0200
Thiosulfan E:0100
Thiosulfan thionel E:0100
Thiosulfil-A forte P:0330
Thiosulfuric acid, diammonium salt A:1270
Thiosulfurous dichloride S:0740
Thio-TEP T:0500
Thiotepa **T:0500**
Thiotepp S:0720
Thiotex T:0520
Thio-1-(thiocarbamoyl)urea D:1600
Thiotox T:0520
Thiourea **T:0510**
2-Thiourea T:0510
Thiourea, *N,N'*-(1,2-ethanediyl)- E:0670
Thiourea, 1-naphthalenyl- A:1500
Thiovanic acid T:0460
Thioxamyl O:0170
Thiram **T:0520**
Thiram 75 .... T:0520
Thiram 80 .... T:0520
Thiramad T:0520
Thiram B T:0520
Thirame (French) T:0520
Thirampa T:0520
Thirasan T:0520
Thireranide D:1570
Thiulin T:0520
Thiulix T:0520
Thiurad T:0520
Thiuram T:0520

Thiuram D T:0520
Thiuram E D:1570
Thiuramin T:0520
Thiuramyl T:0520
Thiuranide D:1570
Thompson's wood fix P:0240
Thoria T:0525
Thorium and compounds **T:0525**
Thorium-232 .... T:0525
Thorium metal, pyrophoric T:0525
Thorium(4+) nitrate T:0525
Thorium(IV) nitrate T:0525
Thorium(IV) oxide T:0525
Thorium oxide (tho2) T:0525
Thorium tetrachloride T:0525
Thorium tetranitrate T:0525
Thorotrast T:0525
Thortrast T:0525
D-Threochloramphenicol C:0620
D-(-)-Threochloramphenicol C:0620
D-(-)-Threo-2-dichloroacetamido-
1-*p*-nitrophenyl-1,3-propanediol C:0620
D-Threo-*N*-dichloroacetyl-1-*p*-nitrophenyl-
2-amino-1,3-propanediol C:0620
D-(−)-Threo-2,2-dichloro-*N*-[b-hydroxy-
α-(hydroxymethyl)]-*p*- nitrophenethylacetamide C:0620
D-Threo-*N*-(1,1'-dihydroxy-
1-*p*-nitrophenylisopropyl)dichloroacetamide C:0620
D-(−)-Threo-1-*p*-nitrophenyl-
2-dichloracetamido-1,3-propanediol C:0620
D-Threo-1-(*p*-nitrophenyl)-
2-(dichloroacetylamino)-1,3-propanediol C:0620
Threthylen T:0740
Threthylene T:0740
THU T:0510
Thylate T:0520
Thylpar M-50 .... M:1070
Tifomycin C:0620
Tifomycine C:0620
Tigrex D:1610
Tiguvon F:0120
Tilcarex P:0230
Tilcarex Q:0110
Tillram D:1570
Timazin F:0370
Tin (elemental) **T:0530**
Tin bifluoride S:0590
Tin(II) chloride S:0580
Tin(II) chloride, dihydrate (1:2:2) S:0580
Tin chloride, fuming T:0550
Tin dichloride S:0580

Totacillin A:1290
Toxalbumin A:0025
Totalciclina A:1290
Totamott D:0460
Totapen A:1290
Totomycin T:0280
TOX 47 .... P:0170
Toxadrin® A:0510
Toxadust T:0650
Toxafeno (Spanish) T:0650
Toxakil T:0650
Toxaphen (German) T:0650
Toxaphene T:0650
Toxaspray T:0650
Toxer total P:0150
Toxic chemical category code N874 .... W:0100
Toxichlor C:0630
Toxilic anhydride M:0210
Toxilic acid M:0200
Toxol (3) P:0170
Toxon 63 .... T:0650
Toxyphen T:0650
Toyo oil yellow G D:1080
TP T:0940
2,4,5-TP S:0300
TPN C:1040
TPN (pesticide) C:1040
TPP T:0940
TPTA T:0950
TPTC T:0950
TPTH T:0540
TPTH T:0950
TPTH technical T:0950
TPTOH T:0950
TPTP T:0800
TPZA T:0950
Trametan T:0520
Tranimul D:0270
Tranqdyn D:0270
Tranquirit D:0270
Transamine D:0100
Transamine T:0100
Transannon C:1350
Trapex M:1030
Trapex-40 .... M:1030
Trapexide M:1030
Travad B:0210
Travex S:0430
Trefanocide T:0840
Treficon T:0840
Treflanocide elancolan T:0840

Treflan® T:0840
Tremolite A:1590
Treomicetina C:0620
Trethylene T:0740
Tri T:0740
TRI-6 .... L:0260
Triacetaldehyde (French) P:0130
Triad T:0740
Triaethylamin (German) T:0810
Trialkylaluminum (general) A:0650
Triamida hexametilfosforica (Spanish) H:0290
2,4,6-Triaminotriazine M:0310
Triammonium tris-(ethanedioato(2-)-o,o′)
ferrate(3-1) F:0150
Triangle C:1390
Trianol direct blue 3B T:0980
Triasol T:0740
Triatomic oxygen O:0230
Triatox® A:0940
1,4,7-Triazaheptane D:0850
Triazine A 384 .... S:0310
Triazine A 1294 .... A:1610
s-Triazine, 2-chloro-4,6-bis(ethylamino)- S:0310
s-Triazine, 2-chloro-4-ethylamino-6-
(1-cyano-1-methyl)ethylamino- C:1580
s-Triazine, 2-chloro-4-(ethylamino)-
6-(isopropylamino)- A:1610
1,3,5-Triazine-2,4-diamine,
6-chloro-N,N′-diethyl- S:0310
1,3,5-Triazine-2,4-diamine,6-chloro-
N-ethyl-N′-(1-methylethyl)- A:1610
1,3,5-Triazine-2,4(1H,3H)-dione,
3-cyclohexyl-6-(dimethylamino)-1-methyl- H:0320
s-Triazine-2,4(1H,3H)-dione, 3-cyclohexyl-6-
(dimethylamino)-1-methyl- H:0320
1,3,5-Triazine-2,4,6-triamine M:0310
s-Triazine-2,4,6(1H,3H,5H)-trione,
dichloro-, potassium derivative P:0920
1,3,5-Triazine-(2,4,6(1H,3H,5H)-trione,
1,3-dichloro-, potassium salt P:0920
s-Triazine, zeazin A:1610
1,2,4-Triazin-5-(4H)-one, 4-Amino-6-
(1,1-dimethylethyl)-3-(methylthio)- M:1330
Triaziridinylphosphine sulfide T:0500
Triazoic acid H:0390
Triazolamine A:0910
1,2,4-Triazol-3-amine A:0910
1H-1,2,4-Triazol-3-amine A:0910
Triazolblau 3B T:0980
s-Triazole, 3-amino- A:0910
δ-2-1,2,2,4-Triazoline, 5-imino- A:0910

U

U 46..D:0100
U 46..D:0610
U 46..C:0900
U 46..M:0290
U 46 DP..D:0100
U 46 DP-fluid..D:0610
U 46 KV-ester..C:0900
U 46 KV-fluid...C:0900
U 46 M-fluid...M:0290
U 1149...F:0490
U 1363...D:1450
U 4224...D:1190
U 4513...D:1460
U 5043...D:0100
U 5965...T:0280
U 6062...C:0620
U 8953...F:0370
U 9889...S:0620
U 25,354 ....D:0650
UC 7744 ....C:0430
UC 7744 (Union Carbide) C:0430
UC 9880..P:1030
UC 10854..P:0350
UC 21149® A:0490
UC 21149® (Union Carbide) A:0490
UCAR 17 ....E:0610
UCAR bisphenol HP.................................B:0550
UCAR butylphenol 4-*t*.............................B:0980
Ucet textile finish 11-74 (obs.)................V:0190
U-Compound...U:0120
Ucon 12...D:0500
Ucon 12/halocarbon 12.............................D:0500
Ucon 22...C:0850
Ucon 22/halocarbon 22.............................C:0850
Ucon 114...D:0680
Ucon fluorocarbon 11 ....F:0360
Ucon refrigerant 11....................................F:0360
U-Dimethylhydrazine D:1200
UDMH D:1200
UL D:0170
Ultamac PR-68 resin.................................E:0290
Ultrabion A:1290
Ultrabron A:1290
Ultracide...M:0540
Ultrafine II A:1480
Ultramac 55 ....E:0280
Ultramac PR-1024 MB-628 resin............B:0810
Ultramac PR-1024 MB-628 resin............E:0290
Ultramac S40 ....D:0460

Ultramac solvent EPA..............................B:0810
Ultramarine green......................................C:1160
Ultra Pure H:0370
Ulup...F:0370
Ulvair..M:1430
Umbrathor...T:0525
Umbethion...C:1420
Umbrium D:0270
UN 1040..E:0660
UN 1086..V:0170
UN 1579..C:0880
UN 1846..C:0510
UN 2862..V:0120
UN 9117..E:0570
UN 9201..A:1480
Unden..P:1180
Unibaryt..B:0210
Unicin..T:0280
Unicrop DNBP...D:1380
Unicrop maneb...M:0240
Unidigin D:0950
Unifos (pesticide).......................................D:0690
Unifume E:0580
Unimoll BB..B:0870
Unimoll DA D:0900
Unimoll DM D:1250
Unimycetin...C:0620
Unimycin T:0280
Union black EM...D:1550
Union Carbide 7,744.................................C:0430
Union Carbide 21,149...............................A:0490
Union Carbide UC 9,880...........................P:1030
Union Carbide UC 10,854.........................P:0350
Union Carbide UC 21,149.........................A:0490
Uniplex 110 ....D:1250
Uniplex 150 ....D:0410
Unipon...D:0670
Unisedil D:0270
United Chemical defoliant No. 1..............S:0430
Unitox...C:0650
Unitox...D:0690
Univerm..C:0510
Unocal 76 RES 6206 ....V:0150
Unocal 76 RES S-55..................................V:0150
Unox epoxide 206......................................V:0190
Upjohn® U-36059......................................A:0940
Uracil, 5-bromo-3-*sec*-butyl-6-methyl B:0640
Uracil, 5-fluoro-...F:0370
Uragan..B:0640
Uragon..B:0640
Uranium and compounds................U:0100

Appendix 6: CAS Number-Cross Index

CAS Number and Record Number

50-00-0 see F:0410
50-07-7 see M:1400
50-14-6 see E:0190
50-18-0 see E:0130
50-28-2 see E:0210
50-29-3 see D:0140
50-32-8 see B:0400
50-55-5 see R:0100
50-76-0 see A:0430
50-78-2 see A:0340
50-99-7 see G:0123
51-21-8 see F:0370
51-28-5 see D:1360
51-75-2 see M:0300
51-79-6 see U:0120
51-83-2 see C:0420
52-24-4 see T:0500
52-68-6 see T:0670
53-16-7 see E:0220
53-70-3 see D:0300
53-86-1 see I:0130
53-96-3 see A:0260
54-11-5 see N:0300
54-62-6 see A:0880
55-18-5 see N:0570
55-21-0 see B:0290
55-38-9 see F:0120
55-63-0 see N:0510
55-91-4 see I:0350
55-98-1 see B:0750
56-23-5 see C:0510
56-25-7 see C:0380
56-38-2 see P:0170
56-53-1 see D:0910
56-55-3 see B:0260
56-72-4 see C:1420
56-75-7 see C:0620
56-81-5 see G:0150
57-06-7 see A:0610
57-12-5 see C:1590
57-13-6 see U:0110
57-14-7 see D:1200
57-24-9 see S:0650
57-41-0 see P:0510
57-47-6 see P:0700
57-50-1 see S:0690
57-55-6 see P:1250
57-57-8 see P:1130
57-63-6 see E:0250
57-74-9 see C:0630

57-97-6 see D:1110
58-22-0 see T:0220
58-36-6 see O:0190
58-89-9 see L:0260
58-89-9 see H:0210
59-05-2 see M:0570
59-88-1 see P:0420
60-00-4 see E:0570
60-09-3 see A:0760
60-11-7 see D:1080
60-29-7 see E:0680
60-34-4 see M:0960
60-35-5 see A:0140
60-41-3 see S:0650
60-51-5 see D:1040
60-54-8 see T:0280
60-57-1 see D:0750
61-82-5 see A:0910
62-38-4 see P:0450
62-44-2 see A:0220
62-50-0 see E:0770
62-53-3 see A:1350
62-55-5 see T:0430
62-56-6 see T:0510
62-73-7 see D:0690
62-74-8 see S:0480
62-75-9 see N:0580
63-25-2 see C:0430
64-00-6 see P:0350
64-02-8 see T:0390
64-17-5 see E:0330
64-18-6 see F:0450
64-19-7 see A:0160
64-67-5 see D:0920
64-75-5 see T:0280
64-86-8 see C:1340
65-30-5 see N:0310
65-85-0 see B:0370
66-56-8 see D:1360
66-81-9 see C:1730
67-56-1 see M:0670
67-63-0 see I:0460
67-64-1 see A:0180
67-66-3 see C:0870
67-68-5 see D:1280
67-72-1 see H:0230
68-11-1 see T:0460
68-12-2 see D:1190
69-53-4 see A:1290
69-72-7 see S:0120
70-30-4 see H:0240
71-23-8 see P:1200
71-36-3 see B:0840

71-41-0 see A:1310
71-43-2 see B:0310
71-55-6 see T:0720
71-63-6 see D:0950
72-20-8 see E:0140
72-43-5 see M:0580
72-54-8 see T:0140
72-57-1 see T:0980
74-82-8 see M:0530
74-83-9 see M:0720
74-84-0 see E:0230
74-85-1 see E:0540
74-86-2 see A:0310
74-87-3 see M:0750
74-88-4 see M:0970
74-89-5 see M:0680
74-90-8 see H:0440
74-93-1 see M:1040
74-95-3 see M:0890
74-96-4 see E:0410
74-97-5 see C:0820
74-98-6 see P:1060
74-99-7 see M:0630
75-00-3 see E:0480
75-01-4 see V:0170
75-02-5 see V:0210
75-04-7 see E:0340
75-05-8 see A:0210
75-07-0 see A:0110
75-08-1 see E:0740
75-09-2 see M:0900
75-12-7 see F:0430
75-15-0 see C:0470
75-18-3 see D:1270
75-19-4 see C:1800
75-20-7 see C:0220
75-21-8 see E:0660
75-24-1 see A:0650
75-25-2 see B:0710
75-26-3 see B:0730
75-27-4 see B:0700
75-28-5 see I:0260
75-28-5 see B:0770
75-31-0 see I:0470
75-34-3 see D:0520
75-35-4 see V:0220
75-36-5 see A:0290
75-38-7 see V:0230
75-39-8 see A:0120
75-43-4 see D:0570
75-44-5 see P:0550
75-45-6 see C:0850
75-47-8 see I:0180

75-50-3 see T:0860
75-52-5 see N:0520
75-54-7 see M:0840
75-55-8 see P:1280
75-56-9 see P:1290
75-60-5 see C:0050
75-61-6 see D:0940
75-63-8 see T:0820
75-64-9 see B:0850
75-65-0 see B:0840
75-68-3 see C:0840
75-69-4 see F:0360
75-71-8 see D:0500
75-73-0 see T:0330
75-74-1 see T:0360
75-77-4 see T:0890
75-78-5 see D:1150
75-79-6 see M:1280
75-84-3 see A:1310
75-83-2 see H:0300
75-85-4 see A:1310
75-86-5 see A:0190
75-87-6 see C:0590
75-99-0 see D:0670
76-01-7 see P:0210
76-02-8 see T:0690
76-03-9 see T:0680
76-06-2 see C:0980
76-11-9 see T:0240
76-12-0 see T:0240
76-13-1 see T:0790
76-14-2 see D:0680
76-15-3 see C:0930
76-16-4 see H:0260
76-22-2 see C:0370
76-44-8 see H:0140
76-87-9 see T:0950
76-87-9 see T:0540
77-47-4 see H:0220
77-73-6 see D:0740
77-78-1 see D:1260
77-81-6 see T:0110
78-00-2 see T:0300
78-10-4 see E:0820
78-11-5 see P:0255
78-30-8 see T:0800
78-32-0 see T:0800
78-34-2 see D:1420
78-53-5 see A:0920
78-59-1 see I:0400
78-62-6 see D:1160
78-63-7 see D:1140
78-67-1 see A:1670

78-76-2 see D:0880	84-66-2 see D:0900	94-78-0 see P:0330	98-51-1 see B:1000
78-78-4 see I:0390	84-74-2 see D:0410	95-06-7 see S:0700	98-54-4 see B:0980
78-79-5 see I:0420	84-80-0 see P:0690	95-13-6 see I:0100	98-56-6 see C:0800
78-81-9 see B:0850	85-00-7 see D:1540	95-47-6 see X:0100	98-82-8 see C:1500
78-82-0 see I:0320	85-01-8 see P:0320	95-48-7 see C:1450	98-83-9 see M:1240
78-83-1 see B:0840	85-44-9 see P:0670	95-49-8 see C:1050	98-86-2 see A:0230
78-84-2 see I:0300	85-68-7 see B:0870	95-50-1 see D:0460	98-87-3 see B:0270
78-86-4 see B:0890	86-30-6 see N:0590	95-51-2 see C:0770	98-88-4 see B:0420
78-87-5 see D:0640	86-50-0 see A:1650	95-53-4 see T:0640	98-92-0 see N:0210
78-88-6 see D:0660	86-73-7 see F:0290	95-54-5 see P:0390	98-95-3 see N:0400
78-89-7 see P:1240	86-88-4 see A:1500	95-55-6 see A:0870	99-08-1 see N:0660
78-92-2 see B:0840	87-59-2 see X:0130	95-57-8 see C:0950	99-35-4 see T:0910
78-93-3 see M:0920	87-60-5 see C:0880	95-63-6 see T:0880	99-55-8 see N:0670
78-94-4 see M:1290	87-62-7 see X:0130	95-64-7 see X:0130	99-59-2 see N:0390
78-97-7 see L:0050	87-63-8 see C:0880	95-68-1 see X:0130	99-65-0 see D:1330
79-00-5 see T:0730	87-65-0 see D:0600	95-69-2 see C:0880	99-71-8 see B:0980
79-01-6 see T:0740	87-68-3 see H:0200	95-70-5 see T:0610	99-98-9 see D:1230
79-04-9 see C:0760	87-86-5 see P:0240	95-74-9 see C:0880	99-99-0 see N:0660
79-06-1 see A:0390	87-90-1 see T:0750	95-78-3 see X:0130	100-00-5 see N:0430
79-09-4 see P:1150	88-05-1 see T:0870	95-79-4 see C:0880	100-01-6 see N:0380
79-10-7 see A:0400	88-06-2 see T:0770	95-80-7 see T:0610	100-14-1 see B:0330
79-11-8 see C:0740	88-10-8 see D:0830	95-81-8 see C:0880	100-21-0 see T:0200
79-19-6 see T:0490	88-18-6 see B:0980	95-83-0 see C:0960	100-25-4 see D:1330
79-20-9 see M:0620	88-72-2 see N:0660	95-85-2 see A:0790	100-37-8 see D:0800
79-21-0 see P:0290	88-75-5 see N:0530	95-95-4 see T:0770	100-39-0 see B:0440
79-22-1 see M:0770	88-85-7 see D:1380	96-09-3 see S:0670	100-40-3 see V:0180
79-24-3 see N:0450	88-89-1 see P:0730	96-10-6 see A:0640	100-41-4 see E:0380
79-27-6 see A:0320	89-72-5 see B:0980	96-12-8 see D:0360	100-42-5 see S:0660
79-29-8 see D:1120	90-04-0 see A:1360	96-14-0 see H:0300	100-44-7 see B:0450
79-31-2 see I:0310	90-13-1 see C:0660	96-18-4 see T:0780	100-47-0 see B:0380
79-31-2 see B:1040	90-43-7 see P:0470	96-22-0 see D:0870	100-48-1 see C:1650
79-34-5 see T:0260	90-94-8 see M:1380	96-23-1 see D:0650	100-52-7 see B:0280
79-41-4 see M:0490	91-20-3 see N:0120	96-33-3 see M:0650	100-54-9 see C:1650
79-43-6 see D:0430	91-22-5 see Q:0050	96-37-7 see M:0830	100-61-8 see M:0700
79-44-7 see D:1130	91-58-7 see C:0600	96-45-7 see E:0670	100-63-0 see P:0420
79-46-9 see N:0550	91-59-8 see N:0170	96-47-9 see M:1250	100-66-3 see A:1370
79-92-5 see C:0360	91-66-7 see D:0810	96-69-5 see T:0440	100-70-9 see C:1650
80-05-7 see B:0550	91-94-1 see D:0470	97-02-9 see D:1320	100-73-2 see A:0380
80-10-4 see D:1480	92-06-8 see T:0210	97-18-7 see B:0560	100-74-3 see E:0780
80-15-9 see C:1510	92-52-4 see B:0480	97-53-0 see E:0850	100-75-4 see N:0640
80-48-8 see T:0630	92-67-1 see A:0780	97-54-1 see I:0335	100-80-1 see V:0240
80-56-8 see T:1000	92-84-2 see P:0360	97-56-3 see A:0770	100-99-2 see A:0650
80-62-6 see M:1060	92-87-5 see B:0350	97-63-2 see E:0760	101-14-4 see M:0850
80-63-7 see M:0760	92-93-3 see N:0410	97-77-8 see D:1570	101-55-3 see P:08100
81-07-2 see S:0100	92-94-4 see T:0210	97-88-1 see B:0970	101-55-3 see B:0720
81-81-2 see W:0100	93-05-0 see D:0890	97-93-8 see A:0650	101-61-1 see M:0870
81-88-9 see C:1250	93-15-2 see M:0945	97-95-0 see E:0430	101-68-8 see M:0880
82-28-0 see A:0850	93-58-3 see M:0710	97-96-1 see E:0460	101-77-9 see D:0250
82-66-6 see D:1450	93-65-2 see C:0900	98-00-0 see F:0520	101-80-4 see O:0180
82-68-8 see P:0230	93-72-1 see S:0300	98-01-1 see F:0510	101-83-7 see D:0720
82-68-8 see Q:0110	93-76-5 see T:0100	98-05-5 see B:0320	101-84-8 see D:1500
83-26-1 see P:0760	94-36-0 see B:0430	98-07-7 see B:0410	102-36-3 see D:0620
83-32-9 see A:0050	94-58-6 see D:0990	98-09-9 see B:0340	102-54-5 see F:0190
83-79-4 see R:0150	94-59-7 see S:0110	98-12-4 see C:1760	102-67-0 see A:0650
84-15-1 see T:0210	94-74-6 see M:0290	98-13-5 see P:0500	102-71-6 see E:0240
84-65-1 see A:1390	94-75-7 see D:0100	98-16-8 see B:0300	102-81-8 see D:0380

102-82-9 see B:0850
103-11-7 see E:0710
103-33-3 see A:1660
103-69-5 see E:0370
103-71-9 see P:0430
103-84-4 see A:0150
103-85-5 see P:0490
104-15-4 see T:0630
104-90-5 see M:0940
104-94-9 see A:1360
105-30-6 see P:1200
105-36-2 see E:0420
105-39-5 see E:0490
105-46-4 see B:0810
105-54-4 see E:0470
105-56-6 see E:0510
105-57-7 see A:0100
105-60-2 see C:0390
105-67-9 see D:1220
106-35-4 see E:0450
106-42-3 see X:0100
106-43-4 see C:1050
106-44-5 see C:1450
106-46-7 see D:0460
106-47-8 see C:0770
106-48-9 see C:0950
106-50-3 see P:0400
106-51-4 see Q:0100
106-63-8 see I:0270
106-87-6 see V:0190
106-88-7 see B:0910
106-89-8 see E:0160
106-92-3 see A:0590
106-93-4 see E:0580
106-94-5 see B:0730
106-95-6 see A:0560
106-96-7 see P:1100
106-97-8 see B:0770
106-99-0 see B:0760
107-00-6 see E:0310
107-02-8 see A:0380
107-05-1 see A:0570
107-06-2 see E:0590
107-07-3 see E:0550
107-10-8 see P:1210
107-11-9 see A:0550
107-12-0 see P:1170
107-13-1 see A:0410
107-15-3 see E:0560
107-16-4 see F:0420
107-18-6 see A:0540
107-19-7 see P:1090
107-20-0 see C:0730
107-21-1 see E:0610
107-27-7 see E:0750
107-29-9 see A:0130

107-30-2 see C:0890
107-31-3 see M:0950
107-37-9 see A:0630
107-41-5 see H:0350
107-44-8 see S:0130
107-49-3 see T:0180
107-66-4 see D:0400
107-72-2 see A:1340
107-83-5 see H:0300
107-87-9 see M:1220
107-89-1 see A:0500
107-92-6 see B:1040
107-98-2 see P:1270
108-01-0 see D:1090
108-03-2 see N:0540
108-05-4 see V:0150
108-10-1 see M:1000
108-11-2 see M:0990
108-18-9 see D:1010
108-20-3 see D:1020
108-21-4 see I:0450
108-23-6 see I:0490
108-24-7 see A:0170
108-31-6 see M:0210
108-38-3 see X:0100
108-39-4 see C:1450
108-41-8 see C:1050
108-42-9 see C:0770
108-43-0 see C:0950
108-45-2 see P:0380
108-46-3 see R:0110
108-57-6 see D:1500
108-60-1 see B:0500
108-62-3 see M:0480
108-67-8 see M:0460
108-67-8 see T:0880
108-68-9 see X:0120
108-69-0 see X:0130
108-71-4 see T:0610
108-71-4 see T:0610
108-78-1 see M:0310
108-83-8 see D:1000
108-84-9 see H:0340
108-86-1 see B:0690
108-87-2 see M:0800
108-88-3 see T:0600
108-89-4 see P:0720
108-90-7 see C:0780
108-91-8 see C:1740
108-93-0 see C:1690
108-94-1 see C:1700
108-95-2 see P:0340
108-98-5 see P:0440
108-99-6 see P:0720
109-06-8 see P:0720
109-19-3 see B:0940

109-59-1 see I:0440
109-60-4 see P:1190
109-61-5 see P:1220
109-63-7 see B:0630
109-65-9 see B:0880
109-66-0 see P:0260
109-69-3 see B:0890
109-73-9 see B:0850
109-77-3 see M:0230
109-79-5 see B:0960
109-86-4 see E:0640
109-87-5 see M:0660
109-89-7 see D:0790
109-90-0 see E:0730
109-92-2 see V:0200
109-93-3 see V:0200
109-94-4 see E:0690
109-99-9 see T:0340
110-00-9 see F:0500
110-12-3 see M:0980
110-13-4 see A:0215
110-16-7 see M:0200
110-17-8 see F:0490
110-19-0 see B:0810
110-43-0 see M:0690
110-46-3 see A:1330
110-49-6 see M:0590
110-54-3 see H:0300
110-57-6 see D:0480
110-62-3 see V:0100
110-68-9 see B:0850
110-75-8 see C:0860
110-78-1 see P:1300
110-80-5 see E:0280
110-82-7 see C:1680
110-83-8 see C:1710
110-85-0 see P:0770
110-86-1 see P:1345
110-88-3 see P:0120
110-89-4 see P:0780
110-91-8 see M:1440
111-15-9 see E:0290
111-27-3 see H:0310
111-30-8 see G:0140
111-34-2 see B:1020
111-40-0 see D:0850
111-41-1 see A:0830
111-42-2 see E:0240
111-42-2 see D:0770
111-44-4 see D:0550
111-65-9 see O:0120
111-69-3 see A:0450
111-76-2 see B:0790
111-84-2 see N:0685
111-91-1 see B:0490
111-92-2 see D:0370

112-57-2 see T:0290
114-26-1 see P:1180
115-07-1 see P:1230
115-09-3 see M:0440
115-10-6 see D:1180
115-11-7 see I:0280
115-21-9 see E:0840
115-25-3 see O:0100
115-26-4 see D:1030
115-29-7 see E:0100
115-32-2 see D:0700
115-77-5 see P:0250
115-86-6 see T:0940
115-90-2 see F:0110
116-06-3 see A:0490
116-14-3 see T:0320
117-52-2 see C:1410
117-79-3 see A:0750
117-81-7 see D:0860
117-84-0 see D:1400
118-52-5 see D:0510
118-74-1 see H:0190
118-96-7 see T:0920
119-34-6 see A:0860
119-38-0 see I:0360
119-61-9 see B:0390
119-90-4 see D:1050
119-93-7 see T:0590
119-94-8 see E:0390
120-12-7 see A:1380
120-36-5 see D:0610
120-61-6 see D:1290
120-71-8 see C:1440
120-80-9 see C:0570
120-82-1 see T:0700
120-83-2 see D:0590
121-14-2 see D:1370
121-21-1 see P:1340
121-29-9 see P:1340
121-44-8 see T:0810
121-45-9 see T:0900
121-69-7 see D:1100
121-75-5 see M:0190
121-82-4 see C:1770
122-14-5 see F:0100
122-34-9 see S:0310
122-39-4 see D:1470
122-42-9 see P:1120
122-60-1 see P:0410
122-66-7 see D:1490
123-05-7 see E:0700
123-19-3 see D:1530
123-30-8 see A:0870
123-31-9 see H:0490
123-33-1 see M:0220
123-38-6 see P:1140

123-12-2 see D:0200
123-51-3 see A:1310
123-51-3 see I:0240
123-54-6 see P:0270
123-62-6 see P:1160
123-63-7 see P:0130
123-72-8 see B:1030
123-73-9 see C:1470
123-86-4 see B:0810
123-91-1 see D:1410
123-92-2 see I:0230
123-92-2 see A:1300
124-04-9 see A:0440
124-09-4 see H:0270
124-38-9 see C:0460
124-40-3 see D:1070
124-48-1 see D:0350
124-65-2 see C:0050
124-65-2 see S:0420
124-87-8 see P:0740
126-33-0 see S:0710
126-72-7 see T:0970
126-73-8 see T:0660
126-75-0 see D:0170
126-85-2 see M:0300
126-98-7 see M:0500
126-99-8 see C:1000
127-00-4 see P:1240
127-18-4 see T:0270
127-19-5 see D:1060
127-91-3 see T:1000
128-37-0 see D:0390
129-00-0 see P:1330
129-06-6 see W:0100
129-67-9 see E:0110
130-15-4 see N:0150
131-11-3 see D:1250
131-52-2 see S:0520
131-74-8 see A:1200
132-32-1 see A:0820
132-64-9 see D:0310
133-06-2 see C:0410
133-90-4 see C:0600
134-29-2 see A:1360
134-32-7 see N:0160
135-19-3 see N:0140
135-20-6 see C:1520
135-88-6 see P:0460
136-40-3 see P:0330
136-78-7 see D:0190
137-05-3 see M:0790
137-26-8 see T:0520
137-32-6 see I:0244
137-32-6 see A:1310
138-22-7 see B:0950
138-86-3 see D:1440

139-13-9 see N:0360
139-40-2 see P:1110
140-29-4 see B:0460
140-31-8 see A:0840
140-57-8 see S:0780
140-76-1 see M:1300
140-80-7 see A:0800
140-88-5 see E:0320
141-32-2 see B:0830
141-43-5 see E:0240
141-66-2 see D:0710
141-78-6 see E:0300
141-79-7 see M:0470
142-04-1 see A:1350
142-64-3 see P:0770
142-71-2 see C:1530
142-82-5 see H:0160
142-84-7 see D:1510
142-96-1 see B:0920
143-33-9 see C:1590
143-33-9 see S:0450
143-50-0 see C:0640
144-49-0 see F:0330
144-62-7 see O:0160
145-73-3 see E:0110
146-84-9 see S:0290
148-01-6 see D:1310
148-82-3 see M:0320
149-74-6 see D:0560
150-76-5 see M:0610
151-38-2 see M:0600
151-50-8 see C:1590
151-50-8 see P:0910
151-56-4 see E:0650
151-67-7 see H:0110
152-16-9 see O:0110
154-93-8 see C:0550
155-25-9 see A:0910
156-10-5 see N:0600
156-59-2 see D:0540
156-60-5 see D:0540
156-62-7 see C:0270
193-39-5 see I:0110
205-99-2 see B:0360
206-44-0 see F:0280
208-96-8 see A:0075
218-01-9 see C:1220
260-94-6 see A:0370
287-92-3 see C:1790
297-78-9 see I:0250
297-97-2 see T:0470
297-99-4 see P:0570
298-00-0 see M:1070
298-02-2 see P:0520
298-03-3 see D:0170
298-04-4 see D:1580

299-84-3 see R:0140
299-86-5 see C:1490
300-62-9 see A:1280
300-76-5 see N:0100
301-03-1 see M:0240
301-04-2 see L:0110
301-05-3 see F:0130
302-01-2 see H:0370
302-17-0 see C:0590
302-70-5 see M:0300
305-03-3 see C:0610
309-00-2 see A:0510
311-45-5 see P:0140
314-40-9 see B:0640
315-18-4 see M:1360
316-42-7 see E:0050
319-84-6 see H:0210
319-85-7 see H:0210
319-86-8 see H:0210
327-98-0 see T:0760
329-71-5 see D:1360
329-99-7 see C:1795
330-54-1 see D:1610
333-41-5 see D:0280
334-88-3 see D:0290
353-42-4 see B:0630
353-50-4 see C:0520
353-59-3 see C:0830
357-57-3 see B:0740
359-06-8 see F:0340
366-70-1 see P:1020
371-62-0 see E:0600
373-57-9 see B:630
379-79-3 see E:0200
402-47-1 see T:0630
409-21-2 see S:0250
420-04-2 see C:1570
420-46-2 see T:0830
430-66-0 see T:0830
431-03-8 see B:0780
439-14-5 see D:0270
443-48-1 see M:1340
446-86-6 see A:1630
460-19-5 see C:1600
462-06-6 see F:0350
463-04-7 see A:1330
463-49-0 see P:1050
463-51-4 see K:0110
463-58-1 see C:0490
463-82-1 see N:0200
465-73-6 see I:0340
470-90-6 see C:0650
471-34-1 see C:0230
479-45-8 see T:0410
485-31-4 see B:0475
492-61-5 see G:0125

492-62-6 see G:0125
492-80-8 see A:1620
494-03-1 see C:0720
497-92-7 see A:0520
498-15-7 see T:1000
502-39-6 see M:1050
503-38-8 see P:0550
504-18-7 see M:0310
504-24-5 see A:0900
504-29-0 see A:0890
505-60-2 see M:1460
506-61-6 see P:1000
506-64-9 see S:0270
506-68-3 see C:1610
506-77-4 see C:1620
506-78-5 see C:1630
506-78-5 see C:1590
506-87-6 see A:1020
506-96-7 see A:0280
507-02-8 see A:0330
509-14-8 see T:0380
510-15-6 see E:0520
513-42-8 see M:0520
513-48-4 see I:0170
513-49-5 see B:0850
514-73-8 see D:1590
518-75-2 see A:1490
522-70-3 see A:1490
526-73-8 see T:0880
528-29-0 see D:1330
532-27-4 see C:0750
532-32-1 see S:0400
534-07-6 see B:0520
534-52-1 see D:1340
535-13-7 see E:0500
535-89-7 see C:1460
536-90-3 see A:1360
538-07-8 see E:0400
540-59-0 see D:0540
540-67-0 see M:0910
540-73-8 see D:1210
540-84-1 see O:0120
540-88-5 see B:0810
541-09-3 see U:0100
541-25-3 see L:0250
541-53-7 see D:1600
541-73-1 see D:0460
541-85-5 see E:0350
542-62-1 see B:0140
542-75-6 see D:0660
542-76-7 see C:1010
542-88-1 see B:0510
542-90-5 see E:0830
542-92-7 see C:1780
543-90-8 see C:0110
544-92-3 see C:1380

548-62-9 see B:0530
552-30-7 see T:0850
554-12-1 see M:1200
554-13-2 see L:0290
554-84-7 see N:0530
555-77-1 see T:0960
556-52-5 see G:0160
556-56-9 see A:0600
556-61-6 see M:1030
556-64-9 see M:1260
557-17-5 see M:1210
557-19-7 see N:0260
557-20-0 see D:0930
557-31-3 see A:0580
558-13-4 see C:0500
563-04-2 see T:0800
563-12-2 see E:0260
563-41-7 see S:0200
563-43-9 see A:0640
563-58-6 see D:0660
563-68-8 see T:0420
563-80-4 see M:1020
573-56-8 see D:1360
573-58-0 see C:1240
577-71-9 see D:1360
578-54-1 see E:0360
583-60-8 see M:0820
584-02-1 see A:1310
584-79-2 see A:0520
584-84-9 see T:0620
586-11-8 see D:1360
590-01-2 see B:0990
591-08-2 see A:0350
591-27-5 see A:0870
591-78-6 see M:0740
591-87-7 see A:0530
592-01-8 see C:1590
592-01-8 see C:0280
592-04-1 see M:0370
592-41-6 see H:0330
592-76-7 see H:0180
592-85-8 see M:0450
592-87-0 see L:0230
593-60-2 see V:0160
593-74-8 see M:0440
594-42-3 see P:0300
594-72-9 see D:0580
597-64-8 see T:0310
598-75-4 see A:1310
600-25-9 see C:0920
602-01-7 see D:1370
602-03-9 see D:1320
602-38-0 see D:1350
602-87-9 see N:0370
603-34-9 see T:0930
605-71-0 see D:1350

606-20-2 see D:1370
606-22-4 see D:1320
606-37-1 see D:1350
608-73-1 see H:0210
608-93-5 see P:0200
609-19-8 see T:0770
610-39-9 see D:1370
612-82-8 see T:0590
612-83-9 see D:0470
615-05-4 see D:0230
615-65-6 see C:0880
616-23-9 see D:0650
618-85-9 see D:1370
618-87-1 see D:1320
620-11-1 see A:1300
621-64-7 see N:0610
624-18-0 see P:0400
624-41-9 see A:1300
624-83-9 see M:1010
624-92-0 see D:1170
625-55-8 see I:0500
626-17-5 see P:0680
626-38-0 see A:1300
627-13-4 see P:1310
628-63-7 see A:1300
628-81-9 see E:0440
628-92-2 see C:1670
628-96-6 see E:0630
629-14-1 see E:0620
630-08-0 see C:0480
630-20-6 see T:0250
630-60-4 see O:0150
630-93-3 see P:0510
631-61-8 see A:0960
633-03-4 see C:1230
636-21-5 see T:0640
638-21-1 see P:0480
639-58-7 see T:0950
640-19-7 see F:0320
642-15-9 see A:1490
644-31-5 see A:0270
644-64-4 see D:1300
645-62-5 see E:0810
646-06-0 see D:1430
671-16-9 see P:1020
675-16-1 see A:1300
676-63-1 see P:1230
676-97-1 see M:1090
680-31-9 see H:0290
681-84-5 see M:1230
684-16-2 see H:0250
696-28-6 see P:0370
709-98-8 see P:1080
732-11-6 see P:0560
752-74-9 see T:0950
759-73-9 see N:0620

760-19-0 see A:0640
764-41-0 see D:0480
765-34-4 see G:0170
766-09-6 see E:0800
768-52-5 see I:0480
786-19-6 see C:0530
814-49-3 see D:0840
814-68-6 see A:0420
814-91-5 see C:1550
822-06-0 see H:0280
823-40-5 see T:0610
834-12-8 see A:0740
900-95-8 see T:0950
920-66-1 see H:0265
924-16-3 see N:0560
928-65-4 see H:0360
929-06-6 see A:0810
930-55-2 see N:0650
933-75-5 see T:0770
933-78-8 see T:0770
944-22-9 see F:0400
947-02-4 see P:0540
950-10-7 see M:0330
950-37-8 see M:0540
957-51-7 see D:1460
959-98-8 see E:0100
991-42-4 see N:0700
992-59-6 see T:0590
993-00-0 see M:0780
999-61-1 see H:0510
999-81-5 see C:0710
1002-16-0 see A:1320
1024-57-3 see H:0150
1066-30-4 see C:1090
1066-33-7 see A:0980
1071-83-6 see G:0180
1072-35-1 see L:0190
1111-78-0 see A:1010
1113-38-8 see A:1150
1116-70-7 see A:0650
1120-71-4 see P:1070
1122-60-7 see N:0440
1125-27-5 see E:0790
1129-41-5 see M:1320
1162-65-8 see A:0470
1162-65-8 see A:0470
1163-19-5 see D:0160
1163-19-5 see P:0810
1165-39-5 see A:0470
1184-66-3 see H:0380
1185-57-5 see F:0140
1189-85-1 see B:0900
1300-71-6 see X:0120
1300-73-8 see X:0130
1302-45-0 see A:0710
1302-52-9 see B:0470

1302-74-5 see A:0660
1302-78-9 see B:0250
1303-28-2 see A:1540
1303-33-9 see A:1560
1303-86-2 see B:0590
1303-96-4 see B:0580
1303-96-4 see B:0580
1304-02-5 see D:0150
1304-28-5 see B:0170
1304-29-6 see B:0200
1304-56-9 see B:0470
1304-76-3 see B:0530
1304-82-1 see B:0540
1304-85-4 see B:0530
1305-62-0 see C:0295
1305-78-8 see C:0320
1305-79-9 see C:0330
1305-99-3 see C:0340
1306-19-0 see C:0140
1306-23-6 see C:0170
1307-82-0 see S:0440
1307-96-6 see C:1300
1308-04-9 see C:1300
1308-13-0 see Z:0130
1308-38-9 see C:1160
1309-32-6 see A:1100
1309-37-1 see I:0210
1309-37-1 see H:0130
1309-45-1 see F:0380
1309-48-4 see M:0140
1309-60-0 see L:0145
1309-63-3 see T:0570
1309-64-4 see A:1480
1310-00-5 see M:0180
1310-02-7 see S:0490
1310-58-3 see P:0950
1310-73-2 see S:0500
1312-43-2 see I:0120
1312-73-8 see P:1010
1313-13-9 see M:0260
1313-27-5 see M:1420
1314-12-1 see T:0420
1314-13-2 see Z:0140
1314-20-1 see T:0525
1314-32-5 see T:0420
1314-36-9 see Y:0100
1314-56-3 see P:0650
1314-61-0 see T:0130
1314-62-1 see V:0120
1314-80-3 see P:0640
1314-84-7 see Z:0150
1314-87-0 see L:0220
1317-39-1 see C:1360
1317-38-0 see C:1360
1317-48-2 see S:0230
1317-60-8 see I:0210

7440-33-7 see T:0985
7440-36-0 see A:1400
7440-37-1 see A:1510
7440-38-2 see A:1520
7440-39-3 see B:0100
7440-41-7 see B:0470
7440-42-8 see B:0580
7440-43-9 see C:0100
7440-44-0 see G:0200
7440-47-3 see C:1130
7440-48-4 see C:1300
7440-50-8 see C:1360
7440-55-3 see G:0050
7440-56-4 see G:0110
7440-58-6 see H:0100
7440-59-7 see H:0120
7440-61-1 see U:0100
7440-62-2 see V:0110
7440-65-5 see Y:0100
7440-66-6 see Z:0100
7440-67-7 see Z:0160
7440-69-9 see B:0530
7440-70-2 see C:0200
7440-74-6 see I:0120
7446-08-4 see S:0150
7446-09-5 see S:0750
7446-11-9 see S:0810
7446-14-2 see L:0210
7446-18-6 see T:0420
7446-27-7 see L:0180
7446-34-6 see S:0180
7446-70-0 see A:0670
7447-39-4 see C:1370
7487-94-7 see M:0360
7488-56-4 see S:0180
7521-80-4 see B:1010
7550-45-0 see T:0580
7553-56-2 see I:0140
7572-29-4 see D:0440
7580-67-8 see L:0310
7616-94-6 see P:0310
7619-62-7 see M:0340
7631-86-9 see D:0260
7631-86-9 see S:0220
7631-89-2 see S:0370
7631-90-5 see S:0410
7632-50-0 see A:1060
7632-51-1 see V:0130
7637-07-2 see B:0620
7645-25-2 see L:0120
7646-78-8 see T:0550
7646-85-7 see Z:0120
7647-01-0 see H:0430
7647-18-9 see A:1420
7664-38-2 see P:0590
7664-39-3 see H:0450

7664-41-7 see A:0950
7664-93-9 see S:0770
7681-49-4 see S:0470
7681-57-4 see S:0510
7693-27-8 see M:0120
7697-37-2 see N:0340
7699-41-4 see S:0220
7699-45-8 see Z:0110
7704-34-9 see S:0730
7705-08-0 see F:0160
7718-54-9 see N:0250
7719-09-7 see T:0480
7719-12-2 see P:0660
7720-78-7 see F:0220
7722-64-7 see P:0980
7722-76-1 see A:1190
7722-84-1 see H:0460
7722-88-5 see T:0400
7723-14-0 see P:0610
7726-95-6 see B:0660
7727-21-1 see P:0990
7727-37-9 see N:0470
7727-43-7 see B:0210
7727-54-0 see A:1180
7738-94-5 see C:1100
7757-74-6 see S:0510
7757-79-1 see P:0960
7757-82-6 see S:0540
7757-83-7 see S:0550
7758-01-2 see P:0870
7758-09-0 see P:0970
7758-89-6 see C:1370
7758-94-3 see F:0210
7758-95-4 see L:0130
7758-97-6 see L:0140
7758-98-7 see C:1390
7758-99-8 see C:1390
7761-88-8 see S:0280
7772-99-8 see S:0580
7773-06-0 see A:1210
7774-29-0 see M:0380
7774-34-7 see C:0250
7775-09-9 see S:0430
7775-11-3 see S:0445
7778-18-9 see C:0350
7778-39-4 see A:1530
7778-43-0 see S:0370
7778-44-1 see C:0210
7778-50-9 see P:0900
7778-54-3 see C:0300
7782-41-4 see F:0310
7782-42-5 see G:0200
7782-44-7 see O:0210
7782-49-2 see S:0140
7782-50-5 see C:0670
7782-63-0 see F:0220

7782-65-2 see G:0120
7782-79-8 see H:0390
7783-00-8 see S:0190
7783-06-4 see H:0480
7783-07-5 see H:0470
7783-18-8 see A:1270
7783-28-0 see A:1190
7783-35-9 see M:0420
7783-41-7 see O:0220
7783-46-2 see L:0160
7783-47-3 see S:0590
7783-54-2 see N:0500
7783-56-4 see A:1470
7783-60-0 see S:0800
7783-66-6 see I:0160
7783-70-2 see A:1430
7783-79-1 see S:0160
7783-80-4 see T:0160
7783-82-6 see T:0990
7783-85-9 see F:0200
7784-18-1 see A:0680
7784-27-2 see A:0690
7784-30-7 see A:0700
7784-34-1 see A:1570
7784-40-9 see L:0120
7784-41-0 see P:0850
7784-42-1 see A:1580
7784-44-3 see A:0970
7784-46-5 see S:0380
7786-34-7 see M:1350
7786-81-4 see N:0290
7787-36-2 see B:0190
7787-47-5 see B:0470
7787-49-7 see B:0470
7787-55-5 see B:0470
7787-56-6 see B:0470
7787-59-9 see B:0530
7787-64-6 see B:0530
7787-71-5 see B:0680
7788-98-9 see A:1050
7788-99-0 see C:1170
7789-00-6 see P:0900
7789-02-8 see C:1150
7789-06-2 see S:0630
7789-09-5 see A:1080
7789-12-0 see S:0445
7789-23-3 see P:0940
7789-30-2 see B:0670
7789-42-6 see C:0120
7789-47-1 see M:0350
7789-61-9 see A:1450
7789-75-5 see C:0290
7789-78-8 see C:0293
7790-69-4 see L:0320
7790-78-5 see C:0130
7790-91-2 see C:0690

7790-94-5 see C:1030
7790-99-0 see I:0150
7791-20-0 see N:0250
7791-23-3 see S:0170
7799-12-0 see T:0420
7803-49-8 see H:0500
7803-51-2 see P:0580
7803-52-3 see S:0600
7803-55-6 see A:1120
7803-57-8 see H:0370
7803-62-5 see S:0210
8000-97-3 see D:0170
8001-35-2 see T:0650
8001-58-9 see C:1290
8002-05-9 see N:0110
8002-74-2 see P:0100
8003-34-7 see P:1340
8003-45-0 see E:0100
8005-38-7 see F:0410
8006-07-3 see F:0410
8006-61-9 see G:0100
8006-64-2 see T:1000
8007-42-9 see L:0260
8007-45-2 see C:1290
8008-20-6 see K:0100
8008-51-3 see C:0370
8011-97-0 see I:0210
8012-01-9 see S:0500
8012-95-1 see M:1385
8013-13-6 see F:0410
8013-54-5 see C:0870
8013-70-5 see I:0460
8013-75-0 see I:0240
8014-94-6 see S:0750
8014-95-7 see S:0770
8022-00-2 see D:0180
8022-04-6 see T:0650
8022-76-2 see D:0660
8023-22-1 see D:0690
8023-53-8 see D:0450
8028-34-0 see M:0400
8030-30-6 see N:0110
8030-64-6 see M:0430
8031-06-9 see N:0110
8032-32-4 see N:0110
8049-47-6 see P:0050
8049-64-7 see L:0140
8051-03-4 see Z:0140
8052-41-3 see S:0610
8052-42-4 see A:1600
8054-98-6 see H:0240
8057-43-0 see P:0050
8057-70-3 see P:0170
8058-73-9 see D:0170
8065-48-3 see D:0170
8072-21-7 see D:0690

8072 39-7 scc D:0690
9000-01-5 see G:0210
9001-37-0 see G:0130
9002-16-8 see P:0050
9004-51-7 see I:0200
9004-66-4 see I:0200
9004-70-0 see N:0420
9009-86-3 see R:0135
9014-01-1 see S:0680
9046-39-3 see P:0050
9061-47-6 see I:0200
9067-26-9 see R:0135
10022-31-8 see B:0160
10024-97-2 see N:0680
10024-97-2 see N:0490
10025-65-7 see P:0800
10025-67-9 see S:0740
10025-69-1 see S:0580
10025-73-7 see C:1110
10025-82-8 see I:0120
10025-87-3 see P:0620
10025-91-9 see A:1460
10025-94-2 see Y:0100
10025-99-7 see P:0890
10026-06-9 see S:0570
10026-08-1 see T:0525
10026-13-8 see P:0630
10028-15-6 see O:0230
10028-22-5 see F:0180
10031-59-1 see T:0420
10034-76-1 see C:0350
10034-82-9 see S:0445
10034-81-8 see M:0150
10034-82-9 see S:0450
10034-85-2 see H:0395
10034-93-2 see H:0380
10035-06-0 see B:0530
10035-10-6 see H:0420
10039-54-0 see H:0380
10042-76-9 see S:0640
10043-01-3 see A:0730
10043-35-3 see B:0580
10043-52-4 see C:0250
10045-89-3 see F:0200
10045-94-0 see M:0390
10048-95-0 see S:0370
10049-04-4 see C:0680
10049-05-5 see C:1200
10049-07-7 see R:0130
10061-01-5 see D:0660
10061-02-6 see D:0660
10101-41-4 see C:0350
10101-53-8 see C:1120
10101-63-0 see L:0170
10101-97-0 see N:0290
10102-03-1 see N:0490

10102-06-4 see U:0100
10102-18-8 see S:0530
10102-20-2 see S:0560
10102-43-9 see N:0490
10102-43-9 see N:0350
10102-44-0 see N:0480
10102-44-0 see N:0490
10102-45-1 see T:0420
10102-53-1 see A:1530
10103-47-6 see C:1150
10108-64-2 see C:0130
10124-36-4 see C:0160
10124-37-5 see C:0310
10124-50-2 see P:0860
10137-69-6 see C:1720
10137-74-3 see C:0240
10140-87-1 see D:0530
10141-00-1 see C:1170
10192-30-0 see A:1230
10192-30-0 see A:1000
10196-04-0 see A:1230
10210-68-1 see C:1310
10213-15-7 see M:0130
10265-92-6 see M:0520
10294-33-4 see B:0600
10294-34-5 see B:0610
10294-56-1 see P:0600
10311-84-9 see D:0210
10326-21-3 see M:0110
10326-27-7 see B:0130
10361-43-0 see B:0530
10361-44-1 see B:0530
10361-92-9 see Y:0100
10361-43-0 see B:0530
10361-44-1 see B:0530
10361-93-0 see Y:0100
10377-60-3 see M:0130
10377-66-9 see M:0270
10421-48-4 see F:0170
10544-72-6 see N:0480
10544-72-6 see N:0490
10544-73-7 see N:0490
10579-83-6 see S:0550
10588-01-9 see S:0445
11004-49-2 see M:0240
11018-89-6 see O:0150
11069-19-5 see D:0480
11095-11-7 see C:0430
11095-17-3 see D:0690
11096-20-1 see D:1040
11096-21-2 see D:0690
11096-32-5 see P:1080
11096-45-0 see S:0430
11096-82-5 see P:0820
11097-69-1 see P:0820
11100-14-4 see P:0820

11103-86-9 see Z:0130
11104-29-3 see P:0820
11107-01-0 see T:0985
11111-91-4 see P:0170
11113-50-1 see B:0576
11115-74-5 see C:1100
11118-72-2 see A:1490
11119-41-8 see T:0120
11119-70-3 see L:0140
11121-31-6 see A:1610
11126-35-5 see A:0340
11126-37-7 see A:0340
11130-21-5 see V:0110
11133-98-5 see B:0470
11138-49-1 see S:0340
11140-60-6 see S:0190
12001-26-2 see M:1370
12001-28-4 see A:1590
12001-29-5 see A:1590
12002-03-8 see P:0180
12002-25-4 see D:1520
12002-35-6 see D:1520
12018-19-8 see Z:0130
12036-35-0 see R:0120
12039-52-0 see T:0420
12040-45-8 see A:1610
12044-50-7 see A:1540
12054-48-7 see N:0270
12055-23-1 see H:0100
12070-10-9 see V:0110
120-7012-1 see T:0985
12075-68-2 see A:0640
12108-13-3 see M:0280
12124-99-1 see A:1220
12125-01-8 see A:1090
12125-02-9 see A:1030
12125-33-6 see M:0240
12125-56-3 see N:0270
12126-59-9 see C:1350
12135-76-1 see A:1220
12136-89-9 see S:0390
12137-27-8 see R:0120
12172-73-5 see A:1590
12185-10-3 see P:0610
12232-99-4 see B:0530
12233-73-7 see B:0530
12324-05-9 see C:1180
12324-08-2 see C:1180
12396-99-5 see S:0750
12414-70-9 see S:0230
12420-12-1 see T:0120
12424-49-6 see G:0200
12426-98-1 see A:1590
12427-38-2 see M:0240
12447-40-4 see B:0580
12504-67-5 see S:0250

12542-85-7 see A:0640
12589-75-2 see B:0530
12604-58-9 see F:0230
12672-29-6 see P:0820
12673-82-4 see S:0730
12679-58-2 see R:0150
12680-07-8 see T:0520
12680-62-5 see T:0520
12718-69-3 see T:0985
12750-99-1 see D:0260
12751-41-6 see G:0200
12767-24-7 see S:0730
12770-50-2 see B:0470
12770-91-1 see C:0930
12772-40-6 see D:0690
12788-93-1 see B:0820
12789-03-6 see C:0630
12797-72-7 see A:1610
13004-56-3 see Q:0120
13007-92-6 see C:1140
13010-47-4 see L:0330
13071-79-9 see T:0190
13106-76-8 see A:1130
13121-70-5 see C:1810
13138-45-9 see N:0280
13171-21-6 see P:0570
13194-48-4 see E:0270
13319-75-0 see B:0620
13327-32-7 see B:0470
13397-24-5 see C: 0350
13426-91-0 see C:1560
13429-07-7 see D:1520
13446-10-1 see A:1170
13450-90-3 see G:0075
13454-96-1 see P:0800
13463-39-3 see N:0240
13463-40-6 see I:0220
13463-67-7 see T:0570
13464-35-2 see P:0860
13464-82-9 see I:0120
13465-08-2 see H:0380
13465-95-7 see B:0180
13466-78-9 see T:1000
13473-90-0 see A:0690
13477-00-4 see B:0130
13477-10-6 see B:0150
13477-34-4 see C:0310
13478-00-7 see N:0280
13494-27-4 see F:0130
13494-80-9 see T:0150
13510-48-0 see B:0470
13510-49-1 see B:0470
13517-17-4 see S:0445
13520-83-7 see U:0100
13530-65-9 see Z:0130
13548-38-4 see C:1150

13552-44-8 see D:0250
13569-65-8 see R:0130
13588-28-8 see D:1520
13597-99-4 see B:0470
13598-00-0 see B:0470
13598-15-7 see B:0470
13598-36-2 see P:0600
13654-09-6 see P:0810
13765-19-0 see C:0260
13770-61-1 see I:0120
13770-96-2 see S:0360
13775-53-6 see S:0350
13814-96-5 see L:0150
13820-41-2 see A:1250
13823-29-5 see T:0525
13825-36-0 see T;0525
13826-63-6 see T:0420
13838-16-9 see E:0150
13943-58-3 see P:0930
13952-84-6 see B:0850
13967-90-3 see B:0120
14018-95-2 see Z:0130
14216-75-2 see N:0280
14221-47-7 see F:0150
14307-35-8 see L:0300
14307-43-8 see A:1240
14452-57-4 see M:0160
14464-46-1 see S:0230
14484-64-1 see F:0130
14489-25-9 see C:1190
14567-73-8 see A:1590
14675-41-3 see Z:0130
14797-55-9 see N:0330
14763-77-0 see C:1380
14807-96-6 see T:0120
14808-60-7 see S:0230
14901-08-7 see C:1660
14977-61-8 see C:1210
15005-90-0 see C:1190
15096-52-3 see S:0350
15347-57-6 see L:0110
15468-32-3 see S:0230
15663-27-1 see C:1260
15699-18-0 see N:0230
15710-66-4 see M:0270
15771-29-6 see S:0510
15892-23-6 see B:0840
15950-66-0 see T:0770
15972-60-8 see A:0480
16045-17-3 see T:0525
16219-75-3 see E:0720
16291-96-6 see G:0200
16752-77-5 see M:0560
16760-37-5 see N:0300
16842-03-8 see C:1320
16853-85-3 see L:0285

16872-11-0 see F:0260
16893-85-9 see S:0490
16919-19-0 see A:1100
16919-58-7 see A:1040
16921-30-5 see P:0890
16940-66-2 see B:0580
16941-12-1 see C:0990
16941-12-1 see P:0800
16949-65-8 see M:0180
16961-83-4 see F:0380
16984-48-8 see F:0300
17068-78-9 see A:1590
17135-66-9 see C:1150
17617-23-1 see F:0390
17702-41-9 see D:0150
17804-35-2 see B:0230
18282-10-5 see T:0530
18396-65-1 see S:0200
18414-36-3 see D:0320
18454-12-1 see L:0140
18810-58-7 see B:0110
18883-66-4 see S:0620
18972-56-0 see M:0180
19034-08-3 see E:0660
19287-45-7 see D:0330
19624-22-7 see P:0190
19701-15-6 see M:0340
20265-96-7 see C:0770
20316-06-7 see M:0240
20324-32-7 see D:1520
20770-05-2 see C:1160
20816-12-0 see O:0140
20830-75-5 see D:0980
20830-81-3 see D:0130
20859-73-8 see A:0710
21059-09-6 see C:1730
21087-64-9 see M:1330
21351-79-1 see C:0580
21548-32-3 see F:0470
21609-90-5 see L:0240
21651-19-4 see T:0530
21725-46-2 see C:1580
21908-53-2 see M:0400
21923-23-9 see C:1080
22194-21-4 see E:0150
22194-22-5 see E:0150
22224-92-6 see F:0050
22306-37-2 see B:0530
22691-02-7 see C:0250
22781-23-3 see B:0220
22831-39-6 see M:0170
22967-92-6 see M:0440
23135-22-0 see O:0170
23214-92-8 see A:0460
23296-15-3 see C:0350
23422-53-9 see F:0440

23505-41-1 see P:0790
32534-81-9 see P:0810
32536-52-0 see P:0810
23720-59-4 see H:0140
23783-98-4 see P:0570
23950-58-5 see P:1040
24678-82-8 see U:0100
24934-91-6 see C:0700
25013-15-4 see V:0240
25057-89-0 see B:0240
25154-54-5 see D:1330
25154-55-6 see N:0530
25167-80-0 see C:0950
25167-82-2 see T:0770
25231-46-3 see T:0630
25265-68-3 see M:1250
25314-61-8 see A:0380
25321-14-6 see D:1370
25321-22-6 see D:0460
25376-45-8 see T:0610
25376-45-8 see T:0610
25377-72-4 see P:0280
25550-49-6 see V:0190
25550-58-7 see D:1360
25551-13-7 see T:0880
25639-42-3 see M:0810
26140-60-3 see T:0210
26249-12-7 see D:0340
26446-77-5 see B:0730
26471-56-7 see D:1320
26545-73-3 see D:0650
26571-79-9 see C:0970
26627-44-1 see N:0360
26627-45-2 see N:0360
26628-22-8 see S:0390
26635-63-2 see A:1310
26764-44-3 see T:0610
26764-44-3 see T:0610
26914-13-6 see A:0340
26952-21-6 see I:0370
27134-26-5 see C:0770
27137-85-5 see D:0630
27156-03-2 see D:0490
27176-87-0 see D:1630
27193-28-8 see O:0130
37211-05-5 see N:0250
27220-59-3 see A:1490
27478-34-8 see D:1350
27497-51-4 see C:0840
27598-85-2 see A:0870
27774-13-6 see V:0140
27858-07-7 see P:0810
27987-06-0 see T:0830
28300-74-5 see A:1440
28355-56-8 see M:0240
28479-22-3 see C:0910

28772-56-7 see B:0650
28805-86-9 see B:0980
29027-17-6 see C:0880
29191-52-4 see A:1360
29371-14-0 see M:0850
29611-03-8 see A:0470
29754-21-0 see A:0410
29847-98-1 see D:0260
29847-98-1 see S:0220
30143-13-6 see T:0610
30143-13-6 see T:0610
30525-89-4 see P:0120
30899-19-5 see A:1310
31012-04-1 see B:0610
31119-53-6 see C:0160
31242-93-0 see C:0655
32215-02-4 see A:0470
33004-01-2 see P:1230
33060-30-9 see E:0540
33089-61-1 see A:0940
33213-65-9 see E:0100
34487-55-3 see A:0510
34590-94-8 see D:1520
34713-94-5 see A:1310
34893-92-0 see D:0620
35400-43-2 see S:0840
36355-01-8 see P:0810
36478-76-9 see U:0100
36483-60-0 see P:0810
37187-22-7 see A:0250
37220-42-1 see I:0220
37224-57-0 see Z:0130
37229-06-4 see H:0140
37232-12-5 see T:0120
37264-96-3 see C: 1310
37293-14-4 see B:0540
37300-23-5 see Z:0130
37324-23-5 see P:0820
37329-49-0 see T:0985
37337-67-0 see D:0260
37337-67-0 see S:0220
39156-41-7 see D:0230
39196-18-4 see T:0450
39287-69-9 see S:0470
39300-53-3 see T:0840
39342-49-9 see Z:0150
39355-35-6 see M:1385
39378-26-2 see P:0230
39378-26-2 see Q:0110
39400-72-1 see A:1610
39404-03-0 see M:0170
39413-34-8 see S:0490
39413-47-3 see B:0470
39456-80-9 see T:0520
39920-37-1 see D:0620
40088-47-9 see P:0810

40334-69-8 see L:0250
40334-70-1 see L:0250
41195-90-8 see D:0620
41766-75-0 see T:0590
42031-19-6 see B:0840
42504-46-1 see I:0430
49690-94-0 see P:0810
50594-66-6 see A:0360
50642-23-4 see S:0130
50723-80-3 see B:0240
50782-69-9 see V:0250
50813-73-5 see N:0110
50888-64-7 see A:0610
50926-93-7 see S:0220
50978-48-8 see A:0610
51005-21-1 see N:0350
51065-07-7 see M:0850
51218-45-2 see M:1310
51235-04-2 see H:0320
51274-07-8 see N:0460
51488-20-1 see C:0910
51887-47-9 see M:0430
52001-89-5 see C:0430
52037-46-4 see T:0740
52110-72-2 see A:1050
52181-51-8 see C:0800
52438-91-2 see B:0500
52627-52-8 see T:0840
53095-31-1 see N:0100
53469-21-9 see P:0820
53558-25-1 see P:1350
53637-13-1 see C:0630
54182-73-9 see A:1650
54322-60-0 see S:0630
54511-18-1 see D:0260
54511-18-1 see S:0220
54841-71-3 see M:0670
54847-97-1 see N:0110
55720-99-5 see C:0655
55819-32-4 see D:0690
55956-21-3 see D:1520
55957-14-7 see S:0370
55963-79-6 see L:0260
56189-09-4 see L:0190
56591-09-4 see S:0730
56645-30-8 see S:0730
56748-40-4 see D:0260
56748-40-4 see S:0220
56795-67-6 see T:0230
56833-73-9 see D:1040
57035-13-9 see S:0730
57158-05-1 see N:0660
57308-10-8 see C:0293

57321-63-8 see C:0655
57360-17-5 see A:0820
57414-01-4 see S:0410
57593-74-5 see P:0150
58128-78-2 see B:0870
58164-88-8 see A:1410
58209-98-6 see A:0470
58391-87-0 see A:0610
59080-40-9 see P:0810
59355-75-8 see M:0640
59536-65-1 see P:0810
60616-74-2 see M:0120
60676-86-0 see S:0220
61028-24-8 see B:0580
61076-97-9 see A:1590
61288-13-9 see P:0810
61373-95-3 see T:0840
61788-32-7 see H:0410
61788-32-7 see T:0210
61789-51-3 see C:1330
61790-53-2 see D:0260
61790-53-2 see S:0220
61840-45-7 see P:0450
61891-42-7 see V:0150
62139-95-1 see D:0690
62476-59-9 see A:0360
62497-63-6 see S:0670
62642-07-3 see C:0160
62869-69-6 see Q:0120
63150-68-5 see M:1310
63705-05-5 see S:0730
63908-52-1 see A:0410
63936-56-1 see P:0810
64070-92-4 see F:0130
64093-79-4 see N:0180
64441-70-9 see D:1250
64475-85-0 see N:0110
64684-45-3 see P:0450
64742-47-8 see K:0100
64742-89-8 see N:0110
64969-34-2 see D:0470
65544-34-5 see M:0310
65931-45-5 see C:1570
65982-50-5 see P:0150
65996-92-1 see C:1290
65996-93-2 see C:1290
65997-15-1 see P:0830
65997-17-3 see F:0240
66039-27-8 see S:0250
66173-72-6 see T:0520
66368-96-5 see H:0280
67016-73-3 see D:0260
67016-73-3 see S:0220

67256-35-3 see S:0220
67757-43-1 see M:0310
67774-32-7 see P:0810
68135-69-3 see S:0550
68379-55-5 see M:0310
68475-76-3 see P:0830
68476-85-7 see L:0270
68855-54-9 see S:0220
68923-44-4 see I:0390
69012-64-2 see S:0220
69020-37-7 see M:1460
69098-86-8 see S:0410
69771-31-9 see A:1610
70371-19-6 see M:0310
71121-36-3 see I:0330
71751-04-7 see A:0710
71808-29-2 see P:0100
72514-83-1 see C:1360
73413-06-6 see D:1460
74552-83-3 see T:0720
75635-23-3 see T:0840
77536-66-4 see A:1590
77536-67-5 see A:1590
78006-92-5 see I:0420
78590-82-6 see Z:0140
78642-65-6 see M:0850
78733-32-1 see E:0710
79637-11-9 see S:0660
79956-36-8 see M:1385
80751-51-5 see N:0360
81133-20-2 see A:1590
81235-32-7 see P:0270
82041-23-4 see V:0150
83008-56-4 see S:0750
83046-05-3 see M:1385
83589-40-6 see F:0360
83730-60-3 see D:1520
84948-57-2 see E:0710
87701-64-2 see E:0540
87701-65-3 see E:0540
89125-89-3 see S:0750
86290-81-5 see G:0100
89830-27-3 see S:0410
90043-99-5 see C:1310
90452-29-2 see N:0350
90640-80-5 see A:1380
90880-94-7 see N:0350
91829-63-9 see S:0410
92046-46-3 see I:0390
92355-34-5 see M:0430
92786-62-4 see M:0430
93196-73-7 see T:0520
93460-77-6 see E:0710

93616-39-8 see A:1610
94449-58-8 see M:1310
94551-77-6 see F:0240
94624-12-1 see A:1310
94977-27-2 see M:0310
95828-55-0 see D:0690
96231-36-6 see E:0280
96638-28-7 see R:0135
98201-60-6 see A:0340
99638-63-8 see T:0120
99932-75-9 see E:0660
100901-72-2 see T:0630
103842-90-6 see P:0255
104512-57-4 see D:1500
105269-70-3 see S:0230
106602-80-6 see P:1260
107231-30-1 see A:0610
107569-51-7 see B:0840
108736-71-6 see P:0255
110540-41-5 see T:0120
110616-89-2 see P:0170
112068-71-0 see F:0410
112388-78-0 see D:1520
112926-00-8 see S:0220
112945-52-5 see S:0220
114213-96-6 see T:0630
116788-91-1 see D:0690
119540-51-1 see S:0770
121448-83-7 see N:0110
123542-79-0 see R:0120
123720-03-6 see M:0430
126033-27-0 see T:0630
127529-01-5 see S:0770
128739-80-0 see T:0630
130565-62-7 see A:0650
132207-33-1 see A:1590
133317-06-3 see M:0240
136338-65-3 see P:0150
137322-20-4 see T:0150
139411-96-4 see H:0240
144647-92-7 see T:0630
154670-12-9 see B:0640
156627-46-2 see T:0630
162744-63-0 see D:1100
168153-21-7 see D:1100
171745-67-8 see D:1100
175446-71-6 see M:1280
205105-68-6 see P:0150
220713-25-7 see B:0840
247050-57-3 see P:0150